CONCISE WORLD ATLAS

DK | Penguin Random House

CONCISE WORLD ATLAS

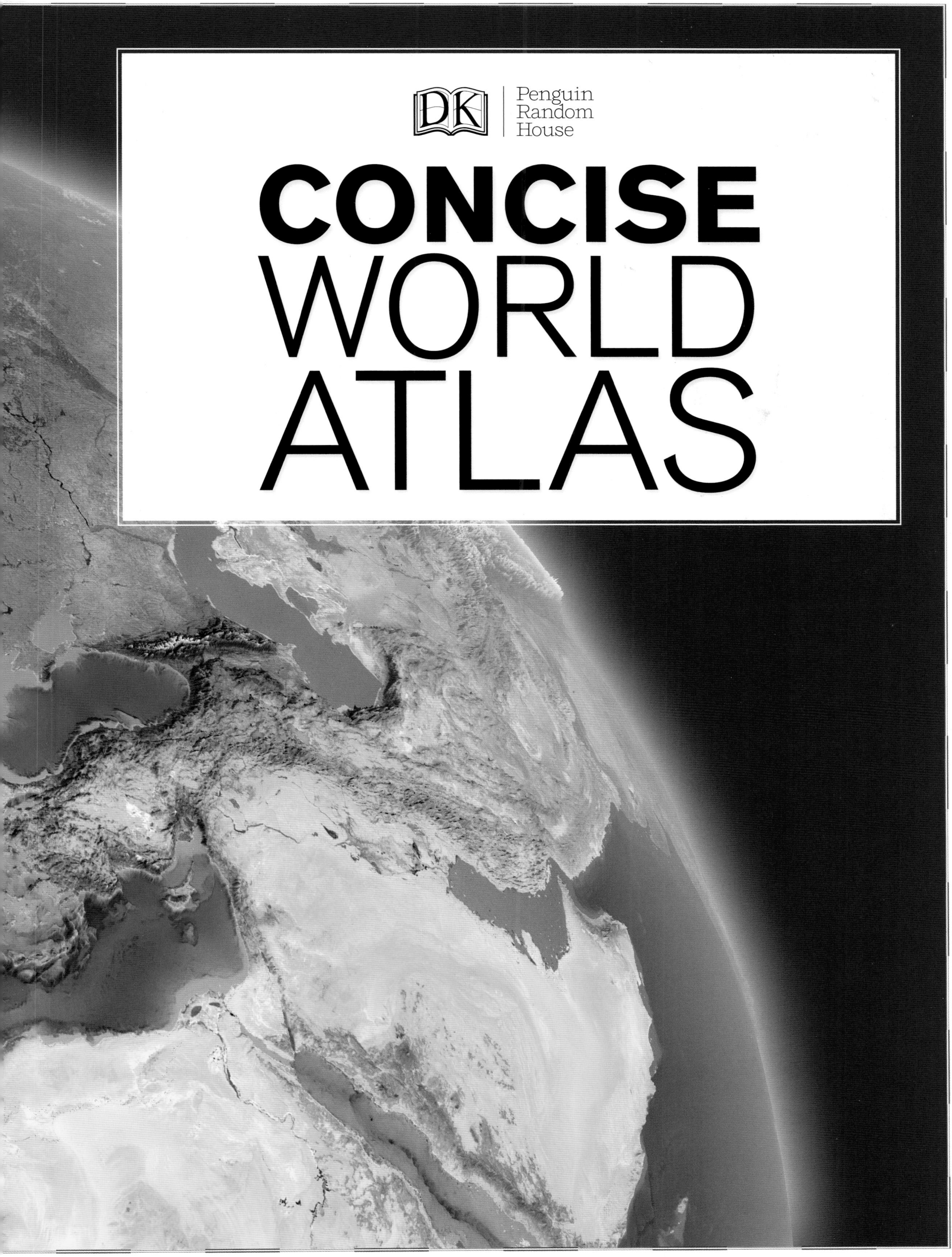

DK | Penguin Random House

FOR THE SEVENTH EDITION

Senior Cartographic Editor Simon Mumford
Producer, Pre-Production Luca Frassinetti **Producer** Vivienne Yong
Jacket Design Development Manager Sophia MTT
Publishing Director Jonathan Metcalf **Associate Publishing Director** Liz Wheeler **Art Director** Karen Self

General Geographical Consultants

Physical Geography Denys Brunsden, Emeritus Professor, Department of Geography, King's College, London
Human Geography Professor J Malcolm Wagstaff, Department of Geography, University of Southampton
Place Names Caroline Burgess, CartoConsulting Ltd, Reading
Boundaries International Boundaries Research Unit, Mountjoy Research Centre, University of Durham

Digital Mapping Consultants

DK Cartopia developed by George Galfalvi and XMap Ltd, London
Professor Jan-Peter Muller, Department of Photogrammetry and Surveying, University College, London
Planets and information on the Solar System provided by Philip Eales and Kevin Tildsley, Planetary Visions Ltd, London

Regional Consultants

North America Dr David Green, Department of Geography, King's College, London • Jim Walsh, Head of Reference, Wessell Library, Tufts University, Medford, Massachussetts
South America Dr David Preston, School of Geography, University of Leeds **Europe** Dr Edward M Yates, formerly of the Department of Geography, King's College, London
Africa Dr Philip Amis, Development Administration Group, University of Birmingham • Dr Ieuan Ll Griffiths, Department of Geography, University of Sussex
Dr Tony Binns, Department of Geography, University of Sussex
Central Asia Dr David Turnock, Department of Geography, University of Leicester **South and East Asia** Dr Jonathan Rigg, Department of Geography, University of Durham
Australasia and Oceania Dr Robert Allison, Department of Geography, University of Durham

Acknowledgments

Digital terrain data created by Eros Data Center, Sioux Falls, South Dakota, USA. Processed by GVS Images Inc, California, USA and Planetary Visions Ltd, London, UK
Cambridge International Reference on Current Affairs (CIRCA), Cambridge, UK • Digitization by Robertson Research International, Swanley, UK • Peter Clark
British Isles maps generated from a dataset supplied by Map Marketing Ltd/European Map Graphics Ltd in combination with DK Cartopia copyright data

DORLING KINDERSLEY CARTOGRAPHY

Editor-in-Chief Andrew Heritage **Managing Cartographer** David Roberts **Senior Cartographic Editor** Roger Bullen
Editorial Direction Louise Cavanagh **Database Manager** Simon Lewis **Art Direction** Chez Picthall

Cartographers
Pamela Alford • James Anderson • Caroline Bowie • Dale Buckton • Tony Chambers • Jan Clark • Bob Croser • Martin Darlison • Damien Demaj • Claire Ellam • Sally Gable
Jeremy Hepworth • Geraldine Horner • Chris Jackson • Christine Johnston • Julia Lunn • Michael Martin • Ed Merritt • James Mills-Hicks • Simon Mumford • John Plumer
John Scott • Ann Stephenson • Gail Townsley • Julie Turner • Sarah Vaughan • Jane Voss • Scott Wallace • Iorwerth Watkins • Bryony Webb • Alan Whitaker • Peter Winfield

Digital Maps Created in DK Cartopia by
Tom Coulson • Thomas Robertshaw
Philip Rowles • Rob Stokes
Managing Editor
Lisa Thomas
Editors
Thomas Heath • Wim Jenkins • Jane Oliver
Siobhan Ryan • Elizabeth Wyse
Editorial Research
Helen Dangerfield • Andrew Rebeiro-Hargrave
Additional Editorial Assistance
Debra Clapson • Robert Damon • Ailsa Heritage
Constance Novis • Jayne Parsons • Chris Whitwell

Placenames Database Team
Natalie Clarkson • Ruth Duxbury • Caroline Falce • John Featherstone • Dan Gardiner
Ciárán Hynes • Margaret Hynes • Helen Rudkin • Margaret Stevenson • Annie Wilson
Senior Managing Art Editor
Philip Lord
Designers
Scott David • Carol Ann Davis • David Douglas • Rhonda Fisher
Karen Gregory • Nicola Liddiard • Paul Williams
Illustrations
Ciárán Hughes • Advanced Illustration, Congleton, UK
Picture Research
Melissa Albany • James Clarke • Anna Lord
Christine Rista • Sarah Moule • Louise Thomas

First published in Great Britain in 2001 by Dorling Kindersley Limited, 80 Strand, London WC2R 0RL.

Second Edition 2003. Reprinted with revisions 2004. Third Edition 2005. Fourth Edition 2008. Fifth Edition 2011. Sixth Edition 2013. Seventh Edition 2016

A Penguin Random House Company

2 4 6 8 10 9 7 5 3 1

001 - 265171 - March 2016

A CIP catalogue record for this book is available from the British Library.

ISBN: 978-0-2412-2634-6

Printed and bound in Hong Kong.

A WORLD OF IDEAS:
SEE ALL THERE IS TO KNOW
www.dk.com

Introduction

EVERYTHING YOU NEED TO KNOW ABOUT OUR PLANET TODAY

For many, the outstanding legacy of the twentieth century was the way in which the Earth shrank. In the third millennium, it is increasingly important for us to have a clear vision of the world in which we live. The human population has increased fourfold since 1900. The last scraps of *terra incognita* – the polar regions and ocean depths – have been penetrated and mapped. New regions have been colonized and previously hostile realms claimed for habitation. The growth of air transport and mass tourism allows many of us to travel further, faster, and more frequently than ever before. In doing so we are given a bird's-eye view of the Earth's surface denied to our forebears.

At the same time, the amount of information about our world has grown enormously. Our multi-media environment hurls uninterrupted streams of data at us, on the printed page, through the airwaves and across our television, computer, and phone screens; events from all corners of the globe reach us instantaneously, and are witnessed as they unfold. Our sense of stability and certainty has been eroded; instead, we are aware that the world is in a constant state of flux and change. Natural disasters, man-made cataclysms, and conflicts between nations remind us daily of the enormity and fragility of our domain. The ongoing threat of international terrorism throws into very stark relief the difficulties that arise when trying to 'know' or 'understand' our planet and its many cultures.

The current crisis in our 'global' culture has made the need greater than ever before for everyone to possess an atlas. The **CONCISE WORLD ATLAS** has been conceived to meet this need.
At its core, like all atlases, it seeks to define where places are, to describe their main characteristics, and to locate them in relation to other places. Every attempt has been made to make the information on the maps as clear, accurate, and accessible as possible using the latest digital cartographic techniques. In addition, each page of the atlas provides a wealth of further information, bringing the maps to life. Using photographs, diagrams, 'at-a-glance' maps, introductory texts, and captions, the atlas builds up a detailed portrait of those features – cultural, political, economic, and geomorphological – that make each region unique and which are also the main agents of change.

This seventh edition of the **CONCISE WORLD ATLAS** incorporates hundreds of revisions and updates affecting every map and every page, distilling the burgeoning mass of information available through modern technology into an extraordinarily detailed and reliable view of our world.

CONTENTS

THE WORLD

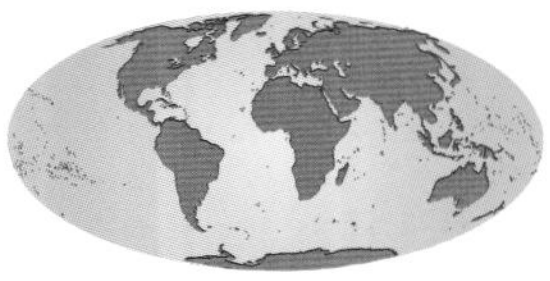

ATLAS OF THE WORLD

North America

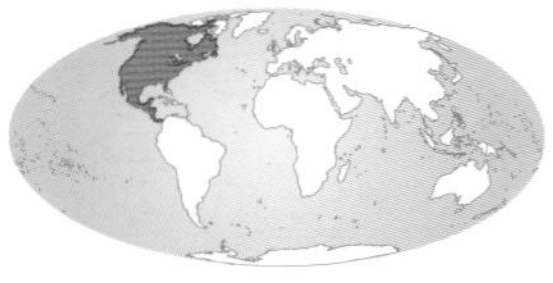

South America

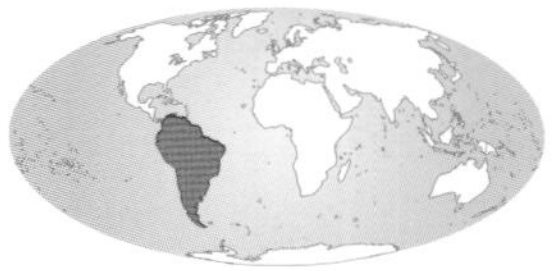

Africa

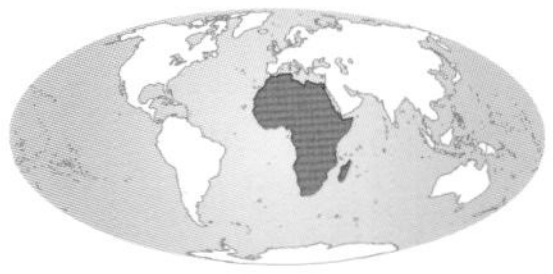

Europe

Asia

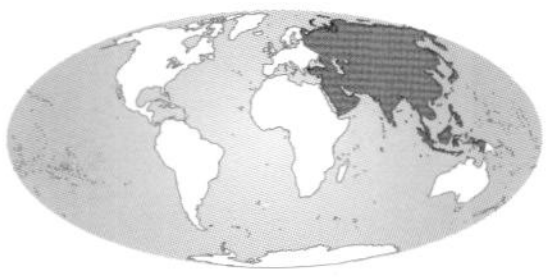

Australasia & Oceania

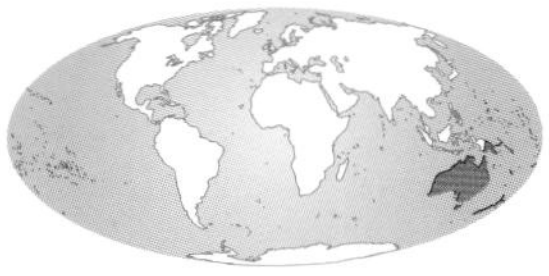

INDEX–GAZETTEER

Key to maps

Regional

Physical features

elevation

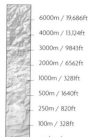

- 6000m / 19,686ft
- 4000m / 13,124ft
- 3000m / 9843ft
- 2000m / 6562ft
- 1000m / 3281ft
- 500m / 1640ft
- 250m / 820ft
- 100m / 328ft
- sea level
- below sea level

- elevation above sea level (mountain height)
- volcano
- pass
- elevation below sea level (depression depth)

- sand desert
- lava flow
- coastline
- reef
- atoll

sea depth

- sea level
- -250m / -820ft
- -500m / -1640ft
- -1000m / -3281ft
- -2000m / -6562ft
- -3000m / -9843ft

- seamount / guyot symbol
- undersea spot depth

Drainage features

- main river
- secondary river
- tertiary river
- minor river
- main seasonal river
- secondary seasonal river
- canal
- waterfall
- rapids
- dam
- perennial lake
- seasonal lake
- perennial salt lake
- seasonal salt lake
- reservoir
- salt flat / salt pan
- marsh / salt marsh
- mangrove
- wadi
- spring / well / waterhole / oasis

Ice features

- ice cap / sheet
- ice shelf
- glacier / snowfield
- summer pack ice limit
- winter pack ice limit

Communications

- motorway / highway
- motorway / highway (under construction)
- major road
- minor road
- tunnel (road)
- main line
- minor line
- tunnel (rail)
- international airport

Borders

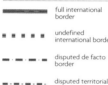

- full international border
- undefined international border
- disputed de facto border
- disputed territorial claim border
- indication of country extent (Pacific only)
- indication of dependent territory extent (Pacific only)
- demarcation/ cease fire line
- autonomous / federal region border
- other 1st order internal administrative border
- 2nd order internal administrative border

Settlements

- built up area

settlement population symbols

- more than 5 million
- 1 million to 5 million
- 500,000 to 1 million
- 100,000 to 500,000
- 50,000 to 100,000
- 10,000 to 50,000
- fewer than 10,000
- country/dependent territory capital city
- autonomous / federal region / other 1st order internal administrative centre
- 2nd order internal administrative centre

Miscellaneous features

- ancient wall
- site of interest
- scientific station

Graticule features

- lines of latitude and longitude / Equator
- Tropics / Polar circles
- 45° degrees of longitude / latitude

Typographic key

Physical features

- landscape features ... *Namib Desert* / *Massif Central* / *ANDES*
- headland ... *Nordkapp*
- elevation / volcano / pass ... Mount Meru 4556 m
- drainage features ... *Lake Geneva*
- rivers / canals / spring / well / waterhole / oasis / waterfall / rapids / dam ... *Mekong*
- ice features ... *Vatnajökull*
- sea features ... *Golfe de Lion* / *Andaman Sea* / *INDIAN OCEAN*
- undersea features ... *Barracuda Fracture Zone*

Regions

- country ... ARMENIA
- dependent territory with parent state ... NIUE (to NZ)
- region outside feature area ... ANGOLA
- autonomous / federal region ... MINAS GERAIS
- other 1st order internal administrative region ... MINSKAYA VOBLASTS'
- 2nd order internal administrative region ... Vaucluse
- cultural region ... New England

Settlements

- capital city ... BEIJING
- dependent territory capital city ... FORT-DE-FRANCE
- other settlements ... Chicago / Adana / Tizi Ozou / Yonezawa / Farnham

Miscellaneous

- sites of interest / miscellaneous ... Valley of the Kings
- Tropics / Polar circles ... *Antarctic Circle*

How to use this Atlas

The atlas is organized by continent, moving eastwards from the International Date Line. The opening section describes the world's structure, systems and its main features. The Atlas of the World which follows, is a continent-by-continent guide to today's world, starting with a comprehensive insight into the physical, political and economic structure of each continent, followed by integrated mapping and descriptions of each region or country.

The world

The introductory section of the Atlas deals with every aspect of the planet, from physical structure to human geography, providing an overall picture of the world we live in. Complex topics such as the landscape of the Earth, climate, oceans, population and economic patterns are clearly explained with the aid of maps, diagrams drawn from the latest information.

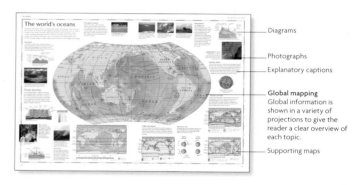

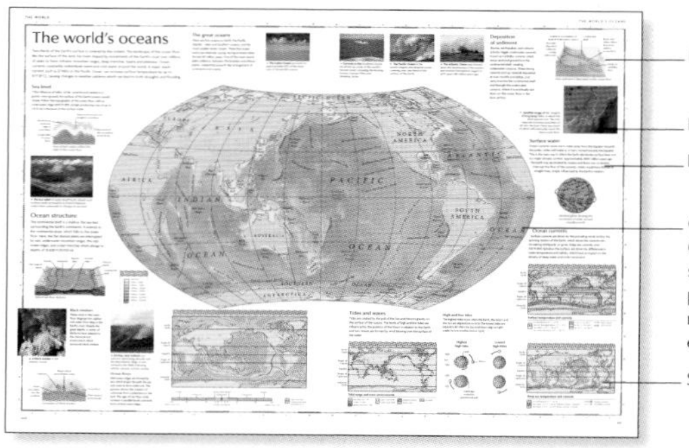
The world's oceans
Diagrams
Photographs
Explanatory captions
Global mapping
Global information is shown in a variety of projections to give the reader a clear overview of each topic.
Supporting maps

The political continent

The political portrait of the continent is a vital reference point for every continental section, showing the position of countries relative to one another, and the relationship between human settlement and geographic location. The complex mosaic of languages spoken in each continent is mapped, as is the effect of communications networks on the pattern of settlement.

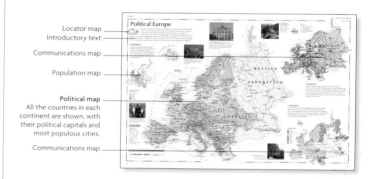

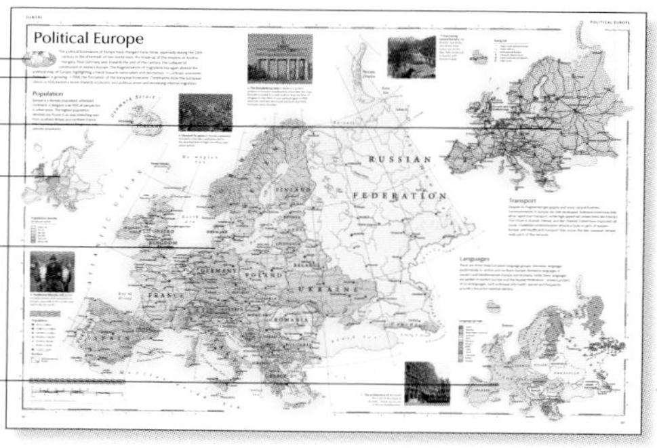
Political Europe
Locator map
Introductory text
Communications map
Population map
Political map
All the countries in each continent are shown, with their political capitals and most populous cities.
Communications map

Continental resources

The Earth's rich natural resources, including oil, gas, minerals and fertile land, have played a key role in the development of society. These pages show the location of minerals and agricultural resources on each continent, and how they have been instrumental in dictating industrial growth and the varieties of economic activity across the continent.

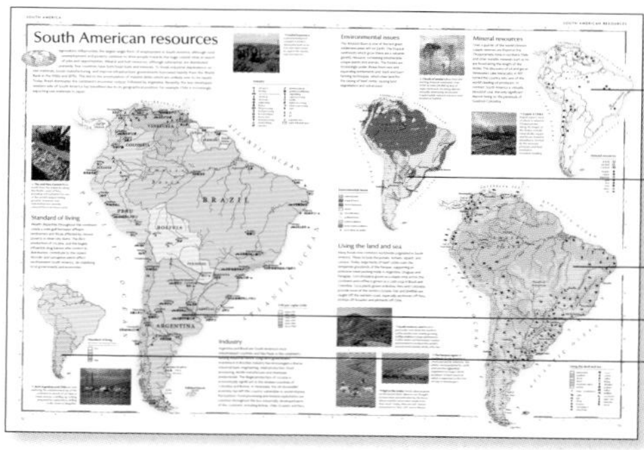
South American resources
Mineral resources map
Environmental issues map
Land use map
Industry map
Comparative wealth map

The physical continent

The astonishing variety of landforms, and the dramatic forces that created and continue to shape the landscape, are explained in the continental physical spread. Cross-sections, illustrations and terrain maps highlight the different parts of the continent, showing how nature's forces have produced the landscapes we see today.

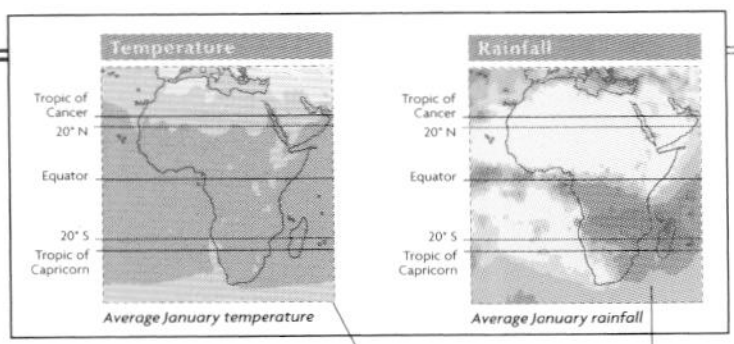

Climate charts
Rainfall and temperature charts clearly show the continental patterns of rainfall and temperature.

Climate map
Climatic regions vary across each continent. The map displays the differing climatic regions, as well as daily hours of sunshine at selected weather stations.

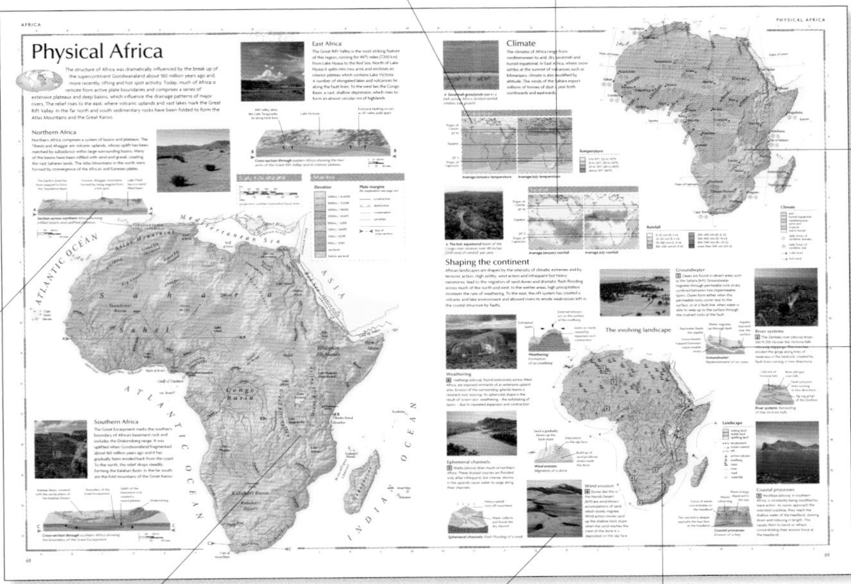

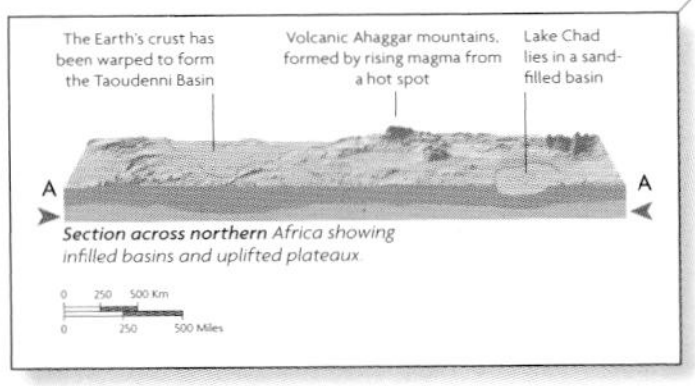

Cross-sections
Detailed cross-sections through selected parts of the continent show the underlying geomorphic structure.

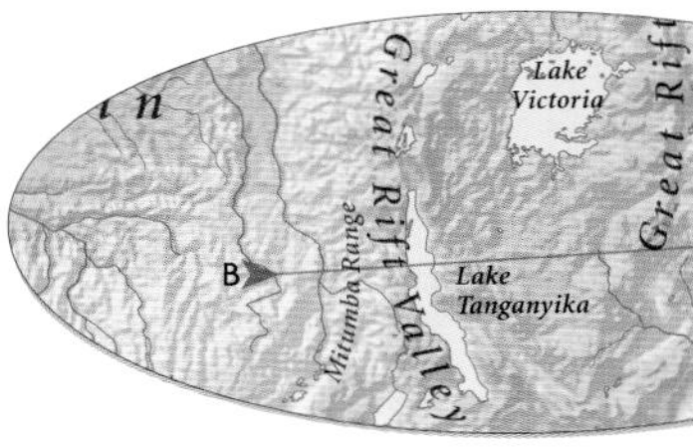

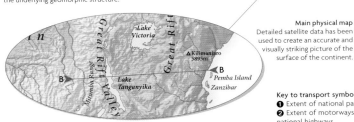

Main physical map
Detailed satellite data has been used to create an accurate and visually striking picture of the surface of the continent.

Photographs
A wide range of beautiful photographs bring the world's regions to life.

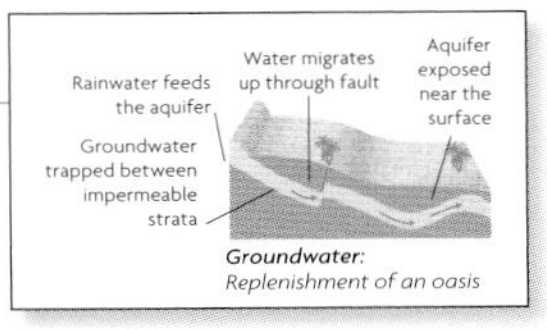

Landform diagrams
The complex formation of many typical landforms is summarized in these easy-to-understand illustrations.

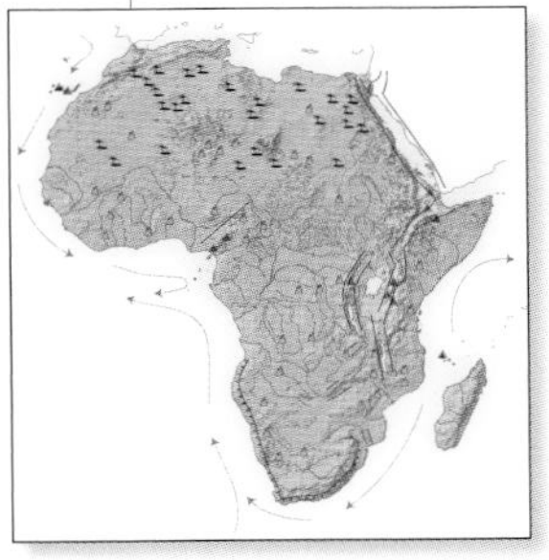

Landscape evolution map
The physical shape of each continent is affected by a variety of forces which continually sculpt and modify the landscape. This map shows the major processes which affect different parts of the continent.

Regional mapping

The main body of the Atlas is a unique regional map set, with detailed information on the terrain, the human geography of the region and its infrastructure. Around the edge of the map, additional 'at-a-glance' maps, give an instant picture of regional industry, land use and agriculture. The detailed terrain map (shown in perspective), focuses on the main physical features of the region, and is enhanced by annotated illustrations, and photographs of the physical structure.

Key to transport symbols
❶ Extent of national paved road network.
❷ Extent of motorways, freeways or major national highways.
❸ Extent of commercial rail network.
❹ Extent of inland waterways navigable by commercial craft.

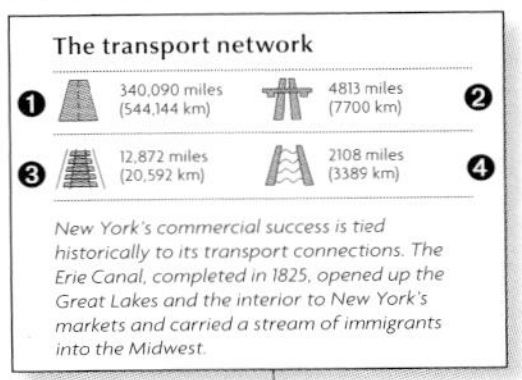

Transport network
The differing extent of the transport network for each region is shown here, along with key facts about the transport system.

World locator
This locates the continent in which the region is found on a small world map.

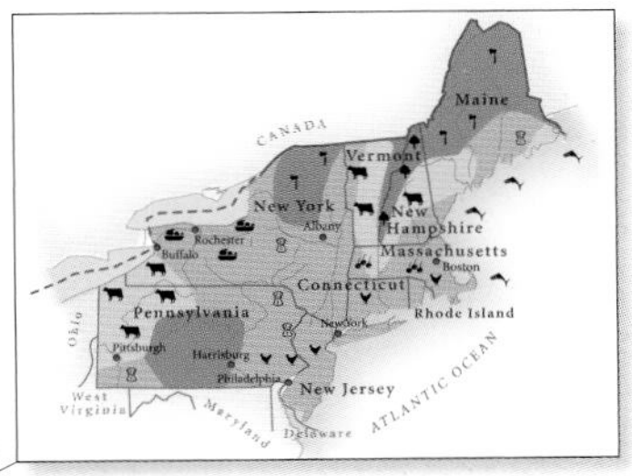

Regional Locator
This small map shows the location of each country in relation to its continent.

Key to main map
A key to the population symbols and land heights accompanies the main map.

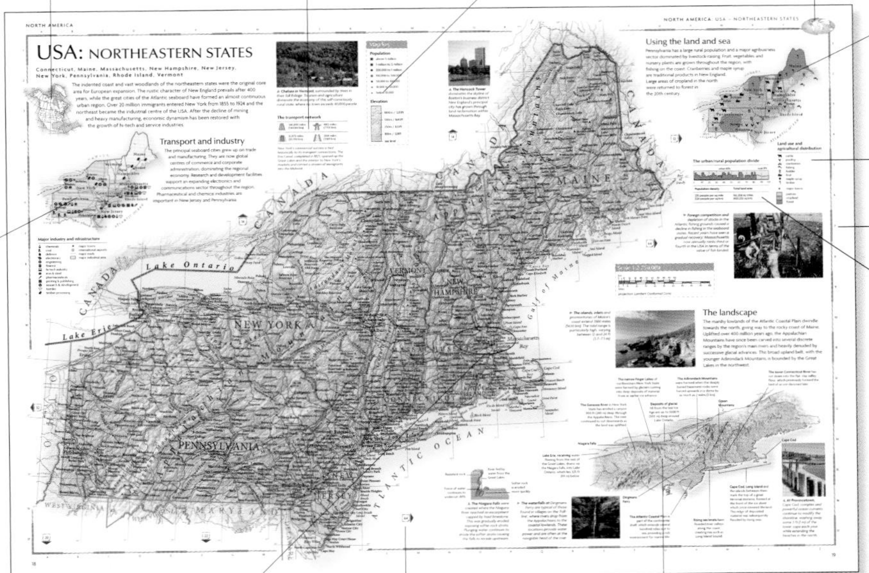

Land use map
This shows the different types of land use which characterize the region, as well as indicating the principal agricultural activities.

Map keys
Each supporting map has its own key.

Grid reference
The framing grid provides a location reference for each place listed in the Index.

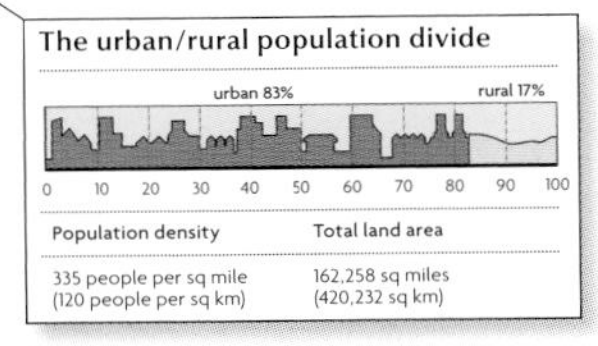

Urban/rural population divide
The proportion of people in the region who live in urban and rural areas, as well as the overall population density and land area are clearly shown in these simple graphics.

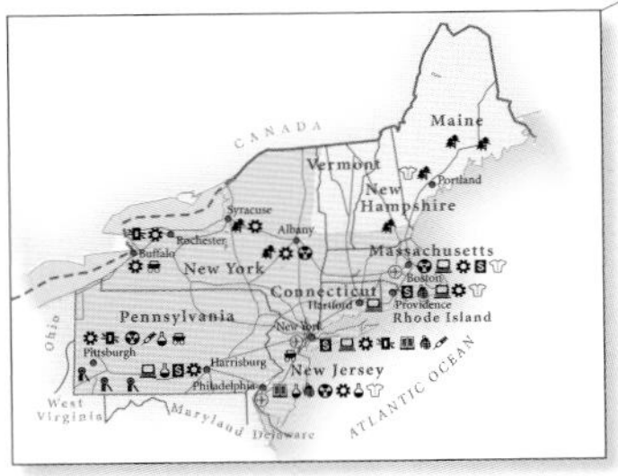

Transport and industry map
The main industrial areas are mapped, and the most important industrial and economic activities of the region are shown.

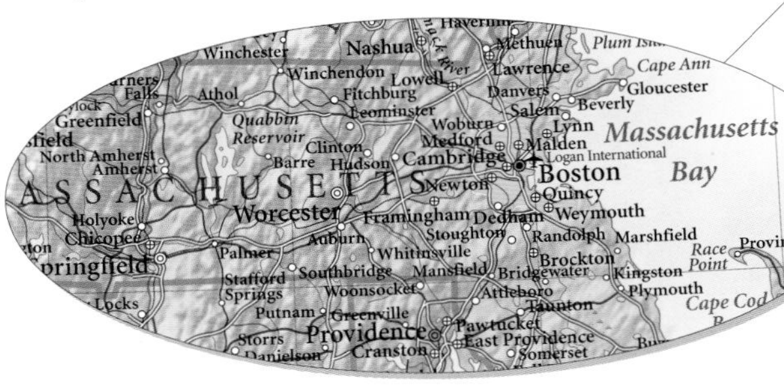

Continuation symbols
These symbols indicate where adjacent maps can be found.

Main regional map
A wealth of information is displayed on the main map, building up a rich portrait of the interaction between the physical landscape and the human and political geography of each region. The key to the regional maps can be found on page viii.

Landscape map
The computer-generated terrain model accurately portrays an oblique view of the landscape. Annotations highlight the most important geographic features of the region.

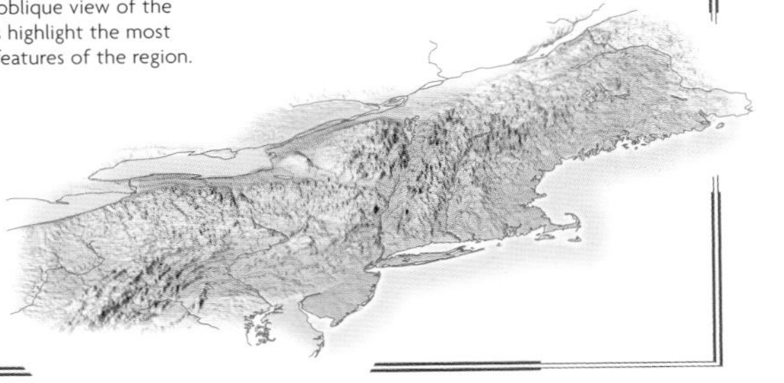

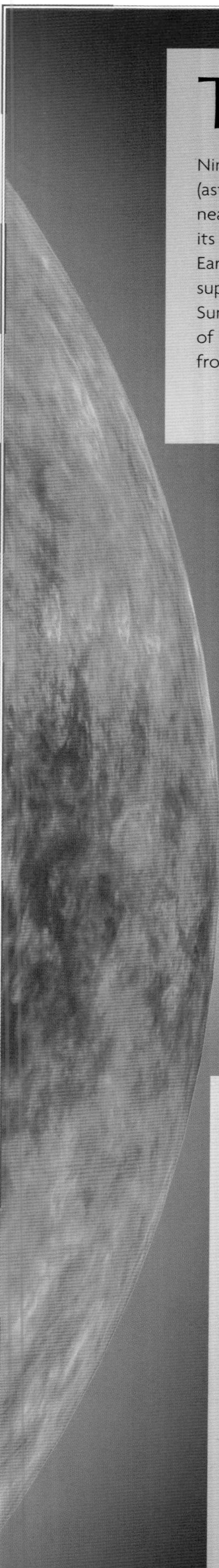

The Solar System

Nine major planets, their satellites and countless minor planets (asteroids) orbit the Sun to form the Solar System. The Sun, our nearest star, creates energy from nuclear reactions deep within its interior, providing all the light and heat which make life on Earth possible. The Earth is unique in the Solar System in that it supports life: its size, gravitational pull and distance from the Sun have all created the optimum conditions for the evolution of life. The planetary images seen here are composites derived from actual spacecraft images (not shown to scale).

Orbits

All the Solar System's planets and dwarf planets orbit the Sun in the same direction and (apart from Pluto) roughly in the same plane. All the orbits have the shapes of ellipses (stretched circles). However in most cases, these ellipses are close to being circular: only Pluto and Eris have very elliptical orbits. Orbital period (the time it takes an object to orbit the Sun) increases with distance from the Sun. The more remote objects not only have further to travel with each orbit, they also move more slowly.

Ceres
(dwarf planet)
Mercury
Venus
Earth
Mars

Jupiter

The Sun

- **Diameter:** *864,948 miles (1,392,000 km)*
- **Mass:** *1990 million million million million tons*

The Sun was formed when a swirling cloud of dust and gas contracted, pulling matter into its centre. When the temperature at the centre rose to 1,000,000°C, nuclear fusion – the fusing of hydrogen into helium, creating energy – occurred, releasing a constant stream of heat and light.

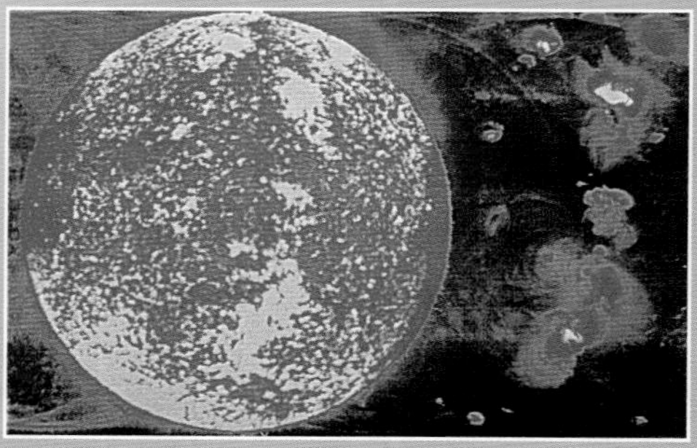

▲ ***Solar flares are** sudden bursts of energy from the Sun's surface. They can be 125,000 miles (200,000 km) long.*

The formation of the Solar System

The cloud of dust and gas thrown out by the Sun during its formation cooled to form the Solar System. The smaller planets nearest the Sun are formed of minerals and metals. The outer planets were formed at lower temperatures, and consist of swirling clouds of gases.

Solar eclipse

A solar eclipse occurs when the Moon passes between Earth and the Sun, casting its shadow on Earth's surface. During a total eclipse *(below)*, viewers along a strip of Earth's surface, called the area of totality, see the Sun totally blotted out for a short time, as the umbra (Moon's full shadow) sweeps over them. Outside this area is a larger one, where the Sun appears only partly obscured, as the penumbra (partial shadow) passes over.

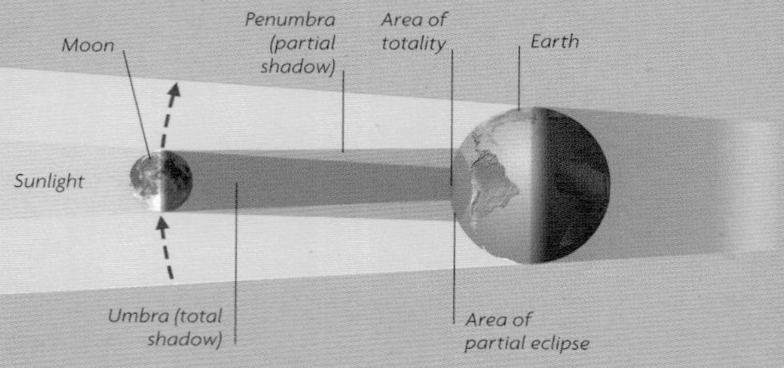
Moon
Penumbra (partial shadow)
Area of totality
Earth
Sunlight
Umbra (total shadow)
Area of partial eclipse

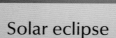

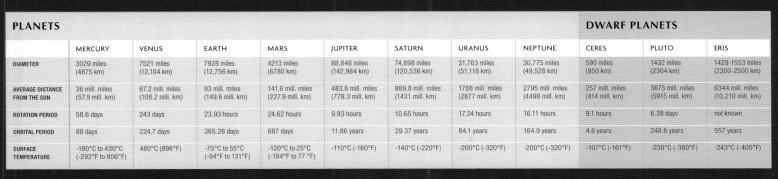

PLANETS								DWARF PLANETS		
	MERCURY	VENUS	EARTH	MARS	JUPITER	SATURN	URANUS	NEPTUNE	CERES	PLUTO	ERIS
DIAMETER	3029 miles (4875 km)	7521 miles (12,104 km)	7928 miles (12,756 km)	4213 miles (6780 km)	88,846 miles (142,984 km)	74,898 miles (120,536 km)	31,763 miles (51,118 km)	30,775 miles (49,528 km)	590 miles (950 km)	1432 miles (2304 km)	1429-1553 miles (2300-2500 km)
AVERAGE DISTANCE FROM THE SUN	36 mill. miles (57.9 mill. km)	67.2 mill. miles (108.2 mill. km)	93 mill. miles (149.6 mill. km)	141.6 mill. miles (227.9 mill. km)	483.6 mill. miles (778.3 mill. km)	889.8 mill. miles (1431 mill. km)	1788 mill. miles (2877 mill. km)	2795 mill. miles (4498 mill. km)	257 mill. miles (414 mill. km)	3675 mill. miles (5915 mill. km)	6344 mill. miles (10,210 mill. km)
ROTATION PERIOD	58.6 days	243 days	23.93 hours	24.62 hours	9.93 hours	10.65 hours	17.24 hours	16.11 hours	9.1 hours	6.38 days	not known
ORBITAL PERIOD	88 days	224.7 days	365.26 days	687 days	11.86 years	29.37 years	84.1 years	164.9 years	4.6 years	248.6 years	557 years
SURFACE TEMPERATURE	-180°C to 430°C (-292°F to 806°F)	480°C (896°F)	-70°C to 55°C (-94°F to 131°F)	-120°C to 25°C (-184°F to 77 °F)	-110°C (-160°F)	-140°C (-220°F)	-200°C (-320°F)	-200°C (-320°F)	-107°C (-161°F)	-230°C (-380°F)	-243°C (-405°F)

AVERAGE DISTANCE FROM THE SUN

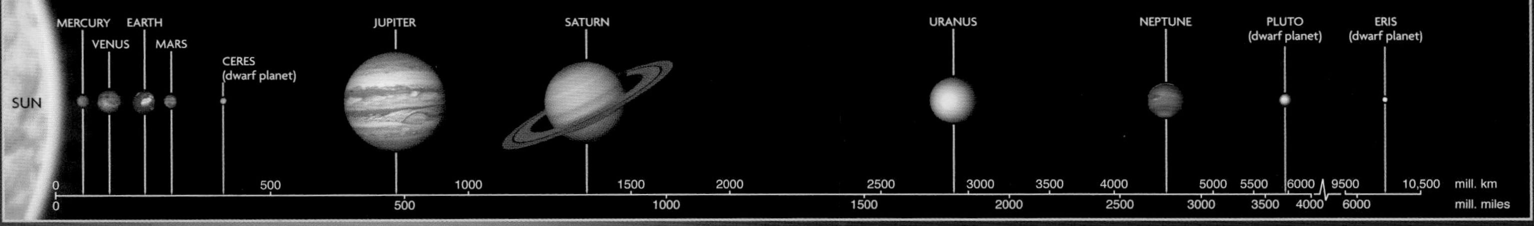
SUN
MERCURY
VENUS
EARTH
MARS
CERES (dwarf planet)
JUPITER
SATURN
URANUS
NEPTUNE
PLUTO (dwarf planet)
ERIS (dwarf planet)
0 500 1000 1500 2000 2500 3000 3500 4000 5000 5500 6000 9500 10,500 mill. km
0 500 1000 1500 2000 2500 3000 3500 4000 6000 mill. miles

Saturn

Uranus
Neptune
Pluto (dwarf planet)
Eris (dwarf planet)

Space debris

Millions of objects, remnants of planetary formation, circle the Sun in a zone lying between Mars and Jupiter: the asteroid belt. Fragments of asteroids break off to form meteoroids, which can reach the Earth's surface. Comets, composed of ice and dust, originated outside our Solar System. Their elliptical orbit brings them close to the Sun and into the inner Solar System.

▲ ***Meteor Crater** in Arizona is 4200 ft (1300 m) wide and 660 ft (200 m) deep. It was formed over 10,000 years ago.*

Possible and actual meteorite craters

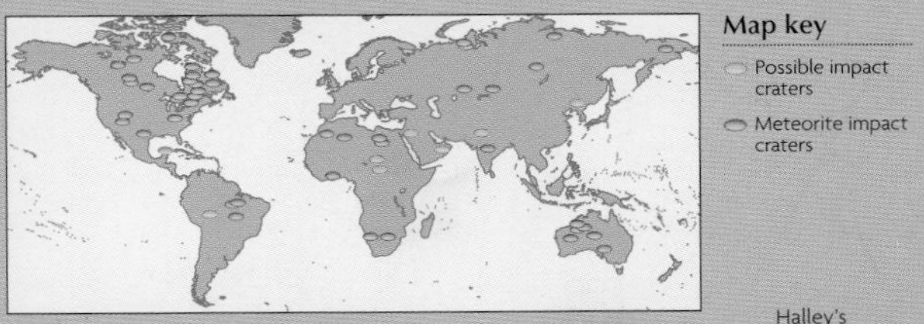
Map key
Possible impact craters
Meteorite impact craters

▲ ***The orbit of** Halley's Comet brings it close to the Earth every 76 years. It last visited in 1986.*

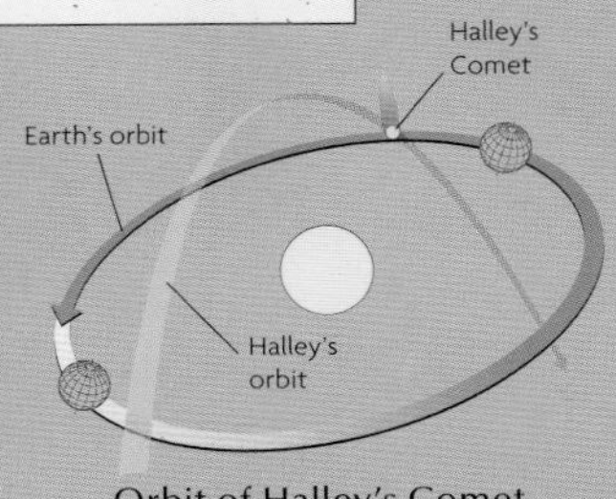
Halley's Comet
Earth's orbit
Halley's orbit

Orbit of Halley's Comet around the Sun

The Earth's atmosphere

During the early stages of the Earth's formation, ash, lava, carbon dioxide and water vapour were discharged onto the surface of the planet by constant volcanic eruptions. The water formed the oceans, while carbon dioxide entered the atmosphere or was dissolved in the oceans. Clouds, formed of water droplets, reflected some of the Sun's radiation back into space. The Earth's temperature stabilized and early life forms began to emerge, converting carbon dioxide into life-giving oxygen.

▲ ***It is thought** that the gases that make up the Earth's atmosphere originated deep within the interior, and were released many millions of years ago during intense volcanic actvity, similar to this eruption at Mount St. Helens.*

The physical world

The Earth's surface is constantly being transformed: it is uplifted, folded and faulted by tectonic forces; weathered and eroded by wind, water and ice. Sometimes change is dramatic, the spectacular results of earthquakes or floods. More often it is a slow process lasting millions of years. A physical map of the world represents a snapshot of the ever-evolving architecture of the Earth. This terrain map shows the whole surface of the Earth, both above and below the sea.

The world in section

These cross-sections around the Earth, one in the northern hemisphere; one straddling the Equator, reveal the limited areas of land above sea level in comparison with the extent of the sea floor. The greater erosive effects of weathering by wind and water limit the upward elevation of land above sea level, while the deep oceans retain their dramatic mountain and trench profiles.

Aleutian Trench
Pacific Ocean
Rocky Mountains
60°N
180°
150°W
120°W

Cross-section: Northern hemisphere

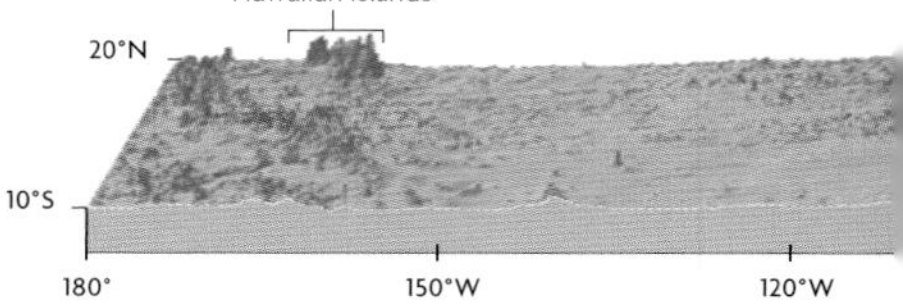
Hawaiian Islands
20°N
10°S
180°
150°W
120°W

Cross-section: Southern hemisphere

Map key

Elevation	Sea depth
6000m / 19,686ft	sea level
4000m / 13,124ft	-250m / -820ft
3000m / 9843ft	-2000m / -6562ft
2000m / 6562ft	-4000m / -13,124ft
1000m / 3281ft	
500m / 1640ft	
250m / 820ft	
100m / 328ft	
sea level	
below sea level	

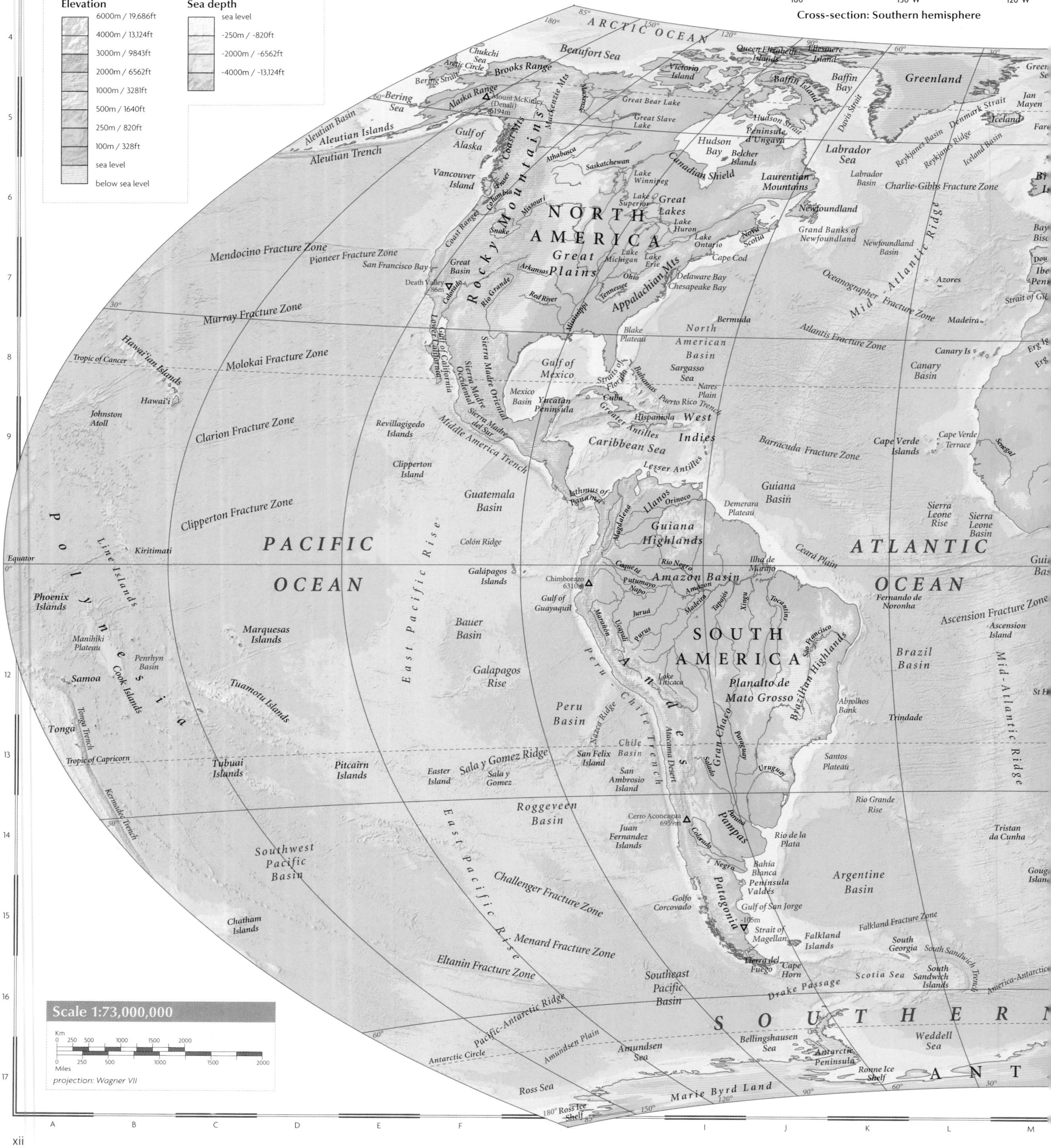
ARCTIC OCEAN
Beaufort Sea
Chukchi Sea
Arctic Circle
Brooks Range
Bering Strait
Bering Sea
Alaska Range
Mount McKinley (Denali) 6194m
Mackenzie Mts
Victoria Island
Queen Elizabeth Islands
Ellesmere Island
Baffin Island
Baffin Bay
Greenland
Great Bear Lake
Great Slave Lake
Aleutian Basin
Aleutian Islands
Aleutian Trench
Gulf of Alaska
Coast Mts
Hudson Strait
Peninsule d'Ungava
Hudson Bay
Belcher Islands
Davis Strait
Denmark Strait
Jan Mayen
Iceland
Labrador Sea
Labrador Basin
Reykjanes Basin
Reykjanes Ridge
Iceland Basin
Athabasca
Saskatchewan
Lake Winnipeg
Canadian Shield
Laurentian Mountains
Charlie-Gibbs Fracture Zone
Vancouver Island
Columbia
Fraser
Missouri
Lake Superior
Great Lakes
Newfoundland
Rocky Mountains
NORTH AMERICA
Lake Huron
Lake Ontario
Nova Scotia
Grand Banks of Newfoundland
Newfoundland Basin
Coast Range
Snake
Great Plains
Lake Michigan
Lake Erie
Cape Cod
Mendocino Fracture Zone
Pioneer Fracture Zone
San Francisco Bay
Great Basin
Arkansas
Ohio
Delaware Bay
Chesapeake Bay
Appalachian Mts
Tennessee
Death Valley -86m
Colorado
Rio Grande
Red River
Mississippi
Oceanographer Fracture Zone
Mid-Atlantic Ridge
Azores
Strait of Gibraltar
Murray Fracture Zone
Bermuda
Blake Plateau
North American Basin
Madeira
Atlantis Fracture Zone
Tropic of Cancer
Hawai'ian Islands
Molokai Fracture Zone
Gulf of California
Lower California
Sierra Madre Occidental
Sierra Madre Oriental
Gulf of Mexico
Straits of Florida
Bahamas
Sargasso Sea
Canary Is
Canary Basin
Hawai'i
Johnston Atoll
Mexico Basin
Yucatan Peninsula
Cuba
Nares Plain
Puerto Rico Trench
Greater Antilles
Hispaniola
West Indies
Revillagigedo Islands
Sierra Madre del Sur
Middle America Trench
Clarion Fracture Zone
Caribbean Sea
Barracuda Fracture Zone
Cape Verde Islands
Cape Verde Terrace
Senegal
Clipperton Island
Lesser Antilles
Guatemala Basin
Isthmus of Panama
Llanos
Orinoco
Guiana Basin
Demerara Plateau
Sierra Leone Rise
Sierra Leone Basin
Clipperton Fracture Zone
Magdalena
Guiana Highlands
Polynesia
Line Islands
Kiritimati
PACIFIC OCEAN
Colón Ridge
East Pacific Rise
Ceará Plain
ATLANTIC OCEAN
Equator
Galápagos Islands
Chimborazo 6310m
Caquetá
Rio Negro
Putumayo
Napo
Amazon Basin
Amazon
Ilha de Marajó
Phoenix Islands
Gulf of Guayaquil
Juruá
Purus
Madeira
Tapajós
Xingu
Tocantins
Fernando de Noronha
Ascension Fracture Zone
Ascension Island
Manihiki Plateau
Marquesas Islands
Bauer Basin
Marañón
Ucayali
SOUTH AMERICA
São Francisco
Brazilian Highlands
Brazil Basin
Penrhyn Basin
Samoa
Cook Islands
Galapagos Rise
Peru-Chile Trench
Andes
Lake Titicaca
Planalto de Mato Grosso
Tuamotu Islands
Abrolhos Bank
Trindade
Tonga
Tonga Trench
Peru Basin
Nazca Ridge
Chile Basin
Atacama Desert
Gran Chaco
Paraguay
Tropic of Capricorn
Tubuai Islands
Pitcairn Islands
Easter Island
Sala y Gomez Ridge
Sala y Gomez
San Felix Island
San Ambrosio Island
Salado
Uruguay
Santos Plateau
Kermadec Trench
Roggeveen Basin
Cerro Aconcagua 6959m
Juan Fernandez Islands
Paraná
Pampas
Rio de la Plata
Rio Grande Rise
Tristan da Cunha
Southwest Pacific Basin
Colorado
Negro
Bahía Blanca
Península Valdés
Argentine Basin
Gough Island
Patagonia
Challenger Fracture Zone
Golfo Corcovado
Gulf of San Jorge
-105m
Chatham Islands
Strait of Magellan
Falkland Islands
Falkland Fracture Zone
Menard Fracture Zone
South Georgia
South Sandwich Trench
Tierra del Fuego
Cape Horn
Eltanin Fracture Zone
Southeast Pacific Basin
Scotia Sea
South Sandwich Islands
America-Antarctic Ridge
Drake Passage
Pacific-Antarctic Ridge
SOUTHERN
Amundsen Plain
Bellingshausen Sea
Weddell Sea
Antarctic Circle
Amundsen Sea
Antarctic Peninsula
Ronne Ice Shelf
ANT
Ross Sea
Marie Byrd Land
Ross Ice Shelf
Scale 1:73,000,000
Km
Miles
projection: Wagner VII

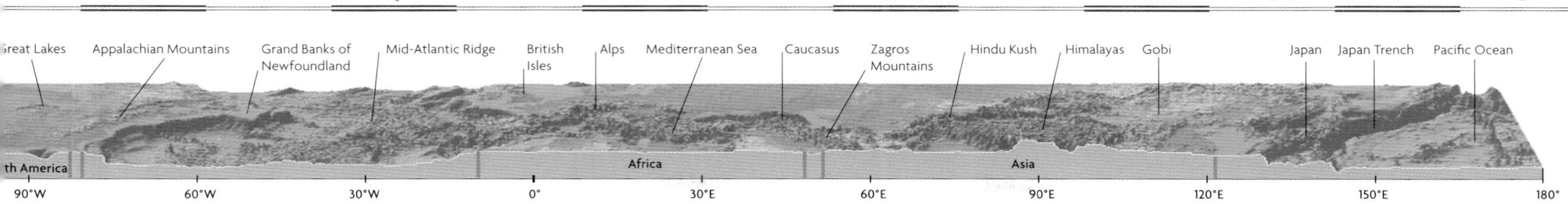

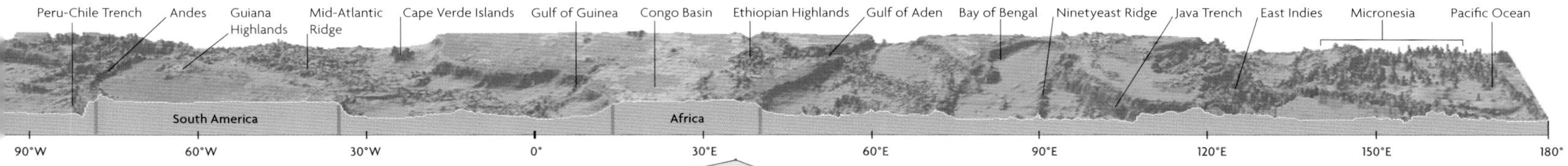

Physical factfile

- **Diameter of Earth at Equator:** *7927 miles (12,756 km)*
- **Equatorial circumference of Earth:** *24,901 miles (40,075 km)*
- **Diameter from Pole to Pole:** *7900 miles (12,714 km)*
- **Polar circumference of Earth:** *24,860 miles (40,008 km)*
- **Mass:** *5988 million million million tons (tonnes)*

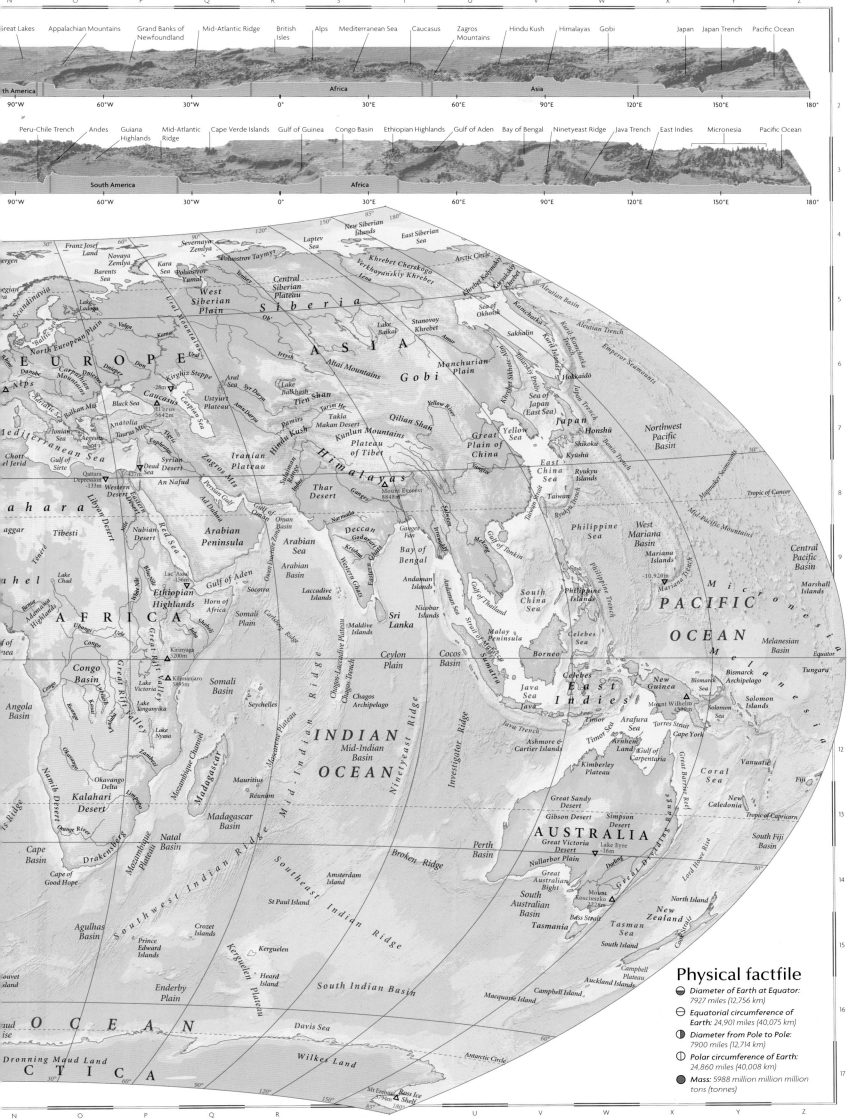

Structure of the Earth

The Earth as it is today is just the latest phase in a constant process of evolution which has occurred over the past 4.5 billion years. The Earth's continents are neither fixed nor stable; over the course of the Earth's history, propelled by currents rising from the intense heat at its centre, the great plates on which they lie have moved, collided, joined together, and separated. These processes continue to mould and transform the surface of the Earth, causing earthquakes and volcanic eruptions and creating oceans, mountain ranges, deep ocean trenches and island chains.

Inside the Earth

The Earth's hot inner core ismade up of solid iron, while the outer core is composed of liquid iron and nickel. The mantle nearest the core is viscous, whereas the rocky upper mantle is fairly rigid. The crust is the rocky outer shell of the Earth. Together, the upper mantle and the crust form the lithosphere.

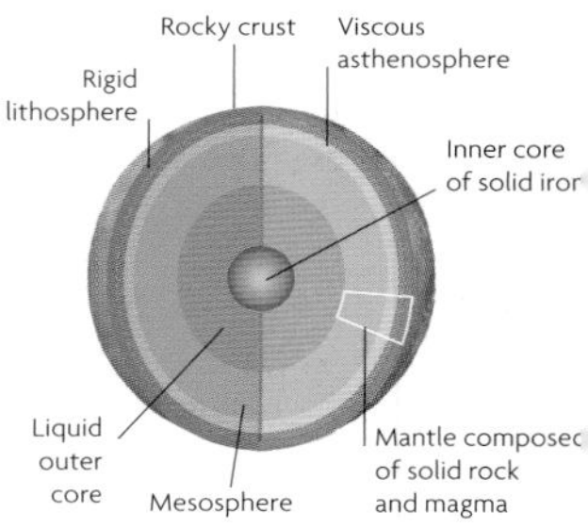
Rocky crust
Viscous asthenosphere
Rigid lithosphere
Inner core of solid iron
Liquid outer core
Mesosphere
Mantle composed of solid rock and magma

The dynamic Earth

The Earth's crust is made up of eight major (and several minor) rigid continental and oceanic tectonic plates, which fit closely together. The positions of the plates are not static. They are constantly moving relative to one another. The type of movement between plates affects the way in which they alter the structure of the Earth. The oldest parts of the plates, known as shields, are the most stable parts of the Earth and little tectonic activity occurs here.

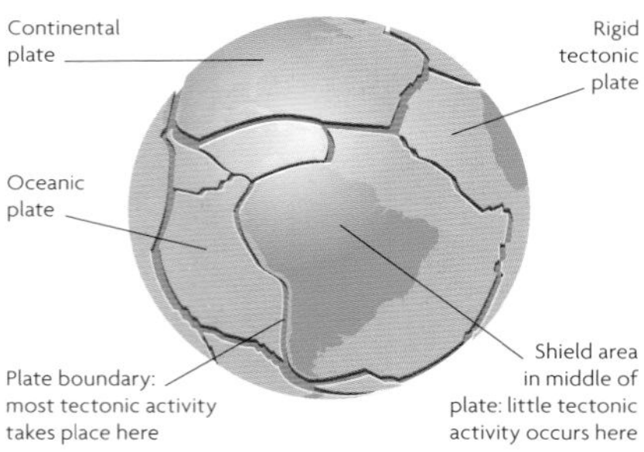
Continental plate
Rigid tectonic plate
Oceanic plate
Plate boundary: most tectonic activity takes place here
Shield area in middle of plate: little tectonic activity occurs here

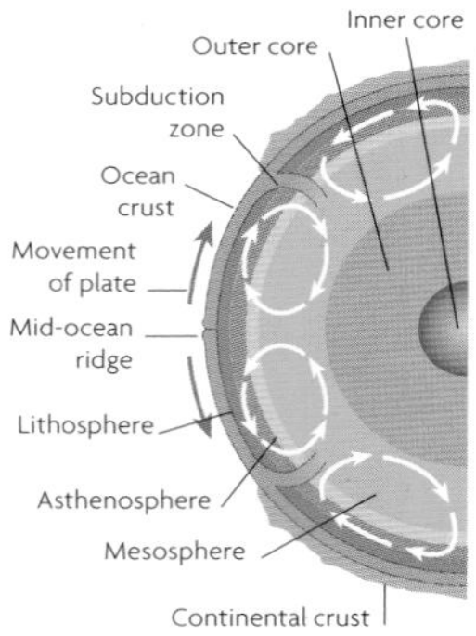
Inner core
Outer core
Subduction zone
Ocean crust
Movement of plate
Mid-ocean ridge
Lithosphere
Asthenosphere
Mesosphere
Continental crust

Convection currents

Deep within the Earth, at its inner core, temperatures may exceed 8100°F (4500°C). This heat warms rocks in the mesosphere which rise through the partially molten mantle, displacing cooler rocks just below the solid crust, which sink, and are warmed again by the heat of the mantle. This process is continually repeated, creating convection currents which form the moving force beneath the Earth's crust.

Plate boundaries

The boundaries between the plates are the areas where most tectonic activity takes place. Three types of movement occur at plate boundaries: the plates can either move towards each other, move apart, or slide past each other. The effect this has on the Earth's structure depends on whether the margin is between two continental plates, two oceanic plates or an oceanic and continental plate.

▲ ***The Mid-Atlantic Ridge** rises above sea level in Iceland, producing geysers and volcanoes.*

Mid-ocean ridges

—— Mid-ocean ridges are formed when two adjacent oceanic plates pull apart, allowing magma to force its way up to the surface, which then cools to form solid rock. Vast amounts of volcanic material are discharged at these mid-ocean ridges which can reach heights of 10,000 ft (3000 m).

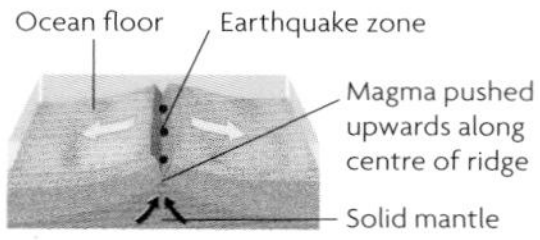
Ocean floor
Earthquake zone
Magma pushed upwards along centre of ridge
Solid mantle

Formation of a mid-ocean ridge

▲ ***Mount Pinatubo** is an active volcano, lying on the Pacific 'Ring of Fire'.*

Ocean plates meeting

△△ Oceanic crust is denser and thinner than continental crust; on average it is 3 miles (5 km) thick, while continental crust averages 18–24 miles (30–40 km). When oceanic plates of similar density meet, the crust is contorted as one plate overrides the other, forming deep sea trenches and volcanic island arcs above sea level.

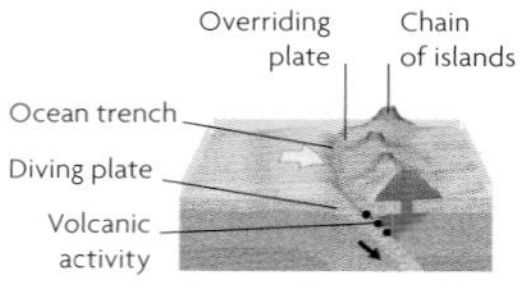
Overriding plate
Chain of islands
Ocean trench
Diving plate
Volcanic activity

Ocean plates meeting to form an island arc

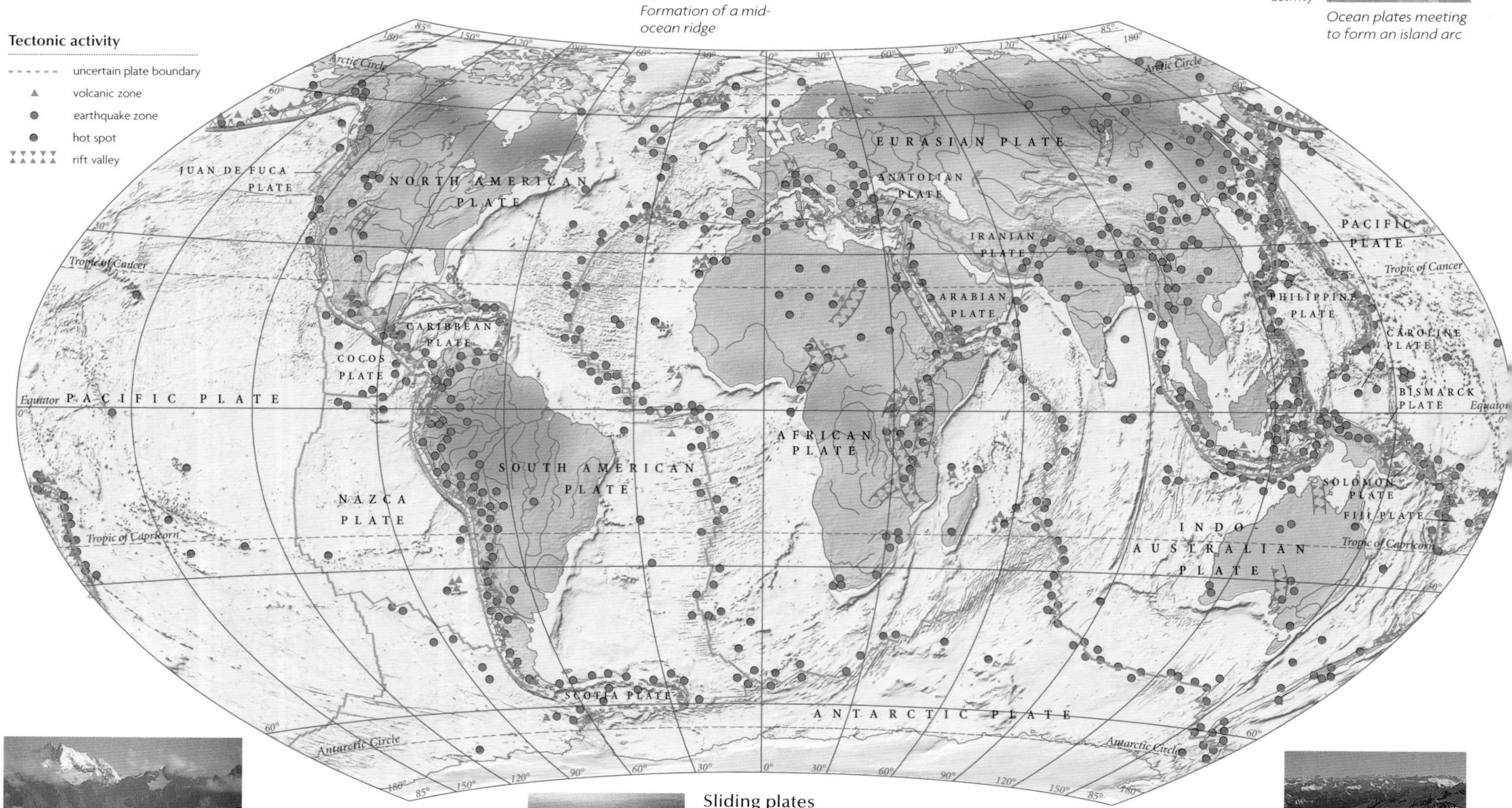
Tectonic activity
uncertain plate boundary
volcanic zone
earthquake zone
hot spot
rift valley
JUAN DE FUCA PLATE
NORTH AMERICAN PLATE
EURASIAN PLATE
ANATOLIAN PLATE
IRANIAN PLATE
ARABIAN PLATE
PACIFIC PLATE
PHILIPPINE PLATE
CAROLINE PLATE
BISMARCK PLATE
CARIBBEAN PLATE
COCOS PLATE
PACIFIC PLATE
AFRICAN PLATE
SOUTH AMERICAN PLATE
NAZCA PLATE
SOLOMON PLATE
FIJI PLATE
INDO-AUSTRALIAN PLATE
SCOTIA PLATE
ANTARCTIC PLATE
Arctic Circle
Tropic of Cancer
Equator
Tropic of Capricorn
Antarctic Circle

◀ ***The Andean mountain** chain is the typical result of the impact of a diving plate.*

Diving plates

△△ When an oceanic and a continental plate meet, the denser oceanic plate is driven underneath the continental plate, which is crumpled by the collision to form mountain ranges. As the ocean plate plunges downward, it heats up, and molten rock (magma) is forced up to the surface.

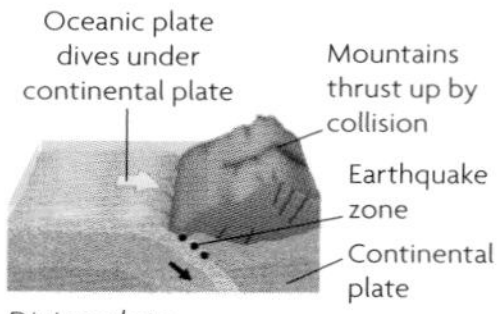
Oceanic plate dives under continental plate
Mountains thrust up by collision
Earthquake zone
Continental plate

Diving plate

▲ ***The deep fracture** caused by the sliding plates of the San Andreas Fault can be clearly seen in parts of California.*

Sliding plates

—— When two plates slide past each other, friction is caused along the fault line which divides them. The plates do not move smoothly, and the uneven movement causes earthquakes.

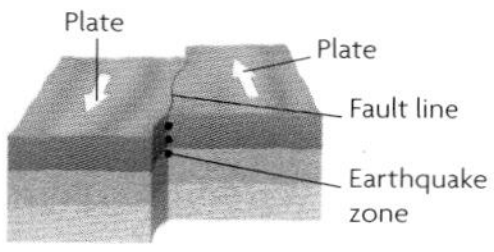
Plate
Plate
Fault line
Earthquake zone

Sliding plates

▶ ***The Alps were** formed when the African Plate collided with the Eurasian Plate, about 65 million years ago.*

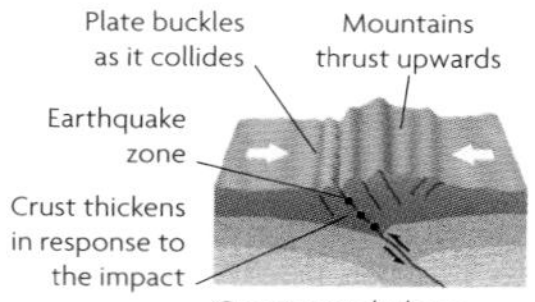
Plate buckles as it collides
Mountains thrust upwards
Earthquake zone
Crust thickens in response to the impact

Continental plates colliding to form a mountain range

Colliding plates

▲▲▲ When two continental plates collide, great mountain chains are thrust upwards as the crust buckles and folds under the force of the impact.

Continental drift

Although the plates which make up the Earth's crust move only a few centimetres in a year, over the millions of years of the Earth's history, its continents have moved many thousands of kilometres, to create new continents, oceans and mountain chains.

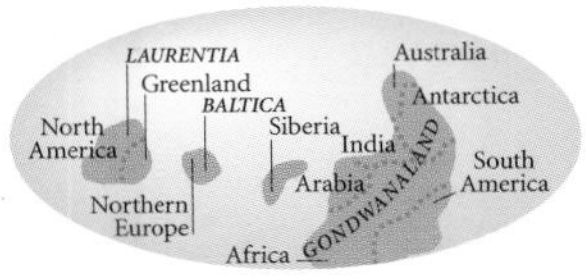

1: Cambrian period

570–510 million years ago. Most continents are in tropical latitudes. The supercontinent of Gondwanaland reaches the South Pole.

2: Devonian period

408–362 million years ago. The continents of Gondwanaland and Laurentia are drifting northwards.

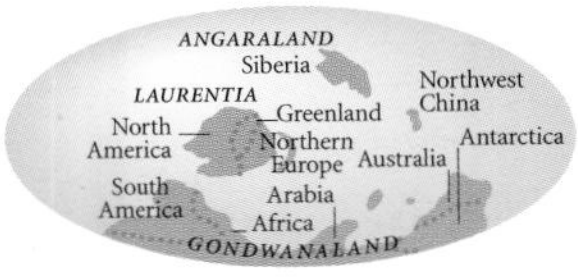

3: Carboniferous period

362–290 million years ago. The Earth is dominated by three continents; Laurentia, Angaraland and Gondwanaland.

4: Triassic period

245–208 million years ago. All three major continents have joined to form the super-continent of Pangea.

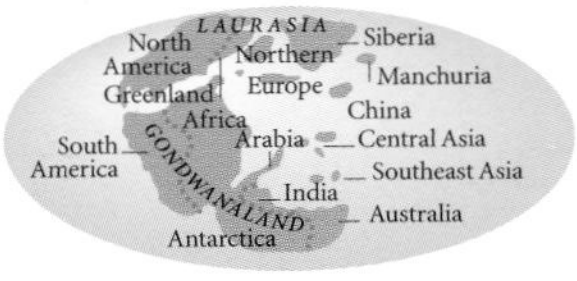

5: Jurassic period

208–145 million years ago. The super-continent of Pangea begins to break up, causing an overall rise in sea levels.

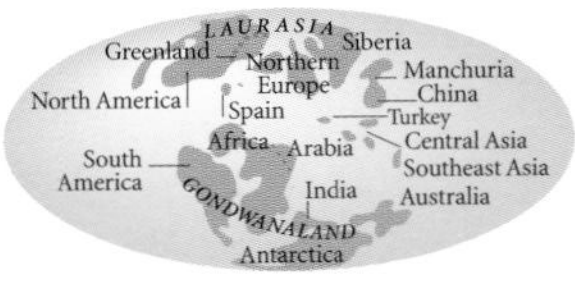

6: Cretaceous period

145–65 million years ago. Warm shallow seas cover much of the land: sea levels are about 80 ft (25 m) above present levels.

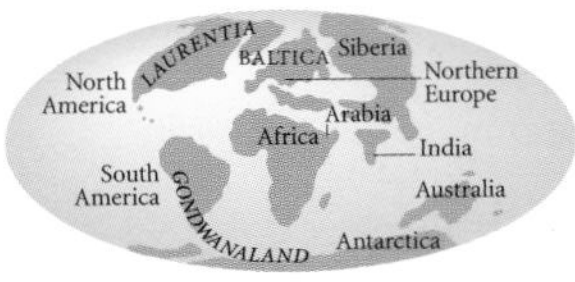

7: Tertiary period

65–2 million years ago. Although the world's geography is becoming more recognizable, major events such as the creation of the Himalayan mountain chain, are still to occur during this period.

Continental shields

The centres of the Earth's continents, known as shields, were established between 2500 and 500 million years ago; some contain rocks over three billion years old. They were formed by a series of turbulent events: plate movements, earthquakes and volcanic eruptions. Since the Pre-Cambrian period, over 570 million years ago, they have experienced little tectonic activity, and today, these flat, low-lying slabs of solidified molten rock form the stable centres of the continents. They are bounded or covered by successive belts of younger sedimentary rock.

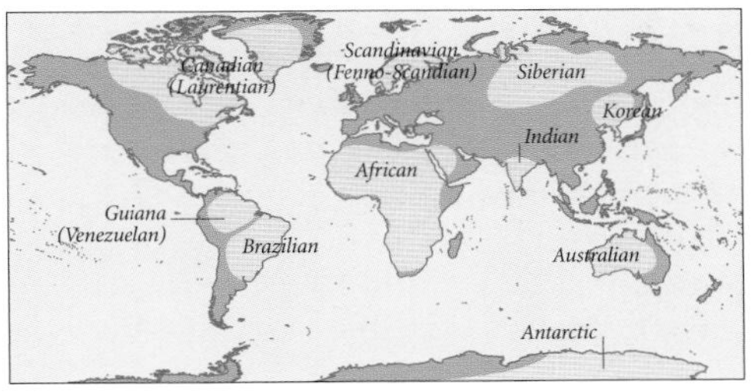

The Hawai'ian island chain

A hot spot lying deep beneath the Pacific Ocean pushes a plume of magma from the Earth's mantle up through the Pacific Plate to form volcanic islands. While the hot spot remains stationary, the plate on which the islands sit is moving slowly. A long chain of islands has been created as the plate passes over the hot spot.

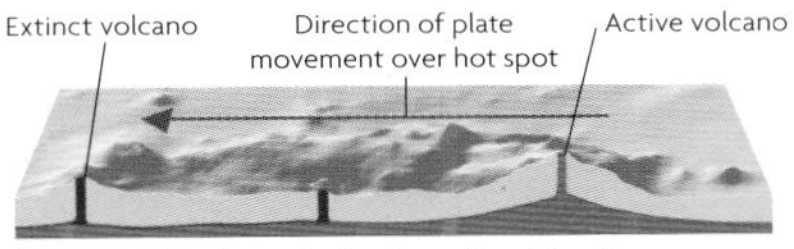

Cross-section through the Hawai'ian Islands

Evolution of the Hawai'ian Islands

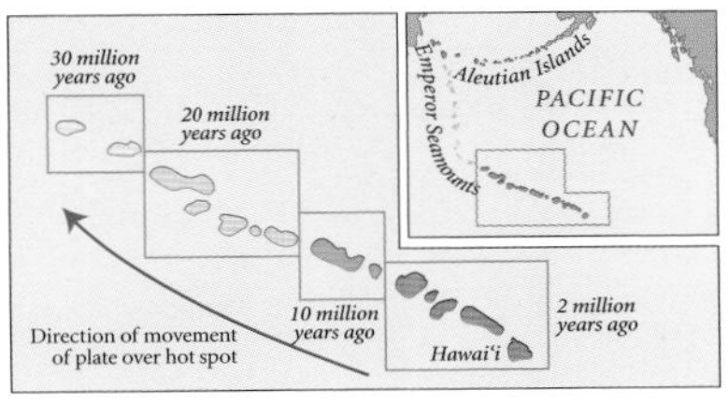

Creation of the Himalayas

Between 10 and 20 million years ago, the Indian subcontinent, part of the ancient continent of Gondwanaland, collided with the continent of Asia. The Indo-Australian Plate continued to move northwards, displacing continental crust and uplifting the Himalayas, the world's highest mountain chain.

Movements of India

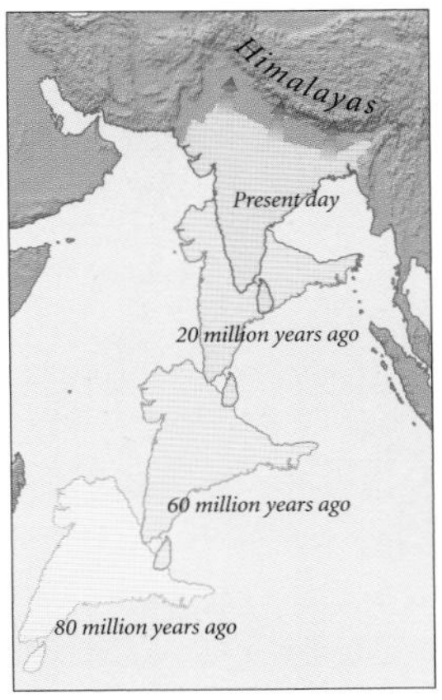

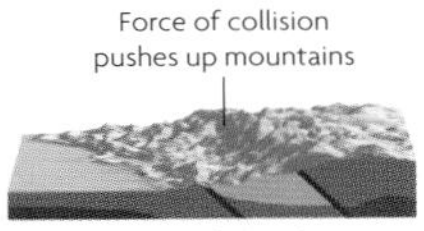

Cross-section through the Himalayas

▲ ***The Himalayas were*** *uplifted when the Indian subcontinent collided with Asia.*

The Earth's geology

The Earth's rocks are created in a continual cycle. Exposed rocks are weathered and eroded by wind, water and chemicals and deposited as sediments. If they pass into the Earth's crust they will be transformed by high temperatures and pressures into metamorphic rocks or they will melt and solidify as igneous rocks.

▲ ***Gneiss formations*** *in Norway's Jotunheimen Mountains.*

Gneiss

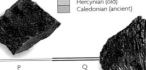

1 Gneiss is a metamorphic rock made at great depth during the formation of mountain chains, when intense heat and pressure transform sedimentary or igneous rocks.

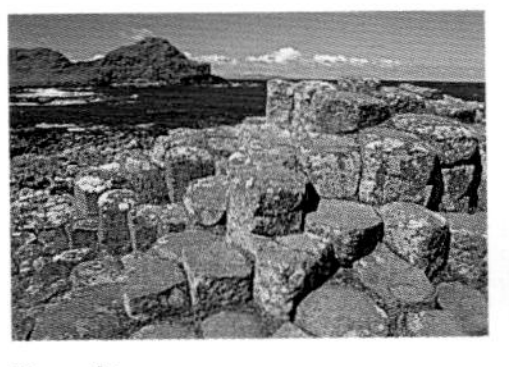

◀ ***Basalt columns*** *at Giant's Causeway, Northern Ireland, UK.*

Basalt

2 Basalt is an igneous rock, formed when small quantities of magma lying close to the Earth's surface cool rapidly.

Limestone

3 Limestone is a sedimentary rock, which is formed mainly from the calcite skeletons of marine animals which have been compressed into rock.

▲ ***Limestone hills, Guilin,*** *China.*

Coral

4 Coral reefs are formed from the skeletons of millions of individual corals.

▲ ***Great Barrier Reef,*** *Australia.*

Sandstone

8 Sandstones are sedimentary rocks formed mainly in deserts, beaches and deltas. Desert sandstones are formed of grains of quartz which have been well rounded by wind erosion.

▲ ***Rock stacks*** *of desert sandstone, at Bryce Canyon National Park, Utah, USA.*

◀ ***Extrusive igneous rocks*** *are formed during volcanic eruptions, as here in Hawai'i.*

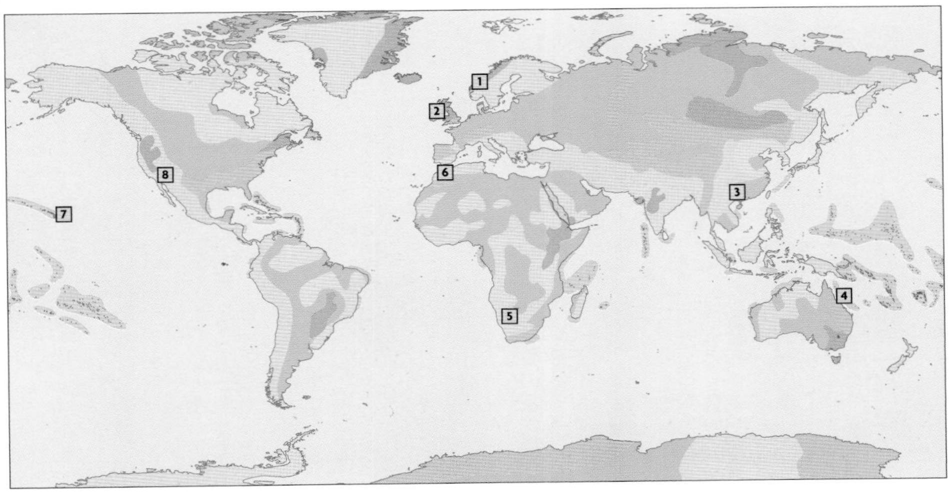

Geological regions

- continental shield
- sedimentary cover
- coral formation
- igneous rock types

Mountain ranges

- Alpine (new)
- Hercynian (old)
- Caledonian (ancient)

Andesite

7 Andesite is an extrusive igneous rock formed from magma which has solidified on the Earth's crust after a volcanic eruption.

Schist

6 Schist is a metamorphic rock formed during mountain building, when temperature and pressure are comparatively high. Both mudstones and shales reform into schist under these conditions.

▶ ***Schist formations*** *in the Atlas Mountains, northwestern Africa.*

Granite

5 Granite is an intrusive igneous rock formed from magma which has solidified deep within the Earth's crust. The magma cools slowly, producing a coarse-grained rock.

▶ ***Namibia's Namaqualand Plateau*** *is formed of granite.*

Shaping the landscape

The basic material of the Earth's surface is solid rock: valleys, deserts, soil, and sand are all evidence of the powerful agents of weathering, erosion, and deposition which constantly shape and transform the Earth's landscapes. Water, either flowing continually in rivers or seas, or frozen and compacted into solid sheets of ice, has the most clearly visible impact on the Earth's surface. But wind can transport fragments of rock over huge distances and strip away protective layers of vegetation, exposing rock surfaces to the impact of extreme heat and cold.

Coastal water

The world's coastlines are constantly changing; every day, tides deposit, sift and sort sand, and gravel on the shoreline. Over longer periods, powerful wave action erodes cliffs and headlands and carves out bays.

► ***A low, wide** sandy beach on South Africa's Cape Peninsula is continually re-shaped by the action of the Atlantic waves.*

▲ ***The sheer chalk** cliffs at Seven Sisters in southern England are constantly under attack from waves.*

Water

Less than 2% of the world's water is on the land, but it is the most powerful agent of landscape change. Water, as rainfall, groundwater and rivers, can transform landscapes through both erosion and deposition. Eroded material carried by rivers forms the world's most fertile soils.

▲ ***Waterfalls such as** the Iguaçu Falls on the border between Argentina and southern Brazil, erode the underlying rock, causing the falls to retreat.*

Groundwater

In regions where there are porous rocks such as chalk, water is stored underground in large quantities; these reservoirs of water are known as aquifers. Rain percolates through topsoil into the underlying bedrock, creating an underground store of water. The limit of the saturated zone is called the water table.

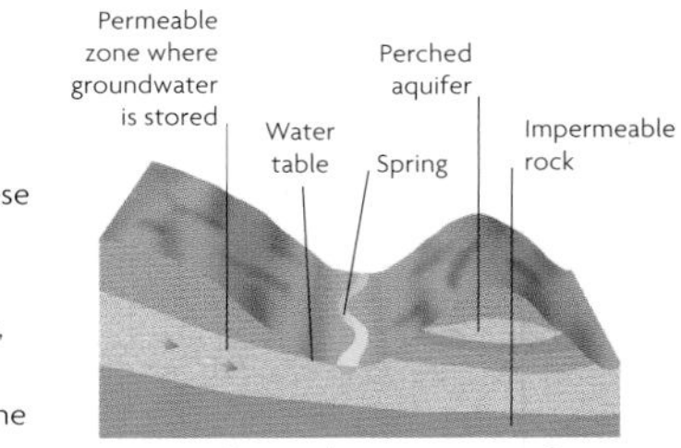
Permeable zone where groundwater is stored
Water table
Perched aquifer
Spring
Impermeable rock

Storage of groundwater in an aquifer

World river systems

drainage basin

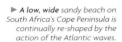

World river systems:
Sediment deposited annually per drainage basin

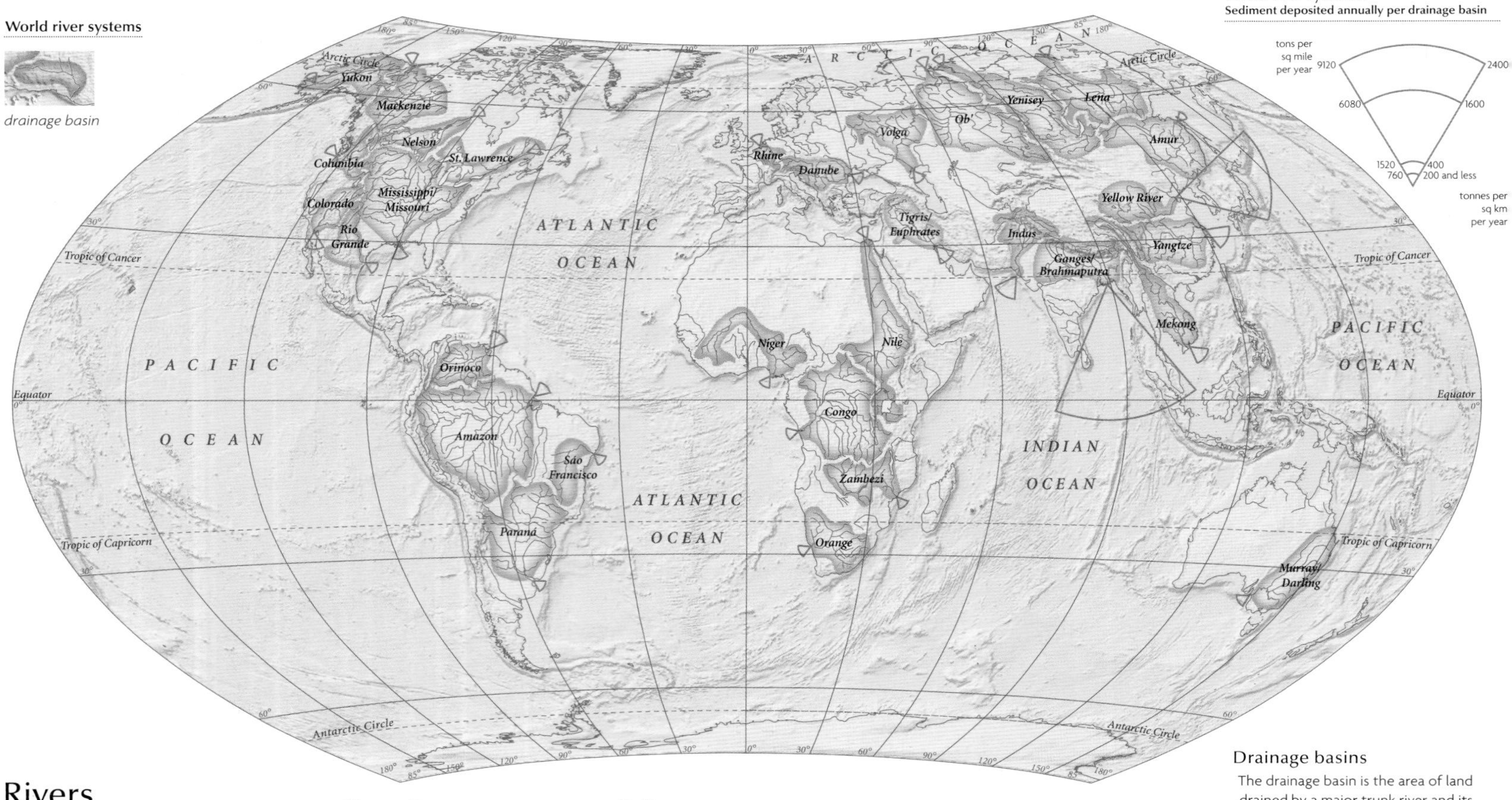
Arctic Circle
Yukon
Mackenzie
Nelson
St. Lawrence
Columbia
Mississippi/ Missouri
Colorado
Rio Grande
Tropic of Cancer
PACIFIC OCEAN
ATLANTIC OCEAN
ARCTIC OCEAN
Orinoco
Amazon
Sao Francisco
Parana
Tropic of Capricorn
Equator
Rhine
Danube
Volga
Ob'
Yenisey
Lena
Amur
Tigris/ Euphrates
Indus
Ganges/ Brahmaputra
Yellow River
Yangtze
Mekong
Niger
Nile
Congo
Zambezi
Orange
INDIAN OCEAN
Murray/ Darling
Antarctic Circle
tons per sq mile per year
9120
6080
1520
760
2400
1600
400
200 and less
tonnes per sq km per year

Rivers

Rivers erode the land by grinding and dissolving rocks and stones. Most erosion occurs in the river's upper course as it flows through highland areas. Rock fragments are moved along the river bed by fast-flowing water and deposited in areas where the river slows down, such as flat plains, or where the river enters seas or lakes.

River valleys

Over long periods of time rivers erode uplands to form characteristic V-shaped valleys with smooth sides.

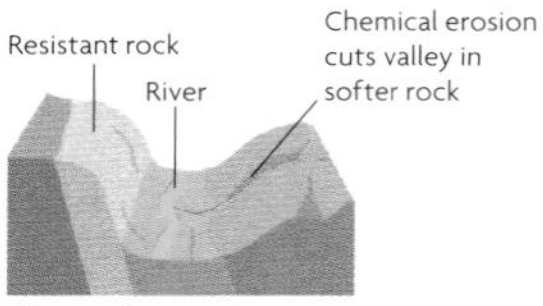
Resistant rock
River
Chemical erosion cuts valley in softer rock

River valley erosion

Deltas

When a river deposits its load of silt and sediment (alluvium) on entering the sea, it may form a delta. As this material accumulates, it chokes the mouth of the river, forcing it to create new channels to reach the sea.

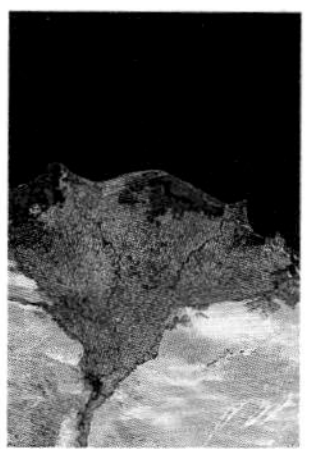

► ***The Nile forms** a broad delta as it flows into the Mediterranean.*

Drainage basins

The drainage basin is the area of land drained by a major trunk river and its smaller branch rivers or tributaries. Drainage basins are separated from one another by natural boundaries known as watersheds.

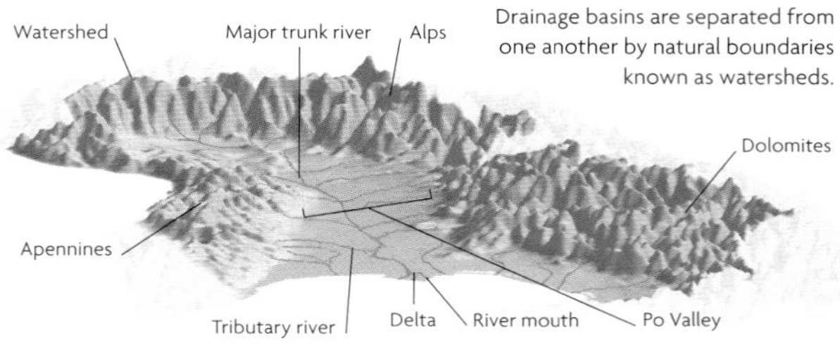
Watershed
Major trunk river
Alps
Dolomites
Apennines
Tributary river
Delta
River mouth
Po Valley

The drainage basin of the Po river, northern Italy.

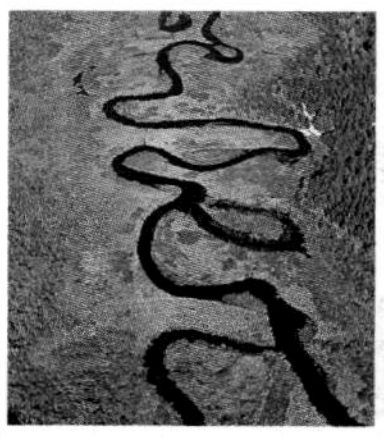

▲ ***The Mississippi River** forms meanders as it flows across the southern US.*

Meanders

In their lower courses, rivers flow slowly. As they flow across the lowlands, they form looping bends called meanders.

▲ ***The meanders of** Utah's San Juan River have become deeply incised.*

◄ ***Mud is deposited** by China's Yellow River in its lower course.*

Deposition

When rivers have deposited large quantities of fertile alluvium, they are forced to find new channels through the alluvium deposits, creating braided river systems.

Landslides

Heavy rain and associated flooding on slopes can loosen underlying rocks, which crumble, causing the top layers of rock and soil to slip.

► ***A huge landslide** in the Swiss Alps has left massive piles of rocks and pebbles called scree.*

▲ ***A deep gully** in the French Alps caused by the scouring of upper layers of turf.*

Gullies

In areas where soil is thin, rainwater is not effectively absorbed, and may flow overland. The water courses downhill in channels, or gullies, and may lead to rapid erosion of soil.

Ice

During its long history, the Earth has experienced a number of glacial episodes when temperatures were considerably lower than today. During the last Ice Age, 18,000 years ago, ice covered an area three times larger than it does today. Over these periods, the ice has left a remarkable legacy of transformed landscapes.

Glaciers

Glaciers are formed by the compaction of snow into 'rivers' of ice. As they move over the landscape, glaciers pick up and carry a load of rocks and boulders which erode the landscape they pass over, and are eventually deposited at the end of the glacier.

▲ ***A massive glacier** advancing down a valley in southern Argentina.*

Post-glacial features

When a glacial episode ends, the retreating ice leaves many features. These include depositional ridges called moraines, which may be eroded into low hills known as drumlins; sinuous ridges called eskers; kames, which are rounded hummocks; depressions known as kettle holes; and windblown loess deposits.

Glacial valleys

Glaciers can erode much more powerfully than rivers. They form steep-sided, flat-bottomed valleys with a typical U-shaped profile. Valleys created by tributary glaciers, whose floors have not been eroded to the same depth as the main glacial valley floor, are called hanging valleys

▲ ***The U-shaped profile** and piles of morainic debris are characteristic of a valley once filled by a glacier.*

▲ ***A series of** hanging valleys high up in the Chilean Andes.*

▲ ***The profile of** the Matterhorn has been formed by three cirques lying 'back-to-back'.*

Past and present world ice-cover and glacial features

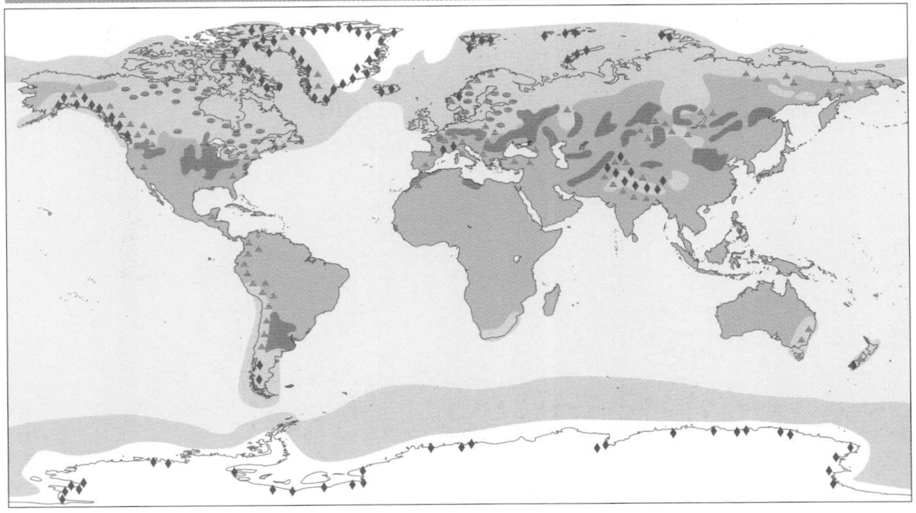

Past and present world ice cover and glacial features

- extent of last Ice Age
- loess deposits
- post-glacial feature
- glacial feature
- present day ice cover
- glacial field

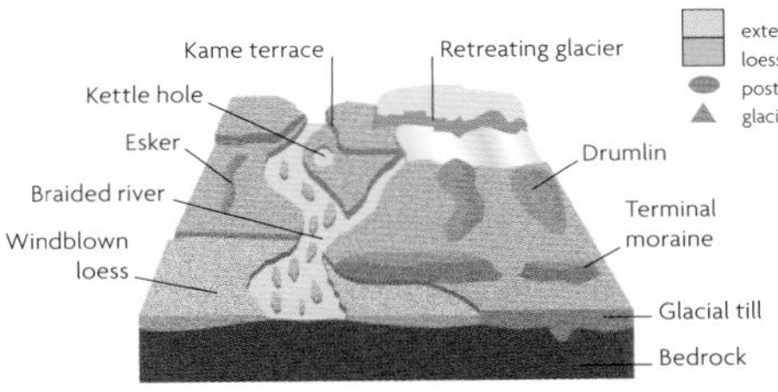

Kame terrace
Retreating glacier
Kettle hole
Esker
Drumlin
Braided river
Terminal moraine
Windblown loess
Glacial till
Bedrock

Post-glacial landscape features

Ice shattering

Water drips into fissures in rocks and freezes, expanding as it does so. The pressure weakens the rock, causing it to crack, and eventually to shatter into polygonal patterns.

▲ ***Irregular polygons show** through the sedge-grass tundra in the Yukon, Canada.*

Cirques

Cirques are basin-shaped hollows which mark the head of a glaciated valley. Where neighboring cirques meet, they are divided by sharp rock ridges called arêtes. It is these arêtes which give the Matterhorn its characteristic profile.

Fjords

Fjords are ancient glacial valleys flooded by the sea following the end of a period of glaciation. Beneath the water, the valley floor can be 4000 ft (1300 m) deep.

▲ ***A fjord fills** a former glacial valley in southern New Zealand.*

Periglaciation

Periglacial areas occur near to the edge of ice sheets. A layer of frozen ground lying just beneath the surface of the land is known as permafrost. When the surface melts in the summer, the water is unable to drain into the frozen ground, and so 'creeps' downhill, a process known as solifluction.

Wind

Strong winds can transport rock fragments great distances, especially where there is little vegetation to protect the rock. In desert areas, wind picks up loose, unprotected sand particles, carrying them over great distances. This powerfully abrasive debris is blasted at the surface by the wind, eroding the landscape into dramatic shapes.

Deposition

The rocky, stony floors of the world's deserts are swept and scoured by strong winds. The smaller, finer particles of sand are shaped into surface ripples, dunes, or sand mountains, which rise to a height of 650 ft (200 m). Dunes usually form single lines, running perpendicular to the direction of the prevailing wind. These long, straight ridges can extend for over 100 miles (160 km).

Dunes

Dunes are shaped by wind direction and sand supply. Where sand supply is limited, crescent-shaped barchan dunes are formed.

Prevailing winds and dust trajectories

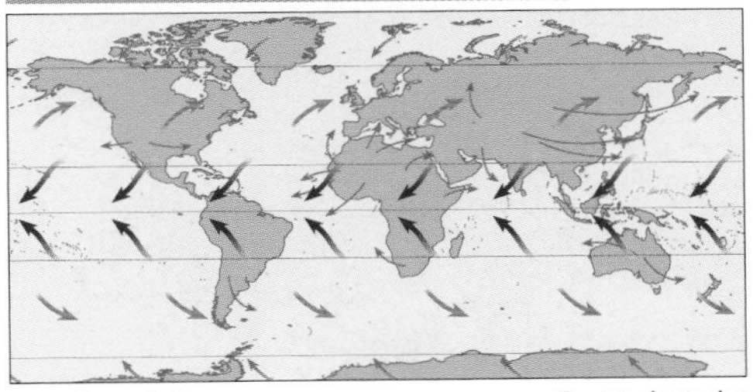

Prevailing winds: northeast trade / southeast trade; westerly / westerly; polar easterly / polar easterly

Dust trajectories: trajectory of aeolian dust

▲ ***Barchan dunes in** the Arabian Desert.*

▲ ***Complex dune system** in the Sahara.*

Types of dune

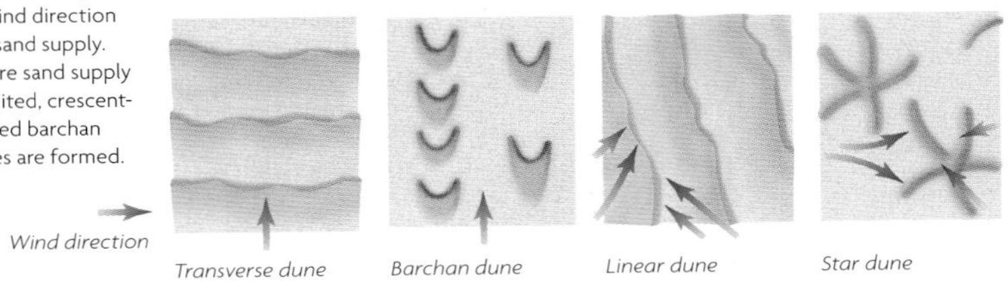

Wind direction
Transverse dune
Barchan dune
Linear dune
Star dune

Hot and cold deserts

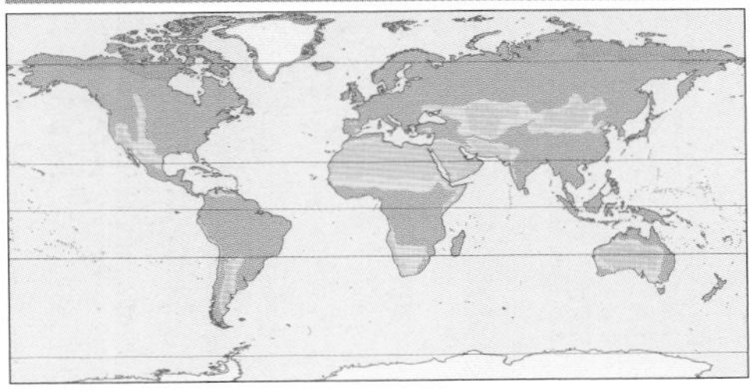

Main desert types: hot arid; semi-arid; cold polar

Heat

Fierce sun can heat the surface of rock, causing it to expand more rapidly than the cooler, underlying layers. This creates tensions which force the rock to crack or break up. In arid regions, the evaporation of water from rock surfaces dissolves certain minerals within the water, causing salt crystals to form in small openings in the rock. The hard crystals force the openings to widen into cracks and fissures.

▲ ***The cracked and** parched floor of Death Valley, California. This is one of the hottest deserts on Earth.*

Temperature

Most of the world's deserts are in the tropics. The cold deserts which occur elsewhere are arid because they are a long way from the rain-giving sea. Rock in deserts is exposed because of lack of vegetation and is susceptible to changes in temperature; extremes of heat and cold can cause both cracks and fissures to appear in the rock.

Desert abrasion

Abrasion creates a wide range of desert landforms from faceted pebbles and wind ripples in the sand, to large-scale features such as yardangs (low, streamlined ridges), and scoured desert pavements.

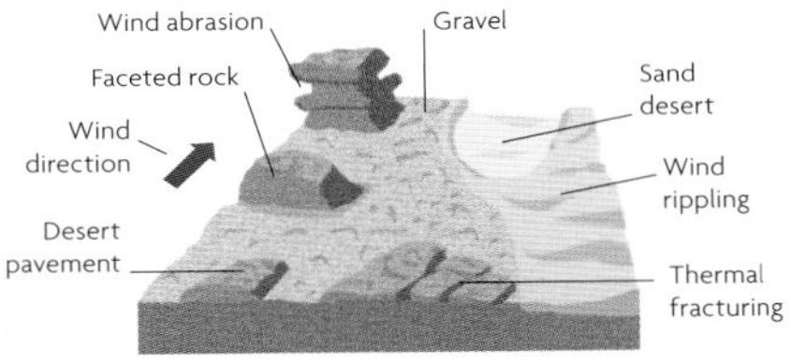

Wind abrasion
Gravel
Faceted rock
Sand desert
Wind direction
Wind rippling
Desert pavement
Thermal fracturing

Features of a desert surface

◀ ***This dry valley** at Ellesmere Island in the Canadian Arctic is an example of a cold desert. The cracked floor and scoured slopes are features also found in hot deserts.*

The world's oceans

Two-thirds of the Earth's surface is covered by the oceans. The landscape of the ocean floor, like the surface of the land, has been shaped by movements of the Earth's crust over millions of years to form volcanic mountain ranges, deep trenches, basins and plateaux. Ocean currents constantly redistribute warm and cold water around the world. A major warm current, such as El Niño in the Pacific Ocean, can increase surface temperature by up to 8°C (10°F), causing changes in weather patterns which can lead to both droughts and flooding.

The great oceans

There are five oceans on Earth: the Pacific, Atlantic, Indian and Southern oceans, and the much smaller Arctic Ocean. These five ocean basins are relatively young, having evolved within the last 80 million years. One of the most recent plate collisions, between the Eurasian and African plates, created the present-day arrangement of continents and oceans.

▲ *The Indian Ocean accounts for approximately 20% of the total area of the world's oceans.*

Sea level

If the influence of tides, winds, currents and variations in gravity were ignored, the surface of the Earth's oceans would closely follow the topography of the ocean floor, with an underwater ridge 3000 ft (915 m) high producing a rise of up to 3 ft (1 m) in the level of the surface water.

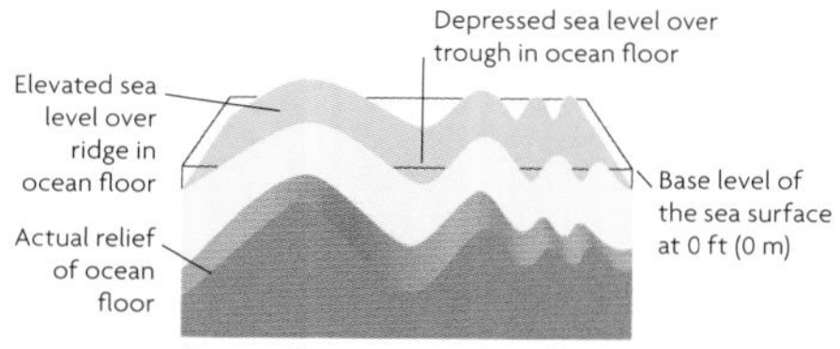
Depressed sea level over trough in ocean floor
Elevated sea level over ridge in ocean floor
Base level of the sea surface at 0 ft (0 m)
Actual relief of ocean floor

How surface waters reflect the relief of the ocean floor

▲ *The low relief of many small Pacific islands such as these atolls at Huahine in French Polynesia makes them vulnerable to changes in sea level.*

Ocean structure

The continental shelf is a shallow, flat sea-bed surrounding the Earth's continents. It extends to the continental slope, which falls to the ocean floor. Here, the flat abyssal plains are interrupted by vast, underwater mountain ranges, the mid-ocean ridges, and ocean trenches which plunge to depths of 35,828 ft (10,920 m).

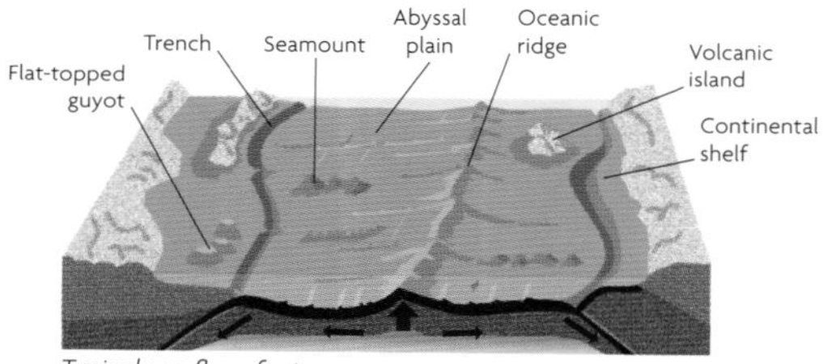
Flat-topped guyot
Trench
Seamount
Abyssal plain
Oceanic ridge
Volcanic island
Continental shelf

Typical sea-floor features

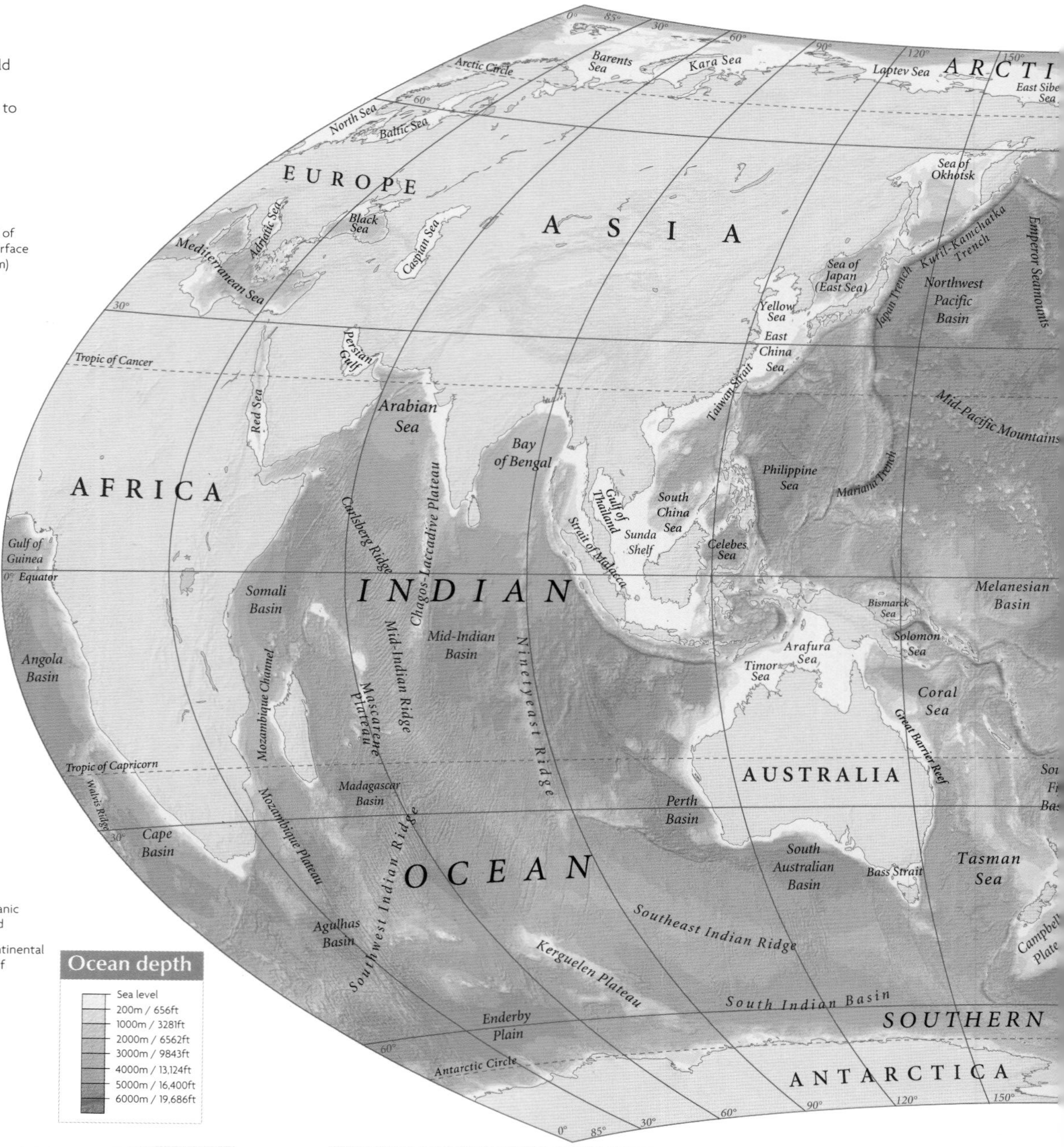
ARCTIC
Arctic Circle
Barents Sea
Kara Sea
Laptev Sea
East Siberian Sea
North Sea
Baltic Sea
EUROPE
ASIA
Sea of Okhotsk
Black Sea
Caspian Sea
Adriatic Sea
Mediterranean Sea
Kuril-Kamchatka Trench
Emperor Seamounts
Sea of Japan (East Sea)
Japan Trench
Northwest Pacific Basin
Yellow Sea
East China Sea
Persian Gulf
Tropic of Cancer
Taiwan Strait
Mid-Pacific Mountains
Red Sea
Arabian Sea
Bay of Bengal
AFRICA
Philippine Sea
Mariana Trench
Gulf of Thailand
South China Sea
Sunda Shelf
Strait of Malacca
Carlsberg Ridge
Chagos-Laccadive Plateau
Gulf of Guinea
Equator
Celebes Sea
Somali Basin
INDIAN
Melanesian Basin
Bismarck Sea
Solomon Sea
Mid-Indian Basin
Ninetyeast Ridge
Mid-Indian Ridge
Angola Basin
Mozambique Channel
Mascarene Plateau
Arafura Sea
Timor Sea
Coral Sea
Great Barrier Reef
Tropic of Capricorn
Walvis Ridge
AUSTRALIA
Madagascar Basin
Perth Basin
Cape Basin
Mozambique Plateau
Southwest Indian Ridge
OCEAN
South Australian Basin
Bass Strait
Tasman Sea
Agulhas Basin
Southeast Indian Ridge
Kerguelen Plateau
Campbell Plateau
South Indian Basin
Enderby Plain
SOUTHERN
Antarctic Circle
ANTARCTICA
Ocean depth
Sea level
200m / 656ft
1000m / 3281ft
2000m / 6562ft
3000m / 9843ft
4000m / 13,124ft
5000m / 16,400ft
6000m / 19,686ft

▲ *A black smoker in the Atlantic Ocean.*

Black smokers

These vents in the ocean floor disgorge hot, sulphur-rich water from deep in the Earth's crust. Despite the great depths, a variety of lifeforms have adapted to the chemical-rich environment which surrounds black smokers.

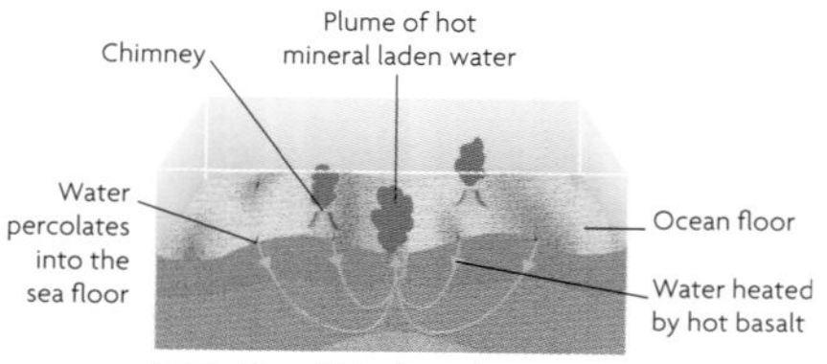
Chimney
Plume of hot mineral laden water
Water percolates into the sea floor
Ocean floor
Water heated by hot basalt

Formation of black smokers

▲ *Surtsey, near Iceland, is a volcanic island lying directly over the Mid-Atlantic Ridge. It was formed in the 1960s following intense volcanic activity nearby.*

Ocean floors

Mid-ocean ridges are formed by lava which erupts beneath the sea and cools to form solid rock. This process mirrors the creation of volcanoes from cooled lava on the land. The ages of sea floor rocks increase in parallel bands outwards from central ocean ridges.

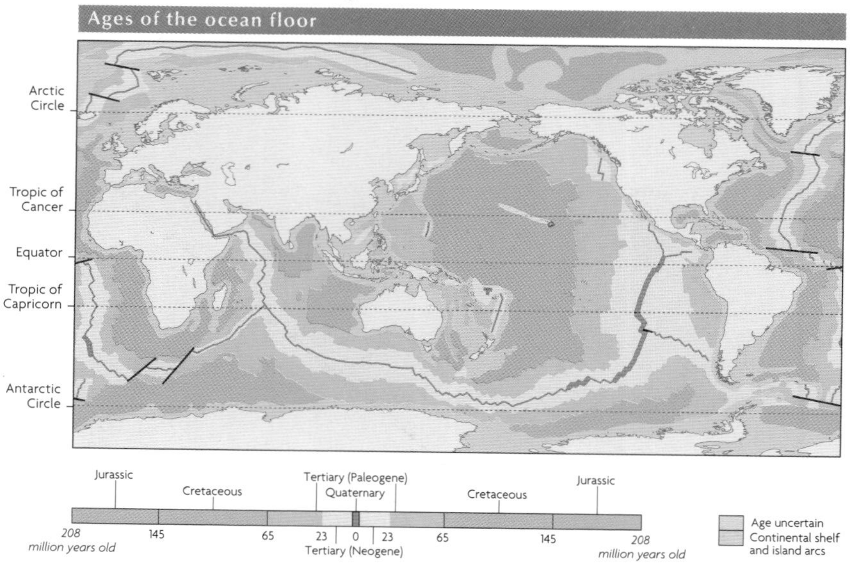
Ages of the ocean floor
Arctic Circle
Tropic of Cancer
Equator
Tropic of Capricorn
Antarctic Circle
Jurassic
Cretaceous
Tertiary (Paleogene)
Quaternary
Cretaceous
Jurassic
208 million years old
145
65
23
0
23
65
145
208 million years old
Tertiary (Neogene)
Age uncertain
Continental shelf and island arcs

▲ ***Currents in the*** *Southern Ocean are driven by some of the world's fiercest winds, including the Roaring Forties, Furious Fifties and Shrieking Sixties.*

▲ ***The Pacific Ocean*** *is the world's largest and deepest ocean, covering over one-third of the surface of the Earth.*

▲ ***The Atlantic Ocean*** *was formed when the landmasses of the eastern and western hemispheres began to drift apart 180 million years ago.*

Deposition of sediment

Storms, earthquakes, and volcanic activity trigger underwater currents known as turbidity currents which scour sand and gravel from the continental shelf, creating underwater canyons. These strong currents pick up material deposited at river mouths and deltas, and carry it across the continental shelf and through the underwater canyons, where it is eventually laid down on the ocean floor in the form of fans.

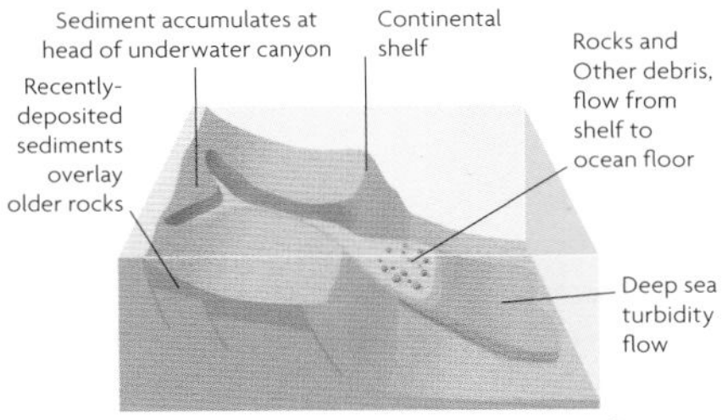
Sediment accumulates at head of underwater canyon
Continental shelf
Rocks and Other debris, flow from shelf to ocean floor
Recently-deposited sediments overlay older rocks
Deep sea turbidity flow

How sediment is deposited on the ocean floor

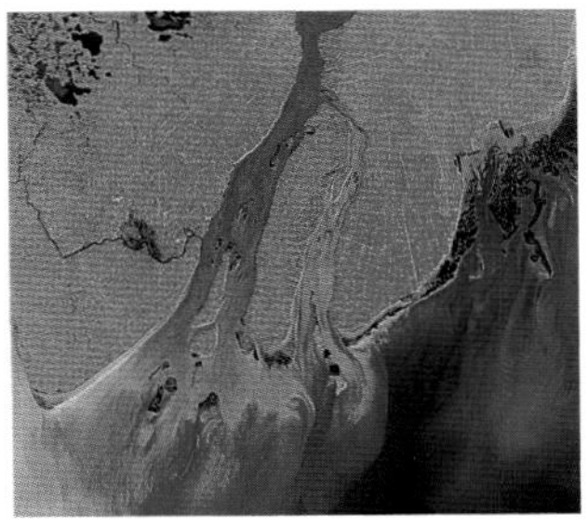

► ***Satellite image of*** *the Yangtze (Chang Jiang) Delta, in which the land appears red. The river deposits immense quantities of silt into the East China Sea, much of which will eventually reach the deep ocean floor.*

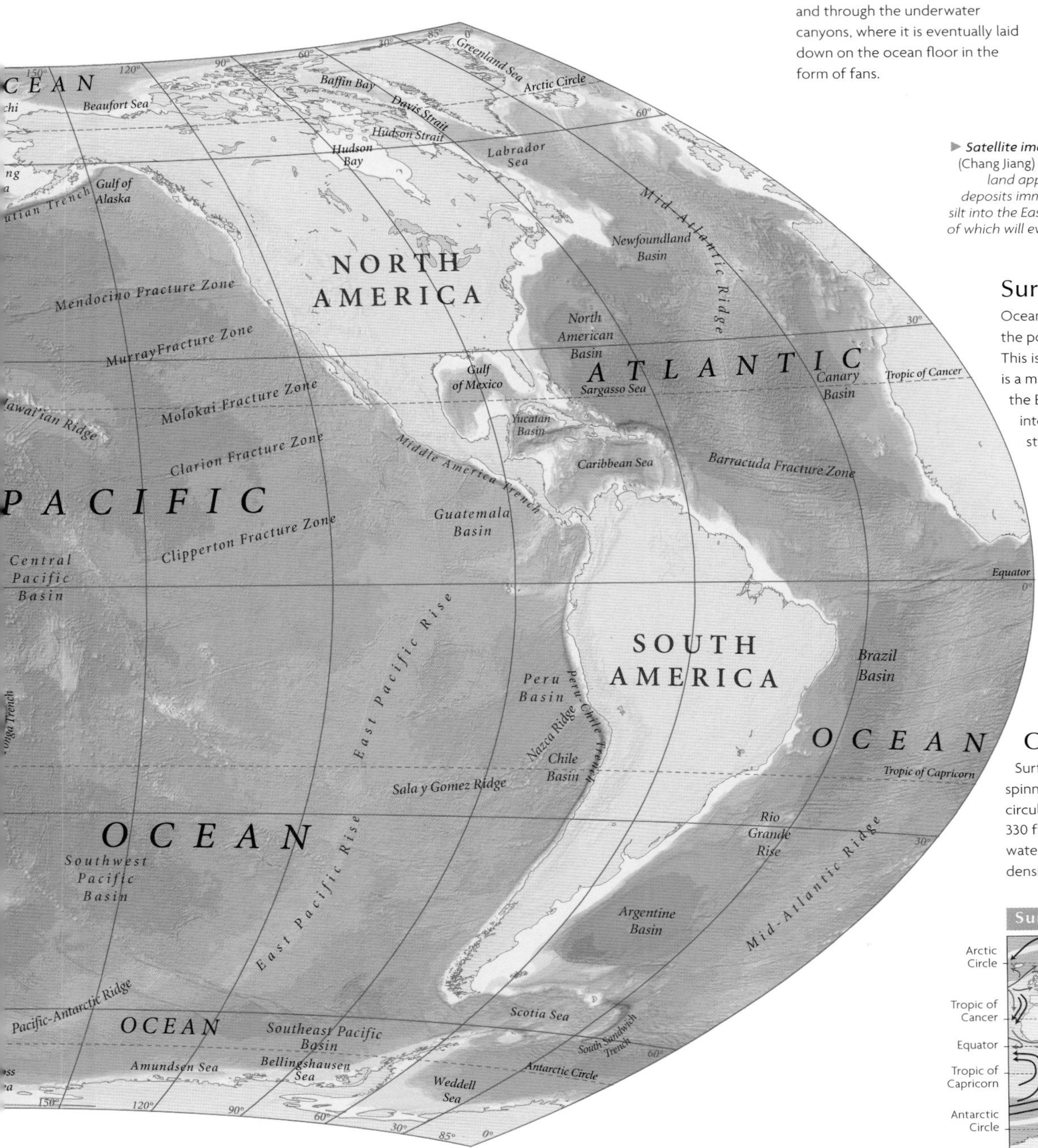
NORTH AMERICA
SOUTH AMERICA
PACIFIC OCEAN
ATLANTIC OCEAN
Beaufort Sea
Baffin Bay
Davis Strait
Hudson Strait
Hudson Bay
Greenland Sea
Arctic Circle
Labrador Sea
Gulf of Alaska
Newfoundland Basin
Mid-Atlantic Ridge
Mendocino Fracture Zone
Murray Fracture Zone
Molokai Fracture Zone
Clarion Fracture Zone
Clipperton Fracture Zone
Hawaiian Ridge
North American Basin
Gulf of Mexico
Sargasso Sea
Canary Basin
Tropic of Cancer
Yucatan Basin
Caribbean Sea
Barracuda Fracture Zone
Middle America Trench
Guatemala Basin
Central Pacific Basin
Equator
East Pacific Rise
Peru Basin
Peru-Chile Trench
Nazca Ridge
Chile Basin
Sala y Gomez Ridge
Brazil Basin
Tropic of Capricorn
Rio Grande Rise
Southwest Pacific Basin
Argentine Basin
Pacific-Antarctic Ridge
SOUTHERN OCEAN
Southeast Pacific Basin
Amundsen Sea
Bellingshausen Sea
Scotia Sea
South Sandwich Trench
Antarctic Circle
Weddell Sea

Surface water

Ocean currents move warm water away from the Equator towards the poles, while cold water is, in turn, moved towards the Equator. This is the main way in which the Earth distributes surface heat and is a major climatic control. Approximately 4000 million years ago, the Earth was dominated by oceans and there was no land to interrupt the flow of the currents, which would have flowed as straight lines, simply influenced by the Earth's rotation.

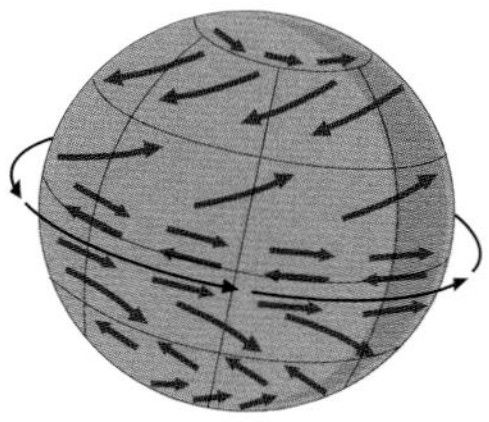

Idealized globe showing the movement of water around a landless Earth.

Ocean currents

Surface currents are driven by the prevailing winds and by the spinning motion of the Earth, which drives the currents into circulating whirlpools, or gyres. Deep sea currents, over 330 ft (100 m) below the surface, are driven by differences in water temperature and salinity, which have an impact on the density of deep water and on its movement.

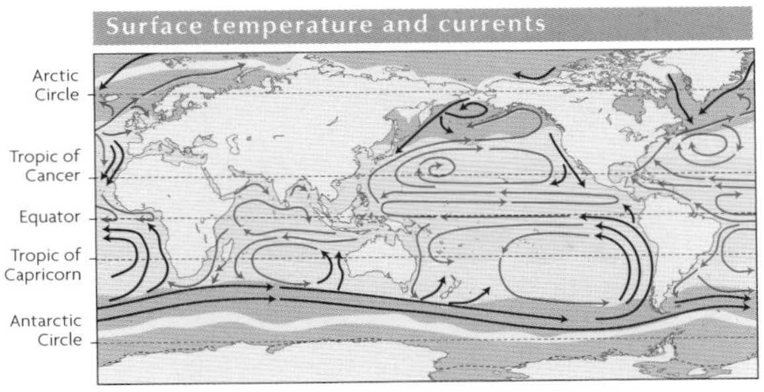
Surface temperature and currents
Arctic Circle
Tropic of Cancer
Equator
Tropic of Capricorn
Antarctic Circle

Surface temperature and currents

- Ice-shelf (below 0°C / 32°F)
- Sea-ice* (average) below -2°C / 28°F
- Sea-water -2–0°C / 28–32°F
- 0–10°C / 32–50°F
- 10–20°C / 50–68°F
- 20–30°C / 68–86°F
- warm current
- cold current

* Sea-water freezes at -1.9°C / 28.4°F

Tides and waves

Tides are created by the pull of the Sun and Moon's gravity on the surface of the oceans. The levels of high and low tides are influenced by the position of the Moon in relation to the Earth and Sun. Waves are formed by wind blowing over the surface of the water.

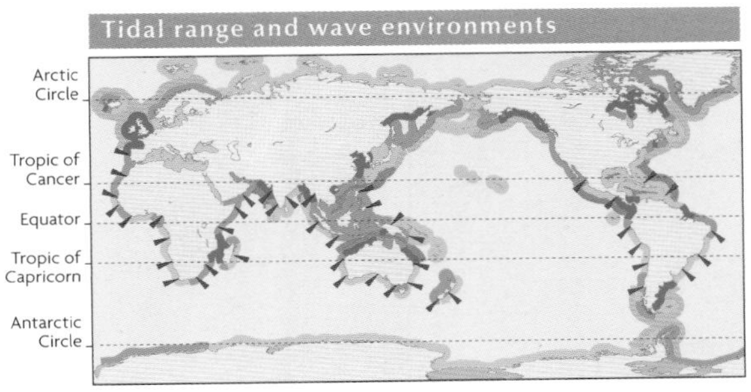
Tidal range and wave environments
Arctic Circle
Tropic of Cancer
Equator
Tropic of Capricorn
Antarctic Circle

Tidal range and wave environments

- less than 2m / 7ft
- 2–4m / 7–13ft
- greater than 4m / 13ft
- east coast swell
- west coast swell
- tropical cyclone storm wave
- ice-shelf

High and low tides

The highest tides occur when the Earth, the Moon and the Sun are aligned *(below left)*. The lowest tides are experienced when the Sun and Moon align at right angles to one another *(below right)*.

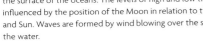

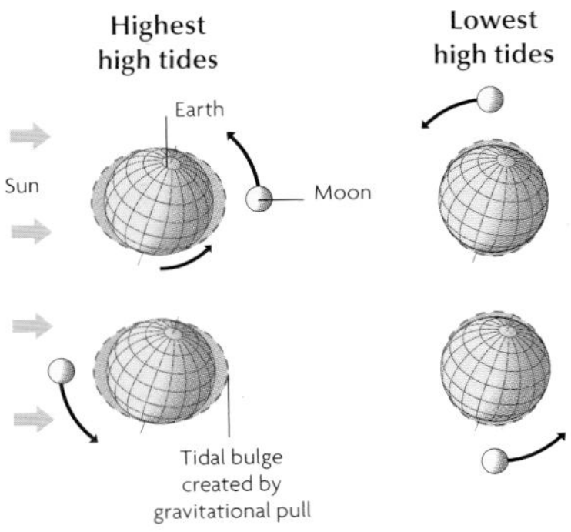
Highest high tides
Lowest high tides
Earth
Sun
Moon
Tidal bulge created by gravitational pull

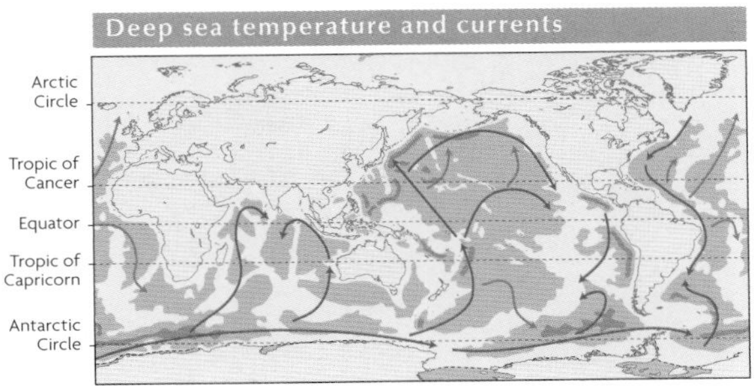
Deep sea temperature and currents
Arctic Circle
Tropic of Cancer
Equator
Tropic of Capricorn
Antarctic Circle

Deep sea temperature and currents

- Ice-shelf (below 0°C / 32°F)
- Sea-water -2–0°C / 28–32°F (below 5000m / 16,400ft)
- Sea-water 0–5°C / 32–41°F (below 4000m / 13,120ft)
- Primary currents
- Secondary currents

The global climate

The Earth's climatic types consist of stable patterns of weather conditions averaged out over a long period of time. Different climates are categorized according to particular combinations of temperature and humidity. By contrast, weather consists of short-term fluctuations in wind, temperature and humidity conditions. Different climates are determined by latitude, altitude, the prevailing wind and circulation of ocean currents. Longer-term changes in climate, such as global warming or the onset of ice ages, are punctuated by shorter-term events which comprise the day-to-day weather of a region, such as frontal depressions, hurricanes and blizzards.

Global air circulation

Air does not simply flow from the Equator to the poles, it circulates in giant cells known as Hadley and Ferrel cells. As air warms it expands, becoming less dense and rising; this creates areas of low pressure. As the air rises it cools and condenses, causing heavy rainfall over the tropics and slight snowfall over the poles. This cool air then sinks, forming high pressure belts. At surface level in the tropics these sinking currents are deflected polewards as the westerlies and towards the equator as the trade winds. At the poles they become the polar easterlies.

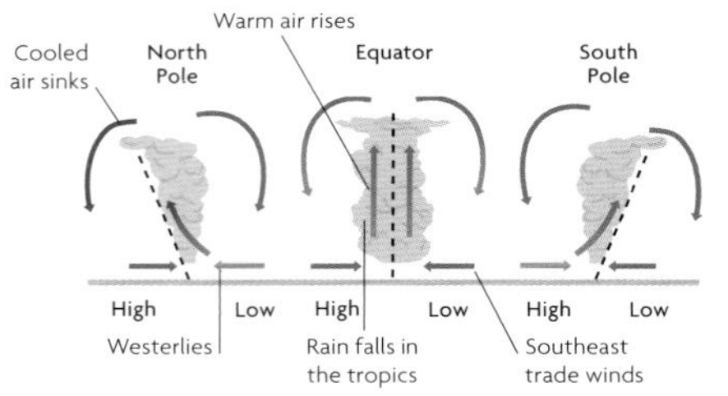
Warm air rises
Cooled air sinks
North Pole
Equator
South Pole
High
Low
High
Low
High
Low
Westerlies
Rain falls in the tropics
Southeast trade winds

▲ *The Antarctic pack ice expands its area by almost seven times during the winter as temperatures drop and surrounding seas freeze.*

The atmosphere, wind and weather

The Earth's atmosphere has been compared to a giant ocean of air which surrounds the planet. Its circulation patterns are similar to the currents in the oceans and are influenced by three factors; the Earth's orbit around the Sun and rotation about its axis, and variations in the amount of heat radiation received from the Sun. If both heat and moisture were not redistributed between the Equator and the poles, large areas of the Earth would be uninhabitable.

◀ *Heavy fogs, as here in southern England, form as moisture-laden air passes over cold ground.*

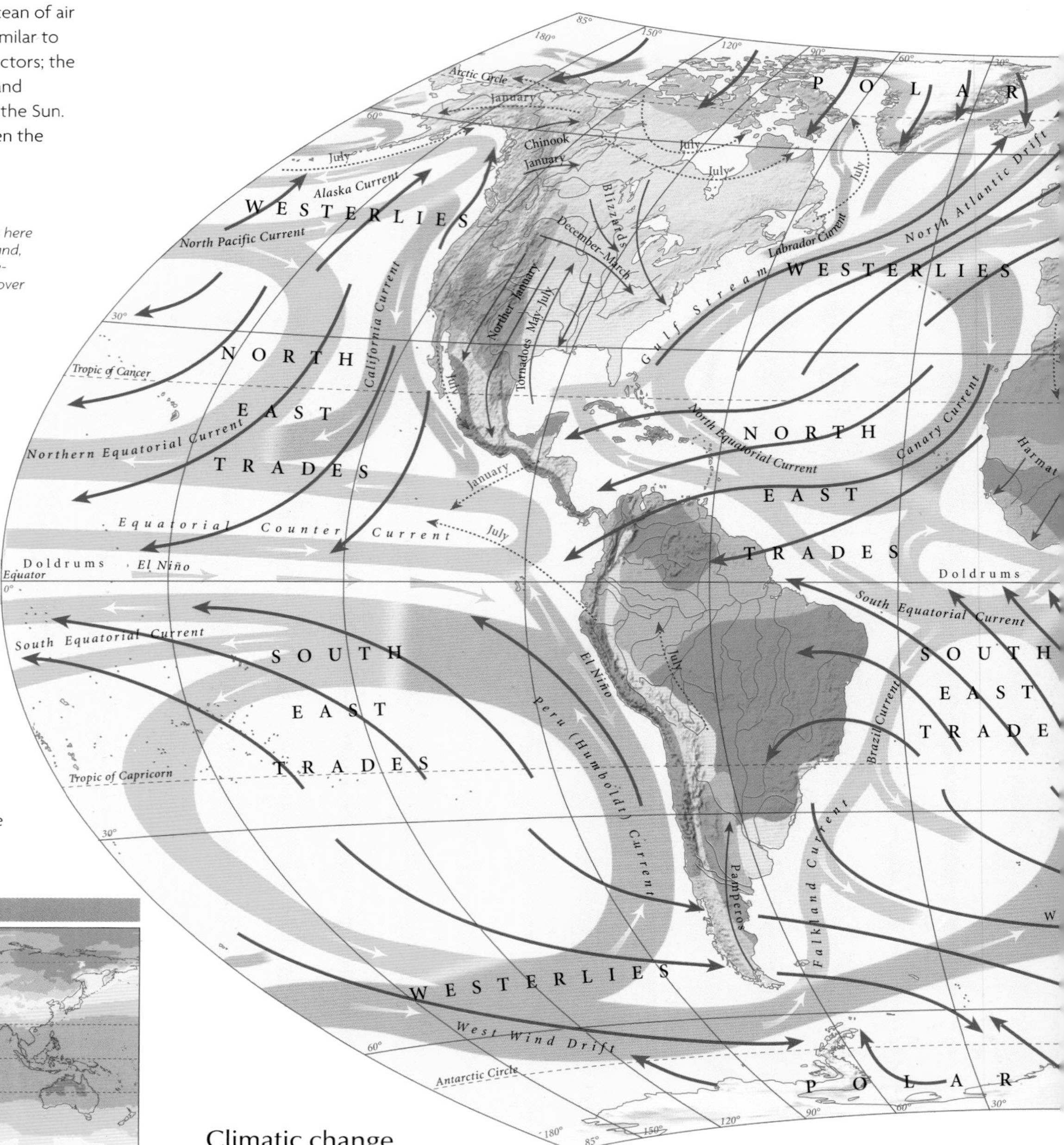
Arctic Circle
January
Chinook
January
July
July
July
July
Alaska Current
W E S T E R L I E S
North Pacific Current
Blizzards
December–March
Northern January
May–July
Tornadoes
California Current
Labrador Current
North Atlantic Drift
P O L A R
W E S T E R L I E S
Gulf Stream
N O R T H
E A S T
T R A D E S
Tropic of Cancer
Northern Equatorial Current
North Equatorial Current
N O R T H
E A S T
T R A D E S
Canary Current
January
Equatorial Counter Current
July
Doldrums
El Niño
Equator
Doldrums
South Equatorial Current
South Equatorial Current
S O U T H
E A S T
T R A D E S
El Niño
Peru (Humboldt) Current
July
S O U T H
E A S T
T R A D E
Brazil Current
Tropic of Capricorn
Falkland Current
Pampero
W E S T E R L I E S
West Wind Drift
Antarctic Circle
P O L A R
Tropic of Capricorn
85°
180°
150°
120°
90°
60°
30°
60°
30°
0°
30°
60°

Temperature

The world can be divided into three major climatic zones, stretching like large belts across the latitudes: the tropics which are warm; the cold polar regions and the temperate zones which lie between them. Temperatures across the Earth range from above 30°C (86°F) in the deserts to as low as -55°C (-70°F) at the poles. Temperature is also controlled by altitude; because air becomes cooler and less dense the higher it gets, mountainous regions are typically colder than those areas which are at, or close to, sea level.

Average January temperatures

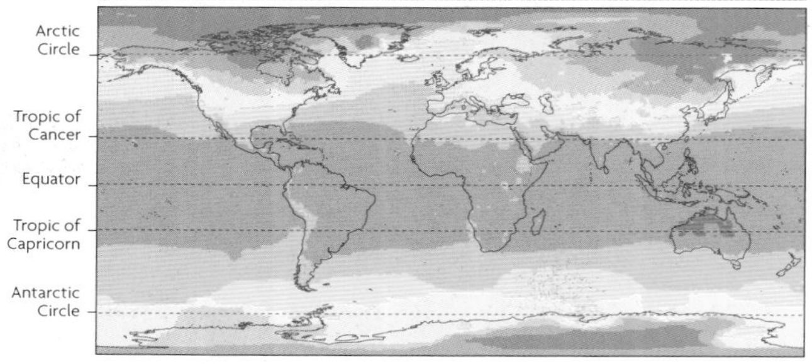
Arctic Circle
Tropic of Cancer
Equator
Tropic of Capricorn
Antarctic Circle

Average July temperatures

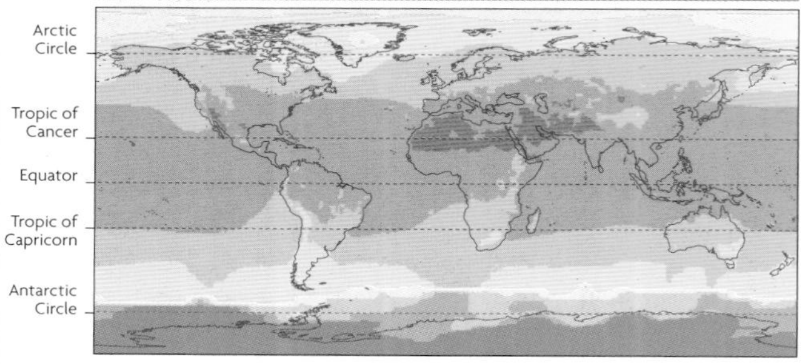
Arctic Circle
Tropic of Cancer
Equator
Tropic of Capricorn
Antarctic Circle

- below - 30°C (-22°F)
- -30 to - 20°C (-22 to -4°F)
- -20 to - 10°C (-4 to 14°F)
- -10 to 0°C (14 to 32°F)
- 0 to 10°C (32 to 50°F)
- 10 to 20°C (50 to 68°F)
- 20 to 30°C (68 to 86°F)
- above 30°C (86°F)

Climatic change

The Earth is currently in a warm phase between ice ages. Warmer temperatures result in higher sea levels as more of the polar ice caps melt. Most of the world's population lives near coasts, so any changes which might cause sea levels to rise, could have a potentially disastrous impact.

▲ *This ice fair, painted by Pieter Brueghel the Younger in the 17th century, shows the Little Ice Age which peaked around 300 years ago.*

The greenhouse effect

Gases such as carbon dioxide are known as 'greenhouse gases' because they allow shortwave solar radiation to enter the Earth's atmosphere, but help to stop longwave radiation from escaping. This traps heat, raising the Earth's temperature. An excess of these gases, such as that which results from the burning of fossil fuels, helps trap more heat and can lead to global warming.

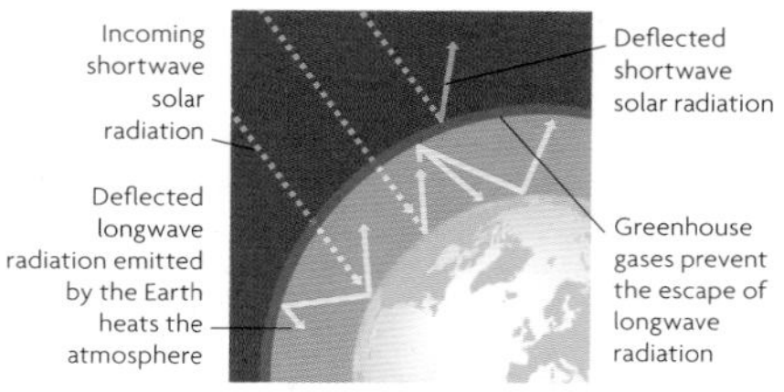
Incoming shortwave solar radiation
Deflected longwave radiation emitted by the Earth heats the atmosphere
Deflected shortwave solar radiation
Greenhouse gases prevent the escape of longwave radiation

◀ *The islands of the Caribbean, Mexico's Gulf coast and the southeastern USA are often hit by hurricanes formed far out in the Atlantic.*

Oceanic water circulation

In general, ocean currents parallel the movement of winds across the Earth's surface. Incoming solar energy is greatest at the Equator and least at the poles. So, water in the oceans heats up most at the Equator and flows polewards, cooling as it moves north or south towards the Arctic or Antarctic. The flow is eventually reversed and cold water currents move back towards the Equator. These ocean currents act as a vast system for moving heat from the Equator towards the poles and are a major influence on the distribution of the Earth's climates.

▲ ***In marginal climatic** zones years of drought can completely dry out the land and transform grassland to desert.*

▲ ***The wide range** of environments found in the Andes is strongly related to their altitude, which modifies climatic influences. While the peaks are snow-capped, many protected interior valleys are semi-tropical.*

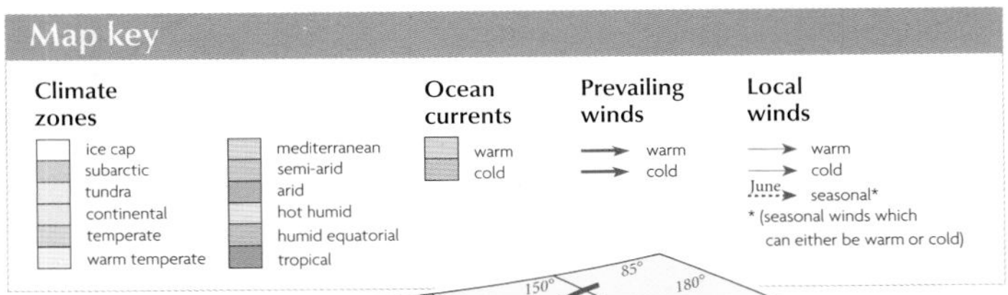
Map key
Climate zones
ice cap
subarctic
tundra
continental
temperate
warm temperate
mediterranean
semi-arid
arid
hot humid
humid equatorial
tropical
Ocean currents
warm
cold
Prevailing winds
warm
cold
Local winds
warm
cold
June seasonal*
* (seasonal winds which can either be warm or cold)

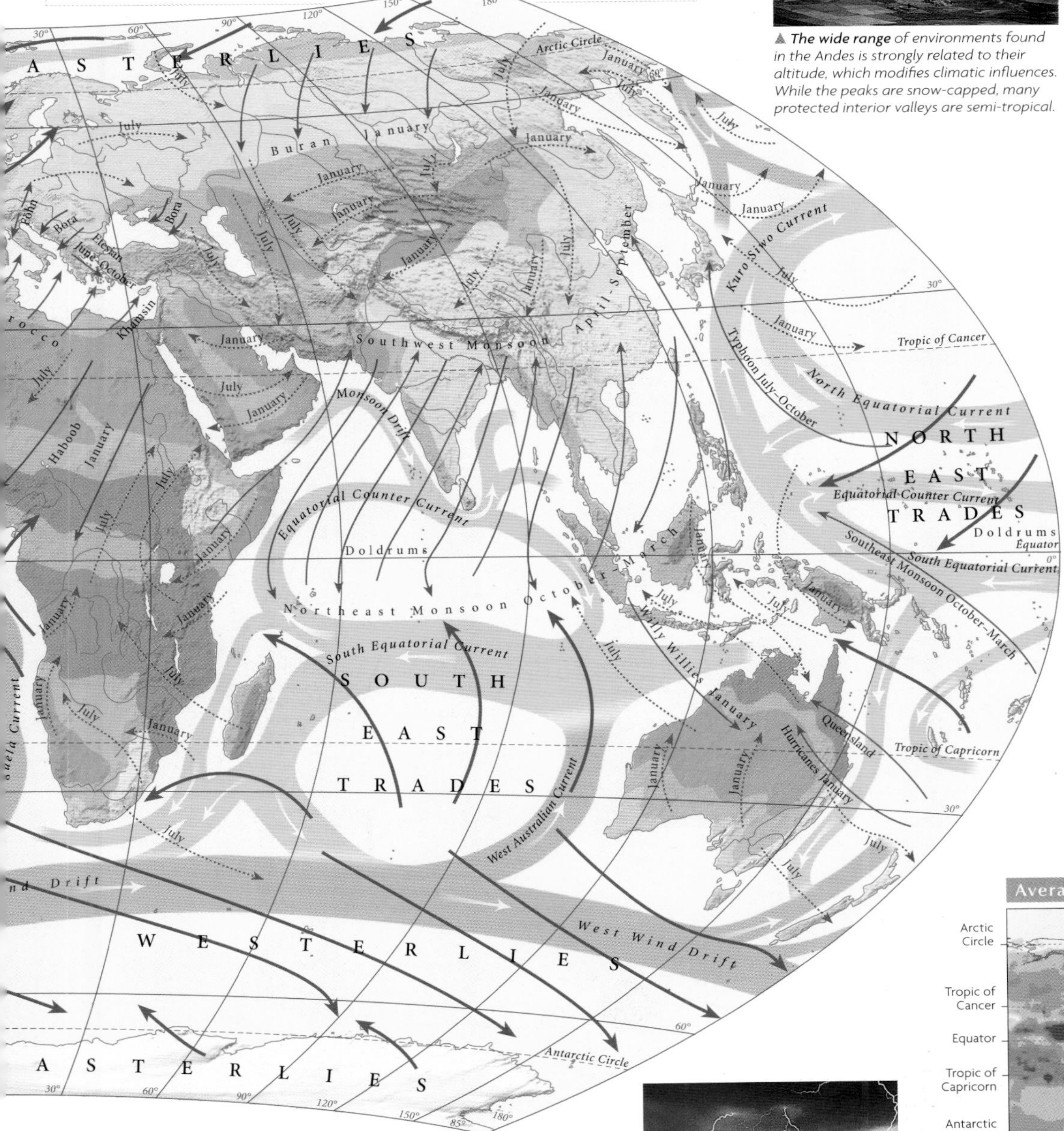
WESTERLIES
Arctic Circle
Buran
Bora
Föhn
Etesian June-October
Khamsin
Southwest Monsoon
Monsoon Drift
Haboob
Equatorial Counter Current
Doldrums
Northeast Monsoon October-March
South Equatorial Current
SOUTH EAST TRADES
West Australian Current
Kuro Siwo Current
Typhoon July-October
Tropic of Cancer
North Equatorial Current
NORTH EAST TRADES
Equatorial Counter Current
Equator
South Equatorial Current
Southeast Monsoon October-March
Willy Willies
Hurricanes January
Queensland
Tropic of Capricorn
West Wind Drift
Antarctic Circle

Tilt and rotation

The tilt and rotation of the Earth during its annual orbit largely control the distribution of heat and moisture across its surface, which correspondingly controls its large-scale weather patterns. As the Earth annually rotates around the Sun, half its surface is receiving maximum radiation, creating summer and winter seasons. The angle of the Earth means that on average the tropics receive two and a half times as much heat from the Sun each day as the poles.

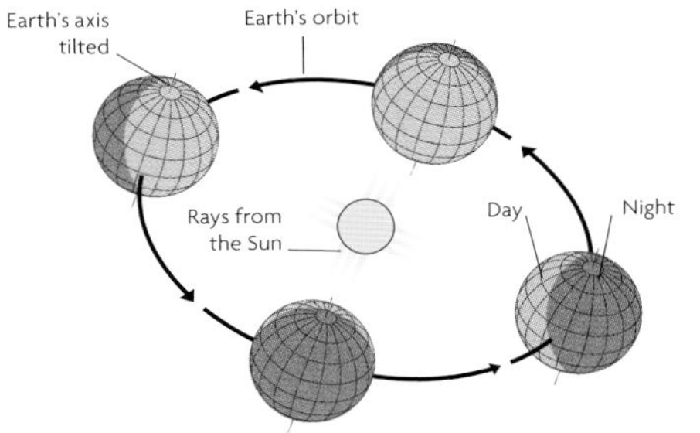
Earth's axis tilted
Earth's orbit
Rays from the Sun
Day
Night

The Coriolis effect

The rotation of the Earth influences atmospheric circulation by deflecting winds and ocean currents. Winds blowing in the northern hemisphere are deflected to the right and those in the southern hemisphere are deflected to the left, creating large-scale patterns of wind circulation, such as the northeast and southeast trade winds and the westerlies. This effect is greatest at the poles and least at the Equator.

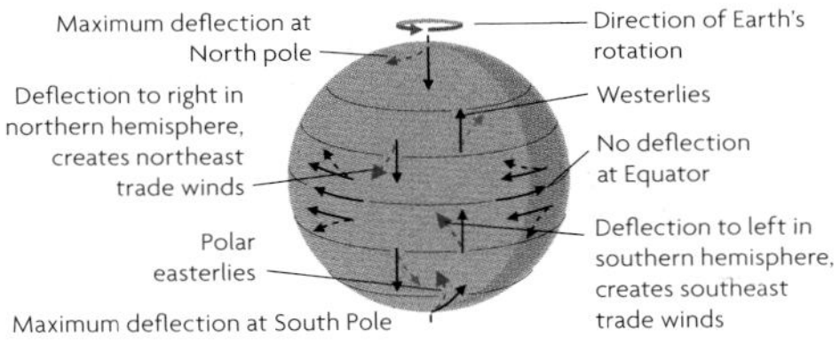
Maximum deflection at North pole
Deflection to right in northern hemisphere, creates northeast trade winds
Polar easterlies
Maximum deflection at South Pole
Direction of Earth's rotation
Westerlies
No deflection at Equator
Deflection to left in southern hemisphere, creates southeast trade winds

Precipitation

When warm air expands, it rises and cools, and the water vapour it carries condenses to form clouds. Heavy, regular rainfall is characteristic of the equatorial region, while the poles are cold and receive only slight snowfall. Tropical regions have marked dry and rainy seasons, while in the temperate regions rainfall is relatively unpredictable.

▲ ***Monsoon rains**, which affect southern Asia from May to September, are caused by sea winds blowing across the warm land.*

▲ ***Heavy tropical rainstorms** occur frequently in Papua New Guinea, often causing soil erosion and landslides in cultivated areas.*

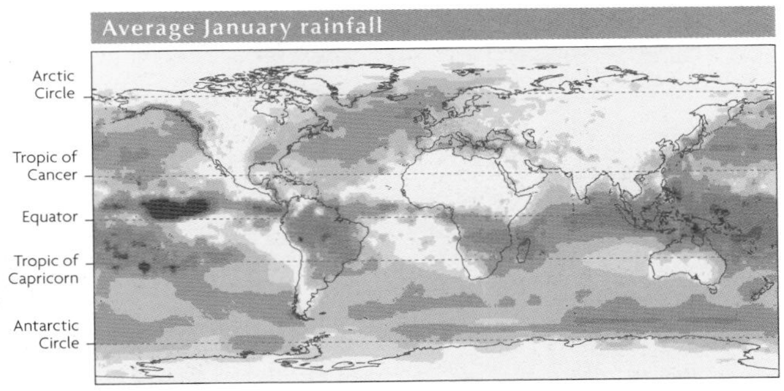
Average January rainfall
Arctic Circle
Tropic of Cancer
Equator
Tropic of Capricorn
Antarctic Circle

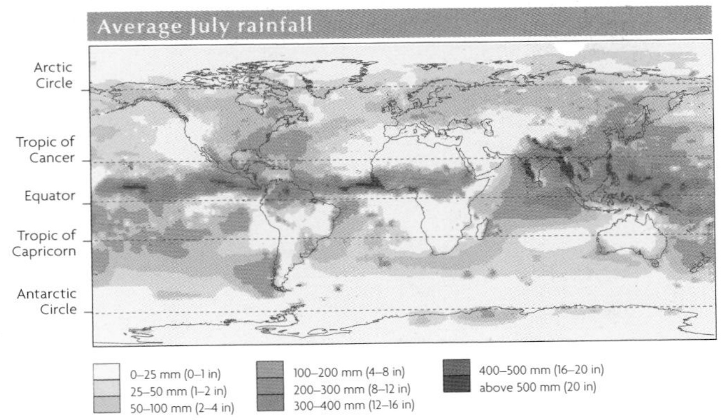
Average July rainfall
Arctic Circle
Tropic of Cancer
Equator
Tropic of Capricorn
Antarctic Circle
0–25 mm (0–1 in)
25–50 mm (1–2 in)
50–100 mm (2–4 in)
100–200 mm (4–8 in)
200–300 mm (8–12 in)
300–400 mm (12–16 in)
400–500 mm (16–20 in)
above 500 mm (20 in)

▲ ***The intensity of** some blizzards in Canada and the northern USA can give rise to snowdrifts as high as 10 ft (3 m).*

▲ ***The Atacama Desert** in Chile is one of the driest places on Earth, with an average rainfall of less than 2 inches (50 mm) per year.*

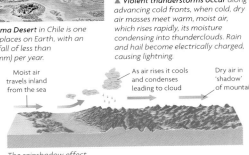

▲ ***Violent thunderstorms occur** along advancing cold fronts, when cold, dry air masses meet warm, moist air, which rises rapidly, its moisture condensing into thunderclouds. Rain and hail become electrically charged, causing lightning.*

The rainshadow effect

When moist air is forced to rise by mountains, it cools and the water vapour falls as precipitation, either as rain or snow. Only the dry, cold air continues over the mountains, leaving inland areas with little or no rain. This is called the rainshadow effect and is one reason for the existence of the Mojave Desert in California, which lies east of the Coast Ranges.

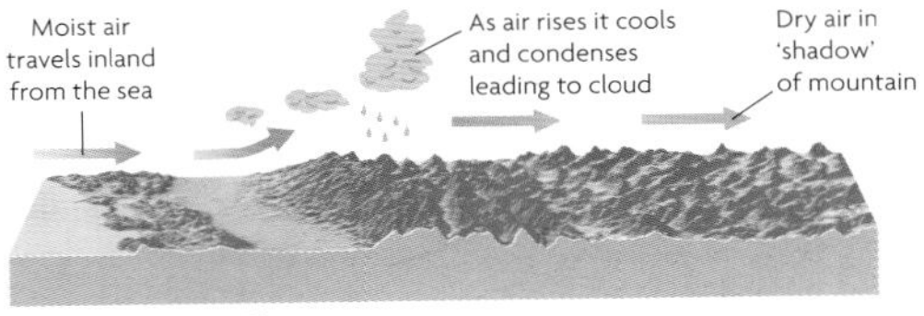
Moist air travels inland from the sea
As air rises it cools and condenses leading to cloud
Dry air in 'shadow' of mountain

The rainshadow effect

Life on Earth

A unique combination of an oxygen-rich atmosphere and plentiful water is the key to life on Earth. Apart from the polar ice caps, there are few areas which have not been colonized by animals or plants over the course of the Earth's history. Plants process sunlight to provide them with their energy, and ultimately all the Earth's animals rely on plants for survival. Because of this reliance, plants are known as primary producers, and the availability of nutrients and temperature of an area is defined as its primary productivity, which affects the quantity and type of animals which are able to live there. This index is affected by climatic factors – cold and aridity restrict the quantity of life, whereas warmth and regular rainfall allow a greater diversity of species.

Biogeographical regions

The Earth can be divided into a series of biogeographical regions, or biomes, ecological communities where certain species of plant and animal co-exist within particular climatic conditions. Within these broad classifications, other factors including soil richness, altitude and human activities such as urbanization, intensive agriculture and deforestation, affect the local distribution of living species within each biome.

Polar regions

A layer of permanent ice at the Earth's poles covers both seas and land. Very little plant and animal life can exist in these harsh regions.

Tundra

A desolate region, with long, dark freezing winters and short, cold summers. With virtually no soil and large areas of permanently frozen ground known as permafrost, the tundra is largely treeless, though it is briefly clothed by small flowering plants in the summer months.

Needleleaf forests

With milder summers than the tundra and less wind, these areas are able to support large forests of coniferous trees.

Broadleaf forests

Much of the northern hemisphere was once covered by deciduous forests, which occurred in areas with marked seasonal variations. Most deciduous forests have been cleared for human settlement.

Temperate rainforests

In warmer wetter areas, such as southern China, temperate deciduous forests are replaced by evergreen forest.

Deserts

Deserts are areas with negligible rainfall. Most hot deserts lie within the tropics; cold deserts are dry because of their distance from the moisture-providing sea.

Mediterranean

Hot, dry summers and short winters typify these areas, which were once covered by evergreen shrubs and woodland, but have now been cleared by humans for agriculture.

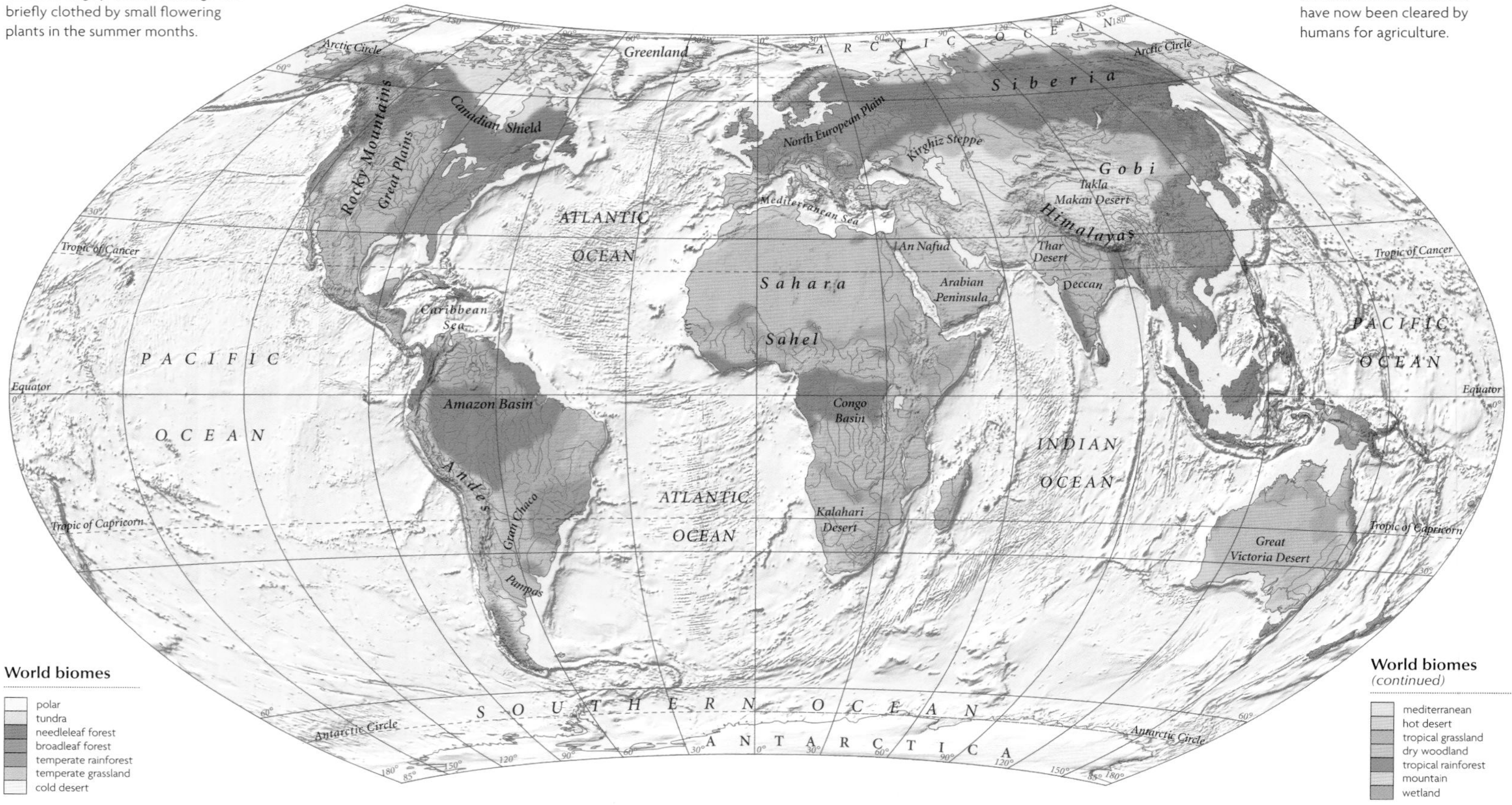

World biomes

- polar
- tundra
- needleleaf forest
- broadleaf forest
- temperate rainforest
- temperate grassland
- cold desert

World biomes *(continued)*

- mediterranean
- hot desert
- tropical grassland
- dry woodland
- tropical rainforest
- mountain
- wetland

Tropical and temperate grasslands

The major grassland areas are found in the centres of the larger continental landmasses. In Africa's tropical savannah regions, seasonal rainfall alternates with drought. Temperate grasslands, also known as steppes and prairies are found in the northern hemisphere, and in South America, where they are known as the pampas.

Dry woodlands

Trees and shrubs, adapted to dry conditions, grow widely spaced from one another, interspersed by savannah grasslands.

Tropical rainforests

Characterized by year-round warmth and high rainfall, tropical rainforests contain the highest diversity of plant and animal species on Earth.

Mountains

Though the lower slopes of mountains may be thickly forested, only ground-hugging shrubs and other vegetation will grow above the tree line which varies according to both altitude and latitude.

Wetlands

Rarely lying above sea level, wetlands are marshes, swamps and tidal flats. Some, with their moist, fertile soils, are rich feeding grounds for fish and breeding grounds for birds. Others have little soil structure and are too acidic to support much plant and animal life.

Biodiversity

The number of plant and animal species, and the range of genetic diversity within the populations of each species, make up the Earth's biodiversity. The plants and animals which are endemic to a region – that is, those which are found nowhere else in the world – are also important in determining levels of biodiversity. Human settlement and intervention have encroached on many areas of the world once rich in endemic plant and animal species. Increasing international efforts are being made to monitor and conserve the biodiversity of the Earth's remaining wild places.

Animal adaptation

The degree of an animal's adaptability to different climates and conditions is extremely important in ensuring its success as a species. Many animals, particularly the largest mammals, are becoming restricted to ever-smaller regions as human development and modern agricultural practices reduce their natural habitats. In contrast, humans have been responsible – both deliberately and accidentally – for the spread of some of the world's most successful species. Many of these introduced species are now more numerous than the indigenous animal populations.

Polar animals

The frozen wastes of the polar regions are able to support only a small range of species which derive their nutritional requirements from the sea. Animals such as the walrus *(left)* have developed insulating fat, stocky limbs and double-layered coats to enable them to survive in the freezing conditions.

Desert animals

Many animals which live in the extreme heat and aridity of the deserts are able to survive for days and even months with very little food or water. Their bodies are adapted to lose heat quickly and to store fat and water. The Gila monster *(above)* stores fat in its tail.

Amazon rainforest

The vast Amazon Basin is home to the world's greatest variety of animal species. Animals are adapted to live at many different levels from the treetops to the tangled undergrowth which lies beneath the canopy. The sloth *(below)* hangs upside down in the branches. Its fur grows from its stomach to its back to enable water to run off quickly.

Diversity of animal species

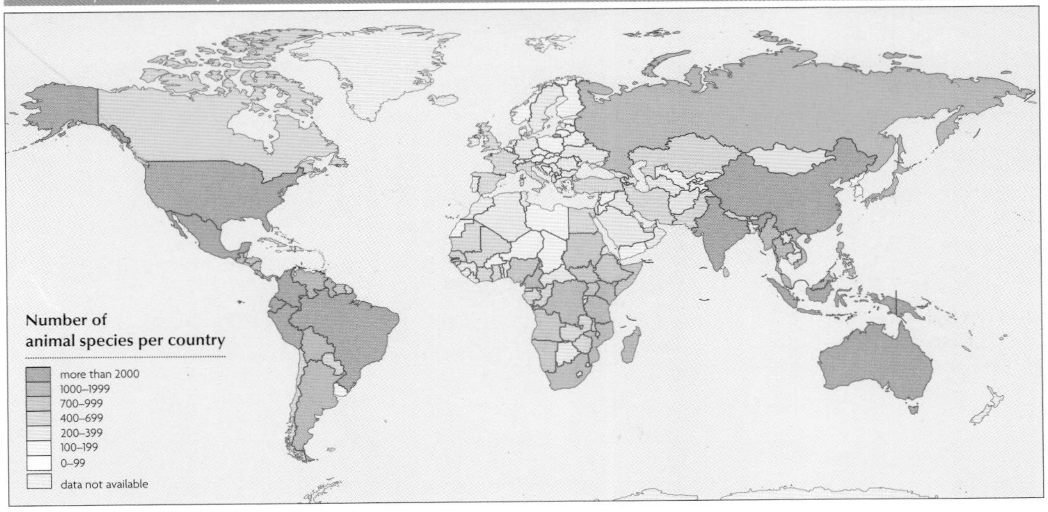
Number of
animal species per country
more than 2000
1000–1999
700–999
400–699
200–399
100–199
0–99
data not available

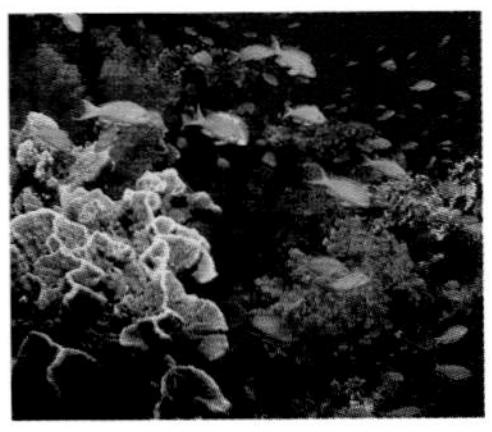

Marine biodiversity

The oceans support a huge variety of different species, from the world's largest mammals like whales and dolphins down to the tiniest plankton. The greatest diversities occur in the warmer seas of continental shelves, where plants are easily able to photosynthesize, and around coral reefs, where complex ecosystems are found. On the ocean floor, nematodes can exist at a depth of more than 10,000 ft (3000 m) below sea level.

High altitudes

Few animals exist in the rarefied atmosphere of the highest mountains. However, birds of prey such as eagles and vultures *(above)*, with their superb eyesight can soar as high as 23,000 ft (7000 m) to scan for prey below.

Urban animals

The growth of cities has reduced the amount of habitat available to many species. A number of animals are now moving closer into urban areas to scavenge from the detritus of the modern city *(left)*. Rodents, particularly rats and mice, have existed in cities for thousands of years, and many insects, especially moths, quickly develop new colouring to provide them with camouflage.

Endemic species

Isolated areas such as Australia and the island of Madagascar, have the greatest range of endemic species. In Australia, these include marsupials such as the kangaroo *(below)*, which carry their young in pouches on their bodies. Destruction of habitat, pollution, hunting, and predators introduced by humans, are threatening this unique biodiversity.

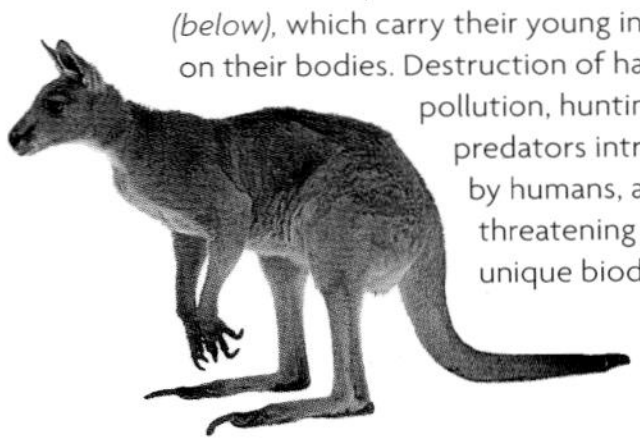

Plant adaptation

Environmental conditions, particularly climate, soil type and the extent of competition with other organisms, influence the development of plants into a number of distinctive forms. Similar conditions in quite different parts of the world create similar adaptations in the plants, which may then be modified by other, local, factors specific to the region.

Cold conditions

In areas where temperatures rarely rise above freezing, plants such as lichens *(left)* and mosses grow densely, close to the ground.

Rainforests

Most of the world's largest and oldest plants are found in rainforests; warmth and heavy rainfall provide ideal conditions for vast plants like the world's largest flower, the rafflesia *(left)*.

Hot, dry conditions

Arid conditions lead to the development of plants whose surface area has been reduced to a minimum to reduce water loss. In cacti *(above)*, which can survive without water for months, leaves are minimal or not present at all.

Ancient plants

Some of the world's most primitive plants still exist today, including algae, cycads and many ferns *(above)*, reflecting the success with which they have adapted to changing conditions.

Resisting predators

A great variety of plants have developed devices including spines *(above)*, poisons, stinging hairs and an unpleasant taste or smell to deter animal predators.

Diversity of plant species

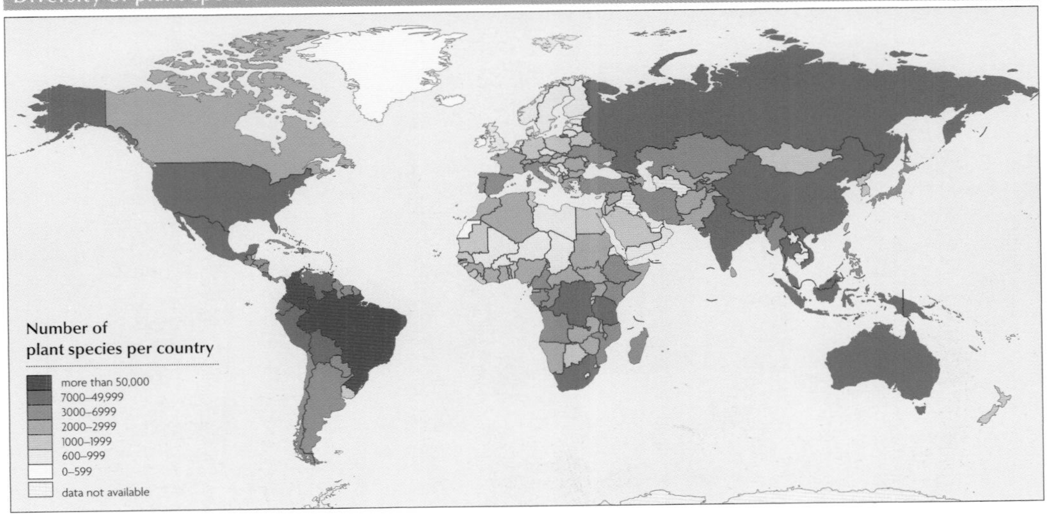
Number of
plant species per country
more than 50,000
7000–49,999
3000–6999
2000–2999
1000–1999
600–999
0–599
data not available

Weeds

Weeds such as bindweed *(above)* are fast-growing, easily dispersed, and tolerant of a number of different environments, enabling them to quickly colonize suitable habitats. They are among the most adaptable of all plants.

Population and settlement

The Earth's population is projected to rise from its current level of about 7.2 billion to reach some 10.5 billion by 2050. The global distribution of this rapidly growing population is very uneven, and is dictated by climate, terrain and natural and economic resources. The great majority of the Earth's people live in coastal zones, and along river valleys. Deserts cover over 20% of the Earth's surface, but support less than 5% of the world's population. It is estimated that over half of the world's population live in cities – most of them in Asia – as a result of mass migration from rural areas in search of jobs. Many of these people live in the so-called 'megacities', some with populations as great as 40 million.

Patterns of settlement

The past 200 years have seen the most radical shift in world population patterns in recorded history.

Nomadic life

All the world's peoples were hunter-gatherers 10,000 years ago. Today nomads, who live by following available food resources, account for less than 0.0001% of the world's population. They are mainly pastoral herders, moving their livestock from place to place in search of grazing land.

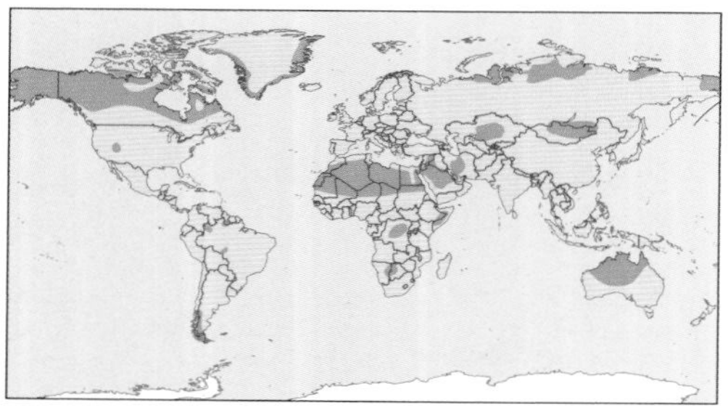

Nomadic population

Nomadic population area

The growth of cities

In 1900 there were only 14 cities in the world with populations of more than a million, mostly in the northern hemisphere. Today, as more and more people in the developing world migrate to towns and cities, there are over 70 cities whose population exceeds 5 million, and around 490 million-cities.

Million-cities in 1900

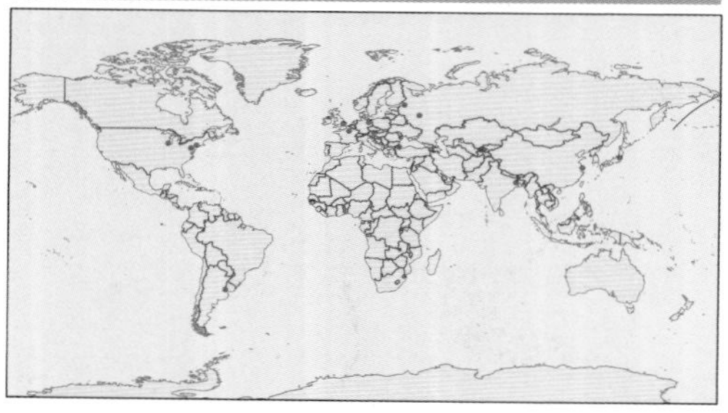

Million-cities in 1900

• Cities over 1 million population

Million-cities in 2005

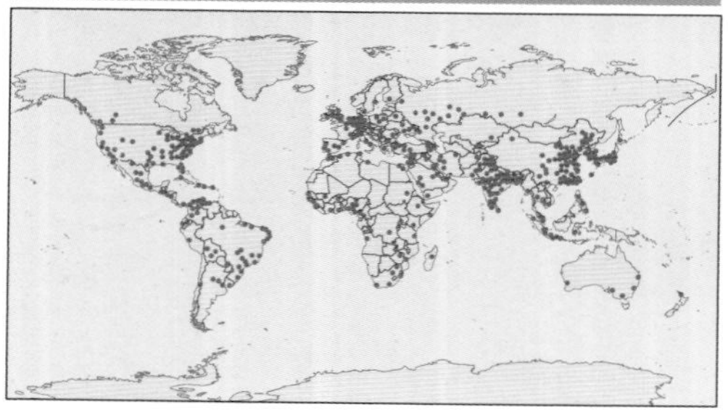

Million-cities in 2005

• Cities over 1 million population

North America

The eastern and western seaboards of the USA, with huge expanses of interconnected cities, towns and suburbs, are vast, densely-populated megalopolises. Central America and the Caribbean also have high population densities. Yet, away from the coasts and in the wildernesses of northern Canada the land is very sparsely settled.

▲ *Vancouver on Canada's west coast, grew up as a port city. In recent years it has attracted many Asian immigrants, particularly from the Pacific Rim.*

▲ *North America's central plains, the continent's agricultural heartland, are thinly populated and highly productive.*

Europe

With its temperate climate, and rich mineral and natural resources, Europe is generally very densely settled. The continent acts as a magnet for economic migrants from the developing world, and immigration is now widely restricted. Birth rates in Europe are generally low, and in some countries, such as Germany, the populations have stabilized at zero growth, with a fast-growing elderly population.

▲ *Many European cities, like Siena, once reflected the 'ideal' size for human settlements. Modern technological advances have enabled them to grow far beyond the original walls.*

▲ *Within the densely-populated Netherlands the reclamation of coastal wetlands is vital to provide much-needed land for agriculture and settlement.*

Population density (inhabitants per sq km)

- 200–1000
- 100–200
- 50–100
- 20–50
- 10–20
- 5–10
- 1–5
- Less than 1

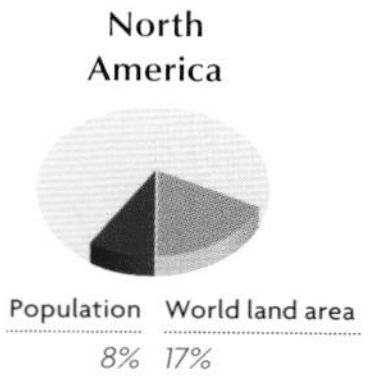
North America
Population
World land area
8%
17%

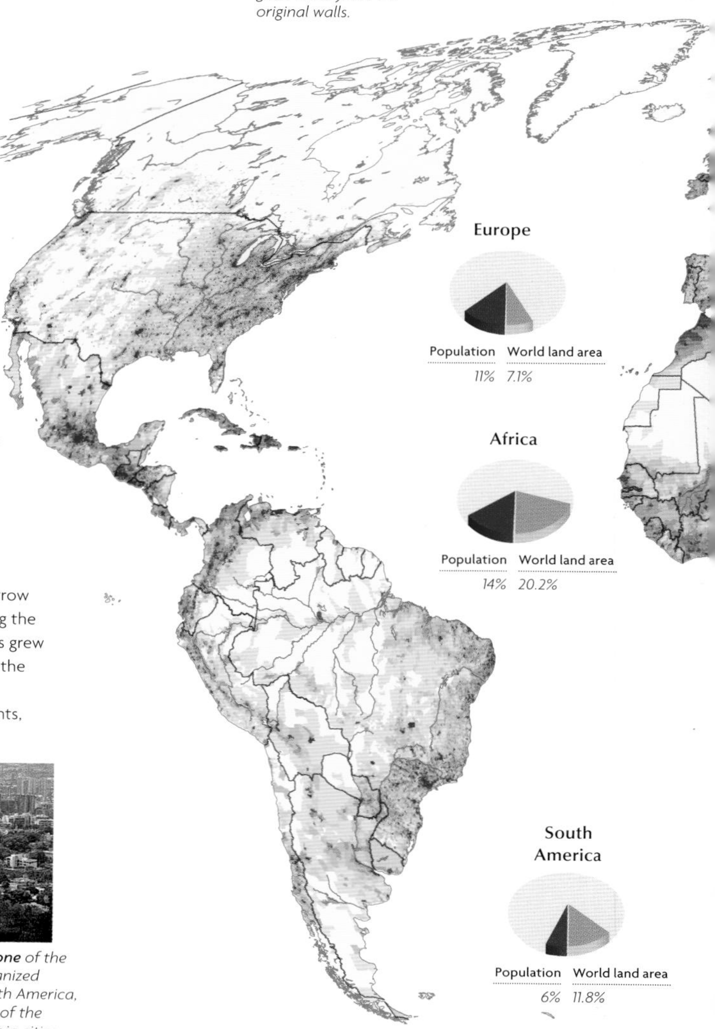
Europe
Population
World land area
11%
7.1%
Africa
Population
World land area
14%
20.2%
South America
Population
World land area
6%
11.8%

South America

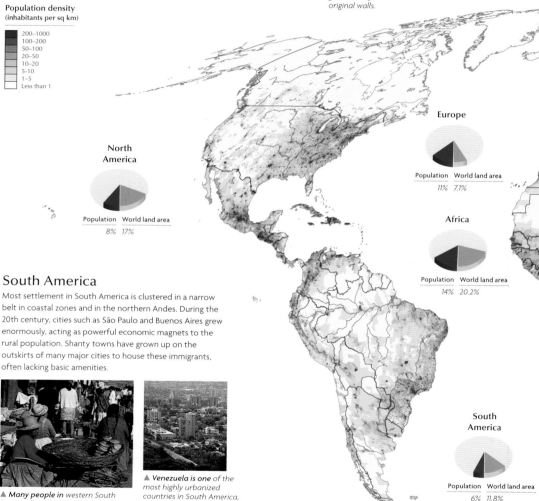

Most settlement in South America is clustered in a narrow belt in coastal zones and in the northern Andes. During the 20th century, cities such as São Paulo and Buenos Aires grew enormously, acting as powerful economic magnets to the rural population. Shanty towns have grown up on the outskirts of many major cities to house these immigrants, often lacking basic amenities.

▲ *Many people in western South America live at high altitudes in the Andes, both in cities and in villages such as this one in Bolivia.*

▲ *Venezuela is one of the most highly urbanized countries in South America, with nearly 90% of the population living in cities such as Caracas.*

Africa

The arid climate of much of Africa means that settlement of the continent is sparse, focusing in coastal areas and fertile regions such as the Nile Valley. Africa still has a high proportion of nomadic agriculturalists, although many are now becoming settled, and the population is predominantly rural.

▲ *Cities such as Nairobi (above), Cairo and Johannesburg have grown rapidly in recent years, although only Cairo has a significant population on a global scale.*

▲ *Traditional lifestyles and homes persist across much of Africa, which has a higher proportion of rural or village-based population than any other continent.*

Asia

Most Asian settlement originally centred around the great river valleys such as the Indus, the Ganges and the Yangtze. Today, almost 60% of the world's population lives in Asia, many in burgeoning cities – particularly in the economically-buoyant Pacific Rim countries. Even rural population densities are high in many countries; practices such as terracing in Southeast Asia making the most of the available land.

▲ *Many of China's cities are now vast urban areas with populations of more than 5 million people.*

▲ *This stilt village in Bangladesh is built to resist the regular flooding. Pressure on land, even in rural areas, forces many people to live in marginal areas.*

Population structures

Population pyramids are an effective means of showing the age structures of different countries, and highlighting changing trends in population growth and decline. The typical pyramid for a country with a growing, youthful population, is broad-based *(left)*, reflecting a high birth rate and a far larger number of young rather than elderly people. In contrast, countries with populations whose numbers are stabilizing have a more balanced distribution of people in each age band, and may even have lower numbers of people in the youngest age ranges, indicating both a high life expectancy, and that the population is now barely replacing itself *(right)*. The Russian Federation *(centre)* shows a marked decline in population due to a combination of a high death rate and low birth rate. The government has taken steps to reverse this trend by providing improved child support and health care. Immigration is also seen as vital to help sustain the population.

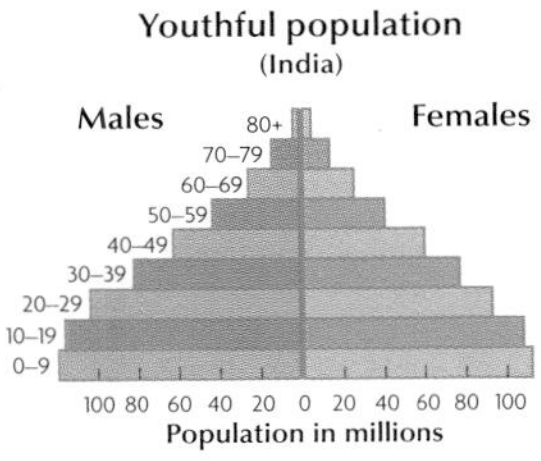
Youthful population
(India)
Males
Females
80+
70–79
60–69
50–59
40–49
30–39
20–29
10–19
0–9
100 80 60 40 20 0 20 40 60 80 100
Population in millions

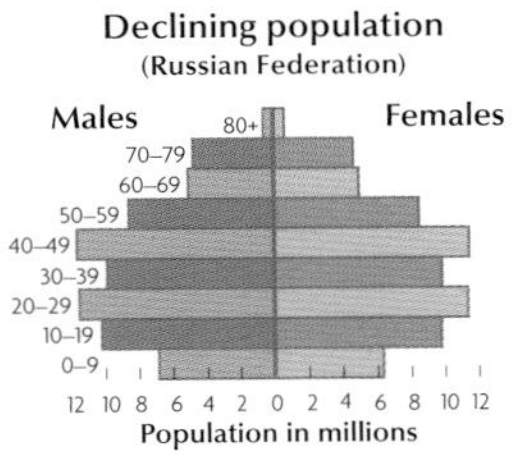
Declining population
(Russian Federation)
Males
Females
80+
70–79
60–69
50–59
40–49
30–39
20–29
10–19
0–9
12 10 8 6 4 2 0 2 4 6 8 10 12
Population in millions

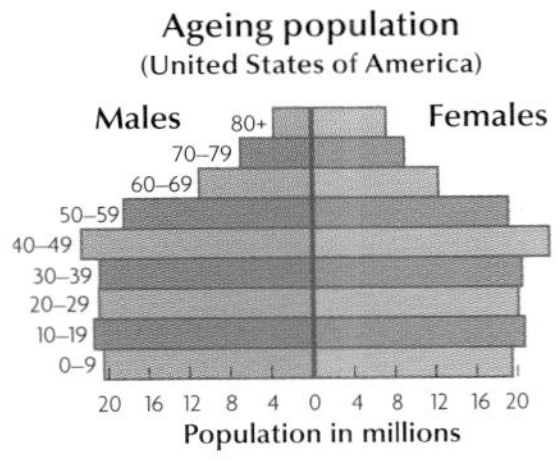
Ageing population
(United States of America)
Males
Females
80+
70–79
60–69
50–59
40–49
30–39
20–29
10–19
0–9
20 16 12 8 4 0 4 8 12 16 20
Population in millions

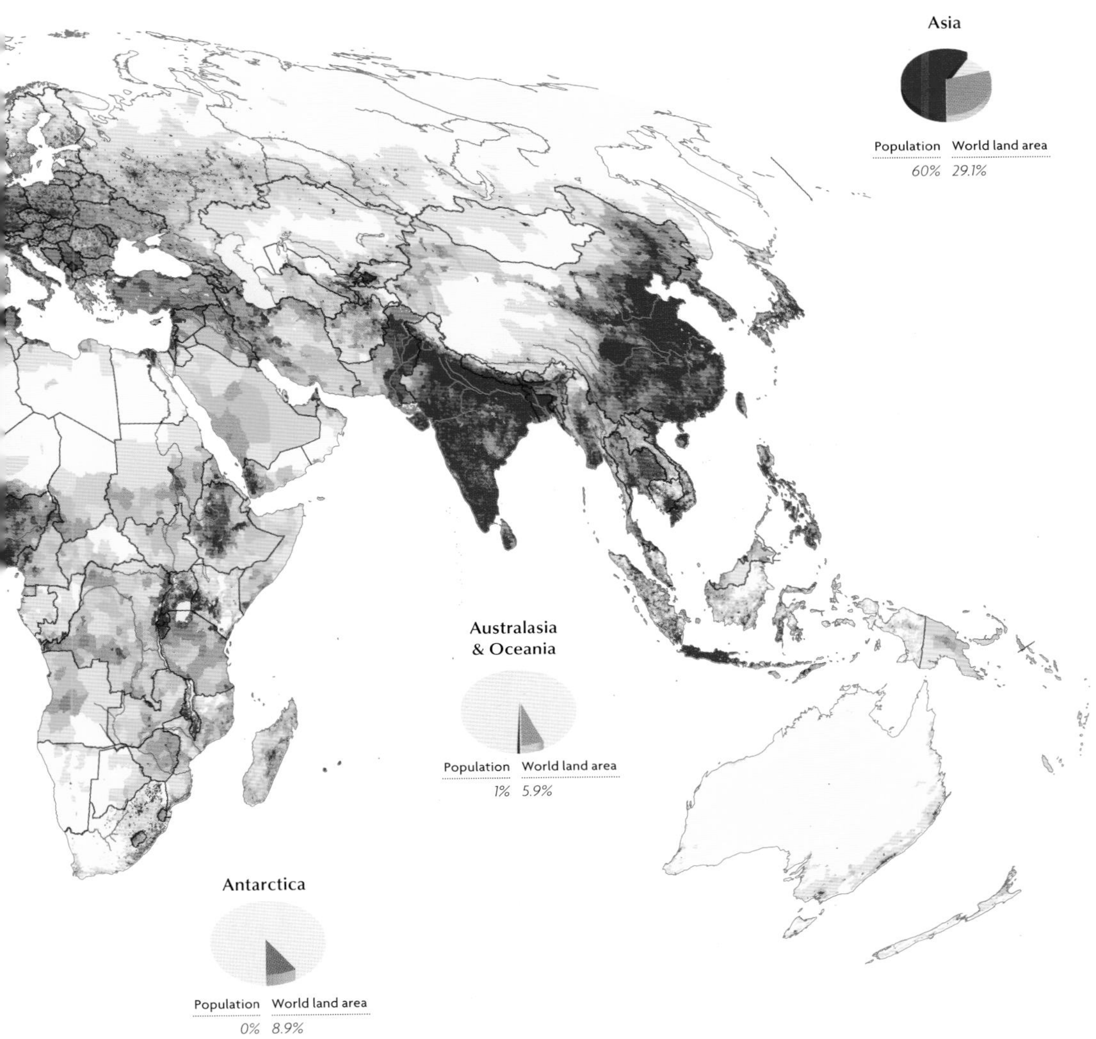
Asia
Population
World land area
60%
29.1%
Australasia & Oceania
Population
World land area
1%
5.9%
Antarctica
Population
World land area
0%
8.9%

Population growth

Improvements in food supply and advances in medicine have both played a major role in the remarkable growth in global population, which has increased five-fold over the last 150 years. Food supplies have risen with the mechanization of agriculture and improvements in crop yields. Better nutrition, together with higher standards of public health and sanitation, have led to increased longevity and higher birth rates.

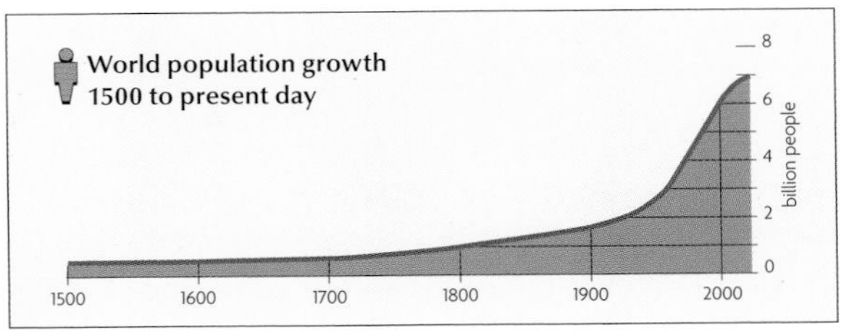
World population growth
1500 to present day
billion people
8
6
4
2
0
1500
1600
1700
1800
1900
2000

World nutrition

Two-thirds of the world's food supply is consumed by the industrialized nations, many of which have a daily calorific intake far higher than is necessary for their populations to maintain a healthy body weight. In contrast, in the developing world, about 800 million people do not have enough food to meet their basic nutritional needs.

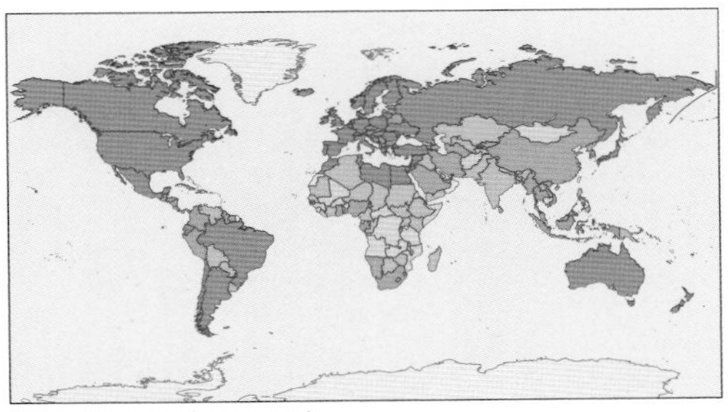

Daily calorie intake per capita

above 3000 | 2500–2999 | 2000–2499 | below 2000 | data not available

World life expectancy

Improved public health and living standards have greatly increased life expectancy in the developed world, where people can now expect to live twice as long as they did 100 years ago. In many of the world's poorest nations, inadequate nutrition and disease, means that the average life expectancy still does not exceed 45 years.

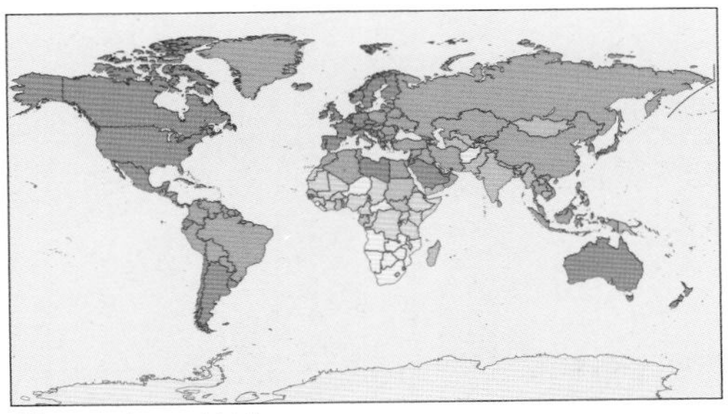

Life expectancy at birth

above 75 years | 65–74 years | 55–64 years | 45–54 years | below 44 years | data not available

Australasia and Oceania

This is the world's most sparsely settled region. The peoples of Australia and New Zealand live mainly in the coastal cities, with only scattered settlements in the arid interior. The Pacific islands can only support limited populations because of their remoteness and lack of resources.

*▶ **Brisbane**, on Australia's Gold Coast is the most rapidly expanding city in the country. The great majority of Australia's population lives in cities near the coasts.*

*◀ **The remote highlands** of Papua New Guinea are home to a wide variety of peoples, many of whom still subsist by traditional hunting and gathering.*

Average world birth rates

Birth rates are much higher in Africa, Asia and South America than in Europe and North America. Increased affluence and easy access to contraception are both factors which can lead to a significant decline in a country's birth rate.

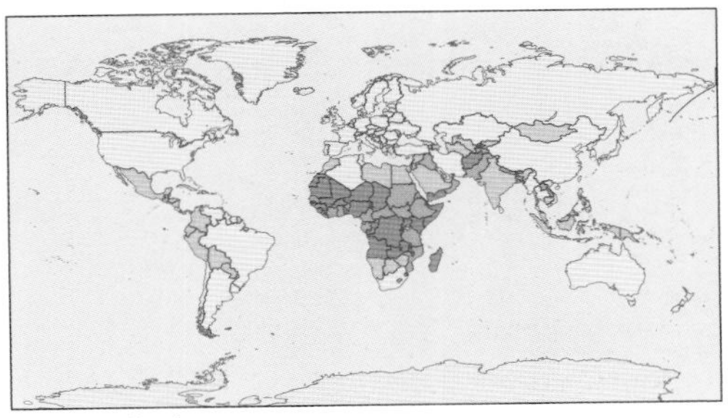

Number of births (per 1000 people)

above 40 | 30–39 | 20–29 | below 20 | data not available

World infant mortality

In parts of the developing world infant mortality rates are still high; access to medical services such as immunization, adequate nutrition and the promotion of breast-feeding have been important in combating infant mortality.

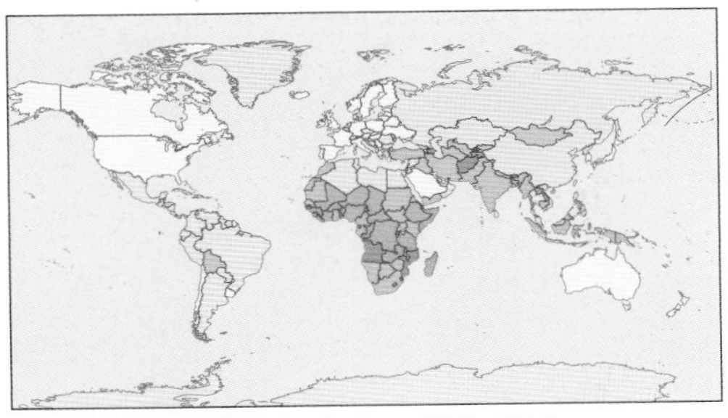

World infant mortality rates (deaths per 1000 live births)

above 125 | 75–124 | 35–74 | 15–34 | below 15 | data not available

The economic system

The wealthy countries of the developed world, with their aggressive, market-led economies and their access to productive new technologies and international markets, dominate the world economic system. At the other extreme, many of the countries of the developing world are locked in a cycle of national debt, rising populations and unemployment. In 2008 a major financial crisis swept the world's banking sector leading to a huge downturn in the global economy. Despite this, China overtook Japan in 2010 to become the world's second largest economy.

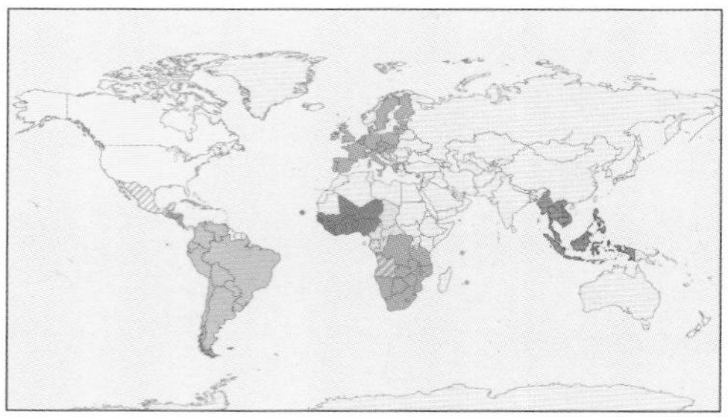

Trade blocs

EU, CACM, NAFTA, SADC, ASEAN, ECOWAS, LAIA, CEEAC

Trade blocs

International trade blocs are formed when groups of countries, often already enjoying close military and political ties, join together to offer mutually preferential terms of trade for both imports and exports. Increasingly, global trade is dominated by three main blocs: the EU, NAFTA, and ASEAN. They are supplanting older trade blocs such as the Commonwealth, a legacy of colonialism.

International trade flows

World trade acts as a stimulus to national economies, encouraging growth. Over the last three decades, as heavy industries have declined, services – banking, insurance, tourism, airlines and shipping – have taken an increasingly large share of world trade. Manufactured articles now account for nearly two-thirds of world trade; raw materials and food make up less than a quarter of the total.

Shipping

Ships carry 80% of international cargo, and extensive container ports, where cargo is stored, are vital links in the international transport network.

Multinationals

Multinational companies are increasingly penetrating inaccessible markets. The reach of many American commodities is now global.

Primary products

Many countries, particularly in the Caribbean and Africa, are still reliant on primary products such as rubber and coffee, which makes them vulnerable to fluctuating prices.

Service industries

Service industries such as banking, tourism and insurance were the fastest-growing industrial sector in the last half of the 20th century. Lloyds of London is the centre of the world insurance market.

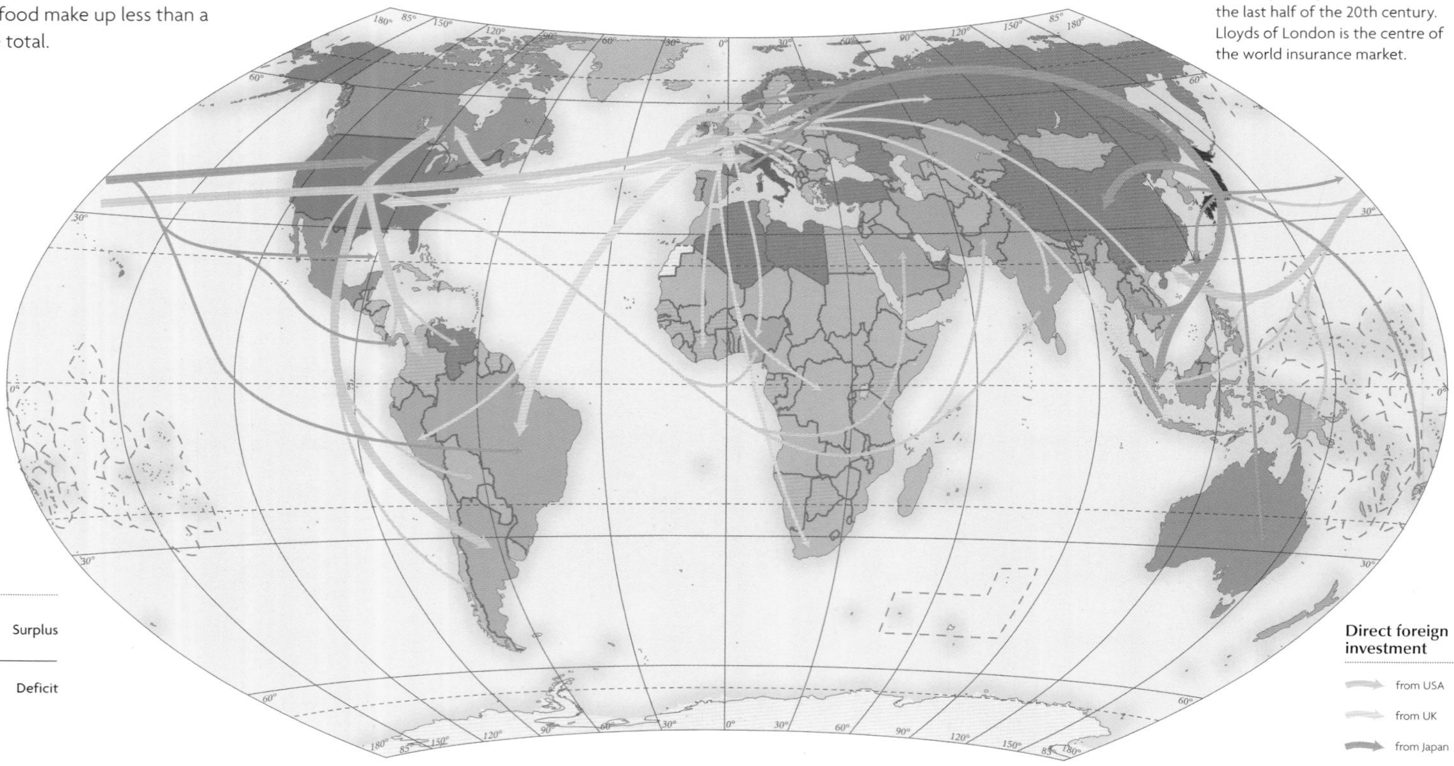

Balance of trade (millions US$)

Surplus: over 30,000; 10,000–29,000; 1000–9999; 0–999

Deficit: 0–999; 1000–9999; 10,000–29,999; over 30,000

data unavailable

Direct foreign investment

from USA

from UK

from Japan

World money markets

The financial world has traditionally been dominated by three major centres – Tokyo, New York and London, which house the headquarters of stock exchanges, multinational corporations and international banks. Their geographic location means that, at any one time in a 24-hour day, one major market is open for trading in shares, currencies and commodities. Since the late 1980s, technological advances have enabled transactions between financial centres to occur at ever-greater speed, and new markets have sprung up throughout the world.

Major money markets

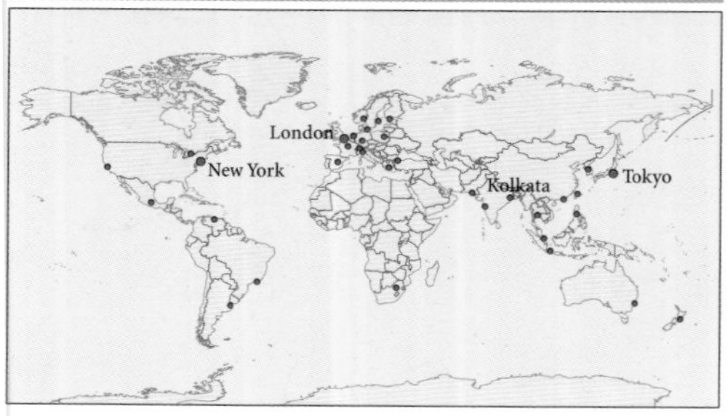

Location of major stock markets

● Major stock markets

▲ ***The Tokyo Stock Market*** *crashed in 1990, leading to slow-down in the growth of the world's most powerful economy, and a refocusing on economic policy away from export-led growth and towards the domestic market.*

New stock markets

New stock markets are now opening in many parts of the world, where economies have recently emerged from state controls. In Moscow and Beijing, and several countries in eastern Europe, newly-opened stock exchanges reflect the transition to market-driven economies.

▲ ***Dealers at the*** *Kolkata Stock Market. The Indian economy has been opened up to foreign investment and many multinationals now have bases there.*

The developing world

International trade in capital and currency is dominated by the rich nations of the northern hemisphere. In parts of Africa and Asia, where exports of any sort are extremely limited, home-produced commodities are simply sold in local markets.

▲ ***Markets have thrived*** *in communist Vietnam since the introduction of a liberal economic policy.*

World wealth disparity

A global assessment of Gross Domestic Product (GDP) by nation reveals great disparities. The developed world, with only a quarter of the world's population, has 80% of the world's manufacturing income. Civil war, conflict and political instability further undermine the economic self-sufficiency of many of the world's poorest nations.

Urban sprawl

Cities are expanding all over the developing world, attracting economic migrants in search of work and opportunities. In cities such as Rio de Janeiro, housing has not kept pace with the population explosion, and squalid shanty towns *(favelas)* rub shoulders with middle-class housing.

▲ ***The favelas of*** *Rio de Janeiro sprawl over the hills surrounding the city.*

▲ ***Cities such as*** *Detroit have been badly hit by the decline in heavy industry.*

Urban decay

Although the USA still dominates the global economy, it faces deficits in both the federal budget and the balance of trade. Vast discrepancies in personal wealth, high levels of unemployment, and the dismantling of welfare provisions throughout the 1980s have led to severe deprivation in several of the inner cities of North America's industrial heartland.

Booming cities

Since the 1980s the Chinese government has set up special industrial zones, such as Shanghai, where foreign investment is encouraged through tax incentives. Migrants from rural China pour into these regions in search of work, creating 'boomtown' economies.

◀ ***Foreign investment has*** *encouraged new infrastructure development in cities like Shanghai.*

Comparative world wealth

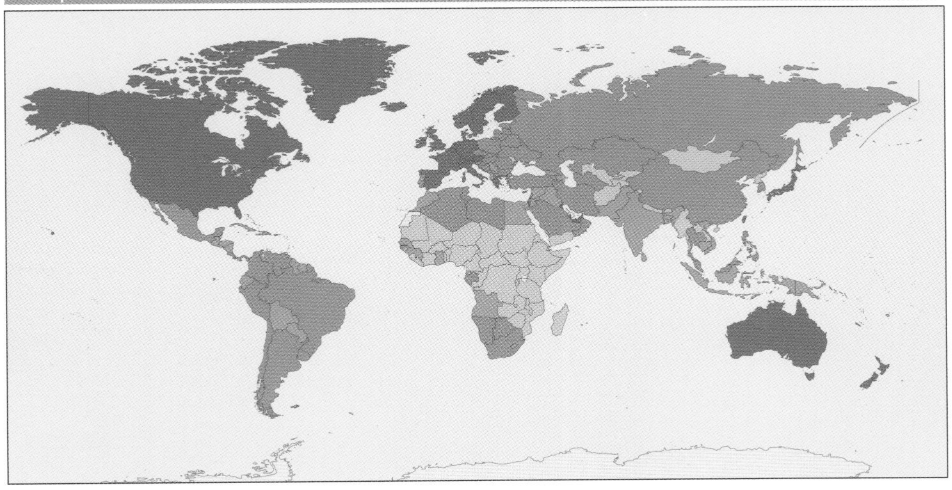

World economies - average GDP per capita (US$)

- above 20,000
- 5000–20,000
- 2000–5000
- below 2000
- data unavailable

Economic 'tigers'

The economic 'tigers' of the Pacific Rim – China, Singapore, and South Korea – have grown faster than Europe and the USA over the last decade. Their export- and service-led economies have benefited from stable government, low labour costs, and foreign investment.

▲ ***Hong Kong, with*** *its fine natural harbour, is one of the most important ports in Asia.*

Agricultural economies

In parts of the developing world, people survive by subsistence farming – only growing enough food for themselves and their families. With no surplus product, they are unable to exchange goods for currency, the only means of escaping the poverty trap. In other countries, farmers have been encouraged to concentrate on growing a single crop for the export market. This reliance on cash crops leaves farmers vulnerable to crop failure and to changes in the market price of the crop.

▲ ***The Ugandan uplands*** *are fertile, but poor infrastructure hampers the export of cash crops.*

▲ ***A shopping arcade*** *in Paris displays a great profusion of luxury goods.*

The affluent West

The capital cities of many countries in the developed world are showcases for consumer goods, reflecting the increasing importance of the service sector, and particularly the retail sector, in the world economy. The idea of shopping as a leisure activity is unique to the western world. Luxury goods and services attract visitors, who in turn generate tourist revenue.

Tourism

In 2004, there were over 940 million tourists worldwide. Tourism is now the world's biggest single industry, employing 130 million people, though frequently in low-paid unskilled jobs. While tourists are increasingly exploring inaccessible and less-developed regions of the world, the benefits of the industry are not always felt at a local level. There are also worries about the environmental impact of tourism, as the world's last wildernesses increasingly become tourist attractions.

▲ ***Botswana's Okavango Delta*** *is an area rich in wildlife. Tourists make safaris to the region, but the impact of tourism is controlled.*

Money flows

In 2008 a global financial crisis swept through the world's economic system. The crisis triggered the failure of several major financial institutions and lead to increased borrowing costs known as the "credit crunch". A consequent reduction in economic activity together with rising inflation forced many governments to introduce austerity measures to reduce borrowing and debt, particulary in Europe where massive "bailouts" were needed to keep some European single currency (Euro) countries solvent.

◀ ***In rural Southeast Asia,*** *babies are given medical checks by UNICEF as part of a global aid programme sponsored by the UN.*

Tourist arrivals

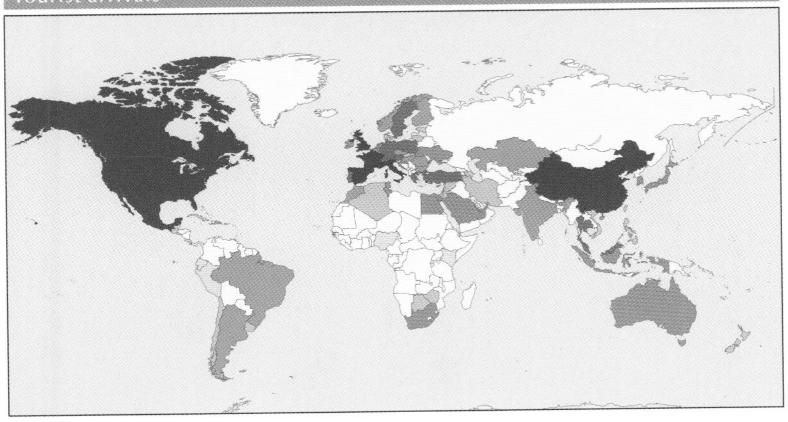

Tourist arrivals

- over 20 million
- 10–20 million
- 5–10 million
- 2.5–5 million
- 1–2.5 million
- 700,000–999,000
- under 700,000
- data unavailable

International debt

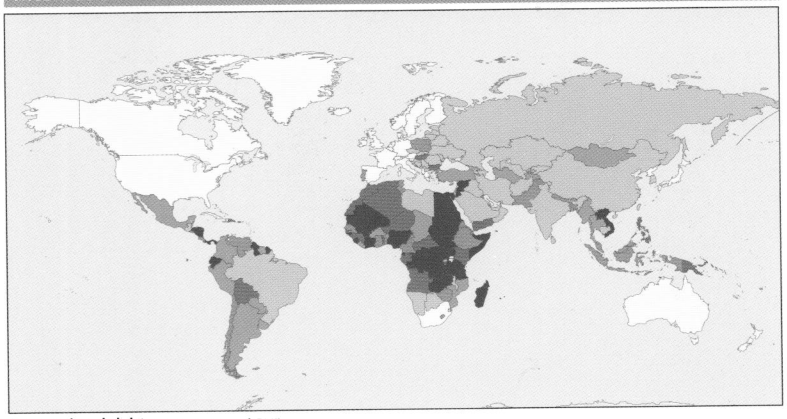

International debt (as percentage of GNI)

- over 100%
- 70–99%
- 50–69%
- 30–49%
- 10–29%
- below 10%
- data unavailable

A B C D E F G H I J K L M

The political world

There are 196 independent countries in the world today. With the exception of Antarctica, where territorial claims have been deferred by international treaty, every land area of the Earth's surface either belongs to, or is claimed by, one country or another. The largest country in the world is the Russian Federation, the smallest is Vatican City. Some 60 overseas dependent territories remain, administered variously by France, Australia, Denmark, New Zealand, Norway, Portugal, the UK, the US and the Netherlands.

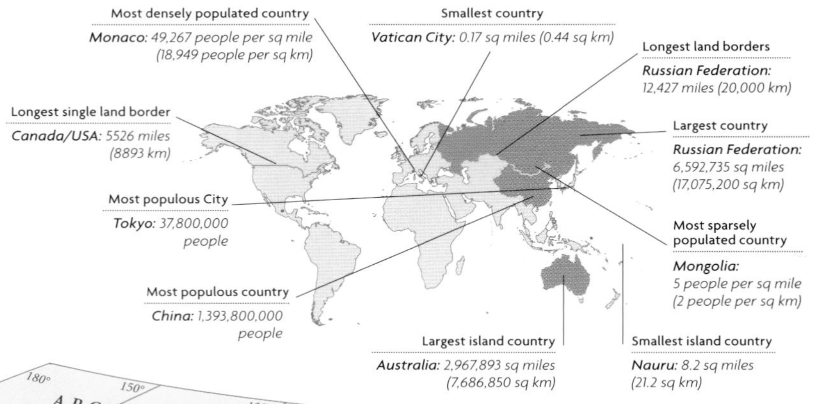
Most densely populated country
Monaco: 49,267 people per sq mile (18,949 people per sq km)
Smallest country
Vatican City: 0.17 sq miles (0.44 sq km)
Longest land borders
Russian Federation: 12,427 miles (20,000 km)
Longest single land border
Canada/USA: 5526 miles (8893 km)
Largest country
Russian Federation: 6,592,735 sq miles (17,075,200 sq km)
Most populous City
Tokyo: 37,800,000 people
Most sparsely populated country
Mongolia: 5 people per sq mile (2 people per sq km)
Most populous country
China: 1,393,800,000 people
Largest island country
Australia: 2,967,893 sq miles (7,686,850 sq km)
Smallest island country
Nauru: 8.2 sq miles (21.2 sq km)

International borders

The map shows three main types of boundary between states. Full borders represent internationally agreed and recognized territorial boundaries. Undefined borders exist where no fixed boundary between states has been demarcated; the boundaries indicated in this way show approximate areas of sovereignty. A disputed border is indicated where a *de facto* territorial boundary exists, which is not agreed or is subject to arbitration.

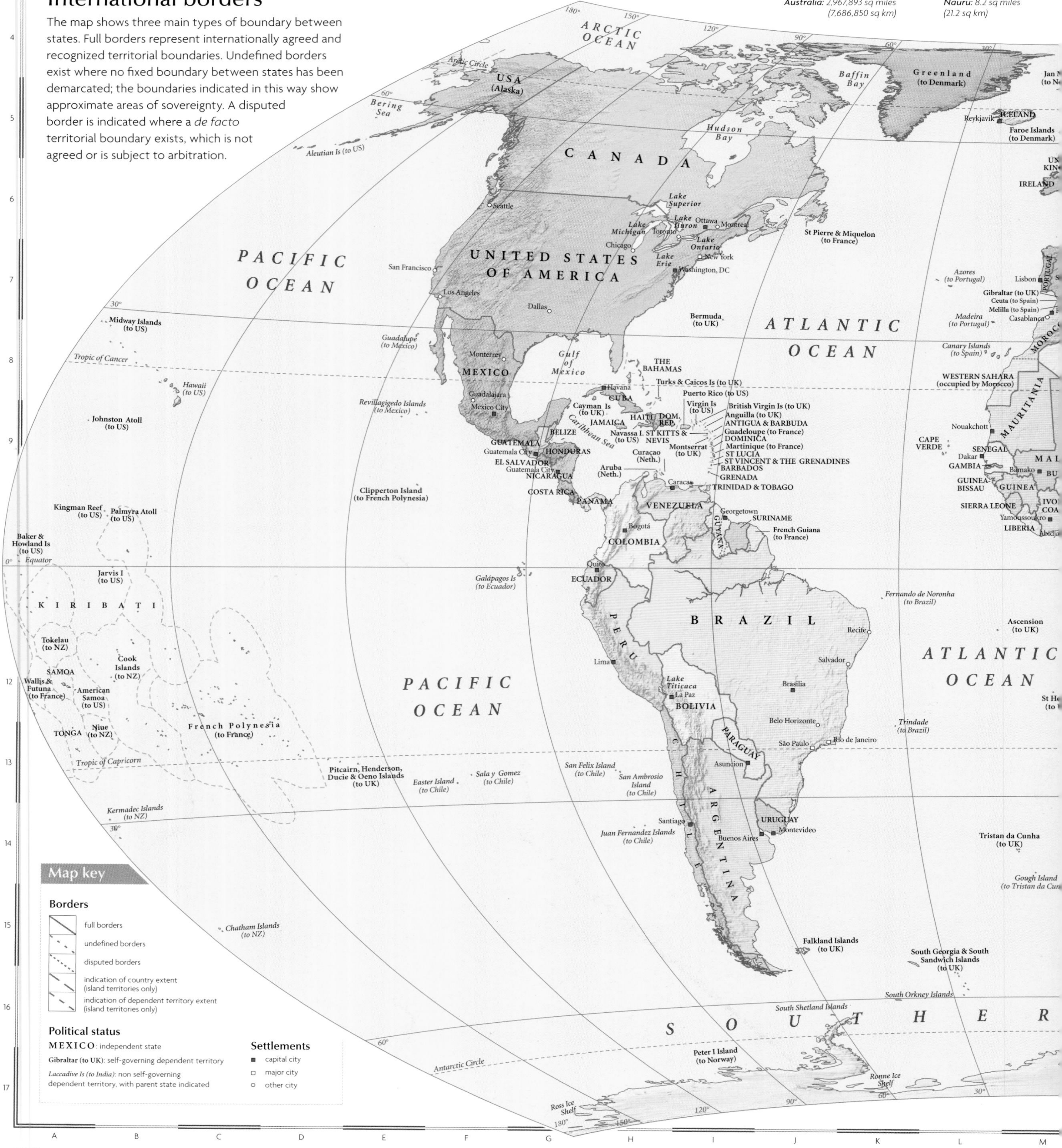
ARCTIC OCEAN
Arctic Circle
USA (Alaska)
Bering Sea
Aleutian Is (to US)
CANADA
Hudson Bay
Baffin Bay
Greenland (to Denmark)
Reykjavik
ICELAND
Faroe Islands (to Denmark)
IRELAND
Seattle
Lake Superior
Lake Michigan
Lake Huron
Ottawa
Montreal
Toronto
Lake Ontario
Chicago
Lake Erie
New York
Washington, DC
St Pierre & Miquelon (to France)
PACIFIC OCEAN
UNITED STATES OF AMERICA
San Francisco
Los Angeles
Dallas
Azores (to Portugal)
Lisbon
Gibraltar (to UK)
Ceuta (to Spain)
Melilla (to Spain)
Casablanca
Madeira (to Portugal)
Midway Islands (to US)
Bermuda (to UK)
ATLANTIC OCEAN
Guadalupe (to Mexico)
Tropic of Cancer
Hawaii (to US)
Monterrey
MEXICO
Gulf of Mexico
THE BAHAMAS
Canary Islands (to Spain)
WESTERN SAHARA (occupied by Morocco)
Havana
CUBA
Turks & Caicos Is (to UK)
Puerto Rico (to US)
Guadalajara
Mexico City
Revillagigedo Islands (to Mexico)
Johnston Atoll (to US)
Cayman Is (to UK)
JAMAICA
HAITI
DOM. REP.
Virgin Is (to US)
British Virgin Is (to UK)
Anguilla (to UK)
ANTIGUA & BARBUDA
Guadeloupe (to France)
DOMINICA
Martinique (to France)
ST LUCIA
ST VINCENT & THE GRENADINES
BARBADOS
GRENADA
TRINIDAD & TOBAGO
Caribbean Sea
BELIZE
Navassa I. (to US)
ST KITTS & NEVIS
Montserrat (to UK)
Curaçao (Neth.)
Aruba (Neth.)
GUATEMALA
Guatemala City
HONDURAS
EL SALVADOR
NICARAGUA
COSTA RICA
PANAMA
Caracas
VENEZUELA
Nouakchott
MAURITANIA
CAPE VERDE
SENEGAL
Dakar
GAMBIA
GUINEA-BISSAU
Bamako
GUINEA
SIERRA LEONE
Yamoussoukro
LIBERIA
Clipperton Island (to French Polynesia)
Kingman Reef (to US)
Palmyra Atoll (to US)
Baker & Howland Is (to US)
Equator
Jarvis I (to US)
Georgetown
GUYANA
SURINAME
French Guiana (to France)
Bogotá
COLOMBIA
Quito
ECUADOR
Galápagos Is (to Ecuador)
KIRIBATI
Fernando de Noronha (to Brazil)
BRAZIL
Recife
Ascension (to UK)
Tokelau (to NZ)
Cook Islands (to NZ)
SAMOA
Wallis & Futuna (to France)
American Samoa (to US)
PERU
Lima
Salvador
Lake Titicaca
La Paz
BOLIVIA
Brasília
PACIFIC OCEAN
Niue (to NZ)
TONGA
French Polynesia (to France)
Belo Horizonte
Trindade (to Brazil)
São Paulo
Rio de Janeiro
PARAGUAY
Asunción
Tropic of Capricorn
Pitcairn, Henderson, Ducie & Oeno Islands (to UK)
Easter Island (to Chile)
Sala y Gomez (to Chile)
San Felix Island (to Chile)
San Ambrosio Island (to Chile)
CHILE
ARGENTINA
Kermadec Islands (to NZ)
Santiago
URUGUAY
Montevideo
Buenos Aires
Juan Fernández Islands (to Chile)
Tristan da Cunha (to UK)
Gough Island (to Tristan da Cunha)
Chatham Islands (to NZ)
Falkland Islands (to UK)
South Georgia & South Sandwich Islands (to UK)
South Orkney Islands
South Shetland Islands
SOUTHER
Peter I Island (to Norway)
Antarctic Circle
Ronne Ice Shelf
Ross Ice Shelf
Map key
Borders
full borders
undefined borders
disputed borders
indication of country extent (island territories only)
indication of dependent territory extent (island territories only)
Political status
MEXICO: independent state
Gibraltar (to UK): self-governing dependent territory
Laccadive Is (to India): non self-governing dependent territory, with parent state indicated
Settlements
capital city
major city
other city

A B C D E F G H I J K L M

The world in 1914

The early years of the 20th century saw the mainly European colonial empires reaching their greatest extents by 1914. Two world wars inaugurated their disintegration, but even in 1950 there were only 82 independent countries. Since then, over 100 have gained their independence, culminating in the breakup of the Soviet Union and former Yugoslavia in the early 1990s.

Percentage of Earth's land surface controlled by colonial empires in 1914

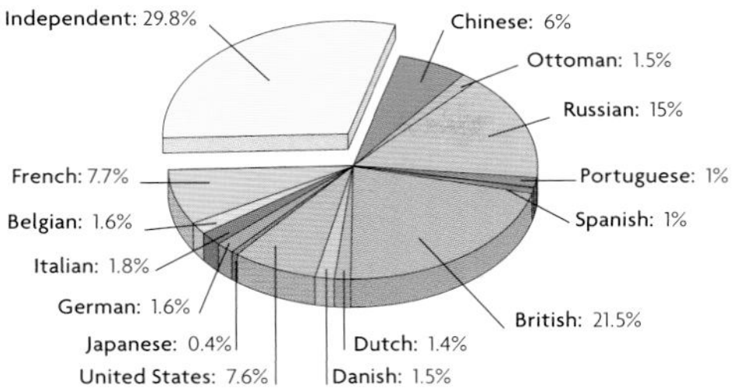

Colonial empires in 1914

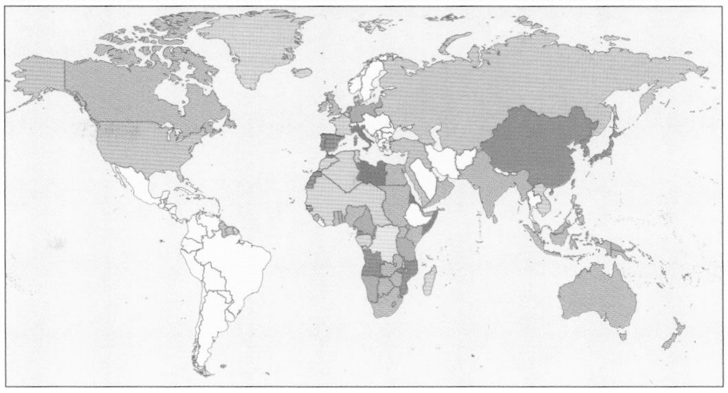

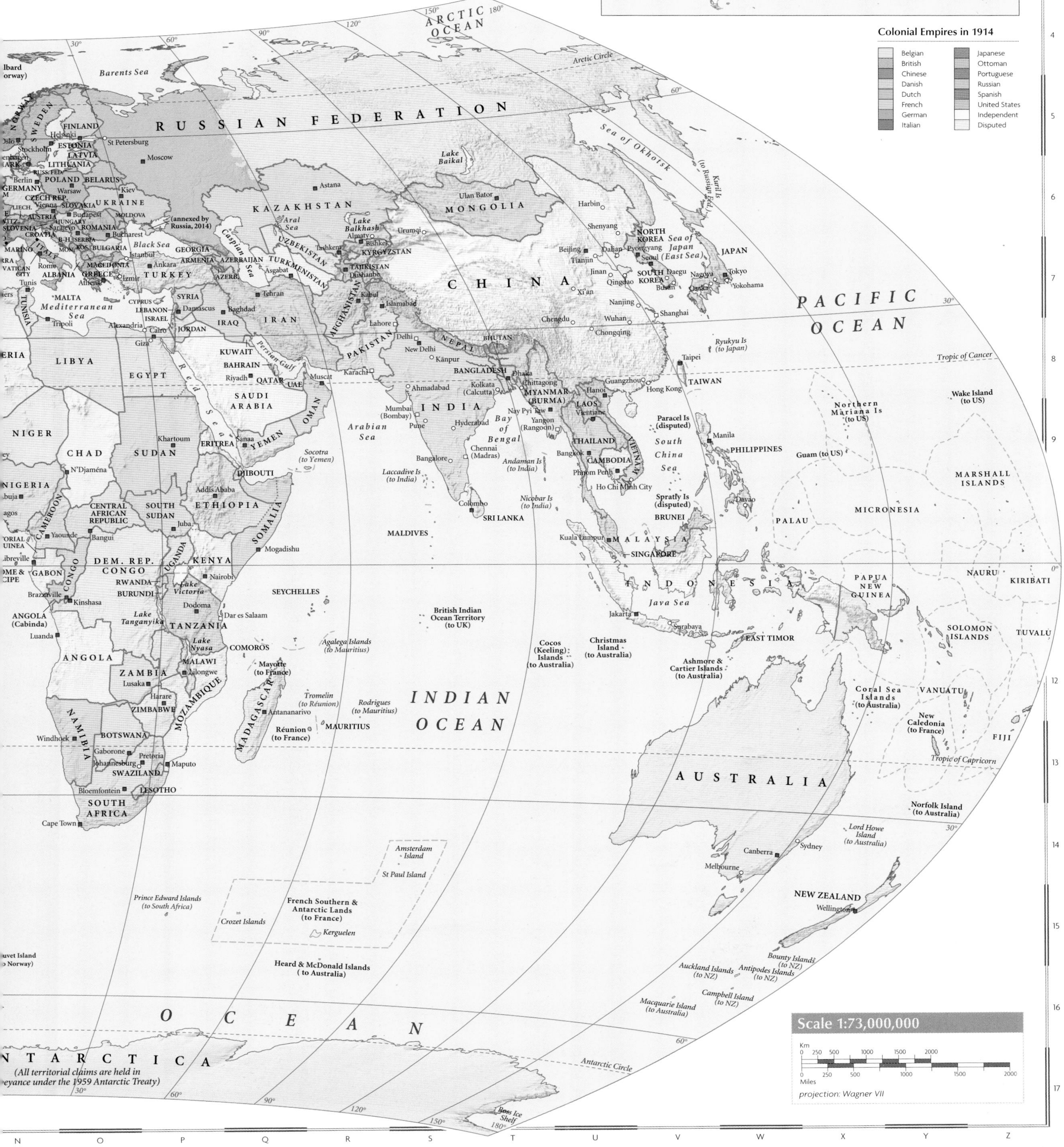

Scale 1:73,000,000

projection: Wagner VII

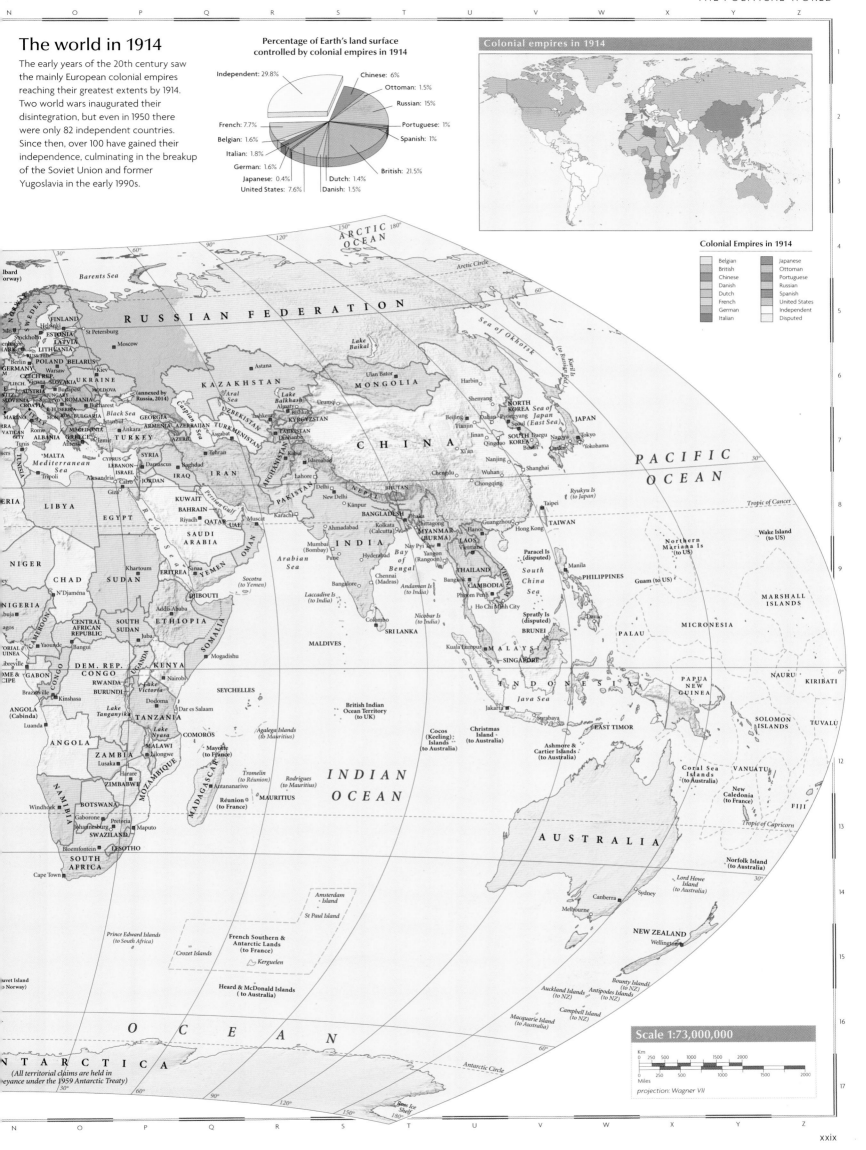

States and boundaries

There are almost 200 sovereign states in the world today; in 1950 there were only 82. Over the last 65 years national self-determination has been a driving force for many states with a history of colonialism and oppression. As more borders have been added to the world map, the number of international border disputes has increased.

In many cases, where the impetus towards independence has been religious or ethnic, disputes with minority groups have also caused violent internal conflict. While many newly-formed states have moved peacefully towards independence, successfully establishing government by multiparty democracy, dictatorship by military regime or individual despot is often the result of the internal power-struggles which characterize the early stages in the lives of new nations.

The nature of politics

Democracy is a broad term: it can range from the ideal of multiparty elections and fair representation to, in countries such as Singapore, a thin disguise for single-party rule. In despotic regimes, on the other hand, a single, often personal authority has total power; institutions such as parliament and the military are mere instruments of the dictator.

◀ ***The stars and** stripes of the US flag are a potent symbol of the country's status as a federal democracy.*

The changing world map

Decolonization

In 1950, large areas of the world remained under the control of a handful of European countries *(page xxix)*. The process of decolonization had begun in Asia, where, following the Second World War, much of south and southeast Asia sought and achieved self-determination. In the 1960s, a host of African states achieved independence, so that by 1965, most of the larger tracts of the European overseas empires had been substantially eroded. The final major stage in decolonization came with the break-up of the Soviet Union and the Eastern bloc after 1990. The process continues today as the last toeholds of European colonialism, often tiny island nations, press increasingly for independence.

▲ ***Icons of communism,** including statues of former leaders such as Lenin and Stalin, were destroyed when the Soviet bloc was dismantled in 1989, creating several new nations.*

New nations 1945–1965

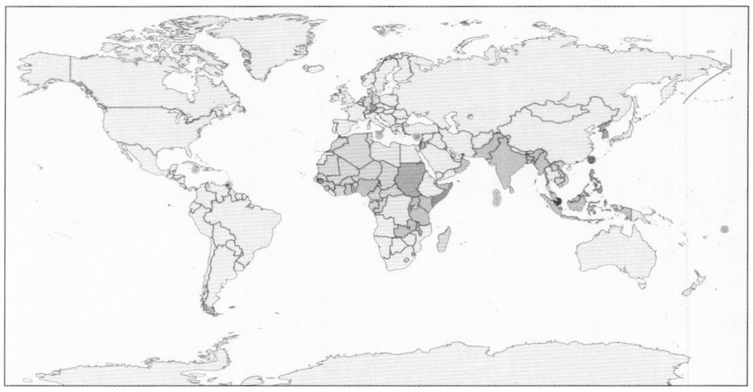

New nations 1965–present

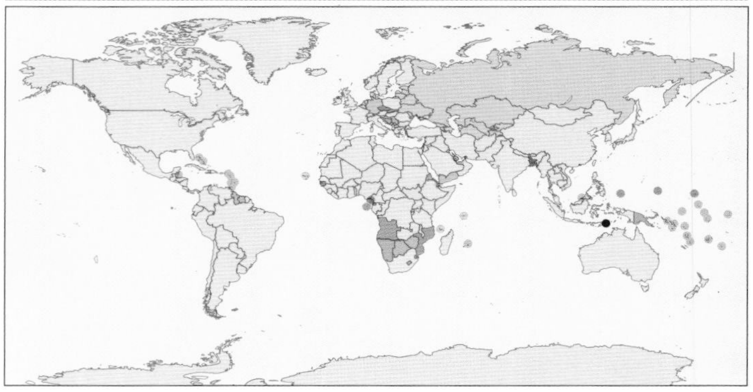

▲ ***Iran has been** one of the modern world's few true theocracies; Islam has an impact on every aspect of political life.*

▲ ***North Korea is** an independent communist republic. Power was transferred directly to Kim Jong-un in 2012 following the death of his father Kim Jong-il.*

Administration at the time of independence

- Australia
- Aust/NZ/UK
- Belgium
- China
- Czechoslovakia
- Egypt/UK
- Ethiopia
- France
- France/UK
- Indonesia
- Italy
- Japan
- Malaysia
- Netherlands
- New Zealand
- Pakistan
- Portugal
- South Africa
- Spain
- Sudan
- UK
- Unified country
- USA
- USSR
- Yugoslavia

◀ ***Afghanistan** has suffered decades of war and occupation resulting in widespread destruction. The hardline Taliban government were ousted by a US-led coalition in 2001 but efforts to stabilise the country are still continuing.*

◀ ***In early 2011,** Egypt underwent a revolution, part of the so called "Arab Spring", which resulted in the ousting of President Hosni Mubarak after nearly 30 years in power.*

▲ ***In Brunei the** Sultan has ruled by decree since 1962; power is closely tied to the royal family. The Sultan's brothers are responsible for finance and foreign affairs.*

Types of government

- Multiparty democracy for more than 10 yrs
- Multiparty democracy within last 10 yrs
- Single-party government
- Military regime
- Theocracy
- Monarchy
- Non-party system
- Transitional regime
- Current civil unrest

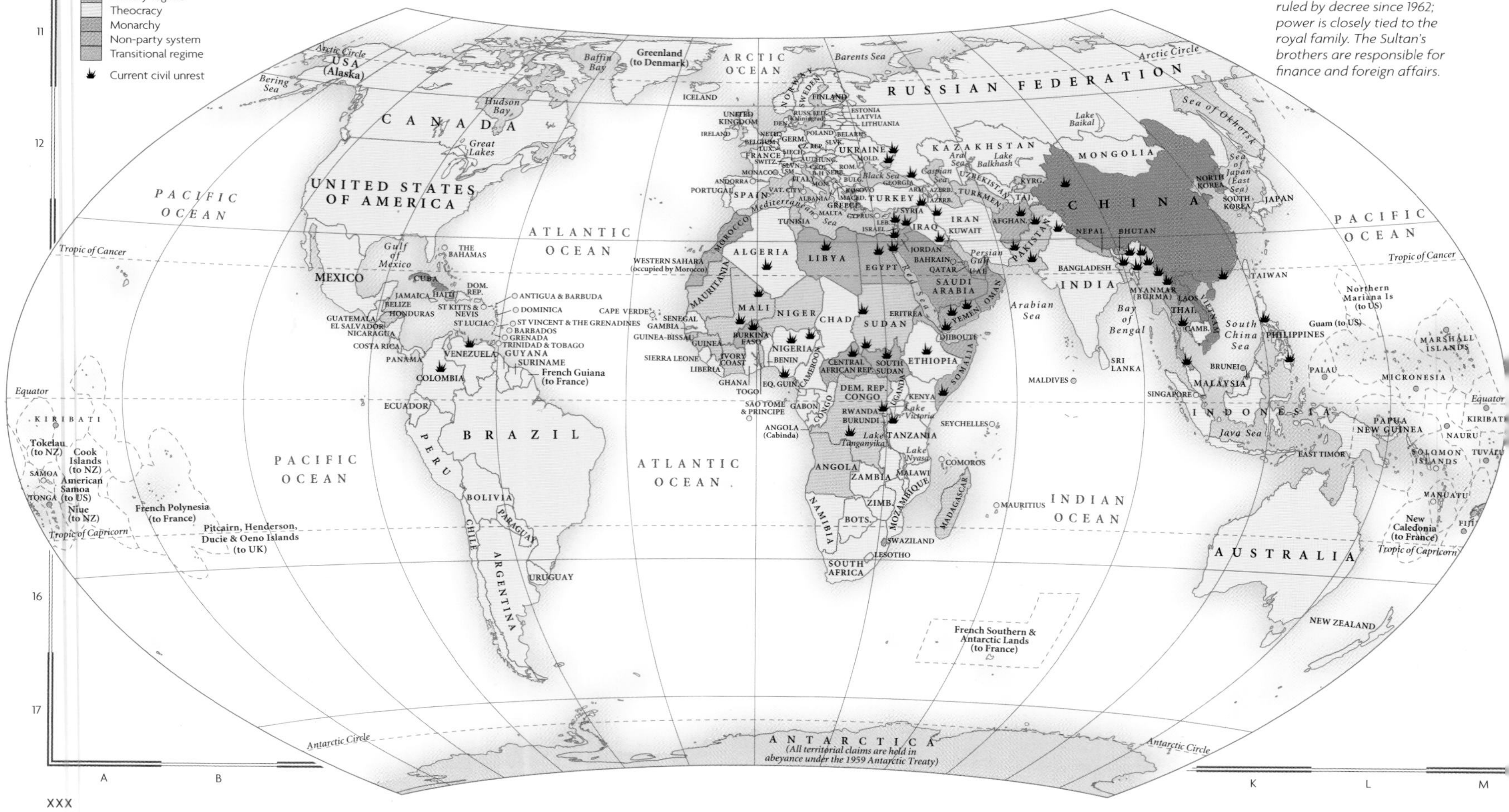

Lines on the map

The determination of international boundaries can use a variety of criteria. Many of the borders between older states follow physical boundaries; some mirror religious and ethnic differences; others are the legacy of complex histories of conflict and colonialism, while others have been imposed by international agreements or arbitration.

Post-colonial borders

When the European colonial empires in Africa were dismantled during the second half of the 20th century, the outlines of the new African states mirrored colonial boundaries. These boundaries had been drawn up by colonial administrators, often based on inadequate geographical knowledge. Such arbitrary boundaries were imposed on people of different languages, racial groups, religions and customs. This confused legacy often led to civil and international war.

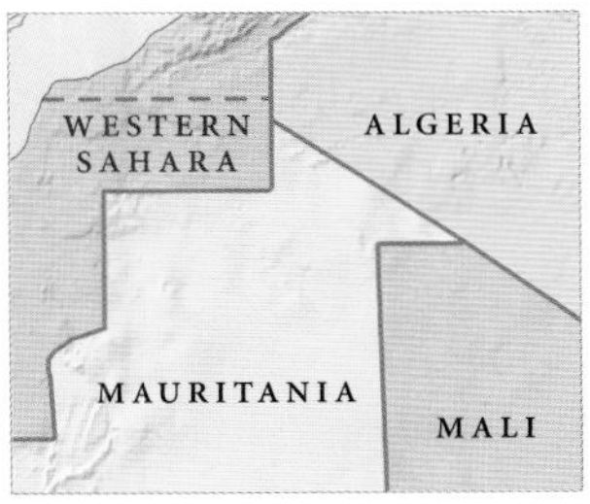

▲ *The conflict that has plagued many African countries since independence has caused millions of people to become refugees.*

Enclaves

The shifting political map over the course of history has frequently led to anomalous situations. Parts of national territories may become isolated by territorial agreement, forming an enclave. The West German part of the city of Berlin, which until 1989 lay a hundred miles (160 km) within East German territory, was a famous example.

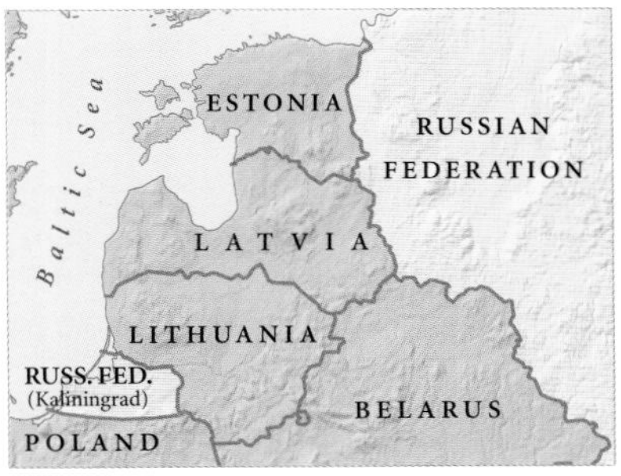

▲ *Since the independence of Lithuania and Belarus, the peoples of the Russian enclave of Kaliningrad have become physically isolated.*

Antarctica

When Antarctic exploration began a century ago, seven nations, Australia, Argentina, Britain, Chile, France, New Zealand and Norway, laid claim to the new territory. In 1961 the Antarctic Treaty, now signed by 45 nations, agreed to hold all territorial claims in abeyance.

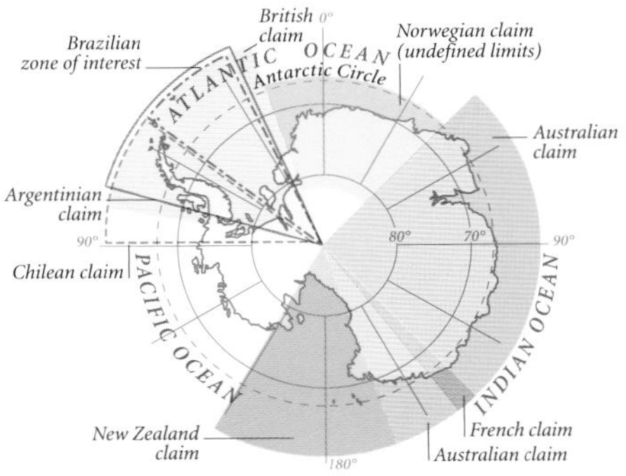

Geometric borders

Straight lines and lines of longitude and latitude have occasionally been used to determine international boundaries; and indeed the world's second longest continuous international boundary, between Canada and the USA, follows the 49th Parallel for over one-third of its course. Many Canadian, American and Australian internal administrative boundaries are similarly determined using a geometric solution.

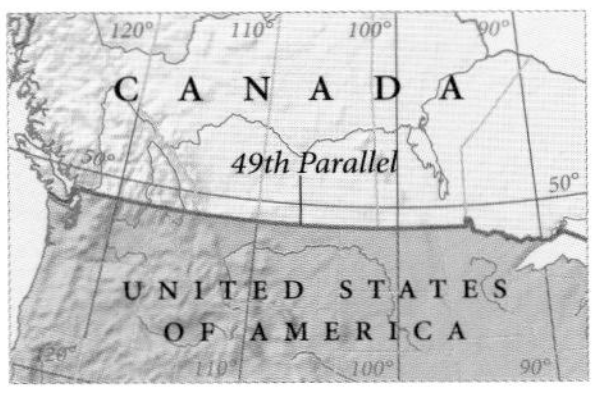

▲ *Different farming techniques in Canada and the USA clearly mark the course of the international boundary in this satellite map.*

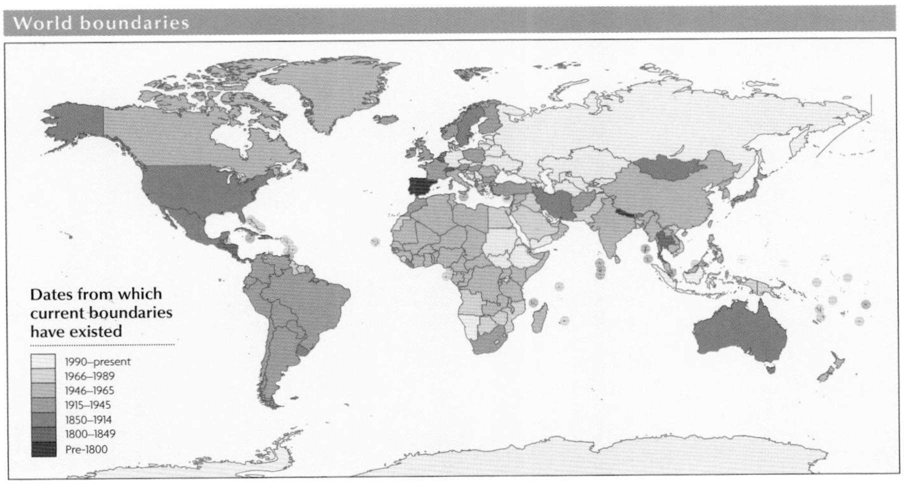

Physical borders

Many of the world's countries are divided by physical borders: lakes, rivers, mountains. The demarcation of such boundaries can, however, lead to disputes. Control of waterways, water supplies and fisheries are frequent causes of international friction.

Lake borders

Countries which lie next to lakes usually fix their borders in the middle of the lake. Unusually the Lake Nyasa border between Malawi and Tanzania runs along Tanzania's shore.

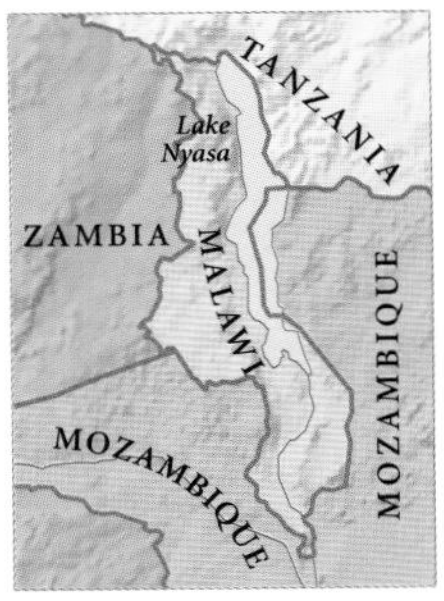

▲ *Complicated agreements between colonial powers led to the awkward division of Lake Nyasa.*

River borders

Rivers alone account for one-sixth of the world's borders. Many great rivers form boundaries between a number of countries. Changes in a river's course and interruptions of its natural flow can lead to disputes, particularly in areas where water is scarce. The centre of the river's course is the nominal boundary line.

▲ *The Danube forms all or part of the border between nine European nations.*

Mountain borders

Mountain ranges form natural barriers and are the basis for many major borders, particularly in Europe and Asia. The watershed is the conventional boundary demarcation line, but its accurate determination is often problematic.

▲ *The Pyrenees form a natural mountain border between France and Spain.*

Shifting boundaries – Poland

Borders between countries can change dramatically over time. The nations of eastern Europe have been particularly affected by changing boundaries. Poland is an example of a country whose boundaries have changed so significantly that it has literally moved around Europe. At the start of the 16th century, Poland was the largest nation in Europe. Between 1772 and 1795, it was absorbed into Prussia, Austria and Russia, and it effectively ceased to exist. After the First World War, Poland became an independent country once more, but its borders changed again after the Second World War following invasions by both Soviet Russia and Nazi Germany.

▲ *In 1634, Poland was the largest nation in Europe, its eastern boundary reaching towards Moscow.*

▲ *From 1772–1795, Poland was gradually partitioned between Austria, Russia and Prussia. Its eastern boundary receded by over 100 miles (160 km).*

▲ *Following the First World War, Poland was reinstated as an independent state, but it was less than half the size it had been in 1634.*

▲ *After the Second World War the Baltic Sea border was extended westwards, but much of the eastern territory was annexed by Russia.*

International disputes

There are more than 60 disputed borders or territories in the world today. Although many of these disputes can be settled by peaceful negotiation, some areas have become a focus for international conflict. Ethnic tensions have been a major source of territorial disagreement throughout history, as has the ownership of, and access to, valuable natural resources. The turmoil of the post-colonial era in many parts of Africa is partly a result of the 19th century 'carve-up' of the continent, which created potential for conflict by drawing often arbitrary lines through linguistic and cultural areas.

North and South Korea

Since 1953, the *de facto* border between North and South Korea has been a ceasefire line which straddles the 38th Parallel and is designated as a demilitarized zone. Both countries have heavy fortifications and troop concentrations behind this zone.

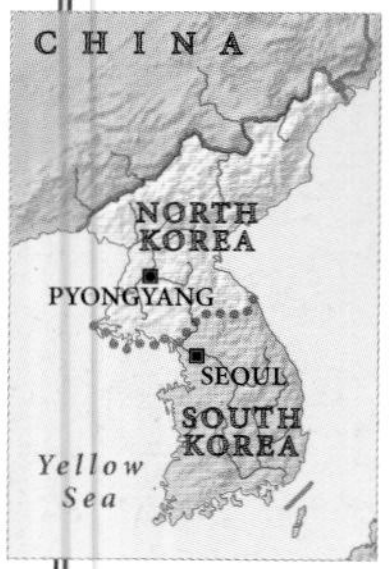

▲ *Heavy fortifications on the border between North and South Korea.*

Cyprus

▲ *The so-called 'green line' divides Cyprus into Greek and Turkish sectors.*

Cyprus was partitioned in 1974, following an invasion by Turkish troops. The south is now the Greek Cypriot Republic of Cyprus, while the self-proclaimed Turkish Republic of Northern Cyprus is recognized only by Turkey.

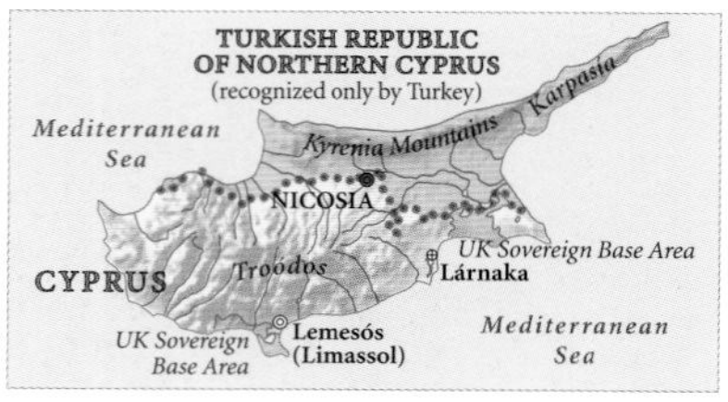

Jammu and Kashmir

Disputes over Jammu and Kashmir have caused three serious wars between India and Pakistan since 1947. Pakistan wishes to annex the largely Muslim territory, while India refuses to cede any territory or to hold a referendum, and also lays claim to the entire territory. Most international maps show the 'line of control' agreed in 1972 as the *de facto* border. In addition India has territorial disputes with neighbouring China. The situation is further complicated by a Kashmiri independence movement, active since the late 1980s.

▲ *Indian army troops maintain their positions in the mountainous terrain of northern Kashmir.*

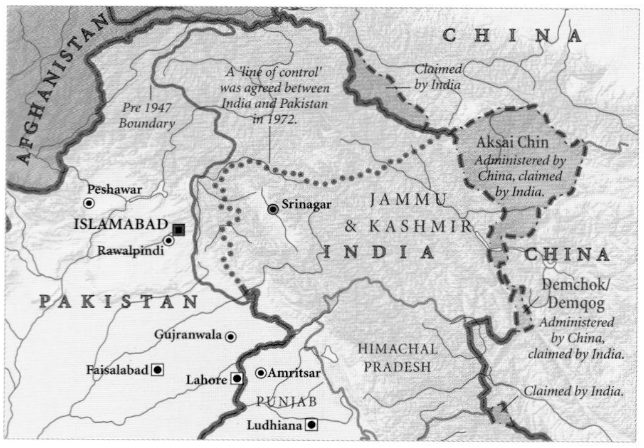

The Falkland Islands

The British dependent territory of the Falkland Islands was invaded by Argentina in 1982, sparking a full-scale war with the UK. Tensions ran high during 2012 in the build up to the thirtieth anniversary of the conflict.

◀ *British warships in Falkland Sound during the 1982 war with Argentina.*

Israel

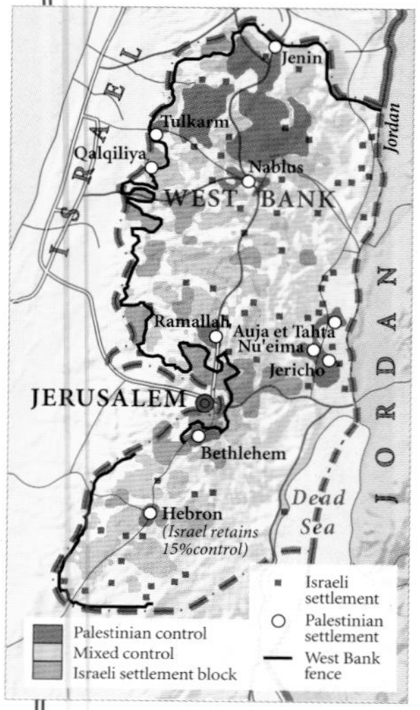

Israel was created in 1948 following the 1947 UN Resolution (147) on Palestine. Until 1979 Israel had no borders, only ceasefire lines from a series of wars in 1948, 1967 and 1973. Treaties with Egypt in 1979 and Jordan in 1994 led to these borders being defined and agreed. Negotiations over Israeli settlements and Palestinian self-government have seen little effective progress since 2000.

▲ *Barbed-wire fences surround a settlement in the Golan Heights.*

Former Yugoslavia

Following the disintegration in 1991 of the communist state of Yugoslavia, the breakaway states of Croatia and Bosnia and Herzegovina came into conflict with the 'parent' state (consisting of Serbia and Montenegro). Warfare focused on ethnic and territorial ambitions in Bosnia. The tenuous Dayton Accord of 1995 sought to recognize the post-1990 borders, whilst providing for ethnic partition and required international peace-keeping troops to maintain the terms of the peace.

▲ *Most claimant states have small military garrisons on the Spratly Islands.*

The Spratly Islands

The site of potential oil and natural gas reserves, the Spratly Islands in the South China Sea have been claimed by China, Vietnam, Taiwan, Malaysia and the Philippines since the Japanese gave up a wartime claim in 1951.

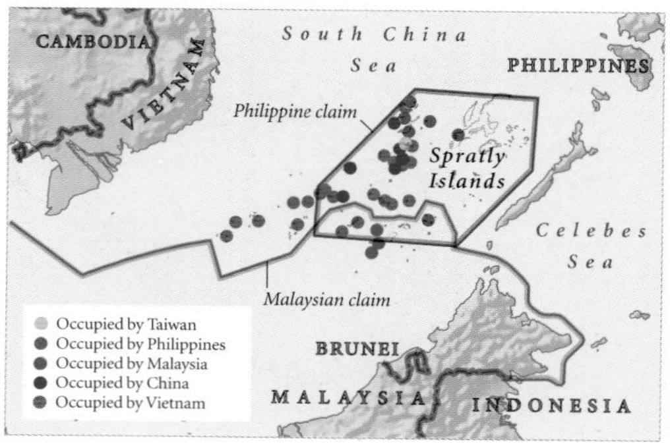

ATLAS
OF THE WORLD

THE MAPS IN THIS ATLAS ARE ARRANGED CONTINENT BY CONTINENT, STARTING FROM THE INTERNATIONAL DATE LINE, AND MOVING EASTWARDS. THE MAPS PROVIDE A UNIQUE VIEW OF TODAY'S WORLD, COMBINING TRADITIONAL CARTOGRAPHIC TECHNIQUES WITH THE LATEST REMOTE-SENSED AND DIGITAL TECHNOLOGY.

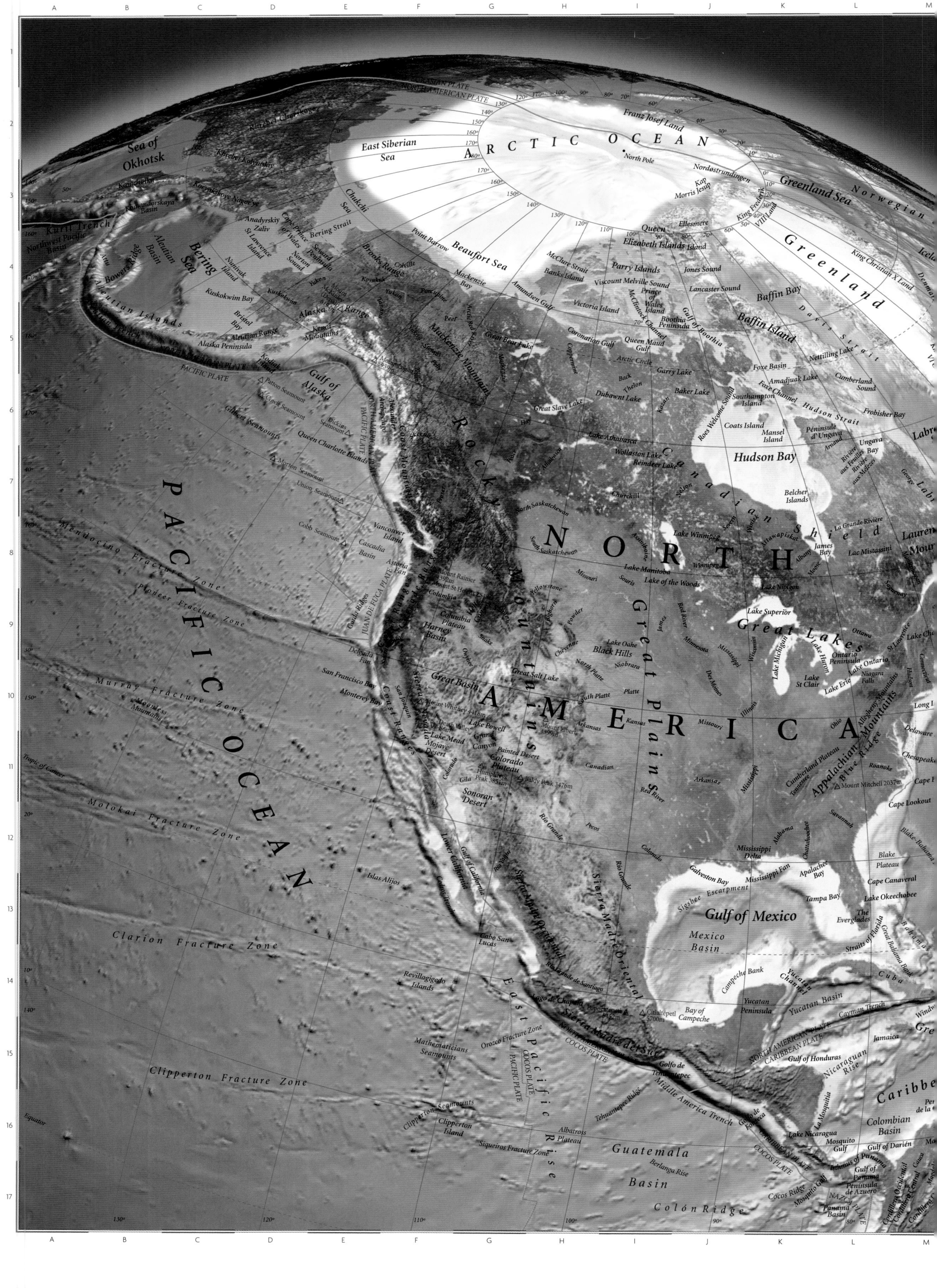
ARCTIC OCEAN
North Pole
Franz Josef Land
East Siberian Sea
Sea of Okhotsk
Chukchi Sea
Bering Strait
Bering Sea
Beaufort Sea
Greenland Sea
Greenland
Baffin Bay
Baffin Island
Davis Strait
Queen Elizabeth Islands
Ellesmere Island
Parry Islands
Banks Island
Victoria Island
Prince of Wales Island
Boothia Peninsula
Gulf of Boothia
Lancaster Sound
Jones Sound
Kuril Trench
Aleutian Islands
Alaska Peninsula
Alaska Range
Brooks Range
Gulf of Alaska
Kodiak Island
Mackenzie Mountains
Great Bear Lake
Great Slave Lake
Lake Athabasca
Hudson Bay
Hudson Strait
James Bay
Canadian Shield
Lake Winnipeg
NORTH AMERICA
Rocky Mountains
Great Plains
Great Lakes
Lake Superior
Lake Michigan
Lake Huron
Lake Erie
Lake Ontario
Appalachian Mountains
Great Basin
Great Salt Lake
Colorado Plateau
Sonoran Desert
Mojave Desert
Grand Canyon
San Francisco Bay
Monterey Bay
Vancouver Island
Queen Charlotte Islands
PACIFIC OCEAN
Mendocino Fracture Zone
Pioneer Fracture Zone
Murray Fracture Zone
Molokai Fracture Zone
Clarion Fracture Zone
Clipperton Fracture Zone
Tropic of Cancer
Equator
Gulf of California
Baja California
Cabo San Lucas
Sierra Madre Occidental
Sierra Madre Oriental
Sierra Madre del Sur
Gulf of Mexico
Mexico Basin
Yucatan Peninsula
Yucatan Channel
Bay of Campeche
Tampa Bay
Cape Canaveral
Straits of Florida
Cuba
Jamaica
Gulf of Honduras
Middle America Trench
East Pacific Rise
Guatemala Basin
Colombian Basin
Isthmus of Panama
Gulf of Panama
Revillagigedo Islands
Clipperton Island
PACIFIC PLATE
NORTH AMERICAN PLATE
JUAN DE FUCA PLATE
COCOS PLATE
CARIBBEAN PLATE
NAZCA PLATE
A B C D E F G H I J K L M
1 2 3 4 5 6 7 8 9 10 11 12 13 14 15 16 17

ATLANTIC OCEAN
MID-ATLANTIC RIDGE
Iceland Basin
Rockall Rise
EURASIAN PLATE
NORTH AMERICAN PLATE
Charlie-Gibbs Fracture Zone
Newfoundland
Flemish Cap
Grand Banks
Newfoundland Basin
Azores
Pico Fracture Zone
Oceanographer Fracture Zone
Georges Bank
Corner Seamounts
Nashville Seamount
Sohm Plain
Atlantis Fracture Zone
Tropic of Cancer
Bermuda
Bermuda Rise
North American Basin
AFRICAN PLATE
Kane Fracture Zone
Sargasso Sea
Nares Plain
Cape Verde Basin
West Indies
Puerto Rico Trench
Barbuda
Antigua
Guadeloupe
Dominica
Martinique
St Lucia
Barbados
SOUTH AMERICAN PLATE
Barracuda Fracture Zone
Lesser Antilles
Venezuelan Basin
Guiana Basin
Demerara Plain
Grenada
Trinidad
Demerara Plateau
CARIBBEAN PLATE
Orinoco
Guiana Highlands
Amazon Fan
Equator

North America

North America is the world's third largest continent with a total area of 9,358,340 sq miles (24,238,000 sq km) including Greenland and the Caribbean islands. It lies wholly within the Northern Hemisphere.

- ○ ***Greatest extent, North–South:*** *4600 miles / 7400 km*
- □ ***Greatest extent, East–West:*** *3500 miles / 5700 km*

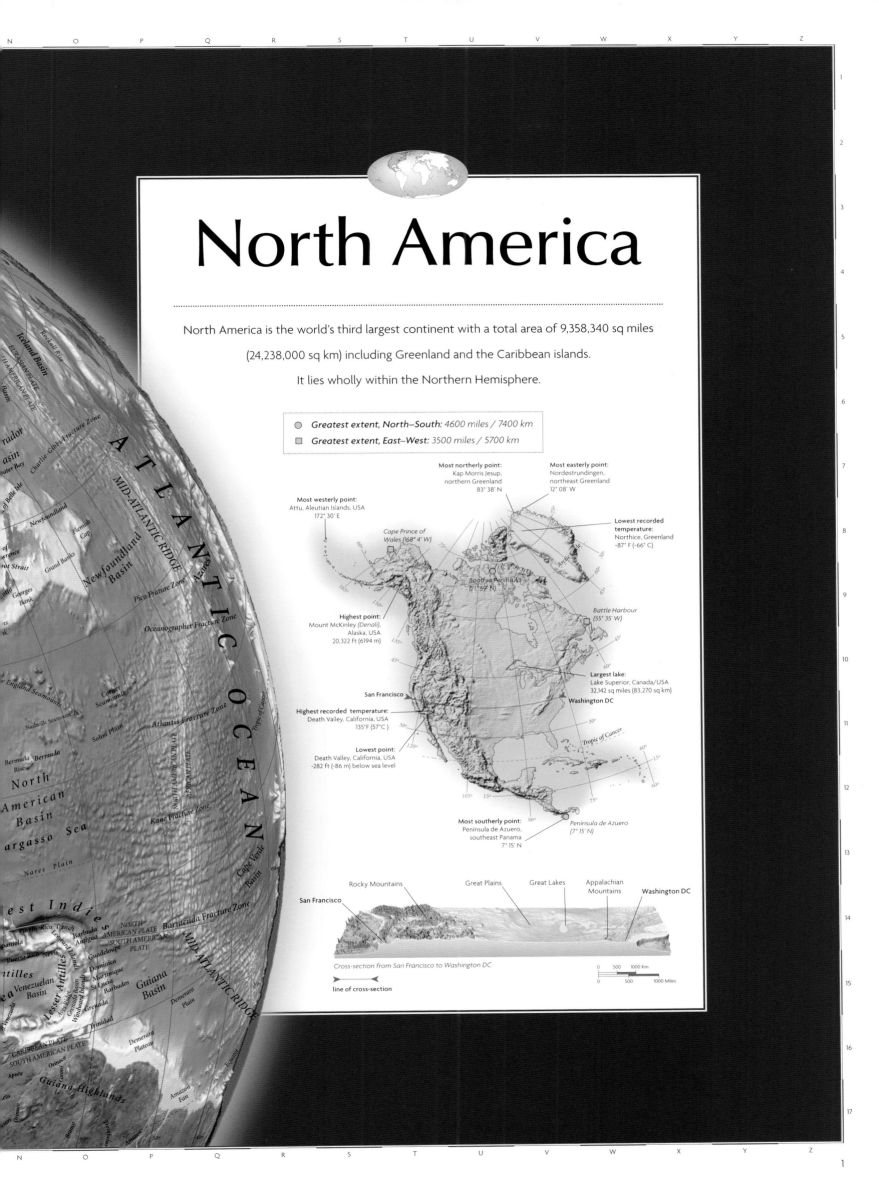

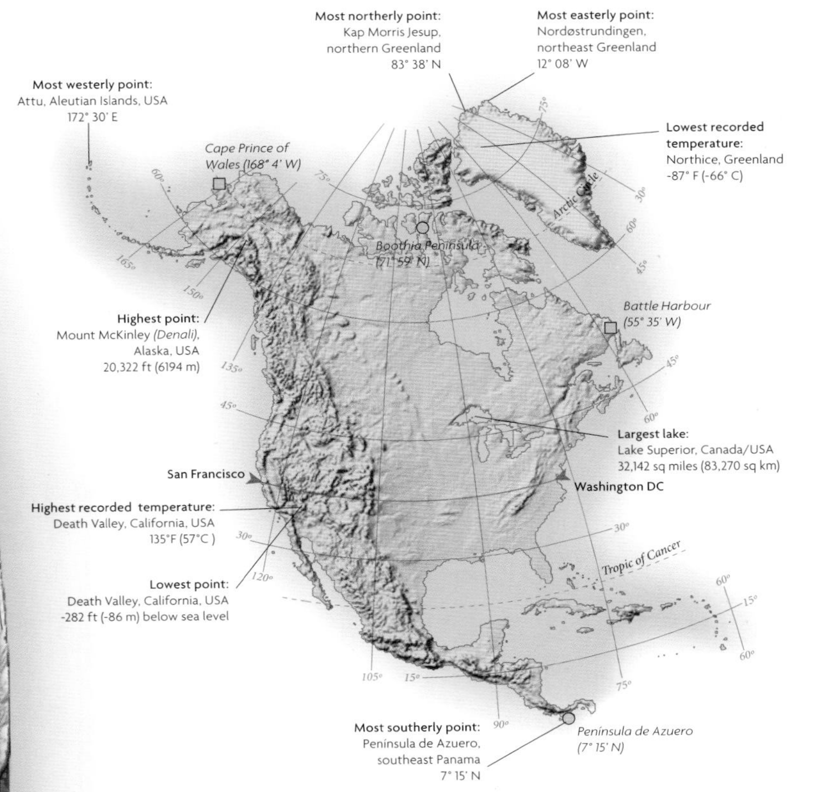
Most northerly point:
Kap Morris Jesup, northern Greenland 83° 38' N
Most easterly point:
Nordøstrundingen, northeast Greenland 12° 08' W
Most westerly point:
Attu, Aleutian Islands, USA 172° 30' E
Cape Prince of Wales (168° 4' W)
Lowest recorded temperature:
Northice, Greenland -87° F (-66° C)
Arctic Circle
Boothia Peninsula (71° 55' N)
Highest point:
Mount McKinley (Denali), Alaska, USA 20,322 ft (6194 m)
Battle Harbour (55° 35' W)
Largest lake:
Lake Superior, Canada/USA 32,142 sq miles (83,270 sq km)
San Francisco
Washington DC
Highest recorded temperature:
Death Valley, California, USA 135°F (57°C)
Lowest point:
Death Valley, California, USA -282 ft (-86 m) below sea level
Tropic of Cancer
Most southerly point:
Peninsula de Azuero, southeast Panama 7° 15' N
Peninsula de Azuero (7° 15' N)

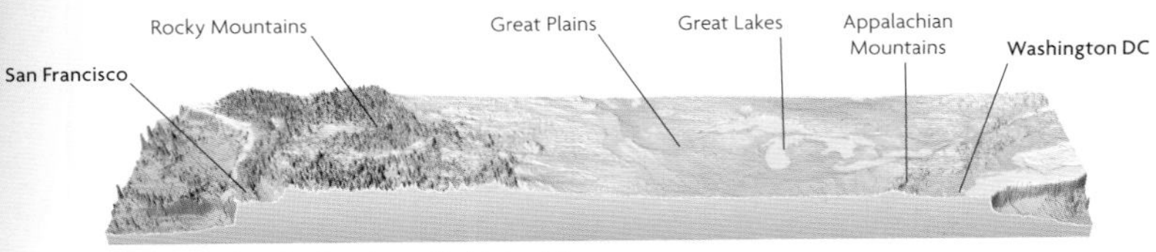
Rocky Mountains
Great Plains
Great Lakes
Appalachian Mountains
Washington DC
San Francisco
0 500 1000 Km
0 500 1000 Miles

Cross-section from San Francisco to Washington DC

line of cross-section

Physical North America

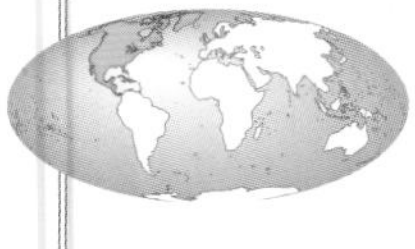

The North American continent can be divided into a number of major structural areas: the Western Cordillera, the Canadian Shield, the Great Plains and Central Lowlands, and the Appalachians. Other smaller regions include the Gulf Atlantic Coastal Plain which borders the southern coast of North America from the southern Appalachians to the Great Plains. This area includes the expanding Mississippi Delta. A chain of volcanic islands, running in an arc around the margin of the Caribbean Plate, lie to the east of the Gulf of Mexico.

The Canadian Shield

Spanning northern Canada and Greenland, this geologically stable plain forms the heart of the continent, containing rocks over two billion years old. A long history of weathering and repeated glaciation has scoured the region, leaving flat plains, gentle hummocks, numerous small basins and lakes, and the bays and islands of the Arctic.

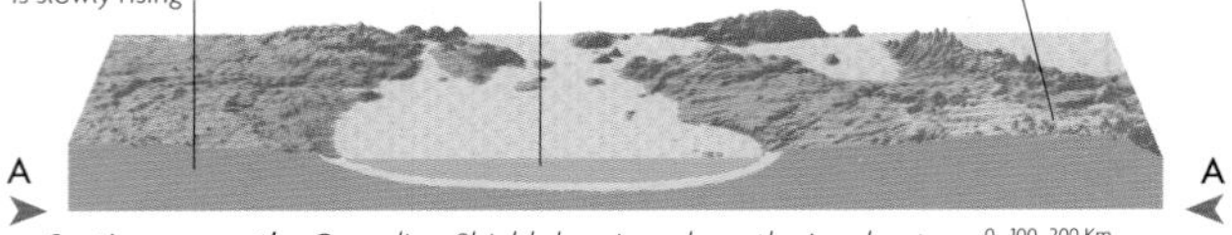
The hard bedrock of the Canadian Shield is slowly rising
Hudson Bay was depressed by the ice sheet to form North America's largest basin
Once overlain by sedimentary rocks, erosion has re-exposed the ancient Laurentian Mountains
A
A
0 100 200 Km
0 100 200 Miles

***Section across the** Canadian Shield showing where the ice sheet has depressed the underlying rock and formed bays and islands.*

The Western Cordillera

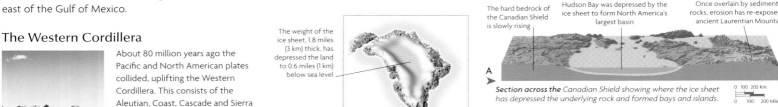

About 80 million years ago the Pacific and North American plates collided, uplifting the Western Cordillera. This consists of the Aleutian, Coast, Cascade and Sierra Nevada mountains, and the inland Rocky Mountains. These run parallel from the Arctic to Mexico.

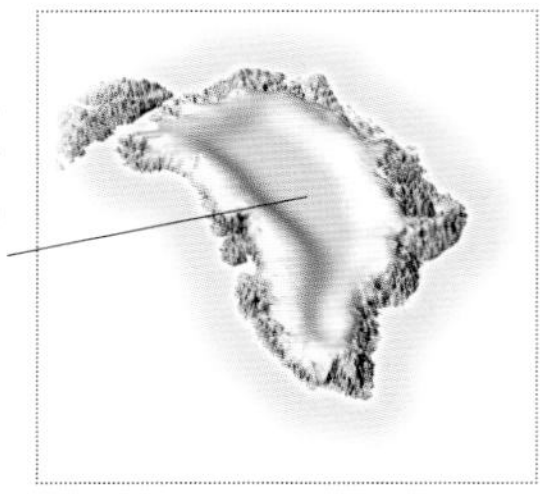
The weight of the ice sheet, 1.8 miles (3 km) thick, has depressed the land to 0.6 miles (1 km) below sea level

▲ ***This computer-generated view** shows the ice-covered island of Greenland without its ice cap.*

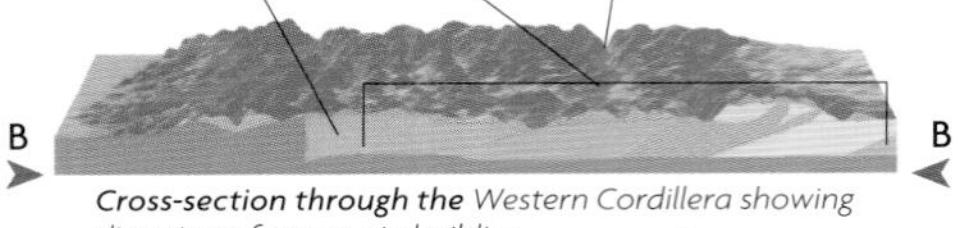
Strata have been thrust eastward along fault lines
The Rocky Mountain Trench is the longest linear fault on the continent
Volcanic rock
B
B
0 50 100 Km
0 50 100 Miles

***Cross-section through the** Western Cordillera showing direction of mountain building.* ⟶

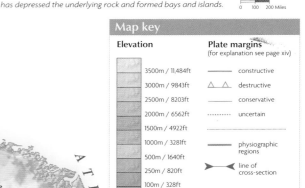

Map key

Elevation

- 3500m / 11,484ft
- 3000m / 9843ft
- 2500m / 8203ft
- 2000m / 6562ft
- 1500m / 4922ft
- 1000m / 3281ft
- 500m / 1640ft
- 250m / 820ft
- 100m / 328ft
- sea level

Plate margins
(for explanation see page xiv)

- constructive
- destructive
- conservative
- uncertain
- physiographic regions
- line of cross-section

Scale 1:42,000,000

Km 0 200 400 600 800 1000
Miles 0 200 400 600 800 1000

projection: Lambert Azimuthal Equal Area

ASIA
Bering Strait
Beaufort Sea
Bering Sea
Aleutian Islands
Aleutian Range
Alaska Range
Mount McKinley 6194m
Brooks Range
Mackenzie Delta
Mackenzie Mountains
Mackenzie
Great Bear Lake
Greenland
Baffin Bay
Baffin Island
Davis Strait
Foxe Basin
Hudson Strait
Labrador Sea
Labrador
Laurentian Mountains
Newfoundland
Hudson Bay
ATLANTIC OCEAN
Gulf of Alaska
NORTH AMERICAN PLATE
PACIFIC PLATE
JUAN DE FUCA PLATE
Coast Mountains
WESTERN CORDILLERA
ROCKY MOUNTAINS
CANADIAN SHIELD
CENTRAL LOWLANDS
GREAT PLAINS
Great Slave Lake
Lake Athabasca
Reindeer Lake
Lake Winnipeg
Lake Manitoba
Lake Superior
Lake Huron
Lake Ontario
Lake Michigan
Lake Erie
Great Lakes
St Lawrence
Nova Scotia
Cape Cod
Mount Rainier 4392m
Mount St Helens 2549m
Cascade Range
Great Basin
Great Salt Lake
Sierra Nevada
San Joaquin Valley
San Andreas Fault
Death Valley -86m
Colorado
Colorado Plateau
Grand Canyon
Mojave Desert
Sonoran Desert
Missouri
Ohio
Arkansas
Mississippi
Appalachian Mountains
APPALACHIANS
GULF ATLANTIC COASTAL PLAIN
Rio Grande
Mississippi Delta
Lower California
Gulf of California
Sierra Madre Occidental
Sierra Madre Oriental
Sierra Madre del Sur
Volcán Pico de Orizaba 5700m
Gulf of Mexico
Yucatan Peninsula
West Indies
Greater Antilles
Lesser Antilles
Caribbean Sea
Lake Nicaragua
Isthmus of Panama
CARIBBEAN PLATE
SOUTH AMERICAN PLATE
SOUTH AMERICA
PACIFIC OCEAN

The Great Plains and Central Lowlands

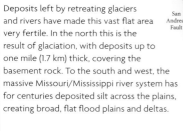

Deposits left by retreating glaciers and rivers have made this vast flat area very fertile. In the north this is the result of glaciation, with deposits up to one mile (1.7 km) thick, covering the basement rock. To the south and west, the massive Missouri/Mississippi river system has for centuries deposited silt across the plains, creating broad, flat flood plains and deltas.

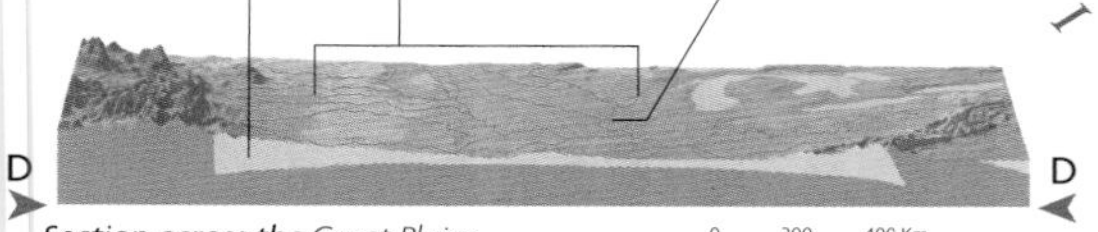
Sedimentary layers overlay domed basement rock
Upland rivers drain south towards the Mississippi Basin
Confluence of the Missouri and Mississippi rivers
D
D
0 200 400 Km
0 200 400 Miles

***Section across the** Great Plains and Central Lowlands showing river systems and structure.*

The Appalachians

The Appalachian Mountains, uplifted about 400 million years ago, are some of the oldest in the world. They have been lowered and rounded by erosion and now slope gently towards the Atlantic across a broad coastal plain.

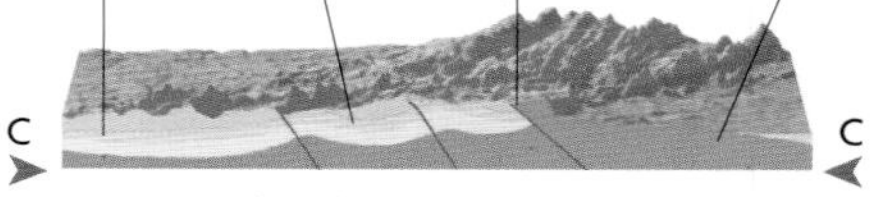
Horizontal strata
Sedimentary strata folded and faulted into ridges and valleys
Softer strata has been crumpled against the harder basement rock
Hard basement rock
C
C
0 25 50 Km
0 25 50 Miles

***Cross-section through the** Appalachians showing the numerous folds, which have subsequently been weathered to create a rounded relief.*

N O P Q R S T U V W X Y Z

Climate

▲ *'Tornado alley' in the Mississippi Valley suffers frequent tornadoes.*

North America's climate includes extremes ranging from freezing Arctic conditions in Alaska and Greenland, to desert in the southwest, and tropical conditions in southeastern Florida, the Caribbean and Central America. Central and southern regions are prone to severe storms including tornadoes and hurricanes.

Climate

- ice cap
- tundra
- subarctic
- cool continental
- warm humid
- semi-arid
- arid
- humid equatorial
- tropical
- daily hours of sunshine, January
- daily hours of sunshine, July
- direction of hurricanes
- tornado zones

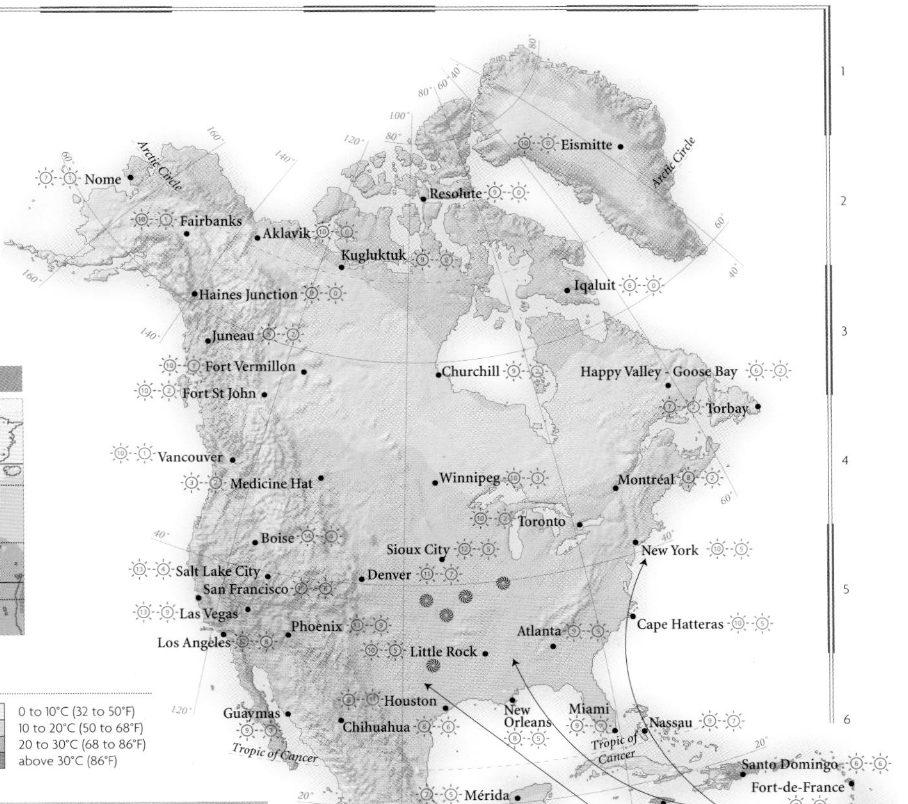

Nome
Arctic Circle
Fairbanks
Aklavik
Resolute
Eismitte
Arctic Circle
Kugluktuk
Iqaluit
Haines Junction
Juneau
Fort Vermillon
Fort St John
Churchill
Happy Valley - Goose Bay
Torbay
Vancouver
Medicine Hat
Winnipeg
Montréal
Toronto
Boise
Sioux City
New York
Salt Lake City
San Francisco
Denver
Las Vegas
Phoenix
Cape Hatteras
Los Angeles
Little Rock
Atlanta
Houston
Guaymas
Chihuahua
New Orleans
Miami
Nassau
Tropic of Cancer
Tropic of Cancer
Santo Domingo
Fort-de-France
Mérida
Kingston
Acapulco
San Salvador
San José

▲ *Much of the southwest is semi-desert; receiving less than 12 inches (300 mm) of rainfall a year.*

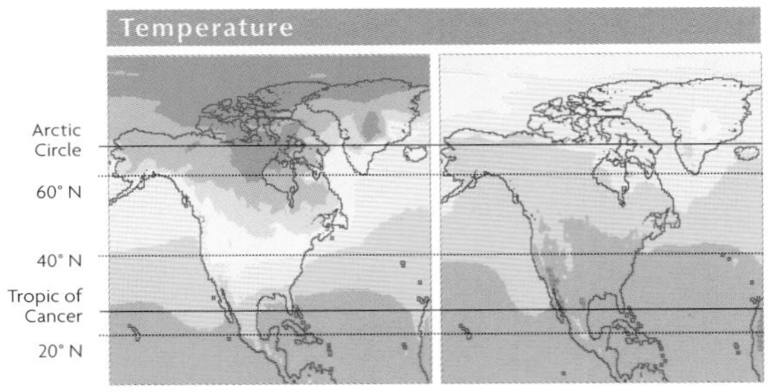

Temperature
Arctic Circle
60° N
40° N
Tropic of Cancer
20° N

Average January temperature *Average July temperature*

Temperature

- below -30°C (-22°F)
- -30 to -20°C (-22 to -4°F)
- -20 to -10°C (-4 to 14°F)
- -10 to 0°C (14 to 32°F)
- 0 to 10°C (32 to 50°F)
- 10 to 20°C (50 to 68°F)
- 20 to 30°C (68 to 86°F)
- above 30°C (86°F)

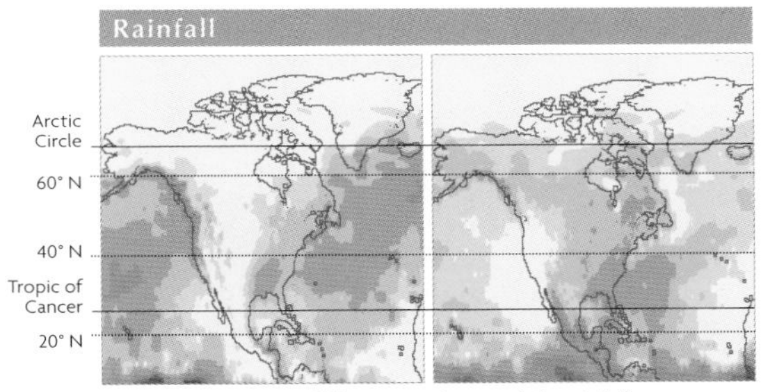

Rainfall
Arctic Circle
60° N
40° N
Tropic of Cancer
20° N

Average January rainfall *Average July rainfall*

Rainfall

- 0–25 mm (0–1 in)
- 25–50 mm (1–2 in)
- 50–100 mm (2–4 in)
- 100–200 mm (4–8 in)
- 200–300 mm (8–12 in)
- 300–400 mm (12–16 in)
- 400–500 mm (16–20 in)
- more than 500 mm (20 in)

◀ *The lush, green mountains of the Lesser Antilles receive annual rainfalls of up to 360 inches (9000 mm).*

Shaping the continent

Glacial processes affect much of northern Canada, Greenland and the Western Cordillera. Along the western coast of North America, Central America and the Caribbean, underlying plates moving together lead to earthquakes and volcanic eruptions. The vast river systems, fed by mountain streams, constantly erode and deposit material along their paths.

Volcanic activity

1 Mount St Helens volcano *(right)* in the Cascade Range erupted violently in May 1980, killing 57 people and levelling large areas of forest. The lateral blast filled a valley for 15 miles (25 km) with debris.

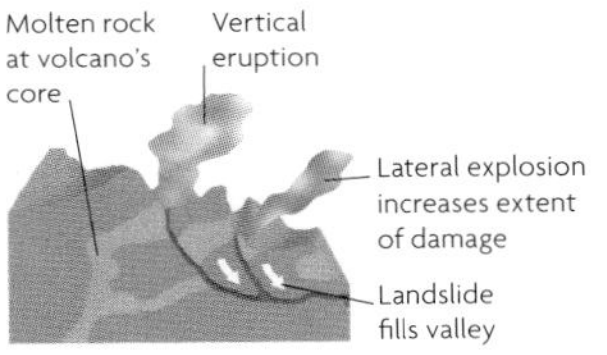

Molten rock at volcano's core
Vertical eruption
Lateral explosion increases extent of damage
Landslide fills valley

***Volcanic activity:** Eruption of Mount St Helens*

Periglaciation

2 The ground in the far north is nearly always frozen: the surface thaws only in summer. This freeze-thaw process produces features such as pingos *(left)*; formed by the freezing of groundwater. With each successive winter ice accumulates producing a mound with a core of ice.

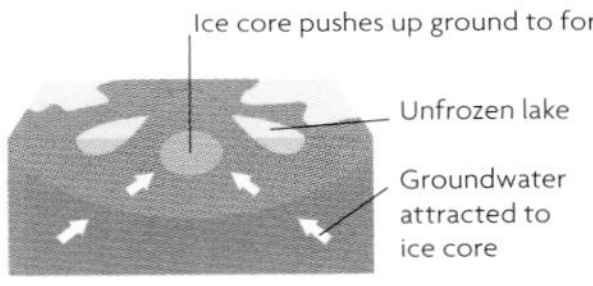

Ice core pushes up ground to form pingo
Unfrozen lake
Groundwater attracted to ice core

***Periglaciation:** Formation of a pingo in the Mackenzie Delta*

Post-glacial lakes

3 A chain of lakes from Great Bear Lake to the Great Lakes *(above)* was created as the ice retreated northwards. Glaciers scoured hollows in the softer lowland rock. Glacial deposits at the lip of the hollows, and ridges of harder rock, trapped water to form lakes.

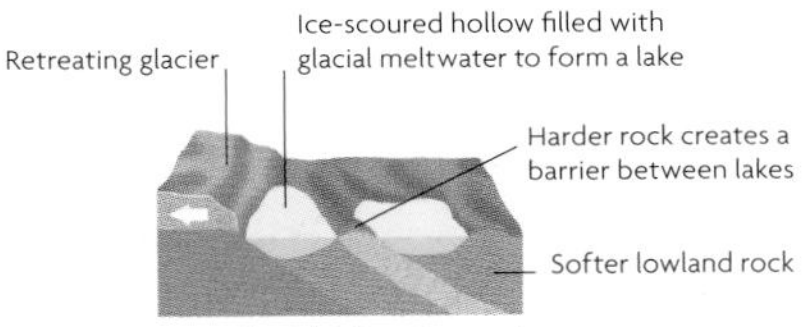

Retreating glacier
Ice-scoured hollow filled with glacial meltwater to form a lake
Harder rock creates a barrier between lakes
Softer lowland rock

***Post-glacial lakes:** Formation of the Great Lakes*

The evolving landscape

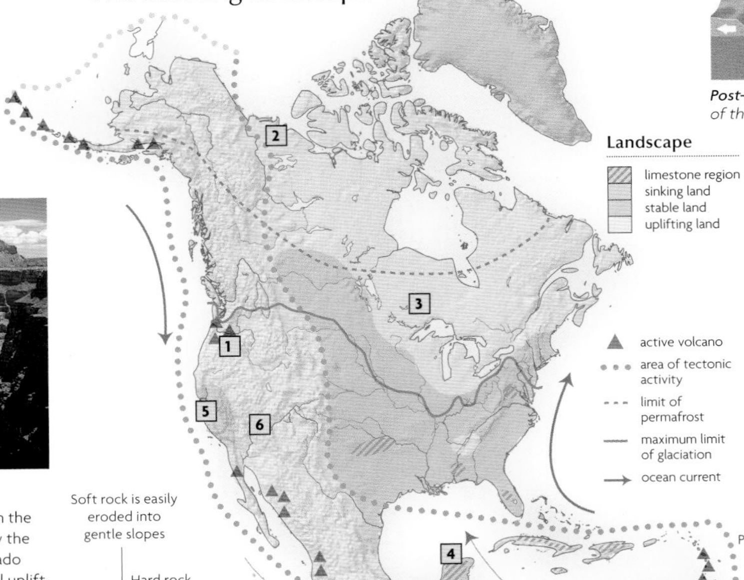

1
2
3
4
5
6

Landscape

- limestone region
- sinking land
- stable land
- uplifting land
- active volcano
- area of tectonic activity
- limit of permafrost
- maximum limit of glaciation
- ocean current

Seismic activity

5 The San Andreas Fault *(above)* places much of the North America's west coast under constant threat from earthquakes. It is caused by the Pacific Plate grinding past the North American Plate at a faster rate, though in the same direction.

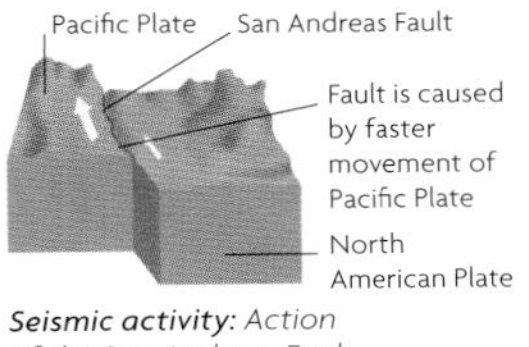

Pacific Plate
San Andreas Fault
Fault is caused by faster movement of Pacific Plate
North American Plate

***Seismic activity:** Action of the San Andreas Fault*

River erosion

6 The Grand Canyon *(above)* in the Colorado Plateau was created by the downward erosion of the Colorado River, combined with the gradual uplift of the plateau, over the past 30 million years. The contours of the canyon formed as the softer rock layers eroded into gentle slopes, and the hard rock layers into cliffs. The depth varies from 3855–6560 ft (1175–2000 m).

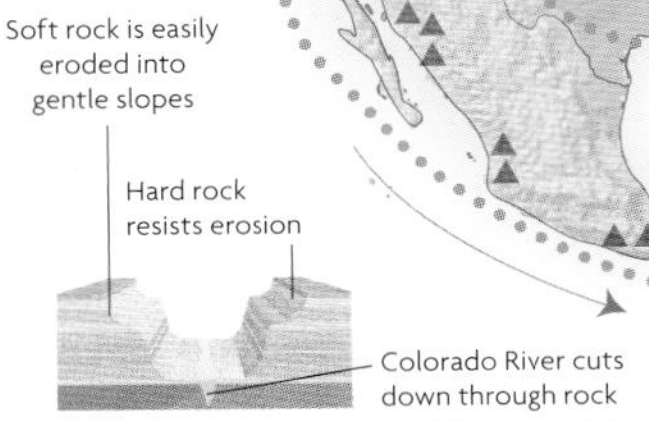

Soft rock is easily eroded into gentle slopes
Hard rock resists erosion
Colorado River cuts down through rock

***River Erosion:** Formation of the Grand Canyon*

Weathering

4 The Yucatan Peninsula is a vast, flat limestone plateau in southern Mexico. Weathering action from both rainwater and underground streams has enlarged fractures in the rock to form caves and hollows, called sinkholes *(above)*.

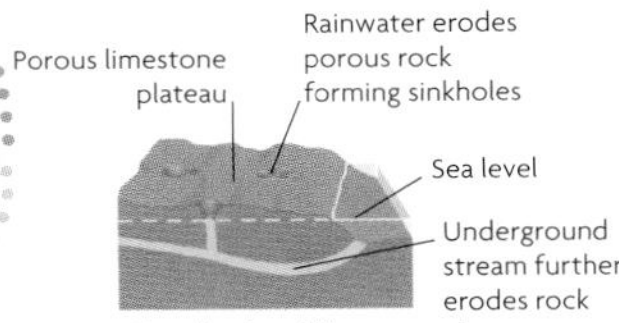

Porous limestone plateau
Rainwater erodes porous rock forming sinkholes
Sea level
Underground stream further erodes rock

***Weathering:** Water erosion on the Yucatan Peninsula*

N O P Q R S T U V W X Y Z

Political North America

Democracy is well established in some parts of the continent but is a recent phenomenon in others. The economically dominant nations of Canada and the USA have a long democratic tradition but elsewhere, notably in the countries of Central America, political turmoil has been more common. In Nicaragua and Haiti, harsh dictatorships have only recently been superseded by democratically-elected governments. North America's largest countries, Canada, Mexico and the USA have federal state systems, sharing political power between national and state governments. The USA has intervened militarily on several occasions in Central America and the Caribbean to protect its strategic interests.

Transport

In the 19th century, railways were used to open up the North American continent. Air transport is now more common for long distance passenger travel, although railways are still extensively used for bulk freight transport. Waterways, like the Mississippi River, are important for the transport of bulk materials, and the Panama Canal is a vital link between the Pacific Ocean and the Caribbean. In the 20th century, road transport increased massively in North America, with the introduction of cheap, mass-produced motor cars and extensive highway construction.

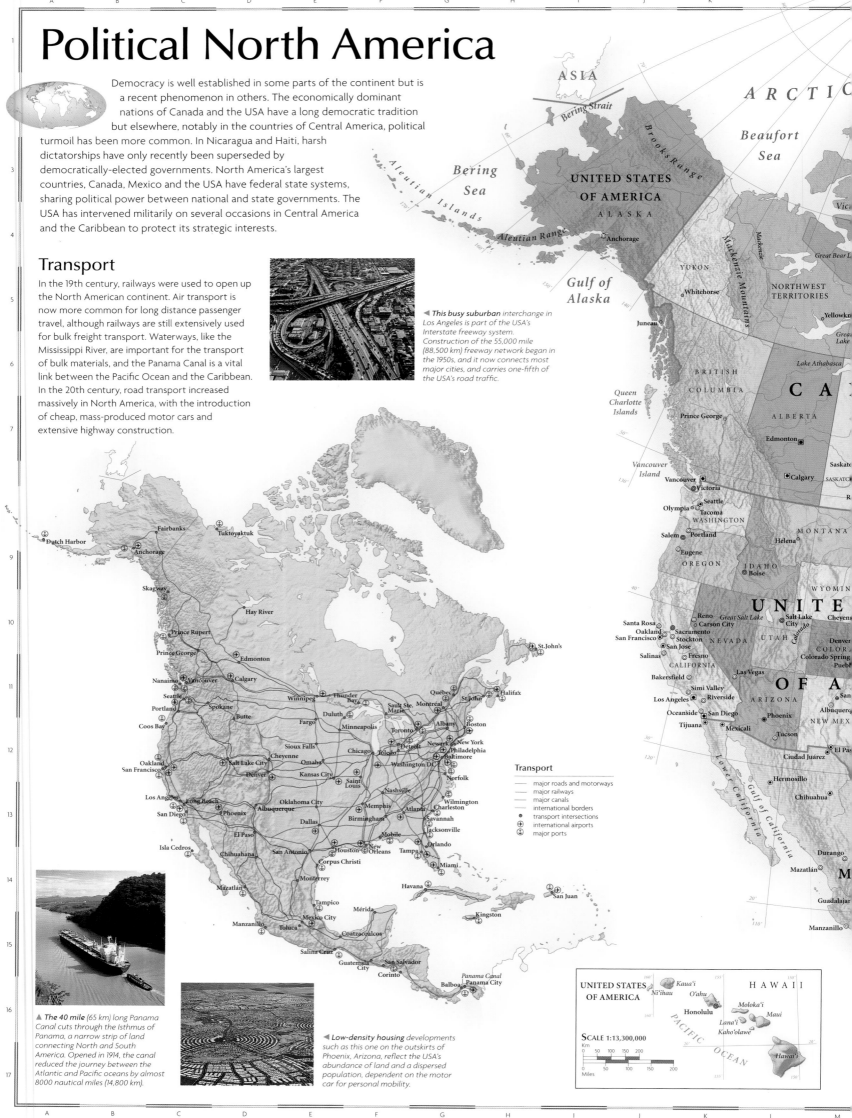

◀ *This busy suburban* interchange in Los Angeles is part of the USA's Interstate freeway system. Construction of the 55,000 mile (88,500 km) freeway network began in the 1950s, and it now connects most major cities, and carries one-fifth of the USA's road traffic.

ASIA
Bering Strait
ARCTIC
Beaufort Sea
Brooks Range
Bering Sea
Aleutian Islands
UNITED STATES OF AMERICA
ALASKA
Aleutian Range
Anchorage
Gulf of Alaska
YUKON
Whitehorse
Mackenzie Mountains
NORTHWEST TERRITORIES
Great Bear L.
Yellowknife
Great Lake
Juneau
Lake Athabasca
BRITISH COLUMBIA
Queen Charlotte Islands
Prince George
ALBERTA
Edmonton
Vancouver Island
Vancouver
Victoria
Calgary
Seattle
Olympia
Tacoma
WASHINGTON
Salem
Portland
Helena
MONTANA
Eugene
OREGON
IDAHO
Boise
WYOMING
Santa Rosa
Oakland
San Francisco
Reno
Carson City
Great Salt Lake
Salt Lake City
Sacramento
Stockton
San Jose
NEVADA
UTAH
Colorado
Denver
Colorado Springs
Salinas
Fresno
CALIFORNIA
Las Vegas
Bakersfield
Simi Valley
Los Angeles
Riverside
ARIZONA
Oceanside
San Diego
Phoenix
NEW MEX
Albuquerque
Tijuana
Mexicali
Tucson
El Paso
Ciudad Juárez
Hermosillo
Chihuahua
Lower California
Gulf of California
Durango
Mazatlán
Guadalajara
Manzanillo

Dutch Harbor
Fairbanks
Anchorage
Tuktoyaktuk
Skagway
Hay River
Prince Rupert
Prince George
Edmonton
St.John's
Nanaimo
Vancouver
Calgary
Seattle
Portland
Spokane
Butte
Coos Bay
Winnipeg
Thunder Bay
Québec
St.John
Halifax
Sault Ste. Marie
Montréal
Duluth
Fargo
Minneapolis
Albany
Boston
Toronto
Detroit
Newark
New York
Philadelphia
Baltimore
Sioux Falls
Chicago
Toledo
Cheyenne
Oakland
San Francisco
Salt Lake City
Omaha
Washington DC
Denver
Kansas City
Saint Louis
Norfolk
Nashville
Wilmington
Charleston
Los Angeles
Long Beach
San Diego
Phoenix
Albuquerque
Oklahoma City
Memphis
Atlanta
Birmingham
Savannah
Dallas
El Paso
Mobile
Jacksonville
Orlando
Isla Cedros
Chihuahua
San Antonio
Houston
New Orleans
Tampa
Corpus Christi
Miami
Monterrey
Mazatlán
Havana
San Juan
Tampico
Mérida
Kingston
Manzanillo
Mexico City
Toluca
Coatzacoalcos
Salina Cruz
Guatemala City
San Salvador
Corinto
Panama Canal
Balboa
Panama City

Transport

- major roads and motorways
- major railways
- major canals
- international borders
- transport intersections
- international airports
- major ports

▲ *The 40 mile* (65 km) long Panama Canal cuts through the Isthmus of Panama, a narrow strip of land connecting North and South America. Opened in 1914, the canal reduced the journey between the Atlantic and Pacific oceans by almost 8000 nautical miles (14,800 km).

◀ *Low-density housing developments* such as this one on the outskirts of Phoenix, Arizona, reflect the USA's abundance of land and a dispersed population, dependent on the motor car for personal mobility.

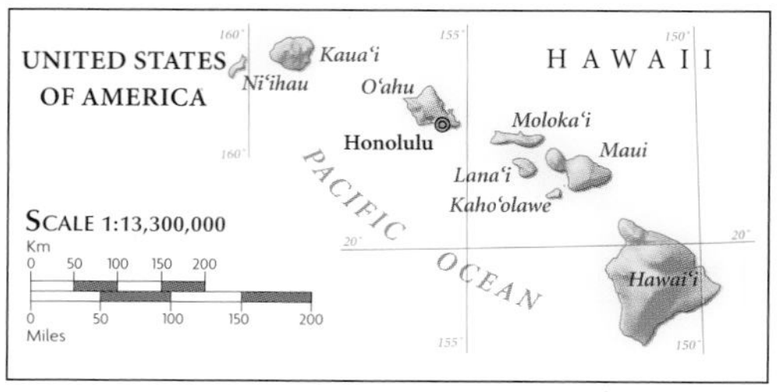

UNITED STATES OF AMERICA
HAWAII
Kaua'i
Ni'ihau
O'ahu
Honolulu
Moloka'i
Maui
Lana'i
Kaho'olawe
PACIFIC OCEAN
Hawai'i
SCALE 1:13,300,000
Km
0 50 100 150 200
Miles
0 50 100 150 200

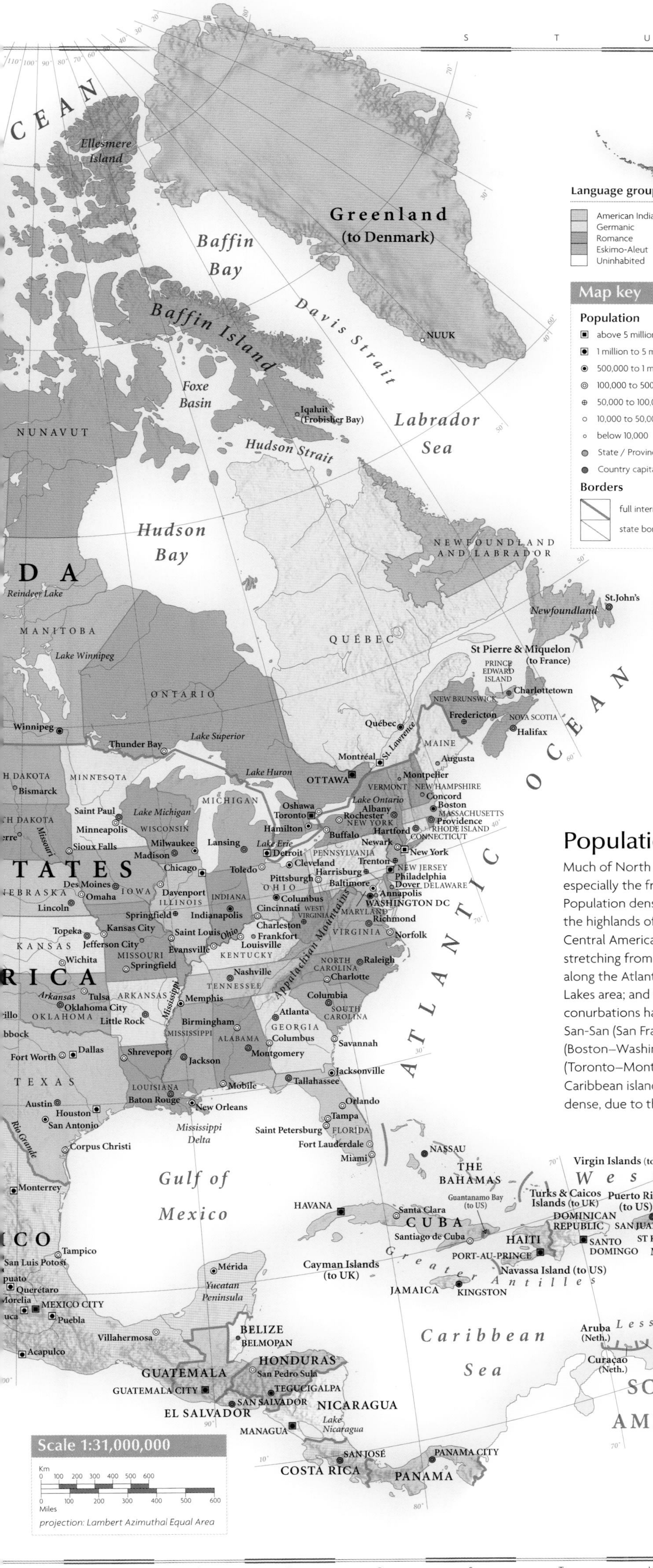

S T U V W Z
1 2 3 8 9 10 11 14 15 16 17
CEAN
Ellesmere Island
Greenland
(to Denmark)
Baffin Bay
Baffin Island
Davis Strait
NUUK
Foxe Basin
Iqaluit (Frobisher Bay)
Labrador Sea
NUNAVUT
Hudson Strait
Hudson Bay
DA
Reindeer Lake
MANITOBA
Lake Winnipeg
ONTARIO
QUÉBEC
NEWFOUNDLAND AND LABRADOR
Newfoundland
St.John's
St Pierre & Miquelon (to France)
PRINCE EDWARD ISLAND
Charlottetown
NEW BRUNSWICK
Fredericton
NOVA SCOTIA
Halifax
Winnipeg
Thunder Bay
Lake Superior
Québec
St. Lawrence
MAINE
Augusta
Montréal
OTTAWA
Lake Huron
H DAKOTA
Bismarck
MINNESOTA
VERMONT
Montpelier
NEW HAMPSHIRE
Concord
MICHIGAN
Oshawa
Toronto
Lake Ontario
Albany
Boston
MASSACHUSETTS
Providence
RHODE ISLAND
Hartford
CONNECTICUT
Rochester
NEW YORK
Buffalo
Saint Paul
Minneapolis
Lake Michigan
WISCONSIN
Hamilton
Lake Erie
Detroit
Sioux Falls
Milwaukee
Lansing
Madison
Chicago
Toledo
Cleveland
PENNSYLVANIA
Newark
New York
Trenton
NEW JERSEY
Philadelphia
Harrisburg
Pittsburgh
Baltimore
Dover
DELAWARE
Annapolis
WASHINGTON DC
TATES
Des Moines
IOWA
Omaha
NEBRASKA
Lincoln
Davenport
ILLINOIS
INDIANA
Indianapolis
Springfield
OHIO
Columbus
Cincinnati
WEST VIRGINIA
MARYLAND
Richmond
Charleston
Frankfort
Louisville
Topeka
Kansas City
Saint Louis
Ohio
VIRGINIA
Norfolk
KANSAS
Jefferson City
Evansville
KENTUCKY
Wichita
MISSOURI
Springfield
Nashville
NORTH CAROLINA
Raleigh
Charlotte
RICA
Tulsa
ARKANSAS
TENNESSEE
Appalachian Mountains
Arkansas
Oklahoma City
Memphis
Columbia
OKLAHOMA
Little Rock
SOUTH CAROLINA
Atlanta
Birmingham
GEORGIA
Columbus
MISSISSIPPI
ALABAMA
Savannah
Fort Worth
Dallas
Shreveport
Montgomery
Jackson
Jacksonville
TEXAS
LOUISIANA
Mobile
Tallahassee
Baton Rouge
New Orleans
Austin
Houston
San Antonio
Orlando
Tampa
FLORIDA
Saint Petersburg
Mississippi Delta
Rio Grande
Fort Lauderdale
Miami
Corpus Christi
ATLANTIC OCEAN
NASSAU
THE BAHAMAS
British Virgin Islands (to UK)
Virgin Islands (to US)
Anguilla (to UK)
Gulf of Mexico
Monterrey
Guantanamo Bay (to US)
Turks & Caicos Islands (to UK)
Puerto Rico (to US)
West Indies
ANTIGUA & BARBUDA
HAVANA
Santa Clara
CUBA
DOMINICAN REPUBLIC
SAN JUAN
Guadeloupe (to France)
Santiago de Cuba
HAITI
SANTO DOMINGO
ST KITTS & NEVIS
DOMINICA
Montserrat (to UK)
Martinique (to France)
ICO
Tampico
PORT-AU-PRINCE
ST LUCIA
San Luis Potosí
Mérida
Cayman Islands (to UK)
Greater Antilles
Navassa Island (to US)
BARBADOS
ST VINCENT & THE GRENADINES
Querétaro
Yucatan Peninsula
JAMAICA
KINGSTON
GRENADA
Morelia
MEXICO CITY
Puebla
Lesser Antilles
TRINIDAD & TOBAGO
PORT-OF-SPAIN
BELIZE
BELMOPAN
Villahermosa
Aruba (Neth.)
Caribbean Sea
Acapulco
HONDURAS
San Pedro Sula
Bonaire (to Neth.)
Curaçao (Neth.)
GUATEMALA
GUATEMALA CITY
TEGUCIGALPA
SOUTH AMERICA
SAN SALVADOR
NICARAGUA
EL SALVADOR
Lake Nicaragua
MANAGUA
SAN JOSÉ
PANAMA CITY
COSTA RICA
PANAMA
Map key
Population
above 5 million
1 million to 5 million
500,000 to 1 million
100,000 to 500,000
50,000 to 100,000
10,000 to 50,000
below 10,000
State / Province capital
Country capital
Borders
full international border
state border
Scale 1:31,000,000
Km
0 100 200 300 400 500 600
0 100 200 300 400 500 600
Miles
projection: Lambert Azimuthal Equal Area
N O P Q R S T U V W X Y Z

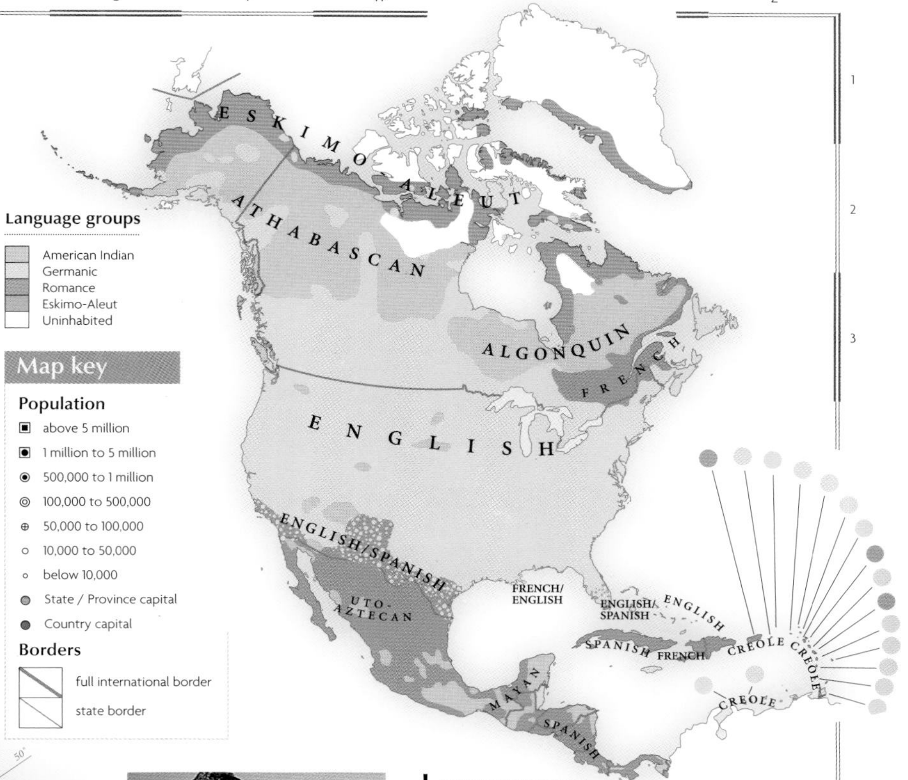

Language groups
American Indian
Germanic
Romance
Eskimo-Aleut
Uninhabited
ESKIMO-ALEUT
ATHABASCAN
ALGONQUIN
FRENCH
ENGLISH
ENGLISH/SPANISH
UTO-AZTECAN
FRENCH/ENGLISH
ENGLISH/SPANISH
ENGLISH
SPANISH
FRENCH
CREOLE
CREOLE
CREOLE
MAYAN
SPANISH

▲ *Land in northern* Canada has been set aside for Inuit reserves, allowing the Inuit and other Native American groups to maintain their traditional practices and culture.

Languages

The three major official languages of North America are of European origin, brought by settlers in the 16th century. In Canada, French and English are spoken; in the USA, English is the main language, with large Spanish-speaking areas in the southwest; Mexicans are Spanish-speaking; while the Caribbean islands use French, English and Spanish as well as the hybrid Creole tongues. In isolated areas, languages of the indigenous peoples still exist, such as Inuit in the far north of the continent.

Population

Much of North America is almost empty, especially the frozen far north. Population densities are highest in the highlands of Mexico and Central America; the coastal plain stretching from the Gulf of Mexico along the Atlantic coast; the Great Lakes area; and the Pacific coast. Large conurbations have developed, notably the San-San (San Francisco–San Diego), Boswash (Boston–Washington) and Main Street (Toronto–Montréal). The populations of the Caribbean islands are small, but settlement is dense, due to the limited amount of land available.

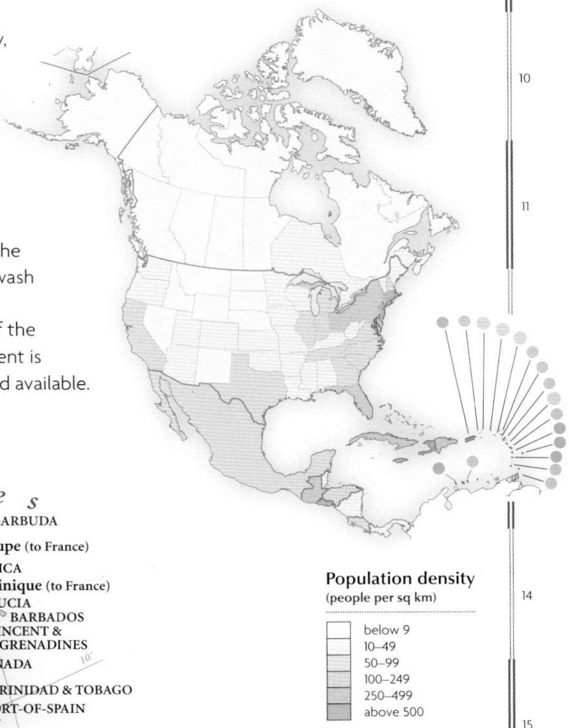

Population density
(people per sq km)
below 9
10–49
50–99
100–249
250–499
above 500

► *Mexico City is* one of the world's largest and highest cities. Fresh water supplies are dwindling, while air pollution regularly creates thick smog.

North American resources

The two northern countries of Canada and the USA are richly endowed with natural resources which have helped to fuel economic development. The USA is the world's largest economy, although today it is facing stiff competition from the Far East. Mexico has relied on oil revenues but there are hopes that the North American Free Trade Agreement (NAFTA), will encourage trade growth with Canada and the USA. The poorer countries of Central America and the Caribbean depend largely on cash crops and tourism.

Industry

The modern, industrialized economies of the USA and Canada contrast sharply with those of Mexico, Central America and the Caribbean. Manufacturing is especially important in the USA; vehicle production is concentrated around the Great Lakes, while electronic and hi-tech industries are increasingly found in the western and southern states. Mexico depends on oil exports and assembly work, taking advantage of cheap labour. Many Central American and Caribbean countries rely heavily on agricultural exports.

◀ *After its purchase* from Russia in 1867, Alaska's frozen lands were largely ignored by the USA. Oil reserves similar in magnitude to those in eastern Texas were discovered in Prudhoe Bay, Alaska in 1968. Freezing temperatures and a fragile environment hamper oil extraction.

Standard of living

The USA and Canada have one of the highest overall standards of living in the world. However, many people still live in poverty, especially in inner city ghettos and some rural areas. Central America and the Caribbean are markedly poorer than their wealthier northern neighbours. Haiti is the poorest country in the western hemisphere.

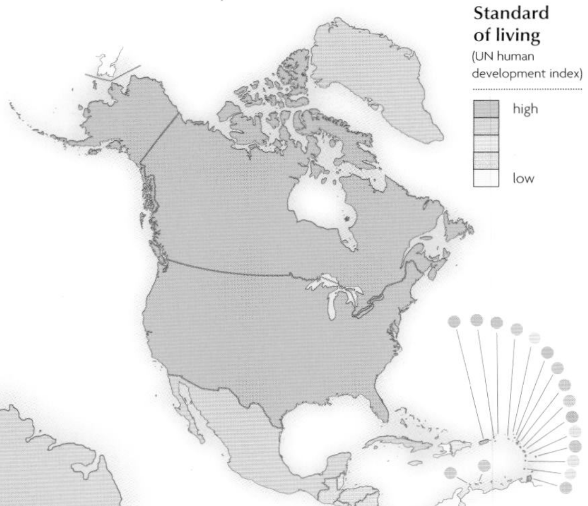

Standard of living
(UN human development index)
high
low

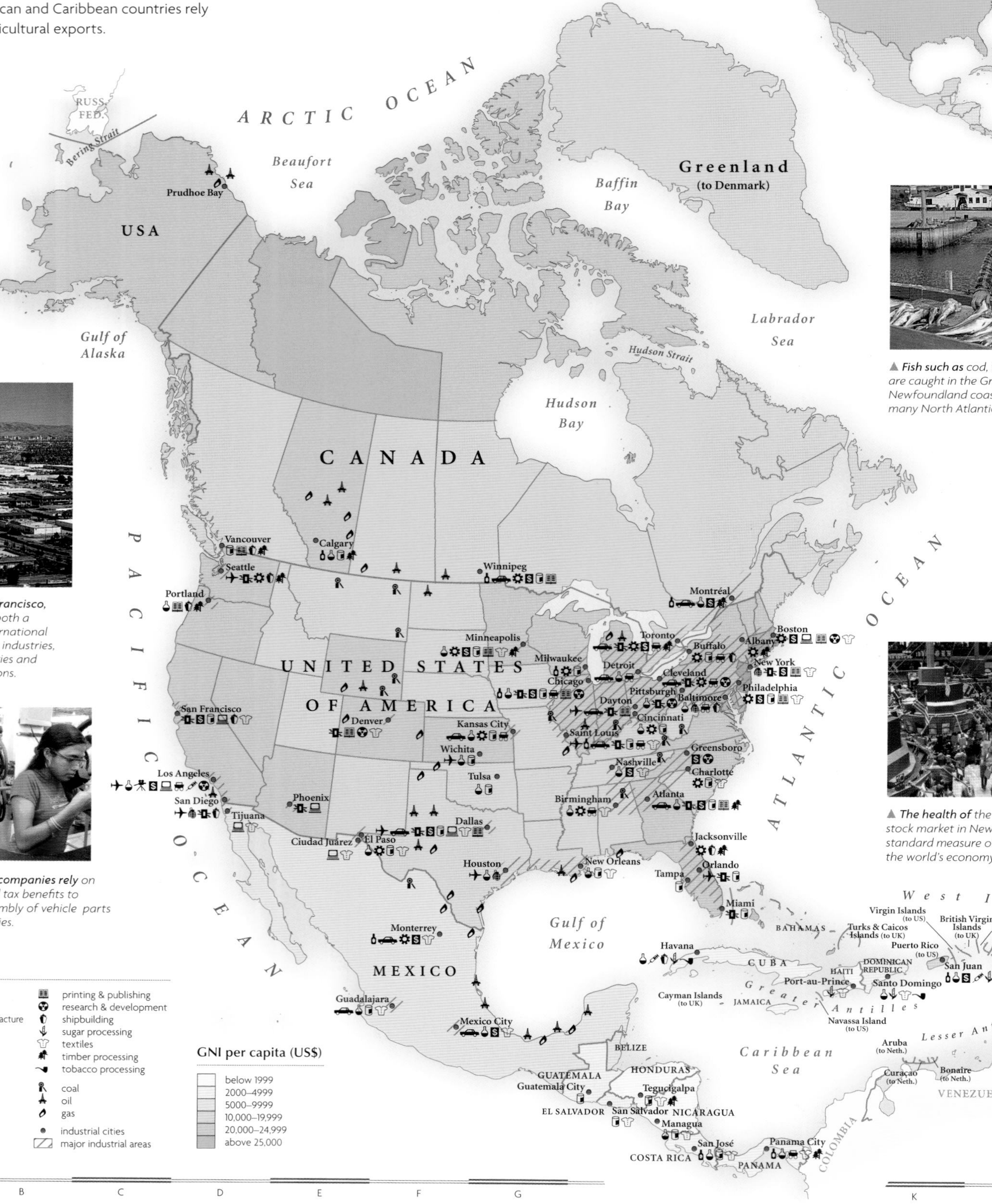

ARCTIC OCEAN
RUSS. FED.
Bering Strait
Bering Sea
Prudhoe Bay
USA
Beaufort Sea
Baffin Bay
Greenland (to Denmark)
Labrador Sea
Hudson Strait
Gulf of Alaska
Hudson Bay
CANADA
Vancouver
Calgary
Seattle
Winnipeg
Portland
Montréal
PACIFIC OCEAN
ATLANTIC OCEAN
Minneapolis
Toronto
Boston
Albany
Buffalo
Milwaukee
Detroit
New York
Chicago
Cleveland
Philadelphia
UNITED STATES OF AMERICA
San Francisco
Pittsburgh
Baltimore
Dayton
Denver
Kansas City
Cincinnati
Saint Louis
Wichita
Greensboro
Nashville
Charlotte
Los Angeles
Tulsa
San Diego
Tijuana
Phoenix
Birmingham
Atlanta
Dallas
Ciudad Juárez
El Paso
Jacksonville
Houston
New Orleans
Orlando
Tampa
Miami
Gulf of Mexico
Monterrey
MEXICO
Guadalajara
Mexico City
West Indies
BAHAMAS
Virgin Islands (to US)
British Virgin Islands (to UK)
Turks & Caicos Islands (to UK)
Anguilla (to UK)
ST KITTS & NEVIS
ANTIGUA & BARBUDA
Montserrat (to UK)
Puerto Rico (to US)
Havana
CUBA
HAITI
DOMINICAN REPUBLIC
San Juan
Guadeloupe (to France)
DOMINICA
Martinique (to France)
Port-au-Prince
Santo Domingo
Cayman Islands (to UK)
JAMAICA
Greater Antilles
Navassa Island (to US)
ST LUCIA
BARBADOS
ST VINCENT & THE GRENADINES
GRENADA
TRINIDAD & TOBAGO
Port-of-Spain
Lesser Antilles
Aruba (to Neth.)
Curaçao (to Neth.)
Bonaire (to Neth.)
VENEZUELA
Caribbean Sea
BELIZE
GUATEMALA
Guatemala City
HONDURAS
Tegucigalpa
EL SALVADOR
San Salvador
NICARAGUA
Managua
San José
COSTA RICA
Panama City
PANAMA
COLOMBIA
Industry
aerospace
brewing
car/vehicle manufacture
chemicals
defence
electronics
engineering
film industry
finance
food processing
hi-tech industry
iron & steel
pharmaceuticals
printing & publishing
research & development
shipbuilding
sugar processing
textiles
timber processing
tobacco processing
coal
oil
gas
industrial cities
major industrial areas
GNI per capita (US$)
below 1999
2000–4999
5000–9999
10,000–19,999
20,000–24,999
above 25,000

▲ *Fish such as* cod, flounder and plaice are caught in the Grand Banks, off the Newfoundland coast, and processed in many North Atlantic coastal settlements.

▲ *South of San Francisco,* 'Silicon Valley' is both a national and international centre for hi-tech industries, electronic industries and research institutions.

▲ *Multinational companies rely* on cheap labour and tax benefits to facilitate the assembly of vehicle parts in Mexican factories.

▲ *The health of* the Wall Street stock market in New York is the standard measure of the state of the world's economy.

N O P Q R S T U V W X Y Z

Environmental issues

Many fragile environments are under threat throughout the region. In Haiti, all the primary rainforest has been destroyed, while air pollution from factories and cars in Mexico City is amongst the worst in the world. Elsewhere, industry and mining pose threats, particularly in the delicate arctic environment of Alaska where oil spills have polluted coastlines and decimated fish stocks.

Environmental issues

- national parks
- risk of acid rain
- tropical forest
- forest destroyed
- desert
- risk of desertification
- polluted rivers
- radioactive contamination
- marine pollution
- heavy marine pollution
- poor urban air quality

▲ *Wild bison* graze in Yellowstone National Park, the world's first national park. Designated in 1872, geothermal springs and boiling mud are among its natural spectacles, making it a major tourist attraction.

Mineral resources

Fossil fuels are exploited in considerable quantities throughout the continent. Coal mining in the Appalachians is declining but vast open pits exist further west in Wyoming. Oil and natural gas are found in Alaska, Texas, the Gulf of Mexico, and the Canadian West. Canada has large quantities of nickel, while Jamaica has considerable deposits of bauxite, and Mexico has large reserves of silver.

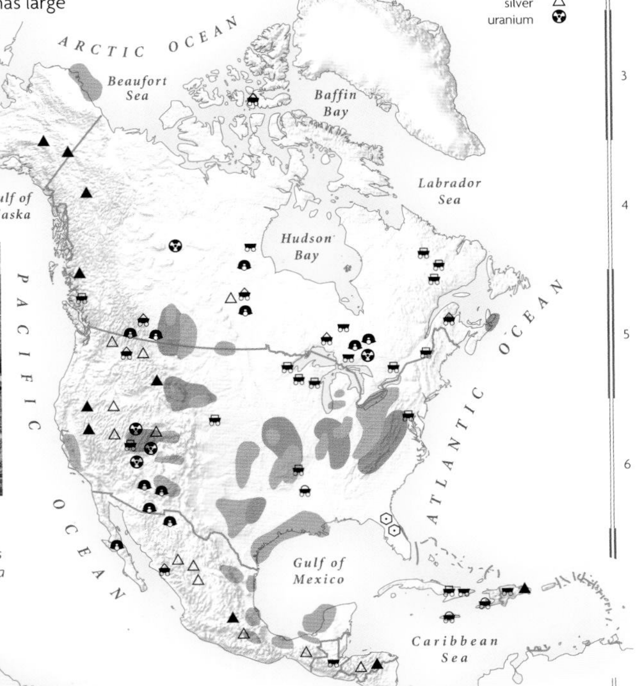

▲ *In addition to* fossil fuels, North America is also rich in exploitable metallic ores. This vast, mile-deep (1.6 km) pit is a copper mine in New Mexico.

Bering Sea

Gulf of Alaska

Using the land and sea

Abundant land and fertile soils stretch from the Canadian prairies to Texas creating North America's agricultural heartland. Cereals and cattle ranching form the basis of the farming economy, with corn and soya beans also important. Fruit and vegetables are grown in California using irrigation, while Florida is a leading producer of citrus fruits. Caribbean and Central American countries depend on cash crops such as bananas, coffee and sugar cane, often grown on large plantations. This reliance on a single crop can leave these countries vulnerable to fluctuating world crop prices.

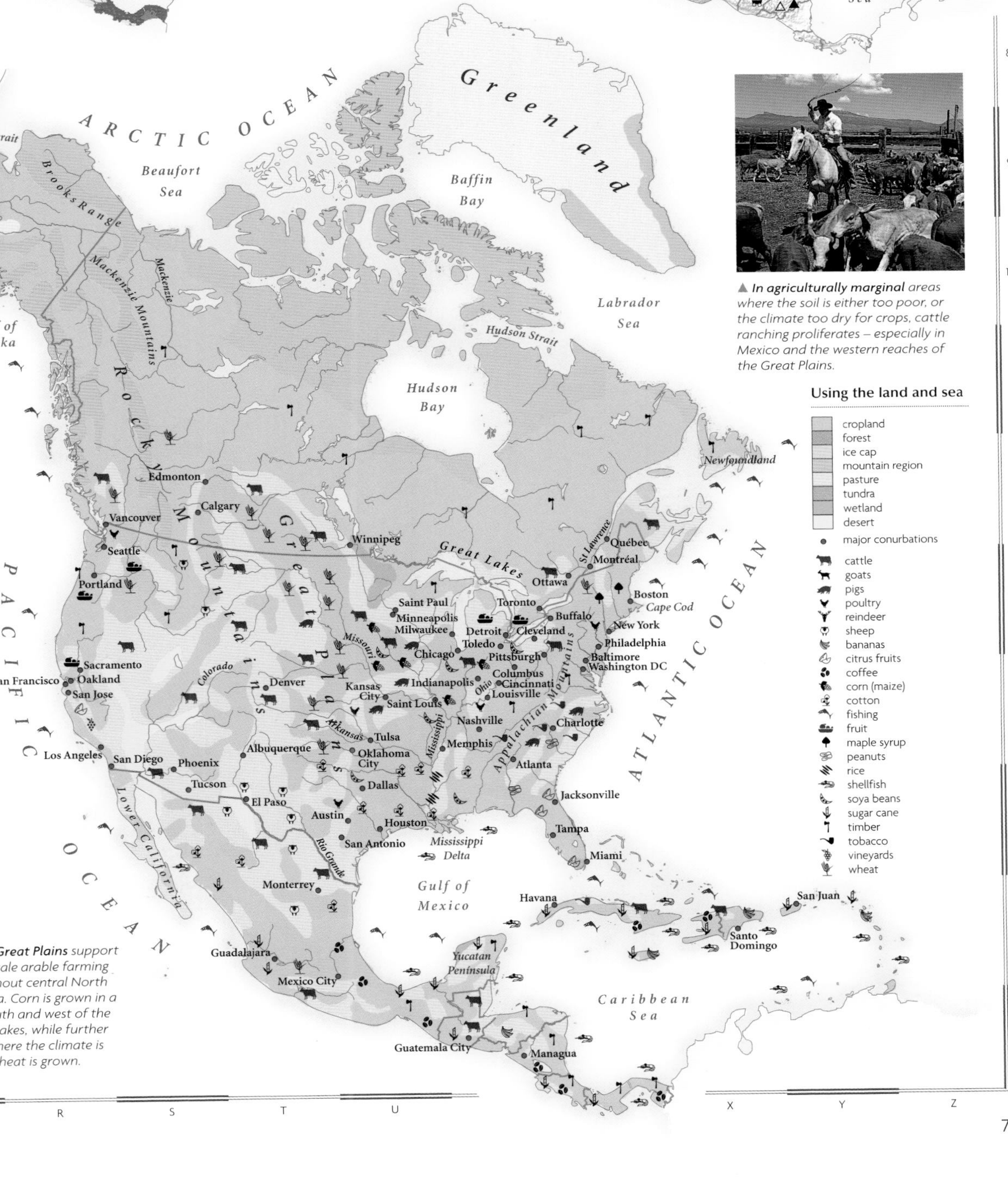

▲ *In agriculturally marginal* areas where the soil is either too poor, or the climate too dry for crops, cattle ranching proliferates – especially in Mexico and the western reaches of the Great Plains.

◀ *Sugar cane is* Cuba's main agricultural crop, and is grown and processed throughout the Caribbean. Fermented sugar is used to make rum.

◀ *The Great Plains* support large-scale arable farming throughout central North America. Corn is grown in a belt south and west of the Great Lakes, while further west where the climate is drier, wheat is grown.

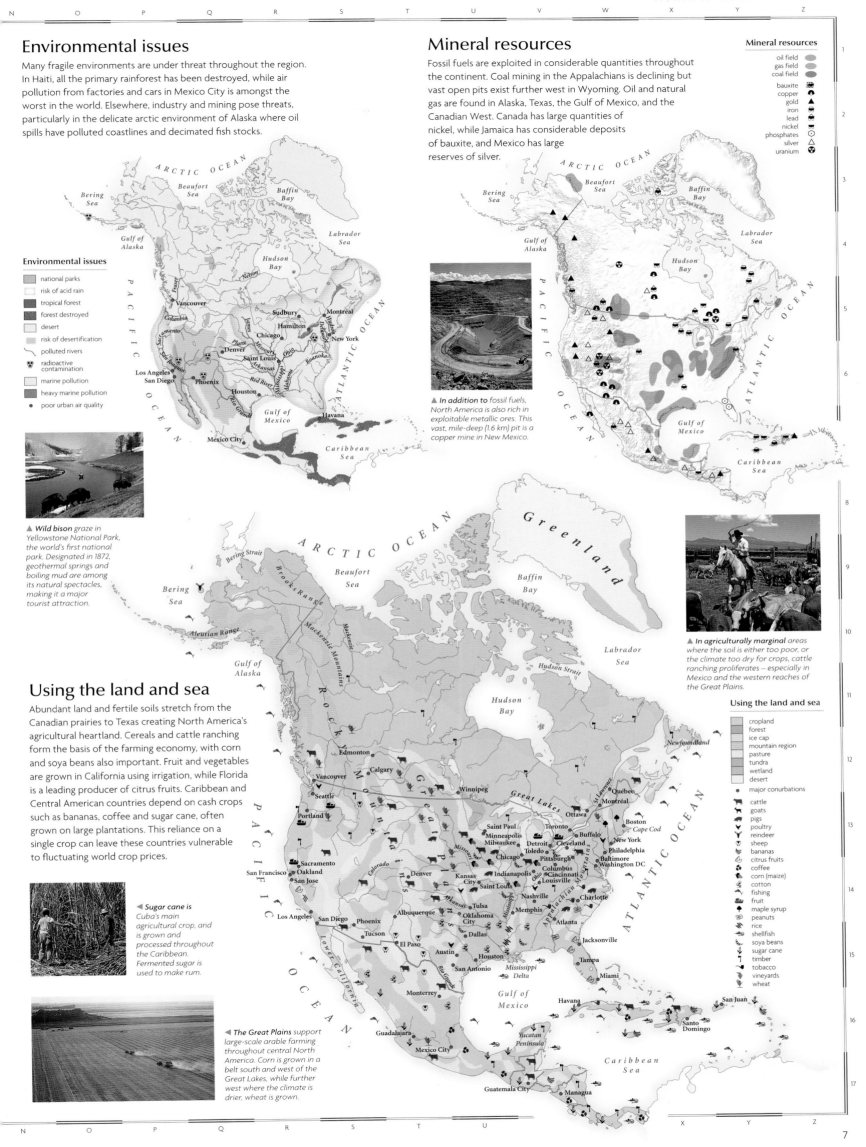

N O P Q R S T U X Y Z

Canada

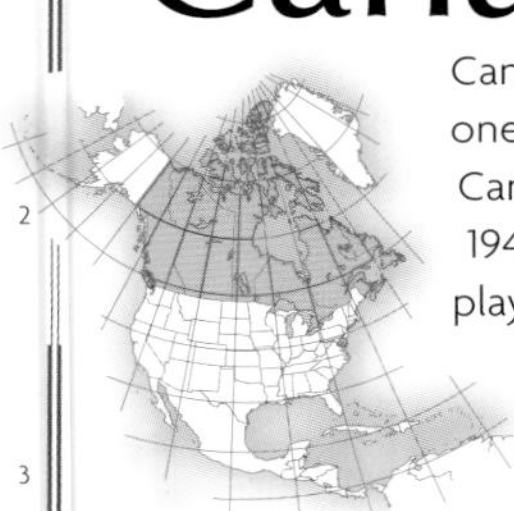

Canada is the second largest country in the world, and with only about one-tenth of its land area inhabited, it is one of the most sparsely populated. Canada became a confederation in 1867, though Newfoundland did not join until 1949. As a founding member of the UN and of the Commonwealth, Canada has played an important role in international affairs. A constitutional crisis, focusing on the French-speaking Québécois, and Inuit and Native American land rights, dominated politics in the 1990s. In 1999, part of the Northwest Territories, Nunavut, became a self-governing homeland for the Inuit.

The Selwyn Mountains in northwestern Canada form part of the Rocky Mountains. The highest point, Keele Peak, rises to 9750 ft (2972 m).

Transport and industry

Abundant energy in the form of coal, oil, natural gas and hydro-electric power underpins Canadian industry. Over 75% of manufacturing is concentrated in the Great Lakes–St. Lawrence region, including prospering aerospace, transport and hi-tech industries. Across Canada as a whole, manufacturing has developed around a diversified, high-quality resource base and a wide range of metallic and non-metallic minerals.

Canada has one of the world's highest rates of energy consumption per person. It is endowed with vast hydro-electric potential from which more than 60% of its electricity requirements are generated.

Major industry and infrastructure

- aerospace
- car manufacture
- chemicals
- engineering
- food processing
- hi-tech industry
- hydro-electric power
- oil & gas
- mining
- timber processing
- capital cities
- major towns
- international airports
- major roads
- major industrial areas

The transport network

309,019 miles (497,375 km)	10,500 miles (16,900 km)
8049 miles (12,995 km)	1864 miles (3000 km)

In recent years the road network has been expanded, especially links to remote areas. Meanwhile, for long-distance travel, air transport now supersedes the declining rail network, which focuses mainly on east–west routes.

Using the land and sea

The majority of Canada's agricultural land is found in the prairies, which cover 140 million acres (57 million ha) and support wheat and grain-fed cattle. More specialized crops, such as fruit and vegetables, are grown in pockets of agricultural land in the east and west. Of Canada's many islands, only Prince Edward Island has notable farmland. Further north, boreal forests, exploited for timber, run in an almost unbroken arc, giving way to uncultivable tundra and ice sheets in the far north.

The urban/rural population divide

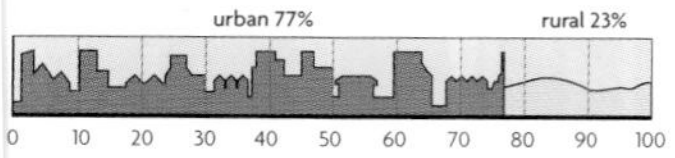

Population density	Total land area
9 people per sq mile (3 people per sq km)	3,559,294 sq miles (9,220,970 sq km)

Land use and agricultural distribution

- cattle
- cereals
- fishing
- fruit
- timber
- capital cities
- major towns
- pasture
- cropland
- forest
- wetland
- mountain region
- barren
- tundra

The climate and topography of the prairies makes them ideally suited to farming. Long summer days, moderate temperatures, limited rainfall and flat plains provide excellent conditions for wheat farming.

Scale 1:14,700,000

Km 0 25 50 100 150 200 250 300 350

Miles 0 25 50 100 150 200 250 300 350

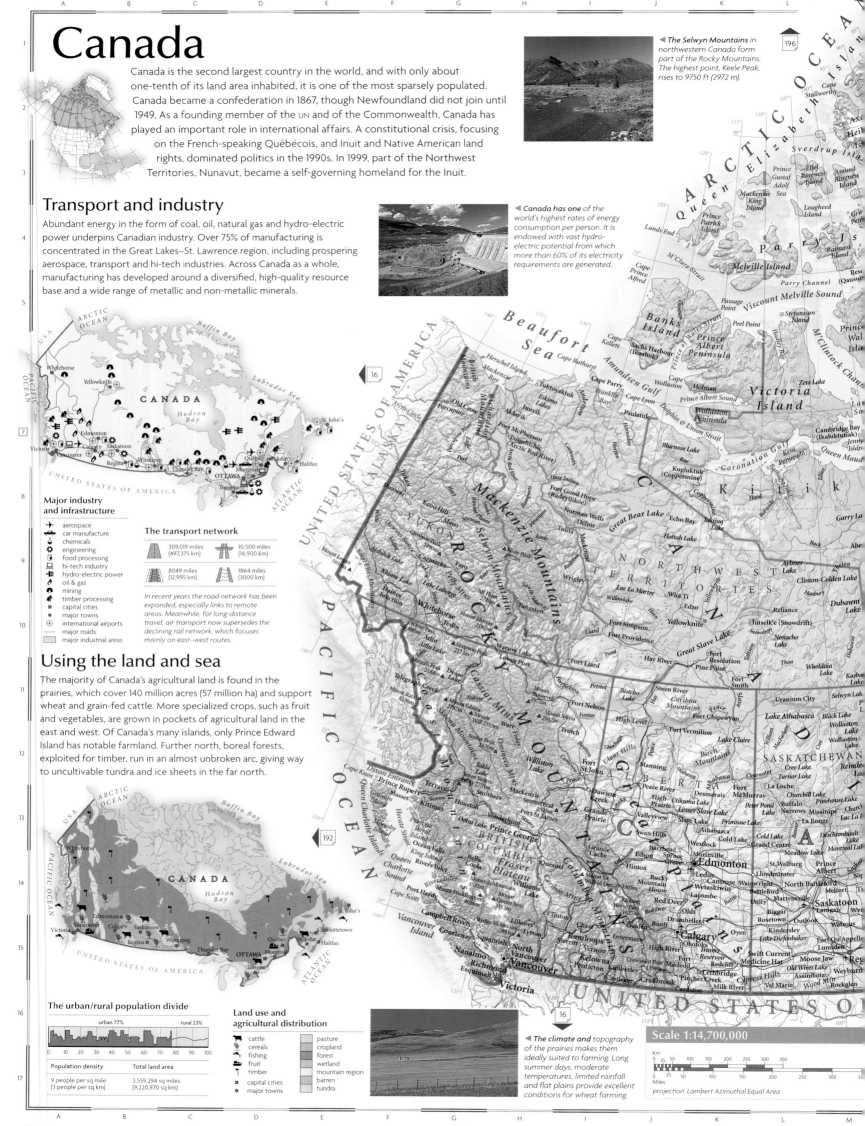

projection: Lambert Azimuthal Equal Area

The landscape

Glaciers on islands in the Arctic Ocean are the last remnants of the ice sheet that once covered and shaped Canada. Hudson Bay is the centre of the Canadian Shield, a huge, eroded plateau marked at its southern extremity by a string of lakes running southeastwards from Great Bear Lake to the Great Lakes. In contrast to the rolling relief of the Shield and the central lowland region, the Rocky Mountains rise to peaks of over 13,000 ft (4000 m), stretching 500 miles (800 km) along the west coast.

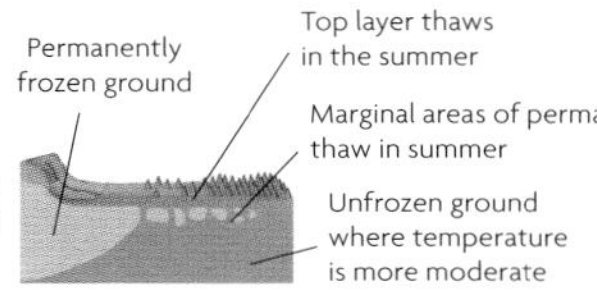

Permanently frozen ground
Top layer thaws in the summer
Marginal areas of permafrost thaw in summer
Unfrozen ground where temperature is more moderate

▶ *Permanently frozen ground known as permafrost is common in Canada's northern tundra. It thickens further north, becoming hundreds of metres deep in parts of the Arctic.*

▲ *Along the northeastern coast of Baffin Island the mountains rise to 8000 ft (2440 m). Glaciers move down through the valleys to the sea, eroding wide U-shaped valleys.*

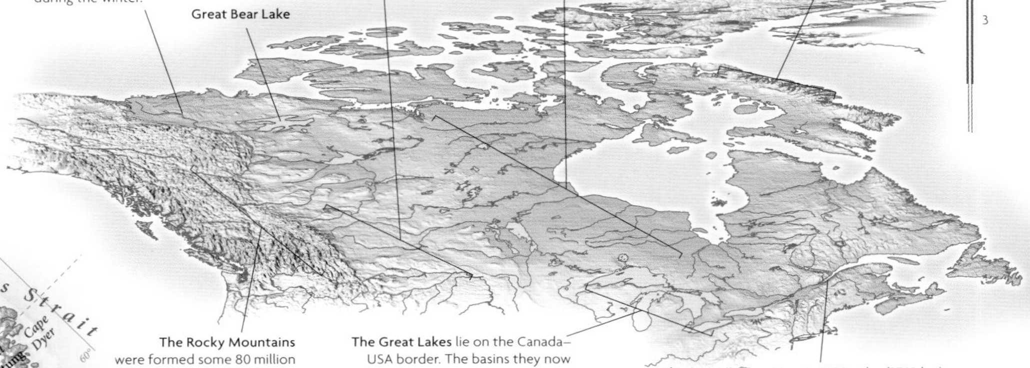

Great Bear Lake

The Mackenzie river, flowing north over the permafrost, forms a wide river channel with many tributaries. Together with the Peel river it has created a long, narrow delta at its mouth. The entire river freezes during the winter.

Fertile prairies stretch from the southern rim of the Canadian Shield, south into the USA.

Exposure to three phases of mountain-building and subsequent erosion over millions of years has moulded the ancient Canadian Shield into a series of basins and ridges.

The Rocky Mountains were formed some 80 million years ago, when the Pacific plate was driven under the North American plate, forcing up the land.

The Great Lakes lie on the Canada–USA border. The basins they now occupy were fashioned by repeated ice advance. At one time, Lakes Superior, Huron and Michigan formed a single large lake, Lake Nipissing.

The St. Lawrence River is 2350 miles (3782 km) long. It flows from the western shore of Lake Superior through the Great Lakes and on to the Atlantic Ocean. From December to April, the St. Lawrence Seaway freezes between Lake Ontario and Montréal.

▶ *The Great Lakes are drained by the St. Lawrence River which flows down through a wide tectonic depression. It forms a broad estuary for much of its course, the width varying from 1.2 miles (1.9 km) in the upper reaches to 90 miles (145 km) at its mouth.*

▶ *Isolated pillars, known as hoodoos near Red Deer river in the badlands of Alberta are a product of wind and water erosion, especially flash floods. The badlands lie in the rain shadow of the Rocky Mountains, which creates a semi-arid climate.*

196
64
16

Map key

Population
- 1 million to 5 million
- 500,000 to 1 million
- 100,000 to 500,000
- 50,000 to 100,000
- 10,000 to 50,000
- below 10,000

Elevation
- 6000m / 19,686ft
- 4000m / 13,124ft
- 3000m / 9843ft
- 2000m / 6562ft
- 1000m / 3281ft
- 500m / 1640ft
- 250m / 820ft
- 100m / 328ft
- sea level

Canada: WESTERN PROVINCES

Alberta, British Columbia, Manitoba, Saskatchewan, Yukon

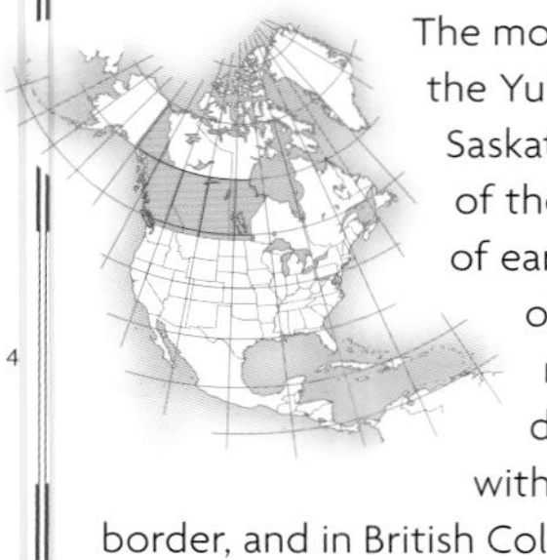

The mountains of the west coast, incorporating British Columbia and the Yukon, descend into the vast, flat prairies of Alberta, Saskatchewan and Manitoba. The empty lands and fertile soils of the prairie provinces attracted migrants, and the descendants of early European immigrants still make up a large proportion of the population. The mechanization of agriculture has reduced the need for labour, and rural population densities remain low. The majority of the people live within 100 miles (160 km) of the southern Canada–USA border, and in British Columbia, one of the leading Canadian provinces in terms of economic wealth. The Yukon, in the far north, remains a relatively unspoilt wilderness, containing large, untapped mineral reserves. This province has a significant population of Native Americans, many of whom maintain a traditional lifestyle.

Using the land and sea

Wheat farming is the economic mainstay of Alberta, Manitoba and Saskatchewan, which contain 82% of farmland in Canada. Cattle are also raised on the prairies. Forestry and fishing are the most prominent resource-based industries in British Columbia. Despite the mountainous terrain, fruit and specialized grains can be grown in the Okanagan and Fraser valleys.

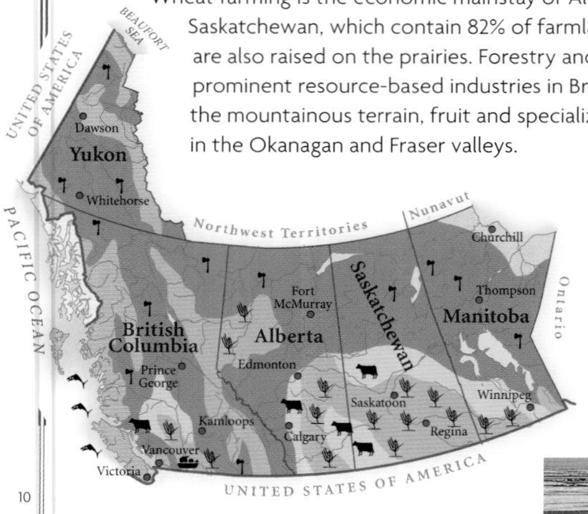

Land use and agricultural distribution

- cattle
- cereals
- fishing
- fruit
- timber
- major towns
- pasture
- cropland
- forest
- wetland
- barren
- tundra

The urban/rural population divide

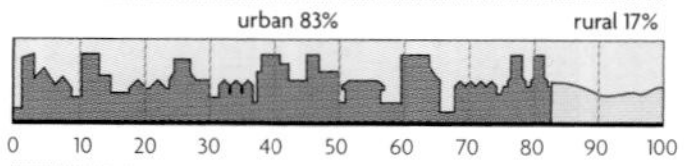

Population density	Total land area
8 people per sq mile (3 people per sq km)	1,230,547 sq miles (3,187,120 sq km)

▲ *Large, highly-mechanized and often very specialized farms, requiring huge investment but little labour, characterize modern farming in the prairies.*

Transport and industry

The western provinces contain a wealth of mineral resources. Alberta holds the bulk of Canada's fossil fuels; the other provinces contain reserves of metallic ores, such as zinc, lead and silver. Isolation from markets has slowed the development of manufacturing, restricting it to the large cities like Vancouver, Winnipeg and Calgary. Hydro-electric power is widely exploited, although there is increasing concern about potential ecological damage.

The transport network

Roads	82,438 miles (135,145 km)
Railways (motorways)	6459 miles (10,401 km)
Rail	24,041 miles (38,694 km)
Waterways	None

The transport network of the western provinces is dominated by east–west routes that weave through mountain passes and the plains. Access to some northern areas is restricted to air travel.

Major industry and infrastructure

- aerospace
- chemicals
- coal
- engineering
- food processing
- hydro-electric power
- mining
- oil & gas
- timber processing
- major towns
- international airports
- major roads
- major industrial areas

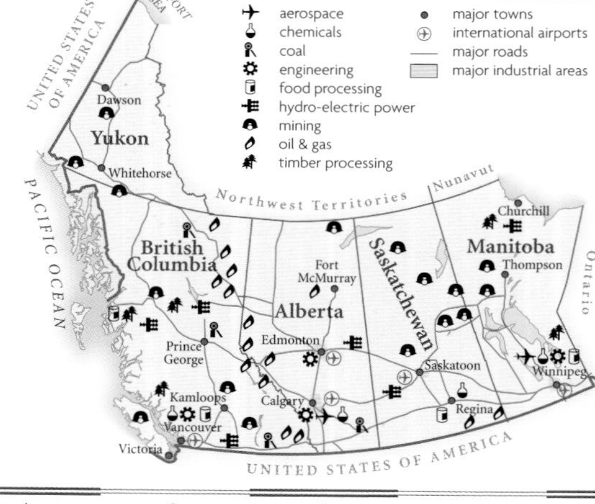

◄ *Much of the Yukon is uninhabited tundra. Industry is based on the extraction of mineral resources, and to a lesser extent, on the scattered forests of the south.*

▲ *The Fraser River valley is a major area of settlement in British Columbia. Railways cross the Rocky Mountains via this valley.*

▲ *Established in 1907, Jasper National Park lies in the heart of the Rocky Mountains. It is noted for its spectacular alpine scenery and contains part of the large Columbia Icefield.*

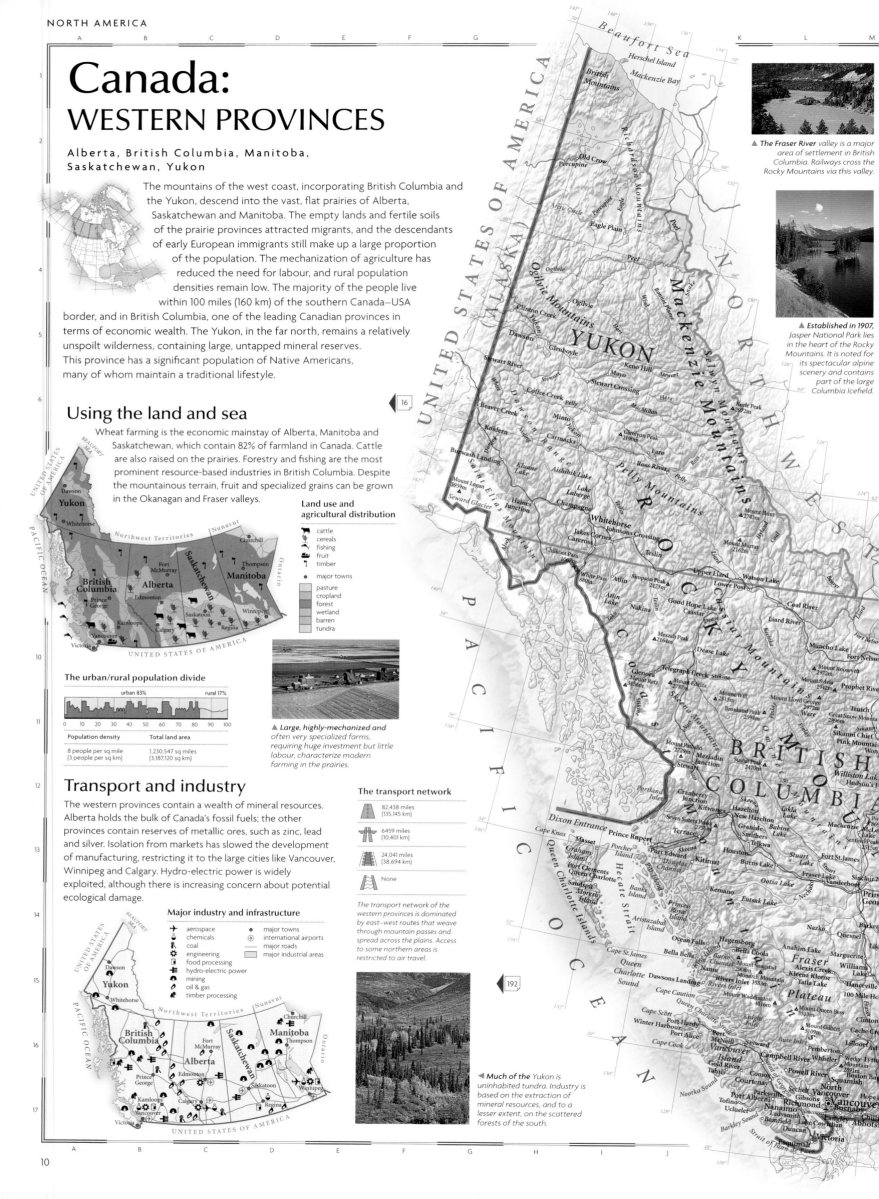

16

192

The landscape

The massive Rocky Mountains form a continental divide between rivers flowing eastward and westward. East of the mountains, stretching from the Arctic Circle south into the USA, lie the interior plains. Covered with glacial deposits from the last Ice Age, these are interspersed with hilly regions and long, steep escarpments.

Mount Logan rises 19,551 ft (5959 m). It is the highest peak in Canada.

The Columbia Icefield in the Rocky Mountains is the source of two major rivers, the Athabasca and the North Saskatchewan.

The badlands of Alberta were created when east-flowing rivers, swollen by meltwater at the end of the last Ice Age, cut deep, wide canyons producing eroded, barren landscapes.

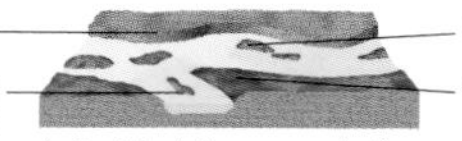

▲ ***Braided rivers are** shallow and fast-flowing. The interlaced branches are formed when excess sediments, which can no longer be transported, are deposited. The sediments collect in the river channel forming bars and sand flats. Islands form when the bars are colonized by vegetation.*

▲ ***Across the tundra** of northern Manitoba, widespread permafrost inhibits water from permeating the soil. This causes rivers like the Churchill to flow in many channels, which can be frozen for up to six months during the winter.*

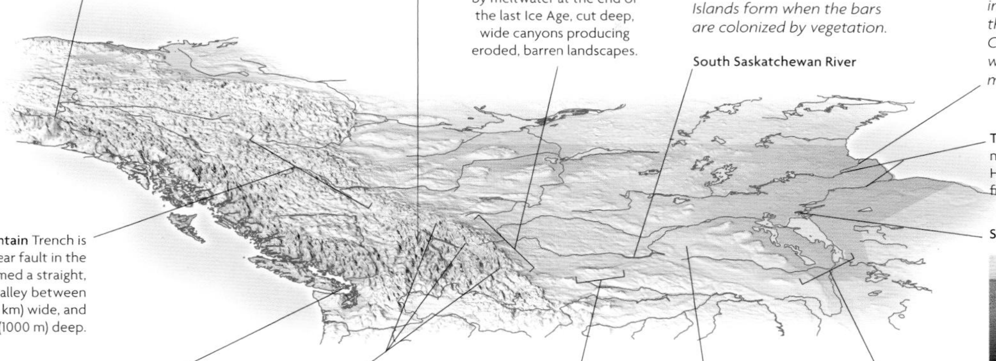

South Saskatchewan River

The Nelson and Churchill rivers drain northward across the Canadian Shield to Hudson Bay. The shield covers three-fifths of Saskatchewan.

Setting Lake

▲ ***Ancient granite outcrops,** part of the Canadian Shield, rise above the surface of Setting Lake, which was initially formed by meltwater from the last Ice Age.*

The Rocky Mountain Trench is the longest linear fault in the world. It has formed a straight, flat-bottomed valley between 2–9 miles (4–15 km) wide, and up to 3280 ft (1000 m) deep.

Hundreds of islands dot the fjord-indented coast of British Columbia; the largest is Vancouver Island.

Three major passes cut through the Rocky Mountains: Yellowhead, Kicking Horse and Crowsnest. They are all used as transport routes through the mountains.

The Cypress Hills rise to 4806 ft (1465 m) above the surrounding plain. Having escaped the last glaciation they contain unique plant and animal life. The silvery lupine, bunchberry and lodgepole pine all grow in the cool, moist climate of the hills.

The Alberta and Saskatchewan plains bear strong testament to past glaciations. The Assiniboine, Saskatchewan and Qu'Appelle rivers occupy flat-bottomed, steep-sided valleys eroded during the last Ice Age by glacial meltwater.

The lowlands of Manitoba are a basin that once held the vast post-glacial Lake Agassiz, remnants of which include Lake Winnipeg, Lake Winnipegosis and Lake Manitoba.

Map key

Population

- ◉ 500,000 to 1 million
- ◎ 100,000 to 500,000
- ⊕ 50,000 to 100,000
- ○ 10,000 to 50,000
- ∘ below 10,000

Elevation

- 6000m / 19,686ft
- 4000m / 13,124ft
- 3000m / 9843ft
- 2000m / 6562ft
- 1000m / 3281ft
- 500m / 1640ft
- 250m / 820ft
- 100m / 328ft
- sea level

Scale 1:8,250,000

Km 0 25 50 100 150 200 250
Miles 0 25 50 100 150 200 250

projection: Lambert Conformal Conic

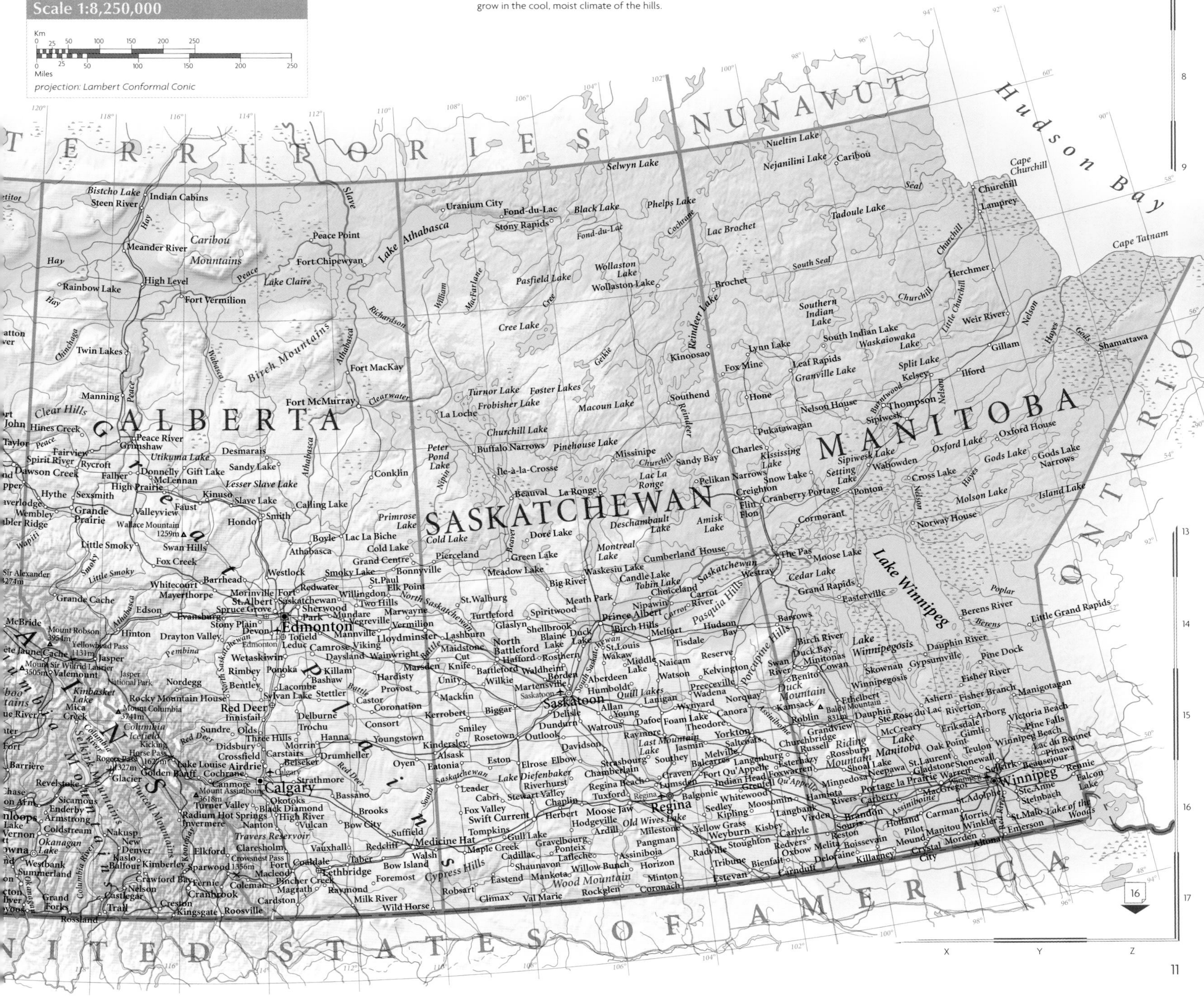

16

Canada: EASTERN PROVINCES

New Brunswick, Newfoundland & Labrador, Nova Scotia, Ontario, Prince Edward Island, Québec, ***St Pierre & Miquelon (to France)***

Colonized by both the English and the French during the 16th century, Canada's eastern provinces are still marked by their dual influences. They contain the last fragment of once-sizeable French territories, the islands of St Pierre and Miquelon. French remains Canada's second official language and Québec's first language. The population of the eastern provinces is highly concentrated in the south, especially along the border with the USA. A recent decline in fishing in the Atlantic provinces has encouraged a steady flow of westerly migration to more prosperous regions. The north, around Hudson Bay, remains snow-covered for most of the year and the indigenous Inuit people make up the bulk of its sparse population.

◀ Rocher Percé, is 290 ft (88 m) high. Lying off the southeastern coast of Québec, it is a sanctuary for sea birds.

Scale 1:7,750,000

Km 0 25 50 100 150 200
Miles 0 25 50 100 150 200

projection: Lambert Conformal Conic

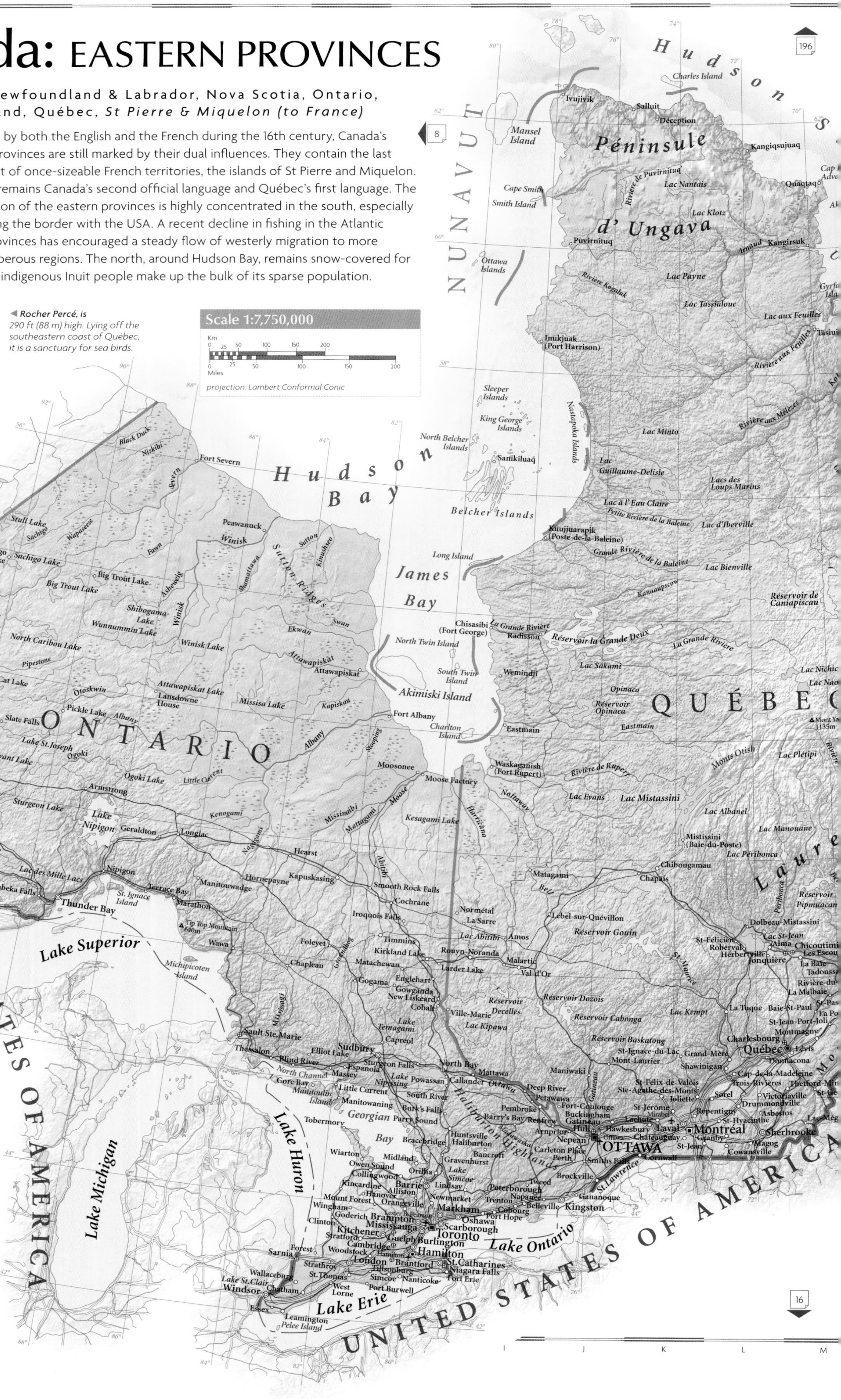

Map key

Population

- ▣ 1 million to 5 million
- ◉ 500,000 to 1 million
- ◎ 100,000 to 500,000
- ⊕ 50,000 to 100,000
- ○ 10,000 to 50,000
- ∘ below 10,000

Elevation

- 500m / 1640ft
- 250m / 820ft
- 100m / 328ft
- sea level

The landscape

Much of eastern Canada is part of the Canadian Shield. Glaciers have scoured the land leaving deposits that have dammed and diverted streams, to create a rocky landscape strewn with lakes and swamps. Much of the ground is subject to permafrost, which further impedes drainage. The uplands in the far east are the most northerly extension of the Appalachian mountain chain.

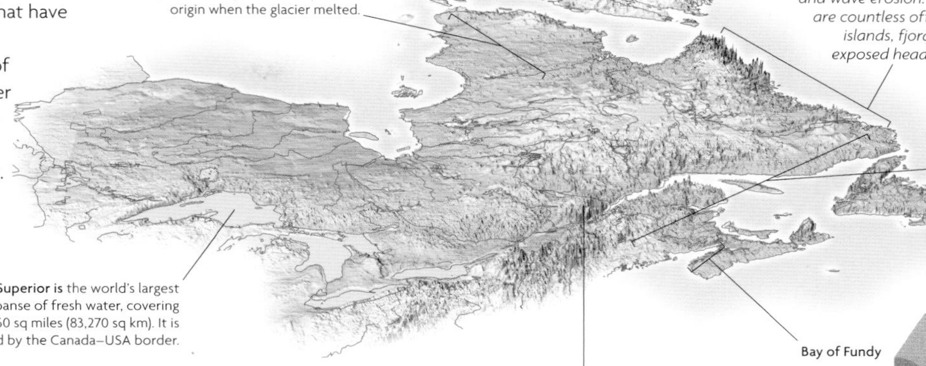

The Péninsule d'Ungava is littered with erratics – isolated rocks which were carried by glaciers and deposited away from their place of origin when the glacier melted.

Lake Superior is the world's largest expanse of fresh water, covering 32,150 sq miles (83,270 sq km). It is crossed by the Canada–USA border.

Laurentides Park

The eroded highlands of New Brunswick, Nova Scotia and Newfoundland are part of the Appalachian mountain chain, formed over 400 million years ago.

Bay of Fundy

Tidal waters are channelled down the bay

▶ *Labrador's indented coast is a product of past glaciations, which caused sea level change, and wave erosion. There are countless offshore islands, fjords and exposed headlands.*

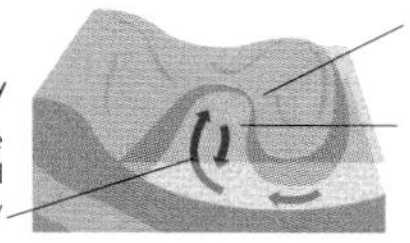

▲ *At the Bay of Fundy, incoming waves are funnelled down the long, narrow, steep-sided bay. These topographical features cause fast-flowing tides which can rise 70 ft (21 m).*

▶ *The forested Laurentides Park incorporates part of the Laurentian Mountains. Within its boundaries are over 1600 lakes.*

▲ *The tides at the Bay of Fundy are among the highest in the world. At low tide the tree-topped rocks have been likened to flowerpots.*

64

▶ *Prince Edward Island is the only Atlantic province with notable agricultural land. The island is Canada's leading producer of potatoes.*

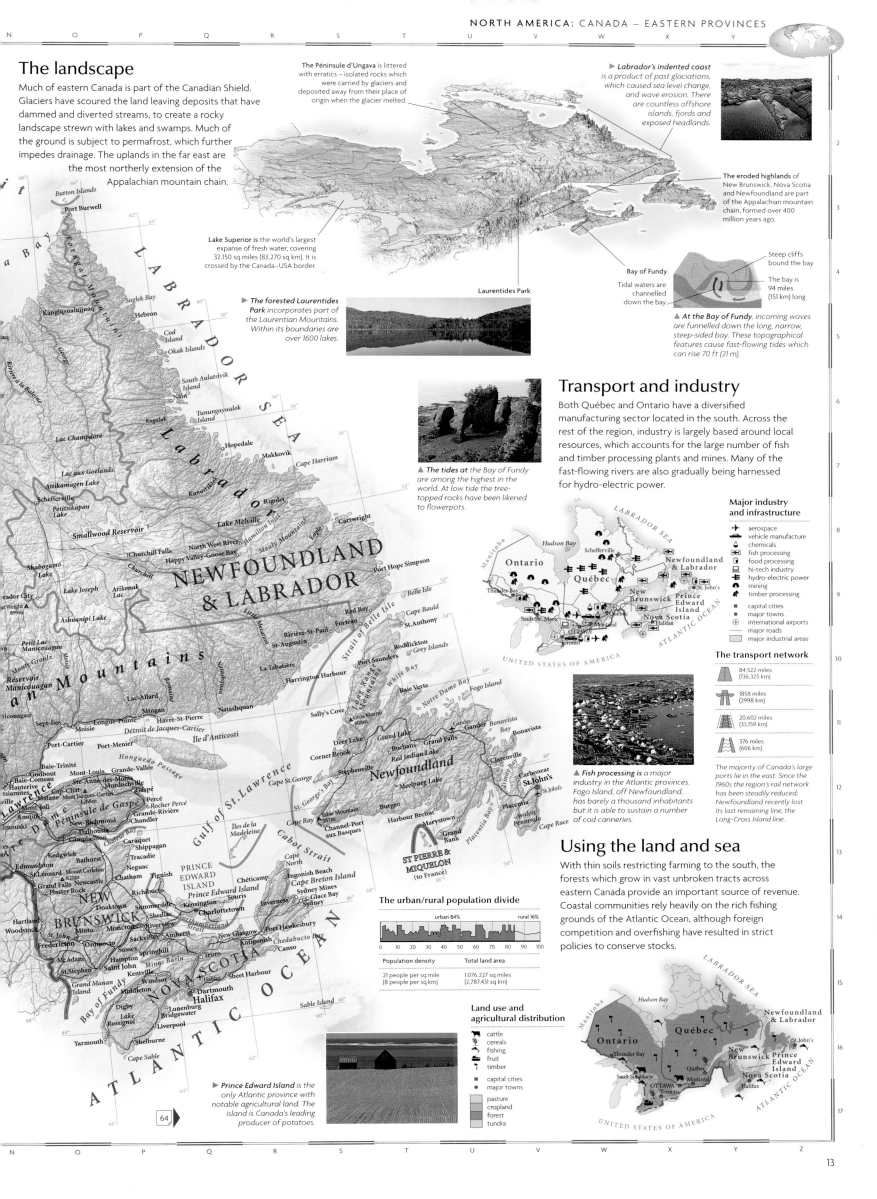

Transport and industry

Both Québec and Ontario have a diversified manufacturing sector located in the south. Across the rest of the region, industry is largely based around local resources, which accounts for the large number of fish and timber processing plants and mines. Many of the fast-flowing rivers are also gradually being harnessed for hydro-electric power.

Major industry and infrastructure

- aerospace
- vehicle manufacture
- chemicals
- fish processing
- food processing
- hi-tech industry
- hydro-electric power
- mining
- timber processing
- capital cities
- major towns
- international airports
- major roads
- major industrial areas

The transport network

- 84,522 miles (136,325 km)
- 1858 miles (2998 km)
- 20,602 miles (33,159 km)
- 376 miles (606 km)

The majority of Canada's large ports lie in the east. Since the 1960s the region's rail network has been steadily reduced; Newfoundland recently lost its last remaining line, the Long-Cross Island line.

▲ *Fish processing is a major industry in the Atlantic provinces. Fogo Island, off Newfoundland, has barely a thousand inhabitants but it is able to sustain a number of cod canneries.*

The urban/rural population divide

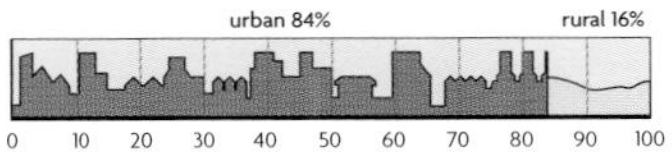

Population density	Total land area
21 people per sq mile (8 people per sq km)	1,076,227 sq miles (2,787,431 sq km)

Using the land and sea

With thin soils restricting farming to the south, the forests which grow in vast unbroken tracts across eastern Canada provide an important source of revenue. Coastal communities rely heavily on the rich fishing grounds of the Atlantic Ocean, although foreign competition and overfishing have resulted in strict policies to conserve stocks.

Land use and agricultural distribution

- cattle
- cereals
- fishing
- fruit
- timber
- capital cities
- major towns
- pasture
- cropland
- forest
- tundra

Southeastern Canada

Southern Ontario, Southern Québec

The southern parts of Québec and Ontario form the economic heart of Canada. The two provinces are divided by their language and culture; in Québec, French is the main language, whereas English is spoken in Ontario. Separatist sentiment in Québec has led to a provincial referendum on the question of a sovereignty association with Canada. The region contains Canada's capital, Ottawa, and its two largest cities: Toronto, the centre of commerce, and Montréal, the cultural and administrative heart of French Canada.

▲ *The port at Montréal is situated on the St. Lawrence Seaway. A network of 16 locks allows sea-going vessels access to routes once plied by fur-trappers and early settlers.*

Transport and industry

The cities of southern Québec and Ontario, and their hinterlands, form the heart of Canadian manufacturing industry. Toronto is Canada's leading financial centre, and Ontario's motor and aerospace industries have developed around the city. A major centre for nickel mining lies to the north of Toronto. Most of Québec's industry is located in Montréal, the oldest port in North America. Chemicals, paper manufacture and the construction of transport equipment are leading industrial activities.

► ***Niagara Falls lies** on the border between Canada and the USA. It comprises a system of two falls: American Falls, in New York, is separated from Horseshoe Falls, in Ontario, by Goat Island. Horseshoe Falls, seen here, plunges 184 ft (56 m) and is 2500 ft (762 m) wide.*

Major industry and infrastructure

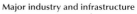

- car manufacture
- chemicals
- engineering
- finance
- food processing
- hi-tech industry
- mining
- iron & steel
- textiles
- paper industry
- timber processing
- capital cities
- major towns
- international airports
- major roads
- major industrial areas

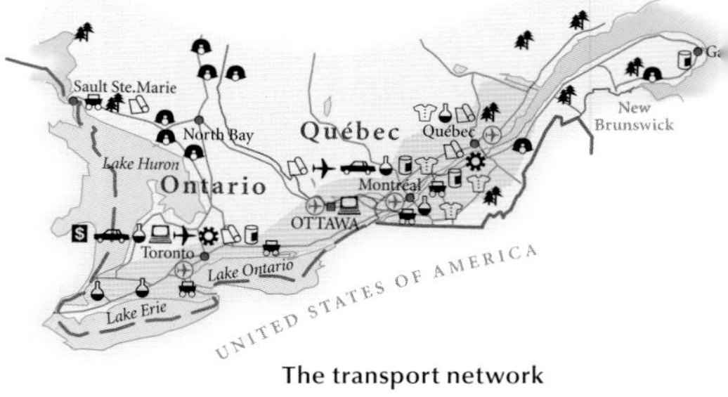

The transport network

The opening of the St. Lawrence Seaway in 1959 finally allowed ocean-going ships (up to 24,000 tons (tonnes)) access to the interior of Canada, creating a vital trading route.

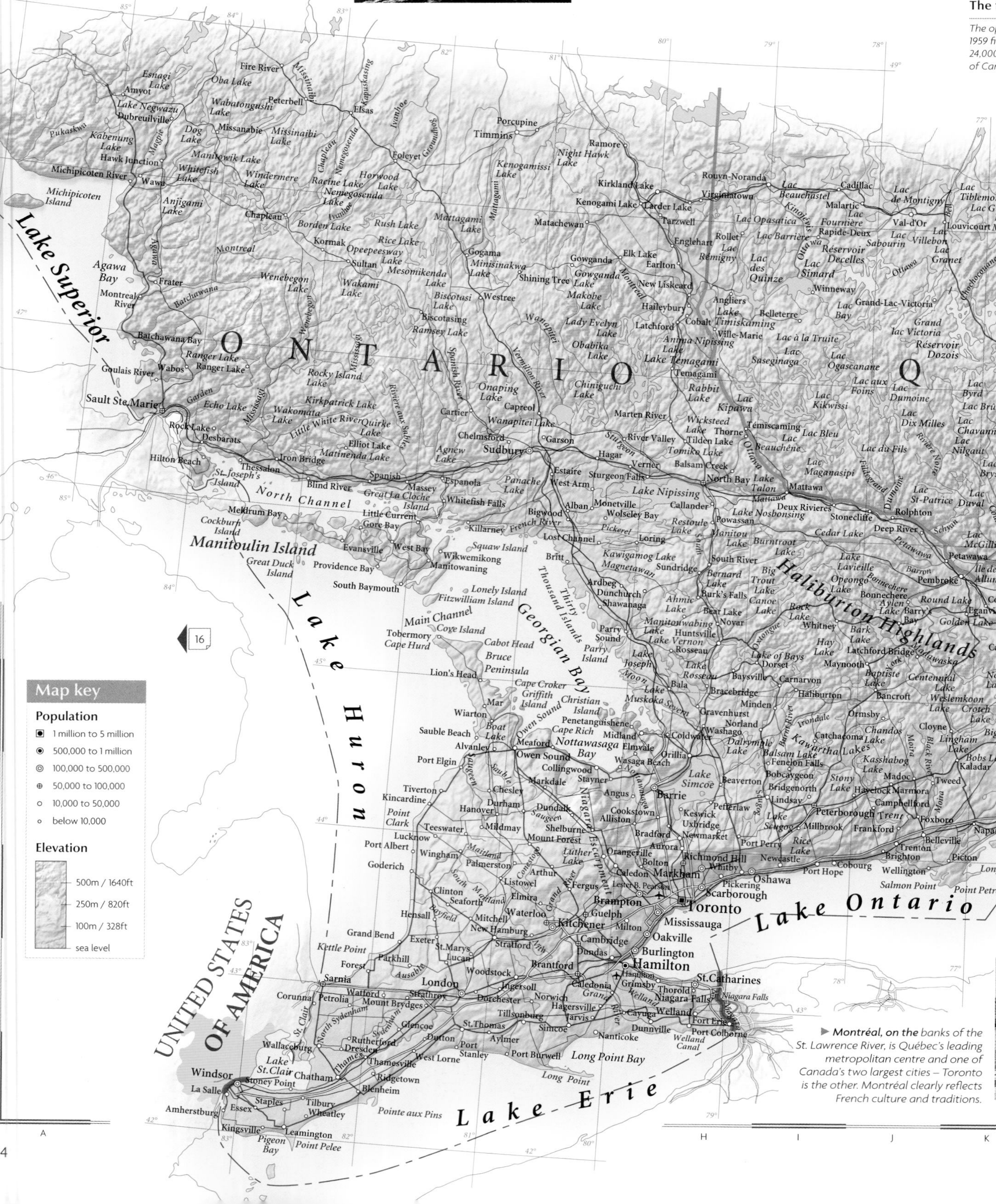

Map key

Population

- 1 million to 5 million
- 500,000 to 1 million
- 100,000 to 500,000
- 50,000 to 100,000
- 10,000 to 50,000
- below 10,000

Elevation

- 500m / 1640ft
- 250m / 820ft
- 100m / 328ft
- sea level

► ***Montréal, on the banks of the** St. Lawrence River, is Québec's leading metropolitan centre and one of Canada's two largest cities – Toronto is the other. Montréal clearly reflects French culture and traditions.*

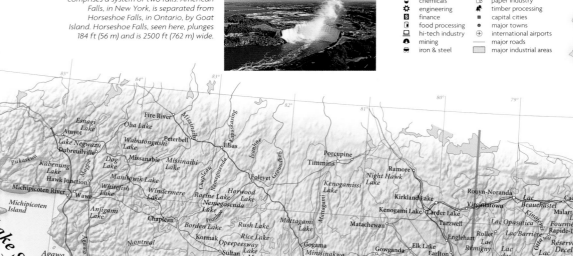

Using the land and sea

The productive Niagara 'fruit belt' on the shores of Lake Erie and Lake Ontario is a major farming region, although available farmland is being challenged by urban expansion. Québec is Canada's leading producer of maple syrup and dairy products. In the north, farmland gives way to extensive areas of forest, partly used for commercial logging. Fishing occurs in Atlantic waters and in the Great Lakes.

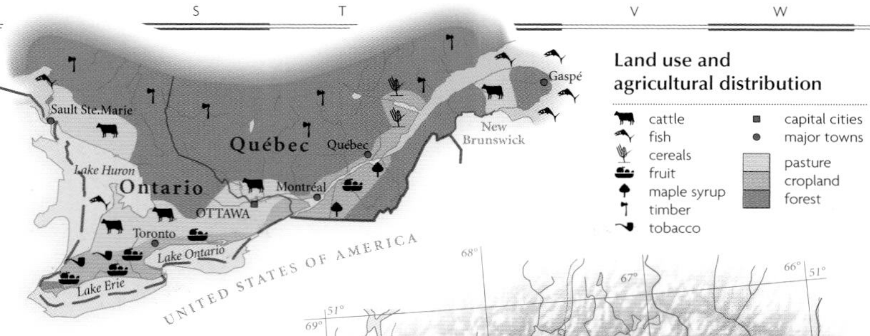

▲ ***Pumpkins are just*** *one of the crops grown in the Niagara 'fruit belt'. The mild climate, moderated by the lakes, allows the cultivation of a wide range of fruit and vegetables, including cherries, apples, peaches, grapes and asparagus. Fruit and vegetable growing is confined to southern Canada, due to the colder climate and short growing season of the northern regions.*

The urban/rural population divide

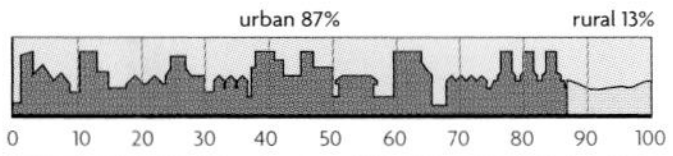

Population density	Total land area
64 people per sq mile (25 people per sq km)	214,230 sq miles (555,000 sq km)

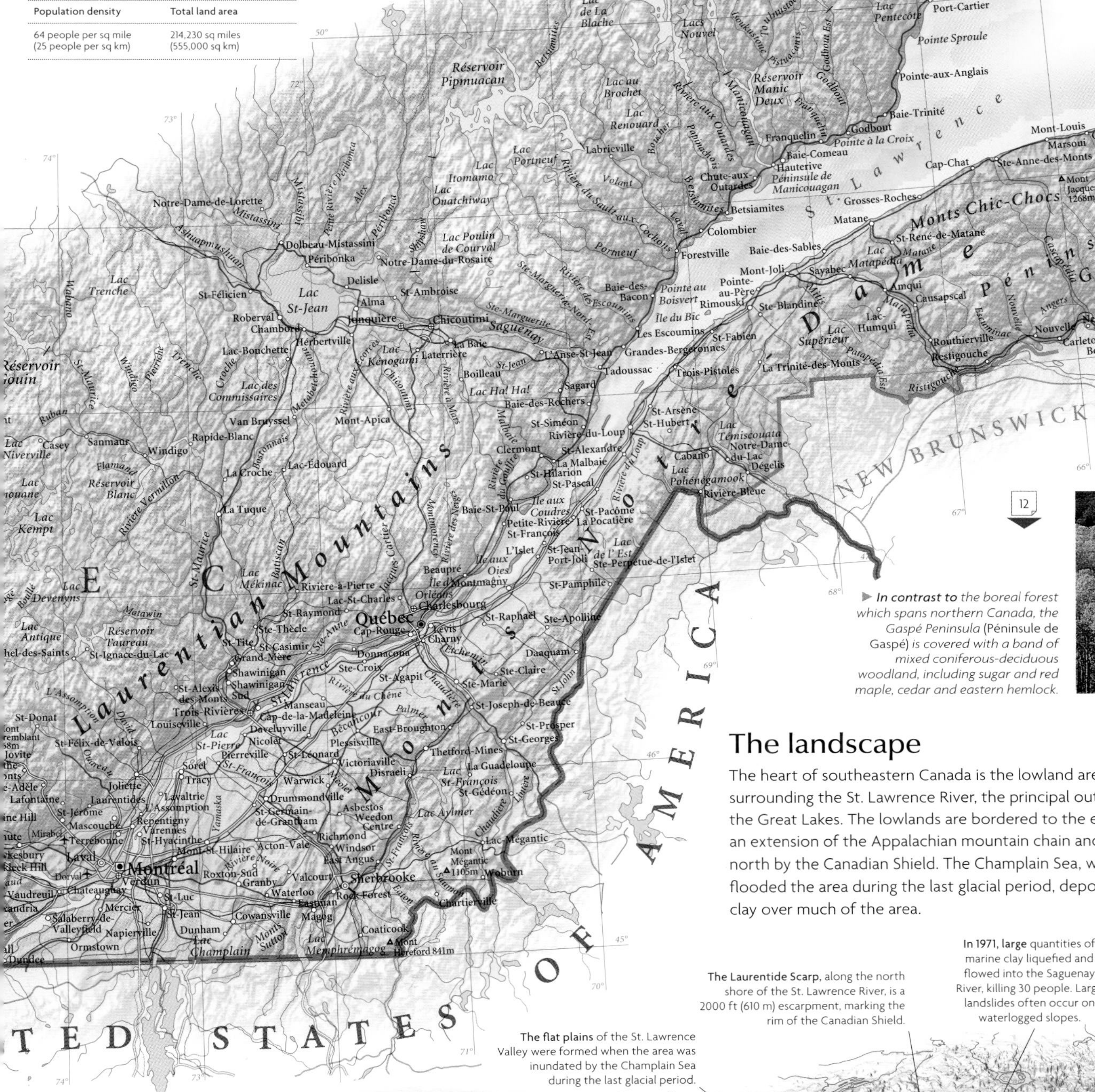

12

▶ ***In contrast to*** *the boreal forest which spans northern Canada, the Gaspé Peninsula* (Péninsule de Gaspé) *is covered with a band of mixed coniferous-deciduous woodland, including sugar and red maple, cedar and eastern hemlock.*

The landscape

The heart of southeastern Canada is the lowland area surrounding the St. Lawrence River, the principal outlet for the Great Lakes. The lowlands are bordered to the east by an extension of the Appalachian mountain chain and to the north by the Canadian Shield. The Champlain Sea, which flooded the area during the last glacial period, deposited clay over much of the area.

▲ ***The wooded Gaspé*** *Peninsula* (Péninsule de Gaspé) *includes the Notre Dame and Shickshock Mountains* (Monts Chic-Chocs)*. These are a northerly outcrop of the Appalachian mountain chain.*

Scale 1:3,250,000

Km 0 5 10 20 30 40 50 60 70
Miles 0 5 10 20 30 40 50 60 70

projection: Lambert Conformal Conic

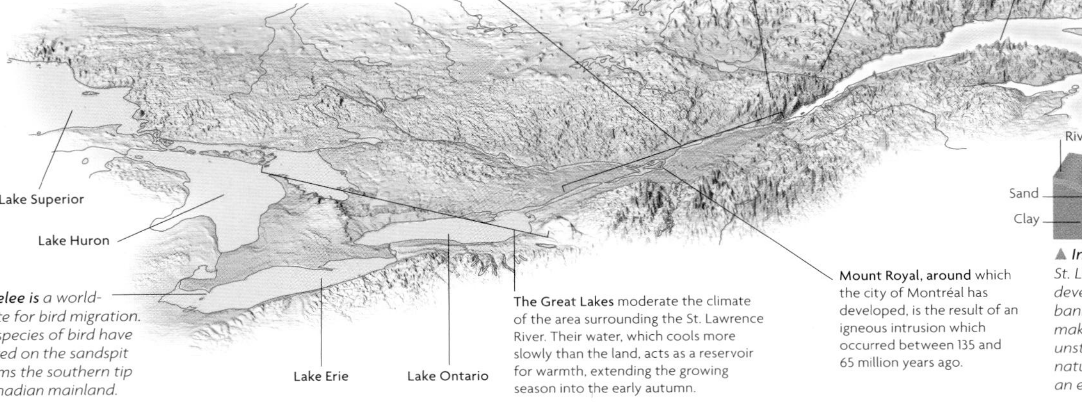

In 1971, large quantities of marine clay liquefied and flowed into the Saguenay River, killing 30 people. Large landslides often occur on waterlogged slopes.

The Laurentide Scarp, along the north shore of the St. Lawrence River, is a 2000 ft (610 m) escarpment, marking the rim of the Canadian Shield.

The flat plains of the St. Lawrence Valley were formed when the area was inundated by the Champlain Sea during the last glacial period.

The Great Lakes moderate the climate of the area surrounding the St. Lawrence River. Their water, which cools more slowly than the land, acts as a reservoir for warmth, extending the growing season into the early autumn.

Mount Royal, around which the city of Montréal has developed, is the result of an igneous intrusion which occurred between 135 and 65 million years ago.

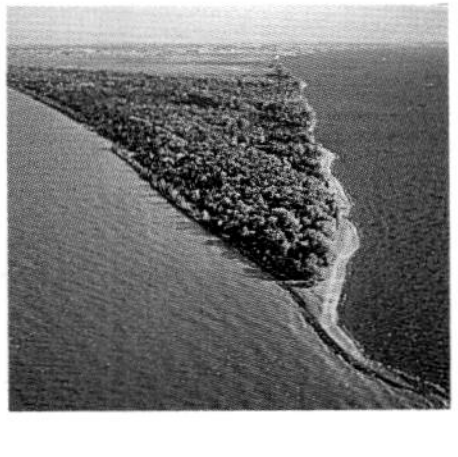

◀ ***Point Pelee is*** *a world-famous site for bird migration. Over 250 species of bird have been sighted on the sandspit which forms the southern tip of the Canadian mainland.*

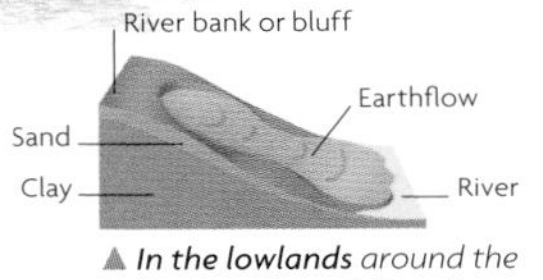

▲ ***In the lowlands*** *around the St. Lawrence, earthflows have developed along gentle river banks where sand overlies clay, making the surface layers very unstable. When the slope's natural equilibrium is disturbed, an earthflow can occur.*

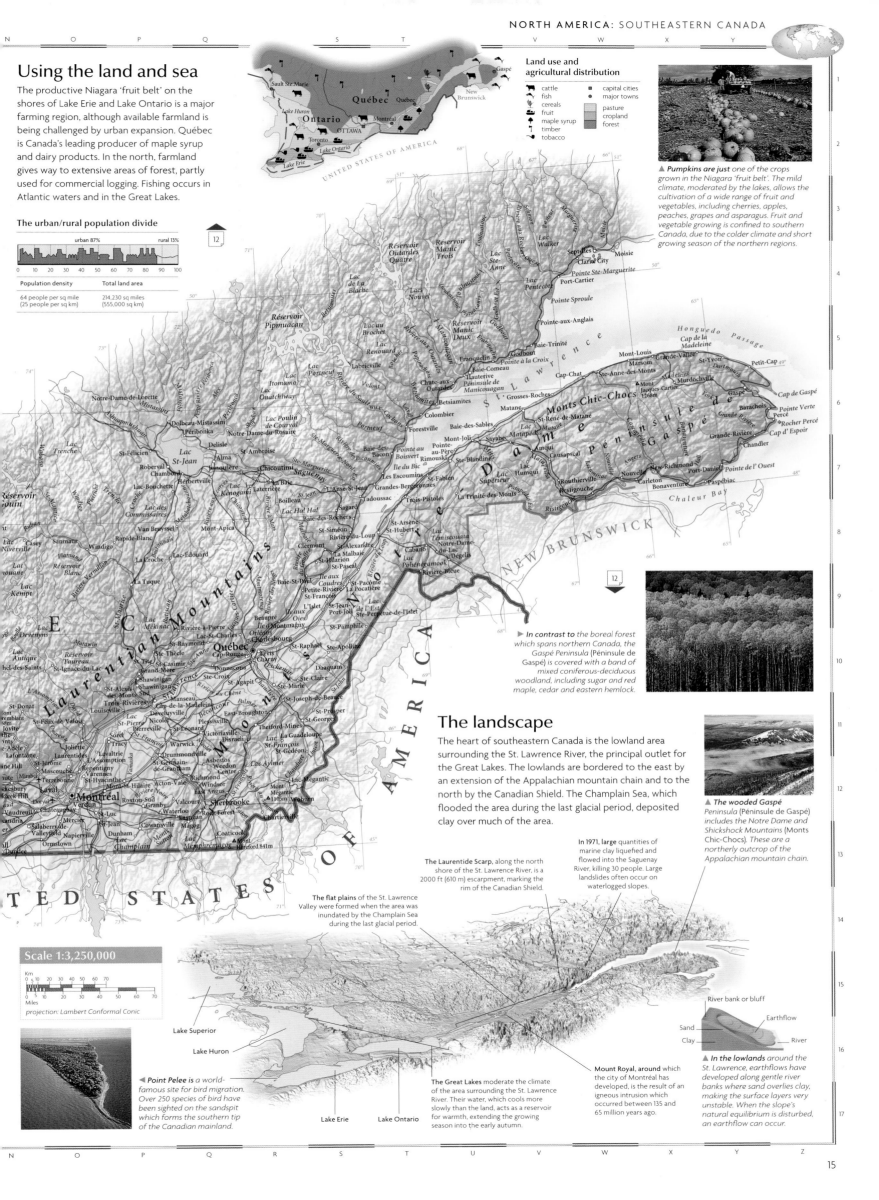

The United States of America

COTERMINOUS USA (FOR ALASKA AND HAWAII SEE PAGES 38-39)

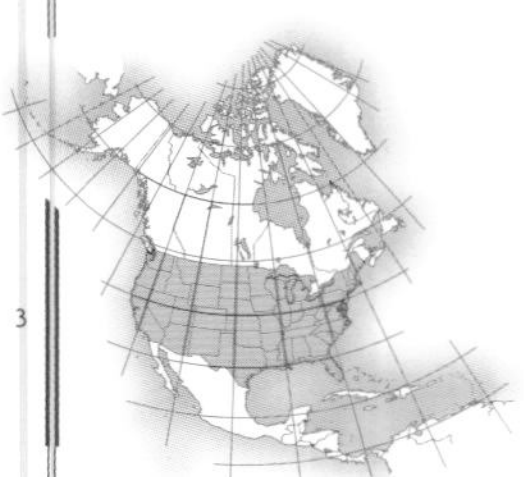

The USA's progression from frontier territory to economic and political superpower has taken less than 200 years. The 48 coterminous states, along with the outlying states of Alaska and Hawaii, are part of a federal union, held together by the guiding principles of the US Constitution, which enshrines the ideals of democracy and liberty for all. Abundant fertile land and a rich resource-base fuelled and sustained the USA's economic development. With the spread of agriculture and the growth of trade and industry came the need for a larger workforce, which was supplied by millions of immigrants, many seeking an escape from poverty and political or religious persecution. Immigration continues today, particularly from Central America and Asia.

▲ *Washington DC was established as the site for the nation's capital in 1790. It is home to the seat of national government, on Capitol Hill, as well as the President's official residence, the White House.*

► *The clear waters of Niagara Falls cascade 190 ft (58 m) into the gorge below. It is one of America's most famous spectacles and a leading tourist attraction. The falls are slowly receding and the gorge may one day stretch from Lake Ontario to Lake Erie.*

▲ *Mount Rainier is a dormant volcano in the Cascade Range, Washington. This 14,090 ft (4392 m) peak is flanked by the most extensive glacier outside Alaska.*

Scale 1:12,700,000

Km 0 25 50 100 150 200 250 300 350 400

Miles 0 25 50 100 150 200 250 300 350 400

projection: Lambert Azimuthal Equal Area

8

192

40

Transport and industry

The USA has been the industrial powerhouse of the world since the Second World War, pioneering mass-production and the consumer lifestyle. Initially, heavy engineering and manufacturing in the northeast led the economy. Today, heavy industry has declined and the USA's economy is driven by service and financial industries, with the most important being defence, hi-tech and electronics.

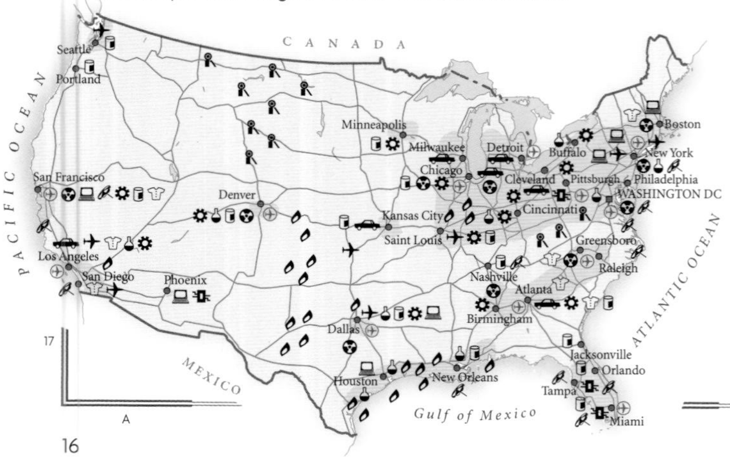

The transport network

3,875,040 miles (6,240,000 km)		52,388 miles (84,361 km)	
148,308 miles (235,238 km)		25,467 miles (41,009 km)	

Transport in the USA is dominated by the car which, with the extensive Interstate Highway system, allows great personal mobility. Today, internal air flights between major cities provide the most rapid cross-country travel.

Major industry and infrastructure

- aerospace
- car manufacture
- chemicals
- coal
- electronics
- engineering
- food processing
- hi-tech industry
- oil & gas
- research & development
- textiles
- tourism
- capital cities
- major towns
- international airports
- major roads
- major industrial areas

The landscape

The high, rugged mountain ranges of the west are about 80 million years old, geologically young compared to the old, eroded, Appalachian mountain chain, which dates from when North America and Europe were joined together as part of the supercontinent Pangaea, 400 million years ago. In contrast, the Great Plains and Mississippi Basin have a low relief and fertile soils.

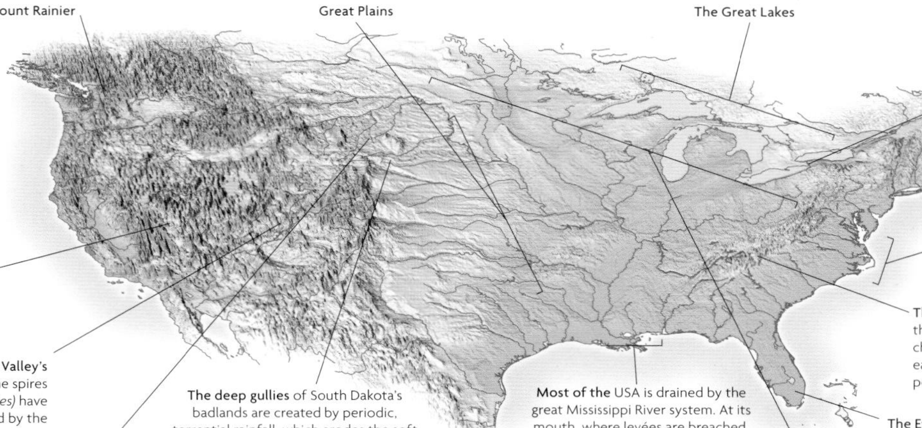

Death Valley, California, 282 ft (86 m) below sea level, is the lowest point in the western hemisphere, and one of the hottest places on Earth. Temperatures of 135° F (57° C) have been recorded here.

Barrier beaches, bars and spits are typical of the Atlantic coast. These sand formations around Cape Hatteras stretch along the coast for 200 miles (320 km).

The Great Smoky Mountains, part of the ancient Appalachian mountain chain, formed a natural barrier to early settlers attempting to penetrate the country's interior.

Monument Valley's striking sandstone spires and pillars *(buttes)* have been formed by the action of wind, water, heat and cold.

The deep gullies of South Dakota's badlands are created by periodic, torrential rainfall, which erodes the soft soils and rocks. Their form has been greatly affected by changes in land use.

Most of the USA is drained by the great Mississippi River system. At its mouth, where levées are breached, floodwaters are carried to the swamps through a series of channels. This region is known as the bayou.

The Everglades are a vast area of saw-grass swamp covering 4000 sq miles (10,300 sq km) of southern Florida.

◀ *Devils Tower, in Wyoming is a 1280 ft (390 m) intrusion of basalt rock, which cooled to form octagonal pillars. In 1906 it became the first US National Monument.*

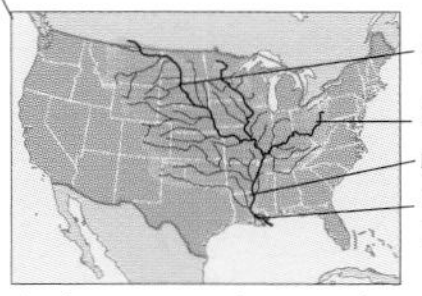

▲ ***The massive drainage** basin of the Mississippi covers 1,250,000 sq miles (3,200,000 sq km). It includes all areas drained by the Mississippi and its chief tributaries, the Missouri and Ohio rivers, and drains the entire region from the Appalachians to the Rockies.*

8

64

64

Map key

Population

- above 5 million
- 1 million to 5 million
- 500,000 to 1 million
- 100,000 to 500,000
- 50,000 to 100,000
- 10,000 to 50,000
- below 10,000

Elevation

- 4000m / 13,124ft
- 3000m / 9843ft
- 2000m / 6562ft
- 1000m / 3281ft
- 500m / 1640ft
- 250m / 820ft
- 100m / 328ft
- sea level

Using the land and sea

Over half of the USA's land area is utilized for agriculture, typified by the large cereal farms and cattle ranches of the Great Plains and Midwest prairie regions. Although wheat and corn are still primary crops, a diverse range of fruits and vegetables are grown in the fertile areas, particularly near the east and west coasts. Despite the abundance of cultivable land, inadequate soil management has resulted in a third of the topsoil being lost through wind and water erosion.

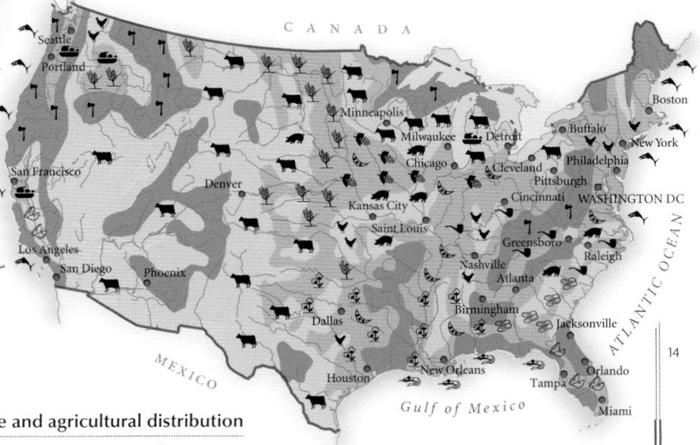

Land use and agricultural distribution

- cattle
- pigs
- poultry
- citrus fruits
- cotton
- fishing
- fruit
- corn (maize)
- peanuts
- shellfish
- soya beans
- timber
- tobacco
- wheat
- capital cities
- major towns
- pasture
- cropland
- forest
- wetland
- desert
- mountain region

The urban/rural population divide

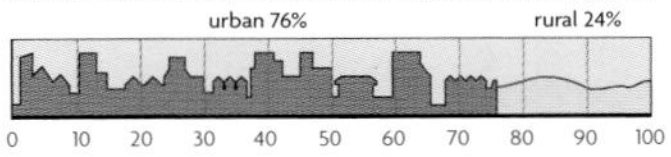

Population density	Total land area
98 people per sq mile (38 people per sq km)	2,959,045 sq miles (7,663,631 sq km)

▶ ***Fakahatchee Strand** is part of the extensive sub-tropical swamps in the Florida Everglades. The swamps support a wide variety of animal life, including many rare birds, fish, alligators and crocodiles.*

◀ ***Farming on the** Great Plains and in the Midwest is characterized by large-scale, mechanized wheat farms.*

USA: NORTHEASTERN STATES

Connecticut, Maine, Massachusetts, New Hampshire, New Jersey, New York, Pennsylvania, Rhode Island, Vermont

The indented coast and vast woodlands of the northeastern states were the original core area for European expansion. The rustic character of New England prevails after 400 years, while the great cities of the Atlantic seaboard have formed an almost continuous urban region. Over 20 million immigrants entered New York from 1855 to 1924 and the northeast became the industrial centre of the USA. After the decline of mining and heavy manufacturing, economic dynamism has been restored with the growth of hi-tech and service industries.

Chelsea in Vermont, surrounded by trees in their fall foliage. Tourism and agriculture dominate the economy of this self-consciously rural state, where no town exceeds 40,000 people.

Map key

Population

- above 5 million
- 1 million to 5 million
- 500,000 to 1 million
- 100,000 to 500,000
- 50,000 to 100,000
- 10,000 to 50,000
- below 10,000

Elevation

- 1000m / 3281ft
- 500m / 1640ft
- 250m / 820ft
- 100m / 328ft
- sea level

The transport network

340,090 miles (544,144 km)	4813 miles (7700 km)
12,872 miles (20,592 km)	2108 miles (3389 km)

New York's commercial success is tied historically to its transport connections. The Erie Canal, completed in 1825, opened up the Great Lakes and the interior to New York's markets and carried a stream of immigrants into the Midwest.

Transport and industry

The principal seaboard cities grew up on trade and manufacturing. They are now global centres of commerce and corporate administration, dominating the regional economy. Research and development facilities support an expanding electronics and communications sector throughout the region. Pharmaceutical and chemical industries are important in New Jersey and Pennsylvania.

Major industry and infrastructure

- chemicals
- coal
- defence
- electronics
- engineering
- finance
- hi-tech industry
- iron & steel
- pharmaceuticals
- printing & publishing
- research & development
- textiles
- timber processing
- major towns
- international airports
- major roads
- major industrial area

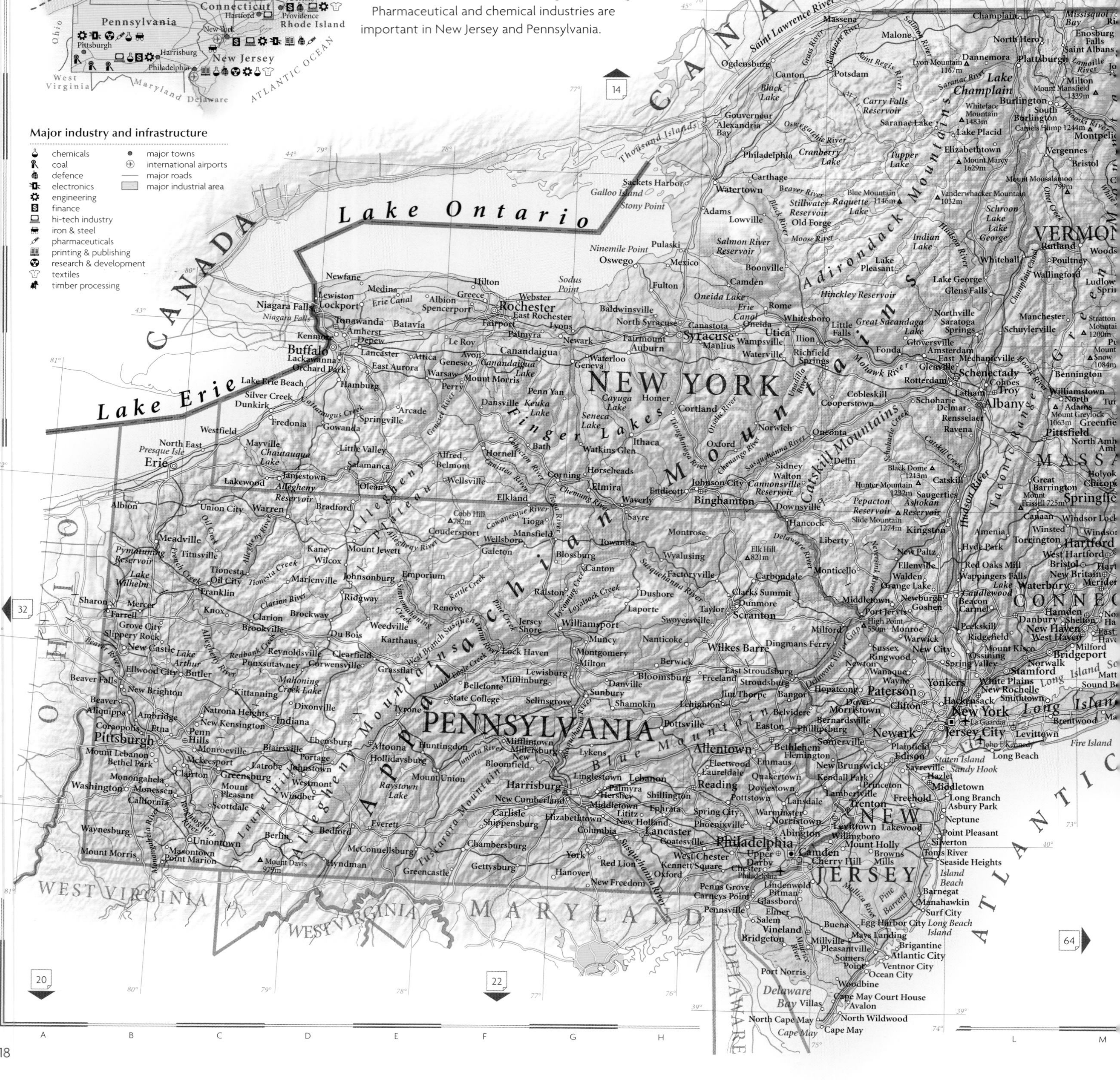

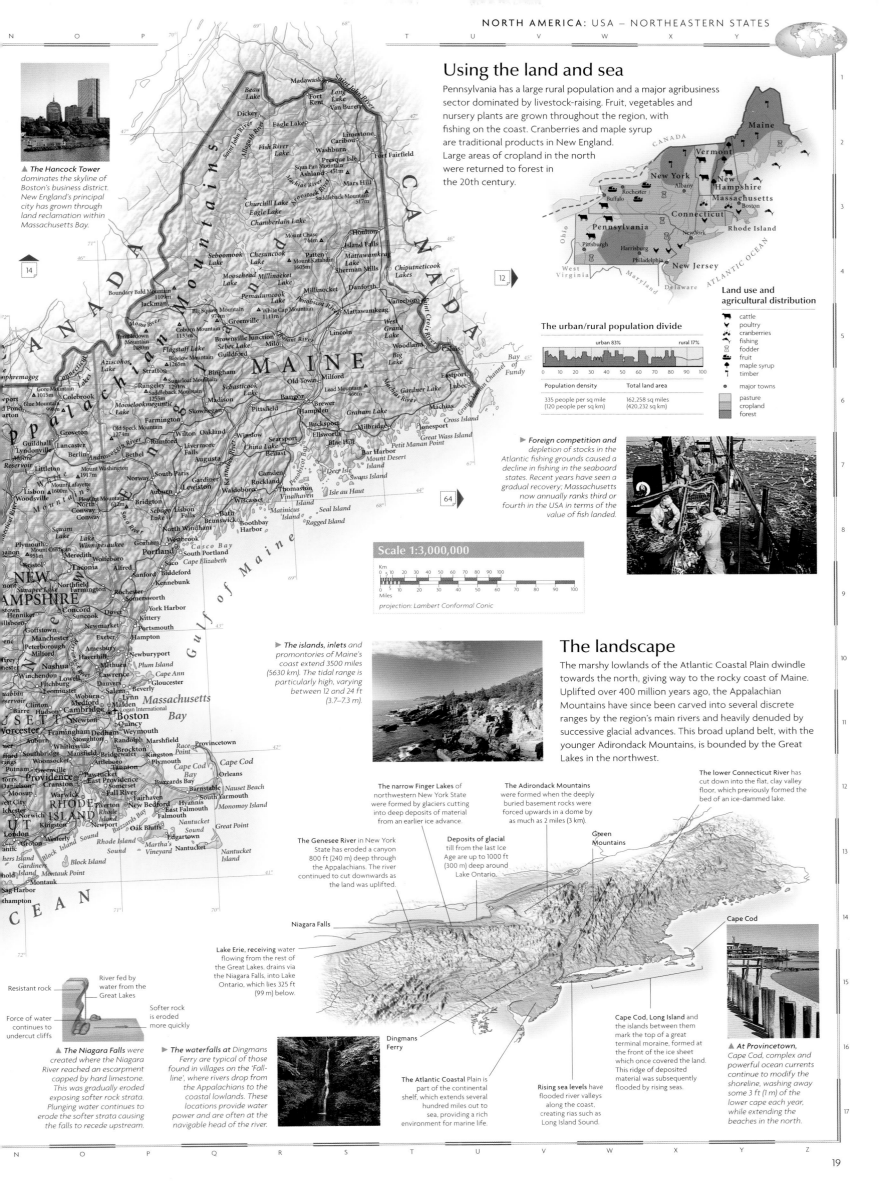

▲ *The Hancock Tower dominates the skyline of Boston's business district. New England's principal city has grown through land reclamation within Massachusetts Bay.*

14

12

64

Using the land and sea

Pennsylvania has a large rural population and a major agribusiness sector dominated by livestock-raising. Fruit, vegetables and nursery plants are grown throughout the region, with fishing on the coast. Cranberries and maple syrup are traditional products in New England. Large areas of cropland in the north were returned to forest in the 20th century.

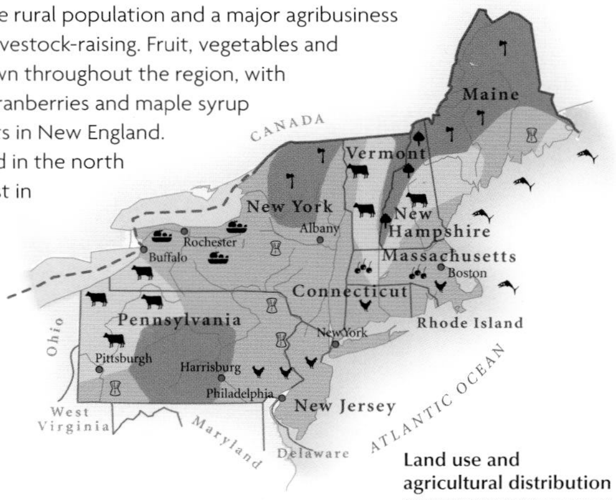

The urban/rural population divide

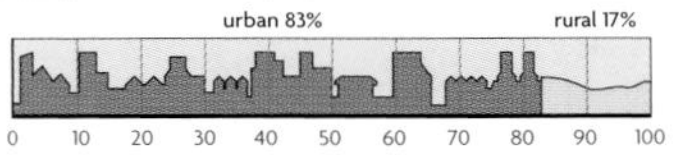

Population density	Total land area
335 people per sq mile (120 people per sq km)	162,258 sq miles (420,232 sq km)

► *Foreign competition and depletion of stocks in the Atlantic fishing grounds caused a decline in fishing in the seaboard states. Gradual recovery; Massachusetts now annually ranks third or fourth in the USA in terms of the value of fish landed.*

Scale 1:3,000,000

Km 0 5 10 20 30 40 50 60 70 80 90 100

Miles 0 5 10 20 30 40 50 60 70 80 90 100

projection: Lambert Conformal Conic

► *The islands, inlets and promontories of Maine's coast extend 3500 miles (5630 km). The tidal range is particularly high, varying between 12 and 24 ft (3.7–7.3 m).*

The landscape

The marshy lowlands of the Atlantic Coastal Plain dwindle towards the north, giving way to the rocky coast of Maine. Uplifted over 400 million years ago, the Appalachian Mountains have since been carved into several discrete ranges by the region's main rivers and heavily denuded by successive glacial advances. This broad upland belt, with the younger Adirondack Mountains, is bounded by the Great Lakes in the northwest.

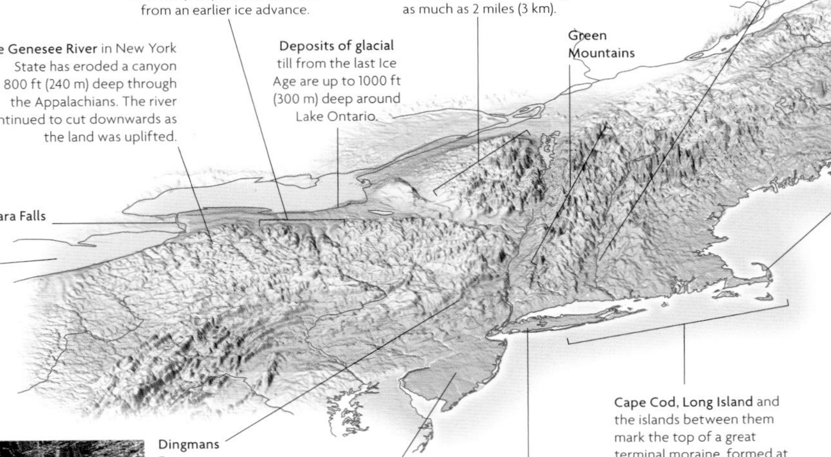

The lower Connecticut River has cut down into the flat, clay valley floor, which previously formed the bed of an ice-dammed lake.

The narrow Finger Lakes of northwestern New York State were formed by glaciers cutting into deep deposits of material from an earlier ice advance.

The Adirondack Mountains were formed when the deeply buried basement rocks were forced upwards in a dome by as much as 2 miles (3 km).

The Genesee River in New York State has eroded a canyon 800 ft (240 m) deep through the Appalachians. The river continued to cut downwards as the land was uplifted.

Deposits of glacial till from the last Ice Age are up to 1000 ft (300 m) deep around Lake Ontario.

Lake Erie, receiving water flowing from the rest of the Great Lakes, drains via the Niagara Falls, into Lake Ontario, which lies 325 ft (99 m) below.

Cape Cod, Long Island and the islands between them mark the top of a great terminal moraine, formed at the front of the ice sheet which once covered the land. This ridge of deposited material was subsequently flooded by rising seas.

The Atlantic Coastal Plain is part of the continental shelf, which extends several hundred miles out to sea, providing a rich environment for marine life.

Rising sea levels have flooded river valleys along the coast, creating rias such as Long Island Sound.

▲ *At Provincetown, Cape Cod, complex and powerful ocean currents continue to modify the shoreline, washing away some 3 ft (1 m) of the lower cape each year, while extending the beaches in the north.*

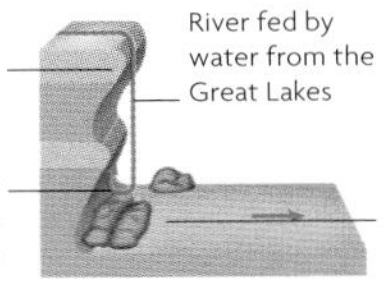

▲ *The Niagara Falls were created where the Niagara River reached an escarpment capped by hard limestone. This was gradually eroded exposing softer rock strata. Plunging water continues to erode the softer strata causing the falls to recede upstream.*

► *The waterfalls at Dingmans Ferry are typical of those found in villages on the 'Fall-line', where rivers drop from the Appalachians to the coastal lowlands. These locations provide water power and are often at the navigable head of the river.*

USA: MID-EASTERN STATES

Delaware, District of Columbia, Kentucky, Maryland, North Carolina, South Carolina, Tennessee, Virginia, West Virginia

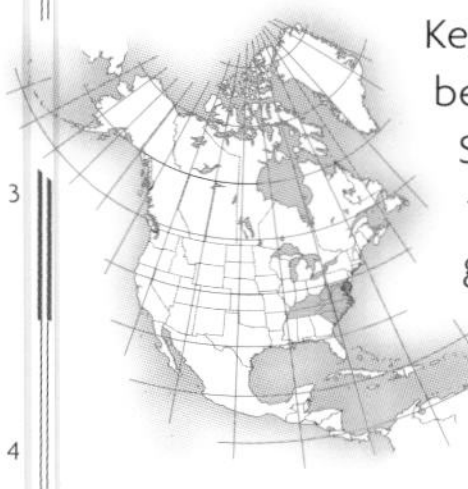

Key events in the history of the USA took place in this diverse region, which became the front line in the Civil War of 1861–65 between North and South. Strong regional contrasts exist between the fertile coastal plains, the isolated upcountry of the Appalachian Mountains and the cotton-growing areas of the Mississippi lowlands to the west. Whilst coal mining, a traditional industry in the Appalachians, has declined in recent years leaving much rural poverty, service industries elsewhere have increased, especially in the US federal capital, Washington DC.

Map key

Population

- 500,000 to 1 million
- 100,000 to 500,000
- 50,000 to 100,000
- 10,000 to 50,000
- below 10,000

Elevation

- 6000m / 19,686ft
- 4000m / 13,124ft
- 3000m / 9843ft
- 2000m / 6562ft
- 1000m / 3281ft
- 500m / 1640ft
- 250m / 820ft
- 100m / 328ft
- sea level

Scale 1:3,250,000

Km 0 5 10 20 30 40 50 60 70 80

Miles 0 5 10 20 30 40 50 60 70 80

projection: Lambert Conformal Conic

▲ *The Bluegrass region of Kentucky centres on the town of Lexington. This exceptionally fertile rolling plain is well known for its thoroughbred horse-breeding ranches.*

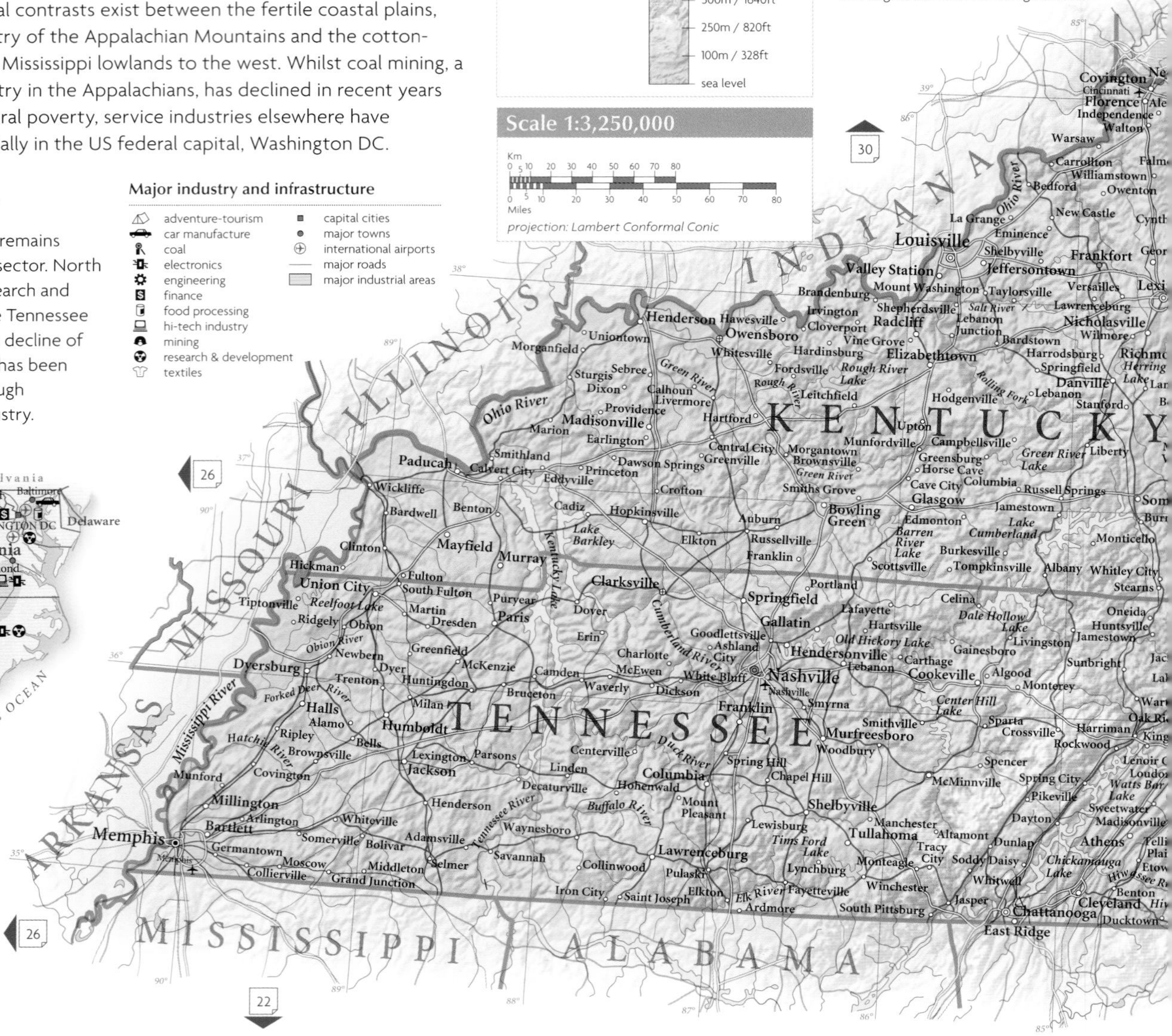

Transport and industry

In the urbanized northeast, manufacturing remains important, alongside a burgeoning service sector. North Carolina is a major centre for industrial research and development. Traditional industries include Tennessee whiskey, and textiles in South Carolina. The decline of open-cast coal mining in the Appalachians has been hastened by environmental controls, although adventure-tourism is a flourishing new industry.

Major industry and infrastructure

- adventure-tourism
- car manufacture
- coal
- electronics
- engineering
- finance
- food processing
- hi-tech industry
- mining
- research & development
- textiles
- capital cities
- major towns
- international airports
- major roads
- major industrial areas

The transport network

	452,218 miles (723,548 km)		5737 miles (8267 km)
	18,336 miles (29,503 km)		4404 miles (7081 km)

Tennessee's rivers are part of an important inland bulk-transport network. Memphis is connected with New Orleans in the south, and with cities as distant as Minneapolis, Sioux City, Chicago and Pittsburgh, via the Mississippi and its tributaries.

The landscape

The eastern tributaries of the Mississippi drain the interior lowlands. The Cumberland Plateau and the parallel ranges of the Appalachians have been successively uplifted and eroded over time, with the eastern side reduced to a series of foothills known as the Piedmont. The broad coastal plain gradually falls away into salt marshes, lagoons and offshore bars, broken by flooded estuaries along the shores of the Atlantic.

◀ *Farmland on the eastern shores of Chesapeake Bay is sustained by artificial drainage. The area also provides refuge for a variety of waterfowl.*

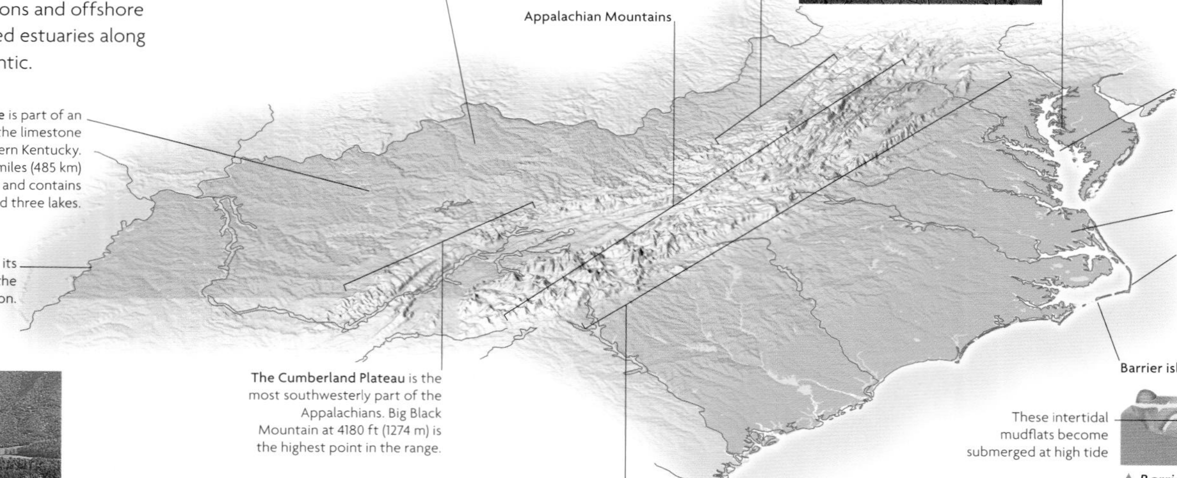

Natural Bridge in eastern Kentucky is an arch 78 ft (26 m) long and 65 ft (20 m) high. It has been shaped from resistant sandstone by gradual weathering processes, which removed the softer rock lying underneath.

The Allegheny Mountains form the northwestern edge of the Appalachian mountain chain. Continuous folding has formed rich seams of bituminous coal.

Appalachian Mountains

The many inlets of Chesapeake Bay are the flooded tributaries of the main river valley, which have been inundated by rising sea levels.

The Mammoth Cave is part of an extensive cave system in the limestone region of southwestern Kentucky. It stretches for over 300 miles (485 km) on five different levels and contains three rivers and three lakes.

Salt marshes such as Great Dismal Swamp, develop where the coast is sheltered. Vast areas of such marshland have been reclaimed for farmland and settlement.

The Mississippi River and its tributary the Ohio River form the western border of the region.

Cape Hatteras is the easternmost point of an offshore barrier island; a wave-deposited sand-bar which has become permanent, establishing its own vegetation.

The Cumberland Plateau is the most southwesterly part of the Appalachians. Big Black Mountain at 4180 ft (1274 m) is the highest point in the range.

Barrier islands

The Blue Ridge mountains are a steep ridge, culminating in Mount Mitchell, the highest point in the Appalachians, at 6684 ft (2037 m).

◀ *The Great Smoky Mountains form the western escarpment of the Appalachians. The region is heavily forested, with over 130 species of tree.*

▲ *Barrier islands are common along the coasts of North and South Carolina. As sea levels rise, wave action builds up ridges of sand and pebbles parallel to the coast, separated by lagoons or intertidal mudflats, which are flooded at high tide.*

◀ *Natural Bridge is* one of Virginia's most popular attractions. The unique 214 ft (65 m) high stone 'bridge' stretches across a 200 ft (60 m) deep gorge.

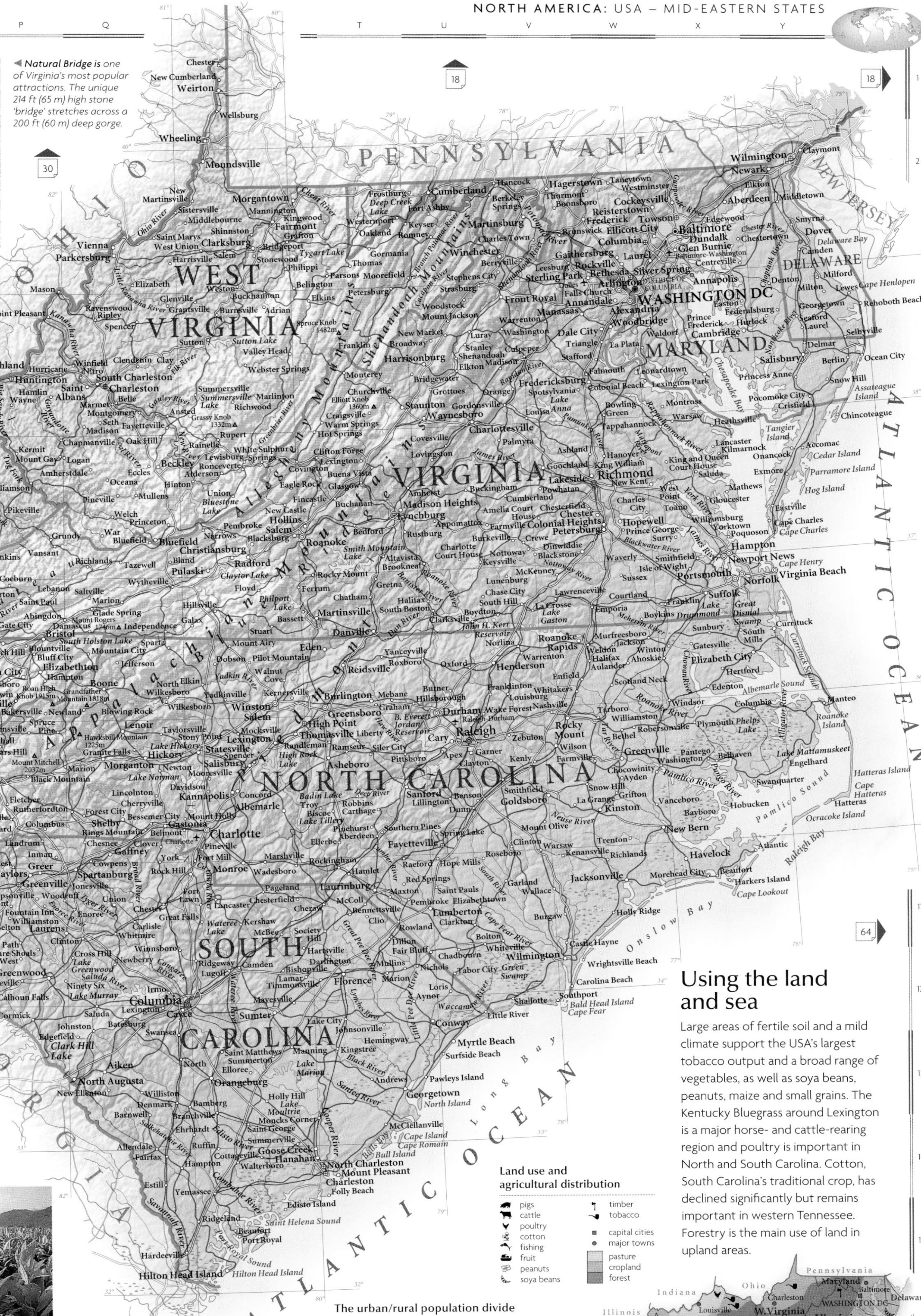

18
18
30
22
64
64

▲ *North Carolina is* the leading grower and processor of tobacco in the USA. The habit of smoking was adopted by Europeans from the native Americans, and tobacco became the main export crop for European colonists.

Using the land and sea

Large areas of fertile soil and a mild climate support the USA's largest tobacco output and a broad range of vegetables, as well as soya beans, peanuts, maize and small grains. The Kentucky Bluegrass around Lexington is a major horse- and cattle-rearing region and poultry is important in North and South Carolina. Cotton, South Carolina's traditional crop, has declined significantly but remains important in western Tennessee. Forestry is the main use of land in upland areas.

Land use and agricultural distribution

- pigs
- cattle
- poultry
- cotton
- fishing
- fruit
- peanuts
- soya beans
- timber
- tobacco
- capital cities
- major towns
- pasture
- cropland
- forest

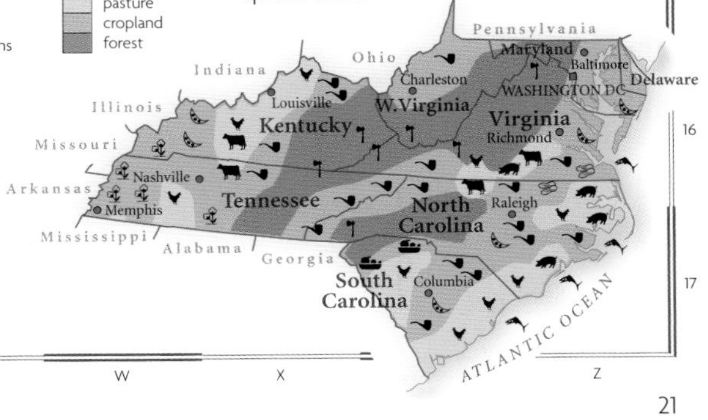

The urban/rural population divide

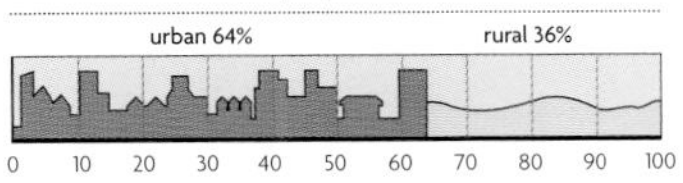

Population density	Total land area
149 people per sq mile (59 people per sq km)	235,226 sq miles (609,212 sq km)

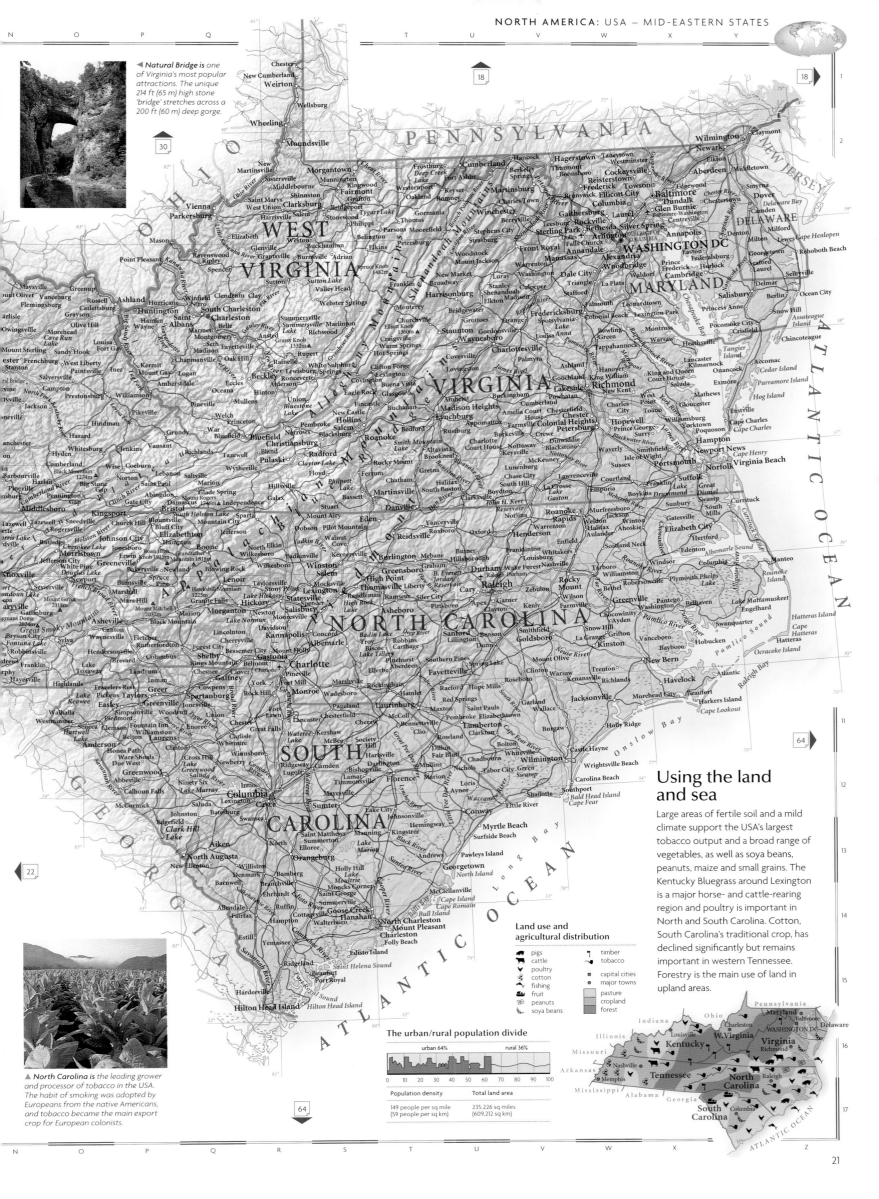

USA: SOUTHERN STATES

Alabama, Florida, Georgia, Louisiana, Mississippi

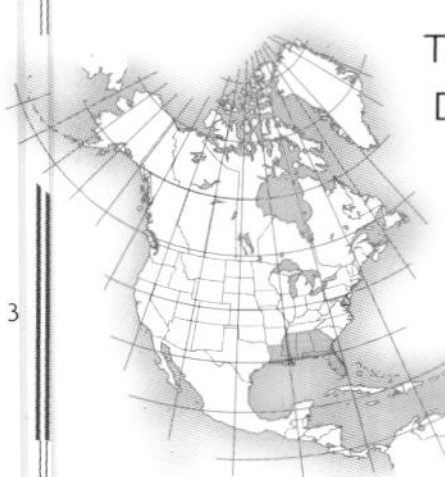

The South has maintained a separate identity and outlook throughout the history of the USA. Defeat in the American Civil War (1861–65) brought chronic poverty to the Confederate states, while the subsequent liberation of four million black slaves began a struggle not resolved until the 1960s, when the Civil Rights movement achieved an end to legal racial segregation. Since then many parts of the region have experienced rapid change: tourism and retirement communities, together with agriculture, have fuelled growth in Florida whilst defence-related industries have boosted the growth of cities such as Miami and Atlanta. Despite these changes, many people retain a strong attachment to their history: in Louisiana, French is still spoken in Cajun communities near the coast.

Transport and industry

Florida's tourist trade is only part of a flourishing service sector, which has swelled the principal cities of the south. Petroleum and mineral extraction has made the Gulf coast a major industrial region. Traditional textile production remains important in Georgia, while advanced new industries have grown from the NASA Space Program.

The transport network

441,625 miles (706,600 km)

5116 miles (8186 km)

16,597 miles (26,555 km)

6179 miles (9942 km)

Atlanta's Hartsfield International airport is one of the busiest in the world. A dramatic rise in the use of regional air transport has helped to integrate the major cities of the southern states.

◄ *The French Quarter is the traditional cultural centre of New Orleans. The city, extensively damaged by Hurricane Katrina in 2005, once thrived on the cotton trade but now relies mainly on tourism and on oil from the Gulf of Mexico.*

Major industry and infrastructure

- aerospace
- car manufacture
- chemicals
- coal
- defence
- electronics
- engineering
- food processing
- oil
- textiles
- tourism
- major towns
- international airports
- major roads
- major industrial areas

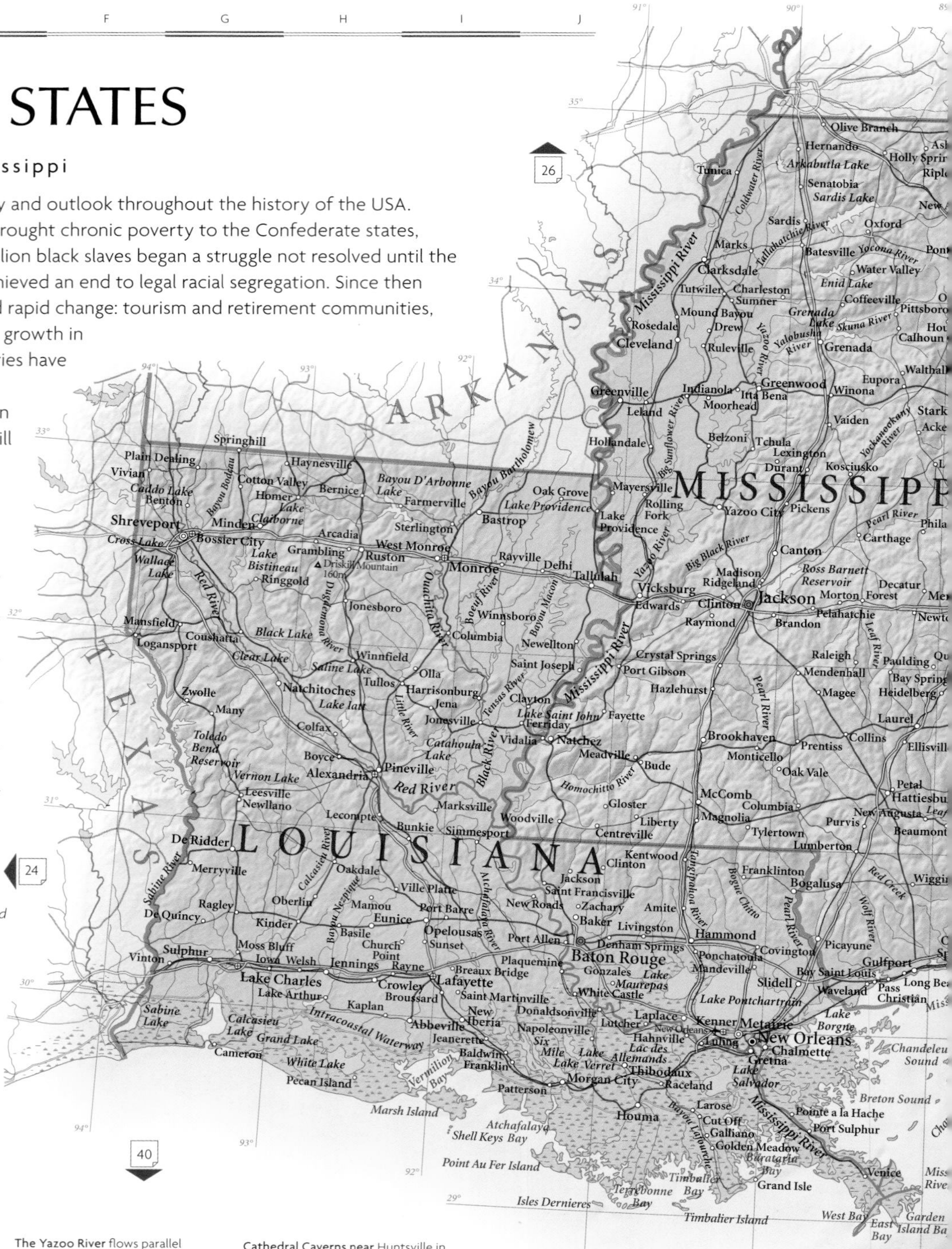

26
24
40

▲ *The cypress swamps of the Mississippi Delta form in the backswamps behind the levees of the river and in the multitude of subsiding delta basins.*

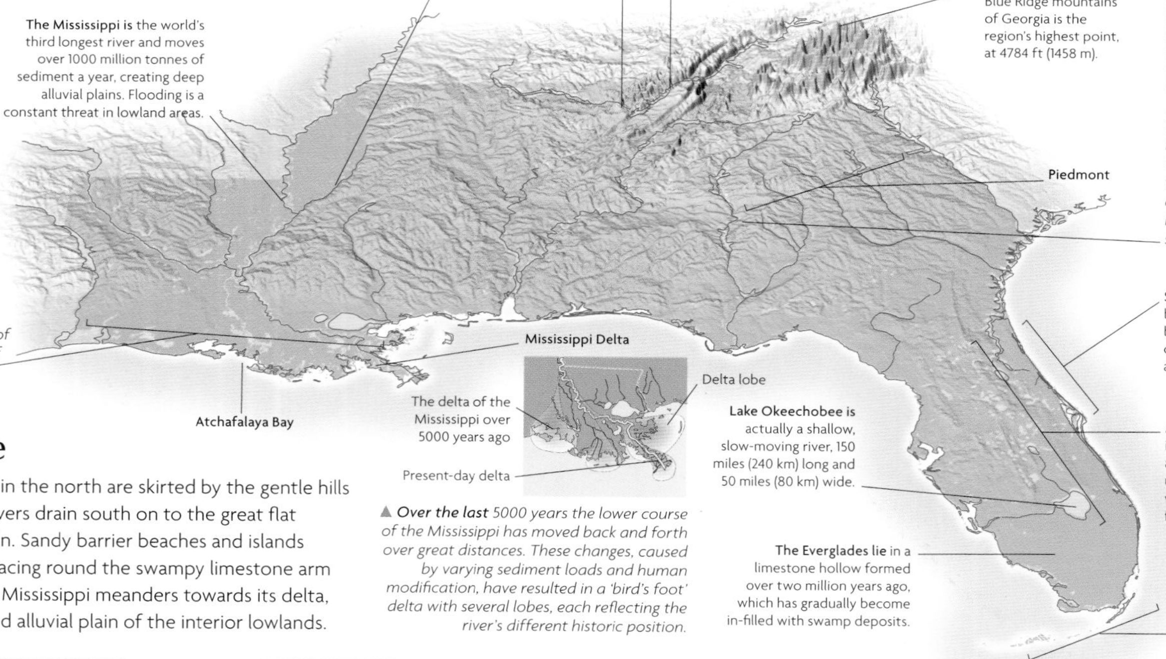

The Yazoo River flows parallel to the Mississippi through a common flood plain. The confluence of the rivers is deferred downstream because flood deposition has built the Mississippi channel up above the level of the Yazoo.

Cathedral Caverns near Huntsville in Alabama is a system of vast limestone caves, with a main opening 1000 ft (300 m) high and 150 ft (50 m) wide.

At De Soto Falls, Alabama, the Little River descends into the deepest canyon east of the Mississippi, with sheer cliff walls up to 700 ft (230 m) high.

Brasstown Bald in the Blue Ridge mountains of Georgia is the region's highest point, at 4784 ft (1458 m).

The Mississippi is the world's third longest river and moves over 1000 million tonnes of sediment a year, creating deep alluvial plains. Flooding is a constant threat in lowland areas.

▲ *In Providence Canyon, Georgia, the Chattahoochee River has cut straight down through the sandy bedrock, to leave sheer rock faces and pinnacles, which have been smoothed by subsequent weathering.*

Sand bars, deposited by waves breaking offshore, form barrier beaches along much of the coastline, creating sheltered lagoons and salt marshes behind them.

Across Florida the coastal plain is mostly less than 75 ft (25 m) above sea level. The land is underlain by limestone, pitted with hollows which have been filled by over 10,000 lakes.

▲ *Over the last 5000 years the lower course of the Mississippi has moved back and forth over great distances. These changes, caused by varying sediment loads and human modification, have resulted in a 'bird's foot' delta with several lobes, each reflecting the river's different historic position.*

Lake Okeechobee is actually a shallow, slow-moving river, 150 miles (240 km) long and 50 miles (80 km) wide.

The Everglades lie in a limestone hollow formed over two million years ago, which has gradually become in-filled with swamp deposits.

The landscape

The Blue Ridge mountains in the north are skirted by the gentle hills of the Piedmont, whose rivers drain south on to the great flat expanse of the coastal plain. Sandy barrier beaches and islands dominate the sea shore, tracing round the swampy limestone arm of Florida. In the west, the Mississippi meanders towards its delta, crossing the thickly mantled alluvial plain of the interior lowlands.

▲ **Mangrove swamps and** islets merge across Whitewater Bay, in the Everglades National Park. Alligators, crocodiles, endangered aquatic mammals such as manatees, and a great variety of birds inhabit the subtropical sanctuary.

◀ **New Orleans was** devastated by Hurricane Katrina in August 2005. Around 1200 lives were lost across the region. Florida and the Gulf coast are prone to hurricanes every autumn.

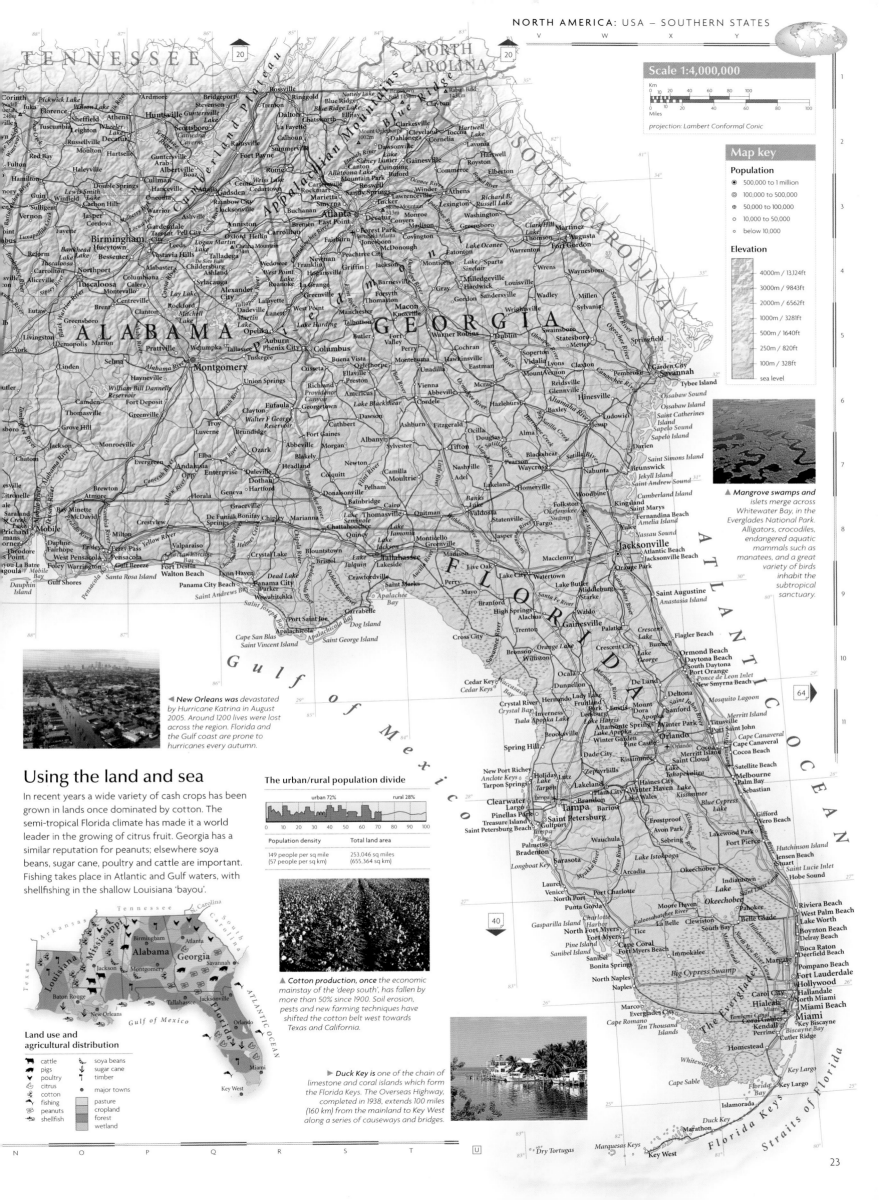

Using the land and sea

In recent years a wide variety of cash crops has been grown in lands once dominated by cotton. The semi-tropical Florida climate has made it a world leader in the growing of citrus fruit. Georgia has a similar reputation for peanuts; elsewhere soya beans, sugar cane, poultry and cattle are important. Fishing takes place in Atlantic and Gulf waters, with shellfishing in the shallow Louisiana 'bayou'.

The urban/rural population divide

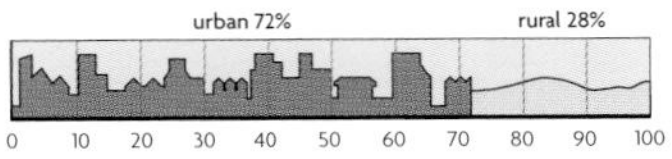

Population density	Total land area
149 people per sq mile (57 people per sq km)	253,046 sq miles (655,364 sq km)

▲ **Cotton production, once** the economic mainstay of the 'deep south', has fallen by more than 50% since 1900. Soil erosion, pests and new farming techniques have shifted the cotton belt west towards Texas and California.

Land use and agricultural distribution

cattle	soya beans
pigs	sugar cane
poultry	timber
citrus	major towns
cotton	pasture
fishing	cropland
peanuts	forest
shellfish	wetland

▶ **Duck Key is** one of the chain of limestone and coral islands which form the Florida Keys. The Overseas Highway, completed in 1938, extends 100 miles (160 km) from the mainland to Key West along a series of causeways and bridges.

USA: Texas

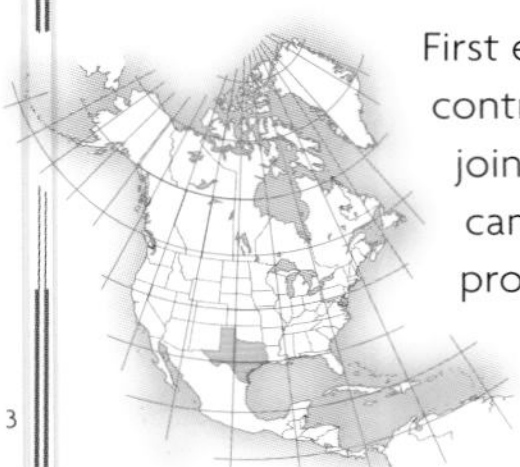

First explored by Spaniards moving north from Mexico in search of gold, Texas was controlled by Spain and then Mexico, before becoming an independent republic in 1836, and joining the Union of States in 1845. During the 19th century, many of the migrants who came to Texas raised cattle on the abundant land; in the 20th century, they were joined by prospectors attracted by the promise of oil riches. Today, although natural resources, especially oil, still form the basis of its wealth, the diversified Texan economy includes thriving hi-tech and finance industries. The major urban centres, home to 80% of the population, lie in the south and east, and include Houston, the 'oil-city', and Dallas–Fort Worth. Hispanic influences remain strong, especially in the south and west.

▲ ***Dallas was founded*** *in 1841 as a prairie trading post and its development was stimulated by the arrival of railroads. Cotton and then oil funded the town's early growth. Today, the modern, high-rise skyline of Dallas reflects the city's position as a leading centre of banking, insurance and the petroleum industry in the southwest.*

Using the land

Cotton production and livestock-raising, particularly cattle, dominate farming, although crop failures and the demands of local markets have led to some diversification. Following the introduction of modern farming techniques, cotton production spread out from the east to the plains of western Texas. Cattle ranches are widespread, while sheep and goats are raised on the dry Edwards Plateau.

▲ ***The huge cattle*** *ranches of Texas developed during the 19th century when land was plentiful and could be acquired cheaply. Today, more cattle and sheep are raised in Texas than in any other state.*

Land use and agricultural distribution

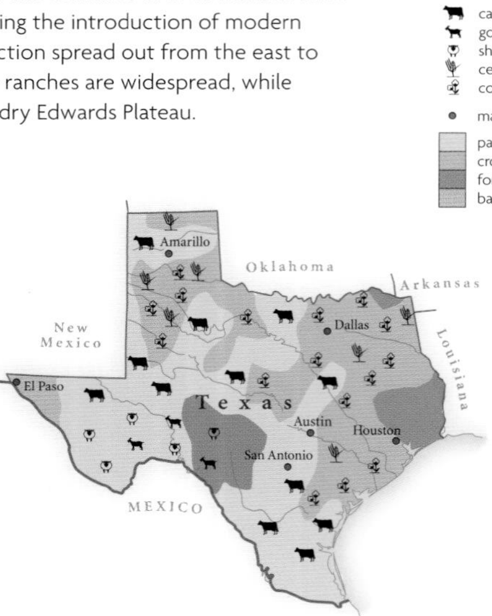

The urban/rural population divide

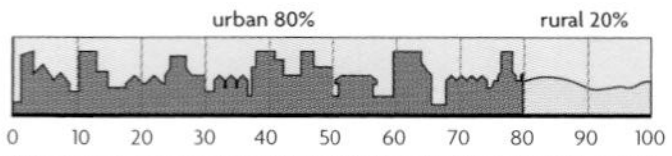

Population density	Total land area
84 people per sq mile (33 people per sq km)	261,797 sq miles (678,028 sq km)

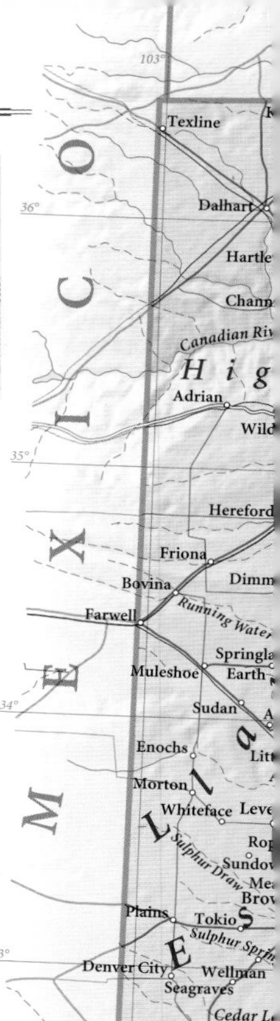

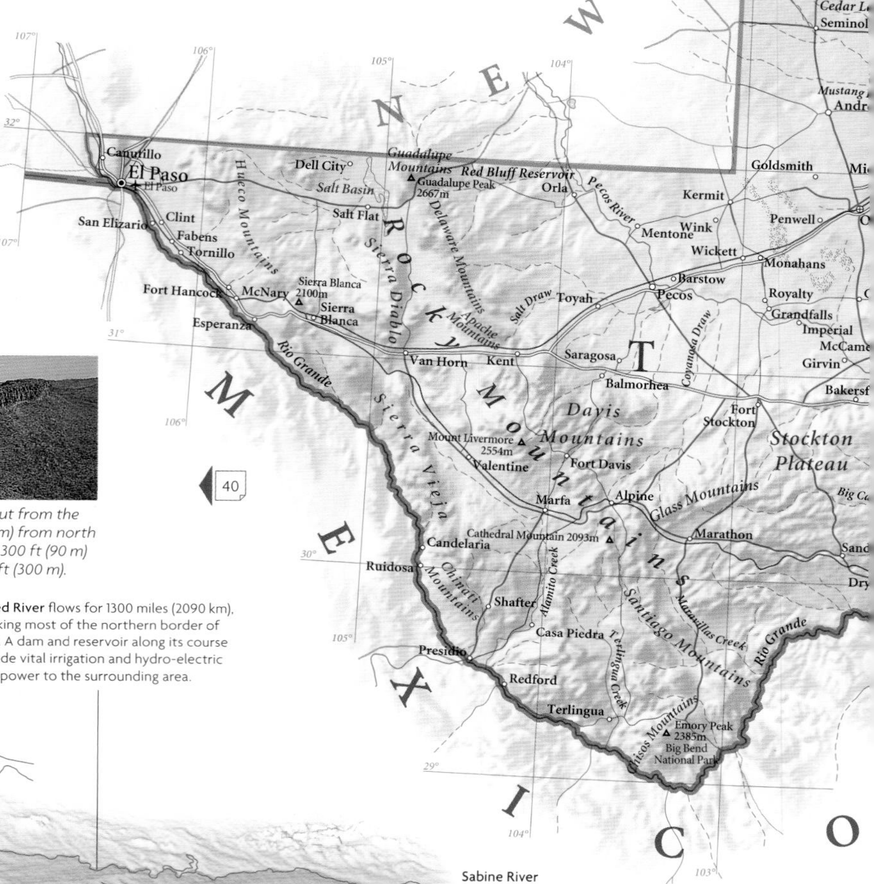

▲ ***Cap Rock Escarpment*** *juts out from the plains, running 200 miles (320 km) from north to south. Its height varies from 300 ft (90 m) rising to sheer cliffs up to 1000 ft (300 m).*

The landscape

Texas is made up of a series of massive steps descending from the mountains and high plains of the west and northwest to the coastal lowlands in the southeast. Many of the state's borders are delineated by water. The Rio Grande flows from the Rocky Mountains to the Gulf of Mexico, marking the border with Mexico.

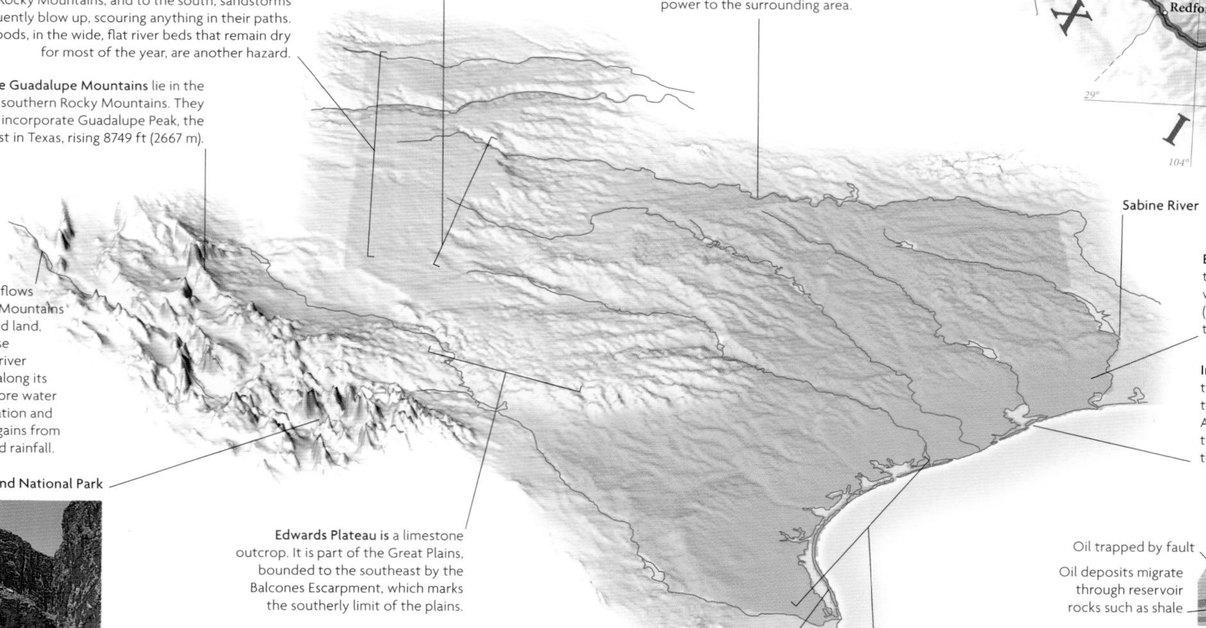

The Llano Estacado or Staked Plain in northern Texas is known for its harsh environment. In the north, freezing winds carrying ice and snow sweep down from the Rocky Mountains, and to the south, sandstorms frequently blow up, scouring anything in their paths. Flash floods, in the wide, flat river beds that remain dry for most of the year, are another hazard.

The Red River flows for 1300 miles (2090 km), marking most of the northern border of Texas. A dam and reservoir along its course provide vital irrigation and hydro-electric power to the surrounding area.

The Guadalupe Mountains lie in the southern Rocky Mountains. They incorporate Guadalupe Peak, the highest in Texas, rising 8749 ft (2667 m).

Sabine River

The Rio Grande flows from the Rocky Mountains through semi-arid land, supporting sparse vegetation. The river actually shrinks along its course, losing more water through evaporation and seepage than it gains from its tributaries and rainfall.

Extensive forests of pine and cypress grow in the eastern corner of the coastal lowlands where the average rainfall is 45 inches (1145 mm) a year. This is higher than the rest of the state and over twice the average in the west.

In the coastal lowlands of southeastern Texas the Earth's crust is warping, causing the land to subside and allowing the sea to invade. Around Galveston, the rate of downward tilting is 6 inches (15 cm) per year. Erosion of the coast is also exacerbated by hurricanes.

Big Bend National Park

Edwards Plateau is a limestone outcrop. It is part of the Great Plains, bounded to the southeast by the Balcones Escarpment, which marks the southerly limit of the plains.

◀ ***Flowing through*** *1500 ft (450 m) high gorges, the shallow, muddy Rio Grande makes a 90° bend, which marks the southern border of Big Bend National Park, giving it its name. The area is a mixture of forested mountains, deserts and canyons.*

Padre Island

Laguna Madre in southern Texas has been almost completely cut off from the sea by Padre Island. This sand bank was created by wave action, carrying and depositing material along the coast. The process is known as longshore drift.

Oil deposits

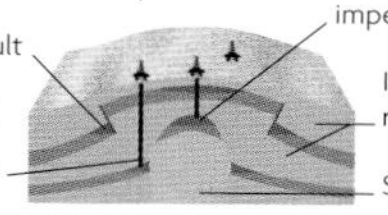

▲ ***Oil deposits are*** *found beneath much of Texas. They collect as oil migrates upwards through porous layers of rock until it is trapped, either by a cap of rock above a salt dome, or by a fault line which exposes impermeable rock through which the oil cannot rise.*

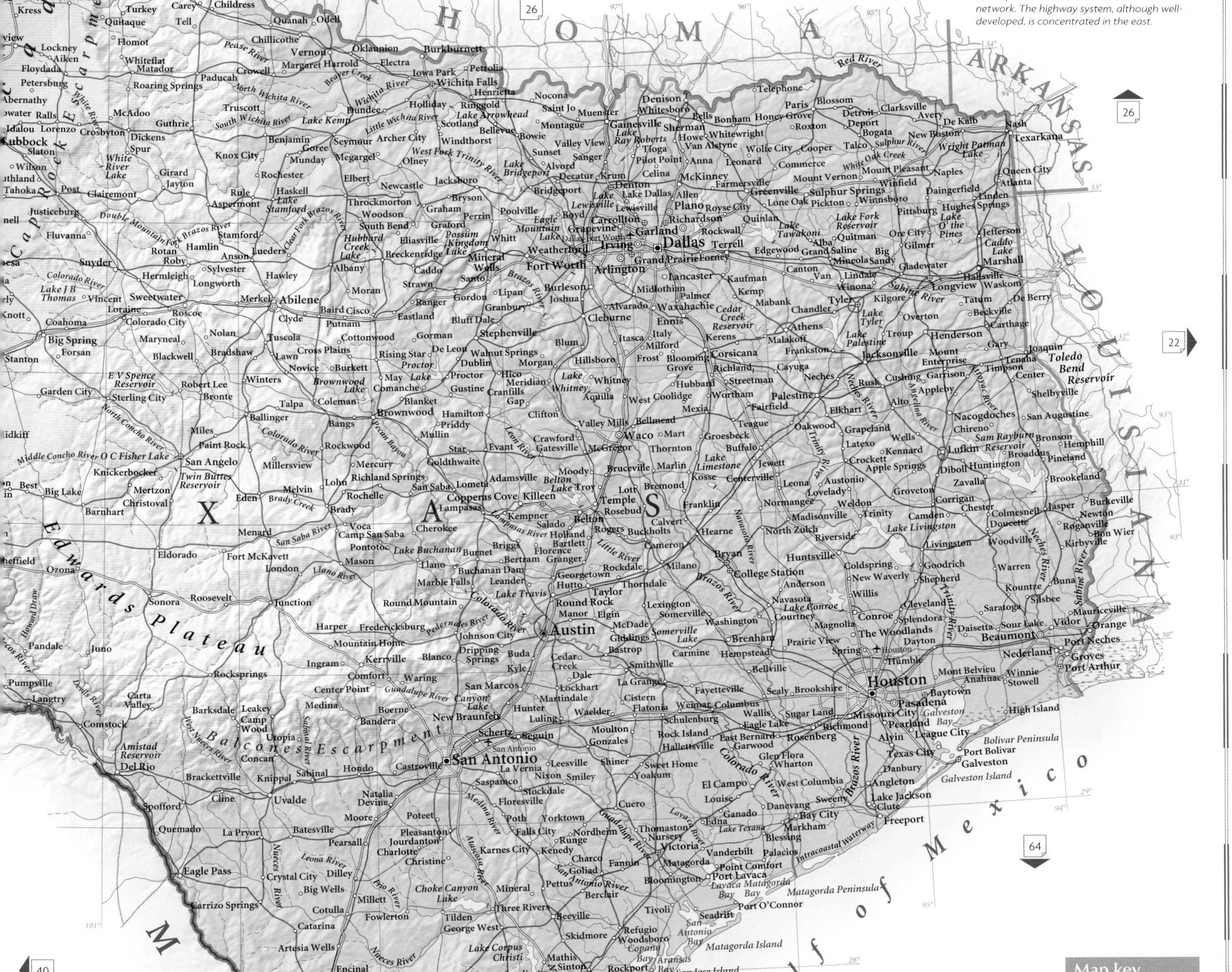

Transport and industry

Industry in the 20th century was largely concentrated on the processing of local raw materials, especially oil – deposits were discovered under 65% of the state's area. The technological demands of the oil industry and defence-related institutions, particularly NASA, have stimulated the development of numerous electronics and hi-tech firms which, alongside many national corporate headquarters, are based in Dallas–Fort Worth and Houston.

Major industry and infrastructure

- chemicals
- defence
- engineering
- finance
- food processing
- gas
- hi-tech industry
- mining
- oil
- textiles
- major towns
- international airports
- major roads
- major industrial areas

The transport network

- 293,509 miles (496,614 km)
- 3229 miles (5166 km)
- 10,681 miles (17,089 km)
- 845 miles (1359 km)

The sheer size of Texas promoted the development of an extensive road and rail network. The highway system, although well-developed, is concentrated in the east.

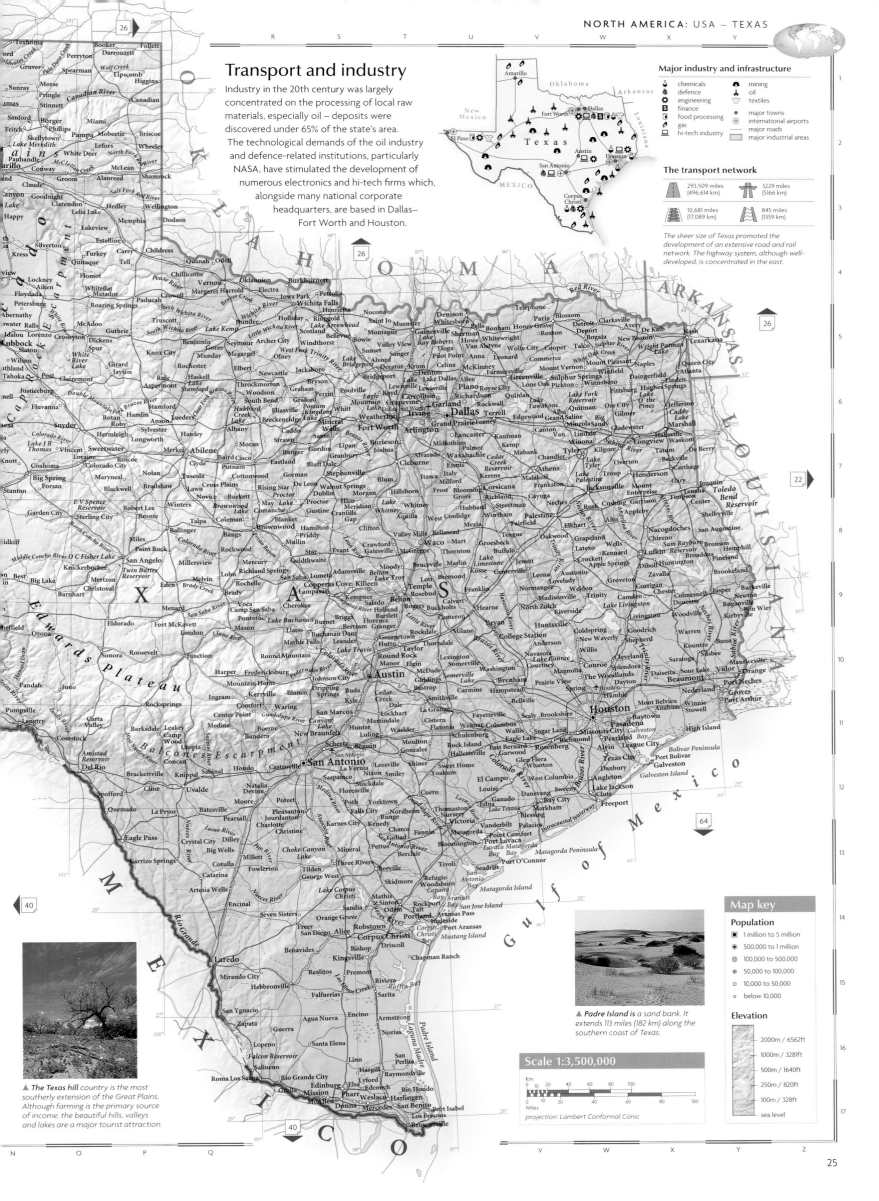

▲ ***The Texas hill** country is the most southerly extension of the Great Plains. Although farming is the primary source of income, the beautiful hills, valleys and lakes are a major tourist attraction.*

▲ ***Padre Island** is a sand bank. It extends 113 miles (182 km) along the southern coast of Texas.*

Scale 1:3,500,000

Km 0 10 20 40 60 80 100
Miles 0 10 20 40 60 80 100

projection: Lambert Conformal Conic

Map key

Population

- 1 million to 5 million
- 500,000 to 1 million
- 100,000 to 500,000
- 50,000 to 100,000
- 10,000 to 50,000
- below 10,000

Elevation

- 2000m / 6562ft
- 1000m / 3281ft
- 500m / 1640ft
- 250m / 820ft
- 100m / 328ft
- sea level

USA: SOUTH MIDWESTERN STATES

Arkansas, Kansas, Missouri, Oklahoma

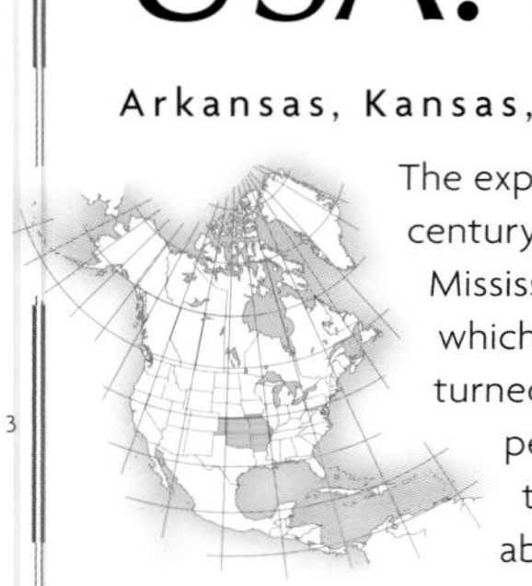

The expansion of the USA focused on this region in the mid-19th century. Settlers spread from the confluence of the Missouri and Mississippi rivers up onto the Great Plains. This treeless expanse, which early explorers had called the 'Great American Desert', was turned into one of the world's richest agricultural regions; but periodic droughts, coupled with over-intensive farming, led to the 'Dustbowl' soil erosion crisis of the 1930s, the abandonment of many farms, and a mass exodus to the west coast. The land has since recovered, although the mechanization of agriculture has led to a decline in the rural population. In recent years, suburban residential development has spread rapidly across the wooded Ozark Plateau in the east of the region.

Transport and industry

The processing of agricultural products, such as brewing and meat packing, has been traditionally important in these states. In Kansas and Oklahoma, diversified manufacturing now supplements income from fossil fuels; Wichita has become a world centre for aeronautical engineering, an industry which also employs many people in neighbouring Missouri.

Major industry and infrastructure

- aerospace
- engineering
- finance
- food processing
- gas
- mining
- oil
- vehicle manufacture
- major towns
- international airports
- major roads
- major industrial areas

▶ *Agricultural produce* **from** *the plains is moved by barges along the Mississippi. The river now carries a far greater tonnage of freight than any other waterway system in the USA.*

The transport network

380,307 miles (608,491 km)	4068 miles (6508 km)
16,185 miles (25,896 km)	1994 miles (3208 km)

The Arkansas River and its tributaries allow access to over half of the USA's navigable inland waterways. A system of locks and dams along the river provides Tulsa in Oklahoma with a navigable water route to the Gulf of Mexico.

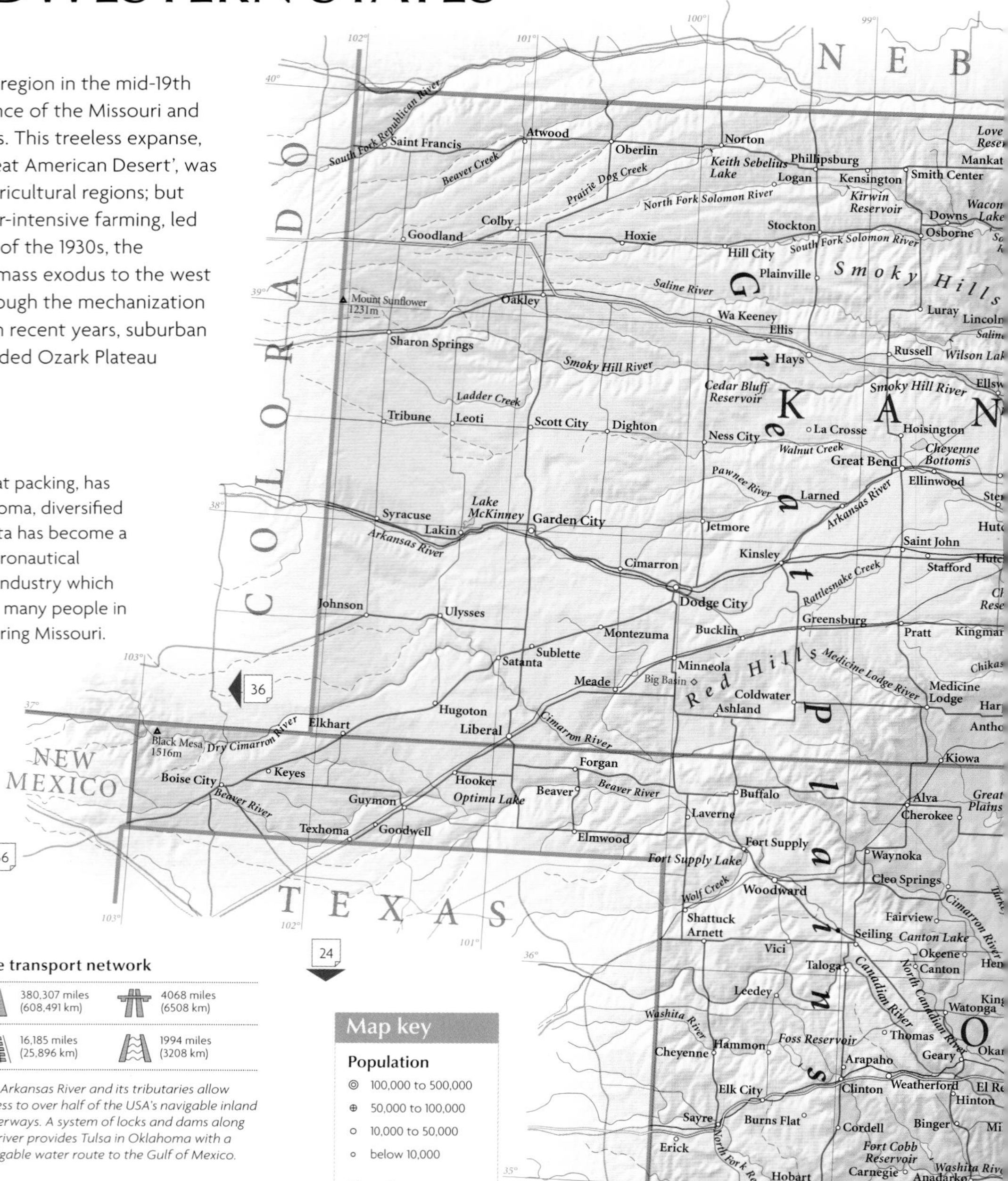

28
36
36
24
24

Map key

Population

- ◎ 100,000 to 500,000
- ⊕ 50,000 to 100,000
- ○ 10,000 to 50,000
- ∘ below 10,000

Elevation

- 1000m / 3281ft
- 500m / 1640ft
- 250m / 820ft
- 100m / 328ft
- sea level

Scale 1:3,250,000

Km 0 5 10 20 30 40 50 60 70
Miles 0 5 10 20 30 40 50 60 70

projection: Lambert Conformal Conic

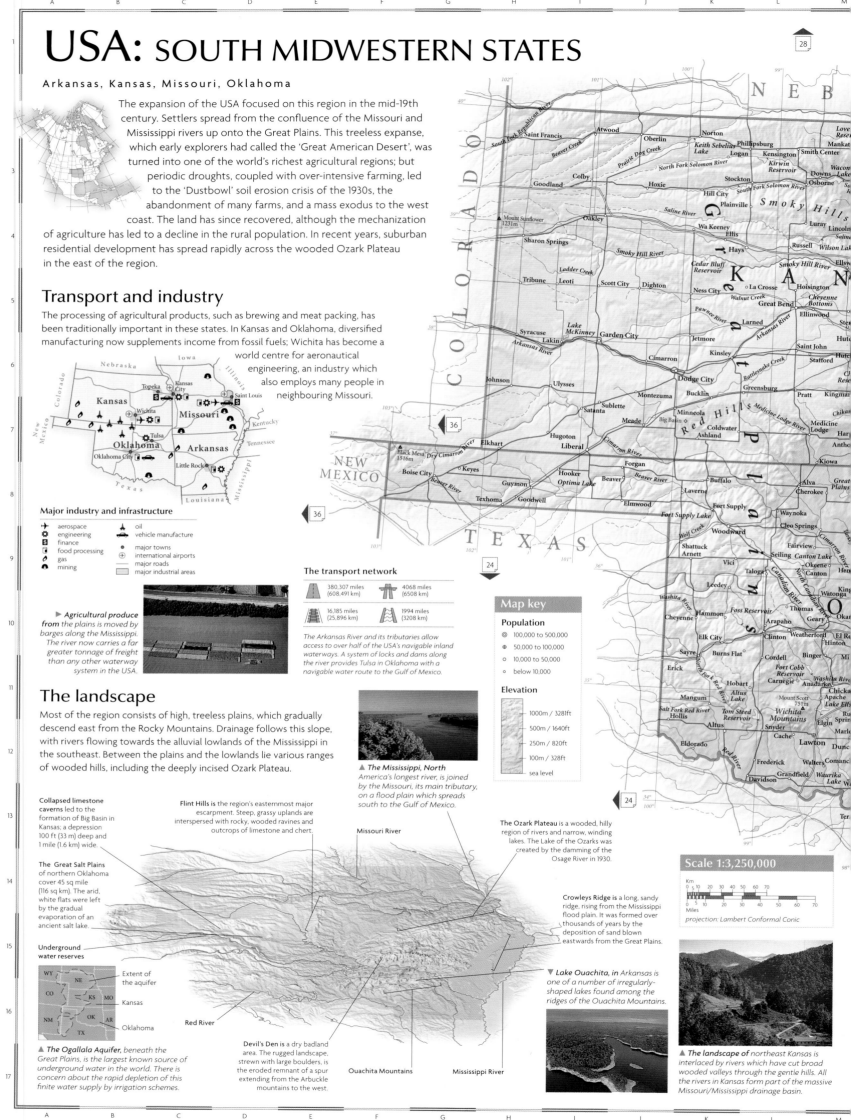

The landscape

Most of the region consists of high, treeless plains, which gradually descend east from the Rocky Mountains. Drainage follows this slope, with rivers flowing towards the alluvial lowlands of the Mississippi in the southeast. Between the plains and the lowlands lie various ranges of wooded hills, including the deeply incised Ozark Plateau.

▲ ***The Mississippi, North*** *America's longest river, is joined by the Missouri, its main tributary, on a flood plain which spreads south to the Gulf of Mexico.*

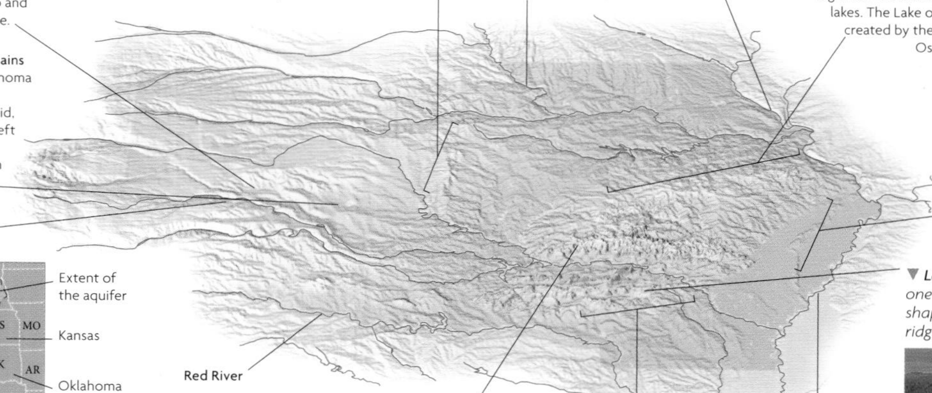

Collapsed limestone caverns led to the formation of Big Basin in Kansas; a depression 100 ft (33 m) deep and 1 mile (1.6 km) wide.

The Great Salt Plains of northern Oklahoma cover 45 sq mile (116 sq km). The arid, white flats were left by the gradual evaporation of an ancient salt lake.

Underground water reserves

Flint Hills is the region's easternmost major escarpment. Steep, grassy uplands are interspersed with rocky, wooded ravines and outcrops of limestone and chert.

Missouri River

The Ozark Plateau is a wooded, hilly region of rivers and narrow, winding lakes. The Lake of the Ozarks was created by the damming of the Osage River in 1930.

Crowleys Ridge is a long, sandy ridge, rising from the Mississippi flood plain. It was formed over thousands of years by the deposition of sand blown eastwards from the Great Plains.

Red River

Devil's Den is a dry badland area. The rugged landscape, strewn with large boulders, is the eroded remnant of a spur extending from the Arbuckle mountains to the west.

Ouachita Mountains

Mississippi River

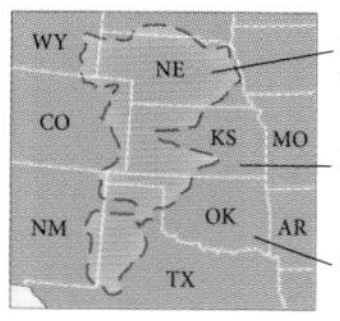

Extent of the aquifer

Kansas

Oklahoma

▲ ***The Ogallala Aquifer,*** *beneath the Great Plains, is the largest known source of underground water in the world. There is concern about the rapid depletion of this finite water supply by irrigation schemes.*

▼ ***Lake Ouachita,*** *in Arkansas is one of a number of irregularly-shaped lakes found among the ridges of the Ouachita Mountains.*

▲ ***The landscape of*** *northeast Kansas is interlaced by rivers which have cut broad wooded valleys through the gentle hills. All the rivers in Kansas form part of the massive Missouri/Mississippi drainage basin.*

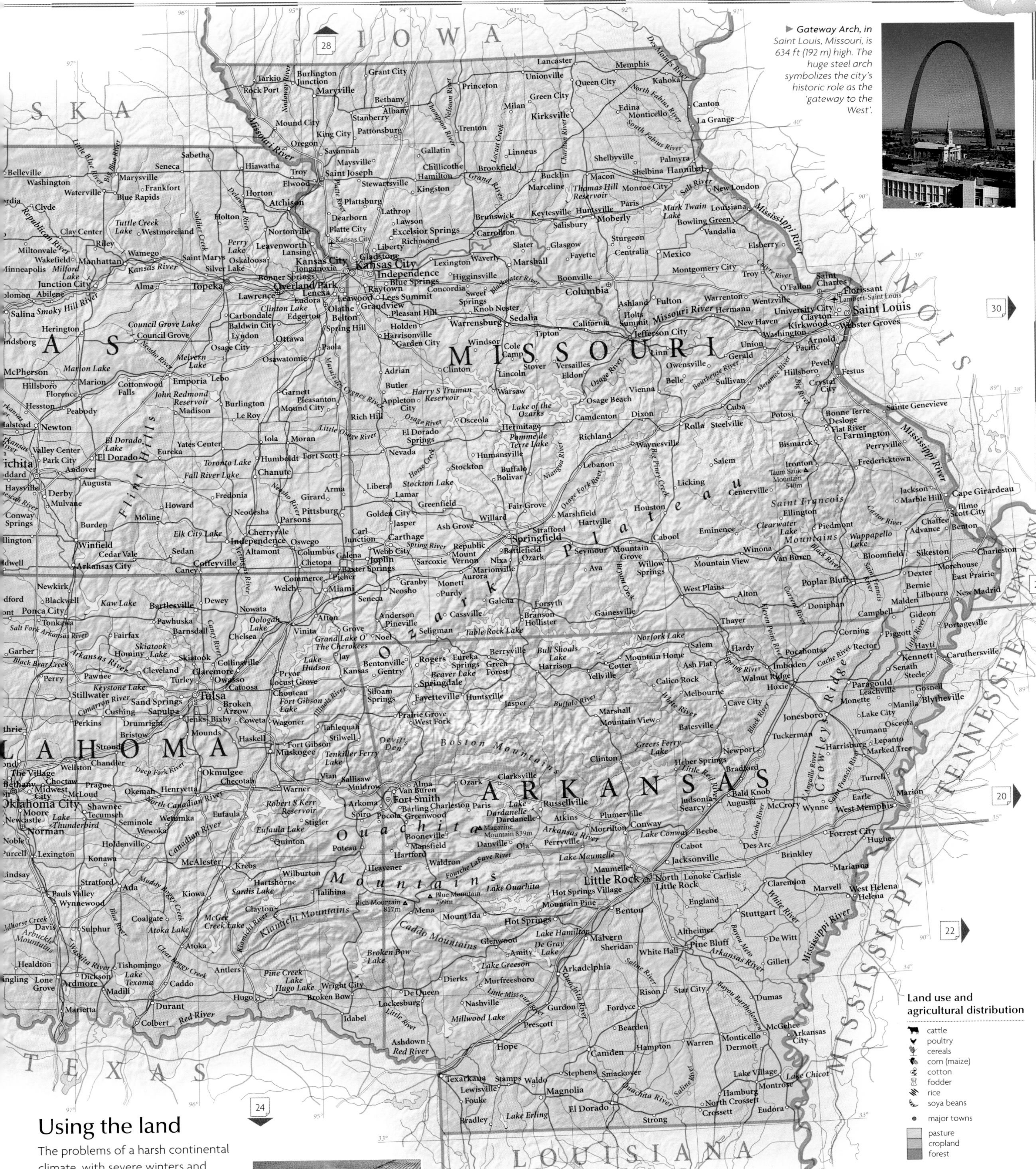

▶ *Gateway Arch, in Saint Louis, Missouri, is 634 ft (192 m) high. The huge steel arch symbolizes the city's historic role as the 'gateway to the West'.*

Using the land

The problems of a harsh continental climate, with severe winters and hot, dry summers, are partially offset by the rich soils of the plains. Kansas is a major cereal producer, ranking first in the USA for the production of wheat and sorghum. Rainfall increases towards the east, favouring the cultivation of soya beans, cotton and rice, with corn concentrated in Missouri. Huge herds of cattle are raised in Oklahoma, Kansas and Missouri.

▲ ***A combine harvester** works the land on the great plains. A hundred years ago this region, also known as the prairies – the French word for pasture – was covered with tall, wild grasses.*

The urban/rural population divide

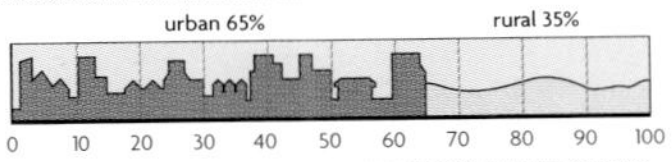

Population density	Total land area
54 people per sq mile (21 people per sq km)	271,436 sq miles (702,992 sq km)

Land use and agricultural distribution

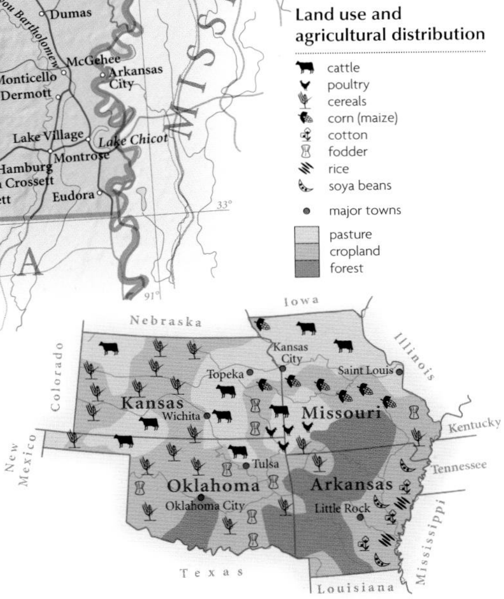

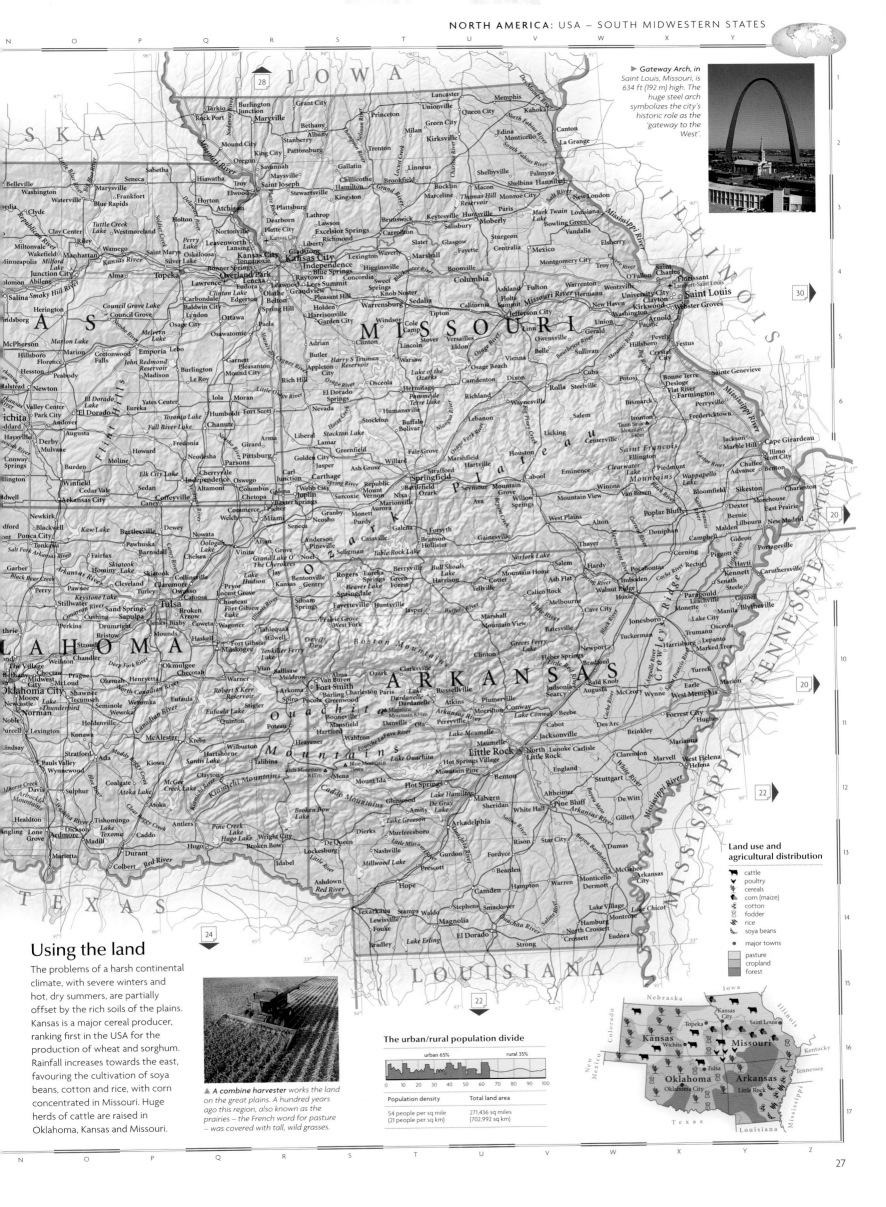

USA: UPPER PLAINS STATES

Iowa, Minnesota, Nebraska, North Dakota, South Dakota

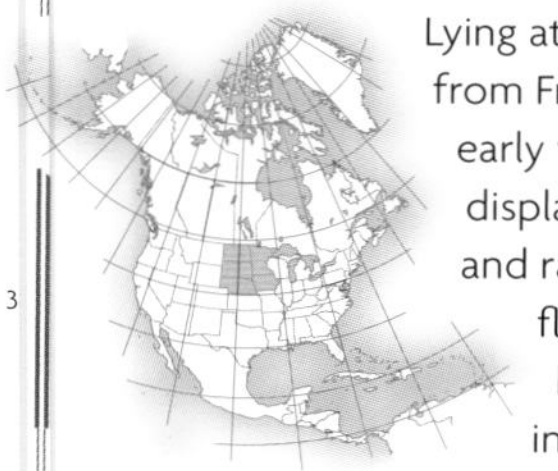

Lying at the very heart of the North American continent, much of this region was acquired from France as part of the Louisiana Purchase in 1803. The area was largely by-passed by the early waves of westward migrants. When Europeans did settle, during the 19th century, they displaced the Native Americans who lived on the plains. The settlers planted arable crops and raised cattle on the immensely fertile prairie land, founding an agrarian tradition which flourishes today. Most of this region remains rural; of the five states, only in Minnesota has there been significant diversification away from agriculture and resource-based industries into the hi-tech and service sectors.

Using the land

The popular image of these states as agricultural is entirely justified; prairies stretch uninterrupted across most of the area. Croplands fall into two regions: the wheat belt of the plains, and the corn belt of the central USA. Cash crops, such as soya beans, are grown to supplement incomes. Livestock, particularly pigs and cattle, are raised throughout this region.

*► **Dark, fertile prairie** soils in the southeast provide Minnesota's most productive farmland. Hot, humid summers create a long growing season for corn cultivation.*

Land use and agricultural distribution

- cattle
- pigs
- corn (maize)
- soya beans
- wheat
- major towns
- pasture
- cropland
- forest
- wetland

The urban/rural population divide

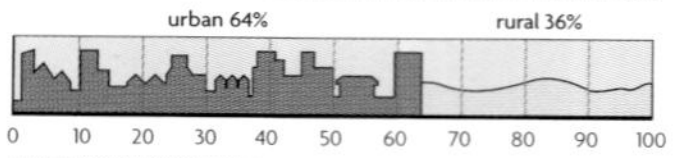

Population density	Total land area
31 people per sq mile (12 people per sq km)	357,212 sq miles (925,143 sq km)

Transport and industry

Food processing and the production of farm machinery are supported by the large agricultural sector. Mineral exploitation is also an important activity: gold is mined in the ore-rich Black Hills of South Dakota, and both North Dakota and Nebraska are emerging as major petroleum producers.

*► **Water erosion along** the Little Missouri River has carried away sedimentary deposits, creating rugged landscapes known as badlands.*

The transport network

504,522 miles (807,235 km)	3422 miles (5475 km)
16,940 miles (27,104 km)	683 miles (1098 km)

Nebraska's central location has made it an important transport artery for east–west traffic. Minnesota's road network radiates out from the hub of the twin cities, Minneapolis–Saint Paul.

Major industry and infrastructure

- coal
- engineering
- electronics
- finance
- food processing
- oil & gas
- mining
- major towns
- international airports
- major roads
- major industrial areas

The landscape

These states straddle the Great Plains and the lowlands of the central USA, with Minnesota lying in a transition zone between the eastern forests and the prairies. The region was shaped by repeated ice advances and retreats, leaving a flat relief, broken only by the numerous lakes and broad river networks which drain the prairies.

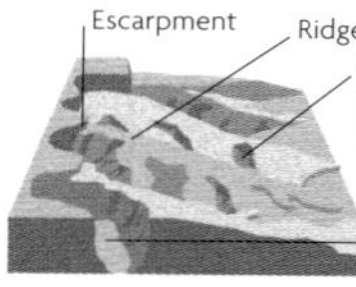

*▲ **Badlands are formed** by stormwater run-off which flows down the impermeable strata of the escarpment and saturates the permeable strata leading to mudslides and the formation of gullies.*

*▲ **In the badlands** of North and South Dakota, horizontal layers of sandstone have been eroded by rivers, leaving a landscape of narrow gullies, sharp crests and pinnacles.*

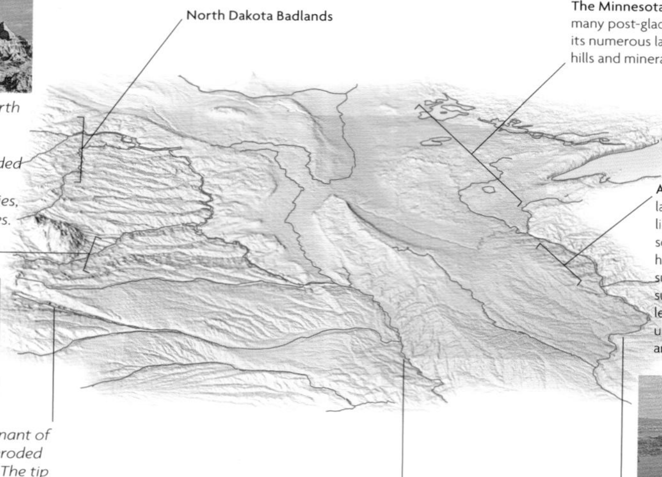

The Minnesota landscape contains many post-glacial features, including its numerous lakes, boulder-strewn hills and mineral-rich deposits.

Although it escaped the last glaciation, the limestone bedrock of southeastern Minnesota has been eroded by surface and subterranean streams, leaving a network of underground caverns and steepsided valleys.

*▲ **Chimney Rock is** a remnant of an ancient land surface, eroded by the North Platte River. The tip of its spire stands 500 ft (150 m) above the plain.*

*◄ **In northeastern Iowa,** the Mississippi and its tributaries have deeply incised the underlying bedrock creating a hilly terrain, with bluffs standing 300 ft (90 m) above the valley.*

32

▶ *Along the shores of Lake Superior in Minnesota, the average number of frost-free days can be as few as 90, and frosts may occur in any month of the year.*

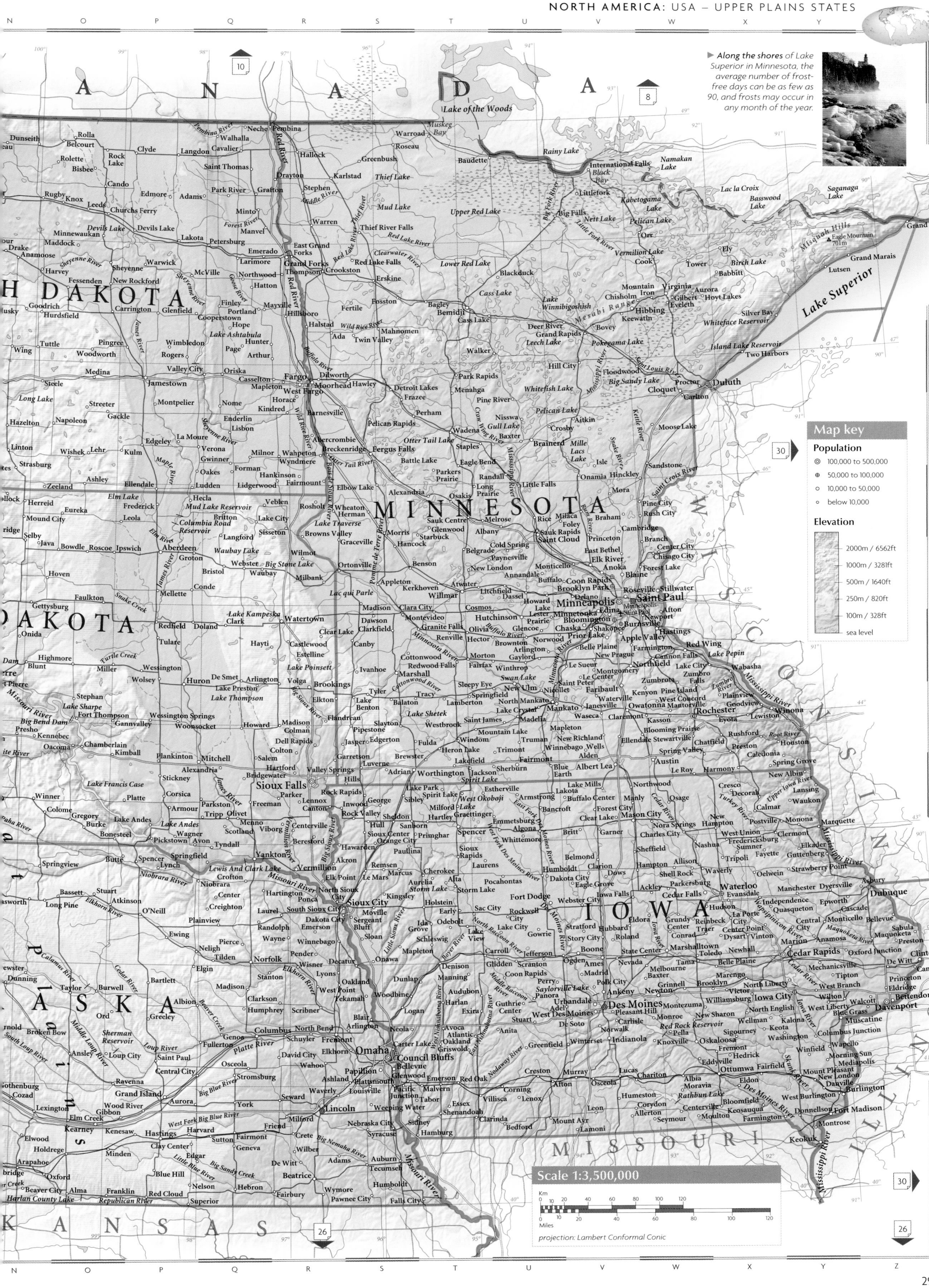

USA: GREAT LAKES STATES

Illinois, Indiana, Michigan, Ohio, Wisconsin

The states bordering the Great Lakes developed rapidly in the second half of the 19th century as a result of improvements in communications: rail to the west and waterways to the south and east. Fertile land and good links with growing eastern seaboard cities encouraged the development of agriculture and food processing. Migrants from Europe and other parts of the USA flooded into the region and for much of the 20th century the region's economy boomed. However, in recent years heavy industry has declined, earning the region the unwanted label the 'Rustbelt'.

Transport and industry

The Great Lakes region is the centre of the USA's car industry. Since the early part of the 20th century, its prosperity has been closely linked to the fortunes of automobile manufacturing. Iron and steel production has expanded to meet demand from this industry. In the 1970s, nationwide recession, cheaper foreign competition in the automobile sector, pollution in and around the Great Lakes and the collapse of the meat-packing industry, centred on Chicago, forced these states to diversify their industrial base. New industries have emerged, notably electronics, service and finance industries.

The transport network

540,682 miles (865,091 km)	6550 miles (10,480 km)
24,928 miles (39,884 km)	2330 miles (3748 km)

Few areas of the USA have a comparable transport system. Chicago is a principal transport terminus with a dense network of roads, railways and Interstate freeways radiating from the city.

Major industry and infrastructure

- car manufacture
- coal
- electronics
- engineering
- finance
- food processing
- iron & steel
- oil
- research & development
- textiles
- major towns
- international airports
- major roads
- major industrial areas

▶ ***Ever since Ransom Olds** and Henry Ford started mass-producing automobiles in Detroit early in the 20th century, the city's name has become synonymous with the American automotive industry.*

The landscape

Much of this region shows the impact of glaciation which lasted until about 10,000 years ago, and extended as far south as Illinois and Ohio. Although the relief of the region slopes towards the Great Lakes, because the ice sheets blocked northerly drainage, most of the rivers today flow southwards, forming part of the massive Mississippi/Missouri drainage basin.

◀ ***The dunes near** Sleeping Bear Point rise 400 ft (120 m) from the banks of Lake Michigan. They are constantly being resculpted by wind action.*

Lake Michigan

The many lakes and marshes of Wisconsin and Michigan are the result of glacial erosion and deposition which occurred during the last Ice Age.

Southwestern Wisconsin is known as a 'driftless' area. Unlike most of the region, low hills protected it from erosion by the advancing ice sheet.

Most of the water used in northern Illinois is pumped from underground reservoirs. Due to increased demand, many areas now face a water shortage. Around Joliet, the water table was lowered by more than 700 ft (210 m) over the last century.

Lake Erie is the shallowest of the five Great Lakes. Its average depth is about 62 ft (19 m). Storms sweeping across from Canada erode its shores and cause the silting of its harbours.

The Appalachian plateau stretches eastward from Ohio. It is dissected by streams flowing west into the Mississippi and Ohio rivers.

Illinois plains

Mississippi River

Ohio River

Relic landforms from the last glaciation, such as shallow basins and ridges, cover all but the south of this region. Ridges, known as moraines, up to 300 ft (100 m) high, lie to the south of Lake Michigan.

Unlike the level prairie to the north, southern Indiana is relatively rugged. Limestone in the hills has been dissolved by water, producing features such as sinkholes and underground caves.

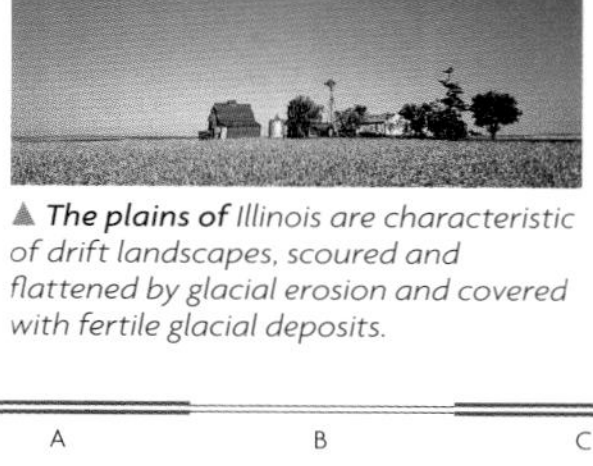

▲ ***The plains of** Illinois are characteristic of drift landscapes, scoured and flattened by glacial erosion and covered with fertile glacial deposits.*

Glacial till

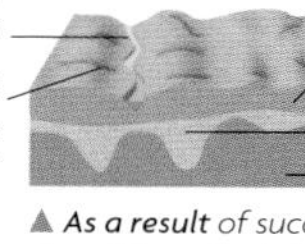

▲ ***As a result** of successive glacial depositions, the total depth of till along the former southern margin of the Laurentide ice sheet can exceed 1300 ft (400 m).*

28
28
26

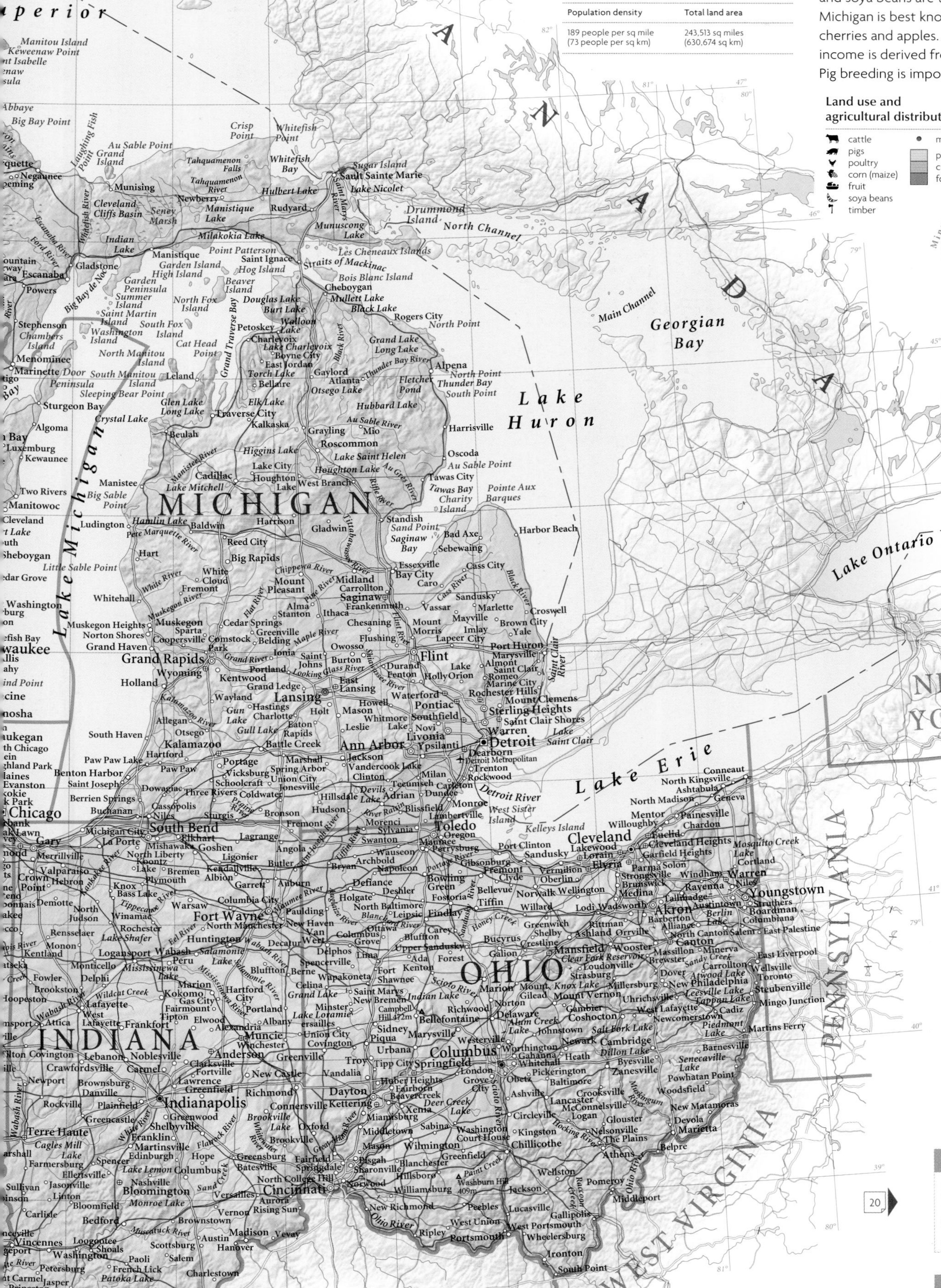

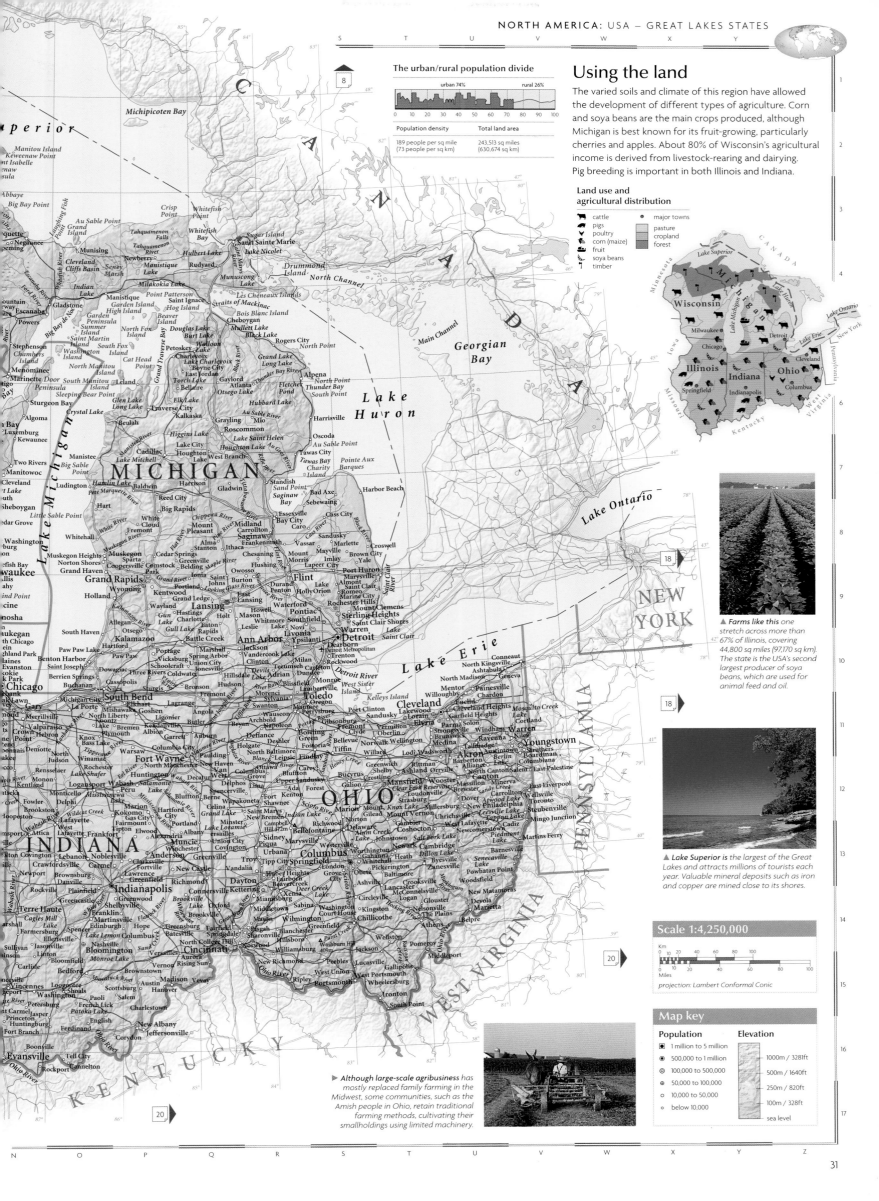

The urban/rural population divide

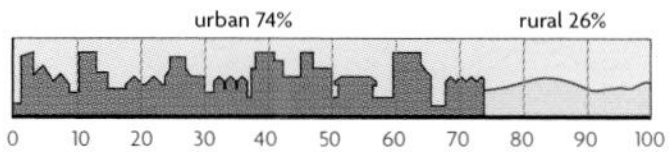

Population density	Total land area
189 people per sq mile (73 people per sq km)	243,513 sq miles (630,674 sq km)

Using the land

The varied soils and climate of this region have allowed the development of different types of agriculture. Corn and soya beans are the main crops produced, although Michigan is best known for its fruit-growing, particularly cherries and apples. About 80% of Wisconsin's agricultural income is derived from livestock-rearing and dairying. Pig breeding is important in both Illinois and Indiana.

Land use and agricultural distribution

▲ *Farms like this one stretch across more than 67% of Illinois, covering 44,800 sq miles (97,170 sq km). The state is the USA's second largest producer of soya beans, which are used for animal feed and oil.*

▲ *Lake Superior is the largest of the Great Lakes and attracts millions of tourists each year. Valuable mineral deposits such as iron and copper are mined close to its shores.*

Scale 1:4,250,000

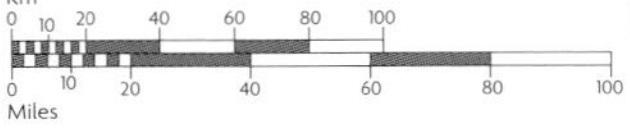

projection: Lambert Conformal Conic

▶ *Although large-scale agribusiness has mostly replaced family farming in the Midwest, some communities, such as the Amish people in Ohio, retain traditional farming methods, cultivating their smallholdings using limited machinery.*

Map key

Population

- 1 million to 5 million
- 500,000 to 1 million
- 100,000 to 500,000
- 50,000 to 100,000
- 10,000 to 50,000
- below 10,000

Elevation

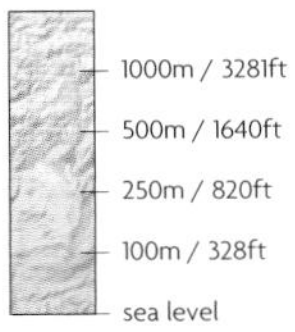

USA: NORTH MOUNTAIN STATES

Idaho, Montana, Oregon, Washington, Wyoming

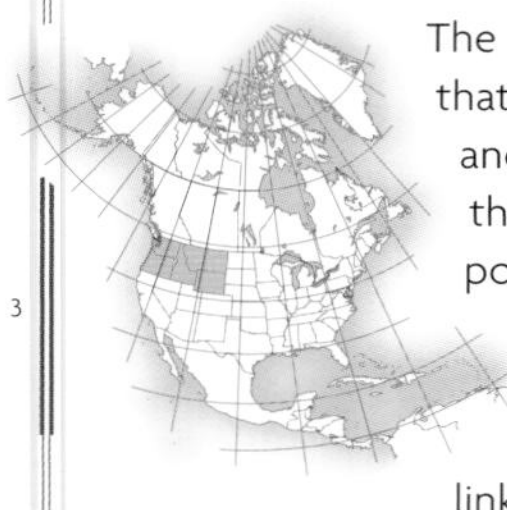

The remoteness of the northwestern states, coupled with the rugged landscape, ensured that this was one of the last areas settled by Europeans in the 19th century. Fur-trappers and gold-prospectors followed the Snake River westwards as it wound its way through the Rocky Mountains. The states of the northwest have pioneered many conservationist policies, with the USA's first national park opened at Yellowstone in 1872. More recently, the Cascades and Rocky Mountains have become havens for adventure tourism. The mountains still serve to isolate the western seaboard from the rest of the continent. This isolation has encouraged west coast cities to expand their trade links with countries of the Pacific Rim.

▲ *The Snake River has cut down into the basalt of the Columbia Basin to form Hells Canyon, the deepest in the USA, with cliffs up to 7900 ft (2408 m) high.*

Map key

Population

- ◉ 500,000 to 1 million
- ◎ 100,000 to 500,000
- ⊕ 50,000 to 100,000
- ○ 10,000 to 50,000
- ○ below 10,000

Elevation

- 4000m / 13,124ft
- 3000m / 9843ft
- 2000m / 6562ft
- 1000m / 3281ft
- 500m / 1640ft
- 250m / 820ft
- 100m / 328ft
- sea level

Using the land

Wheat farming in the east gives way to cattle ranching as rainfall decreases. Irrigated farming in the Snake River valley produces large yields of potatoes and other vegetables. Dairying and fruit-growing take place in the wet western lowlands between the mountain ranges.

The urban/rural population divide

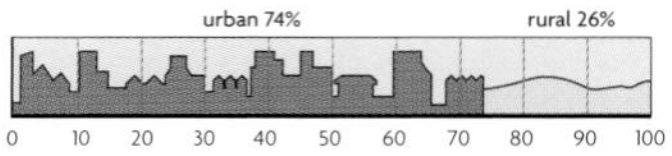

Population density	Total land area
26 people per sq mile (10 people per sq km)	487,970 sq miles (1,263,716 sq km)

Scale 1:4,250,000

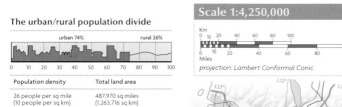

Km 0 10 20 40 60 80 100

Miles 0 10 20 40 60 80 100

projection: Lambert Conformal Conic

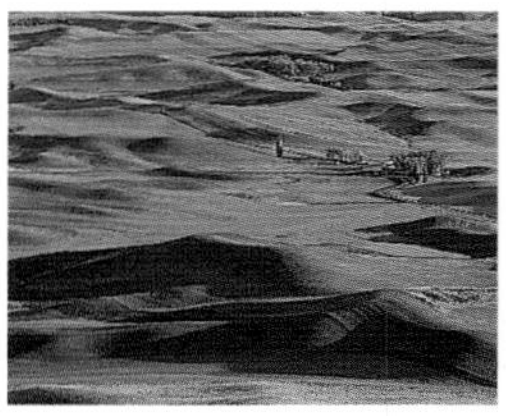

► *Fine-textured, volcanic soils in the hilly Palouse region of eastern Washington are susceptible to erosion.*

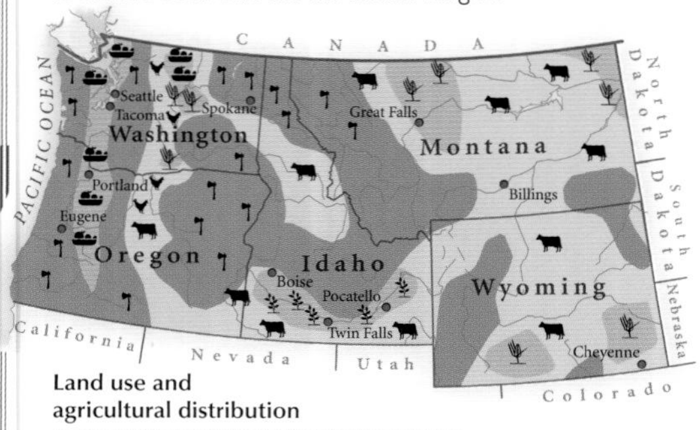

Land use and agricultural distribution

- cattle
- poultry
- cereals
- fruit
- potatoes
- timber
- major towns
- pasture
- cropland
- forest

Transport and industry

Minerals and timber are extremely important in this region. Uranium, precious metals, copper and coal are all mined, the latter in vast open-cast pits in Wyoming; oil and natural gas are extracted further north. Manufacturing, notably related to the aerospace and electronics industries, is important in western cities.

The transport network

- 347,857 miles (556,571 km)
- 4200 miles (6720 km)
- 12,354 miles (19,766 km)
- 1108 miles (1782 km)

The Union Pacific Railroad has been in service across Wyoming since 1867. The route through the Rocky Mountains is now shared with the Interstate 80, a major east–west highway.

Major industry and infrastructure

- adventure tourism
- aerospace
- coal
- chemicals
- electronics
- food processing
- mining
- oil & gas
- timber processing
- major towns
- international airports
- major roads
- major industrial areas

192

8

◄ *Seattle lies in one of Puget Sound's many inlets. The city receives oil and other resources from Alaska, and benefits from expanding trade across the Pacific.*

34

◄ *Crater Lake, Oregon, is 6 miles (10 km) wide and 1800 ft (600 m) deep. It marks the site of a volcanic cone, which collapsed after an eruption within the last 7000 years.*

34

The landscape

The Rocky Mountains are flanked by lower parallel ranges, which spread onto the Great Plains in the east and surmount the broad lava plateau which extends westwards. The Cascade Range divides the Columbia Basin from the coastlands, where the low areas skirting Puget Sound are broken by the steep, volcanic Olympic Mountains and the wooded hills of the Coast Ranges.

Molten rock pools, forming parallel columns
Surrounding strata eroded away
Molten rock wells up from the Earth's core

▲ *Devil's Tower in* Wyoming *is an igneous intrusion, formed below the Earth's surface. Molten rock intruded through cracks in the overlying strata and cooled. Over time, the softer rock layers have been eroded away, leaving only the tower standing.*

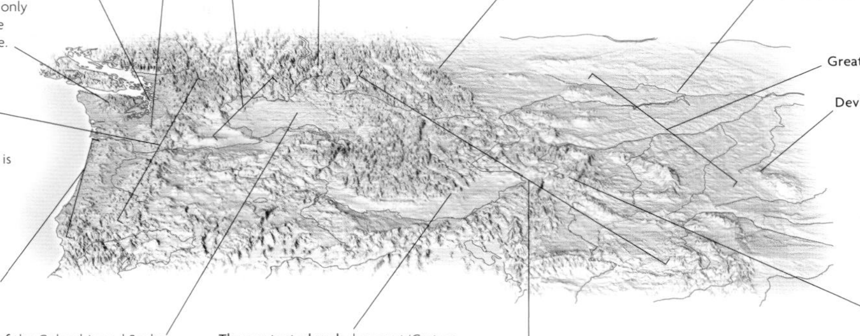
Puget Sound
Mount St Helens erupted in 1980, killing 57 people and devastating a huge area.
Columbia Basin
Grand Coulee and the lesser *coulées* (ravines) were cut by cataclysmic floods, from the release of an ice-dammed lake, at the end of the last Ice Age.
The Continental Divide, or watershed, crosses the Lewis Range. From here, rivers flow east to Hudson Bay, south to the Gulf of Mexico and west to the Pacific Ocean.
Glacial valleys on the seaward side of the Olympic Mountains receive about 142 inches (3600 mm) of rain per year, supporting the only true rainforest of the northern hemisphere.
The Cascades are glacially scoured volcanic mountains, the highest of which is Mount Rainier, a dormant volcano at 14,409 ft (4392 m).
Coast Ranges
Great Plains
Devil's Tower
The plateaus of the Columbia and Snake rivers represent one of the world's largest accumulations of lava. Over 5 million years ago, successive flows of molten basalt buried the existing land surface by up to 450 ft (150 m).
The contorted rock shapes at 'Craters of the Moon' National Monument in Idaho were left 2000 years ago by the sporadic upwelling of viscous lava from fissures in the basalt plateau.
Rocky Mountains

▶ *Piney Buttes are the remnants of an older, higher land surface gradually weathered and eroded into isolated outcrops with flat tops and steep sides.*

▲ *Water from the hot springs in Yellowstone National Park deposits minerals as it cools in rock pools. Long periods of deposition have created these rock terraces.*

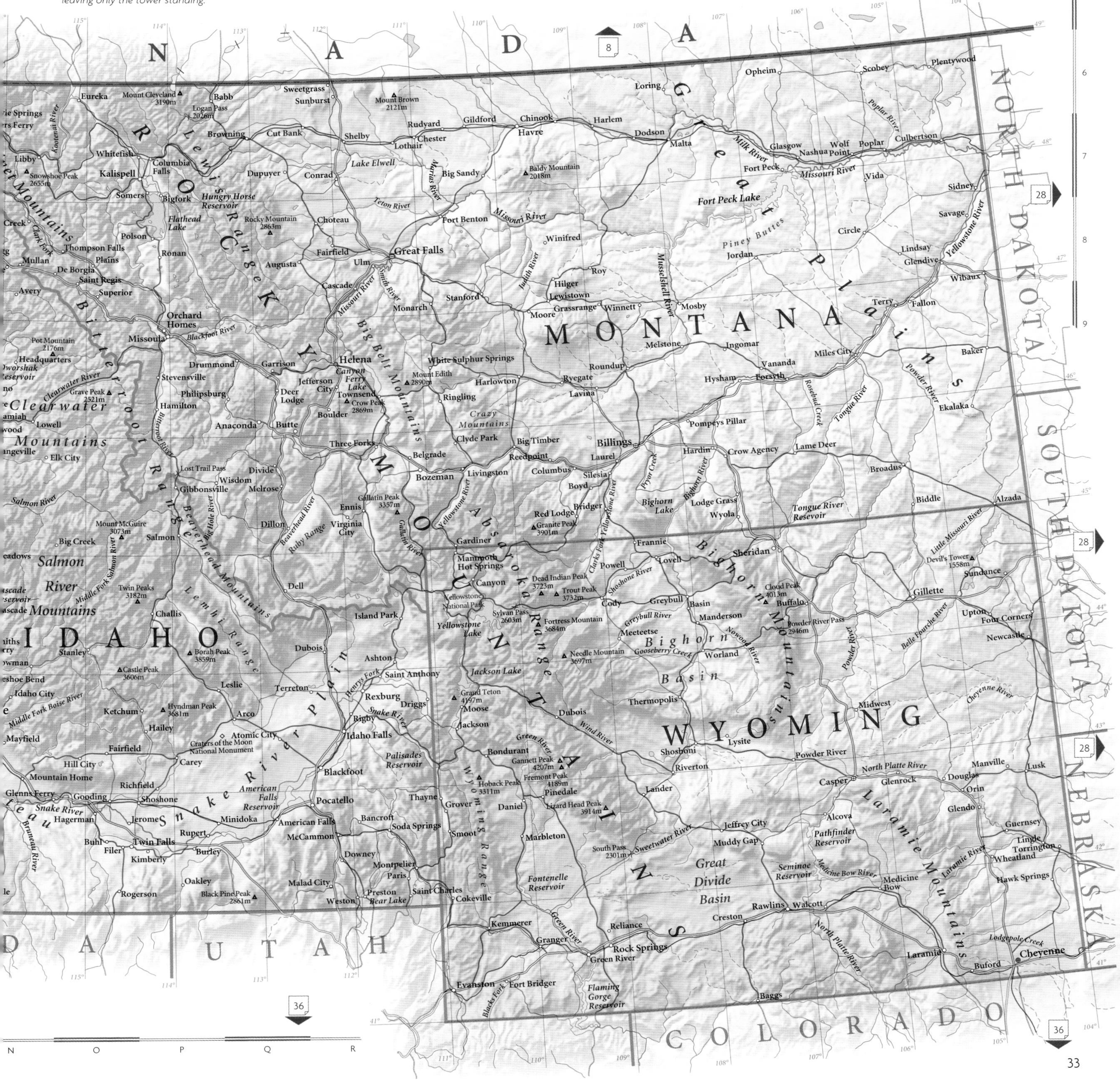
CANADA
MONTANA
IDAHO
WYOMING
UTAH
COLORADO
NORTH DAKOTA
SOUTH DAKOTA
NEBRASKA
ROCKY MOUNTAINS
Great Plains
Lewis Range
Bitterroot Range
Big Belt Mountains
Crazy Mountains
Absaroka Range
Bighorn Mountains
Bighorn Basin
Great Divide Basin
Laramie Mountains
Wind River Range
Wyoming Range
Salmon River Mountains
Clearwater Mountains
Lemhi Range
Snake River Plain
Piney Buttes
Helena
Great Falls
Billings
Missoula
Butte
Bozeman
Cheyenne
Casper
Laramie
Rock Springs
Sheridan
Pocatello
Idaho Falls
Yellowstone Lake
Fort Peck Lake
Missouri River
Yellowstone River
Snake River
8
28
28
28
36
36

USA: CALIFORNIA & NEVADA

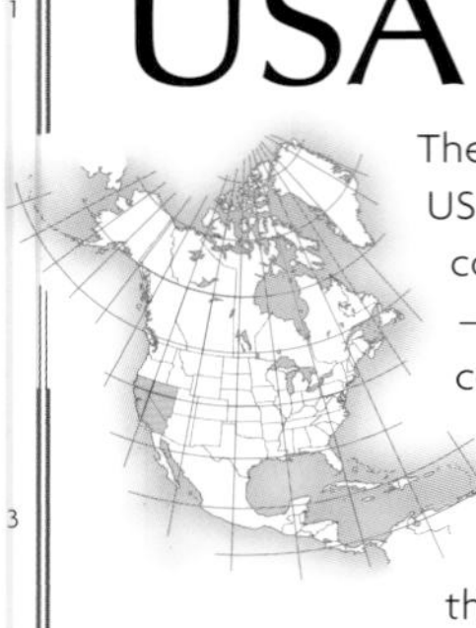

The 'Gold Rush' of 1849 attracted the first major wave of European settlers to the USA's west coast. The pleasant climate, beautiful scenery and dynamic economy continue to attract immigrants – despite the ever-present danger of earthquakes – and California has become the USA's most populous state. The population is concentrated in the vast conurbations of Los Angeles, San Francisco and San Diego; new immigrants include people from South Korea, the Philippines, Vietnam and Mexico. Nevada's arid lands were initially exploited for minerals; in recent years, revenue from mining has been superseded by income from the tourist and gambling centres of Las Vegas and Reno.

Map key

Population

- 1 million to 5 million
- 500,000 to 1 million
- 100,000 to 500,000
- 50,000 to 100,000
- 10,000 to 50,000
- below 10,000

Elevation

- 4000m / 13,124ft
- 3000m / 9843ft
- 2000m / 6562ft
- 1000m / 3281ft
- 500m / 1640ft
- 250m / 820ft
- 100m / 328ft
- sea level

Scale 1:3,250,000

Km 0 5 10 20 30 40 50 60 70 80
Miles 0 5 10 20 30 40 50 60 70 80

projection: Lambert Conformal Conic

Transport and industry

Nevada's rich mineral reserves ushered in a period of mining wealth which has now been replaced by revenue from gambling. California supports a broad set of activities including defence-related industries and research and development facilities. 'Silicon Valley', near San Francisco, is a world leading centre for micro-electronics, while tourism and the Los Angeles film industry also generate large incomes.

Major industry and infrastructure

- aerospace
- car manufacture
- defence
- film industry
- finance
- food processing
- gambling
- hi-tech industry
- mining
- pharmaceuticals
- research & development
- textiles
- tourism
- major towns
- international airports
- major roads
- major industrial areas

◀ *Gambling was **legalized** in Nevada in 1931. Las Vegas has since become the centre of this multi-million dollar industry.*

The transport network

211,459 miles (338,334 km)	2944 miles (4710 km)
7822 miles (12,595 km)	190 miles (360 km)

In California, the motor vehicle is a vital part of daily life, and an extensive freeway system runs throughout the state, cementing its position as the most important mode of transport.

192

The landscape

The broad Central Valley divides California's coastal mountains from the Sierra Nevada. The San Andreas Fault, running beneath much of the state, is the site of frequent earth tremors and sometimes more serious earthquakes. East of the Sierra Nevada, the landscape is characterized by the basin and range topography with stony deserts and many salt lakes.

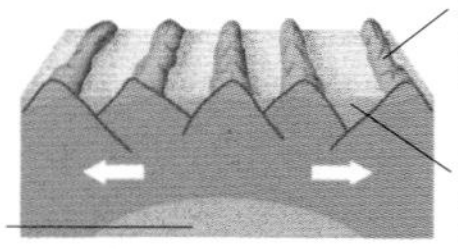

Rising molten rock causes stretching of the Earth's crust

Extensive cracking (faulting) uplifted a series of ridges

As ridges are eroded they fill intervening valleys with sediments

▲ *Molten rock (magma) welling up to form a dome in the Earth's interior, causes the brittle surface rocks to stretch and crack. Some areas were uplifted to form mountains (ranges), while others sunk to form flat valleys (basins).*

◀ *The General Sherman sequoia tree in Sequoia National Park is 2500 years old and at 275 ft (84 m) is one of the largest living things on earth.*

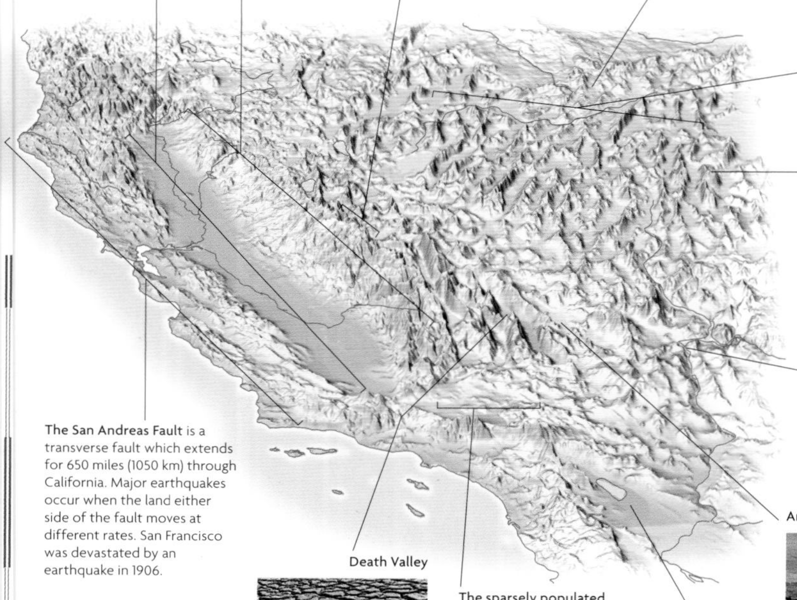

Most of California's agriculture is confined to the fertile and extensively irrigated Central Valley, running between the Coast Ranges and the Sierra Nevada. It incorporates the San Joaquin and Sacramento valleys.

The dramatic granitic rock formations of Half Dome and El Capitan, and the verdant coniferous forests, attract millions of visitors annually to Yosemite National Park in the Sierra Nevada.

Sierra Nevada

The Great Basin dominates most of Nevada's topography containing large open basins, punctuated by eroded features such as *buttes* and *mesas*. River flow tends to be seasonal, dependent upon spring showers and winter snow melt.

Wheeler Peak is home to some of the world's oldest trees, bristlecone pines, which live for up to 5000 years.

When the Hoover Dam across the Colorado River was completed in 1936, it created Lake Mead, one of the largest artificial lakes in the world, extending for 115 miles (285 km) upstream.

The San Andreas Fault is a transverse fault which extends for 650 miles (1050 km) through California. Major earthquakes occur when the land either side of the fault moves at different rates. San Francisco was devastated by an earthquake in 1906.

Death Valley

▶ *Named by migrating settlers in 1849, Death Valley is the driest, hottest place in North America, as well as being the lowest point on land in the western hemisphere, at 282 ft (86 m) below sea level.*

The sparsely populated Mojave Desert receives less than 8 inches (200 mm) of rainfall a year. It is used extensively for weapons-testing and military purposes.

The Salton Sea was created accidentally between 1905 and 1907 when an irrigation channel from the Colorado River broke out of its banks and formed this salty 300 sq mile (777 sq km), land-locked lake.

Amargosa Desert

▲ *The Sierra Nevada create a 'rainshadow', preventing rain from reaching much of Nevada. Pacific air masses, passing over the mountains, are stripped of their moisture.*

Using the land

California is the USA's leading agricultural producer, although low rainfall makes irrigation essential. The long growing season and abundant sunshine allow many crops to be grown in the fertile Central Valley including grapes, citrus fruits, vegetables and cotton. Almost 17 million acres (6.8 million hectares) of California's forests are used commercially. Nevada's arid climate and poor soil are largely unsuitable for agriculture; 85% of its land is state owned and large areas are used for underground testing of nuclear weapons.

Land use and agricultural distribution

- cattle
- citrus fruits
- fruit
- irrigation
- timber
- vineyards
- major towns
- pasture
- cropland
- forest
- desert

▲ *Without considerable irrigation, this fertile valley at Palm Springs would still be part of the Sonoran Desert. California's farmers account for about 80% of the state's total water usage.*

The urban/rural population divide

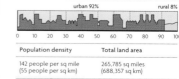

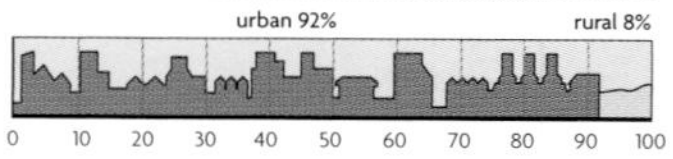

Population density	Total land area
142 people per sq mile (55 people per sq km)	265,785 sq miles (688,357 sq km)

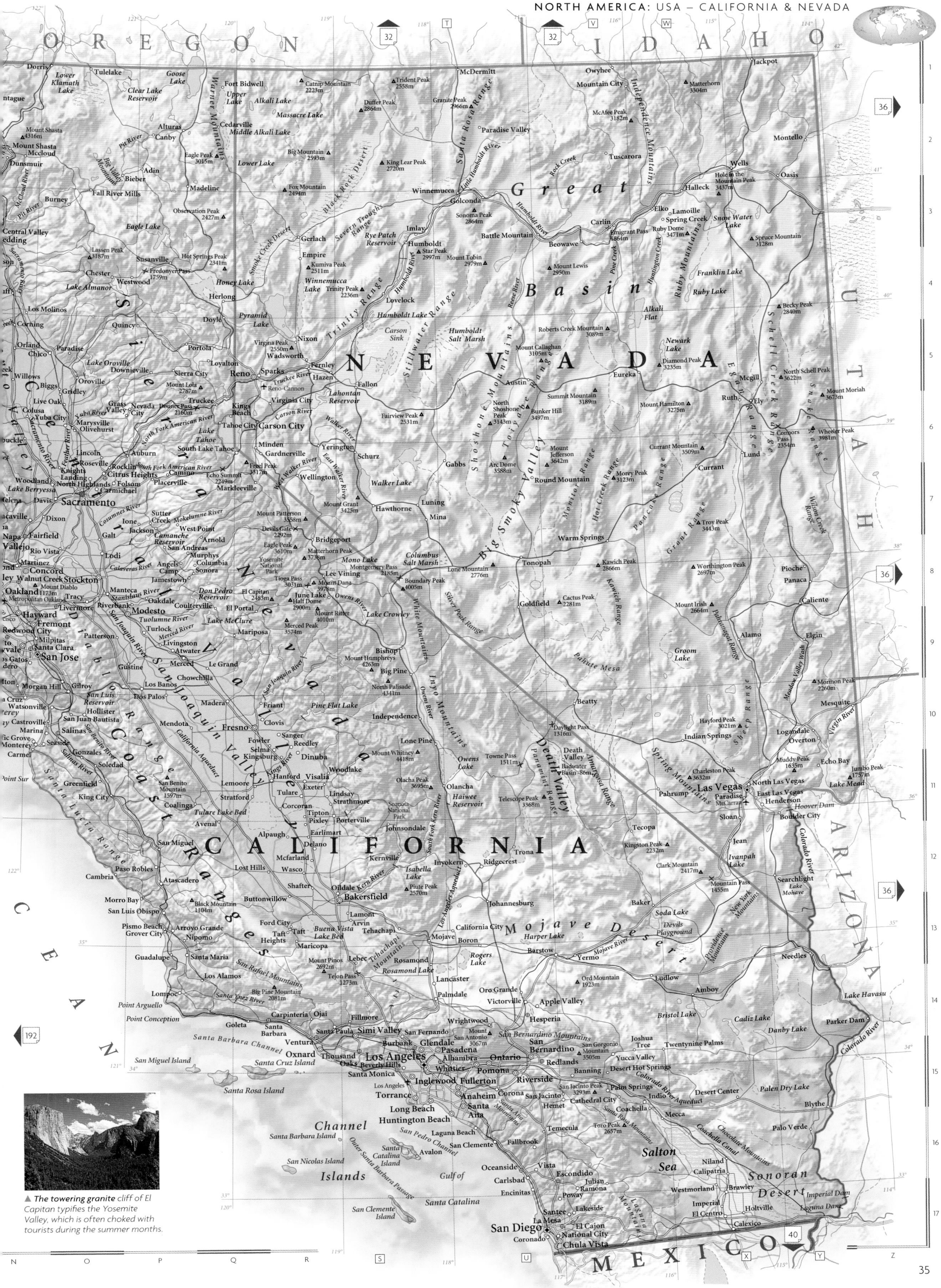

▲ *The towering granite cliff of El Capitan typifies the Yosemite Valley, which is often choked with tourists during the summer months.*

USA: SOUTH MOUNTAIN STATES

Arizona, Colorado, New Mexico, Utah

This arid region, characterized by expansive plateaux and spectacular canyons is home to several distinct peoples. The ruins of cliff dwellings built a thousand years ago by the Anasazi people still exist today, and native Americans own one-third of the land in Arizona. Spanish and Mexican conquest and settlement left a hispanic presence which is strongest in New Mexico. The Mormons, who came to the Great Salt Lake seeking religious freedom in 1847, were among the earliest Anglo-American settlers and now make up over 70% of Utah's population. The region's mineral wealth drove rapid development in the 20th century, yet the constraints of a fragile environment, including widespread water shortages, may limit prospects for growth.

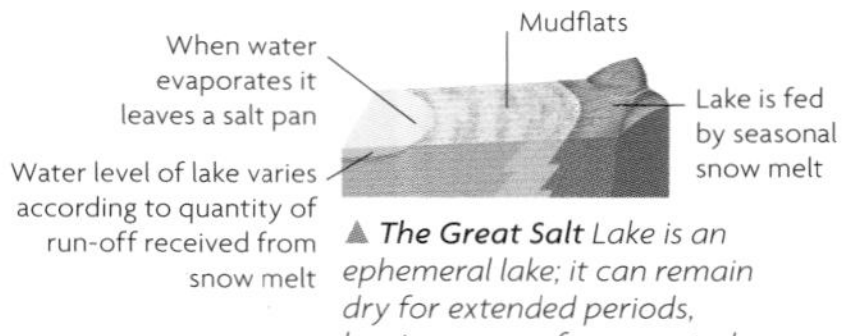
Mudflats
When water evaporates it leaves a salt pan
Lake is fed by seasonal snow melt
Water level of lake varies according to quantity of run-off received from snow melt

▲ ***The Great Salt** Lake is an ephemeral lake; it can remain dry for extended periods, leaving a pan of evaporated mineral salts in its centre.*

The landscape

The arid, rocky expanse of the Colorado Plateau is dissected by immense canyons of the Colorado River. Desert lies to the north and south and branches of the Rocky Mountains run to the east and west. The Great Salt Lake and Desert lie within the Great Basin, a barren region of parallel mountain ranges which extends into Arizona.

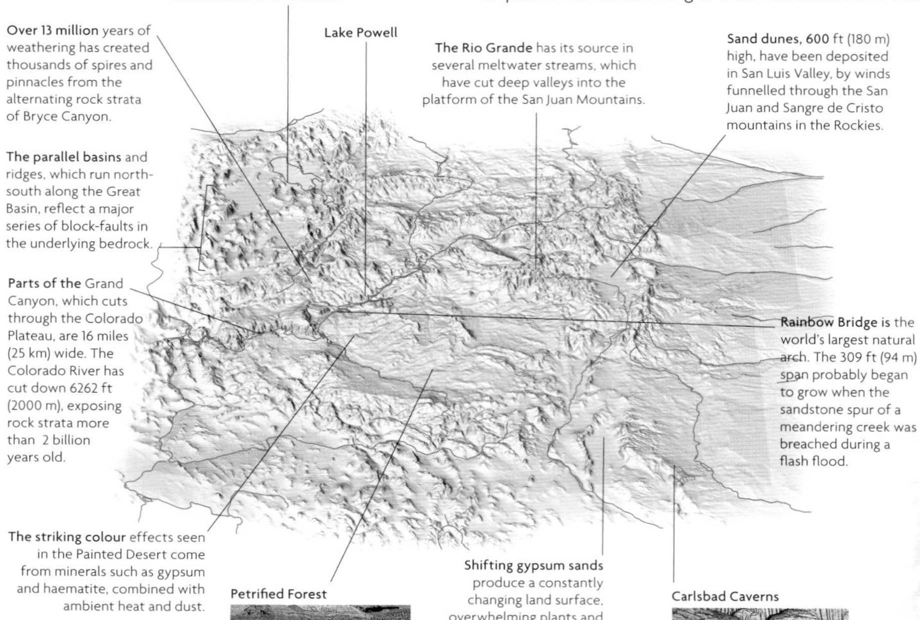
Over 13 million years of weathering has created thousands of spires and pinnacles from the alternating rock strata of Bryce Canyon.
Lake Powell
The Rio Grande has its source in several meltwater streams, which have cut deep valleys into the platform of the San Juan Mountains.
Sand dunes, 600 ft (180 m) high, have been deposited in San Luis Valley, by winds funnelled through the San Juan and Sangre de Cristo mountains in the Rockies.
The parallel basins and ridges, which run north-south along the Great Basin, reflect a major series of block-faults in the underlying bedrock.
Parts of the Grand Canyon, which cuts through the Colorado Plateau, are 16 miles (25 km) wide. The Colorado River has cut down 6262 ft (2000 m), exposing rock strata more than 2 billion years old.
Rainbow Bridge is the world's largest natural arch. The 309 ft (94 m) span probably began to grow when the sandstone spur of a meandering creek was breached during a flash flood.
The striking colour effects seen in the Painted Desert come from minerals such as gypsum and haematite, combined with ambient heat and dust.
Shifting gypsum sands produce a constantly changing land surface, overwhelming plants and any other obstacles in Tularosa Valley.

Petrified Forest

► ***In the arid** landscape of Petrified Forest National Park in Arizona, the grain of prehistoric trees has been preserved as a fossil imprint in the rocks. The bog-preserved trees were gradually turned to stone by seeping mineral-rich water.*

Carlsbad Caverns

► ***The intricate stalactites** of Carlsbad Caverns have grown with the seepage of calcium-rich water, over the last 100,000 years. The huge caves are home to around 100,000 Mexican freetail bats.*

Transport and industry

New industries have helped reduce the region's dependence on the extraction of minerals and fossil fuels. Precision manufacture has grown rapidly, particularly in Arizona and Colorado. Salt Lake City and Denver are well-established financial centres and New Mexico, the USA's main producer of uranium, is a prominent region for nuclear research. Colorado is the USA's most important centre for winter sports.

The transport network

232,434 miles (373,986 km)	4059 miles (6515 km)
8627 miles (13,881 km)	none

The Colorado Rockies are crossed by 32 mountain passes, some as high as 12,183 ft (3713 m). The Eisenhower Tunnel west of Denver carries Interstate Highway 70 straight through the Continental Divide.

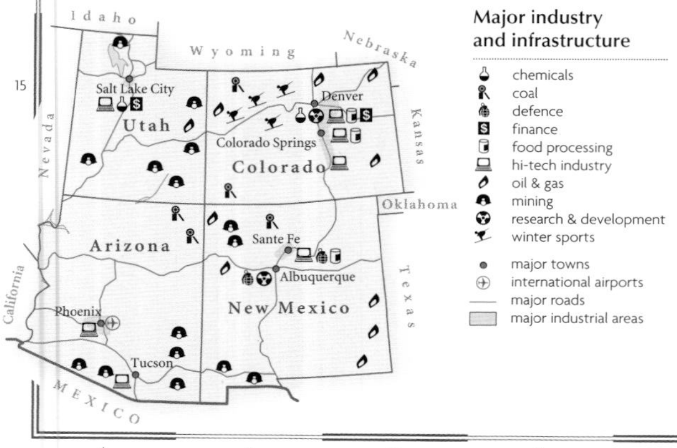
Idaho
Wyoming
Nebraska
Nevada
Salt Lake City
Utah
Denver
Kansas
Colorado Springs
Colorado
Oklahoma
Arizona
Sante Fe
Albuquerque
New Mexico
Texas
California
Phoenix
Tucson
MEXICO
Major industry and infrastructure
chemicals
coal
defence
finance
food processing
hi-tech industry
oil & gas
mining
research & development
winter sports
major towns
international airports
major roads
major industrial areas

▲ ***Glen Canyon Dam** on the Colorado river was completed in 1964. It provides hydro-electric power and irrigation water as part of a long-term federal project to harness the river.*

◄ ***The flat tablelands** (mesas), and the isolated pinnacles (buttes) which rise from the floor of Monument Valley are the resistant remnants of an earlier land surface, gradually cut back by erosion under arid conditions.*

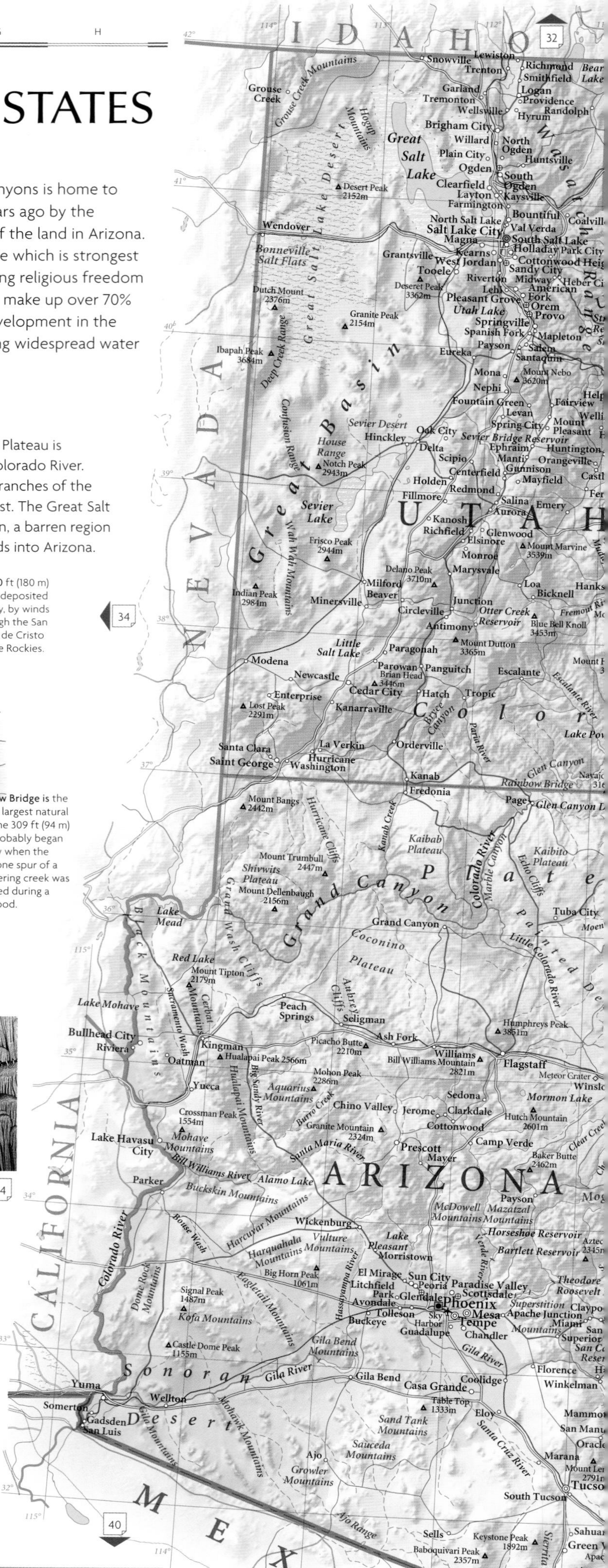
IDAHO
32
NEVADA
UTAH
ARIZONA
CALIFORNIA
MEXICO
Great Salt Lake
Great Salt Lake Desert
Great Basin
Colorado Plateau
Grand Canyon
Coconino Plateau
Kaibab Plateau
Sonoran Desert
Colorado River
Lake Mead
Lake Mohave
Lake Havasu City
Salt Lake City
Ogden
Provo
Cedar City
Saint George
Kanab
Page
Flagstaff
Kingman
Bullhead City
Prescott
Phoenix
Mesa
Tempe
Yuma
Wickenburg
Sells
South Tucson
34
40

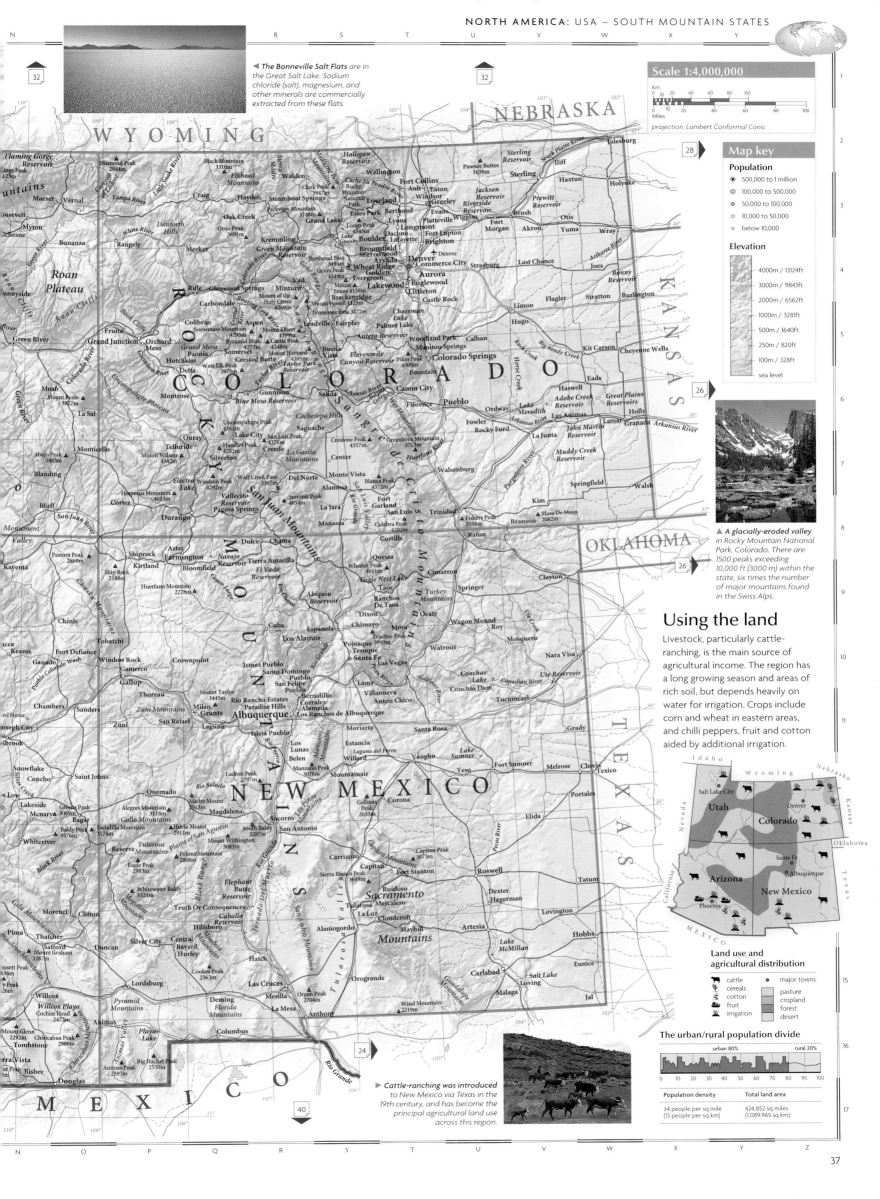

◀ *The Bonneville Salt Flats are in the Great Salt Lake. Sodium chloride (salt), magnesium, and other minerals are commercially extracted from these flats.*

Scale 1:4,000,000

Km 0 10 20 40 60 80 100
Miles 0 10 20 40 60 80 100

projection: Lambert Conformal Conic

Map key

Population

- ◉ 500,000 to 1 million
- ◎ 100,000 to 500,000
- ⊕ 50,000 to 100,000
- ○ 10,000 to 50,000
- ◦ below 10,000

Elevation

- 4000m / 13124ft
- 3000m / 9843ft
- 2000m / 6562ft
- 1000m / 3281ft
- 500m / 1640ft
- 250m / 820ft
- 100m / 328ft
- sea level

▲ *A glacially-eroded valley in Rocky Mountain National Park, Colorado. There are 1500 peaks exceeding 10,000 ft (3000 m) within the state, six times the number of major mountains found in the Swiss Alps.*

Using the land

Livestock, particularly cattle-ranching, is the main source of agricultural income. The region has a long growing season and areas of rich soil, but depends heavily on water for irrigation. Crops include corn and wheat in eastern areas, and chilli peppers, fruit and cotton aided by additional irrigation.

Land use and agricultural distribution

cattle, cereals, cotton, fruit, irrigation, major towns, pasture, cropland, forest, desert

The urban/rural population divide

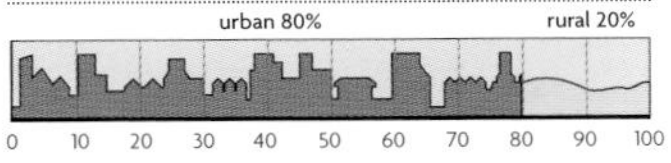

Population density	Total land area
34 people per sq mile (13 people per sq km)	424,852 sq miles (1,089,965 sq km)

▶ *Cattle-ranching was introduced to New Mexico via Texas in the 19th century, and has become the principal agricultural land use across this region.*

USA: HAWAII

The 122 islands of the Hawai'ian archipelago – which are part of Polynesia – are the peaks of the world's largest volcanoes. They rise approximately 6 miles (9.7 km) from the floor of the Pacific Ocean. The largest, the island of Hawai'i, remains highly active. Hawaii became the USA's 50th state in 1959. A tradition of receiving immigrant workers is reflected in the islands' ethnic diversity, with peoples drawn from around the rim of the Pacific. Only 9% of the current population are native Polynesians.

▲ *The island of* Moloka'i is formed from volcanic rock. Mature sand dunes cover the rocks in coastal areas.

Transport and industry

Tourism dominates the economy, with over 90% of the population employed in services. The naval base at Pearl Harbor is also a major source of employment. Industry is concentrated on the island of O'ahu and relies mostly on imported materials, while agricultural produce is processed locally.

The transport network

4102 miles (6600 km)	43 miles (69 km)
none	none

Hawaii relies on ocean-surface transportation. Honolulu is the main focus of this network, bringing foreign trade and the markets of mainland USA to Hawaii's outer islands.

◀ *Haleakala's extinct volcanic* crater is the world's largest. The giant caldera, containing many secondary cones, is 2000 ft (600 m) deep and 20 miles (32 km) in circumference.

Major industry and infrastructure

- food processing
- military base
- textiles
- tourism
- major towns
- international airports
- major roads
- major industrial areas

Using the land and sea

The ice-free coastline of Alaska provides access to salmon fisheries and more than 129 million acres (52.2 million ha) of forest. Most of Alaska is uncultivable, and around 90% of food is imported. Barley, hay and hothouse products are grown around Anchorage, where dairy farming is also concentrated.

The urban/rural population divide

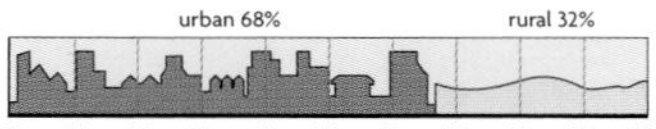

Population density	Total land area
1 person per sq mile (0.4 people per sq km)	571,951 sq miles (1,481,296 sq km)

◀ *A raft of* timber from the Tongass forest is hauled by a tug, bound for the pulp mills of the Alaskan coast between Juneau and Ketchikan.

Scale 1:4,000,000

projection: Lambert Conformal Conic

Map key

Population

- 100,000 to 500,000
- 50,000 to 100,000
- 10,000 to 50,000
- below 10,000

Elevation

- 4000m / 13,124ft
- 3000m / 9843ft
- 2000m / 6562ft
- 1000m / 3281ft
- 500m / 1640ft
- 250m / 820ft
- 100m / 328ft
- sea level

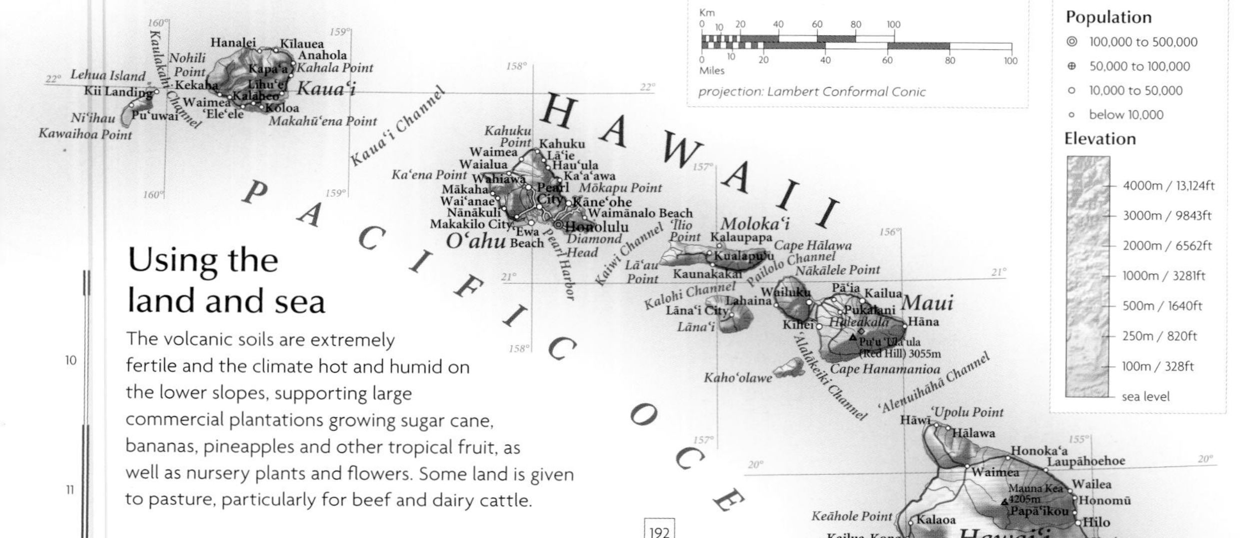

192

Using the land and sea

The volcanic soils are extremely fertile and the climate hot and humid on the lower slopes, supporting large commercial plantations growing sugar cane, bananas, pineapples and other tropical fruit, as well as nursery plants and flowers. Some land is given to pasture, particularly for beef and dairy cattle.

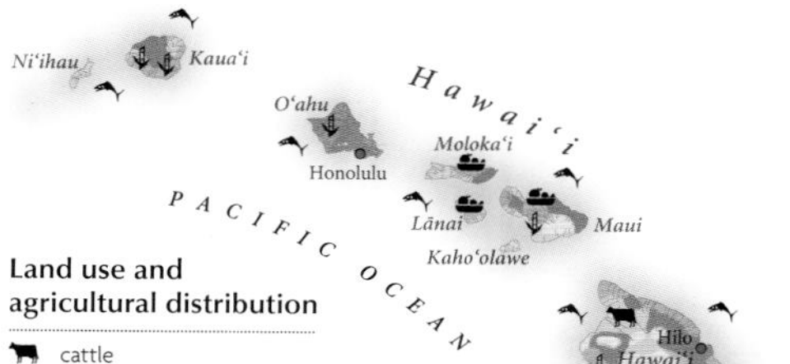

Land use and agricultural distribution

- cattle
- fishing
- fruit
- sugar cane
- major towns
- pasture
- cropland
- forest
- mountain region

► *The island of* Kaua'i is one of the wettest places in the world, receiving some 450 inches (11,500 mm) of rain a year.

The urban/rural population divide

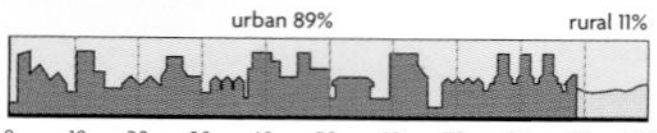

Population density	Total land area
189 people per sq mile (73 people per sq km)	6,423 sq miles (16,636 sq km)

122

192

Map key

Population

- 100,000 to 500,000
- 50,000 to 100,000
- 10,000 to 50,000
- below 10,000

Elevation

- 4000m / 13,124ft
- 3000m / 9843ft
- 2000m / 6562ft
- 1000m / 3281ft
- 500m / 1640ft
- 250m / 820ft
- 100m / 328ft
- sea level

Scale 1:9,000,000

projection: Lambert Conformal Conic

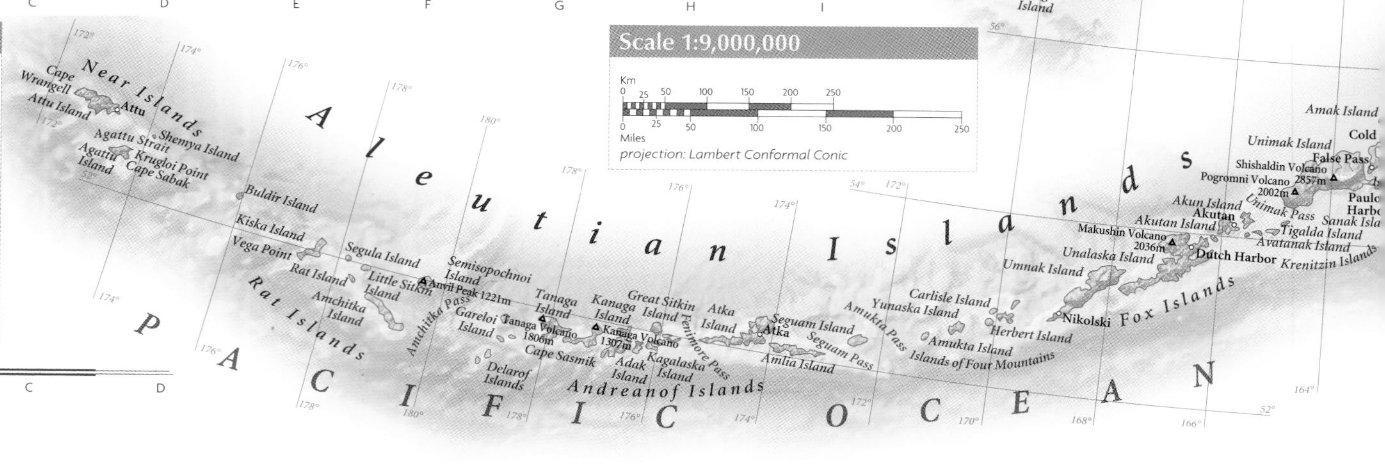

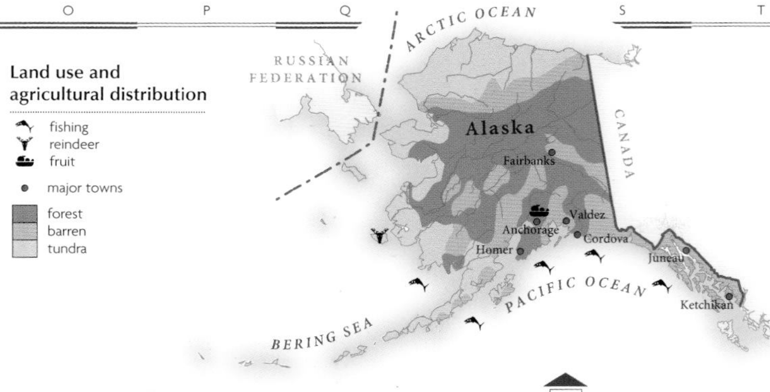
Land use and agricultural distribution
fishing
reindeer
fruit
major towns
forest
barren
tundra
ARCTIC OCEAN
RUSSIAN FEDERATION
CANADA
Alaska
Fairbanks
Valdez
Anchorage
Cordova
Homer
Juneau
Ketchikan
PACIFIC OCEAN
BERING SEA

196

USA: ALASKA

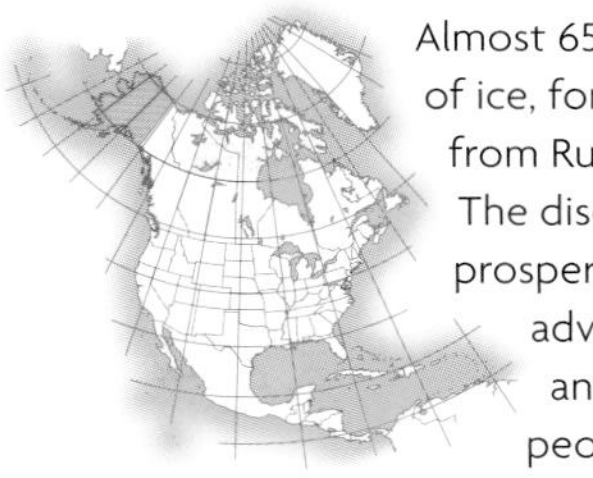

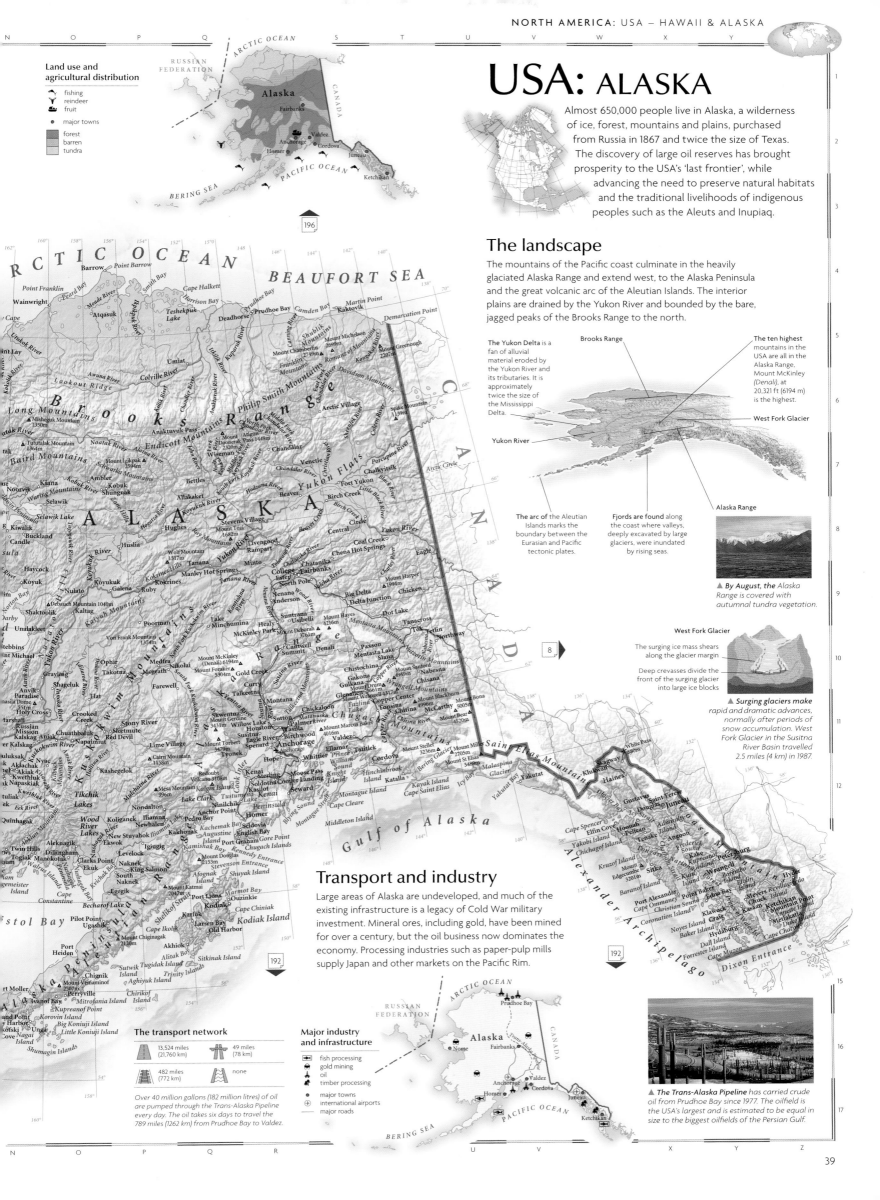

Almost 650,000 people live in Alaska, a wilderness of ice, forest, mountains and plains, purchased from Russia in 1867 and twice the size of Texas. The discovery of large oil reserves has brought prosperity to the USA's 'last frontier', while advancing the need to preserve natural habitats and the traditional livelihoods of indigenous peoples such as the Aleuts and Inupiaq.

The landscape

The mountains of the Pacific coast culminate in the heavily glaciated Alaska Range and extend west, to the Alaska Peninsula and the great volcanic arc of the Aleutian Islands. The interior plains are drained by the Yukon River and bounded by the bare, jagged peaks of the Brooks Range to the north.

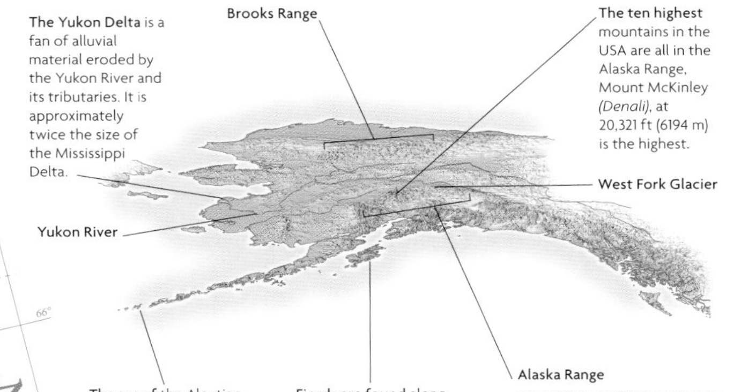
The Yukon Delta is a fan of alluvial material eroded by the Yukon River and its tributaries. It is approximately twice the size of the Mississippi Delta.
Brooks Range
The ten highest mountains in the USA are all in the Alaska Range. Mount McKinley (Denali), at 20,321 ft (6194 m) is the highest.
West Fork Glacier
Yukon River
The arc of the Aleutian Islands marks the boundary between the Eurasian and Pacific tectonic plates.
Fjords are found along the coast where valleys, deeply excavated by large glaciers, were inundated by rising seas.
Alaska Range

▲ *By August, the Alaska Range is covered with autumnal tundra vegetation.*

West Fork Glacier

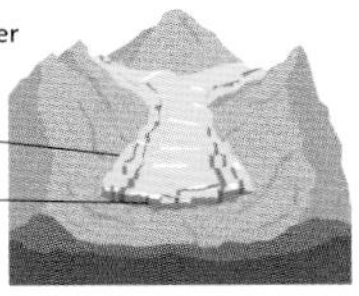
The surging ice mass shears along the glacier margin
Deep crevasses divide the front of the surging glacier into large ice blocks

▲ *Surging glaciers make rapid and dramatic advances, normally after periods of snow accumulation. West Fork Glacier in the Susitna River Basin travelled 2.5 miles (4 km) in 1987.*

ARCTIC OCEAN
BEAUFORT SEA
CANADA
ALASKA
Brooks Range
Alaska Range
Gulf of Alaska
Alexander Archipelago
Alaska Peninsula
Bristol Bay
Kodiak Island
Barrow
Prudhoe Bay
Fairbanks
Anchorage
Juneau
Ketchikan
Valdez
Yukon River
Yukon Flats
Mount McKinley (Denali) 6194m
Saint Elias Mountains
Coast Mountains
Dixon Entrance

8

192

192

Transport and industry

Large areas of Alaska are undeveloped, and much of the existing infrastructure is a legacy of Cold War military investment. Mineral ores, including gold, have been mined for over a century, but the oil business now dominates the economy. Processing industries such as paper-pulp mills supply Japan and other markets on the Pacific Rim.

The transport network

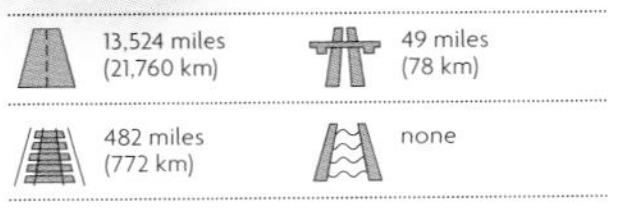
13,524 miles (21,760 km)
49 miles (78 km)
482 miles (772 km)
none

Over 40 million gallons (182 million litres) of oil are pumped through the Trans-Alaska Pipeline every day. The oil takes six days to travel the 789 miles (1262 km) from Prudhoe Bay to Valdez.

Major industry and infrastructure
fish processing
gold mining
oil
timber processing
major towns
international airports
major roads
ARCTIC OCEAN
RUSSIAN FEDERATION
CANADA
Prudhoe Bay
Alaska
Trans-Alaska Pipeline
Nome
Fairbanks
Anchorage
Valdez
Cordova
Homer
Juneau
Ketchikan
PACIFIC OCEAN
BERING SEA

▲ *The Trans-Alaska Pipeline has carried crude oil from Prudhoe Bay since 1977. The oilfield is the USA's largest and is estimated to be equal in size to the biggest oilfields of the Persian Gulf.*

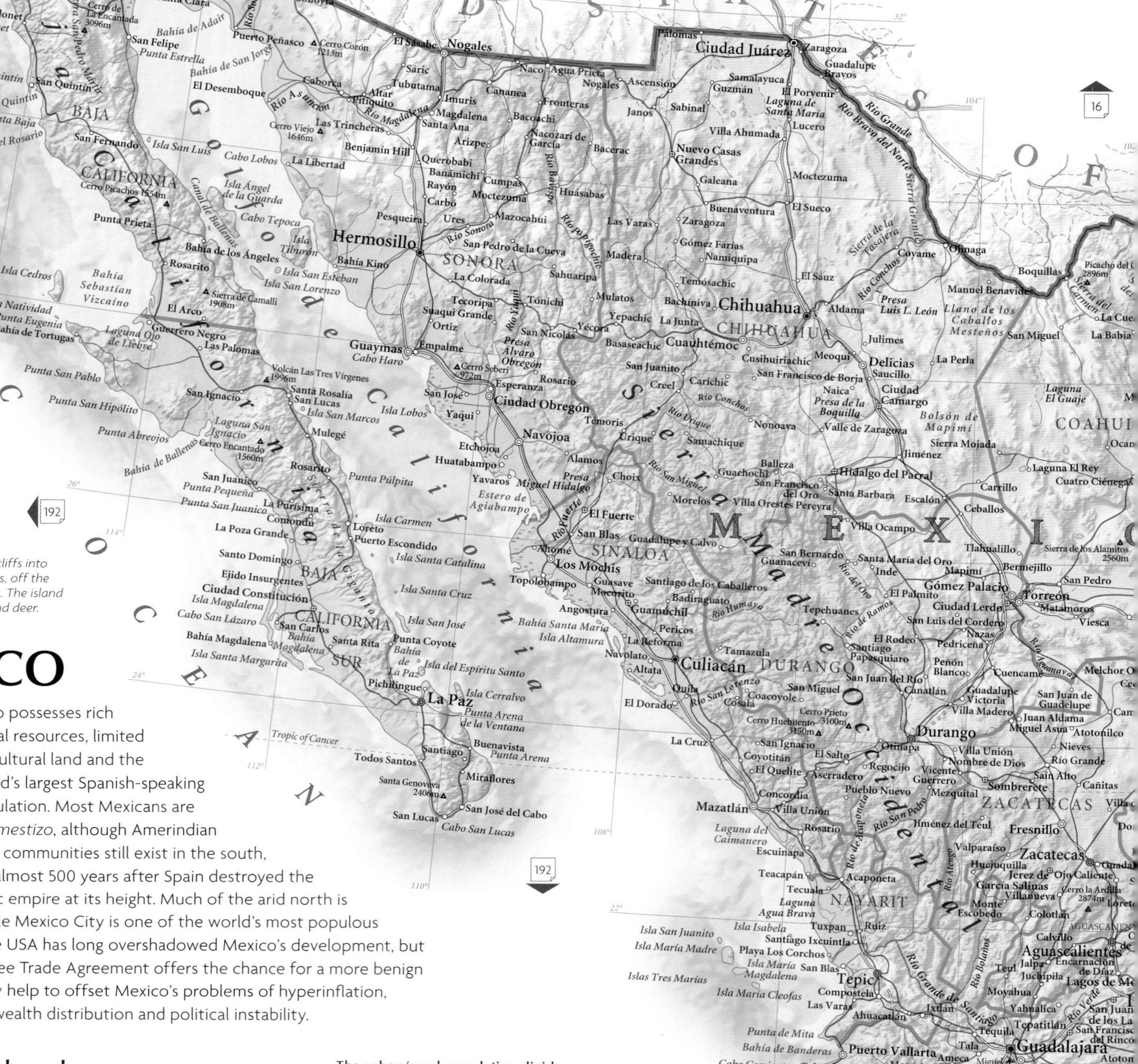

192

Scale 1:7,000,000

Km 0 25 50 100 150 200
Miles 0 25 50 100 150 200
projection: Lambert Conformal Conic

▶ *The rugged, desert landscape of the Sierra Madre del Sur is a product of complex tectonic processes, where the fold mountains in western North America, running north–south, meet the Caribbean mountain arc which runs east–west.*

16

▲ ***Wave action has** cut steep cliffs into the igneous rocks of Isla Cedros, off the Pacific coast of Baja California. The island is home to sea lions, reptiles and deer.*

192

Mexico

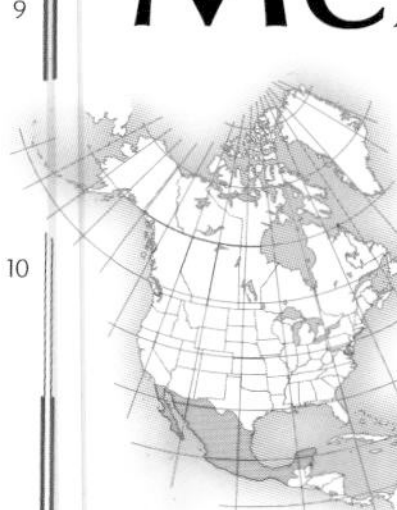

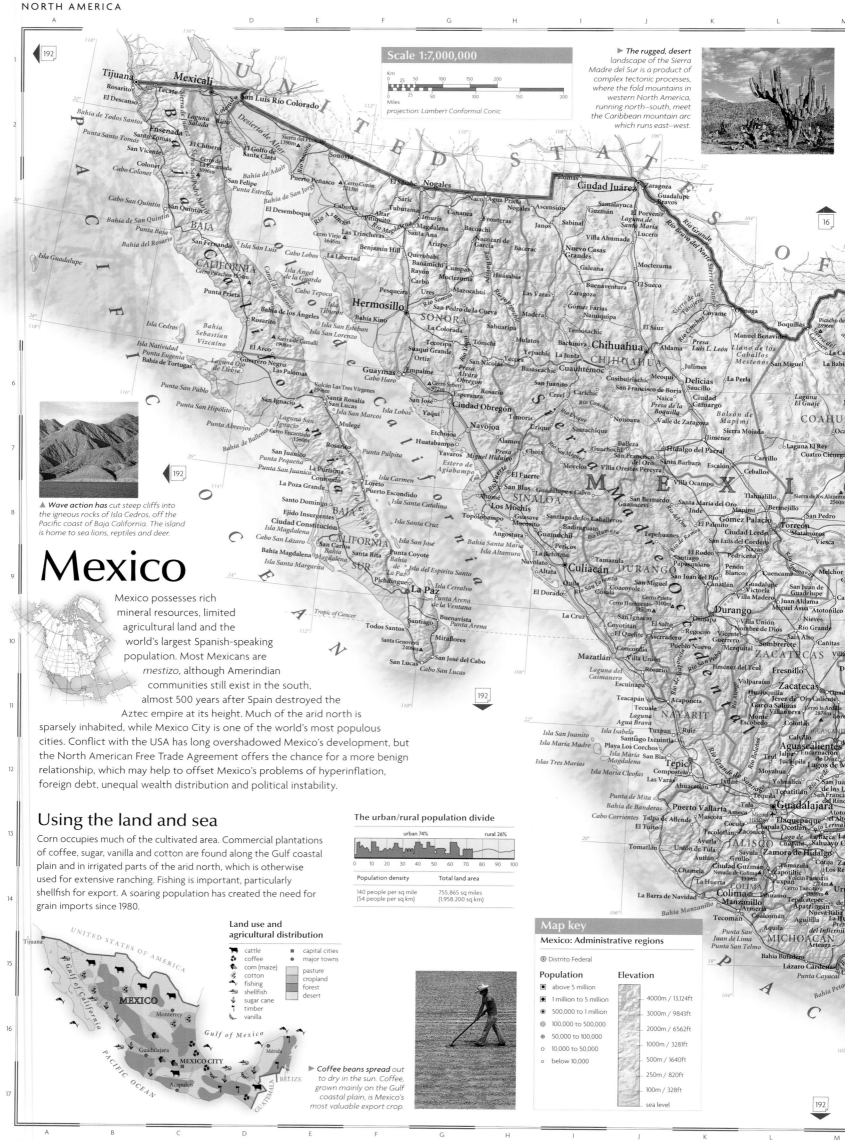

Mexico possesses rich mineral resources, limited agricultural land and the world's largest Spanish-speaking population. Most Mexicans are *mestizo*, although Amerindian communities still exist in the south, almost 500 years after Spain destroyed the Aztec empire at its height. Much of the arid north is sparsely inhabited, while Mexico City is one of the world's most populous cities. Conflict with the USA has long overshadowed Mexico's development, but the North American Free Trade Agreement offers the chance for a more benign relationship, which may help to offset Mexico's problems of hyperinflation, foreign debt, unequal wealth distribution and political instability.

192

Using the land and sea

Corn occupies much of the cultivated area. Commercial plantations of coffee, sugar, vanilla and cotton are found along the Gulf coastal plain and in irrigated parts of the arid north, which is otherwise used for extensive ranching. Fishing is important, particularly shellfish for export. A soaring population has created the need for grain imports since 1980.

The urban/rural population divide

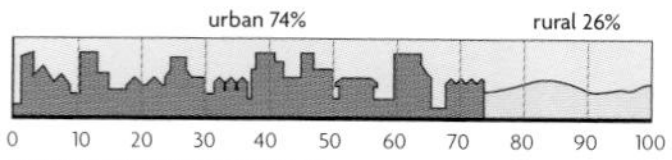

Population density	Total land area
140 people per sq mile (54 people per sq km)	755,865 sq miles (1,958,200 sq km)

Land use and agricultural distribution

▶ ***Coffee beans spread** out to dry in the sun. Coffee, grown mainly on the Gulf coastal plain, is Mexico's most valuable export crop.*

Map key

Mexico: Administrative regions

Ⓓ Distrito Federal

Population

- ▣ above 5 million
- ◙ 1 million to 5 million
- ◉ 500,000 to 1 million
- ◎ 100,000 to 500,000
- ⊕ 50,000 to 100,000
- ○ 10,000 to 50,000
- ○ below 10,000

Elevation

- 4000m / 13,124ft
- 3000m / 9843ft
- 2000m / 6562ft
- 1000m / 3281ft
- 500m / 1640ft
- 250m / 820ft
- 100m / 328ft
- sea level

192

N O P Q R S T U V W X Y

The landscape

The great central plateau rises gently southwards from the Rio Grande, isolated from the coastal plains by the Sierra Madre Oriental and Occidental. The two ranges converge from east and west respectively, culminating in high volcanic peaks around Mexico City. Further ranges of the Sierra Madre rise to the south of the Balsas basin, skirted by the low-lying Isthmus of Tehuantepec *(Istmo de Tehuantepec)* and Yucatan Peninsula.

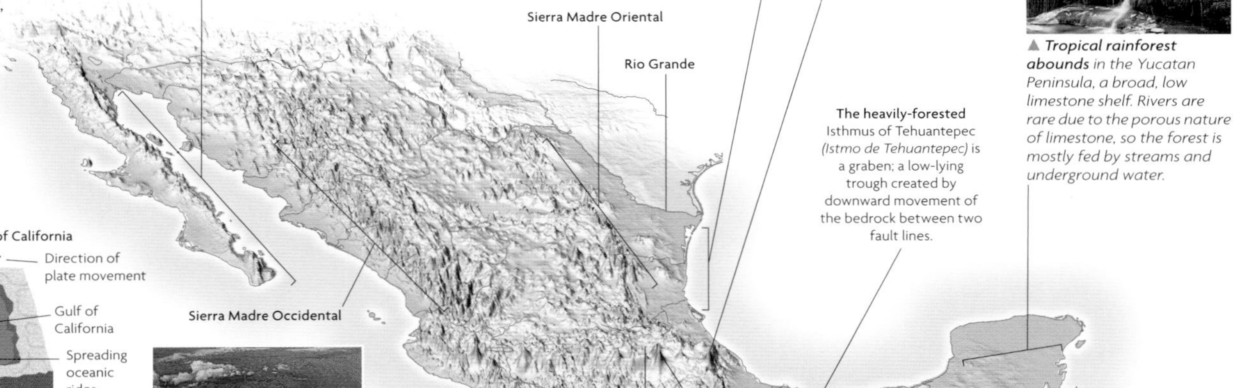

The long, narrow, extremely arid peninsula of Baja (lower) California is an elongated granite block, separated from the mainland by the flooded rift valley of the Gulf of California *(Golfo de California)*.

Wave action has constructed sand bars which shelter lagoons along the shore of the Gulf coastal plain.

The dormant cone of Volcán Pico de Orizaba is, at 18,700 ft (5700 m), the highest peak in Mexico. In North America, only Mount McKinley and Mount Logan are taller.

The heavily-forested Isthmus of Tehuantepec *(Istmo de Tehuantepec)* is a graben; a low-lying trough created by downward movement of the bedrock between two fault lines.

The unstable, earthquake-prone, upland basin around Mexico City was once a region of shallow lakes. Flood control measures and domestic consumption over the last four centuries have caused the virtual disappearance of this surface water.

The highlands of Chiapas are a series of *horsts*, blocks of land thrust upwards between two fault lines. Volcanic cones have developed where lava has flowed out from the faults.

▲ ***Tropical rainforest abounds*** *in the Yucatan Peninsula, a broad, low limestone shelf. Rivers are rare due to the porous nature of limestone, so the forest is mostly fed by streams and underground water.*

Formation of the Gulf of California

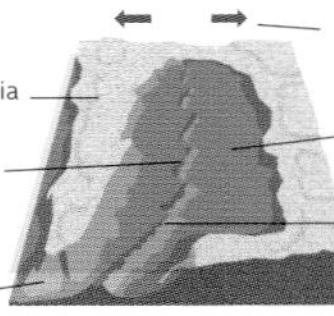

▲ ***The Gulf of California*** (Golfo de California) *began to open out about 4 million years ago as a result of rifting and plate displacement along transform faults.*

▲ ***Popocatépetl is a*** *dormant volcano, part of the Pacific 'Ring of Fire'. The crater is over half a mile (1 km) wide.*

Transport and industry

Oil and gas on the Gulf coast are Mexico's main sources of export income. Metal mining has declined but the country remains a leading global producer of silver. Manufacturing is heavily concentrated around the Mexico City metropolitan area, while the duty-free movement of goods in the USA border region, under the *Maquiladora* (twin plant) scheme, has created new hi-tech and service growth centres.

Major industry and infrastructure

- brewing
- car manufacture
- chemicals
- electronics
- fish processing
- maquiladoras
- mining
- oil & gas
- textiles
- capital cities
- major towns
- international airports
- major roads
- major industrial areas

▲ ***A stone figure*** *reclines by the Temple of Warriors, within the Mayan city of Chichén-Itzá. The Maya civilization flourished across the Yucatan Peninsula between 200 and 900 AD.*

The transport network

- 67,564 miles (108,746 km)
- 3994 miles (6429 km)
- 16,561 miles (26,656 km)
- 1801 miles (2900 km)

Fast, modern highways or autopistas now link Mexico City with Toluca, Puebla and other satellite cities, yet distant centres like Chihuahua are still served by narrow roads and an outdated rail network.

16

58

42

42

N O P X Y Z

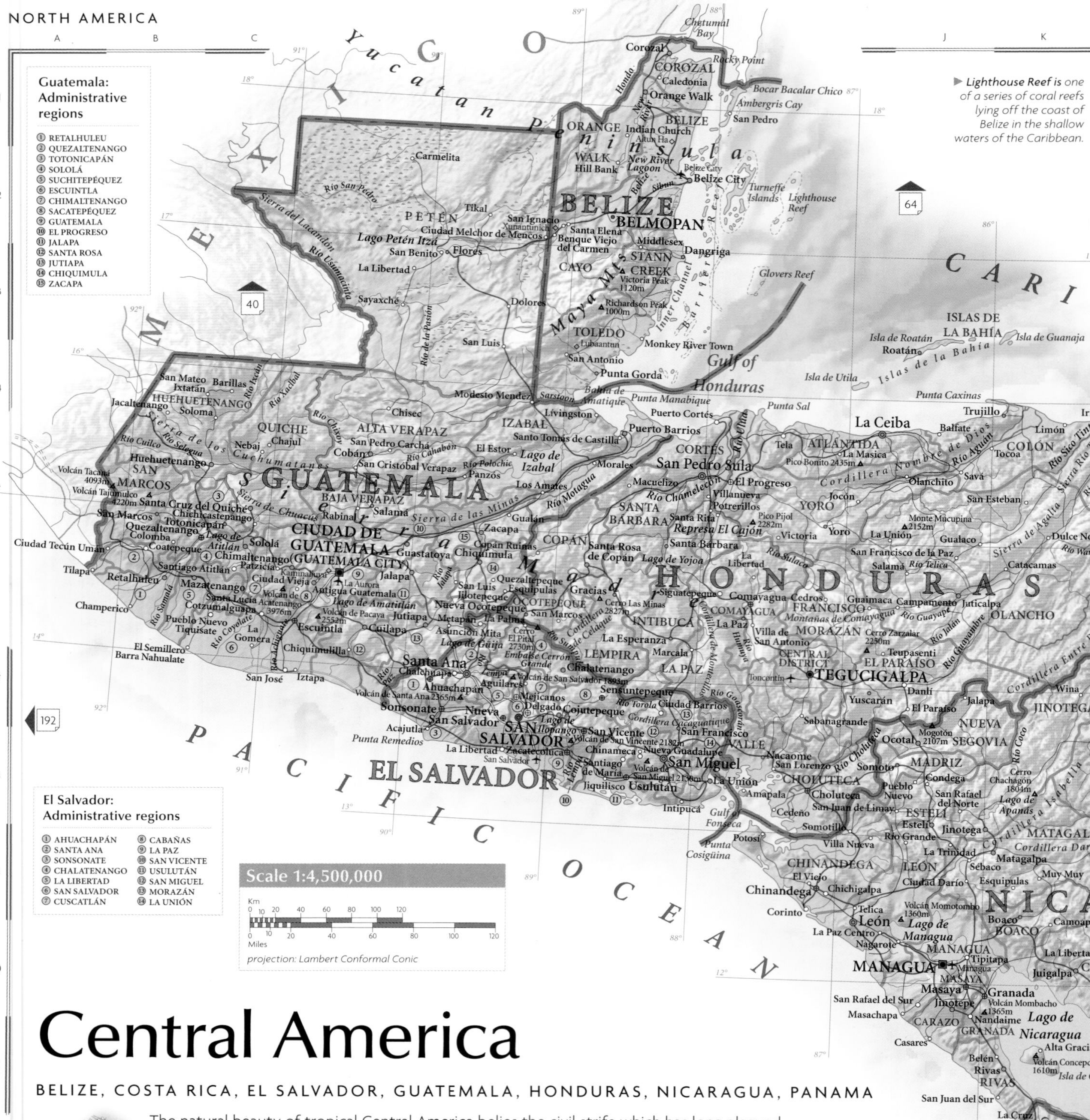

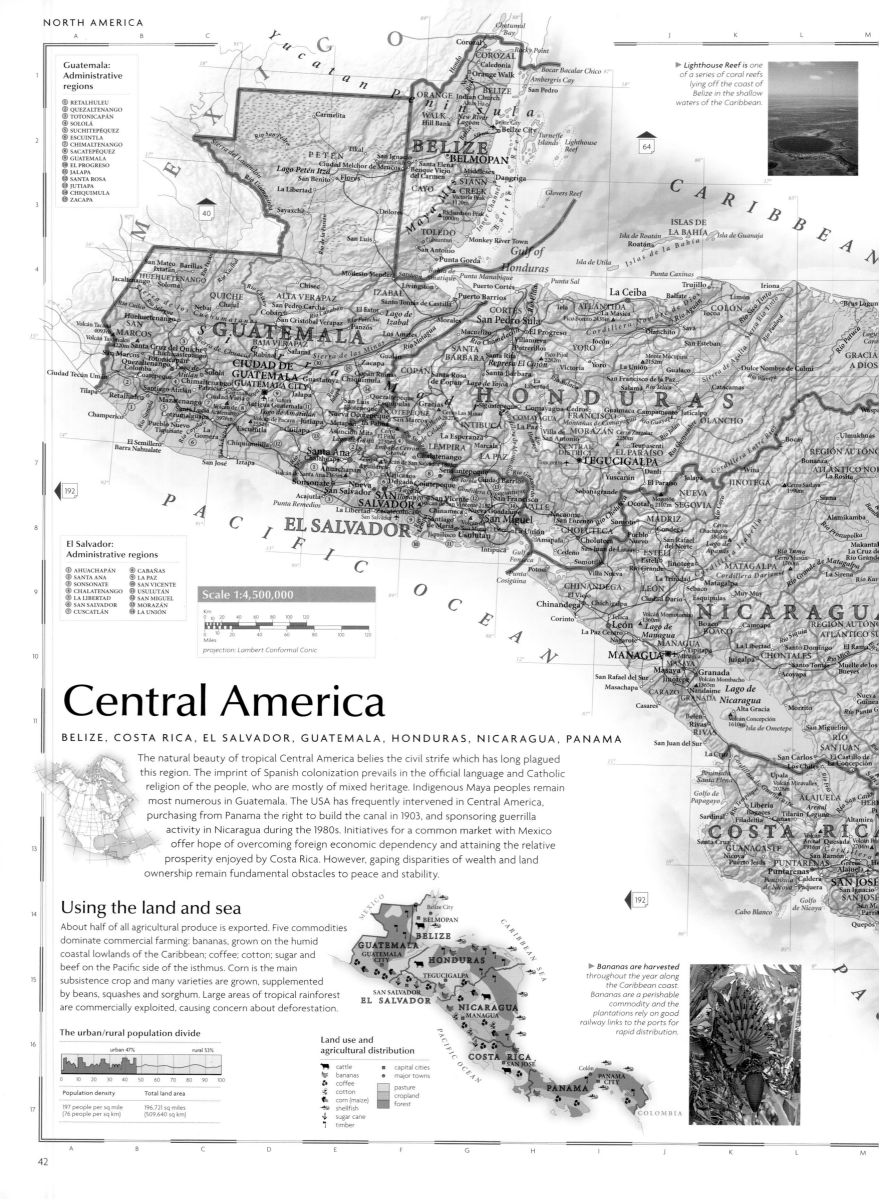

Guatemala: Administrative regions

1. RETALHULEU
2. QUEZALTENANGO
3. TOTONICAPÁN
4. SOLOLÁ
5. SUCHITEPÉQUEZ
6. ESCUINTLA
7. CHIMALTENANGO
8. SACATEPÉQUEZ
9. GUATEMALA
10. EL PROGRESO
11. JALAPA
12. SANTA ROSA
13. JUTIAPA
14. CHIQUIMULA
15. ZACAPA

El Salvador: Administrative regions

1. AHUACHAPÁN
2. SANTA ANA
3. SONSONATE
4. CHALATENANGO
5. LA LIBERTAD
6. SAN SALVADOR
7. CUSCATLÁN
8. CABAÑAS
9. LA PAZ
10. SAN VICENTE
11. USULUTÁN
12. SAN MIGUEL
13. MORAZÁN
14. LA UNIÓN

▶ ***Lighthouse Reef** is one of a series of coral reefs lying off the coast of Belize in the shallow waters of the Caribbean.*

Central America

BELIZE, COSTA RICA, EL SALVADOR, GUATEMALA, HONDURAS, NICARAGUA, PANAMA

The natural beauty of tropical Central America belies the civil strife which has long plagued this region. The imprint of Spanish colonization prevails in the official language and Catholic religion of the people, who are mostly of mixed heritage. Indigenous Maya peoples remain most numerous in Guatemala. The USA has frequently intervened in Central America, purchasing from Panama the right to build the canal in 1903, and sponsoring guerrilla activity in Nicaragua during the 1980s. Initiatives for a common market with Mexico offer hope of overcoming foreign economic dependency and attaining the relative prosperity enjoyed by Costa Rica. However, gaping disparities of wealth and land ownership remain fundamental obstacles to peace and stability.

Using the land and sea

About half of all agricultural produce is exported. Five commodities dominate commercial farming: bananas, grown on the humid coastal lowlands of the Caribbean; coffee; cotton; sugar and beef on the Pacific side of the isthmus. Corn is the main subsistence crop and many varieties are grown, supplemented by beans, squashes and sorghum. Large areas of tropical rainforest are commercially exploited, causing concern about deforestation.

The urban/rural population divide

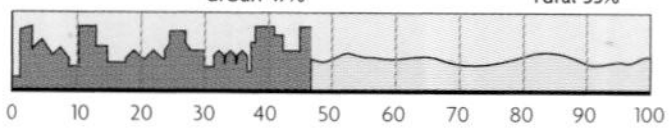

Population density	Total land area
197 people per sq mile (76 people per sq km)	196,721 sq miles (509,640 sq km)

▶ ***Bananas are harvested** throughout the year along the Caribbean coast. Bananas are a perishable commodity and the plantations rely on good railway links to the ports for rapid distribution.*

The landscape

The Sierra Madre range spreads west from Mexico, between the narrow Pacific coastal plain and the limestone lowland of Petén. Parallel hill ranges sweep across Honduras and extend south, past the Caribbean Mosquito Coast, to lakes Managua and Nicaragua. The Cordillera Central rises to the south, gradually descending to Lake Gatún *(Lago Gatún)*. A highly active volcanic belt runs along the Pacific seaboard from Mexico to Costa Rica.

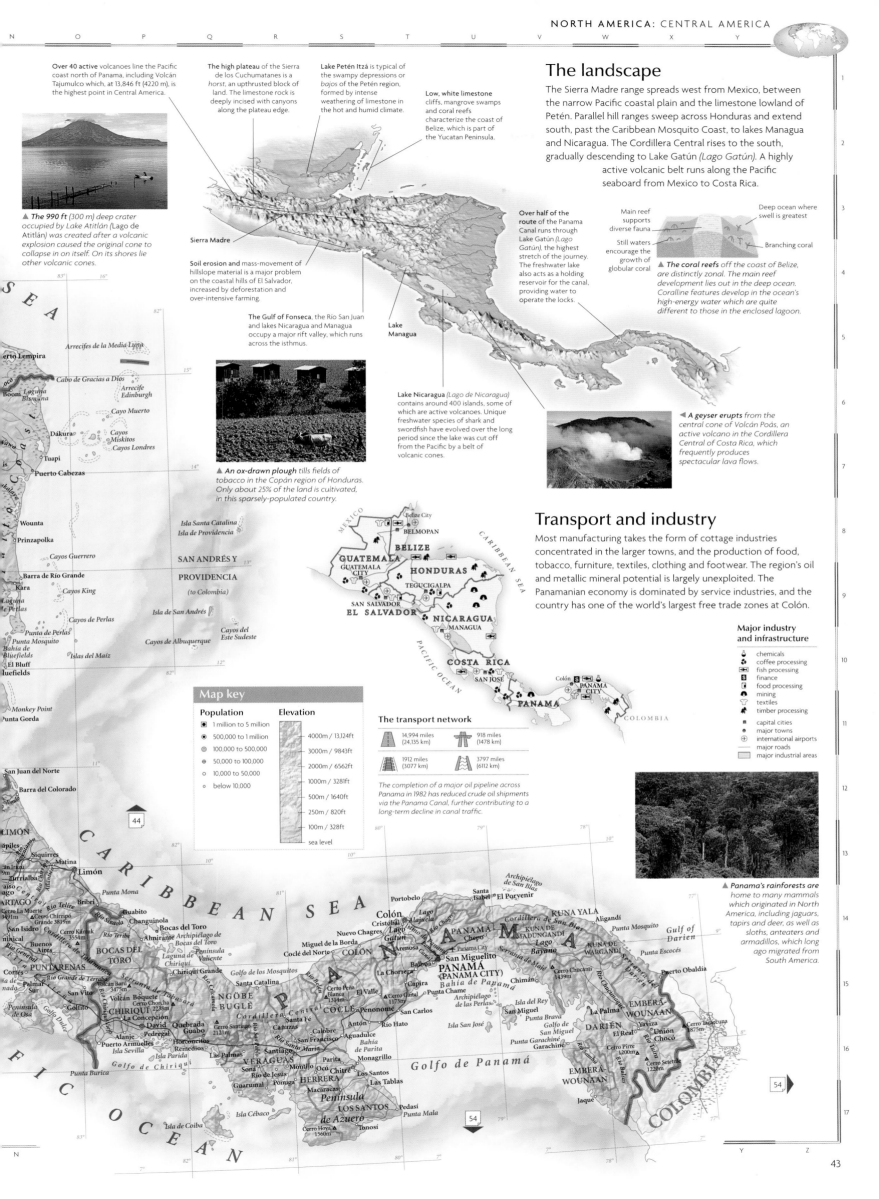

▲ ***The 990 ft** (300 m) deep crater occupied by Lake Atitlán* (Lago de Atitlán) *was created after a volcanic explosion caused the original cone to collapse in on itself. On its shores lie other volcanic cones.*

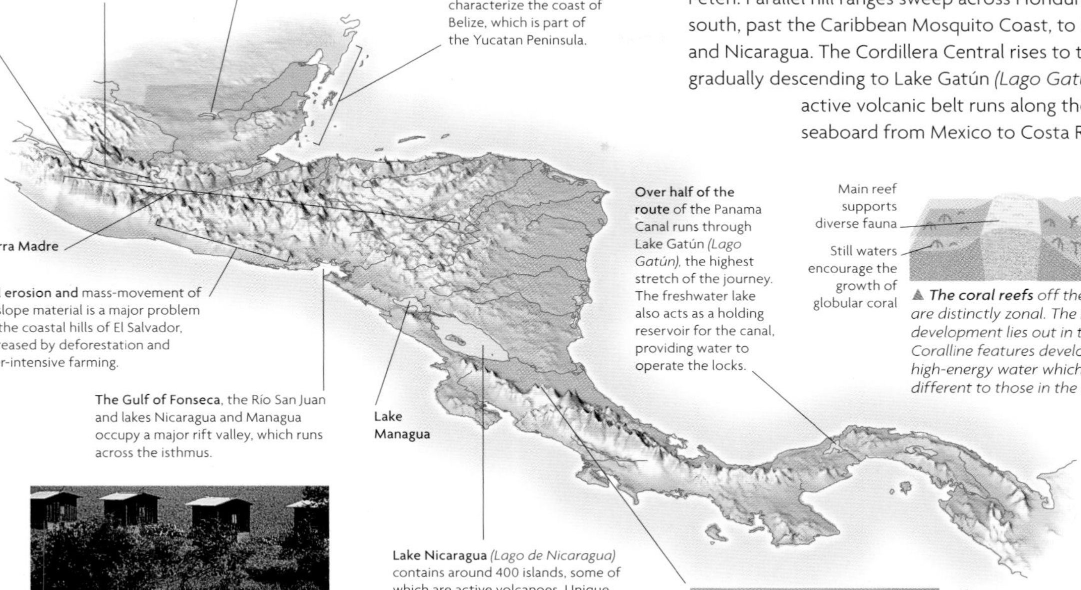

▲ ***The coral reefs** off the coast of Belize, are distinctly zonal. The main reef development lies out in the deep ocean. Coralline features develop in the ocean's high-energy water which are quite different to those in the enclosed lagoon.*

▲ ***An ox-drawn plough** tills fields of tobacco in the Copán region of Honduras. Only about 25% of the land is cultivated, in this sparsely-populated country.*

◀ ***A geyser erupts** from the central cone of Volcán Poás, an active volcano in the Cordillera Central of Costa Rica, which frequently produces spectacular lava flows.*

Transport and industry

Most manufacturing takes the form of cottage industries concentrated in the larger towns, and the production of food, tobacco, furniture, textiles, clothing and footwear. The region's oil and metallic mineral potential is largely unexploited. The Panamanian economy is dominated by service industries, and the country has one of the world's largest free trade zones at Colón.

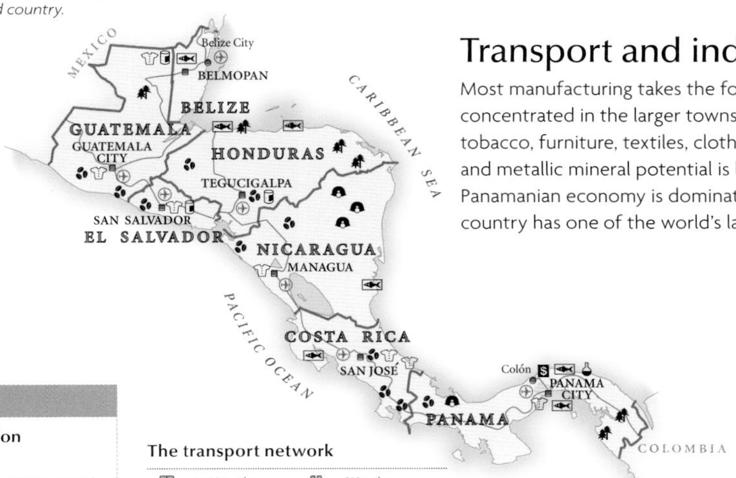

Major industry and infrastructure

- chemicals
- coffee processing
- fish processing
- finance
- food processing
- mining
- textiles
- timber processing
- capital cities
- major towns
- international airports
- major roads
- major industrial areas

The transport network

14,994 miles (24,135 km)	918 miles (1478 km)
1912 miles (3077 km)	3797 miles (6112 km)

The completion of a major oil pipeline across Panama in 1982 has reduced crude oil shipments via the Panama Canal, further contributing to a long-term decline in canal traffic.

▲ *Panama's rainforests are home to many mammals which originated in North America, including jaguars, tapirs and deer, as well as sloths, anteaters and armadillos, which long ago migrated from South America.*

Map key

Population

- 1 million to 5 million
- 500,000 to 1 million
- 100,000 to 500,000
- 50,000 to 100,000
- 10,000 to 50,000
- below 10,000

Elevation

- 4000m / 13,124ft
- 3000m / 9843ft
- 2000m / 6562ft
- 1000m / 3281ft
- 500m / 1640ft
- 250m / 820ft
- 100m / 328ft
- sea level

◀ *The Caribbean's virgin rainforest, seen here in Jamaica, is increasingly at risk from agricultural, industrial and tourist development. On some islands, the rainforest has virtually disappeared.*

▲ *The large bar which lies submerged in front of Marina Cay in the British Virgin Islands, has been built up by waves, depositing a bank of sand which partially encloses the islet.*

The Caribbean

THE BAHAMAS, GREATER ANTILLES, LESSER ANTILLES

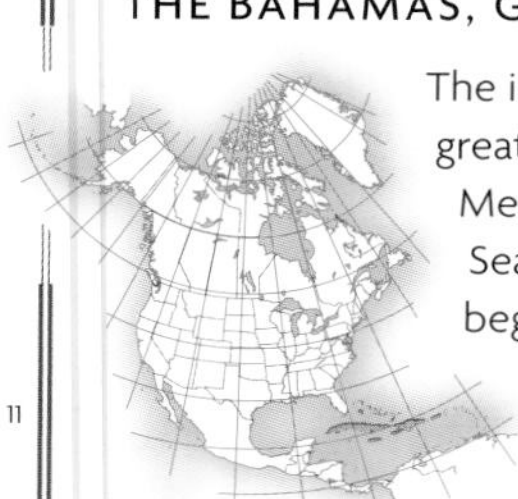

The islands known as the West Indies form a great arc which trails eastwards from the Gulf of Mexico almost to Venezuela, enclosing the Caribbean Sea. During the period of European colonization, which began in the 16th century, Britain, France, Spain and the Netherlands struggled for control of the area. Some countries remained politically tied to their colonial rulers until late in the 20th century, and most islands' economies still bear the legacy of the plantation system. A diverse mix of peoples, with roots drawn from Africa, East Asia and Europe replaced the original Amerindian population, creating a unique and remarkably homogeneous culture, reflected in the various Creole languages and musical forms such as reggae and calypso.

Using the land and sea

Agriculture has long been the basis of most Caribbean economies. Much agricultural land is set aside for cash crops such as sugar, spices, citrus fruits, bananas and cocoa, which are grown for export. Diversification is being encouraged to reduce the islands' reliance on imported grain and vulnerability to price fluctuations.

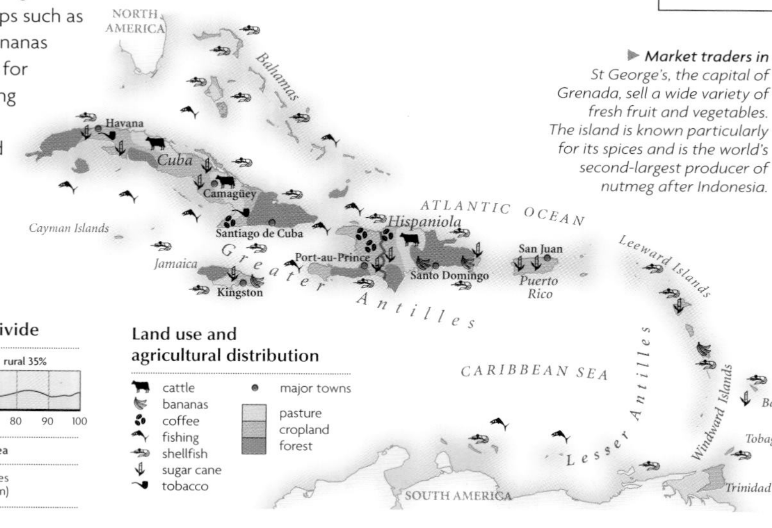

Land use and agricultural distribution

- cattle
- bananas
- coffee
- fishing
- shellfish
- sugar cane
- tobacco
- major towns
- pasture
- cropland
- forest

▶ *Market traders in St George's, the capital of Grenada, sell a wide variety of fresh fruit and vegetables. The island is known particularly for its spices and is the world's second-largest producer of nutmeg after Indonesia.*

The urban/rural population divide

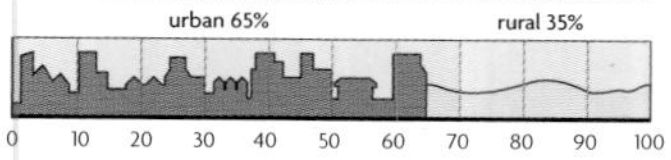

Population density	Total land area
435 people per sq mile (168 people per sq km)	88,396 sq miles (229,005 sq km)

Map key

Population

- 1 million to 5 million
- 500,000 to 1 million
- 100,000 to 500,000
- 50,000 to 100,000
- 10,000 to 50,000
- below 10,000

Elevation

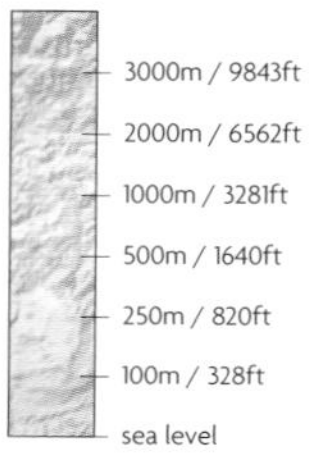

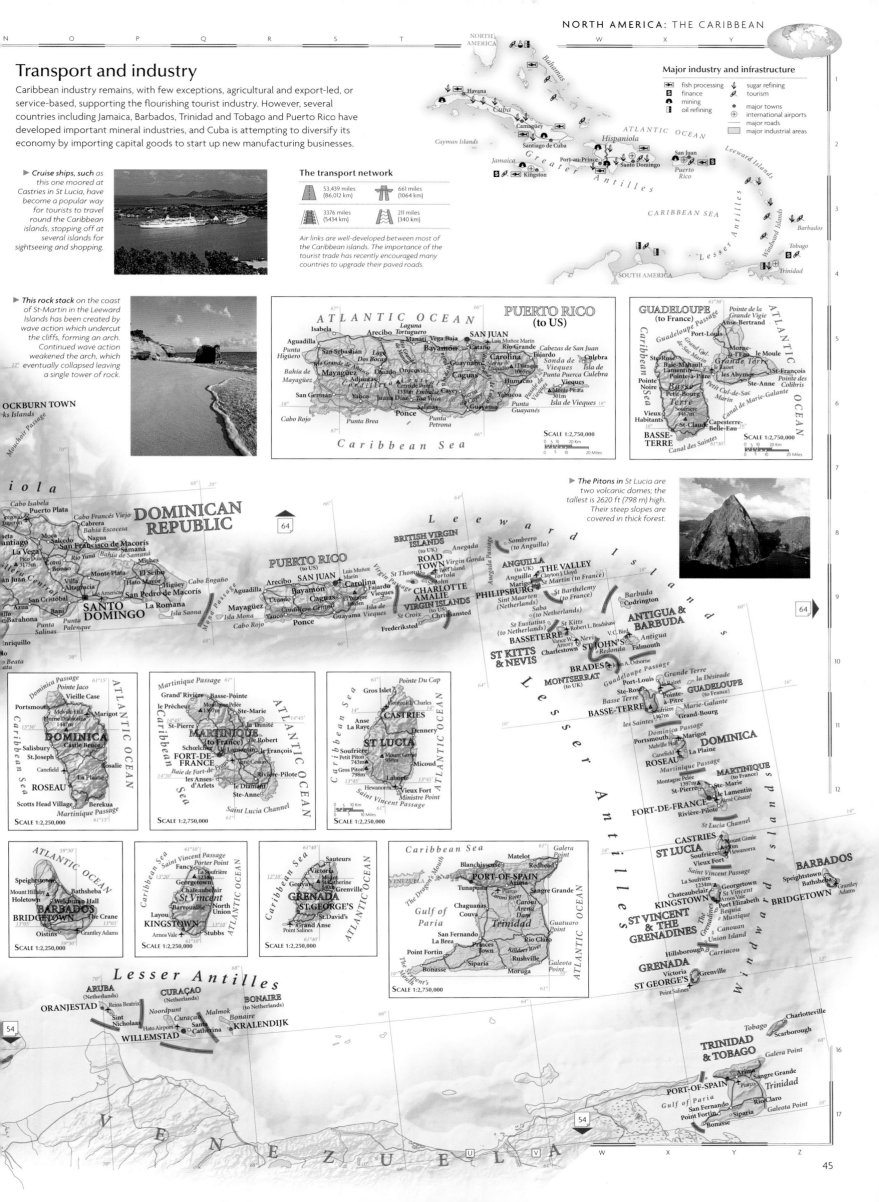

Transport and industry

Caribbean industry remains, with few exceptions, agricultural and export-led, or service-based, supporting the flourishing tourist industry. However, several countries including Jamaica, Barbados, Trinidad and Tobago and Puerto Rico have developed important mineral industries, and Cuba is attempting to diversify its economy by importing capital goods to start up new manufacturing businesses.

▶ *Cruise ships, such as this one moored at Castries in St Lucia, have become a popular way for tourists to travel round the Caribbean islands, stopping off at several islands for sightseeing and shopping.*

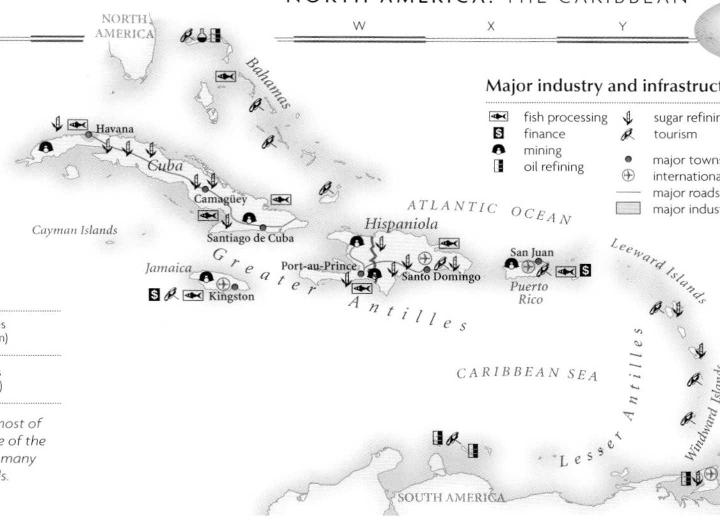

Major industry and infrastructure

- fish processing
- finance
- mining
- oil refining
- sugar refining
- tourism
- major towns
- international airports
- major roads
- major industrial areas

The transport network

53,439 miles (86,012 km)	661 miles (1064 km)	3376 miles (5434 km)	211 miles (340 km)

Air links are well-developed between most of the Caribbean islands. The importance of the tourist trade has recently encouraged many countries to upgrade their paved roads.

▶ *This rock stack on the coast of St-Martin in the Leeward Islands has been created by wave action which undercut the cliffs, forming an arch. Continued wave action weakened the arch, which eventually collapsed leaving a single tower of rock.*

▶ *The Pitons in St Lucia are two volcanic domes; the tallest is 2620 ft (798 m) high. Their steep slopes are covered in thick forest.*

NOR
Rio Grande
Sierra Madre Oriental
Cedros Trench
Gulf of California
Lower California
Sierra Madre Occidental
San Lucas Cape
Lago de Chapala
East Pacific Rise
Tehuantepec Ridge
Clipperton Seamounts
Clipperton Island
PACIFI
COCOS
NAZCA
NAZCA PLATE
PACIFIC PLATE

South America

Reaching from the humid tropics down into the cold south Atlantic, South America has an area of 6,886,000 sq miles (17,835,000 sq km). There are 12 separate countries, with the largest, Brazil, covering almost half the continent.

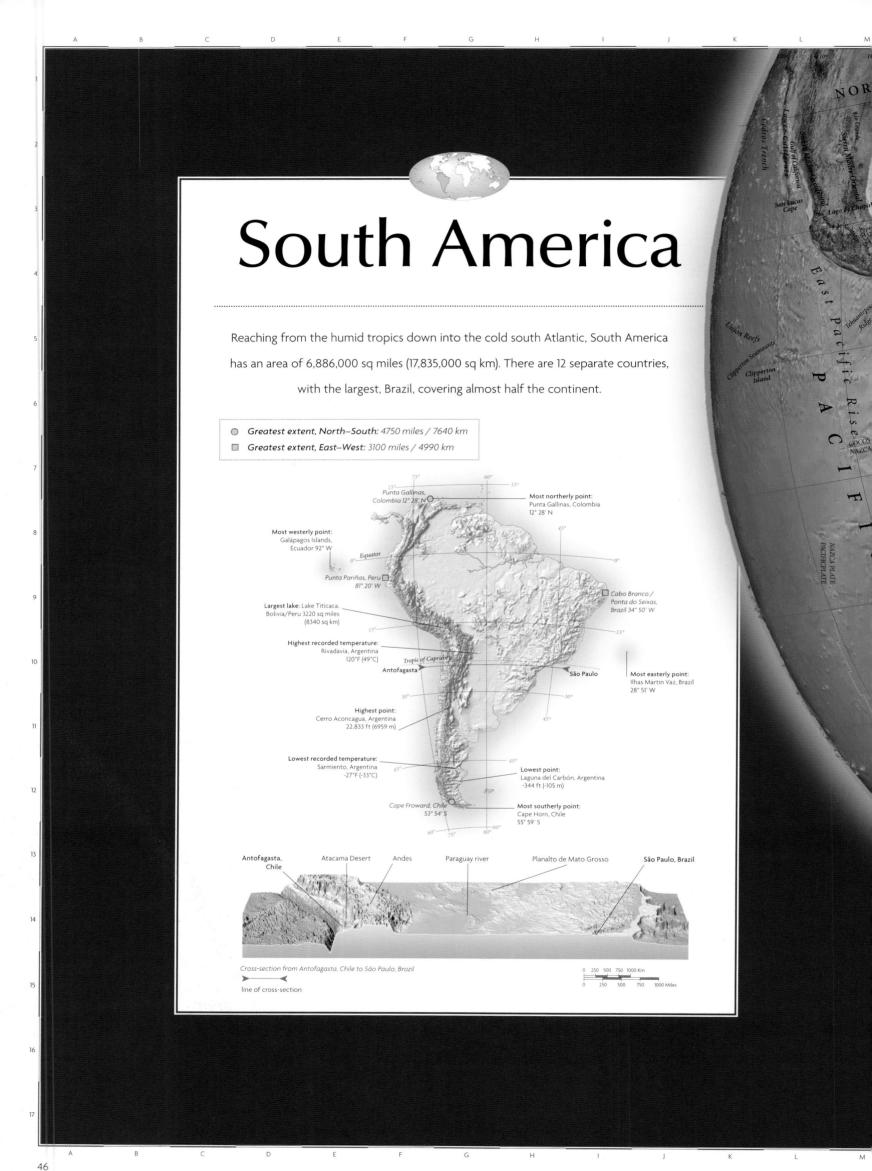

○ ***Greatest extent, North–South:*** *4750 miles / 7640 km*
□ ***Greatest extent, East–West:*** *3100 miles / 4990 km*

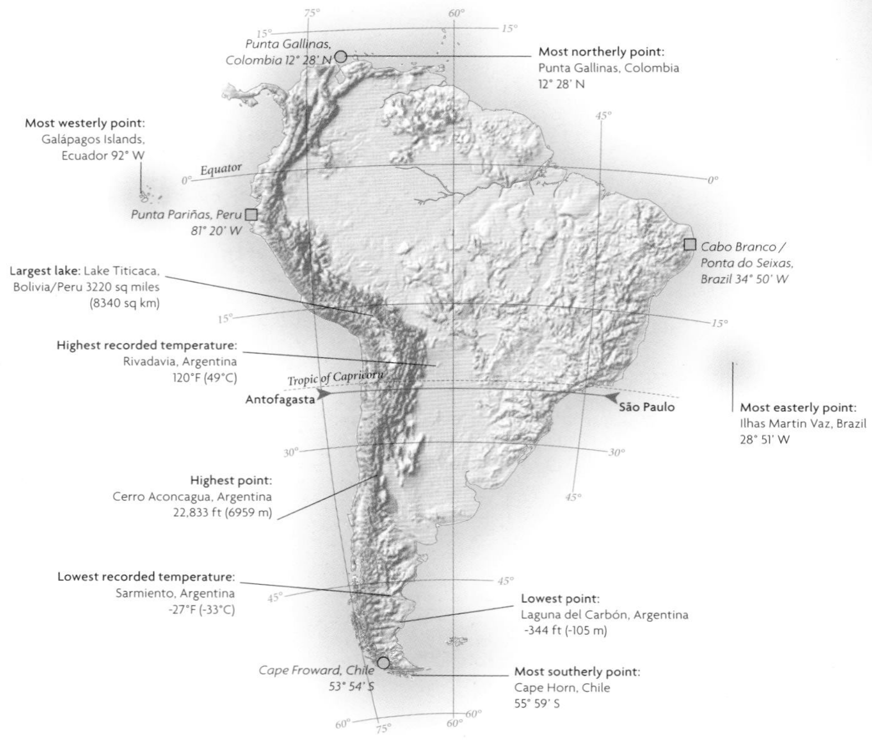

Punta Gallinas, Colombia 12° 28' N
Most northerly point: Punta Gallinas, Colombia 12° 28' N
Most westerly point: Galápagos Islands, Ecuador 92° W
Equator
Punta Pariñas, Peru 81° 20' W
Cabo Branco / Ponta do Seixas, Brazil 34° 50' W
Largest lake: Lake Titicaca, Bolivia/Peru 3220 sq miles (8340 sq km)
Highest recorded temperature: Rivadavia, Argentina 120°F (49°C)
Tropic of Capricorn
Antofagasta
São Paulo
Most easterly point: Ilhas Martin Vaz, Brazil 28° 51' W
Highest point: Cerro Aconcagua, Argentina 22,833 ft (6959 m)
Lowest recorded temperature: Sarmiento, Argentina -27°F (-33°C)
Lowest point: Laguna del Carbón, Argentina -344 ft (-105 m)
Cape Froward, Chile 53° 54' S
Most southerly point: Cape Horn, Chile 55° 59' S

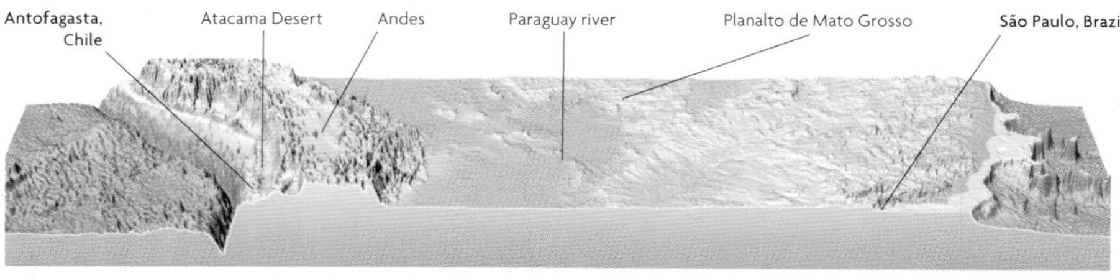

Antofagasta, Chile
Atacama Desert
Andes
Paraguay river
Planalto de Mato Grosso
São Paulo, Brazil

Cross-section from Antofagasta, Chile to São Paulo, Brazil

line of cross-section

0 250 500 750 1000 Km
0 250 500 750 1000 Miles

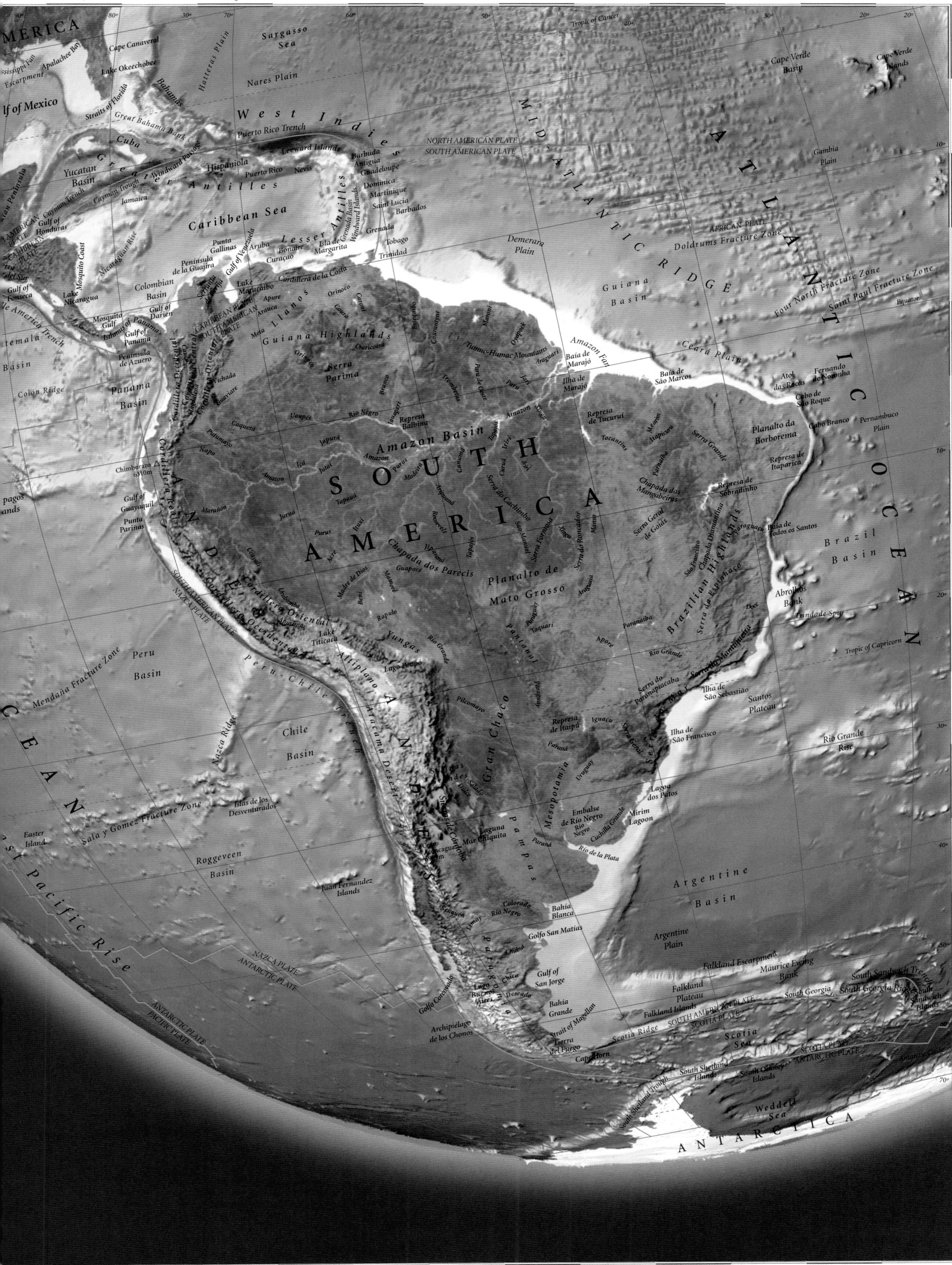
N O P Q R S T U V W X Y Z
1 2 3 4 5 6 7 8 9 10 11 12 13 14 15 16 17
Sargasso Sea
Cape Canaveral
Lake Okeechobee
Bahamas
Hatteras Plain
Nares Plain
Gulf of Mexico
Straits of Florida
Great Bahama Bank
West Indies
Puerto Rico Trench
Cuba
Greater Antilles
Hispaniola
Puerto Rico
Leeward Islands
Nevis
Barbuda
Antigua
Guadeloupe
Dominica
Martinique
Saint Lucia
Barbados
Windward Islands
Windward Passage
Yucatan Basin
Cayman Trough
Jamaica
Gulf of Honduras
Caribbean Sea
Lesser Antilles
Grenada
Tobago
Trinidad
Isla de Margarita
Aruba
Bonaire
Curaçao
Punta Gallinas
Peninsula de la Guajira
Gulf of Venezuela
Colombian Basin
Lake Maracaibo
Cordillera de la Costa
Mosquito Coast
Lake Nicaragua
Gulf of Fonseca
Mosquito Gulf
Gulf of Darien
Isthmus of Panama
Gulf of Panama
Peninsula de Azuero
Panama Basin
Colón Ridge
Tropic of Cancer
NORTH AMERICAN PLATE
SOUTH AMERICAN PLATE
MID-ATLANTIC RIDGE
ATLANTIC OCEAN
Cape Verde Basin
Cape Verde Islands
Gambia Plain
AFRICAN PLATE
Doldrums Fracture Zone
Four North Fracture Zone
Saint Paul Fracture Zone
Equator
Demerara Plain
Guiana Basin
Ceará Plain
Amazon Fan
Baía de Marajó
Ilha de Marajó
Baía de São Marcos
Atol das Rocas
Fernando de Noronha
Cabo de São Roque
Cabo Branco
Pernambuco Plain
Llanos
Apure
Arauca
Meta
Orinoco
Caroni
Guiana Highlands
Serra Parima
Tumuc-Humac Mountains
Rio Negro
Represa Balbina
Amazon
Amazon Basin
SOUTH AMERICA
Japurá
Putumayo
Napo
Purus
Madeira
Juruá
Marañon
Tapajós
Xingu
Tocantins
Represa de Tucuruí
Planalto da Borborema
Serra Grande
Represa de Itaparica
Chapada das Mangabeiras
Represa de Sobradinho
Baía de Todos os Santos
Serra do Cachimbo
Chapada dos Parecis
Planalto de Mato Grosso
Serra do Roncador
Serra Geral de Goiás
Brazilian Highlands
Serra do Espinhaço
São Francisco
Brazil Basin
Abrolhos Bank
Trindade Spur
Tropic of Capricorn
Chimborazo 6310m
Gulf of Guayaquil
Punta Pariñas
Cordillera Occidental
Cordillera Oriental
ANDES
SOUTH AMERICAN PLATE
NAZCA PLATE
Peru Basin
Mendaña Fracture Zone
Peru-Chile Trench
Lake Titicaca
Yungas
Altiplano
Lago Poopó
Atacama Desert
Pantanal
Paraná
Paraguay
Rio Grande
Serra da Mantiqueira
Serra do Paranapiacaba
Ilha de São Sebastião
Santos Plateau
Represa de Itaipú
Iguaçu
Serra Geral
Ilha de São Francisco
Rio Grande Rise
Gran Chaco
Pilcomayo
Chile Basin
Nazca Ridge
Islas de los Desventurados
Sala y Gomez Fracture Zone
Easter Island
Mesopotamia
Uruguay
Lagoa dos Patos
Mirim Lagoon
Embalse de Río Negro
Cuchilla Grande
Laguna Mar Chiquita
Pampas
Río de la Plata
Roggeveen Basin
Juan Fernandez Islands
Argentine Basin
Argentine Plain
Colorado
Río Negro
Bahía Blanca
Golfo San Matías
Patagonia
Gulf of San Jorge
Falkland Escarpment
Maurice Ewing Bank
Falkland Plateau
Falkland Islands
South Georgia
South Sandwich Trench
South Sandwich Islands
Bahía Grande
Golfo Corcovado
Archipiélago de los Chonos
Strait of Magellan
Tierra del Fuego
Cape Horn
Scotia Ridge
SCOTIA PLATE
Scotia Sea
ANTARCTIC PLATE
South Shetland Islands
South Orkney Islands
South Shetland Trough
Weddell Sea
ANTARCTICA
East Pacific Rise
NAZCA PLATE
ANTARCTIC PLATE
PACIFIC PLATE

Physical South America

Three major physiographic regions characterize South America. The oldest, the ancient Brazilian Shield and the smaller Guiana and Patagonian shields, form the stable core of the continent. Stretching along the entire west coast are the younger Andean fold mountains with many summits rising to 20,000 ft (6100 m). These two diverse regions are separated by a number of sedimentary basins carrying South America's large river systems to the sea. These include the massive Amazon Basin and the basin of the Gran Chaco.

The Amazon Basin and Guiana Shield

The Amazon river occupies a large depression in the Earth's crust, formed by the uplift of the Andes. It is covered by thick volcanic deposits and layers of alluvium – these have been laid down by the Amazon's many tributaries. To the north is the smaller Guiana Shield.

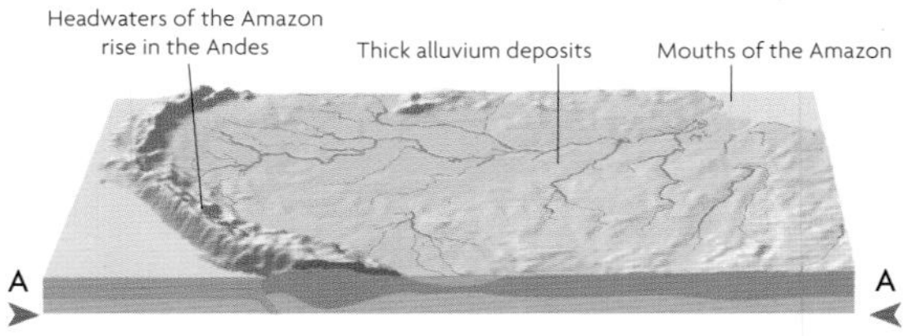

Section across northern South America showing Amazon Basin and its drainage pattern.

The Andean Uplands

The Andean Uplands run along the west coast of South America. They are being uplifted as the Nazca Plate is subducted beneath the South American Plate. They contain some of the world's largest volcanoes, such as Cotopaxi, and Lake Titicaca which occupies a dormant site. The far south has many large ice-sheets and a fragmented coastline.

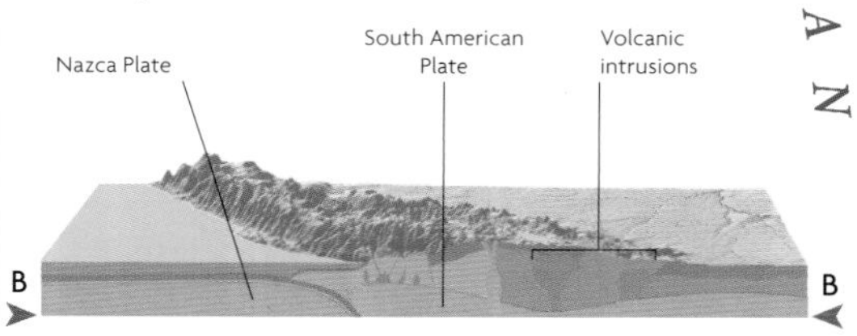

Cross-section through the Andes showing the subduction of the Nazca Plate beneath the South American Plate.

The Brazilian Shield and Gran Chaco

The immense Brazilian Shield underlies more than one-third of South America. It is pitted with numerous volcanic intrusions, and a large basaltic plateau exists between the Paraná river and the Atlantic Ocean. The flat Gran Chaco lies to the west of the shield, covered by sedimentary deposits eroded from the Andes, and transported by South America's mighty rivers.

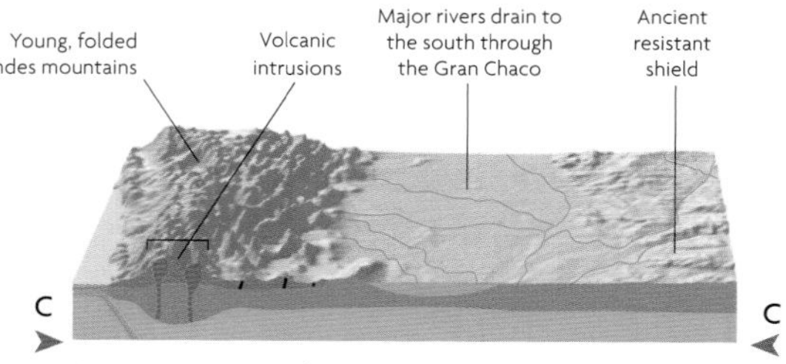

Section across central South America showing the flat basin of the Gran Chaco and the ancient Brazilian Shield.

Map key

Elevation

- 6000m / 19,686ft
- 4000m / 13,124ft
- 3000m / 9843ft
- 2000m / 6562ft
- 1000m / 3281ft
- 500m / 1640ft
- 250m / 820ft
- 100m / 328ft
- sea level

Plate margins (for explanation see page xiv)

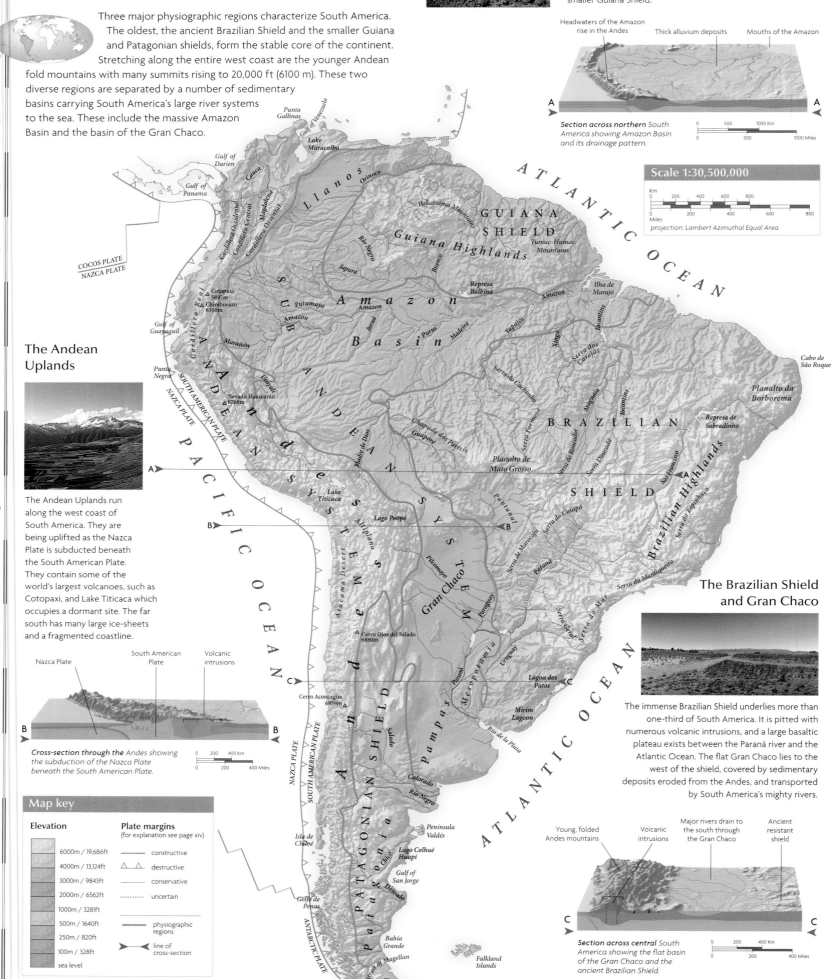

- constructive
- destructive
- conservative
- uncertain
- physiographic regions
- line of cross-section

N O P Q R S T

Climate

The climate of South America is influenced by three principal factors: the seasonal shift of high pressure air masses over the tropics, cold ocean currents along the western coast, affecting temperature and precipitation, and the mountain barrier produced by the Andes, which creates a rain shadow over much of the south.

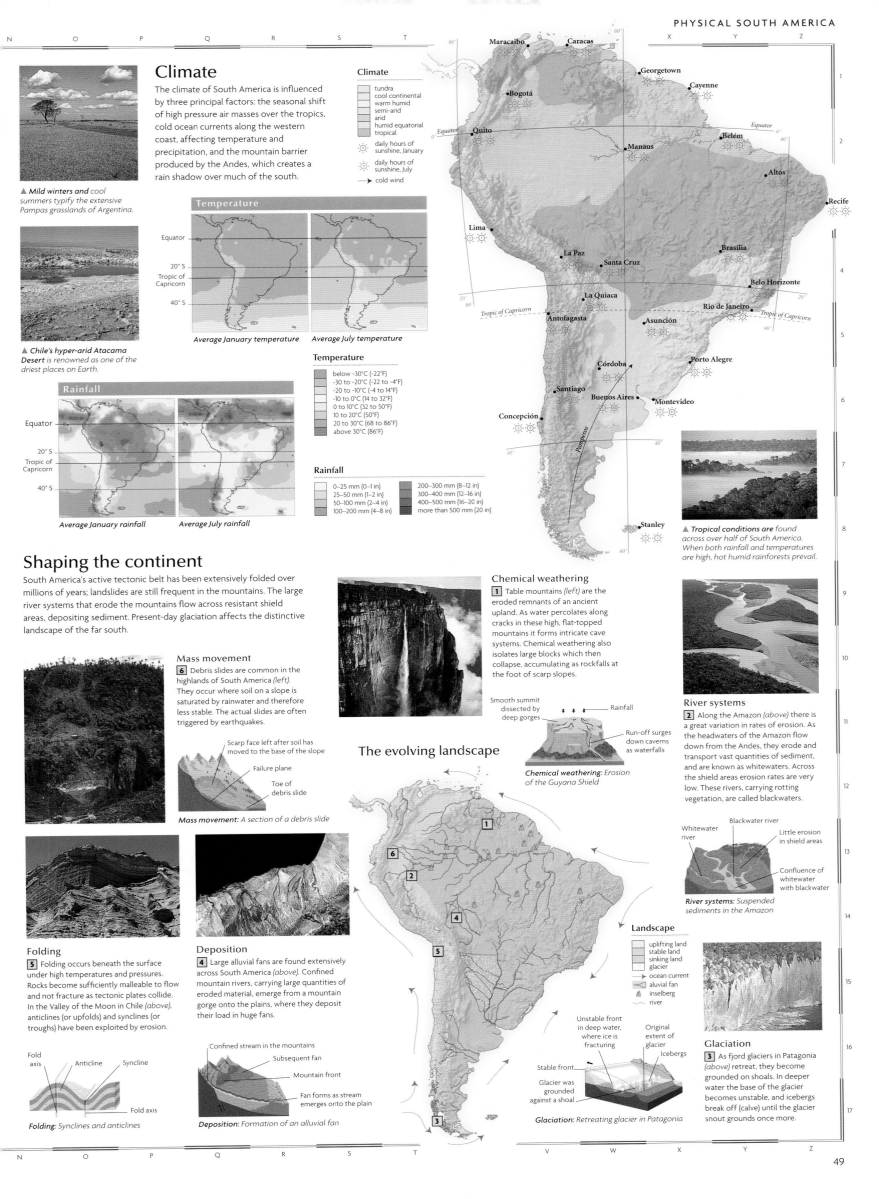

▲ *Mild winters and cool summers typify the extensive Pampas grasslands of Argentina.*

▲ *Chile's hyper-arid Atacama Desert is renowned as one of the driest places on Earth.*

Temperature

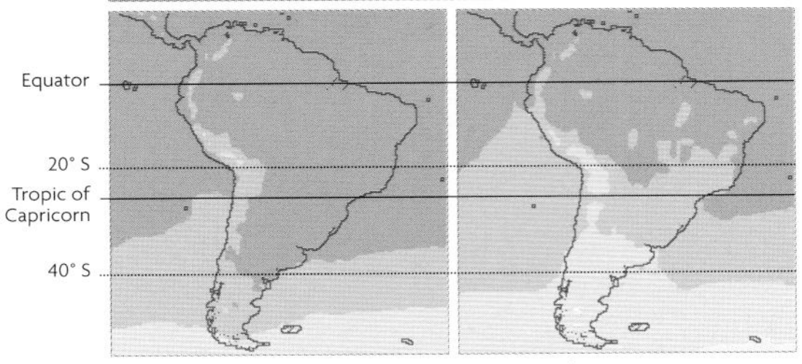

Equator
20° S
Tropic of Capricorn
40° S

Average January temperature *Average July temperature*

Temperature

- below -30°C (-22°F)
- -30 to -20°C (-22 to -4°F)
- -20 to -10°C (-4 to 14°F)
- -10 to 0°C (14 to 32°F)
- 0 to 10°C (32 to 50°F)
- 10 to 20°C (50°F)
- 20 to 30°C (68 to 86°F)
- above 30°C (86°F)

Rainfall

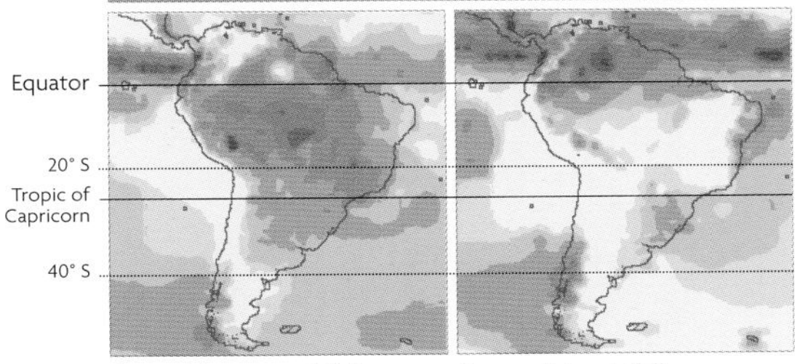

Equator
20° S
Tropic of Capricorn
40° S

Average January rainfall *Average July rainfall*

Rainfall

- 0–25 mm (0–1 in)
- 25–50 mm (1–2 in)
- 50–100 mm (2–4 in)
- 100–200 mm (4–8 in)
- 200–300 mm (8–12 in)
- 300–400 mm (12–16 in)
- 400–500 mm (16–20 in)
- more than 500 mm (20 in)

Climate

- tundra
- cool continental
- warm humid
- semi-arid
- arid
- humid equatorial
- tropical
- daily hours of sunshine, January
- daily hours of sunshine, July
- → cold wind

Maracaibo
Caracas
Georgetown
Cayenne
Bogotá
Equator
Quito
Manaus
Belém
Altos
Recife
Lima
La Paz
Santa Cruz
Brasília
Belo Horizonte
La Quiaca
Tropic of Capricorn
Rio de Janeiro
Antofagasta
Asunción
Córdoba
Porto Alegre
Santiago
Buenos Aires
Montevideo
Concepción
Pamperos
Stanley

▲ *Tropical conditions are found across over half of South America. When both rainfall and temperatures are high, hot humid rainforests prevail.*

Shaping the continent

South America's active tectonic belt has been extensively folded over millions of years; landslides are still frequent in the mountains. The large river systems that erode the mountains flow across resistant shield areas, depositing sediment. Present-day glaciation affects the distinctive landscape of the far south.

Chemical weathering

1 Table mountains *(left)* are the eroded remnants of an ancient upland. As water percolates along cracks in these high, flat-topped mountains it forms intricate cave systems. Chemical weathering also isolates large blocks which then collapse, accumulating as rockfalls at the foot of scarp slopes.

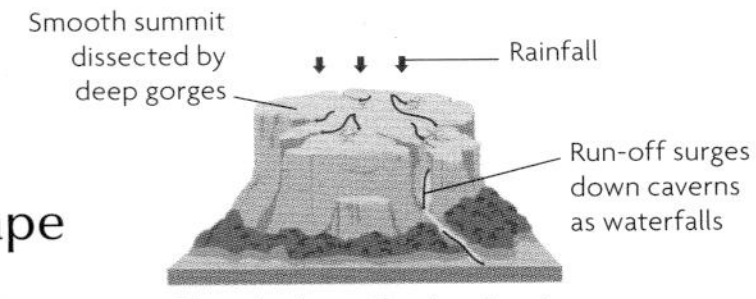

Smooth summit dissected by deep gorges
Rainfall
Run-off surges down caverns as waterfalls

Chemical weathering: *Erosion of the Guyana Shield*

Mass movement

6 Debris slides are common in the highlands of South America *(left)*. They occur where soil on a slope is saturated by rainwater and therefore less stable. The actual slides are often triggered by earthquakes.

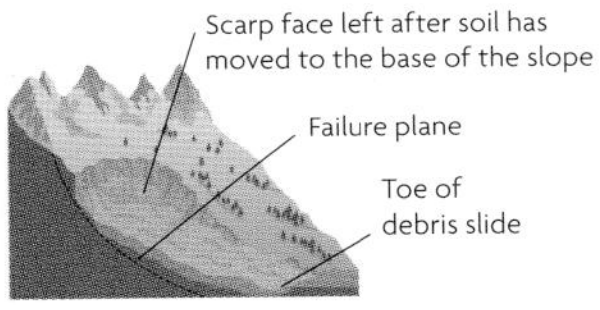

Scarp face left after soil has moved to the base of the slope
Failure plane
Toe of debris slide

Mass movement: *A section of a debris slide*

River systems

2 Along the Amazon *(above)* there is a great variation in rates of erosion. As the headwaters of the Amazon flow down from the Andes, they erode and transport vast quantities of sediment, and are known as whitewaters. Across the shield areas erosion rates are very low. These rivers, carrying rotting vegetation, are called blackwaters.

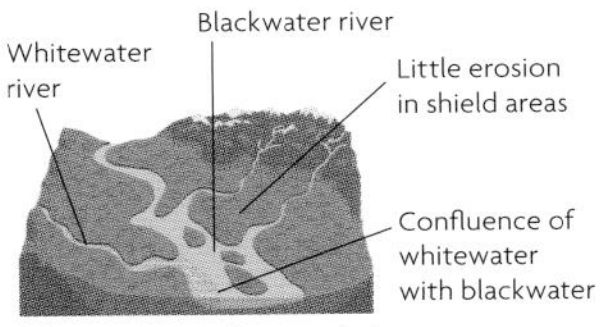

Blackwater river
Whitewater river
Little erosion in shield areas
Confluence of whitewater with blackwater

River systems: *Suspended sediments in the Amazon*

The evolving landscape

1
6
2
4
5
3

Landscape

- uplifting land
- stable land
- sinking land
- glacier
- ocean current
- aluvial fan
- inselberg
- river

Folding

5 Folding occurs beneath the surface under high temperatures and pressures. Rocks become sufficiently malleable to flow and not fracture as tectonic plates collide. In the Valley of the Moon in Chile *(above)*, anticlines (or upfolds) and synclines (or troughs) have been exploited by erosion.

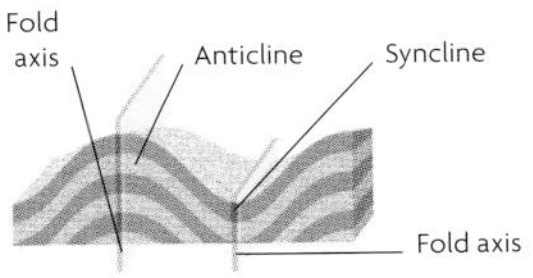

Fold axis
Anticline
Syncline
Fold axis

Folding: *Synclines and anticlines*

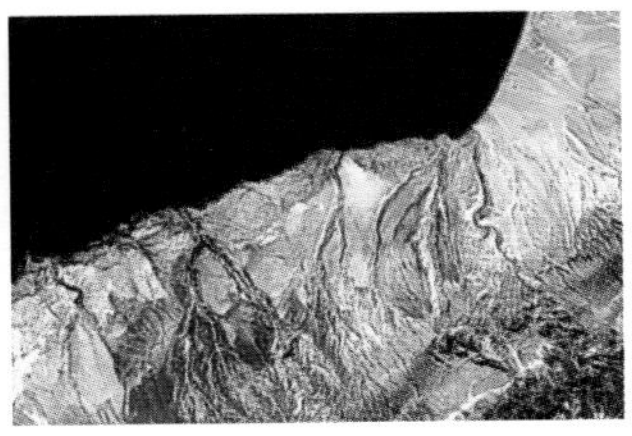

Deposition

4 Large alluvial fans are found extensively across South America *(above)*. Confined mountain rivers, carrying large quantities of eroded material, emerge from a mountain gorge onto the plains, where they deposit their load in huge fans.

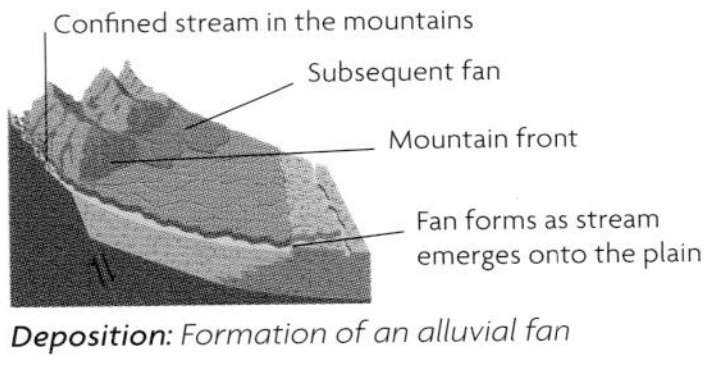

Confined stream in the mountains
Subsequent fan
Mountain front
Fan forms as stream emerges onto the plain

Deposition: *Formation of an alluvial fan*

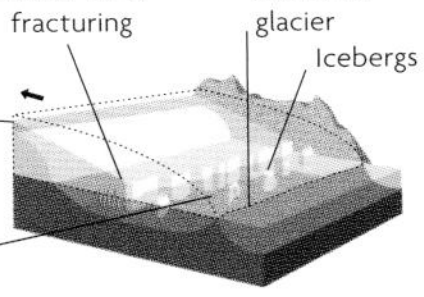

Unstable front in deep water, where ice is fracturing
Original extent of glacier
Icebergs
Stable front
Glacier was grounded against a shoal

Glaciation: *Retreating glacier in Patagonia*

Glaciation

3 As fjord glaciers in Patagonia *(above)* retreat, they become grounded on shoals. In deeper water the base of the glacier becomes unstable, and icebergs break off (calve) until the glacier snout grounds once more.

Political South America

Modern South America's political boundaries have their origins in the territorial endeavours of explorers during the 16th century, who claimed almost the entire continent for Portugal and Spain. The Portuguese land in the east later evolved into the federal states of Brazil, while the Spanish vice-royalties eventually emerged as separate independent nation-states in the early 19th century. South America's growing population has become increasingly urbanized, with the expansion of coastal cities into large conurbations like Rio de Janeiro and Buenos Aires. In Brazil, Argentina, Chile and Uruguay, a succession of military dictatorships has given way to fragile, but strengthening, democracies.

◀ ***Europe retains a*** *small foothold in South America. Kourou in French Guiana was the site chosen by the European Space Agency to launch the Ariane rocket. As a result of its status as a French overseas department, French Guiana is actually part of the European Union.*

Scale 1:24,000,000

Km 0 100 200 300 400 500 600 700 800

0 100 200 300 400 500 600 700 800 Miles

projection: Lambert Azimuthal Equal Area

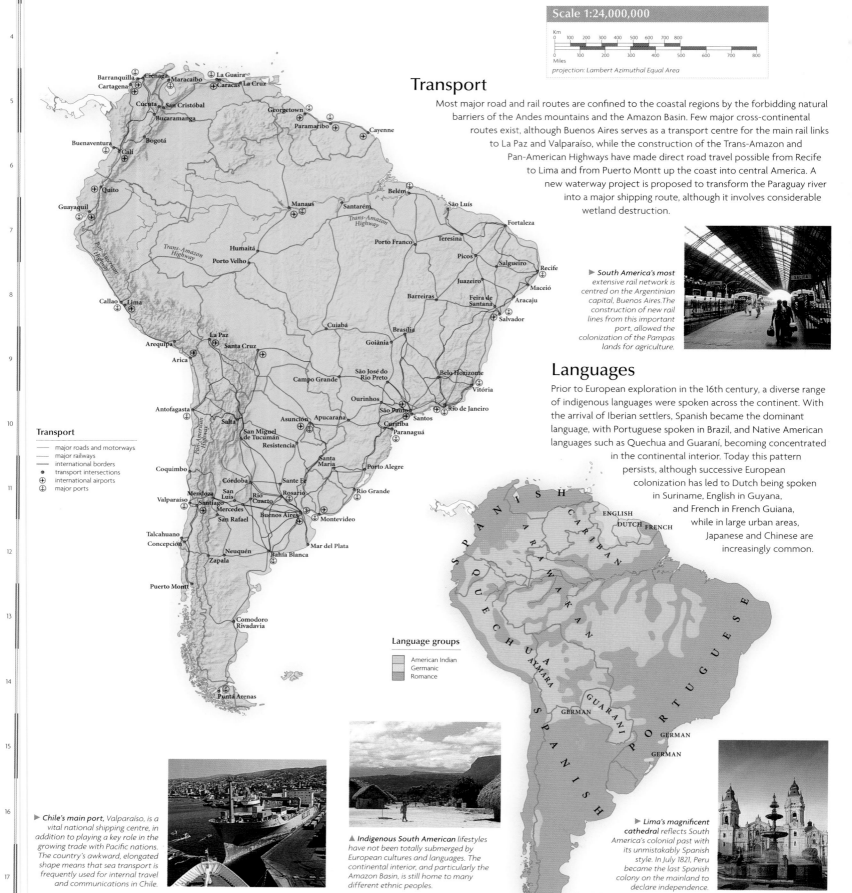

Barranquilla
Cartagena
Ciénaga
Maracaibo
La Guaira
Caracas
La Cruz
Cúcuta
San Cristóbal
Bucaramanga
Georgetown
Paramaribo
Cayenne
Bogotá
Buenaventura
Cali
Quito
Guayaquil
Belém
Manaus
Santarém
São Luís
Trans-Amazon Highway
Fortaleza
Teresina
Porto Franco
Humaitá
Porto Velho
Trans-Amazon Highway
Pan-American Highway
Picos
Salgueiro
Recife
Juazeiro
Maceió
Barreiras
Feira de Santana
Aracaju
Salvador
Callao
Lima
Cuiabá
Brasília
La Paz
Goiânia
Arequipa
Santa Cruz
Arica
Belo Horizonte
São José do Rio Preto
Campo Grande
Vitória
Ourinhos
São Paulo
Rio de Janeiro
Santos
Antofagasta
Asunción
Apucarana
Curitiba
Salta
Paranaguá
San Miguel de Tucumán
Resistencia
Pan-American Highway
Santa Maria
Coquimbo
Porto Alegre
Córdoba
Sante Fe
Rosario
Rio Grande
Mendoza
San Luis
Rio Cuarto
Valparaíso
Santiago
Mercedes
Buenos Aires
Montevideo
San Rafael
Talcahuano
Concepción
Mar del Plata
Neuquén
Bahía Blanca
Zapala
Puerto Montt
Comodoro Rivadavia
Punta Arenas
Transport
major roads and motorways
major railways
international borders
transport intersections
international airports
major ports

Transport

Most major road and rail routes are confined to the coastal regions by the forbidding natural barriers of the Andes mountains and the Amazon Basin. Few major cross-continental routes exist, although Buenos Aires serves as a transport centre for the main rail links to La Paz and Valparaíso, while the construction of the Trans-Amazon and Pan-American Highways have made direct road travel possible from Recife to Lima and from Puerto Montt up the coast into central America. A new waterway project is proposed to transform the Paraguay river into a major shipping route, although it involves considerable wetland destruction.

▶ ***South America's most*** *extensive rail network is centred on the Argentinian capital, Buenos Aires. The construction of new rail lines from this important port, allowed the colonization of the Pampas lands for agriculture.*

Languages

Prior to European exploration in the 16th century, a diverse range of indigenous languages were spoken across the continent. With the arrival of Iberian settlers, Spanish became the dominant language, with Portuguese spoken in Brazil, and Native American languages such as Quechua and Guaraní, becoming concentrated in the continental interior. Today this pattern persists, although successive European colonization has led to Dutch being spoken in Suriname, English in Guyana, and French in French Guiana, while in large urban areas, Japanese and Chinese are increasingly common.

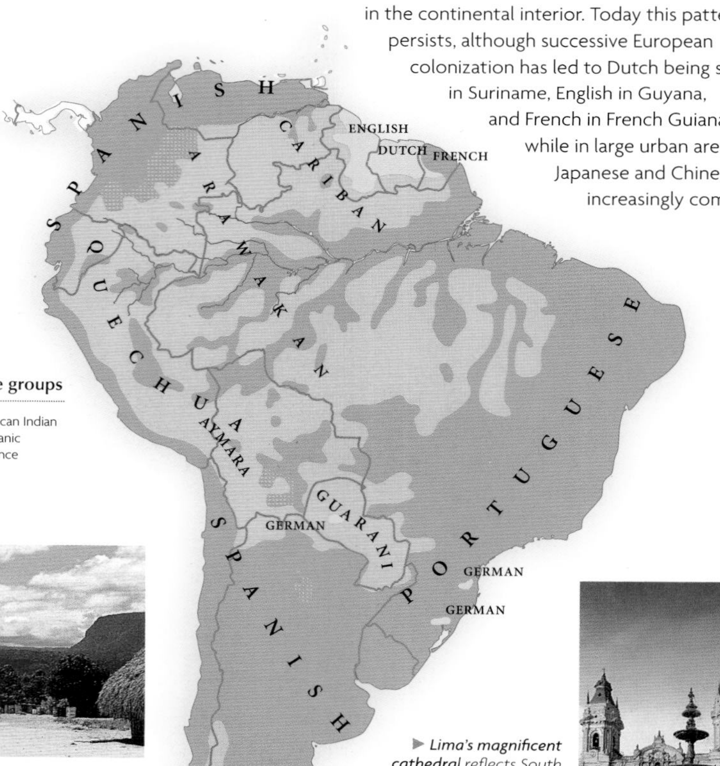
SPANISH
ENGLISH
DUTCH
FRENCH
CARIBAN
ARAWAKAN
QUECHUA
AYMARA
GUARANI
GERMAN
PORTUGUESE
GERMAN
GERMAN
SPANISH
Language groups
American Indian
Germanic
Romance

▶ ***Chile's main port,*** *Valparaíso, is a vital national shipping centre, in addition to playing a key role in the growing trade with Pacific nations. The country's awkward, elongated shape means that sea transport is frequently used for internal travel and communications in Chile.*

▲ ***Indigenous South American*** *lifestyles have not been totally submerged by European cultures and languages. The continental interior, and particularly the Amazon Basin, is still home to many different ethnic peoples.*

▶ ***Lima's magnificent cathedral*** *reflects South America's colonial past with its unmistakably Spanish style. In July 1821, Peru became the last Spanish colony on the mainland to declare independence.*

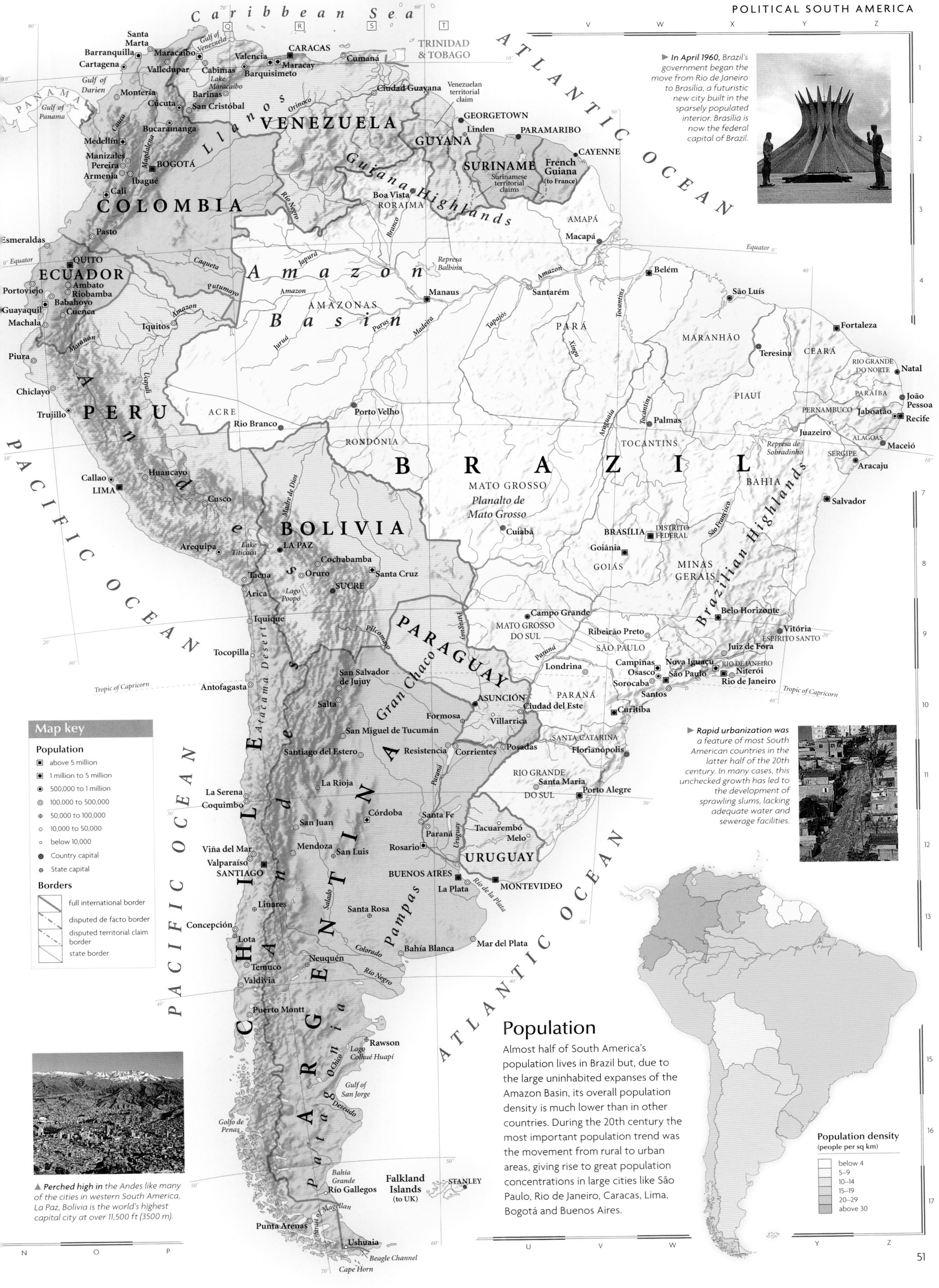
Caribbean Sea
ATLANTIC OCEAN
PACIFIC OCEAN
COLOMBIA
VENEZUELA
GUYANA
SURINAME
French Guiana (to France)
ECUADOR
PERU
BRAZIL
BOLIVIA
PARAGUAY
CHILE
ARGENTINA
URUGUAY
TRINIDAD & TOBAGO
PANAMA
Falkland Islands (to UK)
STANLEY
CARACAS
BOGOTÁ
QUITO
LIMA
LA PAZ
SUCRE
ASUNCIÓN
SANTIAGO
BUENOS AIRES
MONTEVIDEO
GEORGETOWN
PARAMARIBO
CAYENNE
BRASÍLIA
Santa Marta
Barranquilla
Cartagena
Maracaibo
Valledupar
Cabimas
Valencia
Maracay
Barquisimeto
Cumaná
Ciudad Guayana
Venezuelan territorial claim
Surinamese territorial claims
Montería
Cúcuta
San Cristóbal
Barinas
Bucaramanga
Medellín
Manizales
Pereira
Armenia
Ibagué
Cali
Pasto
Esmeraldas
Portoviejo
Ambato
Riobamba
Babahoyo
Guayaquil
Cuenca
Machala
Iquitos
Piura
Chiclayo
Trujillo
Callao
Huancayo
Cusco
Arequipa
Tacna
Arica
Iquique
Tocopilla
Antofagasta
Cochabamba
Oruro
Santa Cruz
Linden
Boa Vista
RORAIMA
AMAPÁ
Macapá
Belém
Santarém
Manaus
São Luís
Fortaleza
Teresina
Natal
João Pessoa
Jaboatão
Recife
Maceió
Aracaju
Salvador
Juazeiro
Palmas
Porto Velho
Rio Branco
ACRE
AMAZONAS
RONDÔNIA
PARÁ
MARANHÃO
CEARÁ
PIAUÍ
RIO GRANDE DO NORTE
PARAÍBA
PERNAMBUCO
ALAGOAS
SERGIPE
BAHIA
TOCANTINS
MATO GROSSO
Planalto de Mato Grosso
Cuiabá
DISTRITO FEDERAL
Goiânia
GOIÁS
MINAS GERAIS
Belo Horizonte
Campo Grande
MATO GROSSO DO SUL
Ribeirão Preto
Vitória
ESPÍRITO SANTO
SÃO PAULO
Juiz de Fora
Londrina
Campinas
Osasco
Nova Iguaçu
RIO DE JANEIRO
Niterói
São Paulo
Sorocaba
Rio de Janeiro
Santos
PARANÁ
Curitiba
Ciudad del Este
Villarrica
SANTA CATARINA
Florianópolis
Posadas
RIO GRANDE DO SUL
Santa Maria
Porto Alegre
Tacuarembó
Melo
San Salvador de Jujuy
Salta
Formosa
San Miguel de Tucumán
Santiago del Estero
Resistencia
Corrientes
La Rioja
La Serena
Coquimbo
San Juan
Córdoba
Santa Fe
Paraná
Mendoza
San Luis
Rosario
Viña del Mar
Valparaíso
La Plata
Linares
Concepción
Lota
Santa Rosa
Bahía Blanca
Mar del Plata
Temuco
Neuquén
Valdivia
Puerto Montt
Rawson
Río Gallegos
Punta Arenas
Ushuaia
Cape Horn
Beagle Channel
Strait of Magellan
Gulf of Venezuela
Lake Maracaibo
Gulf of Darien
Gulf of Panama
Llanos
Guiana Highlands
Amazon Basin
Andes
Brazilian Highlands
Gran Chaco
Pampas
Patagonia
Atacama Desert
Lake Titicaca
Lago Poopó
Lago Colhué Huapí
Gulf of San Jorge
Golfo de Penas
Bahía Grande
Represa Balbina
Represa de Sobradinho
Orinoco
Amazon
Caquetá
Putumayo
Japurá
Juruá
Purus
Madeira
Tapajós
Xingu
Tocantins
Araguaia
São Francisco
Rio Negro
Branco
Ucayali
Marañón
Madre de Dios
Paraguay
Pilcomayo
Paraná
Uruguay
Río de la Plata
Salado
Colorado
Río Negro
Chico
Deseado
Cauca
Magdalena
Equator
Tropic of Capricorn
Map key
Population
above 5 million
1 million to 5 million
500,000 to 1 million
100,000 to 500,000
50,000 to 100,000
10,000 to 50,000
below 10,000
Country capital
State capital
Borders
full international border
disputed de facto border
disputed territorial claim border
state border
Population density
(people per sq km)
below 4
5–9
10–14
15–19
20–29
above 30

▶ *In April 1960, Brazil's government began the move from Rio de Janeiro to Brasília, a futuristic new city built in the sparsely populated interior. Brasília is now the federal capital of Brazil.*

▶ *Rapid urbanization was a feature of most South American countries in the latter half of the 20th century. In many cases, this unchecked growth has led to the development of sprawling slums, lacking adequate water and sewerage facilities.*

▲ *Perched high in the Andes like many of the cities in western South America, La Paz, Bolivia is the world's highest capital city at over 11,500 ft (3500 m).*

Population

Almost half of South America's population lives in Brazil but, due to the large uninhabited expanses of the Amazon Basin, its overall population density is much lower than in other countries. During the 20th century the most important population trend was the movement from rural to urban areas, giving rise to great population concentrations in large cities like São Paulo, Rio de Janeiro, Caracas, Lima, Bogotá and Buenos Aires.

South American resources

Agriculture still provides the largest single form of employment in South America, although rural unemployment and poverty continue to drive people towards the huge coastal cities in search of jobs and opportunities. Mineral and fuel resources, although substantial, are distributed unevenly; few countries have both fossil fuels and minerals. To break industrial dependence on raw materials, boost manufacturing, and improve infrastructure, governments borrowed heavily from the World Bank in the 1960s and 1970s. This led to the accumulation of massive debts which are unlikely ever to be repaid. Today, Brazil dominates the continent's economic output, followed by Argentina. Recently, the less-developed western side of South America has benefited due to its geographical position; for example Chile is increasingly exporting raw materials to Japan.

◀ *Ciudad Guayana is a planned industrial complex in eastern Venezuela, built as an iron and steel centre to exploit the nearby iron ore reserves.*

▲ *The cold Peru Current flows north from the Antarctic along the Pacific coast of Peru, providing rich nutrients for one of the world's largest fishing grounds. However, over-exploitation has severely reduced Peru's anchovy catch.*

Industry
aerospace
brewing
car/vehicle manufacture
chemicals
electronics
engineering
finance
fish processing
food processing
hi-tech industry
iron & steel
meat processing
metal refining
narcotics
pharmaceuticals
printing & publishing
shipbuilding
sugar processing
textiles
timber processing
tobacco processing
wine
oil
gas
industrial cities
major industrial areas
Caribbean Sea
Barranquilla
Cartagena
Maracaibo
Barquisimeto
Caracas
Valencia
Ciudad Guayana
PANAMA
Gulf of Panama
VENEZUELA
Medellín
Bogotá
Cali
COLOMBIA
Georgetown
GUYANA
Paramaribo
SURINAME
French Guiana (to France)
ATLANTIC OCEAN
Quito
ECUADOR
Guayaquil
Iquitos
Amazon Basin
Manaus
Belém
Fortaleza
Natal
Recife
Maceió
Salvador
BRAZIL
Chiclayo
Chimbote
PERU
Lima
Cusco
Arequipa
PACIFIC OCEAN
BOLIVIA
La Paz
Santa Cruz
Sucre
Brasília
Arica
Iquique
Chuquicamata
Antofagasta
Belo Horizonte
PARAGUAY
São Paulo
Rio de Janeiro
Asunción
Ciudad del Este
Curitiba
San Miguel de Tucumán
Corrientes
CHILE
Porto Alegre
Córdoba
Santa Fe
Rosario
URUGUAY
Rio Grande
Valparaíso
Mendoza
Santiago
Buenos Aires
Montevideo
Talca
ARGENTINA
Concepción
Bahía Blanca
Neuquén
Valdivia
Comodoro Rivadavia
Gulf of San Jorge
Bahía Grande
Falkland Islands (to UK)
Punta Arenas
Strait of Magellan
Cape Horn
GNI per capita (US$)
below 999
1000–1999
2000–2999
3000–3999
4000–4999
above 5000

Standard of living

Wealth disparities throughout the continent create a wide gulf between affluent landowners and those afflicted by chronic poverty in inner-city slums. The illicit production of cocaine, and the hugely influential drug barons who control its distribution, contribute to the violent disorder and corruption which affect northwestern South America, de-stabilizing local governments and economies.

Standard of living
(UN human development index)
low
high

▶ *Both Argentina and Chile are now exploring the southernmost tip of the continent in search of oil. Here in Punta Arenas, a drilling rig is being prepared for exploratory drilling in the Strait of Magellan.*

Industry

Argentina and Brazil are South America's most industrialized countries and São Paulo is the continent's leading industrial centre. Long-term government investment in Brazilian industry has encouraged a diverse industrial base; engineering, steel production, food processing, textile manufacture and chemicals predominate. The illegal production of cocaine is economically significant in the Andean countries of Colombia and Bolivia. In Venezuela, the oil-dominated economy has left the country vulnerable to world oil price fluctuations. Food processing and mineral exploitation are common throughout the less industrially developed parts of the continent, including Bolivia, Chile, Ecuador and Peru.

Environmental issues

The Amazon Basin is one of the last great wilderness areas left on Earth. The tropical rainforests which grow there are a valuable genetic resource, containing innumerable unique plants and animals. The forests are increasingly under threat from new and expanding settlements and 'slash and burn' farming techniques, which clear land for the raising of beef cattle, causing land degradation and soil erosion.

▲ *Clouds of smoke billow from the burning Amazon rainforest. Over 11,500 sq miles (30,000 sq km) of virgin rainforest are being cleared annually, destroying an ancient, irreplaceable, natural resource and biodiverse habitat.*

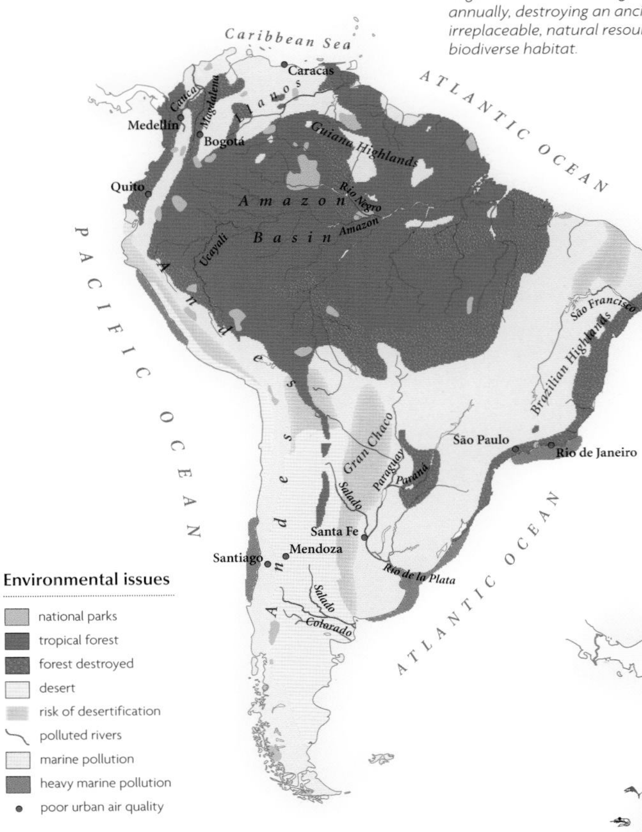
Caribbean Sea
Caracas
Llanos
Medellín
Bogotá
Guiana Highlands
Quito
Amazon Basin
Rio Negro
Amazon
Ucayali
ATLANTIC OCEAN
PACIFIC OCEAN
Andes
São Francisco
Brazilian Highlands
Gran Chaco
Paraguay
Paraná
Salado
São Paulo
Rio de Janeiro
Santa Fe
Mendoza
Santiago
Río de la Plata
Colorado
Environmental issues
national parks
tropical forest
forest destroyed
desert
risk of desertification
polluted rivers
marine pollution
heavy marine pollution
poor urban air quality

Mineral resources

Over a quarter of the world's known copper reserves are found at the Chuquicamata mine in northern Chile, and other metallic minerals such as tin are found along the length of the Andes. The discovery of oil and gas at Venezuela's Lake Maracaibo in 1917 turned the country into one of the world's leading oil producers. In contrast, South America is virtually devoid of coal, the only significant deposit being on the peninsula of Guajira in Colombia.

◀ *Copper is Chile's largest export, most of which is mined at Chuquicamata. Along the length of the Andes, metallic minerals like copper and tin are found in abundance, formed by the excessive pressures and heat involved in mountain-building.*

Caribbean Sea
Llanos
Guiana Highlands
ATLANTIC OCEAN
Amazon Basin
PACIFIC OCEAN
Andes
Brazilian Highlands
Gran Chaco
Mineral resources
oil field
gas field
coal field
bauxite
copper
diamonds
gold
iron
lead
silver
tin

Using the land and sea

Many foods now common worldwide originated in South America. These include the potato, tomato, squash, and cassava. Today, large herds of beef cattle roam the temperate grasslands of the Pampas, supporting an extensive meat-packing trade in Argentina, Uruguay and Paraguay. Corn (maize) is grown as a staple crop across the continent and coffee is grown as a cash crop in Brazil and Colombia. Coca plants grown in Bolivia, Peru and Colombia provide most of the world's cocaine. Fish and shellfish are caught off the western coast, especially anchovies off Peru, shrimps off Ecuador and pilchards off Chile.

Caribbean Sea
Barranquilla
Maracaibo
Caracas
Gulf of Darien
Gulf of Panama
Llanos
Orinoco
Medellín
Bogotá
Cali
Guiana Highlands
ATLANTIC OCEAN
Rio Negro
Amazon Basin
Amazon
Putumayo
Manaus
Ilha de Marajó
Belém
Gulf of Guayaquil
Marañón
Juruá
Purus
Madeira
Tapajós
Xingu
Tocantins
Fortaleza
Recife
Araguaia
Andes
Lima
Planalto de Mato Grosso
Brasília
São Francisco
Salvador
Brazilian Highlands
PACIFIC OCEAN
Belo Horizonte
Pilcomayo
Gran Chaco
Paraguay
Paraná
Rio de Janeiro
São Paulo
Curitiba
Uruguay
Porto Alegre
Córdoba
Rosario
Santiago
Pampas
Buenos Aires
Montevideo
Colorado
Río Negro
Chico
Patagonia
Gulf of San Jorge
Bahía Grande
Tierra del Fuego
Cape Horn
Using the land and sea
barren land
cropland
desert
forest
mountain region
pasture
major conurbations
cattle
pigs
sheep
bananas
corn (maize)
citrus fruits
cocoa
cotton
coffee
fishing
oil palms
peanuts
rubber
shellfish
soya beans
sugar cane
vineyards
wheat

◀ *South America, and Brazil in particular, now leads the world in coffee production, mainly growing Coffea arabica in large plantations. Coffee beans are harvested, roasted and brewed to produce the world's second most popular drink, after tea.*

◀ *High in the Andes, hardy alpacas graze on the barren land. Alpacas are thought to have been domesticated by the Incas, whose nobility wore robes made from their wool. Today, they are still reared and prized for their soft, warm fleeces.*

◀ *The Pampas region of southeast South America is characterized by extensive, flat plains, and populated by cattle and ranchers (gauchos). Argentina is a major world producer of beef, much of which is exported to the USA for use in hamburgers.*

Northern South America

COLOMBIA, GUYANA, SURINAME, VENEZUELA, French Guiana (to France)

Fringed by the Pacific and Atlantic oceans and the Caribbean Sea, South America's northern region has a rich range of natural resources, some exploited for centuries by colonial powers including the Spanish, French, Dutch and British, others still to be fully explored. The prospects for further economic development in Colombia, Guyana and Suriname are blighted by drug-related violence and political instability. Venezuela, despite huge incomes from its oil reserves, remains less developed in other industrial sectors. French Guiana is an overseas *département* of France, now seeking greater autonomy. Most of the major population centres, such as Bogotá, have grown up in the temperate conditions of the high Andes or, like Caracas, at strategic points along the Caribbean coast.

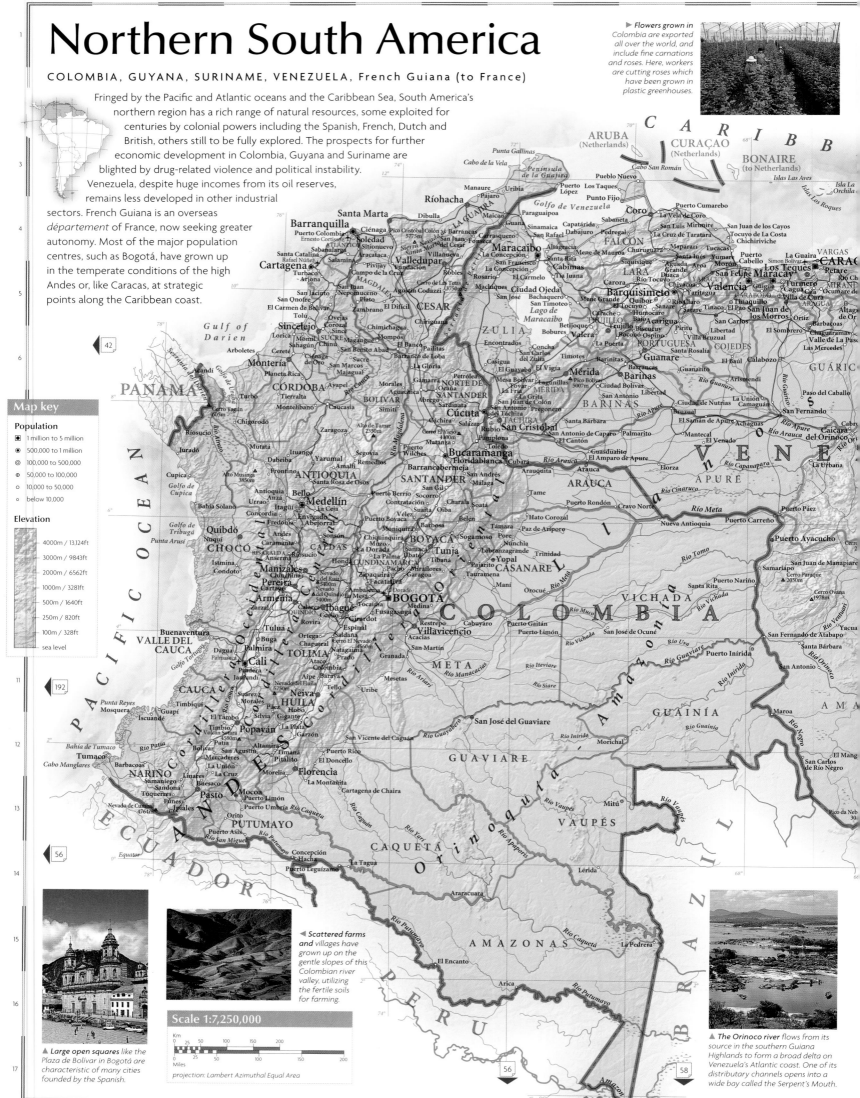

▶ Flowers grown in Colombia are exported all over the world, and include fine carnations and roses. Here, workers are cutting roses which have been grown in plastic greenhouses.

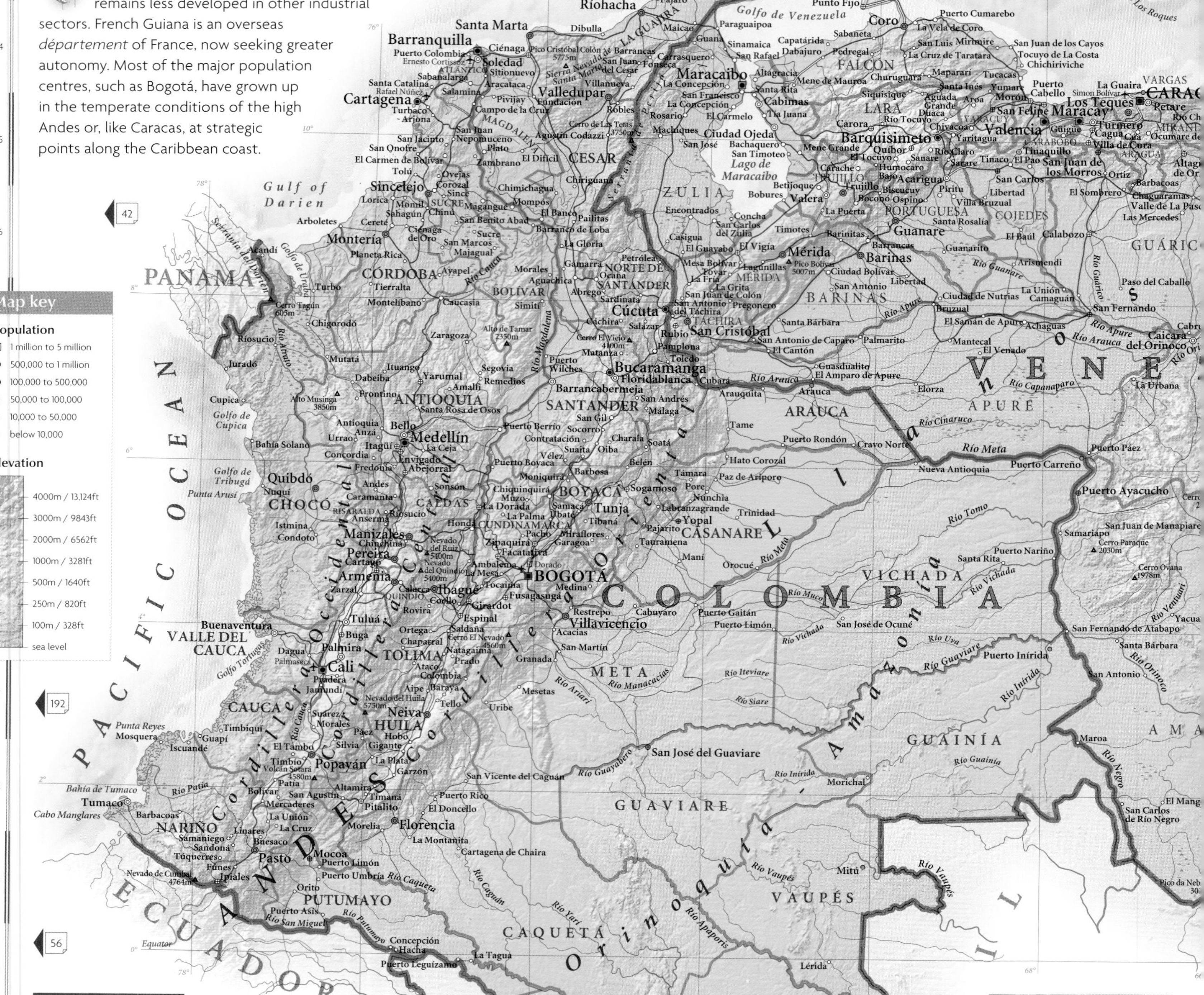

Map key
Population
1 million to 5 million
500,000 to 1 million
100,000 to 500,000
50,000 to 100,000
10,000 to 50,000
below 10,000
Elevation
4000m / 13,124ft
3000m / 9843ft
2000m / 6562ft
1000m / 3281ft
500m / 1640ft
250m / 820ft
100m / 328ft
sea level
PACIFIC OCEAN
COLOMBIA
PANAMA
ECUADOR
PERU
BRAZIL
VENEZUELA
ARUBA (Netherlands)
CURAÇAO (Netherlands)
BONAIRE (to Netherlands)
ANDES
Gulf of Darien
Golfo de Venezuela
Lago de Maracaibo
BOGOTÁ
Medellín
Cali
Barranquilla
Cartagena
Cúcuta
Bucaramanga
Maracaibo
Barquisimeto
Valencia
Maracay
Mérida
Barinas
Leticia
Llanos
Orinoquía
Amazonía
Río Meta
Río Caquetá
Río Putumayo
Río Orinoco
Amazon
Equator
Scale 1:7,250,000
Km
Miles
projection: Lambert Azimuthal Equal Area

42
192
56
56
58

▲ Large open squares like the Plaza de Bolívar in Bogotá are characteristic of many cities founded by the Spanish.

◀ Scattered farms and villages have grown up on the gentle slopes of this Colombian river valley, utilizing the fertile soils for farming.

▲ The Orinoco river flows from its source in the southern Guiana Highlands to form a broad delta on Venezuela's Atlantic coast. One of its distributary channels opens into a wide bay called the Serpent's Mouth.

Transport and industry

Many mineral resources are mined in Colombia, including fuels, gold and precious and semi-precious stones. Revenues from coffee and exports of illegal narcotics are crucial to the economy. Venezuela's major economic activity is the oil industry around Lake Maracaibo *(Lago de Maracaibo)*. Sugar and bauxite are exported from Guyana and Suriname.

The transport network

	31,720 miles (51,054 km)
	3411 miles (5490 km)
	2448 miles (3940 km)
	22,429 miles (36,100 km)

Rivers are an important means of transport in Colombia; many are extensively navigable. The Pan-American Highway runs through Colombia. In Venezuela, much infrastructure investment is linked to the oil industry.

Major industry and infrastructure

- chemicals
- finance
- food processing
- iron & steel
- narcotics
- mining
- oil
- oil refining
- pharmaceuticals
- textiles
- timber processing
- capital cities
- major towns
- international airports
- major roads
- major industrial areas

▲ *Vast oil reserves around Lake Maracaibo* (Lago de Maracaibo) *form the focus of Venezuelan industry. Incomes from oil are used to invest in other industries and in the development of infrastructure.*

Using the land

The Andean basins support cereals and potatoes. Livestock graze at higher altitudes and on the drier tropical grasslands known as the *llanos*; hardy goats are reared in scrubland areas. Grown at higher elevations, coffee is an important cash crop, as is cotton, sugar cane, bananas, citrus fruits, cocoa and rice, farmed on the Caribbean lowlands. Coca is the most widely-grown narcotic plant, with heroin poppies grown in Colombia and marijuana in lowland areas throughout the region.

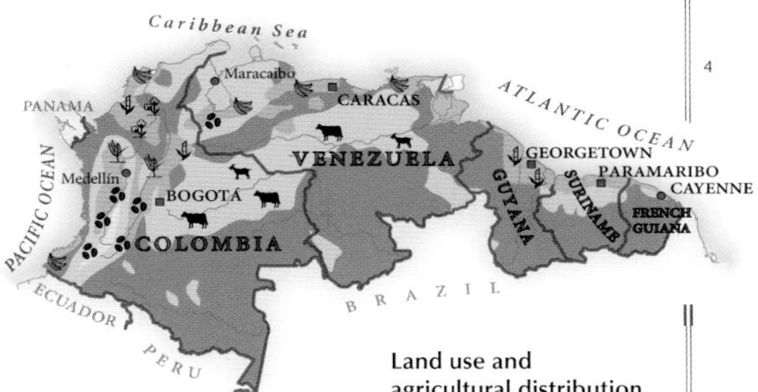

Land use and agricultural distribution

- cattle
- goats
- bananas
- cereals
- coffee
- cotton
- sugar cane
- capital cities
- major towns
- pasture
- cropland
- forest
- wetlands
- mountain region

The urban/rural population divide

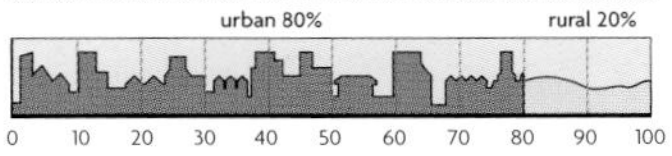

Population density	Total land area
78 people per sq mile (30 people per sq km)	1,111,317 sq miles (2,879,060 sq km)

64

58

▲ *The Sierra Nevada de Santa Marta is a granite massif which rises sharply from the Caribbean lowlands to snow-covered peaks, the tallest of which is 18,947 ft (5775 m) high.*

Guiana Shield

▲ *The Guiana Shield is one of the oldest land surfaces in the world – probably formed more than 4 billion years ago. Chemical weathering over millions of years has created flat-topped table mountains and large numbers of inselbergs.*

The landscape

At its northernmost reaches, in western Colombia and Venezuela, the great Andean mountain chain splits into three distinct ranges: the Cordillera Oriental, Cordillera Central and Cordillera Occidental, intercut by a complex series of lesser ranges and basins. The relief becomes lower toward the coast and the interior plains of the northern Amazon Basin, rising again into the tropical hills of the Guiana Highlands.

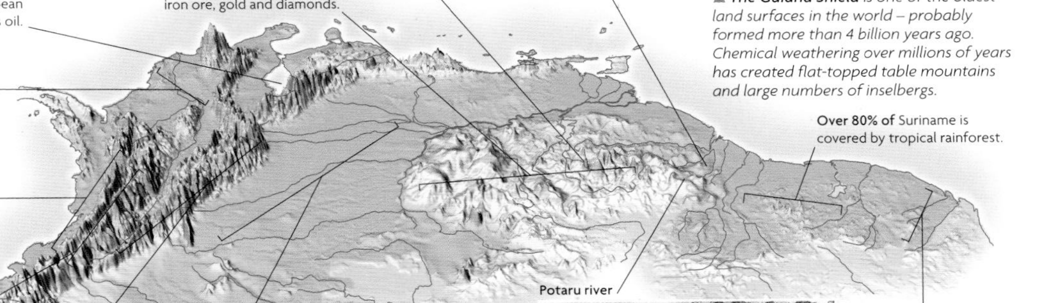

► *The Potaru river descends 741 ft (226 m) over a sandstone ledge at the Kaieteur Falls in Guyana.*

Western South America

BOLIVIA, ECUADOR, PERU

The three states of Western South America share a similar geography and recent history. Dominated by the Inca empire until Spanish conquest in the 16th century, they achieved independence from Spain in the early 19th century. The precipitous terrain of the Andes presents severe difficulties for overland transport and continues to be a barrier to national unity and stability. Although Ecuador is now a relatively stable democracy, the military is highly influential in Peru and Bolivia, while the drug trade and associated corruption discourages external aid and economic progress. Wealth and power are still largely concentrated in the hands of a small elite of families, who attained their position during the Spanish colonial period. Energy resources and political recognition for the indigenous peoples are becoming increasingly important issues, particularly in Bolivia.

▶ *Ecuador's capital city, Quito, lies high in the Andes, nestling between snow-capped peaks. At 9350 ft (2850 m), Quito is the second highest capital in the world – La Paz in Bolivia is the highest.*

The landscape

Bolivia, Peru and Ecuador each possess a high Andean mountain region and an eastern region consisting of tropical lowlands and the Andean slope leading down to them. Towards the south of the region, the mountains widen to form the high plateau of the Altiplano. Peru and Ecuador also have fertile, lowland coastal plains. A wide variety of environments include *selva* (tropical rainforest), *montaña* (mountain forest) and grassland.

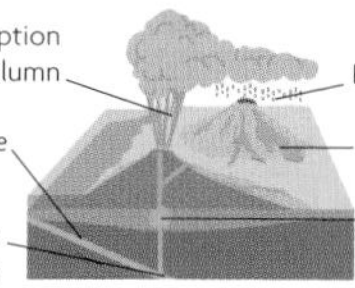

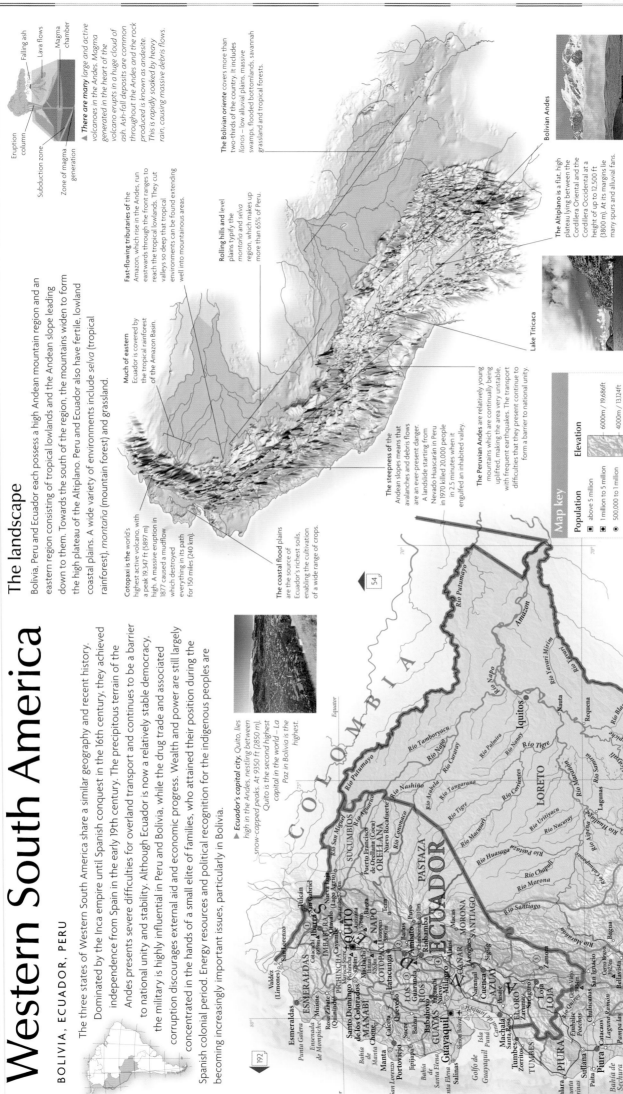

▲ ***There are many** large and active volcanoes in the Andes. Magma generated in the heart of the volcano erupts in a huge cloud of ash. Ash-fall deposits are common throughout the Andes and the rock produced is known as andesite. This is rapidly soaked by heavy rain, causing massive debris flows.*

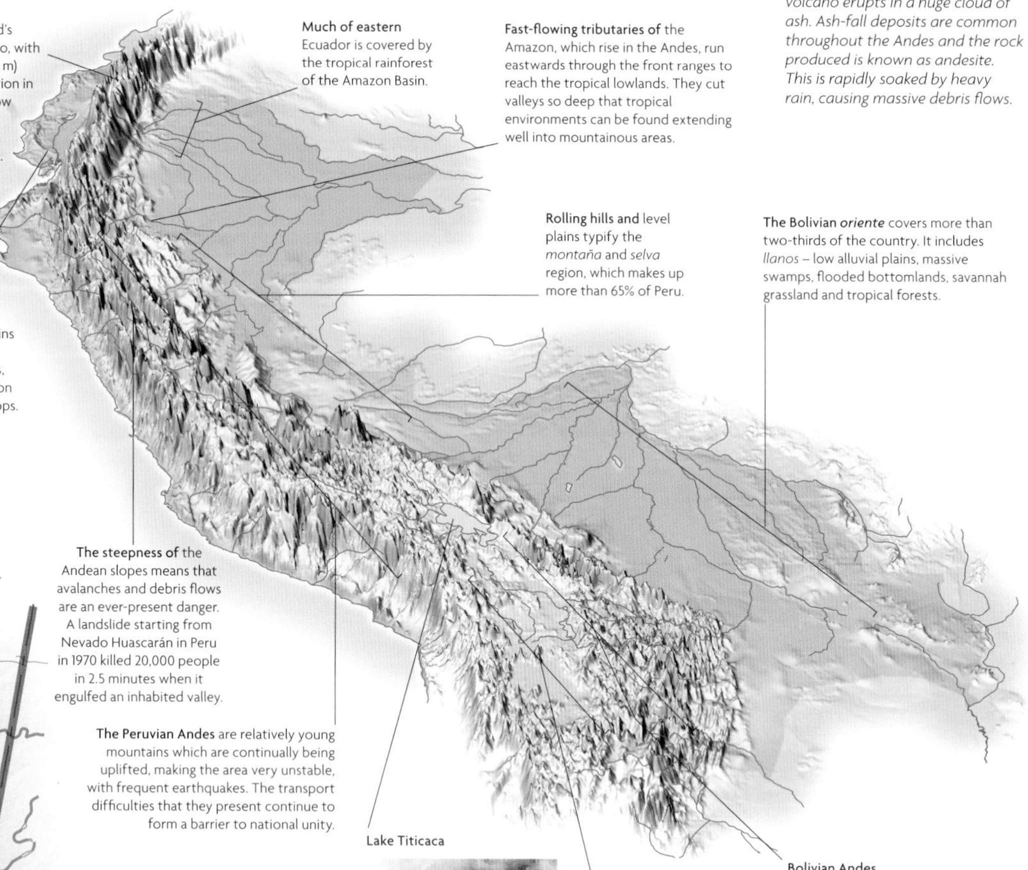

Cotopaxi is the world's highest active volcano, with a peak 19,347 ft (5897 m) high. A massive eruption in 1877 caused a mudflow which destroyed everything in its path for 150 miles (240 km).

Much of eastern Ecuador is covered by the tropical rainforest of the Amazon Basin.

Fast-flowing tributaries of the Amazon, which rise in the Andes, run eastwards through the front ranges to reach the tropical lowlands. They cut valleys so deep that tropical environments can be found extending well into mountainous areas.

Rolling hills and level plains typify the *montaña* and *selva* region, which makes up more than 65% of Peru.

The Bolivian *oriente* covers more than two-thirds of the country. It includes *llanos* – low alluvial plains, massive swamps, flooded bottomlands, savannah grassland and tropical forests.

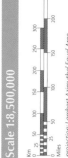

The coastal flood plains are the source of Ecuador's richest soils, enabling the cultivation of a wide range of crops.

The steepness of the Andean slopes means that avalanches and debris flows are an ever-present danger. A landslide starting from Nevado Huascarán in Peru in 1970 killed 20,000 people in 2.5 minutes when it engulfed an inhabited valley.

The Peruvian Andes are relatively young mountains which are continually being uplifted, making the area very unstable, with frequent earthquakes. The transport difficulties that they present continue to form a barrier to national unity.

Lake Titicaca

▲ ***Lake Titicaca, which** forms part of the border between Peru and Bolivia, is the largest lake in South America and the highest significant body of water in the world at an altitude of 12,507 ft (3812 m).*

The Altiplano is a flat, high plateau lying between the Cordillera Oriental and the Cordillera Occidental at a height of up to 12,500 ft (3800 m). At its margins lie many spurs and alluvial fans.

Bolivian Andes

▲ ***Nevado de Illampu** and Nevado de Ancohuma, at 21,275 ft (6485 m) and 21,490 ft (6550 m) respectively, form Illampu, the highest mountain in the Bolivian Andes.*

Map key

Population	
▣	above 5 million
◉	1 million to 5 million
⊙	500,000 to 1 million
◎	100,000 to 500,000
⊕	50,000 to 100,000
○	10,000 to 50,000
○	below 10,000

Elevation
6000m / 19,686ft
4000m / 13,124ft
3000m / 9843ft
2000m / 6562ft
1000m / 3281ft
500m / 1640ft
250m / 820ft
100m / 328ft
sea level

Ecuador: Administrative regions

1. CARCHI
2. TUNGURAHUA
3. BOLÍVAR
4. CHIMBORAZO
5. ZAMORA CHINCHIPE

Scale 1:8,500,000

projection: Lambert Azimuthal Equal Area

▲ **A colony of** marine iguanas basks on the rocks of Isla Fernandina in the Galápagos Islands. Charles Darwin's theory of evolution was inspired by the differences he found between the animal species on neighbouring islands in the Galápagos.

◀ **The Galápagos Islands** are mainly composed of lava, with very little vegetation near to the coasts, although the wetter inland slopes are mantled with forest.

192
58
62

▶ **Llamas, with alpacas** and vicuñas, are indigenous to South America. They thrive in Andean conditions and their wool is both exported and used in the manufacture of local textiles.

Transport and industry

The mountain regions are rich in minerals including lead, copper, silver, gold, zinc and tungsten, though high production and transport costs have meant that they are expensive to extract and vulnerable to price collapses. Foreign debt remains a major burden, hampering industrial development. Manufacturing tends to be small-scale and concentrates on products for local needs, including textiles, food processing and pharmaceuticals. Narcotics are an important, though illegal, export.

▲ **The ancient city** of Machu Picchu, in the Peruvian Andes was built prior to the Inca period. Its impressive ruins reflect a culture which had developed a high degree of sophistication.

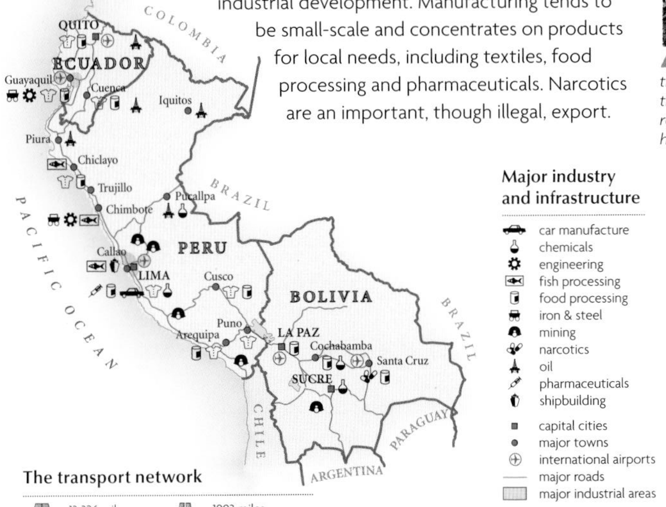

A trans-continental highway is under construction to link Ilo, on Peru's Pacific coast, to Porto Esperança in Brazil, via Puerto Suárez in Bolivia. Establishing port facilities on the Pacific coast is crucial to landlocked Bolivia's further development.

◀ **At Potosí in** Bolivia, silver has been mined for over 400 years.

Bolivia: Capital cities

LA PAZ – seat of government
SUCRE – legal capital

Using the land and sea

The coastal regions support a variety of cash crops including rice, sugar cane, bananas, coffee and cocoa, watered by rainfall or by irrigation schemes. The grasslands of the high *sierra* are used mainly for grazing a wide range of livestock; cattle and sheep are reared, along with pigs, and the indigenous llama and alpaca. Subsistence crops, especially potatoes and cereals, are grown lower down the mountain flanks. Despite government incentives to grow alternative crops, coca, used for cocaine, is the Bolivian and Peruvian *oriente*'s most profitable commercial crop.

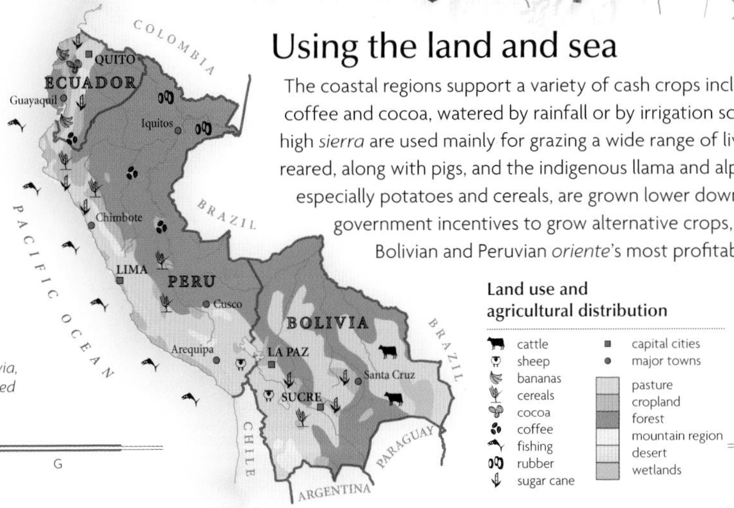

The urban/rural population divide

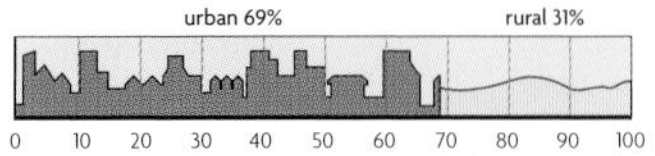

Population density	Total land area
48 people per sq mile (19 people per sq km)	1,019,515 sq miles (2,641,230 sq km)

▶ **Clearance of the** forest in coca-growing regions is encouraged by the Bolivian government. The inaccessible terrain makes policing the growers very difficult. Coca is a popular crop because it is simple to grow and to transport, and is very profitable when illegally processed as cocaine.

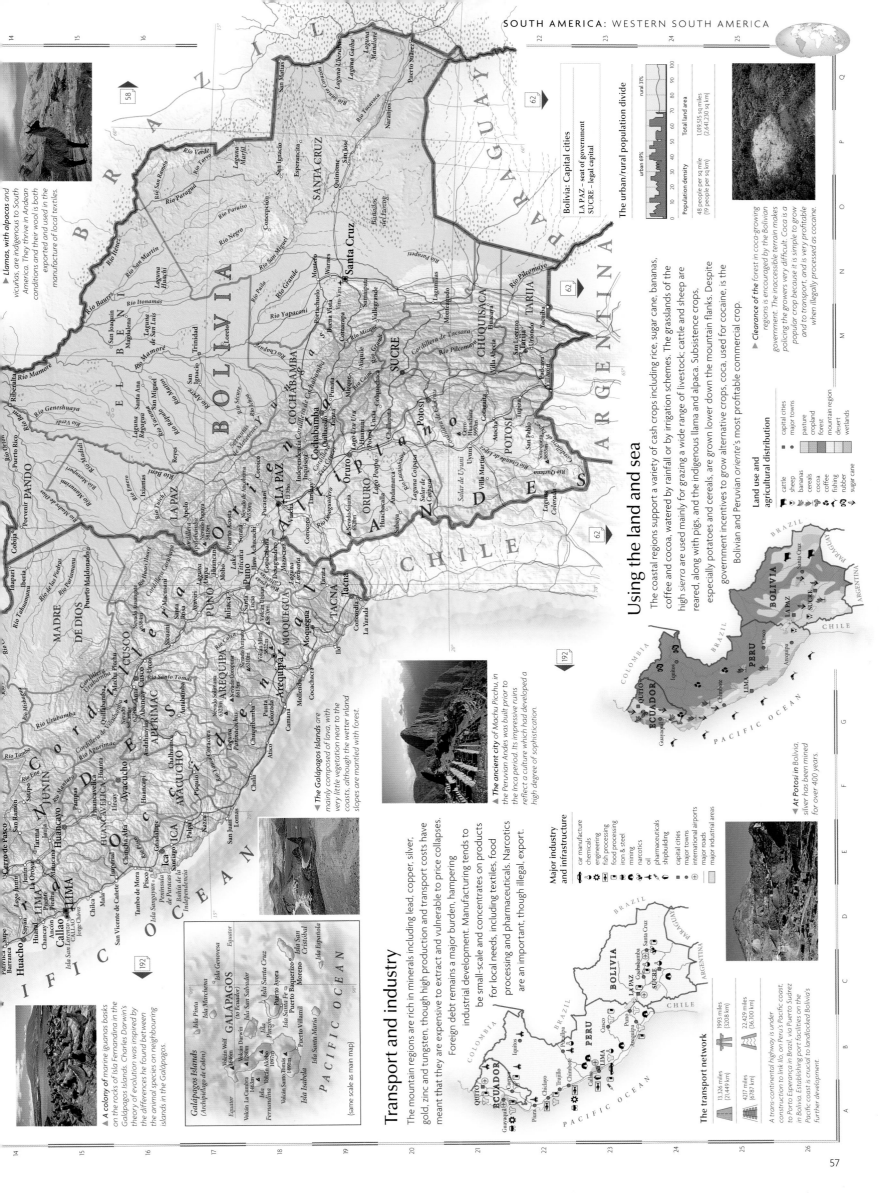

Brazil

Brazil is the largest country in South America, with a population of 191 million – almost half the combined total of the continent. The 26 states which make up the federal republic of Brazil are administered from the purpose-built capital, Brasília. Tropical rainforest, covering more than one-third of the country, contains rich natural resources, but great tracts are sacrificed to agriculture, industry and urban expansion on a daily basis. Most of Brazil's multi-ethnic population now live in cities, some of which are vast areas of urban sprawl; São Paulo is one of the world's biggest conurbations, with more than 20 million inhabitants. Although prosperity is a reality for some, many people still live in great poverty, and mounting foreign debts continue to damage Brazil's prospects of economic advancement.

Using the land

Brazil has immense natural resources, including minerals and hardwoods, many of which are found in the fragile rainforest. Brazil is the world's leading coffee grower and a major producer of livestock, sugar and orange juice concentrate. Soya beans for animal feed, particularly for poultry feed, have become the country's most significant crop.

The urban/rural population divide

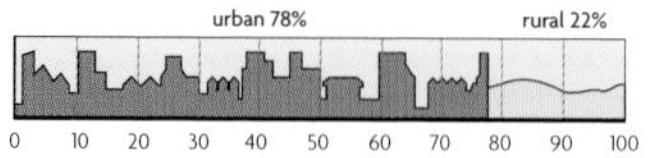

Population density	Total land area
55 people per sq mile (21 people per sq km)	3,286,472 sq miles (8,511,970 sq km)

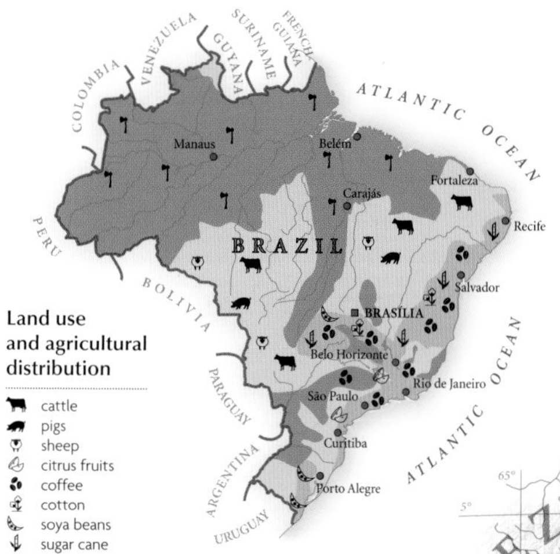

Land use and agricultural distribution

- cattle
- pigs
- sheep
- citrus fruits
- coffee
- cotton
- soya beans
- sugar cane
- timber
- capital cities
- major towns
- pasture
- cropland
- forest

▶ *The fecundity of parts of Brazil's rainforest results from exceptionally high levels of rainfall and the quantities of silt deposited by the Amazon river system.*

The landscape

The Amazon Basin, containing the largest area of tropical rainforest on Earth, covers nearly half of Brazil. It is bordered by two shield areas: in the south by the Brazilian Highlands, and in the north by the Guiana Highlands. The east coast is dominated by a great escarpment which runs for 1600 miles (2565 km).

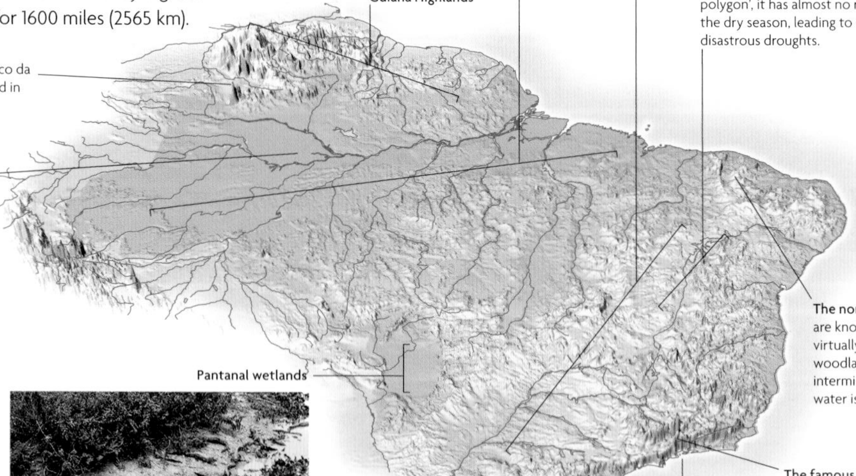

The Amazon Basin is the largest river basin in the world. The Amazon river and over a thousand tributaries drain an area of 2,375,000 sq miles (6,150,000 sq km) and carry one-fifth of the world's fresh water out to sea.

The ancient Brazilian Highlands have a varied topography. Their plateaux, hills and deep valleys are bordered by highly-eroded mountains containing important mineral deposits. They are drained by three great river systems, the Amazon, the Paraguay–Paraná and the São Francisco.

The São Francisco Basin has a climate unique in Brazil. Known as the 'drought polygon', it has almost no rain during the dry season, leading to regular disastrous droughts.

Brazil's highest mountain is the Pico da Neblina which was only discovered in 1962. It is 9888 ft (3014 m) high.

The flood plains which border the Amazon river are made up of a variety of different features including shallow lakes and swamps, mangrove forests in the tidal delta area and fertile levées on river banks and point bars.

The northeastern scrublands are known as the *caatinga*, a virtually impenetrable thorny woodland, sometimes intermixed with cacti where water is scarce.

The famous Sugar Loaf Mountain (*Pão de Açúcar*) which overlooks Rio de Janeiro is a fine example of a volcanic plug a domed core of solidified lava left after the slopes of the original volcano have eroded away.

Deep natural harbours such as Baia de Guanabara were created where the steep slopes of the Serra da Mantiqueira plunge directly into the ocean.

▶ *The Pantanal region in the south of Brazil is an extension of the Gran Chaco plain. The swamps and marshes of this area are renowned for their beauty, and abundant and unique wildlife, including wildfowl and these caimans, a type of crocodile.*

◀ *The Iguaçu river surges over the spectacular Iguaçu Falls* (Saltos do Iguaçu) *towards the Paraná river. Falls like these are increasingly under pressure from large-scale hydro-electric projects such as that at Itaipú.*

Hillslope gullying

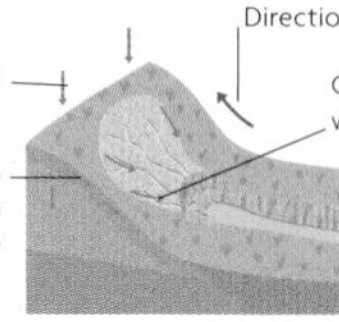

◀ *Large-scale gullies are common in Brazil, particularly on hillslopes from which vegetation has been removed. Gullies grow headwards (up the slope), aided by a combination of erosion through water seepage and rainwater runoff.*

Map key

Population

- above 5 million
- 1 million to 5 million
- 500,000 to 1 million
- 100,000 to 500,000
- 50,000 to 100,000
- 10,000 to 50,000
- below 10,000

Elevation

- 3000m / 9843ft
- 2000m / 6562ft
- 1000m / 3281ft
- 500m / 1640ft
- 250m / 820ft
- 100m / 328ft
- sea level

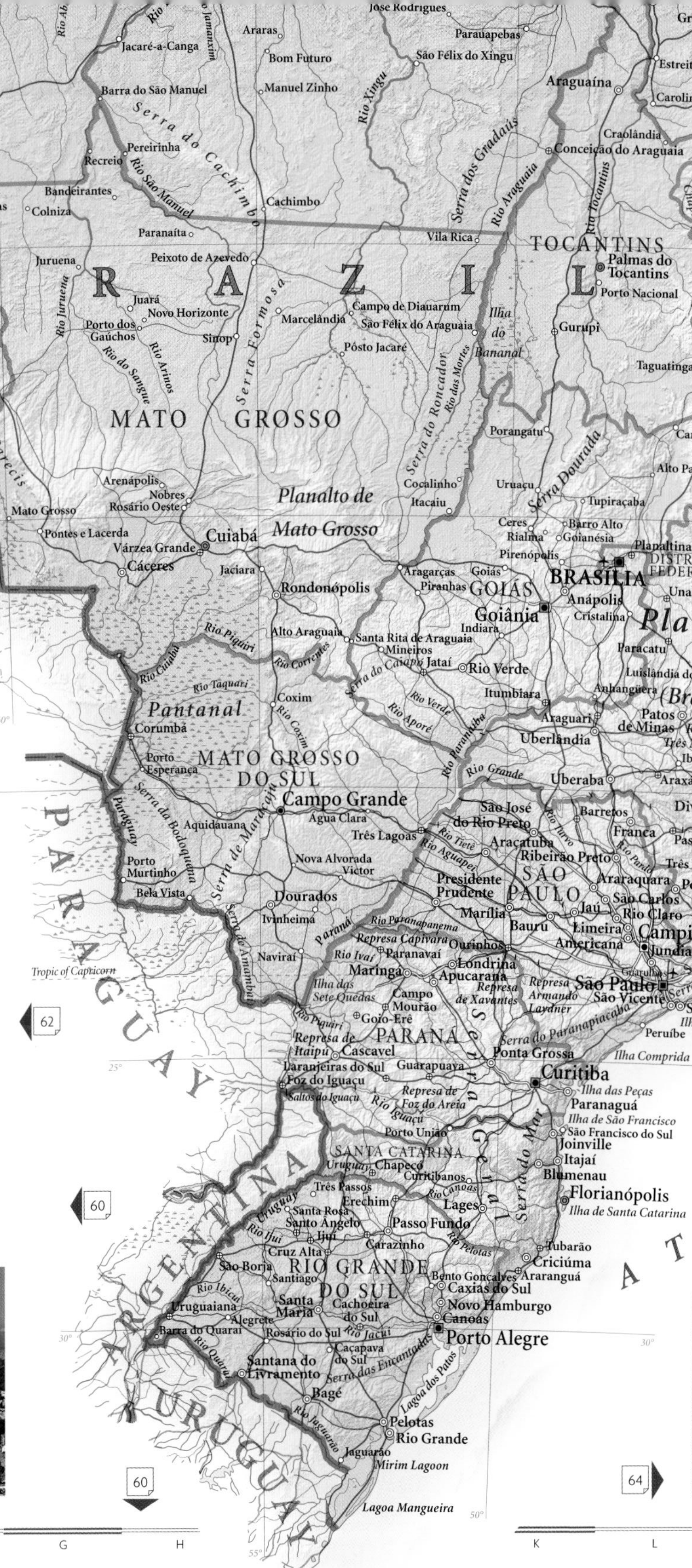

56

Scale 1:14,250,000

projection: Lambert Azimuthal Equal Area

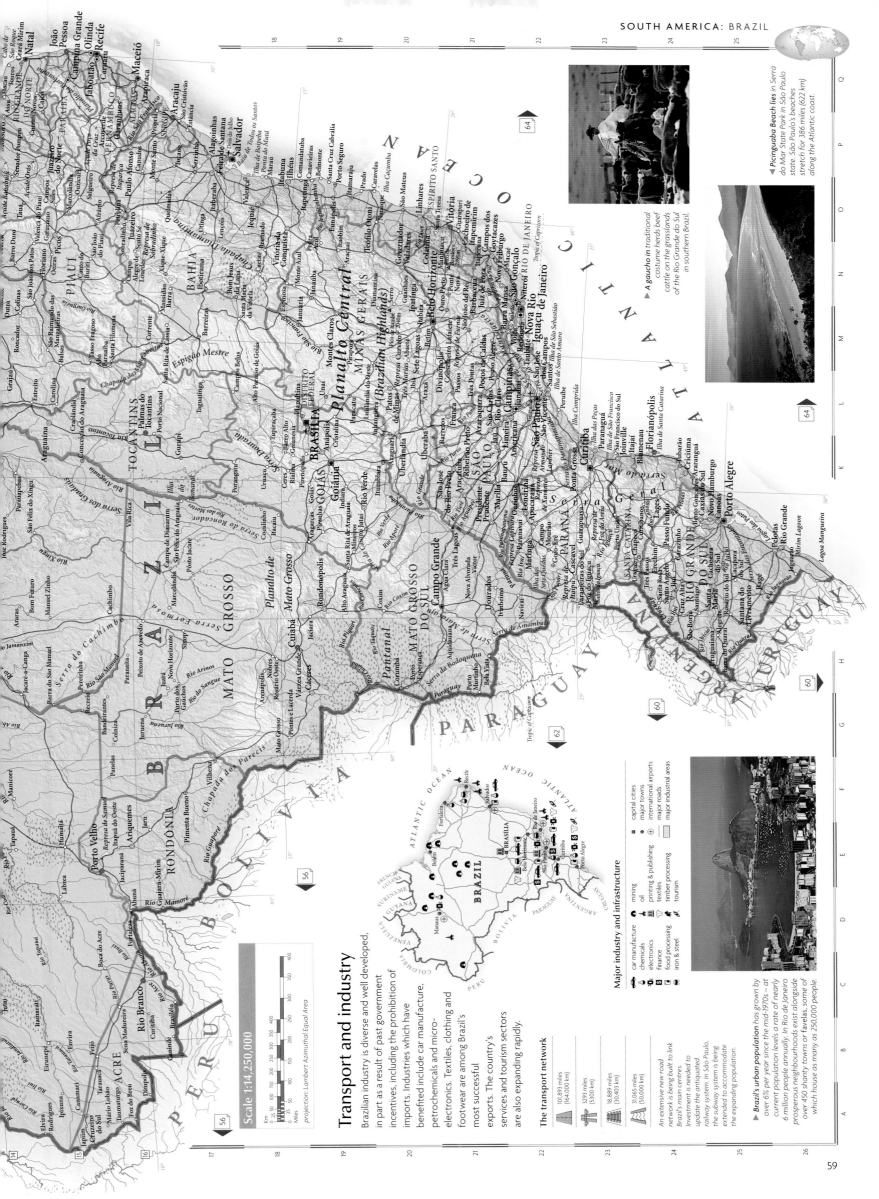

Transport and industry

Brazilian industry is diverse and well developed, in part as a result of past government incentives, including the prohibition of imports. Industries which have benefited include car manufacture, petrochemicals and micro-electronics. Textiles, clothing and footwear are among Brazil's most successful exports. The country's services and tourism sectors are also expanding rapidly.

The transport network

- 101,893 miles (164,000 km)
- 3293 miles (5300 km)
- 18,889 miles (30,403 km)
- 31,065 miles (50,000 km)

An extensive new road network is being built to link Brazil's main centres. Investment is needed to update the antiquated railway system. In São Paulo, the subway system is being extended to accommodate the expanding population.

Major industry and infrastructure

- car manufacture
- chemicals
- electronics
- finance
- food processing
- iron & steel
- mining
- oil
- printing & publishing
- textiles
- timber processing
- tourism
- capital cities
- major towns
- international airports
- major roads
- major industrial areas

▶ ***Brazil's urban population** has grown by over 6% per year since the mid-1970s – at current population levels a rate of nearly 6 million people annually. In Rio de Janeiro prosperous neighbourhoods exist alongside over 450 shanty towns or favelas, some of which house as many as 250,000 people.*

▶ *A **gaucho** in traditional costume herds beef cattle on the grasslands of the Rio Grande do Sul in southern Brazil.*

◀ ***Picinguaba Beach lies** in Serra do Mar State Park in São Paulo state. São Paulo's beaches stretch for 386 miles (622 km) along the Atlantic coast.*

Eastern South America

URUGUAY, NORTHEAST ARGENTINA, SOUTHEAST BRAZIL

The vast conurbations of Rio de Janeiro, São Paulo and Buenos Aires form the core of South America's highly-urbanized eastern region. São Paulo state, with over 40 million inhabitants, is among the world's 20 most powerful economies, and São Paulo is the fastest growing city on the continent. Rio de Janeiro and Buenos Aires, transformed in the last hundred years from port cities to great metropolitan areas each with more than 10 million inhabitants, typify the unstructured growth and wealth disparities of South America's great cities. In Uruguay, two fifths of the population lives in the capital, Montevideo, which faces Buenos Aires across the River Plate *(Río de la Plata)*. Immigration from the countryside has created severe pressure on the urban infrastructure, particularly on available housing, leading to a profusion of crowded shanty settlements *(favelas* or *barrios)*.

Using the land

Most of Uruguay and the Pampas of northern Argentina are devoted to the rearing of livestock, especially cattle and sheep, which are central to both countries' economies. Soya beans, first produced in Brazil's Rio Grande do Sul, are now more widely grown for large-scale export, as are cereals, sugar cane and grapes. Subsistence crops, including potatoes, corn and sugar beet, are grown on the remaining arable land.

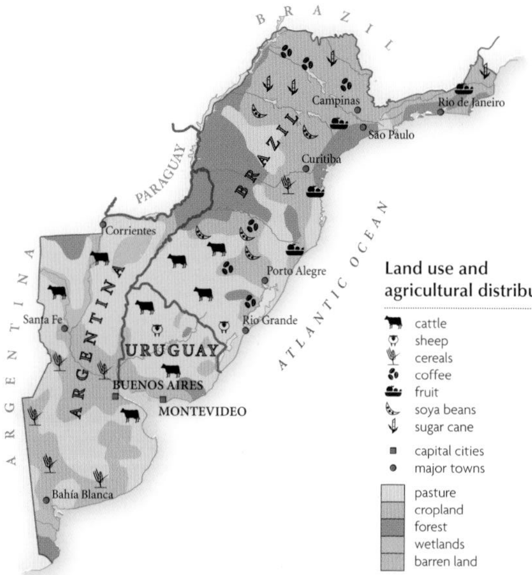

Land use and agricultural distribution

- cattle
- sheep
- cereals
- coffee
- fruit
- soya beans
- sugar cane
- capital cities
- major towns
- pasture
- cropland
- forest
- wetlands
- barren land

◀ *The rolling grasslands of Uruguay are ideally suited to the rearing of cattle. Beef is the country's main export commodity, valued at over one billion US dollars in 2006.*

62

Scale 1:7,000,000

Km 0 25 50 100 150 200
Miles 0 25 50 100 150 200

projection: Lambert Azimuthal Equal Area

▲ ***Soya beans are** harvested, pressed, and processed into soya cake, which is used as animal feed. The cake is fed mainly to chickens on large-scale factory farms, and the growth in soya production has been an important factor in the expansion of the Brazilian poultry trade.*

Transport and industry

Southeast Brazil is home to much of the important motor and capital goods industry, largely based around São Paulo; iron and steel production is also concentrated in this region. Uruguay's economy continues to be based mainly on the export of livestock products including meat and leather goods. Buenos Aires is Argentina's chief port, and the region has a varied and sophisticated economic base including service-based industries such as finance and publishing, as well as primary processing.

Map key

Population	Elevation
above 5 million	2000m / 6562ft
1 million to 5 million	1000m / 3281ft
500,000 to 1 million	500m / 1640ft
100,000 to 500,000	250m / 820ft
50,000 to 100,000	100m / 328ft
10,000 to 50,000	sea level
below 10,000	

58

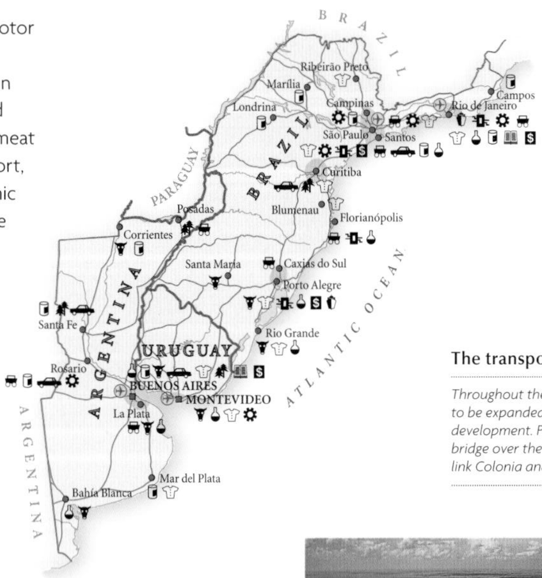

Major industry and infrastructure

- car manufacture
- chemicals
- engineering
- finance
- food processing
- iron & steel
- meat processing
- printing & publishing
- shipbuilding
- textiles
- timber processing
- capital cities
- major towns
- international airports
- major roads
- major industrial areas

The transport network

Throughout the region, road networks need to be expanded to cope with urban development. Plans are underway to build a bridge over the River Plate (Río de la Plata) to link Colonia and Buenos Aires.

▲ ***The Itaipú dam** on the Paraná river is one of the largest hydro-electric projects in the world, jointly financed by Brazil and Paraguay.*

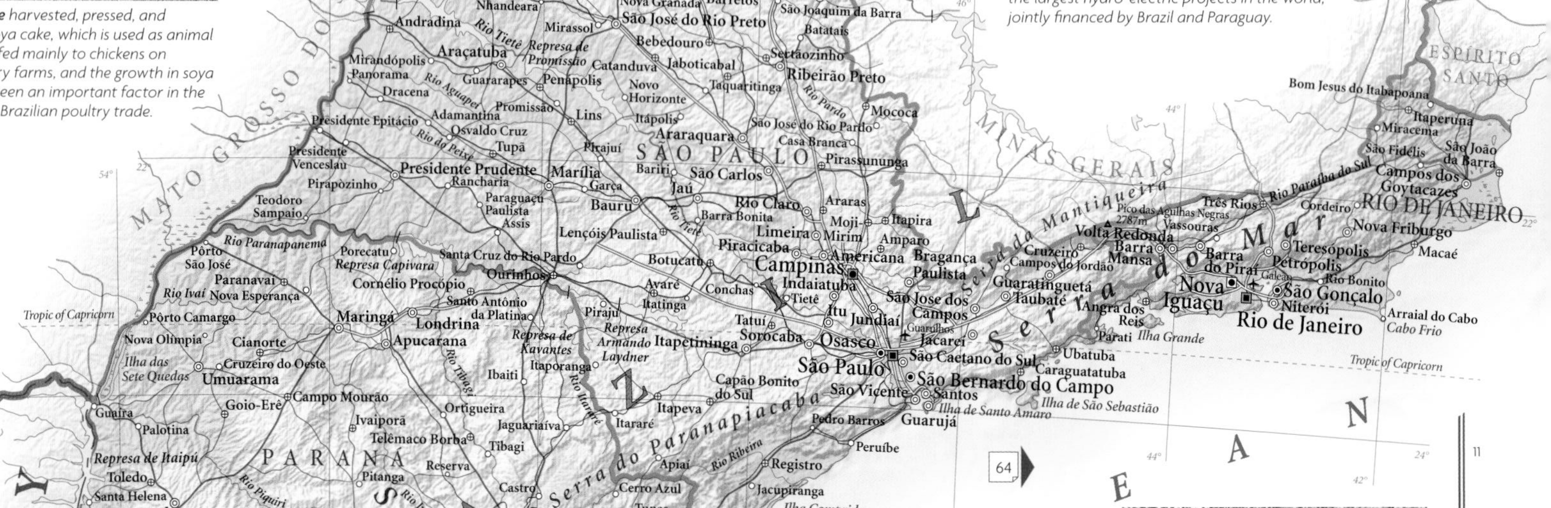

64

▶ *Rio de Janeiro's annual carnival, Mardi Gras, which ushers in the start of Lent, is an extravagant five-day parade through the city, characterized by fantastically decorated floats, exuberant dancing and samba music.*

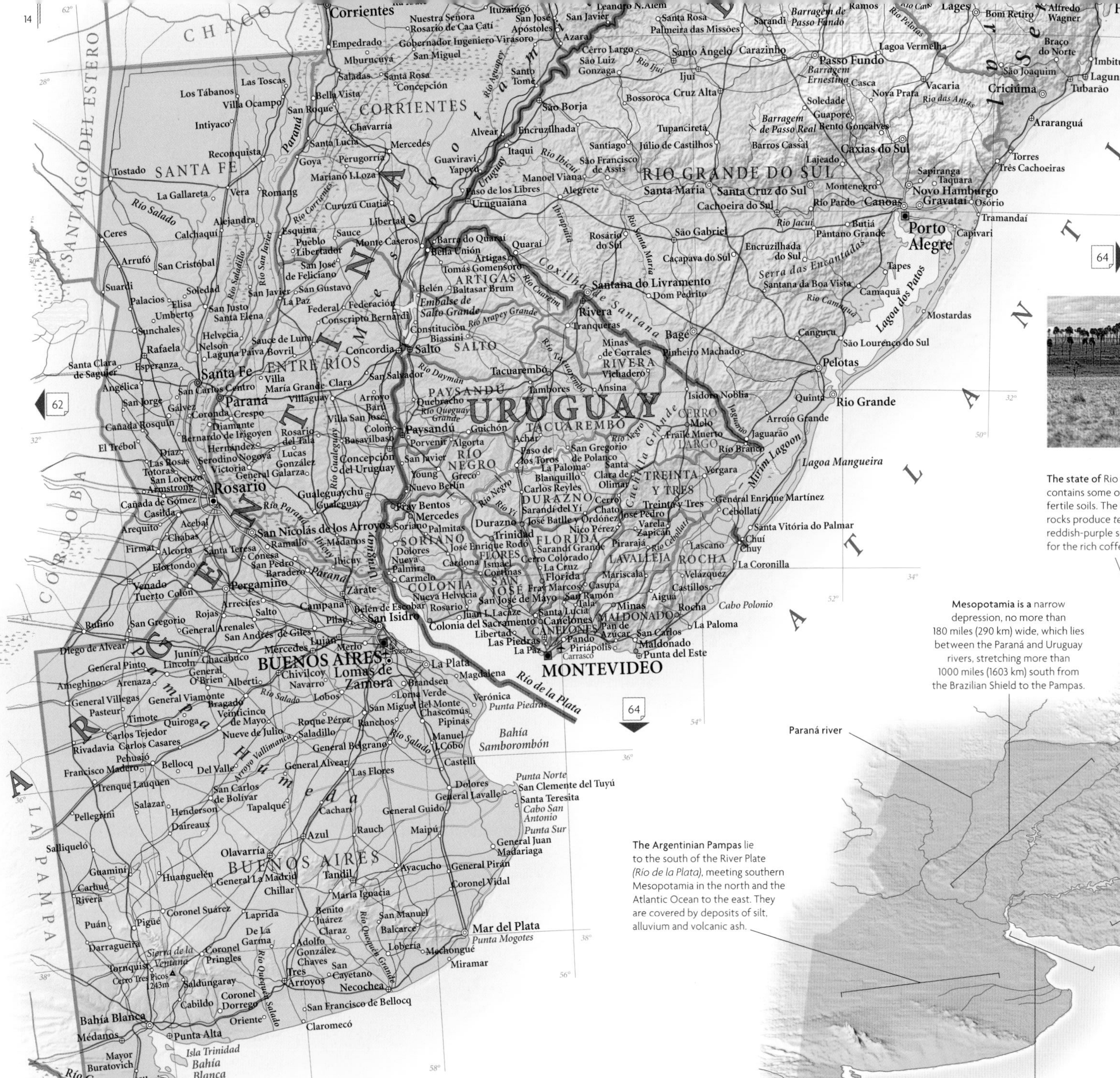

The landscape

The southern reaches of the Brazilian Highlands follow the Atlantic coast to form low, rolling hills in the northeast of Uruguay. Much of South America's mid-eastern region and all of Uruguay has a gentle relief with land rarely rising above 300 ft (100 m). Argentina's northeast comprises two main regions: a long, narrow lowland known as Mesopotamia; and part of the Pampas grasslands.

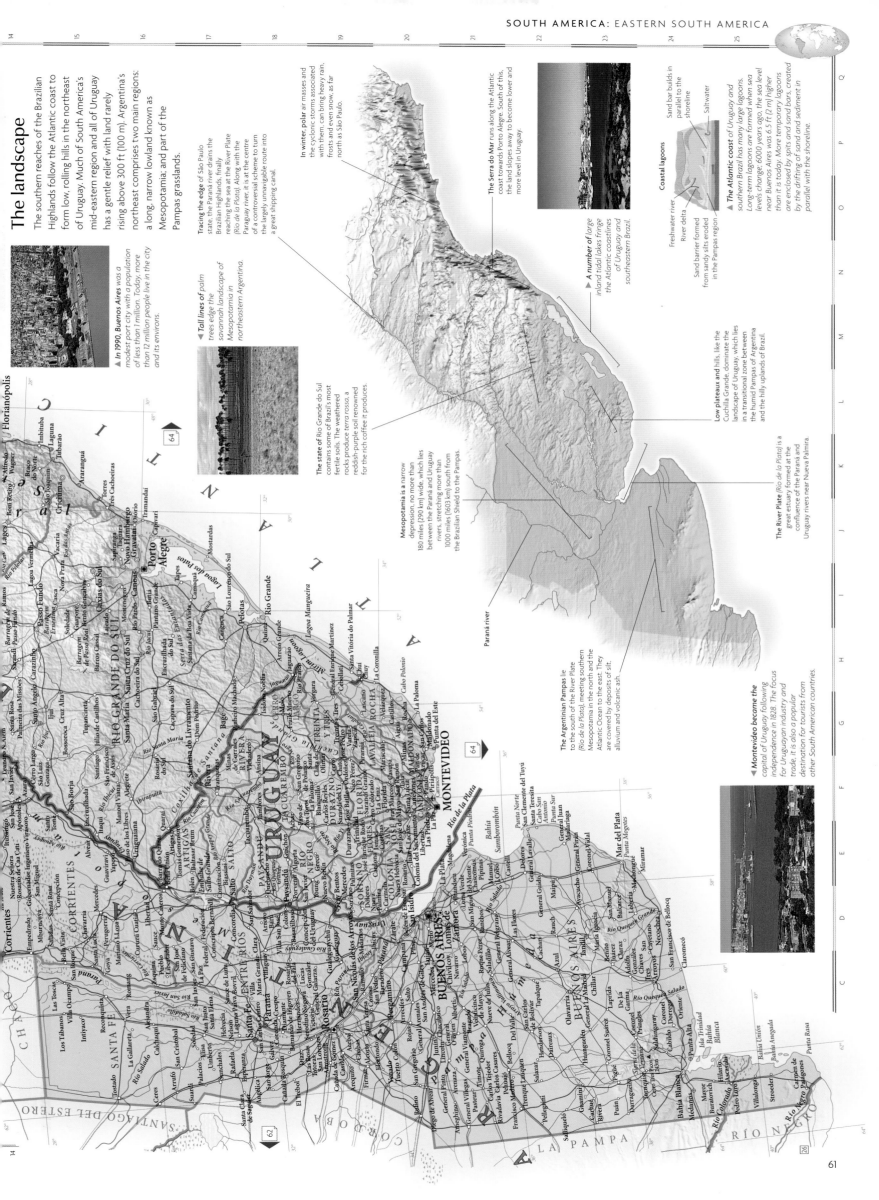

▲ ***In 1990, Buenos Aires** was a modest port city with a population of less than 1 million. Today, more than 12 million people live in the city and its environs.*

◀ ***Tall lines of** palm trees edge the savannah landscape of Mesopotamia in northeastern Argentina.*

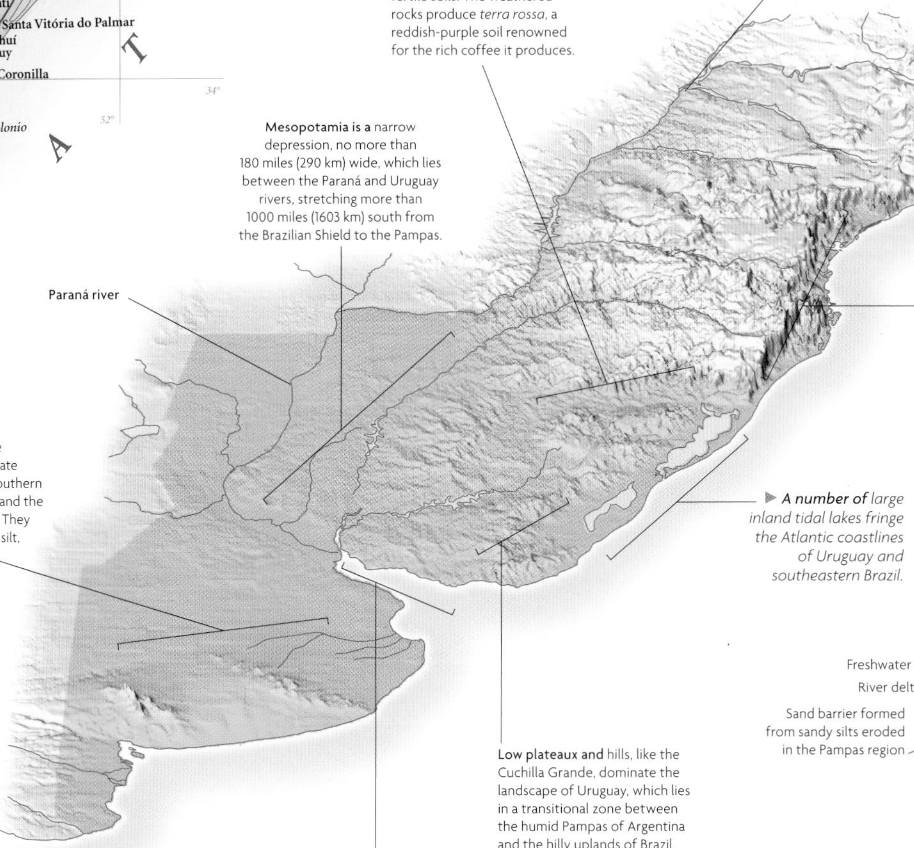

Tracing the edge of São Paulo state, the Paraná river drains the Brazilian Highlands, finally reaching the sea at the River Plate *(Río de la Plata)*. Along with the Paraguay river, it is at the centre of a controversial scheme to turn the largely unavigable route into a great shipping canal.

In winter, polar air masses and the cyclonic storms associated with them, can bring heavy rain, frosts and even snow, as far north as São Paulo.

The state of Rio Grande do Sul contains some of Brazil's most fertile soils. The weathered rocks produce *terra rossa*, a reddish-purple soil renowned for the rich coffee it produces.

Mesopotamia is a narrow depression, no more than 180 miles (290 km) wide, which lies between the Paraná and Uruguay rivers, stretching more than 1000 miles (1603 km) south from the Brazilian Shield to the Pampas.

Paraná river

The Serra do Mar runs along the Atlantic coast towards Porto Alegre. South of this, the land slopes away to become lower and more level in Uruguay.

The Argentinian Pampas lie to the south of the River Plate *(Río de la Plata)*, meeting southern Mesopotamia in the north and the Atlantic Ocean to the east. They are covered by deposits of silt, alluvium and volcanic ash.

Low plateaux and hills, like the Cuchilla Grande, dominate the landscape of Uruguay, which lies in a transitional zone between the humid Pampas of Argentina and the hilly uplands of Brazil.

The River Plate *(Río de la Plata)* is a great estuary formed at the confluence of the Paraná and Uruguay rivers near Nueva Palmira.

▶ ***A number of** large inland tidal lakes fringe the Atlantic coastlines of Uruguay and southeastern Brazil.*

Coastal lagoons

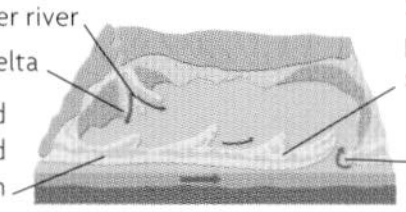
Freshwater river
River delta
Sand barrier formed from sandy silts eroded in the Pampas region
Sand bar builds in parallel to the shoreline
Saltwater

▲ ***The Atlantic coast** of Uruguay and southern Brazil has many large lagoons. Long-term lagoons are formed when sea levels change; 6000 years ago, the sea level near Buenos Aires was 6.5 ft (2 m) higher than it is today. More temporary lagoons are enclosed by spits and sand bars, created by the drifting of sand and sediment in parallel with the shoreline.*

◀ ***Montevideo became the** capital of Uruguay following independence in 1828. The focus for Uruguayan industry and trade, it is also a popular destination for tourists from other South American countries.*

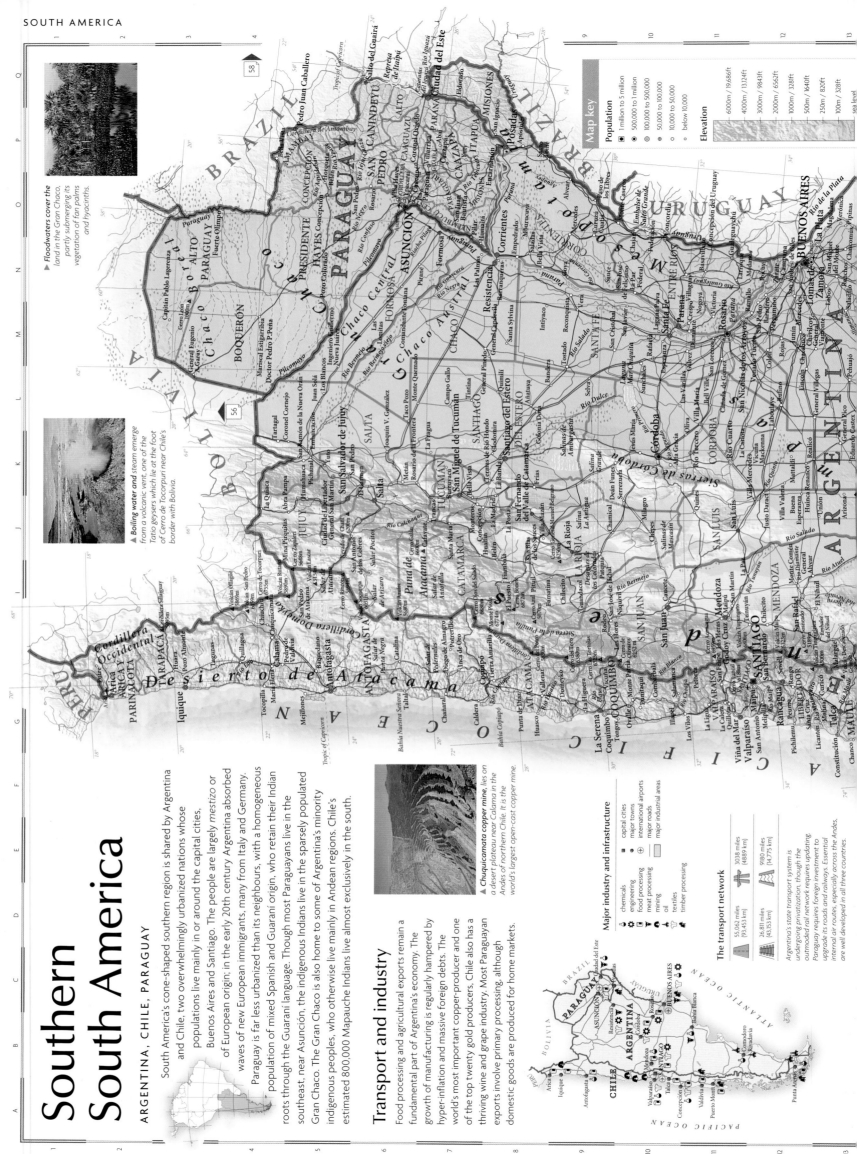

Southern South America

ARGENTINA, CHILE, PARAGUAY

South America's cone-shaped southern region is shared by Argentina and Chile, two overwhelmingly urbanized nations whose populations live mainly in or around the capital cities, Buenos Aires and Santiago. The people are largely *mestizo* or of European origin; in the early 20th century Argentina absorbed waves of new European immigrants, many from Italy and Germany. Paraguay is far less urbanized than its neighbours, with a homogeneous population of mixed Spanish and Guaraní origin, who retain their Indian roots through the Guaraní language. Though most Paraguayans live in the southeast, near Asunción, the indigenous Indians live in the sparsely populated Gran Chaco. The Gran Chaco is also home to some of Argentina's minority indigenous peoples, who otherwise live mainly in Andean regions. Chile's estimated 800,000 Mapauche Indians live almost exclusively in the south.

Transport and industry

Food processing and agricultural exports remain a fundamental part of Argentina's economy. The growth of manufacturing is regularly hampered by hyper-inflation and massive foreign debts. The world's most important copper-producer and one of the top twenty gold producers, Chile also has a thriving wine and grape industry. Most Paraguayan exports involve primary processing, although domestic goods are produced for home markets.

▲ ***Chuquicamata copper mine,*** *lies on a desert plateau near Calama in the Andes of northern Chile. It is the world's largest open-cast copper mine.*

Major industry and infrastructure

- chemicals
- engineering
- food processing
- meat processing
- mining
- oil
- textiles
- timber processing
- capital cities
- major towns
- international airports
- major roads
- major industrial areas

The transport network

- 55,062 miles (93,453 km)
- 3038 miles (4889 km)
- 26,811 miles (43,153 km)
- 9180 miles (14,775 km)

Argentina's state transport system is undergoing privatization, though the outmoded rail network requires updating. Paraguay requires foreign investment to upgrade its roads and railways. Essential internal air routes, especially across the Andes, are well developed in all three countries.

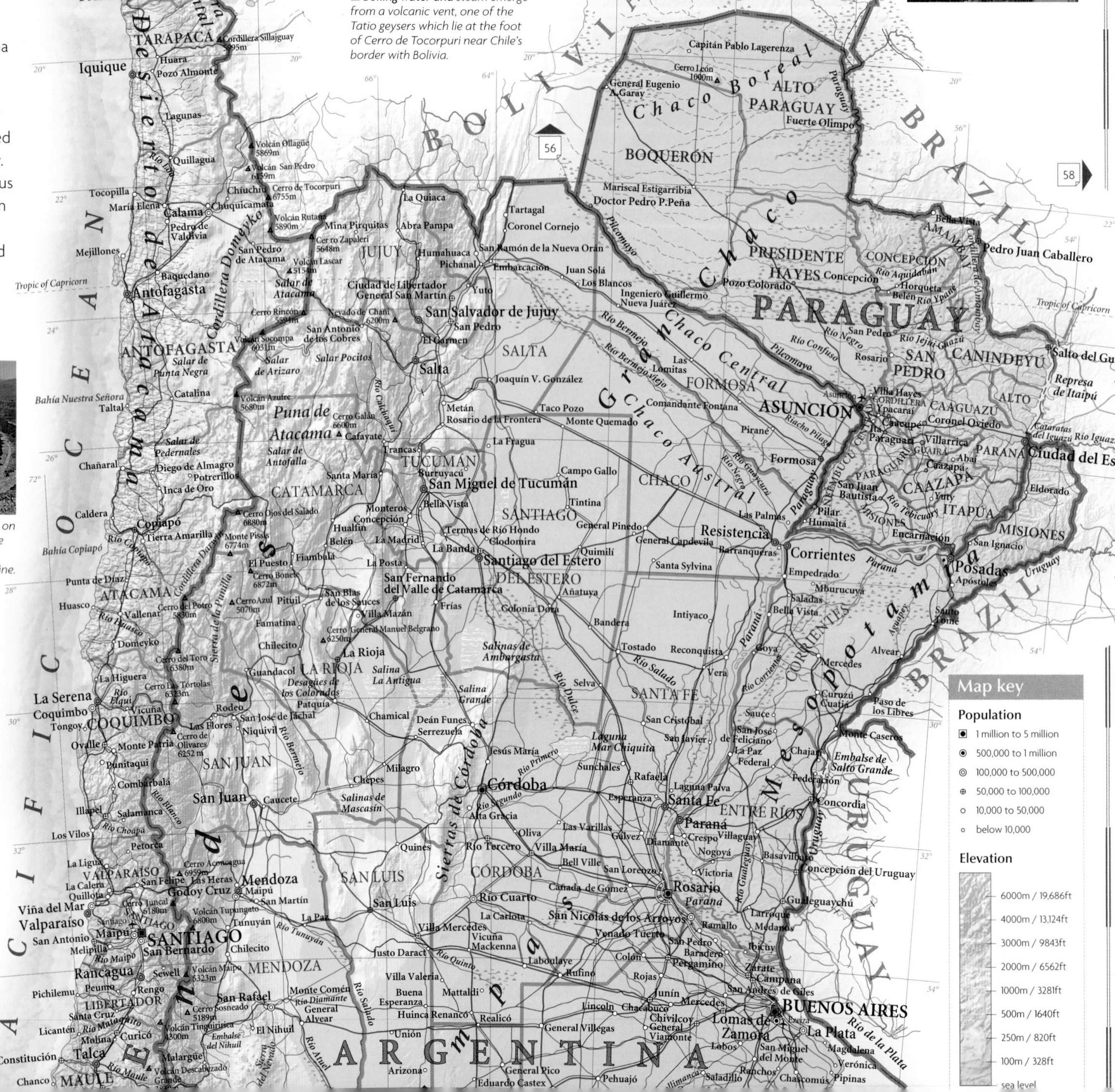

▲ ***Boiling water and*** *steam emerge from a volcanic vent, one of the Tatio geysers which lie at the foot of Cerro de Tocorpuri near Chile's border with Bolivia.*

▶ ***Floodwaters cover the*** *land in the Gran Chaco, partly submerging its vegetation of fan palms and hyacinths.*

56
58

The landscape

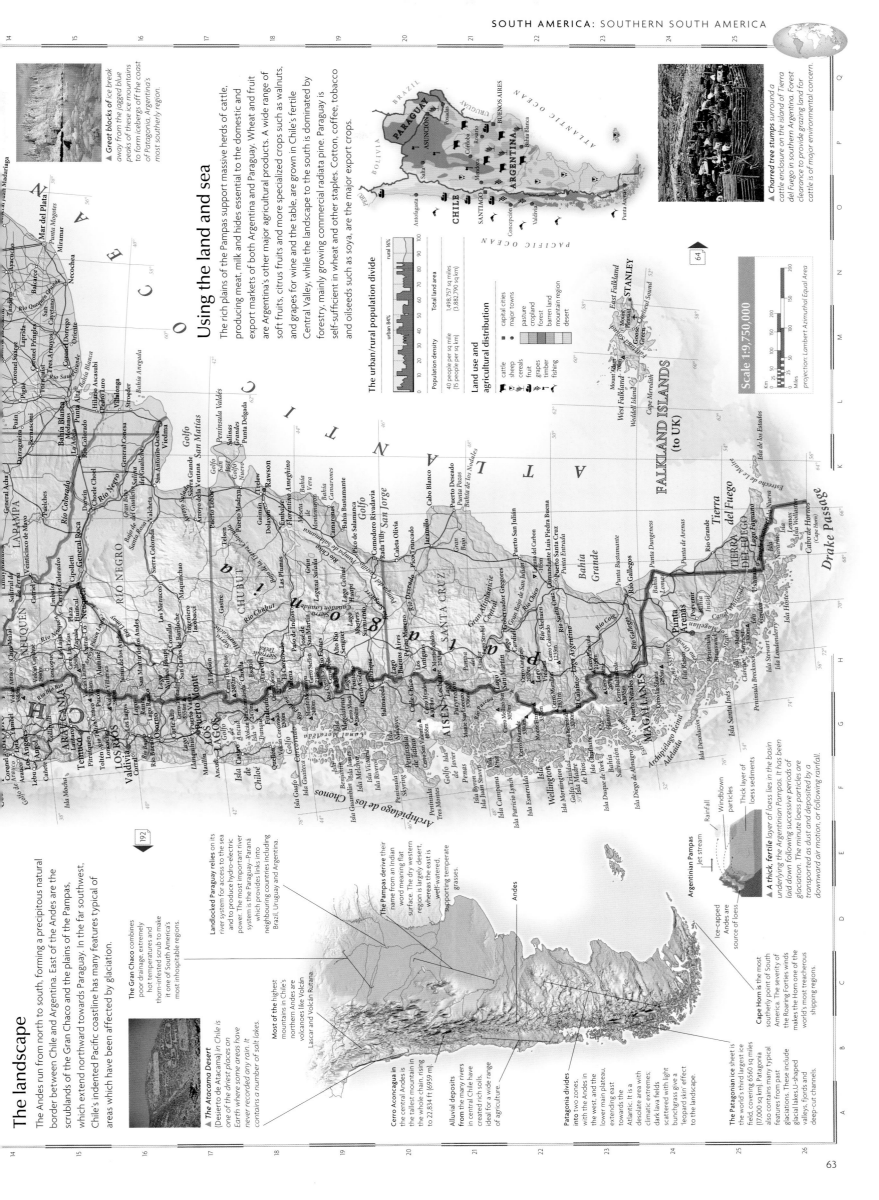

The Andes run from north to south, forming a precipitous natural border between Chile and Argentina. East of the Andes are the scrublands of the Gran Chaco and the plains of the Pampas, which extend northward towards Paraguay. In the far southwest, Chile's indented Pacific coastline has many features typical of areas which have been affected by glaciation.

▲ ***The Atacama Desert*** (Desierto de Atacama) *in Chile is one of the driest places on Earth where some areas have never recorded any rain. It contains a number of salt lakes.*

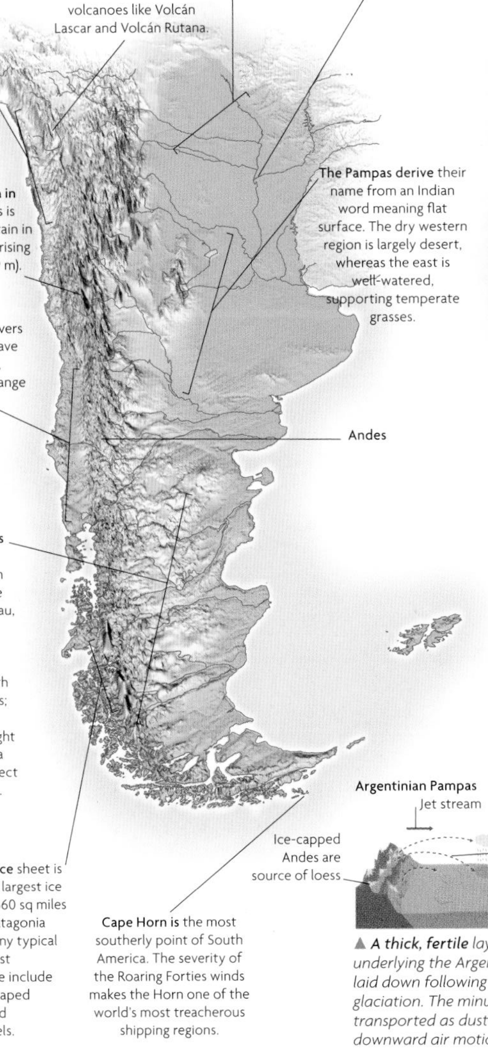

The Gran Chaco combines poor drainage, extremely hot temperatures and thorn-infested scrub to make it one of South America's most inhospitable regions.
Landlocked Paraguay relies on its river system for access to the sea and to produce hydro-electric power. The most important river system is the Paraguay–Paraná which provides links into neighbouring countries including Brazil, Uruguay and Argentina.
Most of the highest mountains in Chile's northern Andes are volcanoes like Volcán Lascar and Volcán Rutana.
The Pampas derive their name from an Indian word meaning flat surface. The dry western region is largely desert, whereas the east is well-watered, supporting temperate grasses.
Cerro Aconcagua in the central Andes is the tallest mountain in the whole chain, rising to 22,834 ft (6959 m).
Alluvial deposits from the many rivers in central Chile have created rich soils, ideal for a wide range of agriculture.
Andes
Patagonia divides into two zones, with the Andes in the west, and the lower main plateau, extending east towards the Atlantic. It is a desolate area with climatic extremes; dark lava fields scattered with light bunchgrass give a 'leopard skin' effect to the landscape.
The Patagonian ice sheet is the world's third largest ice field, covering 6560 sq miles (17,000 sq km). Patagonia also contains many typical features from past glaciations. These include glacial lakes,U-shaped valleys, fjords and deep-cut channels.
Cape Horn is the most southerly point of South America. The severity of the Roaring Forties winds makes the Horn one of the world's most treacherous shipping regions.
Ice-capped Andes are source of loess

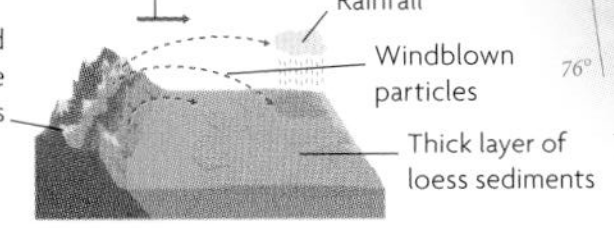

Argentinian Pampas
Jet stream
Rainfall
Windblown particles
Thick layer of loess sediments

▲ ***A thick, fertile*** *layer of loess lies in the basin underlying the Argentinian Pampas. It has been laid down following successive periods of glaciation. The minute loess particles are transported as dust and deposited by a downward air motion, or following rainfall.*

192

NEUQUÉN
LA PAMPA
ARAUCANÍA
LOS RÍOS
LOS LAGOS
RÍO NEGRO
CHUBUT
AISÉN
SANTA CRUZ
MAGALLANES
Patagonia
Puerto Montt
Valdivia
Temuco
Neuquén
General Roca
Bahía Blanca
Mar del Plata
Viedma
Rawson
Comodoro Rivadavia
Río Gallegos
Punta Arenas
Ushuaia
Golfo San Matías
Golfo San Jorge
Bahía Grande
Tierra del Fuego
Archipiélago de los Chonos
FALKLAND ISLANDS (to UK)
STANLEY
West Falkland
East Falkland
Cabo de Hornos (Cape Horn)
Drake Passage
ATLANTIC OCEAN
Scale 1:9,750,000
projection: Lambert Azimuthal Equal Area

64

▲ ***Great blocks of*** *ice break away from the jagged blue peaks of these ice mountains to form icebergs off the coast of Patagonia, Argentina's most southerly region.*

Using the land and sea

The rich plains of the Pampas support massive herds of cattle, producing meat, milk and hides essential to the domestic and export markets of both Argentina and Paraguay. Wheat and fruit are Argentina's other major agricultural products. A wide range of soft fruits, citrus fruits and more specialized crops such as walnuts, and grapes for wine and the table, are grown in Chile's fertile Central Valley, while the landscape to the south is dominated by forestry, mainly growing commercial radiata pine. Paraguay is self-sufficient in wheat and other staples. Cotton, coffee, tobacco and oilseeds such as soya, are the major export crops.

The urban/rural population divide

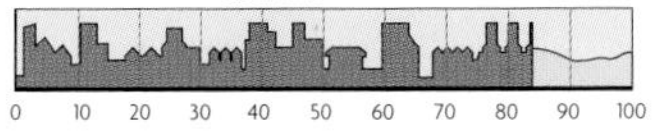

urban 84%
rural 16%
0 10 20 30 40 50 60 70 80 90 100

Population density	Total land area
40 people per sq mile (15 people per sq km)	1,498,757 sq miles (3,882,790 sq km)

Land use and agricultural distribution

cattle
sheep
cereals
fruit
grapes
timber
fishing
capital cities
major towns
pasture
cropland
forest
barren land
mountain region
desert
PERU
BOLIVIA
BRAZIL
PARAGUAY
ASUNCIÓN
URUGUAY
CHILE
ARGENTINA
SANTIAGO
BUENOS AIRES
PACIFIC OCEAN
ATLANTIC OCEAN
Antofagasta
Salta
Posadas
Córdoba
Rosario
Mendoza
Concepción
Valdivia
Bahía Blanca
Punta Arenas

▲ ***Charred tree stumps*** *surround a cattle enclosure on the island of Tierra del Fuego in southern Argentina. Forest clearance to provide grazing land for cattle is of major environmental concern.*

The Atlantic Ocean

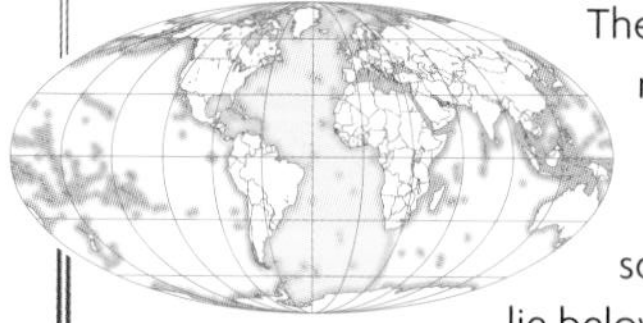

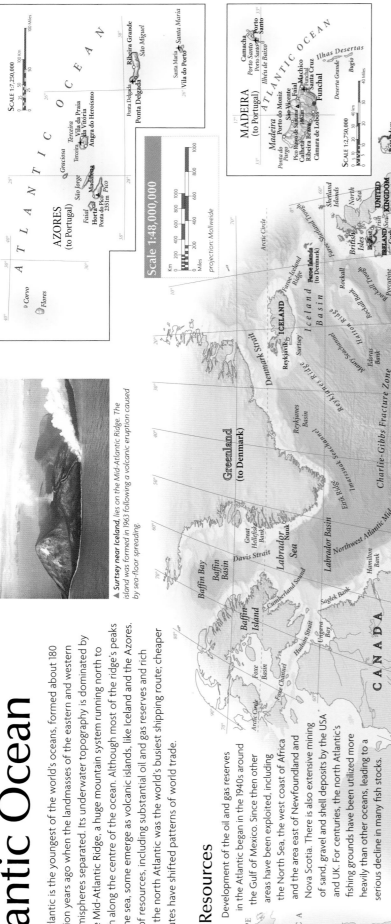

The Atlantic is the youngest of the world's oceans, formed about 180 million years ago when the landmasses of the eastern and western hemispheres separated. Its underwater topography is dominated by the Mid-Atlantic Ridge, a huge mountain system running north to south along the centre of the ocean. Although most of the ridge's peaks lie below the sea, some emerge as volcanic islands, like Iceland and the Azores. The Atlantic contains a wealth of resources, including substantial oil and gas reserves and rich fishing grounds. Until the 1950s, the north Atlantic was the world's busiest shipping route; cheaper air transport and alternative routes have shifted patterns of world trade.

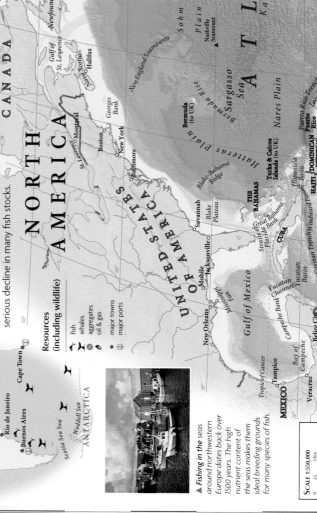

▲ ***Surtsey near Iceland,*** *lies on the Mid-Atlantic Ridge. The island was formed in 1963 following a volcanic eruption caused by sea-floor spreading.*

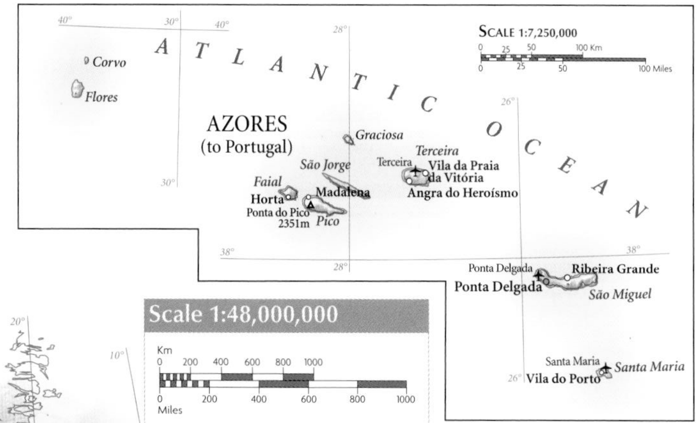

Resources

Development of the oil and gas reserves in the Atlantic began in the 1940s around the Gulf of Mexico. Since then other areas have been exploited, including the North Sea, the west coast of Africa and the area east of Newfoundland and Nova Scotia. There is also extensive mining of sand, gravel and shell deposits by the USA and UK. For centuries, the north Atlantic's fishing grounds have been utilized more heavily than other oceans, leading to a serious decline in many fish stocks.

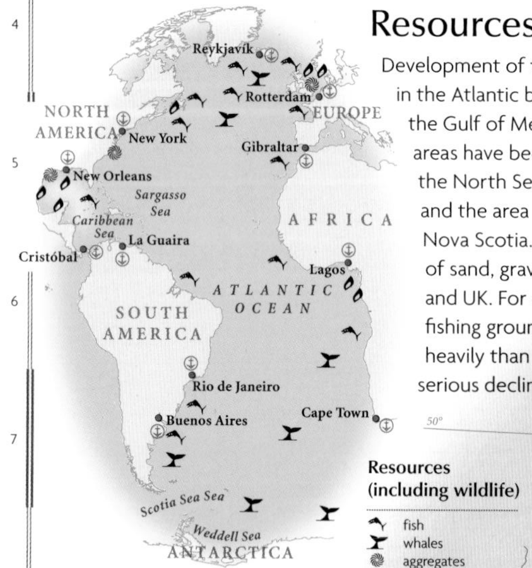

▲ ***Fishing in the*** *seas around northwestern Europe dates back over 1500 years. The high nutrient content of the seas makes them ideal breeding grounds for many species of fish.*

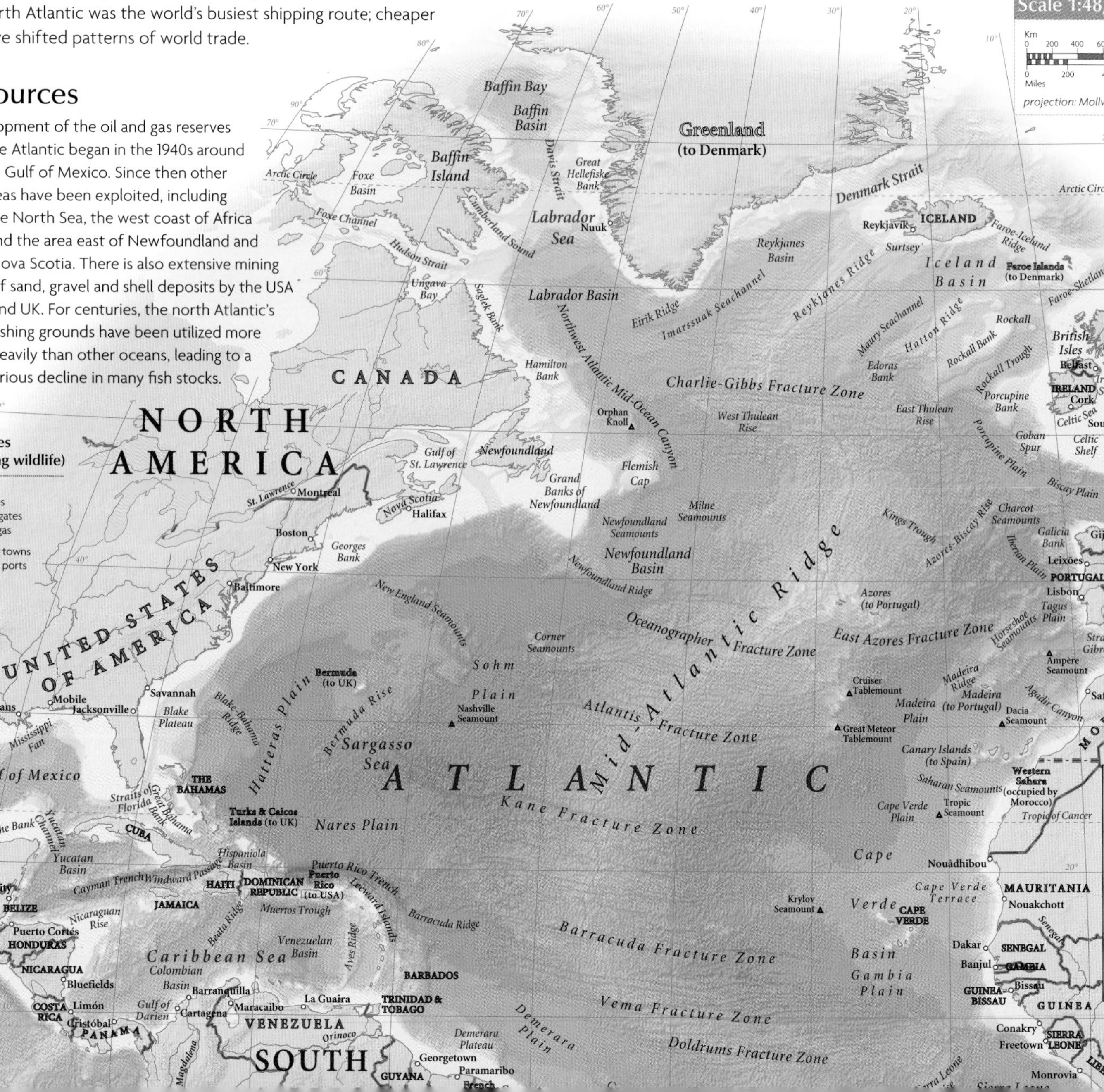

▲ ***On 5 January 1993,*** *the oil tanker Braer ran aground in the Shetland Islands, spilling 83,660 tons (85,000 tonnes) of light crude oil into the ocean, devastating the local marine ecosystem.*

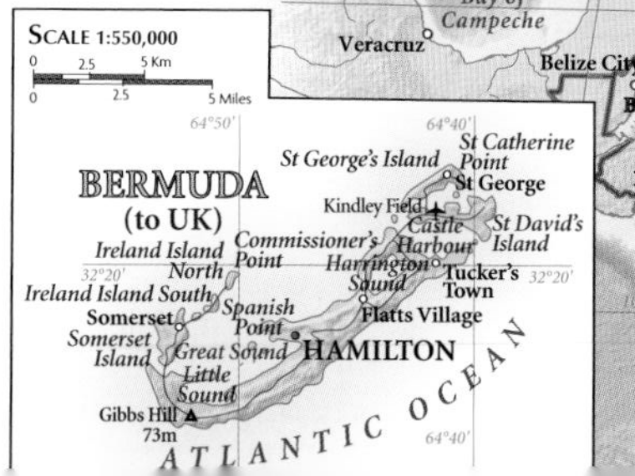

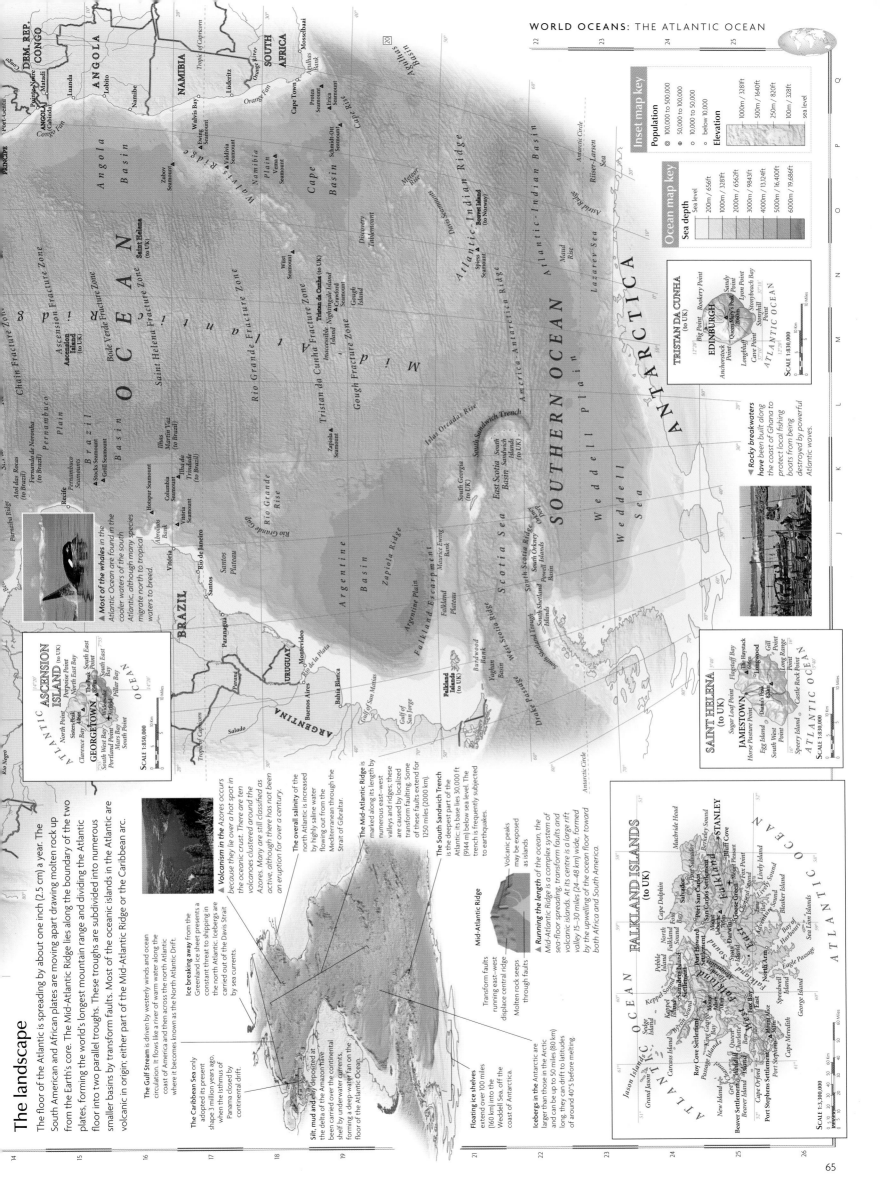

The landscape

The floor of the Atlantic is spreading by about one inch (2.5 cm) a year. The South American and African plates are moving apart drawing molten rock up from the Earth's core. The Mid-Atlantic Ridge lies along the boundary of the two plates, forming the world's longest mountain range and dividing the Atlantic floor into two parallel troughs. These troughs are subdivided into numerous smaller basins by transform faults. Most of the oceanic islands in the Atlantic are volcanic in origin; either part of the Mid-Atlantic Ridge or the Caribbean arc.

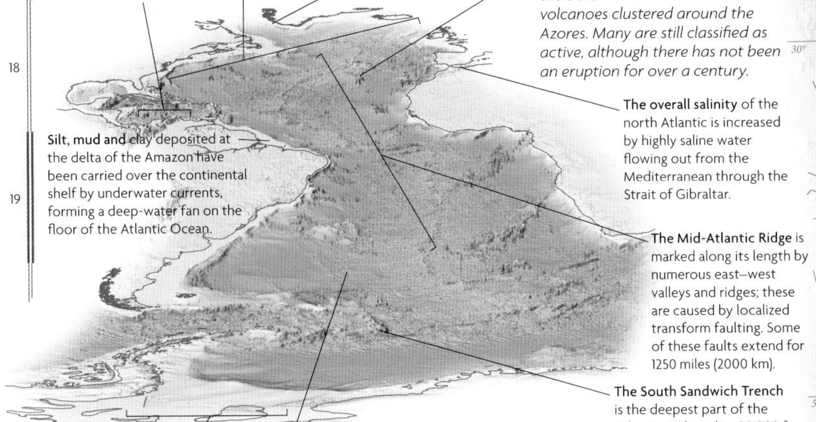

The Gulf Stream is driven by westerly winds and ocean circulation. It flows like a river of warm water along the coast of America and then across the north Atlantic where it becomes known as the North Atlantic Drift.

The Caribbean Sea only adopted its present shape 3 million years ago, when the Isthmus of Panama closed by continental drift.

Ice breaking away from the Greenland ice sheet presents a constant threat to shipping in the north Atlantic. Icebergs are carried out of the Davis Strait by sea currents.

Silt, mud and clay deposited at the delta of the Amazon have been carried over the continental shelf by underwater currents, forming a deep-water fan on the floor of the Atlantic Ocean.

The overall salinity of the north Atlantic is increased by highly saline water flowing out from the Mediterranean through the Strait of Gibraltar.

The Mid-Atlantic Ridge is marked along its length by numerous east–west valleys and ridges; these are caused by localized transform faulting. Some of these faults extend for 1250 miles (2000 km).

The South Sandwich Trench is the deepest part of the Atlantic; its base lies 30,000 ft (9144 m) below sea level. The trench is frequently subjected to earthquakes.

Floating ice shelves extend over 100 miles (160 km) into the Weddell Sea, off the coast of Antarctica.

Icebergs in the Antarctic are larger than those in the Arctic and can be up to 50 miles (80 km) long. they can drift to latitudes of around 40°S before melting.

▲ ***Volcanism in the*** *Azores occurs because they lie over a hot spot in the oceanic crust. There are ten volcanoes clustered around the Azores. Many are still classified as active, although there has not been an eruption for over a century.*

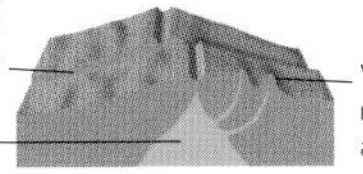
Mid-Atlantic Ridge
Transform faults running east–west displace central ridge
Molten rock seeps through faults
Volcanic peaks may be exposed as islands

▲ ***Running the length*** *of the ocean, the Mid-Atlantic Ridge is a complex system of sea-floor spreading, transform faults and volcanic islands. At its centre is a large rift valley 15–30 miles (24–48 km) wide, formed by the upwelling of the ocean floor toward both Africa and South America.*

▲ ***Most of the whales*** *in the Atlantic Ocean are found in the cooler waters of the south Atlantic, although many species migrate north to tropical waters to breed.*

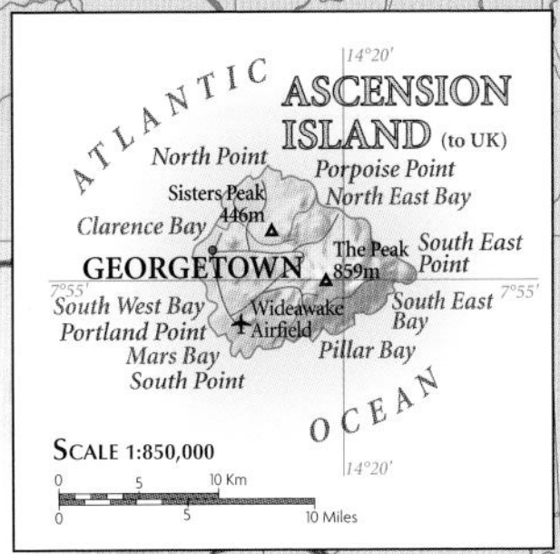
ASCENSION ISLAND (to UK)
ATLANTIC OCEAN
North Point
Porpoise Point
North East Bay
Sisters Peak 446m
Clarence Bay
GEORGETOWN
The Peak 859m
South East Point
South West Bay
Wideawake Airfield
South East Bay
Portland Point
Mars Bay
Pillar Bay
South Point
SCALE 1:850,000

Rio Negro
Parnaíba Ridge
Atol das Rocas (to Brazil)
Fernando de Noronha (to Brazil)
Pernambuco Plain
Recife
Pernambuco Seamounts
Brazil Basin
Stocks Seamount
Gröll Seamount
Chain Fracture Zone
Ascension Fracture Zone
Ascension Island (to UK)
Bode Verde Fracture Zone
OCEAN
Saint Helena Fracture Zone
Saint Helena (to UK)
Mid-Atlantic Ridge
Hotspur Seamount
Abrolhos Bank
Columbia Seamount
Vitória Seamount
Ilhas Martin Vaz (to Brazil)
Ilha da Trindade (to Brazil)
Vitória
BRAZIL
Tropic of Capricorn
Rio de Janeiro
Santos
Santos Plateau
Paranaguá
Paraná
Salado
Rio Grande Gap
Rio Grande Fracture Zone
Rio Grande Rise
URUGUAY
Montevideo
Río de la Plata
ARGENTINA
Buenos Aires
Bahía Blanca
Gulf of San Matías
Gulf of San Jorge
Argentine Basin
Zapiola Ridge
Argentine Plain
Falkland Escarpment
Falkland Plateau
Maurice Ewing Bank
Falkland Islands (to UK)
Burdwood Bank
Yaghan Basin
West Scotia Ridge
Scotia Sea
Drake Passage
South Shetland Trough
South Shetland Islands
South Scotia Ridge
South Orkney Islands
Powell Basin
Orkney Deep
South Georgia (to UK)
East Scotia Basin
South Sandwich Islands (to UK)
South Sandwich Trench
Islas Orcadas Rise
Tristan da Cunha Fracture Zone
Zapiola Seamount
Inaccessible Island
Tristan da Cunha (to UK)
Nightingale Island
Crawford Seamount
Gough Fracture Zone
Gough Island
Wüst Seamount
Discovery Tablemount
America-Antarctica Ridge
SOUTHERN OCEAN
Weddell Plain
Weddell Sea
Antarctic Circle
ANTARCTICA
Lazarev Sea
Maud Rise
Astrid Ridge
Riiser-Larsen Sea
Atlantic-Indian Basin
Atlantic-Indian Ridge
Spiess Seamount
Bouvet Island (to Norway)
Davis Seamounts
Meteor Rise
Agulhas Basin
Cape Rise
Cape Basin
Schmidt-Ott Seamount
Erica Seamount
Protea Seamount
Agulhas Bank
Cape Town
Mosselbaai
SOUTH AFRICA
Orange Fan
Orange River
Lüderitz
Namibia Plain
Vema Seamount
Walvis Ridge
Valdivia Seamount
Ewing Seamount
Walvis Bay
NAMIBIA
Tropic of Capricorn
Zubov Seamount
Angola Basin
Namibe
Lobito
ANGOLA
Luanda
Congo Fan
ANGOLA (Cabinda)
Pointe-Noire
Matadi
DEM. REP. CONGO
Congo
PRINCIPE
Port-Gentil
20

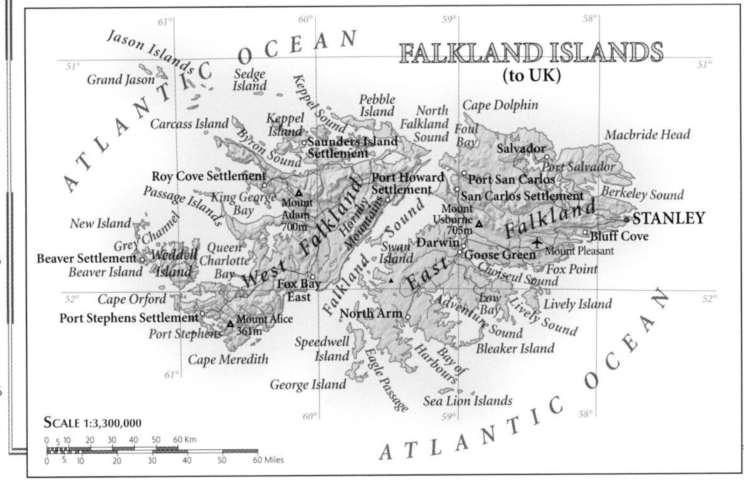
FALKLAND ISLANDS (to UK)
ATLANTIC OCEAN
Jason Islands
Grand Jason
Sedge Island
Keppel Sound
Keppel Island
Pebble Island
North Falkland Sound
Cape Dolphin
Macbride Head
Carcass Island
Byron Sound
Saunders Island Settlement
Foul Bay
Salvador
Port Salvador
Berkeley Sound
STANLEY
Roy Cove Settlement
Port Howard Settlement
Port San Carlos
San Carlos Settlement
Passage Islands
King George Bay
Mount Adam 700m
Mount Usborne 705m
Bluff Cove
Mount Pleasant
New Island
Grey Channel
Queen Charlotte Bay
Weddell Island
West Falkland
Falkland Sound
Hornby Mountains
Darwin
Goose Green
Choiseul Sound
Fox Point
East Falkland
Beaver Settlement
Beaver Island
Swan Island
Fox Bay East
Lively Island
Cape Orford
Port Stephens Settlement
Port Stephens
Mount Alice 361m
North Arm
Low Bay
Adventure Sound
Lively Sound
Bleaker Island
Speedwell Island
Cape Meredith
George Island
Eagle Passage
Bay of Harbours
Sea Lion Islands
SCALE 1:3,300,000

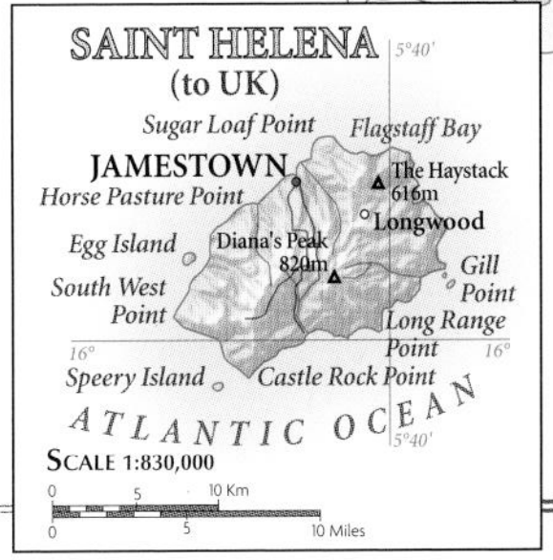
SAINT HELENA (to UK)
Sugar Loaf Point
Flagstaff Bay
JAMESTOWN
The Haystack 616m
Horse Pasture Point
Longwood
Egg Island
Diana's Peak 820m
South West Point
Gill Point
Long Range Point
Speery Island
Castle Rock Point
ATLANTIC OCEAN
SCALE 1:830,000

◀ ***Rocky breakwaters have*** *been built along the coast of Ghana to protect local fishing boats from being destroyed by powerful Atlantic waves.*

TRISTAN DA CUNHA (to UK)
Big Point
Rookery Point
EDINBURGH
Anchorstock Point
Sandy Point
Queen Mary's Peak 2060m
Lyon Point
Longbluff
Stonybeach Bay
Cave Point
Stonyhill Point
ATLANTIC OCEAN
SCALE 1:830,000

Ocean map key

Sea depth
- Sea level
- 200m / 656ft
- 1000m / 3281ft
- 2000m / 6562ft
- 3000m / 9843ft
- 4000m / 13,124ft
- 5000m / 16,400ft
- 6000m / 19,686ft

Inset map key

Population
- ◎ 100,000 to 500,000
- ⊕ 50,000 to 100,000
- ○ 10,000 to 50,000
- ○ below 10,000

Elevation
- 1000m / 3281ft
- 500m / 1640ft
- 250m / 820ft
- 100m / 328ft
- sea level

Africa

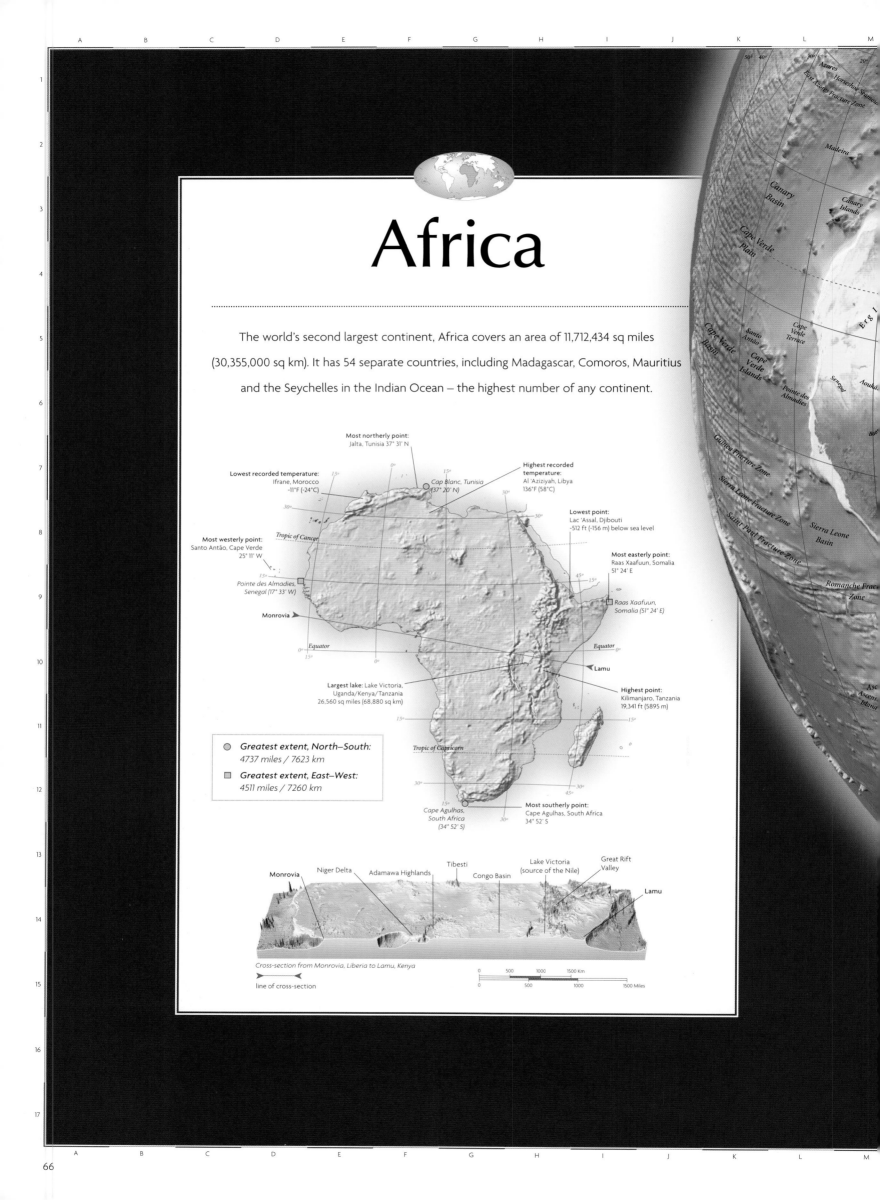

The world's second largest continent, Africa covers an area of 11,712,434 sq miles (30,355,000 sq km). It has 54 separate countries, including Madagascar, Comoros, Mauritius and the Seychelles in the Indian Ocean – the highest number of any continent.

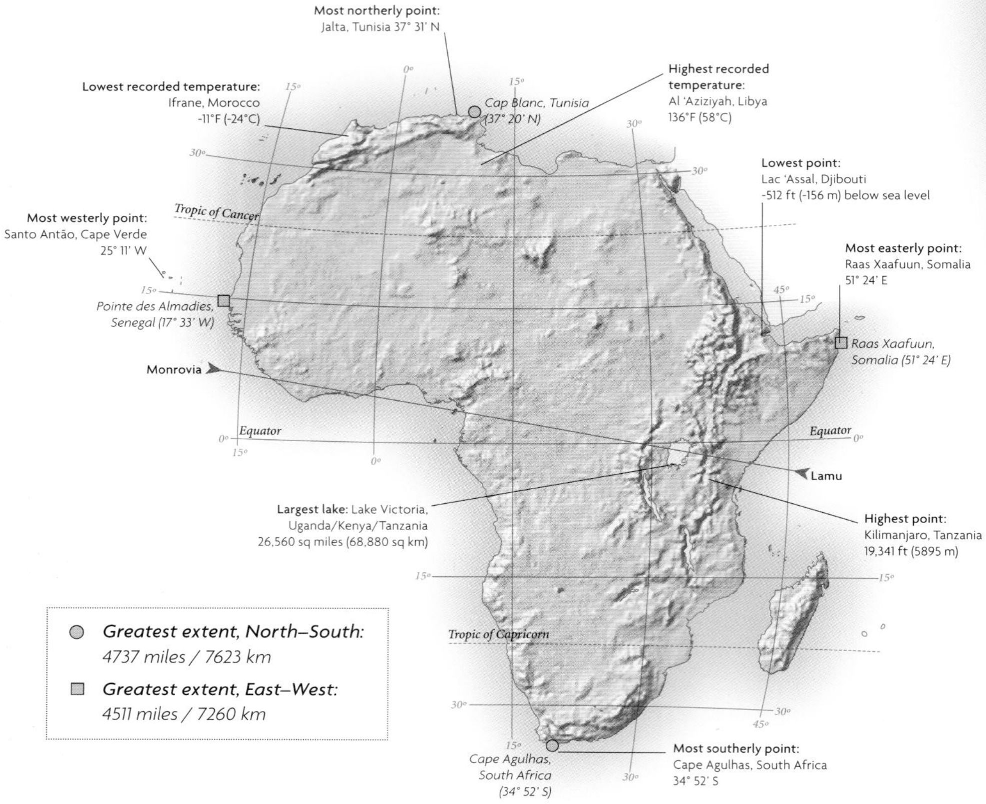

Most northerly point:
Jalta, Tunisia 37° 31' N
Lowest recorded temperature:
Ifrane, Morocco
-11°F (-24°C)
Cap Blanc, Tunisia
(37° 20' N)
Highest recorded temperature:
Al 'Aziziyah, Libya
136°F (58°C)
Lowest point:
Lac 'Assal, Djibouti
-512 ft (-156 m) below sea level
Tropic of Cancer
Most westerly point:
Santo Antão, Cape Verde
25° 11' W
Most easterly point:
Raas Xaafuun, Somalia
51° 24' E
Pointe des Almadies,
Senegal (17° 33' W)
Raas Xaafuun,
Somalia (51° 24' E)
Monrovia
Equator
Equator
Lamu
Largest lake: Lake Victoria,
Uganda/Kenya/Tanzania
26,560 sq miles (68,880 sq km)
Highest point:
Kilimanjaro, Tanzania
19,341 ft (5895 m)
Tropic of Capricorn
Greatest extent, North–South:
4737 miles / 7623 km
Greatest extent, East–West:
4511 miles / 7260 km
Cape Agulhas,
South Africa
(34° 52' S)
Most southerly point:
Cape Agulhas, South Africa
34° 52' S

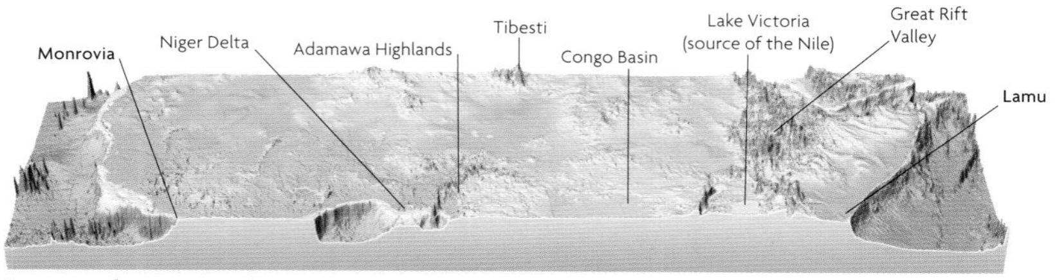

Monrovia
Niger Delta
Adamawa Highlands
Tibesti
Congo Basin
Lake Victoria
(source of the Nile)
Great Rift
Valley
Lamu
0
500
1000
1500 Km
0
500
1000
1500 Miles

Cross-section from Monrovia, Liberia to Lamu, Kenya

line of cross-section

Azores
Horseshoe Seamounts
East Azores Fracture Zone
Madeira
Canary Basin
Canary Islands
Cape Verde Plain
Cape Verde Terrace
Santo Antão
Cape Verde Basin
Cape Verde Islands
Pointe des Almadies
Senegal
Guinea Fracture Zone
Sierra Leone Fracture Zone
Saint Paul Fracture Zone
Sierra Leone Basin
Romanche Fracture Zone

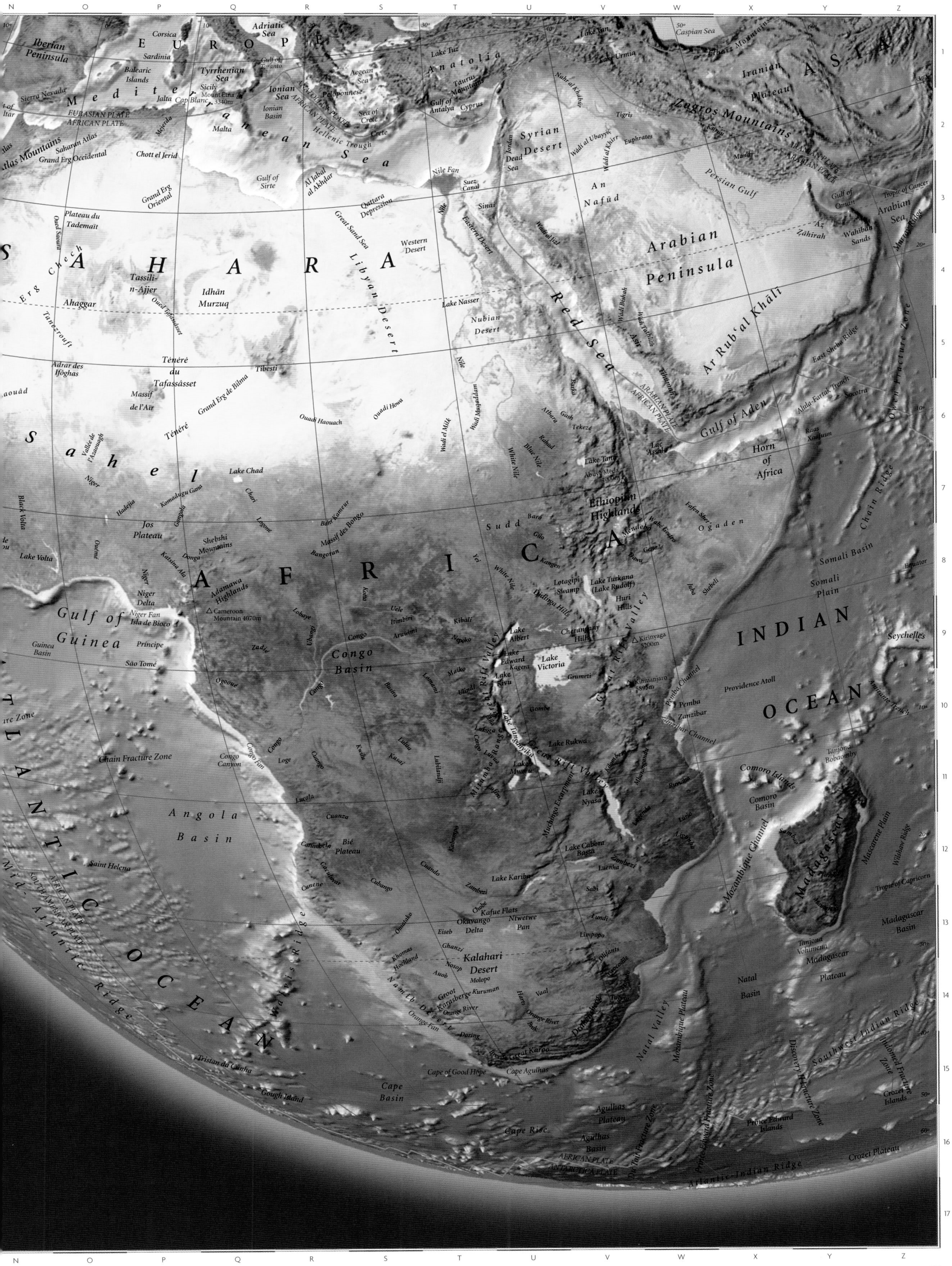

N O P Q R S T U V W X Y Z
1 2 3 4 5 6 7 8 9 10 11 12 13 14 15 16 17
Corsica
Adriatic Sea
Iberian Peninsula
E U R O P E
Sardinia
Balearic Islands
Tyrrhenian Sea
Gulf of Taranto
Sicily
Mount Etna 3340m
Ionian Sea
Ionian Basin
Sierra Nevada
M e d i t e r r a n e a n S e a
Jalta
Cap Blanc
Malta
EURASIAN PLATE
AFRICAN PLATE
Mejerda
Aegean Sea
Peloponnese
Sea of Crete
Crete
Hellenic Trough
Lake Tuz
A n a t o l i a
Taurus Mountains
Gulf of Antalya
Cyprus
Lake Van
Lake Urmia
Caspian Sea
Elburz Mountains
Iranian Plateau
A S I A
Zagros Mountains
Tigris
Euphrates
Saharan Atlas
Atlas Mountains
Grand Erg Occidental
Chott el Jerid
Gulf of Sirte
Al Jabal al Akhdar
Nile Fan
Suez Canal
Sinai
Jordan
Dead Sea
Syrian Desert
Wadi al Ubayyiḍ
An Nafūd
Persian Gulf
Gulf of Oman
Tropic of Cancer
Arabian Sea
Wahiba Sands
Grand Erg Oriental
Qattara Depression
Plateau du Tademaït
Great Sand Sea
Nile
Eastern Desert
Western Desert
Arabian Peninsula
S A H A R A
Erg Chech
Tassili-n-Ajjer
Idhān Murzuq
Ahaggar
Libyan Desert
Lake Nasser
Nubian Desert
Red Sea
Ar Rubʻ al Khālī
Tanezrouft
Adrar des Ifôghas
Ténéré du Tafassâsset
Tibesti
Massif de l'Aïr
Grand Erg de Bilma
Ténéré
Ouadi Haouach
Ouadi Howa
Wadi el Milk
Wadi Muqaddam
ARABIAN PLATE
AFRICAN PLATE
Gulf of Aden
Socotra
Horn of Africa
S a h e l
Lake Chad
Chari
Niger
Black Volta
Lake Volta
Atbara
Tekezē
Blue Nile
White Nile
Lake Tana
Ethiopian Highlands
Lac Assal
Ogaden
Chain Ridge
Owen Fracture Zone
Hadejia
Komadugu Gana
Jos Plateau
Gongola
Logone
Bahr Kameur
Massif des Bongo
Sudd
Baro
Shebshi Mountains
Bangoran
A F R I C A
Somali Basin
Equator
Somali Plain
Juba
Shebeli
Lake Turkana (Lake Rudolf)
Lotagipi Swamp
Huri Hills
Adamawa Highlands
Niger Delta
Cameroon Mountain 4070m
Gulf of Guinea
Niger Fan
Isla de Bioco
Guinea Basin
Principe
São Tomé
Lobaye
Ubangi
Uele
Kibali
Lake Albert
Lake Edward
Lake Kivu
Kagera
Lake Victoria
Cherangany Hills
Kirinyaga 5200m
I N D I A N O C E A N
Seychelles
Congo
Congo Basin
Zadié
Ogooué
Lomami
Grumeti
Kilimanjaro 5895m
Pemba
Zanzibar
Pemba Channel
Zanzibar Channel
Providence Atoll
Lake Tanganyika
Lake Rukwa
Gombe
Chain Fracture Zone
Congo Canyon
Congo Fan
Loge
Kasai
Kwilu
Lulua
Lukuga
Lake Mweru
Comoro Islands
Comoro Basin
Tanjona Bobaomby
Lucala
Cuanza
Angola Basin
Bié Plateau
Lake Nyasa
Muchinga Escarpment
Lake Cabora Bassa
Zambezi
Mozambique Channel
Madagascar
Mascarene Plain
Saint Helena
Cubango
Cunene
Cuando
Lake Kariba
Sabi
Tropic of Capricorn
Kafue Flats
Okavango Delta
Ntwetwe Pan
Limpopo
Madagascar Basin
Tanjona Vohimena
Madagascar Plateau
Eiseb
Ghanzi
Khomas Hochland
Kalahari Desert
Nosop
Molopo
Auob
Kuruman
Groot Karasberge
Orange River
Orange Fan
Namib Desert
Walvis Ridge
Vaal
Natal Basin
Natal Valley
Mozambique Plateau
Southwest Indian Ridge
Great Karoo
Cape of Good Hope
Cape Agulhas
Tristan da Cunha
Cape Basin
Gough Island
Agulhas Plateau
Agulhas Basin
Cape Rise
Prince Edward Islands
Crozet Islands
Crozet Plateau
AFRICAN PLATE
ANTARCTIC PLATE
Atlantic-Indian Ridge
Mid-Atlantic Ridge
AFRICAN PLATE
SOUTH AMERICAN PLATE
A T L A N T I C O C E A N

Physical Africa

The structure of Africa was dramatically influenced by the break up of the supercontinent Gondwanaland about 160 million years ago and, more recently, rifting and hot spot activity. Today, much of Africa is remote from active plate boundaries and comprises a series of extensive plateaux and deep basins, which influence the drainage patterns of major rivers. The relief rises to the east, where volcanic uplands and vast lakes mark the Great Rift Valley. In the far north and south sedimentary rocks have been folded to form the Atlas Mountains and the Great Karoo.

Northern Africa

Northern Africa comprises a system of basins and plateaux. The Tibesti and Ahaggar are volcanic uplands, whose uplift has been matched by subsidence within large surrounding basins. Many of the basins have been infilled with sand and gravel, creating the vast Saharan lands. The Atlas Mountains in the north were formed by convergence of the African and Eurasian plates.

The Earth's crust has been warped to form the Taoudenni Basin
Volcanic Ahaggar mountains, formed by rising magma from a hot spot
Lake Chad lies in a sand-filled basin
A
A

***Section across northern** Africa showing infilled basins and uplifted plateaux.*

0 250 500 Km
0 250 500 Miles

East Africa

The Great Rift Valley is the most striking feature of this region, running for 4475 miles (7200 km) from Lake Nyasa to the Red Sea. North of Lake Nyasa it splits into two arms and encloses an interior plateau which contains Lake Victoria. A number of elongated lakes and volcanoes lie along the fault lines. To the west lies the Congo Basin, a vast, shallow depression, which rises to form an almost circular rim of highlands.

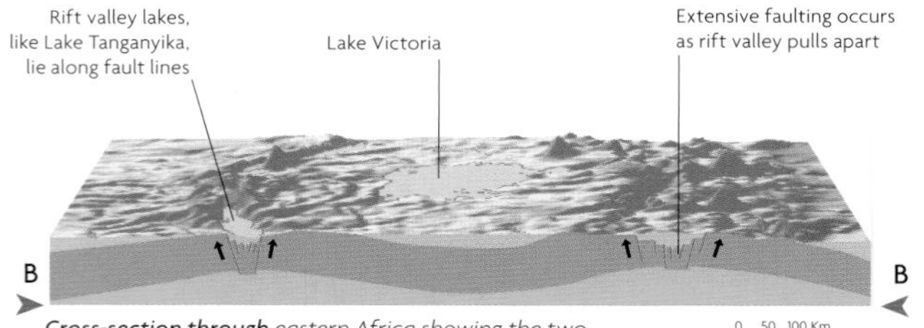

Rift valley lakes, like Lake Tanganyika, lie along fault lines
Lake Victoria
Extensive faulting occurs as rift valley pulls apart
B
B

***Cross-section through** eastern Africa showing the two arms of the Great Rift Valley and its interior plateau.*

0 50 100 Km
0 50 100 Miles

Scale 1:40,000,000

Km 0 200 400 600 800
0 200 400 600 800 Miles

projection: Lambert Azimuthal Equal Area

Map key

Elevation

- 5000m / 16,405ft
- 4000m / 13,124ft
- 3000m / 9843ft
- 2000m / 6562ft
- 1000m / 3281ft
- 500m / 1640ft
- 250m / 820ft
- 100m / 328ft
- sea level
- below sea level

Plate margins
(for explanation see page xiv)

- constructive
- destructive
- conservative
- uncertain
- line of cross-section

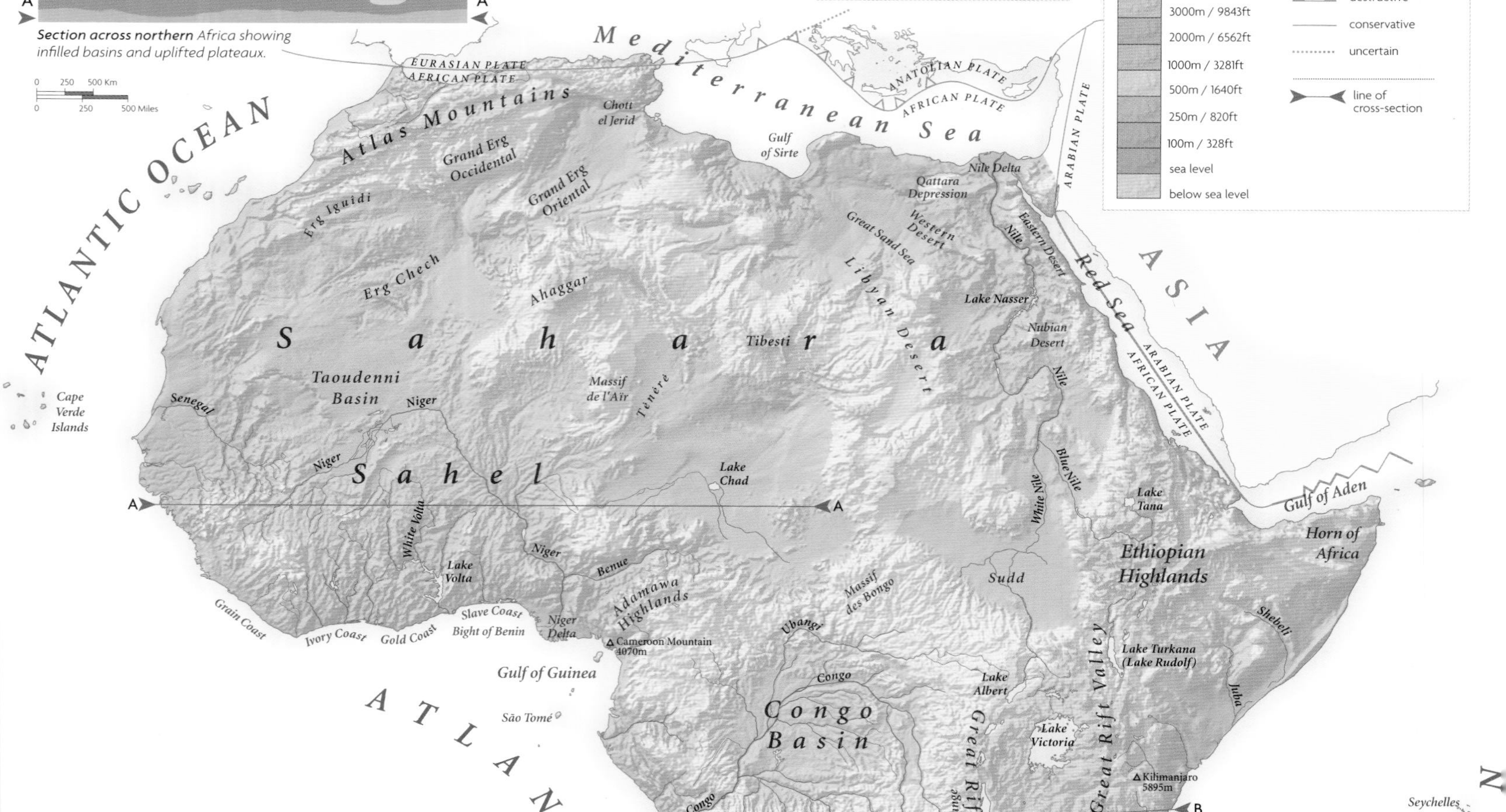
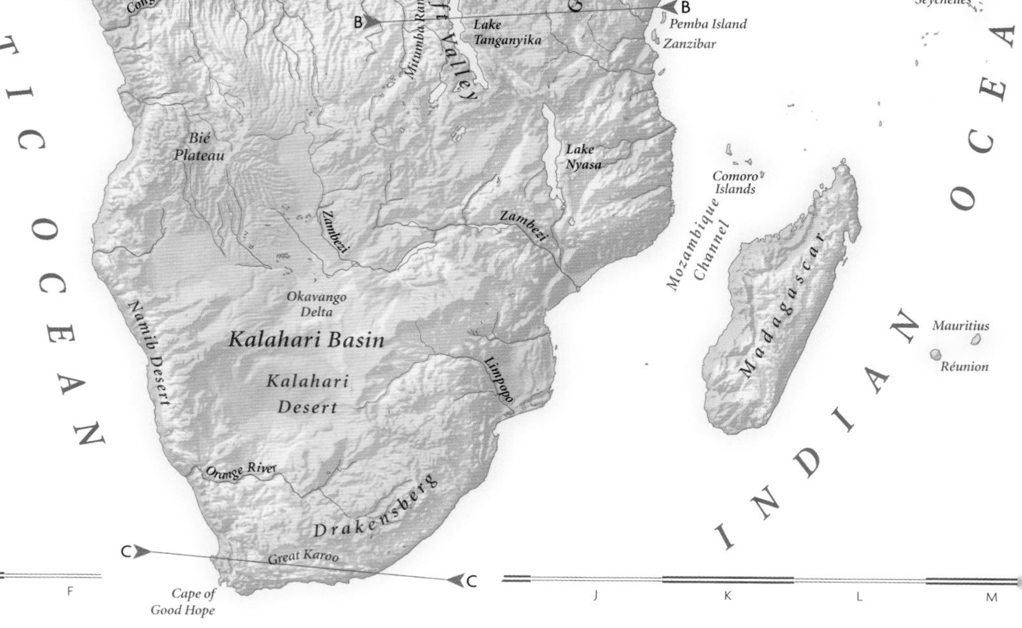

Mediterranean Sea
EURASIAN PLATE
AFRICAN PLATE
ANATOLIAN PLATE
AFRICAN PLATE
ARABIAN PLATE
ATLANTIC OCEAN
Atlas Mountains
Chott el Jerid
Gulf of Sirte
Grand Erg Occidental
Grand Erg Oriental
Erg Iguidi
Erg Chech
Ahaggar
Nile Delta
Qattara Depression
Western Desert
Great Sand Sea
Libyan Desert
Eastern Desert
Nile
Red Sea
ASIA
Lake Nasser
Nubian Desert
ARABIAN PLATE
AFRICAN PLATE
Sahara
Tibesti
Taoudenni Basin
Massif de l'Aïr
Ténéré
Cape Verde Islands
Senegal
Niger
Sahel
Lake Chad
A
A
Blue Nile
White Nile
Lake Tana
Gulf of Aden
Horn of Africa
Ethiopian Highlands
White Volta
Lake Volta
Niger
Benue
Adamawa Highlands
Massif des Bongo
Sudd
Shebeli
Grain Coast
Ivory Coast
Gold Coast
Slave Coast
Bight of Benin
Niger Delta
Cameroon Mountain 4070m
Ubangi
Lake Turkana (Lake Rudolf)
Gulf of Guinea
Congo
Lake Albert
Juba
São Tomé
Congo Basin
Lake Victoria
Great Rift Valley
ATLANTIC OCEAN
Kilimanjaro 5895m
Congo
B
B
Mitumba Range
Lake Tanganyika
Pemba Island
Zanzibar
Seychelles
INDIAN OCEAN
Bié Plateau
Lake Nyasa
Comoro Islands
Zambezi
Zambezi
Mozambique Channel
Madagascar
Okavango Delta
Namib Desert
Kalahari Basin
Kalahari Desert
Limpopo
Mauritius
Réunion
Orange River
Drakensberg
C
Great Karoo
C
Cape of Good Hope

Southern Africa

The Great Escarpment marks the southern boundary of Africa's basement rock and includes the Drakensberg range. It was uplifted when Gondwanaland fragmented about 160 million years ago and it has gradually been eroded back from the coast. To the north, the relief drops steadily, forming the Kalahari Basin. In the far south are the fold mountains of the Great Karoo.

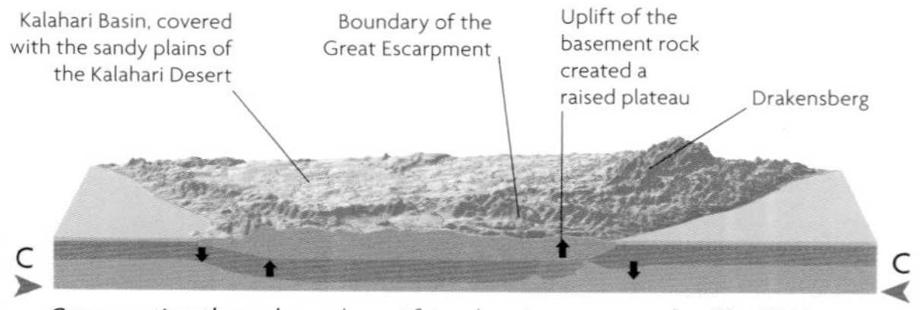

Kalahari Basin, covered with the sandy plains of the Kalahari Desert
Boundary of the Great Escarpment
Uplift of the basement rock created a raised plateau
Drakensberg
C
C

***Cross-section through** southern Africa showing the boundary of the Great Escarpment.*

0 100 200 Km
0 100 200 Miles

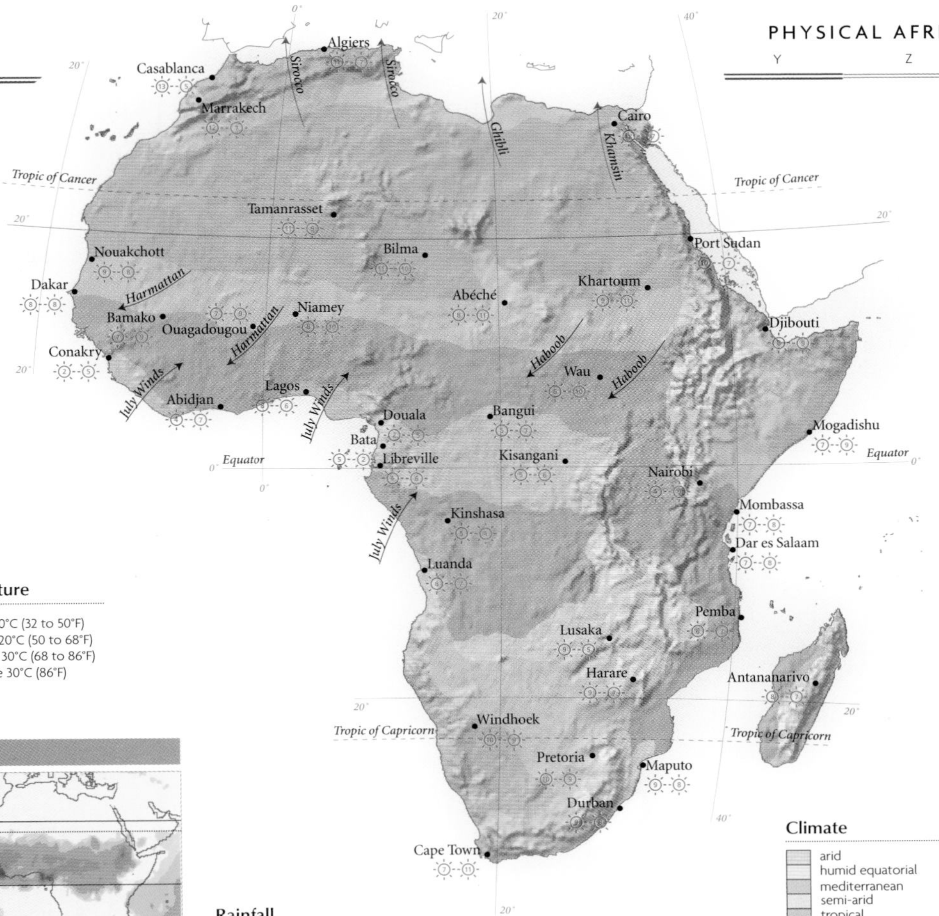
Algiers
Casablanca
Marrakech
Sirocco
Ghibli
Cairo
Khamsin
Tropic of Cancer
Tamanrasset
Nouakchott
Bilma
Port Sudan
Dakar
Harmattan
Bamako
Ouagadougou
Niamey
Abéché
Khartoum
Djibouti
Conakry
Haboob
July Winds
Abidjan
Lagos
Wau
Douala
Bangui
Mogadishu
Bata
Libreville
Kisangani
Equator
Nairobi
Kinshasa
Mombassa
Dar es Salaam
Luanda
Pemba
Lusaka
Harare
Antananarivo
Windhoek
Tropic of Capricorn
Pretoria
Maputo
Durban
Cape Town

Climate

▲ ***Savannah grasslands run** in a belt across Africa; limited rainfall inhibits tree growth.*

The climates of Africa range from mediterranean to arid, dry savannah and humid equatorial. In East Africa, where snow settles at the summit of volcanoes such as Kilimanjaro, climate is also modified by altitude. The winds of the Sahara export millions of tonnes of dust a year both northwards and eastwards.

Temperature

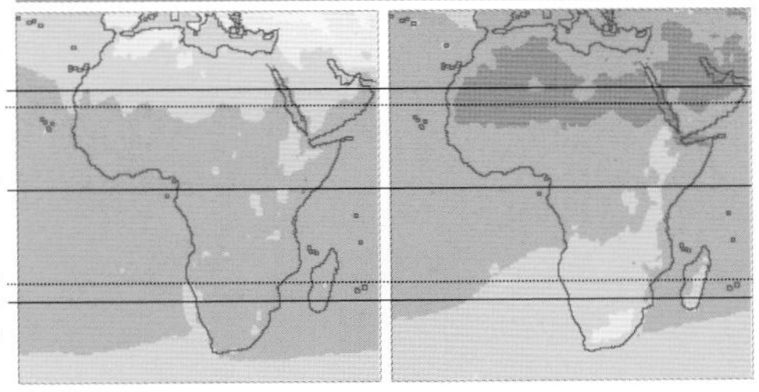
Tropic of Cancer
20° N
Equator
20° S
Tropic of Capricorn

Average January temperature *Average July temperature*

Temperature

- 0 to 10°C (32 to 50°F)
- 10 to 20°C (50 to 68°F)
- 20 to 30°C (68 to 86°F)
- above 30°C (86°F)

▲ ***The hot, equatorial** basin of the Congo river receives over 48 inches (1200 mm) of rainfall per year.*

Rainfall

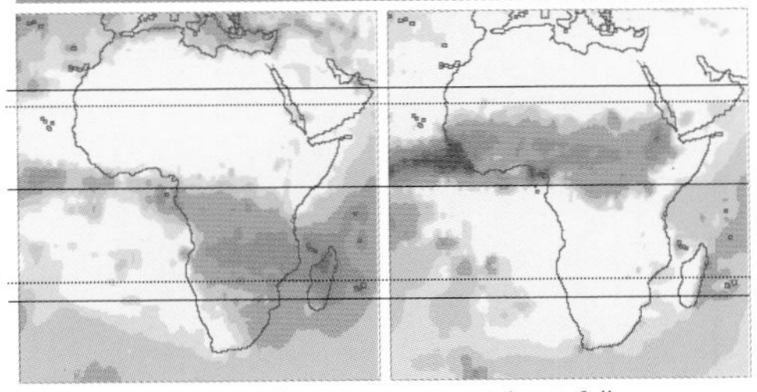
Tropic of Cancer
20° N
Equator
20° S
Tropic of Capricorn

Average January rainfall *Average July rainfall*

Rainfall

- 0–25 mm (0–1 in)
- 25–50 mm (1–2 in)
- 50–100 mm (2–4 in)
- 100–200 mm (4–8 in)
- 200–300 mm (8–12 in)
- 300–400 mm (12–16 in)
- 400–500 mm (16–20 in)
- more than 500 mm (20 in)

Climate

- arid
- humid equatorial
- mediterranean
- semi-arid
- tropical
- warm humid
- daily hours of sunshine, January
- daily hours of sunshine, July
- cold wind
- hot wind

Shaping the continent

African landscapes are shaped by the intensity of climatic extremes and by tectonic action. High aridity, wind action and infrequent but heavy rainstorms, lead to the migration of sand dunes and dramatic flash flooding across much of the north and west. In the wetter areas, high precipitation increases the rate of weathering. To the east, the rift system has created a volcanic and lake environment and allowed rivers to erode weaknesses left in the crustal structure by faults.

Groundwater

1 Oases are found in desert areas such as the Sahara *(left)*. Groundwater migrates through permeable rock strata, confined between two impermeable layers. Oases form either when the permeable rocks come near to the surface, or at a fault line, when water is able to seep up to the surface through the crushed rocks at the fault.

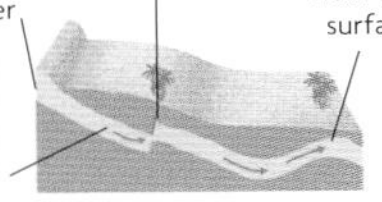
Rainwater feeds the aquifer
Water migrates up through fault
Aquifer exposed near the surface
Groundwater trapped between impermeable strata

***Groundwater:** Replenishment of an oasis*

River systems

2 The Zambezi river *(above)* drops 360 ft (110 m) over the Victoria Falls into a zig-zag gorge. The river has eroded the gorge along lines of weakness in the bedrock, created by fault lines running in two directions.

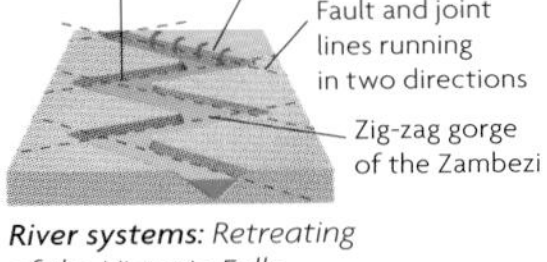
Old site of Victoria Falls
River plunges over falls
Fault and joint lines running in two directions
Zig-zag gorge of the Zambezi

***River systems:** Retreating of the Victoria Falls*

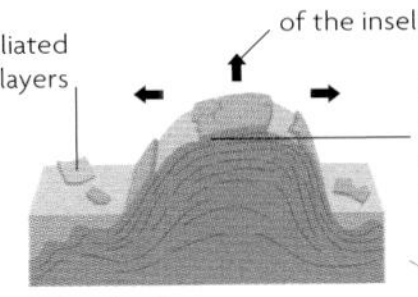
Exfoliated layers
External stresses act on the surface of the inselberg
Joints or cracks caused by expansion and contraction

***Weathering:** Formation of an inselberg*

Weathering

6 Inselbergs *(above)*, found extensively across West Africa, are exposed remnants of an extensive upland area. Erosion of the surrounding uplands leaves a resistant rock outcrop. Its spheroidal shape is the result of 'onion-skin' weathering – the exfoliating of layers – due to repeated expansion and contraction.

The evolving landscape

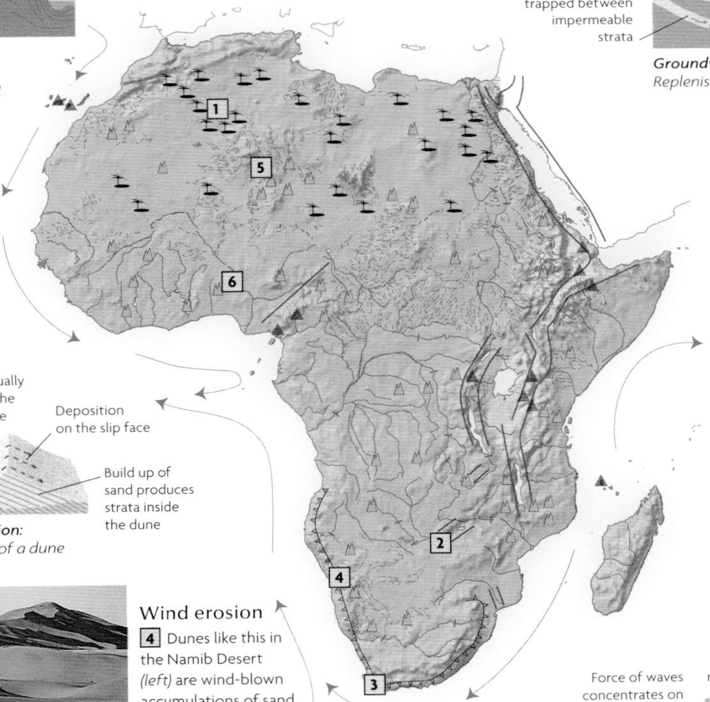
1
5
6
2
4
3

Landscape

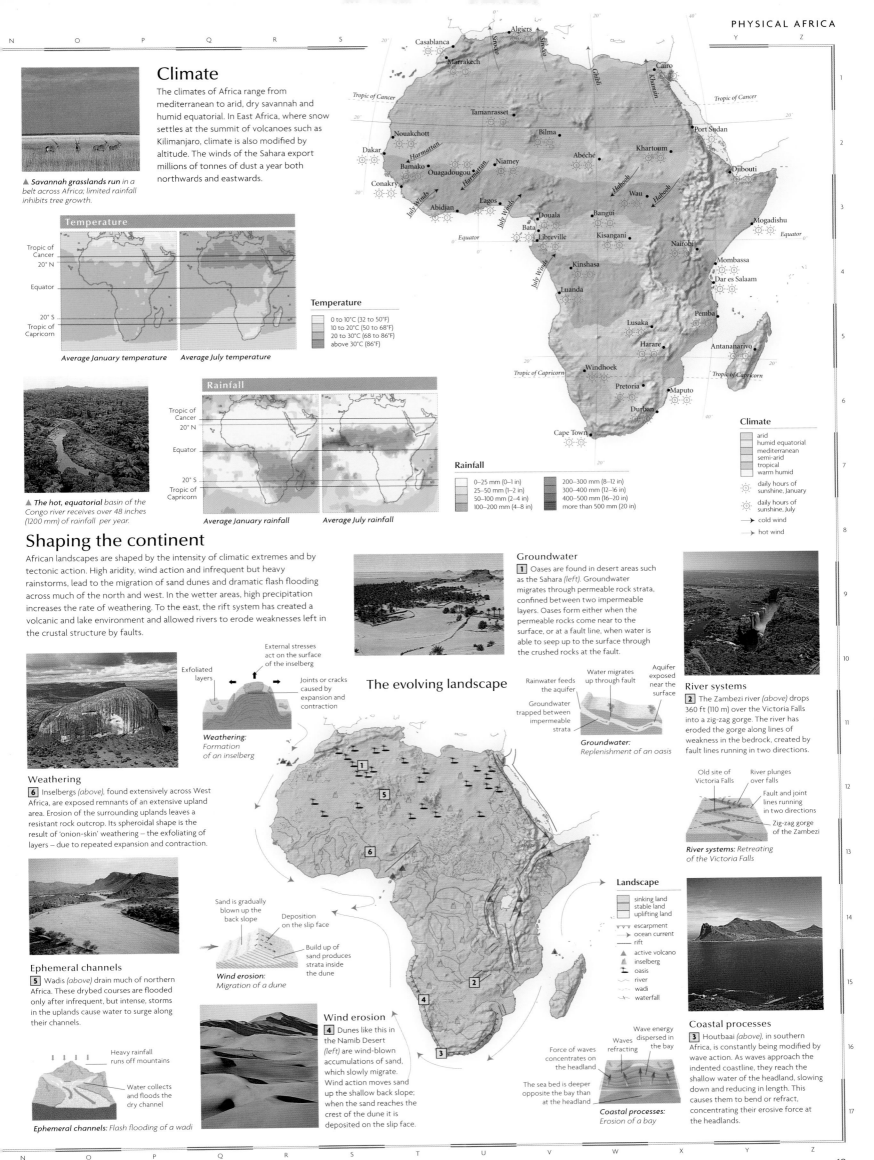

- sinking land
- stable land
- uplifting land
- escarpment
- ocean current
- rift
- active volcano
- inselberg
- oasis
- river
- wadi
- waterfall

Ephemeral channels

5 Wadis *(above)* drain much of northern Africa. These drybed courses are flooded only after infrequent, but intense, storms in the uplands cause water to surge along their channels.

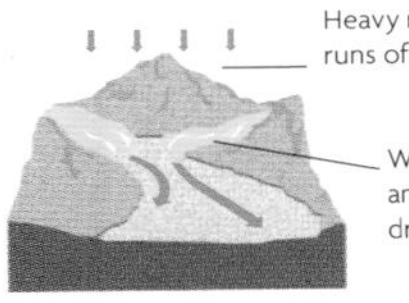
Heavy rainfall runs off mountains
Water collects and floods the dry channel

***Ephemeral channels:** Flash flooding of a wadi*

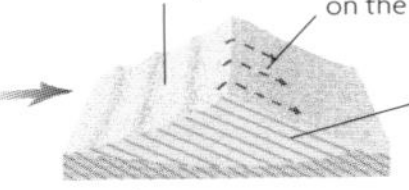
Sand is gradually blown up the back slope
Deposition on the slip face
Build up of sand produces strata inside the dune

***Wind erosion:** Migration of a dune*

Wind erosion

4 Dunes like this in the Namib Desert *(left)* are wind-blown accumulations of sand, which slowly migrate. Wind action moves sand up the shallow back slope; when the sand reaches the crest of the dune it is deposited on the slip face.

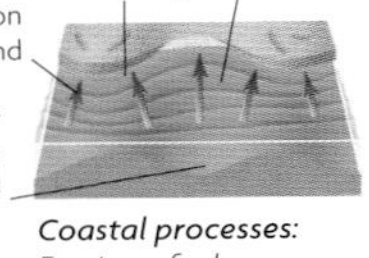
Wave energy dispersed in the bay
Waves refracting
Force of waves concentrates on the headland
The sea bed is deeper opposite the bay than at the headland

***Coastal processes:** Erosion of a bay*

Coastal processes

3 Houtbaai *(above)*, in southern Africa, is constantly being modified by wave action. As waves approach the indented coastline, they reach the shallow water of the headland, slowing down and reducing in length. This causes them to bend or refract, concentrating their erosive force at the headlands.

Political Africa

The political map of modern Africa only emerged following the end of the Second World War. Over the next half-century, all of the countries formerly controlled by European powers gained independence from their colonial rulers – only Liberia and Ethiopia were never colonized.

The post-colonial era has not been an easy period for many countries, but there have been moves towards multi-party democracy across much of the continent. In South Africa, democratic elections replaced the internationally-condemned apartheid system only in 1994. Other countries have still to find political stability; corruption in government and ethnic tensions are serious problems. National infrastructures, based on the colonial transport systems built to exploit Africa's resources, are often inappropriate for independent economic development.

Madeira
(to Portugal)
PORTUG
Casabl
Saf
Marrak
MOROCC
Agadir
Canary Islands
(to Spain)
LAÂYOUNE
Western
Sahara
(Occupied
by Morocco)
Tropic of Cancer
S
MAURITANI
NOUAKCHOTT
CAPE
VERDE
PRAIA
Senegal
SENEGAL
DAKAR
Kaolack
GAMBIA
BANJUL
BAMAKO
GUINEA-
BISSAU
BISSAU
GUINEA
CONAKRY
Koidu
FREETOWN
SIERRA LEONE
YAMOUSSOUK
MONROVIA
LIBERIA

Languages

Three major world languages act as *lingua francas* across the African continent: Arabic in North Africa; English in southern and eastern Africa and Nigeria; and French in Central and West Africa, and in Madagascar. A huge number of African languages are spoken as well – over 2000 have been recorded, with more than 400 in Nigeria alone – reflecting the continuing importance of traditional cultures and values. In the north of the continent, the extensive use of Arabic reflects Middle Eastern influences while Bantu languages are widely spoken across much of southern Africa.

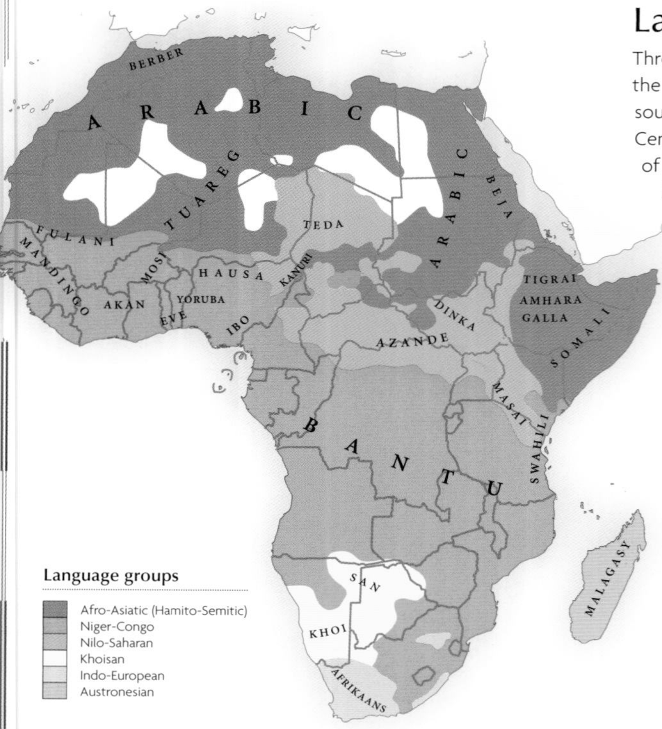
BERBER
ARABIC
TUAREG
TEDA
ARABIC
BEJA
FULANI
MANDINGO
MOSI
HAUSA
KANURI
TIGRAI
AMHARA
GALLA
AKAN
YORUBA
EVE
IBO
DINKA
SOMALI
AZANDE
MASAI
SWAHILI
BANTU
MALAGASY
SAN
KHOI
AFRIKAANS
Language groups
Afro-Asiatic (Hamito-Semitic)
Niger-Congo
Nilo-Saharan
Khoisan
Indo-European
Austronesian

Official African languages

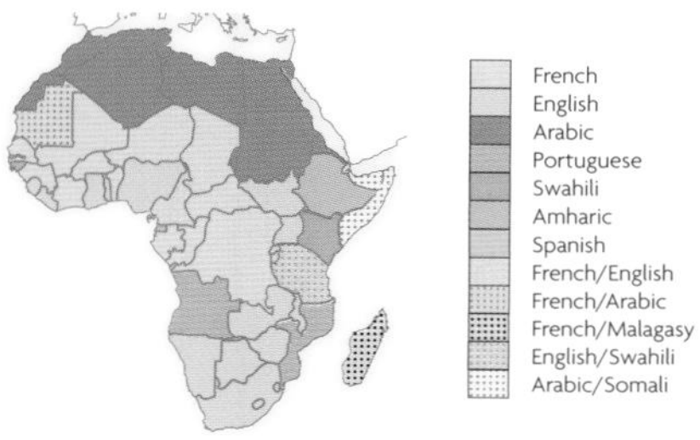
French
English
Arabic
Portuguese
Swahili
Amharic
Spanish
French/English
French/Arabic
French/Malagasy
English/Swahili
Arabic/Somali

▲ *Islamic influences are evident throughout North Africa. The Great Mosque at Kairouan, Tunisia, is Africa's holiest Islamic place.*

▲ *In northeastern Nigeria, people speak Kanuri – a dialect of the Nilo-Saharan language group.*

Transport

African railways were built to aid the exploitation of natural resources, and most offer passage only from the interior to the coastal cities, leaving large parts of the continent untouched – five land-locked countries have no railways at all. The Congo, Nile and Niger river networks offer limited access to land within the continental interior, but have a number of waterfalls and cataracts which prevent navigation from the sea. Many roads were developed in the 1960s and 1970s, but economic difficulties are making the maintenance and expansion of the networks difficult.

▶ *South Africa has the largest concentration of railways in Africa. Over 20,000 miles (32,000 km) of routes have been built since 1870.*

▲ *Traditional means of transport, such as the camel, are still widely used across the less accessible parts of Africa.*

◀ *The Congo river, though not suitable for river transport along its entire length, forms a vital link for people and goods in its navigable inland reaches.*

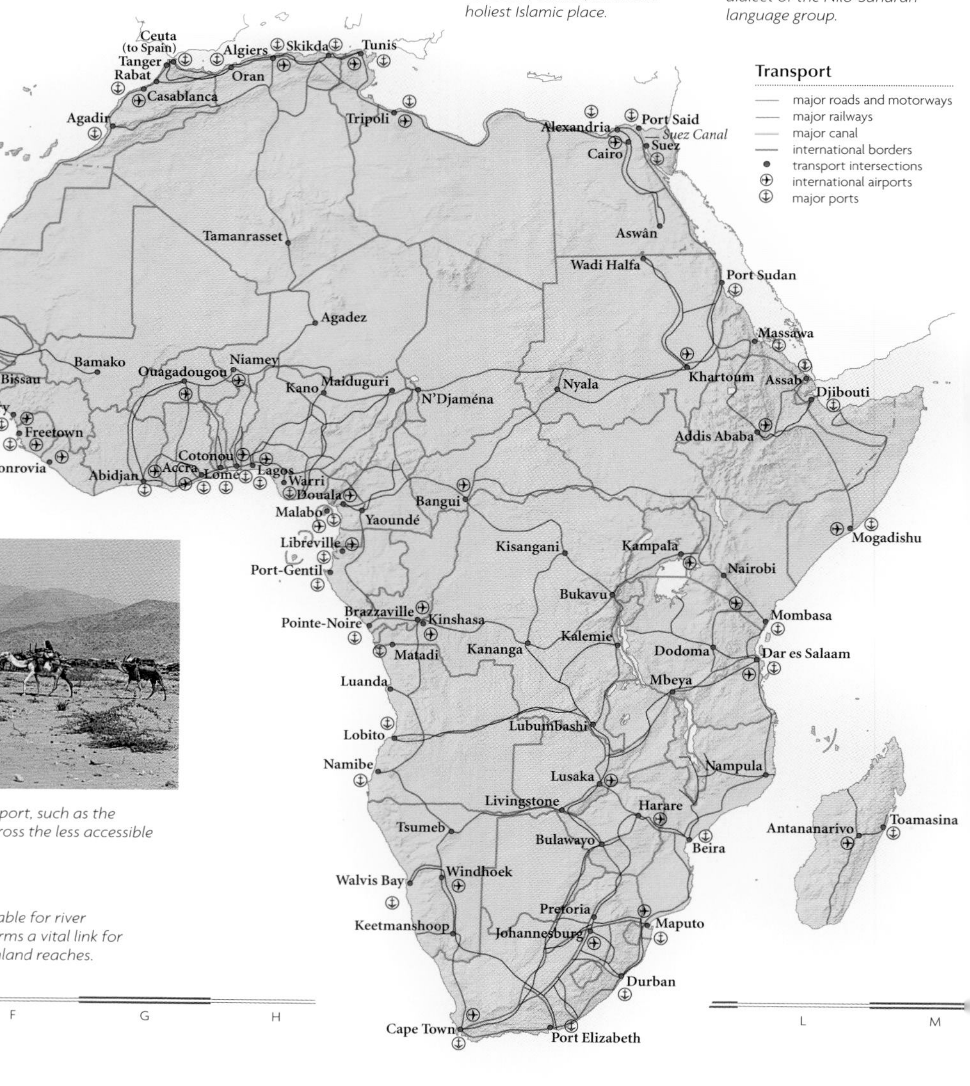
Ceuta
(to Spain)
Tanger
Rabat
Algiers
Skikda
Tunis
Oran
Casablanca
Agadir
Tripoli
Alexandria
Port Said
Suez Canal
Cairo
Suez
Transport
major roads and motorways
major railways
major canal
international borders
transport intersections
international airports
major ports
Tamanrasset
Aswân
Nouâdhibou
Wadi Halfa
Port Sudan
Nouakchott
Agadez
Massawa
Dakar
Banjul
Bamako
Niamey
Ouagadougou
Bissau
Kano
Maiduguri
N'Djaména
Nyala
Khartoum
Assab
Djibouti
Conakry
Freetown
Addis Ababa
Monrovia
Abidjan
Accra
Cotonou
Lomé
Lagos
Warri
Douala
Bangui
Malabo
Yaoundé
Mogadishu
Libreville
Kisangani
Kampala
Nairobi
Port-Gentil
Bukavu
Brazzaville
Kinshasa
Mombasa
Pointe-Noire
Kalemie
Matadi
Kananga
Dodoma
Dar es Salaam
Luanda
Mbeya
Lobito
Lubumbashi
Namibe
Nampula
Lusaka
Livingstone
Harare
Tsumeb
Toamasina
Antananarivo
Bulawayo
Beira
Walvis Bay
Windhoek
Pretoria
Maputo
Keetmanshoop
Johannesburg
Durban
Cape Town
Port Elizabeth

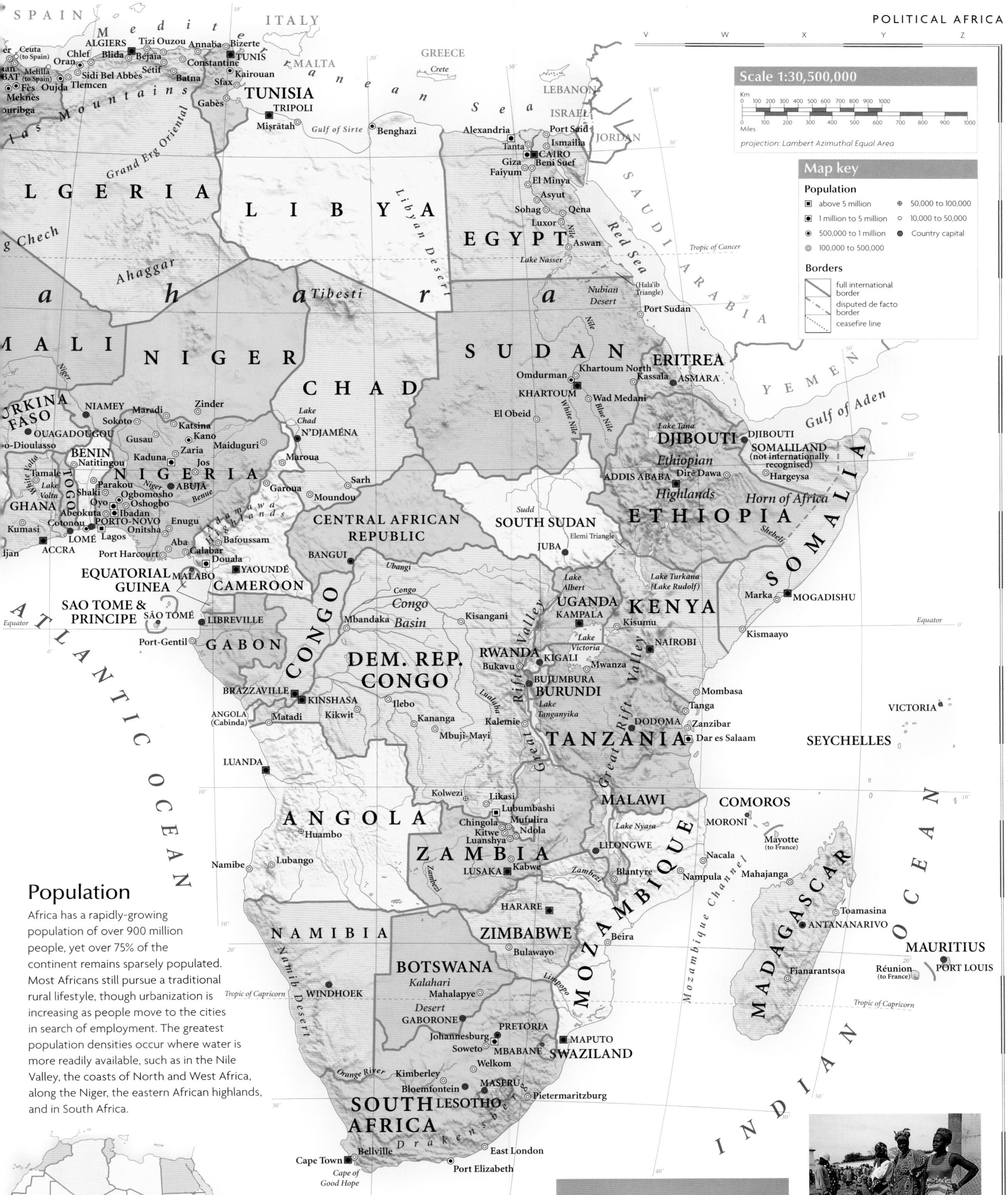

Population

Africa has a rapidly-growing population of over 900 million people, yet over 75% of the continent remains sparsely populated. Most Africans still pursue a traditional rural lifestyle, though urbanization is increasing as people move to the cities in search of employment. The greatest population densities occur where water is more readily available, such as in the Nile Valley, the coasts of North and West Africa, along the Niger, the eastern African highlands, and in South Africa.

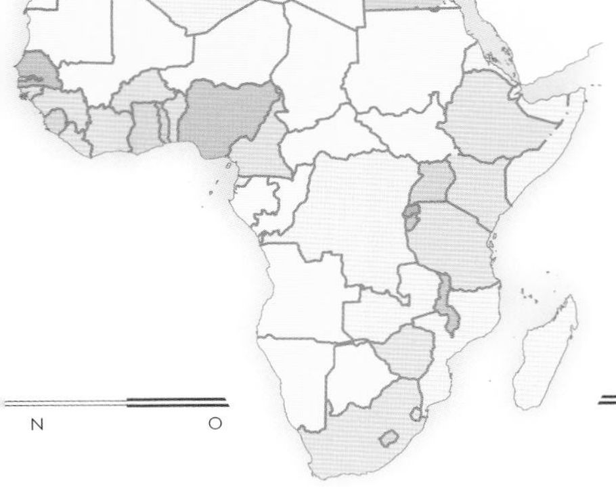

Population density (people per sq km)

- below 49
- 50–99
- 100–149
- 150–199
- 200–299
- above 300

▶ *A thin layer* of smog blankets the dusty streets of Cairo, Africa's most populous city and home to over 15 million people. In the 1990s Cairo grew at a rate of about 1500 people per day.

▲ *Thriving street markets* in Gambia's capital, Banjul, trade a variety of locally-grown produce. Africa's population is still predominantly rural-based.

African resources

The economies of most African countries are dominated by subsistence and cash crop agriculture, with limited industrialization. Manufacturing industry is largely confined to South Africa. Many countries depend on a single resource, such as copper or gold, or a cash crop, such as coffee, for export income, which can leave them vulnerable to fluctuations in world commodity prices. In order to diversify their economies and develop a wider industrial base, investment from overseas is being actively sought by many African governments.

Industry

Many African industries concentrate on the extraction and processing of raw materials. These include the oil industry, food processing, mining and textile production. South Africa accounts for over half of the continent's industrial output with much of the remainder coming from the countries along the northern coast. Over 60% of Africa's workforce is employed in agriculture.

◀ *The unspoilt natural splendour of wildlife reserves, like the Serengeti National Park in Tanzania, attract tourists to Africa from around the globe. The tourist industry in Kenya and Tanzania is particularly well developed, where it accounts for almost 10% of GNI.*

Standard of living

Since the 1960s most countries in Africa have seen significant improvements in life expectancy, healthcare and education. However, 28 of the 30 most deprived countries in the world are African, and the continent as a whole lies well behind the rest of the world in terms of meeting many basic human needs.

Standard of living
(UN human development index)
high
low

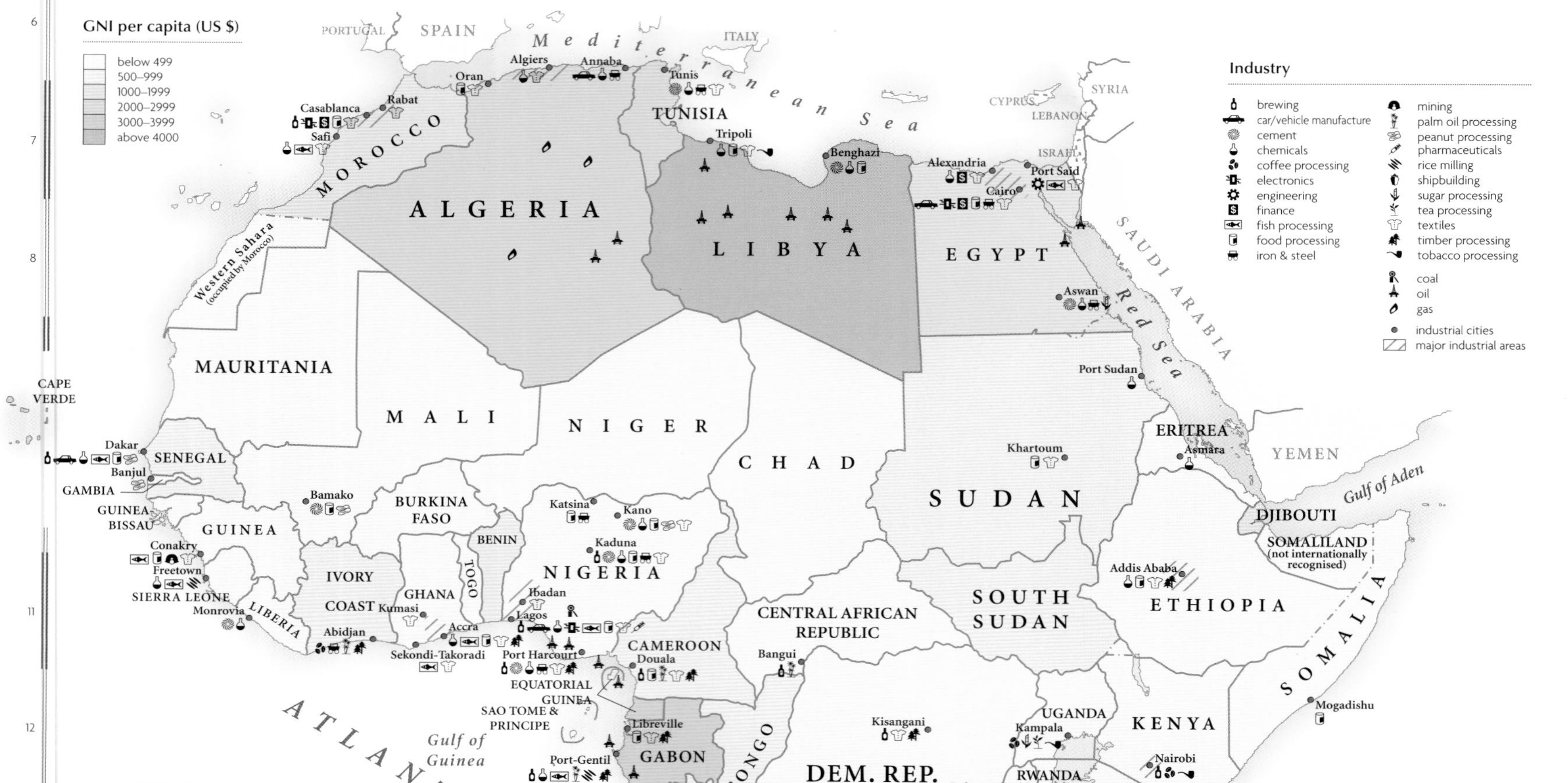

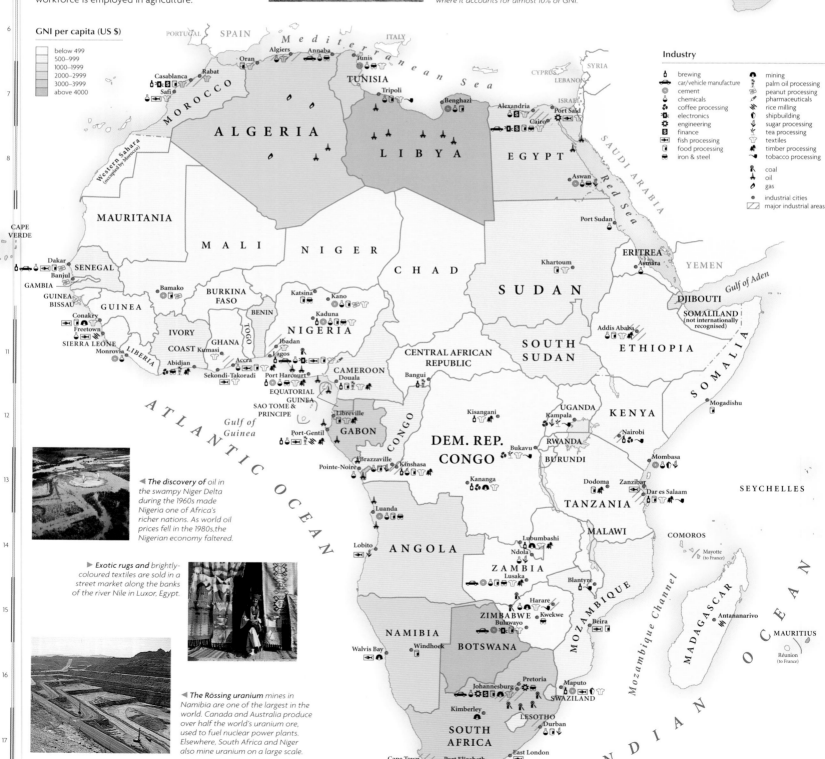

◀ *The discovery of oil in the swampy Niger Delta during the 1960s made Nigeria one of Africa's richer nations. As world oil prices fell in the 1980s, the Nigerian economy faltered.*

▶ *Exotic rugs and brightly-coloured textiles are sold in a street market along the banks of the river Nile in Luxor, Egypt.*

◀ *The Rössing uranium mines in Namibia are one of the largest in the world. Canada and Australia produce over half the world's uranium ore, used to fuel nuclear power plants. Elsewhere, South Africa and Niger also mine uranium on a large scale.*

Environmental issues

One of Africa's most serious environmental problems occurs in marginal areas such as the Sahel where scrub and forest clearance, often for cooking fuel, combined with overgrazing, are causing desertification. Game reserves in southern and eastern Africa have helped to preserve many endangered animals, although the needs of growing populations have led to conflict over land use, and poaching is a serious problem.

Environmental issues

- national parks
- tropical forest
- forest destroyed
- desert
- desertification
- polluted rivers
- radioactive contamination
- marine pollution
- heavy marine pollution
- poor urban air quality

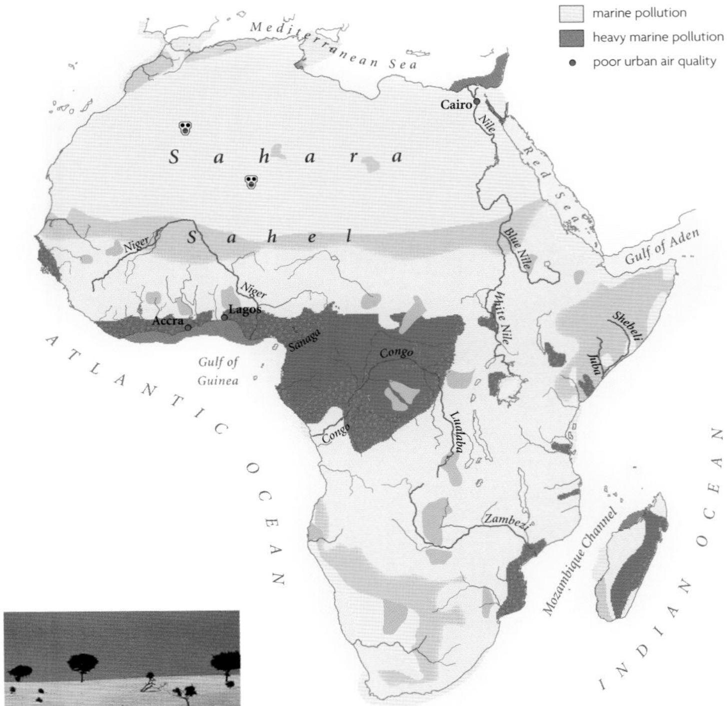

Mineral resources

Africa's ancient plateaux contain some of the world's most substantial reserves of precious stones and metals. About 15% of the world's gold is mined in South Africa; Zambia has great copper deposits; and diamonds are mined in Botswana, Dem. Rep. Congo and South Africa. Oil has brought great economic benefits to Algeria, Libya and Nigeria.

Mineral resources

- oil field
- gas field
- coal field
- bauxite
- copper
- diamonds
- gold
- iron
- phosphates
- tin
- uranium

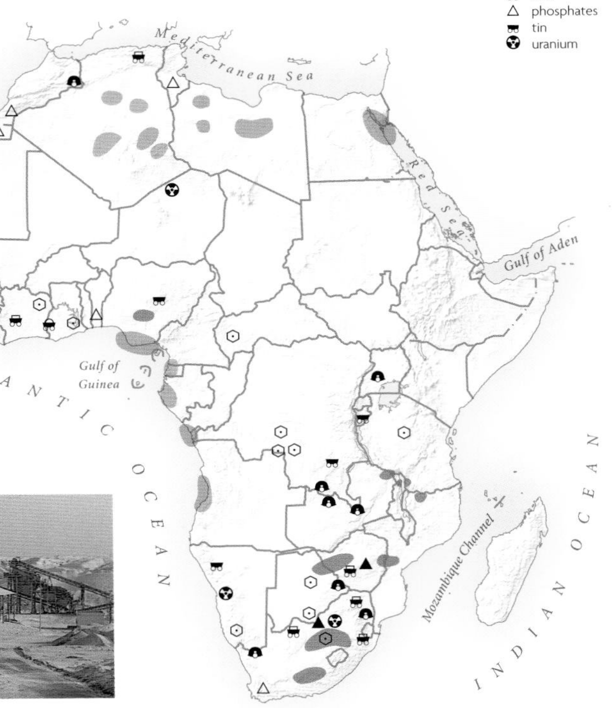

▲ *North and West Africa have large deposits of white phosphate minerals, which are used in making fertilizers. Morocco, Senegal, and Tunisia are among the continent's leading producers.*

▲ ***The Sahel's delicate** natural equilibrium is easily destroyed by the clearing of vegetation, drought and overgrazing. This causes the Sahara to advance south, engulfing the savannah grasslands.*

▲ ***Workers on a tea plantation** gather one of Africa's most important cash crops, providing a valuable source of income. Coffee, rubber, bananas, cotton and cocoa are also widely grown as cash crops.*

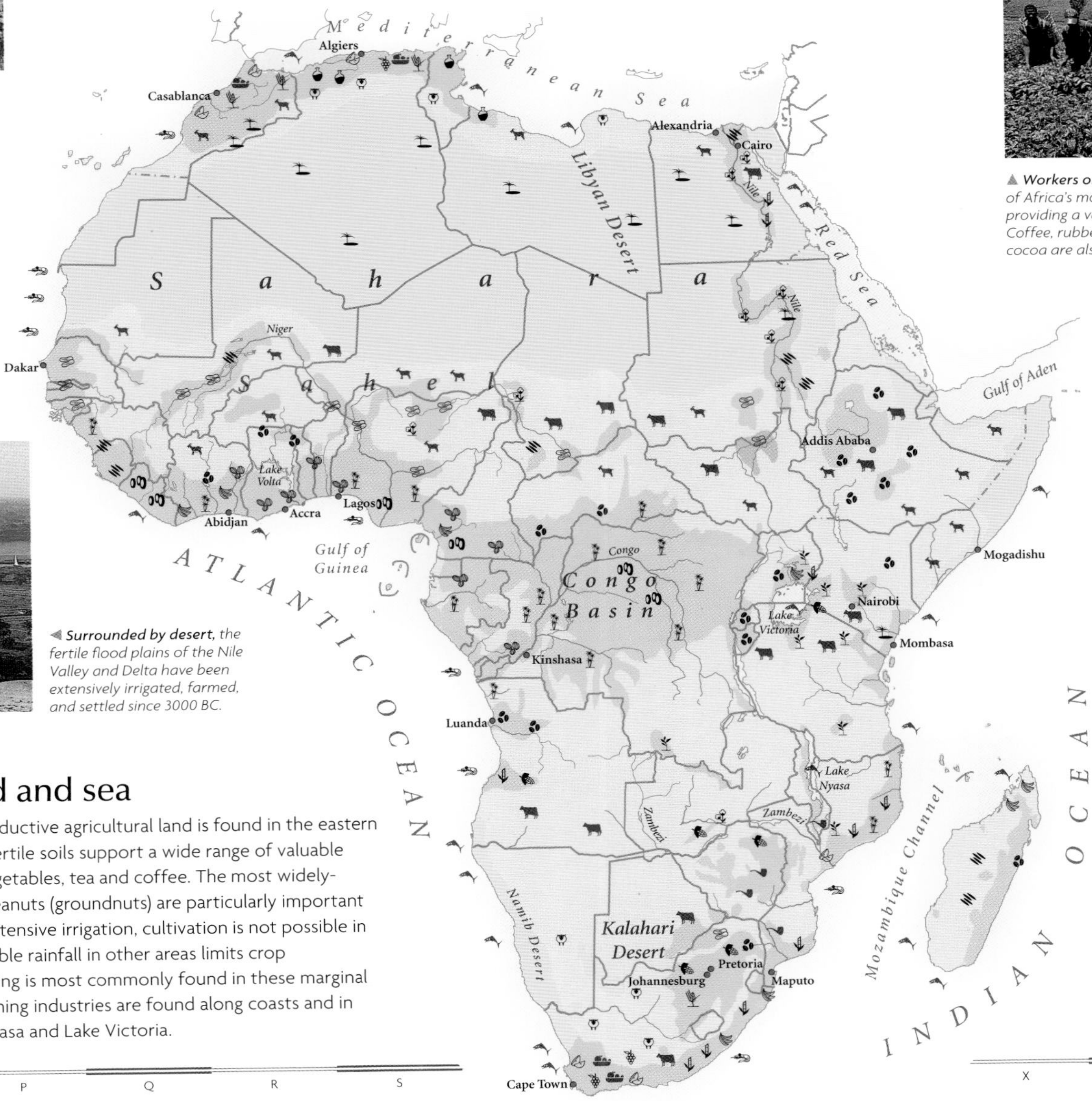

Using the land and sea

- cropland
- desert
- forest
- pasture
- wetland
- major conurbations
- cattle
- goats
- cereals
- sheep
- bananas
- corn (maize)
- citrus fruits
- cocoa
- cotton
- coffee
- dates
- fishing
- fruit
- oil palms
- olives
- peanuts
- rice
- rubber
- shellfish
- sugar cane
- tea
- tobacco
- vineyards
- wheat

◀ ***Surrounded by desert,** the fertile flood plains of the Nile Valley and Delta have been extensively irrigated, farmed, and settled since 3000 BC.*

Using the land and sea

Some of Africa's most productive agricultural land is found in the eastern volcanic uplands, where fertile soils support a wide range of valuable export crops including vegetables, tea and coffee. The most widely-grown grain is corn and peanuts (groundnuts) are particularly important in West Africa. Without intensive irrigation, cultivation is not possible in desert regions and unreliable rainfall in other areas limits crop production. Pastoral herding is most commonly found in these marginal lands. Substantial local fishing industries are found along coasts and in vast lakes such as Lake Nyasa and Lake Victoria.

North Africa

ALGERIA, EGYPT, LIBYA, MOROCCO, TUNISIA, WESTERN SAHARA

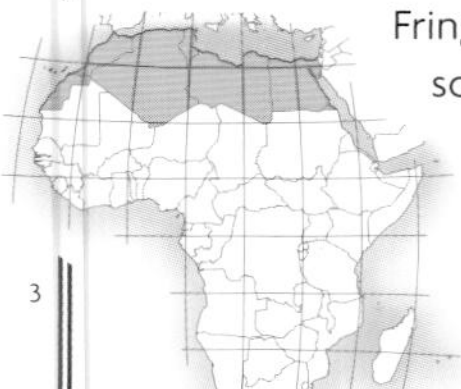

Fringed by the Mediterranean along the northern coast and by the arid Sahara in the south, North Africa reflects the influence of many invaders, both European and, most importantly, Arab, giving the region an almost universal Islamic flavour and a common Arabic language. The countries lying to the west of Egypt are often referred to as the Maghreb, an Arabic term for 'west'. Today, Morocco and Tunisia exploit their culture and landscape for tourism, while rich oil and gas deposits aid development in Libya and Algeria, despite political turmoil. Egypt, with its fertile, Nile-watered agricultural land and varied industrial base, is the most populous nation.

▲ *These rock piles in Algeria's Ahaggar mountains are the result of weathering caused by extremes of temperature. Great cracks or joints appear in the rocks, which are then worn and smoothed by the wind.*

The landscape

The Atlas Mountains, which extend across much of Morocco, northern Algeria and Tunisia, are part of the fold mountain system which also runs through much of southern Europe. They recede to the south and east, becoming a steppe landscape before meeting the Sahara desert which covers more than 90% of the region. The sediments of the Sahara overlie an ancient plateau of crystalline rock, some of which is more than four billion years old.

Map key

Population

- ▣ above 5 million
- ◙ 1 million to 5 million
- ◉ 500,000 to 1 million
- ◎ 100,000 to 500,000
- ⊕ 50,000 to 100,000
- o 10,000 to 50,000
- ∘ below 10,000

Elevation

- 4000m / 13,124ft
- 3000m / 9843ft
- 2000m / 6562ft
- 1000m / 3281ft
- 500m / 1640ft
- 250m / 820ft
- 100m / 328ft
- sea level

Scale 1:12,250,000

Km 0 25 50 100 150 200 250 300

Miles 0 25 50 100 150 200 250 300

projection: Lambert Azimuthal Equal Area

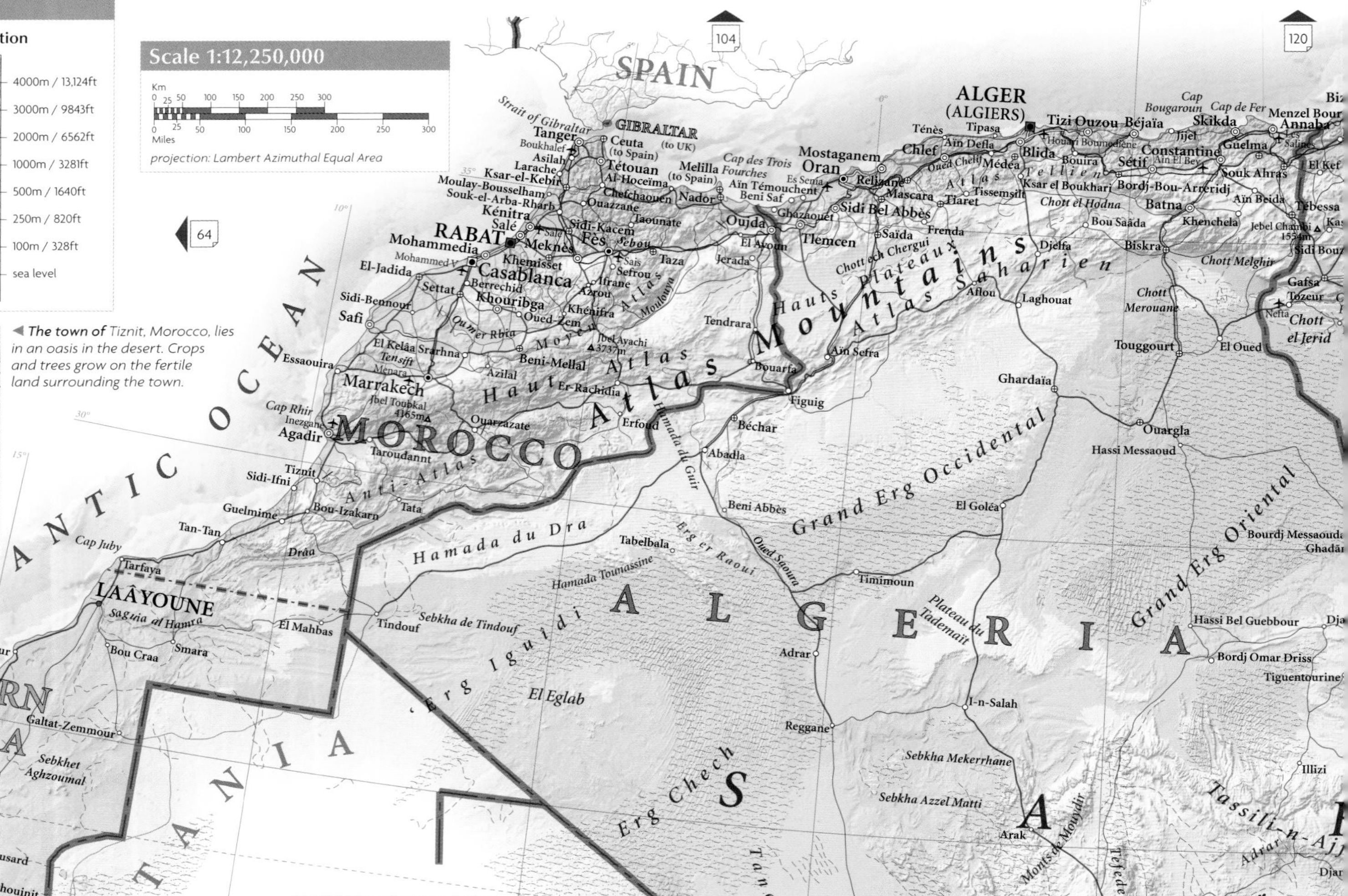

◀ *The town of Tiznit, Morocco, lies in an oasis in the desert. Crops and trees grow on the fertile land surrounding the town.*

▶ ***The Grand Erg Occidental*** *is one of Algeria's great Saharan sand seas. Wind force and direction determines the nature of landforms such as the linear or seif dunes in the foreground.*

Using the land and sea

Sheltered valleys in the Atlas Mountains, the Nile Valley and Delta, and the Mediterranean coast are the main sources of good farming land. A wide variety of valuable crops including cereals, rice and cotton, and woods such as cedar and cork, are grown. Typical Mediterranean crops such as olives, figs, dates and citrus fruits also thrive in these areas. The Nile Valley is particularly fertile, and most of Egypt's population lives close to the river. Elsewhere, irrigation is essential to improve crop yields on the desert margins.

Land use and agricultural distribution

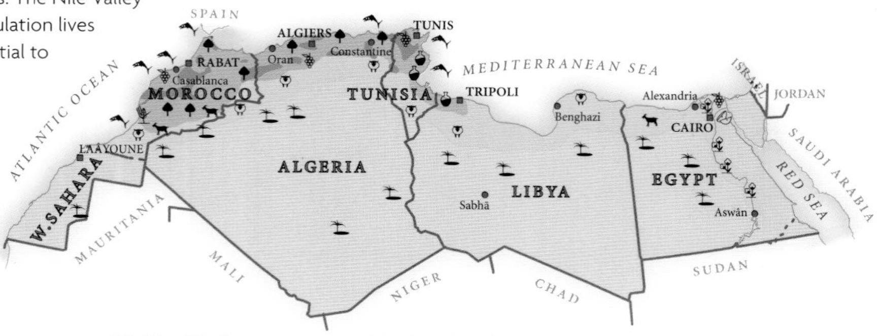

The urban/rural population divide

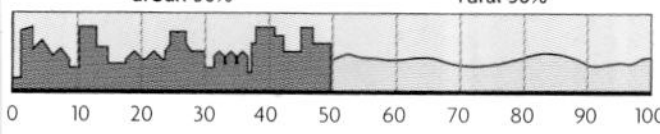

Population density	Total land area
65 people per sq mile (25 people per sq km)	2,215,020 sq miles (5,738,394 sq km)

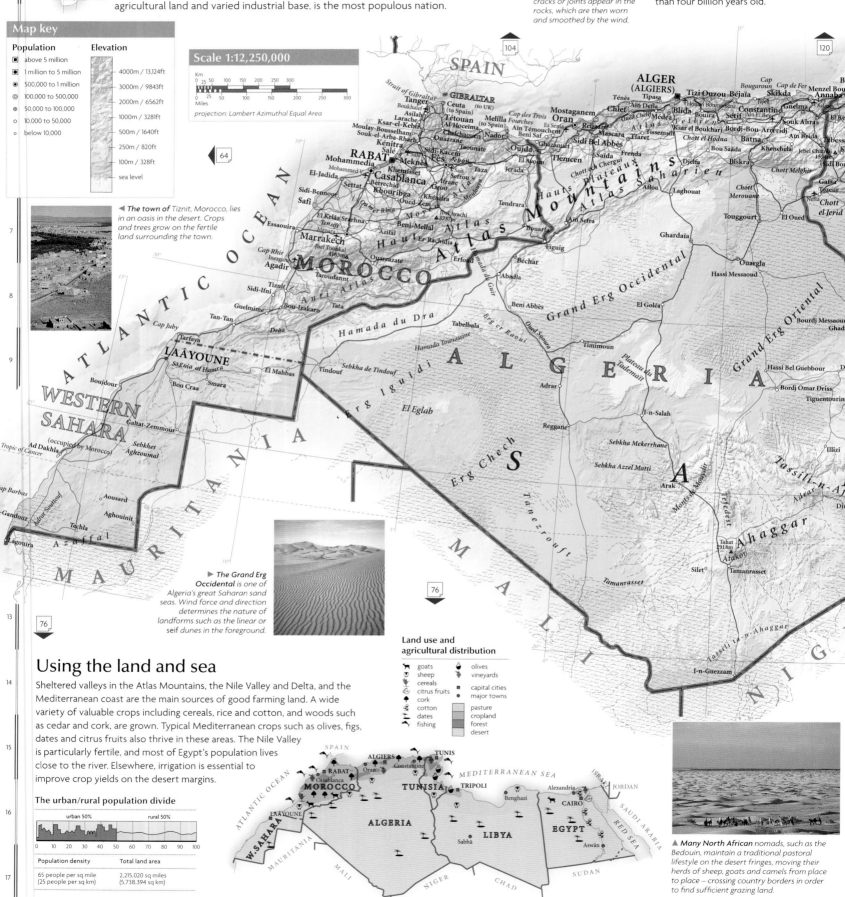

▲ *Many North African nomads, such as the Bedouin, maintain a traditional pastoral lifestyle on the desert fringes, moving their herds of sheep, goats and camels from place to place – crossing country borders in order to find sufficient grazing land.*

N O P Q R S T U V W X Y

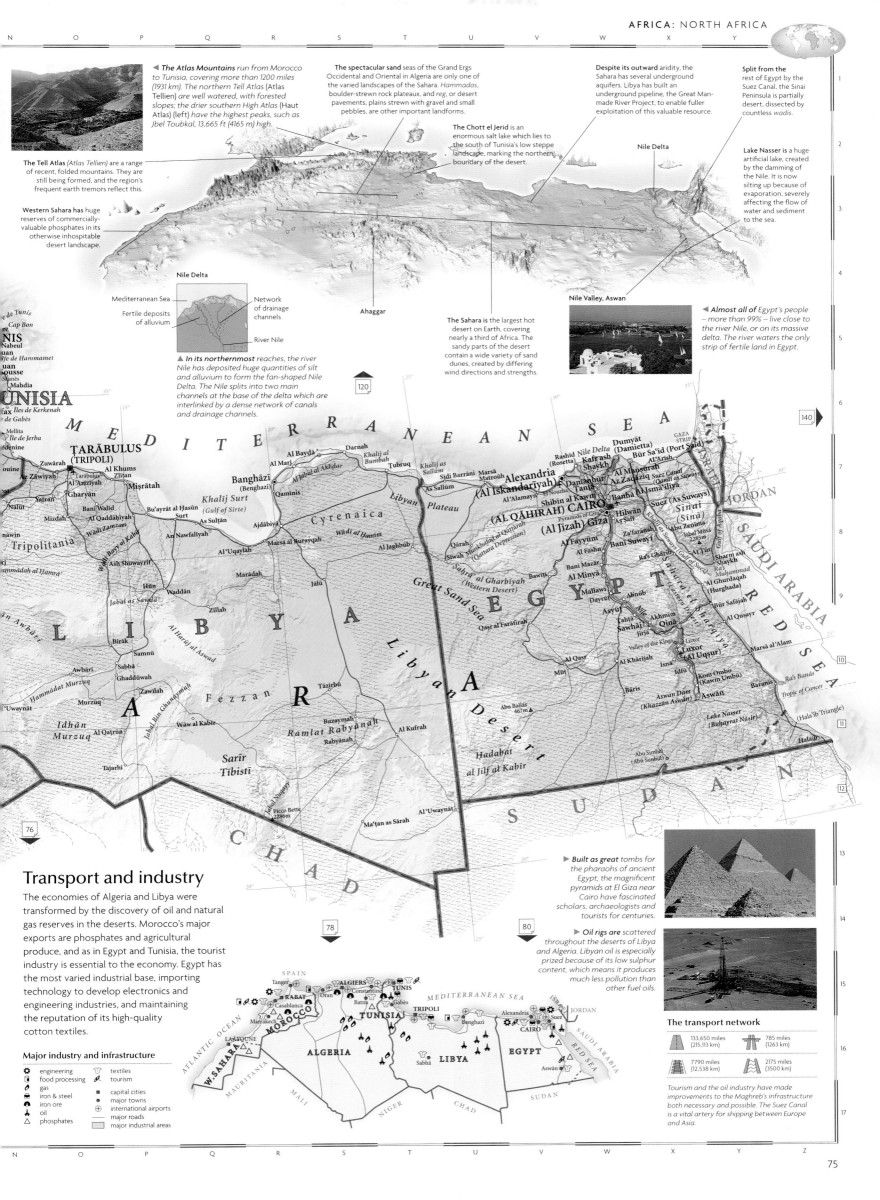

◀ *The Atlas Mountains run from Morocco to Tunisia, covering more than 1200 miles (1931 km). The northern Tell Atlas* (Atlas Tellien) *are well watered, with forested slopes; the drier southern High Atlas* (Haut Atlas) (left) *have the highest peaks, such as Jbel Toubkal, 13,665 ft (4165 m) high.*

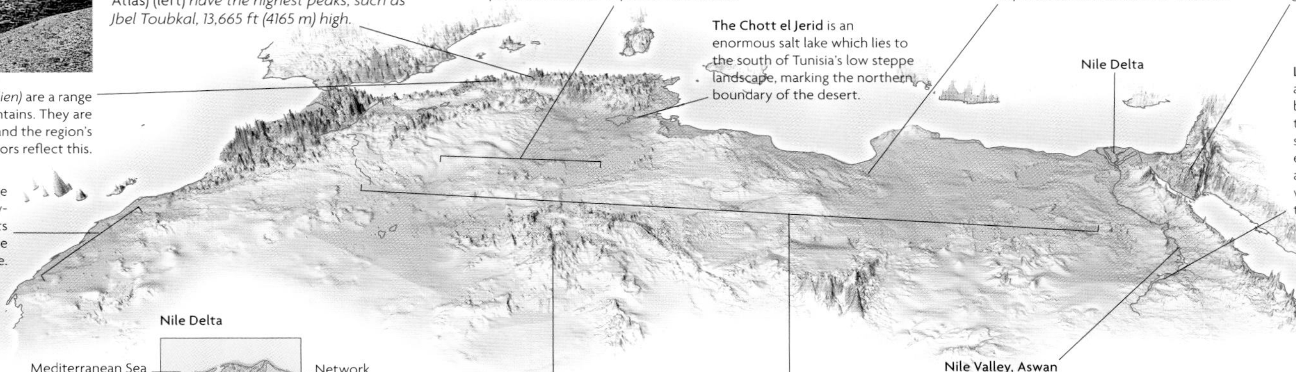

The spectacular sand seas of the Grand Ergs Occidental and Oriental in Algeria are only one of the varied landscapes of the Sahara. *Hammadas*, boulder-strewn rock plateaux, and *reg*, or desert pavements, plains strewn with gravel and small pebbles, are other important landforms.

Despite its outward aridity, the Sahara has several underground aquifers. Libya has built an underground pipeline, the Great Man-made River Project, to enable fuller exploitation of this valuable resource.

Split from the rest of Egypt by the Suez Canal, the Sinai Peninsula is partially desert, dissected by countless *wadis*.

The Chott el Jerid is an enormous salt lake which lies to the south of Tunisia's low steppe landscape, marking the northern boundary of the desert.

The Tell Atlas *(Atlas Tellien)* are a range of recent, folded mountains. They are still being formed, and the region's frequent earth tremors reflect this.

Lake Nasser is a huge artificial lake, created by the damming of the Nile. It is now silting up because of evaporation, severely affecting the flow of water and sediment to the sea.

Western Sahara has huge reserves of commercially-valuable phosphates in its otherwise inhospitable desert landscape.

Nile Delta

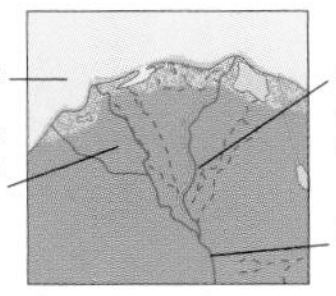

▲ *In its northernmost reaches, the river Nile has deposited huge quantities of silt and alluvium to form the fan-shaped Nile Delta. The Nile splits into two main channels at the base of the delta which are interlinked by a dense network of canals and drainage channels.*

The Sahara is the largest hot desert on Earth, covering nearly a third of Africa. The sandy parts of the desert contain a wide variety of sand dunes, created by differing wind directions and strengths.

Nile Valley, Aswan

◀ *Almost all of Egypt's people – more than 99% – live close to the river Nile, or on its massive delta. The river waters the only strip of fertile land in Egypt.*

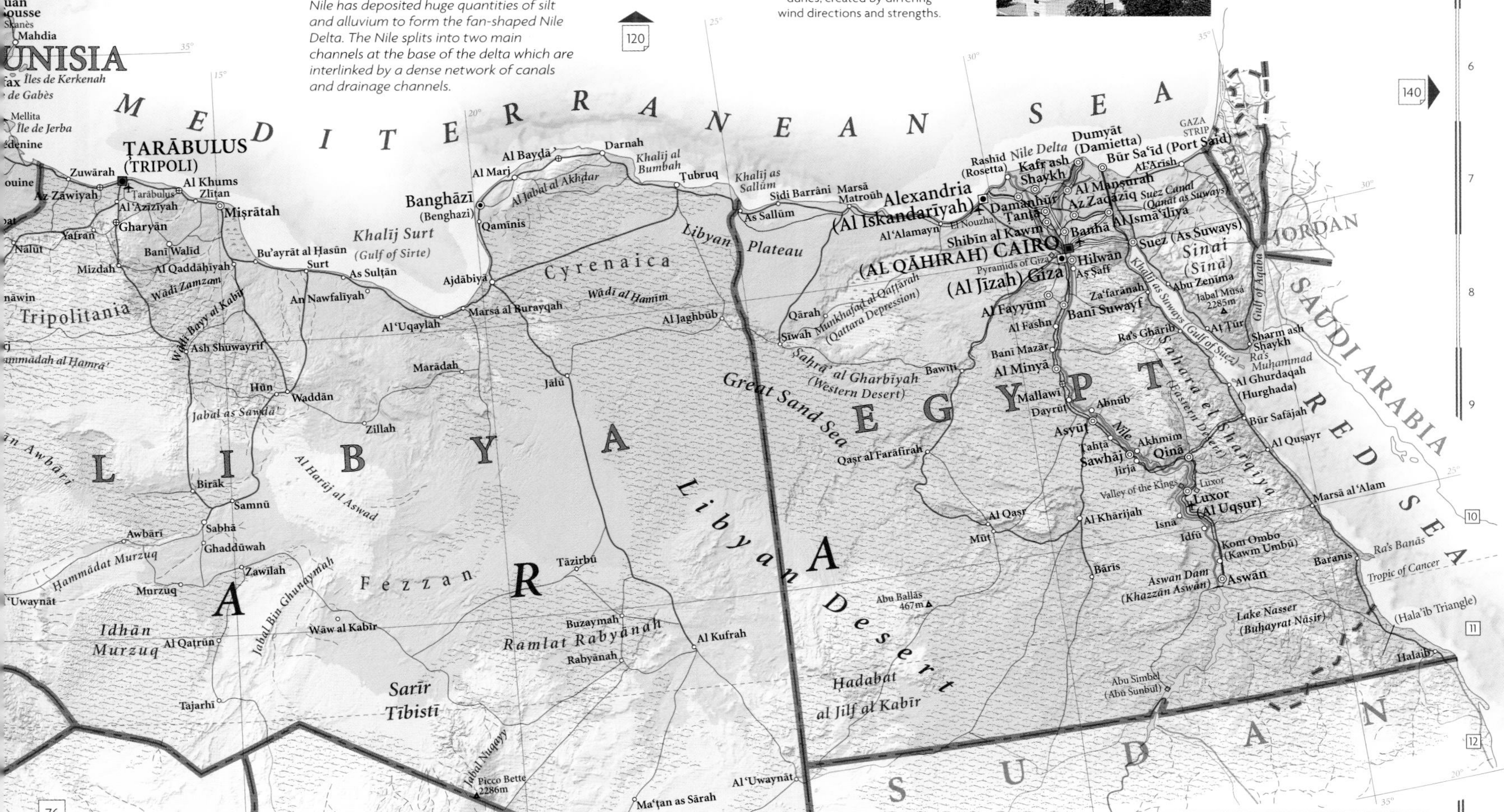

▶ *Built as great tombs for the pharaohs of ancient Egypt, the magnificent pyramids at El Giza near Cairo have fascinated scholars, archaeologists and tourists for centuries.*

▶ *Oil rigs are scattered throughout the deserts of Libya and Algeria. Libyan oil is especially prized because of its low sulphur content, which means it produces much less pollution than other fuel oils.*

Transport and industry

The economies of Algeria and Libya were transformed by the discovery of oil and natural gas reserves in the deserts. Morocco's major exports are phosphates and agricultural produce, and as in Egypt and Tunisia, the tourist industry is essential to the economy. Egypt has the most varied industrial base, importing technology to develop electronics and engineering industries, and maintaining the reputation of its high-quality cotton textiles.

Major industry and infrastructure

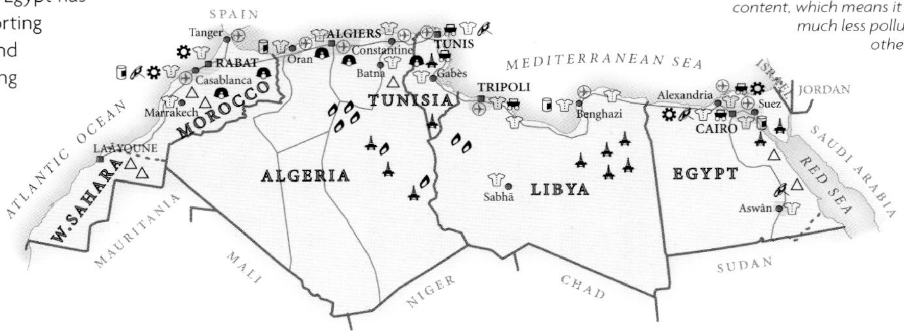

- engineering
- food processing
- gas
- iron & steel
- iron ore
- oil
- phosphates
- textiles
- tourism
- capital cities
- major towns
- international airports
- major roads
- major industrial areas

The transport network

133,650 miles (215,113 km)	785 miles (1263 km)
7790 miles (12,538 km)	2175 miles (3500 km)

Tourism and the oil industry have made improvements to the Maghreb's infrastructure both necessary and possible. The Suez Canal is a vital artery for shipping between Europe and Asia.

N O P Q R S T U V W X Y Z

West Africa

BENIN, BURKINA FASO, CAPE VERDE, GAMBIA, GHANA, GUINEA, GUINEA-BISSAU, IVORY COAST, LIBERIA, MALI, MAURITANIA, NIGER, NIGERIA, SENEGAL, SIERRA LEONE, TOGO

West Africa is an immensely diverse region, encompassing the desert landscapes and mainly Muslim populations of the southern Saharan countries, and the tropical rainforests of the more humid south, with a great variety of local languages and cultures. The rich natural resources and accessibility of the area were quickly exploited by Europeans; most of the Africans taken by slave traders came from this region, causing serious depopulation. The very different influences of West Africa's leading colonial powers, Britain and France, remain today, reflected in the languages and institutions of the countries they once governed.

▶ *The dry scrub of the Sahel is only suitable for grazing herd animals like these cattle in Mali.*

Transport and industry

Abundant natural resources including oil and metallic minerals are found in much of West Africa, although investment is required for their further exploitation. Nigeria experienced an oil boom during the 1970s but subsequent growth has been sporadic. Most industry in other countries has a primary basis, including mining, logging and food processing.

The transport network

62,154 miles (100,038 km)	1037 miles (1669 km)
6752 miles (10,867 km)	10,192 miles (16,405 km)

The road and rail systems are most developed near the coasts. Some of the land-locked countries remain disadvantaged by the difficulty of access to ports, and their poor road networks.

MOROCCO
W. SAHARA
ALGERIA
LIBYA
MAURITANIA
NOUAKCHOTT
MALI
NIGER
CHAD
DAKAR
SENEGAL
BANJUL
GAMBIA
BISSAU
GUINEA-BISSAU
GUINEA
CONAKRY
BAMAKO
BURKINA FASO
NIAMEY
OUAGADOUGOU
Kano
NIGERIA
ABUJA
SIERRA LEONE
FREETOWN
IVORY COAST
GHANA
TOGO
BENIN
MONROVIA
LIBERIA
ACCRA
LOMÉ
PORTO-NOVO
Ibadan
Lagos
Port Harcourt
CAMEROON
YAMOUSSOUKRO
ATLANTIC OCEAN

Major industry and infrastructure

- chemicals
- cotton spinning
- food processing
- mining
- oil
- palm oil processing
- peanut processing
- textiles
- vehicle manufacture
- capital cities
- major towns
- international airports
- major roads
- major industrial areas

CAPE VERDE
Santo Antão
Pombas
Ilhas de Barlavento
Mindelo
São Vicente
Ribeira Brava
Pedra Lume
Amílcar Cabral
Sal
São Nicolau
Boa Vista
João Barrosa
ATLANTIC OCEAN
Tarrafal
Maio
Fogo
São Filipe
Santiago
PRAIA
Ilhas de Sotavento
(same scale as main map)

Map key

Population

- Above 5 million
- 1 million to 5 million
- 500,000 to 1 million
- 100,000 to 500,000
- 50,000 to 100,000
- 10,000 to 50,000
- below 10,000

Elevation

- 2000m / 6562ft
- 1000m / 3281ft
- 500m / 1640ft
- 250m / 820ft
- 100m / 328ft
- sea level

Scale 1:10,000,000

Km 0 25 50 100 150 200 250
Miles 0 25 50 100 150 200 250
projection: Lambert Azimuthal Equal Area

WESTERN SAHARA (occupied by Morocco)
Tropic of Cancer
TIRIS ZEMMOUR
Bir Mogreïn
'Aïn Ben Tili
Yetti
Zouérat
Fdérik
Touâjîl
Tourine
Râs Nouâdhibou
Nouâdhibou
DAKHLET NOUÂDHIBOU
INCHIRI
Akjoujt
Choûm
Atâr
Chinguetti
Ouadâne
ADRAR
Ouarâne
MAURITANIA
NOUAKCHOTT
TRARZA
TAGANT
Tidjikja
Tichit
BRAKNA
Aleg
Boutilimit
Rosso
HODH EL GHARBI
ASSABA
Kiffa
Kaédi
GORGOL
GUIDIMAKA
Sélibabi
ATLANTIC OCEAN
Saint Louis
Louga
DAKAR
Thiès
Rufisque
Mbour
Diourbel
SENEGAL
Kaolack
Tambacounda
BANJUL
GAMBIA
Georgetown
Ziguinchor
Kolda
GUINEA-BISSAU
BISSAU
Bafatá
Arquipélago dos Bijagós
Labé
Fouta Djallon
GUINEA
Kindia
CONAKRY
Kankan
Faranah
Siguiri
Kissidougou
Nzérékoré
FREETOWN
SIERRA LEONE
Bo
Kenema
Makeni
Sherbro Island
MONROVIA
LIBERIA
Buchanan
Greenville
Harper
Cape Palmas
KAYES
Kayes
KOULIKORO
BAMAKO
SIKASSO
Kati
Ségou
Kita
Nioro
Ayoûn el 'Atroûs
Néma
HODH ECH CHARGUI
IVORY COAST

◀ *The southern regions of West Africa still contain great swathes of tropical rainforest, including some of the world's most prized hardwood trees, such as mahogany and iroko.*

Using the land and sea

The humid southern regions are most suitable for cultivation; in these areas, cash crops such as coffee, cotton, cocoa and rubber are grown in large quantities. Peanuts (groundnuts) are grown throughout West Africa. In the north, advancing desertification has made the Sahel increasingly unviable for cultivation, and pastoral farming is more common. Great herds of sheep, cattle and goats are grazed on the savannah grasses, and fishing is important in coastal and delta areas.

▲ *The Gambia, mainland Africa's smallest country, produces great quantities of peanuts (groundnuts). Winnowing is used to separate the nuts from their stalks.*

MOROCCO
W. SAHARA
ALGERIA
LIBYA
SAHARA
MAURITANIA
NOUAKCHOTT
MALI
NIGER
CHAD
Sahel
DAKAR
SENEGAL
BANJUL
GAMBIA
BISSAU
GUINEA-BISSAU
BAMAKO
GUINEA
BURKINA FASO
NIAMEY
OUAGADOUGOU
NIGERIA
ABUJA
CONAKRY
SIERRA LEONE
FREETOWN
IVORY COAST
GHANA
TOGO
BENIN
MONROVIA
LIBERIA
ACCRA
LOMÉ
PORTO-NOVO
Lagos
CAMEROON
YAMOUSSOUKRO
ATLANTIC OCEAN

Land use and agricultural distribution

- goats
- sheep
- cocoa
- coffee
- cotton
- oil palms
- peanuts
- rubber
- shellfish
- capital cities
- major towns
- pasture
- cropland
- forest
- desert

The urban/rural population divide

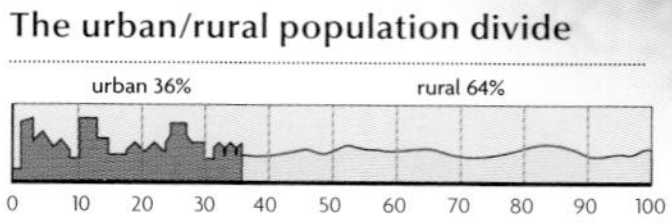
urban 36%
rural 64%
0
10
20
30
40
50
60
70
80
90
100

Population density	Total land area
104 people per sq mile (40 people per sq km)	2,337,137 sq miles (6,054,760 sq km)

N O P Q R S T U V W X Y

The landscape

There are two major topographical areas in West Africa: the northern deserts are part of the Saharan region which stretches across the whole continent; the grasslands of the Sahel and the southern Guinea coast are part of Africa's central plateau. The landscape is generally low, rarely rising above 1500 ft (457 m) and consists mainly of plains, broken by an occasional high plateau or mountain range.

▲ *Inselbergs, found across the Sahel, are isolated hills, or outcrops, formed where the surrounding plain has eroded away, leaving only the more resistant remnants of the original plateau.*

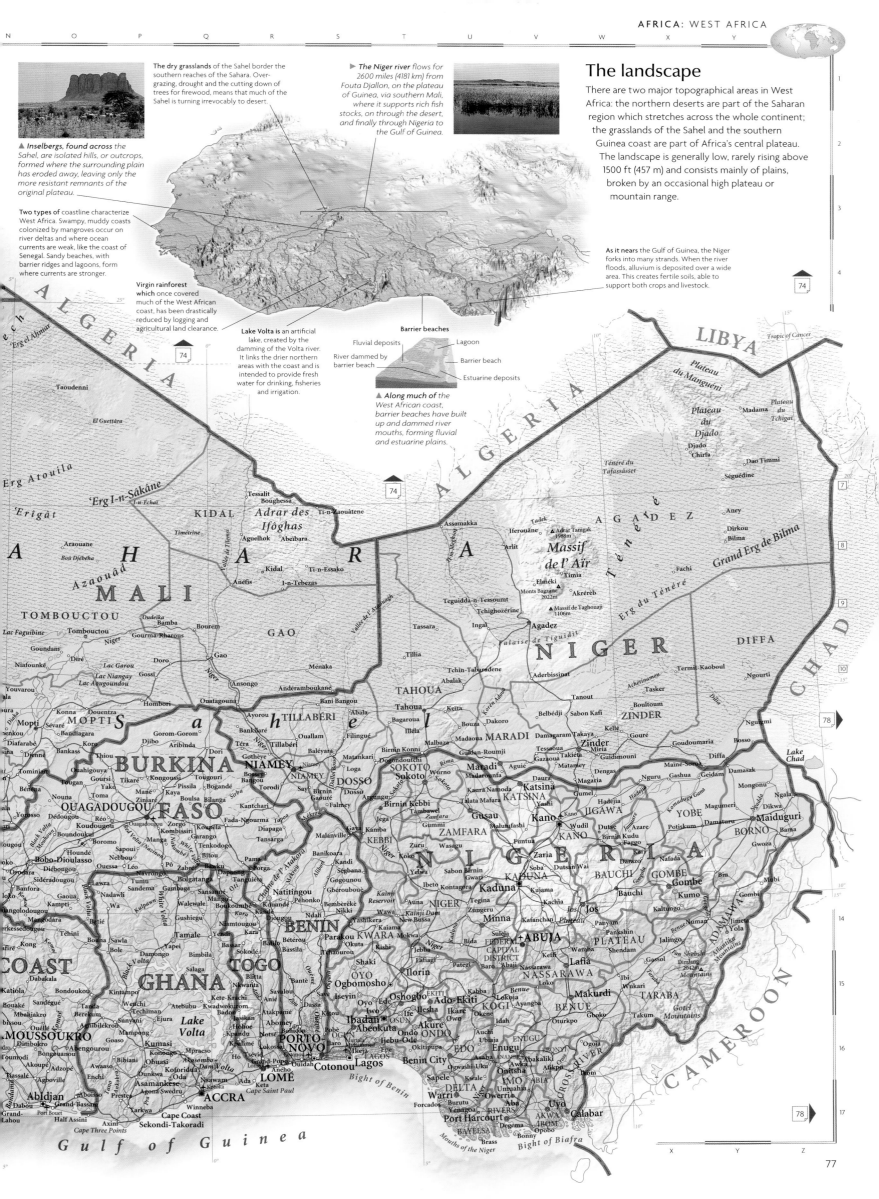

The dry grasslands of the Sahel border the southern reaches of the Sahara. Overgrazing, drought and the cutting down of trees for firewood, means that much of the Sahel is turning irrevocably to desert.

► ***The Niger river** flows for 2600 miles (4181 km) from Fouta Djallon, on the plateau of Guinea, via southern Mali, where it supports rich fish stocks, on through the desert, and finally through Nigeria to the Gulf of Guinea.*

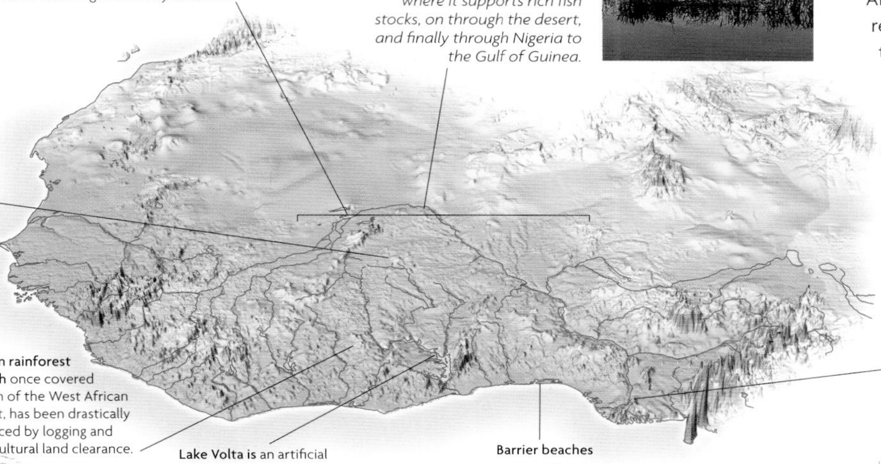

Two types of coastline characterize West Africa. Swampy, muddy coasts colonized by mangroves occur on river deltas and where ocean currents are weak, like the coast of Senegal. Sandy beaches, with barrier ridges and lagoons, form where currents are stronger.

Virgin rainforest which once covered much of the West African coast, has been drastically reduced by logging and agricultural land clearance.

Lake Volta is an artificial lake, created by the damming of the Volta river. It links the drier northern areas with the coast and is intended to provide fresh water for drinking, fisheries and irrigation.

As it nears the Gulf of Guinea, the Niger forks into many strands. When the river floods, alluvium is deposited over a wide area. This creates fertile soils, able to support both crops and livestock.

Barrier beaches

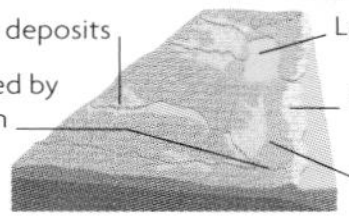

▲ *Along much of the West African coast, barrier beaches have built up and dammed river mouths, forming fluvial and estuarine plains.*

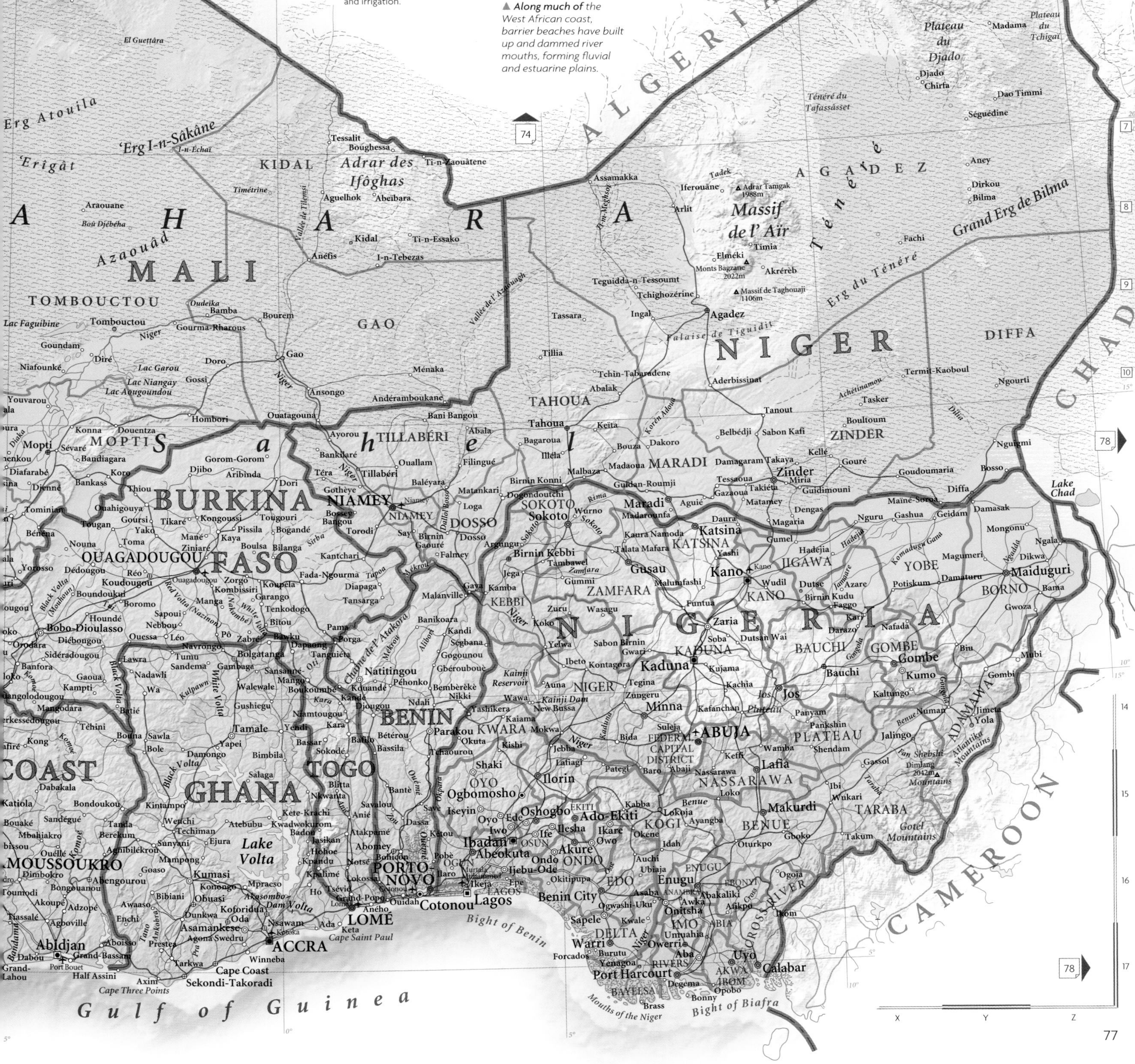

X Y Z

Central Africa

CAMEROON, CENTRAL AFRICAN REPUBLIC, CHAD, CONGO, DEM. REP. CONGO, EQUATORIAL GUINEA, GABON, SAO TOME & PRINCIPE

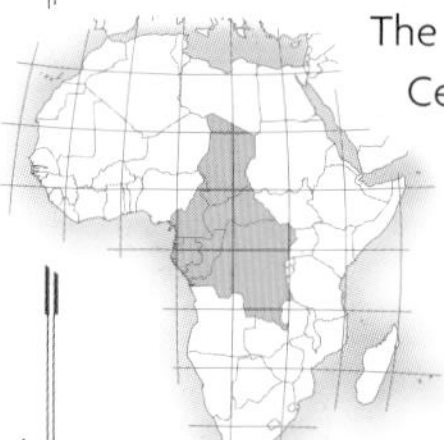

The great rainforest basin of the Congo river embraces most of remote Central Africa. The interior was largely unknown to Europeans until late in the 19th century, when its tribal kingdoms were split – principally between France and Belgium – with Sao Tome and Principe the lone Portuguese territory, and Equatorial Guinea controlled by Spain. Open democracy and regional economic integration are important goals for these nations – several of which have only recently emerged from restrictive regimes – and investment is needed to improve transport infrastructures. Many of the small, but fast-growing and increasingly urban population, speak French, the regional *lingua franca*, along with several hundred Pygmy, Bantu and Sudanic dialects.

Transport and industry

Large reserves of valuable minerals are found in Central Africa: copper, cobalt and diamonds are mined in Dem. Rep. Congo and manganese in Gabon. Congo, Cameroon, Gabon and Equatorial Guinea have oil deposits and oil has also been recently discovered in Chad. Goods such as palm oil and rubber are processed for export.

▲ ***The ancient rocks*** *of Dem. Rep. Congo hold immense and varied mineral reserves. This open pit copper mine is at Kolwezi in the far south.*

Major industry and infrastructure

- brewing
- chemicals
- cobalt
- copper
- diamonds
- food processing
- manganese
- oil
- palm oil processing
- textiles
- tin
- capital cities
- major towns
- international airports
- major roads
- major industrial areas

The transport network

102,747 miles (165,774 km)	37 miles (60 km)
3985 miles (6414 km)	14,110 miles (22,710 km)

The Trans-Gabon railway, which began operating in 1987, has opened up new sources of timber and manganese. Elsewhere, much investment is needed to update and improve road, rail and water transport.

76

The landscape

Lake Chad lies in a desert basin bounded by the volcanic Tibesti mountains in the north, plateaux in the east and, in the south, the broad watershed of the Congo basin. The vast circular depression of the Congo is isolated from the coastal plain by the granite Massif du Chaillu. To the northwest, the volcanoes and fold mountains of the Cameroon Ridge *(Dorsale Camerounaise)* extend as islands into the Gulf of Guinea. The high fold mountains fringing the east of the Congo Basin fall steeply to the lakes of the Great Rift Valley.

74

◀ ***A plug of*** *resistant lava, at the southwestern end of the Cameroon Ridge (Dorsale Camerounaise), is all that remains of an eroded volcano.*

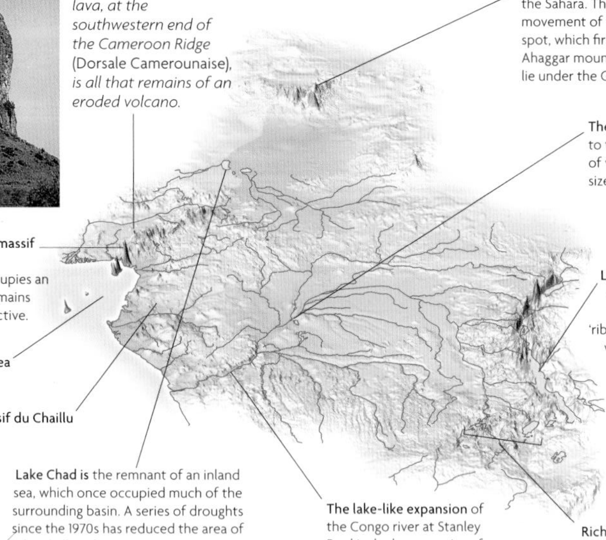

The Tibesti mountains are the highest in the Sahara. They were pushed up by the movement of the African Plate over a hot spot, which first formed the northern Ahaggar mountains and is now thought to lie under the Great Rift Valley.

The Congo river is second only to the Amazon in the volume of water it carries, and in the size of its drainage basin.

Lake Tanganyika, **the** world's second deepest lake, is the largest of a series of linear 'ribbon' lakes occupying a trench within the Great Rift Valley.

The volcanic massif of Cameroon Mountain occupies an area which remains volcanically active.

Gulf of Guinea

Massif du Chaillu

Lake Chad is the remnant of an inland sea, which once occupied much of the surrounding basin. A series of droughts since the 1970s has reduced the area of this shallow freshwater lake to about 1000 sq miles (2599 sq km).

The lake-like expansion of the Congo river at Stanley Pool is the lowest point of the interior basin, although the river still descends more than 1000 ft (300 m) to reach the sea.

Rich mineral deposits in the 'Copper Belt' of Dem. Rep. Congo were formed under intense heat and pressure when the ancient African Shield was uplifted to form the region's mountains.

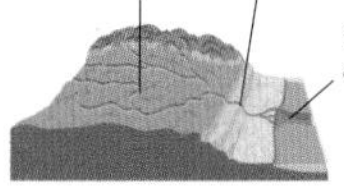

▲ ***The Congo river*** *flows sluggishly through the rainforest of the interior basin. Towards the coast, the river drops steeply in a series of waterfalls and cataracts. At this point, the erosional power of the river becomes so great that it has formed a deep submarine canyon offshore.*

▲ ***Virgin tropical rainforest*** *covers the Ruwenzori range on the borders of Dem. Rep. Congo and Uganda.*

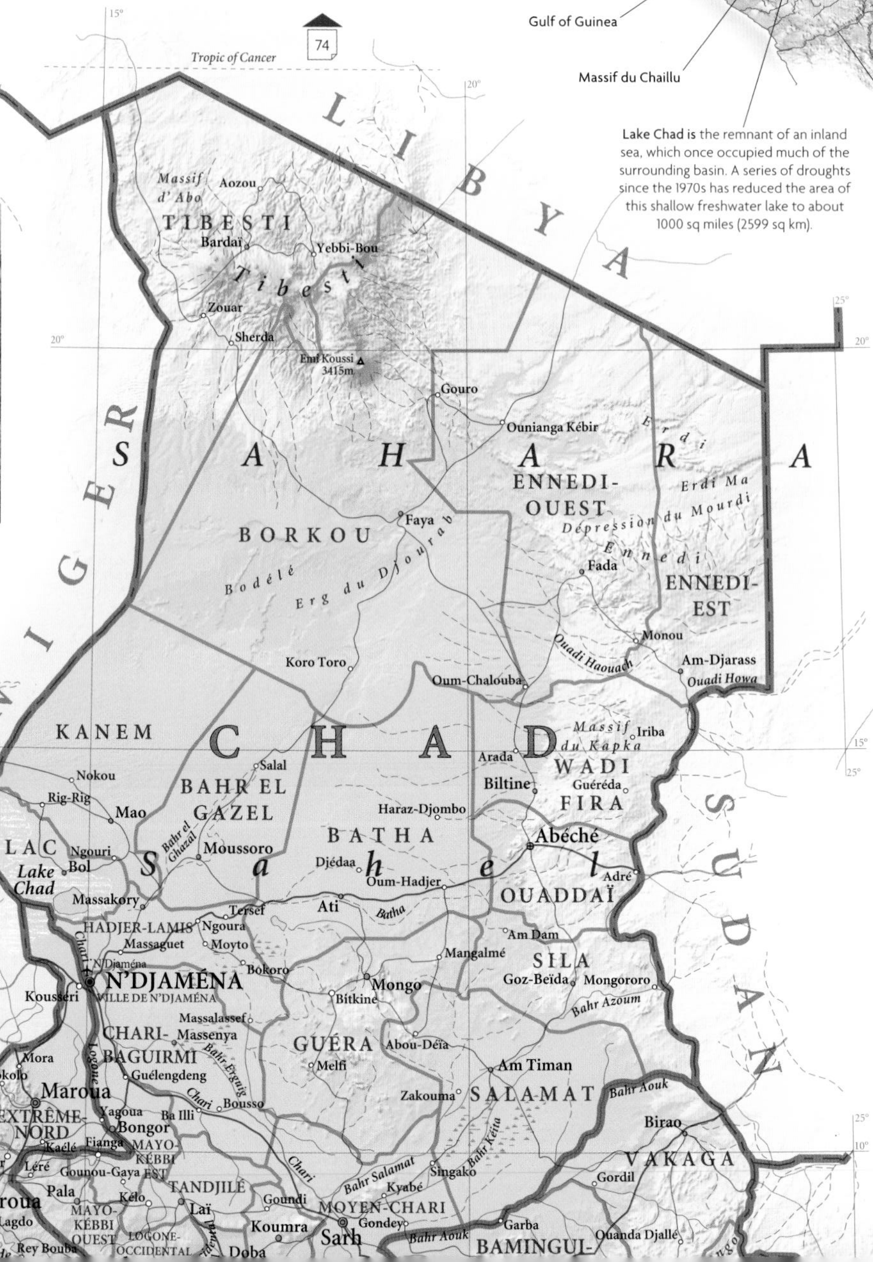

▲ ***The vast sand*** *flats surrounding Lake Chad were once covered by water. Changing climatic patterns caused the lake to shrink, and desert now covers much of its previous area.*

Map key

Population

- 1 million to 5 million
- 500,000 to 1 million
- 100,000 to 500,000
- 50,000 to 100,000
- 10,000 to 50,000
- below 10,000

Elevation

- 4000m / 13,124ft
- 3000m / 9843ft
- 2000m / 6562ft
- 1000m / 3281ft
- 500m / 1640ft
- 250m / 820ft
- 100m / 328ft
- sea level

Scale 1:10,500,000

Km 0 25 50 100 150 200 250

Miles 0 25 50 100 150 200 250

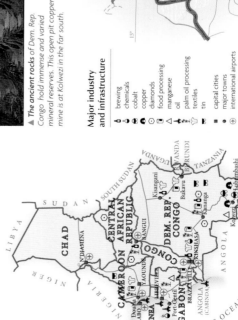

projection: Lambert Azimuthal Equal Area

▲ *The great Congo* river forms part of the border between Congo and Dem. Rep. Congo. The river is fast-flowing, and a series of falls and rapids means that it is only partly navigable.

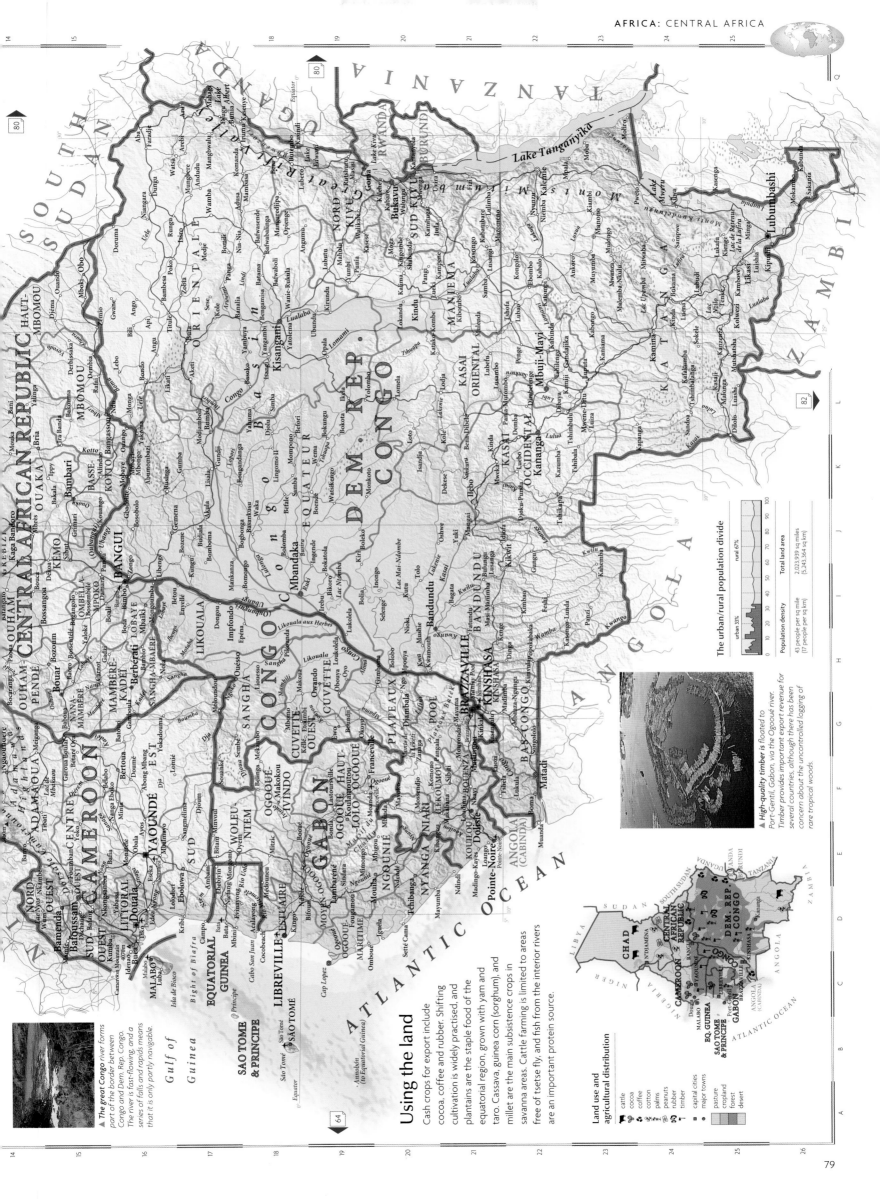

Using the land

Cash crops for export include cocoa, coffee and rubber. Shifting cultivation is widely practised, and plantains are the staple food of the equatorial region, grown with yam and taro. Cassava, guinea corn (sorghum), and millet are the main subsistence crops in savanna areas. Cattle farming is limited to areas free of tsetse fly, and fish from the interior rivers are an important protein source.

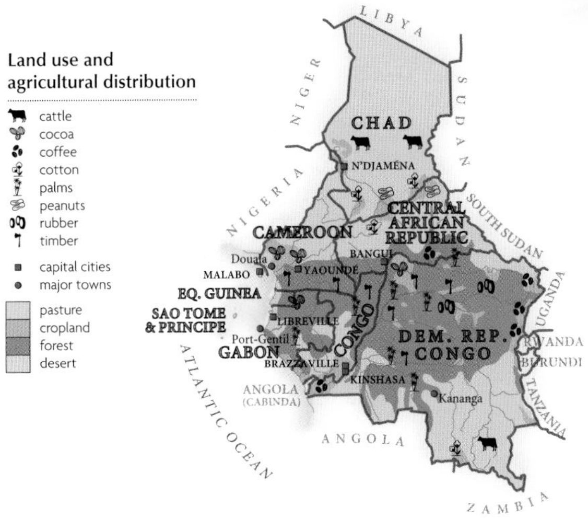

▲ *High-quality timber is* floated to Port-Gentil, Gabon, via the Ogooué river. Timber provides important export revenue for several countries, although there has been concern about the uncontrolled logging of rare tropical woods.

The urban/rural population divide

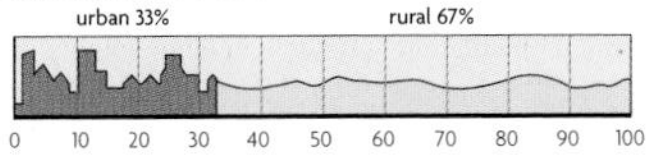

Population density	Total land area
43 people per sq mile (17 people per sq km)	2,023,939 sq miles (5,243,364 sq km)

East Africa

BURUNDI, DJIBOUTI, ERITREA, ETHIOPIA, KENYA, RWANDA, SOMALIA, SOUTH SUDAN, SUDAN, TANZANIA, UGANDA

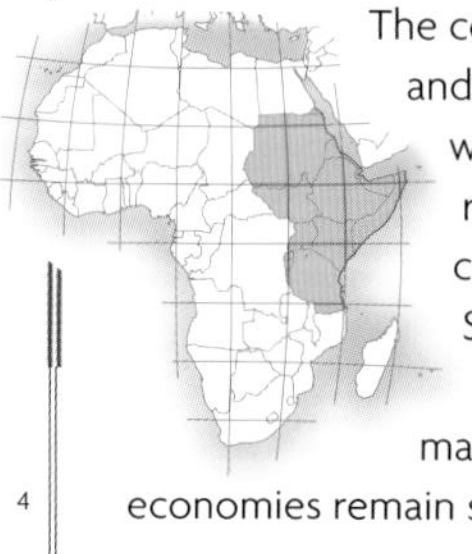

The countries of East Africa divide into two distinct cultural regions. Sudan and the 'Horn' nations have been influenced by the Middle East; Ethiopia was the home of one of the earliest Christian civilizations, and Sudan reflects both Muslim and Christian influences, while the southern countries share a closer cultural affinity with other sub-Saharan nations. Some of Africa's most densely populated countries lie in this region, and the needs of a growing number of people have put pressure on marginal lands and fragile environments. Although most East African economies remain strongly agricultural, Kenya has developed a varied industrial base.

The landscape

East Africa's most significant landscape feature is the Great Rift Valley, which formed during the most recent phase of continental movement when the rigid basement rocks cracked and buckled. Great blocks of land were raised and lowered, creating huge flat-bottomed valleys and steep escarpments, sometimes covered by volcanic extrusions in highland areas.

◀ ***This dome at** Gonder, in Ethiopia, is a volcanic intrusion, formed when molten rock pushed up the surface of the Earth and then solidified, leaving an outcrop of igneous rock.*

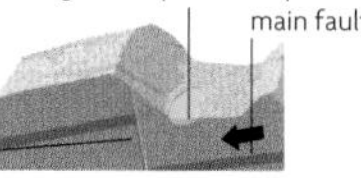

▲ ***The eastern arm** of the Great Rift Valley is gradually being pulled apart; however the forces on one side are greater than the other causing the land to slope. This affects regional drainage which migrates down the slope.*

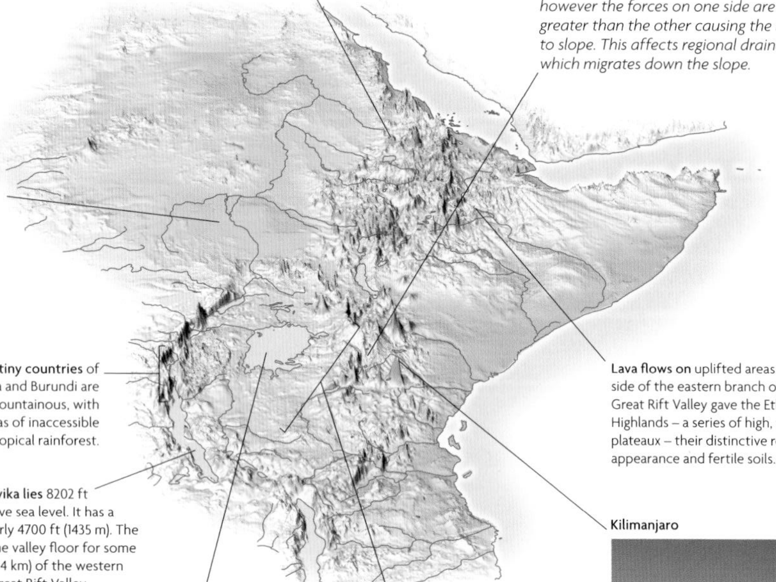

In contrast to the desert conditions that prevail in much of Sudan to the north, annual rainfall in the tropical wetlands of the southern Sudd region in South Sudan, can sometimes exceed 1000 mm (40 inches).

The tiny countries of Rwanda and Burundi are mainly mountainous, with large areas of inaccessible tropical rainforest.

Lake Tanganyika lies 8202 ft (2500 m) above sea level. It has a depth of nearly 4700 ft (1435 m). The lake traces the valley floor for some 400 miles (644 km) of the western arm of the Great Rift Valley.

Lake Victoria occupies a vast basin between the two arms of the Great Rift Valley. It is the world's second largest lake in terms of surface area, extending 26,560 sq miles (68,880 sq km). The lake contains numerous islands and coral reefs.

A vast plateau lies between the eastern and western rift valleys in Kenya, Uganda and western Tanzania. It has been levelled by long periods of erosion to form a peneplain, but is dotted with inselbergs – outcrops of more resistant rocks.

Lava flows on uplifted areas either side of the eastern branch of the Great Rift Valley gave the Ethiopian Highlands – a series of high, wide plateaux – their distinctive rounded appearance and fertile soils.

Kilimanjaro

▲ ***An extinct volcano,** Kilimanjaro is Africa's highest mountain, rising 19,340 ft (5895 m). Once famed for its snow-capped peak, this has almost completely melted due to changing climatic conditions.*

Scale 1:10,500,000

Km 0 25 50 100 150 200 250
Miles 0 25 50 100 150 200 250

projection: Lambert Azimuthal Equal Area

▶ ***The Kassala region** in eastern Sudan is watered by the Atbara river, an important tributary of the Nile. Most of the population is engaged in agriculture, growing cotton and cereals.*

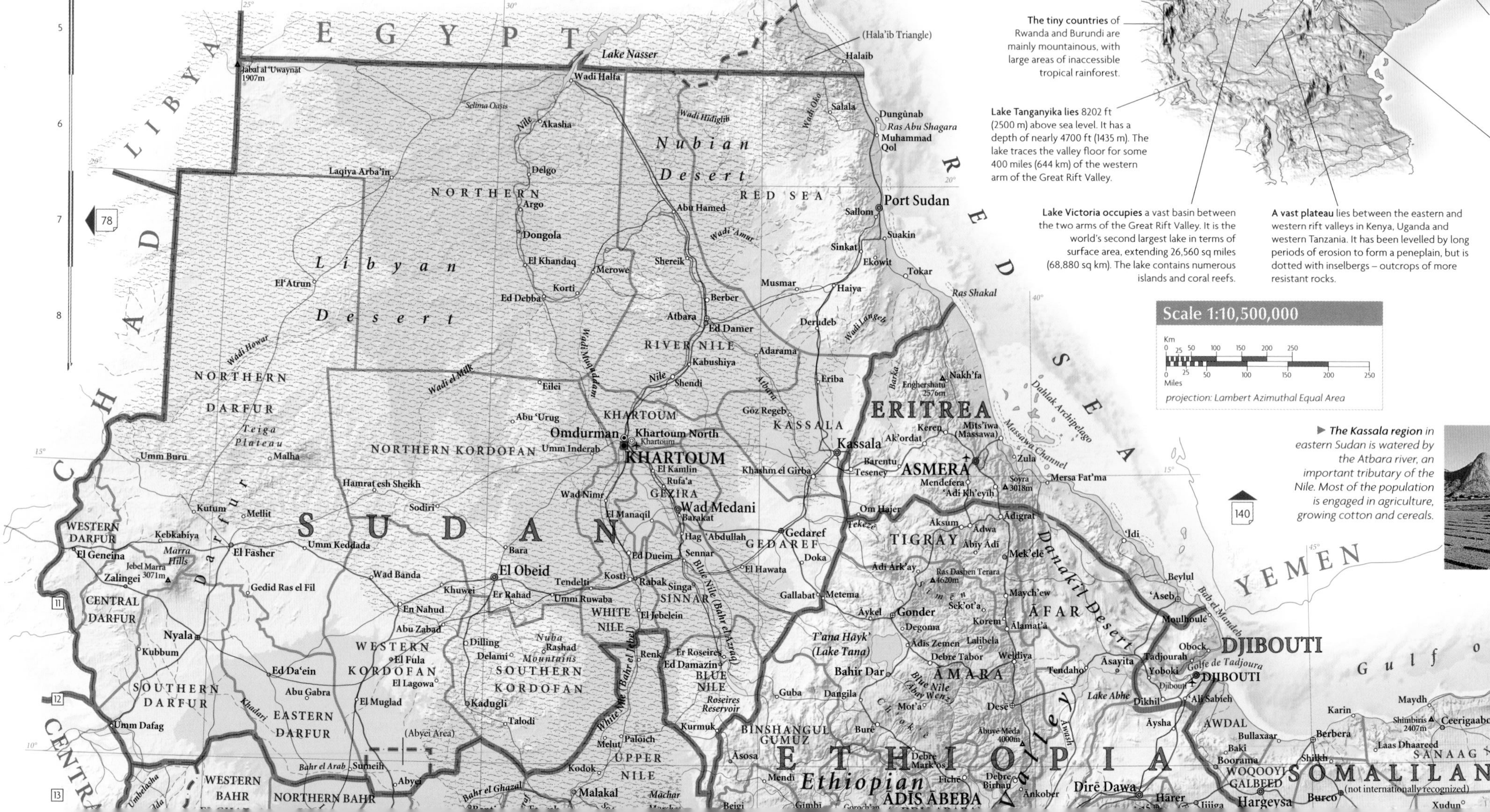

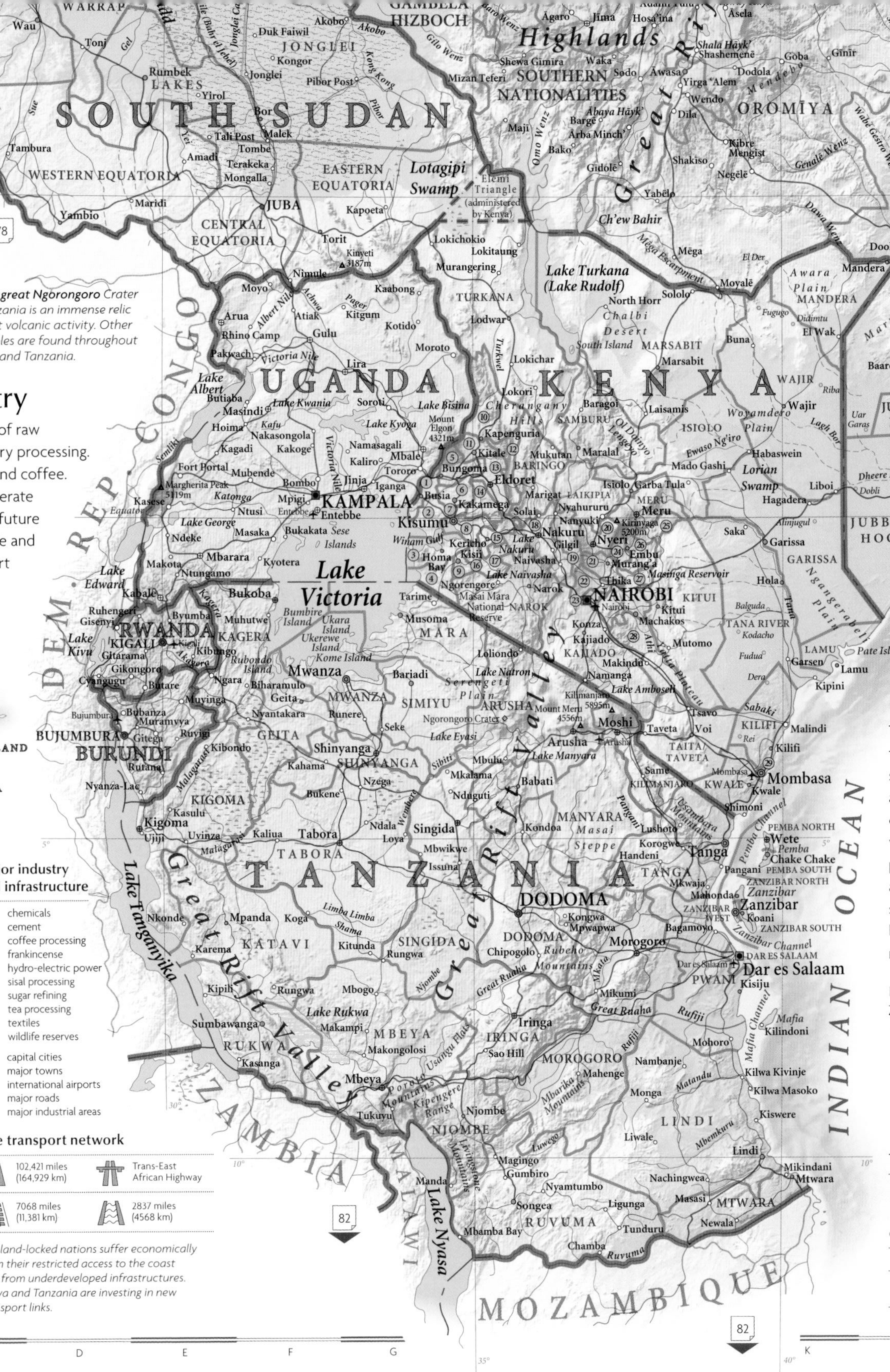

78
78
172
82
82

◀ *The great* **Ngorongoro** *Crater in Tanzania is an immense relic of past volcanic activity. Other examples are found throughout Kenya and Tanzania.*

Transport and industry

Most exports from this region consist of raw materials which have undergone primary processing. These include cotton, sugar, tea, sisal and coffee. Fast-flowing rivers in the highlands generate hydro-electric power, which has great future potential. The appeal of Kenya's wildlife and beaches has made tourism a crucial part of the economy.

Major industry and infrastructure

- chemicals
- cement
- coffee processing
- frankincense
- hydro-electric power
- sisal processing
- sugar refining
- tea processing
- textiles
- wildlife reserves
- capital cities
- major towns
- international airports
- major roads
- major industrial areas

The transport network

- 102,421 miles (164,929 km)
- 7068 miles (11,381 km)
- Trans-East African Highway
- 2837 miles (4568 km)

The land-locked nations suffer economically from their restricted access to the coast and from underdeveloped infrastructures. Kenya and Tanzania are investing in new transport links.

▲ ***The magnificent National** Parks of Kenya and Tanzania provide essential refuges for many of Africa's rarest animals. Tourism brings in much-needed cash to sustain these important conservation projects.*

Map key

Population

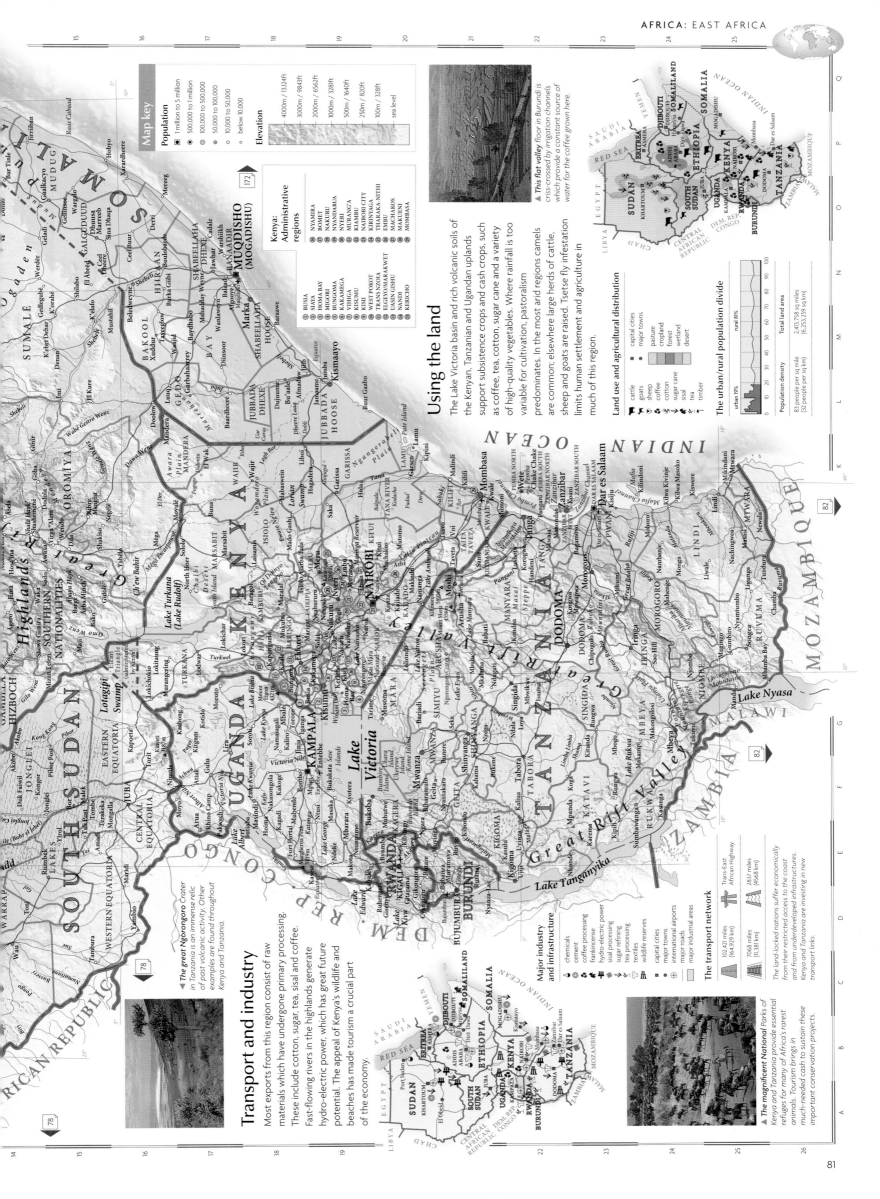

- 1 million to 5 million
- 500,000 to 1 million
- 100,000 to 500,000
- 50,000 to 100,000
- 10,000 to 50,000
- below 10,000

Elevation

- 4000m / 13,124ft
- 3000m / 9843ft
- 2000m / 6562ft
- 1000m / 3281ft
- 500m / 1640ft
- 250m / 820ft
- 100m / 328ft
- sea level

Kenya: Administrative regions

1. BUSIA
2. SIAYA
3. HOMA BAY
4. MIGORI
5. BUNGOMA
6. KAKAMEGA
7. VIHIGA
8. KISUMU
9. KISII
10. WEST POKOT
11. TRANS NZOIA
12. ELGEYO/MARAKWET
13. UASIN GISHU
14. NANDI
15. KERICHO
16. NYAMIRA
17. BOMET
18. NAKURU
19. NYANDARUA
20. NYERI
21. MURANG'A
22. KIAMBU
23. NAIROBI CITY
24. KIRINYAGA
25. THARAKA-NITHI
26. EMBU
27. MACHAKOS
28. MAKUENI
29. MOMBASA

Using the land

The Lake Victoria basin and rich volcanic soils of the Kenyan, Tanzanian and Ugandan uplands support subsistence crops and cash crops, such as coffee, tea, cotton, sugar cane and a variety of high-quality vegetables. Where rainfall is too variable for cultivation, pastoralism predominates. In the most arid regions camels are common; elsewhere large herds of cattle, sheep and goats are raised. Tsetse fly infestation limits human settlement and agriculture in much of this region.

▲ ***This flat valley** floor in Burundi is criss-crossed by irrigation channels which provide a constant source of water for the coffee grown here.*

Land use and agricultural distribution

- cattle
- goats
- sheep
- coffee
- cotton
- sugar cane
- sisal
- tea
- timber
- capital cities
- major towns
- pasture
- cropland
- forest
- wetland
- desert

The urban/rural population divide

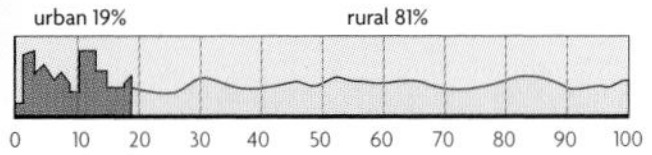

Population density	Total land area
83 people per sq mile (32 people per sq km)	2,413,758 sq miles (6,253,259 sq km)

Southern Africa

ANGOLA, BOTSWANA, LESOTHO, MALAWI, MOZAMBIQUE, NAMIBIA, SOUTH AFRICA, SWAZILAND, ZAMBIA, ZIMBABWE

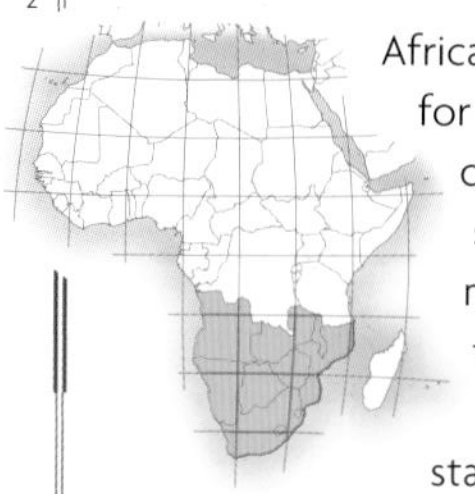

Africa's vast southern plateau has been a contested homeland for disparate peoples for many centuries. The European incursion began with the slave trade and quickened in the 19th century, when the discovery of enormous mineral wealth secured South Africa's regional economic dominance. The struggle against white minority rule led to strife in Namibia, Zimbabwe, and the former Portuguese territories of Angola and Mozambique. South Africa's notorious apartheid laws, which denied basic human rights to more than 75% of the people, led to the state being internationally ostracized until 1994, when the first fully democratic elections inaugurated a new era of racial justice.

Transport and industry

South Africa, the world's largest exporter of gold, has a varied economy which generates about 75% of the region's income and draws migrant labour from neighbouring states. Angola exports petroleum; Botswana and Namibia rely on diamond mining; and Zambia is seeking to diversify its economy to compensate for declining copper reserves.

◀ ***Almost all new** mining ventures in Zimbabwe are now subject to government control. This mine at Bindura in northeastern Zimbabwe produces nickel, one of the country's top three minerals in terms of economic value.*

DEM. REP. CONGO
TANZANIA
LUANDA
Lobito
ANGOLA
ZAMBIA
Ndola
LUSAKA
MALAWI
LILONGWE
Blantyre
HARARE
MOZAMBIQUE
ZIMBABWE
Beira
Bulawayo
NAMIBIA
Walvis Bay
WINDHOEK
BOTSWANA
GABORONE
PRETORIA
MAPUTO
MBABANE
SWAZILAND
Johannesburg
MASERU
LESOTHO
Durban
S.AFRICA
CAPE TOWN
Port Elizabeth
ATLANTIC OCEAN
INDIAN OCEAN

Major industry and infrastructure

- car manufacture
- coal
- copper
- diamonds
- food processing
- gold
- oil
- textiles
- uranium
- wildlife reserves
- capital cities
- major towns
- international airports
- major roads
- major industrial areas

The landscape

Most of southern Africa rests on a concave plateau comprising the Kalahari basin and a mountainous fringe, skirted by a coastal plain which widens out in Mozambique. The plateau extends north, towards the Planalto de Bié in Angola, the Congo Basin and the lake-filled troughs of the Great Rift Valley. The eastern region is drained by the Zambezi and Limpopo rivers, and the Orange is the major western river.

▲ ***Finger Rock**, near Khorixas, Namibia is a remnant of a former land surface, which has been denuded by erosion over the last 5 million years. These occasional stacks of partially weathered rocks interrupt the plains of the dry southern interior.*

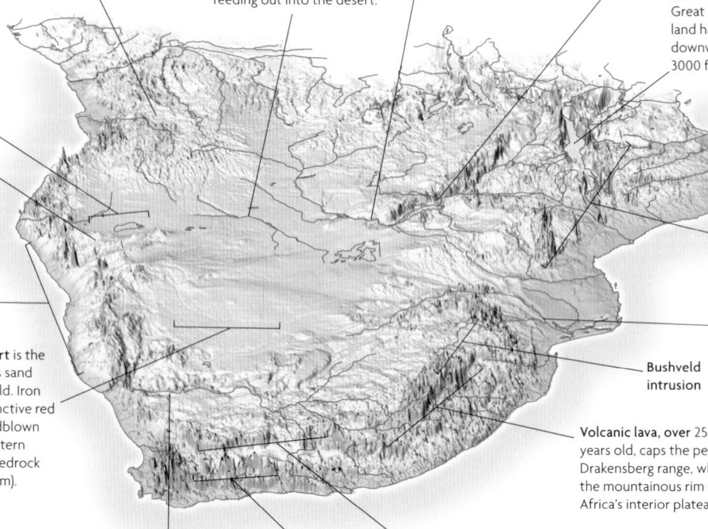
Thousands of years of evaporating water have produced the Etosha Pan, one of the largest salt flats in the world. Lake and river sediments in the area indicate that the region was once less arid.
Planalto de Bié
The Okavango/Cubango river flows from the Planalto de Bié to the swamplands of the Okavango Delta, one of the world's largest inland deltas, where it divides into countless distributary channels, feeding out into the desert.
At Victoria Falls, the Zambezi river has cut a spectacular gorge taking advantage of large joints in the basalt, which were first formed as the lava cooled and contracted.
Lake Nyasa occupies one of the deep troughs of the Great Rift Valley, where the land has been displaced downwards by as much as 3000 ft (920 m).
Khorixas, Namibia
Great Rift Valley
Namib Desert
Limpopo river
The Kalahari Desert is the largest continuous sand surface in the world. Iron oxide gives a distinctive red colour to the windblown sand, which, in eastern areas covers the bedrock by over 200 ft (60 m).
Bushveld intrusion
Volcanic lava, over 250 million years old, caps the peaks of the Drakensberg range, which lie on the mountainous rim of southern Africa's interior plateau.
The Orange River, one of the longest in Africa, rises in Lesotho and is the only major river in the south which flows westward, rather than to the east coast.
The mountains of the Little Karoo are composed of sedimentary rocks which have been substantially folded and faulted.
Broad, flat-topped mountains characterize the Great Karoo, which have been cut from level rock strata under extremely arid conditions.

▶ ***The fast-flowing Zambezi** river cuts a deep, wide channel as it flows along the Zimbabwe/Zambia border.*

The transport network

84,213 miles (135,609 km)	746 miles (1202 km)
23,208 miles (37,372 km)	3815 miles (6144 km)

Southern Africa's Cape-gauge rail network is by far the largest in the continent. About two-thirds of the 20,000 mile (32,000 km) system lies within South Africa. Lines such as the Harare–Bulawayo route have become corridors for industrial growth.

▲ ***Following a series** of droughts, this baobab tree in Zimbabwe now stands alone in a field once filled by sugar cane. The thick trunk and small leaves of the baobab help it to conserve water, enabling it to survive even in drought conditions.*

Bushveld intrusion

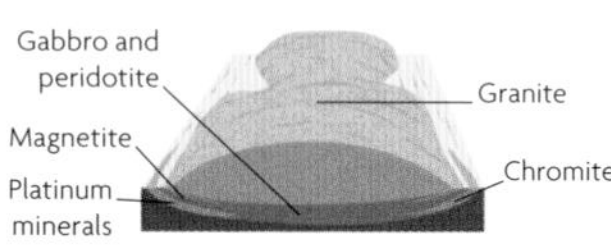
Gabbro and peridotite
Granite
Magnetite
Chromite
Platinum minerals

▲ ***The Bushveld intrusion** lies on South Africa's high 'veld'. Molten magma intruded into the Earth's crust creating a saucer-shaped feature, more than 180 miles (300 km) across, containing regular layers of precious minerals, overlain by a dome of granite.*

Map key

Population

- 1 million to 5 million
- 500,000 to 1 million
- 100,000 to 500,000
- 50,000 to 100,000
- 10,000 to 50,000
- below 10,000

Elevation

- 3000m / 9843ft
- 2000m / 6562ft
- 1000m / 3281ft
- 500m / 1640ft
- 250m / 820ft
- 100m / 328ft
- sea level

Scale 1:10,500,000

Km 0 25 50 100 150 200 250 300
Miles 0 25 50 100 150 200 250 300

projection: Lambert Azimuthal Equal Area

South Africa: Capital cities

PRETORIA – administrative capital
CAPE TOWN – legislative capital
BLOEMFONTEIN – judicial capital

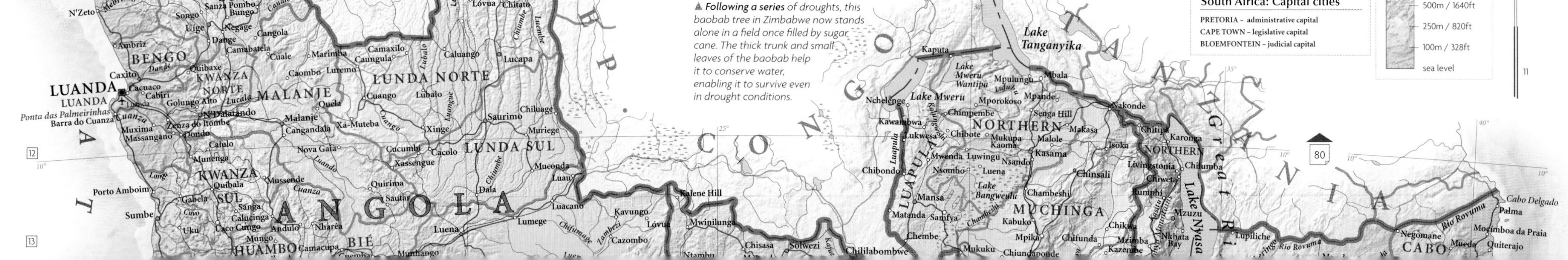
CONGO
CABINDA (To Angola)
Cacongo
Cabinda
Ponta Vermelha
Soyo
ZAIRE
Ponta do Padrão
M'banza Kongo
UIGE
Quinzau
N'Zeto
BENGO
LUANDA
Ponta das Palmeirinhas
Barra do Cuanza
KWANZA NORTE
MALANJE
LUNDA NORTE
LUNDA SUL
KWANZA SUL
Porto Amboim
Sumbe
HUAMBO
BIÉ
ANGOLA
D E M. R E P. C O N G O
Lake Tanganyika
Kaputa
Lake Mweru
NORTHERN
MUCHINGA
LUAPULA
Lake Nyasa
T A N Z A N I A
Great Rift
Cabo Delgado
CABO
Lake Bangweulu

78
78
80

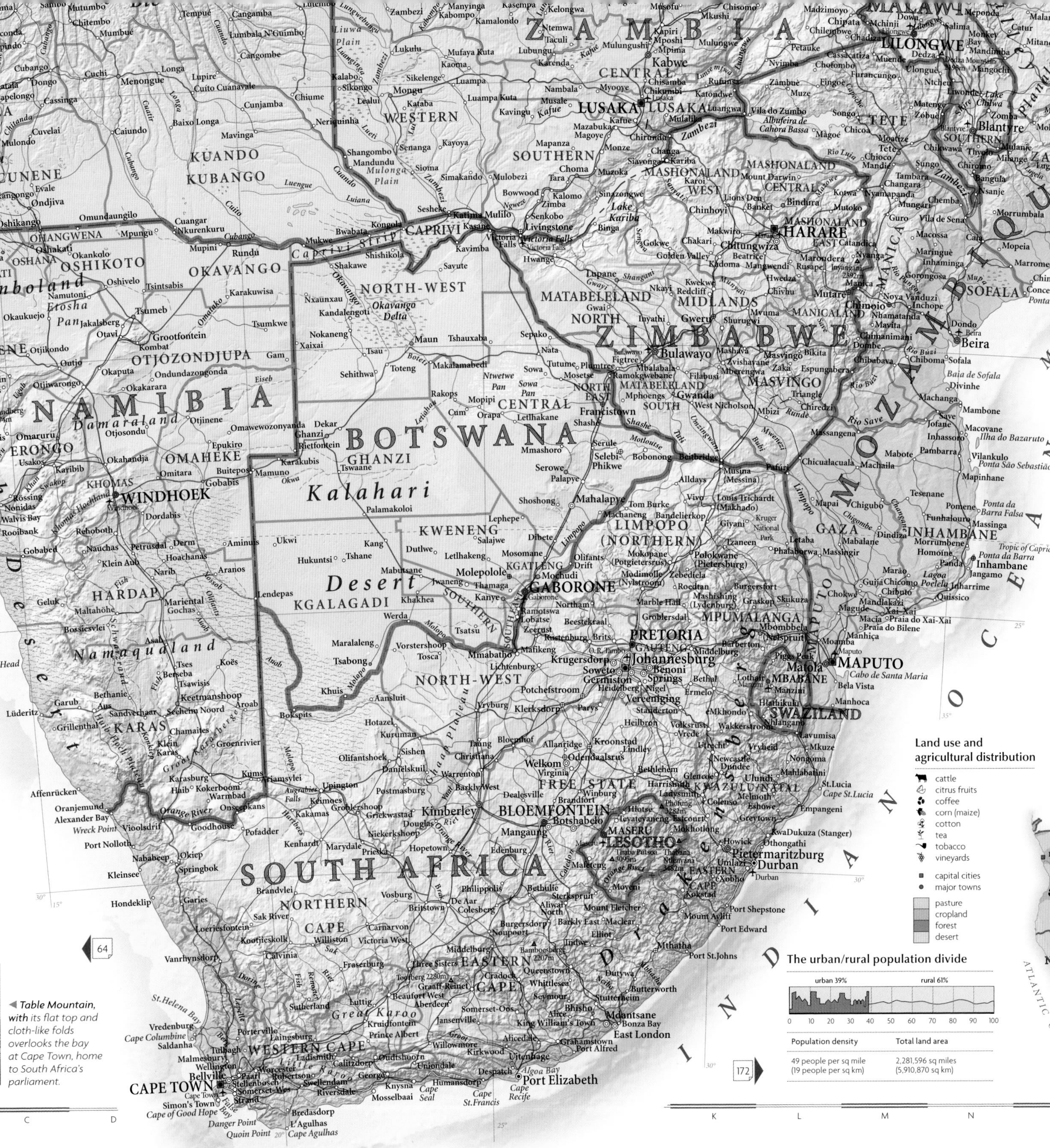

172

64

▲ **A wide range** of crops are grown in South Africa, aided in many areas by irrigation schemes, such as the Orange River Project, which supplement irregular rainfall.

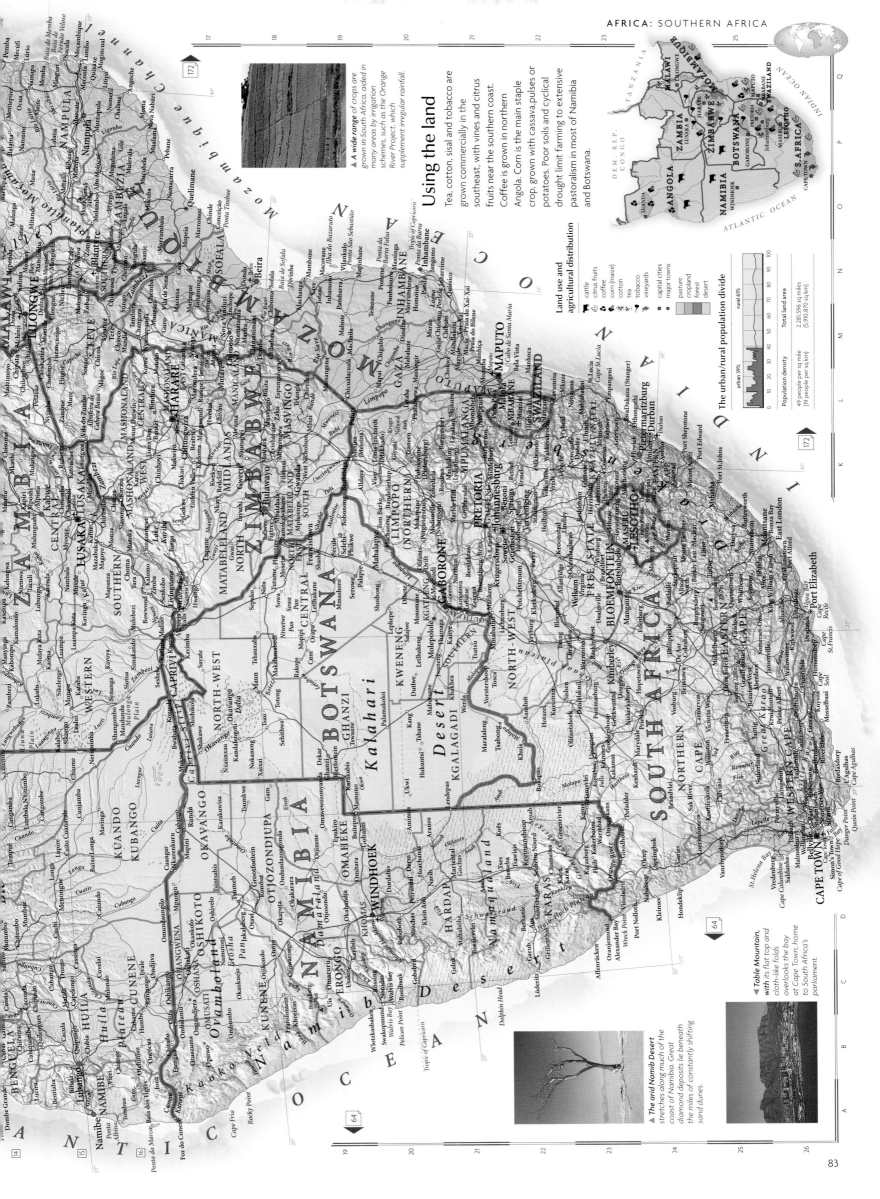

Using the land

Tea, cotton, sisal and tobacco are grown commercially in the southeast, with vines and citrus fruits near the southern coast. Coffee is grown in northern Angola. Corn is the main staple crop, grown with cassava,pulses or potatoes. Poor soils and cyclical drought limit farming to extensive pastoralism in most of Namibia and Botswana.

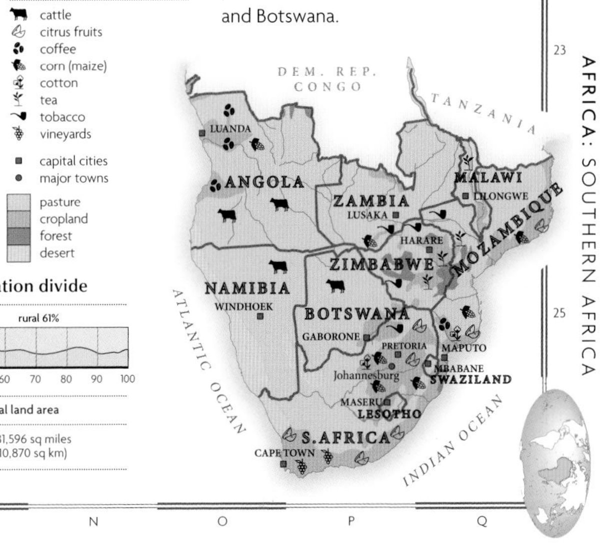

The urban/rural population divide

urban 39% rural 61%

Population density	Total land area
49 people per sq mile (19 people per sq km)	2,281,596 sq miles (5,910,870 sq km)

172

▲ **The arid Namib Desert** stretches along much of the coast of Namibia. Great diamond deposits lie beneath the miles of constantly shifting sand dunes.

64

◀ **Table Mountain, with** its flat top and cloth-like folds overlooks the bay at Cape Town, home to South Africa's parliament.

ARCTIC OCEAN
North Pole
Ellesmere Island
Greenland
King Frederik VIII Land
King Christian X Land
Greenland Sea
Laptev Sea
Severnaya Zemlya
Ostrov Rudol'fa
Franz Josef Land
Spitsbergen
Poluostrov Taymyr
Kara Sea
Mys Flissingskiy
Novaya Zemlya
Barents Sea
Bjørnøya
Barents Trough
Poluostrov Yamal
Baydaratskaya Guba
Gulf of Ob
Kara Strait
Yenisey
Jan Mayen Fracture Zone
Jan Mayen
Kolbeinsey Ridge
Denmark Strait
Arctic Circle
Iceland Plateau
Jan Mayen Ridge
Norwegian Basin
Norwegian Sea
Bjargtangar
Iceland
Vatnajökull
Reykjanes Ridge
Faroe-Iceland Ridge
Voring Plateau
Tromsøflaket
North Cape
Nordkinn
Fugløya Bank
Murmansk Rise
Ostrov Kolguyev
Poluostrov Kanin
Pechora
Vesterålen
Lofoten
Kebnekaise 2117m
Inarijärvi
Kola Peninsula
Ozero Imandra
Torneälven
Luleälven
Kemijoki
White Sea
Onega Bay
Timanskiy Kryazh
Mezen'
West Siberian Plain
Ural Mountains
ASIA
Ob'
Traena Bank
Scandinavia
Kölen
Oulujoki
Umeälven
Gulf of Bothnia
Ozero Vygozero
Lake Onega
Lake Ladoga
Northern Dvina
Iceland Basin
Faroe Islands
Bill Baileys Bank
Faroe-Shetland Trough
Shetland Islands
Hatton Ridge
Viking Bank
Galdhøpiggen 2469m
Ljungan
Ljusnan
Glåma
Åland
Gulf of Finland
Svir
Ozero Beloye
Sukhona
Rybinsk Reservoir
Gor'kiy Reservoir
Kama
Votkinsk Reservoir
Vyatka
Orkney Islands
Outer Hebrides
Ben Nevis 1343m
Grampian Mountains
Rockall Rise
Rockall Trough
Feni Ridge
Norwegian Trench
Skagerrak
Vänern
Vättern
Lake Peipus
Lake Ilmen
Lake Pskov
Gulf of Riga
Gotland
North Sea
Jutland Bank
Great Fisher Bank
Jutland
Kattegat
Sjælland
Baltic Sea
North European Plain
Western Dvina
Moskva
Oka
Sura
Volga Upland
Kuybyshev Reservoir
Belaya
Ural
Volga
Samara
North Channel
British Isles
Pennines
Ireland
Irish Sea
Snowdon 1085m
Shannon
Trent
Severn
Britain
Dogger Bank
Frisian Islands
Elbe
Oder
Warta
Vistula
Neman
Bug
Berezina
Dnieper
Pripet Marshes
Desna
Klyaz'ma
Central Russian Upland
Seym
Don
Khoper
Porcupine Plain
Celtic Sea
Celtic Shelf
St. George's Channel
Bristol Channel
Thames
Strait of Dover
Land's End
English Channel
Channel Islands
Harz
Ardennes
Rhine
Meuse
Moselle
EUROPE
Kiev Reservoir
Kremenchuk Reservoir
Dnieper Lowlands
Donets
Tsimlyansk Reservoir
Volga
Kirghiz Steppe
Dniester
Podil's'ka Vysochina
Pivdennyy Buh
Seine
Marne
Black Forest
Danube
Vosges
Lake Constance
Carpathian Mountains
Tisza
Siret
Prut
Black Sea Lowland
Manych
Yergeni
Loire
Vienne
Cher
Saône
Lake Geneva
Mont Blanc
Bakony
Lake Balaton
Great Hungarian Plain
Drava
Sava
Transylvanian Alps
Sea of Azov
Crimea
Kerch Strait
Kuban'
Charcot Seamounts
Azores-Biscay Rise
Biscay Plain
Bay of Biscay
Garonne
Dordogne
Lot
Massif Central
Cévennes
Rhône
Lake Garda
Po
Adriatic Sea
Dinaric Alps
Black Sea
Galicia Bank
Theta Gap
ATLANTIC OCEAN
Miño
Cordillera Cantábrica
Iberian Peninsula
Aragón
Ebro
Duero
Douro
Sistema Central
Sistema Ibérico
Gulf of Lion
Ligurian Sea
Arno
Apennines
Tiber
Corno Grande 2912m
Balkan Mountains
Danube
Maritsa
Rhodope Mountains
Bosporus
Sea of Marmara
EURASIAN PLATE
ANATOLIAN PLATE
Iberian Plain
Corsica
Strait of Bonifacio
Adriatic Basin
Lake Ohrid
Lake Prespa
Lake Scutari
Anatolia
Lake Tuz
Cabo da Roca
Tagus
Tagus Plain
Guadiana
Júcar
Gulf of Valencia
Balearic Islands
Algerian Basin
Sardinia
Tyrrhenian Sea
Tyrrhenian Basin
Gulf of Taranto
Strait of Otranto
Aegean Sea
Sierra Morena
Guadalquivir
Segura
Gorringe Ridge
Cape Saint Vincent
Sistemas Béticos
Sierra Nevada
Mediterranean Sea
Horseshoe Seamounts
Ampère Seamount
Punta de Tarifa
Strait of Gibraltar
Alboran Sea
Oued Chelif
Mount Etna 3340m
Strait of Messina
Strait of Sicily
Sicily
Ionian Sea
Peloponnese
Mirtoan Sea
Sea of Crete
Karpathos Strait
Crete
Gavdos
Rhodes Basin
Gulf of Antalya
Taurus Mountains
Cyprus
Cyprus Basin
Seine Plain
Seine Seamount
Madeira
Sebou
Rif
Tell Atlas
Atlas Mountains
EURASIAN PLATE
AFRICAN PLATE
Malta
Ionian Basin
Mediterranean Ridge
Dacia Seamount
Agadir Canyon
Oumer Rbia
Middle Atlas
Moulouya
Saharan Atlas
High Atlas
Chott el Jerid
Gulf of Sirte
Levantine Basin
Nile Fan
Suez Canal
Nile
Canary Islands
Grand Erg Occidental
Grand Erg Oriental
Qattara Depression -133m
Western Desert
Libyan Desert
AFRICA
SAHARA
'Erg Iguidi
Erg Chech

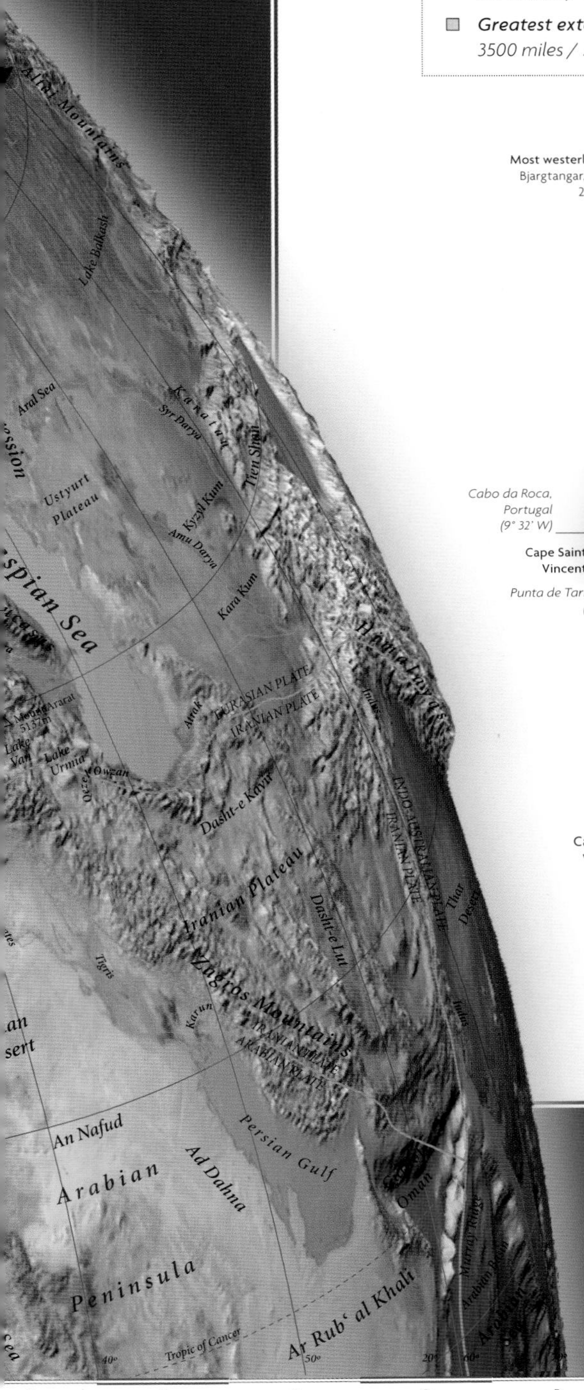
Lake Balkhash
Aral Sea
Syr Darya
Ustyurt Plateau
Kyzyl Kum
Amu Darya
Kara Kum
Tien Shan
EURASIAN PLATE
IRANIAN PLATE
Lake Urmia
Dasht-e Kavir
Iranian Plateau
Dasht-e Lut
Zagros Mountains
ARABIAN PLATE
Tigris
Karun
Persian Gulf
An Nafud
Arabian
Ad Dahna
Peninsula
Ar Rub' al Khali
Tropic of Cancer

Europe

Europe is the world's second smallest continent, covering 4,053,309 sq miles (10,498,000 sq km). It comprises 46 separate countries, including Turkey and the Russian Federation, although the greater parts of these nations lie in Asia.

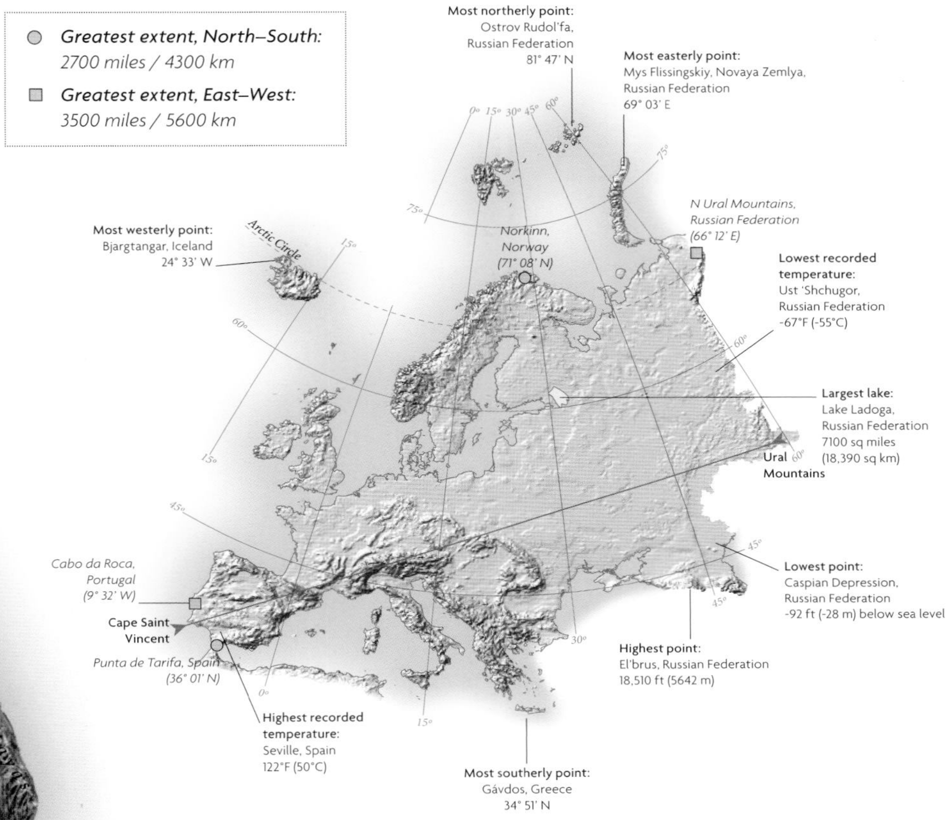
Greatest extent, North–South:
2700 miles / 4300 km
Greatest extent, East–West:
3500 miles / 5600 km
Most northerly point:
Ostrov Rudol'fa, Russian Federation 81° 47' N
Most easterly point:
Mys Flissingskiy, Novaya Zemlya, Russian Federation 69° 03' E
Most westerly point:
Bjargtangar, Iceland 24° 33' W
Arctic Circle
Norkinn, Norway (71° 08' N)
N Ural Mountains, Russian Federation (66° 12' E)
Lowest recorded temperature:
Ust 'Shchugor, Russian Federation -67°F (-55°C)
Largest lake:
Lake Ladoga, Russian Federation 7100 sq miles (18,390 sq km)
Ural Mountains
Cabo da Roca, Portugal (9° 32' W)
Cape Saint Vincent
Punta de Tarifa, Spain (36° 01' N)
Lowest point:
Caspian Depression, Russian Federation -92 ft (-28 m) below sea level
Highest point:
El'brus, Russian Federation 18,510 ft (5642 m)
Highest recorded temperature:
Seville, Spain 122°F (50°C)
Most southerly point:
Gávdos, Greece 34° 51' N

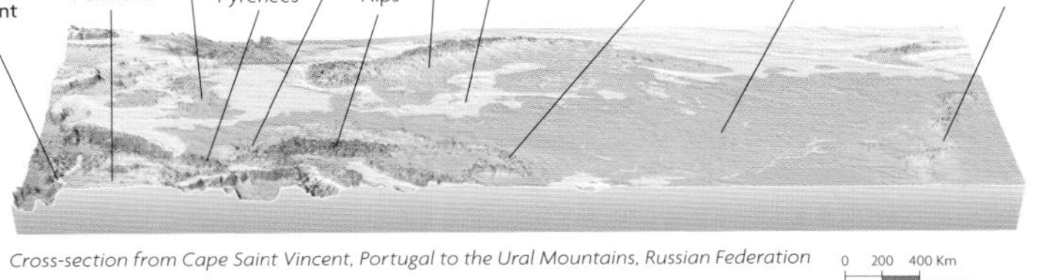
Cape Saint Vincent
Iberian Peninsula
British Isles
Pyrenees
Massif Central
Alps
Scandinavia
Baltic Sea
Carpathian Mountains
North European Plain
Ural Mountains
0 200 400 Km
0 200 400 Miles

Cross-section from Cape Saint Vincent, Portugal to the Ural Mountains, Russian Federation

line of cross-section

Physical Europe

The physical diversity of Europe belies its relatively small size. To the northwest and south it is enclosed by mountains. The older, rounded Atlantic Highlands of Scandinavia and the British Isles lie to the north and the younger, rugged peaks of the Alpine Uplands to the south. In between lies the North European Plain, stretching 2485 miles (4000 km) from The Fens in England to the Ural Mountains in Russia. South of the plain lies a series of gently folded sedimentary rocks separated by ancient plateaux, known as massifs.

The North European Plain

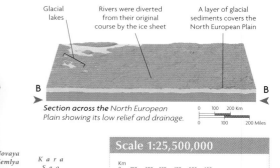

Rising less than 1000 ft (300 m) above sea level, the North European Plain strongly reflects past glaciation. Ridges of both coarse moraine and finer, windblown deposits have accumulated over much of the region. The ice sheet also diverted a number of river channels from their original courses.

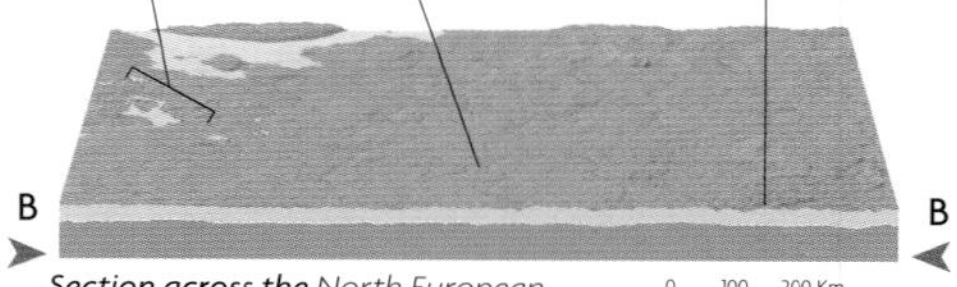

Glacial lakes
Rivers were diverted from their original course by the ice sheet
A layer of glacial sediments covers the North European Plain
B
B
0 100 200 Km
0 100 200 Miles

***Section across the** North European Plain showing its low relief and drainage.*

The Atlantic Highlands

The Atlantic Highlands were formed by compression against the Scandinavian Shield during the Caledonian mountain-building period over 500 million years ago. The highlands were once part of a continuous mountain chain, now divided by the North Sea and a submerged rift valley.

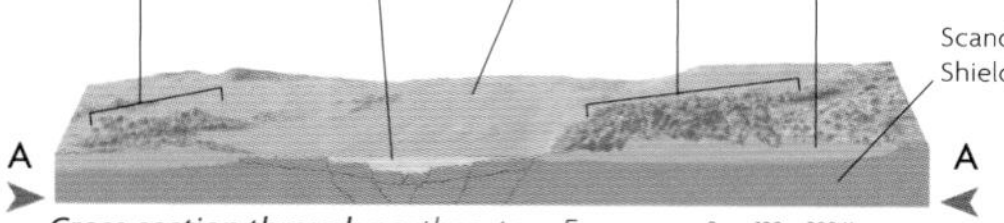

The Atlantic Highlands continue in the British Isles
Rift valley buried by sediments
North Sea
Atlantic Highlands in Norway
Rocks affected by ancient mountain-building
Scandinavian Shield
A
A
0 100 200 Km
0 100 200 Miles

***Cross-section through** northeastern Europe showing the continuous mountain chain and rift valley system.*

Scale 1:25,500,000
Km 0 100 200 300 400 500 600
Miles 0 100 200 300 400 500 600
projection: Lambert Azimuthal Equal Area
Map key
Elevation
4000m / 13,124ft
3000m / 9843ft
2000m / 6562ft
1000m / 3281ft
500m / 1640ft
250m / 820ft
100m / 328ft
sea level
Plate margins
(for explanation see page xiv)
constructive
destructive
conservative
uncertain
physiographic regions
line of cross-section
Novaya Zemlya
Kara Sea
Ostrov Kolguyev
Barents Sea
Iceland
NORTH AMERICAN PLATE
EURASIAN PLATE
Norwegian Sea
Kölen
Kola Peninsula
White Sea
Ural Mountains
Faroe Islands
Shetland Islands
Outer Hebrides
ATLANTIC OCEAN
ATLANTIC HIGHLANDS
SCANDINAVIAN SHIELD
Gulf of Bothnia
Northern Dvina
Lake Onega
Lake Ladoga
British Isles
Ireland
Shannon
Vänern
Vättern
North Sea
Jutland
Baltic Sea
Gulf of Riga
Western Dvina
NORTH EUROPEAN PLAIN
Britain
The Fens
Thames
English Channel
Elbe
Rhine
Oder
Vistula
Harz
Central Russian Upland
Volga Uplands
Volga
Ardennes
Seine
Loire
PLATEAUX AND LOWLANDS
Danube
Carpathian Mountains
Dnieper
Dniester
Don
Bay of Biscay
Massif Central
Mt Blanc 4807m
Garonne
Rhône
ALPS
Great Hungarian Plain
Sea of Azov
Crimea
Caspian Sea
Caucasus
El'brus 5642m
Douro
Iberian Peninsula
Pyrenees
Ebro
ALPINE UPLANDS
Po
Apennines
Dinaric Alps
Balkan Mountains
Black Sea
Corsica
Adriatic Sea
Guadalquivir
Balearic Islands
Sardinia
Vesuvius 1171m
Tyrrhenian Sea
ASIA
EURASIAN PLATE
AFRICAN PLATE
ANATOLIAN PLATE
Mediterranean Sea
Sicily
Etna 3263m
Ionian Sea
Peloponnese
Aegean Sea
Crete

The plateaux and lowlands

The uplifted plateaux or massifs of southern central Europe are the result of long-term erosion, later followed by uplift. They are the source areas of many of the rivers which drain Europe's lowlands. In some of the higher reaches, fractures have enabled igneous rocks from deep in the Earth to reach the surface.

The Alpine uplands

The collision of the African and European continents, which began about 65 million years ago, folded and then uplifted a series of mountain ranges running across southern Europe and into Asia. Two major lines of folding can be traced: one includes the Pyrenees, the Alps and the Carpathian Mountains; the other incorporates the Apennines and the Dinaric Alps.

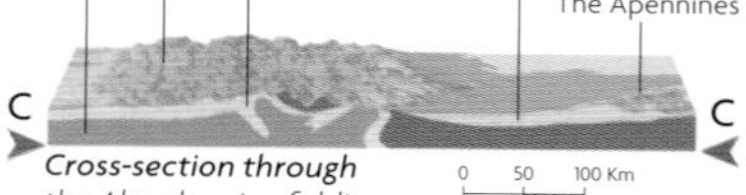

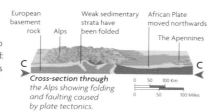

European basement rock
Alps
Weak sedimentary strata have been folded
African Plate moved northwards
The Apennines
C
C
0 50 100 Km
0 50 100 Miles

***Cross-section through** the Alps showing folding and faulting caused by plate tectonics.*

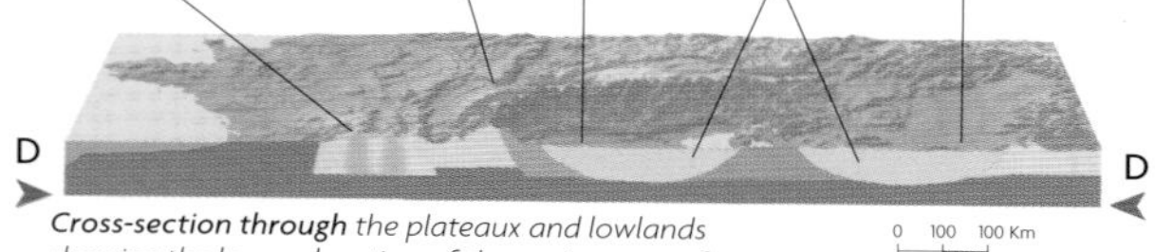

Igneous rocks have intruded into the Massif Central
Older, eroded massifs lie behind the arc of the Alps
Po Valley
Tectonically formed basins
Great Hungarian Plain
D
D
0 100 100 Km
0 100 100 Miles

***Cross-section through** the plateaux and lowlands showing the lower elevation of the ancient massifs.*

▲ ***Frost grips northern*** *and eastern Europe during the long cold winters. Lakes and rivers frequently freeze.*

Climate

Europe experiences few extremes in either rainfall or temperature, with the exception of the far north and south. Along the west coast, the warm currents of the North Atlantic Drift moderate temperatures. Although east–west air movement is relatively unimpeded by relief, the Alpine Uplands halt the progress of north–south air masses, protecting most of the Mediterranean from cold, north winds.

Temperature

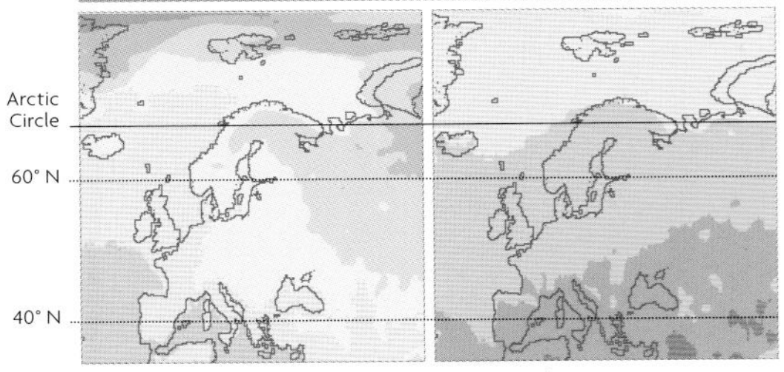
Arctic Circle
60° N
40° N

Average January temperature *Average July temperature*

Temperature

- below -30°C (-22°F)
- -30 to -20°C (-22 to -4°F)
- -20 to -10°C (-4 to 14°F)
- -10 to 0°C (14 to 32°F)
- 0 to 10°C (32 to 50°F)
- 10 to 20°C (50 to 68°F)
- 20 to 30°C (68 to 86°F)
- above 30°C (86°F)

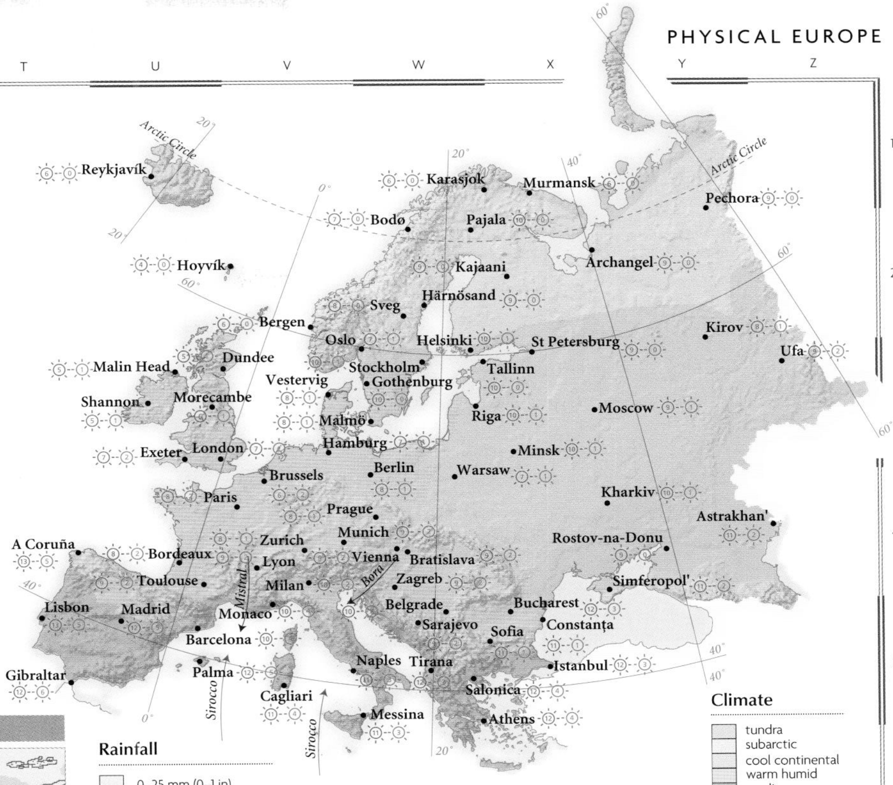
Reykjavík
Karasjok
Murmansk
Pechora
Bodø
Pajala
Hoyvík
Kajaani
Archangel
Sveg
Härnösand
Bergen
Oslo
Helsinki
St Petersburg
Kirov
Ufa
Malin Head
Dundee
Stockholm
Tallinn
Vestervig
Gothenburg
Shannon
Morecambe
Malmö
Riga
Moscow
Hamburg
Exeter
London
Minsk
Berlin
Brussels
Warsaw
Kharkiv
Paris
Prague
Astrakhan'
Munich
A Coruña
Zurich
Bordeaux
Lyon
Vienna
Bratislava
Rostov-na-Donu
Toulouse
Zagreb
Milan
Simferopol'
Mistral
Bora
Lisbon
Madrid
Monaco
Belgrade
Bucharest
Sarajevo
Sofia
Constanța
Barcelona
Naples
Tirana
Istanbul
Gibraltar
Palma
Cagliari
Salonica
Sirocco
Messina
Athens
Arctic Circle

Climate

- tundra
- subarctic
- cool continental
- warm humid
- mediterranean
- semi-arid
- daily hours of sunshine, January
- daily hours of sunshine, July
- cold wind
- hot wind

▲ ***Mild temperatures and*** *frequent rainfall contribute to the fertile farming land found over much of northwestern Europe.*

Rainfall

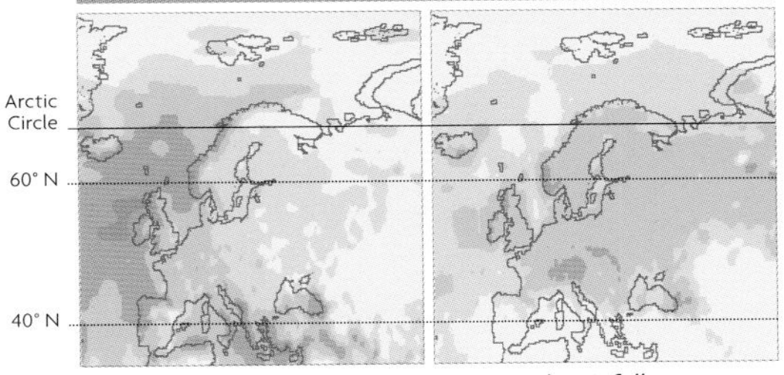
Arctic Circle
60° N
40° N

Average January rainfall *Average July rainfall*

Rainfall

- 0–25 mm (0–1 in)
- 25–50 mm (1–2 in)
- 50–100 mm (2–4 in)
- 100–200 mm (4–8 in)
- 200–300 mm (8–12 in)
- 300–400 mm (12–16 in)
- 400–500 mm (16–20 in)
- more than 500 mm (20 in)

▶ ***Dusty Sirocco winds*** *from Africa help create the semi-arid scrubland common across the Mediterranean coastlands of southern Europe.*

Shaping the continent

Successive Ice Ages have left many relict landforms across Europe. Present glaciers continue to carve peaks and valleys in the northern Atlantic Highlands and Alpine Uplands. Tectonic activity, both past and present, has shaped southern Europe and Iceland. Active volcanoes and earthquakes still occur in Italy and Greece. Europe's extensive coastline, particularly in the northwest, is constantly modified by wave action and fluvial deposits.

Glaciation

1 Valley glaciers, such as this one *(left)* in Iceland, form in hollows at the top of valleys and flow downwards, drawn by gravity. Their growth is dynamic; new snowfall constantly accumulates at the head of the glacier, while the snout melts, depositing material eroded and carried by the glacier.

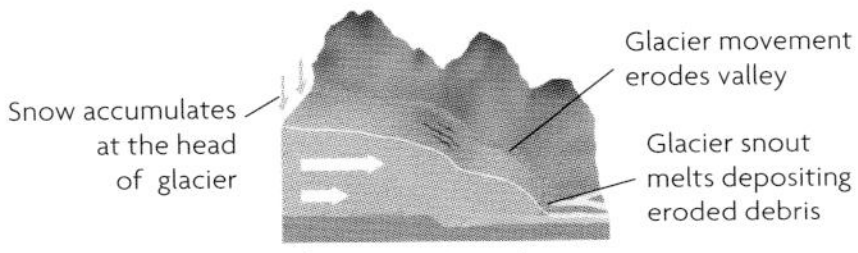
Snow accumulates at the head of glacier
Glacier movement erodes valley
Glacier snout melts depositing eroded debris

Glaciation: *Development of a glacier*

River systems

2 Rivers are continuously transporting eroded material towards the sea. Slow-moving, low-gradient rivers, like this one in western Russia *(above)*, deposit their alluvium load, infilling valleys creating a flood plain. Subsequent climatic and tectonic fluctuations may erode the flood plain to form terraces.

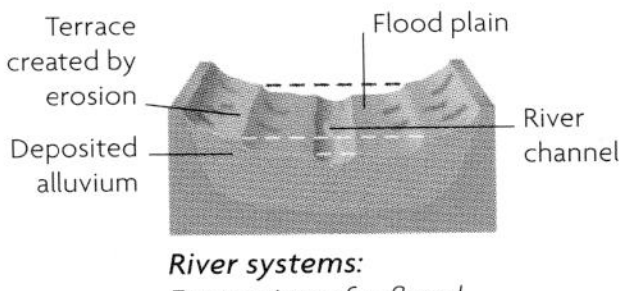
Terrace created by erosion
Flood plain
Deposited alluvium
River channel

River systems: *Formation of a flood plain and terraces*

Coastal processes

5 Spits are narrow bands of sand or shingle, formed by longshore drift; a process whereby waves carry material along the beach. They usually form where the coastline changes direction, and their growth is then halted by an opposing river current, as at Spurn Head, in the British Isles *(left)*. Coastal features such as these are constantly being created and destroyed.

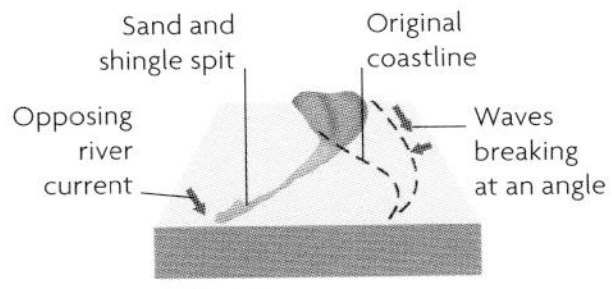
Sand and shingle spit
Original coastline
Opposing river current
Waves breaking at an angle

Coastal processes: *Formation of a spit*

The evolving landscape

Landscape

- uplifting land
- stable land
- sinking land
- limestone region
- glacier
- active volcano
- ocean current
- area of tectonic activity
- maximum limit of glaciation

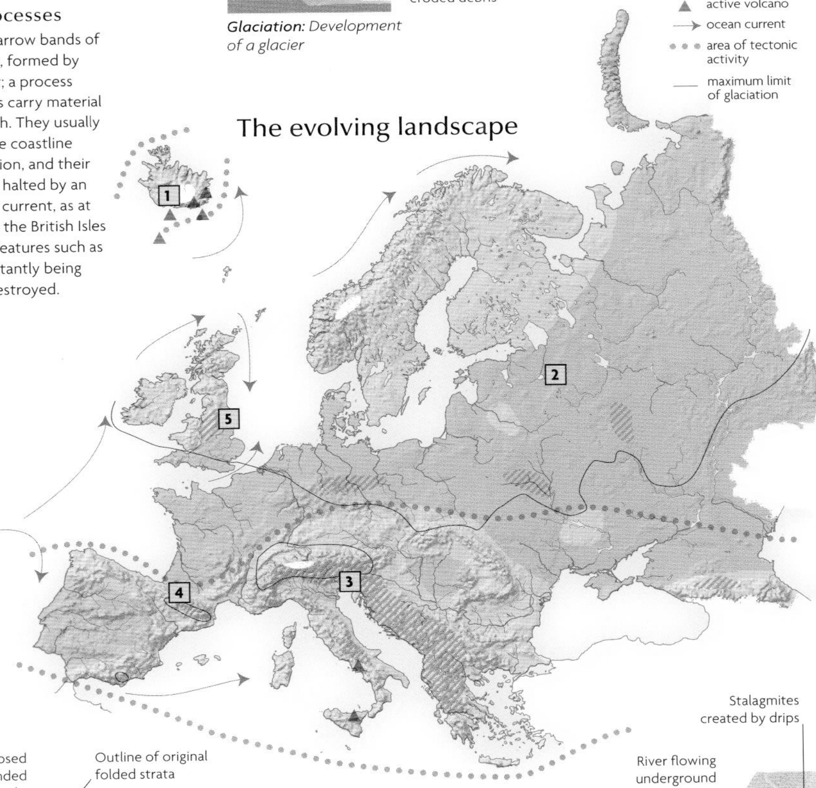
1
2
3
4
5

Erosion and weathering

4 Much of Europe was once subjected to folding and faulting, exposing hard and soft rock layers. Subsequent erosion and weathering has worn away the softer strata, leaving up-ended layers of hard rock as in the French Pyrenees *(above)*.

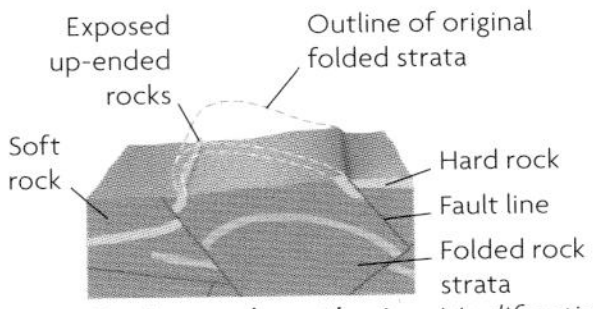
Exposed up-ended rocks
Outline of original folded strata
Soft rock
Hard rock
Fault line
Folded rock strata

Erosion and weathering: *Modification of a fold*

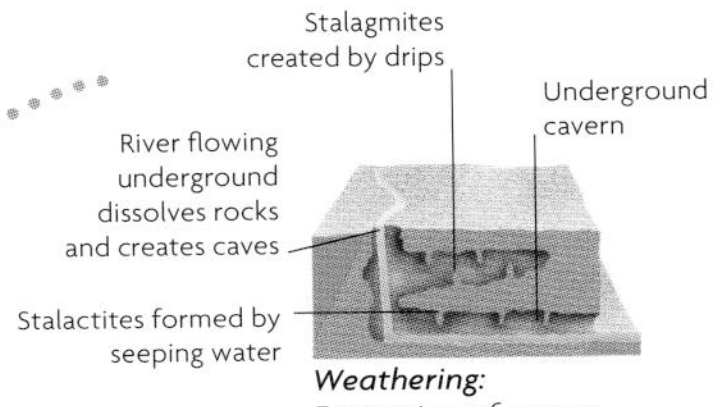
Stalagmites created by drips
Underground cavern
River flowing underground dissolves rocks and creates caves
Stalactites formed by seeping water

Weathering: *Formation of a cave*

Weathering

3 As surface water filters through permeable limestone, the rock dissolves to form underground caves, like Postojna in the Karst region of Slovenia *(above)*. Stalactites grow downwards as lime-enriched water seeps from roof fractures; stalagmites grow upwards where drips splash down.

Political Europe

The political boundaries of Europe have changed many times, especially during the 20th century in the aftermath of two world wars, the break-up of the empires of Austria-Hungary, Nazi Germany and, towards the end of the century, the collapse of communism in eastern Europe. The fragmentation of Yugoslavia has again altered the political map of Europe, highlighting a trend towards nationalism and devolution. In contrast, economic federalism is growing. In 1958, the formation of the European Economic Community (now the European Union or EU) started a move towards economic and political union and increasing internal migration.

▲ ***The Brandenburg Gate** in Berlin is a potent symbol of German reunification. From 1961, the road beneath it ended in a wall, built to stop the flow of refugees to the West. It was opened again in 1989 when the wall was destroyed and East and West Germany were reunited.*

Population

Europe is a densely populated, urbanized continent; in Belgium over 90% of people live in urban areas. The highest population densities are found in an area stretching east from southern Britain and northern France, into Germany. The northern fringes are only sparsely populated.

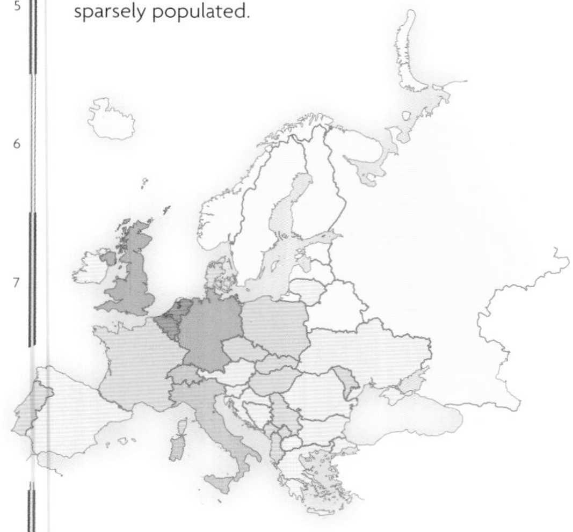

Population density
(people per sq km)

- below 49
- 50–99
- 100–149
- 150–199
- 200–299
- above 300

▲ ***Demand for space** in densely populated European cities like London has led to the development of high-rise offices and urban sprawl.*

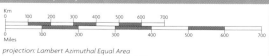

▲ ***Traditional lifestyles still** persist in many remote and rural parts of Europe, especially in the south, east, and in the far north.*

Denmark Strait
Arctic Circle
REYKJAVÍK
ICELAND
Norwegian Sea
Faroe Islands (to Denmark)
ATLANTIC OCEAN
Shetland Islands
Orkney Islands
Outer Hebrides
NORWAY
SWEDEN
FINLAND
Trondheim
Gulf of Bothnia
Bergen
Stavanger
OSLO
Kristiansand
Tampere
Turku
Åland
HELSINKI
Murmansk
St Petersburg
Uppsala
Örebro
STOCKHOLM
Vänern
Vättern
Jönköping
TALLINN
ESTONIA
Gotland
Baltic Sea
Ventspils
LATVIA
RIGA
Western Dvina
Liepāja
LITHUANIA
Vitsyebsk
Kaunas
VILNIUS
MINSK
Babruysk
BELARUS
Brest
RUSS. FED. (Kaliningrad)
Kaliningrad
Gdańsk
SCOTLAND
Aberdeen
Dundee
Glasgow
Edinburgh
North Sea
NORTHERN IRELAND
Belfast
IRELAND
DUBLIN
UNITED KINGDOM
Newcastle upon Tyne
Liverpool
Leeds
Manchester
Sheffield
WALES
Birmingham
ENGLAND
Cardiff
Thames
LONDON
Southampton
English Channel
Channel Islands
Gothenburg
Aalborg
DENMARK
Helsingborg
COPENHAGEN
Malmö
Odense
Groningen
Hamburg
Elbe
Oder
NETH.
AMSTERDAM
THE HAGUE
Rotterdam
Bremen
Hanover
Nijmegen
BERLIN
Bydgoszcz
Vistula
Poznań
WARSAW
Łódź
POLAND
Wrocław
Antwerp
BELGIUM
BRUSSELS
Düsseldorf
GERMANY
Liège
Bonn
Rhine
Leipzig
Dresden
Frankfurt am Main
LUXEMBOURG
PRAGUE
CZECH REPUBLIC
Kraków
Lviv
UK
le Havre
Seine
Rennes
PARIS
St-Nazaire
Nantes
Loire
Orléans
Strasbourg
Nuremberg
Stuttgart
Bay of Biscay
FRANCE
Munich
Danube
SLOVAKIA
BRATISLAVA
VIENNA
Győr
Miskolc
Chernivtsi
Dniester
MOLDOVA
Limoges
BERN
Zurich
Salzburg
SWITZERLAND
Innsbruck
AUSTRIA
BUDAPEST
HUNGARY
Cluj-Napoca
CHIȘINĂU
Bordeaux
Geneva
Lyon
Alps
LIECHTENSTEIN
A Coruña
Bilbao
Milan
LJUBLJANA
SLOVENIA
Trieste
ZAGREB
ROMANIA
Brașov
Porto
Duero
Valladolid
Ebro
Pyrenees
Toulouse
Rhône
Turin
Verona
Venice
Po
Genoa
Bologna
CROATIA
BELGRADE
BUCHAREST
Constanța
PORTUGAL
Tagus
LISBON
Setúbal
MADRID
SPAIN
Zaragoza
ANDORRA LA VELLA
ANDORRA
Marseille
Nice
MONACO
Pisa
Florence
SAN MARINO
Adriatic Sea
BOS. & HERZ.
SARAJEVO
Mostar
SERBIA
Danube
Ruse
Varna
Barcelona
Corsica
ITALY
KOSOVO (disputed)
PRISHTINË
BULGARIA
SOFIA
Stara Zagora
Burgas
Valencia
Seville
Córdoba
Majorca
Minorca
VATICAN CITY
ROME
MONTENEGRO
PODGORICA
SKOPJE
MACEDONIA
Istanbul
Ibiza
Palma
Balearic Islands
Sardinia
Bari
TIRANA
Cádiz
Málaga
Murcia
Naples
ALBANIA
Gibraltar (to UK)
Ceuta (to Spain)
Cagliari
Tyrrhenian Sea
Salonica
Melilla (to Spain)
Mediterranean Sea
Cosenza
Larisa
GREECE
Aegean Sea
Palermo
Messina
Sicily
Catania
ATHENS
Piraeus
Ionian Sea
MALTA
VALLETTA
Iráklein
Crete

Map key

Population

- ■ above 5 million
- ◾ 1 million to 5 million
- ◉ 500,000 to 1 million
- ◎ 100,000 to 500,000
- ⊕ 50,000 to 100,000
- ○ 10,000 to 50,000
- ● Country capital

Borders

- full international border

Scale 1:17,250,000

Km 0 100 200 300 400 500 600 700

Miles 0 100 200 300 400 500 600 700

projection: Lambert Azimuthal Equal Area

Novaya Zemlya
Kara Sea
Barents Sea
Vorkuta
Arctic Circle
Ural Mountains
RUSSIAN FEDERATION
Archangel
Northern Dvina
Lake Onega
Perm'
Kirov
Vologda
Ufa
Yaroslavl'
Kazan'
Nizhniy Novgorod
MOSCOW
Ul'yanovsk
Tol'yatti
Samara
Orenburg
Tula
Kazakhstan
Saratov
Voronezh
Volgograd
Kharkiv
INE
Dnipropetrovs'k
Donets'k
Rostov-na-Donu
Volga
Astrakhan'
Caspian Sea
Dnieper
Sea of Azov
Stavropol'
Crimea
Simferopol'
Novorossiysk
Groznyy
Caucasus
Georgia
Azerbaijan
Black Sea
annexed by Russia, 2014)

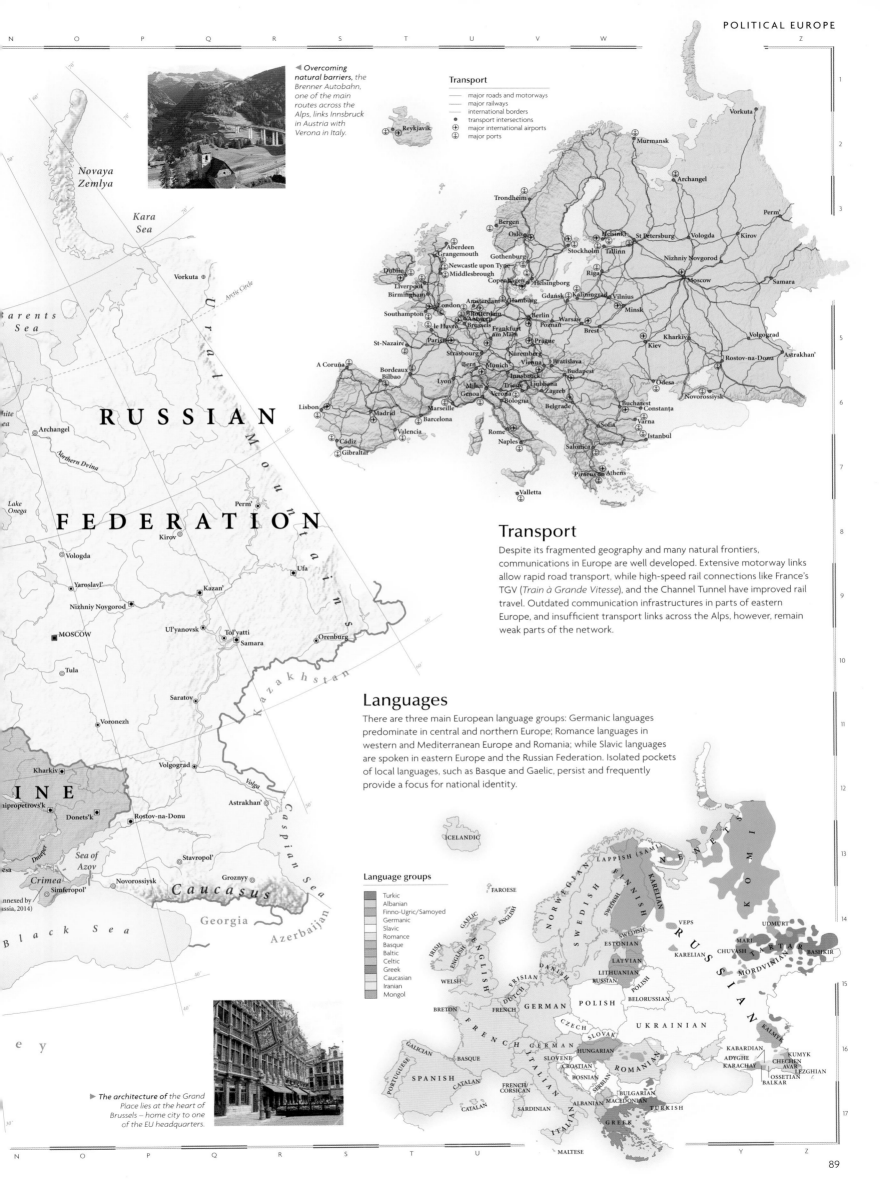

◀ **Overcoming natural barriers,** *the Brenner Autobahn, one of the main routes across the Alps, links Innsbruck in Austria with Verona in Italy.*

Transport
major roads and motorways
major railways
international borders
transport intersections
major international airports
major ports
Reykjavik
Vorkuta
Murmansk
Archangel
Trondheim
Bergen
Oslo
Helsinki
St Petersburg
Vologda
Kirov
Perm'
Aberdeen
Grangemouth
Gothenburg
Stockholm
Tallinn
Nizhniy Novgorod
Newcastle upon Tyne
Middlesbrough
Dublin
Riga
Moscow
Samara
Liverpool
Birmingham
Copenhagen
Helsingborg
Gdańsk
Kaliningrad
Vilnius
Minsk
London
Amsterdam
Hamburg
Rotterdam
Antwerp
Southampton
Le Havre
Brussels
Berlin
Poznań
Warsaw
Brest
Frankfurt am Main
Paris
Prague
St-Nazaire
Kharkiv
Volgograd
Kiev
Strasbourg
Nuremberg
Astrakhan'
A Coruña
Bern
Munich
Vienna
Bratislava
Rostov-na-Donu
Bordeaux
Bilbao
Lyon
Innsbruck
Budapest
Odesa
Milan
Trieste
Ljubljana
Genoa
Verona
Zagreb
Novorossiysk
Bologna
Lisbon
Madrid
Marseille
Belgrade
Bucharest
Constanța
Barcelona
Varna
Valencia
Sofia
Rome
Istanbul
Naples
Cádiz
Gibraltar
Salonica
Piraeus
Athens
Valletta

Transport

Despite its fragmented geography and many natural frontiers, communications in Europe are well developed. Extensive motorway links allow rapid road transport, while high-speed rail connections like France's TGV (*Train à Grande Vitesse*), and the Channel Tunnel have improved rail travel. Outdated communication infrastructures in parts of eastern Europe, and insufficient transport links across the Alps, however, remain weak parts of the network.

Languages

There are three main European language groups: Germanic languages predominate in central and northern Europe; Romance languages in western and Mediterranean Europe and Romania; while Slavic languages are spoken in eastern Europe and the Russian Federation. Isolated pockets of local languages, such as Basque and Gaelic, persist and frequently provide a focus for national identity.

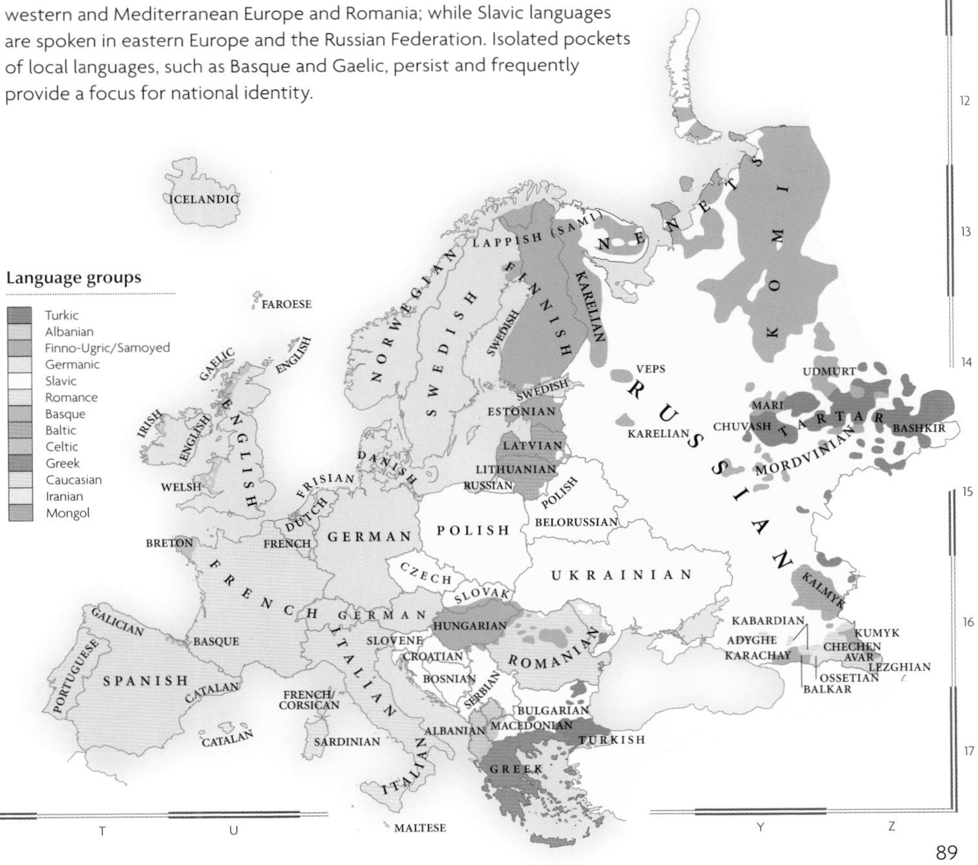
Language groups
Turkic
Albanian
Finno-Ugric/Samoyed
Germanic
Slavic
Romance
Basque
Baltic
Celtic
Greek
Caucasian
Iranian
Mongol
ICELANDIC
FAROESE
LAPPISH (SAMI)
NENETS
KOMI
NORWEGIAN
SWEDISH
FINNISH
KARELIAN
GAELIC
ENGLISH
IRISH
WELSH
VEPS
RUSSIAN
UDMURT
MARI
CHUVASH
TATAR
BASHKIR
MORDVINIAN
ESTONIAN
LATVIAN
LITHUANIAN
DANISH
FRISIAN
DUTCH
POLISH
BELORUSSIAN
BRETON
FRENCH
GERMAN
CZECH
SLOVAK
UKRAINIAN
KALMYK
HUNGARIAN
GALICIAN
BASQUE
SLOVENE
CROATIAN
ROMANIAN
KABARDIAN
ADYGHE
KARACHAY
KUMYK
CHECHEN
AVAR
LEZGHIAN
OSSETIAN
BALKAR
PORTUGUESE
SPANISH
CATALAN
ITALIAN
BOSNIAN
SERBIAN
BULGARIAN
FRENCH/CORSICAN
MACEDONIAN
ALBANIAN
TURKISH
SARDINIAN
GREEK
MALTESE

▶ *The architecture of* the Grand Place *lies at the heart of Brussels – home city to one of the EU headquarters.*

European resources

Europe's large tracts of fertile, accessible land, combined with its generally temperate climate, have allowed a greater percentage of land to be used for agricultural purposes than in any other continent. Extensive coal and iron ore deposits were used to create steel and manufacturing industries during the 19th and 20th centuries. Today, although natural resources have been widely exploited, and heavy industry is of declining importance, the growth of hi-tech and service industries has enabled Europe to maintain its wealth.

Industry

Europe's wealth was generated by the rise of industry and colonial exploitation during the 19th century. The mining of abundant natural resources made Europe the industrial centre of the world. Adaptation has been essential in the changing world economy, and a move to service-based industries has been widespread except in eastern Europe, where heavy industry still dominates.

▲ ***Countries like Hungary*** *are still struggling to modernize inefficient factories left over from extensive, centrally-planned industrialization during the communist era.*

◀ ***Frankfurt am Main*** *is an example of a modern service-based city. The skyline is dominated by headquarters from the worlds of banking and commerce.*

▲ ***Other power sources*** *are becoming more attractive as fossil fuels run out; 16% of Europe's electricity is now provided by hydro-electric power.*

Standard of living

Living standards in western Europe are among the highest in the world, although there is a growing sector of homeless, jobless people. Eastern Europeans have lower overall standards of living – a legacy of stagnated economies.

Standard of living
(UN human development index)

- low
- high
- data not available

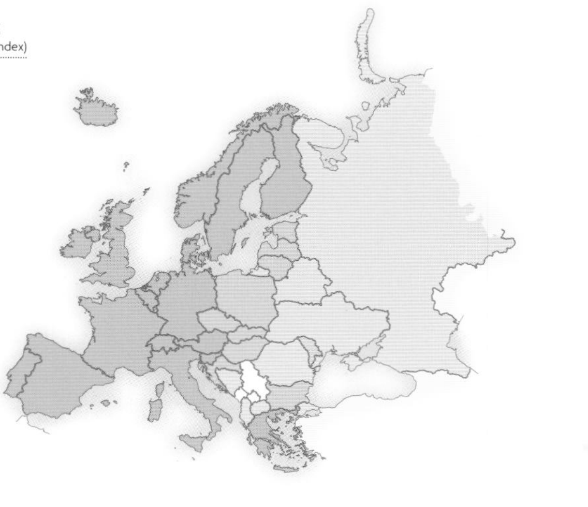

▶ ***Skiing brings millions*** *of tourists to the slopes each year, which means that even unproductive, marginal land is used to create wealth in the French, Swiss, Italian and Austrian Alps.*

Mineral resources

Fossil fuels are Europe's main mineral resource, although fuel demand far outstrips production. Sizeable coal reserves remain in the Donbass in Ukraine, Germany's Ruhr Valley and Poland. Oil and gas reserves are found mainly in the North Sea, the Volga Basin, and the Caucasus.

► *The valuable oil and gas reserves in the North Sea were first discovered in the early 1960s, and are exploited by the UK, Denmark, Germany and Norway.*

Mineral resources
oil field
gas field
coal field
bauxite
iron
lead
mercury
potassium
uranium
zinc

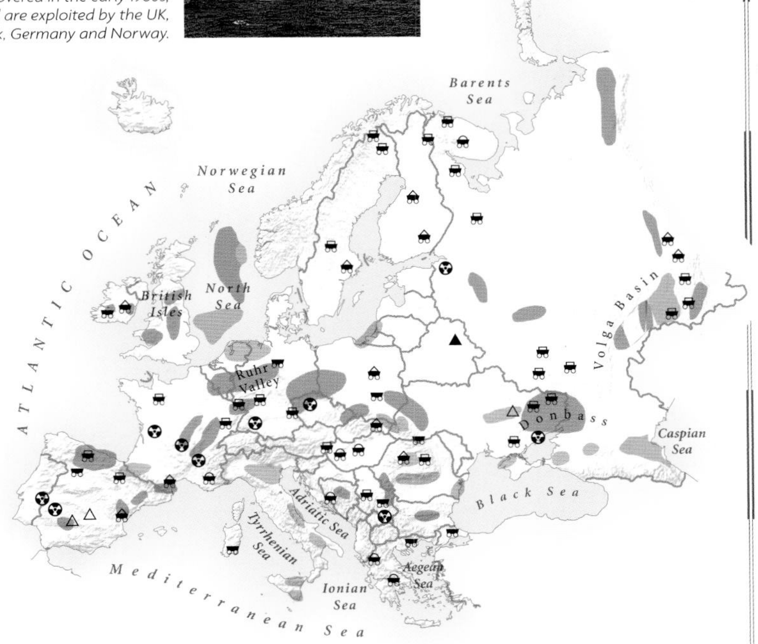

Environmental issues

The partially enclosed waters of the Baltic and Mediterranean seas have become heavily polluted, while the Barents Sea is contaminated with spent nuclear fuel from Russia's navy. During the later stages of the 20th Century acid rain caused by unchecked emissions from factories and power stations was actively destroying northern forests. However, since then international efforts to reduce pollution have brought significant improvements in many areas.

▲ *Coniferous forest covers vast swathes of northern Scandinavia and the Russian Federation. Pollutants from other parts of Europe mixing with rainfall are causing defoliation and serious damage to many forests.*

► *The Camargue in the Rhône Delta, southern France, is a protected wetland area, famous for its native population of white horses, and unique bird and plant life.*

Using the land and sea

Europe's swelling urban population and the outward expansion of many cities has created acute competition for land. Despite this, European resourcefulness has maximized land potential, and over half of Europe's land is still used for a wide variety of agricultural purposes. Land in northern Europe is used for cattle-rearing, pasture, and arable crops. Towards the Mediterranean, the mild climate allows the growing of grapes for wine; olives, sunflowers, tobacco and citrus fruits. EU subsidies, however, have resulted in massive overproduction and a land 'set-aside' policy has been introduced.

Using the land and sea
cropland
forest
ice cap
mountain region
pasture
tundra
wetland
major conurbations
cattle
goats
pigs
poultry
reindeer
sheep
cereals
citrus fruits
cotton
fishing
fodder
fruit
olive oil
potatoes
rice
root crops
roses
shellfish
sunflowers
timber
tobacco
vineyards

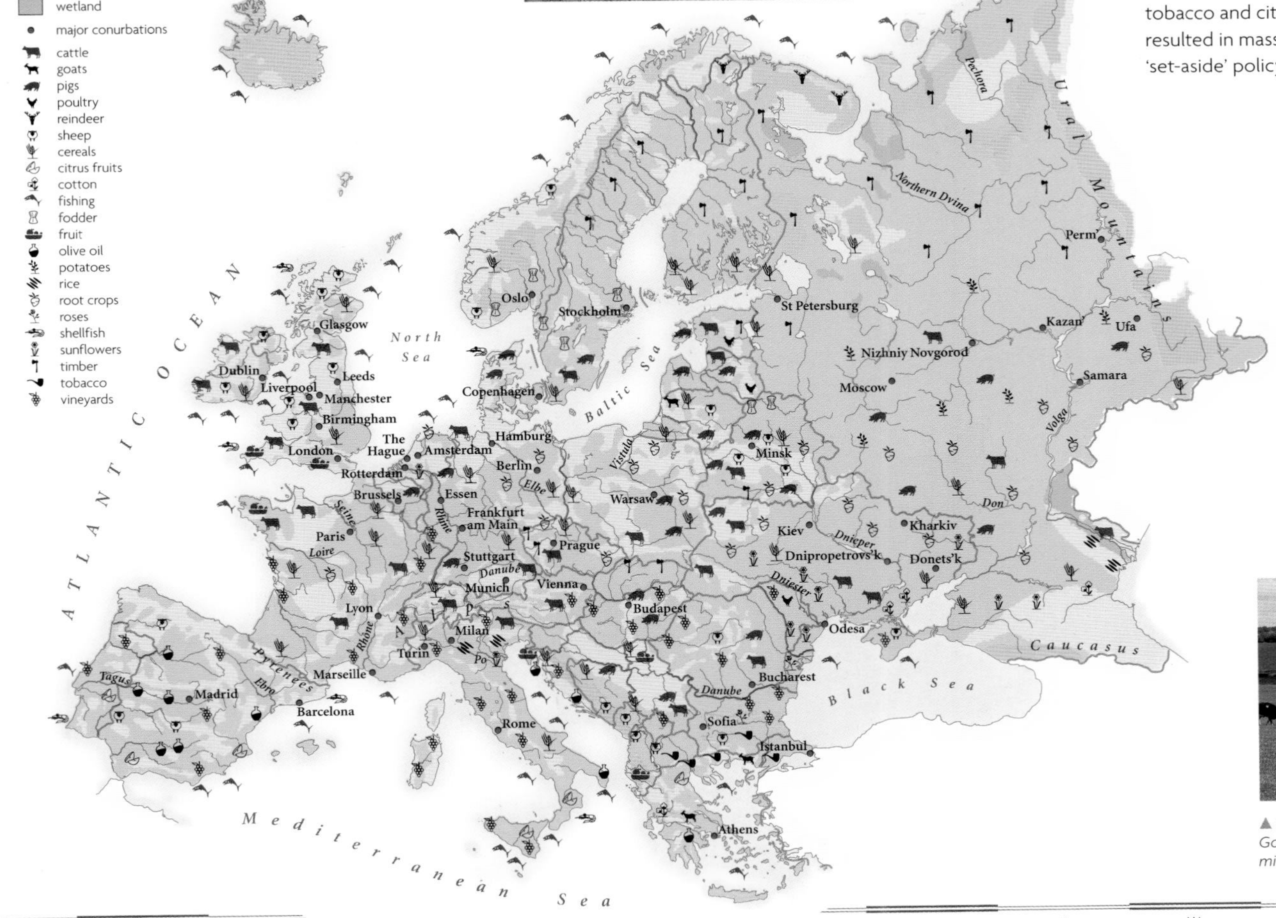

▲ *Bulgarian roses are one of the many diverse crops grown in Europe. Rose oil, extracted from the petals, is used in perfume making.*

▲ *Lowland pastures are used for dairy farming. Good transport links and refrigeration allow fresh milk to be distributed throughout Europe.*

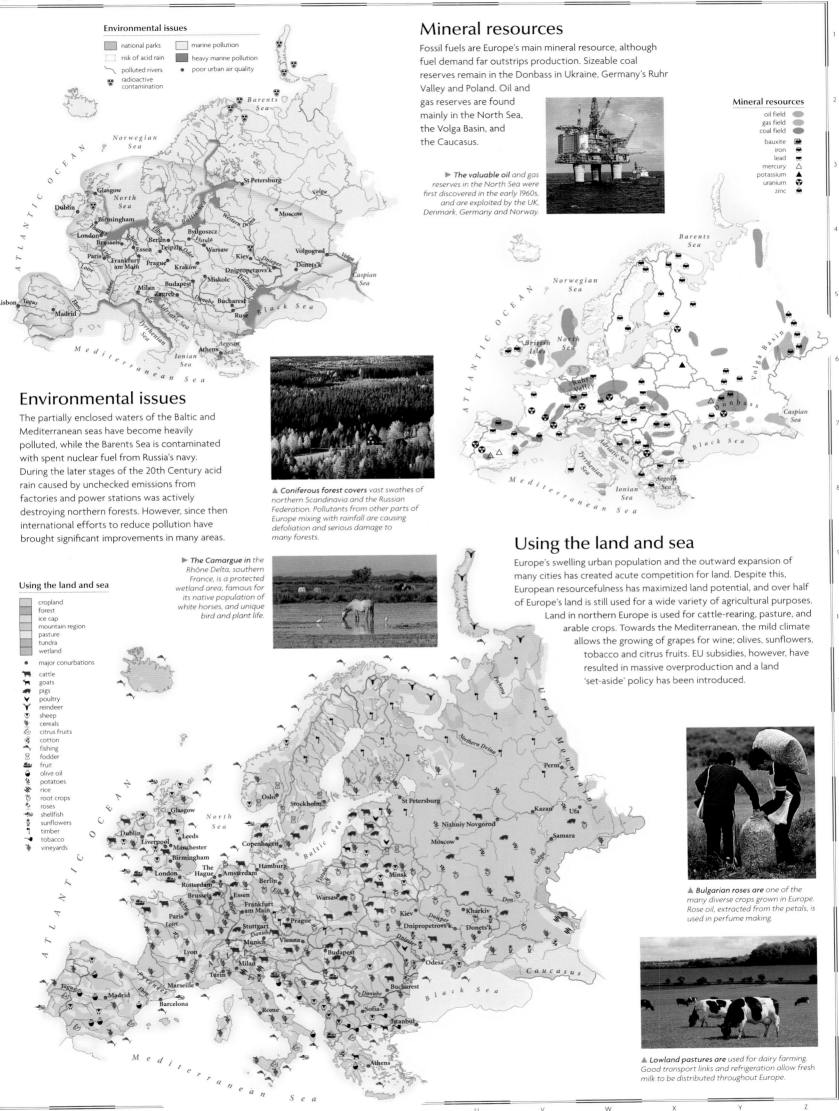

Scandinavia, Finland & Iceland

DENMARK, NORWAY, SWEDEN, FINLAND, ICELAND

Jutting into the Arctic Circle, this northern swathe of Europe has some of the continent's harshest environments, but benefits from great reserves of oil, gas and natural evergreen forests. While most early settlers came from the south, migrants to Finland came from the east, giving it a distinct language and culture. Since the late 19th century, the Scandinavian states have developed strong egalitarian traditions. Today, their welfare benefits systems are among the most extensive in the world, and standards of living are high. The Lapps, or Sami, maintain their traditional lifestyle in the northern regions of Norway, Sweden and Finland.

The landscape

Glaciers up to 10,000 ft (3000 m) deep covered most of Scandinavia and Finland during the last Ice Age. The effects of glaciation mark the entire landscape, from the mountains to the lowlands, across the tundra landscape of Lapland, and the lake districts of Sweden and Finland.

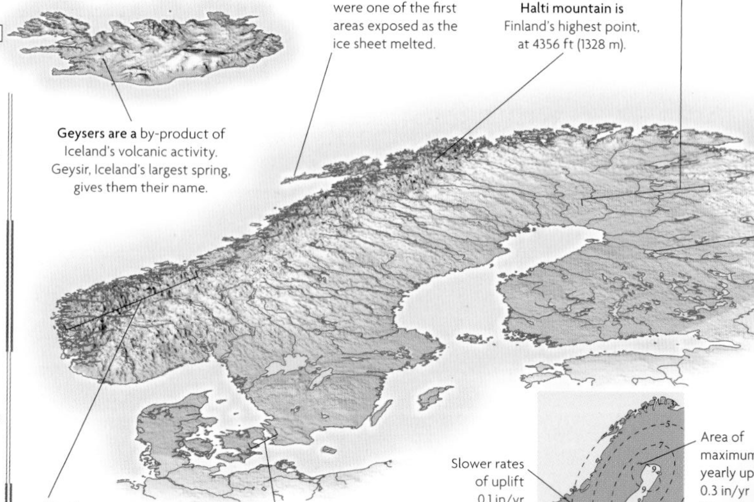

Geysers are a by-product of Iceland's volcanic activity. Geysir, Iceland's largest spring, gives them their name.

The Lofoten Islands were one of the first areas exposed as the ice sheet melted.

Halti mountain is Finland's highest point, at 4356 ft (1328 m).

Lapland, north of the Arctic Circle, is an area of undulating fells and plains known as tundra. The subsoil is permanently frozen and therefore impermeable. There are many peat bogs. Pools reappear in the summer when the surface thaws.

Oulujärvi

◀ *Finland's landscape was fashioned by ice action. Glaciers gouged out its distinctive shallow lake basins, such as Oulujärvi, and left debris called moraines in their wake.*

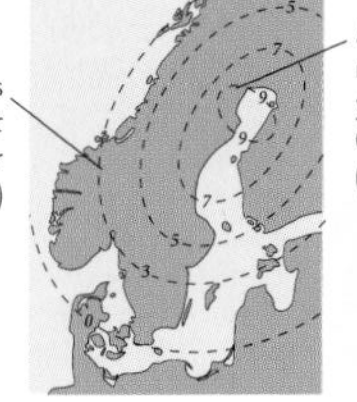

▲ *Scandinavia is still recovering from the last Ice Age, when ice depressed the land by 2000 ft (600 m). This gradual uplift is known as isostatic rebound.*

Fjords

▲ **The fjords on** *the western coast of Norway were once gentle river valleys. Their deep floors and steep sides were carved out by glaciers during the last Ice Age, and they were later flooded by the sea.*

Sjælland coast

▲ **On the coast of** *Sjælland, these cliffs have been eroded by the sea, exposing layers of chalk and limestone.*

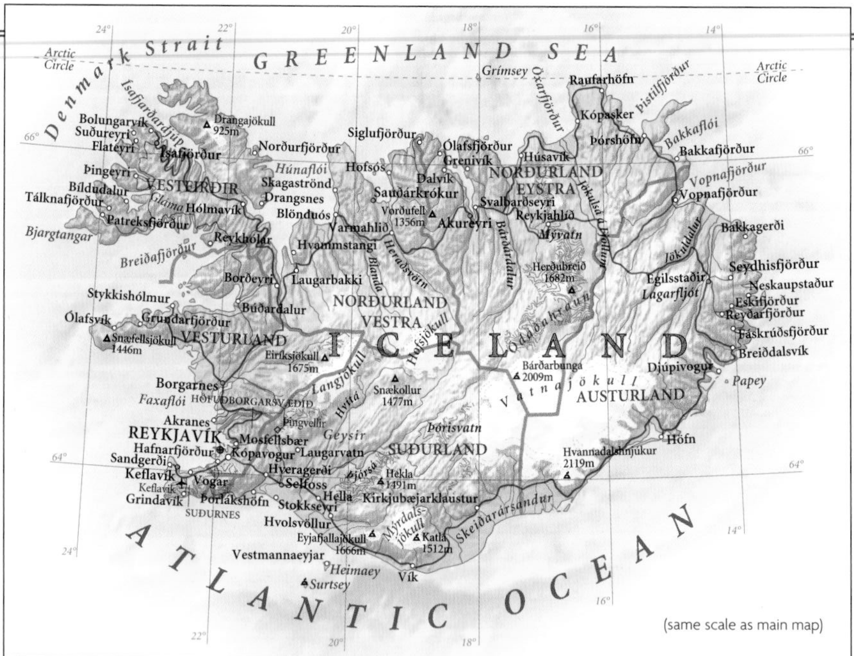

(same scale as main map)

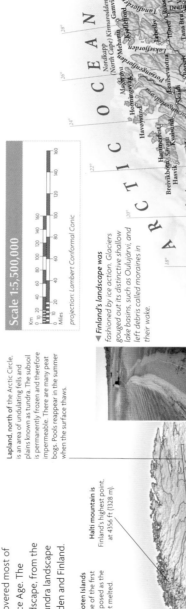

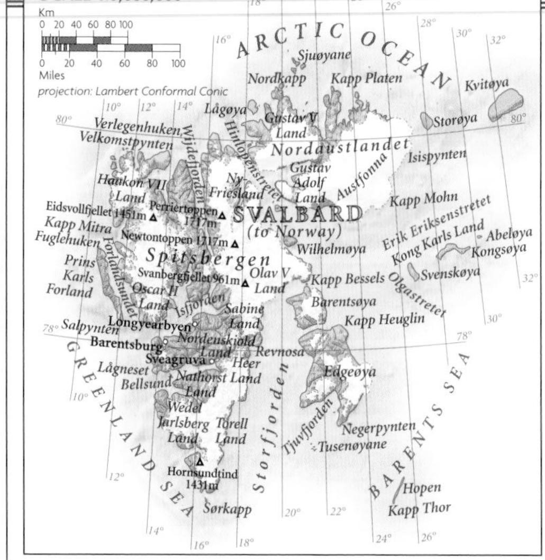

Using the land and sea

The cold climate, short growing season, poorly developed soil, steep slopes, and exposure to high winds across northern regions means that most agriculture is concentrated, with the population, in the south. Most of Finland and much of Norway and Sweden are covered by dense forests of pine, spruce and birch, which supply the timber industries.

Scale 1:5,500,000

projection: Lambert Conformal Conic

122

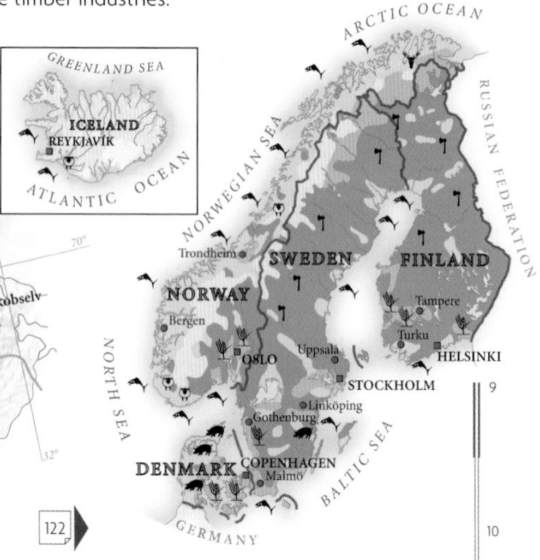

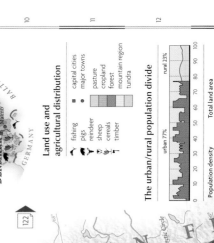

Land use and agricultural distribution

fishing, pigs, reindeer, sheep, cereals, timber; capital cities, major towns; pasture, cropland, forest, mountain region, tundra

The urban/rural population divide

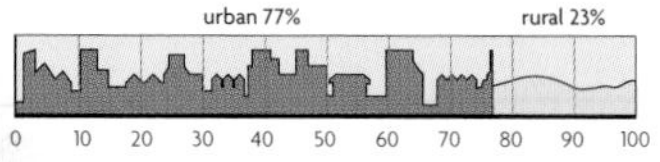

Population density	Total land area
51 people per sq mile	473,970 sq miles

▲ **Many Lappish people,** in addition to traditional reindeer herding, now also make their living from fishing and farming, or working in cities. Tourism provides some with an extra source of income.

▲ **Sweden is one** of the world's largest producers of wood and wood-based products. The traditional movement of logs by floating them down rivers has now been largely replaced by the use of trucks.

Map key

Population

- 1 million to 5 million
- 500,000 to 1 million
- 100,000 to 500,000
- 50,000 to 100,000
- 10,000 to 50,000
- below 10,000

Elevation

- 2000m / 6562ft
- 1000m / 3281ft
- 500m / 1640ft
- 250m / 820ft
- 100m / 328ft
- sea level

Transport and industry

Norway derives its premier industry, the production of oil and gas, from the North Sea, while Denmark exploits its own oil and gas reserves. Hydro-electric power is a major industry, particularly in Sweden and Iceland. Timber processing remains significant in Finland and Sweden, but metal and engineering industries are increasingly important. In Iceland, fish products are the main source of export earnings.

The transport network

- 226,735 miles (364,936 km)
- 2042 miles (3286 km)
- 13,704 miles (22,057 km)
- 6,661 miles (10,721 km)

Although roads now reach most areas, the railways are markedly less developed. Much of the north is not served by rail and must rely on air and sea services for long distance travel and freight transportation.

▶ **The use of** geothermal power in Iceland began half a century ago. Today geothermal power stations supply 89% of the country's domestic heating requirements.

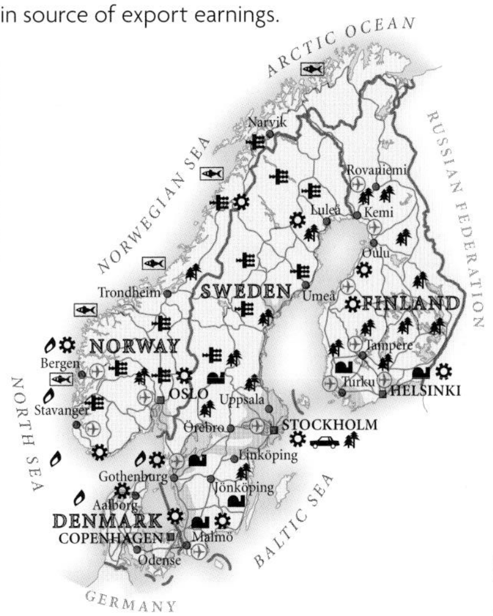

Major industry and infrastructure

- car manufacture
- engineering
- fish processing
- hydro-electric power
- nuclear power
- oil & gas
- timber processing
- capital cities
- major towns
- international airports
- major roads
- major industrial areas

Southern Scandinavia

SOUTHERN NORWAY, SOUTHERN SWEDEN, DENMARK

Scandinavia's economic and political hub is the more habitable and accessible southern region. Many of the area's major cities are on the southern coasts, including Oslo and Stockholm, the capitals of Norway and Sweden. In Denmark, most of the population and the capital, Copenhagen, are located on its many islands. A cultural unity links the three Scandinavian countries. Their main languages, Danish, Swedish and Norwegian, are mutually intelligible, and they all retain their monarchies, although the parliaments have legislative control.

Using the land

Agriculture in southern Scandinavia is highly mechanized although farms are small. Denmark is the most intensively farmed country and its western pastureland is used mainly for pig farming. Cereal crops including wheat, barley and oats, predominate in eastern Denmark and in the far south of Sweden. Southern Norway and Sweden have large tracts of forest which are exploited for logging.

The urban/rural population divide

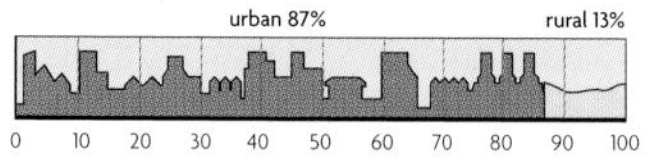
urban 87%
rural 13%
0 10 20 30 40 50 60 70 80 90 100

Population density	Total land area
112 people per sq mile (43 people per sq km)	173,487 sq miles (456,564 sq km)

Land use and agricultural distribution

cattle
pigs
sheep
cereals
fodder
root crops
timber
capital cities
major towns
pasture
cropland
forest
mountain region
NORWEGIAN SEA
NORTH SEA
BALTIC SEA
GERMANY
NORWAY
SWEDEN
DENMARK
Trondheim
Bergen
OSLO
Uppsala
Örebro
STOCKHOLM
Linköping
Gothenburg
Aalborg
COPENHAGEN
Odense
Malmö

▲ **In Norway winters** *are longer and colder inland than in coastal areas, where the warm current of the North Atlantic Drift moderates the climate.*

The landscape

Southern Scandinavia, with the exception of Norway, has a flatter terrain than the rest of the region. Denmark and southern Sweden are both extensions of the North European Plain. In this area, because of glacial deposition rather than erosion, the soils are deeper and more fertile.

Olden

◀ ***In the past,** glaciers such as this one in Olden, Norway, were much larger. Today, many are retreating to yield the spectacular glacial scenery.*

▲ ***Limestone pillars eroded** by the sea dot the coast of Gotland and surrounding islands.*

The peak of Glittertind in the Jotunheimen mountains is 8110 ft (2472 m) high.

Distinctive low ridges, called eskers, are found across southern Sweden. They are formed from sand and gravel deposits left by retreating glaciers.

Acid rain, caused by industrial pollution carried north from elsewhere in Europe, harms plant and animal life in Scandinavian forests and lakes. The region's surface rocks lack lime to neutralize the acid, so making the problem more serious.

Denmark's flat and fertile soils are formed on glacial deposits between 100–160 ft (30–50 m) deep.

Vänern in Sweden is the largest lake in Scandinavia. It covers an area of 2080 sq miles (5390 sq km).

The lakes of southern Sweden remain from a period when the land was completely flooded. As the ice which covered the area melted, the land rose, leaving lakes in shallow, ice-scoured depressions. Sweden has over 90,000 lakes.

Sognefjorden

▲ ***Sognefjorden** is the deepest of Norway's many fjords. It drops to 4291 ft (1308 m) below sea level.*

Erosion by glaciers deepened existing river valleys
Sea level
When the ice retreated the valley was flooded by the sea
Old valley floor
NORWEGIAN SEA
Frohavet
Hustadvika
NORD-TRØNDELAG
SØR-TRØNDELAG
MØRE OG ROMSDAL
SOGN OG FJORDANE
OPPLAND
HEDMARK
JÄMTLAND
VÄSTERNORRLAND
GÄVLEBORG
Gulf of Bothnia
Trondheim
Kristiansund
Molde
Ålesund
Lillehammer
Jotunheimen
Trollheimen
Dovrefjell
Rondane
Glittertind 2472m
Galdhøpiggen 2469m
Snøhetta 2286m
Storsylen 1762m
96
92

Scale 1:3,250,000

projection: Lambert Conformal Conic

Map key

Population

- 1 million to 5 million
- 500,000 to 1 million
- 100,000 to 500,000
- 50,000 to 100,000
- 10,000 to 50,000
- below 10,000

Elevation

- 2000m / 6562ft
- 1000m / 3281ft
- 500m / 1640ft
- 250m / 820ft
- 100m / 328ft

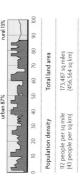

▲ **Shipbuilding in Gothenburg** *has declined in recent years as manufacturers in other sectors have come to the fore. One of these is the car firm, Volvo, a major employer in Gothenburg.*

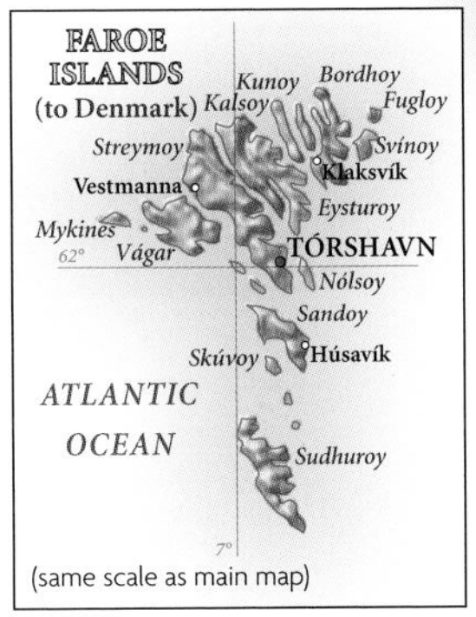

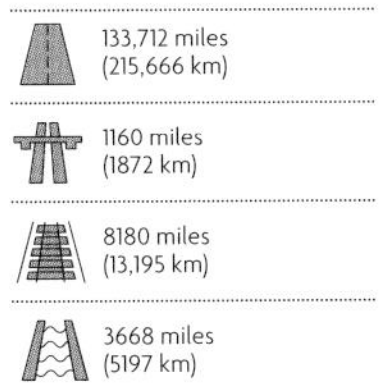

Transport and industry

In Denmark and Norway food processing is a major industry. Swedish iron and steel production supports car manufacturers and component suppliers. Nearly half of Norway's income comes from North Sea oil and gas reserves. Denmark's successful hi-tech, high-profit electronics and light engineering industries largely use imported raw materials.

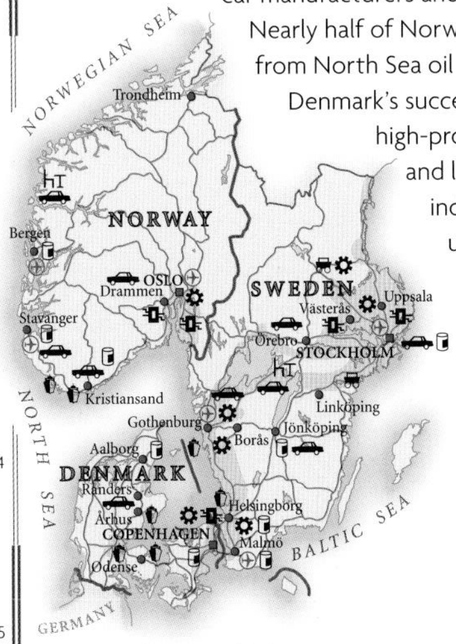

Major industry and infrastructure

- car manufacture
- electronics
- engineering
- furniture industry
- iron & steel
- shipbuilding
- food processing
- capital cities
- major towns
- international airports
- major roads
- major industrial areas

The transport network

- 133,712 miles (215,666 km)
- 1160 miles (1872 km)
- 8180 miles (13,195 km)
- 3668 miles (5197 km)

A major addition to the transport network in this region is the Oresund bridge and tunnel project connecting Copenhagen in Denmark with Malmö in Sweden.

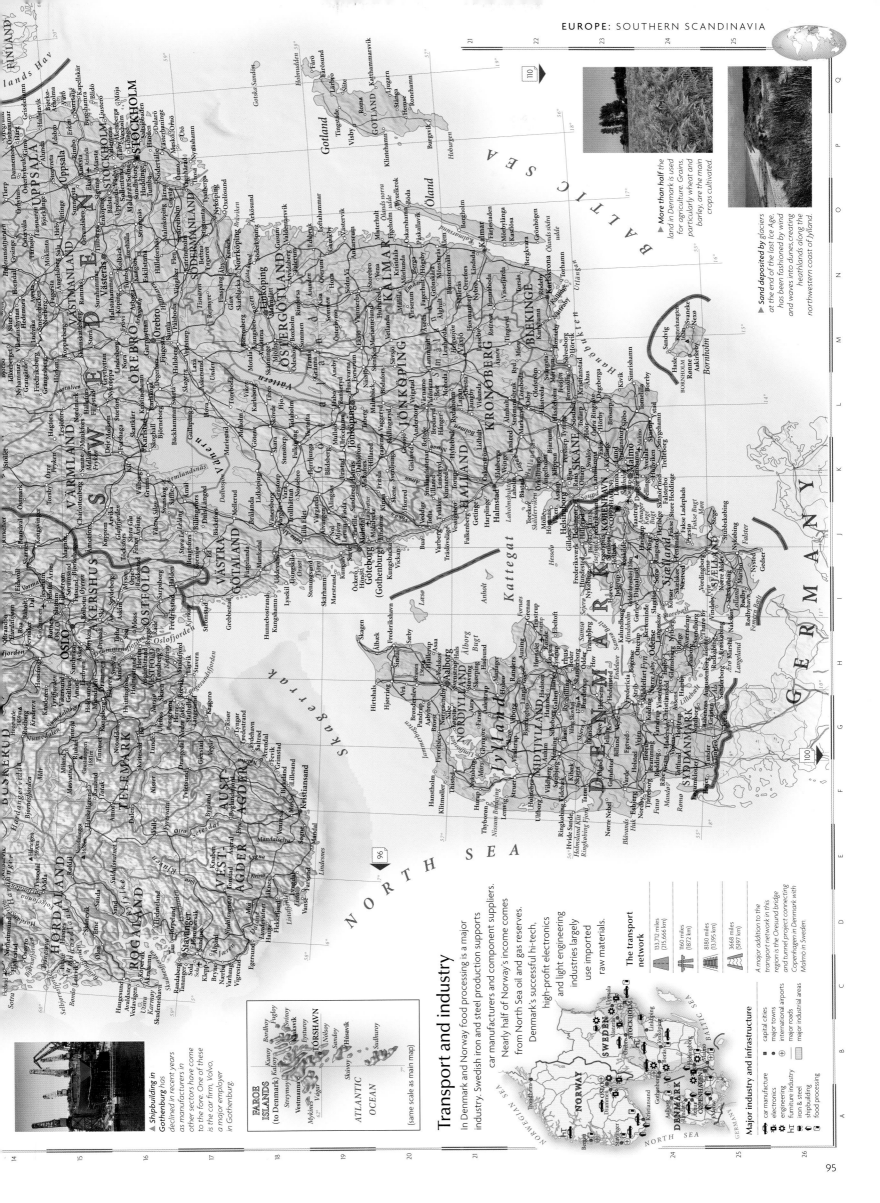

▶ **More than half** *the land in Denmark is used for agriculture. Grains, particularly wheat and barley, are the main crops cultivated.*

▶ **Sand deposited by** *glaciers at the end of the last Ice Age, has been fashioned by wind and waves into dunes, creating heathlands along the northwestern coast of Jylland.*

The British Isles

UNITED KINGDOM, IRELAND

The British Isles have for centuries played a central role in European and world history. England, Wales, Scotland and Northern Ireland together form the United Kingdom (UK), while the southern portion of Ireland is an independent country, self-governing since 1921. Although England has tended to be the politically and economically dominant partner in the UK, the Scots, Welsh and Irish maintain independent cultures, distinct national identities and languages. Southeastern England is the most densely populated part of this crowded region, with over eight million people living in and around the London area.

▲ **The valley of** *Glen Coe in the Scottish Highlands is a U-shaped valley, typical of the north and west of the British Isles, where glaciers shaped much of the landscape.*

Transport and industry

The British Isles' industrial base was founded primarily on coal, iron and textiles, based largely in the north. Today, the most productive sectors include hi-tech industries clustered mainly in southeastern England, chemicals, finance and the service sector, particularly tourism.

Major industry and infrastructure

- car manufacture
- chemicals
- engineering
- hi-tech industry
- iron & steel
- tourism
- capital cities
- major towns
- international airports
- major roads
- major industrial areas

ATLANTIC OCEAN
NORTH SEA
English Channel
UNITED KINGDOM
IRELAND
Aberdeen
Dundee
Edinburgh
Glasgow
Newcastle upon Tyne
Belfast
Leeds
Liverpool
Manchester
DUBLIN
Nottingham
Birmingham
Norwich
Cardiff
Oxford
Bristol
LONDON
Cork

The transport network

285,947 miles (460,240 km)	2023 miles (3578 km)
11,825 miles (19,032km)	3976 miles (6400 km)

The UK's congested roads have become a major focus of environmental concern in recent years. No longer an island, the UK was finally linked to continental Europe by the Channel Tunnel in 1994.

◀ **Clew Bay in** *western Ireland, is characteristic of the heavily indented west coast, where deep wide-mouthed bays separate the mountains of Mayo, Donegal and Kerry as they thrust out into the Atlantic Ocean.*

ATLANTIC OCEAN
NORTH SEA
Shetland Islands
Orkney Islands
Outer Hebrides
Inner Hebrides
Sea of the Hebrides
The Minch
The Little Minch
Isle of Lewis
Isle of Skye
North West Highlands
Grampian Mountains
SCOTLAND
Aberdeen
Dundee
Edinburgh
Glasgow
Inverness
Fort William
Oban
Stornoway
Kirkwall
Lerwick
Ben Nevis 1343m
Cairn Gorm 1245m
Ben Macdui 1309m
St Kilda
Flannan Isles
North Rona
Sula Sgeir
Fair Isle
Cape Wrath
John o'Groats
Moray Firth
Firth of Forth
Firth of Tay
Firth of Lorn

64

92

The landscape

Rugged uplands dominate the landscape of Scotland, Wales and northern England. All the peaks in the British Isles over 4000 ft (1219 m) lie in highland Scotland. Lowland England rises into several ranges of rolling hills, including the older Mendips, and the Cotswolds and the Chilterns, which were formed at the same time as the Alps in southern Europe.

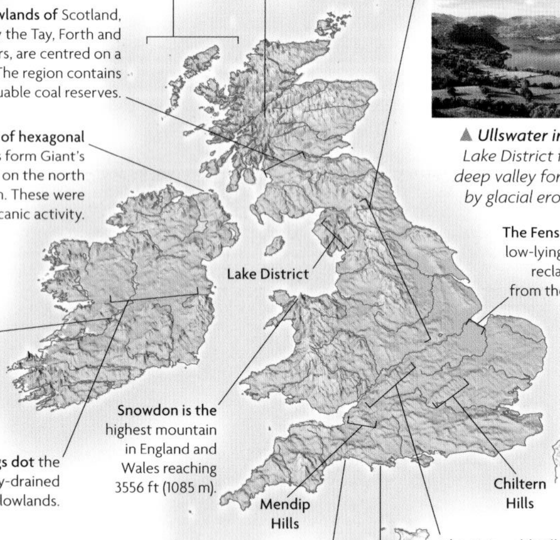

Over 600 islands, mostly uninhabited, lie west and north of the Scottish mainland.
Ben Nevis at 4409 ft (1343 m) is the highest peak in the UK.
The Pennines, sometimes called 'the backbone of England', are formed of limestones and grits.
The lowlands of Scotland, drained by the Tay, Forth and Clyde rivers, are centred on a rift valley. The region contains valuable coal reserves.
Thousands of hexagonal basalt columns form Giant's Causeway on the north coast of Antrim. These were created by volcanic activity.
Lake District
The Fens are a low-lying area reclaimed from the sea.
The British Isles have no large-scale river systems. The Shannon is the longest, at 230 miles (370 km).
Snowdon is the highest mountain in England and Wales reaching 3556 ft (1085 m).
Peat bogs dot the poorly-drained Irish lowlands.
Mendip Hills
Chiltern Hills
The Cotswold Hills are characterized by a series of limestone ridges overlooking clay vales.
Durdle Door
Black Ven, Lyme Regis

▲ **Ullswater in the** *Lake District fills a deep valley formed by glacial erosion.*

◀ **Dartmoor, studded with** *tors, is an exposed part of a vast granite dome, formed when molten rock intruded into the Earth's crust.*

▲ **Coastal erosion around** *the British Isles forms striking features such as this limestone arch, Durdle Door in Dorset.*

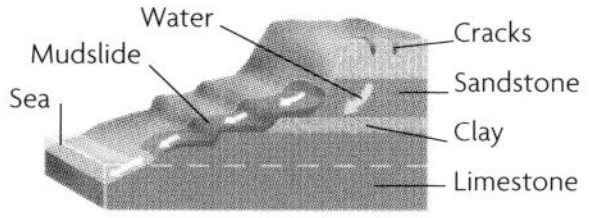

Water
Mudslide
Sea
Cracks
Sandstone
Clay
Limestone

Much of the *south coast is subject to landslides. Following rain, porous sandstones feed water into the underlying, less permeable clays which then crumble and slide into the sea.*

Map key

Population

- above 5 million
- 1 million to 5 million
- 500,000 to 1 million
- 100,000 to 500,000
- 50,000 to 100,000
- 10,000 to 50,000
- below 10,000

Elevation

- 1000m / 3281ft
- 500m / 1640ft
- 250m / 820ft
- 100m / 328ft
- sea level

Using the land

The wetter western parts of the UK suit livestock-rearing and the drier east arable farming, while mountainous areas support sheep farming and forestry. In Ireland and central and southern England, mixed arable, beef and dairy farming predominate, while fruit farming and viticulture are possible in the mild extreme south.

The urban/rural population divide

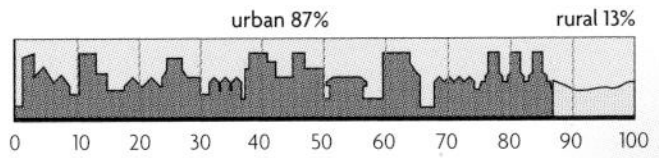

Population density	Total land area
529 people per sq mile (204 people per sq km)	121,684 sq miles (315,160 sq km)

▲ *Exposed highlands, like these in Wales, and in northern England and Scotland are used for grazing sheep.*

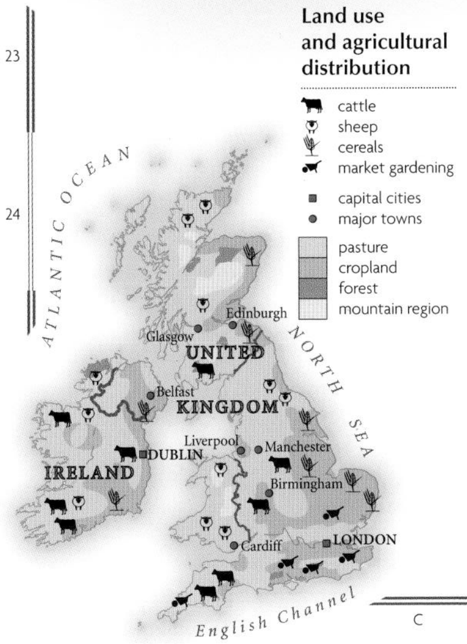

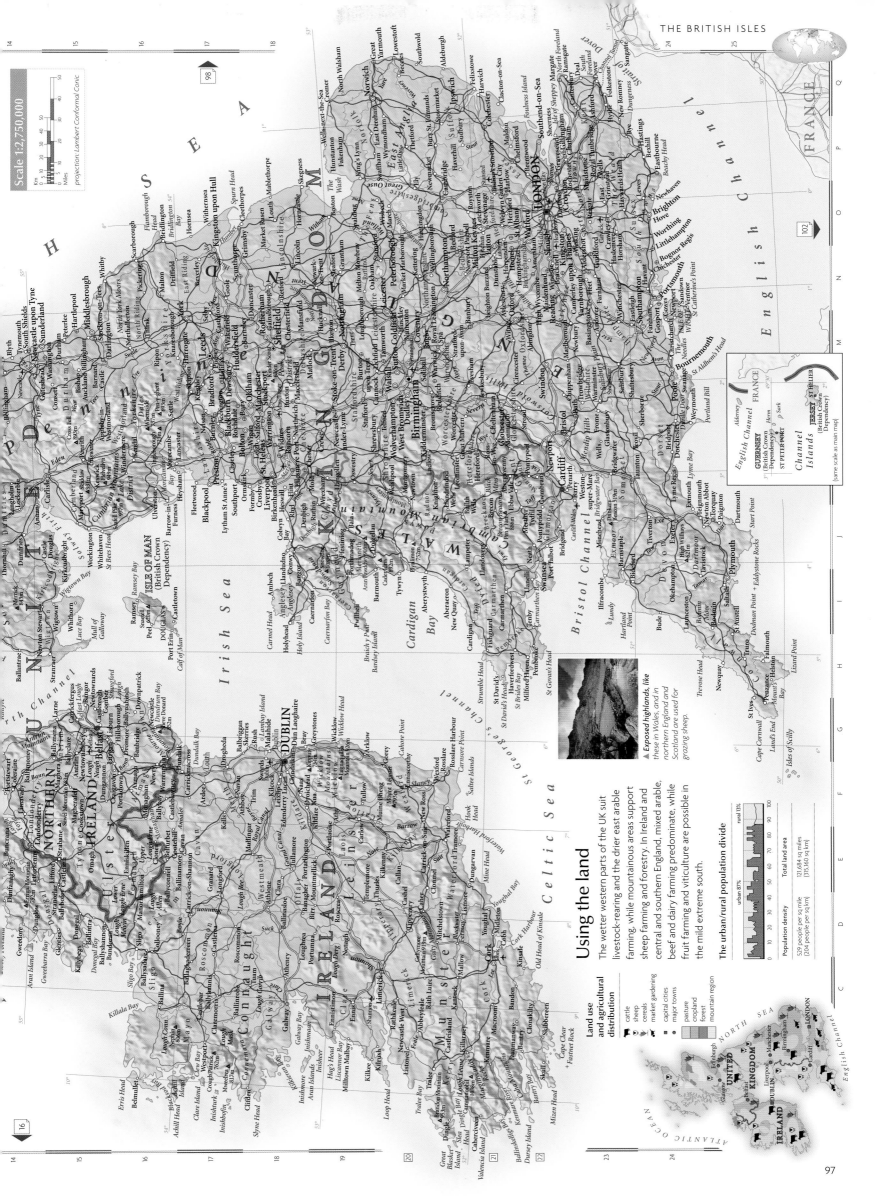

The Low Countries

BELGIUM, LUXEMBOURG, NETHERLANDS

One of northwestern Europe's strategic crossroads, the Low Countries are united by a common history in which they have often been a battleground in European wars. For over a thousand years they were ruled by foreign powers. Even after they achieved independence, the three countries maintained close links, later forming the world's first totally free labour and goods market, the Benelux Economic Union, which became the core of the European Community (now the European Union or EU). These states have remained at the forefront of wider European co-operation; Brussels, The Hague and Luxembourg are hosts to major institutions of the EU.

Major industry and infrastructure

- aerospace
- finance
- engineering
- hi-tech industry
- pharmaceuticals
- textiles
- capital cities
- major towns
- international airports
- major roads
- major industrial areas

Transport and industry

In the western Netherlands, a massive, sprawling industrialized zone encompasses many new hi-tech and service industries. Belgium's central region has emerged as the country's light manufacturing and services centre. Luxembourg city is home to more than 160 banks and the European headquarters of many international companies.

The transport network

140,588 miles (226,281 km)

2565 miles (4129 km)

4099 miles (6598 km)

4134 miles (6653 km)

The Low Countries hold a key position on the North Sea, containing Europe's two largest ports, Rotterdam and Antwerp, which are connected to a comprehensive system of inland waterways.

The landscape

The main geographical regions of the Netherlands are the northern glacial heathlands, the low-lying lands of the Rhine and Maas/Meuse, the reclaimed polders, and the dune coast and islands. Belgium includes part of the Ardennes, together with the coalfields on its northern flanks, and the fertile Flanders plain.

◀ ***Heathlands, like these** at Schoorl, are found along the coast of the Netherlands. Much of the coast was breached by the sea in the 5th century, creating its distinctive inlets and islands.*

▲ ***One-third of the** Netherlands lies below sea level and flooding is a constant threat. Barrages have been built across the mouths of many rivers to contain floodwaters.*

▲ ***Extensive sand dune** systems along the coast have prevented flooding of the land. Behind the dunes, marshy land is drained to form polders, usable land suitable for agriculture.*

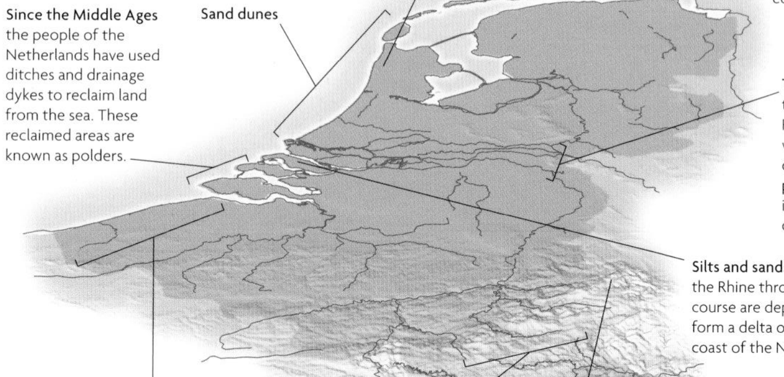

Since the Middle Ages the people of the Netherlands have used ditches and drainage dykes to reclaim land from the sea. These reclaimed areas are known as polders.

The parallel valleys of the Maas/Meuse and Rhine rivers were created when the Rhine was deflected from its previous course by the ice sheet which formed during the last Ice Age.

Silts and sands eroded by the Rhine throughout its course are deposited to form a delta on the west coast of the Netherlands.

The loess soils of the Flanders Plain in western Belgium provide excellent conditions for arable farming.

Hautes Fagnes is the highest part of Belgium. The bogs and streams in this upland region result from high rainfall and low temperatures.

▶ ***Uplifted and folded** 220 million years ago, the Ardennes have since been reduced to relatively level plateaux, then sharply incised by rivers such as the Maas/Meuse.*

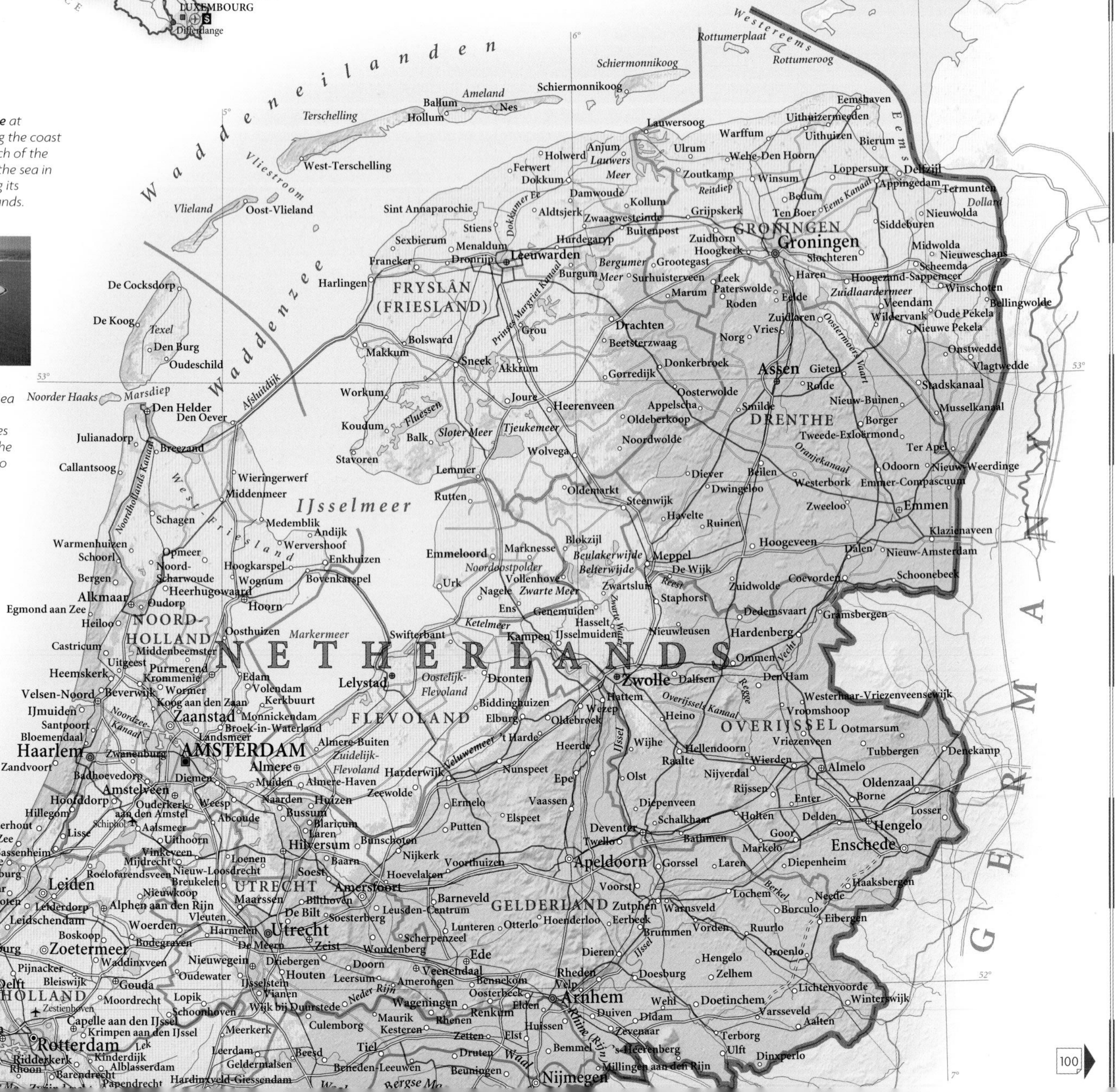

100

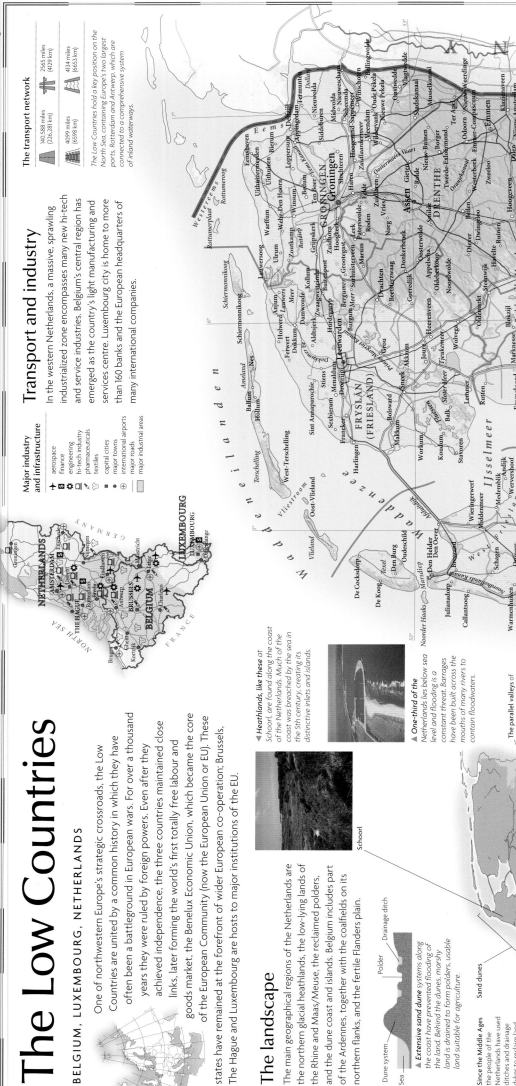

▲ ***The Dutch city*** *of Rotterdam lies within one of the most densely populated and highly industrialized regions in the world, known as 'Randstad Holland'.*

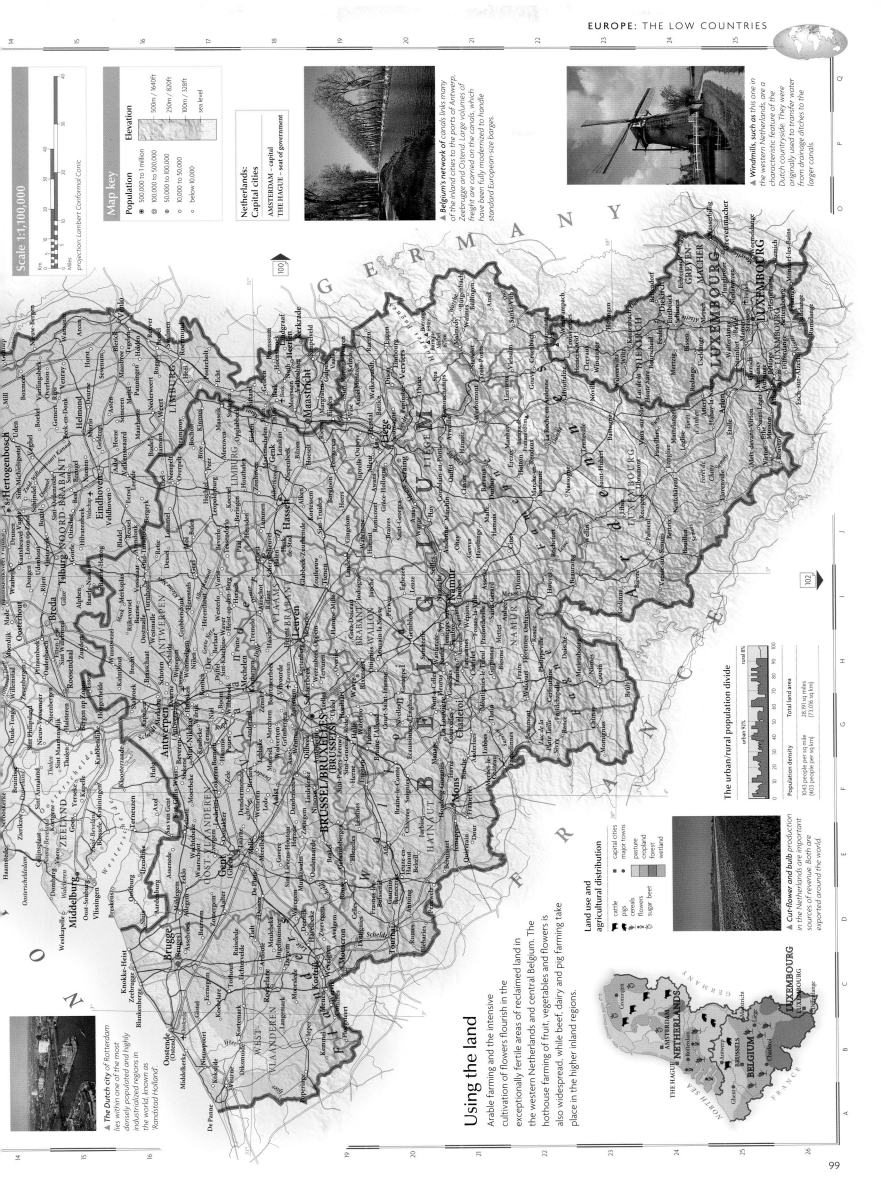

Scale 1:1,100,000

Km 0 5 10 20 30 40

Miles 0 5 10 20 30 40

projection: Lambert Conformal Conic

Map key

Population

- 500,000 to 1 million
- 100,000 to 500,000
- 50,000 to 100,000
- 10,000 to 50,000
- below 10,000

Elevation

- 500m / 1640ft
- 250m / 820ft
- 100m / 328ft
- sea level

Netherlands: Capital cities

AMSTERDAM – capital
THE HAGUE – seat of government

▲ ***Belgium's network of*** *canals links many of the inland cities to the ports of Antwerp, Zeebrugge and Ostend. Large volumes of freight are carried on the canals, which have been fully modernized to handle standard European-size barges.*

▲ ***Windmills, such as*** *this one in the western Netherlands, are a characteristic feature of the Dutch countryside. They were originally used to transfer water from drainage ditches to the larger canals.*

Using the land

Arable farming and the intensive cultivation of flowers flourish in the exceptionally fertile areas of reclaimed land in the western Netherlands and central Belgium. The hothouse farming of fruit, vegetables and flowers is also widespread, while beef, dairy and pig farming take place in the higher inland regions.

Land use and agricultural distribution

- cattle
- pigs
- cereals
- flowers
- sugar beet
- capital cities
- major towns
- pasture
- cropland
- forest
- wetland

▲ ***Cut-flower and bulb*** *production in the Netherlands are important sources of revenue. Both are exported around the world.*

The urban/rural population divide

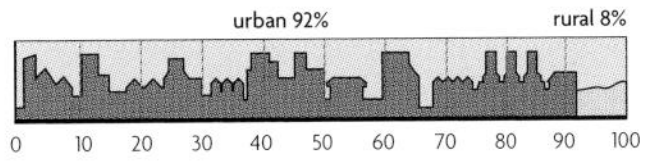

Population density	Total land area
1043 people per sq mile (403 people per sq km)	28,191 sq miles (73,016 sq km)

Germany

Despite the devastation of its industry and infrastructure during the Second World War and its separation from eastern Germany during the Cold War, West Germany made a rapid recovery in the following generation to become Europe's most formidable economic power. When the Berlin Wall was dismantled in 1989, the two halves of Germany were politically united for the first time in 40 years. Complete social and economic unity remain a longer term goal, as East German industry and society adapt to a free market. Germany has been a key player in the creation of the European Union (EU) and in moves toward a single European currency.

DENMARK
BALTIC SEA
NORTH SEA
Hamburg
Bremen
BERLIN
NETHERLANDS
POLAND
Essen
Dortmund
Düsseldorf
GERMANY
Leipzig
Cologne
Dresden
BELG.
LUX.
Frankfurt am Main
CZECH REPUBLIC
Saarbrücken
Nuremberg
FRANCE
Stuttgart
Munich
SWITZ.
AUSTRIA

Using the land

Germany has a large, efficient agricultural sector, and produces more than three-quarters of its own food. The major crops grown are cereals and sugar beet on the more fertile soils, and root crops, rye, oats and fodder on the poorer soils of the northern plains and central uplands. Southern Germany is also a principal producer of high quality wines. Vineyards cover the slopes surrounding the Rhine and its tributaries.

Land use and agricultural distribution

- cattle
- pigs
- cereals
- sugar beet
- vineyards
- capital cities
- major towns
- pasture
- cropland
- forest

The urban/rural population divide

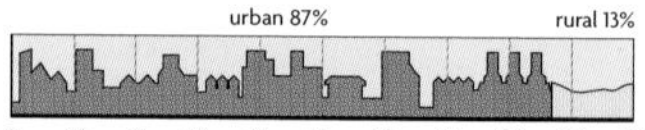

urban 87%
rural 13%
0
10
20
30
40
50
60
70
80
90
100

Population density	Total land area
612 people per sq mile (236 people per sq km)	13,804 sq miles (356,910 sq km)

▲ *The Moselle river flows through the Rhine State Uplands* (Rheinisches Schiefergebirge). *During a period of uplift, pre-existing river meanders were deeply incised, to form its present dramatic contours.*

▶ *The heathlands of northern Germany are covered by glacial deposits of sandy outwash soil which makes them largely infertile. They support only sheep and solitary trees.*

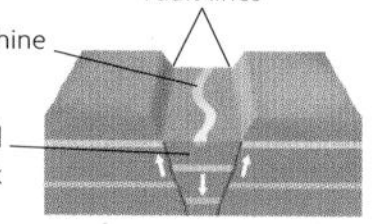

Fault lines
Rhine
Downfaulted block

▲ *Part of the floor of the Rhine Rift Valley was let down between two parallel faults in the Earth's crust.*

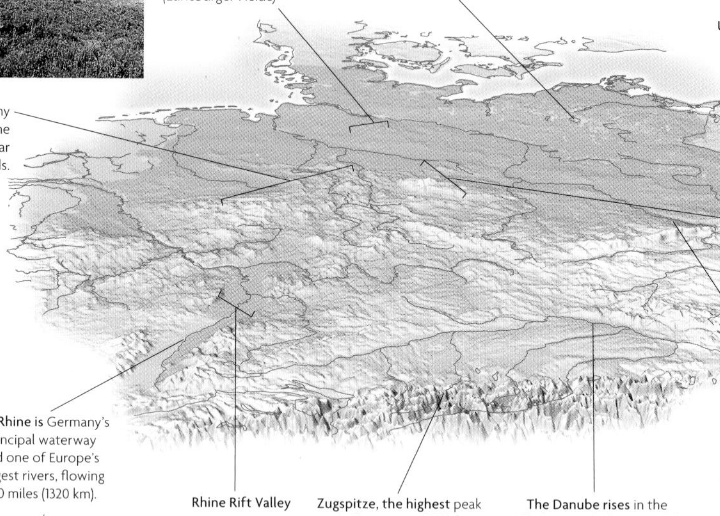

Lüneburg Heath
(Lüneburger Heide)
Müritz lake covers 45 sq miles (117 sq km), but is only 108 ft (33 m) deep. It lies in a shallow valley formed by meltwater flowing out from a retreating ice sheet. These valleys are known as Urstromtäler.
Much of the landscape of northern Germany has been shaped by glaciation. During the last Ice Age, the ice sheet advanced as far the northern slopes of the central uplands.
The Harz Mountains were formed 300 million years ago. They are block-faulted mountains, formed when a section of the Earth's crust was thrust up between two faults.
Elbe river
The Rhine is Germany's principal waterway and one of Europe's longest rivers, flowing 820 miles (1320 km).
Rhine Rift Valley
Zugspitze, the highest peak in Germany at 9719 ft (2962 m), was formed during the Alpine mountain-building period, 30 million years ago.
The Danube rises in the Black Forest (Schwarzwald) and flows east, across a wide valley, on its course to the Black Sea.

The landscape

The plains of northern Germany, the volcanic plateaux and mountains of the central uplands, and the Bavarian Alps are the three principal geographic regions in Germany. North to south the land rises steadily from barely 300 ft (90 m) in the plains to 6500 ft (2000 m) in the Bavarian Alps, which are a small but distinct region in the far south.

◀ *The Elbe flows in wide meanders across the north German plain to the North Sea. At its mouth it is 10 miles (16 km) wide.*

Scale 1:2,500,000

Km 0 5 10 20 30 40 50 60 70
Miles 0 5 10 20 30 40 50 60 70

projection: Lambert Conformal Conic

BALTIC SEA
NORTH SEA
DENMARK
POLAND
NETHERLANDS
North Frisian Islands (Nordfriesische Inseln)
Ostfriesische Inseln
Helgoländer Bucht
Kieler Bucht
Mecklenburger Bucht
Pomeranian Bay
SCHLESWIG-HOLSTEIN
MECKLENBURG-VORPOMMERN
NIEDERSACHSEN
BRANDENBURG
BREMEN
HAMBURG
Hamburg
Bremen
Hannover
BERLIN
Potsdam
Kiel
Lübeck
Rostock
Schwerin
Flensburg
Bremerhaven
Oldenburg
Osnabrück
Wolfsburg
Braunschweig
Stralsund
Greifswald
Neubrandenburg
Lüneburg
Cuxhaven
Wilhelmshaven
Emden
Rügen
Fehmarn
Sylt
Elbe
Weser
Ems
Oder
Havel

92
96
98
110

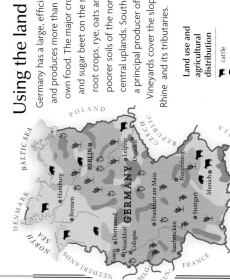

Map key

Population

- 1 million to 5 million
- 500,000 to 1 million
- 100,000 to 500,000
- 50,000 to 100,000
- 10,000 to 50,000
- below 10,000

Elevation

- 2000m / 6562ft
- 1000m / 3281ft
- 500m / 1640ft
- 250m / 820ft
- 100m / 328ft
- sea level

The transport network

- 403,544 miles (649,515 km)
- 7323 miles (11,756 km)
- 22,258 miles (35,868 km)
- 4660 miles (7500 km)

Germany has a complex network of inland waterways. The Rhine and Danube are at the centre of a vast canal system which links central and eastern Europe to the north.

98
102
108
108
110

Transport and industry

Today, the main industries which contribute to Germany's economic power are industrial machine building, electronics, chemicals and car manufacture, including the famous Mercedes and BMW firms. While the introduction of a free market in the east has forced the closure of many less efficient companies there, west German manufacturers have moved in to set up new plants and businesses.

Major industry and infrastructure

- car manufacture
- chemicals
- hi-tech industry
- iron & steel
- mining
- precision engineering
- research & development
- shipbuilding
- capital cities
- major towns
- international airports
- major roads
- major industrial areas

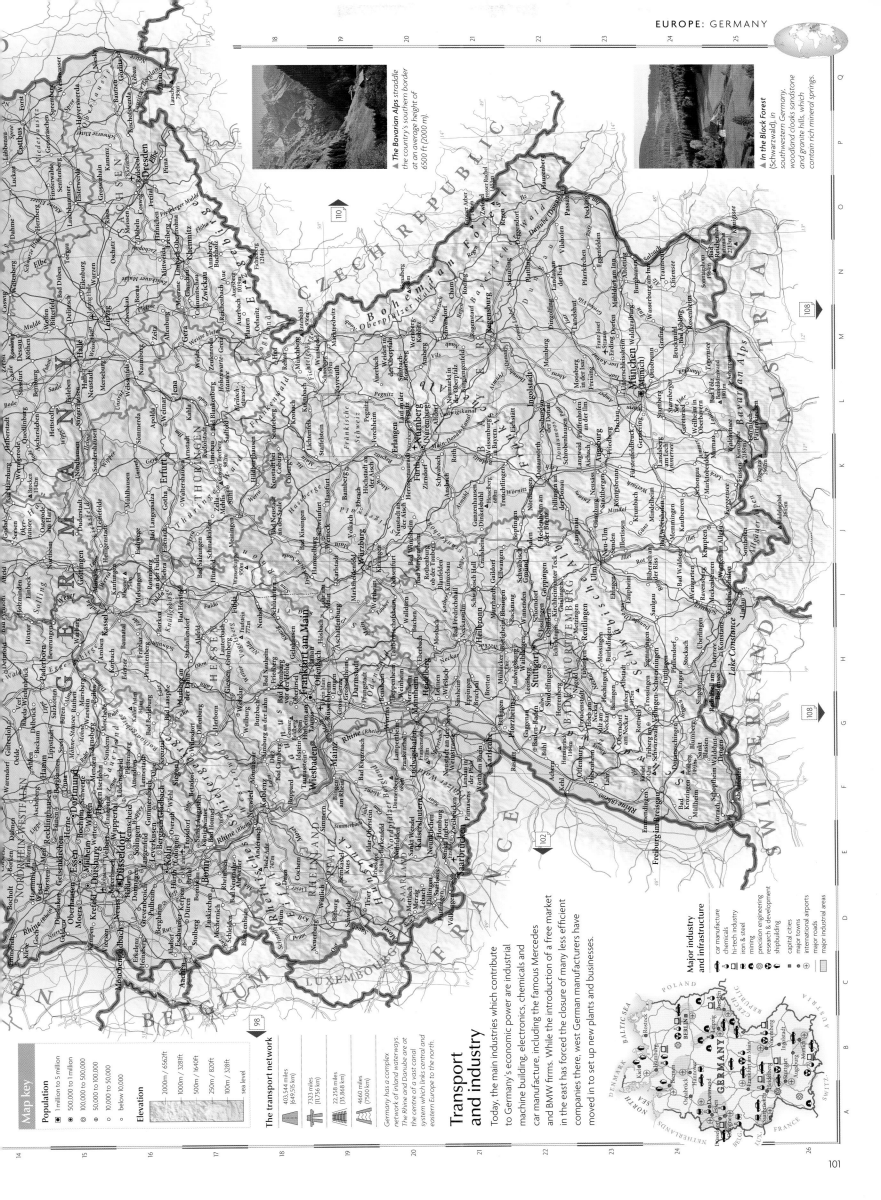

▲ *The Bavarian Alps straddle the country's southern border at an average height of 6500 ft (2000 m).*

▲ *In the Black Forest* (Schwarzwald), *in southwestern Germany, woodland cloaks sandstone and granite hills, which contain rich mineral springs.*

France

FRANCE, MONACO

A major centre of culture and fashion, and a leading producer of both industrial and agricultural goods, France is a key player in the push towards European unity. The founder of modern Republican government in the 18th century, France has been closely involved in European events for many centuries. The Paris Basin is the most highly populated area; Île de France is home to over 11 million people. Large parts of rural France remain thinly populated, particularly the mountainous Massif Central, Pyrenees and southern Alps.

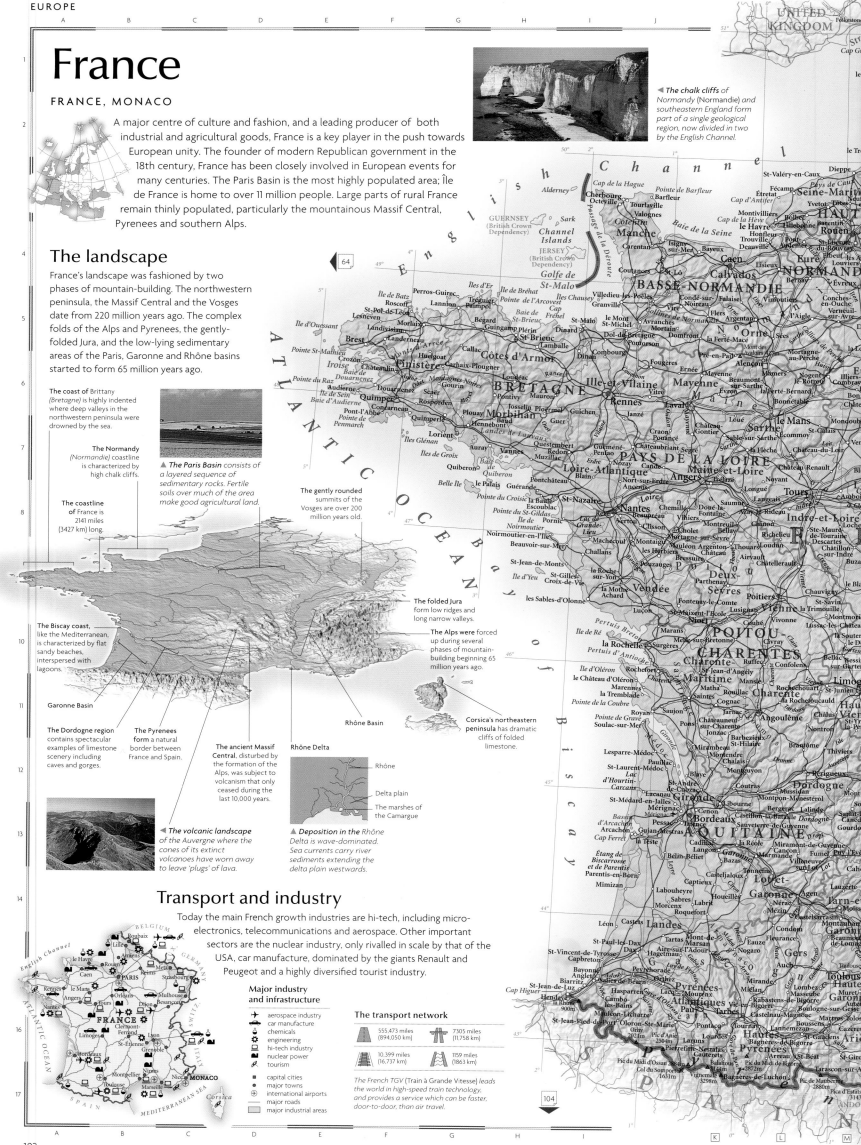

◀ *The chalk cliffs of Normandy (Normandie) and southeastern England form part of a single geological region, now divided in two by the English Channel.*

The landscape

France's landscape was fashioned by two phases of mountain-building. The northwestern peninsula, the Massif Central and the Vosges date from 220 million years ago. The complex folds of the Alps and Pyrenees, the gently-folded Jura, and the low-lying sedimentary areas of the Paris, Garonne and Rhône basins started to form 65 million years ago.

▲ *The Paris Basin consists of a layered sequence of sedimentary rocks. Fertile soils over much of the area make good agricultural land.*

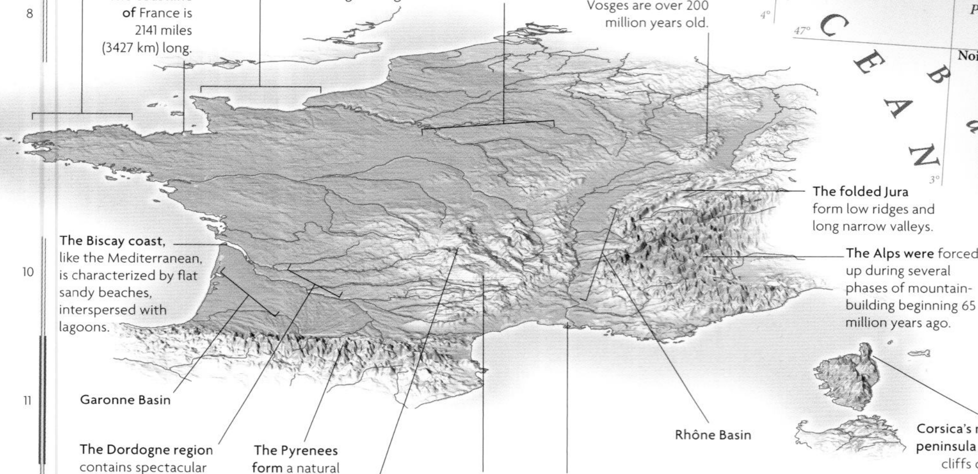
The coast of Brittany (Bretagne) is highly indented where deep valleys in the northwestern peninsula were drowned by the sea.
The Normandy (Normandie) coastline is characterized by high chalk cliffs.
The coastline of France is 2141 miles (3427 km) long.
The gently rounded summits of the Vosges are over 200 million years old.
The folded Jura form low ridges and long narrow valleys.
The Alps were forced up during several phases of mountain-building beginning 65 million years ago.
The Biscay coast, like the Mediterranean, is characterized by flat sandy beaches, interspersed with lagoons.
Garonne Basin
Corsica's northeastern peninsula has dramatic cliffs of folded limestone.
Rhône Basin
The Dordogne region contains spectacular examples of limestone scenery including caves and gorges.
The Pyrenees form a natural border between France and Spain.
The ancient Massif Central, disturbed by the formation of the Alps, was subject to volcanism that only ceased during the last 10,000 years.
Rhône Delta

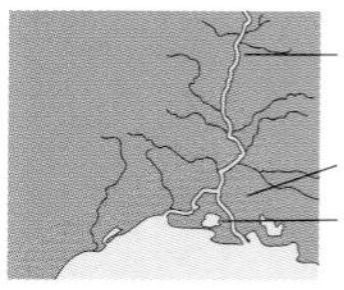
Rhône
Delta plain
The marshes of the Camargue

▲ *Deposition in the Rhône Delta is wave-dominated. Sea currents carry river sediments extending the delta plain westwards.*

◀ *The volcanic landscape of the Auvergne where the cones of its extinct volcanoes have worn away to leave 'plugs' of lava.*

Transport and industry

Today the main French growth industries are hi-tech, including micro-electronics, telecommunications and aerospace. Other important sectors are the nuclear industry, only rivalled in scale by that of the USA, car manufacture, dominated by the giants Renault and Peugeot and a highly diversified tourist industry.

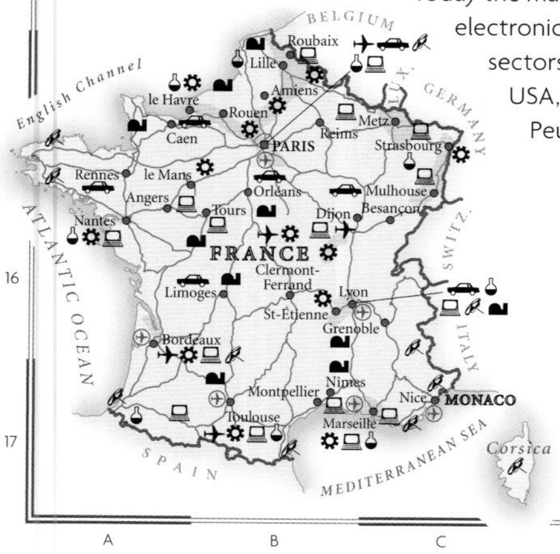
BELGIUM
GERMANY
SWITZ.
ITALY
English Channel
ATLANTIC OCEAN
SPAIN
MEDITERRANEAN SEA
Corsica
MONACO
FRANCE
Roubaix
Lille
Amiens
le Havre
Rouen
Caen
PARIS
Reims
Metz
Strasbourg
Rennes
le Mans
Angers
Orléans
Tours
Nantes
Dijon
Mulhouse
Besançon
Clermont-Ferrand
Lyon
Limoges
St-Étienne
Grenoble
Bordeaux
Montpellier
Nîmes
Nice
Marseille
Toulouse

Major industry and infrastructure

- aerospace industry
- car manufacture
- chemicals
- engineering
- hi-tech industry
- nuclear power
- tourism
- capital cities
- major towns
- international airports
- major roads
- major industrial areas

The transport network

555,473 miles (894,050 km)	7305 miles (11,758 km)
10,399 miles (16,737 km)	1159 miles (1863 km)

The French TGV (Train à Grande Vitesse) leads the world in high-speed train technology, and provides a service which can be faster, door-to-door, than air travel.

UNITED KINGDOM
English Channel
ATLANTIC OCEAN
Bay of Biscay
Channel Islands
GUERNSEY (British Crown Dependency)
JERSEY (British Crown Dependency)
Alderney
Sark
Golfe de St-Malo
Passage de la Déroute
BASSE-NORMANDIE
BRETAGNE
PAYS DE LA LOIRE
POITOU-CHARENTES
AQUITAINE
Cherbourg
Caen
Rennes
Brest
Nantes
Angers
Tours
Le Mans
Poitiers
La Rochelle
Bordeaux
Angoulême
Limoges
Le Havre
Rouen
Dieppe
St-Malo
St-Brieuc
Quimper
Lorient
Vannes
St-Nazaire
Niort
Périgueux
Bayonne
Biarritz
Pau
Tarbes
Lourdes
Pyrénées
Landes
Gironde
Calvados
Manche
Orne
Mayenne
Sarthe
Finistère
Morbihan
Côtes d'Armor
Ille-et-Vilaine
Loire-Atlantique
Maine-et-Loire
Vendée
Deux-Sèvres
Vienne
Charente
Charente-Maritime
Dordogne
Île d'Ouessant
Belle Île
Île de Ré
Île d'Oléron
Île d'Yeu
Noirmoutier-en-l'Île
Pic du Midi d'Ossau 2884m
Vignemale 3298m
SPAIN
ANDORRA

64

104

Scale 1:3,000,000

Km 0 5 10 20 30 40 50 60 70 80
Miles 0 5 10 20 30 40 50 60 70 80

projection: Lambert Conformal Conic

98
100
108
106

Map key

Population

- above 5 million
- 1 million to 5 million
- 500,000 to 1 million
- 100,000 to 500,000
- 50,000 to 100,000
- 10,000 to 50,000
- below 10,000

Elevation

- 4000m / 13,124ft
- 3000m / 9843ft
- 2000m / 6562ft
- 1000m / 3281ft
- 500m / 1640ft
- 250m / 820ft
- 100m / 328ft
- sea level

Using the land

France is western Europe's leading agricultural producer, and benefits from high levels of EU subsidy. The variation in climate and soils across the country provides great potential for agriculture and forestry, reflected in the range of products cultivated, including cereals, olives, herbs, and grapes for its famous wines.

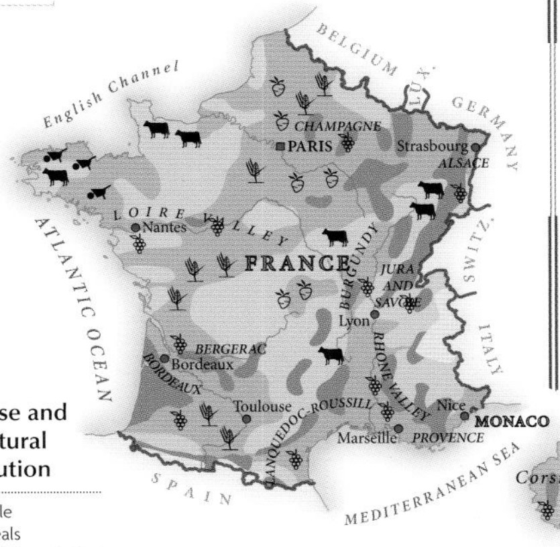

Land use and agricultural distribution

- cattle
- cereals
- market gardening
- sugar beet
- vineyards
- capital cities
- major towns
- pasture
- cropland
- forest
- mountain region

► **The Romans first** introduced wine-making to France when they occupied the region. Traditional vineyards can be found all over France, producing many of the world's classic wines.

The urban/rural population divide

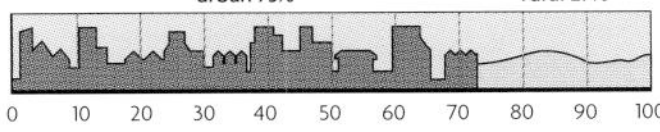

Population density	Total land area
285 people per sq mile (110 people per sq km)	212,930 sq miles (551,500 sq km)

► **The rugged hills** and cliffs of Corsica were uplifted when the African and Eurasian plates collided. Frost action during the Ice Age created their present form.

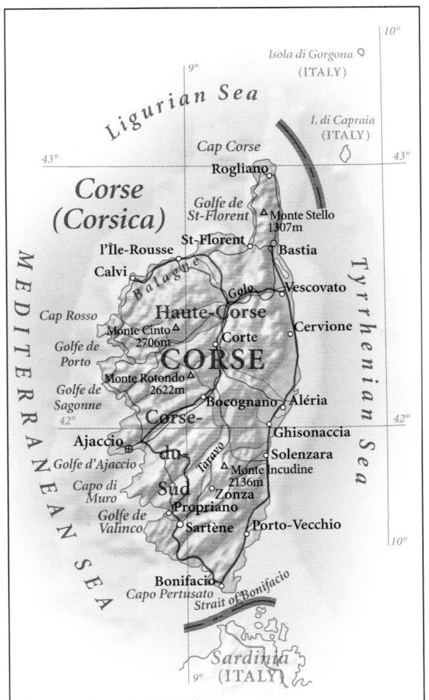

(same scale as main map)

◄ **In the sunny** climate of southern France olives, vines, peppers, garlic and lavender now grow in place of the forests that once covered much of the area.

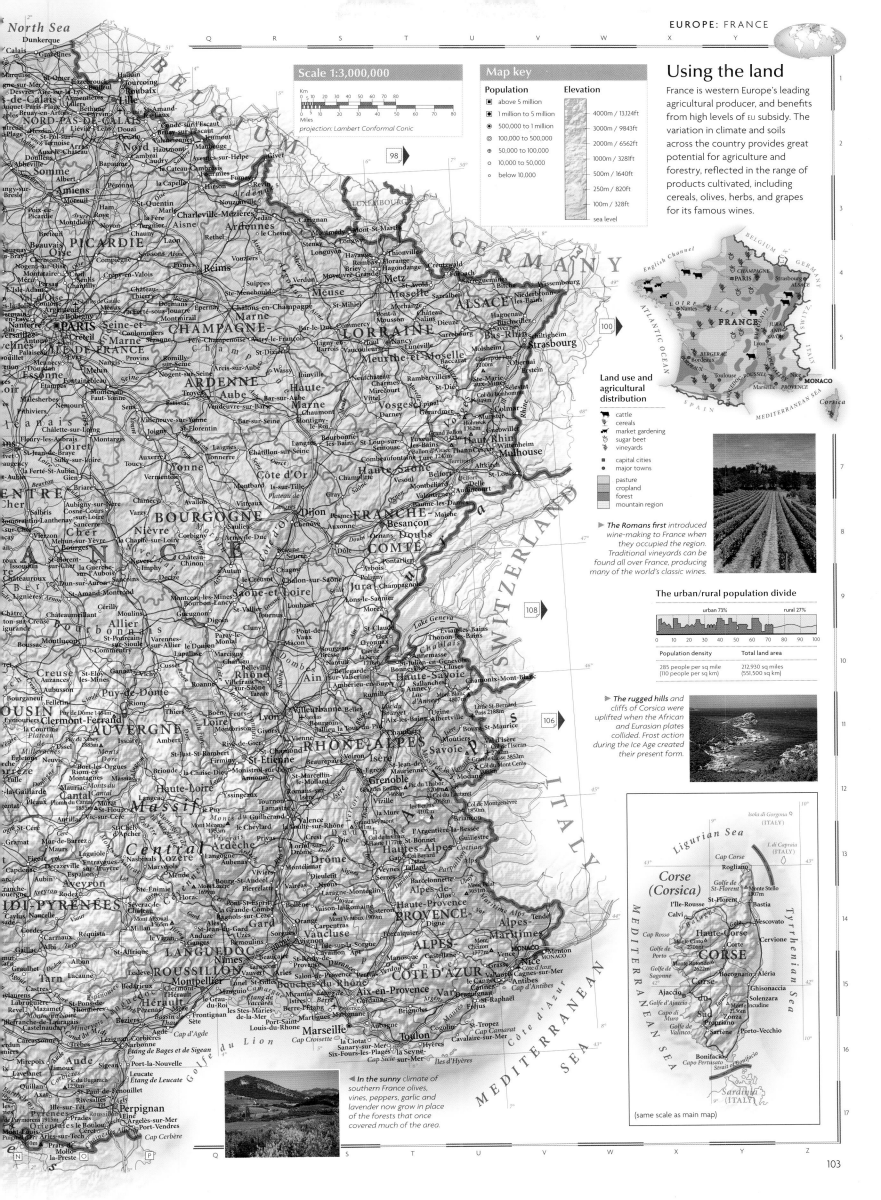

The Iberian peninsula

ANDORRA, GIBRALTAR, PORTUGAL, SPAIN (Azores, Canary Islands, Madeira on p.64)

The Iberian peninsula is separated from the rest of Europe by the Pyrenees, and at its most southerly point is only 5 miles (8 km) from North Africa. The location of Iberia has been central to its diverse history. The Greeks, Carthaginians, Romans, Visigoths and most recently the Moors, invaded Iberia at various times. For much of the 20th century, both Spain and Portugal were governed by right-wing dictators. Since the establishment of democratic governments in the mid-1970s, modernization has been rapid and both countries are now among the most popular of European holiday destinations.

Using the land

The principal crops grown in Iberia are cereals, especially wheat and barley. Both countries are major wine producers, most notably of Rioja, sherry and port. Sheep are kept throughout the region, and citrus fruits thrive on the Mediterranean coast. The successful forest industry in Iberia produces 84% of the world's cork.

▲ ***The steep, terraced*** *slopes of the Douro Valley in northern Portugal, are used to cultivate vines. The grapes harvested produce Portugal's famous port wine.*

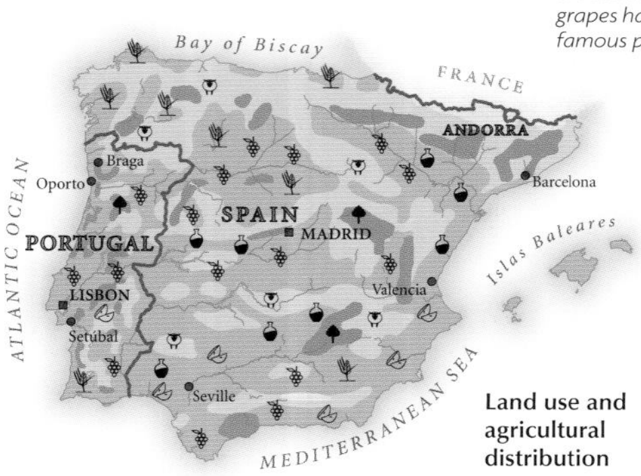

Land use and agricultural distribution

- sheep
- cereals
- citrus fruit
- olives
- vineyards
- cork
- capital cities
- major towns
- pasture
- cropland
- forest
- mountain region

The urban/rural population divide

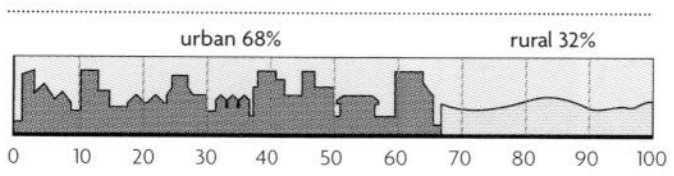

Population density	Total land area
215 people per sq mile (83 people per sq km)	230,569 sq miles (597,170 sq km)

Transport and industry

Since the 1970s, the economies of Spain and Portugal have expanded and diversified. In both countries, tourism has outstripped agriculture in economic importance. Spain's resource base is varied, including coal, iron and the world's largest reserves of mercury. Portugal is a leading producer of tungsten ore.

Major industry and infrastructure

- car manufacture
- chemicals
- engineering
- fish processing
- mining
- textiles
- tourism
- capital cities
- major towns
- international airports
- major roads
- major industrial areas

The transport network

241,720 miles (388,990 km)	1552 miles (2529 km)
11,793 miles (18,979 km)	1159 miles (1865 km)

Radiating from Madrid, the road network in Spain dates from the 18th century, but now includes many motorways. Portugal's road system has been completely modernized in recent years.

◀ ***The eroded cliffs*** *of the Algarve in southern Portugal were carved by Atlantic waves. The numerous rocky bays and beaches, and the region's pleasant climate, have made it a popular tourist destination.*

64

74

N O P Q R S T U V W X Y
1 2 3 4 5 6 10 11 12 13 14 16 17
102
Map key
Population
1 million to 5 million
500,000 to 1 million
100,000 to 500,000
50,000 to 100,000
10,000 to 50,000
below 10,000
Elevation
3000m / 9843ft
2000m / 6562ft
1000m / 3281ft
500m / 1640ft
250m / 820ft
100m / 328ft
sea level
Scale 1:3,000,000
Km
Miles
projection: Lambert Conformal Conic
Biscay
Pyrenees
ANDORRA
ANDORRA LA VELLA
CATALUNA
ARAGON
NAVARRA
PAIS VASCO
LA RIOJA
Donostia/San Sebastián
Bilbao
Vitoria-Gasteiz
Pamplona (Iruña)
Logroño
Burgos
Zaragoza
Huesca
Lleida (Lérida)
Girona
Barcelona
L'Hospitalet de Llobregat
Sabadell
Terrassa
Santa Coloma de Gramenet
Tarragona
Costa Brava
Golf de Sant Jordi
Soria
Guadalajara
Teruel
Castelló
Castellón de la Plana
Valencia
Golfo de Valencia
Cuenca
Albacete
Alicante (Alacant)
Elche (Elx)
Murcia
Cartagena
Costa Blanca
Granada
Almería
Costa del Sol
MEDITERRANEAN SEA
Islas Baleares (Balearic Islands)
Menorca (Minorca)
Mallorca (Majorca)
Palma
Ibiza
Formentera
Melilla (to Spain)
Islas Chafarinas (to Spain)
Peñon de Alhucemas (to Spain)
Isla de Alborán
Alboran Sea
R S T U V W X Y Z

▶ *The climate in northwestern Spain is milder in both summer and winter than in the rest of the country, creating a verdant environment, more commonly associated with northwestern Europe.*

The landscape

A vast plateau, the Meseta dominates the centre of the peninsula, enclosed by the Cordillera Cantábrica to the north and the Sierra Morena to the south. It is drained by three major rivers, the Douro/Duero, the Tagus, and the Guadalquivir. The peninsula experiences great variations in climate and rainfall, both regionally and locally.

▲ ***The Pyrenees form*** *Iberia's northeastern boundary, running for 270 miles (440 km), dividing the peninsula from the rest of Europe.*

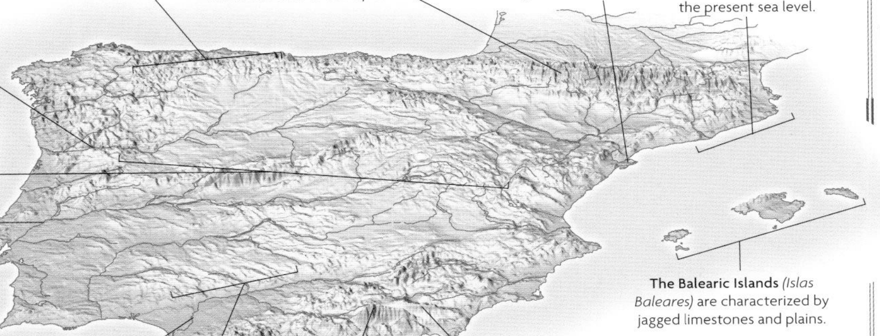
Cordillera Cantábrica
The Ebro river has formed the peninsula's largest delta. Recently, sediment flows have been seriously disturbed by nearby reservoirs.
On the northeastern coast sea level changes are evident from wave-cut beaches which rise up to 200 ft (60 m) above the present sea level.
Douro/Duero river
The Meseta plateau averages 1970 ft (600 m) in height and is now largely dry and treeless.
Tagus River
The Balearic Islands (Islas Baleares) are characterized by jagged limestones and plains.
The Guadalquivir river brings vital irrigation water to the plains, and like many of Iberia's rivers, is prone to flooding.
Sierra Morena
The Sierra Nevada in southern Spain contain Iberia's highest peak, Mulhacén, which rises 11,418 ft (3481 m).

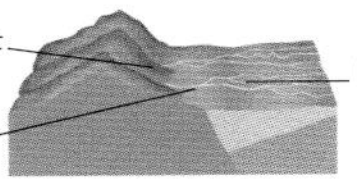
Mountain front
Weathered material
Pediment

▲ ***Pediments are characteristic*** *of semi-arid lands across Iberia. A pediment is a flat, low-lying, eroded platform, cut into the bedrock. Weathered material is transported by streams and deposited in broad fan shapes on the pediment.*

▶ *In the Sierra de los Filabres deforestation and overgrazing, which cause soil erosion, have created semi-desert badlands.*

The Italian peninsula

ITALY, SAN MARINO, VATICAN CITY

The Italian peninsula is a land of great contrasts. Until unification in 1861, Italy was a collection of independent states, whose competitiveness during the Renaissance resulted in the architectural and artistic magnificence of cities such as Rome, Florence and Venice. The majority of Italy's population and economic activity is concentrated in the north, centred on the sophisticated industrial city of Milan. Southern Italy, the *Mezzogiorno*, has a harsh and difficult terrain, and remains far less developed than the north. Attempts to attract industry and investment in the south are frequently deterred by the entrenched network of organized crime and corruption.

The landscape

The mainly mountainous and hilly Italian peninsula took its present form following a collision between the African and Eurasian tectonic plates. The Alps in the northwest rise to a high point of 15,772 ft (4807 m) at Mont Blanc *(Monte Bianco)* on the French border, while the Apennines *(Appennino)* form a rugged backbone, running along the entire length of the country.

▶ ***The island of** Sardinia is an ancient land mass; an uplifted section of very old igneous rocks. Its rugged mountainous regions provide pasture for sheep and goats, while its valleys support some agriculture.*

Mont Blanc *(Monte Bianco)*

The Po Valley once formed part of the Adriatic Sea. Sediments of gravel, sand and clay washed down from the Alps gradually filling the bay and forming a broad, cultivable plain.

Costa Smeralda

The Apennines *(Appennino)* are the source of most of Italy's rivers. They run 823 miles (1324 km) down the length of the peninsula.

The Pontine Marshes *(Agro Pontino)* are bounded by low sand hills which prevent natural drainage.

Vesuvius *(Vesuvio)*

Sardinia is the second largest island in the Mediterranean Sea. The highest point is Punta La Marmora at 6017 ft (1834 m).

The southwestern tip of Sicily lies 95 miles (152 km) from the north African mainland and is part of the same geological region.

The Strait of Messina *(Stretto di Messina)* is between 2 and 12 miles (3–19 km) wide, and is a rich fishing ground.

Sicily is the largest island in the Mediterranean at 9926 sq miles (25,708 sq km).

The distinctive square shape of the Gulf of Taranto *(Golfo di Taranto)* was defined by numerous block faults. Earthquakes are common in this region.

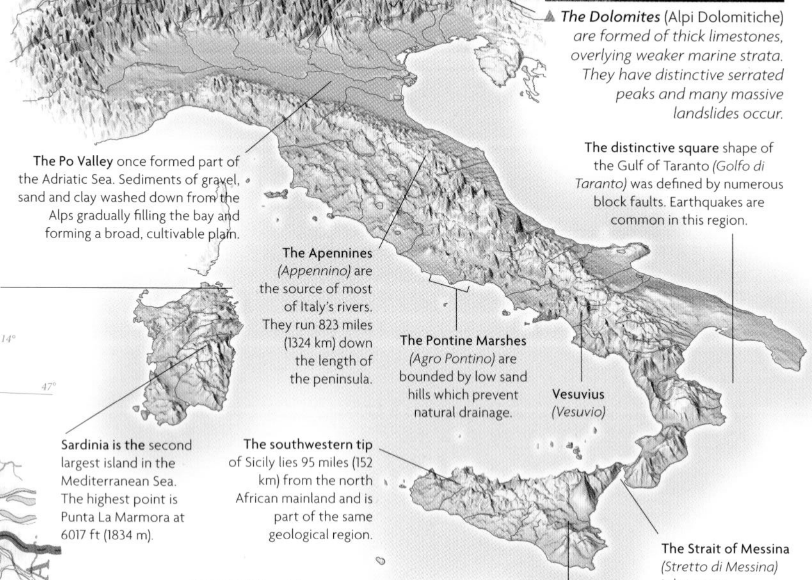

▲ ***The Dolomites** (Alpi Dolomitiche) are formed of thick limestones, overlying weaker marine strata. They have distinctive serrated peaks and many massive landslides occur.*

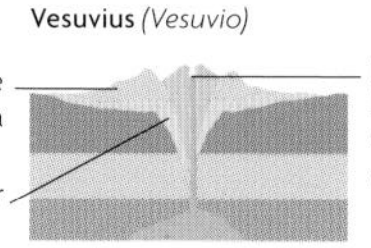
Vesuvius (Vesuvio)
Monte Somma
Old crater
Present-day crater has developed within the old crater of Monte Somma

▲ ***There have been** four volcanoes on the site of Vesuvius since volcanic activity began here more than 10,000 years ago.*

108
108
108
112
120
102

Scale 1:2,750,000

projection: Lambert Conformal Conic

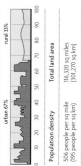

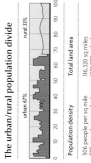

Using the land

Italy produces 95% of its own food. The best farming land is in the Po Valley in northern Italy, where soft wheat and rice are grown. Irrigation is essential to agriculture in much of the south. Italy is a major producer and exporter of citrus fruits, olives, tomatoes and wine.

The urban/rural population divide

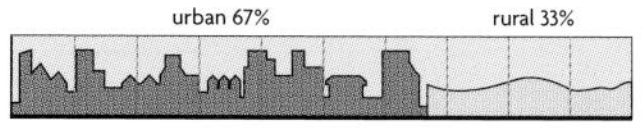
urban 67%
rural 33%
0 10 20 30 40 50 60 70 80 90 100

Population density	Total land area
506 people per sq mile (195 people per sq km)	116,320 sq miles (301,270 sq km)

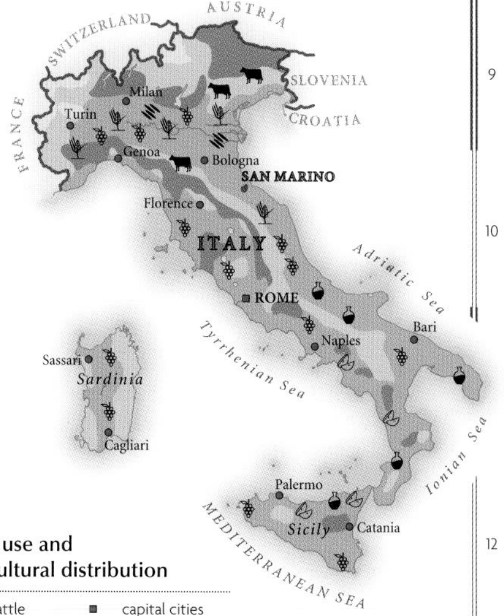
AUSTRIA
SWITZERLAND
SLOVENIA
CROATIA
FRANCE
Turin
Milan
Genoa
Bologna
SAN MARINO
Florence
ITALY
ROME
Adriatic Sea
Tyrrhenian Sea
Naples
Bari
Ionian Sea
Sassari
Sardinia
Cagliari
Palermo
Sicily
Catania
MEDITERRANEAN SEA

Land use and agricultural distribution

- cattle
- cereals
- citrus fruits
- olive oil
- rice
- vineyards
- capital cities
- major towns
- pasture
- cropland
- forest

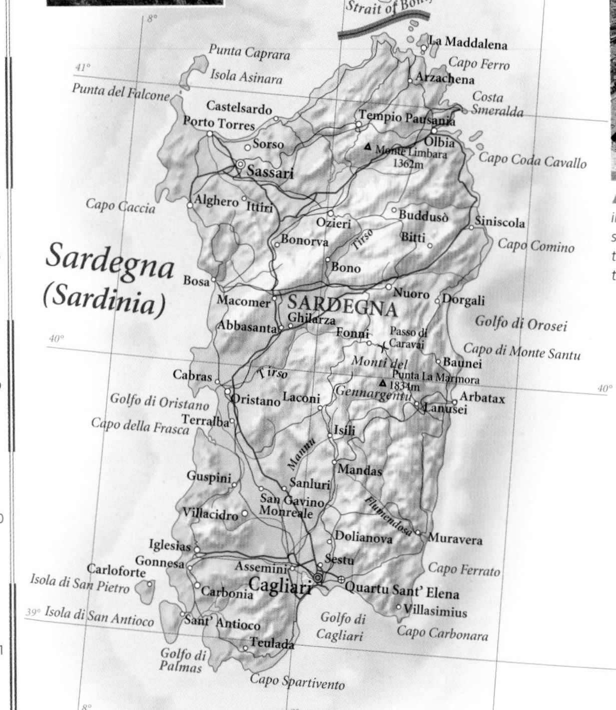

◀ *Tuscany (Toscana) has long produced grapes and olives. Sandstones form its higher reaches, while clays and alluvial soils fill its fertile valleys.*

▲ *Vatican City in Rome is the smallest independent state in the world. As the seat of the Catholic Church it is home to the Pope, spiritual head of 18% of the world's population.*

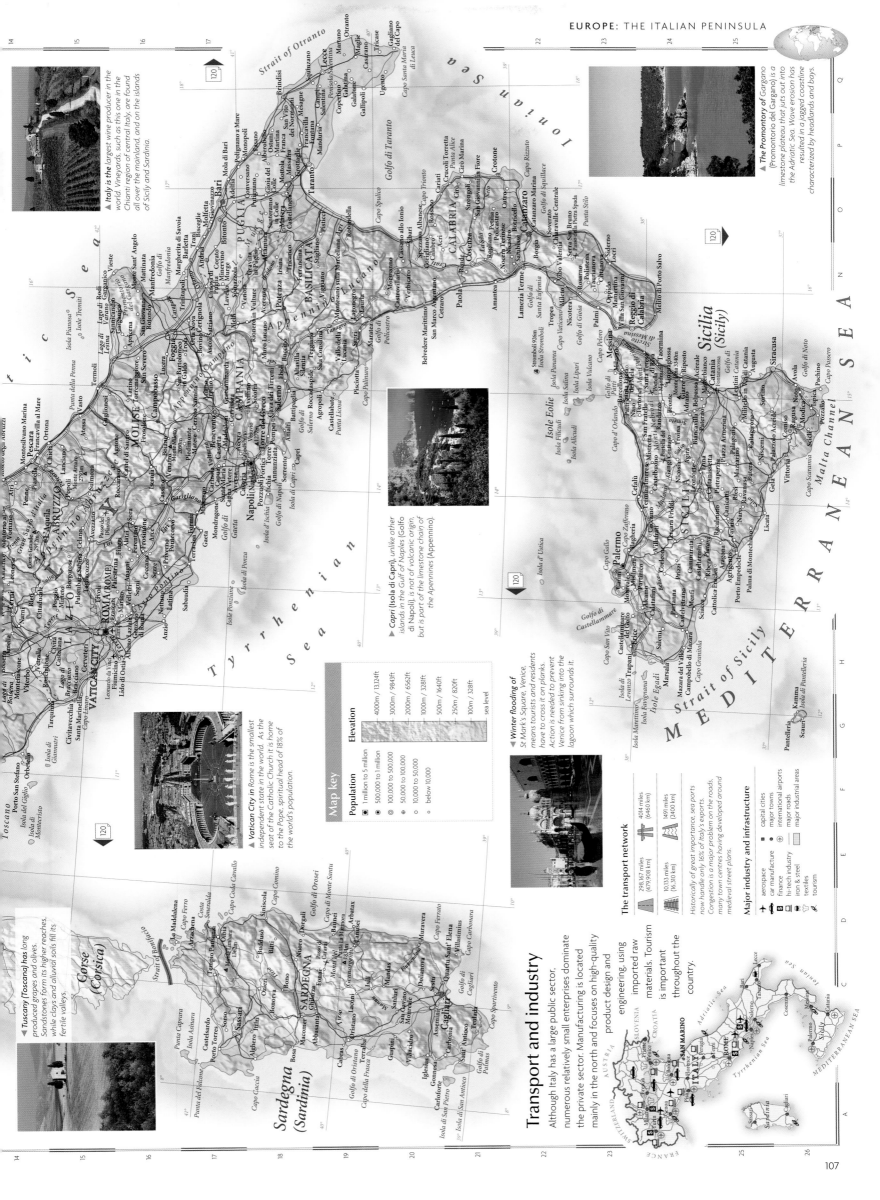

▲ *Italy is the largest wine producer in the world. Vineyards, such as this one in the Chianti region of central Italy, are found all over the mainland, and on the islands of Sicily and Sardinia.*

Map key

Population	Elevation
▣ 1 million to 5 million	4000m / 13,124ft
◉ 500,000 to 1 million	3000m / 9843ft
◎ 100,000 to 500,000	2000m / 6562ft
⊕ 50,000 to 100,000	1000m / 3281ft
○ 10,000 to 50,000	500m / 1640ft
○ below 10,000	250m / 820ft
	100m / 328ft
	sea level

▶ *Capri (Isola di Capri), unlike other islands in the Gulf of Naples (Golfo di Napoli), is not of volcanic origin, but is part of the limestone chain of the Apennines (Appennino).*

◀ *Winter flooding of St Mark's Square, Venice, means tourists and residents have to cross it on planks. Action is needed to prevent Venice from sinking into the lagoon which surrounds it.*

Transport and industry

Although Italy has a large public sector, numerous relatively small enterprises dominate the private sector. Manufacturing is located mainly in the north and focuses on high-quality product design and engineering, using imported raw materials. Tourism is important throughout the country.

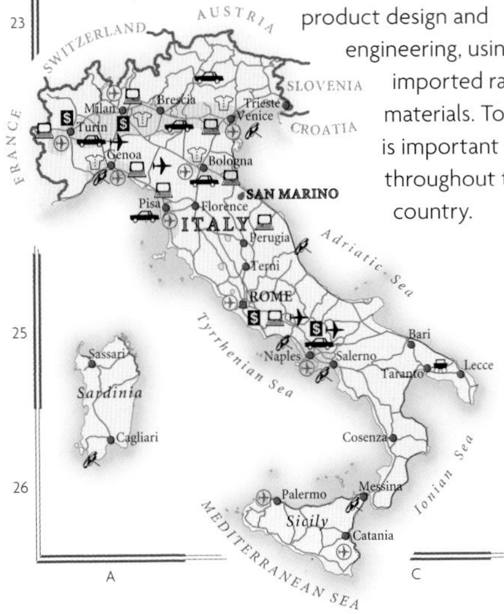

The transport network

298,167 miles (479,908 km)	4014 miles (6460 km)
10,133 miles (16,310 km)	1491 miles (2400 km)

Historically of great importance, sea ports now handle only 16% of Italy's exports. Congestion is a major problem on the roads, many town centres having developed around medieval street plans.

Major industry and infrastructure

aerospace	capital cities
car manufacture	major towns
finance	international airports
hi-tech industry	major roads
iron & steel	major industrial areas
textiles	
tourism	

▲ *The Promontory of Gargano (Promontorio del Gargano) is a limestone plateau that juts out into the Adriatic Sea. Wave erosion has resulted in a jagged coastline characterized by headlands and bays.*

The Alpine states

AUSTRIA, LIECHTENSTEIN, SLOVENIA, SWITZERLAND

The Alpine countries of Austria, Switzerland, Liechtenstein and Slovenia form a narrow strip across western Europe's geographical core, lying on the main north–south trading routes across the Alps. Switzerland, politically neutral since 1815, is an important international meeting place and houses one of the headquarters of the United Nations, although it only became a member in 2002. Austria, once at the heart of the great Habsburg Empire has been a fully independent nation since 1955, and maintains a deserved reputation as an international centre of culture. Slovenia declared independence from the former Yugoslavia in 1991 and despite initial economic hardship, is now starting to achieve the prosperity enjoyed by its Alpine neighbours.

Using the land

The Alpine region's mountainous terrain discourages cultivation over much of the land area. The primary agricultural activity is the raising of dairy and beef cattle on the pasture land of the lower mountain slopes. Austria is self-supporting in grains, and crops such as wheat, barley and grapes are grown on the east Austrian lowlands. Woodlands are more prevalent in the eastern Alps; both Austria and Slovenia have large tracts of forest.

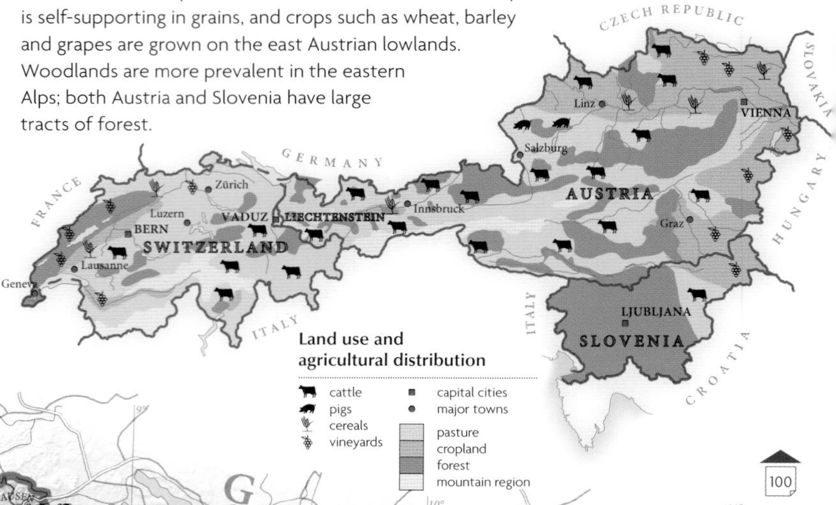

◀ *The Matterhorn, on the Swiss-Italian border, is one of the highest mountains in the Alps, at 14,692 ft (4478 m). The term 'horn' refers to its distinctive peak, formed by three glaciers eroding hollows, known as cirques, in each of its sides.*

100
102
102
106

The landscape

The Alps occupy three-fifths of Switzerland, most of southern Austria and the northwest of Slovenia. They were formed by the collision of the African and Eurasian tectonic plates, which began 65 million years ago. Their complex geology is reflected in the differing heights and rock types of the various ranges. The Rhine flows along Liechtenstein's border with Switzerland, creating a broad flood plain in the north and west of Liechtenstein. In the far northeast and east are a number of lowland regions, including the Vienna Basin, Burgenland and the plain of the Danube. Slovenia's major rivers flow across the lower eastern regions; in the west, the rivers flow largely underground through the limestone Karst region.

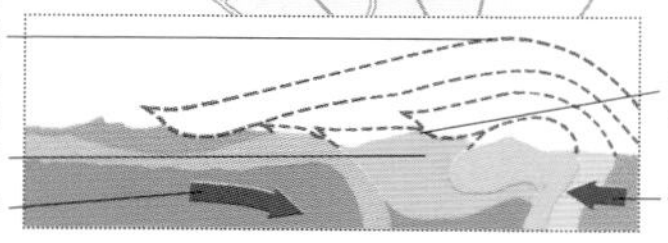

▲ *The convergence of the African and Eurasian plates compressed and folded huge masses of rock strata. As the plates continued to move together, the folded strata were overturned, creating complex nappes. Much of the rock strata has since been eroded, resulting in the current topography of the Alps.*

▲ *Constricted as it cuts through ridges in the Alps, the Danube meanders across the lowlands, where uplift combined with river erosion has deepened meanders.*

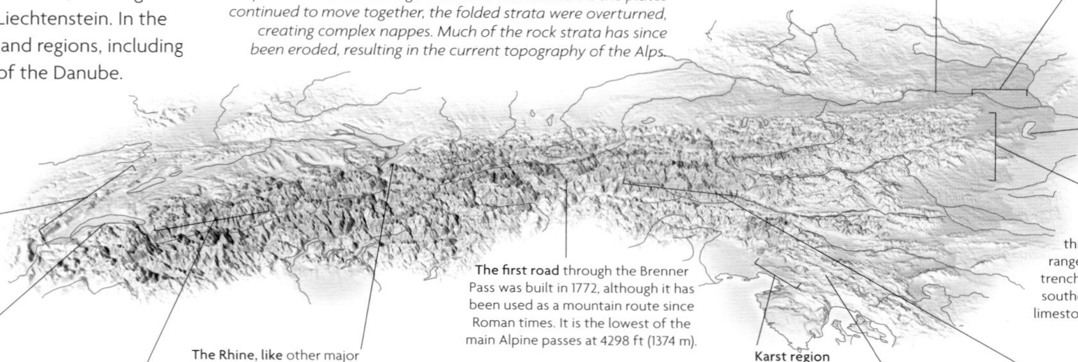

The Vienna Basin lies mainly below 390 ft (120 m). It gradually subsided and filled with sediment as the Alps were uplifted.

Neusiedler See straddles the border of Austria and Hungary; the area around it provides some of the best wine-growing land in Austria.

The Austrian Alps comprise three distinct mountain ranges, separated by deep trenches. The northern and southern ranges are rugged limestones, while the Tauern range is formed of crystalline rocks.

The mountains of the Jura form a natural border between Switzerland and France. Their marine limestones date from over 200 million years ago. When the Alps were formed the Jura were folded into a series of parallel ridges and troughs.

The first road through the Brenner Pass was built in 1772, although it has been used as a mountain route since Roman times. It is the lowest of the main Alpine passes at 4298 ft (1374 m).

Tectonic activity has resulted in dramatic changes in land height over very short distances. Lake Geneva, lying at 1221 ft (372 m) is only 43 miles (70 km) away from the 15,772 ft (4807 m) peak of Mont Blanc, on the France–Italy border.

The Bernese Alps *(Berner Alpen)* contain the Aletsch, which at 15 miles (24 km) is the longest Alpine glacier.

The Rhine, like other major Alpine rivers, follows a broad, flat trough between the mountains. Along part of its course, the Rhine forms the boundary between Switzerland and Liechtenstein.

► *The deep, blue lakes of the Karst region are part of a drainage network which runs largely underground through this limestone area.*

The limestone cave system at Postojna extends for more than 10 miles (16 km) and includes caverns reaching 125 ft (40 m) in height and width.

The Tauern range in the central Austrian Alps contains the highest mountain in Austria, the towering Grossglockner, rising 12,461 ft (3798 m).

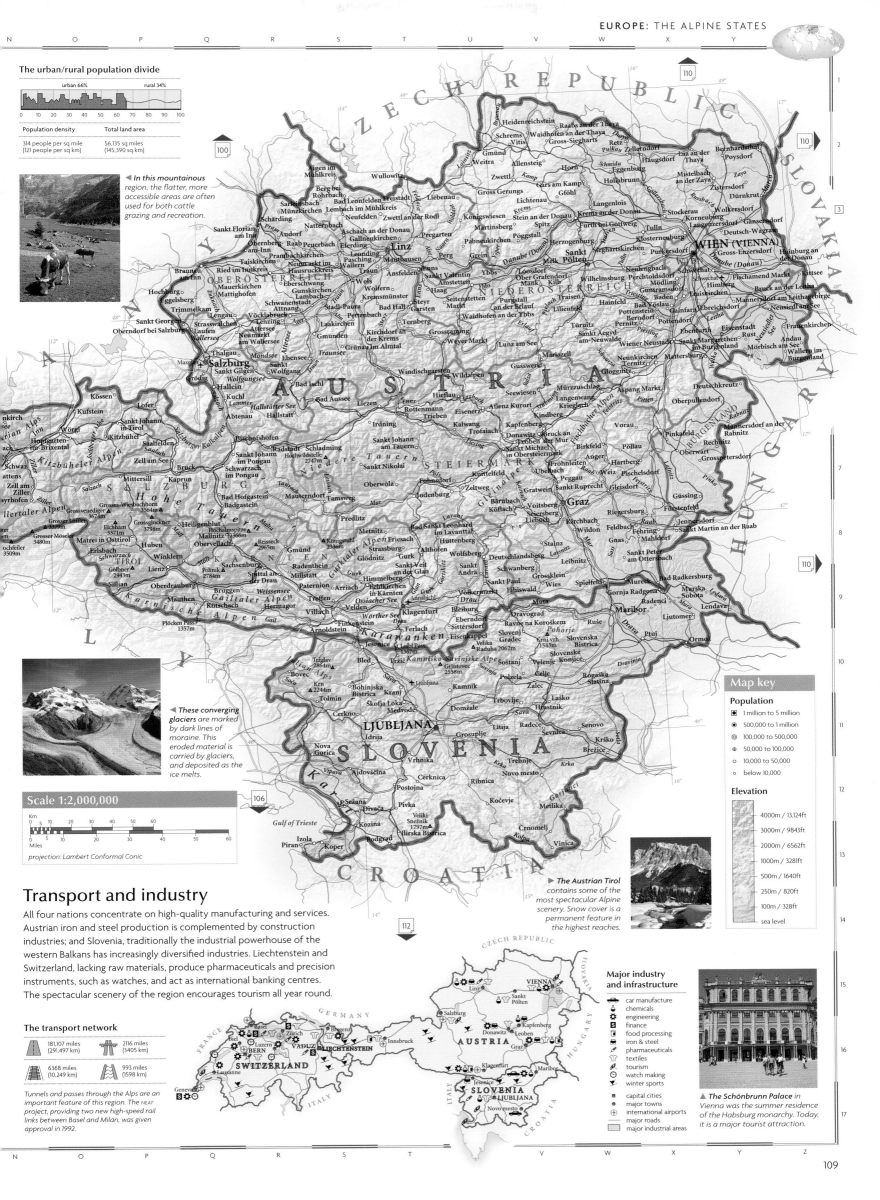

The urban/rural population divide

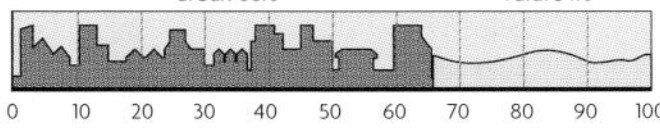

Population density	Total land area
314 people per sq mile (121 people per sq km)	56,135 sq miles (145,390 sq km)

◄ *In this mountainous region, the flatter, more accessible areas are often used for both cattle grazing and recreation.*

◄ *These converging glaciers* are marked by dark lines of moraine. This eroded material is carried by glaciers, and deposited as the ice melts.

Scale 1:2,000,000

Km 0 5 10 20 30 40 50 60
Miles 0 5 10 20 30 40 50 60

projection: Lambert Conformal Conic

Map key

Population

- 1 million to 5 million
- 500,000 to 1 million
- 100,000 to 500,000
- 50,000 to 100,000
- 10,000 to 50,000
- below 10,000

Elevation

- 4000m / 13,124ft
- 3000m / 9843ft
- 2000m / 6562ft
- 1000m / 3281ft
- 500m / 1640ft
- 250m / 820ft
- 100m / 328ft
- sea level

► *The Austrian Tirol* contains some of the most spectacular Alpine scenery. Snow cover is a permanent feature in the highest reaches.

Transport and industry

All four nations concentrate on high-quality manufacturing and services. Austrian iron and steel production is complemented by construction industries; and Slovenia, traditionally the industrial powerhouse of the western Balkans has increasingly diversified industries. Liechtenstein and Switzerland, lacking raw materials, produce pharmaceuticals and precision instruments, such as watches, and act as international banking centres. The spectacular scenery of the region encourages tourism all year round.

The transport network

181,107 miles (291,497 km)	2116 miles (3405 km)
6368 miles (10,249 km)	993 miles (1598 km)

Tunnels and passes through the Alps are an important feature of this region. The NEAT project, providing two new high-speed rail links between Basel and Milan, was given approval in 1992.

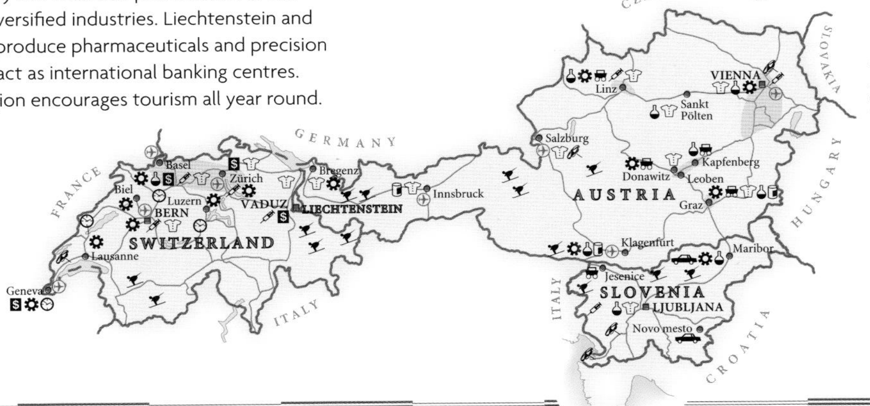

Major industry and infrastructure

- car manufacture
- chemicals
- engineering
- finance
- food processing
- iron & steel
- pharmaceuticals
- textiles
- tourism
- watch making
- winter sports
- capital cities
- major towns
- international airports
- major roads
- major industrial areas

▲ *The Schönbrunn Palace* in Vienna was the summer residence of the Habsburg monarchy. Today, it is a major tourist attraction.

Central Europe

CZECH REPUBLIC, HUNGARY, POLAND, SLOVAKIA

When Slovakia and the Czech Republic became separate countries in 1993, they joined Hungary and Poland in a new role as independent nation states, following centuries of shifting boundaries and imperial strife. This turbulent history bequeathed the region a rich cultural heritage, shared through the works of its many great writers and composers, and celebrated in the vibrant historic capitals of Prague, Budapest and Warsaw. Having shaken off years of Soviet domination in 1989, these states are confronting the challenge of winning commercial investment to modernize outmoded industries as they integrate their economies with those of the European Union.

The landscape

The forested Carpathian Mountains, uplifted with the Alps, lie southeast of the older Bohemian Massif, which contains the Sudeten and Krusné Hory *(Erzgebirge)* ranges. They divide the fertile plains of the Danube to the south and the Vistula *(Wisla)*, which flows north across vast expanses of glacial deposits into the Baltic Sea.

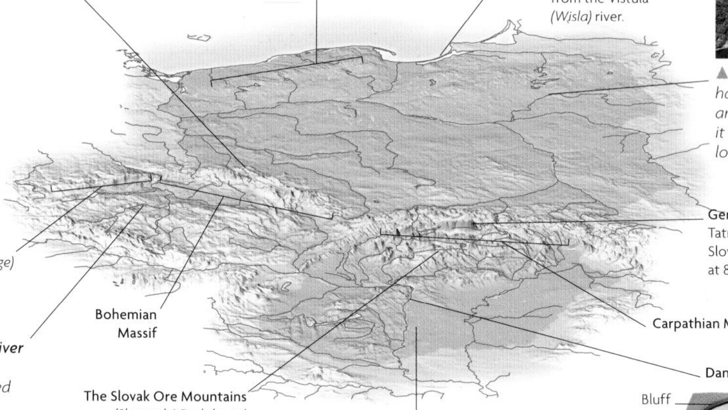

▲ *The Biebrza river has left meanders and oxbow lakes as it flows across low-lying ground.*

◀ *The Berounka river cuts through the precipitous wooded landscape of the Bohemian Massif, banked by a broad flood plain.*

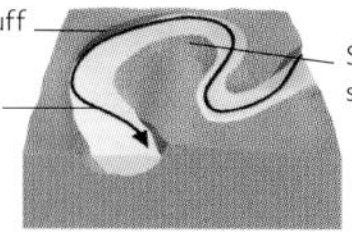

▲ *Meanders form as rivers flow across plains at a low gradient. A steep cliff or bluff, forms on the outside curve, and a gentler slip-off slope on the inside bend.*

Transport and industry

Heavy industry has dominated post-war life in Central Europe. Poland has large coal reserves, having inherited the Silesian coalfield from Germany after the Second World War, allowing the export of large quantities of coal, along with other minerals. Hungary specializes in consumer goods and services, while Slovakia's industrial base is still relatively small. The Czech Republic's traditional glassworks and breweries bring some stability to its precarious Soviet-built manufacturing sector.

Major industry and infrastructure

- car manufacture
- chemicals
- engineering
- food processing
- mining
- shipbuilding
- tourism
- capital cities
- major towns
- international airports
- major roads
- major industrial areas

The transport network

- 213,997 miles (344,600 km)
- 817 miles (1315 km)
- 27,479 miles (44,249 km)
- 3784 miles (6094 km)

The huge growth of tourism and business has prompted major investment in the transport infrastructure, with new road-building schemes within and between the main cities of the region.

▲ ***Budapest, the capital** of Hungary, straddles the Danube. It comprises the historic towns of Buda, on the west bank, and Pest, which contains the Parliament Building, seen here on the far bank.*

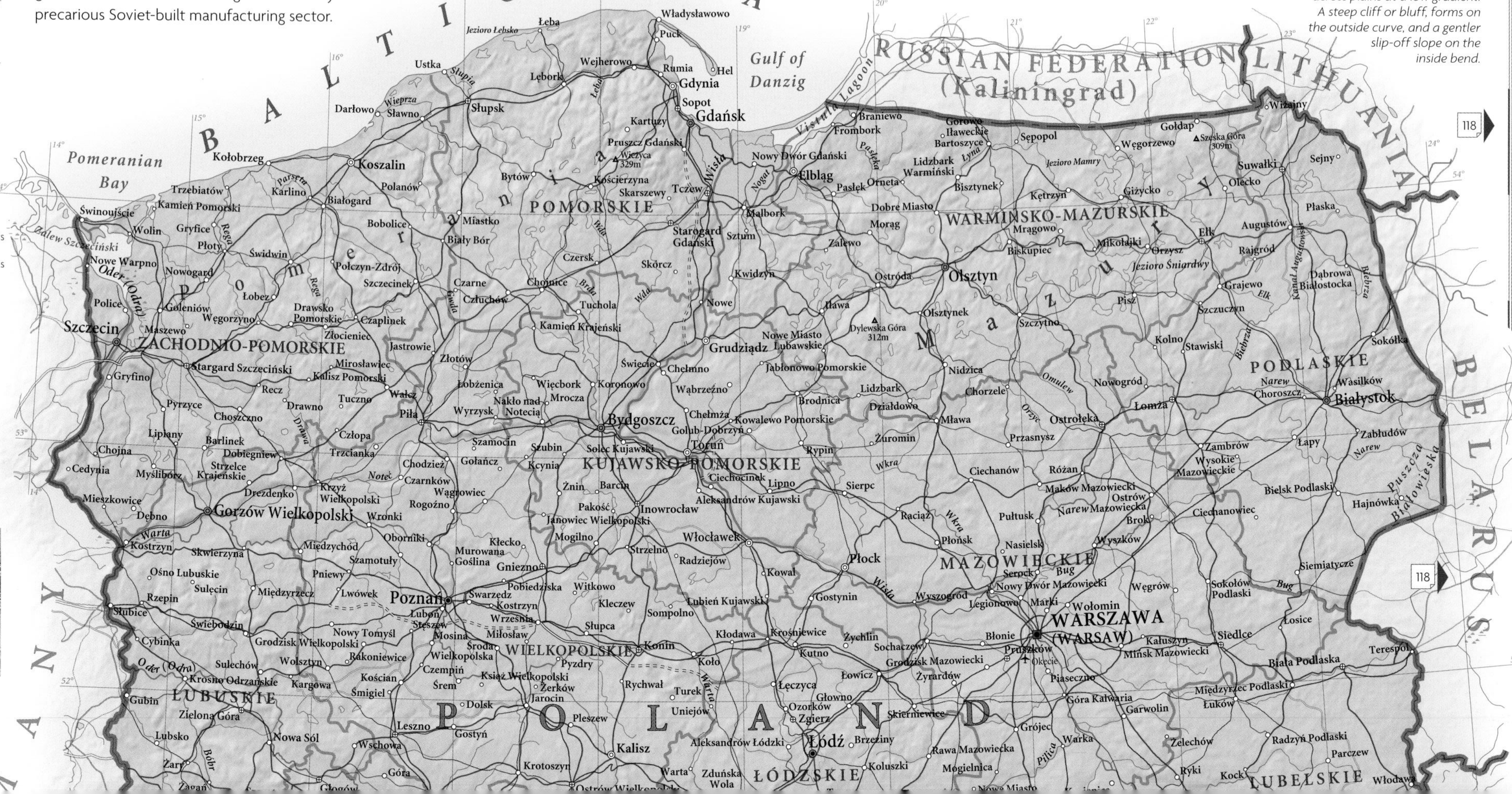

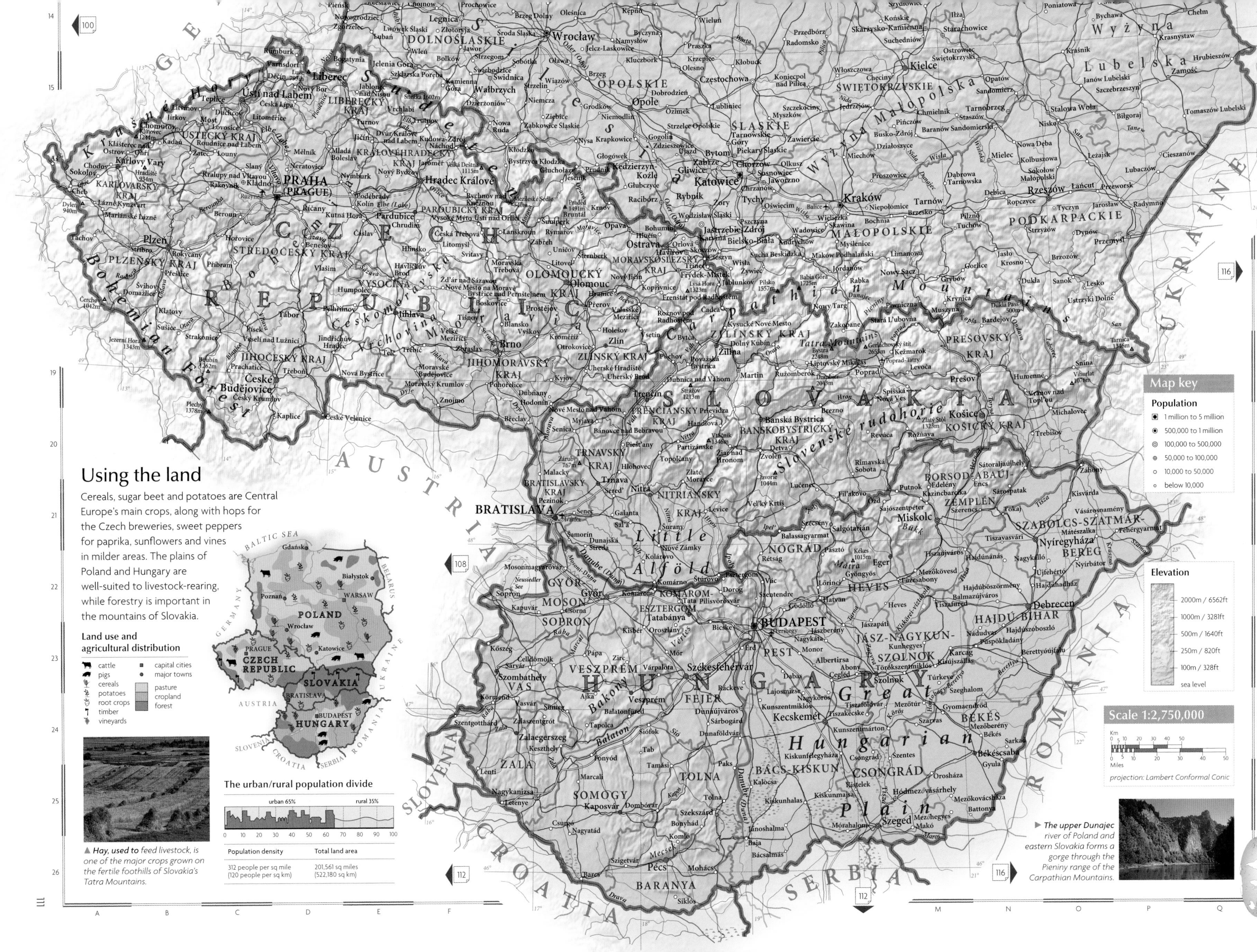

Map key

Population

- 1 million to 5 million
- 500,000 to 1 million
- 100,000 to 500,000
- 50,000 to 100,000
- 10,000 to 50,000
- below 10,000

Elevation

- 2000m / 6562ft
- 1000m / 3281ft
- 500m / 1640ft
- 250m / 820ft
- 100m / 328ft
- sea level

Scale 1:2,750,000

Km 0 5 10 20 30 40 50
Miles 0 5 10 20 30 40 50

projection: Lambert Conformal Conic

Using the land

Cereals, sugar beet and potatoes are Central Europe's main crops, along with hops for the Czech breweries, sweet peppers for paprika, sunflowers and vines in milder areas. The plains of Poland and Hungary are well-suited to livestock-rearing, while forestry is important in the mountains of Slovakia.

Land use and agricultural distribution

- cattle
- pigs
- cereals
- potatoes
- root crops
- timber
- vineyards
- capital cities
- major towns
- pasture
- cropland
- forest

▲ **Hay, used to** feed livestock, *is one of the major crops grown on the fertile foothills of Slovakia's Tatra Mountains.*

The urban/rural population divide

urban 65% · rural 35%

Population density	Total land area
312 people per sq mile (120 people per sq km)	201,561 sq miles (522,180 sq km)

▶ **The upper Dunajec** *river of Poland and eastern Slovakia forms a gorge through the Pieniny range of the Carpathian Mountains.*

Southeast Europe

ALBANIA, BOSNIA & HERZEGOVINA, CROATIA, KOSVOVO, MACEDONIA, MONTENEGRO, SERBIA

For 46 years the federation of Yugoslavia held together the most diverse ethnic region in Europe, along the picturesque mountain hinterland of the Dalmatian coast. Economic collapse resulted in internal tensions. In the early 1990s, civil war broke out in both Croatia and Bosnia as the ethnic populations struggled to establish their own exclusive territories. Peace was only restored by the UN after NATO launched air strikes in 1995. Montenegro voted to split from Serbia in 2006. More recently, Kosovo controversially declared independence from Serbia in 2008, although this may take some time to be fully recognized. Neighbouring Albania is slowly improving its fragile economy but remains one of Europe's poorest nations.

▲ ***Hot, dry summers** and mild winters offer excellent conditions for viticulture in Montenegro. The precipitous Dinaric Alps have kept this region relatively isolated for centuries.*

Scale 1:2,750,000

Km 0 5 10 20 30 40 50 60 70

Miles 0 5 10 20 30 40 50 60 70

projection: Lambert Conformal Conic

The landscape

The Tisza (Tisa), Sava and Drava rivers drain the broad northern lowland, meeting the Danube after it crosses the Hungarian border. In the west, the Dinaric Alps divide the Adriatic Sea from the interior. Mainland valleys and elongated islands run parallel to the steep Dalmatian *(Dalmacija)* coastline, following alternating bands of resistant limestone.

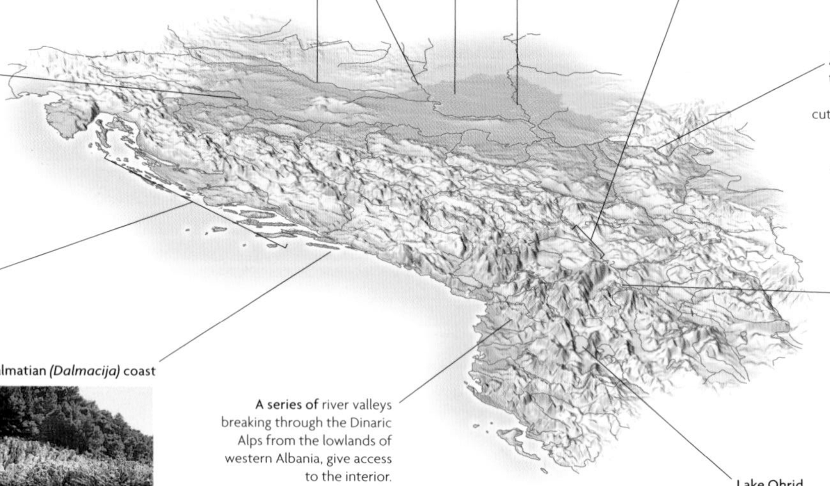
At least 70% of the fresh water in the western Balkans drains eastwards into the Black Sea, mostly via the Danube *(Dunav)*.
The river flood plains of the Pannonian Basin are flanked by terraces of gravel and wind-blown glacial deposits known as loess.
Drava river
Tisza river
Sava river
The elongated islands, promontories and straits of the Dalmatian *(Dalmacija)* coast were formed as the Adriatic Sea rose to flood valleys running parallel to the shore.
Dalmatian *(Dalmacija)* coast
A series of river valleys breaking through the Dinaric Alps from the lowlands of western Albania, give access to the interior.
At Iron Gate *(Derdap)*, on the border with Romania, the Danube narrows and cuts through foothills of the Balkan and Carpathian mountains, forming the deepest gorge in Europe.
A major earthquake at Skopje, Macedonia, in 1963 killed 1000 people. The whole region lies on an active crustal plate margin.
Lake Ohrid

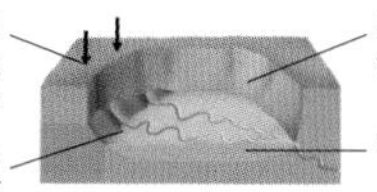
Poljes in the Kosovo region
Underground drainage along joints in the rock
Spring at foot of cliff
Sheer limestone walls enclose all sides
Flat *polje* floor

▲ ***Rain and underground** water dissolve limestone along massive vertical joints (cracks). This creates* poljes: *depressions several miles across with steep walls and broad, flat floors.*

▶ ***Limestone cliffs along** the Dalmatian* (Dalmacija) *shoreline are heavily eroded, as salt water dissolves the rock along existing horizontal cracks, or joints. This tends to form a platform of rock at the foot of the cliff.*

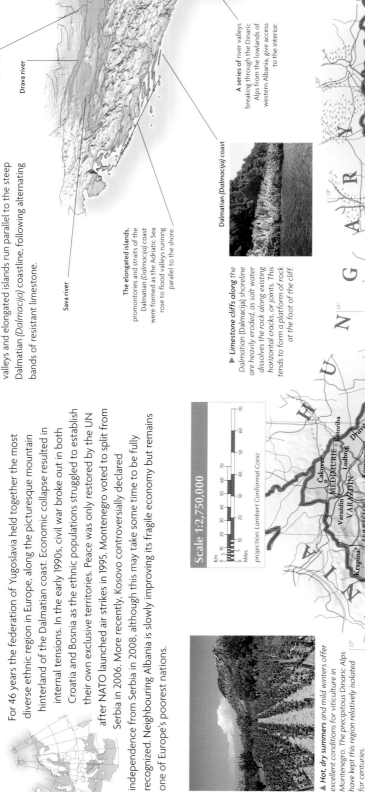

▲ ***Lake Ohrid borders** Albania and Macedonia. Ohrid is the deepest lake in the western Balkans, reaching depths of 938 ft (286 m).*

HUNGARY
ROMANIA
SLOVENIA
CROATIA
BOSNIA & HERZEGOVINA
SERBIA
VOJVODINA
BEOGRAD (BELGRADE)
ZAGREB
Novi Sad
Osijek
Subotica
Rijeka
Zadar
Banja Luka
Karlovac
Gulf of Trieste
Danube (Dunav)
Sava
Drava
Tisa
REPUBLIKA SRPSKA
108
110
114
116
120

◀ *The historic centre* of Mostar in southern Bosnia, with its famous 16th-century Turkish bridge, was destroyed by shelling during 1993. The bridge was rebuilt and opened again in 2004.

114

Transport and industry

Processing industries based on the region's wealth of mineral reserves predominate in Albania and Macedonia. In other regions, industrial plants have been commandeered, if not destroyed in the war and mineral extraction has severely declined. The fast-flowing rivers found throughout the Dinaric Alps are exploited to generate hydro-electric power.

In February 2008, Kosovo (a UN Protectorate within Serbia since 1999) declared independence. Although now recognized by numerous countries, this decision has proved controversial with other states wary of setting a precedent for separatist groups within their own borders. It is therefore likely to be some time before Kosovo becomes universally recognized.

Major industry and infrastructure

- aluminium refining
- car manufacture
- chemicals
- engineering
- food processing
- hydro-electric power
- mining
- shipbuilding
- textiles
- timber processing
- capital cities
- major towns
- international airports
- major roads

The transport network

46,996 miles (75,642 km)		685 miles (1103 km)	
5413 miles (8713 km)		879 miles (1415 km)	

The war resulted in the destruction or disintegration of infrastructure for transport, communications and power supply, though this is now in the process of recovery.

▶ ***Industrial processing plants*** *were established throughout Albania by the Hoxha regime, which collapsed in 1992. They remain incongruous among the villages of one of Europe's most conservative rural societies.*

▲ ***The ancient Croatian*** *port of Dubrovnik was one of the former Yugoslavia's most popular tourist resorts and an important point of access to the sea along the Dalmatian (Dalmacija) coast. Shelling of the old city by Serb forces in 1991 provoked international condemnation.*

Using the land

Crops of wheat, maize, sugar beet, vegetables and fruit are widely grown. The hilly terrain is suited to forestry and livestock farming. The mild, mediterranean climate of the coastal regions provides ideal conditions for growing vines and olives. Albania's largely agricultural economy has been adversely affected by the recent dismantling of state farms.

Land use and agricultural distribution

- pigs
- sheep
- cereals
- fruit
- olives
- sugar beet
- timber
- tobacco
- vineyards
- capital cities
- major towns
- pasture
- cropland
- forest
- mountain region

The urban/rural population divide

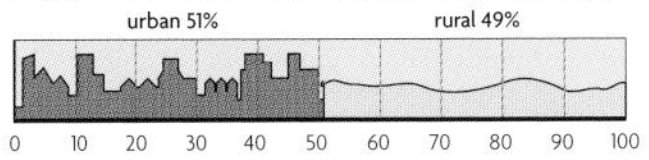

Population density	Total land area
240 people per sq mile (93 people per sq km)	95,038 sq miles (246,278 sq km)

◀ ***Sweet red peppers*** *are dried in the sun, ready to make paprika. Macedonia's economy is mainly agricultural and its fertile soils support a broad range of crops.*

▶ ***The Tara river*** *is one of Montenegro's major rivers. It flows into the Danube via the Drina and Sava rivers. Along its course the Tara has eroded spectacular gorges up to 3280 ft (1000 m) deep.*

Map key

Population

- 1 million to 5 million
- 500,000 to 1 million
- 100,000 to 500,000
- 50,000 to 100,000
- 10,000 to 50,000
- below 10,000

Elevation

- 2000m / 6562ft
- 1000m / 3281ft
- 500m / 1640ft
- 250m / 820ft
- 100m / 328ft
- sea level

Bulgaria & Greece

Including EUROPEAN TURKEY

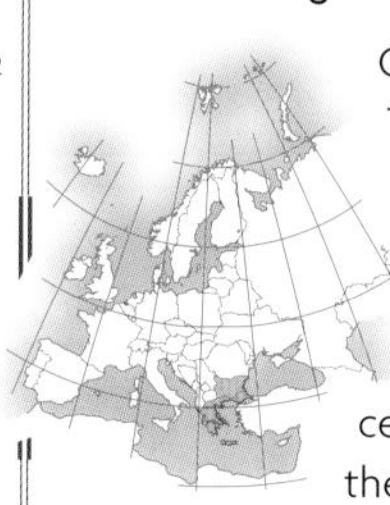

Greece is renowned as the original hearth of western civilization. The rugged terrain and numerous islands have profoundly affected its development, creating a strong agricultural and maritime tradition. In the past 50 years, this formerly rural society has rapidly urbanized, with one third of the population now living in the capital, Athens, and in the northern city of Salonica. Bulgaria, dominated for centuries by the Ottoman Turks, became part of the eastern bloc after the Second World War, only slowly emerging from Soviet influence in 1989. Moves towards democracy led to some instability in Bulgaria and Greece, now outweighed by the challenge of integration with the European Union.

Transport and industry

Soviet investment introduced heavy industry into Bulgaria, and the processing of agricultural produce, such as tobacco, is important throughout the country. Both countries have substantial shipyards and Greece has one of the world's largest merchant fleets. Many small craft workshops, producing textiles and processed foods, are clustered around Greek cities. The service and construction sectors have profited from the successful tourist industry.

112

Major industry and infrastructure

- chemicals
- engineering
- food processing
- shipbuilding
- textiles
- tourism
- capital cities
- major towns
- international airports
- major roads
- major industrial areas

The transport network

- 103,930 miles (167,630 km)
- 345 miles (557 km)
- 4346 miles (6995 km)
- 294 miles (474 km)

Bulgaria's railways require investment to revive an outdated infrastructure. In Greece, despite a developing road network, ferry-boats remain the most effective form of transport in many areas.

▲ *A towering pinnacle* at Metéora in central Greece is home to the monastery of Roussanou. *The 24 rock towers which dominate the plain of* Thessaly (Thessalía) *are remnants of an old plateau. Long-term weathering along fissures in the rock has worn away the rest of the plateau.*

The landscape

Bulgaria's Balkan mountains divide the Danubian Plain (*Dunavska Ravnina*) and Maritsa Basin, meeting the Black Sea in the east along sandy beaches. The steep Rhodope Mountains form a natural barrier with Greece, while the younger Pindus form a rugged central spine which descends into the Aegean Sea to give a vast archipelago of over 2000 islands, the largest of which is Crete.

Mount Olympus is the mythical home of the Greek Gods and, at 9570 ft (2917 m), is the highest mountain in Greece.

▲ **Mount Olympus** *is a composite of rocks formed by two major tectonic events. First the older metamorphic rocks were thrust over the limestones, then two million years ago regional warping and subsequent erosion, re-exposed the limestone.*

The Peloponnese consist of several mountainous peninsulas, linked to the mainland by the Isthmus of Corinth. The Corinth Canal (*Dioryga Korinthou*), built in 1893, cuts through the isthmus, linking the Aegean and Ionian seas.

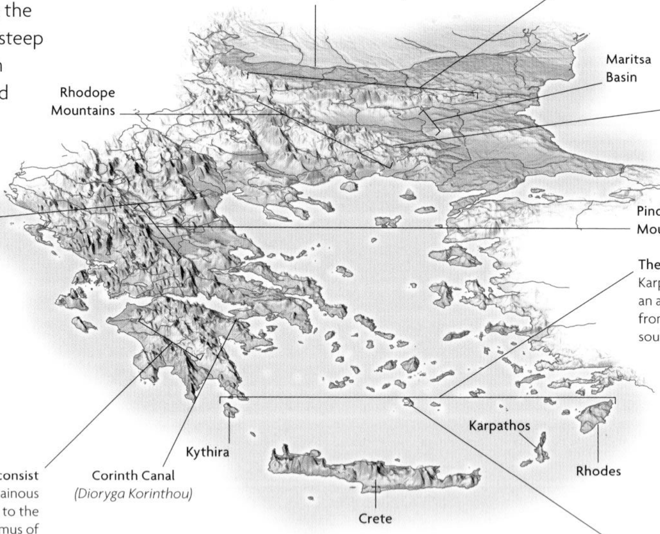

The Danube, Europe's second longest river, forms most of Bulgaria's northern border. The Danubian plain (*Dunavska Ravnina*), extending from the southern bank, is extremely fertile.

The islands of Crete, Kythira, Karpathos and Rhodes are part of an arc which bends southeastwards from the Peloponnese, forming the southern boundary of the Aegean.

▲ *The Arda river cuts through the Rhodope Mountains in rugged, rocky gorges.*

▲ *Layers of black volcanic ash still cover the island of Santorini. This volcano last erupted 3500 years ago, but still shows signs of volcanic activity.*

Scale 1:2,750,000

projection: Lambert Conformal Conic

112
116
112

Map key

Population

- above 5 million
- 1 million to 5 million
- 500,000 to 1 million
- 100,000 to 500,000
- 50,000 to 100,000
- 10,000 to 50,000
- below 10,000

Elevation

- 3000m / 9843ft
- 2000m / 6562ft
- 1000m / 3281ft
- 500m / 1640ft
- 250m / 820ft
- 100m / 328ft
- sea level

▲ **The dry scrubland** *seen here at Vasilikí in Crete, is characteristic of much of southern Greece, and is caused by centuries of forest clearance and soil degradation. Landslides are also common.*

Using the land and sea

The fertile plains of Bulgaria support cattle, fruit, vegetables, tobacco and cereal cultivation, while also providing traditional industries with grapes for wine, sunflowers for oil, and roses for perfume. Over half of Greece is barren upland. Citrus fruit, olives and tobacco are widely exported, yet much of rural life is still characterized by subsistence cropping and goat herding.

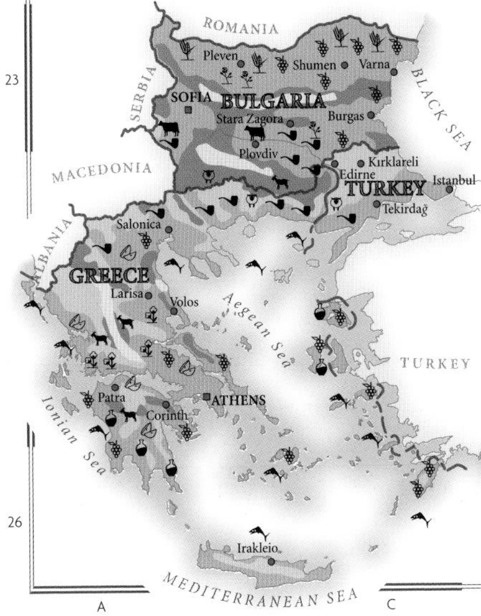

Land use and agricultural distribution

- cattle
- fishing
- goats
- sheep
- cereals
- citrus fruits
- cotton
- olives
- roses
- tobacco
- vineyards
- capital cities
- major towns
- pasture
- cropland
- forest
- mountain region

The urban/rural population divide

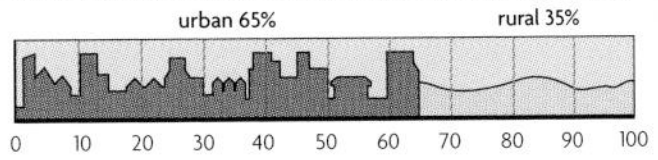

Population density	Total land area
245 people per sq mile (95 people per sq km)	102,353 sq miles (265,164 sq km)

▶ **These terraces,** *built on the hillside at Naxos, an island of the Cyclades group, help to guard against soil erosion.*

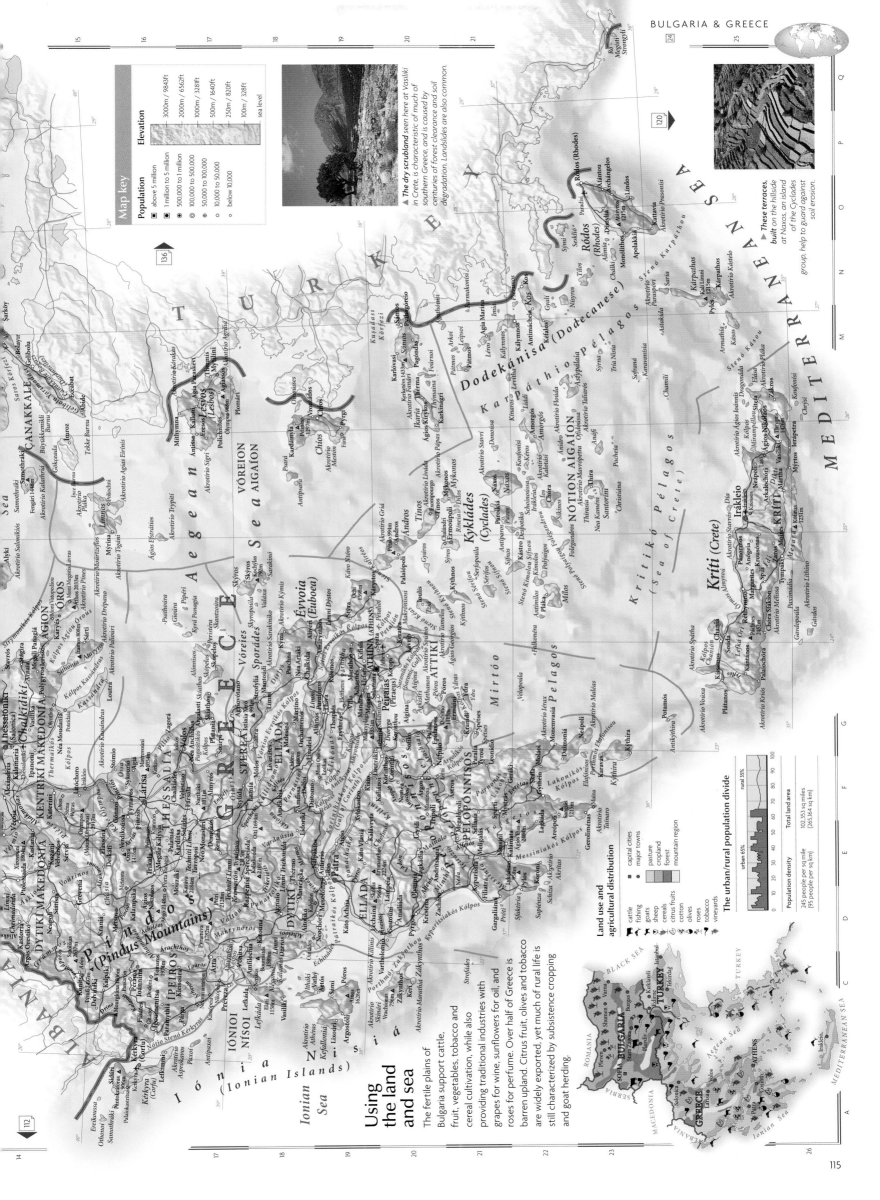

Romania, Moldova & Ukraine

The industrial, social and cultural make-up of Romania and the former Soviet states of Moldova and Ukraine still bear the imprint of their communist past. As part of the USSR, Ukraine was a leading agricultural, industrial and energy producer. These industries, like those in Moldova and Romania, are now being reoriented more firmly towards western markets. As a result of shifting borders, and Soviet policy actively encouraging Russian immigration into other Soviet states like Ukraine and Moldova, all three countries now contain large numbers of foreign nationals. In 2014, the Russian Federation drew international condemnation by annexing the Ukrainian territory of Crimea.

Using the land

The fertile black soils of Ukraine, often called 'the breadbasket of Europe', have enabled the cultivation of a variety of cereals and vegetables, which are widely exported. Romania and Moldova also grow cereals, sunflowers and vegetables, and are noted for the quality of their wines.

◀ *The fertile lands* *and tolerant climate of Moldova are ideally suited to growing grapes for wine.*

Land use and agricultural distribution

- cattle
- pigs
- poultry
- sheep
- cereals
- cotton
- sugar beet
- sunflowers
- vineyards
- capital cities
- major towns
- pasture
- cropland
- forest
- wetland

The urban/rural population divide

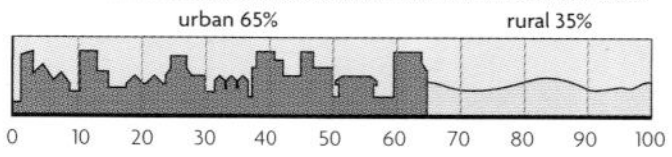

Population density	Total land area
222 people per sq mile (86 people per sq km)	334,947 sq miles (867,740 sq km)

◀ *Glacial lakes are found throughout the Transylvanian Alps (Carpatii Meridionali), although the mountains no longer have any permanent snow cover.*

Transport and industry

Heavy industry using local raw materials characterizes much of this region. The industrial heartland of Ukraine, specializing in metal and machine-building industries, is based around its vast mineral reserves in the Donbass region. In Moldova, food processing draws on produce from its agricultural sector. Romanian industry relies both on local raw materials and imported iron, steel and oil.

Major industry and infrastructure

- car manufacture
- chemicals
- coal
- engineering
- food processing
- mining
- oil & gas
- textiles
- tourism
- capital cities
- major towns
- international airports
- major roads
- major industrial areas

The transport network

170,707 miles (274,757 km)	1170 miles (1883 km)
21,474 miles (34,563 km)	4130 miles (6647 km)

Increased industrialization has necessitated the upgrading of road and rail networks in all three countries. Modernization has tended to focus only on major cities and industrial areas.

▶ *During the 1960s and 1970s, many industries, like this carbon factory, developed using the mineral resources on the flanks of the Transylvanian Alps (Carpatii Meridionali).*

Scale 1:3,500,000
projection: Lambert Conformal Conic
Map key
Population
1 million to 5 million
500,000 to 1 million
100,000 to 500,000
50,000 to 100,000
10,000 to 50,000
below 10,000
Elevation
2000m / 6562ft
1000m / 3281ft
500m / 1640ft
250m / 820ft
100m / 328ft
sea level
RUSSIAN FEDERATION
UKRAINE
BLACK SEA
Sea of Azov
Gulf of Taganrog
Black Sea Lowland
Dnieper Lowland
CHERNIHIVS'KA OBLAST'
SUMS'KA OBLAST'
KYYIVS'KA OBLAST'
POLTAVS'KA OBLAST'
KHARKIVS'KA OBLAST'
LUHANS'KA OBLAST'
CHERKAS'KA OBLAST'
DNIPROPETROVS'KA OBLAST'
DONETS'KA OBLAST'
KIROVOHRADS'KA OBLAST'
MYKOLAYIVS'KA OBLAST'
ODES'KA OBLAST'
KHERSONS'KA OBLAST'
ZAPORIZ'KA OBLAST'
AVTONOMNA RESPUBLIKA KRYM
(annexed by Russian Federation, 2014)
Kryms'kyy Pivostriv (Crimea)
Kyyiv (Kiev)
Chernihiv
Sumy
Kharkiv
Poltava
Luhans'k
Donets'k
Dnipropetrovs'k
Zaporizhzhya
Kryvyy Rih
Kirovohrad
Cherkasy
Kremenchuk
Mykolayiv
Kherson
Odesa
Chişinău
Melitopol'
Mariupol'
Berdyans'k
Simferopol'
Sevastopol'
Yevpatoriya
Yalta
Kerch
Zhytomyr
Bila Tserkva
Nikopol'
Tulcea
Constanţa
Hora Roman-Kosh 1545m
118
120
122
126

▲ ***The Swallow's Nest*** *castle at Yalta is one of many tourist resorts on the Crimean* (Krym) *coast, dubbed the 'Russian Riviera'.*

Old glaciated valley

Water has eroded a new post-glacial valley

▲ ***Balkas are common*** *throughout Ukraine. They are large U-shaped valleys, formed during the last Ice Age, which contain narrower, deep valleys. These were incised by a sudden flow of water, following an ice melt.*

The landscape

Vast flat lowlands and gently rolling hills cover most of southeastern Europe. In the southwest, the Carpathian Mountains form a gentle arc. To the south of the Carpathian Mountains lies the Danube Plain, across which the Danube river flows to the Black Sea. To the north and east, the hills of Moldova level out into low plains, running east to the steppes of Ukraine.

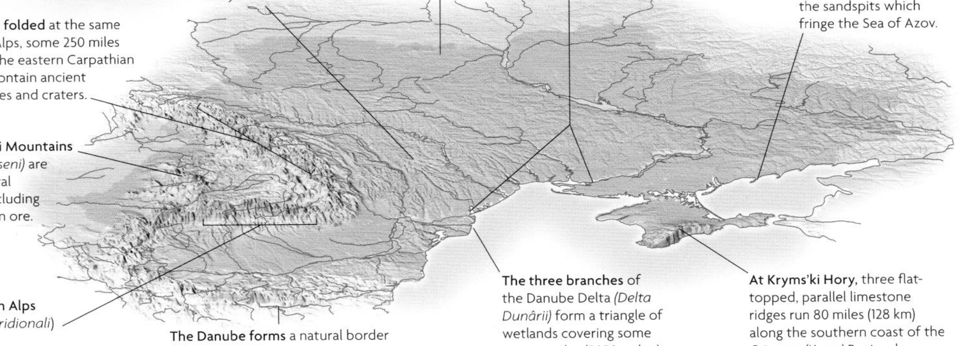

Uplifted and folded at the same time as the Alps, some 250 miles (400 km) of the eastern Carpathian Mountains contain ancient volcanic cones and craters.
The Codrii Hills dominate the landscape of central Moldova; they are intersected by deep, flat valleys and ravines.
Steppe landscape covers two-thirds of Ukraine. These flat, treeless grasslands extend from central Europe to central Asia.
Most of the major rivers in southeastern Europe, like the Danube, the Dniester and Dnieper flow south and east to the Black Sea.
Anti-clockwise currents have created the sandspits which fringe the Sea of Azov.
The Apuseni Mountains *(Muntii Apuseni)* are rich in mineral deposits, including gold and iron ore.
Transylvanian Alps *(Carpatii Meridionali)*
The Danube forms a natural border between Romania and Bulgaria.
The three branches of the Danube Delta *(Delta Dunării)* form a triangle of wetlands covering some 1950 sq miles (5050 sq km).
At Kryms'ki Hory, three flat-topped, parallel limestone ridges run 80 miles (128 km) along the southern coast of the Crimean *(Krym)* Peninsula.

► ***Divided into crystalline*** *massifs, the southern arm of the Carpathian Mountains, the Transylvanian Alps* (Carpatii Meridionali), *extend 170 miles (274 km) across southwestern Romania.*

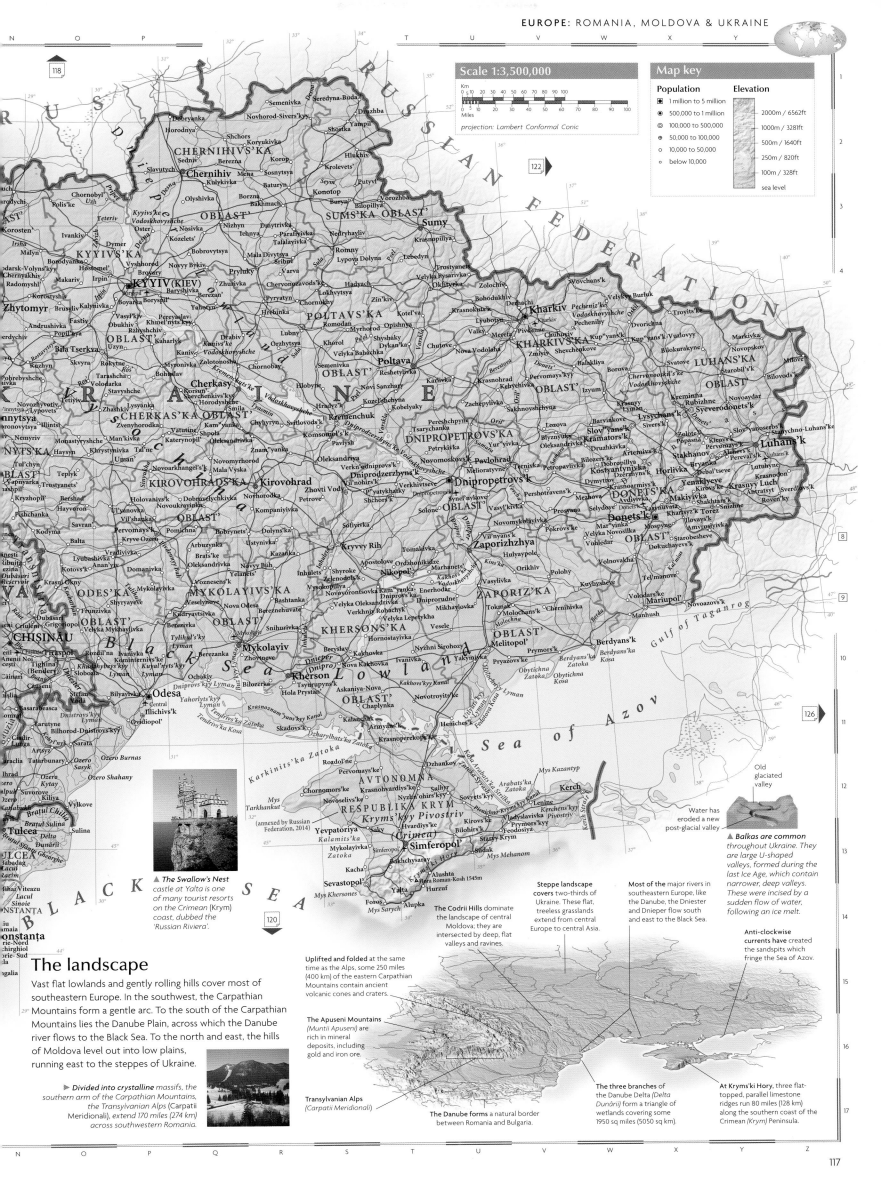

The Baltic states & Belarus

BELARUS, ESTONIA, LATVIA, LITHUANIA, Kaliningrad

Occupying Europe's main corridor to Russia, the four distinct cultures of Estonia, Latvia, Lithuania and Belarus share a history of struggle for nationhood against the interests of more powerful neighbours. As the first republics to declare their independence from the Soviet Union in 1990–91, the Baltic states of Estonia, Latvia and Lithuania sought an economic role in the EU, while reaffirming their European cultural roots through the church and a strong musical tradition. Meanwhile, Belarus has shown economic and political allegiance to Russia by joining the Commonwealth of Independent States.

▲ *The seaport of* Riga *is Latvia's capital and the centre of economic and cultural life. With a 32% Russian minority in Latvia, language and the right to national citizenship are key issues.*

122

122

Using the land

Across the four nations cattle and pig farming are widespread, together with diverse arable crops, including flax for making linen, potatoes used to produce vodka, cereals and other vegetables. Almost a third of the land is forested; demand for timber has increased the importance of forest management.

Land use and agricultural distribution

- cattle
- pigs
- cereals
- flax
- potatoes
- timber
- capital cities
- major towns
- pasture
- cropland
- forest
- wetland

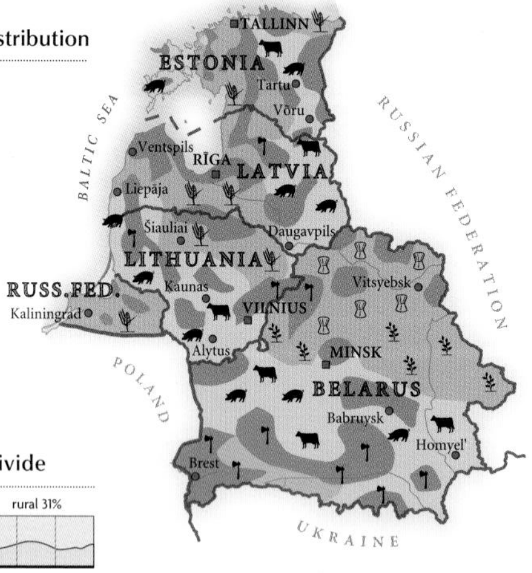

The urban/rural population divide

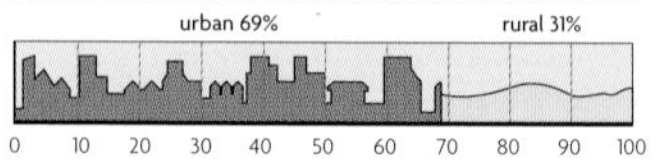

Population density	Total land area
122 people per sq mile (47 people per sq km)	145,006 sq miles (375,656 sq km)

▶ *A pine forest in northern Belarus. Conifers in the north give way to hardwood forest further south. Timber mills are supplied with logs floated along the country's many navigable waterways.*

▲ *The Western Dvina river provides hydro-electric power and, during the summer months, access to the Baltic Sea. The lower course of the river freezes from December to April.*

Map key

Population

- 1 million to 5 million
- 500,000 to 1 million
- 100,000 to 500,000
- 50,000 to 100,000
- 10,000 to 50,000
- below 10,000

Elevation

- 250m / 820ft
- 100m / 328ft
- sea level

BELARUS
RUSSIAN FEDERATION
POLAND
UKRAINE
MINSK
Vilnius
Hrodna/Grodno
Brest
Homyel'/Gomel'
Mahilyow/Mogilev
Babruysk/Bobruysk
Baranavichy/Baranovichi
HRODZYENSKAYA VOBLASTS'
GRODNENSKAYA OBLAST'
MINSKAYA VOBLASTS'
BRESTSKAYA VOBLASTS'
HOMYEL'SKAYA VOBLASTS'
GOMEL'SKAYA OBLAST'
MAHILYOWSKAYA VOBLASTS'
MOGILEVSKAYA OBLAST'
Pripet Marshes
Dnieper
Pripet
110
116
15

Scale 1:2,750,000

projection: Lambert Conformal Conic

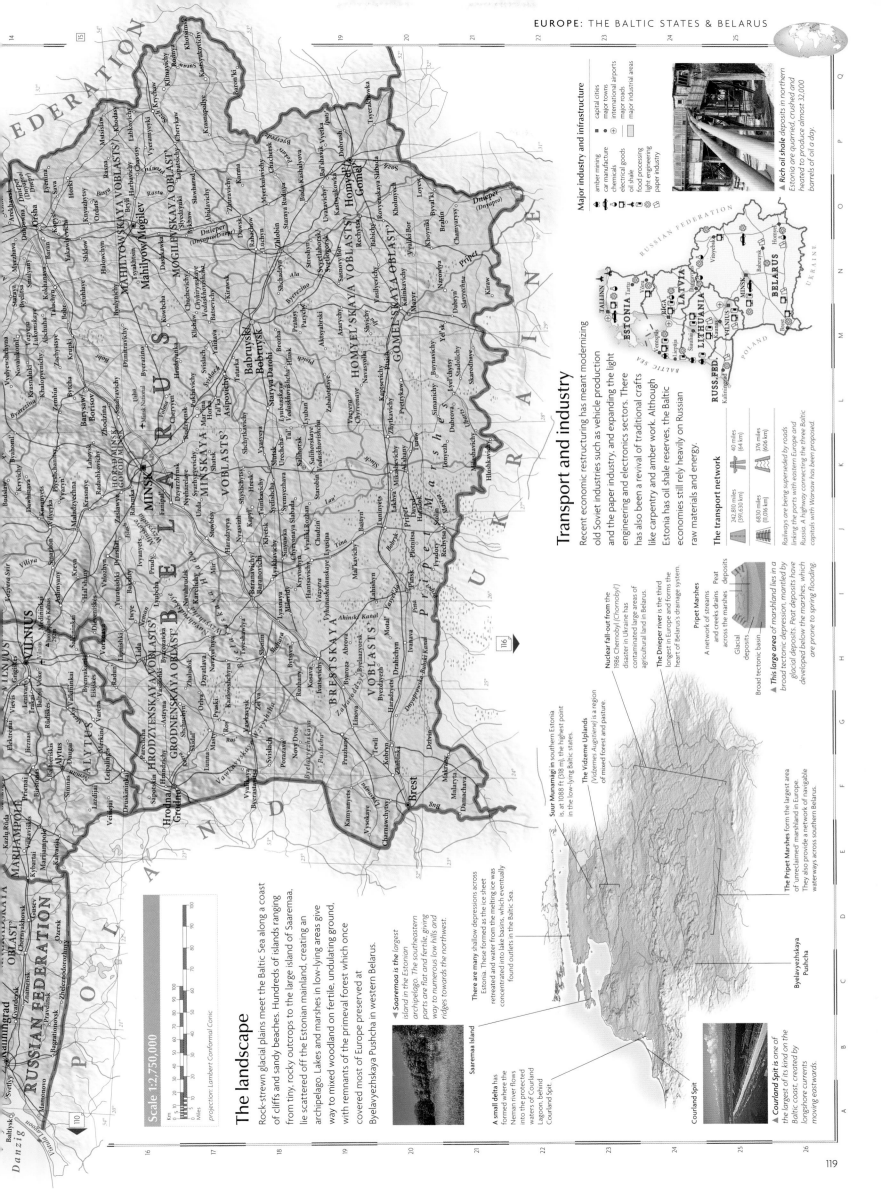

The landscape

Rock-strewn glacial plains meet the Baltic Sea along a coast of cliffs and sandy beaches. Hundreds of islands ranging from tiny, rocky outcrops to the large island of Saaremaa, lie scattered off the Estonian mainland, creating an archipelago. Lakes and marshes in low-lying areas give way to mixed woodland on fertile, undulating ground, with remnants of the primeval forest which once covered most of Europe preserved at Byelavyezhskaya Pushcha in western Belarus.

◀ *Saaremaa is the largest island in the Estonian archipelago. The southeastern parts are flat and fertile, giving way to numerous low hills and ridges towards the northwest.*

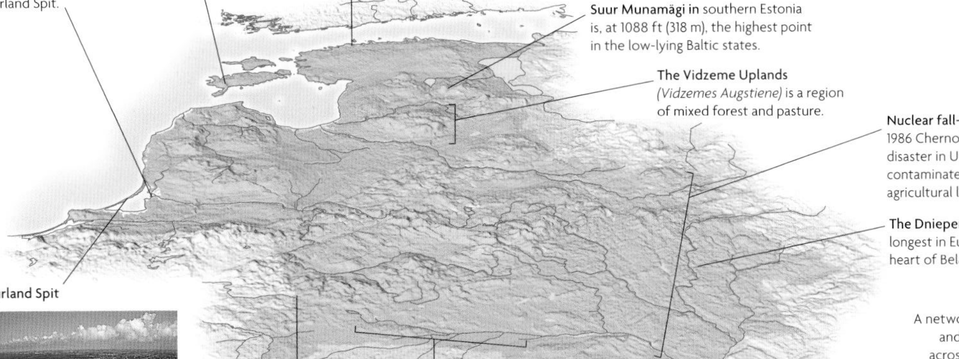
Saaremaa Island
There are many shallow depressions across Estonia. These formed as the ice sheet retreated and water from the melting ice was concentrated into lake basins, which eventually found outlets in the Baltic Sea.
A small delta has formed where the Neman river flows into the protected waters of Courland Lagoon, behind Courland Spit.
Suur Munamägi in southern Estonia is, at 1088 ft (318 m), the highest point in the low-lying Baltic states.
The Vidzeme Uplands (Vidzemes Augstiene) is a region of mixed forest and pasture.
Nuclear fall-out from the 1986 Chernobyl (Chornobyl') disaster in Ukraine has contaminated large areas of agricultural land in Belarus.
The Dnieper river is the third longest in Europe and forms the heart of Belarus's drainage system.
Courland Spit
Byelavyezhskaya Pushcha
The Pripet Marshes form the largest area of 'unreclaimed' marshland in Europe. They also provide a network of navigable waterways across southern Belarus.

▲ *Courland Spit is one of the largest of its kind on the Baltic coast, created by longshore currents moving eastwards.*

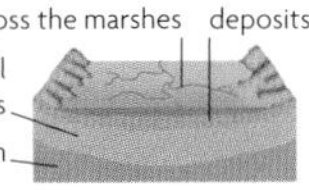
Pripet Marshes
A network of streams and creeks drains across the marshes
Peat deposits
Glacial deposits
Broad tectonic basin

▲ *This large area of marshland lies in a broad tectonic depression, mantled by glacial deposits. Peat deposits have developed below the marshes, which are prone to spring flooding.*

Transport and industry

Recent economic restructuring has meant modernizing old Soviet industries such as vehicle production and the paper industry, and expanding the light engineering and electronics sectors. There has also been a revival of traditional crafts like carpentry and amber work. Although Estonia has oil shale reserves, the Baltic economies still rely heavily on Russian raw materials and energy.

The transport network

242,810 miles (391,630 km)	40 miles (64 km)
6830 miles (11,016 km)	376 miles (606 km)

Railways are being superseded by roads linking the ports with eastern Europe and Russia. A highway connecting the three Baltic capitals with Warsaw has been proposed.

Major industry and infrastructure

amber mining
car manufacture
chemicals
electrical goods
oil shale
food processing
light engineering
paper industry
capital cities
major towns
international airports
major roads
major industrial areas
TALLINN
ESTONIA
Tartu
Võru
Ventspils
Liepāja
RIGA
LATVIA
Šiauliai
Daugavpils
LITHUANIA
Kaunas
VILNIUS
RUSS.FED.
Kaliningrad
Vitsyebsk
MINSK
Babruysk
BELARUS
Homyel'
Brest
BALTIC SEA
RUSSIAN FEDERATION
POLAND
UKRAINE

▲ *Rich oil shale deposits in northern Estonia are quarried, crushed and heated to produce almost 32,000 barrels of oil a day.*

The Mediterranean

The Mediterranean Sea stretches over 2500 miles (4000 km) east to west, separating Europe from Africa. At its most westerly point it is connected to the Atlantic Ocean through the Strait of Gibraltar. In the east, the Suez canal, opened in 1869, gives passage to the Indian Ocean. In the northeast, linked by the Sea of Marmara, lies the Black Sea. The Mediterranean is bordered by almost 30 states and territories, and more than 100 million people live on its shores and islands. Throughout history, the Mediterranean has been a focal area for many great empires and civilizations, reflected in the variety of cultures found on its shores. Since the 1960s, development along the southern coast of Europe has expanded rapidly to accommodate increasing numbers of tourists and to enable the exploitation of oil and gas reserves. This has resulted in rising levels of pollution, threatening the future of the sea.

▲ *Monaco is just one of the luxurious resorts scattered along the Riviera, which stretches along the coast from Cannes in France to La Spezia in Italy. The region's mild winters and hot summers have attracted wealthy tourists since the early 19th century.*

108
102
64

74

The landscape

The Mediterranean Sea is almost totally landlocked, joined to the Atlantic Ocean through the Strait of Gibraltar, which is only 8 miles (13 km) wide. Lying on an active plate margin, sea floor movements have formed a variety of basins, troughs and ridges. A submarine ridge running from Tunisia to the island of Sicily divides the Mediterranean into two distinct basins. The western basin is characterized by broad, smooth abyssal (or ocean) plains. In contrast, the eastern basin is dominated by a large ridge system, running east to west.

The narrow Strait of Gibraltar inhibits water exchange between the Mediterranean Sea and the Atlantic Ocean, producing a high degree of salinity and a low tidal range within the Mediterranean. The lack of tides has encouraged the build-up of pollutants in many semi-enclosed bays.

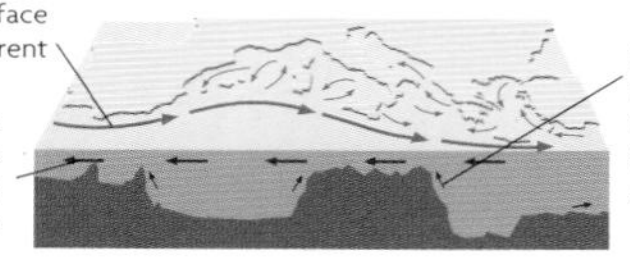

▲ *Because the Mediterranean is almost enclosed by land, its circulation is quite different to the oceans. There is one major current which flows in from the Atlantic and moves east. Currents flowing back to the Atlantic are denser and flow below the main current.*

74

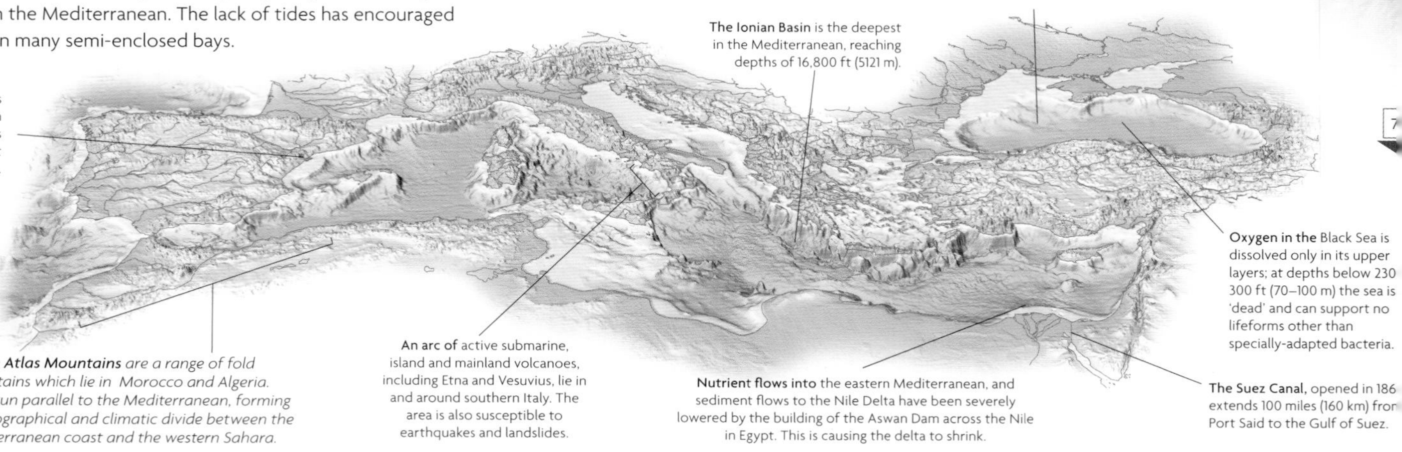

◀ *The Atlas Mountains are a range of fold mountains which lie in Morocco and Algeria. They run parallel to the Mediterranean, forming a topographical and climatic divide between the Mediterranean coast and the western Sahara.*

In 1974 Turkey occupied the northern part of Cyprus while Greek Cypriots remained in control of the south. Cyprus was effectively partitioned and a UN buffer zone currently divides the two areas. In 1983 the north of the island proclaimed itself the Turkish Republic of North Cyprus. It was only recognized by Turkey.

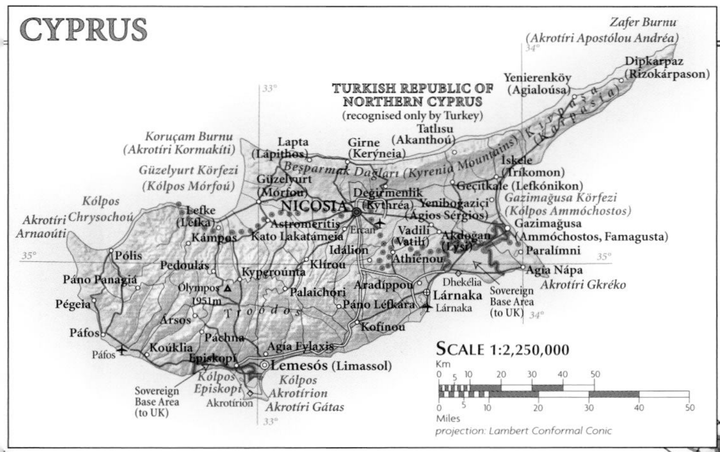

▶ The city of Venice is built on an archipelago of islands and mud-flats in the middle of a lagoon at the head of the Adriatic Sea. The city's numerous canals follow water routes between the original 118 islands.

◀ Cyprus is the third largest Mediterranean island after Sardinia and Sicily. The island is mountainous; containing two main ranges, the Troodos and the Kyrenia mountains .

▲ Beirut is Lebanon's largest city. In the 1960s and 70s it was the chief financial, commercial and transport centre for the Arab states. Devastated by civil war between 1975 and 1990, the city has since been largely rebuilt and has now become a popular tourist destination.

Map key

Population

- above 5 million
- 1 million to 5 million
- 500,000 to 1 million
- 100,000 to 500,000
- 50,000 to 100,000
- 10,000 to 50,000
- below 10,000

Elevation

- 4000m / 13,124ft
- 3000m / 9843ft
- 2000m / 6562ft
- 1000m / 3281ft
- 500m / 1640ft
- 250m / 820ft
- 100m / 328ft
- sea level

Sea depth

- sea level
- 250m / 820ft
- 500m / 1640ft
- 1000m / 3281ft
- 2000m / 6562ft
- 3000m / 9843ft

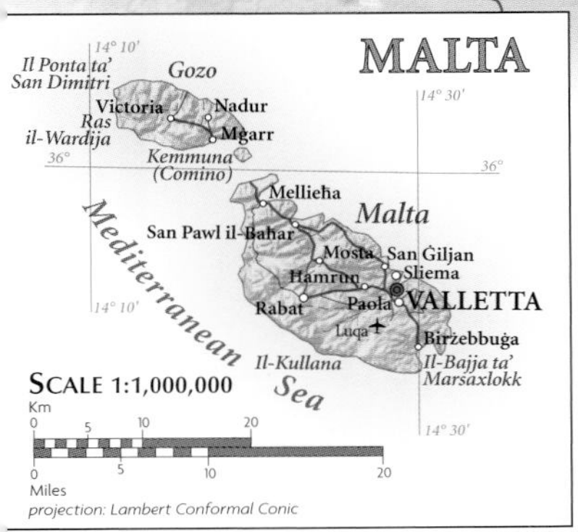

◀ Commercial fisheries are found throughout the Mediterranean. Operations have traditionally been small-scale. As elsewhere, high demand has caused a decline in fish stocks.

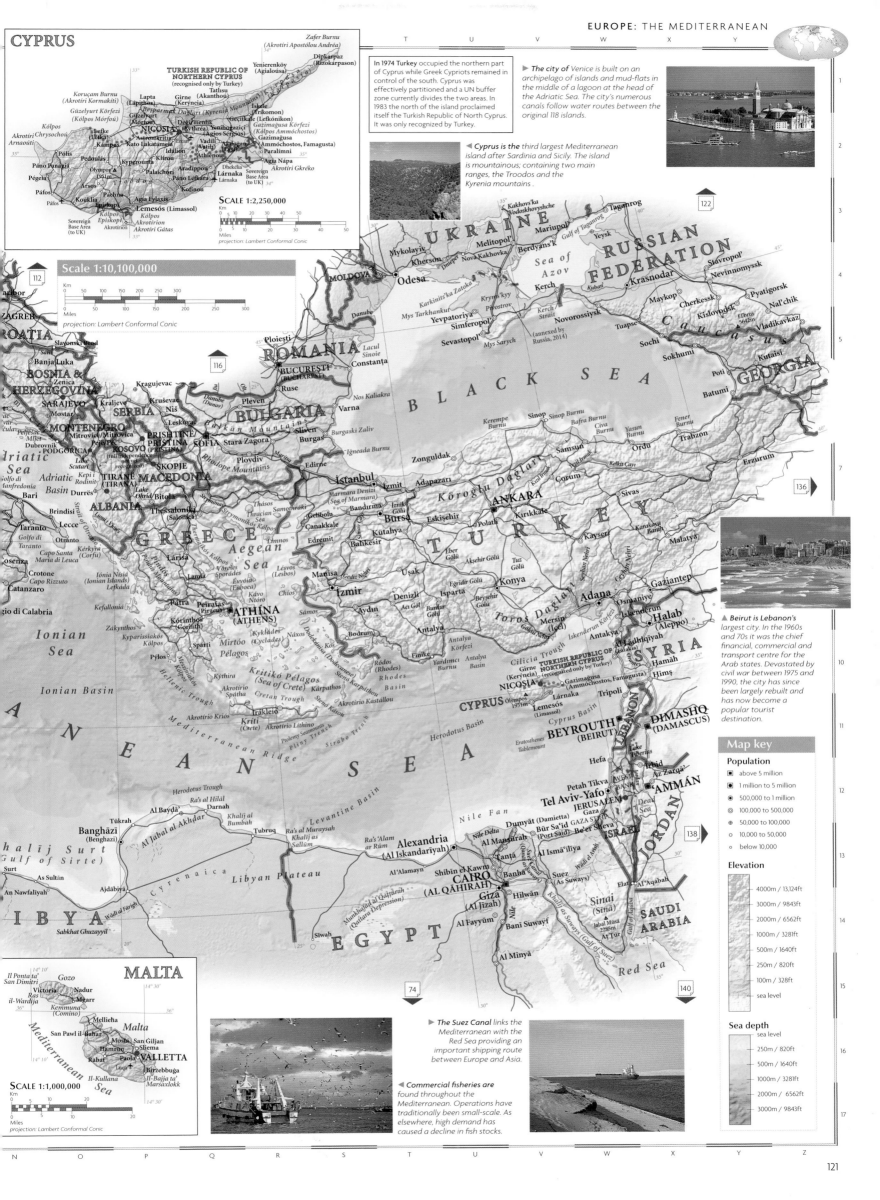

▶ The Suez Canal links the Mediterranean with the Red Sea providing an important shipping route between Europe and Asia.

The Russian Federation

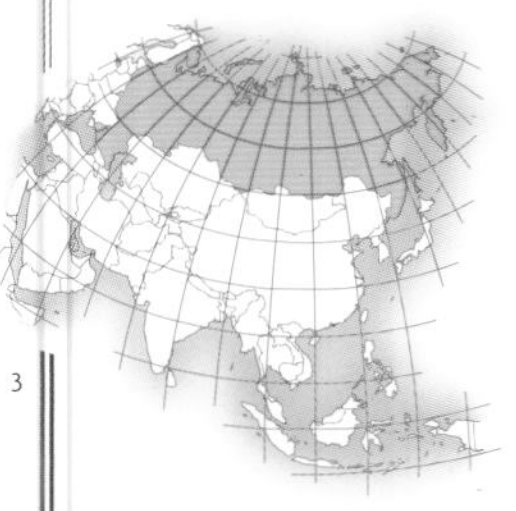

The Cold War era of global relations was concluded in 1991 with the formal dissolution of the Soviet Union. The Russian Federation declared its separate sovereignty from the foundering communist empire following independence declarations from a number of former Soviet republics. As the leading member of the Commonwealth of Independent States, the Russian Federation has a central role in the development of post-Soviet Eurasia. Crossing 11 time zones, the Russian Federation is almost twice the size of the USA, and with more than 150 ethnic minorities and 21 autonomous republics, regionalist dissent within its own territory remains a danger.

▶ *Summer beds of moss and lichen scatter a 90% surface cover of ice across the islands of Franz Josef Land (Zemlya Frantsa-Iosifa), the northernmost land in the eastern hemisphere.*

THE RUSSIAN FEDERATION: ADMINISTRATIVE REGIONS

▶ *The Khatanga river meanders slowly across the Poluostrov Taymyr, a low-lying tundra landscape which floods in the spring thaw, until the water can escape to the sea.*

The landscape

The Ural Mountains *(Ural'skiye Gory)* divide the fertile North European Plain from the West Siberian Plain *(Zapadno-Sibirskaya Ravnina)*, the world's largest area of flat ground, crossed by giant rivers flowing north to the Kara Sea *(Karskoye More)*. The land rises to the Central Siberian Plateau *(Srednesibirskoye Ploskogor'ye)* and becomes more mountainous to the southeast. These immense topographic regions intersect with latitudinal vegetation bands. The tundra of the extreme north gives way to a vast area of coniferous woodland, which is known as *taiga*, larger than the Amazon rainforest. This belt turns to mixed forest and then steppe grasslands towards the south.

The North European Plain is marked by huge moraine ridges left by the Scandinavian Ice Sheet and by longintermoraine drainage channels, known as *Urstromtäler*.

Kara Sea *(Karskoye More)*

Poluostrov Taymyr

The mountains of Verkhoyanskiy Khrebet were formed by movement between the Eurasian and North American plates, during the same period of folding that created the Urals.

Yukagirskoye Ploskogor'ye is a rolling plain with isolated drumlins, dome-like features resulting from glacial deposition.

The Ural Mountains *(Ural'skiye Gory)* extend 1550 miles (2500 km). They were formed over 280 million years ago, folded as the East European and Siberian plates moved closer together.

The Yenisey is one of the world's longest rivers, and also among the most languid, dropping only 500 ft (152 m) over 1200 miles (2000 km).

▶ ***Lake Baikal*** *(Ozero Baykal), occupies a rift valley and is the world's deepest lake, over 1 mile (1.6 km) in depth. It is fed by over 300 rivers and drained by just one, the Angara.*

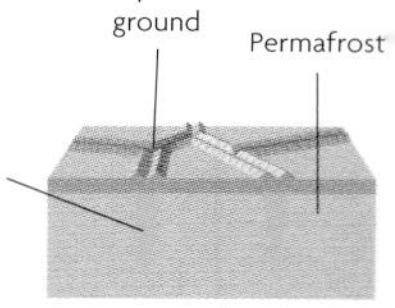

▲ ***Patterned ground*** *is a permafrost feature found extensively across northern Russia. Seasonal contraction of the permafrost creates polygonal cracks, which are filled by ice wedges.*

Transport and industry

Raw materials, particularly fossil fuels, ores and precious metals are abundant, yet often found at sites far from habitation. This inherent 'friction of distance' problem was met from the 1930s by Soviet commitment to heavy industry and the strategic location of plants east of the Urals. It has left a pattern of isolated and often vast industrial complexes, in remote areas from Vladivostok to Murmansk, in the far north and across European Russia, with lighter manufacturing concentrated in urban areas.

Major industry and infrastructure

- aerospace
- car manufacture
- chemicals
- engineering
- gas
- iron & steel
- mining
- oil
- textiles
- timber processing
- capital cities
- major towns
- international airports
- major roads
- major industrial areas

The transport network

218,683 miles (351,976 km)	None
53,147 miles (85,542 km)	59,583 miles (95,900 km)

The recent growth of trade with China and East Asia has put pressure on Siberia's inadequate road and rail network, prompting increased use of the Amur river for freight transport.

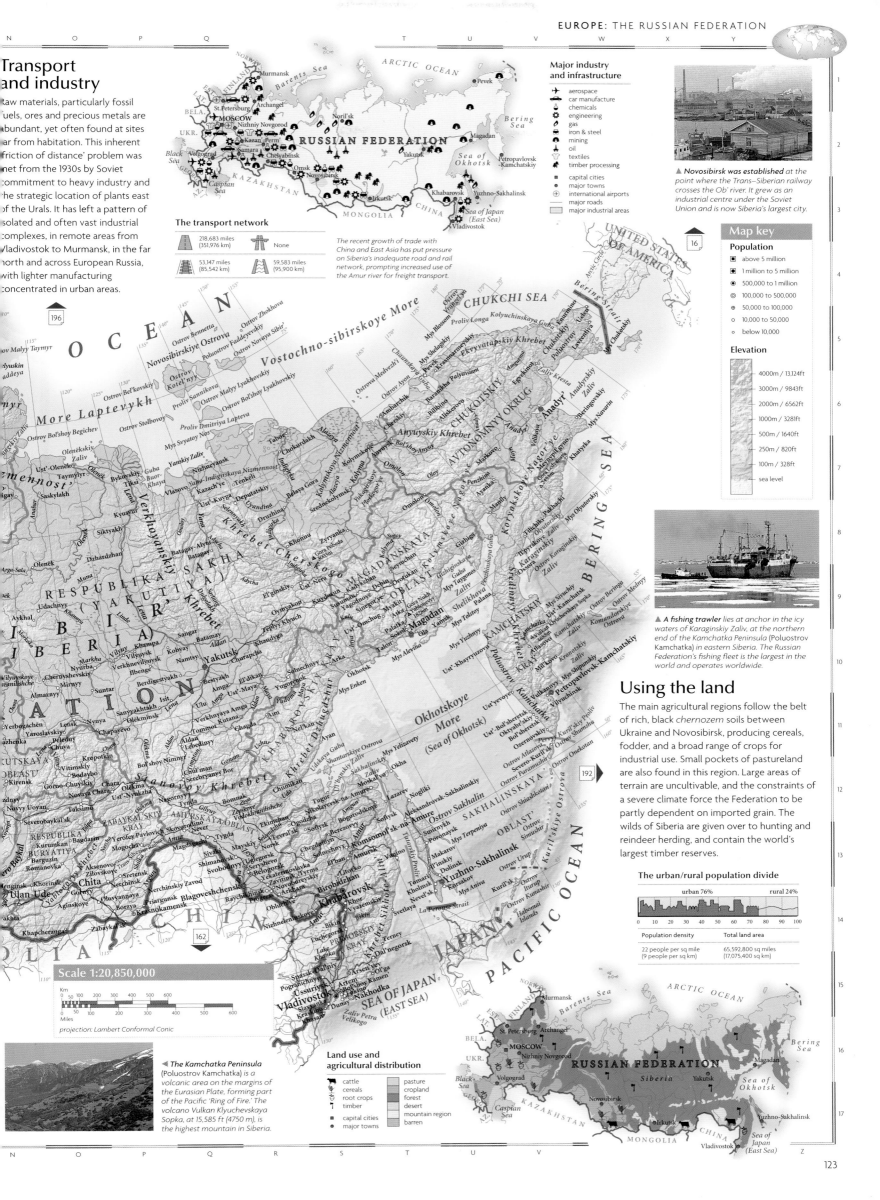

▲ ***Novosibirsk was established** at the point where the Trans–Siberian railway crosses the Ob' river. It grew as an industrial centre under the Soviet Union and is now Siberia's largest city.*

196

16

192

162

Map key

Population

- ▣ above 5 million
- ▣ 1 million to 5 million
- ◉ 500,000 to 1 million
- ◎ 100,000 to 500,000
- ⊕ 50,000 to 100,000
- ○ 10,000 to 50,000
- ○ below 10,000

Elevation

- 4000m / 13,124ft
- 3000m / 9843ft
- 2000m / 6562ft
- 1000m / 3281ft
- 500m / 1640ft
- 250m / 820ft
- 100m / 328ft
- sea level

▲ ***A fishing trawler** lies at anchor in the icy waters of Karaginskiy Zaliv, at the northern end of the Kamchatka Peninsula (Poluostrov Kamchatka) in eastern Siberia. The Russian Federation's fishing fleet is the largest in the world and operates worldwide.*

Using the land

The main agricultural regions follow the belt of rich, black *chernozem* soils between Ukraine and Novosibirsk, producing cereals, fodder, and a broad range of crops for industrial use. Small pockets of pastureland are also found in this region. Large areas of terrain are uncultivable, and the constraints of a severe climate force the Federation to be partly dependent on imported grain. The wilds of Siberia are given over to hunting and reindeer herding, and contain the world's largest timber reserves.

The urban/rural population divide

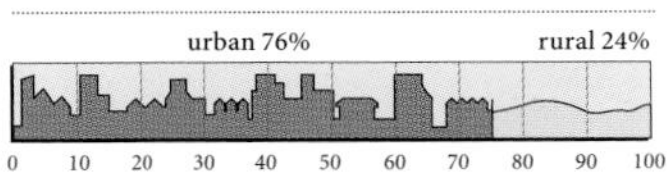

Population density	Total land area
22 people per sq mile (9 people per sq km)	65,592,800 sq miles (17,075,400 sq km)

Scale 1:20,850,000

Km: 0 50 100 200 300 400 500 600

Miles: 0 50 100 200 300 400 500 600

projection: Lambert Conformal Conic

◀ *The Kamchatka Peninsula (Poluostrov Kamchatka) is a volcanic area on the margins of the Eurasian Plate, forming part of the Pacific 'Ring of Fire.' The volcano Vulkan Klyuchevskaya Sopka, at 15,585 ft (4750 m), is the highest mountain in Siberia.*

Land use and agricultural distribution

- cattle
- cereals
- root crops
- timber
- capital cities
- major towns
- pasture
- cropland
- forest
- desert
- mountain region
- barren

Northern European Russia

Reaching into the Arctic Circle, this region of lakeland, forest and tundra is historically bound to Europe by St Petersburg, the old imperial capital of Tsarist Russia and home to a third of the region's population. Communist rule from Moscow left the north politically marginalized, contributing to the present problems of outmoded industry, poor infrastructure and serious environmental neglect. However, with borders embracing Finland, Norway, the Baltic and the northern sea route to the Atlantic, the region's success in foreign trade is now of prime importance to the Russian economy.

▶ *St Peter and Paul Fortress is the oldest building in St Petersburg, founded by Peter the Great in 1703 as a modern, European capital for Russia.*

The landscape

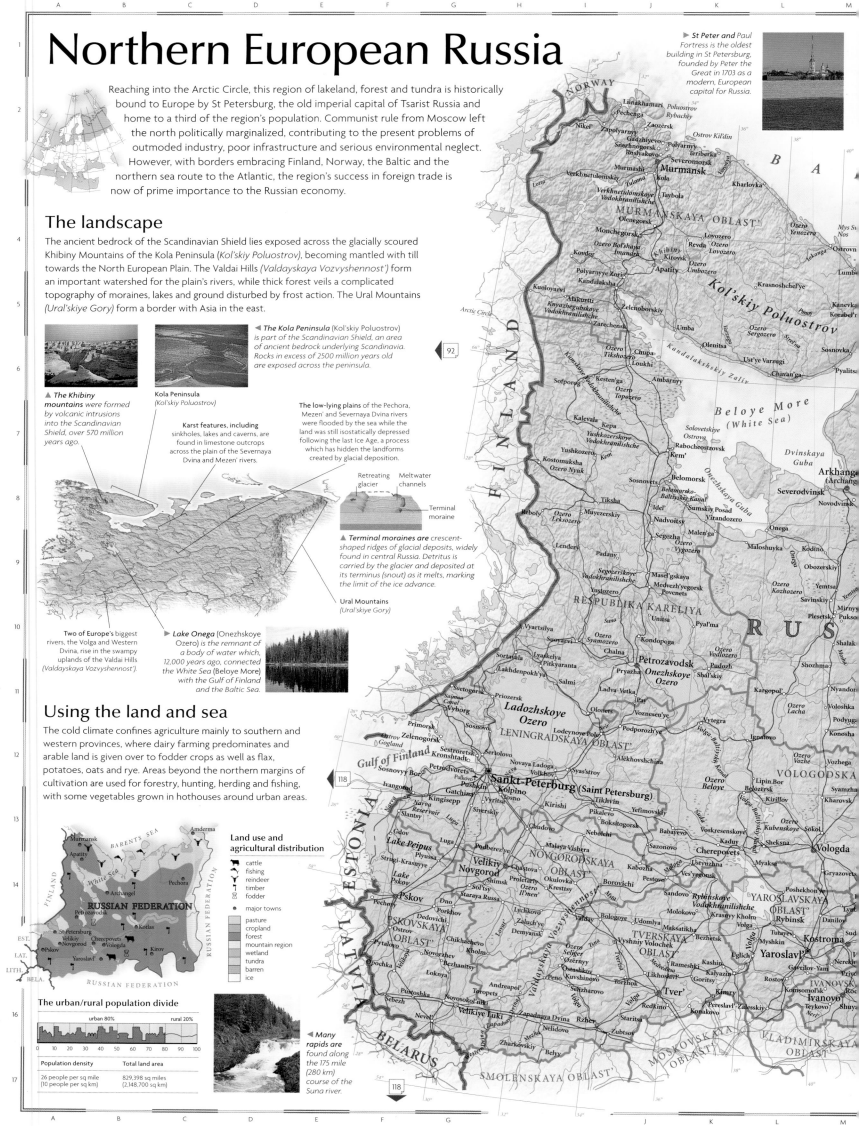

The ancient bedrock of the Scandinavian Shield lies exposed across the glacially scoured Khibiny Mountains of the Kola Peninsula *(Kol'skiy Poluostrov)*, becoming mantled with till towards the North European Plain. The Valdai Hills *(Valdayskaya Vozvyshennost')* form an important watershed for the plain's rivers, while thick forest veils a complicated topography of moraines, lakes and ground disturbed by frost action. The Ural Mountains *(Ural'skiye Gory)* form a border with Asia in the east.

▲ ***The Khibiny mountains** were formed by volcanic intrusions into the Scandinavian Shield, over 570 million years ago.*

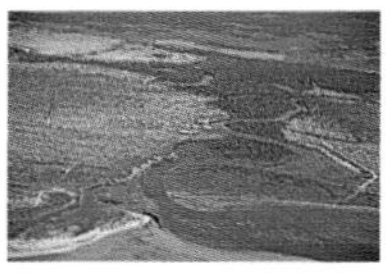

◀ ***The Kola Peninsula** (Kol'skiy Poluostrov) is part of the Scandinavian Shield, an area of ancient bedrock underlying Scandinavia. Rocks in excess of 2500 million years old are exposed across the peninsula.*

Kola Peninsula (*Kol'skiy Poluostrov*)

Karst features, including sinkholes, lakes and caverns, are found in limestone outcrops across the plain of the Severnaya Dvina and Mezen' rivers.

The low-lying plains of the Pechora, Mezen' and Severnaya Dvina rivers were flooded by the sea while the land was still isostatically depressed following the last Ice Age, a process which has hidden the landforms created by glacial deposition.

Ural Mountains (*Ural'skiye Gory*)

Two of Europe's biggest rivers, the Volga and Western Dvina, rise in the swampy uplands of the Valdai Hills (*Valdayskaya Vozvyshennost'*).

▶ ***Lake Onega** (Onezhskoye Ozero) is the remnant of a body of water which, 12,000 years ago, connected the White Sea (Beloye More) with the Gulf of Finland and the Baltic Sea.*

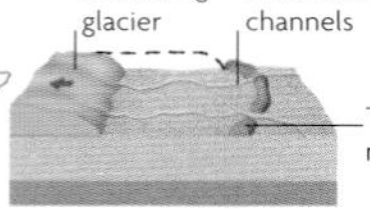

▲ ***Terminal moraines are** crescent-shaped ridges of glacial deposits, widely found in central Russia. Detritus is carried by the glacier and deposited at its terminus (snout) as it melts, marking the limit of the ice advance.*

Using the land and sea

The cold climate confines agriculture mainly to southern and western provinces, where dairy farming predominates and arable land is given over to fodder crops as well as flax, potatoes, oats and rye. Areas beyond the northern margins of cultivation are used for forestry, hunting, herding and fishing, with some vegetables grown in hothouses around urban areas.

Land use and agricultural distribution

- cattle
- fishing
- reindeer
- timber
- fodder
- major towns
- pasture
- cropland
- forest
- mountain region
- wetland
- tundra
- barren
- ice

The urban/rural population divide

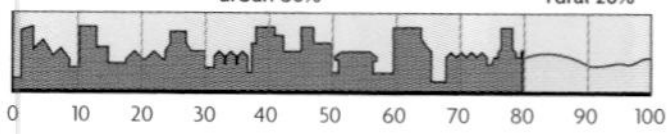

Population density	Total land area
26 people per sq mile (10 people per sq km)	829,398 sq miles (2,148,700 sq km)

◀ ***Many rapids are** found along the 175 mile (280 km) course of the Suna river.*

92
118
118

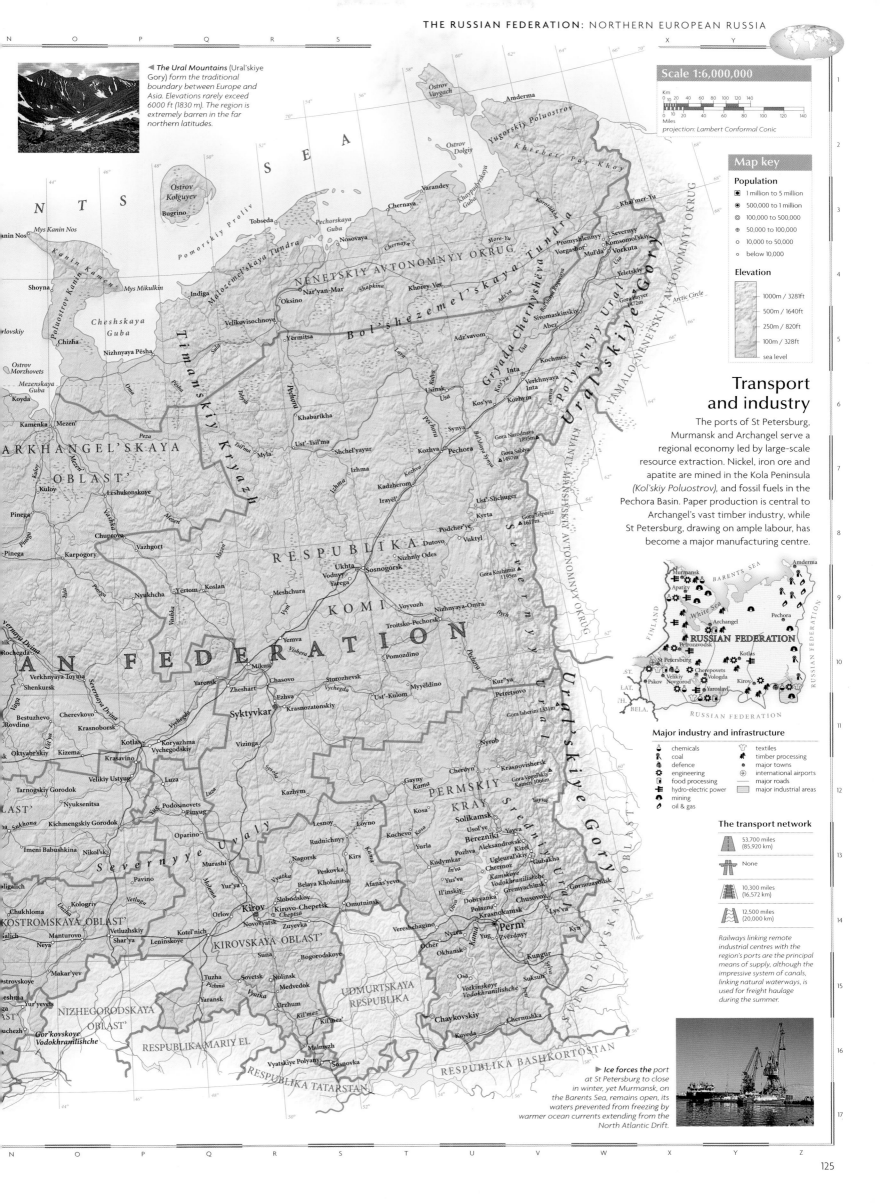

◀ *The Ural Mountains* (Ural'skiye Gory) *form the traditional boundary between Europe and Asia. Elevations rarely exceed 6000 ft (1830 m). The region is extremely barren in the far northern latitudes.*

Scale 1:6,000,000

Km 0 10 20 40 60 80 100 120 140
Miles 0 10 20 40 60 80 100 120 140

projection: Lambert Conformal Conic

Map key

Population

- 1 million to 5 million
- 500,000 to 1 million
- 100,000 to 500,000
- 50,000 to 100,000
- 10,000 to 50,000
- below 10,000

Elevation

- 1000m / 3281ft
- 500m / 1640ft
- 250m / 820ft
- 100m / 328ft
- sea level

Transport and industry

The ports of St Petersburg, Murmansk and Archangel serve a regional economy led by large-scale resource extraction. Nickel, iron ore and apatite are mined in the Kola Peninsula *(Kol'skiy Poluostrov)*, and fossil fuels in the Pechora Basin. Paper production is central to Archangel's vast timber industry, while St Petersburg, drawing on ample labour, has become a major manufacturing centre.

Major industry and infrastructure

- chemicals
- coal
- defence
- engineering
- food processing
- hydro-electric power
- mining
- oil & gas
- textiles
- timber processing
- major towns
- international airports
- major roads
- major industrial areas

The transport network

- 53,700 miles (85,920 km)
- None
- 10,300 miles (16,572 km)
- 12,500 miles (20,000 km)

Railways linking remote industrial centres with the region's ports are the principal means of supply, although the impressive system of canals, linking natural waterways, is used for freight haulage during the summer.

▶ ***Ice forces the** port at St Petersburg to close in winter, yet Murmansk, on the Barents Sea, remains open, its waters prevented from freezing by warmer ocean currents extending from the North Atlantic Drift.*

118

Kaliningrad has been a Russian enclave since 1945. The port is an important centre for the Russian Federation's Baltic fishing fleet.

St Basil's Cathedral, completed in 1561, stands in Moscow's Red Square next to the Kremlin; the original fortified stronghold of the city.

110

Southern European Russia

118

This region, divided from Asia by desert, seas and mountains, has exerted a powerful influence both east and west since the 13th century. Over 70 years of Communist rule produced a highly urbanized, industrial society dominated by Moscow, which was the capital of the Soviet Union until 1991. Almost two-thirds of the Russian Federation's population live in this core area, with a relatively high *per capita* share of its wealth. However, the rapid growth of a market economy has caused great social upheaval, with rising crime and political instability.

The landscape

Ancient folds in the deep sedimentary strata of the North European Plain have created a sequence of high and low regions. The Central Russian Upland *(Srednerusskaya Vozvyshennost')* in the west is deeply incised by rivers draining into the lowland of the Oka and Don rivers. In the east the Volga, Europe's longest river, flows south to the Caspian Sea, dividing the Volga Uplands *(Privolzhskaya Vozvyshennost')* from the foothills of the Ural Mountains *(Ural'skiye Gory)*. The Caucasus mountains and the Black Sea form a natural border to the southwest.

A plantation of *Scots pine helps consolidate the loose sandy soils of the Meshchera Lowland* (Meshcherskaya Nizina), *which lies on the bed of an old glacial lake.*

116

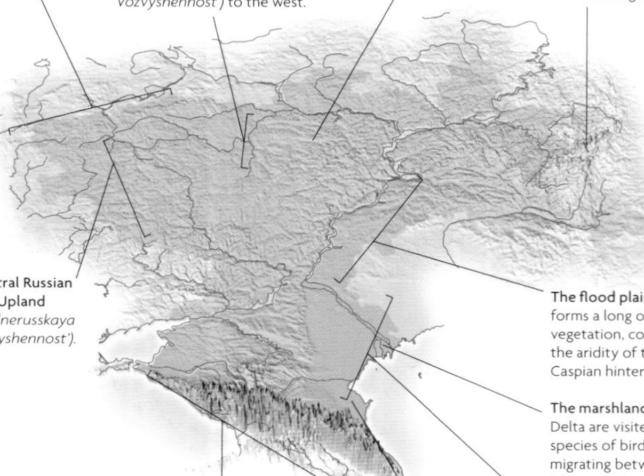

The Smolensk-Moscow Upland *(Smolensko-Moskovskaya Vozvyshennost')* is a series of terminal moraine ridges marking the southern extent of the last glaciation.

Glacial till covers the bedrock to the north of the North European Plain, giving a gentle surface relief.

The lowland of the Oka and Don rivers lies over a broad trough, between the upfolds of the Volga Uplands *(Privolzhskaya Vozvyshennost')* to the east, and the Central Russian Upland *(Srednerusskaya Vozvyshennost')* to the west.

The southern Ural Mountains *(Ural'skiye Gory)* consist of several parallel ranges of ancient fold mountains running from north to south.

Central Russian Upland *(Srednerusskaya Vozvyshennost')*.

The flood plain of the Volga forms a long oasis of verdant vegetation, contrasting with the aridity of the surrounding Caspian hinterland.

The marshlands of the Volga Delta are visited by over 260 species of bird each year, migrating between South Africa and Arctic Siberia.

The Caspian Depression is a large downfold (or syncline) which became flooded, forming the Caspian Sea. The shoreline is 98 ft (30 m) below sea level.

Drifting sand occupies large areas of the south, forming dunes up to 50 ft (15 m) high.

The Caucasus mountains *run from the Black Sea to the Caspian Sea. They include El'brus which, at 18,511 ft (5642 m), is the highest point in Europe. It is still uplifting at a rate of 0.4 inches (10 mm) per year.*

Salt dome

Salt domes, rounded *hills up to 500 ft (150 m) high, are produced as less dense rock salts are displaced under the extreme pressure of denser, overlying strata and forced up towards the surface creating domes. They are widespread in the Caspian Depression.*

Using the land

In the cold, humid north and in the southern Urals *(Ural'skiye Gory)*, small grains, potatoes and flax are commonly rotated with legumes which support livestock farming. The rich chernozem (or black earth) areas support diverse crops such as sugar beet, hemp, sunflowers, millet and vegetables. Further south, aridity restricts husbandry to extensive grazing, with intensive fruit and rice cultivation along the oasis of the Volga.

The urban/rural population divide

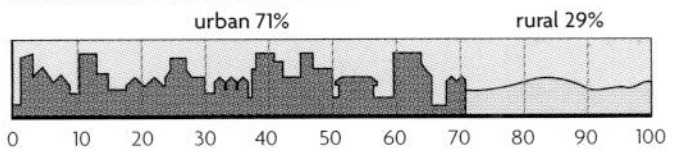

Population density	Total land area
119 people per sq mile (46 people per sq km)	705,916 sq miles (1,828,800 sq km)

Land use and agricultural distribution

- sheep
- flax
- potatoes
- rice
- sunflowers
- sugar beet
- timber
- capital cities
- major towns
- pasture
- cropland
- forest
- wetland
- mountain region
- tundra

Transport and industry

Manufacturing is largely based around Moscow and the Volga region, which became a major industrial area during the Second World War. Both Moscow and Nizhniy Novgorod are centres of skilled labour for light manufacturing and engineering. Most of Russia's main chemical plants are located along the Volga, and one of the world's largest car factories was recently opened in Tol'yatti. Processing and machine construction plants use oil, gas and hydro-electric power from the Volga Basin and metallic minerals from the Urals *(Ural'skiye Gory)* and Kursk.

◀ *Industrial plants are massed along the Volga. Environmental stress from decades of unbridled industrial development has prompted widespread concern about pollution levels.*

The transport network

250,000 miles (402,000 km)	None
28,000 miles (44,800 km)	16,300 miles (26,080 km)

Seventy private and national flag airlines have been created from the reorganization of the state airline Aeroflot, which maintained the world's largest fleet of aircraft during the Soviet era.

Major industry and infrastructure

- aerospace
- car manufacture
- chemicals
- defence
- electronics
- engineering
- gas
- mining
- oil
- textiles
- capital cities
- major towns
- international airports
- major roads
- major industrial areas

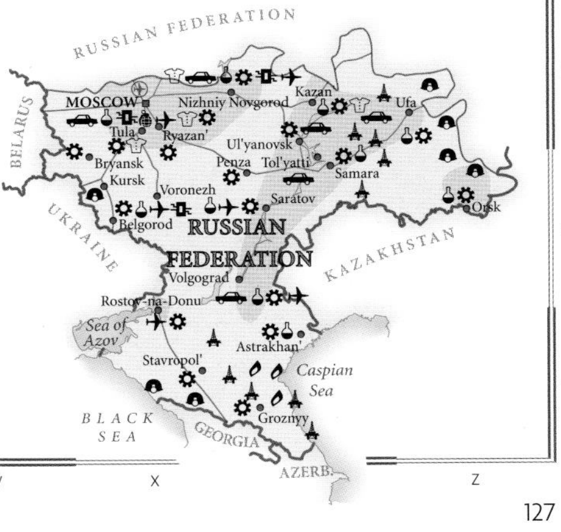

Asia

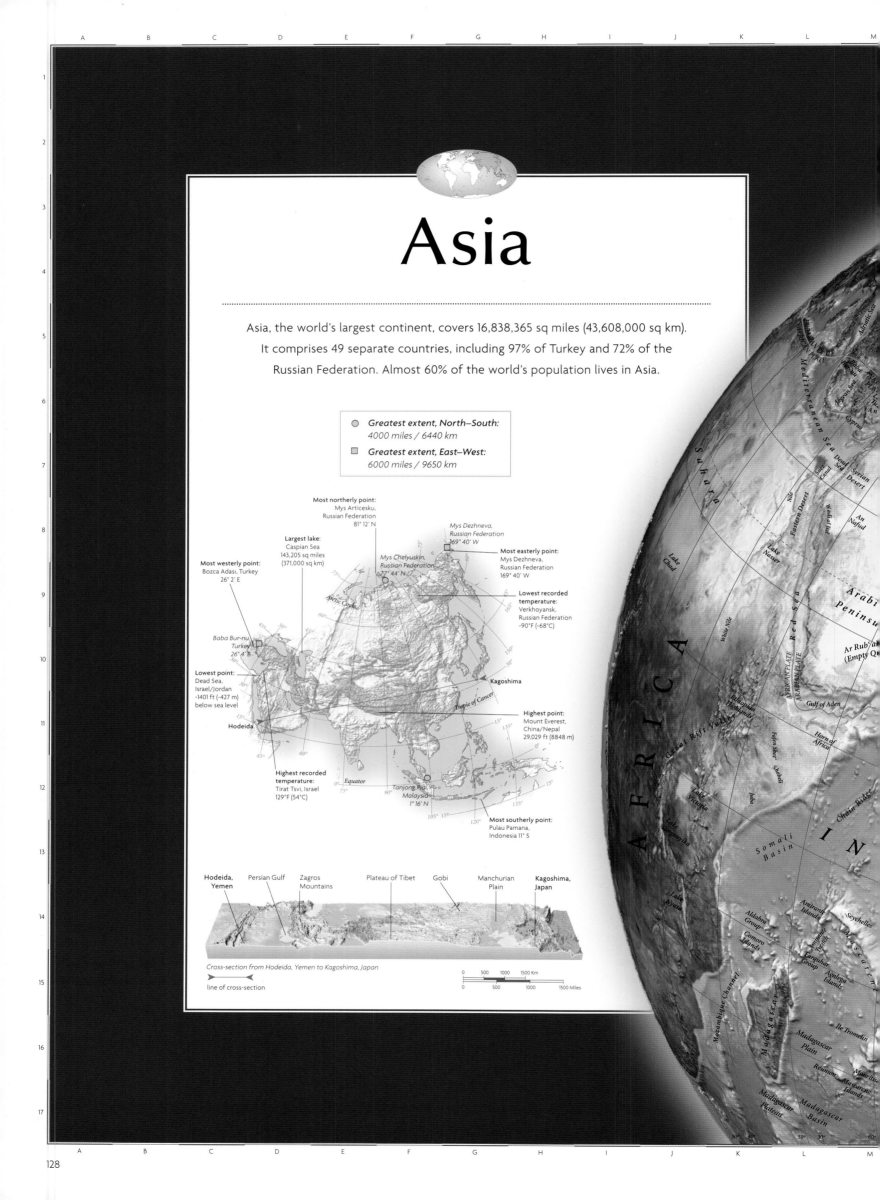

Asia, the world's largest continent, covers 16,838,365 sq miles (43,608,000 sq km). It comprises 49 separate countries, including 97% of Turkey and 72% of the Russian Federation. Almost 60% of the world's population lives in Asia.

○ ***Greatest extent, North–South:***
4000 miles / 6440 km

□ ***Greatest extent, East–West:***
6000 miles / 9650 km

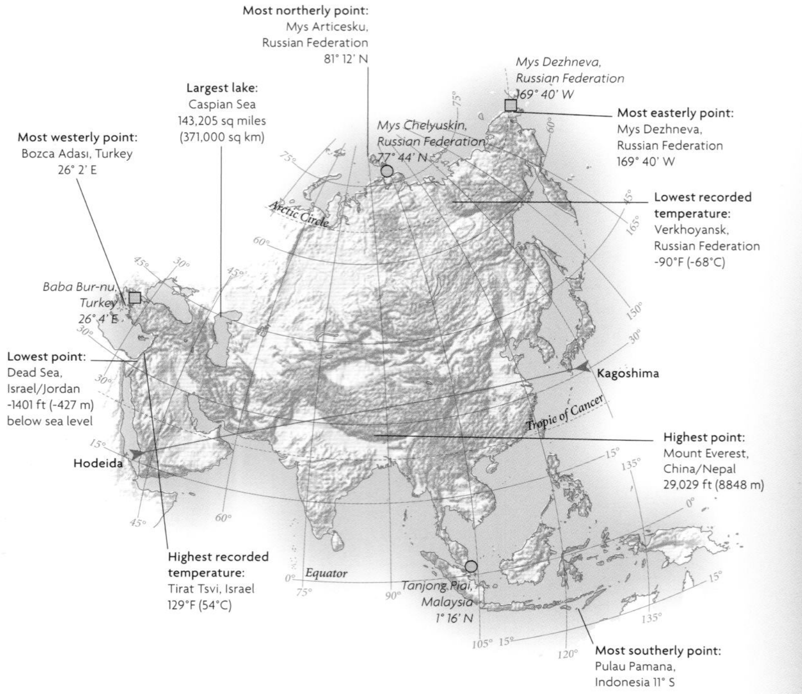

Most northerly point:
Mys Articesku,
Russian Federation
81° 12' N
Largest lake:
Caspian Sea
143,205 sq miles
(371,000 sq km)
Mys Dezhneva,
Russian Federation
169° 40' W
Most easterly point:
Mys Dezhneva,
Russian Federation
169° 40' W
Most westerly point:
Bozca Adası, Turkey
26° 2' E
Mys Chelyuskin,
Russian Federation
77° 44' N
Arctic Circle
Lowest recorded temperature:
Verkhoyansk,
Russian Federation
-90°F (-68°C)
Baba Bur-nu,
Turkey
26° 4' E
Lowest point:
Dead Sea,
Israel/Jordan
-1401 ft (-427 m)
below sea level
Kagoshima
Tropic of Cancer
Highest point:
Mount Everest,
China/Nepal
29,029 ft (8848 m)
Hodeida
Highest recorded temperature:
Tirat Tsvi, Israel
129°F (54°C)
Equator
Tanjong Piai,
Malaysia
1° 16' N
Most southerly point:
Pulau Pamana,
Indonesia 11° S

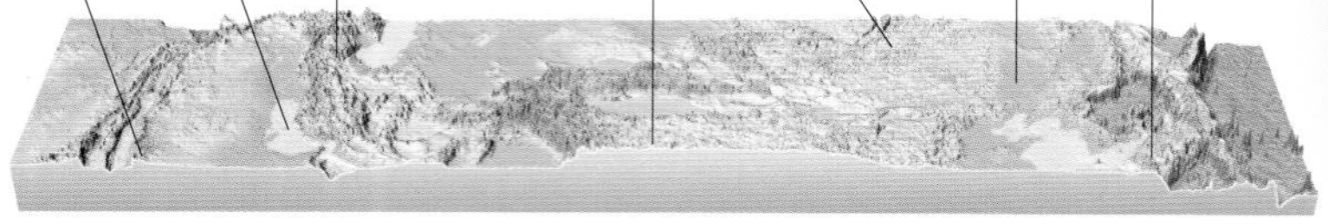

Hodeida, Yemen
Persian Gulf
Zagros Mountains
Plateau of Tibet
Gobi
Manchurian Plain
Kagoshima, Japan

Cross-section from Hodeida, Yemen to Kagoshima, Japan

line of cross-section

0 500 1000 1500 Km
0 500 1000 1500 Miles

Adriatic Sea
Mediterranean Sea
AFRICAN PLATE
Aegean Sea
Cyprus
Sahara
Suez Canal
Dead Sea
Syrian Desert
Nile
Eastern Desert
Wadi al Jarir
An Nafud
Lake Nasser
Lake Chad
Red Sea
Arabian Peninsula
Ar Rub' al (Empty Q
AFRICAN PLATE
ARABIAN PLATE
White Nile
AFRICA
Ethiopian Highlands
Gulf of Aden
Great Rift Valley
Horn of Africa
Fafen Shet'
Shebeli
Juba
Lake Victoria
Lake Tanganyika
Chain Ridge
Somali Basin
Lake Nyasa
Amirante Islands
Seychelles
Aldabra Group
Comoro Islands
Farquhar Group
Agalega Islands
Mozambique Channel
Madagascar
Île Tromelin
Madagascar Plain
Réunion
Mascarene Islands
Madagascar Plateau
Madagascar Basin

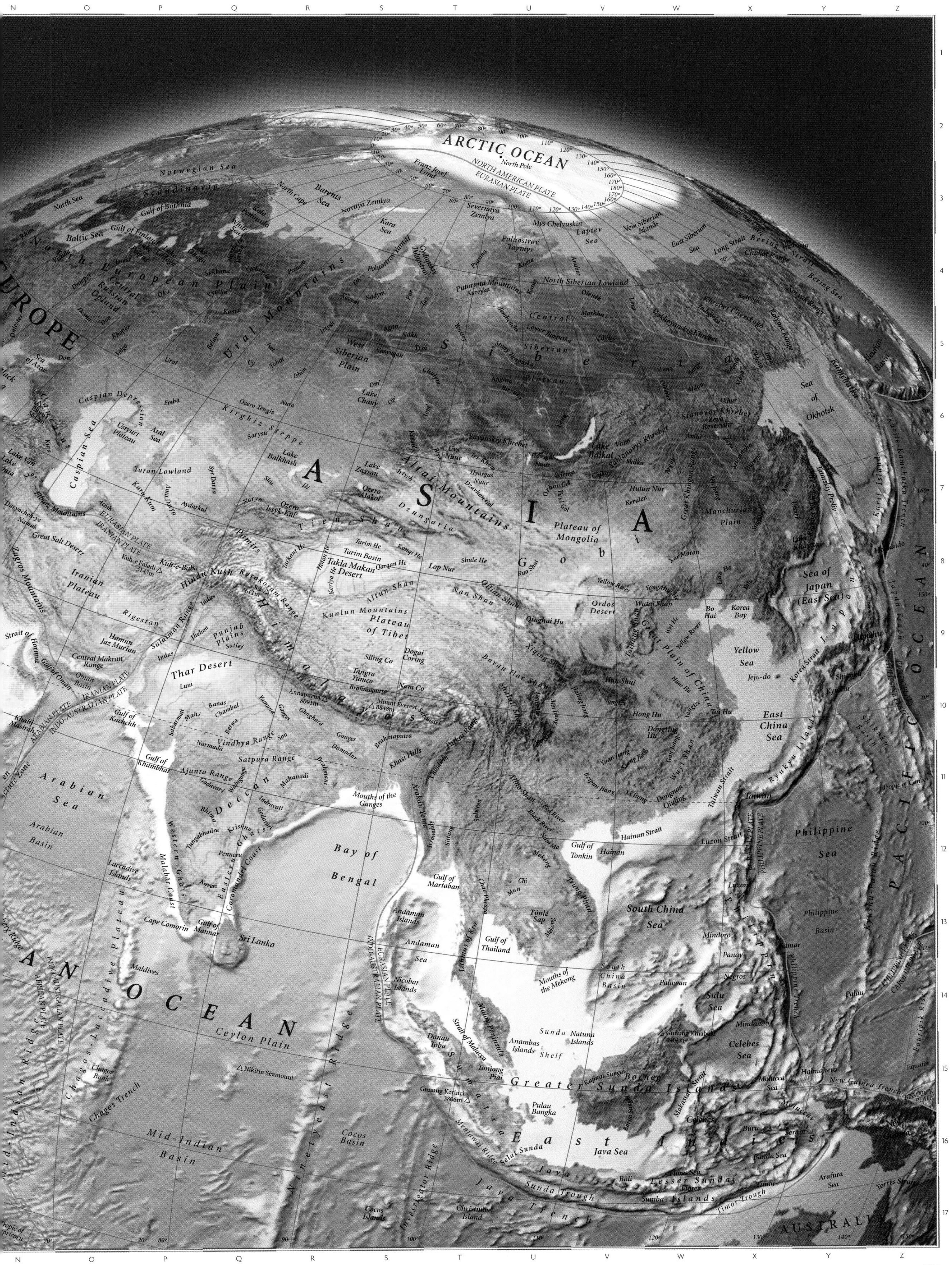
ARCTIC OCEAN
North Pole
NORTH AMERICAN PLATE
EURASIAN PLATE
Norwegian Sea
North Sea
Scandinavia
Gulf of Bothnia
Baltic Sea
Gulf of Finland
Barents Sea
North Cape
Kola Peninsula
White Sea
Franz Josef Land
Novaya Zemlya
Kara Sea
Severnaya Zemlya
Mys Chelyuskin
Laptev Sea
New Siberian Islands
East Siberian Sea
Long Strait
Bering Strait
Chukot Range
Bering Sea
Poluostrov Taymyr
Poluostrov Yamal
North Siberian Lowland
Putorana Mountains
North European Plain
Central Russian Upland
EUROPE
Ural Mountains
West Siberian Plain
Central Siberian Plateau
Siberia
Lower Tunguska
Stony Tunguska
Verkhoyansk Range
Khrebet Cherskogo
Kolyma Range
Koryak Range
Kamchatka
Aleutian Basin
Sea of Okhotsk
Stanovoy Khrebet
Zeya Reservoir
Yablonovyy Khrebet
Lake Baikal
Sayanskiy Khrebet
Caspian Depression
Caspian Sea
Caucasus
Sea of Azov
Lake Van
Aral Sea
Ustyurt Plateau
Kirghiz Steppe
Turan Lowland
Kara Kum
Lake Balkhash
Lake Zaysan
Lake Chany
Ozero Tengiz
Ozero Alakol'
Ozero Issyk-Kul'
Altai Mountains
Dzungaria
Tien Shan
ASIA
Plateau of Mongolia
Gobi
Hulun Nur
Great Khingan Range
Manchurian Plain
Sea of Japan (East Sea)
Japan
Kuril Islands
Kuril-Kamchatka Trench
Hokkaido
Japan Trench
PACIFIC OCEAN
Elburz Mountains
Great Salt Desert
Zagros Mountains
Iranian Plateau
EURASIAN PLATE
IRANIAN PLATE
Kuh-e Fuladi 5143m
Kuh-e-Baba
Pamirs
Hindu Kush
Karakoram Range
Tarim Basin
Takla Makan Desert
Lop Nur
Altun Shan
Nan Shan
Qilian Shan
Kunlun Mountains
Plateau of Tibet
Qinghai Hu
Ordos Desert
Yellow River
Wutai Shan
Bo Hai
Korea Bay
Yellow Sea
Jeju-do
Korea Strait
Great Plain of China
Bayan Har Shan
Himalaya
Rigestan
Sulaiman Range
Punjab Plains
Strait of Hormuz
Gulf of Oman
Hamun Jaz Murian
Central Makran Range
Oman Basin
Thar Desert
Annapurna 8091m
Mount Everest 8848m
Nam Co
Siling Co
Dogai Coring
Tangra Yumco
Brahmaputra
Hong Hu
Dongting Hu
Tai Hu
East China Sea
Ryukyu Islands
Taiwan
Taiwan Strait
Tropic of Cancer
Philippine Sea
ARABIAN PLATE
INDO-AUSTRALIAN PLATE
Gulf of Kachchh
Vindhya Range
Satpura Range
Gulf of Khambhat
Ajanta Range
Arabian Sea
Arabian Basin
Deccan
Western Ghats
Eastern Ghats
Mouths of the Ganges
Khasi Hills
Patkai Range
Arakan Yoma
Bay of Bengal
Gulf of Martaban
Gulf of Tonkin
Hainan
Hainan Strait
Luzon Strait
Luzon
South China Sea
Tônlé Sap
Gulf of Thailand
Isthmus of Kra
Andaman Islands
Andaman Sea
Nicobar Islands
Mouths of the Mekong
South China Basin
Mindoro
Panay
Negros
Palawan
Sulu Sea
Mindanao
Philippine Trench
Philippine Basin
Kyushu-Palau Ridge
Palau
Lakshadweep
Laccadive Islands
Malabar Coast
Coromandel Coast
Cape Comorin
Gulf of Mannar
Sri Lanka
Maldives
INDIAN OCEAN
Chagos-Laccadive Plateau
Chagos Bank
Chagos Trench
Ceylon Plain
Nikitin Seamount
Mid-Indian Basin
Mid-Indian Ridge
Ninetyeast Ridge
Cocos Basin
Investigator Ridge
Danau Toba
Strait of Malacca
Malay Peninsula
Anambas Islands
Natuna Islands
Sunda Shelf
Gunung Kinabalu 4094m
Borneo
Celebes Sea
Celebes
Makassar Strait
Greater Sunda Islands
Sumatra
Gunung Kerinci 3806m
Pulau Bangka
Mentawai Ridge
Selat Sunda
East Indies
Java Sea
Java
Bali
Flores Sea
Flores
Lesser Sunda Islands
Sumba
Timor
Timor Trough
Banda Sea
Buru
Seram
Moluccas
Molucca Sea
Halmahera
New Guinea Trench
Arafura Sea
Torres Strait
Sunda Trough
Java Trench
Cocos Islands
Christmas Island
AUSTRALIA
Tropic of Capricorn
Equator
N O P Q R S T U V W X Y Z
1 2 3 4 5 6 7 8 9 10 11 12 13 14 15 16 17

Physical Asia

The structure of Asia can be divided into two distinct regions. The landscape of northern Asia consists of old mountain chains, shields, plateaux and basins, like the Ural Mountains in the west and the Central Siberian Plateau to the east. To the south of this region, are a series of plateaux and basins, including the vast Plateau of Tibet and the Tarim Basin. In contrast, the landscapes of southern Asia are much younger, formed by tectonic activity beginning about 65 million years ago, leading to an almost continuous mountain chain running from Europe, across much of Asia, and culminating in the mighty Himalayan mountain belt, formed when the Indo-Australian Plate collided with the Eurasian Plate. They are still being uplifted today. North of the mountains lies a belt of deserts, including the Gobi and the Takla Makan. In the far south, tectonic activity has formed narrow island arcs, extending over 4000 miles (7000 km). To the west lies the Arabian Shield, once part of the African Plate. As it was rifted apart from Africa, the Arabian Plate collided with the Eurasian Plate, uplifting the Zagros Mountains.

Coastal Lowlands and Island Arcs

The coastal plains that fringe Southeast Asia contain many large delta systems, caused by high levels of rainfall and erosion of the Himalayas, the Plateau of Tibet and relict loess deposits. To the south is an extensive island archipelago, lying on the drowned Sunda Shelf. Most of these islands are volcanic in origin, caused by the subduction of the Indo-Australian Plate beneath the Eurasian Plate.

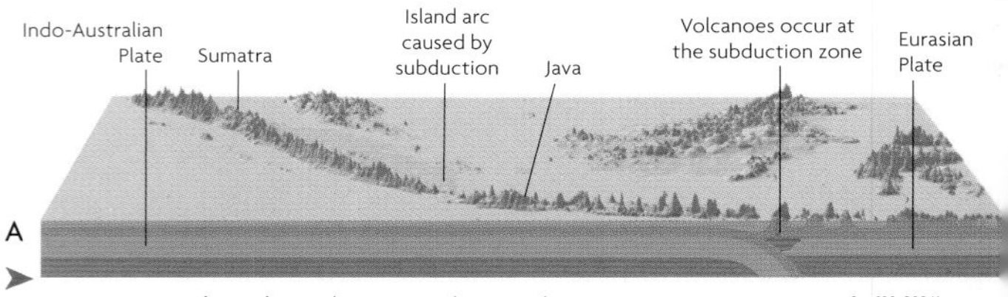

Cross-section through Southeast Asia showing the subduction zone between the Indo-Australian and Eurasian plates and the island arc.

The Indian Shield and Himalayan System

The large shield area beneath the Indian subcontinent is between 2.5 and 3.5 billion years old. As the floor of the southern Indian Ocean spread, it pushed the Indian Shield north. This was eventually driven beneath the Plateau of Tibet. This process closed up the ancient Tethys Sea and uplifted the world's highest mountain chain, the Himalayas. Much of the uplifted rock strata was from the seabed of the Tethys Sea, partly accounting for the weakness of the rocks and the high levels of erosion found in the Himalayas.

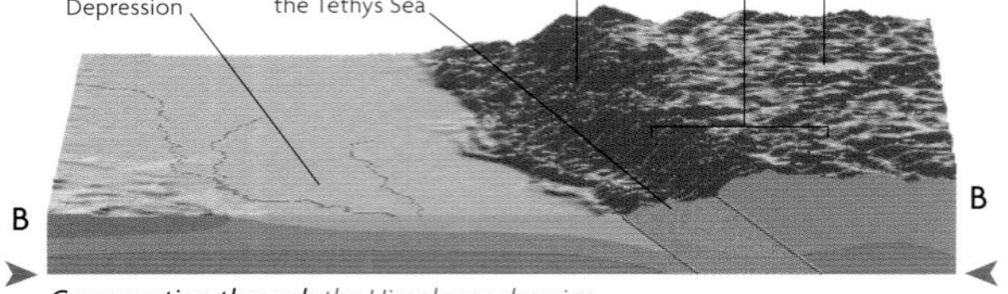

Cross-section through the Himalayas showing thrust faulting of the rock strata.

East Asian Plains and Uplands

Several, small, isolated shield areas, such as the Shandong Peninsula, are found in east Asia. Between these stable shield areas, large river systems like the Yangtze and the Yellow River have deposited thick layers of sediment, forming extensive alluvial plains. The largest of these is the Great Plain of China, the relief of which does not rise above 300 ft (100 m).

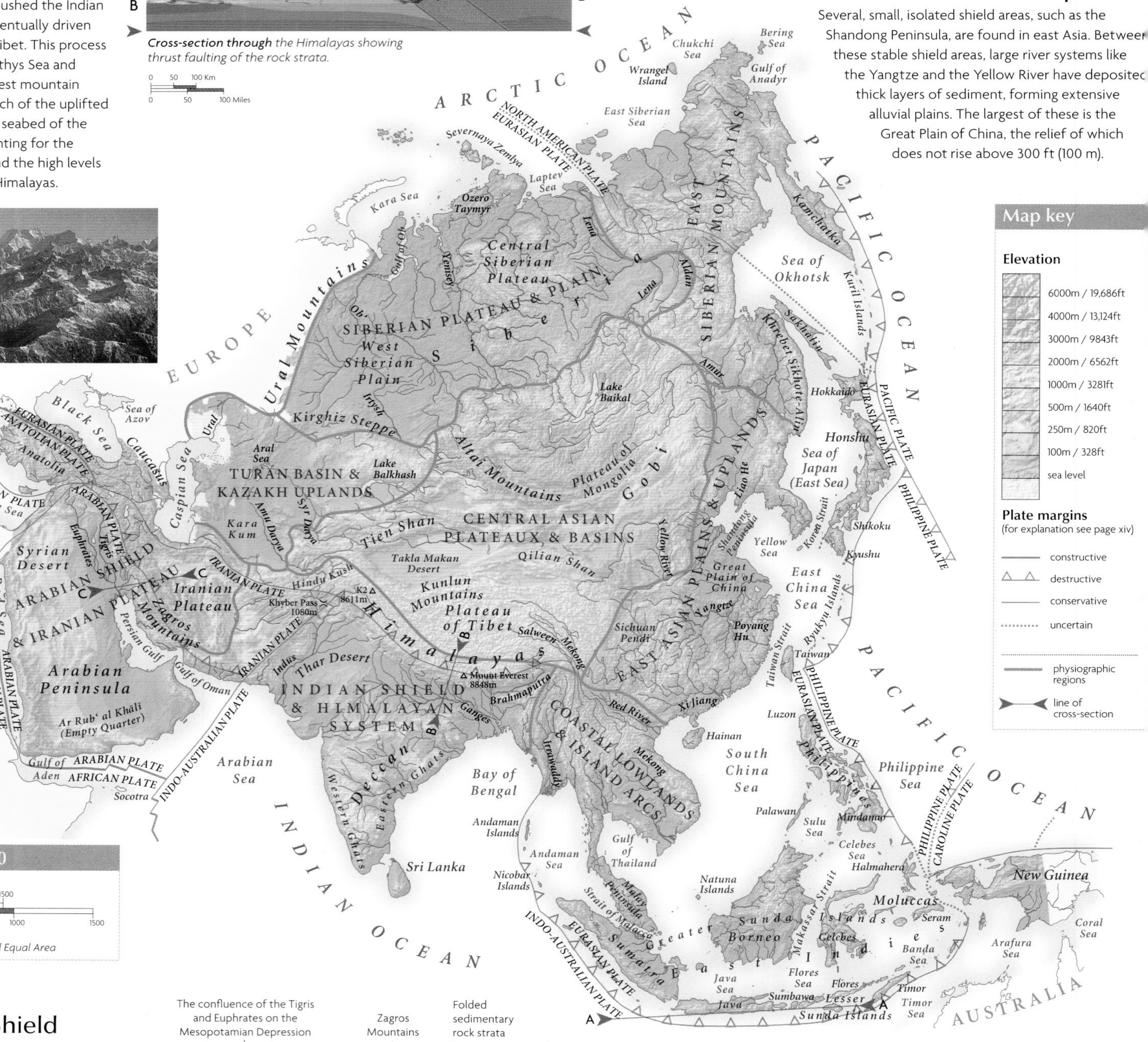

Scale 1:63,000,000

Km 0 250 500 1000 1500

Miles 0 250 500 1000 1500

projection: Lambert Azimuthal Equal Area

The Arabian Shield and Iranian Plateau

Approximately five million years ago, rifting of the continental crust split the Arabian Plate from the African Plate and flooded the Red Sea. As this rift spread, the Arabian Plate collided with the Eurasian Plate, transforming part of the Tethys seabed into the Zagros Mountains which run northwest-southeast across western Iran.

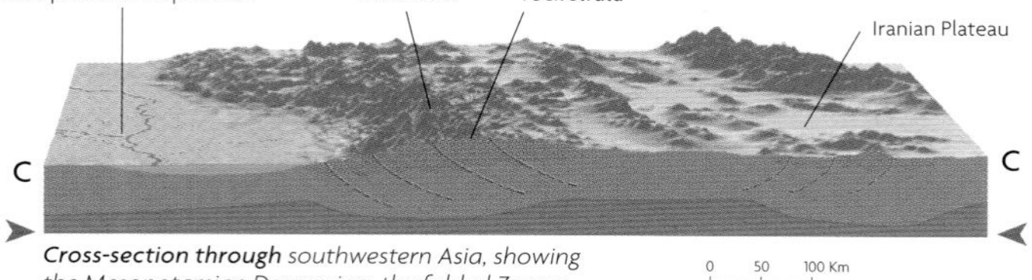

Cross-section through southwestern Asia, showing the Mesopotamian Depression, the folded Zagros Mountains and the Iranian Plateau.

Climate

The climate of Asia exhibits marked differences from region to region, with freezing polar conditions in the north, hot and cold deserts in central regions and subtropical conditions throughout the south. Much of this variation can be attributed to enormous mountain barriers and internal depressions found across the continent. Monsoon winds, which reverse semi-annually, cause alternate wet and dry seasons across southern Asia. These air masses moving north from the ocean are stripped of their moisture over the Himalayas causing arid conditions across the Plateau of Tibet. Both the south and east are susceptible to tropical cyclones or typhoons.

▲ *Tropical cyclones occur principally during late summer and early autumn. The intense winds and heavy rainfall can devastate entire villages.*

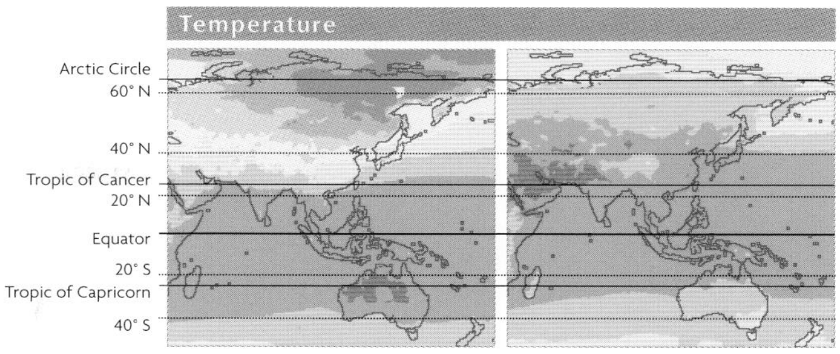
Temperature
Arctic Circle
60° N
40° N
Tropic of Cancer
20° N
Equator
20° S
Tropic of Capricorn
40° S

Average January temperature *Average July temperature*

Temperature

- below -30°C (-22°F)
- -30 to -20°C (-22 to -4°F)
- -20 to -10°C (-4 to 14°F)
- -10 to 0°C (14 to 32°F)
- 0 to 10°C (32 to 50°F)
- 10 to 20°C (50°F)
- 20 to 30°C (68 to 86°F)
- above 30°C (86°F)

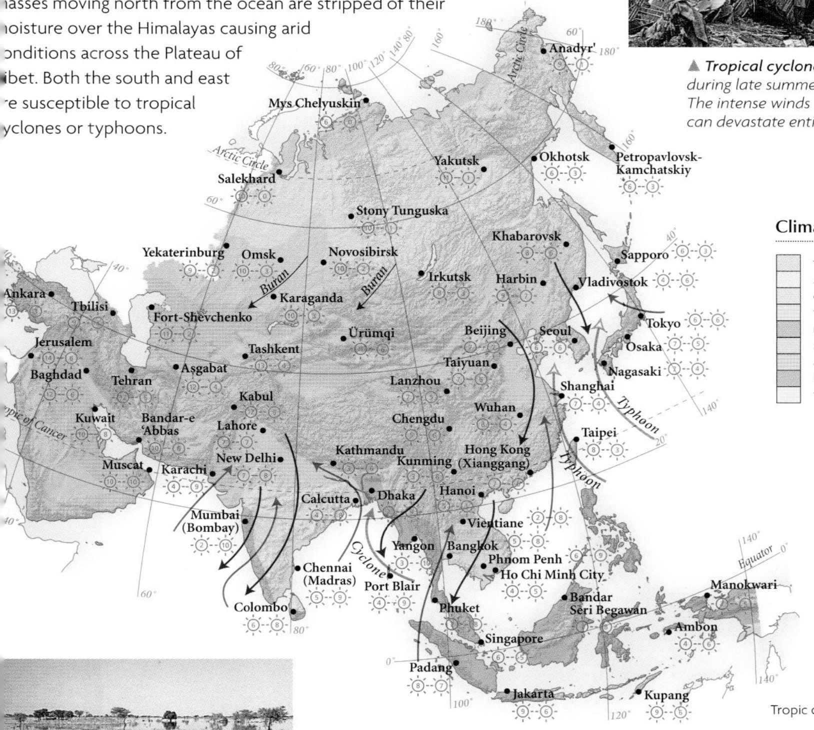
Anadyr'
Mys Chelyuskin
Arctic Circle
Yakutsk
Okhotsk
Petropavlovsk-Kamchatskiy
Salekhard
Stony Tunguska
Khabarovsk
Yekaterinburg
Omsk
Novosibirsk
Sapporo
Irkutsk
Harbin
Vladivostok
Ankara
Tbilisi
Karaganda
Buran
Fort-Shevchenko
Tokyo
Ürümqi
Beijing
Seoul
Osaka
Jerusalem
Tashkent
Taiyuan
Nagasaki
Baghdad
Asgabat
Tehran
Lanzhou
Shanghai
Kabul
Typhoon
Kuwait
Bandar-e 'Abbas
Lahore
Chengdu
Wuhan
Tropic of Cancer
Taipei
Kathmandu
Hong Kong (Xianggang)
New Delhi
Kunming
Muscat
Karachi
Calcutta
Dhaka
Hanoi
Mumbai (Bombay)
Vientiane
Yangon
Bangkok
Cyclone
Phnom Penh
Ho Chi Minh City
Chennai (Madras)
Port Blair
Manokwari
Equator
Colombo
Phuket
Bandar Seri Begawan
Ambon
Singapore
Padang
Jakarta
Kupang

Climate

- tundra
- subarctic
- cool continental
- warm humid
- mediterranean
- semi-arid
- arid
- humid equatorial
- tropical
- daily hours of sunshine, January
- daily hours of sunshine, July
- cyclone
- typhoon
- cold/dry monsoon
- warm/wet monsoon
- cold wind

► *The Gobi Desert experiences major extremes in climate, with winter temperatures sometimes falling below -40°C (-40°F) and summer temperatures exceeding 45°C (113°F).*

◄ *Through India, the southwest monsoon, which brings heavy rainfall from May to September, accounts for 80% of annual precipitation.*

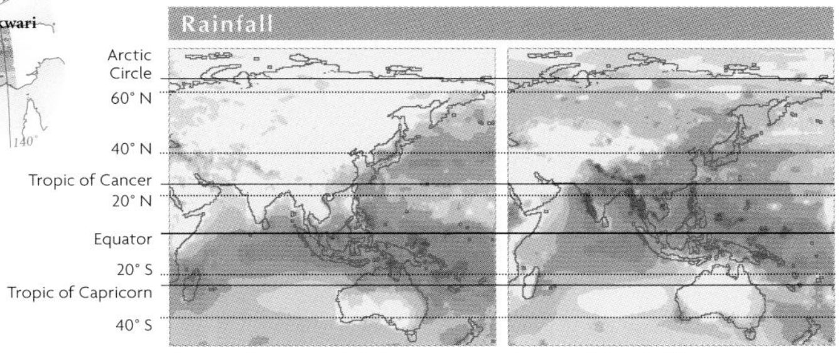
Rainfall
Arctic Circle
60° N
40° N
Tropic of Cancer
20° N
Equator
20° S
Tropic of Capricorn
40° S

Average January rainfall *Average July rainfall*

Rainfall

- 0 –25 mm (0–1 in)
- 25–50 mm (1–2 in)
- 50–100 mm (2–4 in)
- 100–200 mm (4–8 in)
- 200–300 mm (8–12 in)
- 300–400 mm (12–16 in)
- 400–500 mm (16–20 in)
- more than 500 mm (20 in)

Shaping the landscape

In the north, melting of extensive permafrost leads to typical periglacial features such as thermokarst. In the arid areas wind action transports sand creating extensive dune systems. An active tectonic margin in the south causes continued uplift, and volcanic and seismic activity, but also high rates of weathering and erosion. Across the continent, huge rivers erode and transport vast quantities of sediment depositing it on the plains or forming large deltas.

River systems

1 Vast river systems flow across Asia, many originating in the Himalayas and the Plateau of Tibet. Seasonal melting of snow and monsoon rains swell the river flow leading to flooding and erosion. The Yellow River *(right)* gets its colour from the high level of eroded material from the loess plateau.

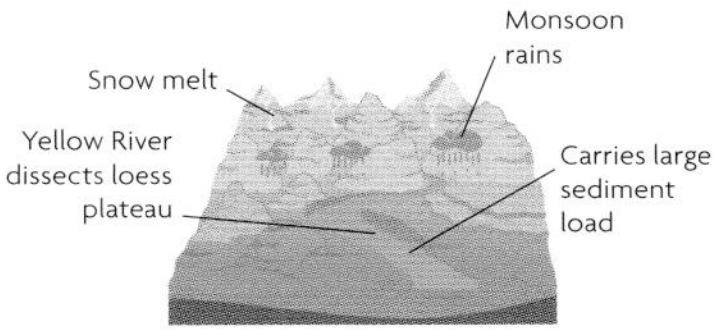
Monsoon rains
Snow melt
Yellow River dissects loess plateau
Carries large sediment load

***River systems:** erosion of the loess plateau by the yellow river*

Chemical weathering

2 Tower karsts are widespread across south China *(left)* and Vietnam. It is thought the karstic towers were formed under a soil cover, where small depressions in the limestone bedrock began to be weathered by soil water acids, eventually creating larger hollows. This process continued over millions of years, deepening the hollows and leaving steep-sided limestone hills.

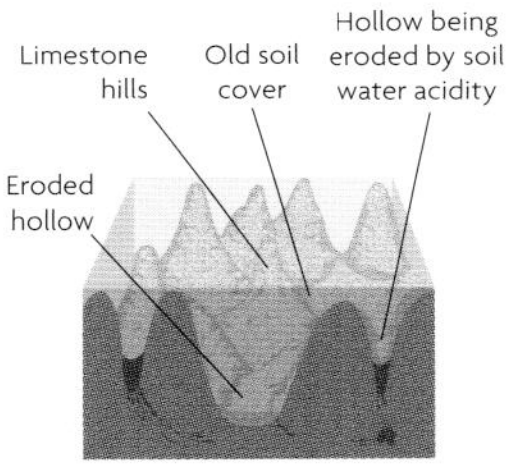
Limestone hills
Old soil cover
Hollow being eroded by soil water acidity
Eroded hollow

***Chemical weathering:** formation of tower karst*

Volcanic activity

3 Volcanic eruptions occur frequently across Southeast Asia's island arcs *(below)*. Low-level eruptions occur when groundwater, superheated by underlying magma, becomes pressurized, forcing hot fluid and rocks up through cracks in the volcanic cone. This is known as a phreatic eruption.

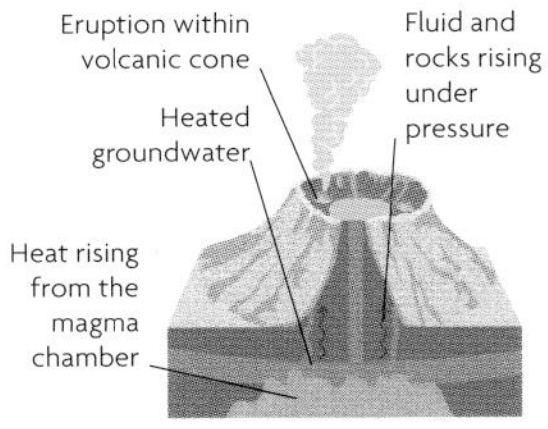
Eruption within volcanic cone
Fluid and rocks rising under pressure
Heated groundwater
Heat rising from the magma chamber

***Volcanic activity:** a phreatic eruption*

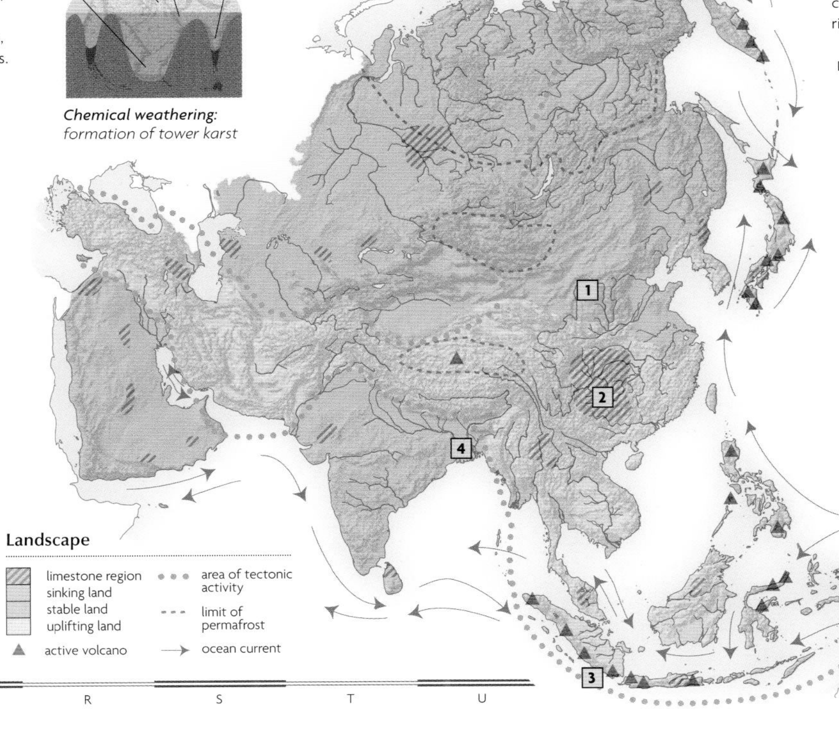
1
2
3
4
Landscape
limestone region
sinking land
stable land
uplifting land
active volcano
area of tectonic activity
limit of permafrost
ocean current

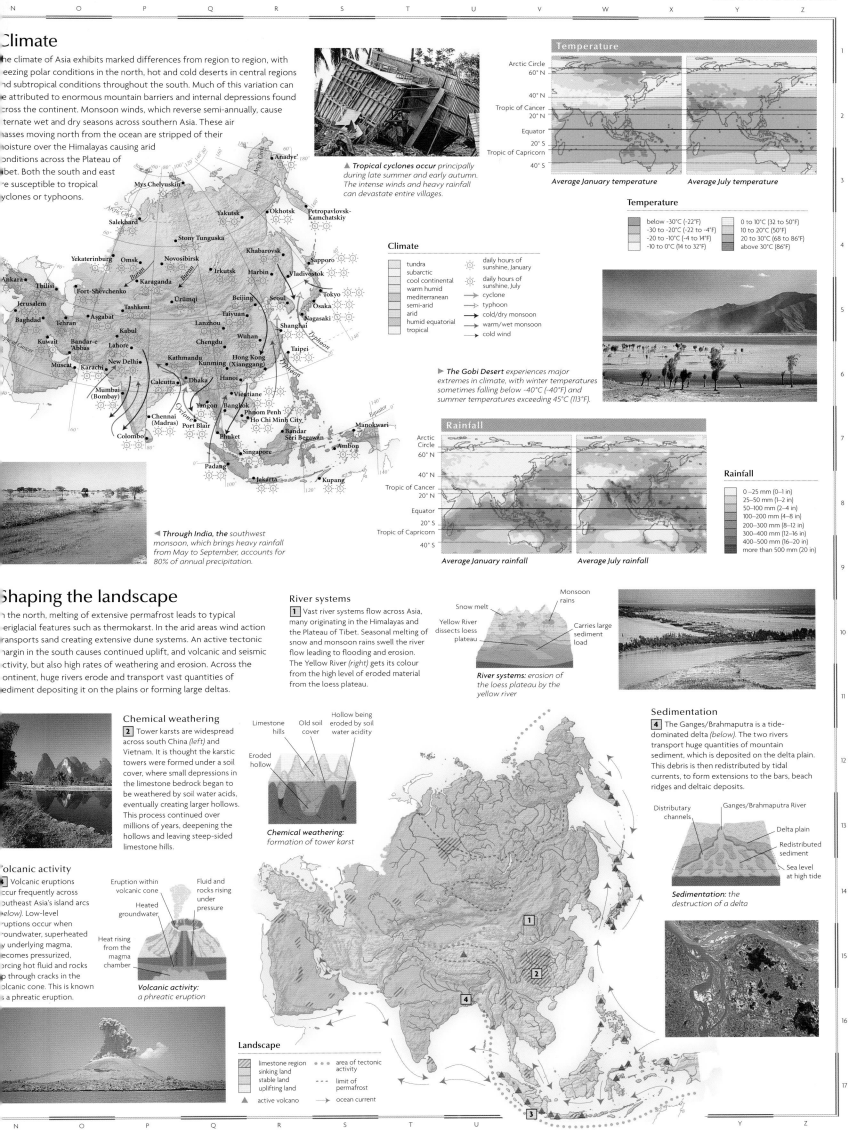

Sedimentation

4 The Ganges/Brahmaputra is a tide-dominated delta *(below)*. The two rivers transport huge quantities of mountain sediment, which is deposited on the delta plain. This debris is then redistributed by tidal currents, to form extensions to the bars, beach ridges and deltaic deposits.

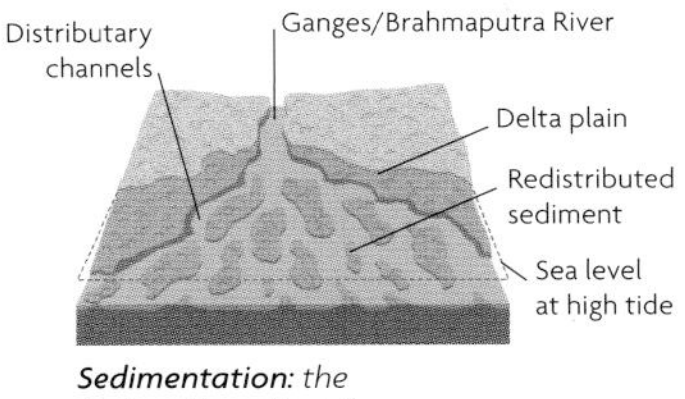
Distributary channels
Ganges/Brahmaputra River
Delta plain
Redistributed sediment
Sea level at high tide

***Sedimentation:** the destruction of a delta*

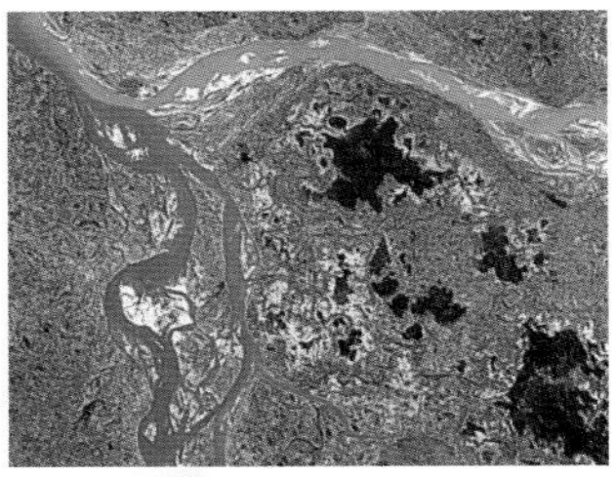

Political Asia

Asia is the world's largest continent, encompassing many different and discrete realms, from the desert Arab lands of the southwest to the subtropical archipelago of Indonesia; from the vast barren wastes of Siberia to the fertile river valleys of China and South Asia, seats of some of the world's most ancient civilizations. The collapse of the Soviet Union has fragmented the north of the continent into the Siberian portion of the Russian Federation, and the new republics of Central Asia. Strong religious traditions heavily influence the politics of South and Southwest Asia. Hindu and Muslim rivalries threaten to upset the political equilibrium in South Asia where India – in terms of population – remains the world's largest democracy. Communist China, another population giant, is reasserting its position as a world political and economic power, while on its doorstep, the dynamic Pacific Rim countries, led by Japan, continue to assert their worldwide economic force.

ARCTIC OCEAN
East Siberian Sea
Laptev Sea
Kara Sea
RUSSIAN FEDERATION
Central Siberian Plateau
West Siberian Plain
Siberia
EUROPE
Ural Mountains
Noril'sk
Yakutsk
Yekaterinburg
Chelyabinsk
Omsk
Tomsk
Novosibirsk
Krasnoyarsk
Novokuznetsk
Irkutsk
Lake Baikal
Arctic Circle
Ural'sk
Rudnyy
ASTANA
Karaganda
Semipalatinsk
Zhezkazgan
KAZAKHSTAN
Aral Sea
Lake Balkhash
Balkhash
Kyzylorda
Aktau
Caspian Sea
Sühbaatar
Choybalsan
Erdenet
ULAN BATOR
MONGOLIA
Altai Mountains
Gobi
Istanbul
Black Sea
ANKARA
TURKEY
Anatolia
Sokhumi
GEORGIA
K'ut'aisi
TBILISI
Bat'umi
Adana
Gaziantep
ARMENIA
YEREVAN
Gancä
AZERB.
BAKU
CYPRUS
NICOSIA
Aleppo
LEBANON
Tripoli
BEIRUT
SYRIA
Haifa
DAMASCUS
Tel Aviv-Yafo
JERUSALEM
Gaza
AMMAN
ISRAEL
JORDAN
Mosul
Kirkuk
Tabriz
TEHRAN
BAGHDAD
Euphrates
Tigris
An Najaf
IRAQ
Qom
Esfahan
Ahvaz
Basra
KUWAIT
IRAN
Iranian Plateau
Shiraz
Kerman
Zahedan
Bandar-e 'Abbas
SAUDI ARABIA
RIYADH
Jedda
At Ta'if
MANAMA
BAHRAIN
QATAR
DOHA
ABU DHABI
UAE
Persian Gulf
Gulf of Oman
MUSCAT
Ar Rustaq
Sur
OMAN
Arabian Peninsula
Ar Rub' al Khali (Empty Quarter)
SANA
YEMEN
Ta'izz
Aden
Gulf of Aden
Socotra (to Yemen)
Red Sea
AFRICA
Tropic of Cancer
Dasoguz
TURKMENISTAN
UZBEKISTAN
Amu Darya
Syr Darya
AŞGABAT
Gorgan
Mashhad
TASHKENT
Taraz
BISHKEK
Almaty
Karakol
KYRGYZSTAN
Osh
Tien Shan
DUSHANBE
TAJIKISTAN
Balkh
Qal'eh-ye Now
Herat
AFGHANISTAN
KABUL
Kandahar
Peshawar
ISLAMABAD
Srinagar
Jammu
Gujranwala
Faisalabad
Lahore
Quetta
Multan
PAKISTAN
Shikarpur
Larkana
Indus
Thar Desert
Karachi
Hyderabad
(line of control)
(claimed by India)
(administered by China, claimed by India)
Ürümqi
Tarim He
Takla Makan Desert
Kunlun Mountains
Plateau of Tibet
CHINA
Himalaya
(Much of Arunachal Pradesh is claimed by China)
Brahmaputra
Ludhiana
Delhi
NEW DELHI
Bareilly
Agra
Jaipur
Kanpur
Lucknow
NEPAL
KATHMANDU
THIMPHU
BHUTAN
Guwahati
Patna
Rangpur
Varanasi
INDIA
Ganges
Rajshahi
BANGLADESH
Brahmanbaria
DHAKA
Khulna
Chittagong
Kolkata (Calcutta)
Jamshedpur
Ahmadabad
Vodadara
Bhopal
Indore
Narmada
Surat
Nagpur
Bhubaneshwar
Mumbai (Bombay)
Pune
Godavari
Solapur
Hyderabad
Krishna
Vijayawada
Hubli
Bangalore
Chennai (Madras)
Mysore
Coimbatore
Kochi
Thiruvananthapuram
Jaffna
SRI LANKA
COLOMBO
SRI JAYEWARDENAPURA KOTTE
Arabian Sea
INDIAN OCEAN
Bay of Bengal
Andaman Islands (to India)
Andaman Sea
Nicobar Islands (to India)
MYANMAR (BURMA)
Pakokku
Mandalay
Taunggyi
NAY PYI TAW
Irrawaddy
Salween
Prome
Pegu
Yangon (Rangoon)
Bassein
Mekong
Salween
Lanzhou
Xi'an
Mianyang
Chengdu
Leshan
Chongqi
Kunming
Liuzhou
Nanning
Yellow River
Baotou
Datong
Inner Mong
Gobi
HANOI
Louangphabang
Vinh
LAOS
Chiang Mai
VIENTIANE
Pakxe
THAILAND
BANGKOK
Batdambang
CAMBODI
PHNOM PENH
Gulf of Thailand
Kota Bharu
Taiping
KUALA LUMPUR
PUTRAJAYA
MAL
Medan
SINGAPORE
Sumatra
Equator
Padang
Jambi
Palembang
JAKA
Yenisey
Ob'
Irtysh
Ishim
Tobol
Chulym
Angara
Lena
Vilyuy
Aldan
Kheta
Anabar
Olenek
Yana
Indigirka
Vitim
Stony Tunguska
Lower Tunguska
Kureyka

Population

Some of the world's most populous and least populous regions are in Asia. The plains of eastern China, the Ganges river plains in India, Japan and the Indonesian island of Java, all have very high population densities; by contrast parts of Siberia and the Plateau of Tibet are virtually uninhabited. China has the world's greatest population – 20% of the globe's total – while India, with the second largest, is likely to overtake China within 30 years.

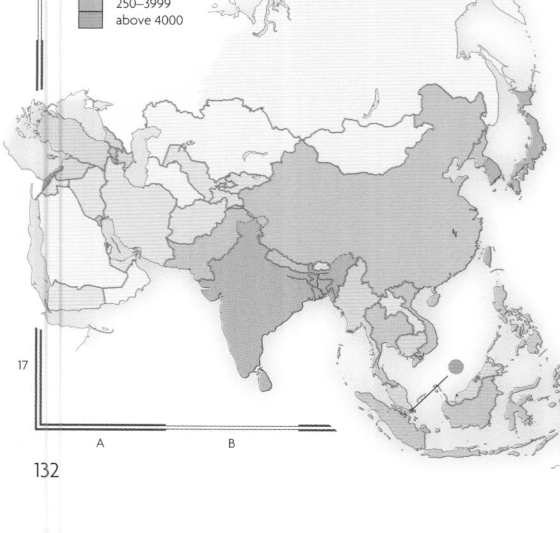
Population density
(people per sq km)
below 9
10–49
50–99
100–249
250–3999
above 4000

◀ *Kolkata's 13 million inhabitants bustle through a maze of crowded, narrow streets. Population densities in India's largest city reach almost 85,000 per sq mile (33,000 per sq km).*

Bering Sea
Anadyr'
PACIFIC OCEAN
Kolyma Range
Magadan
Kamchatka
Petropavlovsk-Kamchatskiy
Sea of Okhotsk
Kuril Islands
Sakhalin
Amur
Yuzhno-Sakhalinsk
Khabarovsk
Hokkaido
Sapporo
Qiqihar
Harbin
Jilin
Vladivostok
Chongjin
Sea of Japan (East Sea)
NORTH KOREA
Shenyang
Hamhung
Wonsan
PYONGYANG
SOUTH KOREA
Dalian
SEOUL
Incheon
Daegu
SEJONG CITY
Busan
Sendai
JAPAN
TOKYO
Yokohama
Kyoto
Nagoya
Kobe
Osaka
Hiroshima
Kitakyushu
Fukuoka
Kumamoto
Kagoshima
Qingdao
Yellow Sea
Xuzhou
Nanjing
Shanghai
East China Sea
Hangzhou
Ningbo
Jinhua
Nanchang
Fuzhou
Ryukyu Islands
Tropic of Cancer
TAIPEI
Taizhong
TAIWAN
Tainan
Gaoxiong
Taiwan Strait
Shantou
Guangzhou
Hong Kong
Philippine Sea
Tuguegarao
Luzon
Baguio
MANILA
Legazpi City
South China Sea
PHILIPPINES
Cebu
Bacolod
Mindanao
Davao
General Santos
Zamboanga
Equator
Jayapura
New Guinea
Papua (Irian Jaya)
BANDAR SERI BEGAWAN
BRUNEI
Manado
Borneo
Moluccas
Balikpapan
Celebes
Ambon
Pontianak
INDONESIA
Banjarmasin
Makassar
Java Sea
Flores Sea
Flores
DILI
EAST TIMOR
Timor
Timor Sea
Semarang
Surabaya
Java
Malang

Map key

Population

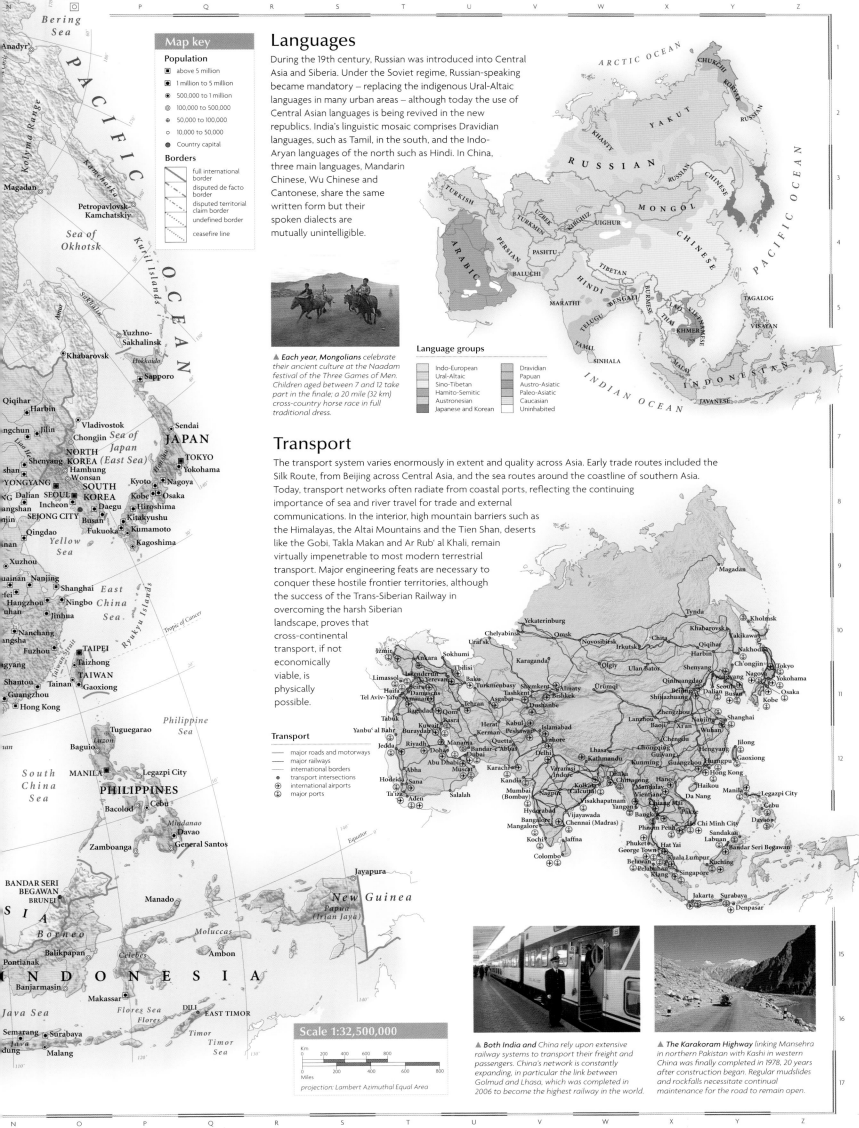

- above 5 million
- 1 million to 5 million
- 500,000 to 1 million
- 100,000 to 500,000
- 50,000 to 100,000
- 10,000 to 50,000
- Country capital

Borders

- full international border
- disputed de facto border
- disputed territorial claim border
- undefined border
- ceasefire line

Languages

During the 19th century, Russian was introduced into Central Asia and Siberia. Under the Soviet regime, Russian-speaking became mandatory – replacing the indigenous Ural-Altaic languages in many urban areas – although today the use of Central Asian languages is being revived in the new republics. India's linguistic mosaic comprises Dravidian languages, such as Tamil, in the south, and the Indo-Aryan languages of the north such as Hindi. In China, three main languages, Mandarin Chinese, Wu Chinese and Cantonese, share the same written form but their spoken dialects are mutually unintelligible.

▲ *Each year, Mongolians celebrate their ancient culture at the Naadam festival of the Three Games of Men. Children aged between 7 and 12 take part in the finale; a 20 mile (32 km) cross-country horse race in full traditional dress.*

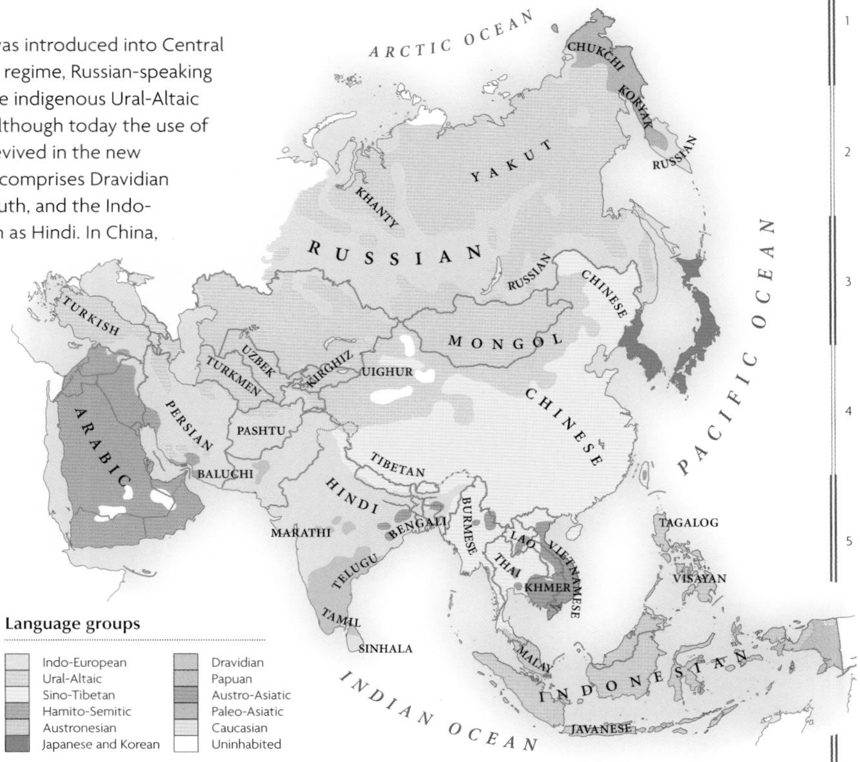
ARCTIC OCEAN
CHUKCHI
KORYAK
RUSSIAN
YAKUT
KHANTY
RUSSIAN
RUSSIAN
CHINESE
TURKISH
MONGOL
UZBEK
TURKMEN
KIRGHIZ
UIGHUR
CHINESE
PACIFIC OCEAN
ARABIC
PERSIAN
PASHTU
BALUCHI
TIBETAN
HINDI
BURMESE
BENGALI
MARATHI
LAO
VIETNAMESE
THAI
TAGALOG
TELUGU
KHMER
VISAYAN
TAMIL
SINHALA
MALAY
INDONESIAN
JAVANESE
INDIAN OCEAN
Language groups
Indo-European
Ural-Altaic
Sino-Tibetan
Hamito-Semitic
Austronesian
Japanese and Korean
Dravidian
Papuan
Austro-Asiatic
Paleo-Asiatic
Caucasian
Uninhabited

Transport

The transport system varies enormously in extent and quality across Asia. Early trade routes included the Silk Route, from Beijing across Central Asia, and the sea routes around the coastline of southern Asia. Today, transport networks often radiate from coastal ports, reflecting the continuing importance of sea and river travel for trade and external communications. In the interior, high mountain barriers such as the Himalayas, the Altai Mountains and the Tien Shan, deserts like the Gobi, Takla Makan and Ar Rub' al Khali, remain virtually impenetrable to most modern terrestrial transport. Major engineering feats are necessary to conquer these hostile frontier territories, although the success of the Trans-Siberian Railway in overcoming the harsh Siberian landscape, proves that cross-continental transport, if not economically viable, is physically possible.

Transport
major roads and motorways
major railways
international borders
transport intersections
international airports
major ports
Magadan
Tynda
Yekaterinburg
Chelyabinsk
Omsk
Novosibirsk
Irkutsk
Chita
Khabarovsk
Kholmsk
Ural'sk
Takikawa
Qiqihar
Harbin
Nakhodka
Izmir
Ankara
Sokhumi
Karaganda
Olgiy
Ulan Bator
Shenyang
Ch'ongjin
Tokyo
Tbilisi
Pyongyang
Nagoya
Yokohama
Iskenderun
Yerevan
Baku
Limassol
Beirut
Damascus
Turkmenbasy
Shymkent
Almaty
Urumqi
Qinhuangdao
Seoul
Beijing
Dalian
Busan
Osaka
Kobe
Haifa
Tel Aviv-Yafo
Amman
Tashkent
Bishkek
Shijiazhuang
Asgabat
Dushanbe
Tehran
Baghdad
Qom
Tabuk
Zhengzhou
Shanghai
Lanzhou
Basra
Herat
Kabul
Islamabad
Baoji
Xi'an
Nanjing
Kuwait
Yanbu' al Bahr
Buraydah
Kerman
Peshawar
Wuhan
Jedda
Riyadh
Manama
Quetta
Lahore
Chengdu
Jilong
Doha
Bandar-e 'Abbas
Lhasa
Chongqing
Hengyang
Dubai
Delhi
Kathmandu
Guiyang
Abu Dhabi
Huangpu
Gaoxiong
Abha
Muscat
Karachi
Varanasi
Kunming
Guangzhou
Hong Kong
Indore
Dhaka
Chittagong
Hanoi
Hodeida
Sana
Kandla
Haikou
Manila
Legazpi City
Kolkata (Calcutta)
Mandalay
Ta'izz
Aden
Salalah
Mumbai (Bombay)
Nagpur
Vientiane
Da Nang
Cebu
Visakhapatnam
Yangon
Chiang Mai
Hyderabad
Vijayawada
Pakxe
Bangkok
Davao
Bangalore
Mangalore
Chennai (Madras)
Phnom Penh
Ho Chi Minh City
Sandakan
Kochi
Jaffna
Labuan
Phuket
Hat Yai
Bandar Seri Begawan
George Town
Colombo
Belawan
Kuala Lumpur
Kuching
Pelabuhan Klang
Singapore
Jakarta
Surabaya
Denpasar

▲ *Both India and China rely upon extensive railway systems to transport their freight and passengers. China's network is constantly expanding, in particular the link between Golmud and Lhasa, which was completed in 2006 to become the highest railway in the world.*

▲ *The Karakoram Highway linking Mansehra in northern Pakistan with Kashi in western China was finally completed in 1978, 20 years after construction began. Regular mudslides and rockfalls necessitate continual maintenance for the road to remain open.*

Scale 1:32,500,000

Km 0 200 400 600 800

Miles 0 200 400 600 800

projection: Lambert Azimuthal Equal Area

Asian resources

Although agriculture remains the economic mainstay of most Asian countries, the number of people employed in agriculture has steadily declined, as new industries have been developed during the past 30 years. China, Indonesia, Malaysia, Thailand and Turkey have all experienced far-reaching structural change in their economies, while the breakup of the Soviet Union has created a new economic challenge in the Central Asian republics. The countries of the Persian Gulf illustrate the rapid transformation from rural nomadism to modern, urban society which oil wealth has brought to parts of the continent. Asia's most economically dynamic countries, Japan, Singapore, South Korea, and Taiwan, fringe the Pacific Ocean and are known as the Pacific Rim. In contrast, other Southeast Asian countries like Laos and Cambodia remain both economically and industrially underdeveloped.

Standard of living

Despite Japan's high standards of living, and Southwest Asia's oil-derived wealth, immense disparities exist across the continent. Afghanistan remains one of the world's most underdeveloped nations, as do the mountain states of Nepal and Bhutan. Further rapid population growth is exacerbating poverty and overcrowding in many parts of India and Bangladesh.

Standard of living
(UN human development index)

low

high

Industry

East Asian industry leads the continent in both productivity and efficiency; electronics, hi-tech industries, car manufacture and shipbuilding are important. The so-called economic 'tigers' of the Pacific Rim are Japan, South Korea and Taiwan and in recent years China has rediscovered its potential as an economic superpower. Heavy industries such as engineering, chemicals, and steel typify the industrial complexes along the corridor created by the Trans-Siberian Railway, the Fergana Valley in Central Asia, and also much of the huge industrial plain of east China. The discovery of oil in the Persian Gulf has brought immense wealth to countries that previously relied on subsistence agriculture on marginal desert land.

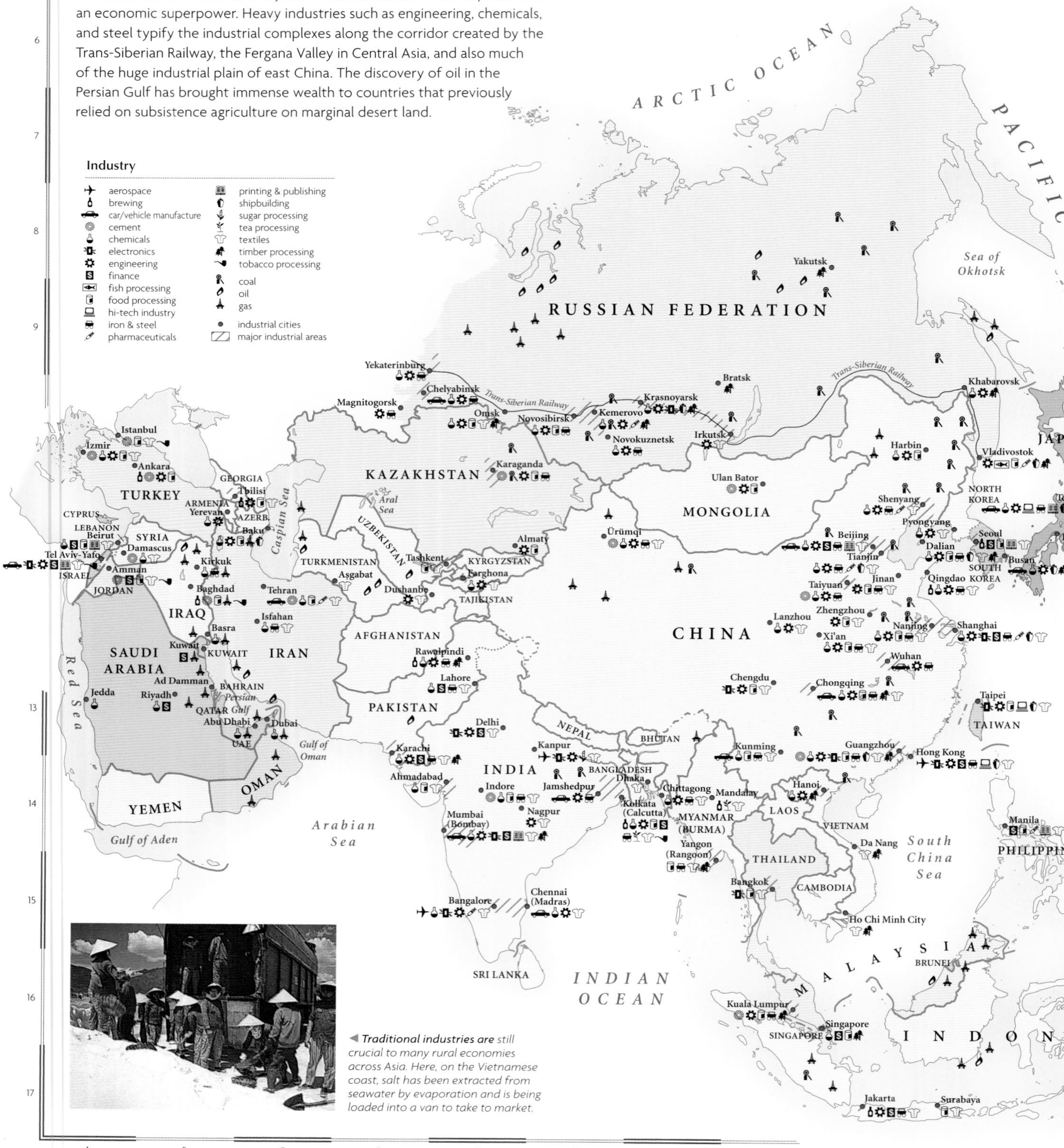

Industry
aerospace
brewing
car/vehicle manufacture
cement
chemicals
electronics
engineering
finance
fish processing
food processing
hi-tech industry
iron & steel
pharmaceuticals
printing & publishing
shipbuilding
sugar processing
tea processing
textiles
timber processing
tobacco processing
coal
oil
gas
industrial cities
major industrial areas
GNI per capita (US$)
below 1999
2000–4999
5000–9999
10,000–19,999
20,000–24,999
above 25,000
ARCTIC OCEAN
PACIFIC OCEAN
INDIAN OCEAN
Sea of Okhotsk
RUSSIAN FEDERATION
KAZAKHSTAN
MONGOLIA
CHINA
JAPAN
NORTH KOREA
SOUTH KOREA
TAIWAN
PHILIPPINES
INDONESIA
MALAYSIA
BRUNEI
SINGAPORE
EAST TIMOR
VIETNAM
LAOS
THAILAND
CAMBODIA
MYANMAR (BURMA)
BANGLADESH
BHUTAN
NEPAL
INDIA
SRI LANKA
PAKISTAN
AFGHANISTAN
IRAN
IRAQ
TURKEY
SYRIA
LEBANON
ISRAEL
JORDAN
CYPRUS
GEORGIA
ARMENIA
AZERB.
TURKMENISTAN
UZBEKISTAN
KYRGYZSTAN
TAJIKISTAN
SAUDI ARABIA
KUWAIT
BAHRAIN
QATAR
UAE
OMAN
YEMEN
Aral Sea
Caspian Sea
Red Sea
Persian Gulf
Gulf of Oman
Gulf of Aden
Arabian Sea
South China Sea
Trans-Siberian Railway
Yekaterinburg
Chelyabinsk
Magnitogorsk
Omsk
Novosibirsk
Kemerovo
Krasnoyarsk
Novokuznetsk
Irkutsk
Bratsk
Yakutsk
Khabarovsk
Vladivostok
Karaganda
Almaty
Tashkent
Farghona
Dushanbe
Asgabat
Ürümqi
Ulan Bator
Harbin
Shenyang
Pyongyang
Seoul
Busan
Dalian
Beijing
Tianjin
Jinan
Qingdao
Taiyuan
Zhengzhou
Lanzhou
Xi'an
Nanjing
Shanghai
Wuhan
Chengdu
Chongqing
Taipei
Guangzhou
Hong Kong
Kunming
Hanoi
Da Nang
Ho Chi Minh City
Bangkok
Yangon (Rangoon)
Mandalay
Chittagong
Dhaka
Kolkata (Calcutta)
Kanpur
Jamshedpur
Delhi
Indore
Nagpur
Mumbai (Bombay)
Ahmadabad
Bangalore
Chennai (Madras)
Karachi
Lahore
Rawalpindi
Tokyo
Nagoya
Kobe
Manila
Kuala Lumpur
Singapore
Jakarta
Surabaya
Istanbul
Izmir
Ankara
Tbilisi
Yerevan
Baku
Tehran
Isfahan
Kirkuk
Baghdad
Basra
Kuwait
Ad Damman
Riyadh
Jedda
Abu Dhabi
Dubai
Damascus
Beirut
Amman
Tel Aviv-Yafo

▲ **On a small** island at the southern tip of the Malay Peninsula lies Singapore, one of the Pacific Rim's most vibrant economic centres. Multinational banking and finance form the core of the city's wealth.

▲ **Iron and steel,** engineering and shipbuilding typify the heavy industry found in eastern China's industrial cities, especially the nation's leading manufacturing centre, Shanghai.

◀ **Traditional industries are** still crucial to many rural economies across Asia. Here, on the Vietnamese coast, salt has been extracted from seawater by evaporation and is being loaded into a van to take to market.

Environmental issues

The transformation of Uzbekistan by the former Soviet Union into the world's fifth largest producer of cotton led to the diversion of several major rivers for irrigation. Starved of this water, the Aral Sea diminished in volume by over 90% since 1960, irreversibly altering the ecology of the area. Heavy industries in eastern China have polluted coastal waters, rivers and urban air, while in Myanmar (Burma), Malaysia and Indonesia, ancient hardwood rainforests are felled faster than they can regenerate.

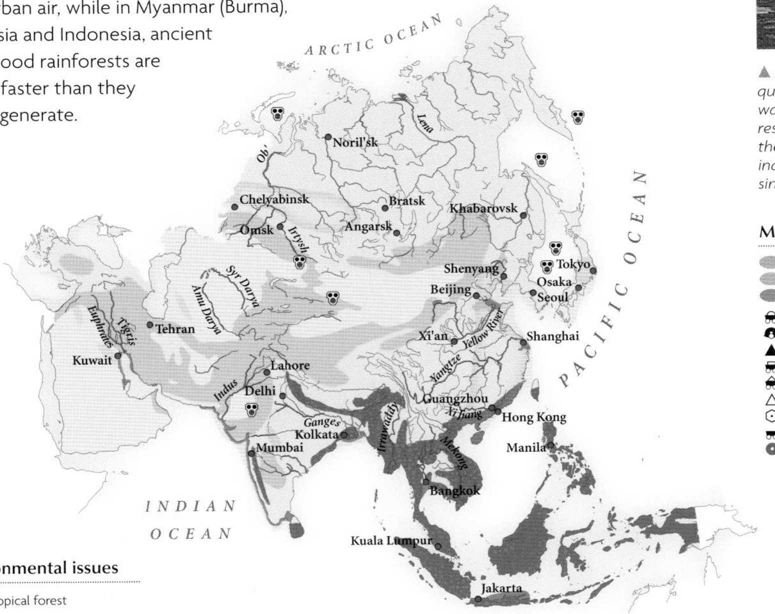
ARCTIC OCEAN
PACIFIC OCEAN
INDIAN OCEAN
Noril'sk
Ob'
Lena
Chelyabinsk
Bratsk
Khabarovsk
Omsk
Irtysh
Angarsk
Syr Darya
Amu Darya
Shenyang
Tokyo
Osaka
Seoul
Beijing
Tehran
Euphrates
Tigris
Kuwait
Xi'an
Yellow River
Yangtze
Shanghai
Lahore
Indus
Delhi
Guangzhou
Xi Jiang
Ganges
Kolkata
Hong Kong
Irrawaddy
Mekong
Mumbai
Manila
Bangkok
Kuala Lumpur
Jakarta

Environmental issues

- tropical forest
- forest destroyed
- desert
- desertification
- acid rain
- polluted rivers
- marine pollution
- heavy marine pollution
- radioactive contamination
- poor urban air quality

▲ *Although Siberia remains a quintessentially frozen, inhospitable wasteland, vast untapped mineral reserves – especially the oil and gas of the West Siberian Plain – have lured industrial development to the area since the 1950s and 1960s.*

Mineral resources

- oil field
- gas field
- coal field
- chromite
- copper
- gold
- iron
- lead
- nickel
- platinum
- tin
- wolfram

◀ ***Commercial logging activities** in Borneo have placed great stress on the rainforest ecosystem. Government attempts to regulate the timber companies and control illegal logging have only been partially successful.*

Mineral resources

At least 60% of the world's known oil and gas deposits are found in Asia; notably the vast oil fields of the Persian Gulf, and the less-exploited oil and gas fields of the Ob' basin in west Siberia. Immense coal reserves in Siberia and China have been utilized to support large steel industries. Southeast Asia has some of the world's largest deposits of tin, found in a belt running down the Malay Peninsula to Indonesia.

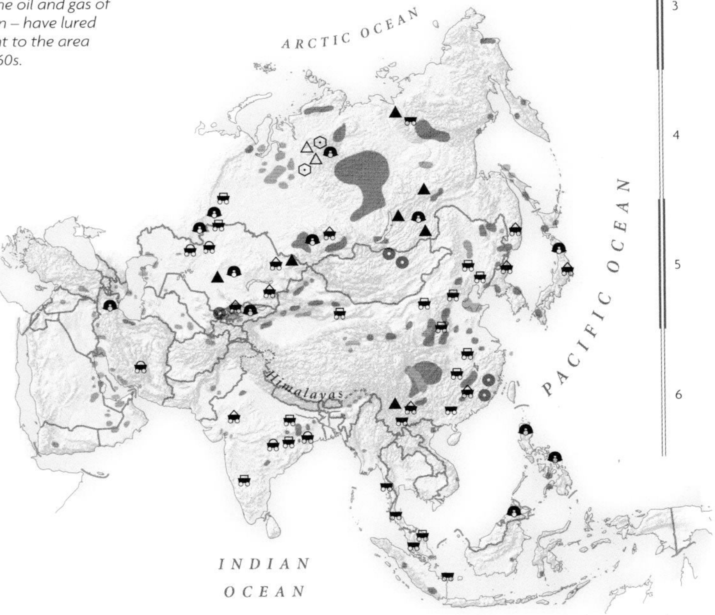
ARCTIC OCEAN
PACIFIC OCEAN
INDIAN OCEAN
Himalayas

Using the land and sea

Vast areas of Asia remain uncultivated as a result of unsuitable climatic and soil conditions. In favourable areas such as river deltas, farming is intensive. Rice is the staple crop of most Asian countries, grown in paddy fields on waterlogged alluvial plains and terraced hillsides, and often irrigated for higher yields. Across the black earth region of the Eurasian steppe in southern Siberia and Kazakhstan, wheat farming is the dominant activity. Cash crops, like tea in Sri Lanka and dates in the Arabian Peninsula, are grown for export, and provide valuable income. The sovereignty of the rich fishing grounds in the South China Sea is disputed by China, Malaysia, Taiwan, the Philippines and Vietnam, because of potential oil reserves.

▲ ***Date palms have** been cultivated in oases throughout the Arabian Peninsula since antiquity. In addition to the fruit, palms are used for timber, fuel, rope, and for making vinegar, syrup and a liquor known as arrack.*

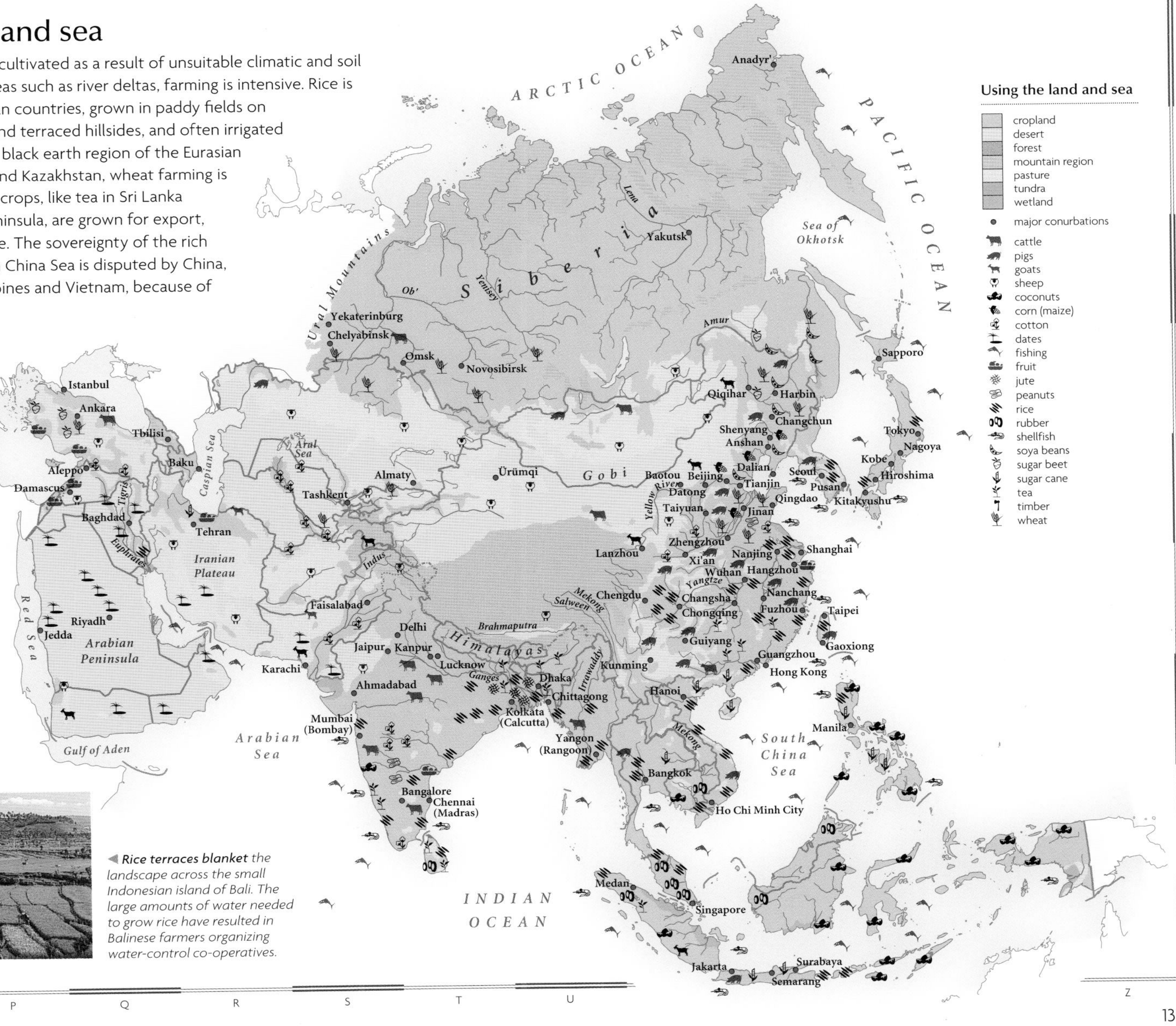
ARCTIC OCEAN
PACIFIC OCEAN
INDIAN OCEAN
Anadyr'
Sea of Okhotsk
Siberia
Lena
Yakutsk
Ural Mountains
Ob'
Yenisey
Yekaterinburg
Chelyabinsk
Omsk
Novosibirsk
Amur
Sapporo
Istanbul
Ankara
Tbilisi
Baku
Caspian Sea
Aral Sea
Qiqihar
Harbin
Changchun
Shenyang
Anshan
Tokyo
Nagoya
Kobe
Hiroshima
Aleppo
Damascus
Tigris
Baghdad
Euphrates
Tashkent
Almaty
Ürümqi
Gobi
Baotou
Beijing
Dalian
Tianjin
Seoul
Pusan
Datong
Qingdao
Kitakyushu
Yellow River
Taiyuan
Jinan
Tehran
Iranian Plateau
Indus
Lanzhou
Zhengzhou
Xi'an
Nanjing
Shanghai
Wuhan
Hangzhou
Yangtze
Red Sea
Faisalabad
Chengdu
Salween
Mekong
Changsha
Nanchang
Chongqing
Fuzhou
Taipei
Riyadh
Jedda
Arabian Peninsula
Delhi
Brahmaputra
Jaipur
Kanpur
Himalayas
Guiyang
Gaoxiong
Karachi
Lucknow
Ganges
Dhaka
Kunming
Guangzhou
Hong Kong
Ahmadabad
Chittagong
Irrawaddy
Hanoi
Kolkata (Calcutta)
Mumbai (Bombay)
Arabian Sea
Yangon (Rangoon)
Manila
South China Sea
Gulf of Aden
Bangkok
Bangalore
Chennai (Madras)
Ho Chi Minh City
Medan
Singapore
Jakarta
Surabaya
Semarang

Using the land and sea

- cropland
- desert
- forest
- mountain region
- pasture
- tundra
- wetland
- major conurbations
- cattle
- pigs
- goats
- sheep
- coconuts
- corn (maize)
- cotton
- dates
- fishing
- fruit
- jute
- peanuts
- rice
- rubber
- shellfish
- soya beans
- sugar beet
- sugar cane
- tea
- timber
- wheat

◀ ***Rice terraces blanket** the landscape across the small Indonesian island of Bali. The large amounts of water needed to grow rice have resulted in Balinese farmers organizing water-control co-operatives.*

Turkey & the Caucasus

ARMENIA, AZERBAIJAN, GEORGIA, TURKEY

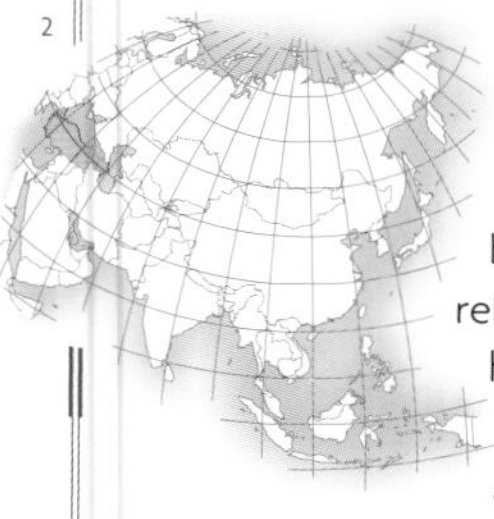

This region occupies the fragmented junction between Europe, Asia and the Russian Federation. Sunni Islam provides a common identity for the secular state of Turkey, which the revered leader Kemal Atatürk established from the remnants of the Ottoman Empire after the First World War. Turkey has a broad resource base and expanding trade links with Europe, but the east is relatively undeveloped and strife between the state and a large Kurdish minority has yet to be resolved. Georgia is similarly challenged by ethnic separatism, while the Christian state of Armenia and the mainly Muslim and oil-rich Azerbaijan are locked in conflict over the territory of Nagorno-Karabakh.

Transport and industry

Turkey leads the region's well-diversified economy. Petrochemicals, textiles, engineering and food processing are the main industries. Azerbaijan is able to export oil, while the other states rely heavily on hydro-electric power and imported fuel. Georgia produces precision machinery. War and earthquake damage have devastated Armenia's infrastructure.

▲ ***Azerbaijan has substantial** oil reserves, located in and around the Caspian Sea. They were some of the earliest oilfields in the world to be exploited.*

Major industry and infrastructure

- carpet weaving
- cement
- chemicals
- coal
- engineering
- food processing
- oil
- textiles
- tourism
- vehicle manufacture
- capital cities
- major towns
- international airports
- major roads
- major industrial areas

The transport network

- 114,867 miles (184,882 km)
- 5778 miles (9300 km)
- 8120 miles (13,069 km)
- 745 miles (1200 km)

Physical and political barriers have severely limited communications between Armenia, Georgia and Azerbaijan. Turkey has a relatively well-developed transport network.

Using the land and sea

Turkey is largely self-sufficient in food. The irrigated Black Sea coastlands have the world's highest yields of hazelnuts. Tobacco, cotton, sultanas, tea and figs are the region's main cash crops and a great range of fruit and vegetables are grown. Wine grapes are among the labour-intensive crops which allow full use of limited agricultural land in the Caucasus. Sturgeon fishing is particularly important in Azerbaijan.

Land use and agricultural distribution

- cattle
- goats
- cotton
- fishing
- fruit
- hazelnuts
- olives
- sugar beet
- tobacco
- vineyards
- capital cities
- major towns
- pasture
- cropland
- forest

The urban/rural population divide

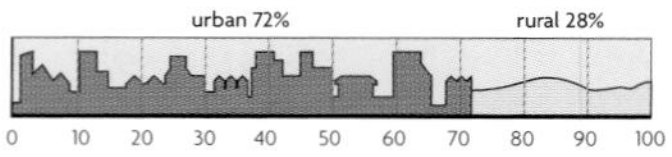

Population density	Total land area
238 people per sq mile (92 people per sq km)	368,912 sq miles (955,730 sq km)

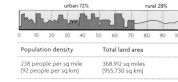

▲ ***For many centuries,** Istanbul has held tremendous strategic importance as a crucial gateway between Europe and Asia. Founded by the Greeks as Byzantium, the city became the centre of the East Roman Empire and was known as Constantinople to the Romans. From the 15th century onwards the city became the centre of the great Ottoman Empire.*

The landscape

The deeply-eroded hills and salty basins of the Anatolian Plateau are bordered by several mountain ranges along the Black Sea coast, and the limestone Taurus Mountains *(Toros Daglari)* in the south. A lowland trough divides the Caucasus and the Lesser Caucasus, which form a formidable barrier of peaks in the north.

Limestone weathering in the Anatolian Plateau

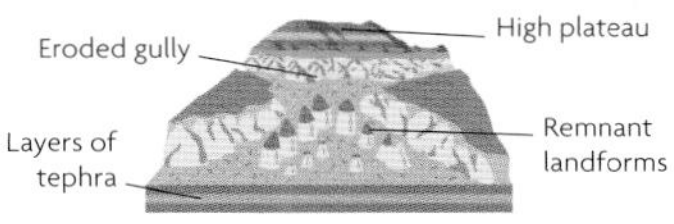

▲ ***In central Turkey,*** *rainwater has chemically weathered away numerous layers of limestone, leaving isolated outcrops and pinnacles and deep eroded gullies.*

▶ ***The Caucasus are*** *fold mountains, which formed around the same time as the Taurus Mountains (Toros Daglari) around 65 million years ago and have since been modified by volcanic eruptions.*

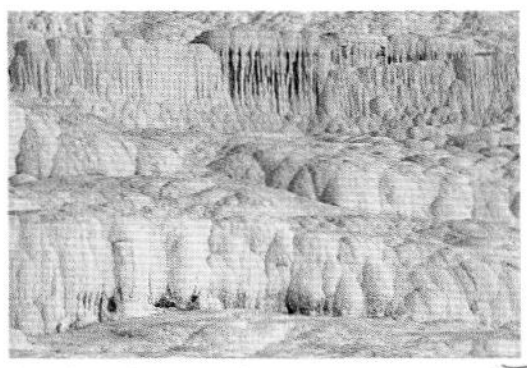

▲ ***The white rock terraces*** *at Pamukkale in western Turkey were formed when underground water, heated by volcanic activity, dissolved minerals in the rocks. When the water reached the surface and evaporated the minerals were left behind in these extraordinary formations.*

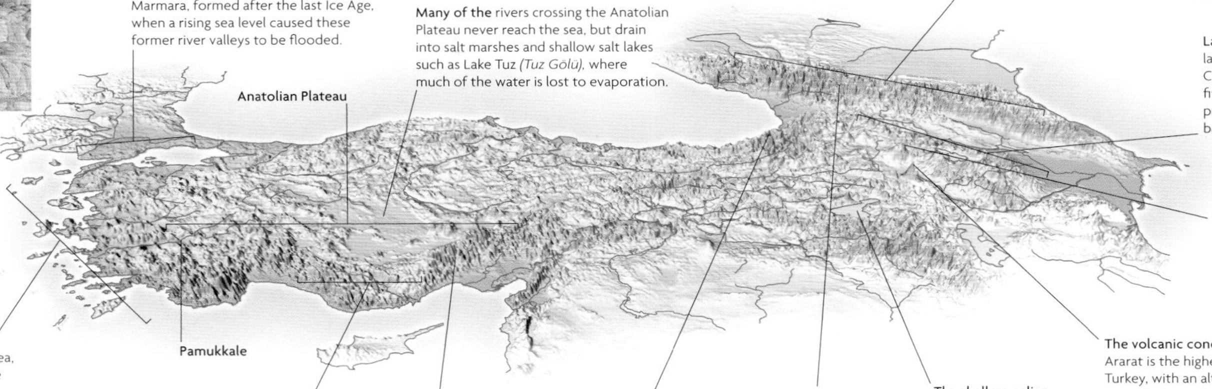

The straits of the Bosporus and the Dardanelles, respectively linking the Black and Mediterranean seas with the Sea of Marmara, formed after the last Ice Age, when a rising sea level caused these former river valleys to be flooded.

Many of the rivers crossing the Anatolian Plateau never reach the sea, but drain into salt marshes and shallow salt lakes such as Lake Tuz *(Tuz Gölü)*, where much of the water is lost to evaporation.

Lava has flowed over large areas of the Lesser Caucasus within the last five million years, producing extensive basalt plateaux.

The earthquake that struck Armenia in 1988 killed over 55,000 people and devastated the country's infrastructure.

Long, parallel mountain ranges run from east to west into the Aegean Sea, which has risen since the last Ice Age to form a drowned coastline of numerous islands and extended inlets.

The volcanic cone of Mount Ararat is the highest peak in Turkey, with an altitude of 16,853 ft (5137 m).

The folded peaks of the Taurus Mountains *(Toros Daglari)* were formed 60–65 million years ago, at the same time as the Alps. The rock is mainly limestone, with deep caves, gorges and underground rivers.

The Cilician Gates *(Gülek Bogazi)*, a major pass through the Taurus Mountains *(Toros Daglari)*, is the point where streams flow from the interior plateau onto the lowland of Adana.

Thick, temperate forest veils the seaward slopes of the Kaçkar Daglari. The southern slopes, which lie in a rainshadow, are dry and barren.

The granite massif near Surami divides the lowlands of Georgia from the oil-rich basin of Azerbaijan's Kura river, which has built a large delta into the Caspian Sea.

The shallow, saline Lake Van *(Van Gölü)* is the largest lake in Turkey. Dry terraces mark a previous shoreline 181 ft (55 m) above the present water level.

▶ ***Since the 6th century BC,*** *the pinnacles and caves of east-central Anatolia have been utilized as dwellings. Many are still inhabited today.*

Map key

Population

- above 5 million
- 1 million to 5 million
- 500,000 to 1 million
- 100,000 to 500,000
- 50,000 to 100,000
- 10,000 to 50,000
- below 10,000

Elevation

- 4000m / 13,124ft
- 3000m / 9843ft
- 2000m / 6562ft
- 1000m / 3281ft
- 500m / 1640ft
- 250m / 820ft
- 100m / 328ft
- sea level

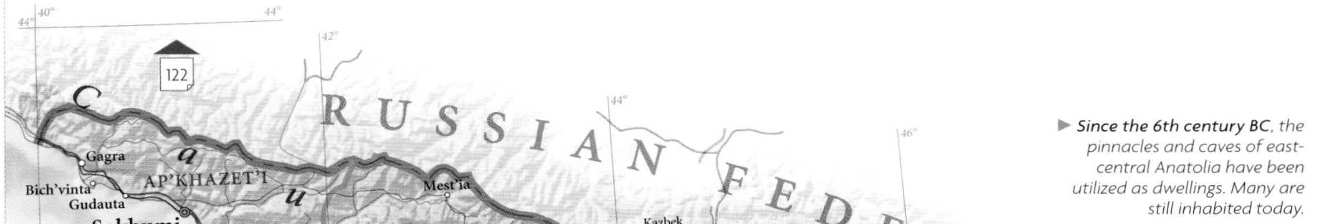

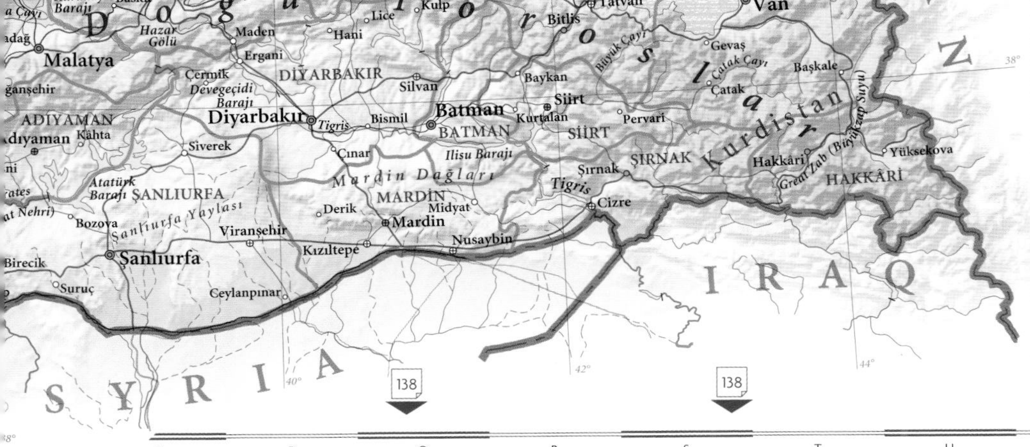

Scale 1:4,500,000

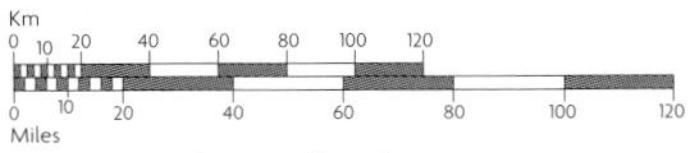

projection: Lambert Conformal Conic

▲ ***The fisheries of*** *Azerbaijan are noted for their hauls of sturgeon, and the Caspian Sea accounts for 80% of the world's total catch. However, stocks are now under serious threat due to overfishing.*

▲ ***Traditional steam baths*** *are found throughout the region, and are used for socializing as well as for bathing.*

The Near East

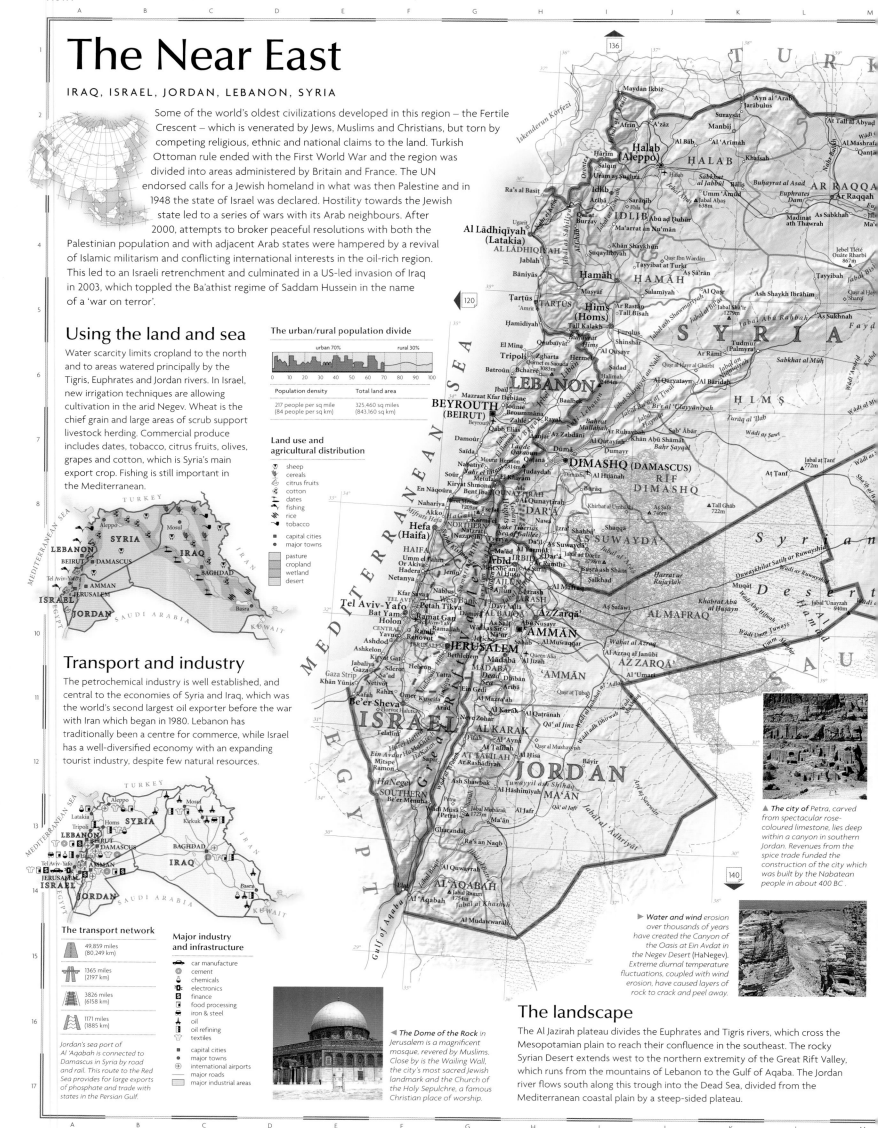

IRAQ, ISRAEL, JORDAN, LEBANON, SYRIA

Some of the world's oldest civilizations developed in this region – the Fertile Crescent – which is venerated by Jews, Muslims and Christians, but torn by competing religious, ethnic and national claims to the land. Turkish Ottoman rule ended with the First World War and the region was divided into areas administered by Britain and France. The UN endorsed calls for a Jewish homeland in what was then Palestine and in 1948 the state of Israel was declared. Hostility towards the Jewish state led to a series of wars with its Arab neighbours. After 2000, attempts to broker peaceful resolutions with both the Palestinian population and with adjacent Arab states were hampered by a revival of Islamic militarism and conflicting international interests in the oil-rich region. This led to an Israeli retrenchment and culminated in a US-led invasion of Iraq in 2003, which toppled the Ba'athist regime of Saddam Hussein in the name of a 'war on terror'.

Using the land and sea

Water scarcity limits cropland to the north and to areas watered principally by the Tigris, Euphrates and Jordan rivers. In Israel, new irrigation techniques are allowing cultivation in the arid Negev. Wheat is the chief grain and large areas of scrub support livestock herding. Commercial produce includes dates, tobacco, citrus fruits, olives, grapes and cotton, which is Syria's main export crop. Fishing is still important in the Mediterranean.

The urban/rural population divide

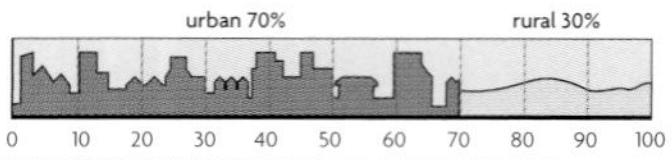

Population density	Total land area
217 people per sq mile (84 people per sq km)	325,460 sq miles (843,160 sq km)

Land use and agricultural distribution

- sheep
- cereals
- citrus fruits
- cotton
- dates
- fishing
- rice
- tobacco
- capital cities
- major towns
- pasture
- cropland
- wetland
- desert

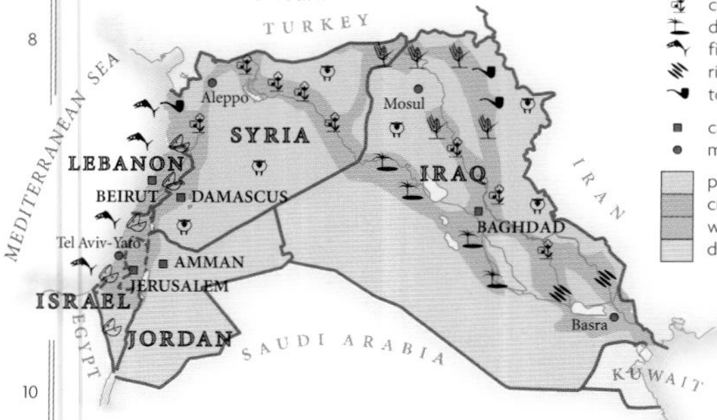

Transport and industry

The petrochemical industry is well established, and central to the economies of Syria and Iraq, which was the world's second largest oil exporter before the war with Iran which began in 1980. Lebanon has traditionally been a centre for commerce, while Israel has a well-diversified economy with an expanding tourist industry, despite few natural resources.

The transport network

- 49,859 miles (80,249 km)
- 1365 miles (2197 km)
- 3826 miles (6158 km)
- 1171 miles (1885 km)

Jordan's sea port of Al 'Aqabah is connected to Damascus in Syria by road and rail. This route to the Red Sea provides for large exports of phosphate and trade with states in the Persian Gulf.

Major industry and infrastructure

- car manufacture
- cement
- chemicals
- electronics
- finance
- food processing
- iron & steel
- oil
- oil refining
- textiles
- capital cities
- major towns
- international airports
- major roads
- major industrial areas

◄ *The Dome of the Rock in Jerusalem is a magnificent mosque, revered by Muslims. Close by is the Wailing Wall, the city's most sacred Jewish landmark and the Church of the Holy Sepulchre, a famous Christian place of worship.*

▲ *The city of Petra, carved from spectacular rose-coloured limestone, lies deep within a canyon in southern Jordan. Revenues from the spice trade funded the construction of the city which was built by the Nabatean people in about 400 BC.*

► *Water and wind erosion over thousands of years have created the Canyon of the Oasis at Ein Avdat in the Negev Desert (HaNegev). Extreme diurnal temperature fluctuations, coupled with wind erosion, have caused layers of rock to crack and peel away.*

The landscape

The Al Jazirah plateau divides the Euphrates and Tigris rivers, which cross the Mesopotamian plain to reach their confluence in the southeast. The rocky Syrian Desert extends west to the northern extremity of the Great Rift Valley, which runs from the mountains of Lebanon to the Gulf of Aqaba. The Jordan river flows south along this trough into the Dead Sea, divided from the Mediterranean coastal plain by a steep-sided plateau.

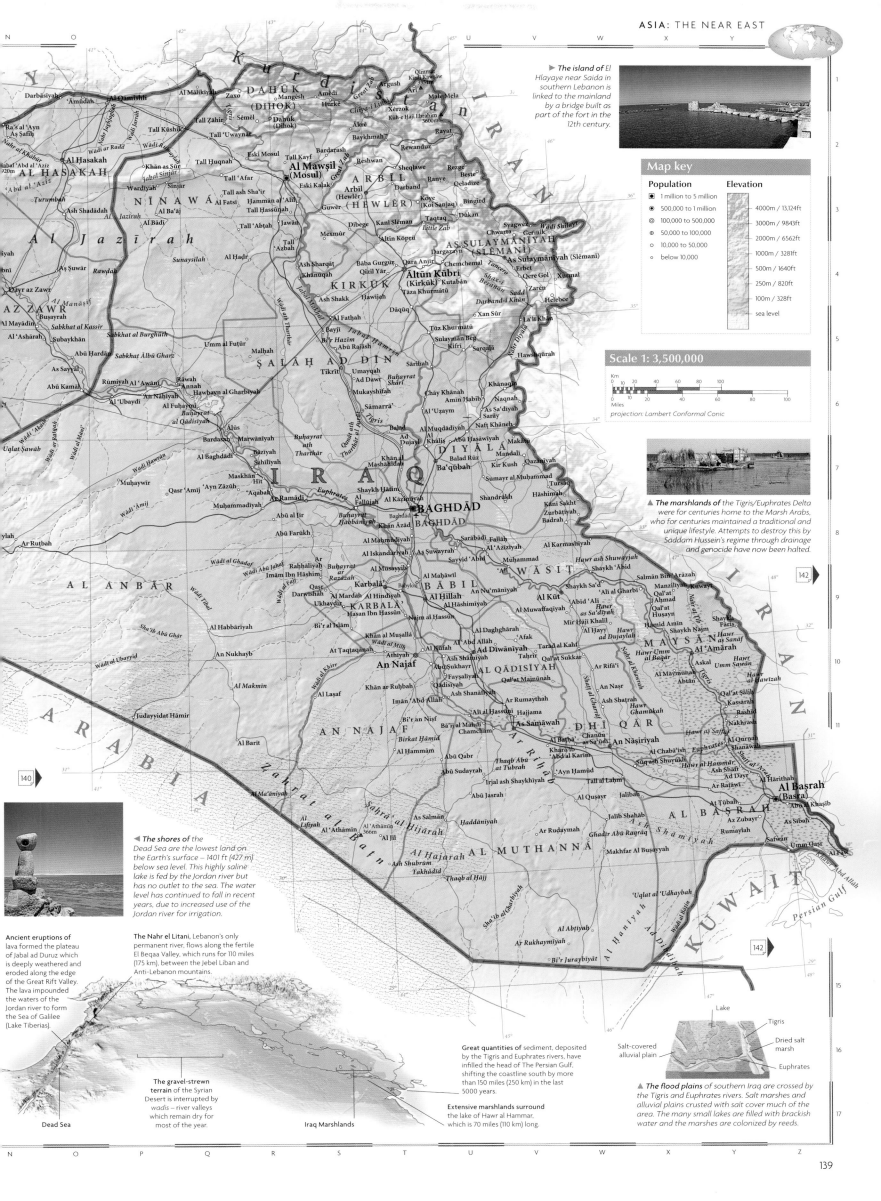

▶ The island of El Hlayaye near Saida in southern Lebanon is linked to the mainland by a bridge built as part of the fort in the 12th century.

Map key

Population

- 1 million to 5 million
- 500,000 to 1 million
- 100,000 to 500,000
- 50,000 to 100,000
- 10,000 to 50,000
- below 10,000

Elevation

- 4000m / 13,124ft
- 3000m / 9843ft
- 2000m / 6562ft
- 1000m / 3281ft
- 500m / 1640ft
- 250m / 820ft
- 100m / 328ft
- sea level

Scale 1: 3,500,000

Km 0 10 20 40 60 80 100

0 10 20 40 60 80 100 Miles

projection: Lambert Conformal Conic

▲ The marshlands of the Tigris/Euphrates Delta were for centuries home to the Marsh Arabs, who for centuries maintained a traditional and unique lifestyle. Attempts to destroy this by Saddam Hussein's regime through drainage and genocide have now been halted.

◀ The shores of the Dead Sea are the lowest land on the Earth's surface – 1401 ft (427 m) below sea level. This highly saline lake is fed by the Jordan river but has no outlet to the sea. The water level has continued to fall in recent years, due to increased use of the Jordan river for irrigation.

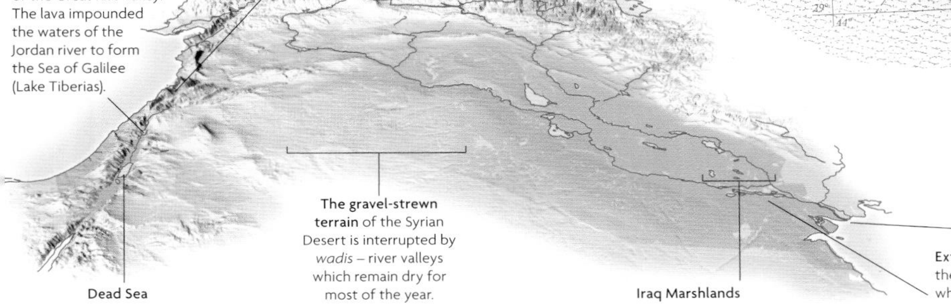

Ancient eruptions of lava formed the plateau of Jabal ad Duruz which is deeply weathered and eroded along the edge of the Great Rift Valley. The lava impounded the waters of the Jordan river to form the Sea of Galilee (Lake Tiberias).

The Nahr el Litani, Lebanon's only permanent river, flows along the fertile El Beqaa Valley, which runs for 110 miles (175 km), between the Jebel Liban and Anti-Lebanon mountains.

The gravel-strewn terrain of the Syrian Desert is interrupted by wadis – river valleys which remain dry for most of the year.

Great quantities of sediment, deposited by the Tigris and Euphrates rivers, have infilled the head of The Persian Gulf, shifting the coastline south by more than 150 miles (250 km) in the last 5000 years.

Extensive marshlands surround the lake of Hawr al Hammar, which is 70 miles (110 km) long.

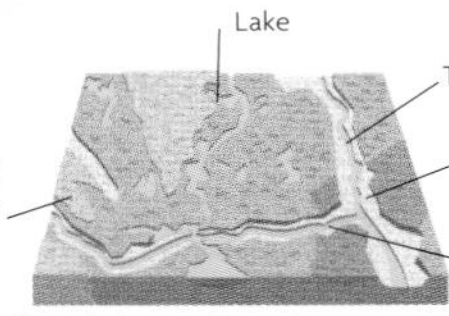

▲ The flood plains of southern Iraq are crossed by the Tigris and Euphrates rivers. Salt marshes and alluvial plains crusted with salt cover much of the area. The many small lakes are filled with brackish water and the marshes are colonized by reeds.

The Arabian Peninsula

BAHRAIN, KUWAIT, OMAN, QATAR, SAUDI ARABIA, UNITED ARAB EMIRATES (UAE), YEMEN

Huge expanses of desert cover much of the Arabian Peninsula, limiting settlement to oases, the mountains along the Red Sea and coastal belts. The most populous area is the fertile highlands of Yemen. The Islamic faith and Arabic language give the region a cultural and religious unity, and the Saudi city of Mecca *(Makkah)* is Islam's most holy place, visited by over two million pilgrims each year. More than half the world's oil reserves are contained in this region, and the exploitation of oil and gas has brought great wealth, particularly to Saudi Arabia. Yemen and Oman are the least developed of the Arabian states, with large rural populations. Within Saudi Arabia over 86% of the people live in urban areas.

Using the land

Most of the Arabian Peninsula is unsuited to settled agriculture, making irrigation and land reclamation projects essential. The narrow coastal plain and isolated oases, commonly amounting to less than 1% of the land area, are used to cultivate grains, coffee and exotic fruits. Goats, sheep and camels are widespread throughout the region.

The urban/rural population divide

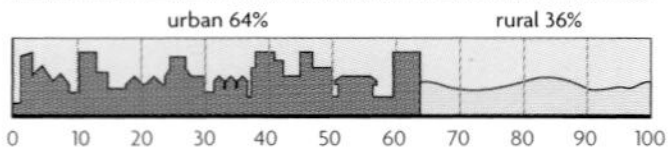

Population density	Total land area
50 people per sq mile (19 people per sq km)	1,147,856 sq miles (2,973,720 sq km)

Land use and agricultural distribution

- goats
- sheep
- cereals
- coffee
- dates
- fruit
- capital cities
- major towns
- pasture
- cropland
- desert

◀ *The fertile soils of Yemen have encouraged settlement of almost all of the land from sea level up to the mountains at 10,000 ft (3050 m). In the higher reaches elaborate terraces have been constructed to facilitate crop cultivation.*

The landscape

A plateau more than 2500 ft (760 m) high extends across much of the Arabian Peninsula. The plateau slopes eastwards from the massive, rifted escarpment along the coast of the Red Sea, to the shallow waters of the Persian Gulf. The interior is characterized by *cuestas* and valleys, drained by a system of *wadis*. A crescent of sand and gravel deserts lies to the east.

The An Nafud Desert is covered with *barchan* dunes varying between 30–100 ft (10–30 m) high. The 'horns' of the crescent-shaped dunes reflect the direction in which they are being moved by the wind.

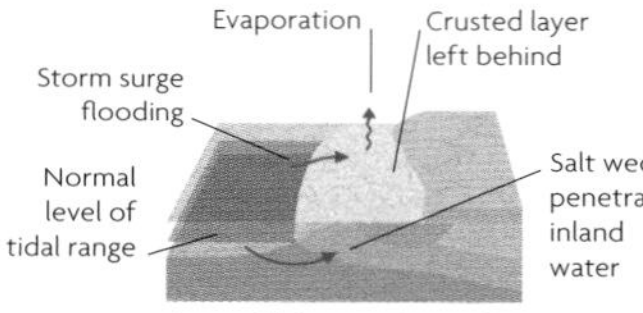

▲ ***A sabkha** is a flat, salt-encrusted plain which occurs near the coast just above the high water mark. Flooding by sea water leads to saturation of the land with saline-rich groundwater. As this evaporates, a cracked layer of sand, cemented together with salt, gypsum and calcium carbonate is left behind.*

Inselbergs are dotted over a wide area of the Najd Plateau. These resistant remnants of the ancient basement rock are left standing when the softer weathered rock has been worn away.

Few areas in the Arabian Peninsula have rivers flowing through them. Most are drained by ephemeral watercourses called *wadis*.

Across the Najd Plateau the flat relief is broken by *mesas*; steep-sided rock plateaux and *cuestas*; ridges with one steep and one gentle slope.

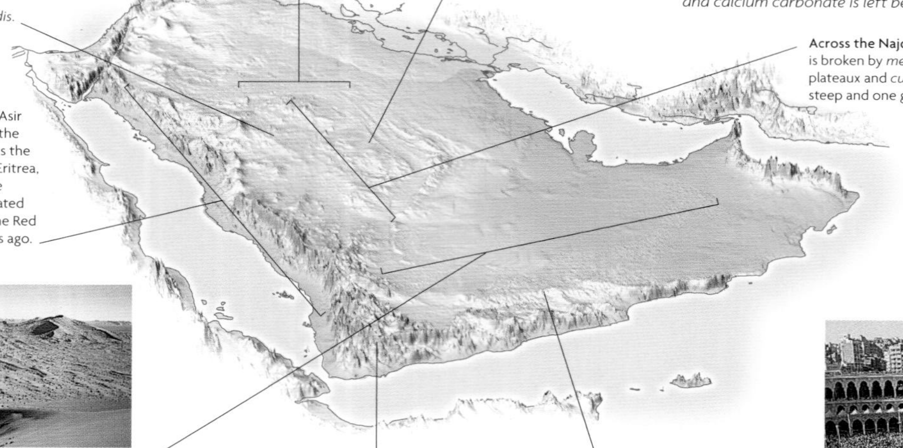

The Hejaz *(Al Hijaz)* and Asir mountains form part of the same geological region as the highlands of Sudan and Eritrea, to which they were once joined. They were separated when faulting opened the Red Sea, over 50 million years ago.

▲ ***Ar Rub' al Khali**, also known as the Empty Quarter, is the most arid part of the Arabian Peninsula. It is the largest uninterrupted sand desert in the world. Ridges of sand up to 25 miles (40 km) long, run northeast–southwest, giving characteristic linear dunes.*

The Jabal an Nabi Shu'ayb in Yemen is the highest point on the peninsula, rising to 12,336 ft (3760 m).

The Arabian Shield underpins the west of the peninsula. It is a fragment of the ancient continent, Gondwanaland, which was separated by rifting millions of years ago.

◀ ***Every Muslim** must make at least one pilgrimage or hajj to Mecca (Makkah), in Saudi Arabia, during their lifetime. The cloth-covered shrine is called the Ka'bah, and is regarded by Muslims as the most sacred place on Earth.*

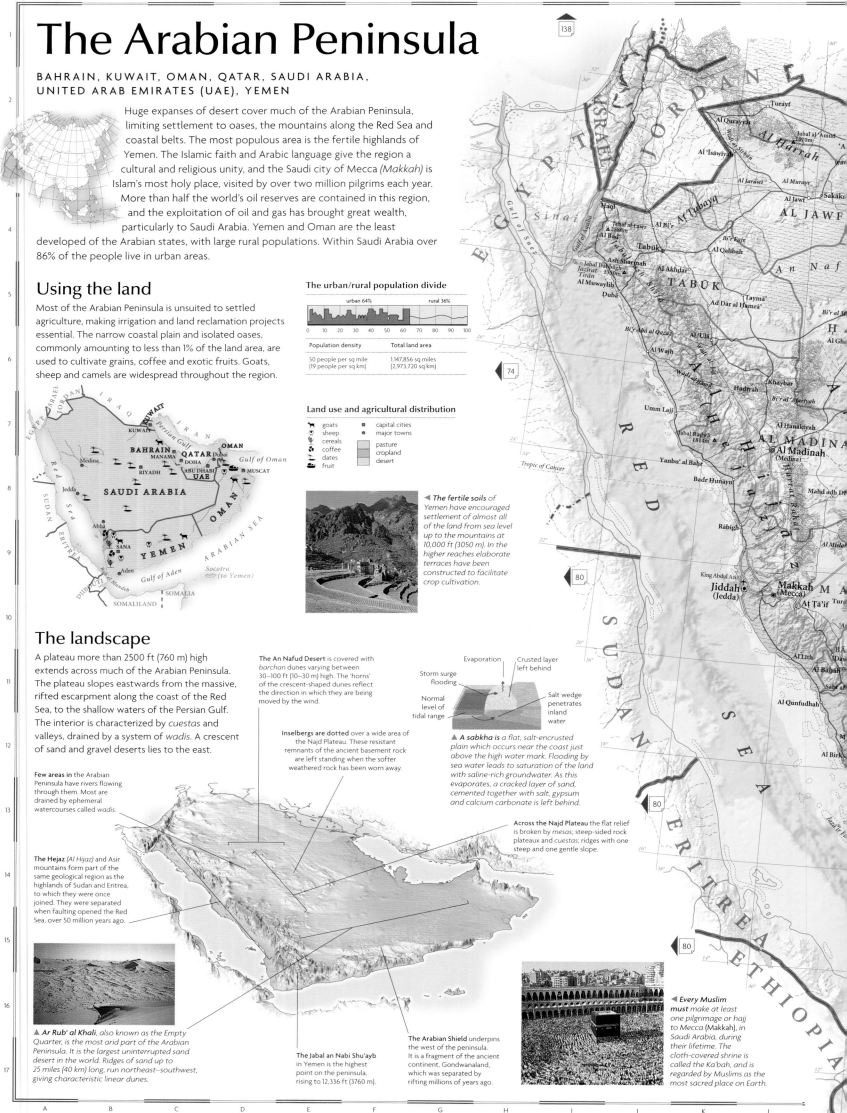

138

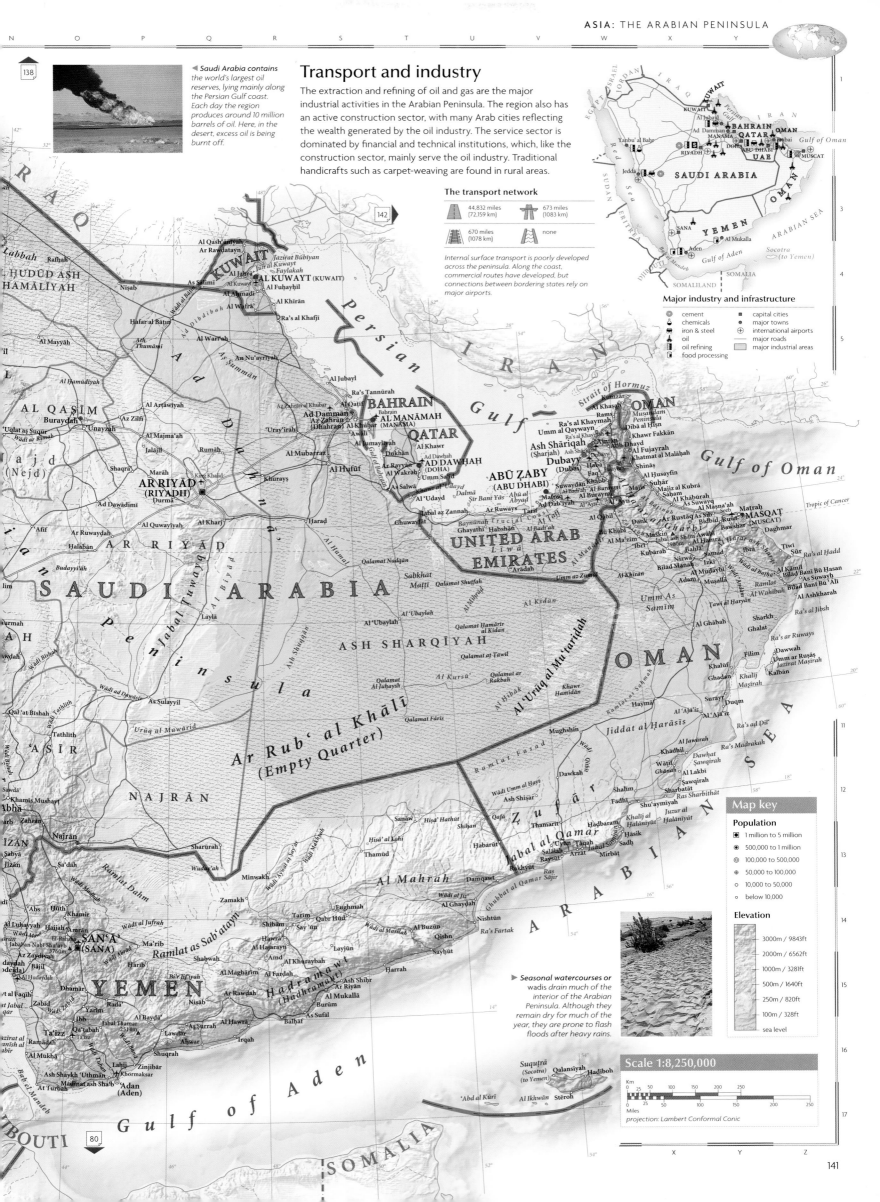

◀ *Saudi Arabia contains the world's largest oil reserves, lying mainly along the Persian Gulf coast. Each day the region produces around 10 million barrels of oil. Here, in the desert, excess oil is being burnt off.*

Transport and industry

The extraction and refining of oil and gas are the major industrial activities in the Arabian Peninsula. The region also has an active construction sector, with many Arab cities reflecting the wealth generated by the oil industry. The service sector is dominated by financial and technical institutions, which, like the construction sector, mainly serve the oil industry. Traditional handicrafts such as carpet-weaving are found in rural areas.

The transport network

44,832 miles (72,159 km)	673 miles (1083 km)
670 miles (1078 km)	none

Internal surface transport is poorly developed across the peninsula. Along the coast, commercial routes have developed, but connections between bordering states rely on major airports.

Major industry and infrastructure

- cement
- chemicals
- iron & steel
- oil
- oil refining
- food processing
- capital cities
- major towns
- international airports
- major roads
- major industrial areas

142

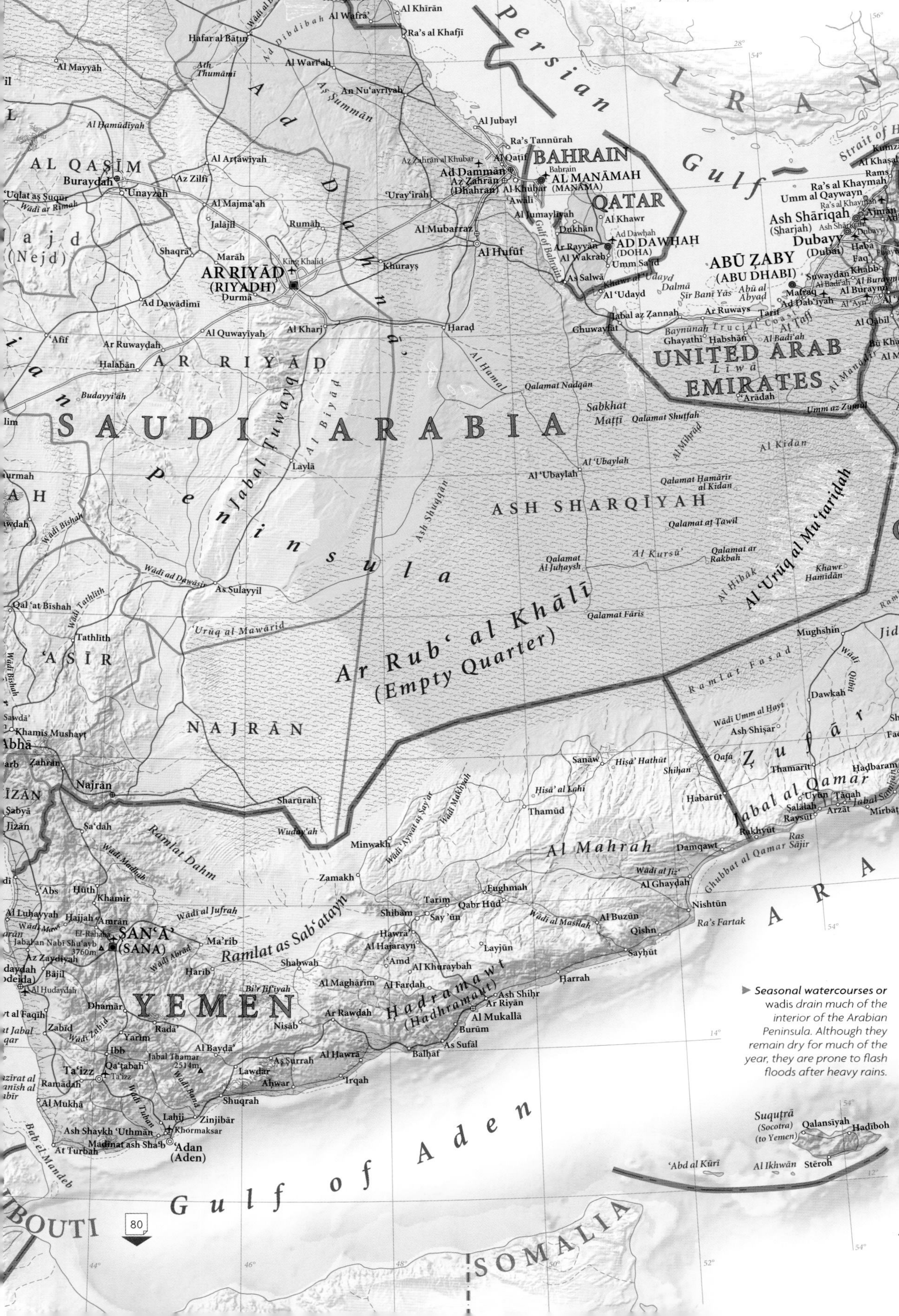

▶ *Seasonal watercourses or wadis drain much of the interior of the Arabian Peninsula. Although they remain dry for much of the year, they are prone to flash floods after heavy rains.*

Map key

Population

- 1 million to 5 million
- 500,000 to 1 million
- 100,000 to 500,000
- 50,000 to 100,000
- 10,000 to 50,000
- below 10,000

Elevation

- 3000m / 9843ft
- 2000m / 6562ft
- 1000m / 3281ft
- 500m / 1640ft
- 250m / 820ft
- 100m / 328ft
- sea level

Scale 1:8,250,000

Km 0 25 50 100 150 200 250
Miles 0 25 50 100 150 200 250

projection: Lambert Conformal Conic

80

Iran & the Gulf states

BAHRAIN, IRAN, KUWAIT, QATAR, UNITED ARAB EMIRATES (UAE)

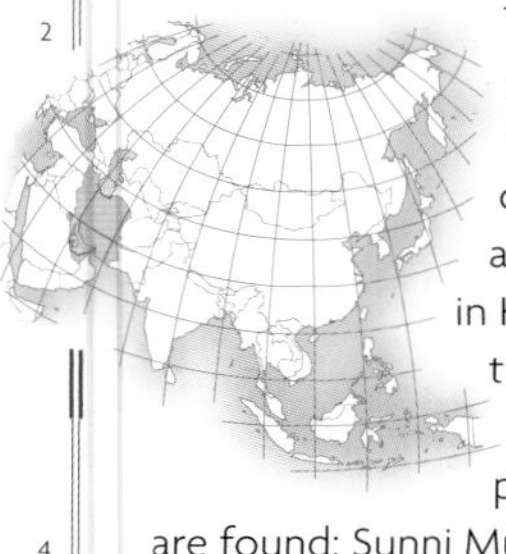

The discovery of oil in the Persian Gulf in the 1930s brought great wealth to the surrounding states. The revenue was largely used to modernize industry and infrastructure, initiating great social change in these formerly agrarian countries. Today, over 90% of the people in the Gulf states live in urban areas, and foreign nationals make up a sizeable proportion of the population in Kuwait, Qatar and the United Arab Emirates. The importance of control of the oil reserves has led to a number of territorial disputes, including most recently the Iran–Iraq War (1980-88) and the First Gulf War (1991). Islam is practised almost exclusively throughout the region and two distinct strands are found; Sunni Muslims in Qatar, Kuwait and UAE, and Shi'a Muslims in Iran and Bahrain. In 1979 Iran became the world's largest theocracy.

The landscape

The land rises steeply from the fragmented coastal lowlands bordering the Persian Gulf, to reach Iran's interior plateau, bounded by heavily-eroded mountain chains. An unstable plate boundary runs northwest to southeast across Iran causing frequent earthquakes. On the sandy west coast of the Persian Gulf, the relief is generally flat, with patches of salt marsh. Bahrain consists of two groups of islands, which are mostly small and rocky.

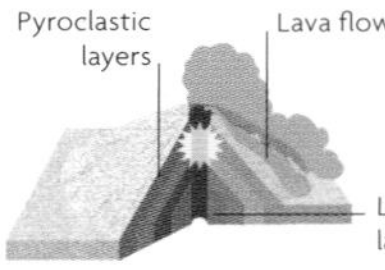

▲ *Qolleh-ye Damavand in the Elburz Mountains is a composite volcano. It comprises layers of lava and pyroclasts – fragmentary rocks which accumulate on the slopes of the volcano after being ejected into the air.*

▲ *Marine sediments from deep beneath the ancient Tethys Sea have been uplifted to form the Elburz Mountains, which stretch along the shores of the Caspian Sea, northern Iran.*

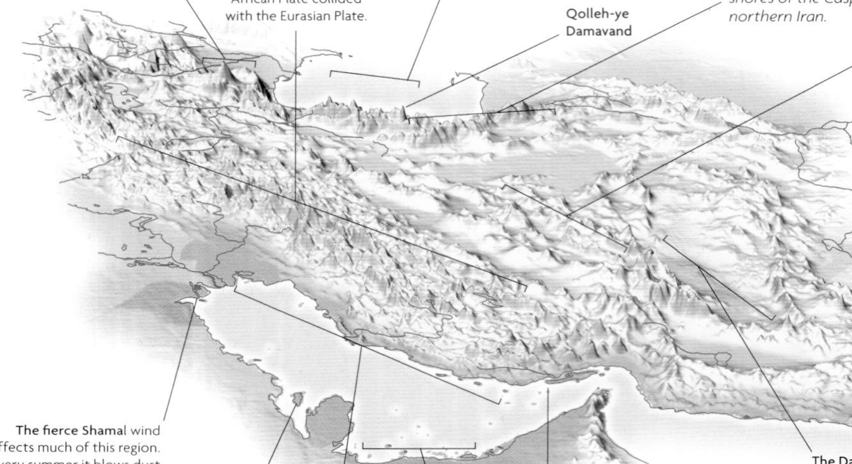

Lava and ash from previous volcanic activity covers a 200 mile (320 km) stretch from the border with Azerbaijan to the Caspian Sea.

Iran's two mountain chains, the Zagros and Elburz, were uplifted at the same time as the Alps in Europe, when the African Plate collided with the Eurasian Plate.

Dominated by a vast, semi-arid interior plateau, most of Iran lies above 1640 ft (500 m). The region is poorly drained with many of its basins remaining dry for months at a time.

The fierce Shamal wind affects much of this region. Every summer it blows dust south from the flood plains of the Tigris and Euphrates, reducing visibility to such **an extent that** Kuwait International Airport is frequently forced to close.

Autumn winds blowing across the Persian Gulf can reach speeds of up to 95 mph (150 kmph) causing severe storms, squalls and waterspouts.

Prolific springs tapping artesian water make cultivation possible across the north of Bahrain's main island. This provides a sharp contrast to the sandy plains in the south and west.

The oilfields of the Persian Gulf are formed from marine shale deposits lying in sedimentary basins at the margins of the Zagros Mountains.

Numerous islands lie along the southern coast of the Persian Gulf. Some of these are salt domes, created when less dense salts were displaced and forced up to the surface by denser, overlying strata.

◄ *The Dasht-e Lut covers a large portion of eastern Iran with its dry, wind-eroded plain of scattered sandstone pillars and salty depressions. During the summer, temperatures soar, making it one of the world's hottest, driest places.*

Using the land and sea

Along the coast of the Caspian Sea, desalinated water allows fruits and vegetables to be produced, although water shortages and desert soils still limit farming. Sheep are the most important livestock raised in Iran and commercial forests cover the northwest of the country. Shrimp stocks were decimated by pollution during the Gulf War, but fishing remains important for domestic and export markets.

Land use and agricultural distribution

- goats
- sheep
- cereals
- citrus fruits
- cotton
- dates
- fishing
- timber
- capital cities
- major towns
- pasture
- cropland
- forest
- desert
- wetland

The urban/rural population divide

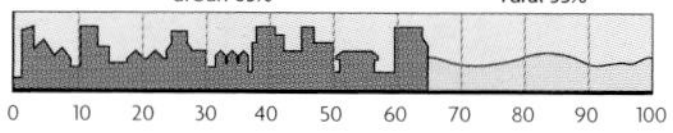

Population density	Total land area
112 people per sq mile (43 people per sq km)	642,883 sq miles (1,665,500 sq km)

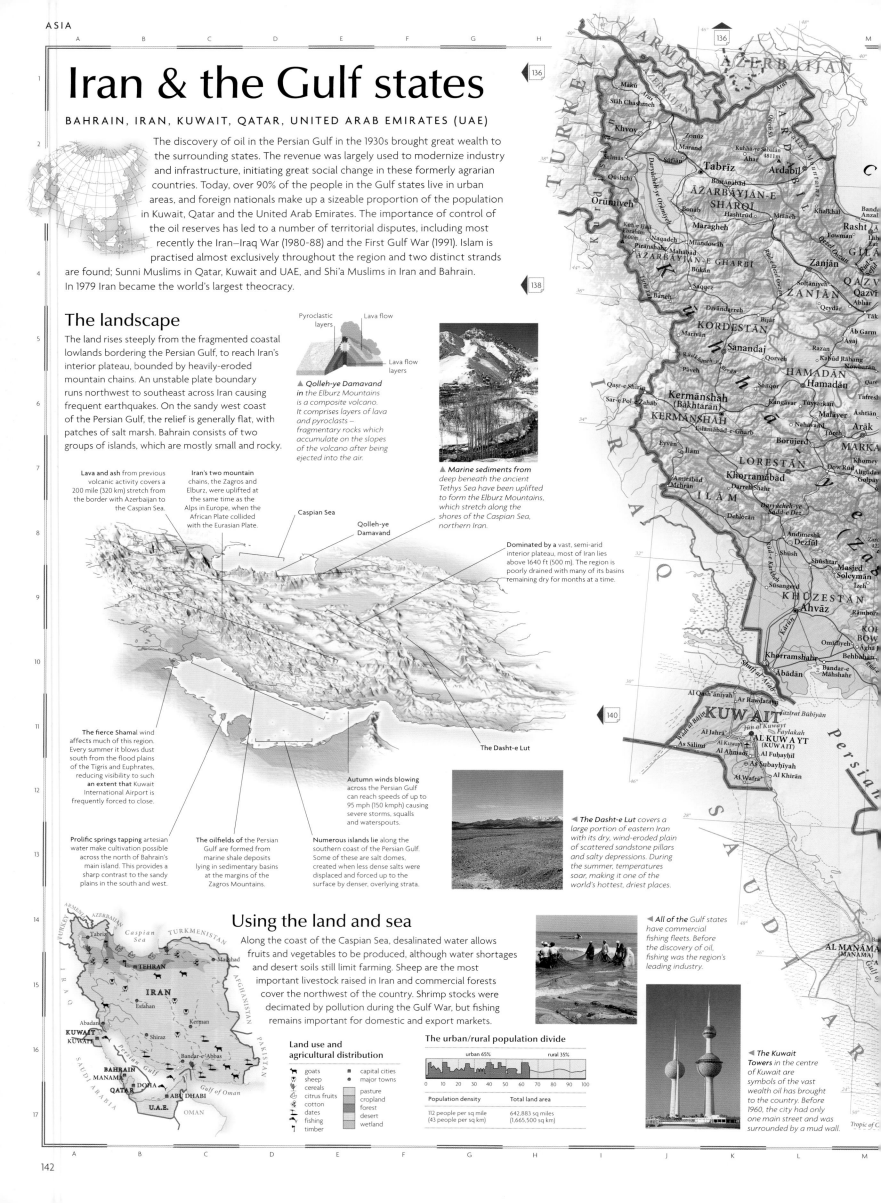

◄ *All of the* Gulf states *have commercial fishing fleets. Before the discovery of oil, fishing was the region's leading industry.*

◄ *The Kuwait Towers in the centre of Kuwait are symbols of the vast wealth oil has brought to the country. Before 1960, the city had only one main street and was surrounded by a mud wall.*

◀ *Many volcanoes lie in Iran's 1200 mile (1930 km) volcanic belt, including the country's highest peak, the now-extinct Qolleh-ye Damavand at 18,600 ft (5671 m).*

146

▶ *Extensive oil and gas exploitation in the Gulf region has allowed the economic transformation of the Gulf states. Consequently, many of these states have a hugely improved per capita income compared to the 1960's.*

Transport and industry

Both onshore and offshore oil reserves are exploited throughout the region. Kuwait not only extracts but also refines 80% of its oil. Bahrain has diversified its economy to become the main commercial and financial centre in the Persian Gulf. Iran produces a wide range of products: textile mills are widespread and carpet-weaving is an important export industry.

Major industry and infrastructure

carpet manufacture	capital city
chemicals	major towns
finance	international airports
food processing	major roads
oil	major industrial areas
oil refining	
textiles	

The transport network

63,543 miles (102,274 km)	884 miles (1423 km)
3822 miles (6151 km)	562 miles (904 km)

Major towns and neighbouring countries are linked by adequate road networks, although rural areas are less well served. Bahrain is linked to the mainland by a 15 mile (25 km) long causeway.

Map key

Population

- above 5 million
- 1 million to 5 million
- 500,000 to 1 million
- 100,000 to 500,000
- 50,000 to 100,000
- 10,000 to 50,000
- below 10,000

Elevation

- 4000m / 13,124ft
- 3000m / 9843ft
- 2000m / 6562ft
- 1000m / 3281ft
- 500m / 1640ft
- 250m / 820ft
- 100m / 328ft
- sea level

Scale 1:6,000,000

Km 0 10 20 40 60 80 100 120 140 160 180 200

Miles 0 10 20 40 60 80 100 120 140 160 180 200

projection: Lambert Conformal Conic

Kazakhstan

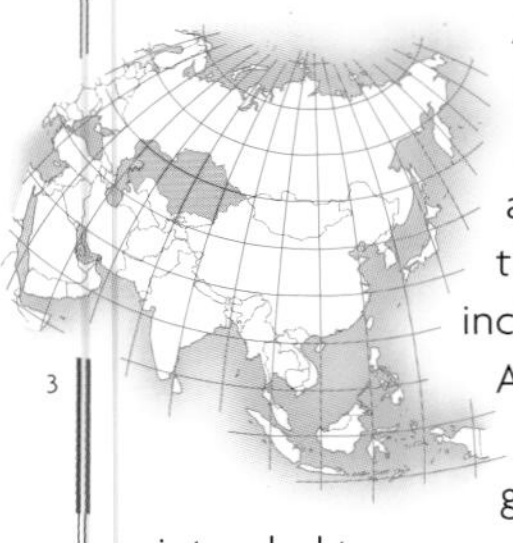

Abundant natural resources lie in the immense steppe grasslands, deserts and central plateau of the former Soviet republic of Kazakhstan. An intensive programme of industrial and agricultural development to exploit these resources during the Soviet era resulted in catastrophic industrial pollution, including fallout from nuclear testing and the shrinkage of the Aral Sea. Since independence, the government has encouraged foreign investment and liberalized the economy to promote growth. The adoption of Kazakh as the national language is intended to encourage a new sense of national identity in a state where living conditions for the majority remain harsh, both in cramped urban centres and impoverished rural areas.

Transport and industry

The single most important industry in Kazakhstan is mining, based around extensive oil deposits near the Caspian Sea, the world's largest chromium mine, and vast reserves of iron ore. Recent foreign investment has helped to develop industries including food processing and steel manufacture, and to expand the exploitation of mineral resources. The Russian space programme is still based at Baykonyr, near Kyzylorda in central Kazakhstan.

Major industry and infrastructure

- chemicals
- engineering
- fish processing
- food processing
- iron & steel
- metallurgy
- mining
- oil
- capital cities
- major towns
- international airports
- major roads
- major industrial areas

The transport network

- 48,263 miles (77,680 km)
- none
- 8483 miles (13,660 km)
- 3900 miles (2423 km)

Industrial areas in the north and east are well-connected to Russia. Air and rail links with Germany and China have been established through foreign investment. Better access to Baltic ports is being sought.

An open-cast coal mine in Kazakhstan. Foreign investment is being actively sought by the Kazakh government in order to fully exploit the potential of the country's rich mineral reserves.

Using the land and sea

The rearing of large herds of sheep and goats on the steppe grasslands forms the core of Kazakh agriculture. Arable cultivation and cotton-growing in pasture and desert areas was encouraged during the Soviet era, but relative yields are low. The heavy use of fertilizers and the diversion of natural water sources for irrigation has degraded much of the land.

The urban/rural population divide

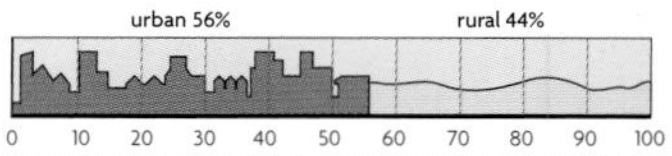

Population density	Total land area
16 people per sq mile (6 people per sq km)	1,048,878 sq miles (2,717,300 sq km)

Land use and agricultural distribution

- cattle
- goats
- sheep
- cotton
- fishing
- wheat
- capital cities
- major towns
- pasture
- cropland
- forest
- mountain region
- desert

The nomadic peoples who moved their herds around the steppe grasslands are now largely settled, although echoes of their traditional lifestyle, in particular their superb riding skills, remain.

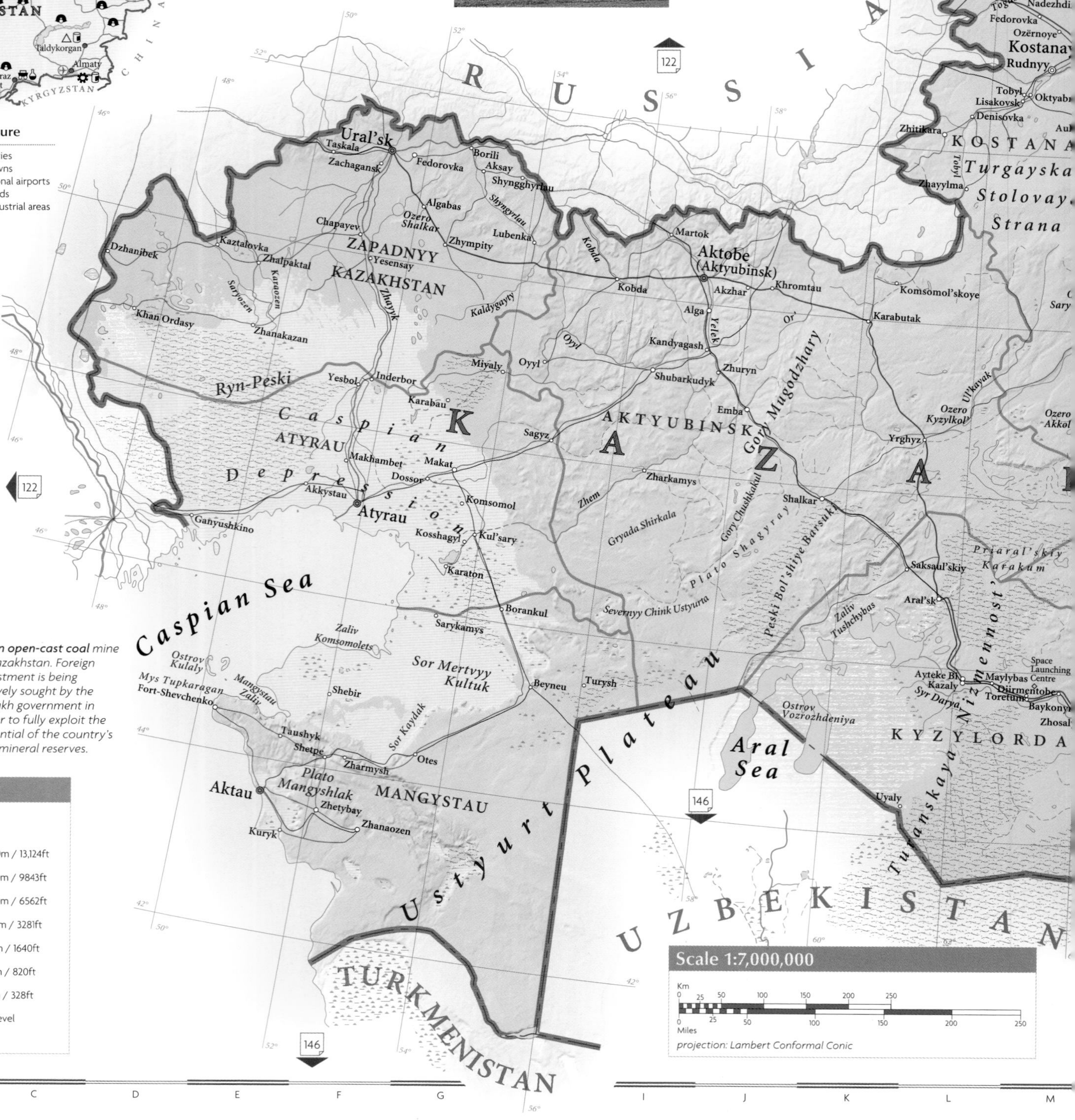

Map key

Population

- 1 million to 5 million
- 500,000 to 1 million
- 100,000 to 500,000
- 50,000 to 100,000
- 10,000 to 50,000
- below 10,000

Elevation

- 4000m / 13,124ft
- 3000m / 9843ft
- 2000m / 6562ft
- 1000m / 3281ft
- 500m / 1640ft
- 250m / 820ft
- 100m / 328ft
- sea level

Scale 1:7,000,000

Km 0 25 50 100 150 200 250

Miles 0 25 50 100 150 200 250

projection: Lambert Conformal Conic

122
122
146
146

N O P Q R S T U V W X Y

The landscape

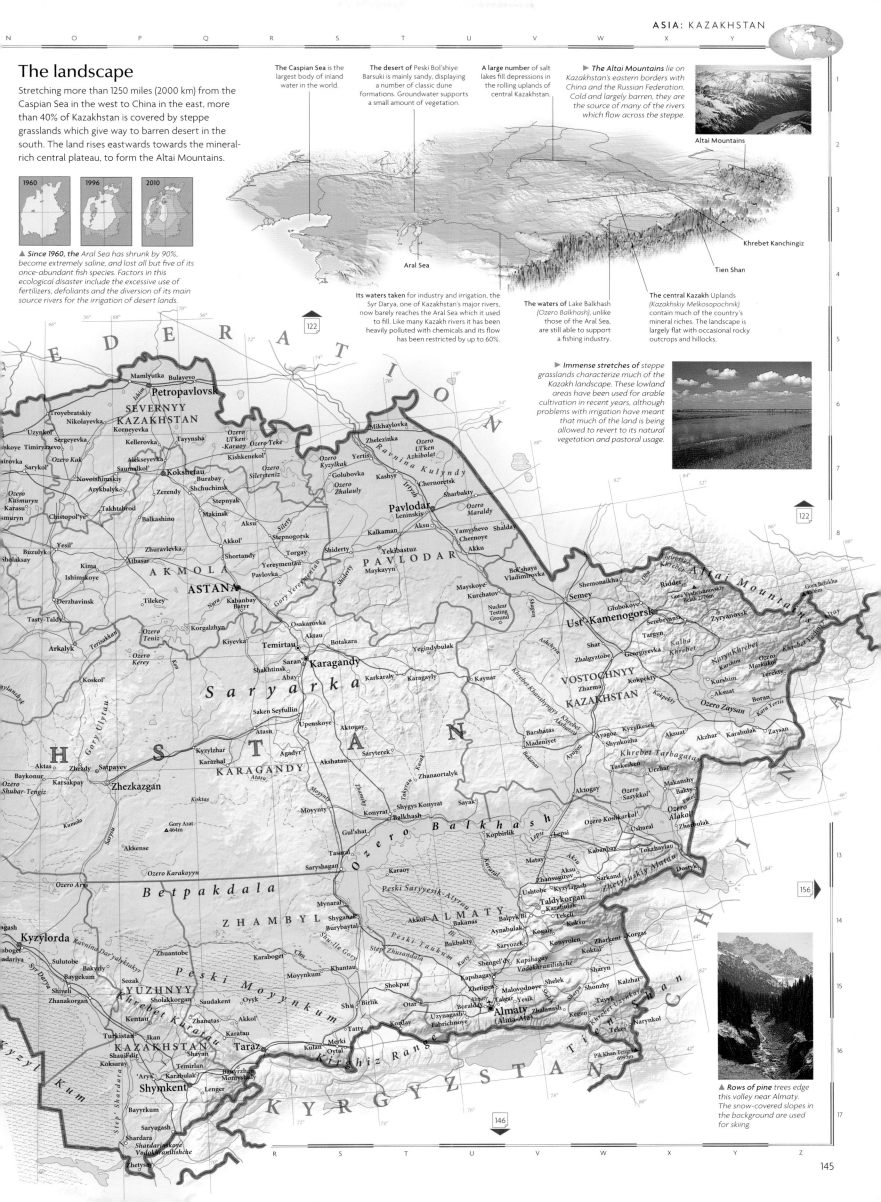

Stretching more than 1250 miles (2000 km) from the Caspian Sea in the west to China in the east, more than 40% of Kazakhstan is covered by steppe grasslands which give way to barren desert in the south. The land rises eastwards towards the mineral-rich central plateau, to form the Altai Mountains.

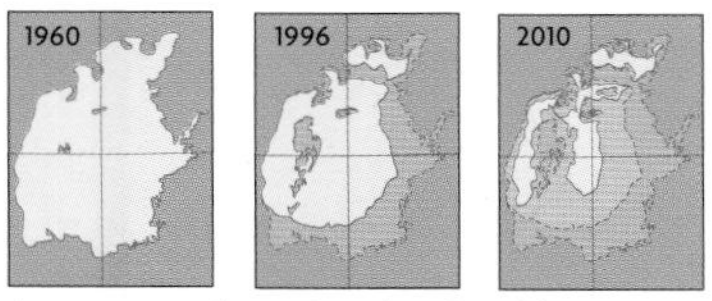

▲ *Since 1960, the Aral Sea has shrunk by 90%, become extremely saline, and lost all but five of its once-abundant fish species. Factors in this ecological disaster include the excessive use of fertilizers, defoliants and the diversion of its main source rivers for the irrigation of desert lands.*

The Caspian Sea is the largest body of inland water in the world.

The desert of Peski Bol'shiye Barsuki is mainly sandy, displaying a number of classic dune formations. Groundwater supports a small amount of vegetation.

A large number of salt lakes fill depressions in the rolling uplands of central Kazakhstan.

▶ ***The Altai Mountains** lie on Kazakhstan's eastern borders with China and the Russian Federation. Cold and largely barren, they are the source of many of the rivers which flow across the steppe.*

Altai Mountains

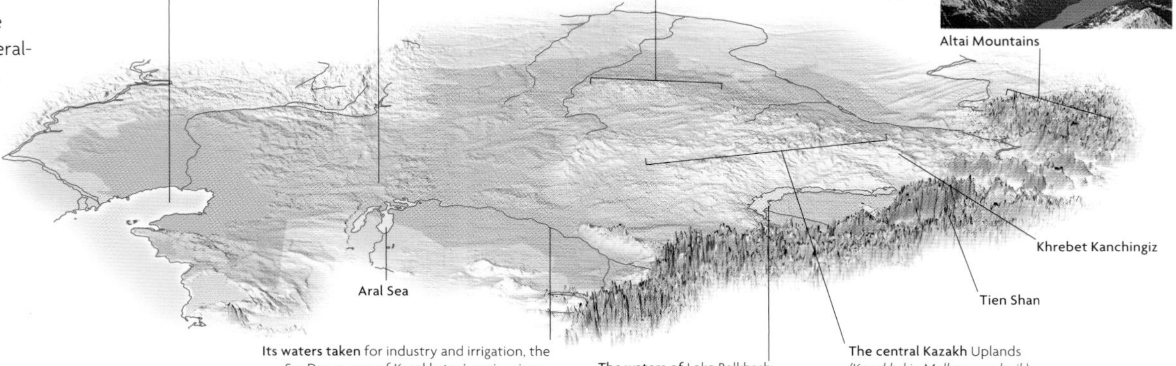

Its waters taken for industry and irrigation, the Syr Darya, one of Kazakhstan's major rivers, now barely reaches the Aral Sea which it used to fill. Like many Kazakh rivers it has been heavily polluted with chemicals and its flow has been restricted by up to 60%.

The waters of Lake Balkhash (*Ozero Balkhash*), unlike those of the Aral Sea, are still able to support a fishing industry.

The central Kazakh Uplands (*Kazakhskiy Melkosopochnik*) contain much of the country's mineral riches. The landscape is largely flat with occasional rocky outcrops and hillocks.

▶ ***Immense stretches of** steppe grasslands characterize much of the Kazakh landscape. These lowland areas have been used for arable cultivation in recent years, although problems with irrigation have meant that much of the land is being allowed to revert to its natural vegetation and pastoral usage.*

122
122
156
146

▲ *Rows of pine trees edge this valley near Almaty. The snow-covered slopes in the background are used for skiing.*

R S T U V W X Y Z

Central Asia

KYRGYZSTAN, TAJIKISTAN, TURKMENISTAN, UZBEKISTAN

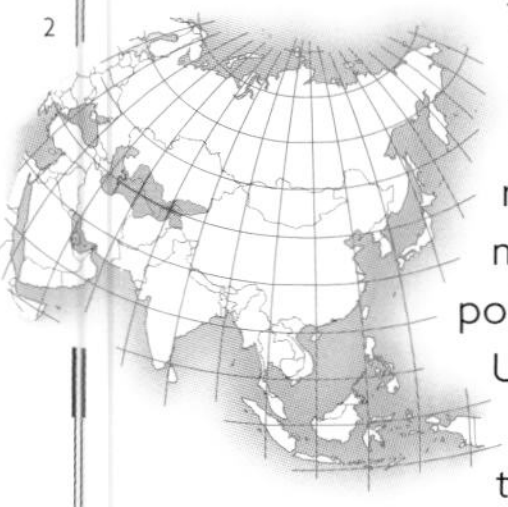

The four republics that declared independence in 1991 were created in the early years of the Soviet Union, promoting ethnic divisions in a region whose common focus, since the 8th century, has been Islam. Traditional rural and nomadic ways of life have survived the Soviet era, while the benefits of modern industry and grand irrigation schemes have resulted in severe pollution in the delicate, arid environment of the steppe, particularly in Uzbekistan. Many ethnic minority groups are scattered among the four republics, with isolated communities in the mountains of Kyrgyzstan.

The current Islamic revival has brought hope of greater regional unity, in spite of religious factionalism which, in 1992, plunged Tajikistan into civil war.

▲ *The southern shoreline of the Aral Sea has retreated over 30 miles (48 km) since 1960. A major cause is the diversion of water from the Amu Darya river for irrigation via the Kara Kum Canal* (Garagum Kanaly).

◄ ***The desert of*** *the Kara Kum (Garagum) occupies over 70% of Turkmenistan; its wind-scoured surface of dune ridges and depressions severely limits human settlement.*

Transport and industry

Fossil fuels are extracted and processed in all four states, with scope for further exploitation. Agriculture provides raw materials for many industries, including food and textiles processing, and the manufacture of leather goods, clothing and carpets. Farm machinery is also produced.

The transport network

73,658 miles (118,555 km)	87 miles (140 km)
4773 miles (7683 km)	1180 miles (1900 km)

The Kara Kum Canal (Garagumskiy Kanal) runs for 870 miles (1400 km) from the Amu Darya river to the Caspian Sea. The canal is principally used for irrigation but is navigable for 280 miles (450 km).

Major industry and infrastructure

- carpet weaving
- chemicals
- engineering
- food processing
- oil & gas
- textiles
- capital cities
- major towns
- international airports
- major roads
- major industrial areas

N O P Q R S T U V W X Y

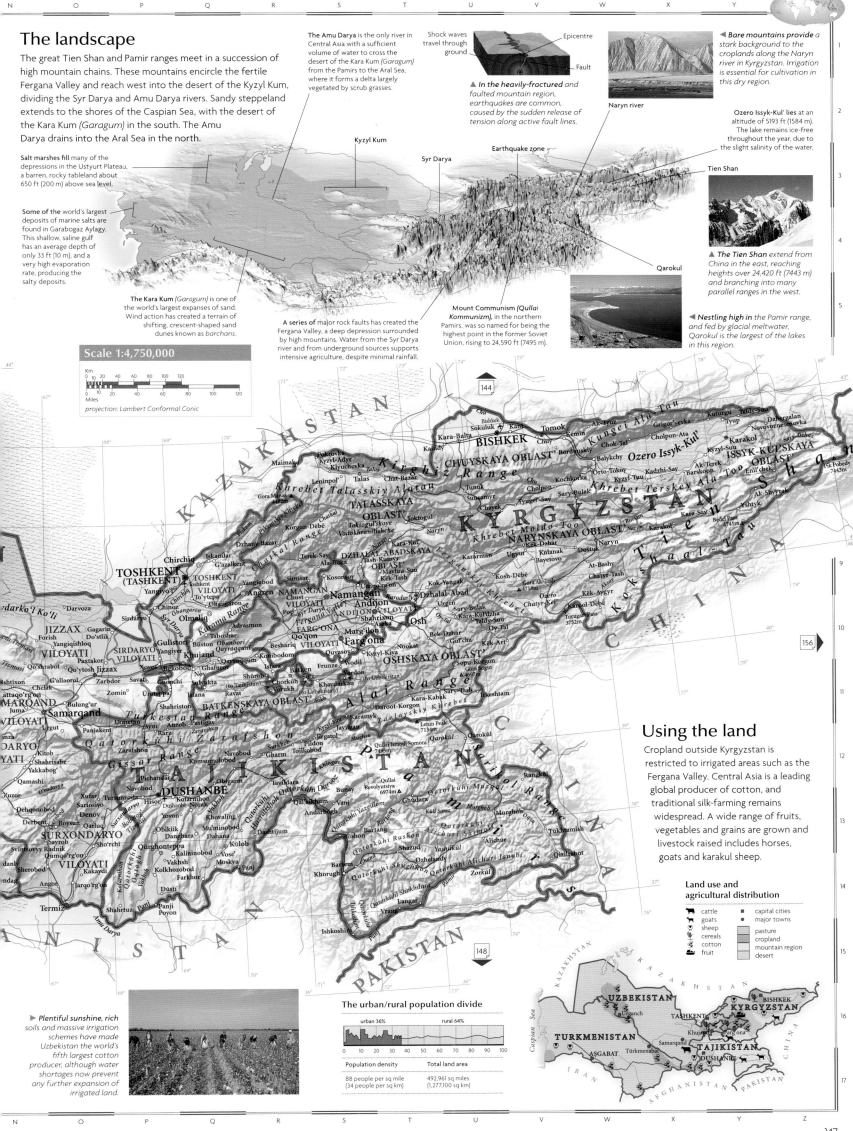

The landscape

The great Tien Shan and Pamir ranges meet in a succession of high mountain chains. These mountains encircle the fertile Fergana Valley and reach west into the desert of the Kyzyl Kum, dividing the Syr Darya and Amu Darya rivers. Sandy steppeland extends to the shores of the Caspian Sea, with the desert of the Kara Kum *(Garagum)* in the south. The Amu Darya drains into the Aral Sea in the north.

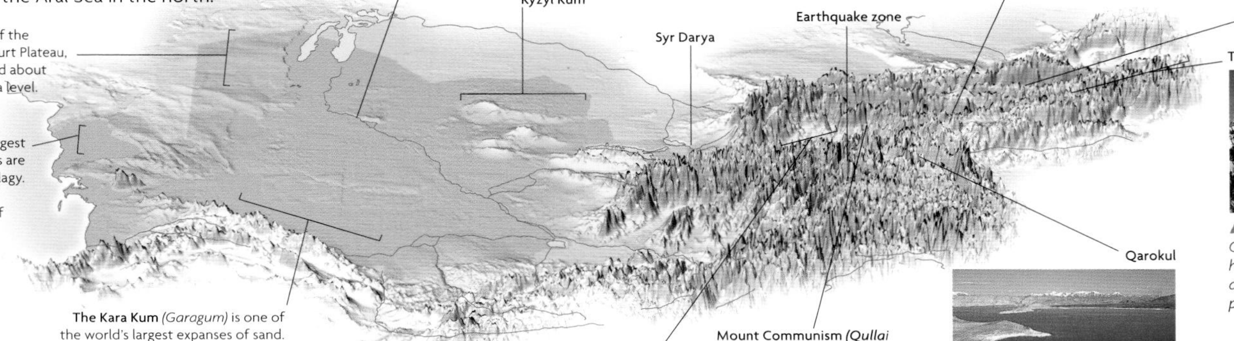

Salt marshes fill many of the depressions in the Ustyurt Plateau, a barren, rocky tableland about 650 ft (200 m) above sea level.

Some of the world's largest deposits of marine salts are found in Garabogaz Aylagy. This shallow, saline gulf has an average depth of only 33 ft (10 m), and a very high evaporation rate, producing the salty deposits.

The Kara Kum *(Garagum)* is one of the world's largest expanses of sand. Wind action has created a terrain of shifting, crescent-shaped sand dunes known as *barchans*.

A series of major rock faults has created the Fergana Valley, a deep depression surrounded by high mountains. Water from the Syr Darya river and from underground sources supports intensive agriculture, despite minimal rainfall.

Mount Communism *(Qullai Kommunizm)*, in the northern Pamirs, was so named for being the highest point in the former Soviet Union, rising to 24,590 ft (7495 m).

The Amu Darya is the only river in Central Asia with a sufficient volume of water to cross the desert of the Kara Kum *(Garagum)* from the Pamirs to the Aral Sea, where it forms a delta largely vegetated by scrub grasses.

▲ ***In the heavily-fractured** and faulted mountain region, earthquakes are common, caused by the sudden release of tension along active fault lines.*

◀ ***Bare mountains provide** a stark background to the croplands along the Naryn river in Kyrgyzstan. Irrigation is essential for cultivation in this dry region.*

Ozero Issyk-Kul' lies at an altitude of 5193 ft (1584 m). The lake remains ice-free throughout the year, due to the slight salinity of the water.

▲ ***The Tien Shan** extend from China in the east, reaching heights over 24,420 ft (7443 m) and branching into many parallel ranges in the west.*

◀ ***Nestling high in** the Pamir range, and fed by glacial meltwater, Qarokul is the largest of the lakes in this region.*

Scale 1:4,750,000

Km 0 10 20 40 60 80 100 120
Miles 0 10 20 40 60 80 100 120
projection: Lambert Conformal Conic

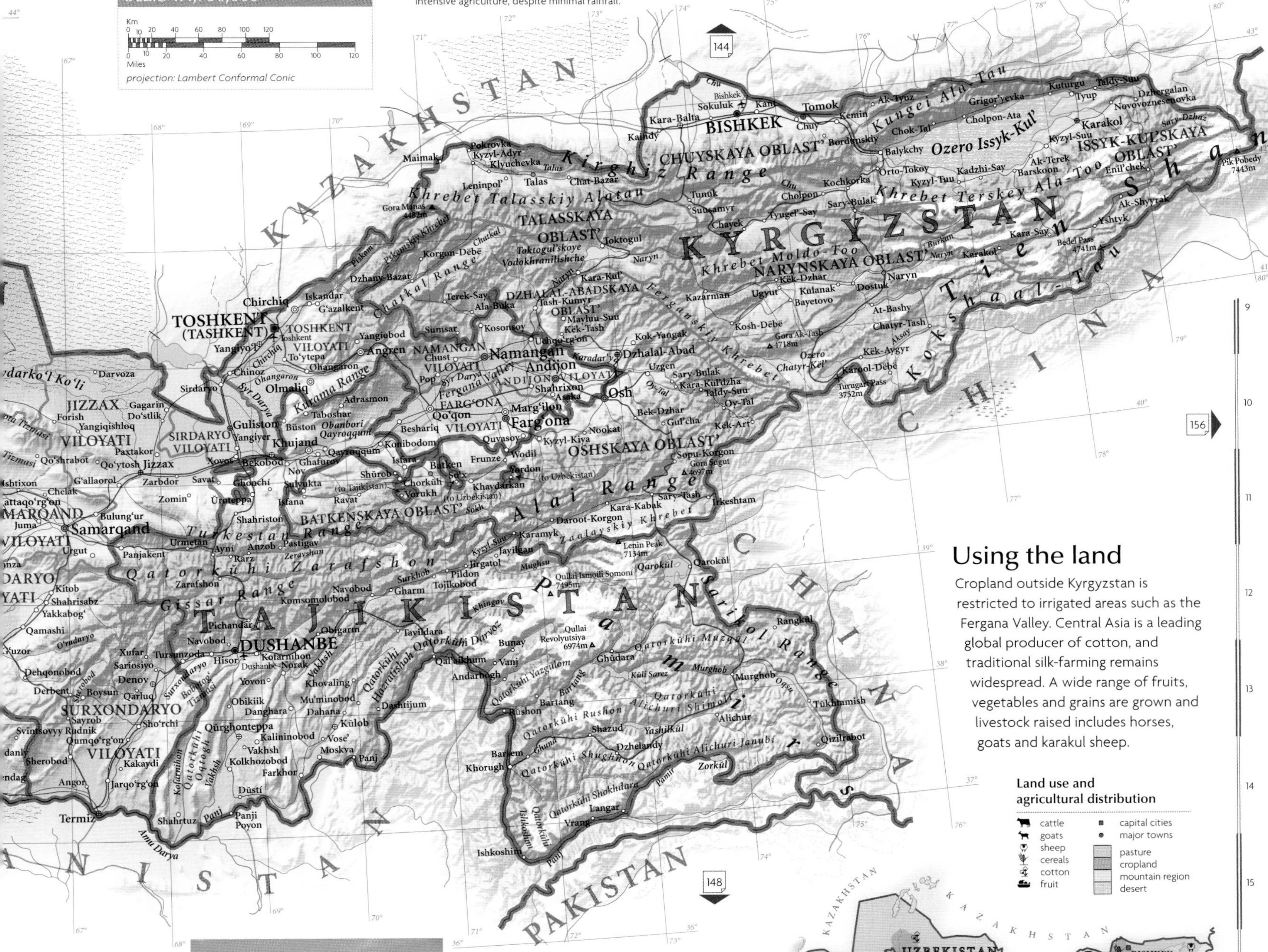

Using the land

Cropland outside Kyrgyzstan is restricted to irrigated areas such as the Fergana Valley. Central Asia is a leading global producer of cotton, and traditional silk-farming remains widespread. A wide range of fruits, vegetables and grains are grown and livestock raised includes horses, goats and karakul sheep.

Land use and agricultural distribution

cattle
goats
sheep
cereals
cotton
fruit
capital cities
major towns
pasture
cropland
mountain region
desert

► ***Plentiful sunshine, rich** soils and massive irrigation schemes have made Uzbekistan the world's fifth largest cotton producer, although water shortages now prevent any further expansion of irrigated land.*

The urban/rural population divide

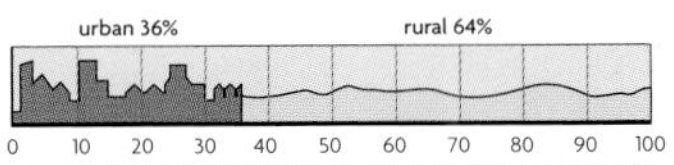

Population density	Total land area
88 people per sq mile (34 people per sq km)	492,961 sq miles (1,277,100 sq km)

N O P Q R S T U V W X Y Z

Afghanistan & Pakistan

Pakistan was created by the partition of British India in 1947, becoming the western arm of a new Islamic state for Indian Muslims; the eastern sector, in Bengal, seceded to become the separate country of Bangladesh in 1971. Over half of Pakistan's 158 million people live in the Punjab, at the fertile head of the great Indus Basin. The river sustains a national economy based on irrigated agriculture, including cotton for the vital textiles industry. Afghanistan, a mountainous, landlocked country, with an ancient and independent culture, has been wracked by war since 1979. Factional strife escalated into an international conflict in late 2001, as US-led troops ousted the militant and fundamentally Islamist *taliban* regime as part of their 'war on terror'.

◀ **The town of** Bamian lies high in the Hindu Kush west of Kabul. Between the 2nd and 5th centuries two huge statues of Buddha were carved into the nearby rock, the largest of which stood 125 ft (38 m) high. The statues were destroyed by the taliban regime in March 2001.

Transport and industry

Pakistan is highly dependent on the cotton textiles industry, although diversified manufacture is expanding around cities such as Karachi and Lahore. Afghanistan's limited industry is based mainly on the processing of agricultural raw materials and includes traditional crafts such as carpet-making.

Major industry and infrastructure

carpet weaving	■ capital cities
✿ chemicals	● major towns
✿ engineering	✈ international airports
S finance	— major roads
▮ food processing	major industrial areas
iron & steel	
♨ oil & gas	
⟁ textiles	

The transport network

🛣	96,154 miles (154,763 km)
🛣	211 miles (340 km)
🚃	4852 miles (7814 km)
🚃	745 miles (1200 km)

The Karakoram Highway was completed after 20 years of construction in 1978. It breaches the Himalayan mountain barrier providing a commercial motor route linking lowland Pakistan and China.

▶ **The Karakoram Highway** is one of the highest major roads in the world. It took over 24,000 workers almost 20 years to complete.

The landscape

Afghanistan's topography is dominated by the mountains of the Hindu Kush, which spread south and west into numerous mountain spurs. The dry plateau of southwestern Afghanistan extends into Pakistan and the hills which overlook the great Indus Basin. In northern Pakistan the Hindu Kush, Himalayan and Karakoram ranges meet to form one of the world's highest mountain regions.

Hunza river

◀ **The Hunza river** rises in the northern Karakoram Range, running for 120 miles (193 km) before joining the Gilgit river.

▶ **The arid Hindu Kush** makes much of Afghanistan uninhabitable, with over 50% of the land lying above 6500 ft (2000 m).

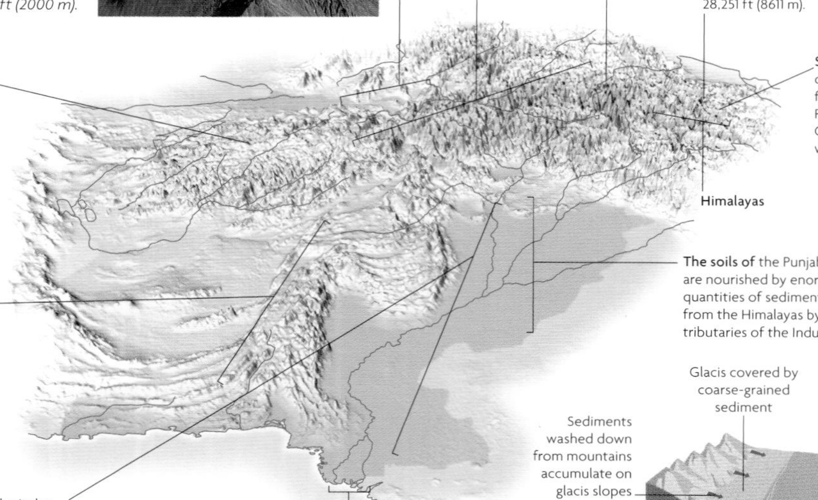

The plains and foothills which extend from the northern slopes of the Hindu Kush are part of the great grassy steppe lands of Central Asia.

Hindu Kush

K2 (Mount Godwin Austen), in the Karakoram Range, is the second highest mountain in the world, at an altitude of 28,251 ft (8611 m).

Some of the largest glaciers outside the polar regions are found in the Karakoram Range, including Siachen Glacier *(Siachen Muztagh)*, which is 40 miles (72 km) long.

Frequent earthquakes mean that mountain-building processes are continuing in this region, as the Indo-Australian Plate drifts northwards, colliding with the Eurasian Plate.

Himalayas

Mountain chains running southwest from the Hindu Kush into Pakistan form a barrier to the humid winds which blow from the Indian Ocean, creating arid conditions across southern Afghanistan.

The soils of the Punjab plain are nourished by enormous quantities of sediment, carried from the Himalayas by the five tributaries of the Indus river.

Glacis covered by coarse-grained sediment

Sediments washed down from mountains accumulate on glacis slopes

Fine sediments deposited on salt flats are removed by wind erosion.

Bedrock

▲ **Glacis are gentle**, debris-covered slopes which lead into salt flats or deserts. They typically occur at the base of mountains in arid regions such as Afghanistan.

The Indus Basin is part of the Indus-Ganges lowland, a vast depression which has been filled with layers of sediment over the last 50 million years. These deposits are estimated to be over 16,400 ft (5000 m) deep.

The Indus Delta is prone to heavy flooding and high levels of salinity. It remains a largely uncultivated wilderness area.

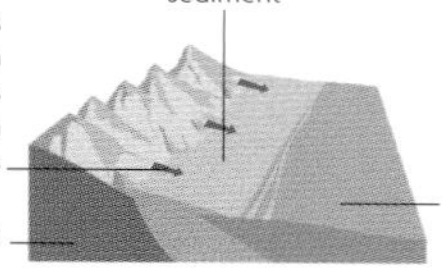

Scale 1:5,000,000

Km
0 20 40 60 80 100 120 140 160
Miles
0 10 20 40 60 80 100 120 140 160

projection: Lambert Conformal Conic

Map key

Population

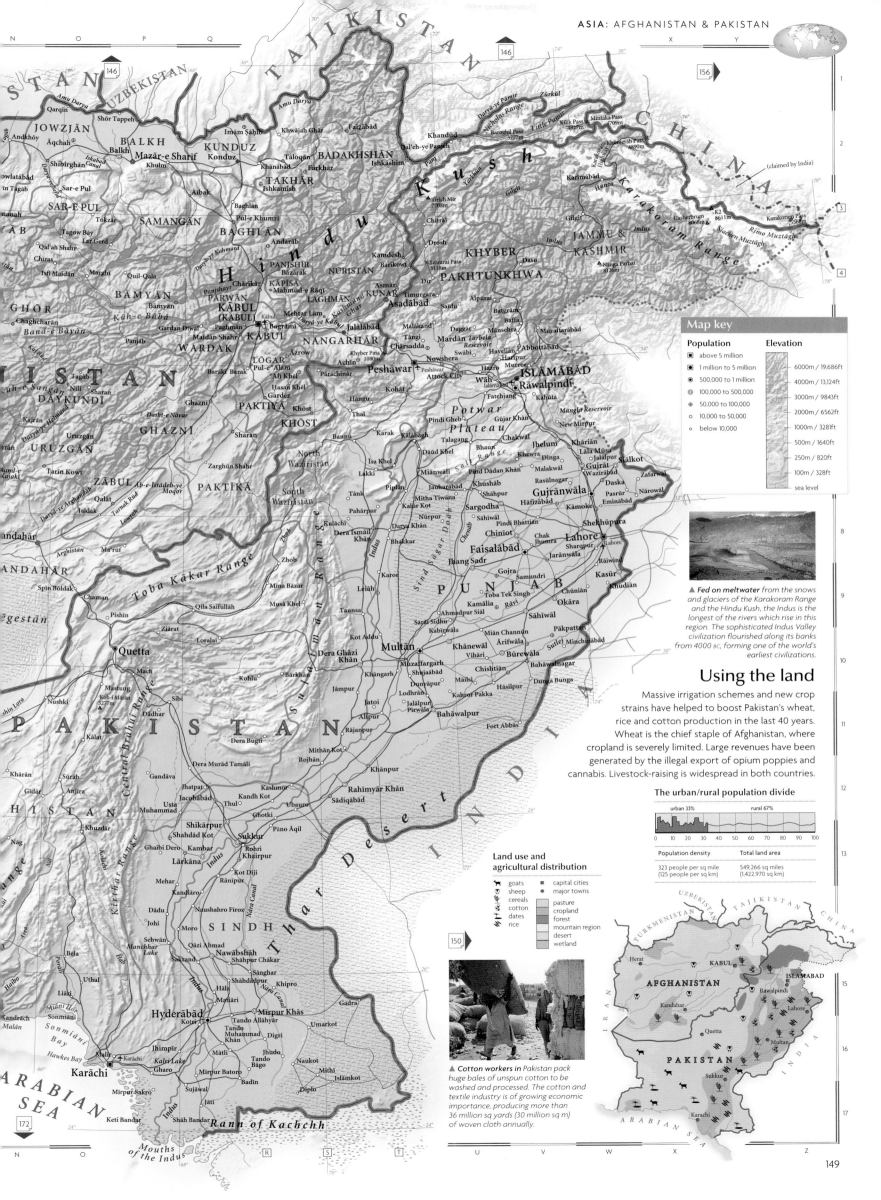

- above 5 million
- 1 million to 5 million
- 500,000 to 1 million
- 100,000 to 500,000
- 50,000 to 100,000
- 10,000 to 50,000
- below 10,000

Elevation

- 6000m / 19,686ft
- 4000m / 13,124ft
- 3000m / 9843ft
- 2000m / 6562ft
- 1000m / 3281ft
- 500m / 1640ft
- 250m / 820ft
- 100m / 328ft
- sea level

▲ *Fed on meltwater from the snows and glaciers of the Karakoram Range and the Hindu Kush, the Indus is the longest of the rivers which rise in this region. The sophisticated Indus Valley civilization flourished along its banks from 4000 BC, forming one of the world's earliest civilizations.*

Using the land

Massive irrigation schemes and new crop strains have helped to boost Pakistan's wheat, rice and cotton production in the last 40 years. Wheat is the chief staple of Afghanistan, where cropland is severely limited. Large revenues have been generated by the illegal export of opium poppies and cannabis. Livestock-raising is widespread in both countries.

The urban/rural population divide

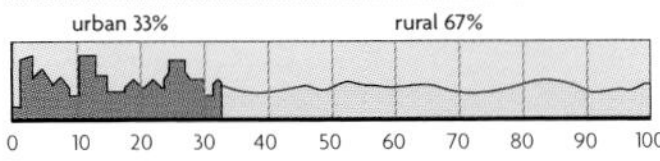

Population density	Total land area
323 people per sq mile (125 people per sq km)	549,266 sq miles (1,422,970 sq km)

Land use and agricultural distribution

▲ *Cotton workers in Pakistan pack huge bales of unspun cotton to be washed and processed. The cotton and textile industry is of growing economic importance, producing more than 36 million sq yards (30 million sq m) of woven cloth annually.*

South Asia

BANGLADESH, BHUTAN, INDIA, MALDIVES, NEPAL, PAKISTAN, SRI LANKA

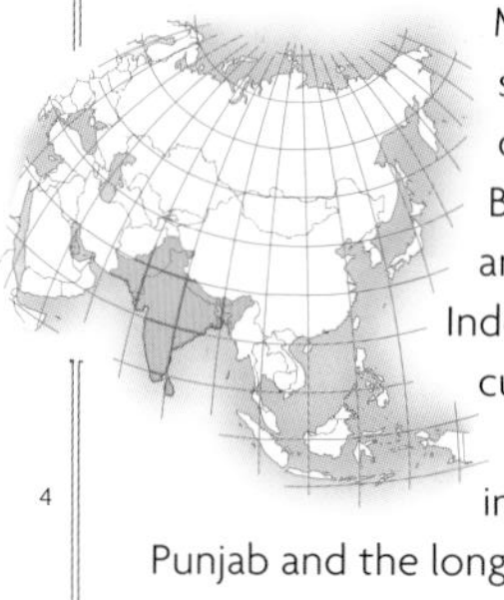

More than one-fifth of the world's population lives in the south Asian subcontinent. Great cultural diversity has come from a long succession of foreign invaders, including Hindu Aryans, Islamic Moguls and the British, whose empire incorporated the princely states of the Maharajas and extended to the borders of Nepal and Bhutan in the Himalayas. Independent since 1947, India is the world's largest democracy, and at the current rate of growth, may overtake China as the world's most populous country during the 21st century. There are points of tension in the region over claims for independence by the Sikhs in the Indian Punjab and the long-standing dispute with Pakistan over Jammu and Kashmir in the north.

◀ *The towering Karakoram and Hindu Kush ranges, formed at the same time as the Himalayas, dominate Pakistan's northern borders. K2 on the border of northern Pakistan is the second highest mountain on Earth, at 28,251 ft (8611 m).*

The landscape

South Asia is effectively isolated from the rest of Asia by desert along the western flank of Pakistan, and a continuous wall of mountains, dominated by the Himalayas, to the north and east. The great basins of the Indus and Ganges separate this mountain fringe from the rolling plateau of the Indian peninsula, which is bordered by a line of coastal hills, the Eastern and Western Ghats.

◀ ***The Indus valley** near Skardu in northern Pakistan has been partially infilled by great quantities of eroded sediment. Most of this is carried from the region's bare slopes by swollen rivers during the spring thaw and mass movement activity.*

The Himalayas are the highest and most extensive mountain system in the world. They were formed when the Indo-Australian Plate collided with the Eurasian Plate about 40 million years ago, thrusting up huge masses of land and creating a 'ripple' effect, which formed lesser mountain ranges in Tibet and Southeast Asia. Mount Everest is the world's tallest mountain at 29,029 ft (8848 m).

Almost all of Bangladesh lies in the immense delta formed by the Ganges and the Brahmaputra which merge and flow out into the Bay of Bengal.

The Indus river flows more than 1970 miles (3180 km) from southwestern Tibet to its mouth on the Arabian Sea. It has an estimated catchment area of 450,000 sq miles (1,165,500 sq km).

The coast of western Pakistan is a staircase of folded rock strata caused by successive periods of rapid uplift.

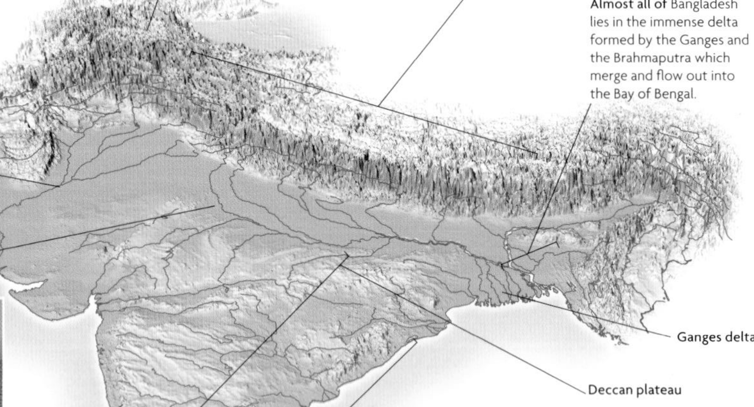

▲ ***Rivers flowing from** the Himalayas into a broad depression in northern India have formed marshes around Bharatpur. They are now a sanctuary for numerous bird species.*

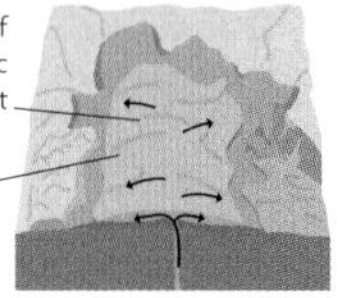

▲ ***The Deccan plateau** covers an area of more than 123,553 sq miles (320,000 sq km). It is formed of deep layers of volcanic basalt, reaching thicknesses of more than 9800 ft (3000 m) towards the coast. Distinctive stepped valleys cut in the basalt plateau by rivers are known as 'traps'.*

The Western Ghats are formed by a fault scarp which runs unbroken for more than 930 miles (1500 km). They reach their highest point at the southern Cardamom Hills.

Trivandrum in southern India normally receives the first of the monsoon rains, which are essential to south Asian agriculture and moderate the extreme summer heat. The monsoon then moves northwards over a period of about two months.

Coastal deposition has formed many typical features along the western coast of Sri Lanka. These include spits and bars, sometimes enclosing lagoons.

146
148
156
142

Map key

Population
- above 5 million
- 1 million to 5 million
- 500,000 to 1 million
- 100,000 to 500,000
- 50,000 to 100,000
- 10,000 to 50,000
- below 10,000

Elevation
- 6000m / 19,686ft
- 4000m / 13,124ft
- 3000m / 9843ft
- 2000m / 6562ft
- 1000m / 3281ft
- 500m / 1640ft
- 250m / 820ft
- 100m / 328ft
- sea level

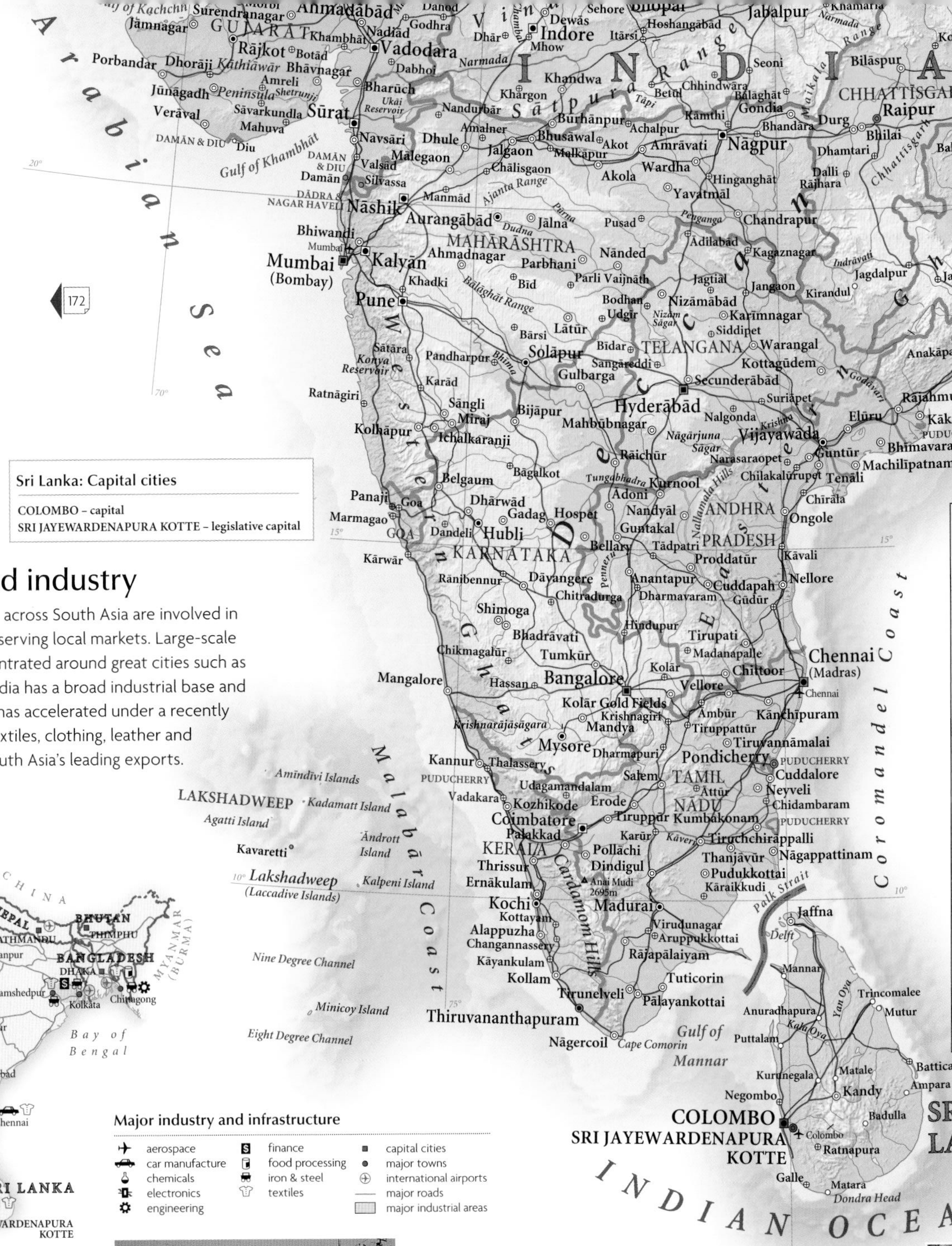

172

Sri Lanka: Capital cities

COLOMBO – capital
SRI JAYEWARDENAPURA KOTTE – legislative capital

Transport and industry

Most industrial workers across South Asia are involved in small-scale production serving local markets. Large-scale industry remains concentrated around great cities such as Kolkata and Mumbai. India has a broad industrial base and manufacturing growth has accelerated under a recently liberalized economy. Textiles, clothing, leather and jewellery are among South Asia's leading exports.

Major industry and infrastructure

- aerospace
- car manufacture
- chemicals
- electronics
- engineering
- finance
- food processing
- iron & steel
- textiles
- capital cities
- major towns
- international airports
- major roads
- major industrial areas

The transport network

- 1,068,996 miles (1,720,579 km)
- 21,015 miles (33,840 km)
- 46,724 miles (75,204 km)
- 15,319 miles (24,656 km)

India's railway network, established under British colonial rule, is the fifth most extensive in the world and continues to play a unique role in integrating the country's disparate regions.

◀ ***Religion and commerce** sit side by side in the Nepalese capital, Kathmandu. Nepal is a Hindu state and these small, highly decorated shrines are commonplace. As in India, cows are venerated, and allowed free rein throughout the city.*

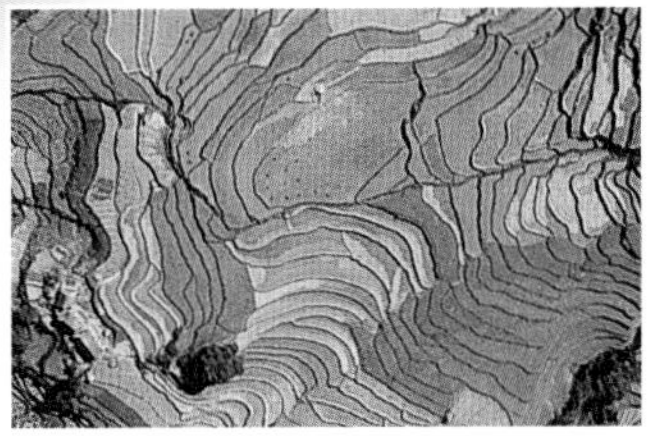

▶ ***Terracing allows steep** hillslopes to be cultivated in Nepal, a country where agricultural land is very limited. Because of poor soil quality, these terraces are often abandoned within a few years.*

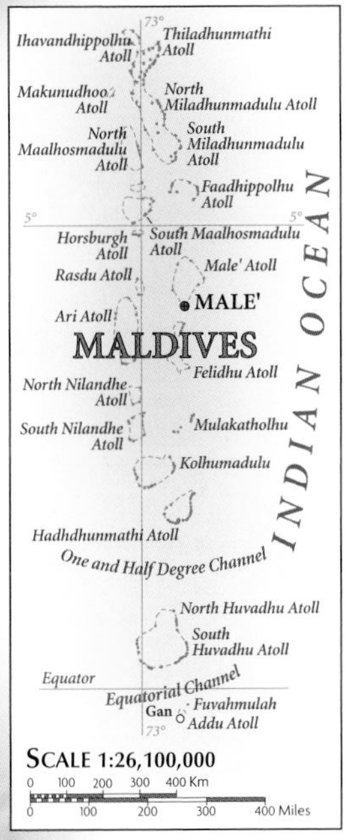

Using the land and sea

Over 60% of South Asia's population is involved in agriculture. Traditional subsistence farming prevails and productivity is generally low. The monsoon region of the east is the world's most extensive rice-growing area. Corn, millet and groundnuts are staple crops in drier areas, with wheat towards the north. Terracing increases cultivable land in the mountains. Livestock-raising is widespread throughout the subcontinent and fishing is common along the entire coast, although because few fishing craft are mechanized, total fish catches are low.

The urban/rural population divide

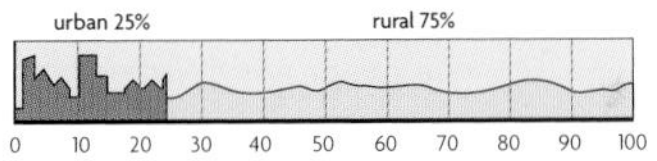

Population density	Total land area
888 people per sq mile (343 people per sq km)	1,573,285 sq miles (4,075,868 sq km)

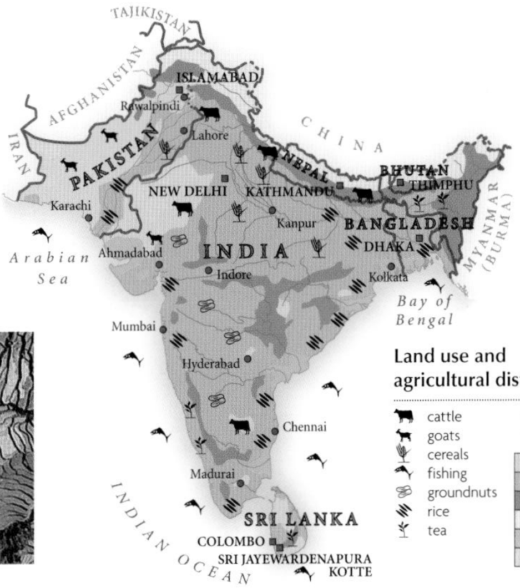

Land use and agricultural distribution

- cattle
- goats
- cereals
- fishing
- groundnuts
- rice
- tea
- capital cities
- major towns
- pasture
- cropland
- forest
- mountain region
- wetland
- desert

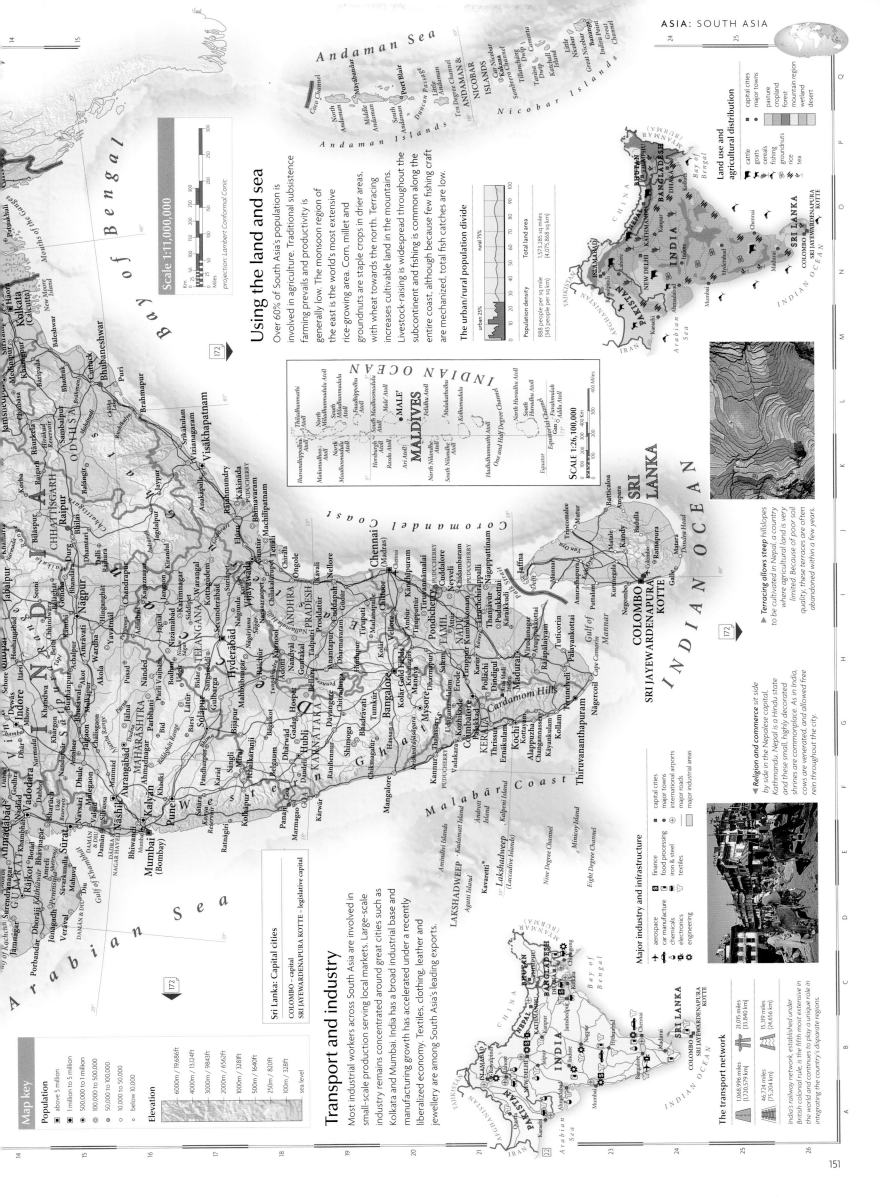

Northern India & the Himalayan states

BANGLADESH, BHUTAN, NEPAL, Arunachal Pradesh, Assam, Bihar, Chandigarh, Delhi, Haryana, Himachal Pradesh, Jammu & Kashmir, Jharkhand, Manipur, Meghalaya, Mizoram, Nagaland, Punjab, Rajasthan, Sikkim, Tripura, Uttarakhand, Uttar Pradesh, West Bengal

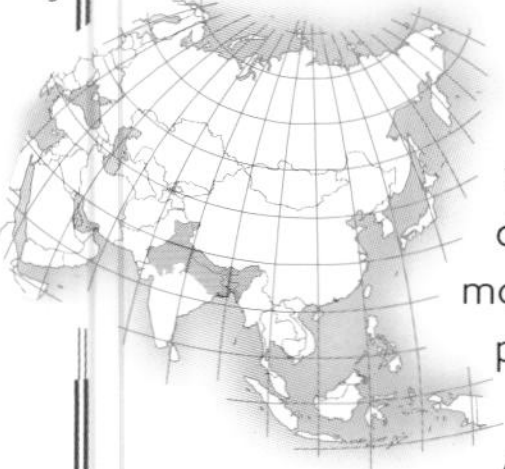

The Ganges and Brahmaputra river basins and the massive mountain barrier of the Himalayas define this region's landscape and have served to reinforce potent cultural and religious differences among its people. Hinduism pervades most aspects of national life and is a growing political force within India, a secular country which also encompasses the centre of Sikhism at Amritsar and the world's largest Muslim minority. Nepal is a crowded mountain state, which faces severe ecological problems from deforestation, while the tiny Himalayan Buddhist kingdom of Bhutan is emerging from long-term isolation, to welcome selected visitors. The Muslim state of Bangladesh, formerly East Pakistan, is one of the world's most densely populated countries and one of the poorest, with more than 145 million people living largely on the massive Ganges/Brahmaputra delta. Many Bangladeshis live under threat of repeated, catastrophic floods.

156
148
154

◀ *The Golden Temple in Amritsar, the most sacred shrine of the Sikh religion, was the scene of violent clashes between Sikh separatists and government forces in 1984.*

Map key

Population

- 1 million to 5 million
- 500,000 to 1 million
- 100,000 to 500,000
- 50,000 to 100,000
- 10,000 to 50,000
- below 10,000

Elevation

- 6000m / 19,686ft
- 4000m / 13,124ft
- 3000m / 9843ft
- 2000m / 6562ft
- 1000m / 3281ft
- 500m / 1640ft
- 250m / 820ft
- 100m / 328ft
- sea level

Transport and industry

Textiles, engineering, chemicals and electronics are leading industries in north India. The plateau of Chota Nagpur provides ore for iron and steel production in the major industrial region northeast of Kolkata. Bangladesh processes jute and Nepal has a small manufacturing sector based on agricultural produce, while Bhutan's limited industry is concentrated in the southern lowland area.

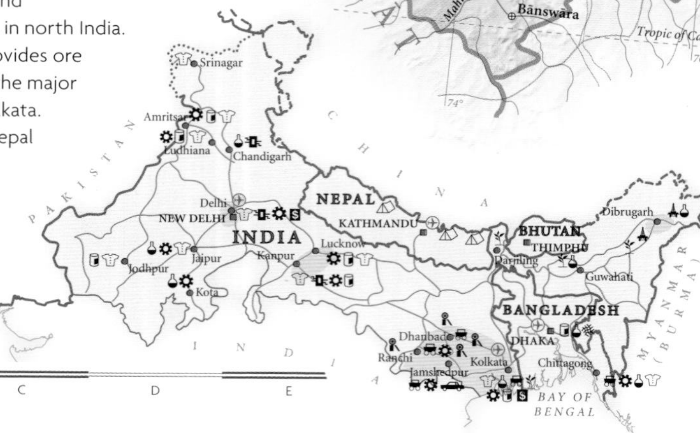

Major industry and infrastructure

The transport network

Over 60% of Bangladesh's internal trade is carried by boat. The country has a very disjointed land transport network, with no bridges over the Brahmaputra and few road crossings on the Ganges river.

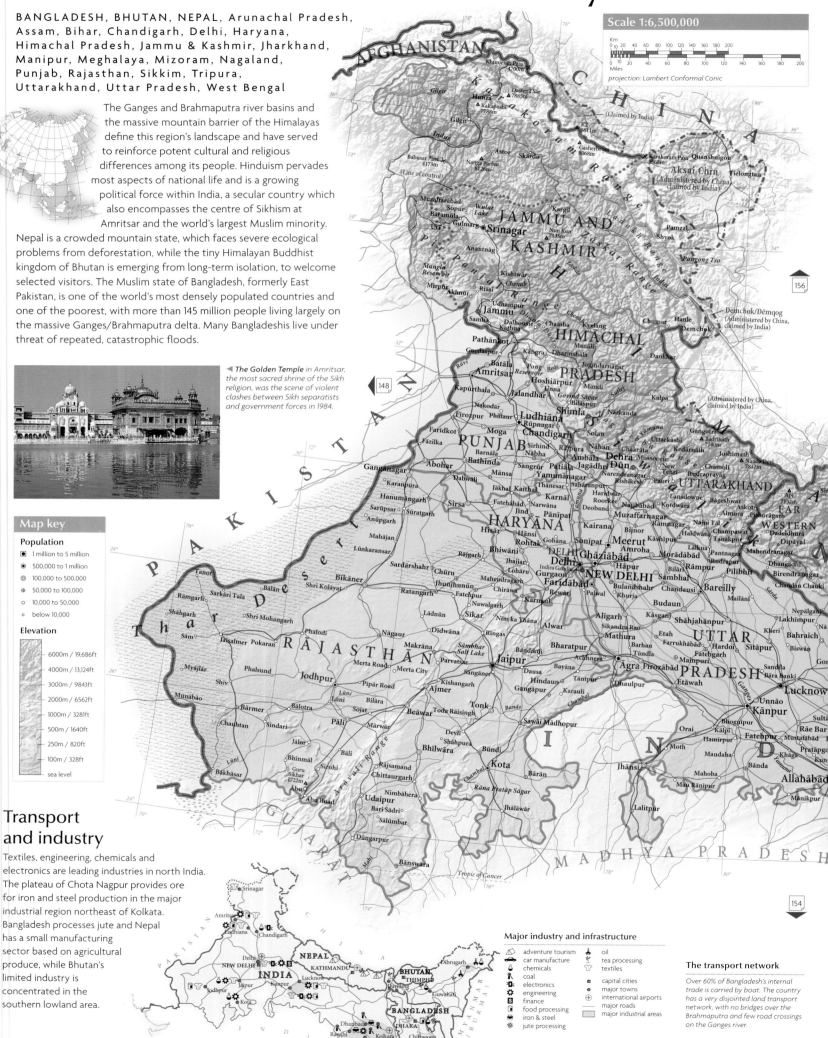

The landscape

Most of the region is drained by the Ganges river, which meets the Brahmaputra in Bangladesh to form an immense delta before flowing into the Bay of Bengal. The Himalayas extend eastwards over 1500 miles (2400 km), from the parallel ranges running through Jammu and Kashmir. The Thar Desert occupies the southwest.

► *The Pir Panjal range in southwestern Kashmir rises to elevations of 12,500 ft (3810 m). Despite the freezing conditions, settlements and extensive pastures are found above the tree line.*

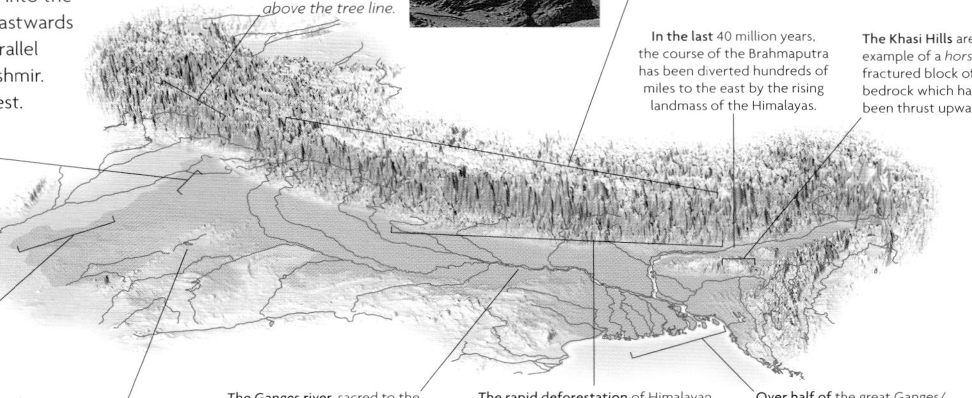

The Indian Punjab lies mainly to the west of the Ganges watershed and its rivers flow into the Indus. Control of this water resource has been a source of great friction with neighbouring Pakistan.

The border between India and Pakistan runs through the Thar Desert, an area of sandy *seif* dunes 50–100 ft (15–30 m) in height. Fossils found in the desert indicate that the dunes, stabilized by vegetation, have been in their current position for about 3000 years.

Sambhar Salt Lake in Rajasthan is India's largest lake. Unlike most of the Himalayan lakes which are glacial in origin – formed in ice-scoured basins or as the result of depositional damming – it is an ephemeral salt lake filled periodically by flash flooding.

The Ganges river, sacred to the Hindu people, drains a vast lowland area at the base of the Himalayas. The northern plains are covered by sandy deposits, broken by mud-banks formed when the river floods.

The rapid deforestation of Himalayan valleys has led to acute soil erosion and increased rates of rainwater run-off, both cited as possible causes of the worsening floods downstream in the Ganges/Brahmaputra delta, although natural rates are high and may be the real cause.

Over half of the great Ganges/Brahmaputra delta floods each year during the monsoon as rivers, swollen by meltwater from the Himalayas and by excess rainwater, break their banks and fertilize the land with nutrient-rich sediment.

The northern ranges of the Himalayas contain the highest mountains in the world, with average heights of more than 23,000 ft (7000 m) and many peaks higher than 26,000 ft (8000 m).

In the last 40 million years, the course of the Brahmaputra has been diverted hundreds of miles to the east by the rising landmass of the Himalayas.

The Khasi Hills are an example of a *horst*, a fractured block of bedrock which has been thrust upwards.

▲ ***The summit of** Machhapuchhre rises to 22,942 ft (6993 m). It is also known as the 'Fish's Tail' because of its distinctive peak.*

Debris slides in the middle Himalayas

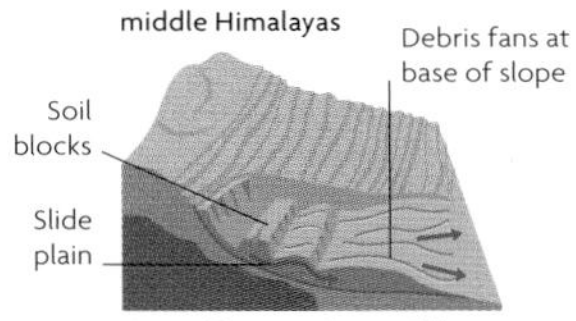

▲ ***Soil loss** in the middle Himalayas has largely been attributed to debris slides, where large blocks of soil are mobilized by saturation along a slide plane. Once mobile, the soil slides down the slope, gaining speed and thinning to form a fan at the base of the slope.*

Using the land

Grain production dominates land use. Rice is most widely grown in the east. Irrigation and new crop strains have dramatically increased yields in the Punjab, a major wheat-producing area. River flood plains are intensively farmed and livestock-herding is widespread, particularly in Bhutan. Regional crops include jute in Bangladesh, tea in Assam, cardamom in Sikkim and saffron in Kashmir.

Land use and agricultural distribution

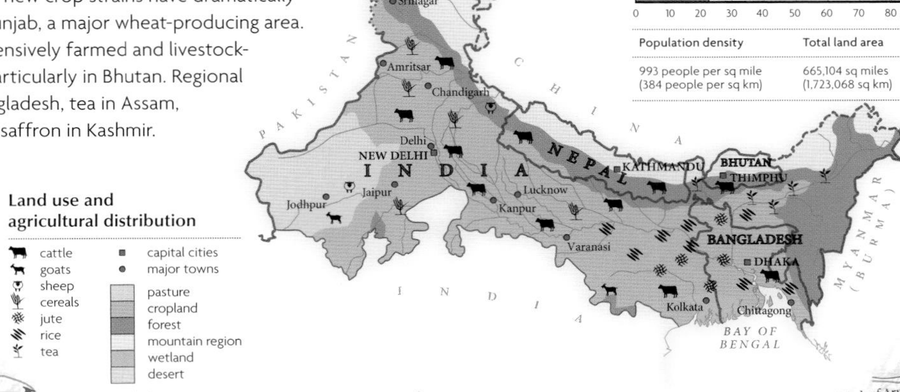

The urban/rural population divide

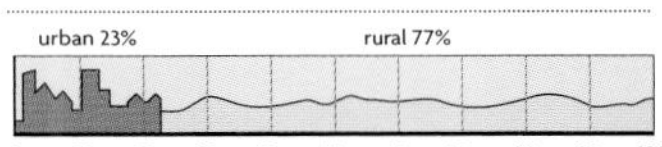

Population density	Total land area
993 people per sq mile (384 people per sq km)	665,104 sq miles (1,723,068 sq km)

▲ ***An adverse climate**, steep slopes and poor soils limit crop cultivation in Bhutan, which is a largely agrarian economy. Rice, corn and wheat are the main staples, although orchards are being established as the soil and climate suit this type of farming.*

▲ ***Flooded streets** in Dhaka, Bangladesh are a testament to the region's vulnerability to flooding. In 1988 alone, 75% of the country was flooded, leaving thousands of people dead and over 25 million homeless.*

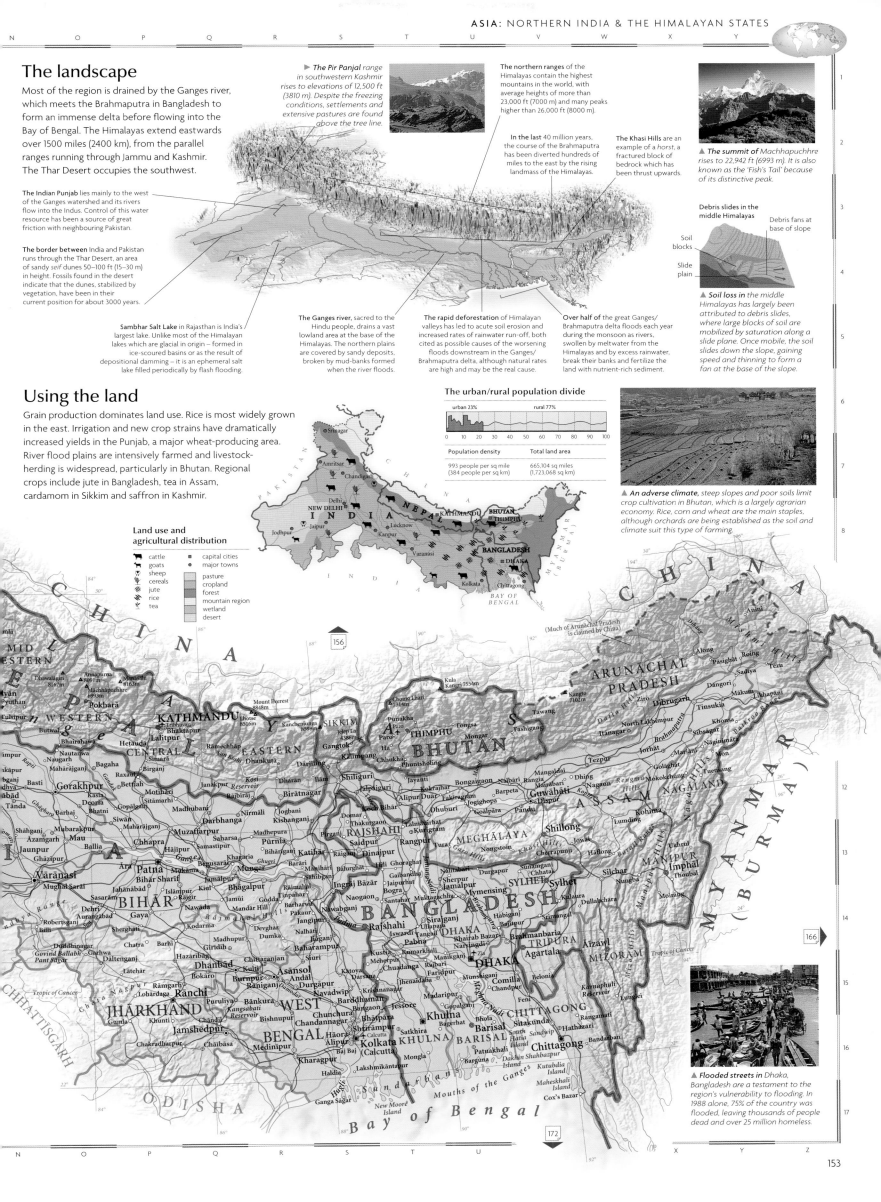

Southern India & Sri Lanka

SRI LANKA, Andhra Pradesh, Chhattisgarh, Dadra & Nagar Haveli, Daman & Diu, Goa, Gujarat, Karnataka, Kerala, Lakshadweep, Madhya Pradesh, Maharashtra, Odisha, Puducherry, Tamil Nadu, Telangana

The unique and highly independent southern states reflect the diverse and decentralized nature of India, which has fourteen official languages. The southern half of the peninsula lay beyond the reach of early invaders from the north and retained the distinct and ancient culture of Dravidian peoples such as the Tamils, whose language is spoken in preference to Hindi throughout southern India. The interior plateau of southern India is less densely populated than the coastal lowlands, where the European colonial imprint is strongest. Urban and industrial growth is accelerating, but southern India's vast population remains predominantly rural. The island of Sri Lanka has two distinct cultural groups; the mainly Buddhist Sinhalese majority, and the Tamil minority whose struggle for a homeland in the northeast led to prolonged civil war.

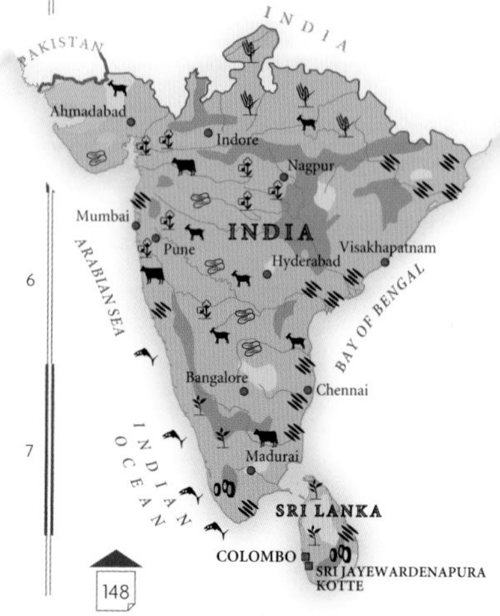

PAKISTAN
INDIA
Ahmadabad
Indore
Nagpur
Mumbai
Pune
Hyderabad
Visakhapatnam
ARABIAN SEA
BAY OF BENGAL
Bangalore
Chennai
Madurai
INDIAN OCEAN
SRI LANKA
COLOMBO
SRI JAYEWARDENAPURA KOTTE

Using the land and sea

Rice is the main staple in the east, in Sri Lanka and along the humid Malabar Coast. Groundnuts are grown on the Deccan plateau, with wheat, corn and chickpeas, towards the north. Sri Lanka is a leading exporter of tea, coconuts and rubber. Cotton plantations supply local mills around Nagpur and Mumbai. Fishing supports many communities in Kerala and the Laccadive Islands.

Land use and agricultural distribution

cattle	rice	pasture
goats	rubber	cropland
cereals	tea	forest
cotton		wetland
fishing	capital cities	
groundnuts	major towns	

The urban/rural population divide

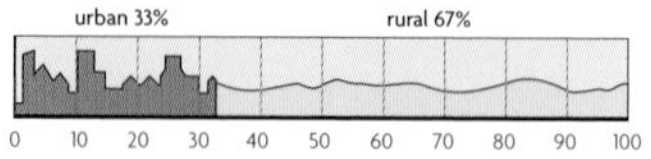

urban 33%
rural 67%
0
10
20
30
40
50
60
70
80
90
100

Population density	Total land area
730 people per sq mile (282 people per sq km)	698,295 sq miles (1,809,054 sq km)

The landscape

The undulating Deccan plateau underlies most of southern India; it slopes gently down towards the east and is largely enclosed by the Ghats coastal hill ranges. The Western Ghats run continuously along the Arabian Sea coast, while the Eastern Ghats are interrupted by rivers which follow the slope of the plateau and flow across broad lowlands into the Bay of Bengal. The plateaux and basins of Sri Lanka's central highlands are surrounded by a broad plain.

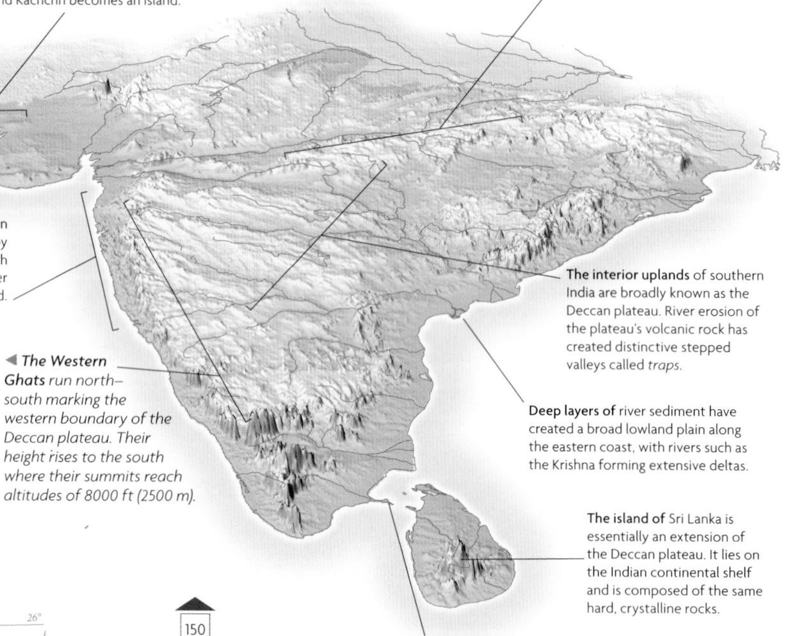

The Rann of Kachchh tidal marshes encircle the low-lying Kachchh peninsula. For several months during the rainy season the water level of the marshes rises and Kachchh becomes an island.
Along the northern boundary of the Deccan plateau, old basement rocks are interspersed with younger sedimentary strata. This creates spectacular scarplands, cut by numerous waterfalls along the softer sedimentary strata.
The Konkan coast, which runs between Daman and Goa, is characterized by rocky headlands, and bays with crescent-shaped beaches. Flooded river valleys known as *rias* extend inland.
The interior uplands of southern India are broadly known as the Deccan plateau. River erosion of the plateau's volcanic rock has created distinctive stepped valleys called *traps*.
Deep layers of river sediment have created a broad lowland plain along the eastern coast, with rivers such as the Krishna forming extensive deltas.
The island of Sri Lanka is essentially an extension of the Deccan plateau. It lies on the Indian continental shelf and is composed of the same hard, crystalline rocks.
Adam's Bridge

◀ *The Western Ghats run north–south marking the western boundary of the Deccan plateau. Their height rises to the south where their summits reach altitudes of 8000 ft (2500 m).*

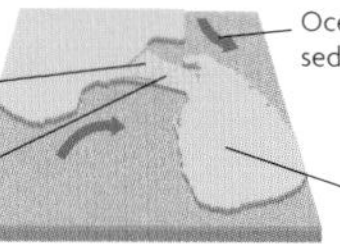

Relict of ancient tombolo
Adam's Bridge
Ocean currents cause sediment build up
Sri Lanka

▲ **Adam's Bridge (Rama's Bridge)** *is a chain of sandy shoals lying about 4 ft (1.2 m) under the sea between India and Sri Lanka. They once formed the world's longest tombolo, or land bridge, before the sea level began to rise several thousand years ago.*

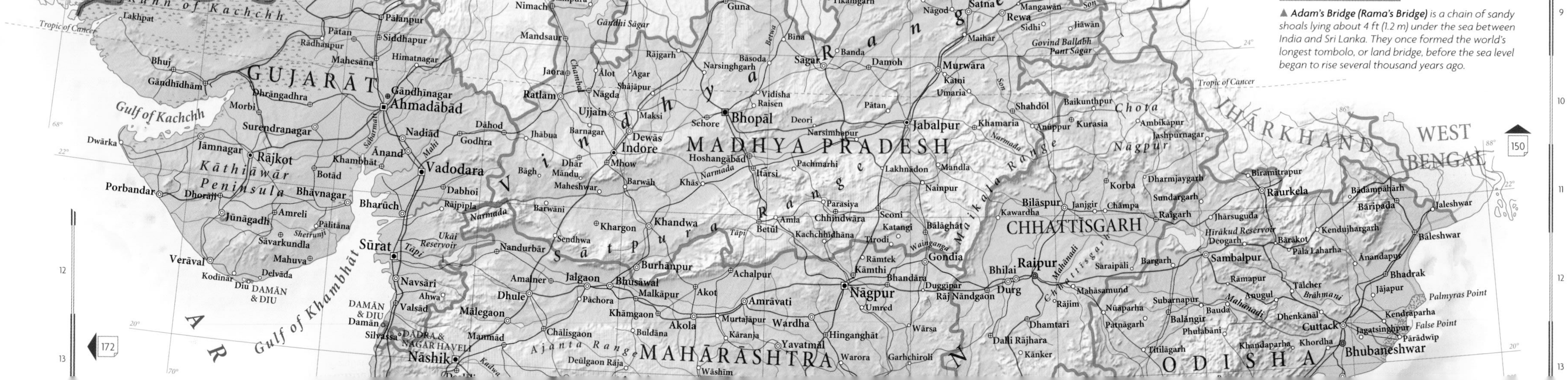

PAKISTAN
RĀJASTHĀN
UTTAR PRADESH
GUJARĀT
MADHYA PRADESH
CHHATTĪSGARH
JHĀRKHAND
WEST BENGAL
MAHĀRĀSHTRA
ODISHA
Gulf of Kachchh
Gulf of Khambhāt
Rann of Kachchh
Kāthiāwār Peninsula
Tropic of Cancer
Ahmadābād
Vadodara
Sūrat
Rājkot
Indore
Bhopāl
Jabalpur
Gwalior
Nāgpur
Raipur
Bhubaneshwar
Cuttack
Nāshik

148
150
150
150
172

Map key

Population

- ■ above 5 million
- ■ 1 million to 5 million
- ● 500,000 to 1 million
- ◎ 100,000 to 500,000
- ⊕ 50,000 to 100,000
- ○ 10,000 to 50,000
- ○ below 10,000

Elevation

- 2000m / 6562ft
- 1000m / 3281ft
- 500m / 1640ft
- 250m / 820ft
- 100m / 328ft
- sea level

◀ *Sea pencils thrive on* the coral reefs around the coast of the Laccadive Islands and Sri Lanka. The reefs support an amazing diversity of marine life, but are increasingly under threat from growing coastal populations.

Sri Lanka: Capital cities

COLOMBO – capital
SRI JAYEWARDENAPURA KOTTE – legislative capital

▲ *Local fisheries around* Sri Lanka afford great potential for exploitation. However, many fishermen living on the coastal fringes saw their livelihoods destroyed by the devastating effects of the Asian tsunami in 2004.

Scale 1:7,000,000

Km 0 25 50 100 150 200
Miles 0 25 50 100 150 200

projection: Lambert Conformal Conic

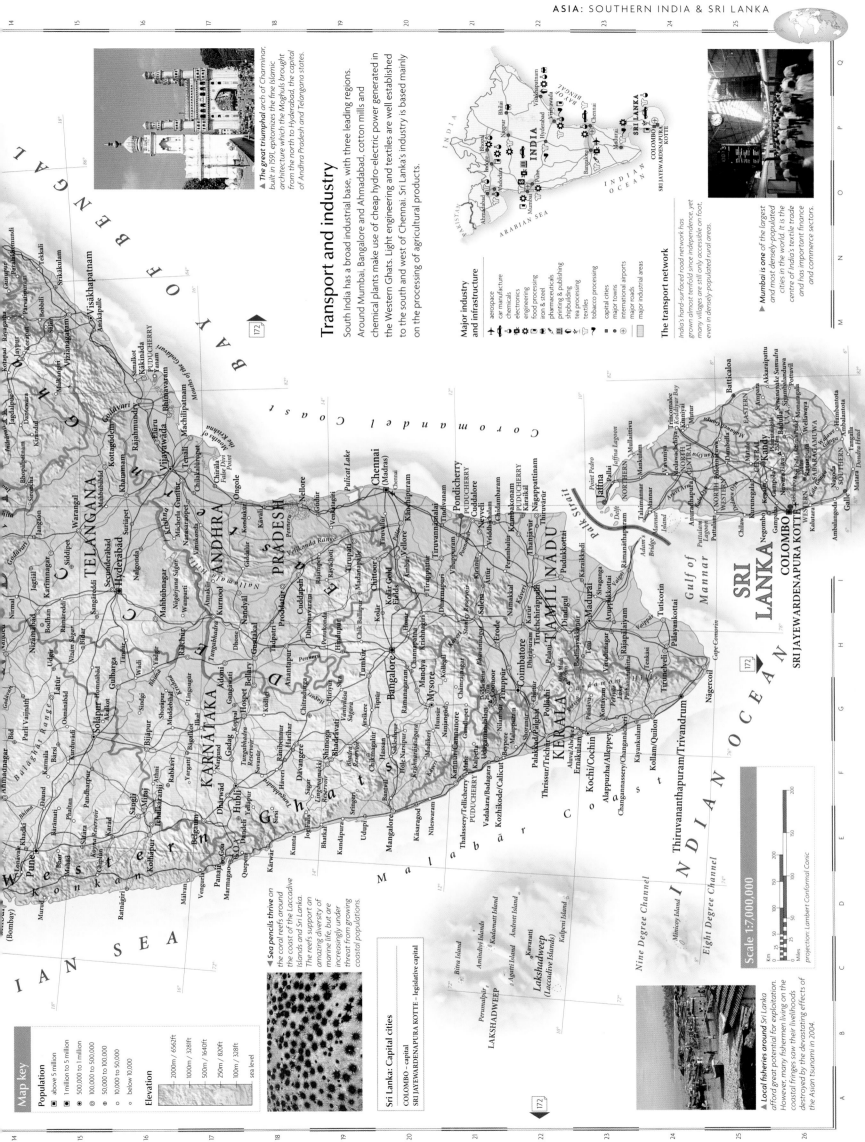

▲ *The great triumphal arch of Charminar,* built in 1591, epitomizes the fine Islamic architecture which the Moghuls brought from the north to Hyderabad, the capital of Andhra Pradesh and Telangana states.

Transport and industry

South India has a broad industrial base, with three leading regions. Around Mumbai, Bangalore and Ahmadabad, cotton mills and chemical plants make use of cheap hydro-electric power generated in the Western Ghats. Light engineering and textiles are well established to the south and west of Chennai. Sri Lanka's industry is based mainly on the processing of agricultural products.

Major industry and infrastructure

- aerospace
- car manufacture
- chemicals
- electronics
- engineering
- food processing
- iron & steel
- pharmaceuticals
- printing & publishing
- shipbuilding
- tea processing
- textiles
- tobacco processing
- capital cities
- major towns
- international airports
- major roads
- major industrial areas

The transport network

India's hard-surfaced road network has grown almost tenfold since independence, yet many villages are still only accessible on foot, even in densely-populated rural areas.

▶ *Mumbai is one* of the largest and most densely-populated cities in the world. It is the centre of India's textile trade and has important finance and commerce sectors.

Mainland East Asia

CHINA, MONGOLIA, NORTH KOREA, SOUTH KOREA, TAIWAN

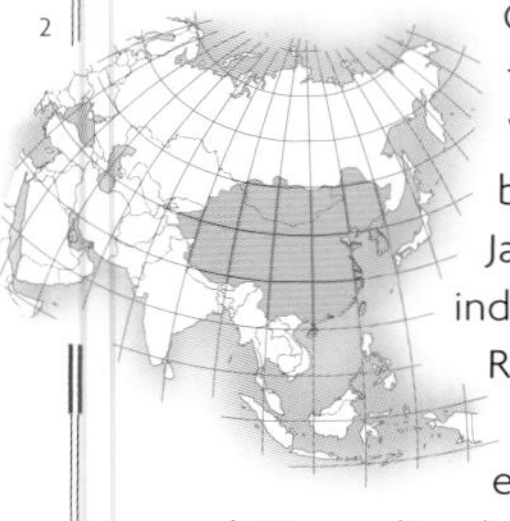

China, the world's most populous nation, has an unbroken cultural history, longer than that of any other country, and is rapidly emerging as a leading world power. When Mao Zedong established Communist rule in 1949, China had become a backward feudal empire, stricken by civil war and over a century of European and Japanese incursions. The closed regime withstood the traumas of rapid industrialization, communalized farming and the brutal purges of the Cultural Revolution but, since the 1980s has introduced economic reforms, led by expanded foreign trade. China's population is heavily concentrated in the east and, despite accelerating urban growth, remains predominantly rural. One cultural group, the Han, make up over 90% of the people, while five 'Autonomous Regions' have been established in the south and west for the main ethnic minorities.

◀ ***Gansu province, through** which the ancient Silk Route passes on its way to the west, is characterized by extensive loess deposits which are terraced and used for crop cultivation.*

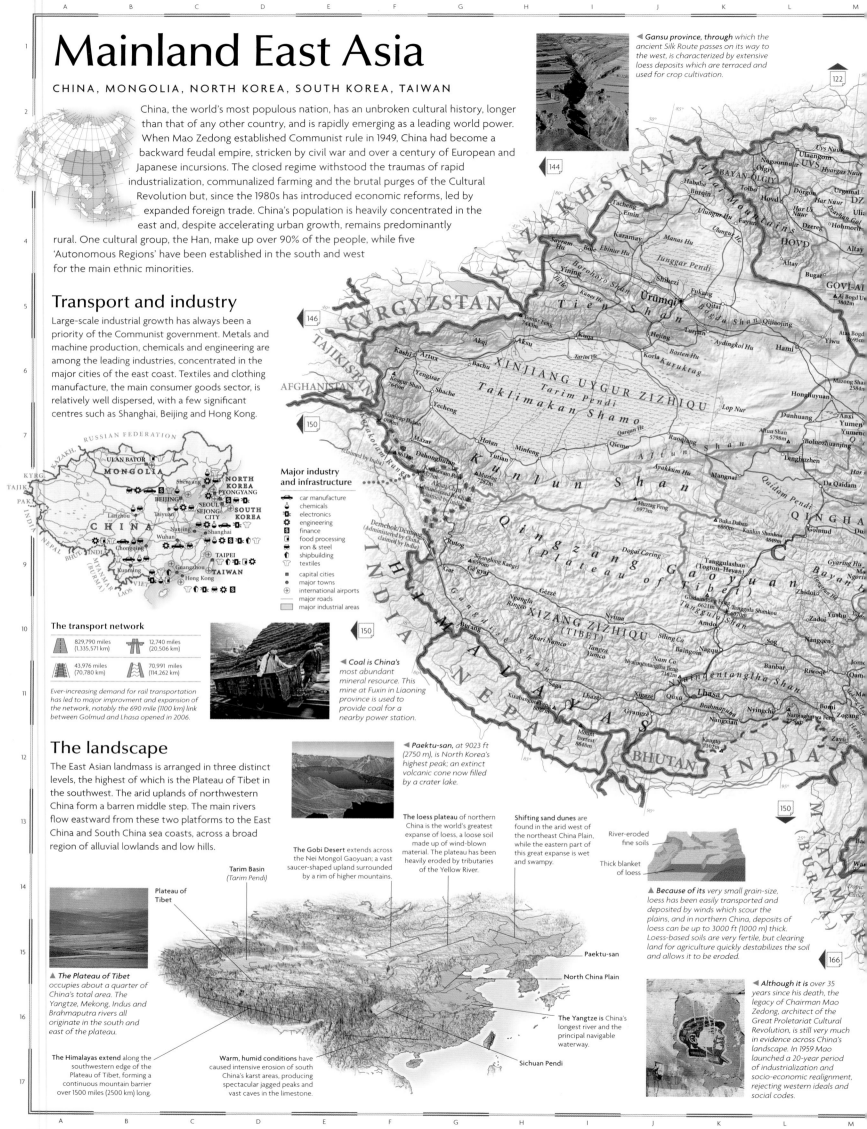

122

144

146

150

Transport and industry

Large-scale industrial growth has always been a priority of the Communist government. Metals and machine production, chemicals and engineering are among the leading industries, concentrated in the major cities of the east coast. Textiles and clothing manufacture, the main consumer goods sector, is relatively well dispersed, with a few significant centres such as Shanghai, Beijing and Hong Kong.

The transport network

829,790 miles (1,335,571 km)	12,740 miles (20,506 km)
43,976 miles (70,780 km)	70,991 miles (114,262 km)

Ever-increasing demand for rail transportation has led to major improvment and expansion of the network, notably the 690 mile (1100 km) link between Golmud and Lhasa opened in 2006.

◀ ***Coal is China's** most abundant mineral resource. This mine at Fuxin in Liaoning province is used to provide coal for a nearby power station.*

150

The landscape

The East Asian landmass is arranged in three distinct levels, the highest of which is the Plateau of Tibet in the southwest. The arid uplands of northwestern China form a barren middle step. The main rivers flow eastward from these two platforms to the East China and South China sea coasts, across a broad region of alluvial lowlands and low hills.

◀ ***Paektu-san,** at 9023 ft (2750 m), is North Korea's highest peak; an extinct volcanic cone now filled by a crater lake.*

150

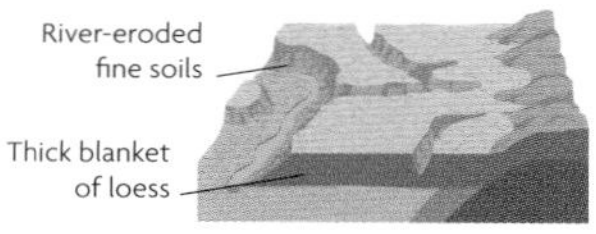

▲ ***Because of its** very small grain-size, loess has been easily transported and deposited by winds which scour the plains, and in northern China, deposits of loess can be up to 3000 ft (1000 m) thick. Loess-based soils are very fertile, but clearing land for agriculture quickly destabilizes the soil and allows it to be eroded.*

166

▲ ***The Plateau of Tibet** occupies about a quarter of China's total area. The Yangtze, Mekong, Indus and Brahmaputra rivers all originate in the south and east of the plateau.*

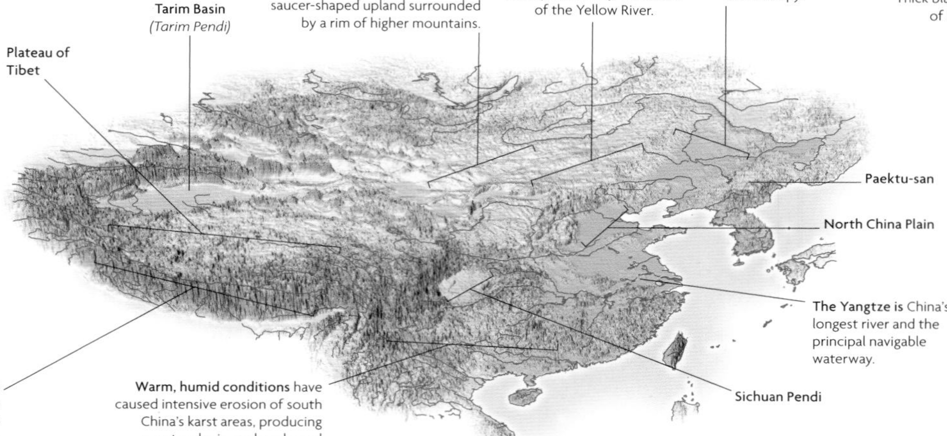

The Gobi Desert extends across the Nei Mongol Gaoyuan; a vast saucer-shaped upland surrounded by a rim of higher mountains.

The loess plateau of northern China is the world's greatest expanse of loess, a loose soil made up of wind-blown material. The plateau has been heavily eroded by tributaries of the Yellow River.

Shifting sand dunes are found in the arid west of the northeast China Plain, while the eastern part of this great expanse is wet and swampy.

The Yangtze is China's longest river and the principal navigable waterway.

The Himalayas extend along the southwestern edge of the Plateau of Tibet, forming a continuous mountain barrier over 1500 miles (2500 km) long.

Warm, humid conditions have caused intensive erosion of south China's karst areas, producing spectacular jagged peaks and vast caves in the limestone.

◀ ***Although it is** over 35 years since his death, the legacy of Chairman Mao Zedong, architect of the Great Proletariat Cultural Revolution, is still very much in evidence across China's landscape. In 1959 Mao launched a 20-year period of industrialization and socio-economic realignment, rejecting western ideals and social codes.*

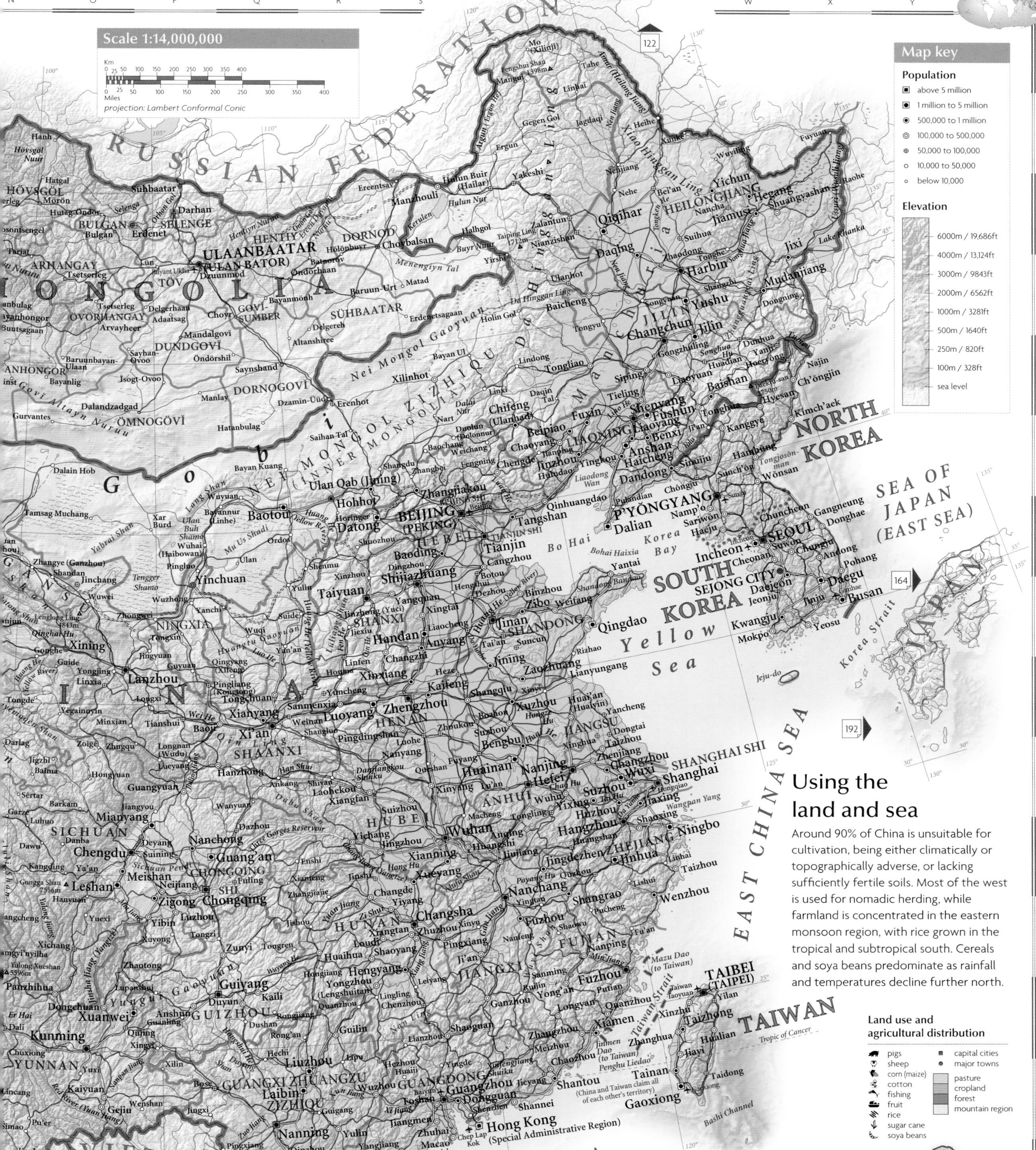

Using the land and sea

Around 90% of China is unsuitable for cultivation, being either climatically or topographically adverse, or lacking sufficiently fertile soils. Most of the west is used for nomadic herding, while farmland is concentrated in the eastern monsoon region, with rice grown in the tropical and subtropical south. Cereals and soya beans predominate as rainfall and temperatures decline further north.

◀ *The Great Wall* of China remains one of the world's largest-ever construction projects, and is so vast that it is visible from space. Sections were added as late as 1640 and it runs for over 4000 miles (6400 km) from the Yellow Sea to Central Asia.

The urban/rural population divide

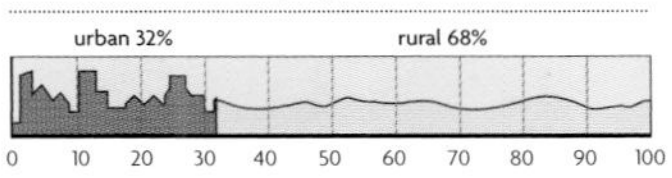

Population density	Total land area
325 people per sq mile (125 people per sq km)	4,288,672 sq miles (11,110,550 sq km)

Land use and agricultural distribution

A B C D E F G H I J

Western China

Gansu, Ningxia, Qinghai, Tibet, Xinjiang

The plateaux and basins of China's dry, desolate western domain are sparsely populated and largely undeveloped, although they have rich mineral reserves; they also form a critical buffer zone for China, in a geographically important and culturally sensitive part of the Asian continent. Across most of the west, the Han Chinese are outnumbered by a range of cultural groups, including the Uygur, the largest group of the various semi-nomadic Muslim peoples from Central Asia. The remote, inhospitable Plateau of Tibet is the world's coldest and highest plateau. It has been occupied by the Chinese since 1950. Tibet is one of western China's five 'Autonomous Regions', but its reclusive Buddhist culture has been systematically undermined by the Chinese government.

Map key

Population

- 1 million to 5 million
- 500,000 to 1 million
- 100,000 to 500,000
- 50,000 to 100,000
- 10,000 to 50,000
- below 10,000

Elevation

- 6000m / 19,686ft
- 4000m / 13,124ft
- 3000m / 9843ft
- 2000m / 6562ft
- 1000m / 3281ft
- 500m / 1640ft
- 250m / 820ft
- 100m / 328ft
- sea level

Scale 1:7,750,000

Km 0 25 50 100 150 200

Miles 0 25 50 100 150 200

projection: Lambert Conformal Conic

▲ ***The Lhasa He** is one of the many rivers which drain the vast Plateau of Tibet. From its source in the Nyainqêntanglha Shan range and fed by the spring meltwater, it eventually joins the upper Brahmaputra 40 miles (65 km) southwest of Lhasa.*

Using the land

Agriculture is constrained by the cold, dry climate and lack of fertile soils in the region, although irrigation and glasshouse farming are increasing agricultural potential. Large quantities of fruit, like melons and grapes, are grown at the oases of Hami and Turpan in Xinjiang, and new irrigation schemes have greatly increased cotton and wheat production in the Tarim Basin *(Tarim Pendi)*. Most of the great area of Tibet and Qinghai is devoted to pastoralism. Sheep are the principal livestock.

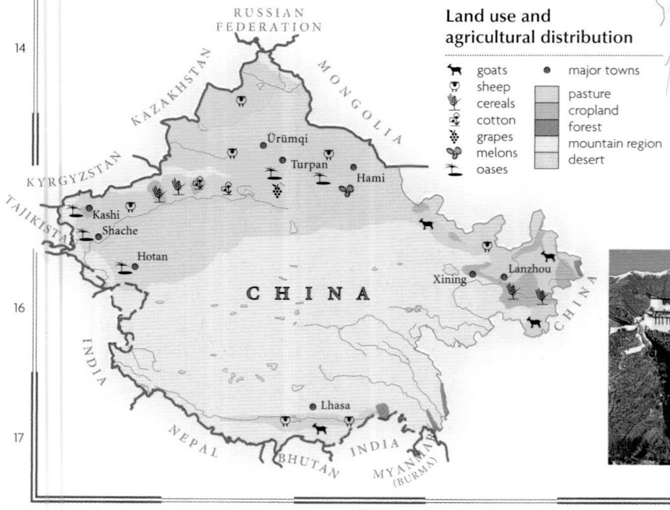

◀ ***The Potala Palace**, in Tibet's capital, Lhasa, was the former residence of the Dalai Lama, Tibetan Buddhism's spiritual leader. Tibet remains only sparsely populated; forming over 20% of China's landmass, it supports fewer than 1% of its population.*

144 122 146 146 148 150 150 150

A B C D E F G H I J K L M

The landscape

The Himalayas mark the southwestern edge of the Plateau of Tibet, an extreme mountain wilderness which occupies nearly a quarter of China's total area. A large structural depression, the Qaidam Pendi, lies at its northeastern edge. The Kunlun mountain chain isolates the plateau from the desert to the north, where the Tien Shan range forms a spur between the Tarim Basin *(Tarim Pendi)* and Dzungarian Basin *(Junggar Pendi)*.

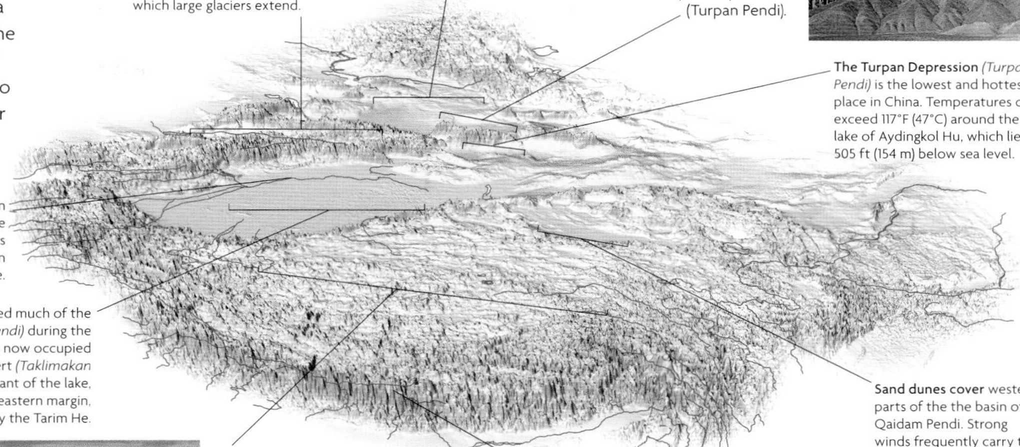

The Tien Shan reach elevations of over 24,419 ft (7443 m) and have permanent ice fields, from which large glaciers extend.

Northwestern China is largely a region of internal drainage. The Tarim He flows only as far as Lop Nur, where its water is lost by evapotranspiration from the lake and land surface.

A vast glacial lake filled much of the Tarim Basin *(Tarim Pendi)* during the last Ice Age. This area is now occupied by the Takla Makan Desert *(Taklimakan Shamo)*. A remnant of the lake, Lop Nur, forms the eastern margin, where it is fed by the Tarim He.

► ***The Bogda Shan,** an eastward arm of the Tien Shan range, rise high above the Turpan Depression* (Turpan Pendi).

The Turpan Depression *(Turpan Pendi)* is the lowest and hottest place in China. Temperatures can exceed 117°F (47°C) around the lake of Aydingkol Hu, which lies 505 ft (154 m) below sea level.

Sand dunes cover western parts of the the basin of Qaidam Pendi. Strong winds frequently carry the sands east, threatening the agricultural areas around the lake of Qinghai Hu.

Mount Everest is the world's highest peak, at 29,029 ft (8848 m). The summit marks the border between China and Nepal.

◄ ***The terrain of** the Plateau of Tibet consists of mountain peaks and open plateaux, dotted with brackish lakes. These are probably remnants of the Tethys Sea, which covered the area before it was uplifted following the collision of the Indo-Australian and Eurasian plates.*

▲ ***The Tarim Basin** (Tarim Pendi) has no permanent rivers. Rainfall from the surrounding Plateau of Tibet and Tien Shan ranges drains into the basin's sand and gravel floor.*

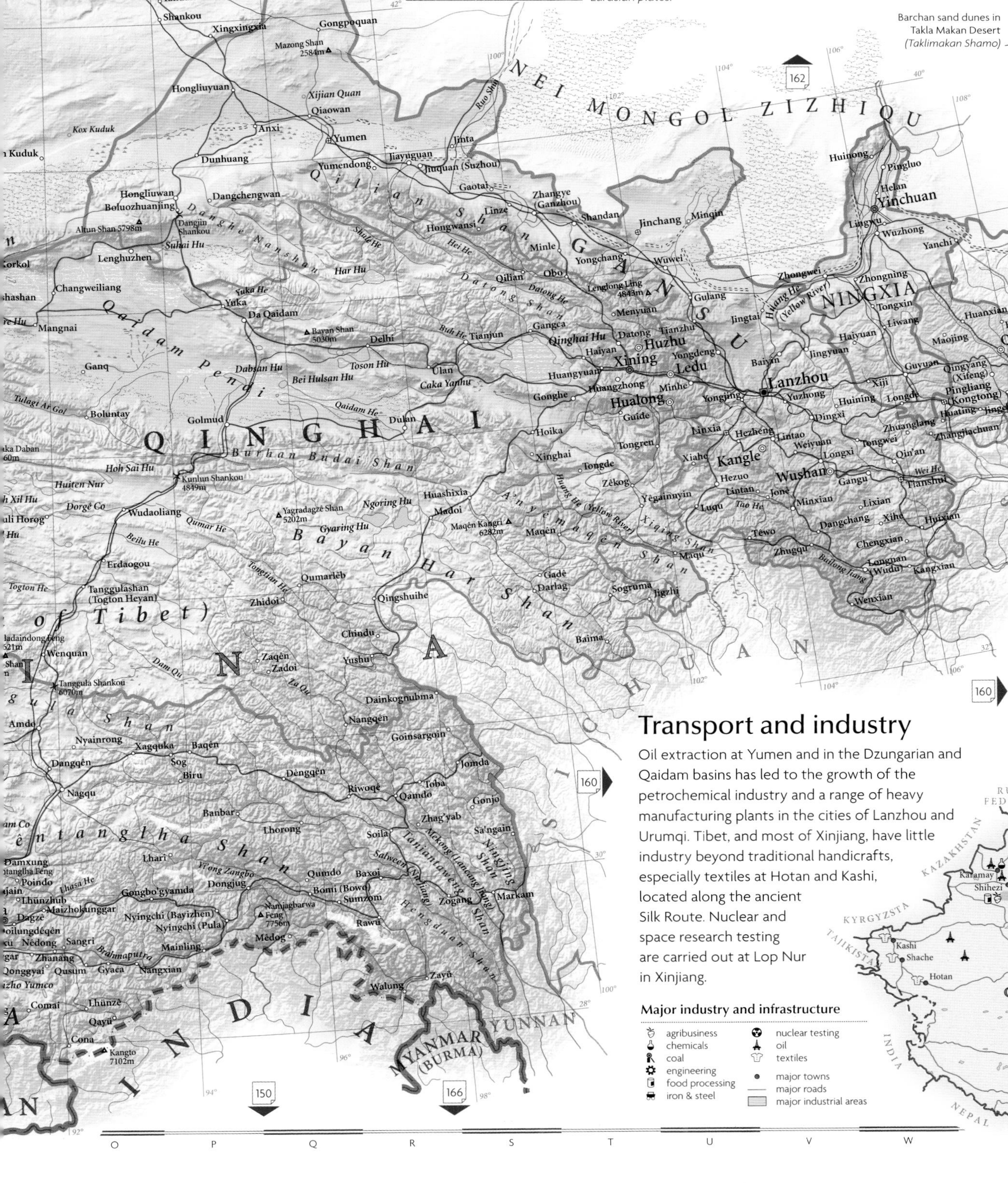

▲ ***From its source,** high in eastern Qinghai, the Yellow River starts on a 3395 mile (5464 km) journey to the Yellow Sea.*

Transport and industry

Oil extraction at Yumen and in the Dzungarian and Qaidam basins has led to the petrochemical industry and a range of heavy manufacturing plants in the cities of Lanzhou and Urumqi. Tibet, and most of Xinjiang, have little industry beyond traditional handicrafts, especially textiles at Hotan and Kashi, located along the ancient Silk Route. Nuclear and space research testing are carried out at Lop Nur in Xinjiang.

The transport network

The construction of roads connecting Lhasa in Tibet with Sichuan, Qinghai and Xinjiang was achieved in the 1950s, in spite of the extreme physical conditions of the Plateau of Tibet.

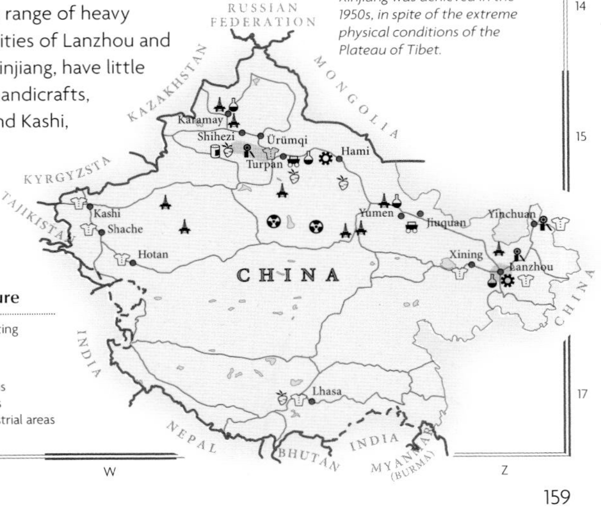

Major industry and infrastructure

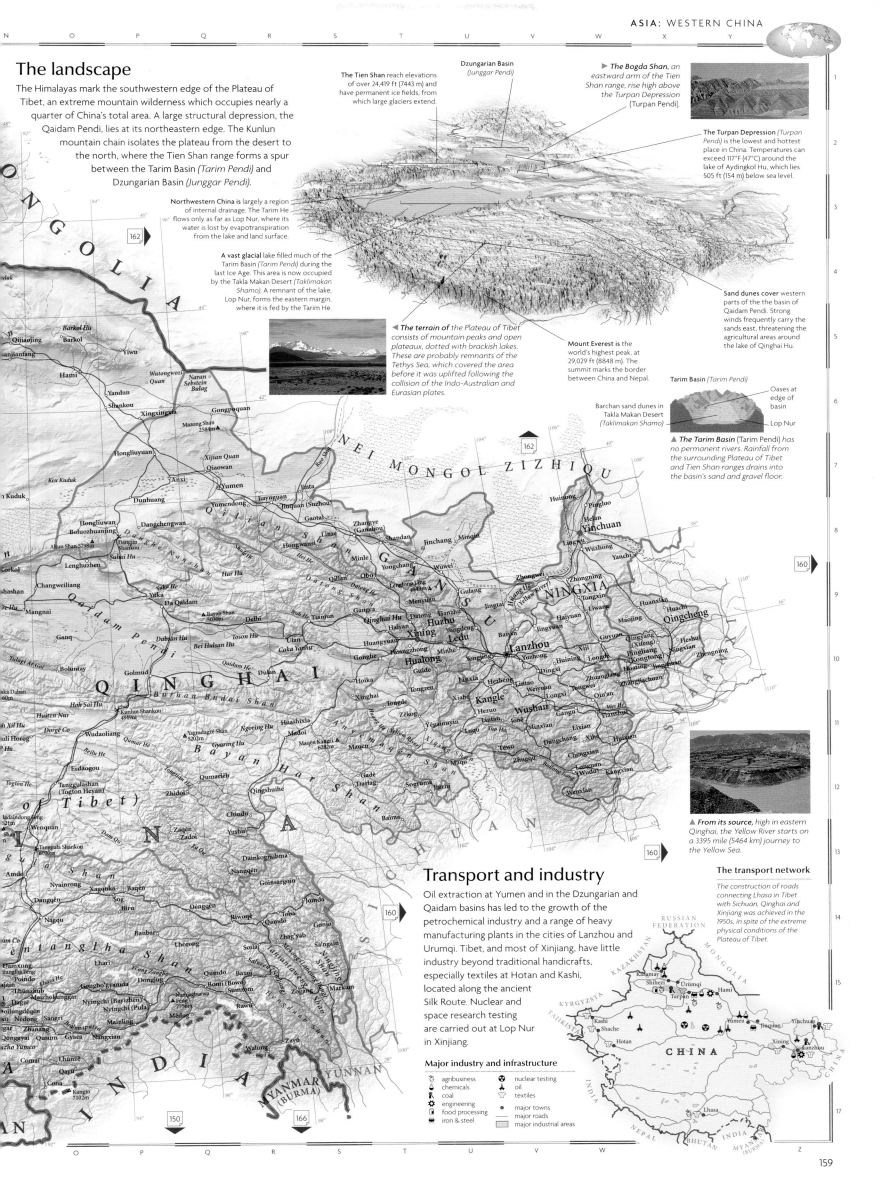

- agribusiness
- chemicals
- coal
- engineering
- food processing
- iron & steel
- nuclear testing
- oil
- textiles
- major towns
- major roads
- major industrial areas

Eastern China

TAIWAN, Anhui, Beijing, Chongqing, Fujian, Guangdong, Guangxi, Guizhou, Hainan, Hebei, Henan, Hubei, Hunan, Jiangsu, Jiangxi, Shaanxi, Shandong, Shanghai, Shanxi, Sichuan, Tianjin, Yunnan, Zhejiang

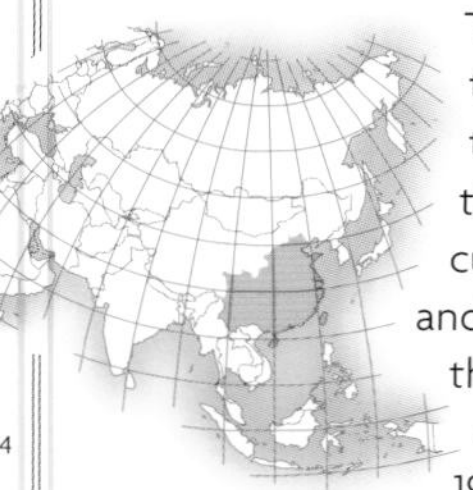

The east is China's heartland. Massive industrial development since 1949 has transformed much of the densely populated rural landscape, in a region still prone to flooding and drought. Over 30 cities have populations of over a million, including the giant metropolis of Shanghai and the capital Beijing, which has been China's cultural and political centre since the 13th century. The ethnically diverse southwest and the oil-rich interior provinces of Sichuan and Shaanxi have largely missed out on the remarkable economic growth occurring in designated free-trade areas along the coasts of the South and East China seas. The republic of Taiwan was established in 1949 by Chinese nationalists ousted from the mainland by the victorious Communist forces. Taiwan now has one of the strongest economies in the world but its sovereignty is not recognized by China. Hong Kong provides a major international trade link for China; a 99-year 'lease' period of British control was concluded in 1997.

▲ ***North of the** Qin Ling range in Shaanxi province, is an agriculturally fertile region covered with fine, wind-blown deposits and known as the loess plateau. The loose sediments are vulnerable to water erosion.*

Using the land and sea

This is a region of intensive cultivation. Wheat, millet, sorghum and cotton are the main crops of the Yellow River basin. South from Sichuan, rice becomes the principal crop, grown with wheat, corn and cotton along the Yangtze river. Tea is produced in the hills and sugar cane along the coast of the southeast, where flat land is limited. Pigs and poultry are raised in great numbers.

Land use and agricultural distribution

- cattle
- pigs
- cereals
- corn (maize)
- cotton
- fishing
- peanuts
- rice
- sugar cane
- tea
- capital cities
- major towns
- pasture
- cropland
- forest
- mountain region

▲ ***On the hills** above the North China Plain, slopes are terraced to utilize the rich loess soils of the Taihang Shan range.*

◀ ***The former Portuguese** territory of Macao, with its colonial architecture, bars and casinos, reverted to Chinese rule in 1999.*

Map key

Population

- above 5 million
- 1 million to 5 million
- 500,000 to 1 million
- 100,000 to 500,000
- 50,000 to 100,000
- 10,000 to 50,000
- below 10,000

Elevation

- 6000m / 19,686ft
- 4000m / 13,124ft
- 3000m / 9843ft
- 2000m / 6562ft
- 1000m / 3281ft
- 500m / 1640ft
- 250m / 820ft
- 100m / 328ft
- sea level

Scale 1:8,500,000

Km 0 25 50 100 150 200 250 300

Miles 0 25 50 100 150 200 250 300

projection: Lambert Conformal Conic

158 162 166

The landscape

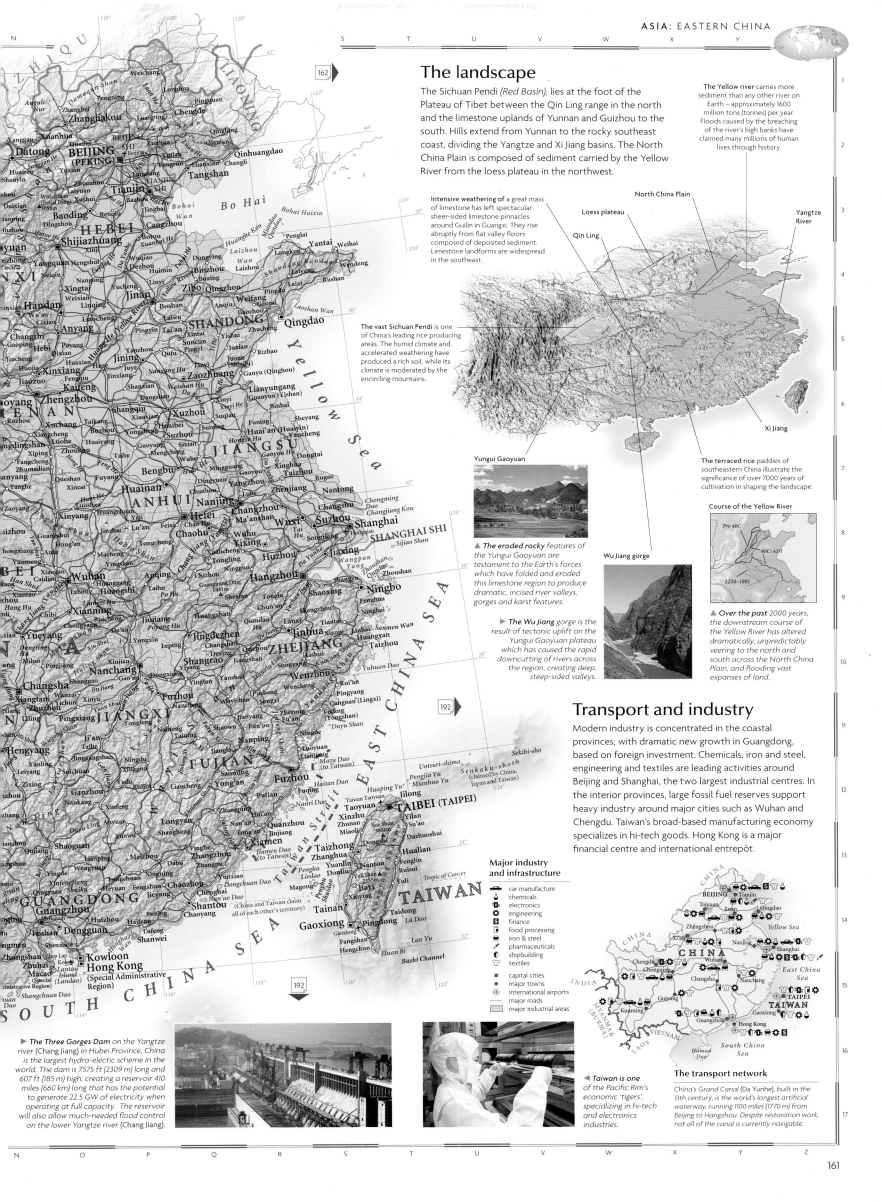

The Sichuan Pendi *(Red Basin)*, lies at the foot of the Plateau of Tibet between the Qin Ling range in the north and the limestone uplands of Yunnan and Guizhou to the south. Hills extend from Yunnan to the rocky southeast coast, dividing the Yangtze and Xi Jiang basins. The North China Plain is composed of sediment carried by the Yellow River from the loess plateau in the northwest.

The Yellow river carries more sediment than any other river on Earth – approximately 1600 million tons (tonnes) per year. Floods caused by the breaching of the river's high banks have claimed many millions of human lives through history.

Intensive weathering of a great mass of limestone has left spectacular sheer-sided limestone pinnacles around Guilin in Guangxi. They rise abruptly from flat valley floors composed of deposited sediment. Limestone landforms are widespread in the southeast.

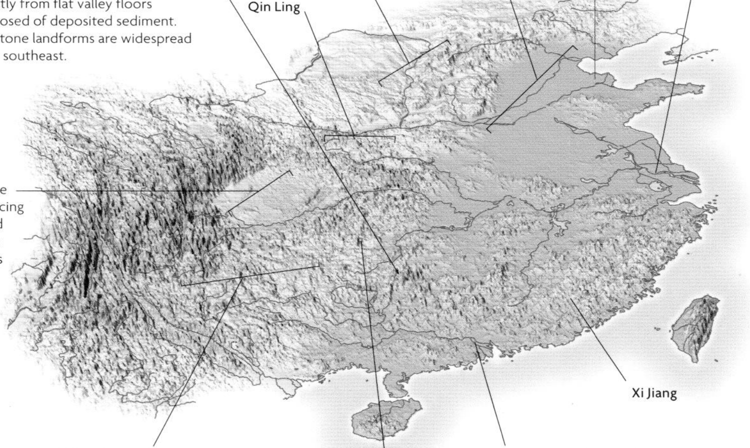

The vast Sichuan Pendi is one of China's leading rice producing areas. The humid climate and accelerated weathering have produced a rich soil, while its climate is moderated by the encircling mountains.

The terraced rice paddies of southeastern China illustrate the significance of over 7000 years of cultivation in shaping the landscape.

▲ ***The eroded rocky** features of the Yungui Gaoyuan are testament to the Earth's forces which have folded and eroded this limestone region to produce dramatic, incised river valleys, gorges and karst features.*

Wu Jiang gorge

▶ ***The Wu Jiang** gorge is the result of tectonic uplift on the Yungui Gaoyuan plateau which has caused the rapid downcutting of rivers across the region, creating deep, steep-sided valleys.*

Course of the Yellow River

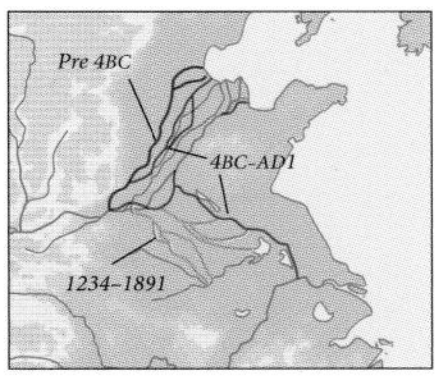

▲ ***Over the past** 2000 years, the downstream course of the Yellow River has altered dramatically, unpredictably veering to the north and south across the North China Plain, and flooding vast expanses of land.*

Transport and industry

Modern industry is concentrated in the coastal provinces, with dramatic new growth in Guangdong, based on foreign investment. Chemicals, iron and steel, engineering and textiles are leading activities around Beijing and Shanghai, the two largest industrial centres. In the interior provinces, large fossil fuel reserves support heavy industry around major cities such as Wuhan and Chengdu. Taiwan's broad-based manufacturing economy specializes in hi-tech goods. Hong Kong is a major financial centre and international entrepôt.

The transport network

China's Grand Canal (Da Yunhe), built in the 13th century, is the world's longest artificial waterway, running 1100 miles (1770 m) from Beijing to Hangzhou. Despite restoration work, not all of the canal is currently navigable.

▶ ***The Three Gorges Dam** on the Yangtze river (Chang Jiang) in Hubei Province, China is the largest hydro-electic scheme in the world. The dam is 7575 ft (2309 m) long and 607 ft (185 m) high, creating a reservoir 410 miles (660 km) long that has the potential to generate 22.5 GW of electricity when operating at full capacity. The reservoir will also allow much-needed flood control on the lower Yangtze river (Chang Jiang).*

◀ ***Taiwan is one** of the Pacific Rim's economic 'tigers', specializing in hi-tech and electronics industries.*

Northeastern China, Mongolia & Korea

MONGOLIA, NORTH KOREA, SOUTH KOREA, Heilongjiang, Inner Mongolia, Jilin, Liaoning

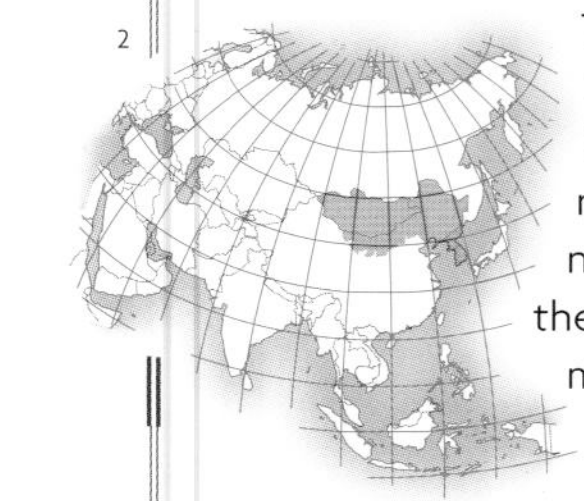

This northerly region has for centuries been a domain of shifting borders and competing colonial powers. Mongolia was the heartland of Chinghiz Khan's vast Mongol empire in the 13th century, while northeastern China was home to the Manchus, China's last ruling dynasty (1644–1911). The mineral and forest wealth of the northeast helped make this China's principal region of heavy industry, although the outdated state factories now face decline. South Korea's state-led market economy has grown dramatically and Seoul is now one of the world's largest cities. The austere communist regime of North Korea has isolated itself from the expanding markets of the Pacific Rim and faces continuing economic stagnation.

▲ *The Eurasian steppe* stretches from the mouth of the Danube in Europe, to Mongolia. In Mongolia, nomadic people have lived in felt huts called yurts or gers, for thousands of years.

Map key

Population

- ▣ above 5 million
- ◙ 1 million to 5 million
- ◉ 500,000 to 1 million
- ◎ 100,000 to 500,000
- ⊕ 50,000 to 100,000
- ○ 10,000 to 50,000
- ○ below 10,000

Elevation

- 4000m / 13,124ft
- 3000m / 9843ft
- 2000m / 6562ft
- 1000m / 3281ft
- 500m / 1640ft
- 250m / 820ft
- 100m / 328ft
- sea level

Scale 1:7,750,000

Km 0 25 50 100 150 200

Miles 0 25 50 100 150 200

projection: Lambert Conformal Conic

The landscape

The great North China Plain is largely enclosed by mountain ranges including the Great and Lesser Khingan Ranges (*Da Hinggan Ling* and *Xiao Hinggan Ling*) in the north, and the Changbai Shan, which extend south into the rugged peninsula of Korea. The broad steppeland plateau of Nei Mongol Gaoyuan borders the southeastern edge of the great cold desert of the Gobi which extends west across the southern reaches of Mongolia. In northwest Mongolia the Altai Mountains and various lesser ranges are interspersed with lakeland basins.

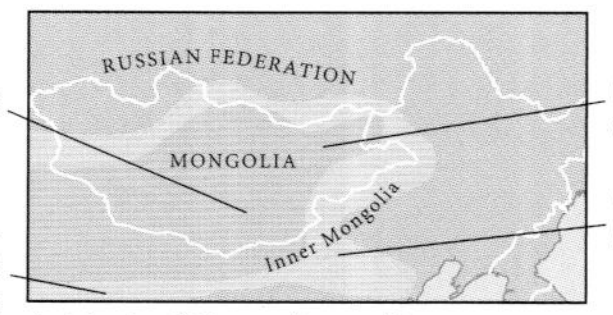

▲ *Much of Mongolia* and Inner Mongolia is a vast desert area. To the south and east, a semi-arid region extends into China proper.

▲ *The Gobi desert* stretches from Central Asia, through Mongolia and into China. Bare rock surfaces, rather than sand dunes, typify the cold desert landscape of the Gobi.

Tributaries of the Amur river follow U-shaped valleys through the Great Khingan Range (*Da Hinggan Ling*). These were cut by ice-age glaciers between 3 and 10 million years ago.

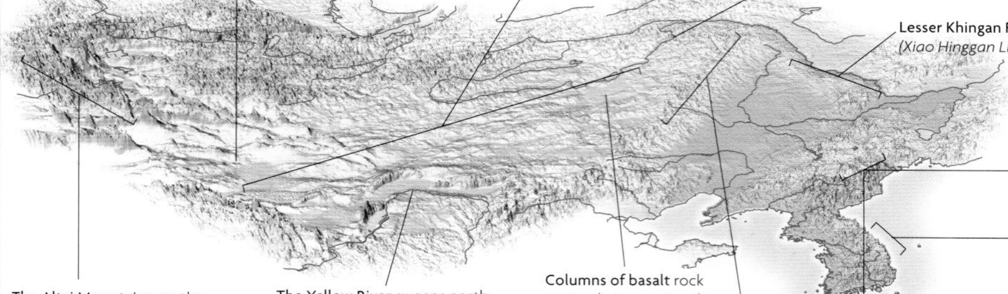

The Altai Mountains are the highest and longest of the mountain ranges which extend into Mongolia from the northwest. These mountains provide one of the last refuges for the endangered snow leopard.

The Yellow River sweeps north around the Ordos Desert (*Mu Us Shadi*), bringing water to an otherwise barren region.

Columns of basalt rock protrude in occasional clusters from the flat surface of the eastern Gobi. Their regular, six-sided form was produced when the rock cooled and contracted from its molten state.

Great Khingan Range (*Da Hinggan Ling*)

A crater lake occupies the 9023 ft (2750 m) snowy summit of the extinct volcano Paektu-san, the highest peak in the mountains of the Changbai Shan.

◄ *The wooded mountain* range of T'aebaek-sanmaek forms the backbone of the Korean peninsula, running north–south along the eastern coastline.

Transport and industry

North Korea's centrally-planned economy is strongly oriented towards heavy industry, while South Korea has a broad manufacturing base which includes textiles, steel, electronics, and one of the world's largest shipbuilding industries. Mongolia and Inner Mongolia's great mineral resource potential is largely undeveloped. The heavy industrial region around Shenyang produces iron, steel, chemicals and cement on a massive scale.

▲ *Ulan Bator, the Mongolian capital bears many of the hallmarks of Soviet-style central planning, the result of economic and industrial assistance from the Soviet Union following Mongolian independence in 1921.*

The transport network

Liaoning has China's most comprehensive railway network, the legacy of the Japanese occupation of Manchuria in the 20th century. The railways are used primarily for freight transport.

▶ *While North Korea has remained politically and economically isolated from the rest of the world, South Korea has enjoyed immense economic growth. It has benefited considerably from US economic aid in the aftermath of the Korean war of 1950–1953.*

South Korea: Capital cities

SEOUL – capital
SEJONG CITY – administrative capital

Using the land and sea

Mongolia and Inner Mongolia rely heavily on livestock farming, with only about 1% of the land area cultivated. Northeastern China produces wheat, corn, soya beans and sugar beet. The cool climate limits the range of crops and large upland areas of the northeast remain forested. Rice is the staple food of North and South Korea. The latter has become a leading ocean-fishing nation.

Japan

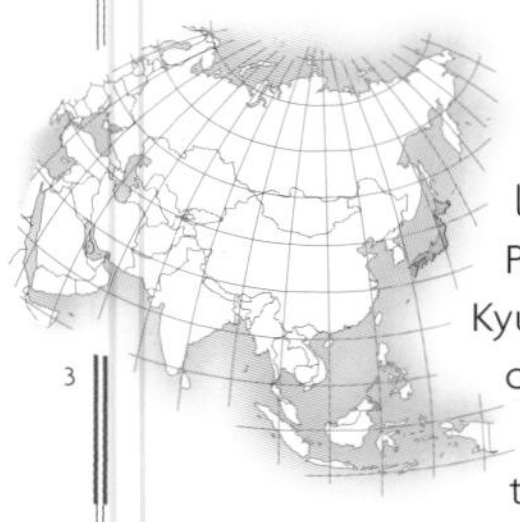

In the years since the end of the Second World War, Japan has become the world's most dynamic industrial nation. The country comprises a string of over 4000 islands which lie in a great northeast to southwest arc in the northwest Pacific. Four major islands: Hokkaido, Honshu, Shikoku and Kyushu are home to the great majority of Japan's population of 128 million people, although the mountainous terrain of the central region means that most cities are situated on the coast. A densely populated industrial belt stretches along much of Honshu's southern coast, including Japan's crowded capital, Tokyo. Alongside its spectacular economic growth and the increasing westernization of its cities, Japan still maintains a most singular culture, reflected in its traditional food, formal behavioural codes, unique Shinto religion and a deep reverence for the emperor.

Using the land and sea

Although only about 11% of Japan is suitable for cultivation, substantial government support, a favourable climate and intensive farming methods enable the country to be virtually self-sufficient in rice production. Northern Hokkaido, the largest and most productive farming region, has an open terrain and climate similar to that of the US Midwest, and produces over half of Japan's cereal requirements. Farmers are being encouraged to diversify by growing fruit, vegetables and wheat, as well as raising livestock.

Land use and agricultural distribution

The urban/rural population divide

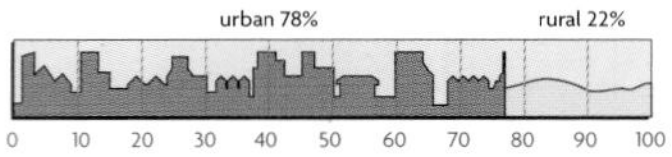

urban 78%
rural 22%
0 10 20 30 40 50 60 70 80 90 100

Population density	Total land area
885 people per sq mile (342 people per sq km)	145,869 sq miles (377,800 sq km)

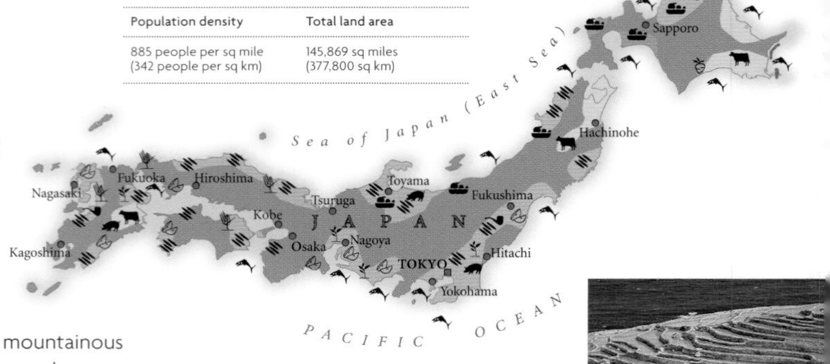

cattle
pigs
fishing
cereals
citrus fruits
fruit
herbs
rice
root crops
tobacco
capital cities
major towns
pasture
cropland
forest
Sapporo
Hachinohe
Sea of Japan (East Sea)
Fukuoka
Hiroshima
Nagasaki
Kagoshima
Kobe
Osaka
Tsuruga
Toyama
Nagoya
JAPAN
Fukushima
TOKYO
Hitachi
Yokohama
PACIFIC OCEAN

▶ *Cutting terraces maximizes the limited agricultural land, enabling Japan to produce large quantities of rice.*

The landscape

The islands of Japan lie on the Pacific 'Ring of Fire', and form a series of clearly defined arcs. The largely mountainous landscape was formed very recently in geological terms. Volcanic eruptions and earthquakes continue to reshape the terrain and to shake the country's complex infrastructure. There is no one continuous mountain range; the mountains divide into many small land blocks separated by lowlands and dissected by numerous river valleys.

Sea of Japan (East Sea)
Active volcanic island
Japan Trench (subduction zone)

▲ *Japan is part of an arc of volcanic islands, formed by the Pacific Plate diving under the Eurasian Plate. This process generates intense stress which is periodically released as earthquakes.*

◀ *Mount Fuji is Japan's highest mountain, rising 12,388 ft (3776 m) above the Kanto Plain in the central region of Honshu. The flat land below is suitable for growing crops such as tea. Like many Japanese mountains, it is revered as a sacred site.*

▶ *Trees cling to the sheer slopes of the waterfalls on the northern island of Hokkaido. The island's climate is similar to that in northern Europe, with long, cold winters and short, warm summers.*

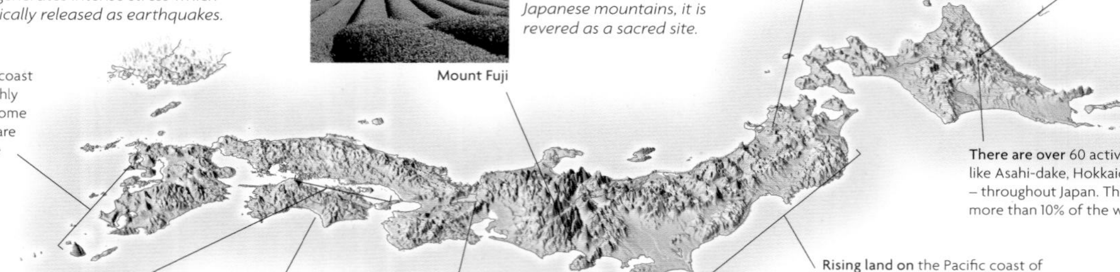

Mount Fuji
A number of rivers which emerge from the volcanic parts of northwestern Honshu are so highly acidic that their water is unsuitable for irrigation and consumption.
In much of Kyushu the coast is subsiding, giving a highly indented coastline. In some places, former hilltops are barely visible above the current sea level.
There are over 60 active volcanoes – like Asahi-dake, Hokkaido's highest peak – throughout Japan. This accounts for more than 10% of the world's total.
The Inland Sea (Seto-naikai) has resulted from the depression of faulted blocks which has allowed sea water to invade the region between northern Shikoku and western Honshu.
Strong southeasterly winds blowing onshore during the winter create sand dunes which extend for miles along the eastern coasts.
Biwa-ko is the largest lake in Japan, covering 260 sq miles (673 sq km) in central Honshu. The depression in which it lies was created by recent faulting of the underlying rocks.
Rising land on the Pacific coast of Honshu leads to typical features such as raised beaches, some lying over 1000 ft (300 m) above sea level.

Sea of Japan
East China Sea
Philippine Sea
Korea Strait
JAPAN
Honshu
Shikoku
Kyūshū
Kitakyūshū
Fukuoka
Nagasaki
Kumamoto
Kagoshima
Miyazaki
Ōita
Hiroshima
Okayama
Kōbe
Ōsaka
Kyoto
Nagoya
Matsuyama
Takamatsu
Kōchi
Tokushima
Wakayama
Tottori
Matsue
Kanazawa
Fukui
Gifu
Toyama
Shizuoka
Hamamatsu
Yamaguchi
Shimonoseki
Tsushima
Iki
Gotō-rettō
Yaku-shima
Tanega-shima
Oki-shotō
Chūgoku-sanchi
Shikoku-sanchi
Kii-sanchi
Kii-suidō
Bungo-suidō
Ōsumi-kaikyō

157

▼ *Autumnal trees near Gifu, on central Honshu, create a spectacular display. Native trees on this island include camphor, pasania, Japanese evergreen oak, camellia and holly.*

▶ *The Kobe earthquake in January 1995 highlighted Japan's vulnerability to earthquakes, despite technological advances. It shattered much of the infrastructure of this important port. More than 5000 people died as buildings and overhead highways collapsed and fires broke out.*

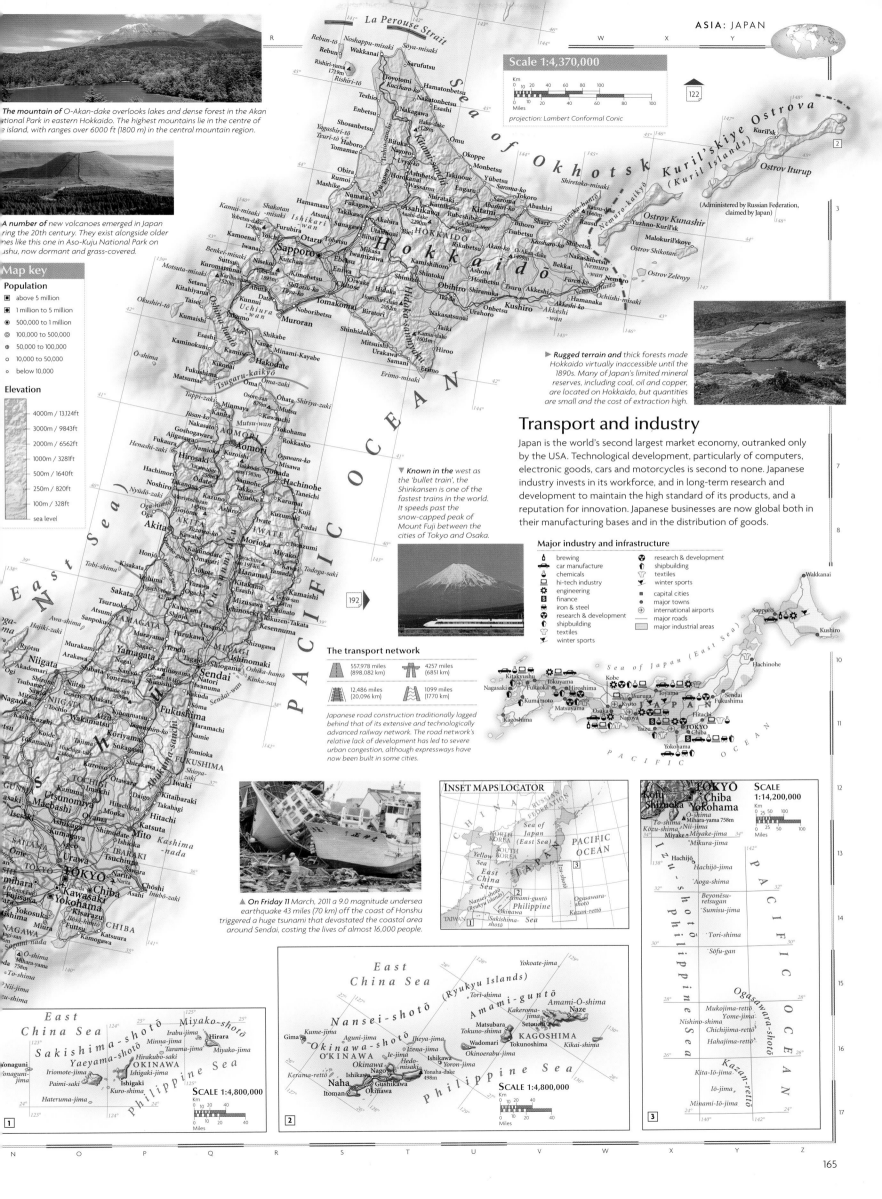

The mountain of O-Akan-dake overlooks lakes and dense forest in the Akan National Park in eastern Hokkaido. The highest mountains lie in the centre of the island, with ranges over 6000 ft (1800 m) in the central mountain region.

A number of new volcanoes emerged in Japan during the 20th century. They exist alongside older ones like this one in Aso-Kuju National Park on Kyushu, now dormant and grass-covered.

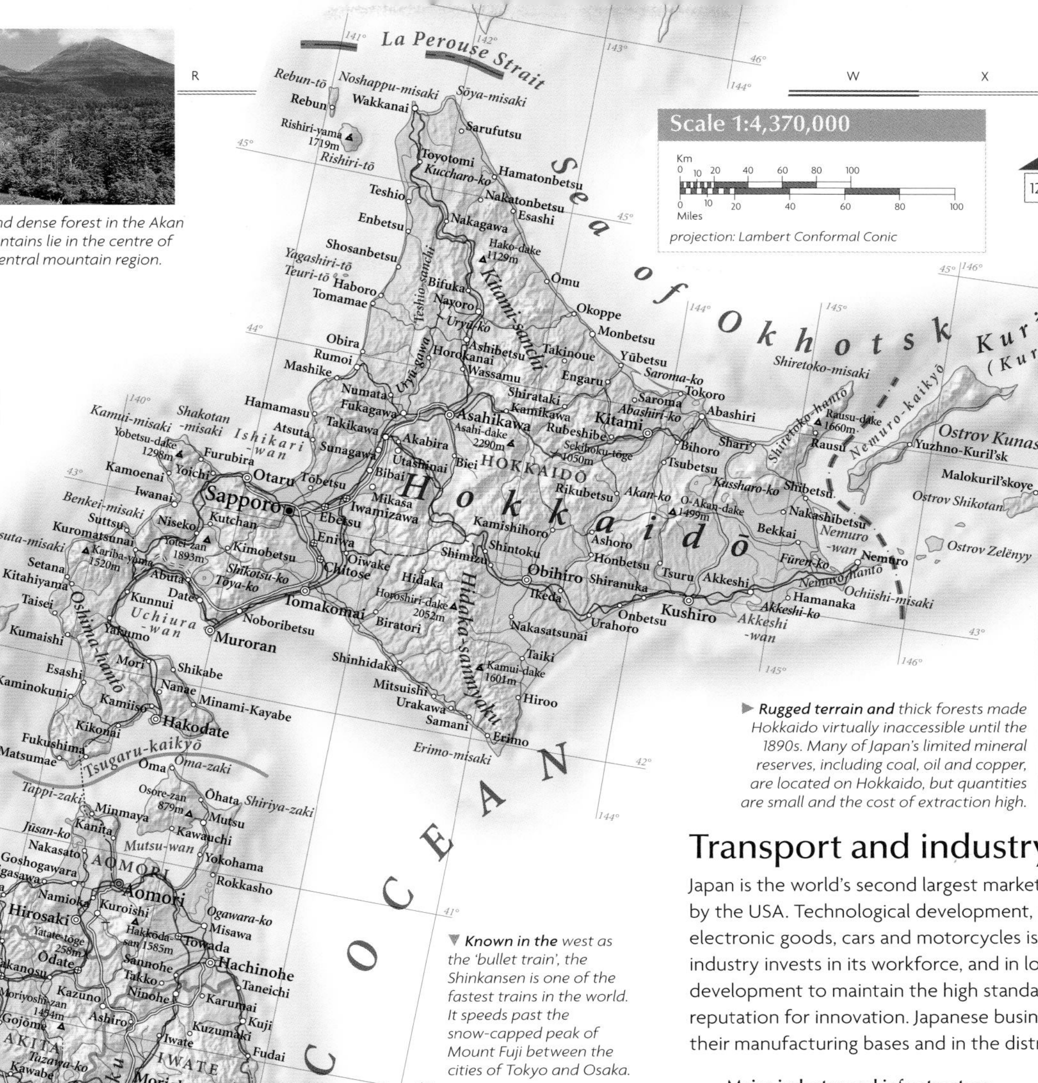

Map key

Population

- above 5 million
- 1 million to 5 million
- 500,000 to 1 million
- 100,000 to 500,000
- 50,000 to 100,000
- 10,000 to 50,000
- below 10,000

Elevation

- 4000m / 13,124ft
- 3000m / 9843ft
- 2000m / 6562ft
- 1000m / 3281ft
- 500m / 1640ft
- 250m / 820ft
- 100m / 328ft
- sea level

122

192

► *Rugged terrain and* thick forests made Hokkaido virtually inaccessible until the 1890s. Many of Japan's limited mineral reserves, including coal, oil and copper, are located on Hokkaido, but quantities are small and the cost of extraction high.

Transport and industry

Japan is the world's second largest market economy, outranked only by the USA. Technological development, particularly of computers, electronic goods, cars and motorcycles is second to none. Japanese industry invests in its workforce, and in long-term research and development to maintain the high standard of its products, and a reputation for innovation. Japanese businesses are now global both in their manufacturing bases and in the distribution of goods.

▼ *Known in the* west as the 'bullet train', the Shinkansen is one of the fastest trains in the world. It speeds past the snow-capped peak of Mount Fuji between the cities of Tokyo and Osaka.

Major industry and infrastructure

- brewing
- car manufacture
- chemicals
- hi-tech industry
- engineering
- finance
- iron & steel
- research & development
- shipbuilding
- textiles
- winter sports
- capital cities
- major towns
- international airports
- major roads
- major industrial areas

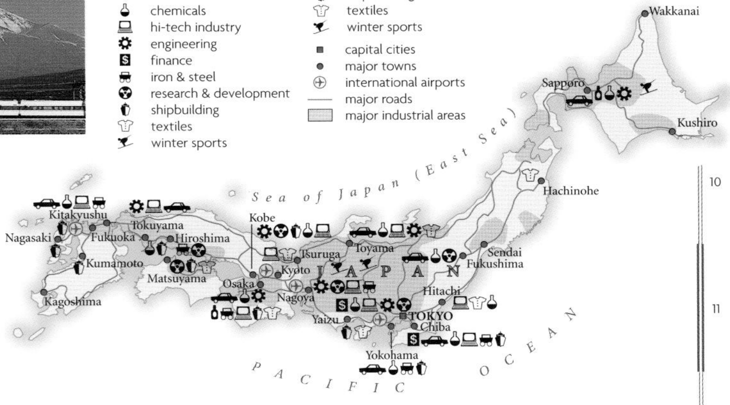

The transport network

557,978 miles (898,082 km)		4257 miles (6851 km)	
12,486 miles (20,096 km)		1099 miles (1770 km)	

Japanese road construction traditionally lagged behind that of its extensive and technologically advanced railway network. The road network's relative lack of development has led to severe urban congestion, although expressways have now been built in some cities.

▲ *On Friday 11* March, 2011 a 9.0 magnitude undersea earthquake 43 miles (70 km) off the coast of Honshu triggered a huge tsunami that devastated the coastal area around Sendai, costing the lives of almost 16,000 people.

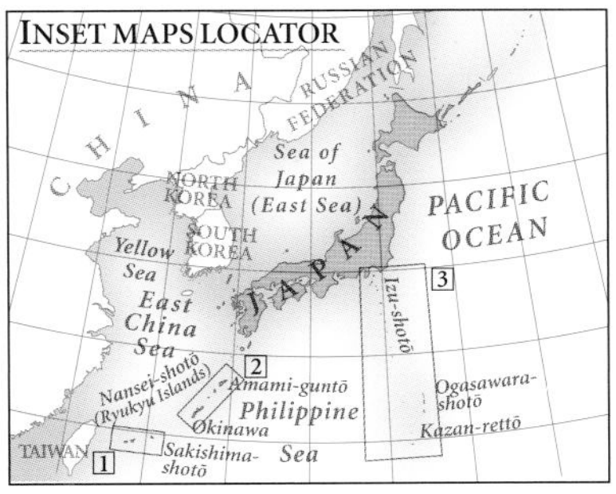

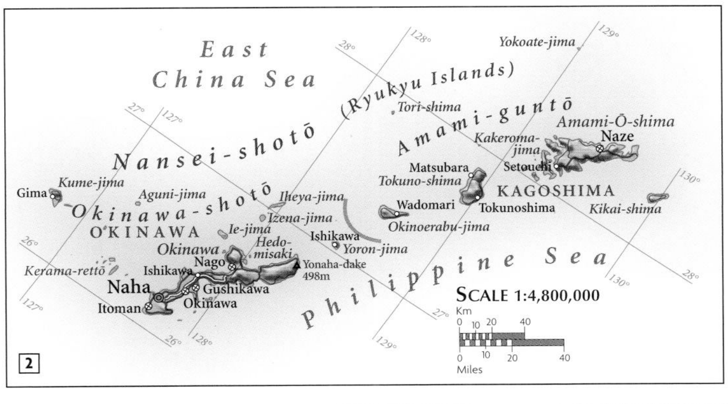

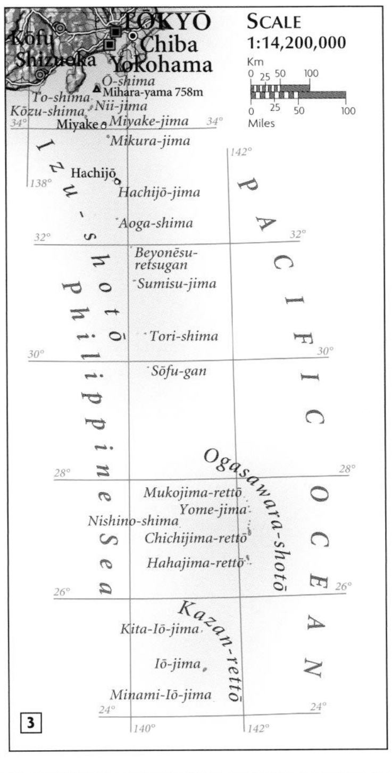

Mainland Southeast Asia

CAMBODIA, LAOS, MYANMAR (BURMA), THAILAND, VIETNAM

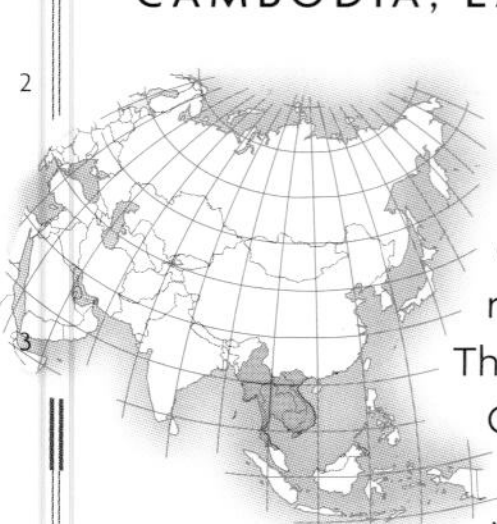

Thickly forested mountains, intercut by the broad valleys of five great rivers characterize the landscape of Southeast Asia's mainland countries. Agriculture remains the main activity for much of the population, which is concentrated in the river flood plains and deltas. Linked ethnic and cultural roots give the region a distinct identity. Most people on the mainland are Theravada Buddhists, and the Philippines is the only predominantly Christian country in Southeast Asia. Foreign intervention began in the 16th century with the opening of the spice trade; Cambodia, Laos and Vietnam were French colonies until the end of the Second World War, Myanmar (Burma) was under British control. Only Thailand was never colonized. Today, Thailand is poised to play a leading role in the economic development of the Pacific Rim, and Laos and Vietnam continue to mend the devastation of the Vietnam War, and to develop their economies. With ongoing political instability and a shattered infrastructure, Cambodia faces an uncertain future, while Myanmar (Burma) is seeking investment and the ending of its long isolation from the world community.

▲ *The Irrawaddy river is Myanmar's (Burma) vital central artery, watering the ricefields and providing a rich source of fish, as well as an important transport link, particularly for local traffic.*

The landscape

A series of mountain ranges runs north–south through the mainland, formed as the result of the collision between the Eurasian Plate and the Indian subcontinent, which created the Himalayas. They are interspersed by the valleys of a number of great rivers. On their passage to the sea these rivers have deposited sediment, forming huge, fertile flood plains and deltas.

The coastline of the Isthmus of Kra

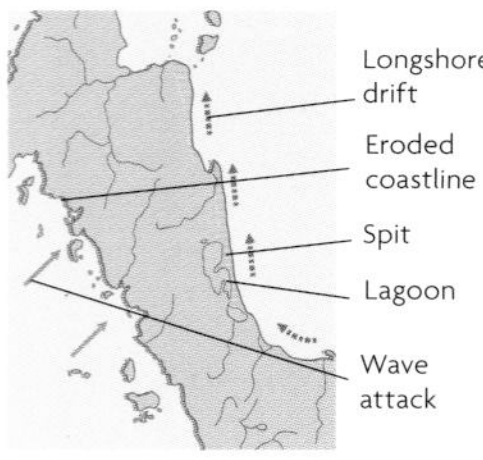

Longshore drift
Eroded coastline
Spit
Lagoon
Wave attack

◀ *The east and west coasts of the Isthmus of Kra differ greatly. The tectonically uplifting west coast is exposed to the harsh south-westerly monsoon and is heavily eroded. On the east coast, longshore currents produce depositional features such as spits and lagoons.*

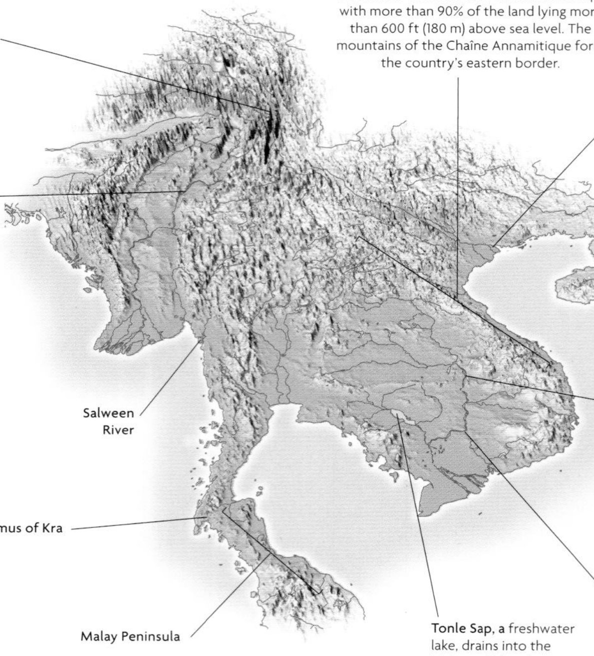

Mountains dominate the Laotian landscape with more than 90% of the land lying more than 600 ft (180 m) above sea level. The mountains of the Chaîne Annamitique form the country's eastern border.
Hkakabo Razi is the highest point in mainland Southeast Asia. It rises 19,300 ft (5885 m) at the border between China and Myanmar (Burma).
The Red River delta in northern Vietnam is fringed to the north by steep-sided, round-topped limestone hills, typical of karst scenery.
The Irrawaddy river runs virtually north–south, draining the plains of northern Myanmar (Burma). The Irrawaddy delta is the country's main rice-growing area.
Salween River
Isthmus of Kra
Malay Peninsula
Tonle Sap, a freshwater lake, drains into the Mekong delta via the Mekong river. It is the largest lake in Southeast Asia.
The Mekong river flows through southern China and Myanmar (Burma), then for much of its length forms the border between Laos and Thailand, flowing through Cambodia before terminating in a vast delta on the southern Vietnamese coast.

▲ *The coast of the Isthmus of Kra, in southeast Thailand has many small, precipitous islands like these, formed by chemical erosion on limestone, which is weathered along vertical cracks. The humidity of the climate in Southeast Asia increases the rate of weathering.*

◀ *The fast-flowing waters of the Mekong river cascade over this waterfall in Champasak province in Laos. The force of the water erodes rocks at the base of the fall.*

Using the land and sea

The fertile flood plains of rivers such as the Mekong and Salween, and the humid climate, enable the production of rice throughout the region. Cambodia, Myanmar (Burma) and Laos still have substantial forests, producing hardwoods such as teak and rosewood. Cash crops include tropical fruits such as coconuts, bananas and pineapples, rubber, oil palm, sugar cane and the jute substitute, kenaf. Pigs and cattle are the main livestock raised. Large quantities of marine and freshwater fish are caught throughout the region.

▲ *Commercial logging – still widespread in Myanmar (Burma) – has now been stopped in Thailand because of over-exploitation of the tropical rainforest.*

The urban/rural population divide

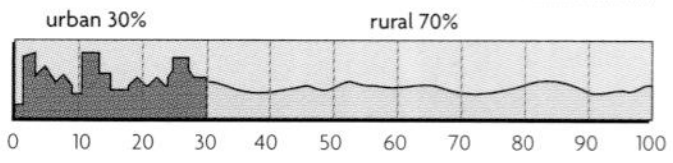

urban 30%
rural 70%
0
10
20
30
40
50
60
70
80
90
100

Population density	Total land area
345 people per sq mile (133 people per sq km)	733,828 sq miles (1,901,110 sq km)

INDIA
BANGLADESH
CHINA
MYANMAR (BURMA)
Mandalay
NAY PYI TAW
HANOI
Gulf of Tonkin
Bay of Bengal
VIENTIANE
LAOS
Hue
Yangon
THAILAND
BANGKOK
VIETNAM
CAMBODIA
PHNOM PENH
Andaman Sea
South China Sea
Gulf of Thailand
Ho Chi Minh City
MALAYSIA

Land use and agricultural distribution

- cattle
- pigs
- bananas
- coconuts
- fishing
- oil palms
- rice
- rubber
- sugar cane
- timber
- capital cities
- major towns
- pasture
- cropland
- forest
- wetland

152
172
168
BANGLADESH
Tropic of Cancer
Bay of Bengal
CHIN STATE
Mount Victoria 3053m
Maungdaw
Buthidaung
Rathedaung
Sittwe
Kyaukpyu
Ramree Island
Cheduba Island
Thandwe
Pathein
Yangon (Rangoon)
Mouths of the Irrawaddy
Preparis Island
Great Coco Island
Little Coco Island
Gulf of Mottama
Andaman Sea
Myeik Archipelago
Strait of Malacca
SAGAING
MAGWAY
BAGO
AYEYARWADY
Mandalay
NAY PYI TAW
Pakokku
Sagaing
Monywa
Kalemyo
Hakha
Falam
Magway
Minbu
Pyapon
Bago
Taungoo
Prome
Phuket
Thalang
Ko Phuket
Ban Kam Phuang
Ko Phra Thong
Zadetkyi

Transport and industry

Industrial manufacturing has become increasingly important in Thailand and Vietnam in recent years. The assembling of component-based electrical and electronic goods is becoming more common throughout this region, with foreign companies benefiting from low labour costs and the upgrading of technology. The economies of Myanmar (Burma) and Cambodia are still based on agricultural produce and the processing of raw materials. Tin is the region's most important metal, and nickel, copper and chromite are also mined, although the quantities produced are not significant on a global scale. Thailand's successful tourist industry is the country's highest earner of foreign exchange.

The transport network

82,958 miles (133,524 km)	267 miles (430 km)
7500 miles (12,071 km)	28,585 miles (46,008 km)

Transport development has concentrated on the building of road networks. Water and sea transport remain important, although air links have improved, particularly in Thailand and the Philippines.

Major industry and infrastructure

chemicals	oil & gas	capital cities
electronics	mining	major towns
engineering	shipbuilding	international airports
finance	textiles	major roads
food processing	timber processing	major industrial areas
iron & steel		

160

160

170

168

▶ ***Opium poppies are*** *destroyed under army supervision in Thailand. This action is part of a government-sponsored initiative to reduce the trade in drugs such as heroin, which is derived from these plants. Drug trafficking is a major problem throughout the region; the area is known as the 'Golden Triangle', and Laos is the third-largest producer of opium poppies in the world.*

The Paracel Islands are a strategically sensitive island group, disputed by several surrounding countries. The Paracels are claimed by China, Taiwan and Vietnam, though only China has actually occupied them.

Map key

Population

- above 5 million
- 1 million to 5 million
- 500,000 to 1 million
- 100,000 to 500,000
- 50,000 to 100,000
- 10,000 to 50,000
- below 10,000

Elevation

- 4000m / 13,124ft
- 3000m / 9843ft
- 2000m / 6562ft
- 1000m / 3281ft
- 500m / 1640ft
- 250m / 820ft
- 100m / 328ft
- sea level

Scale 1:8,600,000

Km 0 25 50 100 150 200
Miles 0 25 50 100 150 200

projection: Lambert Conformal Conic

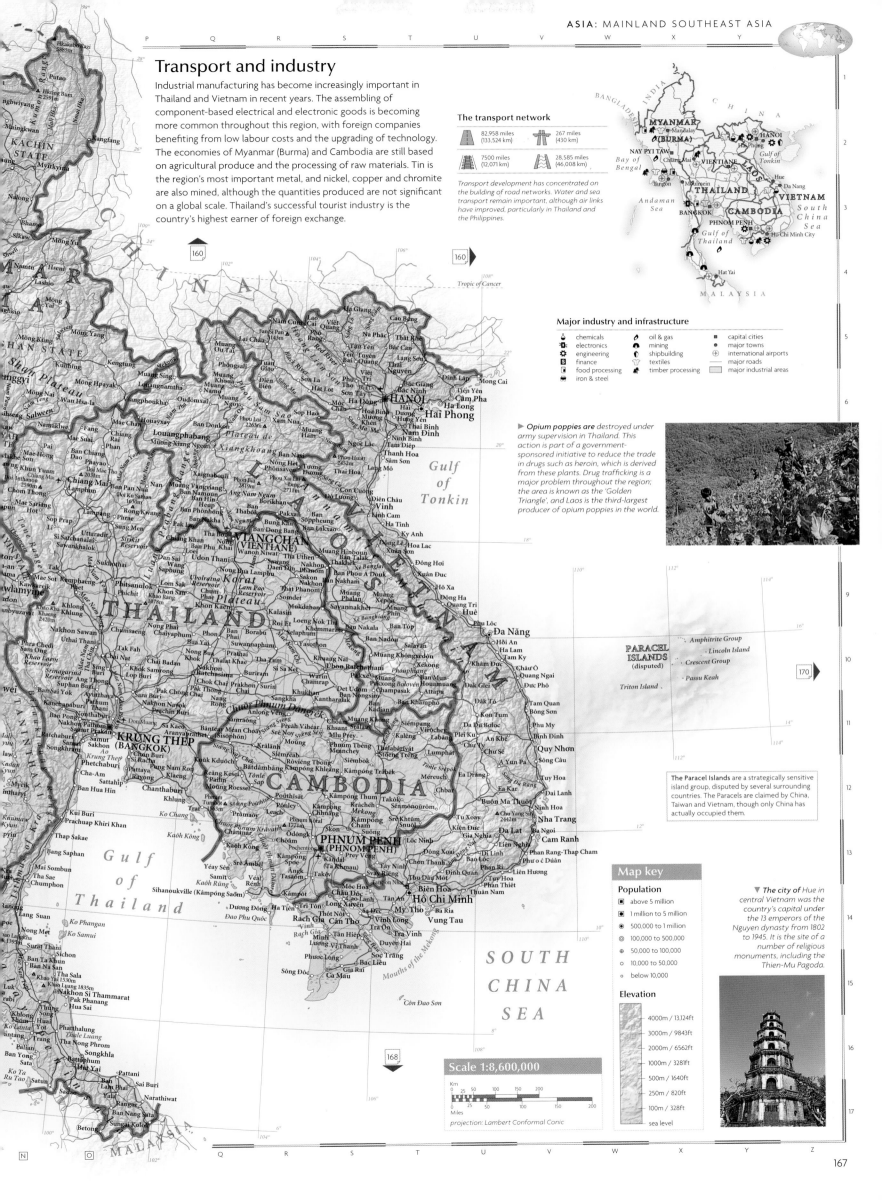

▼ ***The city of*** *Hue in central Vietnam was the country's capital under the 13 emperors of the Nguyen dynasty from 1802 to 1945. It is the site of a number of religious monuments, including the Thien-Mu Pagoda.*

Western Maritime Southeast Asia

BRUNEI, INDONESIA, MALAYSIA, SINGAPORE

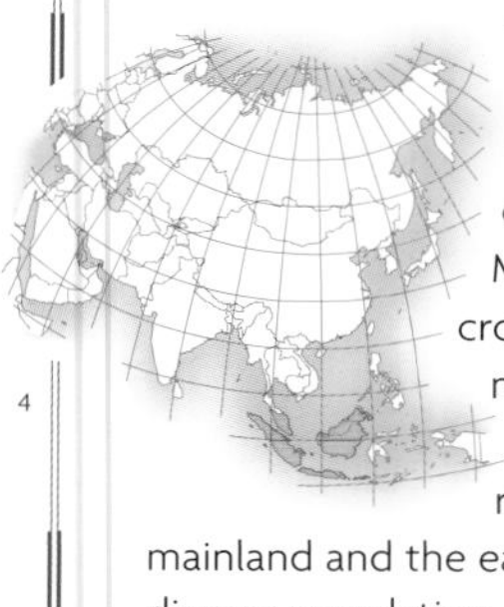

The world's largest archipelago, Indonesia's myriad islands stretch 3100 miles (5000 km) eastwards across the Pacific, from the Malay Peninsula to western New Guinea. Only about 1500 of the 13,677 islands are inhabited and the huge, predominently Muslim population is unevenly distributed, with some two-thirds crowded onto the western islands of Java, Madura and Bali. The national government is trying to resettle large numbers of people from these islands to other parts of the country to reduce population pressure there. Malaysia, split between the mainland and the east Malaysian states of Sabah and Sarawak on Borneo, has a diverse population, as well as a fast-growing economy, although the pace of its development is still far outstripped by that of Singapore. This small island nation is the financial and commercial capital of Southeast Asia. The Sultanate of Brunei in northern Borneo, one of the world's last princely states, has an extremely high standard of living, based on its oil revenues.

The landscape

Indonesia's western islands are characterized by rugged volcanic mountains cloaked with dense tropical forest, which slope down to coastal plains covered by thick alluvial swamps. The Sunda Shelf, an extension of the Eurasian Plate, lies between Java, Bali, Sumatra and Borneo. These islands' mountains rise from a base below the sea, and they were once joined together by dry land, which has since been submerged by rising sea levels.

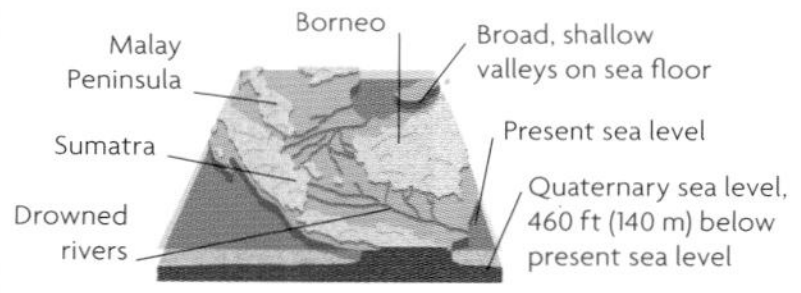

▲ *The Sunda Shelf underlies this whole region. It is one of the largest submarine shelves in the world, covering an area of 714,285 sq miles (1,850,000 sq km). During the early Quaternary period, when sea levels were lower, the shelf was exposed.*

◀ *On January 24, 2005 a 9.2 magnitude earthquake off the coast of Sumatra triggered a devastating tsunami that was up to 90 ft (30 m) high in places. The death toll was estimated to be around 230,000 people from fourteen different countries around the Indian Ocean.*

Transport and industry

Singapore has a thriving economy based on international trade and finance. Annual trade through the port is among the highest of any in the world. Indonesia's western islands still depend on natural resources, particularly petroleum, gas and wood, although the economy is rapidly diversifying with manufactured exports including garments, consumer electronics and footwear. A high-profile aircraft industry has developed in Bandung on Java. Malaysia has a fast-growing and varied manufacturing sector, although oil, gas and timber remain important resource-based industries.

▶ *Ranks of gleaming skyscrapers, new motorways and infrastructure construction reflect the investment which is pouring into Southeast Asian cities like the Malaysian capital, Kuala Lumpur. Traditional housing and markets still exist amidst the new developments. Many of the city's inhabitants subsist at a level far removed from the prosperity implied by its outward modernity.*

Malaysia: Capital cities
KUALA LUMPUR – capital
PUTRAJAYA – administrative capital

166
172
172

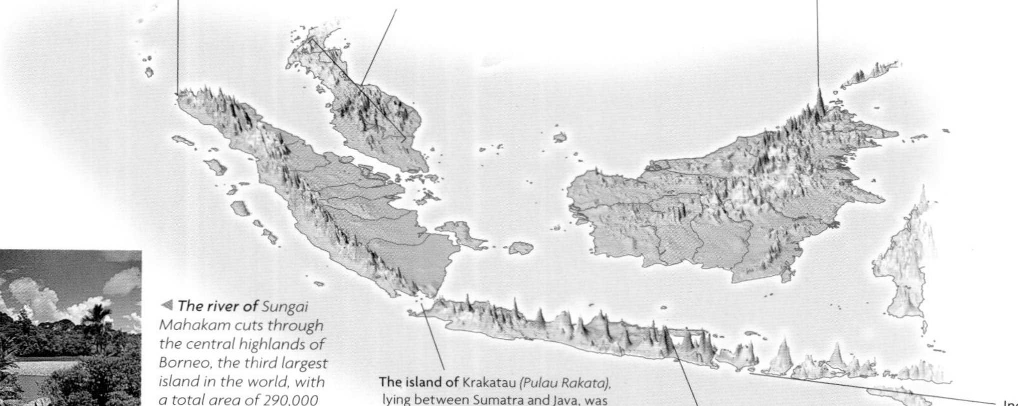

Malay Peninsula has a rugged east coast, but the west coast, fronting the Strait of Malacca, has many sheltered beaches and bays. The two coasts are divided by the Banjaran Titiwangsa, which run the length of the peninsula.

Gunung Kinabalu is the highest peak in Malaysia, rising 13,455 ft (4101 m).

The island of Krakatau *(Pulau Rakata)*, lying between Sumatra and Java, was all but destroyed in 1883, when the volcano erupted. The release of gas and dust into the atmosphere disrupted cloud cover and global weather patterns for several years.

Indonesia has more than 220 volcanoes, most of which are still active. They are strung out along the island arc from Sumatra through the Lesser Sunda Islands, into the Moluccas and Celebes.

◀ *The river of Sungai Mahakam cuts through the central highlands of Borneo, the third largest island in the world, with a total area of 290,000 sq miles (757,050 sq km). Although mountainous, Borneo is one of the most stable of the Indonesian islands, with little volcanic activity.*

N O P Q R S T U V W X Y

he transport network

	165,272 miles (266,010 km)
	958 miles (1,542 km)
	5,061 miles (8,146 km)
	18,070 miles (29,084 km)

ngapore's metro system, ompleted in 1991, is among he most efficient in the vorld. Malaysia has several ıst, modern highways and nost roads are paved. ndonesia's many islands nake improvement of the hipping infrastructure priority.

Major industry and infrastructure

- aerospace
- copra processing
- chemicals
- electronics
- engineering
- finance
- food processing
- iron & steel
- oil
- ship building
- timber processing
- textiles
- capital cities
- major towns
- international airports
- major roads
- major industrial areas

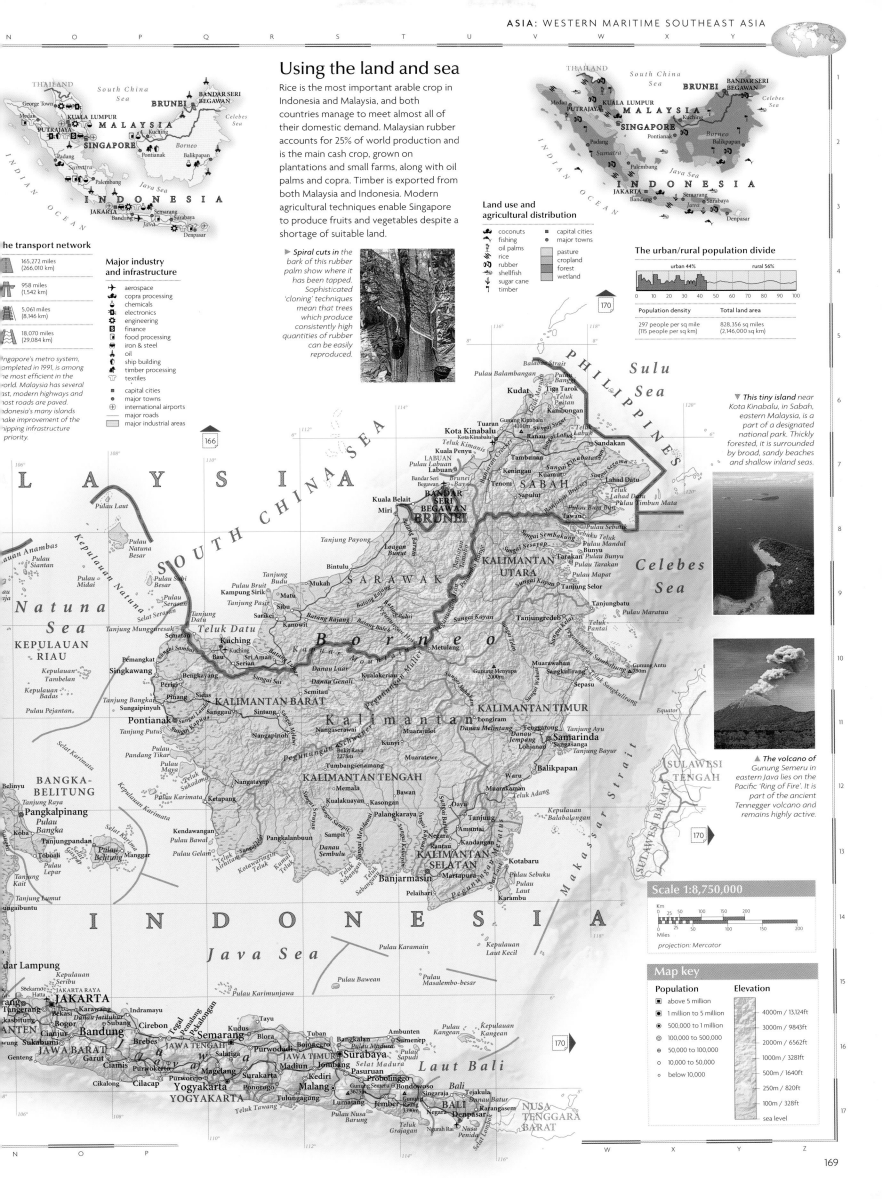

Using the land and sea

Rice is the most important arable crop in Indonesia and Malaysia, and both countries manage to meet almost all of their domestic demand. Malaysian rubber accounts for 25% of world production and is the main cash crop, grown on plantations and small farms, along with oil palms and copra. Timber is exported from both Malaysia and Indonesia. Modern agricultural techniques enable Singapore to produce fruits and vegetables despite a shortage of suitable land.

▶ *Spiral cuts in the bark of this rubber palm show where it has been tapped. Sophisticated 'cloning' techniques mean that trees which produce consistently high quantities of rubber can be easily reproduced.*

Land use and agricultural distribution

- coconuts
- fishing
- oil palms
- rice
- rubber
- shellfish
- sugar cane
- timber
- capital cities
- major towns
- pasture
- cropland
- forest
- wetland

The urban/rural population divide

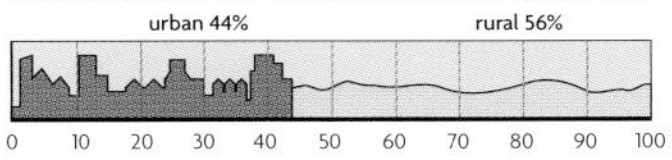

Population density	Total land area
297 people per sq mile (115 people per sq km)	828,356 sq miles (2,146,000 sq km)

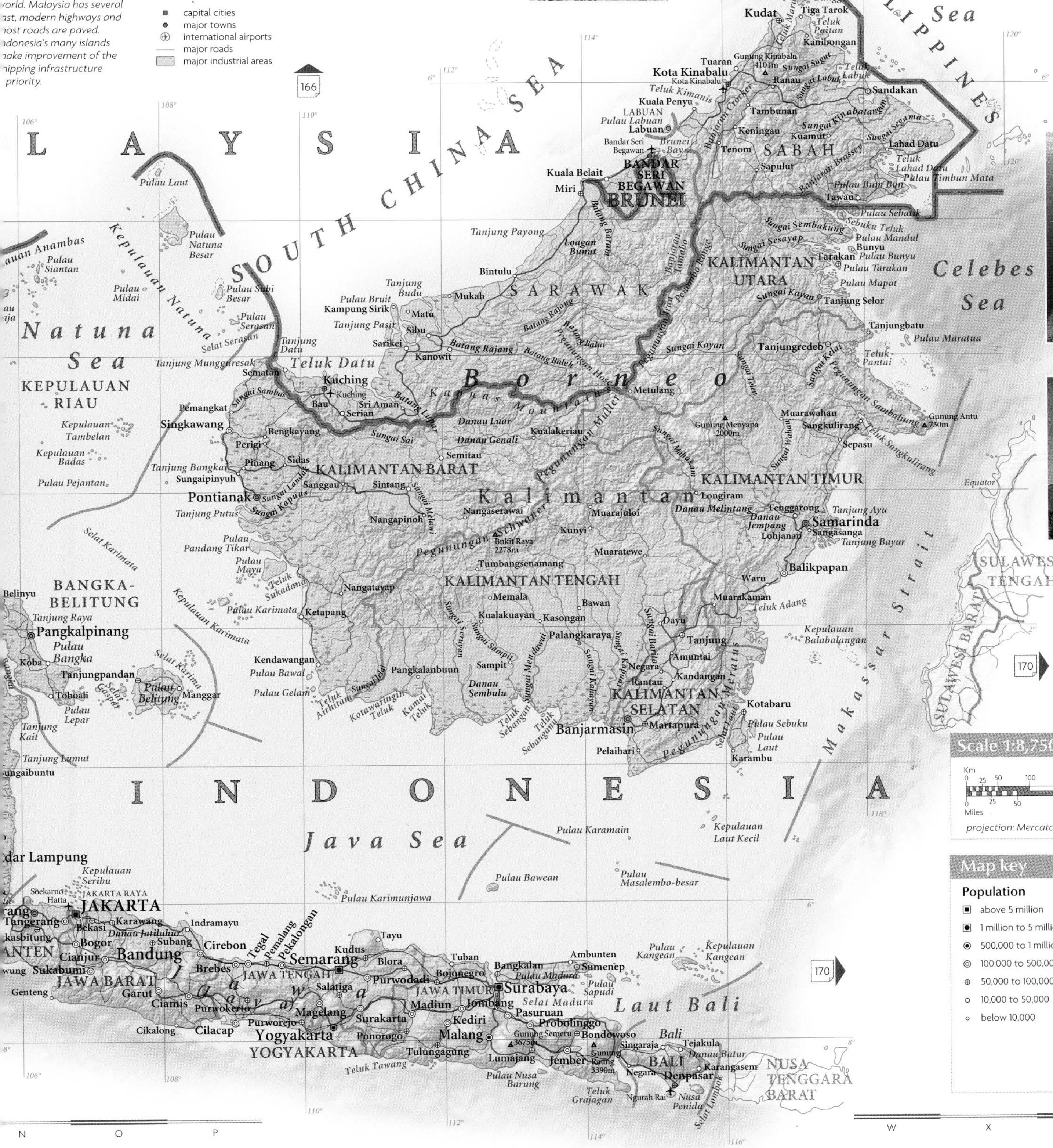

▼ *This tiny island near Kota Kinabalu, in Sabah, eastern Malaysia, is a part of a designated national park. Thickly forested, it is surrounded by broad, sandy beaches and shallow inland seas.*

▲ *The volcano of Gunung Semeru in eastern Java lies on the Pacific 'Ring of Fire'. It is part of the ancient Tennegger volcano and remains highly active.*

Scale 1:8,750,000

Km 0 25 50 100 150 200
Miles 0 25 50 100 150 200

projection: Mercator

Map key

Population

- above 5 million
- 1 million to 5 million
- 500,000 to 1 million
- 100,000 to 500,000
- 50,000 to 100,000
- 10,000 to 50,000
- below 10,000

Elevation

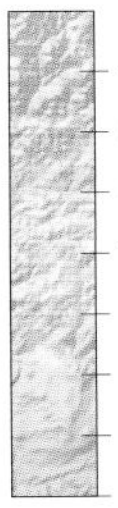

Eastern Maritime Southeast Asia

EAST TIMOR, INDONESIA, PHILIPPINES

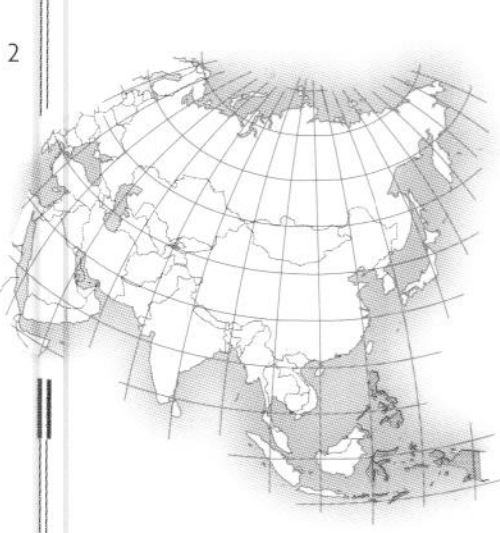

The Philippines takes its name from Philip II of Spain who was king when the islands were colonized during the 16th century. Almost 400 years of Spanish, and later US, rule have left their mark on the country's culture; English is widely spoken and over 90% of the population is Christian. The Philippines' economy is agriculturally based – inadequate infrastructure and electrical power shortages have so far hampered faster industrial growth. Indonesia's eastern islands are less economically developed than the rest of the country. Papua, which constitutes the western portion of New Guinea, is one of the world's last great wildernesses. After a long struggle, East Timor gained full autonomy from Indonesia in 2002.

▲ ***The traditional boat-shaped** houses of the Toraja people in Sulawesi. Although now Christian, the Toraja still practice the animist traditions and rituals of their ancestors. They are famous for their elaborate funeral ceremonies and burial sites in cliffside caves.*

The landscape

Located on the Pacific 'Ring of Fire' the Philippines' 7100 islands are subject to frequent earthquakes and volcanic activity. Their terrain is largely mountainous, with narrow coastal plains and interior valleys and plains. Luzon and Mindanao are by far the largest islands and comprise roughly 66% of the country's area. Indonesia's eastern islands are mountainous and dotted with volcanoes, both active and dormant.

► ***Lake Taal** on the Philippines island of Luzon lies within the crater of an immense volcano that erupted twice in the 20th century, first in 1911 and again in 1965, causing the deaths of more than 3200 people.*

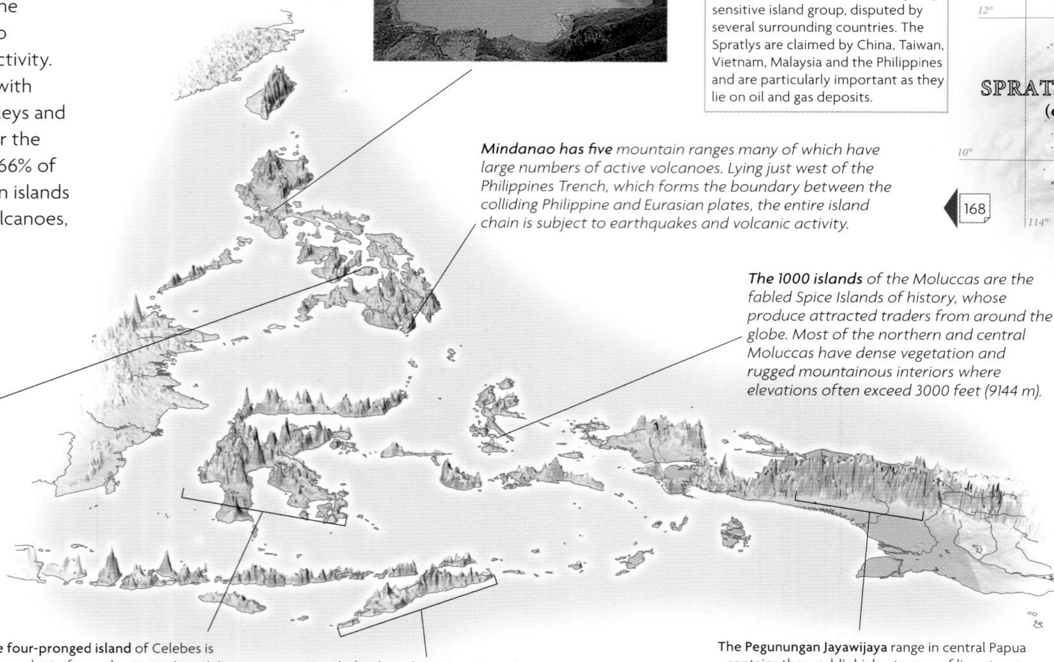

***Mindanao has five** mountain ranges many of which have large numbers of active volcanoes. Lying just west of the Philippines Trench, which forms the boundary between the colliding Philippine and Eurasian plates, the entire island chain is subject to earthquakes and volcanic activity.*

***The 1000 islands** of the Moluccas are the fabled Spice Islands of history, whose produce attracted traders from around the globe. Most of the northern and central Moluccas have dense vegetation and rugged mountainous interiors where elevations often exceed 3000 feet (9144 m).*

The four-pronged island of Celebes is the product of complex tectonic activity which ruptured and then reattached small fragments of the Earth's crust to form the island's many peninsulas.

Coral islands such as Timor in eastern Indonesia show evidence of very recent and dramatic movements of the Earth's plates. Reefs in Timor have risen by as much as 4000 ft (1300 m) in the last million years.

The Pegunungan Jayawijaya range in central Papua contains the world's highest range of limestone mountains, some with peaks more than 16,400 ft (5000 m) high. Heavy rainfall and high temperatures, which promote rapid weathering, have led to the creation of large underground caves and river systems such as the river of Sungai Baliem.

▲ ***Bohol in the** southern Philippines is famous for its so-called 'chocolate hills'. There are more than 1000 of these regular mounds on the island. The hills are limestone in origin, the smoothed remains of an earlier cycle of erosion. Their brown appearance in the dry season gives them their name.*

The Spratly Islands are a strategically sensitive island group, disputed by several surrounding countries. The Spratlys are claimed by China, Taiwan, Vietnam, Malaysia and the Philippines and are particularly important as they lie on oil and gas deposits.

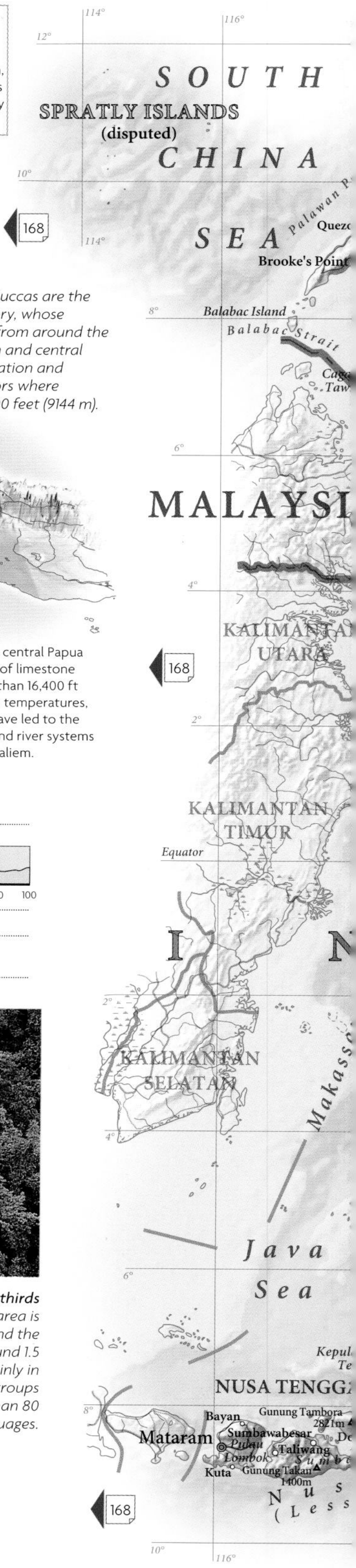

SOUTH CHINA SEA
SPRATLY ISLANDS
(disputed)
Brooke's Point
Balabac Island
MALAYSI
KALIMANTAN UTARA
KALIMANTAN TIMUR
Equator
KALIMANTAN SELATAN
Java Sea
NUSA TENGGA
Bayan
Gunung Tambora
Mataram
Lombok
Kuta

168
168
168

Using the land and sea

Indonesia's eastern islands are less intensively cultivated than those in the west. Coconuts, coffee and spices such as cloves and nutmeg are the major commercial crops while rice, corn and soya beans are grown for local consumption. The Philippines' rich, fertile soils support year-round production of a wide range of crops. The country is one of the world's largest producers of coconuts and a major exporter of coconut products, including one-third of the world's copra. Although much of the arable land is given over to rice and corn, the main staple food crops, tropical fruits such as bananas, pineapples and mangos,and sugar cane are also grown for export.

Luzon Strait
Luzon
Baguio
MANILA
Philippine Sea
South China Sea
PHILIPPINES
Cebu
Sulu Sea
Butuan
Mindanao
Zamboanga
Davao
MALAYSIA
Celebes Sea
PACIFIC OCEAN
Manado
Halmahera
Maluku (Moluccas)
Celebes
Ceram
Ambon
Jayapura
Makassar
Banda Sea
New Guinea
PAPUA NEW GUINEA
INDONESIA
Arafura Sea
Lombok
Sumbawa
Flores
DILI
EAST TIMOR
Sumba
Timor
Kupang
Timor Sea
INDIAN OCEAN

Land use and agricultural distribution

- coconuts
- fishing
- rice
- rubber
- shellfish
- sugar cane
- capital cities
- major towns
- pasture
- cropland
- forest
- wetland

The urban/rural population divide

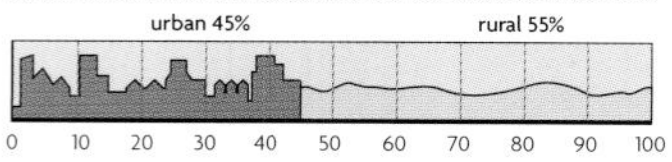

urban 45%
rural 55%
0 10 20 30 40 50 60 70 80 90 100

Population density	Total land area
258 people per sq mile (160 people per sq km)	654,771 sq miles (1,053,755 sq km)

▲ ***More than two-thirds** of Papua's land area is heavily forested and the population of around 1.5 million live mainly in isolated tribal groups using more than 80 distinct languages.*

◄ ***The terracing of** land to restrict soil erosion and create flat surfaces for agriculture is a common practice throughout Southeast Asia, particularly where land is scarce. These terraces are on Luzon in the Philippines.*

Transport and industry

The Philippines' economy is primarily a mixture of agriculture and light industry. The manufacturing sector is still developing; many factories are licensees of foreign companies producing finished goods for export. Mining is also important – the country's chromite, nickel and copper deposits are among the largest in the world. Agriculture is the main activity in eastern Indonesia. Most industry has a primary basis, including logging, food-processing and mining. Nickel, the most important metal, is produced on Sulawesi, in Papua, and in the Moluccas.

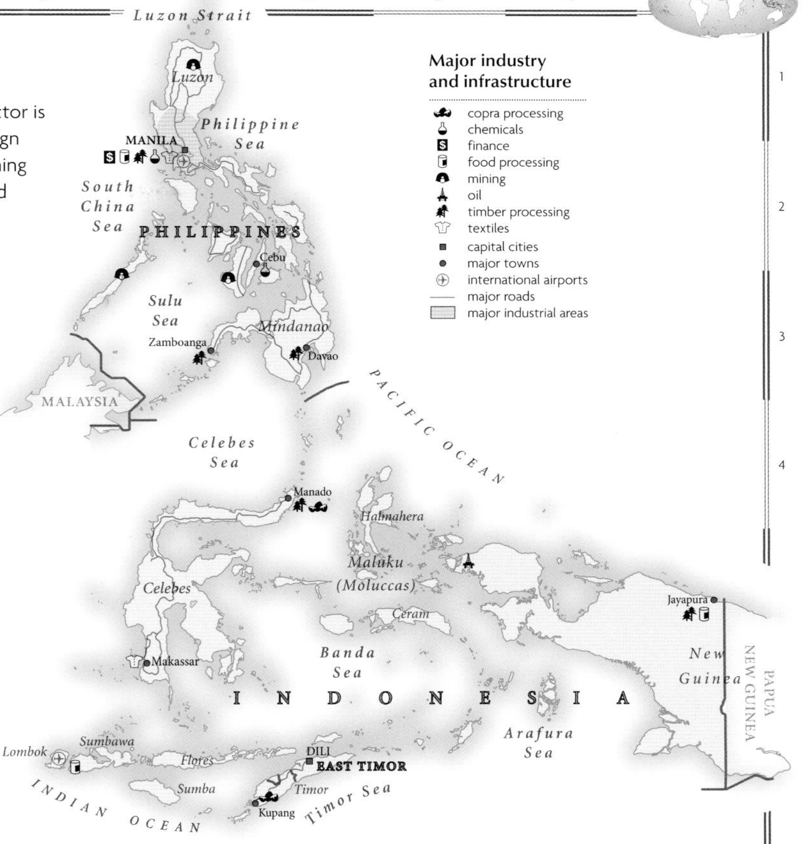
Luzon Strait
Luzon
MANILA
Philippine Sea
South China Sea
PHILIPPINES
Cebu
Sulu Sea
Zamboanga
Mindanao
Davao
MALAYSIA
PACIFIC OCEAN
Celebes Sea
Manado
Halmahera
Maluku (Moluccas)
Celebes
Ceram
Jayapura
Makassar
Banda Sea
INDONESIA
New Guinea
PAPUA NEW GUINEA
Lombok
Sumbawa
Flores
Sumba
DILI
EAST TIMOR
Timor
Kupang
Timor Sea
Arafura Sea
INDIAN OCEAN

Major industry and infrastructure

- copra processing
- chemicals
- finance
- food processing
- mining
- oil
- timber processing
- textiles
- capital cities
- major towns
- international airports
- major roads
- major industrial areas

The transport network

- 16,652 miles (26,800 km)
- None
- 500 miles (805 km)
- 8704 miles (14,008 km)

Sulawesi has some good roads, but on Papua and the Moluccas there are few road interconnections between major settled areas. Water and sea transport remain important although air links have improved in the Philippines.

crissa
EGO

▲ ***Manila is the** Philippines' chief port and transport centre, and the focus of the country's commercial, industrial and cultural activities. Much of the city lies below sea level, and it suffers from floods during the rainy summer season.*

Map key

Population

- above 5 million
- 1 million to 5 million
- 500,000 to 1 million
- 100,000 to 500,000
- 50,000 to 100,000
- 10,000 to 50,000
- below 10,000

Elevation

- 4000m / 13,124ft
- 3000m / 9843ft
- 2000m / 6562ft
- 1000m / 3281ft
- 500m / 1640ft
- 250m / 820ft
- 100m / 328ft
- sea level

Scale 1:,11,800,000

Km 0 50 100 200 300 400

0 50 100 200 300 400 Miles

projection: Mercator

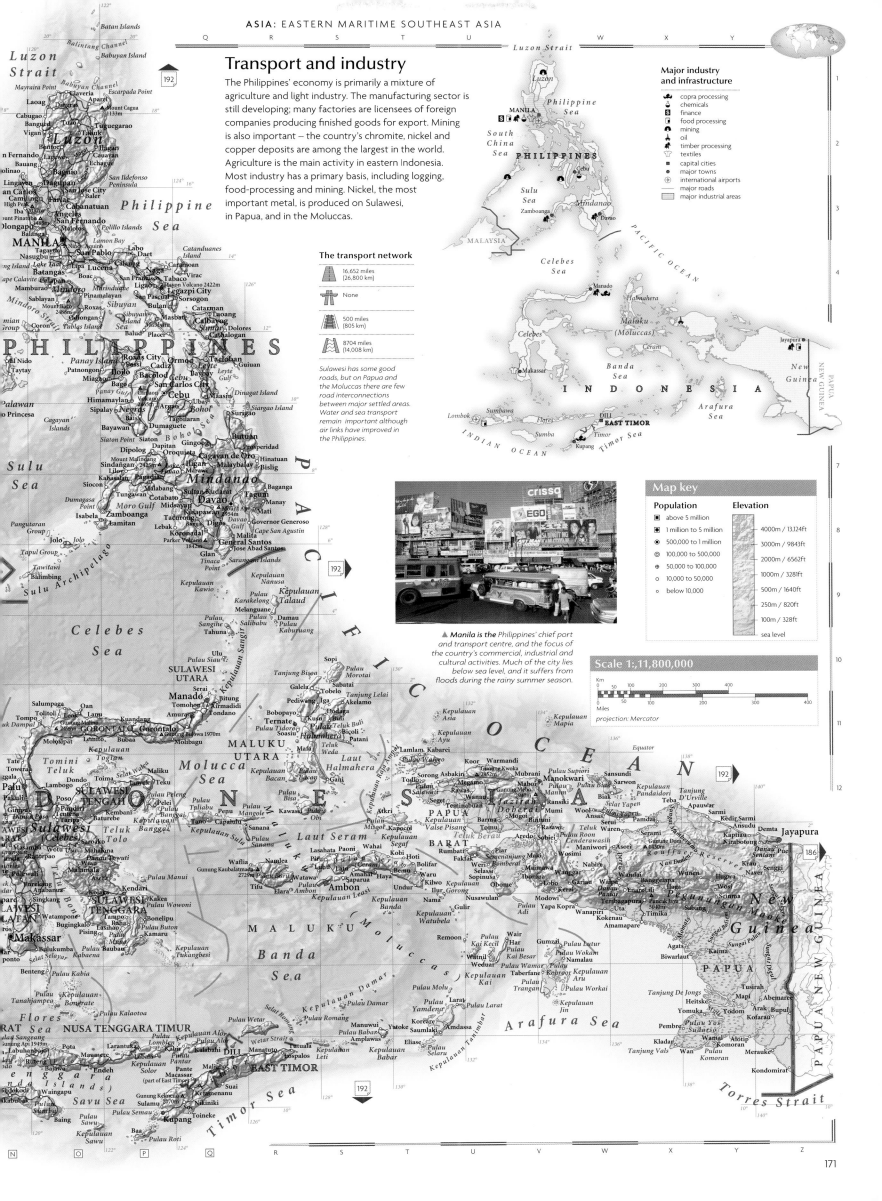

Q R S T U V W X Y
Luzon Strait
Batan Islands
Balintang Channel
Babuyan Island
Babuyan Channel
Luzon
Philippine Sea
MANILA
Mindoro
Sibuyan Sea
PHILIPPINES
Panay Island
Negros
Cebu
Bohol
Leyte
Samar
Dinagat Island
Siargao Island
Palawan
Cagayan Islands
Sulu Sea
Mindanao
Moro Gulf
Davao
Zamboanga
Sulu Archipelago
Celebes Sea
PACIFIC OCEAN
SULAWESI UTARA
Manado
Gorontalo
MALUKU UTARA
Molucca Sea
Halmahera
SULAWESI TENGAH
Sulawesi (Celebes)
Teluk Tomini
Teluk Tolo
Laut Halmahera
Laut Seram
Ambon
Ceram
MALUKU (Moluccas)
Banda Sea
SULAWESI TENGGARA
SULAWESI SELATAN
Makassar
Flores Sea
NUSA TENGGARA TIMUR
DILI
EAST TIMOR
Savu Sea
Kupang
Timor Sea
PAPUA
Teluk Cenderawasih
Jayapura
New Guinea
PAPUA NEW GUINEA
Arafura Sea
Kepulauan Aru
Torres Strait
Equator
192
186
N O P Q R S T U V W X Y Z

The Indian Ocean

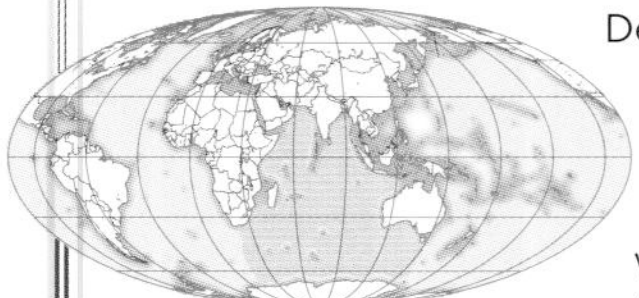

Despite being the smallest of the three major oceans, the evolution of the Indian Ocean was the most complex. The ocean basin was formed during the break up of the supercontinent Gondwanaland, when the Indian subcontinent moved northeast, Africa moved west and Australia separated from Antarctica. Like the Pacific Ocean, the warm waters of the Indian Ocean are punctuated by coral atolls and islands. About one-fifth of the world's population – over 1000 million people – live on its shores. Those people living along the northern coasts are constantly threatened by flooding and typhoons caused by the monsoon winds.

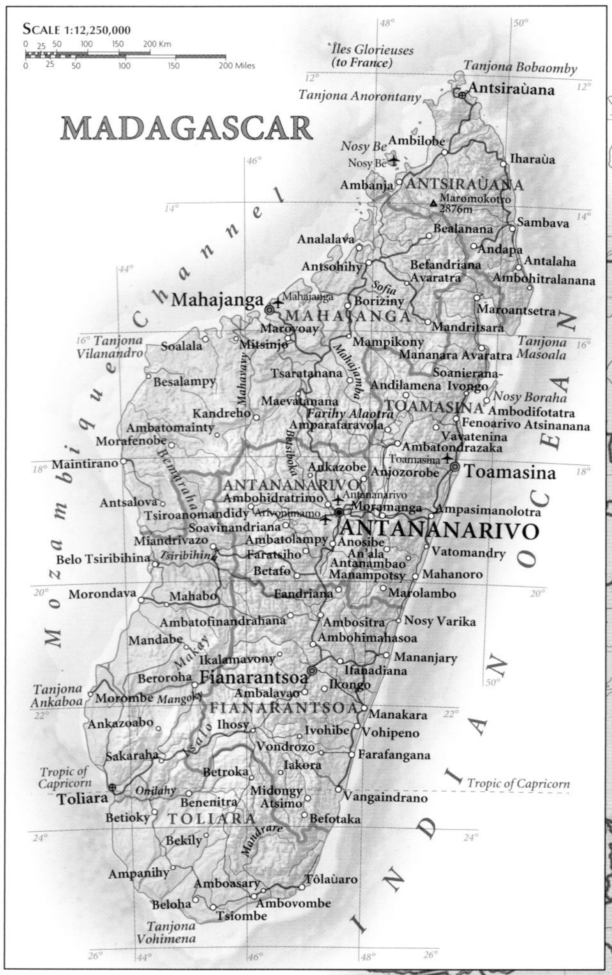

The landscape

The Indian Ocean began forming about 150 million years ago, but in its present form it is relatively young, only about 36 million years old. Along the three subterranean mountain chains of its mid-ocean ridge the seafloor is still spreading. The Indian Ocean has fewer trenches than other oceans and only a narrow continental shelf around most of its surrounding land.

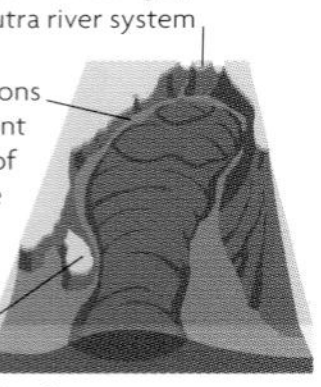

▲ ***The Ganges Fan*** *is one of the world's largest submarine accumulations of sediment, extending far beyond Sri Lanka. It is fed by the Ganges/Brahmaputra river system, whose sediment is carried through a network of underwater canyons at the edge of the continental shelf.*

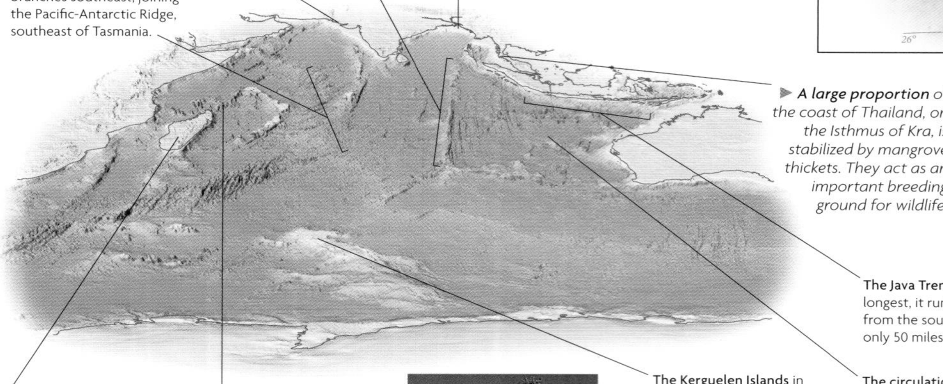

The mid-oceanic ridge runs from the Arabian Sea. It diverges east of Madagascar, one arm runs southwest to join the Mid-Atlantic Ridge, the other branches southeast, joining the Pacific-Antarctic Ridge, southeast of Tasmania.

The Ninetyeast Ridge takes its name from the line of longitude it follows. It is the world's longest and straightest under-sea ridge.

Two of the world's largest rivers flow into the Indian Ocean; the Indus and the Ganges/Brahmaputra. Both have deposited enormous fans of sediment.

The relief of Madagascar rises from a low-lying coastal strip in the east, to the central plateau. The plateau is also a major watershed separating Madagascar's three main river basins.

The Kerguelen Islands in the Southern Ocean were created by a hot spot in the Earth's crust. The islands were formed in succession as the Antarctic Plate moved slowly over the hot spot.

The Java Trench is the world's longest, it runs 1600 miles (2570 km) from the southwest of Java, but is only 50 miles (80 km) wide.

The circulation in the northern Indian Ocean is controlled by the monsoon winds. Biannually these winds reverse their pattern, causing a reversal in the surface currents and alternative high and low pressure conditions over Asia and Australia.

► ***A large proportion*** *of the coast of Thailand, on the Isthmus of Kra, is stabilized by mangrove thickets. They act as an important breeding ground for wildlife.*

► ***The central group*** *of the Seychelles are mountainous, granite islands. They have a narrow coastal belt and lush, tropical vegetation cloaks the highlands.*

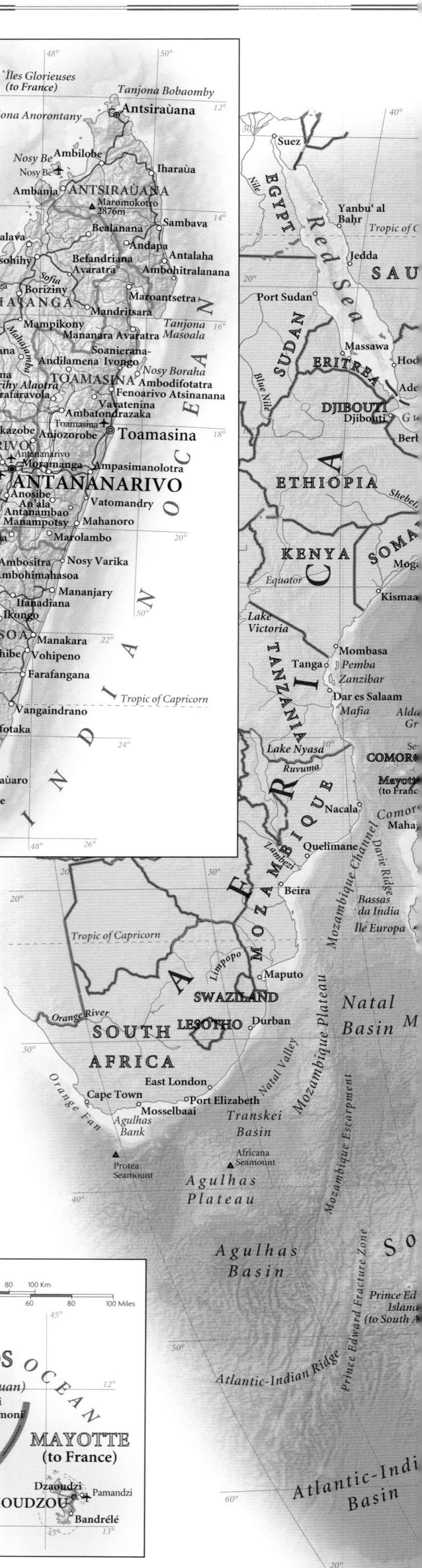

Resources

Many of the small islands in the Indian Ocean rely exclusively on tuna-fishing and tourism to maintain their economies. Most fisheries are artisanal, although large-scale tuna-fishing does take place in the Seychelles, Mauritius and the western Indian Ocean. Other resources include oil in The Gulf, pearls in the Red Sea and tin from deposits off the shores of Burma, Thailand and Indonesia.

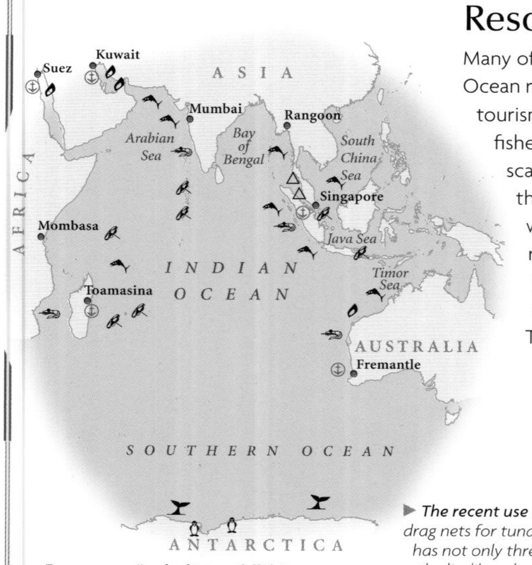

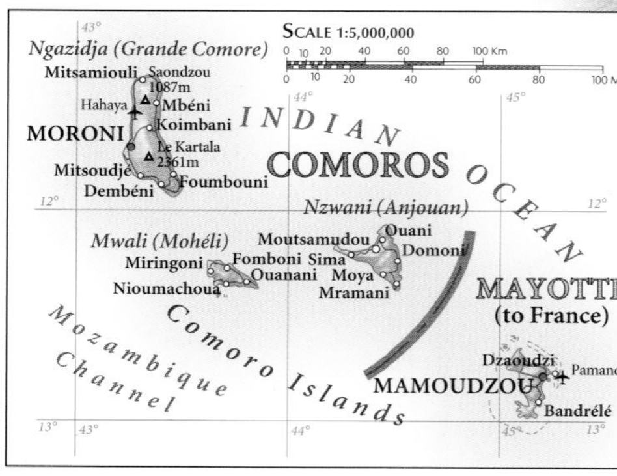

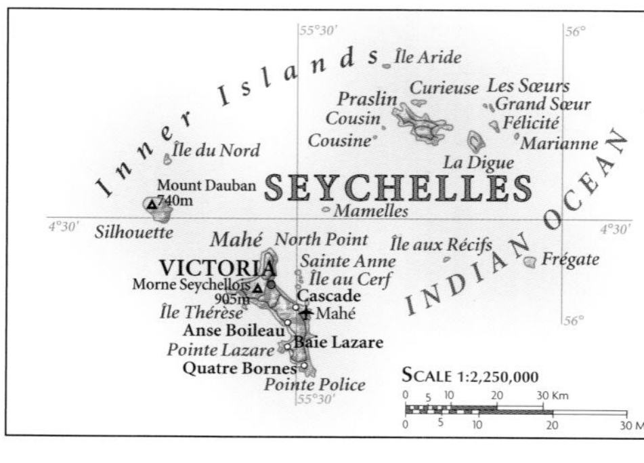

► ***The recent use*** *of large drag nets for tuna-fishing has not only threatened the livelihoods of many small-scale fisheries, but also caused widespread environmental concern about the potential impact on other marine species.*

▲ ***Coral reefs support*** *an enormous diversity of animal and plant life. Many species of tropical fish, like these squirrel fish, live and feed around the profusion of reefs and atolls in the Indian Ocean.*

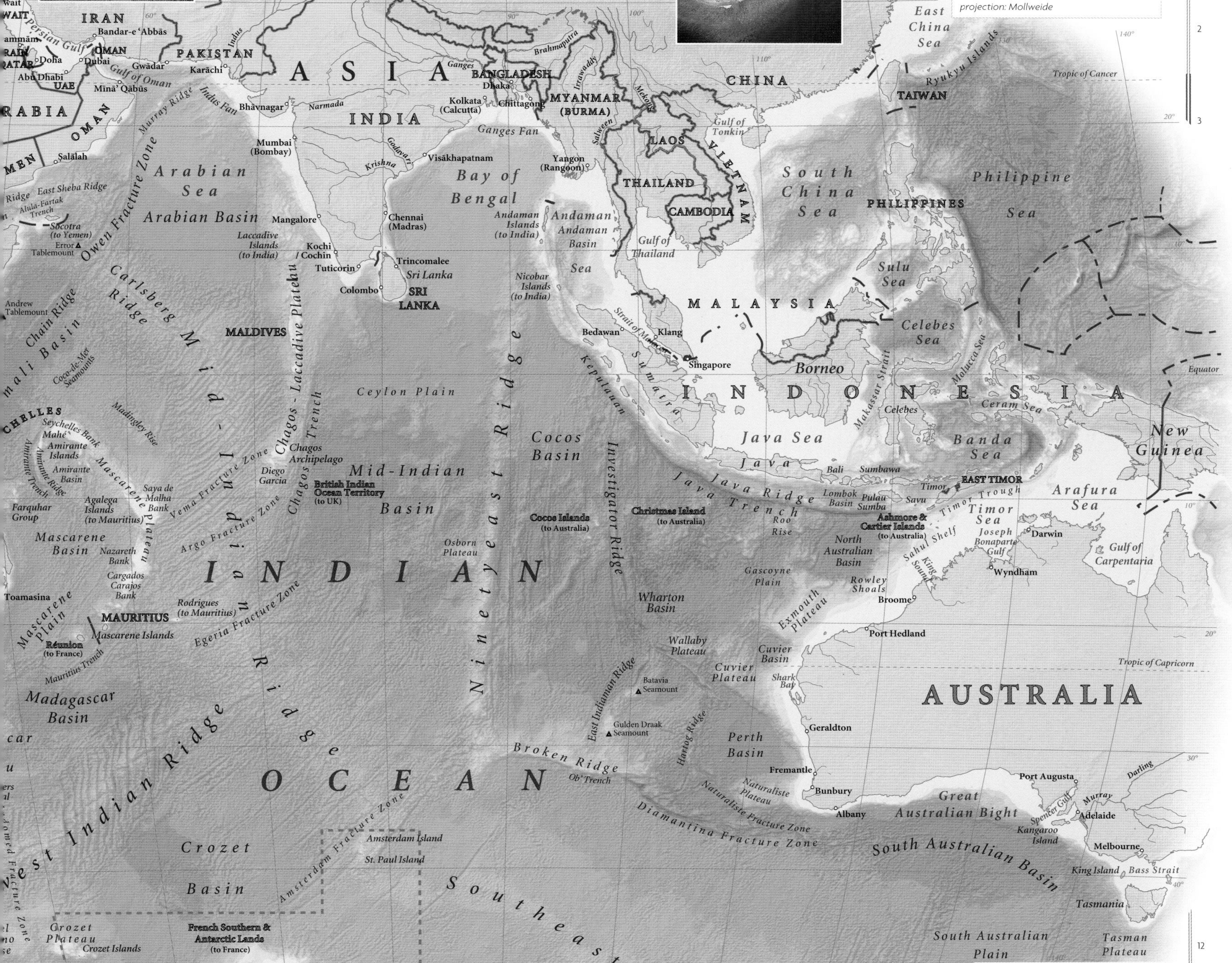

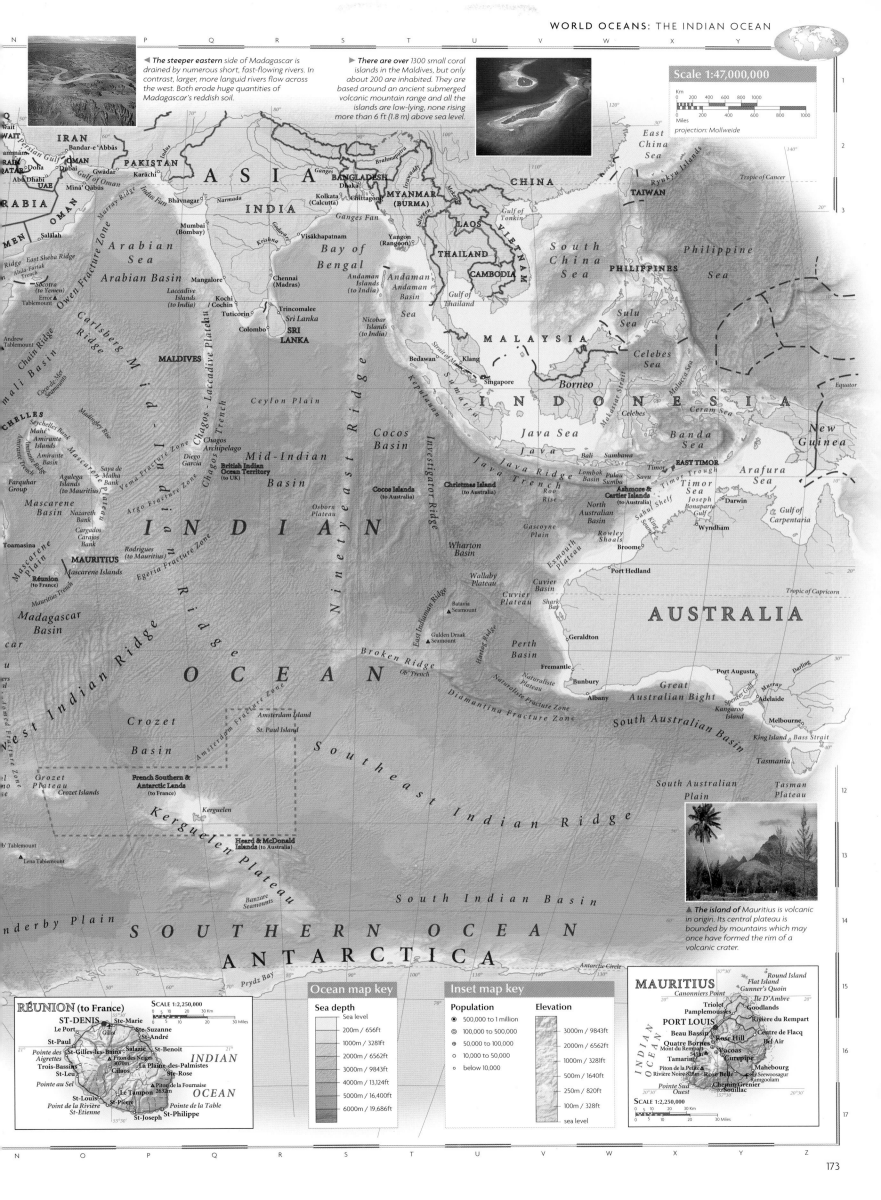

◀ *The steeper eastern* side of Madagascar is drained by numerous short, fast-flowing rivers. In contrast, larger, more languid rivers flow across the west. Both erode huge quantities of Madagascar's reddish soil.

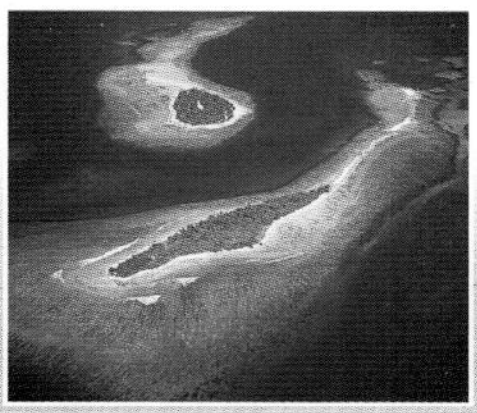

▶ *There are over* 1300 small coral islands in the Maldives, but only about 200 are inhabited. They are based around an ancient submerged volcanic mountain range and all the islands are low-lying, none rising more than 6 ft (1.8 m) above sea level.

▲ *The island of* Mauritius is volcanic in origin. Its central plateau is bounded by mountains which may once have formed the rim of a volcanic crater.

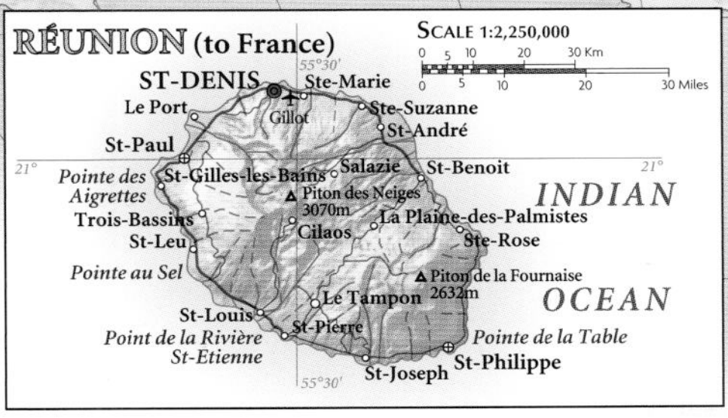

Ocean map key

Sea depth

- Sea level
- 200m / 656ft
- 1000m / 3281ft
- 2000m / 6562ft
- 3000m / 9843ft
- 4000m / 13,124ft
- 5000m / 16,400ft
- 6000m / 19,686ft

Inset map key

Population

- 500,000 to 1 million
- 100,000 to 500,000
- 50,000 to 100,000
- 10,000 to 50,000
- below 10,000

Elevation

- 3000m / 9843ft
- 2000m / 6562ft
- 1000m / 3281ft
- 500m / 1640ft
- 250m / 820ft
- 100m / 328ft
- sea level

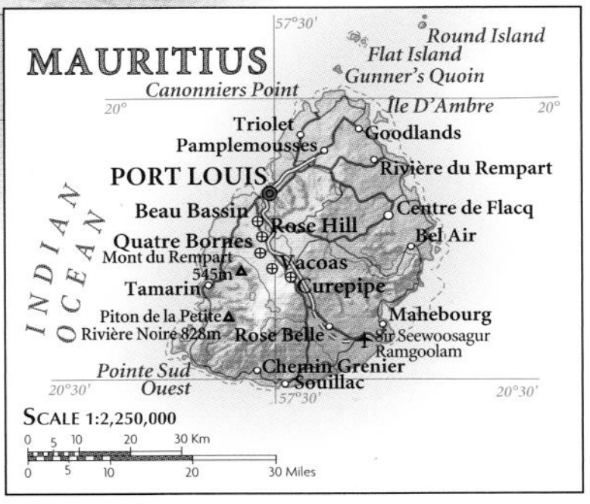

Australasia & Oceania

Australasia and Oceania, covering a land area of 3,285,048 sq miles (8,508,238 sq km), takes in 14 countries including the continent of Australia, New Zealand, Papua New Guinea and many island groups scattered across the Pacific Ocean.

- *Greatest extent, North–South:* 2000 miles / 3200 km
- *Greatest extent, East–West:* 2500 miles /4000 km

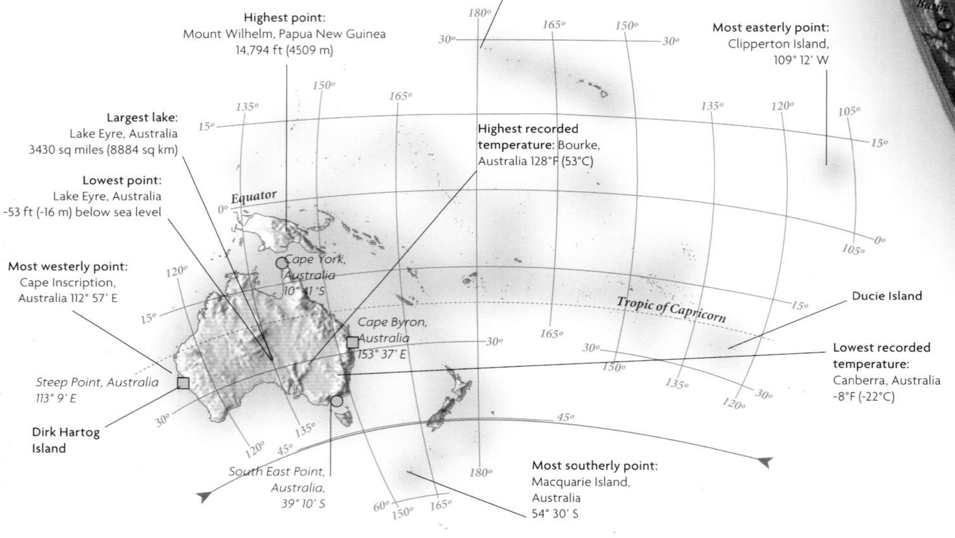

Most northerly point:
Eastern Island, Midway Islands 28° 15′ N
Highest point:
Mount Wilhelm, Papua New Guinea 14,794 ft (4509 m)
Most easterly point:
Clipperton Island, 109° 12′ W
Largest lake:
Lake Eyre, Australia 3430 sq miles (8884 sq km)
Highest recorded temperature: Bourke, Australia 128°F (53°C)
Lowest point:
Lake Eyre, Australia -53 ft (-16 m) below sea level
Equator
Most westerly point:
Cape Inscription, Australia 112° 57′ E
Cape York, Australia 10° 41′ S
Cape Byron, Australia 153° 37′ E
Tropic of Capricorn
Ducie Island
Lowest recorded temperature:
Canberra, Australia -8°F (-22°C)
Steep Point, Australia 113° 9′ E
Dirk Hartog Island
South East Point, Australia, 39° 10′ S
Most southerly point:
Macquarie Island, Australia 54° 30′ S

Dirk Hartog Island, Australia
Great Dividing Range
New Caledonia
New Zealand
Tonga
Tuamoto Islands
Ducie Island, Pitcairn Islands
0 500 1000 1500 Km
0 500 1000 1500 Miles

Cross-section from Dirk Hartog Island, Australia to Ducie Island, Pitcairn Islands

line of cross-section

Hainan Dao
South China Sea
Luzon
Philippine Sea
Philippine Basin
Palawan
Sulu Sea
Philippines
Palau Islands
Mindanao
Celebes Sea
Halmahera
West Caroline Basin
Molucca Sea
New Guinea Trench
Greater Sunda Islands
Java Sea
Flores Sea
Banda Sea
Lesser Sunda Islands
INDO-AUSTRALIAN PLATE
Arafura Sea
Timor Sea
Gulf of Papua
Torres Strait
Cape York
Arnhem Land
Kimberley Plateau
Gulf of Carpentaria
Exmouth Plateau
Great Sandy Desert
Tanami Desert
Australia
Gibson Desert
Lake Carnegie
Simpson Desert
Lake Eyre
Great Victoria Desert
Nullarbor Plain
Lake Gairdner
Lake Frome
Lake Torrens
Great Australian Bight
Kangaroo Island
South Australian Basin
Mount Kosciuszko
South East Point
INDIAN OCEAN

N O P Q R S T U V W X Y Z

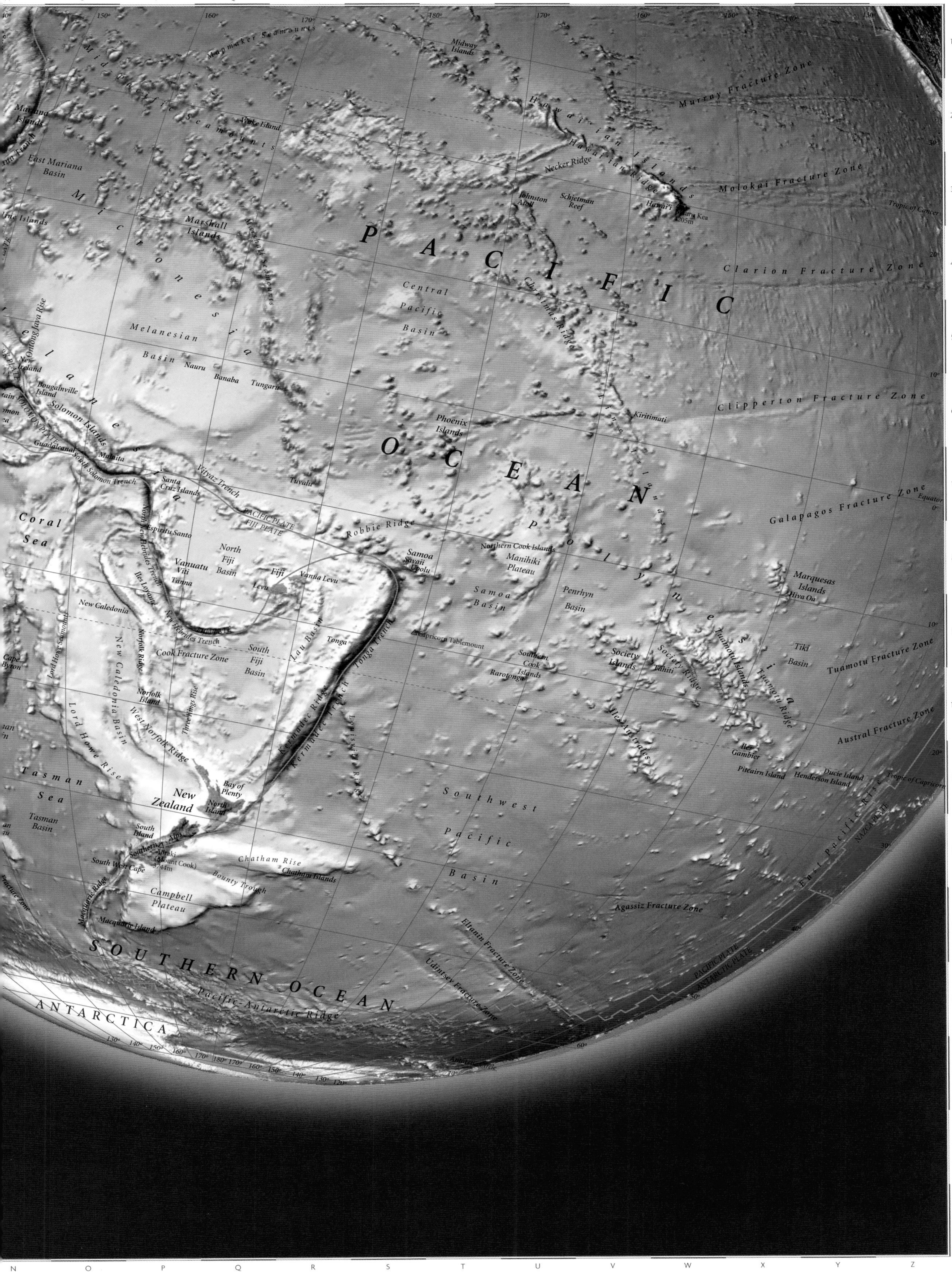
PACIFIC OCEAN
SOUTHERN OCEAN
ANTARCTICA
Musician Seamounts
Midway Islands
Mid Pacific Seamounts
Mariana Islands
Wake Island
Murray Fracture Zone
Hawaiian Islands
Hawaiian Ridge
Necker Ridge
Molokai Fracture Zone
East Mariana Basin
Johnston Atoll
Schjetman Reef
Hawai'i
Tropic of Cancer
Micronesia
Marshall Islands
Clarion Fracture Zone
Central Pacific Basin
Christmas Ridge
Ontong Java Rise
Melanesian Basin
Nauru
Banaba
Tungaru
New Ireland
Bougainville Island
Melanesia
Solomon Islands
Clipperton Fracture Zone
Kiritimati
Phoenix Islands
Line Islands
Guadalcanal
Malaita
South Solomon Trench
Santa Cruz Islands
Vityaz Trench
Tuvalu
Galapagos Fracture Zone
Equator
Coral Sea
Espiritu Santo
PACIFIC PLATE
FIJI PLATE
Robbie Ridge
North Fiji Basin
Vanuatu
Fiji
Samoa
Savai'i
Upolu
Northern Cook Islands
Manihiki Plateau
Polynesia
Marquesas Islands
Hiva Oa
Viti Levu
Vanua Levu
Tanna
Iles Loyauté
New Caledonia
Samoa Basin
Penrhyn Basin
New Hebrides Trench
Lau Basin
Tonga
Tonga Trench
Capricorn Tablemount
Cook Fracture Zone
South Fiji Basin
Southern Cook Islands
Rarotonga
Society Islands
Tahiti
Society Ridge
Tuamotu Islands
Tiki Basin
Tuamotu Fracture Zone
Tuamotu Ridge
New Caledonia Basin
Norfolk Ridge
Norfolk Island
West Norfolk Ridge
Three Kings Rise
Lord Howe Rise
Kermadec Ridge
Kermadec Trench
Louisville Ridge
Iles Australes
Iles Gambier
Austral Fracture Zone
Pitcairn Island
Henderson Island
Ducie Island
Tropic of Capricorn
Tasman Sea
New Zealand
Bay of Plenty
North Island
Southwest Pacific Basin
East Pacific Rise
NAZCA PLATE
Tasman Basin
South Island
Southern Alps
Aoraki (Mount Cook) 3,744m
South West Cape
Chatham Rise
Chatham Islands
Bounty Trough
Campbell Plateau
Macquarie Ridge
Macquarie Island
Agassiz Fracture Zone
Eltanin Fracture Zone
Udintsev Fracture Zone
Pacific-Antarctic Ridge
PACIFIC PLATE
ANTARCTIC PLATE

N O P Q R S T U V W X Y Z

Political Australasia & Oceania

◀ ***Western Australia's mineral*** *wealth has transformed its state capital, Perth, into one of Australia's major cities. Perth is one of the world's most isolated cities – over 2500 miles (4000 km) from the population centres of the eastern seaboard.*

Scale 1:35,500,000

Km 0 200 400 600 800

Miles 0 200 400 600 800

projection: Lambert Azimuthal Equal Area

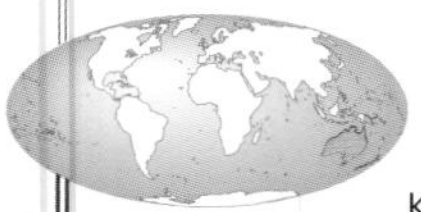

Vast expanses of ocean separate this geographically fragmented realm, characterized more by each country's isolation than by any political unity. Australia's and New Zealand's traditional ties with the United Kingdom, as members of the Commonwealth, are now being called into question as Australasian and Oceanian nations are increasingly looking to forge new relationships with neighbouring Asian countries like Japan. External influences have featured strongly in the politics of the Pacific Islands; the various territories of Micronesia were largely under US control until the late 1980s, and France, New Zealand, the USA and the UK still have territories under colonial rule in Polynesia. Nuclear weapons-testing by Western superpowers was widespread during the Cold War period, but has now been discontinued.

Population

Density of settlement in the region is generally low. Australia is one of the least densely populated countries on Earth with over 80% of its population living within 25 miles (40 km) of the coast – mostly in the southeast of the country. New Zealand, and the island groups of Melanesia, Micronesia and Polynesia, are much more densely populated, although many of the smaller islands remain uninhabited.

Population density
(people per sq km)

- below 4
- 5-24
- 25-49
- 50-99
- 100-199
- 200-299
- above 300

▲ ***The myriad of*** *small coral islands which are scattered across the Pacific Ocean are often uninhabited, as they offer little shelter from the weather, often no fresh water, and only limited food supplies.*

◀ ***The planes of*** *the Australian Royal Flying Doctor Service are able to cover large expanses of barren land quickly, bringing medical treatment to the most inaccessible and far-flung places.*

Northern Mariana Islands
(to US)
Mariana Islands
Saipan
Philippine Sea
Guam
(to US)
HAGÅTÑA
Micronesia
Bikini Atoll
Yap
Caroline Islands
Chuuk
Pohnpei
PALIKIR
Ralik
MELEKEOK
Babeldaob
Kosrae
MICRONESIA
PALAU
Melanesia
NAURU
YAREN
PAPUA NEW GUINEA
Equator
Bismarck Sea
New Ireland
Wewak
New Britain
Rabaul
Madang
Solomon Islands
New Guinea
Mount Hagen
Lae
Ubai
Arawa
Bougainville Island
Solomon Sea
New Georgia Islands
HONIARA
SOLOMON ISLANDS
Tapini
Guadalcanal
PORT MORESBY
Santa Cruz Islands
Arafura Sea
Torres Strait
VANUATU
Espiritu Santo
Malekula
PORT-VILA
Erromango
Coral Sea
Cape York Peninsula
Darwin
Arnhem Land
Gulf of Carpentaria
Timor Sea
Joseph Bonaparte Gulf
Katherine
Great Barrier Reef
Cairns
Coral Sea Islands
(to Australia)
New Caledonia
(to France)
NOUMÉA
Wyndham
Normanton
Townsville
Kimberley Plateau
NORTHERN TERRITORY
Hughenden
Mackay
Derby
Tennant Creek
Tanami Desert
Mount Isa
Great Dividing Range
Rockhampton
Broome
QUEENSLAND
Barcaldine
INDIAN OCEAN
Great Sandy Desert
Alice Springs
Simpson Desert
Charleville
Miles
Brisbane
Port Hedland
Toowoomba
Norfolk Island
(to Australia)
AUSTRALIA
Cunnamulla
Grafton
Hamersley Range
Gibson Desert
Lake Eyre North
Grey Range
Bourke
Barwon
Lord Howe Island
(to Australia)
WESTERN AUSTRALIA
SOUTH AUSTRALIA
Flinders Ranges
Darling
NEW SOUTH WALES
Wilcannia
Dubbo
Newcastle
Lake Torrens
Sydney
Carnarvon
Tropic of Capricorn
Great Victoria Desert
Lake Everard
Lake Gairdner
Port Augusta
Campbelltown
Wollongong
CANBERRA
AUSTRALIAN CAPITAL TERRITORY
Mount Magnet
Ceduna
Whyalla
Murray
Wagga Wagga
Nullarbor Plain
Kalgoorlie
Great Australian Bight
Adelaide
Bendigo
Geraldton
Kangaroo Island
Horsham
VICTORIA
Ballarat
Melbourne
Tasman Sea
Geelong
Mount Gambier
Perth
Esperance
Bass Strait
Launceston
TASMANIA
Albany
Tasmania
Hobart

N O P Q R V W X Y Z

Languages

English is spoken throughout Australia and New Zealand. In Australia, English has been superimposed on a mosaic of Aboriginal languages. In New Zealand, the indigenous language, Maori, is the official language besides English. In Papua New Guinea, Melanesian Pidgin has become a *lingua franca* alongside several hundred indigenous languages. Across the region, the indigenous languages can be grouped into (1) the Aboriginal languages of Australia, (2) the Papuan languages spoken mostly inland in Papua New Guinea, and (3) the widely dispersed Austronesian, which includes coastal languages of Papua New Guinea, New Zealand Maori and languages of Oceania.

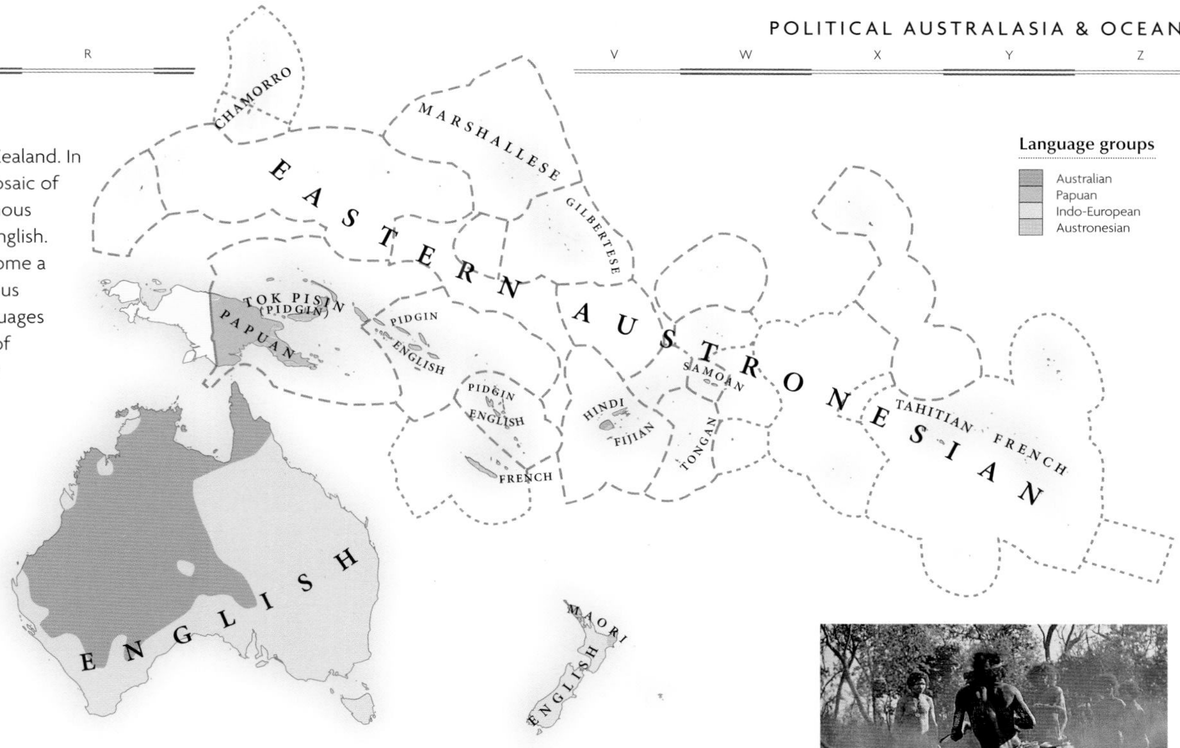

▲ ***Aboriginal languages and*** *cultures are preserved in the central and northern regions of Australia. Ever since the arrival of European settlers, Australia's indigenous peoples have been marginalized. Recently, both their culture and land rights have been increasingly recognized.*

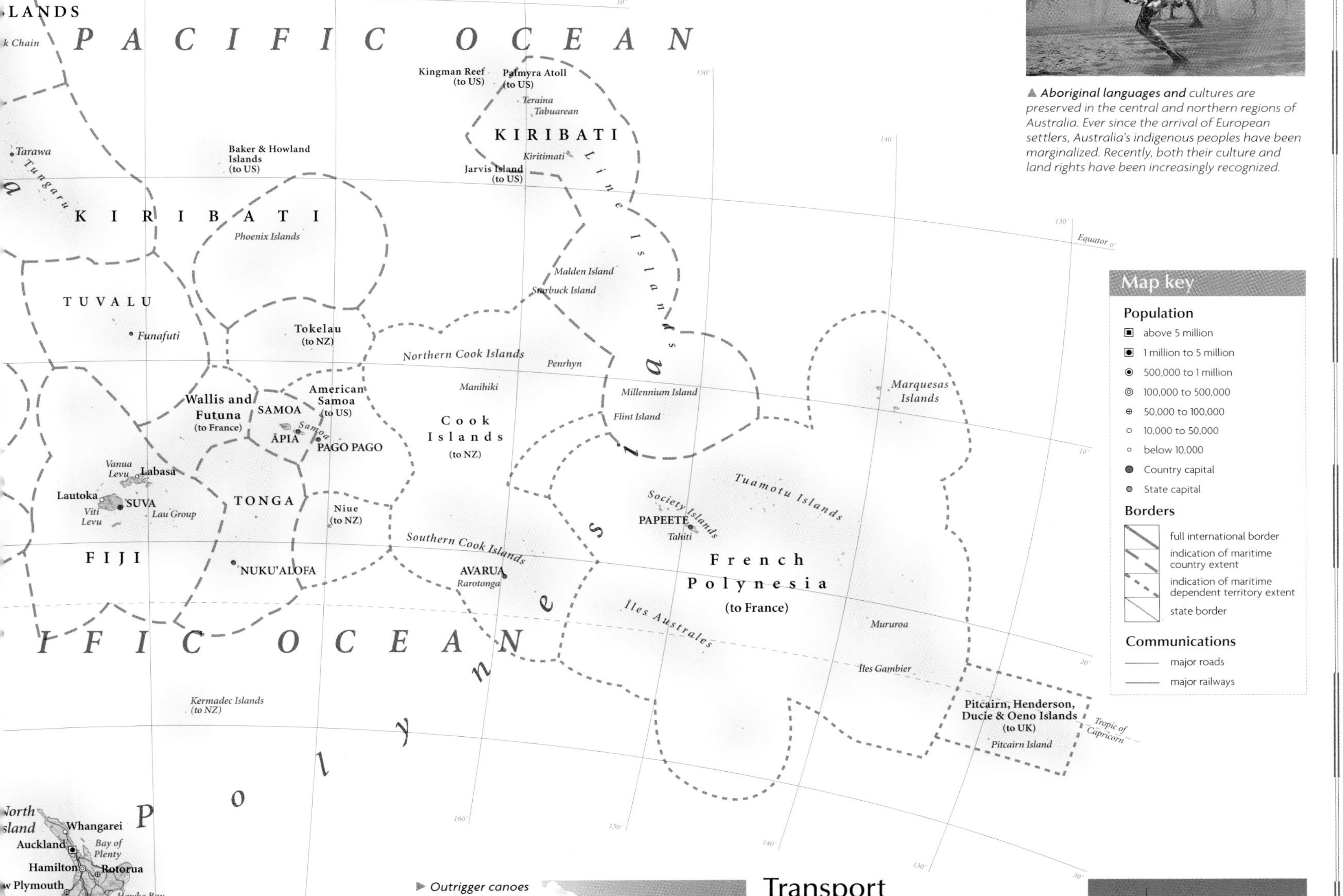

▶ ***Outrigger canoes have*** *been used for centuries throughout the Pacific islands, especially in Micronesia. Hunting and fishing expeditions traditionally required several nights spent at sea, and stronger canoes were built for this purpose.*

Transport

While sea travel remains of paramount importance throughout the continent, well-developed regional and international air travel has reduced the region's global isolation. Internal air travel is particularly important in Australia, where distances are great and road systems are poorly developed or in some areas non-existent. Australia's rail system, still operating on three different gauges, a legacy of its piecemeal development, is being upgraded, particularly in the north-south links.

▲ ***Australia's vast interior*** *is traversed by a limited number of vital roads, linking the major coastal cities to one another. Bulk freight crosses the country along these roads in huge articulated trucks known as 'road trains'.*

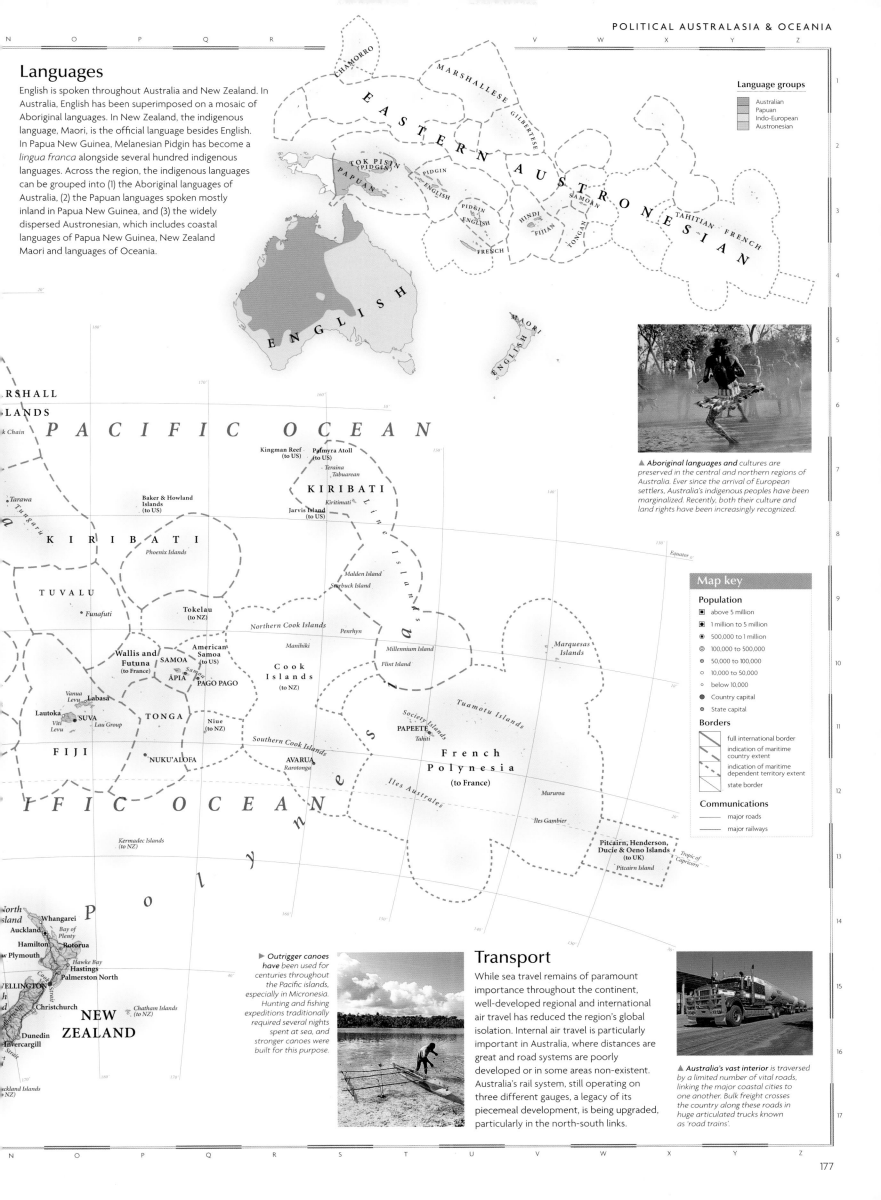

N O P Q R S T U V W X Y Z

Australasian & Oceanian resources

Natural resources are of major economic importance throughout Australasia and Oceania. Australia in particular is a major world exporter of raw materials such as coal, iron ore and bauxite, while New Zealand's agricultural economy is dominated by sheep-raising. Trade with western Europe has declined significantly in the last 20 years, and the Pacific Rim countries of Southeast Asia are now the main trading partners, as well as a source of new settlers to the region. Australasia and Oceania's greatest resources are its climate and environment; tourism increasingly provides a vital source of income for the whole continent.

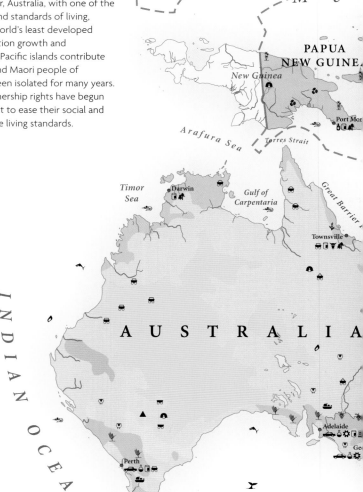

▲ *The largely unpolluted waters of the Pacific Ocean support rich and varied marine life, much of which is farmed commercially. Here, oysters are gathered for market off the coast of New Zealand's South Island.*

▶ *Huge flocks of sheep are a common sight in New Zealand, where they outnumber people by 12 to 1. New Zealand is one of the world's largest exporters of wool and frozen lamb.*

Standard of living

In marked contrast to its neighbour, Australia, with one of the world's highest life expectancies and standards of living, Papua New Guinea is one of the world's least developed countries. In addition, high population growth and urbanization rates throughout the Pacific islands contribute to overcrowding. The Aboriginal and Maori people of Australia and New Zealand have been isolated for many years. Recently, their traditional land ownership rights have begun to be legally recognized in an effort to ease their social and economic isolation, and to improve living standards.

Environmental issues

The prospect of rising sea levels poses a threat to many low-lying islands in the Pacific. Nuclear weapons-testing, once common throughout the region, was finally discontinued in 1996. Australia's ecological balance has been irreversibly altered by the introduction of alien species. Although it has the world's largest underground water reserve, the Great Artesian Basin, the availability of fresh water in Australia remains critical. Periodic droughts combined with over-grazing lead to desertification and increase the risk of devastating bush fires, and occasional flash floods.

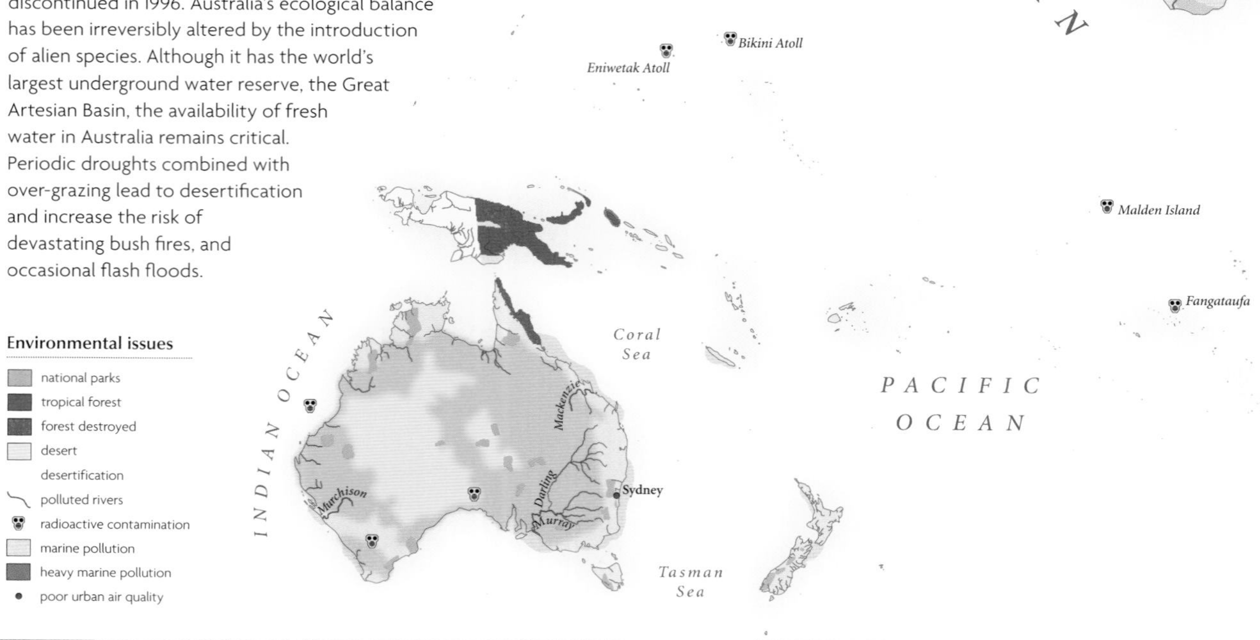

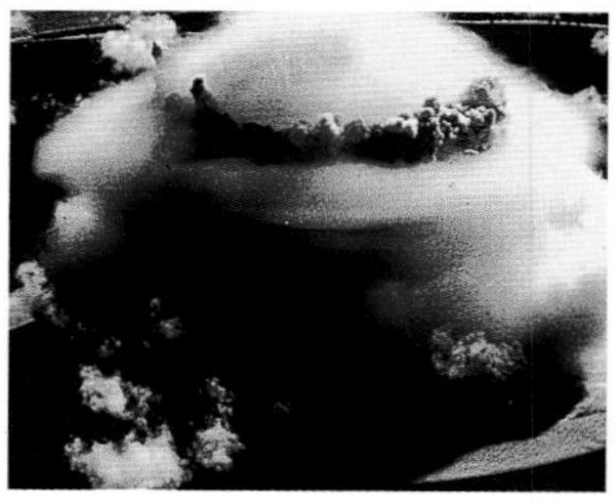

▲ *In 1946 Bikini Atoll, in the Marshall Islands, was chosen as the site for Operation Crossroads – investigating the effects of atomic bombs upon naval vessels. Further nuclear tests continued until the early 1990s. The long-term environmental effects are unknown.*

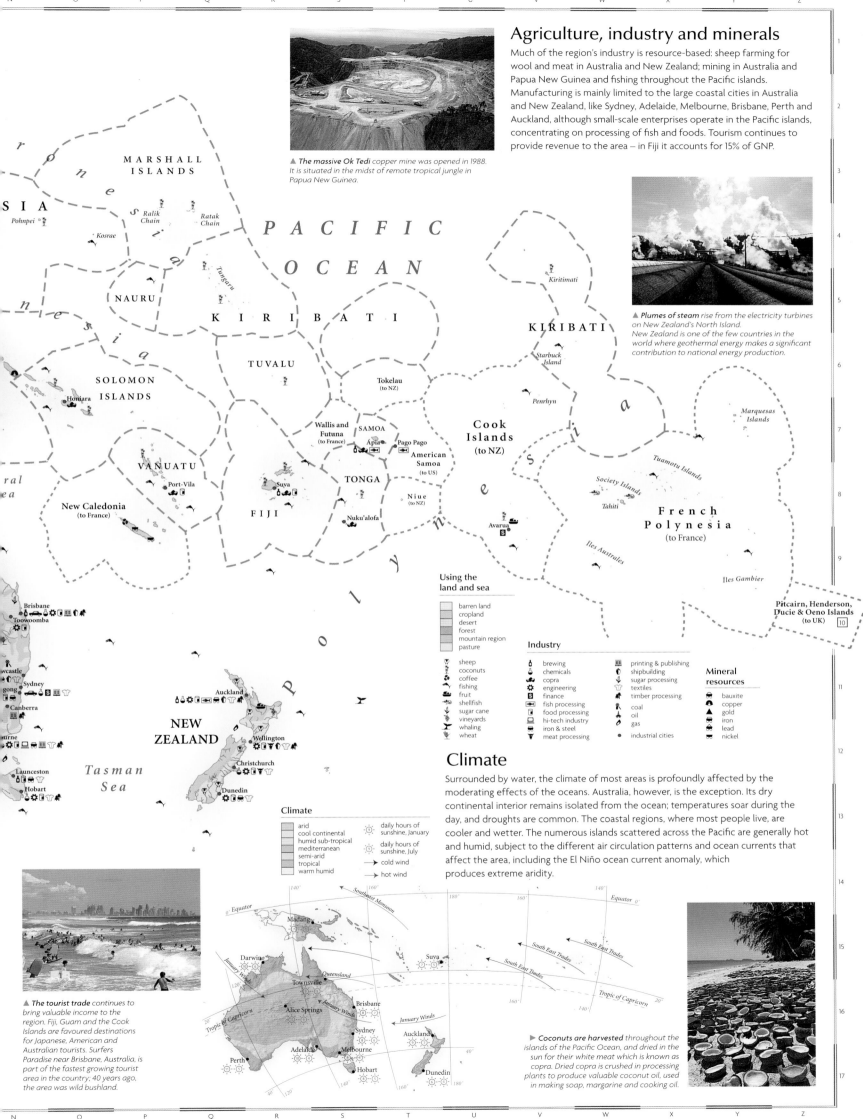

▲ ***The massive Ok Tedi*** *copper mine was opened in 1988. It is situated in the midst of remote tropical jungle in Papua New Guinea.*

Agriculture, industry and minerals

Much of the region's industry is resource-based: sheep farming for wool and meat in Australia and New Zealand; mining in Australia and Papua New Guinea and fishing throughout the Pacific islands. Manufacturing is mainly limited to the large coastal cities in Australia and New Zealand, like Sydney, Adelaide, Melbourne, Brisbane, Perth and Auckland, although small-scale enterprises operate in the Pacific islands, concentrating on processing of fish and foods. Tourism continues to provide revenue to the area – in Fiji it accounts for 15% of GNP.

▲ ***Plumes of steam*** *rise from the electricity turbines on New Zealand's North Island. New Zealand is one of the few countries in the world where geothermal energy makes a significant contribution to national energy production.*

Climate

Surrounded by water, the climate of most areas is profoundly affected by the moderating effects of the oceans. Australia, however, is the exception. Its dry continental interior remains isolated from the ocean; temperatures soar during the day, and droughts are common. The coastal regions, where most people live, are cooler and wetter. The numerous islands scattered across the Pacific are generally hot and humid, subject to the different air circulation patterns and ocean currents that affect the area, including the El Niño ocean current anomaly, which produces extreme aridity.

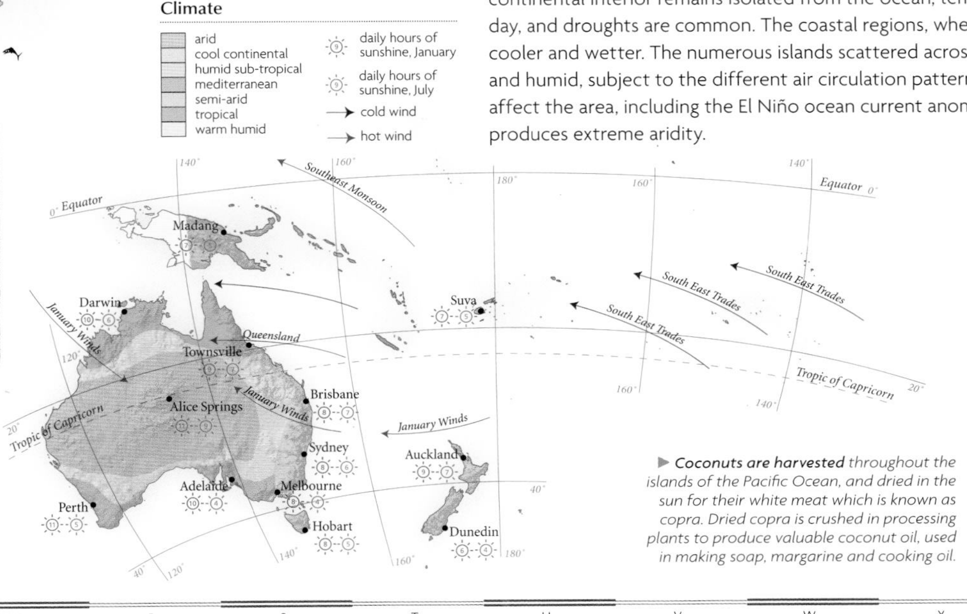

▲ ***The tourist trade*** *continues to bring valuable income to the region. Fiji, Guam and the Cook Islands are favoured destinations for Japanese, American and Australian tourists. Surfers Paradise near Brisbane, Australia, is part of the fastest growing tourist area in the country; 40 years ago, the area was wild bushland.*

▶ ***Coconuts are harvested*** *throughout the islands of the Pacific Ocean, and dried in the sun for their white meat which is known as copra. Dried copra is crushed in processing plants to produce valuable coconut oil, used in making soap, margarine and cooking oil.*

Australia

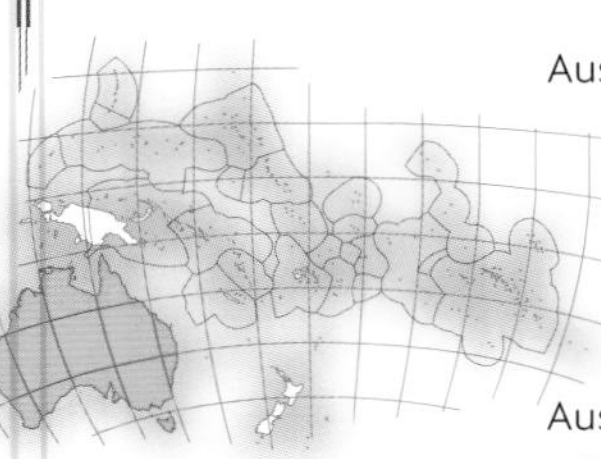

Australia is the world's smallest continent, a stable landmass lying between the Indian and Pacific oceans. Previously home to its aboriginal peoples only, since the end of the 18th century immigration has transformed the face of the country. Initially settlers came mainly from western Europe, particularly the UK, and for years Australia remained wedded to its British colonial past. More recent immigrants have come from eastern Europe, and from Asian countries such as Japan, South Korea and Indonesia. Australia is now forging strong trading links with these 'Pacific Rim' countries and its economic future seems to lie with Asia and the Americas, rather than Europe, its traditional partner.

◀ ***Uluru (Ayers Rock),*** *the world's largest free-standing rock, is a massive outcrop of red sandstone in Australia's desert centre. Wind and sandstorms have ground the rock into the smooth curves seen here. Uluru is revered as a sacred site by many aboriginal peoples.*

Scale 1:11,500,000

Km: 0 25 50 100 150 200 250 300 350
Miles: 0 25 50 100 150 200 250 300 350

projection: Lambert Conformal Conic

Map key

Population	Elevation
1 million to 5 million	2000m / 6562ft
500,000 to 1 million	1000m / 3281ft
100,000 to 500,000	500m / 1640ft
50,000 to 100,000	250m / 820ft
10,000 to 50,000	100m / 328ft
below 10,000	sea level

Using the land

Over 104 million sheep are dispersed in vast herds around the country, contributing to a major export industry. Cattle-ranching is important, particularly in the west. Wheat, and grapes for Australia's wine industry, are grown mainly in the south. Much of the country is desert, unsuitable for agriculture unless irrigation is used.

The urban/rural population divide

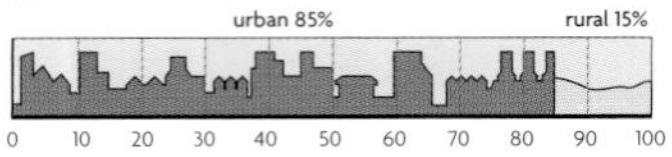

Population density	Total land area
6 people per sq mile (2 people per sq km)	2,967,893 sq miles (7,686,850 sq km)

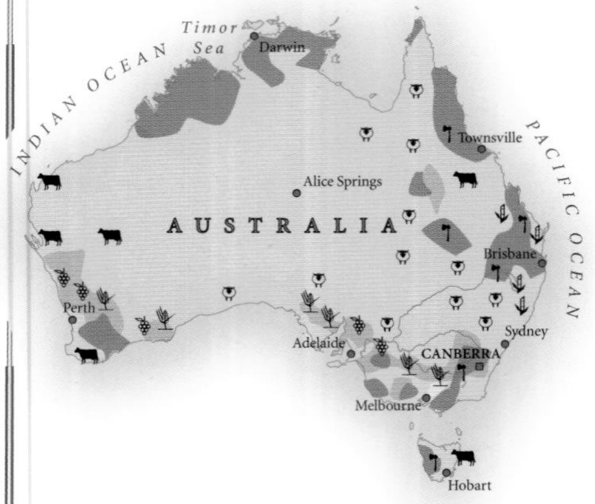

Land use and agricultural distribution

cattle; sheep; cereals; sugar cane; timber; vineyards; capital cities; major towns; pasture; cropland; forest; desert; mountain region

▲ ***Lines of ripening*** *vines stretch for miles in Barossa Valley, a major wine-growing region near Adelaide.*

The landscape

Australia consists of many eroded plateaux, lying firmly in the middle of the Indo-Australian Plate. It is the world's flattest continent, and the driest, after Antarctica. The coasts tend to be more hilly and fertile, especially in the east. The mountains of the Great Dividing Range form a natural barrier between the eastern coastal areas and the flat, dry plains and desert regions of the Australian 'outback.'

▲ ***The Great Barrier Reef*** *is the world's largest area of coral islands and reefs. It runs for about 1240 miles (2000 km) along the Queensland coast.*

▲ ***The Pinnacles are*** *a series of rugged sandstone pillars. Their strange shapes have been formed by water and wind erosion.*

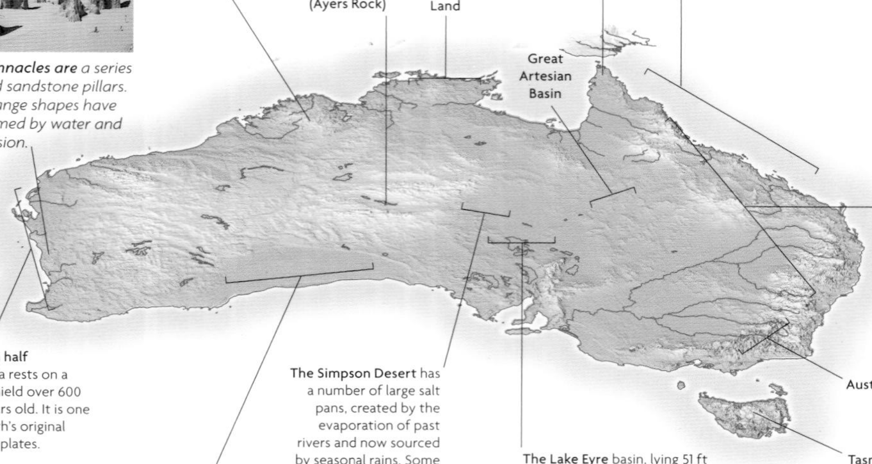

The ancient Kimberley Plateau is the source of some of Australia's richest mineral deposits, including diamonds.

Uluru (Ayers Rock)

Arnhem Land

The tropical rain forest of the Cape York Peninsula contains more than 600 different varieties of tree.

Great Artesian Basin

The Great Dividing Range forms a watershed between east- and west-flowing rivers. Erosion has created deep valleys, gorges and waterfalls where rivers tumble over escarpments on their way to the sea.

Australian Alps

More than half of Australia rests on a uniform shield over 600 million years old. It is one of the Earth's original geological plates.

The Nullarbor Plain is a low-lying limestone plateau which is so flat that the Trans-Australian Railway runs through it in a straight line for more than 300 miles (483 km).

The Simpson Desert has a number of large salt pans, created by the evaporation of past rivers and now sourced by seasonal rains. Some are crusted with gypsum, but most are covered by common salt crystals.

The Lake Eyre basin, lying 51 ft (16 m) below sea level, is one of the largest inland drainage systems in the world, covering an area of more than 500,000 sq miles (1,300,000 sq km).

Tasmania has the same geological structure as the Australian Alps. During the last period of glaciation, 18,000 years ago, sea levels were some 300 ft (100 m) lower and it was joined to the mainland.

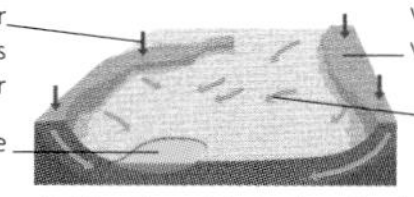

▲ ***The Great Artesian Basin*** *underlies nearly 20% of the total area of Australia, providing a valuable store of underground water, essential to Australian agriculture. The ephemeral rivers which drain the northern part of the basin have highly braided courses and, in consequence, the area is known as 'channel country.'*

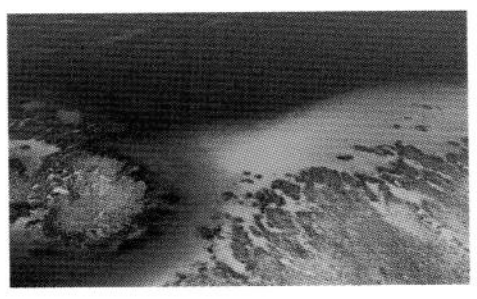

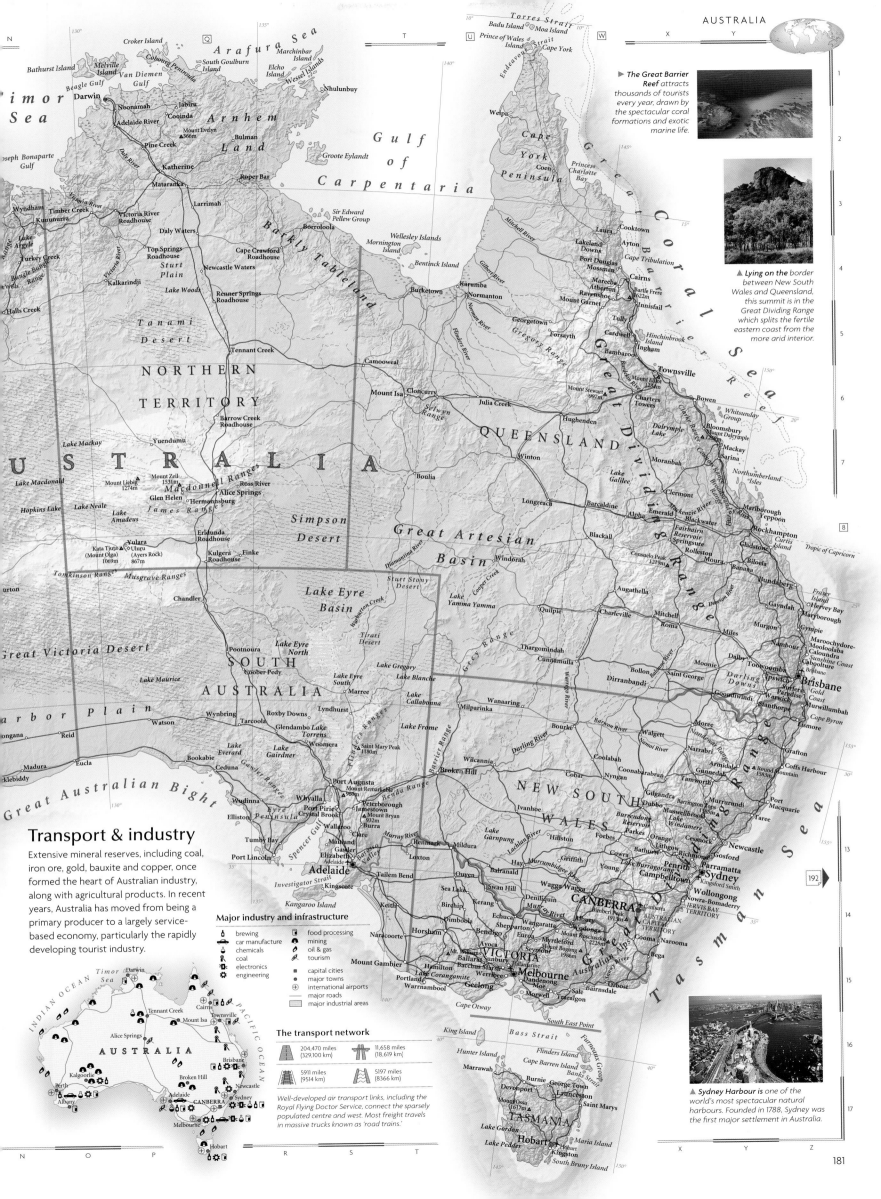

▶ **The Great Barrier Reef** attracts thousands of tourists every year, drawn by the spectacular coral formations and exotic marine life.

▲ **Lying on the border** between New South Wales and Queensland, this summit is in the Great Dividing Range which splits the fertile eastern coast from the more arid interior.

Transport & industry

Extensive mineral reserves, including coal, iron ore, gold, bauxite and copper, once formed the heart of Australian industry, along with agricultural products. In recent years, Australia has moved from being a primary producer to a largely service-based economy, particularly the rapidly developing tourist industry.

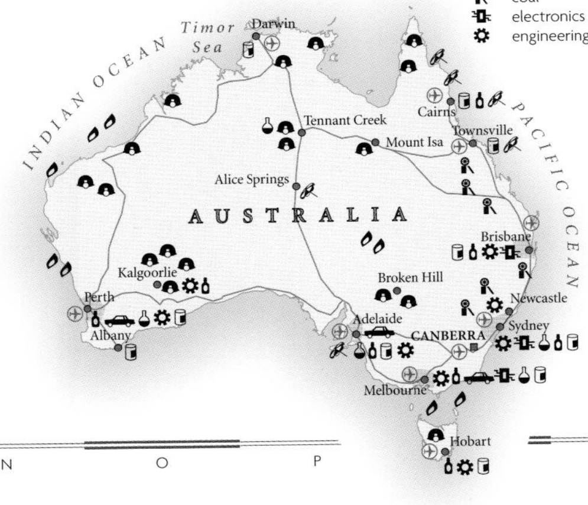

Major industry and infrastructure

- brewing
- car manufacture
- chemicals
- coal
- electronics
- engineering
- food processing
- mining
- oil & gas
- tourism
- capital cities
- major towns
- international airports
- major roads
- major industrial areas

The transport network

204,470 miles (329,100 km)	11,658 miles (18,619 km)
5911 miles (9514 km)	5197 miles (8366 km)

Well-developed air transport links, including the Royal Flying Doctor Service, connect the sparsely populated centre and west. Most freight travels in massive trucks known as 'road trains.'

▲ **Sydney Harbour is** one of the world's most spectacular natural harbours. Founded in 1788, Sydney was the first major settlement in Australia.

192

180

192

192

Southeast Australia

New South Wales, South Australia, Tasmania, Victoria

The southeast of Australia is the most industrialized, economically stable, urbanized and ethnically diverse region, centred on the states of Victoria and New South Wales. The first area to be extensively settled, the southeast remains the country's focus, with the four states which comprise this region containing more than 70% of the population in only 27% of the land area. The southeast – the cultural and artistic heartland of Australia – takes in five of the country's great cities: Sydney, the largest city; Adelaide; Melbourne; Hobart; and Canberra, the centre of federal government.

▲ *Bondi Beach in Sydney is a famous 'surf beach'; its rolling waves and sandy beaches draw locals, tourists and surf enthusiasts from all over the world.*

Transport and industry

Most manufacturing and service industry is based in the southeast. A thriving tourist industry contributes to 5% of GDP. The manufacture of electronic equipment, chemicals and vehicles is complemented by the more traditional fishing, agricultural and mining industries; iron ore and brown coal (lignite) are particularly important.

The transport network

The region's road links are well developed. A high-speed train service linking Melbourne, Sydney and Canberra is under discussion. High levels of air traffic, servicing the expanding tourist industry, is causing increased congestion.

Major industry and infrastructure

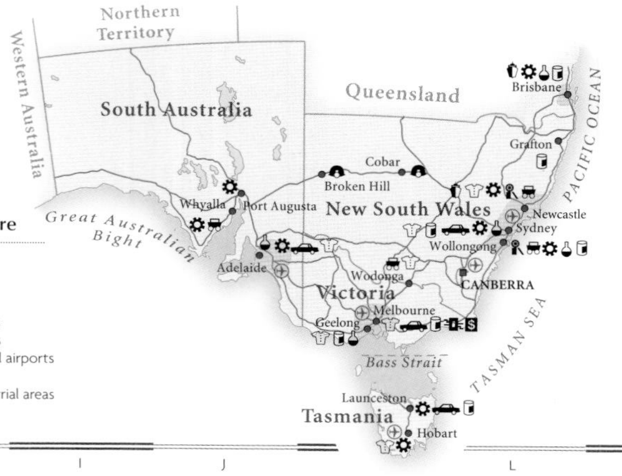

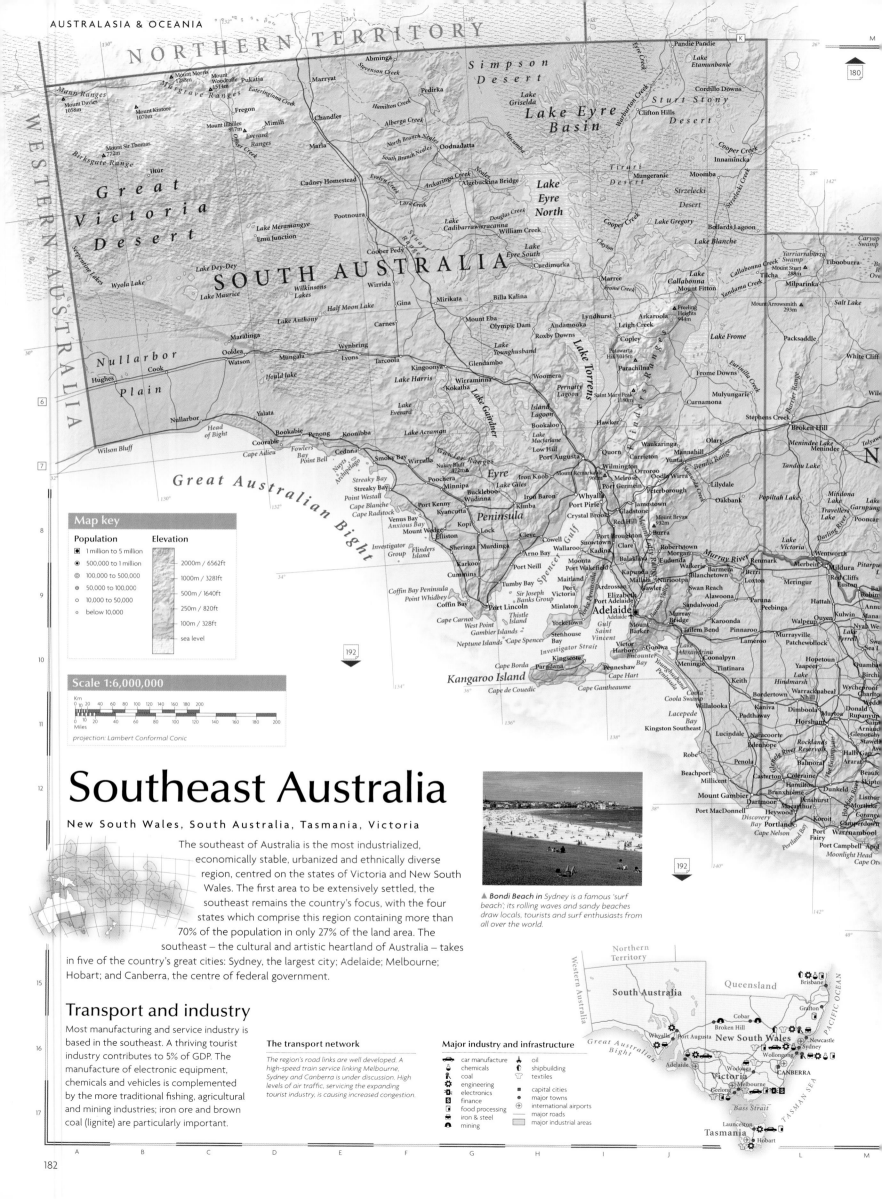

▲ *The fertile Darling Downs, known as the 'breadbasket of Australia', support a wide range of crops including cereals, sugar cane and fruit.*

Using the land and sea

The western flanks of the Great Dividing Range and the northern deserts of South Australia support massive herds of sheep and cattle, while more intensive stock-rearing occurs near the cities. Sugar cane is the most important industrial crop, and cereals including wheat, maize, barley and sorghum are also grown. Grapes, citrus and orchard fruits are among the wide range of fruit and vegetables cultivated in this region. Tasmania's forestry and fishing contributes to over one-third of the state's exports.

► *The Murray River has its source in the eastern uplands of the Great Dividing Range. Fed by melting snow, it runs for 1609 miles (2589 km), and has sufficient volume to reach the ocean southeast of Adelaide despite a minimal gradient for most of its lower reaches.*

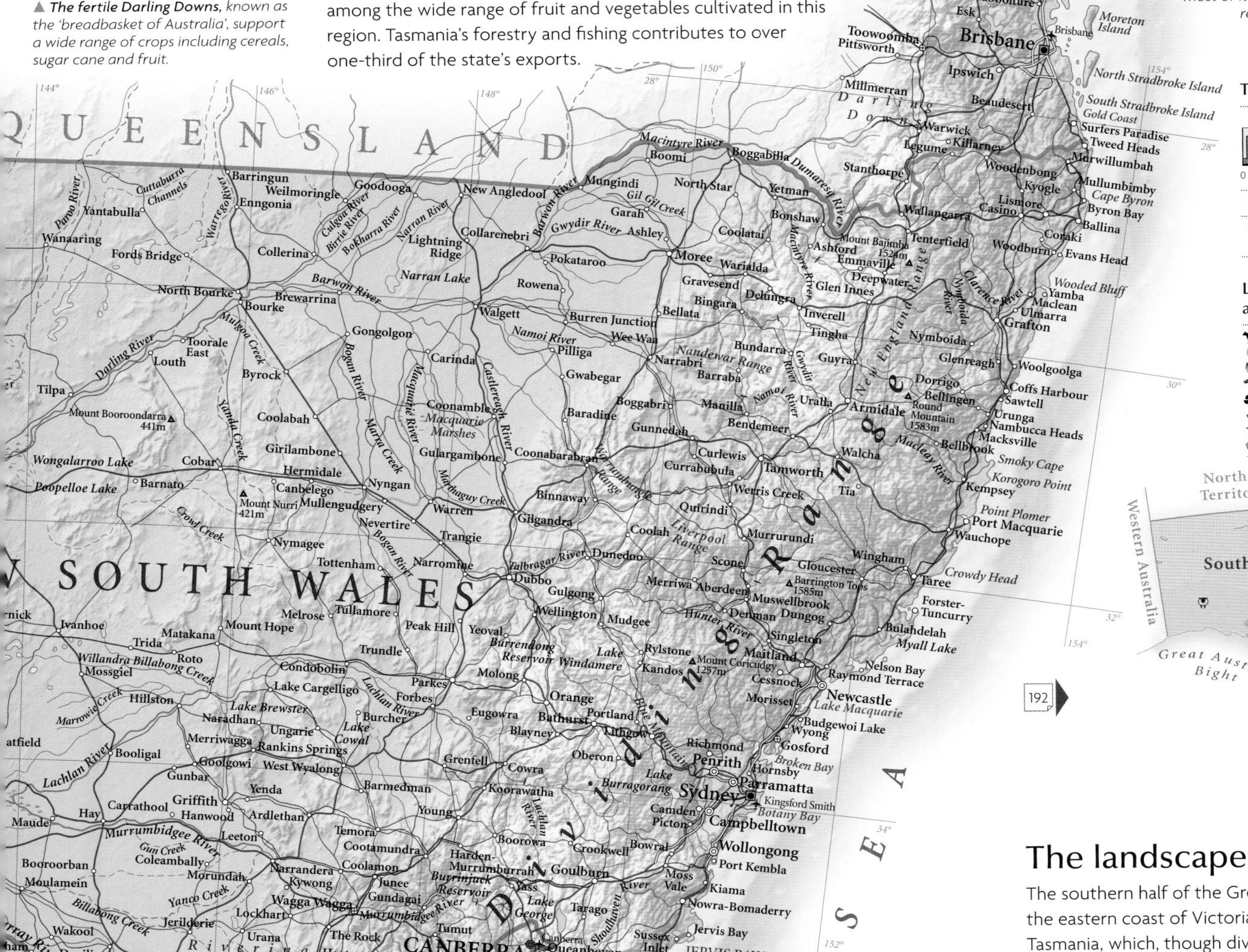

The urban/rural population divide

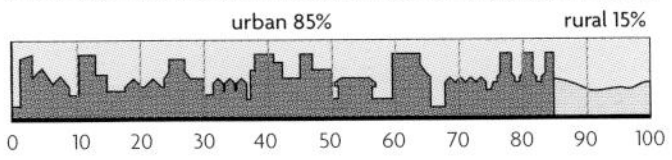

Population density	Total land area
18 people per sq mile (7 people per sq km)	778,022 sq miles (2,015,600 sq km)

Land use and agricultural distribution

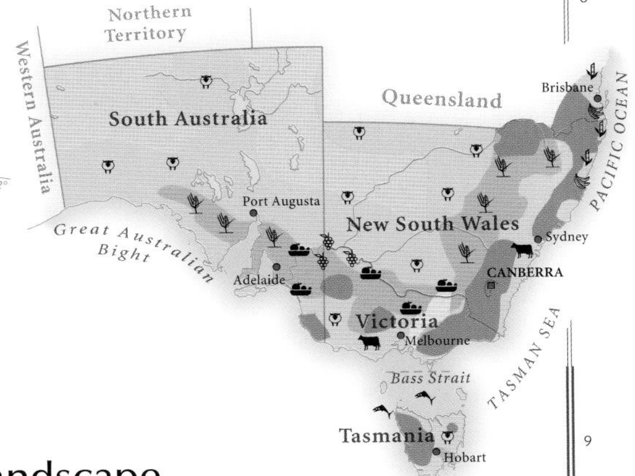

192

The landscape

The southern half of the Great Dividing Range runs parallel to the eastern coast of Victoria and New South Wales as far as Tasmania, which, though divided from the mainland is part of the same mountain chain. South Australia comprises the Australian shield and half of the dry, flat Nullarbor Plain. The Murray/Darling river basin is the only major river system.

◄ *The heavily folded Flinders Ranges is part of an arc of sedimentary rocks reaching northward from Kangaroo Island.*

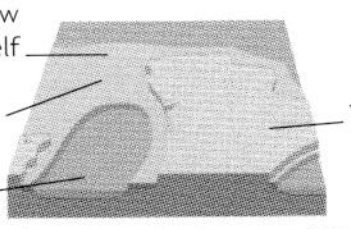

▲ *Tasmania is part of Australia's eastern highlands, separated from the mainland by 155 miles (250 km) of the Bass Strait. In the recent geological past, dry land links between Tasmania and Victoria would have been possible during periods of world-wide glaciation, when the sea level was more than 180 ft (55 m) below that of present sea levels.*

The Musgrave and Everard ranges form bare, rounded hills made up of ancient granite and gneiss.

Lake Eyre is the largest of southern Australia's dry lakes. Lying -51 ft (-16 m) below sea level, it has flooded only three times in the last century.

The Murray/Darling is Australia's longest river at 1703 miles (2739 km).

Great Dividing Range

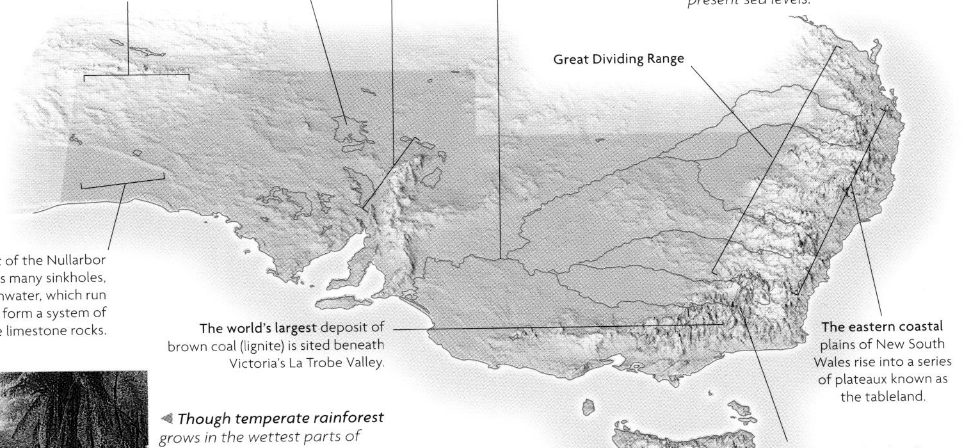

The eastern part of the Nullarbor Plain has many sinkholes, eroded by rainwater, which run underground to form a system of long caves in the limestone rocks.

The world's largest deposit of brown coal (lignite) is sited beneath Victoria's La Trobe Valley.

The eastern coastal plains of New South Wales rise into a series of plateaux known as the tableland.

The glaciated central plateau of Tasmania has many lakes, including Lake St Clair, a piedmont lake more than 700 ft (200 m) deep.

Mount Kosciuszko, the highest point in the Snowy Mountains, is the tallest mountain in Australia at 7316 ft (2228 m).

◄ *Though temperate rainforest grows in the wettest parts of Tasmania, extreme variations in the levels of rainfall over the island mean that some drier areas may experience forest fires.*

New Zealand

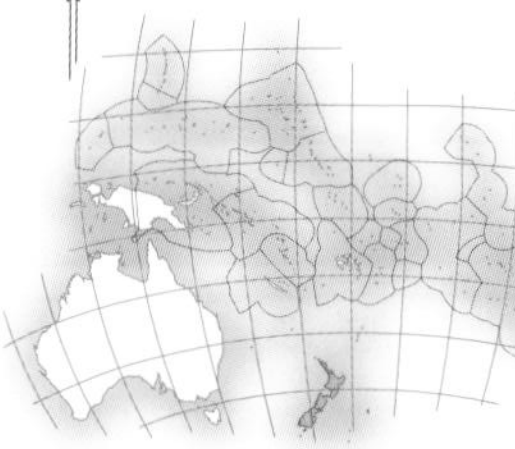

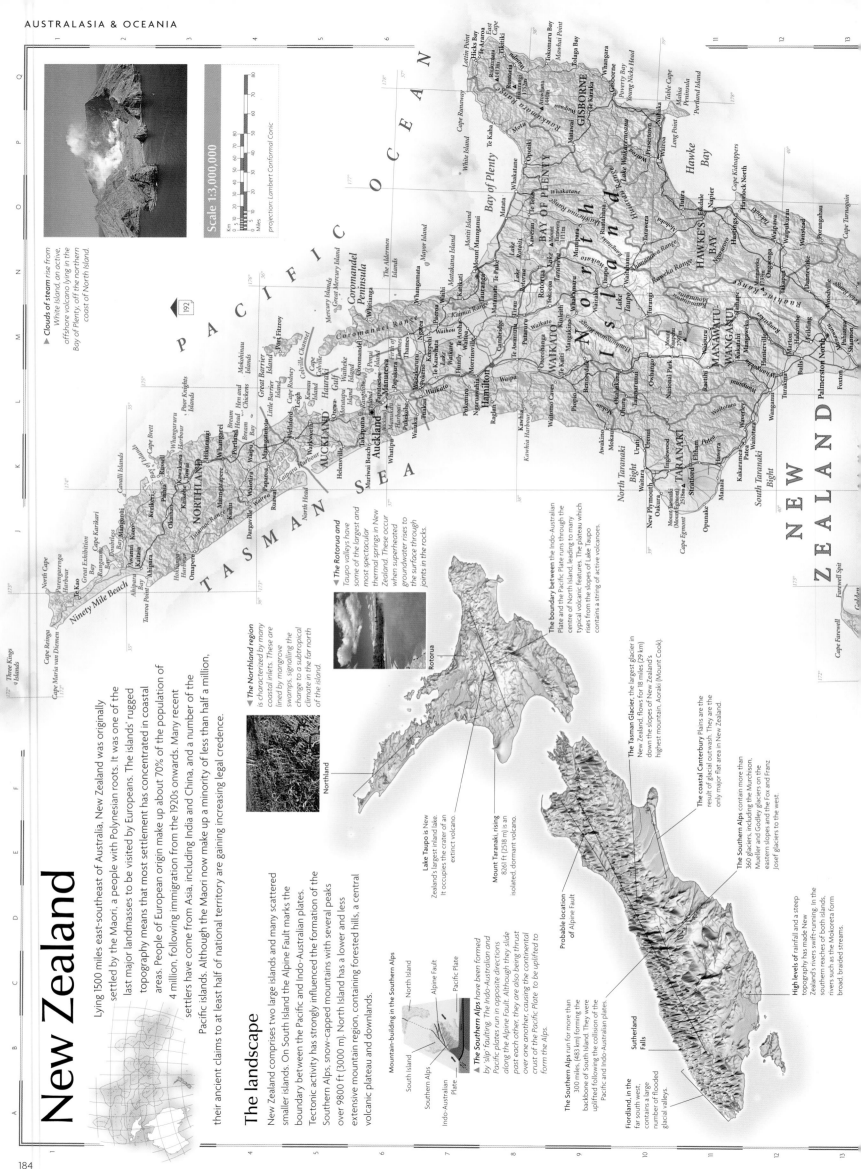

Lying 1500 miles east-southeast of Australia, New Zealand was originally settled by the Maori, a people with Polynesian roots. It was one of the last major landmasses to be visited by Europeans. The islands' rugged topography means that most settlement has concentrated in coastal areas. People of European origin make up about 70% of the population of 4 million, following immigration from the 1920s onwards. Many recent settlers have come from Asia, including India and China, and a number of the Pacific islands. Although the Maori now make up a minority of less than half a million, their ancient claims to at least half of national territory are gaining increasing legal credence.

The landscape

New Zealand comprises two large islands and many scattered smaller islands. On South Island the Alpine Fault marks the boundary between the Pacific and Indo-Australian plates. Tectonic activity has strongly influenced the formation of the Southern Alps, snow-capped mountains with several peaks over 9800 ft (3000 m). North Island has a lower and less extensive mountain region, containing forested hills, a central volcanic plateau and downlands.

◀ *The Northland region is characterized by many coastal inlets. These are lined by mangrove swamps, signalling the change to a subtropical climate in the far north of the island.*

◀ *The Rotorua and Taupo valleys have some of the largest and most spectacular thermal springs in New Zealand. These occur when superheated groundwater rises to the surface through joints in the rocks.*

Mountain-building in the Southern Alps

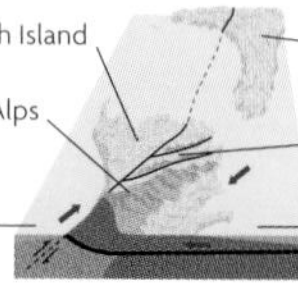

▲ **The Southern Alps** *have been formed by 'slip' faulting. The Indo-Australian and Pacific plates run in opposite directions along the Alpine Fault. Although they slide past each other, they are also being thrust over one another, causing the continental crust of the Pacific Plate to be uplifted to form the Alps.*

Lake Taupo is New Zealand's largest inland lake. It occupies the crater of an extinct volcano.

Mount Taranaki, rising 8261 ft (2518 m) is an isolated, dormant volcano.

The boundary between the Indo-Australian Plate and the Pacific Plate runs through the centre of North Island, leading to many typical volcanic features. The plateau which rises from the slopes of Lake Taupo contains a string of active volcanoes.

The Southern Alps run for more than 300 miles, (483 km) forming the backbone of South Island. They were uplifted following the collision of the Pacific and Indo-Australian plates.

Fiordland, in the far south west, contains a large number of flooded glacial valleys.

The Tasman Glacier, the largest glacier in New Zealand, flows for 18 miles (29 km) down the slopes of New Zealand's highest mountain, Aoraki (Mount Cook).

The coastal Canterbury Plains are the result of glacial outwash. They are the only major flat area in New Zealand.

The Southern Alps contain more than 360 glaciers, including the Murchison, Mueller and Godley glaciers on the eastern slopes and the Fox and Franz Josef glaciers to the west.

High levels of rainfall and a steep topography has made New Zealand's rivers swift-running. In the southern reaches of both islands, rivers such as the Mokoreta form broad, braided streams.

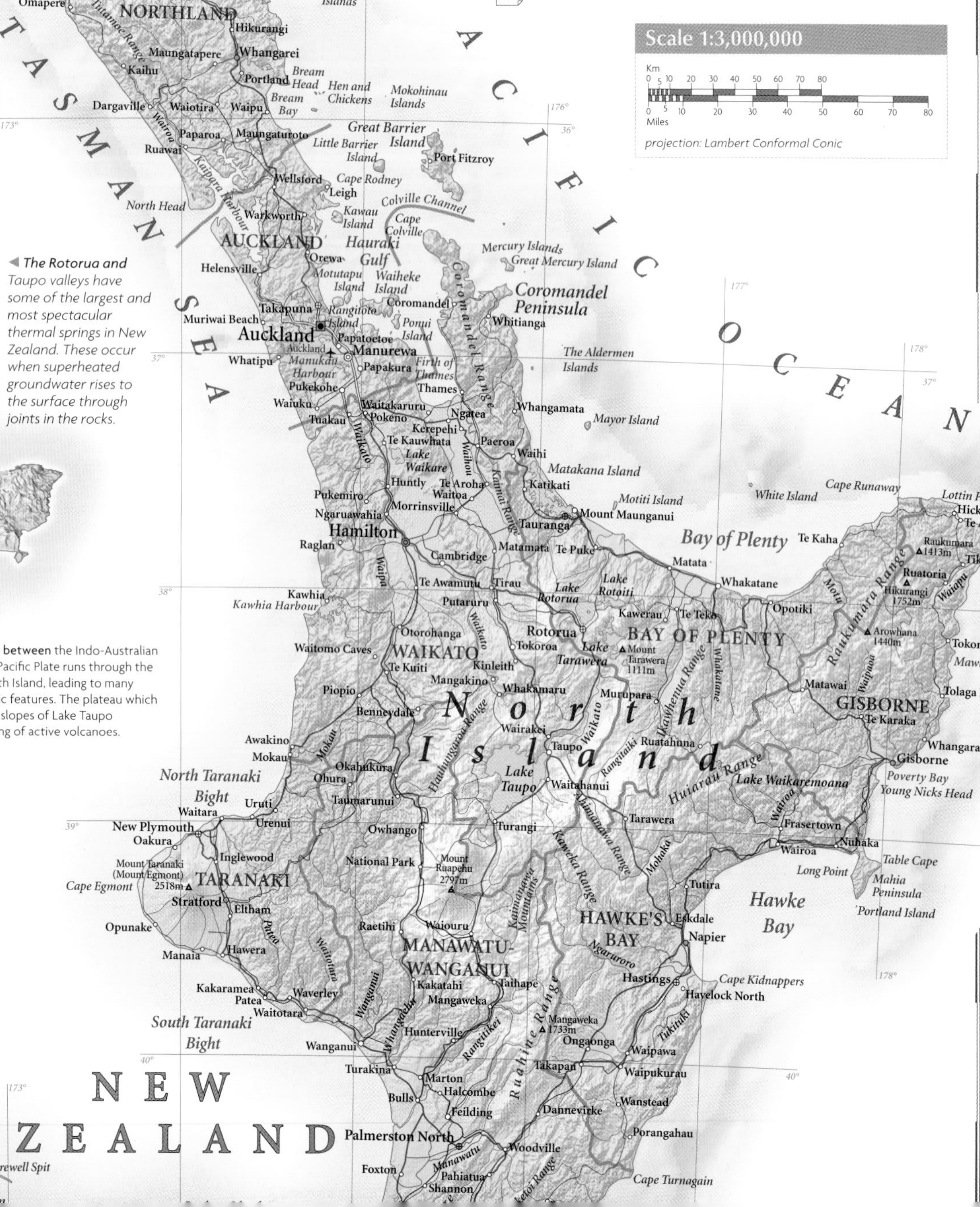

▶ *Clouds of steam rise from White Island, an active, offshore volcano lying in the Bay of Plenty, off the northern coast of North Island.*

Scale 1:3,000,000

projection: Lambert Conformal Conic

Map key

Population

- 1 million to 5 million
- 500,000 to 1 million
- 100,000 to 500,000
- 50,000 to 100,000
- 10,000 to 50,000
- below 10,000

Elevation

- 3000m / 9843ft
- 2000m / 6562ft
- 1000m / 3281ft
- 500m / 1640ft
- 250m / 820ft
- 100m / 328ft
- sea level

▲ ***The snow-capped peak*** *of Aoraki (Mount Cook), on the west coast of South Island, overlooks a heath strewn with foxgloves. Though still the highest peak in New Zealand, at 12,349 ft (3744 m), a massive rock fall in 1991 reduced the height of the mountain by 66 ft (20 m).*

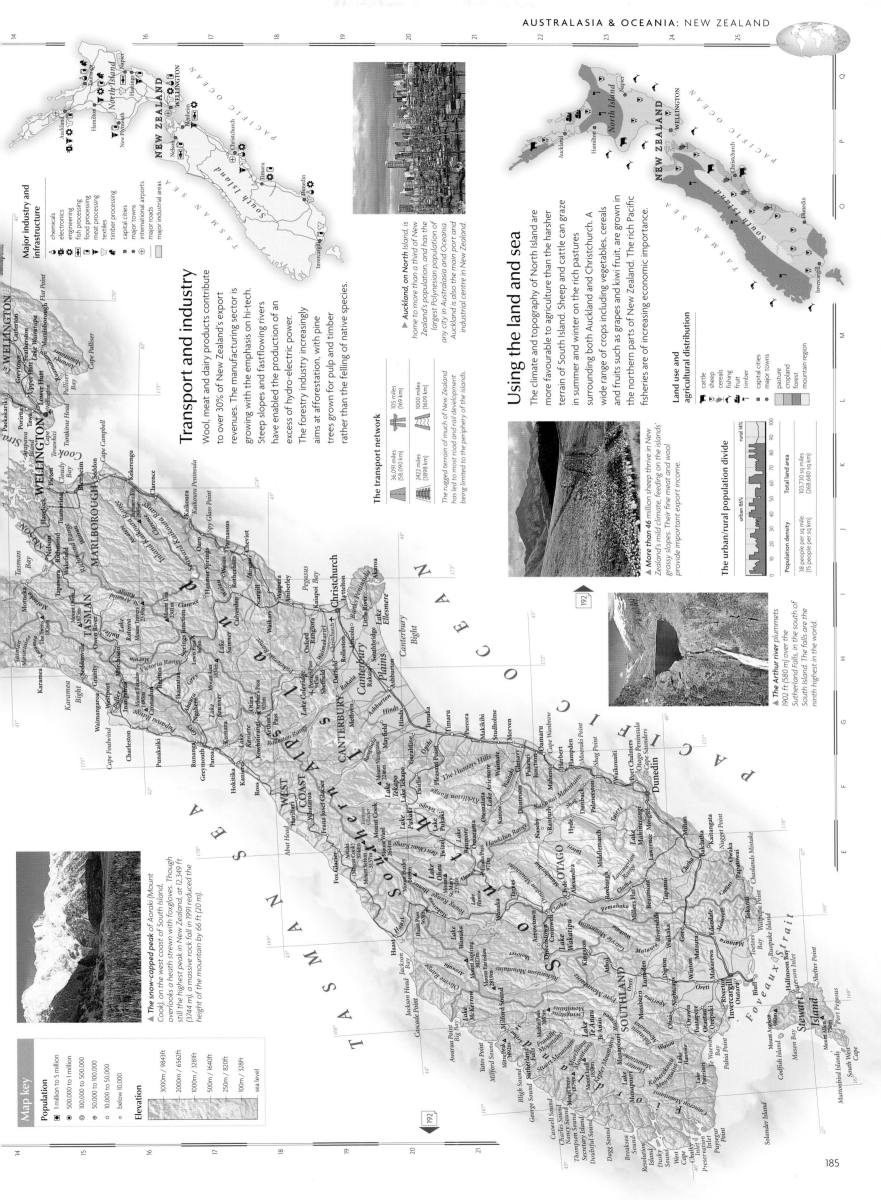

Transport and industry

Wool, meat and dairy products contribute to over 30% of New Zealand's export revenues. The manufacturing sector is growing with the emphasis on hi-tech. Steep slopes and fastflowing rivers have enabled the production of an excess of hydro-electric power. The forestry industry increasingly aims at afforestation, with pine trees grown for pulp and timber rather than the felling of native species.

Major industry and infrastructure

- chemicals
- electronics
- engineering
- fish processing
- food processing
- meat processing
- textiles
- timber processing
- capital cities
- major towns
- international airports
- major roads
- major industrial areas

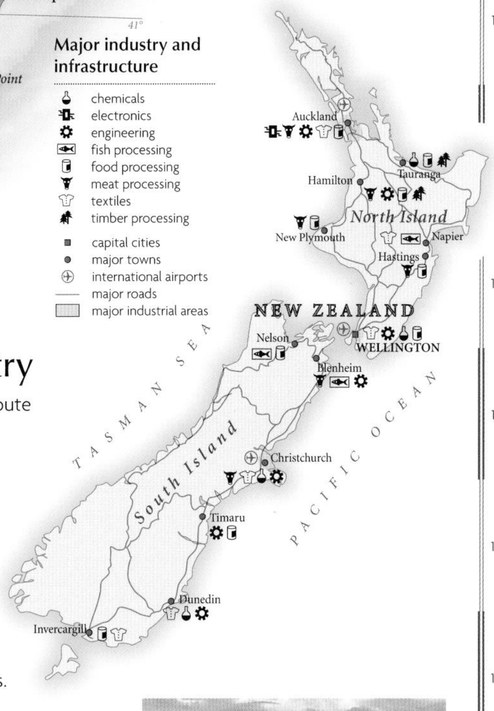

The transport network

36,091 miles (58,090 km)	105 miles (169 km)
2422 miles (3898 km)	1000 miles (1609 km)

The rugged terrain of much of New Zealand has led to most road and rail development being limited to the periphery of the islands.

▶ ***Auckland, on North Island,*** *is home to more than a third of New Zealand's population, and has the largest Polynesian population of any city in Australasia and Oceania. Auckland is also the main port and industrial centre in New Zealand.*

▲ ***The Arthur river*** *plummets 1902 ft (580 m) over the Sutherland Falls, in the south of South Island. The falls are the ninth highest in the world.*

192

▲ ***More than 46*** *million sheep thrive in New Zealand's mild climate, feeding on the islands' grassy slopes. Their fine meat and wool provide important export income.*

Using the land and sea

The climate and topography of North Island are more favourable to agriculture than the harsher terrain of South Island. Sheep and cattle can graze in summer and winter on the rich pastures surrounding both Auckland and Christchurch. A wide range of crops including vegetables, cereals and fruits such as grapes and kiwi fruit, are grown in the northern parts of New Zealand. The rich Pacific fisheries are of increasing economic importance.

The urban/rural population divide

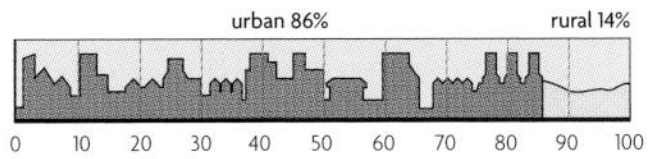

Population density	Total land area
38 people per sq mile (15 people per sq km)	103,730 sq miles (268,680 sq km)

Land use and agricultural distribution

- cattle
- sheep
- cereals
- fishing
- fruit
- timber
- capital cities
- major towns
- pasture
- cropland
- forest
- mountain region

Melanesia

FIJI, New Caledonia *(to France)*, PAPUA NEW GUINEA, SOLOMON ISLANDS,VANUATU

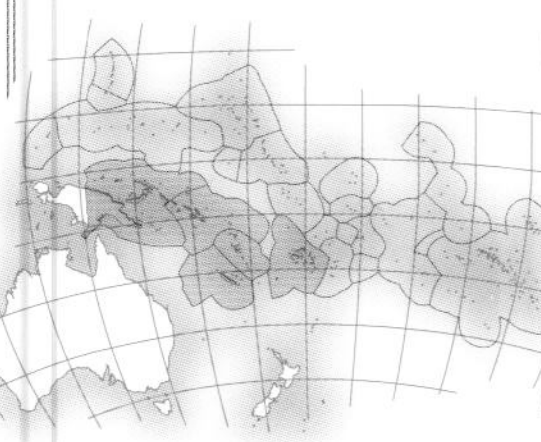

Lying in the southwest Pacific Ocean, northeast of Australia and south of the Equator, the islands of Melanesia form one of the three geographic divisions (along with Polynesia and Micronesia) of Oceania. Melanesia's name derives from the Greek melas, 'black', and nesoi, 'islands'. Most of the larger islands are volcanic in origin. The smaller islands tend to be coral atolls and are mainly uninhabited. Rugged mountains, covered by dense rainforest, take up most of the land area. Melanesian's cultivate yams, taro, and sweet potatoes for local consumption and live in small, usually dispersed, homesteads.

▲ ***Huli tribesmen from** Southern Highlands Province in Papua New Guinea parade in ceremonial dress, their powdered wigs decorated with exotic plumage and their faces and bodies painted with coloured pigments.*

Map key

Population

- ◎ 100,000 to 500,000
- ⊕ 50,000 to 100,000
- ○ 10,000 to 50,000
- ○ below 10,000

Elevation

- 4000m / 13,124ft
- 3000m / 9843ft
- 2000m / 6562ft
- 1000m / 3281ft
- 500m / 1640ft
- 250m / 820ft
- 100m / 328ft
- sea level

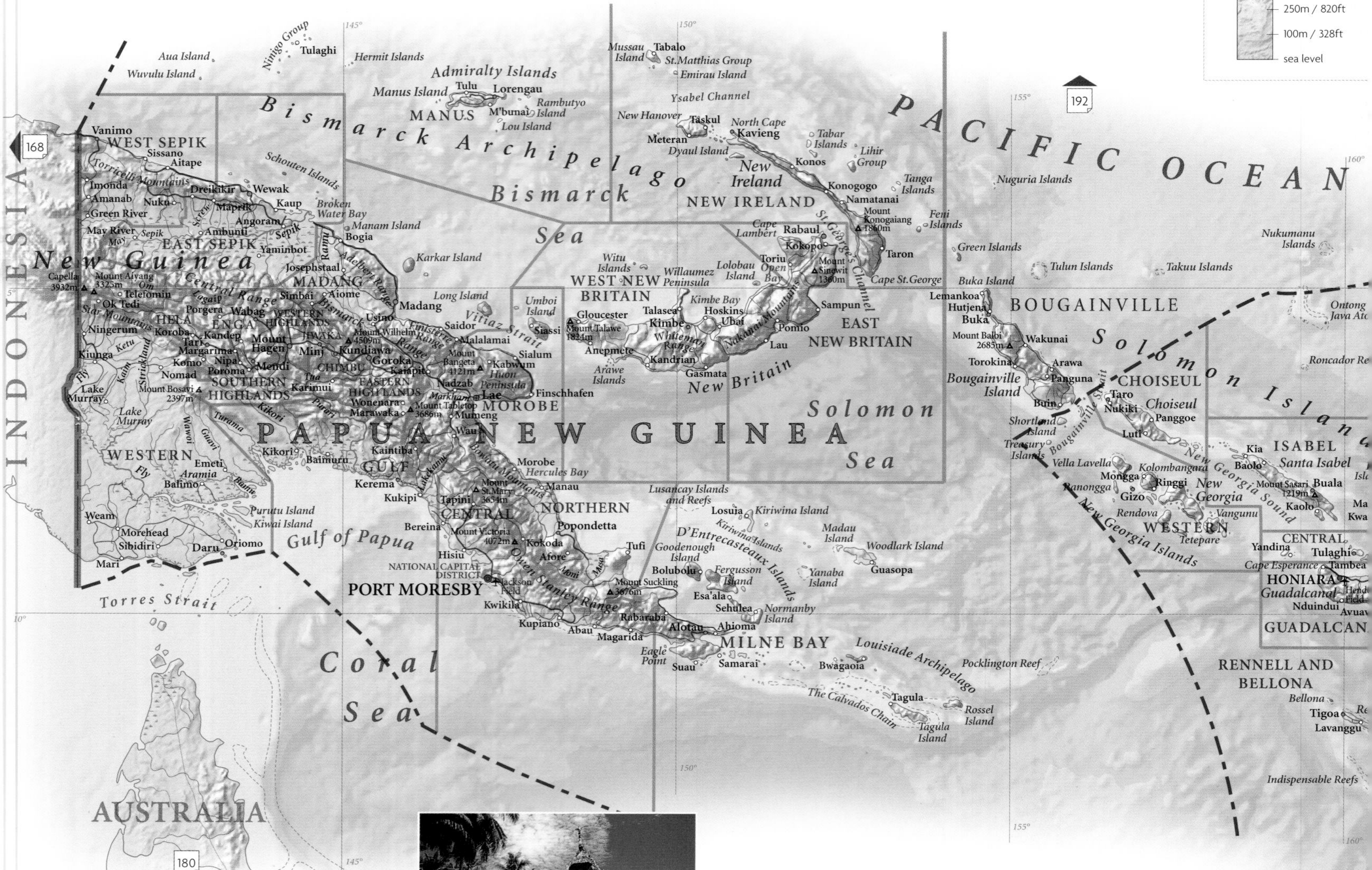

◀ ***Lying close to** the banks of the Sepik river in northern Papua New Guinea, this building is known as the Spirit House. It is constructed from leaves and twigs, ornately woven and trimmed into geometric patterns. The house is decorated with a mask and topped by a carved statue.*

▲ ***On one of** Vanuatu's many islands, beach houses stand at the water's edge, surrounded by coconut palms and other tropical vegetation. The unspoilt beaches and tranquillity of its islands are drawing ever-larger numbers of tourists to Vanuatu.*

Transport and Industry

The processing of natural resources generates significant export revenue for the countries of Melanesia. The region relies mainly on copra, tuna and timber exports, with some production of cocoa and palm oil. The islands have substantial mineral resources including the world's largest copper reserves on Bougainville Island; gold, and potential oil and natural gas. Tourism has become the fastest growing sector in most of the countries' economies.

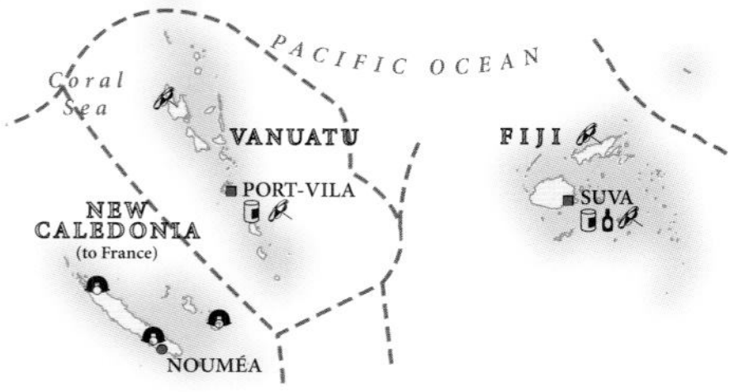

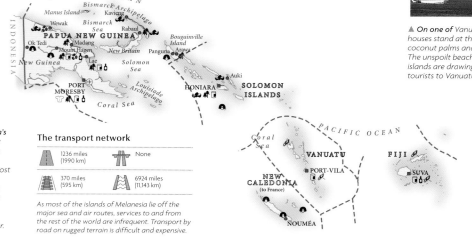

Major industry and infrastructure

- beverages
- coffee processing
- copra processing
- food processing
- mining
- textiles
- timber processing
- tourism
- capital cities
- major towns
- international airports
- major roads

◀ ***On New Caledonia's** main island, relatively high interior plateaux descend to coastal plains. Nickel is the most important mineral resource, but the hills also harbour metallic deposits including chrome, cobalt, iron, gold, silver and copper.*

The transport network

1236 miles (1990 km)	None
370 miles (595 km)	6924 miles (11,143 km)

As most of the islands of Melanesia lie off the major sea and air routes, services to and from the rest of the world are infrequent. Transport by road on rugged terrain is difficult and expensive.

N O P Q R S T U V W X Y

The Landscape

Melanesia comprises high, volcanic islands, low coral islands and continental islands. New Guinea is part of the Australian continental platform, and is separated from it only by the shallow flooding of the Torres Strait. The plate margin of the Pacific and Indo-Australian plates cuts through mainland Papua New Guinea. Volcanic activity, resulting from the collision of these plates, has sculpted much of Melanesia's landscape.

◄ *The slopes of this extinct volcano near Talasea on the island of New Britain have been almost entirely colonized by rainforest vegetation.*

▲ ***A series of** coral reefs can be seen in the clear waters off Cape Esperance on the island of Guadalcanal in the Solomons.*

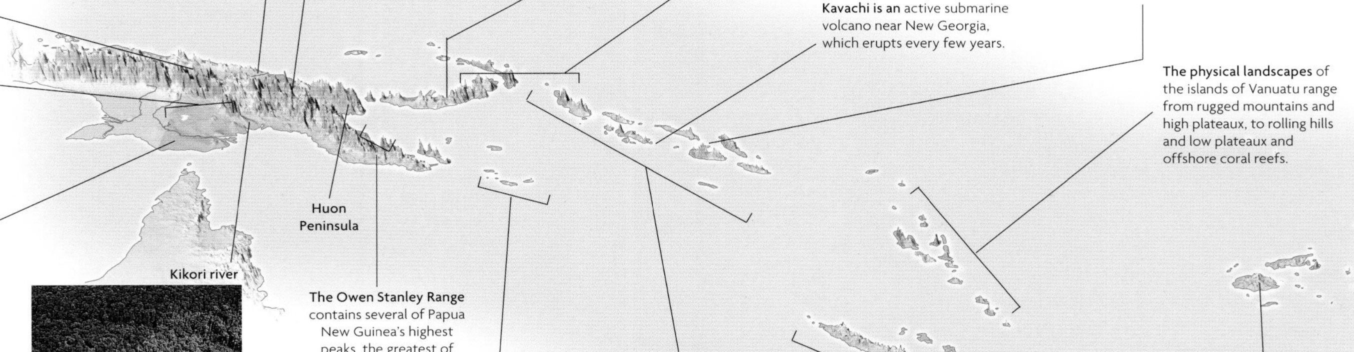

The Sepik river drains the lowlands north of the Central Range, flowing eastward into the Bismarck Sea.
The Bismarck Range is precipitous, rugged and covered in dense vegetation, rising to 14,793 ft (4509-m) at Mount Wilhelm in central Papua New Guinea.
Most of Papua New Guinea's outlying islands, including New Britain, Bougainville Island and New Ireland, are precipitous and of volcanic origin.
Kavachi is an active submarine volcano near New Georgia, which erupts every few years.
The Star Mountains include some of the most remote terrain on Earth. The area is rich in gold and copper.
The lowland plains in the south and north of Papua New Guinea's main island are swampy, and contain some fertile alluvial soils. This contrasts with the mountainous islands in the rest of the country where soils are generally thin and nutrients are retained in the existing vegetation.
The physical landscapes of the islands of Vanuatu range from rugged mountains and high plateaux, to rolling hills and low plateaux and offshore coral reefs.
Southern Papua New Guinea is part of the Indo-Australian Plate. New Guinea only became separated physically from Australia about 8000-years ago following the flooding of the Torres Strait.
Huon Peninsula
Kikori river
The Owen Stanley Range contains several of Papua New Guinea's highest peaks, the greatest of which is Mount Victoria at 13,200 ft (4035 m).
The Solomon Islands are mountainous continental-type islands with largely andesitic volcanoes.
New Caledonia's main island is surrounded by coral reef that extends from the Huon island group in the north, to Île des Pins in the south.
Viti Levu, the largest of Fiji's islands, contains the country's highest mountain, Mount Victoria at 4339 ft (1323 m).
The Louisiade Archipelago contains 10 volcanic islands and numerous coral islets. Tagula Island is the largest of the islands, containing the archipelago's highest peak at 2645 ft (806 m).

► ***Papua New Guinea's** rivers, though fairly short, carry extremely high sediment loads, largely due to soil erosion. This is caused by a combination of very steep slopes and heavy rainfall, and is made worse by forest clearance, particularly 'slash and burn' techniques and road or mine operations.*

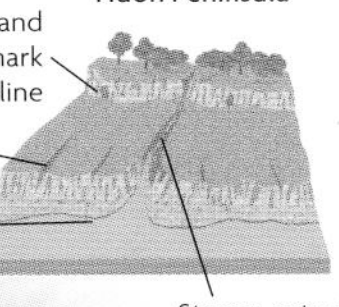

Huon Peninsula
Caves and undercut cliffs mark former shoreline
Former level of beach
Current beach
Stream cuts down through recently exposed land

Uplift of the land in tectonically active regions can lead to former coastlines being lifted beyond the reach of the sea. New cliffs and caves are formed at a lower level, and rivers cut down through the lower land to reach sea level once more.

Using the land and sea

Almost 60% of the population of Melanesia is engaged in agriculture and animal husbandry at a subsistence level. Coconuts and cocoa are grown for export revenue. Over 80% of the land area is cloaked by tropical forest and woodlands, which have proved to be a rich timber source. In coastal areas, fishing, mainly for tuna, is a staple industry.

PACIFIC OCEAN
Manus Island
Bismarck Archipelago
Bismarck Sea
Wewak
Rabaul
PAPUA NEW GUINEA
Madang
INDONESIA
New Guinea
New Britain
Lae
Solomon Sea
Bougainville Island
Arawa
PORT MORESBY
Louisiade Archipelago
Coral Sea
HONIARA
SOLOMON ISLANDS
PACIFIC OCEAN
Coral Sea
VANUATU
PORT-VILA
FIJI
SUVA
NEW CALEDONIA (to France)
NOUMÉA
Land use and agricultural distribution
bananas
cocoa
coconuts
fishing
oil palms
rubber
timber
capital cities
major towns
cropland
forest
wetland

The urban/rural population divide

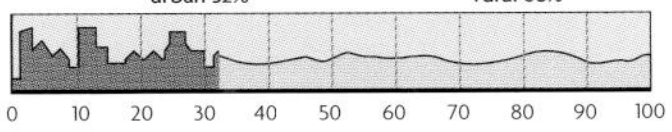

urban 32%
rural 68%
0 10 20 30 40 50 60 70 80 90 100

Population density	Total land area
32 people per sq mile (12 people per sq km)	205,354 sq miles (332,008 sq km)

◄ ***Abaca Eco-tourist Park** near Lautoka on the island of Viti Levu in western Fiji is one of a number of projects aimed at combining tourism with awareness about the environment. The government and people of Fiji are keen to protect the unique ecology of the islands and prevent further damage to the coral reefs. Until the recent ending of nuclear testing in the Pacific by Western nations, Fiji lay downwind of some of the main testing sites.*

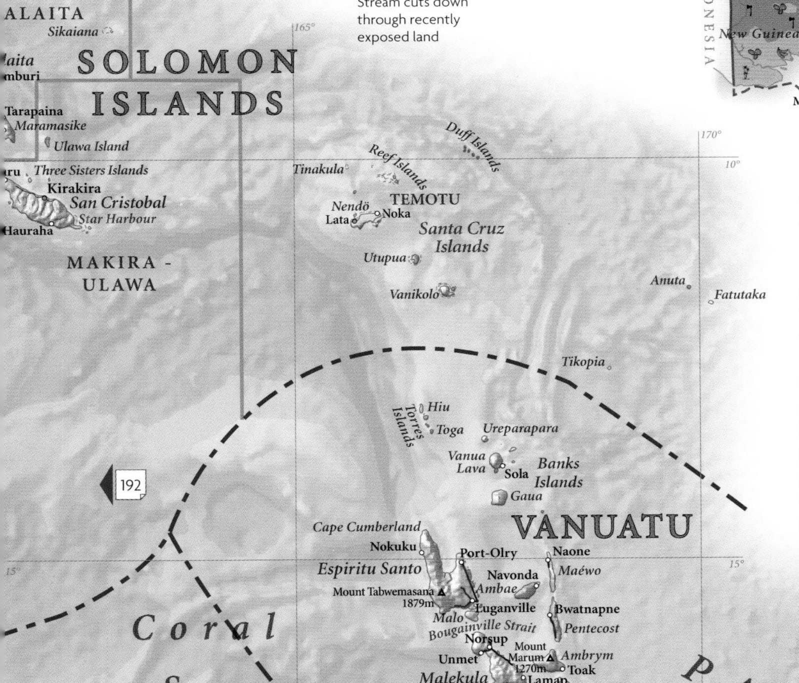

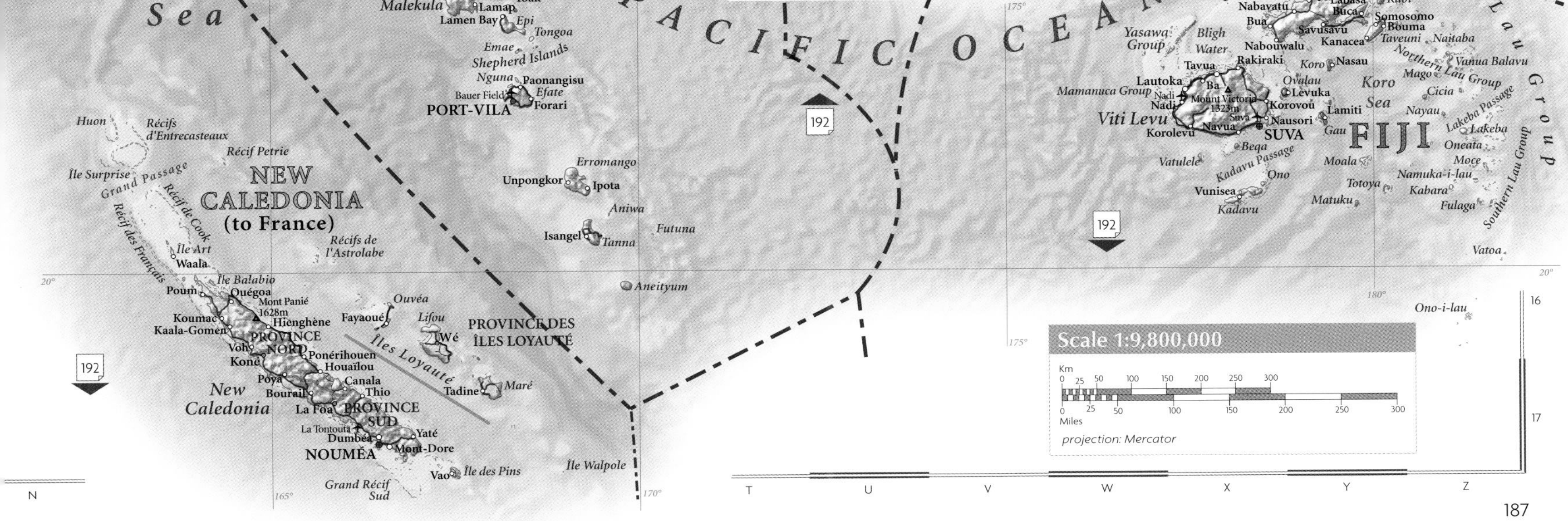

MALAITA
Sikaiana
SOLOMON ISLANDS
Tarapaina
Maramasike
Ulawa Island
Three Sisters Islands
Kirakira
San Cristobal
Star Harbour
Hauraha
MAKIRA - ULAWA
Duff Islands
Reef Islands
Tinakula
TEMOTU
Nendö
Noka
Lata
Santa Cruz Islands
Utupua
Vanikolo
Anuta
Fatutaka
Tikopia
Hiu
Torres Islands
Toga
Ureparapara
Vanua Lava
Sola
Banks Islands
Gaua
VANUATU
Cape Cumberland
Nokuku
Port-Olry
Espiritu Santo
Naone
Maéwo
Navonda
Ambae
Mount Tabwemasana 1879m
Luganville
Bwatnapne
Malo
Bougainville Strait
Pentecost
Norsup
Mount Marum 1270m
Ambrym
Unmet
Toak
Malekula
Lamap
Lamen Bay
Epi
Tongoa
Emae
Shepherd Islands
Nguna
Paonangisu
Bauer Field
Efate
PORT-VILA
Forari
Erromango
Unpongkor
Ipota
Aniwa
Isangel
Tanna
Futuna
Aneityum
Coral Sea
PACIFIC OCEAN
Huon
Récifs d'Entrecasteaux
Récif Petrie
Île Surprise
Grand Passage
Récif de Cook
Récif des Français
NEW CALEDONIA (to France)
Récifs de l'Astrolabe
Île Art
Waala
Île Balabio
Poum
Ouégoa
Mont Panié 1628m
Ouvéa
Koumac
Hienghène
Fayaoué
Lifou
Kaala-Gomen
PROVINCE NORD
PROVINCE DES ÎLES LOYAUTÉ
Voh
Wé
Koné
Ponérihouen
Houaïlou
Îles Loyauté
Poya
Canala
Bourail
Thio
Tadine
Maré
New Caledonia
La Foa
PROVINCE SUD
La Tontouta
Yaté
Dumbéa
Mont-Dore
NOUMÉA
Vao
Île des Pins
Île Walpole
Grand Récif Sud
Great Sea Reef
Vanua Levu
Cikobia
Qelelevu Lagoon
Navoalevu
Nabuna
Naduri
Labasa
Nabavatu
Rabi
Bua
Buca
Somosomo
Savusavu
Bouma
Yasawa Group
Bligh Water
Nabouwalu
Kanacea
Taveuni
Naitaba
Tavua
Rakiraki
Koro
Nasau
Northern Lau Group
Vanua Balavu
Lautoka
Ba
Ovalau
Mago
Mamanuca Group
Nadi
Mount Victoria 1323m
Levuka
Koro Sea
Cicia
Viti Levu
Navua
Korovou
Nayau
Lakeba Passage
Korolevu
Nausori
Lamiti
Lakeba
Vatulele
SUVA
Gau
FIJI
Oneata
Beqa
Kadavu Passage
Moala
Moce
Ono
Namuka-i-lau
Lau Group
Vunisea
Totoya
Kabara
Kadavu
Matuku
Fulaga
Southern Lau Group
Vatoa
Ono-i-lau
Scale 1:9,800,000
Km 0 25 50 100 150 200 250 300
0 25 50 100 150 200 250 300 Miles
projection: Mercator

192

T U V W X Y Z

Micronesia

MARSHALL ISLANDS, MICRONESIA, NAURU, PALAU, Guam, Northern Mariana Islands, Wake Island

The Micronesian islands lie in the western reaches of the Pacific Ocean and are all part of the same volcanic zone. The Federated States of Micronesia is the largest group, with more than 600 atolls and forested volcanic islands in an area of more than 1120 sq miles (2900 sq km). Micronesia is a mixture of former colonies, overseas territories and dependencies. Most of the region still relies on aid and subsidies to sustain economies limited by resources, isolation, and an emigrating population, drawn to New Zealand and Australia by the attractions of a western lifestyle.

Palau

Palau is an archipelago of over 200 islands, only eight of which are inhabited. It was the last remaining UN trust territory in the Pacific, controlled by the USA until 1994, when it became independent. The economy operates on a subsistence level, with coconuts and cassava the principal crops. Fishing licences and tourism provide foreign currency.

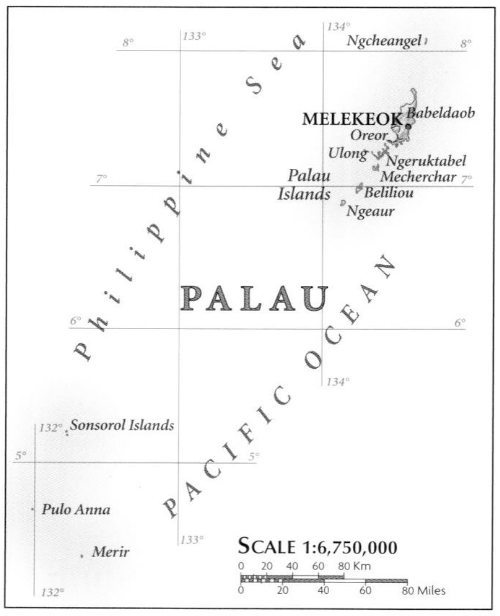

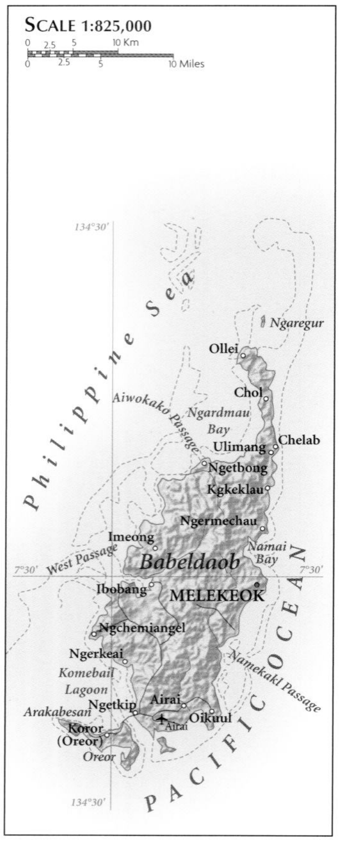

Northern Mariana Islands *(to US)*

A US Commonwealth territory, the Northern Marianas comprise the whole of the Mariana archipelago except for Guam. The islands retain their close links with the United States and continue to receive US aid. Tourism, though bringing in much-needed revenue, has speeded the decline of the traditional subsistence economy. Most of the population lives on Saipan.

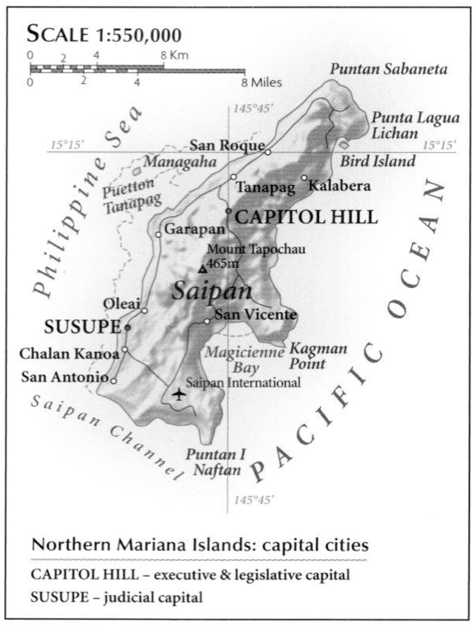

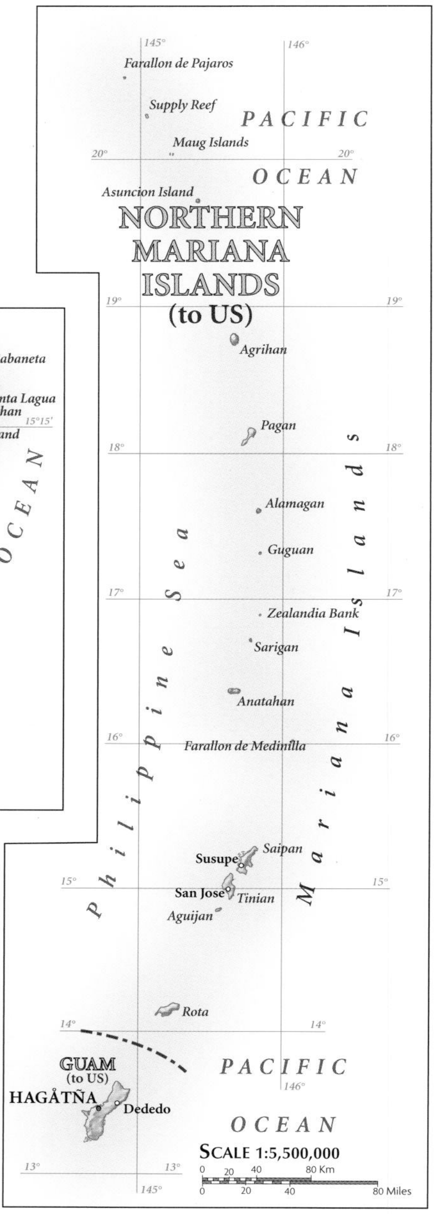

▲ *The Palau Islands* have numerous hidden lakes and lagoons. These sustain their own ecosystems which have developed in isolation. This has produced adaptations in the animals and plants which are often unique to each lake.

Micronesia

A mixture of high volcanic islands and low-lying coral atolls, the Federated States of Micronesia include all the Caroline Islands except Palau. Pohnpei, Kosrae, Chuuk and Yap are the four main island cluster states, each of which has its own language, with English remaining the official language. Nearly half the population is concentrated on Pohnpei, the largest island. Independent since 1986, the islands continue to receive considerable aid from the USA which supplements an economy based primarily on fishing and copra processing.

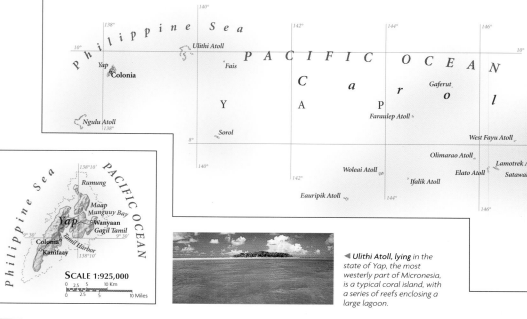

Guam *(to US)*

Lying at the southern end of the Mariana Islands, Guam is an important US military base and tourist destination. Social and political life is dominated by the indigenous Chamorro, who make up just under half the population, although the increasing prevalence of western culture threatens Guam's traditional social stability.

◀ *The tranquillity of* these coastal lagoons, at Inarajan in southern Guam, belies the fact that the island lies in a region where typhoons are common.

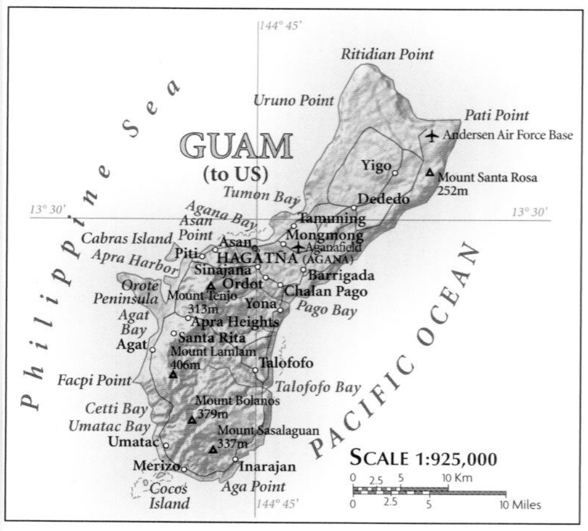

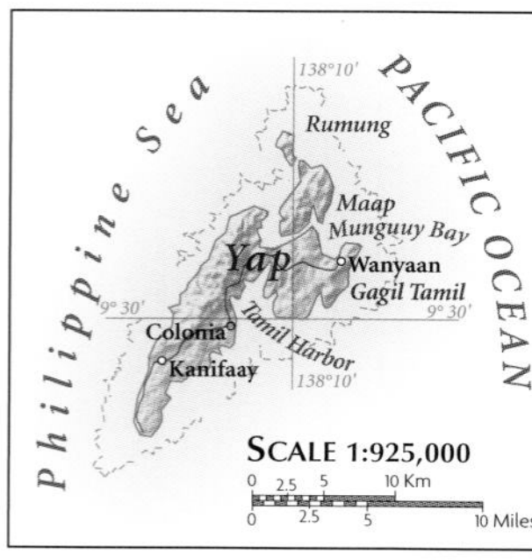

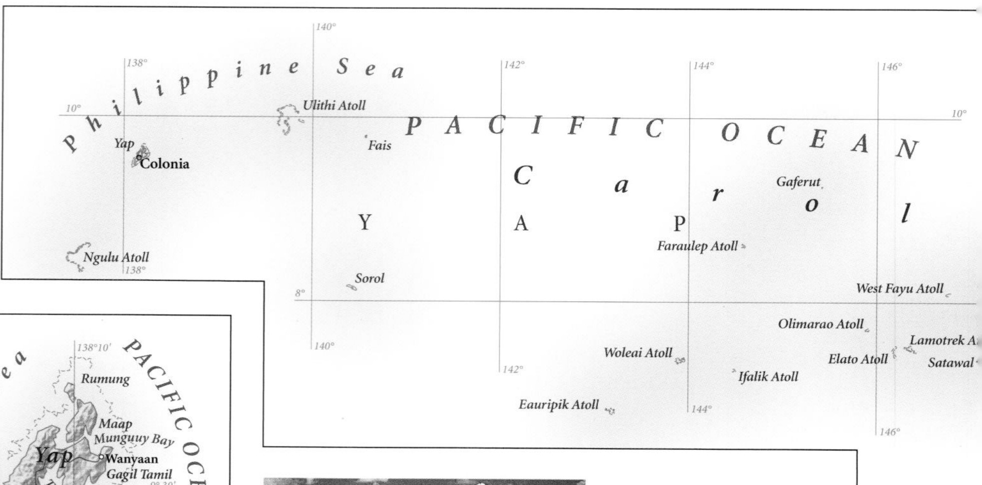

◀ *Ulithi Atoll, lying* in the state of Yap, the most westerly part of Micronesia, is a typical coral island, with a series of reefs enclosing a large lagoon.

Marshall Islands

A group of 34 widely-scattered atolls in the central Pacific Ocean, the Marshall Islands include some of the largest atolls in the world, formed from low coral islands with sandy beaches and enclosing vast lagoons. Formerly under US protection as part of the UN Trust Territory of the Pacific Islands, and including the former US nuclear testing sites of Bikini atoll and Enewetak Atoll, the Marshall Islands became self-governing in 1979. The economy is reliant on US aid and on the rent paid by the USA for its missile base on Kwajalein atoll.

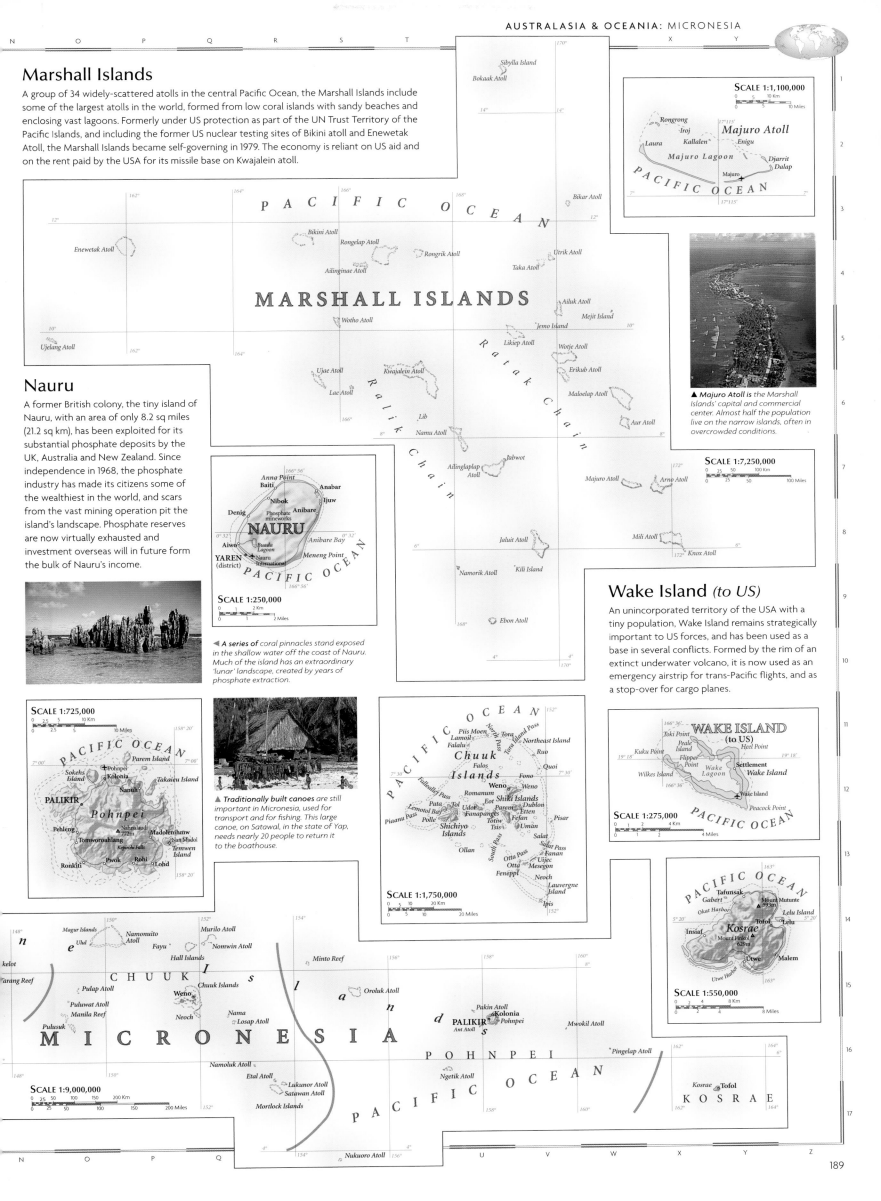

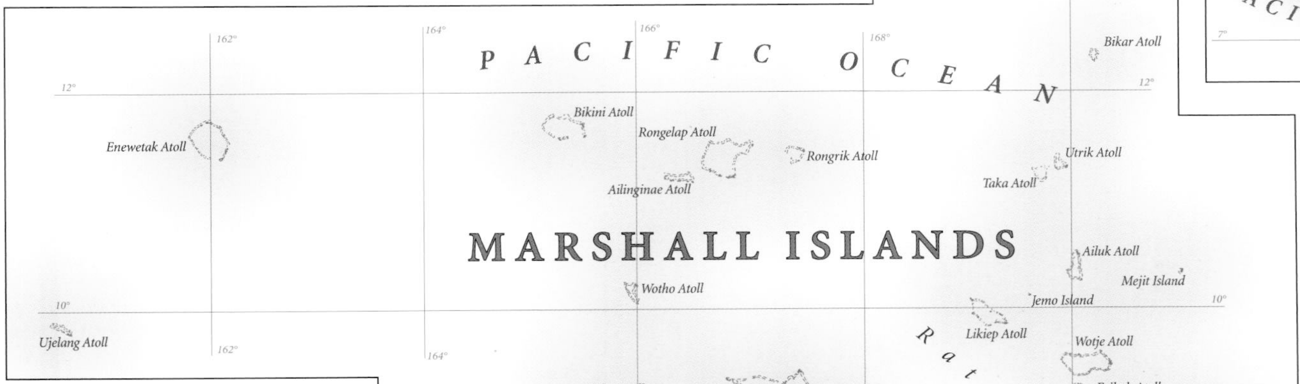

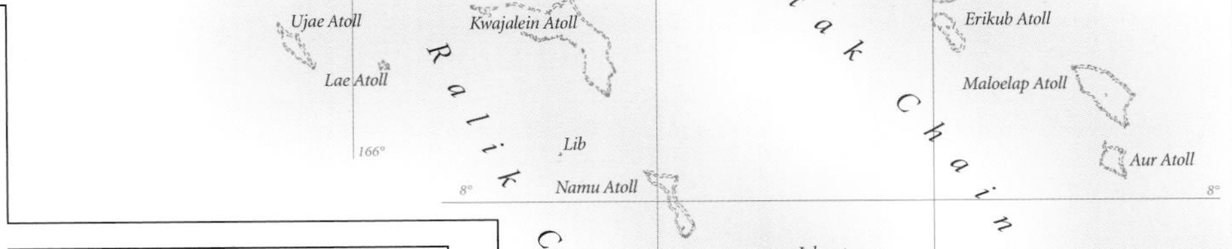

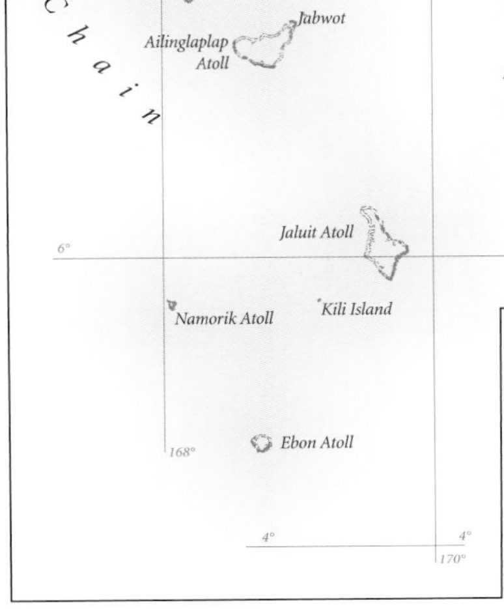

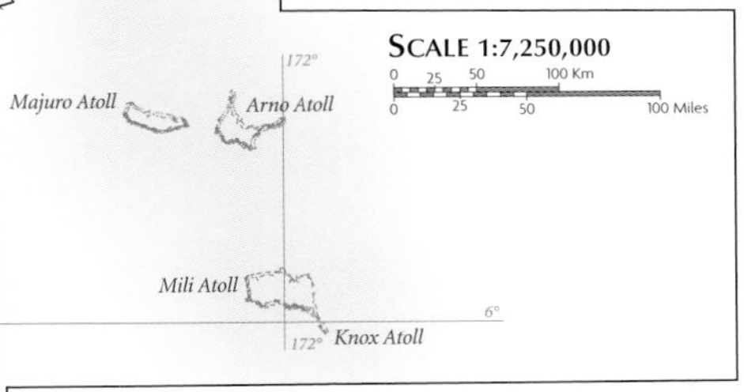

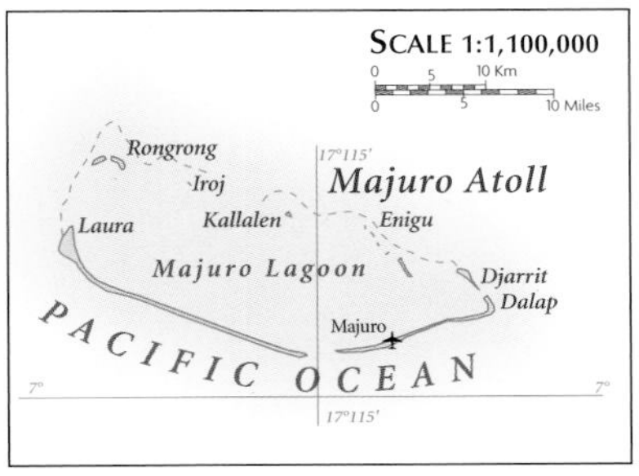

▲ **Majuro Atoll is** the Marshall Islands' capital and commercial center. Almost half the population live on the narrow islands, often in overcrowded conditions.

Nauru

A former British colony, the tiny island of Nauru, with an area of only 8.2 sq miles (21.2 sq km), has been exploited for its substantial phosphate deposits by the UK, Australia and New Zealand. Since independence in 1968, the phosphate industry has made its citizens some of the wealthiest in the world, and scars from the vast mining operation pit the island's landscape. Phosphate reserves are now virtually exhausted and investment overseas will in future form the bulk of Nauru's income.

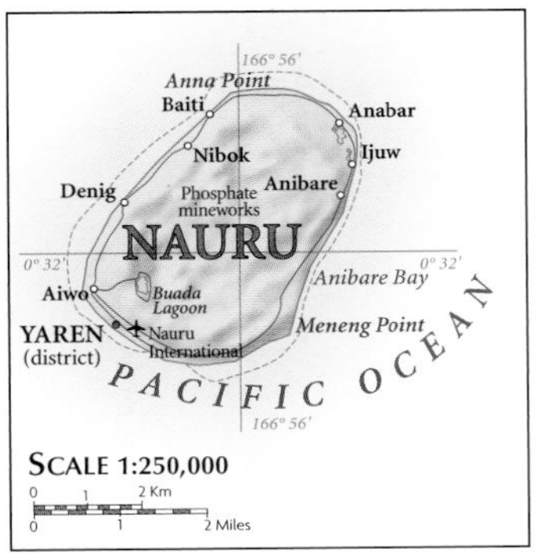

◀ **A series of** coral pinnacles stand exposed in the shallow water off the coast of Nauru. Much of the island has an extraordinary 'lunar' landscape, created by years of phosphate extraction.

Wake Island (to US)

An unincorporated territory of the USA with a tiny population, Wake Island remains strategically important to US forces, and has been used as a base in several conflicts. Formed by the rim of an extinct underwater volcano, it is now used as an emergency airstrip for trans-Pacific flights, and as a stop-over for cargo planes.

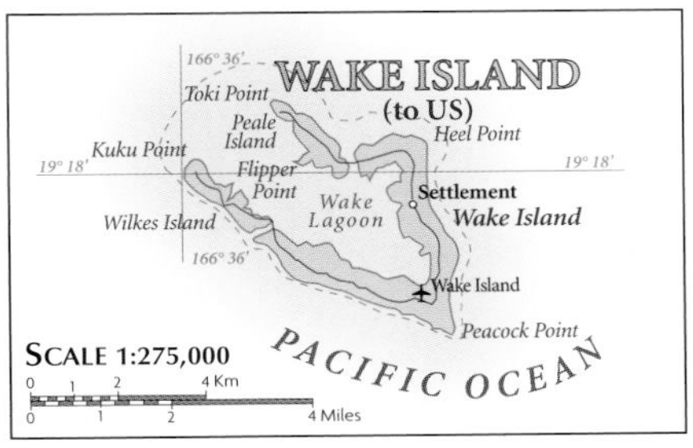

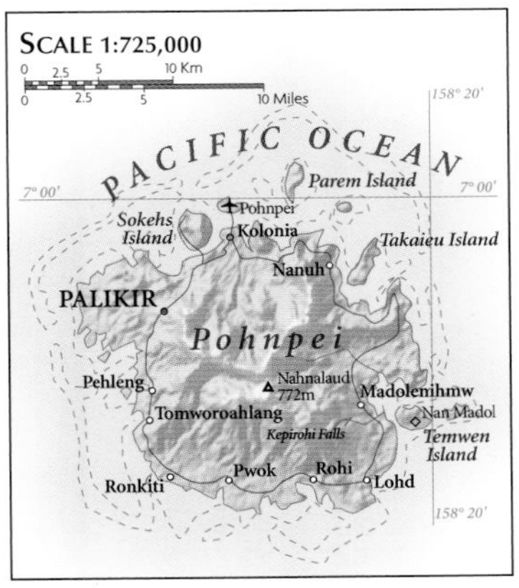

▲ **Traditionally built canoes** are still important in Micronesia, used for transport and for fishing. This large canoe, on Satawal, in the state of Yap, needs nearly 20 people to return it to the boathouse.

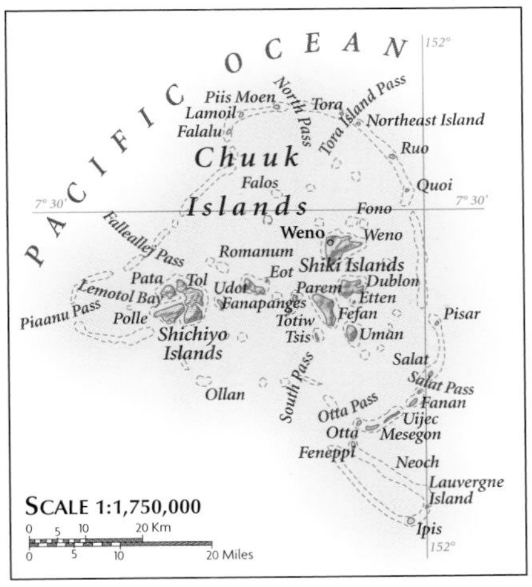

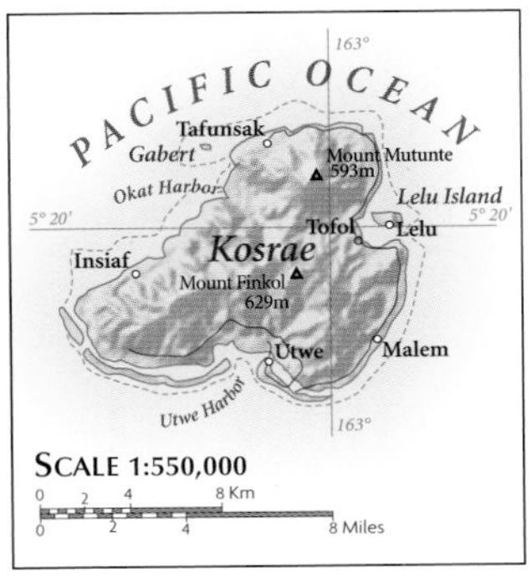

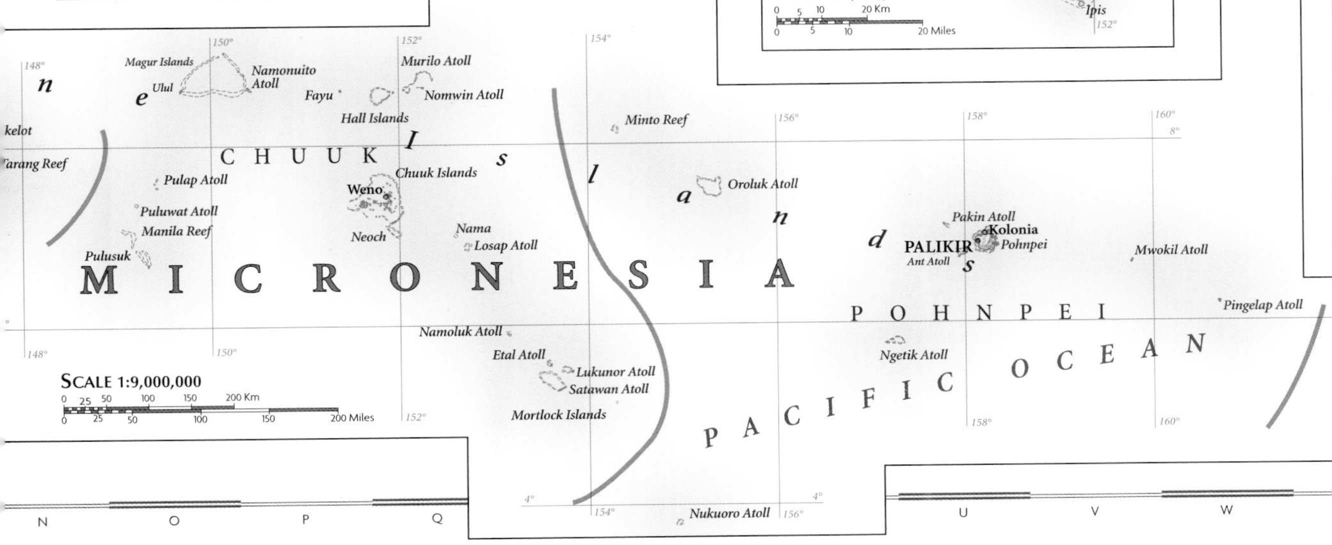

Polynesia

KIRIBATI, TUVALU, Cook Islands, Easter Island, French Polynesia, Niue, Pitcairn Islands, Tokelau, Wallis & Futuna

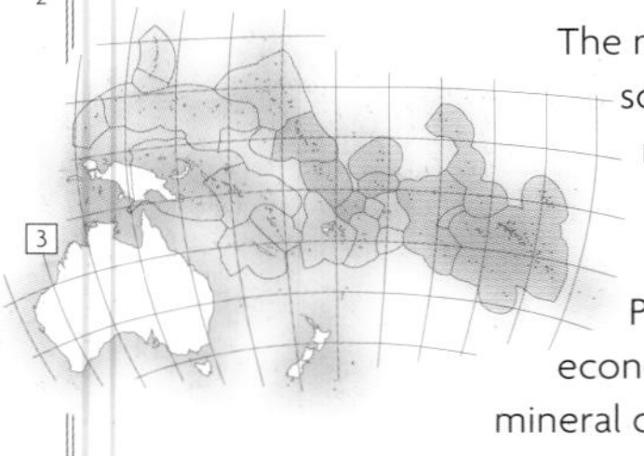

The numerous island groups of Polynesia lie to the east of Australia, scattered over a vast area in the south Pacific. The islands are a mixture of low-lying coral atolls, some of which enclose lagoons, and the tips of great underwater volcanoes. The populations on the islands are small, and most people are of Polynesian origin, as are the Maori of New Zealand. Local economies remain simple, relying mainly on subsistence crops, mineral deposits – many now exhausted – fishing and tourism.

Kiribati

A former British colony, Kiribati became independent in 1979. Banaba's phosphate deposits ran out in 1980, following decades of exploitation by the British. Economic development remains slow and most agriculture is at a subsistence level, though coconuts provide export income, and underwater agriculture is being developed.

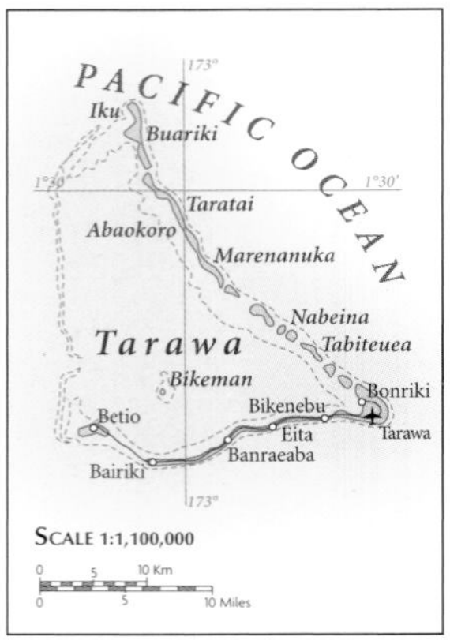

► *With the exception of Banaba all the islands in Kiribati's three groups are low-lying, coral atolls. This aerial view shows the sparsely vegetated islands, intercut by many small lagoons.*

Tuvalu

A chain of nine coral atolls, 360 miles (579 km) long with a land area of just over 9 sq miles (23 sq km), Tuvalu is one of the world's smallest and most isolated states. As the Ellice Islands, Tuvalu was linked to the Gilbert Islands (now part of Kiribati) as a British colony until independence in 1978. Politically and socially conservative, Tuvaluans live by fishing and subsistence farming.

▲ ***Funafuti Atoll contains** more than 40% of Tuvalu's people, giving it an extremely high population density.*

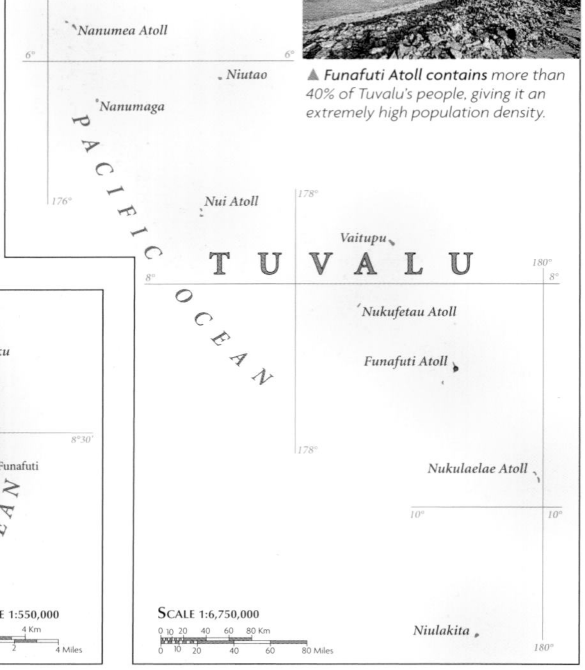

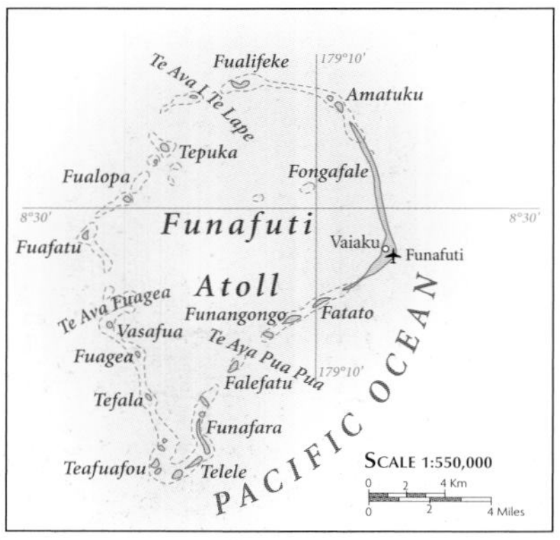

Tokelau *(to New Zealand)*

A low-lying coral atoll, Tokelau is a dependent territory of New Zealand with few natural resources. Although a 1990 cyclone destroyed crops and infrastructure, a tuna cannery and the sale of fishing licences have raised revenue and a catamaran link between the islands has increased their tourism potential. Tokelau's small size and economic weakness makes independence from New Zealand unlikely.

▲ ***Fishermen cast their** nets to catch small fish in the shallow waters off Atafu Atoll, the most westerly island in Tokelau.*

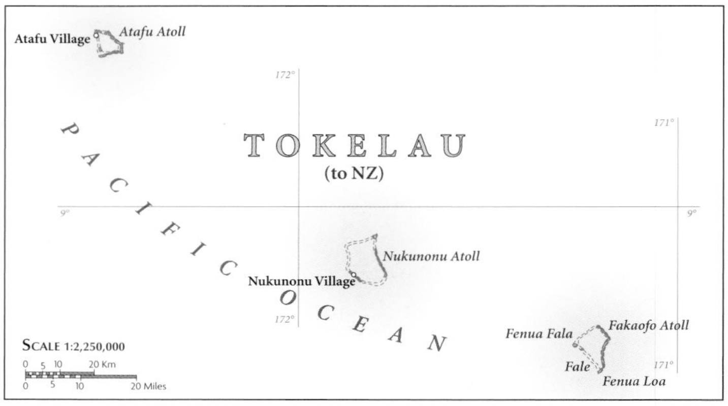

Wallis & Futuna *(to France)*

In contrast to other French overseas territories in the south Pacific, the inhabitants of Wallis and Futuna have shown little desire for greater autonomy. A subsistence economy produces a variety of tropical crops, while foreign currency remittances come from expatriates and from the sale of licences to Japanese and Korean fishing fleets.

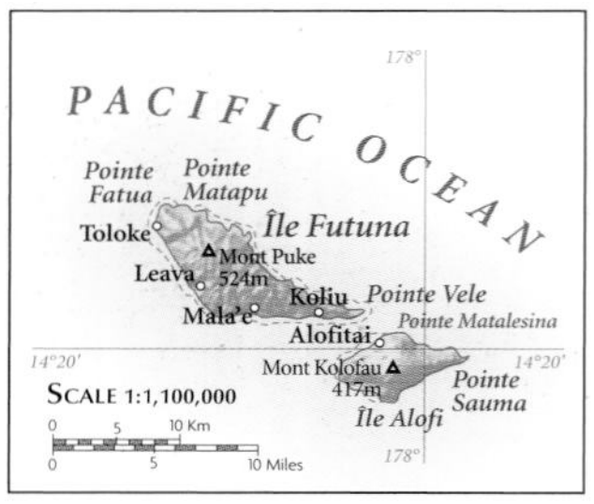

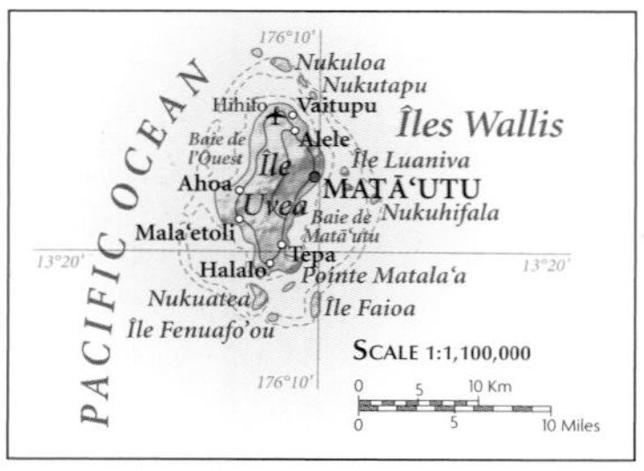

Cook Islands *(to New Zealand)*

A mixture of coral atolls and volcanic peaks, the Cook Islands achieved self-government in 1965 but exist in free association with New Zealand. A diverse economy includes pearl and giant clam farming, and an ostrich farm, plus tourism and banking. A 1991 friendship treaty with France provides for French surveillance of territorial waters.

▲ ***Palm trees fringe** the white sands of a beach on Aitutaki in the Southern Cook Islands, where tourism is of increasing economic importance.*

Niue *(to New Zealand)*

Niue, the world's largest coral island, is self-governing but exists in free association with New Zealand. Tropical fruits are grown for local consumption; tourism and the sale of postage stamps provide foreign currency. The lack of local job prospects has led more than 10,000 Niueans to emigrate to New Zealand, which has now invested heavily in Niue's economy in the hope of reversing this trend.

▲ ***Waves have cut** back the original coastline, exposing a sandy beach, near Mutalau in the northeast corner of Niue.*

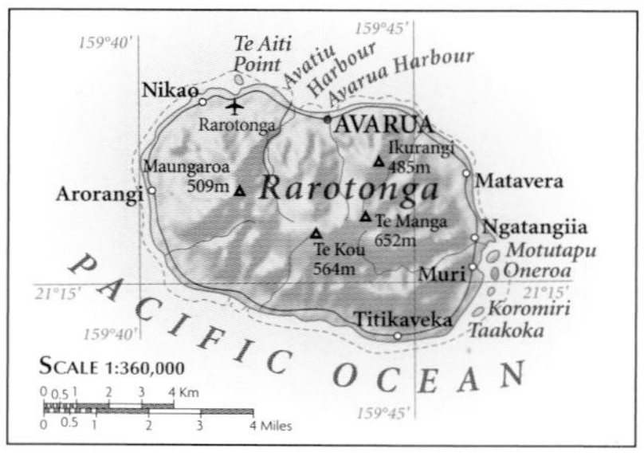

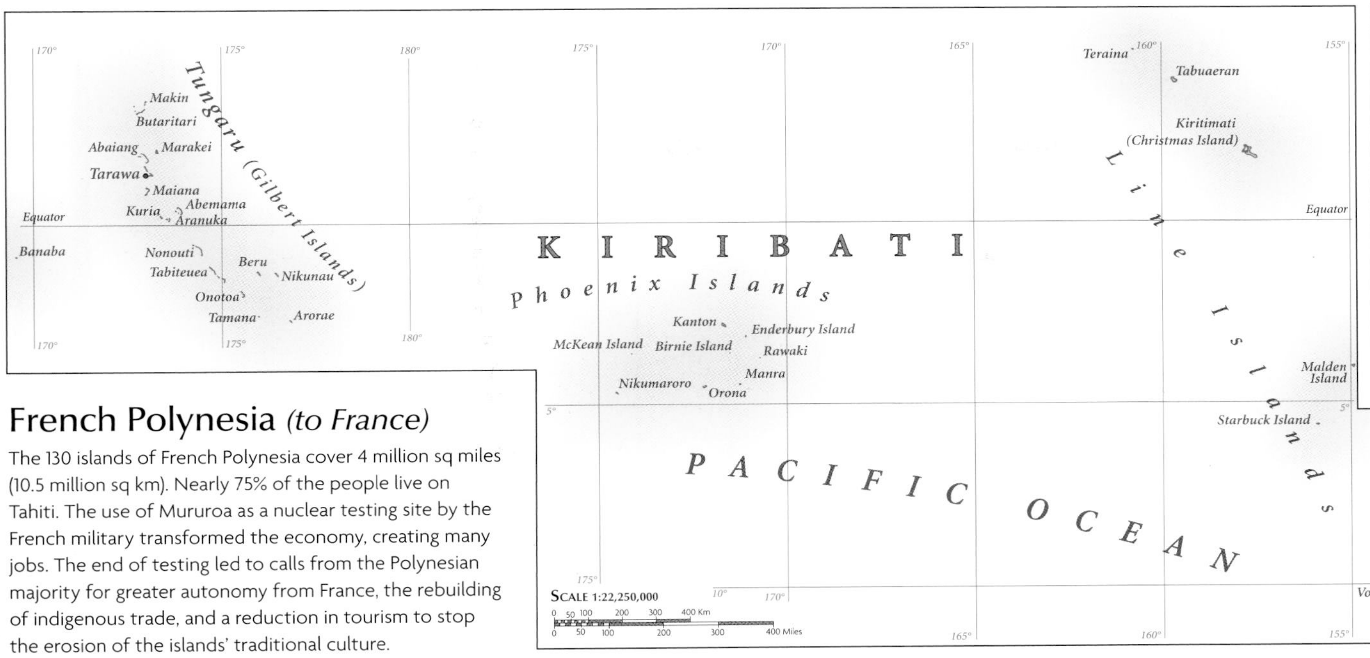

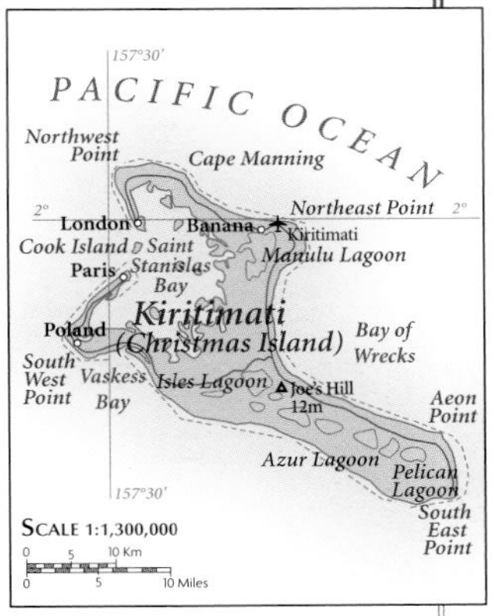

French Polynesia *(to France)*

The 130 islands of French Polynesia cover 4 million sq miles (10.5 million sq km). Nearly 75% of the people live on Tahiti. The use of Mururoa as a nuclear testing site by the French military transformed the economy, creating many jobs. The end of testing led to calls from the Polynesian majority for greater autonomy from France, the rebuilding of indigenous trade, and a reduction in tourism to stop the erosion of the islands' traditional culture.

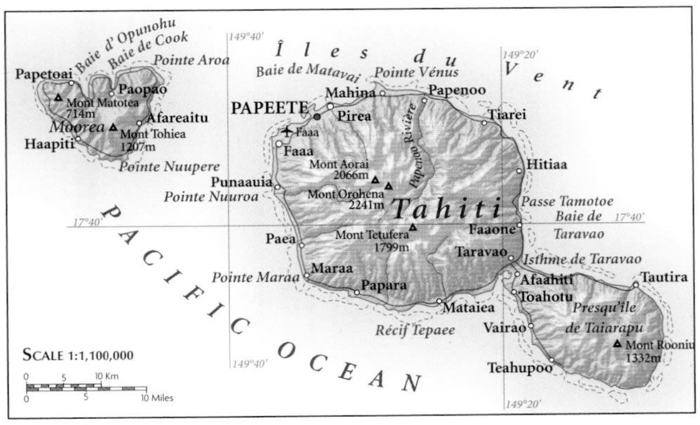

◀ **The traditional Tahitian** *welcome for visitors, who are greeted by parties of canoes, has become a major tourist attraction.*

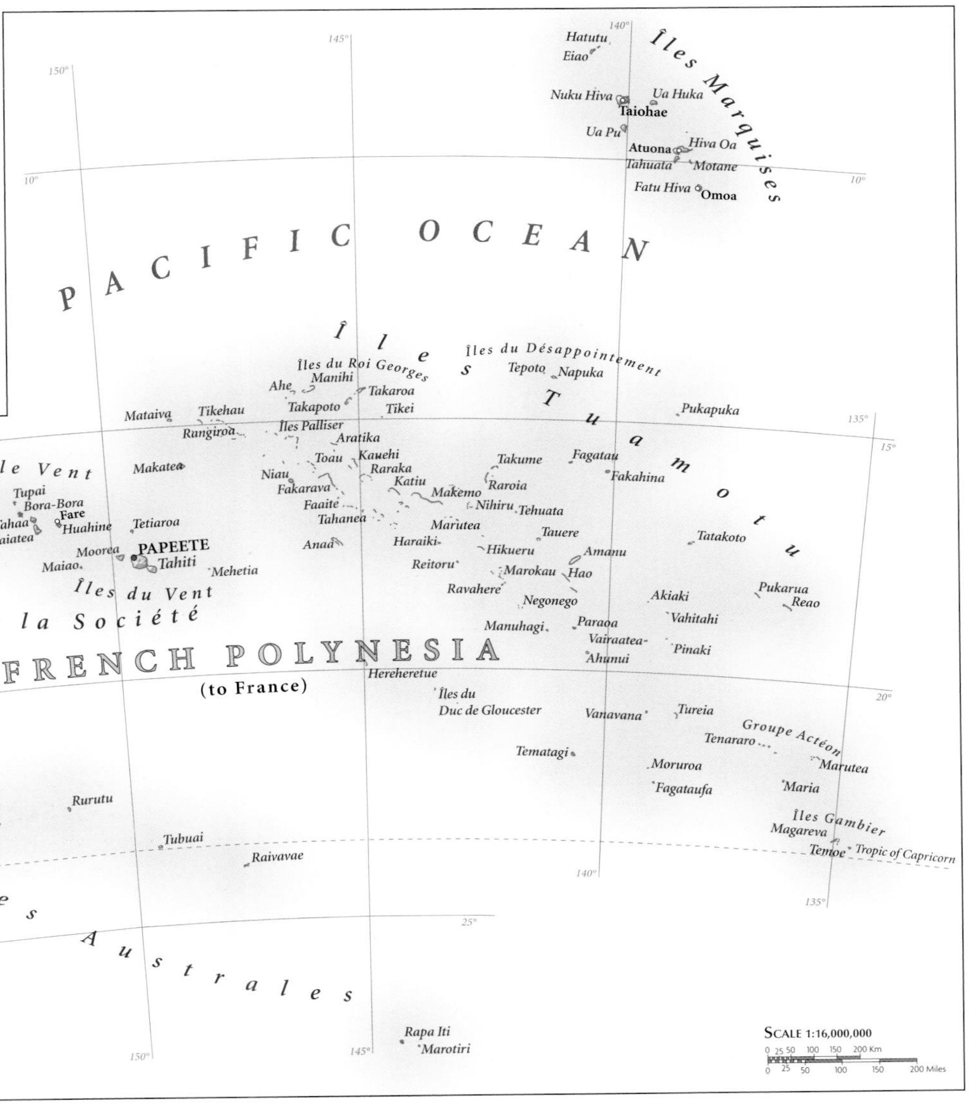

Pitcairn Group of Islands
(to UK)

Britain's most isolated dependency, Pitcairn Island was first populated by mutineers from the HMS *Bounty* in 1790. Emigration is further depleting the already limited gene pool of the island's inhabitants, with associated social and health problems. Barter, fishing and subsistence farming form the basis of the economy whilst offshore mineral exploitation may boost the economy in future.

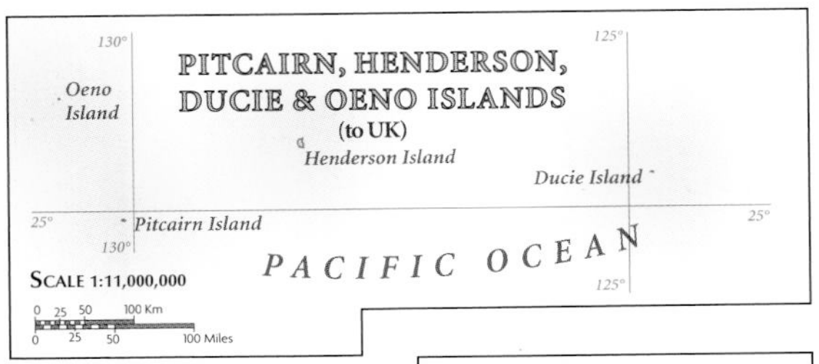

◀ **The Pitcairn Islanders** *rely on regular airdrops from New Zealand and periodic visits by supply vessels to provide them with basic commodities.*

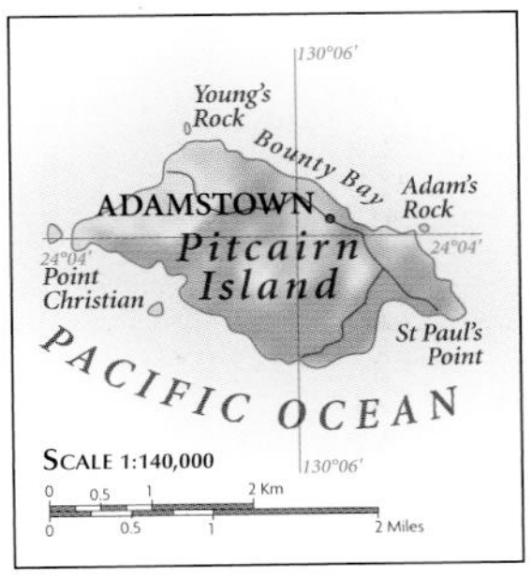

Easter Island *(to Chile)*

One of the most easterly islands in Polynesia, Easter Island *(Isla de Pascua)* – also known as Rapa Nui, is part of Chile. The mainly Polynesian inhabitants support themselves by farming, which is mainly of a subsistence nature, and includes cattle rearing and crops such as sugar cane, bananas, corn, gourds and potatoes. In recent years, tourism has become the most important source of income and the island sustains a small commercial airport.

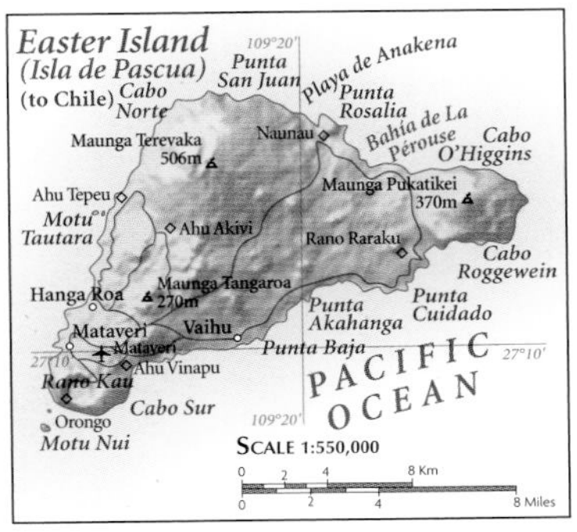

▲ **The Naunau,** *a series of huge stone statues overlook Playa de Anakena, on Easter Island. Carved from a soft volcanic rock, they were erected between 400 and 900 years ago.*

The Pacific Ocean

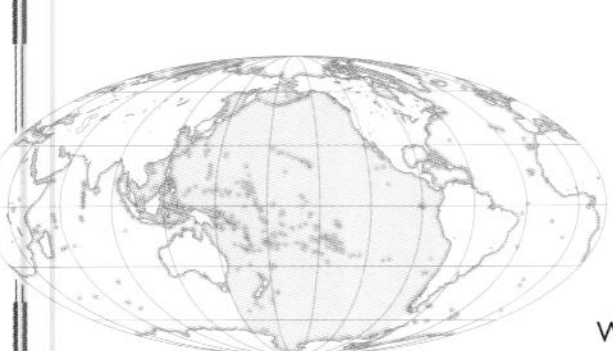

The Pacific is the world's largest and deepest ocean. It is nearly twice the area of the Atlantic and contains almost three times as much water. The ocean is dotted with islands and surrounded by some of the world's most populous states; over half the world's population lives on its shores. The Pacific is bordered by active plate margins known as the 'Ring of Fire', causing earthquakes and tsunamis, and creating volcanic islands and subterranean mountain chains. The largest underwater mountains break the surface as island arcs. The fisheries of the Pacific are some of the most productive in the world and provide a vital resource for many of the Pacific islands. Since the Second World War there has been a shift in trading patterns, with a considerable growth in trade between the United States and the countries of the Pacific Rim.

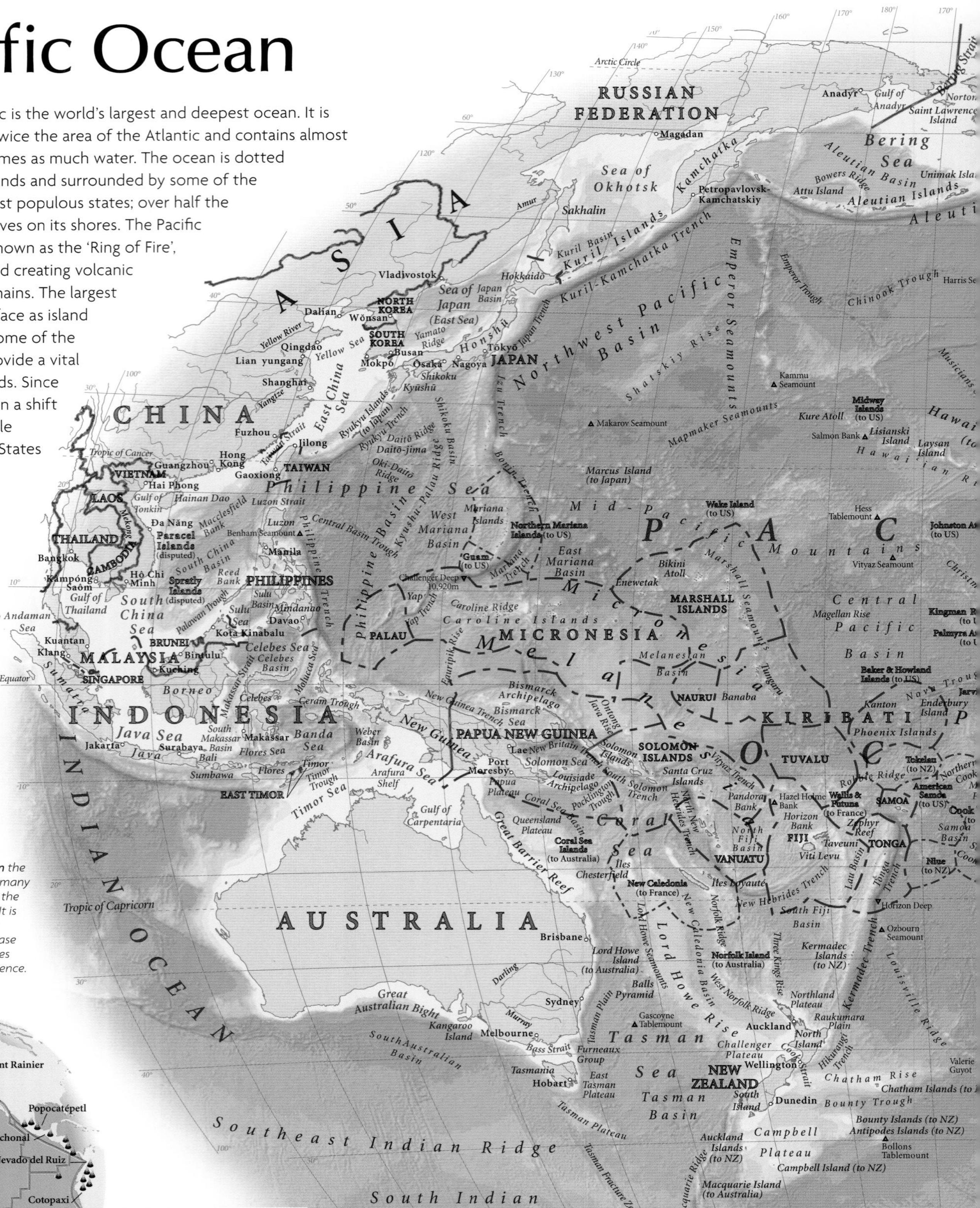

The Ring of Fire

The active plate margins surrounding the Pacific have created numerous land and island volcanoes along its border. The actual basin of the Pacific is made up of a number of separate tectonic plates which move away from each other, colliding with other plates. When they collide, the oceanic plates, being thinner, are forced beneath the thicker continental plates, forming deep ocean trenches and high ridges. These collision zones are known as subduction zones and are characterized by intense seismic and volcanic activity.

◀ *Mayon Volcano in the Philippines is one of many active volcanoes on the Pacific 'Ring of Fire'. It is noted for its perfect conical shape; the base of the cone is 80 miles (130 km) in circumference.*

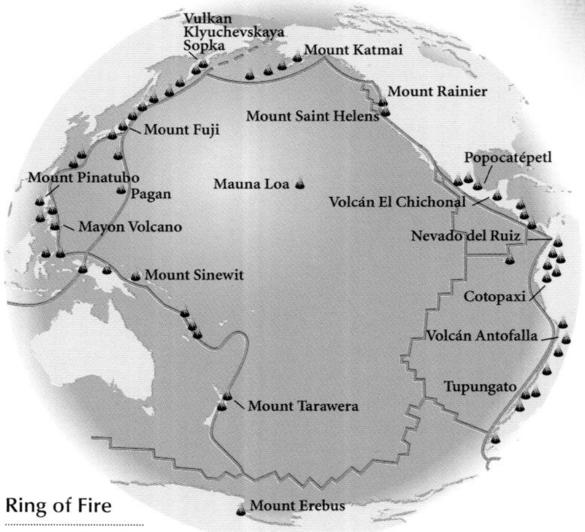

Ring of Fire
— plate boundaries
▲ major volcanoes

◀ *The Hawaiian volcanoes lie in the centre of a plate, not on a plate margin, and are known as intraplate volcanoes. They are associated with hot spots, whereby a plume of hot molten rock rises to the surface as the plate moves over it.*

American Samoa and Samoa

American Samoa and Samoa are part of the island archipelago of Polynesia. The two most populous islands are Tutuila in American Samoa and 'Upolu in Samoa. Although the economies of both these states remain predominantly resource-based, both are expanding their light manufacturing sectors, and the US administration is the primary employer in American Samoa. Tuna fishing is particularly important: 25% of all tuna consumed in the USA is processed and canned in Pago Pago.

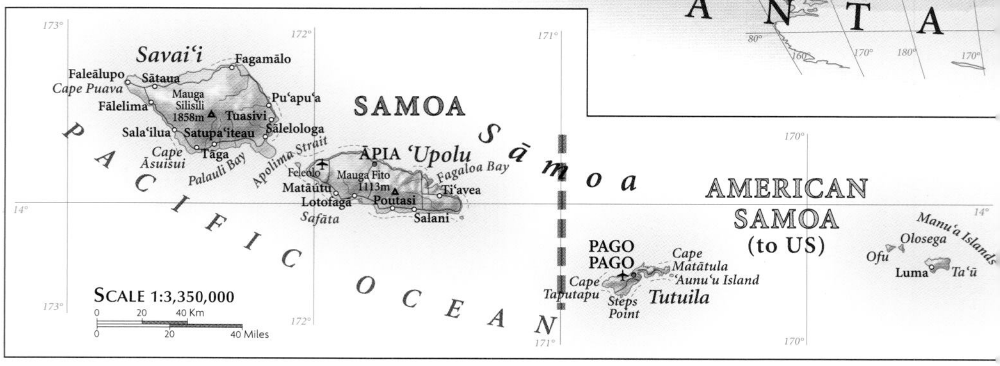

▶ *Many of the buildings in Samoa reflect the country's colonial past. Once a colony of New Zealand, Samoa is now an independent state; American Samoa remains an unincorporated territory of the United States.*

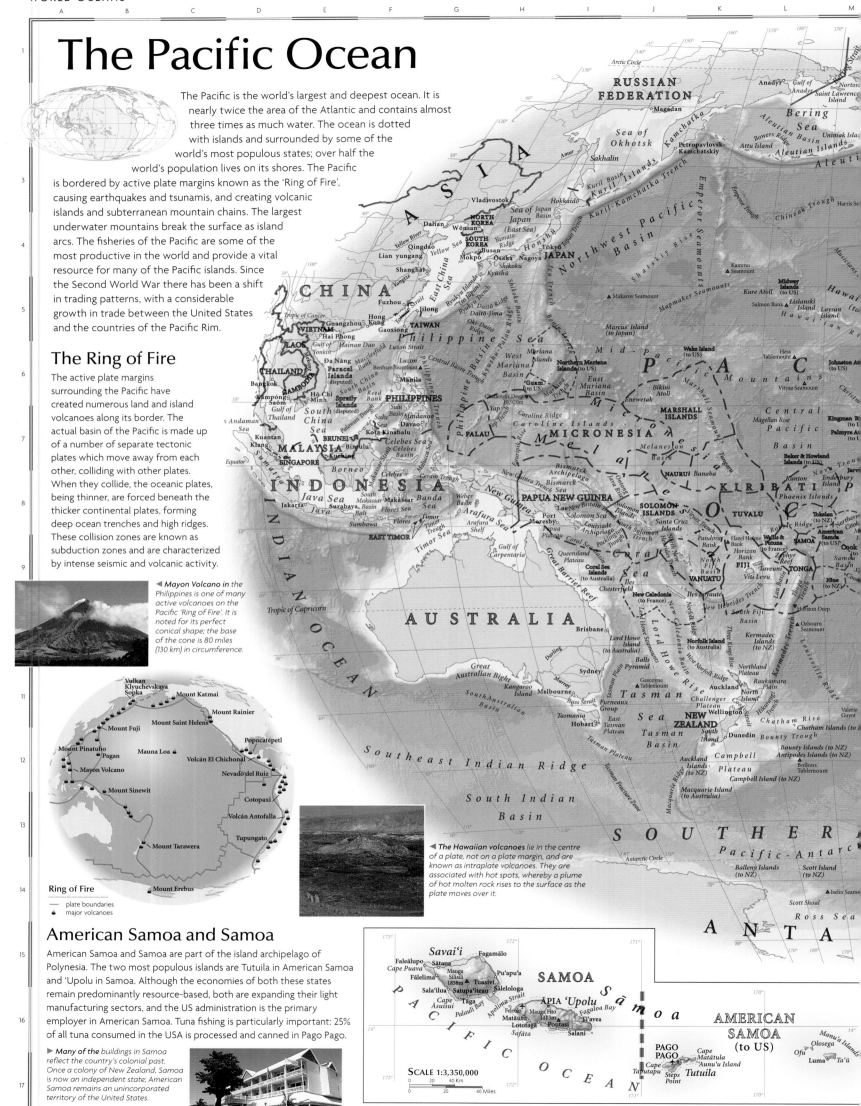

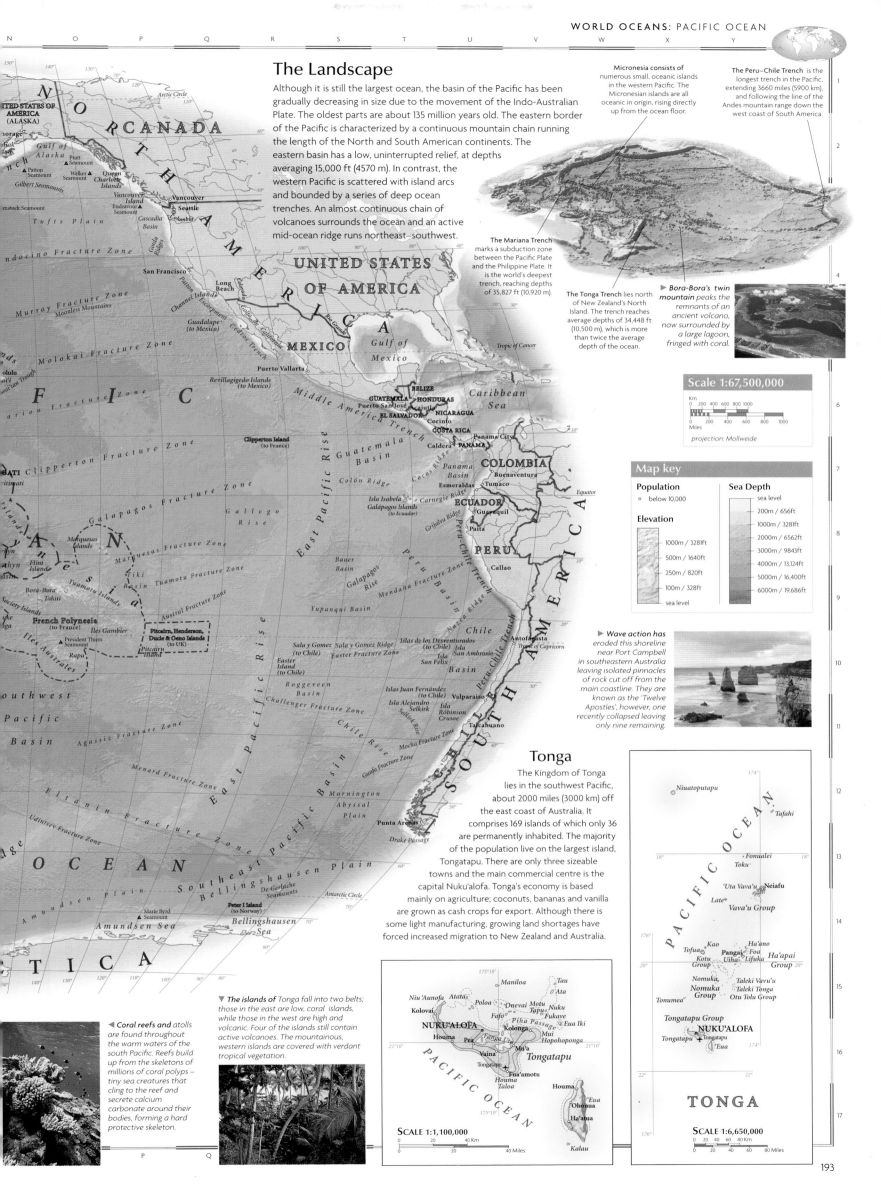

The Landscape

Although it is still the largest ocean, the basin of the Pacific has been gradually decreasing in size due to the movement of the Indo-Australian Plate. The oldest parts are about 135 million years old. The eastern border of the Pacific is characterized by a continuous mountain chain running the length of the North and South American continents. The eastern basin has a low, uninterrupted relief, at depths averaging 15,000 ft (4570 m). In contrast, the western Pacific is scattered with island arcs and bounded by a series of deep ocean trenches. An almost continuous chain of volcanoes surrounds the ocean and an active mid-ocean ridge runs northeast–southwest.

Micronesia consists of numerous small, oceanic islands in the western Pacific. The Micronesian islands are all oceanic in origin, rising directly up from the ocean floor.

The Peru–Chile Trench is the longest trench in the Pacific, extending 3660 miles (5900 km), and following the line of the Andes mountain range down the west coast of South America.

The Mariana Trench marks a subduction zone between the Pacific Plate and the Philippine Plate. It is the world's deepest trench, reaching depths of 35,827 ft (10,920 m).

The Tonga Trench lies north of New Zealand's North Island. The trench reaches average depths of 34,448 ft (10,500 m), which is more than twice the average depth of the ocean.

▶ *Bora-Bora's twin mountain peaks the remnants of an ancient volcano, now surrounded by a large lagoon, fringed with coral.*

Scale 1:67,500,000

Km 0 200 400 600 800 1000

Miles 0 200 400 600 800 1000

projection: Mollweide

Map key

Population

○ below 10,000

Elevation

1000m / 3281ft
500m / 1640ft
250m / 820ft
100m / 328ft
sea level

Sea Depth

sea level
200m / 656ft
1000m / 3281ft
2000m / 6562ft
3000m / 9843ft
4000m / 13,124ft
5000m / 16,400ft
6000m / 19,686ft

▶ *Wave action has eroded this shoreline near Port Campbell in southeastern Australia leaving isolated pinnacles of rock cut off from the main coastline. They are known as the 'Twelve Apostles', however, one recently collapsed leaving only nine remaining.*

Tonga

The Kingdom of Tonga lies in the southwest Pacific, about 2000 miles (3000 km) off the east coast of Australia. It comprises 169 islands of which only 36 are permanently inhabited. The majority of the population live on the largest island, Tongatapu. There are only three sizeable towns and the main commercial centre is the capital Nuku'alofa. Tonga's economy is based mainly on agriculture; coconuts, bananas and vanilla are grown as cash crops for export. Although there is some light manufacturing, growing land shortages have forced increased migration to New Zealand and Australia.

◀ *Coral reefs and atolls are found throughout the warm waters of the south Pacific. Reefs build up from the skeletons of millions of coral polyps – tiny sea creatures that cling to the reef and secrete calcium carbonate around their bodies, forming a hard protective skeleton.*

▼ *The islands of Tonga fall into two belts; those in the east are low, coral islands, while those in the west are high and volcanic. Four of the islands still contain active volcanoes. The mountainous, western islands are covered with verdant tropical vegetation.*

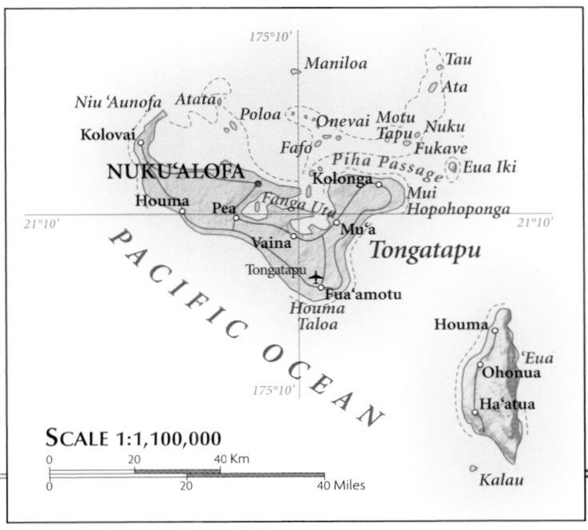

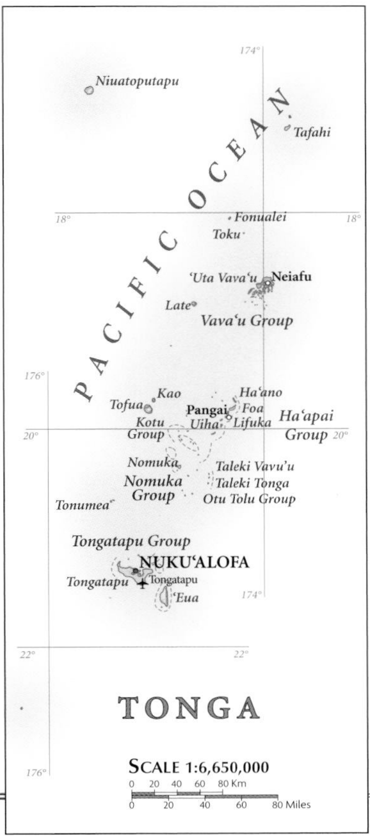

Antarctica

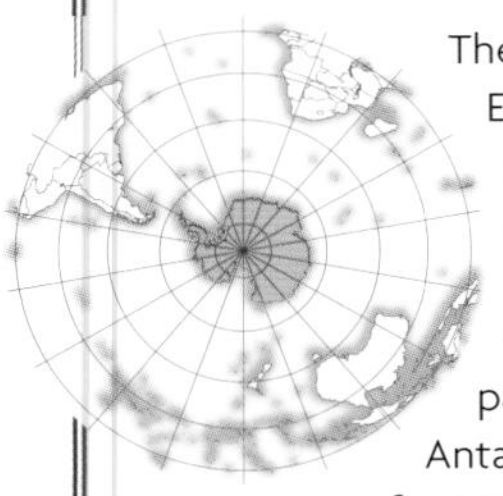

The ice-covered continent of Antarctica, which is the Earth's most southerly region, has for over 200 years drawn explorers and entrepreneurs seeking challenge and riches in its wintry lands. The extreme climate has deterred any large-scale settlement of the continent, and though commercial hunters built outposts in the past, habitation is now limited to scientific bases. The Antarctic Treaty, which came into force in 1961, provides for international governance and scientific co-operation in place of potential territorial conflict.

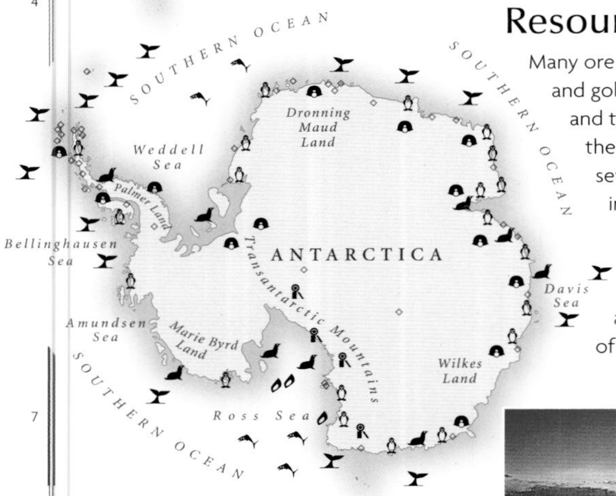
SOUTHERN OCEAN
Dronning Maud Land
Weddell Sea
Palmer Land
Bellinghausen Sea
ANTARCTICA
Transantarctic Mountains
Davis Sea
Amundsen Sea
Marie Byrd Land
Wilkes Land
Ross Sea

Resources (including wildlife)

- coal
- fish
- minerals
- oil & gas
- penguins
- seals
- whales
- polar research base

Resources

Many ore minerals, including iron and gold, are found in the Antarctic, and there are also coal reserves in the Transantarctic Mountains. The severe conditions and environmental importance of the region mean that exploitation of potential mineral resources is both uneconomic and undesirable. The unique wildlife and landscape draw a small number of tourists annually.

◀ *Most settlements in Antarctica are research bases such as this one at Rothera on Adelaide Island, although there is a small Chilean settlement on King George Island.*

The landscape

There are two distinct parts to Antarctica: West Antarctica, a series of ice-covered, mountainous islands, joined together by the ice; and the high plateau of East Antarctica. The Ross Sea and the Weddell Sea are outliers of the Southern Ocean – deep bays partially covered by thick ice shelves.

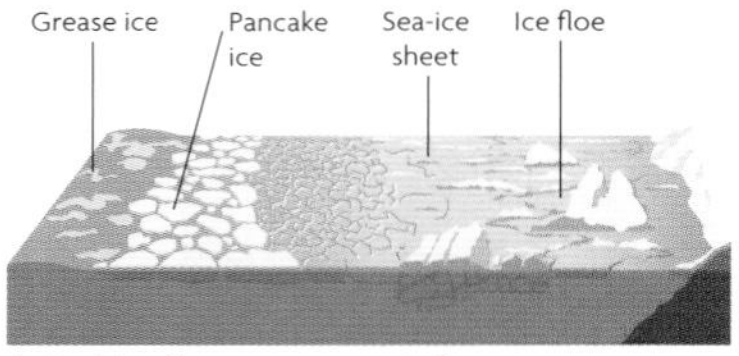
Grease ice
Pancake ice
Sea-ice sheet
Ice floe

▲ *Pack ice forms out at sea in freezing temperatures. At the outer limits, grease ice congeals on the surface of the ocean. This is then spun around by wind and waves into irregular 'pancakes', freezing and breaking up several times before bonding together again to form sea-ice sheets, which finally cement into enormous ice floes.*

◀ *On Elephant Island, the coast is edged by glaciers, although the land is not permanently covered by ice.*

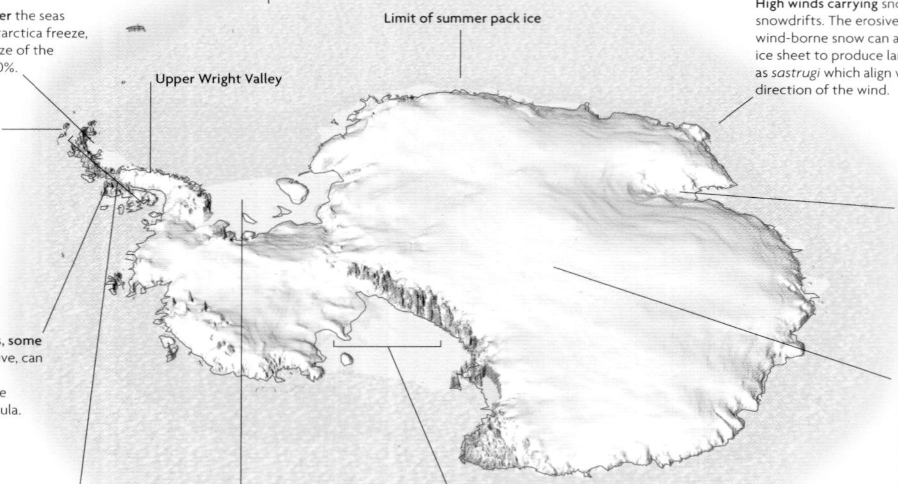
Limit of winter pack ice
Limit of summer pack ice
Upper Wright Valley
Elephant Island

During the winter the seas surrounding Antarctica freeze, increasing the size of the continent by 100%.

High winds carrying snow form huge snowdrifts. The erosive power of the wind-borne snow can also sculpt the ice sheet to produce landforms known as *sastrugi* which align with the direction of the wind.

The Lambert Glacier is the largest glacier system in the world, up to 50 miles (80 km) wide at its seaward limit, and reaching 180 miles (300 km) into the interior by way of the Prince Charles Mountains.

Many volcanoes, some of them still active, can be found in the mountains of the Antarctic Peninsula.

Antarctica is the highest continent on Earth, because of the great thickness of ice which overlays the land. In places the ice alone can reach up to 15,700 ft (4800 m) thick. Much of the basement rock of west Antarctica lies below sea level, pushed down by the weight of the ice.

The mountainous Antarctic Peninsula is formed of rocks 65–225 million years old, overlain by more recent rocks and glacial deposits. It is connected to the Andes in South America by a submarine ridge.

Nearly half – 44% – of the Antarctic coastline is bounded by ice shelves, like the Ronne Ice Shelf, which float on the Ocean. These are joined to the inland ice sheet by dome-shaped ice 'rises'.

More than 30% of Antarctic ice is contained in the Ross Ice Shelf.

◀ *The barren, flat-bottomed Upper Wright Valley was once filled by a glacier, but is now dry, strewn with boulders and pebbles. In some dry valleys, there has been no rain for over 2 million years.*

▲ *Large colonies of seabirds live in the extremely harsh Antarctic climate. The Emperor penguins seen here, the smaller Adélie penguin, the Antarctic petrel and the South Polar skua are the only birds which breed exclusively on the continent.*

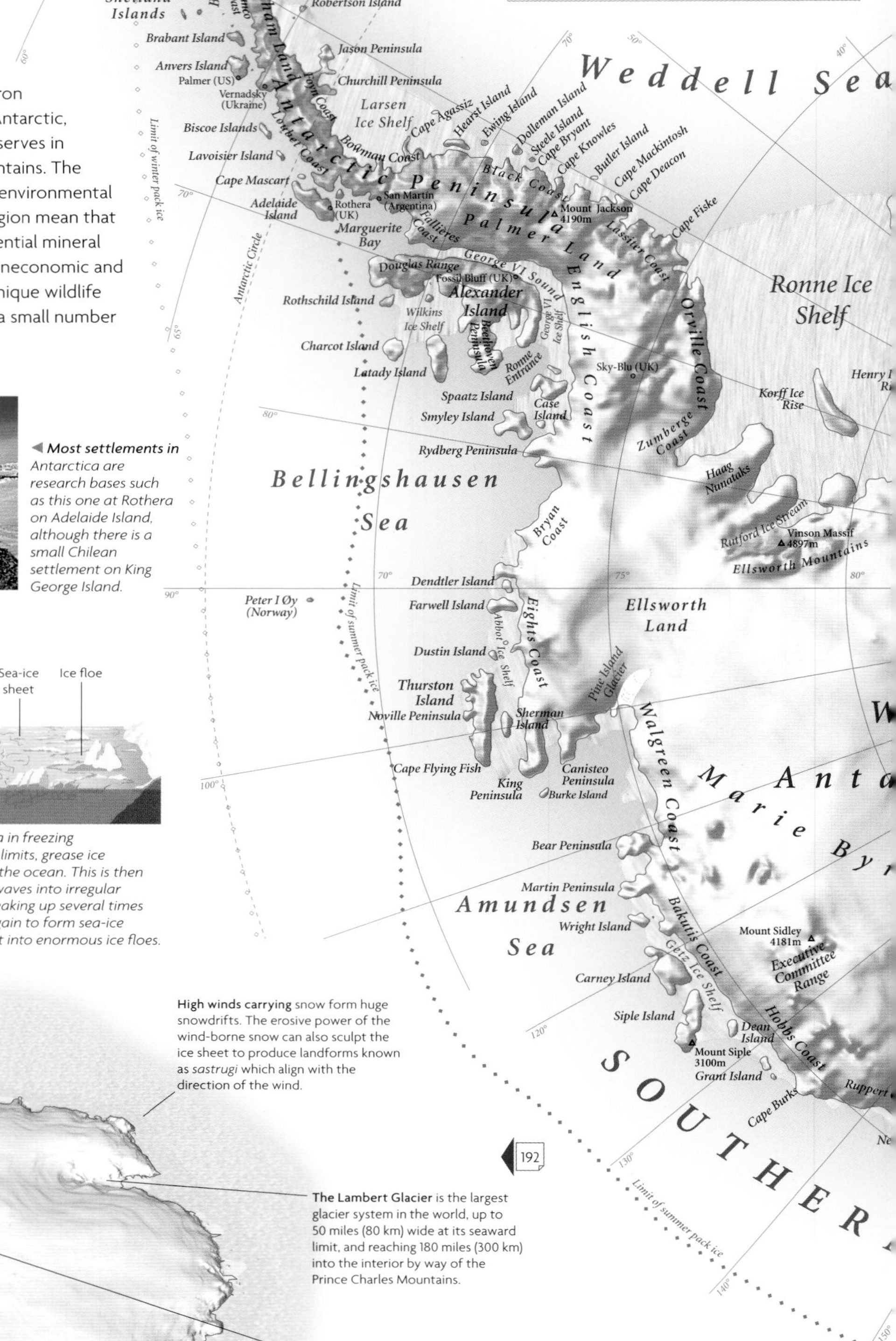
South Orkney Islands
Laurie Island
Orcadas (Argentina)
Coronation Island
Signy (UK)
Scotia Sea
Research Stations on King George Island
Arctowski (Poland)
Artigas (Uruguay)
Bellingshausen (Russian Federation)
Comandante Ferraz (Brazil)
Great Wall (China)
Jubany (Argentina)
King Sejong (South Korea)
Teniente Rodolfo Marsh (Chile)
Clarence Island
Elephant Island
Drake Passage
King George Island
Capitán Arturo Prat (Chile)
Livingston Island
South Shetland Islands
Bransfield Strait
Joinville Island
Dundee Island
General Bernardo O'Higgins (Chile)
Esperanza (Argentina)
Marambio (Argentina)
Snowhill Island
James Ross Island
Robertson Island
Brabant Island
Anvers Island
Palmer (US)
Vernadsky (Ukraine)
Jason Peninsula
Churchill Peninsula
Larsen Ice Shelf
Graham Land
Antarctic Peninsula
Biscoe Islands
Lavoisier Island
Cape Mascart
Adelaide Island
Rothera (UK)
San Martin (Argentina)
Marguerite Bay
Weddell Sea
Cape Agassiz
Hearst Island
Ewing Island
Dolleman Island
Steele Island
Cape Bryant
Cape Knowles
Butler Island
Cape Mackintosh
Cape Deacon
Black Coast
Palmer Land
Mount Jackson 4190m
Lassiter Coast
Cape Fiske
Douglas Range
Fossil Bluff (UK)
George VI Sound
Alexander Island
Wilkins Ice Shelf
Rothschild Island
English Coast
Orville Coast
Ronne Ice Shelf
Charcot Island
Latady Island
Sky-Blu (UK)
Korff Ice Rise
Spaatz Island
Smyley Island
Case Island
Zumberge Coast
Rydberg Peninsula
Bellingshausen Sea
Haag Nunataks
Rutford Ice Stream
Vinson Massif 4897m
Ellsworth Mountains
Bryan Coast
Peter I Øy (Norway)
Dendtler Island
Farwell Island
Ellsworth Land
Eights Coast
Dustin Island
Thurston Island
Noville Peninsula
Abbot Ice Shelf
Pine Island Glacier
Sherman Island
Cape Flying Fish
King Peninsula
Canisteo Peninsula
Burke Island
Walgreen Coast
Marie Byrd Land
Bear Peninsula
Martin Peninsula
Amundsen Sea
Wright Island
Bakutis Coast
Getz Ice Shelf
Mount Sidley 4181m
Executive Committee Range
Carney Island
Siple Island
Dean Island
Hobbs Coast
Mount Siple 3100m
Grant Island
Cape Burks
Ruppert Coast
Southern Ocean
Antarctic Circle
Limit of winter pack ice
Limit of summer pack ice

192

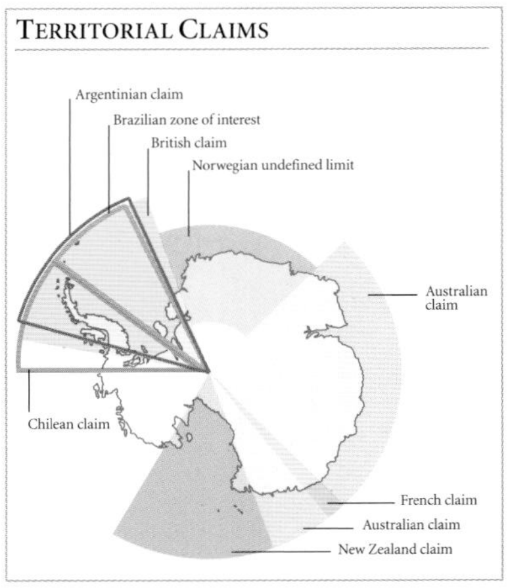

Territorial Claims

Argentinian claim
Brazilian zone of interest
British claim
Norwegian undefined limit
Australian claim
Chilean claim
French claim
Australian claim
New Zealand claim

◀ ***The sun sets*** *over the Antarctic Peninsula for more than six months during the winter. However, there are more hours of sunshine during the brief Antarctic summer than most equatorial countries experience in a whole year.*

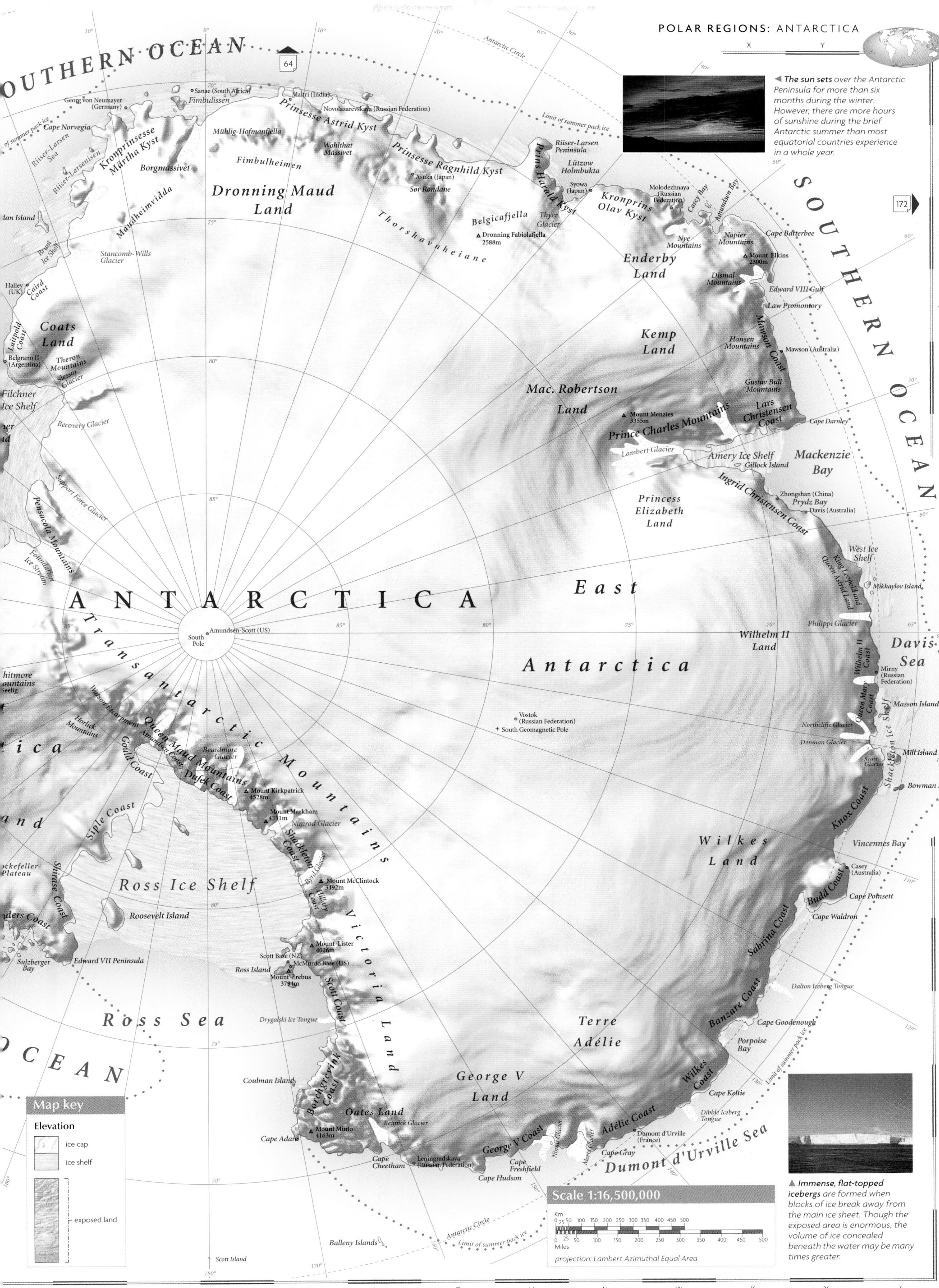
SOUTHERN OCEAN
64
172
SOUTHERN OCEAN
Antarctic Circle
Limit of summer pack ice
Sanae (South Africa)
Fimbulisen
Georg von Neumayer (Germany)
Maitri (India)
Novolazarevskaya (Russian Federation)
Prinsesse Astrid Kyst
Cape Norvegia
Riiser-Larsen Sea
Riiser-Larsenisen
Kronprinsesse Märtha Kyst
Mühlig-Hofmanfjella
Wohlthat Massivet
Prinsesse Ragnhild Kyst
Riiser-Larsen Peninsula
Lützow Holmbukta
Prins Harald Kyst
Borgmassivet
Fimbulheimen
Asuka (Japan)
Sør Rondane
Syowa (Japan)
Molodezhnaya (Russian Federation)
Casey Bay
Amundsen Bay
Kronprins Olav Kyst
Dronning Maud Land
Maudheimvidda
Thorshavnheiane
Belgicafjella
Dronning Fabiolafjella 2588m
Thyer Glacier
Nye Mountains
Napier Mountains
Cape Batterbee
Mount Elkins 2300m
Enderby Land
Stancomb-Wills Glacier
Brunt Ice Shelf
Halley (UK)
Caird Coast
Dismal Mountains
Edward VIII Gulf
Law Promontory
Coats Land
Luitpold Coast
Belgrano II (Argentina)
Theron Mountains
Slessor Glacier
Kemp Land
Hansen Mountains
Mawson Coast
Mawson (Australia)
Filchner Ice Shelf
Recovery Glacier
Mac. Robertson Land
Gustav Bull Mountains
Mount Menzies 3355m
Prince Charles Mountains
Lars Christensen Coast
Cape Darnley
Lambert Glacier
Amery Ice Shelf
Gillock Island
Mackenzie Bay
Support Force Glacier
Ingrid Christensen Coast
Zhongshan (China)
Prydz Bay
Davis (Australia)
Princess Elizabeth Land
Pensacola Mountains
Foundation Ice Stream
West Ice Shelf
King Leopold and Queen Astrid Land
Mikhaylov Island
ANTARCTICA
East Antarctica
South Pole
Amundsen-Scott (US)
Philippi Glacier
Wilhelm II Land
Wilhelm II Coast
Davis Sea
Mirny (Russian Federation)
Queen Mary Coast
Masson Island
Transantarctic Mountains
Whitmore Mountains
Watson Escarpment
Horlick Mountains
Queen Maud Mountains
Amundsen Coast
Gould Coast
Dufek Coast
Beardmore Glacier
Vostok (Russian Federation)
South Geomagnetic Pole
Northcliffe Glacier
Denman Glacier
Shackleton Ice Shelf
Scott Glacier
Mill Island
Bowman Island
Mount Kirkpatrick 4528m
Mount Markham 4351m
Nimrod Glacier
Shackleton Coast
Siple Coast
Knox Coast
Vincennes Bay
Wilkes Land
Casey (Australia)
Byrd Glacier
Mount McClintock 3492m
Hillary Coast
Ross Ice Shelf
Rockefeller Plateau
Siple Coast
Budd Coast
Cape Poinsett
Cape Waldron
Roosevelt Island
Saunders Coast
Sabrina Coast
Victoria Land
Mount Lister 4026m
Scott Base (NZ)
McMurdo Base (US)
Ross Island
Mount Erebus 3794m
Sulzberger Bay
Edward VII Peninsula
Dalton Iceberg Tongue
Banzare Coast
Scott Coast
Drygalski Ice Tongue
Ross Sea
Cape Goodenough
Terre Adélie
Porpoise Bay
OCEAN
Borchgrevink Coast
Coulman Island
George V Land
Wilkes Coast
Cape Keltie
Oates Land
Rennick Glacier
Dibble Iceberg Tongue
Adélie Coast
Dumont d'Urville (France)
Mount Minto 4163m
Cape Adare
George V Coast
Ninnis Glacier
Mertz Glacier
Cape Gray
Dumont d'Urville Sea
Cape Cheetham
Leningradskaya (Russian Federation)
Cape Freshfield
Cape Hudson
Balleny Islands
Scott Island
Map key
Elevation
ice cap
ice shelf
exposed land
Scale 1:16,500,000
Km
Miles
projection: Lambert Azimuthal Equal Area
X
Y
N
O
P
Q
R
S
T
U
V
W
X
Y
Z
1
2
3
4
5
6
12
13
14
15
16
17

▲ ***Immense, flat-topped icebergs*** *are formed when blocks of ice break away from the main ice sheet. Though the exposed area is enormous, the volume of ice concealed beneath the water may be many times greater.*

A B C D E F G H I J K L M

The Arctic

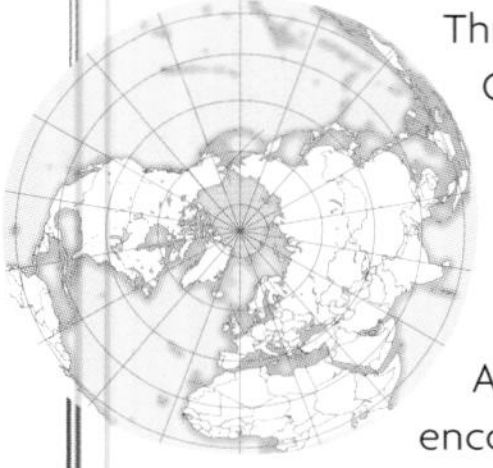

Three continents, Asia, North America and Europe, reach into the Arctic Circle at their northernmost limits, almost entirely encircling the Arctic Ocean. Despite the region's extraordinarily harsh climate, it has been inhabited for thousands of years by peoples such as the European Lapps, the Russian Nenet, and the North American Inuit, who draw a living from fishing, herding and hunting. More recently, particularly in the Russian Arctic, opportunities to exploit oil and other mineral reserves have encouraged immigration. Pollution of the Arctic's unique ecology and damage to the traditional lifestyles of many native peoples have been the unfortunate results of this activity, and international co-operation is needed to safeguard the future of the region.

Map key

Population

- above 5 million
- 1 million to 5 million
- 500,000 to 1 million
- 100,000 to 500,000
- 50,000 to 100,000
- 10,000 to 50,000
- below 10,000

Sea depth

- Sea level
- 200m / 656ft
- 1000m / 3281ft
- 2000m / 6562ft
- 3000m / 9843ft
- 4000m / 13,124ft
- 5000m / 16,400ft
- 6000m / 19,686ft

Scale 1:23,500,000

Km 0 100 200 300 400 500 600

Miles 0 100 200 300 400 500 600

projection: Lambert Azimuthal Equal Area

192

▲ *Wind-blown snow etches deep patterns in the ice sheet known as sastrugi. They align with the direction of the wind.*

Resources

Large quantities of coal, oil and natural gas are to be found in the basins of the Arctic Ocean, and in northern Canada, Alaska and the Russian Federation. The cost and difficulty of extraction and, more recently, awareness of damage to the environment, have limited exploitation to coastal regions. The unfrozen waters have stocks of fish including cod, plaice and haddock. Quotas have now been put in place to restrict the number of fish caught annually. Reindeer are herded in large numbers by many of the native Arctic peoples. Most grain and vegetables are imported from elsewhere.

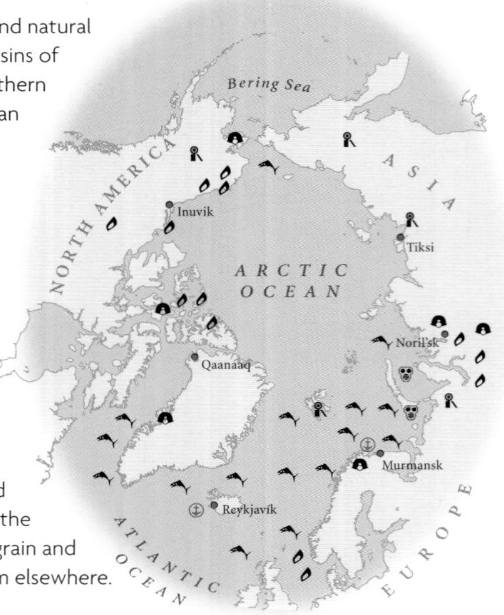

Bering Sea
NORTH AMERICA
ASIA
Inuvik
Tiksi
ARCTIC OCEAN
Noril'sk
Qaanaaq
Murmansk
Reykjavík
ATLANTIC OCEAN
EUROPE

▲ *Icebreakers, ships with specially strengthened hulls, designed to break a path through the ice, are used to keep important routes open during the winter, when falling temperatures cause much of the Arctic Ocean to freeze over.*

Resources

- coal
- fish
- mining
- oil & gas
- radioactive contamination
- major towns
- major ports

8

8

The landscape

The Arctic Ocean comprises two large ocean basins divided by three submarine ridges, the greatest of which, the Lomonosov Ridge, is a huge underwater mountain range which has an average height of more than 10,000 ft (3000 m). The lands which encircle the Arctic Ocean are underlain by great shield areas of ancient rocks, which were heavily glaciated during the last Ice Age.

◀ *Icebergs are constantly broken up and re-shaped by wind and the oceans. This flat-topped iceberg has been undercut, leaving a craggy ice cliff.*

The Canadian Shield underlies almost all of the Canadian Arctic. It is a very stable plateau of ancient rock, now covered by glacial lakes and sediment, which supports tundra vegetation.

The Arctic Ocean is the world's smallest ocean with a total area of 5,440,000 sq miles (15,100,000 sq km).

At a latitude of more than 75° N, the Arctic Ocean is almost permanently covered by pack-ice, though high winds and the movement of the seas may cause the ice to crack and break up.

In the more southerly reaches of the Arctic, like Siberia, much of the land is covered by permafrost. In the summer, higher temperatures warm the frozen ground, causing a number of typical phenomena. These include solifluction, the fast downhill movement of top soil layers; freeze/thaw activity, which patterns the ground into regular polygonal shapes, and the formation of large domes with a frozen ice core, known as pingos.

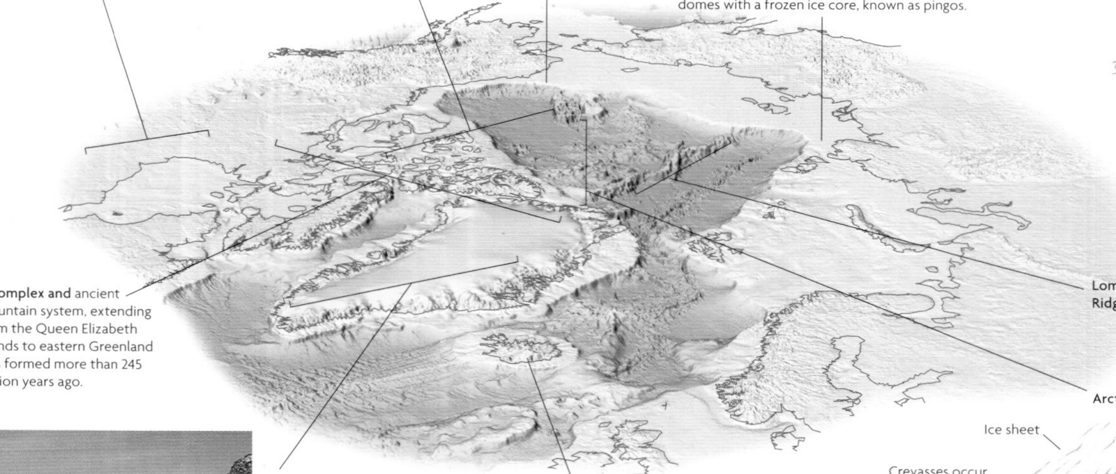

A complex and ancient mountain system, extending from the Queen Elizabeth Islands to eastern Greenland was formed more than 245 million years ago.

Lomonosov Ridge

Arctic ice shelf

◀ *Much of Greenland is covered by a massive ice sheet more than 650,000 sq miles (1,683,400 sq km) in extent. The weight of the ice has depressed the central land area to form a basin lying more than 1000 ft (300 m) below sea level. Only at the edges of the island is bare rock visible.*

Iceland has five major glaciers, sustained by heavy snowfall. Parts of the ice cap cover active volcanoes, such as Bárdharbunga, which periodically erupt causing the melted ice to form a great lake at the glacier margins.

Ice sheet
Iceberg
Crevasses occur at the edge of the ice sheet
Sea water melts the edge of the ice sheet

▲ *At the boundary of the Arctic ice shelves, sea water flows under the ice causing melting and forming crevasses on the surface. This eventually weakens blocks of ice which break away as icebergs. This process is known as calving.*

CANADA
NORTH AMERICA
Great Bear Lake
Great Slave Lake
Kugluktuk (Coppermine)
Bathurst Inlet
Cambridge Bay (Ikaluktutiak)
Nelson
Churchill
Back
Hudson Bay
Southampton Island
Coats Island
Repulse Bay
Melville Peninsula
Mansel Island
Ivujivik
Foxe Basin
Prince Charles Island
Foxe Peninsula
Inukjuak (Port Harrison)
Hudson Strait
Baffin Isla
Kimmirut (Lake Harbour)
Iqaluit (Frobisher Bay)
Frobisher Bay
Cumberland Sound
Ungava Bay
Cape Chidley
Davis Str
Nain
Labrador Sea
Maniits
NUUK
Paamiut
Ivittuut
Qaqortoq
Narsarsuaq
Nanortalik
Nunap Isua (Kap Farvel)
Eirik Ridge
Labrador Basin
ATLAN

64

A B C D E F G H I J K L M

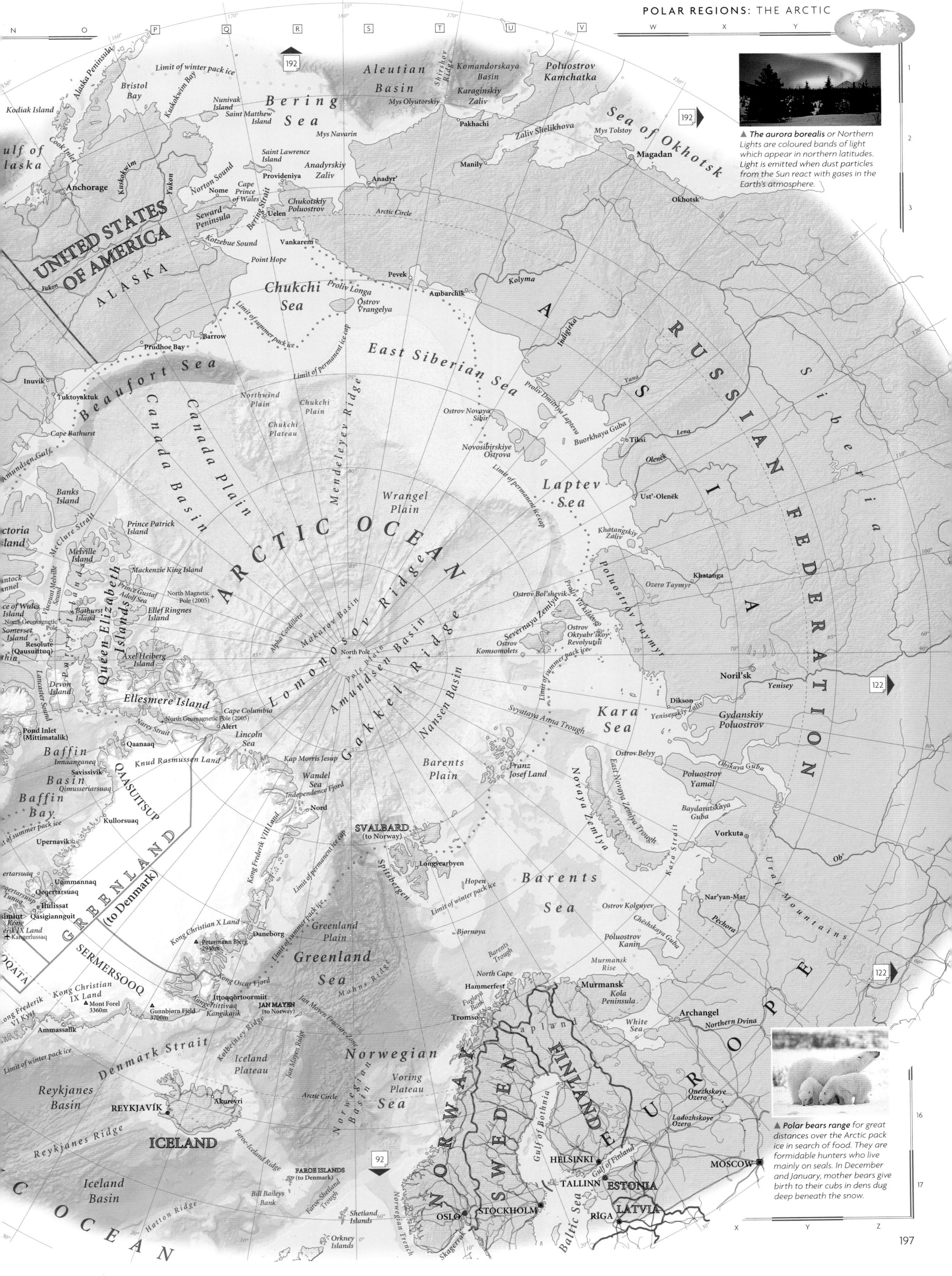

▲ ***The aurora borealis*** *or Northern Lights are coloured bands of light which appear in northern latitudes. Light is emitted when dust particles from the Sun react with gases in the Earth's atmosphere.*

▲ ***Polar bears range*** *for great distances over the Arctic pack ice in search of food. They are formidable hunters who live mainly on seals. In December and January, mother bears give birth to their cubs in dens dug deep beneath the snow.*

192
122
92

Geographical comparisons

Largest countries

Russian Federation	6,592,735 sq miles	(17,075,200 sq km)
Canada	3,855,171 sq miles	(9,984,670 sq km)
USA	3,794,100 sq miles	(9,826,675 sq km)
China	3,705,386 sq miles	(9,596,960 sq km)
Brazil	3,286,470 sq miles	(8,511,965 sq km)
Australia	2,967,893 sq miles	(7,686,850 sq km)
India	1,269,339 sq miles	(3,287,590 sq km)
Argentina	1,068,296 sq miles	(2,766,890 sq km)
Kazakhstan	1,049,150 sq miles	(2,717,300 sq km)
Algeria	919,590 sq miles	(2,381,740 sq km)

Smallest countries

Vatican City	0.17 sq miles	(0.44 sq km)
Monaco	0.75 sq miles	(1.95 sq km)
Nauru	8.2 sq miles	(21.2 sq km)
Tuvalu	10 sq miles	(26 sq km)
San Marino	24 sq miles	(61 sq km)
Liechtenstein	62 sq miles	(160 sq km)
Marshall Islands	70 sq miles	(181 sq km)
St. Kitts & Nevis	101 sq miles	(261 sq km)
Maldives	116 sq miles	(300 sq km)
Malta	124 sq miles	(320 sq km)

Largest islands

To the nearest 1000 – or 100,000 for the largest

Greenland	849,400 sq miles	(2,200,000 sq km)
New Guinea	312,000 sq miles	(808,000 sq km)
Borneo	292,222 sq miles	(757,050 sq km)
Madagascar	229,300 sq miles	(594,000 sq km)
Sumatra	202,300 sq miles	(524,000 sq km)
Baffin Island	183,800 sq miles	(476,000 sq km)
Honshu	88,800 sq miles	(230,000 sq km)
Britain	88,700 sq miles	(229,800 sq km)
Victoria Island	81,900 sq miles	(212,000 sq km)
Ellesmere Island	75,700 sq miles	(196,000 sq km)

Richest countries

GNI per capita, in US$

Monaco	186,950
Liechtenstein	136,770
Norway	102,610
Switzerland	90,760
Qatar	86,790
Luxembourg	69,900
Australia	65,390
Sweden	61,760
Denmark	61,680
Singapore	54,040

Poorest countries

GNI per capita, in US$

Burundi	260
Malawi	270
Somalia	288
Central African Republic	320
Niger	400
Liberia	410
Dem. Rep. Congo	430
Madagascar	440
Guinea	460
Ethiopia	470
Eritrea	490
Gambia	500

Most populous countries

China	1,393,800,000
India	1,267,400,000
USA	322,600,000
Indonesia	252,800,000
Brazil	202,120,000
Pakistan	185,100,000
Nigeria	178,500,000
Bangladesh	159,000,000
Russian Federation	142,500,000
Japan	127,000,000

Least populous countries

Vatican City	842
Nauru	9488
Tuvalu	10,782
Palau	21,186
San Marino	32,742
Monaco	36,950
Liechtenstein	37,313
St Kitts & Nevis	51,538
Marshall Islands	70,983
Dominica	73,449
Andorra	85,458
Antigua & Barbuda	91,295

Most densely populated countries

Monaco	49,267 people per sq mile	(18,949 per sq km)
Singapore	23,305 people per sq mile	(9016 per sq km)
Vatican City	4953 people per sq mile	(1914 per sq km)
Bahrain	4762 people per sq mile	(1841 per sq km)
Maldives	3448 people per sq mile	(1333 per sq km)
Malta	3226 people per sq mile	(1250 per sq km)
Bangladesh	3066 people per sq mile	(1184 per sq km)
Taiwan	1879 people per sq mile	(725 per sq km)
Barbados	1807 people per sq mile	(698 per sq km)
Mauritius	1671 people per sq mile	(645 per sq km)

Most sparsely populated countries

Mongolia	5 people per sq mile	(2 per sq km)
Namibia	7 people per sq mile	(3 per sq km)
Australia	8 people per sq mile	(3 per sq km)
Suriname	8 people per sq mile	(3 per sq km)
Iceland	8 people per sq mile	(3 per sq km)
Botswana	9 people per sq mile	(4 per sq km)
Libya	9 people per sq mile	(4 per sq km)
Mauriania	10 people per sq mile	(4 per sq km)
Canada	10 people per sq mile	(4 per sq km)
Guyana	11 people per sq mile	(4 per sq km)

Most widely spoken languages

1. Chinese (Mandarin)	6. Arabic
2. English	7. Bengali
3. Hindi	8. Portuguese
4. Spanish	9. Malay-Indonesian
5. Russian	10. French

Largest conurbations

Urban area population

Tokyo	37,800,000
Jakarta	30,500,000
Manila	24,100,000
Delhi	24,000,000
Karachi	23,500,000
Seoul	23,500,000
Shanghai	23,400,000
Beijing	21,000,000
New York City	20,600,000
Guangzhou	20,600,000
São Paulo	20,300,000
Mexico City	20,000,000
Mumbai	17,700,000
Osaka	17,400,000
Lagos	17,000,000
Moscow	16,100,000
Dhaka	15,700,000
Lahore	15,600,000
Los Angeles	15,000,000
Bangkok	15,000,000
Kolkatta	14,700,000
Buenos Aires	14,100,000
Tehran	13,500,000
Istanbul	13,300,000
Shenzhen	12,000,000

Countries with the most land borders

14: China	(Afghanistan, Bhutan, India, Kazakhstan, Kyrgyzstan, Laos, Mongolia, Myanmar (Burma), Nepal, North Korea, Pakistan, Russian Federation, Tajikistan, Vietnam)
14: Russian Federation	(Azerbaijan, Belarus, China, Estonia, Finland, Georgia, Kazakhstan, Latvia, Lithuania, Mongolia, North Korea, Norway, Poland, Ukraine)
10: Brazil	(Argentina, Bolivia, Colombia, French Guiana, Guyana, Paraguay, Peru, Suriname, Uruguay, Venezuela)
9: Congo, Dem. Rep.	(Angola, Burundi, Central African Republic, Congo, Rwanda, South Sudan, Tanzania, Uganda, Zambia)
9: Germany	(Austria, Belgium, Czech Republic, Denmark, France, Luxembourg, Netherlands, Poland, Switzerland)
8: Austria	(Czech Republic, Germany, Hungary, Italy, Liechtenstein, Slovakia, Slovenia, Switzerland)
8: France	(Andorra, Belgium, Germany, Italy, Luxembourg, Monaco, Spain, Switzerland)
8: Tanzania	(Burundi, Dem. Rep. Congo, Kenya, Malawi, Mozambique, Rwanda, Uganda, Zambia)
8: Turkey	(Armenia, Azerbaijan, Bulgaria, Georgia, Greece, Iran, Iraq, Syria)
8: Zambia	(Angola, Botswana, Dem. Rep.Congo, Malawi, Mozambique, Namibia, Tanzania, Zimbabwe)

Longest rivers

Nile (NE Africa)	4160 miles	(6695 km)
Amazon (South America)	4049 miles	(6516 km)
Yangtze (China)	3915 miles	(6299 km)
Mississippi/Missouri (USA)	3710 miles	(5969 km)
Ob'-Irtysh (Russian Federation)	3461 miles	(5570 km)
Yellow River (China)	3395 miles	(5464 km)
Congo (Central Africa)	2900 miles	(4667 km)
Mekong (Southeast Asia)	2749 miles	(4425 km)
Lena (Russian Federation)	2734 miles	(4400 km)
Mackenzie (Canada)	2640 miles	(4250 km)
Yenisey (Russian Federation)	2541 miles	(4090km)

Highest mountains

Height above sea level

Everest	29,029 ft	(8848 m)
K2	28,253 ft	(8611 m)
Kangchenjunga I	28,210 ft	(8598 m)
Makalu I	27,767 ft	(8463 m)
Cho Oyu	26,907 ft	(8201 m)
Dhaulagiri I	26,796 ft	(8167 m)
Manaslu I	26,783 ft	(8163 m)
Nanga Parbat I	26,661 ft	(8126 m)
Annapurna I	26,547 ft	(8091 m)
Gasherbrum I	26,471 ft	(8068 m)

Largest bodies of inland water

With area and depth

Caspian Sea	143,243 sq miles (371,000 sq km)	3215 ft (980 m)
Lake Superior	31,151 sq miles (83,270 sq km)	1289 ft (393 m)
Lake Victoria	26,828 sq miles (69,484 sq km)	328 ft (100 m)
Lake Huron	23,436 sq miles (60,700 sq km)	751 ft (229 m)
Lake Michigan	22,402 sq miles (58,020 sq km)	922 ft (281 m)
Lake Tanganyika	12,703 sq miles (32,900 sq km)	4700 ft (1435 m)
Great Bear Lake	12,274 sq miles (31,790 sq km)	1047 ft (319 m)
Lake Baikal	11,776 sq miles (30,500 sq km)	5712 ft (1741 m)
Great Slave Lake	10,981 sq miles (28,440 sq km)	459 ft (140 m)
Lake Erie	9,915 sq miles (25,680 sq km)	197 ft (60 m)

Deepest ocean features

Challenger Deep, Mariana Trench (Pacific)	35,827 ft	(10,920 m)
Vityaz III Depth, Tonga Trench (Pacific)	35,704 ft	(10,882 m)
Vityaz Depth, Kuril-Kamchatka Trench (Pacific)	34,588 ft	(10,542 m)
Cape Johnson Deep, Philippine Trench (Pacific)	34,441 ft	(10,497 m)
Kermadec Trench (Pacific)	32,964 ft	(10,047 m)
Ramapo Deep, Japan Trench (Pacific)	32,758 ft	(9984 m)
Milwaukee Deep, Puerto Rico Trench (Atlantic)	30,185 ft	(9200 m)
Argo Deep, Torres Trench (Pacific)	30,070 ft	(9165 m)
Meteor Depth, South Sandwich Trench (Atlantic)	30,000 ft	(9144 m)
Planet Deep, New Britain Trench (Pacific)	29,988 ft	(9140 m)

Greatest waterfalls

Mean flow of water

Boyoma (Dem. Rep. Congo)	600,400 cu. ft/sec	(17,000 cu.m/sec)
Khône (Laos/Cambodia)	410,000 cu. ft/sec	(11,600 cu.m/sec)
Niagara (USA/Canada)	195,000 cu. ft/sec	(5500 cu.m/sec)
Grande, Salto (Uruguay)	160,000 cu. ft/sec	(4500 cu.m/sec)
Paulo Afonso (Brazil)	100,000 cu. ft/sec	(2800 cu.m/sec)
Urubupungá, Salto do (Brazil)	97,000 cu. ft/sec	(2750 cu.m/sec)
Iguaçu (Argentina/Brazil)	62,000 cu. ft/sec	(1700 cu.m/sec)
Maribondo, Cachoeira do (Brazil)	53,000 cu. ft/sec	(1500 cu.m/sec)
Victoria (Zimbabwe)	39,000 cu. ft/sec	(1100 cu.m/sec)
Murchison Falls (Uganda)	42,000 cu. ft/sec	(1200 cu.m/sec)
Churchill (Canada)	35,000 cu. ft/sec	(1000 cu.m/sec)
Kaveri Falls (India)	33,000 cu. ft/sec	(900 cu.m/sec)

Highest waterfalls

* Indicates that the total height is a single leap

Angel (Venezuela)	3212 ft	(979 m)
Tugela (South Africa)	3110 ft	(948 m)
Utigard (Norway)	2625 ft	(800 m)
Mongefossen (Norway)	2539 ft	(774 m)
Mtarazi (Zimbabwe)	2500 ft	(762 m)
Yosemite (USA)	2425 ft	(739 m)
Ostre Mardola Foss (Norway)	2156 ft	(657 m)
Tyssestrengane (Norway)	2119 ft	(646 m)
***Cuquenan** (Venezuela)	2001 ft	(610 m)
Sutherland (New Zealand)	1903 ft	(580 m)
***Kjellfossen** (Norway)	1841 ft	(561 m)

Largest deserts

NB – Most of Antarctica is a polar desert, with only 50mm of precipitation annually

Sahara	3,450,000 sq miles	(9,065,000 sq km)
Gobi	500,000 sq miles	(1,295,000 sq km)
Ar Rub al Khali	289,600 sq miles	(750,000 sq km)
Great Victorian	249,800 sq miles	(647,000 sq km)
Sonoran	120,000 sq miles	(311,000 sq km)
Kalahari	120,000 sq miles	(310,800 sq km)
Kara Kum	115,800 sq miles	(300,000 sq km)
Takla Makan	100,400 sq miles	(260,000 sq km)
Namib	52,100 sq miles	(135,000 sq km)
Thar	33,670 sq miles	(130,000 sq km)

Hottest inhabited places

Djibouti (Djibouti)	86° F	(30 °C)
Tombouctou (Mali)	84.7° F	(29.3 °C)
Tirunelveli (India)		
Tuticorin (India)		
Nellore (India)	84.5° F	(29.2 °C)
Santa Marta (Colombia)		
Aden (Yemen)	84° F	(28.9 °C)
Madurai (India)		
Niamey (Niger)		
Hodeida (Yemen)	83.8° F	(28.8 °C)
Ouagadougou (Burkina Faso)		
Thanjavur (India)		
Tiruchchirappalli (India)		

Driest inhabited places

Aswân (Egypt)	0.02 in	(0.5 mm)
Luxor (Egypt)	0.03 in	(0.7 mm)
Arica (Chile)	0.04 in	(1.1 mm)
Ica (Peru)	0.1 in	(2.3 mm)
Antofagasta (Chile)	0.2 in	(4.9 mm)
Al Minya (Egypt)	0.2 in	(5.1 mm)
Asyut (Egypt)	0.2 in	(5.2 mm)
Callao (Peru)	0.5 in	(12.0 mm)
Trujillo (Peru)	0.55 in	(14.0 mm)
Al Fayyum (Egypt)	0.8 in	(19.0 mm)

Wettest inhabited places

Mawsynram (India)	467 in	(11,862 mm)
Mount Waialeale (Hawaii, USA)	460 in	(11,684 mm)
Cherrapunji (India)	450 in	(11,430 mm)
Cape Debundsha (Cameroon)	405 in	(10,290 mm)
Quibdo (Colombia)	354 in	(8892 mm)
Buenaventura (Colombia)	265 in	(6743 mm)
Monrovia (Liberia)	202 in	(5131 mm)
Pago Pago (American Samoa)	196 in	(4990 mm)
Mawlamyine (Myanmar [Burma])	191 in	(4852 mm)
Lae (Papua New Guinea)	183 in	(4645 mm)

Standard time zones

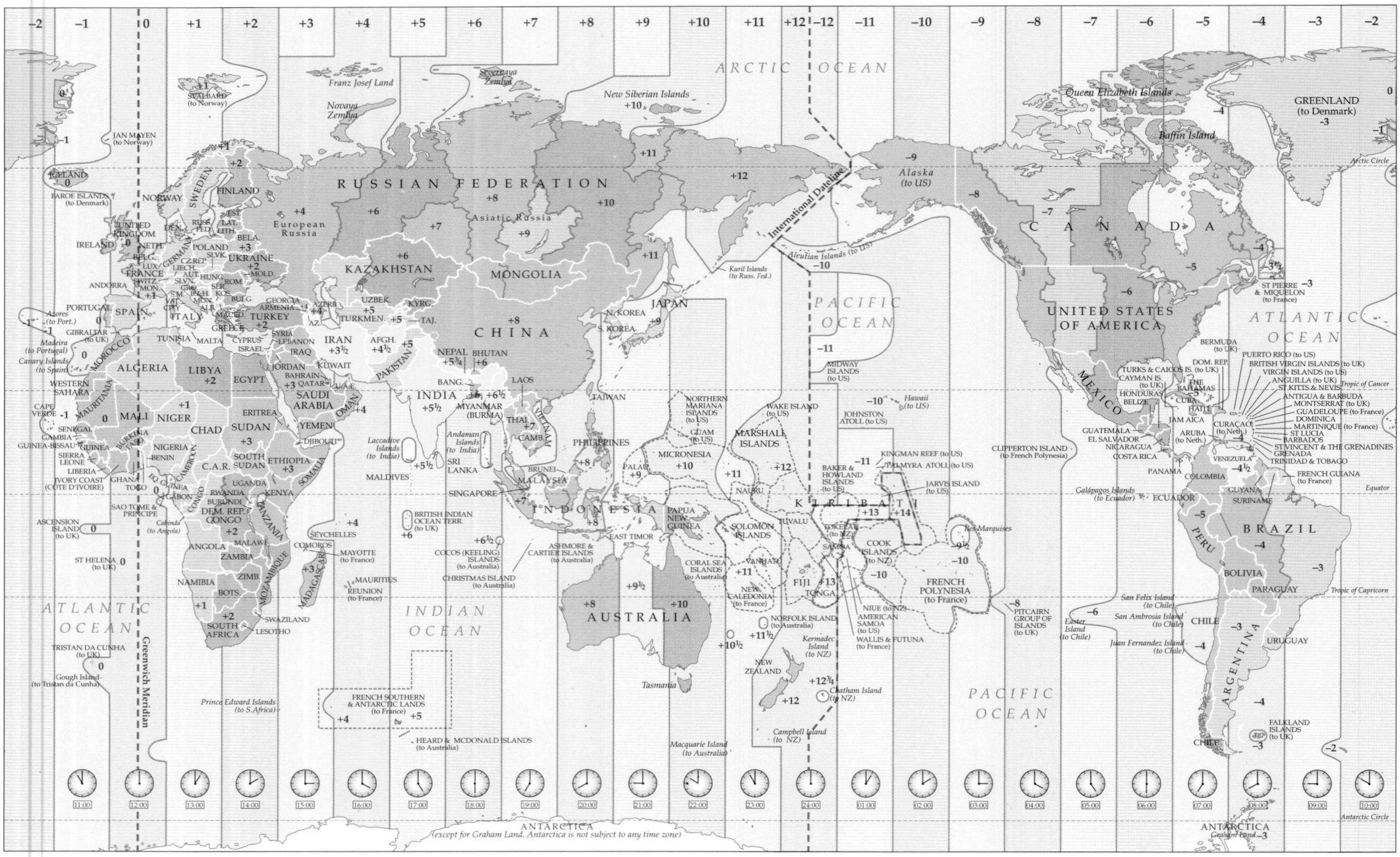

The numbers at the top of the map indicate the number of hours each time zone is ahead or behind Coordinated Universal Time (UTC). The clocks and 24-hour times given at the bottom of the map show the time in each time zone when it is 12:00 hours noon (UTC)

Time Zones

Because Earth is a rotating sphere, the Sun shines on only half of its surface at any one time. Thus, it is simultaneously morning, evening and night time in different parts of the world *(see diagram below)*. Because of these disparities, each country or part of a country adheres to a local time.

A region of Earth's surface within which a single local time is used is called a time zone. There are 24 one hour time zones around the world, arranged roughly in longitudinal bands.

Standard Time

Standard time is the official local time in a particular country or part of a country. It is defined by the time zone or zones associated with that country or region. Although time zones are arranged roughly in longitudinal bands, in many places the borders of a zone do not fall exactly on longitudinal meridians, as can be seen on the map *(above)*, but are determined by geographical factors or by borders between countries or parts of countries. Most countries have just one time zone and one standard time, but some large countries (such as the USA, Canada and Russia) are split between several time zones, so standard time varies across those countries. For example, the coterminous United States straddles four time zones and so has four standard times, called the Eastern, Central, Mountain and Pacific standard times. China is unusual in that just one standard time is used for the whole country, even though it extends across 60° of longitude from west to east.

Day and night around the world

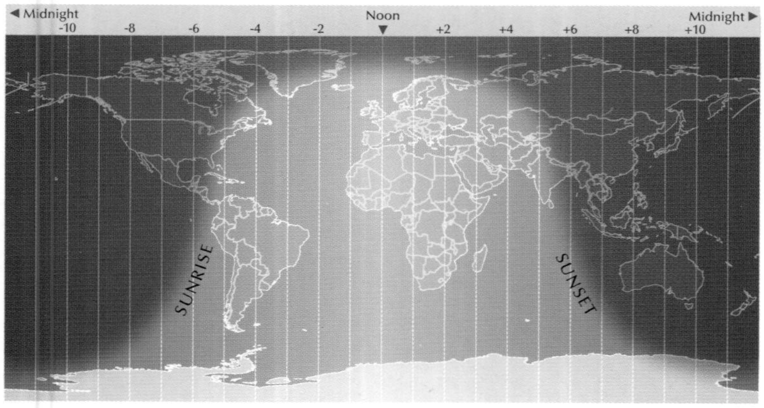

Coordinated Universal Time (UTC)

Coordinated Universal Time (UTC) is a reference by which the local time in each time zone is set. For example, Australian Western Standard Time (the local time in Western Australia) is set 8 hours ahead of UTC (it is UTC+8) whereas Eastern Standard Time in the United States is set 5 hours behind UTC (it is UTC-5). UTC is a successor to, and closely approximates, Greenwich Mean Time (GMT). However, UTC is based on an atomic clock, whereas GMT is determined by the Sun's position in the sky relative to the 0° longitudinal meridian, which runs through Greenwich, UK.

The International Dateline

The International Dateline is an imaginary line from pole to pole that roughly corresponds to the 180° longitudinal meridian. It is an arbitrary marker between calendar days. The dateline is needed because of the use of local times around the world rather than a single universal time. When moving from west to east across the dateline, travellers have to set their watches back one day. Those travelling in the opposite direction, from east to west, must add a day.

Daylight Saving Time

Daylight saving is a summertime adjustment to the local time in a country or region, designed to cause a higher proportion of its citizens' waking hours to pass during daylight. To follow the system, timepieces are advanced by an hour on a pre-decided date in spring and reverted back in autumn. About half of the world's nations use daylight saving.

Countries of the World

There are currently 196 independent countries in the world and almost 60 dependencies. Antarctica is the only land area on Earth that is not officially part of, and does not belong to, any single country.

In 1950, the world comprised 82 countries. In the decades following, many more states came into being as they achieved independence from their former colonial rulers. Most recent additions were caused by the breakup of the former Soviet Union in 1991, and the former Yugoslavia in 1992, which swelled the ranks of independent states. In July 2011, South Sudan became the latest country to be formed after declaring independence from Sudan.

Country factfile key

Formation Date of political origin or independence/ date current borders were established

Population Total population / population density – based on total *land* area

Languages An asterisk (*) denotes the official language(s)

Calorie consumption Average number of kilocalories consumed daily per person

AFGHANISTAN
Central Asia

Official name Islamic Republic of Afghanistan
Formation 1919 / 1919
Capital Kabul
Population 31.3 million / 124 people per sq mile (48 people per sq km)
Total area 250,000 sq. miles (647,500 sq. km)
Languages Pashtu*, Tajik, Dari*, Farsi, Uzbek, Turkmen
Religions Sunni Muslim 80%, Shi'a Muslim 19%, Other 1%
Ethnic mix Pashtun 38%, Tajik 25%, Hazara 19%, Uzbek and Turkmen 15%, Other 3%
Government Nonparty system
Currency Afghani = 100 puls
Literacy rate rate 32%
Calorie consumption 2090 kilocalories

ALBANIA
Southeast Europe

Official name Republic of Albania
Formation 1912 / 1921
Capital Tirana
Population 3.2 million / 302 people per sq mile (117 people per sq km)
Total area 11,100 sq. miles (28,748 sq. km)
Languages Albanian*, Greek
Religions Sunni Muslim 70%, Albanian Orthodox 20%, Roman Catholic 10%
Ethnic mix Albanian 98%, Greek 1%, Other 1%
Government Parliamentary system
Currency Lek = 100 qindarka (qintars)
Literacy rate 97%
Calorie consumption 3023 kilocalories

ALGERIA
North Africa

Official name People's Democratic Republic of Algeria
Formation 1962 / 1962
Capital Algiers
Population 39.9 million / 43 people per sq mile (17 people per sq km)
Total area 919,590 sq. miles (2,381,740 sq. km)
Languages Arabic*, Tamazight (Kabyle, Shawia, Tamashek), French
Religions Sunni Muslim 99%, Christian and Jewish 1%
Ethnic mix Arab 75%, Berber 24%, European and Jewish 1%
Government Presidential system
Currency Algerian dinar = 100 centimes
Literacy rate 73%
Calorie consumption 3296 kilocalories

ANDORRA
Southwest Europe

Official name Principality of Andorra
Formation 1278 / 1278
Capital Andorra la Vella
Population 85,485 / 475 people per sq mile (184 people per sq km)
Total area 181 sq. miles (468 sq. km)
Languages Spanish, Catalan*, French, Portuguese
Religions Roman Catholic 94%, Other 6%
Ethnic mix Spanish 46%, Andorran 28%, Other 18%, French 8%
Government Parliamentary system
Currency Euro = 100 cents
Literacy rate 99%
Calorie consumption Not available

ANGOLA
Southern Africa

Official name Republic of Angola
Formation 1975 / 1975
Capital Luanda
Population 22.1 million / 46 people per sq mile (18 people per sq km)
Total area 481,351 sq. miles (1,246,700 sq. km)
Languages Portuguese*, Umbundu, Kimbundu, Kikongo
Religions Roman Catholic 68%, Protestant 20%, Indigenous beliefs 12%
Ethnic mix Ovimbundu 37%, Kimbundu 25%, Other 25%, Bakongo 13%
Government Presidential system
Currency Readjusted kwanza = 100 lwei
Literacy rate 71%
Calorie consumption 2473 kilocalories

ANTIGUA & BARBUDA
West Indies

Official name Antigua and Barbuda
Formation 1981 / 1981
Capital St. John's
Population 91,295 / 537 people per sq mile (207 people per sq km)
Total area 170 sq. miles (442 sq. km)
Languages English*, English patois
Religions Anglican 45%, Other Protestant 42%, Roman Catholic 10%, Other 2%, Rastafarian 1%
Ethnic mix Black African 95%, Other 5%
Government Parliamentary system
Currency East Caribbean dollar = 100 cents
Literacy rate 99%
Calorie consumption 2396 kilocalories

ARGENTINA
South America

Official name Argentine Republic
Formation 1816 / 1816
Capital Buenos Aires
Population 41.8 million / 40 people per sq mile (15 people per sq km)
Total area 1,068,296 sq. miles (2,766,890 sq. km)
Languages Spanish*, Italian, Amerindian languages
Religions Roman Catholic 70%, Other 18%, Protestant 9%, Muslim 2%, Jewish 1%
Ethnic mix Indo-European 97%, Mestizo 2%, Amerindian 1%
Government Presidential system
Currency Argentine peso = 100 centavos
Literacy rate 98%
Calorie consumption 3155 kilocalories

ARMENIA
Southwest Asia

Official name Republic of Armenia
Formation 1991 / 1991
Capital Yerevan
Population 3 million / 261 people per sq mile (101 people per sq km)
Total area 11,506 sq. miles (29,800 sq. km)
Languages Armenian*, Azeri, Russian
Religions Armenian Apostolic Church (Orthodox) 88%, Armenian Catholic Church 6%, Other 6%
Ethnic mix Armenian 98%, Other 1%, Yezidi 1%
Government Parliamentary system
Currency Dram = 100 luma
Literacy rate 99%
Calorie consumption 2809 kilocalories

AUSTRALIA
Australasia & Oceania

Official name Commonwealth of Australia
Formation 1901 / 1901
Capital Canberra
Population 23.6 million / 8 people per sq mile (3 people per sq km)
Total area 2,967,893 sq. miles (7,686,850 sq. km)
Languages English*, Italian, Cantonese, Greek, Arabic, Vietnamese, Aboriginal languages
Religions Roman Catholic 26%, Nonreligious 19%, Anglican 19%, Other 17%, Other Christian 13%, United Church 6%
Ethnic mix European origin 50%, Australian 25.5%, other 19%, Asian 5%, Aboriginal 0.5%
Government Parliamentary system
Currency Australian dollar = 100 cents
Literacy rate 99%
Calorie consumption 3265 kilocalories

AUSTRIA
Central Europe

Official name Republic of Austria
Formation 1918 / 1919
Capital Vienna
Population 8.5 million / 266 people per sq mile (103 people per sq km)
Total area 32,378 sq. miles (83,858 sq. km)
Languages German*, Croatian, Slovenian, Hungarian (Magyar)
Religions Roman Catholic 78%, Nonreligious 9%, Other (including Jewish and Muslim) 8%, Protestant 5%
Ethnic mix Austrian 93%, Croat, Slovene, and Hungarian 6%, Other 1%
Government Parliamentary system
Currency Euro = 100 cents
Literacy rate 99%
Calorie consumption 3784 kilocalories

AZERBAIJAN
Southwest Asia

Official name Republic of Azerbaijan
Formation 1991 / 1991
Capital Baku
Population 9.5 million / 284 people per sq mile (110 people per sq km)
Total area 33,436 sq. miles (86,600 sq. km)
Languages Azeri*, Russian
Religions Shi'a Muslim 68%, Sunni Muslim 26%, Russian Orthodox 3%, Armenian Apostolic Church (Orthodox) 2%, Other 1%
Ethnic mix Azeri 91%, Other 3%, Lazs 2%, Armenian 2%, Russian 2%
Government Presidential system
Currency New manat = 100 gopik
Literacy rate 99%
Calorie consumption 2952 kilocalories

THE BAHAMAS
West Indies

Official name Commonwealth of The Bahamas
Formation 1973 / 1973
Capital Nassau
Population 400,000 / 103 people per sq mile (40 people per sq km)
Total area 5382 sq. miles (13,940 sq. km)
Languages English*, English Creole, French Creole
Religions Baptist 32%, Anglican 20%, Roman Catholic 19%, Other 17%, Methodist 6%, Church of God 6%
Ethnic mix Black African 85%, European 12%, Asian and Hispanic 3%
Government Parliamentary system
Currency Bahamian dollar = 100 cents
Literacy rate 96%
Calorie consumption 2575 kilocalories

BAHRAIN
Southwest Asia

Official name Kingdom of Bahrain
Formation 1971 / 1971
Capital Manama
Population 1.3 million / 4762 people per sq mile (1841 people per sq km)
Total area 239 sq. miles (620 sq. km)
Languages Arabic*
Religions Muslim (mainly Shi'a) 99%, Other 1%
Ethnic mix Bahraini 63%, Asian 19%, Other Arab 10%, Iranian 8%
Government Mixed monarchical–parliamentary system
Currency Bahraini dinar = 1000 fils
Literacy rate 95%
Calorie consumption Not available

BANGLADESH
South Asia

Official name People's Republic of Bangladesh
Formation 1971 / 1971
Capital Dhaka
Population 159 million / 3066 people per sq mile (1184 people per sq km)
Total area 55,598 sq. miles (144,000 sq. km)
Languages Bengali*, Urdu, Chakma, Marma (Magh), Garo, Khasi, Santhali, Tripuri, Mro
Religions Muslim (mainly Sunni) 88%, Hindu 11%, Other 1%
Ethnic mix Bengali 98%, Other 2%
Government Parliamentary system
Currency Taka = 100 poisha
Literacy rate 59%
Calorie consumption 2450 kilocalories

BARBADOS
West Indies

Official name Barbados
Formation 1966 / 1966
Capital Bridgetown
Population 300,000 / 1807 people per sq mile (698 people per sq km)
Total area 166 sq. miles (430 sq. km)
Languages Bajan (Barbadian English), English*
Religions Anglican 40%, Other 24%, Nonreligious 17%, Pentecostal 8%, Methodist 7%, Roman Catholic 4%
Ethnic mix Black African 92%, White 3%, Other 3%, Mixed race 2%
Government Parliamentary system
Currency Barbados dollar = 100 cents
Literacy rate 99%
Calorie consumption 3047 kilocalories

BELARUS
Eastern Europe

Official name Republic of Belarus
Formation 1991 / 1991
Capital Minsk
Population 9.3 million / 116 people per sq mile (45 people per sq km)
Total area 80,154 sq. miles (207,600 sq. km)
Languages Belarussian*, Russian*
Religions Orthodox Christian 80%, Roman Catholic 14%, Other 4%, Protestant 2%
Ethnic mix Belarussian 81%, Russian 11%, Polish 4%, Ukrainian 2%, Other 2%
Government Presidential system
Currency Belarussian rouble = 100 kopeks
Literacy rate 99%
Calorie consumption 3253 kilocalories

BELGIUM
Northwest Europe

Official name Kingdom of Belgium
Formation 1830 / 1919
Capital Brussels
Population 11.1 million / 876 people per sq mile (338 people per sq km)
Total area 11,780 sq. miles (30,510 sq. km)
Languages Dutch*, French*, German*
Religions Roman Catholic 88%, Other 10%, Muslim 2%
Ethnic mix Fleming 58%, Walloon 33%, Other 6%, Italian 2%, Moroccan 1%
Government Parliamentary system
Currency Euro = 100 cents
Literacy rate 99%
Calorie consumption 3793 kilocalories

BELIZE
Central America

Official name Belize
Formation 1981 / 1981
Capital Belmopan
Population 300,000 / 34 people per sq mile (13 people per sq km)
Total area 8867 sq. miles (22,966 sq. km)
Languages English Creole, Spanish, English*, Mayan, Garifuna (Carib)
Religions Roman Catholic 62%, Other 13%, Anglican 12%, Methodist 6%, Mennonite 4%, Seventh-day Adventist 3%
Ethnic mix Mestizo 49%, Creole 25%, Maya 11%, Garifuna 6%, Other 6%, Asian Indian 3%
Government Parliamentary system
Currency Belizean dollar = 100 cents
Literacy rate 75%
Calorie consumption 2751 kilocalories

BENIN
West Africa

Official name Republic of Benin
Formation 1960 / 1960
Capital Porto-Novo
Population 10.6 million / 248 people per sq mile (96 people per sq km)
Total area 43,483 sq. miles (112,620 sq. km)
Languages Fon, Bariba, Yoruba, Adja, Houeda, Somba, French*
Religions Indigenous beliefs and Voodoo 50%, Christian 30%, Muslim 20%
Ethnic mix Fon 41%, Other 21%, Adja 16%, Yoruba 12%, Bariba 10%
Government Presidential system
Currency CFA franc = 100 centimes
Literacy rate 29%
Calorie consumption 2594 kilocalories

BHUTAN
South Asia

Official name Kingdom of Bhutan
Formation 1656 / 1865
Capital Thimphu
Population 800,000 / 44 people per sq mile (17 people per sq km)
Total area 18,147 sq. miles (47,000 sq. km)
Languages Dzongkha*, Nepali, Assamese
Religions Mahayana Buddhist 75%, Hindu 25%
Ethnic mix Drukpa 50%, Nepalese 35%, Other 15%
Government Mixed monarchical–parliamentary system
Currency Ngultrum = 100 chetrum
Literacy rate 53%
Calorie consumption Not available

BOLIVIA
South America

Official name Plurinational State of Bolivia
Formation 1825 / 1938
Capital La Paz (administrative); Sucre (judicial)
Population 10.8 million / 26 people per sq mile (10 people per sq km)
Total area 424,162 sq. miles (1,098,580 sq. km)
Languages Aymara*, Quechua*, Spanish*
Religions Roman Catholic 93%, Other 7%
Ethnic mix Quechua 37%, Aymara 32%, Mixed race 13%, European 10%, Other 8%
Government Presidential system
Currency Boliviano = 100 centavos
Literacy rate 94%
Calorie consumption 2254 kilocalories

BOSNIA & HERZEGOVINA
Southeast Europe

Official name Bosnia and Herzegovina
Formation 1992 / 1992
Capital Sarajevo
Population 3.8 million / 192 people per sq mile (74 people per sq km)
Total area 19,741 sq. miles (51,129 sq. km)
Languages Bosnian*, Serbian*, Croatian*
Religions Muslim (mainly Sunni) 40%, Orthodox Christian 31%, Roman Catholic 15%, Other 10%, Protestant 4%
Ethnic mix Bosniak 48%, Serb 34%, Croat 16%, Other 2%
Government Parliamentary system
Currency Marka = 100 pfeninga
Literacy rate 98%
Calorie consumption 3130 kilocalories

BOTSWANA
Southern Africa

Official name Republic of Botswana
Formation 1966 / 1966
Capital Gaborone
Population 2 million / 9 people per sq mile (4 people per sq km)
Total area 231,803 sq. miles (600,370 sq. km)
Languages Setswana, English*, Shona, San, Khoikhoi, isiNdebele
Religions Christian (mainly Protestant) 70%, Nonreligious 20%, Traditional beliefs 6%, Other (including Muslim) 4%
Ethnic mix Tswana 79%, Kalanga 11%, Other 10%
Government Presidential system
Currency Pula = 100 thebe
Literacy rate 87%
Calorie consumption 2285 kilocalories

BRAZIL
South America

Official name Federative Republic of Brazil
Formation 1822 / 1828
Capital Brasília
Population 202 million / 62 people per sq mile (24 people per sq km)
Total area 3,286,470 sq. miles (8,511,965 sq. km)
Languages Portuguese*, German, Italian, Spanish, Polish, Japanese, Amerindian languages
Religions Roman Catholic 74%, Protestant 15%, Atheist 7%, Other 3%, Afro-American Spiritist 1%
Ethnic mix White 54%, Mixed race 38%, Black 6%, Other 2%
Government Presidential system
Currency Real = 100 centavos
Literacy rate 91%
Calorie consumption 3263 kilocalories

BRUNEI
Southeast Asia

Official name Brunei Darussalam
Formation 1984 / 1984
Capital Bandar Seri Begawan
Population 400,000 / 197 people per sq mile (76 people per sq km)
Total area 2228 sq. miles (5770 sq. km)
Languages Malay*, English, Chinese
Religions Muslim (mainly Sunni) 66%, Buddhist 14%, Other 10%, Christian 10%
Ethnic mix Malay 67%, Chinese 16%, Other 11%, Indigenous 6%
Government Monarchy
Currency Brunei dollar = 100 cents
Literacy rate 95%
Calorie consumption 2949 kilocalories

BULGARIA
Southeast Europe

Official name Republic of Bulgaria
Formation 1908 / 1947
Capital Sofia
Population 7.2 million / 169 people per sq mile (65 people per sq km)
Total area 42,822 sq. miles (110,910 sq. km)
Languages Bulgarian*, Turkish, Romani
Religions Bulgarian Orthodox 83%, Muslim 12%, Other 4%, Roman Catholic 1%
Ethnic mix Bulgarian 84%, Turkish 9%, Roma 5%, Other 2%
Government Parliamentary system
Currency Lev = 100 stotinki
Literacy rate 98%
Calorie consumption 2877 kilocalories

BURKINA FASO
West Africa

Official name Burkina Faso
Formation 1960 / 1960
Capital Ouagadougou
Population 17.4 million / 165 people per sq mile (64 people per sq km)
Total area 105,869 sq. miles (274,200 sq. km)
Languages Mossi, Fulani, French*, Tuare g, Dyula, Songhai
Religions Muslim 55%, Christian 25%, Traditional beliefs 20%
Ethnic mix Mossi 48%, Other 21%, Peul 10%, Lobi 7%, Bobo 7%, Mandé 7%
Government Transitional regime
Currency CFA franc = 100 centimes
Literacy rate 29%
Calorie consumption 2720 kilocalories

BURUNDI
Central Africa

Official name Republic of Burundi
Formation 1962 / 1962
Capital Bujumbura
Population 10.5 million / 1060 people per sq mile (409 people per sq km)
Total area 10,745 sq. miles (27,830 sq. km)
Languages Kirundi*, French*, Kiswahili
Religions Roman Catholic 62%, Traditional beliefs 23%, Muslim 10%, Protestant 5%
Ethnic mix Hutu 85%, Tutsi 14%, Twa 1%
Government Presidential system
Currency Burundian franc = 100 centimes
Literacy rate 87%
Calorie consumption 1604 kilocalories

CAMBODIA
Southeast Asia

Official name Kingdom of Cambodia
Formation 1953 / 1953
Capital Phnom Penh
Population 15.4 million / 226 people per sq mile (87 people per sq km)
Total area 69,900 sq. miles (181,040 sq. km)
Languages Khmer*, French, Chinese, Vietnamese, Cham
Religions Buddhist 93%, Muslim 6%, Christian 1%
Ethnic mix Khmer 90%, Vietnamese 5%, Other 4%, Chinese 1%
Government Parliamentary system
Currency Riel = 100 sen
Literacy rate 74%
Calorie consumption 2411 kilocalories

CAMEROON
Central Africa

Official name Republic of Cameroon
Formation 1960 / 1961
Capital Yaoundé
Population 22.8 million / 127 people per sq mile (49 people per sq km)
Total area 183,567 sq. miles (475,400 sq. km)
Languages Bamileke, Fang, Fulani, French*, English*
Religions Roman Catholic 35%, Traditional beliefs 25%, Muslim 22%, Protestant 18%
Ethnic mix Cameroon highlanders 31%, Other 21%, Equatorial Bantu 19%, Kirdi 11%, Fulani 10%, Northwestern Bantu 8%
Government Presidential system
Currency CFA franc = 100 centimes
Literacy rate 71%
Calorie consumption 2586 kilocalories

CANADA
North America

Official name Canada
Formation 1867 / 1949
Capital Ottawa
Population 35.5 million / 10 people per sq mile (4 people per sq km)
Total area 3,855,171 sq. miles (9,984,670 sq. km)
Languages English*, French*, Chinese, Italian, German, Ukrainian, Portuguese, Inuktitut, Cree
Religions Roman Catholic 44%, Protestant 29%, Other and nonreligious 27%
Ethnic mix European origin 66%, other 27%, Asian 5%, Amerindian 2%
Government Parliamentary system
Currency Canadian dollar = 100 cents
Literacy rate 99%
Calorie consumption 3419 kilocalories

CAPE VERDE
Atlantic Ocean

Official name Republic of Cape Verde
Formation 1975 / 1975
Capital Praia
Population 500,000 / 321 people per sq mile (124 people per sq km)
Total area 1557 sq. miles (4033 sq. km)
Languages Portuguese Creole, Portuguese*
Religions Roman Catholic 97%, Other 2%, Protestant (Church of the Nazarene) 1%
Ethnic mix Mestiço 71%, African 28%, European 1%
Government Mixed presidential–parliamentary system
Currency Escudo = 100 centavos
Literacy rate 85%
Calorie consumption 2716 kilocalories

CENTRAL AFRICAN REPUBLIC
Central Africa

Official name Central African Republic
Formation 1960 / 1960
Capital Bangui
Population 4.7 million / 20 people per sq mile (8 people per sq km)
Total area 240,534 sq. miles (622,984 sq. km)
Languages Sango, Banda, Gbaya, French*
Religions Traditional beliefs 35%, Roman Catholic 25%, Protestant 25%, Muslim 15%
Ethnic mix Baya 33%, Banda 27%, Other 17%, Mandjia 13%, Sara 10%
Government Transitional regime
Currency CFA franc = 100 centimes
Literacy rate 37%
Calorie consumption 2154 kilocalories

CHAD
Central Africa

Official name Republic of Chad
Formation 1960 / 1960
Capital N'Djaména
Population 13.2 million / 27 people per sq mile (10 people per sq km)
Total area 495,752 sq. miles (1,284,000 sq. km)
Languages French*, Sara, Arabic*, Maba
Religions Muslim 51%, Christian 35%, Animist 7%, Traditional beliefs 7%
Ethnic mix Other 30%, Sara 28%, Mayo-Kebbi 12%, Arab 12%, Ouaddai 9%, Kanem-Bornou 9%
Government Presidential system
Currency CFA franc = 100 centimes
Literacy rate 37%
Calorie consumption 2110 kilocalories

CHILE
South America

Official name Republic of Chile
Formation 1818 / 1883
Capital Santiago
Population 17.8 million / 62 people per sq mile (24 people per sq km)
Total area 292,258 sq. miles (756,950 sq. km)
Languages Spanish*, Amerindian languages
Religions Roman Catholic 89%, Other and nonreligious 11%
Ethnic mix Mestizo and European 90%, Other Amerindian 9%, Mapuche 1%
Government Presidential system
Currency Chilean peso = 100 centavos
Literacy rate 99%
Calorie consumption 2989 kilocalories

CHINA
East Asia

Official name People's Republic of China
Formation 960 / 1999
Capital Beijing
Population 1.39 billion / 387 people per sq mile (149 people per sq km)
Total area 3,705,386 sq. miles (9,596,960 sq. km)
Languages Mandarin*, Wu, Cantonese, Hsiang, Min, Hakka, Kan
Religions Nonreligious 59%, Traditional beliefs 20%, Other 13%, Buddhist 6%, Muslim 2%
Ethnic mix Han 92%, Other 4%, Hui 1%, Miao 1%, Manchu 1%, Zhuang 1%
Government One-party state
Currency Renminbi (known as yuan) = 10 jiao = 100 fen
Literacy rate 95%
Calorie consumption 3108 kilocalories

COLOMBIA
South America

Official name Republic of Colombia
Formation 1819 / 1903
Capital Bogotá
Population 48.9 million / 122 people per sq mile (47 people per sq km)
Total area 439,733 sq. miles (1,138,910 sq. km)
Languages Spanish*, Wayuu, Páez, and other Amerindian languages
Religions Roman Catholic 95%, Other 5%
Ethnic mix Mestizo 58%, White 20%, European–African 14%, African 4%, African–Amerindian 3%, Amerindian 1%
Government Presidential system
Currency Colombian peso = 100 centavos
Literacy rate 94%
Calorie consumption 2804 kilocalories

COMOROS
Indian Ocean

Official name Union of the Comoros
Formation 1975 / 1975
Capital Moroni
Population 800,000 / 929 people per sq mile (359 people per sq km)
Total area 838 sq. miles (2170 sq. km)
Languages Arabic*, Comoran*, French*
Religions Muslim (mainly Sunni) 98%, Other 1%, Roman Catholic 1%
Ethnic mix Comoran 97%, Other 3%
Government Presidential system
Currency Comoros franc = 100 centimes
Literacy rate 76%
Calorie consumption 2139 kilocalories

CONGO
Central Africa

Official name Republic of the Congo
Formation 1960 / 1960
Capital Brazzaville
Population 4.6 million / 35 people per sq mile (13 people per sq km)
Total area 132,046 sq. miles (342,000 sq. km)
Languages Kongo, Teke, Lingala, French*
Religions Traditional beliefs 50%, Roman Catholic 35%, Protestant 13%, Muslim 2%
Ethnic mix Bakongo 51%, Teke 17%, Other 16%, Mbochi 11%, Mbédé 5%
Government Presidential system
Currency CFA franc = 100 centimes
Literacy rate 79%
Calorie consumption 2195 kilocalories

CONGO, DEM. REP.
Central Africa

Official name Democratic Republic of the Congo
Formation 1960 / 1960
Capital Kinshasa
Population 69.4 million / 79 people per sq mile (31 people per sq km)
Total area 905,563 sq. miles (2,345,410 sq. km)
Languages Kiswahili, Tshiluba, Kikongo, Lingala, French*
Religions Roman Catholic 50%, Protestant 20%, Traditional beliefs and other 10%, Muslim 10%, Kimbanguist 10%
Ethnic mix Other 55%, Mongo, Luba, Kongo, and Mangbetu-Azande 45%
Government Presidential system
Currency Congolese franc = 100 centimes
Literacy rate 61%
Calorie consumption 1585 kilocalories

COSTA RICA
Central America

Official name Republic of Costa Rica
Formation 1838 / 1838
Capital San José
Population 4.9 million / 249 people per sq mile (96 people per sq km)
Total area 19,730 sq. miles (51,100 sq. km)
Languages Spanish*, English Creole, Bribri, Cabecar
Religions Roman Catholic 71%, Evangelical 14%, Nonreligious 11%, Other 4%
Ethnic mix Mestizo and European 94%, Black 3%, Other 1%, Chinese 1%, Amerindian 1%
Government Presidential system
Currency Costa Rican colón = 100 céntimos
Literacy rate 97%
Calorie consumption 2898 kilocalories

CROATIA
Southeast Europe

Official name Republic of Croatia
Formation 1991 / 1991
Capital Zagreb
Population 4.3 million / 197 people per sq mile (76 people per sq km)
Total area 21,831 sq. miles (56,542 sq. km)
Languages Croatian*
Religions Roman Catholic 88%, Other 7%, Orthodox Christian 4%, Muslim 1%
Ethnic mix Croat 90%, Other 5%, Serb 5%
Government Parliamentary system
Currency Kuna = 100 lipa
Literacy rate 99%
Calorie consumption 3052 kilocalories

CUBA
West Indies

Official name Republic of Cuba
Formation 1902 / 1902
Capital Havana
Population 11.3 million / 264 people per sq mile (102 people per sq km)
Total area 42,803 sq. miles (110,860 sq. km)
Languages Spanish*
Religions Nonreligious 49%, Roman Catholic 40%, Atheist 6%, Other 4%, Protestant 1%
Ethnic mix Mulatto (mixed race) 51%, White 37%, Black 11%, Chinese 1%
Government One-party state
Currency Cuban peso = 100 centavos
Literacy rate 99%
Calorie consumption 3277 kilocalories

CYPRUS
Southeast Europe

Official name Republic of Cyprus
Formation 1960 / 1960
Capital Nicosia
Population 1.2 million / 336 people per sq mile (130 people per sq km)
Total area 3571 sq. miles (9250 sq. km)
Languages Greek*, Turkish*
Religions Orthodox Christian 78%, Muslim 18%, Other 4%
Ethnic mix Greek 81%, Turkish 11%, Other 8%
Government Presidential system
Currency Euro = 100 cents; (TRNC: new Turkish lira = 100 kurus)
Literacy rate 99%
Calorie consumption 2661 kilocalories

CZECH REPUBLIC
Central Europe

Official name Czech Republic
Formation 1993 / 1993
Capital Prague
Population 10.7 million / 351 people per sq mile (136 people per sq km)
Total area 30,450 sq. miles (78,866 sq. km)
Languages Czech*, Slovak, Hungarian (Magyar)
Religions Roman Catholic 39%, Atheist 38%, Other 18%, Protestant 3%, Hussite 2%
Ethnic mix Czech 90%, Moravian 4%, Other 4%, Slovak 2%
Government Parliamentary system
Currency Czech koruna = 100 haleru
Literacy rate 99%
Calorie consumption 3292 kilocalories

DENMARK
Northern Europe

Official name Kingdom of Denmark
Formation 950 / 1944
Capital Copenhagen
Population 5.6 million / 342 people per sq mile (132 people per sq km)
Total area 16,639 sq. miles (43,094 sq. km)
Languages Danish*
Religions Evangelical Lutheran 95%, Roman Catholic 3%, Muslim 2%
Ethnic mix Danish 96%, Other (including Scandinavian and Turkish) 3%, Faeroese and Inuit 1%
Government Parliamentary system
Currency Danish krone = 100 øre
Literacy rate 99%
Calorie consumption 3363 kilocalories

DJIBOUTI
East Africa

Official name Republic of Djibouti
Formation 1977 / 1977
Capital Djibouti
Population 900,000 / 101 people per sq mile (39 people per sq km)
Total area 8494 sq. miles (22,000 sq. km)
Languages Somali, Afar, French*, Arabic*
Religions Muslim (mainly Sunni) 94%, Christian 6%
Ethnic mix Issa 60%, Afar 35%, Other 5%
Government Presidential system
Currency Djibouti franc = 100 centimes
Literacy rate 70%
Calorie consumption 2526 kilocalories

DOMINICA
West Indies

Official name Commonwealth of Dominica
Formation 1978 / 1978
Capital Roseau
Population 73,449 / 253 people per sq mile (98 people per sq km)
Total area 291 sq. miles (754 sq. km)
Languages French Creole, English*
Religions Roman Catholic 77%, Protestant 15%, Other 8%
Ethnic mix Black 87%, Mixed race 9%, Carib 3%, Other 1%
Government Parliamentary system
Currency East Caribbean dollar = 100 cents
Literacy rate 88%
Calorie consumption 3047 kilocalories

DOMINICAN REPUBLIC
West Indies

Official name Dominican Republic
Formation 1865 / 1865
Capital Santo Domingo
Population 10.5 million / 562people per sq mile (217 people per sq km)
Total area 18,679 sq. miles (48,380 sq. km)
Languages Spanish*, French Creole
Religions Roman Catholic 95%, Other and nonreligious 5%
Ethnic mix Mixed race 73%, European 16%, Black African 11%
Government Presidential system
Currency Dominican Republic peso = 100 centavos
Literacy rate 91%
Calorie consumption 2614 kilocalories

EAST TIMOR
Southeast Asia

Official name Democratic Republic of Timor-Leste
Formation 2002 / 2002
Capital Dili
Population 1.2 million / 213 people per sq mile (82 people per sq km)
Total area 5756 sq. miles (14,874 sq. km)
Languages Tetum (Portuguese/Austronesian)*, Bahasa Indonesia, Portuguese*
Religions Roman Catholic 95%, Other (including Muslim and Protestant) 5%
Ethnic mix Papuan groups approx 85%, Indonesian approx 13%, Chinese 2%
Government Parliamentary system
Currency US dollar = 100 cents
Literacy rate 58%
Calorie consumption 2083 kilocalories

ECUADOR
South America

Official name Republic of Ecuador
Formation 1830 / 1942
Capital Quito
Population 16 million / 150 people per sq mile (58 people per sq km)
Total area 109,483 sq. miles (283,560 sq. km)
Languages Spanish*, Quechua, other Amerindian languages
Religions Roman Catholic 95%, Protestant, Jewish, and other 5%
Ethnic mix Mestizo 77%, White 11%, Amerindian 7%, Black African 5%
Government Presidential system
Currency US dollar = 100 cents
Literacy rate 93%
Calorie consumption 2477 kilocalories

EGYPT
North Africa

Official name Arab Republic of Egypt
Formation 1936 / 1982
Capital Cairo
Population 83.4 million / 217 people per sq mile (84 people per sq km)
Total area 386,660 sq. miles (1,001,450 sq. km)
Languages Arabic*, French, English, Berber
Religions Muslim (mainly Sunni) 90%, Coptic Christian and other 9%, Other Christian 1%
Ethnic mix Egyptian 99%, Nubian, Armenian, Greek, and Berber 1%
Government Transitional regime
Currency Egyptian pound = 100 piastres
Literacy rate 74%
Calorie consumption 3557 kilocalories

EL SALVADOR
Central America

Official name Republic of El Salvador
Formation 1841 / 1841
Capital San Salvador
Population 6.4 million / 800 people per sq mile (309 people per sq km)
Total area 8124 sq. miles (21,040 sq. km)
Languages Spanish*
Religions Roman Catholic 80%, Evangelical 18%, Other 2%
Ethnic mix Mestizo 90%, White 9%, Amerindian 1%
Government Presidential system
Currency Salvadorean colón = 100 centavos; and US dollar = 100 cents
Literacy rate 86%
Calorie consumption 2513 kilocalories

EQUATORIAL GUINEA
Central Africa

Official name Republic of Equatorial Guinea
Formation 1968 / 1968
Capital Malabo
Population 800,000 / 74 people per sq mile (29 people per sq km)
Total area 10,830 sq. miles (28,051 sq. km)
Languages Spanish*, Fang, Bubi, French*
Religions Roman Catholic 90%, Other 10%
Ethnic mix Fang 85%, Other 11%, Bubi 4%
Government Presidential system
Currency CFA franc = 100 centimes
Literacy rate 94%
Calorie consumption Not available

ERITREA
East Africa

Official name State of Eritrea
Formation 1993 / 2002
Capital Asmara
Population 6.5 million / 143 people per sq mile (55 people per sq km)
Total area 46,842 sq. miles (121,320 sq. km)
Languages Tigrinya*, English*, Tigre, Afar, Arabic*, Saho, Bilen, Kunama, Nara, Hadareb
Religions Christian 50%, Muslim 48%, Other 2%
Ethnic mix Tigray 50%, Tigre 31%, Other 9%, Afar 5%, Saho 5%
Government Mixed presidential–parliamentary system
Currency Nakfa = 100 cents
Literacy rate 70%
Calorie consumption 1640 kilocalories

ESTONIA
Northeast Europe

Official name Republic of Estonia
Formation 1991 / 1991
Capital Tallinn
Population 1.3 million / 75 people per sq mile (29 people per sq km)
Total area 17,462 sq. miles (45,226 sq. km)
Languages Estonian*, Russian
Religions Evangelical Lutheran 56%, Orthodox Christian 25%, Other 19%
Ethnic mix Estonian 69%, Russian 25%, Other 4%, Ukrainian 2%
Government Parliamentary system
Currency Euro = 100 cents
Literacy rate 99%
Calorie consumption 3214 kilocalories

ETHIOPIA
East Africa

Official name Federal Democratic Republic of Ethiopia
Formation 1896 / 2002
Capital Addis Ababa
Population 96.5 million / 225 people per sq mile (87 people per sq km)
Total area 435,184 sq. miles (1,127,127 sq. km)
Languages Amharic*, Tigrinya, Galla, Sidamo, Somali, English, Arabic
Religions Orthodox Christian 40%, Muslim 40%, Traditional beliefs 15%, Other 5%
Ethnic mix Oromo 40%, Amhara 25%, Other 13%, Sidama 9%, Tigray 7%, Somali 6%
Government Parliamentary system
Currency Birr = 100 cents
Literacy rate 39%
Calorie consumption 2131 kilocalories

FIJI
Australasia & Oceania

Official name Republic of Fiji
Formation 1970 / 1970
Capital Suva
Population 900,000 / 128 people per sq mile (49 people per sq km)
Total area 7054 sq. miles (18,270 sq. km)
Languages Fijian, English*, Hindi, Urdu, Tamil, Telugu
Religions Hindu 38%, Methodist 37%, Roman Catholic 9%, Muslim 8%, Other 8%
Ethnic mix Melanesian 51%, Indian 44%, Other 5%
Government Parliamentary system
Currency Fiji dollar = 100 cents
Literacy rate 94%
Calorie consumption 2930 kilocalories

FINLAND
Northern Europe

Official name Republic of Finland
Formation 1917 / 1947
Capital Helsinki
Population 5.4 million / 46 people per sq mile (18 people per sq km)
Total area 130,127 sq. miles (337,030 sq. km)
Languages Finnish*, Swedish*, Sámi
Religions Evangelical Lutheran 83%, Other 15%, Orthodox Christian 1%, Roman Catholic 1%
Ethnic mix Finnish 93%, Other (including Sámi) 7%
Government Parliamentary system
Currency Euro = 100 cents
Literacy rate 99%
Calorie consumption 3285 kilocalories

FRANCE
Western Europe

Official name French Republic
Formation 987 / 1919
Capital Paris
Population 64.6 million / 304 people per sq mile (117 people per sq km)
Total area 211,208 sq. miles (547,030 sq. km)
Languages French*, Provençal, German, Breton, Catalan, Basque
Religions Roman Catholic 88%, Muslim 8%, Protestant 2%, Buddhist 1%, Jewish 1%
Ethnic mix French 90%, North African (mainly Algerian) 6%, German (Alsace) 2%, Breton 1%, Other (including Corsicans) 1%
Government Mixed presidential–parliamentary system
Currency Euro = 100 cents
Literacy rate 99%
Calorie consumption 3524 kilocalories

GABON
Central Africa

Official name Gabonese Republic
Formation 1960 / 1960
Capital Libreville
Population 1.7 million / 17 people per sq mile (7 people per sq km)
Total area 103,346 sq. miles (267,667 sq. km)
Languages Fang, French*, Punu, Sira, Nzebi, Mpongwe
Religions Christian (mainly Roman Catholic) 55%, Traditional beliefs 40%, Other 4%, Muslim 1%
Ethnic mix Fang 26%, Shira-punu 24%, Other 16%, Foreign residents 15%, Nzabi-duma 11%, Mbédé-Teke 8%
Government Presidential system
Currency CFA franc = 100 centimes
Literacy rate 82%
Calorie consumption 2781 kilocalories

GAMBIA
West Africa

Official name Republic of the Gambia
Formation 1965 / 1965
Capital Banjul
Population 1.9 million / 492 people per sq mile (190 people per sq km)
Total area 4363 sq. miles (11,300 sq. km)
Languages Mandinka, Fulani, Wolof, Jola, Soninke, English*
Religions Sunni Muslim 90%, Christian 8%, Traditional beliefs 2%
Ethnic mix Mandinka 42%, Fulani 18%, Wolof 16%, Jola 10%, Serahuli 9%, Other 5%
Government Presidential system
Currency Dalasi = 100 butut
Literacy rate 52%
Calorie consumption 2849 kilocalories

GEORGIA
Southwest Asia

Official name Georgia
Formation 1991 / 1991
Capital Tbilisi
Population 4.3 million / 160 people per sq mile (62 people per sq km)
Total area 26,911 sq. miles (69,700 sq. km)
Languages Georgian*, Russian, Azeri, Armenian, Mingrelian, Ossetian, Abkhazian* (in Abkhazia)
Religions Georgian Orthodox 74%, Muslim 10%, Russian Orthodox 10%, Armenian Apostolic Church (Orthodox) 4%, Other 2%
Ethnic mix Georgian 84%, Azeri 6%, Armenian 6%, Russian 2%, Ossetian 1%, Other 1%
Government Presidential system
Currency Lari = 100 tetri
Literacy rate 99%
Calorie consumption 2731 kilocalories

GERMANY
Northern Europe

Official name Federal Republic of Germany
Formation 1871 / 1990
Capital Berlin
Population 82.7 million / 613 people per sq mile (237 people per sq km)
Total area 137,846 sq. miles (357,021 sq. km)
Languages German*, Turkish
Religions Protestant 34%, Roman Catholic 33%, Other 30%, Muslim 3%
Ethnic mix German 92%, Other European 3%, Other 3%, Turkish 2%
Government Parliamentary system
Currency Euro = 100 cents
Literacy rate 99%
Calorie consumption 3539 kilocalories

GHANA
West Africa

Official name Republic of Ghana
Formation 1957 / 1957
Capital Accra
Population 26.4 million / 297 people per sq mile (115 people per sq km)
Total area 92,100 sq. miles (238,540 sq. km)
Languages Twi, Fanti, Ewe, Ga, Adangbe, Gurma, Dagomba (Dagbani), English*
Religions Christian 69%, Muslim 16%, Traditional beliefs 9%, Other 6%
Ethnic mix Akan 49%, Mole-Dagbani 17%, Ewe 13%, Other 9%, Ga and Ga-Adangbe 8%, Guan 4%
Government Presidential system
Currency Cedi = 100 pesewas
Literacy rate 72%
Calorie consumption 3003 kilocalories

GREECE
Southeast Europe

Official name Hellenic Republic
Formation 1829 / 1947
Capital Athens
Population 11.1 million / 220 people per sq mile (85 people per sq km)
Total area 50,942 sq. miles (131,940 sq. km)
Languages Greek*, Turkish, Macedonian, Albanian
Religions Orthodox Christian 98%, Muslim 1%, Other 1%
Ethnic mix Greek 98%, Other 2%
Government Parliamentary system
Currency Euro = 100 cents
Literacy rate 97%
Calorie consumption 3433 kilocalories

GRENADA
West Indies

Official name Grenada
Formation 1974 / 1974
Capital St. George's
Population 110,152 / 841 people per sq mile (324 people per sq km)
Total area 131 sq. miles (340 sq. km)
Languages English*, English Creole
Religions Roman Catholic 68%, Anglican 17%, Other 15%
Ethnic mix Black African 82%, Mulatto (mixed race) 13%, East Indian 3%, Other 2%
Government Parliamentary system
Currency East Caribbean dollar = 100 cents
Literacy rate 96%
Calorie consumption 2453 kilocalories

GUATEMALA
Central America

Official name Republic of Guatemala
Formation 1838 / 1838
Capital Guatemala City
Population 15.9 million / 380 people per sq mile (147 people per sq km)
Total area 42,042 sq. miles (108,890 sq. km)
Languages Quiché, Mam, Cakchiquel, Kekchi, Spanish*
Religions Roman Catholic 65%, Protestant 33%, Other and nonreligious 2%
Ethnic mix Amerindian 60%, Mestizo 30%, Other 10%
Government Presidential system
Currency Quetzal = 100 centavos
Literacy rate 78%
Calorie consumption 2419 kilocalories

GUINEA
West Africa

Official name Republic of Guinea
Formation 1958 / 1958
Capital Conakry
Population 12 million / 126 people per sq mile (49 people per sq km)
Total area 94,925 sq. miles (245,857 sq. km)
Languages Pulaar, Malinké, Soussou, French*
Religions Muslim 85%, Christian 8%, Traditional beliefs 7%
Ethnic mix Peul 40%, Malinké 30%, Soussou 20%, Other 10%
Government Presidential system
Currency Guinea franc = 100 centimes
Literacy rate 25%
Calorie consumption 2553 kilocalories

GUINEA-BISSAU
West Africa

Official name Republic of Guinea-Bissau
Formation 1974 / 1974
Capital Bissau
Population 1.7 million / 157 people per sq mile (60 people per sq km)
Total area 13,946 sq. miles (36,120 sq. km)
Languages Portuguese Creole, Balante, Fulani, Malinké, Portuguese*
Religions Traditional beliefs 50%, Muslim 40%, Christian 10%
Ethnic mix Balante 30%, Fulani 20%, Other 16%, Mandyako 14%, Mandinka 13%, Papel 7%
Government Presidential system
Currency CFA franc = 100 centimes
Literacy rate 57%
Calorie consumption 2304 kilocalories

GUYANA
South America

Official name Cooperative Republic of Guyana
Formation 1966 / 1966
Capital Georgetown
Population 800,000 / 11 people per sq mile (4 people per sq km)
Total area 83,000 sq. miles (214,970 sq. km)
Languages English Creole, Hindi, Tamil, Amerindian languages, English*
Religions Christian 57%, Hindu 28%, Muslim 10%, Other 5%
Ethnic mix East Indian 43%, Black African 30%, Mixed race 17%, Amerindian 9%, Other 1%
Government Presidential system
Currency Guyanese dollar = 100 cents
Literacy rate 85%
Calorie consumption 2648 kilocalories

HAITI
West Indies

Official name Republic of Haiti
Formation 1804 / 1844
Capital Port-au-Prince
Population 10.5 million / 987 people per sq mile (381 people per sq km)
Total area 10,714 sq. miles (27,750 sq. km)
Languages French Creole*, French*
Religions Roman Catholic 55%, Protestant 28%, Other (including Voodoo) 16%, Nonreligious 1%
Ethnic mix Black African 95%, Mulatto (mixed race) and European 5%
Government Presidential system
Currency Gourde = 100 centimes
Literacy rate 49%
Calorie consumption 2091 kilocalories

HONDURAS
Central America

Official name Republic of Honduras
Formation 1838 / 1838
Capital Tegucigalpa
Population 8.3 million / 192 people per sq mile (74 people per sq km)
Total area 43,278 sq. miles (112,090 sq. km)
Languages Spanish*, Garifuna (Carib), English Creole
Religions Roman Catholic 97%, Protestant 3%
Ethnic mix Mestizo 90%, Black African 5%, Amerindian 4%, White 1%
Government Presidential system
Currency Lempira = 100 centavos
Literacy rate 85%
Calorie consumption 2651 kilocalories

HUNGARY
Central Europe

Official name Hungary
Formation 1918 / 1947
Capital Budapest
Population 9.9 million / 278 people per sq mile (107 people per sq km)
Total area 35,919 sq. miles (93,030 sq. km)
Languages Hungarian (Magyar)*
Religions Roman Catholic 52%, Calvinist 16%, Other 15%, Nonreligious 14%, Lutheran 3%
Ethnic mix Magyar 90%, Roma 4%, German 3%, Serb 2%, Other 1%
Government Parliamentary system
Currency Forint = 100 fillér
Literacy rate 99%
Calorie consumption 2968 kilocalories

ICELAND
Northwest Europe

Official name Republic of Iceland
Formation 1944 / 1944
Capital Reykjavík
Population 300,000 / 8 people per sq mile (3 people per sq km)
Total area 39,768 sq. miles (103,000 sq. km)
Languages Icelandic*
Religions Evangelical Lutheran 84%, Other (mostly Christian) 10%, Roman Catholic 3%, Nonreligious 3%
Ethnic mix Icelandic 94%, Other 5%, Danish 1%
Government Parliamentary system
Currency Icelandic króna = 100 aurar
Literacy rate 99%
Calorie consumption 3339 kilocalories

INDIA
South Asia

Official name Republic of India
Formation 1947 / 1947
Capital New Delhi
Population 1.27 billion / 1104 people per sq mile (426 people per sq km)
Total area 1,269,339 sq. miles (3,287,590 sq. km)
Languages Hindi*, English*, Urdu, Bengali, Marathi, Telugu, Tamil, Bihari, Gujarati, Kanarese
Religions Hindu 81%, Muslim 13%, Christian 2%, Sikh 2%, Buddhist 1%, Other 1%
Ethnic mix Indo-Aryan 72%, Dravidian 25%, Mongoloid and other 3%
Government Parliamentary system
Currency Indian rupee = 100 paise
Literacy rate 63%
Calorie consumption 2459 kilocalories

INDONESIA
Southeast Asia

Official name Republic of Indonesia
Formation 1949 / 1999
Capital Jakarta
Population 253 million / 364 people per sq mile (141 people per sq km)
Total area 741,096 sq. miles (1,919,440 sq. km)
Languages Javanese, Sundanese, Madurese, Bahasa Indonesia*, Dutch
Religions Sunni Muslim 86%, Protestant 6%, Roman Catholic 3%, Hindu 2%, Other 2%, Buddhist 1%
Ethnic mix Javanese 41%, Other 29%, Sundanese 15%, Coastal Malays 12%, Madurese 3%
Government Presidential system
Currency Rupiah = 100 sen
Literacy rate 93%
Calorie consumption 2777 kilocalories

IRAN
Southwest Asia

Official name Islamic Republic of Iran
Formation 1502 / 1990
Capital Tehran
Population 78.5 million / 124 people per sq mile (48 people per sq km)
Total area 636,293 sq. miles (1,648,000 sq. km)
Languages Farsi*, Azeri, Luri, Gilaki, Mazanderani, Kurdish, Turkmen, Arabic, Baluchi
Religions Shi'a Muslim 89%, Sunni Muslim 9%, Other 2%
Ethnic mix Persian 51%, Azari 24%, Other 10%, Lur and Bakhtiari 8%, Kurdish 7%
Government Islamic theocracy
Currency Iranian rial = 100 dinars
Literacy rate 84%
Calorie consumption 3058 kilocalories

IRAQ
Southwest Asia

Official name Republic of Iraq
Formation 1932 / 1990
Capital Baghdad
Population 34.8 million / 206 people per sq mile (80 people per sq km)
Total area 168,753 sq. miles (437,072 sq. km)
Languages Arabic*, Kurdish*, Turkic languages, Armenian, Assyrian
Religions Shi'a Muslim 60%, Sunni Muslim 35%, Other (including Christian) 5%
Ethnic mix Arab 80%, Kurdish 15%, Turkmen 3%, Other 2%
Government Parliamentary system
Currency New Iraqi dinar = 1000 fils
Literacy rate 79%
Calorie consumption 2489 kilocalories

IRELAND
Northwest Europe

Official name Ireland
Formation 1922 / 1922
Capital Dublin
Population 4.7 million / 177 people per sq mile (68 people per sq km)
Total area 27,135 sq. miles (70,280 sq. km)
Languages English*, Irish*
Religions Roman Catholic 87%, Other and nonreligious 10%, Anglican 3%
Ethnic mix Irish 99%, Other 1%
Government Parliamentary system
Currency Euro = 100 cents
Literacy rate 99%
Calorie consumption 3591 kilocalories

ISRAEL
Southwest Asia

Official name State of Israel
Formation 1948 / 1994
Capital Jerusalem (not internationally recognized)
Population 7.8 million / 994 people per sq mile (384 people per sq km)
Total area 8019 sq. miles (20,770 sq. km)
Languages Hebrew*, Arabic*, Yiddish, German, Russian, Polish, Romanian, Persian
Religions Jewish 76%, Muslim (mainly Sunni) 16%, Other 4%, Druze 2%, Christian 2%
Ethnic mix Jewish 76%, Arab 20%, Other 4%
Government Parliamentary system
Currency Shekel = 100 agorot
Literacy rate 98%
Calorie consumption 3619 kilocalories

ITALY
Southern Europe

Official name Italian Republic
Formation 1861 / 1947
Capital Rome
Population 61.1 million / 538 people per sq mile (208 people per sq km)
Total area 116,305 sq. miles (301,230 sq. km)
Languages Italian*, German, French, Rhaeto-Romanic, Sardinian
Religions Roman Catholic 85%, Other and nonreligious 13%, Muslim 2%
Ethnic mix Italian 94%, Other 4%, Sardinian 2%
Government Parliamentary system
Currency Euro = 100 cents
Literacy rate 99%
Calorie consumption 3539 kilocalories

IVORY COAST
West Africa

Official name Republic of Côte d'Ivoire
Formation 1960 / 1960
Capital Yamoussoukro
Population 20.8 million / 169 people per sq mile (65 people per sq km)
Total area 124,502 sq. miles (322,460 sq. km)
Languages Akan, French*, Krou, Voltaique
Religions Muslim 38%, Traditional beliefs 25%, Roman Catholic 25%, Other 6%, Protestant 6%
Ethnic mix Akan 42%, Voltaique 18%, Mandé du Nord 17%, Krou 11%, Mandé du Sud 10%, Other 2%
Government Presidential system
Currency CFA franc = 100 centimes
Literacy rate 41%
Calorie consumption 2799 kilocalories

JAMAICA
West Indies

Official name Jamaica
Formation 1962 / 1962
Capital Kingston
Population 2.8 million / 670 people per sq mile (259 people per sq km)
Total area 4243 sq. miles (10,990 sq. km)
Languages English Creole, English*
Religions Other and nonreligious 45%, Other Protestant 20%, Church of God 18%, Baptist 10%, Anglican 7%
Ethnic mix Black 91%, Mulatto (mixed race) 7%, European and Chinese 1%, East Indian 1%
Government Parliamentary system
Currency Jamaican dollar = 100 cents
Literacy rate 88%
Calorie consumption 2746 kilocalories

JAPAN
East Asia

Official name Japan
Formation 1590 / 1972
Capital Tokyo
Population 127 million / 874 people per sq mile (337 people per sq km)
Total area 145,882 sq. miles (377,835 sq. km)
Languages Japanese*, Korean, Chinese
Religions Shinto and Buddhist 76%, Buddhist 16%, Other (including Christian) 8%
Ethnic mix Japanese 99%, Other (mainly Korean) 1%
Government Parliamentary system
Currency Yen = 100 sen
Literacy rate 99%
Calorie consumption 2719 kilocalories

JORDAN
Southwest Asia

Official name Hashemite Kingdom of Jordan
Formation 1946 / 1967
Capital Amman
Population 7.5 million / 218 people per sq mile (84 people per sq km)
Total area 35,637 sq. miles (92,300 sq. km)
Languages Arabic*
Religions Sunni Muslim 92%, Christian 6%, Other 2%
Ethnic mix Arab 98%, Circassian 1%, Armenian 1%
Government Monarchy
Currency Jordanian dinar = 1000 fils
Literacy rate 98%
Calorie consumption 3149 kilocalories

KAZAKHSTAN
Central Asia

Official name Republic of Kazakhstan
Formation 1991 / 1991
Capital Astana
Population 16.6 million / 16 people per sq mile (6 people per sq km)
Total area 1,049,150 sq. miles (2,717,300 sq. km)
Languages Kazakh*, Russian, Ukrainian, German, Uzbek, Tatar, Uighur
Religions Muslim (mainly Sunni) 47%, Orthodox Christian 44%, Other 7%, Protestant 2%
Ethnic mix Kazakh 57%, Russian 27%, Other 8%, Uzbek 3%, Ukrainian 2%, German 2%
Government Presidential system
Currency Tenge = 100 tiyn
Literacy rate 99%
Calorie consumption 3107 kilocalories

KENYA
East Africa

Official name Republic of Kenya
Formation 1963 / 1963
Capital Nairobi
Population 45.5 million / 208 people per sq mile (80 people per sq km)
Total area 224,961 sq. miles (582,650 sq. km)
Languages Kiswahili*, English*, Kikuyu, Luo, Kalenjin, Kamba
Religions Christian 80%, Muslim 10%, Traditional beliefs 9%, Other 1%
Ethnic mix Other 28%, Kikuyu 22%, Luo 14%, Luhya 14%, Kalenjin 11%, Kamba 11%
Government Presidential system
Currency Kenya shilling = 100 cents
Literacy rate 72%
Calorie consumption 2206 kilocalories

KIRIBATI
Australasia & Oceania

Official name Republic of Kiribati
Formation 1979 / 1979
Capital Tarawa Atoll
Population 104,488 / 381 people per sq mile (147 people per sq km)
Total area 277 sq. miles (717 sq. km)
Languages English*, Kiribati
Religions Roman Catholic 55%, Kiribati Protestant Church 36%, Other 9%
Ethnic mix Micronesian 99%, Other 1%
Government Presidential system
Currency Australian dollar = 100 cents
Literacy rate 99%
Calorie consumption 3022 kilocalories

KOSOVO (not yet recognised)
Southeast Europe

Official name Republic of Kosovo
Formation 2008 / 2008
Capital Pristina
Population 1.9 million / 451 people per sq mile (174 people per sq km)
Total area 4212 sq. miles (10,908 sq. km)
Languages Albanian*, Serbian*, Bosniak, Gorani, Roma, Turkish
Religions Muslim 92%, Roman Catholic 4%, Orthodox Christian 4%
Ethnic mix Albanian 92%, Serb 4%, Bosniak and Gorani 2%, Turkish 1%, Roma 1%
Government Parliamentary system
Currency Euro = 100 cents
Literacy rate 92%
Calorie consumption Not available

KUWAIT
Southwest Asia

Official name State of Kuwait
Formation 1961 / 1961
Capital Kuwait City
Population 3.5 million / 509 people per sq mile (196 people per sq km)
Total area 6880 sq. miles (17,820 sq. km)
Languages Arabic*, English
Religions Sunni Muslim 45%, Shi'a Muslim 40%, Christian, Hindu, and other 15%
Ethnic mix Kuwaiti 45%, Other Arab 35%, South Asian 9%, Other 7%, Iranian 4%
Government Monarchy
Currency Kuwaiti dinar = 1000 fils
Literacy rate 96%
Calorie consumption 3471 kilocalories

KYRGYZSTAN
Central Asia

Official name Kyrgyz Republic
Formation 1991 / 1991
Capital Bishkek
Population 5.6 million / 73 people per sq mile (28 people per sq km)
Total area 76,641 sq. miles (198,500 sq. km)
Languages Kyrgyz*, Russian*, Uzbek, Tatar, Ukrainian
Religions Muslim (mainly Sunni) 70%, Orthodox Christian 30%
Ethnic mix Kyrgyz 69%, Uzbek 14%, Russian 9%, Other 6%, Dungan 1%, Uighur 1%
Government Presidential system
Currency Som = 100 tyiyn
Literacy rate 99%
Calorie consumption 2828 kilocalories

LAOS
Southeast Asia

Official name Lao People's Democratic Republic
Formation 1953 / 1953
Capital Vientiane
Population 6.9 million / 77 people per sq mile (30 people per sq km)
Total area 91,428 sq. miles (236,800 sq. km)
Languages Lao*, Mon-Khmer, Yao, Vietnamese, Chinese, French
Religions Buddhist 65%, Other (including animist) 34%, Christian 1%
Ethnic mix Lao Loum 66%, Lao Theung 30%, Lao Soung 2%, Other 2%
Government One-party state
Currency Kip = 100 at
Literacy rate 73%
Calorie consumption 2356 kilocalories

LATVIA
Northeast Europe

Official name Republic of Latvia
Formation 1991 / 1991
Capital Riga
Population 2 million / 80 people per sq mile (31 people per sq km)
Total area 24,938 sq. miles (64,589 sq. km)
Languages Latvian*, Russian
Religions Other 43%, Lutheran 24%, Roman Catholic 18%, O rthodox Christian 15%
Ethnic mix Latvian 62%, Russian 27%, Other 4%, Belarussian 3%, Ukrainian 2%, Polish 2%
Government Parliamentary system
Currency Euro = 100 cents
Literacy rate 99%
Calorie consumption 3293 kilocalories

LEBANON
Southwest Asia

Official name Lebanese Republic
Formation 1941 / 1941
Capital Beirut
Population 5 million / 1266 people per sq mile (489 people per sq km)
Total area 4015 sq. miles (10,400 sq. km)
Languages Arabic*, French, Armenian, Assyrian
Religions Muslim 60%, Christian 39%, Other 1%
Ethnic mix Arab 95%, Armenian 4%, Other 1%
Government Parliamentary system
Currency Lebanese pound = 100 piastres
Literacy rate 90%
Calorie consumption 3181 kilocalories

LESOTHO
Southern Africa

Official name Kingdom of Lesotho
Formation 1966 / 1966
Capital Maseru
Population 2.1 million / 179 people per sq mile (69 people per sq km)
Total area 11,720 sq. miles (30,355 sq. km)
Languages English*, Sesotho*, isiZulu
Religions Christian 90%, Traditional beliefs 10%
Ethnic mix Sotho 99%, European and Asian 1%
Government Parliamentary system
Currency Loti = 100 lisente; and South African rand = 100 cents
Literacy rate 76%
Calorie consumption 2595 kilocalories

LIBERIA
West Africa

Official name Republic of Liberia
Formation 1847 / 1847
Capital Monrovia
Population 4.4 million / 118 people per sq mile (46 people per sq km)
Total area 43,000 sq. miles (111,370 sq. km)
Languages Kpelle, Vai, Bassa, Kru, Grebo, Kissi, Gola, Loma, English*
Religions Christian 40%, Traditional beliefs 40%, Muslim 20%
Ethnic mix Indigenous tribes (12 groups) 49%, Kpellé 20%, Bassa 16%, Gio 8%, Krou 7%
Government Presidential system
Currency Liberian dollar = 100 cents
Literacy rate 43%
Calorie consumption 2251 kilocalories

LIBYA
North Africa

Official name State of Libya
Formation 1951 / 1951
Capital Tripoli
Population 6.3 million / 9 people per sq mile (4 people per sq km)
Total area 679,358 sq. miles (1,759,540 sq. km)
Languages Arabic*, Tuareg
Religions Muslim (mainly Sunni) 97%, Other 3%
Ethnic mix Arab and Berber 97%, Other 3%
Government Transitional regime
Currency Libyan dinar = 1000 dirhams
Literacy rate 90%
Calorie consumption 3211 kilocalories

LIECHTENSTEIN
Central Europe

Official name Principality of Liechtenstein
Formation 1719 / 1719
Capital Vaduz
Population 37,313 / 602 people per sq mile (233 people per sq km)
Total area 62 sq. miles (160 sq. km)
Languages German*, Alemannish dialect, Italian
Religions Roman Catholic 79%, Other 13%, Protestant 8%
Ethnic mix Liechtensteiner 66%, Other 12%, Swiss 10%, Austrian 6%, German 3%, Italian 3%
Government Parliamentary system
Currency Swiss franc = 100 rappen/centimes
Literacy rate 99%
Calorie consumption Not available

LITHUANIA
Northeast Europe

Official name Republic of Lithuania
Formation 1991 / 1991
Capital Vilnius
Population 3 million / 119 people per sq mile (46 people per sq km)
Total area 25,174 sq. miles (65,200 sq. km)
Languages Lithuanian*, Russian
Religions Roman Catholic 77%, Other 17%, Russian Orthodox 4%, Protestant 1%, Old believers 1%
Ethnic mix Lithuanian 85%, Polish 7%, Russian 6%, Belarussian 1%, Other 1%
Government Parliamentary system
Currency Euro = 100 cents
Literacy rate 99%
Calorie consumption 3463 kilocalories

LUXEMBOURG
Northwest Europe

Official name Grand Duchy of Luxembourg
Formation 1867 / 1867
Capital Luxembourg-Ville
Population 500,000 / 501 people per sq mile (193 people per sq km)
Total area 998 sq. miles (2586 sq. km)
Languages Luxembourgish*, German*, French*
Religions Roman Catholic 97%, Protestant, Orthodox Christian, and Jewish 3%
Ethnic mix Luxembourger 62%, Foreign residents 38%
Government Parliamentary system
Currency Euro = 100 cents
Literacy rate 99%
Calorie consumption 3568 kilocalories

MACEDONIA
Southeast Europe

Official name Republic of Macedonia
Formation 1991 / 1991
Capital Skopje
Population 2.1 million / 212 people per sq mile (82 people per sq km)
Total area 9781 sq. miles (25,333 sq. km)
Languages Macedonian*, Albanian*, Turkish, Romani, Serbian
Religions Orthodox Christian 65%, Muslim 29%, Roman Catholic 4%, Other 2%
Ethnic mix Macedonian 64%, Albanian 25%, Turkish 4%, Roma 3%, Serb 2%, Other 2%
Government Mixed presidential–parliamentary system
Currency Macedonian denar = 100 deni
Literacy rate 98%
Calorie consumption 2923 kilocalories

MADAGASCAR
Indian Ocean

Official name Republic of Madagascar
Formation 1960 / 1960
Capital Antananarivo
Population 23.6 million / 105 people per sq mile (41 people per sq km)
Total area 226,656 sq. miles (587,040 sq. km)
Languages Malagasy*, French*, English*
Religions Traditional beliefs 52%, Christian (mainly Roman Catholic) 41%, Muslim 7%
Ethnic mix Other Malay 46%, Merina 26%, Betsimisaraka 15%, Betsileo 12%, Other 1%
Government Mixed presidential–parliamentary system
Currency Ariary = 5 iraimbilanja
Literacy rate 64%
Calorie consumption 2052 kilocalories

MALAWI
Southern Africa

Official name Republic of Malawi
Formation 1964 / 1964
Capital Lilongwe
Population 16.8 million / 463 people per sq mile (179 people per sq km)
Total area 45,745 sq. miles (118,480 sq. km)
Languages Chewa, Lomwe, Yao, Ngoni, English*
Religions Protestant 55%, Roman Catholic 20%, Muslim 20%, Traditional beliefs 5%
Ethnic mix Bantu 99%, Other 1%
Government Presidential system
Currency Malawi kwacha = 100 tambala
Literacy rate 61%
Calorie consumption 2334 kilocalories

MALAYSIA
Southeast Asia

Official name Malaysia
Formation 1963 / 1965
Capital Kuala Lumpur; Putrajaya (administrative)
Population 30.2 million / 238 people per sq mile (92 people per sq km)
Total area 127,316 sq. miles (329,750 sq. km)
Languages Bahasa Malaysia*, Malay, Chinese, Tamil, English
Religions Muslim (mainly Sunni) 61%, Buddhist 19%, Christian 9%, Hindu 6%, Other 5%
Ethnic mix Malay 53%, Chinese 26%, Indigenous tribes 12%, Indian 8%, Other 1%
Government Parliamentary system
Currency Ringgit = 100 sen
Literacy rate 93%
Calorie consumption 2855 kilocalories

MALDIVES
Indian Ocean

Official name Republic of Maldives
Formation 1965 / 1965
Capital Male'
Population 400,000 / 3448 people per sq mile (1333 people per sq km)
Total area 116 sq. miles (300 sq. km)
Languages Dhivehi (Maldivian), Sinhala, Tamil, Arabic
Religions Sunni Muslim 100%
Ethnic mix Arab–Sinhalese–Malay 100%
Government Presidential system
Currency Rufiyaa = 100 laari
Literacy rate 98%
Calorie consumption 2722 kilocalories

MALI
West Africa

Official name Republic of Mali
Formation 1960 / 1960
Capital Bamako
Population 15.8 million / 34 people per sq mile (13 people per sq km)
Total area 478,764 sq. miles (1,240,000 sq. km)
Languages Bambara, Fulani, Senufo, Soninke, French*
Religions Muslim (mainly Sunni) 90%, Traditional beliefs 6%, Christian 4%
Ethnic mix Bambara 52%, Other 14%, Fulani 11%, Saracolé 7%, Soninka 7%, Tuareg 5%, Mianka 4%
Government Presidential system
Currency CFA franc = 100 centimes
Literacy rate 34%
Calorie consumption 2833 kilocalories

MALTA
Southern Europe

Official name Republic of Malta
Formation 1964 / 1964
Capital Valletta
Population 400,000 / 3226 people per sq mile (1250 people per sq km)
Total area 122 sq. miles (316 sq. km)
Languages Maltese*, English*
Religions Roman Catholic 98%, Other and nonreligious 2%
Ethnic mix Maltese 96%, Other 4%
Government Parliamentary system
Currency Euro = 100 cents
Literacy rate 92%
Calorie consumption 3389 kilocalories

MARSHALL ISLANDS
Australasia & Oceania

Official name Republic of the Marshall Islands
Formation 1986 / 1986
Capital Majuro
Population 70,983 / 1014 people per sq mile (392 people per sq km)
Total area 70 sq. miles (181 sq. km)
Languages Marshallese*, English*, Japanese, German
Religions Protestant 90%, Roman Catholic 8%, Other 2%
Ethnic mix Micronesian 90%, Other 10%
Government Presidential system
Currency US dollar = 100 cents
Literacy rate 91%
Calorie consumption Not available

MAURITANIA
West Africa

Official name Islamic Republic of Mauritania
Formation 1960 / 1960
Capital Nouakchott
Population 4 million / 10 people per sq mile (4 people per sq km)
Total area 397,953 sq. miles (1,030,700 sq. km)
Languages Arabic*, Hassaniyah Arabic, Wolof, French
Religions Sunni Muslim 100%
Ethnic mix Maure 81%, Wolof 7%, Tukolor 5%, Other 4%, Soninka 3%
Government Presidential system
Currency Ouguiya = 5 khoums
Literacy rate 46%
Calorie consumption 2791 kilocalories

MAURITIUS
Indian Ocean

Official name Republic of Mauritius
Formation 1968 / 1968
Capital Port Louis
Population 1.2 million / 1671 people per sq mile (645 people per sq km)
Total area 718 sq. miles (1860 sq. km)
Languages French Creole, Hindi, Urdu, Tamil, Chinese, English*, French
Religions Hindu 48%, Roman Catholic 24%, Muslim 17%, Protestant 9%, Other 2%
Ethnic mix Indo-Mauritian 68%, Creole 27%, Sino-Mauritian 3%, Franco-Mauritian 2%
Government Parliamentary system
Currency Mauritian rupee = 100 cents
Literacy rate 89%
Calorie consumption 3055 kilocalories

MEXICO
North America

Official name United Mexican States
Formation 1836 / 1848
Capital Mexico City
Population 124 million / 168 people per sq mile (65 people per sq km)
Total area 761,602 sq. miles (1,972,550 sq. km)
Languages Spanish*, Nahuatl, Mayan, Zapotec, Mixtec, Otomi, Totonac, Tzotzil, Tzeltal
Religions Roman Catholic 77%, Other 14%, Protestant 6%, Nonreligious 3%
Ethnic mix Mestizo 60%, Amerindian 30%, European 9%, Other 1%
Government Presidential system
Currency Mexican peso = 100 centavos
Literacy rate 94%
Calorie consumption 3072 kilocalories

MICRONESIA
Australasia & Oceania

Official name Federated States of Micronesia
Formation 1986 / 1986
Capital Palikir (Pohnpei Island)
Population 105,681 / 390 people per sq mile (151 people per sq km)
Total area 271 sq. miles (702 sq. km)
Languages Trukese, Pohnpeian, Kosraean, Yapese, English*
Religions Roman Catholic 50%, Protestant 47%, Other 3%
Ethnic mix Chuukese 49%, Pohnpeian 24%, Other 14%, Kosraean 6%, Yapese 5%, Asian 2%
Government Nonparty system
Currency US dollar = 100 cents
Literacy rate 81%
Calorie consumption Not available

MOLDOVA
Southeast Europe

Official name Republic of Moldova
Formation 1991 / 1991
Capital Chisinau
Population 3.5 million / 269 people per sq mile (104 people per sq km)
Total area 13,067 sq. miles (33,843 sq. km)
Languages Moldovan*, Ukrainian, Russian
Religions Orthodox Christian 93%, Other 6%, Baptist 1%
Ethnic mix Moldovan 84%, Ukrainian 7%, Gagauz 5%, Russian 2%, Bulgarian 1%, Other 1%
Government Parliamentary system
Currency Moldovan leu = 100 bani
Literacy rate 99%
Calorie consumption 2837 kilocalories

MONACO
Southern Europe

Official name Principality of Monaco
Formation 1861 / 1861
Capital Monaco-Ville
Population 36,950 / 49,267 people per sq mile (18,949 people per sq km)
Total area 0.75 sq. miles (1.95 sq. km)
Languages French*, Italian, Monégasque, English
Religions Roman Catholic 89%, Protestant 6%, Other 5%
Ethnic mix French 47%, Other 21%, Italian 16%, Monégasque 16%
Government Mixed monarchical–parliamentary system
Currency Euro = 100 cents
Literacy rate 99%
Calorie consumption Not available

MONGOLIA
East Asia

Official name Mongolia
Formation 1924 / 1924
Capital Ulan Bator
Population 2.9 million / 5 people per sq mile (2 people per sq km)
Total area 604,247 sq. miles (1,565,000 sq. km)
Languages Khalkha Mongolian, Kazakh, Chinese, Russian
Religions Tibetan Buddhist 50%, Nonreligious 40%, Shamanist and Christian 6%, Muslim 4%
Ethnic mix Khalkh 95%, Kazakh 4%, Other 1%
Government Mixed presidential–parliamentary system
Currency Tugrik (tögrög) = 100 möngö
Literacy rate 98%
Calorie consumption 2463 kilocalories

MONTENEGRO
Southeast Europe

Official name Montenegro
Formation 2006 / 2006
Capital Podgorica
Population 600,000 / 113 people per sq mile (43 people per sq km)
Total area 5332 sq. miles (13,812 sq. km)
Languages Montenegrin*, Serbian, Albanian, Bosniak, Croatian
Religions Orthodox Christian 74%, Muslim 18%, Roman Catholic 4%, Other 4%
Ethnic mix Montenegrin 43%, Serb 32%, Other 12%, Bosniak 8%, Albanian 5%
Government Parliamentary system
Currency Euro = 100 cents
Literacy rate 98%
Calorie consumption 3568 kilocalories

MOROCCO
North Africa

Official name Kingdom of Morocco
Formation 1956 / 1969
Capital Rabat
Population 35.5 million / 194 people per sq mile (75 people per sq km)
Total area 172,316 sq. miles (446,300 sq. km)
Languages Arabic*, Tamazight (Berber), French, Spanish
Religions Muslim (mainly Sunni) 99%, Other (mostly Christian) 1%
Ethnic mix Arab 70%, Berber 29%, European 1%
Government Mixed monarchical–parliamentary system
Currency Moroccan dirham = 100 centimes
Literacy rate 67%
Calorie consumption 3334 kilocalories

MOZAMBIQUE
Southern Africa

Official name Republic of Mozambique
Formation 1975 / 1975
Capital Maputo
Population 26.5 million / 88 people per sq mile (34 people per sq km)
Total area 309,494 sq. miles (801,590 sq. km)
Languages Makua, Xitsonga, Sena, Lomwe, Portuguese*
Religions Traditional beliefs 56%, Christian 30%, Muslim 14%
Ethnic mix Makua Lomwe 47%, Tsonga 23%, Malawi 12%, Shona 11%, Yao 4%, Other 3%
Government Presidential system
Currency New metical = 100 centavos
Literacy rate 51%
Calorie consumption 2283 kilocalories

MYANMAR (BURMA)
Southeast Asia

Official name Republic of the Union of Myanmar
Formation 1948 / 1948
Capital Nay Pyi Taw
Population 53.7 million / 212 people per sq mile (82 people per sq km)
Total area 261,969 sq. miles (678,500 sq. km)
Languages Myanmar (Burmese)*, Shan, Karen, Rakhine, Chin, Yangbye, Kachin, Mon
Religions Buddhist 89%, Christian 4%, Muslim 4%, Other 2%, Animist 1%
Ethnic mix Burman (Bamah) 68%, Other 12%, Shan 9%, Karen 7%, Rakhine 4%
Government Presidential system
Currency Kyat = 100 pyas
Literacy rate 93%
Calorie consumption 2571 kilocalories

NAMIBIA
Southern Africa

Official name Republic of Namibia
Formation 1990 / 1994
Capital Windhoek
Population 2.3 million / 7 people per sq mile (3 people per sq km)
Total area 318,694 sq. miles (825,418 sq. km)
Languages Ovambo, Kavango, English*, Bergdama, German, Afrikaans
Religions Christian 90%, Traditional beliefs 10%
Ethnic mix Ovambo 50%, Other tribes 22%, Kavango 9%, Damara 7%, Herero 7%, Other 5%
Government Presidential system
Currency Namibian dollar = 100 cents; and South African rand = 100 cents
Literacy rate 76%
Calorie consumption 2086 kilocalories

NAURU
Australasia & Oceania

Official name Republic of Nauru
Formation 1968 / 1968
Capital None
Population 9488 / 1171 people per sq mile (452 people per sq km)
Total area 8.1 sq. miles (21 sq. km)
Languages Nauruan*, Kiribati, Chinese, Tuvaluan, English
Religions Nauruan Congregational Church 60%, Roman Catholic 35%, Other 5%
Ethnic mix Nauruan 93%, Chinese 5%, European 1%, Other Pacific islanders 1%
Government Nonparty system
Currency Australian dollar = 100 cents
Literacy rate 95%
Calorie consumption Not available

NEPAL
South Asia

Official name Federal Democratic Republic of Nepal
Formation 1769 / 1769
Capital Kathmandu
Population 28.1 million / 532 people per sq mile (205 people per sq km)
Total area 54,363 sq. miles (140,800 sq. km)
Languages Nepali*, Maithili, Bhojpuri
Religions Hindu 81%, Buddhist 11%, Muslim 4%, Other (including Christian) 4%
Ethnic mix Other 52%, Chhetri 16%, Hill Brahman 13%, Tharu 7%, Magar 7%, Tamang 5%
Government Transitional regime
Currency Nepalese rupee = 100 paisa
Literacy rate 57%
Calorie consumption 2673 kilocalories

NETHERLANDS
Northwest Europe

Official name Kingdom of the Netherlands
Formation 1648 / 1839
Capital Amsterdam; The Hague (administrative)
Population 16.8 million / 1283 people per sq mile (495 people per sq km)
Total area 16,033 sq. miles (41,526 sq. km)
Languages Dutch*, Frisian
Religions Roman Catholic 36%, Other 34%, Protestant 27%, Muslim 3%
Ethnic mix Dutch 82%, Other 12%, Surinamese 2%, Turkish 2%, Moroccan 2%
Government Parliamentary system
Currency Euro = 100 cents
Literacy rate 99%
Calorie consumption 3147 kilocalories

NEW ZEALAND
Australasia & Oceania

Official name New Zealand
Formation 1947 / 1947
Capital Wellington
Population 4.6 million / 44 people per sq mile (17 people per sq km)
Total area 103,737 sq. miles (268,680 sq. km)
Languages English*, Maori*
Religions Anglican 24%, Other 22%, Presbyterian 18%, Nonreligious 16%, Roman Catholic 15%, Methodist 5%
Ethnic mix European 75%, Maori 15%, Other 7%, Samoan 3%
Government Parliamentary system
Currency New Zealand dollar = 100 cents
Literacy rate 99%
Calorie consumption 3170 kilocalories

NICARAGUA
Central America

Official name Republic of Nicaragua
Formation 1838 / 1838
Capital Managua
Population 6.2 million / 135 people per sq mile (52 people per sq km)
Total area 49,998 sq. miles (129,494 sq. km)
Languages Spanish*, English Creole, Miskito
Religions Roman Catholic 80%, Protestant Evangelical 17%, Other 3%
Ethnic mix Mestizo 69%, White 17%, Black 9%, Amerindian 5%
Government Presidential system
Currency Córdoba oro = 100 centavos
Literacy rate 78%
Calorie consumption 2564 kilocalories

NIGER
West Africa

Official name Republic of Niger
Formation 1960 / 1960
Capital Niamey
Population 18.5 million / 38 people per sq mile (15 people per sq km)
Total area 489,188 sq. miles (1,267,000 sq. km)
Languages Hausa, Djerma, Fulani, Tuareg, Teda, French*
Religions Muslim 99%, Other (including Christian) 1%
Ethnic mix Hausa 53%, Djerma and Songhai 21%, Tuareg 11%, Fulani 7%, Kanuri 6%, Other 2%
Government Presidential system
Currency CFA franc = 100 centimes
Literacy rate 16%
Calorie consumption 2546 kilocalories

NIGERIA
West Africa

Official name Federal Republic of Nigeria
Formation 1960 / 1961
Capital Abuja
Population 179 million / 508 people per sq mile (196 people per sq km)
Total area 356,667 sq. miles (923,768 sq. km)
Languages Hausa, English*, Yoruba, Ibo
Religions Muslim 50%, Christian 40%, Traditional beliefs 10%
Ethnic mix Other 29%, Hausa 21%, Yoruba 21%, Ibo 18%, Fulani 11%
Government Presidential system
Currency Naira = 100 kobo
Literacy rate 51%
Calorie consumption 2700 kilocalories

NORTH KOREA
East Asia

Official name Democratic People's Republic of Korea
Formation 1948 / 1953
Capital Pyongyang
Population 25 million / 538 people per sq mile (208 people per sq km)
Total area 46,540 sq. miles (120,540 sq. km)
Languages Korean*
Religions Atheist 100%
Ethnic mix Korean 100%
Government One-party state
Currency North Korean won = 100 chon
Literacy rate 99%
Calorie consumption 2094 kilocalories

NORWAY
Northern Europe

Official name Kingdom of Norway
Formation 1905 / 1905
Capital Oslo
Population 5.1 million / 43 people per sq mile (17 people per sq km)
Total area 125,181 sq. miles (324,220 sq. km)
Languages Norwegian* (Bokmål "book language" and Nynorsk "new Norsk"), Sámi
Religions Evangelical Lutheran 88%, Other and nonreligious 8%, Muslim 2%, Pentecostal 1%, Roman Catholic 1%
Ethnic mix Norwegian 93%, Other 6%, Sámi 1%
Government Parliamentary system
Currency Norwegian krone = 100 øre
Literacy rate 99%
Calorie consumption 3484 kilocalories

OMAN
Southwest Asia

Official name Sultanate of Oman
Formation 1951 / 1951
Capital Muscat
Population 3.9 million / 48 people per sq mile (18 people per sq km)
Total area 82,031 sq. miles (212,460 sq. km)
Languages Arabic*, Baluchi, Farsi, Hindi, Punjabi
Religions Ibadi Muslim 75%, Other Muslim and Hindu 25%
Ethnic mix Arab 88%, Baluchi 4%, Persian 3%, Indian and Pakistani 3%, African 2%
Government Monarchy
Currency Omani rial = 1000 baisa
Literacy rate 87%
Calorie consumption 3143 kilocalories

PAKISTAN
South Asia

Official name Islamic Republic of Pakistan
Formation 1947 / 1971
Capital Islamabad
Population 185 million / 622 people per sq mile (240 people per sq km)
Total area 310,401 sq. miles (803,940 sq. km)
Languages Punjabi, Sindhi, Pashtu, Urdu*, Baluchi, Brahui
Religions Sunni Muslim 77%, Shi'a Muslim 20%, Hindu 2%, Christian 1%
Ethnic mix Punjabi 56%, Pathan (Pashtun) 15%, Sindhi 14%, Mohajir 7%, Baluchi 4%, Other 4%
Government Parliamentary system
Currency Pakistani rupee = 100 paisa
Literacy rate 55%
Calorie consumption 2440 kilocalories

PALAU
Australasia & Oceania

Official name Republic of Palau
Formation 1994 / 1994
Capital Ngerulmud
Population 21,186 / 108 people per sq mile (42 people per sq km)
Total area 177 sq. miles (458 sq. km)
Languages Palauan*, English*, Japanese, Angaur, Tobi, Sonsorolese
Religions Christian 66%, Modekngei 34%
Ethnic mix Palauan 74%, Filipino 16%, Other 6%, Chinese and other Asian 4%
Government Nonparty system
Currency US dollar = 100 cents
Literacy rate 99%
Calorie consumption Not available

PANAMA
Central America

Official name Republic of Panama
Formation 1903 / 1903
Capital Panama City
Population 3.9 million / 133 people per sq mile (51 people per sq km)
Total area 30,193 sq. miles (78,200 sq. km)
Languages English Creole, Spanish*, Amerindian languages, Chibchan languages
Religions Roman Catholic 84%, Protestant 15%, Other 1%
Ethnic mix Mestizo 70%, Black 14%, White 10%, Amerindian 6%
Government Presidential system
Currency Balboa = 100 centésimos; and US dollar = 100 cents
Literacy rate 94%
Calorie consumption 2733 kilocalories

PAPUA NEW GUINEA
Australasia & Oceania

Official name Independent State of Papua New Guinea
Formation 1975 / 1975
Capital Port Moresby
Population 7.5 million / 43 people per sq mile (17 people per sq km)
Total area 178,703 sq. miles (462,840 sq. km)
Languages Pidgin English, Papuan, English*, Motu, 800 (est.) native languages
Religions Protestant 60%, Roman Catholic 37%, Other 3%
Ethnic mix Melanesian and mixed race 100%
Government Parliamentary system
Currency Kina = 100 toea
Literacy rate 63%
Calorie consumption 2193 kilocalories

PARAGUAY
South America

Official name Republic of Paraguay
Formation 1811 / 1938
Capital Asunción
Population 6.9 million / 45 people per sq mile (17 people per sq km)
Total area 157,046 sq. miles (406,750 sq. km)
Languages Guaraní*, Spanish*, German
Religions Roman Catholic 90%, Protestant (including Mennonite) 10%
Ethnic mix Mestizo 91%, Other 7%, Amerindian 2%
Government Presidential system
Currency Guaraní = 100 céntimos
Literacy rate 94%
Calorie consumption 2589 kilocalories

PERU
South America

Official name Republic of Peru
Formation 1824 / 1941
Capital Lima
Population 30.8 million / 62 people per sq mile (24 people per sq km)
Total area 496,223 sq. miles (1,285,200 sq. km)
Languages Spanish*, Quechua*, Aymara
Religions Roman Catholic 81%, Other 19%
Ethnic mix Amerindian 45%, Mestizo 37%, White 15%, Other 3%
Government Presidential system
Currency New sol = 100 céntimos
Literacy rate 94%
Calorie consumption 2700 kilocalories

PHILIPPINES
Southeast Asia

Official name Republic of the Philippines
Formation 1946 / 1946
Capital Manila
Population 100 million / 870 people per sq mile (336 people per sq km)
Total area 115,830 sq. miles (300,000 sq. km)
Languages Filipino*, English*, Tagalog, Cebuano, Ilocano, Hiligaynon, many other local languages
Religions Roman Catholic 81%, Protestant 9%, Muslim 5%, Other (including Buddhist) 5%
Ethnic mix Other 34%, Tagalog 28%, Cebuano 13%, Ilocano 9%, Hiligaynon 8%, Bisaya 8%
Government Presidential system
Currency Philippine peso = 100 centavos
Literacy rate 95%
Calorie consumption 2570 kilocalories

POLAND
Northern Europe

Official name Republic of Poland
Formation 1918 / 1945
Capital Warsaw
Population 38.2 million / 325 people per sq mile (125 people per sq km)
Total area 120,728 sq. miles (312,685 sq. km)
Languages Polish*
Religions Roman Catholic 93%, Other and nonreligious 5%, Orthodox Christian 2%
Ethnic mix Polish 98%, Other 2%
Government Parliamentary system
Currency Zloty = 100 groszy
Literacy rate 99%
Calorie consumption 3485 kilocalories

PORTUGAL
Southwest Europe

Official name Portuguese Republic
Formation 1139 / 1640
Capital Lisbon
Population 10.6 million / 299 people per sq mile (115 people per sq km)
Total area 35,672 sq. miles (92,391 sq. km)
Languages Portuguese*
Religions Roman Catholic 92%, Protestant 4%, Nonreligious 3%, Other 1%
Ethnic mix Portuguese 98%, African and other 2%
Government Parliamentary system
Currency Euro = 100 cents
Literacy rate 94%
Calorie consumption 3456 kilocalories

QATAR
Southwest Asia

Official name State of Qatar
Formation 1971 / 1971
Capital Doha
Population 2.3 million / 542 people per sq mile (209 people per sq km)
Total area 4416 sq. miles (11,437 sq. km)
Languages Arabic*
Religions Muslim (mainly Sunni) 95%, Other 5%
Ethnic mix Qatari 20%, Indian 20%, Other Arab 20%, Nepalese 13%, Filipino 10%, Other 10%, Pakistani 7%
Government Monarchy
Currency Qatar riyal = 100 dirhams
Literacy rate 97%
Calorie consumption Not available

ROMANIA
Southeast Europe

Official name Romania
Formation 1878 / 1947
Capital Bucharest
Population 21.6 million / 243 people per sq mile (94 people per sq km)
Total area 91,699 sq. miles (237,500 sq. km)
Languages Romanian*, Hungarian (Magyar), Romani, German
Religions Romanian Orthodox 87%, Protestant 5%, Roman Catholic 5%, Greek Orthodox 1%, Greek Catholic (Uniate) 1%, Other 1%
Ethnic mix Romanian 89%, Magyar 7%, Roma 3%, Other 1%
Government Presidential system
Currency New Romanian leu = 100 bani
Literacy rate 99%
Calorie consumption 3363 kilocalories

RUSSIAN FEDERATION
Europe / Asia

Official name Russian Federation
Formation 1480 / 1991
Capital Moscow
Population 143 million / 22 people per sq mile (8 people per sq km)
Total area 6,592,735 sq. miles (17,075,200 sq. km)
Languages Russian*, Tatar, Ukrainian, Chavash, various other national languages
Religions Orthodox Christian 75%, Muslim 14%, Other 11%
Ethnic mix Russian 80%, Other 12%, Tatar 4%, Ukrainian 2%, Bashkir 1%, Chavash 1%
Government Mixed Presidential–Parliamentary system
Currency Russian rouble = 100 kopeks
Literacy rate 99%
Calorie consumption 3358 kilocalories

RWANDA
Central Africa

Official name Republic of Rwanda
Formation 1962 / 1962
Capital Kigali
Population 12.1 million / 1256 people per sq mile (485 people per sq km)
Total area 10,169 sq. miles (26,338 sq. km)
Languages Kinyarwanda*, French*, Kiswahili, English*
Religions Christian 94%, Muslim 5%, Traditional beliefs 1%
Ethnic mix Hutu 85%, Tutsi 14%, Other (including Twa) 1%
Government Presidential system
Currency Rwanda franc = 100 centimes
Literacy rate 66%
Calorie consumption 2148 kilocalories

ST KITTS & NEVIS
West Indies

Official name Federation of Saint Christopher and Nevis
Formation 1983 / 1983
Capital Basseterre
Population 51,538 / 371 people per sq mile (143 people per sq km)
Total area 101 sq. miles (261 sq. km)
Languages English*, English Creole
Religions Anglican 33%, Methodist 29%, Other 22%, Moravian 9%, Roman Catholic 7%
Ethnic mix Black 95%, Mixed race 3%, White 1%, Other and Amerindian 1%
Government Parliamentary system
Currency East Caribbean dollar = 100 cents
Literacy rate 98%
Calorie consumption 2507 kilocalories

ST LUCIA
West Indies

Official name Saint Lucia
Formation 1979 / 1979
Capital Castries
Population 200,000 / 847 people per sq mile (328 people per sq km)
Total area 239 sq. miles (620 sq. km)
Languages English*, French Creole
Religions Roman Catholic 90%, Other 10%
Ethnic mix Black 83%, Mulatto (mixed race) 13%, Asian 3%, Other 1%
Government Parliamentary system
Currency East Caribbean dollar = 100 cents
Literacy rate 95%
Calorie consumption 2629 kilocalories

ST VINCENT & THE GRENADINES
West Indies

Official name Saint Vincent and the Grenadines
Formation 1979 / 1979
Capital Kingstown
Population 102,918 / 786 people per sq mile (303 people per sq km)
Total area 150 sq. miles (389 sq. km)
Languages English*, English Creole
Religions Anglican 47%, Methodist 28%, Roman Catholic 13%, Other 12%
Ethnic mix Black 66%, Mulatto (mixed race) 19%, Other 12%, Carib 2%, Asian 1%
Government Parliamentary system
Currency East Caribbean dollar = 100 cents
Literacy rate 88%
Calorie consumption 2960 kilocalories

SAMOA
Australasia & Oceania

Official name Independent State of Samoa
Formation 1962 / 1962
Capital Apia
Population 200,000 / 183 people per sq mile (71 people per sq km)
Total area 1104 sq. miles (2860 sq. km)
Languages Samoan*, English*
Religions Christian 99%, Other 1%
Ethnic mix Polynesian 91%, Euronesian 7%, Other 2%
Government Parliamentary system
Currency Tala = 100 sene
Literacy rate 99%
Calorie consumption 2872 kilocalories

SAN MARINO
Southern Europe

Official name Republic of San Marino
Formation 1631 / 1631
Capital San Marino
Population 32,742 / 1364 people per sq mile (537 people per sq km)
Total area 23.6 sq. miles (61 sq. km)
Languages Italian*
Religions Roman Catholic 93%, Other and nonreligious 7%
Ethnic mix Sammarinese 88%, Italian 10%, Other 2%
Government Parliamentary system
Currency Euro = 100 cents
Literacy rate 99%
Calorie consumption Not available

SAO TOME & PRINCIPE
West Africa

Official name Democratic Republic of Sao Tome and Principe
Formation 1975 / 1975
Capital São Tomé
Population 200,000 / 539 people per sq mile (208 people per sq km)
Total area 386 sq. miles (1001 sq. km)
Languages Portuguese Creole, Portuguese*
Religions Roman Catholic 84%, Other 16%
Ethnic mix Black 90%, Portuguese and Creole 10%
Government Presidential system
Currency Dobra = 100 céntimos
Literacy rate 70%
Calorie consumption 2676 kilocalories

SAUDI ARABIA
Southwest Asia

Official name Kingdom of Saudi Arabia
Formation 1932 / 1932
Capital Riyadh
Population 29.4 million / 36 people per sq mile (14 people per sq km)
Total area 756,981 sq. miles (1,960,582 sq. km)
Languages Arabic*
Religions Sunni Muslim 85%, Shi'a Muslim 15%
Ethnic mix Arab 72%, Foreign residents (mostly south and southeast Asian) 20%, Afro-Asian 8%
Government Monarchy
Currency Saudi riyal = 100 halalat
Literacy rate 94%
Calorie consumption 3122 kilocalories

SENEGAL
West Africa

Official name Republic of Senegal
Formation 1960 / 1960
Capital Dakar
Population 14.5 million / 195 people per sq mile (75 people per sq km)
Total area 75,749 sq. miles (196,190 sq. km)
Languages Wolof, Pulaar, Serer, Diola, Mandinka, Malinké, Soninké, French*
Religions Sunni Muslim 95%, Christian (mainly Roman Catholic) 4%, Traditional beliefs 1%
Ethnic mix Wolof 43%, Serer 15%, Peul 14%, Other 14%, Toucouleur 9%, Diola 5%
Government Presidential system
Currency CFA franc = 100 centimes
Literacy rate 52%
Calorie consumption 2426 kilocalories

SERBIA
Southeast Europe

Official name Republic of Serbia
Formation 2006 / 2008
Capital Belgrade
Population 9.5 million / 318 people per sq mile (123 people per sq km)
Total area 29,905 sq. miles (77,453 sq. km)
Languages Serbian*, Hungarian (Magyar)
Religions Orthodox Christian 85%, Roman Catholic 6%, Other 6%, Muslim 3%
Ethnic mix Serb 83%, Other 10%, Magyar 4%, Bosniak 2%, Roma 1%
Government Parliamentary system
Currency Serbian dinar = 100 para
Literacy rate 98%
Calorie consumption 2724 kilocalories

SEYCHELLES
Indian Ocean

Official name Republic of Seychelles
Formation 1976 / 1976
Capital Victoria
Population 91,650 / 881 people per sq mile (339 people per sq km)
Total area 176 sq. miles (455 sq. km)
Languages French Creole*, English*, French*
Religions Roman Catholic 82%, Anglican 6%, Other (including Muslim) 6%, Other Christian 3%, Hindu 2%, Seventh-day Adventist 1%
Ethnic mix Creole 89%, Indian 5%, Other 4%, Chinese 2%
Government Presidential system
Currency Seychelles rupee = 100 cents
Literacy rate 92%
Calorie consumption 2426 kilocalories

SIERRA LEONE
West Africa

Official name Republic of Sierra Leone
Formation 1961 / 1961
Capital Freetown
Population 6.2 million / 224 people per sq mile (87 people per sq km)
Total area 27,698 sq. miles (71,740 sq. km)
Languages Mende, Temne, Krio, English*
Religions Muslim 60%, Christian 30%, Traditional beliefs 10%
Ethnic mix Mende 35%, Temne 32%, Other 21%, Limba 8%, Kuranko 4%
Government Presidential system
Currency Leone = 100 cents
Literacy rate 44%
Calorie consumption 2333 kilocalories

SINGAPORE
Southeast Asia

Official name Republic of Singapore
Formation 1965 / 1965
Capital Singapore
Population 5.5 million / 23,305 people per sq mile (9016 people per sq km)
Total area 250 sq. miles (648 sq. km)
Languages Mandarin*, Malay*, Tamil*, English*
Religions Buddhist 55%, Taoist 22%, Muslim 16%, Hindu, Christian, and Sikh 7%
Ethnic mix Chinese 74%, Malay 14%, Indian 9%, Other 3%
Government Parliamentary system
Currency Singapore dollar = 100 cents
Literacy rate 96%
Calorie consumption Not available

SLOVAKIA
Central Europe

Official name Slovak Republic
Formation 1993 / 1993
Capital Bratislava
Population 5.5 million / 290 people per sq mile (112 people per sq km)
Total area 18,859 sq. miles (48,845 sq. km)
Languages Slovak*, Hungarian (Magyar), Czech
Religions Roman Catholic 69%, Nonreligious 13%, Other 13%, Greek Catholic (Uniate) 4%, Orthodox Christian 1%
Ethnic mix Slovak 86%, Magyar 10%, Roma 2%, Czech 1%, Other 1%
Government Parliamentary system
Currency Euro = 100 cents
Literacy rate 99%
Calorie consumption 2902 kilocalories

SLOVENIA
Central Europe

Official name Republic of Slovenia
Formation 1991 / 1991
Capital Ljubljana
Population 2.1 million / 269 people per sq mile (104 people per sq km)
Total area 7820 sq. miles (20,253 sq. km)
Languages Slovenian*
Religions Roman Catholic 58%, Other 28%, Atheist 10%, Orthodox Christian 2%, Muslim 2%
Ethnic mix Slovene 83%, Other 12%, Serb 2%, Croat 2%, Bosniak 1%
Government Parliamentary system
Currency Euro = 100 cents
Literacy rate 99%
Calorie consumption 3173 kilocalories

SOLOMON ISLANDS
Australasia & Oceania

Official name Solomon Islands
Formation 1978 / 1978
Capital Honiara
Population 600,000 / 56 people per sq mile (21 people per sq km)
Total area 10,985 sq. miles (28,450 sq. km)
Languages English*, Pidgin English, Melanesian Pidgin, 120 (est.) native languages
Religions Church of Melanesia (Anglican) 34%, Roman Catholic 19%, South Seas Evangelical Church 17%, Methodist 11%, Seventh-day Adventist 10%, Other 9%
Ethnic mix Melanesian 93%, Polynesian 4%, Micronesian 2%, Other 1%
Government Parliamentary system
Currency Solomon Islands dollar = 100 cents
Literacy rate 77%
Calorie consumption 2473 kilocalories

SOMALIA
East Africa

Official name Federal Republic of Somalia
Formation 1960 / 1960
Capital Mogadishu
Population 10.8 million / 45 people per sq mile (17 people per sq km)
Total area 246,199 sq. miles (637,657 sq. km)
Languages Somali*, Arabic*, English, Italian
Religions Sunni Muslim 99%, Christian 1%
Ethnic mix Somali 85%, Other 15%
Government Non-party system
Currency Somali shilin = 100 senti
Literacy rate 24%
Calorie consumption 1696 kilocalories

SOUTH AFRICA
Southern Africa

Official name Republic of South Africa
Formation 1934 / 1994
Capital Pretoria; Cape Town; Bloemfontein
Population 53.1 million / 113 people per sq mile (43 people per sq km)
Total area 471,008 sq. miles (1,219,912 sq. km)
Languages English, isiZulu, isiXhosa, Afrikaans, Sepedi, Setswana, Sesotho, Xitsonga, siSwati, Tshivenda, isiNdebele
Religions Christian 68%, Traditional beliefs and animist 29%, Muslim 2%, Hindu 1%
Ethnic mix Black 80%, Mixed race 9%, White 9%, Asian 2%
Government Presidential system
Currency Rand = 100 cents
Literacy rate 94%
Calorie consumption 3007 kilocalories

SOUTH KOREA
East Asia

Official name Republic of Korea
Formation 1948 / 1953
Capital Seoul; Sejong City (administrative)
Population 49.5 million / 1299 people per sq mile (501 people per sq km)
Total area 38,023 sq. miles (98,480 sq. km)
Languages Korean*
Religions Mahayana Buddhist 47%, Protestant 38%, Roman Catholic 11%, Confucianist 3%, Other 1%
Ethnic mix Korean 100%
Government Presidential system
Currency South Korean won = 100 chon
Literacy rate 99%
Calorie consumption 3329 kilocalories

SOUTH SUDAN
East Africa

Official name Republic of South Sudan
Formation 2011 / 2011
Capital Juba
Population 11.7 million / 47 people per sq mile (18 people per sq km)
Total area 248,777 sq. miles (644,329 sq. km)
Languages Arabic, Dinka, Nuer, Zande, Bari, Shilluk, Lotuko, English*
Religions Over half of the population follow Christian or traditional beliefs.
Ethnic mix Dinka 40%, Nuer 15%, Bari 10%, Shilluk/Anwak 10%, Azande 10%, Arab 10%, Other 5%
Government Transitional regime
Currency South Sudan pound = 100 piastres
Literacy rate 37%
Calorie consumption Not available

SPAIN
Southwest Europe

Official name Kingdom of Spain
Formation 1492 / 1713
Capital Madrid
Population 47.1 million / 244 people per sq mile (94 people per sq km)
Total area 194,896 sq. miles (504,782 sq. km)
Languages Spanish*, Catalan*, Galician*, Basque*
Religions Roman Catholic 96%, Other 4%
Ethnic mix Castilian Spanish 72%, Catalan 17%, Galician 6%, Basque 2%, Other 2%, Roma 1%
Government Parliamentary system
Currency Euro = 100 cents
Literacy rate 98%
Calorie consumption 3183 kilocalories

SRI LANKA
South Asia

Official name Democratic Socialist Republic of Sri Lanka
Formation 1948 / 1948
Capital Colombo; Sri Jayewardenapura Kotte
Population 21.4 million / 856 people per sq mile (331 people per sq km)
Total area 25,332 sq. miles (65,610 sq. km)
Languages Sinhala*, Tamil*, Sinhala-Tamil, English
Religions Buddhist 69%, Hindu 15%, Muslim 8%, Christian 8%
Ethnic mix Sinhalese 74%, Tamil 18%, Moor 7%, Other 1%
Government Mixed presidential–parliamentary system
Currency Sri Lanka rupee = 100 cents
Literacy rate 91%
Calorie consumption 2539 kilocalories

SUDAN
East Africa

Official name Republic of the Sudan
Formation 1956 / 2011
Capital Khartoum
Population 38.8 million / 54 people per sq mile (21 people per sq km)
Total area 718,722 sq. miles (1,861,481 sq. km)
Languages Arabic, Nubian, Beja, Fur
Religions Nearly the whole population is Muslim (mainly Sunni)
Ethnic mix Arab 60%, Other 18%, Nubian 10%, Beja 8%, Fur 3%, Zaghawa 1%
Government Presidential system
Currency New Sudanese pound = 100 piastres
Literacy rate 73%
Calorie consumption 2346 kilocalories

SURINAME
South America

Official name Republic of Suriname
Formation 1975 / 1975
Capital Paramaribo
Population 500,000 / 8 people per sq mile (3 people per sq km)
Total area 63,039 sq. miles (163,270 sq. km)
Languages Sranan (creole), Dutch*, Javanese, Sarnami Hindi, Saramaccan, Chinese, Carib
Religions Hindu 27%, Protestant 25%, Roman Catholic 23%, Muslim 20%, Traditional beliefs 5%
Ethnic mix East Indian 27%, Creole 18%, Black 15%, Javanese 15%, Mixed race 13%, Other 6%, Amerindian 4%, Chinese 2%
Government Mixed presidential–parliamentary system
Currency Surinamese dollar = 100 cents
Literacy rate 95%
Calorie consumption 2727 kilocalories

SWAZILAND
Southern Africa

Official name Kingdom of Swaziland
Formation 1968 / 1968
Capital Mbabane
Population 1.3 million / 196 people per sq mile (76 people per sq km)
Total area 6704 sq. miles (17,363 sq. km)
Languages English*, siSwati*, isiZulu, Xitsonga
Religions Traditional beliefs 40%, Other 30%, Roman Catholic 20%, Muslim 10%
Ethnic mix Swazi 97%, Other 3%
Government Monarchy
Currency Lilangeni = 100 cents
Literacy rate 83%
Calorie consumption 2275 kilocalories

SWEDEN
Northern Europe

Official name Kingdom of Sweden
Formation 1523 / 1921
Capital Stockholm
Population 9.6 million / 60 people per sq mile (23 people per sq km)
Total area 173,731 sq. miles (449,964 sq. km)
Languages Swedish*, Finnish, Sámi
Religions Evangelical Lutheran 75%, Other 13%, Muslim 5%, Other Protestant 5%, Roman Catholic 2%
Ethnic mix Swedish 86%, Foreign-born or first-generation immigrant 12%, Finnish and Sámi 2%
Government Parliamentary system
Currency Swedish krona = 100 öre
Literacy rate 99%
Calorie consumption 3160 kilocalories

SWITZERLAND
Central Europe

Official name Swiss Confederation
Formation 1291 / 1857
Capital Bern
Population 8.2 million / 534 people per sq mile (206 people per sq km)
Total area 15,942 sq. miles (41,290 sq. km)
Languages German*, Swiss-German, French*, Italian*, Romansch
Religions Roman Catholic 42%, Protestant 35%, Other and nonreligious 19%, Muslim 4%
Ethnic mix German 64%, French 20%, Other 9.5%, Italian 6%, Romansch 0.5%
Government Parliamentary system
Currency Swiss franc = 100 rappen/centimes
Literacy rate 99%
Calorie consumption 3487 kilocalories

SYRIA
Southwest Asia

Official name Syrian Arab Republic
Formation 1941 / 1967
Capital Damascus
Population 22 million / 310 people per sq mile (120 people per sq km)
Total area 71,498 sq. miles (184,180 sq. km)
Languages Arabic*, French, Kurdish, Armenian, Circassian, Turkic languages, Assyrian, Aramaic
Religions Sunni Muslim 74%, Alawi 12%, Christian 10%, Druze 3%, Other 1%
Ethnic mix Arab 90%, Kurdish 9%, Armenian, Turkmen, and Circassian 1%
Government Presidential system
Currency Syrian pound = 100 piastres
Literacy rate 85%
Calorie consumption 3106 kilocalories

TAIWAN
East Asia

Official name Republic of China (ROC)
Formation 1949 / 1949
Capital Taibei (Taipei)
Population 23.4 million / 1879 people per sq mile (725 people per sq km)
Total area 13,892 sq. miles (35,980 sq. km)
Languages Amoy Chinese, Mandarin Chinese*, Hakka Chinese
Religions Buddhist, Confucianist, and Taoist 93%, Christian 5%, Other 2%
Ethnic mix Han Chinese (pre-20th-century migration) 84%, Han Chinese (20th-century migration) 14%, Aboriginal 2%
Government Presidential system
Currency Taiwan dollar = 100 cents
Literacy rate 98%
Calorie consumption 2997 kilocalories

TAJIKISTAN
Central Asia

Official name Republic of Tajikistan
Formation 1991 / 1991
Capital Dushanbe
Population 8.4 million / 152 people per sq mile (59 people per sq km)
Total area 55,251 sq. miles (143,100 sq. km)
Languages Tajik*, Uzbek, Russian
Religions Sunni Muslim 95%, Shi'a Muslim 3%, Other 2%
Ethnic mix Tajik 80%, Uzbek 15%, Other 3%, Russian 1%, Kyrgyz 1%
Government Presidential system
Currency Somoni = 100 diram
Literacy rate 99%
Calorie consumption 2101 kilocalories

TANZANIA
East Africa

Official name United Republic of Tanzania
Formation 1964 / 1964
Capital Dodoma
Population 50.8 million / 148 people per sq mile (57 people per sq km)
Total area 364,898 sq. miles (945,087 sq. km)
Languages Kiswahili*, Sukuma, Chagga, Nyamwezi, Hehe, Makonde, Yao, Sandawe, English*
Religions Christian 63%, Muslim 35%, Other 2%
Ethnic mix Native African (over 120 tribes) 99%, European, Asian, and Arab 1%
Government Presidential system
Currency Tanzanian shilling = 100 cents
Literacy rate 68%
Calorie consumption 2208 kilocalories

THAILAND
Southeast Asia

Official name Kingdom of Thailand
Formation 1238 / 1907
Capital Bangkok
Population 67.2 million / 341 people per sq mile (132 people per sq km)
Total area 198,455 sq. miles (514,000 sq. km)
Languages Thai*, Chinese, Malay, Khmer, Mon, Karen, Miao
Religions Buddhist 95%, Muslim 4%, Other (including Christian) 1%
Ethnic mix Thai 83%, Chinese 12%, Malay 3%, Khmer and Other 2%
Government Transitional regime
Currency Baht = 100 satang
Literacy rate 96%
Calorie consumption 2784 kilocalories

TOGO
West Africa

Official name Togolese Republic
Formation 1960 / 1960
Capital Lomé
Population 7 million / 333 people per sq mile (129 people per sq km)
Total area 21,924 sq. miles (56,785 sq. km)
Languages Ewe, Kabye, Gurma, French*
Religions Christian 47%, Traditional beliefs 33%, Muslim 14%, Other 6%
Ethnic mix Ewe 46%, Other African 41%, Kabye 12%, European 1%
Government Presidential system
Currency CFA franc = 100 centimes
Literacy rate 60%
Calorie consumption 2366 kilocalories

TONGA
Australasia & Oceania

Official name Kingdom of Tonga
Formation 1970 / 1970
Capital Nuku'alofa
Population 106,440 / 383 people per sq mile (148 people per sq km)
Total area 289 sq. miles (748 sq. km)
Languages English*, Tongan*
Religions Free Wesleyan 41%, Other 17%, Roman Catholic 16%, Church of Jesus Christ of Latter-day Saints 14%, Free Church of Tonga 12%
Ethnic mix Tongan 98%, Other 2%
Government Monarchy
Currency Pa'anga (Tongan dollar) = 100 seniti
Literacy rate 99%
Calorie consumption Not available

TRINIDAD & TOBAGO
West Indies

Official name Republic of Trinidad and Tobago
Formation 1962 / 1962
Capital Port-of-Spain
Population 1.3 million / 656 people per sq mile (253 people per sq km)
Total area 1980 sq. miles (5128 sq. km)
Languages English Creole, English*, Hindi, French, Spanish
Religions Roman Catholic 26%, Hindu 23%, Other and nonreligious 23%, Anglican 8%, Baptist 7%, Pentecostal 7%, Muslim 6%
Ethnic mix East Indian 40%, Black 38%, Mixed race 20%, White and Chinese 1%, other 1%
Government Parliamentary system
Currency Trinidad and Tobago dollar = 100 cents
Literacy rate 99%
Calorie consumption 2889 kilocalories

TUNISIA
North Africa

Official name Tunisian Republic
Formation 1956 / 1956
Capital Tunis
Population 11.1 million / 185 people per sq mile (71 people per sq km)
Total area 63,169 sq. miles (163,610 sq. km)
Languages Arabic*, French
Religions Muslim (mainly Sunni) 98%, Christian 1%, Jewish 1%
Ethnic mix Arab and Berber 98%, Jewish 1%, European 1%
Government Mixed presidential–parliamentary system
Currency Tunisian dinar = 1000 millimes
Literacy rate 80%
Calorie consumption 3362 kilocalories

TURKEY
Asia / Europe

Official name Republic of Turkey
Formation 1923 / 1939
Capital Ankara
Population 75.8 million / 255 people per sq mile (98 people per sq km)
Total area 301,382 sq. miles (780,580 sq. km)
Languages Turkish*, Kurdish, Arabic, Circassian, Armenian, Greek, Georgian, Ladino
Religions Muslim (mainly Sunni) 99%, Other 1%
Ethnic mix Turkish 70%, Kurdish 20%, Other 8%, Arab 2%
Government Parliamentary system
Currency Turkish lira = 100 kurus
Literacy rate 95%
Calorie consumption 3680 kilocalories

TURKMENISTAN
Central Asia

Official name Turkmenistan
Formation 1991 / 1991
Capital Ashgabat
Population 5.3 million / 28 people per sq mile (11 people per sq km)
Total area 188,455 sq. miles (488,100 sq. km)
Languages Turkmen*, Uzbek, Russian, Kazakh, Tatar
Religions Sunni Muslim 89%, Orthodox Christian 9%, Other 2%
Ethnic mix Turkmen 85%, Other 6%, Uzbek 5%, Russian 4%
Government Presidential system
Currency New manat = 100 tenge
Literacy rate 99%
Calorie consumption 2883 kilocalories

TUVALU
Australasia & Oceania

Official name Tuvalu
Formation 1978 / 1978
Capital Funafuti Atoll
Population 10,782 / 1078 people per sq mile (415 people per sq km)
Total area 10 sq. miles (26 sq. km)
Languages Tuvaluan, Kiribati, English*
Religions Church of Tuvalu 97%, Baha'i 1%, Seventh-day Adventist 1%, Other 1%
Ethnic mix Polynesian 96%, Micronesian 4%
Government Nonparty system
Currency Australian dollar = 100 cents; and Tuvaluan dollar = 100 cents
Literacy rate 95%
Calorie consumption Not available

UGANDA
East Africa

Official name Republic of Uganda
Formation 1962 / 1962
Capital Kampala
Population 38.8 million / 504 people per sq mile (194 people per sq km)
Total area 91,135 sq. miles (236,040 sq. km)
Languages Luganda, Nkole, Chiga, Lango, Acholi, Teso, Lugbara, English*
Religions Christian 85%, Muslim (mainly Sunni) 12%, Other 3%
Ethnic mix Other 50%, Baganda 17%, Banyakole 10%, Basoga 9%, Iteso 7%, Bakiga 7%
Government Presidential system
Currency Uganda shilling = 100 cents
Literacy rate 74%
Calorie consumption 2279 kilocalories

UKRAINE
Eastern Europe

Official name Ukraine
Formation 1991 / 1991
Capital Kiev
Population 44.9 million / 193 people per sq mile (74 people per sq km)
Total area 223,089 sq. miles (603,700 sq. km)
Languages Ukrainian*, Russian, Tatar
Religions Christian (mainly Orthodox) 95%, Other 5%
Ethnic mix Ukrainian 78%, Russian 17%, Other 5%
Government Presidential system
Currency Hryvna = 100 kopiykas
Literacy rate 99%
Calorie consumption 3142 kilocalories

UNITED ARAB EMIRATES
Southwest Asia

Official name United Arab Emirates
Formation 1971 / 1972
Capital Abu Dhabi
Population 9.4 million / 291 people per sq mile (112 people per sq km)
Total area 32,000 sq. miles (82,880 sq. km)
Languages Arabic*, Farsi, Indian and Pakistani languages, English
Religions Muslim (mainly Sunni) 96%, Christian, Hindu, and other 4%
Ethnic mix Asian 60%, Emirian 25%, Other Arab 12%, European 3%
Government Monarchy
Currency UAE dirham = 100 fils
Literacy rate 90%
Calorie consumption 3215 kilocalories

UNITED KINGDOM
Northwest Europe

Official name United Kingdom of Great Britain and Northern Ireland
Formation 1707 / 1922
Capital London
Population 63.5 million / 681 people per sq mile (263 people per sq km)
Total area 94,525 sq. miles (244,820 sq. km)
Languages English*, Welsh*, Scottish Gaelic, Irish
Religions Anglican 45%, Other and nonreligious 36%, Roman Catholic 9%, Presbyterian 4%, Muslim 3%, Methodist 2%, Hindu 1%
Ethnic mix English 80%, Scottish 9%, West Indian, Asian, and other 5%, Northern Irish 3%, Welsh 3%
Government Parliamentary system
Currency Pound sterling = 100 pence
Literacy rate 99%
Calorie consumption 3414 kilocalories

UNITED STATES
North America

Official name United States of America
Formation 1776 / 1959
Capital Washington D.C.
Population 323 million / 91 people per sq mile (35 people per sq km)
Total area 3,794,100 sq. miles (9,826,675 sq. km)
Languages English*, Spanish, Chinese, French, German, Tagalog, Vietnamese, Italian, Korean, Russian, Polish
Religions Protestant 52%, Roman Catholic 25%, Other and nonreligious 20%, Jewish 2%, Muslim 1%
Ethnic mix White 60%, Hispanic 17%, Black American/African 14%, Asian 6%, American Indians & Alaksa Natives 2%, Pacific Islanders 1%
Government Presidential system
Currency US dollar = 100 cents
Literacy rate 99%
Calorie consumption 3639 kilocalories

URUGUAY
South America

Official name Oriental Republic of Uruguay
Formation 1828 / 1828
Capital Montevideo
Population 3.4 million / 50 people per sq mile (19 people per sq km)
Total area 68,039 sq. miles (176,220 sq. km)
Languages Spanish*
Religions Roman Catholic 66%, Other and nonreligious 30%, Jewish 2%, Protestant 2%
Ethnic mix White 90%, Mestizo 6%, Black 4%
Government Presidential system
Currency Uruguayan peso = 100 centésimos
Literacy rate 98%
Calorie consumption 2939 kilocalories

UZBEKISTAN
Central Asia

Official name Republic of Uzbekistan
Formation 1991 / 1991
Capital Tashkent
Population 29.3 million / 170 people per sq mile (65 people per sq km)
Total area 172,741 sq. miles (447,400 sq. km)
Languages Uzbek*, Russian, Tajik, Kazakh
Religions Sunni Muslim 88%, Orthodox Christian 9%, Other 3%
Ethnic mix Uzbek 80%, Russian 6%, Other 6%, Tajik 5%, Kazakh 3%
Government Presidential system
Currency Som = 100 tiyin
Literacy rate 99%
Calorie consumption 2675 kilocalories

VANUATU
Australasia & Oceania

Official name Republic of Vanuatu
Formation 1980 / 1980
Capital Port Vila
Population 300,000 / 64 people per sq mile (25 people per sq km)
Total area 4710 sq. miles (12,200 sq. km)
Languages Bislama (Melanesian pidgin)*, English*, French*, other indigenous languages
Religions Presbyterian 37%, Other 19%, Anglican 15%, Roman Catholic 15%, Traditional beliefs 8%, Seventh-day Adventist 6%
Ethnic mix ni-Vanuatu 94%, European 4%, Other 2%
Government Parliamentary system
Currency Vatu = 100 centimes
Literacy rate 83%
Calorie consumption 2820 kilocalories

VATICAN CITY
Southern Europe

Official name State of the Vatican City
Formation 1929 / 1929
Capital Vatican City
Population 842 / 4953 people per sq mile (1914 people per sq km)
Total area 0.17 sq. miles (0.44 sq. km)
Languages Italian*, Latin*
Religions Roman Catholic 100%
Ethnic mix The current pope is Argentinian, though most popes for the last 500 years have been Italian. Cardinals are from many nationalities, but Italians form the largest group. Most of the resident lay persons are Italian.
Government Papal state
Currency Euro = 100 cents
Literacy rate 99%
Calorie consumption Not available

VENEZUELA
South America

Official name Bolivarian Republic of Venezuela
Formation 1830 / 1830
Capital Caracas
Population 30.9 million / 91 people per sq mile (35 people per sq km)
Total area 352,143 sq. miles (912,050 sq. km)
Languages Spanish*, Amerindian languages
Religions Roman Catholic 96%, Protestant 2%, Other 2%
Ethnic mix Mestizo 69%, White 20%, Black 9%, Amerindian 2%
Government Presidential system
Currency Bolívar fuerte = 100 céntimos
Literacy rate 96%
Calorie consumption 2880 kilocalories

VIETNAM
Southeast Asia

Official name Socialist Republic of Vietnam
Formation 1976 / 1976
Capital Hanoi
Population 92.5 million / 736 people per sq mile (284 people per sq km)
Total area 127,243 sq. miles (329,560 sq. km)
Languages Vietnamese*, Chinese, Thai, Khmer, Muong, Nung, Miao, Yao, Jarai
Religions Other 74%, Buddhist 14%, Roman Catholic 7%, Cao Dai 3%, Protestant 2%
Ethnic mix Vietnamese 86%, Other 8%, Muong 2%, Tay 2%, Thai 2%
Government One-party state
Currency Dông = 10 hao = 100 xu
Literacy rate 94%
Calorie consumption 2745 kilocalories

YEMEN
Southwest Asia

Official name Republic of Yemen
Formation 1990 / 1990
Capital Sana
Population 25 million / 115 people per sq mile (44 people per sq km)
Total area 203,849 sq. miles (527,970 sq. km)
Languages Arabic*
Religions Sunni Muslim 55%, Shi'a Muslim 42%, Christian, Hindu, and Jewish 3%
Ethnic mix Arab 99%, Afro-Arab, Indian, Somali, and European 1%
Government Transitional regime
Currency Yemeni rial = 100 fils
Literacy rate 66%
Calorie consumption 2223 kilocalories

ZAMBIA
Southern Africa

Official name Republic of Zambia
Formation 1964 / 1964
Capital Lusaka
Population 15 million / 52 people per sq mile (20 people per sq km)
Total area 290,584 sq. miles (752,614 sq. km)
Languages Bemba, Tonga, Nyanja, Lozi, Lala-Bisa, Nsenga, English*
Religions Christian 63%, Traditional beliefs 36%, Muslim and Hindu 1%
Ethnic mix Bemba 34%, Other African 26%, Tonga 16%, Nyanja 14%, Lozi 9%, European 1%
Government Presidential system
Currency New Zambian kwacha = 100 ngwee
Literacy rate 61%
Calorie consumption 1930 kilocalories

ZIMBABWE
Southern Africa

Official name Republic of Zimbabwe
Formation 1980 / 1980
Capital Harare
Population 14.6 million / 98 people per sq mile (38 people per sq km)
Total area 150,803 sq. miles (390,580 sq. km)
Languages Shona, isiNdebele, English*
Religions Syncretic (Christian/traditional beliefs) 50%, Christian 25%, Traditional beliefs 24%, Other (including Muslim) 1%
Ethnic mix Shona 71%, Ndebele 16%, Other African 11%, White 1%, Asian 1%
Government Presidential system
Currency US $, South African rand, Euro, UK £, Botswana pula, Australian $, Chinese yuan, Indian rupee, and Japanese yen are legal tender
Literacy rate 84%
Calorie consumption 2110 kilocalories

GLOSSARY

This glossary lists all geographical, technical and foreign language terms which appear in the text, followed by a brief definition of the term. Any acronyms used in the text are also listed in full. Terms in italics are for cross-reference and indicate that the word is separately defined in the glossary.

A

Aboriginal The original (indigenous) inhabitants of a country or continent. Especially used with reference to Australia.

Abyssal plain A broad plain found in the depths of the ocean, more than 10,000 ft (3000 m) below sea level.

Acid rain Rain, sleet, snow or mist which has absorbed waste gases from fossil-fuelled power stations and vehicle exhausts, becoming more acid. It causes severe environmental damage.

Adaptation The gradual evolution of plants and animals so that they become better suited to survive and reproduce in their *environment*.

Afforestation The planting of new forest in areas which were once forested but have been cleared.

Agribusiness A term applied to activities such as the growing of crops, rearing of animals or the manufacture of farm machinery, which eventually leads to the supply of agricultural produce at market.

Air mass A huge, homogeneous mass of air, within which horizontal patterns of temperature and *humidity* are consistent. Air masses are separated by *fronts*.

Alliance An agreement between two or more states, to work together to achieve common purposes.

Alluvial fan A large fan-shaped deposit of fine sediments deposited by a river as it emerges from a narrow, mountain valley onto a broad, open *plain*.

Alluvium Material deposited by rivers. Nowadays usually only applied to finer particles of silt and clay.

Alpine Mountain *environment*, between the *treeline* and the level of permanent snow cover.

Alpine mountains Ranges of mountains formed between 30 and 65 million years ago, by *folding*, in west and central Europe.

Amerindian A term applied to people *indigenous* to North, Central and South America.

Animal husbandry The business of rearing animals.

Antarctic circle The parallel which lies at *latitude* of 66° 32′ S.

Anticline A geological *fold* that forms an arch shape, curving upwards in the rock *strata*.

Anticyclone An area of relatively high atmospheric pressure.

Aquaculture Collective term for the farming of produce derived from the sea, including fish-farming, the cultivation of shellfish, and plants such as seaweed.

Aquifer A body of rock which can absorb water. Also applied to any rock strata that have sufficient porosity to yield *groundwater* through wells or springs.

Arable Land which has been ploughed and is being used, or is suitable, for growing crops.

Archipelago A group or chain of islands.

Arctic Circle The parallel which lies at a *latitude* of 66° 32′ N.

Arête A thin, jagged mountain ridge which divides two adjacent *cirques*, found in regions where *glaciation* has occurred.

Arid Dry. An area of low rainfall, where the rate of *evaporation* may be greater than that of *precipitation*. Often defined as those areas that receive less than one inch (25 mm) of rain a year. In these areas only drought-resistant plants can survive.

Artesian well A naturally occurring source of underground water, stored in an *aquifer*.

Artisanal Small-scale, manual operation, such as fishing, using little or no machinery.

ASEAN Association of Southeast Asian Nations. Established in 1967 to promote economic, social and cultural co-operation. Its members include Brunei, Indonesia, Malaysia, Philippines, Singapore and Thailand.

Aseismic A region where *earthquake* activity has ceased.

Asteroid A minor planet circling the Sun, mainly between the orbits of Mars and Jupiter.

Asthenosphere A zone of hot, partially melted rock, which underlies the *lithosphere*, within the Earth's *crust*.

Atmosphere The envelope of odourless, colourless and tasteless gases surrounding the Earth, consisting of *oxygen* (23%), *nitrogen* (75%), argon (1%), *carbon dioxide* (0.03%), as well as tiny proportions of other gases.

Atmospheric pressure The pressure created by the action of gravity on the gases surrounding the Earth.

Atoll A ring-shaped island or *coral reef* often enclosing a *lagoon* of sea water.

Avalanche The rapid movement of a mass of snow and ice down a steep slope. Similar movements of other materials are described as *rock avalanches* or *landslides* and *sand avalanches*.

B

Badlands A landscape that has been heavily eroded and dissected by rainwater, and which has little or no vegetation.

Back slope The gentler windward slope of a sand *dune* or gentler slope of a *cuesta*.

Bajos An *alluvial fan* deposited by a river at the base of mountains and hills which encircle *desert* areas.

Bar, coastal An offshore strip of sand or shingle, either above or below the water. Usually parallel to the shore but sometimes crescent-shaped or at an oblique angle.

Barchan A crescent-shaped sand *dune*, formed where wind direction is very consistent. The horns of the crescent point downwind and where there is enough sand the barchan is mobile.

Barrio A Spanish term for the shanty towns – self-built settlements – which are clustered around many South and Central American cities (*see also Favela*).

Basalt Dark, fine-grained *igneous rock*. Formed near the Earth's surface from fast-cooling *lava*.

Base level The level below which flowing water cannot erode the land.

Basement rock A mass of ancient rock often of *Pre-Cambrian age*, covered by a layer of more recent *sedimentary rocks*. Commonly associated with *shield* areas.

Beach Lake or sea shore where waves break and there is an accumulation of loose material – mud, sand, shingle or pebbles.

Bedrock Solid, consolidated and relatively unweathered rock, found on the surface of the land or just below a layer of soil or *weathered* rock.

Biodiversity The quantity of animal or plant species in a given area.

Biomass The total mass of organic matter – plants and animals – in a given area. It is usually measured in kilogrammes per square metre. Plant biomass is proportionally greater than that of animals, except in cities.

Biosphere The zone just above and below the Earth's surface, where all plants and animals live.

Blizzard A severe windstorm with snow and sleet. Visibility is often severely restricted.

Bluff The steep bank of a *meander*, formed by the erosive action of a river.

Boreal forest Tracts of mainly coniferous forest found in northern *latitudes*.

Breccia A type of rock composed of sharp fragments, cemented by a fine-grained material such as clay.

Butte An isolated, flat-topped hill with steep or vertical sides, buttes are the eroded remnants of a former land surface.

C

Caatinga Portuguese (Brazilian) term for thorny woodland growing in areas of pale granitic soils.

CACM Central American Common Market. Established in 1960 to further economic ties between its members, which are Costa Rica, El Salvador, Guatemala, Honduras and Nicaragua.

Calcite Hexagonal crystals of calcium carbonate.

Caldera A huge volcanic vent, often containing a number of smaller vents, and sometimes a crater lake.

Carbon cycle The transfer of carbon to and from the *atmosphere*. This occurs on land through *photosynthesis*. In the sea, *carbon dioxide* is absorbed, some returning to the air and some taken up into the bodies of sea creatures.

Carbon dioxide A colourless, odourless gas (CO_2) which makes up 0.03% of the *atmosphere*.

Carbonation The process whereby rocks are broken down by carbonic acid. Carbon dioxide in the air dissolves in rainwater, forming carbonic acid. *Limestone* terrain can be rapidly eaten away.

Cash crop A single crop grown specifically for export sale, rather than for local use. Typical examples include coffee, tea and citrus fruits.

Cassava A type of grain meal, used to produce tapioca. A staple crop in many parts of Africa.

Castle kopje Hill or rock outcrop, especially in southern Africa, where steep sides, and a summit composed of blocks, give a castle-like appearance.

Cataracts A series of stepped waterfalls created as a river flows over a band of hard, resistant rock.

Causeway A raised route through marshland or a body of water.

CEEAC Economic Community of Central African States. Established in 1983 to promote regional co-operation and if possible, establish a common market between 16 Central African nations.

Chemical weathering The chemical reactions leading to the decomposition of rocks. Types of chemical weathering include *carbonation*, *hydrolysis* and *oxidation*.

Chernozem A fertile soil, also known as 'black earth' consisting of a layer of dark topsoil, rich in decaying vegetation, overlying a lighter chalky layer.

Cirque Armchair-shaped basin, found in mountain regions, with a steep back, or rear, wall and a raised rock lip, often containing a lake (or *tarn*). The cirque floor has been eroded by a *glacier*, while the back wall is eroded both by the *glacier* and by *weathering*.

Climate The average weather conditions in a given area over a period of years, sometimes defined as 30 years or more.

Cold War A period of hostile relations between the USA and the Soviet Union and their allies after the Second World War.

Composite volcano Also known as a strato-volcano, the volcanic cone is composed of alternating deposits of *lava* and *pyroclastic* material.

Compound A substance made up of *elements* chemically combined in a consistent way.

Condensation The process whereby a gas changes into a liquid. For example, water vapour in the *atmosphere* condenses around tiny airborne particles to form droplets of water.

Confluence The point at which two rivers meet.

Conglomerate Rock composed of large, water-worn or rounded pebbles, held together by a natural cement.

Coniferous forest A forest type containing trees which are generally, but not necessarily, *evergreen* and have slender, needle-like leaves and which reproduce by means of seeds contained in a cone.

Continental drift The theory that the continents of today are fragments of one or more prehistoric *supercontinents* which have moved across the Earth's surface, creating ocean basins. The theory has been superseded by a more sophisticated one – *plate tectonics*.

Continental shelf An area of the continental crust, below sea level, which slopes gently. It is separated from the deep ocean by a much more steeply inclined *continental slope*.

Continental slope A steep slope running from the edge of the *continental shelf* to the ocean floor.

Conurbation A vast metropolitan area created by the expansion of towns and cities into a virtually continuous urban area.

Cool continental A rainy *climate* with warm summers [warmest month below 76°F (22°C)] and often severe winters [coldest month below 32°F (0°C)].

Copra The dried, white kernel of a coconut, from which coconut oil is extracted.

Coral reef An underwater barrier created by colonies of the coral polyp. Polyps secrete a protective skeleton of calcium carbonate, and reefs develop as live polyps build on the skeletons of dead generations.

Core The centre of the Earth, consisting of a dense mass of iron and nickel. It is thought that the outer core is molten or liquid, and that the hot inner core is solid due to extremely high pressures.

Coriolis effect A deflecting force caused by the rotation of the Earth. In the northern hemisphere a body, such as an *air mass* or ocean current, is deflected to the right, and in the southern hemisphere to the left. This prevents winds from blowing straight from areas of high to low pressure.

Coulées A US / Canadian term for a ravine formed by river *erosion*.

Craton A large block of the Earth's *crust* which has remained stable for a long period of *geological time*. It is made up of ancient *shield* rocks.

Cretaceous A period of *geological time* beginning about 145 million years ago and lasting until about 65 million years ago.

Crevasse A deep crack in a *glacier*.

Crust The hard, thin outer shell of the Earth. The crust floats on the *mantle*, which is softer and more dense. Under the oceans (oceanic crust) the crust is 3.7–6.8 miles (6–11 km) thick. Continental crust averages 18–24 miles (30–40 km).

Crystalline rock Rocks formed when molten *magma* crystallizes (*igneous rocks*) or when heat or pressure cause re-crystallization (*metamorphic rocks*). Crystalline rocks are distinct from *sedimentary rocks*.

Cuesta A hill which rises into a steep slope on one side but has a gentler gradient on its other slope.

Cyclone An area of low *atmospheric pressure*, occurring where the air is warm and relatively low in density, causing low level winds to spiral. *Hurricanes* and *typhoons* are tropical cyclones.

D

De facto
1 Government or other activity that takes place, or exists in actuality if not by right.
2 A border, which exists in practice, but which is not officially recognized by all the countries it adjoins.

Deciduous forest A forest of trees which shed their leaves annually at a particular time or season. In *temperate* climates the fall of leaves occurs in the Autumn. Some *coniferous* trees, such as the larch, are deciduous. Deciduous vegetation contrasts with *evergreen*, which keeps its leaves for more than a year.

Defoliant Chemical spray used to remove foliage (leaves) from trees.

Deforestation The act of cutting down and clearing large areas of forest for human activities, such as agricultural land or urban development.

Delta Low-lying, fan-shaped area at a river mouth, formed by the *deposition* of successive layers of *sediment*. Slowing as it enters the sea, a river deposits sediment and may, as a result, split into numerous smaller channels, known as *distributaries*.

Denudation The combined effect of *weathering*, *erosion* and *mass movement*, which, over long periods, exposes underlying rocks.

Deposition The laying down of material that has accumulated:
(1) after being *eroded* and then transported by physical forces such as wind, ice or water;
(2) as organic remains, such as coal and coral;
(3) as the result of *evaporation* and chemical *precipitation*.

Depression
1 In climatic terms it is a large low pressure system.
2 A complex *fold*, producing a large valley, which incorporates both a *syncline* and an *anticline*.

Desert An *arid* region of low rainfall, with little vegetation or animal life, which is adapted to the dry conditions. The term is now applied not only to hot tropical and subtropical regions, but to arid areas of the continental interiors and to the ice deserts of the *Arctic* and *Antarctic*.

Desertification The gradual extension of *desert* conditions in *arid* or *semi-arid* regions, as a result of climatic change or human activity, such as over-grazing and *deforestation*.

Despot A ruler with absolute power. Despots are often associated with oppressive regimes.

Detritus Piles of rock deposited by an erosive agent such as a river or *glacier*.

Distributary A minor branch of a river, which does not rejoin the main stream, common at *deltas*.

Diurnal Daily, something that occurs each day. Diurnal temperature refers to the variation in temperature over the course of a full day and night.

Divide A US term describing the area of high ground separating two *drainage basins*.

Donga A steep-sided *gully*, resulting from *erosion* by a river or by floods.

Dormant A term used to describe a *volcano* which is not currently erupting. They differ from extinct volcanoes as dormant volcanoes are still considered likely to erupt in the future.

Drainage basin The area drained by a single river system, its boundary is marked by a *watershed* or *divide*.

Drought A long period of continuously low rainfall.

Drumlin A long, streamlined hillock composed of material deposited by a *glacier*. They often occur in groups known as swarms.

Dune A mound or ridge of sand, shaped, and often moved, by the wind. They are found in hot *deserts* and on low-lying coasts where onshore winds blow across sandy beaches.

Dyke A wall constructed in low-lying areas to contain floodwaters or protect from high tides.

E

Earthflow The rapid movement of soil and other loose surface material down a slope, when saturated by water. Similar to a mudflow but not as fast-flowing, due to a lower percentage of water.

Earthquake Sudden movements of the Earth's *crust*, causing the ground to shake. Frequently occurring at *tectonic plate* margins. The shock, or series of shocks, spreads out from an *epicentre*.

EC The European Community (*see* EU).

Ecosystem A system of living organisms – plants and animals – interacting with their *environment*.

ECOWAS Economic Community of West African States. Established in 1975, it incorporates 16 West African states and aims to promote closer regional and economic co-operation.

Element
1 A constituent of the *climate* – *precipitation*, *humidity*, temperature, *atmospheric pressure* or wind.
2 A substance that cannot be separated into simpler substances by chemical means.

El Niño A climatic phenomenon, the El Niño effect occurs about 14 times each century and leads to major shifts in global air circulation. It is associated with unusually warm currents off the coasts of Peru, Ecuador and Chile. The anomaly can last for up to two years.

Environment The conditions created by the surroundings (both natural and artificial) within which an organism lives. In human geography the word includes the surrounding economic, cultural and social conditions.

Eon (aeon) Traditionally a long, but indefinite, period of *geological time*.

Ephemeral A non-permanent feature, often used in connection with seasonal rivers or lakes in dry areas.

Epicentre The point on the Earth's surface directly above the underground origin – or focus – of an *earthquake*.

Equator The line of *latitude* which lies equidistant between the North and South Poles.

Erg An extensive area of sand *dunes*, particularly in the Sahara Desert.

Erosion The processes which wear away the surface of the land. *Glaciers*, wind, rivers, waves and currents all carry debris which causes *erosion*. Some definitions also include *mass movement* due to gravity as an agent of erosion.

Escarpment A steep slope at the margin of a level, upland surface. In a landscape created by *folding*, escarpments (or scarps) frequently lie behind a more gentle backward slope.

Esker A narrow, winding ridge of sand and gravel deposited by streams of water flowing beneath or at the edge of a *glacier*.

Erratic A rock transported by a *glacier* and deposited some distance from its place of origin.

Eustacy A world-wide fall or rise in ocean levels.

EU The European Union. Established in 1965, it was formerly known as the EEC (European Economic Community) and then the EC (European Community). Its members are Austria, Belgium, Denmark, Finland, France, Germany, Greece, Ireland, Italy, Luxembourg, Netherlands, Portugal, Spain, Sweden and UK. It seeks to establish an integrated European common market and eventual federation.

Evaporation The process whereby a liquid or solid is turned into a gas or vapour. Also refers to the diffusion of water vapour into the *atmosphere* from exposed water surfaces such as lakes and seas.

Evapotranspiration The loss of moisture from the Earth's surface through a combination of *evaporation*, and *transpiration* from the leaves of plants.

Evergreen Plants with long-lasting leaves, which are not shed annually or seasonally.

Exfoliation A kind of *weathering* whereby scale-like flakes of rock are peeled or broken off by the development of salt crystals in water within the rocks. *Groundwater*, which contains dissolved salts, seeps to the surface and evaporates, precipitating a film of salt crystals, which expands causing fine cracks. As these grow, flakes of rock break off.

Extrusive rock *Igneous* rock formed when molten material (*magma*) pours forth at the Earth's surface and cools rapidly. It usually has a glassy texture.

F

Factionalism The actions of one or more minority political group acting against the interests of the majority government.

Fault A fracture or crack in rock, where strains (*tectonic* movement) have caused blocks to move, vertically or laterally, relative to each other.

Fauna Collective name for the animals of a particular period of time, or region.

Favela Brazilian term for the shanty towns or self-built, temporary dwellings which have grown up around the edge of many South and Central American cities.

Ferrel cell A component in the global pattern of air circulation, which rises in the colder *latitudes* (60° N and S) and descends in warmer *latitudes* (30° N and S). The Ferrel cell forms part of the world's three-cell air circulation pattern, with the *Hadley* and Polar cells.

Fissure A deep crack in a rock or a *glacier*.

Fjord A deep, narrow inlet, created when the sea inundates the *U-shaped valley* created by a *glacier*.

Flash flood A sudden, short-lived rise in the water level of a river or stream, or surge of water down a dry river channel, or *wadi*, caused by heavy rainfall.

Flax A plant used to make linen.

Flood plain The broad, flat part of a river valley, adjacent to the river itself, formed by *sediment* deposited during flooding.

Flora The collective name for the plants of a particular period of time or region.

Flow The movement of a river within its banks, particularly in terms of the speed and volume of water.

Fold A bend in the rock *strata* of the Earth's *crust*, resulting from compression.

Fossil The remains, or traces, of a dead organism preserved in the Earth's *crust*.

Fossil dune A *dune* formed in a once-*arid* region which is now wetter. *Dunes* normally move with the wind, but in these cases vegetation makes them stable.

Fossil fuel Fuel – coal, natural gas or oil – composed of the fossilized remains of plants and animals.

Front The boundary between two *air masses*, which contrast sharply in temperature and *humidity*.

Frontal depression An area of low pressure caused by rising warm air. They are generally 600–1200 miles (1000–2000 km) in diameter. Within *depressions* there are both warm and cold fronts.

Frost shattering A form of *weathering* where water freezes in cracks, causing expansion. As temperatures fluctuate and the ice melts and refreezes, it eventually causes the rocks to shatter and fragments of rock to break off.

G

Gaucho South American term for a stock herder or cowboy who works on the grassy *plains* of Paraguay, Uruguay and Argentina.

Geological time-scale The chronology of the Earth's history as revealed in its rocks. Geological time is divided into a number of periods: *eon*, era, period, epoch, age and chron (the shortest). These units are not of uniform length.

Geosyncline A concave fold (*syncline*) or large depression in the Earth's *crust*, extending hundreds of kilometres. This basin contains a deep layer of sediment, especially at its centre, from the land masses around it.

Geothermal energy Heat derived from hot rocks within the Earth's *crust* and resulting in hot springs, steam or hot rocks at the surface. The energy is generated by rock movements, and from the breakdown of radioactive elements occurring under intense pressure.

GDP Gross Domestic Product. The total value of goods and services produced by a country excluding income from foreign countries.

Geyser A jet of steam and hot water that intermittently erupts from vents in the ground in areas that are, or were, *volcanic*. Some geysers occasionally reach heights of 196 ft (60 m).

Ghetto An area of a city or region occupied by an overwhelming majority of people from one racial or religious group, who may be subject to persecution or containment.

Glaciation The growth of *glaciers* and *ice sheets*, and their impact on the landscape.

Glacier A body of ice moving downslope under the influence of gravity and consisting of compacted and frozen snow. A glacier is distinct from an *ice sheet*, which is wider and less confined by features of the landscape.

Glacio-eustacy A world-wide change in the level of the oceans, caused when the formation of *ice sheets* takes up water or when their melting returns water to the ocean. The formation of ice sheets in the *Pleistocene* epoch, for example, caused sea level to drop by about 320 ft (100 m).

Glaciofluvial To do with glacial *meltwater*, the landforms it creates and its processes; *erosion*, transportation and *deposition*. Glaciofluvial effects are more powerful and rapid where they occur within or beneath the *glacier*, rather than beyond its edge.

Glacis A gentle slope or *pediment*.

Global warming An increase in the average temperature of the Earth. At present the *greenhouse effect* is thought to contribute to this.

GNP Gross National Product. The total value of goods and services produced by a country.

Gondwanaland The *supercontinent* thought to have existed over 200 million years ago in the southern hemisphere. Gondwanaland is believed to have comprised today's Africa, Madagascar, Australia, parts of South America, *Antarctica* and the Indian subcontinent.

Graben A block of rock let down between two parallel *faults*. Where the graben occurs within a valley, the structure is known as a *rift valley*.

Grease ice Slicks of ice which form in *Antarctic* seas, when ice crystals are bonded together by wind and wave action.

Greenhouse effect A change in the temperature of the *atmosphere*. Short-wave solar radiation travels through the *atmosphere* unimpeded to the Earth's surface, whereas outgoing, long-wave terrestrial radiation is absorbed by materials that re-radiate it back to the Earth. Radiation trapped in this way, by water vapour, carbon dioxide and other 'greenhouse gases', keeps the Earth warm. As more *carbon dioxide* is released into the atmosphere by the burning of *fossil fuels*, the greenhouse effect may cause a global increase in temperature.

Groundwater Water that has seeped into the pores, cavities and cracks of rocks or into soil and water held in an *aquifer*.

Gully A deep, narrow channel eroded in the landscape by *ephemeral* streams.

Guyot A small, flat-topped submarine mountain, formed as a result of subsidence which occurs during *sea-floor spreading*.

Gypsum A soft mineral *compound* (hydrated calcium sulphate), used as the basis of many forms of plaster, including plaster of Paris.

H

Hadley cell A large-scale component in the global pattern of air circulation. Warm air rises over the *Equator* and blows at high altitude towards the poles, sinking in subtropical regions (30° N and 30° S) and creating high pressure. The air then flows at the surface towards the *Equator* in the form of trade winds. There is one cell in each hemisphere. Named after G Hadley, who published his theory in 1735.

Hamada An Arabic word for a plateau of bare rock in a *desert*.

Hanging valley A tributary valley which ends suddenly, high above the bed of the main valley. The effect is found where the main valley has been more deeply eroded by a *glacier*, than has the tributary valley. A stream in a hanging valley will descend to the floor of the main valley as a waterfall or *cataract*.

Headwards The action of a river eroding back upstream, as opposed to the normal process of downstream *erosion*. Headwards erosion is often associated with *gullying*.

Hoodos Pinnacles of rock which have been worn away by *weathering* in *semi-arid* regions.

Horst A block of the Earth's *crust* which has been left upstanding by the sinking of adjoining blocks along fault lines.

Hot spot A region of the Earth's *crust* where high thermal activity occurs, often leading to volcanic eruptions. Hot spots often occur far from plate boundaries, but their movement is associated with *plate tectonics*.

Humid equatorial Rainy *climate* with no winter, where the coolest month is generally above 64°F (18°C).

Humidity The relative amount of moisture held in the Earth's *atmosphere*.

Hurricane
1 A tropical *cyclone* occurring in the Caribbean and western North Atlantic.
2 A wind of more than 65 knots (75 kmph).

Hydro-electric power Energy produced by harnessing the rapid movement of water down steep mountain slopes to drive turbines to generate electricity.

Hydrolysis The chemical breakdown of rocks in reaction with water, forming new compounds.

I

Ice Age A period in the Earth's history when surface temperatures in the temperate *latitudes* were much lower and *ice sheets* expanded considerably. There have been *ice ages* from *Pre-Cambrian* times onwards. The most recent began two million years ago and ended 10,000 years ago.

Ice cap A permanent dome of ice in highland areas. The term ice cap is often seen as distinct from *ice sheet*, which denotes a much wider covering of ice; and is also used refer to the very extensive polar and Greenland ice caps.

Ice floe A large, flat mass of ice floating free on the ocean surface. It is usually formed after the break-up of winter ice by heavy storms.

Ice sheet A continuous, very thick layer of ice and snow. The term is usually used of ice masses which are continental in extent.

Ice shelf A floating mass of ice attached to the edge of a coast. The seaward edge is usually a sheer cliff up to 100 ft (30 m) high.

Ice wedge Massive blocks of ice up to 6.5 ft (2 m) wide at the top and extending 32 ft (10 m) deep. They are found in cracks in *polygonally-patterned* ground in *periglacial* regions.

Iceberg A large mass of ice in a lake or a sea, which has broken off from a floating *ice sheet* (an *ice shelf*) or from a *glacier*.

Igneous rock Rock formed when molten material, *magma*, from the hot, lower layers of the Earth's *crust*, cools, solidifies and crystallizes, either within the Earth's *crust* (*intrusive*) or on the surface (*extrusive*).

IMF International Monetary Fund. Established in 1944 as a UN agency, it contains 182 members around the world and is concerned with world monetary stability and economic development.

Incised meander A *meander* where the river, following its original course, cuts deeply into *bedrock*. This may occur when a mature, meandering river begins to erode its bed much more vigorously after the surrounding land has been uplifted.

Indigenous People, plants or animals native to a particular region.

Infrastructure The communications and services – roads, railways and telecommunications – necessary for the functioning of a country or region.

Inselberg An isolated, steep-sided hill, rising from a low *plain* in *semi-arid* and *savannah* landscapes. Inselbergs are usually composed of a rock, such as granite, which resists *erosion*.

Interglacial A period of global *climate*, between two *ice ages*, when temperatures rise and *ice sheets* and *glaciers* retreat.

Intraplate volcano A *volcano* which lies in the centre of one of the Earth's *tectonic plates*, rather than, as is more common, at its edge. They are thought to have been formed by a *hot spot*.

Intrusion (intrusive igneous rock) Rock formed when molten material, *magma*, penetrates existing rocks below the Earth's surface before cooling and solidifying. These rocks cool more slowly than extrusive rock and therefore tend to have coarser grains.

Irrigation The artificial supply of agricultural water to dry areas, often involving the creation of canals and the diversion of natural watercourses.

Island arc A curved chain of islands. Typically, such an arc fringes an ocean trench, formed at the margin between two *tectonic plates*. As one plate overrides another, *earthquakes* and volcanic activity are common and the islands themselves are often volcanic cones.

Isostasy The state of equilibrium which the Earth's *crust* maintains as its lighter and heavier parts float on the denser underlying mantle.

Isthmus A narrow strip of land connecting two larger landmasses or islands.

J

Jet stream A narrow belt of westerly winds in the *troposphere*, at altitudes above 39,000 ft (12,000 m). Jet streams tend to blow more strongly in winter and include: the subtropical jet stream; the *polar* front jet stream in mid-*latitudes*; the *Arctic* jet stream; and the polar-night jet stream.

Joint A crack in a rock, formed where blocks of rock have not shifted relative to each other, as is the case with a *fault*. Joints are created by *folding*; by shrinkage in *igneous rock* as it cools or *sedimentary rock* as it dries out; and by the release of pressure in a rock mass when overlying materials are removed by *erosion*.

Jute A plant fibre used to make coarse ropes, sacks and matting.

K

Kame A mound of stratified sand and gravel with steep sides, deposited in a *crevasse* by *meltwater* running over a *glacier*. When the ice retreats, this forms an undulating terrain of hummocks.

Karst A barren *limestone* landscape created by carbonic acid in streams and rainwater, in areas where *limestone* is close to the surface. Typical features include caverns, tower-like hills, *sinkholes* and flat limestone pavements.

Kettle hole A round hollow formed in a glacial deposit by a detached block of glacial ice, which later melted. They can fill with water to form kettle-lakes.

L

Lagoon A shallow stretch of coastal salt-water behind a partial barrier such as a sandbank or *coral reef*. Lagoon is also used to describe the water encircled by an *atoll*.

LAIA Latin American Integration Association. Established in 1980, its members are Argentina, Bolivia, Brazil, Chile, Colombia, Ecuador, Mexico, Paraguay, Peru, Uruguay and Venezuela. It aims to promote economic co-operation between member states.

Landslide The sudden downslope movement of a mass of rock or earth on a slope, caused either by heavy rain; the impact of waves; an *earthquake* or human activity.

Laterite A hard red deposit left by *chemical weathering* in tropical conditions, and consisting mainly of oxides of iron and aluminium.

Latitude The angular distance from the *Equator*, to a given point on the Earth's surface. Imaginary lines of *latitude* running parallel to the Equator encircle the Earth, and are measured in degrees north or south of the Equator. The Equator is 0°, the poles 90° South and North respectively. Also called parallels.

Laurasia In the theory of *continental drift*, the northern part of the great *supercontinent* of *Pangaea*. Laurasia is said to consist of N America, Greenland and all of Eurasia north of the Indian subcontinent.

Lava The molten rock, *magma*, which erupts onto the Earth's surface through a *volcano*, or through a *fault* or crack in the Earth's *crust*. Lava refers to the rock both in its molten and in its later, solidified form.

Leaching The process whereby water dissolves minerals and moves them down through layers of soil or rock.

Levée A raised bank alongside the channel of a river. Levées are either human-made or formed in times of flood when the river overflows its channel, slows and deposits much of its *sediment* load.

Lichen An organism which is the symbiotic product of an algae and a fungus. Lichens form in tight crusts on stones and trees, and are resistant to extreme cold. They are often found in tundra regions.

Lignite Low-grade coal, also known as brown coal. Found in large deposits in eastern Europe.

Limestone A porous *sedimentary* rock formed from carbonate materials.

Lingua franca The language adopted as the common language between speakers whose native languages are different. This is common in former colonial states.

Lithosphere The rigid upper layer of the Earth, comprising the *crust* and the upper part of the *mantle*.

Llanos Vast grassland *plains* of northern South America.

Loess Fine-grained, yellow deposits of unstratified silts and sands. Loess is believed to be wind-carried *sediment* created in the last *Ice Age*. Some deposits may later have been redistributed by rivers. Loess-derived soils are of high quality, fertile and easy to work.

Longitude A division of the Earth which pinpoints how far east or west a given place is from the Prime Meridian (0°) which runs through the Royal Observatory at Greenwich, England (UK). Imaginary lines of longitude are drawn around the world from pole to pole. The world is divided into 360 degrees.

Longshore drift The transport of sand and silt along the coast, carried by waves hitting the beach at an angle.

M

Magma Underground, molten rock, which is very hot and highly charged with gas. It is generated at great pressure, at depths 10 miles (16 km) or more below the Earth's surface. It can issue as *lava* at the Earth's surface or, more often, solidify below the surface as *intrusive igneous rock*.

Mantle The layer of the Earth between the *crust* and the *core*. It is about 1800 miles (2900 km) thick. The uppermost layer of the mantle is the soft, 125 mile (200 km) thick *asthenosphere* on which the more rigid *lithosphere* floats.

Maquiladoras Factories on the Mexico side of the Mexico/US border, which are allowed to import raw materials and components duty-free and use low-cost labour to assemble the goods, finally exporting them for sale in the US.

Market gardening The intensive growing of fruit and vegetables close to large local markets.

Mass movement Downslope movement of weathered materials such as rock, often helped by rainfall or glacial *meltwater*. Mass movement may be a gradual process or rapid, as in a *landslide* or rockfall.

Massif A single very large mountain or an area of mountains with uniform characteristics and clearly-defined boundaries.

Meander A loop-like bend in a river, which is found typically in the lower, mature reaches of a river but can form wherever the valley is wide and the slope gentle.

Mediterranean climate A temperate *climate* of hot, dry summers and warm, damp winters. This is typical of the western fringes of the world's continents in the warm temperate regions between *latitudes* of 30° and 40° (north and south).

Meltwater Water resulting from the melting of a *glacier* or *ice sheet*.

Mesa A broad, flat-topped hill, characteristic of *arid* regions.

Mesosphere A layer of the Earth's *atmosphere*, between the *stratosphere* and the *thermosphere*. Extending from about 25–50 miles (40–80 km) above the surface of the Earth.

Mestizo A person of mixed *Amerindian* and European origin.

Metallurgy The refining and working of metals.

Metamorphic rocks Rocks which have been altered from their original form, in terms of texture, composition and structure by intense heat, pressure, or by the introduction of new chemical substances – or a combination of more than one of these.

Meteor A body of rock, metal or other material, which travels through space at great speeds. Meteors are visible as they enter the Earth's *atmosphere* as shooting stars and fireballs.

Meteorite The remains of a *meteor* that has fallen to Earth.

Meteoroid A *meteor* which is still travelling in space, outside the Earth's *atmosphere*.

Mezzogiorno A term applied to the southern portion of Italy.

Milankovitch hypothesis A theory suggesting that there are a series of cycles which slightly alter the Earth's position when rotating about the Sun. The cycles identified all affect the amount of *radiation* the Earth receives at different *latitudes*. The theory is seen as a key factor in the cause of *ice ages*.

Millet A grain-crop, forming part of the staple diet in much of Africa.

Mistral A strong, dry, cold northerly or north-westerly wind, which blows from the Massif Central of France to the Mediterranean Sea. It is common in winter and its cold blasts can cause crop damage in the Rhône Delta, in France.

Mohorovicic discontinuity (Moho) The structural divide at the margin between the Earth's *crust* and the *mantle*. On average it is 20 miles (35 km) below the continents and 6 miles (10 km) below the oceans. The different densities of the *crust* and the mantle cause *earthquake* waves to accelerate at this point.

Monarchy A form of government in which the head of state is a single hereditary monarch. The monarch may be a mere figurehead, or may retain significant authority.

Monsoon A wind which changes direction bi-annually. The change is caused by the reversal of pressure over landmasses and the adjacent oceans. Because the inflowing moist winds bring rain, the term monsoon is also used to refer to the rains themselves. The term is derived from and most commonly refers to the seasonal winds of south and east Asia.

Montaña Mountain areas along the west coast of South America.

Moraine Debris, transported and deposited by a *glacier* or *ice sheet* in unstratified, mixed, piles of rock, boulders, pebbles and clay.

Mountain-building The formation of *fold* mountains by tectonic activity. Also known as orogeny, mountain-building often occurs on the margin where two *tectonic plates* collide. The periods when most mountain-building occurred are known as orogenic phases and lasted many millions of years.

Mudflow An *avalanche* of mud which occurs when a mass of soil is drenched by rain or melting snow. It is a type of *mass movement*, faster than an *earthflow* because it is lubricated by water.

N

Nappe A mass of rocks which has been overfolded by repeated thrust *faulting*.

NAFTA The North American Free Trade Association. Established in 1994 between Canada, Mexico and the US to set up a free-trade zone.

NASA The National Aeronautics and Space Administration. It is a US government agency established in 1958 to develop manned and unmanned space programmes.

NATO The North Atlantic Treaty Organization. Established in 1949 to promote mutual defence and co-operation between its members, which are Belgium, Canada, Czech Republic, Denmark, France, Germany, Greece, Iceland, Italy, Luxembourg, the Netherlands, Norway, Portugal, Poland, Spain, Turkey, UK, and US.

Nitrogen The odourless, colourless gas which makes up 78% of the atmosphere. Within the soil, it is a vital nutrient for plants.

Nomads (nomadic) Wandering communities who move around in search of suitable pasture for their herds of animals.

Nuclear fusion A technique used to create a new nucleus by the merging of two lighter ones, resulting in the release of large quantities of energy.

O

Oasis A fertile area in the midst of a *desert*, usually watered by an underground *aquifer*.

Oceanic ridge A mid-ocean ridge formed, according to the theory of *plate tectonics*, when plates drift apart and hot *magma* pours through to form new oceanic *crust*.

Oligarchy The government of a state by a small, exclusive group of people – such as an elite class or a family group.

Onion-skin weathering The *weathering* away or *exfoliation* of a rock or outcrop by the peeling off of surface layers.

Oriente A flatter region lying to the east of the Andes in South America.

Outwash plain *Glaciofluvial* material (typically clay, sand and gravel) carried beyond an ice sheet by *meltwater* streams, forming a broad, flat deposit.

Oxbow lake A crescent-shaped lake formed on a river *flood plain* when a river erodes the outside bend of a *meander*, making the neck of the *meander* narrower until the river cuts across the neck. The meander is cut off and is dammed off with sediment, creating an oxbow lake. Also known as a cut-off or mortlake.

Oxidation A form of *chemical weathering* where oxygen dissolved in water reacts with minerals in rocks – particularly iron – to form oxides. Oxidation causes brown or yellow staining on rocks, and eventually leads to the break down of the rock.

Oxygen A colourless, odourless gas which is one of the main constituents of the Earth's *atmosphere* and is essential to life on Earth.

Ozone layer A layer of enriched *oxygen* (O_3) within the stratosphere, mostly between 18–50 miles (30–80 km) above the Earth's surface. It is vital to the existence of life on Earth because it absorbs harmful shortwave ultraviolet radiation, while allowing beneficial longer wave ultraviolet radiation to penetrate to the Earth's surface.

P

Pacific Rim The name given to the economically-dynamic countries bordering the Pacific Ocean.

Pack ice Ice masses more than 10 ft (3 m) thick which form on the sea surface and are not attached to a landmass.

Pancake ice Thin discs of ice, up to 8 ft (2.4 m) wide which form when slicks of *grease ice* are tossed together by winds and stormy seas.

Pangaea In the theory of continental drift, Pangaea is the original great land mass which, about 190 million years ago, began to split into Gondwanaland in the south and Laurasia in the north, separated by the Tethys Sea.

Pastoralism Grazing of livestock– usually sheep, goats or cattle. Pastoralists in many drier areas have traditionally been *nomadic*.

Parallel *see Latitude.*

Peat Ancient, partially-decomposed vegetation found in wet, boggy conditions where there is little *oxygen*. It is the first stage in the development of coal and is often dried for use as fuel. It is also used to improve soil quality.

Pediment A gently-sloping ramp of *bedrock* below a steeper slope, often found at mountain edges in *desert* areas, but also in other climatic zones. Pediments may include depositional elements such as *alluvial fans*.

Peninsula A thin strip of land surrounded on three of its sides by water. Large examples include Florida and Korea.

Per capita Latin term meaning 'for each person'.

Periglacial Regions on the edges of *ice sheets* or *glaciers* or, more commonly, cold regions experiencing intense frost action, *permafrost* or both. Periglacial climates bring long, freezing winters and short, mild summers.

Permafrost Permanently frozen ground, typical of *Arctic* regions. Although a layer of soil above the permafrost melts in summer, the melted water does not drain through the permafrost.

Permeable rocks Rocks through which water can seep, because they are either porous or cracked.

Pharmaceuticals The manufacture of medicinal drugs.

Phreatic eruption A volcanic eruption which occurs when *lava* combines with *groundwater*, superheating the water and causing a sudden emission of steam at the surface.

Physical weathering (mechanical weathering) The breakdown of rocks by physical, as opposed to chemical, processes. Examples include: changes in pressure or temperature; the effect of windblown sand; the pressure of growing salt crystals in cracks within rock; and the expansion and contraction of water within rock as it freezes and thaws.

Pingo A dome of earth with a core of ice, found in *tundra* regions. Pingos are formed either when *groundwater* freezes and expands, pushing up the land surface, or when trapped, freezing water in a lake expands and pushes up lake *sediments* to form the pingo dome.

Placer A belt of mineral-bearing rock *strata* lying at or close to the Earth's surface, from which minerals can be easily extracted.

Plain A flat, level region of land, often relatively low-lying.

Plateau A highland tract of flat land.

Plate *see Tectonic plates.*

Plate tectonics The study of *tectonic plates*, which helps to explain *continental drift*, mountain formation and volcanic activity. The movement of tectonic plates may be explained by the currents of rock rising and falling from within the Earth's *mantle*, as it heats up and then cools. The boundaries of the plates are known as plate margins and most mountains, *earthquakes* and *volcanoes* occur at these margins. Constructive margins are moving apart; destructive margins are crunching together and conservative margins are sliding past one another.

Pleistocene A period of *geological time* spanning from about 5.2 million years ago to 1.6 million years ago.

Plutonic rock *Igneous* rocks found deep below the surface. They are coarse-grained because they cooled and solidified slowly.

Polar The zones within the *Arctic* and *Antarctic* circles.

Polje A long, broad *depression* found in *karst* (*limestone*) regions.

Polygonal patterning Typical ground patterning, found in areas where the soil is subject to severe frost action, often in *periglacial* regions.

Porosity A measure of how much water can be held within a rock or a soil. Porosity is measured as the percentage of holes or pores in a material, compared to its total volume. For example, the porosity of slate is less than 1%, whereas that of gravel is 25–35%.

Prairies Originally a French word for grassy *plains* with few or no trees.

Pre-Cambrian The earliest period of *geological time* dating from over 570 million years ago.

Precipitation The fall of moisture from the *atmosphere* onto the surface of the Earth, whether as dew, hail, rain, sleet or snow.

Pyramidal peak A steep, isolated mountain summit, formed when the back walls of three or more *cirques* are cut back and move towards each other. The cliffs around such a horned peak, or horn, are divided by sharp *arêtes*. The Matterhorn in the Swiss Alps is an example.

Pyroclasts Fragments of rock ejected during volcanic eruptions.

Q

Quaternary The current period of *geological time*, which started about 1.6 million years ago.

R

Radiation The emission of energy in the form of particles or waves. Radiation from the sun includes heat, light, ultraviolet rays, gamma rays and X-rays. Only some of the solar energy radiated into space reaches the Earth.

Rainforest Dense forests in tropical zones with high rainfall, temperature and *humidity*. Strictly, the term applies to the equatorial rainforest in tropical lowlands with constant rainfall and no seasonal change. The Congo and Amazon basins are examples. The term is applied more loosely to lush forest in other climates. Within rainforests organic life is dense and varied: at least 40% of all plant and animal species are found here and there may be as many as 100 tree species per hectare.

Rainshadow An area which experiences low rainfall, because of its position on the leeward side of a mountain range.

Reg A large area of stony *desert*, where tightly-packed gravel lies on top of clayey sand. A reg is formed where the wind blows away the finer sand.

Remote-sensing Method of obtaining information about the *environment* using unmanned equipment, such as a satellite, which relays the information to a point where it is collected and used.

Resistance The capacity of a rock to resist *denudation*, by processes such as *weathering* and *erosion*.

Ria A flooded *V-shaped river valley* or estuary, flooded by a rise in sea level (*eustacy*) or sinking land. It is shorter than a *fjord* and gets deeper as it meets the sea.

Rift valley A long, narrow depression in the Earth's *crust*, formed by the sinking of rocks between two *faults*.

River channel The trough which contains a river and is moulded by the flow of water within it.

Roche moutonée A rock found in a glaciated valley. The side facing the flow of the *glacier* has been smoothed and rounded, while the other side has been left more rugged because the *glacier*, as it flows over it, has plucked out frozen fragments and carried them away.

Runoff Water draining from a land surface by flowing across it.

S

Sabkha The floor of an isolated *depression* which occurs in an *arid environment* – usually covered by salt deposits and devoid of vegetation.

SADC Southern African Development Community. Established in 1992 to promote economic integration between its member states, which are Angola, Botswana, Lesotho, Malawi, Mauritius, Mozambique, Namibia, South Africa, Swaziland, Tanzania, Zambia and Zimbabwe.

Salt plug A rounded hill produced by the upward doming of rock *strata* caused by the movement of salt or other evaporite deposits under intense pressure.

Sastrugi Ice ridges formed by wind action. They lie parallel to the direction of the wind.

Savannah Open grassland found between the zone of *deserts*, and that of tropical *rainforests* in the tropics and subtropics. Scattered trees and shrubs are found in some kinds of savannah. A savannah *climate* usually has wet and dry seasons.

Scarp *see Escarpment.*

Scree Piles of rock fragments beneath a cliff or rock face, caused by mechanical *weathering*, especially *frost shattering*, where the expansion and contraction of freezing and thawing water within the rock, gradually breaks it up.

Sea-floor spreading The process whereby *tectonic plates* move apart, allowing hot *magma* to erupt and solidify. This forms a new sea floor and, ultimately, widens the ocean.

Seamount An isolated, submarine mountain or hill, probably of volcanic origin.

Season A period of time linked to regular changes in the weather, especially the intensity of solar *radiation*.

Sediment Grains of rock transported and deposited by rivers, sea, ice or wind.

Sedimentary rocks Rocks formed from the debris of pre-existing rocks or of organic material. They are found in many *environments* – on the ocean floor, on beaches, rivers and *deserts*. Organically-formed sedimentary rocks include coal and chalk. Other sedimentary rocks, such as flint, are formed by chemical processes. Most of these rocks contain *fossils*, which can be used to date them.

Seif A sand *dune* which lies parallel to the direction of the prevailing wind. Seifs form steep-sided ridges, sometimes extending for miles.

Seismic activity Movement within the Earth, such as an *earthquake* or *tremor*.

Selva A region of wet forest found in the Amazon Basin.

Semi-arid, semi-desert The *climate* and landscape which lies between *savannah* and *desert* or between savannah and a *mediterranean* climate. In semi-arid conditions there is a little more moisture than in a true *desert*; and more patches of drought-resistant vegetation can survive.

Shale (marine shale) A compacted *sedimentary rock*, with fine-grained particles. Marine shale is formed on the seabed. Fuel such as oil may be extracted from it.

Sheetwash Water which runs downhill in thin sheets without forming channels. It can cause *sheet erosion*.

Sheet erosion The washing away of soil by a thin film or sheet of water, known as *sheetwash*.

Shield A vast stable block of the Earth's *crust*, which has experienced little or no *mountain-building*.

Sierra The Spanish word for mountains.

Sinkhole A circular *depression* in a *limestone* region. They are formed by the collapse of an underground cave system or the *chemical weathering* of the *limestone*.

Sisal A plant-fibre used to make matting.

Slash and burn A farming technique involving the cutting down and burning of scrub forest, to create agricultural land. After a number of seasons this land is abandoned and the process is repeated. This practice is common in Africa and South America.

Slip face The steep leeward side of a sand *dune* or slope. Opposite side to a *back slope*.

Soil A thin layer of rock particles mixed with the remains of dead plants and animals. This occurs naturally on the surface of the Earth and provides a medium for plants to grow.

Soil creep The very gradual downslope movement of rock debris and soil, under the influence of gravity. This is a type of *mass movement*.

Soil erosion The wearing away of soil more quickly than it is replaced by natural processes. Soil can be carried away by wind as well as by water. Human activities, such as over-grazing and the clearing of land for farming, accelerate the process in many areas.

Solar energy Energy derived from the Sun. Solar energy is converted into other forms of energy. For example, the wind and waves, as well as the creation of plant material in photosynthesis, depend on solar energy.

Solifluction A kind of *soil creep*, where water in the surface layer has saturated the soil and rock debris which slips slowly downhill. It often happens where frozen top-layer deposits thaw, leaving frozen layers below them.

Sorghum A type of grass found in South America, similar to sugar cane. When refined it is used to make molasses.

Spit A thin linear deposit of sand or shingle extending from the sea shore. Spits are formed as angled waves shift sand along the beach, eventually extending a ridge of sand beyond a change in the angle of the coast. Spits are common where the coastline bends, especially at estuaries.

Squash A type of edible gourd.

Stack A tall, isolated pillar of rock near a coastline, created as wave action erodes away the adjacent rock.

Stalactite A tapering cylinder of mineral deposit, hanging from the roof of a cave in a *karst* area. It is formed by calcium carbonate, dissolved in water, which drips through the roof of a *limestone* cavern.

Stalagmite A cone of calcium carbonate, similar to a *stalactite*, rising from the floor of a *limestone* cavern and formed when drops of water fall from the roof of a *limestone* cave. If the water has dripped from a *stalactite* above the stalagmite, the two may join to form a continuous pillar.

Staple crop The main crop on which a country is economically and or physically reliant. For example, the major crop grown for large-scale local consumption in South Asia is rice.

Steppe Large areas of dry grassland in the northern hemisphere – particularly found in southeast Europe and central Asia.

Strata The plural of stratum, a distinct, virtually horizontal layer of deposited material, lying parallel to other layers.

Stratosphere A layer of the *atmosphere*, above the *troposphere*, extending from about 7–30 miles (11–50 km) above the Earth's surface. In the lower part of the stratosphere, the temperature is relatively stable and there is little moisture.

Strike-slip fault Occurs where plates move sideways past each other and blocks of rocks move horizontally in relation to each other, not up or down as in normal *faults*.

Subduction zone A region where two *tectonic plates* collide, forcing one beneath the other. Typically, a dense oceanic plate dips below a lighter continental plate, melting in the heat of the *asthenosphere*. This is why the zone is also called a destructive margin (*see Plate tectonics*). These zones are characterized by *earthquakes*, volcanoes, *mountain-building* and the development of oceanic trenches and *island arcs*.

Submarine canyon A steep-sided valley, which extends along the *continental shelf* to the ocean floor. Often formed by *turbidity currents*.

Submarine fan Deposits of silt and *alluvium*, carried by large rivers forming great fan-shaped deposits on the ocean floor.

Subsistence agriculture An agricultural practice, whereby enough food is produced to support the farmer and his dependents, but not providing any surplus to generate an income.

Subtropical A term applied loosely to *climates* which are nearly tropical or tropical for a part of the year – areas north or south of the *tropics* but outside the *temperate zone*.

Supercontinent A large continent that breaks up to form smaller continents or which forms when smaller continents merge. In the theory of *continental drift*, the supercontinents are *Pangaea*, *Gondwanaland* and *Laurasia*.

Sustainable development An approach to development, applied to economies across the world which exploit natural resources without destroying them or the *environment*.

Syncline A basin-shaped downfold in rock *strata*, created when the *strata* are compressed, for example where *tectonic plates* collide.

T

Tableland A highland area with a flat or gently undulating surface.

Taiga The belt of *coniferous* forest found in the north of Asia and North America. The conifers are adapted to survive low temperatures and long periods of snowfall.

Tarn A Scottish term for a small mountain lake, usually found at the head of a *glacier*.

Tectonic plates Plates, or tectonic plates, are the rigid slabs which form the Earth's outer shell, the *lithosphere*. Eight big plates and several smaller ones have been identified.

Temperate A moderate *climate* without extremes of temperature, typical of the mid-*latitudes* between the *tropics* and the *polar* circles.

Theocracy A state governed by religious laws – today Iran is the world's largest theocracy.

Thermokarst Subsidence created by the thawing of ground ice in *periglacial* areas, creating depressions.

Thermosphere A layer of the Earth's *atmosphere* which lies above the *mesophere*, about 60–300 miles (100–500 km) above the Earth

Terraces Steps cut into steep slopes to create flat surfaces for cultivating crops. They also help reduce soil *erosion* on unconsolidated slopes. They are most common in heavily-populated parts of Southeast Asia.

Till Unstratified glacial deposits or drift left by a *glacier* or *ice sheet*. Till includes mixtures of clay, sand, gravel and boulders.

Topography The typical shape and features of a given area such as land height and terrain.

Tombolo A large sand *spit* which attaches part of the mainland to an island.

Tornado A violent, spiralling windstorm, with a centre of very low pressure. Wind speeds reach 200 mph (320 kmph) and there is often thunder and heavy rain.

Transform fault In *plate tectonics*, a *fault* of continental scale, occurring where two plates slide past each other, staying close together for example, the San Andreas Fault, USA. The jerky, uneven movement creates *earthquakes* but does not destroy or add to the Earth's *crust*

Transpiration The loss of water vapour through the pores (or stomata) of plants. The process helps to return moisture to the *atmosphere*.

Trap An area of fine-grained *igneous rock* which has been extruded and cooled on the Earth's surface in stages, forming a series of steps or terraces.

Treeline The line beyond which trees cannot grow, dependent on *latitude* and altitude, as well as local factors such as soil.

Tremor A slight *earthquake*.

Trench (oceanic trench) A long, deep trough in the ocean floor, formed, according to the theory of *plate tectonics*, when two plates collide and one dives under the other, creating a *subduction zone*.

Tropics The zone between the *Tropic of Cancer* and the *Tropic of Capricorn* where the *climate* is hot. Tropical climate is also applied to areas rather further north and south of the *Equator* where the climate is similar to that of the true tropics.

Tropic of Cancer A line of *latitude* or imaginary circle round the Earth, lying at 23° 28' N.

Tropic of Capricorn A line of *latitude* or imaginary circle round the Earth, lying at 23° 28' S.

Troposphere The lowest layer of the Earth's *atmosphere*. From the surface, it reaches a height of between 4–10 miles (7–16 km). It is the most turbulent zone of the atmosphere and accounts for the generation of most of the world's weather. The layer above it is called the *stratosphere*.

Tsunami A huge wave created by shock waves from an *earthquake* under the sea. Reaching speeds of up to 600 mph (960 kmph), the wave may increase to heights of 50 ft (15 m) on entering coastal waters; and it can cause great damage.

Tundra The treeless *plains* of the *Arctic Circle*, found south of the *polar* region of permanent ice and snow, and north of the belt of *coniferous* forests known as *taiga*. In this region of long, very cold winters, vegetation is usually limited to mosses, *lichens*, sedges and rushes, although flowers and dwarf shrubs blossom in the brief summer.

Turbidity current An oceanic feature. A turbidity current is a mass of *sediment*-laden water which has substantial erosive power. Turbidity currents are thought to contribute to the formation of *submarine canyons*.

Typhoon A kind of *hurricane* (or tropical cyclone) bringing violent winds and heavy rain, a typhoon can do great damage. They occur in the South China Sea, especially around the Philippines.

U

U-shaped valley A river valley that has been deepened and widened by a *glacier*. They are characteristically flat-bottomed and steep-sided and generally much deeper than river valleys.

UN United Nations. Established in 1945, it contains 188 nations and aims to maintain international peace and security, and promote co-operation over economic, social, cultural and humanitarian problems.

UNICEF United Nations Children's Fund. A UN organization set up to promote family and child related programmes.

Urstromtäler A German word used to describe *meltwater* channels which flowed along the front edge of the advancing *ice sheet* during the last Ice Age, 18,000–20,000 years ago.

V

V-shaped valley A typical valley eroded by a river in its upper course.

Virgin rainforest Tropical *rainforest* in its original state, untouched by human activity such as logging, clearance for agriculture, settlement or road building.

Viticulture The cultivation of grapes for wine.

Volcano An opening or vent in the Earth's *crust* where molten rock, *magma*, erupts. Volcanoes tend to be conical but may also be a crack in the Earth's surface or a hole blasted through a mountain. The magma is accompanied by other materials such as gas, steam and fragments of rock, or *pyroclasts*. They tend to occur on destructive or constructive *tectonic plate* margins.

W-Z

Wadi The dry bed left by a torrent of water. Also classified as a *ephemeral* stream, found in *arid* and *semi-arid* regions, which are subject to sudden and often severe flash flooding.

Warm humid climate A rainy climate with warm summers and mild winters.

Water cycle The continuous circulation of water between the Earth's surface and the *atmosphere*. The processes include *evaporation* and *transpiration* of moisture into the atmosphere, and its return as *precipitation*, some of which flows into lakes and oceans.

Water table The upper level of *groundwater* saturation in permeable rock *strata*.

Watershed The dividing line between one *drainage basin* – an area where all streams flow into a single river system – and another. In the US, watershed also means the whole drainage basin of a single river system – its catchment area.

Waterspout A rotating column of water in the form of cloud, mist and spray which form on open water. Often has the appearance of a small *tornado*.

Weathering The decay and break-up of rocks at or near the Earth's surface, caused by water, wind, heat or ice, organic material or the *atmosphere*. *Physical weathering* includes the effects of frost and temperature changes. Biological weathering includes the effects of plant roots, burrowing animals and the acids produced by animals, especially as they decay after death. *Carbonation* and *hydrolysis* are among many kinds of *chemical weathering*.

Geographical names

The following glossary lists all geographical terms occurring on the maps and in main-entry names in the Index-Gazetteer. These terms may precede, follow or be run together with the proper element of the name; where they precede it the term is reversed for indexing purposes - thus Poluostrov Yamal is indexed as Yamal, Poluostrov.

Key

Geographical term
Language, Term

A

Å *Danish, Norwegian*, River
Āb *Persian*, River
Adrar *Berber*, Mountains
Agía, Ágios *Greek*, Saint
Air *Indonesian*, River
Akrotírio *Greek*, Cape, point
Alpen *German*, Alps
Alt- *German*, Old
Altiplanicie *Spanish*, Plateau
Älv, -älven *Swedish*, River
-ån *Swedish*, River
Anse *French*, Bay
'Aqabat *Arabic*, Pass
Archipiélago *Spanish*, Archipelago
Arcipelago *Italian*, Archipelago
Arquipélago *Portuguese*, Archipelago
Arrecife(s) *Spanish*, Reef(s)
Aru *Tamil*, River
Augstiene *Latvian*, Upland
Aukštuma *Lithuanian*, Upland
Aust- *Norwegian*, Eastern
Avtonomnyy Okrug *Russian*, Autonomous district
Āw *Kurdish*, River
'Ayn *Arabic*, Spring, well
'Ayoûn *Arabic*, Wells

B

Baelt *Danish*, Strait
Bahía *Spanish*, Bay
Baḥr *Arabic*, River
Baía *Portuguese*, Bay
Baie *French*, Bay
Bañado *Spanish*, Marshy land
Bandao *Chinese*, Peninsula
Banjaran *Malay*, Mountain range
Barajı *Turkish*, Dam
Barragem *Portuguese*, Reservoir
Bassin *French*, Basin
Batang *Malay*, Stream
Beinn, Ben *Gaelic*, Mountain
-berg *Afrikaans, Norwegian*, Mountain
Besar *Indonesian, Malay*, Big
Birkat, Birket *Arabic*, Lake, well, cistern
Boğazı *Turkish*, Strait, defile
Boka *Serbo-Croatian*, Bay
Bol'sh-aya, -iye, -oy, -oye *Russian*, Big
Botigh(i) *Uzbek*, Depression basin
-bre(en) *Norwegian*, Glacier
Bredning *Danish*, Bay
Bucht *German*, Bay
Bugt(en) *Danish*, Bay
Buḥayrat *Arabic*, Lake, reservoir
Buḥeirat *Arabic*, Lake
Bukit *Malay*, Mountain
-bukta *Norwegian*, Bay
bukten *Swedish*, Bay
Bulag *Mongolian*, Spring
Bulak *Uighur*, Spring
Burnu *Turkish*, Cape, point
Buuraha *Somali*, Mountains

C

Cabo *Portuguese*, Cape
Caka *Tibetan*, Salt lake
Canal *Spanish*, Channel
Cap *French*, Cape
Capo *Italian*, Cape, headland
Cascada *Portuguese*, Waterfall
Cayo(s) *Spanish*, Islet(s), rock(s)
Cerro *Spanish*, Hill
Chaîne *French*, Mountain range
Chapada *Portuguese*, Hills, upland
Chau *Cantonese*, Island
Chay *Turkish*, River
Chhâk *Cambodian*, Bay
Chhu *Tibetan*, River
-chŏsuji *Korean*, Reservoir
Chott *Arabic*, Depression, salt lake
Chŭli *Uzbek*, Grassland, steppe
Ch'ün-tao *Chinese*, Island group
Chuŏr Phnum *Cambodian*, Mountains
Ciudad *Spanish*, City, town
Co *Tibetan*, Lake
Colline(s) *French*, Hill(s)
Cordillera *Spanish*, Mountain range
Costa *Spanish*, Coast
Côte *French*, Coast
Coxilha *Portuguese*, Mountains
Cuchilla *Spanish*, Mountains

D

Daban *Mongolian, Uighur*, Pass
Dağı *Azerbaijani, Turkish*, Mountain
Dağları *Azerbaijani, Turkish*, Mountains
-dake *Japanese*, Peak
-dal(en) *Norwegian*, Valley
Danau *Indonesian*, Lake
Dao *Chinese*, Island
Ðao *Vietnamese*, Island
Daryā *Persian*, River
Daryācheh *Persian*, Lake
Dasht *Persian*, Desert, plain
Dawḥat *Arabic*, Bay
Denizi *Turkish*, Sea
Dere *Turkish*, Stream
Desierto *Spanish*, Desert
Dili *Azerbaijani*, Spit
-do *Korean*, Island
Dooxo *Somali*, Valley
Düzü *Azerbaijani*, Steppe
-dwīp *Bengali*, Island

E

-eilanden *Dutch*, Islands
Embalse *Spanish*, Reservoir
Ensenada *Spanish*, Bay
Erg *Arabic*, Dunes
Estany *Catalan*, Lake
Estero *Spanish*, Inlet
Estrecho *Spanish*, Strait
Étang *French*, Lagoon, lake
-ey *Icelandic*, Island
Ezero *Bulgarian, Macedonian*, Lake
Ezers *Latvian*, Lake

F

Feng *Chinese*, Peak
-fjella *Norwegian*, Mountain
Fjord *Danish*, Fjord
-fjord(en) *Danish, Norwegian, Swedish*, fjord
-fjördhur *Icelandic*, Fjord
Fleuve *French*, River
Fliegu *Maltese*, Channel
-fljór *Icelandic*, River
-flói *Icelandic*, Bay
Forêt *French*, Forest

G

-gan *Japanese*, Rock
-gang *Korean*, River
Ganga *Hindi, Nepali, Sinhala*, River
Gaoyuan *Chinese*, Plateau
Garagumy *Turkmen*, Sands
-gawa *Japanese*, River
Gebel *Arabic*, Mountain
-gebirge *German*, Mountain range
Ghadīr *Arabic*, Well
Ghubbat *Arabic*, Bay
Gjiri *Albanian*, Bay
Gol *Mongolian*, River
Golfe *French*, Gulf
Golfo *Italian, Spanish*, Gulf
Göl(ü) *Turkish*, Lake
Golyam, -a *Bulgarian*, Big
Gora *Russian, Serbo-Croatian*, Mountain
Góra *Polish*, mountain
Gory *Russian*, Mountain
Gryada *Russian*, ridge
Guba *Russian*, Bay
-gundo *Korean*, island group
Gunung *Malay*, Mountain

H

Ḥadd *Arabic*, Spit
-haehyŏp *Korean*, Strait
Haff *German*, Lagoon
Hai *Chinese*, Bay, lake, sea
Haixia *Chinese*, Strait
Ḥammādah *Arabic*, Desert
Ḥammādat *Arabic*, Rocky plateau
Hāmūn *Persian*, Lake
-hantō *Japanese*, Peninsula
Har, Haré *Hebrew*, Mountain
Ḥarrat *Arabic*, Lava-field
Hav(et) *Danish, Swedish*, Sea
Hawr *Arabic*, Lake
Hāyk' *Amharic*, Lake
He *Chinese*, River
-hegység *Hungarian*, Mountain range
Heide *German*, Heath, moorland
Helodrano *Malagasy*, Bay
Higashi- *Japanese*, East(ern)
Ḥiṣā' *Arabic*, Well
Hka *Burmese*, River
-ho *Korean*, Lake
Ḥolot *Hebrew*, Dunes
Hora *Belarussian, Czech*, Mountain
Hrada *Belarussian*, Mountain, ridge
Hsi *Chinese*, River
Hu *Chinese*, Lake
Huk *Danish*, Point

I

Île(s) *French*, Island(s)
Ilha(s) *Portuguese*, Island(s)
Ilhéu(s) *Portuguese*, Islet(s)
-isen *Norwegian*, Ice shelf
Imeni *Russian*, In the name of
Inish- *Gaelic*, Island
Insel(n) *German*, Island(s)
Irmağı, Irmak *Turkish*, River
Isla(s) *Spanish*, Island(s)
Isola (Isole) *Italian*, Island(s)

J

Jabal *Arabic*, Mountain
Jāl *Arabic*, Ridge
-järv *Estonian*, Lake
-järvi *Finnish*, Lake
Jazā'ir *Arabic*, Islands
Jazīrat *Arabic*, Island
Jazīreh *Persian*, Island
Jebel *Arabic*, Mountain
Jezero *Serbo-Croatian*, Lake
Jezioro *Polish*, Lake
Jiang *Chinese*, River
-jima *Japanese*, Island
Jižní *Czech*, Southern
-jõgi *Estonian*, River
-joki *Finnish*, River
-jökull *Icelandic*, Glacier
Jūn *Arabic*, Bay
Juzur *Arabic*, Islands

K

Kaikyō *Japanese*, Strait
-kaise *Lappish*, Mountain
Kali *Nepali*, River
Kalnas *Lithuanian*, Mountain
Kalns *Latvian*, Mountain
Kang *Chinese*, Harbour
Kangri *Tibetan*, Mountain(s)
Kaôh *Cambodian*, Island
Kapp *Norwegian*, Cape
Káto *Greek*, Lower
Kavīr *Persian*, Desert
K'edi *Georgian*, Mountain range
Kediet *Arabic*, Mountain
Kepi *Albanian*, Cape, point
Kepulauan *Indonesian, Malay*, Island group
Khalig, Khalīj *Arabic*, Gulf
Khawr *Arabic*, Inlet
Khola *Nepali*, River
Khrebet *Russian*, Mountain range
Ko *Thai*, Island
-ko *Japanese*, Inlet, lake
Kólpos *Greek*, Bay
-kopf *German*, Peak
Körfäzi *Azerbaijani*, Bay
Körfezi *Turkish*, Bay
Kõrgustik *Estonian*, Upland
Kosa *Russian, Ukrainian*, Spit
Koshi *Nepali*, River
Kou *Chinese*, River-mouth
Kowtal *Persian*, Pass
Kray *Russian*, Region, territory
Kryazh *Russian*, Ridge
Kuduk *Uighur*, Well
Kūh(hā) *Persian*, Mountain(s)
-kul' *Russian*, Lake
Kŭl(i) *Tajik, Uzbek*, Lake
-kundo *Korean*, Island group
-kysten *Norwegian*, Coast
Kyun *Burmese*, Island

L

Laaq *Somali*, Watercourse
Lac *French*, Lake
Lacul *Romanian*, Lake
Lagh *Somali*, Stream
Lago *Italian, Portuguese, Spanish*, Lake
Lagoa *Portuguese*, Lagoon
Laguna *Italian, Spanish*, Lagoon, lake
Laht *Estonian*, Bay
Laut *Indonesian*, Bay
Lembalemba *Malagasy*, Plateau
Lerr *Armenian*, Mountain
Lerrnashghtʻa *Armenian*, Mountain range
Les *Czech*, Forest
Lich *Armenian*, Lake
Liehtao *Chinese*, Island group
Liqeni *Albanian*, Lake
Límni *Greek*, Lake
Ling *Chinese*, Mountain range
Llano *Spanish*, Plain, prairie
Lumi *Albanian*, River
Lyman *Ukrainian*, Estuary

M

Madīnat *Arabic*, City, town
Mae Nam *Thai*, River
-mägi *Estonian*, Hill
Maja *Albanian*, Mountain
Mal *Albanian*, Mountains
Mal-aya, -oye, -yy, *Russian*, Small
-man *Korean*, Bay
Mar *Spanish*, Sea
Marios *Lithuanian*, Lake
Massif *French*, Mountains
Meer *German*, Lake
-meer *Dutch*, Lake
Melkosopochnik *Russian*, Plain
-meri *Estonian*, Sea
Mifraẕ *Hebrew*, Bay
Minami- *Japanese*, South(ern)
-misaki *Japanese*, Cape, point
Monkhafad *Arabic*, Depression
Montagne(s) *French*, Mountain(s)
Montañas *Spanish*, Mountains
Mont(s) *French*, Mountain(s)
Monte *Italian, Portuguese*, Mountain
More *Russian*, Sea
Mörön *Mongolian*, River
Mys *Russian*, Cape, point

N

-nada *Japanese*, Open stretch of water
Nadi *Bengali*, River
Nagor'ye *Russian*, Upland
Naḥal *Hebrew*, River
Nahr *Arabic*, River
Nam *Laotian*, River
Namakzār *Persian*, Salt desert
Né-a, -on, -os *Greek*, New
Nedre- *Norwegian*, Lower
-neem *Estonian*, Cape, point
Nehri *Turkish*, River
-nes *Norwegian*, Cape, point
Nevado *Spanish*, Mountain (snow-capped)
Nieder- *German*, Lower
Nishi- *Japanese*, West(ern)
-nísi *Greek*, Island
Nisoi *Greek*, Islands
Nizhn-eye, -iy, -iye, -yaya *Russian*, Lower
Nizmennost' *Russian*, Lowland, plain
Nord *Danish, French, German*, North
Norte *Portuguese, Spanish*, North
Nos *Bulgarian*, Point, spit
Nosy *Malagasy*, Island
Nov-a, -i, *Bulgarian, Serbo-Croatian*, New
Nov-aya, -o, -oye, -yy, -yye *Russian*, New
Now-a, -e, -y *Polish*, New
Nur *Mongolian*, Lake
Nuruu *Mongolian*, Mountains
Nuur *Mongolian*, Lake
Nyzovyna *Ukrainian*, Lowland, plain

O

-ø *Danish*, Island
Ober- *German*, Upper
Oblast' *Russian*, Province
Órmos *Greek*, Bay
Orol(i) *Uzbek*, Island
Øster- *Norwegian*, Eastern
Ostrov(a) *Russian*, Island(s)
Otok *Serbo-Croatian*, Island
Oued *Arabic*, Watercourse
-oy *Faeroese*, Island
-øy(a) *Norwegian*, Island
Oya *Sinhala*, River
Ozero *Russian, Ukrainian*, Lake

P

Passo *Italian*, Pass
Pegunungan *Indonesian, Malay*, Mountain range
Pélagos *Greek*, Sea
Pendi *Chinese*, Basin
Penisola *Italian*, Peninsula
Pertuis *French*, Strait
Peski *Russian*, Sands
Phanom *Thai*, Mountain
Phou *Laotian*, Mountain
Pi *Chinese*, Point
Pic *Catalan, French*, Peak
Pico *Portuguese, Spanish*, Peak
-piggen *Danish*, Peak
Pik *Russian*, Peak
Pivostriv *Ukrainian*, Peninsula
Planalto *Portuguese*, Plateau
Planina, Planini *Bulgarian, Macedonian, Serbo-Croatian*, Mountain range
Plato *Russian*, Plateau
Ploskogor'ye *Russian*, Upland
Poluostrov *Russian*, Peninsula
Ponta *Portuguese*, Point
Porthmós *Greek*, Strait
Pótamos *Greek*, River
Presa *Spanish*, Dam
Prokhod *Bulgarian*, Pass
Proliv *Russian*, Strait
Pulau *Indonesian Malay*, Island
Pulu *Malay*, Island
Punta *Spanish*, Point
Pushcha *Belorussian*, Forest
Puszcza *Polish*, Forest

Q

Qā' *Arabic*, Depression
Qalamat *Arabic*, Well
Qatorkŭh(i) *Tajik*, Mountain
Qiuling *Chinese*, Hills
Qolleh *Persian*, Mountain
Qu *Tibetan*, Stream
Quan *Chinese*, Well
Qulla(i) *Tajik*, Peak
Qundao *Chinese*, Island group

R

Raas *Somali*, Cape
-rags *Latvian*, Cape
Ramlat *Arabic*, Sands
Ra's *Arabic*, Cape, headland, point
Ravnina *Bulgarian, Russian*, Plain
Récif *French*, Reef
Recife *Portuguese*, Reef
Reka *Bulgarian*, River
Represa (Rep.) *Portuguese, Spanish*, Reservoir
Reshteh *Persian*, Mountain range
Respublika *Russian*, Republic, first-order administrative division
Respublika(si) *Uzbek*, Republic, first-order administrative division
-retsugan *Japanese*, Chain of rocks
-rettō *Japanese*, Island chain
Riacho *Spanish*, Stream
Riban' *Malagasy*, Mountains
Rio *Portuguese*, River
Río *Spanish*, River
Riu *Catalan*, River
Rivier *Dutch*, River
Rivière *French*, River
Rowd *Pashtu*, River
Rt *Serbo-Croatian*, Point
Rūd *Persian*, River
Rūdkhāneh *Persian*, River
Rudohorie *Slovak*, Mountains
Ruisseau *French*, Stream

S

-saar *Estonian*, Island
-saari *Finnish*, Island
Sabkhat *Arabic*, Salt marsh
Sāgar(a) *Hindi*, Lake, reservoir
Ṣaḥrā' *Arabic*, Desert
Saint, Sainte *French*, Saint
Salar *Spanish*, Salt-pan
Salto *Portuguese, Spanish*, Waterfall
Samudra *Sinhala*, Reservoir
-san *Japanese, Korean*, Mountain
-sanchi *Japanese*, Mountains
-sandur *Icelandic*, Beach
Sankt *German, Swedish*, Saint
-sanmaek *Korean*, Mountain range
-sanmyaku *Japanese*, Mountain range
San, Santa, Santo *Italian, Portuguese, Spanish*, Saint
São *Portuguese*, Saint
Sarīr *Arabic*, Desert
Sebkha, Sebkhet *Arabic*, Depression, salt marsh
Sedlo *Czech*, Pass
See *German*, Lake
Selat *Indonesian*, Strait
Selatan *Indonesian*, Southern
-selkä *Finnish*, Lake, ridge
Selseleh *Persian*, Mountain range
Serra *Portuguese*, Mountain
Serranía *Spanish*, Mountain
-seto *Japanese*, Channel, strait
Sever-naya, -noye, -nyy, -o *Russian*, Northern
Sha'ib *Arabic*, Watercourse
Shākh *Kurdish*, Mountain
Shamo *Chinese*, Desert
Shan *Chinese*, Mountain(s)
Shankou *Chinese*, Pass
Shanmo *Chinese*, Mountain range
Shaṭṭ *Arabic*, Distributary
Shet' *Amharic*, River
Shi *Chinese*, Municipality
-shima *Japanese*, Island
Shiqqat *Arabic*, Depression
-shotō *Japanese*, Group of islands
Shuiku *Chinese*, Reservoir
Shŭrkhog(i) *Uzbek*, Salt marsh
Sierra *Spanish*, Mountains
Sint *Dutch*, Saint
-sjø(en) *Norwegian*, Lake
-sjön *Swedish*, Lake
Solonchak *Russian*, Salt lake
Solonchakovyye Vpadiny *Russian*, Salt basin, wetlands
Sơn *Vietnamese*, Mountain
Sông *Vietnamese*, River
Sør- *Norwegian*, Southern
-spitze *German*, Peak
Star-á, -é *Czech*, Old
Star-aya, -oye, -yy, -yye *Russian*, Old
Stenó *Greek*, Strait
Step' *Russian*, Steppe
Štít *Slovak*, Peak
Stœng *Cambodian*, River
Stolovaya Strana *Russian*, Plateau
Stredné *Slovak*, Middle
Střední *Czech*, Middle
Stretto *Italian*, Strait
Su Anbarı *Azerbaijani*, Reservoir
-suidō *Japanese*, Channel, strait
Sund *Swedish*, Sound, strait
Sungai *Indonesian, Malay*, River
Suu *Turkish*, River

T

Tal *Mongolian*, Plain
Tandavan' *Malagasy*, Mountain range
Tangorombohitr' *Malagasy*, Mountain massif
Tanjung *Indonesian, Malay*, Cape, point
Tao *Chinese*, Island
Ṭaraq *Arabic*, Hills
Tassili *Berber*, Mountain, plateau
Tau *Russian*, Mountain(s)
Taungdan *Burmese*, Mountain range
Techníti Límni *Greek*, Reservoir
Tekojärvi *Finnish*, Reservoir
Teluk *Indonesian, Malay*, Bay
Tengah *Indonesian*, Middle
Terara *Amharic*, Mountain
Timur *Indonesian*, Eastern
-tind(an) *Norwegian*, Peak
Tizma(si) *Uzbek*, Mountain range, ridge
-tō *Japanese*, island
Tog *Somali*, Valley
-tōge *Japanese*, pass
Togh(i) *Uzbek*, mountain
Tônlé *Cambodian*, Lake
Top *Dutch*, Peak
-tunturi *Finnish*, Mountain
Ṭurāq *Arabic*, hills
Tur'at *Arabic*, Channel

U

Udde(n) *Swedish*, Cape, point
'Uqlat *Arabic*, Well
Utara *Indonesian*, Northern
Uul *Mongolian*, Mountains

V

Väin *Estonian*, Strait
Vallée *French*, Valley
Varful *Romanian*, Peak
-vatn *Icelandic*, Lake
-vatnet *Norwegian*, Lake
Velayat *Turkmen*, Province
-vesi *Finnish*, Lake
Vestre- *Norwegian*, Western
-vidda *Norwegian*, Plateau
-vík *Icelandic*, Bay
-viken *Swedish*, Bay, inlet
Vinh *Vietnamese*, Bay
Víztárloló *Hungarian*, Reservoir
Vodaskhovishcha *Belarussian*, Reservoir
Vodokhranilishche (Vdkhr.) *Russian*, Reservoir
Vodoskhovyshche (Vdskh.) *Ukrainian*, Reservoir
Volcán *Spanish*, Volcano
Vostochn-o, -yy *Russian*, Eastern
Vozvyshennost' *Russian*, Upland, plateau
Vozyera *Belarussian*, Lake
Vpadina *Russian*, Depression
Vrchovina *Czech*, Mountains
Vrh *Croat, Slovene*, Peak
Vychodné *Slovak*, Eastern
Vysochyna *Ukrainian*, Upland
Vysočina *Czech*, Upland

W

Waadi *Somali*, Watercourse
Wādī *Arabic* Watercourse
Wāḥat, Wâhat *Arabic*, Oasis
Wald *German*, Forest
Wan *Chinese*, Bay
Way *Indonesian*, River
Webi *Somali*, River
Wenz *Amharic*, River
Wiloyat(i) *Uzbek*, Province
Wyżyna *Polish*, Upland
Wzgórza *Polish*, Upland
Wzvyshsha *Belarussian*, Upland

X

Xé *Laotian*, River
Xi *Chinese*, Stream

Y

-yama *Japanese*, Mountain
Yanchi *Chinese*, Salt lake
Yanhu *Chinese*, Salt lake
Yarımadası *Azerbaijani, Turkish*, Peninsula
Yaylası *Turkish*, Plateau
Yazovir *Bulgarian*, Reservoir
Yoma *Burmese*, Mountains
Ytre- *Norwegian*, Outer
Yu *Chinese*, Islet
Yunhe *Chinese*, Canal
Yuzhn-o, -yy *Russian*, Southern

Z

-zaki *Japanese*, Cape, point
Zaliv *Bulgarian, Russian*, Bay
-zan *Japanese*, Mountain
Zangbo *Tibetan*, River
Zapadn-aya, -o, -yy *Russian*, Western
Západné *Slovak*, Western
Západní *Czech*, Western
Zatoka *Polish, Ukrainian*, Bay
-zee *Dutch*, Sea
Zemlya *Russian*, Earth, land
Zizhiqu *Chinese*, Autonomous region

Index

Glossary of Abbreviations

This glossary provides a comprehensive guide to the abbreviations used in this Atlas, and in the Index.

A
abbrev. abbreviated
AD Anno Domini
Afr. Afrikaans
Alb. Albanian
Amh. Amharic
anc. ancient
approx. approximately
Ar. Arabic
Arm. Armenian
ASEAN Association of South East Asian Nations
ASSR Autonomous Soviet Socialist Republic
Aust. Australian
Az. Azerbaijani
Azerb. Azerbaijan

B
Basq. Basque
BC before Christ
Bel. Belorussian
Ben. Bengali
Ber. Berber
B-H Bosnia-Herzegovina
bn billion (one thousand million)
BP British Petroleum
Bret. Breton
Brit. British
Bul. Bulgarian
Bur. Burmese

C
C central
C. Cape
°C degrees Centigrade
CACM Central America Common Market
Cam. Cambodian
Cant. Cantonese
CAR Central African Republic
Cast. Castilian
Cat. Catalan
CEEAC Central America Common Market
Chin. Chinese
CIS Commonwealth of Independent States
cm centimetre(s)
Cro. Croat
Cz. Czech
Czech Rep. Czech Republic

D
Dan. Danish
Div. Divehi
Dom. Rep. Dominican Republic
Dut. Dutch

E
E east
EC *see* EU
EEC *see* EU
ECOWAS Economic Community of West African States
ECU European Currency Unit
EMS European Monetary System
Eng. English
est estimated
Est. Estonian
EU European Union (previously European Community [EC], European Economic Community [EEC])

F
°F degrees Fahrenheit
Faer. Faeroese
Fij. Fijian
Fin. Finnish
Fr. French
Fris. Frisian
ft foot/feet
FYROM Former Yugoslav Republic of Macedonia

G
g gram(s)
Gael. Gaelic
Gal. Galician
GDP Gross Domestic Product (the total value of goods and services produced by a country excluding income from foreign countries)
Geor. Georgian
Ger. German
Gk Greek
GNP Gross National Product (the total value of goods and services produced by a country)

H
Heb. Hebrew
HEP hydro-electric power
Hind. Hindi
hist. historical
Hung. Hungarian

I
I. Island
Icel. Icelandic
in inch(es)
In. Inuit (Eskimo)
Ind. Indonesian
Intl International
Ir. Irish
Is Islands
It. Italian

J
Jap. Japanese

K
Kaz. Kazakh
kg kilogram(s)
Kir. Kirghiz
km kilometre(s)
km² square kilometre (singular)
Kor. Korean
Kurd. Kurdish

L
L. Lake
LAIA Latin American Integration Association
Lao. Laotian
Lapp. Lappish
Lat. Latin
Latv. Latvian
Liech. Liechtenstein
Lith. Lithuanian
Lus. Lusatian
Lux. Luxembourg

M
m million/metre(s)
Mac. Macedonian
Maced. Macedonia
Mal. Malay
Malg. Malagasy
Malt. Maltese
mi. mile(s)
Mong. Mongolian
Mt. Mountain
Mts Mountains

N
N north
NAFTA North American Free Trade Agreement
Nep. Nepali
Neth. Netherlands
Nic. Nicaraguan
Nor. Norwegian
NZ New Zealand

P
Pash. Pashtu
PNG Papua New Guinea
Pol. Polish
Poly. Polynesian
Port. Portuguese
prev. previously

R
Rep. Republic
Res. Reservoir
Rmsch Romansch
Rom. Romanian
Rus. Russian
Russ. Fed. Russian Federation

S
S south
SADC Southern Africa Development Community
SCr. Serbian, Croatian
Sinh. Sinhala
Slvk Slovak
Slvn. Slovene
Som. Somali
Sp. Spanish
St., St Saint
Strs Straits
Swa. Swahili
Swe. Swedish
Switz. Switzerland

T
Taj. Tajik
Th. Thai
Thai. Thailand
Tib. Tibetan
Turk. Turkish
Turkm. Turkmenistan

U
UAE United Arab Emirates
Uigh. Uighur
UK United Kingdom
Ukr. Ukrainian
UN United Nations
Urd. Urdu
US/USA United States of America
USSR Union of Soviet Socialist Republics
Uzb. Uzbek

V
var. variant
Vdkhr. Vodokhranilishche (Russian for reservoir)
Vdskh. Vodoskhovyshche (Ukrainian for reservoir)
Vtn. Vietnamese

W
W west
Wel. Welsh

This index lists all the placenames and features shown on the regional and continental maps in this Atlas. Placenames are referenced to the largest scale map on which they appear. The policy followed throughout the Atlas is to use the local spelling or local name at regional level; commonly-used English language names may occasionally be added (in parentheses) where this is an aid to identification e.g. Firenze (Florence). English names, where they exist, have been used for all international features e.g. oceans and country names; they are also used on the continental maps and in the introductory World Today section; these are then fully cross-referenced to the local names found on the regional maps. The index also contains commonly-found alternative names and variant spellings, which are also fully cross-referenced.

All main entry names are those of settlements unless otherwise indicated by the use of italicized definitions or representative symbols, which are keyed at the foot of each page.

1

A

◆ Country
● Country Capital
◇ Dependent Territory
○ Dependent Territory Capital
◈ Administrative Regions
✈ International Airport
▲ Mountain
▲▲ Mountain Range
Volcano
≈ River
◎ Lake
⊠ Reservoir

80 D11 **Abu Zabad** Western Kordofan, C Sudan *12°21′N 29°16′E*
143 P16 **Abū Ẓabī** *var.* Abū Ẓabī, *Eng.* Abu Dhabi. ● (United Arab Emirates) Abū Ẓaby, C United Arab Emirates *24°30′N 54°20′E*
Abū Ẓabī *see* Abū Ẓabī
75 X8 **Abu Zenīma** E Egypt *29°01′N 33°08′E*
95 N17 **Åby** Östergötland, S Sweden *58°40′N 16°10′E*
Abyaḍ, Al Baḥr al *see* White Nile
Åbybro *see* Aabybro
80 D13 **Abyei** Southern Kordofan, S Sudan *09°35′N 28°28′E*
Abyla *see* Ávila
Abymes *see* les Abymes
Abyssinia *see* Ethiopia
Acāba *see* Assaba
54 F11 **Acacías** Meta, C Colombia *03°59′N 73°46′W*
58 L13 **Açailândia** Maranhão, E Brazil *04°51′S 47°26′W*
Acaill *see* Achill Island
42 E8 **Acajutla** Sonsonate, W El Salvador *13°34′N 89°50′W*
79 D17 **Acalayong** SW Equatorial Guinea *01°05′N 09°34′E*
41 N13 **Acámbaro** Guanajuato, C Mexico *20°01′N 100°42′W*
54 C6 **Acandí** Chocó, NW Colombia *08°32′N 77°20′W*
104 H4 **A Cañiza** *var.* La Cañiza. Galicia, NW Spain *42°13′N 08°16′W*
40 J11 **Acaponeta** Nayarit, C Mexico *22°30′N 105°21′W*
40 J11 **Acaponeta, Río de** ≈ C Mexico
41 O16 **Acapulco** *var.* Acapulco de Juárez. Guerrero, S Mexico *16°51′N 99°53′W*
Acapulco de Juárez *see* Acapulco
55 T13 **Acarai Mountains** *Sp.* Serra Acaraí. ▲▲ Brazil/Guyana
Acaraí, Serra *see* Acarai Mountains
58 O13 **Acaraú** Ceará, NE Brazil *04°35′S 37°37′W*
54 J6 **Acarigua** Portuguesa, N Venezuela *09°35′N 69°12′W*
104 H2 **A Carreira** Galicia, NW Spain *43°21′N 08°12′W*
42 C6 **Acatenango, Volcán de** 🌋 S Guatemala *14°30′N 90°52′W*
41 Q15 **Acatlán** *var.* Acatlán de Osorio. Puebla, S Mexico *18°12′N 98°02′W*
Acatlán de Osorio *see* Acatlán
41 S15 **Acayucan** *var.* Acayucán. Veracruz-Llave, E Mexico *17°59′N 94°58′W*
Accho *see* Akko
21 Y5 **Accomac** Virginia, NE USA *37°43′N 75°41′N*
77 Q17 **Accra** ● (Ghana)SE Ghana *05°33′N 00°15′W*
97 L17 **Accrington** NW England, United Kingdom *53°46′N 02°21′W*
61 B19 **Acebal** Santa Fe, C Argentina *33°14′S 60°50′W*
168 H8 **Aceh** *off.* Daerah Istimewa Aceh, *var.* Acheen, Achin, Atchin, Atjeh. ◆ *autonomous district* NW Indonesia
107 M18 **Acerenza** Basilicata, S Italy *40°46′N 15°51′E*
107 K17 **Acerra** *anc.* Acerrae. Campania, S Italy *40°56′N 14°22′E*
Acerrae *see* Acerra
57 J17 **Achacachi** La Paz, W Bolivia *16°01′S 68°44′W*
54 K7 **Achaguas** Apure, C Venezuela *07°46′N 68°14′W*
154 H12 **Achalpur** *prev.* Elichpur, Ellichpur. Mahārāshtra, C India *21°19′N 77°30′E*
61 F18 **Achar** Tacuarembó, C Uruguay *32°20′S 56°15′W*
137 R10 **Ach'ara** *prev.* Achara, *var.* Ajaria. ◆ *autonomous republic* SW Georgia
Achara *see* Ach'ara
115 H19 **Acharnés** *var.* Aharnes; *prev.* Akharnaí. Attikí, C Greece *38°09′N 23°58′E*
Ach'asar Lerr *see* Achk'asari, Mta
Acheen *see* Aceh
99 K16 **Achel** Limburg, NE Belgium *51°15′N 05°31′E*
115 D16 **Achelóos** *var.* Akhelóos, Aspropótamos; *anc.* Achelous. ≈ W Greece
Achelous *see* Achelóos
163 W8 **Acheng** Heilongjiang, NE China *45°32′N 126°56′E*
109 N6 **Achenkirch** Tirol, W Austria *47°31′N 11°42′E*
101 L24 **Achenpass** *pass* Austria/Germany
109 N7 **Achensee** ◎ W Austria
101 F22 **Achern** Baden-Württemberg, SW Germany *48°37′N 08°04′E*
115 C16 **Acherón** ≈ W Greece
77 W11 **Achétinamou** ≈ S Niger
152 J12 **Achhnera** Uttar Pradesh, N India *27°10′N 77°45′E*
42 C7 **Achiguate, Río** ≈ S Guatemala
97 A16 **Achill Head** *Ir.* Ceann Acla. *headland* W Ireland *53°58′N 10°14′W*
97 A16 **Achill Island** *Ir.* Acaill. *island* W Ireland
100 H11 **Achim** Niedersachsen, NW Germany *53°01′N 09°01′E*
149 S5 **Achīn** Nangarhār, E Afghanistan *34°04′N 70°41′E*
Achin *see* Aceh
122 K12 **Achinsk** Krasnoyarskiy Kray, S Russian Federation *56°21′N 90°25′E*
162 E5 **Achit Nuur** ◎ NW Mongolia
137 T11 **Achk'asari, Mta** *Arm.* Ach'asar Lerr. ▲ Armenia/Georgia *41°09′N 43°55′E*
126 K13 **Achuyevo** Krasnodarskiy Kray, SW Russian Federation *46°00′N 38°01′E*
81 F16 **Achwa** *var.* Aswa. ≈ N Uganda
136 E15 **Acıgöl** *salt lake* SW Turkey
107 L24 **Acireale** Sicilia, Italy, C Mediterranean Sea *37°36′N 15°10′E*
Aciris *see* Agri
25 N7 **Ackerly** Texas, SW USA *32°31′N 101°43′W*
22 M4 **Ackerman** Mississippi, S USA *33°18′N 89°10′W*
29 W13 **Ackley** Iowa, C USA *42°33′N 93°03′W*
44 J5 **Acklins Island** *island* SE The Bahamas
Acla, Ceann *see* Achill Head
62 H11 **Aconcagua, Cerro** ▲ W Argentina *32°36′S 69°53′W*
Açores/Açores, Arquipélago dos/Açores, Ilhas dos *see* Azores
104 H2 **A Coruña** *Cast.* La Coruña, *Eng.* Corunna; *anc.* Caronium. Galicia, NW Spain *43°22′N 08°24′W*
104 G2 **A Coruña** *Cast.* La Coruña. ◆ *province* Galicia, NW Spain
42 L10 **Acoyapa** Chontales, S Nicaragua *11°58′N 85°10′W*
106 H13 **Acquapendente** Lazio, C Italy *42°44′N 11°52′E*
106 J13 **Acquasanta Terme** Marche, C Italy *42°46′N 13°24′E*
106 I13 **Acquasparta** Lazio, C Italy *42°41′N 12°31′E*
106 C9 **Acqui Terme** Piemonte, NW Italy *44°41′N 08°28′E*
Acrae *see* Palazzolo Acreide
182 F7 **Acraman, Lake** *salt lake* South Australia
59 A15 **Acre** *off.* Estado do Acre. ◆ *state* W Brazil
Acre *see* Akko
59 C16 **Acre, Rio** ≈ W Brazil
107 N20 **Acri** Calabria, SW Italy *39°30′N 16°22′E*
Acte *see* Ágion Óros
191 Y12 **Actéon, Groupe** *island group* Îles Tuamotu, SE French Polynesia
15 P12 **Acton-Vale** Québec, SE Canada *45°39′N 72°31′W*
41 P13 **Actopan** *var.* Actopán. Hidalgo, C Mexico *20°19′N 98°59′W*
Açu *see* Assu
Acunum Acusio *see* Montélimar
77 Q17 **Ada** SE Ghana *05°47′N 00°42′E*
112 L8 **Ada** Vojvodina, N Serbia *45°48′N 20°08′E*
29 R5 **Ada** Minnesota, N USA *47°18′N 96°31′W*
31 R12 **Ada** Ohio, N USA *40°46′N 83°49′W*
27 O12 **Ada** Oklahoma, C USA *34°47′N 96°41′W*
162 L8 **Adaatsag** *var.* Tavin. Dundgovĭ, C Mongolia *46°27′N 105°43′E*
Ada Bazar *see* Adapazarı
40 D3 **Adair, Bahía de** *bay* NW Mexico
104 M7 **Adaja** ≈ N Spain
38 H17 **Adak Island** *island* Aleutian Islands, Alaska, USA
Adalia *see* Antalya
Adalia, Gulf of *see* Antalya Körfezi
141 X9 **Adam** N Oman *22°22′N 57°30′E*
Adama *see* Nazrēt
60 I8 **Adamantina** São Paulo, S Brazil *21°41′S 51°04′W*
79 E14 **Adamaoua** *Eng.* Adamawa. ◆ *province* N Cameroon
68 F11 **Adamaoua, Massif d'** *Eng.* Adamawa Highlands. *plateau* NW Cameroon
77 Y14 **Adamawa** ◆ *state* E Nigeria
Adamawa *see* Adamaoua
Adamawa Highlands *see* Adamaoua, Massif d'
106 F6 **Adamello** ▲ N Italy *46°09′N 10°33′E*
81 J14 **Ādamī Tulu** Oromīya, C Ethiopia *07°52′N 38°39′E*
63 M23 **Adam, Mount** *var.* Monte Independencia. ▲ West Falkland, Falkland Islands *51°36′S 60°00′W*
29 R16 **Adams** Nebraska, C USA *40°25′N 96°30′W*
18 H8 **Adams** New York, NE USA *43°48′N 75°57′W*
29 Q3 **Adams** North Dakota, N USA *48°23′N 98°01′W*
155 I23 **Adam's Bridge** *chain of shoals* NW Sri Lanka
32 H10 **Adams, Mount** ▲ Washington, NW USA *46°12′N 121°29′W*
Adam's Peak *see* Sri Pada
191 R16 **Adam's Rock** *island* Pitcairn Island, Pitcairn Islands
191 P16 **Adamstown** ○ (Pitcairn Islands)Pitcairn Island, Pitcairn Islands *25°04′S 130°05′W*
20 G10 **Adamsville** Tennessee, S USA *35°14′N 88°23′W*
25 S9 **Adamsville** Texas, SW USA *31°15′N 98°09′W*
141 O17 **'Adan** *Eng.* Aden. SW Yemen *12°51′N 45°05′E*
136 K16 **Adana** *var.* Seyhan. Adana, S Turkey *37°N 35°19′E*
136 K16 **Adana** *var.* Seyhan. ◆ *province* S Turkey
Ādancata *see* Horlivka
169 V12 **Adang, Teluk** *bay* Borneo, C Indonesia
136 F11 **Adapazarı** *prev.* Ada Bazar. Sakarya, NW Turkey *40°49′N 30°24′E*
80 H8 **Adarama** River Nile, NE Sudan *17°04′N 34°57′E*
195 Q16 **Adare, Cape** *cape* Antarctica
Ādavāni *see* Ādoni
106 E6 **Adda** *anc.* Addua. ≈ N Italy
80 A13 **Adda** ≈ W South Sudan
143 Q17 **Aḍ Ḍab'iyah** Abū Ẓaby, C United Arab Emirates *24°17′N 54°08′E*
143 O18 **Aḍ Ḍafrah** *desert* S United Arab Emirates
141 Q6 **Ad Dahnā'** *desert* E Saudi Arabia
74 A11 **Ad Dakhla** *var.* Dakhla. SW Western Sahara *23°46′N 15°56′W*
Ad Dalanj *see* Dilling
Ad Damar *see* Ed Damer
Ad Damazīn *see* Ed Damazin
Ad Dāmir *see* Ed Damer
173 N2 **Ad Dammām** *desert* NE Saudi Arabia
141 R6 **Ad Dammān** *var.* Dammām. Ash Sharqīyah, NE Saudi Arabia *26°23′N 50°05′E*
Ad Dāmūr *see* Damoûr
140 K5 **Ad Dār al Ḥamrā'** Tabūk, NW Saudi Arabia *27°22′N 37°46′E*
140 M13 **Ad Darb** Jīzān, SW Saudi Arabia *17°45′N 42°15′E*
141 O8 **Ad Dawādimī** Ar Riyāḍ, C Saudi Arabia *24°32′N 44°21′E*
143 N16 **Ad Dawḥah** *Eng.* Doha. ● (Qatar) C Qatar *25°15′N 51°36′E*
143 N16 **Ad Dawḥah** *Eng.* Doha. ✈ C Qatar *25°11′N 51°37′E*
139 S6 **Ad Dawr** Ṣalāḥ ad Dīn, N Iraq *34°30′N 43°49′E*
139 Y12 **Ad Dayr** *var.* Dayr, Shahbān. Al Baṣrah, E Iraq *30°45′N 47°36′E*
Addi Arkay *see* Ādī Ārk'ay
Ad Dibakah *see* Dibege
139 X15 **Ad Dibdibah** *physical region* Iraq/Kuwait
Aḍ Diffah *see* Libyan Plateau
Addis Ababa *see* Ādīs Ābeba
Addison *see* Webster Springs
139 U10 **Ad Dīwānīyah** *var.* Diwaniyah. C Iraq *32°00′N 44°57′E*
Ad Dīwānīyah *see* Al Qādisīyah
Addoo Atoll *see* Addu Atoll
Addua *see* Adda
151 K22 **Addu Atoll** *var.* Addoo Atoll, Seenu Atoll. *atoll* S Maldives
Ad Dujail *see* Ad Dujayl
139 T7 **Ad Dujayl** *var.* Ad Dujail. Ṣalāḥ ad Dīn, N Iraq *33°49′N 44°16′E*
Ad Dulaym *see* Al Anbār
Ad Duwaym/Ad Duwēm *see* Ed Dueim
99 D16 **Adegem** Oost-Vlaanderen, NW Belgium *51°12′N 03°31′E*
23 U7 **Adel** Georgia, SE USA *31°08′N 83°25′W*
29 U14 **Adel** Iowa, C USA *41°36′N 94°01′W*
182 I9 **Adelaide** *state capital* South Australia *34°56′S 138°36′E*
44 H2 **Adelaide** New Providence, N The Bahamas *24°59′N 77°30′W*
182 I9 **Adelaide** ✈ South Australia *34°55′S 138°31′E*
194 H6 **Adelaide Island** *island* Antarctica
181 P2 **Adelaide River** Northern Territory, N Australia *13°12′S 131°06′E*
76 M10 **'Adel Bagrou** Hodh ech Chargui, SE Mauritania *15°33′N 07°04′W*
186 D6 **Adelbert Range** ▲▲ N Papua New Guinea
180 K3 **Adele Island** *island* Western Australia
107 O17 **Adelfia** Puglia, SE Italy *41°01′N 16°52′E*
195 V16 **Adélie Coast** *physical region* Antarctica
195 V14 **Adélie, Terre** *physical region* Antarctica
Adelnau *see* Odolanów
Adelsberg *see* Postojna
Aden *see* 'Adan
Aden *see* Khormaksar
141 Q17 **Aden, Gulf of** *gulf* SW Arabian Sea
77 V10 **Aderbissinat** Agadez, C Niger *15°30′N 07°57′E*
Adhaim *see* Al 'Uẓaym
143 R16 **Adh Dhayd** *var.* Al Dhaid. Ash Shāriqah, NE United Arab Emirates *25°19′N 55°51′E*
140 M4 **'Adhfā'** *spring/well* NW Saudi Arabia *29°15′N 41°24′E*
138 I13 **'Ādhriyāt, Jabāl al** ▲ S Jordan
80 I10 **Ādī Ārk'ay** *var.* Addi Arkay. Āmara, N Ethiopia *13°18′N 37°56′E*
182 C7 **Adieu, Cape** *headland* South Australia *32°01′S 132°12′E*
106 H8 **Adige** *Ger.* Etsch. ≈ N Italy
80 J10 **Ādīgrat** Tigray, N Ethiopia *14°17′N 39°27′E*
154 I13 **Ādilābād** *var.* Ādilābād. Telangana, C India *19°40′N 78°31′E*
35 P2 **Adin** California, W USA *41°10′N 120°57′W*
171 V14 **Adi, Pulau** *island* E Indonesia
18 K8 **Adirondack Mountains** ▲▲ New York, NE USA
80 J13 **Ādīs Ābeba** *Eng.* Addis Ababa. ● (Ethiopia) Ādīs Ābeba, C Ethiopia *8°59′N 38°43′E*
80 J13 **Ādīs Ābeba** ✈ Ādīs Ābeba, C Ethiopia *08°58′N 38°53′E*
80 I11 **Ādīs Zemen** Āmara, N Ethiopia *12°00′N 37°43′E*
Adi Ugri *see* Mendefera
137 N15 **Adıyaman** Adıyaman, SE Turkey *37°46′N 38°15′E*
137 N15 **Adıyaman** ◆ *province* S Turkey
116 L11 **Adjud** Vrancea, E Romania *46°07′N 27°10′E*
45 T6 **Adjuntas** C Puerto Rico *18°10′N 66°42′W*
Adjuntas, Presa de las *see* Vicente Guerrero, Presa
Adkup *see* Erikub Atoll
126 L15 **Adler** Krasnodarskiy Kray, SW Russian Federation *43°25′N 39°58′E*
Adler *see* Orlice
108 G7 **Adliswil** Zürich, NW Switzerland *47°19′N 08°32′E*
32 G7 **Admiralty Inlet** *inlet* Washington, NW USA
39 X13 **Admiralty Island** *island* Alexander Archipelago, Alaska, USA
186 E5 **Admiralty Islands** *island group* N Papua New Guinea
136 B14 **Adnan Menderes** ✈ (İzmir) İzmir, W Turkey *38°16′N 27°09′E*
37 V6 **Adobe Creek Reservoir** ⊠ Colorado, C USA
77 T16 **Ado-Ekiti** Ekiti, SW Nigeria *07°42′N 05°13′E*
Adola *see* Kibre Mengist
61 C23 **Adolfo González Chaues** Buenos Aires, E Argentina *38°13′S 60°05′W*
155 H17 **Ādoni** *var.* Ādavāni. Andhra Pradesh, C India *15°38′N 77°16′E*
102 K15 **Adour** *anc.* Aturus. ≈ SW France
Adowa *see* Ādwa
105 O15 **Adra** Andalucía, S Spain *36°45′N 03°01′W*
107 L24 **Adrano** Sicilia, Italy, C Mediterranean Sea *37°39′N 14°49′E*
74 I9 **Adrar** C Algeria *27°56′N 00°12′W*
76 K7 **Adrar** ◆ *region* C Mauritania
74 L11 **Adrar** ▲▲ SE Algeria
74 A12 **Adrar Souttouf** ▲▲ SW Western Sahara
Adrasman *see* Adrasmon
147 Q10 **Adrasmon** *Rus.* Adrasman. NW Tajikistan *40°38′N 69°56′E*
78 K10 **Adré** Ouaddaï, E Chad *13°26′N 22°14′E*
106 H9 **Adria** *anc.* Atria, Hadria, Hatria. Veneto, NE Italy *45°03′N 12°04′E*
31 R10 **Adrian** Michigan, N USA *41°54′N 84°02′W*
29 S11 **Adrian** Minnesota, N USA *43°38′N 95°55′W*
27 R5 **Adrian** Missouri, C USA *38°24′N 94°21′W*
24 M2 **Adrian** Texas, SW USA *35°16′N 102°39′W*
21 S4 **Adrian** West Virginia, NE USA *38°53′N 80°14′W*
Adrianople/Adrianopolis *see* Edirne
121 P7 **Adriatic Basin** *undersea feature* Adriatic Sea, N Mediterranean Sea *42°00′N 17°30′E*
Adriatico, Mare *see* Adriatic Sea
106 L13 **Adriatic Sea** *Alb.* Deti Adriatik, *It.* Mare Adriatico, *SCr.* Jadransko More, *Slvn.* Jadransko Morje. *sea* N Mediterranean Sea
Adriatik, Deti *see* Adriatic Sea
Adua *see* Ādwa
Aduana del Sásabe *see* El Sásabe
79 O17 **Adusa** Orientale, NE Dem. Rep. Congo *01°25′N 28°05′E*
118 J13 **Adutiškis** Vilnius, E Lithuania *55°09′N 26°34′E*
27 Y7 **Advance** Missouri, C USA *37°06′N 89°54′W*
65 D25 **Adventure Sound** *bay* East Falkland, Falkland Islands
80 J10 **Ādwa** *var.* Adowa, *It.* Adua. Tigray, N Ethiopia *14°08′N 38°51′E*
123 Q8 **Adycha** ≈ NE Russian Federation
126 L14 **Adygeya, Respublika** ◆ *autonomous republic* SW Russian Federation
Adzhikui *see* Ajyguýy
77 N17 **Adzopé** SE Ivory Coast *06°07′N 03°49′W*
125 U4 **Adz'va** ≈ NW Russian Federation
125 U5 **Adz'vavom** Respublika Komi, NW Russian Federation *66°35′N 59°13′E*
Ædua *see* Autun
115 K19 **Aegean Islands** *island group* Greece/Turkey
Aegean North *see* Vóreion Aigaíon
115 I17 **Aegean Sea** *Gk.* Aigaíon Pelagos, Aigaío Pélagos, *Turk.* Ege Denizi. *sea* NE Mediterranean Sea
Aegean South *see* Nótion Aigaíon
118 H3 **Aegviidu** *Ger.* Charlottenhof. Harjumaa, NW Estonia *59°17′N 25°37′E*
Aegyptus *see* Egypt
Aelana *see* Al 'Aqabah
Aelok *see* Ailuk Atoll
Aelōninae *see* Ailinginae Atoll
Aelōnlaplap *see* Ailinglaplap Atoll
Æmilia *see* Emilia-Romagna
Æmilianum *see* Millau
Aemona *see* Ljubljana
Aenaria *see* Ischia
Aeolian Islands *see* Eolie, Isole
191 Z3 **Aeon Point** *headland* Kiritimati, NE Kiribati *01°46′N 157°11′W*
Æsernia *see* Isernia
104 G3 **A Estrada** Galicia, NW Spain *42°41′N 08°29′W*
115 C18 **Aetós** Itháki, Iónia Nísoi, Greece, C Mediterranean Sea *38°21′N 20°40′E*
191 Q8 **Afaahiti** Tahiti, W French Polynesia *17°43′S 149°18′W*
139 U10 **'Afak** Al Qādisīyah, C Iraq *32°04′N 45°17′E*
Afanasjevo *see* Afanas'yevo
125 T14 **Afanas'yevo** *var.* Afanasjevo. Kirovskaya Oblast', NW Russian Federation *58°55′N 53°13′E*
Afándou *see* Afántou
115 O23 **Afántou** *var.* Afándou. Ródos, Dodekánisa, Greece, Aegean Sea *36°17′N 28°10′E*
80 K11 **Āfar** ◆ *region* NE Ethiopia
Afar Depression *see* Danakil Desert
191 O7 **Afareaitu** Moorea, W French Polynesia *17°33′S 149°47′W*
140 L7 **'Afarīyah, Bi'r al** *well* NW Saudi Arabia
Afars et des Issas, Territoire Français des *see* Djibouti
83 D22 **Affenrücken** Karas, SW Namibia *28°05′S 15°49′E*
Afghānestān, Dowlat-e Eslāmī-ye *see* Afghanistan
148 M6 **Afghanistan** *off.* Islamic Republic of Afghanistan, *Per.* Dowlat-e Eslāmī-ye Afghānestān; *prev.* Republic of Afghanistan. ◆ *islamic state* C Asia
Afgoi *see* Afgooye
81 N17 **Afgooye** *It.* Afgoi. Shabeellaha Hoose, S Somalia *02°09′N 45°07′E*
141 N8 **'Afīf** Ar Riyāḍ, C Saudi Arabia *23°57′N 42°57′E*
77 V17 **Afikpo** Ebonyi, S Nigeria *05°52′N 07°58′E*
Afiun Karahissar *see* Afyon
109 V6 **Aflenz Kurort** Steiermark, E Austria *47°33′N 15°14′E*
74 J6 **Aflou** N Algeria *34°09′N 02°06′E*
81 L18 **Afmadow** Jubbada Hoose, S Somalia *0°24′N 42°04′E*
39 Q14 **Afognak Island** *island* Alaska, USA
104 J2 **A Fonsagrada** Galicia, NW Spain *43°09′N 07°03′W*
186 E9 **Afore** Northern, S Papua New Guinea *09°10′S 148°30′E*
59 O15 **Afrânio** Pernambuco, E Brazil *08°32′S 40°54′W*
66-67 **Africa** *continent*
68 L11 **Africa, Horn of** *physical region* Ethiopia/Somalia
172 K11 **Africana Seamount** *undersea feature* SW Indian Ocean *37°10′S 29°10′E*
86 A14 **African Plate** *tectonic plate*
138 I2 **'Afrīn** Ḥalab, N Syria *36°31′N 36°51′E*
136 M15 **Afşin** Kahramanmaraş, C Turkey *38°14′N 36°54′E*
98 J7 **Afsluitdijk** *dam* N Netherlands
29 U15 **Afton** Iowa, C USA *41°01′N 94°12′W*
29 W9 **Afton** Minnesota, N USA *44°54′N 92°46′W*
27 R8 **Afton** Oklahoma, C USA *36°41′N 94°57′W*
136 F14 **Afyon** *prev.* Afyonkarahisar. Afyon, W Turkey *38°46′N 30°32′E*
136 F14 **Afyon** *var.* Afiun Karahissar, Afyonkarahisar. ◆ *province* W Turkey
Afyonkarahisar *see* Afyon
Agadès *see* Agadez
77 V10 **Agadez** *prev.* Agadès. Agadez, C Niger *16°57′N 07°56′E*
77 W8 **Agadez** ◆ *department* N Niger
74 E8 **Agadir** SW Morocco *30°30′N 09°37′W*
64 M9 **Agadir Canyon** *undersea feature* SE Atlantic Ocean *32°30′N 12°50′W*
145 R12 **Agadyr'** Karaganda, C Kazakhstan *48°15′N 72°55′E*
173 O7 **Agalega Islands** *island group* N Mauritius
42 K6 **Agalta, Sierra de** ▲▲ E Honduras
122 I10 **Agan** ≈ C Russian Federation
Agana/Agaña *see* Hagåtña
188 B15 **Agana Bay** *bay* NW Guam
171 Kk13 **Agano-gawa** ≈ Honshū, C Japan
188 B17 **Aga Point** *headland* S Guam
154 G9 **Agar** Madhya Pradesh, C India *23°44′N 76°01′E*
81 I14 **Āgaro** Oromīya, C Ethiopia *07°52′N 36°36′E*
153 V15 **Agartala** *state capital* Tripura, NE India *23°49′N 91°15′E*
194 I5 **Agassiz, Cape** *headland* Antarctica *68°29′S 62°59′W*
175 V13 **Agassiz Fracture Zone** *tectonic feature* S Pacific Ocean
9 N2 **Agassiz Ice Cap** *Ice feature* Nunavut, N Canada
188 B16 **Agat** W Guam *13°20′N 144°38′E*
188 B16 **Agat Bay** *bay* W Guam
Agatha *see* Agde
115 M20 **Agathónisi** *island* Dodekánisa, Greece, Aegean Sea
171 X14 **Agats** Papua, E Indonesia *05°33′S 138°07′E*
155 C21 **Agatti Island** *island* Lakshadweep, India, N Indian Ocean
38 D16 **Agattu Island** *island* Aleutian Islands, Alaska, USA
38 D16 **Agattu Strait** *strait* Aleutian Islands, Alaska, USA
14 B8 **Agawa** ≈ Ontario, S Canada
14 B8 **Agawa Bay** *lake bay* Ontario, S Canada
77 N17 **Agboville** SE Ivory Coast *05°55′N 04°15′W*
137 V12 **Ağdam** *Rus.* Agdam. SW Azerbaijan *40°04′N 46°00′E*
Agdam *see* Ağdam
103 P16 **Agde** *anc.* Agatha. Hérault, S France *43°19′N 03°29′E*
103 P16 **Agde, Cap d'** *headland* S France *43°17′N 03°30′E*
Agedabia *see* Ajdābiyā
102 L14 **Agen** *anc.* Aginnum. Lot-et-Garonne, SW France *44°12′N 00°37′E*
Agendicum *see* Sens
165 O13 **Ageo** Saitama, Honshū, S Japan *35°58′N 139°36′E*
109 R5 **Ager** ≈ N Austria
Ager Hiywet *see* Hägere Hiywet
108 G8 **Ägerisee** ◎ W Switzerland
142 M10 **Āghā Jārī** Khūzestān, SW Iran *30°48′N 49°45′E*
39 P15 **Aghiyuk Island** *island* Alaska, USA
74 B12 **Aghouinit** SE Western Sahara *22°14′N 13°10′W*
Aghri Dagh *see* Büyükağrı Dağı
74 B10 **Aghzoumal, Sebkhet** *var.* Sebjet Agsumal. *salt lake* E Western Sahara
115 F15 **Agiá** *var.* Ayiá. Thessalía, C Greece *39°43′N 22°45′E*
40 G7 **Agiabampo, Estero de** *estuary* NW Mexico
121 P3 **Agía Fýlaxis** *var.* Ayia Phyla. S Cyprus *34°43′N 33°02′E*
Agialoúsa *see* Yenierenköy
115 M21 **Agía Marína** Léros, Dodekánisa, Greece, Aegean Sea *37°09′N 26°51′E*
121 Q2 **Agía Nápa** *var.* Ayia Napa. E Cyprus *34°59′N 34°00′E*
115 L16 **Agía Paraskeví** Lésvos, E Greece *39°13′N 26°19′E*
115 J15 **Agías Eirínis, Akrotírio** *headland* Límnos, E Greece *39°47′N 25°21′E*
Agiasós *see* Agiassós
115 L17 **Agiassós** *var.* Agiasós, Ayiásos, Ayiássos. Lésvos, E Greece *39°05′N 26°23′E*
Aginnum *see* Agen
123 O14 **Aginskoye** Zabaykal'skiy Kray, S Russian Federation *51°10′N 114°32′E*
115 I14 **Ágion Óros** *Eng.* Mount Athos. ◆ *monastic republic* NE Greece
115 H14 **Ágion Óros** *var.* Akte, Aktí; *anc.* Acte. *peninsula* NE Greece
114 D13 **Ágios Achílleios** *religious building* Dytikí Makedonía, N Greece
115 J16 **Ágios Efstrátios** *var.* Áyios Evstrátios, Hagios Evstrátios. *island* E Greece *39°30′N 25°00′E*
115 H20 **Ágios Geórgios** *island* Kykládes, Greece, Aegean Sea
Ágios Geórgios *see* Ro
115 E21 **Ágios Ilías** ▲ S Greece *36°57′N 22°19′E*
115 K25 **Ágios Ioánnis, Akrotírio** *headland* Kríti, Greece, E Mediterranean Sea *35°19′N 25°46′E*
115 L20 **Ágios Kírykos** *var.* Áyios Kírikos. Ikaría, Dodekánisa, Greece, Aegean Sea *37°34′N 26°15′E*
115 K25 **Ágios Nikólaos** *var.* Áyios Nikólaos. Kríti, Greece, E Mediterranean Sea *35°12′N 25°43′E*
115 D16 **Ágios Nikolaos** Thessalía, C Greece *39°33′N 21°21′E*
115 H14 **Agíou Órous, Kólpos** *gulf* N Greece
107 K24 **Agira** *anc.* Agyrium. Sicilia, Italy, C Mediterranean Sea *37°39′N 14°32′E*
115 G20 **Agkístri** *island* S Greece
114 G12 **Agkístro** *var.* Angistro. ▲ NE Greece *41°21′N 23°29′E*
103 O17 **Agly** ≈ S France
Agnetheln *see* Agnita
14 E10 **Agnew Lake** ◎ Ontario, S Canada
77 O16 **Agnibilékrou** E Ivory Coast *07°10′N 03°11′W*
116 I11 **Agnita** *Ger.* Agnetheln, *Hung.* Szentágota. Sibiu, SW Romania *45°59′N 24°40′E*
107 K15 **Agnone** Molise, C Italy *41°49′N 14°21′E*
164 K14 **Ago** Mie, Honshū, SW Japan *34°18′N 136°50′E*
106 C8 **Agogna** ≈ N Italy
Agoitz *see* Aoiz
Agoiz *see* Aoiz
77 P17 **Agona Swedru** *var.* Swedru. SE Ghana *05°31′N 00°42′W*
Agordat *see* Ak'ordat
Agosta *see* Augusta
103 N15 **Agout** ≈ S France
152 J12 **Āgra** Uttar Pradesh, N India *27°09′N 78°E*
Agra and Oudh, United Provinces of *see* Uttar Pradesh
Agram *see* Zagreb
105 U5 **Agramunt** Cataluña, NE Spain *41°48′N 01°07′E*
105 Q5 **Agreda** Castilla y León, N Spain *41°51′N 01°55′W*
137 S13 **Ağrı** *var.* Karaköse; *prev.* Karakılısse. Ağrı, NE Turkey *39°44′N 43°04′E*
137 S13 **Ağrı** ◆ *province* NE Turkey
107 N19 **Agri** *anc.* Aciris. ≈ S Italy
Agri Dagi *see* Büyükağrı Dağı
107 J24 **Agrigento** *Gk.* Akragas; *prev.* Girgenti. Sicilia, Italy, C Mediterranean Sea *37°19′N 13°33′E*
188 K4 **Agrihan** *island* N Northern Mariana Islands
115 L17 **Agriliá, Akrotírio** *prev.* Ákra Maléas. *cape* Lésvos, E Greece
115 D18 **Agrínio** *prev.* Agrinion. Dytikí Elláda, W Greece *38°38′N 21°25′E*
Agrinion *see* Agrínio
115 G17 **Agriovótano** Évvoia, C Greece *39°00′N 23°18′E*
107 L18 **Agropoli** Campania, S Italy *40°22′N 14°59′E*
127 T3 **Agryz** Udmurtskaya Respublika, NW Russian Federation *56°27′N 52°58′E*
137 U11 **Ağstafa** *Rus.* Akstafa. NW Azerbaijan *41°06′N 45°28′E*
137 X11 **Ağsu** *Rus.* Akhsu. C Azerbaijan
Agsumal, Sebjet *see* Aghzoumal, Sebkhet
40 J11 **Agua Brava, Laguna** *lagoon* W Mexico
54 F7 **Aguachica** Cesar, N Colombia *08°16′N 73°35′W*
59 J20 **Água Clara** Mato Grosso do Sul, SW Brazil *20°21′S 52°58′W*
44 D5 **Aguada de Pasajeros** Cienfuegos, C Cuba *22°23′N 80°51′W*
54 J5 **Aguada Grande** Lara, N Venezuela *10°38′N 69°29′W*
45 S5 **Aguadilla** W Puerto Rico *18°27′N 67°08′W*
43 S16 **Aguadulce** Coclé, S Panama *08°16′N 80°31′W*
104 L14 **Aguadulce** Andalucía, S Spain *37°15′N 04°59′W*
41 O8 **Agualeguas** Nuevo León, NE Mexico *26°17′N 99°30′W*
40 L9 **Aguanaval, Río** ≈ C Mexico
42 J5 **Aguán, Río** ≈ N Honduras
25 R16 **Agua Nueva** Texas, SW USA *26°57′N 98°34′W*
60 J8 **Aguapeí, Rio** ≈ S Brazil
61 E14 **Aguapey, Río** ≈ NE Argentina
40 G3 **Agua Prieta** Sonora, NW Mexico *31°16′N 109°33′W*
104 G5 **A Guarda** *var.* A Guardia, Laguardia, La Guardia. Galicia, NW Spain *41°54′N 08°53′W*
A Guardia *see* A Guarda
56 E6 **Aguarico, Río** ≈ Ecuador/Peru
55 O6 **Aguasay** Monagas, NE Venezuela *09°25′N 63°44′W*
40 M12 **Aguascalientes** Aguascalientes, C Mexico *21°54′N 102°17′W*
40 L12 **Aguascalientes** ◆ *state* C Mexico
57 I18 **Aguas Calientes, Río** ≈ S Peru
105 R7 **Aguasvivas** ≈ NE Spain
60 J7 **Água Vermelha, Represa de** ⊠ S Brazil
56 E12 **Aguaytía** Ucayali, C Peru *09°02′S 75°30′W*
104 I5 **A Gudiña** *var.* La Gudiña. Galicia, NW Spain *42°04′N 07°08′W*
104 G7 **Águeda** Aveiro, N Portugal *40°34′N 08°28′W*
104 J8 **Águeda** ≈ Portugal/Spain
77 Q8 **Aguelhok** Kidal, NE Mali *19°18′N 00°50′E*
77 V12 **Aguié** Maradi, S Niger *13°28′N 07°43′E*
188 K8 **Aguijan** *island* S Northern Mariana Islands
104 M14 **Aguilar** *var.* Aguilar de la Frontera. Andalucía, S Spain *37°31′N 04°40′W*
104 M3 **Aguilar de Campóo** Castilla y León, N Spain *42°47′N 04°15′W*
Aguilar de la Frontera *see* Aguilar
42 F7 **Aguilares** San Salvador, C El Salvador *13°56′N 89°09′W*
105 Q14 **Águilas** Murcia, SE Spain *37°24′N 01°36′W*
40 L15 **Aguililla** Michoacán, SW Mexico *18°43′N 102°45′W*
Agulhas *see* L'Agulhas
172 J11 **Agulhas Bank** *undersea feature* SW Indian Ocean *35°30′S 21°00′E*
172 K11 **Agulhas Basin** *undersea feature* SW Indian Ocean *47°00′S 27°00′E*
83 F26 **Agulhas, Cape** *Afr.* Kaap Agulhas. *headland* SW South Africa *34°51′S 19°59′E*
Agulhas, Kaap *see* Agulhas, Cape
60 O9 **Agulhas Negras, Pico das** ▲ SE Brazil *22°21′S 44°39′W*
172 K11 **Agulhas Plateau** *undersea feature* SW Indian Ocean *40°00′S 26°00′E*
165 S16 **Aguni-jima** *island* Nansei-shotō, SW Japan
Agurain *see* Salvatierra
54 G5 **Agustín Codazzi** *var.* Codazzi. Cesar, N Colombia *10°02′N 73°15′W*
Agyrium *see* Agira
74 L12 **Ahaggar** *high plateau region* SE Algeria
146 E12 **Ahal Welaýaty** *Rus.* Akhalskiy Velayat. ◆ *province* C Turkmenistan
142 K2 **Ahar** Āẕarbāyjān-e Sharqī, NW Iran *38°25′N 47°07′E*
Aharnes *see* Acharnés
138 J3 **Aḥaṣ, Jabal** ▲ NW Syria
138 J3 **Aḥaṣ, Jebal** ▲ W Syria
185 G16 **Ahaura** ≈ South Island, New Zealand
100 E13 **Ahaus** Nordrhein-Westfalen, NW Germany *52°04′N 07°01′E*
191 U9 **Ahe** *atoll* Îles Tuamotu, C French Polynesia
184 N10 **Ahimanawa Range** ▲▲ North Island, New Zealand
119 I19 **Ahinski Kanal** *Rus.* Oginskiy Kanal. *canal* SW Belarus
186 G10 **Ahioma** SE Papua New Guinea *10°20′S 150°35′E*
184 I2 **Ahipara** Northland, North Island, New Zealand *35°11′S 173°07′E*
184 I2 **Ahipara Bay** *bay* SE Tasman Sea
Áhkájávrre *see* Akkajaure
Áhkká *see* Akka
39 N13 **Ahklun Mountains** ▲▲ Alaska, USA
137 R14 **Ahlat** Bitlis, E Turkey *38°45′N 42°28′E*
101 F14 **Ahlen** Nordrhein-Westfalen, W Germany *51°46′N 07°53′E*
154 D10 **Ahmadābād** *var.* Ahmedabad. Gujarāt, W India *23°03′N 72°40′E*
143 R10 **Aḥmadābād** Kermān, C Iran *35°51′N 59°36′E*
Ahmadi *see* Al Aḥmadī
Ahmad Khel *see* Ḥasan Khēl
155 F14 **Ahmadnagar** *var.* Ahmednagar. Mahārāshtra, W India *19°08′N 74°48′E*
149 T9 **Ahmadpur Siāl** Punjab, E Pakistan *30°40′N 71°52′E*
77 N5 **Aḥmar, 'Erg el** *desert* N Mali
80 K13 **Ahmar Mountains** ▲▲ C Ethiopia
Ahmedabad *see* Ahmadābād
Ahmednagar *see* Ahmadnagar
114 N12 **Ahmetbey** Kırklareli, NW Turkey *41°26′N 27°35′E*
14 H12 **Ahmic Lake** ◎ Ontario, S Canada
190 G12 **Ahoa** Île Uvea, E Wallis and Futuna *13°17′S 176°12′W*
40 G8 **Ahomé** Sinaloa, C Mexico *25°55′N 109°10′W*
21 X8 **Ahoskie** North Carolina, SE USA *36°17′N 76°59′W*
101 D17 **Ahr** ≈ W Germany
143 N12 **Ahram** *var.* Ahrom. Būshehr, S Iran *28°52′N 51°18′E*
100 J9 **Ahrensburg** Schleswig-Holstein, N Germany *53°41′N 10°14′E*
Ahrom *see* Ahram
93 L17 **Ähtäri** Etelä-Pohjanmaa, W Finland *62°34′N 24°08′E*
114 N10 **Ahtopol** *var.* Akhtopol. Burgas, E Bulgaria *42°06′N 27°57′E*
40 K12 **Ahuacatlán** Nayarit, C Mexico *21°02′N 104°30′W*
42 E7 **Ahuachapán** Ahuachapán, W El Salvador *13°55′N 89°51′W*
42 A9 **Ahuachapán** ◆ *department* W El Salvador
191 V16 **Ahu Akivi** *var.* Siete Moai. *ancient monument* Easter Island, Chile, E Pacific Ocean
191 W11 **Ahunui** *atoll* Îles Tuamotu, C French Polynesia
185 E20 **Ahuriri** ≈ South Island, New Zealand
95 L22 **Åhus** Skåne, S Sweden *55°55′N 14°18′E*
Ahu Tahira *see* Ahu Vinapu
191 V16 **Ahu Tepeu** *ancient monument* Easter Island, Chile, E Pacific Ocean
191 V17 **Ahu Vinapu** *var.* Ahu Tahira. *ancient monument* Easter Island, Chile, E Pacific Ocean
142 L9 **Ahvāz** *var.* Ahwāz; *prev.* Nāsiri. Khūzestān, SW Iran *31°20′N 48°38′E*
Ahvenanmaa *see* Åland
141 Q16 **Aḥwar** SW Yemen *13°34′N 46°41′E*
Ahwāz *see* Ahvāz
94 H7 **Åi Åfjord** *var.* Åi Åfjord, Årnes. Sør-Trøndelag, C Norway *63°57′N 10°12′E*
see also Åi Åfjord
Åi Åfjord *see* Åi Åfjord
149 P3 **Aībak** *var.* Haibak; *prev.* Āybak, Samangān. Samangān, NE Afghanistan *36°16′N 68°04′E*
101 K22 **Aichach** Bayern, SE Germany *48°26′N 11°06′E*
164 L14 **Aichi** *off.* Aichi-ken, *var.* Aiti. ◆ *prefecture* Honshū, SW Japan
Aïdin *see* Aydın
Aidussina *see* Ajdovščina
Aifir, Clochán an *see* Giant's Causeway
Aigaíon Pelagos/Aigaío Pélagos *see* Aegean Sea
109 S3 **Aigen im Mülkreis** Oberösterreich, N Austria *48°39′N 13°57′E*
115 G20 **Aígina** *var.* Aíyina, Egina. Aígina, C Greece *37°45′N 23°26′E*
115 G20 **Aígina** *island* S Greece
115 E18 **Aígio** *var.* Egio; *prev.* Aíyion. Dytikí Elláda, S Greece *38°15′N 22°05′E*
108 C10 **Aigle** Vaud, SW Switzerland *46°20′N 06°58′E*
103 P14 **Aigoual, Mont** ▲ S France *44°09′N 03°34′E*
173 O16 **Aigrettes, Pointe des** *headland* W Réunion *21°02′S 55°14′E*
61 G19 **Aiguá** *var.* Aigua. Maldonado, S Uruguay *34°13′S 54°46′W*
103 S13 **Aigues** ≈ SE France
103 N10 **Aigurande** Indre, C France *46°26′N 01°49′E*
Ai-hun *see* Heihe
165 N10 **Aikawa** Niigata, Sado, C Japan *38°04′N 138°15′E*
21 Q13 **Aiken** South Carolina, SE USA *33°34′N 81°44′W*
25 N4 **Aiken** Texas, SW USA *34°06′N 101°31′W*

◆ Country
● Country Capital
◇ Dependent Territory
○ Dependent Territory Capital
◆ Administrative Regions
✈ International Airport
▲ Mountain
▲▲ Mountain Range
🌋 Volcano
≈ River
◎ Lake
⊠ Reservoir

160 F13 **Ailao Shan** ▲▲ SW China
189 R4 **Ailinginae Atoll** *var.* Aelōninae. *atoll* Ralik Chain, SW Marshall Islands
189 T7 **Ailinglaplap Atoll** *var.* Aelōnlaplap. *atoll* Ralik Chain, S Marshall Islands
Aillionn, Loch *see* Allen, Lough
96 H13 **Ailsa Craig** *island* SW Scotland, United Kingdom
189 V5 **Ailuk Atoll** *var.* Aelok. *atoll* Ratak Chain, NE Marshall Islands
123 R11 **Aim** Khabarovskiy Kray, E Russian Federation 58°45′N 134°08′E
45 Q12 **Aimé Césaire** ✕ (Fort-de-France) C Martinique 14°34′N 61°00′W
103 R11 **Ain** ⬖ *department* E France
103 S10 **Ain** ≈ E France
118 G7 **Ainaži** *Est.* Heinaste, *Ger.* Hainasch. N Latvia 57°51′N 24°24′E
74 L6 **Aïn Beida** NE Algeria 35°52′N 07°25′E
76 K4 **'Aïn Ben Tili** Tiris Zemmour, N Mauritania 25°58′N 09°30′W
74 J5 **Aïn Defla** *var.* Aïn Eddefla. N Algeria 36°16′N 01°58′E
Aïn Eddefla *see* Aïn Defla
74 L5 **Ain El Bey** ✕ (Constantine) NE Algeria 36°15′N 06°36′E
115 C19 **Aínos** ▲ Kefallonía, Iónia Nísoi, Greece, C Mediterranean Sea 38°08′N 20°39′E
105 T4 **Ainsa** Aragón, NE Spain 42°25′N 00°08′E
74 I7 **Aïn Sefra** NW Algeria 32°45′N 00°32′W
29 N13 **Ainsworth** Nebraska, C USA 42°33′N 99°51′W
Aintab *see* Gaziantep
74 H5 **Aïn Témouchent** N Algeria 35°18′N 01°09′W
186 C6 **Aiome** Madang, N Papua New Guinea 05°08′S 144°45′E
Aïoun el Atrous/Aïoun el Atroûss *see* 'Ayoûn el 'Atroûs
54 E11 **Aipe** Huila, C Colombia 03°15′N 75°17′W
56 D9 **Aipena, Río** ≈ N Peru
57 L19 **Aiquile** Cochabamba, C Bolivia 18°10′S 65°10′W
Aïr *see* Aïr, Massif de l'
188 E10 **Airai** Babeldaob, C Palau
188 E10 **Airai** ✕ (Oreor) Babeldaob, N Palau 07°22′N 134°34′E
168 I11 **Airbangis** Sumatera, NW Indonesia 0°12′N 99°22′E
11 Q16 **Airdrie** Alberta, SW Canada 51°20′N 114°00′W
96 I12 **Airdrie** S Scotland, United Kingdom 55°52′N 03°59′W
Air du Azbine *see* Aïr, Massif de l'
97 M17 **Aire** ≈ N England, United Kingdom
102 K15 **Aire-sur-l'Adour** Landes, SW France 43°43′N 00°16′W
103 O1 **Aire-sur-la-Lys** Pas-de-Calais, N France 50°39′N 02°24′E
9 Q6 **Air Force Island** *island* Baffin Island, Nunavut, NE Canada
169 Q13 **Airhitam, Teluk** *bay* Borneo, C Indonesia
171 Q11 **Airmadidi** Sulawesi, N Indonesia 01°25′N 124°59′E
77 V8 **Aïr, Massif de l'** *var.* Aïr, Air du Azbine, Asben. ▲▲ NC Niger
108 G10 **Airolo** Ticino, S Switzerland 46°32′N 08°38′E
102 K9 **Airvault** Deux-Sèvres, W France 46°51′N 00°07′W
101 K19 **Aisch** ≈ S Germany
63 G20 **Aisén** *off.* Región Aisén del General Carlos Ibañez del Campo, *var.* Aysen. ⬖ *region* S Chile
10 H7 **Aishihik Lake** ◎ Yukon, W Canada
103 P3 **Aisne** ⬖ *department* N France
103 R4 **Aisne** ≈ NE France
109 T4 **Aist** ≈ N Austria
114 K13 **Aisými** Anatolikí Makedonía kai Thráki, NE Greece 41°00′N 25°55′E
105 S11 **Aitana** ▲ E Spain 38°39′N 00°15′W
186 B5 **Aitape** *var.* Eitape. West Sepik, NW Papua New Guinea 03°10′S 142°17′E
Aiti *see* Aichi
29 V6 **Aitkin** Minnesota, N USA 46°31′N 93°42′W
115 D18 **Aitolikó** *var.* Etoliko; *prev.* Aitolikón. Dytikí Elláda, C Greece 38°26′N 21°21′E
Aitolikón *see* Aitolikó
190 L15 **Aitutaki** *island* S Cook Islands
116 H11 **Aiud** *Ger.* Strassburg, *Hung.* Nagyenyed; *prev.* Engeten. Alba, SW Romania 46°19′N 23°43′E
118 I9 **Aiviekste** ≈ C Latvia
189 Q8 **Aiwo** SW Nauru 0°32′S 166°54′E
188 E8 **Aiwokako Passage** *passage* Babeldaob, N Palau
Aix *see* Aix-en-Provence
103 S15 **Aix-en-Provence** *var.* Aix; *anc.* Aquae Sextiae. Bouches-du-Rhône, SE France 43°31′N 05°27′E
Aix-la-Chapelle *see* Aachen
103 T11 **Aix-les-Bains** Savoie, E France 45°40′N 05°55′E
186 A6 **Aiyang, Mount** ▲ NW Papua New Guinea 05°03′S 141°15′E
Aíyina *see* Aígina
Aíyion *see* Aígio
153 W15 **Āīzawl** *state capital* Mizoram, NE India 23°41′N 92°45′E
118 H9 **Aizkraukle** S Latvia 56°39′N 25°07′E
118 C9 **Aizpute** W Latvia 56°43′N 21°32′E
165 O11 **Aizuwakamatsu** Fukushima, Honshū, C Japan 37°30′N 139°58′E
103 X15 **Ajaccio** Corse, France, C Mediterranean Sea 41°54′N 08°43′E
103 X15 **Ajaccio, Golfe d'** *gulf* Corse, France, C Mediterranean Sea
41 Q15 **Ajalpán** Puebla, S Mexico 18°26′N 97°20′W
154 F13 **Ajanta Range** ▲▲ C India
Ajaria *see* Ach'ara
Ajastan *see* Armenia
93 G14 **Ajaureforsen** Västerbotten, N Sweden 65°31′N 15°44′E
185 H17 **Ajax, Mount** ▲ South Island, New Zealand 42°34′S 172°06′E
162 F9 **Aj Bogd Uul** ▲ SW Mongolia 44°49′N 95°01′E
75 R8 **Ajdābiyā** *var.* Agedabia, Ajdābiyah. NE Libya 30°46′N 20°14′E
Ajdābiyah *see* Ajdābiyā
109 S12 **Ajdovščina** *Ger.* Haidenschaft, *It.* Aidussina. W Slovenia 45°52′N 13°55′E
165 Q7 **Ajigasawa** Aomori, Honshū, C Japan 40°45′N 140°11′E
Ajjinena *see* El Geneina
111 H23 **Ajka** Veszprém, W Hungary 47°18′N 17°32′E
138 G9 **'Ajlūn** 'Ajlūn, N Jordan 32°20′N 35°45′E
138 G9 **'Ajlūn** *off.* Muḩāfa at 'Ajlūn. ⬖ *governorate* N Jordan
138 H9 **'Ajlūn, Jabal** ▲▲ W Jordan
'Ajlūn, Muḩāfa at *see* 'Ajlūn
Ájluokta *see* Drag
143 R15 **'Ajmān** *var.* Ajman, 'Ujmān. 'Ajmān, NE United Arab Emirates 25°36′N 55°42′E
152 G12 **Ajmer** *var.* Ajmere. Rājasthān, N India 26°29′N 74°40′E
36 J15 **Ajo** Arizona, SW USA 32°22′N 112°51′W
105 N2 **Ajo, Cabo de** *headland* N Spain 43°31′N 03°36′W
36 J16 **Ajo Range** ▲▲ Arizona, SW USA
146 C14 **Ajyguýy** *Rus.* Adzhikui. Balkan Welaýaty, W Turkmenistan 39°46′N 53°57′E
Akaba *see* Al 'Aqabah
165 T3 **Akabira** Hokkaidō, NE Japan 43°30′N 142°04′E
165 N10 **Akadomari** Niigata, Sado, C Japan 37°54′N 138°24′E
81 E20 **Akagera** ≈ Rwanda/Tanzania
191 W16 **Akahanga, Punta** *headland* Easter Island, Chile, E Pacific Ocean
80 J13 **Āk'ak'i** Oromīya, C Ethiopia 08°50′N 38°51′E
155 G15 **Akalkot** Mahārāshtra, W India 17°36′N 76°10′E
Akamagaseki *see* Shimonoseki
165 V3 **Akan** Hokkaidō, NE Japan 44°06′N 144°03′E
165 U4 **Akan** Hokkaidō, NE Japan 43°09′N 144°08′E
165 U4 **Akan-ko** ◎ Hokkaidō, NE Japan
Akanthoú *see* Tatlısu
185 I19 **Akaroa** Canterbury, South Island, New Zealand 43°48′S 172°58′E
80 E6 **Akasha** Northern, N Sudan 21°03′N 30°46′E
164 I13 **Akashi** *var.* Akasi. Hyōgo, Honshū, SW Japan 34°39′N 135°00′E
139 N7 **'Akāsh, Wādī** *var.* Wādī 'Ukash. *dry watercourse* W Iraq
Akasi *see* Akashi
92 K11 **Äkäsjokisuu** Lappi, N Finland 67°28′N 23°44′E
137 S11 **Akbaba Dağı** ▲ Armenia/Turkey 41°04′N 43°28′E
Akbük Limanı *see* Güllük Körfezi
127 V8 **Akbulak** Orenburgskaya Oblast', W Russian Federation 51°01′N 55°35′E
137 O11 **Akçaabat** Trabzon, NE Turkey 41°00′N 39°36′E
137 N15 **Akçadağ** Malatya, C Turkey 38°21′N 37°59′E
136 G11 **Akçakoca** Düzce, NW Turkey 41°05′N 31°08′E
Akchakaya, Vpadina *see* Akdzhakaya, Vpadina
76 H7 **Akchâr** *desert* W Mauritania
Akchatau *see* Akshatau
136 L13 **Akdağlar** ▲▲ C Turkey
136 E17 **Ak Dağları** ▲▲ SW Turkey
136 K13 **Akdağmadeni** Yozgat, C Turkey 39°40′N 35°52′E
146 G8 **Akdepe** *prev.* Ak-Tepe, Leninsk, *Turkm.* Lenin. Daşoguz Welaýat, N Turkmenistan 42°10′N 59°17′E
Ak-Dere *see* Byala
121 P2 **Akdoğan** *Gk.* Lýsi. C Cyprus 35°06′N 33°42′E
122 J14 **Ak-Dovurak** Respublika Tyva, S Russian Federation 51°09′N 90°36′E
146 F9 **Akdzhakaya, Vpadina** *var.* Vpadina Akchakaya. *depression* N Turkmenistan
171 S11 **Akelamo** Pulau Halmahera, E Indonesia 01°27′N 128°39′E
Aken *see* Aachen
Akermanceaster *see* Bath
95 P15 **Åkersberga** Stockholm, C Sweden 59°28′N 18°19′E
95 H15 **Akershus** ⬖ *county* S Norway
79 L16 **Aketi** Orientale, N Dem. Rep. Congo 02°44′N 23°46′E
146 C10 **Akgyr Erezi** *Rus.* Gryada Akkyr. *hill range* NW Turkmenistan
Akhalskiy Velayat *see* Ahal Welaýaty
137 S10 **Akhaltsikhe** *prev.* Akhalts'ikhe. SW Georgia 41°39′N 43°04′E
Akhalts'ikhe *see* Akhaltsikhe
Akhangaran *see* Ohangaron
Akharnaí *see* Acharnés
75 R7 **Akhḑar, Al Jabal al** *hill range* NE Libya
Akhelóös *see* Achelóos
39 Q15 **Akhiok** Kodiak Island, Alaska, USA 56°57′N 154°12′W
136 C13 **Akhisar** Manisa, W Turkey 38°54′N 27°50′E
75 X10 **Akhmīm** *var.* Akhmîm; *anc.* Panopolis. C Egypt 26°35′N 31°48′E
152 H6 **Akhnūr** Jammu and Kashmir, NW India 32°53′N 74°46′E
Akhsu *see* Ağsu
Akhtopol *see* Ahtopol
127 P11 **Akhtuba** ≈ SW Russian Federation
127 P11 **Akhtubinsk** Astrakhanskaya Oblast', SW Russian Federation 48°17′N 46°14′E
Akhtyrka *see* Okhtyrka
164 H14 **Aki** Kōchi, Shikoku, SW Japan 33°30′N 133°53′E
39 N12 **Akiachak** Alaska, USA 60°54′N 161°25′W
39 N12 **Akiak** Alaska, USA 60°54′N 161°12′W
191 X11 **Akiaki** *atoll* Îles Tuamotu, E French Polynesia
12 H9 **Akimiski Island** *island* Nunavut, C Canada
136 K17 **Akıncı Burnu** *headland* S Turkey 36°21′N 35°47′E
Akıncılar *see* Selçuk
117 U10 **Akinovka** Zaporiz'ka Oblast', S Ukraine
Åkirkeby *see* Aakirkeby
165 P8 **Akita** Akita, Honshū, C Japan 39°44′N 140°06′E
165 Q8 **Akita** *off.* Akita-ken. ⬖ *prefecture* Honshū, C Japan
76 H8 **Akjoujt** *prev.* Fort-Repoux. Inchiri, W Mauritania 19°42′N 14°28′W
92 H11 **Akka** *Lapp.* Áhkká. ▲ N Sweden 67°33′N 17°27′E
92 H11 **Akkajaure** *Lapp.* Áhkájávrre. ◎ N Sweden
Akkala *see* Oqqal'a
155 L25 **Akkaraipattu** Eastern Province, E Sri Lanka 07°13′N 81°51′E
145 P13 **Akkense** *Kaz.* Aqkengse. Karaganda, C Kazakhstan 46°39′N 68°06′E
Akkerman *see* Bilhorod-Dnistrovs'kyy
127 W8 **Akkermanovka** Orenburgskaya Oblast', W Russian Federation 51°11′N 58°03′E
165 V4 **Akkeshi** Hokkaidō, NE Japan 43°03′N 144°49′E
165 V4 **Akkeshi-ko** ◎ Hokkaidō, NE Japan
165 V5 **Akkeshi-wan** *bay* NW Pacific Ocean
138 F8 **Akko** *Eng.* Acre, *Fr.* Saint-Jean-d'Acre, *Bibl.* Accho, Ptolemaïs. Northern, N Israel 32°55′N 35°04′E
145 Q8 **Akkol'** *Kaz.* Aqköl; *prev.* Alekseyevka, *Kaz.* Alekseevka. Akmola, C Kazakhstan 51°58′N 70°58′E
145 T14 **Akkol'** *Kaz.* Aqköl. Almaty, SE Kazakhstan 45°01′N 75°38′E
145 Q16 **Akkol'** *Kaz.* Aqköl. Zhambyl, C Kazakhstan 43°25′N 70°47′E
144 M11 **Akkol', Ozero** *prev.* Ozero Zhaman-Akkol. ◎ C Kazakhstan
98 L6 **Akkrum** Fryslân, N Netherlands 53°01′N 05°52′E
145 U8 **Akku** *Kaz.* Aqqū; *prev.* Lebyazh'ye. Pavlodar, NE Kazakhstan 51°29′N 77°48′E
144 F12 **Akkystau** *Kaz.* Aqqystaū. Atyrau, SW Kazakhstan 47°17′N 51°03′E
8 G6 **Aklavik** Northwest Territories, NW Canada 68°15′N 135°02′W
118 B9 **Akmenrags** *prev.* Akmeņrags. *headland* W Latvia 56°49′N 21°03′E
Akmeņrags *see* Akmenrags
158 E9 **Akmeqit** Xinjiang Uygur Zizhiqu, NW China 37°10′N 76°59′E
146 J14 **Akmeydan** Mary Welaýaty, C Turkmenistan 37°50′N 62°08′E
145 P9 **Akmola** *off.* Akmolinskaya Oblast', *Kaz.* Aqmola Oblysy; *prev.* Tselinogradskaya Oblast. ⬖ *province* C Kazakhstan
Akmola *see* Astana
Akmolinsk *see* Astana
Akmolinskaya Oblast' *see* Akmola
Aknavásár *see* Târgu Ocna
118 I11 **Aknīste** S Latvia 56°09′N 25°43′E
81 G14 **Akobo** Jonglei, E South Sudan 07°50′N 33°05′E
81 G14 **Akobo** *var.* Ākobowenz. ≈ Ethiopia/Sudan
Ākobowenz *see* Akobo
154 H12 **Akola** Mahārāshtra, C India 20°44′N 77°00′E
80 J9 **Ak'ordat** *var.* Agordat, Akurdet. C Eritrea 15°33′N 38°01′E
77 Q16 **Akosombo Dam** *dam* SE Ghana
154 H12 **Akot** Mahārāshtra, C India 20°45′N 77°00′E
77 N16 **Akoupé** SE Ivory Coast 06°19′N 03°54′W
12 M3 **Akpatok Island** *island* Nunavut, E Canada
158 G7 **Akqi** Xinjiang Uygur Zizhiqu, NW China 40°51′N 78°20′E
138 I2 **Akrād, Jabal al** ▲▲ N Syria
Akragas *see* Agrigento
92 H3 **Akranes** Vesturland, W Iceland 64°19′N 22°01′W
139 S2 **Ākrē** *Ar.* 'Aqrah. Dahūk, N Iraq 36°46′N 43°52′E
95 C16 **Åkrehamn** Rogaland, S Norway 59°15′N 15°13′E
77 V9 **Akrérèb** Agadez, C Niger 17°45′N 09°01′E
115 D22 **Akrítas, Akrotírio** *headland* S Greece 36°43′N 21°52′E
37 V3 **Akron** Colorado, C USA 40°09′N 103°12′W
29 R12 **Akron** Iowa, C USA 42°49′N 96°33′W
31 U12 **Akron** Ohio, N USA 41°05′N 81°31′W
Akrotiri *see* Akrotírion
Akrotiri Bay *see* Akrotírion, Kólpos
Akrotírio Kastállou *see* Kástelo, Akrotírio
121 P3 **Akrotírion** *var.* Akrotiri. *UK air base* S Cyprus34°36′N 32°57′E
121 P3 **Akrotírion, Kólpos** *var.* Akrotiri Bay. *bay* S Cyprus
121 O3 **Akrotiri Sovereign Base Area** *UK military installation* S Cyprus
158 F11 **Aksai Chin** *Chin.* Aksayqin. *disputed region* China/India
Aksaj *see* Aksay
136 I15 **Aksaray** Aksaray, C Turkey 38°23′N 33°50′E
136 I15 **Aksaray** ⬖ *province* C Turkey
144 G8 **Aksay** *var.* Aksaj, *Kaz.* Aqsay. Zapadnyy Kazakhstan, NW Kazakhstan 51°11′N 53°00′E
127 O11 **Aksay** Volgogradskaya Oblast', SW Russian Federation 47°59′N 43°54′E
147 W10 **Ak-say** *var.* Toxkan He. ≈ China/Kyrgyzstan
Aksay/Aksay Kazakzu Zizhixian *see* Boluozhuanjing/Hongliuwan
Aksayqin *see* Aksai Chin
158 G11 **Aksayqin Hu** ◎ NW China
136 G14 **Akşehir** Konya, W Turkey 38°22′N 31°24′E
136 G14 **Akşehir Gölü** ◎ C Turkey
136 G16 **Akseki** Antalya, SW Turkey 37°03′N 31°46′E
123 P13 **Aksenovo-Zilovskoye** Zabaykal'skiy Kray, S Russian Federation 53°01′N 117°26′E
145 S12 **Akshatau** *Kaz.* Aqshataū; *prev.* Akchatau. Karaganda, C Kazakhstan 47°59′N 74°02′E
145 V11 **Akshatau, Khrebet** ▲▲ E Kazakhstan
147 Y8 **Ak-Shyyrak** Issyk-Kul'skaya Oblast', E Kyrgyzstan 41°46′N 78°34′E
Akstafa *see* Ağstafa
158 H7 **Aksu** Xinjiang Uygur Zizhiqu, NW China 41°17′N 80°15′E
145 R8 **Aksu** *Kaz.* Aqsū. Akmola, N Kazakhstan 52°31′N 72°00′E
145 W13 **Aksu** *Kaz.* Aqsū. Almaty, SE Kazakhstan 45°31′N 79°28′E
145 T8 **Aksu** *var.* Jermak, *Kaz.* Ermak; *prev.* Yermak. Pavlodar, NE Kazakhstan 52°03′N 76°55′E
145 V13 **Aksu** *Kaz.* Aqsū. ≈ SE Kazakhstan
145 Y11 **Aksuat** *Kaz.* Aqsūat. Vostochnyy Kazakhstan, SE Kazakhstan 48°16′N 83°39′E
145 X11 **Aksuat** *Kaz.* Aqsūat. Vostochnyy Kazakhstan, E Kazakhstan 47°48′N 82°51′E
127 S4 **Aksubayevo** Respublika Tatarstan, W Russian Federation 54°52′N 50°50′E
158 H7 **Aksu He** *Rus.* Sary-Dzhaz. ≈ China/Kyrgyzstan *see also* Sary-Dzhaz
80 J10 **Āksum** Tigray, N Ethiopia 14°06′N 38°42′E
145 O12 **Aktas** *Kaz.* Aqtas. Karaganda, C Kazakhstan 48°03′N 66°21′E
Aktash *see* Oqtosh
147 V9 **Ak-Tash, Gora** ▲ C Kyrgyzstan 40°53′N 74°39′E
145 R10 **Aktau** *Kaz.* Aqtaū. Karaganda, C Kazakhstan 50°13′N 73°06′E
144 E11 **Aktau** *Kaz.* Aqtaū; *prev.* Shevchenko. Mangistau, W Kazakhstan 43°37′N 51°14′E
Aktau, Khrebet *see* Oqtogh, Qatorkŭhi. SW Tajikistan
Aktau, Khrebet *see* Oqtov Tizmasi, C Uzbekistan
Akte *see* Ágion Óros
Ak-Tepe *see* Akdepe
147 X7 **Ak-Terek** Issyk-Kul'skaya Oblast', E Kyrgyzstan 42°14′N 77°46′E
Aktí *see* Ágion Óros
Aktjubinsk/Aktyubinsk *see* Aktobe
158 E8 **Akto** Xinjiang Uygur Zizhiqu, NW China 39°07′N 75°43′E
144 I10 **Aktobe** *Kaz.* Aqtöbe; *prev.* Aktjubinsk, Aktyubinsk. Aktyubinsk, NW Kazakhstan 50°18′N 57°10′E
145 V12 **Aktogay** *Kaz.* Aqtoghay. Vostochnyy Kazakhstan, E Kazakhstan 46°56′N 79°40′E
119 M18 **Aktsyabrski** *Rus.* Oktyabr'skiy; *prev.* Karpilovka. Homyel'skaya Voblasts', SE Belarus 52°38′N 28°53′E
144 H11 **Aktyubinsk** *off.* Aktyubinskaya Oblast', *Kaz.* Aqtöbe Oblysy. ⬖ *province* W Kazakhstan
Aktyubinsk *see* Aktobe
147 W7 **Ak-Tyuz** *var.* Aktyuz. Chuyskaya Oblast', N Kyrgyzstan 42°50′N 76°05′E
79 J17 **Akula** Equateur, NW Dem. Rep. Congo 02°21′N 20°13′E
164 C15 **Akune** Kagoshima, Kyūshū, SW Japan 32°00′N 130°12′E
38 L16 **Akun Island** *island* Aleutian Islands, Alaska, USA
Akurdet *see* Ak'ordat
77 T16 **Akure** Ondo, SW Nigeria 07°18′N 05°13′E
92 J2 **Akureyri** Norðurland Eystra, N Iceland 65°40′N 18°07′W
38 L17 **Akutan** Akutan Island, Alaska, USA 54°08′N 165°47′W
38 K17 **Akutan Island** *island* Aleutian Islands, Alaska, USA
77 V17 **Akwa Ibom** ⬖ *state* SE Nigeria
Akyab *see* Sittwe
127 W7 **Ak"yar** Respublika Bashkortostan, W Russian Federation 51°51′N 58°13′E
144 J10 **Akzhar** *prev.* Novorossiyskiy, Novorossiyskoye. Aktyubinsk, NW Kazakhstan 50°14′N 57°57′E
145 Y11 **Akzhar** *Kaz.* Aqzhar. Vostochnyy Kazakhstan, E Kazakhstan 47°36′N 83°38′E
94 F13 **Ål** Buskerud, S Norway 60°37′N 08°33′E
119 N18 **Ala** *Rus.* Ola. ≈ SE Belarus
23 P5 **Alabama** *off.* State of Alabama, *also known as* Camellia State, Heart of Dixie, The Cotton State, Yellowhammer State. ⬖ *state* S USA
23 P6 **Alabama River** ≈ Alabama, S USA
23 P4 **Alabaster** Alabama, S USA 33°14′N 86°49′W
139 U10 **Al 'Abd Allāh** *var.* Al Abdullah. Al Qādisīyah, S Iraq 32°06′N 45°08′E
139 W14 **Al Abṭīyah** *well* S Iraq
147 S9 **Ala-Buka** Dzhalal-Abadskaya Oblast', W Kyrgyzstan 41°22′N 71°27′E
136 J12 **Alaca** Çorum, N Turkey 40°10′N 34°52′E
136 K10 **Alaçam** Samsun, N Turkey 41°36′N 35°35′E
Alacant *see* Alicante
23 V9 **Alachua** Florida, SE USA 29°48′N 82°29′W
137 S13 **Aladağlar** ▲▲ W Turkey
136 K15 **Ala Dağları** ▲▲ C Turkey
162 I5 **Alag-Erdene** *var.* Manhan. Hövsgöl, N Mongolia 50°05′N 100°01′E
127 O16 **Alagir** Respublika Severnaya Osetiya, SW Russian Federation 43°02′N 44°10′E
106 B6 **Alagna Valsesia** Valle d'Aosta, NW Italy 45°51′N 07°50′E
103 P12 **Alagnon** ≈ C France
59 P16 **Alagoas** *off.* Estado de Alagoas. ⬖ *state* E Brazil
59 P17 **Alagoinhas** Bahia, E Brazil 12°09′S 38°21′W
105 R5 **Alagón** Aragón, NE Spain 41°46′N 01°07′W
104 J9 **Alagón** ≈ W Spain
93 K16 **Alahärmä** Etelä-Pohjanmaa, W Finland 63°15′N 22°50′E
al Ahdar *see* Al Akhḑar
Álaheaieatnu *see* Altaelva
142 K12 **Al Aḩmadī** *var.* Ahmadi. E Kuwait 29°02′N 48°01′E
Al Ain *see* Al 'Ayn
105 Z8 **Alaior** *prev.* Alayor. Menorca, Spain, W Mediterranean Sea 39°56′N 04°08′E
147 T11 **Alai Range** *Rus.* Alayskiy Khrebet. ▲▲ Kyrgyzstan/Tajikistan
Alais *see* Alès
141 X11 **Al 'Ajā'iz** E Oman 19°33′N 57°12′E
141 X11 **Al 'Ajā'iz** *oasis* SE Oman
93 L16 **Alajärvi** Etelä-Pohjanmaa, W Finland 63°00′N 23°50′E
118 K4 **Alajõe** Ida-Virumaa, NE Estonia 59°00′N 27°26′E
42 M13 **Alajuela** Alajuela, C Costa Rica 10°00′N 84°12′W
42 L12 **Alajuela** *off.* Provincia de Alajuela. ⬖ *province* N Costa Rica
43 T14 **Alajuela, Lago** ◎ C Panama
38 M11 **Alakanuk** Alaska, USA 62°41′N 164°37′W
140 K5 **Al Akhḑar** *var.* al Ahdar. Tabūk, NW Saudi Arabia 28°04′N 37°13′E
Alaköl *see* Alakol', Ozero
145 X13 **Alakol', Ozero** *Kaz.* Alaköl. ◎ SE Kazakhstan
124 I5 **Alakurtti** Murmanskaya Oblast', NW Russian Federation 66°57′N 30°27′E
38 F10 **Alalākeiki Channel** *var.* Alalakeiki Channel. *channel* Hawaii, USA, C Pacific Ocean
75 U8 **Al 'Alamayn** *var.* El 'Alamein. N Egypt 30°50′N 28°57′E
Al 'Amādiyah *see* Amêdî
188 K5 **Alamagan** *island* C Northern Mariana Islands
139 X10 **Al 'Amārah** *var.* Amara. Maysān, E Iraq 31°51′N 47°10′E
80 J11 **Ālamat'ā** Tigray, N Ethiopia 12°22′N 39°32′E
37 R11 **Alameda** New Mexico, SW USA 35°09′N 106°37′W
121 T13 **'Alam el Rûm, Râs** *headland* N Egypt 31°21′N 27°23′E
Alamícamba *see* Alamikamba
42 M8 **Alamikamba** *var.* Alamícamba. Región Autónoma Atlántico Norte, NE Nicaragua 13°26′N 84°09′W
24 K11 **Alamito Creek** ≈ Texas, SW USA
40 M8 **Alamitos, Sierra de los** ▲ NE Mexico 26°15′N 102°14′W
41 Q12 **Álamo** Veracruz-Llave, C Mexico 20°55′N 97°41′W
35 X9 **Alamo** Nevada, W USA 37°21′N 115°08′W
20 F9 **Alamo** Tennessee, S USA 35°47′N 89°09′W
37 S14 **Alamogordo** New Mexico, SW USA 32°52′N 105°57′W
36 J12 **Alamo Lake** ⊠ Arizona, SW USA
40 H7 **Alamos** Sonora, NW Mexico 26°59′N 108°53′W
37 S7 **Alamosa** Colorado, C USA 37°25′N 105°51′W
139 P9 **Al Anbār** *off.* Muḩāfa at al Anbār, *var.* Ad Dulaym. ⬖ *governorate* SW Iraq
93 J20 **Åland** *off.* Provincial Autonomy of the Åland Islands, *Fin.* Ahvenanmaa. ⬖ *autonomous region* SW Finland
88 K9 **Åland** *var.* Aland Islands, *Fin.* Ahvenanmaa. *island group* SW Finland
Aland Islands *see* Åland
Åland Islands, Provincial Autonomy of the *see* Åland
Aland Sea *see* Ålands Hav
95 Q14 **Ålands Hav** *var.* Aland Sea. *strait* Baltic Sea/Gulf of Bothnia
43 P16 **Alanje** Chiriquí, SW Panama 08°26′N 82°33′W
25 O2 **Alanreed** Texas, SW USA 35°12′N 100°45′W
136 G17 **Alanya** Antalya, S Turkey 36°32′N 32°02′E
23 U7 **Alapaha River** ≈ Florida/Georgia, SE USA
122 G10 **Alapayevsk** Sverdlovskaya Oblast', C Russian Federation 57°48′N 61°50′E
138 F14 **Al 'Aqabah** *var.* Akaba, Aqaba, 'Aqaba; *anc.* Aelana, Elath. Al 'Aqabah, SW Jordan 29°32′N 35°00′E
138 G14 **Al 'Aqabah** *off.* Muḩāfa at al 'Aqabah. ⬖ *governorate* SW Jordan
Al 'Arabīyah as Su'ūdīyah *see* Saudi Arabia
al Araïch *see* Larache
105 Q10 **Alarcón** Castilla-La Mancha, C Spain 39°32′N 02°05′W
105 Q9 **Alarcón, Embalse de** ⊠ C Spain
138 J2 **Al 'Arīmah** *Fr.* Arime. Ḩalab, N Syria 36°27′N 37°41′E
75 X7 **Al 'Arīsh** *var.* El 'Arîsh. NE Egypt 33°00′N 31°00′E
141 P6 **Al Arṭāwīyah** Ar Riyāḑ, N Saudi Arabia 26°34′N 45°20′E
Alasca, Golfo de *see* Alaska, Gulf of
136 D14 **Alaşehir** Manisa, W Turkey 38°19′N 28°30′E
139 N5 **Al 'Ashārah** *var.* Ashara. Dayr az Zawr, E Syria 34°51′N 40°36′E
Al Ashkhara *see* Al Ashkharah
141 Z9 **Al Ashkharah** *var.* Al Ashkhara. NE Oman 21°47′N 59°30′E
Al 'Aşimah *see* 'Ammān
39 P8 **Alaska** *off.* State of Alaska, *also known as* Land of the Midnight Sun, The Last Frontier, Seward's Folly; *prev.* Russian America. ⬖ *state* NW USA
39 T13 **Alaska, Gulf of** *var.* Golfo de Alasca. *gulf* Canada/USA
39 O15 **Alaska Peninsula** *peninsula* Alaska, USA
39 Q11 **Alaska Range** ▲▲ Alaska, USA
Al-Asnam *see* Chlef
106 B10 **Alassio** Liguria, NW Italy 44°01′N 08°12′E
137 Y12 **Älät** *Rus.* Alyat; *prev.* Alyaty-Pristan'. SE Azerbaijan 39°57′N 49°24′E
Alat *see* Olot
139 S13 **Al 'Athāmīn** An Najaf, S Iraq 30°27′N 43°41′E
39 P7 **Alatna River** ≈ Alaska, USA
107 J15 **Alatri** Lazio, C Italy 41°43′N 13°21′E
Alattio *see* Alta
127 P5 **Alatyr'** Chuvashskaya Respublika, W Russian Federation 54°50′N 46°28′E
56 C7 **Alausí** Chimborazo, C Ecuador 02°11′S 78°52′W
105 O3 **Álava** *Basq.* Araba. ⬖ *province* País Vasco, N Spain
137 T11 **Alaverdi** N Armenia 41°06′N 44°37′E
Alavo *see* Alavus
93 N14 **Ala-Vuokki** Kainuu, E Finland 64°46′N 29°29′E
93 K17 **Alavus** *Swe.* Alavo. Etelä-Pohjanmaa, W Finland 62°33′N 23°38′E
Al 'Awābī *see* Awābī
139 P6 **Al 'Awānī** Al Anbār, W Iraq 34°28′N 41°43′E
75 U12 **Al Awaynāt** SE Libya 21°46′N 24°51′E
Al Awaynāt *see* Al 'Uwaynāt
182 K9 **Alawoona** South Australia 34°45′S 140°28′E
Alaykel'/Alay-Kuu *see* Këk-Art
143 R17 **Al 'Ayn** *var.* Al Ain. Abū Ẓaby, E United Arab Emirates 24°13′N 55°44′E
143 R17 **Al 'Ayn** *var.* Al Ain. ✕ Abū Ẓaby, E United Arab Emirates 24°16′N 55°31′E
138 G12 **Al 'Aynā** Al Karak, W Jordan 30°59′N 35°43′E
Alayor *see* Alaior
Alayskiy Khrebet *see* Alai Range
123 S6 **Alazeya** ≈ NE Russian Federation
139 U8 **Al 'Azīzīyah** *var.* Aziziya. Wāsiţ, E Iraq 32°54′N 45°05′E
120 M12 **Al 'Azīzīyah** NW Libya 32°32′N 13°01′E
138 I10 **Al Azraq al Janūbī** Az Zarqā', N Jordan 31°49′N 36°48′E
106 B9 **Alba** *anc.* Alba Pompeia. Piemonte, NW Italy 44°42′N 08°02′E
25 V6 **Alba** Texas, SW USA 32°47′N 95°37′W
116 G11 **Alba** ⬖ *county* W Romania
139 P3 **Al Ba'āj** Nīnawá, N Iraq 36°02′N 41°43′E
138 J2 **Al Bāb** Ḩalab, N Syria 36°24′N 37°32′E
116 G10 **Albac** *Hung.* Fehérvölgy; *prev.* Álbák. Alba, SW Romania 46°27′N 22°58′E
105 Q11 **Albacete** Castilla-La Mancha, C Spain 39°N 01°52′W
105 P11 **Albacete** ⬖ *province* Castilla-La Mancha, C Spain
140 I4 **Al Bad'** Tabūk, NW Saudi Arabia 28°28′N 35°00′E
104 L7 **Alba de Tormes** Castilla y León, N Spain 40°50′N 05°30′W
139 P3 **Al Bādī** Nīnawá, N Iraq 35°57′N 41°37′E
141 V8 **Al Badī'ah** ✕ (Abū Ẓaby) Abū Ẓaby, C United Arab Emirates 24°27′N 54°39′E
143 P17 **Al Badī'ah** *var.* Al Bedei'ah. *spring/well* C United Arab Emirates 23°44′N 53°50′E
95 H19 **Ålbæk** Nordjylland, N Denmark 57°33′N 10°24′E
139 Q7 **Al Baghdādī** *var.* Khān al Baghdādī. Al Anbār, SW Iraq 33°60′N 42°50′E
Al Bāha *see* Al Bāḩah
140 M11 **Al Bāḩah** *var.* Al Bāha. Al BāḨah, SW Saudi Arabia 20°01′N 41°29′E
140 M11 **Al Bāḩah** *var.* Minţaqat al Bāḩah. ⬖ *province* W Saudi Arabia
Al Baḩrayn *see* Bahrain
105 S11 **Albaida** Valenciana, E Spain 38°51′N 00°31′W
116 H11 **Alba Iulia** *Ger.* Weissenburg, *Hung.* Gyulafehérvár; *prev.* Bălgrad, Karlsburg, Károly-Fehérvár. Alba, W Romania 46°06′N 23°33′E
Álbák *see* Albac
138 G10 **Al Balqā'** *off.* Muḩāfaẓat al Balqā', *var.* Balqā'. ⬖ *governorate* NW Jordan
14 F11 **Alban** Ontario, S Canada 46°07′N 80°37′W
103 O15 **Alban** Tarn, S France 43°52′N 02°30′E
12 K11 **Albanel, Lac** ◎ Québec, SE Canada
113 L20 **Albania** *off.* Republic of Albania, *Alb.* Republika e Shqipqërisë, Shqipëria; *prev.* People's Socialist Republic of Albania. ◆ *republic* SE Europe
Albania *see* Aubagne
107 H15 **Albano Laziaie** Lazio, C Italy 41°44′N 12°40′E
180 J14 **Albany** Western Australia 35°03′S 117°54′E
23 S7 **Albany** Georgia, SE USA 31°35′N 84°09′W
31 P13 **Albany** Indiana, N USA 40°18′N 85°14′W
20 L8 **Albany** Kentucky, S USA 36°42′N 85°08′W
29 U7 **Albany** Minnesota, N USA 45°39′N 94°33′W
27 R2 **Albany** Missouri, C USA 40°15′N 94°15′W
18 L10 **Albany** *state capital* New York, NE USA 42°39′N 73°45′W
32 G12 **Albany** Oregon, NW USA 44°38′N 123°06′W
25 Q6 **Albany** Texas, SW USA 32°44′N 99°18′W
12 F10 **Albany** ≈ Ontario, S Canada
Alba Pompeia *see* Alba
Alba Regia *see* Székesfehérvár
138 J6 **Al Bāridah** *var.* Bāridah. Ḩimş, C Syria 34°15′N 37°39′E
139 Q11 **Al Barīt** Al Anbār, S Iraq 31°16′N 42°28′E
105 R8 **Albarracín** Aragón, NE Spain 40°24′N 01°25′W
139 Y12 **Al Başrah** *Eng.* Basra, *hist.* Busra, Bussora. Al Başrah, SE Iraq 30°30′N 47°50′E
139 Y13 **Al Başrah** *off.* Muḩāfa at al Başrah, *Eng.* Basra. ⬖ *governorate* SE Iraq
139 V11 **Al Baṭḩā'** Dhī Qār, SE Iraq 31°06′N 45°54′E
141 X8 **Al Bāṭinah** *var.* Batinah. *coastal region* N Oman
0 H16 **Albatross Plateau** *undersea feature* E Pacific Ocean 10°00′N 103°00′W
Al Batrūn *see* Batroûn
121 Q12 **Al Bayḑā'** *var.* Beida. NE Libya 32°46′N 21°43′E
141 P16 **Al Bayḑā'** *var.* Al Beida. SW Yemen 13°58′N 45°38′E
Al Bedei'ah *see* Al Badī'ah
Al Beida *see* Al Bayḑā'
21 S10 **Albemarle** *var.* Albermarle. North Carolina, SE USA 35°21′N 80°12′W
Albemarle Island *see* Isabela, Isla
21 N8 **Albemarle Sound** *inlet* W Atlantic Ocean
106 B10 **Albenga** Liguria, NW Italy 44°04′N 08°13′E
104 L8 **Alberche** ≈ C Spain
103 O17 **Albères, Chaîne des** *var.* les Albères, Montes Albères. ▲▲ France/Spain
Albères, Montes *see* Albères, Chaîne des
182 F2 **Alberga Creek** *seasonal river* South Australia
104 G7 **Albergaria-a-Velha** Aveiro, N Portugal 40°42′N 08°28′W
105 S10 **Alberic** Valenciana, E Spain 39°07′N 00°31′W
Albermarle *see* Albemarle
107 P18 **Alberobello** Puglia, SE Italy 40°47′N 17°14′E
108 J7 **Alberschwende** Vorarlberg, W Austria 47°28′N 09°50′E
103 O3 **Albert** Somme, N France 50°N 02°38′E
11 O12 **Alberta** ⬖ *province* SW Canada
Albert Edward Nyanza *see* Edward, Lake
61 C20 **Alberti** Buenos Aires, E Argentina 35°03′S 60°15′W
111 K23 **Albertirsa** Pest, C Hungary 47°15′N 19°36′E
99 I16 **Albertkanaal** *canal* N Belgium
79 P17 **Albert, Lake** *var.* Albert Nyanza, Lac Mobutu Sese Seko. ◎ Uganda/Dem. Rep. Congo
29 V11 **Albert Lea** Minnesota, N USA 43°39′N 93°22′W
81 F16 **Albert Nile** ≈ NW Uganda
Albert Nyanza *see* Albert, Lake
103 T11 **Albertville** Savoie, E France 45°41′N 06°24′E
23 Q2 **Albertville** Alabama, S USA 34°16′N 86°12′W
Albertville *see* Kalemie
103 N15 **Albi** *anc.* Albiga. Tarn, S France 43°55′N 02°09′E
29 W15 **Albia** Iowa, C USA 41°01′N 92°48′W
Albiga *see* Albi
55 X9 **Albina** Marowijne, NE Suriname 05°29′N 54°08′W
83 A15 **Albina, Ponta** *headland* SW Angola 15°52′S 11°45′E
30 M16 **Albion** Illinois, N USA 38°22′N 88°03′W
31 P11 **Albion** Indiana, N USA 41°23′N 85°26′W
29 P14 **Albion** Nebraska, C USA 41°41′N 98°00′W
18 E9 **Albion** New York, NE USA 43°13′N 78°09′W
18 B12 **Albion** Pennsylvania, NE USA 41°52′N 80°18′W
Al Biqā' *see* El Beqaa
140 J4 **Al Bi'r** *var.* Bi'r Ibn Hirmās. Tabūk, NW Saudi Arabia 28°52′N 36°16′E
140 M12 **Al Birk** Makkah, SW Saudi Arabia 18°13′N 41°36′E
141 Q9 **Al Biyāḑ** *desert* C Saudi Arabia
98 H13 **Alblasserdam** Zuid-Holland, SW Netherlands 51°52′N 04°40′E
Albocácer *see* Albocàsser
105 T8 **Albocàsser** *Cast.* Albocácer. Valenciana, E Spain 40°21′N 00°01′E
Albona *see* Labin
105 O17 **Alborán, Isla de** *island* S Spain
Alborán, Mar de *see* Alboran Sea
105 N17 **Alboran Sea** *Sp.* Mar de Alborán. *sea* SW Mediterranean Sea
Ålborg *see* Aalborg
95 H21 **Ålborg Bugt** *var.* Aalborg Bugt. *bay* N Denmark
Ålborg-Nørresundby *see* Aalborg
143 N5 **Alborz** *var.* Ostān-e Alborz. ⬖ *province* N Iran
Alborz, Ostān-e *see* Alborz
143 O5 **Alborz, Reshteh-ye Kūhhā-ye** *Eng.* Elburz Mountains. ▲▲ N Iran
105 Q14 **Albox** Andalucía, S Spain 37°22′N 02°08′W
101 H23 **Albstadt** Baden-Württemberg, SW Germany 48°13′N 09°01′E
104 G14 **Albufeira** Beja, S Portugal 37°05′N 08°15′W
139 P5 **Ālbū Gharz, Sabkhat** ◎ W Iraq
105 O15 **Albuñol** Andalucía, S Spain 36°48′N 03°11′W
37 Q11 **Albuquerque** New Mexico, SW USA 35°05′N 106°38′W
141 W8 **Al Buraymī** *var.* Buraimi. N Oman 24°16′N 55°48′E
143 R17 **Al Buraymī** *var.* Buraimi. *spring/well* Oman/United Arab Emirates 24°27′N 55°33′E
Al Burayqah *see* Marsá al Burayqah
Alburgum *see* Aalborg
104 I10 **Alburquerque** Extremadura, W Spain 39°12′N 07°00′W
181 V14 **Albury** New South Wales, SE Australia 36°03′S 146°53′E
141 T14 **Al Buzūn** SE Yemen 15°40′N 50°53′E
93 G17 **Alby** Västernorrland, C Sweden 62°30′N 15°25′E
Albyn, Glen *see* Mor, Glen
104 G12 **Alcácer do Sal** Setúbal, W Portugal 38°22′N 08°29′W
Alcalá de Chisvert/Alcalá de Chivert *see* Alcalà de Xivert
104 K14 **Alcalá de Guadaira** Andalucía, S Spain 37°20′N 05°50′W
105 O8 **Alcalá de Henares** *Ar.* Alkal'a; *anc.* Complutum. Madrid, C Spain 40°28′N 03°22′W
104 K16 **Alcalá de los Gazules** Andalucía, S Spain 36°29′N 05°43′W
105 T8 **Alcalà de Xivert** *var.* Alcalá de Chisvert, *Cast.* Alcalá de Chivert. Valenciana, E Spain 40°19′N 00°13′E
105 N14 **Alcalá La Real** Andalucía, S Spain 37°28′N 03°55′W
107 I23 **Alcamo** Sicilia, Italy, C Mediterranean Sea 37°58′N 12°58′E
105 T4 **Alcanadre** ≈ NE Spain
105 T8 **Alcanar** Cataluña, NE Spain 40°33′N 00°28′E

◆ Country
● Country Capital
◇ Dependent Territory
○ Dependent Territory Capital
⬖ Administrative Regions
✕ International Airport
▲ Mountain
▲▲ Mountain Range
⛰ Volcano
≈ River
◎ Lake
⊠ Reservoir

104 J5 **Alcañices** Castilla y León, N Spain 41°41′N 06°21′W
105 T7 **Alcañiz** Aragón, NE Spain 41°03′N 00°09′W
104 I9 **Alcántara** Extremadura, W Spain 39°42′N 06°54′W
104 J9 **Alcántara, Embalse de** ⊠ W Spain
105 R13 **Alcantarilla** Murcia, SE Spain 37°59′N 01°12′W
105 P11 **Alcaraz** Castilla-La Mancha, C Spain 38°40′N 02°29′W
105 P12 **Alcaraz, Sierra de** ▲▲ C Spain
104 I12 **Alcarrache** ≈ SW Spain
105 T6 **Alcarràs** Cataluña, NE Spain 41°34′N 00°31′E
105 N14 **Alcaudete** Andalucía, S Spain 37°35′N 04°05′W
Alcázar *see* Ksar-el-Kebir
105 O10 **Alcázar de San Juan** *anc.* Alce. Castilla-La Mancha, C Spain 39°24′N 03°12′W
Alcazarquivir *see* Ksar-el-Kebir
Alce *see* Alcázar de San Juan
57 B17 **Alcebo, Volcán** 🌋 Galapagos Islands, Ecuador, E Pacific Ocean 0°25′S 91°06′W
139 X12 **Al Chabā'ish** *var.* Al Kaba'ish. Dhī Qār, SE Iraq 30°58′N 47°02′E
117 Y7 **Alchevs'k** *prev.* Kommunarsk, Voroshilovsk. Luhans'ka Oblast', E Ukraine 48°29′N 38°52′E
Alcira *see* Alzira
21 N9 **Alcoa** Tennessee, S USA 35°47′N 83°58′W
104 F9 **Alcobaça** Leiria, C Portugal 39°32′N 08°59′W
105 N8 **Alcobendas** Madrid, C Spain 40°32′N 03°38′W
Alcoi *see* Alcoy
105 P7 **Alcolea del Pinar** Castilla-La Mancha, C Spain 41°02′N 02°28′W
104 I11 **Alconchel** Extremadura, W Spain 38°31′N 07°04′W
Alcora *see* L'Alcora
105 N8 **Alcorcón** Madrid, C Spain 40°20′N 03°52′W
105 S7 **Alcorisa** Aragón, NE Spain 40°53′N 00°23′W
61 B19 **Alcorta** Santa Fe, C Argentina 33°32′S 61°07′W
104 H14 **Alcoutim** Faro, S Portugal 37°28′N 07°29′W
33 W15 **Alcova** Wyoming, C USA 42°33′N 106°40′W
105 S11 **Alcoy** *Cat.* Alcoi. Valenciana, E Spain 38°42′N 00°29′W
105 Y9 **Alcúdia** Mallorca, Spain, W Mediterranean Sea 39°51′N 03°06′E
105 Y9 **Alcúdia, Badia d'** *bay* Mallorca, Spain, W Mediterranean Sea
172 M7 **Aldabra Group** *island group* SW Seychelles
139 U10 **Al Daghghārah** Bābil, C Iraq 32°10′N 44°57′E
40 J5 **Aldama** Chihuahua, N Mexico 28°50′N 105°52′W
41 P11 **Aldama** Tamaulipas, C Mexico 22°54′N 98°05′W
123 Q11 **Aldan** Respublika Sakha (Yakutiya), NE Russian Federation 58°31′N 125°15′E
123 Q10 **Aldan** ≈ NE Russian Federation
Aldar *see* Aldarhaan
al Dar al Baida *see* Rabat
162 G7 **Aldarhaan** *var.* Aldar. Dzavhan, W Mongolia 47°43′N 96°36′E
97 Q20 **Aldeburgh** E England, United Kingdom 52°12′N 01°36′E
105 P5 **Aldehuela de Calatañazor** Castilla y León, N Spain 41°42′N 02°46′W
Aldeia Nova *see* Aldeia Nova de São Bento
104 H13 **Aldeia Nova de São Bento** *var.* Aldeia Nova. Beja, S Portugal 37°55′N 07°24′W
29 V11 **Alden** Minnesota, N USA 43°40′N 93°34′W
184 N6 **Aldermen Islands, The** *island group* N New Zealand
97 L25 **Alderney** *island* Channel Islands
97 N22 **Aldershot** S England, United Kingdom 51°15′N 00°47′W
21 R6 **Alderson** West Virginia, NE USA 37°43′N 80°38′W
Al Dhaid *see* Adh Dhayd
98 L5 **Aldtsjerk** *Dutch.* Oudkerk. Fryslân, N Netherlands 53°16′N 05°52′E
30 J11 **Aledo** Illinois, N USA 41°12′N 90°45′W
76 H9 **Aleg** Brakna, SW Mauritania 17°03′N 13°53′W
64 Q10 **Alegranza** *island* Islas Canarias, Spain, NE Atlantic Ocean
37 P12 **Alegres Mountain** ▲ New Mexico, SW USA 34°09′N 108°11′W
61 F15 **Alegrete** Rio Grande do Sul, S Brazil 29°46′S 55°46′W
61 C16 **Alejandra** Santa Fe, C Argentina 29°54′S 59°50′W
193 T11 **Alejandro Selkirk, Isla** *island* Islas Juan Fernández, Chile, E Pacific Ocean
124 I12 **Alëkhovshchina** Leningradskaya Oblast', NW Russian Federation 60°22′N 33°57′E
39 O13 **Aleknagik** Alaska, USA 59°16′N 158°37′W
Aleksandriya *see* Oleksandriya
Aleksandropol' *see* Gyumri
126 L3 **Aleksandrov** Vladimirskaya Oblast', W Russian Federation 56°24′N 38°42′E
113 N14 **Aleksandrovac** Serbia, C Serbia 43°28′N 21°05′E
127 R9 **Aleksandrov Gay** Saratovskaya Oblast', W Russian Federation 50°08′N 48°34′E
127 U6 **Aleksandrovka** Orenburgskaya Oblast', W Russian Federation 52°47′N 54°14′E
Aleksandrovka *see* Oleksandrivka
125 V13 **Aleksandrovsk** Permskiy Kray, NW Russian Federation 59°12′N 57°27′E
Aleksandrovsk *see* Zaporizhzhya
127 N14 **Aleksandrovskoye** Stavropol'skiy Kray, SW Russian Federation 44°43′N 42°56′E
123 T12 **Aleksandrovsk-Sakhalinskiy** Ostrov Sakhalin, Sakhalinskaya Oblast', SE Russian Federation 50°55′N 142°12′E
110 J10 **Aleksandrów Kujawski** Kujawskio-pormorskie, C Poland 52°52′N 18°40′E
110 K12 **Aleksandrów Łódzki** Łódzkie, C Poland 51°49′N 19°19′E
114 J8 **Aleksandŭr Stamboliyski, Yazovir** ⊠ N Bulgaria
Alekseevka *see* Akkol', Akmola, Kazakhstan
Alekseevka *see* Terekty
145 P7 **Alekseyevka** *Kaz.* Alekseevka. Akmola, N Kazakhstan 53°32′N 69°30′E
126 L9 **Alekseyevka** Belgorodskaya, W Russian Federation 50°35′N 38°41′E
127 S7 **Alekseyevka** Samarskaya Oblast', W Russian Federation 52°37′N 51°20′E
Alekseyevka *see* Akkol', Akmola, Kazakhstan
Alekseyevka *see* Terekty, Vostochnyy Kazakhstan, Kazakhstan
127 R4 **Alekseyevskoye** Respublika Tatarstan, W Russian Federation 55°18′N 50°11′E
126 K5 **Aleksin** Tul'skaya Oblast', W Russian Federation 54°30′N 37°08′E
113 O14 **Aleksinac** Serbia, SE Serbia 43°33′N 21°43′E
190 G11 **Alele** Île Uvea, E Wallis and Futuna 13°14′S 176°09′W
95 N20 **Ålem** Kalmar, S Sweden 56°57′N 16°25′E
102 L6 **Alençon** Orne, N France 48°26′N 00°04′E
58 I12 **Alenquer** Pará, NE Brazil 01°58′S 54°45′W
38 G10 **'Alenuihaha Channel** *var.* Alenuihaha Channel. *channel* Hawai'i, USA, C Pacific Ocean
Alep/Aleppo *see* Ḥalab
103 Y15 **Aléria** Corse, France, C Mediterranean Sea 42°06′N 09°29′E
197 Q11 **Alert** Ellesmere Island, Nunavut, N Canada 82°28′N 62°13′W
103 Q14 **Alès** *prev.* Alais. Gard, S France 44°08′N 04°05′E
116 G9 **Aleşd** *Hung.* Élesd. Bihor, SW Romania 47°03′N 22°22′E
106 C9 **Alessandria** *Fr.* Alexandrie. Piemonte, N Italy 44°54′N 08°37′E
Ålestrup *see* Aalestrup
94 D9 **Ålesund** Møre og Romsdal, S Norway 62°28′N 06°11′E
108 E10 **Aletschhorn** ▲ SW Switzerland 46°33′N 08°01′E
197 S1 **Aleutian Basin** *undersea feature* Bering Sea 57°00′N 177°00′E
38 H17 **Aleutian Islands** *island group* Alaska, USA
39 P14 **Aleutian Range** ▲▲ Alaska, USA
0 B5 **Aleutian Trench** *undersea feature* S Bering Sea 57°00′N 177°00′W
123 T10 **Alevina, Mys** *cape* E Russian Federation
15 Q6 **Alex** ≈ Québec, SE Canada
28 J3 **Alexander** North Dakota, N USA 47°48′N 103°38′W
39 W14 **Alexander Archipelago** *island group* Alaska, USA
Alexanderbaai *see* Alexander Bay
83 D23 **Alexander Bay** *Afr.* Alexanderbaai. Northern Cape, W South Africa 28°40′S 16°30′E
23 Q5 **Alexander City** Alabama, S USA 32°56′N 85°57′W
194 J6 **Alexander Island** *island* Antarctica
Alexander Range *see* Kirghiz Range
183 O12 **Alexandra** Victoria, SE Australia 37°12′S 145°43′E
185 D22 **Alexandra** Otago, South Island, New Zealand 45°15′S 169°25′E
115 F14 **Alexándreia** *var.* Alexándria. Kentrikí Makedonía, N Greece 40°38′N 22°27′E
Alexandretta *see* İskenderun
Alexandretta, Gulf of *see* İskenderun Körfezi
15 N13 **Alexandria** Ontario, SE Canada 45°19′N 74°37′W
121 U13 **Alexandria** *Ar.* Al Iskandarīyah. N Egypt 31°07′N 29°51′E
44 J12 **Alexandria** C Jamaica 18°18′N 77°21′W
116 J15 **Alexandria** Teleorman, S Romania 43°58′N 25°18′E
31 P13 **Alexandria** Indiana, N USA 40°15′N 85°40′W
20 M4 **Alexandria** Kentucky, S USA 38°59′N 84°22′W
22 H7 **Alexandria** Louisiana, S USA 31°19′N 92°27′W
29 T7 **Alexandria** Minnesota, N USA 45°54′N 95°22′W
29 Q11 **Alexandria** South Dakota, N USA 43°39′N 97°46′W
21 W4 **Alexandria** Virginia, NE USA 38°49′N 77°06′W
Alexándria *see* Alexándreia
18 I7 **Alexandria Bay** New York, NE USA 44°20′N 75°54′W
Alexandrie *see* Alessandria
182 J10 **Alexandrina, Lake** ◎ South Australia
114 K13 **Alexandroúpoli** *var.* Alexandroúpolis, *Turk.* Dedeagaç, Dedeagach. Anatolikí Makedonía kai Thráki, NE Greece 40°52′N 25°53′E
Alexandroúpolis *see* Alexandroúpoli
10 L15 **Alexis Creek** British Columbia, SW Canada 52°06′N 123°25′W
122 I13 **Aleysk** Altayskiy Kray, S Russian Federation 52°32′N 82°46′E
139 S8 **Al Fallūjah** *var.* Falluja. Al Anbār, C Iraq 33°21′N 43°46′E
105 R8 **Alfambra** ≈ E Spain
Al Faqa *see* Faq'
141 R15 **Al Farḍah** C Yemen 14°51′N 48°33′E
105 Q4 **Alfaro** La Rioja, N Spain 42°13′N 01°45′W
105 U5 **Alfarràs** Cataluña, NE Spain 41°50′N 00°34′E
Al Fāshir *see* El Fasher
75 W8 **Al Fashn** *var.* El Fashn. C Egypt 28°49′N 30°54′E
114 M7 **Alfatar** Silistra, NE Bulgaria 43°56′N 27°17′E
139 S5 **Al Fatḥah** Ṣalāḥ ad Dīn, C Iraq 35°06′N 43°34′E
139 Q3 **Al Fatsī** Nīnawá, N Iraq 36°04′N 42°39′E
139 Z13 **Al Fāw** *var.* Fao. Al Başrah, SE Iraq 29°55′N 48°26′E
75 W8 **Al Fayyūm** *var.* El Faiyûm. N Egypt 29°19′N 30°50′E
115 D20 **Alfeiós** *prev.* Alfiós; *anc.* Alpheius, Alpheus. ≈ S Greece
100 I13 **Alfeld** Niedersachsen, C Germany 51°58′N 09°49′E
Alfiós *see* Alfeiós
Alföld *see* Great Hungarian Plain
94 C11 **Ålfotbreen** *glacier* S Norway
19 P9 **Alfred** Maine, NE USA 43°29′N 70°44′W
18 F11 **Alfred** New York, NE USA 42°15′N 77°47′W
61 K14 **Alfredo Wagner** Santa Catarina, S Brazil 27°40′S 49°22′W
94 M12 **Alfta** Gävleborg, C Sweden 61°20′N 16°05′E
140 K12 **Al Fuḥayḥīl** *var.* Fahaheel. SE Kuwait 29°01′N 48°05′E
139 Q6 **Al Fuḥaymī** Al Anbār, C Iraq 34°18′N 42°09′E
143 S16 **Al Fujayrah** *Eng.* Fujairah. Al Fujayrah, NE United Arab Emirates 25°09′N 56°18′E
143 S16 **Al Fujayrah** *Eng.* Fujairah. ✈ Al Fujayrah, NE United Arab Emirates 25°04′N 56°12′E
Al-Furāt *see* Euphrates
144 I10 **Alga** *Kaz.* Algha. Aktyubinsk, NW Kazakhstan 49°56′N 57°19′E
144 G9 **Algabas** *Kaz.* Alghabas. Zapadnyy Kazakhstan, NW Kazakhstan 50°43′N 52°09′E
95 C17 **Ålgård** Rogaland, S Norway 58°45′N 05°52′E
104 G14 **Algarve** *cultural region* S Portugal
182 G3 **Algebuckina Bridge** South Australia 28°03′S 135°48′E
104 K16 **Algeciras** Andalucía, SW Spain 36°08′N 05°27′W
105 S10 **Algemesí** Valenciana, E Spain 39°11′N 00°27′W
Al-Genain *see* El Geneina
120 F9 **Alger** *var.* Algiers, El Djazaïr, Al Jazair. ● (Algeria) N Algeria 36°47′N 02°58′E
74 H9 **Algeria** *off.* Democratic and Popular Republic of Algeria. ◆ *republic* N Africa
Algeria, Democratic and Popular Republic of *see* Algeria
120 J8 **Algerian Basin** *var.* Balearic Plain. *undersea feature* W Mediterranean Sea
Algha *see* Alga
138 I4 **Al Ghāb** *Valley* NW Syria
141 X10 **Al Ghābah** *var.* Ghaba. C Oman 21°22′N 57°14′E
Alghabas *see* Algabas
141 U14 **Al Ghayḍah** E Yemen 16°15′N 52°13′E
140 M6 **Al Ghazālah** Ḥā'il, NW Saudi Arabia 26°55′N 41°23′E
107 B17 **Alghero** Sardegna, Italy, C Mediterranean Sea 40°34′N 08°19′E
95 M20 **Älghult** Kronoberg, S Sweden 57°00′N 15°34′E
75 X9 **Al Ghurdaqah** *var.* Ghurdaqah, Hurghada. E Egypt 27°17′N 33°47′E
Algiers *see* Alger
105 S10 **Alginet** Valenciana, E Spain 39°16′N 00°28′W
83 I26 **Algoa Bay** *bay* S South Africa
104 L15 **Algodonales** Andalucía, S Spain 36°54′N 05°24′W
105 N9 **Algodor** ≈ C Spain
Al Golea *see* El Goléa
31 N6 **Algoma** Wisconsin, N USA 44°41′N 87°24′W
29 U12 **Algona** Iowa, C USA 43°04′N 94°13′W
20 L8 **Algood** Tennessee, S USA 36°12′N 85°27′W
105 O2 **Algorta** País Vasco, N Spain 43°20′N 03°00′W
61 E18 **Algorta** Río Negro, W Uruguay 32°26′S 57°18′W
Al Haba *see* Haba
139 Q10 **Al Habbārīyah** Al Anbār, S Iraq 32°16′N 42°12′E
Al Hadhar *see* Al Ḥaḍr
139 Q4 **Al Ḥaḍr** *var.* Al Hadhar; *anc.* Hatra. Nīnawýa, NW Iraq 35°34′N 42°44′E
139 T13 **Al Ḥajarah** *desert* S Iraq
141 W8 **Al Ḥajar al Gharbī** ▲▲ N Oman
141 Y8 **Al Ḥajar ash Sharqī** ▲▲ NE Oman
141 R15 **Al Hajarayn** C Yemen 15°29′N 48°24′E
138 L10 **Al Ḥamād** *desert* Jordan/Saudi Arabia
Al Hamad *see* Syrian Desert
75 N9 **Al Ḥamādah al Ḥamrā'** *var.* Al Ḥamrā'. *desert* NW Libya
105 N15 **Alhama de Granada** Andalucía, S Spain 37°00′N 03°59′W
105 R13 **Alhama de Murcia** Murcia, SE Spain 37°51′N 01°25′W
35 T15 **Alhambra** California, W USA 34°08′N 118°06′W
139 T12 **Al Ḥammām** An Najaf, S Iraq 31°09′N 44°04′E
141 X8 **Al Ḥamrā'** NE Oman 23°07′N 57°23′E
Al Ḥamrā' *see* Al Ḥamādah al Ḥamrā'
141 O6 **Al Ḥamūdīyah** *spring/well* N Saudi Arabia 27°05′N 44°24′E
140 M7 **Al Ḥanākīyah** Al Madīnah, W Saudi Arabia 24°55′N 40°31′E
139 W14 **Al Ḥanīyah** *escarpment* Iraq/Saudi Arabia
139 Y12 **Al Hārithah** Al Başrah, SE Iraq 30°43′N 47°44′E
140 L3 **Al Ḥarrah** *desert* NW Saudi Arabia
75 Q10 **Al Harūj al Aswad** *desert* C Libya
Al Hasaifin *see* Al Ḥusayfin
139 N2 **Al Ḥasakah** *var.* Al Hasijah, El Haseke, *Fr.* Hassetché. Al Ḥasakah, NE Syria 36°22′N 40°44′E
139 O2 **Al Ḥasakah** *off.* Muḥāfaẓat al Ḥasakah, *var.* Al Hasakah, Āl Hasakah, Hasakah, Hassakeh. ⬖ *governorate* NE Syria
Al Hasakah *see* Al Ḥasakah
Al Hasakah *see* 'Āmūdah
139 T9 **Al Hāshimīyah** Bābil, C Iraq 32°24′N 44°39′E
138 G13 **Al Hāshimīyah** Ma'ān, S Jordan 30°31′N 35°46′E
Al Hasijah *see* Al Ḥasakah
104 M15 **Alhaurín el Grande** Andalucía, S Spain 36°39′N 04°41′W
141 Q16 **Al Ḥawrā** S Yemen 13°54′N 47°36′E
139 V10 **Al Ḥayy** *var.* Kut al Hai, Kūt al Ḥayy. Wāsiţ, E Iraq 32°11′N 46°03′E
141 U11 **Al Ḥibāk** *desert* E Saudi Arabia
138 H8 **Al Hījānah** *var.* Hejanah, Hijanah. Rīf Dimashq, W Syria 33°23′N 36°34′E
140 K7 **Al Ḥijāz** *Eng.* Hejaz. *physical region* NW Saudi Arabia
Al Hilbeh *see* 'Ulayyānīyah, Bi'r al
139 T9 **Al Ḥillah** *var.* Hilla. Bābil, C Iraq 32°28′N 44°29′E
139 T9 **Al Hindīyah** *var.* Hindiya. Bābil, C Iraq 32°32′N 44°14′E
138 G12 **Al Ḥisā** Aţ Ţafilah, W Jordan 30°49′N 35°58′E
74 G5 **Al-Hoceïma** *var.* al Hoceima, Al-Hoceima, Alhucemas; *prev.* Villa Sanjurjo. N Morocco 35°14′N 03°56′W
Alhucemas *see* Al-Hoceïma
105 N17 **Alhucemas, Peñon de** *island group* S Spain
141 N15 **Al Ḥudaydah** *Eng.* Hodeida. W Yemen 15°N 42°50′E
141 N15 **Al Ḥudaydah** *var.* Hodeida. ✈ W Yemen 14°45′N 43°01′E
140 M4 **Al Ḥudūd ash Shamālīyah** *var.* Minţaqat al Ḥudūd ash Shamālīyah, *Eng.* Northern Border Region. ⬖ *province* N Saudi Arabia
141 S7 **Al Hufūf** *var.* Hofuf. Ash Sharqīyah, NE Saudi Arabia 25°21′N 49°34′E
al-Hurma *see* Al Khurmah
141 X7 **Al Ḥusayfin** *var.* Al Hasaifin. N Oman 24°33′N 56°33′E
138 G9 **Al Ḥuşn** *var.* Husn. Irbid, N Jordan 32°29′N 35°53′E
139 U9 **'Alī** Wāsiţ, E Iraq 32°43′N 45°21′E
104 L10 **Alia** Extremadura, W Spain 39°25′N 05°12′W
143 P9 **'Alīābād** Yazd, C Iran 36°55′N 54°33′E
'Alīābad *see* Qā'emshahr
105 S7 **Aliaga** Aragón, NE Spain 40°40′N 00°42′W
136 B13 **Aliağa** İzmir, W Turkey 38°49′N 26°59′E
Aliákmon *see* Aliákmonas
115 F14 **Aliákmonas** *prev.* Aliákmon; *anc.* Haliacmon. ≈ N Greece
139 W9 **'Alī al Gharbī** Maysān, E Iraq 32°28′N 46°42′E
139 U11 **'Alī al Ḥassūnī** Al Qādisīyah, S Iraq 31°25′N 44°50′E
115 G18 **Alíartos** Stereá Elláda, C Greece 38°23′N 23°06′E
Äli-Bayramli *see* Şirvan
114 P12 **Alibey Barajī** ⊠ NW Turkey
77 S13 **Alibori** ≈ N Benin
112 M10 **Alibunar** Vojvodina, NE Serbia 45°06′N 20°59′E
105 S12 **Alicante** *Cat.* Alacant, *Lat.* Lucentum. Valenciana, SE Spain 38°21′N 00°29′W
105 S12 **Alicante** ⬖ *province* Valenciana, SE Spain
105 S12 **Alicante** ✈ Murcia, E Spain 38°21′N 00°29′W
83 I25 **Alice** Eastern Cape, S South Africa 32°47′S 26°50′E
25 S14 **Alice** Texas, SW USA 27°45′N 98°06′W
83 I25 **Alicedale** Eastern Cape, S South Africa 33°19′S 26°05′E
65 B25 **Alice, Mount** *hill* West Falkland, Falkland Islands
107 P20 **Alice, Punta** *headland* S Italy 39°24′N 17°09′E
181 Q7 **Alice Springs** Northern Territory, C Australia 23°42′S 133°52′E
23 N4 **Aliceville** Alabama, S USA 33°07′N 88°09′W
147 U13 **Alichur** SE Tajikistan 37°49′N 73°45′E
147 U14 **Alichuri Janubí, Qatorkŭhi** *Rus.* Yuzhno-Alichurskiy Khrebet. ▲▲ SE Tajikistan
147 U13 **Alichuri Shimolí, Qatorkŭhi** *Rus.* Severo-Alichurskiy Khrebet. ▲▲ SE Tajikistan
107 K22 **Alicudi, Isola** *island* Isole Eolie, S Italy
Alifu Atoll *see* Ari Atoll
43 W14 **Aligandí** Kuna Yala, NE Panama 09°15′N 78°05′W
152 J11 **Alīgarh** Uttar Pradesh, N India 27°54′N 78°04′E
142 M7 **Alīgūdarz** Lorestān, W Iran 33°24′N 49°19′E
163 U5 **Alihe** *var.* Oroqen Zizhiqi. Nei Mongol Zizhiqu, N China 50°34′N 123°40′E
0 F12 **Alijos, Islas** *islets* California, SW USA
149 R6 **'Alī Kbel** *Pash.* 'Alī Khēl. Paktīkā, E Afghanistan 33°55′N 69°49′E
149 R6 **'Alī Khēl** *var.* Ali Khel, Jaji; *prev.* 'Alī Kheyl. Paktiyā, SE Afghanistan 33°55′N 69°46′E
'Alī Khēl *see* 'Alī Kbel, Paktīkā, Afghanistan
Ali Khel *see* 'Alī Kheyl, Paktīā, Afghanistan
'Alī Kheyl *see* 'Alī Khēl
141 V17 **Al Ikhwān** *island group* SE Yemen
Aliki *see* Alykí
79 H19 **Alima** ≈ C Congo
Al Imārāt al 'Arabīyahal Muttaḥidah *see* United Arab Emirates
115 N23 **Alimía** *island* Dodekánisa, Greece, Aegean Sea
55 V12 **Alimimuni Piek** ▲ S Suriname 02°26′N 55°46′W
79 K15 **Alindao** Basse-Kotto, S Central African Republic 04°58′N 21°16′E
95 J18 **Alingsås** Västra Götaland, S Sweden 57°55′N 12°30′E
81 K18 **Alinjugul** *spring/well* E Kenya 0°03′S 40°31′E
149 S11 **Alīpur** Punjab, E Pakistan 29°22′N 70°59′E
153 T12 **Alīpur Duār** West Bengal, NE India 26°28′N 89°25′E
18 B14 **Aliquippa** Pennsylvania, NE USA 40°36′N 80°15′W
80 L12 **'Ali Sabieh** *var.* 'Alī Sabīḥ. S Djibouti 11°07′N 42°44′E
'Alī Sabīḥ *see* 'Ali Sabieh
140 K3 **Al 'Īsāwīyah** Al Jawf, NW Saudi Arabia 30°41′N 37°58′E
104 J10 **Aliseda** Extremadura, W Spain 39°25′N 06°42′W
139 T8 **Al Iskandarīyah** Bābil, C Iraq 32°53′N 44°22′E
Al Iskandarīyah *see* Alexandria
123 T6 **Aliskerovo** Chukotskiy Avtonomnyy Okrug, NE Russian Federation 67°40′N 167°37′E
75 W7 **Al Ismā'īlīya** *var.* Ismailia, Ismâ'ilîya. N Egypt 30°32′N 32°13′E
114 H13 **Alistráti** Kentrikí Makedonía, NE Greece 41°03′N 23°58′E
39 P15 **Alitak Bay** *bay* Kodiak Island, Alaska, USA
Al Ittiḥād *see* Madīnat ash Sha'b
115 H18 **Alivéri** *var.* Alivérion. Évvoia, C Greece 38°24′N 24°02′E
Alivérion *see* Alivéri
Aliwal-Noord *see* Aliwal North
83 I24 **Aliwal North** *Afr.* Aliwal-Noord. Eastern Cape, SE South Africa 30°42′S 26°43′E
121 Q13 **Al Jabal al Akhḍar** ▲▲ NE Libya
138 H13 **Al Jafr** Ma'ān, S Jordan 30°18′N 36°13′E
75 T8 **Al Jaghbūb** NE Libya 29°45′N 24°31′E
142 K11 **Al Jahrā'** *var.* Al Jahrah, Jahra. C Kuwait 29°18′N 47°36′E
Al Jahrah *see* Al Jahrā'
Al Jamāhīrīyah al 'Arabīyah al Lībīyah ash Sha'bīyah al Ishtirākīy *see* Libya
140 K3 **Al Jarāwī** *spring/well* NW Saudi Arabia 30°12′N 38°48′E
141 X11 **Al Jawārah** *oasis* SE Oman
140 L3 **Al Jawf** *off.* Jauf. Al Jawf, NW Saudi Arabia 29°51′N 39°49′E
140 L4 **Al Jawf** *var.* Minţaqat al Jawf. ⬖ *province* N Saudi Arabia
Al Jawlān *see* Golan Heights
Al Jazair *see* Alger
139 N4 **Al Jazīrah** *physical region* Iraq/Syria
104 F14 **Aljezur** Faro, S Portugal 37°18′N 08°49′W
139 S13 **Al Jīl** An Najaf, S Iraq 30°28′N 43°57′E
138 G11 **Al Jīzah** *var.* Jiza. 'Ammān, N Jordan 31°42′N 35°57′E
Al Jīzah *see* Giza
Al Jubail *see* Al Jubayl
141 S6 **Al Jubayl** *var.* Al Jubail. Ash Sharqīyah, NE Saudi Arabia 27°N 49°36′E
141 T10 **Āl Juḥaysh, Qalamat** *well* SE Saudi Arabia
143 N15 **Al Jumaylīyah** N Qatar 25°37′N 51°05′E
Al Junaynah *see* El Geneina
104 G13 **Aljustrel** Beja, S Portugal 37°52′N 08°10′W
Al Kaba'ish *see* Al Chabā'ish
Al-Kadhimain *see* Al Kāẓimīyah
Al Kāf *see* El Kef
Alkal'a *see* Alcalá de Henares
35 W4 **Alkali Flat** *salt flat* Nevada, W USA
35 Q1 **Alkali Lake** ◎ Nevada, W USA
141 Z9 **Al Kāmil** NE Oman 22°14′N 58°15′E
138 G11 **Al Karak** *var.* El Kerak, Karak, Kerak; *anc.* Kir Moab, Kir of Moab. Al Karak, W Jordan 31°11′N 35°42′E
138 G12 **Al Karak** *off.* Muḥāfaẓat al Karak. ⬖ *governorate* W Jordan
139 W8 **Al Karmashīyah** Wāsiţ, E Iraq 32°57′N 46°10′E
Al-Kashaniya *see* Al Qash'ānīyah
Al-Kasr al-Kebir *see* Ksar-el-Kebir
139 T8 **Al Kāẓimīyah** *var.* Al-Kadhimain, Kadhimain. Baghdād, C Iraq 33°22′N 44°20′E
99 J18 **Alken** Limburg, NE Belgium 50°52′N 05°19′E
141 X8 **Al Khābūrah** *var.* Khabura. N Oman 23°57′N 57°06′E
Al Khalīl *see* Hebron
139 T7 **Al Khāliş** Diyālá, C Iraq 33°51′N 44°33′E
Al Khaluf *see* Khalūf
75 W10 **Al Khārijah** *var.* El Khârga. C Egypt 25°31′N 30°36′E
141 Q8 **Al Kharj** Ar Riyāḍ, C Saudi Arabia 24°12′N 47°12′E
141 W6 **Al Khaşab** *var.* Khasab. N Oman 26°11′N 56°18′E
Al Khaur *see* Al Khawr
143 N15 **Al Khawr** *var.* Al Khaur, Al Khor. N Qatar 25°40′N 51°33′E
142 K12 **Al Khīrān** *var.* Al Khiran. SE Kuwait 28°34′N 48°21′E
141 W9 **Al Khīrān** *spring/well* NW Oman 22°31′N 55°42′E
Al Khiyām *see* El Khiyam
Al-Khobar *see* Al Khubar
Al Khor *see* Al Khawr
141 S6 **Al Khubar** *var.* Al-Khobar. Ash Sharqīyah, NE Saudi Arabia 26°00′N 50°00′E
120 M12 **Al Khums** *var.* Homs, Khoms, Khums. NW Libya 32°39′N 14°16′E
141 R15 **Al Khuraybah** C Yemen 15°05′N 48°17′E
140 M9 **Al Khurmah** *var.* al-Hurma. Makkah, W Saudi Arabia 21°59′N 42°00′E
141 V9 **Al Kidan** *desert* NE Saudi Arabia
127 V4 **Alkino-2** Respublika Bashkortostan, W Russian Federation 54°30′N 55°40′E
98 H9 **Alkmaar** Noord-Holland, NW Netherlands 52°37′N 04°45′E
139 T10 **Al Kūfah** *var.* Kufa. An Najaf, S Iraq 32°02′N 44°25′E
75 T11 **Al Kufrah** SE Libya 24°11′N 23°19′E
141 T10 **Al Kursū'** *desert* E Saudi Arabia
139 V9 **Al Kūt** *var.* Kūt al 'Amārah, Kut al Imara. Wāsiţ, E Iraq 32°30′N 45°51′E
Al-Kuwait *see* Al Kuwayt
Al Kuwayr *see* Guwēr
142 K11 **Al Kuwayt** *var.* Al-Kuwait, *Eng.* Kuwait, Kuwait City; *prev.* Qurein. ● (Kuwait) E Kuwait 29°23′N 48°00′E
142 K11 **Al Kuwayt** ✈ C Kuwait 29°13′N 47°57′E
115 G19 **Alkyonidón, Kólpos** *gulf* C Greece
141 N4 **Al Labbah** *physical region* N Saudi Arabia
138 G4 **Al Lādhiqīyah** *Eng.* Latakia, *Fr.* Lattaquié; *anc.* Laodicea, Laodicea ad Mare. Al Lādhiqīyah, W Syria 35°31′N 35°47′E
138 H4 **Al Lādhiqīyah** *off.* Muḥāfaẓat al Ļādhiqīyah, *var.* Al Lathqiyah, Latakia, Lattakia. ⬖ *governorate* W Syria
19 R2 **Allagash River** ≈ Maine, NE USA
152 M13 **Allahābād** Uttar Pradesh, N India 25°27′N 81°50′E
143 S3 **Allāh Dāgh, Reshteh-ye** ▲▲ NE Iran
39 Q8 **Allakaket** Alaska, USA 66°34′N 152°39′W
141 X12 **Al Lakbī** S Oman 18°27′N 56°37′E
11 T15 **Allan** Saskatchewan, S Canada 51°50′N 105°59′W
Allanmyo *see* Aunglan
83 I22 **Allanridge** Free State, C South Africa 27°45′S 26°40′E
104 H4 **Allariz** Galicia, NW Spain 42°11′N 07°48′W
139 R11 **Al Laşaf** *var.* Al Lussuf. An Najaf, S Iraq 31°38′N 43°16′E
Al Lathqiyah *see* Al Lādhiqīyah
23 S2 **Allatoona Lake** ⊠ Georgia, SE USA
83 J19 **Alldays** Limpopo, NE South Africa 22°39′S 29°04′E
Alle *see* Łyna
31 P10 **Allegan** Michigan, N USA 42°31′N 85°51′W
18 E14 **Allegheny Mountains** ▲▲ NE USA
18 E12 **Allegheny Plateau** ▲▲ New York/Pennsylvania, NE USA
18 D11 **Allegheny Reservoir** ⊠ New York/Pennsylvania, NE USA
18 E12 **Allegheny River** ≈ New York/Pennsylvania, NE USA
22 K9 **Allemands, Lac des** ◎ Louisiana, S USA
25 U6 **Allen** Texas, SW USA 33°06′N 96°40′W
21 R14 **Allendale** South Carolina, SE USA 33°01′N 81°19′W
41 N6 **Allende** Coahuila, NE Mexico 28°22′N 100°50′W
41 O9 **Allende** Nuevo León, NE Mexico 25°20′N 100°01′W
97 D16 **Allen, Lough** *Ir.* Loch Aillionn. ◎ NW Ireland
185 B26 **Allen, Mount** ▲ Stewart Island, Southland, SW New Zealand 47°05′S 167°49′E
109 V2 **Allensteig** Niederösterreich, N Austria 48°40′N 15°24′E
Allenstein *see* Olsztyn
18 I14 **Allentown** Pennsylvania, NE USA 40°37′N 75°30′W
155 G23 **Alleppey** *var.* Alappuzha. Kerala, SW India 09°30′N 76°22′E *see also* Alappuzha
100 J12 **Aller** ≈ NW Germany
29 V16 **Allerton** Iowa, C USA 40°42′N 93°22′W
99 K19 **Alleur** Liège, E Belgium 50°40′N 05°33′E
101 J25 **Allgäuer Alpen** ▲▲ Austria/Germany
28 J13 **Alliance** Nebraska, C USA 42°08′N 102°54′W
31 U12 **Alliance** Ohio, N USA 40°55′N 81°06′W
103 O10 **Allier** ⬖ *department* N France
139 R13 **Al Līfīyah** *well* An Najaf, S Iraq
44 J13 **Alligator Pond** C Jamaica 17°52′N 77°34′W
21 Y9 **Alligator River** ≈ North Carolina, SE USA
29 W12 **Allison** Iowa, C USA 42°45′N 92°48′W
14 G14 **Alliston** Ontario, S Canada 44°09′N 79°51′W
140 L11 **Al Līth** Makkah, SW Saudi Arabia 21°N 41°E
Al Liwā' *see* Līwā
96 J12 **Alloa** C Scotland, United Kingdom 56°07′N 03°49′W
103 U14 **Allos** Alpes-de-Haute-Provence, SE France 44°16′N 06°37′E
108 D6 **Allschwil** Basel Landschaft, NW Switzerland 47°34′N 07°32′E
Al Lubnān *see* Lebanon
141 N14 **Al Luḥayyah** W Yemen 15°44′N 42°45′E
14 K12 **Allumettes, Île des** *island* Québec, SE Canada
Al Lussuf *see* Al Laşaf
109 S5 **Alm** ≈ N Austria
15 Q7 **Alma** Québec, SE Canada 48°32′N 71°41′W
27 S10 **Alma** Arkansas, C USA 35°28′N 94°13′W
23 V7 **Alma** Georgia, SE USA 31°32′N 82°27′W
27 P4 **Alma** Kansas, C USA 39°01′N 96°17′W
31 Q8 **Alma** Michigan, N USA 43°22′N 84°39′W
29 O17 **Alma** Nebraska, C USA 40°06′N 99°21′W
30 I7 **Alma** Wisconsin, N USA 44°21′N 91°54′W
139 R12 **Al Ma'ānīyah** *well* An Najaf, S Iraq
Alma-Ata *see* Almaty
Alma-Atinskaya Oblast' *see* Almaty
105 T5 **Almacelles** Cataluña, NE Spain 41°44′N 00°26′E
104 F11 **Almada** Setúbal, W Portugal 38°40′N 09°09′W
104 L11 **Almadén** Castilla-La Mancha, C Spain 38°47′N 04°50′W
66 L6 **Almadies, Pointe des** *headland* W Senegal 14°44′N 17°31′W
140 L7 **Al Madīnah** *Eng.* Medina. Al Madīnah, W Saudi Arabia 24°25′N 39°29′E
140 L7 **Al Madīnah** *off.* Minţaqat al Madīnah. ⬖ *province* W Saudi Arabia
138 H9 **Al Mafraq** *var.* Mafraq. Al Mafraq, N Jordan 32°20′N 36°12′E
138 J10 **Al Mafraq** *off.* Muḥāfaẓat al Mafraq. ⬖ *governorate* NW Jordan
141 R15 **Al Maghārīm** C Yemen 15°00′N 47°49′E
105 N11 **Almagro** Castilla-La Mancha, C Spain 38°54′N 03°43′W
Al Maḥallah al Kubrá *see* El Maḥalla el Kubra
139 T9 **Al Maḥāwīl** *var.* Khān al Maḥāwīl. Bābil, C Iraq 32°39′N 44°28′E
Al Mahdīyah *see* Mahdia
139 T8 **Al Maḥmūdīyah** *var.* Mahmudiya. Baghdād, C Iraq 33°04′N 44°22′E
141 T14 **Al Mahrah** ▲▲ E Yemen
141 P7 **Al Majma'ah** Ar Riyāḍ, C Saudi Arabia 25°55′N 45°19′E
139 Q11 **Al Makmīn** *well* S Iraq
139 Q1 **Al Mālikīyah** *var.* Malkiye. Al Ḥasakah, N Syria 37°12′N 42°13′E
Almalyk *see* Olmaliq
Al Mamlakah *see* Morocco
Al Mamlaka al Urdunīya al Hashemīyah *see* Jordan
143 Q18 **Al Manādir** *var.* Al Manadir. *desert* Oman/United Arab Emirates
142 L15 **Al Manāmah** *Eng.* Manama. ● (Bahrain) N Bahrain 26°13′N 50°33′E
139 O5 **Al Manāşif** ▲▲ E Syria
35 O4 **Almanor, Lake** ⊠ California, W USA
105 R11 **Almansa** Castilla-La Mancha, C Spain 38°52′N 01°06′W
75 W7 **Al Manşūrah** *var.* Manşûra, El Manşûra. N Egypt 31°03′N 31°23′E
104 L3 **Almanza** Castilla y León, N Spain 42°40′N 05°01′W
104 L8 **Almanzor** ▲ W Spain 40°13′N 05°18′W
105 P14 **Almanzora** ≈ SE Spain
139 S9 **Al Mardah** Karbalā', C Iraq 32°35′N 43°30′E
Al-Mariyya *see* Almería
75 R7 **Al Marj** *var.* Barka, *It.* Barce. NE Libya 32°30′N 20°54′E
138 L2 **Al Mashrafah** Ar Raqqah, N Syria 36°25′N 39°07′E
141 X8 **Al Maşna'ah** *var.* Al Muşana'a. NE Oman 23°45′N 57°38′E
Almassora *see* Almazora
Almatinskaya Oblast' *see* Almaty
145 U15 **Almaty** *var.* Alma-Ata. Almaty, SE Kazakhstan 43°19′N 76°55′E
145 S14 **Almaty** *off.* Almatinskaya Oblast', *Kaz.* Almaty Oblysy; *prev.* Alma-Atinskaya Oblast'. ⬖ *province* SE Kazakhstan
145 U15 **Almaty** ✈ Almaty, SE Kazakhstan 43°15′N 76°57′E
Almaty Oblysy *see* Almaty
al-Mawailih *see* Al Muwayliḥ
139 R3 **Al Mawşil** *Eng.* Mosul. Nīnawá, N Iraq 36°21′N 43°08′E
139 N5 **Al Mayādīn** *var.* Mayadin, *Fr.* Meyadine. Dayr az Zawr, E Syria 35°00′N 40°31′E
139 X10 **Al Maymūnah** *var.* Maimuna. Maysān, SE Iraq 31°43′N 46°55′E
141 N5 **Al Mayyāh** Ḥā'il, N Saudi Arabia 27°57′N 42°53′E
Al Ma'zam *see* Al Ma'zim
105 P6 **Almazán** Castilla y León, N Spain 41°29′N 02°31′W
141 W8 **Al Ma'zim** *var.* Al Ma'zam. NW Oman 23°22′N 56°16′E
123 N11 **Almaznyy** Respublika Sakha (Yakutiya), NE Russian Federation 62°19′N 114°14′E
105 T9 **Almazora** *Cat.* Almassora. Valenciana, E Spain 39°55′N 00°02′W
Al Mazra' *see* Al Mazra'ah
138 G11 **Al Mazra'ah** *var.* Al Mazra', Mazra'a. Al Karak, W Jordan 31°18′N 35°32′E
101 G15 **Alme** ≈ W Germany
104 I7 **Almeida** Guarda, N Portugal 40°43′N 06°53′W
104 G10 **Almeirim** Santarém, C Portugal 39°12′N 08°37′W
98 O10 **Almelo** Overijssel, E Netherlands 52°22′N 06°42′E
105 S9 **Almenara** Valenciana, E Spain 39°46′N 00°14′W
105 P12 **Almenaras** ▲ S Spain 38°31′N 02°27′W
105 P5 **Almenar de Soria** Castilla y León, N Spain 41°41′N 02°12′W
104 J6 **Almendra, Embalse de** ⊠ Castilla y León, NW Spain
104 J11 **Almendralejo** Extremadura, W Spain 38°41′N 06°25′W
98 J10 **Almere** *var.* Almere-stad. Flevoland, C Netherlands 52°22′N 05°12′E
98 J10 **Almere-Buiten** Flevoland, C Netherlands 52°24′N 05°15′E
98 J10 **Almere-Haven** Flevoland, C Netherlands 52°20′N 05°13′E
Almere-stad *see* Almere
105 P15 **Almería** *Ar.* Al-Mariyya; *anc.* Unci, *Lat.* Portus Magnus. Andalucía, S Spain 36°50′N 02°26′W
105 P14 **Almería** ⬖ *province* Andalucía, S Spain
105 P15 **Almería, Golfo de** *gulf* S Spain
127 S5 **Al'met'yevsk** Respublika Tatarstan, W Russian Federation 54°53′N 52°20′E
95 L21 **Älmhult** Kronoberg, S Sweden 56°32′N 14°10′E
141 U9 **Al Miḥrāḍ** *desert* NE Saudi Arabia
Al Mīnā' *see* El Mina
104 L17 **Almina, Punta** *headland* Ceuta, Spain, N Africa 35°54′N 05°16′W
75 W9 **Al Minyā** *var.* El Minya, Minya. C Egypt 28°06′N 30°40′E
Al Miqdādīyah *see* Al Muqdādīyah
43 P14 **Almirante** Bocas del Toro, NW Panama 09°20′N 82°22′W
Almirós *see* Almyrós
140 M9 **Al Mislaḥ** *spring/well* W Saudi Arabia 22°46′N 40°47′E
Almissa *see* Omiš
104 G13 **Almodôvar** *var.* Almodóvar. Beja, S Portugal 37°31′N 08°03′W
104 M11 **Almodóvar del Campo** Castilla-La Mancha, C Spain 38°43′N 04°10′W
105 Q9 **Almodóvar del Pinar** Castilla-La Mancha, C Spain 39°44′N 01°55′W
31 S9 **Almont** Michigan, N USA 42°53′N 83°02′W
14 L13 **Almonte** Ontario, SE Canada 45°13′N 76°12′W

◆ Country
● Country Capital
◇ Dependent Territory
○ Dependent Territory Capital
⬖ Administrative Regions
✈ International Airport
▲ Mountain
▲▲ Mountain Range
🌋 Volcano
≈ River
◎ Lake
⊠ Reservoir

104 J14 **Almonte** Andalucía, S Spain *37°16′N 06°31′W*
104 K9 **Almonte** ≈ W Spain
152 K9 **Almora** Uttarakhand, N India *29°36′N 79°40′E*
104 M8 **Almorox** Castilla-La Mancha, C Spain *40°13′N 04°22′W*
141 S7 **Al Mubarraz** Ash Sharqīyah, E Saudi Arabia *25°28′N 49°34′E*
Al Muḍaibī *see* Al Muḍaybī
138 G15 **Al Mudawwarah** Ma'ān, SW Jordan *29°20′N 36°E*
141 Y9 **Al Muḍaybī** *var.* Al Muḍaibī. NE Oman *22°35′N 58°08′E*
Almudébar *see* Almudévar
105 S5 **Almudévar** *var.* Almudébar. Aragón, NE Spain *42°03′N 00°34′W*
141 S15 **Al Mukallā** *var.* Mukalla. SE Yemen *14°36′N 49°07′E*
141 N16 **Al Mukhā** *Eng.* Mocha. SW Yemen *13°18′N 43°17′E*
105 N15 **Almuñécar** Andalucía, S Spain *36°44′N 03°41′W*
139 U7 **Al Muqdādīyah** *var.* Al Miqdādīyah. Diyālá, C Iraq *33°58′N 44°58′E*
140 L3 **Al Murayr** *spring/well* NW Saudi Arabia *30°06′N 39°54′E*
136 M12 **Almus** Tokat, N Turkey *40°22′N 36°54′E*
Al Muṣana'a *see* Al Maṣna'ah
139 T9 **Al Musayyib** *var.* Musaiyib. Bābil, C Iraq *32°47′N 44°20′E*
139 V13 **Al Muthannā** *off.* Muḥāfa at al Muthannā, *var.* As Samāwah. ◈ *governorate* S Iraq
139 V9 **Al Muwaffaqīyah** Wāsiṭ, S Iraq *32°19′N 45°22′E*
138 H10 **Al Muwaqqar** *var.* El Muwaqqar. 'Ammān, W Jordan *31°49′N 36°06′E*
140 J5 **Al Muwayliḥ** *var.* al-Muwaileh. Tabūk, NW Saudi Arabia *27°39′N 35°33′E*
115 F17 **Almyrós** *var.* Almirós. Thessalía, C Greece *39°11′N 22°45′E*
115 I24 **Almyroú, Órmos** *bay* Kríti, Greece, E Mediterranean Sea
Al Nūwfalīyah *see* An Nawfalīyah
96 L13 **Alnwick** N England, United Kingdom *55°27′N 01°44′W*
Al Obayyid *see* El Obeid
Al Odaid *see* Al 'Udayd
190 B16 **Alofi** ○ (Niue) W Niue *19°01′S 169°55′E*
190 A16 **Alofi Bay** *bay* W Niue, C Pacific Ocean
190 E13 **Alofi, Île** *island* S Wallis and Futuna
190 E13 **Alofitai** Île Alofi, W Wallis and Futuna *14°21′S 178°03′W*
Aloha State *see* Hawai'i
118 G7 **Aloja** Latvia *57°47′N 24°53′E*
153 X10 **Along** Arunāchel Pradesh, NE India *28°15′N 94°56′E*
115 H16 **Alónnisos** *island* Vóreies Sporádes, Greece, Aegean Sea
104 M15 **Álora** Andalucía, S Spain *36°50′N 04°43′W*
171 Q16 **Alor, Kepulauan** *island group* E Indonesia
171 Q16 **Alor, Pulau** *prev.* Ombai. *island* Kepulauan Alor, E Indonesia
171 O16 **Alor, Selat** *strait* Flores Sea/Savu Sea
168 I7 **Alor Setar** *var.* Alor Star, Alur Setar. Kedah, Peninsular Malaysia *06°06′N 100°23′E*
Alor Star *see* Alor Setar
Alost *see* Aalst
154 F9 **Ālot** Madhya Pradesh, C India *23°56′N 75°40′E*
186 G10 **Alotau** Milne Bay, SE Papua New Guinea *10°20′S 150°23′E*
171 Y16 **Alotip** Papua, E Indonesia *08°07′S 140°06′E*
Al Oued *see* El Oued
35 R12 **Alpaugh** California, W USA *35°52′N 119°29′W*
Alpen *see* Alps
31 R6 **Alpena** Michigan, N USA *45°04′N 83°27′W*
Alpes *see* Alps
103 S14 **Alpes-de-Haute-Provence** ◈ *department* SE France
103 U14 **Alpes-Maritimes** ◈ *department* SE France
181 W8 **Alpha** Queensland, E Australia *23°40′S 146°38′E*
197 R9 **Alpha Cordillera** *var.* Alpha Ridge. *undersea feature* Arctic Ocean *85°30′N 125°00′W*
Alpha Ridge *see* Alpha Cordillera
Alpheius *see* Alfeiós
99 I15 **Alphen** Noord-Brabant, S Netherlands *51°29′N 04°57′E*
Alphen *see* Alphen aan den Rijn
98 H11 **Alphen aan den Rijn** *var.* Alphen. Zuid-Holland, C Netherlands *52°08′N 04°40′E*
Alpheus *see* Alfeiós
Alpi *see* Alps
104 G10 **Alpiarça** Santarém, C Portugal *39°15′N 08°35′W*
24 K10 **Alpine** Texas, SW USA *30°22′N 103°40′W*
108 F8 **Alpnach** Unterwalden, W Switzerland *46°56′N 08°17′E*
84 D11 **Alps** *Fr.* Alpes, *Ger.* Alpen, *It.* Alpi. ▲▲ C Europe
141 W8 **Al Qābil** *var.* Qabil. N Oman *23°55′N 55°50′E*
Al Qaḍārif *see* Gedaref
75 P8 **Al Qaddāḥīyah** N Libya *31°21′N 15°16′E*
139 V10 **Al Qādisīyah** *off.* Muḥāfa at al Qādisīyah, *var.* Ad Diwānīyah. ◈ *governorate* S Iraq
Al Qāhirah *see* Cairo
140 K4 **Al Qalībah** Tabūk, NW Saudi Arabia *28°29′N 37°40′E*
139 O1 **Al Qāmishlī** *var.* Kamishli, Qamishly. Al Ḥasakah, NE Syria *37°N 41°E*
138 I6 **Al Qaryatayn** *var.* Qaryatayn, *Fr.* Qariateïne. Ḥimṣ, C Syria *34°13′N 37°13′E*
142 K11 **Al Qash'ānīyah** *var.* Al-Kashaniya. NE Kuwait *29°59′N 47°42′E*
141 N7 **Al Qaṣīm** *var.* Minṭaqat Qaṣīm, Qassim. ◈ *province* C Saudi Arabia
75 V10 **Al Qaşr** *var.* Al Qaşr *var.* El Qaşr. C Egypt *25°43′N 28°54′E*
138 J5 **Al Qaşr** Ḥimṣ, C Syria *35°06′N 37°39′E*
Al Qaşr *see* Al Qaşr
Al Qaşrayn *see* Kasserine
141 S6 **Al Qaṭīf** Ash Sharqīyah, NE Saudi Arabia *26°27′N 50°01′E*

138 G11 **Al Qaṭrānah** *var.* El Qatrani, Qatrana. Al Karak, W Jordan *31°14′N 36°03′E*
75 P11 **Al Qaṭrūn** SW Libya *24°57′N 14°40′E*
Al Qayrawān *see* Kairouan
Al-Qsar al-Kbir *see* Ksar-el-Kebir
Al Qubayyāt *see* Qoubaïyât
Al Quds/Al Quds ash Sharīf *see* Jerusalem
104 H12 **Alqueva, Barragem do** ▣ Portugal/Spain
138 G8 **Al Qunayṭirah** *var.* El Kuneitra, El Quneitra, Kuneitra, Qunaytra. Al Qunayṭirah, SW Syria *33°08′N 35°49′E*
138 G8 **Al Qunayṭirah** *off.* Muḥāfaẓat al Qunayṭirah, *var.* El Q'unayṭirah, Qunaytirah, *Fr.* Kuneitra. ◈ *governorate* SW Syria
140 M11 **Al Qunfudhah** Makkah, SW Saudi Arabia *19°19′N 41°03′E*
140 K2 **Al Qurayyāt** Al Jawf, NW Saudi Arabia *31°25′N 37°26′E*
139 Y11 **Al Qurnah** *var.* Kurna. Al Baṣrah, SE Iraq *31°01′N 47°27′E*
75 Y10 **Al Quşayr** *var.* Al Quşayr *var.* Quşair, Quseir. E Egypt *26°05′N 34°16′E*
139 V12 **Al Quşayr** Al Muthannā, S Iraq *30°36′N 45°52′E*
138 I6 **Al Quşayr** *var.* El Quseir, Quşayr, *Fr.* Kousseir. Ḥimṣ, W Syria *34°36′N 36°36′E*
Al Quşayr *see* Al Quşayr
138 H7 **Al Quṭayfah** *var.* Quṭayfah, Qutayfe, Quteife, *Fr.* Kouteifé. Rif Dimashq, W Syria *33°44′N 36°33′E*
141 P8 **Al Quwayʻīyah** Ar Riyāḍ, C Saudi Arabia *24°06′N 45°18′E*
Al Quwayr *see* Guwēr
138 F14 **Al Quwayrah** *var.* El Quweira. Al 'Aqabah, SW Jordan *29°47′N 35°18′E*
Al Rayyan *see* Ar Rayyān
Al Ruweis *see* Ar Ruways
95 G24 **Als** *Ger.* Alsen. *island* SW Denmark
103 U5 **Alsace** *Ger.* Elsass; *anc.* Alsatia. ◈ *region* NE France
11 R16 **Alsask** Saskatchewan, S Canada *51°24′N 109°55′W*
Alsasua *see* Altsasu
Alsatia *see* Alsace
101 C16 **Alsdorf** Nordrhein-Westfalen, W Germany *50°52′N 06°09′E*
10 G8 **Alsek** ≈ Canada/USA
Alsen *see* Als
101 F19 **Alsenz** ≈ W Germany
101 H17 **Alsfeld** Hessen, C Germany *50°45′N 09°14′E*
119 K20 **Al'shany** *Rus.* Ol'shany. Brestskaya Voblasts', SW Belarus *52°05′N 27°21′E*
Alsókubin *see* Dolný Kubín
118 C9 **Alsunga** W Latvia *56°59′N 21°31′E*
Alt *see* Olt
92 K9 **Alta** *Fin.* Alattio. Finnmark, N Norway *69°58′N 23°17′E*
29 T12 **Alta** Iowa, C USA *42°40′N 95°17′W*
108 I7 **Altach** Vorarlberg, W Austria *47°22′N 09°39′E*
92 K9 **Altaelva** *Lapp.* Álaheaieatnu. ≈ N Norway
92 J8 **Altafjorden** *fjord* NE Norwegian Sea
62 K10 **Alta Gracia** Córdoba, C Argentina *31°42′S 64°25′W*
42 K11 **Alta Gracia** Rivas, SW Nicaragua *11°35′N 85°38′W*
54 H4 **Altagracia** Zulia, NW Venezuela *10°44′N 71°30′W*
54 M5 **Altagracia de Orituco** Guárico, N Venezuela *09°54′N 66°24′W*
Altai *see* Altai Mountains
129 T7 **Altai Mountains** *var.* Altai, *Chin.* Altay Shan, *Rus.* Altay. ▲▲ Asia/Europe
23 V6 **Altamaha River** ≈ Georgia, SE USA
58 J13 **Altamira** Pará, NE Brazil *03°13′S 52°15′W*
54 D12 **Altamira** Huila, S Colombia *02°04′N 75°47′W*
42 M13 **Altamira** Alajuela, N Costa Rica *10°25′N 84°21′W*
41 Q11 **Altamira** Tamaulipas, C Mexico *22°25′N 97°55′W*
30 L15 **Altamont** Illinois, N USA *39°03′N 88°45′W*
27 Q7 **Altamont** Kansas, C USA *37°11′N 95°18′W*
32 H16 **Altamont** Oregon, NW USA *42°12′N 121°44′W*
20 K10 **Altamont** Tennessee, S USA *35°28′N 85°42′W*
23 X11 **Altamonte Springs** Florida, SE USA *28°39′N 81°22′W*
107 O17 **Altamura** *anc.* Lupatia. Puglia, SE Italy *40°50′N 16°33′E*
40 H9 **Altamura, Isla** *island* C Mexico
Altan *see* Erdenehayrhan
Altanbulag *see* Bayanhayrhan
163 Q7 **Altan Emel** *var.* Xin Barag Youqi. Nei Mongol Zizhiqu, N China *48°37′N 116°40′E*
Altan-Ovoo *see* Tsenher
163 N9 **Altanshiree** *var.* Chamdmanī. Dornigovi, SE Mongolia *45°36′N 110°30′E*
Altanteel *see* Dzereg
162 D5 **Altantsögts** *var.* Tsagaantüngi. Bayan-Ölgiy, NW Mongolia *49°06′N 90°26′E*
40 F3 **Altar** Sonora, NW Mexico *30°41′N 111°53′W*
40 D2 **Altar, Desierto de** *var.* Sonoran Desert. *desert* Mexico/USA *see also* Sonoran Desert
Altar, Desierto de *see* Sonoran Desert
105 Q8 **Alta, Sierra** ▲ N Spain *40°29′N 01°36′W*
40 H9 **Altata** Sinaloa, C Mexico *24°40′N 107°54′W*
42 D4 **Alta Verapaz** *off.* Departamento de Alta Verapaz. ◈ *department* C Guatemala
Alta Verapaz, Departamento de *see* Alta Verapaz
107 L18 **Altavilla Silentia** Campania, S Italy *40°32′N 15°06′E*
21 T7 **Altavista** Virginia, NE USA *37°06′N 79°17′W*
158 L2 **Altay** Xinjiang Uygur Zizhiqu, NW China *47°51′N 88°06′E*

162 D6 **Altay** *var.* Chihertey. Bayan-Ölgiy, W Mongolia *48°10′N 89°35′E*
162 G8 **Altay** *prev.* Yösönbulag. Govĭ-Altay, W Mongolia *46°23′N 96°17′E*
162 E8 **Altay** *var.* Bor-Üdzüür. Hovd, W Mongolia *45°46′N 92°13′E*
Altay *see* Altai Mountains, Asia/Europe
Altay *see* Bayantes, Mongolia
122 J14 **Altay, Respublika** *var.* Gornyy Altay; *prev.* Gorno-Altayskaya Respublika. ◈ *autonomous republic* S Russian Federation
Altay Shan *see* Altai Mountains
123 I13 **Altayskiy Kray** ◈ *territory* S Russian Federation
Altbetsche *see* Bečej
101 L20 **Altdorf** Bayern, SE Germany *49°23′N 11°22′E*
108 G8 **Altdorf** *var.* Altorf. Uri, C Switzerland *46°53′N 08°38′E*
105 T11 **Altea** Valenciana, E Spain *38°37′N 00°03′W*
100 L10 **Alte Elde** ≈ N Germany
101 M16 **Altenburg** Thüringen, E Germany *50°59′N 12°27′E*
Altenburg *see* București, Romania
Altenburg *see* Baia de Criș, Romania
100 P12 **Alte Oder** ≈ NE Germany
104 H10 **Alter do Chão** Portalegre, C Portugal *39°12′N 07°40′W*
92 I10 **Altevatnet** *Lapp.* Álttesjávri. ◎ N Norway
27 V12 **Altheimer** Arkansas, C USA *34°19′N 91°51′W*
109 T9 **Althofen** Kärnten, S Austria *46°52′N 14°27′E*
114 H7 **Altimir** Vratsa, NW Bulgaria *43°33′N 23°48′E*
136 K11 **Altınkaya Barajı** ▣ N Turkey
139 S3 **Altin Köprü** *var.* Altun Kupri. Kirkūk, N Iraq *35°50′N 44°10′E*
Altin Köprü *see* Āltūn Kūbrī
136 E13 **Altıntaş** Kütahya, W Turkey *39°05′N 30°07′E*
57 K18 **Altiplano** *physical region* W South America
Altkanischa *see* Kanjiža
103 U7 **Altkirch** Haut-Rhin, NE France *47°37′N 07°14′E*
Altlublau *see* Stará L'ubovňa
100 L12 **Altmark** *cultural region* N Germany
Altmoldowa *see* Moldova Veche
25 W8 **Alto** Texas, SW USA *31°39′N 95°04′W*
104 H11 **Alto Alentejo** *physical region* S Portugal
59 I19 **Alto Araguaia** Mato Grosso, C Brazil *17°19′S 53°10′W*
58 L12 **Alto Bonito** Pará, NE Brazil *01°48′S 46°18′W*
83 O15 **Alto Molócuè** Zambézia, NE Mozambique *15°38′S 37°42′E*
30 K15 **Alton** Illinois, N USA *38°53′N 90°10′W*
27 W8 **Alton** Missouri, C USA *36°41′N 91°25′W*
11 X17 **Altona** Manitoba, S Canada *49°06′N 97°35′W*
18 E14 **Altoona** Pennsylvania, NE USA *40°32′N 78°23′W*
30 J6 **Altoona** Wisconsin, N USA *44°49′N 91°22′W*
62 N3 **Alto Paraguay** *off.* Departamento del Alto Paraguay. ◈ *department* N Paraguay
Alto Paraguay, Departamento del *see* Alto Paraguay
59 L17 **Alto Paraíso de Goiás** Goiás, S Brazil *14°04′S 47°15′W*
62 P6 **Alto Paraná** *off.* Departamento del Alto Paraná. ◈ *department* E Paraguay
Alto Paraná *see* Paraná
Alto Paraná, Departamento del *see* Alto Paraná
59 L15 **Alto Parnaíba** Maranhão, E Brazil *09°08′S 45°56′W*
56 H13 **Alto Purús, Río** ≈ E Peru
Altorf *see* Altdorf
63 H19 **Alto Río Senguer** *var.* Alto Río Senguerr. Chubut, S Argentina *45°01′S 70°55′W*
Alto Río Senguerr *see* Alto Río Senguer
41 Q13 **Altotonga** Veracruz-Llave, E Mexico *19°46′N 97°14′W*
101 N23 **Altötting** Bayern, SE Germany *48°12′N 12°37′E*
Altpasua *see* Stara Pazova
Altraga *see* Bayandzürh
105 P3 **Altsasu** *Cast.* Alsasua. Navarra, N Spain *42°54′N 02°10′W*
Alt-Schwanenburg *see* Gulbene
Altsohl *see* Zvolen
108 I7 **Altstätten** Sankt Gallen, NE Switzerland *47°22′N 09°33′E*
Álttesjávri *see* Altevatnet
42 G1 **Altun Ha** *ruins* Belize, N Belize
139 T4 **Āltūn Kūbrī** *prev.* Kirkūk, *var.* Altin Köprü, Karkūk, Kerkuk. Kirkūk, N Iraq *35°28′N 44°26′E*
Altun Kupri *see* Altin Köprü
158 D8 **Altun Shan** ▲ C China *39°19′N 93°37′E*
158 L9 **Altun Shan** *var.* Altyn Tagh. ▲▲ NW China
35 P2 **Alturas** California, W USA *41°28′N 120°32′W*
26 K12 **Altus** Oklahoma, C USA *34°39′N 99°21′W*
26 K11 **Altus Lake** ▣ Oklahoma, C USA
Altvater *see* Praděd
Altyn Tagh *see* Altun Shan
al-'Ubaila *see* Al 'Ubaylah
139 O6 **Al 'Ubaydī** Al Anbār, W Iraq *34°22′N 41°15′E*
141 T9 **Al 'Ubaylah** *var.* al-'Ubaila. Ash Sharqīyah, E Saudi Arabia *22°02′N 50°57′E*
141 T9 **Al 'Ubaylah** *spring/well* E Saudi Arabia *22°02′N 50°56′E*
Al Ubayyiḍ *see* El Obeid
141 T7 **Al 'Udayd** *var.* Al Odaid. Abū Ẓaby, W United Arab Emirates *24°34′N 51°27′E*
118 J8 **Alūksne** *Ger.* Marienburg. NE Latvia *57°26′N 27°02′E*

140 K6 **Al 'Ulā** Al Madīnah, NW Saudi Arabia *26°39′N 37°55′E*
173 N4 **Alula-Fartak Trench** *var.* Illaue Fartak Trench. *undersea feature* W Indian Ocean *14°04′N 51°47′E*
138 I11 **Al 'Umarī** 'Ammān, E Jordan *37°30′N 31°30′E*
31 S13 **Alum Creek Lake** ▣ Ohio, N USA
63 H15 **Aluminé** Neuquén, C Argentina *39°15′S 71°00′W*
95 O14 **Alunda** Uppsala, C Sweden *60°04′N 18°04′E*
117 T14 **Alupka** Avtonomna Respublika Krym, S Ukraine *44°24′N 34°01′E*
75 P8 **Al 'Uqaylah** N Libya *30°13′N 19°12′E*
Al Uqṣur *see* Luxor
Al Urdunn *see* Jordan
168 J9 **Alur Panal** *bay* Sumatera, W Indonesia
Alur Setar *see* Alor Setar
141 V10 **Al 'Urūq al Mu'tariḍah** *salt lake* SE Saudi Arabia
139 Q7 **Ālūs** Al Anbār, C Iraq *34°05′N 42°27′E*
117 T13 **Alushta** Avtonomna Respublika Krym, S Ukraine *44°41′N 34°24′E*
151 G22 **Aluva** *var.* Alwaye. Kerala, SW India *10°06′N 76°23′E* *see also* Alwaye
75 N11 **Al 'Uwaynāt** *var.* Al Awaynāt. SW Libya *25°47′N 10°34′E*
139 T6 **Al 'Uẓaym** *var.* Adhaim. Diyālá, E Iraq *34°12′N 44°31′E*
26 L8 **Alva** Oklahoma, C USA *36°48′N 98°40′W*
104 H8 **Alva** ≈ N Portugal
95 J18 **Älvängen** Västra Götaland, S Sweden *57°56′N 12°09′E*
14 F14 **Alvanley** Ontario, S Canada *44°33′N 81°05′W*
41 S14 **Alvarado** Veracruz-Llave, E Mexico *18°47′N 95°45′W*
25 T7 **Alvarado** Texas, SW USA *32°24′N 97°12′W*
58 D13 **Alvarães** Amazonas, NW Brazil *03°13′S 64°53′W*
40 G6 **Álvaro Obregón, Presa** ▣ W Mexico
94 H10 **Alvdal** Hedmark, S Norway *62°07′N 10°39′E*
94 K12 **Älvdalen** Dalarna, C Sweden *61°13′N 14°04′E*
61 E15 **Alvear** Corrientes, NE Argentina *29°05′S 56°35′W*
104 F10 **Alverca do Ribatejo** Lisboa, C Portugal *38°56′N 09°01′W*
95 L20 **Alvesta** Kronoberg, S Sweden *56°52′N 14°34′E*
25 W12 **Alvin** Texas, SW USA *29°25′N 95°14′W*
95 O13 **Älvkarleby** Uppsala, C Sweden *60°34′N 17°30′E*
25 S5 **Alvord** Texas, SW USA *33°22′N 97°39′W*
93 G18 **Älvros** Jämtland, C Sweden *62°04′N 14°30′E*
92 J13 **Älvsbyn** Norrbotten, N Sweden *65°41′N 21°00′E*
142 K12 **Al Wafrā'** SE Kuwait *28°38′N 47°57′E*
140 J6 **Al Wajh** Tabūk, NW Saudi Arabia *26°16′N 36°30′E*
143 N16 **Al Wakrah** *var.* Wakra. C Qatar *25°09′N 51°36′E*
138 M8 **al Walaj, Sha'īb** *dry watercourse* W Iraq
152 I11 **Alwar** Rājasthān, N India *27°32′N 76°35′E*
141 Q5 **Al Wari'ah** Ash Sharqīyah, N Saudi Arabia *27°54′N 47°23′E*
155 G22 **Alwaye** *var.* Aluva. Kerala, SW India *10°06′N 76°23′E* *see also* Aluva
Alxa Zuoqi *see* Bayan Hot
Alx Youqi *see* Ehen Hudag
Al Yaman *see* Yemen
138 G9 **Al Yarmūk** Irbid, N Jordan *32°41′N 35°55′E*
Alyat/Alyaty-Pristan' *see* Ālāt
115 I14 **Alykí** *var.* Aliki. Thásos, N Greece *40°36′N 24°45′E*
119 F14 **Alytus** *Pol.* Olita. Alytus, S Lithuania *54°24′N 24°02′E*
119 F15 **Alytus** ◈ *province* S Lithuania
101 N23 **Alz** ≈ SE Germany
33 Y11 **Alzada** Montana, NW USA *45°00′N 104°24′W*
122 L12 **Alzamay** Irkutskaya Oblast', S Russian Federation *55°33′N 98°36′E*
99 M25 **Alzette** ≈ S Luxembourg
105 S10 **Alzira** *var.* Alcira; *anc.* Saetabicula, Suero. Valenciana, E Spain *39°10′N 00°27′W*
Al Zubair *see* Az Zubayr
181 O8 **Amadeus, Lake** *seasonal lake* Northern Territory, C Australia
81 E15 **Amadi** Western Equatoria, SW South Sudan *05°32′N 30°20′E*
9 R7 **Amadjuak Lake** ◎ Baffin Island, Nunavut, N Canada
95 J23 **Amager** *island* E Denmark
165 N14 **Amagi-san** ▲ Honshū, S Japan *34°51′N 138°57′E*
171 S13 **Amahai** *var.* Masohi. Palau Seram, E Indonesia *03°19′S 128°56′E*
38 M16 **Amak Island** *island* Alaska, USA
164 C15 **Amakusa** *prev.* Hondo. Kumamoto, Shimo-jima, SW Japan *32°28′N 130°12′E*
164 B14 **Amakusa-nada** *gulf* SW Japan
95 J16 **Åmål** Västra Götaland, S Sweden *59°04′N 12°41′E*
54 E8 **Amalfi** Antioquia, N Colombia *06°54′N 75°04′W*
107 L18 **Amalfi** Campania, S Italy *40°37′N 14°35′E*
115 D19 **Amaliáda** *var.* Amaliás. Dytikí Elláda, S Greece *37°48′N 21°21′E*
Amaliás *see* Amaliáda
154 F12 **Amalner** Mahārāshtra, C India *21°03′N 75°04′E*
171 W14 **Amamapare** Papua, E Indonesia *04°51′S 136°44′E*
59 H21 **Amambaí, Serra de** *var.* Cordillera de Amambay, Serra de Amambay. ▲▲ Brazil/Paraguay *see also* Amambay, Cordillera de
62 P4 **Amambay** *off.* Departamento del Amambay. ◈ *department* E Paraguay
62 P5 **Amambay, Cordillera de** *var.* Serra de Amambaí, Serra de Amambay. ▲▲ Brazil/Paraguay *see also* Amambaí, Serra de

Amambay, Departamento del *see* Amambay
Amambay, Serra de *see* Amambaí, Serra de/Amambay, Cordillera de
165 U16 **Amami-guntō** *island group* SW Japan
165 V15 **Amami-Ō-shima** *island* S Japan
186 A5 **Amanab** West Sepik, NW Papua New Guinea *03°38′S 141°16′E*
Amānat al 'Āṣimah *see* Baghdād
106 J13 **Amandola** Marche, C Italy *42°58′N 13°22′E*
107 N21 **Amantea** Calabria, SW Italy *39°06′N 16°05′E*
191 W10 **Amanu** *island* Îles Tuamotu, C French Polynesia
58 J10 **Amapá** Amapá, NE Brazil *02°00′N 50°50′W*
58 J11 **Amapá** *off.* Estado de Amapá; *prev.* Território de Amapá. ◈ *state* NE Brazil
Amapá, Estado de *see* Amapá
42 H8 **Amapala** Valle, S Honduras *13°16′N 87°39′W*
Amapá, Território de *see* Amapá
80 J12 **Āmara** *var.* Amhara. ◈ N Ethiopia
Amara *see* Al 'Amārah
'Amāra, Al *see* Maysān
104 H6 **Amarante** Porto, N Portugal *41°16′N 08°05′W*
166 M5 **Amarapura** Mandalay, C Myanmar (Burma) *21°54′N 96°01′E*
Amardalay *see* Delgertsogt
104 I12 **Amareleja** Beja, S Portugal *38°12′N 07°13′W*
35 V11 **Amargosa Range** ▲▲ California, W USA
25 N2 **Amarillo** Texas, SW USA *35°13′N 101°50′W*
Amarinthos *see* Amárynthos
107 K15 **Amaro, Monte** ▲ C Italy *42°03′N 14°06′E*
115 H18 **Amárynthos** *var.* Amarinthos. Évvoia, C Greece *38°24′N 23°53′E*
Amasia *see* Amasya
136 K12 **Amasya** *anc.* Amasia. Amasya, N Turkey *40°37′N 35°50′E*
136 K11 **Amasya** ◈ *province* N Turkey
42 F4 **Amatique, Bahía de** *bay* Gulf of Honduras, W Caribbean Sea
42 D6 **Amatitlán, Lago de** ◎ S Guatemala
107 J14 **Amatrice** Lazio, C Italy *42°38′N 13°19′E*
190 C8 **Amatuku** *atoll* C Tuvalu
99 J20 **Amay** Liège, E Belgium *50°33′N 05°19′E*
48 F7 **Amazon** *Sp.* Amazonas. ≈ Brazil/Peru
59 C14 **Amazonas** *off.* Estado do Amazonas. ◈ *state* N Brazil
54 G15 **Amazonas** *off.* Comisaria del Amazonas. ◈ *province* SE Colombia
56 C10 **Amazonas** *off.* Departamento de Amazonas. ◈ *department* N Peru
54 M12 **Amazonas** *off.* Territorio Amazonas. ◈ *federal territory* S Venezuela
Amazonas *see* Amazon
Amazonas, Comisaria del *see* Amazonas
Amazonas, Departamento de *see* Amazonas
Amazonas, Estado do *see* Amazonas
Amazonas, Territorio *see* Amazonas
48 F7 **Amazon Basin** *basin* N South America
47 V5 **Amazon Fan** *undersea feature* W Atlantic Ocean *05°00′N 47°30′W*
58 K11 **Amazon, Mouths of the** *delta* NE Brazil
187 R13 **Ambae** *var.* Aoba, Omba. *island* C Vanuatu
152 I9 **Ambāla** Haryāna, NW India *30°19′N 76°49′E*
155 J26 **Ambalangoda** Southern Province, SW Sri Lanka *06°14′N 80°03′E*
155 K26 **Ambalantota** Southern Province, S Sri Lanka *06°07′N 81°01′E*
172 I6 **Ambalavao** Fianarantsoa, C Madagascar *21°50′S 46°56′E*
54 E10 **Ambalema** Tolima, C Colombia *04°49′N 74°48′W*
79 E17 **Ambam** Sud, S Cameroon *02°23′N 11°17′E*
172 J2 **Ambanja** Antsirañana, N Madagascar *13°40′S 48°27′E*
123 T6 **Ambarchik** Respublika Sakha (Yakutiya), NE Russian Federation *69°33′N 162°08′E*
62 K9 **Ambargasta, Salinas de** *salt lake* C Argentina
124 J6 **Ambarnyy** Respublika Kareliya, NW Russian Federation *65°53′N 33°44′E*
56 C7 **Ambato** Tungurahua, C Ecuador *01°18′S 78°39′W*
172 I6 **Ambato Finandrahana** Fianarantsoa, SE Madagascar *20°33′S 46°48′E*
172 I5 **Ambatolampy** Antananarivo, C Madagascar *19°21′S 47°27′E*
172 H4 **Ambatomainty** Mahajanga, W Madagascar *17°40′S 45°39′E*
172 J4 **Ambatondrazaka** Toamasina, C Madagascar *17°49′S 48°28′E*
101 L20 **Amberg** *var.* Amberg in der Oberpfalz. Bayern, SE Germany *49°26′N 11°52′E*
Amberg in der Oberpfalz *see* Amberg
42 H1 **Ambergris Cay** *island* NE Belize
103 S11 **Ambérieu-en-Bugey** Ain, E France *45°57′N 05°21′E*
185 I18 **Amberley** Canterbury, South Island, New Zealand *43°09′S 172°43′E*
103 P11 **Ambert** Puy-de-Dôme, C France *45°33′N 03°45′E*
Ambianum *see* Amiens
76 J11 **Ambidédi** Kayes, SW Mali *14°37′N 11°49′W*
154 M10 **Ambikāpur** Chhattīsgarh, C India *23°09′N 83°12′E*
172 J2 **Ambilobe** Antsirañana, N Madagascar *13°10′S 49°03′E*
39 O7 **Ambler** Alaska, USA *67°05′N 157°51′W*
Amblève *see* Amel
Ambo *see* Hāgere Hiywet
172 I8 **Amboasary** Toliara, S Madagascar *25°01′S 46°23′E*

172 J4 **Ambodifotatra** *var.* Ambodifototra. Toamasina, E Madagascar *16°59′S 49°51′E*
Amboenten *see* Ambunten
172 I5 **Ambohidratrimo** Antananarivo, C Madagascar *18°48′S 47°26′E*
172 I6 **Ambohimahasoa** Fianarantsoa, SE Madagascar *21°07′S 47°13′E*
172 K3 **Ambohitralanana** Antsirañana, NE Madagascar *15°13′S 50°28′E*
Amboina *see* Ambon
102 M8 **Amboise** Indre-et-Loire, C France *47°25′N 01°00′E*
171 S13 **Ambon** *prev.* Amboina, Amboyna. Pulau Ambon, E Indonesia *03°41′S 128°10′E*
171 S13 **Ambon, Pulau** *island* E Indonesia
81 I20 **Amboseli, Lake** ◎ Kenya/Tanzania
172 I6 **Ambositra** Fianarantsoa, SE Madagascar *20°31′S 47°15′E*
172 I8 **Ambovombe** Toliara, S Madagascar *25°10′S 46°06′E*
35 W14 **Amboy** California, W USA *34°33′N 115°44′W*
30 L11 **Amboy** Illinois, N USA *41°42′N 89°19′W*
Amboyna *see* Ambon
Ambracia *see* Árta
Ambre, Cap d' *see* Bobaomby, Tanjona
18 B14 **Ambridge** Pennsylvania, NE USA *40°33′N 80°11′W*
Ambrim *see* Ambrym
82 A11 **Ambriz** Bengo, NW Angola *07°55′S 13°11′E*
Ambrizete *see* N'Zeto
187 R13 **Ambrym** *var.* Ambrim. *island* C Vanuatu
169 T16 **Ambunten** *prev.* Amboenten. Pulau Madura, E Indonesia *06°55′S 113°45′E*
186 B6 **Ambunti** East Sepik, NW Papua New Guinea *04°12′S 142°49′E*
155 I20 **Āmbūr** Tamil Nādu, SE India *12°48′N 78°44′E*
38 E17 **Amchitka Island** *island* Aleutian Islands, Alaska, USA
38 F17 **Amchitka Pass** *strait* Aleutian Islands, Alaska, USA
141 R15 **'Amd** C Yemen *15°10′N 47°58′E*
78 J10 **Am Dam** Sila, E Chad *12°46′N 20°29′E*
171 U16 **Amdassa** Pulau Yamdena, E Indonesia *07°40′S 131°24′E*
125 U1 **Amderma** Nenetskiy Avtonomnyy Okrug, NW Russian Federation *69°45′N 61°36′E*
78 L9 **Am Djarass** Ennedi-Est, NE Chad *17°14′N 21°32′E*
159 N14 **Amdo** Xizang Zizhiqu, W China *32°15′N 91°43′E*
40 K13 **Ameca** Jalisco, SW Mexico *20°34′N 104°03′W*
41 P14 **Amecameca** *var.* Amecameca de Juárez. México, C Mexico *19°08′N 98°48′W*
Amecameca de Juárez *see* Amecameca
139 R1 **Amēdī** *Ar.* Al 'Amādīyah. Dahūk, N Iraq *37°09′N 43°27′E*
61 A20 **Ameghino** Buenos Aires, E Argentina *34°51′S 62°28′W*
99 M21 **Amel** *Fr.* Amblève. Liège, E Belgium *50°20′N 06°13′E*
98 K4 **Ameland** *Fris.* It Amelân. *island* Waddeneilanden, N Netherlands
Amelân, It *see* Ameland
107 H14 **Amelia** Umbria, C Italy *42°33′N 12°26′E*
21 V6 **Amelia Court House** Virginia, NE USA *37°20′N 77°59′W*
23 W8 **Amelia Island** *island* Florida, SE USA
18 L12 **Amenia** New York, NE USA *41°51′N 73°31′W*
America *see* United States of America
65 M21 **America-Antarctica Ridge** *undersea feature* S Atlantic Ocean
America in Miniature *see* Maryland
60 L9 **Americana** São Paulo, S Brazil *22°44′S 47°19′W*
33 Q15 **American Falls** Idaho, NW USA *42°47′N 112°51′W*
33 Q15 **American Falls Reservoir** ▣ Idaho, NW USA
36 L3 **American Fork** Utah, W USA *40°24′N 111°47′W*
192 K16 **American Samoa** ◇ US *unincorporated territory* W Polynesia
23 S6 **Americus** Georgia, SE USA *32°04′N 84°13′W*
98 K12 **Amerongen** Utrecht, C Netherlands *52°00′N 05°30′E*
98 K11 **Amersfoort** Utrecht, C Netherlands *52°09′N 05°23′E*
97 N21 **Amersham** SE England, United Kingdom *51°40′N 00°37′W*
30 I5 **Amery** Wisconsin, N USA *45°18′N 92°20′W*
195 W6 **Amery Ice Shelf** *ice shelf* Antarctica
29 V13 **Ames** Iowa, C USA *42°01′N 93°37′W*
19 P10 **Amesbury** Massachusetts, NE USA *42°51′N 70°55′W*
Amestratus *see* Mistretta
115 F18 **Amfíkleia** *var.* Amfiklia. Stereá Elláda, C Greece *38°38′N 22°35′E*
Amfiklia *see* Amfíkleia
115 D17 **Amfilochía** *var.* Amfilokhía. Dytikí Elláda, C Greece *38°52′N 21°09′E*
Amfilokhía *see* Amfilochía
114 H13 **Amfípoli** *anc.* Amphipolis. *site of ancient city* Kentrikí Makedonía, NE Greece
115 F18 **Ámfissa** Stereá Elláda, C Greece *38°32′N 22°22′E*
123 Q10 **Amga** Respublika Sakha (Yakutiya), NE Russian Federation *60°55′N 131°45′E*
123 Q11 **Amga** ≈ NE Russian Federation
163 R7 **Amgalang** *var.* Xin Barag Zuoqi. Nei Mongol Zizhiqu, N China *48°12′N 118°15′E*
123 V5 **Amguema** ≈ NE Russian Federation
123 S12 **Amgun'** ≈ SE Russian Federation
Amhara *see* Āmara
13 P15 **Amherst** Nova Scotia, SE Canada *45°50′N 64°14′W*
18 M11 **Amherst** Massachusetts, NE USA *42°22′N 72°31′W*

18 D10 **Amherst** New York, NE USA *42°57′N 78°47′W*
24 M4 **Amherst** Texas, SW USA *33°59′N 102°24′W*
21 U6 **Amherst** Virginia, NE USA *37°35′N 79°04′W*
Amherst *see* Kyaikkami
14 C18 **Amherstburg** Ontario, S Canada *42°05′N 83°06′W*
21 Q6 **Amherstdale** West Virginia, NE USA *37°46′N 81°46′W*
14 K15 **Amherst Island** *island* Ontario, SE Canada
Amida *see* Diyarbakır
28 J6 **Amidon** North Dakota, N USA *46°29′N 103°19′W*
103 O3 **Amiens** *anc.* Ambianum, Samarobriva. Somme, N France *49°54′N 02°18′E*
139 P8 **'Āmij, Wādī** *var.* Wadi 'Amīq. *dry watercourse* W Iraq
136 L17 **Amik Ovasi** ◎ S Turkey
76 E9 **Amilcar Cabral** ✈ Sal, NE Cape Verde
Amīlḥayt, Wādī *see* Umm al Ḥayt, Wādī
Amíndaion/Amindeo *see* Amýntaio
155 C21 **Amīndīvi Islands** *island group* Lakshadweep, India, N Indian Ocean
139 U6 **Āmīn Ḥabīb** Diyālá, E Iraq *34°17′N 45°10′E*
83 E20 **Aminuis** Omaheke, E Namibia *23°43′S 19°21′E*
'Amīq, Wadi *see* 'Āmij, Wādī
142 J7 **Amīrābād** Īlām, NW Iran *33°20′N 46°16′E*
Amirante Bank *see* Amirante Ridge
173 N6 **Amirante Basin** *undersea feature* W Indian Ocean *07°00′S 54°00′E*
173 N6 **Amirante Islands** *var.* Amirantes Group. *island group* C Seychelles
173 N7 **Amirante Ridge** *var.* Amirante Bank. *undersea feature* W Indian Ocean *06°00′S 53°10′E*
Amirantes Group *see* Amirante Islands
173 N7 **Amirante Trench** *undersea feature* W Indian Ocean *08°00′S 52°30′E*
11 U13 **Amisk Lake** ◎ Saskatchewan, C Canada
Amistad, Presa de la *see* Amistad Reservoir
25 O12 **Amistad Reservoir** *var.* Presa de la Amistad. ▣ Mexico/USA
Amisus *see* Samsun
22 K8 **Amite** *var.* Amite City. Louisiana, S USA *30°40′N 90°30′W*
Amite City *see* Amite
27 T12 **Amity** Arkansas, C USA *34°15′N 93°27′W*
154 H11 **Amla** *prev.* Amulla. Madhya Pradesh, C India *21°53′N 78°10′E*
38 I17 **Amlia Island** *island* Aleutian Islands, Alaska, USA
97 I18 **Amlwch** NW Wales, United Kingdom *53°25′N 04°23′W*
Ammaia *see* Portalegre
138 H10 **'Ammān** *var.* Amman; *anc.* Philadelphia, *Bibl.* Rabbah Ammon, Rabbath Ammon. ● (Jordan) 'Ammān, NW Jordan *31°57′N 35°56′E*
138 H10 **'Ammān** *off.* Muḥāfaẓat 'Ammān; *prev.* Al 'Āṣimah. ◈ *governorate* NW Jordan
'Ammān, Muḥāfaẓat *see* 'Ammān
93 N14 **Ämmänsaari** Kainuu, E Finland *64°51′N 28°58′E*
92 H13 **Ammarnäs** Västerbotten, N Sweden *65°58′N 16°10′E*
197 O15 **Ammassalik** *var.* Angmagssalik. Sermersooq, S Greenland *65°51′N 37°30′W*
101 K24 **Ammer** ≈ SE Germany
101 K24 **Ammersee** ◎ SE Germany
98 J13 **Ammerzoden** Gelderland, C Netherlands *51°46′N 05°07′E*
Ammóchostos *see* Gazimağusa
Ammóchostos, Kólpos *see* Gazimağusa Körfezi
Amnok-kang *see* Yalu
Amoea *see* Portalegre
Amoentai *see* Amuntai
Amoerang *see* Amurang
143 O4 **Āmol** *var.* Amul. Māzandarān, N Iran *36°31′N 52°24′E*
115 K21 **Amorgós** Amorgós, Kykládes, Greece, Aegean Sea *36°49′N 25°54′E*
115 K22 **Amorgós** *island* Kykládes, Greece, Aegean Sea
23 N3 **Amory** Mississippi, S USA *33°58′N 88°29′W*
12 I13 **Amos** Québec, SE Canada *48°34′N 78°08′W*
95 G15 **Åmot** Buskerud, S Norway *59°52′N 09°55′E*
95 E15 **Åmot** Telemark, S Norway *59°34′N 07°59′E*
95 J15 **Åmotfors** Värmland, C Sweden *59°46′N 12°24′E*
76 L10 **Amourj** Hodh ech Chargui, SE Mauritania *16°04′N 07°12′W*
Amoy *see* Xiamen
172 H7 **Ampanihy** Toliara, SW Madagascar *24°40′S 44°45′E*
155 L25 **Ampara** *var.* Amparai. Eastern Province, E Sri Lanka *07°17′N 81°41′E*
172 J4 **Amparafaravola** Toamasina, E Madagascar *17°33′S 48°13′E*
Amparai *see* Ampara
60 M9 **Amparo** São Paulo, S Brazil *22°40′S 46°49′W*
172 J5 **Ampasimanolotra** Toamasina, E Madagascar *18°49′S 49°04′E*
57 H17 **Ampato, Nevado** ▲ S Peru *15°52′S 71°51′W*
101 L23 **Amper** ≈ SE Germany
64 M9 **Ampère Seamount** *undersea feature* E Atlantic Ocean *35°05′N 13°00′W*
Amphipolis *see* Amfípoli
167 X10 **Amphitrite Group** *Chin.* Xuande Qundao, *Viet.* Nhom An Vinh. *island group* N Paracel Islands
171 T16 **Amplawas** *var.* Emplawas. Pulau Babar, E Indonesia *08°01′S 129°42′E*
105 U7 **Amposta** Cataluña, NE Spain *40°43′N 00°34′E*
15 V7 **Amqui** Québec, SE Canada *48°28′N 67°27′W*
141 O14 **'Amrān** W Yemen *15°39′N 43°59′E*
Amraoti *see* Amrāvati

154 H12 **Amrāvati** *prev.* Amraoti. Mahārāshtra, C India 20°56′N 77°45′E
154 C11 **Amreli** Gujarāt, W India 21°36′N 71°20′E
108 H6 **Amriswil** Thurgau, NE Switzerland 47°33′N 09°18′E
138 H5 **'Amrīt** *ruins* Tarṭūs, W Syria
152 H7 **Amritsar** Punjab, N India 31°38′N 74°55′E
152 J10 **Amroha** Uttar Pradesh, N India 28°54′N 78°29′E
100 G7 **Amrum** *island* NW Germany
93 I15 **Åmsele** Västerbotten, N Sweden 64°31′N 19°24′E
98 I10 **Amstelveen** Noord-Holland, C Netherlands 52°18′N 04°50′E
98 I10 **Amsterdam** ● (Netherlands) Noord-Holland, C Netherlands 52°22′N 04°54′E
18 K10 **Amsterdam** New York, NE USA 42°56′N 74°11′W
173 Q11 **Amsterdam Fracture Zone** *tectonic feature* S Indian Ocean
173 R11 **Amsterdam Island** *island* NE French Southern and Antarctic Territories
109 U4 **Amstetten** Niederösterreich, N Austria 48°08′N 14°52′E
78 J11 **Am Timan** Salamat, SE Chad 11°02′N 20°17′E
146 L12 **Amu-Buxoro Kanali** *var.* Aral-Bukhorskiy Kanal. *canal* C Uzbekistan
139 O1 **'Āmūdah** *var.* Amude. Al Ḥasakah, N Syria 37°06′N 40°56′E
147 O15 **Amu Darya** *Rus.* Amudar'ya, *Taj.* Dar"yoi Amu, *Turkm.* Amyderya, *Uzb.* Amudaryo; *anc.* Oxus. ≈ C Asia
Amu-Dar'ya *see* Amyderýa
Amudar'ya/Amudaryo/Amu, Dar"yoi *see* Amu Darya
Amude *see* 'Āmūdah
140 L3 **'Amūd, Jabal al** ▲ NW Saudi Arabia 30°59′N 39°17′E
38 J17 **Amukta Island** *island* Aleutian Islands, Alaska, USA
38 I17 **Amukta Pass** *strait* Aleutian Islands, Alaska, USA
Amul *see* Āmol
Amulla *see* Amla
197 S10 **Amundsen Basin** *var.* Fram Basin. *undersea basin* Arctic Ocean
195 X3 **Amundsen Bay** *bay* Antarctica
195 P10 **Amundsen Coast** *physical region* Antarctica
193 O14 **Amundsen Plain** *undersea feature* S Pacific Ocean
195 Q9 **Amundsen-Scott** *US research station* Antarctica 89°59′S 10°00′E
194 J11 **Amundsen Sea** *sea* S Pacific Ocean
94 M12 **Amungen** ⊛ C Sweden
169 U13 **Amuntai** *prev.* Amoentai. Borneo, C Indonesia 02°24′S 115°14′E
129 W6 **Amur** *Chin.* Heilong Jiang. ≈ China/Russian Federation
171 Q11 **Amurang** *prev.* Amoerang. Sulawesi, C Indonesia 01°12′N 124°37′E
105 O3 **Amurrio** País Vasco, N Spain 43°03′N 03°00′W
123 S13 **Amursk** Khabarovskiy Kray, SE Russian Federation 50°13′N 136°34′E
123 Q12 **Amurskaya Oblast'** ◈ *province* SE Russian Federation
80 G7 **'Amur, Wadi** ≈ NE Sudan
115 C17 **Amvrakikós Kólpos** *gulf* W Greece
Amvrosiyevka *see* Amvrosiyivka
117 X8 **Amvrosiyivka** *Rus.* Amvrosiyevka. Donets'ka Oblast', SE Ukraine 47°46′N 38°30′E
146 M14 **Amyderýa** *Rus.* Amu-Dar'ya. Lebap Welaýaty, NE Turkmenistan 37°58′N 65°14′E
Amyderya *see* Amu Darya
114 E13 **Amýntaio** *var.* Amindeo; *prev.* Amíndaion. Dytikí Makedonía, N Greece 40°42′N 21°42′E
14 B6 **Amyot** Ontario, S Canada 48°28′N 84°58′W
191 U10 **Anaa** *atoll* Îles Tuamotu, C French Polynesia
Anabanoea *see* Anabanua
171 N14 **Anabanua** *prev.* Anabanoea. Sulawesi, C Indonesia 03°58′S 120°07′E
189 R8 **Anabar** NE Nauru 0°30′S 166°56′E
123 N8 **Anabar** ≈ NE Russian Federation
An Abhainn Mhór *see* Blackwater
55 O6 **Anaco** Anzoátegui, NE Venezuela 09°30′N 64°28′W
33 Q10 **Anaconda** Montana, NW USA 46°09′N 112°56′W
32 H7 **Anacortes** Washington, NW USA 48°30′N 122°36′W
26 M11 **Anadarko** Oklahoma, C USA 35°04′N 98°16′W
114 N12 **Ana Dere** ≈ NW Turkey
104 G8 **Anadia** Aveiro, N Portugal 40°26′N 08°27′W
Anadolu Dağları *see* Doğu Karadeniz Dağları
123 V6 **Anadyr'** Chukotskiy Avtonomnyy Okrug, NE Russian Federation 64°41′N 177°22′E
123 V6 **Anadyr'** ≈ NE Russian Federation
Anadyr, Gulf of *see* Anadyrskiy Zaliv
129 X4 **Anadyrskiy Khrebet** *var.* Chukot Range. ▲▲ NE Russian Federation
123 W6 **Anadyrskiy Zaliv** *Eng.* Gulf of Anadyr. *gulf* NE Russian Federation
115 K22 **Anáfi** *anc.* Anaphe. *island* Kykládes, Greece, Aegean Sea
107 J15 **Anagni** Lazio, C Italy 41°43′N 13°12′E
'Ānah *see* 'Annah
35 T15 **Anaheim** California, W USA 33°50′N 117°54′W
10 L15 **Anahim Lake** British Columbia, SW Canada 52°26′N 125°13′W
38 B8 **Anahola** Kaua'i, Hawai'i, USA, C Pacific Ocean 22°09′N 159°19′W
41 O7 **Anáhuac** Nuevo León, NE Mexico 27°13′N 100°09′W
25 X11 **Anahuac** Texas, SW USA 29°44′N 94°41′W
155 G22 **Anai Mudi** ▲ S India 10°16′N 77°08′E
Anaiza *see* 'Unayzah
155 M15 **Anakāpalle** Andhra Pradesh, E India 17°42′N 83°06′E
191 W15 **Anakena, Playa de** *beach* Easter Island, Chile, E Pacific Ocean
39 Q7 **Anaktuvuk Pass** Alaska, USA 68°08′N 151°44′W
39 Q6 **Anaktuvuk River** ≈ Alaska, USA
172 J3 **Analalava** Mahajanga, NW Madagascar 14°38′S 47°46′E
44 F6 **Ana Maria, Golfo de** *gulf* N Caribbean Sea
Anambas Islands *see* Anambas, Kepulauan
169 N8 **Anambas, Kepulauan** *var.* Anambas Islands. *island group* W Indonesia
77 U17 **Anambra** ◈ *state* SE Nigeria
29 N4 **Anamoose** North Dakota, N USA 47°50′N 100°14′W
29 Y13 **Anamosa** Iowa, C USA 42°06′N 91°17′W
136 H17 **Anamur** İçel, S Turkey 36°06′N 32°49′E
136 H17 **Anamur Burnu** *headland* S Turkey 36°03′N 32°49′E
154 O12 **Ānandapur** *var.* Anandpur. Odisha, E India 21°14′N 86°10′E
Anandpur *see* Ānandapur
155 H18 **Anantapur** Andhra Pradesh, S India 14°41′N 77°36′E
152 H5 **Anantnāg** *var.* Islamabad. Jammu and Kashmir, NW India 33°44′N 75°11′E
Ananyev *see* Anan'yiv
117 O9 **Anan'yiv** *Rus.* Ananyev. Odes'ka Oblast', SW Ukraine 47°43′N 29°51′E
126 J14 **Anapa** Krasnodarskiy Kray, SW Russian Federation 44°55′N 37°20′E
Anaphe *see* Anáfi
59 K18 **Anápolis** Goiás, C Brazil 16°19′S 48°58′W
143 R10 **Anār** Kermān, C Iran 30°49′N 55°18′E
Anár *see* Inari
143 P7 **Anārak** Eṣfahān, C Iran 33°21′N 53°43′E
Anar Dara *see* Anār Darreh
148 J7 **Anār Darreh** *var.* Anar Dara. Farāh, W Afghanistan 32°45′N 61°38′E
Anárjohka *see* Inarijoki
23 X9 **Anastasia Island** *island* Florida, SE USA
188 K7 **Anatahan** *island* C Northern Mariana Islands
128 M6 **Anatolia** *plateau* C Turkey
86 F14 **Anatolian Plate** *tectonic feature* Asia/Europe
114 H13 **Anatolikí Makedonía kai Thráki** *Eng.* Macedonia East and Thrace. ◈ *region* NE Greece
Anatom *see* Aneityum
62 L8 **Añatuya** Santiago del Estero, N Argentina 28°28′S 62°52′W
An Baile Meánach *see* Ballymena
Anbār, Muḥāfa at al *see* Al Anbār
An Bhearú *see* Barrow
An Bhóinn *see* Boyne
An Blascaod Mór *see* Great Blasket Island
An Cabhán *see* Cavan
An Caisleán Nua *see* Newcastle
An Caisleán Riabhach *see* Castlerea, Ireland
An Caisleán Riabhach *see* Castlereagh
56 C13 **Ancash** *off.* Departamento de Ancash. ◈ *department* W Peru
Ancash, Departamento de *see* Ancash
An Cathair *see* Caher
102 J8 **Ancenis** Loire-Atlantique, NW France 47°23′N 01°10′W
An Chanáil Ríoga *see* Royal Canal
An Cheacha *see* Caha Mountains
39 R11 **Anchorage** Alaska, USA 61°13′N 149°52′W
39 R12 **Anchorage** ✈ Alaska, USA 61°08′N 150°00′W
39 Q13 **Anchor Point** Alaska, USA 59°46′N 151°49′W
An Chorr Chríochach *see* Cookstown
65 M24 **Anchorstock Point** *headland* W Tristan da Cunha 37°07′S 12°21′W
An Clár *see* Clare
An Clochán *see* Clifden
An Clochán Liath *see* Dunglow
23 U12 **Anclote Keys** *island group* Florida, SE USA
An Cóbh *see* Cobh
57 J17 **Ancohuma, Nevado de** ▲ W Bolivia 15°51′S 68°33′W
An Comar *see* Comber
57 D14 **Ancón** Lima, W Peru 11°45′S 77°08′W
106 J12 **Ancona** Marche, C Italy 43°38′N 13°30′E
Ancuabe *see* Ancuabi
82 Q13 **Ancuabi** *var.* Ancuabe. Cabo Delgado, NE Mozambique 13°00′S 39°50′E
63 F17 **Ancud** *prev.* San Carlos de Ancud. Los Lagos, S Chile 41°53′S 73°50′W
63 G17 **Ancud, Golfo de** *gulf* S Chile
Ancyra *see* Ankara
163 V8 **Anda** Heilongjiang, NE China 46°25′N 125°20′E
57 G16 **Andahuaylas** Apurímac, S Peru 13°39′S 73°24′W
An Daingean *see* Dingle
153 R15 **Andāl** West Bengal, NE India 23°35′N 87°14′E
94 E9 **Åndalsnes** Møre og Romsdal, S Norway 62°33′N 07°42′E
104 K13 **Andalucía** *Eng.* Andalusia. ◈ *autonomous community* S Spain
23 P7 **Andalusia** Alabama, S USA 31°18′N 86°29′W
Andalusia *see* Andalucía
151 Q21 **Andaman and Nicobar Islands** *var.* Andamans and Nicobars. ◈ *union territory* India, NE Indian Ocean
173 T4 **Andaman Basin** *undersea feature* NE Indian Ocean 10°00′N 94°00′E
151 P19 **Andaman Islands** *island group* India, NE Indian Ocean
Andamans and Nicobars *see* Andaman and Nicobar Islands
173 T4 **Andaman Sea** *sea* NE Indian Ocean
57 K19 **Andamarca** Oruro, C Bolivia 18°46′S 67°31′W
182 H5 **Andamooka** South Australia 30°26′S 137°12′E
141 Y9 **'Andām, Wādī** *seasonal river* NE Oman
172 J3 **Andapa** Antsirañana, NE Madagascar 14°39′S 49°40′E
149 R4 **Andarāb** *var.* Banow. Baghlān, NE Afghanistan 35°36′N 69°18′E
Andarbag *see* Andarbogh
147 S13 **Andarbogh** *Rus.* Andarbag, Anderbak. S Tajikistan 37°51′N 71°45′E
109 Z5 **Andau** Burgenland, E Austria 47°47′N 17°02′E
108 I10 **Andeer** Graubünden, S Switzerland 46°36′N 09°24′E
92 H9 **Andenes** Nordland, C Norway 69°18′N 16°10′E
99 J20 **Andenne** Namur, SE Belgium 50°29′N 05°06′E
77 S11 **Andéramboukane** Gao, E Mali 15°24′N 03°03′E
Anderbak *see* Andarbogh
99 G18 **Anderlecht** Brussels, C Belgium 50°50′N 04°18′E
99 G21 **Anderlues** Hainaut, S Belgium 50°24′N 04°16′E
108 G9 **Andermatt** Uri, C Switzerland 46°39′N 08°36′E
101 E17 **Andernach** *anc.* Antunnacum. Rheinland-Pfalz, SW Germany 50°26′N 07°24′E
188 D15 **Andersen Air Force Base** *air base* NE Guam 13°34′N 144°55′E
39 R9 **Anderson** Alaska, USA 64°20′N 149°11′W
35 N4 **Anderson** California, W USA 40°26′N 122°21′W
31 P13 **Anderson** Indiana, N USA 40°06′N 85°40′W
27 R8 **Anderson** Missouri, C USA 36°39′N 94°26′W
21 P11 **Anderson** South Carolina, SE USA 34°30′N 82°39′W
25 V10 **Anderson** Texas, SW USA 30°29′N 96°00′W
95 K20 **Anderstorp** Jönköping, S Sweden 57°17′N 13°38′E
54 D9 **Andes** Antioquia, W Colombia 05°40′N 75°56′W
47 P7 **Andes** ▲▲ W South America
29 P12 **Andes, Lake** ⊛ South Dakota, N USA
92 H9 **Andfjorden** *fjord* E Norwegian Sea
155 H16 **Andhra Pradesh** ◈ *state* E India
98 J8 **Andijk** Noord-Holland, NW Netherlands 52°38′N 05°00′E
147 S10 **Andijon** *Rus.* Andizhan. Andijon Viloyati, E Uzbekistan 40°46′N 72°19′E
147 S10 **Andijon Viloyati** *Rus.* Andizhanskaya Oblast'. ◈ *province* E Uzbekistan
Andikíthira *see* Antikýthira
172 J4 **Andilamena** Toamasina, C Madagascar 17°00′S 48°35′E
142 L8 **Andīmeshk** *var.* Andimishk; *prev.* Salehābād. Khūzestān, SW Iran 32°30′N 48°26′E
Andimishk *see* Andīmeshk
Andipáros *see* Antíparos
Andipaxi *see* Antípaxoi
Andípsara *see* Antípsara
136 L16 **Andırın** Kahramanmaraş, S Turkey 37°33′N 36°18′E
158 J8 **Andirlangar** Xinjiang Uygur Zizhiqu, NW China 37°38′N 83°40′E
Andírrion *see* Antírrio
Ándissa *see* Ántissa
Andizhan *see* Andijon
Andizhanskaya Oblast' *see* Andijon Viloyati
149 N2 **Andkhvōy** *prev.* Andkhvoy. Fāryāb, N Afghanistan 36°56′N 65°08′E
105 Q2 **Andoain** País Vasco, N Spain 43°13′N 02°02′W
163 Y15 **Andong** *Jap.* Antō. E South Korea 36°34′N 128°44′E
109 R4 **Andorf** Oberösterreich, N Austria 48°22′N 13°33′E
105 S7 **Andorra** Aragón, NE Spain 40°59′N 00°27′W
105 V4 **Andorra** *off.* Principality of Andorra, *Cat.* Valls d'Andorra, *Fr.* Vallée d'Andorre. ◆ *monarchy* SW Europe
Andorra *see* Andorra la Vella
105 V4 **Andorra la Vella** *var.* Andorra, *Fr.* Andorre la Vielle, *Sp.* Andorra la Vieja. ● (Andorra) C Andorra 42°30′N 01°30′E
Andorra la Vieja *see* Andorra la Vella
Andorra, Principality of *see* Andorra
Andorra, Valls d'/Andorra, Vallée d' *see* Andorra
Andorre la Vielle *see* Andorra la Vella
97 M22 **Andover** S England, United Kingdom 51°13′N 01°28′W
27 N6 **Andover** Kansas, C USA 37°42′N 97°08′W
92 G10 **Andøya** *island* C Norway
60 I8 **Andradina** São Paulo, S Brazil 20°54′S 51°19′W
105 X9 **Andratx** Mallorca, Spain, W Mediterranean Sea 39°35′N 02°25′E
39 N10 **Andreafsky River** ≈ Alaska, USA
38 H17 **Andreanof Islands** *island group* Aleutian Islands, Alaska, USA
124 H16 **Andreapol'** Tverskaya Oblast', W Russian Federation 56°38′N 32°17′E
Andreas, Cape *see* Zafer Burnu
Andreevka *see* Kabanbay
21 N10 **Andrews** North Carolina, SE USA 35°19′N 84°01′W
21 T13 **Andrews** South Carolina, SE USA 33°27′N 79°33′W
24 M7 **Andrews** Texas, SW USA 32°19′N 102°34′W
173 N5 **Andrew Tablemount** *var.* Gora Andryu. *undersea feature* W Indian Ocean 06°45′N 50°30′E
Andreyevka *see* Kabanbay
107 N17 **Andria** Puglia, SE Italy 41°13′N 16°17′E
113 K16 **Andrijevica** E Montenegro 42°45′N 19°45′E
115 E20 **Andrítsaina** Pelopónnisos, S Greece 37°29′N 21°52′E
An Droichead Nua *see* Newbridge
Andropov *see* Rybinsk
115 J19 **Ándros** Ándros, Kykládes, Greece, Aegean Sea 37°49′N 24°54′E
115 J20 **Ándros** *island* Kykládes, Greece, Aegean Sea
19 O7 **Androscoggin River** ≈ Maine/New Hampshire, NE USA
44 F3 **Andros Island** *island* NW The Bahamas
127 R7 **Androsovka** Samarskaya Oblast', W Russian Federation 52°41′N 49°34′E
44 G3 **Andros Town** Andros Island, NW The Bahamas 24°40′N 77°47′W
155 D21 **Āndrott Island** *island* Lakshadweep, India, N Indian Ocean
117 N5 **Andrushivka** Zhytomyrs'ka Oblast', N Ukraine 50°01′N 29°02′E
111 K17 **Andrychów** Małopolskie, S Poland 49°51′N 19°18′E
Andryu, Gora *see* Andrew Tablemount
92 I10 **Andselv** Troms, N Norway 69°05′N 18°30′E
79 O17 **Andudu** Orientale, NE Dem. Rep. Congo 02°26′N 28°39′E
105 N13 **Andújar** *anc.* Illiturgis. Andalucía, SW Spain 38°02′N 04°03′W
82 C12 **Andulo** Bié, W Angola 11°29′S 16°43′E
103 Q14 **Anduze** Gard, S France 44°03′N 03°59′E
An Earagail *see* Errigal Mountain
95 L19 **Aneby** Jönköping, S Sweden 57°50′N 14°45′E
Anécho *see* Aného
77 Q9 **Anéfis** Kidal, NE Mali 18°05′N 00°38′E
45 U8 **Anegada** *island* NE British Virgin Islands
61 B25 **Anegada, Bahía** *bay* E Argentina
45 U9 **Anegada Passage** *passage* Anguilla/British Virgin Islands
77 R17 **Aného** *var.* Anécho; *prev.* Petit-Popo. S Togo 06°14′N 01°36′E
197 D17 **Aneityum** *var.* Anatom; *prev.* Kéamu. *island* S Vanuatu
117 N10 **Anenii Noi** *Rus.* Novyye Aneny. C Moldova 46°52′N 29°10′E
186 F7 **Anepmete** New Britain, E Papua New Guinea 05°47′S 148°37′E
105 U4 **Aneto** ▲ NE Spain 42°36′N 00°37′E
146 F13 **Änew** *Rus.* Annau. Ahal Welaýaty, C Turkmenistan 37°51′N 58°22′E
Änewetak *see* Enewetak Atoll
77 Y8 **Aney** Agadez, NE Niger 19°22′N 13°00′E
An Fheoir *see* Nore
122 L12 **Angara** ≈ C Russian Federation
122 M13 **Angarsk** Irkutskaya Oblast', S Russian Federation 52°31′N 103°55′E
Angaur *see* Ngeaur
93 G17 **Ånge** Västernorrland, C Sweden 62°31′N 15°40′E
Angel *see* Úhlava
40 D4 **Ángel de la Guarda, Isla** *island* NW Mexico
171 O3 **Angeles** *off.* Angeles City. Luzon, N Philippines 15°16′N 120°37′E
Angeles City *see* Angeles
Angel Falls *see* Ángel, Salto
95 J22 **Ängelholm** Skåne, S Sweden 56°14′N 12°52′E
61 A17 **Angélica** Santa Fe, C Argentina 31°33′S 61°33′W
25 W8 **Angelina River** ≈ Texas, SW USA
55 Q9 **Ángel, Salto** *Eng.* Angel Falls. *waterfall* E Venezuela
95 M15 **Ängelsberg** Västmanland, C Sweden 59°57′N 16°01′E
35 P8 **Angels Camp** California, W USA 38°03′N 120°31′W
109 W7 **Anger** Steiermark, SE Austria 47°16′N 15°41′E
Angerapp *see* Ozersk
Angerburg *see* Węgorzewo
93 H15 **Ångermanälven** ≈ N Sweden
100 P11 **Angermünde** Brandenburg, NE Germany 53°02′N 13°59′E
102 K7 **Angers** *anc.* Juliomagus. Maine-et-Loire, NW France 47°30′N 00°33′W
15 W7 **Angers** ≈ Québec, SE Canada
93 J16 **Ängesön** *island* N Sweden
Angistro *see* Ágkistro
114 H13 **Angítis** ≈ NE Greece
167 R13 **Ângk Tasaôm** *prev.* Angtassom. Takêv, S Cambodia
185 C25 **Anglem, Mount** ▲ Stewart Island, Southland, SW New Zealand 46°44′S 167°56′E
97 I18 **Anglesey** *cultural region* NW Wales, United Kingdom
97 I18 **Anglesey** *island* NW Wales, United Kingdom
102 I15 **Anglet** Pyrénées-Atlantiques, SW France 43°29′N 01°30′W
25 W12 **Angleton** Texas, SW USA 29°10′N 95°27′W
Anglia *see* England
14 H9 **Angliers** Québec, SE Canada 47°33′N 79°17′W
Anglo-Egyptian Sudan *see* Sudan
Angmagssalik *see* Ammassalik
167 Q7 **Ang Nam Ngum** ⊛ C Laos
79 N16 **Ango** Orientale, N Dem. Rep. Congo 04°01′N 25°52′E
83 Q15 **Angoche** Nampula, E Mozambique 16°10′S 39°58′E
63 G14 **Angol** Araucanía, C Chile 37°47′S 72°45′W
31 Q11 **Angola** Indiana, N USA 41°37′N 85°00′W
82 A9 **Angola** *off.* Republic of Angola; *prev.* People's Republic of Angola, Portuguese West Africa. ◆ *republic* SW Africa
65 P15 **Angola Basin** *undersea feature* E Atlantic Ocean 15°00′S 03°00′E
Angola, People's Republic of *see* Angola
Angola, Republic of *see* Angola
39 X13 **Angoon** Admiralty Island, Alaska, USA 57°33′N 134°30′W
147 O14 **Angor** Surkhondaryo Viloyati, S Uzbekistan 37°30′N 67°06′E
Angora *see* Ankara
186 C6 **Angoram** East Sepik, NW Papua New Guinea 04°04′S 144°04′E
40 H8 **Angostura** Sinaloa, C Mexico 25°18′N 108°10′W
Angostura *see* Ciudad Bolívar
41 U17 **Angostura, Presa de la** ⊠ SE Mexico
28 J11 **Angostura Reservoir** ⊠ South Dakota, N USA
102 L11 **Angoulême** *anc.* Iculisma. Charente, W France 45°39′N 00°10′E
102 K11 **Angoumois** *cultural region* W France
64 O2 **Angra do Heroísmo** Terceira, Azores, Portugal, NE Atlantic Ocean 38°40′N 27°14′W
60 O10 **Angra dos Reis** Rio de Janeiro, SE Brazil 22°59′S 44°17′W
Angra Pequena *see* Lüderitz
147 Q10 **Angren** Toshkent Viloyati, E Uzbekistan 41°05′N 70°18′E
Angtassom *see* Ângk Tasaôm
167 O10 **Ang Thong** *var.* Angthong. Ang Thong, C Thailand 14°35′N 100°25′E
Angthong *see* Ang Thong
79 M16 **Angu** Orientale, N Dem. Rep. Congo 03°38′N 24°14′E
105 S5 **Angües** Aragón, NE Spain 42°07′N 00°10′W
45 U9 **Anguilla** ◇ *UK dependent territory* E West Indies
45 V9 **Anguilla** *island* E West Indies
44 F4 **Anguilla Cays** *islets* SW The Bahamas
Angul *see* Anugul
161 N1 **Anguli Nur** ⊛ E China
79 O18 **Angumu** Orientale, E Dem. Rep. Congo 0°10′S 27°42′E
14 G14 **Angus** Ontario, S Canada 44°19′N 79°52′W
96 J10 **Angus** *cultural region* E Scotland, United Kingdom
59 K19 **Anhangüera** Goiás, S Brazil 18°12′S 48°19′W
99 I21 **Anhée** Namur, S Belgium 50°18′N 04°52′E
95 I21 **Anholt** *island* C Denmark
160 M11 **Anhua** *var.* Dongping. Hunan, S China 28°25′N 111°10′E
161 P8 **Anhui** *var.* Anhui Sheng, Anhwei, Wan. ◈ *province* E China
AnhuiSheng/Anhwei Wan *see* Anhui
39 O11 **Aniak** Alaska, USA 61°34′N 159°31′W
39 O12 **Aniak River** ≈ Alaska, USA
An Iarmhí *see* Westmeath
189 R8 **Anibare** E Nauru 0°31′S 166°56′E
189 R8 **Anibare Bay** *bay* E Nauru, W Pacific Ocean
Anicium *see* le Puy
115 K22 **Ánidro** *island* Kykládes, Greece, Aegean Sea
77 R15 **Anié** C Togo 07°48′N 01°12′E
77 Q15 **Anié** ≈ C Togo
102 J16 **Anie, Pic d'** ▲ SW France 42°56′N 00°44′W
127 Y7 **Anikhovka** Orenburgskaya Oblast', W Russian Federation 51°27′N 60°17′E
14 G9 **Anima Nipissing Lake** ⊛ Ontario, S Canada
37 O16 **Animas** New Mexico, SW USA 31°55′N 108°49′W
37 P16 **Animas Peak** ▲ New Mexico, SW USA 31°34′N 108°46′W
37 P16 **Animas Valley** *valley* New Mexico, SW USA
116 F13 **Anina** *Ger.* Steierdorf, *Hung.* Stájerlakanina; *prev.* Ştaierdorf-Anina, Steierdorf-Anina, Steyerlak-Anina. Caraş-Severin, SW Romania 45°05′N 21°51′E
29 U14 **Anita** Iowa, C USA 41°27′N 94°45′W
123 U14 **Aniva, Mys** *headland* Ostrov Sakhalin, SE Russian Federation 46°02′N 143°25′E
187 S15 **Aniwa** *island* S Vanuatu
93 M19 **Anjalankoski** Kymenlaakso, S Finland 60°39′N 26°54′E
'Anjar *see* Aanjar
Anjiangying *see* Luanping
14 B8 **Anjigami Lake** ⊛ Ontario, S Canada
149 N13 **Anjīra** Baluchistān, SW Pakistan 28°19′N 66°19′E
164 K14 **Anjō** *var.* Anzyô. Aichi, Honshū, SW Japan 34°56′N 137°05′E
102 J8 **Anjou** *cultural region* NW France
Anjouan *see* Nzwani
172 J4 **Anjozorobe** Antananarivo, C Madagascar 18°22′S 47°52′E
163 W13 **Anju** W North Korea 39°36′N 125°44′E
98 M5 **Anjum** *Fris.* Eanjum. Fryslân, N Netherlands 53°22′N 06°09′E
172 G6 **Ankaboa, Tanjona** *prev./Fr.* Cap Saint-Vincent. *headland* W Madagascar 21°57′S 43°16′E
160 L7 **Ankang** *prev.* Xing'an. Shaanxi, C China 32°45′N 109°00′E
136 I12 **Ankara** *prev.* Angora; *anc.* Ancyra. ● (Turkey) Ankara, C Turkey 39°55′N 32°50′E
136 H12 **Ankara** ◈ *province* C Turkey
95 N19 **Ankarsrum** Kalmar, S Sweden 57°40′N 16°19′E
172 H6 **Ankazoabo** Toliara, SW Madagascar 22°18′S 44°30′E
172 I4 **Ankazobe** Antananarivo, C Madagascar 18°20′S 47°07′E
29 V14 **Ankeny** Iowa, C USA 41°43′N 93°37′W
167 V11 **An Khê** Gia Lai, C Vietnam 13°57′N 108°39′E
100 O9 **Anklam** Mecklenburg-Vorpommern, NE Germany 53°51′N 13°42′E
80 K13 **Ānkober** Āmara, N Ethiopia 09°36′N 39°44′E
77 O17 **Ankobra** ≈ S Ghana
79 N22 **Ankoro** Katanga, SE Dem. Rep. Congo 06°45′S 26°58′E
99 L24 **Anlier, Forêt d'** *forest* SE Belgium
160 I13 **Anlong** *var.* Xin'an. Guizhou, S China 25°05′N 105°29′E
An Longfort *see* Longford
167 R11 **Ânlong Vêng** Siĕmréab, NW Cambodia 14°16′N 104°08′E
An Lorgain *see* Lurgan
161 N8 **Anlu** Hubei, C China 31°15′N 113°41′E
An Mhí *see* Meath
An Mhuir Cheilteach *see* Celtic Sea
An Muileann gCearr *see* Mullingar
93 F16 **Ånn** Jämtland, C Sweden 63°19′N 12°34′E
126 M8 **Anna** Voronezhskaya Oblast', W Russian Federation 51°31′N 40°23′E
30 L17 **Anna** Illinois, N USA 37°27′N 89°15′W
25 U5 **Anna** Texas, SW USA 33°21′N 96°33′W
74 L5 **Annaba** *prev.* Bône. NE Algeria 36°55′N 07°47′E
An Nabatīyah at Taḥtā *see* Nabatîyé
101 N17 **Annaberg-Buchholz** Sachsen, E Germany 50°35′N 13°01′E
109 T9 **Annabichl** ✈ (Klagenfurt) Kärnten, S Austria 46°39′N 14°21′E
140 M5 **An Nafūd** *desert* NW Saudi Arabia
139 P6 **'Annah** *var.* 'Ānah. Al Anbār, NW Iraq 34°50′N 42°00′E
139 P6 **An Nāḥiyah** Al Anbār, W Iraq 34°24′N 41°33′E
139 T10 **An Najaf** *off.* Muḥāfa at an Najaf, *var.* Najaf. An Najaf, S Iraq 31°59′N 44°19′E
21 V5 **Anna, Lake** ⊠ Virginia, NE USA
97 F16 **Annalee** ≈ N Ireland
167 S9 **Annamite Mountains** *var.* annamsescordillera, *Fr.* Chaîne Annamitique, *Lao.* Phou Louang. ▲▲ C Laos
Annamitique, Chaîne *see* Annamite Mountains
Annamses Cordillera *see* Annamite Mountains
97 J14 **Annan** S Scotland, United Kingdom 55°N 03°20′W
29 U8 **Annandale** Minnesota, N USA 45°15′N 94°07′W
21 W4 **Annandale** Virginia, NE USA 38°48′N 77°10′W
189 Q7 **Anna Point** *headland* N Nauru 0°30′S 166°56′E
21 X3 **Annapolis** *state capital* Maryland, NE USA 38°59′N 76°30′W
188 A10 **Anna, Pulo** *island* S Palau
153 O10 **Annapurṇa** ▲ C Nepal 28°30′N 83°50′E
An Nāqūrah *see* En Nâqoûra
31 R10 **Ann Arbor** Michigan, N USA 42°17′N 83°45′W
An Nás *see* Naas
139 W12 **An Nāṣirīyah** *var.* Nasiriya. Dhī Qār, SE Iraq 31°04′N 46°17′E
139 W11 **An Naṣr** Dhī Qār, E Iraq 31°34′N 46°08′E
Annau *see* Änew
121 O13 **An Nawfalīyah** *var.* Al Nūwfalīyah. N Libya 30°46′N 17°48′E
19 P10 **Ann, Cape** *headland* Massachusetts, NE USA 42°39′N 70°35′W
180 I10 **Annean, Lake** ⊛ Western Australia
Anneciacum *see* Annecy
103 T11 **Annecy** *anc.* Anneciacum. Haute-Savoie, E France 45°53′N 06°09′E
103 T11 **Annecy, Lac d'** ⊛ E France
103 T10 **Annemasse** Haute-Savoie, E France 46°10′N 06°13′E
39 Z14 **Annette Island** *island* Alexander Archipelago, Alaska, USA
An Nhon *see* Binh Đinh
An Nīl al Abyaḍ *see* White Nile
An Nīl al Azraq *see* Blue Nile
23 Q3 **Anniston** Alabama, S USA 33°39′N 85°49′W
79 A19 **Annobón** *island* W Equatorial Guinea
103 R12 **Annonay** Ardèche, E France 45°15′N 04°40′E
44 K12 **Annotto Bay** C Jamaica 18°16′N 76°47′W
141 R5 **An Nu'ayrīyah** *var.* Narıya. Ash Sharqīyah, NE Saudi Arabia 27°30′N 48°30′E
182 M9 **Annuello** Victoria, SE Australia 34°54′S 142°50′E
139 Q10 **An Nukhayb** Al Anbār, S Iraq 32°02′N 42°15′E
139 U9 **An Nu'mānīyah** Wāsiṭ, E Iraq 32°34′N 45°23′E
Áno Arkhánai *see* Archánes
115 J25 **Anógeia** *var.* Anogia, Anóyia. Kríti, Greece, E Mediterranean Sea 35°17′N 24°53′E
Anogia *see* Anógeia
29 V8 **Anoka** Minnesota, N USA 45°15′N 93°26′W
An Ómaigh *see* Omagh
172 I1 **Anorontany, Tanjona** *Fr.* Cap Saint-Sébastien. *headland* N Madagascar
172 J5 **Anosibe An'Ala** Toamasina, E Madagascar 19°24′S 48°11′E
Anóyia *see* Anógeia
An Pointe *see* Warrenpoint
161 P9 **Anqing** Anhui, E China 30°32′N 116°59′E
161 Q5 **Anqiu** Shandong, E China 36°25′N 119°10′E
An Ráth *see* Ráth Luirc
An Ribhéar *see* Kenmare River
An Ros *see* Rush
99 K19 **Ans** Liège, E Belgium 50°39′N 05°32′E
Anṣāb *see* Niṣāb
171 W12 **Ansas** Papua, E Indonesia 01°44′S 135°52′E
101 J20 **Ansbach** Bayern, SE Germany 49°18′N 10°36′E
An Sciobairín *see* Skibbereen
An Scoil *see* Skull
An Seancheann *see* Old Head of Kinsale
45 Y5 **Anse-Bertrand** Grande Terre, N Guadeloupe 16°28′N 61°31′W
172 H17 **Anse Boileau** Mahé, NE Seychelles 04°43′S 55°29′E
45 S11 **Anse La Raye** NW Saint Lucia 13°57′N 61°01′W
54 D9 **Anserma** Caldas, W Colombia 05°15′N 75°47′W
109 T4 **Ansfelden** Oberösterreich, N Austria 48°12′N 14°17′E
163 U12 **Anshan** Liaoning, NE China 41°06′N 122°55′E
160 J12 **Anshun** Guizhou, S China 26°15′N 105°58′E
61 F17 **Ansina** Tacuarembó, C Uruguay 31°58′S 55°28′W
29 O15 **Ansley** Nebraska, C USA 41°16′N 99°22′W
25 P6 **Anson** Texas, SW USA 32°45′N 99°55′W
77 Q10 **Ansongo** Gao, E Mali 15°39′N 00°33′E
An Srath Bán *see* Strabane
21 R5 **Ansted** West Virginia, NE USA 38°08′N 81°06′W
171 Y13 **Ansudu** Papua, E Indonesia 02°09′S 139°19′E
57 G15 **Anta** Cusco, S Peru 13°30′S 72°08′W
57 G16 **Antabamba** Apurímac, C Peru 14°23′S 72°54′W
Antafalva *see* Kovačica
136 L17 **Antakya** *anc.* Antioch, Antiochia. Hatay, S Turkey 36°12′N 36°10′E
172 K3 **Antalaha** Antsirañana, NE Madagascar 14°53′S 50°16′E
136 F17 **Antalya** *prev.* Adalia; *anc.* Attaleia, *Bibl.* Attalia. Antalya, SW Turkey 36°53′N 30°42′E
136 F17 **Antalya** ◈ *province* SW Turkey
136 F16 **Antalya** ✈ Antalya, SW Turkey 36°53′N 30°45′E
121 U10 **Antalya Basin** *undersea feature* E Mediterranean Sea
Antalya, Gulf of *see* Antalya Körfezi
136 F16 **Antalya Körfezi** *var.* Gulf of Adalia, *Eng.* Gulf of Antalya. *gulf* SW Turkey
172 J5 **Antanambao Manampotsy** Toamasina, E Madagascar 19°30′S 48°36′E
172 I5 **Antananarivo** *prev.* Tananarive. ● (Madagascar) Antananarivo, C Madagascar 18°52′S 47°30′E
172 I4 **Antananarivo** ◈ *province* C Madagascar
172 J5 **Antananarivo** ✈ Antananarivo, C Madagascar 18°52′S 47°30′E
An tAonach *see* Nenagh
194-195 **Antarctica** *continent*
194 I5 **Antarctic Peninsula** *peninsula* Antarctica
61 J15 **Antas, Rio das** ≈ S Brazil
189 U16 **Ant Atoll** *atoll* Caroline Islands, E Micronesia
An Teampall Mór *see* Templemore
Antep *see* Gaziantep
104 M15 **Antequera** *anc.* Anticaria, Antiquaria. Andalucía, S Spain 37°01′N 04°34′W
Antequera *see* Oaxaca
37 S5 **Antero Reservoir** ⊠ Colorado, C USA
26 M7 **Anthony** Kansas, C USA 37°10′N 98°02′W
37 R16 **Anthony** New Mexico, SW USA 32°00′N 106°36′W
182 D5 **Anthony, Lake** *salt lake* South Australia
74 E8 **Anti-Atlas** ▲▲ SW Morocco
103 U15 **Antibes** *anc.* Antipolis. Alpes-Maritimes, SE France 43°35′N 07°07′E
103 U15 **Antibes, Cap d'** *headland* SE France 43°33′N 07°08′E
Anticaria *see* Antequera
13 Q11 **Anticosti, Île d'** *Eng.* Anticosti Island. *island* Québec, E Canada
Anticosti Island *see* Anticosti, Île d'
102 K3 **Antifer, Cap d'** *headland* N France 49°43′N 00°10′E
30 L6 **Antigo** Wisconsin, N USA 45°10′N 89°10′W
13 Q15 **Antigonish** Nova Scotia, SE Canada 45°39′N 62°00′W
64 P11 **Antigua** Fuerteventura, Islas Canarias, NE Atlantic Ocean
45 X10 **Antigua** *island* S Antigua and Barbuda, Leeward Islands
Antigua *see* Antigua Guatemala
45 W9 **Antigua and Barbuda** ◆ *commonwealth republic* E West Indies
42 C6 **Antigua Guatemala** *var.* Antigua. Sacatepéquez, SW Guatemala 14°33′N 90°42′W
41 P11 **Antiguo Morelos** *var.* Antiguo-Morelos. Tamaulipas, C Mexico 22°35′N 99°08′W
115 F19 **Antíkyras, Kólpos** *gulf* C Greece
115 G24 **Antikýthira** *var.* Andikíthira. *island* S Greece
138 I7 **Anti-Lebanon** *var.* Jebel esh Sharqi, *Ar.* Al Jabal ash Sharqī, *Fr.* Anti-Liban. ▲▲ Lebanon/Syria
Anti-Liban *see* Anti-Lebanon
115 M22 **Antimácheia** Kos, Dodekánisa, Greece 47°19′N 07°54′E
115 I22 **Antímilos** *island* Kykládes, Greece, Aegean Sea
36 L6 **Antimony** Utah, W USA 38°07′N 112°00′W
An tInbhear Mór *see* Arklow
30 M10 **Antioch** Illinois, N USA 42°28′N 88°06′W
Antioch *see* Antakya
102 I10 **Antioche, Pertuis d'** *inlet* W France
Antiochia *see* Antakya
54 D8 **Antioquia** Antioquia, C Colombia 06°36′N 75°53′W
54 E8 **Antioquia** *off.* Departamento de Antioquia. ◈ *province* C Colombia
Antioquia, Departamento de *see* Antioquia
115 J21 **Antíparos** *var.* Andipáros. *island* Kykládes, Greece, Aegean Sea
115 B17 **Antípaxoi** *var.* Andipaxi. *island* Iónia Nísiá, Greece, C Mediterranean Sea
122 J8 **Antipayuta** Yamalo-Nenetskiy Avtonomnyy Okrug, N Russian Federation 69°08′N 76°43′E
192 L12 **Antipodes Islands** *island group* S New Zealand
Antipolis *see* Antibes
115 J18 **Antípsara** *var.* Andípsara. *island* E Greece
Antiquaria *see* Antequera
15 N10 **Antique, Lac** ⊛ Québec, SE Canada
115 E18 **Antírrio** *var.* Andírrion. Dytikí Elláda, C Greece 38°20′N 21°46′E
115 K16 **Ántissa** *var.* Ándissa. Lésvos, E Greece 39°15′N 26°00′E
An tIúr *see* Newry

◆ Country
● Country Capital
◇ Dependent Territory
○ Dependent Territory Capital
◈ Administrative Regions
✈ International Airport
▲ Mountain
▲▲ Mountain Range
Volcano
≈ River
⊛ Lake
⊠ Reservoir

Antivari *see* Bar
56 *C6* **Antizana** ▲ N Ecuador *0°29′S 78°08′W*
27 *Q13* **Antlers** Oklahoma, C USA *34°15′N 95°38′W*
93 *J14* **Antnäs** Norrbotten, N Sweden *65°32′N 21°53′E*
Antō *see* Andong
62 *G5* **Antofagasta** Antofagasta, N Chile *23°40′S 70°23′W*
62 *G6* **Antofagasta** ***off.*** Región de Antofagasta. ◈ *region* N Chile
Antofagasta, Región de *see* Antofagasta
62 *I7* **Antofalla, Salar de** *salt lake* NW Argentina
99 *D20* **Antoing** Hainaut, SW Belgium *50°34′N 03°26′E*
43 *S16* **Antón** Coclé, C Panama *08°23′N 80°15′W*
24 *M5* **Anton** Texas, SW USA *33°48′N 102°09′W*
37 *T11* **Anton Chico** New Mexico, SW USA *35°12′N 105°09′W*
60 *K12* **Antonina** Paraná, S Brazil *25°28′S 48°43′W*
188 *C16* **Antonio B. Won Pat International** ✈ (Agana) C Guam *13°28′N 144°48′E*
103 *O5* **Antony** Hauts-de-Seine, N France *48°45′N 02°17′E*
Antratsit *see* Antratsyt
117 *Y8* **Antratsyt** ***Rus.*** Antratsit. Luhans'ka Oblast', E Ukraine *48°07′N 39°05′E*
97 *G15* **Antrim** ***Ir.*** Aontroim. NE Northern Ireland, United Kingdom *54°43′N 06°13′W*
97 *G14* **Antrim** ***Ir.*** Aontroim. *cultural region* NE Northern Ireland, United Kingdom
97 *G14* **Antrim Mountains** ⛰ NE Northern Ireland, United Kingdom
172 *H5* **Antsalova** Mahajanga, W Madagascar *18°40′S 44°37′E*
Antserana *see* Antsiraňana
An tSionainn *see* Shannon
172 *J2* **Antsiraňana** ***var.*** Antserana; ***prev.*** Antsirane, Diégo-Suarez. Antsiraňana, N Madagascar *12°19′S 49°17′E*
172 *J2* **Antsiraňana** ◈ *province* N Madagascar
Antsirane *see* Antsiraňana
An tSiúir *see* Suir
118 *I7* **Antsla** ***Ger.*** Anzen. Võrumaa, SE Estonia *57°52′N 26°33′E*
An tSláine *see* Slaney
172 *J3* **Antsohihy** Mahajanga, NW Madagascar *14°50′S 47°58′E*
63 *G14* **Antuco, Volcán** 🌋 C Chile *37°29′S 71°25′W*
169 *W10* **Antu, Gunung** ▲ Borneo, N Indonesia *0°57′N 118°51′E*
An Tullach *see* Tullow
An-tung *see* Dandong
Antunnacum *see* Andernach
Antwerp *see* Antwerpen
99 *G16* **Antwerpen** ***Eng.*** Antwerp, ***Fr.*** Anvers. Antwerpen, N Belgium *51°13′N 04°25′E*
99 *H16* **Antwerpen** ***Eng.*** Antwerp. ◈ *province* N Belgium
An Uaimh *see* Navan
154 *N12* **Anugul** ***var.*** Angul. Odisha, E India *20°51′N 84°59′E*
152 *F9* **Anūpgarh** Rājasthān, NW India *29°10′N 73°14′E*
154 *K10* **Anūppur** Madhya Pradesh, C India *23°05′N 81°45′E*
155 *K24* **Anuradhapura** North Central Province, C Sri Lanka *08°20′N 80°25′E*
Anvers *see* Antwerpen
194 *G4* **Anvers Island** *island* Antarctica
39 *N11* **Anvik** Alaska, USA *62°39′N 160°12′W*
39 *N10* **Anvik River** ≈ Alaska, USA
38 *F17* **Anvil Peak** ▲ Semisopochnoi Island, Alaska, USA *51°59′N 179°36′E*
An Vinh, Nhom *see* Amphitrite Group
159 *P7* **Anxi** ***var.*** Yuanquan. Gansu, N China *40°32′N 95°50′E*
182 *F8* **Anxious Bay** *bay* South Australia
161 *O5* **Anyang** Henan, C China *36°11′N 114°18′E*
159 *S11* **A'nyêmaqên Shan** ⛰ C China
118 *H12* **Anykščiai** Utena, E Lithuania *55°30′N 25°34′E*
161 *P13* **Anyuan** ***var.*** Xinshan. Jiangxi, S China *25°10′N 115°25′E*
123 *T7* **Anyuysk** Chukotskiy Avtonomnyy Okrug, NE Russian Federation *68°22′N 161°33′E*
123 *T7* **Anyuyskiy Khrebet** ⛰ NE Russian Federation
54 *D8* **Anzá** Antioquia, C Colombia *06°18′N 75°54′W*
Anzen *see* Antsla
107 *I16* **Anzio** Lazio, C Italy *41°28′N 12°38′E*
55 *O6* **Anzoátegui** ***off.*** Estado Anzoátegui. ◈ *state* NE Venezuela
Anzoátegui, Estado *see* Anzoátegui
147 *P12* **Anzob** W Tajikistan *39°24′N 68°55′E*
Anzyô *see* Anjō
Aoba *see* Ambae
165 *X13* **Aoga-shima** *island* Izu-shotō, SE Japan
Aohan Qi *see* Xinhui
105 *R3* **Aoiz** ***Bas.*** Agoitz, ***var.*** Agoiz. Navarra, N Spain *42°47′N 01°23′W*
167 *O11* **Ao Krung Thep** ***var.*** Krung Thep Mahanakhon, ***Eng.*** Bangkok. ● (Thailand) Bangkok, C Thailand *13°44′N 100°30′E*
186 *M9* **Aola** ***var.*** Tenaghau. Guadalcanal, C Solomon Islands *09°32′S 160°28′E*
166 *M15* **Ao Luk Nua** Krabi, SW Thailand *08°21′N 98°43′E*
Aomen Tebie Xingzhengqu *see* Macao
172 *N8* **Aomori** Aomori, Honshū, C Japan *40°50′N 140°43′E*
172 *N8* **Aomori** ***off.*** Aomori-ken. ◈ *prefecture* Honshū, C Japan
Aomori-ken *see* Aomori
Aontroim *see* Antrim
115 *C15* **Aóos** ***var.*** Vijosa, Vijosë, ***Alb.*** Lumi i Vjosës. ≈ Albania/Greece ***see also*** Vjosës, Lumi i
Aóos *see* Vjosës, Lumi i
191 *Q7* **Aorai, Mont** 🌋 Tahiti, W French Polynesia *17°36′S 149°29′W*
185 *E19* **Aoraki** ***prev.*** Aorangi, Mount Cook. ▲ South Island, New Zealand *43°38′S 170°05′E*
167 *R13* **Aôral, Phnum** ***prev.*** Phnom Aural. ▲ W Cambodia *12°01′N 104°10′E*
Aorangi *see* Aoraki
185 *L15* **Aorangi Mountains** ⛰ North Island, New Zealand
184 *H13* **Aorere** ≈ South Island, New Zealand
106 *A7* **Aosta** ***anc.*** Augusta Praetoria. Valle d'Aosta, NW Italy *45°43′N 07°20′E*
77 *O11* **Aougoundou, Lac** ◎ S Mali
76 *K9* **Aoukâr** ***var.*** Aouker. *plateau* C Mauritania
78 *J13* **Aouk, Bahr** ≈ Central African Republic/Chad
Aouker *see* Aoukâr
74 *B11* **Aousard** SE Western Sahara *22°42′N 14°22′W*
164 *H12* **Aoya** Tottori, Honshū, SW Japan *35°31′N 134°01′E*
Aoyang *see* Shanggao
78 *H5* **Aozou** Tibesti, N Chad *22°01′N 17°11′E*
26 *M11* **Apache** Oklahoma, C USA *34°57′N 98°21′W*
36 *L14* **Apache Junction** Arizona, SW USA *33°25′N 111°33′W*
24 *J9* **Apache Mountains** ⛰ Texas, SW USA
36 *M16* **Apache Peak** ▲ Arizona, SW USA *31°50′N 110°25′W*
116 *H10* **Apahida** Cluj, NW Romania *46°49′N 23°45′E*
23 *T9* **Apalachee Bay** *bay* Florida, SE USA
23 *T3* **Apalachee River** ≈ Georgia, SE USA
23 *S10* **Apalachicola** Florida, SE USA *29°43′N 84°58′W*
23 *S10* **Apalachicola Bay** *bay* Florida, SE USA
23 *R9* **Apalachicola River** ≈ Florida, SE USA
Apam *see* Apan
Apamama *see* Abemama
41 *P14* **Apan** ***var.*** Apam. Hidalgo, C Mexico *19°48′N 98°25′W*
42 *J8* **Apanás, Lago de** ◎ NW Nicaragua
54 *H14* **Apaporis, Río** ≈ Brazil/Colombia
185 *C23* **Aparima** ≈ South Island, New Zealand
171 *O1* **Aparri** Luzon, N Philippines *18°16′N 121°42′E*
112 *J9* **Apatin** Vojvodina, NW Serbia *45°40′N 19°01′E*
124 *J4* **Apatity** Murmanskaya Oblast', NW Russian Federation *67°34′N 33°26′E*
55 *X9* **Apatou** NW French Guiana *05°10′N 54°22′W*
40 *M14* **Apatzingán** ***var.*** Apatzingán de la Constitución. Michoacán, SW Mexico *19°05′N 102°20′W*
Apatzingán de la Constitución *see* Apatzingán
171 *X12* **Apauwar** Papua, E Indonesia *01°36′S 138°10′E*
Apaxtla *see* Apaxtla de Castrejón
41 *O15* **Apaxtla de Castrejón** ***var.*** Apaxtla. Guerrero, S Mexico *18°06′N 99°55′W*
118 *J7* **Ape** NE Latvia *57°32′N 26°42′E*
98 *L11* **Apeldoorn** Gelderland, E Netherlands *52°13′N 05°57′E*
Apennines *see* Appennino
Apenrade *see* Aabenraa
57 *L17* **Apere, Río** ≈ C Bolivia
55 *W11* **Apetina** Sipaliwini, SE Suriname *03°30′N 55°03′W*
21 *U9* **Apex** North Carolina, SE USA *35°43′N 78°51′W*
79 *M16* **Api** Orientale, N Dem. Rep. Congo *03°40′N 25°26′E*
152 *M9* **Api** ▲ NW Nepal *30°07′N 80°57′E*
192 *H16* **Āpia** ● (Samoa) Upolu, SE Samoa *13°50′S 171°47′W*
Apia *see* Abaiang
60 *K11* **Apiaí** São Paulo, S Brazil *24°29′S 48°51′W*
170 *M16* **Api, Gunung** ▲ Pulau Sangeang, S Indonesia *08°09′S 119°03′E*
187 *N9* **Apio** Maramasike Island, N Solomon Islands *09°36′S 161°25′E*
41 *O15* **Apipilulco** Guerrero, S Mexico *18°11′N 99°40′W*
41 *P14* **Apizaco** Tlaxcala, S Mexico *19°26′N 98°09′W*
137 *Q8* **Apkhazeti** ***var.*** Abkhazia; ***prev.*** Ap'khazet'i. ◈ *autonomous republic* NW Georgia
Ap'khazet'i *see* Apkhazeti
104 *I4* **A Pobla de Trives** ***Cast.*** Puebla de Trives. Galicia, NW Spain *42°21′N 07°16′W*
55 *U9* **Apoera** Sipaliwini, NW Suriname *05°12′N 57°13′W*
115 *O23* **Apolakkiá** Ródos, Dodekánisa, Greece, Aegean Sea *36°02′N 27°48′E*
101 *L16* **Apolda** Thüringen, C Germany *51°02′N 11°31′E*
192 *H16* **Apolima Strait** *strait* C Pacific Ocean
182 *M13* **Apollo Bay** Victoria, SE Australia *38°40′S 143°44′E*
Apollonia *see* Sozopol
57 *J16* **Apolo** La Paz, W Bolivia *14°48′S 68°31′W*
57 *J16* **Apolobamba, Cordillera** ⛰ Bolivia/Peru
171 *Q8* **Apo, Mount** 🌋 Mindanao, S Philippines *06°54′N 125°16′E*
23 *W11* **Apopka** Florida, SE USA *28°40′N 81°30′W*
23 *W11* **Apopka, Lake** ◎ Florida, SE USA
59 *J19* **Aporé, Rio** ≈ SW Brazil
30 *K2* **Apostle Islands** *island group* Wisconsin, N USA
Apostolas Andreas, Cape *see* Zafer Burnu
61 *F14* **Apóstoles** Misiones, NE Argentina *27°54′S 55°45′W*
Apostólou Andréa, Akrotíri *see* Zafer Burnu
117 *S9* **Apostolove** ***Rus.*** Apostolovo. Dnipropetrovs'ka Oblast', E Ukraine *47°40′N 33°45′E*
Apostolovo *see* Apostolove
17 *S10* **Appalachian Mountains** ⛰ E USA
95 *K14* **Äppelbo** Dalarna, C Sweden *60°30′N 14°00′E*
98 *N7* **Appelscha** ***Fris.*** Appelskea. Fryslân, N Netherlands *52°57′N 06°19′E*
Appelskea *see* Appelscha
106 *G11* **Appennino** ***Eng.*** Apennines. ⛰ Italy/San Marino
107 *L17* **Appennino Campano** ⛰ C Italy
108 *I7* **Appenzell** Inner-Rhoden, NW Switzerland *47°20′N 09°25′E*
55 *V12* **Appikalo** Sipaliwini, S Suriname *02°07′N 56°16′W*
98 *O5* **Appingedam** Groningen, NE Netherlands *53°18′N 06°52′E*
25 *X8* **Appleby** Texas, SW USA *31°43′N 94°36′W*
97 *L15* **Appleby-in-Westmorland** Cumbria, NW England, United Kingdom *54°35′N 02°26′W*
30 *K10* **Apple River** ≈ Illinois, N USA
30 *I5* **Apple River** ≈ Wisconsin, N USA
25 *W9* **Apple Springs** Texas, SW USA *31°13′N 94°57′W*
29 *S8* **Appleton** Minnesota, N USA *45°12′N 96°01′W*
30 *M7* **Appleton** Wisconsin, N USA *44°17′N 88°24′W*
27 *S5* **Appleton City** Missouri, C USA *38°11′N 94°01′W*
35 *U14* **Apple Valley** California, W USA *34°30′N 117°11′W*
29 *V9* **Apple Valley** Minnesota, N USA *44°43′N 93°13′W*
21 *U6* **Appomattox** Virginia, NE USA *37°21′N 78°51′W*
188 *B16* **Apra Harbor** *harbour* W Guam
188 *B16* **Apra Heights** W Guam
106 *F6* **Aprica, Passo dell'** *pass* N Italy
107 *M15* **Apricena** ***anc.*** Hadria Picena. Puglia, SE Italy *41°47′N 15°27′E*
114 *I9* **Apriltsi** Lovech, N Bulgaria *42°50′N 24°54′E*
126 *L14* **Apsheronsk** Krasnodarskiy Kray, SW Russian Federation *44°27′N 39°45′E*
Apsheronskiy Poluostrov *see* Abşeron Yarımadası
103 *S15* **Apt** ***anc.*** Apta Julia. Vaucluse, SE France *43°54′N 05°24′E*
Apta Julia *see* Apt
38 *H12* **'Āpua Point** ***var.*** Apua Point. *headland* Hawai'i, USA, C Pacific Ocean *19°15′N 155°13′W*
60 *I10* **Apucarana** Paraná, S Brazil *23°34′S 51°28′W*
Apulia *see* Puglia
54 *K8* **Apure** ***off.*** Estado Apure. ◈ *state* C Venezuela
Apure, Estado *see* Apure
54 *J7* **Apure, Río** ≈ W Venezuela
57 *F16* **Apurímac** ***off.*** Departamento de Apurímac. ◈ *department* C Peru
Apurímac, Departamento de *see* Apurímac
57 *F15* **Apurímac, Río** ≈ S Peru
116 *G10* **Apuseni, Munții** ⛰ W Romania
Aqaba/'Aqaba *see* Al 'Aqabah
138 *F15* **Aqaba, Gulf of** ***var.*** Gulf of Elat, ***Ar.*** Khalīj al 'Aqabah; ***anc.*** Sinus Aelaniticus. *gulf* NE Red Sea
139 *R7* **'Aqabah** Al Anbār, C Iraq *33°33′N 42°55′E*
'Aqabah, Khalīj al *see* Aqaba, Gulf of
'Aqabah, Muḩāfa at al *see* Al 'Aqabah
149 *O2* **Āqchah** ***var.*** Āqcheh. Jowzjān, N Afghanistan *37°N 66°07′E*
Āqcheh *see* Āqchah
Aqkengse *see* Akkense
Aqköl *see* Akkol'
Aqmola *see* Astana
Aqmola Oblysy *see* Akmola
158 *L10* **Aqqikkol Hu** ◎ NW China
Aqqū *see* Akku
Aqqystaū *see* Akkystau
'Aqrah *see* Ākrē
Aqsay *see* Aksay
Aqshataū *see* Akshatau
Aqsū *see* Aksu
Aqsūat *see* Aksuat
Aqtas *see* Aktas
Aqtaū *see* Aktau
Aqtöbe *see* Aktobe
Aqtöbe Oblysy *see* Aktyubinsk
Aqtoghay *see* Aktogay
Aquae Augustae *see* Dax
Aquae Calidae *see* Bath
Aquae Flaviae *see* Chaves
Aquae Grani *see* Aachen
Aquae Panoniae *see* Baden
Aquae Sextiae *see* Aix-en-Provence
Aquae Solis *see* Bath
Aquae Tarbelicae *see* Dax
36 *J11* **Aquarius Mountains** ⛰ Arizona, SW USA
62 *O5* **Aquidabán, Río** ≈ E Paraguay
59 *H20* **Aquidauana** Mato Grosso do Sul, S Brazil *20°27′S 55°45′W*
40 *L15* **Aquila** Michoacán, S Mexico *18°36′N 103°32′W*
Aquila/Aquila degli Abruzzi *see* L'Aquila
25 *T8* **Aquilla** Texas, SW USA *31°51′N 97°13′W*
44 *L9* **Aquin** S Haiti *18°16′N 73°24′W*
Aquisgranum *see* Aachen
102 *J13* **Aquitaine** ◈ *region* SW France
Aqzhar *see* Akzhar
153 *P13* **Āra** ***prev.*** Arrah. Bihār, N India *25°34′N 84°40′E*
105 *S4* **Ara** ≈ NE Spain
23 *P2* **Arab** Alabama, S USA *34°19′N 86°30′W*
Araba *see* Álava
138 *G12* **'Arabah, Wādī al** ***Heb.*** Ha'Arava. *dry watercourse* Israel/Jordan
117 *U12* **Arabats'ka Strilka, Kosa** *spit* S Ukraine
117 *U12* **Arabats'ka Zatoka** *gulf* S Ukraine
'Arab, Baḩr al *see* Arab, Bahr el
80 *C12* **Arab, Bahr el** ***var.*** Baḩr al 'Arab. ≈ S Sudan
56 *E7* **Arabela, Río** ≈ N Peru
173 *O4* **Arabian Basin** *undersea feature* N Arabian Sea *11°30′N 65°00′E*
Arabian Desert *see* Sahara el Sharqîya
141 *N9* **Arabian Peninsula** *peninsula* SW Asia
85 *P15* **Arabian Plate** *tectonic feature* Africa/Asia/Europe
141 *W14* **Arabian Sea** *sea* NW Indian Ocean
Arabicus, Sinus *see* Red Sea
'Arabī, Khalīj al *see* Persian Gulf
Arabistan *see* Khūzestān
'Arabīyah as Su'ūdīyah, Al Mamlakah al *see* Saudi Arabia
'Arabīyah Jumhūrīyah, Mişr al *see* Egypt
138 *I9* **'Arab, Jabal al** ⛰ S Syria
Arab Republic of Egypt *see* Egypt
139 *Y12* **'Arab, Shaţţ al** ***Eng.*** Shatt al Arab, ***Per.*** Arvand Rūd. ≈ Iran/Iraq
136 *I11* **Araç** Kastamonu, N Turkey *41°14′N 33°20′E*
59 *P16* **Aracaju** *state capital* Sergipe, E Brazil *10°45′S 37°07′W*
54 *F5* **Aracataca** Magdalena, N Colombia *10°38′N 74°09′W*
58 *P13* **Aracati** Ceará, E Brazil *04°32′S 37°45′W*
60 *J8* **Araçatuba** São Paulo, S Brazil *21°12′S 50°24′W*
136 *I11* **Araç Çayı** ≈ N Turkey
104 *J13* **Aracena** Andalucía, S Spain *37°54′N 06°33′W*
115 *F20* **Arachnaío** ⛰ S Greece
115 *D16* **Árachthos** ***var.*** Arta, ***prev.*** Árakhthos; ***anc.*** Arachthus. ≈ W Greece
Arachthus *see* Árachthos
59 *N19* **Araçuai** Minas Gerais, SE Brazil *16°52′S 42°03′W*
138 *F11* **Arad** Southern, S Israel *31°16′N 35°09′E*
116 *F11* **Arad** Arad, W Romania *46°12′N 21°20′E*
116 *F11* **Arad** ◈ *county* W Romania
78 *J9* **Arada** Wadi Fira, NE Chad *15°00′N 20°38′E*
143 *P18* **'Arādah** Abū Z̧aby, S United Arab Emirates *22°57′N 53°24′E*
Aradhippou *see* Aradíppou
121 *Q3* **Aradíppou** ***var.*** Aradhippou. SE Cyprus *34°57′N 33°37′E*
174 *K6* **Arafura Sea** ***Ind.*** Laut Arafuru. *sea* W Pacific Ocean
174 *L6* **Arafura Shelf** *undersea feature* C Arafura Sea
Arafuru, Laut *see* Arafura Sea
59 *J18* **Aragarças** Goiás, C Brazil *15°55′S 52°12′W*
Aragats, Gora *see* Aragats Lerr
137 *T12* **Aragats Lerr** ***Rus.*** Gora Aragats. ▲ W Armenia *40°31′N 44°06′E*
32 *E14* **Arago, Cape** *headland* Oregon, NW USA *43°17′N 124°25′W*
105 *R6* **Aragón** *autonomous community* E Spain
105 *Q4* **Aragón** ≈ NE Spain
107 *I24* **Aragona** Sicilia, Italy, C Mediterranean Sea *37°25′N 13°37′E*
105 *Q7* **Aragoncillo** ▲ C Spain *40°59′N 02°01′W*
54 *L5* **Aragua** ***off.*** Estado Aragua. ◈ *state* N Venezuela
55 *N6* **Aragua de Barcelona** Anzoátegui, NE Venezuela *09°30′N 64°51′W*
55 *O5* **Aragua de Maturín** Monagas, NE Venezuela *09°58′N 63°30′W*
Aragua, Estado *see* Aragua
59 *K15* **Araguaia, Río** ***var.*** Araguaya. ≈ C Brazil
59 *K19* **Araguari** Minas Gerais, SE Brazil *18°38′S 48°13′W*
58 *J11* **Araguari, Rio** ≈ SW Brazil
Araguaya *see* Araguaia, Río
104 *K14* **Arahal** Andalucía, S Spain *37°15′N 05°33′W*
165 *N11* **Arai** Niigata, Honshū, C Japan *37°02′N 138°17′E*
Árainn *see* Inishmore
Árainn Mhór *see* Aran Island
Ara Jovis *see* Aranjuez
74 *J11* **Arak** C Algeria *25°17′N 03°45′E*
171 *Y15* **Arak** Papua, E Indonesia *07°14′S 139°40′E*
142 *M7* **Arāk** ***prev.*** Sultānābād. Markazī, W Iran *34°07′N 49°39′E*
188 *D10* **Arakabesan** *island* Palau Islands, N Palau
55 *S7* **Arakaka** NW Guyana *07°37′N 59°58′W*
Arakan State *see* Rakhine State
166 *K5* **Arakan Yoma** ⛰ W Myanmar (Burma)
165 *O10* **Arakawa** Niigata, Honshū, C Japan *38°06′N 139°25′E*
Árakhthos *see* Árachthos
Araks/Arak's *see* Aras
158 *H7* **Aral** Xinjiang Uygur Zizhiqu, NW China *40°40′N 81°19′E*
Aral *see* Aralsk, Kazakhstan
Aral *see* Vose', Tajikistan
Aral-Bukhorskiy Kanal *see* Amu-Buxoro Kanali
137 *T12* **Aralik** Iğdır, E Turkey *39°54′N 44°28′E*
146 *H5* **Aral Sea** ***Kaz.*** Aral Tengizi, ***Rus.*** Aral'skoye More, ***Uzb.*** Orol Dengizi. *inland sea* Kazakhstan/Uzbekistan
144 *L13* **Aral'sk** ***Kaz.*** Aral. Kzylorda, SW Kazakhstan *46°48′N 61°40′E*
Aral'skoye More/Aral Tengizi *see* Aral Sea
41 *O10* **Aramberri** Nuevo León, NE Mexico *24°05′N 99°52′W*
186 *B8* **Aramia** ≈ SW Papua New Guinea
105 *N5* **Aranda de Duero** Castilla y León, N Spain *41°40′N 03°41′W*
112 *M12* **Aranđelovac** ***prev.*** Arandjelovac. Serbia, C Serbia *44°18′N 20°32′E*
Arandjelovac *see* Aranđelovac
97 *J19* **Aran Fawddwy** ▲ NW Wales, United Kingdom *52°48′N 03°42′W*
97 *C14* **Aran Island** ***Ir.*** Árainn Mhór. *island* NW Ireland
97 *A18* **Aran Islands** *island group* W Ireland
105 *N9* **Aranjuez** ***anc.*** Ara Jovis. Madrid, C Spain *40°02′N 03°37′W*
83 *E20* **Aranos** Hardap, SE Namibia *24°10′S 19°08′E*
25 *U14* **Aransas Bay** *inlet* Texas, SW USA
25 *T14* **Aransas Pass** Texas, SW USA *27°54′N 97°09′W*
191 *O3* **Aranuka** ***prev.*** Nanouki. *atoll* Tungaru, W Kiribati
143 *N6* **Ārān-va-Bidgol** ***var.*** Golārā. Eşfahān, C Iran *34°03′N 51°30′E*
167 *Q11* **Aranyaprathet** Prachin Buri, S Thailand *13°42′N 102°32′E*
Aranyosasztal *see* Zlatý Stôl
Aranyosgyéres *see* Câmpia Turzii
Aranyosmarót *see* Zlaté Moravce
164 *C14* **Arao** Kumamoto, Kyūshū, SW Japan *32°58′N 130°26′E*
77 *O8* **Araouane** Tombouctou, N Mali *18°53′N 03°31′W*
26 *L10* **Arapaho** Oklahoma, C USA *35°34′N 98°57′W*
29 *N16* **Arapahoe** Nebraska, C USA *40°18′N 99°54′W*
57 *I16* **Arapa, Laguna** ◎ SE Peru
185 *K14* **Arapawa Island** *island* C New Zealand
61 *E17* **Arapey Grande, Río** ≈ N Uruguay
59 *P16* **Arapiraca** Alagoas, E Brazil *09°45′S 36°40′W*
140 *M3* **'Ar'ar** Al Ḩudūd ash Shamālīyah, NW Saudi Arabia *31°N 41°E*
54 *G15* **Araracuara** Caquetá, S Colombia *0°36′S 72°24′W*
61 *K15* **Araranguá** Santa Catarina, S Brazil *28°56′S 49°30′W*
60 *L8* **Araraquara** São Paulo, S Brazil *21°46′S 48°08′W*
59 *O13* **Araras** Ceará, E Brazil *04°08′S 40°30′W*
58 *I14* **Araras** Pará, N Brazil *06°04′S 54°34′W*
60 *L9* **Araras** São Paulo, S Brazil *22°21′S 47°21′W*
60 *H11* **Araras, Serra das** ⛰ S Brazil
137 *U12* **Ararat** S Armenia *39°49′N 44°45′E*
182 *M12* **Ararat** Victoria, SE Australia *37°20′S 143°00′E*
Ararat, Mount *see* Büyükağrı Dağı
140 *M3* **'Ar'ar, Wādī** *dry watercourse* Iraq/Saudi Arabia
129 *N7* **Aras** ***Arm.*** Arak's, ***Az.*** Araz Nehri, ***Per.*** Rūd-e Aras, ***Rus.*** Araks; ***prev.*** Araxes. ≈ SW Asia
Aras de Alpuente *see* Aras de los Olmos
105 *R9* **Aras de los Olmos** ***prev.*** Aras de Alpuente. Valenciana, E Spain *39°55′N 01°09′W*
137 *S13* **Aras Güneyi Dağları** ⛰ NE Turkey
Aras, Rūd-e *see* Aras
191 *U9* **Aratika** *atoll* Îles Tuamotu, C French Polynesia
Aratürük *see* Yiwu
54 *I8* **Arauca** Arauca, NE Colombia *07°03′N 70°47′W*
54 *I8* **Arauca** ***off.*** Intendencia de Arauca. ◈ *province* NE Colombia
Arauca, Intendencia de *see* Arauca
63 *G15* **Araucanía** ***off.*** Región de la Araucanía. ◈ *region* C Chile
Araucanía, Región de la *see* Araucanía
54 *L7* **Arauca, Río** ≈ Colombia/Venezuela
63 *F14* **Arauco** Bío Bío, C Chile *37°15′S 73°22′W*
63 *F14* **Arauco, Golfo de** *gulf* S Chile
54 *H8* **Arauquita** Arauca, C Colombia *06°57′N 71°19′W*
Arausio *see* Orange
152 *F13* **Arāvali Range** ⛰ N India
186 *J7* **Arawa** Bougainville, NE Papua New Guinea *06°15′S 155°35′E*
185 *C20* **Arawata** ≈ South Island, New Zealand
186 *F7* **Arawe Islands** *island group* E Papua New Guinea
59 *L20* **Araxá** Minas Gerais, SE Brazil *19°37′S 46°50′W*
Araxes *see* Aras
55 *O5* **Araya** Sucre, N Venezuela *10°34′N 64°15′W*
Araz Nehri *see* Aras
105 *R4* **Arba** ≈ N Spain
81 *I15* **Ārba Minch'** Southern Nationalities, S Ethiopia *06°02′N 37°34′E*
Arbat *see* Erbet
'Arbat *see* Erbet
107 *D19* **Arbatax** Sardegna, Italy, C Mediterranean Sea *39°57′N 09°42′E*
Arbe *see* Rab
Arbela *see* Arbīl
139 *S3* **Arbīl** ***Kurd.*** Hewlêr, ***var.*** Erbil, Irbīl, Hawlēr; ***anc.*** Arbela. Arbīl, N Iraq *36°12′N 44°01′E*
139 *S3* **Arbīl** ***off.*** Muḩāfa at Arbīl, ***var.*** Irbil, ***off. Kurd.*** Parêzga-î Hewlêr, ***Kurd.*** Hewlêr. ◈ *governorate* N Iraq
Arbīl, Muḩāfa at *see* Arbīl
95 *M16* **Arboga** Västmanland, C Sweden *59°24′N 15°50′E*
95 *M16* **Arbogaån** ≈ C Sweden
103 *S9* **Arbois** Jura, E France *46°54′N 05°45′E*
54 *D6* **Arboletes** Antioquia, NW Colombia *08°52′N 76°25′W*
11 *X15* **Arborg** Manitoba, S Canada *50°52′N 97°20′W*
94 *N12* **Arbrå** Gävleborg, C Sweden *61°27′N 16°21′E*
96 *K10* **Arbroath** ***anc.*** Aberbrothock. E Scotland, United Kingdom *56°34′N 02°35′W*
35 *N6* **Arbuckle** California, W USA *39°00′N 122°05′W*
27 *N12* **Arbuckle Mountains** ⛰ Oklahoma, C USA
162 *I5* **Arbulag** ***var.*** Mandal. Hövsgöl, N Mongolia *49°55′N 99°21′E*
Arbuzinka *see* Arbuzynka
117 *Q8* **Arbuzynka** ***Rus.*** Arbuzinka. Mykolayivs'ka Oblast', S Ukraine *47°54′N 31°19′E*
103 *U12* **Arc** ≈ E France
102 *J13* **Arcachon** Gironde, SW France *44°40′N 01°11′W*
102 *J13* **Arcachon, Bassin d'** *inlet* SW France
18 *E10* **Arcade** New York, NE USA *42°32′N 78°19′W*
23 *W14* **Arcadia** Florida, SE USA *27°13′N 81°51′W*
22 *H5* **Arcadia** Louisiana, S USA *32°33′N 92°55′W*
30 *J7* **Arcadia** Wisconsin, N USA *44°15′N 91°30′W*
Arcae Remorum *see* Châlons-en-Champagne
44 *L9* **Arcahaie** C Haiti *18°46′N 72°32′W*
34 *K3* **Arcata** California, W USA *40°51′N 124°06′W*
35 *U6* **Arc Dome** ▲ Nevada, W USA *38°52′N 117°20′W*
107 *J16* **Arce** Lazio, C Italy *41°35′N 13°34′E*
41 *O15* **Arcelia** Guerrero, S Mexico *18°20′N 100°16′W*
99 *M15* **Arcen** Limburg, SE Netherlands *51°28′N 06°10′E*
115 *J25* **Archánes** ***var.*** Áno Arkhánai, Epáno Archánes; ***prev.*** Epáno Arkhánai. Kríti, Greece, E Mediterranean Sea *35°12′N 25°10′E*
Archangel *see* Arkhangel'sk
Archangel Bay *see* Chëshskaya Guba
115 *O23* **Archángelos** ***var.*** Arhangelos, Arkhángelos. Ródos, Dodekánisa, Greece, Aegean Sea *36°13′N 28°07′E*
114 *F7* **Archar** ≈ NW Bulgaria
31 *R11* **Archbold** Ohio, N USA *41°31′N 84°18′W*
105 *R12* **Archena** Murcia, SE Spain *38°07′N 01°17′W*
25 *R5* **Archer City** Texas, SW USA *33°36′N 98°37′W*
104 *M14* **Archidona** Andalucía, S Spain *37°06′N 04°23′W*
65 *B25* **Arch Islands** *island group* SW Falkland Islands
106 *G13* **Arcidosso** Toscana, C Italy *42°52′N 11°30′E*
103 *Q5* **Arcis-sur-Aube** Aube, N France *48°32′N 04°09′E*
182 *F3* **Arckaringa Creek** *seasonal river* South Australia
106 *G7* **Arco** Trentino-Alto Adige, N Italy *45°53′N 10°52′E*
33 *Q14* **Arco** Idaho, NW USA *43°38′N 113°18′W*
30 *M14* **Arcola** Illinois, N USA *39°39′N 88°19′W*
105 *P6* **Arcos de Jalón** Castilla y León, N Spain *41°12′N 02°13′W*
104 *K15* **Arcos de la Frontera** Andalucía, S Spain *36°45′N 05°49′W*
104 *G5* **Arcos de Valdevez** Viana do Castelo, N Portugal *41°51′N 08°25′W*
59 *P15* **Arcoverde** Pernambuco, E Brazil *08°23′S 37°W*
102 *H5* **Arcovest, Pointe de l'** *headland* NW France *48°49′N 02°58′W*
Arctic Mid Oceanic Ridge *see* Gakkel Ridge
197 *R8* **Arctic Ocean** *ocean*
8 *G7* **Arctic Red River** ≈ Northwest Territories/Yukon, NW Canada
Arctic Red River *see* Tsiigehtchic
39 *S6* **Arctic Village** Alaska, USA *68°07′N 145°32′W*
194 *H1* **Arctowski** *Polish research station* South Shetland Islands, Antarctica *61°57′S 58°23′W*
114 *I12* **Arda** ***var.*** Ardhas, ***Gk.*** Ardas. ≈ Bulgaria/Greece ***see also*** Ardas
142 *L2* **Ardabīl** ***var.*** Ardebil. Ardabīl, NW Iran *38°15′N 48°18′E*
142 *L2* **Ardabīl** ***off.*** Ostān-e Ardabīl. ◈ *province* NW Iran
Ardabīl, Ostān-e *see* Ardabīl
137 *R11* **Ardahan** Ardahan, NE Turkey *41°08′N 42°41′E*
137 *S11* **Ardahan** ◈ *province* Ardahan, Turkey
143 *P8* **Ardakān** Yazd, C Iran *32°20′N 53°59′E*
94 *E12* **Årdalstangen** Sogn Og Fjordane, S Norway *61°14′N 07°45′E*
137 *R11* **Ardanuç** Artvin, NE Turkey *41°07′N 42°04′E*
114 *L12* **Ardas** ***var.*** Ardhas, ***Bul.*** Arda. ≈ Bulgaria/Greece ***see also*** Arda
138 *I13* **Arḑ aş Şawwān** ***var.*** Ardh es Suwwān. *plain* S Jordan
127 *P5* **Ardatov** Respublika Mordoviya, W Russian Federation *54°49′N 46°13′E*
14 *G12* **Ardbeg** Ontario, S Canada *45°38′N 80°05′W*
Ardeal *see* Transylvania
Ardebil *see* Ardabīl
103 *Q13* **Ardèche** ◈ *department* E France
103 *Q13* **Ardèche** ≈ C France
97 *F17* **Ardee** ***Ir.*** Baile Átha Fhirdhia. Louth, NE Ireland *53°52′N 06°33′W*
103 *Q3* **Ardennes** ◈ *department* NE France
99 *J23* **Ardennes** *physical region* Belgium/France
137 *Q11* **Ardeşen** Rize, NE Turkey *41°14′N 41°00′E*
143 *O7* **Ardestān** ***var.*** Ardistan. Eşfahān, C Iran *33°29′N 52°17′E*
108 *J9* **Ardez** Graubünden, SE Switzerland *46°47′N 10°09′E*
Ardhas *see* Arda/Ardas
Ardh es Suwwān *see* Arḑ aş Şawwān
Ardila *see* Ardila, Ribeira de
104 *I12* **Ardila, Ribeira de** ***Sp.*** Ardilla. ≈ Portugal/Spain ***see also*** Ardilla
Ardila, Ribeira de *see* Ardilla
11 *T17* **Ardill** Saskatchewan, S Canada *49°56′N 105°49′W*
104 *I12* **Ardilla** ***Port.*** Ribeira de Ardila. ≈ Portugal/Spain ***see also*** Ardila, Ribeira de
40 *M11* **Ardilla, Cerro la** ▲ C Mexico *22°15′N 102°33′W*
114 *J12* **Ardino** Kardzhali, S Bulgaria *41°38′N 25°22′E*
Ardistan *see* Ardestān
183 *P9* **Ardlethan** New South Wales, SE Australia *34°24′S 146°52′E*
Ard Mhacha *see* Armagh
23 *P1* **Ardmore** Alabama, S USA *34°59′N 86°51′W*
27 *N13* **Ardmore** Oklahoma, C USA *34°11′N 97°08′W*
20 *J10* **Ardmore** Tennessee, S USA *35°00′N 86°48′W*
96 *G10* **Ardnamurchan, Point of** *headland* N Scotland, United Kingdom *56°42′N 06°15′W*
Árdni *see* Arnøya
99 *C17* **Ardooie** West-Vlaanderen, W Belgium *50°59′N 03°10′E*
182 *I9* **Ardrossan** South Australia *34°27′S 137°54′E*
116 *H9* **Ardusat** ***Hung.*** Erdőszáda. Maramureş, N Romania *47°36′N 23°25′E*
93 *F16* **Åre** Jämtland, C Sweden *63°25′N 13°04′E*
79 *P16* **Arebi** Orientale, NE Dem. Rep. Congo *02°47′N 29°35′E*
45 *T5* **Arecibo** C Puerto Rico *18°29′N 66°44′W′N*
171 *V13* **Aredo** Papua Barat, E Indonesia *02°27′S 133°59′E*
59 *P14* **Areia Branca** Rio Grande do Norte, E Brazil *04°53′S 37°03′W*
119 *O14* **Arekhawsk** ***Rus.*** Orekhovsk. Vitsyebskaya Voblasts', N Belarus *54°42′N 30°30′E*
Arel *see* Arlon
Arelas/Arelate *see* Arles
Arenal, Embalse de *see* Arenal Laguna
42 *L12* **Arenal Laguna** ***var.*** Embalse de Arenal. ◎ NW Costa Rica
42 *L13* **Arenal, Volcán** 🌋 NW Costa Rica *10°21′N 84°42′W*
34 *K6* **Arena, Point** *headland* California, W USA *38°57′N 123°44′W*
59 *H17* **Arenápolis** Mato Grosso, W Brazil *14°25′S 56°52′W*
40 *G10* **Arena, Punta** *headland* NW Mexico *23°28′N 109°24′W*
104 *L8* **Arenas de San Pedro** Castilla y León, N Spain *40°12′N 05°05′W*
63 *I24* **Arenas, Punta de** *headland* S Argentina *53°10′S 68°15′W*
61 *B20* **Arenaza** Buenos Aires, E Argentina *34°55′S 61°45′W*
95 *F17* **Arendal** Aust-Agder, S Norway *58°27′N 08°45′E*
99 *J16* **Arendonk** Antwerpen, N Belgium *51°18′N 05°06′E*
43 *T15* **Arenosa** Panamá, N Panama *09°02′N 79°57′W*
Arensburg *see* Kuressaare
105 *W5* **Arenys de Mar** Cataluña, NE Spain *41°35′N 02°33′E*
106 *C9* **Arenzano** Liguria, NW Italy *44°25′N 08°43′E*
115 *F22* **Areópoli** ***prev.*** Areópolis. Pelopónnisos, S Greece *36°40′N 22°24′E*
Areópolis *see* Areópoli
57 *H18* **Arequipa** Arequipa, SE Peru *16°24′S 71°33′W*
57 *G17* **Arequipa** ***off.*** Departamento de Arequipa. ◈ *department* Arequia SW Peru
Arequipa, Departamento de *see* Arequipa
61 *B19* **Arequito** Santa Fe, C Argentina *33°09′S 61°28′W*
104 *M7* **Arévalo** Castilla y León, N Spain *41°04′N 04°44′W*
106 *H12* **Arezzo** ***anc.*** Arretium. Toscana, C Italy *43°28′N 11°50′E*
105 *Q4* **Arga** ≈ N Spain
Argaeus *see* Erciyes Dağı
115 *G17* **Argalastí** Thessalía, C Greece *39°13′N 23°13′E*
105 *O10* **Argamasilla de Alba** Castilla-La Mancha, C Spain *39°08′N 03°05′W*
158 *L8* **Argan** Xinjiang Uygur Zizhiqu, NW China *40°09′N 88°16′E*
105 *O8* **Arganda** Madrid, C Spain *40°19′N 03°26′W*
104 *H8* **Arganil** Coimbra, N Portugal *40°13′N 08°03′W*
171 *P6* **Argao** Cebu, C Philippines *09°52′N 123°33′E*
153 *V15* **Argartala** Tripura, NE India
123 *N9* **Arga-Sala** ≈ Respublika Sakha (Yakutiya),NE Russian Federation
103 *P17* **Argelès-sur-Mer** Pyrénées-Orientales, S France *42°33′N 03°01′E*
103 *T15* **Argens** ≈ SE France
106 *H9* **Argenta** Emilia-Romagna, N Italy *44°37′N 11°49′E*
102 *K5* **Argentan** Orne, N France *48°45′N 00°01′W*
103 *N12* **Argentat** Corrèze, C France *45°06′N 01°57′E*
106 *A9* **Argentera** Piemonte, NE Italy *44°25′N 06°57′E*
103 *N5* **Argenteuil** Val-d'Oise, N France *48°57′N 02°13′E*
62 *K13* **Argentina** ***off.*** Argentine Republic. ◆ *republic* S South America
Argentina Basin *see* Argentine Basin
Argentine Abyssal Plain *see* Argentine Plain
65 *I19* **Argentine Basin** ***var.*** Argentina Basin. *undersea feature* SW Atlantic Ocean *45°00′S 45°00′W*
65 *I20* **Argentine Plain** ***var.*** Argentine Abyssal Plain. *undersea feature* SW Atlantic Ocean *47°31′S 50°00′W*
Argentine Republic *see* Argentina
Argentine Rise *see* Falkland Plateau
63 *H22* **Argentino, Lago** ◎ S Argentina
102 *K8* **Argenton-Château** Deux-Sèvres, W France *46°59′N 00°22′W*
102 *M9* **Argenton-sur-Creuse** Indre, C France *46°34′N 01°32′E*
Argentoratum *see* Strasbourg
116 *I12* **Argeş** ◈ *county* S Romania
116 *K14* **Argeş** ≈ S Romania
149 *O8* **Arghandāb, Daryā-ye** ≈ SE Afghanistan
Arghastān *see* Arghistān
Arghestān *see* Arghistān
149 *O8* **Arghistān** ***Pash.*** Arghastān; ***prev.*** Arghestān. ≈ SE Afghanistan
Argirocastro *see* Gjirokastër
80 *E7* **Argo** Northern, N Sudan *19°31′N 30°25′E*
173 *P7* **Argo Fracture Zone** *tectonic feature* C Indian Ocean
115 *F20* **Argolikós Kólpos** *gulf* S Greece
103 *R4* **Argonne** *physical region* NE France
115 *F20* **Árgos** Pelopónnisos, S Greece *37°38′N 22°43′E*
Argōsh *see* Argush
115 *D14* **Árgos Orestikó** Dytikí Makedonía, N Greece *40°27′N 21°15′E*
115 *B19* **Argostóli** ***var.*** Argostólion. Kefallonía, Iónia Nísiá, Greece, C Mediterranean Sea *38°13′N 20°29′E*
Argostólion *see* Argostóli
Argovie *see* Aargau
35 *O14* **Arguello, Point** *headland* California, W USA *34°34′N 120°39′W*
127 *P16* **Argun** Chechenskaya Respublika, SW Russian Federation *43°16′N 45°53′E*
157 *T2* **Argun** ***Chin.*** Ergun He, ***Rus.*** Argun'. ≈ China/Russian Federation
77 *T12* **Argungu** Kebbi, NW Nigeria *12°45′N 04°24′E*
139 *S1* **Argush** ***Ar.*** Argūsh, ***var.*** Argōsh. Dahūk, N Iraq *37°07′N 44°13′E*
Argūsh *see* Argush
Arguut *see* Guchin-Us

◆ Country
● Country Capital
◇ Dependent Territory
○ Dependent Territory Capital
◈ Administrative Regions
✈ International Airport
▲ Mountain
⛰ Mountain Range
🌋 Volcano
≈ River
◎ Lake
⊠ Reservoir

181 *N3* **Argyle, Lake** *salt lake* Western Australia
96 *G12* **Argyll** *cultural region* W Scotland, United Kingdom
Argyrokastron *see* Gjirokastër
162 *I7* **Arhangay** ⬖ *province* C Mongolia
Arhangelos *see* Archángelos
95 *G22* **Århus** ***var.*** Aarhus. Midtjylland, C Denmark *56°09′N 10°11′E*
139 *T1* **Arî** ***Ar.*** Ārī. Arbīl, E Iraq *37°07′N 44°34′E*
Ārī *see* Arî
Aria *see* Herāt
83 *F22* **Ariamsvlei** Karas, SE Namibia *28°08′S 19°50′E*
107 *L17* **Ariano Irpino** Campania, S Italy *41°08′N 15°00′E*
54 *F11* **Ariari, Río** ≈ C Colombia
151 *K19* **Ari Atoll** ***var.*** Alifu Atoll. *atoll* C Maldives
77 *P11* **Aribinda** N Burkina Faso *14°12′N 00°50′W*
62 *G2* **Arica** ***hist.*** San Marcos de Arica. Arica y Parinacota, N Chile *18°31′S 70°18′W*
54 *H16* **Arica** Amazonas, S Colombia *02°09′S 71°48′W*
62 *G2* **Arica** ✈ Arica y Parinacota, N Chile *18°30′S 70°20′W*
62 *H2* **Arica y Parinacota** ⬖ *region* N Chile
114 *E13* **Aridaía** ***var.*** Aridea, Aridhaía. Dytikí Makedonía, N Greece *40°59′N 22°04′E*
Aridea *see* Aridaía
172 *I15* **Aride, Île** *island* Inner Islands, NE Seychelles
Aridhaía *see* Aridaía
103 *N17* **Ariège** ⬖ *department* S France
102 *M16* **Ariège** ***var.*** la Riege. ≈ Andorra/France
116 *H11* **Arieş** ≈ W Romania
149 *U10* **Ārifwāla** Punjab, E Pakistan *30°15′N 73°08′E*
Ariguaní *see* El Difícil
138 *G11* **Arīhā** Al Karak, W Jordan *31°25′N 35°47′E*
138 *I3* **Arīḩā** ***var.*** Arīhā. Idlib, W Syria *35°50′N 36°36′E*
Arīhā *see* Arīḩā
Arīḩā *see* Jericho
37 *W4* **Arikaree River** ≈ Colorado/Nebraska, C USA
112 *L13* **Arilje** Serbia, W Serbia *43°45′N 20°06′E*
45 *U14* **Arima** Trinidad, Trinidad and Tobago *10°38′N 61°17′W*
Arime *see* Al 'Arīmah
Ariminum *see* Rimini
59 *H16* **Arinos, Rio** ≈ W Brazil
40 *M14* **Ario de Rosales** ***var.*** Ario de Rosáles. Michoacán, SW Mexico *19°12′N 101°42′W*
Ario de Rosáles *see* Ario de Rosales
118 *F12* **Ariogala** Kaunas, C Lithuania *55°16′N 23°30′E*
47 *T7* **Aripuanã** ≈ W Brazil
59 *E15* **Ariquemes** Rondônia, W Brazil *09°55′S 63°06′W*
121 *W13* **'Arīsh, Wādi el** ≈ NE Egypt
54 *K6* **Arismendi** Barinas, C Venezuela *08°29′N 68°22′W*
10 *J14* **Aristazabal Island** *island* SW Canada
60 *F13* **Aristóbulo del Valle** Misiones, NE Argentina *27°09′S 54°54′W*
172 *I5* **Arivonimamo** ✈ (Antananarivo) Antananarivo, C Madagascar *19°00′S 47°11′E*
Arixang *see* Wenquan
105 *Q6* **Ariza** Aragón, NE Spain *41°19′N 02°03′W*
62 *I6* **Arizaro, Salar de** *salt lake* NW Argentina
105 *O2* **Arizgoiti** ***var.*** Basauri. País Vasco, N Spain *43°13′N 02°54′W*
62 *K13* **Arizona** San Luis, C Argentina *35°44′S 65°16′W*
36 *J12* **Arizona** ***off.*** State of Arizona, ***also known as*** Copper State, Grand Canyon State. ⬖ *state* SW USA
40 *G4* **Arizpe** Sonora, NW Mexico *30°20′N 110°11′W*
95 *J16* **Årjäng** Värmland, C Sweden *59°24′N 12°09′E*
143 *P8* **Arjenān** Yazd, C Iran *32°19′N 53°48′E*
92 *I13* **Arjeplog** ***Lapp.*** Árjepluovve. Norrbotten, N Sweden *66°04′N 18°E*
Árjepluovve *see* Arjeplog
54 *E5* **Arjona** Bolívar, N Colombia *10°14′N 75°22′W*
105 *N13* **Arjona** Andalucía, S Spain *37°56′N 04°04′W*
123 *S10* **Arka** Khabarovskiy Kray, E Russian Federation *60°04′N 142°17′E*
22 *L2* **Arkabutla Lake** ▣ Mississippi, S USA
127 *O7* **Arkadak** Saratovskaya Oblast', W Russian Federation *51°55′N 43°29′E*
27 *T13* **Arkadelphia** Arkansas, C USA *34°07′N 93°06′W*
115 *J25* **Arkalochóri** ; ***prev.*** Arkalokhórion. Kríti, Greece, E Mediterranean Sea *35°09′N 25°15′E*
Arkalohori/Arkalokhórion *see* Arkalochóri
145 *O10* **Arkalyk** ***Kaz.*** Arqalyq. Kostanay, N Kazakhstan *50°17′N 66°51′E*
27 *U10* **Arkansas** ***off.*** State of Arkansas, ***also known as*** The Land of Opportunity. ⬖ *state* S USA
27 *W14* **Arkansas City** Arkansas, C USA *33°36′N 91°12′W*
27 *O7* **Arkansas City** Kansas, C USA *37°03′N 97°02′W*
16 *K11* **Arkansas River** ≈ C USA
182 *J5* **Arkaroola** South Australia *30°21′S 139°20′E*
Arkhángelos *see* Archángelos
124 *L8* **Arkhangel'sk** ***Eng.*** Archangel. Arkhangel'skaya Oblast', NW Russian Federation *64°32′N 40°40′E*
124 *L9* **Arkhangel'skaya Oblast'** ⬖ *province* NW Russian Federation
127 *O14* **Arkhangel'skoye** Stavropol'skiy Kray, SW Russian Federation *44°37′N 44°03′E*
123 *R14* **Arkhara** Amurskaya Oblast', S Russian Federation *49°20′N 130°04′E*
97 *G19* **Arklow** ***Ir.*** An tInbhear Mór. SE Ireland *52°48′N 06°09′W*
115 *M20* **Arkoí** *island* Dodekánisa, Greece, Aegean Sea
27 *R11* **Arkoma** Oklahoma, C USA *35°19′N 94°27′W*
100 *O7* **Arkona, Kap** *headland* NE Germany *54°40′N 13°24′E*
95 *N17* **Arkösund** Östergötland, S Sweden *58°28′N 16°55′E*
122 *J6* **Arkticheskogo Instituta, Ostrova** *island* N Russian Federation
95 *O15* **Arlanda** ✈ (Stockholm) Stockholm, C Sweden *59°40′N 17°58′E*
146 *C11* **Arlandag** ***Rus.*** Gora Arlan. ▲ W Turkmenistan *39°39′N 54°28′E*
Arlan, Gora *see* Arlandag
105 *O5* **Arlanza** ≈ N Spain
105 *N5* **Arlanzón** ≈ N Spain
103 *R15* **Arles** ***var.*** Arles-sur-Rhône; ***anc.*** Arelas, Arelate. Bouches-du-Rhône, SE France *43°41′N 04°38′E*
Arles-sur-Rhône *see* Arles
103 *O17* **Arles-sur-Tech** Pyrénées-Orientales, S France *42°27′N 02°37′E*
29 *U9* **Arlington** Minnesota, N USA *44°36′N 94°04′W*
29 *R15* **Arlington** Nebraska, C USA *41°27′N 96°21′W*
32 *J11* **Arlington** Oregon, NW USA *45°43′N 120°10′W*
29 *R10* **Arlington** South Dakota, N USA *44°21′N 97°07′W*
20 *E10* **Arlington** Tennessee, S USA *35°17′N 89°40′W*
25 *T6* **Arlington** Texas, SW USA *32°44′N 97°05′W*
21 *W4* **Arlington** Virginia, NE USA *38°54′N 77°09′W*
32 *H7* **Arlington** Washington, NW USA *48°12′N 122°07′W*
30 *M10* **Arlington Heights** Illinois, N USA *42°08′N 88°03′W*
77 *U8* **Arlit** Agadez, C Niger *18°54′N 07°25′E*
99 *L24* **Arlon** ***Dut.*** Aarlen, ***Ger.*** Arel, ***Lat.*** Orolaunum. Luxembourg, SE Belgium *49°41′N 05°49′E*
27 *R7* **Arma** Kansas, C USA *37°32′N 94°42′W*
97 *F16* **Armagh** ***Ir.*** Ard Mhacha. S Northern Ireland, United Kingdom *54°15′N 06°33′W*
97 *F16* **Armagh** *cultural region* S Northern Ireland, United Kingdom
102 *K15* **Armagnac** *cultural region* S France
103 *Q7* **Armançon** ≈ C France
60 *K10* **Armando Laydner, Represa** ▣ S Brazil
115 *M24* **Armathía** *island* SE Greece
137 *T12* **Armavir** ***prev.*** Hoktemberyan, ***Rus.*** Oktemberyan. SW Armenia *40°09′N 43°58′E*
126 *M14* **Armavir** Krasnodarskiy Kray, SW Russian Federation *44°59′N 41°07′E*
54 *E10* **Armenia** Quindío, W Colombia *04°32′N 75°40′W*
137 *T12* **Armenia** ***off.*** Republic of Armenia, ***var.*** Ajastan, ***Arm.*** Hayastani Hanrapetut'yun; ***prev.*** Armenian Soviet Socialist Republic. ◆ *republic* SW Asia
Armenian Soviet Socialist Republic *see* Armenia
Armenia, Republic of *see* Armenia
Armenierstadt *see* Gherla
103 *O1* **Armentières** Nord, N France *50°41′N 02°53′E*
40 *K14* **Armería** Colima, SW Mexico *18°55′N 103°59′W*
183 *T5* **Armidale** New South Wales, SE Australia *30°32′S 151°40′E*
29 *P11* **Armour** South Dakota, N USA *43°19′N 98°21′W*
61 *B18* **Armstrong** Santa Fe, C Argentina *32°46′S 61°39′W*
11 *N16* **Armstrong** British Columbia, SW Canada *50°27′N 119°14′W*
12 *D11* **Armstrong** Ontario, S Canada *50°20′N 89°02′W*
29 *U11* **Armstrong** Iowa, C USA *43°24′N 94°28′W*
25 *S16* **Armstrong** Texas, SW USA *26°55′N 97°47′W*
117 *S11* **Armyans'k** ***Rus.*** Armyansk. Avtonomna Respublika Krym, S Ukraine *46°05′N 33°43′E*
115 *H14* **Arnaía** ***Cont.*** Arnea. Kentrikí Makedonía, N Greece *40°30′N 23°36′E*
121 *N2* **Arnaoúti, Akrotíri** ***var.*** Arnaoútis, Cape Arnaouti. *headland* W Cyprus *35°06′N 32°16′E*
Arnaouti, Cape/Arnaoútis *see* Arnaoúti, Akrotíri
12 *L4* **Arnaud** ≈ Québec, E Canada
103 *Q8* **Arnay-le-Duc** Côte d'Or, C France *47°08′N 04°27′E*
Arnea *see* Arnaía
105 *Q4* **Arnedo** La Rioja, N Spain *42°14′N 02°05′W*
95 *I14* **Årnes** Akershus, S Norway *60°07′N 11°28′E*
Årnes *see* Åi Åfjord
26 *K9* **Arnett** Oklahoma, C USA *36°08′N 99°46′W*
98 *L12* **Arnhem** Gelderland, SE Netherlands *51°59′N 05°54′E*
181 *Q2* **Arnhem Land** *physical region* Northern Territory, N Australia
106 *F11* **Arno** ≈ C Italy
Arņo *see* Arno Atoll
189 *W7* **Arno Atoll** ***var.*** Arņo. *atoll* Ratak Chain, NE Marshall Islands
182 *H8* **Arno Bay** South Australia *33°55′S 136°31′E*
35 *Q8* **Arnold** California, W USA *38°15′N 120°19′W*
27 *X5* **Arnold** Missouri, C USA *38°25′N 90°22′W*
29 *N15* **Arnold** Nebraska, C USA *41°25′N 100°11′W*
109 *R10* **Arnoldstein** ***Slvn.*** Pod Klošter. Kärnten, S Austria *46°34′N 13°43′E*
103 *N9* **Arnon** ≈ C France
45 *P14* **Arnos Vale** ✈ (Kingstown) Saint Vincent, SE Saint Vincent and the Grenadines *13°08′N 61°13′W*
92 *I8* **Arnøya** ***Lapp.*** Árdni. *island* N Norway
14 *L12* **Arnprior** Ontario, SE Canada *45°31′N 76°11′W*
101 *G15* **Arnsberg** Nordrhein-Westfalen, W Germany *51°24′N 08°04′E*
101 *K16* **Arnstadt** Thüringen, C Germany *50°50′N 10°57′E*
Arnswalde *see* Choszczno
54 *K5* **Aroa** Yaracuy, N Venezuela *10°26′N 68°54′W*
83 *E21* **Aroab** Karas, SE Namibia *26°47′S 19°40′E*
Ároania *see* Chelmós
191 *O6* **Aroa, Pointe** *headland* Moorea, W French Polynesia *17°27′S 149°45′W*
Aroe Islands *see* Aru, Kepulauan
101 *H15* **Arolsen** Niedersachsen, C Germany *51°23′N 09°00′E*
106 *C7* **Arona** Piemonte, NE Italy *45°45′N 08°33′E*
19 *R3* **Aroostook River** ≈ Canada/USA
Arop Island *see* Long Island
38 *M12* **Aropuk Lake** ◎ Alaska, USA
191 *P4* **Arorae** *atoll* Tungaru, W Kiribati
190 *G16* **Arorangi** Rarotonga, S Cook Islands *21°13′S 159°49′W*
108 *I9* **Arosa** Graubünden, S Switzerland *46°48′N 09°42′E*
104 *F4* **Arousa, Ría de** *estuary* E Atlantic Ocean
184 *P8* **Arowhana** ▲ North Island, New Zealand *38°07′S 177°52′E*
137 *V12* **Arp'a** ***Az.*** Arpaçay. ≈ Armenia/Azerbaijan
137 *S11* **Arpaçay** Kars, NE Turkey *40°51′N 43°20′E*
Arpaçay *see* Arp'a
Arqalyq *see* Arkalyk
149 *N14* **Arra** ≈ SW Pakistan
Arrabona *see* Győr
Arrah *see* Āra
Ar Rahad *see* Er Rahad
139 *R9* **Ar Raḩḩālīyah** Al Anbār, C Iraq *32°53′N 43°21′E*
60 *Q10* **Arraial do Cabo** Rio de Janeiro, SE Brazil *22°57′S 42°00′W*
104 *H11* **Arraiolos** Évora, S Portugal *38°44′N 07°59′W*
139 *R8* **Ar Ramādī** ***var.*** Ramadi, Rumadiya. Al Anbār, SW Iraq *33°27′N 43°19′E*
138 *J6* **Ar Rāmī** Ḩimş, C Syria *34°32′N 37°54′E*
Ar Rams *see* Rams
138 *H9* **Ar Ramthā** ***var.*** Ramtha. Irbid, N Jordan *32°34′N 36°00′E*
96 *H13* **Arran, Isle of** *island* SW Scotland, United Kingdom
138 *L3* **Ar Raqqah** ***var.*** Rakka; ***anc.*** Nicephorium. Ar Raqqah, N Syria *35°57′N 39°03′E*
138 *L3* **Ar Raqqah** ***off.*** Muḩāfaẓat al Raqqah, ***var.*** Raqqah, ***Fr.*** Rakka. ⬖ *governorate* N Syria
103 *O2* **Arras** ***anc.*** Nemetocenna. Pas-de-Calais, N France *50°17′N 02°46′E*
105 *P3* **Arrasate** ***Cast.*** Mondragón. País Vasco, N Spain *43°04′N 02°30′W*
138 *G12* **Ar Rashādīyah** Aţ ţafilah, W Jordan *30°42′N 35°38′E*
138 *I5* **Ar Rastān** ***var.*** Rastâne. Ḩimş, W Syria *34°57′N 36°43′E*
139 *X12* **Ar Raṭāwī** Al Başrah, E Iraq *30°37′N 47°12′E*
102 *L15* **Arrats** ≈ S France
141 *N10* **Ar Rawḑah** Makkah, S Saudi Arabia *21°19′N 42°48′E*
141 *Q15* **Ar Rawḑah** S Yemen *14°26′N 47°14′E*
142 *K11* **Ar Rawḑatayn** ***var.*** Raudhatain. N Kuwait *29°80′N 47°50′E*
143 *N16* **Ar Rayyān** ***var.*** Al Rayyan. C Qatar *25°18′N 51°29′E*
102 *L17* **Arreau** Hautes-Pyrénées, S France *42°55′N 00°21′E*
64 *Q11* **Arrecife** ***var.*** Arrecife de Lanzarote, Puerto Arrecife. Lanzarote, Islas Canarias, NE Atlantic Ocean *28°57′N 13°33′W*
Arrecife de Lanzarote *see* Arrecife
43 *P6* **Arrecife Edinburgh** *reef* NE Nicaragua
61 *C19* **Arrecifes** Buenos Aires, E Argentina *34°06′S 60°09′W*
102 *F6* **Arrée, Monts d'** ▲▲ NW France
Ar Refā'ī *see* Ar Rifā'ī
Arretium *see* Arezzo
Arriaca *see* Guadalajara
109 *S9* **Arriach** Kärnten, S Austria *46°43′N 13°52′E*
41 *T16* **Arriaga** Chiapas, SE Mexico *16°14′N 93°54′W*
41 *N12* **Arriaga** San Luis Potosí, C Mexico *21°55′N 101°23′W*
139 *W10* **Ar Rifā'ī** ***var.*** Ar Refā'ī. Dhī Qār, SE Iraq *31°47′N 46°07′E*
139 *V12* **Ar Riḩāb** *salt flat* S Iraq
Arriondas *see* Les Arriondes
141 *Q7* **Ar Riyāḑ** ***Eng.*** Riyadh. ● (Saudi Arabia) Ar Riyāḑ, C Saudi Arabia *24°50′N 46°50′E*
141 *O8* **Ar Riyāḑ** ***off.*** Minţaqat ar Riyāḑ. ⬖ *province* C Saudi Arabia
141 *S15* **Ar Riyān** S Yemen *14°43′N 49°18′E*
Arrö *see* Ærø
61 *H18* **Arroio Grande** Rio Grande do Sul, S Brazil *32°15′S 53°02′W*
102 *K15* **Arros** ≈ S France
103 *Q9* **Arroux** ≈ C France
25 *R5* **Arrowhead, Lake** ▣ Texas, SW USA
182 *L5* **Arrowsmith, Mount** *hill* New South Wales, SE Australia
185 *D21* **Arrowtown** Otago, South Island, New Zealand *44°57′S 168°51′E*
61 *D17* **Arroyo Barú** Entre Ríos, E Argentina *31°52′S 58°26′W*
104 *J10* **Arroyo de la Luz** Extremadura, W Spain *39°28′N 06°36′W*
63 *J16* **Arroyo de la Ventana** Río Negro, SE Argentina *41°41′S 66°03′W*
35 *P13* **Arroyo Grande** California, W USA *35°07′N 120°35′W*
Ar Ru'ays *see* Ar Ruways
141 *R11* **Ar Rub' al Khālī** ***Eng.*** Empty Quarter, Great Sandy Desert. *desert* SW Asia
139 *V13* **Ar Ruḑaymah** Al Muthanná, S Iraq *30°20′N 45°26′E*
61 *A16* **Arrufó** Santa Fe, C Argentina *30°15′S 61°45′W*
138 *I7* **Ar Ruḩaybah** ***var.*** Ruhaybeh, ***Fr.*** Rouhaïbé. Rīf Dimashq, W Syria *33°45′N 36°40′E*
139 *V15* **Ar Rukhaymīyah** *well* S Iraq
139 *U11* **Ar Rumaythah** ***var.*** Rumaitha. Al Muthanná, S Iraq *31°31′N 45°15′E*
141 *X8* **Ar Rustāq** ***var.*** Rostak, Rustaq. N Oman *23°34′N 57°25′E*
139 *N8* **Ar Ruţbah** ***var.*** Rutba. Al Anbār, SW Iraq *33°03′N 40°16′E*
140 *M3* **Ar Rūthīyah** *spring/well* NW Saudi Arabia *31°18′N 41°23′E*
ar-Ruwaida *see* Ar Ruwayḑah
141 *O8* **Ar Ruwayḑah** ***var.*** ar-Ruwaida. Jīzān, C Saudi Arabia *23°48′N 44°44′E*
143 *N15* **Ar Ruways** ***var.*** Al Ruweis, Ar Ru'ays, Ruwais. N Qatar *26°08′N 51°13′E*
143 *O17* **Ar Ruways** ***var.*** Ar Ru'ays. Abū Ẓaby, W United Arab Emirates *24°09′N 52°57′E*
Års *see* Aars
Arsanias *see* Murat Nehri
123 *S15* **Arsen'yev** Primorskiy Kray, SE Russian Federation *44°09′N 133°28′E*
155 *G19* **Arsikere** Karnātaka, W India *13°20′N 76°13′E′N*
127 *R3* **Arsk** Respublika Tatarstan, W Russian Federation *56°07′N 49°54′E*
94 *N10* **Årskogen** Gävleborg, C Sweden *62°07′N 17°19′E*
121 *O3* **Ársos** C Cyprus *34°51′N 32°46′E*
94 *N13* **Årsunda** Gävleborg, C Sweden *60°31′N 16°45′E*
115 *C17* **Árta** ***anc.*** Ambracia. Ípeiros, W Greece *39°08′N 20°59′E*
105 *Y9* **Artà** Mallorca, Spain, W Mediterranean Sea *39°42′N 03°20′E*
Arta *see* Árachthos
137 *T12* **Artashat** S Armenia *39°57′N 44°34′E*
40 *M15* **Arteaga** Michoacán, SW Mexico *18°22′N 102°18′W*
123 *S15* **Artem** Primorskiy Kray, SE Russian Federation *43°24′N 132°20′E*
44 *C4* **Artemisa** La Habana, W Cuba *22°49′N 82°47′W*
117 *W7* **Artemivs'k** Donets'ka Oblast', E Ukraine *48°35′N 37°58′E*
122 *K13* **Artemovsk** Krasnoyarskiy Kray, S Russian Federation *54°22′N 93°24′E*
105 *U5* **Artesa de Segre** Cataluña, NE Spain *41°54′N 01°03′E*
37 *U14* **Artesia** New Mexico, SW USA *32°50′N 104°24′W*
25 *Q14* **Artesia Wells** Texas, SW USA *28°13′N 99°18′W*
108 *G8* **Arth** Schwyz, C Switzerland *47°05′N 08°39′E*
14 *F15* **Arthur** Ontario, S Canada *43°49′N 80°31′W*
30 *M14* **Arthur** Illinois, N USA *39°42′N 88°28′W*
28 *L14* **Arthur** Nebraska, C USA *41°35′N 101°42′W*
29 *Q5* **Arthur** North Dakota, N USA *47°03′N 97°12′W*
185 *B21* **Arthur** ≈ South Island, New Zealand
18 *B13* **Arthur, Lake** ▣ Pennsylvania, NE USA
183 *N15* **Arthur River** ≈ Tasmania, SE Australia
185 *G18* **Arthur's Pass** Canterbury, South Island, New Zealand *42°59′S 171°33′E*
185 *G17* **Arthur's Pass** *pass* South Island, New Zealand
44 *I3* **Arthur's Town** Cat Island, C The Bahamas *24°34′N 75°39′W*
44 *M9* **Artibonite, Rivière de l'** ≈ C Haiti
61 *E16* **Artigas** ***prev.*** San Eugenio, San Eugenio del Cuareim. Artigas, N Uruguay *30°25′S 56°28′W*
61 *E16* **Artigas** ⬖ *department* N Uruguay
194 *H1* **Artigas** *Uruguayan research station* Antarctica *61°57′S 58°23′W*
137 *T11* **Art'ik** W Armenia *40°38′N 43°58′E*
187 *O16* **Art, Île** *island* Îles Belep, W New Caledonia
103 *O2* **Artois** *cultural region* N France
136 *L12* **Artova** Tokat, N Turkey *40°04′N 36°17′E*
105 *Y9* **Artrutx, Cap d'** ***var.*** Cabo Dartuch. *cape* Menorca, Spain, W Mediterranean Sea
Artsiz *see* Artsyz
117 *N11* **Artsyz** ***Rus.*** Artsiz. Odes'ka Oblast', SW Ukraine *45°59′N 29°26′E*
158 *E7* **Artux** Xinjiang Uygur Zizhiqu, NW China *39°40′N 76°10′E*
137 *R11* **Artvin** Artvin, NE Turkey *41°12′N 41°48′E*
137 *R11* **Artvin** ⬖ *province* NE Turkey
146 *G14* **Artyk** Ahal Welaýaty, C Turkmenistan *37°32′N 59°16′E*
79 *Q16* **Aru** Orientale, NE Dem. Rep. Congo *02°53′N 30°50′E*
81 *E17* **Arua** NW Uganda *03°02′N 30°56′E*
104 *I4* **A Rúa de Valdeorras** ***var.*** La Rúa. Galicia, NW Spain *42°22′N 07°12′W*
Aruângua *see* Luangwa
45 *O15* **Aruba** ***var.*** Oruba. ◇ *Dutch self-governing territory* S West Indies
47 *Q4* **Aruba** *island* Aruba, Lesser Antilles
Aru Islands *see* Aru, Kepulauan
171 *W15* **Aru, Kepulauan** ***Eng.*** Aru Islands; ***prev.*** Aroe Islands. *island group* E Indonesia
153 *W10* **Arunāchal Pradesh** ***prev.*** North East Frontier Agency, North East Frontier Agency of Assam. ⬖ *state* NE India
Arun Qi *see* Naji
155 *H23* **Aruppukkottai** Tamil Nādu, SE India *09°31′N 78°03′E*
81 *I20* **Arusha** Arusha, N Tanzania *03°23′S 36°40′E*
81 *I21* **Arusha** ⬖ *region* E Tanzania
81 *I20* **Arusha** ✈ Arusha, N Tanzania *03°26′S 37°07′E*
54 *C9* **Arusí, Punta** *headland* NW Colombia *05°36′N 77°30′W*
155 *J23* **Aruvi Aru** ≈ NW Sri Lanka
79 *M17* **Aruwimi** ***var.*** Ituri (upper course). ≈ NE Dem. Rep. Congo
Árva *see* Orava
37 *T4* **Arvada** Colorado, C USA *39°48′N 105°06′W*
Arvand Rūd *see* 'Arab, Shaţţ al
162 *J8* **Arvayheer** Övörhangay, C Mongolia *46°13′N 102°47′E*
9 *O10* **Arviat** ***prev.*** Eskimo Point. Nunavut, C Canada *61°10′N 94°15′W*
93 *I14* **Arvidsjaur** Norrbotten, N Sweden *65°34′N 19°12′E*
95 *J15* **Arvika** Värmland, C Sweden *59°41′N 12°38′E*
92 *J8* **Årviksand** Troms, N Norway *70°10′N 20°30′E*
35 *S13* **Arvin** California, W USA *35°12′N 118°52′W*
163 *S8* **Arxan** Nei Mongol Zizhiqu, N China *47.11N 119.58′E*
145 *P7* **Arykbalyk** ***Kaz.*** Aryqbalyq. Severnyy Kazakhstan, N Kazakhstan *53°00′N 68°11′E*
Aryqbalyq *see* Arykbalyk
145 *P17* **'Arys'** ***prev.*** Arys'. Yuzhnyy Kazakhstan, S Kazakhstan *42°26′N 68°49′E*
Arys' *see* 'Arys'
Arys *see* Orzysz
Arys Köli *see* Arys, Ozero
145 *O14* **Arys, Ozero** ***Kaz.*** Arys Köli. ◎ C Kazakhstan
107 *D16* **Arzachena** Sardegna, Italy, C Mediterranean Sea *41°05′N 09°21′E*
127 *O4* **Arzamas** Nizhegorodskaya Oblast', W Russian Federation *55°25′N 43°51′E*
141 *V13* **Arzāt** S Oman *17°00′N 54°18′E*
104 *H3* **Arzúa** Galicia, NW Spain *42°55′N 08°10′W*
111 *A16* **Aš** ***Ger.*** Asch. Karlovarský Kraj, W Czech Republic *50°18′N 12°12′E*
95 *H15* **Ås** Akershus, S Norway *59°40′N 10°50′E*
Åsa *see* Asaa
95 *H20* **Asaa** ***var.*** Åsa. Nordjylland, N Denmark *57°07′N 10°24′E*
83 *E21* **Asab** Karas, S Namibia *25°29′S 17°59′E*
77 *U16* **Asaba** Delta, S Nigeria *06°10′N 06°44′E*
76 *J10* **Assaba** ***var.*** Açâba. ⬖ *region* S Mauritania
149 *S4* **Asadābād** ***var.*** Asadābād; ***prev.*** Chaghasarāy. Kunar, E Afghanistan *34°52′N 71°09′E*
Asadābād *see* Asadābād
138 *K3* **Asad, Buḩayrat al** ***Eng.*** Lake Assad. ◎ N Syria
63 *H20* **Asador, Pampa del** *plain* S Argentina
165 *P14* **Asahi** Chiba, Honshū, S Japan *35°43′N 140°38′E*
164 *M11* **Asahi** Toyama, Honshū, SW Japan *36°56′N 137°34′E*
165 *T13* **Asahi-dake** ▲ Hokkaidō, N Japan *43°42′N 142°50′E*
165 *T3* **Asahikawa** Hokkaidō, N Japan *43°46′N 142°23′E*
147 *S10* **Asaka** ***Rus.*** Assake; ***prev.*** Leninsk. Andijon Viloyati, E Uzbekistan *40°39′N 72°16′E*
77 *P17* **Asamankese** SE Ghana *05°47′N 00°41′W*
188 *B15* **Asan** W Guam *13°28′N 144°43′E*
188 *B15* **Asan Point** *headland* W Guam
153 *R15* **Āsānsol** West Bengal, NE India *23°40′N 86°59′E*
80 *K12* **Āsayita** Āfar, NE Ethiopia *11°35′S 41°23′E*
171 *T12* **Asbakin** Papua Barat, E Indonesia *0°45′S 131°40′E*
Asben *see* Aïr, Massif de l'
15 *Q12* **Asbestos** Québec, SE Canada *45°46′N 71°56′W*
29 *Y13* **Asbury** Iowa, C USA *42°30′N 90°45′W*
18 *K15* **Asbury Park** New Jersey, NE USA *40°13′N 74°00′W*
41 *Z12* **Ascención, Bahiá de la** *bay* NW Caribbean Sea
40 *I3* **Ascensión** Chihuahua, N Mexico *31°07′N 107°59′W*
Ascension *see* Saint Helena, Ascension and Tristan da Cunha
65 *M14* **Ascension Fracture Zone** *tectonic feature* C Atlantic Ocean
Ascension Island ◇ *dependency of St.Helena* C Atlantic Ocean
65 *N16* **Ascension Island** *island* C Atlantic Ocean
Asch *see* Aš
109 *S3* **Aschach an der Donau** Oberösterreich, N Austria *48°22′N 14°00′E*
101 *H18* **Aschaffenburg** Bayern, SW Germany *49°58′N 09°10′E*
101 *F14* **Ascheberg** Nordrhein-Westfalen, W Germany *51°46′N 07°36′E*
101 *L14* **Aschersleben** Sachsen-Anhalt, C Germany *51°46′N 11°28′E*
106 *G12* **Asciano** Toscana, C Italy *43°15′N 11°32′E*
106 *J13* **Ascoli Piceno** ***anc.*** Asculum Picenum. Marche, C Italy *42°51′N 13°34′E*
107 *M17* **Ascoli Satriano** ***anc.*** Asculub, Ausculum Apulum. Puglia, SE Italy *41°13′N 15°32′E*
108 *G11* **Ascona** Ticino, S Switzerland *46°10′N 08°45′E*
Asculub *see* Ascoli Satriano
Asculum Picenum *see* Ascoli Piceno
80 *L11* **'Aseb** ***var.*** Assab, ***Amh.*** Āseb. SE Eritrea *13°04′N 42°36′E*
95 *M20* **Åseda** Kronoberg, S Sweden *57°10′N 15°20′E*
127 *T6* **Asekeyevo** Orenburgskaya Oblast', W Russian Federation *53°36′N 52°53′E*
81 *J14* **Āsela** ***var.*** Asella, Aselle, Asselle. Oromīya, C Ethiopia *07°55′N 39°08′E*
93 *H15* **Åsele** Västerbotten, N Sweden *64°10′N 17°20′E*
Asella/Aselle *see* Āsela
98 *N7* **Assen** Drenthe, NE Netherlands *53°N 06°34′E*
94 *K12* **Åsen** Dalarna, C Sweden *61°18′N 13°49′E*
114 *J11* **Asenovgrad** ***prev.*** Stanimaka. Plovdiv, C Bulgaria *42°00′N 24°53′E*
171 *O13* **Asera** Sulawesi, C Indonesia *03°24′S 121°42′E*
95 *E17* **Åseral** Vest-Agder, S Norway *58°37′N 07°27′E*
118 *J3* **Aseri** ***var.*** Asserien, ***Ger.*** Asserin. Ida-Virumaa, NE Estonia *59°29′N 26°51′E*
104 *G3* **A Serra de Outes** Galicia, NW Spain *42°50′N 08°54′W*
40 *J10* **Aserradero** Durango, W Mexico
146 *F13* **Aşgabat** ***prev.*** Ashgabat, Ashkhabad, Poltoratsk. ● (Turkmenistan) Ahal Welaýaty, C Turkmenistan *37°58′N 58°22′E*
146 *F13* **Aşgabat** ✈ Ahal Welaýaty, C Turkmenistan *38°06′N 58°10′E*
95 *H16* **Åsgårdstrand** Vestfold, S Norway *59°22′N 10°27′E*
Ashara *see* Al 'Ashārah
23 *T6* **Ashburn** Georgia, SE USA *31°42′N 83°39′W*
185 *G19* **Ashburton** Canterbury, South Island, New Zealand *43°55′S 171°47′E*
185 *G19* **Ashburton** ≈ South Island, New Zealand
180 *H8* **Ashburton River** ≈ Western Australia
145 *V10* **Ashchysu** ≈ E Kazakhstan
10 *M16* **Ashcroft** British Columbia, SW Canada *50°41′N 121°17′W*
138 *E10* **Ashdod** ***anc.*** Azotos, ***Lat.*** Azotus. Central, W Israel *31°48′N 34°38′E*
27 *S14* **Ashdown** Arkansas, C USA *33°40′N 94°09′W*
21 *T9* **Asheboro** North Carolina, SE USA *35°42′N 79°50′W*
11 *X15* **Ashern** Manitoba, S Canada *51°10′N 98°22′W*
21 *P10* **Asheville** North Carolina, SE USA *35°36′N 82°33′W*
12 *E8* **Asheweig** ≈ Ontario, C Canada
27 *V9* **Ash Flat** Arkansas, C USA *36°13′N 91°36′W*
183 *T4* **Ashford** New South Wales, SE Australia *29°18′S 151°09′E*
97 *P22* **Ashford** SE England, United Kingdom *51°09′N 00°52′E*
36 *K11* **Ash Fork** Arizona, SW USA *35°12′N 112°31′W*
Ashgabat *see* Aşgabat
27 *T7* **Ash Grove** Missouri, C USA *37°19′N 93°35′W*
165 *O12* **Ashikaga** ***var.*** Asikaga. Tochigi, Honshū, S Japan *36°21′N 139°26′E*
165 *Q8* **Ashiro** Iwate, Honshū, C Japan *40°04′N 141°00′E*
164 *F15* **Ashizuri-misaki** Shikoku, SW Japan
138 *E10* **Ashkelon** ***prev.*** Ashqelon. Southern, C Israel *31°40′N 34°35′E*
Ashkhabad *see* Aşgabat
23 *Q4* **Ashland** Alabama, S USA *33°16′N 85°50′W*
26 *K7* **Ashland** Kansas, C USA *37°12′N 99°47′W*
21 *P5* **Ashland** Kentucky, S USA *38°28′N 82°40′W*
19 *S2* **Ashland** Maine, NE USA *46°36′N 68°24′W*
22 *M1* **Ashland** Mississippi, S USA *34°51′N 89°10′W*
27 *U4* **Ashland** Missouri, C USA *38°46′N 92°15′W*
29 *S15* **Ashland** Nebraska, C USA *41°01′N 96°21′W*
31 *T12* **Ashland** Ohio, N USA *40°52′N 82°19′W*
32 *G15* **Ashland** Oregon, NW USA *42°11′N 122°42′W*
21 *W6* **Ashland** Virginia, NE USA *37°45′N 77°28′W*
30 *K3* **Ashland** Wisconsin, N USA *46°34′N 90°54′W*
20 *I8* **Ashland City** Tennessee, S USA *36°16′N 87°05′W*
183 *S4* **Ashley** New South Wales, SE Australia *29°21′S 149°49′E*
29 *O7* **Ashley** North Dakota, N USA *46°00′N 99°22′W*
173 *W7* **Ashmore and Cartier Islands** ◇ *Australian external territory* E Indian Ocean
119 *I14* **Ashmyany** ***Rus.*** Oshmyany. Hrodzyenskaya Voblasts', W Belarus *54°24′N 25°57′E*
18 *K12* **Ashokan Reservoir** ▣ New York, NE USA
165 *U4* **Ashoro** Hokkaidō, NE Japan *43°16′N 143°33′E*
Ashqelon *see* Ashkelon
Ashraf *see* Behshahr
139 *O3* **Ash Shadādah** ***var.*** Ash Shaddādah, Jisr ash Shadadi, Shaddādī, Shedadi, Tell Shedadi. Al Ḩasakah, NE Syria *36°00′N 40°42′E*
Ash Shaddādah *see* Ash Shadādah
139 *Y12* **Ash Shāfī** Al Başrah, E Iraq *30°49′N 47°30′E*
139 *R4* **Ash Shakk** ***var.*** Shaykh. Şalāḩ ad Dīn, C Iraq *35°15′N 43°27′E*
Ash Sham/Ash Shām *see* Rīf Dimashq
139 *T10* **Ash Shāmīyah** ***var.*** Shamiya. Al Qādisīyah, C Iraq *31°56′N 44°37′E*
139 *Y13* **Ash Shāmīyah** ***var.*** Al Bādiyah al Janūbīyah. *desert* S Iraq
139 *T11* **Ash Shanāfīyah** ***var.*** Ash Shināfīyah. Al Qādisīyah, S Iraq *31°35′N 44°38′E*
138 *G13* **Ash Sharāh** ***var.*** Esh Sharā. ▲▲ W Jordan
143 *R16* **Ash Shāriqah** ***Eng.*** Sharjah. Ash Shāriqah, NE United Arab Emirates *25°22′N 55°28′E*
143 *R16* **Ash Shāriqah** ***var.*** Sharjah. ✈ Ash Shāriqah, NE United Arab Emirates *25°19′N 55°37′E*
140 *I4* **Ash Sharmah** ***var.*** Šarma. Tabūk, NW Saudi Arabia *28°02′N 35°16′E*
139 *R4* **Ash Sharqāţ** Nīnawýa, NW Iraq *35°31′N 43°15′E*
141 *S10* **Ash Sharqīyah** ***off.*** Al Minţaqah ash Sharqīyah, ***Eng.*** Eastern Region. ⬖ *province* E Saudi Arabia
Ash Sharqīyah *see* Al 'Ubaylah
139 *W11* **Ash Shaţrah** ***var.*** Shatra. Dhī Qār, SE Iraq *31°26′N 46°10′E*
138 *G13* **Ash Shawbak** Ma'ān, W Jordan *30°32′N 35°34′E*
138 *L5* **Ash Shaykh Ibrāhīm** Ḩimş, C Syria *35°03′N 38°50′E*
141 *O17* **Ash Shaykh 'Uthmān** SW Yemen *12°53′N 45°00′E*
141 *S15* **Ash Shiḩr** SE Yemen *14°45′N 49°24′E*
Ash Shināfīyah *see* Ash Shanāfīyah
141 *V12* **Ash Shişar** ***var.*** Shisur. SW Oman *18°13′N 53°35′E*
139 *S13* **Ash Shubrūm** *well* S Iraq
141 *R10* **Ash Shuqqān** *desert* E Saudi Arabia
75 *O9* **Ash Shuwayrif** ***var.*** Ash Shwayrif. N Libya *29°54′N 14°16′E*
Ash Shwayrif *see* Ash Shuwayrif
31 *U10* **Ashtabula** Ohio, N USA *41°54′N 80°46′W*
29 *Q5* **Ashtabula, Lake** ▣ North Dakota, N USA
137 *T12* **Ashtarak** W Armenia *40°18′N 44°22′E*
142 *M6* **Āshtīān** ***var.*** Āshtiyān. Markazī, W Iran *34°23′N 49°55′E*
Āshtiyān *see* Āshtīān
33 *R13* **Ashton** Idaho, NW USA *44°04′N 111°27′W*
13 *O10* **Ashuanipi Lake** ◎ Newfoundland and Labrador, E Canada
15 *P6* **Ashuapmushuan** ≈ Québec, SE Canada
23 *Q3* **Ashville** Alabama, S USA *33°50′N 86°15′W*
31 *S14* **Ashville** Ohio, N USA *39°43′N 82°57′W*
30 *K3* **Ashwabay, Mount** *hill* Wisconsin, N USA
128-129 **Asia** *continent*
171 *T11* **Asia, Kepulauan** *island group* E Indonesia
154 *N13* **Āsika** Odisha, E India *19°38′N 84°41′E*
Asikaga *see* Ashikaga
93 *M18* **Asikkala** ***var.*** Vääksy. Päijät-Häme, S Finland *61°09′N 25°36′E*
74 *G5* **Asilah** N Morocco *35°32′N 06°04′W*
'Aşī, Nahr al *see* Orontes
107 *B16* **Asinara, Isola** *island* W Italy
122 *J12* **Asino** Tomskaya Oblast', C Russian Federation *56°56′N 86°02′E*
119 *O14* **Asintorf** ***Rus.*** Osintorf. Vitsyebskaya Voblasts', N Belarus *54°43′N 30°35′E*
119 *L17* **Asipovichy** ***Rus.*** Osipovichi. Mahilyowskaya Voblasts', C Belarus *53°18′N 28°40′E*
141 *N12* **'Asīr** ***off.*** Minţaqat 'Asīr. ⬖ *province* SW Saudi Arabia
140 *M11* **'Asīr** ***Eng.*** Asir. ▲▲ SW Saudi Arabia
'Asīr, Minţaqat *see* 'Asīr
139 *X10* **Askal** Maysān, E Iraq *31°45′N 47°07′E*
137 *P13* **Aşkale** Erzurum, NE Turkey *39°56′N 40°39′E*
117 *T11* **Askaniya-Nova** Khersons'ka Oblast', S Ukraine *46°27′N 33°54′E*
95 *H15* **Asker** Akershus, S Norway *59°52′N 10°26′E*
95 *L17* **Askersund** Örebro, C Sweden *58°55′N 14°55′E*
Askī Kalak *see* Eski Kaļak
95 *I15* **Askim** Østfold, S Norway *59°15′N 11°10′E*
127 *V3* **Askino** Respublika Bashkortostan, W Russian Federation *56°07′N 56°39′E*
115 *D14* **Áskio** ▲▲ N Greece
152 *L9* **Askot** Uttarakhand, N India *29°44′N 80°20′E*
94 *C12* **Askvoll** Sogn Og Fjordane, S Norway *61°21′N 05°04′E*
136 *A13* **Aslan Burnu** *headland* W Turkey *38°44′N 26°43′E*
136 *L16* **Aslantaş Barajı** ▣ S Turkey
149 *S4* **Asmār** ***var.*** Bar Kunar. Kunar, E Afghanistan *34°59′N 71°29′E*
Asmara *see* Asmera
80 *I9* **Asmera** ***var.*** Asmara. ● (Eritrea) C Eritrea *15°15′N 38°58′E*
95 *L21* **Åsnen** ◎ S Sweden
115 *F19* **Asopós** ≈ S Greece
171 *W13* **Asori** Papua, E Indonesia *02°37′S 136°06′E*
80 *G12* **Āsosa** Bīnishangul Gumuz, W Ethiopia *10°06′N 34°27′E*
32 *M10* **Asotin** Washington, NW USA *46°18′N 117°03′W*
Aspadana *see* Eşfahān
Aspang *see* Aspang Markt
109 *X6* **Aspang Markt** ***var.*** Aspang. Niederösterreich, E Austria *47°34′N 16°06′E*
105 *S12* **Aspe** Valenciana, E Spain *38°21′N 00°43′W*
37 *R5* **Aspen** Colorado, C USA *39°12′N 106°49′W*
25 *P6* **Aspermont** Texas, SW USA *33°08′N 100°14′W*
Asphaltites, Lacus *see* Dead Sea
Aspinwall *see* Colón
185 *C20* **Aspiring, Mount** ▲ South Island, New Zealand *44°21′S 168°47′E*
115 *B16* **Asprókavos, Akrotírio** *headland* Kérkyra, Iónia Nisiá, Greece, C Mediterranean Sea *39°22′N 20°07′E*
Aspropótamos *see* Achelóos
Assab *see* 'Aseb
138 *L4* **As Sabkhah** ***var.*** Sabkha. Ar Raqqah, NE Syria *35°30′N 35°54′E*
139 *U6* **As Sa'dīyah** Diyālá, E Iraq *34°11′N 45°09′E*
Assad, Lake *see* Asad, Buḩayrat al
138 *I8* **Aş Şafā** ▲ S Syria *33°03′N 37°07′E*
138 *I10* **Aş Şafāwī** Al Mafraq, N Jordan *37°12′N 32°30′E*
75 *W8* **Aş Şaff** ***var.*** El Şaff. N Egypt *29°34′N 31°16′E*
139 *N2* **Aş Şafīḩ** Al Ḩasakah, N Syria *36°42′N 40°12′E*
Aş Şaḩrā' ash Sharqīyah *see* Sahara el Sharqîya
Assake *see* Asaka
As Salamīyah *see* Salamīyah
141 *Q4* **As Sālimī** ***var.*** Salemy. SW Kuwait *29°07′N 46°41′E*
67 *W7* **'Assal, Lac** ◎ C Djibouti
75 *T7* **As Sallūm** ***var.*** Salûm. NW Egypt *31°31′N 25°09′E*
139 *T13* **As Salmān** Al Muthanná, S Iraq *30°29′N 44°34′E*
138 *G10* **As Salţ** ***var.*** Salt. Al Balqā', NW Jordan *32°03′N 35°44′E*
142 *M16* **As Salwā** ***var.*** Salwa, Salwah. S Qatar *24°44′N 50°52′E*
153 *V12* **Assam** ⬖ *state* NE India
Assamaka *see* Assamakka
77 *T8* **Assamakka** ***var.*** Assamaka. Agadez, NW Niger *19°24′N 05°53′E*
139 *U11* **As Samāwah** ***var.*** Samawa. Al Muthanná, S Iraq *31°17′N 45°06′E*

◆ Country
● Country Capital
◇ Dependent Territory
○ Dependent Territory Capital
⬖ Administrative Regions
✈ International Airport
▲ Mountain
▲▲ Mountain Range
Volcano
≈ River
◎ Lake
▣ Reservoir

As Samāwah *see* Al Muthannā
As Saqia al Hamra *see* Saguia al Hamra
138 J4 **Aş Şā'rān** Ḩamāh, C Syria 35°15′N 37°28′E
138 G9 **Aş Şarīḩ** Irbid, N Jordan 32°31′N 35°54′E
21 Z5 **Assateague Island** *island* Maryland, NE USA
139 O6 **As Sayyāl** *var.* Sayyāl. Dayr az Zawr, E Syria 34°37′N 40°52′E
99 G18 **Asse** Vlaams Brabant, C Belgium 50°55′N 04°12′E
99 D16 **Assebroek** West-Vlaanderen, NW Belgium 51°12′N 03°16′E
Asselle *see* Āsela
107 C20 **Assemini** Sardegna, Italy, C Mediterranean Sea 39°16′N 08°58′E
99 E16 **Assenede** Oost-Vlaanderen, NW Belgium 51°15′N 03°43′E
95 G24 **Assens** Syddtjylland, C Denmark 55°16′N 09°54′N
Asserien/Asserin *see* Aseri
99 I21 **Assesse** Namur, SE Belgium 50°22′N 05°01′E
141 Y8 **As Sib** *var.* Seeb. NE Oman 23°40′N 58°03′E
139 Z13 **As Sibah** *var.* Sibah. Al Başrah, SE Iraq 30°13′N 47°24′E
11 T17 **Assiniboia** Saskatchewan, S Canada 49°39′N 105°59′W
11 V15 **Assiniboine** ⚟ Manitoba, S Canada
11 P16 **Assiniboine, Mount** ▲ Alberta/British Columbia, SW Canada 50°54′N 115°43′W
Assiout *see* Asyūţ
60 J9 **Assis** São Paulo, S Brazil 22°37′S 50°25′W
106 I13 **Assisi** Umbria, C Italy 43°04′N 12°36′E
Assiut *see* Asyūţ
Assling *see* Jesenice
Assouan *see* Aswān
59 P14 **Assu** *var.* Açu. Rio Grande do Norte, E Brazil 05°33′S 36°55′W
Assuan *see* Aswān
142 K12 **Aş Şubayḩīyah** *var.* Subiyah. S Kuwait 28°55′N 47°57′E
141 R16 **As Sufāl** S Yemen 14°06′N 48°42′E
138 L5 **As Sukhnah** *var.* Sukhne, *Fr.* Soukhné. Ḩimş, C Syria 34°56′N 38°52′E
139 U4 **As Sulaymānīyah** *var.* Sulaimaniya, *Kurd.* Slēmānī. As Sulaymānīyah, NE Iraq 35°32′N 45°27′E
139 U4 **As Sulaymānīyah** *off.* Muḩāfa at as Sulaymānīyah, *off. Kurd.* Parēzga-i Slēmānī, *Kurd.* Slēmānī. ◆ *governorate* N Iraq
as Sulaymānīyah, Muḩāfa at *see* As Sulaymānīyah
141 P11 **As Sulayyil** Ar Riyāḑ, S Saudi Arabia 20°29′N 45°33′E
121 O13 **As Sulţān** N Libya 31°01′N 17°21′E
141 Q5 **Aş Şummān** *desert* N Saudi Arabia
141 Q16 **Aş Şurrah** SW Yemen 13°56′N 46°23′E
139 N4 **Aş Şuwār** *var.* Şuwār. Dayr az Zawr, E Syria 35°31′N 40°37′E
138 H9 **As Suwaydā'** *var.* El Suweida, Es Suweida, Suweida, *Fr.* Soueida. As Suwaydā', SW Syria 32°43′N 36°33′E
138 H9 **As Suwaydā'** *off.* Muḩāfaẓat as Suwaydā', *var.* As Suwaydā, Suwaydá, Suweida. ◆ *governorate* S Syria
141 Z9 **As Suwayḩ** NE Oman 22°07′N 59°42′E
141 X8 **As Suwayq** *var.* Suwaik. N Oman 23°49′N 57°30′E
139 T8 **Aş Şuwayrah** *var.* Suwaira. Wāsiţ, E Iraq 32°57′N 44°47′E
As Suways *see* Suez
Asta Colonia *see* Asti
Astacus *see* İzmit
115 M23 **Astakída** *island* SE Greece
145 Q9 **Astana** *prev.* Akmola, Akmolinsk, Tselinograd, Aqmola. ● (Kazakhstan) Akmola, N Kazakhstan 51°13′N 71°25′E
142 M3 **Āstāneh** *var.* Āstāneh-ye Ashrafīyeh. Gīlān, NW Iran 37°17′N 49°58′E
Āstāneh-ye Ashrafīyeh *see* Āstāneh
Asta Pompeia *see* Asti
137 Y14 **Astara** S Azerbaijan 38°28′N 48°51′E
Astarabad *see* Gorgān
99 L15 **Asten** Noord-Brabant, SE Netherlands 51°24′N 05°45′E
Asterābād *see* Gorgān
106 C8 **Asti** *anc.* Asta Colonia, Asta Pompeia, Hasta Colonia, Hasta Pompeia. Piemonte, NW Italy 44°54′N 08°11′E
Astigi *see* Ecija
Astipálaia *see* Astypálaia
148 L16 **Astola Island** *island* SW Pakistan
152 H4 **Astor** Jammu and Kashmir, NW India 35°21′N 74°52′E
104 K4 **Astorga** *anc.* Asturica Augusta. Castilla y León, N Spain 42°27′N 06°04′W
32 F10 **Astoria** Oregon, NW USA 46°12′N 123°50′W
0 F8 **Astoria Fan** *undersea feature* E Pacific Ocean 45°15′N 126°15′W
95 J22 **Åstorp** Skåne, S Sweden 56°09′N 12°57′E
Astrabad *see* Gorgān
127 Q13 **Astrakhan'** Astrakhanskaya Oblast', SW Russian Federation 46°20′N 48°01′E
Astrakhan-Bazar *see* Cälilabad
127 Q11 **Astrakhanskaya Oblast'** ◆ *province* SW Russian Federation
93 J15 **Åsträsk** Västerbotten, N Sweden 64°38′N 20°00′E
Astrida *see* Butare
65 O22 **Astrid Ridge** *undersea feature* S Atlantic Ocean
187 P15 **Astrolabe, Récifs de l'** *reef* C New Caledonia
121 P2 **Astromerítis** N Cyprus 35°09′N 33°02′E
115 F20 **Ástros** Pelopónnisos, S Greece 37°24′N 22°43′E
119 G16 **Astryna** *Rus.* Ostryna. Hrodzyenskaya Voblasts', W Belarus 53°44′N 24°33′E
104 J2 **Asturias** ◆ *autonomous community* NW Spain
Asturias *see* Oviedo
Asturica Augusta *see* Astorga
115 L22 **Astypálaia** *var.* Astipálaia, *It.* Stampalia. *island* Kykládes, Greece, Aegean Sea
192 G16 **Āsuisui, Cape** *headland* Savai'i, W Samoa 13°44′S 172°29′W
195 S2 **Asuka** *Japanese research station* Antarctica 71°49′S 23°52′E
62 O6 **Asunción** ● (Paraguay) Central, S Paraguay 25°17′S 57°36′W
62 O6 **Asunción** ✕ Central, S Paraguay 25°15′S 57°40′W
188 K3 **Asuncion Island** *island* N Northern Mariana Islands
42 E6 **Asunción Mita** Jutiapa, SE Guatemala 14°20′N 89°42′W
Asunción Nochixtlán *see* Nochixtlán
40 E3 **Asunción, Río** ⚟ NW Mexico
95 M18 **Åsunden** ◎ S Sweden
118 K11 **Asvyeya** *Rus.* Osveya. Vitsyebskaya Voblasts', N Belarus 56°00′N 28°05′E
Aswa *see* Achwa
75 X11 **Aswān** *var.* Assouan, Assuan, Aswân; *anc.* Syene. SE Egypt 24°03′N 32°59′E
Aswân *see* Aswān
Aswan Dam *see* Khazzān Aswān
75 W9 **Asyūţ** *var.* Assiout, Assiut, Asyût, Siut; *anc.* Lycopolis. C Egypt 27°06′N 31°11′E
Asyût *see* Asyūţ
193 W15 **Ata** *island* Tongatapu Group, SW Tonga
62 G8 **Atacama** *off.* Región de Atacama. ◆ *region* C Chile
Atacama Desert *see* Atacama, Desierto de
62 H4 **Atacama, Desierto de** *Eng.* Atacama Desert. *desert* N Chile
62 I6 **Atacama, Puna de** ▲ NW Argentina
Atacama, Región de *see* Atacama
62 I5 **Atacama, Salar de** *salt lake* N Chile
54 E11 **Ataco** Tolima, C Colombia 03°36′N 75°23′W
190 H8 **Atafu Atoll** *island* NW Tokelau
190 H8 **Atafu Village** Atafu Atoll, NW Tokelau 08°40′S 172°40′W
74 K12 **Atakor** ▲ SE Algeria
77 R14 **Atakora, Chaîne de l'** *var.* Atakora Mountains. ▲ N Benin
Atakora Mountains *see* Atakora, Chaîne de l'
77 R16 **Atakpamé** C Togo 07°32′N 01°08′E
146 F11 **Atakui** Ahal Welaýaty, C Turkmenistan 40°04′N 58°03′E
58 B13 **Atalaia do Norte** Amazonas, N Brazil 04°22′S 70°10′W
146 M14 **Atamyrat** *prev.* Kerki. Lebap Welaýaty, E Turkmenistan 37°52′N 65°06′E
76 I7 **Aţâr** Adrar, W Mauritania 20°30′N 13°03′W
162 G10 **Atas Bogd** ▲ SW Mongolia 43°17′N 96°47′E
35 P12 **Atascadero** California, W USA 35°28′N 120°40′W
25 S13 **Atascosa River** ⚟ Texas, SW USA
145 R11 **Atasu** Karaganda, C Kazakhstan 48°42′N 71°38′E
145 R12 **Atasu** ⚟ Karaganda, C Kazakhstan
193 V15 **Atata** *island* Tongatapu Group, S Tonga
136 H10 **Atatürk** ✕ (İstanbul) İstanbul, NW Turkey 40°58′N 28°50′E
137 N16 **Atatürk Barajı** ⊠ S Turkey
115 O23 **Atavýros** *prev.* Attavyros. Ródos, Dodekánisa, Aegean Sea 36°10′N 27°50′E
115 O23 **Atávyros** *prev.* Attávyros. ▲ Ródos, Dodekánisa, Greece, Aegean Sea 36°10′N 27°50′E
Atax *see* Aude
80 G8 **Atbara** *var.* 'Aţbārah. River Nile, NE Sudan 17°42′N 34°E
80 H8 **Atbara** *var.* Nahr 'Aţbarah. ⚟ Eritrea/Sudan
'Aţbārah/'Aţbarah, Nahr *see* Atbara
145 P9 **Atbasar** Akmola, N Kazakhstan 51°49′N 68°18′E
At-Bashi *see* At-Bashy
147 W9 **At-Bashy** *var.* At-Bashi. Narynskaya Oblast', C Kyrgyzstan 41°07′N 75°48′E
22 I10 **Atchafalaya Bay** *bay* Louisiana, S USA
22 I8 **Atchafalaya River** ⚟ Louisiana, S USA
Atchin *see* Aceh
27 Q3 **Atchison** Kansas, C USA 39°31′N 95°07′W
77 P16 **Atebubu** C Ghana 07°47′N 01°00′W
105 Q6 **Ateca** Aragón, NE Spain 41°20′N 01°49′W
40 K11 **Atengo, Río** ⚟ C Mexico
Aternum *see* Pescara
107 K15 **Atessa** Abruzzo, C Italy 42°03′N 14°25′E
Ateste *see* Este
99 E19 **Ath** *var.* Aat. Hainaut, SW Belgium 50°38′N 03°47′E
11 Q13 **Athabasca** Alberta, SW Canada 54°44′N 113°15′W
11 Q12 **Athabasca** *var.* Athabaska. ⚟ Alberta, SW Canada
11 R10 **Athabasca, Lake** ◎ Alberta/ Saskatchewan, SW Canada
Athabaska *see* Athabasca
115 C16 **Athamánon** ▲ C Greece
97 F17 **Athboy** *Ir.* Baile Átha Buí. E Ireland 53°38′N 06°55′W
Athenae *see* Athína
97 C18 **Athenry** *Ir.* Baile Átha an Rí. W Ireland 53°19′N 08°49′W
23 P2 **Athens** Alabama, S USA 34°48′N 86°58′W
23 T3 **Athens** Georgia, SE USA 33°57′N 83°24′W
31 T14 **Athens** Ohio, N USA 39°20′N 82°06′W
20 M10 **Athens** Tennessee, S USA 35°27′N 84°38′W
25 V7 **Athens** Texas, SW USA 32°12′N 95°51′W
Athens *see* Athína
115 B18 **Athéras, Akrotírio** *headland* Kefallonía, Iónia Nisiá, Greece, C Mediterranean Sea 38°20′N 20°24′E
181 W4 **Atherton** Queensland, NE Australia 17°18′S 145°29′E
81 I19 **Athi** ⚟ S Kenya
121 Q2 **Athiénou** SE Cyprus 35°01′N 33°31′E
115 H19 **Athína** *Eng.* Athens, *prev.* Athínai; *anc.* Athenae. ● (Greece) Attikí, C Greece 37°59′N 23°44′E
Athínai *see* Athína
139 S10 **Athiyah** An Najaf, C Iraq 32°01′N 44°04′E
97 D18 **Athlone** *Ir.* Baile Átha Luain. C Ireland 53°25′N 07°56′W
155 F16 **Athni** Karnātaka, W India 16°43′N 75°04′E
185 C23 **Athol** Southland, South Island, New Zealand 45°30′S 168°35′E
19 N11 **Athol** Massachusetts, NE USA 42°35′N 72°11′W
115 I15 **Áthos** ▲ NE Greece 40°10′N 24°21′E
Athos, Mount *see* Ágion Óros
Ath Thawrah *see* Madīnat ath Thawrah
141 P5 **Ath Thumāmī** *spring/well* N Saudi Arabia 27°56′N 45°06′E
99 L25 **Athus** Luxembourg, SE Belgium 49°34′N 05°50′E
97 E19 **Athy** *Ir.* Baile Átha Í. C Ireland 52°59′N 06°59′W
78 I10 **Ati** Batha, C Chad 13°11′N 18°20′E
81 F16 **Atiak** NW Uganda 03°14′N 32°05′E
57 G17 **Atico** Arequipa, SW Peru 16°13′S 73°13′W
105 O6 **Atienza** Castilla-La Mancha, C Spain 41°12′N 02°52′W
39 Q6 **Atigun Pass** *pass* Alaska, USA
12 B12 **Atikokan** Ontario, S Canada 48°45′N 91°38′W
13 O9 **Atikonak Lac** ◎ Newfoundland and Labrador, E Canada
42 C6 **Atitlán, Lago de** ◎ W Guatemala
190 L16 **Atiu** *island* S Cook Islands
Atjeh *see* Aceh
123 T9 **Atka** Magadanskaya Oblast', E Russian Federation 60°45′N 151°35′E
38 H17 **Atka** Atka Island, Alaska, USA 52°12′N 174°14′W
38 H17 **Atka Island** *island* Aleutian Islands, Alaska, USA
127 O7 **Atkarsk** Saratovskaya Oblast', W Russian Federation 52°15′N 43°48′E
27 U11 **Atkins** Arkansas, C USA 35°15′N 92°56′W
29 O13 **Atkinson** Nebraska, C USA 42°31′N 98°57′W
171 T12 **Atkri** Papua Barat, E Indonesia 01°45′S 130°04′E
41 O13 **Atlacomulco** *var.* Atlacomulco de Fabela. México, C Mexico 19°49′N 99°54′W
Atlacomulco de Fabela *see* Atlacomulco
23 S3 **Atlanta** *state capital* Georgia, SE USA 33°45′N 84°23′W
31 R6 **Atlanta** Michigan, N USA 45°01′N 84°07′W
25 X6 **Atlanta** Texas, SW USA 33°06′N 94°09′W
29 T15 **Atlantic** Iowa, C USA 41°24′N 95°00′W
21 Y10 **Atlantic** North Carolina, SE USA 34°52′N 76°20′W
23 W8 **Atlantic Beach** Florida, SE USA 30°19′N 81°24′W
18 J17 **Atlantic City** New Jersey, NE USA 39°23′N 74°27′W
172 L14 **Atlantic-Indian Basin** *undersea feature* SW Indian Ocean 60°00′S 15°00′E
172 K13 **Atlantic-Indian Ridge** *undersea feature* SW Indian Ocean 53°00′S 15°00′E
54 E4 **Atlántico** *off.* Departamento del Atlántico. ◆ *province* NW Colombia
64-65 **Atlantic Ocean** *ocean*
Atlántico, Departamento del *see* Atlántico
42 K7 **Atlántico Norte, Región Autónoma** *prev.* Zelaya Norte. ◆ *autonomous region* NE Nicaragua
42 L10 **Atlántico Sur, Región Autónoma** *prev.* Zelaya Sur. ◆ *autonomous region* SE Nicaragua
42 I5 **Atlántida** ◆ *department* N Honduras
77 Y15 **Atlantika Mountains** ▲ E Nigeria
64 J10 **Atlantis Fracture Zone** *tectonic feature* NW Atlantic Ocean
74 H7 **Atlas Mountains** ▲ NW Africa
123 V11 **Atlasova, Ostrov** *island* SE Russian Federation
123 V10 **Atlasovo** Kamchatskiy Kray, E Russian Federation 55°42′N 159°35′E
120 G11 **Atlas Saharien** *var.* Saharan Atlas. ▲ Algeria/Morocco
Atlas, Tell *see* Atlas Tellien
120 H10 **Atlas Tellien** *Eng.* Tell Atlas. ▲ N Algeria
10 I9 **Atlin** British Columbia, W Canada 59°31′N 133°41′W
10 I9 **Atlin Lake** ◎ British Columbia, W Canada
41 P14 **Atlixco** Puebla, S Mexico 18°55′N 98°26′W
94 B11 **Atløyna** *island* S Norway
155 I17 **Ātmakūr** Andhra Pradesh, C India 15°52′N 78°42′E
23 O8 **Atmore** Alabama, S USA 31°01′N 87°29′W
101 I20 **Atmühl** ⚟ S Germany
94 H11 **Atna** ⚟ S Norway
164 E13 **Atō** Yamaguchi, Honshū, SW Japan 34°24′N 131°42′E
57 L21 **Atocha** Potosí, S Bolivia 20°55′S 66°14′W
27 P12 **Atoka** Oklahoma, C USA 34°22′N 96°08′W
27 O12 **Atoka Lake** *var.* Atoka Reservoir. ⊠ Oklahoma, C USA
Atoka Reservoir *see* Atoka Lake
33 Q14 **Atomic City** Idaho, NW USA 43°26′N 112°48′W
40 L10 **Atotonilco** Zacatecas, C Mexico 24°12′N 102°46′W
Atotonilco *see* Atotonilco el Alto
40 M13 **Atotonilco el Alto** *var.* Atotonilco. Jalisco, SW Mexico 20°35′N 102°30′W
77 N7 **Atouila, 'Erg** *desert* N Mali
41 N16 **Atoyac** *var.* Atoyac de Alvarez. Guerrero, S Mexico 17°12′N 100°28′W
Atoyac de Alvarez *see* Atoyac
41 P15 **Atoyac, Río** ⚟ S Mexico
39 O5 **Atqasuk** Alaska, USA 70°28′N 157°24′W
Atrak/Atrak, Rūd-e *see* Etrek
95 J20 **Ätran** ⚟ S Sweden
54 C7 **Atrato, Río** ⚟ NW Colombia
Atrek *see* Etrek
107 K14 **Atri** Abruzzo, C Italy 42°35′N 13°59′E
Atria *see* Adria
165 P9 **Atsumi** Yamagata, Honshū, C Japan 38°38′N 139°36′E
165 S3 **Atsuta** Hokkaidō, NE Japan 43°28′N 141°24′E
143 Q17 **Aţ Ţaff** *desert* C United Arab Emirates
138 G12 **Aţ Ţafīlah** *var.* Et Tafila, Tafila. Aţ Ţafīlah, W Jordan 30°52′N 35°36′E
138 G12 **Aţ Ţafīlah** *off.* Muḩāfaẓat aţ Ţafīlah. ◆ *governorate* W Jordan
140 L10 **Aţ Ţā'if** Makkah, W Saudi Arabia 21°50′N 40°50′E
Attaleia/Attalia *see* Antalya
23 Q3 **Attalla** Alabama, S USA 34°01′N 86°05′W
138 L2 **At Tall al Abyaḑ** *var.* Tall al Abyaḑ, Tell Abyad, *Fr.* Tell Abiad. Ar Raqqah, N Syria 36°36′N 34°00′E
At Ta'mīm *see* Kirkūk
138 L7 **Aţ Ţanf** Ḩimş, S Syria 33°29′N 38°39′E
163 N9 **Attanshiree** Dornogovĭ, SE Mongolia 45°36′N 110°30′E
167 T10 **Attapu** *var.* Attopeu, Samakhixai. Attapu, S Laos 14°48′N 106°51′E
139 S10 **Aţ Ţaqţaqānah** An Najaf, C Iraq 32°03′N 43°54′E
Attávyros *see* Atávyros
12 G9 **Attawapiskat** Ontario, C Canada 52°55′N 82°26′W
12 F9 **Attawapiskat** ⚟ Ontario, S Canada
12 D9 **Attawapiskat Lake** ◎ Ontario, C Canada
At Taybé *see* Ţayyibah
101 F16 **Attendorn** Nordrhein-Westfalen, W Germany 51°07′N 07°54′E
109 R5 **Attersee** Salzburg, NW Austria 47°55′N 13°31′E
109 R5 **Attersee** ◎ N Austria
99 L24 **Attert** Luxembourg, SE Belgium 49°45′N 05°47′E
138 M4 **At Tibnī** *var.* Tibnī. Dayr az Zawr, NE Syria 35°30′N 39°48′E
31 N13 **Attica** Indiana, N USA 40°17′N 87°15′W
18 E10 **Attica** New York, NE USA 42°51′N 78°13′W
Attica *see* Attikí
13 N7 **Attikamagen Lake** ◎ Newfoundland and Labrador, E Canada
115 H20 **Attikí** *Eng.* Attica. ◆ *region* C Greece
19 O12 **Attleboro** Massachusetts, NE USA 41°55′N 71°15′W
109 R5 **Attnang** Oberösterreich, N Austria 48°02′N 13°44′E
149 U6 **Attock City** Punjab, E Pakistan 33°52′N 72°20′E
Attopeu *see* Attapu
25 X8 **Attoyac River** ⚟ Texas, SW USA
38 D16 **Attu** Attu Island, Alaska, USA 52°53′N 173°18′E
139 Y12 **Aţ Ţūbah** Al Başrah, E Iraq 30°29′N 47°28′E
140 K4 **Aţ Ţubayq** *plain* Jordan/ Saudi Arabia
38 C16 **Attu Island** *island* Aleutian Islands, Alaska, USA
75 X8 **Aţ Ţūr** *var.* El Ţûr. NE Egypt 28°14′N 33°36′E
155 I21 **Āttūr** Tamil Nādu, SE India 11°34′N 78°39′E
141 N17 **At Turbah** SW Yemen 12°42′N 43°31′E
62 I12 **Atuel, Río** ⚟ C Argentina
191 X7 **Atuona** Hiva Oa, NE French Polynesia 09°47′S 139°03′W
Aturus *see* Adour
95 M18 **Åtvidaberg** Östergötland, S Sweden 58°12′N 16°00′E
35 P9 **Atwater** California, W USA 37°19′N 120°33′W
29 T8 **Atwater** Minnesota, N USA 45°08′N 94°48′W
26 I2 **Atwood** Kansas, C USA 39°48′N 101°03′W
31 U12 **Atwood Lake** ⊠ Ohio, N USA
127 P5 **Atyashevo** Respublika Mordoviya, W Russian Federation 54°36′N 46°04′E
144 F12 **Atyrau** *prev.* Gur'yev. Atyrau, W Kazakhstan 47°07′N 51°56′E
144 E11 **Atyrau** *off.* Atyrauskaya Oblast', *var.Kaz.* Atyraū Oblysy; *prev.* Gur'yevskaya Oblast'. ◆ *province* W Kazakhstan
Atyraū Oblysy/ Atyrauskaya Oblast' *see* Atyrau
Atyrauskaya Oblast' *see* Atyrau
108 J7 **Au** Vorarlberg, NW Austria 47°19′N 10°01′E
186 B4 **Aua Island** *island* NW Papua New Guinea
103 S16 **Aubagne** *anc.* Albania. Bouches-du-Rhône, SE France 43°17′N 05°35′E
99 L25 **Aubange** Luxembourg, SE Belgium 49°35′N 05°49′E
103 Q6 **Aube** ◆ *department* N France
103 R6 **Aube** ⚟ N France
99 L19 **Aubel** Liège, E Belgium 50°45′N 05°49′E
103 Q13 **Aubenas** Ardèche, E France 44°37′N 04°24′E
103 O8 **Aubigny-sur-Nère** Cher, C France 47°30′N 02°27′E
103 O13 **Aubin** Aveyron, S France 44°30′N 02°18′E
103 O13 **Aubrac, Monts d'** ▲ S France
36 J10 **Aubrey Cliffs** *cliff* Arizona, SW USA
23 R5 **Auburn** Alabama, S USA 32°37′N 85°30′W
35 P6 **Auburn** California, W USA 38°53′N 121°03′W
30 K14 **Auburn** Illinois, N USA 39°35′N 89°45′W
31 Q11 **Auburn** Indiana, N USA 41°22′N 85°03′W
20 J7 **Auburn** Kentucky, S USA 36°52′N 86°42′W
19 P8 **Auburn** Maine, NE USA 44°05′N 70°15′W
19 N11 **Auburn** Massachusetts, NE USA 42°11′N 71°47′W
29 S16 **Auburn** Nebraska, C USA 40°23′N 95°50′W
18 H10 **Auburn** New York, NE USA 42°55′N 76°31′W
32 H8 **Auburn** Washington, NW USA 47°18′N 122°13′W
103 N11 **Aubusson** Creuse, C France 45°58′N 02°10′E
118 E10 **Auce** *Ger.* Autz. SW Latvia 56°28′N 22°54′E
102 L15 **Auch** *Lat.* Augusta Auscorum, Elimberrum. Gers, S France 43°40′N 00°37′E
77 U16 **Auchi** Edo, S Nigeria 07°01′N 06°17′E
23 T9 **Aucilla River** ⚟ Florida/ Georgia, SE USA
184 L6 **Auckland** Auckland, North Island, New Zealand 36°53′S 174°46′E
184 K5 **Auckland** *off.* Auckland Region. ◆ *region* North Island, New Zealand
184 L6 **Auckland** ✕ Auckland, North Island, New Zealand 37°01′S 174°49′E
192 J12 **Auckland Islands** *island group* S New Zealand
Auckland Region *see* Auckland
103 O16 **Aude** ◆ *department* S France
103 N16 **Aude** *anc.* Atax. ⚟ S France
Audenarde *see* Oudenaarde
Audern *see* Audru
102 E6 **Audierne** Finistère, NW France 48°01′N 04°30′W
102 E6 **Audierne, Baie d'** *bay* NW France
103 U7 **Audincourt** Doubs, E France 47°29′N 06°50′E
118 G5 **Audru** *Ger.* Audern. Pärnumaa, SW Estonia 58°24′N 24°22′E
29 T14 **Audubon** Iowa, C USA 41°44′N 94°56′W
101 N17 **Aue** Sachsen, E Germany 50°35′N 12°42′E
100 H12 **Aue** ⚟ NW Germany
100 L9 **Auerbach** Bayern, SE Germany 49°41′N 11°41′E
101 M17 **Auerbach** Sachsen, E Germany 50°30′N 12°24′E
108 I10 **Auererrhein** ⚟ SW Switzerland
101 N17 **Auersberg** ▲ E Germany 50°30′N 12°42′E
181 W9 **Augathella** Queensland, E Australia 25°54′S 146°38′E
31 Q12 **Auglaize River** ⚟ Ohio, N USA
83 F22 **Augrabies Falls** *waterfall* W South Africa
31 R7 **Au Gres River** ⚟ Michigan, N USA
Augsbourg *see* Augsburg
101 K22 **Augsburg** *Fr.* Augsbourg; *anc.* Augusta Vindelicorum. Bayern, S Germany 48°22′N 10°54′E
180 I14 **Augusta** Western Australia 34°18′S 115°10′E
107 L25 **Augusta** *It.* Agosta. Sicilia, Italy, C Mediterranean Sea 37°14′N 15°14′E
27 W11 **Augusta** Arkansas, C USA 35°16′N 91°21′W
23 V3 **Augusta** Georgia, SE USA 33°29′N 81°58′W
27 O6 **Augusta** Kansas, C USA 37°40′N 96°59′W
19 Q7 **Augusta** *state capital* Maine, NE USA 44°20′N 69°44′W
33 Q8 **Augusta** Montana, NW USA 47°28′N 112°23′W
Augusta *see* London
Augusta Auscorum *see* Auch
Augusta Emerita *see* Mérida
Augusta Praetoria *see* Aosta
Augusta Suessionum *see* Soissons
Augusta Trajana *see* Stara Zagora
Augusta Treverorum *see* Trier
Augusta Vangionum *see* Worms
Augusta Vindelicorum *see* Augsburg
95 G24 **Augustenborg** *Ger.* Augustenburg. Syddanmark, SW Denmark 54°57′N 09°53′E
Augustenburg *see* Augustenborg
39 Q13 **Augustine Island** *island* Alaska, USA
14 L9 **Augustines, Lac des** ◎ Québec, SE Canada
Augustobona Tricassium *see* Troyes
Augustodunum *see* Autun
Augustodurum *see* Bayeux
Augustoritum Lemovicensium *see* Limoges
110 O8 **Augustów** *Rus.* Avgustov. Podlaskie, NE Poland 53°52′N 22°58′E
Augustow Canal *see* Augustowski, Kanał
110 O8 **Augustowski, Kanał** *Eng.* Augustow Canal, *Rus.* Avgustovskiy Kanal. *canal* NE Poland
180 I9 **Augustus, Mount** ▲ Western Australia 24°42′S 117°42′E
Aujuittuq *see* Grise Fiord
186 M9 **Auki** Malaita, N Solomon Islands 08°48′S 160°45′E
21 W8 **Aulander** North Carolina, SE USA 36°15′N 77°16′W
180 L7 **Auld, Lake** *salt lake* C Western Australia
Aulie Ata/Auliye-Ata *see* Taraz
Äūliekōl *see* Auliyekol'
144 M8 **Auliyekol'** *Kaz.* Äūliekōl; *prev.* Semiozernoye. Kostanay, N Kazakhstan 52°22′N 64°06′E
106 E10 **Aulla** Toscana, C Italy 44°15′N 10°00′E
102 F6 **Aulne** ⚟ NW France
Aulong *see* Ulong
37 T3 **Ault** Colorado, C USA 40°34′N 104°43′W
95 F22 **Aulum** *var.* avlum. Midtjylland, C Denmark 56°16′N 08°48′E
103 N3 **Aumale** Seine-Maritime, N France 49°45′N 01°43′E
Auminzatau, Gory *see* Ovminzatovo Tog'lari
77 T14 **Auna** Niger, W Nigeria 10°13′N 04°43′E
166 L6 **Aunglan** *var.* Allanmyo, Myaydo. Magway, C Myanmar (Burma) 19°25′N 95°13′E
95 H21 **Auning** Midtjylland, C Denmark 56°26′N 10°23′E
192 K17 **'Aunu'u Island** *island* W American Samoa
83 E20 **Auob** *var.* Oup. ⚟ Namibia/South Africa
93 K19 **Aura** Varsinais-Suomi, SW Finland 60°37′N 22°35′E
109 R5 **Aurach** ⚟ N Austria
Aural, Phnom *see* Aôral, Phnum
153 O14 **Aurangābād** Bihār, N India 24°46′N 84°23′E
154 F13 **Aurangābād** Mahārāshtra, C India 19°52′N 75°22′E
189 V7 **Aur Atoll** *atoll* E Marshall Islands
102 G7 **Auray** Morbihan, NW France 47°40′N 02°59′W
94 G13 **Aurdal** Oppland, S Norway 60°51′N 09°25′E
94 F8 **Aure** Møre og Romsdal, S Norway 63°16′N 08°31′E
29 T12 **Aurelia** Iowa, C USA 42°42′N 95°26′W
Aurelia Aquensis *see* Baden-Baden
Aurelianum *see* Orléans
120 J10 **Aurès, Massif de l'** ▲ NE Algeria
100 F10 **Aurich** Niedersachsen, NW Germany 53°28′N 07°28′E
103 O13 **Aurillac** Cantal, C France 44°56′N 02°26′E
Aurine, Alpi *see* Zillertaler Alpen
Aurium *see* Ourense
14 H15 **Aurora** Ontario, S Canada 44°00′N 79°26′W
55 S8 **Aurora** NW Guyana 06°46′N 59°45′W
37 T4 **Aurora** Colorado, C USA 39°42′N 104°51′W
30 M11 **Aurora** Illinois, N USA 41°46′N 88°19′W
31 Q15 **Aurora** Indiana, N USA 39°01′N 84°55′W
29 W4 **Aurora** Minnesota, N USA 47°31′N 92°14′W
27 S8 **Aurora** Missouri, C USA 36°58′N 93°43′W
29 P16 **Aurora** Nebraska, C USA 40°52′N 98°00′W
36 J5 **Aurora** Utah, W USA 38°55′N 111°55′W
Aurora *see* San Francisco, Philippines
Aurora *see* Maéwo, Vanuatu
94 F10 **Aursjøen** ◎ S Norway
94 I9 **Aursunden** ◎ S Norway
83 D21 **Aus** Karas, SW Namibia 26°38′S 16°19′E
Ausa *see* Vic
14 E16 **Ausable** ⚟ Ontario, S Canada
31 O3 **Au Sable Point** *headland* Michigan, N USA 46°40′N 86°08′W
31 S7 **Au Sable Point** *headland* Michigan, N USA 44°19′N 83°20′W
31 R6 **Au Sable River** ⚟ Michigan, N USA
57 H16 **Ausangate, Nevado** ▲ C Peru 13°47′S 71°13′W
Auschwitz *see* Oświęcim
Ausculum Apulum *see* Ascoli Satriano
105 Q4 **Ausejo** La Rioja, N Spain 42°21′N 02°10′W
108 H7 **Ausser Rhoden** ◆ *canton* NE Switzerland
Aussig *see* Ústí nad Labem
95 F17 **Aust-Agder** ◆ *county* S Norway
92 P2 **Austfonna** *glacier* NE Svalbard
31 P15 **Austin** Indiana, N USA 38°45′N 85°48′W
29 W11 **Austin** Minnesota, N USA 43°40′N 92°58′W
35 U5 **Austin** Nevada, W USA 39°30′N 117°05′W
25 S10 **Austin** *state capital* Texas, SW USA 30°16′N 97°45′W
180 J10 **Austin, Lake** *salt lake* Western Australia
31 V11 **Austintown** Ohio, N USA 41°06′N 80°45′W
25 V9 **Austonio** Texas, SW USA 31°09′N 95°39′W
Australes, Archipel des *see* Australes, Îles
Australes et Antarctiques Françaises, Terres *see* French Southern and Antarctic Lands
191 T14 **Australes, Îles** *var.* Archipel des Australes, Îles Tubuai, Tubuai Islands, *Eng.* Austral Islands. *island group* SW French Polynesia
175 Y11 **Austral Fracture Zone** *tectonic feature* S Pacific Ocean
174 M8 **Australia** *continent*
181 O7 **Australia** *off.* Commonwealth of Australia. ◆ *commonwealth republic*
Australia, Commonwealth of *see* Australia
183 Q12 **Australian Alps** ▲ SE Australia
183 R11 **Australian Capital Territory** *prev.* Federal Capital Territory. ◆ *territory* SE Australia
Australie, Bassin Nord de l' *see* North Australian Basin
Austral Islands *see* Australes, Îles
Austrava *see* Ostrov
109 T6 **Austria** *off.* Republic of Austria, *Ger.* Österreich. ◆ *republic* C Europe
Austria, Republic of *see* Austria
92 K3 **Austurland** ◆ *region* SE Iceland
92 G10 **Austvågøya** *island* C Norway
Ausuitoq *see* Grise Fiord
58 G13 **Autazes** Amazonas, N Brazil 03°37′S 59°08′W
102 M16 **Auterive** Haute-Garonne, S France 43°22′N 01°28′E
Autesiodorum *see* Auxerre
103 N2 **Authie** ◆ N France
Autissiodorum *see* Auxerre
40 K14 **Autlán** *var.* Autlán de Navarro. Jalisco, SW Mexico 19°48′N 104°20′W
Autlán de Navarro *see* Autlán
Autricum *see* Chartres
103 Q9 **Autun** *anc.* Ædua, Augustodunum. Saône-et-Loire, C France 46°58′N 04°18′E
Autz *see* Auce
99 H20 **Auvelais** Namur, S Belgium 50°27′N 04°38′E
103 P11 **Auvergne** ◆ *region* C France
102 M12 **Auvézère** ⚟ W France
103 P7 **Auxerre** *anc.* Autesiodorum, Autissiodorum. Yonne, C France 47°48′N 03°35′E
103 N2 **Auxi-le-Château** Pas-de-Calais, N France 50°14′N 02°06′E
103 S8 **Auxonne** Côte d'Or, C France 47°12′N 05°22′E
55 P9 **Auyan Tebuy** ▲ SE Venezuela 05°48′N 62°27′W
103 O10 **Auzances** Creuse, C France 46°01′N 02°29′E
27 U8 **Ava** Missouri, C USA 36°57′N 92°39′W
142 M5 **Āvaj** Qazvin, N Iran
95 C15 **Avaldsnes** Rogaland, S Norway 59°21′N 05°16′E
103 Q8 **Avallon** Yonne, C France 47°30′N 03°54′E
102 K6 **Avaloirs, Mont des** ▲ NW France 48°27′N 00°11′W
35 S16 **Avalon** Santa Catalina Island, California, W USA 33°20′N 118°19′W
18 J17 **Avalon** New Jersey, NE USA 39°04′N 74°42′W
13 V13 **Avalon Peninsula** *peninsula* Newfoundland and Labrador, E Canada
Avanersuaq *see* Avannaarsua
Avannaarsua *var.* Avanersuaq, *Dan.* Nordgrønland. ◆ *province* N Greenland
60 K10 **Avaré** São Paulo, S Brazil 23°06′S 48°57′W
Avaricum *see* Bourges
190 H16 **Avarua** ○ (Cook Islands) Rarotonga, S Cook Islands 21°12′S 159°46′E
190 H16 **Avarua Harbour** *harbour* Rarotonga, S Cook Islands
Avasfelsőfalu *see* Negrești-Oaș
38 L17 **Avatanak Island** *island* Aleutian Islands, Alaska, USA
190 B16 **Avatele** S Niue 19°06′S 169°55′E
190 H15 **Avatiu Harbour** *harbour* Rarotonga, S Cook Islands
Avdeyevka *see* Avdiyivka
114 J13 **Ávdira** Anatolikí Makedonía kai Thráki, NE Greece 40°58′N 24°58′E
117 X8 **Avdiyivka** *Rus.* Avdeyevka. Donets'ka Oblast', SE Ukraine 48°06′N 37°46′E
Avdzaga *see* Gurvanbulag
104 G6 **Ave** ⚟ N Portugal
104 G7 **Aveiro** *anc.* Talabriga. Aveiro, W Portugal 40°38′N 08°40′W
104 G7 **Aveiro** ◆ *district* N Portugal
Avela *see* Ávila
99 D18 **Avelgem** West-Vlaanderen, W Belgium 50°46′N 03°25′E
61 D20 **Avellaneda** Buenos Aires, E Argentina 34°43′S 58°23′W
107 L17 **Avellino** *anc.* Abellinum. Campania, S Italy 40°54′N 14°46′E
35 Q12 **Avenal** California, W USA 36°00′N 120°07′W
Avenio *see* Avignon
94 E8 **Averøya** *island* S Norway
107 K17 **Aversa** Campania, S Italy 40°58′N 14°13′E
33 N9 **Avery** Idaho, NW USA 47°14′N 115°48′W
25 W5 **Avery** Texas, SW USA 33°33′N 94°46′W
Aves, Islas de *see* Las Aves, Islas
Avesnes *see* Avesnes-sur-Helpe
103 Q2 **Avesnes-sur-Helpe** *var.* Avesnes. Nord, N France 50°08′N 03°57′E
64 G12 **Aves Ridge** *undersea feature* SE Caribbean Sea 14°00′N 63°30′W
95 M14 **Avesta** Dalarna, C Sweden 60°09′N 16°10′E
103 O14 **Aveyron** ◆ *department* S France
103 N14 **Aveyron** ⚟ S France
107 J15 **Avezzano** Abruzzo, C Italy 42°02′N 13°26′E
115 D16 **Avgó** ▲ C Greece 39°31′N 21°24′E
Avgustov *see* Augustów
Avgustovskiy Kanal *see* Augustowski, Kanał
96 J9 **Aviemore** N Scotland, United Kingdom 57°06′N 04°01′W
185 F21 **Aviemore, Lake** ◎ South Island, New Zealand
103 R15 **Avignon** *anc.* Avenio. Vaucluse, SE France 43°57′N 04°49′E
104 M7 **Ávila** *var.* Avila; *anc.* Abela, Abula, Abyla, Avela. Castilla y León, C Spain 40°39′N 04°42′W
104 L8 **Ávila** ◆ *province* Castilla y León, C Spain
104 K2 **Avilés** Asturias, NW Spain 43°33′N 05°55′W
118 J4 **Avinurme** *Ger.* Awwinorm. Ida-Virumaa, NE Estonia 58°58′N 26°53′E
104 H10 **Avis** Portalegre, C Portugal 39°03′N 07°53′W
Avlum *see* Aulum
182 M11 **Avoca** Victoria, SE Australia 37°09′S 143°34′E
29 T14 **Avoca** Iowa, C USA 41°27′N 95°20′W
182 M11 **Avoca River** ⚟ Victoria, SE Australia
107 L25 **Avola** Sicilia, Italy, C Mediterranean Sea 36°54′N 15°08′E
18 F10 **Avon** New York, NE USA 42°53′N 77°41′W
29 P12 **Avon** South Dakota, N USA 43°00′N 98°03′W
97 M23 **Avon** ⚟ S England, United Kingdom
97 L20 **Avon** ⚟ C England, United Kingdom
36 K13 **Avondale** Arizona, SW USA 33°25′N 112°20′W
23 X13 **Avon Park** Florida, SE USA 27°36′N 81°30′W
102 J5 **Avranches** Manche, N France 48°42′N 01°21′W
Avveel *see* Ivalojoki, Finland
186 M6 **Avuavu** *var.* Kolotambu. Guadalcanal, C Solomon Islands 09°52′S 160°25′E
103 O3 **Avure** ⚟ N France
Avveel *see* Ivalo, Finland
Avvil *see* Ivalo
77 O17 **Awaaso** *var.* Awaso. SW Ghana 06°10′N 02°18′W
141 X8 **Awābī** *var.* Al 'Awābī. NE Oman 23°20′N 57°35′E
184 L9 **Awakino** Waikato, North Island, New Zealand 38°40′S 174°37′E

◆ Country
● Country Capital
◇ Dependent Territory
○ Dependent Territory Capital
◆ Administrative Regions
✕ International Airport
▲ Mountain
▲ Mountain Range
⊼ Volcano
⚟ River
◎ Lake
⊠ Reservoir

142 M15 'Awālī C Bahrain 26°07′N 50°33′E
99 K19 Awans Liège, E Belgium 50°39′N 05°30′E
184 I2 Awanui Northland, North Island, New Zealand 35°01′S 173°16′E
148 M14 Awārān Baluchistān, SW Pakistan 26°31′N 65°10′E
81 K16 Awara Plain *plain* NE Kenya
80 M13 Āwarē Sumalē, E Ethiopia 08°12′N 44°09′E
138 M6 'Awāriḍ, Wādī *dry watercourse* E Syria
185 B20 Awarua Point *headland* South Island, New Zealand 44°15′S 168°03′E
81 J14 Āwasa Southern Nationalities, S Ethiopia 06°54′N 38°26′E
80 K13 Āwash Āfar, NE Ethiopia 08°59′N 40°16′E
80 K12 Āwash *var.* Hawash. ≈ C Ethiopia
Awaso *see* Awaaso
158 H7 Awat Xinjiang Uygur Zizhiqu, NW China 40°36′N 80°22′E
185 J15 Awatere ≈ South Island, New Zealand
75 O10 Awbārī SW Libya 26°35′N 12°46′E
75 N9 Awbāri, Idhān *var.* Edeyen d'Oubari. *desert* Algeria/Libya
80 M12 Awdal *off.* Gobolka Awdal. ◆ N Somalia
80 C13 Aweil Northern Bahr el Ghazal, NW South Sudan 08°42′N 27°20′E
96 H11 Awe, Loch ◎ W Scotland, United Kingdom
77 U16 Awka Anambra, SW Nigeria 06°12′N 07°04′E
39 O6 Awuna River ≈ Alaska, USA
Awwinorm *see* Avinurme
Ax *see* Dax
Axarfjördhur *see* Öxarfjörður
103 N17 Axat Aude, S France 42°47′N 02°14′E
99 F16 Axel Zeeland, SW Netherlands 51°16′N 03°55′E
197 P9 Axel Heiberg Island *var.* Axel Heiburg. *island* Nunavut, N Canada
Axel Heiburg *see* Axel Heiberg Island
77 O17 Axim S Ghana 04°53′N 02°14′W
114 F13 Axiós *var.* Vardar. ≈ Greece/FYR Macedonia *see also* Vardar
Axiós *see* Vardar
103 N17 Ax-les-Thermes Ariège, S France 42°43′N 01°49′E
120 D11 Ayachi, Jbel ▲ C Morocco 32°30′N 05°00′W
61 D22 Ayacucho Buenos Aires, E Argentina 37°09′S 58°30′W
57 F15 Ayacucho Ayacucho, S Peru 13°10′S 74°15′W
57 E16 Ayacucho *off.* Departamento de Ayacucho. ◆ *department* SW Peru
Ayacucho, Departamento de *see* Ayacucho
145 W11 Ayagoz *var.* Ayaguz, *Kaz.* Ayakōz; *prev.* Sergiopol. Vostochnyy Kazakhstan, E Kazakhstan 47°54′N 80°25′E
145 V12 Ayagoz *var.* Ayaguz, *Kaz.* Ayakoz. ≈ E Kazakhstan
Ayaguz *see* Ayagoz
Ayakagytma *see* Oyoqog'itma
Ayakkuduk *see* Oyoqquduq
158 L10 Ayakkum Hu ◎ NW China
Ayaköz *see* Ayagoz
104 H14 Ayamonte Andalucía, S Spain 37°13′N 07°24′W
123 S11 Ayan Khabarovskiy Kray, E Russian Federation 56°27′N 138°09′E
136 J10 Ayancık Sinop, N Turkey 41°56′N 34°35′E
55 S9 Ayanganna Mountain ▲ C Guyana 05°21′N 59°54′W
77 U16 Ayangba Kogi, C Nigeria 07°36′N 07°10′E
123 U7 Ayanka Krasnoyarskiy Kray, E Russian Federation 63°42′N 167°31′E
54 E7 Ayapel Córdoba, NW Colombia 08°16′N 75°10′W
136 H12 Ayaş Ankara, N Turkey 40°02′N 32°21′E
57 I16 Ayaviri Puno, S Peru 14°53′S 70°35′W
Āybak *see* Aibak
147 N10 Aydarkoʻl Koʻli *Rus.* Ozero Aydarkul'. ◎ C Uzbekistan
Aydarkul', Ozero *see* Aydarkoʻl Koʻli
21 W10 Ayden North Carolina, SE USA 35°28′N 77°25′W
136 C15 Aydın *var.* Aïdin; *anc.* Tralles Aydin. Aydın, SW Turkey 37°51′N 27°51′E
136 C15 Aydın *var.* Aïdin. ◆ *province* SW Turkey
136 I17 Aydıncık İçel, S Turkey 36°08′N 33°17′E
136 C15 Aydın Dağları ▲ W Turkey
158 L6 Aydingkol Hu ◎ NW China
127 X7 Aydyrlinskiy Orenburgskaya Oblast', W Russian Federation 52°03′N 59°54′E
105 S4 Ayerbe Aragón, NE Spain 42°16′N 00°41′W
Ayers Rock *see* Uluru
Ayeyarwady *see* Ayeyawady
Ayeyarwady *see* Irrawaddy
166 K8 Ayeyawady *prev.* Ayeyarwady *var.* Irrawaddy. ◆ *region* SW Myanmar (Burma)
Ayiá *see* Agiá
Ayia Napa *see* Agía Nápa
Ayia Phyla *see* Agía Fýlaxis
Ayiásos/Ayiássos *see* Agiassos
Áyios Evstrátios *see* Ágios Efstrátios
Áyios Kírikos *see* Ágios Kírykos
Áyios Nikólaos *see* Ágios Nikólaos
80 I11 Āykel Āmara, N Ethiopia 12°33′N 37°01′E
123 N9 Aykhal Respublika Sakha (Yakutiya), NE Russian Federation 66°07′N 110°25′E
14 J12 Aylen Lake ◎ Ontario, SE Canada
97 N21 Aylesbury SE England, United Kingdom 51°50′N 00°50′W
105 O6 Ayllón Castilla y León, N Spain 41°25′N 03°23′W
14 F17 Aylmer Ontario, S Canada 42°46′N 80°57′W
14 L12 Aylmer Québec, SE Canada 45°23′N 75°51′W
15 R12 Aylmer, Lac ◎ Québec, SE Canada
8 L9 Aylmer Lake ◎ Northwest Territories, NW Canada
145 V14 Aynabulak *Kaz.* Aynabulaq. Almaty, SE Kazakhstan 44°37′N 77°59′E
Aynabulaq *see* Aynabulak
138 K2 'Ayn al 'Arab *Kurd.* Kobanî. Ḩalab, N Syria 36°55′N 38°21′E
Aynayn *see* 'Aynīn
139 V12 'Ayn Ḩamūd Dhī Qār, S Iraq 30°51′N 45°37′E
147 P12 Ayní *prev.* Varzimanor Ayni. W Tajikistan 39°24′N 68°30′E
140 M10 'Aynīn *var.* Aynayn. *spring/well* SW Saudi Arabia 20°52′N 41°41′E
21 U12 Aynor South Carolina, SE USA 33°59′N 79°11′W
139 Q7 'Ayn Zāzūh Al Anbār, C Iraq 33°29′N 42°34′E
153 N12 Ayodhya Uttar Pradesh, N India 26°47′N 82°12′E
123 S6 Ayon, Ostrov *island* NE Russian Federation
105 R11 Ayora Valenciana, E Spain 39°04′N 01°04′W
77 Q11 Ayorou Tillabéri, W Niger 14°45′N 00°54′E
79 E16 Ayos Centre, S Cameroon 03°53′N 12°31′E
76 L5 'Ayoûn 'Abd el Mâlek *well* N Mauritania
76 K10 'Ayoûn el 'Atroûs *var.* Aioun el Atrous, Aïoun el Atroûss. Hodh el Gharbi, SE Mauritania 16°38′N 09°36′W
96 I13 Ayr W Scotland, United Kingdom 55°28′N 04°38′W
96 I13 Ayr ≈ W Scotland, United Kingdom
96 I13 Ayrshire *cultural region* SW Scotland, United Kingdom
Aysen *see* Aisén
80 L12 Āysha Sumalē, E Ethiopia 10°36′N 42°31′E
144 L14 Ayteke Bi *Kaz.* Zhangaqazaly; *prev.* Novokazalinsk. Kzylorda, SW Kazakhstan 45°53′N 62°10′E
146 K8 Aytim Navoiy Viloyati, N Uzbekistan 42°15′N 63°25′E
181 W4 Ayton Queensland, NE Australia 15°54′S 145°19′E
114 M9 Aytos Burgas, E Bulgaria 42°43′N 27°14′E
171 T11 Ayu, Kepulauan *island group* E Indonesia
167 U12 A Yun Pa *prev.* Cheo Reo. Gia Lai, S Vietnam 13°19′N 108°27′E
169 V11 Ayu, Tanjung *headland* Borneo, N Indonesia 0°25′N 117°34′E
41 P16 Ayutlá *var.* Ayutla de los Libres. Guerrero, S Mexico 16°51′N 99°16′W
40 K13 Ayutla Jalisco, C Mexico 20°07′N 104°18′W
Ayutla de los Libres *see* Ayutlá
167 O11 Ayutthaya *var.* Phra Nakhon Si Ayutthaya. Phra Nakhon Si Ayutthaya, C Thailand 14°20′N 100°35′E
136 B13 Ayvalık Balıkesir, W Turkey 39°18′N 26°42′E
99 L20 Aywaille Liège, E Belgium 50°28′N 05°40′E
141 R13 'Aywat aş Şay'ar, Wādī *seasonal river* N Yemen
Azaffal *see* Azeffâl
105 T9 Azahar, Costa del *coastal region* E Spain
105 S6 Azaila Aragón, NE Spain 41°17′N 00°20′W
104 F10 Azambuja Lisboa, C Portugal 39°04′N 08°52′W
153 N13 Āzamgarh Uttar Pradesh, N India 26°03′N 83°10′E
77 O9 Azaouâd *desert* C Mali
77 S10 Azaouagh, Vallée de l' *var.* Azaouak. ≈ W Niger
Azaouak *see* Azaouagh, Vallée de l'
61 F14 Azara Misiones, NE Argentina 28°03′S 55°42′W
Azaran *see* Hashtrūd
Azärbaycan/Azärbaycan Respublikasi *see* Azerbaijan
Āzarbāyjān-e Bākhtarī *see* Āzarbāyjān-e Gharbī
142 I4 Āzarbāyjān-e Gharbī *off.* Ostān-e Āzarbāyjān-e Gharbī, *Eng.* West Azerbaijan; *prev.* Āzarbāyjān-e Bākhtarī. ◆ *province* NW Iran
Āzarbāyjān-e Gharbī, Ostān-e *see* Āzarbāyjān-e Gharbī
Āzarbāyjān-e Khāvarī *see* Āzarbāyjān-e Sharqī
142 J3 Āzarbāyjān-e Sharqī *off.* Ostān-e Āzarbāyjān-e Sharqī, *Eng.* East Azerbaijan; *prev.* Āzarbāyjān-e Khāvarī. ◆ *province* NW Iran
Āzarbāyjān-e Sharqī, Ostān-e *see* Āzarbāyjān-e Sharqī
77 W13 Azare Bauchi, N Nigeria 11°41′N 10°09′E
119 M19 Azarychy *Rus.* Ozarichi. Homyel'skaya Voblasts', SE Belarus 52°31′N 29°19′E
145 P13 Azat, Gory *hill* C Kazakhstan
102 L8 Azay-le-Rideau Indre-et-Loire, C France 47°16′N 00°25′E
138 I2 A'zāz Ḩalab, NW Syria 36°35′N 37°03′E
76 H7 Azeffâl *var.* Azaffal. *desert* Mauritania/Western Sahara
137 V12 Azerbaijan *off.* Azerbaijani Republic, *Az.* Azärbaycan, Azärbaycan Respublikasi; *prev.* Azerbaijan SSR. ◆ *republic* SE Asia
Azerbaijani Republic *see* Azerbaijan
Azerbaijan SSR *see* Azerbaijan
74 F7 Azilal C Morocco 31°58′N 06°53′W
Azimabad *see* Patna
19 O6 Azischos Lake ◎ Maine, NE USA
Azizbekov *see* Vayk'
Azizie *see* Telish
Aziziya *see* Al 'Azīzīyah
127 T4 Aznakayevo Respublika Tatarstan, W Russian Federation 54°55′N 53°15′E
56 C8 Azogues Cañar, S Ecuador 02°44′S 78°48′W
64 N2 Azores *var.* Açores, Ilhas dos Açores, *Port.* Arquipélago dos Açores. *island group* Portugal, NE Atlantic Ocean
64 L8 Azores-Biscay Rise *undersea feature* E Atlantic Ocean 19°00′W 42°40′N
Azotos/Azotus *see* Ashdod
78 K11 Azoum, Bahr *seasonal river* SE Chad
126 L12 Azov Rostovskaya Oblast', SW Russian Federation 47°07′N 39°26′E
126 J13 Azov, Sea of *Rus.* Azovskoye More, *Ukr.* Azovs'ke More. *sea* NE Black Sea
Azovs'ke More/Azovskoye More *see* Azov, Sea of
138 I10 Azraq, Wāḩat al *oasis* N Jordan
Āzro *see* Āzrow
74 G6 Azrou C Morocco 33°30′N 05°12′W
149 R5 Āzrow *var.* Āzro. Lōgar, E Afghanistan 34°11′N 69°39′E
37 P8 Aztec New Mexico, SW USA 36°49′N 107°59′W
36 M13 Aztec Peak ▲ Arizona, SW USA 33°48′N 110°54′W
45 N9 Azua *var.* Azua de Compostela. S Dominican Republic 18°29′N 70°44′W
Azua de Compostela *see* Azua
104 K12 Azuaga Extremadura, W Spain 38°16′N 05°40′W
56 B8 Azuay ◆ *province* W Ecuador
164 C13 Azuchi-Ō-shima *island* SW Japan
105 O11 Azuer ≈ C Spain
43 S17 Azuero, Península de *peninsula* S Panama
62 I6 Azufre, Volcán *var.* Volcán Lastarria. ⛰ N Chile 25°16′S 68°35′W
116 J12 Azuga Prahova, SE Romania 45°27′N 25°34′E
61 C22 Azul Buenos Aires, E Argentina 36°46′S 59°50′W
62 I8 Azul, Cerro ▲ NW Argentina 28°28′S 68°43′W
56 E12 Azul, Cordillera ▲ C Peru
165 P11 Azuma-san ▲ Honshū, C Japan 37°44′N 140°05′E
103 V15 Azur, Côte d' *coastal region* SE France
191 Z3 Azur Lagoon ◎ Kiritimati, E Kiribati
'Azza *see* Gaza
Az Zāb al Kabīr *see* Great Zab
138 H7 Az Zabdānī *var.* Zabadani. Rīf Dimashq, W Syria 33°45′N 36°07′E
141 W8 Az Zāhirah *desert* NW Oman
141 S6 Az Zahrān *Eng.* Dhahran. Ash Sharqīyah, NE Saudi Arabia 26°18′N 50°02′E
141 R6 Az Zahrān al Khubar *var.* Dhahran Al Khobar. ✈ Ash Sharqīyah, NE Saudi Arabia 26°28′N 49°42′E
75 W7 Az Zaqāzīq *var.* Zagazig. N Egypt 30°36′N 31°32′E
138 H10 Az Zarqā' *var.* Zarqa. Az Zarqā', NW Jordan 32°04′N 36°06′E
138 I11 Az Zarqā' *off.* Muḩāfazat az Zarqā', *var.* Zarqa. ◆ *governorate* N Jordan
75 O7 Az Zāwiyah *var.* Zawia. NW Libya 32°45′N 12°44′E
141 N15 Az Zaydīyah W Yemen 15°20′N 43°03′E
74 I11 Azzel Matti, Sebkha *var.* Sebkra Azz el Matti. *salt flat* C Algeria
141 P6 Az Zilfī Ar Riyāḑ, N Saudi Arabia 26°17′N 44°48′E
139 Y13 Az Zubayr *var.* Al Zubair. Al Başrah, SE Iraq 30°24′N 47°45′E
Az Zuqur *see* Jabal Zuqar, Jazīrat

B

187 X15 Ba *prev.* Mba. Viti Levu, W Fiji 17°35′S 177°40′E
Ba *see* Đa Răng, Sông
171 P17 Baa Pulau Rote, C Indonesia 10°44′S 123°06′E
138 H7 Baalbek *var.* Ba'labakk; *anc.* Heliopolis. E Lebanon 34°00′N 36°15′E
108 G8 Baar Zug, N Switzerland 47°12′N 08°32′E
81 L17 Baardheere *var.* Bardere, *It.* Bardera. Gedo, SW Somalia 02°13′N 42°19′E
Baargaal *see* Bargaal
99 I15 Baarle-Hertog Antwerpen, N Belgium 51°26′N 04°56′E
99 I15 Baarle-Nassau Noord-Brabant, S Netherlands 51°27′N 04°56′E
98 J11 Baarn Utrecht, C Netherlands 52°13′N 05°16′E
162 H9 Baatsagaan *var.* Bayansayr. Bayanhongor, C Mongolia 45°36′N 99°27′E
114 D13 Baba *var.* Buševa, *Gk.* Varnoús. ▲ FYR Macedonia/Greece
76 H10 Bababé Brakna, W Mauritania 16°22′N 13°57′W
136 G10 Baba Burnu *headland* NW Turkey 41°18′N 31°24′E
117 N13 Babadag Tulcea, SE Romania 44°53′N 28°47′E
137 X10 Babadağ Dağı ▲ NE Azerbaijan 41°02′N 48°04′E
146 H14 Babadayhan *Rus.* Babadaykhan; *prev.* Kirovsk. Ahal Welaýaty, C Turkmenistan 37°39′N 60°17′E
Babadaykhan *see* Babadayhan
146 G14 Babadurmaz Ahal Welaýaty, C Turkmenistan 37°39′N 59°03′E
114 M12 Babaeski Kırklareli, NW Turkey 41°26′N 27°06′E
139 T4 Bāba Gurgur Kirkūk, N Iraq 35°34′N 44°18′E
56 B7 Babahoyo *prev.* Bodegas. Los Ríos, C Ecuador 01°53′S 79°31′W
149 P5 Bābā, Kūh-e ▲ C Afghanistan
171 N12 Babana Sulawesi, C Indonesia 02°03′S 119°13′E
Babao *see* Qilian
171 Q12 Babar, Kepulauan *island group* E Indonesia
171 T12 Babar, Pulau *island* Kepulauan Babar, E Indonesia
Bābāsar Pass *see* Babusar Pass
Babashy, Gory *see* Başgyr
146 C9 Babaşy *Rus.* Gory Babashy. ▲ W Turkmenistan
168 M13 Babat Sumatera, W Indonesia 02°45′S 104°01′E
Babatag, Khrebet *see* Bototog', Tizmasi
81 H21 Babati Manyara, NE Tanzania 04°12′S 35°45′E
124 J13 Babayevo Vologodskaya Oblast', NW Russian Federation 59°23′N 35°52′E
127 Q15 Babayurt Respublika Dagestan, SW Russian Federation 43°38′N 46°49′E
33 P6 Babb Montana, NW USA 48°51′N 113°26′W
29 X4 Babbitt Minnesota, N USA 47°42′N 91°56′W
188 E9 Babeldaob *var.* Babeldaop, Babelthuap. *island* N Palau
Babeldaop *see* Babeldaob
141 N17 Bab el Mandeb *strait* Gulf of Aden/Red Sea
Babelthuap *see* Babeldaob
111 K17 Babia Góra *var.* Babia Hora. ▲ Poland/Slovakia 49°33′N 19°32′E
Babia Hora *see* Babia Góra
Babian Jiang *see* Black River
Babichi *see* Babichy
119 N19 Babichy *Rus.* Babichi. Homyel'skaya Voblasts', SE Belarus 52°17′N 30°00′E
139 U9 Bābil *off.* Muḩāfa at Bābil, *var.* Babylon, Al Ḩillah. ◆ *governorate* S Iraq
Bābil, Muḩāfa at at *see* Bābil
112 I10 Babina Greda Vukovar-Srijem, E Croatia 45°09′N 18°33′E
10 K13 Babine Lake ◎ British Columbia, SW Canada
143 O4 Bābol *var.* Babul, Balfrush, Barfrush; *prev.* Barfurush. Māzandarān, N Iran 36°34′N 52°39′E
143 O4 Bābolsar *var.* Babulsar; *prev.* Meshed-i-Sar. Māzandarān, N Iran 36°41′N 52°39′E
36 L16 Baboquivari Peak ▲ Arizona, SW USA 31°46′N 111°36′W
79 G15 Baboua Nana-Mambéré, W Central African Republic 05°46′N 14°47′E
119 M17 Babruysk *Rus.* Bobruysk. Mahilyowskaya Voblasts', E Belarus 53°07′N 29°13′E
Babu *see* Hezhou
Babul *see* Bābol
Babulsar *see* Bābolsar
113 O19 Babuna ▲ C FYR Macedonia
113 O19 Babuna ≈ C FYR Macedonia
152 G4 Babusar Pass *prev.* Bābāsar Pass. *pass* India/Pakistan
148 K7 Bābūs, Dasht-e *Pash.* Bebas, Dasht-i. ▲ W Afghanistan
171 O1 Babuyan Channel *channel* N Philippines
171 O1 Babuyan Islands *island* N Philippines
139 T9 Babylon *site of ancient city* C Iraq
Babylon *see* Bābil
112 J9 Bač *Ger.* Batsch. Vojvodina, NW Serbia 45°24′N 19°17′E
58 M13 Bacabal Maranhão, E Brazil 04°15′S 44°45′W
41 Y14 Bacalar Quintana Roo, SE Mexico 18°38′N 88°17′W
41 Y14 Bacalar Chico, Boca *strait* SE Mexico
171 Q12 Bacan, Kepulauan *island group* E Indonesia
171 S12 Bacan, Pulau *prev.* Batjan. *island* Maluku, E Indonesia
116 L10 Bacău *Hung.* Bákó. Bacău, NE Romania 46°36′N 26°56′E
116 K11 Bacău ◆ *county* E Romania
Bắc Bộ, Vinh *see* Tonkin, Gulf of
167 T5 Bắc Can *var.* Bach Thong. Bắc Thai, N Vietnam 22°07′N 105°50′E
103 T5 Baccarat Meurthe-et-Moselle, NE France 48°27′N 06°46′E
183 N12 Bacchus Marsh Victoria, SE Australia 37°41′S 144°30′E
40 H4 Bacerac Sonora, NW Mexico 30°27′N 108°55′W
116 L10 Băceşti Vaslui, E Romania 46°50′N 27°14′E
167 T6 Bắc Giang Ha Băc, N Vietnam 21°17′N 106°12′E
54 I5 Bachaquero Zulia, NW Venezuela 09°57′N 71°09′W
Bacher *see* Pohorje
118 M13 Bacheykava *Rus.* Bocheykovo. Vitsyebskaya Voblasts', N Belarus 55°01′N 29°09′E
40 I5 Bachíniva Chihuahua, N Mexico 28°41′N 107°13′W
Bach Quy, Đao *see* Passu Keah
Bach Thong *see* Bắc Can
158 G8 Bachu Xinjiang Uygur Zizhiqu, NW China 39°46′N 78°30′E
9 N8 Back ≈ Nunavut, N Canada
112 K10 Bačka Palanka *prev.* Palanka. Serbia, NW Serbia 44°22′N 20°57′E
112 K8 Bačka Topola *Hung.* Topolya; *prev. Hung.* Bácstopolya. Vojvodina, N Serbia 45°48′N 19°39′E
95 J17 Bäckefors Västra Götaland, S Sweden 58°49′N 12°07′E
Bäckermühle Schulzenmühle *see* Żywiec
95 L16 Bäckhammar Värmland, C Sweden 59°09′N 14°13′E
112 K9 Bački Petrovac *Hung.* Petrőcz; *prev.* Petrovac, Petrovácz. Vojvodina, NW Serbia 45°22′N 19°34′E
101 I21 Backnang Baden-Württemberg, SW Germany 48°57′N 09°26′E
167 S15 Bac Liêu *var.* Vinh Loi. Minh Hai, S Vietnam 09°17′N 105°44′E
167 T6 Bắc Ninh Ha Băc, N Vietnam 21°10′N 106°04′E
40 G4 Bacoachi Sonora, NW Mexico 30°36′N 110°00′W
171 P6 Bacolod *off.* Bacolod City. Negros, C Philippines 10°43′N 122°58′E
Bacolod City *see* Bacolod
171 O4 Baco, Mount ▲ Mindoro, N Philippines 12°50′N 121°08′E
111 K25 Bácsalmás Bács-Kiskun, S Hungary 46°07′N 19°20′E
Bácsjózseffalva *see* Žednik
111 J24 Bács-Kiskun *off.* Bács-Kiskun Megye. ◆ *county* S Hungary
Bács-Kiskun Megye *see* Bács-Kiskun
Bácsszenttamás *see* Srbobran
Bácstopolya *see* Bačka Topola
Bactra *see* Balkh
Bada *see* Xilin
155 F21 Badagara *var.* Vadakara. Kerala, SW India 11°36′N 75°34′E *see also* Vadakara
101 M24 Bad Aibling Bayern, SE Germany 47°52′N 12°00′E
162 I13 Badain Jaran Shamo *desert* N China
104 I11 Badajoz *anc.* Pax Augusta. Extremadura, W Spain 38°53′N 06°58′W
104 J11 Badajoz ◆ *province* Extremadura, W Spain
149 S2 Badakhshān ◆ *province* NE Afghanistan
105 W6 Badalona *anc.* Baetulo. Cataluña, E Spain 41°27′N 02°15′E
154 O11 Bādāmpāhārh *var.* Bādāmapāhārh. Odisha, E India 22°04′N 86°06′E
169 O10 Badas, Kepulauan *island group* W Indonesia
109 S6 Bad Aussee Salzburg, E Austria 47°35′N 13°44′E
31 S8 Bad Axe Michigan, N USA 43°48′N 83°00′W
101 G16 Bad Berleburg Nordrhein-Westfalen, W Germany 51°03′N 08°24′E
101 L17 Bad Blankenburg Thüringen, C Germany 50°43′N 11°19′E
Bad Borseck *see* Borsec
101 G18 Bad Camberg Hessen, W Germany 50°18′N 08°15′E
100 L8 Bad Doberan Mecklenburg-Vorpommern, N Germany 54°06′N 11°55′E
101 N14 Bad Düben Sachsen, E Germany 51°35′N 12°34′E
109 X4 Baden *var.* Baden bei Wien; *anc.* Aquae Panoniae, Thermae Pannonicae. Niederösterreich, NE Austria 48°01′N 16°14′E
108 F9 Baden Aargau, N Switzerland 47°28′N 08°19′E
101 G21 Baden-Baden *anc.* Aurelia Aquensis. Baden-Württemberg, SW Germany 48°46′N 08°14′E
Baden bei Wien *see* Baden
101 G22 Baden-Württemberg *Fr.* Bade-Wurtemberg. ◆ *state* SW Germany
112 A10 Baderna Istra, NW Croatia 45°12′N 13°45′E
Bade-Wurtemberg *see* Baden-Württemberg
101 H20 Bad Fredrichshall Baden-Württemberg, S Germany 49°13′N 09°15′E
100 P11 Bad Freienwalde Brandenburg, NE Germany 52°47′N 14°04′E
109 Q8 Badgastein *var.* Gastein. Salzburg, NW Austria 47°07′N 13°09′E
Badger State *see* Wisconsin
148 L4 Bādghīs ◆ *province* NW Afghanistan
109 T5 Bad Hall Oberösterreich, N Austria 48°03′N 14°13′E
101 J14 Bad Harzburg Niedersachsen, C Germany 51°52′N 10°34′E
101 I16 Bad Hersfeld Hessen, C Germany 50°52′N 09°42′E
98 I10 Badhoevedorp Noord-Holland, C Netherlands 52°21′N 04°46′E
109 Q8 Bad Hofgastein Salzburg, NW Austria 47°11′N 13°07′E
Bad Homburg *see* Bad Homburg vor der Höhe
101 G18 Bad Homburg vor der Höhe *var.* Bad Homburg. Hessen, W Germany 50°14′N 08°37′E
101 E17 Bad Honnef Nordrhein-Westfalen, W Germany 50°39′N 07°13′E
149 Q17 Badin Sind, SE Pakistan 24°38′N 68°53′E
21 S10 Badin Lake ◎ North Carolina, SE USA
40 I8 Badiraguato Sinaloa, C Mexico 25°21′N 107°31′W
109 R6 Bad Ischl Oberösterreich, N Austria 47°43′N 13°36′E
Badjawa *see* Bajawa
Badje-Sohppar *see* Övre Soppero
101 J18 Bad Kissingen Bayern, SE Germany 50°12′N 10°04′E
Bad Königswart *see* Lázně Kynžvart
101 F19 Bad Kreuznach Rheinland-Pfalz, SW Germany 49°51′N 07°52′E
101 F24 Bad Krozingen Baden-Württemberg, SW Germany 47°55′N 07°43′E
101 G16 Bad Laasphe Nordrhein-Westfalen, W Germany 50°57′N 08°24′E
28 J6 Badlands *physical region* North Dakota/South Dakota, N USA
101 K16 Bad Langensalza Thüringen, C Germany 51°05′N 10°40′E
109 T3 Bad Leonfelden Oberösterreich, N Austria 48°31′N 14°17′E
101 I20 Bad Mergentheim Baden-Württemberg, SW Germany 49°30′N 09°46′E
101 H17 Bad Nauheim Hessen, W Germany 50°22′N 08°45′E
101 E17 Bad Neuenahr-Ahrweiler Rheinland-Pfalz, W Germany 50°33′N 07°07′E
Bad Neustadt *see* Bad Neustadt an der Saale
101 J18 Bad Neustadt an der Saale *var.* Bad Neustadt. Berlin, C Germany 50°21′N 10°13′E
Badnur *see* Betūl
100 H13 Bad Oeynhausen Nordrhein-Westfalen, NW Germany 52°12′N 08°48′E
100 J9 Bad Oldesloe Schleswig-Holstein, N Germany 53°49′N 10°22′E
77 Q16 Badou C Togo 07°37′N 00°37′E
100 H13 Bad Pyrmont Niedersachsen, C Germany 51°58′N 09°16′E
109 X9 Bad Radkersburg Steiermark, SE Austria 46°40′N 16°02′E
139 V8 Badrah Wāsit, E Iraq 33°06′N 45°58′E
Badrah *see* Tarialan
101 N24 Bad Reichenhall Bayern, SE Germany 47°43′N 12°52′E
140 K8 Badr Ḩunayn Al Madīnah, W Saudi Arabia 23°46′N 38°45′E
152 K8 Badrīnāth ▲ N India 30°44′N 79°29′E
28 M10 Bad River ≈ South Dakota, N USA
30 K4 Bad River ≈ Wisconsin, N USA
100 H13 Bad Salzuflen Nordrhein-Westfalen, NW Germany 52°06′N 08°45′E
101 J16 Bad Salzungen Thüringen, C Germany 50°48′N 10°15′E
109 V8 Bad Sankt Leonhard im Lavanttal Kärnten, S Austria 46°55′N 14°51′E
100 K9 Bad Schwartau Schleswig-Holstein, N Germany 53°55′N 10°42′E
101 L24 Bad Tölz Bayern, SE Germany 47°44′N 11°34′E
181 U1 Badu Island *island* Queensland, NE Australia
155 K25 Badulla Uva Province, C Sri Lanka 06°59′N 81°03′E
109 X5 Bad Vöslau Niederösterreich, NE Austria 47°58′N 16°13′E
101 I24 Bad Waldsee Baden-Württemberg, S Germany 47°54′N 09°44′E
35 U11 Badwater Basin *depression* California, W USA
101 J20 Bad Windsheim Bayern, C Germany 49°30′N 10°25′E
101 J23 Bad Wörishofen Bayern, S Germany 48°00′N 10°36′E
100 G10 Bad Zwischenahn Niedersachsen, NW Germany 53°10′N 08°01′E
104 M13 Baena Andalucía, S Spain 36°37′N 04°20′W
163 V15 Baengnyong-do *prev.* Paengnyŏng. *island* NW South Korea
Baeterrae/Baeterrae Septimanorum *see* Béziers
Baetic Cordillera/Baetic Mountains *see* Béticos, Sistemas
Baetulo *see* Badalona
57 K18 Baeza Napo, NE Ecuador 0°30′S 77°52′W
105 N13 Baeza Andalucía, S Spain 38°00′N 03°28′W
79 D15 Bafang Ouest, W Cameroon 05°10′N 10°11′E
76 H12 Bafatá C Guinea-Bissau 12°09′N 14°38′W
149 U5 Baffa Khyber Pakhtunkhwa, NW Pakistan 34°30′N 73°18′E
197 O11 Baffin Basin *undersea feature* N Labrador Sea
197 N12 Baffin Bay *bay* Canada/Greenland
25 T15 Baffin Bay *inlet* Texas, SW USA
196 M12 Baffin Island *island* Nunavut, NE Canada
79 E15 Bafia Centre, C Cameroon 04°49′N 11°14′E
77 R14 Bafilo NE Togo 09°23′N 01°20′E
76 J12 Bafing ≈ W Africa
76 J12 Bafoulabé Kayes, W Mali 13°43′N 10°49′W
79 D15 Bafoussam Ouest, W Cameroon 05°31′N 10°25′E
143 R9 Bāfq Yazd, C Iran 31°35′N 55°21′E
136 L10 Bafra Samsun, N Turkey 41°34′N 35°56′E
136 L10 Bafra Burnu *headland* N Turkey 41°42′N 36°02′E
143 S12 Bāft Kermān, S Iran 29°12′N 56°36′E
79 N18 Bafwabalinga Orientale, NE Dem. Rep. Congo 0°52′N 26°55′E
79 N18 Bafwaboli Orientale, NE Dem. Rep. Congo 0°36′N 26°08′E
79 N17 Bafwasende Orientale, NE Dem. Rep. Congo 01°00′N 27°09′E
42 K13 Bagaces Guanacaste, NW Costa Rica 10°31′N 85°18′W
153 O12 Bagaha Bihār, N India 27°08′N 84°04′E
155 F16 Bāgalkot Karnātaka, W India 16°11′N 75°42′E
81 J22 Bagamoyo Pwani, E Tanzania 06°26′S 38°55′E
Bagan Datok *see* Bagan Datuk
168 J8 Bagan Datuk *var.* Bagan Datok. Perak, Peninsular Malaysia 03°58′N 100°47′E
171 R7 Baganga Mindanao, S Philippines 07°31′N 126°34′E
168 J9 Bagansiapiapi *var.* Pasirpangarayan. Sumatera, W Indonesia 02°06′N 100°52′E
162 M8 Baganuur *var.* Nüürst. Töv, C Mongolia 47°44′N 108°22′E
Bagaria *see* Bagheria
77 T11 Bagaroua Tahoua, W Niger 14°34′N 04°24′E
79 I20 Bagata Bandundu, W Dem. Rep. Congo 03°47′S 17°57′E
Bagdad *see* Baghdad
123 O13 Bagdarin Respublika Buryatiya, S Russian Federation 54°27′N 113°34′E
61 G17 Bagé Rio Grande do Sul, S Brazil 31°22′S 54°06′W
Bagenalstown *see* Muine Bheag
153 T16 Bagerhat *var.* Bagherhat. Khulna, S Bangladesh 22°40′N 89°48′E
103 P16 Bages et de Sigean, Étang de ◎ S France
33 W17 Baggs Wyoming, C USA 41°02′N 107°39′W
154 F11 Bāgh Madhya Pradesh, C India 22°22′N 74°49′E
139 T8 Baghdād *var.* Bagdad, *Eng.* Baghdad. ● Baghdād, C Iraq 33°20′N 44°26′E
139 T8 Baghdād *off.* Muḩāfa at Baghdād, *var.* Amānat al 'Āşimah. ◆ *governorate* C Iraq
139 T8 Baghdād ✈ (Baghdād) Baghdād, C Iraq 33°20′N 44°26′E
Baghdad *see* Baghdād
Baghdād, Muḩāfa at at *see* Baghdād
Bagherhat *see* Bagerhat
107 J23 Bagheria *var.* Bagaria. Sicilia, Italy, C Mediterranean Sea 38°05′N 13°31′E
143 S10 Bāghīn Kermān, C Iran 30°50′N 57°00′E
149 Q3 Baghlān Baghlān, NE Afghanistan 36°11′N 68°44′E
149 Q3 Baghlān *var.* Bāghlān. ◆ *province* NE Afghanistan
Bāghlān *see* Baghlān
148 M7 Bāghrān Helmand, S Afghanistan 32°55′N 64°57′E
29 T4 Bagley Minnesota, N USA 47°31′N 95°24′W
106 H10 Bagnacavallo Emilia-Romagna, C Italy 44°00′N 12°59′E
102 K16 Bagnères-de-Bigorre Hautes-Pyrénées, S France 43°04′N 00°09′E
102 L17 Bagnères-de-Luchon Hautes-Pyrénées, S France 42°46′N 00°34′E
106 F11 Bagni di Lucca Toscana, C Italy 44°01′N 10°38′E
106 H11 Bagno di Romagna Emilia-Romagna, C Italy 43°51′N 11°57′E
103 R14 Bagnols-sur-Cèze Gard, S France 44°10′N 04°37′E
162 M14 Bag Nur ◎ N China
166 L8 Bago *var.* Pegu. Bago, SW Myanmar (Burma) 17°18′N 96°31′E
171 P6 Bago *off.* Bago City. Negros, C Philippines 10°30′N 122°49′E
166 L7 Bago *var.* Pegu. ◆ *region* S Myanmar (Burma)
Bago City *see* Bago
76 M13 Bagoé ≈ Ivory Coast/Mali
Bagrāmē *see* Bagrāmī
149 R5 Bagrāmī *var.* Bagrāmē. Kābōl, E Afghanistan 34°29′N 69°16′E
119 B14 Bagrationovsk *Ger.* Preussisch Eylau. Kaliningradskaya Oblast', W Russian Federation 54°24′N 20°39′E
Bagrax *see* Bohu
Bagrax Hu *see* Bosten Hu
56 C10 Bagua Amazonas, NE Peru 05°37′S 78°36′W
171 O2 Baguio *off.* Baguio City. Luzon, N Philippines 16°25′N 120°36′E
Baguio City *see* Baguio
77 V9 Bagzane, Monts ▲ N Niger 17°48′N 08°43′E
Bāḩah, Minţaqat al *see* Al Bāḩah
Bahama Islands *see* Bahamas, The
Bahamas *see* Bahamas, The
Bahamas, Commonwealth of The *see* Bahamas, The
0 L13 Bahamas, The *var.* Bahama Islands, Bahamas. ◆ *island group* N West Indies
44 H3 Bahamas, The *off.* Commonwealth of The Bahamas. ◆ *commonwealth republic* N West Indies
153 S15 Baharampur *prev.* Berhampore. West Bengal, NE India 24°06′N 88°19′E
146 E12 Baharly *var.* Bäherden, *Rus.* Bakharden; *prev.* Bakherden. Ahal Welaýaty, C Turkmenistan 38°30′N 57°18′E
149 U10 Bahāwalnagar Punjab, E Pakistan 30°00′N 73°03′E
149 T11 Bahāwalpur Punjab, E Pakistan 29°25′N 71°40′E
136 L16 Bahçe Osmaniye, S Turkey 37°14′N 36°34′E
160 J8 Ba He ≈ C China
Bäherden *see* Baharly
59 N16 Bahia *off.* Estado da Bahia. ◆ *state* E Brazil
61 B24 Bahía Blanca Buenos Aires, E Argentina 38°43′S 62°19′W
40 L15 Bahía Bufadero Michoacán, SW Mexico
63 J19 Bahía Bustamante Chubut, SE Argentina 45°06′S 66°30′W
40 D5 Bahía de los Ángeles Baja California Norte, NW Mexico
40 C6 Bahía de Tortugas Baja California Sur, NW Mexico 27°42′N 114°54′W
Bahia, Estado da *see* Bahia
42 J4 Bahía, Islas de la *Eng.* Bay Islands. *island group* N Honduras
40 E5 Bahía Kino Sonora, NW Mexico 28°48′N 111°55′W
40 E9 Bahía Magdalena *var.* Puerto Magdalena. Baja California Sur, NW Mexico 24°34′N 112°07′W
54 C8 Bahía Solano *var.* Ciudad Mutis, Solano. Chocó, W Colombia 06°13′N 77°27′W
80 I11 Bahir Dar *var.* Bahr Dar, Bahrdar Giyorgis. Āmara, N Ethiopia 11°34′N 37°23′E
141 X8 Bahlā' *var.* Bahlah, Bahlat. NW Oman 22°58′N 57°16′E
Bāhla *see* Bālān
Bahlah/Bahlat *see* Bahlā'
152 M11 Bahraich Uttar Pradesh, N India 27°35′N 81°36′E
143 M14 Bahrain *off.* State of Bahrain, Dawlat al Bahrayn, *Ar.* Al Baḩrayn, *prev.* Bahrein; *anc.* Tylos, Tyros. ◆ *monarchy* SW Asia
142 M14 Bahrain ✈ C Bahrain 26°15′N 50°39′E
142 M15 Bahrain, Gulf of *gulf* Persian Gulf, NW Arabian Sea
Bahrain, State of *see* Bahrain
138 I7 Bahrat Mallāḩah ◎ W Syria
Bahrayn, Dawlat al *see* Bahrain
Bahr Dar/Bahrdar Giyorgis *see* Bahir Dar
Bahrein *see* Bahrain
Bahr el, Azraq *see* Blue Nile
Bahr el Gebel *see* Central Equatoria
Bahr el Jebel *see* Central Equatoria
80 E13 Bahr ez Zaref ≈ Jonglei, E South Sudan
67 R8 Bahr Kameur ≈ N Central African Republic
Bahr Tabariya, Sea of *see* Tiberias, Lake
143 W15 Bāhū Kalāt Sīstān va Balūchestān, SE Iran 25°42′N 61°28′E
118 N13 Bahushewsk *Rus.* Bogushëvsk. Vitsyebskaya Voblasts', NE Belarus 54°51′N 30°13′E
Bai *see* Tagow Bāy

◆ Country
● Country Capital
◇ Dependent Territory
○ Dependent Territory Capital
◆ Administrative Regions
✈ International Airport
▲ Mountain
▲ Mountain Range
⛰ Volcano
≈ River
◎ Lake
Reservoir

116 G13 **Baia de Aramă** Mehedinți, SW Romania 45°00′N 22°43′E
116 G11 **Baia de Criș** *Ger.* Altenburg, *Hung.* Körösbánya. Hunedoara, SW Romania 46°10′N 22°41′E
83 A16 **Baia dos Tigres** Namibe, SW Angola 16°36′S 11°44′E
82 A13 **Baia Farta** Benguela, W Angola 12°38′S 13°12′E
116 H9 **Baia Mare** *Ger.* Frauenbach, *Hung.* Nagybánya; *prev.* Neustadt. Maramureș, NW Romania 47°40′N 23°35′E
116 H8 **Baia Sprie** *Ger.* Mittelstadt, *Hung.* Felsőbánya. Maramureș, NW Romania 47°40′N 23°42′E
78 G13 **Baïbokoum** Logone-Oriental, SW Chad 07°46′N 15°43′E
160 F12 **Baicao Ling** ▲▲ SW China
163 U9 **Baicheng** *var.* Pai-ch'eng; *prev.* T'aon-an. Jilin, NE China 45°32′N 122°51′E
158 I6 **Baicheng** *var.* Bay. Xinjiang Uygur Zizhiqu, NW China 41°49′N 81°45′E
116 J13 **Băicoi** Prahova, SE Romania 45°02′N 25°51′E
Baidoa *see* Baydhabo
15 U6 **Baie-Comeau** Québec, SE Canada 49°12′N 68°10′W
15 U6 **Baie-des-Sables** Québec, SE Canada 48°41′N 67°55′W
15 T7 **Baie-des-Bacon** Québec, SE Canada 48°31′N 69°17′W
15 S8 **Baie-des-Rochers** Québec, SE Canada 47°57′N 69°50′W
Baie-du-Poste *see* Mistissini
172 H17 **Baie Lazare** Mahé, NE Seychelles 04°45′S 55°29′E
45 Y5 **Baie-Mahault** Basse Terre, C Guadeloupe 16°16′N 61°35′W
15 R9 **Baie-St-Paul** Québec, SE Canada 47°27′N 70°30′W
15 V5 **Baie-Trinité** Québec, SE Canada 49°25′N 67°20′W
13 T11 **Baie Verte** Newfoundland and Labrador, SE Canada 49°55′N 56°12′W
Baiguan *see* Shangyu
Baihe *see* Erdaobaihe
139 U11 **Bā'ij al Mahdī** Al Muthanná, S Iraq 31°21′N 44°57′E
Baiji *see* Bayjī
Baikal, Lake *see* Baykal, Ozero
Bailādila *see* Kirandul
Baile an Chaistil *see* Ballycastle
Baile an Róba *see* Ballinrobe
Baile an tSratha *see* Ballintra
Baile Átha an Rí *see* Athenry
Baile Átha Buí *see* Athboy
Baile Átha Cliath *see* Dublin
Baile Átha Fhirdhia *see* Ardee
Baile Átha Í *see* Athy
Baile Átha Luain *see* Athlone
Baile Átha Troim *see* Trim
Baile Brigín *see* Balbriggan
Baile Easa Dara *see* Ballysadare
116 I13 **Băile Govora** Vâlcea, SW Romania 45°00′N 24°08′E
116 F13 **Băile Herculane** *Ger.* Herkulesbad, *Hung.* Herkulesfürdő. Caraș-Severin, SW Romania 44°51′N 22°24′E
Baile Locha Riach *see* Loughrea
Baile Mhistéala *see* Mitchelstown
Baile Monaidh *see* Ballymoney
105 N12 **Bailén** Andalucía, S Spain 38°06′N 03°46′W
Baile na hInse *see* Ballynahinch
Baile na Lorgan *see* Castleblayney
Baile na Mainistreach *see* Newtownabbey
Baile Nua na hArda *see* Newtownards
116 I12 **Băile Olănești** Vâlcea, SW Romania 45°14′N 24°18′E
116 H14 **Băilești** Dolj, SW Romania 44°01′N 23°20′E
163 N12 **Bailingmiao** *var.* Darhan Muminggan Lianheqi. Nei Mongol Zizhiqu, N China 41°41′N 110°25′E
58 K11 **Bailique, Ilha** *island* NE Brazil
103 O1 **Bailleul** Nord, N France 50°43′N 02°43′E
78 H12 **Ba Illi** Chari-Baguirmi, SW Chad 10°31′N 16°29′E
159 V12 **Bailong Jiang** ≋ C China
82 C13 **Bailundo** *Port.* Vila Teixeira da Silva. Huambo, C Angola 12°12′S 15°52′E
159 T13 **Baima** *var.* Sêraitang. Qinghai, C China 32°55′N 100°44′E
Baima *see* Baxoi
186 C8 **Baimuru** Gulf, S Papua New Guinea 07°34′S 144°49′E
158 M16 **Bainang** Xizang Zizhiqu, W China 28°57′N 89°31′E
23 S8 **Bainbridge** Georgia, SE USA 30°54′N 84°33′W
171 O17 **Baing** Pulau Sumba, SE Indonesia 10°09′S 120°34′E
158 M14 **Baingoin** *var.* Pubao. Xizang Zizhiqu, W China 31°22′N 90°00′E
104 G2 **Baio** Galicia, NW Spain 43°08′N 08°58′W
104 G4 **Baiona** Galicia, NW Spain 42°06′N 08°49′W
163 V7 **Baiquan** Heilongjiang, NE China 47°37′N 126°04′E
Bā'ir *see* Bāyir
158 I11 **Bairab Co** ◎ W China
25 Q7 **Baird** Texas, SW USA 32°23′N 99°24′W
39 N7 **Baird Mountains** ▲▲ Alaska, USA
Baireuth *see* Bayreuth
190 H3 **Bairiki** Tarawa, NW Kiribati 01°20′N 173°01′E
Bairin Youqi *see* Daban
Bairin Zuoqi *see* Lindong
Bairkum *see* Bayyrkum
183 P12 **Bairnsdale** Victoria, SE Australia 37°51′S 147°38′E
171 P6 **Bais** Negros, S Philippines 09°36′N 123°07′E
102 L15 **Baïse** *var.* Baise. ≋ S France
Baise *see* Baïse
163 W11 **Baishan** *prev.* Hunjiang. Jilin, NE China 41°57′N 126°31′E
Baishan *see* Mashan
118 F12 **Baisogala** Šiauliai, C Lithuania 55°38′N 23°44′E
189 Q7 **Baiti** N Nauru 0°30′S 166°55′E
Baitou Shan *see* Paektu-san
104 G13 **Baixo Alentejo** *physical region* S Portugal
64 P5 **Baixo, Ilhéu de** *island* Madeira, Portugal, NE Atlantic Ocean
83 E15 **Baixo Longa** Kuando Kubango, SE Angola 15°39′S 18°39′E
159 V10 **Baiyin** Gansu, C China 36°33′N 104°11′E
160 E8 **Baiyü** *var.* Jianshe. Sichuan, C China 30°37′N 97°15′E
161 N14 **Baiyun** ✈ (Guangzhou) Guangdong, S China 23°12′N 113°19′E
160 K4 **Baiyu Shan** ▲▲ C China
111 J25 **Baja** Bács-Kiskun, S Hungary 46°13′N 18°56′E
40 C4 **Baja California** *Eng.* Lower California. *peninsula* NW Mexico
40 C4 **Baja California Norte** ◈ *state* NW Mexico
40 E9 **Baja California Sur** ◈ *state* NW Mexico
Bājah *see* Béja
Bajan *see* Bayan
191 V16 **Baja, Punta** *headland* Easter Island, Chile, E Pacific Ocean 27°10′S 109°21′W
40 B4 **Baja, Punta** *headland* NW Mexico 29°57′N 115°48′W
55 R5 **Baja, Punta** *headland* NE Venezuela
42 D5 **Baja Verapaz** *off.* Departamento de Baja Verapaz. ◈ *department* C Guatemala
Baja Verapaz, Departamento de *see* Baja Verapaz
171 N16 **Bajawa** *prev.* Badjawa. Flores, S Indonesia 08°46′S 120°59′E
153 S16 **Baj Baj** *prev.* Budge-Budge. West Bengal, E India 22°29′N 88°11′E
141 N15 **Bājil** W Yemen 15°05′N 43°16′E
183 U4 **Bajimba, Mount** ▲ New South Wales, SE Australia 29°19′S 152°04′E
112 K13 **Bajina Bašta** Serbia, W Serbia 43°58′N 19°33′E
153 U14 **Bajitpur** Dhaka, E Bangladesh 24°12′N 90°57′E
112 K8 **Bajmok** Vojvodina, NW Serbia 45°59′N 19°25′E
Bajo Boquete *see* Boquete
113 L17 **Bajram Curri** Kukës, N Albania 42°23′N 20°06′E
79 J14 **Bakala** Ouaka, C Central African Republic 06°03′N 20°31′E
127 T4 **Bakaly** Respublika Bashkortostan, W Russian Federation 55°10′N 53°46′E
Bakan *see* Shimonoseki
145 U14 **Bakanas** *Kaz.* Baqanas. Almaty, SE Kazakhstan 44°50′N 76°13′E
145 V12 **Bakanas** *Kaz.* Baqanas. ≋ E Kazakhstan
149 R4 **Bākārak** Panjshīr, NE Afghanistan 35°16′N 69°28′E
145 U14 **Bakbakty** *Kaz.* Baqbaqty. Almaty, SE Kazakhstan 44°36′N 76°41′E
122 J12 **Bakchar** Tomskaya Oblast', C Russian Federation 56°58′N 81°59′E
76 I11 **Bakel** E Senegal 14°54′N 12°26′W
35 W13 **Baker** California, W USA 35°15′N 116°04′W
22 J8 **Baker** Louisiana, S USA 30°35′N 91°10′W
33 Y9 **Baker** Montana, NW USA 46°22′N 104°16′W
32 L12 **Baker** Oregon, NW USA 44°46′N 117°50′W
192 L7 **Baker and Howland Islands** ◇ *US unincorporated territory* W Polynesia
36 L12 **Baker Butte** ▲ Arizona, SW USA 34°24′N 111°22′W
39 X15 **Baker Island** *island* Alexander Archipelago, Alaska, USA
9 N9 **Baker Lake** *var.* Qamanittuaq. Nunavut, N Canada 64°20′N 96°10′W
9 N9 **Baker Lake** ◎ Nunavut, N Canada
32 H6 **Baker, Mount** ▲ Washington, NW USA 48°46′N 121°48′W
35 R13 **Bakersfield** California, W USA 35°22′N 119°01′W
24 M9 **Bakersfield** Texas, SW USA 30°54′N 102°21′W
21 P9 **Bakersville** North Carolina, SE USA 36°01′N 82°09′W
Bakhābī *see* Bū Khābī
Bakharden *see* Baharly
Bakhardok *see* Bokurdak
143 U5 **Bākharz, Kuhhā-ye** ▲▲ NE Iran
152 D13 **Bākhāsar** Rājasthān, NW India 24°42′N 71°11′E
Bakhchisaray *see* Bakhchysaray
117 T13 **Bakhchysaray** *Rus.* Bakhchisaray. Avtonomna Respublika Krym, S Ukraine 44°44′N 33°53′E
Bakherden *see* Baharly
117 R3 **Bakhmach** Chernihivs'ka Oblast', N Ukraine 51°10′N 32°48′E
Bākhtarān *see* Kermānshāh
143 Q11 **Bakhtegān, Daryācheh-ye** ◎ C Iran
Bakhty *see* Bakty
137 Z11 **Bakı** *Eng.* Baku. ● (Azerbaijan) E Azerbaijan 40°24′N 49°51′E
80 M12 **Baki** Awdal, N Somalia 10°10′N 43°45′E
137 Z11 **Bakı** ✈ E Azerbaijan 40°26′N 49°55′E
136 C13 **Bakır Çayı** ≋ W Turkey
92 L1 **Bakkafjörður** Austurland, NE Iceland 66°01′N 14°49′W
92 L1 **Bakkaflói** *sea area* W Norwegian Sea
92 L2 **Bakkagerði** Austurland, NE Iceland 65°32′N 13°46′W
81 I15 **Bako** Southern Nationalities, S Ethiopia 05°45′N 36°39′E
76 L15 **Bako** NW Ivory Coast 09°08′N 07°40′W
Bákó *see* Bacău
111 H23 **Bakony** *Eng.* Bakony Mountains, *Ger.* Bakonywald. ▲▲ W Hungary
Bakony Mountains/Bakonywald *see* Bakony
81 M16 **Bakool** *off.* Gobolka Bakool. ◈ *region* W Somalia
Bakool, Gobolka *see* Bakool
79 L15 **Bakouma** Mbomou, SE Central African Republic 05°42′N 22°43′E
127 N15 **Baksan** Kabardino-Balkarskaya Respublika, SW Russian Federation 43°43′N 43°31′E
119 I16 **Bakshty** Hrodzyenskaya Voblasts', W Belarus 53°56′N 26°11′E
145 X12 **Bakty** *prev.* Bakhty. Vostochnyy Kazakhstan, E Kazakhstan 46°41′N 82°45′E
Baku *see* Bakı
194 K12 **Bakutis Coast** *physical region* Antarctica
Bakwanga *see* Mbuji-Mayi
145 O15 **Bakyrly** Yuzhnyy Kazakhstan, S Kazakhstan 44°30′N 67°41′E
14 H13 **Bala** Ontario, S Canada 45°01′N 79°37′W
136 I13 **Balâ** Ankara, C Turkey 39°34′N 33°07′E
97 J19 **Bala** NW Wales, United Kingdom 52°54′N 03°31′W
170 L7 **Balabac Island** *island* W Philippines
Balabac, Selat *see* Balabac Strait
169 V5 **Balabac Strait** *var.* Selat Balabac. *strait* Malaysia/Philippines
Ba'labakk *see* Baalbek
187 P16 **Balabio, Île** *island* Province Nord, W New Caledonia
116 I14 **Balaci** Teleorman, S Romania 44°21′N 24°55′E
139 S7 **Balad** Şalāḩ ad Dīn, N Iraq 34°00′N 44°07′E
139 U7 **Balad Rūz** Diyālá, E Iraq 33°42′N 45°04′E
154 J11 **Bālāghāt** Madhya Pradesh, C India 21°48′N 80°11′E
155 F14 **Bālāghāt Range** ▲▲ W India
103 X14 **Balagne** *physical region* Corse, France, C Mediterranean Sea
105 U5 **Balaguer** Cataluña, NE Spain 41°48′N 00°48′E
105 S3 **Balaïtous** *var.* Pic de Balaïtous. ▲ France/Spain 42°51′N 00°17′W
Balaitous, Pic de *see* Balaïtous
Bálák *see* Ballangen
127 O3 **Balakhna** Nizhegorodskaya Oblast', W Russian Federation 56°26′N 43°43′E
122 L12 **Balakhta** Krasnoyarskiy Kray, S Russian Federation 55°22′N 91°24′E
182 I9 **Balaklava** South Australia 34°10′S 138°22′E
Balakleya *see* Balakliya
117 V6 **Balakliya** *Rus.* Balakleya. Kharkivs'ka Oblast', E Ukraine 49°27′N 36°53′E
127 Q7 **Balakovo** Saratovskaya Oblast', W Russian Federation 52°03′N 47°47′E
83 P14 **Balama** Cabo Delgado, N Mozambique 13°18′S 38°39′E
169 U6 **Balambangan, Pulau** *island* East Malaysia
Bālā Morghāb *see* Bālā Murghāb
148 L3 **Bālā Murghāb** *prev.* Bālā Morghāb. Laghmān, NW Afghanistan 35°38′N 63°21′E
152 E11 **Bālān** *prev.* Bāhla. Rājasthān, NW India 27°45′N 71°32′E
116 J10 **Bălan** *Hung.* Balánbánya. Harghita, C Romania 46°39′N 25°47′E
Balánbánya *see* Bălan
171 O3 **Balanga** Luzon, N Philippines 14°40′N 120°32′E
154 M12 **Balāngīr** *prev.* Bolangir. Odisha, E India 20°41′N 83°30′E
127 N8 **Balashov** Saratovskaya Oblast', W Russian Federation 51°32′N 43°14′E
Balasore *see* Bāleshwar
111 K21 **Balassagyarmat** Nógrád, N Hungary 48°06′N 19°17′E
29 S10 **Balaton** Minnesota, N USA 44°13′N 95°52′W
111 H24 **Balaton** *var.* Lake Balaton, *Ger.* Plattensee. ◎ W Hungary
111 I23 **Balatonfüred** *var.* Füred. Veszprém, W Hungary 46°59′N 17°53′E
Balaton, Lake *see* Balaton
116 I11 **Bălăușeri** *Ger.* Bladenmarkt, *Hung.* Balavásár. Mureș, C Romania 46°24′N 24°41′E
Balavásár *see* Bălăușeri
105 Q11 **Balazote** Castilla-La Mancha, C Spain 38°54′N 02°09′W
Balázsfalva *see* Blaj
119 F14 **Balbieriškis** Kaunas, S Lithuania 54°29′N 23°52′E
186 J7 **Balbi, Mount** ▲ Bougainville Island, NE Papua New Guinea 05°51′S 154°58′E
58 F11 **Balbina, Represa** ⊠ NW Brazil
43 T15 **Balboa** Panamá, C Panama 08°55′N 79°36′W
97 G17 **Balbriggan** *Ir.* Baile Brigín. E Ireland 53°37′N 06°11′W
Balbunar *see* Kubrat
81 N17 **Balcad** Shabeellaha Dhexe, C Somalia 02°19′N 45°19′E
61 D23 **Balcarce** Buenos Aires, E Argentina 37°51′S 58°16′W
11 U16 **Balcarres** Saskatchewan, S Canada 50°49′N 103°31′W
114 O8 **Balchik** Dobrich, NE Bulgaria 43°25′N 28°11′E
185 E24 **Balclutha** Otago, South Island, New Zealand 46°15′S 169°45′E
25 Q12 **Balcones Escarpment** *escarpment* Texas, SW USA
18 F14 **Bald Eagle Creek** ≋ Pennsylvania, NE USA
Baldenburg *see* Biały Bór
21 V12 **Bald Head Island** *island* North Carolina, SE USA
27 W10 **Bald Knob** Arkansas, C USA 35°18′N 91°34′W
30 K17 **Bald Knob** *hill* Illinois, N USA
Baldohn *see* Baldone
118 G9 **Baldone** *Ger.* Baldohn. W Latvia 56°46′N 24°18′E
22 I9 **Baldwin** Louisiana, S USA 29°50′N 91°32′W
31 P7 **Baldwin** Michigan, N USA 43°54′N 85°50′W
27 Q4 **Baldwin City** Kansas, C USA 38°43′N 95°12′W
39 N8 **Baldwin Peninsula** *headland* Alaska, USA 66°45′N 162°19′W
18 H9 **Baldwinsville** New York, NE USA 43°09′N 76°19′W
23 N2 **Baldwyn** Mississippi, S USA 34°30′N 88°38′W
11 W15 **Baldy Mountain** ▲ Manitoba, S Canada 51°29′N 100°46′W
33 T7 **Baldy Mountain** ▲ Montana, NW USA 48°09′N 109°39′W
37 O13 **Baldy Peak** ▲ Arizona, SW USA 33°56′N 109°37′W
Bâle *see* Basel
Balearic Plain *see* Algerian Basin
Baleares *see* Illes Baleares
105 X11 **Baleares, Islas** *Eng.* Balearic Islands. *island group* Spain, W Mediterranean Sea
Baleares Major *see* Mallorca
Balearic Islands *see* Baleares, Islas
Balearis Minor *see* Menorca
169 S9 **Baleh, Batang** ≋ East Malaysia
12 J8 **Baleine, Grande Rivière de la** ≋ Québec, E Canada
12 K7 **Baleine, Petite Rivière de la** ≋ Québec, SE Canada
12 K7 **Baleine, Petite Rivière de la** ≋ Québec, NE Canada
13 N6 **Baleine, Rivière à la** ≋ Québec, E Canada
99 J16 **Balen** Antwerpen, N Belgium 51°12′N 05°12′E
171 O3 **Baler** Luzon, N Philippines 15°47′N 121°30′E
154 P11 **Bāleshwar** *prev.* Balasore. Odisha, E India 21°31′N 86°59′E
77 S12 **Baléyara** Tillabéri, W Niger 13°48′N 02°57′E
127 T1 **Balezino** Udmurtskaya Respublika, NW Russian Federation 57°57′N 53°03′E
42 J4 **Balfate** Colón, N Honduras 15°47′N 86°24′W
11 O17 **Balfour** British Columbia, SW Canada 49°39′N 116°57′W
29 N3 **Balfour** North Dakota, N USA 47°55′N 100°34′W
Balfrush *see* Bābol
114 L9 **Balgarka** , Bŭlgarka. E Bulgaria 42°43′N 26°19′E
122 L14 **Balgazyn** Respublika Tyva, S Russian Federation 50°53′N 95°12′E
11 U16 **Balgonie** Saskatchewan, S Canada 50°30′N 104°12′W
Bălgrad *see* Alba Iulia
81 J19 **Balguda** *spring/well* S Kenya 01°28′S 39°50′E
158 K6 **Balguntay** Xinjiang Uygur Zizhiqu, NW China 42°45′N 86°18′E
141 R16 **Balḩāf** S Yemen 14°02′N 48°16′E
152 F13 **Bāli** Rājasthān, N India 25°10′N 73°20′E
169 U17 **Bali** ◈ *province* S Indonesia
169 T17 **Bali** *island* C Indonesia
111 K16 **Balice** ✈ (Kraków) Małopolskie, S Poland 49°57′N 19°49′E
171 Y14 **Baliem, Sungai** ≋ Papua, E Indonesia
136 C12 **Balıkesir** Balıkesir, W Turkey 39°38′N 27°52′E
136 C12 **Balıkesir** ◈ *province* NW Turkey
138 L3 **Balīkh, Nahr** ≋ N Syria
169 V12 **Balikpapan** Borneo, C Indonesia 01°15′S 116°50′E
169 T16 **Bali, Laut** *Eng.* Bali Sea. *sea* C Indonesia
171 N9 **Balimbing** Tawitawi, SW Philippines 05°10′N 120°00′E
186 B8 **Balimo** Western, SW Papua New Guinea 08°00′S 143°00′E
Bálinc *see* Balinț
101 H23 **Balingen** Baden-Württemberg, SW Germany 48°16′N 08°51′E
116 F11 **Balinț** *Hung.* Bálinc. Timiș, W Romania 45°48′N 21°54′E
171 O1 **Balintang Channel** *channel* N Philippines
138 K3 **Bālis** Ḩalab, N Syria 36°01′N 38°03′E
Bali Sea *see* Bali, Laut
98 K7 **Balk** Fryslân, N Netherlands 52°54′N 05°34′E
146 B11 **Balkanabat** *Rus.* Nebitdag. Balkan Welaýaty, W Turkmenistan 39°33′N 54°19′E
121 R6 **Balkan Mountains** *Bul./SCr.* Stara Planina. ▲▲ Bulgaria/Serbia
Balkanskiy Welayat *see* Balkan Welaýaty
146 B9 **Balkan Welaýaty** *Rus.* Balkanskiy Velayat. ◈ *province* W Turkmenistan
Balkash *see* Balkhash
145 P8 **Balkashino** Akmola, N Kazakhstan 52°36′N 68°46′E
Balkash, Ozero *see* Balkhash, Ozero
149 O2 **Balkh** *anc.* Bactra. Balkh, N Afghanistan 36°46′N 66°54′E
149 P2 **Balkh** ◈ *province* N Afghanistan
145 T13 **Balkhash** *Kaz.* Balqash, *var.* Balkash. Karaganda, SE Kazakhstan 46°52′N 74°55′E
Balkhash, Lake *see* Balkhash, Ozero
145 T13 **Balkhash, Ozero** *var.* Ozero Balkash, *Eng.* Lake Balkhash, *Kaz.* Balqash. ◎ SE Kazakhstan
Balla Balla *see* Mbalabala
96 H10 **Ballachulish** N Scotland, United Kingdom 56°40′N 05°10′W
180 M12 **Balladonia** Western Australia 32°21′S 123°32′E
97 C16 **Ballaghaderreen** *Ir.* Bealach an Doirín. C Ireland 53°51′N 08°29′W
92 H10 **Ballangen** *Lapp.* Bálák. Nordland, NW Norway 68°18′N 16°48′E
97 H14 **Ballantrae** W Scotland, United Kingdom 55°05′N 05°W
183 N12 **Ballarat** Victoria, SE Australia 37°36′S 143°51′E
180 K11 **Ballard, Lake** *salt lake* Western Australia
Ballari *see* Bellary
76 L11 **Ballé** Koulikoro, W Mali 15°18′N 08°31′W
40 D7 **Ballenas, Bahía de** *bay* NW Mexico
40 D5 **Ballenas, Canal de** *channel* NW Mexico
195 R17 **Balleny Islands** *island group* Antarctica
40 J7 **Balleza** *var.* San Pablo Balleza. Chihuahua, N Mexico 26°55′N 106°21′W
114 M13 **Ballı** Tekirdağ, NW Turkey 40°48′N 27°03′E
153 O13 **Ballia** Uttar Pradesh, N India 25°45′N 84°09′E
183 V4 **Ballina** New South Wales, SE Australia 28°50′S 153°37′E
97 C16 **Ballina** *Ir.* Béal an Átha. W Ireland 54°07′N 09°09′W
97 D16 **Ballinamore** *Ir.* Béal an Átha Móir. NW Ireland 54°03′N 07°47′W
97 D18 **Ballinasloe** *Ir.* Béal Átha na Sluaighe. W Ireland 53°20′N 08°13′W
25 P8 **Ballinger** Texas, SW USA 31°44′N 99°57′W
97 C17 **Ballinrobe** *Ir.* Baile an Róba. W Ireland 53°37′N 09°14′W
97 A21 **Ballinskelligs Bay** *Ir.* Bá na Scealg. *inlet* SW Ireland
97 D15 **Ballintra** *Ir.* Baile an tSratha. NW Ireland 54°35′N 08°07′W
103 T7 **Ballon d'Alsace** ▲ NE France 47°50′N 06°54′E
Ballon de Guebwiller *see* Grand Ballon
113 K21 **Ballsh** *var.* Ballshi. Fier, SW Albania 40°35′N 19°45′E
Ballshi *see* Ballsh
98 K4 **Ballum** Fryslân, N Netherlands 53°27′N 05°40′E
97 F16 **Ballybay** *Ir.* Béal Átha Beithe. N Ireland 54°08′N 06°54′W
97 E14 **Ballybofey** *Ir.* Bealach Féich. NW Ireland 54°49′N 07°47′W
97 G14 **Ballycastle** *Ir.* Baile an Chaistil. N Northern Ireland, United Kingdom 55°12′N 06°14′W
97 G15 **Ballyclare** *Ir.* Bealach Cláir. E Northern Ireland, United Kingdom 54°45′N 06°W
97 E16 **Ballyconnell** *Ir.* Béal Átha Conaill. N Ireland 54°07′N 07°35′W
97 C17 **Ballyhaunis** *Ir.* Beál Átha hAmhnais. W Ireland 53°45′N 08°45′W
97 G14 **Ballymena** *Ir.* An Baile Meánach. NE Northern Ireland, United Kingdom 54°52′N 06°17′W
97 F14 **Ballymoney** *Ir.* Baile Monaidh. NE Northern Ireland, United Kingdom 55°10′N 06°30′W
97 G15 **Ballynahinch** *Ir.* Baile na hInse. SE Northern Ireland, United Kingdom 54°24′N 05°54′W
97 D16 **Ballysadare** *Ir.* Baile Easa Dara. NW Ireland 54°13′N 08°30′W
97 D15 **Ballyshannon** *Ir.* Béal Átha Seanaidh. NW Ireland 54°30′N 08°11′W
63 H19 **Balmaceda** Aisén, S Chile 45°52′S 72°43′W
63 G23 **Balmaceda, Cerro** ▲ S Chile 51°27′S 73°26′W
111 N22 **Balmazújváros** Hajdú-Bihar, E Hungary 47°36′N 21°18′E
108 E10 **Balmhorn** ▲ SW Switzerland 46°27′N 07°41′E
182 L12 **Balmoral** Victoria, SE Australia 37°16′S 141°38′E
24 K9 **Balmorhea** Texas, SW USA 30°58′N 103°44′W
Balneario Claromecó *see* Claromecó
148 M12 **Balochistan** *prev.* Baluchistan, Beluchistan. ◈ *province* SW Pakistan
82 B13 **Balombo** *Port.* Norton de Matos, Vila Norton de Matos. Benguela, W Angola 12°21′S 14°46′E
82 B13 **Balombo** ≋ W Angola
181 X10 **Balonne River** ≋ Queensland, E Australia
152 E13 **Bālotra** Rājasthān, N India 25°50′N 72°21′E
145 V14 **Balpyk Bi** *prev.* Kirovskiy, *Kaz.* Kīrov. Almaty, SE Kazakhstan 44°54′N 78°13′E
Balqā'/Balqā', Muḩāfaẓat al *see* Al Balqā'
Balqash *see* Balkhash/Balkhash, Ozero
152 M12 **Balrāmpur** Uttar Pradesh, N India 27°25′N 82°10′E
182 M9 **Balranald** New South Wales, SE Australia 34°39′S 143°33′E
116 H14 **Balș** Olt, S Romania 44°20′N 24°06′E
14 H11 **Balsam Creek** Ontario, S Canada 46°26′N 79°10′W
30 I5 **Balsam Lake** Wisconsin, N USA 45°27′N 92°28′W
14 I14 **Balsam Lake** ◎ Ontario, SE Canada
59 M14 **Balsas** Maranhão, E Brazil 07°30′S 46°W
40 M15 **Balsas, Río** *var.* Río Mexcala. ≋ S Mexico
43 W16 **Balsas, Río** ≋ SE Panama
119 O18 **Bal'shavik** *Rus.* Bol'shevik. Homyel'skaya Voblasts', SE Belarus 52°34′N 30°53′E
95 O15 **Bålsta** Uppsala, C Sweden 59°31′N 17°32′E
108 E7 **Balsthal** Solothurn, NW Switzerland 47°18′N 07°42′E
117 O8 **Balta** Odes'ka Oblast', SW Ukraine 47°58′N 29°39′E
105 N5 **Baltanás** Castilla y León, N Spain 41°56′N 04°12′W
61 E16 **Baltasar Brum** Artigas, N Uruguay 30°44′S 57°19′W
116 M9 **Bălți** *Rus.* Bel'tsy. N Moldova 47°45′N 27°57′E
Baltic Port *see* Paldiski
118 B10 **Baltic Sea** *Ger.* Ostee, *Rus.* Baltiskoye More. *sea* N Europe
21 X3 **Baltimore** Maryland, NE USA 39°17′N 76°37′W
31 T13 **Baltimore** Ohio, N USA 39°48′N 82°33′W
21 X3 **Baltimore-Washington** ✈ Maryland, E USA 39°10′N 76°40′W
Baltischport/Baltiski *see* Paldiski
Baltiskoye More *see* Baltic Sea
119 A14 **Baltiysk** *Ger.* Pillau. Kaliningradskaya Oblast', W Russian Federation 54°39′N 19°54′E
Baltkrievija *see* Belarus
119 H14 **Baltoji Voke** Vilnius, SE Lithuania 54°35′N 25°13′E
Balūchestān va Sīstān *see* Sīstān va Balūchestān
Baluchistan *see* Balochistan
171. P5 **Balud** Masbate, N Philippines 12°03′N 123°12′E
169 T9 **Balui, Batang** ≋ East Malaysia
153 S13 **Bālurghat** West Bengal, NE India 25°12′N 88°50′E
118 J8 **Balvi** NE Latvia 57°07′N 27°14′E
147 W7 **Balykchy** *Kir.* Ysyk-Köl; *prev.* Issyk-Kul', Rybach'ye. Issyk-Kul'skaya Oblast', NE Kyrgyzstan 42°29′N 76°08′E
56 B7 **Balzar** Guayas, W Ecuador 01°25′S 79°54′W
108 I8 **Balzers** S Liechtenstein 47°04′N 09°32′E
143 T12 **Bam** Kermān, SE Iran 29°08′N 58°27′E
77 Y13 **Bama** Borno, NE Nigeria 11°28′N 13°46′E
76 L12 **Bamako** ● (Mali) Capital District, SW Mali 12°39′N 08°02′W
77 P10 **Bamba** Gao, C Mali 17°03′N 01°19′W
42 M8 **Bambana, Río** ≋ NE Nicaragua
79 J15 **Bambari** Ouaka, C Central African Republic 05°45′N 20°37′E
181 W5 **Bambaroo** Queensland, NE Australia 19°00′S 146°16′E
101 K19 **Bamberg** Bayern, SE Germany 49°54′N 10°53′E
21 R14 **Bamberg** South Carolina, SE USA 33°16′N 81°02′W
79 M16 **Bambesa** Orientale, N Dem. Rep. Congo 03°25′N 25°43′E
76 G11 **Bambey** W Senegal 14°43′N 16°26′W
79 H16 **Bambio** Sangha-Mbaéré, SW Central African Republic 03°57′N 16°54′E
83 I24 **Bamboesberge** ▲ S South Africa 31°24′S 26°10′E
79 D14 **Bamenda** Nord-Ouest, W Cameroon 05°55′N 10°09′E
10 K17 **Bamfield** Vancouver Island, British Columbia, SW Canada 48°48′N 125°05′W
Bami *see* Bamy
79 J14 **Bamingui** Bamingui-Bangoran, C Central African Republic 07°38′N 20°06′E
78 J13 **Bamingui** ≋ N Central African Republic
78 J13 **Bamingui-Bangoran** ◈ *prefecture* N Central African Republic
143 V13 **Bampūr** Sīstān va Balūchestān, SE Iran 27°13′N 60°28′E
186 C8 **Bamu** ≋ SW Papua New Guinea
146 E12 **Bamy** *Rus.* Bami. Ahal Welaýaty, C Turkmenistan 38°42′N 56°47′E
149 P4 **Bāmyān** *prev.* Bāmīān. Bāmyān, NE Afghanistan 34°50′N 67°50′E
149 O4 **Bāmyān** *prev.* Bāmīān. ◈ *province* C Afghanistan
Bán *see* Bánovce nad Bebravou
81 N17 **Banaadir** *off.* Gobolka Banaadir. ◈ *region* S Somalia
Banaadir, Gobolka *see* Banaadir
191 N3 **Banaba** *var.* Ocean Island. *island* Tungaru, W Kiribati
59 O14 **Banabuiú, Açude** ⊠ NE Brazil
57 O19 **Bañados del Izozog** *salt lake* SE Bolivia
97 D18 **Banagher** *Ir.* Beannchar. C Ireland 53°12′N 07°56′W
79 M17 **Banalia** Orientale, N Dem. Rep. Congo 01°33′N 25°23′E
76 L12 **Banamba** Koulikoro, W Mali 13°29′N 07°22′W
40 G4 **Banámichi** Sonora, NW Mexico 30°00′N 110°14′W
181 Y9 **Banana** Queensland, E Australia 24°33′S 150°07′E
191 Z2 **Banana** *prev.* Main Camp. Kiritimati, E Kiribati 02°00′N 157°25′W
59 K16 **Bananal, Ilha do** *island* C Brazil
23 Y12 **Banana River** *lagoon* Florida, SE USA
151 Q22 **Bananga** Andaman and Nicobar Islands, India, NE Indian Ocean 06°57′N 93°54′E
Banaras *see* Vārānasi
114 N13 **Banarlı** Tekirdağ, NW Turkey 41°04′N 27°21′E
152 H12 **Banās** ≋ N India
75 Z11 **Banās, Rās** *headland* E Egypt 23°55′N 35°47′E
112 N10 **Banatski Karlovac** Vojvodina, NE Serbia 45°03′N 21°02′E
141 P16 **Banā, Wādī** *dry watercourse* SW Yemen
136 E14 **Banaz** Uşak, W Turkey 38°47′N 29°46′E
136 E14 **Banaz Çayı** ≋ W Turkey
159 P14 **Banbar** *var.* Coka. Xizang Zizhiqu, W China 31°01′N 94°43′E
97 G15 **Banbridge** *Ir.* Droichead na Banna. SE Northern Ireland, United Kingdom 54°21′N 06°16′W
Ban Bua Yai *see* Bua Yai
97 M21 **Banbury** S England, United Kingdom 52°04′N 01°20′W
167 O7 **Ban Chiang Dao** Chiang Mai, NW Thailand 19°22′N 98°59′E
96 K9 **Banchory** NE Scotland, United Kingdom 58°05′N 00°35′W
14 J13 **Bancroft** Ontario, SE Canada 45°03′N 77°52′W
33 R15 **Bancroft** Idaho, NW USA 42°43′N 111°54′W
29 U11 **Bancroft** Iowa, C USA 43°17′N 94°13′W
154 I9 **Bānda** Madhya Pradesh, C India 24°01′N 78°59′E
152 L13 **Bānda** Uttar Pradesh, N India 25°28′N 80°20′E
168 F7 **Banda Aceh** *var.* Banda Atjeh; *prev.* Koetaradja, Kutaradja, Kutaraja. Sumatera, W Indonesia 05°30′N 95°20′E
Banda Atjeh *see* Banda Aceh
171 S14 **Banda, Kepulauan** *island group* E Indonesia
Banda, Laut *see* Banda Sea
77 N17 **Bandama** *var.* Bandama Fleuve. ≋ S Ivory Coast
77 N15 **Bandama Blanc** ≋ C Ivory Coast
Bandama Fleuve *see* Bandama
Bandar 'Abbās *see* Bandar-e 'Abbās
153 W16 **Bandarban** Chittagong, SE Bangladesh 22°13′N 92°13′E
80 Q13 **Bandarbeyla** *var.* Bender Beila, Bender Beyla. Bari, NE Somalia 09°28′N 50°48′E
143 R14 **Bandar-e 'Abbās** *var.* Bandar 'Abbās; *prev.* Gombroon. Hormozgān, S Iran 27°11′N 56°11′E
142 M3 **Bandar-e Anzalī** Gīlān, NW Iran 37°26′N 49°29′E
143 N12 **Bandar-e Būshehr** *var.* Būshehr, *Eng.* Bushire. Būshehr, S Iran 28°59′N 50°50′E
143 O13 **Bandar-e Dayyer** *var.* Deyyer. Būshehr, SE Iran 27°50′N 51°55′E
142 M11 **Bandar-e Gonāveh** *var.* Ganāveh; *prev.* Gonāveh. Būshehr, SW Iran 29°33′N 50°39′E
143 T15 **Bandar-e Jāsk** *var.* Jāsk. Hormozgān, SE Iran 25°35′N 58°06′E
143 O13 **Bandar-e Kangān** *var.* Kangān. Būshehr, S Iran 25°50′N 57°30′E
143 R14 **Bandar-e Khamīr** Hormozgān, S Iran 27°00′N 55°50′E
Bandar-e Langeh *see* Bandar-e Lengeh
143 Q14 **Bandar-e Lengeh** *var.* Bandar-e Langeh, Lingeh. Hormozgān, S Iran 26°34′N 54°52′E
142 L10 **Bandar-e Māhshahr** *var.* Māh-Shahr; *prev.* Bandar-e Ma'shūr. Khūzestān, SW Iran 30°34′N 49°10′E
Bandar-e Ma'shūr *see* Bandar-e Māhshahr
143 O14 **Bandar-e Nakhīlū** Hormozgān, S Iran
Bandar-e Shāh *see* Bandar-e Torkaman
143 P4 **Bandar-e Torkaman** *var.* Bandar-e Torkeman, Bandar-e Torkman; *prev.* Bandar-e Shāh. Golestán, N Iran 36°55′N 54°05′E
Bandar-e Torkeman/Bandar-e Torkman *see* Bandar-e Torkaman
Bandar Kassim *see* Boosaaso
168 M15 **Bandar Lampung** *var.* Bandarlampung, Tanjungkarang-Telukbetung; *prev.* Tandjoengkarang, Tanjungkarang, Teloekbetoeng, Telukbetung. Sumatera, W Indonesia 05°28′S 105°16′E
Bandarlampung *see* Bandar Lampung
Bandar Maharani *see* Muar
Bandar Masulipatnam *see* Machilīpatnam
Bandar Penggaram *see* Batu Pahat
169 T7 **Bandar Seri Begawan** *prev.* Brunei Town. ● (Brunei) N Brunei 04°56′N 114°58′E
169 T7 **Bandar Seri Begawan** ✈ N Brunei 04°56′N 114°58′E
171 R15 **Banda Sea** *var.* Laut Banda. *sea* E Indonesia
104 H5 **Bande** Galicia, NW Spain 42°01′N 07°58′W
59 G15 **Bandeirantes** Mato Grosso, W Brazil 09°04′S 57°53′W
59 N20 **Bandeira, Pico da** ▲ SE Brazil 20°25′S 41°45′W
83 K19 **Bandelierkop** Limpopo, NE South Africa 23°21′S 29°46′E
62 L8 **Bandera** Santiago del Estero, N Argentina 28°53′S 62°15′W
25 Q11 **Bandera** Texas, SW USA 29°44′N 99°06′W
40 J13 **Banderas, Bahía de** *bay* W Mexico
77 O11 **Bandiagara** Mopti, C Mali 14°20′N 03°37′W
152 I12 **Bāndīkūi** Rājasthān, N India 27°01′N 76°33′E
136 C11 **Bandırma** *var.* Penderma. Balıkesir, NW Turkey 40°21′N 27°58′E
Bandjarmasin *see* Banjarmasin
Bandoeng *see* Bandung
97 C21 **Bandon** *Ir.* Droicheadna Bandan. SW Ireland 51°44′N 08°44′W
32 E14 **Bandon** Oregon, NW USA 43°07′N 124°24′W
167 R8 **Ban Dong Bang** Nong Khai, E Thailand 18°00′N 104°08′E
167 Q6 **Ban Donkon** Oudômxai, N Laos 20°20′N 101°37′E
172 J14 **Bandrélé** SE Mayotte
79 H20 **Bandundu** *prev.* Banningville. Bandundu, W Dem. Rep. Congo 03°19′S 17°24′E
79 I21 **Bandundu** *off.* Région de Bandundu. ◈ *region* W Dem. Rep. Congo
Bandundu, Région de *see* Bandundu
169 O16 **Bandung** *prev.* Bandoeng. Jawa, C Indonesia 06°47′S 107°28′E
116 L15 **Băneasa** Constanța, SW Romania 45°56′N 27°55′E
142 J4 **Bāneh** Kordestān, N Iran 35°58′N 45°54′E
44 I7 **Banes** Holguín, E Cuba 20°58′N 75°43′W
11 P16 **Banff** Alberta, SW Canada 51°10′N 115°34′W
96 K8 **Banff** NE Scotland, United Kingdom 57°39′N 02°33′W
96 K8 **Banff** *cultural region* NE Scotland, United Kingdom
Bánffyhunyad *see* Huedin
77 N14 **Banfora** SW Burkina Faso 10°36′N 04°45′W
155 H19 **Bangalore** *var.* Bengalooru, Bengaluru. *state capital* Karnātaka, S India 12°58′N 77°35′E
153 S16 **Bangaon** West Bengal, NE India 23°01′N 88°50′E
79 L15 **Bangassou** Mbomou, SE Central African Republic 04°51′N 22°55′E
186 D7 **Bangeta, Mount** ▲ C Papua New Guinea 06°11′S 147°02′E
171 P12 **Banggai, Kepulauan** *island group* C Indonesia
171 Q12 **Banggai, Pulau** *island* Kepulauan Banggai, N Indonesia
171 X13 **Banggelapa** Papua, E Indonesia 03°47′S 136°53′E
Banggi *see* Banggi, Pulau
169 V6 **Banggi, Pulau** *var.* Banggi. *island* East Malaysia
152 K5 **Banggong Co** *var.* Pangong Tso. ◎ China/India *see also* Pangong Tso
121 P13 **Banghāzī** *Eng.* Bengazi, Benghazi, *It.* Bengasi. NE Libya 32°07′N 20°04′E

◆ Country
● Country Capital
◇ Dependent Territory
○ Dependent Territory Capital
◈ Administrative Regions
✈ International Airport
▲ Mountain
▲▲ Mountain Range
Volcano
≋ River
◎ Lake
⊠ Reservoir

Bang Hieng see Xé Banghiang
169 O13 **Bangka-Belitung** off. Propinsi Bangka-Belitung. ◐ province W Indonesia
169 P11 **Bangkai, Tanjung** var. Bankai. headland Borneo, N Indonesia 0°21′N 108°53′E
169 S16 **Bangkalan** Pulau Madura, C Indonesia 07°05′S 112°44′E
169 N12 **Bangka, Pulau** island W Indonesia
169 N13 **Bangka, Selat** strait Sumatera, W Indonesia
169 N13 **Bangka, Selat** var. Selat Likupang. strait Sulawesi, N Indonesia
168 J11 **Bangkinang** Sumatera, W Indonesia 0°21′N 100°52′E
168 K12 **Bangko** Sumatera, W Indonesia 02°05′S 102°20′E
Bangkok see Ao Krung Thep
Bangkok, Bight of see Krung Thep, Ao
153 T14 **Bangladesh** off. People's Republic of Bangladesh; prev. East Pakistan. ◆ republic S Asia
Bangladesh, People's Republic of see Bangladesh
167 V13 **Ba Ngoi** Khanh Hoa, S Vietnam 11°56′N 109°07′E
Ba Ngoi see Cam Ranh
Bangong Co see Pangong Tso
97 I18 **Bangor** NW Wales, United Kingdom 53°13′N 04°08′W
97 G15 **Bangor** Ir. Beannchar. E Northern Ireland, United Kingdom 54°40′N 05°40′W
19 R6 **Bangor** Maine, NE USA 44°48′N 68°47′W
18 I14 **Bangor** Pennsylvania, NE USA 40°52′N 75°12′W
67 R8 **Bangoran** ≈ S Central African Republic
Bang Phra see Trat
Bang Pla Soi see Chon Buri
25 Q8 **Bangs** Texas, SW USA 31°43′N 99°07′W
167 N13 **Bang Saphan** var. Bang Saphan Yai. Prachuap Khiri Khan, SW Thailand 11°10′N 99°33′E
Bang Saphan Yai see Bang Saphan
36 I8 **Bangs, Mount** ▲ Arizona, SW USA 36°47′N 113°51′W
93 E15 **Bangsund** Nord-Trøndelag, C Norway 64°22′N 11°22′E
171 O2 **Bangued** Luzon, N Philippines 17°36′N 120°40′E
79 I15 **Bangui** ● (Central African Republic) Ombella-Mpoko, SW Central African Republic 04°21′N 18°32′E
79 I15 **Bangui** ✈ Ombella-Mpoko, SW Central African Republic 04°19′N 18°34′E
83 N16 **Bangula** Southern, S Malawi 16°38′S 35°04′E
Bangwaketse see Southern
82 K12 **Bangweulu, Lake** var. Lake Bengweulu. ⊛ N Zambia
121 V13 **Banhā** var. Benha. N Egypt 30°28′N 31°11′E
Ban Hat Yai see Hat Yai
167 Q7 **Ban Hin Heup** Viangchan, C Laos 18°37′N 102°19′E
Ban Houayxay/Ban Houei Sai see Houayxay
167 O12 **Ban Hua Hin** var. Hua Hin. Prachuap Khiri Khan, SW Thailand 12°34′N 99°58′E
79 L14 **Bani** Haute-Kotto, E Central African Republic 07°06′N 22°51′E
45 O9 **Baní** S Dominican Republic 18°19′N 70°21′W
77 N12 **Bani** ≈ S Mali
Bāmiān see Bāmyān
Banias see Bāniyās
77 S11 **Bani Bangou** Tillabéri, SW Niger 15°04′N 02°40′E
76 M12 **Banifing** var. Ngorolaka. ≈ Burkina Faso/Mali
Banijska Palanka see Glina
77 R13 **Banikoara** N Benin 11°18′N 02°26′E
75 W9 **Banī Mazār** var. Beni Mazâr. C Egypt 28°29′N 30°48′E
114 K8 **Baniski Lom** ≈ N Bulgaria
21 U7 **Banister River** ≈ Virginia, NE USA
121 V14 **Banī Suwayf** var. Beni Suef. N Egypt 29°09′N 31°04′E
75 O8 **Banī Walīd** NW Libya 31°46′N 13°59′E
138 H5 **Bāniyās** var. Banias, Baniyas, Paneas. Tarṭūs, W Syria 35°12′N 35°57′E
113 K14 **Banja** Serbia, W Serbia 43°33′N 19°35′E
Banjak, Kepulauan see Banyak, Kepulauan
112 J12 **Banja Koviljača** Serbia, W Serbia 44°31′N 19°11′E
112 G11 **Banja Luka** ◐ Republika Srpska, NW Bosnia and Herzegovina
169 T13 **Banjarmasin** prev. Bandjarmasin. Borneo, C Indonesia 03°22′S 114°33′E
76 F11 **Banjul** prev. Bathurst. ● (Gambia) W Gambia 13°26′N 16°43′W
76 F11 **Banjul** ✈ W Gambia 13°18′N 16°39′W
Bank see Bankä
137 Y13 **Bankä** Rus. Bank. SE Azerbaijan 39°25′N 49°13′E
167 S11 **Ban Kadian** var. Ban Kadiene. Champasak, S Laos 14°25′N 105°42′E
Ban Kadiene see Ban Kadian
Bankai see Bangkai, Tanjung
166 M14 **Ban Kam Phuam** Phangnga, SW Thailand 09°16′N 98°24′E
Ban Kantang see Kantang
77 O11 **Bankass** Mopti, S Mali 14°05′N 03°30′W
95 L19 **Bankeryd** Jönköping, S Sweden 57°51′N 14°07′E
83 K16 **Banket** Mashonaland West, N Zimbabwe 17°23′S 30°24′E
167 T11 **Ban Khamphô** Attapu, S Laos 14°35′N 106°18′E
23 O4 **Bankhead Lake** ⊠ Alabama, S USA
77 Q11 **Bankilaré** Tillabéri, SW Niger 14°34′N 00°41′E
Banks, Îles see Banks Islands
10 I14 **Banks Island** island British Columbia, SW Canada
187 R12 **Banks Islands** Fr. Îles Banks. island group N Vanuatu
23 U8 **Banks Lake** ⊛ Georgia, SE USA
32 K8 **Banks Lake** ⊠ Washington, NW USA
185 I19 **Banks Peninsula** peninsula South Island, New Zealand
183 Q15 **Banks Strait** strait SW Tasman Sea
Ban Kui Nua see Kui Buri
153 R16 **Bānkura** West Bengal, NE India 23°14′N 87°05′E
167 S8 **Ban Lakxao** var. Lak Sao. Bolikhamxai, C Laos 18°10′N 104°58′E
167 O16 **Ban Lam Phai** Songkhla, SW Thailand 06°43′N 100°57′E
Ban Mae Sot see Mae Sot
Ban Mae Suai see Mae Suai
Ban Mak Khaeng see Udon Thani
166 M3 **Banmauk** Sagaing, N Myanmar (Burma) 24°26′N 95°54′E
Banmo see Bhamo
167 T10 **Ban Mun-Houamuang** S Laos 15°11′N 106°44′E
97 F14 **Bann** var. Lower Bann, Upper Bann. ≈ N Northern Ireland, United Kingdom
167 S10 **Ban Nadou** Salavan, S Laos 15°51′N 105°38′E
167 S9 **Ban Nakala** Savannakhét, S Laos 16°14′N 105°09′E
167 Q8 **Ban Nakha** Viangchan, C Laos 18°13′N 102°29′E
167 S9 **Ban Nakham** Khammouan, S Laos 17°10′N 105°25′E
167 P7 **Ban Namoun** Xaignabouli, N Laos 18°40′N 101°34′E
167 O17 **Ban Nang Sata** Yala, SW Thailand 06°15′N 101°13′E
167 N15 **Ban Na San** Surat Thani, SW Thailand 08°53′N 99°17′E
167 R7 **Ban Nasi** Xiangkhoang, N Laos 19°37′N 103°33′E
44 I3 **Bannerman Town** Eleuthera Island, C The Bahamas 24°38′N 76°09′W
35 V15 **Banning** California, W USA 33°55′N 116°52′W
Banningville see Bandundu
167 S11 **Ban Nongsim** Champasak, S Laos 14°45′N 106°00′E
149 S7 **Bannu** prev. Edwardesabad. Khyber Pakhtunkhwa, NW Pakistan 33°00′N 70°36′E
Bañolas see Banyoles
56 C7 **Baños** Tungurahua, C Ecuador 01°26′S 78°24′W
Bánovce see Bánovce nad Bebravou
111 I19 **Bánovce nad Bebravou** var. Bánovce, Hung. Bán. Trenčiansky Kraj, W Slovakia 48°43′N 18°15′E
112 I12 **Banovići** ◐ Federacija Bosne I Hercegovine, E Bosnia and Herzegovina
Banow see Andarāb
Ban Pak Phanang see Pak Phanang
167 O7 **Ban Pan Nua** Lampang, NW Thailand 18°51′N 99°57′E
167 Q9 **Ban Phai** Khon Kaen, E Thailand 16°00′N 102°42′E
167 Q8 **Ban Phônhông** var. Phônhông. C Laos 18°29′N 102°26′E
167 T9 **Ban Phou A Douk** Khammouan, C Laos 17°12′N 106°07′E
167 Q8 **Ban Phu** Uthai Thani, W Thailand
167 O11 **Ban Pong** Ratchaburi, W Thailand 13°49′N 99°53′E
190 I3 **Banraeaba** Tarawa, W Kiribati 01°20′N 173°02′E
167 N10 **Ban Sai Yok** Kanchanaburi, W Thailand 14°24′N 98°54′E
Ban Sattahip/Ban Sattahipp see Sattahip
Ban Sichon see Sichon
Ban Si Racha see Si Racha
111 J19 **Banská Bystrica** Ger. Neusohl, Hung. Besztercebánya. Banskobystrický Kraj, C Slovakia 48°46′N 19°08′E
111 K20 **Banskobystrický Kraj** ◐ region C Slovakia
167 R8 **Ban Sôppheung** Bolikhamxai, C Laos 18°33′N 104°18′E
Ban Sop Prap see Sop Prap
152 G15 **Bānswāra** Rājasthān, N India 23°32′N 74°28′E
167 N15 **Ban Ta Khun** Surat Thani, SW Thailand 08°53′N 98°52′E
Ban Takua Pa see Takua Pa
167 S8 **Ban Talak** Khammouan, C Laos 17°33′N 105°40′E
77 R15 **Bantè** W Benin 08°25′N 01°58′E
167 Q11 **Bânteay Méan Choăy** var. Sisŏphŏn. Bătdâmbâng, NW Cambodia 13°37′N 102°58′E
167 N16 **Banten** off. Propinsi Banten. ◐ province W Indonesia
Propinsi Banten see Banten
167 Q8 **Ban Thabôk** Bolikhamxai, C Laos 18°21′N 103°12′E
167 T9 **Ban Tôp** Savannakhét, S Laos 16°07′N 106°07′E
97 B21 **Bantry** Ir. Beanntraí. Cork, SW Ireland 51°41′N 09°27′W
97 A21 **Bantry Bay** Ir. Bá Bheanntraí. bay SW Ireland
155 F19 **Bantvāl** var. Bantwāl. Karnātaka, E India 12°57′N 75°04′E
Bantwāl see Bantvāl
114 N9 **Banya** Burgas, E Bulgaria 42°46′N 27°49′E
168 G10 **Banyak, Kepulauan** prev. Kepulauan Banjak. island group NW Indonesia
105 U8 **Banya, La** headland E Spain 40°34′N 00°37′E
79 E14 **Banyo** Adamaoua, NW Cameroon 06°47′N 11°50′E
105 X4 **Banyoles** var. Bañolas. Cataluña, NE Spain 42°07′N 02°46′E
167 N16 **Ban Yong Sata** Trang, SW Thailand 07°09′N 99°42′E
195 X14 **Banzare Coast** physical region Antarctica
173 Q14 **Banzare Seamounts** undersea feature S Indian Ocean
Banzart see Bizerte
163 Q12 **Baochang** var. Taibus Qi. Nei Mongol Zizhiqu, N China 41°55′N 115°22′E
161 O3 **Baoding** var. Pao-ting; prev. Tsingyuan. Hebei, E China 38°47′N 115°30′E
Baoebaoe see Baubau
Baoi, Oileán see Dursey Island
160 J6 **Baoji** var. Pao-chi, Paoki. Shaanxi, C China 34°23′N 107°16′E
163 U9 **Baokang** var. Hoqin Zuoyi Zhongji. Nei Mongol Zizhiqu, N China 44°08′N 123°18′E
186 L8 **Baolo** Santa Isabel, N Solomon Islands 07°41′S 158°47′E
167 U13 **Bao Lôc** Lâm Đông, S Vietnam 11°33′N 107°48′E
163 Z7 **Baoqing** Heilongjiang, NE China 46°15′N 132°12′E
Baoqing see Shaoyang
79 H15 **Baoro** Nana-Mambéré, W Central African Republic 05°40′N 16°00′E
160 E12 **Baoshan** var. Pao-shan. Yunnan, SW China 25°05′N 99°07′E
163 N13 **Baotou** var. Pao-t'ou, Paotow. Nei Mongol Zizhiqu, N China 40°38′N 109°59′E
76 L14 **Baoulé** ≈ S Mali
76 K12 **Baoulé** ≈ W Mali
Bao Yèn see Phô Rang
Ba-Pahalaborwa see Phalaborwa
103 O2 **Bapaume** Pas-de-Calais, N France 50°06′N 02°50′E
14 J13 **Baptiste Lake** ⊛ Ontario, SE Canada
Bapu see Meigu
Baqanas see Bakanas
Baqbaqty see Bakbakty
159 P14 **Baqên** var. Dartang. Xizang Zizhiqu, W China 31°50′N 94°08′E
138 F14 **Bāqir, Jabal** ▲ S Jordan
139 T7 **Ba'qūbah** var. Qubba. Diyālá, C Iraq 33°45′N 44°40′E
Ba'qūbah see Diyālá
62 H5 **Baquedano** Antofagasta, N Chile 23°20′S 69°50′W
Baquerizo Moreno see Puerto Baquerizo Moreno
113 J18 **Bar** It. Antivari. S Montenegro 42°02′N 19°09′E
116 M6 **Bar** Vinnyts'ka Oblast', C Ukraine 49°05′N 27°40′E
80 E10 **Bara** Northern Kordofan, C Sudan 13°42′N 30°21′E
81 M18 **Baraawe** It. Brava. Shabeellaha Hoose, S Somalia 01°10′N 43°59′E
152 M12 **Bāra Banki** Uttar Pradesh, N India 26°56′N 81°11′E
30 L8 **Baraboo** Wisconsin, N USA 43°27′N 89°45′W
30 K8 **Baraboo Range** hill range Wisconsin, N USA
Baracaldo see San Vicente de Barakaldo
15 Y6 **Barachois** Québec, SE Canada 48°37′N 64°14′W
44 J7 **Baracoa** Guantánamo, E Cuba 20°23′N 74°31′W
61 C19 **Baradero** Buenos Aires, E Argentina 33°50′S 59°30′W
183 R6 **Baradine** New South Wales, SE Australia 30°55′S 149°03′E
Baraf Daja Islands see Damar, Kepulauan
Baragarh see Bargarh
81 I17 **Baragoi** Samburu, W Kenya 01°39′N 36°46′E
45 N9 **Barahona** SW Dominican Republic 18°13′N 71°07′W
153 W13 **Barail Range** ▲▲ NE India
Baraka see Barka
80 G10 **Barakat** Gezira, C Sudan 14°18′N 33°32′E
149 Q6 **Barakī Barak** var. Barakī, Baraki Rajan. Lōgar, E Afghanistan 33°58′N 68°58′E
Baraki Rajan see Barakī Barak
154 N11 **Bārākot** Odisha, E India 21°35′N 85°00′E
55 S7 **Barama River** ≈ N Guyana
155 E14 **Bārāmati** Mahārāshtra, W India 18°12′N 74°39′E
152 H5 **Bārāmūla** Jammu and Kashmir, NW India 34°15′N 74°24′E
119 N14 **Baran'** Vitsyebskaya Voblasts', NE Belarus 54°29′N 30°18′E
152 I14 **Bārān** Rājasthān, N India 25°08′N 76°32′E
Barānān, Shākh-i see Beranan, Shax-i
119 I17 **Baranavichy** Pol. Baranowicze, Rus. Baranovichi. Brestskaya Voblasts', SW Belarus 53°08′N 26°02′E
123 T6 **Baranikha** Chukotskiy Avtonomnyy Okrug, NE Russian Federation 68°29′N 168°13′E
75 Y11 **Baranīs** var. Berenice, Mînâ Baranîs. SE Egypt 23°58′N 35°29′E
116 M4 **Baranivka** Zhytomyrs'ka Oblast', N Ukraine 50°16′N 27°40′E
39 W14 **Baranof Island** island Alexander Archipelago, Alaska, USA
Baranovichi/Baranowicze see Baranavichy
111 N15 **Baranów Sandomierski** Podkarpackie, SE Poland 50°28′N 21°31′E
111 I26 **Baranya** off. Baranya Megye. ◐ county S Hungary
Baranya Megye see Baranya
153 R13 **Barāri** Bihār, NE India 25°31′N 87°23′E
22 L10 **Barataria Bay** bay Louisiana, S USA
Barat Daya, Kepulauan see Damar, Kepulauan
118 L12 **Baravukha** Rus. Borovukha. Vitsyebskaya Voblasts', N Belarus 55°34′N 28°36′E
54 E11 **Baraya** Huila, C Colombia 03°11′N 75°04′W
59 M21 **Barbacena** Minas Gerais, SE Brazil 21°13′S 43°47′W
54 B13 **Barbacoas** Nariño, SW Colombia 01°38′N 78°08′W
54 L6 **Barbacoas** Aragua, N Venezuela 09°29′N 66°58′W
45 Z13 **Barbados** ◆ commonwealth republic SE West Indies
47 S3 **Barbados** island Barbados
105 U11 **Barbaria, Cap de** var. Cabo de Berberia. headland Formentera, E Spain 38°39′N 01°24′E
114 N13 **Barbaros** Tekirdağ, NW Turkey 40°55′N 27°28′E
74 A11 **Barbas, Cap** headland S Western Sahara 22°14′N 16°45′W
105 T5 **Barbastro** Aragón, NE Spain 42°02′N 00°07′E
104 K16 **Barbate** ≈ SW Spain
104 K16 **Barbate de Franco** Andalucía, S Spain 36°11′N 05°55′W
83 K21 **Barberton** Mpumalanga, NE South Africa 25°48′S 31°03′E
31 U12 **Barberton** Ohio, N USA 41°02′N 81°37′W
102 K12 **Barbezieux-St-Hilaire** Charente, W France 45°28′N 00°09′W
54 G9 **Barbosa** Boyacá, C Colombia 05°57′N 73°37′W
21 N7 **Barbourville** Kentucky, S USA 36°52′N 83°54′W
45 W9 **Barbuda** island N Antigua and Barbuda
181 W8 **Barcaldine** Queensland, E Australia 23°33′S 145°21′E
Barcarozsnyó see Râşnov
104 I11 **Barcarrota** Extremadura, W Spain 38°31′N 06°51′W
Barcău see Berettyó
Barce see Al Marj
107 L23 **Barcellona** var. Barcellona Pozzo di Gotto. Sicilia, Italy, C Mediterranean Sea 38°10′N 15°15′E
Barcellona Pozzo di Gotto see Barcellona
105 W6 **Barcelona** anc. Barcino, Barcinona. Cataluña, E Spain 41°25′N 02°10′E
55 N5 **Barcelona** Anzoátegui, NE Venezuela 10°08′N 64°43′W
105 S5 **Barcelona** ◐ province Cataluña, NE Spain
105 W6 **Barcelona** ✈ Cataluña, E Spain 41°25′N 02°10′E
103 U14 **Barcelonnette** Alpes-de-Haute-Provence, SE France 44°24′N 06°37′E
58 E12 **Barcelos** Amazonas, N Brazil 0°59′S 62°58′W
104 G5 **Barcelos** Braga, N Portugal 41°32′N 08°37′W
110 I10 **Barcin** Ger. Bartschin. Kujawski-pomorskie, C Poland 52°51′N 17°55′E
Barcino/Barcinona see Barcelona
Barcoo see Cooper Creek
111 H26 **Barcs** Somogy, SW Hungary 45°58′N 17°26′E
137 W11 **Bärdä** Rus. Barda. C Azerbaijan 40°25′N 47°07′E
Barda see Bärdä
78 H5 **Bardaï** Tibesti, N Chad 21°21′N 16°56′E
139 R2 **Bardarash** Dahūk, N Iraq 36°32′N 43°36′E
92 K3 **Bárðarbunga** ▲ C Iceland 64°39′N 17°30′W
92 K3 **Bárðardalur** valley C Iceland
139 Q7 **Bardasah** Al Anbār, SW Iraq 34°02′N 42°28′E
153 S16 **Barddhamān** West Bengal, NE India 23°10′N 88°03′E
111 N18 **Bardejov** Ger. Bartfeld, Hung. Bártfa. Presovský Kraj, E Slovakia 49°17′N 21°18′E
105 R4 **Bárdenas Reales** physical region N Spain
Bardera/Bardere see Baardheere
Bardesīr see Bardsīr
106 E9 **Bardi** Emilia-Romagna, C Italy 44°39′N 09°44′E
106 A8 **Bardonecchia** Piemonte, W Italy 45°04′N 06°40′E
97 H19 **Bardsey Island** island NW Wales, United Kingdom
143 S11 **Bardsīr** var. Bardesīr, Mashīz. Kermān, C Iran 29°58′N 56°29′E
20 L6 **Bardstown** Kentucky, S USA 37°49′N 85°29′W
Barduli see Barletta
20 G7 **Bardwell** Kentucky, S USA 36°52′N 89°01′W
152 K11 **Bareilly** var. Bareli. Uttar Pradesh, N India 28°20′N 79°24′E
Bareli see Bareilly
Baren see Jiashi
98 H13 **Barendrecht** Zuid-Holland, SW Netherlands 51°52′N 04°31′E
102 M3 **Barentin** Seine-Maritime, N France 49°33′N 00°57′E
92 N3 **Barentsburg** Spitsbergen, W Svalbard 78°01′N 14°19′E
Barentsevo More/Barents Havet see Barents Sea
92 O3 **Barentsøya** island E Svalbard
197 T11 **Barents Plain** undersea feature N Barents Sea
125 P3 **Barents Sea** Nor. Barents Havet, Rus. Barentsevo More. sea Arctic Ocean
197 U14 **Barents Trough** undersea feature SW Barents Sea 75°00′N 29°00′E
80 I9 **Barentu** W Eritrea 15°08′N 37°35′E
102 J3 **Barfleur** Manche, N France 49°41′N 01°18′W
102 J3 **Barfleur, Pointe de** headland N France 49°46′N 01°09′W
Barfrush/Barfurush see Bābol
158 H14 **Barga** Xizang Zizhiqu, W China 30°51′N 81°20′E
80 Q12 **Bargaal** prev. Baargaal. Bari, NE Somalia 11°12′N 51°04′E
154 M12 **Bargarh** var. Baragarh. Odisha, E India 21°25′N 83°35′E
105 N9 **Bargas** Castilla-La Mancha, C Spain 39°56′N 04°00′W
81 I15 **Bargē** Southern Nationalities, S Ethiopia 06°11′N 37°04′E
106 A9 **Barge** Piemonte, NE Italy 44°49′N 07°21′E
153 U16 **Barguna** Barisal, S Bangladesh 22°09′N 90°07′E
Bārgusad see Vorotan
123 N13 **Barguzin** Respublika Buryatiya, S Russian Federation 53°37′N 109°47′E
153 O13 **Barhaj** Uttar Pradesh, N India 26°16′N 83°44′E
183 N10 **Barham** New South Wales, SE Australia 35°39′S 144°09′E
152 J12 **Barhan** Uttar Pradesh, N India 27°21′N 78°11′E
19 S7 **Bar Harbor** Mount Desert Island, Maine, NE USA 44°23′N 68°14′W
153 R14 **Barharwa** Jhārkhand, NE India 24°52′N 87°47′E
153 P15 **Barhi** Jhārkhand, N India 24°19′N 85°25′E
107 O17 **Bari** var. Bari delle Puglie; anc. Barium. Puglia, SE Italy 41°06′N 16°52′E
80 P12 **Bari** Off. Gobolka Bari. ◐ region NE Somalia
167 T14 **Ba Ria** var. Châu Thanh. Ba Ria-Vung Tau, S Vietnam 10°30′N 107°10′E
81 G21 **Bariadi** Simiyu, NE Tanzania 02°48′S 33°59′E
Bāridah see Al Bāridah
Bari delle Puglie see Bari
Bari, Gobolka see Bari
Barikot see Barīkowṭ
149 T4 **Barīkowṭ** var. Barikot. Kunar, NE Afghanistan 35°18′N 71°36′E
42 C4 **Barillas** var. Santa Cruz Barillas. Huehuetenango, NW Guatemala 15°50′N 91°20′W
54 J6 **Barinas** Barinas, W Venezuela 08°36′N 70°15′W
54 I7 **Barinas** off. Estado Barinas; prev. Zamora. ◐ state C Venezuela
Barinas, Estado see Barinas
81 H18 **Baringo** ◐ county C Kenya
54 I6 **Barinitas** Barinas, NW Venezuela 08°45′N 70°26′W
154 P11 **Bāripada** Odisha, E India 21°56′N 86°43′E
60 K9 **Bariri** São Paulo, S Brazil 22°04′S 48°46′W
75 W11 **Bārīs** var. Bârîs. S Egypt 24°28′N 30°39′E
152 G14 **Bari Sādri** Rājasthān, N India 24°25′N 74°28′E
153 U16 **Barisal** Barisal, S Bangladesh 22°41′N 90°20′E
153 U16 **Barisal** ◐ division S Bangladesh
168 I10 **Barisan, Pegunungan** ▲▲ Sumatera, W Indonesia
169 T12 **Barito, Sungai** ≈ Borneo, C Indonesia
Barium see Bari
Bārjās see Porjus
80 I9 **Barka** var. Baraka, Ar. Khawr Barakah. seasonal river Eritrea/Sudan
Barka see Al Marj
160 H8 **Barkam** Sichuan, C China 31°56′N 102°22′E
118 J9 **Barkava** C Latvia 56°43′N 26°34′E
10 M15 **Barkerville** British Columbia, SW Canada 53°06′N 121°35′W
14 J12 **Bark Lake** ⊛ Ontario, SE Canada
20 H7 **Barkley, Lake** ⊠ Kentucky/Tennessee, S USA
10 K17 **Barkley Sound** inlet British Columbia, W Canada
83 J24 **Barkly East** Afr. Barkly-Oos. Eastern Cape, SE South Africa 30°58′S 27°35′E
Barkly-Oos see Barkly East
181 S4 **Barkly Tableland** plateau Northern Territory/Queensland, N Australia
Barkly-Wes see Barkly West
83 H22 **Barkly West** Afr. Barkly-Wes. Northern Cape, N South Africa 28°32′S 24°32′E
159 O5 **Barkol** var. Barkol Kazak Zizhixian. Xinjiang Uygur Zizhiqu, NW China 43°37′N 93°01′E
159 O5 **Barkol Hu** ⊛ NW China
Barkol Kazak Zizhixian see Barkol
30 J3 **Bark Point** headland Wisconsin, N USA 46°53′N 91°11′W
25 P11 **Barksdale** Texas, SW USA 29°43′N 100°03′W
Bar Kunar see Asmār
116 L11 **Bârlad** prev. Bîrlad. Vaslui, E Romania 46°12′N 27°39′E
116 M11 **Bârlad** prev. Bîrlad. ≈ E Romania
76 D9 **Barlavento, Ilhas de** var. Windward Islands. island group N Cape Verde
103 R5 **Bar-le-Duc** var. Bar-sur-Ornain. Meuse, NE France 48°46′N 05°10′E
180 K11 **Barlee, Lake** ⊛ Western Australia
180 H8 **Barlee Range** ▲▲ Western Australia
107 N16 **Barletta** anc. Barduli. Puglia, SE Italy 41°20′N 16°17′E
110 E10 **Barlinek** Ger. Berlinchen. Zachodnio-pomorskie, NW Poland 53°N 15°11′E
27 S11 **Barling** Arkansas, C USA 35°19′N 94°18′W
171 U12 **Barma** Papua Barat, E Indonesia 01°55′S 132°57′E
183 Q9 **Barmedman** New South Wales, SE Australia 34°09′S 147°21′E
Barmen-Elberfeld see Wuppertal
152 D12 **Bārmer** Rājasthān, NW India 25°43′N 71°25′E
182 K9 **Barmera** South Australia 34°14′S 140°26′E
97 I19 **Barmouth** NW Wales, United Kingdom 52°44′N 04°04′W
154 F10 **Barnagar** Madhya Pradesh, C India 23°01′N 75°28′E
152 H9 **Barnāla** Punjab, NW India 30°26′N 75°33′E
97 L15 **Barnard Castle** N England, United Kingdom 54°35′N 01°55′W
183 O6 **Barnato** New South Wales, SE Australia 31°39′S 145°01′E
122 I13 **Barnaul** Altayskiy Kray, C Russian Federation 53°21′N 83°45′E
109 V8 **Bärnbach** Steiermark, SE Austria 47°05′N 15°07′E
18 K16 **Barnegat** New Jersey, NE USA 39°43′N 74°12′W
23 S4 **Barnesville** Georgia, SE USA 33°03′N 84°09′W
29 R6 **Barnesville** Minnesota, N USA 46°39′N 96°25′W
31 U13 **Barnesville** Ohio, N USA 39°59′N 81°10′W
98 K11 **Barneveld** var. Barnveld. Gelderland, C Netherlands 52°08′N 05°34′E
25 O9 **Barnhart** Texas, SW USA 31°07′N 101°09′W
27 P8 **Barnsdall** Oklahoma, C USA 36°33′N 96°09′W
97 M17 **Barnsley** N England, United Kingdom 53°34′N 01°28′W
19 Q12 **Barnstable** Massachusetts, NE USA 41°42′N 70°17′W
97 I23 **Barnstaple** SW England, United Kingdom 51°05′N 04°04′W
Barnveld see Barneveld
21 Q14 **Barnwell** South Carolina, SE USA 33°14′N 81°21′W
77 U15 **Baro** Niger, C Nigeria 08°35′N 06°28′E
67 U8 **Baro** var. Baro Wenz. ≈ Ethiopia/Sudan
Baro see Baro Wenz
Baroda see Vadodara
149 U2 **Baroghil Pass** var. Kowtal-e Barowghil. pass Afghanistan/Pakistan
119 Q17 **Baron'ki** Rus. Boron'ki. Mahilyowskaya Voblasts', E Belarus 53°09′N 32°08′E
182 J9 **Barossa Valley** valley South Australia
Baroui see Salisbury
81 H14 **Baro Wenz** var. Baro, Nahr Barū. ≈ Ethiopia/Sudan
Baro Wenz see Baro
Barowghīl, Kowtal-e see Baroghil Pass
153 U12 **Barpeta** Assam, NE India 26°19′N 91°05′E
31 S7 **Barques, Pointe Aux** headland Michigan, N USA 44°04′N 82°57′W
54 I5 **Barquisimeto** Lara, NW Venezuela 10°03′N 69°18′W
59 N16 **Barra** Bahia, E Brazil 11°06′S 43°15′W
96 E9 **Barra** island NW Scotland, United Kingdom
183 T5 **Barraba** New South Wales, SE Australia 30°24′S 150°37′E
60 L9 **Barra Bonita** São Paulo, S Brazil 22°30′S 48°35′W
64 J12 **Barracuda Fracture Zone** var. Fifteen Twenty Fracture Zone. tectonic feature SW Atlantic Ocean
64 G11 **Barracuda Ridge** undersea feature N Atlantic Ocean
43 N12 **Barra del Colorado** Limón, NE Costa Rica 10°44′N 83°35′W
43 N9 **Barra de Río Grande** Región Autónoma Atlántico Sur, E Nicaragua 12°56′N 83°30′W
82 A11 **Barra do Cuanza** Luanda, NW Angola 09°12′S 13°08′E
60 O9 **Barra do Piraí** Rio de Janeiro, SE Brazil 22°30′S 43°47′W
61 D16 **Barra do Quaraí** Rio Grande do Sul, SE Brazil 31°03′S 58°10′W
59 G14 **Barra do São Manuel** Pará, N Brazil 07°12′S 58°03′W
83 N19 **Barra Falsa, Ponta da** headland S Mozambique 22°57′S 35°36′E
96 E10 **Barra Head** headland NW Scotland, United Kingdom 56°46′N 07°37′W
60 O9 **Barra Mansa** Rio de Janeiro, SE Brazil 22°25′S 44°03′W
57 D14 **Barranca** Lima, W Peru 10°46′S 77°46′W
54 F8 **Barrancabermeja** Santander, N Colombia 07°01′N 73°51′W
54 H4 **Barrancas** La Guajira, N Colombia 10°59′N 72°46′W
54 J6 **Barrancas** Barinas, NW Venezuela 08°47′N 70°07′W
55 Q6 **Barrancas** Monagas, NE Venezuela 08°45′N 62°12′W
54 F6 **Barranco de Loba** Bolívar, N Colombia 08°56′N 74°07′W
104 I12 **Barrancos** Beja, S Portugal 38°08′N 06°59′W
62 N7 **Barranqueras** Chaco, N Argentina 27°29′S 58°54′W
54 E4 **Barranquilla** Atlántico, N Colombia 10°59′N 74°48′W
83 N20 **Barra, Ponta da** headland S Mozambique 23°46′S 35°33′E
105 P11 **Barrax** Castilla-La Mancha, C Spain 39°04′N 02°12′W
19 N11 **Barre** Massachusetts, NE USA 42°24′N 72°06′W
18 M7 **Barre** Vermont, NE USA 44°09′N 72°25′W
59 M17 **Barreiras** Bahia, E Brazil 12°09′S 44°58′W
104 F11 **Barreiro** Setúbal, W Portugal 38°40′N 09°05′W
65 C26 **Barren Island** island S Falkland Islands
20 K7 **Barren River Lake** ⊠ Kentucky, S USA
60 L7 **Barretos** São Paulo, S Brazil 20°33′S 48°33′W
11 P14 **Barrhead** Alberta, SW Canada 54°10′N 114°22′W
14 G14 **Barrie** Ontario, S Canada 44°22′N 79°42′W
11 N16 **Barrière** British Columbia, SW Canada 51°10′N 120°06′W
14 H8 **Barrière, Lac** ⊛ Québec, SE Canada
182 L6 **Barrier Range** hill range New South Wales, SE Australia
42 G3 **Barrier Reef** reef E Belize
188 C16 **Barrigada** C Guam 13°27′N 144°48′E
Barrington Island see Santa Fe, Isla
183 T7 **Barrington Tops** ▲ New South Wales, SE Australia 32°06′S 151°18′E
183 O4 **Barringun** New South Wales, SE Australia 29°02′S 145°45′E
59 K18 **Barro Alto** Goiás, S Brazil 15°07′S 48°56′W
59 N14 **Barro Duro** Piauí, NE Brazil 05°49′S 42°30′W
30 I5 **Barron** Wisconsin, N USA 45°24′N 91°50′W
14 J12 **Barron** ≈ Ontario, SE Canada
61 H15 **Barros Cassal** Rio Grande do Sul, S Brazil 29°12′S 52°33′W
45 P14 **Barrouallie** Saint Vincent, W Saint Vincent and the Grenadines 13°14′N 61°17′W
39 O4 **Barrow** Alaska, USA 71°17′N 156°47′W
97 E20 **Barrow** Ir. An Bhearú. ≈ SE Ireland
181 Q6 **Barrow Creek Roadhouse** Northern Territory, N Australia 21°30′S 133°52′E
97 J16 **Barrow-in-Furness** NW England, United Kingdom 54°07′N 03°14′W
180 G7 **Barrow Island** island Western Australia
39 O4 **Barrow, Point** headland Alaska, USA 71°23′N 156°28′W
11 V14 **Barrows** Manitoba, S Canada 52°49′N 101°36′W
97 J22 **Barry** S Wales, United Kingdom 51°24′N 03°18′W
14 J12 **Barry's Bay** Ontario, SE Canada 45°30′N 77°41′W
144 K14 **Barsakel'mes, Ostrov** island SW Kazakhstan
Barsč Łužyca see Forst
147 S14 **Barsem** S Tajikistan 37°36′N 71°43′E
145 V11 **Barshatas** Vostochnyy Kazakhstan, E Kazakhstan 48°13′N 78°33′E
155 F14 **Bārsi** Mahārāshtra, W India 18°14′N 75°42′E
100 I13 **Barsinghausen** Niedersachsen, C Germany 53°19′N 09°30′E
147 X8 **Barskoon** Issyk-Kul'skaya Oblast', E Kyrgyzstan 42°07′N 77°34′E
100 F10 **Barssel** Niedersachsen, NW Germany 53°10′N 07°46′E
35 U14 **Barstow** California, W USA 34°52′N 117°00′W
24 L8 **Barstow** Texas, SW USA 31°27′N 103°23′W
103 R6 **Bar-sur-Aube** Aube, NE France 48°13′N 04°43′E
Bar-sur-Ornain see Bar-le-Duc
103 Q6 **Bar-sur-Seine** Aube, N France 48°06′N 04°22′E
147 S13 **Bartang** S Tajikistan 38°06′N 71°48′E
147 T13 **Bartang** ≈ SE Tajikistan
Bartenstein see Bartoszyce
Bártfa/Bartfeld see Bardejov
100 N7 **Barth** Mecklenburg-Vorpommern, NE Germany 54°21′N 12°43′E
27 W13 **Bartholomew, Bayou** ≈ Arkansas/Louisiana, S USA
55 T8 **Bartica** N Guyana 06°24′N 58°36′W
136 H10 **Bartın** Bartın, NW Turkey 41°37′N 32°20′E
136 H10 **Bartın** ◐ province NW Turkey
181 W4 **Bartle Frere** ▲ Queensland, E Australia 17°15′S 145°43′E
27 P8 **Bartlesville** Oklahoma, C USA 36°44′N 95°59′W
29 P14 **Bartlett** Nebraska, C USA 41°51′N 98°32′W
20 E10 **Bartlett** Tennessee, S USA 35°12′N 89°52′W
25 T9 **Bartlett** Texas, SW USA 30°47′N 97°25′W
36 L13 **Bartlett Reservoir** ⊠ Arizona, SW USA
19 N6 **Barton** Vermont, NE USA 44°44′N 72°09′W
110 L7 **Bartoszyce** Ger. Bartenstein. Warmińsko-mazurskie, NE Poland 54°16′N 20°49′E
23 W12 **Bartow** Florida, SE USA 27°54′N 81°50′W
Bartschin see Barcin
168 J10 **Barumun, Sungai** ≈ Sumatera, W Indonesia
Barū, Nahr see Baro Wenz
169 S17 **Barung, Nusa** island S Indonesia
168 H9 **Barus** Sumatera, NW Indonesia 02°02′N 98°20′E
162 I9 **Baruunbayan-Ulaan** var. Hövör. Övörhangay, C Mongolia 45°10′N 101°19′E
Baruunsuu see Tsogttsetsiy
163 P8 **Baruun-Urt** Sühbaatar, E Mongolia 46°40′N 113°17′E
43 P15 **Barú, Volcán** var. Volcán de Chiriquí. ✹ W Panama 08°49′N 82°32′W
99 K21 **Barvaux** Luxembourg, SE Belgium 50°21′N 05°30′E
42 M13 **Barva, Volcán** ✹ NW Costa Rica 10°07′N 84°08′W
117 W6 **Barvinkove** Kharkivs'ka Oblast', E Ukraine 48°54′N 37°03′E
154 G11 **Barwāh** Madhya Pradesh, C India 22°17′N 76°01′E
Bärwalde Neumark see Mieszkowice
154 F11 **Barwāni** Madhya Pradesh, C India 22°02′N 74°56′E
183 P5 **Barwon River** ≈ New South Wales, SE Australia
119 L15 **Barysaw** Rus. Borisov. Minskaya Voblasts', NE Belarus 54°14′N 28°30′E
127 Q6 **Barysh** Ul'yanovskaya Oblast', W Russian Federation 53°32′N 47°06′E
117 Q4 **Baryshivka** Kyyivs'ka Oblast', N Ukraine 50°21′N 31°21′E
114 G8 **Barzia** var. Bŭrziya. NW Bulgaria
79 J17 **Basankusu** Equateur, NW Dem. Rep. Congo 01°12′N 19°50′E
117 N11 **Basarabeasca** Rus. Bessarabka. SE Moldova 46°22′N 28°56′E
116 M14 **Basarabi** Constanța, SW Romania 44°10′N 28°26′E
40 H6 **Basaseachic** Chihuahua, NW Mexico 28°18′N 108°13′W
Basauri see Arizgoiti
61 D18 **Basavilbaso** Entre Ríos, E Argentina 32°23′S 58°55′W
79 F21 **Bas-Congo** off. Région du Bas-Congo; prev. Bas-Zaïre. ◐ region SW Dem. Rep. Congo
108 E6 **Basel** Eng. Basle, Fr. Bâle. Basel Stadt, NW Switzerland 47°33′N 07°36′E
Baselland see Basel Landschaft
108 E7 **Basel Landschaft** prev. Baselland. ◐ canton NW Switzerland
108 E6 **Basel Stadt** former canton Basel. ◐ canton NW Switzerland
143 T14 **Bashākerd, Kūhhā-ye** ▲▲ SE Iran
11 Q15 **Bashaw** Alberta, SW Canada 52°40′N 112°53′W
146 K16 **Bashbedeng** Mary Welaýaty, S Turkmenistan 35°44′N 63°07′E
161 T15 **Bashi Channel** Chin. Pa-shih Hai-hsia. channel Philippines/Taiwan
Bashkiria see Bashkortostan, Respublika
122 F11 **Bashkortostan, Respublika** prev. Bashkiria. ◐ autonomous republic W Russian Federation
127 N6 **Bashmakovo** Penzenskaya Oblast', W Russian Federation 53°13′N 43°00′E
146 J10 **Bashsakarba** Lebap Welaýaty, NE Turkmenistan 40°25′N 62°16′E
117 R9 **Bashtanka** Mykolayivs'ka Oblast', S Ukraine 47°24′N 32°27′E
22 H8 **Basile** Louisiana, S USA 30°29′N 92°36′W
107 M18 **Basilicata** ◐ region S Italy
33 V13 **Basin** Wyoming, C USA 44°22′N 108°02′W
97 N22 **Basingstoke** S England, United Kingdom 51°16′N 01°08′W
143 U8 **Bāsīrān** Khorāsān-e Janūbī, E Iran 31°57′N 59°10′E
112 B10 **Baška** It. Bescanuova. Primorje-Gorski Kotar, NW Croatia 44°58′N 14°46′E
137 T15 **Başkale** Van, SE Turkey 38°03′N 43°59′E
14 L10 **Baskatong, Réservoir** ⊠ Québec, SE Canada
137 O14 **Baskil** Elazığ, E Turkey 38°38′N 38°47′E
Basle see Basel

◆ Country
● Country Capital
◇ Dependent Territory
○ Dependent Territory Capital
◐ Administrative Regions
✈ International Airport
▲ Mountain
▲▲ Mountain Range
✹ Volcano
≈ River
⊛ Lake
⊠ Reservoir

154 H9 Bāsoda Madhya Pradesh, C India 23°54′N 77°58′E
79 L17 Basoko Orientale, N Dem. Rep. Congo 01°14′N 23°26′E
Basque Country, The see País Vasco
Basra see Al Başrah
Basra see Al Başrah
Başrah, Muḥāfa at al see Al Başrah
103 U5 Bas-Rhin ◈ department NE France
Bassam see Grand-Bassam
11 Q16 Bassano Alberta, SW Canada 50°48′N 112°28′W
106 H7 Bassano del Grappa Veneto, NE Italy 45°45′N 11°45′E
77 Q15 Bassar var. Bassari. NW Togo 09°15′N 00°47′E
Bassari see Bassar
172 L9 Bassas da India island group W Madagascar
108 D7 Bassecourt Jura, W Switzerland 47°20′N 07°16′E
Bassein see Pathein
79 J15 Basse-Kotto ◈ prefecture S Central African Republic
102 J5 Basse-Normandie Eng. Lower Normandy. ◈ region N France
45 Q11 Basse-Pointe N Martinique 14°52′N 61°07′W
76 H12 Basse Santa Su E Gambia 13°18′N 14°10′W
Basse-Saxe see Niedersachsen
45 X6 Basse-Terre ○ (Guadeloupe) Basse Terre, SW Guadeloupe 16°08′N 61°40′W
45 V10 Basseterre ● (Saint Kitts and Nevis) Saint Kitts, Saint Kitts and Nevis 17°16′N 62°45′W
45 X6 Basse Terre island W Guadeloupe
29 O13 Bassett Nebraska, C USA 42°34′N 99°32′W
21 S7 Bassett Virginia, NE USA 36°45′N 79°59′W
37 N15 Bassett Peak ▲ Arizona, SW USA 32°30′N 110°16′W
76 M10 Bassikounou Hodh ech Chargui, SE Mauritania 15°55′N 05°59′W
77 R15 Bassila W Benin 08°25′N 01°58′E
Bass, Îlots de see Marotiri
31 O11 Bass Lake Indiana, N USA 41°12′N 86°35′W
183 O14 Bass Strait strait SE Australia
100 H11 Bassum Niedersachsen, NW Germany 52°52′N 08°44′E
29 X3 Basswood Lake ◎ Canada/USA
95 J21 Båstad Skåne, S Sweden 56°25′N 12°50′E
Basṭah see Beste
153 N12 Basti Uttar Pradesh, N India 26°48′N 82°44′E
103 X14 Bastia Corse, France, C Mediterranean Sea 42°42′N 09°27′E
99 L23 Bastogne Luxembourg, SE Belgium 50°N 05°43′E
22 I5 Bastrop Louisiana, S USA 32°46′N 91°54′W
25 T11 Bastrop Texas, SW USA 30°07′N 97°21′W
93 J15 Bastuträsk Västerbotten, N Sweden 64°47′N 20°05′E
119 J19 Bastyn' Rus. Bostyn'. Brestskaya Voblasts', SW Belarus 52°23′N 26°45′E
Basuo see Dongfang
Basutoland see Lesotho
119 O15 Basya ≈ E Belarus
Bas-Zaïre see Bas-Congo
79 D17 Bata NW Equatorial Guinea 01°51′N 09°48′E
79 D17 Bata ✈ S Equatorial Guinea 01°55′N 09°48′E
Batae Coritanorum see Leicester
123 Q8 Batagay Respublika Sakha (Yakutiya), NE Russian Federation 67°36′N 134°44′E
123 P8 Batagay-Alyta Respublika Sakha (Yakutiya), NE Russian Federation 67°48′N 130°15′E
112 L10 Batajnica Vojvodina, N Serbia 44°55′N 20°17′E
136 H15 Batakık Gölü ◎ S Turkey
114 H11 Batak, Yazovir ◎ SW Bulgaria
152 H7 Batāla Punjab, N India 31°48′N 75°12′E
104 F9 Batalha Leiria, C Portugal 39°40′N 08°50′W
79 N17 Batama Orientale, NE Dem. Rep. Congo 0°54′N 26°25′E
123 Q10 Batamay Respublika Sakha (Yakutiya), NE Russian Federation 63°28′N 129°33′E
160 F9 Batang var. Bazhong. Sichuan, C China 30°04′N 99°10′E
79 I14 Batangafo Ouham, NW Central African Republic 07°19′N 18°22′E
171 P8 Batangas off. Batangas City. Luzon, N Philippines 13°47′N 121°03′E
Batangas City see Batangas
Bătania see Battonya
171 Q10 Batan Islands island group N Philippines
60 L8 Batatais São Paulo, S Brazil 20°54′S 47°37′W
18 E10 Batavia New York, NE USA 43°00′N 78°11′W
Batavia see Jakarta
173 T9 Batavia Seamount undersea feature E Indian Ocean 27°42′S 100°36′E
126 L12 Bataysk Rostovskaya Oblast', SW Russian Federation 47°10′N 39°46′E
14 B9 Batchawana ≈ Ontario, S Canada
14 B9 Batchawana Bay Ontario, S Canada 46°55′N 84°36′W
167 Q12 Bătdâmbâng prev. Battambang. Bătdâmbâng, NW Cambodia 13°06′N 103°13′E
79 G20 Batéké, Plateaux plateau S Congo
183 S11 Batemans Bay New South Wales, SE Australia 35°45′S 150°09′E
21 Q13 Batesburg South Carolina, SE USA 33°54′N 81°33′W
28 K12 Batesland South Dakota, N USA 43°05′N 102°07′W
27 V10 Batesville Arkansas, C USA 35°45′N 91°39′W
31 Q14 Batesville Indiana, N USA 39°18′N 85°13′W
22 L2 Batesville Mississippi, S USA 34°18′N 89°56′W
25 Q13 Batesville Texas, SW USA 28°56′N 99°38′W
44 L13 Bath E Jamaica 17°57′N 76°22′W
97 L22 Bath hist. Akermanceaster; anc. Aquae Calidae, Aquae Solis. SW England, United Kingdom 51°23′N 02°22′W
19 Q8 Bath Maine, NE USA 43°54′N 69°49′W
18 F11 Bath New York, NE USA 42°20′N 77°16′W
Bath see Berkeley Springs
78 I10 Batha off. Région du Batha. ◈ region C Chad
78 I10 Batha seasonal river C Chad
Batha, Région du see Batha
141 Y8 Baṭḥā', Wādī al dry watercourse NE Oman
152 H9 Bathinda Punjab, NW India 30°14′N 74°54′E
98 M11 Bathmen Overijssel, E Netherlands 52°15′N 06°16′E
45 Z14 Bathsheba E Barbados 13°13′N 59°31′W
183 R8 Bathurst New South Wales, SE Australia 33°32′S 149°35′E
13 O13 Bathurst New Brunswick, SE Canada 47°37′N 65°40′W
Bathurst see Banjul
8 H6 Bathurst, Cape headland Northwest Territories, NW Canada 70°33′N 128°00′W
196 L8 Bathurst Inlet Nunavut, N Canada 66°23′N 107°00′W
196 L8 Bathurst Inlet inlet Nunavut, N Canada
181 N1 Bathurst Island island Northern Territory, N Australia
197 O9 Bathurst Island island Parry Islands, Nunavut, N Canada
77 O14 Batié SW Burkina Faso 09°53′N 02°53′W
Batinah see Al Bāṭinah
141 Y9 Bāṭin, Wādī al dry watercourse SW Asia
15 P9 Batiscan ≈ Québec, SE Canada
136 F16 Batı Toroslar ▲▲ SW Turkey
Batjan see Bacan, Pulau
147 R11 Batken Batkenskaya Oblast', SW Kyrgyzstan 40°03′N 70°50′E
Batken Oblasty see Batkenskaya Oblast'
147 Q11 Batkenskaya Oblast' Kir. Batken Oblasty. ◈ province SW Kyrgyzstan
Batlle y Ordóñez see José Batlle y Ordóñez
183 Q10 Batlow New South Wales, SE Australia 35°32′S 148°09′E
137 Q15 Batman var. Iluh. Batman, SE Turkey 37°52′N 41°06′E
137 Q15 Batman ◈ province SE Turkey
74 L6 Batna NE Algeria 35°34′N 06°10′E
163 O7 Batnorov var. Dundbürd. Hentiy, E Mongolia 47°55′N 111°37′E
Batoe see Batu, Kepulauan
162 K7 Bat-Öldziy var. Övt. Övörhangay, C Mongolia 46°50′N 102°15′E
Bat-Öldziyt see Dzaamar
22 J8 Baton Rouge state capital Louisiana, S USA 30°28′N 91°09′W
79 G15 Batouri Est, E Cameroon 04°26′N 14°27′E
138 G14 Baṭrā', Jibāl al ▲ S Jordan
138 G6 Batroûn var. Al Batrūn. N Lebanon 34°15′N 35°42′E
Batsch see Bač
Batsevichi see Batsevichy
119 M17 Batsevichy Rus. Batsevichi. Mahilyowskaya Voblasts', E Belarus 53°24′N 29°14′E
92 M7 Båtsfjord Finnmark, N Norway 70°37′N 29°42′E
Batshireet see Hentiy
162 L7 Batsümber var. Mandal. Töv, C Mongolia 48°24′N 106°47′E
Battambang see Bătdâmbâng
195 X3 Batterbee, Cape headland Antarctica
155 L24 Batticaloa Eastern Province, E Sri Lanka 07°44′N 81°43′E
99 L19 Battice Liège, E Belgium 50°39′N 05°50′E
107 L18 Battipaglia Campania, S Italy 40°36′N 14°59′E
11 R15 Battle ≈ Alberta/Saskatchewan, SW Canada
Battle Born State see Nevada
31 Q10 Battle Creek Michigan, N USA 42°20′N 85°10′W
27 T7 Battlefield Missouri, C USA 37°07′N 93°22′W
11 S15 Battleford Saskatchewan, S Canada 52°45′N 108°20′W
29 S6 Battle Lake Minnesota, N USA 46°16′N 95°42′W
35 U3 Battle Mountain Nevada, W USA 40°37′N 116°55′W
111 M25 Battonya Rom. Bătania. Békés, SE Hungary 46°16′N 21°00′E
162 J7 Battsengel var. Jargalant. Arhangay, C Mongolia 47°46′N 101°56′E
168 D11 Batu, Kepulauan prev. Batoe. island group W Indonesia
137 Q10 Batumi W Georgia 41°39′N 41°38′E
168 K10 Batu Pahat prev. Bandar Penggaram. Johor, Peninsular Malaysia 01°51′N 102°56′E
171 O12 Baturebe Sulawesi, N Indonesia 01°43′S 121°43′E
122 J12 Baturino Tomskaya Oblast', C Russian Federation 57°46′N 85°08′E
117 R3 Baturyn Chernihivs'ka Oblast', N Ukraine 51°20′N 32°54′E
138 F10 Bat Yam Tel Aviv, C Israel 32°01′N 34°45′E
127 Q4 Batyrevo Chuvashskaya Respublika, W Russian Federation 55°04′N 47°34′E
Batys Qazaqstan Oblysy see Zapadnyy Kazakhstan
102 F5 Batz, Île de island NW France
169 Q10 Bau Sarawak, East Malaysia 01°25′N 110°10′E
171 N2 Bauang Luzon, N Philippines 16°33′N 120°19′E
171 P14 Baubau var. Baoebaoe. Pulau Buton, C Indonesia 05°30′S 122°37′E
77 W14 Bauchi Bauchi, NE Nigeria 10°18′N 09°46′E
77 W14 Bauchi ◈ state C Nigeria
102 H7 Baud Morbihan, NW France 47°52′N 02°59′W
29 T2 Baudette Minnesota, N USA 48°42′N 94°36′W
193 S9 Bauer Basin undersea feature E Pacific Ocean 10°00′S 101°45′W
187 R14 Bauer Field var. Port Vila. ✈ (Port-Vila) Éfaté, C Vanuatu 17°42′S 168°21′E
13 T9 Bauld, Cape headland Newfoundland and Labrador, E Canada 51°35′N 55°22′W
103 T8 Baume-les-Dames Doubs, E France 47°22′N 06°20′E
101 I15 Baunatal Hessen, C Germany 51°15′N 09°25′E
107 D18 Baunei Sardegna, Italy, C Mediterranean Sea 40°04′N 09°36′E
57 M15 Baures, Río ≈ N Bolivia
60 K9 Bauru São Paulo, S Brazil 22°19′S 49°07′W
Baushar see Bawshar
118 G10 Bauska Ger. Bauske. S Latvia 56°25′N 24°11′E
Bauske see Bauska
101 Q15 Bautzen Lus. Budyšin. Sachsen, E Germany 51°11′N 14°29′E
145 Q16 Bauyrzhan Momyshuly Kaz. Baüyrzhan Momyshuly; prev. Burnoye. Zhambyl, S Kazakhstan 42°36′N 70°46′E
Bauzanum see Bolzano
Bavaria see Bayern
109 N7 Bavarian Alps Ger. Bayrische Alpen. ▲▲ Austria/Germany
Bavière see Bayern
40 H4 Bavispe, Río ≈ NW Mexico
127 T5 Bavly Respublika Tatarstan, W Russian Federation 54°20′N 53°21′E
169 P13 Bawal, Pulau island N Indonesia
169 T12 Bawan Borneo, C Indonesia 01°36′S 113°55′E
183 O12 Baw Baw, Mount ▲ Victoria, SE Australia 37°49′S 146°16′E
169 S15 Bawean, Pulau island S Indonesia
75 V9 Bawītī var. Bawīṭī. N Egypt 28°19′N 28°53′E
Bawīṭī see Bawītī
77 Q13 Bawku N Ghana 11°00′N 00°12′W
Bawlake see Bawlakhe
167 N7 Bawlakhe var. Bawlake. Kayah State, C Myanmar (Burma) 19°10′N 97°19′E
169 H11 Bawo Ofuloa Pulau Tanahmasa, W Indonesia 0°10′S 98°24′E
141 Y8 Bawshar var. Baushar. NE Oman 23°32′N 58°24′E
Baxian see Bazhou
Ba Xian see Bazhou
158 M8 Baxkorgan Xinjiang Uygur Zizhiqu, W China 39°05′N 90°00′E
23 V6 Baxley Georgia, SE USA 31°46′N 82°21′W
159 R15 Baxoi var. Baima. Xizang Zizhiqu, W China 30°01′N 96°53′E
29 W14 Baxter Iowa, C USA 41°49′N 93°09′W
29 U6 Baxter Minnesota, N USA 46°21′N 94°18′W
27 R8 Baxter Springs Kansas, C USA 37°01′N 94°45′W
81 M17 Bay off. Gobolka Bay. ◈ region SW Somalia
Bay see Baicheng
44 H7 Bayamo Granma, E Cuba 20°21′N 76°38′W
45 U5 Bayamón E Puerto Rico 18°24′N 66°09′W
163 W8 Bayan Heilongjiang, NE China 46°05′N 127°24′E
170 L16 Bayan prev. Bajan. Pulau Lombok, C Indonesia 08°16′S 116°28′E
162 M8 Bayan var. Maanīt. Töv, C Mongolia 47°14′N 107°34′E
Bayan see Hölönbuyr, Dornod, Mongolia
Bayan see Ihhet, Dornogovĭ, Mongolia
Bayan see Bayan-Uul, Govĭ-Altay, Mongolia
Bayan see Bayanhutag, Hentiy, Mongolia
Bayan see Bürentogtoh, Hövsgol, Mongolia
152 I12 Bayāna Rājasthān, N India 26°55′N 77°18′E
149 N5 Bāyān, Band-e ▲▲ C Afghanistan
162 H8 Bayanbulag Bayanhongor, C Mongolia 46°46′N 98°07′E
Bayanbulag see Ömnödelger
158 J5 Bayanbulak Xinjiang Uygur Zizhiqu, W China 43°05′N 84°05′E
162 L7 Bayanchandmanī var. Ihsüüj. Töv, C Mongolia 48°12′N 106°23′E
162 J11 Bayandalay var. Dalay. Ömnögovĭ, S Mongolia 43°27′N 103°30′E
163 O9 Bayandelger var. Shireet. Sühbaatar, SE Mongolia 45°33′N 112°19′E
162 I5 Bayandzürh var. Altraga. Hövsgöl, N Mongolia 50°08′N 98°54′E
Bayan Gol see Dengkou, China
Bayangol see Bugat, Mongolia
162 I9 Bayangovĭ var. Örgön. Bayanhongor, C Mongolia 44°43′N 100°23′E
159 R12 Bayan Har Shan var. Bayan Khar. ▲▲ C China
162 G6 Bayanhayrhan var. Altanbulag. Dzavhan, N Mongolia 49°16′N 96°22′E
162 I8 Bayanhongor Bayanhongor, C Mongolia 46°08′N 100°42′E
162 H9 Bayanhongor ◈ province C Mongolia
162 K14 Bayan Hot var. Alxa Zuoqi. Nei Mongol Zizhiqu, N China 38°49′N 105°40′E
163 O8 Bayanhutag var. Bayan. Hentiy, C Mongolia 47°13′N 110°57′E
163 T9 Bayan Huxu var. Horqin Zuoyi Zhongqi. Nei Mongol Zizhiqu, N China 45°02′N 121°28′E
Bayan Khar see Bayan Har Shan
163 N12 Bayan Kuang prev. Bayan Obo. Nei Mongol Zizhiqu, N China 41°45′N 109°58′E
168 J7 Bayan Lepas ✈ (George Town) Pinang, Peninsular Malaysia 05°18′N 100°15′E
162 I10 Bayanlig var. Hatansuudal. Bayanhongor, C Mongolia 44°34′N 100°41′E
162 K13 Bayan Mod Nei Mongol Zizhiqu, N China 40°45′N 104°29′E
163 N8 Bayanmönh var. Ulaan-Ereg. Hentiy, E Mongolia 46°50′N 109°39′E
162 L12 Bayannur var. Linhe. Nei Mongol Zizhiqu, N China 40°46′N 107°27′E
Bayan Nuru see Xar Burd
162 E5 Bayannuur var. Tsul-Ulaan. Bayan-Ölgiy, W Mongolia 48°51′N 91°13′E
Bayan Obo see Bayan Kuang
43 V15 Bayano, Lago ◎ E Panama
162 C5 Bayan-Ölgiy ◈ province NW Mongolia
162 H9 Bayan-Öndör var. Bulgan. Bayanhongor, C Mongolia 44°48′N 98°39′E
162 K8 Bayan-Öndör var. Bumbat. Övörhangay, C Mongolia 46°30′N 104°08′E
162 L8 Bayan-Önjüül var. Ihhayrhan. Töv, C Mongolia 46°57′N 105°51′E
163 O7 Bayan-Ovoo var. Javhlant. Hentiy, E Mongolia 47°46′N 112°06′E
162 L11 Bayan-Ovoo var. Erdenetsogt. Ömnögovĭ, S Mongolia 42°54′N 106°16′E
Bayansayr see Baatsagaan
159 Q9 Bayan Shan ▲ C China 37°36′N 96°23′E
162 J9 Bayanteeg Övörhangay, C Mongolia 45°39′N 101°30′E
162 G5 Bayantes var. Altay. Dzavhan, N Mongolia 49°40′N 96°21′E
Bayantöhöm see Büren
162 M8 Bayantsagaan var. Dzogsool. Töv, C Mongolia 46°46′N 107°18′E
163 P7 Bayantümen var. Tsagaanders. Dornod, NE Mongolia 48°03′N 114°16′E
Bayan-Uhaa see Ih-Uul
163 R10 Bayan Ul var. Xi Ujimqin Qi. Nei Mongol Zizhiqu, N China 44°31′N 117°36′E
Bayan-Ulaan see Dzüünbayan-Ulaan
163 O7 Bayan-Uul var. Javarthushuu. Dornod, NE Mongolia 49°05′N 112°40′E
162 F7 Bayan-Uul var. Bayan. Govĭ-Altay, W Mongolia 47°05′N 95°13′E
162 M8 Bayanuur var. Tsul-Ulaan. Töv, C Mongolia 47°44′N 108°22′E
28 J14 Bayard Nebraska, C USA 41°45′N 103°19′W
37 P15 Bayard New Mexico, SW USA 32°45′N 108°07′W
103 T13 Bayard, Col pass SE France
Bayasgalant see Mönhhaan
136 J12 Bayat Çorum, N Turkey 40°34′N 34°07′E
171 P6 Bayawan Negros, C Philippines 09°22′N 122°50′E
143 R10 Bayāz Kermān, C Iran 30°41′N 55°29′E
171 Q6 Baybay Leyte, C Philippines 10°41′N 124°49′E
21 X10 Bayboro North Carolina, SE USA 35°08′N 76°49′W
137 P12 Bayburt Bayburt, NE Turkey 40°15′N 40°16′E
137 P12 Bayburt ◈ province NE Turkey
31 R8 Bay City Michigan, N USA 43°35′N 83°52′W
25 V12 Bay City Texas, SW USA 28°59′N 96°00′W
Baydarata Bay see Baydaratskaya Guba
122 J7 Baydaratskaya Guba var. Baydarata Bay. bay N Russian Federation
81 M16 Baydhabo var. Baydhowa, Isha Baydhabo, It. Baidoa. Bay, SW Somalia 03°08′N 43°39′E
Baydhowa see Baydhabo
101 N21 Bayerischer Wald ▲▲ SE Germany
101 K21 Bayern Eng. Bavaria, Fr. Bavière. ◈ state SE Germany
147 V9 Bayetovo Narynskaya Oblast', C Kyrgyzstan 41°14′N 74°55′E
102 K4 Bayeux anc. Augustodurum. Calvados, N France 49°16′N 00°42′W
14 E15 Bayfield ≈ Ontario, S Canada
145 O15 Baygekum Kaz. Bäygequm. Kzylorda, S Kazakhstan 44°15′N 66°34′E
Bäygequm see Baygekum
136 C14 Bayındır Izmir, SW Turkey 38°12′N 27°40′E
138 H12 Bāyir var. Bā'ir. Ma'ān, S Jordan 30°46′N 36°40′E
Bay Islands see Bahía, Islas de la
Bayizhen see Nyingchi
139 R5 Bayjī var. Baiji. Ṣalāḥ ad Dīn, N Iraq 34°56′N 43°29′E
Baykadam see Saudakent
123 N13 Baykal, Ozero Eng. Lake Baikal. ◎ S Russian Federation
123 M14 Baykal'sk Irkutskaya Oblast', S Russian Federation 51°30′N 104°03′E
137 R15 Baykan Siirt, SE Turkey 38°08′N 41°43′E
139 S2 Baykhmah var. Bēkma. Arbīl, E Iraq 36°40′N 44°15′E
123 L11 Baykit Krasnoyarskiy Kray, C Russian Federation 61°37′N 96°23′E
145 N12 Baykonur var. Baykonyr. Karaganda, C Kazakhstan 47°50′N 75°33′E
Baykonur see Baykonyr
144 M14 Baykonyr var. Baykonur, Kaz. Bayqongyr; prev. Leninsk. Kzylorda, S Kazakhstan 45°38′N 63°20′E
Baykonyr see Baykonur
158 E7 Baykurt Xinjiang Uygur Zizhiqu, W China 39°56′N 75°33′E
14 I9 Bay, Lac ◎ Québec, SE Canada
127 W6 Baymak Respublika Bashkortostan, W Russian Federation 52°34′N 58°20′E
23 O8 Bay Minette Alabama, S USA 30°52′N 87°46′W
143 O17 Baynūnah desert W United Arab Emirates
184 O8 Bay of Plenty off. Bay of Plenty Region. ◈ region North Island, New Zealand
Bay of Plenty Region see Bay of Plenty
191 Z3 Bay of Wrecks bay Kiritimati, E Kiribati
Bayonnaise Rocks see Beyonēsu-retsugan
102 I15 Bayonne anc. Lapurdum. Pyrénées-Atlantiques, SW France 43°30′N 01°28′W
22 H4 Bayou D'Arbonne Lake ▣ Louisiana, S USA
23 N9 Bayou La Batre Alabama, S USA 30°24′N 88°15′W
Bayou State see Mississippi
Bayqadam see Saudakent
Bayqongyr see Baykonyr
Bayram-Ali see Bayramaly
146 J14 Bayramaly var. Bayramaly; prev. Bayram-Ali. Mary Welaýaty, S Turkmenistan 37°33′N 62°08′E
101 L19 Bayreuth var. Baireuth. Bayern, SE Germany 49°57′N 11°34′E
Bayrische Alpen see Bavarian Alps
Bayrūt see Beyrouth
22 L9 Bay Saint Louis Mississippi, S USA 30°18′N 89°19′W
Baysān see Beit She'an
Bayshint see Öndörshireet
14 H13 Bays, Lake of ◎ Ontario, S Canada
22 M6 Bay Springs Mississippi, S USA 31°58′N 89°17′W
Bay State see Massachusetts
Baysun see Boysun
14 H13 Baysville Ontario, S Canada 45°10′N 79°03′W
141 N15 Bayt al Faqīh W Yemen 14°30′N 43°20′E
158 M4 Baytik Shan ▲▲ China/Mongolia
Bayt Laḥm see Bethlehem
25 W11 Baytown Texas, SW USA 29°43′N 94°59′W
169 V11 Bayur, Tanjung headland Borneo, N Indonesia 0°43′S 117°32′E
121 N14 Bayy al Kabīr, Wādī dry watercourse NW Libya
145 P17 Bayyrkum Kaz. Bayyrqum; prev. Bairkum. Yuzhnyy Kazakhstan, S Kazakhstan 41°57′N 68°05′E
Bayyrqum see Bayyrkum
105 P14 Baza Andalucía, S Spain 37°30′N 02°45′W
137 X10 Bazardüzü Dağı Rus. Gora Bazardyuzyu. ▲ N Azerbaijan 41°13′N 47°50′E
Bazardyuzyu, Gora see Bazardüzü Dağı
Bazargic see Dobrich
83 N18 Bazaruto, Ilha do island SE Mozambique
102 K14 Bazas Gironde, SW France 44°27′N 00°11′W
105 O14 Baza, Sierra de ▲▲ S Spain
160 J8 Bazhong var. Bazhou. Sichuan, C China 31°55′N 106°44′E
Bazhong see Batang
161 P3 Bazhou prev. Baxian, Ba Xian. Hebei, E China 39°05′N 116°24′E
Bazhou see Bazhong
14 M9 Bazin ≈ Québec, SE Canada
Bazin see Pezinok
139 Q7 Bāzīyah Al Anbār, C Iraq 33°50′N 42°41′E
138 H6 Bcharré var. Bcharreh, Bsharrī, Bsherri. NE Lebanon 34°16′N 36°01′E
Bcharreh see Bcharré
28 J5 Beach North Dakota, N USA 46°55′N 104°00′W
182 K12 Beachport South Australia 37°29′S 140°03′E
97 O23 Beachy Head headland SE England, United Kingdom 50°44′N 00°16′E
18 K13 Beacon New York, NE USA 41°30′N 73°54′W
63 J25 Beagle Channel channel Argentina/Chile
181 O1 Beagle Gulf gulf Northern Territory, N Australia
Bealach an Doirín see Ballaghaderreen
Bealach Cláir see Ballyclare
Bealach Féich see Ballybofey
172 J3 Bealanana Mahajanga, NE Madagascar 14°33′S 48°44′E
Béal an Átha see Ballina
Béal an Átha Móir see Ballinamore
Béal an Mhuirhead see Belmullet
Béal Átha Beithe see Ballybay
Béal Átha Conaill see Ballyconnell
Béal Átha hAmhnais see Ballyhaunis
Béal Átha na Sluaighe see Ballinasloe
Béal Átha Seanaidh see Ballyshannon
Bealdovuopmi see Peltovuoma
Béal Feirste see Belfast
Béal Tairbirt see Belturbet
Beanna Boirche see Mourne Mountains
Beannchar see Banagher, Ireland
Beannchar see Bangor, Northern Ireland, UK
Beanntraí see Bantry
Bearalváhki see Berlevåg
23 N2 Bear Creek ≈ Alabama/Mississippi, S USA
30 J13 Bear Creek ≈ Illinois, N USA
27 U13 Bearden Arkansas, C USA 33°43′N 92°37′W
195 Q10 Beardmore Glacier glacier Antarctica
30 K13 Beardstown Illinois, N USA 40°01′N 90°25′W
28 L14 Bear Hill ▲ Nebraska, C USA 41°24′N 101°49′W
Bear Island see Bjørnøya
14 H12 Bear Lake Ontario, S Canada 45°28′N 79°31′W
36 M1 Bear Lake ◎ Idaho/Utah, NW USA
39 U11 Bear, Mount ▲ Alaska, USA 61°16′N 141°09′W
102 J16 Béarn cultural region SW France
194 J11 Bear Peninsula peninsula Antarctica
152 I7 Beās ≈ India/Pakistan
105 P3 Beasain País Vasco, N Spain 43°03′N 02°11′W
105 O12 Beas de Segura Andalucía, S Spain 38°16′N 02°54′W
45 N10 Beata, Cabo headland SW Dominican Republic 17°34′N 71°25′W
45 N10 Beata, Isla island SW Dominican Republic
64 F11 Beata Ridge undersea feature N Caribbean Sea 16°00′N 72°30′W
29 R17 Beatrice Nebraska, C USA 40°14′N 96°43′W
83 L16 Beatrice Mashonaland East, NE Zimbabwe 18°15′S 30°55′E
11 N11 Beatton ≈ British Columbia, W Canada
11 N11 Beatton River British Columbia, W Canada 57°35′N 121°45′W
35 V10 Beatty Nevada, W USA 36°53′N 116°44′W
21 N6 Beattyville Kentucky, S USA 37°33′N 83°44′W
173 X16 Beau Bassin W Mauritius 20°13′S 57°27′E
103 R15 Beaucaire Gard, S France 43°49′N 04°37′E
14 I8 Beauchastel, Lac ◎ Québec, SE Canada
14 I10 Beauchêne, Lac ◎ Québec, SE Canada
183 V3 Beaudesert Queensland, E Australia 28°00′S 152°27′E
182 M12 Beaufort Victoria, SE Australia 37°27′S 143°24′E
21 X11 Beaufort North Carolina, SE USA 34°44′N 76°41′W
21 R15 Beaufort South Carolina, SE USA 32°23′N 80°40′W
38 M11 Beaufort Sea sea Arctic Ocean
Beaufort-Wes see Beaufort West
83 G25 Beaufort West Afr. Beaufort-Wes. Western Cape, SW South Africa 32°21′S 22°35′E
103 N7 Beaugency Loiret, C France 47°46′N 01°38′E
19 R1 Beau Lake ▣ Maine, NE USA
96 I8 Beauly N Scotland, United Kingdom 57°29′N 04°29′W
99 G21 Beaumont Hainaut, S Belgium 50°12′N 04°13′E
185 E23 Beaumont Otago, South Island, New Zealand 45°48′S 169°32′E
22 M7 Beaumont Mississippi, S USA 31°10′N 88°55′W
25 X10 Beaumont Texas, SW USA 30°05′N 94°06′W
102 M15 Beaumont-de-Lomagne Tarn-et-Garonne, S France 43°54′N 01°00′E
102 L6 Beaumont-sur-Sarthe Sarthe, NW France 48°15′N 00°07′E
103 R8 Beaune Côte d'Or, C France 47°02′N 04°50′E
15 R9 Beaupré Québec, SE Canada 47°03′N 70°52′W
102 J8 Beaupréau Maine-et-Loire, NW France 47°13′N 00°57′W
99 I22 Beauraing Namur, SE Belgium 50°07′N 04°57′E
103 R12 Beaurepaire Isère, E France 45°20′N 05°03′E
11 Y16 Beausejour Manitoba, S Canada 50°04′N 96°30′W
103 N4 Beauvais anc. Bellovacum, Caesaromagus. Oise, N France 49°27′N 02°04′E
11 S13 Beauval Saskatchewan, C Canada 55°10′N 107°37′W
102 I9 Beauvoir-sur-Mer Vendée, NW France 46°54′N 02°03′W
39 R8 Beaver Alaska, USA 66°22′N 147°31′W
26 J8 Beaver Oklahoma, C USA 36°48′N 100°32′W
18 B14 Beaver Pennsylvania, NE USA 40°39′N 80°19′W
36 K6 Beaver Utah, W USA 38°16′N 112°38′W
10 L9 Beaver ≈ British Columbia/Yukon, W Canada
11 S13 Beaver ≈ Saskatchewan, C Canada
29 N17 Beaver City Nebraska, C USA 40°08′N 99°49′W
10 G6 Beaver Creek Yukon, W Canada 62°20′N 140°45′W
31 R14 Beavercreek Ohio, N USA 39°42′N 83°58′W
39 S8 Beaver Creek ≈ Alaska, USA
26 H3 Beaver Creek ≈ Kansas/Nebraska, C USA
28 J5 Beaver Creek ≈ Montana/North Dakota, N USA
29 Q14 Beaver Creek ≈ Nebraska, C USA
25 Q4 Beaver Creek ≈ Texas, SW USA
30 M8 Beaver Dam Wisconsin, N USA 43°28′N 88°49′W
30 M8 Beaver Dam Lake ◎ Wisconsin, N USA
18 B14 Beaver Falls Pennsylvania, NE USA 40°45′N 80°20′W
33 P12 Beaverhead Mountains ▲▲ Idaho/Montana, NW USA
33 Q12 Beaverhead River ≈ Montana, NW USA
65 A25 Beaver Island island W Falkland Islands
31 P5 Beaver Island island Michigan, N USA
27 S9 Beaver Lake ▣ Arkansas, C USA
11 N13 Beaverlodge Alberta, W Canada 55°11′N 119°29′W
18 I8 Beaver River ≈ New York, NE USA
26 J8 Beaver River ≈ Oklahoma, C USA
18 B13 Beaver River ≈ Pennsylvania, NE USA
65 A25 Beaver Settlement Beaver Island, W Falkland Islands 51°30′S 61°15′W
Beaver State see Oregon
14 H14 Beaverton Ontario, S Canada 44°24′N 79°07′W
32 G11 Beaverton Oregon, NW USA 45°29′N 122°49′W
152 G12 Beāwar Rājasthān, N India 26°08′N 74°22′E
Bebas, Dasht-i see Bābūs, Dasht-e
60 L8 Bebedouro São Paulo, S Brazil 20°58′S 48°28′W
101 I16 Bebra Hessen, C Germany 50°59′N 09°46′E
41 W12 Becal Campeche, SE Mexico 19°49′N 90°28′W
15 Q11 Bécancour ≈ Québec, SE Canada
97 Q19 Beccles E England, United Kingdom 52°27′N 01°32′E
112 L9 Bečej Ger. Altbetsche, Hung. Óbecse, Rácz-Becse; prev. Magyar-Becse, Stari Bečej. Vojvodina, N Serbia 45°36′N 20°02′E
104 I3 Becerreá Galicia, NW Spain 42°51′N 07°10′W
74 H7 Béchar prev. Colomb-Béchar. W Algeria 31°38′N 02°11′W
39 O14 Becharof Lake ◎ Alaska, USA
116 H15 Bechet var. Bechetu. Dolj, SW Romania 43°45′N 23°57′E
Bechetu see Bechet
21 R6 Beckley West Virginia, NE USA 37°46′N 81°12′W
101 G14 Beckum Nordrhein-Westfalen, W Germany 51°45′N 08°03′E
25 X7 Beckville Texas, SW USA 32°14′N 94°27′W
35 X4 Becky Peak ▲ Nevada, W USA 39°59′N 114°33′W
116 I9 Beclean Hung. Bethlen; prev. Betlen. Bistriţa-Năsăud, N Romania 47°10′N 24°11′E
Bécs see Wien
111 H18 Bečva Ger. Betschau, Pol. Beczwa. ≈ E Czech Republic
Beczwa see Bečva
103 P15 Bédarieux Hérault, S France 43°37′N 03°10′E
120 B10 Beddouza, Cap headland W Morocco 32°35′N 09°16′W
80 I13 Bedelē Oromīya, C Ethiopia 08°25′N 36°21′E
147 Y8 Bedel Pass Rus. Pereval Bedel. pass China/Kyrgyzstan
Bedel, Pereval see Bedel Pass
95 H22 Beder Midtjylland, C Denmark 56°03′N 10°13′E
97 N20 Bedford E England, United Kingdom 52°08′N 00°29′W
31 O15 Bedford Indiana, N USA 38°51′N 86°29′W
29 U16 Bedford Iowa, C USA 40°40′N 94°43′W
20 L4 Bedford Kentucky, S USA 38°36′N 85°18′W
18 D15 Bedford Pennsylvania, NE USA 40°00′N 78°29′W
21 T6 Bedford Virginia, NE USA 37°20′N 79°31′W
97 N20 Bedfordshire cultural region E England, United Kingdom
127 N5 Bednodem'yanovsk Penzenskaya Oblast', W Russian Federation 53°55′N 43°14′E
98 N5 Bedum Groningen, NE Netherlands 53°18′N 06°36′E
27 V11 Beebe Arkansas, C USA 35°04′N 91°52′W
Beechy Group see Chichijima-rettō
45 T9 Beef Island ✈ (Road Town) Tortola, E British Virgin Islands 18°25′N 64°31′W
Beehive State see Utah
99 L18 Beek Limburg, SE Netherlands 50°56′N 05°47′E
99 L18 Beek ✈ (Maastricht) Limburg, SE Netherlands 50°55′N 05°47′E
99 K14 Beek-en-Donk Noord-Brabant, S Netherlands 51°31′N 05°37′E
138 F13 Be'er Menuha prev. Be'ér Menuḥa. Southern, S Israel 30°22′N 35°09′E
Be'ér Menuḥa see Be'er Menuha
99 D16 Beernem West-Vlaanderen, NW Belgium 51°09′N 03°18′E
99 I16 Beerse Antwerpen, N Belgium 51°20′N 04°52′E
Beersheba see Be'er Sheva
138 E11 Be'er Sheva var. Beersheba, Ar. Bir es Saba; prev. Be'ér Sheva'. Southern, S Israel 31°15′N 34°47′E
Be'ér Sheva' see Be'er Sheva
98 J13 Beesd Gelderland, C Netherlands 51°52′N 05°12′E
99 M16 Beesel Limburg, SE Netherlands 51°16′N 06°02′E
83 J21 Beestekraal North-West, N South Africa 25°21′S 27°40′E
194 J7 Beethoven Peninsula peninsula Alexander Island, Antarctica
Beetstersweach see Beetsterzwaag
98 M6 Beetsterzwaag Fris. Beetstersweach. Fryslân, N Netherlands 53°03′N 06°04′E
25 S13 Beeville Texas, SW USA 28°25′N 97°47′W
79 J18 Befale Equateur, NW Dem. Rep. Congo 0°25′N 20°48′E
Befandriana see Befandriana Avaratra
172 J3 Befandriana Avaratra var. Befandriana, Befandriana Nord. Mahajanga, NW Madagascar 15°14′S 48°33′E
Befandriana Nord see Befandriana Avaratra
79 K18 Befori Equateur, N Dem. Rep. Congo 0°09′N 22°18′E
172 I7 Befotaka Fianarantsoa, S Madagascar 23°49′S 47°00′E
183 R11 Bega New South Wales, SE Australia 36°43′S 149°50′E
102 G5 Bégard Côtes d'Armor, NW France 48°37′N 03°18′W
112 M9 Begejski Kanal canal Vojvodina, NE Serbia
94 G13 Begna ≈ S Norway
Begoml' see Byahoml'
Begovat see Bekobod
153 Q13 Begusarai Bihār, NE India 25°25′N 86°08′E
143 R9 Behābād Yazd, C Iran 32°23′N 59°50′E
Behagle see Laï
55 Z10 Béhague, Pointe headland E French Guiana 04°38′N 51°52′W
Behar see Bihār
142 M10 Behbahān var. Behbehān. Khūzestān, SW Iran 30°38′N 50°07′E
Behbehān see Behbahān
44 G3 Behring Point Andros Island, W The Bahamas 24°28′N 77°44′W
143 P4 Behshahr prev. Ashraf. Māzandarān, N Iran 36°42′N 53°36′E
163 V6 Bei'an Heilongjiang, NE China 48°16′N 126°29′E
Beibunar see Sredishte
Beibu Wan see Tonkin, Gulf of
Beida see Al Bayḍā'
80 H13 Beigi Oromīya, C Ethiopia 09°13′N 34°48′E
160 L16 Beihai Guangxi Zhuangzu Zizhiqu, S China 21°29′N 109°10′E
159 Q10 Bei Hulsan Hu ◎ C China
161 N13 Bei Jiang ≈ S China
161 O2 Beijing var. Pei-ching, Eng. Peking; prev. Pei-p'ing. ● (China) Beijing Shi, E China 39°58′N 116°23′E
161 P2 Beijing ✈ Beijing Shi, N China 39°54′N 116°22′E
Beijing see Beijing Shi, China

◆ Country
● Country Capital
◇ Dependent Territory
○ Dependent Territory Capital
◈ Administrative Regions
✈ International Airport
▲ Mountain
▲▲ Mountain Range
▲ Volcano
≈ River
◎ Lake
▣ Reservoir

161 O2 **Beijing Shi** var. Beijing, Jing, Pei-ching, Eng. Peking; prev. Pei-p'ing. ⬖ municipality E China
76 G8 **Beïla** Trarza, W Mauritania 18°07′N 15°56′W
98 N7 **Beilen** Drenthe, NE Netherlands 52°52′N 06°27′E
160 L15 **Beiliu** var. Lingcheng. Guangxi Zhuangzu Zizhiqu, S China 22°50′N 110°22′E
159 O12 **Beilu He** ≈ W China
Beilul see Beylul
163 U12 **Beining** prev. Beizhen. Liaoning, NE China 41°34′N 121°51′E
96 H8 **Beinn Dearg** ▲ N Scotland, United Kingdom 57°47′N 04°52′W
Beinn MacDuibh see Ben Macdui
160 I12 **Beipan Jiang** ≈ S China
163 T12 **Beipiao** Liaoning, NE China 41°49′N 120°45′E
83 N17 **Beira** Sofala, C Mozambique 19°45′S 34°56′E
83 N17 **Beira** ✈ Sofala, C Mozambique 19°39′S 35°05′E
104 I7 **Beira Alta** former province N Portugal
104 H9 **Beira Baixa** former province C Portugal
104 G8 **Beira Litoral** former province N Portugal
Beirut see Beyrouth
Beisān see Beit She'an
11 Q16 **Beiseker** Alberta, SW Canada 51°20′N 113°34′W
Beitai Ding see Wutai Shan
83 K19 **Beitbridge** Matabeleland South, S Zimbabwe 22°10′S 30°02′E
Beit Lekhem see Bethlehem
138 G9 **Beit She'an** Ar. Baysān, Beisān; anc. Scythopolis, prev. Bet She'an. Northern, N Israel 32°30′N 35°30′E
116 G10 **Beiuş** Hung. Belényes. Bihor, NW Romania 46°40′N 22°21′E
Beizhen see Beining
104 H12 **Beja** anc. Pax Julia. Beja, SE Portugal 38°01′N 07°52′W
74 M5 **Béja** var. Bājah. N Tunisia 36°45′N 09°04′E
104 G13 **Beja** ⬖ district S Portugal
120 I9 **Béjaïa** var. Bejaïa, Fr. Bougie; anc. Saldae. NE Algeria 36°49′N 05°03′E
Bejaïa see Béjaïa
104 K8 **Béjar** Castilla y León, N Spain 40°24′N 05°45′W
Bejraburi see Phetchaburi
Bekaa Valley see El Beqaa
Bekabad see Bekobod
Békás see Bicaz
169 O15 **Bekasi** Jawa, C Indonesia 06°14′S 106°595′E
Bek-Budi see Qarshi
Bekdaş/Bekdash see Garabogaz
147 T10 **Bek-Dzhar** Oshskaya Oblast', SW Kyrgyzstan 40°22′N 73°08′E
111 N24 **Békés** Rom. Bichiş. Békés, SE Hungary 46°45′N 21°09′E
111 M24 **Békés** off. Békés Megye. ⬖ county SE Hungary
111 M24 **Békéscsaba** Rom. Bichiş-Ciaba. Békés, SE Hungary 46°40′N 21°05′E
Békés Megye see Békés
172 H7 **Bekily** Toliara, S Madagascar 24°12′S 45°20′E
165 W4 **Bekkai** var. Betsukai. Hokkaidō, NE Japan 43°23′N 145°07′E
Bēkma see Baykhmah
147 Q11 **Bekobod** Rus. Bekabad; prev. Begovat. Toshkent Viloyati, E Uzbekistan 40°17′N 69°11′E
127 O7 **Bekovo** Penzenskaya Oblast', W Russian Federation 52°27′N 43°41′E
Bel see Beliu
152 M13 **Bela** Uttar Pradesh, N India 25°55′N 82°00′E
149 N15 **Bela** Baluchistān, SW Pakistan 26°12′N 66°20′E
79 F15 **Bélabo** Est, C Cameroon 04°54′N 13°10′E
112 N10 **Bela Crkva** Ger. Weisskirchen, Hung. Fehértemplom. Vojvodina, W Serbia 44°55′N 21°28′E
173 Y16 **Bel Air** var. Rivière Sèche. E Mauritius
104 L12 **Belalcázar** Andalucía, S Spain 38°33′N 05°07′W
113 P15 **Bela Palanka** Serbia, SE Serbia 43°13′N 22°19′E
119 H16 **Belarus** off. Republic of Belarus, var. Belorussia, Latv. Baltkrievija; prev. Belorussian SSR, Rus. Belorusskaya SSR. ◆ republic E Europe
Belarus, Republic of see Belarus
Belau see Palau
59 H21 **Bela Vista** Mato Grosso do Sul, SW Brazil 22°04′S 56°25′W
83 L21 **Bela Vista** Maputo, S Mozambique 26°20′S 32°40′E
168 I8 **Belawan** Sumatera, W Indonesia 03°46′N 98°44′E
Běla Woda see Weisswasser
127 U4 **Belaya** ≈ W Russian Federation
123 R7 **Belaya Gora** Respublika Sakha (Yakutiya), NE Russian Federation 68°25′N 146°12′E
126 M11 **Belaya Kalitiva** Rostovskaya Oblast', SW Russian Federation 48°09′N 40°43′E
125 R14 **Belaya Kholunitsa** Kirovskaya Oblast', NW Russian Federation 58°54′N 50°52′E
Belaya Tserkov' see Bila Tserkva
77 V11 **Belbédji** Zinder, S Niger 14°35′N 08°00′E
111 K14 **Bełchatów** var. Belchatow. Łódzski, C Poland 51°23′N 19°20′E
Belchatow see Bełchatów
Belcher, Îles see Belcher Islands
12 H7 **Belcher Islands** Fr. Îles Belcher. island group Nunavut, SE Canada
105 S6 **Belchite** Aragón, NE Spain 41°18′N 00°45′W
29 O2 **Belcourt** North Dakota, N USA 48°50′N 99°44′W
31 P9 **Belding** Michigan, N USA 43°06′N 85°13′W
127 U5 **Belebey** Respublika Bashkortostan, W Russian Federation 54°04′N 54°13′E
81 N16 **Beledweyne** var. Belet Huen, It. Belet Uen. Hiiraan, C Somalia 04°39′N 45°12′E
146 B10 **Belek** Balkan Welaýaty, W Turkmenistan 39°57′N 53°51′E
58 L12 **Belém** var. Pará. state capital Pará, N Brazil 01°27′S 48°29′W
65 I14 **Belém Ridge** undersea feature C Atlantic Ocean
62 I7 **Belén** Catamarca, NW Argentina 27°36′N 67°00′W
54 G9 **Belén** Boyacá, C Colombia 06°01′N 72°55′W
42 J11 **Belén** Rivas, SW Nicaragua 11°30′N 85°55′W
62 O5 **Belén** Concepción, C Paraguay 23°25′S 57°14′W
61 D16 **Belén** Salto, N Uruguay 30°47′S 57°47′W
37 R12 **Belen** New Mexico, SW USA 34°37′N 106°46′W
61 D20 **Belén de Escobar** Buenos Aires, E Argentina 34°21′S 58°47′W
114 J7 **Belene** Pleven, N Bulgaria 43°39′N 25°09′E
114 J7 **Belene, Ostrov** island N Bulgaria
43 R15 **Belén, Río** ≈ C Panama
Belényes see Beiuş
Embalse de Belesar see Belesar, Encoro de
104 H3 **Belesar, Encoro de** Sp. Embalse de Belesar. ▣ NW Spain
Belet Huen/Belet Uen see Beledweyne
126 J5 **Belëv** Tul'skaya Oblast', W Russian Federation 53°48′N 36°07′E
19 R7 **Belfast** Maine, NE USA 44°25′N 69°02′W
97 G15 **Belfast** Ir. Béal Feirste. ◐ E Northern Ireland, United Kingdom 54°35′N 05°55′W
97 G15 **Belfast Aldergrove** ✈ E Northern Ireland, United Kingdom 54°37′N 06°11′W
97 G15 **Belfast Lough** Ir. Loch Lao. inlet E Northern Ireland, United Kingdom
28 K5 **Belfield** North Dakota, N USA 46°53′N 103°12′W
103 U7 **Belfort** Territoire-de-Belfort, E France 47°38′N 06°52′E
Belgard see Białogard
155 E17 **Belgaum** Karnātaka, W India 15°52′N 74°30′E
Belgian Congo see Congo (Democratic Republic of)
België/Belgique see Belgium
99 F20 **Belgium** off. Kingdom of Belgium, Dut. België, Fr. Belgique. ◆ monarchy NW Europe
Belgium, Kingdom of see Belgium
126 J8 **Belgorod** Belgorodskaya Oblast', W Russian Federation 50°38′N 36°37′E
Belgorod-Dnestrovskiy see Bilhorod-Dnistrovs'kyy
126 J8 **Belgorodskaya Oblast'** ⬖ province W Russian Federation
Belgrad see Beograd
29 T8 **Belgrade** Minnesota, N USA 45°27′N 94°59′W
33 S11 **Belgrade** Montana, NW USA 45°46′N 111°10′W
Belgrade see Beograd
Belgrano, Cabo see Meredith, Cape
195 N5 **Belgrano II** Argentinian research station Antarctica 77°50′S 35°25′W
21 X9 **Belhaven** North Carolina, SE USA 35°36′N 76°50′W
107 I23 **Belice** anc. Hypsas. ≈ Sicilia, Italy, C Mediterranean Sea
Belice see Belize/Belize City
Beli Drim see Drini i Bardhë
Beligrad see Berat
188 C8 **Beliliou** prev. Peleliu. island S Palau
114 L8 **Beli Lom, Yazovir** ▣ NE Bulgaria
112 I8 **Beli Manastir** Hung. Pélmonostor; prev. Monostor. Osijek-Baranja, NE Croatia 45°46′N 16°38′E
102 J13 **Bélin-Béliet** Gironde, SW France 44°30′N 00°48′W
79 F17 **Bélinga** Ogooué-Ivindo, NE Gabon 01°05′N 13°12′E
21 S4 **Belington** West Virginia, NE USA 39°01′N 79°57′W
127 O6 **Belinskiy** Penzenskaya Oblast', W Russian Federation 52°58′N 43°25′E
169 N12 **Belinyu** Pulau Bangka, W Indonesia 01°37′S 105°45′E
169 O13 **Belitung, Pulau** island W Indonesia
116 F10 **Beliu** Hung. Bel. Arad, W Romania 46°31′N 21°57′E
114 I9 **Beli Vit** ≈ NW Bulgaria
42 G2 **Belize** Sp. Belice; prev. British Honduras, Colony of Belize. ◆ commonwealth republic Central America
42 F2 **Belize** Sp. Belice. ⬖ district NE Belize
42 G2 **Belize** ≈ Belize/Guatemala
Belize see Belize City
42 G2 **Belize City** var. Belize, Sp. Belice. Belize, NE Belize 17°29′N 88°10′W
42 G2 **Belize City** ✈ Belize, NE Belize 17°31′N 88°15′W
Belize, Colony of see Belize
Beljak see Villach
39 N16 **Belkofski** Alaska, USA 55°07′N 162°04′W
123 O6 **Bel'kovskiy, Ostrov** island Novosibirskiye Ostrova, NE Russian Federation
14 J8 **Bell** ≈ Québec, SE Canada
10 J15 **Bella Bella** British Columbia, SW Canada 52°04′N 128°07′W
102 M10 **Bellac** Haute-Vienne, C France 46°07′N 01°04′E
10 K15 **Bella Coola** British Columbia, SW Canada 52°23′N 126°46′W
106 D6 **Bellagio** Lombardia, N Italy 45°58′N 09°15′E
31 P6 **Bellaire** Michigan, N USA 44°59′N 85°12′W
106 D6 **Bellano** Lombardia, N Italy 46°06′N 09°21′E
155 G17 **Bellary** var. Ballari. Karnātaka, S India 15°11′N 76°54′E
183 S5 **Bellata** New South Wales, SE Australia 29°58′S 149°49′E
61 D16 **Bella Unión** Artigas, N Uruguay 30°18′S 57°35′W
61 C14 **Bella Vista** Corrientes, NE Argentina 28°30′S 59°03′W
62 J7 **Bella Vista** Tucumán, N Argentina 27°05′S 65°19′W
62 P4 **Bella Vista** Amambay, C Paraguay 22°08′S 56°20′W
56 B10 **Bellavista** Cajamarca, N Peru 05°43′S 78°48′W
56 D11 **Bellavista** San Martín, N Peru 07°04′S 76°35′W
183 U6 **Bellbrook** New South Wales, SE Australia 30°48′S 152°32′E
27 V5 **Belle** Missouri, C USA 38°17′N 91°43′W
21 Q5 **Belle** West Virginia, NE USA 38°13′N 81°32′W
31 R13 **Bellefontaine** Ohio, N USA 40°22′N 83°45′W
18 F14 **Bellefonte** Pennsylvania, NE USA 40°54′N 77°43′W
28 J9 **Belle Fourche** South Dakota, N USA 44°40′N 103°50′W
28 J9 **Belle Fourche Reservoir** ▣ South Dakota, N USA
28 K9 **Belle Fourche River** ≈ South Dakota/Wyoming, N USA
103 S10 **Bellegarde-sur-Valserine** Ain, E France 46°06′N 05°49′E
23 Y14 **Belle Glade** Florida, SE USA 26°40′N 80°40′W
102 G8 **Belle Île** island NW France
13 T9 **Belle Isle** island Belle Isle, Newfoundland and Labrador, E Canada
13 S10 **Belle Isle, Strait of** strait Newfoundland and Labrador, E Canada
Bellenz see Bellinzona
29 W14 **Belle Plaine** Iowa, C USA 41°54′N 92°16′W
29 V9 **Belle Plaine** Minnesota, N USA 44°39′N 93°47′W
14 I9 **Belleterre** Québec, SE Canada 47°24′N 78°40′W
14 J15 **Belleville** Ontario, SE Canada 44°10′N 77°22′W
103 R10 **Belleville** Rhône, E France 46°09′N 04°42′E
30 K15 **Belleville** Illinois, N USA 38°31′N 89°58′W
27 N3 **Belleville** Kansas, C USA 39°51′N 97°38′W
29 Z13 **Bellevue** Iowa, C USA 42°15′N 90°25′W
29 S15 **Bellevue** Nebraska, C USA 41°08′N 95°53′W
31 S11 **Bellevue** Ohio, N USA 41°16′N 82°50′W
25 S5 **Bellevue** Texas, SW USA 33°38′N 98°00′W
32 H8 **Bellevue** Washington, NW USA 47°36′N 122°12′W
55 Y11 **Bellevue de l'Inini, Montagnes** ▲▲ S French Guiana
103 S11 **Belley** Ain, E France 45°46′N 05°41′E
Bellin see Kangirsuk
183 V6 **Bellingen** New South Wales, SE Australia 30°27′S 152°53′E
97 L14 **Bellingham** N England, United Kingdom 55°09′N 02°16′W
32 H6 **Bellingham** Washington, NW USA 48°46′N 122°29′W
Belling Hausen Mulde see Southeast Pacific Basin
194 H2 **Bellingshausen** Russian research station South Shetland Islands, Antarctica 61°57′S 58°23′W
Bellingshausen see Motu One
Bellingshausen Abyssal Plain see Bellingshausen Plain
196 R14 **Bellingshausen Plain** var. Bellingshausen Abyssal Plain. undersea feature SE Pacific Ocean 64°00′S 90°00′W
194 I8 **Bellingshausen Sea** sea Antarctica
98 P6 **Bellingwolde** Groningen, NE Netherlands 53°07′N 07°10′E
108 H11 **Bellinzona** Ger. Bellenz. Ticino, S Switzerland 46°12′N 09°02′E
25 T8 **Bellmead** Texas, SW USA 31°36′N 97°02′W
54 E8 **Bello** Antioquia, W Colombia 06°19′N 75°34′W
61 B21 **Bellocq** Buenos Aires, E Argentina 35°55′S 61°32′W
Bello Horizonte see Belo Horizonte
186 L10 **Bellona** var. Mungiki. island S Solomon Islands
Bellovacum see Beauvais
182 D7 **Bell, Point** headland South Australia 32°13′S 133°08′E
20 F9 **Bells** Tennessee, S USA 35°42′N 89°05′W
25 U5 **Bells** Texas, SW USA 33°36′N 96°24′W
92 N3 **Bellsund** inlet SW Svalbard
106 H6 **Belluno** Veneto, NE Italy 46°08′N 12°13′E
62 L11 **Bell Ville** Córdoba, C Argentina 32°35′S 62°41′W
83 E26 **Bellville** Western Cape, SW South Africa 33°50′S 18°43′E
25 U11 **Bellville** Texas, SW USA 29°57′N 96°15′W
104 L12 **Belméz** Andalucía, S Spain 38°16′N 05°12′W
29 V12 **Belmond** Iowa, C USA 42°51′N 93°36′W
18 E11 **Belmont** New York, NE USA 42°14′N 78°02′W
21 R10 **Belmont** North Carolina, SE USA 35°13′N 81°01′W
59 O18 **Belmonte** Bahia, E Brazil 15°53′S 38°54′W
104 I8 **Belmonte** Castelo Branco, C Portugal 40°21′N 07°20′W
105 P10 **Belmonte** Castilla-La Mancha, C Spain 39°34′N 02°43′W
42 G2 **Belmopan** ● (Belize) Cayo, C Belize 17°13′N 88°48′W
97 B16 **Belmullet** Ir. Béal an Mhuirhead. Mayo, W Ireland 54°14′N 09°59′W
99 E20 **Belœil** Hainaut, SW Belgium 50°33′N 03°45′E
123 R13 **Belogorsk** Amurskaya Oblast', SE Russian Federation 50°53′N 128°24′E
Belogorsk see Bilohirs'k
114 F7 **Belogradchik** Vidin, NW Bulgaria 43°37′N 22°42′E
172 H8 **Beloha** Toliara, S Madagascar 25°09′S 45°04′E
59 M20 **Belo Horizonte** prev. Bello Horizonte. state capital Minas Gerais, SE Brazil 19°54′S 43°54′W
26 M3 **Beloit** Kansas, C USA 39°27′N 98°06′W
30 L9 **Beloit** Wisconsin, N USA 42°31′N 89°01′W
Belokorovichi see Novi Bilokorovychi
124 J8 **Belomorsk** Respublika Kareliya, NW Russian Federation 64°30′N 34°43′E
124 J8 **Belomorsko-Baltiyskiy Kanal** Eng. White Sea-Baltic Canal, White Sea Canal. canal NW Russian Federation
153 V15 **Belonia** Tripura, NE India 23°15′N 91°25′E
Beloozersk see Byelaazyorsk
Belopol'ye see Bilopillya
105 O4 **Belorado** Castilla y León, N Spain 42°25′N 03°11′W
126 L14 **Belorechensk** Krasnodarskiy Kray, SW Russian Federation 44°46′N 39°53′E
127 W5 **Beloretsk** Respublika Bashkortostan, W Russian Federation 53°56′N 58°26′E
Belorussia/Belorussian SSR see Belarus
Belorusskaya Gryada see Byelaruskaya Hrada
Belorusskaya SSR see Belarus
Beloshchel'ye see Nar'yan-Mar
114 N8 **Beloslav** Varna, E Bulgaria 43°13′N 27°42′E
Belostok see Białystok
Belo-sur-Tsiribihina see Belo Tsiribihina
172 H5 **Belo Tsiribihina** var. Belo-sur-Tsiribihina. Toliara, W Madagascar 19°40′S 44°30′E
Belovár see Bjelovar
Belovezhskaya, Pushcha see Białowieska, Puszcza/Byelavyezhskaya, Pushcha
114 H10 **Belovo** Pazardzhik, C Bulgaria 42°10′N 24°01′E
Belovodsk see Bilovods'k
122 H9 **Beloyarskiy** Khanty-Mansiyskiy Avtonomnyy Okrug-Yugra, N Russian Federation 63°40′N 66°31′E
124 K7 **Beloye More** Eng. White Sea. sea NW Russian Federation
124 K13 **Beloye, Ozero** ◎ NW Russian Federation
114 J10 **Belozem** Plovdiv, C Bulgaria 42°11′N 25°00′E
124 K13 **Belozërsk** Vologodskaya Oblast', NW Russian Federation 59°59′N 37°49′E
108 D8 **Belp** Bern, W Switzerland 46°54′N 07°31′E
108 D8 **Belp** ✈ (Bern) Bern, C Switzerland 46°55′N 07°29′E
107 L24 **Belpasso** Sicilia, Italy, C Mediterranean Sea 37°35′N 14°59′E
31 U14 **Belpre** Ohio, N USA 39°14′N 81°34′W
98 M8 **Belterwijde** ◎ N Netherlands
27 R4 **Belton** Missouri, C USA 38°48′N 94°31′W
21 P11 **Belton** South Carolina, SE USA 34°31′N 82°29′W
25 T9 **Belton** Texas, SW USA 31°04′N 97°30′W
25 S9 **Belton Lake** ▣ Texas, SW USA
Bel'tsy see Bălţi
97 E16 **Belturbet** Ir. Béal Tairbirt. Cavan, N Ireland 54°06′N 07°26′W
Beluchistan see Balochistan
145 Z9 **Belukha, Gora** ▲ Kazakhstan/Russian Federation 49°50′N 86°44′E
107 M20 **Belvedere Marittimo** Calabria, SW Italy 39°37′N 15°52′E
30 L10 **Belvidere** Illinois, N USA 42°15′N 88°50′W
18 J14 **Belvidere** New Jersey, NE USA 40°50′N 75°05′W
Bely see Belyy
127 V8 **Belyayevka** Orenburgskaya Oblast', W Russian Federation 51°25′N 56°26′E
Belynichi see Byalynichy
124 H17 **Belyy** var. Bely, Beyj. Tverskaya Oblast', W Russian Federation 55°51′N 32°57′E
126 I6 **Belyye Berega** Bryanskaya Oblast', W Russian Federation 53°11′N 34°42′E
122 J6 **Belyy, Ostrov** island N Russian Federation
122 J11 **Belyy Yar** Tomskaya Oblast', C Russian Federation 58°26′N 84°57′E
100 N13 **Belzig** Brandenburg, NE Germany 52°09′N 12°37′E
22 K4 **Belzoni** Mississippi, S USA 33°10′N 90°29′W
172 H4 **Bemaraha** var. Plateau du Bemaraha. ▲▲ W Madagascar
Bemaraha, Plateau du see Bemaraha
82 B10 **Bembe** Uíge, NW Angola 07°03′S 14°25′E
77 S14 **Bembèrèkè** var. Bimbéréké. N Benin 10°10′N 02°41′E
104 K12 **Bembézar** ≈ SW Spain
104 J3 **Bembibre** Castilla y León, N Spain 42°37′N 06°25′W
29 T4 **Bemidji** Minnesota, N USA 47°27′N 94°53′W
98 L12 **Bemmel** Gelderland, SE Netherlands 51°53′N 05°54′E
171 T13 **Bemu** Pulau Seram, E Indonesia 03°21′S 129°58′E
Benāb see Bonāb
105 T5 **Benabarre** var. Benavarn. Aragón, NE Spain 42°06′N 00°28′E
Benaco see Garda, Lago di
79 L20 **Bena-Dibele** Kasai-Oriental, C Dem. Rep. Congo 04°01′S 22°50′E
105 R9 **Benagéber, Embalse de** E Spain
183 O11 **Benalla** Victoria, SE Australia 36°33′S 146°00′E
104 M14 **Benamejí** Andalucía, S Spain 37°16′N 04°33′W
Benares see Vārānasi
Benavarn see Benabarre
104 F10 **Benavente** Santarém, C Portugal 38°59′N 08°49′W
104 K5 **Benavente** Castilla y León, N Spain 42°N 05°40′W
25 S15 **Benavides** Texas, SW USA 27°36′N 98°24′W
96 F8 **Benbecula** island NW Scotland, United Kingdom
Bencovazzo see Benkovac
32 H13 **Bend** Oregon, NW USA 44°04′N 121°19′W
182 K7 **Benda Range** ▲▲ South Australia
183 T6 **Bendemeer** New South Wales, SE Australia 30°54′S 151°12′E
Bender see Tighina
Bender Beila/Bender Beyla see Bandarbeyla
Bender Cassim/Bender Qaasim see Boosaaso
Bendery see Tighina
183 N11 **Bendigo** Victoria, SE Australia 36°46′S 144°19′E
118 E10 **Bēne** SW Latvia 56°30′N 23°04′E
98 K13 **Beneden-Leeuwen** Gelderland, C Netherlands 51°52′N 05°32′E
101 L24 **Benediktenwand** ▲ S Germany 47°39′N 11°28′E
Benemérita de San Cristóbal see San Cristóbal
77 N12 **Bénéna** Ségou, S Mali 13°04′N 04°20′W
172 I7 **Benenitra** Toliara, S Madagascar 23°25′S 45°06′E
Beneschau see Benešov
Beneški Zaliv see Venice, Gulf of
111 D17 **Benešov** Ger. Beneschau. Středočeský Kraj, W Czech Republic 49°48′N 14°41′E
107 L17 **Benevento** anc. Beneventum, Malventum. Campania, S Italy 41°07′N 14°45′E
Beneventum see Benevento
173 S3 **Bengal, Bay of** bay N Indian Ocean
Bengalooru see Bangalore
Bengaluru see Bangalore
79 M17 **Bengamisa** Orientale, N Dem. Rep. Congo 0°58′N 25°11′E
Bengasi see Banghāzī
Bengazi see Banghāzī
161 P7 **Bengbu** var. Peng-pu. Anhui, E China 32°57′N 117°17′E
32 L9 **Benge** Washington, NW USA 46°55′N 118°01′W
Benghazi see Banghāzī
168 K10 **Bengkalis** Pulau Bengkalis, W Indonesia 01°27′N 102°10′E
168 K10 **Bengkalis, Pulau** island W Indonesia
169 Q10 **Bengkayang** Borneo, C Indonesia 0°45′N 109°28′E
168 K14 **Bengkulu** prev. Bengkoeloe, Benkoelen, Benkulen. Sumatera, W Indonesia 03°46′S 102°16′E
168 K13 **Bengkulu** off. Propinsi Bengkulu; prev. Bengkoeloe, Benkoelen, Benkulen. ⬖ province W Indonesia
Bengkulu, Propinsi see Bengkulu
82 A11 **Bengo** ⬖ province W Angola
95 J16 **Bengtsfors** Västra Götaland, S Sweden 59°03′N 12°14′E
82 B13 **Benguela** var. Benguella. Benguela, W Angola 12°35′S 13°30′E
83 A14 **Benguela** ⬖ province W Angola
Benguella see Benguela
138 F10 **Ben Gurion** ✈ Tel Aviv, C Israel 32°04′N 34°45′E
Bengweulu, Lake see Bangweulu, Lake
Benha see Banhā
192 F6 **Benham Seamount** undersea feature W Philippine Sea 15°48′N 124°15′E
96 H6 **Ben Hope** ▲ N Scotland, United Kingdom 58°25′N 04°36′W
79 P18 **Beni** Nord-Kivu, NE Dem. Rep. Congo 0°31′N 29°30′E
74 H8 **Beni Abbès** W Algeria 30°07′N 02°09′W
105 T8 **Benicarló** Valenciana, E Spain 40°25′N 00°25′E
105 T9 **Benicasim** Cat. Benicàssim. Valenciana, E Spain 40°03′N 00°03′E
Benicàssim see Benicasim
105 T12 **Benidorm** Valenciana, SE Spain 38°33′N 00°09′W
Beni Mazâr see Banī Mazār
121 C11 **Beni-Mellal** C Morocco 32°20′N 06°21′W
77 R14 **Benin** off. Republic of Benin; prev. Dahomey. ◆ republic W Africa
77 S17 **Benin, Bight of** gulf W Africa
77 U16 **Benin City** Edo, SW Nigeria 06°23′N 05°40′E
Benin, Republic of see Benin
57 K16 **Beni, Río** ≈ N Bolivia
120 F10 **Beni Saf** var. Beni-Saf. NW Algeria 35°19′N 01°23′W
Beni-Saf see Beni Saf
Benishangul see Bīnshangul Gumuz
105 T11 **Benissa** Valenciana, E Spain 38°43′N 00°03′E
Beni Suef see Banī Suwayf
11 V15 **Benito** Manitoba, S Canada 51°57′N 101°24′W
Benito see Uolo, Río
61 C23 **Benito Juárez** Buenos Aires, E Argentina 37°43′S 59°48′W
41 P14 **Benito Juárez Internacional** ✈ (México) México, S Mexico 19°24′N 99°02′W
25 P5 **Benjamin** Texas, SW USA 33°35′N 99°49′W
58 B13 **Benjamin Constant** Amazonas, N Brazil 04°22′S 70°02′W
40 F4 **Benjamín Hill** Sonora, NW Mexico 30°13′N 111°08′W
63 F19 **Benjamín, Isla** island Archipiélago de los Chonos, S Chile
164 Q4 **Benkei-misaki** headland Hokkaidō, NE Japan 42°49′N 140°10′E
28 L17 **Benkelman** Nebraska, C USA 40°04′N 101°30′W
96 I7 **Ben Klibreck** ▲ N Scotland, United Kingdom 58°15′N 04°23′W
Benkoelen/Bengkoeloe see Bengkulu
112 D13 **Benkovac** It. Bencovazzo. Zadar, SW Croatia 44°02′N 15°36′E
Benkulen see Bengkulu
96 I11 **Ben Lawers** ▲ C Scotland, United Kingdom 56°33′N 04°13′W
96 J9 **Ben Macdui** var. Beinn MacDuibh. ▲ C Scotland, United Kingdom 57°02′N 03°42′W
96 G11 **Ben More** ▲ W Scotland, United Kingdom 56°26′N 06°00′W
96 I11 **Ben More** ▲ C Scotland, United Kingdom 56°22′N 04°31′W
96 H7 **Ben More Assynt** ▲ N Scotland, United Kingdom 58°09′N 04°51′W
185 E20 **Benmore, Lake** ◎ South Island, New Zealand
98 L12 **Bennekom** Gelderland, SE Netherlands 52°00′N 05°40′E
123 Q5 **Bennetta, Ostrov** island Novosibirskiye Ostrova, NE Russian Federation
21 T11 **Bennettsville** South Carolina, SE USA 34°36′N 79°40′W
96 H10 **Ben Nevis** ▲ N Scotland, United Kingdom 56°80′N 05°00′W
184 M9 **Benneydale** Waikato, North Island, New Zealand 38°31′S 175°22′E
Bennichab see Bennichchâb
76 H8 **Bennichchâb** var. Bennichab. Inchiri, W Mauritania 19°26′N 15°21′W
18 L10 **Bennington** Vermont, NE USA 42°51′N 73°09′W
185 E20 **Ben Ohau Range** ▲▲ South Island, New Zealand
83 J21 **Benoni** Gauteng, NE South Africa 26°04′S 28°18′E
172 J2 **Be, Nosy** var. Nossi-Bé. island NW Madagascar
Bénoué see Benue
42 F2 **Benque Viejo del Carmen** Cayo, W Belize 17°04′N 89°08′W
101 G19 **Bensheim** Hessen, W Germany 49°41′N 08°38′E
37 N16 **Benson** Arizona, SW USA 31°55′N 110°16′W
29 S8 **Benson** Minnesota, N USA 45°19′N 95°36′W
21 U10 **Benson** North Carolina, SE USA 35°22′N 78°33′W
171 N15 **Benteng** Pulau Selayar, C Indonesia 06°07′S 120°28′E
83 A14 **Bentiaba** Namibe, SW Angola 14°18′S 12°27′E
181 T4 **Bentinck Island** island Wellesley Islands, Queensland, N Australia
80 E13 **Bentiu** Unity, N South Sudan 09°14′N 29°49′E
138 G8 **Bent Jbaïl** var. Bint Jubayl. S Lebanon 33°07′N 35°26′E
11 Q15 **Bentley** Alberta, SW Canada 52°27′N 114°02′W
61 I15 **Bento Gonçalves** Rio Grande do Sul, S Brazil 29°12′S 51°34′W
27 U12 **Benton** Arkansas, C USA 34°34′N 92°35′W
30 L16 **Benton** Illinois, N USA 38°00′N 88°55′W
20 H7 **Benton** Kentucky, S USA 36°51′N 88°21′W
22 G5 **Benton** Louisiana, S USA 32°41′N 93°44′W
27 Y7 **Benton** Missouri, C USA 37°05′N 89°34′W
20 M10 **Benton** Tennessee, S USA 35°10′N 84°39′W
31 O10 **Benton Harbor** Michigan, N USA 42°07′N 86°27′W
27 S9 **Bentonville** Arkansas, C USA 36°23′N 94°13′W
77 V16 **Benue** ⬖ state SE Nigeria
78 F13 **Benue** Fr. Bénoué. ≈ Cameroon/Nigeria
163 V12 **Benxi** prev. Pen-ch'i, Penhsihu, Penki. Liaoning, NE China 41°20′N 123°45′E
Benyakoni see Byenyakoni
112 K10 **Beočin** Vojvodina, N Serbia 45°13′N 19°43′E
Beodericsworth see Bury St Edmunds
112 M11 **Beograd** Eng. Belgrade, Ger. Belgrad; anc. Singidunum. ● (Serbia) Serbia, N Serbia 44°48′N 20°27′E
112 L11 **Beograd** Eng. Belgrade. ✈ Serbia, N Serbia 44°45′N 20°21′E
76 M16 **Béoumi** C Ivory Coast 07°40′N 05°34′W
35 V3 **Beowawe** Nevada, W USA 40°33′N 116°31′W
164 E14 **Beppu** Ōita, Kyūshū, SW Japan 33°18′N 131°30′E
187 X15 **Beqa** prev. Mbengga. island W Fiji
45 Y14 **Bequia** island C Saint Vincent and the Grenadines
139 U4 **Beranan, Shax-î** var. Shākh-i Barānān. ▲▲ E Iraq
113 L16 **Berane** prev. Ivangrad. E Montenegro 42°51′N 19°51′E
113 L21 **Berat** var. Berati, SCr. Beligrad. Berat, C Albania 40°43′N 19°58′E
113 L21 **Berat** ⬖ district C Albania
Berătău see Berettyó
Berati see Berat
Beraun see Berounka, Czech Republic
Beraun see Beroun, Czech Republic
171 U13 **Berau, Teluk** var. MacCluer Gulf. bay Papua, E Indonesia
80 G8 **Berber** River Nile, NE Sudan 18°01′N 34°00′E
80 N12 **Berbera** Woqooyi Galbeed, NW Somalia 10°24′N 45°02′E
79 H16 **Berbérati** Mambéré-Kadéï, SW Central African Republic 04°14′N 15°50′E
Berberia, Cabo de see Barbaria, Cap de
55 T9 **Berbice River** ≈ NE Guyana
Berchid see Berrechid
103 N2 **Berck-Plage** Pas-de-Calais, N France 50°24′N 01°35′E
25 T13 **Berclair** Texas, SW USA 28°33′N 97°32′W
117 W10 **Berda** ≈ SE Ukraine
Berdichev see Berdychiv
123 P10 **Berdigestyakh** Respublika Sakha (Yakutiya), NE Russian Federation 62°02′N 127°03′E
122 J12 **Berdsk** Novosibirskaya Oblast', C Russian Federation 54°42′N 82°56′E
117 W10 **Berdyans'k** Rus. Berdyansk; prev. Osipenko. Zaporiz'ka Oblast', SE Ukraine 46°46′N 36°49′E
117 W10 **Berdyans'ka Kosa** spit SE Ukraine
117 V10 **Berdyans'ka Zatoka** gulf S Ukraine
117 N5 **Berdychiv** Rus. Berdichev. Zhytomyrs'ka Oblast', N Ukraine 49°54′N 28°39′E
20 M6 **Berea** Kentucky, S USA 37°34′N 84°18′W
Beregovo/Beregszász see Berehove
116 G8 **Berehove** Cz. Berehovo, Hung. Beregszász, Rus. Beregovo. Zakarpats'ka Oblast', W Ukraine 48°13′N 22°39′E
Berehovo see Berehove
186 D9 **Bereina** Central, S Papua New Guinea 08°29′S 146°30′E
146 C11 **Bereket** prev. Rus. Gazandzhyk, Kazandzhik, Turkm. Gazanjyk. Balkan Welaýaty, W Turkmenistan 39°17′N 55°27′E
45 O12 **Berekua** S Dominica 15°14′N 61°19′W
77 O16 **Berekum** W Ghana 07°27′N 02°35′W
Berenice see Baranīs
11 O14 **Berens** ≈ Manitoba/Ontario, C Canada
11 X14 **Berens River** Manitoba, C Canada 52°22′N 97°00′W
29 R12 **Beresford** South Dakota, N USA 43°02′N 96°45′W
116 J4 **Berestechko** Volyns'ka Oblast', NW Ukraine 50°21′N 25°06′E
116 M11 **Bereşti** Galaţi, E Romania 46°04′N 27°54′E
117 U6 **Berestova** ≈ E Ukraine
Beretău see Berettyó
111 N23 **Berettyó** Rom. Barcău; prev. Berătău, Beretău. ≈ Hungary/Romania
111 N23 **Berettyóújfalu** Hajdú-Bihar, E Hungary 47°15′N 21°33′E
Berëza/Bereza Kartuska see Byaroza
117 Q4 **Berezan'** Kyyivs'ka Oblast', N Ukraine 50°18′N 31°30′E
117 Q10 **Berezanka** Mykolayivs'ka Oblast', S Ukraine 46°51′N 31°24′E
116 J6 **Berezhany** Pol. Brzeżany. Ternopil's'ka Oblast', W Ukraine 49°29′N 25°00′E
Berezina see Byarezina
Berezino see Byerazino
117 P10 **Berezivka** Rus. Berezovka. Odes'ka Oblast', SW Ukraine 47°12′N 30°56′E
117 Q2 **Berezna** Chernihivs'ka Oblast', NE Ukraine 51°35′N 31°50′E
116 L3 **Berezne** Rivnens'ka Oblast', NW Ukraine 51°00′N 26°46′E
117 R9 **Bereznehuvate** Mykolayivs'ka Oblast', S Ukraine 47°18′N 32°52′E
125 N10 **Bereznik** Arkhangel'skaya Oblast', NW Russian Federation 62°50′N 42°40′E
125 U13 **Berezniki** Permskiy Kray, NW Russian Federation 59°26′N 56°49′E
Berëzovka see Byarozawka, Belarus
Berezovka see Berezivka, Ukraine
122 H9 **Berezovo** Khanty-Mansiyskiy Avtonomnyy Okrug-Yugra, N Russian Federation 63°48′N 64°38′E
127 O9 **Berezovskaya** Volgogradskaya Oblast', SW Russian Federation 50°17′N 43°58′E
123 S13 **Berezovyy** Khabarovskiy Kray, E Russian Federation 51°42′N 135°39′E
83 E25 **Berg** ≈ W South Africa
Berg see Berg bei Rohrbach
105 V4 **Berga** Cataluña, NE Spain 42°06′N 01°41′E
95 N20 **Berga** Kalmar, S Sweden 57°13′N 16°03′E
136 B13 **Bergama** İzmir, W Turkey 39°08′N 27°10′E
106 E7 **Bergamo** anc. Bergomum. Lombardia, N Italy 45°42′N 09°40′E
105 P3 **Bergara** País Vasco, N Spain 43°05′N 02°25′W
109 S3 **Berg bei Rohrbach** var. Berg. Oberösterreich, N Austria 48°34′N 14°02′E
100 O6 **Bergen** Mecklenburg-Vorpommern, NE Germany 54°25′N 13°25′E
101 I11 **Bergen** Niedersachsen, NW Germany 52°49′N 09°57′E
98 H8 **Bergen** Noord-Holland, NW Netherlands 52°40′N 04°42′E
94 C13 **Bergen** Hordaland, S Norway 60°24′N 05°19′E
Bergen see Mons
55 W9 **Berg en Dal** Brokopondo, C Suriname 05°15′N 55°20′W
99 G15 **Bergen op Zoom** Noord-Brabant, S Netherlands 51°30′N 04°17′E
102 L13 **Bergerac** Dordogne, SW France 44°51′N 00°30′E
99 J16 **Bergeyk** Noord-Brabant, S Netherlands 51°19′N 05°21′E
101 D16 **Bergheim** Nordrhein-Westfalen, W Germany 50°57′N 06°39′E
55 X10 **Bergi** Sipaliwini, E Suriname 04°36′N 54°24′W
101 E16 **Bergisch Gladbach** Nordrhein-Westfalen, W Germany 50°59′N 07°09′E
101 F14 **Bergkamen** Nordrhein-Westfalen, W Germany 51°32′N 07°41′E
95 N21 **Bergkvara** Kalmar, S Sweden 56°22′N 16°04′E
Bergomum see Bergamo
98 K13 **Bergse Maas** ≈ S Netherlands
95 P15 **Bergshamra** Stockholm, C Sweden 59°37′N 18°40′E
94 N10 **Bergsjö** Gävleborg, C Sweden 62°00′N 17°10′E
93 J14 **Bergsviken** Norrbotten, N Sweden 65°16′N 21°24′E
Bergum see Burgum
98 M6 **Bergumer Meer** ◎ N Netherlands
94 N12 **Bergviken** ◎ C Sweden
168 M11 **Berhala, Selat** strait Sumatera, W Indonesia
Berhampore see Baharampur
99 J17 **Beringen** Limburg, NE Belgium 51°04′N 05°14′E
39 T12 **Bering Glacier** glacier Alaska, USA
Beringov Proliv see Bering Strait
192 L2 **Bering Sea** sea N Pacific Ocean
38 L9 **Bering Strait** Rus. Beringov Proliv. strait Bering Sea/Chukchi Sea
Berislav see Beryslav
105 O15 **Berja** Andalucía, S Spain 36°51′N 02°56′W
94 H9 **Berkåk** Sør-Trøndelag, S Norway 62°50′N 10°01′E

◆ Country
● Country Capital
◇ Dependent Territory
○ Dependent Territory Capital
⬖ Administrative Regions
✈ International Airport
▲ Mountain
▲▲ Mountain Range
Volcano
≈ River
◎ Lake
▣ Reservoir

98 N11 **Berkel** ≈ Germany/Netherlands
35 N8 **Berkeley** California, W USA 37°52′N 122°16′W
65 E24 **Berkeley Sound** sound NE Falkland Islands
21 V2 **Berkeley Springs** var. Bath. West Virginia, NE USA 39°38′N 78°14′W
195 N6 **Berkner Island** island Antarctica
114 G8 **Berkovitsa** Montana, NW Bulgaria 43°15′N 23°05′E
97 M22 **Berkshire** former county S England, United Kingdom
99 H17 **Berlaar** Antwerpen, N Belgium 51°08′N 04°39′E
Berlanga see Berlanga de Duero
105 P6 **Berlanga de Duero** var. Berlanga. Castilla y León, N Spain 41°28′N 02°51′W
0 I16 **Berlanga Rise** undersea feature E Pacific Ocean 08°30′N 93°30′W
99 F17 **Berlare** Oost-Vlaanderen, NW Belgium 51°02′N 04°01′E
104 E9 **Berlenga, Ilha da** island C Portugal
92 M7 **Berlevåg** Lapp. Bearalváhki. Finnmark, N Norway 70°51′N 29°04′E
100 O12 **Berlin** ● (Germany) Berlin, NE Germany 52°31′N 13°26′E
21 Z4 **Berlin** Maryland, NE USA 38°19′N 75°13′W
19 O7 **Berlin** New Hampshire, NE USA 44°27′N 71°13′W
18 D16 **Berlin** Pennsylvania, NE USA 39°54′N 78°57′W
30 L7 **Berlin** Wisconsin, N USA 43°57′N 88°59′W
100 O12 **Berlin** ◈ state NE Germany
Berlinchen see Berlinek
31 U12 **Berlin Lake** ◎ Ohio, N USA
183 R11 **Bermagui** New South Wales, SE Australia 36°26′S 150°01′E
40 L8 **Bermejillo** Durango, C Mexico 25°55′N 103°39′W
62 L5 **Bermejo, Río** ≈ N Argentina
62 I10 **Bermejo, Río** ≈ W Argentina
62 M6 **Bermejo viejo, Río** ≈ N Argentina
105 P2 **Bermeo** País Vasco, N Spain 43°25′N 02°44′W
104 K6 **Bermillo de Sayago** Castilla y León, N Spain 41°22′N 06°08′W
106 E6 **Bermina, Pizzo** Rmsch. Piz Bernina. ▲ Italy/Switzerland 46°22′N 09°52′E see also Bernina, Piz
64 A12 **Bermuda** var. Bermuda Islands, Bermudas; prev. Somers Islands. ◇ UK crown colony NW Atlantic Ocean
1 N11 **Bermuda** var. Great Bermuda, Long Island, Main Island. island Bermuda
Bermuda Islands see Bermuda
Bermuda-New England Seamount Arc see New England Seamounts
1 N11 **Bermuda Rise** undersea feature C Sargasso Sea 32°30′N 65°00′W
Bermudas see Bermuda
108 D8 **Bern** Fr. Berne. ● (Switzerland) Bern, W Switzerland 46°57′N 07°26′E
108 D9 **Bern** Fr. Berne. ◈ canton W Switzerland
37 R11 **Bernalillo** New Mexico, SW USA 35°18′N 106°33′W
14 H12 **Bernard Lake** ◎ Ontario, S Canada
61 B18 **Bernardo de Irigoyen** Santa Fe, NE Argentina 32°09′S 61°06′W
18 J14 **Bernardsville** New Jersey, NE USA 40°43′N 74°34′W
63 K14 **Bernasconi** La Pampa, C Argentina 37°55′S 63°44′W
100 O12 **Bernau** Brandenburg, NE Germany 52°41′N 13°36′E
102 L4 **Bernay** Eure, N France 49°05′N 00°36′E
101 L14 **Bernburg** Sachsen-Anhalt, C Germany 51°47′N 11°45′E
109 X5 **Berndorf** Niederösterreich, NE Austria 47°58′N 16°08′E
31 Q12 **Berne** Indiana, N USA 40°39′N 84°57′W
Berne see Bern
108 D10 **Berner Alpen** var. Berner Oberland, Eng. Bernese Oberland. ▲▲ SW Switzerland
Berner Oberland/Bernese Oberland see Berner Alpen
109 Y2 **Bernhardsthal** Niederösterreich, N Austria 48°40′N 16°51′E
22 H4 **Bernice** Louisiana, S USA 32°49′N 92°39′W
27 Y8 **Bernie** Missouri, C USA 36°40′N 89°58′W
180 G9 **Bernier Island** island Western Australia
Bernina Pass see Bernina, Passo del
108 J10 **Bernina, Passo del** Eng. Bernina Pass. pass SE Switzerland
108 J10 **Bernina, Piz** It. Pizzo Bernina. ▲ Italy/Switzerland 46°22′N 09°55′E see also Bermina, Pizzo
Bernina, Piz see Bernina, Pizzo
99 E20 **Bernissart** Hainaut, SW Belgium 50°29′N 03°37′E
101 E18 **Bernkastel-Kues** Rheinland-Pfalz, W Germany 49°55′N 07°04′E
Beroea see Ḥalab
172 H6 **Beroroha** Toliara, SW Madagascar 21°40′S 45°10′E
Béroubouay see Gbéroubouè
111 C17 **Beroun** Ger. Beraun. Středočeský Kraj, W Czech Republic 49°58′N 14°05′E
111 C16 **Berounka** Ger. Beraun. ≈ W Czech Republic
113 Q18 **Berovo** E FYR Macedonia 41°45′N 22°50′E
74 F6 **Berrechid** var. Berchid. W Morocco 33°16′N 07°32′W
103 R15 **Berre, Étang de** ◎ SE France
103 S15 **Berre-l'Étang** Bouches-du-Rhône, SE France 43°28′N 05°10′E
182 K9 **Berri** South Australia 34°16′S 140°35′E
31 O10 **Berrien Springs** Michigan, N USA 41°57′N 86°20′W
183 O10 **Berrigan** New South Wales, SE Australia 35°41′S 145°50′E
103 N9 **Berry** cultural region C France
35 N7 **Berryessa, Lake** ◎ California, W USA
44 G2 **Berry Islands** island group N The Bahamas
27 T9 **Berryville** Arkansas, C USA 36°22′N 93°35′W
21 V3 **Berryville** Virginia, NE USA 39°08′N 77°59′W
83 D21 **Berseba** Karas, S Namibia 26°00′S 17°46′E
117 O8 **Bershad'** Vinnyts'ka Oblast', C Ukraine 48°20′N 29°30′E
28 L3 **Berthold** North Dakota, N USA 48°16′N 101°48′W
37 T3 **Berthoud** Colorado, C USA 40°18′N 105°04′W
37 S4 **Berthoud Pass** pass Colorado, C USA
79 F15 **Bertoua** Est, E Cameroon 04°34′N 13°42′E
25 S10 **Bertram** Texas, SW USA 30°44′N 98°03′W
63 G22 **Bertrand, Cerro** ▲ S Argentina 50°00′S 73°27′W
99 J23 **Bertrix** Luxembourg, SE Belgium 49°52′N 05°15′E
191 P3 **Beru** var. Peru. atoll Tungaru, W Kiribati
Beruni see Beruniy
146 I9 **Beruniy** var. Biruni, Rus. Beruni. Qoraqalpog'iston Respublikasi, W Uzbekistan 41°48′N 60°39′E
58 F13 **Beruri** Amazonas, NW Brazil 03°44′S 61°13′W
18 H14 **Berwick** Pennsylvania, NE USA 41°03′N 76°13′W
96 K12 **Berwick** cultural region SE Scotland, United Kingdom
96 L12 **Berwick-upon-Tweed** N England, United Kingdom 55°46′N 02°W
117 S10 **Beryslav** Rus. Berislav. Khersons'ka Oblast', S Ukraine 46°51′N 33°26′E
Berytus see Beyrouth
172 H4 **Besalampy** Mahajanga, W Madagascar 16°43′S 44°29′E
103 T8 **Besançon** anc. Besontium, Vesontio. Doubs, E France 47°14′N 06°01′E
103 P10 **Besbre** ≈ C France
Bescanuova see Baška
Besdan see Bezdan
Besed' see Byesyedz'
147 R10 **Besharik** Rus. Besharyk; prev. Kirovo. Farg'ona Viloyati, E Uzbekistan 40°26′N 70°33′E
Besharyk see Besharik
146 L9 **Beshbuloq** Rus. Beshulak. Navoiy Viloyati, N Uzbekistan 41°55′N 64°13′E
Beshenkovichi see Byeshankovichy
146 M13 **Beshkent** Qashqadaryo Viloyati, S Uzbekistan 38°47′N 65°42′E
Beshulak see Beshbuloq
112 L10 **Beška** Vojvodina, N Serbia 45°09′N 20°04′E
Beskra see Biskra
127 O16 **Beslan** Respublika Severnaya Osetiya, SW Russian Federation 43°12′N 44°33′E
113 P16 **Besna Kobila** ▲ SE Serbia 42°30′N 22°16′E
137 N16 **Besni** Adıyaman, S Turkey 37°42′N 37°53′E
Besontium see Besançon
121 Q2 **Beşparmak Dağları** Eng. Kyrenia Mountains. ▲▲ N Cyprus
Bessarabka see Basarabeasca
92 O2 **Bessels, Kapp** headland C Svalbard 78°36′N 21°43′E
23 P4 **Bessemer** Alabama, S USA 33°24′N 86°57′W
30 K3 **Bessemer** Michigan, N USA 46°28′N 90°03′W
21 Q10 **Bessemer City** North Carolina, SE USA 35°16′N 81°16′W
102 M10 **Bessines-sur-Gartempe** Haute-Vienne, C France 46°06′N 01°22′E
99 K15 **Best** Noord-Brabant, S Netherlands 51°31′N 05°24′E
25 N9 **Best** Texas, SW USA 31°13′N 101°34′W
139 U2 **Beste** Ar. Baṣṭah. As Sulaymānīyah, E Iraq 36°20′N 45°14′E
125 O11 **Bestuzhevo** Arkhangel'skaya Oblast', NW Russian Federation 61°36′N 43°54′E
123 M11 **Bestyakh** Respublika Sakha (Yakutiya), NE Russian Federation 61°25′N 129°05′E
Beszterce see Bistriţa
Besztercebánya see Banská Bystrica
172 I5 **Betafo** Antananarivo, C Madagascar 19°50′S 46°50′E
104 H2 **Betanzos** Galicia, NW Spain 43°17′N 08°17′W
104 G2 **Betanzos, Ría de** estuary NW Spain
79 G15 **Bétaré Oya** Est, E Cameroon 05°34′N 14°09′E
105 S9 **Bétera** Valenciana, E Spain 39°35′N 00°28′W
77 R15 **Bétérou** C Benin 09°13′N 02°18′E
83 K21 **Bethal** Mpumalanga, NE South Africa 26°27′S 29°28′E
30 K15 **Bethalto** Illinois, N USA 38°54′N 90°02′W
83 D21 **Bethanie** var. Bethanien, Bethany. Karas, S Namibia 26°32′S 17°11′E
Bethanien see Bethanie
27 S2 **Bethany** Missouri, C USA 40°15′N 94°03′W
27 N10 **Bethany** Oklahoma, C USA 35°31′N 97°37′W
Bethany see Bethanie
39 N12 **Bethel** Alaska, USA 60°47′N 161°45′W
19 P7 **Bethel** Maine, NE USA 44°24′N 70°47′W
21 W9 **Bethel** North Carolina, SE USA 35°46′N 77°21′W
18 B15 **Bethel Park** Pennsylvania, NE USA 40°21′N 80°03′W
21 W3 **Bethesda** Maryland, NE USA 39°00′N 77°05′W
83 J22 **Bethlehem** Free State, C South Africa 28°12′S 28°16′E
18 I14 **Bethlehem** Pennsylvania, NE USA 40°36′N 75°22′W
138 F10 **Bethlehem** var. Beit Lekhem, Ar. Bayt Laḥm, Heb. Bet Leḥem. C West Bank 31°43′N 35°12′E
Bethlen see Beclean
83 I24 **Bethulie** Free State, C South Africa 30°30′S 25°59′E
103 O1 **Béthune** Pas-de-Calais, N France 50°32′N 02°38′E
102 M3 **Béthune** ≈ N France
104 M14 **Béticos, Sistemas** var. Sistema Penibético, Eng. Baetic Cordillera, Baetic Mountains. ▲▲ S Spain
54 I6 **Betijoque** Trujillo, NW Venezuela 09°25′N 70°45′W
59 M20 **Betim** Minas Gerais, SE Brazil 19°56′S 44°10′W
190 H3 **Betio** Tarawa, W Kiribati 01°21′N 172°56′E
172 H7 **Betioky** Toliara, S Madagascar 23°42′S 44°22′E
Bet Leḥem see Bethlehem
Betlen see Beclean
167 O17 **Betong** Yala, SW Thailand 05°45′N 101°05′E
79 I16 **Bétou** Likouala, N Congo 03°08′N 18°31′E
145 P14 **Betpakdala** Kaz. Betpaqdala prev. Betpak-Dala. plateau S Kazakhstan
Betpak-Dala see Betpakdala
Betpaqdala see Betpakdala
172 H7 **Betroka** Toliara, S Madagascar 23°15′S 46°07′E
Betschau see Bečva
Bet She'an see Beit She'an
15 T6 **Betsiamites** Québec, SE Canada 48°56′N 68°40′W
15 T6 **Betsiamites** ≈ Québec, SE Canada
172 I4 **Betsiboka** ≈ N Madagascar
Betsukai see Bekkai
99 M25 **Bettembourg** Luxembourg, S Luxembourg 49°31′N 06°06′E
99 M23 **Bettendorf** Diekirch, NE Luxembourg 49°53′N 06°13′E
29 Z14 **Bettendorf** Iowa, C USA 41°31′N 90°31′W
Bette, Pic see Bette, Picco
75 R13 **Bette, Picco** var. Bikkū Bittī, Pic Bette. ▲ S Libya 22°02′N 19°07′E
153 P12 **Bettiah** Bihār, N India 26°49′N 84°30′E
39 Q7 **Bettles** Alaska, USA 66°54′N 151°40′W
95 N17 **Bettna** Södermanland, C Sweden 58°52′N 16°40′E
154 H11 **Betūl** prev. Badnur. Madhya Pradesh, C India 21°55′N 77°54′E
154 H9 **Betwa** ≈ C India
101 F16 **Betzdorf** Rheinland-Pfalz, W Germany 50°47′N 07°50′E
82 C9 **Béu** Uíge, NW Angola 06°15′S 15°32′E
31 P6 **Beulah** Michigan, N USA 43°35′N 83°52′W
28 L5 **Beulah** North Dakota, N USA 47°16′N 101°48′W
98 M8 **Beulakerwijde** ◎ N Netherlands
98 L13 **Beuningen** Gelderland, SE Netherlands 51°52′N 05°47′E
Beuthen see Bytom
103 N7 **Beuvron** ≈ C France
99 F16 **Beveren** Oost-Vlaanderen, N Belgium 51°13′N 04°15′E
21 T9 **B. Everett Jordan Reservoir** var. Jordan Lake. ⊠ North Carolina, SE USA
97 N17 **Beverley** E England, United Kingdom 53°51′N 00°26′W
Beverley see Beverly
99 J17 **Beverlo** Limburg, NE Belgium 51°06′N 05°14′E
19 P11 **Beverly** Massachusetts, NE USA 42°33′N 70°49′W
32 J9 **Beverly** var. Beverley. Washington, NW USA 46°50′N 119°57′W
35 S15 **Beverly Hills** California, W USA 34°02′N 118°25′W
101 I14 **Beverungen** Nordrhein-Westfalen, C Germany 51°39′N 09°22′E
98 H9 **Beverwijk** Noord-Holland, W Netherlands 52°29′N 04°40′E
108 C10 **Bex** Vaud, W Switzerland 46°15′N 07°00′E
97 P23 **Bexhill** var. Bexhill-on-Sea. SE England, United Kingdom 50°50′N 00°28′E
Bexhill-on-Sea see Bexhill
136 E17 **Bey Dağları** ▲▲ SW Turkey
Beyj see Belyy
136 E10 **Beykoz** İstanbul, NW Turkey 41°09′N 29°06′E
76 K15 **Beyla** SE Guinea 08°43′N 08°41′W
137 X12 **Beyläqan** prev. Zhdanov. SW Azerbaijan 39°43′N 47°38′E
80 L10 **Beylul** var. Beilul. SE Eritrea 13°10′N 42°27′E
144 H14 **Beyneu** Kaz. Beyneū. Mangistau, SW Kazakhstan 45°20′N 55°11′E
Beyneū see Beyneu
165 X14 **Beyonēsu-retsugan** Eng. Bayonnaise Rocks. island group SE Japan
136 G12 **Beypazarı** Ankara, NW Turkey 40°10′N 31°56′E
155 F21 **Beypore** Kerala, SW India 11°10′N 75°49′E
138 G7 **Beyrouth** var. Bayrūt, Eng. Beirut; anc. Berytus. ● (Lebanon) W Lebanon 33°55′N 35°31′E
138 G7 **Beyrouth** ✈ W Lebanon 33°52′N 35°30′E
136 G15 **Beyşehir** Konya, SW Turkey 37°40′N 31°43′E
136 G15 **Beyşehir Gölü** ◎ C Turkey
108 J7 **Bezau** Vorarlberg, NW Austria 47°24′N 09°55′E
112 J8 **Bezdan** Ger. Besdan, Hung. Bezdán. Vojvodina, NW Serbia 45°51′N 19°00′E
Bezdezh see Byezdzyezh
124 G15 **Bezhanitsy** Pskovskaya Oblast', W Russian Federation 56°57′N 29°53′E
124 K15 **Bezhetsk** Tverskaya Oblast', W Russian Federation 57°47′N 36°42′E
103 P16 **Béziers** anc. Baeterrae, Baeterrae Septimanorum, Julia Beterrae. Hérault, S France 43°21′N 03°13′E
Bezmein see Abadan
Bezwada see Vijayawāda
154 P12 **Bhadrak** var. Bhadrakh. Odisha, E India 21°04′N 86°30′N
Bhadrakh see Bhadrak
155 F19 **Bhadra Reservoir** ⊠ SW India
155 F18 **Bhadrāvati** Karnātaka, SW India 13°52′N 75°43′E
153 R14 **Bhāgalpur** Bihār, NE India 25°14′N 86°59′E
Bhairab see Bhairab Bazar
153 U14 **Bhairab Bazar** var. Bhairab. Dhaka, C Bangladesh 24°04′N 91°00′E
153 O11 **Bhairahawā** Western, S Nepal 27°31′N 83°27′E
149 S8 **Bhakkar** Punjab, E Pakistan 31°40′N 71°08′E
153 P11 **Bhaktapur** Central, C Nepal 27°47′N 85°21′E
167 N3 **Bhamo** var. Banmo. Kachin State, N Myanmar (Burma) 24°15′N 97°15′E
Bhāmragad see Bhāmragarh
154 K13 **Bhāmragarh** var. Bhāmragad. Mahārāshtra, C India 19°28′N 80°39′E
154 J12 **Bhandāra** Mahārāshtra, C India 21°10′N 79°41′E
Bhārat see India
152 J12 **Bharatpur** prev. Bhurtpore. Rājasthān, N India 27°14′N 77°29′E
154 D11 **Bharūch** Gujarāt, W India 21°48′N 72°55′E
155 E18 **Bhatkal** Karnātaka, W India 13°59′N 74°34′E
153 O13 **Bhatni** var. Bhatni Junction. Uttar Pradesh, N India 26°23′N 83°56′E
Bhatni Junction see Bhatni
153 S16 **Bhātpāra** West Bengal, NE India 22°55′N 88°30′E
149 U7 **Bhaun** Punjab, E Pakistan 32°53′N 72°48′E
Bhaunagar see Bhāvnagar
Bhavānipātna see Bhawānipatna
155 H21 **Bhavānisāgar** ⊠ S India
154 D11 **Bhāvnagar** prev. Bhaunagar. Gujarāt, W India 21°46′N 72°14′E
154 M13 **Bhawānipatna** var. Bhavānipātna. Odisha, E India 19°56′N 83°09′E
Bheanntraí, Bá see Bantry Bay
Bheara, Béal an see Gweebarra Bay
154 K12 **Bhilai** Chhattīsgarh, C India 21°12′N 81°26′E
152 G13 **Bhīlwāra** Rājasthān, N India 25°23′N 74°39′E
155 E14 **Bhīma** ≈ S India
155 K16 **Bhīmavaram** Andhra Pradesh, E India 16°34′N 81°35′E
154 I7 **Bhind** Madhya Pradesh, C India 26°33′N 78°47′E
152 E13 **Bhīnmāl** Rājasthān, N India 25°01′N 72°22′E
Bhir see Bīd
83 J25 **Bhisho** prev. Bisho. Eastern Cape, S South Africa 32°46′S 27°21′E see also Bhisho
154 D13 **Bhiwandi** Mahārāshtra, W India 19°21′N 73°08′E
152 H10 **Bhiwāni** Haryāna, N India 28°50′N 76°10′E
152 L13 **Bhognīpur** Uttar Pradesh, N India 26°12′N 79°48′E
153 U16 **Bhola** Barisal, S Bangladesh 22°40′N 91°36′E
154 H10 **Bhopāl** state capital Madhya Pradesh, C India 23°17′N 77°25′E
155 J14 **Bhopālpatnam** Chhattīsgarh, C India 18°51′N 80°22′E
155 E14 **Bhor** Mahārāshtra, W India 18°10′N 73°55′E
154 O12 **Bhubaneshwar** prev. Bhubaneswar, Bhuvaneshwar. state capital Odisha, E India 20°16′N 85°51′E
Bhubaneswar see Bhubaneshwar
154 B9 **Bhuj** Gujarāt, W India 23°16′N 69°40′E
Bhuket see Phuket
Bhurtpore see Bharatpur
Bhusaval see Bhusāwal
154 G12 **Bhusāwal** prev. Bhusaval. Mahārāshtra, C India 21°01′N 75°50′E
153 T12 **Bhutan** off. Kingdom of Bhutan, var. Druk-yul. ◆ monarchy S Asia
Bhutan, Kingdom of see Bhutan
Bhuvaneshwar see Bhubaneshwar
143 T15 **Bīābān, Kūh-e** ▲▲ S Iran
77 V18 **Biafra, Bight of** var. Bight of Bonny. bay W Africa
171 W12 **Biak** Papua, E Indonesia 01°10′S 136°05′E
171 W12 **Biak, Pulau** island E Indonesia
110 P12 **Biała Podlaska** Lubelskie, E Poland 52°03′N 23°08′E
110 F7 **Białogard** Ger. Belgard. Zachodnio-pomorskie, NW Poland 54°01′N 15°59′E
110 P10 **Białowieska, Puszcza** Bel. Byelavyezhskaya Pushcha, Rus. Belovezhskaya Pushcha. physical region Belarus/Poland see also Byelavyezhskaya, Pushcha
Białowieska, Puszcza see Byelavyezhskaya Pushcha
110 G8 **Biały Bór** Ger. Baldenburg. Zachodnio-pomorskie, NW Poland 53°53′N 16°49′E
110 P9 **Białystok** Rus. Belostok, Bielostok. Podlaskie, NE Poland 53°08′N 23°09′E
107 L24 **Biancavilla** prev. Inessa. Sicilia, Italy, C Mediterranean Sea 37°38′N 14°52′E
Bianco, Monte see Blanc, Mont
Bianjiang see Xunke
76 L15 **Biankouma** W Ivory Coast 07°44′N 07°37′W
167 R7 **Bia, Phou** var. Pou Bia. ▲ C Laos 18°59′N 103°09′E
Bia, Pou see Bia, Phou
143 R5 **Bīārjmand** Semnān, N Iran 36°05′N 55°50′E
105 P4 **Biarra** ≈ NE Spain
102 I15 **Biarritz** Pyrénées-Atlantiques, SW France 43°25′N 01°40′W
108 H10 **Biasca** Ticino, S Switzerland 46°22′N 08°59′E
61 E17 **Biassini** Salto, N Uruguay 31°18′S 57°05′W
Biasteri see Laguardia
165 S3 **Bibai** Hokkaidō, NE Japan 43°21′N 141°53′E
83 B15 **Bibala** Port. Vila Arriaga. Namibe, SW Angola 14°46′S 13°21′E
104 J4 **Bibéi** ≈ NW Spain
Biberach see Biberach an der Riss
101 I23 **Biberach an der Riss** var. Biberach, Ger. Biberach an der Riß. Baden-Württemberg, S Germany 48°06′N 09°48′E
108 E7 **Biberist** Solothurn, NW Switzerland 47°11′N 07°34′E
77 O16 **Bibiani** SW Ghana 06°28′N 02°20′W
112 C13 **Bibinje** Zadar, SW Croatia 44°04′N 15°17′E
Biblical Gebal see Jbaïl
116 I5 **Bibrka** Pol. Bóbrka, Rus. Bobrka. L'vivs'ka Oblast', NW Ukraine 49°39′N 24°16′E
117 N10 **Bîc** ≈ S Moldova
113 M18 **Bicaj** Kukës, NE Albania 42°00′N 20°24′E
116 K10 **Bicaz** Hung. Békás. Neamţ, NE Romania 46°53′N 26°05′E
183 Q16 **Bicheno** Tasmania, SE Australia 41°56′S 148°15′E
Bichiş see Békés
Bichiş-Ciaba see Békéscsaba
Bichitra see Phichit
137 P8 **Bich'vinta** Prev. Bichvint'a, Rus. Pitsunda. NW Georgia 43°12′N 40°21′E
Bichvint'a see Bich'vinta
15 T7 **Bic, Île du** island Québec, SE Canada
32 J10 **Bickleton** Washington, NW USA 46°00′N 120°16′W
36 L6 **Bicknell** Utah, W USA 38°20′N 111°32′W
171 S11 **Bicoli** Pulau Halmahera, E Indonesia 0°34′N 128°33′E
111 J22 **Bicske** Fejér, C Hungary 47°30′N 18°36′E
155 F14 **Bīd** prev. Bhir. Mahārāshtra, W India 19°17′N 75°22′E
77 U15 **Bida** Niger, C Nigeria 09°06′N 06°02′E
155 H15 **Bīdar** Karnātaka, C India 17°56′N 77°35′E
141 Y8 **Bidbid** NE Oman 23°25′N 58°08′E
19 P9 **Biddeford** Maine, NE USA 43°29′N 70°27′W
98 L9 **Biddinghuizen** Flevoland, C Netherlands 52°28′N 05°41′E
33 X11 **Biddle** Montana, NW USA 45°04′N 105°21′W
97 J23 **Bideford** SW England, United Kingdom 51°01′N 04°12′W
82 D13 **Bié** ◈ province C Angola
35 O2 **Bieber** California, W USA 41°07′N 121°09′W
110 O9 **Biebrza** ≈ NE Poland
165 T3 **Biei** Hokkaidō, NE Japan 43°33′N 142°28′E
108 D8 **Biel** Fr. Bienne. Bern, W Switzerland 47°09′N 07°16′E
100 G13 **Bielefeld** Nordrhein-Westfalen, NW Germany 52°01′N 08°32′E
108 D8 **Bieler See** Fr. Lac de Bienne. ◎ W Switzerland
Bielitz/Bielitz-Biala see Bielsko-Biała
106 C7 **Biella** Piemonte, N Italy 45°34′N 08°04′E
Bielostok see Białystok
111 J17 **Bielsko-Biała** Ger. Bielitz, Bielitz-Biala. Śląskie, S Poland 49°49′N 19°01′E
110 P10 **Bielsk Podlaski** Białystok, E Poland 52°45′N 23°11′E
Bien Bien see Điện Biên Phu
Biên Đông see South China Sea
11 V17 **Bienfait** Saskatchewan, S Canada 49°06′N 102°47′W
167 T14 **Biên Hoa** Đông Nai, S Vietnam 10°58′N 106°50′E
Bienne see Biel
Bienne, Lac de see Bieler See
12 K8 **Bienville, Lac** ◎ Québec, C Canada
82 D13 **Bié, Planalto do** var. Bié Plateau. plateau C Angola
Bié Plateau see Bié, Planalto do
108 B9 **Bière** Vaud, W Switzerland 46°32′N 06°19′E
98 O4 **Bierum** Groningen, NE Netherlands 53°25′N 06°51′E
98 I13 **Biesbos** var. Biesbosch. wetland S Netherlands
Biesbosch see Biesbos
99 H21 **Biesme** Namur, S Belgium 50°19′N 04°43′E
101 H21 **Bietigheim-Bissingen** Baden-Württemberg, SW Germany 48°57′N 09°07′E
99 I23 **Bièvre** Namur, SE Belgium 49°56′N 05°01′E
79 D18 **Bifoun** Moyen-Ogooué, NW Gabon 0°15′S 10°24′E
165 T2 **Bifuka** Hokkaidō, NE Japan 44°28′N 142°20′E
136 C11 **Biga** Çanakkale, NW Turkey 40°13′N 27°14′E
136 C13 **Bigadiç** Balıkesir, W Turkey 39°24′N 28°08′E
26 J7 **Big Basin** basin Kansas, C USA
185 B20 **Big Bay** bay South Island, New Zealand
31 O5 **Big Bay de Noc** ◎ Michigan, N USA
31 N3 **Big Bay Point** headland Michigan, N USA 46°51′N 87°40′W
33 R10 **Big Belt Mountains** ▲▲ Montana, NW USA
29 N10 **Big Bend Dam** dam South Dakota, N USA
24 K12 **Big Bend National Park** national park Texas, S USA
22 K5 **Big Black River** ≈ Mississippi, S USA
27 O3 **Big Blue River** ≈ Kansas/Nebraska, C USA
24 M10 **Big Canyon** ≈ Texas, SW USA
33 N12 **Big Creek** Idaho, NW USA 45°05′N 115°20′W
23 N8 **Big Creek Lake** ⊠ Alabama, S USA
23 X15 **Big Cypress Swamp** wetland Florida, SE USA
39 S9 **Big Delta** Alaska, USA 64°09′N 145°50′W
30 K6 **Big Eau Pleine Reservoir** ⊠ Wisconsin, N USA
19 P5 **Bigelow Mountain** ▲ Maine, NE USA 45°09′N 70°17′W
162 G9 **Biger** var. Jargalant. Govĭ-Altay, W Mongolia 45°39′N 97°10′E
29 U3 **Big Falls** Minnesota, N USA 48°13′N 93°48′W
33 P8 **Bigfork** Montana, NW USA 48°03′N 114°04′W
29 U3 **Big Fork River** ≈ Minnesota, N USA
11 S15 **Biggar** Saskatchewan, S Canada 52°03′N 107°59′W
180 L3 **Bigge Island** island Western Australia
35 O5 **Biggs** California, W USA 39°24′N 121°44′W
32 I11 **Biggs** Oregon, NW USA 45°39′N 120°49′W
14 K13 **Big Gull Lake** ◎ Ontario, SE Canada
37 P16 **Big Hachet Peak** ▲ New Mexico, SW USA 31°38′N 108°24′W
33 P11 **Big Hole River** ≈ Montana, NW USA
33 V13 **Bighorn Basin** basin Wyoming, C USA
33 U11 **Bighorn Lake** ◎ Montana/Wyoming, N USA
33 W13 **Bighorn Mountains** ▲▲ Wyoming, C USA
36 J13 **Big Horn Peak** ▲ Arizona, SW USA 33°40′N 113°01′W
33 V11 **Bighorn River** ≈ Montana/Wyoming, NW USA
9 S7 **Big Island** island Nunavut, NE Canada
39 O16 **Big Koniuji Island** island Shumagin Islands, Alaska, USA
25 N9 **Big Lake** Texas, SW USA 31°12′N 101°29′W
19 T5 **Big Lake** ◎ Maine, NE USA
30 I3 **Big Manitou Falls** waterfall Wisconsin, N USA
35 R2 **Big Mountain** ▲ Nevada, W USA 41°18′N 119°03′W
108 G10 **Bignasco** Ticino, S Switzerland 46°21′N 08°37′E
29 R16 **Big Nemaha River** ≈ Nebraska, C USA
76 G12 **Bignona** SW Senegal 12°49′N 16°16′W
Bigorra see Tarbes
Bigosovo see Bihosava
35 S10 **Big Pine** California, W USA 37°09′N 118°18′W
35 Q14 **Big Pine Mountain** ▲ California, W USA 34°41′N 119°37′W
27 V6 **Big Piney Creek** ≈ Missouri, C USA
65 M24 **Big Point** headland N Tristan da Cunha 37°10′S 12°18′W
31 P8 **Big Rapids** Michigan, N USA 43°42′N 85°28′W
30 K6 **Big Rib River** ≈ Wisconsin, N USA
14 L14 **Big Rideau Lake** ◎ Ontario, SE Canada
11 T14 **Big River** Saskatchewan, C Canada 53°48′N 106°55′W
27 X5 **Big River** ≈ Missouri, C USA
31 N7 **Big Sable Point** headland Michigan, N USA 44°03′N 86°30′W
33 S7 **Big Sandy** Montana, NW USA 48°08′N 110°09′W
25 W6 **Big Sandy** Texas, SW USA 32°34′N 95°06′W
37 V5 **Big Sandy Creek** ≈ Colorado, C USA
29 Q16 **Big Sandy Creek** ≈ Nebraska, C USA
29 V5 **Big Sandy Lake** ◎ Minnesota, N USA
36 J11 **Big Sandy River** ≈ Arizona, SW USA
21 P5 **Big Sandy River** ≈ S USA
23 V6 **Big Satilla Creek** ≈ Georgia, SE USA
29 R12 **Big Sioux River** ≈ Iowa/South Dakota, N USA
35 U7 **Big Smoky Valley** valley Nevada, W USA
25 N7 **Big Spring** Texas, SW USA 32°15′N 101°30′W
19 Q5 **Big Squaw Mountain** ▲ Maine, NE USA 45°28′N 69°42′W
21 O7 **Big Stone Gap** Virginia, NE USA 36°52′N 82°45′W
29 Q8 **Big Stone Lake** ◎ Minnesota/South Dakota, N USA
22 K4 **Big Sunflower River** ≈ Mississippi, S USA
33 T11 **Big Timber** Montana, NW USA 45°50′N 109°57′W
12 D8 **Big Trout Lake** Ontario, C Canada 53°40′N 90°00′W
14 I12 **Big Trout Lake** ◎ Ontario, SE Canada
35 O2 **Big Valley Mountains** ▲▲ California, W USA
25 Q13 **Big Wells** Texas, SW USA 28°34′N 99°34′W
14 F11 **Bigwood** Ontario, S Canada 46°03′N 80°37′W
112 D11 **Bihać** ◈ Federacija Bosne I Hercegovine, NW Bosnia and Herzegovina
153 P14 **Bihār** prev. Behar. ◈ state N India
Bihār see Bihār Sharīf
81 F20 **Biharamulo** Kagera, NW Tanzania 02°37′S 31°20′E
153 R13 **Bihārīganj** Bihār, NE India 25°44′N 86°59′E
153 P14 **Bihār Sharīf** var. Bihār. Bihār, N India 25°13′N 85°31′E
116 F10 **Bihor** ◈ county NW Romania
165 V3 **Bihoro** Hokkaidō, NE Japan 43°50′N 144°05′E
118 K11 **Bihosava** Rus. Bigosovo. Vitsyebskaya Voblasts', NW Belarus 55°50′N 27°46′E
Bijagos Archipelago see Bijagós, Arquipélago dos
76 G13 **Bijagós, Arquipélago dos** var. Bijagos Archipelago. island group W Guinea-Bissau
155 F16 **Bījāpur** Karnātaka, C India 16°50′N 75°42′E
142 K5 **Bījār** Kordestān, W Iran 35°52′N 47°39′E
112 J11 **Bijeljina** Republika Srpska, NE Bosnia and Herzegovina 44°46′N 19°13′E
113 K15 **Bijelo Polje** E Montenegro 43°03′N 19°44′E
160 I11 **Bijie** Guizhou, S China 27°15′N 105°16′E
152 J10 **Bijnor** Uttar Pradesh, N India 29°22′N 78°09′E
152 F11 **Bīkāner** Rājasthān, NW India 28°01′N 73°22′E
189 V3 **Bikar Atoll** var. Pikaar. atoll Ratak Chain, N Marshall Islands
190 H3 **Bikeman** atoll Tungaru, W Kiribati
190 I3 **Bikenebu** Tarawa, W Kiribati
123 S14 **Bikin** Khabarovskiy Kray, SE Russian Federation 46°45′N 134°06′E
123 S14 **Bikin** ≈ SE Russian Federation
189 R3 **Bikini Atoll** var. Pikinni. atoll Ralik Chain, NW Marshall Islands
83 L17 **Bikita** Masvingo, E Zimbabwe 20°06′S 31°41′E
Bikkū Bittī see Bette, Picco
79 I19 **Bikoro** Equateur, W Dem. Rep. Congo 0°45′S 18°09′E
141 Z9 **Bilād Banī Bū 'Alī** NE Oman 22°02′N 59°18′E
141 Z9 **Bilād Banī Bū Ḥasan** NE Oman 22°09′N 59°14′E
141 X9 **Bilād Manaḥ** var. Manaḥ. NE Oman 22°44′N 57°36′E
77 Q12 **Bilanga** C Burkina Faso 12°35′N 00°08′W
152 F12 **Bilāra** Rājasthān, N India 26°10′N 73°48′E
152 K10 **Bilāri** Uttar Pradesh, N India 28°37′N 78°48′E
138 J5 **Bil'ās, Jabal al** ▲ C Syria
154 L11 **Bilāspur** Chhattīsgarh, C India 22°06′N 82°08′E
152 I8 **Bilāspur** Himāchal Pradesh, N India 31°18′N 76°48′E
168 J9 **Bila, Sungai** ≈ Sumatera, W Indonesia
137 Y13 **Biläsuvar** Rus. Bilyasuvar; prev. Pushkino. SE Azerbaijan 39°26′N 48°34′E
117 O5 **Bila Tserkva** Rus. Belaya Tserkov'. Kyyivs'ka Oblast', N Ukraine 49°49′N 30°08′E
167 N11 **Bilauktaung Range** var. Thanintari Taungdan. ▲▲ Myanmar (Burma)/Thailand
105 O2 **Bilbao** Basq. Bilbo. País Vasco, N Spain 43°15′N 02°56′W
Bilbo see Bilbao
92 H2 **Bíldudalur** Vestfirðir, NW Iceland 65°40′N 23°35′W
113 I16 **Bileća** ◈ Republika Srpska, S Bosnia and Herzegovina
136 E12 **Bilecik** Bilecik, NW Turkey 39°59′N 29°54′E
136 F12 **Bilecik** ◈ province NW Turkey
116 E11 **Biled** Ger. Billed, Hung. Billéd. Timiş, W Romania 45°55′N 20°55′E
111 O15 **Biłgoraj** Lubelskie, E Poland 50°31′N 22°41′E
117 P11 **Bilhorod-Dnistrovs'kyy** Rus. Belgorod-Dnestrovskiy, Rom. Cetatea Albă, prev. Akkerman; anc. Tyras. Odes'ka Oblast', SW Ukraine 46°10′N 30°19′E
79 M16 **Bili** Orientale, N Dem. Rep. Congo 04°07′N 25°09′E
123 T6 **Bilibino** Chukotskiy Avtonomnyy Okrug, NE Russian Federation 67°56′N 166°45′E
166 M8 **Bilin** Mon State, S Myanmar (Burma) 17°14′N 97°12′E
113 N21 **Bilisht** var. Bilishti. Korçë, SE Albania 40°36′N 21°00′E
Bilishti see Bilisht
183 N10 **Billabong Creek** var. Moulamein Creek. seasonal river New South Wales, SE Australia
182 G4 **Billa Kalina** South Australia 29°57′S 136°13′E
197 Q17 **Bill Baileys Bank** undersea feature N Atlantic Ocean 60°35′N 10°15′W
Billed/Billéd see Biled
153 N14 **Billi** Uttar Pradesh, N India 24°30′N 82°59′E
97 M15 **Billingham** N England, United Kingdom 54°36′N 01°17′W
33 U11 **Billings** Montana, NW USA 45°47′N 108°32′W
95 J16 **Billingsfors** Västra Götaland, S Sweden 58°57′N 12°14′E
Bill of Cape Clear, The see Clear, Cape
28 L9 **Billsburg** South Dakota, N USA 44°22′N 101°40′W
95 F23 **Billund** Syddtjylland, W Denmark 55°44′N 09°07′E
36 L11 **Bill Williams Mountain** ▲ Arizona, SW USA 35°12′N 112°12′W
36 I12 **Bill Williams River** ≈ Arizona, SW USA
77 Y8 **Bilma** Agadez, NE Niger 18°22′N 13°01′E
77 Y8 **Bilma, Grand Erg de** desert NE Niger
181 Y9 **Biloela** Queensland, E Australia 24°27′S 150°31′E
112 G8 **Bilo Gora** ▲▲ N Croatia
117 U13 **Bilohirs'k** Rus. Belogorsk; prev. Karasubazar. Avtonomna Respublika Krym, S Ukraine 45°04′N 34°35′E
Bilokorovychi see Novi Bilokorovychi
Bilokurakine see Bilokurakyne
117 X5 **Bilokurakyne** var. Bilokurakine. Luhans'ka Oblast', E Ukraine 49°32′N 38°44′E
117 T3 **Bilopillya** Rus. Belopol'ye. Sums'ka Oblast', NE Ukraine 51°09′N 34°17′E
117 Y6 **Bilovods'k** Rus. Belovodsk. Luhans'ka Oblast', E Ukraine 49°11′N 39°34′E
22 M9 **Biloxi** Mississippi, S USA 30°24′N 88°53′W
117 R10 **Bilozerka** Khersons'ka Oblast', S Ukraine 46°36′N 32°23′E
117 W7 **Bilozers'ke** Donets'ka Oblast', E Ukraine 48°29′N 37°03′E
98 J11 **Bilthoven** Utrecht, C Netherlands 52°07′N 05°12′E
78 K9 **Biltine** Wadi Fira, E Chad 14°30′N 20°53′E
Biltine, Préfecture de see Wadi Fira
Bilüü see Ulaanhus
Bilwi see Puerto Cabezas
Bilyasuvar see Biläsuvar
117 O11 **Bilyayivka** Odes'ka Oblast', SW Ukraine 46°28′N 30°11′E
99 K18 **Bilzen** Limburg, NE Belgium 50°52′N 05°31′E
Bimbéréké see Bembèrèkè
183 R10 **Bimberi Peak** ▲ New South Wales, SE Australia 35°42′S 148°46′E
77 Q15 **Bimbila** E Ghana 08°54′N 00°05′E
79 I15 **Bimbo** Ombella-Mpoko, SW Central African Republic 04°19′N 18°27′E
44 F2 **Bimini Islands** island group W The Bahamas
154 I9 **Bina** Madhya Pradesh, C India 24°09′N 78°10′E
143 T4 **Bīnālūd, Kūh-e** ▲▲ NE Iran
99 F20 **Binche** Hainaut, S Belgium 50°25′N 04°10′E
Bindloe Island see Marchena, Isla
83 L16 **Bindura** Mashonaland Central, NE Zimbabwe 17°20′S 31°21′E
105 T5 **Binefar** Aragón, NE Spain 41°51′N 00°17′E

◆ Country
● Country Capital
◇ Dependent Territory
○ Dependent Territory Capital
◈ Administrative Regions
✈ International Airport
▲ Mountain
▲▲ Mountain Range
Volcano
≈ River
◎ Lake
⊠ Reservoir

83 J16 **Binga** Matabeleland North, W Zimbabwe 17°40′S 27°22′E
183 T5 **Bingara** New South Wales, SE Australia 29°54′S 150°36′E
101 F18 **Bingen am Rhein** Rheinland-Pfalz, SW Germany 49°58′N 07°54′E
26 M11 **Binger** Oklahoma, C USA 35°19′N 98°19′W
Bingerau see Węgrów
Bin Ghalfān, Jazā'ir see Ḩalāniyāt, Juzur al
19 Q6 **Bingham** Maine, NE USA 45°01′N 69°51′W
18 H11 **Binghamton** New York, NE USA 42°06′N 75°55′W
Bin Ghanīmah, Jabal see Bin Ghunaymah, Jabal
75 P11 **Bin Ghunaymah, Jabal** var. Jabal Bin Ghanīmah. ▲ C Libya
139 U3 **Bingird** As Sulaymānīyah, NE Iraq 36°03′N 45°03′E
Bingmei see Congjiang
137 P14 **Bingöl** Bingöl, E Turkey 38°54′N 40°29′E
137 P14 **Bingöl** ◆ province E Turkey
161 R6 **Binhai** var. Dongkan. Jiangsu, E China 34°00′N 119°51′E
167 V11 **Binh Đinh** var. An Nhon. Binh Đinh, C Vietnam 13°53′N 109°07′E
Binh Sơn see Châu Ô
Binimani see Bintimani
168 I8 **Binjai** Sumatera, W Indonesia 03°37′N 98°30′E
183 R6 **Binnaway** New South Wales, SE Australia 31°34′S 149°24′E
108 E6 **Binningen** Basel Landschaft, NW Switzerland 47°32′N 07°35′E
80 H12 **Binshangul Gumuz** var. Benishangul. ◆ W Ethiopia
168 J8 **Bintang, Banjaran** ▲ Peninsular Malaysia
168 M10 **Bintan, Pulau** island Kepulauan Riau, W Indonesia
76 J14 **Bintimani** var. Binimani. ▲ NE Sierra Leone 09°21′N 11°09′W
Bint Jubayl see Bent Jbaïl
169 S9 **Bintulu** Sarawak, East Malaysia 03°12′N 113°01′E
169 S9 **Bintuni** prev. Steenkool. Papua Barat, E Indonesia 02°03′S 133°45′E
163 W8 **Binxian** prev. Binzhou. Heilongjiang, NE China 45°44′N 127°27′E
160 K14 **Binyang** var. Binzhou. Guangxi Zhuangzu Zizhiqu, S China 23°15′N 108°40′E
161 Q4 **Binzhou** Shandong, E China 37°23′N 118°03′E
Binzhou see Binyang
Binzhou see Binxian
63 G14 **Bío Bío** var. Región del Bío Bío. ◆ region C Chile
Bío Bío, Región del see Bío Bío
63 G14 **Bío Bío, Río** ↝ C Chile
79 C16 **Bioco, Isla de** var. Bioko, Eng. Fernando Po, Sp. Fernando Póo; prev. Macías Nguema Biyogo. island NW Equatorial Guinea
112 D13 **Biograd na Moru** It. Zaravecchia. Zadar, SW Croatia 43°57′N 15°27′E
Bioko see Bioco, Isla de
113 F14 **Biokovo** ▲ S Croatia
Biorra see Birr
Bipontium see Zweibrücken
143 W13 **Birag, Kūh-e** ▲ SE Iran
75 O10 **Birāk** var. Brak. C Libya 27°32′N 14°17′E
139 S10 **Bi'r al Islām** Karbalā', C Iraq 32°15′N 43°40′E
154 N11 **Biramitrapur** var. Birmitrapur. Odisha, E India 22°24′N 84°42′E
139 T11 **Bi'r an Nişf** An Najaf, S Iraq 31°22′N 44°07′E
78 L12 **Birao** Vakaga, NE Central African Republic 10°14′N 22°49′E
146 J10 **Birata** Rus. Darganata, Dargan-Ata. Lebap Welaýaty, NE Turkmenistan 40°30′N 62°09′E
158 M6 **Biratar Bulak** well NW China
153 R12 **Birātnagar** Eastern, SE Nepal 26°28′N 87°16′E
165 R5 **Biratori** Hokkaidō, NE Japan 42°35′N 142°07′E
39 S8 **Birch Creek** Alaska, USA 66°17′N 145°54′W
38 M11 **Birch Creek** ↝ Alaska, USA
11 T14 **Birch Hills** Saskatchewan, S Canada 52°58′N 105°22′W
182 M10 **Birchip** Victoria, SE Australia 36°01′S 142°55′E
29 X4 **Birch Lake** ◎ Minnesota, N USA
11 Q11 **Birch Mountains** ▲ Alberta, W Canada
11 V15 **Birch River** Manitoba, S Canada 52°22′N 101°03′W
44 H12 **Birchs Hill** hill W Jamaica
39 R11 **Birchwood** Alaska, USA 61°24′N 149°28′W
188 I5 **Bird Island** island S Northern Mariana Islands
137 N16 **Birecik** Şanlıurfa, S Turkey 37°03′N 37°59′E
152 M10 **Birendranagar** var. Surkhet. Mid Western, W Nepal 28°35′N 81°36′E
Bir es Saba see Be'er Sheva
74 A12 **Bir-Gandouz** SW Western Sahara 21°35′N 16°27′W
153 P12 **Birganj** Central, C Nepal 27°03′N 84°53′E
81 B14 **Biri** ↝ W South Sudan
Bi'r Ibn Hirmās see Al Bi'r
143 U8 **Bīrjand** Khorāsān-e Janūbī, E Iran 32°54′N 59°14′E
Birkaland see Pirkanmaa
139 T11 **Birkat Ḩāmid** well S Iraq
95 F18 **Birkeland** Aust-Agder, S Norway 58°18′N 08°13′E
101 E19 **Birkenfeld** Rheinland-Pfalz, SW Germany 49°39′N 07°10′E
97 K18 **Birkenhead** NW England, United Kingdom 53°24′N 03°02′W
109 W7 **Birkfeld** Steiermark, SE Austria 47°21′N 15°40′E
182 A2 **Birksgate Range** ▲ South Australia
Birlad see Bârlad
145 S15 **Birlik** var. Novotroickoje; prev. Brlik. Zhambyl, SE Kazakhstan 43°39′N 73°45′E
97 K20 **Birmingham** C England, United Kingdom 52°30′N 01°50′W
23 P4 **Birmingham** Alabama, S USA 33°30′N 86°47′W
97 M20 **Birmingham** ✈ C England, United Kingdom 52°27′N 01°46′W
Birmitrapur see Biramitrapur
Bir Moghrein see Bir Mogreïn
76 J4 **Bir Mogreïn** var. Bir Moghrein; prev. Fort-Trinquet. Tiris Zemmour, N Mauritania 25°10′N 11°35′W
191 S4 **Birnie Island** atoll Phoenix Islands, C Kiribati
77 S12 **Birnin Gaouré** var. Birni-Ngaouré. Dosso, SW Niger 12°59′N 03°02′E
Birni-Ngaouré see Birnin Gaouré
77 S12 **Birnin Kebbi** Kebbi, NW Nigeria 12°28′N 04°08′E
77 T12 **Birnin Konni** var. Birni-Nkonni. Tahoua, SW Niger 13°51′N 05°15′E
Birni-Nkonni see Birnin Konni
77 W13 **Birnin Kudu** Jigawa, N Nigeria 11°28′N 09°29′E
123 S16 **Birobidzhan** Yevreyskaya Avtonomnaya Oblast', SE Russian Federation 48°42′N 132°55′E
97 D18 **Birr** var. Parsonstown, Ir. Biorra. C Ireland 53°06′N 07°55′W
183 P4 **Birrie River** ↝ New South Wales/Queensland, SE Australia
108 D7 **Birse** ↝ NW Switzerland
Birsen see Biržai
108 E6 **Birsfelden** Basel Landschaft, NW Switzerland 47°33′N 07°37′E
127 U4 **Birsk** Respublika Bashkortostan, W Russian Federation 55°24′N 55°33′E
119 F14 **Birštonas** Kaunas, C Lithuania 54°37′N 24°00′E
159 P14 **Biru** Xinjiang Uygur Zizhiqu, W China 31°30′N 93°56′E
Biruni see Beruniy
122 L12 **Biryusa** ↝ C Russian Federation
122 L12 **Biryusinsk** Irkutskaya Oblast', C Russian Federation 55°52′N 97°48′E
118 G10 **Biržai** Ger. Birsen. Panevėžys, NE Lithuania 56°12′N 24°47′E
121 P16 **Birżebbuġa** SE Malta 35°50′N 14°32′E
Bisanthe see Tekirdağ
171 R12 **Bisa, Pulau** island Maluku, E Indonesia
37 N17 **Bisbee** Arizona, SW USA 31°27′N 109°55′W
29 O2 **Bisbee** North Dakota, N USA 48°36′N 99°21′W
Biscaia, Baía de see Biscay, Bay of
102 I13 **Biscarrosse et de Parentis, Étang de** ◎ SW France
104 M1 **Biscay, Bay of** Sp. Golfo de Vizcaya, Port. Baía de Biscaia. bay France/Spain
23 Z16 **Biscayne Bay** bay Florida, SE USA
64 M7 **Biscay Plain** undersea feature SE Bay of Biscay 07°15′W 45°00′N
107 N17 **Bisceglie** Puglia, SE Italy 41°14′N 16°31′E
Bischoflack see Škofja Loka
Bischofsburg see Biskupiec
109 Q7 **Bischofshofen** Salzburg, NW Austria 47°25′N 13°13′E
101 P15 **Bischofswerda** Sachsen, E Germany 51°07′N 14°13′E
103 V5 **Bischwiller** Bas-Rhin, NE France 48°46′N 07°52′E
21 T10 **Biscoe** North Carolina, SE USA 35°20′N 79°46′W
194 G5 **Biscoe Islands** island group Antarctica
14 E9 **Biscotasi Lake** ◎ Ontario, S Canada
14 E9 **Biscotasing** Ontario, S Canada 47°16′N 82°04′W
54 J6 **Biscucuy** Portuguesa, NW Venezuela 09°22′N 69°59′W
99 M24 **Bissen** Luxembourg, C Luxembourg 49°47′N 06°04′E
114 K11 **Biser** Haskovo, S Bulgaria 41°52′N 25°59′E
113 D15 **Biševo** It. Busi. island SW Croatia
141 N12 **Bishah, Wādi** dry watercourse C Saudi Arabia
147 U7 **Bishkek** var. Pishpek; prev. Frunze. ● (Kyrgyzstan) Chuyskaya Oblast', N Kyrgyzstan 42°54′N 74°27′E
147 U7 **Bishkek** ✈ Chuyskaya Oblast', N Kyrgyzstan 42°55′N 74°37′E
153 R16 **Bishnupur** West Bengal, NE India 23°05′N 87°20′E
Bisho see Bhisho
35 S9 **Bishop** California, W USA 37°22′N 118°24′W
25 S15 **Bishop** Texas, SW USA 27°36′N 97°49′W
97 L15 **Bishop Auckland** N England, United Kingdom 54°41′N 01°41′W
Bishop's Lynn see King's Lynn
97 O21 **Bishop's Stortford** E England, United Kingdom 51°45′N 00°11′E
21 S12 **Bishopville** South Carolina, SE USA 34°18′N 80°15′W
138 M5 **Bishrī, Jabal** ▲ E Syria
163 U4 **Bishui** Heilongjiang, NE China 52°06′N 123°42′E
81 G17 **Bisina, Lake** prev. Lake Salisbury. ◎ E Uganda
Biskara see Biskra
74 L6 **Biskra** var. Beskra, Biskara. NE Algeria 34°51′N 05°44′E
110 M8 **Biskupiec** Ger. Bischofsburg. Warmińsko-Mazurskie, NE Poland 53°52′N 20°57′E
171 R7 **Bislig** Mindanao, S Philippines 08°10′N 126°19′E
27 X6 **Bismarck** Missouri, C USA 37°46′N 90°37′W
28 M5 **Bismarck** state capital North Dakota, N USA 46°49′N 100°48′W
186 D5 **Bismarck Archipelago** island group NE Papua New Guinea
129 Z16 **Bismarck Plate** tectonic feature W Pacific Ocean
186 D7 **Bismarck Range** ▲ N Papua New Guinea
186 E6 **Bismarck Sea** sea W Pacific Ocean
137 P15 **Bismil** Diyarbakır, SE Turkey 37°53′N 40°38′E
43 N6 **Bismuna, Laguna** lagoon NE Nicaragua
Bisnulok see Phitsanulok
171 R10 **Bisoa, Tanjung** headland Pulau Halmahera, N Indonesia 02°15′N 127°57′E
28 K7 **Bison** South Dakota, N USA 45°31′N 102°27′W
93 H17 **Bispgården** Jämtland, C Sweden 63°00′N 16°40′E
76 G13 **Bissau** ● (Guinea-Bissau) W Guinea-Bissau 11°52′N 15°39′W
76 G13 **Bissau** ✈ W Guinea-Bissau 11°53′N 15°41′W
Bissojohka see Børselv
76 G12 **Bissorã** W Guinea-Bissau 12°16′N 15°35′W
11 O10 **Bistcho Lake** ◎ Alberta, W Canada
22 G5 **Bistineau, Lake** ⊠ Louisiana, S USA
Bistrica see Ilirska Bistrica
116 I9 **Bistrița** Ger. Bistritz, Hung. Beszterce; prev. Nösen. Bistrița-Năsăud, N Romania 47°10′N 24°31′E
116 K10 **Bistrița** Ger. Bistritz. ↝ NE Romania
116 I9 **Bistrița-Năsăud** ◆ county N Romania
Bistritz see Bistrița
Bistritz ober Pernstein see Bystřice nad Pernštejnem
152 L11 **Biswān** Uttar Pradesh, N India 27°30′N 81°00′E
110 M7 **Bisztynek** Warmińsko-Mazurskie, NE Poland 54°05′N 20°53′E
79 E17 **Bitam** Woleu-Ntem, N Gabon 02°05′N 11°30′E
101 D18 **Bitburg** Rheinland-Pfalz, SW Germany 49°58′N 06°31′E
103 U4 **Bitche** Moselle, NE France 49°01′N 07°27′E
78 I11 **Bitkine** Guéra, C Chad 11°59′N 18°13′E
137 R15 **Bitlis** Bitlis, SE Turkey 38°23′N 42°04′E
137 R14 **Bitlis** ◆ province E Turkey
Bitoeng see Bitung
113 N20 **Bitola** Turk. Monastir; prev. Bitolj. S FYR Macedonia 41°01′N 21°22′E
Bitolj see Bitola
107 O17 **Bitonto** anc. Butuntum. Puglia, SE Italy 41°07′N 16°41′E
77 Q13 **Bitou** var. Bittou. SE Burkina Faso 11°19′N 00°17′W
155 C20 **Bitra Island** island Lakshadweep, India, N Indian Ocean
101 M14 **Bitterfeld** Sachsen-Anhalt, E Germany 51°37′N 12°18′E
32 O9 **Bitterroot Range** ▲ Idaho/Montana, NW USA
33 P10 **Bitterroot River** ↝ Montana, NW USA
107 D18 **Bitti** Sardegna, Italy, C Mediterranean Sea 40°30′N 09°31′E
Bittou see Bitou
171 Q11 **Bitung** prev. Bitoeng. Sulawesi, C Indonesia 01°28′N 125°13′E
60 I12 **Bituruna** Paraná, S Brazil 26°11′S 51°34′W
77 Y13 **Biu** Borno, E Nigeria 10°35′N 12°13′E
Biumba see Byumba
164 J13 **Biwa-ko** ◎ Honshū, SW Japan
171 X14 **Biwarlaut** Papua, E Indonesia 05°44′S 138°14′E
27 P10 **Bixby** Oklahoma, C USA 35°56′N 95°52′W
122 J13 **Biya** ↝ S Russian Federation
122 J13 **Biysk** Altayskiy Kray, S Russian Federation 52°34′N 85°09′E
164 H13 **Bizen** Okayama, Honshū, SW Japan 34°45′N 134°10′E
Bizerta see Bizerte
120 K10 **Bizerte** Ar. Banzart, Eng. Bizerta. N Tunisia 37°18′N 09°48′E
105 O2 **Bizkaia** Cast. Vizcaya. ◆ province País Vasco, N Spain
92 G2 **Bjargtangar** headland W Iceland 63°30′N 24°29′W
Bjärnå see Perniö
95 K22 **Bjärnum** Skåne, S Sweden 56°15′N 13°45′E
93 I16 **Bjästa** Västernorrland, C Sweden 63°12′N 18°30′E
113 I14 **Bjelašnica** ▲ SE Bosnia and Herzegovina 43°13′N 18°18′E
112 C10 **Bjelolasica** ▲ NW Croatia 45°13′N 14°56′E
112 F8 **Bjelovar** Hung. Belovár. Bjelovar-Bilogora, N Croatia 45°54′N 16°49′E
112 F8 **Bjelovar-Bilogora** off. Bjelovarsko-Bilogorska Županija. ◆ province NE Croatia
Bjelovarsko-Bilogorska Županija see Bjelovar-Bilogora
92 H10 **Bjerkvik** Nordland, C Norway 68°31′N 16°08′E
95 G21 **Bjerringbro** Midtjylland, NW Denmark 56°23′N 09°40′E
Bjeshkët e Namuna see North Albanian Alps
95 L14 **Björbo** Dalarna, C Sweden 60°28′N 14°44′E
95 I15 **Bjørkelangen** Akershus, S Norway 59°54′N 11°33′E
95 O14 **Björklinge** Uppsala, C Sweden 60°03′N 17°33′E
95 P14 **Björkö-Arholma** Stockholm, C Sweden 59°51′N 19°01′E
93 I14 **Björksele** Västerbotten, N Sweden 64°58′N 18°30′E
93 I16 **Björna** Västernorrland, C Sweden 63°34′N 18°38′E
95 C14 **Bjørnafjorden** fjord S Norway
95 L16 **Björneborg** Värmland, C Sweden 59°13′N 14°15′E
Björneborg see Pori
95 E14 **Bjørnesfjorden** ◎ S Norway
92 M9 **Bjørnevatn** Finnmark, N Norway 69°40′N 29°57′E
197 T13 **Bjørnøya** Eng. Bear Island. island N Norway
93 I15 **Bjurholm** Västerbotten, N Sweden 63°57′N 19°16′E
95 J22 **Bjuv** Skåne, S Sweden 56°05′N 12°57′E
76 M12 **Bla** Ségou, W Mali 12°58′N 05°45′W
181 W8 **Blackall** Queensland, E Australia 24°26′S 145°32′E
29 V2 **Black Bay** lake bay Minnesota, N USA
27 N9 **Black Bear Creek** ↝ Oklahoma, C USA
97 K17 **Blackburn** NW England, United Kingdom 53°45′N 02°29′W
39 T11 **Blackburn, Mount** ▲ Alaska, USA 61°43′N 143°25′W
35 N5 **Black Butte Lake** ⊠ California, W USA
194 J5 **Black Coast** physical region Antarctica
11 Q16 **Black Diamond** Alberta, SW Canada 50°42′N 114°09′W
18 K11 **Black Dome** ▲ New York, NE USA 42°16′N 74°07′W
113 L18 **Black Drin** Alb. Lumi i Drinit të Zi, SCr. Crni Drim. ↝ Albania/FYR Macedonia
29 U4 **Blackduck** Minnesota, N USA 47°45′N 94°33′W
12 D6 **Black Duck** ↝ Ontario, C Canada
33 R14 **Blackfoot** Idaho, NW USA 43°11′N 112°20′W
33 P9 **Blackfoot River** ↝ Montana, NW USA
Black Forest see Schwarzwald
28 J10 **Blackhawk** South Dakota, N USA 44°09′N 103°18′W
28 I10 **Black Hills** ▲ South Dakota/Wyoming, N USA
11 T10 **Black Lake** ◎ Saskatchewan, C Canada
22 G6 **Black Lake** ⊠ Louisiana, S USA
31 Q5 **Black Lake** ◎ Michigan, N USA
18 I7 **Black Lake** ◎ New York, NE USA
26 F7 **Black Mesa** ▲ Oklahoma, C USA 37°00′N 103°07′W
21 P10 **Black Mountain** North Carolina, SE USA 35°37′N 82°19′W
35 P13 **Black Mountain** ▲ California, W USA 35°22′N 120°21′W
37 Q2 **Black Mountain** ▲ Colorado, C USA 40°47′N 107°23′W
96 K1 **Black Mountains** ▲ SE Wales, United Kingdom
36 H10 **Black Mountains** ▲ Arizona, SW USA
21 O7 **Black Mountains** ▲ Kentucky, E USA 36°54′N 82°53′W
33 Q16 **Black Pine Peak** ▲ Idaho, NW USA 42°07′N 113°07′W
97 K17 **Blackpool** NW England, United Kingdom 53°50′N 03°03′W
37 Q14 **Black Range** ▲ New Mexico, SW USA
44 I12 **Black River** W Jamaica 18°02′N 77°52′W
14 J14 **Black River** ↝ Ontario, SE Canada
129 U12 **Black River** Chin. Babian Jiang, Lixian Jiang, Fr. Rivière Noire, Vtn. Sông Đa. ↝ China/Vietnam
44 I12 **Black River** ↝ W Jamaica
39 T7 **Black River** ↝ Alaska, USA
37 N13 **Black River** ↝ Arizona, SW USA
27 X7 **Black River** ↝ Arkansas/Missouri, C USA
22 I7 **Black River** ↝ Louisiana, S USA
31 S8 **Black River** ↝ Michigan, N USA
31 Q5 **Black River** ↝ Michigan, N USA
18 I8 **Black River** ↝ New York, NE USA
21 T13 **Black River** ↝ South Carolina, SE USA
30 J7 **Black River** ↝ Wisconsin, N USA
30 J7 **Black River Falls** Wisconsin, N USA 44°18′N 90°51′W
35 R3 **Black Rock Desert** desert Nevada, W USA
Black Sand Desert see Garagum
21 S7 **Blacksburg** Virginia, NE USA 37°15′N 80°25′W
136 H10 **Black Sea** var. Euxine Sea, Bul. Cherno More, Rom. Marea Neagră, Rus. Chernoye More, Turk. Karadeniz, Ukr. Chorne More. sea Asia/Europe
117 Q10 **Black Sea Lowland** Ukr. Prychornomor'ska Nyzovyna. depression SE Europe
33 S17 **Blacks Fork** ↝ Wyoming, C USA
23 V7 **Blackshear** Georgia, SE USA 31°18′N 82°14′W
23 S6 **Blackshear, Lake** ⊠ Georgia, SE USA
97 A16 **Blacksod Bay** Ir. Cuan an Fhóid Duibh. inlet W Ireland
21 V7 **Blackstone** Virginia, NE USA 37°04′N 78°00′W
77 O14 **Black Volta** var. Borongo, Mouhoun, Moun Hou, Fr. Volta Noire. ↝ W Africa
23 O5 **Black Warrior River** ↝ Alabama, S USA
181 X8 **Blackwater** Queensland, E Australia 23°34′S 148°51′E
97 D20 **Blackwater** Ir. An Abhainn Mhór. ↝ S Ireland
27 T4 **Blackwater River** ↝ Missouri, C USA
21 W7 **Blackwater River** ↝ Virginia, NE USA
Blackwater State see Nebraska
27 N8 **Blackwell** Oklahoma, C USA 36°48′N 97°16′W
25 P7 **Blackwell** Texas, SW USA 32°05′N 100°19′W
99 J15 **Bladel** Noord-Brabant, S Netherlands 51°22′N 05°13′E
Bladenmarkt see Bălăușeri
114 G11 **Blagoevgrad** prev. Gorna Dzhumaya. Blagoevgrad, W Bulgaria 42°01′N 23°05′E
114 G11 **Blagoevgrad** ◆ province SW Bulgaria
123 Q14 **Blagoveshchensk** Amurskaya Oblast', SE Russian Federation 50°19′N 127°30′E
127 V4 **Blagoveshchensk** Respublika Bashkortostan, W Russian Federation 55°03′N 56°01′E
102 I7 **Blain** Loire-Atlantique, NW France 47°26′N 01°47′W
29 V8 **Blaine** Minnesota, N USA 45°09′N 93°13′W
32 H6 **Blaine** Washington, NW USA 48°59′N 122°45′W
11 T15 **Blaine Lake** Saskatchewan, S Canada 52°49′N 106°48′W
29 S14 **Blair** Nebraska, C USA 41°32′N 96°07′W
96 J10 **Blairgowrie** C Scotland, United Kingdom 56°19′N 03°25′W
18 C15 **Blairsville** Pennsylvania, NE USA 40°25′N 79°12′W
116 H11 **Blaj** Ger. Blasendorf, Hung. Balázsfalva. Alba, SW Romania 46°10′N 23°57′E
64 F9 **Blake-Bahama Ridge** undersea feature W Atlantic Ocean 29°00′N 73°30′W
23 S7 **Blakely** Georgia, SE USA 31°22′N 84°55′W
64 E10 **Blake Plateau** var. Blake Terrace. undersea feature W Atlantic Ocean 31°00′N 79°00′W
30 M1 **Blake Point** headland Michigan, N USA 48°11′N 88°25′W
Blake Terrace see Blake Plateau
61 B24 **Blanca, Bahía** bay E Argentina
56 C12 **Blanca, Cordillera** ▲ W Peru
105 T12 **Blanca, Costa** physical region SE Spain
37 S7 **Blanca Peak** ▲ Colorado, C USA 37°34′N 105°29′W
24 I9 **Blanca, Sierra** ▲ Texas, SW USA 31°15′N 105°26′W
120 K9 **Blanc, Cap** headland N Tunisia 37°20′N 09°41′E
Blanc, Cap see Nouâdhibou, Râs
31 R12 **Blanchard River** ↝ Ohio, N USA
182 E8 **Blanche, Cape** headland South Australia 33°03′S 134°10′E
182 J4 **Blanche, Lake** ◎ South Australia
31 R14 **Blanchester** Ohio, N USA 39°17′N 83°59′W
182 J9 **Blanchetown** South Australia 34°21′S 139°36′E
45 U13 **Blanchisseuse** Trinidad, Trinidad and Tobago 10°47′N 61°18′W
103 T11 **Blanc, Mont** It. Monte Bianco. ▲ France/Italy 45°45′N 06°51′E
25 R11 **Blanco** Texas, SW USA 30°06′N 98°25′W
42 K14 **Blanco, Cabo** headland NW Costa Rica 09°34′N 85°06′W
32 D14 **Blanco, Cape** headland Oregon, NW USA 42°49′N 124°33′W
62 H10 **Blanco, Río** ↝ W Argentina
56 F10 **Blanco, Río** ↝ NE Peru
15 O9 **Blanc, Réservoir** ⊠ Québec, SE Canada
21 R7 **Bland** Virginia, NE USA 37°06′N 81°08′W
92 I2 **Blanda** ↝ N Iceland
37 O7 **Blanding** Utah, W USA 37°37′N 109°28′W
105 X5 **Blanes** Cataluña, NE Spain 41°41′N 02°48′E
103 N3 **Blangy-sur-Bresle** Seine-Maritime, N France 49°55′N 01°37′E
111 C18 **Blanice** Ger. Blanitz. ↝ SE Czech Republic
Blanitz see Blanice
99 C16 **Blankenberge** West-Vlaanderen, NW Belgium 51°19′N 03°08′E
101 D17 **Blankenheim** Nordrhein-Westfalen, W Germany 50°26′N 06°41′E
25 R8 **Blanket** Texas, SW USA 31°49′N 98°47′W
55 O3 **Blanquilla, Isla** var. La Blanquilla. island N Venezuela
Blanquilla, La see Blanquilla, Isla
61 F18 **Blanquillo** Durazno, C Uruguay 32°53′S 55°37′W
111 G18 **Blansko** Ger. Blanz. Jihomoravský Kraj, SE Czech Republic 49°22′N 16°39′E
83 N15 **Blantyre** var. Blantyre-Limbe. Southern, S Malawi 15°45′S 35°04′E
83 N15 **Blantyre** ✈ Southern, S Malawi 15°34′S 35°03′E
Blantyre-Limbe see Blantyre
Blanz see Blansko
98 J10 **Blaricum** Noord-Holland, C Netherlands 52°16′N 05°15′E
Blasendorf see Blaj
Blatnitsa see Durankulak
113 F15 **Blato** It. Blatta. Dubrovnik-Neretva, S Croatia 42°57′N 16°47′E
Blatta see Blato
108 E10 **Blatten** Valais, SW Switzerland 46°22′N 08°00′E
101 J20 **Blaufelden** Baden-Württemberg, SW Germany 49°21′N 10°01′E
95 E23 **Blåvands Huk** headland W Denmark 55°33′N 08°04′E
102 G6 **Blavet** ↝ NW France
102 J12 **Blaye** Gironde, SW France 45°08′N 00°40′W
183 R8 **Blayney** New South Wales, SE Australia 33°33′S 149°13′E
65 D25 **Bleaker Island** island SE Falkland Islands
109 T10 **Bled** Ger. Veldes. NW Slovenia 46°23′N 14°06′E
99 D20 **Bléharies** Hainaut, SW Belgium 50°31′N 03°25′E
109 U9 **Bleiburg** Slvn. Pliberk. Kärnten, S Austria 46°36′N 14°49′E
101 L17 **Bleiloch-stausee** ⊠ C Germany
98 H12 **Bleiswijk** Zuid-Holland, W Netherlands 52°01′N 04°32′E
95 L22 **Blekinge** ◆ county S Sweden
14 D17 **Blenheim** Ontario, S Canada 42°20′N 81°59′W
185 K15 **Blenheim** Marlborough, South Island, New Zealand 41°32′S 174°E
99 M15 **Blerick** Limburg, SE Netherlands 51°22′N 06°10′E
Blesae see Blois
25 V13 **Blessing** Texas, SW USA 28°52′N 96°12′W
14 I10 **Bleu, Lac** ◎ Québec, SE Canada
Blibba see Blitta
120 H10 **Blida** var. El Boulaida, El Boulaïda. N Algeria 36°30′N 02°50′E
95 P15 **Blidö** Stockholm, C Sweden 59°37′N 18°55′E
95 K18 **Blidsberg** Västra Götaland, S Sweden 57°55′N 13°30′E
185 A21 **Bligh Sound** sound South Island, New Zealand
187 X14 **Bligh Water** strait NW Fiji
14 D11 **Blind River** Ontario, S Canada 46°12′N 82°59′W
31 R11 **Blissfield** Michigan, N USA 41°49′N 83°51′W
77 R15 **Blitta** prev. Blibba. C Togo 08°19′N 00°59′E
19 O13 **Block Island** island Rhode Island, NE USA
19 O13 **Block Island Sound** sound Rhode Island, NE USA
98 H10 **Bloemendaal** Noord-Holland, W Netherlands 52°23′N 04°39′E
83 H23 **Bloemfontein** var. Mangaung. ● (South Africa-judicial capital) Free State, C South Africa 29°07′S 26°14′E
83 I22 **Bloemhof** North-West, NW South Africa 27°39′S 25°37′E
102 M7 **Blois** anc. Blesae. Loir-et-Cher, C France 47°36′N 01°20′E
98 L8 **Blokzijl** Overijssel, N Netherlands 52°46′N 05°58′E
95 N20 **Blomstermåla** Kalmar, S Sweden 56°58′N 16°19′E
92 I2 **Blönduós** Norðurland Vestra, N Iceland 65°39′N 20°15′W
110 L11 **Błonie** Mazowieckie, C Poland 52°13′N 20°36′E
97 C14 **Bloody Foreland** Ir. Cnoc Fola. headland NW Ireland 55°09′N 08°18′W
31 N15 **Bloomfield** Indiana, N USA 39°01′N 86°58′W
29 X16 **Bloomfield** Iowa, C USA 40°45′N 92°24′W
27 Y8 **Bloomfield** Missouri, C USA 36°54′N 89°58′W
37 P9 **Bloomfield** New Mexico, SW USA 36°42′N 108°00′W
25 U7 **Blooming Grove** Texas, SW USA 32°05′N 96°43′W
29 W10 **Blooming Prairie** Minnesota, N USA 43°52′N 93°03′W
30 L13 **Bloomington** Illinois, N USA 40°28′N 88°59′W
31 O15 **Bloomington** Indiana, N USA 39°10′N 86°31′W
29 V9 **Bloomington** Minnesota, N USA 44°50′N 93°18′W
25 U13 **Bloomington** Texas, SW USA 28°39′N 96°53′W
18 H14 **Bloomsburg** Pennsylvania, NE USA 40°59′N 76°27′W
181 X7 **Bloomsbury** Queensland, NE Australia 20°47′S 148°35′E
169 R16 **Blora** Jawa, C Indonesia 06°55′S 111°29′E
18 G12 **Blossburg** Pennsylvania, NE USA 41°38′N 77°00′W
25 V5 **Blossom** Texas, SW USA 33°39′N 95°23′W
123 T5 **Blossom, Mys** headland Ostrov Vrangelya, NE Russian Federation 70°49′N 178°49′E
23 R8 **Blountstown** Florida, SE USA 30°26′N 85°03′W
21 P8 **Blountville** Tennessee, S USA 36°31′N 82°19′W
21 Q9 **Blowing Rock** North Carolina, SE USA 36°15′N 81°53′W
108 J8 **Bludenz** Vorarlberg, W Austria 47°10′N 09°50′E
36 L6 **Blue Bell Knoll** ▲ Utah, W USA 38°11′N 111°31′W
23 Y12 **Blue Cypress Lake** ◎ Florida, SE USA
29 U11 **Blue Earth** Minnesota, N USA 43°38′N 94°06′W
21 Q7 **Bluefield** Virginia, NE USA 37°15′N 81°16′W
21 R7 **Bluefield** West Virginia, NE USA 37°16′N 81°13′W
43 N10 **Bluefields** Región Autónoma Atlántico Sur, SE Nicaragua 12°01′N 83°47′W
43 N10 **Bluefields, Bahía de** bay W Caribbean Sea
29 Z14 **Blue Grass** Iowa, C USA 41°30′N 90°46′W
Bluegrass State see Kentucky
Blue Hen State see Delaware
19 S7 **Blue Hill** Maine, NE USA 44°25′N 68°36′W
29 P16 **Blue Hill** Nebraska, C USA 40°19′N 98°27′W
30 J5 **Blue Hills** hill range Wisconsin, N USA
34 L3 **Blue Lake** California, W USA 40°52′N 124°00′W
Blue Law State see Connecticut
37 Q6 **Blue Mesa Reservoir** ⊠ Colorado, C USA
27 S12 **Blue Mountain** ▲ Arkansas, C USA 34°42′N 94°04′W
19 O6 **Blue Mountain** ▲ New Hampshire, NE USA 44°48′N 71°26′W
18 K8 **Blue Mountain** ▲ New York, NE USA 43°52′N 74°24′W
18 H15 **Blue Mountain** ridge Pennsylvania, NE USA
44 H10 **Blue Mountain Peak** ▲ E Jamaica 18°02′N 76°34′W
183 S8 **Blue Mountains** ▲ New South Wales, SE Australia
32 L11 **Blue Mountains** ▲ Oregon/Washington, NW USA
80 G12 **Blue Nile** ◆ state E Sudan
80 H12 **Blue Nile** var. Abai, Bahr el, Azraq, Amh. Ābay Wenz, Ar. An Nīl al Azraq. ↝ Ethiopia/Sudan
8 J7 **Bluenose Lake** ◎ Nunavut, NW Canada
27 O3 **Blue Rapids** Kansas, C USA 39°39′N 96°38′W
23 S1 **Blue Ridge** Georgia, SE USA 34°51′N 84°19′W
17 S11 **Blue Ridge** var. Blue Ridge Mountains. ▲ North Carolina/Virginia, USA
23 S1 **Blue Ridge Lake** ⊠ Georgia, SE USA
Blue Ridge Mountains see Blue Ridge
11 N15 **Blue River** British Columbia, SW Canada 52°03′N 119°21′W
27 O12 **Blue River** ↝ Oklahoma, C USA
27 R4 **Blue Springs** Missouri, C USA 39°01′N 94°16′W
21 R6 **Bluestone Lake** ⊠ West Virginia, NE USA
185 C25 **Bluff** Southland, South Island, New Zealand 46°36′S 168°22′E
37 O8 **Bluff** Utah, W USA 37°15′N 109°33′W
21 P8 **Bluff City** Tennessee, S USA 36°28′N 82°15′W
65 E24 **Bluff Cove** East Falkland, Falkland Islands 51°45′S 58°11′W
25 S7 **Bluff Dale** Texas, SW USA 32°18′N 98°01′W
183 N15 **Bluff Hill Point** headland Tasmania, SE Australia 41°03′S 144°35′E
31 Q12 **Bluffton** Indiana, N USA 40°44′N 85°10′W
31 R12 **Bluffton** Ohio, N USA 40°54′N 83°53′W
25 T7 **Blum** Texas, SW USA 32°08′N 97°24′W
101 G24 **Blumberg** Baden-Württemberg, SW Germany 47°48′N 08°31′E
60 K13 **Blumenau** Santa Catarina, S Brazil 26°55′S 49°07′W
29 N9 **Blunt** South Dakota, N USA 44°30′N 99°58′E
32 H15 **Bly** Oregon, NW USA 42°22′N 121°04′W
39 R13 **Blying Sound** sound Alaska, USA
97 M14 **Blyth** N England, United Kingdom 55°07′N 01°30′W
35 Y16 **Blythe** California, W USA 33°35′N 114°36′W
27 Y9 **Blytheville** Arkansas, C USA 35°56′N 89°55′W
117 V7 **Blyznyuky** Kharkivs'ka Oblast', E Ukraine 48°51′N 36°32′E
95 G16 **Bø** Telemark, S Norway 59°24′N 09°04′E
76 I15 **Bo** S Sierra Leone 07°58′N 11°45′W
171 O4 **Boac** Marinduque, N Philippines 13°26′N 121°50′E
42 K10 **Boaco** Boaco, S Nicaragua 12°28′N 85°45′W
42 J10 **Boaco** ◆ department C Nicaragua
79 I15 **Boali** Ombella-Mpoko, SW Central African Republic 04°52′N 18°00′E
Boalsert see Bolsward
31 V12 **Boardman** Ohio, N USA 41°01′N 80°39′W
32 J11 **Boardman** Oregon, NW USA 45°50′N 119°42′W
14 F13 **Boat Lake** ◎ Ontario, S Canada
58 F10 **Boa Vista** state capital Roraima, NW Brazil 02°51′N 60°43′W
76 D9 **Boa Vista** island Ilhas de Barlavento, E Cape Verde
23 Q2 **Boaz** Alabama, S USA 34°12′N 86°10′W
160 L15 **Bobai** Guangxi Zhuangzu Zizhiqu, S China 22°09′N 109°57′E
172 J1 **Bobaomby, Tanjona** Fr. Cap d'Ambre. headland N Madagascar 11°58′S 49°13′E
155 M14 **Bobbili** Andhra Pradesh, E India 18°32′N 83°29′E
106 D9 **Bobbio** Emilia-Romagna, C Italy 44°48′N 09°27′E
14 I14 **Bobcaygeon** Ontario, SE Canada 44°32′N 78°33′W
Bober see Bóbr
103 O5 **Bobigny** Seine-St-Denis, N France 48°55′N 02°27′E
77 N13 **Bobo-Dioulasso** SW Burkina Faso 11°12′N 04°21′W
110 G8 **Bobolice** Ger. Bublitz. Zachodnio-pomorskie, NW Poland 53°56′N 16°37′E
83 J19 **Bobonong** Central, E Botswana 21°58′S 28°26′E
171 R11 **Bobopayo** Pulau Halmahera, E Indonesia 01°07′N 127°26′E
113 J15 **Bobotov Kuk** ▲ N Montenegro 43°06′N 19°00′E
114 G10 **Bobov Dol** var. Bobovdol. Kyustendil, W Bulgaria 42°21′N 22°59′E
Bobovdol see Bobov Dol
119 M15 **Bobr** Minskaya Voblasts', NW Belarus 54°20′N 29°16′E
119 M15 **Bobr** ↝ C Belarus
111 E14 **Bóbr** Eng. Bobrawa, Ger. Bober. ↝ SW Poland
Bobrawa see Bóbr
Bobrik see Bobryk
Bobrinets see Bobrynets'
Bobrka/Bóbrka see Bibrka
126 L8 **Bobrov** Voronezhskaya Oblast', W Russian Federation 51°10′N 40°03′E
117 Q4 **Bobrovytsya** Chernihivs'ka Oblast', N Ukraine 50°43′N 31°24′E
Bobruysk see Babruysk
119 J19 **Bobryk** Rus. Bobrik. ↝ SW Belarus
117 Q8 **Bobrynets'** Rus. Bobrinets. Kirovohrads'ka Oblast', C Ukraine 48°02′N 32°10′E
14 K14 **Bobs Lake** ◎ Ontario, SE Canada
54 I6 **Bobures** Zulia, NW Venezuela 09°15′N 71°10′W
42 H1 **Boca Bacalar Chico** headland N Belize 15°05′N 82°12′W
112 G11 **Bočac** ◆ Republika Srpska, NW Bosnia and Herzegovina
41 R14 **Boca del Río** Veracruz-Llave, S Mexico 19°08′N 96°08′W
55 O4 **Boca de Pozo** Nueva Esparta, NE Venezuela 11°00′N 64°23′W
59 C15 **Boca do Acre** Amazonas, N Brazil 08°45′S 67°23′W
55 N12 **Boca Mavaca** Amazonas, S Venezuela 02°30′N 65°11′W
79 G14 **Bocaranga** Ouham-Pendé, W Central African Republic 07°07′N 15°40′E
23 Z15 **Boca Raton** Florida, SE USA 26°22′N 80°05′W
43 P14 **Bocas del Toro** Bocas del Toro, NW Panama 09°20′N 82°15′W
43 P15 **Bocas del Toro** off. Provincia de Bocas del Toro. ◆ province NW Panama
43 P15 **Bocas del Toro, Archipiélago de** island group NW Panama
Bocas del Toro, Provincia de see Bocas del Toro
42 L7 **Bocay** Jinotega, N Nicaragua 14°19′N 85°08′W
105 N6 **Boceguillas** Castilla y León, N Spain 41°20′N 03°39′W
Bocheykovo see Bacheykava
111 L17 **Bochnia** Małopolskie, SE Poland 49°58′N 20°27′E
99 K16 **Bocholt** Limburg, NE Belgium 51°10′N 05°37′E
101 D14 **Bocholt** Nordrhein-Westfalen, W Germany 51°50′N 06°37′E
101 E15 **Bochum** Nordrhein-Westfalen, W Germany 51°29′N 07°13′E
103 Y15 **Bocognano** Corse, France, C Mediterranean Sea 42°04′N 09°03′E
54 I6 **Boconó** Trujillo, NW Venezuela 09°17′N 70°17′W

◆ Country
● Country Capital
◇ Dependent Territory
○ Dependent Territory Capital
◆ Administrative Regions
✈ International Airport
▲ Mountain
▲ Mountain Range
Volcano
↝ River
◎ Lake
⊠ Reservoir

116 F12 **Bocşa** *Ger.* Bokschen, *Hung.* Boksánbánya. Caraş-Severin, SW Romania 45°23′N 21°47′E
79 H15 **Boda** Lobaye, SW Central African Republic 04°17′N 17°25′E
94 L12 **Boda** Dalarna, C Sweden 61°00′N 15°15′E
95 O20 **Böda** Kalmar, S Sweden 57°16′N 17°04′E
95 L19 **Bodafors** Jönköping, S Sweden 57°30′N 14°40′E
123 O12 **Bodaybo** Irkutskaya Oblast', E Russian Federation 57°52′N 114°05′E
22 G5 **Bodcau, Bayou** *var.* Bodcau Creek. ≈ Louisiana, S USA
Bodcau Creek *see* Bodcau, Bayou
44 D8 **Bodden Town** *var.* Boddentown. Grand Cayman, SW Cayman Islands 19°20′N 81°14′W
Boddentown *see* Bodden Town
101 K14 **Bode** ≈ C Germany
34 L7 **Bodega Head** *headland* California, W USA 38°16′N 123°04′W
Bodegas *see* Babahoyo
98 H11 **Bodegraven** Zuid-Holland, C Netherlands 52°05′N 04°45′E
78 H8 **Bodélé** *depression* W Chad
92 J13 **Boden** Norrbotten, N Sweden 65°50′N 21°44′E
Bodensee *see* Constance, Lake, C Europe
65 M15 **Bode Verde Fracture Zone** *tectonic feature* E Atlantic Ocean
155 H14 **Bodhan** Telangana, C India 18°40′N 77°51′E
Bodī *see* Jinst
155 H22 **Bodināyakkanūr** Tamil Nādu, SE India 10°02′N 77°18′E
108 H10 **Bodio** Ticino, S Switzerland 46°23′N 08°55′E
Bodjonegoro *see* Bojonegoro
97 I24 **Bodmin** SW England, United Kingdom 50°29′N 04°43′W
97 I24 **Bodmin Moor** *moorland* SW England, United Kingdom
92 G12 **Bodø** Nordland, C Norway 67°17′N 14°22′E
59 H20 **Bodoquena, Serra da** ▲▲ SW Brazil
136 B16 **Bodrum** Muğla, SW Turkey 37°01′N 27°28′E
Bodzafordulő *see* Întorsura Buzăului
99 L14 **Boekel** Noord-Brabant, SE Netherlands 51°35′N 05°42′E
Boeloekoemba *see* Bulukumba
103 Q11 **Boën** Loire, E France 45°45′N 04°01′E
79 K18 **Boende** Equateur, C Dem. Rep. Congo 0°12′S 20°54′E
25 R11 **Boerne** Texas, SW USA 29°47′N 98°44′W
Boeroe *see* Buru, Pulau
Boetoeng *see* Buton, Pulau
22 I5 **Boeuf River** ≈ Arkansas/Louisiana, S USA
76 H14 **Boffa** W Guinea 10°12′N 14°02′W
Bó Finne, Inis *see* Inishbofin
Boga *see* Bogë
166 L9 **Bogale** Ayeyawady, SW Myanmar (Burma) 16°16′N 95°21′E
22 L8 **Bogalusa** Louisiana, S USA 30°47′N 89°51′W
77 Q12 **Bogandé** C Burkina Faso 13°02′N 00°08′W
79 I15 **Bogangolo** Ombella-Mpoko, C Central African Republic 05°36′N 18°17′E
183 Q7 **Bogan River** ≈ New South Wales, SE Australia
25 W5 **Bogata** Texas, SW USA 33°28′N 95°12′W
111 D14 **Bogatynia** *Ger.* Reichenau. Dolnośląskie, SW Poland 50°53′N 14°55′E
136 K13 **Boğazlıyan** Yozgat, C Turkey 39°13′N 35°17′E
79 J17 **Bogbonga** Equateur, NW Dem. Rep. Congo 01°36′N 19°24′E
158 J14 **Bogcang Zangbo** ≈ W China
162 I9 **Bogd** *var.* Horiult. Bayanhongor, C Mongolia 45°09′N 100°50′E
162 J10 **Bogd** *var.* Hovd. Övörhangay, C Mongolia 44°43′N 102°08′E
158 L5 **Bogda Feng** ▲ NW China 43°51′N 88°14′E
114 I9 **Bogdan** ▲ C Bulgaria 42°37′N 24°28′E
113 Q20 **Bogdanci** SE FYR Macedonia 41°12′N 22°34′E
158 M5 **Bogda Shan** *var.* Po-ko-to Shan. ▲▲ NW China
113 K17 **Bogë** *var.* Boga. Shkodër, N Albania 42°25′N 19°38′E
Bogeda'er *see* Wenquan
Bogendorf *see* Łuków
95 G23 **Bogense** Syddtjylland, C Denmark 55°34′N 10°06′E
183 T3 **Boggabilla** New South Wales, SE Australia 28°37′S 150°21′E
183 S6 **Boggabri** New South Wales, SE Australia 30°44′S 150°00′E
186 D6 **Bogia** Madang, N Papua New Guinea 04°16′S 144°56′E
97 N23 **Bognor Regis** SE England, United Kingdom 50°47′N 00°41′W
Bogodukhov *see* Bohodukhiv
181 V15 **Bogong, Mount** ▲ Victoria, SE Australia 36°43′S 147°19′E
169 O16 **Bogor** *Dut.* Buitenzorg. Jawa, C Indonesia 06°34′S 106°45′E
126 L5 **Bogoroditsk** Tul'skaya Oblast', W Russian Federation 53°46′N 38°09′E
127 O3 **Bogorodsk** Nizhegorodskaya Oblast', W Russian Federation 56°06′N 43°29′E
Bogorodskoje *see* Bogorodskoye
123 S12 **Bogorodskoye** Khabarovskiy Kray, SE Russian Federation 52°22′N 140°33′E
125 R15 **Bogorodskoye** *var.* Bogorodskoje. Kirovskaya Oblast', NW Russian Federation 57°50′N 50°41′E
54 F10 **Bogotá** *prev.* Santa Fe, Santa Fe de Bogotá. ● (Colombia) Cundinamarca, C Colombia 04°38′N 74°05′W
153 T14 **Bogra** Rajshahi, N Bangladesh 24°52′N 89°28′E
Bogschan *see* Boldu
122 L12 **Boguchany** Krasnoyarskiy Kray, C Russian Federation 58°20′N 97°20′E
126 M9 **Boguchar** Voronezhskaya Oblast', W Russian Federation 49°57′N 40°34′E
76 H10 **Bogué** Brakna, SW Mauritania 16°36′N 14°15′W
22 K8 **Bogue Chitto** ≈ Louisiana/Mississippi, S USA
Bogushëvsk *see* Bahushewsk
Boguslav *see* Bohuslav
44 K12 **Bog Walk** C Jamaica 18°06′N 77°01′W
161 Q3 **Bo Hai** *var.* Gulf of Chihli. *gulf* NE China
161 R3 **Bohai Haixia** *strait* NE China
161 Q3 **Bohai Wan** *bay* NE China
111 C17 **Bohemia** *Cz.* Čechy, *Ger.* Böhmen. W Czech Republic
111 B18 **Bohemian Forest** *Cz.* Ceský Les, Šumava, *Ger.* Böhmerwald. ▲▲ C Europe
Bohemian-Moravian Highlands *see* Českomoravská Vrchovina
77 R16 **Bohicon** S Benin 07°14′N 02°04′E
109 S11 **Bohinjska Bistrica** *Ger.* Wocheiner Feistritz. NW Slovenia 46°16′N 13°55′E
Bohkká *see* Pokka
Böhmen *see* Bohemia
Böhmerwald *see* Bohemian Forest
Böhmisch-Krumau *see* Český Krumlov
Böhmisch-Leipa *see* Česká Lípa
Böhmisch-Mährische Höhe *see* Českomoravská Vrchovina
Böhmisch-Trübau *see* Česká Třebová
117 U5 **Bohodukhiv** *Rus.* Bogodukhov. Kharkivs'ka Oblast', E Ukraine 50°10′N 35°32′E
171 Q6 **Bohol** *island* C Philippines
171 Q7 **Bohol Sea** *var.* Mindanao Sea. *sea* S Philippines
116 I7 **Bohorodchany** Ivano-Frankivs'ka Oblast', W Ukraine 48°46′N 24°31′E
Böhöt *see* Öndörshil
158 K6 **Bohu** *var.* Bagrax. Xinjiang Uygur Zizhiqu, NW China 42°00′N 86°28′E
111 I17 **Bohumín** *Ger.* Oderberg; *prev.* Neuoderberg, Nový Bohumín. Moravskoslezský Kraj, E Czech Republic 49°55′N 18°20′E
117 P6 **Bohuslav** *Rus.* Boguslav. Kyyivs'ka Oblast', N Ukraine 49°33′N 30°53′E
58 F11 **Boiaçu** Roraima, N Brazil 0°27′S 61°46′W
107 K16 **Boiano** Molise, C Italy 41°28′N 14°28′E
15 R8 **Boilleau** Québec, SE Canada 48°06′N 70°49′W
59 O17 **Boipeba, Ilha de** *island* SE Brazil
104 G3 **Boiro** Galicia, NW Spain 42°39′N 08°53′W
31 Q5 **Bois Blanc Island** *island* Michigan, N USA
29 R7 **Bois de Sioux River** ≈ Minnesota, N USA
33 N14 **Boise** *var.* Boise City. *state capital* Idaho, NW USA 43°39′N 116°14′W
26 G8 **Boise City** Oklahoma, C USA 36°44′N 102°31′W
Boise City *see* Boise
33 N14 **Boise River, Middle Fork** ≈ Idaho, NW USA
Bois, Lac des *see* Woods, Lake of the
Bois-le-Duc *see* 's-Hertogenbosch
11 W17 **Boissevain** Manitoba, S Canada 49°14′N 100°02′W
15 T7 **Boisvert, Pointe au** *headland* Québec, SE Canada 48°34′N 69°07′W
100 K10 **Boizenburg** Mecklenburg-Vorpommern, N Germany 53°23′N 10°43′E
Bojador *see* Boujdour
113 K18 **Bojana** *Alb.* Bunë. ≈ Albania/Montenegro *see also* Bunë
Bojana *see* Bunë, Lumi i
143 S3 **Bojnūrd** *var.* Bujnurd. Khorāsān-e Shemālī, N Iran 37°31′N 57°24′E
169 R16 **Bojonegoro** *prev.* Bodjonegoro. Jawa, C Indonesia 07°06′S 111°50′E
189 T1 **Bokaak Atoll** *var.* Bokak, Taongi. *atoll* Ratak Chain, NE Marshall Islands
Bokak *see* Bokaak Atoll
146 K8 **Boʻkantov Togʻlari** *Rus.* Gory Bukantau. ▲▲ N Uzbekistan
153 Q15 **Bokāro** Jhārkhand, N India 23°46′N 85°55′E
79 I18 **Bokatola** Equateur, NW Dem. Rep. Congo 0°37′S 18°45′E
76 H13 **Boké** W Guinea 10°56′N 14°18′W
Bokhara *see* Buxoro
183 Q4 **Bokharra River** ≈ New South Wales/Queensland, SE Australia
147 X8 **Bokonbayevo** *Kir.* Kajisay; *prev.* Kadzhi-Say. Issyk-Kul'skaya Oblast', NE Kyrgyzstan 42°07′N 76°59′E
78 H11 **Bokoro** Hadjer-Lamis, W Chad 12°23′N 17°03′E
79 K19 **Bokota** Equateur, NW Dem. Rep. Congo 0°56′S 22°24′E
167 N13 **Bokpyin** Taninthayi, S Myanmar (Burma) 11°16′N 98°47′E
Boksánbánya/Bokschen *see* Bocşa
83 F21 **Bokspits** Kgalagadi, SW Botswana 26°50′S 20°41′E
79 K18 **Bokungu** Equateur, C Dem. Rep. Congo 0°41′S 22°19′E
146 F12 **Bokurdak** *Rus.* Bakhardok. Ahal Welaýaty, C Turkmenistan 38°51′N 58°34′E
78 G10 **Bol** Lac, W Chad 13°27′N 14°40′E
76 G13 **Bolama** SW Guinea-Bissau 11°35′N 15°30′W
Bolangir *see* Balāngīr
Bolanos *see* Bolanos, Mount, Guam
Bolaños *see* Bolaños de Calatrava, Spain
105 N11 **Bolaños de Calatrava** *var.* Bolaños. Castilla-La Mancha, C Spain 38°55′N 03°39′W
188 B17 **Bolanos, Mount** *var.* Bolanos. ▲ S Guam 13°18′N 144°41′E
40 L12 **Bolaños, Río** ≈ C Mexico
115 M14 **Bolayır** Çanakkale, NW Turkey 40°31′N 26°46′E
102 L3 **Bolbec** Seine-Maritime, N France 49°34′N 00°31′E
116 L13 **Boldu** *var.* Bogschan. Buzău, SE Romania 45°18′N 27°15′E
146 H8 **Boldumsaz** *prev.* Kalinin, Kalininsk, Porsy. Daşoguz Welaýaty, N Turkmenistan 42°12′N 59°33′E
158 I4 **Bole** *var.* Bortala. Xinjiang Uygur Zizhiqu, NW China 44°52′N 82°06′E
77 O15 **Bole** NW Ghana 09°02′N 02°29′W
79 J19 **Boleko** Equateur, W Dem. Rep. Congo 01°27′S 19°52′E
111 E14 **Bolesławiec** *Ger.* Bunzlau. Dolnośląskie, SW Poland 51°16′N 15°34′E
127 R4 **Bolgar** *prev.* Kuybyshev. Respublika Tatarstan, W Russian Federation 54°58′N 49°03′E
77 P14 **Bolgatanga** N Ghana 10°45′N 00°52′W
Bolgrad *see* Bolhrad
117 N12 **Bolhrad** *Rus.* Bolgrad. Odes'ka Oblast', SW Ukraine 45°42′N 28°35′E
163 Y8 **Boli** Heilongjiang, NE China 45°45′N 130°32′E
79 I19 **Bolia** Bandundu, W Dem. Rep. Congo 01°34′S 18°24′E
93 J14 **Boliden** Västerbotten, N Sweden 64°52′N 20°20′E
171 T13 **Bolifar** Pulau Seram, E Indonesia 03°08′S 130°34′E
171 N2 **Bolinao** Luzon, N Philippines 16°22′N 119°52′E
54 C12 **Bolívar** Cauca, SW Colombia 01°52′N 76°56′W
27 T6 **Bolivar** Missouri, C USA 37°37′N 93°25′W
20 F10 **Bolivar** Tennessee, S USA 35°17′N 88°59′W
54 F7 **Bolívar** *off.* Departmento de Bolívar. ◈ *province* N Colombia
56 A13 **Bolívar** ◈ *province* C Ecuador
55 N9 **Bolívar** *off.* Estado Bolívar. ◈ *state* SE Venezuela
Bolívar, Departmento de *see* Bolívar
Bolívar, Estado *see* Bolívar
25 X12 **Bolivar Peninsula** *headland* Texas, SW USA 29°26′N 94°41′W
54 I6 **Bolívar, Pico** ▲ W Venezuela 08°33′N 71°05′W
57 K17 **Bolivia** *off.* Plurinational State of Bolivia; *prev.* Republic of Bolivia. ◆ *republic* W South America
Bolivia, Plurinatinoal State of *see* Bolivia
Bolivia, Republic of *see* Bolivia
112 O13 **Boljevac** Serbia, E Serbia 43°50′N 21°57′E
Bolkenhain *see* Bolków
126 J5 **Bolkhov** Orlovskaya Oblast', W Russian Federation 53°28′N 36°00′E
111 F14 **Bolków** *Ger.* Bolkenhain. Dolnośląskie, SW Poland 50°55′N 15°49′E
182 K3 **Bollards Lagoon** South Australia 28°58′S 140°52′E
103 R14 **Bollène** Vaucluse, SE France 44°16′N 04°45′E
94 N12 **Bollnäs** Gävleborg, C Sweden 61°18′N 16°27′E
181 W10 **Bollon** Queensland, C Australia 28°07′S 147°28′E
192 L12 **Bollons Tablemount** *undersea feature* S Pacific Ocean 49°40′S 176°10′W
93 H17 **Bollstabruk** Västernorrland, C Sweden 63°00′N 17°41′E
Bolluilos de Par del Condado *see* Bollulos Par del Condado
104 J14 **Bollulos Par del Condado** *var.* Bolluilos de Par del Condado. Andalucía, S Spain 37°20′N 06°32′W
95 K21 **Bolmen** ◎ S Sweden
137 T10 **Bolnisi** S Georgia 41°28′N 44°34′E
79 H19 **Bolobo** Bandundu, W Dem. Rep. Congo 02°10′S 16°17′E
106 G10 **Bologna** Emilia-Romagna, N Italy 44°30′N 11°20′E
124 I15 **Bologoye** Tverskaya Oblast', W Russian Federation 57°54′N 34°04′E
79 J18 **Bolomba** Equateur, NW Dem. Rep. Congo 0°27′N 19°13′E
41 X13 **Bolónchén de Rejón** *var.* Bolonchén de Rejón. Campeche, SE Mexico 20°00′N 89°34′W
114 J13 **Boloústra, Akrotírio** *headland* NE Greece 40°56′N 24°58′E
167 L8 **Bolovén, Phouphiang** *Fr.* Plateau des Bolovens. *plateau* S Laos
Bolovens, Plateau des *see* Bolovén, Phouphiang
106 H13 **Bolsena** Lazio, C Italy 42°39′N 11°59′E
107 G14 **Bolsena, Lago di** ◎ C Italy
126 B3 **Bol'shakovo** *Ger.* Kreuzingen; *prev.* Gross-Skaisgirren. Kaliningradskaya Oblast', W Russian Federation 54°53′N 21°38′E
Bol'shaya Berëstovitsa *see* Vyalikaya Byerastavitsa
127 S7 **Bol'shaya Chernigovka** Samarskaya Oblast', W Russian Federation 52°07′N 50°49′E
127 S7 **Bol'shaya Glushitsa** Samarskaya Oblast', W Russian Federation 52°22′N 50°29′E
124 J4 **Bol'shaya Imandra, Ozero** ◎ NW Russian Federation
Bol'shaya Khobda *see* Kobda
126 M12 **Bol'shaya Martynovka** Rostovskaya Oblast', SW Russian Federation 47°19′N 41°40′E
122 K12 **Bol'shaya Murta** Krasnoyarskiy Kray, C Russian Federation 56°51′N 93°10′E
125 V4 **Bol'shaya Rogovaya** ≈ NW Russian Federation
125 U7 **Bol'shaya Synya** ≈ NW Russian Federation
145 V9 **Bol'shaya Vladimirovka** Vostochnyy Kazakhstan, E Kazakhstan 50°53′N 79°29′E
123 V11 **Bol'sheretsk** Kamchatskiy Kray, E Russian Federation 52°20′N 156°24′E
127 W3 **Bol'sheust'ikinskoye** Respublika Bashkortostan, W Russian Federation 56°00′N 58°13′E
Bol'shevik *see* Bal'shavik
122 L5 **Bol'shevik, Ostrov** *island* Severnaya Zemlya, N Russian Federation
125 U4 **Bol'shezemel'skaya Tundra** *physical region* NW Russian Federation
144 J13 **Bol'shiye Barsuki, Peski** *desert* SW Kazakhstan
123 T7 **Bol'shoy Anyuy** ≈ NE Russian Federation
123 N7 **Bol'shoy Begichev, Ostrov** *island* NE Russian Federation
123 S15 **Bol'shoye Kamen'** Primorskiy Kray, SE Russian Federation 43°06′N 132°21′E
127 O4 **Bol'shoye Murashkino** Nizhegorodskaya Oblast', W Russian Federation 55°46′N 44°48′E
127 W4 **Bol'shoy Iremel'** ▲ W Russian Federation 54°31′N 58°47′E
127 R7 **Bol'shoy Irgiz** ≈ W Russian Federation
123 Q6 **Bol'shoy Lyakhovskiy, Ostrov** *island* NE Russian Federation
123 Q11 **Bol'shoy Nimnyr** Respublika Sakha (Yakutiya), NE Russian Federation 57°55′N 125°34′E
Bol'shoy Rozhan *see* Vyaliki Rozhan
Bol'shoy Uzen' *see* Karaozen
40 K6 **Bolsón de Mapimí** ▲▲ NW Mexico
98 K6 **Bolsward** *Fris.* Boalsert. Fryslân, N Netherlands 53°04′N 05°31′E
105 T4 **Boltaña** Aragón, NE Spain 42°28′N 00°02′E
14 G15 **Bolton** Ontario, S Canada 43°52′N 79°45′W
97 K17 **Bolton** *prev.* Bolton-le-Moors. NW England, United Kingdom 53°35′N 02°26′W
21 V12 **Bolton** North Carolina, SE USA 34°22′N 78°26′W
Bolton-le-Moors *see* Bolton
136 G11 **Bolu** Bolu, NW Turkey 40°45′N 31°38′E
136 G11 **Bolu** ◈ *province* NW Turkey
186 G9 **Bolubolu** Goodenough Island, S Papua New Guinea 09°22′S 150°22′E
92 H1 **Bolungarvík** Vestfirðir, NW Iceland 66°09′N 23°17′W
159 O10 **Boluntay** Qinghai, W China 36°30′N 92°11′E
159 P8 **Boluozhuanjing**, Aksay Kazakzu Zizhixian. Gansu, N China 39°25′N 94°09′E
136 F14 **Bolvadin** Afyon, W Turkey 38°43′N 31°02′E
114 M10 **Bolyarovo** *prev.* Pashkeni. Yambol, E Bulgaria 42°09′N 26°49′E
106 G6 **Bolzano** *Ger.* Bozen; *anc.* Bauzanum. Trentino-Alto Adige, N Italy 46°30′N 11°22′E
79 F22 **Boma** Bas-Congo, W Dem. Rep. Congo 05°42′S 13°05′E
183 R12 **Bombala** New South Wales, SE Australia 36°54′S 149°15′E
104 F10 **Bombarral** Leiria, C Portugal 39°15′N 09°09′W
Bombay *see* Mumbai
171 U13 **Bomberai, Semenanjung** *cape* Papua Barat, E Indonesia
81 F18 **Bombo** S Uganda 0°36′N 32°33′E
162 I8 **Bömbögör** *var.* Dzadgay. Bayanhongor, C Mongolia 46°12′N 99°29′E
79 I17 **Bomboma** Equateur, NW Dem. Rep. Congo 02°23′N 19°03′E
81 I19 **Bomet** ◈ *county* W Kenya
59 I14 **Bom Futuro** Pará, N Brazil 06°27′S 54°44′W
159 Q15 **Bomi** *var.* Bowo, Zhamo. Xizang Zizhiqu, W China 29°43′N 96°12′E
79 N17 **Bomili** Orientale, NE Dem. Rep. Congo 01°45′N 27°01′E
59 N17 **Bom Jesus da Lapa** Bahia, E Brazil 13°16′S 43°23′W
60 Q8 **Bom Jesus do Itabapoana** Rio de Janeiro, SE Brazil 21°07′S 41°43′W
95 C15 **Bømlafjorden** *fjord* S Norway
95 B15 **Bømlo** *island* S Norway
12 Q12 **Bomnak** Amurskaya Oblast', SE Russian Federation 54°43′N 128°50′E
79 I17 **Bomongo** Equateur, NW Dem. Rep. Congo 01°22′N 18°21′E
61 K14 **Bom Retiro** Santa Catarina, S Brazil 27°45′S 49°31′W
79 L15 **Bomu** *var.* Mbomou, Mbomu, M'Bomu. ≈ Central African Republic/Dem. Rep. Congo
142 J3 **Bonāb** *var.* Benāb, Bunab. Āžarbāyjān-e Sharqī, N Iran 37°30′N 46°03′E
45 S9 **Bonaire** ◈ *Dutch special municipality* S Caribbean Sea
45 Q16 **Bonaire** *island* Lesser Antilles
39 U11 **Bona, Mount** ▲ Alaska, USA 61°22′N 141°45′W
183 Q12 **Bonang** Victoria, SE Australia 37°13′S 148°43′E
42 L7 **Bonanza** Región Autónoma Atlántico Norte, NE Nicaragua 13°59′N 84°30′W
37 O4 **Bonanza** Utah, W USA 40°01′N 109°12′W
45 O9 **Bonao** C Dominican Republic 18°55′N 70°25′W
180 L3 **Bonaparte Archipelago** *island group* Western Australia
32 K6 **Bonaparte, Mount** ▲ Washington, NW USA 48°47′N 119°07′W
39 N11 **Bonasila Dome** ▲ Alaska, USA 62°24′N 160°28′W
45 T15 **Bonasse** Trinidad, Trinidad and Tobago 10°02′N 61°48′W
15 X7 **Bonaventure** Québec, SE Canada 48°03′N 65°30′W
15 X7 **Bonaventure** ≈ Québec, SE Canada
13 V11 **Bonavista** Newfoundland, Newfoundland and Labrador, SE Canada 48°38′N 53°08′W
13 U11 **Bonavista Bay** *inlet* NW Atlantic Ocean
79 E19 **Bonda** Ogooué-Lolo, C Gabon 0°50′S 12°28′E
127 N6 **Bondari** Tambovskaya Oblast', W Russian Federation 52°58′N 42°02′E
106 G9 **Bondeno** Emilia-Romagna, C Italy 44°53′N 11°24′E
30 L4 **Bond Falls Flowage** ◎ Michigan, N USA
79 L16 **Bondo** Orientale, N Dem. Rep. Congo 03°52′N 23°41′E
171 N17 **Bondokodi** Pulau Sumba, S Indonesia 09°36′S 119°01′E
77 O15 **Bondoukou** E Ivory Coast 08°03′N 02°45′W
Bondoukui/Bondoukuy *see* Boundoukui
169 T17 **Bondowoso** Jawa, C Indonesia 07°54′S 113°50′E
33 S14 **Bondurant** Wyoming, C USA 43°14′N 110°26′W
Bône *see* Annaba, Algeria
Bone *see* Watampone, Indonesia
30 I5 **Bone Lake** ◎ Wisconsin, N USA
171 P14 **Bonelipu** Pulau Buton, C Indonesia 04°42′S 123°09′E
171 O15 **Bonerate, Kepulauan** *var.* Macan. *island group* C Indonesia
29 O12 **Bonesteel** South Dakota, N USA 43°01′N 98°55′W
62 I8 **Bonete, Cerro** ▲ N Argentina 27°58′S 68°22′W
171 O14 **Bone, Teluk** *bay* Sulawesi, C Indonesia
108 D6 **Bonfol** Jura, NW Switzerland 47°28′N 07°08′E
153 U12 **Bongaigaon** Assam, NE India 26°30′N 90°31′E
79 K17 **Bongandanga** Equateur, NW Dem. Rep. Congo 01°28′N 21°03′E
78 L13 **Bongo, Massif des** *var.* Chaîne des Mongos. ▲▲ NE Central African Republic
78 G12 **Bongor** Mayo-Kébbi Est, SW Chad 10°18′N 15°20′E
77 N16 **Bongouanou** E Ivory Coast 06°39′N 04°12′W
167 V11 **Bông Sơn** *var.* Hoai Nhơn. Binh Ðinh, C Vietnam 14°28′N 109°00′E
25 U5 **Bonham** Texas, SW USA 33°36′N 96°12′W
Bonhard *see* Bonyhád
103 U6 **Bonhomme, Col du** *pass* NE France
103 Y16 **Bonifacio** Corse, France, C Mediterranean Sea 41°24′N 09°09′E
Bonifacio, Bocche de/Bonifacio, Bouches de *see* Bonifacio, Strait of
103 Y16 **Bonifacio, Strait of** *Fr.* Bouches de Bonifacio, *It.* Bocche di Bonifacio. *strait* C Mediterranean Sea
23 Q8 **Bonifay** Florida, SE USA 30°49′N 85°42′W
Bonin Islands *see* Ogasawara-shotō
192 H5 **Bonin Trench** *undersea feature* NW Pacific Ocean
23 W15 **Bonita Springs** Florida, SE USA 26°19′N 81°48′W
42 I5 **Bonito, Pico** ▲ N Honduras 15°33′N 86°55′W
101 E17 **Bonn** Nordrhein-Westfalen, W Germany 50°44′N 07°06′E
92 H11 **Bonnåsjøen** Nordland, C Norway 67°35′N 15°39′E
14 J12 **Bonnechere** Ontario, SE Canada 45°39′N 77°36′W
14 J12 **Bonnechere** ≈ Ontario, SE Canada
33 N7 **Bonners Ferry** Idaho, NW USA 48°41′N 116°19′W
27 R4 **Bonner Springs** Kansas, C USA 39°03′N 94°52′W
102 L6 **Bonnétable** Sarthe, NW France 48°09′N 00°24′E
27 X6 **Bonne Terre** Missouri, C USA 37°55′N 90°33′W
10 J5 **Bonnet Plume** ≈ Yukon, NW Canada
102 M6 **Bonneval** Eure-et-Loir, C France 48°12′N 01°23′E
103 T10 **Bonneville** Haute-Savoie, E France 46°05′N 06°25′E
36 J3 **Bonneville Salt Flats** *salt flat* Utah, W USA
77 U18 **Bonny** Rivers, S Nigeria 04°25′N 07°13′E
Bonny, Bight of *see* Biafra, Bight of
37 W4 **Bonny Reservoir** ▦ Colorado, C USA
11 R14 **Bonnyville** Alberta, SW Canada 54°16′N 110°46′W
107 C18 **Bono** Sardegna, Italy, C Mediterranean Sea 40°24′N 09°01′E
Bononia *see* Vidin, Bulgaria
Bononia *see* Boulogne-sur-Mer, France
107 B18 **Bonorva** Sardegna, Italy, C Mediterranean Sea 40°27′N 08°46′E
30 M15 **Bonpas Creek** ≈ Illinois, N USA
190 I3 **Bonriki** Tarawa, W Kiribati 01°23′N 173°09′E
183 T4 **Bonshaw** New South Wales, SE Australia 29°06′S 151°15′E
76 I16 **Bonthe** SW Sierra Leone 07°32′N 12°30′W
171 N2 **Bontoc** Luzon, N Philippines 17°04′N 120°58′E
25 Y9 **Bon Wier** Texas, SW USA 30°43′N 93°40′W
111 J25 **Bonyhád** *Ger.* Bonhard. Tolna, S Hungary 46°20′N 18°31′E
Bonzabaai *see* Bonza Bay
83 J25 **Bonza Bay** *Afr.* Bonzabaai. Eastern Cape, S South Africa 32°58′S 27°58′E
182 D7 **Bookabie** South Australia 31°49′S 132°41′E
182 H6 **Bookaloo** South Australia 31°56′S 137°21′E
37 P5 **Book Cliffs** *cliff* Colorado/Utah, W USA
25 P1 **Booker** Texas, SW USA 36°27′N 100°32′W
76 K15 **Boola** SE Guinea 08°22′N 08°41′W
183 O8 **Booligal** New South Wales, SE Australia 33°56′S 144°54′E
99 G17 **Boom** Antwerpen, N Belgium 51°05′N 04°24′E
43 N6 **Boom** *var.* Boon. Región Autónoma Atlántico Norte, NE Nicaragua 14°52′N 83°36′W
183 S3 **Boomi** New South Wales, SE Australia 28°43′S 149°35′E
Boon *see* Boom
162 H9 **Bööncagaan Nuur** ◎ S Mongolia
29 V13 **Boone** Iowa, C USA 42°04′N 93°52′W
21 Q8 **Boone** North Carolina, SE USA 36°13′N 81°41′W
27 S11 **Booneville** Arkansas, C USA 35°09′N 93°57′W
21 N6 **Booneville** Kentucky, S USA 37°26′N 83°45′W
23 N2 **Booneville** Mississippi, S USA 34°39′N 88°34′W
21 V3 **Boonsboro** Maryland, NE USA 39°30′N 77°39′W
34 L6 **Boonville** California, W USA 38°58′N 123°21′W
31 N16 **Boonville** Indiana, N USA 38°03′N 87°16′W
27 U4 **Boonville** Missouri, C USA 38°58′N 92°43′W
18 I9 **Boonville** New York, NE USA 43°28′N 75°17′W
80 M12 **Boorama** Awdal, NW Somalia 09°58′N 43°15′E
183 O6 **Booroondarra, Mount** *hill* New South Wales, SE Australia
183 N9 **Booroorban** New South Wales, SE Australia 34°55′S 144°45′E
183 R9 **Boorowa** New South Wales, SE Australia 34°26′S 148°42′E
99 H17 **Boortmeerbeek** Vlaams Brabant, C Belgium 50°58′N 04°27′E
80 P11 **Boosaaso** *var.* Bandar Kassim, Bender Qaasim, Bosaso, *It.* Bender Cassim. Bari, N Somalia 11°26′N 49°37′E
19 Q8 **Boothbay Harbor** Maine, NE USA 43°52′N 69°35′W
Boothia Felix *see* Boothia Peninsula
9 N6 **Boothia, Gulf of** *gulf* Nunavut, NE Canada
9 N6 **Boothia Peninsula** *prev.* Boothia Felix. *peninsula* Nunavut, NE Canada
79 E18 **Booué** Ogooué-Ivindo, NE Gabon 0°03′S 11°58′E
101 J21 **Bopfingen** Baden-Württemberg, S Germany 48°51′N 10°21′E
101 F18 **Boppard** Rheinland-Pfalz, W Germany 50°13′N 07°36′E
62 M4 **Boquerón** *off.* Departamento de Boquerón. ◈ *department* W Paraguay
Boquerón, Departamento de *see* Boquerón
43 P15 **Boquete** *var.* Bajo Boquete. Chiriquí, W Panama 08°45′N 82°26′W
40 J6 **Boquilla, Presa de la** ▦ N Mexico
40 L5 **Boquillas** *var.* Boquillas del Carmen. Coahuila, NE Mexico 29°10′N 102°55′W
Boquillas del Carmen *see* Boquillas
112 P12 **Bor** Serbia, E Serbia 44°05′N 22°07′E
81 F15 **Bor** Jonglei, E South Sudan 06°12′N 31°33′E
95 L20 **Bor** Jönköping, S Sweden 57°04′N 14°10′E
136 J15 **Bor** Niğde, S Turkey 37°49′N 35°00′E
191 S10 **Bora-Bora** *island* Îles Sous le Vent, W French Polynesia
167 Q9 **Borabu** Maha Sarakham, E Thailand 16°01′N 103°06′E
172 K4 **Boraha, Nosy** *island* E Madagascar
33 P13 **Borah Peak** ▲ Idaho, NW USA 44°21′N 113°53′W
145 U16 **Boralday** *prev.* Burunday. Almaty, SE Kazakhstan 43°21′N 76°48′E
145 Y11 **Boran** *prev.* Buran. Vostochnyy Kazakhstan, E Kazakhstan 48°00′N 85°09′E
144 G13 **Borankul** *prev.* Opornyy. Mangistau, SW Kazakhstan 46°09′N 54°32′E
95 J19 **Borås** Västra Götaland, S Sweden 57°44′N 12°55′E
143 N11 **Borāzjān** *var.* Borazjān. Būshehr, S Iran 29°19′N 51°12′E
Borazjān *see* Borāzjān
58 G13 **Borba** Amazonas, N Brazil 04°39′S 59°35′W
104 H11 **Borba** Évora, S Portugal 38°48′N 07°28′W
Borbetomagus *see* Worms
55 O7 **Borbón** Bolívar, E Venezuela 07°55′N 64°03′W
59 Q15 **Borborema, Planalto da** *plateau* NE Brazil
116 M14 **Borcea, Braţul** ≈ S Romania
Borchalo *see* Marneuli
195 R15 **Borchgrevink Coast** *physical region* Antarctica
137 Q11 **Borçka** Artvin, NE Turkey 41°24′N 41°38′E
98 N11 **Borculo** Gelderland, E Netherlands 52°07′N 06°31′E
182 G10 **Borda, Cape** *headland* South Australia 35°45′S 136°34′E
102 K13 **Bordeaux** *anc.* Burdigala. Gironde, SW France 44°49′N 00°33′W
11 T15 **Borden** Saskatchewan, S Canada 52°23′N 107°10′W
14 D8 **Borden Lake** ◎ Ontario, S Canada
9 N4 **Borden Peninsula** *peninsula* Baffin Island, Nunavut, NE Canada
182 K11 **Bordertown** South Australia 36°21′S 140°48′E
92 H2 **Borðeyri** Norðurland Vestra, NW Iceland 65°12′N 21°09′W
95 B18 **Bordhoy** *Dan.* Bordø. *island* NE Faroe Islands
106 B11 **Bordighera** Liguria, NW Italy 43°48′N 07°40′E
74 K5 **Bordj-Bou-Arreridj** *var.* Bordj Bou Arreridj, Bordj Bou Arréridj. N Algeria 36°02′N 04°49′E
74 L10 **Bordj Omar Driss** E Algeria 28°09′N 06°52′E
143 N13 **Bord Khūn** Hormozgān, S Iran
Bordø *see* Bordhoy
147 V7 **Bordunskiy** Chuyskaya Oblast', N Kyrgyzstan 42°37′N 75°31′E
95 M17 **Borensberg** Östergötland, S Sweden 58°33′N 15°15′E
Borgå *see* Porvoo
92 H3 **Borgarnes** Vesturland, W Iceland 64°33′N 21°55′W
93 G14 **Børgefjell** ▲▲ C Norway
98 O7 **Borger** Drenthe, NE Netherlands 52°54′N 06°48′E
25 N2 **Borger** Texas, SW USA 35°40′N 101°24′W
95 N20 **Borgholm** Kalmar, S Sweden 56°50′N 16°41′E
107 N22 **Borgia** Calabria, SW Italy 38°48′N 16°28′E
99 J18 **Borgloon** Limburg, NE Belgium 50°48′N 05°21′E
Borg Massif *see* Borgmassivet
195 P2 **Borgmassivet** *Eng.* Borg Massif. ▲▲ Antarctica
22 L9 **Borgne, Lake** ◎ Louisiana, S USA
106 C7 **Borgomanero** Piemonte, NE Italy 45°42′N 08°33′E
106 G10 **Borgo Panigale** ✖ (Bologna) Emilia-Romagna, N Italy 44°33′N 11°16′E
107 J15 **Borgorose** Lazio, C Italy 42°10′N 13°15′E
106 A9 **Borgo San Dalmazzo** Piemonte, N Italy 44°19′N 07°29′E
106 G11 **Borgo San Lorenzo** Toscana, C Italy 43°58′N 11°22′E
106 C7 **Borgosesia** Piemonte, NE Italy 45°41′N 08°21′E
106 E9 **Borgo Val di Taro** Emilia-Romagna, C Italy 44°29′N 09°48′E
106 G6 **Borgo Valsugana** Trentino-Alto Adige, N Italy 46°04′N 11°31′E
Borhoyn Tal *see* Dzamïn-Üüd
167 R8 **Borikhan** *var.* Borikhane. Bolikhamxai, C Laos 18°36′N 103°43′E
Borikhane *see* Borikhan
144 G8 **Borili** *prev.* Burlin. Zapadnyy Kazakhstan, NW Kazakhstan 51°25′N 52°42′E
Borislav *see* Boryslav
127 N8 **Borisoglebsk** Voronezhskaya Oblast', W Russian Federation 51°23′N 42°00′E
Borisov *see* Barysaw
Borisovgrad *see* Parvomay
Borispol' *see* Boryspil'
172 I3 **Boriziny** *prev./Fr.* Port-Bergé. Mahajanga, NW Madagascar 15°31′S 47°40′E
105 Q5 **Borja** Aragón, NE Spain 41°50′N 01°32′W
Borjas Blancas *see* Les Borges Blanques
137 S10 **Borjomi** *Rus.* Borzhomi. C Georgia 41°50′N 43°24′E
118 L12 **Borkavichy** *Rus.* Borkovichi. Vitsyebskaya Voblasts', N Belarus 55°40′N 28°20′E
101 H16 **Borken** Hessen, C Germany 51°01′N 09°16′E
101 E14 **Borken** Nordrhein-Westfalen, W Germany 51°51′N 06°51′E
92 H10 **Borkenes** Troms, N Norway 68°46′N 16°10′E
78 H7 **Borkou** *off.* Région du Borkou. ◈ *region* N Chad
Borkou, Région du *see* Borkou
Borkovichi *see* Borkavichy
100 E9 **Borkum** *island* NW Germany
81 K17 **Bor, Lagh** *var.* Lak Bor. *dry watercourse* NE Kenya
Bor, Lak *see* Bor, Lagh
95 M14 **Borlänge** Dalarna, C Sweden 60°29′N 15°25′E
106 C9 **Bormida** ≈ NW Italy
106 F6 **Bormio** Lombardia, N Italy 46°27′N 10°24′E
101 M16 **Borna** Sachsen, E Germany 51°07′N 12°30′E
98 O10 **Borne** Overijssel, E Netherlands 52°18′N 06°45′E
99 F17 **Bornem** Antwerpen, N Belgium 51°06′N 04°14′E
169 S10 **Borneo** *island* Brunei/Indonesia/Malaysia
101 E16 **Bornheim** Nordrhein-Westfalen, W Germany 50°46′N 06°58′E
95 L24 **Bornholm** ◈ *county* E Denmark
95 L24 **Bornholm** *island* E Denmark
77 Y13 **Borno** ◈ *state* NE Nigeria
104 K15 **Bornos** Andalucía, S Spain 36°50′N 05°42′W
162 L7 **Bornuur** Töv, C Mongolia 48°28′N 106°15′E
117 O4 **Borodyanka** Kyyivs'ka Oblast', N Ukraine 50°40′N 29°54′E
158 I5 **Borohoro Shan** ▲▲ NW China
77 O13 **Boromo** SW Burkina Faso 11°47′N 02°54′W
35 T13 **Boron** California, W USA 35°00′N 117°42′W
Borongo *see* Black Volta
Boron'ki *see* Baron'ki
Borosjenő *see* Ineu
Borossebes *see* Sebiş
76 L15 **Borotou** NW Ivory Coast 08°46′N 07°30′W
117 W6 **Borova** Kharkivs'ka Oblast', E Ukraine 49°22′N 37°39′E
114 H8 **Borovan** Vratsa, NW Bulgaria 43°25′N 23°45′E
124 I14 **Borovichi** Novgorodskaya Oblast', W Russian Federation 58°24′N 33°56′E
Borovlje *see* Ferlach
114 K8 **Borovo** Ruse, N Bulgaria 43°28′N 25°46′E
112 J9 **Borovo** Vukovar-Srijem, NE Croatia 45°22′N 18°57′E
Borovoye *see* Burabay
126 K4 **Borovsk** Kaluzhskaya Oblast', W Russian Federation 55°12′N 36°22′E
145 N7 **Borovskoye** Kostanay, N Kazakhstan 53°48′N 64°17′E
Borovukha *see* Baravukha
95 L23 **Borrby** Skåne, S Sweden 55°27′N 14°10′E
105 T9 **Borriana** *var.* Burriana. Valenciana, E Spain 39°54′N 00°05′W
181 R3 **Borroloola** Northern Territory, N Australia 16°09′S 136°18′E
116 F9 **Borş** Bihor, NW Romania 47°07′N 21°49′E
116 I9 **Borşa** *Hung.* Borsa. Maramureş, N Romania 47°40′N 24°37′E
116 J10 **Borsec** *Ger.* Bad Borseck, *Hung.* Borszék. Harghita, C Romania 46°58′N 25°32′E
92 K8 **Børselv** *Lapp.* Bissojohka. Finnmark, N Norway 70°18′N 25°35′E
113 L23 **Borsh** *var.* Borshi. Vlorë, S Albania 40°04′N 19°51′E
Borshchev *see* Borshchiv
116 K7 **Borshchiv** *Pol.* Borszczów, *Rus.* Borshchev. Ternopil's'ka Oblast', W Ukraine 48°48′N 26°00′E
Borshi *see* Borsh
111 L20 **Borsod-Abaúj-Zemplén** *off.* Borsod-Abaúj-Zemplén Megye. ◈ *county* NE Hungary
Borsod-Abaúj-Zemplén Megye *see* Borsod-Abaúj-Zemplén

◆ Country
● Country Capital
◇ Dependent Territory
○ Dependent Territory Capital
◈ Administrative Regions
✖ International Airport
▲ Mountain
▲▲ Mountain Range
Volcano
≈ River
◎ Lake
▦ Reservoir

99 E15 **Borssele** Zeeland, SW Netherlands *51°26′N 03°45′E*
Borszczów *see* Borshchiv
Borszék *see* Borsec
Bortala *see* Bole
103 O12 **Bort-les-Orgues** Corrèze, C France *45°28′N 02°31′E*
Bor u České Lípy *see* Nový Bor
Bor-Üdzüür *see* Altay
143 N9 **Borūjen** Chahār MaḨall va Bakhtīārī, C Iran *32°N 51°09′E*
142 L7 **Borūjerd** *var.* Burujird. Lorestān, W Iran *33°55′N 48°46′E*
116 H6 **Boryslav** *Pol.* Borysław, *Rus.* Borislav. L'vivs'ka Oblast', NW Ukraine *49°18′N 23°28′E*
Borysław *see* Boryslav
117 P4 **Boryspil'** *Rus.* Borispol'. Kyyivs'ka Oblast', N Ukraine *50°21′N 30°59′E*
117 P4 **Boryspil'** *Rus.* Borispol'. ✈ (Kyyiv) Kyyivs'ka Oblast', N Ukraine *50°21′N 30°46′E*
Borzhomi *see* Borjomi
117 R3 **Borzna** Chernihivs'ka Oblast', NE Ukraine *51°15′N 32°25′E*
123 O14 **Borzya** Zabaykal'skiy Kray, S Russian Federation *50°18′N 116°24′E*
107 B18 **Bosa** Sardegna, Italy, C Mediterranean Sea *40°18′N 08°28′E*
112 F10 **Bosanska Dubica** *var.* Kozarska Dubica. ◈ Republika Srpska, NW Bosnia and Herzegovina
112 G10 **Bosanska Gradiška** *var.* Gradiška. ◈ Republika Srpska, N Bosnia and Herzegovina
112 F10 **Bosanska Kostajnica** *var.* Srpska Kostajnica. ◈ Republika Srpska, NW Bosnia and Herzegovina
112 E11 **Bosanska Krupa** *var.* Krupa, Krupa na Uni. ◈ Federacija Bosne I Hercegovine, NW Bosnia and Herzegovina
112 H10 **Bosanski Brod** *var.* Srpski Brod. ◈ Republika Srpska, N Bosnia and Herzegovina
112 E10 **Bosanski Novi** *var.* Novi Grad. Republika Srpska, NW Bosnia and Herzegovina *45°03′N 16°23′E*
112 E11 **Bosanski Petrovac** *var.* Petrovac. Federacija Bosne I Hercegovine, NW Bosnia and Herzegovina *44°34′N 16°21′E*
112 I10 **Bosanski Šamac** *var.* Šamac. Republika Srpska, N Bosnia and Herzegovina *45°03′N 18°27′E*
112 E12 **Bosansko Grahovo** *var.* Grahovo, Hrvatsko Grahovi. Federacija Bosne I Hercegovine, W Bosnia and Herzegovina *44°10′N 16°22′E*
Bosaso *see* Boosaaso
186 B7 **Bosavi, Mount** ⛰ W Papua New Guinea *06°33′S 142°50′E*
160 J14 **Bose** Guangxi Zhuangzu Zizhiqu, S China *23°55′N 106°32′E*
161 Q5 **Boshan** Shandong, E China *36°32′N 117°47′E*
113 P16 **Bosilegrad** *prev.* Bosiljgrad. Serbia, SE Serbia *42°30′N 22°30′E*
Bosiljgrad *see* Bosilegrad
Bösing *see* Pezinok
98 H12 **Boskoop** Zuid-Holland, C Netherlands *52°04′N 04°40′E*
111 G18 **Boskovice** *Ger.* Boskowitz. Jihomoravský Kraj, SE Czech Republic *49°30′N 16°39′E*
Boskowitz *see* Boskovice
112 I10 **Bosna** ≈ N Bosnia and Herzegovina
113 G14 **Bosne I Hercegovine, Federacija** ◈ *republic* Bosnia and Herzegovina
112 H12 **Bosnia and Herzegovina** *off.* Republic of Bosnia and Herzegovina. ◆ *republic* SE Europe
Bosnia and Herzegovina, Republic of *see* Bosnia and Herzegovina
79 J16 **Bosobolo** Equateur, NW Dem. Rep. Congo *04°11′N 19°55′E*
165 O14 **Bōsō-hantō** *peninsula* Honshū, S Japan
Bosora *see* Buşrá ash Shām
Bosphorus/Bosporus *see* İstanbul Boğazı
Bosporus Cimmerius *see* Kerch Strait
Bosporus Thracius *see* İstanbul Boğazı
Bosra *see* Buşrá ash Shām
79 H14 **Bossangoa** Ouham, C Central African Republic *06°32′N 17°25′E*
Bossé Bangou *see* Bossey Bangou
79 I15 **Bossembélé** Ombella-Mpoko, C Central African Republic *05°13′N 17°39′E*
79 H15 **Bossentélé** Ouham-Pendé, W Central African Republic *05°36′N 16°37′E*
77 R12 **Bossey Bangou** *var.* Bossé Bangou. Tillabéri, SW Niger *13°22′N 01°18′E*
22 G5 **Bossier City** Louisiana, S USA *32°31′N 93°43′W*
83 D20 **Bossiesvlei** Hardap, S Namibia *25°02′S 16°48′E*
77 Y11 **Bosso** Diffa, SE Niger *13°42′N 13°18′E*
61 F15 **Bossoroca** Rio Grande do Sul, S Brazil *28°45′S 54°54′W*
158 J10 **Bostan** Xinjiang Uygur Zizhiqu, W China *41°20′N 83°15′E*
142 K3 **Bostānābād** Āẕarbāyjān-e Sharqī, N Iran *37°52′N 46°51′E*
158 K6 **Bosten Hu** *var.* Bagrax Hu. ◎ NW China
97 O18 **Boston** *prev.* St.Botolph's Town. E England, United Kingdom *52°59′N 00°01′W*
19 O11 **Boston** *state capital* Massachusetts, NE USA *42°22′N 71°04′W*
146 I9 **Bo'ston** *Rus.* Bustan. Qoraqalpog'iston Respublikasi, W Uzbekistan *41°49′N 60°51′E*
10 M17 **Boston Bar** British Columbia, SW Canada *49°54′N 121°22′W*
27 T10 **Boston Mountains** ▲▲ Arkansas, C USA
15 P8 **Bostonnais** ≈ Québec, SE Canada
Bostyn' *see* Bastyn'
112 J10 **Bosut** ≈ E Croatia
154 C11 **Botād** Gujarāt, W India *22°12′N 71°44′E*
145 S10 **Botakara** *Kas.* Botaqara; *prev.* Ul'yanovskiy. Karaganda, C Kazakhstan *50°05′N 73°45′E*
183 T9 **Botany Bay** *inlet* New South Wales, SE Australia
Botaqara *see* Botakara
83 G18 **Boteti** *var.* Botletle. ≈ N Botswana
114 J9 **Botev** ▲ C Bulgaria *42°45′N 24°57′E*
114 H9 **Botevgrad** *prev.* Orkhaniye. Sofia, W Bulgaria *42°55′N 23°47′E*
93 J16 **Bothnia, Gulf of** *Fin.* Pohjanlahti, *Swe.* Bottniska Viken. *gulf* N Baltic Sea
183 P17 **Bothwell** Tasmania, SE Australia *42°24′S 147°01′E*
104 H5 **Boticas** Vila Real, N Portugal *41°41′N 07°40′W*
55 W10 **Boti-Pasi** Sipaliwini, C Suriname *04°15′N 55°27′W*
Botletle *see* Boteti
127 P16 **Botlikh** Chechenskaya Respublika, SW Russian Federation *42°39′N 46°12′E*
117 N10 **Botna** ≈ E Moldova
116 I9 **Botoşani** *Hung.* Botosány. Botoşani, NE Romania *47°44′N 26°41′E*
116 K8 **Botoşani** ◈ *county* NE Romania
Botosány *see* Botoşani
147 P12 **Bototog', Tizmasi** *Rus.* Khrebet Babatag. ▲▲ Tajikistan/Uzbekistan
161 P4 **Botou** *prev.* Bozhen. Hebei, E China *38°09′N 116°37′E*
99 M20 **Botrange** ▲ E Belgium *50°30′N 06°03′E*
107 O21 **Botricello** Calabria, SW Italy *38°56′N 16°51′E*
83 I23 **Botshabelo** Free State, C South Africa *29°15′S 26°51′E*
93 J15 **Botsmark** Västerbotten, N Sweden *64°15′N 20°15′E*
83 G19 **Botswana** *off.* Republic of Botswana. ◆ *republic* S Africa
Botswana, Republic of *see* Botswana
29 N2 **Bottineau** North Dakota, N USA *48°50′N 100°28′W*
Bottniska Viken *see* Bothnia, Gulf of
60 L9 **Botucatu** São Paulo, S Brazil *22°52′S 48°30′W*
76 M16 **Bouaflé** C Ivory Coast *06°59′N 05°45′W*
77 N16 **Bouaké** *var.* Bwake. C Ivory Coast *07°42′N 05°00′W*
79 G14 **Bouar** Nana-Mambéré, W Central African Republic *05°58′N 15°38′E*
74 H7 **Bouarfa** NE Morocco *32°33′N 01°54′W*
111 B19 **Boubín** ▲ SW Czech Republic *49°00′N 13°51′E*
79 I14 **Bouca** Ouham, W Central African Republic *06°57′N 18°18′E*
15 T5 **Boucher** ≈ Québec, SE Canada
103 R15 **Bouches-du-Rhône** ◈ *department* SE France
74 C9 **Bou Craa** *var.* Bu Craa. NW Western Sahara *26°32′N 12°52′W*
77 O9 **Boû Djébéha** *oasis* C Mali
108 C8 **Boudry** Neuchâtel, W Switzerland *46°57′N 06°46′E*
79 F21 **Bouenza** ◈ *province* S Congo
186 J7 **Bougainville** *off.* Autonomous Region of Bougainville; *prev.* North Solomons. ◈ *autonomous region* Bougainville, NE Papua New Guinea Oceania
Bougainville, Autonomous Region of *see* Bougainville
180 L2 **Bougainville, Cape** *cape* Western Australia
65 E24 **Bougainville, Cape** *headland* East Falkland, Falkland Islands *51°18′S 58°28′W*
Bougainville, Détroit de *see* Bougainville Strait
186 J7 **Bougainville Island** *island* NE Papua New Guinea
186 I8 **Bougainville Strait** *strait* N Solomon Islands
187 Q13 **Bougainville Strait** *Fr.* Détroit de Bougainville. *strait* C Vanuatu
120 I9 **Bougaroun, Cap** *headland* NE Algeria *37°07′N 06°18′E*
77 R8 **Boughessa** Kidal, NE Mali *20°05′N 02°13′E*
Bougie *see* Béjaïa
76 L13 **Bougouni** Sikasso, SW Mali *11°25′N 07°28′W*
99 J24 **Bouillon** Luxembourg, SE Belgium *49°47′N 05°04′E*
74 K5 **Bouira** *var.* Bouïra. N Algeria *36°22′N 03°55′E*
74 D8 **Bou-Izakarn** SW Morocco *29°12′N 09°43′W*
74 B9 **Boujdour** *var.* Bojador. W Western Sahara *26°06′N 14°29′W*
74 G5 **Boukhalef** ✈ (Tanger) N Morocco *35°45′N 05°53′W*
Boukombé *see* Boukoumbé
77 R14 **Boukoumbé** *var.* Boukombé. C Benin *10°13′N 01°09′E*
76 G6 **Boû Lanouâr** Dakhlet Nouâdhibou, W Mauritania *21°17′N 16°29′W*
37 T4 **Boulder** Colorado, C USA *40°02′N 105°18′W*
33 R10 **Boulder** Montana, NW USA *46°14′N 112°07′W*
35 X12 **Boulder City** Nevada, W USA *35°58′N 114°49′W*
181 T7 **Boulia** Queensland, C Australia *23°02′S 139°58′E*
15 N10 **Boullé** ≈ Québec, SE Canada
102 J9 **Boulogne** ≈ NW France
Boulogne *see* Boulogne-sur-Mer
102 L16 **Boulogne-sur-Gesse** Haute-Garonne, S France *43°18′N 00°38′E*
103 N1 **Boulogne-sur-Mer** *var.* Boulogne; *anc.* Bononia, Gesoriacum, Gessoriacum. Pas-de-Calais, N France *50°43′N 01°36′E*
77 Q12 **Boulsa** C Burkina Faso *12°41′N 00°29′W*
77 W11 **Boultoum** Zinder, C Niger *14°43′N 10°22′E*
187 Y14 **Bouma** Taveuni, N Fiji *16°49′S 179°50′W*
79 G16 **Boumba** ≈ SE Cameroon
76 J9 **Boûmdeïd** *var.* Boumdeït. Assaba, S Mauritania *17°26′N 11°21′W*
Boumdeït *see* Boûmdeïd
115 C17 **Boumistós** ▲ W Greece *38°48′N 20°59′E*
77 O15 **Bouna** NE Ivory Coast *09°16′N 03°00′W*
19 P4 **Boundary Bald Mountain** ▲ Maine, NE USA *45°45′N 70°10′W*
35 S8 **Boundary Peak** ▲ Nevada, W USA *37°50′N 118°21′W*
76 M14 **Boundiali** N Ivory Coast *09°30′N 06°31′W*
79 G19 **Boundji** Cuvette, C Congo *01°05′S 15°18′E*
77 O13 **Boundoukui** *var.* Bondoukui, Bondoukuy. W Burkina Faso *11°51′N 03°47′W*
36 L2 **Bountiful** Utah, W USA *40°53′N 111°52′W*
Bounty Basin *see* Bounty Trough
191 Q16 **Bounty Bay** *bay* Pitcairn Island, C Pacific Ocean
192 L12 **Bounty Islands** *island group* S New Zealand
175 Q13 **Bounty Trough** *var.* Bounty Basin. *undersea feature* S Pacific Ocean
187 P17 **Bourail** Province Sud, C New Caledonia *21°35′S 165°29′E*
27 V5 **Bourbeuse River** ≈ Missouri, C USA
103 Q9 **Bourbon-Lancy** Saône-et-Loire, C France *46°39′N 03°48′E*
31 N11 **Bourbonnais** Illinois, N USA *41°08′N 87°52′W*
103 O10 **Bourbonnais** *cultural region* C France
103 S7 **Bourbonne-les-Bains** Haute-Marne, N France *48°00′N 05°43′E*
Bourbon Vendée *see* la Roche-sur-Yon
74 M8 **Bourdj Messaouda** E Algeria *30°18′N 09°19′E*
77 Q10 **Bourem** Gao, C Mali *16°56′N 00°21′W*
Bourg *see* Bourg-en-Bresse
103 N11 **Bourganeuf** Creuse, C France *45°57′N 01°47′E*
Bourgas *see* Burgas
Bourge-en-Bresse *see* Bourg-en-Bresse
103 S10 **Bourg-en-Bresse** *var.* Bourg, Bourge-en-Bresse. Ain, E France *46°12′N 05°13′E*
103 O8 **Bourges** *anc.* Avaricum. Cher, C France *47°06′N 02°24′E*
103 T11 **Bourget, Lac du** ◎ E France
103 P8 **Bourgogne** *Eng.* Burgundy. ◈ *region* E France
103 S11 **Bourgoin-Jallieu** Isère, E France *45°34′N 05°16′E*
103 R14 **Bourg-St-Andéol** Ardèche, E France *44°24′N 04°36′E*
103 U11 **Bourg-St-Maurice** Savoie, E France *45°37′N 06°49′E*
108 C11 **Bourg St. Pierre** Valais, SW Switzerland *45°54′N 07°10′E*
76 H8 **Boû Rjeimât** *well* W Mauritania
183 P5 **Bourke** New South Wales, SE Australia *30°08′S 145°57′E*
97 M24 **Bournemouth** S England, United Kingdom *50°43′N 01°54′W*
99 M23 **Bourscheid** Diekirch, NE Luxembourg *49°55′N 06°04′E*
74 K6 **Bou Saâda** *var.* Bou Saada. N Algeria *35°10′N 04°09′E*
36 I13 **Bouse Wash** ≈ Arizona, SW USA
103 N10 **Boussac** Creuse, C France *46°20′N 02°12′E*
102 M16 **Boussens** Haute-Garonne, S France *43°11′N 00°58′E*
78 H12 **Bousso** *prev.* Fort-Bretonnet. Chari-Baguirmi, S Chad *10°32′N 16°45′E*
76 H9 **Boutilimit** Trarza, SW Mauritania *17°33′N 14°42′W*
65 D21 **Bouvet Island** ◇ *Norwegian dependency* S Atlantic Ocean
77 U11 **Bouza** Tahoua, SW Niger *14°25′N 06°09′E*
109 R10 **Bovec** *Ger.* Flitsch, *It.* Plezzo. NW Slovenia *46°21′N 13°33′E*
98 J8 **Bovenkarspel** Noord-Holland, NW Netherlands *52°33′N 05°03′E*
29 V5 **Bovey** Minnesota, N USA *47°18′N 93°25′W*
32 M9 **Bovill** Idaho, NW USA *46°50′N 116°24′W*
24 L4 **Bovina** Texas, SW USA *34°30′N 102°52′W*
107 M17 **Bovino** Puglia, SE Italy *41°14′N 15°19′E*
61 C17 **Bovril** Entre Ríos, E Argentina *31°24′S 59°25′W*
28 L2 **Bowbells** North Dakota, N USA *48°48′N 102°15′W*
11 Q16 **Bow City** Alberta, SW Canada *50°27′N 112°16′W*
29 O8 **Bowdle** South Dakota, N USA *45°27′N 99°39′W*
181 X6 **Bowen** Queensland, NE Australia *20°S 148°10′E*
192 L2 **Bowers Ridge** *undersea feature* S Bering Sea *50°00′N 180°00′W*
25 S5 **Bowie** Texas, SW USA *33°33′N 97°51′W*
11 R17 **Bow Island** Alberta, SW Canada *49°53′N 111°24′W*
Bowkān *see* Būkān
20 J7 **Bowling Green** Kentucky, S USA *37°00′N 86°29′W*
27 V3 **Bowling Green** Missouri, C USA *39°21′N 91°11′W*
31 R11 **Bowling Green** Ohio, N USA *41°22′N 83°40′W*
21 W5 **Bowling Green** Virginia, NE USA *38°02′N 77°22′W*
28 J6 **Bowman** North Dakota, N USA *46°11′N 103°26′W*
9 Q7 **Bowman Bay** *bay* NW Atlantic Ocean
194 I5 **Bowman Coast** *physical region* Antarctica
28 J7 **Bowman-Haley Lake** ⊠ North Dakota, N USA
195 Z11 **Bowman Island** *island* Antarctica
Bowo *see* Bomi
183 S9 **Bowral** New South Wales, SE Australia *34°28′S 150°52′E*
186 E8 **Bowutu Mountains** ▲▲ C Papua New Guinea
83 I16 **Bowwood** Southern, S Zambia *17°09′S 26°16′E*
28 I12 **Box Butte Reservoir** ⊠ Nebraska, C USA
28 J10 **Box Elder** South Dakota, N USA *44°06′N 103°04′W*
95 M18 **Boxholm** Östergötland, S Sweden *58°12′N 15°05′E*
Bo Xian/Boxian *see* Bozhou
161 Q4 **Boxing** Shandong, E China *37°06′N 118°05′E*
99 L14 **Boxmeer** Noord-Brabant, SE Netherlands *51°39′N 05°57′E*
99 J14 **Boxtel** Noord-Brabant, S Netherlands *51°36′N 05°20′E*
136 J10 **Boyabat** Sinop, N Turkey *41°27′N 34°45′E*
54 F9 **Boyacá** *off.* Departamento de Boyacá. ◈ *province* C Colombia
Boyacá, Departamento de *see* Boyacá
117 O4 **Boyarka** Kyyivs'ka Oblast', N Ukraine *50°19′N 30°20′E*
22 H7 **Boyce** Louisiana, S USA *31°23′N 92°40′W*
114 H8 **Boychinovtsi** Montana, NW Bulgaria *43°28′N 23°20′E*
33 U11 **Boyd** Montana, NW USA *45°27′N 109°03′W*
25 S6 **Boyd** Texas, SW USA *33°01′N 97°33′W*
21 V8 **Boydton** Virginia, NE USA *36°40′N 78°26′W*
Boyer Ahmadī va Kohkīlūyeh *see* Kohgīlūyeh va Bowyer Aḩmad
29 T13 **Boyer River** ≈ Iowa, C USA
21 W8 **Boykins** Virginia, NE USA *36°35′N 77°11′W*
11 Q13 **Boyle** Alberta, SW Canada *54°38′N 112°45′W*
97 D16 **Boyle** *Ir.* Mainistirna Búille. C Ireland *53°58′N 08°18′W*
97 F17 **Boyne** *Ir.* An Bhóinn. ≈ E Ireland
31 Q5 **Boyne City** Michigan, N USA *45°13′N 85°00′W*
23 Z14 **Boynton Beach** Florida, SE USA *26°31′N 80°04′W*
147 O13 **Boysun** *Rus.* Baysun. Surkhondaryo Viloyati, S Uzbekistan *38°14′N 67°08′E*
Bozau *see* Întorsura Buzăului
136 B12 **Bozcaada** *Island* Çanakkale, NW Turkey
136 C14 **Boz Dağları** ▲▲ W Turkey
33 S11 **Bozeman** Montana, NW USA *45°40′N 111°02′W*
Bozen *see* Bolzano
79 J16 **Bozene** Equateur, NW Dem. Rep. Congo *02°56′N 19°15′E*
Bozhen *see* Botou
161 P7 **Bozhou** *var.* Boxian, Bo Xian. Anhui, E China *33°46′N 115°44′E*
136 H16 **Bozkır** Konya, S Turkey *37°10′N 32°15′E*
136 K13 **Bozok Yaylasī** *plateau* C Turkey
79 H14 **Bozoum** Ouham-Pendé, W Central African Republic *06°17′N 16°26′E*
137 N16 **Bozova** Sanlıurfa, S Turkey *37°23′N 38°33′E*
Bozrah *see* Buşrá ash Shām
136 E12 **Bozüyük** Bilecik, NW Turkey *39°55′N 30°02′E*
106 B9 **Bra** Piemonte, NW Italy *44°42′N 07°51′E*
194 G4 **Brabant Island** *island* Antarctica
99 I20 **Brabant Walloon** ◈ *province* C Belgium
113 F15 **Brač** *var.* Brach, *It.* Brazza; *anc.* Brattia. *island* S Croatia
Bracara Augusta *see* Braga
107 H15 **Bracciano** Lazio, C Italy *42°04′N 12°12′E*
107 H14 **Bracciano, Lago di** ◎ C Italy
14 H13 **Bracebridge** Ontario, S Canada *45°02′N 79°19′W*
Brach *see* Brač
93 G17 **Bräcke** Jämtland, C Sweden *62°43′N 15°30′E*
25 P12 **Brackettville** Texas, SW USA *29°19′N 100°27′W*
97 N22 **Bracknell** S England, United Kingdom *51°26′N 00°46′W*
61 K14 **Braço do Norte** Santa Catarina, S Brazil *28°16′S 49°11′W*
116 G11 **Brad** *Hung.* Brád. Hunedoara, SW Romania *45°52′N 23°00′E*
107 N18 **Bradano** ≈ S Italy
23 V13 **Bradenton** Florida, SE USA *27°30′N 82°34′W*
45 V10 **Brades** ***de facto capital, de jure capital. Previous capital destroyed by volcano in 1995*** Plymouth. ● (Montserrat) SW Montserrat *16°45′N 62°12′W*
14 H14 **Bradford** Ontario, S Canada *44°09′N 79°34′W*
97 L17 **Bradford** N England, United Kingdom *53°48′N 01°45′W*
27 W10 **Bradford** Arkansas, C USA *35°25′N 91°27′W*
18 D12 **Bradford** Pennsylvania, NE USA *41°57′N 78°38′W*
27 T15 **Bradley** Arkansas, C USA *33°06′N 93°39′W*
25 P7 **Bradshaw** Texas, SW USA *32°06′N 99°52′W*
25 Q9 **Brady** Texas, SW USA *31°08′N 99°22′W*
25 Q9 **Brady Creek** ≈ Texas, SW USA
95 G22 **Brædstrup** Syddanmark, C Denmark *55°58′N 09°38′E*
96 J10 **Braemar** NE Scotland, United Kingdom *57°12′N 02°52′W*
116 K8 **Brăeşti** Botoşani, NW Romania *47°50′N 26°26′E*
104 G5 **Braga** *anc.* Bracara Augusta. Braga, NW Portugal *41°32′N 08°26′W*
104 G5 **Braga** ◈ *district* N Portugal
116 J15 **Bragadiru** Teleorman, S Romania *43°43′N 25°32′E*
61 C20 **Bragado** Buenos Aires, E Argentina *35°10′S 60°29′W*
104 J5 **Bragança** *Eng.* Braganza; *anc.* Julio Briga. Bragança, NE Portugal *41°47′N 06°46′W*
104 I5 **Bragança** ◈ *district* N Portugal
60 N9 **Bragança Paulista** São Paulo, S Brazil *22°55′S 46°30′W*
Braganza *see* Bragança
Bragin *see* Brahin
29 V7 **Braham** Minnesota, N USA *45°43′N 93°10′W*
Brahe *see* Brda
Brahestad *see* Raahe
119 O20 **Brahin** *Rus.* Bragin. Homyel'skaya Voblasts', SE Belarus *51°47′N 30°16′E*
153 U15 **Brahmanbaria** Chittagong, E Bangladesh *23°58′N 91°04′E*
154 O12 **Brāhmani** ≈ E India
154 N13 **Brahmapur** Odisha, E India *19°21′N 84°51′E*
129 S10 **Brahmaputra** *var.* Padma, Tsangpo, *Ben.* Jamuna, *Chin.* Yarlung Zangbo Jiang, *Ind.* Bramaputra, Dihang, Siang. ≈ S Asia
97 H19 **Braich y Pwll** *headland* NW Wales, United Kingdom *52°47′N 04°46′W*
183 R10 **Braidwood** New South Wales, SE Australia *35°26′S 149°48′E*
30 M11 **Braidwood** Illinois, N USA *41°16′N 88°12′W*
116 M13 **Brăila** Brăila, E Romania *45°17′N 27°57′E*
116 L13 **Brăila** ◈ *county* SE Romania
99 G19 **Braine-l'Alleud** Brabant Walloon, C Belgium *50°41′N 04°22′E*
99 F19 **Braine-le-Comte** Hainaut, SW Belgium *50°37′N 04°08′E*
29 U6 **Brainerd** Minnesota, N USA *46°22′N 94°10′W*
99 J19 **Braives** Liège, E Belgium *50°37′N 05°09′E*
83 H23 **Brak** ≈ C South Africa
Brak *see* Birāk
99 E18 **Brakel** Oost-Vlaanderen, SW Belgium *50°50′N 03°48′E*
98 J13 **Brakel** Gelderland, C Netherlands *51°49′N 05°05′E*
76 H9 **Brakna** ◈ *region* S Mauritania
95 J17 **Brålanda** Västra Götaland, S Sweden *58°32′N 12°18′E*
Bramaputra *see* Brahmaputra
95 F23 **Bramming** Syddtjylland, W Denmark *55°28′N 08°42′E*
14 G15 **Brampton** Ontario, S Canada *43°42′N 79°46′W*
100 F12 **Bramsche** Niedersachsen, NW Germany *52°25′N 07°58′E*
116 J12 **Bran** *Ger.* Törzburg, *Hung.* Törcsvár. Braşov, S Romania *45°31′N 25°23′E*
29 W8 **Branch** Minnesota, N USA *45°29′N 92°57′W*
21 R14 **Branchville** South Carolina, SE USA *33°15′N 80°49′W*
47 Y6 **Branco, Cabo** *headland* E Brazil *07°08′S 34°45′W*
58 F11 **Branco, Rio** ≈ N Brazil
108 J8 **Brand** Vorarlberg, W Austria *47°07′N 09°45′E*
83 B18 **Brandberg** ▲ NW Namibia *21°20′S 14°22′E*
95 H14 **Brandbu** Oppland, S Norway *60°24′N 10°30′E*
95 F22 **Brande** Midtjylland, W Denmark *55°57′N 09°08′E*
Brandebourg *see* Brandenburg
100 M12 **Brandenburg** *var.* Brandenburg an der Havel. Brandenburg, NE Germany *52°25′N 12°34′E*
20 K5 **Brandenburg** Kentucky, S USA *38°00′N 86°11′W*
100 N12 **Brandenburg** *off.* Freie und Hansestadt Hamburg, *Fr.* Brandebourg. ◈ *state* NE Germany
Brandenburg an der Havel *see* Brandenburg
83 I23 **Brandfort** Free State, C South Africa *28°42′S 26°28′E*
11 W16 **Brandon** Manitoba, S Canada *49°50′N 99°57′W*
23 V12 **Brandon** Florida, SE USA *27°56′N 82°17′W*
22 L6 **Brandon** Mississippi, S USA *32°16′N 90°01′W*
97 A20 **Brandon Mountain** *Ir.* Cnoc Bréanainn. ▲ SW Ireland *52°13′N 10°16′W*
Brandsen *see* Coronel Brandsen
95 I14 **Brandval** Hedmark, S Norway *60°18′N 12°01′E*
83 F24 **Brandvlei** Northern Cape, W South Africa *30°27′S 20°29′E*
23 U9 **Branford** Florida, SE USA *29°57′N 82°54′W*
110 K7 **Braniewo** *Ger.* Braunsberg. Warmińsko-mazurskie, N Poland *54°24′N 19°50′E*
194 H3 **Bransfield Strait** *strait* Antarctica
37 U8 **Branson** Colorado, C USA *37°01′N 103°52′W*
27 T8 **Branson** Missouri, C USA *36°38′N 93°13′W*
14 G16 **Brantford** Ontario, S Canada *43°09′N 80°17′W*
102 L12 **Brantôme** Dordogne, SW France *45°21′N 00°37′E*
182 L12 **Branxholme** Victoria, SE Australia *37°51′S 141°48′E*
Brasil *see* Brazil
59 C16 **Brasiléia** Acre, W Brazil *10°59′S 48°45′W*
59 K18 **Brasília** ● (Brazil) Distrito Federal, C Brazil *15°45′S 47°57′W*
Brasil, Repùblica Federativa do *see* Brazil
Braslav *see* Braslaw
118 J12 **Braslaw** *Pol.* Brasław, *Rus.* Braslav. Vitsyebskaya Voblasts', N Belarus *55°38′N 27°02′E*
116 J12 **Braşov** *Ger.* Kronstadt, *Hung.* Brassó; *prev.* Oraşul Stalin. Braşov, C Romania *45°40′N 25°35′E*
116 I12 **Braşov** ◈ *county* C Romania
77 U18 **Brass** Bayelsa, S Nigeria *04°19′N 06°21′E*
99 H16 **Brasschaat** *var.* Brasschaet. Antwerpen, N Belgium *51°17′N 04°30′E*
Brasschaet *see* Brasschaat
169 V8 **Brassey, Banjaran** *var.* Brassey Range. ▲▲ East Malaysia
Brassey Range *see* Brassey, Banjaran
Brassó *see* Braşov
23 T1 **Brasstown Bald** ▲ Georgia, SE USA *34°52′N 83°48′W*
113 K22 **Brataj** Vlorë, SW Albania *40°18′N 19°37′E*
114 J10 **Bratan** *var.* Morozov. ▲ C Bulgaria *42°31′N 25°08′E*
114 H10 **Bratia** *var.* Bratiya. C Bulgaria *42°36′N 24°08′E*
111 F21 **Bratislava** *Ger.* Pressburg, *Hung.* Pozsony. ● (Slovakia) Bratislavský Kraj, W Slovakia *48°10′N 17°10′E*
111 H21 **Bratislavský Kraj** ◈ *region* W Slovakia
Bratiya *see* Bratia
122 M12 **Bratsk** Irkutskaya Oblast', C Russian Federation *56°20′N 101°50′E*
117 Q8 **Brats'ke** Mykolayivs'ka Oblast', S Ukraine *47°52′N 31°34′E*
122 M13 **Bratskoye Vodokhranilishche** *Eng.* Bratsk Reservoir. ⊠ S Russian Federation
Bratsk Reservoir *see* Bratskoye Vodokhranilishche
Brattia *see* Brač
94 D9 **Brattvåg** Møre og Romsdal, S Norway *62°36′N 06°21′E*
112 K12 **Bratunac** ◈ Republika Srpska, E Bosnia and Herzegovina
114 J10 **Bratya Daskalovi** *prev.* Grozdovo. Stara Zagora, C Bulgaria *42°13′N 25°21′E*
109 U2 **Braunau** ≈ N Austria
Braunau *see* Braunau am Inn
109 Q4 **Braunau am Inn** *var.* Braunau. Oberösterreich, N Austria *48°16′N 13°03′E*
Braunsberg *see* Braniewo
100 J13 **Braunschweig** *Eng./Fr.* Brunswick. Niedersachsen, N Germany *52°16′N 10°32′E*
Brava *see* Baraawe
105 Y6 **Brava, Costa** *coastal region* NE Spain
43 V16 **Brava, Punta** *headland* E Panama *08°21′N 78°22′W*
95 N17 **Bråviken** *inlet* S Sweden
56 B10 **Bravo, Cerro** ▲ N Peru *05°33′S 79°10′W*
Bravo del Norte, Río/ Bravo, Río *see* Grande, Rio
35 X17 **Brawley** California, W USA *32°58′N 115°31′W*
97 G18 **Bray** *Ir.* Bré. E Ireland *53°12′N 06°06′W*
59 G16 **Brazil** *off.* Federative Republic of Brazil, *Port.* Repùblica Federativa do Brasil, *Sp.* Brasil; *prev.* United States of Brazil. ◆ *federal republic* South America
65 K15 **Brazil Basin** *var.* Brazilian Basin, Brazil'skaya Kotlovina. *undersea feature* W Atlantic Ocean *15°00′S 25°00′W*
Brazil, Federative Republic of *see* Brazil
Brazilian Basin *see* Brazil Basin
Brazilian Highlands *see* Central, Planalto
Brazil'skaya Kotlovina *see* Brazil Basin
Brazil, United States of *see* Brazil
25 U10 **Brazos River** ≈ Texas, SW USA
Brazza *see* Brač
79 G21 **Brazzaville** ● (Congo) Capital District, S Congo *04°14′S 15°14′E*
79 G21 **Brazzaville** ✈ Pool, S Congo *04°15′S 15°15′E*
112 J11 **Brčko** Brčko Distrikt, NE Bosnia and Herzegovina *44°52′N 18°49′E*
112 J11 **Brčko Distrikt** ◈ *self-governing district* Bosnia and Herzegovina
110 H8 **Brda** *Ger.* Brahe. ≈ N Poland
Bré *see* Bray
185 A23 **Breaksea Sound** *sound* South Island, New Zealand
184 L4 **Bream Bay** *bay* North Island, New Zealand
184 L4 **Bream Head** *headland* North Island, New Zealand *35°51′S 174°35′E*
Bréanainn, Cnoc *see* Brandon Mountain
45 S6 **Brea, Punta** *headland* W Puerto Rico *17°56′N 66°55′W*
22 I9 **Breaux Bridge** Louisiana, S USA *30°16′N 91°54′W*
116 J13 **Breaza** Prahova, SE Romania *45°13′N 25°38′E*
169 P16 **Brebes** Jawa, C Indonesia *06°54′S 109°00′E*
96 K10 **Brechin** E Scotland, United Kingdom *56°45′N 02°38′W*
99 H15 **Brecht** Antwerpen, N Belgium *51°21′N 04°33′E*
37 S4 **Breckenridge** Colorado, C USA *39°28′N 106°02′W*
29 R6 **Breckenridge** Minnesota, N USA *46°15′N 96°35′W*
25 R6 **Breckenridge** Texas, SW USA *32°45′N 98°56′W*
97 J21 **Brecknock** *cultural region* SE Wales, United Kingdom
63 G25 **Brecknock, Península** *headland* S Chile *54°39′S 71°48′W*
111 G19 **Břeclav** *Ger.* Lundenburg. Jihomoravský Kraj, SE Czech Republic *49°05′N 16°51′E*
97 J21 **Brecon** E Wales, United Kingdom *51°58′N 03°26′W*
97 J21 **Brecon Beacons** ▲▲ S Wales, United Kingdom
99 I14 **Breda** Noord-Brabant, S Netherlands *51°35′N 04°46′E*
95 K20 **Bredaryd** Jönköping, S Sweden *57°10′N 13°45′E*
83 F26 **Bredasdorp** Western Cape, SW South Africa *34°32′S 20°02′E*
93 H16 **Bredbyn** Västernorrland, N Sweden *63°28′N 18°04′E*
122 F11 **Bredy** Chelyabinskaya Oblast', C Russian Federation *52°23′N 60°24′E*
99 K17 **Bree** Limburg, NE Belgium *51°08′N 05°36′E*
67 T15 **Breede** ≈ S South Africa
98 I7 **Breezand** Noord-Holland, NW Netherlands *52°52′N 04°47′E*
113 P18 **Bregalnica** ≈ E FYR Macedonia
108 I6 **Bregenz** *anc.* Brigantium. Vorarlberg, W Austria *47°31′N 09°46′E*
108 J7 **Bregenzer Wald** ▲▲ W Austria
114 F6 **Bregovo** Vidin, NW Bulgaria *44°07′N 22°40′E*
102 H5 **Bréhat, Île de** *island* NW France
92 H2 **Breiðafjörður** *bay* W Iceland
92 L3 **Breiðdalsvík** Austurland, E Iceland *64°48′N 14°02′W*
108 H9 **Breil** *Ger.* Brigels. Graubünden, S Switzerland *46°46′N 09°04′E*
92 J8 **Breivikbotn** Finnmark, N Norway *70°36′N 22°19′E*
94 I9 **Brekken** Sør-Trøndelag, S Norway *62°39′N 11°49′E*
94 G7 **Brekstad** Sør-Trøndelag, S Norway *63°42′N 09°40′E*
94 B10 **Bremangerlandet** *island* S Norway
Brême *see* Bremen
100 H11 **Bremen** *Fr.* Brême. Bremen, NW Germany *53°06′N 08°48′E*
23 R3 **Bremen** Georgia, SE USA *33°43′N 85°09′W*
31 O11 **Bremen** Indiana, N USA *41°24′N 86°07′W*
100 H10 **Bremen** *off.* Freie Hansestadt Bremen, *Fr.* Brême. ◈ *state* N Germany
100 G9 **Bremerhaven** Bremen, NW Germany *53°33′N 08°35′E*
Bremersdorp *see* Manzini
32 G8 **Bremerton** Washington, NW USA *47°34′N 122°37′W*
100 H10 **Bremervörde** Niedersachsen, NW Germany *53°29′N 09°06′E*
25 U9 **Bremond** Texas, SW USA *31°10′N 96°40′W*
25 U10 **Brenham** Texas, SW USA *30°09′N 96°24′W*
108 M8 **Brenner** Tirol, W Austria *47°10′N 11°51′E*
Brenner, Col du/Brennero, Passo del *see* Brenner Pass
108 M8 **Brenner Pass** *var.* Brenner Sattel, *Fr.* Col du Brenner, *Ger.* Brennerpass, *It.* Passo del Brennero. *pass* Austria/Italy
Brennerpass *see* Brenner Pass
Brenner Sattel *see* Brenner Pass
108 G10 **Brenno** ≈ SW Switzerland
106 F7 **Breno** Lombardia, N Italy *45°58′N 10°18′E*
23 O5 **Brent** Alabama, S USA *32°54′N 87°10′W*
106 H7 **Brenta** ≈ NE Italy
97 P21 **Brentwood** E England, United Kingdom *51°38′N 00°21′E*
18 L14 **Brentwood** Long Island, New York, NE USA *40°46′N 73°12′W*
106 F7 **Brescia** *anc.* Brixia. Lombardia, N Italy *45°33′N 10°13′E*
99 D15 **Breskens** Zeeland, SW Netherlands *51°24′N 03°33′E*
Breslau *see* Wrocław
106 H5 **Bressanone** *Ger.* Brixen. Trentino-Alto Adige, N Italy *46°44′N 11°41′E*
96 M2 **Bressay** *island* NE Scotland, United Kingdom
102 K9 **Bressuire** Deux-Sèvres, W France *46°50′N 00°29′W*
119 F20 **Brest** *Pol.* Brześć nad Bugiem, *Rus.* Brest-Litovsk; *prev.* Brześć Litewski. Brestskaya Voblasts', SW Belarus *52°06′N 23°42′E*
102 F5 **Brest** Finistère, NW France *48°24′N 04°31′W*
Brest-Litovsk *see* Brest
112 A10 **Brestova** Istra, NW Croatia *45°09′N 14°13′E*
Brestskaya Oblast' *see* Brestskaya Voblasts'
119 G19 **Brestskaya Voblasts'** *prev.* *Rus.* Brestskaya Oblast'. ◈ *province* SW Belarus
102 G6 **Bretagne** *Eng.* Brittany, *Lat.* Britannia Minor. ◈ *region* NW France
116 G12 **Bretea-Română** *Hung.* Oláhbrettye; *prev.* Bretea-Romînă. Hunedoara, W Romania *45°39′N 23°00′E*
Bretea-Romînă *see* Bretea-Română
103 O3 **Breteuil** Oise, N France *49°37′N 02°18′E*
102 I10 **Breton, Pertuis** *inlet* W France
22 L10 **Breton Sound** *sound* Louisiana, S USA
184 K2 **Brett, Cape** *headland* North Island, New Zealand *35°11′S 174°21′E*
101 G21 **Bretten** Baden-Württemberg, SW Germany *49°01′N 08°42′E*
99 K15 **Breugel** Noord-Brabant, S Netherlands *51°30′N 05°30′E*
106 B6 **Breuil-Cervinia** *It.* Cervinia. Valle d'Aosta, NW Italy *45°57′N 07°37′E*
98 I11 **Breukelen** Utrecht, C Netherlands *52°11′N 05°01′E*
21 P10 **Brevard** North Carolina, SE USA *35°13′N 82°46′W*
38 L9 **Brevig Mission** Alaska, USA *65°19′N 166°29′W*
95 G16 **Brevik** Telemark, S Norway *59°05′N 09°42′E*
183 P5 **Brewarrina** New South Wales, SE Australia *30°01′S 146°50′E*
19 R6 **Brewer** Maine, NE USA *44°46′N 68°44′W*
29 T11 **Brewster** Minnesota, N USA *43°43′N 95°28′W*
29 N14 **Brewster** Nebraska, C USA *41°57′N 99°52′W*
31 U12 **Brewster** Ohio, N USA *40°42′N 81°36′W*
Brewster, Kap *see* Kangikajik
183 O8 **Brewster, Lake** ◎ New South Wales, SE Australia
23 P7 **Brewton** Alabama, S USA *31°06′N 87°04′W*
Brezhnev *see* Naberezhnyye Chelny
109 W12 **Brežice** *Ger.* Rann. E Slovenia *45°54′N 15°35′E*
114 G9 **Breznik** Pernik, W Bulgaria *42°45′N 22°54′E*
111 K19 **Brezno** *Ger.* Bries, Briesen, *Hung.* Breznóbánya; *prev.* Brezno nad Hronom. Banskobystrický Kraj, C Slovakia *48°49′N 19°40′E*
Breznóbánya/Brezno nad Hronom *see* Brezno
116 I12 **Brezoi** Vâlcea, SW Romania *45°18′N 24°15′E*
114 J10 **Brezovo** *prev.* Abrashlare. Plovdiv, C Bulgaria *42°19′N 25°05′E*
79 K14 **Bria** Haute-Kotto, C Central African Republic *06°30′N 22°00′E*
103 U13 **Briançon** *anc.* Brigantio. Hautes-Alpes, SE France *44°55′N 06°37′E*
36 K7 **Brian Head** ▲ Utah, W USA *37°40′N 112°49′W*
103 O7 **Briare** Loiret, C France *47°35′N 02°46′E*
183 V2 **Bribie Island** *island* Queensland, E Australia
43 O14 **Bribrí** Limón, E Costa Rica *09°37′N 82°51′W*
116 L8 **Briceni** *var.* Brinceni, *Rus.* Brichany. N Moldova *48°21′N 27°02′E*
Bricgstow *see* Bristol
Brichany *see* Briceni
99 M24 **Bridel** Luxembourg, C Luxembourg *49°40′N 06°03′E*

- ◆ Country
- ● Country Capital
- ◇ Dependent Territory
- ○ Dependent Territory Capital
- ◈ Administrative Regions
- ✈ International Airport
- ▲ Mountain
- ▲▲ Mountain Range
- ⛰ Volcano
- ≈ River
- ◎ Lake
- ⊠ Reservoir

97 J22 **Bridgend** S Wales, United Kingdom 51°30′N 03°37′W
14 I14 **Bridgenorth** Ontario, SE Canada 44°21′N 78°22′W
23 Q1 **Bridgeport** Alabama, S USA 34°57′N 85°42′W
35 R8 **Bridgeport** California, W USA 38°14′N 119°15′W
18 L13 **Bridgeport** Connecticut, NE USA 41°10′N 73°12′W
31 N15 **Bridgeport** Illinois, N USA 38°42′N 87°45′W
28 J14 **Bridgeport** Nebraska, C USA 41°37′N 103°07′W
25 S6 **Bridgeport** Texas, SW USA 33°12′N 97°45′W
21 S3 **Bridgeport** West Virginia, NE USA 39°17′N 80°15′W
25 S5 **Bridgeport, Lake** ⊠ Texas, SW USA
33 U11 **Bridger** Montana, NW USA 45°16′N 108°55′W
18 I17 **Bridgeton** New Jersey, NE USA 39°24′N 75°10′W
180 J14 **Bridgetown** Western Australia 34°01′S 116°07′E
45 Y14 **Bridgetown** ● (Barbados) SW Barbados 13°05′N 59°36′W
183 P17 **Bridgewater** Tasmania, SE Australia 42°47′S 147°15′E
13 P16 **Bridgewater** Nova Scotia, SE Canada 44°19′N 64°30′W
19 P12 **Bridgewater** Massachusetts, NE USA 41°59′N 70°58′W
29 Q11 **Bridgewater** South Dakota, N USA 43°33′N 97°30′W
21 U5 **Bridgewater** Virginia, NE USA 38°22′N 78°58′W
19 P8 **Bridgton** Maine, NE USA 44°04′N 70°43′W
97 K23 **Bridgwater** SW England, United Kingdom 51°08′N 03°W
97 K22 **Bridgwater Bay** bay SW England, United Kingdom
97 O16 **Bridlington** E England, United Kingdom 54°05′N 00°12′W
97 O16 **Bridlington Bay** bay E England, United Kingdom
183 P15 **Bridport** Tasmania, SE Australia 41°03′S 147°26′E
97 K24 **Bridport** S England, United Kingdom 50°44′N 02°43′W
103 O5 **Brie** cultural region N France
Brieg see Brzeg
Briel see Brielle
98 G12 **Brielle** var. Briel, Bril, Eng. The Brill. Zuid-Holland, SW Netherlands 51°54′N 04°10′E
108 E9 **Brienz** Bern, C Switzerland 46°45′N 08°00′E
108 E9 **Brienzer See** ◎ SW Switzerland
Bries/Briesen see Brezno
Brietzig see Brzesko
103 S4 **Briey** Meurthe-et-Moselle, NE France 49°15′N 05°57′E
108 E10 **Brig** Fr. Brigue, It. Briga. Valais, SW Switzerland 46°19′N 08°E
Briga see Brig
101 G24 **Brigach** ≈ S Germany
18 K17 **Brigantine** New Jersey, NE USA 39°23′N 74°21′W
Brigantio see Briançon
Brigantium see Bregenz
Brigels see Breil
25 S9 **Briggs** Texas, SW USA 30°52′N 97°55′W
36 L1 **Brigham City** Utah, W USA 41°30′N 112°00′W
14 J15 **Brighton** Ontario, SE Canada 44°01′N 77°44′W
97 O23 **Brighton** SE England, United Kingdom 50°50′N 00°10′W
37 T4 **Brighton** Colorado, C USA 39°58′N 104°46′W
30 K15 **Brighton** Illinois, N USA 39°01′N 90°09′W
103 T16 **Brignoles** Var, W France 43°25′N 06°03′E
Brigue see Brig
105 O7 **Brihuega** Castilla-La Mancha, C Spain 40°45′N 02°52′W
112 A10 **Brijuni** It. Brioni. island group NW Croatia
76 G12 **Brikama** W Gambia 13°13′N 16°37′W
Bril see Brielle
Brill, The see Brielle
101 G15 **Brilon** Nordrhein-Westfalen, W Germany 51°24′N 08°34′E
Brinceni see Briceni
107 Q18 **Brindisi** anc. Brundisium, Brundusium. Puglia, SE Italy 40°39′N 17°55′E
27 W11 **Brinkley** Arkansas, C USA 34°53′N 91°11′W
Brioni see Brijuni
103 P12 **Brioude** anc. Brivas. Haute-Loire, C France 45°18′N 03°23′E
Briovera see St-Lô
183 U2 **Brisbane** state capital Queensland, E Australia 27°30′S 153°E
183 V2 **Brisbane** ✈ Queensland, E Australia 27°30′S 153°00′E
25 P2 **Briscoe** Texas, SW USA 35°34′N 100°17′W
106 H10 **Brisighella** Emilia-Romagna, C Italy 44°12′N 11°45′E
108 G11 **Brissago** Ticino, S Switzerland 46°07′N 08°40′E
97 K22 **Bristol** anc. Bricgstow. SW England, United Kingdom 51°27′N 02°35′W
18 M12 **Bristol** Connecticut, NE USA 41°40′N 72°56′W
23 R9 **Bristol** Florida, SE USA 30°25′N 84°58′W
19 N9 **Bristol** New Hampshire, NE USA 43°34′N 71°42′W
29 Q8 **Bristol** South Dakota, N USA 45°18′N 97°45′W
21 P8 **Bristol** Tennessee, S USA 36°36′N 82°11′W
18 M8 **Bristol** Vermont, NE USA 44°07′N 73°00′W
39 N14 **Bristol Bay** bay Alaska, USA
97 I22 **Bristol Channel** inlet England/Wales, United Kingdom
35 W14 **Bristol Lake** ◎ California, W USA
27 P10 **Bristow** Oklahoma, C USA 35°49′N 96°23′W
86 C10 **Britain** var. Great Britain. island United Kingdom
Britannia Minor see Bretagne
10 L12 **British Columbia** Fr. Colombie-Britannique. ⬖ province SW Canada
British Guiana see Guyana
British Honduras see Belize
173 Q7 **British Indian Ocean Territory** ◇ UK dependent territory C Indian Ocean
86 B9 **British Isles** island group NW Europe
10 I1 **British Mountains** ▲▲ Yukon, NW Canada
British North Borneo see Sabah
British Solomon Islands Protectorate see Solomon Islands
45 S8 **British Virgin Islands** var. Virgin Islands. ◇ UK dependent territory E West Indies
83 J21 **Brits** North-West, N South Africa 25°39′S 27°47′E
83 H24 **Britstown** Northern Cape, W South Africa 30°36′S 23°30′E
14 F12 **Britt** Ontario, S Canada 45°46′N 80°34′W
29 V12 **Britt** Iowa, C USA 43°06′N 93°48′W
Brittany see Bretagne
29 Q7 **Britton** South Dakota, N USA 45°47′N 97°45′W
Briva Curretia see Brive-la-Gaillarde
Briva Isarae see Pontoise
Brivas see Brioude
Brive see Brive-la-Gaillarde
102 M12 **Brive-la-Gaillarde** prev. Brive; anc. Briva Curretia. Corrèze, C France 45°09′N 01°31′E
105 O4 **Briviesca** Castilla y León, N Spain 42°33′N 03°19′W
Brixen see Bressanone
Brixia see Brescia
Brlik see Birlik
Brněnský Kraj see Jihomoravský Kraj
111 G18 **Brno** Ger. Brünn. Jihomoravský Kraj, SE Czech Republic 49°11′N 16°35′E
96 G7 **Broad Bay** bay NW Scotland, United Kingdom
25 X8 **Broaddus** Texas, SW USA 31°18′N 94°16′W
183 O12 **Broadford** Victoria, SE Australia 37°07′S 145°04′E
96 G9 **Broadford** N Scotland, United Kingdom 57°14′N 05°54′W
96 J13 **Broad Law** ▲ S Scotland, United Kingdom 55°30′N 03°22′W
23 U3 **Broad River** ≈ Georgia, SE USA
21 N8 **Broad River** ≈ North Carolina/South Carolina, SE USA
181 Y8 **Broadsound Range** ▲▲ Queensland, E Australia
33 X11 **Broadus** Montana, NW USA 45°28′N 105°22′W
21 U4 **Broadway** Virginia, NE USA 38°36′N 78°48′W
118 E9 **Broceni** SW Latvia 56°41′N 22°31′E
11 U11 **Brochet** Manitoba, C Canada 57°55′N 101°40′W
11 U10 **Brochet, Lac** ◎ Manitoba, C Canada
15 S5 **Brochet, Lac au** ◎ Québec, SE Canada
101 K14 **Brocken** ▲ C Germany 51°48′N 10°38′E
19 O12 **Brockton** Massachusetts, NE USA 42°04′N 71°01′W
14 L14 **Brockville** Ontario, SE Canada 44°35′N 75°44′W
18 D13 **Brockway** Pennsylvania, NE USA 41°14′N 78°45′W
Brod/Bród see Slavonski Brod
9 N5 **Brodeur Peninsula** peninsula Baffin Island, Nunavut, NE Canada
96 H13 **Brodick** W Scotland, United Kingdom 55°34′N 05°10′W
Brod na Savi see Slavonski Brod
110 K9 **Brodnica** Ger. Buddenbrock. Kujawski-pomorskie, C Poland 53°15′N 19°23′E
Brod-Posavina see Slavonski Brod-Posavina
Brodsko-Posavska Županija see Slavonski Brod-Posavina
116 J5 **Brody** L'vivs'ka Oblast', NW Ukraine 50°05′N 25°08′E
98 I10 **Broek-in-Waterland** Noord-Holland, C Netherlands 52°27′N 04°59′E
32 L13 **Brogan** Oregon, NW USA 44°15′N 117°34′W
110 N10 **Brok** Mazowieckie, C Poland 52°42′N 21°53′E
27 P9 **Broken Arrow** Oklahoma, C USA 36°03′N 95°47′W
183 T9 **Broken Bay** bay New South Wales, SE Australia
29 N15 **Broken Bow** Nebraska, C USA 41°24′N 99°38′W
27 R13 **Broken Bow** Oklahoma, C USA 34°01′N 94°44′W
27 R12 **Broken Bow Lake** ⊠ Oklahoma, C USA
182 L6 **Broken Hill** New South Wales, SE Australia 31°58′S 141°27′E
173 S10 **Broken Ridge** undersea feature S Indian Ocean 31°30′S 95°00′E
186 C6 **Broken Water Bay** bay W Bismarck Sea
55 W10 **Brokopondo** Brokopondo, NE Suriname 05°04′N 55°00′W
55 W10 **Brokopondo** ⬖ district C Suriname
Bromberg see Bydgoszcz
95 L22 **Bromölla** Skåne, S Sweden 56°04′N 14°28′E
97 L20 **Bromsgrove** W England, United Kingdom 52°20′N 02°03′W
95 G20 **Brønderslev** Nordjylland, N Denmark 57°16′N 09°58′E
106 D8 **Broni** Lombardia, N Italy 45°04′N 09°18′E
10 K11 **Bronlund Peak** ▲ British Columbia, W Canada 57°27′N 126°43′W
93 F14 **Brønnøysund** Nordland, C Norway 65°38′N 12°15′E
23 V10 **Bronson** Florida, SE USA 29°25′N 82°38′W
31 Q11 **Bronson** Michigan, N USA 41°52′N 85°11′W
25 X8 **Bronson** Texas, SW USA 31°20′N 94°00′W
107 L24 **Bronte** Sicilia, Italy, C Mediterranean Sea 37°47′N 14°50′E
25 P8 **Bronte** Texas, SW USA 31°53′N 100°17′W
25 Y9 **Brookeland** Texas, SW USA 31°05′N 93°57′W
170 M7 **Brooke's Point** Palawan, W Philippines 08°54′N 117°54′E
27 T3 **Brookfield** Missouri, C USA 39°46′N 93°04′W
22 K7 **Brookhaven** Mississippi, S USA 31°34′N 90°26′W
32 E16 **Brookings** Oregon, NW USA 42°03′N 124°16′W
29 R10 **Brookings** South Dakota, N USA 44°15′N 96°46′W
29 W14 **Brooklyn** Iowa, C USA 41°43′N 92°27′W
29 U8 **Brooklyn Park** Minnesota, N USA 45°06′N 93°18′W
21 U7 **Brookneal** Virginia, NE USA 37°03′N 78°56′W
11 R16 **Brooks** Alberta, SW Canada 50°35′N 111°54′W
25 V11 **Brookshire** Texas, SW USA 29°47′N 95°57′W
38 L8 **Brooks Mountain** ▲ Alaska, USA 65°31′N 167°24′W
38 M11 **Brooks Range** ▲▲ Alaska, USA
31 O12 **Brookston** Indiana, N USA 40°34′N 86°53′W
23 V11 **Brooksville** Florida, SE USA 28°33′N 82°23′W
23 N4 **Brooksville** Mississippi, S USA 33°13′N 88°34′W
180 J13 **Brookton** Western Australia 32°24′S 117°04′E
31 Q14 **Brookville** Indiana, N USA 39°25′N 85°00′W
18 D13 **Brookville** Pennsylvania, NE USA 41°07′N 79°05′W
31 Q14 **Brookville Lake** ⊠ Indiana, N USA
180 K5 **Broome** Western Australia 17°58′S 122°15′E
37 S4 **Broomfield** Colorado, C USA 39°55′N 105°05′W
Broos see Orăștie
96 J7 **Brora** N Scotland, United Kingdom 57°59′N 04°00′W
96 I7 **Brora** ≈ N Scotland, United Kingdom
95 F23 **Brørup** Syddtjylland, W Denmark 55°29′N 09°01′E
95 L23 **Brösarp** Skåne, S Sweden 55°43′N 14°10′E
116 J9 **Broșteni** Suceava, NE Romania 47°14′N 25°43′E
102 M6 **Brou** Eure-et-Loir, C France 48°12′N 01°10′E
Broucsella see Brussel/Bruxelles
Broughton Bay see Tongjosŏn-man
Broughton Island see Qikiqtarjuaq
138 G7 **Broummâna** C Lebanon 33°53′N 35°39′E
22 I9 **Broussard** Louisiana, S USA 30°09′N 91°57′W
98 E13 **Brouwersdam** dam SW Netherlands
98 E13 **Brouwershaven** Zeeland, SW Netherlands 51°44′N 03°50′E
117 P4 **Brovary** Kyyivs'ka Oblast', N Ukraine 50°30′N 30°45′E
95 G20 **Brovst** Nordjylland, N Denmark 57°06′N 09°32′E
31 S8 **Brown City** Michigan, N USA 43°12′N 82°50′W
24 M6 **Brownfield** Texas, SW USA 33°11′N 102°16′W
33 Q7 **Browning** Montana, NW USA 48°33′N 113°00′W
33 R6 **Brown, Mount** ▲ Montana, NW USA 48°52′N 111°08′W
0 M9 **Browns Bank** undersea feature NW Atlantic Ocean 42°40′N 66°05′W
31 O14 **Brownsburg** Indiana, N USA 39°50′N 86°24′W
18 J16 **Browns Mills** New Jersey, NE USA 39°58′N 74°33′W
44 J12 **Browns Town** C Jamaica 18°28′N 77°22′W
31 P15 **Brownstown** Indiana, N USA 38°52′N 86°02′W
29 R8 **Browns Valley** Minnesota, N USA 45°36′N 96°49′W
20 K7 **Brownsville** Kentucky, S USA 37°10′N 86°18′W
20 F9 **Brownsville** Tennessee, S USA 35°35′N 89°15′W
25 T17 **Brownsville** Texas, SW USA 25°56′N 97°28′W
55 W10 **Brownsweg** Brokopondo, C Suriname
29 U9 **Brownton** Minnesota, N USA 44°43′N 94°21′W
19 R5 **Brownville Junction** Maine, NE USA 45°20′N 69°04′W
25 R8 **Brownwood** Texas, SW USA 31°42′N 98°59′W
25 R8 **Brownwood Lake** ⊠ Texas, SW USA
104 I9 **Brozas** Extremadura, W Spain 39°37′N 06°48′W
119 M18 **Brozha** Mahilyowskaya Voblasts', E Belarus 52°57′N 29°07′E
103 O2 **Bruay-en-Artois** Pas-de-Calais, N France 50°31′N 02°30′E
103 P2 **Bruay-sur-l'Escaut** Nord, N France 50°24′N 03°33′E
14 F13 **Bruce Peninsula** peninsula Ontario, S Canada
20 H9 **Bruceton** Tennessee, S USA 36°02′N 88°14′W
25 T9 **Bruceville** Texas, SW USA 31°17′N 97°15′W
101 G21 **Bruchsal** Baden-Württemberg, SW Germany 49°07′N 08°35′E
109 Q7 **Bruck** Salzburg, NW Austria 47°18′N 12°51′E
Bruck see Bruck an der Mur
109 Y4 **Bruck an der Leitha** Niederösterreich, NE Austria 48°02′N 16°47′E
109 V7 **Bruck an der Mur** var. Bruck. Steiermark, C Austria 47°25′N 15°17′E
101 M24 **Bruckmühl** Bayern, SE Germany 47°52′N 11°54′E
168 E7 **Brueuh, Pulau** island NW Indonesia
Bruges see Brugge
108 F6 **Brugg** Aargau, NW Switzerland 47°29′N 08°13′E
99 C16 **Brugge** Fr. Bruges. West-Vlaanderen, NW Belgium 51°13′N 03°14′E
109 R9 **Bruggen** Kärnten, S Austria 46°46′N 13°13′E
101 E16 **Brühl** Nordrhein-Westfalen, W Germany 50°50′N 06°55′E
99 F14 **Bruinisse** Zeeland, SW Netherlands 51°40′N 04°04′E
169 R9 **Bruit, Pulau** island East Malaysia
14 K10 **Brûlé, Lac** ◎ Québec, SE Canada
30 M4 **Brule River** ≈ Michigan/Wisconsin, N USA
99 H23 **Brûly** Namur, S Belgium 49°55′N 04°31′E
59 N17 **Brumado** Bahia, E Brazil 14°14′S 41°38′W
98 M11 **Brummen** Gelderland, E Netherlands 52°05′N 06°10′E
94 H13 **Brumunddal** Hedmark, S Norway 60°54′N 11°00′E
23 Q6 **Brundidge** Alabama, S USA 31°43′N 85°49′W
Brundisium/Brundusium see Brindisi
33 N15 **Bruneau River** ≈ Idaho, NW USA
Bruneck see Brunico
169 T8 **Brunei** off. Brunei Darussalam, Mal. Negara Brunei Darussalam. ◆ monarchy SE Asia
169 T7 **Brunei Bay** var. Teluk Brunei. bay N Brunei
Brunei Darussalam see Brunei
Brunei, Teluk see Brunei Bay
Brunei Town see Bandar Seri Begawan
106 H5 **Brunico** Ger. Bruneck. Trentino-Alto Adige, N Italy 46°49′N 11°57′E
Brünn see Brno
185 G17 **Brunner, Lake** ◎ South Island, New Zealand
99 M18 **Brunssum** Limburg, SE Netherlands 50°57′N 05°59′E
23 W7 **Brunswick** Georgia, SE USA 31°09′N 81°30′W
19 Q8 **Brunswick** Maine, NE USA 43°54′N 69°58′W
21 V3 **Brunswick** Maryland, NE USA 39°18′N 77°37′W
27 T3 **Brunswick** Missouri, C USA 39°25′N 93°07′W
31 T11 **Brunswick** Ohio, N USA 41°14′N 81°50′W
Brunswick see Braunschweig
63 H24 **Brunswick, Península** headland S Chile 53°30′S 71°27′W
111 H17 **Bruntál** Ger. Freudenthal. Moravskoslezský Kraj, E Czech Republic 50°00′N 17°27′E
195 N3 **Brunt Ice Shelf** ice shelf Antarctica
Brusa see Bursa
114 G7 **Brusartsi** Montana, NW Bulgaria 43°39′N 23°04′E
37 U3 **Brush** Colorado, C USA 40°15′N 103°37′W
42 M5 **Brus Laguna** Gracias a Dios, E Honduras 15°46′N 84°32′W
60 K13 **Brusque** Santa Catarina, S Brazil 27°07′S 48°54′W
Brussa see Bursa
99 E18 **Brussel** var. Brussels, Fr. Bruxelles, Ger. Brüssel; anc. Broucsella. ● (Belgium) Brussels, C Belgium 50°52′N 04°21′E see also Bruxelles
Brussel see Bruxelles
Brüssel/Brussels see Brussel/Bruxelles
117 O5 **Brusyliv** Zhytomyrs'ka Oblast', N Ukraine 50°16′N 29°31′E
183 Q12 **Bruthen** Victoria, SE Australia 37°43′S 147°49′E
Bruttium see Calabria
Brüx see Most
99 E18 **Bruxelles** var. Brussels, Dut. Brussel, Ger. Brüssel; anc. Broucsella. ● Brussels, C Belgium 50°52′N 04°21′E see also Brussel
Bruxelles see Brussel
54 J7 **Bruzual** Apure, W Venezuela 07°59′N 69°18′W
31 Q11 **Bryan** Ohio, N USA 41°30′N 84°34′W
25 U10 **Bryan** Texas, SW USA 30°40′N 96°23′W
194 J4 **Bryan Coast** physical region Antarctica
122 L11 **Bryanka** Krasnoyarskiy Kray, C Russian Federation 59°01′N 93°13′E
117 Y7 **Bryanka** Luhans'ka Oblast', E Ukraine 48°30′N 38°45′E
182 J8 **Bryan, Mount** ▲ South Australia 33°25′S 138°59′E
126 I6 **Bryansk** Bryanskaya Oblast', W Russian Federation 53°16′N 34°07′E
126 H6 **Bryanskaya Oblast'** ⬖ province W Russian Federation
194 J5 **Bryant, Cape** headland Antarctica
27 U8 **Bryant Creek** ≈ Missouri, C USA
36 K8 **Bryce Canyon** canyon Utah, W USA
119 O15 **Bryli** Mahilyowskaya Voblasts', E Belarus 53°54′N 30°33′E
95 C17 **Bryne** Rogaland, S Norway 58°43′N 05°40′E
25 R6 **Bryson** Texas, SW USA 33°09′N 98°23′W
21 N10 **Bryson City** North Carolina, SE USA 35°26′N 83°27′W
14 K11 **Bryson, Lac** ◎ Québec, SE Canada
126 K13 **Bryukhovetskaya** Krasnodarskiy Kray, SW Russian Federation 45°49′N 38°01′E
111 H15 **Brzeg** Ger. Brieg; anc. Civitas Altae Ripae. Opolskie, S Poland 50°52′N 17°27′E
111 G14 **Brzeg Dolny** Ger. Dyhernfurth. Dolnośląskie, SW Poland 51°15′N 16°40′E
Brześć Litewski/Brześć nad Bugiem see Brest
111 L17 **Brzesko** Ger. Brietzig. Małopolskie, SE Poland 49°59′N 20°34′E
Brzeżany see Berezhany
110 K12 **Brzeziny** Łódzkie, C Poland 51°49′N 19°41′E
Brzostowica Wielka see Vyalikaya Byerastavitsa
111 O17 **Brzozów** Podkarpackie, SE Poland 49°38′N 22°00′E
Bsharrī/Bsherri see Bcharré
187 X14 **Bua** Vanua Levu, N Fiji 16°48′S 178°36′E
95 J20 **Bua** Halland, S Sweden 57°14′N 12°07′E
82 M13 **Bua** ≈ C Malawi
Bua see Čiovo
81 L18 **Bu'aale** It. Buale. Jubbada Dhexe, SW Somalia 0°52′N 42°37′E
Buache, Mount see Mutunte, Mount
189 Q8 **Buada Lagoon** lagoon Nauru, C Pacific Ocean
186 M8 **Buala** Santa Isabel, E Solomon Islands 08°06′S 159°31′E
Buale see Bu'aale
190 H1 **Buariki** atoll Tungaru, W Kiribati
167 Q10 **Bua Yai** var. Ban Bua Yai. Nakhon Ratchasima, E Thailand 15°35′N 102°25′E
75 P8 **Bu'ayrāt al Ḥasūn** var. Buwayrāt al Hasūn. C Libya 31°22′N 15°41′E
76 H13 **Buba** S Guinea-Bissau 11°36′N 14°55′W
171 P11 **Bubaa** Sulawesi, N Indonesia 0°32′N 122°27′E
81 D20 **Bubanza** NW Burundi 03°04′S 29°22′E
83 K18 **Bubi** prev. Bubye. ≈ S Zimbabwe
142 L11 **Būbīyan, Jazīrat** island E Kuwait
Bublitz see Bobolice
Bubye see Bubi
187 Y13 **Buca** prev. Mbutha. Vanua Levu, N Fiji 16°39′S 179°51′E
136 F16 **Bucak** Burdur, SW Turkey 37°28′N 30°37′E
54 G8 **Bucaramanga** Santander, N Colombia 07°08′N 73°10′W
107 M18 **Buccino** Campania, S Italy 40°37′N 15°25′E
116 K9 **Bucecea** Botoșani, NE Romania 47°45′N 26°30′E
116 J6 **Buchach** Pol. Buczacz. Ternopil's'ka Oblast', W Ukraine 49°04′N 25°23′E
183 Q12 **Buchan** Victoria, SE Australia 37°26′S 148°11′E
76 J17 **Buchanan** prev. Grand Bassa. SW Liberia 05°53′N 10°03′W
23 R3 **Buchanan** Georgia, SE USA 33°48′N 85°11′W
31 O11 **Buchanan** Michigan, N USA 41°49′N 86°21′W
21 T6 **Buchanan** Virginia, NE USA 37°31′N 79°40′W
25 R10 **Buchanan Dam** Texas, SW USA 30°42′N 98°24′W
25 R10 **Buchanan, Lake** ⊠ Texas, SW USA
96 L8 **Buchan Ness** headland NE Scotland, United Kingdom 57°28′N 01°46′W
13 T12 **Buchans** Newfoundland and Labrador, SE Canada 48°49′N 56°53′W
Bucharest see București
101 H20 **Buchen** Baden-Württemberg, SW Germany 49°31′N 09°18′E
100 I10 **Buchholz in der Nordheide** Niedersachsen, NW Germany 53°19′N 09°52′E
108 F7 **Buchs** Aargau, N Switzerland 47°24′N 08°04′E
108 I8 **Buchs** Sankt Gallen, NE Switzerland 47°10′N 09°28′E
100 H13 **Bückeburg** Niedersachsen, NW Germany 52°16′N 09°03′E
36 K14 **Buckeye** Arizona, SW USA 33°22′N 112°34′W
Buckeye State see Ohio
21 S4 **Buckhannon** West Virginia, NE USA 38°59′N 80°14′W
25 T9 **Buckholts** Texas, SW USA 30°52′N 97°07′W
96 K8 **Buckie** NE Scotland, United Kingdom 57°39′N 02°56′W
14 M12 **Buckingham** Québec, SE Canada 45°35′N 75°25′W
21 U6 **Buckingham** Virginia, NE USA 37°33′N 78°34′W
97 N21 **Buckinghamshire** cultural region SE England, United Kingdom
39 N8 **Buckland** Alaska, USA 65°58′N 161°07′W
182 G7 **Buckleboo** South Australia 32°55′S 136°11′E
26 K7 **Bucklin** Kansas, C USA 37°33′N 99°37′W
27 T3 **Bucklin** Missouri, C USA 39°46′N 92°53′W
36 I12 **Buckskin Mountains** ▲▲ Arizona, SW USA
19 R7 **Bucksport** Maine, NE USA 44°34′N 68°46′W
82 A9 **Buco Zau** Cabinda, NW Angola 04°45′S 12°34′E
Bu Craa see Bou Craa
116 K14 **București** Eng. Bucharest, Ger. Bukarest, prev. Altenburg; anc. Cetatea Damboviței. ● (Romania) București, S Romania 44°27′N 26°06′E
31 S12 **Bucyrus** Ohio, N USA 40°47′N 82°57′W
Buczacz see Buchach
94 E9 **Bud** Møre og Romsdal, S Norway 62°55′N 06°55′E
25 S11 **Buda** Texas, SW USA 30°05′N 97°50′W
119 O18 **Buda-Kashalyova** Rus. Buda-Koshelëvo. Homyel'skaya Voblasts', SE Belarus 52°43′N 30°34′E
Buda-Koshelëvo see Buda-Kashalyova
166 L4 **Budalin** Sagaing, C Myanmar (Burma) 22°24′N 95°08′E
111 J22 **Budapest** off. Budapest Főváros, SCr. Budimpešta. ● (Hungary) Pest, N Hungary 47°30′N 19°03′E
Budapest Főváros see Budapest
92 I2 **Búðardalur** Vesturland, W Iceland 65°07′N 21°45′W
152 K11 **Budaun** Uttar Pradesh, N India 28°02′N 79°07′E
141 O9 **Budayyi'ah** oasis C Saudi Arabia
195 Y12 **Budd Coast** physical region Antarctica
Buddenbrock see Brodnica
107 C17 **Budduso** Sardegna, Italy, C Mediterranean Sea 40°37′N 09°19′E
97 I23 **Bude** SW England, United Kingdom 50°50′N 04°33′W
22 J7 **Bude** Mississippi, S USA 31°27′N 90°51′W
Budějovický Kraj see Jihočeský Kraj
99 K16 **Budel** Noord-Brabant, SE Netherlands 51°17′N 05°35′E
100 I8 **Büdelsdorf** Schleswig-Holstein, N Germany 54°20′N 09°40′E
127 O14 **Budënnovsk** Stavropol'skiy Kray, SW Russian Federation 44°46′N 44°07′E
116 K14 **Budești** Călărași, SE Romania 44°13′N 26°30′E
Budge-Budge see Baj Baj
Budgewoi see Budgewoi Lake
183 T8 **Budgewoi Lake** var. Budgewoi. New South Wales, SE Australia 33°14′S 151°34′E
Budimpešta see Budapest
79 J16 **Budjala** Equateur, NW Dem. Rep. Congo 02°39′N 19°42′E
106 G10 **Budrio** Emilia-Romagna, C Italy 44°33′N 11°34′E
Budslav see Budslaw
119 K14 **Budslaw** Rus. Budslav. Minskaya Voblasts', N Belarus 54°47′N 27°27′E
Budua see Budva
169 R9 **Budu, Tanjung** headland East Malaysia 02°51′N 111°42′E
113 J17 **Budva** It. Budua. W Montenegro 42°17′N 18°49′E
Budweis see České Budějovice
Budyšin see Bautzen
79 D16 **Buea** Sud-Ouest, SW Cameroon 04°09′N 09°13′E
103 S13 **Buëch** ≈ SE France
18 J17 **Buena** New Jersey, NE USA 39°30′N 74°55′W
62 K12 **Buena Esperanza** San Luis, C Argentina 34°45′S 65°15′W
54 C11 **Buenaventura** Valle del Cauca, W Colombia 03°54′N 77°02′W
40 I4 **Buenaventura** Chihuahua, N Mexico 29°50′N 107°30′W
57 M18 **Buena Vista** Santa Cruz, C Bolivia 17°28′S 63°37′W
40 G10 **Buenavista** Baja California Sur, NW Mexico 23°39′N 109°41′W
37 S5 **Buena Vista** Colorado, C USA 38°50′N 106°07′W
23 S5 **Buena Vista** Georgia, SE USA 32°19′N 84°31′W
21 T6 **Buena Vista** Virginia, NE USA 37°44′N 79°22′W
44 F5 **Buena Vista, Bahia de** bay N Cuba
35 R13 **Buena Vista Lake Bed** ◎ California, W USA
105 P8 **Buendía, Embalse de** ⊠ C Spain
63 F16 **Bueno, Río** ≈ S Chile
62 N12 **Buenos Aires** hist. Santa Maria del Buen Aire. ● (Argentina) Buenos Aires, E Argentina 34°40′S 58°30′W
43 O15 **Buenos Aires** Puntarenas, SE Costa Rica 09°10′N 83°23′W
61 C20 **Buenos Aires** off. Provincia de Buenos Aires. ⬖ province E Argentina
63 H19 **Buenos Aires, Lago** var. Lago General Carrera. ◎ Argentina/Chile
Buenos Aires, Provincia de see Buenos Aires
54 C13 **Buesaco** Nariño, SW Colombia 01°22′N 77°07′W
29 U8 **Buffalo** Minnesota, N USA 45°11′N 93°50′W
26 T6 **Buffalo** Missouri, C USA 37°38′N 93°05′W
18 D10 **Buffalo** New York, NE USA 42°53′N 78°53′W
27 K8 **Buffalo** Oklahoma, C USA 36°51′N 99°38′W
28 J7 **Buffalo** South Dakota, N USA 45°25′N 103°35′W
25 V8 **Buffalo** Texas, SW USA 31°25′N 96°04′W
33 W12 **Buffalo** Wyoming, C USA 44°21′N 106°40′W
29 U11 **Buffalo Center** Iowa, C USA 43°23′N 93°57′W
24 M3 **Buffalo Lake** ⊠ Texas, SW USA
30 K7 **Buffalo Lake** ⊠ Wisconsin, N USA
11 S12 **Buffalo Narrows** Saskatchewan, C Canada 55°52′N 108°28′W
27 U9 **Buffalo River** ≈ Arkansas, C USA
29 R5 **Buffalo River** ≈ Minnesota, N USA
20 I10 **Buffalo River** ≈ Tennessee, S USA
30 J6 **Buffalo River** ≈ Wisconsin, N USA
44 L12 **Buff Bay** E Jamaica 18°18′N 76°40′W
23 T3 **Buford** Georgia, SE USA 34°07′N 84°00′W
28 J3 **Buford** North Dakota, N USA 48°00′N 103°58′W
33 Y17 **Buford** Wyoming, C USA 41°05′N 105°17′W
116 J14 **Buftea** Ilfov, S Romania 44°34′N 25°58′E
84 I9 **Bug** Bel. Zakhodni Buh, Eng. Western Bug, Rus. Zapadnyy Bug, Ukr. Zakhidnyy Buh. ≈ E Europe
54 D11 **Buga** Valle del Cauca, W Colombia 03°53′N 76°17′W
Buga see Dörvöljin
103 O17 **Bugarach, Pic du** ▲ S France 42°52′N 02°23′E
162 F8 **Bugat** var. Bayangol. Govĭ-Altay, SW Mongolia 45°33′N 94°22′E
146 B12 **Bugdaýly** Rus. Bugdayly. Balkan Welaýaty, W Turkmenistan 38°42′N 54°14′E
Bugdayly see Bugdaýly
Buggs Island Lake see John H. Kerr Reservoir
Bughotu see Santa Isabel
171 O14 **Bugingkalo** Sulawesi, C Indonesia 04°49′S 121°42′E
64 P6 **Bugio** island Madeira, Portugal, NE Atlantic Ocean
92 M8 **Bugøynes** Finnmark, N Norway 69°57′N 29°34′E
125 Q3 **Bugrino** Nenetskiy Avtonomnyy Okrug, NW Russian Federation 68°48′N 49°12′E
127 T5 **Bugul'ma** Respublika Tatarstan, W Russian Federation 54°31′N 52°45′E
Bügür see Luntai
127 T6 **Buguruslan** Orenburgskaya Oblast', W Russian Federation 53°38′N 52°30′E
159 R9 **Buh He** ≈ C China
101 F22 **Bühl** Baden-Württemberg, SW Germany 48°42′N 08°07′E
33 O15 **Buhl** Idaho, NW USA 42°36′N 114°45′W
116 K10 **Buhuși** Bacău, E Romania 46°41′N 26°45′E
Buie d'Istria see Buje
97 J20 **Builth Wells** E Wales, United Kingdom 52°07′N 03°28′W
186 J8 **Buin** Bougainville Island, NE Papua New Guinea 06°52′S 155°42′E
108 J9 **Buin, Piz** ▲ Austria/Switzerland 46°51′N 10°07′E
127 Q4 **Buinsk** Chuvashskaya Respublika, W Russian Federation 55°09′N 47°00′E
127 Q4 **Buinsk** Respublika Tatarstan, W Russian Federation 54°58′N 48°16′E
163 R8 **Buir Nur** Mong. Buyr Nuur. ◎ China/Mongolia see also Buyr Nuur
Buir Nur see Buyr Nuur
98 M5 **Buitenpost** Fris. Bûtenpost. Fryslân, N Netherlands 53°15′N 06°09′E
Buitenzorg see Bogor
83 F19 **Buitepos** Omaheke, E Namibia 22°17′S 19°59′E
105 N7 **Buitrago del Lozoya** Madrid, C Spain 41°00′N 03°38′W
Buj see Buy
104 M13 **Bujalance** Andalucía, S Spain 37°54′N 04°23′W
113 O17 **Bujanovac** SE Serbia 42°29′N 21°44′E
105 S6 **Bujaraloz** Aragón, NE Spain 41°29′N 00°10′W
112 A9 **Buje** It. Buie d'Istria. Istra, NW Croatia 45°23′N 13°40′E
Bujnurd see Bojnūrd
81 D21 **Bujumbura** prev. Usumbura. ● (Burundi) W Burundi 03°25′S 29°24′E
81 D20 **Bujumbura** ✈ W Burundi 03°21′S 29°19′E
186 J6 **Buka** Bougainville, Papua New Guinea 05°24′S 154°40′E
159 N11 **Buka Daban** var. Bukadaban Feng. ▲ C China 36°09′N 90°52′E
Bukadaban Feng see Buka Daban
186 J6 **Buka Island** island NE Papua New Guinea
81 F18 **Bukakata** S Uganda 0°18′S 31°57′E
79 N24 **Bukama** Katanga, SE Dem. Rep. Congo 09°13′S 25°52′E
142 J4 **Būkān** var. Bowkān. Āžarbāyjān-e Gharbī, NW Iran 36°31′N 46°10′E
Bukantau, Gory see Bo'kantov Tog'lari
Bukarest see București
79 O19 **Bukavu** prev. Costermansville. Sud-Kivu, E Dem. Rep. Congo 02°19′S 28°49′E
81 F21 **Bukene** Tabora, NW Tanzania 04°15′S 32°51′E
141 W8 **Bū Khābī** var. Bakhābī. NW Oman 23°29′N 56°06′E
Bukhara see Buxoro
Bukharskaya Oblast' see Buxoro Viloyati
168 M14 **Bukitkemuning** Sumatera, W Indonesia 04°43′S 104°27′E
168 I11 **Bukittinggi** prev. Fort de Kock. Sumatera, W Indonesia 0°18′S 100°20′E
111 L21 **Bükk** ▲▲ NE Hungary
81 F19 **Bukoba** Kagera, NW Tanzania 01°19′S 31°49′E
113 N20 **Bukovo** S FYR Macedonia 40°59′N 21°20′E
108 G6 **Bülach** Zürich, NW Switzerland 47°31′N 08°30′E
Būlaevo see Bulayevo
Bulag see Tünel, Hövsgöl, Mongolia
Bulag see Möngönmorĭt, Töv, Mongolia
Bulagiyn Denj see Bulgan
183 U7 **Bulahdelah** New South Wales, SE Australia 32°24′S 152°13′E
171 P4 **Bulan** Luzon, N Philippines 12°40′N 123°55′E
137 N11 **Bulancak** Giresun, N Turkey 40°57′N 38°14′E
152 J10 **Bulandshahr** Uttar Pradesh, N India 28°30′N 77°49′E
137 R14 **Bulanık** Muş, E Turkey 39°04′N 42°16′E
127 V7 **Bulanovo** Orenburgskaya Oblast', W Russian Federation 52°27′N 55°08′E
83 J17 **Bulawayo** var. Buluwayo. Bulawayo, SW Zimbabwe 20°08′S 28°37′E
83 J17 **Bulawayo** ✈ Matabeleland North, SW Zimbabwe 20°00′S 28°36′E
145 Q6 **Bulayevo** Kaz. Būlaevo. Severnyy Kazakhstan, N Kazakhstan 54°55′N 70°29′E
136 D15 **Buldan** Denizli, SW Turkey 38°03′N 28°50′E
154 G12 **Buldāna** Mahārāshtra, C India 20°31′N 76°18′E
38 E16 **Buldir Island** island Aleutian Islands, Alaska, USA
Buldur see Burdur
162 I8 **Bulgan** var. Bulagiyn Denj. Arhangay, C Mongolia 47°14′N 100°56′E
162 D7 **Bulgan** var. Jargalant. Bayan-Ölgiy, W Mongolia 46°56′N 91°07′E
162 K6 **Bulgan** Bulgan, N Mongolia 50°31′N 101°30′E
162 F7 **Bulgan** var. Bürenhayrhan. Hovd, W Mongolia 46°04′N 91°34′E
162 J10 **Bulgan** Ömnögovĭ, S Mongolia 44°07′N 103°28′E
162 J7 **Bulgan** ⬖ province N Mongolia
Bulgan see Bayan-Öndör, Bayanhongor, C Mongolia
Bulgan see Darvi, Hovd, Mongolia
Bulgan see Tsagaan-Üür, Hövsgöl, Mongolia
114 H10 **Bulgaria** off. Republic of Bulgaria, Bul. Bŭlgariya; prev. People's Republic of Bulgaria. ◆ republic SE Europe
Bulgaria, People's Republic of see Bulgaria
Bulgaria, Republic of see Bulgaria
Bŭlgariya see Bulgaria
Bŭlgarka see Balgarka
171 S11 **Buli** Pulau Halmahera, E Indonesia 0°56′N 128°17′E
171 S11 **Buli, Teluk** bay Pulau Halmahera, E Indonesia
160 J13 **Buliu He** ≈ S China
Bullange see Büllingen
Bulla, Ostrov see Xärä Zirä Adasi
104 M11 **Bullaque** ≈ C Spain
105 Q13 **Bullas** Murcia, SE Spain 38°02′N 01°40′W
80 M12 **Bullaxaar** Woqooyi Galbeed, NW Somalia 10°28′N 44°15′E
108 C9 **Bulle** Fribourg, SW Switzerland 46°37′N 07°04′E
185 G15 **Buller** ≈ South Island, New Zealand
183 P12 **Buller, Mount** ▲ Victoria, SE Australia 37°10′S 146°31′E
36 H11 **Bullhead City** Arizona, SW USA 35°07′N 114°32′W
99 N21 **Büllingen** Fr. Bullange. Liège, E Belgium 50°23′N 06°15′E
Bullion State see Missouri
21 T14 **Bull Island** island South Carolina, SE USA

◆ Country
● Country Capital
◇ Dependent Territory
○ Dependent Territory Capital
⬖ Administrative Regions
✈ International Airport
▲ Mountain
▲▲ Mountain Range
Volcano
≈ River
◎ Lake
⊠ Reservoir

182 M4 **Bulloo River Overflow** *wetland* New South Wales, SE Australia
184 M12 **Bulls** Manawatu-Wanganui, North Island, New Zealand *40°10′S 175°22′E*
21 T14 **Bulls Bay** *bay* South Carolina, SE USA
27 U9 **Bull Shoals Lake** ⊠ Arkansas/Missouri, C USA
181 Q2 **Bulman** Northern Territory, N Australia *13°39′S 134°21′E*
162 I6 **Bulnayn Nuruu** ▲ N Mongolia
171 O11 **Bulowa, Gunung** ▲ Sulawesi, N Indonesia *0°33′N 123°39′E*
Bulqiza *see* Bulqizë
113 L19 **Bulqizë** *var.* Bulqiza. Dibër, C Albania *41°30′N 20°16′E*
Bulsar *see* Valsād
171 N14 **Bulukumba** *prev.* Boeloekoemba. Sulawesi, C Indonesia *05°35′S 120°13′E*
147 O11 **Bulungh'ur** *Rus.* Bulungur; *prev.* Krasnogvardeysk. Samarqand Viloyati, C Uzbekistan *39°46′N 67°18′E*
79 I21 **Bulungu** Bandundu, SW Dem. Rep. Congo *04°36′S 18°34′E*
Bulungur *see* Bulungh'ur
Buluwayo *see* Bulawayo
79 K17 **Bumba** Equateur, N Dem. Rep. Congo *02°14′N 22°25′E*
121 R12 **Bumbah, Khalij al** *gulf* N Libya
Bumbat *see* Bayan-Öndör
81 F19 **Bumbire Island** *island* N Tanzania
169 V8 **Bum Bun, Pulau** *island* East Malaysia
81 J17 **Buna** Wajir, NE Kenya *02°40′N 39°34′E*
25 Y10 **Buna** Texas, SW USA *30°25′N 94°00′W*
Bunab *see* Bonāb
Bunai *see* M'bunai
147 S13 **Bunay** S Tajikistan *38°29′N 71°41′E*
180 I13 **Bunbury** Western Australia *33°24′S 115°44′E*
97 E14 **Buncrana** *Ir.* Bun Cranncha. NW Ireland *55°08′N 07°27′W*
Bun Cranncha *see* Buncrana
181 Z9 **Bundaberg** Queensland, E Australia *24°50′S 152°16′E*
183 T5 **Bundarra** New South Wales, SE Australia *30°12′S 151°06′E*
100 G13 **Bünde** Nordrhein-Westfalen, NW Germany *52°12′N 08°34′E*
152 H13 **Būndi** Rājasthān, N India *25°28′N 75°42′E*
Bun Dobhráin *see* Bundoran
97 D15 **Bundoran** *Ir.* Bun Dobhráin. NW Ireland *54°30′N 08°11′W*
Bunë *see* Bojana
113 K18 **Bunë, Lumi i** *SCr.* Bojana. ≈ Albania/Montenegro *see also* Bojana
171 Q8 **Bunga** ≈ Mindanao, S Philippines
168 I12 **Bungalaut, Selat** *strait* W Indonesia
167 R8 **Bung Kan** Nong Khai, E Thailand *18°19′N 103°39′E*
181 N4 **Bungle Bungle Range** ▲ Western Australia
82 C10 **Bungo** Uíge, NW Angola *07°30′S 15°24′E*
81 G18 **Bungoma** Bungoma, W Kenya *0°34′N 34°34′E*
81 G18 **Bungoma** ◆ *county* W Kenya
164 F15 **Bungo-suidō** *strait* SW Japan
164 E14 **Bungo-Takada** Ōita, Kyūshū, SW Japan *33°34′N 131°28′E*
100 K8 **Bungsberg** *hill* N Germany
Bungur *see* Bunyu
79 P17 **Bunia** Orientale, NE Dem. Rep. Congo *01°33′N 30°16′E*
35 U6 **Bunker Hill** ▲ Nevada, W USA *39°16′N 117°06′W*
22 I7 **Bunkie** Louisiana, S USA *30°58′N 92°12′W*
23 X10 **Bunnell** Florida, SE USA *29°28′N 81°15′W*
105 S10 **Buñol** Valenciana, E Spain *39°25′N 00°47′W*
98 K11 **Bunschoten** Utrecht, C Netherlands *52°15′N 05°23′E*
136 K14 **Bünyan** Kayseri, C Turkey *38°51′N 35°50′E*
169 W8 **Bunyu** *var.* Bungur. Borneo, N Indonesia *03°33′N 117°50′E*
169 W8 **Bunyu, Pulau** *island* N Indonesia
Bunzlau *see* Bolesławiec
Buoddobohki *see* Patoniva
123 P7 **Buor-Khaya, Guba** *bay* N Russian Federation
123 P7 **Buor-Khaya, Guba** *bay* N Russian Federation
171 Z15 **Bupul** Papua, E Indonesia *07°24′S 140°57′E*
80 P12 **Buraan** Bari, N Somalia *10°03′N 49°08′E*
145 Q7 **Burabay** *prev.* Borovoye. Akmola, N Kazakhstan *53°07′N 70°20′E*
Buraida *see* Buraydah
Buraimi *see* Al Buraymī
Buran *see* Boran
158 G15 **Burang** Xizang Zizhiqu, W China *30°28′N 81°13′E*
Burao *see* Burco
138 H8 **Burāq** Dar'ā, S Syria *33°11′N 36°28′E*
141 O6 **Buraydah** *var.* Buraida. Al Qaṣīm, N Saudi Arabia *26°50′N 44°E*
35 S15 **Burbank** California, W USA *34°10′N 118°25′W*
31 N11 **Burbank** Illinois, N USA *41°45′N 87°48′W*
183 Q8 **Burcher** New South Wales, SE Australia *33°29′S 147°16′E*
80 N13 **Burco** *var.* Burao, Bur'o. Togdheer, NW Somalia *09°29′N 45°31′E*
162 K8 **Bürd** *var.* Ongon. Övörhangay, C Mongolia *46°58′N 103°45′E*
146 L13 **Burdalyk** Lebap Welaýaty, E Turkmenistan *38°31′N 64°21′E*
181 W6 **Burdekin River** ≈ Queensland, NE Australia
27 O7 **Burden** Kansas, C USA *37°18′N 96°45′W*
Burdigala *see* Bordeaux
136 E15 **Burdur** *var.* Buldur. Burdur, SW Turkey *37°44′N 30°17′E*
136 E15 **Burdur** *var.* Buldur. ◆ *province* SW Turkey
136 E15 **Burdur Gölü** *salt lake* SW Turkey
65 H21 **Burdwood Bank** *undersea feature* SW Atlantic Ocean
80 I12 **Burē** Āmara, N Ethiopia *10°43′N 37°09′E*
80 H13 **Burē** Oromīya, C Ethiopia *08°13′N 35°09′E*
93 J15 **Bureå** Västerbotten, N Sweden *64°36′N 21°15′E*
162 K7 **Büreghangay** *var.* Darhan. Bulgan, C Mongolia *48°07′N 103°54′E*
101 G14 **Büren** Nordrhein-Westfalen, W Germany *51°34′N 08°34′E*
162 L8 **Büren** *var.* Bayantöhöm. Töv, C Mongolia *46°57′N 105°09′E*
162 K6 **Bürengiyn Nuruu** ▲ N Mongolia
Bürenhayrhan *see* Bulgan
162 I6 **Bürentogtoh** *var.* Bayan. Hövsgol, C Mongolia *49°36′N 99°36′E*
149 U10 **Būrewāla** *var.* Mandi Būrewāla. Punjab, E Pakistan *30°05′N 72°47′E*
92 J9 **Burfjord** Troms, N Norway *69°55′N 21°54′E*
100 L13 **Burg** *var.* Burg an der Ihle, Burg bei Magdeburg. Sachsen-Anhalt, C Germany *52°17′N 11°51′E*
Burg an der Ihle *see* Burg
114 N10 **Burgas** *var.* Bourgas. Burgas, E Bulgaria *42°31′N 27°30′E*
114 M10 **Burgas** ◆ *province* E Bulgaria
114 N9 **Burgas** ✈ Burgas, E Bulgaria *42°35′N 27°33′E*
114 N10 **Burgaski Zaliv** *gulf* E Bulgaria
114 M10 **Burgasko Ezero** *lagoon* E Bulgaria
21 V11 **Burgaw** North Carolina, SE USA *34°33′N 77°56′W*
Burg bei Magdeburg *see* Burg
108 E8 **Burgdorf** Bern, NW Switzerland *47°03′N 07°38′E*
109 Y7 **Burgenland** *off.* Land Burgenland. ◆ *state* SE Austria
13 S13 **Burgeo** Newfoundland, Newfoundland and Labrador, SE Canada *47°37′N 57°38′W*
83 I24 **Burgersdorp** Eastern Cape, SE South Africa *31°00′S 26°20′E*
83 K20 **Burgersfort** Mpumalanga, NE South Africa *24°39′S 30°18′E*
101 N23 **Burghausen** Bayern, SE Germany *48°10′N 12°48′E*
139 O5 **Burghūth, Sabkhat al** ⊚ E Syria
101 M20 **Burglengenfeld** Bayern, SE Germany *49°11′N 12°01′E*
41 P9 **Burgos** Tamaulipas, C Mexico *24°55′N 98°46′W*
105 N4 **Burgos** Castilla y León, N Spain *42°21′N 03°41′W*
105 N4 **Burgos** ◆ *province* Castilla y León, N Spain
Burgstadlberg *see* Hradiště
95 P20 **Burgsvik** Gotland, SE Sweden *57°01′N 18°18′E*
98 L6 **Burgum** *Dutch.* Bergum. Fryslân, N Netherlands *53°12′N 05°59′E*
Burgundy *see* Bourgogne
159 Q11 **Burhan Budai Shan** ▲ C China
136 B12 **Burhaniye** Balıkesir, W Turkey *39°29′N 26°59′E*
154 G12 **Burhānpur** Madhya Pradesh, C India *21°18′N 76°14′E*
127 W7 **Buribay** Respublika Bashkortostan, W Russian Federation *51°57′N 58°11′E*
43 O17 **Burica, Punta** *headland* Costa Rica/Panama *08°02′N 82°53′W*
167 Q10 **Buriram** *var.* Buri Ram, Puriramya. Buri Ram, E Thailand *15°01′N 103°06′E*
Buri Ram *see* Buriram
105 S10 **Burjassot** Valenciana, E Spain *39°33′N 00°26′W*
81 N16 **Burka Giibi** Hiiraan, C Somalia *03°52′N 45°07′E*
147 X8 **Burkan** ≈ E Kyrgyzstan
25 R4 **Burkburnett** Texas, SW USA *34°06′N 98°34′W*
29 O12 **Burke** South Dakota, N USA *43°09′N 99°18′W*
10 K15 **Burke Channel** *channel* British Columbia, W Canada
194 J10 **Burke Island** *island* Antarctica
20 L7 **Burkesville** Kentucky, S USA *36°48′N 85°21′W*
181 T4 **Burketown** Queensland, NE Australia *17°49′S 139°28′E*
25 Q8 **Burkett** Texas, SW USA *32°01′N 99°17′W*
25 Y9 **Burkeville** Texas, SW USA *30°58′N 93°41′W*
21 V7 **Burkeville** Virginia, NE USA *37°11′N 78°12′W*
Burkina *see* Burkina Faso
77 O12 **Burkina Faso** *off.* Burkina Faso; *prev.* Upper Volta. ◆ *republic* W Africa
Burkina Faso *see* Burkina Faso
194 L13 **Burks, Cape** *headland* Antarctica
14 H12 **Burk's Falls** Ontario, S Canada *45°38′N 79°25′W*
101 H23 **Burladingen** Baden-Württemberg, S Germany *48°18′N 09°05′E*
25 T7 **Burleson** Texas, SW USA *32°32′N 97°19′W*
33 P15 **Burley** Idaho, NW USA *42°32′N 113°47′W*
Burlin *see* Borili
14 G16 **Burlington** Ontario, S Canada *43°19′N 79°48′W*
37 W4 **Burlington** Colorado, C USA *39°17′N 102°17′W*
29 Y15 **Burlington** Iowa, C USA *40°48′N 91°05′W*
27 P5 **Burlington** Kansas, C USA *38°11′N 95°46′W*
21 T9 **Burlington** North Carolina, SE USA *36°05′N 79°27′W*
28 M3 **Burlington** North Dakota, N USA *48°16′N 101°25′W*
18 L7 **Burlington** Vermont, NE USA *44°28′N 73°14′W*
30 M9 **Burlington** Wisconsin, N USA *42°38′N 88°12′W*
27 Q1 **Burlington Junction** Missouri, C USA *40°27′N 95°04′W*
Burma *see* Myanmar (Burma)
10 L17 **Burnaby** British Columbia, SW Canada *49°16′N 122°58′W*
117 O12 **Burnas, Ozero** ⊚ SW Ukraine
25 S10 **Burnet** Texas, SW USA *30°46′N 98°14′W*
35 O3 **Burney** California, W USA *40°52′N 121°42′W*
183 O16 **Burnie** Tasmania, SE Australia *41°07′S 145°52′E*
97 L17 **Burnley** NW England, United Kingdom *53°48′N 02°14′W*
Burnoye *see* Bauyrzhan Momyshuly
153 R15 **Burnpur** West Bengal, NE India *23°39′N 86°55′E*
32 K14 **Burns** Oregon, NW USA *43°35′N 119°03′W*
26 K11 **Burns Flat** Oklahoma, C USA *35°21′N 99°10′W*
20 M7 **Burnside** Kentucky, S USA *36°55′N 84°34′W*
8 K8 **Burnside** ≈ Nunavut, NW Canada
32 L15 **Burns Junction** Oregon, NW USA *42°46′N 117°51′W*
10 L13 **Burns Lake** British Columbia, SW Canada *54°14′N 125°45′W*
29 V9 **Burnsville** Minnesota, N USA *44°49′N 93°14′W*
21 P9 **Burnsville** North Carolina, SE USA *35°56′N 82°18′W*
21 R4 **Burnsville** West Virginia, NE USA *38°50′N 80°39′W*
14 I13 **Burnt River** ≈ Ontario, SE Canada
14 I11 **Burntroot Lake** ⊚ Ontario, SE Canada
11 W12 **Burntwood** ≈ Manitoba, C Canada
Bur'o *see* Burco
158 L2 **Burqin** Xinjiang Uygur Zizhiqu, NW China *47°42′N 86°50′E*
182 J8 **Burra** South Australia *33°41′S 138°54′E*
183 S9 **Burragorang, Lake** ⊠ New South Wales, SE Australia
96 K5 **Burray** *island* NE Scotland, United Kingdom
113 L19 **Burrel** *var.* Burreli. Dibër, C Albania *41°36′N 20°00′E*
Burreli *see* Burrel
183 R8 **Burrendong Reservoir** ⊠ New South Wales, SE Australia
183 R5 **Burren Junction** New South Wales, SE Australia *30°06′S 149°01′E*
Burriana *see* Borriana
183 R10 **Burrinjuck Reservoir** ⊠ New South Wales, SE Australia
36 J12 **Burro Creek** ≈ Arizona, SW USA
40 M5 **Burro, Serranías del** ▲ NW Mexico
62 K7 **Burruyacú** Tucumán, N Argentina *26°30′S 64°45′W*
136 E12 **Bursa** *var.* Brussa, *prev.* Brusa; *anc.* Prusa. Bursa, NW Turkey *40°12′N 29°04′E*
136 D12 **Bursa** *var.* Brusa, Brussa. ◆ *province* NW Turkey
75 Y9 **Būr Safājah** *var.* Būr Safājah. E Egypt *26°43′N 33°55′E*
Būr Safājah *see* Būr Safājah
75 W7 **Būr Sa'īd** *var.* Port Said. N Egypt *31°17′N 32°18′E*
81 O14 **Bur Tinle** Nugaal, C Somalia *07°50′N 48°01′E*
31 Q5 **Burt Lake** ⊚ Michigan, N USA
118 H7 **Burtnieks** *var.* Burtnieks Ezers. ⊚ N Latvia
Burtnieks Ezers *see* Burtnieks
31 Q9 **Burton** Michigan, N USA *43°00′N 84°16′W*
Burton on Trent *see* Burton upon Trent
97 M19 **Burton upon Trent** *var.* Burton on Trent, Burton-upon-Trent. C England, United Kingdom *52°48′N 01°36′W*
93 J15 **Burträsk** Västerbotten, N Sweden *64°31′N 20°40′E*
Burubaytal *see* Burybaytal
Burujird *see* Borūjerd
Burultokay *see* Fuhai
141 R15 **Burūm** SE Yemen *14°22′N 48°53′E*
Burunday *see* Boralday
81 D21 **Burundi** *off.* Republic of Burundi; *prev.* Kingdom of Burundi, Urundi. ◆ *republic* C Africa
Burundi, Kingdom of *see* Burundi
Burundi, Republic of *see* Burundi
171 R13 **Buru, Pulau** *prev.* Boeroe. *island* E Indonesia
77 T17 **Burutu** Delta, S Nigeria *05°18′N 05°32′E*
10 G7 **Burwash Landing** Yukon, W Canada *61°26′N 139°12′W*
29 O14 **Burwell** Nebraska, C USA *41°46′N 99°04′W*
97 L17 **Bury** NW England, United Kingdom *53°36′N 02°17′W*
123 N13 **Buryatiya, Respublika** *prev.* Buryatskaya ASSR. ◆ *autonomous republic* S Russian Federation
Buryatskaya ASSR *see* Buryatiya, Respublika
145 S14 **Burybaytal** *prev.* Burubaytal. Zhambyl, SE Kazakhstan *44°56′N 73°59′E*
117 S3 **Buryn'** Sums'ka Oblast', NE Ukraine *51°13′N 33°50′E*
97 P20 **Bury St Edmunds** *hist.* Beodericsworth. E England, United Kingdom *52°15′N 00°43′E*
Būrziya *see* Barzia
106 D9 **Busalla** Liguria, NW Italy *44°35′N 08°55′E*
163 Z16 **Busan** *off.* Pusan-gwangyŏksi, *var.* Vusan; *prev.* Pusan, *Jap.* Fusan. SE South Korea *35°11′N 129°04′E*
139 N5 **Buşayrah** Dayr az Zawr, E Syria *35°10′N 40°25′E*
Buševa *see* Baba
143 N12 **Būshehr** *off.* Ostān-e Būshehr. ◆ *province* SW Iran
Būshehr/Bushire *see* Bandar-e Būshehr
Būshehr, Ostān-e *see* Būshehr
25 N2 **Bushland** Texas, SW USA *35°11′N 102°04′W*
30 J12 **Bushnell** Illinois, N USA *40°33′N 90°30′W*
Busi *see* Biševo
81 H18 **Busia** Western, W Kenya *0°28′N 34°07′E*
81 G18 **Busia** SE Uganda *01°20′N 34°48′E*
81 G18 **Busia** ◆ *county* W Kenya
Busiasch *see* Buziaş
79 K16 **Businga** Equateur, NW Dem. Rep. Congo *03°20′N 20°53′E*
79 J18 **Busira** ≈ NW Dem. Rep. Congo
116 I5 **Bus'k** *Rus.* Busk. L'vivs'ka Oblast', W Ukraine *49°59′N 24°34′E*
95 E14 **Buskerud** ◆ *county* S Norway
113 F14 **Buško Jezero** ⊠ SW Bosnia and Herzegovina
111 M15 **Busko-Zdrój** Świętokrzyskie, C Poland *50°28′N 20°44′E*
Busra *see* Al Başrah, Iraq
Buşrá *see* Buşrá ash Shām, Syria
138 H9 **Buşrá ash Shām** *var.* Bosora, Bosra, Bozrah, Buşrá. Dar'ā, S Syria *32°31′N 36°29′E*
180 I13 **Busselton** Western Australia *33°43′S 115°15′E*
81 C14 **Busseri** ≈ W South Sudan
106 E9 **Busseto** Emilia-Romagna, C Italy *45°00′N 10°06′E*
106 A8 **Bussoleno** Piemonte, NE Italy *45°11′N 07°07′E*
Bussora *see* Al Başrah
41 N7 **Bustamante** Nuevo León, NE Mexico *26°29′N 100°30′W*
63 I23 **Bustamante, Punta** *headland* S Argentina *51°35′S 68°58′W*
Bustan *see* Bo'ston
116 J12 **Buşteni** Prahova, SE Romania *45°23′N 25°32′E*
106 D7 **Busto Arsizio** Lombardia, N Italy *45°37′N 08°51′E*
147 Q10 **Büston** *Rus.* Buston. NW Tajikistan *40°31′N 69°21′E*
100 H8 **Büsum** Schleswig-Holstein, N Germany *54°08′N 08°52′E*
98 J10 **Bussum** Noord-Holland, C Netherlands *52°17′N 05°10′E*
79 M16 **Buta** Orientale, N Dem. Rep. Congo *02°50′N 24°41′E*
81 E20 **Butare** *prev.* Astrida. S Rwanda *02°39′S 29°44′E*
191 O2 **Butaritari** *atoll* Tungaru, W Kiribati
Butawal *see* Butwal
96 H13 **Bute** *cultural region* SW Scotland, United Kingdom
162 K6 **Büteeliyn Nuruu** ▲ N Mongolia
10 L16 **Bute Inlet** *fjord* British Columbia, W Canada
96 H12 **Bute, Island of** *island* SW Scotland, United Kingdom
79 P18 **Butembo** Nord-Kivu, NE Dem. Rep. Congo *0°09′N 29°17′E*
Būtenpost *see* Buitenpost
107 K25 **Butera** Sicilia, Italy, C Mediterranean Sea *37°12′N 14°12′E*
99 M20 **Bütgenbach** Liège, E Belgium *50°25′N 06°12′E*
Butha Qi *see* Zalantun
166 J5 **Buthidaung** Rakhine State, W Myanmar (Burma) *20°50′N 92°25′E*
61 I16 **Butiá** Rio Grande do Sul, S Brazil *30°09′S 51°55′W*
81 F17 **Butiaba** NW Uganda *01°48′N 31°21′E*
23 N6 **Butler** Alabama, S USA *32°05′N 88°13′W*
23 S5 **Butler** Georgia, SE USA *32°33′N 84°14′W*
31 Q11 **Butler** Indiana, N USA *41°25′N 84°52′W*
27 R5 **Butler** Missouri, C USA *38°17′N 94°21′W*
18 B14 **Butler** Pennsylvania, NE USA *40°51′N 79°52′W*
194 K5 **Butler Island** *island* Antarctica
21 U8 **Butner** North Carolina, SE USA *36°07′N 78°45′W*
171 P14 **Buton, Pulau** *var.* Pulau Butung; *prev.* Boetoeng. *island* C Indonesia
Bütow *see* Bytów
113 L23 **Butrintit, Liqeni i** ⊚ S Albania
23 N3 **Buttahatchee River** ≈ Alabama/Mississippi, S USA
33 Q10 **Butte** Montana, NW USA *46°01′N 112°33′W*
29 O12 **Butte** Nebraska, C USA *42°54′N 98°51′W*
168 J7 **Butterworth** Pinang, Peninsular Malaysia *05°24′N 100°22′E*
83 J25 **Butterworth** *var.* Gcuwa. Eastern Cape, SE South Africa *32°20′S 28°09′E*
13 O3 **Button Islands** *island group* Nunavut, NE Canada
35 R13 **Buttonwillow** California, W USA *35°24′N 119°26′W*
171 Q7 **Butuan** *off.* Butuan City. Mindanao, S Philippines *08°57′N 125°33′E*
Butuan City *see* Butuan
Butung, Pulau *see* Buton, Pulau
Butuntum *see* Bitonto
126 M8 **Buturlinovka** Voronezhskaya Oblast', W Russian Federation *50°48′N 40°33′E*
153 O11 **Butwal** *var.* Butawal. Western, C Nepal *27°41′N 83°28′E*
101 G17 **Butzbach** Hessen, W Germany *50°26′N 08°40′E*
100 L9 **Bützow** Mecklenburg-Vorpommern, N Germany *53°49′N 11°58′E*
80 N13 **Buuhoodle** Togdheer, N Somalia *08°18′N 46°15′E*
81 N16 **Buulobarde** *var.* Buulo Berde. Hiiraan, C Somalia *03°52′N 45°37′E*
Buulo Berde *see* Buulobarde
80 P12 **Buuraha Cal Miskaat** ▲ NE Somalia
81 L19 **Buur Gaabo** Jubbada Hoose, S Somalia *01°14′S 41°48′E*
99 M22 **Buurgplaatz** ▲ N Luxembourg *50°09′N 06°02′E*
162 H8 **Buutsagaan** *var.* Buyant. Bayanhongor, C Mongolia *46°07′N 98°45′E*
Buwayrāt al Hasūn *see* Bu'ayrāt al Ḩasūn
146 L11 **Buxoro** *var.* Bokhara, *Rus.* Bukhara. Buxoro Viloyati, C Uzbekistan *39°51′N 64°23′E*
146 J11 **Buxoro Viloyati** *Rus.* Bukharskaya Oblast'. ◆ *province* C Uzbekistan
100 I10 **Buxtehude** Niedersachsen, NW Germany *53°29′N 09°42′E*
97 L18 **Buxton** C England, United Kingdom *53°18′N 01°52′W*
124 M14 **Buy** *var.* Buj. Kostromskaya Oblast', NW Russian Federation *58°27′N 41°31′E*
Buyanbat *see* Jargalant
162 D6 **Buyant** Bayan-Ölgiy, W Mongolia *48°31′N 89°36′E*
Buyant *see* Buutsagaan, Bayanhongor, Mongolia
Buyant *see* Otgon, Dzavhan, Mongolia
Buyant *see* Galshar, Hentiy, Mongolia
163 N10 **Buyant-Uhaa** Dornogovi, SE Mongolia *44°52′N 110°12′E*
162 M7 **Buyant Ukha** ✈ (Ulaanbaatar) Töv, N Mongolia
127 Q16 **Buynaksk** Respublika Dagestan, SW Russian Federation *42°53′N 47°03′E*
119 L20 **Buynavichy** *Rus.* Buynovichi. Homyel'skaya Voblasts', SE Belarus *51°52′N 28°33′E*
Buynovichi *see* Buynavichy
76 L16 **Buyo** SW Ivory Coast *06°16′N 07°03′W*
76 L16 **Buyo, Lac de** ⊠ W Ivory Coast
163 R7 **Buyr Nuur** *var.* Buir Nur. ⊚ China/Mongolia *see also* Buir Nur
Buyr Nuur *see* Buir Nur
137 T13 **Büyükağrı Dağı** *var.* Aghri Dagh, Agri Dagi, Koh I Noh, Masis, *Eng.* Great Ararat, Mount Ararat. ▲ E Turkey *39°43′N 44°19′E*
137 R15 **Büyük Çayı** ≈ NE Turkey
114 O13 **Büyük Çekmece** İstanbul, NW Turkey *41°02′N 28°35′E*
114 N12 **Büyükkarıştıran** Kırklareli, NW Turkey *41°17′N 27°33′E*
115 L14 **Büyükkemikli Burnu** *cape* NW Turkey
136 E15 **Büyükmenderes Nehri** ≈ SW Turkey
Büyükzap Suyu *see* Great Zab
102 M9 **Buzançais** Indre, C France *46°53′N 01°25′E*
116 K13 **Buzău** Buzău, SE Romania *45°08′N 26°51′E*
116 K13 **Buzău** ◆ *county* SE Romania
116 L12 **Buzău** ≈ E Romania
75 S11 **Buzaymah** *var.* Bzīmah. SE Libya *24°53′N 22°01′E*
164 E13 **Buzen** Fukuoka, Kyūshū, SW Japan *33°37′N 131°06′E*
116 F12 **Buziaş** *Ger.* Busiasch, *Hung.* Buziásfürdő; *prev.* Buziás. Timiş, W Romania *45°38′N 21°36′E*
Buziás *see* Buziaş
Buziásfürdő *see* Buziaş
83 M18 **Búzi, Rio** ≈ C Mozambique
117 Q10 **Buz'kyy Lyman** *bay* S Ukraine
Büzmeýin *see* Abadan
127 T6 **Buzuluk** Orenburgskaya Oblast', W Russian Federation *52°47′N 52°16′E*
127 N8 **Buzuluk** ≈ SW Russian Federation
Buzuluk *see* Buzylyk
145 O8 **Buzylyk** *prev.* Buzuluk. Akmola, C Kazakhstan *51°53′N 66°09′E*
19 P12 **Buzzards Bay** Massachusetts, NE USA *41°45′N 70°37′W*
19 P13 **Buzzards Bay** *bay* Massachusetts, NE USA
83 G16 **Bwabata** Caprivi, NE Namibia *17°52′S 22°39′E*
186 H10 **Bwagaoia** Misima Island, SE Papua New Guinea *10°39′S 152°48′E*
Bwake *see* Bouaké
187 R13 **Bwatnapne** Pentecost, C Vanuatu *15°42′S 168°07′E*
119 K14 **Byahoml'** *Rus.* Begoml'. Vitsyebskaya Voblasts', N Belarus *54°44′N 28°04′E*
114 K8 **Byala** Ruse, N Bulgaria *43°27′N 25°44′E*
114 N9 **Byala** *prev.* Ak-Dere. Varna, E Bulgaria *42°52′N 27°53′E*
Byala Reka *see* Erythropótamos
114 H8 **Byala Slatina** Vratsa, NW Bulgaria *43°28′N 23°56′E*
119 N15 **Byalynichy** *Rus.* Belynichi. Mahilyowskaya Voblasts', E Belarus *54°00′N 29°42′E*
Byan Tumen *see* Choybalsan
119 L14 **Byarezina** *prev.* Byerezino, *Rus.* Berezina. ≈ C Belarus
119 G19 **Byaroza** *Pol.* Bereza Kartuska, *Rus.* Berëza. Brestskaya Voblasts', SW Belarus *52°32′N 24°59′E*
119 H16 **Byarozawka** *Rus.* Berëzovka. Hrodzyenskaya Voblasts', W Belarus *53°45′N 25°30′E*
Byblos *see* Jbail
111 O14 **Bychawa** Lubelskie, SE Poland *51°06′N 22°34′E*
Bychikha *see* Bychykha
118 N11 **Bychykha** *Rus.* Bychikha. Vitsyebskaya Voblasts', NE Belarus *55°41′N 29°59′E*
111 I14 **Byczyna** *Ger.* Pitschen. Opolskie, S Poland *51°06′N 18°13′E*
110 I10 **Bydgoszcz** *Ger.* Bromberg. Kujawski-pomorskie, C Poland *53°06′N 18°00′E*
119 H19 **Byelaazyorsk** *Rus.* Beloozersk. Brestskaya Voblasts', SW Belarus *52°28′N 25°10′E*
119 I17 **Byelaruskaya Hrada** *Rus.* Belorusskaya Gryada. *ridge* N Belarus
119 G18 **Byelavyezhskaya Pushcha** *Pol.* Puszcza Białowieska, *Rus.* Belovezhskaya Pushcha. *forest* Belarus/Poland *see also* Białowieska, Puszcza
Byelavyezhskaya, Pushcha *see* Białowieska, Puszcza
119 H15 **Byenyakoni** *Rus.* Benyakoni. Hrodzyenskaya Voblasts', W Belarus *54°15′N 25°22′E*
119 M16 **Byerazino** *Rus.* Berezino. Minskaya Voblasts', C Belarus *53°50′N 29°00′E*
118 L13 **Byerazino** *Rus.* Berezino. Vitsyebskaya Voblasts', N Belarus *54°54′N 28°12′E*
Byerezino *see* Byarezina
118 M13 **Byeshankovichy** *Rus.* Beshenkovichi. Vitsyebskaya Voblasts', N Belarus *55°03′N 29°27′E*
31 U13 **Byesville** Ohio, N USA *39°58′N 81°32′W*
119 P18 **Byesyedz'** *Rus.* Besed'. ≈ SE Belarus
119 H19 **Byezdzyezh** *Rus.* Bezdezh. Brestskaya Voblasts', SW Belarus *52°19′N 25°18′E*
93 J15 **Bygdeå** Västerbotten, N Sweden *64°03′N 20°49′E*
94 F12 **Bygdin** ⊚ S Norway
93 J15 **Bygdsiljum** Västerbotten, N Sweden *64°20′N 20°31′E*
95 E17 **Bygland** Aust-Agder, S Norway *58°46′N 07°50′E*
95 E17 **Byglandsfjord** Aust-Agder, S Norway *58°42′N 07°51′E*
119 N16 **Bykhaw** *Rus.* Bykhov. Mahilyowskaya Voblasts', E Belarus *53°31′N 30°15′E*
Bykhov *see* Bykhaw
127 P9 **Bykovo** Volgogradskaya Oblast', SW Russian Federation *49°52′N 45°24′E*
123 P7 **Bykovskiy** Respublika Sakha (Yakutiya), NE Russian Federation *71°57′N 129°07′E*
195 R12 **Byrd Glacier** *glacier* Antarctica
14 K10 **Byrd, Lac** ⊚ Québec, SE Canada
183 P5 **Byrock** New South Wales, SE Australia *30°40′S 146°24′E*
30 L10 **Byron** Illinois, N USA *42°06′N 89°15′W*
183 V4 **Byron Bay** New South Wales, SE Australia *28°39′S 153°34′E*
183 V4 **Byron, Cape** *headland* New South Wales, E Australia *28°37′S 153°40′E*
63 F21 **Byron, Isla** *island* S Chile
Byron Island *see* Nikunau
65 B24 **Byron Sound** *sound* NW Falkland Islands
122 M6 **Byrranga, Gory** ▲ N Russian Federation
93 J14 **Byske** Västerbotten, N Sweden *64°58′N 21°10′E*
111 K18 **Bystrá** ▲ N Slovakia *49°10′N 19°49′E*
111 F18 **Bystřice nad Pernštejnem** *Ger.* Bistritz ober Pernstein. Vysočina, C Czech Republic *49°32′N 16°16′E*
Bystrovka *see* Kemin
111 G16 **Bystrzyca Kłodzka** *Ger.* Habelschwerdt. Wałbrzych, SW Poland *50°19′N 16°39′E*
111 I18 **Bytča** Žilinský Kraj, N Slovakia *49°15′N 18°32′E*
119 L15 **Bytcha** Minskaya Voblasts', NE Belarus *54°19′N 28°24′E*
Byten/Byten' *see* Bytsyen'
111 J16 **Bytom** *Ger.* Beuthen. Śląskie, S Poland *50°21′N 18°51′E*
110 H7 **Bytów** *Ger.* Bütow. Pomorskie, N Poland *54°10′N 17°30′E*
119 H18 **Bytsyen'** *Pol.* Byteń, *Rus.* Byten'. Brestskaya Voblasts', SW Belarus *52°53′N 25°30′E*
81 E19 **Byumba** *var.* Biumba. N Rwanda *01°37′S 30°05′E*
Byuzmeyin *see* Abadan
119 O20 **Byval'ki** Homyel'skaya Voblasts', SE Belarus *51°51′N 30°38′E*
95 O20 **Byxelkrok** Kalmar, S Sweden *57°18′N 17°01′E*
Byzantium *see* İstanbul
Bzīmah *see* Buzaymah

C

62 O6 **Caacupé** Cordillera, S Paraguay *25°23′S 57°05′W*
62 P6 **Caaguazú** *off.* Departamento de Caaguazú. ◆ *department* C Paraguay
Caaguazú, Departamento de *see* Caaguazú
82 C13 **Caála** *var.* Kaala, Robert Williams, *Port.* Vila Robert Williams. Huambo, C Angola *12°51′S 15°33′E*
62 P7 **Caazapá** Caazapá, S Paraguay *26°09′S 56°21′W*
62 P7 **Caazapá** *off.* Departamento de Caazapá. ◆ *department* SE Paraguay
Caazapá, Departamento de *see* Caazapá
81 P15 **Cabaad, Raas** *headland* C Somalia *06°13′N 49°01′E*
55 N10 **Cabadisocaña** Amazonas, S Venezuela *04°28′N 64°45′W*
44 F5 **Cabaiguán** Sancti Spíritus, C Cuba *22°04′N 79°32′W*
Caballeria, Cabo *see* Cavalleria, Cap de
37 Q14 **Caballo Reservoir** ⊠ New Mexico, SW USA
40 L6 **Caballos Mesteños, Llano de los** *plain* N Mexico
104 L2 **Cabanaquinta** Asturias, N Spain *43°10′N 05°37′W*
42 B9 **Cabañas** ◆ *department* E El Salvador
171 O3 **Cabanatuan** *off.* Cabanatuan City. Luzon, N Philippines *15°27′N 120°57′E*
Cabanatuan City *see* Cabanatuan
15 T8 **Cabano** Québec, SE Canada *47°40′N 68°56′W*
104 L11 **Cabeza del Buey** Extremadura, W Spain *38°44′N 05°13′W*
45 V5 **Cabezas de San Juan** *headland* E Puerto Rico *18°23′N 65°37′W*
105 N2 **Cabezón de la Sal** Cantabria, N Spain *43°19′N 04°14′W*
Cabhán *see* Cavan
61 B23 **Cabildo** Buenos Aires, E Argentina *38°28′S 61°50′W*
Cabillonum *see* Chalon-sur-Saône
54 H5 **Cabimas** Zulia, NW Venezuela *10°26′N 71°27′W*
82 A9 **Cabinda** *var.* Kabinda. Cabinda, NW Angola *05°34′S 12°12′E*
82 A9 **Cabinda** *var.* Kabinda. ◆ *province* NW Angola
33 N7 **Cabinet Mountains** ▲ Idaho/Montana, NW USA
82 B11 **Cabiri** Bengo, NW Angola *08°50′S 13°42′E*
63 J20 **Cabo Blanco** Santa Cruz, SE Argentina *47°13′S 65°43′W*
82 P13 **Cabo Delgado** *off.* Província de Cabo Delgado. ◆ *province* NE Mozambique
14 L9 **Cabonga, Réservoir** ⊠ Québec, SE Canada
27 V7 **Cabool** Missouri, C USA *37°07′N 92°06′W*
183 V2 **Caboolture** Queensland, E Australia *27°05′S 152°50′E*
Cabora Bassa, Lake *see* Cahora Bassa, Albufeira de
40 F3 **Caborca** Sonora, NW Mexico *30°44′N 112°06′W*
Cabo San Lucas *see* San Lucas
27 V11 **Cabot** Arkansas, C USA *34°58′N 92°01′W*
14 F12 **Cabot Head** *headland* Ontario, S Canada *45°13′N 81°17′W*
13 R13 **Cabot Strait** *strait* E Canada
Cabo Verde, Ilhas do *see* Cape Verde
104 M14 **Cabra** Andalucía, S Spain *37°28′N 04°28′W*
107 B19 **Cabras** Sardegna, Italy, C Mediterranean Sea *39°55′N 08°30′E*
188 A15 **Cabras Island** *island* W Guam
45 O8 **Cabrera** N Dominican Republic *19°40′N 69°54′W*
104 J4 **Cabrera** ≈ NW Spain
105 X10 **Cabrera, Illa de** *anc.* Capraria. *island* Islas Baleares, Spain, W Mediterranean Sea
105 Q15 **Cabrera, Sierra** ▲ S Spain
11 S16 **Cabri** Saskatchewan, S Canada *50°38′N 108°28′W*
105 R10 **Cabriel** ≈ E Spain
54 M7 **Cabruta** Guárico, C Venezuela *07°39′N 66°19′W*
171 N2 **Cabugao** Luzon, N Philippines *17°55′N 120°29′E*
54 G10 **Cabuyaro** Meta, C Colombia *04°21′N 72°47′W*
60 I13 **Caçador** Santa Catarina, S Brazil *26°47′S 51°00′W*
42 G8 **Cacaguatique, Cordillera** *var.* Cordillera. ▲ NE El Salvador
112 L13 **Čačak** Serbia, C Serbia *43°52′N 20°23′E*
55 Y10 **Cacao** NE French Guiana *04°37′N 52°29′W*
61 H16 **Caçapava do Sul** Rio Grande do Sul, S Brazil *30°28′S 53°29′W*
21 U3 **Cacapon River** ≈ West Virginia, NE USA
107 J23 **Caccamo** Sicilia, Italy, C Mediterranean Sea *37°56′N 13°40′E*
107 A17 **Caccia, Capo** *headland* Sardegna, Italy, C Mediterranean Sea *40°34′N 08°09′E*
146 H15 **Çäçe** *var.* Chäche, *Rus.* Chaacha. Ahal Welaýaty, S Turkmenistan *36°49′N 60°33′E*
59 G18 **Cáceres** Mato Grosso, W Brazil *16°05′S 57°40′W*
104 J10 **Cáceres** *Ar.* Qazris. Extremadura, W Spain *39°29′N 06°23′W*
104 J9 **Cáceres** ◆ *province* Extremadura, W Spain
Cachacrou *see* Scotts Head Village
61 C21 **Cacharí** Buenos Aires, E Argentina *36°24′S 59°32′W*
26 L12 **Cache** Oklahoma, C USA *34°37′N 98°37′W*
10 M16 **Cache Creek** British Columbia, SW Canada *50°49′N 121°20′W*
35 N6 **Cache Creek** ≈ California, W USA
37 S3 **Cache La Poudre River** ≈ Colorado, C USA
Cacheo *see* Cacheu
27 W11 **Cache River** ≈ Arkansas, C USA
30 L17 **Cache River** ≈ Illinois, N USA
76 G12 **Cacheu** *var.* Cacheo. W Guinea-Bissau *12°12′N 16°10′W*
59 I15 **Cachimbo** Pará, NE Brazil *09°21′S 54°58′W*
59 H15 **Cachimbo, Serra do** ▲ C Brazil
82 D13 **Cachingues** Bié, C Angola *13°05′S 16°48′E*
54 G7 **Cáchira** Norte de Santander, N Colombia *07°44′N 73°07′W*
61 H16 **Cachoeira do Sul** Rio Grande do Sul, S Brazil *29°58′S 52°54′W*
59 O20 **Cachoeiro de Itapemirim** Espírito Santo, SE Brazil *20°51′S 41°07′W*
82 E12 **Cacolo** Lunda Sul, NE Angola *10°09′S 19°21′E*
83 C14 **Caconda** Huíla, C Angola *13°43′S 15°03′E*
82 A9 **Cacongo** Cabinda, NW Angola *05°13′S 12°08′E*
35 U9 **Cactus Peak** ▲ Nevada, W USA *37°42′N 116°51′W*
82 A11 **Cacuaco** Luanda, NW Angola *08°47′S 13°21′E*
83 B14 **Cacula** Huíla, SW Angola *14°33′S 14°04′E*
67 R12 **Caculuvar** ≈ SW Angola
59 O19 **Caçumba, Ilha** *island* SE Brazil
55 N10 **Cacuri** Amazonas, S Venezuela
81 N17 **Cadale** Shabeellaha Dhexe, E Somalia *02°48′N 46°19′E*
105 X4 **Cadaqués** Cataluña, NE Spain *42°17′N 03°16′E*
111 J18 **Čadca** *Hung.* Csaca. Žilinský Kraj, N Slovakia *49°27′N 18°46′E*
27 P13 **Caddo** Oklahoma, C USA *34°07′N 96°15′W*
25 R6 **Caddo** Texas, SW USA *32°42′N 98°40′W*
25 X6 **Caddo Lake** ⊠ Louisiana/Texas, SW USA
27 S12 **Caddo Mountains** ▲ Arkansas, C USA
41 O8 **Cadereyta** Nuevo León, NE Mexico *25°35′N 99°54′W*
97 J19 **Cader Idris** ▲ NW Wales, United Kingdom *52°43′N 03°57′W*
182 F3 **Cadibarrawirracanna, Lake** *salt lake* South Australia
14 I7 **Cadillac** Québec, SE Canada *48°12′N 78°23′W*
11 T17 **Cadillac** Saskatchewan, S Canada *49°43′N 107°41′W*
102 K13 **Cadillac** Gironde, SW France *44°37′N 00°16′W*
31 P7 **Cadillac** Michigan, N USA *44°15′N 85°23′W*
105 V4 **Cadí, Torreta de** *prev.* Torre de Cadí. ▲ NE Spain *42°16′N 01°38′E*
Torre de Cadí *see* Cadí, Toretta de
171 P5 **Cadiz** *off.* Cadiz City. Negros, C Philippines *10°58′N 123°18′E*
104 J15 **Cádiz** *anc.* Gades, Gadier, Gadir, Gadire. Andalucía, SW Spain *36°32′N 06°18′W*
20 H7 **Cadiz** Kentucky, S USA *36°52′N 87°50′W*
31 U13 **Cadiz** Ohio, N USA *40°16′N 81°00′W*
104 K15 **Cádiz** ◆ *province* Andalucía, SW Spain
104 I15 **Cádiz, Bahía de** *bay* SW Spain
Cadiz City *see* Cadiz
104 H15 **Cádiz, Golfo de** *Eng.* Gulf of Cadiz. *gulf* Portugal/Spain
Cadiz, Gulf of *see* Cádiz, Golfo de
35 X14 **Cadiz Lake** ⊚ California, W USA

◆ Country
● Country Capital
◇ Dependent Territory
○ Dependent Territory Capital
◆ Administrative Regions
✈ International Airport
▲ Mountain
▲ Mountain Range
▲ Volcano
≈ River
⊚ Lake
⊠ Reservoir

◆ Country ● Country Capital ◇ Dependent Territory ○ Dependent Territory Capital ⬖ Administrative Regions ✈ International Airport ▲ Mountain ▲▲ Mountain Range ▲ Volcano ~ River ◎ Lake ⊠ Reservoir

27 V2 **Canton** Missouri, C USA 40°07′N 91°31′W
18 J7 **Canton** New York, NE USA 44°36′N 75°10′W
21 O10 **Canton** North Carolina, SE USA 35°31′N 82°50′W
31 U12 **Canton** Ohio, N USA 40°48′N 81°23′W
26 L9 **Canton** Oklahoma, C USA 36°03′N 98°35′W
18 G12 **Canton** Pennsylvania, NE USA 41°38′N 76°49′W
29 R11 **Canton** South Dakota, N USA 43°19′N 96°33′W
25 V7 **Canton** Texas, SW USA 32°33′N 95°51′W
Canton *see* Guangzhou
Canton Island *see* Kanton
26 L9 **Canton Lake** ⊠ Oklahoma, C USA
106 D7 **Cantù** Lombardia, N Italy 45°44′N 09°98′E
Cantuaria/Cantwaraburh *see* Canterbury
39 R10 **Cantwell** Alaska, USA 63°23′N 148°57′W
59 O16 **Canudos** Bahia, E Brazil 09°51′S 39°08′W
47 T7 **Canumã, Rio** ≈ N Brazil
Canusium *see* Puglia, Canosa di
24 G7 **Canutillo** Texas, SW USA 31°53′N 106°34′W
25 N3 **Canyon** Texas, SW USA 34°58′N 101°56′W
33 S12 **Canyon** Wyoming, C USA 44°44′N 110°30′W
32 K13 **Canyon City** Oregon, NW USA 44°24′N 118°58′W
33 R10 **Canyon Ferry Lake** ◎ Montana, NW USA
25 S11 **Canyon Lake** ⊠ Texas, SW USA
167 T5 **Cao Băng** *var.* Caobang. Cao Băng, N Vietnam 22°40′N 106°16′E
Caobang *see* Cao Băng
160 J12 **Caodu He** ≈ S China
Caohai *see* Weining
167 S14 **Cao Lanh** Đông Thap, S Vietnam 10°35′N 105°25′E
82 C11 **Caombo** Malanje, NW Angola 08°42′S 16°33′E
Caorach, Cuan na g *see* Sheep Haven
Caozhou *see* Heze
171 Q12 **Capalulu** Pulau Mangole, E Indonesia 01°51′S 125°53′E
54 K8 **Capanaparo, Río** ≈ Colomiba/Venezuela
58 L12 **Capanema** Pará, NE Brazil 01°08′S 47°07′W
60 L10 **Capão Bonito do Sul** São Paulo, S Brazil 24°01′S 48°23′W
60 I13 **Capão Doce, Morro do** ▲ S Brazil 26°37′S 51°28′W
54 I4 **Capatárida** Falcón, N Venezuela 11°11′N 70°37′W
102 I15 **Capbreton** Landes, SW France 43°40′N 01°25′W
Cap-Breton, Île du *see* Cape Breton Island
15 W6 **Cap-Chat** Québec, SE Canada 49°04′N 66°43′W
15 P11 **Cap-de-la-Madeleine** Québec, SE Canada 46°22′N 72°31′W
103 N13 **Capdenac** Aveyron, S France 44°35′N 02°06′E
Cap des Palmès *see* Palmas, Cape
183 Q15 **Cape Barren Island** *island* Furneaux Group, Tasmania, SE Australia
65 O18 **Cape Basin** *undersea feature* S Atlantic Ocean 37°00′S 07°00′E
13 R14 **Cape Breton Island** *Fr.* Île du Cap-Breton. *island* Nova Scotia, SE Canada
23 Y11 **Cape Canaveral** Florida, SE USA 28°24′N 80°36′W
21 Y6 **Cape Charles** Virginia, NE USA 37°16′N 76°01′W
77 P17 **Cape Coast** *prev.* Cape Coast Castle. S Ghana 05°10′N 01°13′W
Cape Coast Castle *see* Cape Coast
19 Q12 **Cape Cod Bay** *bay* Massachusetts, NE USA
23 W15 **Cape Coral** Florida, SE USA 26°33′N 81°57′W
181 R4 **Cape Crawford Roadhouse** Northern Territory, N Australia 16°39′S 135°44′E
9 Q7 **Cape Dorset** *var.* Kingait. Baffin Island, Nunavut, NE Canada 76°14′N 76°32′W
21 N8 **Cape Fear River** ≈ North Carolina, SE USA
27 Y7 **Cape Girardeau** Missouri, C USA 37°19′N 89°31′W
21 T14 **Cape Island** *island* South Carolina, SE USA
186 A6 **Capella** ▲ NW Papua New Guinea 05°00′S 141°09′E
98 H12 **Capelle aan den IJssel** Zuid-Holland, SW Netherlands 51°56′N 04°36′E
83 C15 **Capelongo** Huíla, C Angola 14°45′S 15°02′E
18 J17 **Cape May** New Jersey, NE USA 38°54′N 74°54′W
18 J17 **Cape May Court House** New Jersey, NE USA 39°03′N 74°46′W
Cape Palmas *see* Harper
8 I16 **Cape Parry** Northwest Territories, N Canada 70°10′N 124°33′W
65 P19 **Cape Rise** *undersea feature* SW Indian Ocean 42°00′S 15°00′E
Cape Saint Jacques *see* Vung Tau
Capesterre *see* Capesterre-Belle-Eau
45 Y6 **Capesterre-Belle-Eau** *var.* Capesterre. Basse Terre, S Guadeloupe 16°03′N 61°34′W
83 D26 **Cape Town** *var.* Ekapa, *Afr.* Kaapstad, Kapstad. ● (South Africa-legislative capital) Western Cape, SW South Africa 33°56′S 18°28′E
83 E26 **Cape Town** ✈ Western Cape, SW South Africa 31°51′S 21°06′E
76 D9 **Cape Verde** *off.* Republic of Cape Verde, *Port.* Cabo Verde, Ilhas do Cabo Verde. ◆ *republic* E Atlantic Ocean
64 L11 **Cape Verde Basin** *undersea feature* E Atlantic Ocean 15°00′N 30°00′W
66 K5 **Cape Verde Islands** *island group* E Atlantic Ocean
64 L10 **Cape Verde Plain** *undersea feature* E Atlantic Ocean 23°00′N 26°00′W
Cape Verde Plateau/Cape Verde Rise *see* Cape Verde Terrace
Cape Verde, Republic of *see* Cape Verde
64 L11 **Cape Verde Terrace** *var.* Cape Verde Plateau, Cape Verde Rise. *undersea feature* E Atlantic Ocean 18°00′N 20°00′W
181 V2 **Cape York Peninsula** *peninsula* Queensland, N Australia
44 M8 **Cap-Haïtien** *var.* Le Cap. N Haiti 19°44′N 72°12′W
43 T15 **Capira** Panamá, C Panama 08°48′N 79°51′W
14 K8 **Capitachouane** ≈ Québec, SE Canada
14 L8 **Capitachouane, Lac** ◎ Québec, SE Canada
37 T13 **Capitan** New Mexico, SW USA 33°33′N 105°34′W
194 G3 **Capitán Arturo Prat** *Chilean research station* South Shetland Islands, Antarctica 62°24′S 59°42′W
37 S13 **Capitan Mountains** ▲▲ New Mexico, SW USA
62 M3 **Capitán Pablo Lagerenza** *var.* Mayor Pablo Lagerenza. Chaco, N Paraguay 19°55′S 60°46′W
37 T13 **Capitan Peak** ▲ New Mexico, SW USA 33°35′N 105°15′W
188 H5 **Capitol Hill** ● (Northern Mariana Islands-legislative capital) Saipan, S Northern Mariana Islands
60 I9 **Capivara, Represa** ⊠ S Brazil
61 J16 **Capivari** Rio Grande do Sul, S Brazil 30°08′S 50°32′W
113 H15 **Čapljina** Federicija Bosna I Hercegovina, S Bosnia and Herzegovina 43°07′N 17°42′E
83 M15 **Capoche** *var.* Kapoche. ≈ Mozambique/Zambia
Capo Delgado, Província de *see* Cabo Delgado
107 K17 **Capodichino** ✈ (Napoli) Campania, S Italy 40°53′N 14°15′E
Capodistria *see* Koper
106 E12 **Capraia, Isola di** *island* Arcipelago Toscano, C Italy
107 B16 **Caprara, Punta** *var.* Punta dello Scorno. *headland* Isola Asinara, W Italy 41°07′N 08°19′E
Capraria *see* Cabrera, Illa de
14 F10 **Capreol** Ontario, S Canada 46°43′N 80°56′W
107 K18 **Capri** Campania, S Italy 40°33′N 14°14′E
175 S9 **Capricorn Tablemount** *undersea feature* W Pacific Ocean 18°34′S 172°12′W
107 J18 **Capri, Isola di** *island* S Italy
83 G16 **Caprivi** ◈ *district* NE Namibia
Caprivi Concession *see* Caprivi Strip
83 F16 **Caprivi Strip** *Ger.* Caprivizipfel; *prev.* Caprivi Concession. *cultural region* NE Namibia
Caprivizipfel *see* Caprivi Strip
25 O5 **Cap Rock Escarpment** *cliffs* Texas, SW USA
15 R10 **Cap-Rouge** Québec, SE Canada 46°45′N 71°18′W
Cap Saint-Jacques *see* Vung Tau
38 F12 **Captain Cook** Hawaii, USA, C Pacific Ocean 19°30′N 155°55′W
183 R10 **Captains Flat** New South Wales, SE Australia 35°37′S 149°28′E
102 K14 **Captieux** Gironde, SW France 44°16′N 00°15′W
107 K17 **Capua** Campania, S Italy 41°06′N 14°13′E
54 F14 **Caquetá** *off.* Departamanto del Caquetá. ◈ *province* S Colombia
Caquetá, Departamanto del *see* Caquetá
54 E13 **Caquetá, Río** *var.* Rio Japurá, Yapurá. ≈ Brazil/Colombia *see also* Japurá, Rio
Caquetá, Río *see* Japurá, Rio
CAR *see* Central African Republic
Cara *see* Kara
57 I16 **Carabaya, Cordillera** ▲▲ E Peru
54 K5 **Carabobo** *off.* Estado Carabobo. ◈ *state* N Venezuela
Carabobo, Estado *see* Carabobo
116 I14 **Caracal** Olt, S Romania 44°07′N 24°18′E
58 F10 **Caracaraí** Rondônia, W Brazil 01°47′N 61°11′W
54 L5 **Caracas** ● (Venezuela) Distrito Federal, N Venezuela 10°29′N 66°54′W
54 I5 **Carache** Trujillo, N Venezuela 09°40′N 70°15′W
60 N10 **Caraguatatuba** São Paulo, S Brazil 23°37′S 45°24′W
48 I7 **Carajás, Serra dos** ▲▲ N Brazil
Caralis *see* Cagliari
54 E9 **Caramanta** Antioquia, W Colombia 05°36′N 75°38′W
171 P4 **Caramoan** Catanduanes Island, N Philippines 13°47′N 123°49′E
Caramurat *see* Mihail Kogălniceanu
116 F12 **Caransebeş** *Ger.* Karansebesch, *Hung.* Karánsebes. Caraş-Severin, SW Romania 45°23′N 22°13′E
Carapella *see* Carapelle
107 M16 **Carapelle** *var.* Carapella. ≈ SE Italy
55 O9 **Carapo** Bolívar, SE Venezuela
13 P13 **Caraquet** New Brunswick, SE Canada 47°48′N 64°59′W
Caras *see* Caraz
116 F12 **Caraşova** *Hung.* Krassóvár. Caraş-Severin, SW Romania 45°11′N 21°51′E
116 F12 **Caraş-Severin** ◈ *county* SW Romania
42 M5 **Caratasca, Laguna de** *lagoon* NE Honduras
58 C13 **Carauari** Amazonas, NW Brazil 04°55′S 66°57′W
Caravaca *see* Caravaca de la Cruz
105 Q12 **Caravaca de la Cruz** *var.* Caravaca. Murcia, SE Spain 38°06′N 01°51′W
106 E7 **Caravaggio** Lombardia, N Italy 45°31′N 09°39′E
107 C18 **Caravai, Passo di** *pass* Sardegna, Italy, C Mediterranean Sea
59 O19 **Caravelas** Bahia, E Brazil 17°45′S 39°15′W
56 C12 **Caraz** *var.* Caras. Ancash, W Peru 09°03′S 77°47′W
61 H14 **Carazinho** Rio Grande do Sul, S Brazil 28°16′S 52°46′W
42 J11 **Carazo** ◈ *department* SW Nicaragua
Carballiño *see* O Carballiño
104 G2 **Carballo** Galicia, NW Spain 43°13′N 08°41′W
11 W16 **Carberry** Manitoba, S Canada 49°52′N 99°20′W
40 F4 **Carbó** Sonora, NW Mexico 29°41′N 111°00′W
107 C20 **Carbonara, Capo** *headland* Sardegna, Italy, C Mediterranean Sea 39°06′N 09°31′E
37 Q5 **Carbondale** Colorado, C USA 39°24′N 107°12′W
30 L17 **Carbondale** Illinois, N USA 37°43′N 89°13′W
27 Q4 **Carbondale** Kansas, C USA 38°49′N 95°41′W
18 I13 **Carbondale** Pennsylvania, NE USA 41°34′N 75°30′W
13 V12 **Carbonear** Newfoundland, Newfoundland and Labrador, SE Canada 47°45′N 53°16′W
105 Q9 **Carboneras de Guadazaón** *var.* Carboneras de Guadazón. Castilla-La Mancha, C Spain 39°54′N 01°50′W
Carboneras de Guadazón *see* Carboneras de Guadazaón
23 O3 **Carbon Hill** Alabama, S USA 33°53′N 87°31′W
107 B20 **Carbonia** *var.* Carbonia Centro. Sardegna, Italy, C Mediterranean Sea 39°11′N 08°31′E
Carbonia Centro *see* Carbonia
63 I22 **Carbon, Laguna del** *depression* S Argentina
105 S10 **Carcaixent** Valenciana, E Spain 39°08′N 00°28′W
Carcaso *see* Carcassonne
65 B24 **Carcass Island** *island* NW Falkland Islands
103 O16 **Carcassonne** *anc.* Carcaso. Aude, S France 43°13′N 02°21′E
105 R12 **Carche** ▲ S Spain 38°24′N 01°11′W
56 A13 **Carchi** ◈ *province* N Ecuador
10 I8 **Carcross** Yukon, W Canada 60°11′N 134°41′W
Cardamomes, Chaîne des *see* Krâvanh, Chuŏr Phnum
155 G22 **Cardamom Hills** ▲▲ SW India
Cardamom Mountains *see* Krâvanh, Chuŏr Phnum
104 M12 **Cardeña** Andalucía, S Spain 38°16′N 04°20′W
44 D4 **Cárdenas** Matanzas, W Cuba 23°02′N 81°12′W
41 O11 **Cárdenas** San Luis Potosí, C Mexico 22°03′N 99°30′W
41 U15 **Cárdenas** Tabasco, SE Mexico 18°00′N 93°21′W
63 H21 **Cardiel, Lago** ◎ S Argentina
97 K22 **Cardiff** *Wel.* Caerdydd. ◐ S Wales, United Kingdom 51°30′N 03°13′W
97 J22 **Cardiff-Wales** ✈ S Wales, United Kingdom 51°24′N 03°22′W
97 I21 **Cardigan** *Wel.* Aberteifi. SW Wales, United Kingdom 52°06′N 04°40′W
97 I20 **Cardigan** *cultural region* W Wales, United Kingdom
97 I20 **Cardigan Bay** *bay* W Wales, United Kingdom
19 N8 **Cardigan, Mount** ▲ New Hampshire, NE USA 43°39′N 71°52′W
14 M13 **Cardinal** Ontario, SE Canada 44°48′N 75°22′W
105 V5 **Cardona** Cataluña, NE Spain 41°55′N 01°41′E
61 E19 **Cardona** Soriano, SW Uruguay 33°53′S 57°18′W
105 V4 **Cardoner** ≈ NE Spain
11 Q17 **Cardston** Alberta, SW Canada 49°14′N 113°19′W
181 W5 **Cardwell** Queensland, NE Australia 18°24′S 146°06′E
116 G8 **Carei** *Ger.* Gross-Karol, Karol, *Hung.* Nagykároly; *prev.* Careii-Mari. Satu Mare, NW Romania 47°40′N 22°28′E
Careii-Mari *see* Carei
58 F13 **Careiro** Amazonas, NW Brazil 03°40′S 60°23′W
102 J4 **Carentan** Manche, N France 49°18′N 01°15′W
104 M2 **Cares** ≈ N Spain
33 P14 **Carey** Idaho, NW USA 43°17′N 113°58′W
31 S12 **Carey** Ohio, N USA 40°57′N 83°22′W
25 P4 **Carey** Texas, SW USA 34°28′N 100°18′W
180 L11 **Carey, Lake** ◎ Western Australia
173 O8 **Cargados Carajos Bank** *undersea feature* C Indian Ocean
102 G6 **Carhaix-Plouguer** Finistère, NW France 48°16′N 03°35′W
61 A22 **Carhué** Buenos Aires, E Argentina 37°10′S 62°45′W
55 O5 **Cariaco** Sucre, NE Venezuela 10°33′N 63°37′W
107 O20 **Cariati** Calabria, SW Italy 39°30′N 16°57′E
2 H17 **Caribbean Plate** *tectonic feature*
44 I11 **Caribbean Sea** *sea* W Atlantic Ocean
11 N15 **Cariboo Mountains** ▲▲ British Columbia, SW Canada
11 W9 **Caribou** Manitoba, C Canada 59°27′N 97°43′W
19 S2 **Caribou** Maine, NE USA 46°51′N 68°00′W
11 P10 **Caribou Mountains** ▲▲ Alberta, SW Canada
Caribrod *see* Dimitrovgrad
40 I6 **Carichic** Chihuahua, N Mexico 27°57′N 107°01′W
103 R3 **Carignan** Ardennes, N France 49°38′N 05°08′E
183 Q5 **Carinda** New South Wales, SE Australia 30°26′S 147°45′E
105 R6 **Cariñena** Aragón, NE Spain 41°20′N 01°13′W
107 I23 **Carini** Sicilia, Italy, C Mediterranean Sea 38°06′N 13°09′E
107 K17 **Carinola** Campania, S Italy 41°14′N 14°03′E
Carinthi *see* Kärnten
55 O5 **Caripe** Monagas, NE Venezuela 10°13′N 63°30′W
55 P5 **Caripito** Monagas, NE Venezuela 10°03′N 63°05′W
15 W7 **Carleton** Québec, SE Canada 48°07′N 66°07′W
31 S10 **Carleton** Michigan, N USA 42°03′N 83°23′W
13 O14 **Carleton, Mount** ▲ New Brunswick, SE Canada 47°10′N 66°54′W
14 L13 **Carleton Place** Ontario, SE Canada 45°08′N 76°09′W
35 V3 **Carlin** Nevada, W USA 40°40′N 116°09′W
30 K14 **Carlinville** Illinois, N USA 39°16′N 89°52′W
97 K14 **Carlisle** *anc.* Caer Luel, Luguvallium, Luguvallum. NW England, United Kingdom 54°54′N 02°55′W
27 V11 **Carlisle** Arkansas, C USA 34°46′N 91°45′W
31 N15 **Carlisle** Indiana, N USA 38°57′N 87°23′W
29 V14 **Carlisle** Iowa, C USA 41°30′N 93°29′W
21 N5 **Carlisle** Kentucky, S USA 38°19′N 84°02′W
18 F15 **Carlisle** Pennsylvania, NE USA 40°10′N 77°10′W
21 Q11 **Carlisle** South Carolina, SE USA 34°35′N 81°27′W
38 J17 **Carlisle Island** *island* Aleutian Islands, Alaska, USA
27 R7 **Carl Junction** Missouri, C USA 37°10′N 94°34′W
107 A20 **Carloforte** Sardegna, Italy, C Mediterranean Sea 39°10′N 08°17′E
Carlopago *see* Karlobag
61 B21 **Carlos Casares** Buenos Aires, E Argentina 35°39′S 61°28′W
61 E18 **Carlos Reyles** Durazno, C Uruguay 33°00′S 56°30′W
61 A21 **Carlos Tejedor** Buenos Aires, E Argentina 35°25′S 62°25′W
97 F19 **Carlow** *Ir.* Ceatharlach. SE Ireland 52°50′N 06°55′W
97 F19 **Carlow** *Ir.* Cheatharlach. *cultural region* SE Ireland
96 F7 **Carloway** NW Scotland, United Kingdom 58°17′N 06°48′W
35 U17 **Carlsbad** California, W USA 33°09′N 117°21′W
37 U15 **Carlsbad** New Mexico, SW USA 32°24′N 104°15′W
Carlsbad *see* Karlovy Vary
129 N13 **Carlsberg Ridge** *undersea feature* S Arabian Sea 06°00′N 61°00′E
Carlsruhe *see* Karlsruhe
29 W6 **Carlton** Minnesota, N USA 46°39′N 92°25′W
11 V17 **Carlyle** Saskatchewan, S Canada 49°39′N 102°18′W
30 L15 **Carlyle** Illinois, N USA 38°36′N 89°22′W
30 L15 **Carlyle Lake** ⊠ Illinois, N USA
10 H7 **Carmacks** Yukon, W Canada 62°04′N 136°21′W
106 B9 **Carmagnola** Piemonte, NW Italy 44°50′N 07°43′E
11 X16 **Carman** Manitoba, S Canada 49°32′N 97°59′W
Carmana/Carmania *see* Kermān
97 I21 **Carmarthen** SW Wales, United Kingdom 51°52′N 04°19′W
97 I21 **Carmarthen** *cultural region* SW Wales, United Kingdom
97 I22 **Carmarthen Bay** *inlet* SW Wales, United Kingdom
103 N14 **Carmaux** Tarn, S France 44°03′N 02°09′E
35 N11 **Carmel** California, W USA 36°32′N 121°54′W
31 O13 **Carmel** Indiana, N USA 39°58′N 86°07′W
18 L13 **Carmel** New York, NE USA 41°25′N 73°40′W
97 H18 **Carmel Head** *headland* NW Wales, United Kingdom 53°24′N 04°35′W
42 E2 **Carmelita** Petén, N Guatemala 17°33′N 90°11′W
61 D19 **Carmelo** Colonia, SW Uruguay 34°00′S 58°20′W
41 V14 **Carmen** *var.* Ciudad del Carmen. Campeche, SE Mexico 18°38′N 91°50′W
61 A25 **Carmen de Patagones** Buenos Aires, E Argentina 40°45′S 63°00′W
40 F8 **Carmen, Isla** *island* NW Mexico
40 M5 **Carmen, Sierra del** ▲▲ NW Mexico
30 M16 **Carmi** Illinois, N USA 38°05′N 88°09′W
35 O7 **Carmichael** California, W USA 38°36′N 121°21′W
Carmiel *see* Karmi'el
25 U11 **Carmine** Texas, SW USA 30°07′N 96°40′W
104 K14 **Carmona** Andalucía, S Spain 37°28′N 05°38′W
Carmona *see* Uíge
Carnaro *see* Kvarner
180 G9 **Carnarvon** Western Australia 24°57′S 113°38′E
14 I13 **Carnarvon** Ontario, SE Canada 45°03′N 78°41′W
83 G24 **Carnarvon** Northern Cape, W South Africa 30°59′S 22°08′E
Carnarvon *see* Caernarfon
180 K9 **Carnarvon Range** ▲▲ Western Australia
Carn Domhnach *see* Carndonagh
96 E13 **Carndonagh** *Ir.* Carn Domhnach. NW Ireland 55°15′N 07°15′W
11 V17 **Carnduff** Saskatchewan, S Canada 49°11′N 101°50′W
26 L11 **Carnegie** Oklahoma, C USA 35°06′N 98°36′W
180 L9 **Carnegie, Lake** *salt lake* Western Australia
193 U8 **Carnegie Ridge** *undersea feature* E Pacific Ocean 01°00′S 85°00′W
96 H9 **Carn Eige** ▲ N Scotland, United Kingdom 57°18′N 05°04′W
182 F5 **Carnes** South Australia 30°12′S 134°31′E
194 J12 **Carney Island** *island* Antarctica
18 H16 **Carneys Point** New Jersey, NE USA 39°38′N 75°29′W
Carniche, Alpi *see* Karnische Alpen
151 Q21 **Car Nicobar** *island* Nicobar Islands, India, NE Indian Ocean
79 H15 **Carnot** Mambéré-Kadéï, W Central African Republic 04°58′N 15°55′E
182 F10 **Carnot, Cape** *headland* South Australia 34°57′S 135°39′E
96 K11 **Carnoustie** E Scotland, United Kingdom 56°30′N 02°42′W
97 F20 **Carnsore Point** *Ir.* Ceann an Chairn. *headland* SE Ireland 52°10′N 06°22′W
8 H7 **Carnwath** ≈ Northwest Territories, NW Canada
31 R8 **Caro** Michigan, N USA 43°29′N 83°24′W
23 Z15 **Carol City** Florida, SE USA 25°56′N 80°15′W
59 L14 **Carolina** Maranhão, E Brazil 07°20′S 47°25′W
45 U5 **Carolina** E Puerto Rico 18°22′N 65°57′W
21 V12 **Carolina Beach** North Carolina, SE USA 34°02′N 77°53′W
Caroline Island *see* Millennium Island
189 N15 **Caroline Islands** *island group* C Micronesia
129 Z14 **Caroline Plate** *tectonic feature*
192 H7 **Caroline Ridge** *undersea feature* E Philippine Sea 08°00′N 150°00′E
Carolopois *see* Châlons-en-Champagne
45 V14 **Caroni Arena Dam** ⊠ Trinidad, Trinidad and Tobago
Caronie, Monti *see* Nebrodi, Monti
55 P7 **Caroní, Río** ≈ E Venezuela
45 U14 **Caroni River** ≈ Trinidad, Trinidad and Tobago
Caronium *see* A Coruña
54 J5 **Carora** Lara, N Venezuela 10°12′N 70°07′W
86 F12 **Carpathian Mountains** *var.* Carpathians, *Cz./Pol.* Karpaty, *Ger.* Karpaten. ▲▲ E Europe
Carpathians *see* Carpathian Mountains
Carpathos/Carpathus *see* Kárpathos
116 H12 **Carpaţii Meridionalii** *var.* Alpi Transilvaniei, Carpaţii Sudici, *Eng.* South Carpathians, Transylvanian Alps, *Ger.* Südkarpaten, Transsylvanische Alpen, *Hung.* Déli-Kárpátok, Erdélyi-Havasok. ▲▲ C Romania
Carpaţii Sudici *see* Carpaţii Meridionalii
174 L7 **Carpentaria, Gulf of** *gulf* N Australia
Carpentoracte *see* Carpentras
103 R14 **Carpentras** *anc.* Carpentoracte. Vaucluse, SE France 44°03′N 05°03′E
106 F9 **Carpi** Emilia-Romagna, N Italy 44°47′N 10°53′E
116 E11 **Cărpiniş** *Hung.* Gyertyámos. Timiş, W Romania 45°46′N 20°53′E
35 R14 **Carpinteria** California, W USA 34°24′N 119°30′W
23 S9 **Carrabelle** Florida, SE USA 29°51′N 84°39′W
Carraig Aonair *see* Fastnet Rock
Carraig Fhearghais *see* Carrickfergus
Carraig Mhachaire Rois *see* Carrickmacross
Carraig na Siúire *see* Carrick-on-Suir
Carrantual *see* Carrauntoohil
106 E10 **Carrara** Toscana, C Italy 44°05′N 10°07′E
61 F20 **Carrasco** ✈ (Montevideo) Canelones, S Uruguay 34°51′S 56°00′W
105 P9 **Carrascosa del Campo** Castilla-La Mancha, C Spain 40°02′N 02°35′W
54 H4 **Carrasquero** Zulia, NW Venezuela 11°00′N 72°01′W
183 O9 **Carrathool** New South Wales, SE Australia 34°25′S 145°30′E
Carrauntohil *see* Carrauntoohil
97 B21 **Carrauntoohil** *Ir.* Carrantual, Carrauntohil, Corrán Tuathail. ▲ SW Ireland 51°98′N 09°53′W
45 Y15 **Carriacou** *island* N Grenada
97 G15 **Carrickfergus** *Ir.* Carraig Fhearghais. NE Northern Ireland, United Kingdom 54°43′N 05°49′W
97 F16 **Carrickmacross** *Ir.* Carraig Mhachaire Rois. N Ireland 53°58′N 06°43′W
97 D16 **Carrick-on-Shannon** *Ir.* Cora Droma Rúisc. NW Ireland 53°37′N 08°05′W
97 E20 **Carrick-on-Suir** *Ir.* Carraig na Siúire. S Ireland 52°21′N 07°25′W
182 I7 **Carrieton** South Australia 32°27′S 138°33′E
40 L7 **Carrillo** Chihuahua, N Mexico 25°53′N 103°54′W
29 O4 **Carrington** North Dakota, N USA 47°27′N 99°07′W
104 M4 **Carrión** ≈ N Spain
104 M4 **Carrión de los Condes** Castilla y León, N Spain 42°20′N 04°37′W
25 P13 **Carrizo Springs** Texas, SW USA 28°33′N 99°54′W
37 S13 **Carrizozo** New Mexico, SW USA 33°38′N 105°52′W
29 T13 **Carroll** Iowa, C USA 42°04′N 94°52′W
23 N4 **Carrollton** Alabama, S USA 33°13′N 88°05′W
23 R3 **Carrollton** Georgia, SE USA 33°33′N 85°04′W
30 K14 **Carrollton** Illinois, N USA 39°18′N 90°24′W
20 L4 **Carrollton** Kentucky, S USA 38°41′N 85°09′W
31 R8 **Carrollton** Michigan, N USA 43°27′N 83°55′W
27 T3 **Carrollton** Missouri, C USA 39°20′N 93°30′W
31 U12 **Carrollton** Ohio, N USA 40°34′N 81°05′W
25 T6 **Carrollton** Texas, SW USA 32°57′N 96°53′W
11 U14 **Carrot** ≈ Saskatchewan, S Canada
11 U14 **Carrot River** Saskatchewan, C Canada 53°18′N 103°32′W
18 J7 **Carry Falls Reservoir** ⊠ New York, NE USA
136 L11 **Çarşamba** Samsun, N Turkey 41°13′N 36°43′E
28 L6 **Carson** North Dakota, N USA 46°26′N 101°34′W
35 Q6 **Carson City** *state capital* Nevada, W USA 39°10′N 119°46′W
35 R6 **Carson River** ≈ Nevada, W USA
35 S5 **Carson Sink** *salt flat* Nevada, W USA
11 Q16 **Carstairs** Alberta, SW Canada 51°35′N 114°02′W
Carstensz, Puntjak *see* Jaya, Puncak
54 E5 **Cartagena** *var.* Cartagena de los Indes. Bolívar, NW Colombia 10°24′N 75°33′W
105 R13 **Cartagena** *anc.* Carthago Nova. Murcia, SE Spain 37°36′N 00°59′W
54 E13 **Cartagena de Chaira** Caquetá, S Colombia 01°19′N 74°52′W
Cartagena de los Indes *see* Cartagena
54 D10 **Cartago** Valle del Cauca, W Colombia 04°45′N 75°55′W
43 N14 **Cartago** Cartago, C Costa Rica 09°50′N 83°54′W
42 M14 **Cartago** *off.* Provincia de Cartago. ◈ *province* C Costa Rica
Cartago, Provincia de *see* Cartago
25 O11 **Carta Valley** Texas, SW USA 29°46′N 100°37′W
104 F10 **Cartaxo** Santarém, C Portugal 39°10′N 08°47′W
104 I14 **Cartaya** Andalucía, S Spain 37°16′N 07°09′W
Carteret Islands *see* Tulun Islands
29 S15 **Carter Lake** Iowa, C USA 41°17′N 95°55′W
23 S3 **Cartersville** Georgia, SE USA 34°10′N 84°48′W
185 M14 **Carterton** Wellington, North Island, New Zealand 41°01′S 175°30′E
30 J13 **Carthage** Illinois, N USA 40°25′N 91°09′W
22 L5 **Carthage** Mississippi, S USA 32°43′N 89°31′W
27 R7 **Carthage** Missouri, C USA 37°10′N 94°20′W
18 I8 **Carthage** New York, NE USA 43°58′N 75°36′W
21 T10 **Carthage** North Carolina, SE USA 35°21′N 79°27′W
20 K8 **Carthage** Tennessee, S USA 36°14′N 85°59′W
25 X7 **Carthage** Texas, SW USA 32°10′N 94°21′W
74 M5 **Carthage** ✈ (Tunis) N Tunisia 36°51′N 10°12′E
Carthago Nova *see* Cartagena
14 E10 **Cartier** Ontario, S Canada 46°40′N 81°31′W
13 S8 **Cartwright** Newfoundland and Labrador, E Canada 53°40′N 57°W
55 P9 **Caruana de Montaña** Bolívar, SE Venezuela 05°16′N 63°12′W
59 Q15 **Caruaru** Pernambuco, E Brazil 08°15′S 35°55′W
55 P5 **Carúpano** Sucre, NE Venezuela 10°39′N 63°14′W
Carusbur *see* Cherbourg
58 M12 **Carutapera** Maranhão, E Brazil 01°12′S 45°57′W
27 Y9 **Caruthersville** Missouri, C USA 36°11′N 89°40′W
103 O1 **Carvin** Pas-de-Calais, N France 50°31′N 03°00′E
58 E12 **Carvoeiro** Amazonas, NW Brazil 01°24′S 61°59′W
104 E10 **Carvoeiro, Cabo** *headland* C Portugal 39°19′N 09°27′W
21 U9 **Cary** North Carolina, SE USA 35°47′N 78°46′W
182 M3 **Caryapundy Swamp** *wetland* New South Wales/Queensland, SE Australia
65 E24 **Carysfort, Cape** *headland* East Falkland, Falkland Islands 51°26′S 57°50′W
74 F6 **Casablanca** *Ar.* Dar-el-Beida. NW Morocco 33°39′N 07°31′W
60 M8 **Casa Branca** São Paulo, S Brazil 21°47′S 47°05′W
36 L14 **Casa Grande** Arizona, SW USA 32°52′N 111°45′W
106 C8 **Casale Monferrato** Piemonte, NW Italy 45°08′N 08°27′E
106 E8 **Casalpusterlengo** Lombardia, N Italy 45°10′N 09°37′E
54 H10 **Casanare** *off.* Intendencia de Casanare. ◈ *province* C Colombia
Casanare, Intendencia de *see* Casanare
55 P5 **Casanay** Sucre, NE Venezuela 10°30′N 63°25′W
24 K11 **Casa Piedra** Texas, SW USA 29°43′N 104°03′W
107 Q19 **Casarano** Puglia, SE Italy 40°01′N 18°10′E
42 J11 **Casares** Carazo, W Nicaragua 11°37′N 86°19′W
105 R10 **Casas Ibáñez** Castilla-La Mancha, C Spain 39°17′N 01°28′W
61 I14 **Casca** Rio Grande do Sul, S Brazil 28°39′S 51°55′W
172 I17 **Cascade** Mahé, NE Seychelles 04°39′S 55°29′E
33 N13 **Cascade** Idaho, NW USA 44°31′N 116°02′W
29 Y13 **Cascade** Iowa, C USA 42°18′N 91°00′W
33 R9 **Cascade** Montana, NW USA 47°15′N 111°46′W
185 B20 **Cascade Point** *headland* South Island, New Zealand 44°00′S 168°23′E
32 G13 **Cascade Range** ▲▲ Oregon/Washington, NW USA
33 N12 **Cascade Reservoir** ⊠ Idaho, NW USA
0 E8 **Cascadia Basin** *undersea feature* NE Pacific Ocean 47°00′N 127°30′W
104 E11 **Cascais** Lisboa, C Portugal 38°41′N 09°25′W
15 W7 **Cascapédia** ≈ Québec, SE Canada
59 I22 **Cascavel** Ceará, E Brazil 04°10′S 38°15′W
60 G11 **Cascavel** Paraná, S Brazil 24°56′S 53°28′W
106 I13 **Cascia** Umbria, C Italy 42°45′N 13°01′E
106 F11 **Cascina** Toscana, C Italy 43°40′N 10°33′E
19 Q8 **Casco Bay** *bay* Maine, NE USA
194 J7 **Case Island** *island* Antarctica
106 B8 **Caselle** ✈ (Torino) Piemonte, NW Italy 45°06′N 07°41′E
107 K17 **Caserta** Campania, S Italy 41°05′N 14°20′E
15 N8 **Casey** Québec, SE Canada 47°50′N 74°09′W
30 M14 **Casey** Illinois, N USA 39°18′N 87°59′W
195 Y12 **Casey** *Australian research station* Antarctica 65°58′S 111°04′E
195 W3 **Casey Bay** *bay* Antarctica
80 Q11 **Caseyr, Raas** *headland* NE Somalia 11°51′S 51°16′E
97 D20 **Cashel** *Ir.* Caiseal. S Ireland 52°31′N 07°53′W
54 G6 **Casigua** Zulia, W Venezuela 08°46′N 72°30′W
61 B19 **Casilda** Santa Fe, C Argentina 33°05′S 61°10′W
Casim *see* General Toshevo
183 V4 **Casino** New South Wales, SE Australia 28°50′S 153°02′E
107 J16 **Cassino** *prev.* San Germano; *anc.* Casinum. Lazio, C Italy 41°29′N 13°50′E
Casinum *see* Cassino
111 E17 **Čáslav** *Ger.* Tschaslau. Střední Čechy, C Czech Republic 49°54′N 15°23′E
56 C13 **Casma** Ancash, C Peru 09°30′S 78°18′W
167 S7 **Ca, Sông** ≈ N Vietnam
107 K17 **Casoria** Campania, S Italy 40°54′N 14°28′E
105 T6 **Caspe** Aragón, NE Spain 41°14′N 00°03′W
33 X15 **Casper** Wyoming, C USA 42°48′N 106°22′W
84 M10 **Caspian Depression** *Kaz.* Kaspīy Mangy Oypaty, *Rus.* Prikaspiyskaya Nizmennost'. *depression* Kazakhstan/Russian Federation
130 D10 **Caspian Sea** *Az.* Xäzär Dänizi, *Kaz.* Kaspiy Tengizi, *Per.* Baḥr-e Khazar, Daryā-ye Khazar, *Rus.* Kaspiyskoye More. *inland sea* Asia/Europe
83 L14 **Cassacatiza** Tete, NW Mozambique 14°20′S 32°24′E
Cassai *see* Kasai
82 F13 **Cassamba** Moxico, E Angola 13°07′S 20°22′E
107 N20 **Cassano allo Ionio** Calabria, SE Italy 39°46′N 16°16′E
31 S8 **Cass City** Michigan, N USA 43°36′N 83°10′W
Cassel *see* Kassel
14 M13 **Casselman** Ontario, SE Canada 45°18′N 75°05′W
29 R5 **Casselton** North Dakota, N USA 46°53′N 97°10′W
Cássia *see* Santa Rita de Cassia
10 J9 **Cassiar** British Columbia, W Canada 59°16′N 129°40′W
10 K10 **Cassiar Mountains** ▲▲ British Columbia, W Canada
83 C15 **Cassinga** Huíla, SW Angola 15°08′S 16°05′E
29 T4 **Cass Lake** Minnesota, N USA 47°22′N 94°36′W
29 T4 **Cass Lake** ◎ Minnesota, N USA
31 P10 **Cassopolis** Michigan, N USA 41°56′N 86°00′W
31 S8 **Cass River** ≈ Michigan, N USA
27 S8 **Cassville** Missouri, C USA 36°42′N 93°52′W
Castamoni *see* Kastamonu
58 L12 **Castanhal** Pará, NE Brazil 01°16′S 47°55′W
104 G8 **Castanheira de Pêra** Leiria, C Portugal 40°01′N 08°12′W
41 N7 **Castaños** Coahuila, NE Mexico 26°48′N 101°26′W
108 I10 **Castasegna** Graubünden, SE Switzerland 46°21′N 09°30′E
106 D8 **Casteggio** Lombardia, N Italy 45°02′N 09°10′E
107 K23 **Castelbuono** Sicilia, Italy, C Mediterranean Sea 37°56′N 14°05′E
107 K15 **Castel di Sangro** Abruzzo, C Italy 41°46′N 14°03′E
106 H7 **Castelfranco Veneto** Veneto, NE Italy 45°40′N 11°55′E
102 K14 **Casteljaloux** Lot-et-Garonne, SW France 44°19′N 00°03′E
107 L18 **Castellabate** *var.* Santa Maria di Castellabate. Campania, S Italy 40°16′N 14°57′E
107 I23 **Castellammare del Golfo** Sicilia, Italy, C Mediterranean Sea 38°02′N 12°53′E
107 H22 **Castellammare, Golfo di** *gulf* Sicilia, Italy, C Mediterranean Sea
103 U15 **Castellane** Alpes-de-Haute-Provence, SE France 43°49′N 06°34′E
107 O18 **Castellaneta** Puglia, SE Italy 40°38′N 16°57′E
106 E9 **Castel l'Arquato** Emilia-Romagna, C Italy 44°52′N 09°51′E
61 E21 **Castelli** Buenos Aires, E Argentina 36°07′S 57°47′W
105 S8 **Castelló de la Plana** *var.* Castellón de la Plana. ◈ *province* Valenciana, E Spain
Castelló de la Plana *see* Castellón de la Plana
Castellón *see* Castellón de la Plana
105 T9 **Castellón de la Plana** *var.* Castellón, *Cat.* Castelló de la Plana. Valenciana, E Spain 39°59′N 00°03′W
Castellón de la Plana *see* Castelló de la Plana
105 S7 **Castellote** Aragón, NE Spain 40°46′N 00°18′W
103 N16 **Castelnaudary** Aude, S France 43°18′N 01°57′E
102 L16 **Castelnau-Magnoac** Hautes-Pyrénées, S France 43°18′N 00°30′E
106 F10 **Castelnovo ne' Monti** Emilia-Romagna, C Italy 44°26′N 10°24′E
Castelnuovo *see* Herceg-Novi
104 H9 **Castelo Branco** Castelo Branco, C Portugal 39°50′N 07°30′W
104 H8 **Castelo Branco** ◈ *district* C Portugal
104 I10 **Castelo de Vide** Portalegre, C Portugal 39°25′N 07°27′W
104 G9 **Castelo do Bode, Barragem do** ⊠ C Portugal

◆ Country
● Country Capital
◇ Dependent Territory
○ Dependent Territory Capital
◈ Administrative Regions
✈ International Airport
▲ Mountain
▲▲ Mountain Range
Volcano
≈ River
◎ Lake
⊠ Reservoir

106 G10 **Castel San Pietro Terme** Emilia-Romagna, C Italy 44°22′N 11°34′E
107 B17 **Castelsardo** Sardegna, Italy, C Mediterranean Sea 40°54′N 08°42′E
102 M14 **Castelsarrasin** Tarn-et-Garonne, S France 44°02′N 01°06′E
107 I24 **Casteltermini** Sicilia, Italy, C Mediterranean Sea 37°33′N 13°38′E
107 H24 **Castelvetrano** Sicilia, Italy, C Mediterranean Sea 37°40′N 12°46′E
182 L12 **Casterton** Victoria, SE Australia 37°37′S 141°22′E
102 J15 **Castets** Landes, SW France 43°55′N 01°08′W
106 H12 **Castiglione del Lago** Umbria, C Italy 43°07′N 12°02′E
106 F13 **Castiglione della Pescaia** Toscana, C Italy 42°46′N 10°53′E
106 F8 **Castiglione delle Stiviere** Lombardia, N Italy 45°24′N 10°31′E
104 M9 **Castilla-La Mancha** ◐ *autonomous community* NE Spain
105 N10 **Castilla Nueva** *cultural region* C Spain
105 N6 **Castilla Vieja** ◐ *cultural region* N Spain
104 L5 **Castilla y León** *var.* Castillia Leon. ◐ *autonomous community* NW Spain
Castilla Leon *see* Castilla y León
Castillo de Locubim *see* Castillo de Locubín
105 N14 **Castillo de Locubín** *var.* Castillo de Locubim. Andalucía, S Spain 37°32′N 03°56′W
102 K13 **Castillon-la-Bataille** Gironde, SW France 44°51′N 00°01′W
63 I19 **Castillo, Pampa del** *plain* S Argentina
61 G19 **Castillos** Rocha, SE Uruguay 34°12′S 53°52′W
97 B16 **Castlebar** *Ir.* Caisleán an Bharraigh. W Ireland 53°52′N 09°17′W
97 F16 **Castleblayney** *Ir.* Baile na Lorgan. N Ireland 54°07′N 06°44′W
45 O11 **Castle Bruce** E Dominica 15°24′N 61°26′W
36 M5 **Castle Dale** Utah, W USA 39°10′N 111°02′W
36 I14 **Castle Dome Peak** ▲ Arizona, SW USA 33°04′N 114°08′W
97 J14 **Castle Douglas** S Scotland, United Kingdom 54°56′N 03°56′W
97 E14 **Castlefinn** *Ir.* Caisleán na Finne. NW Ireland 54°47′N 07°35′W
97 M17 **Castleford** N England, United Kingdom 53°44′N 01°21′W
11 O17 **Castlegar** British Columbia, SW Canada 49°18′N 117°48′W
64 B12 **Castle Harbour** *inlet* Bermuda, NW Atlantic Ocean
21 V12 **Castle Hayne** North Carolina, SE USA 34°23′N 78°07′W
97 B20 **Castleisland** *Ir.* Oileán Ciarral. SW Ireland 52°12′N 09°30′W
183 N12 **Castlemaine** Victoria, SE Australia 37°06′S 144°13′E
37 R5 **Castle Peak** ▲ Colorado, C USA 39°00′N 106°51′W
33 O13 **Castle Peak** ▲ Idaho, NW USA 44°02′N 114°42′W
184 N13 **Castlepoint** Wellington, North Island, New Zealand 40°54′S 176°13′E
97 D17 **Castlerea** *Ir.* An Caisleán Riabhach. W Ireland 53°45′N 08°32′W
97 G15 **Castlereagh** *Ir.* An Caisleán Riabhach. N Northern Ireland, United Kingdom 54°33′N 05°53′W
183 R6 **Castlereagh River** ∼ New South Wales, SE Australia
37 T5 **Castle Rock** Colorado, C USA 39°22′N 104°51′W
30 K7 **Castle Rock Lake** ⊠ Wisconsin, N USA
65 G25 **Castle Rock Point** *headland* S Saint Helena 16°02′S 05°45′W
97 I16 **Castletown** SE Isle of Man 54°05′N 04°39′W
29 R9 **Castlewood** South Dakota, N USA 44°43′N 97°01′W
11 R15 **Castor** Alberta, SW Canada 52°14′N 111°54′W
14 M13 **Castor** ∼ Ontario, SE Canada
27 X7 **Castor River** ∼ Missouri, C USA
Castra Albiensium *see* Castres
Castra Regina *see* Regensburg
103 N15 **Castres** *anc.* Castra Albiensium. Tarn, S France 43°36′N 02°15′E
98 H9 **Castricum** Noord-Holland, W Netherlands 52°33′N 04°40′E
45 S11 **Castries** ● (Saint Lucia) N Saint Lucia 14°01′N 60°59′W
60 J11 **Castro** Paraná, S Brazil 24°46′S 50°03′W
63 F17 **Castro** Los Lagos, W Chile 42°27′S 73°48′W
104 H7 **Castro Daire** Viseu, N Portugal 40°54′N 07°55′W
104 M13 **Castro del Río** Andalucía, S Spain 37°41′N 04°29′W
Castrogiovanni *see* Enna
104 H14 **Castro Marim** Faro, S Portugal 37°13′N 07°26′W
104 J2 **Castropol** Asturias, N Spain 43°30′N 07°01′W
105 O2 **Castro-Urdiales** *var.* Castro Urdiales. Cantabria, N Spain 43°23′N 03°11′W
104 G13 **Castro Verde** Beja, S Portugal 37°42′N 08°05′W
107 N19 **Castrovillari** Calabria, SW Italy 39°48′N 16°12′E
35 N10 **Castroville** California, W USA 36°46′N 121°46′W
25 R12 **Castroville** Texas, SW USA 29°21′N 98°52′W
104 K11 **Castuera** Extremadura, W Spain 38°44′N 05°33′W
61 F19 **Casupá** Florida, S Uruguay 34°09′S 55°38′W
185 A22 **Caswell Sound** *sound* South Island, New Zealand
137 Q13 **Çat** Erzurum, NE Turkey 39°40′N 41°03′E
42 K6 **Catacamas** Olancho, C Honduras 14°55′N 85°54′W
56 A10 **Catacaos** Piura, NW Peru 05°22′S 80°40′W
22 I7 **Catahoula Lake** ⊚ Louisiana, S USA
137 S15 **Çatak** Van, SE Turkey 38°02′N 43°05′E
137 S15 **Çatak Çayi** ∼ SE Turkey
114 O12 **Çatalca** İstanbul, NW Turkey 41°09′N 28°28′E
114 O12 **Çatalca Yarimadasi** *physical region* NW Turkey
62 H6 **Catalina** Antofagasta, N Chile 25°19′S 69°37′W
Catalonia *see* Cataluña
105 U5 **Cataluña** *Cat.* Catalunya, *Eng.* Catalonia. ◐ *autonomous community* N Spain
Catalunya *see* Cataluña
62 I7 **Catamarca** *off.* Provincia de Catamarca. ◐ *province* NW Argentina
Catamarca *see* San Fernando del Valle de Catamarca
Catamarca, Provincia de *see* Catamarca
83 M16 **Catandica** Manica, C Mozambique 18°05′S 33°10′E
171 P4 **Catanduanes Island** *island* N Philippines
60 K8 **Catanduva** São Paulo, S Brazil 21°05′S 49°00′W
107 L24 **Catania** Sicilia, Italy, C Mediterranean Sea 37°31′N 15°04′E
107 M24 **Catania, Golfo di** *gulf* Sicilia, Italy, C Mediterranean Sea
45 U5 **Cataño** *var.* Cantaño. E Puerto Rico 18°26′N 66°06′W
107 O21 **Catanzaro** Calabria, SW Italy 38°53′N 16°36′E
107 O22 **Catanzaro Marina** *var.* Marina di Catanzaro. Calabria, S Italy 38°48′N 16°33′E
Marina di Catanzaro *see* Catanzaro Marina
25 Q14 **Catarina** Texas, SW USA 28°19′N 99°36′W
171 Q5 **Catarman** Samar, C Philippines 12°29′N 124°34′E
105 S10 **Catarroja** Valenciana, E Spain 39°24′N 00°24′W
21 R11 **Catawba River** ∼ North Carolina/South Carolina, SE USA
171 Q5 **Catbalogan** Samar, C Philippines 11°49′N 124°55′E
14 I14 **Catchacoma** Ontario, SE Canada 44°43′N 78°19′W
41 S15 **Catemaco** Veracruz-Llave, SE Mexico 18°28′N 95°10′W
Cathair na Mart *see* Westport
Cathair Saidhbhín *see* Caherciveen
31 P5 **Cat Head Point** *headland* Michigan, N USA 45°11′N 85°37′W
23 Q2 **Cathedral Caverns** *cave* Alabama, S USA
35 V16 **Cathedral City** California, W USA 33°45′N 116°27′W
24 K10 **Cathedral Mountain** ▲ Texas, SW USA 30°10′N 103°39′W
32 G10 **Cathlamet** Washington, NW USA 46°12′N 123°24′W
76 G13 **Catió** S Guinea-Bissau 11°13′N 15°10′W
55 O10 **Catisimiña** Bolívar, SE Venezuela 04°07′N 63°40′W
44 J3 **Cat Island** *island* C The Bahamas
12 B9 **Cat Lake** Ontario, S Canada 51°47′N 91°52′W
21 P5 **Catlettsburg** Kentucky, S USA 38°24′N 82°37′W
185 D24 **Catlins** ∼ South Island, New Zealand
35 R1 **Catnip Mountain** ▲ Nevada, W USA 41°53′N 119°19′W
41 Z11 **Catoche, Cabo** *headland* SE Mexico 21°36′N 87°04′W
27 P9 **Catoosa** Oklahoma, C USA 36°11′N 95°45′W
41 N10 **Catorce** San Luis Potosí, C Mexico 23°42′N 100°49′W
63 I14 **Catriel** Río Negro, C Argentina 37°54′S 67°52′W
62 K13 **Catriló** La Pampa, C Argentina 36°28′S 63°20′W
58 F11 **Catrimani** Roraima, N Brazil 0°24′N 61°30′W
58 E10 **Catrimani, Rio** ∼ N Brazil
18 K11 **Catskill** New York, NE USA 42°13′N 73°52′W
18 K11 **Catskill Creek** ∼ New York, NE USA
18 J11 **Catskill Mountains** ▲▲ New York, NE USA
18 D11 **Cattaraugus Creek** ∼ New York, NE USA
Cattaro *see* Kotor
Cattaro, Bocche di *see* Kotorska, Boka
107 I24 **Cattolica Eraclea** Sicilia, Italy, C Mediterranean Sea 37°27′N 13°24′E
83 B14 **Catumbela** ∼ W Angola
83 N14 **Catur** Niassa, N Mozambique 13°50′S 35°43′E
82 C10 **Cauale** ∼ NE Angola
171 O2 **Cauayan** Luzon, N Philippines 16°55′N 121°46′E
54 C12 **Cauca** *off.* Departamento del Cauca. ◐ *province* SW Colombia
47 P5 **Cauca** ∼ SE Venezuela
Cauca, Departamento del *see* Cauca
58 P13 **Caucaia** Ceará, E Brazil 03°44′S 38°45′W
54 E7 **Cauca, Río** ∼ N Colombia
54 E7 **Caucasia** Antioquia, NW Colombia 07°59′N 75°13′W
137 Q8 **Caucasus** *Rus.* Kavkaz. ▲▲ Georgia/Russian Federation
62 I10 **Caucete** San Juan, W Argentina 31°38′S 68°16′W
105 R11 **Caudete** Castilla-La Mancha, C Spain 38°42′N 01°00′W
103 P2 **Caudry** Nord, N France 50°07′N 03°24′E
82 D11 **Caungula** Lunda Norte, NE Angola 08°22′S 18°37′E
62 G13 **Cauquenes** Maule, C Chile 35°58′S 72°22′W
55 N8 **Caura, Río** ∼ C Venezuela
15 V7 **Causapscal** Québec, SE Canada 48°22′N 67°14′W
117 N10 **Căuşeni** *Rus.* Kaushany. E Moldova 46°37′N 29°21′E
102 M14 **Caussade** Tarn-et-Garonne, S France 44°10′N 01°31′E
102 K17 **Cauterets** Hautes-Pyrénées, S France 42°53′N 00°08′W
10 J15 **Caution, Cape** *headland* British Columbia, SW Canada 51°10′N 127°43′W
44 H7 **Cauto** ∼ E Cuba
Cauvery *see* Kāveri
102 L3 **Caux, Pays de** *physical region* N France
107 L18 **Cava de' Tirreni** Campania, S Italy 40°42′N 14°42′E
104 G6 **Cávado** ∼ N Portugal
Cavaia *see* Kavajë
103 R15 **Cavaillon** Vaucluse, SE France 43°51′N 05°01′E
103 U16 **Cavalaire-sur-Mer** Var, SE France 43°10′N 06°31′E
106 G6 **Cavalese** *Ger.* Gablös. Trentino-Alto Adige, N Italy 46°18′N 11°29′E
29 Q2 **Cavalier** North Dakota, N USA 48°47′N 97°37′W
76 L17 **Cavalla** *var.* Cavally, Cavally Fleuve. ∼ Ivory Coast/Liberia
105 Y8 **Cavalleria, Cap de** *var.* Cabo Caballeria. *headland* Menorca, Spain, W Mediterranean Sea 40°04′N 04°06′E
184 K2 **Cavalli Islands** *island group* N New Zealand
Cavally/Cavally Fleuve *see* Cavalla
97 E16 **Cavan** *Ir.* Cabhán. N Ireland 54°N 07°21′W
97 E16 **Cavan** *Ir.* An Cabhán. *cultural region* N Ireland
106 H8 **Cavarzere** Veneto, NE Italy 45°08′N 12°05′E
27 W9 **Cave City** Arkansas, C USA 35°56′N 91°33′W
20 K7 **Cave City** Kentucky, S USA 37°08′N 85°57′W
65 M25 **Cave Point** *headland* S Tristan da Cunha
21 N5 **Cave Run Lake** ⊠ Kentucky, S USA
58 K11 **Caviana de Fora, Ilha** *var.* Ilha Caviana. *island* N Brazil
Caviana, Ilha *see* Caviana de Fora, Ilha
113 I16 **Cavtat** *It.* Ragusavecchia. Dubrovnik-Neretva, SE Croatia 42°36′N 18°13′E
Cawnpore *see* Kānpur
Caxamarca *see* Cajamarca
58 A13 **Caxias** Amazonas, W Brazil 04°27′S 71°27′W
58 N13 **Caxias** Maranhão, E Brazil 04°53′S 43°20′W
61 I15 **Caxias do Sul** Rio Grande do Sul, S Brazil 29°14′S 51°10′W
42 J4 **Caxinas, Punta** *headland* N Honduras 16°01′N 86°02′W
82 B11 **Caxito** Bengo, NW Angola 08°34′S 13°38′E
136 F14 **Çay** Afyon, W Turkey 38°35′N 31°01′E
40 L15 **Cayacal, Punta** *var.* Punta Mongrove. *headland* S Mexico 17°55′N 102°09′W
56 C6 **Cayambe** Pichincha, N Ecuador 0°02′N 78°08′W
56 C6 **Cayambe** ▲ N Ecuador 0°00′S 77°58′W
21 R12 **Cayce** South Carolina, SE USA 33°58′N 81°04′W
55 Y10 **Cayenne** ○ (French Guiana) NE French Guiana 04°55′N 52°18′W
55 Y10 **Cayenne** ✈ NE French Guiana 04°55′N 52°18′W
44 K10 **Cayes** *var.* Les Cayes. SW Haiti 18°10′N 73°48′W
45 U6 **Cayey** C Puerto Rico 18°06′N 66°11′W
45 U6 **Cayey, Sierra de** ▲ E Puerto Rico
103 N14 **Caylus** Tarn-et-Garonne, S France 44°13′N 01°42′E
44 E8 **Cayman Brac** *island* E Cayman Islands
44 D8 **Cayman Islands** ◇ *UK dependent territory* W West Indies
64 D11 **Cayman Trench** *undersea feature* NW Caribbean Sea 19°00′N 80°00′W
47 O3 **Cayman Trough** *undersea feature* NW Caribbean Sea 18°00′N 81°00′W
80 O13 **Caynabo** Togdheer, N Somalia 08°55′N 46°28′E
42 F3 **Cayo** ◐ *district* SW Belize
Cayo *see* San Ignacio
43 N9 **Cayos Guerrero** *reef* E Nicaragua
43 O9 **Cayos King** *reef* E Nicaragua
44 E4 **Cay Sal** *islet* SW The Bahamas
14 G16 **Cayuga** Ontario, S Canada 42°57′N 79°49′W
25 V8 **Cayuga** Texas, SW USA 31°55′N 95°57′W
18 G10 **Cayuga Lake** ⊚ New York, NE USA
104 K13 **Cazalla de la Sierra** Andalucía, S Spain 37°56′N 05°46′W
116 L14 **Căzăneşti** Ialomiţa, SE Romania 44°36′N 27°03′E
102 M16 **Cazères** Haute-Garonne, S France 43°15′N 01°11′E
112 E10 **Cazin** ◐ Federacija Bosne I Hercegovine, NW Bosnia and Herzegovina
82 G13 **Cazombo** Moxico, E Angola 11°54′S 22°56′E
105 O13 **Cazorla** Andalucía, S Spain 37°55′N 03°W
Cazza *see* Sušac
104 L4 **Cea** ∼ NW Spain
Ceadâr-Lunga *see* Ciadîr-Lunga
Ceanannas *see* Kells
Ceann Toirc *see* Kanturk
58 O13 **Ceará** *off.* Estado do Ceará. ◐ *state* C Brazil
Ceará *see* Fortaleza
Ceara Abyssal Plain *see* Ceará Plain
Ceará, Estado do *see* Ceará
59 Q14 **Ceará Mirim** Rio Grande do Norte, E Brazil 05°30′S 35°51′W
64 J13 **Ceará Plain** *var.* Ceara Abyssal Plain. *undersea feature* W Atlantic Ocean 0°00′36°30′W
64 I13 **Ceará Ridge** *undersea feature* C Atlantic Ocean
Ceatharlach *see* Carlow
43 Q17 **Cébaco, Isla** *island* SW Panama
40 K7 **Ceballos** Durango, C Mexico 26°33′N 104°07′W
61 G19 **Cebollatí** Rocha, E Uruguay 33°15′S 53°46′W
61 G19 **Cebollatí, Río** ∼ E Uruguay
105 P5 **Cebollera** ▲ N Spain 42°01′N 02°40′W
104 M8 **Cebreros** Castilla y León, N Spain 40°27′N 04°28′W
171 P6 **Cebu** *off.* Cebu City. Cebu, C Philippines 10°17′N 123°46′E
171 P6 **Cebu** *island* C Philippines
Cebu City *see* Cebu
107 J16 **Ceccano** Lazio, C Italy 41°34′N 13°20′E
Čechy *see* Bohemia
106 F12 **Cecina** Toscana, C Italy 43°19′N 10°31′E
26 K4 **Cedar Bluff Reservoir** ⊠ Kansas, C USA
30 M8 **Cedarburg** Wisconsin, N USA 43°18′N 87°58′W
36 J7 **Cedar City** Utah, W USA 37°40′N 113°03′W
25 T11 **Cedar Creek** Texas, SW USA 30°04′N 97°30′W
28 L7 **Cedar Creek** ∼ North Dakota, N USA
25 U7 **Cedar Creek Reservoir** ⊠ Texas, SW USA
29 W13 **Cedar Falls** Iowa, C USA 42°31′N 92°27′W
31 N8 **Cedar Grove** Wisconsin, N USA 43°31′N 87°48′W
21 Y6 **Cedar Island** *island* Virginia, NE USA
23 U11 **Cedar Key** Cedar Keys, Florida, SE USA 29°08′N 83°03′W
23 U11 **Cedar Keys** *island group* Florida, SE USA
11 V14 **Cedar Lake** ⊚ Manitoba, C Canada
14 I11 **Cedar Lake** ⊚ Ontario, SE Canada
24 M6 **Cedar Lake** ⊚ Texas, SW USA
29 X13 **Cedar Rapids** Iowa, C USA 41°58′N 91°40′W
29 X14 **Cedar River** ∼ Iowa/Minnesota, C USA
29 O14 **Cedar River** ∼ Nebraska, C USA
31 P8 **Cedar Springs** Michigan, N USA 43°13′N 85°33′W
23 R3 **Cedartown** Georgia, SE USA 34°00′N 85°16′W
27 O7 **Cedar Vale** Kansas, C USA 37°06′N 96°30′W
35 Q2 **Cedarville** California, W USA 41°30′N 120°10′W
104 H1 **Cedeira** Galicia, NW Spain 43°40′N 08°03′W
42 H8 **Cedeño** Choluteca, S Honduras 13°10′N 87°25′W
41 N10 **Cedral** San Luis Potosí, C Mexico 23°47′N 100°40′W
42 I6 **Cedros** Francisco Morazán, C Honduras 14°38′N 86°42′W
40 M9 **Cedros** Zacatecas, C Mexico 24°39′N 101°47′W
40 B5 **Cedros, Isla** *island* W Mexico
193 R5 **Cedros Trench** *undersea feature* E Pacific Ocean 27°45′N 115°45′W
182 E7 **Ceduna** South Australia 32°09′S 133°43′E
110 D10 **Cedynia** *Ger.* Zehden. Zachodnio-pomorskie, W Poland 52°54′N 14°15′E
80 P12 **Ceelaayo** Sanaag, N Somalia 11°18′N 49°20′E
81 O16 **Ceel Buur** *It.* El Bur. Galguduud, C Somalia 04°36′N 46°33′E
81 N15 **Ceel Dheere** *var.* Ceel Dher, *It.* El Dere. Galguduud, C Somalia 05°18′N 46°07′E
Ceel Dher *see* Ceel Dheere
80 O12 **Ceerigaabo** *var.* Erigabo, Erigavo. Sanaag, N Somalia 10°34′N 47°22′E
107 J23 **Cefalù** *anc.* Cephaloedium. Sicilia, Italy, C Mediterranean Sea 38°02′N 14°02′E
105 N6 **Cega** ∼ Castilla y León, N Spain
111 K23 **Cegléd** *prev.* Czegléd. Pest, C Hungary 47°10′N 19°47′E
113 N18 **Čegrane** W FYR Macedonia 41°50′N 20°59′E
105 Q13 **Cehegín** Murcia, SE Spain 38°04′N 01°48′W
136 K12 **Çekerek** Yozgat, N Turkey 40°04′N 35°30′E
146 B13 **Çekiçler** *Rus.* Chekishlyar, *Turkm.* Chekichler. Balkan Welaýaty, W Turkmenistan 37°35′N 53°52′E
107 J15 **Celano** Abruzzo, C Italy 42°06′N 13°33′E
104 H4 **Celanova** Galicia, NW Spain 42°09′N 07°58′W
42 F6 **Celaque, Cordillera de** ▲▲ W Honduras
41 N13 **Celaya** Guanajuato, C Mexico 20°32′N 100°48′W
Celebes *see* Sulawesi
192 F7 **Celebes Basin** *undersea feature* SE South China Sea 04°00′N 122°00′E
192 F7 **Celebes Sea** *Ind.* Laut Sulawesi. *sea* Indonesia/Philippines
41 W12 **Celestún** Yucatán, E Mexico 20°50′N 90°22′W
31 Q12 **Celina** Ohio, N USA 40°34′N 84°35′W
20 L8 **Celina** Tennessee, S USA 36°32′N 85°30′W
25 U5 **Celina** Texas, SW USA 33°19′N 96°46′W
112 G11 **Čelinac Donji** Republika Srpska, N Bosnia and Herzegovina 44°43′N 17°19′E
109 V10 **Celje** *Ger.* Cilli. C Slovenia 46°16′N 15°14′E
111 G23 **Celldömölk** Vas, W Hungary 47°16′N 17°10′E
100 J12 **Celle** *var.* Zelle. Niedersachsen, N Germany 52°38′N 10°05′E
99 D19 **Celles** Hainaut, SW Belgium 50°42′N 03°25′E
104 I7 **Celorico da Beira** Guarda, N Portugal 40°38′N 07°24′W
Celovec *see* Klagenfurt
64 M7 **Celtic Sea** *Ir.* An Mhuir Cheilteach. *sea* SW British Isles
64 N7 **Celtic Shelf** *undersea feature* E Atlantic Ocean 07°00′W 49°15′N
114 L13 **Çeltik Gölu** ⊚ NW Turkey
146 J17 **Çemenibit** *prev. Rus.* Chemenibit. Mary Welaýaty, S Turkmenistan 35°27′N 62°19′E
113 M14 **Čemerno** ▲ C Serbia
105 Q12 **Cenajo, Embalse del** ⊠ S Spain
171 V13 **Cenderawasih, Teluk** *var.* Teluk Irian, Teluk Sarera. *bay* W Pacific Ocean
105 P4 **Cenicero** La Rioja, N Spain 42°29′N 02°38′W
106 E9 **Ceno** ∼ NW Italy
102 K13 **Cenon** Gironde, SW France 44°51′N 00°33′W
14 K13 **Centennial Lake** ⊚ Ontario, SE Canada
Centennial State *see* Colorado
37 S7 **Center** Colorado, C USA 37°45′N 106°06′W
29 Q13 **Center** Nebraska, C USA 42°33′N 97°51′W
28 M5 **Center** North Dakota, N USA 47°07′N 101°18′W
25 X8 **Center** Texas, SW USA 31°49′N 94°10′W
29 W8 **Center City** Minnesota, N USA 45°25′N 92°48′W
36 L5 **Centerfield** Utah, W USA 39°07′N 111°49′W
20 K9 **Center Hill Lake** ⊠ Tennessee, S USA
29 X13 **Center Point** Iowa, C USA 42°11′N 91°47′W
25 R11 **Center Point** Texas, SW USA 29°56′N 99°01′W
29 W16 **Centerville** Iowa, C USA 40°43′N 92°51′W
27 W7 **Centerville** Missouri, C USA 37°27′N 91°01′W
29 R12 **Centerville** South Dakota, N USA 43°07′N 96°57′W
20 I9 **Centerville** Tennessee, S USA 35°45′N 87°29′W
25 V9 **Centerville** Texas, SW USA 31°17′N 95°59′W
40 M5 **Centinela, Picacho** ▲ NE Mexico 29°07′N 102°40′W
106 G9 **Cento** Emilia-Romagna, N Italy 44°43′N 11°16′E
Centrafricaine, République *see* Central African Republic
39 S8 **Central** Alaska, USA 65°34′N 144°48′W
37 P15 **Central** New Mexico, SW USA 32°46′N 108°09′W
83 H18 **Central** ◐ *district* E Botswana
138 E10 **Central** ◐ *district* C Israel
82 M13 **Central** ◐ *region* C Malawi
153 P12 **Central** ◐ *zone* C Nepal
186 E9 **Central** ◐ *province* S Papua New Guinea
63 I21 **Central** ◐ *department* C Paraguay
155 K25 **Central** ◐ *province* C Sri Lanka
83 J14 **Central** ◐ *province* C Zambia
117 P11 **Central** ✈ (Odesa) Odes'ka Oblast', SW Ukraine 46°26′N 30°41′E
Central *see* Centre
Central *see* Rennell and Bellona
79 H14 **Central African Republic** *var.* République Centrafricaine, *abbrev.* CAR; *prev.* Ubangi-Shari, Oubangui-Chari, Territoire de l'Oubangui-Chari. ◆ *republic* C Africa
192 C6 **Central Basin Trough** *undersea feature* W Pacific Ocean 16°45′N 130°00′E
Central Borneo *see* Kalimantan Tengah
149 P12 **Central Brāhui Range** ▲▲ W Pakistan
Central Celebes *see* Sulawesi Tengah
29 Y13 **Central City** Iowa, C USA 42°12′N 91°31′W
20 I6 **Central City** Kentucky, S USA 37°17′N 87°07′W
29 P15 **Central City** Nebraska, C USA 41°04′N 97°59′W
48 D6 **Central, Cordillera** ▲▲ W Bolivia
54 D11 **Central, Cordillera** ▲▲ W Colombia
42 M13 **Central, Cordillera** ▲▲ C Costa Rica
45 N9 **Central, Cordillera** ▲▲ C Dominican Republic
43 R16 **Central, Cordillera** ▲▲ C Panama
45 S6 **Central, Cordillera** ▲▲ Puerto Rico
80 A11 **Central Darfur** ◐ *state* SW Sudan
42 H7 **Central District** *var.* Tegucigalpa. ◐ *district* C Honduras
81 E16 **Central Equatoria** *var.* Bahr el Gebel, Bahr el Jebel. ◐ *state* S South Sudan
Central Finland *see* Keski-Suomi
Central Group *see* Inner Islands
30 L15 **Centralia** Illinois, N USA 38°31′N 89°07′W
27 U4 **Centralia** Missouri, C USA 39°12′N 92°08′W
32 G9 **Centralia** Washington, NW USA 46°43′N 122°57′W
Central Indian Ridge *see* Mid-Indian Ridge
Central Java *see* Jawa Tengah
Central Kalimantan *see* Kalimantan Tengah
148 L14 **Central Makrān Range** ▲▲ W Pakistan
Central Ostrobothnia *see* Keski-Pohjanmaa
192 K7 **Central Pacific Basin** *undersea feature* C Pacific Ocean 05°00′N 175°00′W
59 M19 **Central, Planalto** *var.* Brazilian Highlands. ▲▲ E Brazil
32 F15 **Central Point** Oregon, NW USA 42°22′N 122°55′W
Central Provinces and Berar *see* Madhya Pradesh
186 B6 **Central Range** ▲▲ NW Papua New Guinea
Central Russian Upland *see* Srednerusskaya Vozvyshennost'
Central Siberian Plateau/Central Siberian Uplands *see* Srednesibirskoye Ploskogor'ye
104 K8 **Central, Sistema** ▲▲ C Spain
Central Sulawesi *see* Sulawesi Tengah
35 N3 **Central Valley** California, W USA 40°39′N 122°21′W
35 P8 **Central Valley** *valley* California, W USA
23 Q3 **Centre** Alabama, S USA 34°09′N 85°40′W
79 E15 **Centre** *Eng.* Central. ◐ *province* C Cameroon
102 M8 **Centre** ◐ *region* N France
173 Y16 **Centre de Flacq** E Mauritius 20°12′S 57°43′E
55 Y9 **Centre Spatial Guyanais** *space station* N French Guiana
23 O5 **Centreville** Alabama, S USA 32°58′N 87°08′W
21 X3 **Centreville** Maryland, NE USA 39°03′N 76°04′W
22 J7 **Centreville** Mississippi, S USA 31°05′N 91°04′W
Centum Cellae *see* Civitavecchia
160 M14 **Cenxi** Guangxi Zhuangzu Zizhiqu, S China 22°58′N 111°00′E
Ceos *see* Tziá
Cephaloedium *see* Cefalù
112 I9 **Čepin** *Hung.* Csepén. Osijek-Baranja, E Croatia 45°32′N 18°33′E
Ceram *see* Seram, Pulau
Ceram Sea *see* Seram, Laut
192 G8 **Ceram Trough** *undersea feature* W Pacific Ocean
Cerasus *see* Giresun
36 I10 **Cerbat Mountains** ▲▲ Arizona, SW USA
103 P17 **Cerbère, Cap** *headland* S France 42°28′N 03°15′E
104 F13 **Cercal do Alentejo** Setúbal, S Portugal 37°48′N 08°40′W
111 A18 **Čerchov** *Ger.* Czerkow. ▲ W Czech Republic 49°24′N 12°47′E
103 O13 **Cère** ∼ C France
61 A16 **Ceres** Santa Fe, C Argentina 29°55′S 61°55′W
59 K18 **Ceres** Goiás, C Brazil 15°21′S 49°34′W
Ceresio *see* Lugano, Lago di
103 O17 **Céret** Pyrénées-Orientales, S France 42°30′N 02°44′E
54 E6 **Cereté** Córdoba, NW Colombia 08°54′N 75°51′W
172 I17 **Cerf, Île au** *island* Inner Islands, NE Seychelles
99 G22 **Cerfontaine** Namur, S Belgium 50°08′N 04°25′E
Cergy-Pontoise *see* Pontoise
107 N16 **Cerignola** Puglia, SE Italy 41°17′N 15°53′E
Cerigo *see* Kýthira
103 O9 **Cérilly** Allier, C France 46°38′N 02°51′E
136 I11 **Çerkeş** Çankırı, N Turkey 40°51′N 32°52′E
136 D10 **Çerkezköy** Tekirdağ, NW Turkey 41°17′N 28°00′E
109 T12 **Cerknica** *Ger.* Zirknitz. SW Slovenia 45°48′N 14°21′E
109 S11 **Cerkno** W Slovenia 46°07′N 13°58′E
116 F10 **Cermei** *Hung.* Csermő. Arad, W Romania 46°33′N 21°51′E
137 O15 **Çermik** Diyarbakır, SE Turkey 38°09′N 39°27′E
112 I10 **Cerna** Vukovar-Srijem, E Croatia 45°10′N 18°36′E
Cernăuţi *see* Chernivtsi
116 M14 **Cernavodă** Constanţa, SW Romania 44°20′N 28°03′E
103 U7 **Cernay** Haut-Rhin, NE France 47°49′N 07°11′E
Černice *see* Schwarzach
41 O8 **Cerralvo** Nuevo León, NE Mexico 26°10′N 99°40′W
40 G9 **Cerralvo, Isla** *island* NW Mexico
107 L16 **Cerreto Sannita** Campania, S Italy 41°17′N 14°39′E
113 L20 **Cërrik** *var.* Cerriku. Elbasan, C Albania 41°01′N 19°55′E
Cerriku *see* Cërrik
41 O11 **Cerritos** San Luis Potosí, C Mexico 22°25′N 100°16′W
60 K11 **Cerro Azul** Paraná, S Brazil 24°48′S 49°14′W
61 F18 **Cerro Chato** Treinta y Tres, E Uruguay 33°04′S 55°08′W
61 F19 **Cerro Colorado** Florida, S Uruguay 33°52′S 55°33′W
56 E13 **Cerro de Pasco** Pasco, C Peru 10°43′S 76°15′W
61 G14 **Cêrro Largo** Rio Grande do Sul, S Brazil 28°10′S 54°43′W
61 G18 **Cerro Largo** ◐ *department* NE Uruguay
42 E7 **Cerrón Grande, Embalse** ⊠ N El Salvador
63 I14 **Cerros Colorados, Embalse** ⊚ W Argentina
105 V5 **Cervera** Cataluña, NE Spain 41°40′N 01°16′E
104 M3 **Cervera del Pisuerga** Castilla y León, N Spain 42°51′N 04°30′W
105 Q5 **Cervera del Río Alhama** La Rioja, N Spain 42°01′N 01°58′W
107 H15 **Cerveteri** Lazio, C Italy 42°00′N 12°06′E
106 H10 **Cervia** Emilia-Romagna, N Italy 44°14′N 12°22′E
106 J7 **Cervignano del Friuli** Friuli-Venezia Giulia, NE Italy 45°49′N 13°18′E
107 L17 **Cervinara** Campania, S Italy 41°02′N 14°36′E
Cervinia *see* Breuil-Cervinia
106 B6 **Cervino, Monte** *var.* Matterhorn. ▲ Italy/Switzerland 46°00′N 07°39′E
see also Matterhorn
Cervino, Monte *see* Matterhorn
103 Y14 **Cervione** Corse, France, C Mediterranean Sea 42°22′N 09°28′E
104 I1 **Cervo** Galicia, NW Spain 43°39′N 07°25′W
54 F5 **Cesar** *off.* Departamento del Cesar. ◐ *province* N Colombia
Cesar, Departamento del *see* Cesar
106 H10 **Cesena** *anc.* Caesena. Emilia-Romagna, N Italy 44°09′N 12°14′E
106 I10 **Cesenatico** Emilia-Romagna, N Italy 44°12′N 12°24′E
118 H8 **Cēsis** *Ger.* Wenden. C Latvia 57°19′N 25°17′E
111 D15 **Česká Lípa** *Ger.* Böhmisch-Leipa. Liberecký Kraj, N Czech Republic 50°43′N 14°35′E
Česká Republika *see* Czech Republic
111 F17 **Česká Třebová** *Ger.* Böhmisch-Trübau. Pardubický Kraj, C Czech Republic 49°54′N 16°27′E
111 D19 **České Budějovice** *Ger.* Budweis. Jihočeský Kraj, S Czech Republic 48°58′N 14°29′E
111 D19 **České Velenice** Jihočeský Kraj, S Czech Republic 48°49′N 14°57′E
111 E18 **Českomoravská Vrchovina** *var.* Českomoravská Vysočina, *Eng.* Bohemian-Moravian Highlands, *Ger.* Böhmisch-Mährische Höhe. ▲▲ S Czech Republic
Českomoravská Vysočina *see* Českomoravská Vrchovina
111 C19 **Český Krumlov** *var.* Böhmisch-Krumau, *Ger.* Krummau. Jihočeský Kraj, S Czech Republic 48°48′N 14°18′E
Český Les *see* Bohemian Forest
112 F8 **Česma** ∼ N Croatia
136 A14 **Çeşme** İzmir, W Turkey 38°19′N 26°20′E
Cess *see* Cestos
183 T8 **Cessnock** New South Wales, SE Australia 32°51′S 151°21′E
76 K17 **Cestos** *var.* Cess. ∼ S Liberia
118 I9 **Cesvaine** E Latvia 56°58′N 26°15′E
116 G14 **Cetate** Dolj, SW Romania 44°06′N 23°01′E
Cetatea Albă *see* Bilhorod-Dnistrovs'kyy
Cetatea Damboviţei *see* Bucureşti
113 J17 **Cetinje** *It.* Cettigne. S Montenegro 42°23′N 18°55′E
107 N20 **Cetraro** Calabria, S Italy 39°30′N 15°59′E
Cette *see* Sète
188 A17 **Cetti Bay** *bay* SW Guam
Cettigne *see* Cetinje
104 L17 **Ceuta** *var.* Sebta. Ceuta, Spain, N Africa 35°53′N 05°19′W
88 C16 **Ceuta** *enclave* Spain, N Africa
106 B9 **Ceva** Piemonte, NE Italy 44°24′N 08°01′E
103 P14 **Cévennes** ▲▲ S France
108 G10 **Cevio** Ticino, S Switzerland 46°18′N 08°36′E
136 K16 **Ceyhan** Adana, S Turkey 37°02′N 35°48′E
136 K17 **Ceyhan Nehri** ∼ S Turkey
137 P17 **Ceylanpınar** Şanlıurfa, SE Turkey 36°53′N 40°02′E
Ceylon *see* Sri Lanka
173 R6 **Ceylon Plain** *undersea feature* N Indian Ocean 04°00′S 82°00′E
Ceyre to the Caribs *see* Marie-Galante
103 Q14 **Cèze** ∼ S France
Chaacha *see* Çäçe
127 P6 **Chaadayevka** Penzenskaya Oblast', W Russian Federation 53°07′N 45°55′E
167 O12 **Cha-Am** Phetchaburi, SW Thailand 12°48′N 99°58′E
143 W15 **Chābahār** *var.* Chāh Bahār, Chahbar. Sīstān va Balūchestān, SE Iran 25°21′N 60°38′E
Chabarikha *see* Khabarikha
61 B19 **Chabas** Santa Fe, C Argentina 33°16′S 61°23′W
103 T10 **Chablais** *physical region* E France
61 B20 **Chacabuco** Buenos Aires, E Argentina 34°40′S 60°27′W
42 K8 **Chachagón, Cerro** ▲ N Nicaragua 13°18′N 85°39′W
56 C10 **Chachapoyas** Amazonas, NW Peru 06°13′S 77°54′W
Chäche *see* Çäçe
119 O18 **Chachersk** *Rus.* Chechersk. Homyel'skaya Voblasts', SE Belarus 52°54′N 30°54′E
119 N16 **Chachevichy** *Rus.* Chechevichi. Mahilyowskaya Voblasts', E Belarus 53°31′N 29°51′E
61 B14 **Chaco** *off.* Provincia de Chaco. ◐ *province* NE Argentina
Chaco *see* Gran Chaco
62 M6 **Chaco Austral** *physical region* N Argentina
62 M3 **Chaco Boreal** *physical region* N Paraguay
62 M6 **Chaco Central** *physical region* N Argentina
39 Y15 **Chacon, Cape** *headland* Prince of Wales Island, Alaska, USA 54°41′N 132°00′W
Chaco, Provincia de *see* Chaco
78 H9 **Chad** *off.* Republic of Chad, *Fr.* Tchad. ◆ *republic* C Africa
122 K14 **Chadan** Respublika Tyva, S Russian Federation 51°16′N 91°25′E
21 U12 **Chadbourn** North Carolina, SE USA 34°19′N 78°49′W
83 L14 **Chadiza** Eastern, E Zambia 14°04′S 32°27′E
67 Q7 **Chad, Lake** *Fr.* Lac Tchad. ⊚ C Africa
Chad, Republic of *see* Chad
28 J12 **Chadron** Nebraska, C USA 42°48′N 102°57′W
Chadyr-Lunga *see* Ciadîr-Lunga
163 W14 **Chaeryŏng** SW North Korea 38°22′N 125°35′E
105 P17 **Chafarinas, Islas** *island group* S Spain
27 Y7 **Chaffee** Missouri, C USA 37°10′N 89°39′W
148 L12 **Chāgai Hills** *var.* Chāh Gay. ▲▲ Afghanistan/Pakistan
123 Q11 **Chagda** Respublika Sakha (Yakutiya), NE Russian Federation 58°43′N 130°38′E
Chaghasarāy *see* Asadābād
149 N5 **Chaghcharān** *var.* Chakhcharan, Cheghcheran, Qala Āhangarān. Gōwr, C Afghanistan 34°28′N 65°18′E
103 R9 **Chagny** Saône-et-Loire, C France 46°54′N 04°45′E
173 Q7 **Chagos Archipelago** *var.* Oil Islands. *island group* British Indian Ocean Territory
129 O15 **Chagos Bank** *undersea feature* C Indian Ocean 06°15′S 72°00′E
129 O14 **Chagos-Laccadive Plateau** *undersea feature* N Indian Ocean 03°00′S 73°00′E
173 Q7 **Chagos Trench** *undersea feature* N Indian Ocean 07°00′S 73°30′E
43 T14 **Chagres, Río** ∼ C Panama
45 U14 **Chaguanas** Trinidad, Trinidad and Tobago 10°31′N 61°25′W
54 M6 **Chaguaramas** Guárico, N Venezuela 09°23′N 66°18′W
Chagyl *see* Çagyl
Chahār Maḩāll and Bakhtiyāri *see* Chahār Maḩāll va Bakhtīārī
Chahār Maḩāll va Bakhtīārī, Ostān-e *see* Chahār Maḩāll va Bakhtīārī
142 M9 **Chahār Maḩāll va Bakhtīārī** *off.* Ostān-e Chahār Maḩāll va Bakhtīārī, *var.* Chahar Mahall and Bakhtiyari. ◐ *province* SW Iran
Chāh Bahār/Chahbar *see* Chābahār
143 V13 **Chāh Derāz** Sīstān va Balūchestān, SE Iran 27°07′N 60°01′E
Chāh Gay *see* Chāgai Hills
167 P10 **Chai Badan** Lop Buri, C Thailand 15°08′N 101°03′E

◆ Country ● Country Capital ◇ Dependent Territory ○ Dependent Territory Capital ◐ Administrative Regions ✈ International Airport ▲ Mountain ▲▲ Mountain Range Volcano ∼ River ⊚ Lake ⊠ Reservoir

153 Q16 **Chāībāsa** Jhārkhand, N India *22°31′N 85°50′E*
79 E19 **Chaillu, Massif du** ▲ C Gabon
167 O10 **Chai Nat** *var.* Chainat, Jainat, Jayanath. Chai Nat, C Thailand *15°10′N 100°10′E*
Chainat *see* Chai Nat
65 M14 **Chain Fracture Zone** *tectonic feature* E Atlantic Ocean
173 N5 **Chain Ridge** *undersea feature* W Indian Ocean *06°00′N 54°00′E*
Chairn, Ceann an *see* Carnsore Point
158 L5 **Chaiwopu** Xinjiang Uygur Zizhiqu, W China *43°32′N 87°55′E*
167 Q10 **Chaiyaphum** *var.* Jayabum. Chaiyaphum, C Thailand *15°46′N 101°55′E*
62 N10 **Chajarí** Entre Ríos, E Argentina *30°45′S 57°57′W*
42 C5 **Chajul** Quiché, W Guatemala *15°28′N 91°02′W*
83 K16 **Chakari** Mashonaland West, N Zimbabwe *18°05′S 29°51′E*
148 J9 **Chakhānsūr** Nīmrōz, SW Afghanistan *31°11′N 62°06′E*
Chakhānsūr *see* Nīmrōz
Chakhcharan *see* Chaghcharān
149 V8 **Chak Jhumra** *var.* Jhumra. Punjab, E Pakistan *31°33′N 73°14′E*
146 I16 **Chaknakdysonga** Ahal Welaýaty, S Turkmenistan *35°39′N 61°24′E*
153 P16 **Chakradharpur** Jhārkhand, N India *22°42′N 85°38′E*
152 J8 **Chakrāta** Uttarakhand, N India *30°42′N 77°52′E*
149 U7 **Chakwāl** Punjab, NE Pakistan *32°56′N 72°53′E*
57 F17 **Chala** Arequipa, SW Peru *15°52′S 74°13′W*
102 K12 **Chalais** Charente, W France *45°16′N 00°02′E*
108 D10 **Chalais** Valais, SW Switzerland *46°18′N 07°37′E*
115 J20 **Chalándri** *var.* Halandri; *prev.* Khalándrion. *prehistoric site* Sýros, Kykládes, Greece, Aegean Sea
188 H6 **Chalan Kanoa** Saipan, S Northern Mariana Islands *15°08′S 145°43′E*
188 C16 **Chalan Pago** C Guam
Chalap Dalam/Chalap Dalan *see* Chehel Abdālān, Kūh-e
42 F7 **Chalatenango** Chalatenango, N El Salvador *14°04′N 88°53′W*
42 A9 **Chalatenango** ◆ *department* NW El Salvador
83 P15 **Chalaua** Nampula, NE Mozambique *16°04′S 39°08′E*
81 I16 **Chalbi Desert** *desert* N Kenya
42 D7 **Chalchuapa** Santa Ana, W El Salvador *13°59′N 89°41′W*
Chalcidice *see* Chalkidikí
Chalcis *see* Chalkída
Chālderān *see* Siāh Chashmeh
103 N6 **Châlette-sur-Loing** Loiret, C France *48°01′N 02°45′E*
15 X8 **Chaleur Bay** *Fr.* Baie des Chaleurs. *bay* New Brunswick/Québec, E Canada
Chaleurs, Baie des *see* Chaleur Bay
57 G16 **Chalhuanca** Apurímac, S Peru *14°17′S 73°15′W*
154 F12 **Chālisgaon** Mahārāshtra, C India *20°29′N 75°10′E*
115 N23 **Chálki** *island* Dodekánisa, Greece, Aegean Sea
115 F16 **Chalkiádes** Thessalía, C Greece *39°24′N 22°25′E*
115 H18 **Chalkída** *var.* Halkida, *prev.* Khalkís; *anc.* Chalcis. Évvoia, E Greece *38°27′N 23°38′E*
115 G14 **Chalkidikí** *var.* Khalkidhikí; *anc.* Chalcidice. *peninsula* NE Greece
185 A24 **Chalky Inlet** *inlet* South Island, New Zealand
39 S7 **Chalkyitsik** Alaska, USA *66°39′N 143°43′W*
102 I9 **Challans** Vendée, NW France *46°51′N 01°52′W*
57 K19 **Challapata** Oruro, SW Bolivia *18°50′S 66°45′W*
192 H6 **Challenger Deep** *undersea feature* W Pacific Ocean *11°20′N 142°12′E*
Challenger Deep *see* Mariana Trench
193 S11 **Challenger Fracture Zone** *tectonic feature* SE Pacific Ocean
192 K11 **Challenger Plateau** *undersea feature* E Tasman Sea
33 P13 **Challis** Idaho, NW USA *44°31′N 114°14′W*
22 L9 **Chalmette** Louisiana, S USA *29°56′N 89°57′W*
124 J11 **Chalna** Respublika Kareliya, NW Russian Federation *61°53′N 33°59′E*
103 Q5 **Châlons-en-Champagne** *prev.* Châlons-sur-Marne, *hist.* Arcae Remorum; *anc.* Carolopois. Marne, NE France *48°58′N 04°22′E*
Châlons-sur-Marne *see* Châlons-en-Champagne
103 R9 **Chalon-sur-Saône** *anc.* Cabillonum. Saône-et-Loire, C France *46°47′N 04°51′E*
Chaltel, Cerro *see* Fitzroy, Monte
102 M11 **Châlus** Haute-Vienne, C France *45°38′N 01°00′E*
143 N4 **Chālūs** Māzandarān, N Iran *36°40′N 51°25′E*
101 N20 **Cham** Bayern, SE Germany *49°13′N 12°40′E*
108 F7 **Cham** Zug, N Switzerland *47°11′N 08°28′E*
37 R8 **Chama** New Mexico, SW USA *36°54′N 106°34′W*
Cha Mai *see* Thung Song
83 E22 **Chamaites** Karas, S Namibia *27°15′S 17°52′E*
149 O9 **Chaman** Baluchistān, SW Pakistan *30°55′N 66°27′E*
37 R9 **Chama, Rio** ≈ New Mexico, SW USA
152 I6 **Chamba** Himāchal Pradesh, N India *32°33′N 76°10′E*
81 I25 **Chamba** Ruvuma, S Tanzania *11°33′S 37°01′E*
150 H12 **Chambal** ≈ C India
11 U16 **Chamberlain** Saskatchewan, S Canada *50°49′N 105°29′W*
29 O11 **Chamberlain** South Dakota, N USA *43°48′N 99°19′W*
19 R3 **Chamberlain Lake** ◎ Maine, NE USA
39 S5 **Chamberlin, Mount** ▲ Alaska, USA *69°16′N 144°54′W*
37 O11 **Chambers** Arizona, SW USA *35°11′N 109°25′W*
18 F16 **Chambersburg** Pennsylvania, NE USA *39°54′N 77°39′W*
31 N5 **Chambers Island** *island* Wisconsin, N USA
103 T11 **Chambéry** *anc.* Camberia. Savoie, E France *45°34′N 05°56′E*
82 L12 **Chambeshi** Muchinga, NE Zambia *10°55′S 31°07′E*
82 L12 **Chambeshi** ≈ NE Zambia
74 M6 **Chambi, Jebel** *var.* Jabal ash Sha'nabi. ▲ W Tunisia *35°16′N 08°39′E*
15 Q7 **Chambord** Québec, SE Canada *48°25′N 72°02′W*
139 U11 **Chamcham** Al Muthannā, S Iraq *31°17′N 45°05′E*
Chamchamāl *see* Chemchemal
Chamdmanī *see* Altanshiree
40 J14 **Chamela** Jalisco, SW Mexico *19°31′N 105°02′W*
42 G5 **Chamelecón, Río** ≈ NW Honduras
62 J9 **Chamical** La Rioja, C Argentina *30°21′S 66°19′W*
115 L23 **Chamilí** *island* Kykládes, Greece, Aegean Sea
167 Q13 **Châmnar** Kaôh Kŏng, SW Cambodia *11°45′N 103°32′E*
152 K9 **Chamoli** Uttarakhand, N India *30°22′N 79°19′E*
103 U11 **Chamonix-Mont-Blanc** Haute-Savoie, E France *45°55′N 06°52′E*
154 L11 **Chāmpa** Chhattisgarh, C India *22°02′N 82°42′E*
10 H8 **Champagne** Yukon, W Canada *60°48′N 136°22′W*
103 Q5 **Champagne** *cultural region* N France
Champagne *see* Campania
103 Q5 **Champagne-Ardenne** ◆ *region* N France
103 S9 **Champagnole** Jura, E France *46°44′N 05°55′E*
30 M13 **Champaign** Illinois, N USA *40°07′N 88°15′W*
167 S10 **Champasak** Champasak, S Laos *14°50′N 105°51′E*
103 U6 **Champ de Feu** ▲ NE France *48°24′N 07°15′E*
13 O7 **Champdoré, Lac** ◎ Québec, NE Canada
42 B6 **Champerico** Retalhuleu, SW Guatemala *14°18′N 91°54′W*
108 C11 **Champéry** Valais, SW Switzerland *46°12′N 06°52′E*
18 L6 **Champlain** New York, NE USA *44°58′N 73°25′W*
18 L9 **Champlain Canal** *canal* New York, NE USA
15 P13 **Champlain, Lac** ◎ Québec, Canada/USA *see also* Champlain, Lake
18 L7 **Champlain, Lake** ◎ Canada/USA *see also* Champlain, Lac
103 S7 **Champlitte** Haute-Saône, E France *47°36′N 05°31′E*
41 W13 **Champotón** Campeche, SE Mexico *19°18′N 90°43′W*
104 G10 **Chamusca** Santarém, C Portugal *39°21′N 08°29′W*
119 O20 **Chamyarysy** *Rus.* Chemerisy. Homyel'skaya Voblasts', SE Belarus *51°42′N 30°27′E*
127 P5 **Chamzinka** Respublika Mordoviya, W Russian Federation *54°22′N 45°22′E*
Chanáil Mhór, An *see* Grand Canal
Chanak *see* Çanakkale
62 G7 **Chañaral** Atacama, N Chile *26°19′S 70°34′W*
104 H13 **Chança, Rio** *var.* Chanza. ≈ Portugal/Spain
57 D14 **Chancay** Lima, W Peru *11°36′S 77°14′W*
Chan-chiang/Chanchiang *see* Zhanjiang
62 G13 **Chanco** Maule, C Chile *35°43′S 72°35′W*
39 R7 **Chandalar** Alaska, USA *67°30′N 148°29′W*
39 R6 **Chandalar River** ≈ Alaska, USA
152 L10 **Chandan Chauki** Uttar Pradesh, N India *28°32′N 80°43′E*
153 S16 **Chandannagar** *prev.* Chandernagore. West Bengal, E India *22°52′N 88°21′E*
152 K10 **Chandausi** Uttar Pradesh, N India *28°27′N 78°43′E*
22 M10 **Chandeleur Islands** *island group* Louisiana, S USA
22 M9 **Chandeleur Sound** *sound* N Gulf of Mexico
Chandernagore *see* Chandannagar
152 I8 **Chandīgarh** *state capital* Punjab, N India *30°41′N 76°51′E*
153 Q16 **Chāndil** Jhārkhand, NE India *22°58′N 86°04′E*
182 D2 **Chandler** South Australia *26°59′S 133°22′E*
15 Y7 **Chandler** Québec, SE Canada *48°21′N 64°41′W*
36 L14 **Chandler** Arizona, SW USA *33°18′N 111°50′W*
27 O10 **Chandler** Oklahoma, C USA *35°43′N 96°54′W*
25 V7 **Chandler** Texas, SW USA *32°18′N 95°28′W*
39 Q6 **Chandler River** ≈ Alaska, USA
56 H13 **Chandles, Río** ≈ E Peru
162 H9 **Chandmanī** *var.* Talshand. Govĭ-Altayĭ, C Mongolia *45°21′N 98°00′E*
162 E7 **Chandmanī** *var.* Urdgol. Hovd, W Mongolia *47°39′N 92°46′E*
14 J13 **Chandos Lake** ◎ Ontario, SE Canada
153 U15 **Chandpur** Chittagong, C Bangladesh *23°13′N 90°43′E*
154 I13 **Chandrapur** Mahārāshtra, C India *19°58′N 79°21′E*
83 J15 **Changa** Southern, S Zambia *16°24′S 28°27′E*
Chang'an *see* Rong'an, Guangxi Zhuangzu Zizhiqu, S China
Changan *see* Xi'an, Shaanxi, C China
155 G23 **Changanācheri** *var.* Changannassery. Kerala, SW India *09°26′N 76°31′E* *see also* Changannassery
83 M19 **Changane** ≈ S Mozambique
83 M16 **Changara** Tete, NW Mozambique *16°54′S 33°15′E*
163 X11 **Changbai** *var.* Changbai Chaoxianzu Zizhixian. Jilin, NE China *41°25′N 128°08′E*
Changbai Chaoxianzu Zizhixian *see* Changbai
163 X11 **Changbai Shan** ▲▲ NE China
163 V10 **Changchun** *var.* Ch'angch'un, Ch'ang-ch'un; *prev.* Hsinking. *province capital* Jilin, NE China *43°53′N 125°18′E*
Ch'angch'un/Ch'ang-ch'un *see* Changchun
160 M10 **Changde** Hunan, S China *29°04′N 111°42′E*
Changhua *see* Zhanghua
168 L10 **Changi** ✈ (Singapore) E Singapore *01°22′N 103°58′E*
158 L5 **Changji** Xinjiang Uygur Zizhiqu, NW China *44°02′N 87°12′E*
160 L17 **Changjiang** *var.* Changjiang Lizu Zizhixian, Shiliu. Hainan, S China *19°16′N 109°09′E*
157 R11 **Chang Jiang** *var.* Yangtze Kiang, *Eng.* Yangtze. ≈ C China
157 N12 **Chang Jiang** *Eng.* Yangtze. ≈ SW China
161 S8 **Changjiang Kou** *delta* E China
Changjiang Lizu Zizhixian *see* Changjiang
Changkiakow *see* Zhangjiakou
167 F12 **Chang, Ko** *island* S Thailand
161 Q2 **Changli** Hebei, E China *39°44′N 119°13′E*
163 V10 **Changling** Jilin, NE China *44°15′N 124°03′E*
Changning *see* Xunwu
161 N11 **Changsha** *var.* Ch'angsha, Ch'ang-sha. *province capital* Hunan, S China *28°10′N 113°E*
Ch'angsha/Ch'ang-sha *see* Changsha
161 Q10 **Changshan** Zhejiang, SE China *28°54′N 118°30′E*
163 V14 **Changshan Qundao** *island group* NE China
161 S8 **Changshu** *var.* Ch'ang-shu. Jiangsu, E China *31°39′N 120°45′E*
Ch'ang-shu *see* Changshu
163 V11 **Changtu** Liaoning, NE China *42°50′N 123°59′E*
43 P14 **Changuinola** Bocas del Toro, NW Panama *09°28′N 82°31′W*
159 N9 **Changweiliang** Qinghai, W China *38°24′N 92°08′E*
160 K6 **Changwu** *var.* Zhaoren. Shaanxi, C China *35°12′N 107°46′E*
163 U13 **Changxing Dao** *island* N China
160 M9 **Changyang** *var.* Longzhouping. Hubei, C China *30°45′N 111°13′E*
163 W14 **Changyŏn** SW North Korea *38°19′N 125°15′E*
161 N5 **Changzhi** Shanxi, C China *36°10′N 113°02′E*
161 R8 **Changzhou** Jiangsu, E China *31°45′N 119°58′E*
115 H24 **Chaniá** *var.* Hania, Khaniá, *Eng.* Canea; *anc.* Cydonia. Kríti, Greece, E Mediterranean Sea *35°31′N 24°00′E*
62 J5 **Chañi, Nevado de** ▲ NW Argentina *24°09′S 65°44′W*
115 H24 **Chanión, Kólpos** *gulf* Kríti, Greece, E Mediterranean Sea
Chankiri *see* Çankırı
30 M11 **Channahon** Illinois, N USA *41°25′N 88°13′W*
155 H20 **Channapatna** Karnātaka, E India *12°43′N 77°14′E*
97 K26 **Channel Islands** *Fr.* Iles Normandes. *island group* S English Channel
35 R16 **Channel Islands** *island group* California, W USA
13 S13 **Channel-Port aux Basques** Newfoundland and Labrador, SE Canada *47°35′N 59°02′W*
Channel, The *see* English Channel
97 Q23 **Channel Tunnel** *tunnel* France/United Kingdom
24 M2 **Channing** Texas, SW USA *35°41′N 102°21′W*
Chantabun/Chantaburi *see* Chanthaburi
104 H3 **Chantada** Galicia, NW Spain *42°36′N 07°46′W*
167 P12 **Chanthaburi** *var.* Chantabun, Chantaburi. Chantaburi, S Thailand *12°35′N 102°08′E*
103 O4 **Chantilly** Oise, N France *49°12′N 02°28′E*
139 V12 **Chanūn as Sa'ūdī** Dhī Qār, S Iraq *31°04′N 46°00′E*
27 Q6 **Chanute** Kansas, C USA *37°40′N 95°27′W*
Chanza *see* Chança, Rio
Ch'ao-an/Chaochow *see* Chaozhou
161 P8 **Chao Hu** ◎ E China
167 P11 **Chao Phraya, Mae Nam** ≈ C Thailand
163 T8 **Chaor He** *prev.* Qulin Gol. ≈ NE China
Chaouèn *see* Chefchaouen
161 P14 **Chaoyang** Guangdong, S China *23°17′N 116°33′E*
163 T12 **Chaoyang** Liaoning, NE China *41°34′N 120°29′E*
Chaoyang *see* Jiayin, Heilongjiang, China
Chaoyang *see* Huinan, Jilin China
161 Q14 **Chaozhou** *var.* Chaoan, Chao'an, Ch'ao-an; *prev.* Chaochow. Guangdong, SE China *23°42′N 116°36′E*
58 N13 **Chapadinha** Maranhão, E Brazil *03°45′S 43°23′W*
12 K12 **Chapais** Québec, SE Canada *49°47′N 74°54′W*
40 L13 **Chapala** Jalisco, SW Mexico *20°20′N 103°10′W*
40 L13 **Chapala, Lago de** ◎ C Mexico
146 F13 **Chapan, Gora** ▲ C Turkmenistan *37°48′N 58°03′E*
57 M18 **Chapare, Río** ≈ C Bolivia
54 E11 **Chaparral** Tolima, C Colombia *03°45′N 75°30′W*
144 F9 **Chapayev** Zapadnyy Kazakhstan, NW Kazakhstan *50°12′N 51°09′E*
123 O11 **Chapayevo** Respublika Sakha (Yakutiya), NE Russian Federation *60°03′N 117°19′E*
127 R6 **Chapayevsk** Samarskaya Oblast', W Russian Federation *52°57′N 49°42′E*
60 H13 **Chapecó** Santa Catarina, S Brazil *27°14′S 52°41′W*
60 I13 **Chapecó, Rio** ≈ S Brazil
20 J9 **Chapel Hill** Tennessee, S USA *35°38′N 86°40′W*
44 J12 **Chapelton** C Jamaica *18°05′N 77°16′W*
14 C8 **Chapleau** Ontario, SE Canada *47°50′N 83°24′W*
14 D7 **Chapleau** ≈ Ontario, S Canada
11 T16 **Chaplin** Saskatchewan, S Canada *50°27′N 106°37′W*
126 M6 **Chaplygin** Lipetskaya Oblast', W Russian Federation *53°13′N 39°58′E*
117 S11 **Chaplynka** Khersons'ka Oblast', S Ukraine *46°20′N 33°34′E*
9 O6 **Chapman, Cape** *headland* Nunavut, NE Canada *69°15′N 89°09′W*
25 T15 **Chapman Ranch** Texas, SW USA *27°32′N 97°25′W*
Chapman's *see* Okwa
21 P5 **Chapmanville** West Virginia, NE USA *37°58′N 82°01′W*
28 K15 **Chappell** Nebraska, C USA *41°05′N 102°28′W*
Chapra *see* Chhapra
56 D9 **Chapuli, Río** ≈ N Peru
76 I6 **Châr** *well* N Mauritania
123 P12 **Chara** Zabaykal'skiy Kray, S Russian Federation *56°57′N 118°05′E*
123 O11 **Chara** ≈ C Russian Federation
54 G8 **Charala** Santander, C Colombia *06°17′N 73°09′W*
41 N10 **Charcas** San Luis Potosí, C Mexico *23°09′N 101°10′W*
25 T13 **Charco** Texas, SW USA *28°42′N 97°35′E*
194 H7 **Charcot Island** *island* Antarctica
64 M8 **Charcot Seamounts** *undersea feature* E Atlantic Ocean *11°30′W 45°00′N*
Chardara *see* Shardara
Chardarinskoye Vodokhranilishche *see* Shardarinskoye Vodokhranilishche
31 U11 **Chardon** Ohio, N USA *41°34′N 81°12′W*
44 K9 **Chardonnières** SW Haiti *18°16′N 74°10′W*
Chardzhev *see* Türkmenabat
Chardzhevskaya Oblast *see* Lebap Welaýaty
Chardzhou/Chardzhui *see* Türkmenabat
102 L11 **Charente** ◆ *department* W France
102 J11 **Charente** ≈ W France
102 J10 **Charente-Maritime** ◆ *department* W France
137 U12 **Ch'arents'avan** C Armenia *40°23′N 44°41′E*
78 I12 **Chari** *var.* Shari. ≈ Central African Republic/Chad
78 G11 **Chari-Baguirmi** *off.* Région du Chari-Baguirmi. ◆ *region* SW Chad
Chari-Baguirmi, Région du *see* Chari-Baguirmi
149 Q4 **Chārīkār** Parwān, NE Afghanistan *35°01′N 69°11′E*
29 V15 **Chariton** Iowa, C USA *41°00′N 93°18′W*
27 U3 **Chariton River** ≈ Missouri, C USA
55 T7 **Charity** NW Guyana *07°22′N 58°34′W*
31 R7 **Charity Island** *island* Michigan, N USA
Chärjew *see* Türkmenabat
Chärjew Oblasty *see* Lebap Welaýaty
Charkhlik/Charkhliq *see* Ruoqiang
99 G20 **Charleroi** Hainaut, S Belgium *50°25′N 04°27′E*
11 V12 **Charles** Manitoba, C Canada *55°27′N 100°58′W*
15 R10 **Charlesbourg** Québec, SE Canada *46°50′N 71°15′W*
21 Y7 **Charles, Cape** *headland* Virginia, NE USA *37°09′N 75°57′W*
29 W12 **Charles City** Iowa, C USA *43°04′N 92°40′W*
21 W6 **Charles City** Virginia, NE USA *37°21′N 77°05′W*
103 O5 **Charles de Gaulle** ✈ (Paris) Seine-et-Marne, N France *49°04′N 02°36′E*
12 K1 **Charles Island** *island* Nunavut, NE Canada
Charles Island *see* Santa María, Isla
30 K9 **Charles Mound** *hill* Illinois, N USA
185 A22 **Charles Sound** *sound* South Island, New Zealand
185 G15 **Charleston** West Coast, South Island, New Zealand *41°54′S 171°25′E*
27 S11 **Charleston** Arkansas, C USA *35°19′N 94°02′W*
30 M14 **Charleston** Illinois, N USA *39°30′N 88°10′W*
22 L3 **Charleston** Mississippi, S USA *34°00′N 90°03′W*
27 Z7 **Charleston** Missouri, C USA *36°54′N 89°22′W*
21 T15 **Charleston** South Carolina, SE USA *32°48′N 79°57′W*
21 Q5 **Charleston** *state capital* West Virginia, NE USA *38°21′N 81°38′W*
14 L14 **Charleston Lake** ◎ Ontario, SE Canada
35 W11 **Charleston Peak** ▲ Nevada, W USA *36°16′N 115°40′W*
45 W10 **Charlestown** Nevis, Saint Kitts and Nevis *17°08′N 62°37′W*
31 P16 **Charlestown** Indiana, N USA *38°27′N 85°40′W*
18 M9 **Charlestown** New Hampshire, NE USA *43°14′N 72°23′W*
21 V3 **Charles Town** West Virginia, NE USA *39°18′N 77°54′W*
181 W9 **Charleville** Queensland, E Australia *26°25′S 146°18′E*
103 R3 **Charleville-Mézières** Ardennes, N France *49°45′N 04°43′E*
31 P5 **Charlevoix** Michigan, N USA *45°19′N 85°15′W*
31 Q6 **Charlevoix, Lake** ◎ Michigan, N USA
39 T9 **Charley River** ≈ Alaska, USA
64 J6 **Charlie-Gibbs Fracture Zone** *tectonic feature* N Atlantic Ocean
103 Q10 **Charlieu** Loire, E France *46°11′N 04°10′E*
31 Q9 **Charlotte** Michigan, N USA *42°33′N 84°50′W*
21 R10 **Charlotte** North Carolina, SE USA *35°14′N 80°51′W*
20 I8 **Charlotte** Tennessee, S USA *36°11′N 87°18′W*
25 R13 **Charlotte** Texas, SW USA *28°51′N 98°42′W*
21 R10 **Charlotte** ✈ North Carolina, SE USA *35°12′N 80°54′W*
45 T9 **Charlotte Amalie** *prev.* Saint Thomas. ○ (Virgin Islands (US)) Saint Thomas, N Virgin Islands (US) *18°22′N 64°56′W*
21 U7 **Charlotte Court House** Virginia, NE USA *37°04′N 78°37′W*
23 W14 **Charlotte Harbor** *inlet* Florida, SE USA
Charlotte Island *see* Abaiang
95 J15 **Charlottenberg** Värmland, C Sweden *59°53′N 12°17′E*
Charlottenhof *see* Aegviidu
21 U5 **Charlottesville** Virginia, NE USA *38°02′N 78°29′W*
13 Q14 **Charlottetown** *province capital* Prince Edward Island, Prince Edward Island, SE Canada *46°14′N 63°09′W*
Charlotte Town *see* Roseau, Dominica
Charlotte Town *see* Gouyave, Grenada
45 Z16 **Charlotteville** Tobago, Trinidad and Tobago *11°16′N 60°33′W*
182 M11 **Charlton** Victoria, SE Australia *36°18′S 143°19′E*
12 H10 **Charlton Island** *island* Northwest Territories, C Canada
103 T6 **Charmes** Vosges, NE France *48°19′N 06°19′E*
119 F19 **Charnawchytsy** *Rus.* Chernavchitsy. Brestskaya Voblasts', SW Belarus *52°13′N 23°44′E*
15 R10 **Charny** Québec, SE Canada *46°43′N 71°15′W*
149 T5 **Chārsadda** Khyber Pakhtunkhwa, NW Pakistan *34°12′N 71°46′E*
Charshanga/Charshangngy/Charshangy *see* Köýtendag
Charshy *see* Shar
181 W6 **Charters Towers** Queensland, NE Australia *20°02′S 146°20′E*
15 R12 **Chartierville** Québec, SE Canada *45°19′N 71°13′W*
102 M6 **Chartres** *anc.* Autricum, Civitas Carnutum. Eure-et-Loir, C France *48°27′N 01°27′E*
Charyn *see* Sharyn
Charyn *see* Sharyn
61 D21 **Chascomús** Buenos Aires, E Argentina *35°34′S 58°01′W*
11 N16 **Chase** British Columbia, SW Canada *50°49′N 119°41′W*
21 U7 **Chase City** Virginia, NE USA *36°48′N 78°27′W*
19 S4 **Chase, Mount** ▲ Maine, NE USA *46°06′N 68°30′W*
118 M13 **Chashniki** Vitsyebskaya Voblasts', N Belarus *54°52′N 29°10′E*
115 D15 **Chásia** ▲ C Greece
29 V9 **Chaska** Minnesota, N USA *44°47′N 93°36′W*
185 D25 **Chaslands Mistake** *headland* South Island, New Zealand *46°37′S 169°21′E*
125 R11 **Chasovo** Respublika Komi, NW Russian Federation *61°58′N 50°34′E*
Chasovo *see* Vazhgort
124 H14 **Chastova** Novgorodskaya Oblast', NW Russian Federation *58°37′N 32°05′E*
143 R3 **Chāt** Golestān, N Iran *37°52′N 55°27′E*
Chatak *see* Chhatak
Chatang *see* Zhanang
39 R9 **Chatanika** Alaska, USA *65°06′N 147°28′W*
39 R9 **Chatanika River** ≈ Alaska, USA
147 T8 **Chat-Bazar** Talasskaya Oblast', NW Kyrgyzstan *42°29′N 72°37′E*
45 Y14 **Chateaubelair** Saint Vincent, W Saint Vincent and the Grenadines *13°15′N 61°05′W*
102 J7 **Châteaubriant** Loire-Atlantique, NW France *47°43′N 01°22′W*
103 Q8 **Château-Chinon** Nièvre, C France *47°04′N 03°50′E*
108 C10 **Château d'Oex** Vaud, W Switzerland *46°28′N 07°09′E*
102 L7 **Château-du-Loir** Sarthe, NW France *47°40′N 00°25′E*
102 M6 **Châteaudun** Eure-et-Loir, C France *48°04′N 01°20′E*
102 K7 **Château-Gontier** Mayenne, NW France *47°49′N 00°42′W*
15 O13 **Châteauguay** Québec, SE Canada *45°22′N 73°44′W*
102 F6 **Châteaulin** Finistère, NW France *48°12′N 04°07′W*
103 N9 **Châteaumeillant** Cher, C France *46°33′N 02°10′E*
102 K11 **Châteauneuf-sur-Charente** Charente, W France *45°34′N 00°33′W*
102 M7 **Château-Renault** Indre-et-Loire, C France *47°34′N 00°52′E*
103 N9 **Châteauroux** *prev.* Indreville. Indre, C France *46°50′N 01°43′E*
103 T5 **Château-Salins** Moselle, NE France *48°50′N 06°29′E*
103 P4 **Château-Thierry** Aisne, N France *49°03′N 03°24′E*
99 H21 **Châtelet** Hainaut, S Belgium *50°24′N 04°32′E*
Châtelherault *see* Châtellerault
102 L9 **Châtellerault** *var.* Châtelherault. Vienne, W France *46°49′N 00°33′E*
29 X10 **Chatfield** Minnesota, N USA *43°51′N 92°11′W*
13 O14 **Chatham** New Brunswick, SE Canada *47°02′N 65°30′W*
14 D17 **Chatham** Ontario, S Canada *42°24′N 82°11′W*
97 P22 **Chatham** SE England, United Kingdom *51°23′N 00°31′E*
30 K14 **Chatham** Illinois, N USA *39°40′N 89°42′W*
21 T7 **Chatham** Virginia, NE USA *36°49′N 79°26′W*
63 F22 **Chatham, Isla** *island* S Chile
175 R12 **Chatham Island** *island* Chatham Islands, New Zealand
Chatham Island *see* San Cristóbal, Isla
Chatham Island Rise *see* Chatham Rise
175 R12 **Chatham Islands** *island group* New Zealand, SW Pacific Ocean
175 Q12 **Chatham Rise** *var.* Chatham Island Rise. *undersea feature* S Pacific Ocean
39 X13 **Chatham Strait** *strait* Alaska, USA
Chathóir, Rinn *see* Cahore Point
102 M9 **Châtillon-sur-Indre** Indre, C France *46°58′N 01°10′E*
103 Q7 **Châtillon-sur-Seine** Côte d'Or, C France *47°51′N 04°30′E*
147 S8 **Chatkal** *Uzb.* Chotqol. ≈ Kyrgyzstan/Uzbekistan
147 R9 **Chatkal Range** *Rus.* Chatkal'skiy Khrebet. ▲▲ Kyrgyzstan/Uzbekistan
Chatkal'skiy Khrebet *see* Chatkal Range
23 N7 **Chatom** Alabama, S USA *31°28′N 88°15′W*
Chatrapur *see* Chhatrapur
143 S10 **Chatrūd** Kermān, C Iran *30°39′N 56°57′E*
23 S2 **Chatsworth** Georgia, SE USA *34°46′N 84°46′W*
Chāttagām *see* Chittagong
23 S8 **Chattahoochee** Florida, SE USA *30°40′N 84°51′W*
23 R8 **Chattahoochee River** ≈ SE USA
20 L10 **Chattanooga** Tennessee, S USA *35°05′N 85°16′W*
147 V10 **Chatyr-Kël', Ozero** ◎ C Kyrgyzstan
147 W9 **Chatyr-Tash** Narynskaya Oblast', C Kyrgyzstan *40°54′N 76°22′E*
15 R12 **Chaudière** ≈ Québec, SE Canada
167 S14 **Châu Đốc** *var.* Chauphu, Chau Phu. An Giang, S Vietnam *10°53′N 105°07′E*
152 D13 **Chauhtan** *prev.* Chohtan. Rājasthān, NW India *25°27′N 71°08′E*
166 L5 **Chauk** Magway, W Myanmar (Burma) *20°52′N 94°50′E*
103 R6 **Chaumont** *prev.* Chaumont-en-Bassigny. Haute-Marne, N France *48°07′N 05°08′E*
Chaumont-en-Bassigny *see* Chaumont
123 T5 **Chaunskaya Guba** *bay* NE Russian Federation
103 P3 **Chauny** Aisne, N France *49°37′N 03°13′E*
167 U10 **Châu Ô** *var.* Bình Sơn. Quang Ngai, C Vietnam *15°18′N 108°45′E*
Chau Phu *see* Châu Đốc
102 I5 **Chausey, Îles** *island group* N France
Chausy *see* Chavusy
18 C11 **Chautauqua Lake** ◎ New York, NE USA
ChâuThanh *see* Ba Ria
102 L9 **Chauvigny** Vienne, W France *46°35′N 00°37′E*
124 L6 **Chavan'ga** Murmanskaya Oblast', NW Russian Federation *66°07′N 37°44′E*
14 K10 **Chavannes, Lac** ◎ Québec, SE Canada
Chavantes, Represa de *see* Xavantes, Represa de
61 D15 **Chavarría** Corrientes, NE Argentina *28°57′S 58°35′W*
Chavash Respubliki *see* Chuvashskaya Respublika
104 I5 **Chaves** *anc.* Aquae Flaviae. Vila Real, N Portugal *41°44′N 07°28′W*
Chávez, Isla *see* Santa Cruz, Isla
82 G13 **Chavuma** North Western, NW Zambia *13°04′S 22°43′E*
119 O16 **Chavusy** *Rus.* Chausy. Mahilyowskaya Voblasts', E Belarus *53°48′N 30°58′E*
Chayan *see* Shayan
147 U8 **Chayek** Narynskaya Oblast', C Kyrgyzstan *41°54′N 74°28′E*
139 T6 **Chāy Khānah** Diyālá, E Iraq *34°19′N 44°33′E*
125 T16 **Chaykovskiy** Permskiy Kray, NW Russian Federation *56°45′N 54°09′E*
167 T12 **Chbar** Mŏndŏl Kiri, E Cambodia *12°46′N 107°10′E*
23 Q4 **Cheaha Mountain** ▲ Alabama, S USA *33°29′N 85°48′W*
Cheatharlach *see* Carlow
21 S2 **Cheat River** ≈ NE USA
111 A16 **Cheb** *Ger.* Eger. Karlovarský Kraj, W Czech Republic *50°05′N 12°23′E*
127 Q3 **Cheboksary** Chuvashskaya Respublika, W Russian Federation *56°06′N 47°15′E*
31 Q5 **Cheboygan** Michigan, N USA *45°40′N 84°28′W*
Chechaouèn *see* Chefchaouen
Chechenia *see* Chechenskaya Respublika
127 O15 **Chechenskaya Respublika** *Eng.* Chechenia, Chechnia, *Rus.* Chechnya. ◆ *autonomous republic* SW Russian Federation
67 N4 **Chech, Erg** *desert* Algeria/Mali
Chechevichi *see* Chachevichy
Che-chiang *see* Zhejiang
Chechnia/Chechnya *see* Chechenskaya Respublika
Ch'ech'on *see* Jecheon
111 L15 **Chęciny** Świętokrzyskie, S Poland *50°51′N 20°31′E*
27 Q10 **Checotah** Oklahoma, C USA *35°28′N 95°31′W*
13 R15 **Chedabucto Bay** *inlet* Nova Scotia, E Canada
166 J7 **Cheduba Island** *island* W Myanmar (Burma)
37 T5 **Cheesman Lake** ⊠ Colorado, C USA
195 S16 **Cheetham, Cape** *headland* Antarctica *70°26′S 162°40′E*
74 G5 **Chefchaouen** *var.* Chaouèn, Chechaouèn, *Sp.* Xauen. N Morocco *35°10′N 05°16′W*
Chefoo *see* Yantai
38 M12 **Chefornak** Alaska, USA *60°09′N 164°09′W*
123 R13 **Chegdomyn** Khabarovskiy Kray, SE Russian Federation *51°09′N 132°58′E*
76 M4 **Chegga** Tiris Zemmour, NE Mauritania *25°27′N 05°49′W*
Cheghcheran *see* Chaghcharān
32 G9 **Chehalis** Washington, NW USA *46°39′N 122°57′W*
32 G9 **Chehalis River** ≈ Washington, NW USA
148 M6 **Chehel Abdālān, Kūh-e** *var.* Chalap Dalam, *Pash.* Chalap Dalan. ▲▲ C Afghanistan
115 D14 **Cheimadítis, Límni** *var.* Límni Cheimadítis. ◎ N Greece
Cheimadítis, Límni *see* Cheimadítis, Límni
103 U15 **Cheiron, Mont** ▲ SE France *43°49′N 07°00′E*
163 Y17 **Cheju** ✈ S South Korea *33°31′N 126°29′E*
Cheju *see* Jeju
Cheju-do *see* Jeju-do
Cheju-haehyeop *see* Jeju-haehyeop
Cheju Strait *see* Jeju-haehyeop
Chekiang *see* Zhejiang
Chekichler/Chekishlyar *see* Çekiçler
188 F8 **Chelab** Babeldaob, N Palau
147 N11 **Chelak** *Rus.* Chelek. Samarqand Viloyati, C Uzbekistan *39°55′N 66°45′E*
32 J7 **Chelan, Lake** ◎ Washington, NW USA
Chelek *see* Chelak
Cheleken *see* Hazar
Chélif/Chéliff *see* Chelif, Oued
74 J5 **Chelif, Oued** *var.* Chélif, Chéliff, Chellif, Shellif. ≈ N Algeria
Chelkar *see* Shalkar
Chelkar Ozero *see* Shalkar, Ozero
Chellif *see* Chelif, Oued
111 P14 **Chełm** *Rus.* Kholm. Lubelskie, SE Poland *51°08′N 23°29′E*
110 I9 **Chełmno** *Ger.* Culm, Kulm. Kujawski-pomorskie, C Poland *53°21′N 18°27′E*
115 E19 **Chelmós** *var.* Aroania. ▲ S Greece
14 F10 **Chelmsford** Ontario, S Canada *46°33′N 81°16′W*
97 P21 **Chelmsford** E England, United Kingdom *51°44′N 00°28′E*
110 J9 **Chełmża** *Ger.* Culmsee, Kulmsee. Kujawski-pomorskie, C Poland *53°11′N 18°34′E*
27 Q8 **Chelsea** Oklahoma, C USA *36°32′N 95°25′W*
18 M8 **Chelsea** Vermont, NE USA *43°58′N 72°29′W*
97 L21 **Cheltenham** C England, United Kingdom *51°54′N 02°04′W*
105 R9 **Chelva** Valenciana, E Spain *39°45′N 01°00′W*
122 G11 **Chelyabinsk** Chelyabinskaya Oblast', C Russian Federation *55°12′N 61°25′E*
122 F11 **Chelyabinskaya Oblast'** ◆ *province* C Russian Federation
123 N5 **Chelyuskin, Mys** *headland* N Russian Federation *77°42′N 104°13′E*
41 Y12 **Chemax** Yucatán, SE Mexico *20°41′N 87°54′W*
83 N16 **Chemba** Sofala, C Mozambique *17°11′S 34°53′E*
82 J13 **Chembe** Luapula, NE Zambia *11°58′S 28°45′E*
139 T4 **Chemchemal** *Ar.* Juwārtā, *var.* Chamchamāl. At Ta'mím, N Iraq *35°32′N 44°50′E*
Chemenibit *see* Çemenibit
Chemerisy *see* Chamyarysy
116 K7 **Chemerivtsi** Khmel'nyts'ka Oblast', W Ukraine *49°00′N 26°21′E*
102 J8 **Chemillé** Maine-et-Loire, NW France *47°15′N 00°42′W*
173 X17 **Chemin Grenier** S Mauritius *20°29′S 57°28′E*
101 N16 **Chemnitz** *prev.* Karl-Marx-Stadt. Sachsen, E Germany *50°50′N 12°55′E*
Chemulpo *see* Incheon
32 H14 **Chemult** Oregon, NW USA *43°14′N 121°48′W*
18 G12 **Chemung River** ≈ New York/Pennsylvania, NE USA
149 U8 **Chenāb** ≈ India/Pakistan
39 S9 **Chena Hot Springs** Alaska, USA *65°06′N 146°02′W*
18 I11 **Chenango River** ≈ New York, NE USA
168 J7 **Chenderoh, Tasik** ◎ Peninsular Malaysia
15 Q11 **Chêne, Rivière du** ≈ Québec, SE Canada
32 L8 **Cheney** Washington, NW USA *47°29′N 117°34′W*
26 M6 **Cheney Reservoir** ⊠ Kansas, C USA
Chengchiatun *see* Liaoyuan
Ch'eng-chou/Chengchow *see* Zhengzhou
161 P1 **Chengde** *var.* Jehol. Hebei, E China *41°N 117°57′E*
160 I9 **Chengdu** *var.* Chengtu, Ch'eng-tu. *province capital* Sichuan, C China *30°41′N 104°03′E*
161 Q14 **Chenghai** Guangdong, S China *23°30′N 116°42′E*
Chenghsien *see* Zhengzhou
160 H13 **Chengjiang** Yunnan, SW China *24°40′N 102°55′E*
Chengjiang *see* Taihe
160 L17 **Chengmai** *var.* Jinjiang. Hainan, S China *19°45′N 109°59′E*
Chengtu/Ch'eng-tu *see* Chengdu
Chengwen *see* Chindu
159 W12 **Chengxian** *var.* Cheng Xiang. Gansu, C China *33°42′N 105°45′E*
Cheng Xiang *see* Chengxian
Chengyang *see* Juxian
Chengzhong *see* Ningming
Chenkiang *see* Zhenjiang
155 J19 **Chennai** *prev.* Madras. *state capital* Tamil Nādu, S India *13°05′N 80°18′E*
155 J19 **Chennai** ✈ Tamil Nādu, S India *13°07′N 80°13′E*
103 R8 **Chenôve** Côte d'Or, C France *47°16′N 05°00′E*
Chenstokhov *see* Częstochowa

◆ Country ● Country Capital ◇ Dependent Territory ○ Dependent Territory Capital ◆ Administrative Regions ✈ International Airport ▲ Mountain ▲▲ Mountain Range Volcano ≈ River ◎ Lake ⊠ Reservoir

160 L11 **Chenxi** var. Chenyang. Hunan, S China 28°02′N 110°15′E
Chen Xian/Chenxian/Chen Xiang see Chenzhou
Chenyang see Chenxi
161 N12 **Chenzhou** var. Chenxian, Chen Xian, Chen Xiang. Hunan, S China 25°51′N 113°01′E
163 X15 **Cheonan** Jap. Tenan; prev. Ch'ŏnan. W South Korea 36°51′N 127°11′E
163 W13 **Cheongju** prev. Chŏngju. W North Korea 39°44′N 125°13′E
Cheo Reo see A Yun Pa
114 I11 **Chepelare** Smolyan, S Bulgaria 41°44′N 24°41′E
114 I11 **Chepelarska Reka** ≋ S Bulgaria
56 B11 **Chepén** La Libertad, C Peru 07°15′S 79°23′W
62 J10 **Chepes** La Rioja, C Argentina 31°19′S 66°40′W
161 O15 **Chep Lap Kok** ✕ S China 22°23′N 114°11′E
43 U14 **Chepo** Panamá, C Panama 09°09′N 79°03′W
Chepping Wycombe see High Wycombe
125 R14 **Cheptsa** ≋ NW Russian Federation
30 K3 **Chequamegon Point** headland Wisconsin, N USA 46°42′N 90°45′W
103 O8 **Cher** ◈ department C France
102 M8 **Cher** ≋ C France
Cherangani Hills see Cherangany Hills
81 H17 **Cherangany Hills** var. Cherangani Hills. ▲ W Kenya
21 S11 **Cheraw** South Carolina, SE USA 34°42′N 79°52′W
102 J3 **Cherbourg** anc. Carusbur. Manche, N France 49°40′N 01°36′W
127 R5 **Cherdakly** Ul'yanovskaya Oblast', W Russian Federation 54°21′N 48°54′E
125 U12 **Cherdyn'** Permskiy Kray, NW Russian Federation 60°21′N 56°39′E
124 J14 **Cherekha** ≋ W Russian Federation
122 M13 **Cheremkhovo** Irkutskaya Oblast', S Russian Federation 53°16′N 102°44′E
Cheren see Keren
124 K14 **Cherepovets** Vologodskaya Oblast', NW Russian Federation 59°09′N 37°50′E
125 O11 **Cherevkovo** Arkhangel'skaya Oblast', NW Russian Federation 61°45′N 45°16′E
74 I6 **Chergui, Chott ech** salt lake NW Algeria
Cherikov see Cherykaw
117 P6 **Cherkas'ka Oblast'** var. Cherkasy, Rus. Cherkasskaya Oblast'. ◈ province C Ukraine
Cherkasskaya Oblast' see Cherkas'ka Oblast'
Cherkassy see Cherkasy
117 Q6 **Cherkasy** Rus. Cherkassy. Cherkas'ka Oblast', C Ukraine 49°26′N 32°05′E
Cherkasy see Cherkas'ka Oblast'
126 M15 **Cherkessk** Karachayevo-Cherkesskaya Respublika, SW Russian Federation 44°12′N 42°06′E
122 H12 **Cherlak** Omskaya Oblast', C Russian Federation 54°06′N 74°59′E
122 H12 **Cherlakskoye** Omskaya Oblast', C Russian Federation 53°42′N 74°23′E
125 U13 **Chermoz** Permskiy Kray, NW Russian Federation 58°49′N 56°07′E
Chernavchitsy see Charnawchytsy
125 T3 **Chernaya** Nenetskiy Avtonomnyy Okrug, NW Russian Federation 68°36′N 56°34′E
125 T4 **Chernaya** ≋ NW Russian Federation
Chernigov see Chernihiv
Chernigovskaya Oblast' see Chernihivs'ka Oblast'
117 Q2 **Chernihiv** Rus. Chernigov. Chernihivs'ka Oblast', NE Ukraine 51°28′N 31°19′E
Chernihiv see Chernihivs'ka Oblast'
117 V9 **Chernihivka** Zaporiz'ka Oblast', SE Ukraine 47°11′N 36°10′E
117 P2 **Chernihivs'ka Oblast'** var. Chernihiv, Rus. Chernigovskaya Oblast'. ◈ province NE Ukraine
114 I9 **Cherni Osŭm** ≋ N Bulgaria
116 J8 **Chernivets'ka Oblast'** var. Chernivtsi, Rus. Chernovitskaya Oblast'. ◈ province W Ukraine
114 I9 **Cherni Vit** ≋ NW Bulgaria
114 G10 **Cherni Vrah**, Cherni Vrŭkh. ▲ W Bulgaria 42°33′N 23°18′E
Cherni Vrŭkh see Cherni Vrah
116 K8 **Chernivtsi** Ger. Czernowitz, Rom. Cernăuţi, Rus. Chernovtsy. Chernivets'ka Oblast', W Ukraine 48°18′N 25°55′E
116 M7 **Chernivtsi** Vinnyts'ka Oblast', C Ukraine 48°33′N 28°06′E
Chernivtsi see Chernivets'ka Oblast'
Chernobyl' see Chornobyl'
Cherno More see Black Sea
Chernomorskoye see Chornomors'ke
145 T7 **Chernoretsk** prev. Chernoretskoye. Pavlodar, NE Kazakhstan 52°51′N 76°37′E
Chernoretskoye see Chernoretsk
Chernovitskaya Oblast' see Chernivets'ka Oblast'
Chernovtsy see Chernivtsi
145 U8 **Chernoye** Pavlodar, NE Kazakhstan 51°40′N 77°33′E
Chernoye More see Black Sea
125 U16 **Chernushka** Permskiy Kray, NW Russian Federation 56°30′N 56°07′E
117 N4 **Chernyakhiv** Rus. Chernyakhov. Zhytomyrs'ka Oblast', N Ukraine 50°30′N 28°38′E
Chernyakhov see Chernyakhiv
119 C14 **Chernyakhovsk** Ger. Insterburg. Kaliningradskaya Oblast', W Russian Federation 54°36′N 21°49′E
126 K8 **Chernyanka** Belgorodskaya Oblast', W Russian Federation 50°59′N 37°54′E
125 V5 **Chernyshëva, Gryada** ▲ NW Russian Federation
144 J14 **Chernyshëva, Zaliv** gulf SW Kazakhstan
123 O10 **Chernyshevskiy** Respublika Sakha (Yakutiya), NE Russian Federation 62°57′N 112°29′E
127 P13 **Chernyye Zemli** plain SW Russian Federation
Chërnyy Irtysh see Ertix He, China/Kazakhstan
Chërnyy Irtysh see Kara Irtysh, Kazakhstan
127 V7 **Chernyy Otrog** Orenburgskaya Oblast', W Russian Federation 52°03′N 56°09′E
29 T12 **Cherokee** Iowa, C USA 42°45′N 95°33′W
26 M8 **Cherokee** Oklahoma, C USA 36°45′N 98°22′W
25 R9 **Cherokee** Texas, SW USA 30°56′N 98°42′W
21 O8 **Cherokee Lake** ⊠ Tennessee, S USA
Cherokees, Lake O' The see Grand Lake O' The Cherokees
44 H1 **Cherokee Sound** Great Abaco, N The Bahamas 26°16′N 77°03′W
153 V13 **Cherrapunji** Meghālaya, NE India 25°16′N 91°42′E
28 L9 **Cherry Creek** ≋ South Dakota, N USA
18 J16 **Cherry Hill** New Jersey, NE USA 39°55′N 75°01′W
27 Q7 **Cherryvale** Kansas, C USA 37°16′N 95°33′W
21 Q10 **Cherryville** North Carolina, SE USA 35°22′N 81°22′W
Cherski Range see Cherskogo, Khrebet
123 T6 **Cherskiy** Respublika Sakha (Yakutiya), NE Russian Federation 68°45′N 161°15′E
123 R8 **Cherskogo, Khrebet** var. Cherski Range. ▲ NE Russian Federation
Cherso see Cres
126 L10 **Chertkovo** Rostovskaya Oblast', SW Russian Federation 49°22′N 40°10′E
Cherven' see Chervyen'
114 H8 **Cherven Bryag** Pleven, N Bulgaria 43°16′N 24°06′E
116 M4 **Chervonoarmiys'k** Zhytomyrs'ka Oblast', N Ukraine 50°27′N 28°15′E
Chervonograd see Chervonohrad
116 I4 **Chervonohrad** Rus. Chervonograd. L'vivs'ka Oblast', NW Ukraine 50°25′N 24°10′E
117 W6 **Chervonooskil's'ke Vodoskhovyshche** Rus. Krasnooskol'skoye Vodokhranilishche. ⊠ NE Ukraine
Chervonoye, Ozero see Chyrvonaye, Vozyera
117 S4 **Chervonozavods'ke** Poltavs'ka Oblast', C Ukraine 50°24′N 33°22′E
119 L16 **Chervyen'** Rus. Cherven'. Minskaya Voblasts', C Belarus 53°42′N 28°26′E
119 P16 **Cherykaw** Rus. Cherikov. Mahilyowskaya Voblasts', E Belarus 53°34′N 31°23′E
31 R9 **Chesaning** Michigan, N USA 43°10′N 84°07′W
21 X5 **Chesapeake Bay** inlet NE USA
Chesha Bay see Chëshskaya Guba
Cheshevlya see Tsyeshawlya
97 K18 **Cheshire** cultural region C England, United Kingdom
125 P5 **Chëshskaya Guba** var. Archangel Bay, Chesha Bay, Dvina Bay. bay NW Russian Federation
14 F14 **Chesley** Ontario, S Canada 44°17′N 81°06′W
21 Q10 **Chesnee** South Carolina, SE USA 35°09′N 81°51′W
97 K18 **Chester** Wel. Caerleon, hist. Legaceaster, Lat. Deva, Devana Castra. C England, United Kingdom 53°12′N 02°54′W
35 O4 **Chester** California, W USA 40°18′N 121°14′W
30 K16 **Chester** Illinois, N USA 37°54′N 89°49′W
33 S7 **Chester** Montana, NW USA 48°30′N 110°59′W
18 I16 **Chester** Pennsylvania, NE USA 39°51′N 75°21′W
21 R11 **Chester** South Carolina, SE USA 34°43′N 81°14′W
25 X9 **Chester** Texas, SW USA 30°55′N 94°36′W
21 W6 **Chester** Virginia, NE USA 37°22′N 77°27′W
21 R11 **Chester** West Virginia, NE USA 40°34′N 80°33′W
97 M18 **Chesterfield** C England, United Kingdom 53°15′N 01°25′W
21 S11 **Chesterfield** South Carolina, SE USA 34°44′N 80°04′W
21 W6 **Chesterfield** Virginia, NE USA 37°22′N 77°31′W
192 J9 **Chesterfield, Îles** island group NW New Caledonia
9 O9 **Chesterfield Inlet** Nunavut, NW Canada 63°19′N 90°57′W
9 O9 **Chesterfield Inlet** inlet Nunavut, N Canada
21 Y3 **Chester River** ≋ Delaware/Maryland, NE USA
21 X3 **Chestertown** Maryland, NE USA 39°13′N 76°04′W
19 R4 **Chesuncook Lake** ◎ Maine, NE USA
30 J5 **Chetek** Wisconsin, N USA 45°19′N 91°37′W
13 R14 **Chéticamp** Nova Scotia, SE Canada 46°14′N 61°19′W
27 Q8 **Chetopa** Kansas, C USA 37°02′N 95°05′W
41 Y14 **Chetumal** var. Payo Obispo. Quintana Roo, SE Mexico 18°32′N 88°16′W
Chetumal, Bahia/Chetumal, Bahía de see Chetumal Bay
42 G1 **Chetumal Bay** var. Bahia Chetumal, Bahía de Chetumal. bay Belize/Mexico
10 M13 **Chetwynd** British Columbia, W Canada 55°42′N 121°36′W
38 M11 **Chevak** Alaska, USA 61°31′N 165°35′W
36 M12 **Chevelon Creek** ≋ Arizona, SW USA
185 J17 **Cheviot** Canterbury, South Island, New Zealand 42°48′S 173°17′E
96 L13 **Cheviot Hills** hill range England/Scotland, United Kingdom
96 L13 **Cheviot, The** ▲ NE England, United Kingdom 55°28′N 02°10′W
14 M11 **Chevreuil, Lac du** ◎ Québec, SE Canada
81 I16 **Ch'ew Bahir** var. Lake Stefanie. ◎ Ethiopia/Kenya
32 L7 **Chewelah** Washington, NW USA 48°16′N 117°42′W
26 K10 **Cheyenne** Oklahoma, C USA 35°37′N 99°43′W
33 Z17 **Cheyenne** state capital Wyoming, C USA 41°08′N 104°46′W
26 L5 **Cheyenne Bottoms** ◎ Kansas, C USA
16 J8 **Cheyenne River** ≋ South Dakota/Wyoming, N USA
37 W5 **Cheyenne Wells** Colorado, C USA 38°49′N 102°21′W
108 C9 **Cheyres** Vaud, W Switzerland 46°48′N 06°48′E
Chezdi-Oşorheiu see Târgu Secuiesc
153 P13 **Chhapra** prev. Chapra. Bihār, N India 25°50′N 84°42′E
153 V13 **Chhatak** var. Chatak. Sylhet, NE Bangladesh 25°02′N 91°33′E
154 J9 **Chhatarpur** Madhya Pradesh, C India 24°54′N 79°35′E
154 N13 **Chhatrapur** prev. Chatrapur. Odisha, E India 19°26′N 85°02′E
154 K12 **Chhattīsgarh** ◈ state E India
154 L12 **Chhattīsgarh** plain C India
154 I11 **Chhindwāra** Madhya Pradesh, C India 22°04′N 78°58′E
153 T12 **Chhukha** SW Bhutan 27°02′N 89°36′E
Chiai see Jiayi
Chia-i see Jiayi
Chia-mu-ssu see Jiamusi
83 B15 **Chiange** Port. Vila de Almoster. Huíla, SW Angola 15°44′S 13°54′E
Chiang-hsi see Jiangxi
Chiang Kai-shek see Taiwan Taoyuan
167 P8 **Chiang Khan** Loei, E Thailand 17°51′N 101°43′E
167 O7 **Chiang Mai** var. Chiangmai, Chiengmai, Kiangmai. Chiang Mai, NW Thailand 18°48′N 98°59′E
167 O7 **Chiang Mai** ✕ Chiang Mai, NW Thailand 18°44′N 98°53′E
Chiangmai see Chiang Mai
167 O6 **Chiang Rai** var. Chianpai, Chienrai, Muang Chiang Rai. Chiang Rai, NW Thailand 19°56′N 99°51′E
Chiang-su see Jiangsu
Chianning/Chian-ning see Nanjing
Chianpai see Chiang Rai
106 G12 **Chianti** cultural region C Italy
Chiapa see Chiapa de Corzo
41 U16 **Chiapa de Corzo** var. Chiapa. Chiapas, SE Mexico 16°42′N 92°59′W
41 V16 **Chiapas** ◈ state SE Mexico
106 J12 **Chiaravalle** Marche, C Italy 43°36′N 13°19′E
107 N22 **Chiaravalle Centrale** Calabria, SW Italy 38°40′N 16°25′E
106 E7 **Chiari** Lombardia, N Italy 45°33′N 10°00′E
108 H12 **Chiasso** Ticino, S Switzerland 45°51′N 09°02′E
137 S9 **Chiatura** prev. Chiat'ura. C Georgia 42°13′N 43°11′E
Chiat'ura see Chiatura
41 P15 **Chiautla** var. Chiautla de Tapia. Puebla, S Mexico 18°16′N 98°31′W
Chiautla de Tapia see Chiautla
106 D10 **Chiavari** Liguria, NW Italy 44°19′N 09°19′E
106 E6 **Chiavenna** Lombardia, N Italy 46°19′N 09°22′E
Chiayi see Jiayi
Chiazza see Piazza Armerina
165 O14 **Chiba** var. Tiba. Chiba, Honshū, S Japan 35°37′N 140°06′E
165 O13 **Chiba** off. Chiba-ken, var. Tiba. ◈ prefecture Honshū, S Japan
83 M18 **Chibabava** Sofala, C Mozambique 20°17′S 33°39′E
Chiba-ken see Chiba
161 O10 **Chibi** prev. Puqi. Hubei, C China 29°45′N 113°55′E
83 B15 **Chibia** Port. João de Almeida, Vila João de Almeida. Huíla, SW Angola 15°11′S 13°41′E
83 M18 **Chiboma** Sofala, C Mozambique 20°06′S 33°54′E
82 J12 **Chibondo** Luapula, N Zambia 10°42′S 28°42′E
82 K11 **Chibote** Luapula, NE Zambia 09°52′S 29°33′E
12 K12 **Chibougamau** Québec, SE Canada 49°56′N 74°24′W
164 H11 **Chiburi-jima** island Oki-shotō, SW Japan
83 M20 **Chibuto** Gaza, S Mozambique 24°40′S 33°33′E
31 N11 **Chicago** Illinois, N USA 41°51′N 87°39′W
31 N11 **Chicago Heights** Illinois, N USA 41°30′N 87°38′W
15 W6 **Chic-Chocs, Monts** Eng. Shickshock Mountains. ▲ Québec, SE Canada
39 W13 **Chichagof Island** island Alexander Archipelago, Alaska, USA
57 K20 **Chichas, Cordillera de** ▲ SW Bolivia
41 X12 **Chichén-Itzá, Ruinas** ruins Yucatán, SE Mexico
97 N23 **Chichester** SE England, United Kingdom 50°50′N 00°48′W
42 C5 **Chichicastenango** Quiché, W Guatemala 14°55′N 91°06′W
42 I9 **Chichigalpa** Chinandega, NW Nicaragua 12°35′N 87°04′W
Ch'i-ch'i-ha-erh see Qiqihar
165 X16 **Chichijima-rettō** Eng. Beechy Group. island group SE Japan
54 K4 **Chichiriviche** Falcón, N Venezuela 10°58′N 68°17′W
39 R11 **Chickaloon** Alaska, USA 61°48′N 148°27′W
20 L10 **Chickamauga Lake** ⊠ Tennessee, S USA
23 N7 **Chickasawhay River** ≋ Mississippi, S USA
26 M11 **Chickasha** Oklahoma, C USA 35°03′N 97°57′W
39 T9 **Chicken** Alaska, USA 64°04′N 141°56′W
104 J16 **Chiclana de la Frontera** Andalucía, S Spain 36°26′N 06°09′W
56 B11 **Chiclayo** Lambayeque, NW Peru 06°47′S 79°47′W
35 N5 **Chico** California, W USA 39°42′N 121°51′W
83 L15 **Chicoa** Tete, NW Mozambique 15°45′S 32°25′E
83 M20 **Chicomo** Gaza, S Mozambique 24°29′S 34°15′E
18 M11 **Chicopee** Massachusetts, NE USA 42°08′N 72°34′W
63 I19 **Chico, Río** ≋ SE Argentina
63 I21 **Chico, Río** ≋ S Argentina
27 W14 **Chicot, Lake** ◎ Arkansas, C USA
15 R7 **Chicoutimi** Québec, SE Canada 48°24′N 71°04′W
15 Q8 **Chicoutimi** ≋ Québec, SE Canada
83 L19 **Chicualacuala** Gaza, SW Mozambique 22°06′S 31°42′E
83 B14 **Chicuma** Benguela, C Angola 13°33′S 14°41′E
155 J21 **Chidambaram** Tamil Nādu, SE India 11°25′N 79°42′E
196 K13 **Chidley, Cape** headland Newfoundland and Labrador, E Canada 60°25′N 64°39′W
101 N24 **Chiemsee** ◎ SE Germany
Chiengmai see Chiang Mai
Chienrai see Chiang Rai
106 B8 **Chieri** Piemonte, NW Italy 45°01′N 07°49′E
106 F8 **Chiese** ≋ N Italy
107 K14 **Chieti** var. Teate. Abruzzo, C Italy 42°22′N 14°10′E
99 E19 **Chièvres** Hainaut, SW Belgium 50°34′N 03°49′E
163 S12 **Chifeng** var. Ulanhad. Nei Mongol Zizhiqu, N China 42°17′N 118°56′E
82 F13 **Chifumage** ≋ E Angola
82 M13 **Chifunda** Muchinga, NE Zambia 11°57′S 32°36′E
Chiganak see Shyganak
39 P15 **Chiginagak, Mount** ▲ Alaska, USA 57°10′N 157°00′W
Chigirin see Chyhyryn
Chigirinskoye Vodokhranilishche see Chyhyrynskaya Vodaskhovishcha
41 P13 **Chignahuapan** Puebla, S Mexico 19°52′N 98°03′W
39 O15 **Chignik** Alaska, USA 56°18′N 158°24′W
83 M19 **Chigombe** ≋ S Mozambique
54 D7 **Chigorodó** Antioquia, NW Colombia 07°42′N 76°45′W
83 M19 **Chigubo** Gaza, S Mozambique 22°50′S 33°30′E
Chihertey see Altay
Chih-fu see Yantai
Chihli see Hebei
Chihli, Gulf of see Bo Hai
40 J6 **Chihuahua** Chihuahua, NW Mexico 28°40′N 106°06′W
40 I6 **Chihuahua** ◈ state N Mexico
Chiili see Shiyeli
26 M7 **Chikaskia River** ≋ Kansas/Oklahoma, C USA
155 H19 **Chik Ballāpur** Karnātaka, W India 13°28′N 77°42′E
124 G15 **Chikhachevo** Pskovskaya Oblast', W Russian Federation 57°17′N 29°51′E
155 F19 **Chikmagalūr** Karnātaka, W India 13°20′N 75°46′E
129 V7 **Chikoy** ≋ C Russian Federation
83 J15 **Chikumbi** Lusaka, C Zambia 15°11′S 28°20′E
82 M13 **Chikwa** Muchinga, NE Zambia 11°39′S 32°45′E
Chikwana see Chikwawa
83 N15 **Chikwawa** var. Chikwana. Southern, S Malawi 16°03′S 34°48′E
155 J16 **Chilakalūrupet** Andhra Pradesh, E India 16°09′N 80°13′E
146 L14 **Chilan** Lebap Welaýaty, E Turkmenistan 37°57′N 64°58′E
Chilapa see Chilapa de Alvarez
41 P16 **Chilapa de Alvarez** var. Chilapa. Guerrero, S Mexico 17°38′N 99°11′W
155 J25 **Chilaw** North Western Province, W Sri Lanka 07°34′N 79°48′E
57 D15 **Chilca** Lima, W Peru 12°35′S 76°41′W
23 Q4 **Childersburg** Alabama, S USA 33°16′N 86°21′W
25 P4 **Childress** Texas, SW USA 34°25′N 100°14′W
63 G14 **Chile** off. Republic of Chile. ◆ republic SW South America
47 R10 **Chile Basin** undersea feature E Pacific Ocean 33°00′S 80°00′W
63 H20 **Chile Chico** Aisén, W Chile 46°34′S 71°44′W
62 I9 **Chilecito** La Rioja, NW Argentina 29°10′S 67°30′W
62 H12 **Chilecito** Mendoza, W Argentina 33°53′S 69°03′W
83 L14 **Chilembwe** Eastern, E Zambia 13°54′S 31°38′E
Chile, Republic of see Chile
193 S11 **Chile Rise** undersea feature SE Pacific Ocean 40°00′S 90°00′W
117 N13 **Chilia, Braţul** ≋ SE Romania
Chilia-Nouă see Kiliya
Chilik see Shelek
Chilik see Shelek
154 O13 **Chilika Lake** var. Chilka Lake. ◎ E India
82 J13 **Chililabombwe** Copperbelt, C Zambia 12°20′S 27°52′E
Chi-lin see Jilin
Chilka Lake see Chilika Lake
10 H9 **Chilkoot Pass** pass British Columbia, W Canada
Chill Ala, Cuan see Killala Bay
62 G13 **Chillán** Bío Bío, C Chile 36°37′S 72°10′W
61 C22 **Chillar** Buenos Aires, E Argentina 37°16′S 59°58′W
Chill Chiaráin, Cuan see Kilkieran Bay
30 K12 **Chillicothe** Illinois, N USA 40°55′N 89°29′W
27 S3 **Chillicothe** Missouri, C USA 39°47′N 93°33′W
31 S14 **Chillicothe** Ohio, N USA 39°20′N 83°00′W
25 Q4 **Chillicothe** Texas, SW USA 34°15′N 99°31′W
10 M17 **Chilliwack** British Columbia, SW Canada 49°09′N 121°54′W
Chill Mhantáin, Ceann see Wicklow Head
Chill Mhantáin, Sléibhte see Wicklow Mountains
108 C10 **Chillon** Vaud, W Switzerland 46°24′N 06°56′E
Chil'mamedkum, Peski/Chilmämetgum see Çilmämmetgum
63 F17 **Chiloé, Isla de** var. Isla Grande de Chiloé. island W Chile
32 H15 **Chiloquin** Oregon, NW USA 42°33′N 121°33′W
41 O16 **Chilpancingo** var. Chilpancingo de los Bravos. Guerrero, S Mexico 17°33′N 99°30′W
Chilpancingo de los Bravos see Chilpancingo
97 N21 **Chiltern Hills** hill range S England, United Kingdom
30 M7 **Chilton** Wisconsin, N USA 44°04′N 88°10′W
82 F11 **Chiluage** Lunda Sul, NE Angola 09°32′S 21°48′E
82 N12 **Chilumba** prev. Deep Bay. Northern, N Malawi 10°27′S 34°12′E
Chilung see Jilong
83 N15 **Chilwa, Lake** var. Lago Chirua, Lake Shirwa. ◎ SE Malawi
167 R10 **Chi, Mae Nam** ≋ E Thailand
42 C6 **Chimaltenango** Chimaltenango, C Guatemala 14°40′N 90°48′W
42 A2 **Chimaltenango** off. Departamento de Chimaltenango. ◈ department S Guatemala
Chimaltenango, Departamento de see Chimaltenango
43 V15 **Chimán** Panamá, E Panama 08°42′N 78°35′W
83 M17 **Chimanimani** prev. Mandidzudzure, Melsetter. Manicaland, E Zimbabwe 19°48′S 32°52′E
99 G22 **Chimay** Hainaut, S Belgium 50°03′N 04°20′E
37 S10 **Chimayo** New Mexico, SW USA 36°00′N 105°55′W
Chimbay see Chimboy
56 A13 **Chimborazo** ◈ province C Ecuador
56 C7 **Chimborazo** ⛰ C Ecuador 01°29′S 78°50′W
56 C12 **Chimbote** Ancash, W Peru 09°04′S 78°34′W
146 H7 **Chimboy** Rus. Chimbay. Qoraqalpog'iston Respublikasi, NW Uzbekistan 43°03′N 59°52′E
186 D7 **Chimbu** var. Simbu. ◈ province C Papua New Guinea
54 F6 **Chimichagua** Cesar, N Colombia 09°19′N 73°51′W
Chimishliya see Cimişlia
Chimkent see Shymkent
Chimkentskaya Oblast' see Yuzhnyy Kazakhstan
28 I14 **Chimney Rock** rock Nebraska, C USA
83 M17 **Chimoio** Manica, C Mozambique 19°08′S 33°29′E
82 K11 **Chimpembe** Northern, NE Zambia 09°31′S 29°33′E
41 O8 **China** Nuevo León, NE Mexico 25°42′N 99°15′W
156 M9 **China** off. People's Republic of China, Chin. Chung-hua Jen-min Kung-ho-kuo, Zhonghua Renmin Gongheguo; prev. Chinese Empire. ◆ republic E Asia
19 Q7 **China Lake** ◎ Maine, NE USA
42 F8 **Chinameca** San Miguel, E El Salvador 13°30′N 88°20′W*
Chi-nan/Chinan see Jinan
42 H9 **Chinandega** Chinandega, NW Nicaragua 12°37′N 87°08′W
42 H9 **Chinandega** ◈ department NW Nicaragua
China, People's Republic of see China
China, Republic of see Taiwan
24 J11 **Chinati Mountains** ▲ Texas, SW USA
Chinaz see Chinoz
57 E15 **Chincha Alta** Ica, SW Peru 13°25′S 76°07′W
11 N11 **Chinchaga** ≋ Alberta, SW Canada
Chin-chiang see Quanzhou
Chinchilla see Chinchilla de Monte Aragón
105 Q11 **Chinchilla de Monte Aragón** var. Chinchilla. Castilla-La Mancha, C Spain 38°56′N 01°44′W
54 D10 **Chinchiná** Caldas, W Colombia 04°59′N 75°37′W
105 O8 **Chinchón** Madrid, C Spain 40°08′N 03°26′W
41 Z14 **Chinchorro, Banco** island SE Mexico
Chin-chou/Chinchow see Jinzhou
21 Z5 **Chincoteague** Assateague Island, Virginia, NE USA 37°55′N 75°22′W
83 O17 **Chinde** Zambézia, NE Mozambique 18°35′S 36°28′E
159 R13 **Chindu** var. Chengwen; prev. Chuqung. Qinghai, C China 33°19′N 97°08′E
Chindwin see Chindwinn
166 M2 **Chindwinn** var. Chindwin. ≋ N Myanmar (Burma)
Chinese Empire see China
Chinghai see Qinghai
Ch'ing Hai see Qinghai Hu, China
Chingildi see Shengeldi
Chingirlau see Shyngghyrlau
82 J13 **Chingola** Copperbelt, C Zambia 12°31′S 27°53′E
Ching-Tao/Ch'ing-tao see Qingdao
82 C13 **Chinguar** Huambo, C Angola 12°33′S 16°25′E
76 I7 **Chinguețți** var. Chinguetti. Adrar, C Mauritania 20°25′N 12°24′W
Chinhae see Jinhae
166 K4 **Chin Hills** ▲ W Myanmar (Burma)
83 K16 **Chinhoyi** prev. Sinoia. Mashonaland West, N Zimbabwe 17°22′S 30°12′E
Chinhsien see Jinzhou
39 Q14 **Chiniak, Cape** headland Kodiak Island, Alaska, USA 57°37′N 152°10′W
14 G10 **Chiniguchi Lake** ◎ Ontario, S Canada
149 U8 **Chiniot** Punjab, NE Pakistan 31°40′N 73°00′E
Chinju see Jinju
Chinkai see Jinhae
78 M13 **Chinko** ≋ E Central African Republic
37 O9 **Chinle** Arizona, SW USA 36°09′N 109°33′W
Chinmen Tao see Jinmen Dao
Chinnchâr see Shinshār
Chinnereth see Tiberias, Lake
164 C12 **Chino** var. Tino. Nagano, Honshū, S Japan 36°00′N 138°10′E
102 L8 **Chinon** Indre-et-Loire, C France 47°10′N 00°15′E
33 T7 **Chinook** Montana, NW USA 48°35′N 109°13′W
Chinook State see Washington
192 L4 **Chinook Trough** undersea feature N Pacific Ocean
36 K11 **Chino Valley** Arizona, SW USA 34°45′N 112°27′W
147 P10 **Chinoz** Rus. Chinaz. Toshkent Viloyati, E Uzbekistan 40°58′N 68°46′E
82 L12 **Chinsali** Muchinga, NE Zambia 10°33′S 32°05′E
166 K5 **Chin State** ◈ state W Myanmar (Burma)
Chinsura see Chunchura
Chin-tō see Jin-do
54 E6 **Chinú** Córdoba, NW Colombia 09°07′N 75°25′W
99 K24 **Chiny, Forêt de** forest SE Belgium
83 M15 **Chioco** Tete, NW Mozambique 16°22′S 32°50′E
106 H8 **Chioggia** anc. Fossa Claudia. Veneto, NE Italy 45°14′N 12°17′E
114 H12 **Chionótrypa** ▲ NE Greece 41°16′N 24°06′E
115 L18 **Chíos** var. Hios, Khíos, It. Scio, Turk. Sakiz-Adasi. Chíos, E Greece 38°23′N 26°07′E
115 K18 **Chíos** var. Khíos. island E Greece
83 M14 **Chipata** prev. Fort Jameson. Eastern, E Zambia 13°40′S 32°42′E
83 C14 **Chipindo** Huíla, C Angola 13°53′S 15°47′E
23 R8 **Chipley** Florida, SE USA 30°46′N 85°32′W
155 D15 **Chiplūn** Mahārāshtra, W India 17°32′N 73°32′E
81 H22 **Chipogolo** Dodoma, C Tanzania 06°52′S 36°03′E
23 R8 **Chipola River** ≋ Florida, SE USA
97 L22 **Chippenham** S England, United Kingdom 51°28′N 02°07′W
30 J6 **Chippewa Falls** Wisconsin, N USA 44°56′N 91°25′W
30 J4 **Chippewa, Lake** ⊠ Wisconsin, N USA
31 Q8 **Chippewa River** ≋ Michigan, N USA
30 I6 **Chippewa River** ≋ Wisconsin, N USA
Chipping Wycombe see High Wycombe
114 G8 **Chiprovtsi** Montana, NW Bulgaria 43°23′N 22°53′E
19 T4 **Chiputneticook Lakes** lakes Canada/USA
56 D13 **Chiquián** Ancash, W Peru 10°09′S 78°08′W
41 Y11 **Chiquilá** Quintana Roo, SE Mexico 21°25′N 87°20′W
42 E6 **Chiquimula** Chiquimula, SE Guatemala 14°46′N 89°32′W
42 A3 **Chiquimula** off. Departamento de Chiquimula. ◈ department SE Guatemala
Chiquimula, Departamento de see Chiquimula
42 D7 **Chiquimulilla** Santa Rosa, S Guatemala 14°06′N 90°23′W
54 F9 **Chiquinquirá** Boyacá, C Colombia 05°37′N 73°51′W
155 J17 **Chīrāla** Andhra Pradesh, E India 15°49′N 80°21′E
149 N4 **Chīras** Gōwr, N Afghanistan 35°15′N 65°39′E
152 H11 **Chirāwa** Rājasthān, N India 28°12′N 75°42′E
Chirchik see Chirchiq
147 Q9 **Chirchiq** Rus. Chirchik. Toshkent Viloyati, E Uzbekistan 41°30′N 69°32′E
147 P10 **Chirchiq** ≋ E Uzbekistan
Chire see Shire
83 L18 **Chiredzi** Masvingo, SE Zimbabwe 21°00′S 31°38′E
25 X8 **Chireno** Texas, SW USA 31°30′N 94°21′W
77 X7 **Chirfa** Agadez, NE Niger 21°01′N 12°41′E
37 O16 **Chiricahua Mountains** ▲ Arizona, SW USA
37 O16 **Chiricahua Peak** ▲ Arizona, SW USA 31°51′N 109°17′W
54 F6 **Chiriguaná** Cesar, N Colombia 09°24′N 73°38′W
39 P15 **Chirikof Island** island Alaska, USA
43 P16 **Chiriquí** off. Provincia de Chiriquí. ◈ province SW Panama
43 P17 **Chiriquí, Golfo de** Eng. Chiriqui Gulf. gulf SW Panama
43 P15 **Chiriquí Grande** Bocas del Toro, W Panama 08°58′N 82°08′W
Chiriqui Gulf see Chiriquí, Golfo de
43 P15 **Chiriquí, Laguna de** lagoon NW Panama
Chiriquí, Provincia de see Chiriquí
43 O16 **Chiriquí Viejo, Río** ≋ W Panama
Chiriquí, Volcán de see Barú, Volcán
83 N15 **Chiromo** Southern, S Malawi 16°32′S 35°07′E
114 J10 **Chirpan** Stara Zagora, C Bulgaria 42°12′N 25°20′E
43 N14 **Chirripó Atlántico, Río** ≋ E Costa Rica
Chirripó, Cerro see Chirripó Grande, Cerro
Chirripó del Pacífico, Río see Chirripó, Río
43 N14 **Chirripó Grande, Cerro** var. Cerro Chirripó. ▲ SE Costa Rica 09°31′N 83°28′W
43 N13 **Chirripó, Río** var. Río Chirripó del Pacífico. ≋ NE Costa Rica
Chirua, Lago see Chilwa, Lake
83 J15 **Chirundu** Southern, S Zambia 16°03′S 28°50′E
29 W8 **Chisago City** Minnesota, N USA 45°22′N 92°53′W
83 J14 **Chisamba** Central, C Zambia 15°00′S 28°22′E
39 T10 **Chisana** Alaska, USA 62°09′N 142°07′W
82 I13 **Chisasa** North Western, NW Zambia 12°09′S 25°30′E
12 I9 **Chisasibi** prev. Fort George. Québec, C Canada 53°50′N 79°01′W
42 D4 **Chisec** Alta Verapaz, C Guatemala 15°50′N 90°18′W
127 U5 **Chishmy** Respublika Bashkortostan, W Russian Federation 54°33′N 55°21′E
29 V4 **Chisholm** Minnesota, N USA 47°29′N 92°52′W
149 U10 **Chishtiān** var. Chishtiān Mandi. Punjab, E Pakistan 29°44′N 72°54′E
Chishtiān Mandi see Chishtiān
160 I11 **Chishui He** ≋ C China
Chisimaio/Chisimayu see Kismaayo
117 N10 **Chişinău** Rus. Kishinev. ● (Moldova) C Moldova 47°N 28°51′E
117 N10 **Chişinău** ✕ S Moldova 46°54′N 28°56′E
Chişinău-Criş see Chişineu-Criş
116 F10 **Chişineu-Criş** Hung. Kisjenő; prev. Chişinău-Criş. Arad, W Romania 46°33′N 21°30′E
83 K14 **Chisomo** Central, C Zambia 13°30′S 30°37′E
106 A8 **Chisone** ≋ NW Italy
24 K12 **Chisos Mountains** ▲ Texas, SW USA
39 T10 **Chistochina** Alaska, USA 62°34′N 144°39′W
127 R4 **Chistopol'** Respublika Tatarstan, W Russian Federation 55°20′N 50°39′E
145 O8 **Chistopol'ye** Severnyy Kazakhstan, N Kazakhstan 52°37′N 67°14′E
123 O13 **Chita** Zabaykal'skiy Kray, S Russian Federation 52°03′N 113°35′E
83 B16 **Chitado** Cunene, SW Angola 17°16′S 13°54′E
Chitaldroog/Chitaldrug see Chitradurga
83 C15 **Chitanda** ≋ S Angola
Chitangwiza see Chitungwiza
82 F10 **Chitato** Lunda Norte, NE Angola 07°23′S 20°46′E
83 C14 **Chitembo** Bié, C Angola 13°33′S 16°47′E
39 T11 **Chitina** Alaska, USA 61°31′N 144°26′W
39 T11 **Chitina River** ≋ Alaska, USA
82 M11 **Chitipa** Northern, NW Malawi 09°41′S 33°19′E
165 S4 **Chitose** var. Titose. Hokkaidō, NE Japan 42°51′N 141°40′E
155 G18 **Chitradurga** prev. Chitaldroog, Chitaldrug. Karnātaka, W India 14°16′N 76°23′E
149 T3 **Chitrāl** Khyber Pakhtunkhwa, NW Pakistan 35°51′N 71°47′E
43 S16 **Chitré** Herrera, S Panama 07°57′N 80°26′W
153 V16 **Chittagong** Ben. Chāttagām. Chittagong, SE Bangladesh 22°20′N 91°48′E
153 U16 **Chittagong** ◈ division E Bangladesh
153 Q15 **Chittaranjan** West Bengal, NE India 23°52′N 86°40′E
152 G14 **Chittaurgarh** var. Chittorgarh. Rājasthān, N India 24°54′N 74°42′E
155 I19 **Chittoor** Andhra Pradesh, E India 13°13′N 79°06′E
Chittorgarh see Chittaurgarh
155 G21 **Chittūr** Kerala, SW India 10°42′N 76°46′E
83 K16 **Chitungwiza** prev. Chitangwiza. Mashonaland East, NE Zimbabwe 18°S 31°06′E
62 H4 **Chiuchíu** Antofagasta, N Chile 22°13′S 68°34′W
82 F12 **Chiumbe** var. Tshiumbe. ≋ Angola/Dem. Rep. Congo
83 F15 **Chiume** Moxico, E Angola 15°08′S 21°09′E
82 K13 **Chiundaponde** Muchinga, NE Zambia 12°14′S 30°40′E
106 H13 **Chiusi** Toscana, C Italy 43°00′N 11°56′E
54 J5 **Chivacoa** Yaracuy, N Venezuela 10°10′N 68°54′W
106 B8 **Chivasso** Piemonte, NW Italy 45°13′N 07°54′E
83 L17 **Chivhu** prev. Enkeldoorn. Midlands, C Zimbabwe 19°01′S 30°54′E
61 C20 **Chivilcoy** Buenos Aires, E Argentina 34°55′S 60°00′W
Chiwei Yu see Sekibi-sho
82 N12 **Chiweta** Northern, N Malawi 10°36′S 34°09′E
42 D4 **Chixoy, Río** var. Río Negro, Río Salinas. ≋ Guatemala/Mexico
82 H13 **Chizela** North Western, NW Zambia 13°11′S 24°59′E
125 O5 **Chizha** Nenetskiy Avtonomnyy Okrug, NW Russian Federation 67°04′N 44°19′E
161 Q9 **Chizhou** var. Guichi. Anhui, E China 30°39′N 117°29′E
164 I12 **Chizu** Tottori, Honshū, SW Japan 35°15′N 134°14′E
Chkalov see Orenburg
74 J5 **Chlef** var. Ech Chleff, Ech Chleliff; prev. Al-Asnam, El Asnam, Orléansville. NW Algeria 36°11′N 01°21′E
115 G18 **Chlómo** ▲ C Greece 38°36′N 22°57′E
111 M15 **Chmielnik** Świętokrzyskie, C Poland 50°37′N 20°43′E

◆ Country ● Country Capital ◇ Dependent Territory ○ Dependent Territory Capital ◈ Administrative Regions ✕ International Airport ▲ Mountain ▲▲ Mountain Range ⛰ Volcano ≋ River ◎ Lake ⊠ Reservoir

167 S11 **Chŏâm Khsant** Preăh Vihéar, N Cambodia 14°13′N 104°56′E
62 G10 **Choapa, Río** var. Choapo. ≈ C Chile
Choapas see Las Choapas
Choapo see Choapa, Río
Choarta see Chwarta
67 T13 **Chobe** ≈ N Botswana
14 K8 **Chochocouane** ≈ Québec, SE Canada
110 E13 **Chocianów** Ger. Kotzenan. Dolnośląskie, SW Poland 51°23′N 15°55′E
54 C9 **Chocó** off. Departamento del Chocó. ◈ province W Colombia
Chocó, Departamento del see Chocó
35 X16 **Chocolate Mountains** ⛰ California, W USA
21 W9 **Chocowinity** North Carolina, SE USA 35°33′N 77°03′W
27 N10 **Choctaw** Oklahoma, C USA 35°30′N 97°16′W
23 Q8 **Choctawhatchee Bay** bay Florida, SE USA
23 Q8 **Choctawhatchee River** ≈ Florida, SE USA
Chodau see Chodov
163 V14 **Ch'o-do** island SW North Korea
Chodorów see Khodoriv
111 A16 **Chodov** Ger. Chodau. Karlovarský Kraj, W Czech Republic 50°15′N 12°45′E
110 G10 **Chodzież** Wielkopolskie, C Poland 53°N 16°55′E
63 J15 **Choele Choel** Río Negro, C Argentina 39°19′S 65°42′W
83 L14 **Chofombo** Tete, NW Mozambique 14°43′S 31°48′E
Chohtan see Chauhtan
11 U14 **Choiceland** Saskatchewan, C Canada 53°28′N 104°26′W
186 K8 **Choiseul** ◈ province NW Solomon Islands
186 K8 **Choiseul** var. Lauru. island NW Solomon Islands
63 M23 **Choiseul Sound** sound East Falkland, Falkland Islands
40 H7 **Choix** Sinaloa, C Mexico 26°43′N 108°20′W
110 D10 **Chojna** Zachodnio-pomorskie, W Poland 52°56′N 14°25′E
110 H8 **Chojnice** Ger. Konitz. Pomorskie, N Poland 53°41′N 17°34′E
111 F14 **Chojnów** Ger. Hainau, Haynau. Dolnośląskie, SW Poland 51°16′N 15°55′E
167 Q10 **Chok Chai** Nakhon Ratchasima, C Thailand 14°45′N 102°10′E
80 I12 **Ch'ok'ē** var. Choke Mountains. ⛰ NW Ethiopia
25 R13 **Choke Canyon Lake** ⊠ Texas, SW USA
Choke Mountains see Ch'ok'ē
Chokpar see Shokpar
147 W7 **Chok-Tal** var. Choktal. Issyk-Kul'skaya Oblast', E Kyrgyzstan 42°37′N 76°45′E
Choktal see Chok-Tal
Chókué see Chokwè
123 R7 **Chokurdakh** Respublika Sakha (Yakutiya), NE Russian Federation 70°38′N 148°18′E
83 L20 **Chokwè** var. Chókué. Gaza, S Mozambique 24°27′S 32°55′E
188 F8 **Chol** Babeldaob, N Palau
160 E8 **Chola Shan** ⛰ C China
102 J8 **Cholet** Maine-et-Loire, NW France 47°03′N 00°52′W
63 H17 **Cholila** Chubut, W Argentina 42°33′S 71°28′W
Cholo see Thyolo
147 V8 **Cholpon** Narynskaya Oblast', C Kyrgyzstan 42°07′N 75°25′E
147 X7 **Cholpon-Ata** Issyk-Kul'skaya Oblast', E Kyrgyzstan 42°39′N 77°05′E
41 P14 **Cholula** Puebla, S Mexico 19°03′N 98°19′W
42 I8 **Choluteca** Choluteca, S Honduras 13°15′N 87°10′W
42 H8 **Choluteca** ◈ department S Honduras
42 H8 **Choluteca, Río** ≈ SW Honduras
83 I15 **Choma** Southern, S Zambia 16°48′S 26°58′E
153 T11 **Chomo Lhari** ▲ NW Bhutan 27°59′N 89°24′E
167 N7 **Chom Thong** Chiang Mai, NW Thailand 18°25′N 98°44′E
111 B15 **Chomutov** Ger. Komotau. Ústecký Kraj, NW Czech Republic 50°28′N 13°24′E
123 N11 **Chona** ≈ C Russian Federation
Ch'ŏnan see Cheonan
167 P11 **Chon Buri** prev. Bang Pla Soi. Chon Buri, S Thailand 13°24′N 100°59′E
56 B6 **Chone** Manabí, W Ecuador 0°44′S 80°04′W
Chong'an see Wuyishan
163 W13 **Ch'ŏngch'ŏn-gang** ≈ W North Korea
163 Y11 **Ch'ŏngjin** NE North Korea 41°48′N 129°44′E
Chŏngju see Cheongju
161 S8 **Chongming Dao** island E China
160 J10 **Chongqing** var. Ch'ung-ching, Ch'ung-ch'ing, Chungking, Pahsien, Tchongking, Yuzhou. Chongqing Shi, C China 29°34′N 106°27′E
Chŏngup see Chŏnju
161 O10 **Chongyang** var. Tiancheng. Hubei, C China 29°35′N 114°03′E
160 J15 **Chongzuo** prev. Taiping. Guangxi Zhuangzu Zizhiqu, S China 22°18′N 107°23′E
163 Y16 **Chŏnju** prev. Chŏnju, Chŏngup, Jap. Seiyu. SW South Korea 35°51′N 127°08′E
Chŏnju see Chŏnju
Chŏnju see Jeonju
Chonnacht see Connaught
Chonogol see Erdenetsagaan
63 F19 **Chonos, Archipiélago de los** island group S Chile
42 K10 **Chontales** ◈ department S Nicaragua
167 T13 **Chơn Thanh** Sông Be, S Vietnam 11°25′N 106°38′E
158 K17 **Cho Oyu** var. Qowowuyag. ▲ China/Nepal 28°07′N 86°37′E
116 G7 **Chop** Cz. Čop, Hung. Csap. Zakarpats'ka Oblast', W Ukraine 48°26′N 22°13′E
21 Y3 **Choptank River** ≈ Maryland, NE USA
115 J22 **Chóra** prev. Íos. Íos, Kykládes, Greece, Aegean Sea 36°42′N 25°16′E
115 H25 **Chóra Sfakíon** var. Sfákia. Kríti, Greece, E Mediterranean Sea 35°12′N 24°05′E
Chorcaí, Cuan see Cork Harbour
43 P15 **Chorcha, Cerro** ▲ W Panama 08°39′N 82°07′W
Chorku see Chorkŭh
147 R11 **Chorkŭh** Rus. Chorku. N Tajikistan 40°04′N 70°30′E
97 K17 **Chorley** NW England, United Kingdom 53°40′N 02°38′W
Chorne More see Black Sea
117 R5 **Chornobay** Cherkas'ka Oblast', C Ukraine 49°40′N 32°20′E
117 O3 **Chornobyl'** Rus. Chernobyl'. Kyyivs'ka Oblast', N Ukraine 51°17′N 30°15′E
117 R12 **Chornomors'ke** Rus. Chernomorskoye. Avtonomna Respublika Krym, S Ukraine 45°29′N 32°45′E
117 R4 **Chornukhy** Poltavs'ka Oblast', C Ukraine 50°15′N 32°57′E
Chorokh/Chorokhi see Çoruh Nehri
110 O9 **Choroszcz** Podlaskie, NE Poland 53°10′N 23°E
116 K6 **Chortkiv** Rus. Chortkov. Ternopil's'ka Oblast', W Ukraine 49°01′N 25°46′E
Chortkov see Chortkiv
Chorum see Çorum
110 M9 **Chorzele** Mazowieckie, C Poland 53°16′N 20°53′E
111 J16 **Chorzów** Ger. Königshütte; prev. Królewska Huta. Śląskie, S Poland 50°17′N 18°58′E
163 W12 **Ch'osan** N North Korea 40°45′N 125°52′E
Chośebuz see Cottbus
Chōsen-kaikyō see Korea Strait
164 P14 **Chōshi** var. Tyôsi. Chiba, Honshū, S Japan 35°44′N 140°48′E
63 H14 **Chos Malal** Neuquén, W Argentina 37°23′S 70°16′W
Chosŏn-minjujuŭi-inmin-kanghwaguk see North Korea
110 E9 **Choszczno** Ger. Arnswalde. Zachodnio-pomorskie, NW Poland 53°10′N 15°24′E
153 O15 **Chota Nāgpur** plateau N India
33 R8 **Choteau** Montana, NW USA 47°48′N 112°40′W
Chotqol see Chatkal
14 M8 **Chouart** ≈ Québec, SE Canada
76 I7 **Choûm** Adrar, C Mauritania 21°19′N 12°59′W
27 Q9 **Chouteau** Oklahoma, C USA 36°11′N 95°20′W
21 X8 **Chowan River** ≈ North Carolina, SE USA
35 Q10 **Chowchilla** California, W USA 37°06′N 120°15′W
163 Q7 **Choybalsan** var. Hulstay. Dornod, NE Mongolia 48°25′N 114°56′E
163 P7 **Choybalsan** prev. Byan Tumen. Dornod, E Mongolia 48°03′N 114°32′E
162 M9 **Choyr** Govĭ Sumber, C Mongolia 46°20′N 108°21′E
185 I19 **Christchurch** Canterbury, South Island, New Zealand 43°31′S 172°39′E
97 M24 **Christchurch** S England, United Kingdom 50°44′N 01°45′W
185 I18 **Christchurch** ✈ Canterbury, South Island, New Zealand 43°28′S 172°33′E
44 J12 **Christiana** C Jamaica 18°13′N 77°29′W
83 H22 **Christiana** Free State, C South Africa 27°55′S 25°10′E
115 J23 **Christiána** var. Christiani. island Kykládes, Greece, Aegean Sea
Christiani see Christiána
Christiania see Oslo
14 G13 **Christian Island** island Ontario, S Canada
191 P16 **Christian, Point** headland Pitcairn Island, Pitcairn Islands 25°04′S 130°08′E
38 M11 **Christian River** ≈ Alaska, USA
Christiansand see Kristiansand
21 S7 **Christiansburg** Virginia, NE USA 37°07′N 80°26′W
95 G23 **Christiansfeld** Syddanmark, SW Denmark 55°21′N 09°30′E
Christianshåb see Qasigiannguit
39 X14 **Christian Sound** inlet Alaska, USA
45 T9 **Christiansted** Saint Croix, S Virgin Islands (US) 17°43′N 64°42′W
Christiansund see Kristiansund
25 R13 **Christine** Texas, SW USA 28°47′N 98°30′W
173 U7 **Christmas Island** ◇ Australian external territory E Indian Ocean
129 T17 **Christmas Island** island E Indian Ocean
Christmas Island see Kiritimati
192 M7 **Christmas Ridge** undersea feature C Pacific Ocean
30 L16 **Christopher** Illinois, N USA 37°58′N 89°03′W
25 P9 **Christoval** Texas, SW USA 31°09′N 100°30′W
111 F17 **Chrudim** Pardubický Kraj, C Czech Republic 49°58′N 15°49′E
115 K25 **Chrýsi** island SE Greece
121 N2 **Chrysochoú, Kólpos** var. Khrysokhou Bay. bay E Mediterranean Sea
114 I13 **Chrysoúpoli** var. Hrisoupoli; prev. Khrisoúpolis. Anatolikí Makedonía kai Thráki, NE Greece 40°59′N 24°42′E
111 K16 **Chrzanów** var. Chrzanow, Ger. Zaumgarten. Śląskie, S Poland 50°10′N 19°21′E
Chu see Shu
42 C5 **Chuacús, Sierra de** ⛰ W Guatemala
153 S15 **Chuadanga** Khulna, W Bangladesh 23°38′N 88°52′E
Chuan see Sichuan
Ch'uan-chou see Quanzhou
Chuankou see Minhe
39 O11 **Chuathbaluk** Alaska, USA 61°36′N 159°14′W
Chubek see Moskva
63 I17 **Chubut** off. Provincia de Chubut. ◈ province S Argentina
Chubut, Provincia de see Chubut
63 I17 **Chubut, Río** ≈ SE Argentina
43 V15 **Chucanti, Cerro** ▲ E Panama 08°48′N 78°27′W
Ch'u-chiang see Shaoguan
43 W15 **Chucunaque, Río** ≈ E Panama
Chudin see Chudzin
116 M5 **Chudniv** Zhytomyrs'ka Oblast', N Ukraine 50°02′N 28°06′E
124 H13 **Chudovo** Novgorodskaya Oblast', W Russian Federation 59°07′N 31°42′E
Chudskoye Ozero see Peipus, Lake
119 J18 **Chudzin** Rus. Chudin. Brestskaya Voblasts', SW Belarus 52°44′N 26°59′E
39 Q13 **Chugach Islands** island group Alaska, USA
39 S11 **Chugach Mountains** ⛰ Alaska, USA
164 G12 **Chūgoku-sanchi** ⛰ Honshū, SW Japan
Chugqênsumdo see Jigzhi
Chuguyev see Chuhuyiv
117 V5 **Chuhuyiv** var. Chuguyev. Kharkivs'ka Oblast', E Ukraine 49°51′N 36°44′E
61 H19 **Chuí** Rio Grande do Sul, S Brazil 33°45′S 53°23′W
Chuí see Chuy
Chu-Iliyskiye Gory see Gory Shu-Ile
Chukai see Cukai
Chukchi Avtonomnyy Okrug see Chukotskiy Avtonomnyy Okrug
Chukchi Peninsula see Chukotskiy Poluostrov
197 R6 **Chukchi Plain** undersea feature Arctic Ocean
197 R6 **Chukchi Plateau** undersea feature Arctic Ocean
197 R4 **Chukchi Sea** Rus. Chukotskoye More. sea Arctic Ocean
125 N14 **Chukhloma** Kostromskaya Oblast', NW Russian Federation 58°42′N 42°39′E
Chukotka see Chukotskiy Avtonomnyy Okrug
Chukot Range see Anadyrskiy Khrebet
123 V6 **Chukotskiy Avtonomnyy Okrug** var. Chukchi Avtonomnyy Okrug, Chukotka. ◈ autonomous district NE Russian Federation
123 W5 **Chukotskiy, Mys** headland NE Russian Federation 64°15′N 173°03′W
123 V5 **Chukotskiy Poluostrov** Eng. Chukchi Peninsula. peninsula NE Russian Federation
Chukotskoye More see Chukchi Sea
Chukurkak see Chuqurqoq
Chulakkurgan see Sholakkorgan
35 U17 **Chula Vista** California, W USA 32°38′N 117°04′W
123 Q12 **Chul'man** Respublika Sakha (Yakutiya), NE Russian Federation 56°50′N 124°47′E
56 B9 **Chulucanas** Piura, NW Peru 05°08′S 80°10′W
122 J12 **Chulym** ≈ C Russian Federation
152 K6 **Chumar** Jammu and Kashmir, N India 32°38′N 78°36′E
114 K9 **Chumerna** ▲ C Bulgaria 42°45′N 25°58′E
123 R12 **Chumikan** Khabarovskiy Kray, E Russian Federation 54°41′N 135°12′E
167 Q9 **Chum Phae** Khon Kaen, C Thailand 16°31′N 102°09′E
167 N13 **Chumphon** var. Jumporn. Chumphon, SW Thailand 10°30′N 99°11′E
167 O9 **Chumsaeng** var. Chum Saeng. Nakhon Sawan, C Thailand 15°52′N 100°18′E
Chum Saeng see Chumsaeng
122 L12 **Chuna** ≈ C Russian Federation
161 R9 **Chun'an** var. Qiandaohu; prev. Pailing. Zhejiang, SE China 29°37′N 118°59′E
Chunan see Zhunan
Chuncheng see Yangchun
163 Y14 **Chuncheon** Jap. Shunsen; prev. Ch'unch'ŏn. N South Korea 37°52′N 127°48′E
Ch'unch'ŏn see Chuncheon
153 S16 **Chunchura** prev. Chinsura. West Bengal, NE India 22°54′N 88°20′E
Chundzha see Shonzhy
Ch'ung-ch'ing/Ch'ung-ching see Chongqing
Chung-hua Jen-min Kung-ho-kuo see China
163 Y15 **Chungju** Jap. Chūshū; prev. Ch'ungju. C South Korea 36°57′N 127°50′E
Ch'ungju see Chungju
Chungking see Chongqing
161 T14 **Chungyang Shanmo** Chin. Taiwan Shan. ⛰ C Taiwan
149 V9 **Chūniān** Punjab, E Pakistan 30°57′N 74°01′E
122 L12 **Chunskiy** Irkutskaya Oblast', S Russian Federation 56°10′N 99°15′E
122 M11 **Chunya** ≈ C Russian Federation
124 J6 **Chupa** Respublika Kareliya, NW Russian Federation 66°15′N 33°02′E
125 P8 **Chuprovo** Respublika Komi, NW Russian Federation 64°16′N 46°27′E
57 G17 **Chuquibamba** Arequipa, SW Peru 15°47′S 72°44′W
62 H4 **Chuquicamata** Antofagasta, N Chile 22°20′S 68°56′W
57 L21 **Chuquisaca** ◈ department S Bolivia
Chuquisaca see Sucre
Chuqung see Chindu
146 I8 **Chuqurqoq** Rus. Chukurkak. Qoraqalpog'iston Respublikasi, NW Uzbekistan 42°44′N 61°33′E
127 T2 **Chur** Udmurtskaya Respublika, NW Russian Federation 57°06′N 52°57′E
108 I9 **Chur** Fr. Coire, It. Coira, Rmsch. Cuera, Quera; anc. Curia Rhaetorum. Graubünden, E Switzerland 46°52′N 09°32′E
123 Q10 **Churapcha** Respublika Sakha (Yakutiya), NE Russian Federation 61°59′N 132°06′E
11 V16 **Churchbridge** Saskatchewan, S Canada 50°55′N 101°53′W
21 O8 **Church Hill** Tennessee, S USA 36°31′N 82°42′W
11 X9 **Churchill** Manitoba, C Canada 58°46′N 94°10′W
11 X10 **Churchill** ≈ Manitoba/Saskatchewan, C Canada
13 P9 **Churchill** ≈ Newfoundland and Labrador, E Canada
11 Y9 **Churchill, Cape** headland Manitoba, C Canada 58°42′N 93°12′W
13 P9 **Churchill Falls** Newfoundland and Labrador, E Canada 53°38′N 64°00′W
11 S12 **Churchill Lake** ◎ Saskatchewan, C Canada
19 Q3 **Churchill Lake** ⊠ Maine, NE USA
194 I5 **Churchill Peninsula** peninsula Antarctica
22 H8 **Church Point** Louisiana, S USA 30°24′N 92°13′W
29 O3 **Churchs Ferry** North Dakota, N USA 48°15′N 99°12′W
146 G12 **Churchuri** Ahal Welaýaty, C Turkmenistan 38°55′N 59°13′E
21 T5 **Churchville** Virginia, NE USA 38°13′N 79°10′W
152 G10 **Chūru** Rājasthān, NW India 28°18′N 75°00′E
54 J4 **Churuguara** Falcón, N Venezuela 10°52′N 69°35′W
167 U11 **Chư Sê** Gia Lai, C Vietnam 13°38′N 108°06′E
144 J12 **Chushkakul, Gory** ⛰ SW Kazakhstan
Chūshū see Chungju
37 O9 **Chuska Mountains** ⛰ Arizona/New Mexico, SW USA
Chu, Sông see Sam, Nam
125 V14 **Chusovoy** Permskiy Kray, NW Russian Federation 58°17′N 57°54′E
147 R10 **Chust** Namangan Viloyati, E Uzbekistan 40°58′N 71°12′E
Chust see Khust
15 U6 **Chute-aux-Outardes** Québec, SE Canada 49°07′N 68°25′W
117 U5 **Chutove** Poltavs'ka Oblast', C Ukraine 49°45′N 35°11′E
167 U11 **Chư Ty** var. Đưc Cơ. Gia Lai, C Vietnam 13°48′N 107°41′E
189 O15 **Chuuk** var. Truk. ◈ state C Micronesia
189 P15 **Chuuk Islands** var. Hogoley Islands; prev. Truk Islands. island group Caroline Islands, C Micronesia
Chuvashia see Chuvashskaya Respublika
Chuvashiya see Chuvashskaya Respublika
127 P4 **Chuvashskaya Respublika** var. Chuvashiya, Eng. Chuvashia. ◈ autonomous republic W Russian Federation
Chuwārtah see Chwarta
Chu Xian/Chuxian see Chuzhou
160 G13 **Chuxiong** Yunnan, SW China 25°02′N 101°32′E
147 V7 **Chuy** Chuyskaya Oblast', N Kyrgyzstan 42°45′N 75°11′E
61 H19 **Chuy** var. Chuí. Rocha, E Uruguay 33°42′S 53°27′W
123 O11 **Chuya** Respublika Sakha (Yakutiya), NE Russian Federation 59°30′N 112°26′E
Chüy Oblasty see Chuyskaya Oblast'
147 U8 **Chuyskaya Oblast'** Kir. Chüy Oblasty. ◈ province N Kyrgyzstan
161 Q7 **Chuzhou** var. Chuxian, Chu Xian. Anhui, E China 32°20′N 118°18′E
139 U3 **Chwarta** Ar. Juwārtā, var. Chwārtā, Choarta, Chuwārtah. As Sulaymānīyah, NE Iraq 35°11′N 45°59′E
Chwārtā see Chwarta
119 N16 **Chyhirynskaye Vodaskhovishcha** Rus. Chigirinskoye Vodokhranilishche. ⊠ E Belarus
117 R6 **Chyhyryn** Rus. Chigirin. Cherkas'ka Oblast', N Ukraine 49°03′N 32°40′E
119 J18 **Chyrvonaya Slabada** Rus. Krasnaya Slabada, Krasnaya Sloboda. Minskaya Voblasts', S Belarus 52°51′N 27°10′E
119 L19 **Chyrvonaye, Vozyera** Rus. Ozero Chervonoye. ◎ SE Belarus
117 N11 **Ciadîr-Lunga** var. Ceadâr-Lunga, Rus. Chadyr-Lunga. S Moldova 46°03′N 28°50′E
169 P16 **Ciamis** prev. Tjiamis. Jawa, C Indonesia 07°20′S 108°21′E
107 I16 **Ciampino** ✈ Lazio, C Italy 41°48′N 12°36′E
169 N16 **Cianjur** prev. Tjiandjoer. Jawa, C Indonesia 06°50′S 107°09′E
60 H10 **Cianorte** Paraná, S Brazil 23°42′S 52°31′W
Ciarraí see Kerry
112 N13 **Ćićevac** Serbia, E Serbia 43°44′N 21°25′E
187 Z14 **Cicia** prev. Thithia. island Lau Group, E Fiji
105 P4 **Cidacos** ≈ N Spain
136 I10 **Cide** Kastamonu, N Turkey 41°53′N 33°01′E
110 L10 **Ciechanów** prev. Zichenau. Mazowieckie, C Poland 52°53′N 20°37′E
110 O10 **Ciechanowiec** Ger. Rudelstadt. Podlaskie, E Poland 52°43′N 22°30′E
110 J10 **Ciechocinek** Kujawskio-pomorskie, C Poland 52°53′N 18°49′E
44 F6 **Ciego de Ávila** Ciego de Ávila, C Cuba 21°50′N 78°44′W
54 F4 **Ciénaga** Magdalena, N Colombia 11°01′N 74°15′W
54 E6 **Ciénaga de Oro** Córdoba, NW Colombia 08°54′N 75°39′W
44 E5 **Cienfuegos** Cienfuegos, C Cuba 22°10′N 80°27′W
104 F4 **Cíes, Illas** island group NW Spain
111 P16 **Cieszanów** Podkarpackie, SE Poland 50°15′N 23°09′E
111 J17 **Cieszyn** Cz. Těšín, Ger. Teschen. Śląskie, S Poland 49°45′N 18°35′E
105 R12 **Cieza** Murcia, SE Spain 38°14′N 01°25′W
136 F13 **Çifteler** Eskişehir, W Turkey 39°25′N 31°00′E
105 P7 **Cifuentes** Castilla-La Mancha, C Spain 40°47′N 02°37′W
Čiganak see Shyganak
105 P9 **Cigüela** ≈ C Spain
136 H14 **Cihanbeyli** Konya, C Turkey 38°40′N 32°55′E
136 H14 **Cihanbeyli Yaylası** plateau C Turkey
104 L10 **Cíjara, Embalse de** ⊠ C Spain
169 P16 **Cikalong** Jawa, S Indonesia 07°46′S 108°13′E
169 N16 **Cikawung** Jawa, S Indonesia 06°49′S 105°29′E
187 Y13 **Cikobia** prev. Thikombia. island N Fiji
169 P17 **Cilacap** prev. Tjilatjap. Jawa, C Indonesia 07°44′S 109°E
173 O16 **Cilaos** C Réunion 21°08′S 55°28′E
137 S11 **Çıldır** Ardahan, NE Turkey 41°08′N 43°08′E
137 S11 **Çıldır Gölü** ◎ NE Turkey
160 M10 **Cili** Hunan, S China 29°24′N 110°59′E
Cilician Gates see Gülek Boğazı
121 V10 **Cilicia Trough** undersea feature E Mediterranean Sea
Cill Airne see Killarney
Cill Chainnigh see Kilkenny
Cill Chaoi see Kilkee
Cill Choca see Kilcock
Cill Dara see Kildare
105 N3 **Cilleruelo de Bezana** Castilla y León, N Spain 42°58′N 03°50′W
Cilli see Celje
Cill Mhantáin see Wicklow
Cill Rois see Kilrush
146 C11 **Çilmämmetgum** Rus. Peski Chil'mamedkum,, Turkm. Chilmämetgum. desert Balkan Welaýaty, W Turkmenistan
137 Z11 **Çilov Adası** Rus. Ostrov Zhiloy. island E Azerbaijan
26 J6 **Cimarron** Kansas, C USA 37°49′N 100°20′W
37 T9 **Cimarron** New Mexico, SW USA 36°30′N 104°55′W
26 M9 **Cimarron River** ≈ Kansas/Oklahoma, C USA
117 N11 **Cimişlia** Rus. Chimishliya. S Moldova 46°31′N 28°45′E
Cîmpia Turzii see Câmpia Turzii
Cîmpina see Câmpina
Cîmpulung see Câmpulung
Cîmpulung Moldovenesc see Câmpulung Moldovenesc
137 P15 **Çınar** Dıyarbakır, SE Turkey 37°45′N 40°22′E
54 J8 **Cinaruco, Río** ≈ Colombia/Venezuela
Cina Selatan, Laut see South China Sea
105 T5 **Cinca** ≈ NE Spain
112 G13 **Cincar** ▲ SW Bosnia and Herzegovina 43°55′N 17°05′E
31 Q15 **Cincinnati** Ohio, N USA 39°04′N 84°34′W
21 M4 **Cincinnati** ✈ Kentucky, S USA 39°03′N 84°39′W
Cinco de Outubro see Xá-Muteba
136 C15 **Çine** Aydın, SW Turkey 37°37′N 28°03′E
99 J21 **Ciney** Namur, SE Belgium 50°17′N 05°06′E
104 H6 **Cinfães** Viseu, N Portugal 41°04′N 08°06′W
106 J12 **Cingoli** Marche, C Italy 43°25′N 13°09′E
41 U16 **Cintalapa** var. Cintalapa de Figueroa. Chiapas, SE Mexico 16°42′N 93°40′W
Cintalapa de Figueroa see Cintalapa
103 X14 **Cinto, Monte** ▲ Corse, France, C Mediterranean Sea 42°22′N 08°57′E
Cintra see Sintra
105 Q5 **Cintruénigo** Navarra, N Spain 42°05′N 01°50′W
Cionn tSáile see Kinsale
116 K13 **Ciorani** Prahova, SE Romania 44°49′N 26°25′E
113 E14 **Čiovo** It. Bua. island S Croatia
Cipiúr see Kippure
63 I15 **Cipolletti** Río Negro, C Argentina 38°55′S 68°W
120 L7 **Circeo, Capo** headland C Italy 41°15′N 13°03′E
39 S8 **Circle** var. Circle City. Alaska, USA 65°51′N 144°04′W
33 X8 **Circle** Montana, NW USA 47°25′N 105°32′W
Circle City see Circle
31 S14 **Circleville** Ohio, N USA 39°36′N 82°57′W
36 K6 **Circleville** Utah, W USA 38°10′N 112°16′W
169 P16 **Cirebon** prev. Tjirebon. Jawa, S Indonesia 06°46′S 108°33′E
97 L21 **Cirencester** anc. Corinium, Corinium Dobunorum. C England, United Kingdom 51°44′N 01°59′W
Cirkvenica see Crikvenica
107 O20 **Cirò** Calabria, SW Italy 39°22′N 17°02′E
107 O20 **Cirò Marina** Calabria, S Italy 39°21′N 17°07′E
102 K14 **Ciron** ≈ SW France
Cirquenizza see Crikvenica
25 R7 **Cisco** Texas, SW USA 32°23′N 98°58′W
116 I12 **Cisnădie** Ger. Heltau, Hung. Nagydisznód. Sibiu, SW Romania 45°42′N 24°09′E
63 G18 **Cisnes, Río** ≈ S Chile
25 T11 **Cistern** Texas, SW USA 29°46′N 97°12′W
104 L3 **Cistierna** Castilla y León, N Spain 42°47′N 05°08′W
Citharista see la Ciotat
Citlaltépetl see Orizaba, Volcán Pico de
55 X10 **Citron** NW French Guiana 04°49′N 53°55′W
23 N7 **Citronelle** Alabama, S USA 31°05′N 88°13′W
35 O7 **Citrus Heights** California, W USA 38°42′N 121°18′W
106 H7 **Cittadella** Veneto, NE Italy 45°37′N 11°46′E
106 H13 **Città della Pieve** Umbria, C Italy 42°57′N 12°01′E
106 H12 **Città di Castello** Umbria, C Italy 43°27′N 12°13′E
107 I14 **Cittaducale** Lazio, C Italy 42°24′N 12°55′E
107 N22 **Cittanova** Calabria, SW Italy 38°21′N 16°05′E
Cittavecchia see Stari Grad
116 G10 **Ciucea** Hung. Csucsa. Cluj, NW Romania 46°58′N 22°50′E
116 M13 **Ciucurova** Tulcea, SE Romania 44°57′N 28°24′E
Ciudad Acuña see Villa Acuña
41 N15 **Ciudad Altamirano** Guerrero, S Mexico 18°20′N 100°40′W
42 G7 **Ciudad Barrios** San Miguel, NE El Salvador 13°46′N 88°13′W
54 I7 **Ciudad Bolívar** Barinas, NW Venezuela 08°22′N 70°37′W
55 N7 **Ciudad Bolívar** prev. Angostura. Bolívar, E Venezuela 08°08′N 63°31′W
40 K6 **Ciudad Camargo** Chihuahua, N Mexico 27°42′N 105°10′W
40 E8 **Ciudad Constitución** Baja California Sur, NW Mexico 25°09′N 111°43′W
Ciudad Cortés see Cortés
41 V17 **Ciudad Cuauhtémoc** Chiapas, SE Mexico 15°38′N 91°59′W
42 J9 **Ciudad Darío** var. Dario. Matagalpa, W Nicaragua 12°42′N 86°10′W
Ciudad de Dolores Hidalgo see Dolores Hidalgo
42 C6 **Ciudad de Guatemala** Eng. Guatemala City; prev. Santiago de los Caballeros. ● (Guatemala) Guatemala, C Guatemala 14°38′N 90°29′W
Ciudad del Carmen see Carmen
62 Q6 **Ciudad del Este** prev. Ciudad Presidente Stroessner, Presidente Stroessner, Puerto Presidente Stroessner. Alto Paraná, SE Paraguay 25°34′S 54°40′W
62 K5 **Ciudad de Libertador General San Martín** var. Libertador General San Martín. Jujuy, C Argentina 23°50′S 64°45′W
Ciudad Delicias see Delicias
41 O11 **Ciudad del Maíz** San Luis Potosí, C Mexico 22°26′N 99°36′W
Ciudad de México see México
54 J7 **Ciudad de Nutrias** Barinas, NW Venezuela 08°03′N 69°17′W
Ciudad de Panama see Panamá
55 P7 **Ciudad Guayana** prev. San Tomé de Guayana, Santo Tomé de Guayana. Bolívar, NE Venezuela 08°22′N 62°37′W
40 K14 **Ciudad Guzmán** Jalisco, SW Mexico 19°40′N 103°30′W
41 V17 **Ciudad Hidalgo** Chiapas, SE Mexico 14°40′N 92°11′W
41 N14 **Ciudad Hidalgo** Michoacán, SW Mexico 19°40′N 100°34′W
40 J3 **Ciudad Juárez** Chihuahua, N Mexico 31°39′N 106°26′W
40 L8 **Ciudad Lerdo** Durango, C Mexico 25°34′N 103°30′W
41 Q11 **Ciudad Madero** var. Villa Cecilia. Tamaulipas, C Mexico 22°18′N 97°56′W
41 P11 **Ciudad Mante** Tamaulipas, C Mexico 22°43′N 99°02′W
42 F2 **Ciudad Melchor de Mencos** var. Melchor de Mencos. Petén, NE Guatemala 17°03′N 89°12′W
41 P8 **Ciudad Miguel Alemán** Tamaulipas, C Mexico 26°20′N 98°56′W
Ciudad Mutis see Bahía Solano
40 G6 **Ciudad Obregón** Sonora, NW Mexico 27°32′N 109°53′W
54 I5 **Ciudad Ojeda** Zulia, NW Venezuela 10°12′N 71°17′W
55 P7 **Ciudad Piar** Bolívar, E Venezuela 07°25′N 63°19′W
Ciudad Porfirio Díaz see Piedras Negras
Ciudad Presidente Stroessner see Ciudad del Este
Ciudad Quesada see Quesada
105 N11 **Ciudad Real** Castilla-La Mancha, C Spain 38°59′N 03°55′W
105 N11 **Ciudad Real** ◈ province Castilla-La Mancha, C Spain
104 J7 **Ciudad-Rodrigo** Castilla y León, N Spain 40°36′N 06°33′W
42 A6 **Ciudad Tecún Umán** San Marcos, SW Guatemala 14°40′N 92°06′W
Ciudad Trujillo see Santo Domingo
41 P12 **Ciudad Valles** San Luis Potosí, C Mexico 21°59′N 99°01′W
41 O10 **Ciudad Victoria** Tamaulipas, C Mexico 23°44′N 99°07′W
42 C6 **Ciudad Vieja** Suchitepéquez, S Guatemala 14°30′N 90°46′W
116 L8 **Ciuhuru** var. Reuţel. ≈ N Moldova
105 Z8 **Ciutadella** var. Ciutadella de Menorca. Menorca, Spain, W Mediterranean Sea 40°N 03°50′E
Ciutadella Ciutadella de Menorca see Ciutadella
136 L11 **Civa Burnu** headland N Turkey 41°22′N 36°39′E
106 J7 **Cividale del Friuli** Friuli-Venezia Giulia, NE Italy 46°06′N 13°25′E
107 H14 **Civita Castellana** Lazio, C Italy 42°16′N 12°24′E
106 J12 **Civitanova Marche** Marche, C Italy 43°18′N 13°41′E
Civitas Altae Ripae see Brzeg
Civitas Carnutum see Chartres
Civitas Eburovicum see Évreux
Civitas Nemetum see Speyer
107 G15 **Civitavecchia** anc. Centum Cellae, Trajani Portus. Lazio, C Italy 42°05′N 11°47′E
102 L10 **Civray** Vienne, W France 46°10′N 00°18′E
136 E14 **Çivril** Denizli, W Turkey 38°18′N 29°43′E
161 O5 **Cixian** Hebei, E China 36°19′N 114°22′E
137 R16 **Cizre** Şırnak, SE Turkey 37°21′N 42°11′E
Clacton see Clacton-on-Sea
97 Q21 **Clacton-on-Sea** var. Clacton. E England, United Kingdom 51°48′N 01°09′E
22 H5 **Claiborne, Lake** ⊠ Louisiana, S USA
102 L10 **Clain** ≈ W France
11 Q11 **Claire, Lake** ◎ Alberta, C Canada
25 O6 **Clairemont** Texas, SW USA 33°09′N 100°45′W
34 M3 **Clair Engle Lake** ⊠ California, W USA
18 B15 **Clairton** Pennsylvania, NE USA 40°17′N 79°52′W
32 F7 **Clallam Bay** Washington, NW USA 48°13′N 124°16′W
103 P8 **Clamecy** Nièvre, C France 47°28′N 03°30′E
23 P5 **Clanton** Alabama, S USA 32°50′N 86°37′W
61 D17 **Clara** Entre Ríos, E Argentina 31°50′S 58°48′W
97 E18 **Clara** Ir. Clóirtheach. C Ireland 53°20′N 07°36′W
29 T9 **Clara City** Minnesota, N USA 44°57′N 95°22′W
61 D23 **Claraz** Buenos Aires, E Argentina 37°56′S 59°18′W
Clár Chlainne Mhuiris see Claremorris
182 I8 **Clare** South Australia 33°49′S 138°35′E
97 C19 **Clare** Ir. An Clár. cultural region W Ireland
97 C18 **Clare** ≈ W Ireland
97 A16 **Clare Island** Ir. Cliara. island W Ireland
44 J12 **Claremont** C Jamaica 18°23′N 77°11′W
29 W10 **Claremont** Minnesota, N USA 44°01′N 93°00′W
19 N9 **Claremont** New Hampshire, NE USA 43°21′N 72°18′W
27 Q9 **Claremore** Oklahoma, C USA 36°20′N 95°37′W
97 C17 **Claremorris** Ir. Clár Chlainne Mhuiris. W Ireland 53°47′N 09°W
185 J16 **Clarence** Canterbury, South Island, New Zealand 42°08′S 173°54′E
185 J16 **Clarence** ≈ South Island, New Zealand
65 F15 **Clarence Bay** bay Ascension Island, C Atlantic Ocean
63 H25 **Clarence, Isla** island S Chile
194 H2 **Clarence Island** island South Shetland Islands, Antarctica
183 V5 **Clarence River** ≈ New South Wales, SE Australia
44 J5 **Clarence Town** Long Island, C The Bahamas 23°03′N 74°57′W
27 W12 **Clarendon** Arkansas, C USA 34°41′N 91°19′W
25 O3 **Clarendon** Texas, SW USA 34°57′N 100°54′W
13 U12 **Clarenville** Newfoundland, Newfoundland and Labrador, SE Canada 48°10′N 54°00′W
11 Q17 **Claresholm** Alberta, SW Canada 50°02′N 113°33′W
29 T16 **Clarinda** Iowa, C USA 40°44′N 95°02′W
55 N5 **Clarines** Anzoátegui, NE Venezuela 09°56′N 65°11′W
29 V12 **Clarion** Iowa, C USA 42°43′N 93°43′W
18 C13 **Clarion** Pennsylvania, NE USA 41°11′N 79°21′W
193 O6 **Clarion Fracture Zone** tectonic feature NE Pacific Ocean
18 D13 **Clarion River** ≈ Pennsylvania, NE USA
29 Q9 **Clark** South Dakota, N USA 44°50′N 97°44′W
36 K11 **Clarkdale** Arizona, SW USA 34°46′N 112°03′W
15 W4 **Clarke City** Québec, SE Canada 50°09′N 66°36′W
183 Q15 **Clarke Island** island Furneaux Group, Tasmania, SE Australia
181 X6 **Clarke Range** ⛰ Queensland, E Australia
23 T2 **Clarkesville** Georgia, SE USA 34°36′N 83°31′W
29 S9 **Clarkfield** Minnesota, N USA 44°48′N 95°49′W
33 N7 **Clark Fork** Idaho, NW USA 48°06′N 116°10′W
33 N8 **Clark Fork** ≈ Idaho/Montana, NW USA
21 P13 **Clark Hill Lake** var. J.Storm Thurmond Reservoir. ⊠ Georgia/South Carolina, SE USA
39 Q12 **Clark, Lake** ◎ Alaska, USA
35 W12 **Clark Mountain** ▲ California, W USA 35°30′N 115°34′W
37 S3 **Clark Peak** ▲ Colorado, C USA 40°36′N 105°57′W
14 D14 **Clark, Point** headland Ontario, S Canada 44°04′N 81°45′W
21 S3 **Clarksburg** West Virginia, NE USA 39°16′N 80°22′W
22 K2 **Clarksdale** Mississippi, S USA 34°12′N 90°34′W
33 U12 **Clarks Fork Yellowstone River** ≈ Montana/Wyoming, NW USA
29 R14 **Clarkson** Nebraska, C USA 41°42′N 97°07′W
39 O13 **Clarks Point** Alaska, USA 58°50′N 158°33′W
18 I13 **Clarks Summit** Pennsylvania, NE USA 41°29′N 75°42′W
32 M10 **Clarkston** Washington, NW USA 46°25′N 117°02′W
44 J12 **Clark's Town** C Jamaica 18°25′N 77°32′W
27 T10 **Clarksville** Arkansas, C USA 35°29′N 93°29′W
31 P13 **Clarksville** Indiana, N USA 40°01′N 85°54′W
20 I8 **Clarksville** Tennessee, S USA 36°32′N 87°22′W
25 W5 **Clarksville** Texas, SW USA 33°37′N 95°04′W
21 U8 **Clarksville** Virginia, NE USA 36°36′N 78°36′W
21 U11 **Clarkton** North Carolina, SE USA 34°28′N 78°39′W
61 C24 **Claromecó** var. Balneario Claromecó. Buenos Aires, E Argentina 38°51′S 60°01′W
25 N3 **Claude** Texas, SW USA 35°06′N 101°22′W
Clausentum see Southampton
171 O1 **Claveria** Luzon, N Philippines 18°36′N 121°04′E
99 J20 **Clavier** Liège, E Belgium 50°27′N 05°21′E

◆ Country
● Country Capital
◇ Dependent Territory
○ Dependent Territory Capital
◈ Administrative Regions
✈ International Airport
▲ Mountain
⛰ Mountain Range
Volcano
≈ River
◎ Lake
⊠ Reservoir

23 W6 Claxton Georgia, SE USA 32°09′N 81°54′W
21 R4 Clay West Virginia, NE USA 38°28′N 81°17′W
27 N3 Clay Center Kansas, C USA 39°22′N 97°08′W
29 P16 Clay Center Nebraska, C USA 40°31′N 98°03′W
21 Y2 Claymont Delaware, NE USA 39°48′N 75°27′W
36 M14 Claypool Arizona, SW USA 33°24′N 110°50′W
23 R6 Clayton Alabama, S USA 31°52′N 85°27′W
23 T1 Clayton Georgia, SE USA 34°52′N 83°24′W
22 J5 Clayton Louisiana, S USA 31°43′N 91°32′W
27 X5 Clayton Missouri, C USA 38°39′N 90°21′W
37 V9 Clayton New Mexico, SW USA 36°27′N 103°12′W
21 V9 Clayton North Carolina, SE USA 35°39′N 78°27′W
27 Q12 Clayton Oklahoma, C USA 34°35′N 95°21′W
45 V9 Clayton J. Lloyd ✈ (The Valley) C Anguilla 18°12′N 63°02′W
182 I4 Clayton River seasonal river South Australia
21 R7 Claytor Lake ⊠ Virginia, NE USA
27 P13 Clear Boggy Creek ≈ Oklahoma, C USA
97 B22 Clear, Cape var. The Bill of Cape Clear, Ir. Ceann Cléire. headland SW Ireland 51°25′N 09°31′W
36 M12 Clear Creek ≈ Arizona, SW USA
39 S12 Cleare, Cape headland Montague Island, Alaska, USA 59°46′N 147°54′W
18 E13 Clearfield Pennsylvania, NE USA 41°02′N 78°27′W
36 L2 Clearfield Utah, W USA 41°06′N 112°03′W
25 Q6 Clear Fork Brazos River ≈ Texas, SW USA
31 T12 Clear Fork Reservoir ⊠ Ohio, N USA
11 N12 Clear Hills ▲ Alberta, SW Canada
34 M6 Clearlake California, W USA 38°57′N 122°38′W
29 V12 Clear Lake Iowa, C USA 43°07′N 93°27′W
29 R9 Clear Lake South Dakota, N USA 44°45′N 96°40′W
34 M6 Clear Lake ◎ California, W USA
22 G6 Clear Lake ⊠ Louisiana, S USA
35 P1 Clear Lake Reservoir ⊠ California, W USA
11 N16 Clearwater British Columbia, SW Canada 51°38′N 120°02′W
23 U12 Clearwater Florida, SE USA 27°58′N 82°46′W
11 R12 Clearwater ≈ Alberta/ Saskatchewan, C Canada
27 W7 Clearwater Lake ⊠ Missouri, C USA
33 N10 Clearwater Mountains ▲ Idaho, NW USA
33 N10 Clearwater River ≈ Idaho, NW USA
29 S4 Clearwater River ≈ Minnesota, N USA
25 T7 Cleburne Texas, SW USA 32°21′N 97°24′W
32 I9 Cle Elum Washington, NW USA 47°12′N 120°56′W
97 O17 Cleethorpes E England, United Kingdom 53°34′N 00°02′W
Cléire, Ceann see Clear, Cape
21 O11 Clemson South Carolina, SE USA 34°40′N 82°50′W
21 Q4 Clendenin West Virginia, NE USA 38°29′N 81°21′W
26 M9 Cleo Springs Oklahoma, C USA 36°25′N 98°25′W
Clerk Island see Onotoa
181 X8 Clermont Queensland, E Australia 22°47′S 147°41′E
15 S8 Clermont Québec, SE Canada 47°41′N 70°15′W
103 O4 Clermont Oise, N France 49°23′N 02°26′E
29 X12 Clermont Iowa, C USA 43°00′N 91°39′W
103 P11 Clermont-Ferrand Puy-de-Dôme, C France 45°47′N 03°05′E
103 Q15 Clermont-l'Hérault Hérault, S France 43°37′N 03°25′E
99 M22 Clervaux Diekirch, N Luxembourg 50°03′N 06°02′E
106 G6 Cles Trentino-Alto Adige, N Italy 46°22′N 11°04′E
182 H8 Cleve South Australia 33°43′S 136°30′E
Cleve see Kleve
23 T2 Cleveland Georgia, SE USA 34°36′N 83°45′W
22 K3 Cleveland Mississippi, S USA 33°45′N 90°43′W
31 T11 Cleveland Ohio, N USA 41°30′N 81°42′W
27 O9 Cleveland Oklahoma, C USA 36°18′N 96°27′W
20 L10 Cleveland Tennessee, S USA 35°10′N 84°51′W
25 W10 Cleveland Texas, SW USA 30°19′N 95°06′W
31 N7 Cleveland Wisconsin, N USA 43°58′N 87°45′W
31 O4 Cleveland Cliffs Basin ◎ Michigan, N USA
31 U11 Cleveland Heights Ohio, N USA 41°30′N 81°34′W
33 P6 Cleveland, Mount ▲ Montana, NW USA 48°55′N 113°51′W
Cleves see Kleve
97 B16 Clew Bay Ir. Cuan Mó. inlet W Ireland
23 Y14 Clewiston Florida, SE USA 26°45′N 80°55′W
Cliara see Clare Island
97 A17 Clifden Ir. An Clochán. Galway, W Ireland 53°29′N 10°14′W
37 O14 Clifton Arizona, SW USA 33°03′N 109°18′W
18 K14 Clifton New Jersey, NE USA 40°50′N 74°28′W
25 S8 Clifton Texas, SW USA 31°43′N 97°36′W
21 S6 Clifton Forge Virginia, NE USA 37°49′N 79°50′W
182 I1 Clifton Hills South Australia 27°03′S 138°49′E
11 S17 Climax Saskatchewan, S Canada 49°12′N 108°22′W
21 O8 Clinch River ≈ Tennessee/ Virginia, S USA
25 P12 Cline Texas, SW USA 29°14′N 100°07′W
21 N10 Clingmans Dome ▲ North Carolina/Tennessee, SE USA 35°33′N 83°30′W
24 H8 Clint Texas, SW USA 31°35′N 106°13′W
10 M16 Clinton British Columbia, SW Canada 51°06′N 121°31′W
14 E15 Clinton Ontario, S Canada 43°36′N 81°33′W
27 U10 Clinton Arkansas, C USA 35°34′N 92°28′W
30 L10 Clinton Illinois, N USA 40°09′N 88°57′W
29 Z14 Clinton Iowa, C USA 41°50′N 90°11′W
20 G7 Clinton Kentucky, S USA 36°39′N 89°00′W
22 J8 Clinton Louisiana, S USA 30°52′N 91°01′W
19 N11 Clinton Massachusetts, NE USA 42°25′N 71°40′W
31 R10 Clinton Michigan, N USA 42°04′N 83°58′W
22 K5 Clinton Mississippi, S USA 32°22′N 90°22′W
27 S5 Clinton Missouri, C USA 38°22′N 93°51′W
21 V10 Clinton North Carolina, SE USA 35°00′N 78°19′W
26 L10 Clinton Oklahoma, C USA 35°31′N 98°58′W
21 Q12 Clinton South Carolina, SE USA 34°28′N 81°52′W
21 M9 Clinton Tennessee, S USA 36°07′N 84°08′W
8 L9 Clinton-Colden Lake ◎ Northwest Territories, NW Canada
10 H5 Clinton Creek Yukon, NW Canada 64°24′N 140°35′W
30 L13 Clinton Lake ⊠ Illinois, N USA
27 Q4 Clinton Lake ⊠ Kansas, N USA
21 T11 Clio South Carolina, SE USA 34°34′N 79°33′W
193 O7 Clipperton Fracture Zone tectonic feature E Pacific Ocean
193 Q7 Clipperton Island ◇ French overseas territory E Pacific Ocean
46 K6 Clipperton Island island E Pacific Ocean
0 F16 Clipperton Seamounts undersea feature E Pacific Ocean 08°00′N 111°00′W
102 J8 Clisson Loire-Atlantique, NW France 47°06′N 01°19′W
62 K7 Clodomira Santiago del Estero, N Argentina 27°35′S 64°14′W
Cloich na Coillte see Clonakilty
Clóirtheach see Clara
97 C21 Clonakilty Ir. Cloich na Coillte. SW Ireland 51°37′N 08°54′W
181 T6 Cloncurry Queensland, C Australia 20°45′S 140°30′E
97 F18 Clondalkin Ir. Cluain Dolcáin. E Ireland 53°19′N 06°24′W
97 E16 Clones Ir. Cluain Eois. N Ireland 54°11′N 07°14′W
97 D20 Clonmel Ir. Cluain Meala. S Ireland 52°21′N 07°42′W
100 G11 Cloppenburg Niedersachsen, NW Germany 52°51′N 08°03′E
29 W6 Cloquet Minnesota, N USA 46°43′N 92°27′W
37 S14 Cloudcroft New Mexico, SW USA 32°57′N 105°44′W
33 W12 Cloud Peak ▲ Wyoming, C USA 44°22′N 107°10′W
185 K14 Cloudy Bay inlet South Island, New Zealand
21 R10 Clover South Carolina, SE USA 35°06′N 81°13′W
34 M6 Cloverdale California, W USA 38°49′N 123°03′W
20 J5 Cloverport Kentucky, S USA 37°50′N 86°37′W
35 Q10 Clovis California, W USA 36°48′N 119°43′W
37 W12 Clovis New Mexico, SW USA 34°22′N 103°12′W
14 K13 Cloyne Ontario, SE Canada 44°48′N 77°09′W
Cluain Dolcáin see Clondalkin
Cluain Eois see Clones
Cluainín see Manorhamilton
Cluain Meala see Clonmel
116 H10 Cluj ◆ county NW Romania
Cluj see Cluj-Napoca
116 H10 Cluj-Napoca Ger. Klausenburg, Hung. Kolozsvár; prev. Cluj. Cluj, NW Romania 46°47′N 23°36′E
Clunia see Feldkirch
103 R10 Cluny Saône-et-Loire, C France 46°25′N 04°38′E
103 T10 Cluses Haute-Savoie, E France 46°04′N 06°34′E
106 E7 Clusone Lombardia, N Italy 45°56′N 10°00′E
25 W12 Clute Texas, SW USA 29°01′N 95°24′W
185 D23 Clutha ≈ South Island, New Zealand
97 J18 Clwyd cultural region NE Wales, United Kingdom
185 D22 Clyde Otago, South Island, New Zealand 45°12′S 169°21′E
27 N3 Clyde Kansas, C USA 39°35′N 97°24′W
29 P2 Clyde North Dakota, N USA 48°44′N 98°51′W
31 S11 Clyde Ohio, N USA 41°18′N 82°58′W
25 Q7 Clyde Texas, SW USA 32°24′N 99°29′W
14 K13 Clyde ≈ Ontario, SE Canada
96 J13 Clyde ≈ W Scotland, United Kingdom
96 H12 Clydebank S Scotland, United Kingdom 55°54′N 04°24′W
96 H13 Clyde, Firth of inlet S Scotland, United Kingdom
33 S11 Clyde Park Montana, NW USA 45°56′N 110°39′W
35 W16 Coachella California, W USA 33°38′N 116°10′W
35 W16 Coachella Canal canal California, W USA
40 I9 Coacoyole Durango, C Mexico 24°30′N 106°33′W
25 N7 Coahoma Texas, SW USA 32°18′N 101°18′W
10 K8 Coal ≈ Yukon, NW Canada
40 L14 Coalcomán var. Coalcomán de Matamoros. Michoacán, S Mexico 18°49′N 103°13′W
Coalcomán de Matamoros see Coalcomán
39 T8 Coal Creek Alaska, USA 65°21′N 143°08′W
11 Q17 Coaldale Alberta, SW Canada 49°42′N 112°36′W
27 P12 Coalgate Oklahoma, C USA 34°33′N 96°15′W
35 P11 Coalinga California, W USA 36°08′N 120°21′W
10 L9 Coal River British Columbia, W Canada 59°38′N 126°45′W
21 Q6 Coal River ≈ West Virginia, NE USA
36 M2 Coalville Utah, W USA 40°56′N 111°22′W
58 E13 Coari Amazonas, N Brazil 04°08′S 63°07′W
104 I7 Côa, Rio ≈ N Portugal
59 D14 Coari, Rio ≈ NW Brazil
Coast see Pwani
10 G12 Coast Mountains Fr. Chaîne Côtière. ▲ Canada/USA
16 C7 Coast Ranges ▲ W USA
96 I12 Coatbridge S Scotland, United Kingdom 55°52′N 04°01′W
42 B6 Coatepeque Quezaltenango, SW Guatemala 14°42′N 91°50′W
18 H16 Coatesville Pennsylvania, NE USA 39°58′N 75°47′W
15 Q13 Coaticook Québec, SE Canada 45°07′N 71°46′W
9 P9 Coats Island island Nunavut, NE Canada
195 O4 Coats Land physical region Antarctica
41 T14 Coatzacoalcos var. Quetzalcoalco; prev. Puerto México. Veracruz-Llave, E Mexico 18°06′N 94°26′W
41 S14 Coatzacoalcos, Río ≈ SE Mexico
116 M15 Cobadin Constanța, SW Romania 44°05′N 28°13′E
14 H9 Cobalt Ontario, S Canada 47°24′N 79°41′W
42 D5 Cobán Alta Verapaz, C Guatemala 15°28′N 90°20′W
183 O6 Cobar New South Wales, SE Australia 31°31′S 145°51′E
18 F12 Cobb Hill ▲ Pennsylvania, NE USA 41°52′N 77°52′W
0 D8 Cobb Seamount undersea feature E Pacific Ocean 47°00′N 131°00′W
14 K12 Cobden Ontario, SE Canada 45°36′N 76°54′W
97 D21 Cobh Ir. An Cóbh; prev. Cove of Cork, Queenstown. SW Ireland 51°51′N 08°17′W
57 J14 Cobija Pando, NW Bolivia 11°04′S 68°49′W
Coblence/Coblenz see Koblenz
18 J10 Cobleskill New York, NE USA 42°40′N 74°29′W
14 I15 Cobourg Ontario, SE Canada 43°57′N 78°06′W
181 P1 Cobourg Peninsula headland Northern Territory, N Australia 11°27′S 132°33′E
183 O10 Cobram Victoria, SE Australia 35°56′S 145°36′E
82 N13 Cóbuè Niassa, N Mozambique 12°08′S 34°46′E
101 K18 Coburg Bayern, SE Germany 50°16′N 10°58′E
19 Q5 Coburn Mountain ▲ Maine, NE USA 45°28′N 70°07′W
57 H18 Cocachacra Arequipa, SW Peru 17°05′S 71°45′W
59 J17 Cocalinho Mato Grosso, W Brazil 14°22′S 51°00′W
Cocanada see Kākināda
105 S11 Cocentaina Valenciana, E Spain 38°44′N 00°27′W
57 L18 Cochabamba hist. Oropeza. Cochabamba, C Bolivia 17°23′S 66°10′W
57 K18 Cochabamba ◆ department C Bolivia
57 L18 Cochabamba, Cordillera de ▲ C Bolivia
101 E18 Cochem Rheinland-Pfalz, W Germany 50°09′N 07°09′E
37 R6 Cochetopa Hills ▲ Colorado, C USA
155 G22 Cochin var. Kochchi, Kochi. Kerala, SW India 09°56′N 76°15′E see also Kochi
44 D5 Cochinos, Bahía de Eng. Bay of Pigs. bay SE Cuba
37 O16 Cochise Head ▲ Arizona, SW USA 32°03′N 109°19′W
23 U5 Cochran Georgia, SE USA 32°23′N 83°21′W
11 P16 Cochrane Alberta, SW Canada 51°15′N 114°25′W
12 G12 Cochrane Ontario, S Canada 49°04′N 81°02′W
63 G20 Cochrane Aisén, S Chile 47°16′S 72°33′W
11 U10 Cochrane ≈ Manitoba/ Saskatchewan, C Canada
Cochrane, Lago see Pueyrredón, Lago
Cocibolca see Nicaragua, Lago de
Cockade State see Maryland
44 M6 Cockburn Harbour South Caicos, S Turks and Caicos Islands 21°28′N 71°30′W
14 C11 Cockburn Island island Ontario, S Canada
44 J3 Cockburn Town San Salvador, E The Bahamas 24°01′N 74°31′W
21 X2 Cockeysville Maryland, NE USA 39°29′N 76°34′W
181 N12 Cocklebiddy Western Australia 32°02′S 125°54′E
44 I12 Cockpit Country, The physical region W Jamaica
43 S16 Coclé off. Provincia de Coclé. ◆ province C Panama
43 S15 Coclé del Norte Colón, C Panama 09°04′N 80°32′W
Coclé, Provincia de see Coclé
23 Y12 Cocoa Florida, SE USA 28°23′N 80°44′W
23 Y12 Cocoa Beach Florida, SE USA 28°19′N 80°36′W
79 D17 Cocobeach Estuaire, NW Gabon 00°59′N 09°34′E
44 G5 Coco, Cayo island C Cuba
151 Q19 Coco Channel strait Andaman Sea/Bay of Bengal
173 N6 Coco-de-Mer Seamounts undersea feature W Indian Ocean 0°30′N 56°00′E
36 K10 Coconino Plateau plain Arizona, SW USA
43 N6 Coco, Río var. Río Wanki, Segoviao Wangkí. ≈ Honduras/Nicaragua
173 T7 Cocos Basin undersea feature E Indian Ocean 05°00′S 94°00′E
188 B17 Cocos Island island S Guam
Cocos Island Ridge see Cocos Ridge
129 S17 Cocos Islands island group E Indian Ocean
173 T8 Cocos (Keeling) Islands ◇ Australian external territory E Indian Ocean
0 G15 Cocos Plate tectonic feature
193 T7 Cocos Ridge var. Cocos Island Ridge. undersea feature E Pacific Ocean 05°30′N 86°00′W
40 K13 Cocula Jalisco, SW Mexico 20°22′N 103°50′W
107 D17 Coda Cavallo, Capo headland Sardegna, Italy, C Mediterranean Sea 40°49′N 09°43′E
58 E13 Codajás Amazonas, N Brazil 03°50′S 62°12′W
Codazzi see Agustín Codazzi
19 Q12 Cod, Cape headland Massachusetts, NE USA 41°50′N 69°56′W
185 B25 Codfish Island island SW New Zealand
106 H9 Codigoro Emilia-Romagna, N Italy 44°50′N 12°07′E
13 P5 Cod Island island Newfoundland and Labrador, E Canada
116 J12 Codlea Ger. Zeiden, Hung. Feketehalom. Brașov, C Romania 45°43′N 25°27′E
58 M13 Codó Maranhão, E Brazil 04°28′S 43°51′W
106 E8 Codogno Lombardia, N Italy 45°10′N 09°42′E
116 M10 Codrii hill range C Moldova
45 W9 Codrington Barbuda, Antigua and Barbuda 17°43′N 61°49′W
106 J7 Codroipo Friuli-Venezia Giulia, NE Italy 45°58′N 13°00′E
28 M12 Cody Nebraska, C USA 42°54′N 101°13′W
33 U12 Cody Wyoming, C USA 43°31′N 109°04′W
21 P7 Coeburn Virginia, NE USA 36°56′N 82°27′W
54 E10 Coello Tolima, W Colombia 04°15′N 74°52′W
Coemba see Cuemba
181 V2 Coen Queensland, NE Australia 14°03′S 143°16′E
101 E14 Coesfeld Nordrhein-Westfalen, W Germany 51°55′N 07°10′E
32 M8 Coeur d'Alene Idaho, NW USA 47°40′N 116°46′W
32 M8 Coeur d'Alene Lake ⊠ Idaho, NW USA
98 O8 Coevorden Drenthe, NE Netherlands 52°39′N 06°45′E
10 H6 Coffee Creek Yukon, W Canada 62°52′N 139°05′W
30 L15 Coffeen Lake ⊠ Illinois, N USA
22 L3 Coffeeville Mississippi, S USA 33°58′N 89°40′W
27 Q8 Coffeyville Kansas, C USA 37°02′N 95°37′W
182 G9 Coffin Bay South Australia 34°39′S 135°30′E
182 F9 Coffin Bay Peninsula peninsula South Australia
183 V5 Coffs Harbour New South Wales, SE Australia 30°18′S 153°08′E
105 R10 Cofrentes Valenciana, E Spain 39°14′N 01°04′W
117 N10 Cogîlnic Ukr. Kohyl'nyk. ≈ Moldova/Ukraine
102 K11 Cognac anc. Compniacum. Charente, W France 45°42′N 00°19′W
106 B7 Cogne Valle d'Aosta, NW Italy 45°37′N 07°27′E
103 U16 Cogolin Var, SE France 43°15′N 06°30′E
105 O7 Cogolludo Castilla-La Mancha, C Spain 40°58′N 03°05′W
Cohalm see Rupea
92 K8 Čohkarášša var. Cuokkaraš ša. ▲ N Norway 69°57′N 24°38′E
Čohkkiras see Jukkasjärvi
18 F11 Cohocton River ≈ New York, NE USA
18 L10 Cohoes New York, NE USA 42°46′N 73°42′W
183 N10 Cohuna Victoria, SE Australia 35°51′S 144°15′E
43 P17 Coiba, Isla de island SW Panama
63 H23 Coig, Río ≈ S Argentina
63 G19 Coihaique var. Coyhaique. Aisén, S Chile 45°32′S 72°00′W
155 G21 Coimbatore Tamil Nādu, S India 11°N 76°57′E
104 G8 Coimbra anc. Conimbria, Conimbriga. Coimbra, W Portugal 40°12′N 08°25′W
104 G8 Coimbra ◆ district N Portugal
104 L15 Coín Andalucía, S Spain 36°40′N 04°45′W
Coin de Mire see Gunner's Quoin
57 J20 Coipasa, Laguna ◎ W Bolivia
57 J20 Coipasa, Salar de salt lake W Bolivia
Coira/Coire see Chur
Coirib, Loch see Corrib, Lough
54 K6 Cojedes off. Estado Cojedes. ◆ state N Venezuela
Cojedes, Estado see Cojedes
42 F7 Cojutepeque Cuscatlán, C El Salvador 13°43′N 88°56′W
Coka see Banbar
33 S16 Cokeville Wyoming, C USA 42°03′N 110°55′W
182 M13 Colac Victoria, SE Australia 38°22′S 143°38′E
59 O20 Colatina Espírito Santo, SE Brazil 19°35′S 40°37′W
27 O13 Colbert Oklahoma, C USA 33°51′N 96°30′W
100 L12 Colbitz-Letzlinger Heide heathland N Germany
26 I3 Colby Kansas, C USA 39°24′N 101°04′W
57 H17 Colca, Río ≈ SW Peru
97 P21 Colchester hist. Colneceaste; anc. Camulodunum. E England, United Kingdom 51°54′N 00°54′E
19 N13 Colchester Connecticut, NE USA 41°34′N 72°17′W
38 M16 Cold Bay Alaska, USA 55°11′N 162°43′W
11 R14 Cold Lake Alberta, SW Canada 54°26′N 110°16′W
11 R13 Cold Lake ◎ Alberta/ Saskatchewan, S Canada
29 U8 Cold Spring Minnesota, N USA 45°27′N 94°25′W
25 W10 Coldspring Texas, SW USA 30°34′N 95°10′W
11 N17 Coldstream British Columbia, SW Canada 50°13′N 119°09′W
96 L13 Coldstream SE Scotland, United Kingdom 55°39′N 02°19′W
14 H13 Coldwater Ontario, S Canada 44°43′N 79°36′W
26 K7 Coldwater Kansas, C USA 37°16′N 99°20′W
31 Q10 Coldwater Michigan, N USA 41°56′N 85°00′W
25 N1 Coldwater Creek ≈ Oklahoma/Texas, SW USA
22 K2 Coldwater River ≈ Mississippi, S USA
183 O9 Coleambally New South Wales, SE Australia 34°48′S 145°54′E
19 O6 Colebrook New Hampshire, NE USA 44°52′N 71°27′W
27 T5 Cole Camp Missouri, C USA 38°27′N 93°12′W
39 T6 Coleen River ≈ Alaska, USA
11 P17 Coleman Alberta, SW Canada 49°36′N 114°26′W
25 Q8 Coleman Texas, SW USA 31°50′N 99°27′W
Çölemerik see Hakkâri
83 K22 Colenso KwaZulu/Natal, E South Africa 28°44′S 29°50′E
182 L12 Coleraine Victoria, SE Australia 37°39′S 141°42′E
97 F14 Coleraine Ir. Cúil Raithin. N Northern Ireland, United Kingdom 55°08′N 06°40′W
185 G18 Coleridge, Lake ◎ South Island, New Zealand
83 H24 Colesberg Northern Cape, C South Africa 30°41′S 25°08′E
22 H7 Colfax Louisiana, S USA 31°31′N 92°42′W
32 L9 Colfax Washington, NW USA 46°52′N 117°21′W
30 J6 Colfax Wisconsin, N USA 45°00′N 91°44′W
63 I19 Colhué Huapí, Lago ◎ S Argentina
45 Z6 Colibris, Pointe des headland Grande Terre, E Guadeloupe 16°15′N 61°10′W
106 D6 Colico Lombardia, N Italy 46°08′N 09°24′E
99 E14 Colijnsplaat Zeeland, SW Netherlands 51°36′N 03°47′E
40 L14 Colima Colima, S Mexico 19°13′N 103°46′W
40 K14 Colima ◆ state SW Mexico
40 L14 Colima, Nevado de ⛰ C Mexico 19°36′N 103°36′W
59 M14 Colinas Maranhão, E Brazil 06°02′S 44°15′W
96 F10 Coll island W Scotland, United Kingdom
105 N7 Collado Villalba var. Villalba. Madrid, C Spain 40°38′N 04°00′W
183 R4 Collarenebri New South Wales, SE Australia 29°31′S 148°33′E
37 P5 Collbran Colorado, C USA 39°14′N 107°57′W
106 G12 Colle di Val d'Elsa Toscana, C Italy 43°26′N 11°06′E
39 R9 College Alaska, USA 64°49′N 148°06′W
32 K10 College Place Washington, NW USA 46°03′N 118°23′W
25 U10 College Station Texas, SW USA 30°38′N 96°21′W
183 P4 Collerina New South Wales, SE Australia 29°41′S 146°36′E
180 I13 Collie Western Australia 33°20′S 116°06′E
180 L4 Collier Bay bay Western Australia
21 F10 Collierville Tennessee, S USA 35°02′N 89°39′W
106 F11 Collina, Passo della pass C Italy
14 G14 Collingwood Ontario, S Canada 44°30′N 80°14′W
184 I13 Collingwood Tasman, South Island, New Zealand 40°40′S 172°40′E
22 L7 Collins Mississippi, S USA 31°39′N 89°33′W
30 K15 Collinsville Illinois, N USA 38°40′N 89°58′W
27 P9 Collinsville Oklahoma, C USA 36°21′N 95°50′W
20 H10 Collinwood Tennessee, S USA 35°10′N 87°44′W
Collipo see Leiria
63 G14 Collipulli Araucanía, C Chile 37°55′S 72°30′W
97 D16 Collooney Ir. Cúil Mhuine. NW Ireland 54°11′N 08°29′W
29 R10 Colman South Dakota, N USA 43°58′N 96°48′W
103 U6 Colmar Ger. Kolmar. Haut-Rhin, NE France 48°05′N 07°21′E
104 M15 Colmenar Andalucía, S Spain 36°54′N 04°20′W
Colmenar see Colmenar de Oreja
105 O9 Colmenar de Oreja var. Colmenar. Madrid, C Spain 40°06′N 03°25′W
105 N7 Colmenar Viejo Madrid, C Spain 40°39′N 03°46′W
25 X9 Colmesneil Texas, SW USA 30°54′N 94°25′W
Cöln see Köln
Colneceaste see Colchester
59 G15 Colniza Mato Grosso, W Brazil 09°16′S 59°25′W
Cologne see Köln
42 B6 Colomba Quezaltenango, SW Guatemala 14°45′N 91°39′W
Colomb-Béchar see Béchar
54 E11 Colombia Huila, C Colombia 03°24′N 74°49′W
54 G10 Colombia off. Republic of Colombia. ◆ republic N South America
64 E12 Colombian Basin undersea feature SW Caribbean Sea 13°00′N 76°00′W
Colombia, Republic of see Colombia
Colombie-Britannique see British Columbia
15 T6 Colombier Québec, SE Canada 48°51′N 68°52′W
155 J25 Colombo ● (Sri Lanka) Western Province, W Sri Lanka 06°55′N 79°52′E
155 J25 Colombo ✈ Western Province, SW Sri Lanka 06°50′N 79°59′E
29 N11 Colome South Dakota, N USA 43°13′N 99°42′W
61 B19 Colón Buenos Aires, E Argentina 33°53′S 61°06′W
61 D18 Colón Entre Ríos, E Argentina 32°10′S 58°16′W
44 D5 Colón Matanzas, C Cuba 22°43′N 80°54′W
43 T14 Colón prev. Aspinwall. Colón, C Panama 09°04′N 80°33′W
42 K5 Colón ◆ department NE Honduras
43 S15 Colón off. Provincia de Colón. ◆ province N Panama
57 A16 Colón, Archipiélago de var. Islas de los Galápagos, Eng. Galapagos Islands, Tortoise Islands. island group Ecuador, E Pacific Ocean
44 K5 Colonel Hill Crooked Island, SE The Bahamas 22°43′N 74°12′W
40 C3 Colonet Baja California Norte, NW Mexico 31°00′N 116°11′W
40 B3 Colonet, Cabo headland NW Mexico 30°57′N 116°19′W
188 G14 Colonia Yap, W Micronesia 09°29′N 138°06′E
61 D19 Colonia ◆ department SW Uruguay
Colonia see Kolonia, Micronesia
Colonia see Colonia del Sacramento, Uruguay
Colonia Agrippina see Köln
61 D20 Colonia del Sacramento var. Colonia. Colonia, SW Uruguay 34°29′S 57°48′W
62 L8 Colonia Dora Santiago del Estero, N Argentina 28°34′S 62°59′W
Colonia Julia Fanestris see Fano
21 W5 Colonial Beach Virginia, NE USA 38°15′N 76°57′W
21 V6 Colonial Heights Virginia, NE USA 37°15′N 77°24′W
Colón, Provincia de see Colón
193 S7 Colón Ridge undersea feature E Pacific Ocean 02°00′N 96°00′W
96 F12 Colonsay island W Scotland, United Kingdom
57 K22 Colorada, Laguna ◎ SW Bolivia
37 R6 Colorado off. State of Colorado, also known as Centennial State, Silver State. ◆ state C USA
63 H22 Colorado, Cerro ▲ S Argentina 49°58′S 71°38′W
25 O7 Colorado City Texas, SW USA 32°24′N 100°51′W
36 M7 Colorado Plateau plateau W USA
61 A24 Colorado, Río ≈ E Argentina
43 N12 Colorado, Río ≈ NE Costa Rica
Colorado, Río see Colorado River
16 F12 Colorado River var. Río Colorado. ≈ Mexico/USA
16 K14 Colorado River ≈ Texas, SW USA
35 W15 Colorado River Aqueduct aqueduct California, W USA
44 A4 Colorados, Archipiélago de los island group NW Cuba
62 J9 Colorados, Desagües de los ◎ W Argentina
37 T5 Colorado Springs Colorado, C USA 38°50′N 104°47′W
40 L11 Colotlán Jalisco, SW Mexico 22°08′N 103°15′W
57 L19 Colquechaca Potosí, C Bolivia 18°40′S 66°00′W
23 S7 Colquitt Georgia, SE USA 31°10′N 84°43′W
29 R11 Colton South Dakota, N USA 43°47′N 96°55′W
32 M10 Colton Washington, NW USA 46°34′N 117°10′W
35 P8 Columbia California, W USA 38°01′N 120°22′W
30 K16 Columbia Illinois, N USA 38°26′N 90°12′W
20 L7 Columbia Kentucky, S USA 37°05′N 85°19′W
22 I6 Columbia Louisiana, S USA 32°05′N 92°03′W
21 W3 Columbia Maryland, NE USA 39°13′N 76°51′W
22 L7 Columbia Mississippi, S USA 31°15′N 89°50′W
27 U4 Columbia Missouri, C USA 38°56′N 92°19′W
21 Y9 Columbia North Carolina, SE USA 35°55′N 76°15′W
18 G16 Columbia Pennsylvania, NE USA 40°01′N 76°30′W
21 Q12 Columbia state capital South Carolina, SE USA 34°00′N 81°02′W
20 I9 Columbia Tennessee, S USA 35°37′N 87°02′W
0 F9 Columbia ≈ Canada/USA
32 K9 Columbia Basin basin Washington, NW USA
197 Q10 Columbia, Cape headland Ellesmere Island, Nunavut, NE Canada
31 Q12 Columbia City Indiana, N USA 41°09′N 85°29′W
21 W3 Columbia, District of ◆ federal district NE USA
33 P7 Columbia Falls Montana, NW USA 48°22′N 114°10′W
11 O15 Columbia Icefield ice field Alberta/British Columbia, S Canada
11 O15 Columbia, Mount ▲ Alberta/British Columbia, SW Canada 52°07′N 117°30′W
11 N15 Columbia Mountains ▲ British Columbia, SW Canada
23 P4 Columbiana Alabama, S USA 33°10′N 86°36′W
31 V12 Columbiana Ohio, N USA 40°53′N 80°41′W
32 M14 Columbia Plateau plateau Idaho/Oregon, NW USA
29 P7 Columbia Road Reservoir ⊠ South Dakota, N USA
65 K16 Columbia Seamount undersea feature C Atlantic Ocean 20°30′S 32°00′W
83 D25 Columbine, Cape headland SW South Africa 32°50′S 17°39′E
105 U9 Columbretes, Illes prev. Islas Columbretes. island group E Spain
Columbretes, Islas see Columbretes, Illes
23 R5 Columbus Georgia, SE USA 32°29′N 84°58′W
31 P14 Columbus Indiana, N USA 39°12′N 85°55′W
27 R7 Columbus Kansas, C USA 37°09′N 94°52′W
23 N4 Columbus Mississippi, S USA 33°30′N 88°25′W
33 U11 Columbus Montana, NW USA 45°38′N 109°15′W
29 Q15 Columbus Nebraska, C USA 41°25′N 97°22′W
37 Q16 Columbus New Mexico, SW USA 31°49′N 107°38′W
21 P10 Columbus North Carolina, SE USA 35°15′N 82°09′W
28 K2 Columbus North Dakota, N USA 48°52′N 102°47′W
31 S13 Columbus state capital Ohio, N USA 39°58′N 83°W
25 U11 Columbus Texas, SW USA 29°42′N 96°35′W
30 L8 Columbus Wisconsin, N USA 43°21′N 89°00′W
31 R12 Columbus Grove Ohio, N USA 40°55′N 84°03′W
29 Y15 Columbus Junction Iowa, C USA 41°16′N 91°21′W
44 J3 Columbus Point headland Cat Island, C The Bahamas 24°07′N 75°19′W
35 T8 Columbus Salt Marsh salt marsh Nevada, W USA
35 N6 Colusa California, W USA 39°10′N 122°03′W
32 L7 Colville Washington, NW USA 48°33′N 117°54′W
184 M5 Colville, Cape headland North Island, New Zealand 36°28′S 175°20′E
184 M5 Colville Channel channel North Island, New Zealand
39 P6 Colville River ≈ Alaska, USA
97 J18 Colwyn Bay N Wales, United Kingdom 53°18′N 03°43′W
106 H9 Comacchio var. Commachio; anc. Comactium. Emilia-Romagna, N Italy 44°41′N 12°10′E
106 H9 Comacchio, Valli di lagoon Adriatic Sea, N Mediterranean Sea
Comactium see Comacchio
41 V17 Comalapa Chiapas, SE Mexico 15°42′N 92°06′W
41 U15 Comalcalco Tabasco, SE Mexico 18°16′N 93°05′W
63 H16 Comallo Río Negro, SW Argentina 40°58′S 70°13′W
26 M12 Comanche Oklahoma, C USA 34°22′N 97°57′W
25 R8 Comanche Texas, SW USA 31°55′N 98°36′W
194 H2 Comandante Ferraz Brazilian research station Antarctica 61°57′S 58°23′W
62 N6 Comandante Fontana Formosa, N Argentina 25°19′S 59°42′W
63 I22 Comandante Luis Peidra Buena Santa Cruz, S Argentina 50°04′S 68°55′W
59 O18 Comandatuba Bahia, SE Brazil 15°13′S 39°00′W
116 K11 Comănești Hung. Kománfalva. Bacău, SW Romania 46°25′N 26°29′E
57 M19 Comarapa Santa Cruz, C Bolivia 17°53′S 64°30′W
116 J13 Comarnic Prahova, SE Romania 45°18′N 25°37′E
42 H6 Comayagua Comayagua, W Honduras 14°30′N 87°39′W
42 H6 Comayagua ◆ department W Honduras
42 I6 Comayagua, Montañas de ▲ C Honduras
21 R15 Combahee River ≈ South Carolina, SE USA
62 G10 Combarbalá Coquimbo, C Chile 31°15′S 71°03′W
103 S7 Combeaufontaine Haute-Saône, E France 47°43′N 05°52′E
97 G15 Comber Ir. An Comar. E Northern Ireland, United Kingdom 54°33′N 05°45′W
99 K20 Comblain-au-Pont Liège, E Belgium 50°29′N 05°36′E
102 I6 Combourg Ille-et-Vilaine, NW France 48°21′N 01°44′W
44 M9 Comendador prev. Elías Piña. W Dominican Republic 18°53′N 71°42′W
Comer See see Como, Lago di
25 R11 Comfort Texas, SW USA 29°58′N 98°54′W
153 V15 Comilla Ben. Kumillā. Chittagong, E Bangladesh 23°28′N 91°10′E
99 B18 Comines Hainaut, W Belgium 50°46′N 02°58′E
Comino see Kemmuna
107 D18 Comino, Capo headland Sardegna, Italy, C Mediterranean Sea 40°32′N 09°49′E
107 K25 Comiso Sicilia, Italy, C Mediterranean Sea 36°57′N 14°37′E
41 V16 Comitán var. Comitán de Domínguez. Chiapas, SE Mexico 16°15′N 92°06′W
Comitán de Domínguez see Comitán
Commachio see Comacchio
Commander Islands see Komandorskiye Ostrova
103 O10 Commentry Allier, C France 46°18′N 02°46′E
23 T2 Commerce Georgia, SE USA 34°12′N 83°27′W
27 R8 Commerce Oklahoma, C USA 36°55′N 94°52′W
25 V5 Commerce Texas, SW USA 33°16′N 95°52′W
37 T4 Commerce City Colorado, C USA 39°45′N 104°54′W
103 S5 Commercy Meuse, NE France 48°46′N 05°36′E
55 W9 Commewijne var. Commewyne. ◆ district NE Suriname
Commewyne see Commewijne
15 P8 Commissaires, Lac des ◎ Québec, SE Canada
64 A12 Commissioner's Point headland W Bermuda
9 O7 Committee Bay bay Nunavut, N Canada
106 D7 Como anc. Comum. Lombardia, N Italy 45°48′N 09°05′E
63 J19 Comodoro Rivadavia Chubut, SE Argentina 45°50′S 67°30′W
106 D6 Como, Lago di var. Lario, Eng. Lake Como, Ger. Comer See. ◎ N Italy
Como, Lake see Como, Lago di
40 E7 Comondú Baja California Sur, NW Mexico 26°01′N 111°50′W
116 F12 Comorâște Hung. Komornok. Caraș-Severin, SW Romania 45°13′N 21°34′E
Comores, République Fédérale Islamique des see Comoros
155 G24 Comorin, Cape headland SE India 08°00′N 77°10′E
172 M8 Comoro Basin undersea feature SW Indian Ocean 14°00′S 44°00′E

◆ Country ● Country Capital ◇ Dependent Territory ○ Dependent Territory Capital ◆ Administrative Regions ✈ International Airport ▲ Mountain ▲ Mountain Range ⛰ Volcano ≈ River ◎ Lake ⊠ Reservoir

172 K14 **Comoro Islands** island group W Indian Ocean
172 H13 **Comoros** off. Federal Islamic Republic of the Comoros, Fr. République Fédérale Islamique des Comores. ◆ republic W Indian Ocean
Comoros, Federal Islamic Republic of the see Comoros
10 L17 **Comox** Vancouver Island, British Columbia, SW Canada 49°40′N 124°55′W
103 O4 **Compiègne** Oise, N France 49°25′N 02°50′E
Complutum see Alcalá de Henares
Compniacum see Cognac
40 K12 **Compostela** Nayarit, C Mexico 21°12′N 104°52′W
Compostella see Santiago de Compostela
60 L11 **Comprida, Ilha** island S Brazil
117 N11 **Comrat** Rus. Komrat. S Moldova 46°18′N 28°40′E
25 O11 **Comstock** Texas, SW USA 29°39′N 101°10′W
31 P9 **Comstock Park** Michigan, N USA 43°00′N 85°40′W
193 N3 **Comstock Seamount** undersea feature N Pacific Ocean 48°15′N 156°55′W
Comum see Como
159 N17 **Cona** Xizang Zizhiqu, W China 27°59′N 91°54′E
76 H14 **Conakry** ● (Guinea) SW Guinea 09°31′N 13°43′W
76 H14 **Conakry** ✕ SW Guinea 09°37′N 13°32′W
Conamara see Connemara
Conca see Cuenca
25 Q12 **Concan** Texas, SW USA 29°27′N 99°43′W
102 F6 **Concarneau** Finistère, NW France 47°53′N 03°55′W
83 O17 **Conceição** Sofala, C Mozambique 18°47′S 36°18′E
59 K15 **Conceição do Araguaia** Pará, NE Brazil 08°15′S 49°15′W
58 F10 **Conceição do Maú** Roraima, W Brazil 03°35′N 59°52′W
61 D14 **Concepción** var. Concepcion. Corrientes, NE Argentina 28°25′S 57°54′W
62 J8 **Concepción** Tucumán, N Argentina 27°20′S 65°35′W
57 O17 **Concepción** Santa Cruz, E Bolivia 16°15′S 62°08′W
62 G13 **Concepción** Bío Bío, C Chile 36°47′S 73°01′W
54 E14 **Concepción** Putumayo, S Colombia 0°03′N 75°35′W
62 O5 **Concepción** var. Villa Concepción. Concepción, C Paraguay 23°26′S 57°24′W
62 O5 **Concepción** off. Departamento de Concepción. ◈ department E Paraguay
Concepción see La Concepción
Concepción de la Vega see La Vega
41 N9 **Concepción del Oro** Zacatecas, C Mexico 24°38′N 101°25′W
61 D18 **Concepción del Uruguay** Entre Ríos, E Argentina 32°30′S 58°15′W
Concepción, Departamento de see Concepción
42 K11 **Concepción, Volcán** ⛰ SW Nicaragua 11°31′N 85°37′W
44 J4 **Conception Island** island C The Bahamas
35 P14 **Conception, Point** headland California, W USA 34°27′N 120°28′W
54 H6 **Concha** Zulia, W Venezuela 09°02′N 71°45′W
60 L9 **Conchas** São Paulo, S Brazil 23°00′S 47°58′W
37 U11 **Conchas Dam** New Mexico, SW USA 35°21′N 104°11′W
37 U10 **Conchas Lake** ⊠ New Mexico, SW USA
102 M5 **Conches-en-Ouche** Eure, N France 49°00′N 01°00′E
37 N12 **Concho** Arizona, SW USA 34°28′N 109°33′W
40 J5 **Conchos, Río** ≈ NW Mexico
41 O8 **Conchos, Río** ≈ C Mexico
108 C8 **Concise** Vaud, W Switzerland 46°52′N 06°40′E
35 N8 **Concord** California, W USA 37°58′N 122°01′W
19 O9 **Concord** state capital New Hampshire, NE USA 43°10′N 71°32′W
21 R10 **Concord** North Carolina, SE USA 35°25′N 80°34′W
61 D17 **Concordia** Entre Ríos, E Argentina 31°25′S 58°W
60 I13 **Concórdia** Santa Catarina, S Brazil 27°14′S 52°01′W
54 D9 **Concordia** Antioquia, W Colombia 06°03′N 75°57′W
40 J10 **Concordia** Sinaloa, C Mexico 23°18′N 106°02′W
57 I19 **Concordia** Tacna, SW Peru 18°12′S 70°19′W
27 N3 **Concordia** Kansas, C USA 39°35′N 97°39′W
27 S4 **Concordia** Missouri, C USA 38°58′N 93°34′W
167 S7 **Con Cuông** Nghệ An, N Vietnam 19°02′N 104°54′E
167 T15 **Côn Đao Sơn** var. Con Son. island S Vietnam
Condate see Rennes, Ille-et-Vilaine, France
Condate see St-Claude, Jura, France
Condate see Montereau-Faut-Yonne, Seine-St-Denis, France
29 P8 **Conde** South Dakota, N USA 45°08′N 98°07′W
42 J8 **Condega** Estelí, NW Nicaragua 13°19′N 86°26′W
103 P2 **Condé-sur-l'Escaut** Nord, N France 50°27′N 03°36′E
102 K5 **Condé-sur-Noireau** Calvados, N France 48°52′N 00°31′W
Condivincum see Nantes
183 P8 **Condobolin** New South Wales, SE Australia 33°04′S 147°08′E
102 L15 **Condom** Gers, S France 43°56′N 00°23′E
32 J11 **Condon** Oregon, NW USA 45°15′N 120°10′W
54 D9 **Condoto** Chocó, W Colombia 05°06′N 76°37′W
23 P7 **Conecuh River** ≈ Alabama/Florida, SE USA
106 H7 **Conegliano** Veneto, NE Italy 45°53′N 12°18′E
61 C19 **Conesa** Buenos Aires, E Argentina 33°36′S 60°21′W
14 F15 **Conestogo** ≈ Ontario, S Canada
Confluentes see Koblenz
102 L10 **Confolens** Charente, W France 46°00′N 00°40′E
36 J4 **Confusion Range** ▲▲ Utah, W USA
62 N6 **Confuso, Río** ≈ C Paraguay
21 R12 **Congaree River** ≈ South Carolina, SE USA
Công Hoa Xa Hôi Chu Nghia Viêt Nam see Vietnam
160 K12 **Congjiang** var. Bingmei. Guizhou, S China 25°48′N 108°55′E
79 G18 **Congo** off. Republic of the Congo, Fr. Moyen-Congo; prev. Middle Congo. ◆ republic C Africa
79 K19 **Congo** off. Democratic Republic of Congo; prev. Zaire, Belgian Congo, Congo (Kinshasa). ◆ republic C Africa
67 T11 **Congo** var. Kongo, Fr. Zaire. ≈ C Africa
Congo see Zaire (province) Angola
68 G12 **Congo Basin** drainage basin W Dem. Rep. Congo
67 Q11 **Congo Canyon** var. Congo Seavalley, Congo Submarine Canyon. undersea feature E Atlantic Ocean 06°00′S 11°50′E
Congo Cone see Congo Fan
Congo/Congo (Kinshasa) see Congo (Democratic Republic of)
65 P15 **Congo Fan** var. Congo Cone. undersea feature E Atlantic Ocean 06°00′S 09°00′E
Congo Seavalley see Congo Canyon
Congo Submarine Canyon see Congo Canyon
Coni see Cuneo
63 H18 **Cónico, Cerro** ▲ SW Argentina 43°12′S 71°42′W
Conimbria/Conimbriga see Coimbra
Conjeeveram see Kānchipuram
11 R13 **Conklin** Alberta, C Canada 55°36′N 111°06′W
24 M1 **Conlen** Texas, SW USA 36°16′N 102°10′W
Con, Loch see Conn, Lough
Connacht see Connaught
97 B17 **Connaught** var. Connacht, Ir. Chonnacht, Cúige. cultural region W Ireland
31 V10 **Conneaut** Ohio, N USA 41°56′N 80°32′W
18 L13 **Connecticut** off. State of Connecticut, also known as Blue Law State, Constitution State, Land of Steady Habits, Nutmeg State. ◈ state NE USA
19 N8 **Connecticut** ≈ Canada/USA
19 O6 **Connecticut Lakes** lakes New Hampshire, NE USA
32 K9 **Connell** Washington, NW USA 46°39′N 118°51′W
97 B17 **Connemara** Ir. Conamara. physical region W Ireland
31 Q14 **Connersville** Indiana, N USA 39°38′N 85°15′W
97 B16 **Conn, Lough** Ir. Loch Con. ◎ W Ireland
35 X6 **Connors Pass** pass Nevada, W USA
181 X7 **Connors Range** ▲▲ Queensland, E Australia
56 E7 **Cononaco, Río** ≈ E Ecuador
29 W13 **Conrad** Iowa, C USA 42°13′N 92°52′W
33 R7 **Conrad** Montana, NW USA 48°10′N 111°58′W
25 W10 **Conroe** Texas, SW USA 30°19′N 95°28′W
25 V10 **Conroe, Lake** ⊠ Texas, SW USA
61 C17 **Conscripto Bernardi** Entre Ríos, E Argentina 31°03′S 59°05′W
59 M20 **Conselheiro Lafaiete** Minas Gerais, SE Brazil 20°40′S 43°48′W
Consentia see Cosenza
97 L14 **Consett** N England, United Kingdom 54°50′N 01°53′W
44 B5 **Consolación del Sur** Pinar del Río, W Cuba 22°32′N 83°32′W
Con Son see Côn Đao Sơn
11 R15 **Consort** Alberta, SW Canada 51°58′N 110°44′W
Constance see Konstanz
108 I6 **Constance, Lake** Ger. Bodensee. ◎ C Europe
104 G9 **Constância** Santarém, C Portugal 39°29′N 08°22′W
117 N14 **Constanța** var. Küstendje, Eng. Constanza, Ger. Konstanza, Turk. Köstence. Constanța, SE Romania 44°09′N 28°37′E
116 L14 **Constanța** ◈ county SE Romania
Constantia see Coutances
Constantia see Konstanz
104 K13 **Constantina** Andalucía, S Spain 37°54′N 05°36′W
74 L5 **Constantine** var. Qacentina, Ar. Qoussantîna. NE Algeria 36°23′N 06°44′E
39 O14 **Constantine, Cape** headland Alaska, USA 58°23′N 158°53′W
Constantinople see İstanbul
Constantiola see Oltenița
Constanz see Konstanz
Constanza see Constanța
62 G13 **Constitución** Maule, C Chile 35°20′S 72°28′W
61 D17 **Constitución** Salto, N Uruguay 31°05′S 57°51′W
Constitution State see Connecticut
105 N10 **Consuegra** Castilla-La Mancha, C Spain 39°28′N 03°36′W
181 X9 **Consuelo Peak** ▲ Queensland, E Australia 24°45′S 148°01′E
56 E11 **Contamana** Loreto, N Peru 07°19′S 75°04′W
Contrasto, Colle del see Contrasto, Portella del
107 K23 **Contrasto, Portella del** var. Colle del Contrasto. pass Sicilia, Italy, C Mediterranean Sea
54 G8 **Contratación** Santander, C Colombia 06°18′N 73°27′W
102 M8 **Contres** Loir-et-Cher, C France 47°24′N 01°30′E
107 O17 **Conversano** Puglia, SE Italy 40°58′N 17°07′E
27 U11 **Conway** Arkansas, C USA 35°05′N 92°27′W
19 O8 **Conway** New Hampshire, NE USA 43°58′N 71°05′W
21 U13 **Conway** South Carolina, SE USA 33°51′N 79°04′W
25 N2 **Conway** Texas, SW USA 35°10′N 101°23′W
27 U11 **Conway, Lake** ⊠ Arkansas, C USA
27 N7 **Conway Springs** Kansas, C USA 37°23′N 97°38′W
97 J18 **Conwy** N Wales, United Kingdom 53°17′N 03°51′W
23 T3 **Conyers** Georgia, SE USA 33°40′N 84°01′W
Coo see Kos
182 F4 **Coober Pedy** South Australia 29°01′S 134°47′E
181 P2 **Cooinda** Northern Territory, N Australia 12°54′S 132°31′E
182 B6 **Cook** South Australia 30°37′S 130°26′E
29 W4 **Cook** Minnesota, N USA 47°51′N 92°41′W
191 N6 **Cook, Baie de** bay Moorea, W French Polynesia
10 J16 **Cook, Cape** headland Vancouver Island, British Columbia, SW Canada 50°04′N 127°52′W
37 Q15 **Cookes Peak** ▲ New Mexico, SW USA 32°32′N 107°43′W
20 L8 **Cookeville** Tennessee, S USA 36°10′N 85°30′W
175 P9 **Cook Fracture Zone** tectonic feature S Pacific Ocean
Cook, Grand Récif de see Cook, Récif de
39 Q12 **Cook Inlet** inlet Alaska, USA
191 X2 **Cook Island** island Line Islands, E Kiribati
190 J14 **Cook Islands** ◇ self-governing entity in free association with New Zealand S Pacific Ocean
Cook, Mount see Aoraki
187 O15 **Cook, Récif de** var. Grand Récif de Cook. reef S New Caledonia
14 G14 **Cookstown** Ontario, S Canada 44°12′N 79°39′W
97 F15 **Cookstown** Ir. An Chorr Chríochach. C Northern Ireland, United Kingdom 54°39′N 06°45′W
185 K14 **Cook Strait** var. Raukawa. strait New Zealand
181 W3 **Cooktown** Queensland, NE Australia 15°28′S 145°15′E
183 P6 **Coolabah** New South Wales, SE Australia 31°03′S 146°42′E
182 J11 **Coola Coola Swamp** wetland South Australia
183 S7 **Coolah** New South Wales, SE Australia 31°49′S 149°43′E
183 P9 **Coolamon** New South Wales, SE Australia 34°49′S 147°13′E
183 T4 **Coolatai** New South Wales, SE Australia 29°16′S 150°45′E
180 K12 **Coolgardie** Western Australia 31°01′S 121°12′E
36 L14 **Coolidge** Arizona, SW USA 32°58′N 111°29′W
25 U8 **Coolidge** Texas, SW USA 31°45′N 96°39′W
183 Q11 **Cooma** New South Wales, SE Australia 36°16′S 149°09′E
Coomassie see Kumasi
183 R6 **Coonabarabran** New South Wales, SE Australia 31°19′S 149°18′E
182 J10 **Coonalpyn** South Australia 35°43′S 139°50′E
183 R6 **Coonamble** New South Wales, SE Australia 30°56′S 148°22′E
Coondapoor see Kundāpura
155 G21 **Coonoor** Tamil Nādu, SE India 11°21′N 76°46′E
29 U14 **Coon Rapids** Iowa, C USA 41°52′N 94°40′W
29 V8 **Coon Rapids** Minnesota, N USA 45°12′N 93°18′W
25 V5 **Cooper** Texas, SW USA 33°23′N 95°42′W
181 U9 **Cooper Creek** var. Barcoo, Cooper's Creek. seasonal river Queensland/South Australia
39 R12 **Cooper Landing** Alaska, USA 60°27′N 149°59′W
21 T14 **Cooper River** ≈ South Carolina, SE USA
Cooper's Creek see Cooper Creek
44 H1 **Coopers Town** Great Abaco, N The Bahamas 26°46′N 77°27′W
18 J10 **Cooperstown** New York, NE USA 42°43′N 74°56′W
29 P4 **Cooperstown** North Dakota, N USA 47°26′N 98°07′W
31 P9 **Coopersville** Michigan, N USA 43°03′N 85°55′W
182 D7 **Coorabie** South Australia 31°57′S 132°18′E
23 Q3 **Coosa River** ≈ Alabama/Georgia, S USA
32 E14 **Coos Bay** Oregon, NW USA 43°22′N 124°13′W
183 Q9 **Cootamundra** New South Wales, SE Australia 34°41′S 148°03′E
97 E16 **Cootehill** Ir. Muinchille. N Ireland 54°04′N 07°05′W
Cop see Chop
57 J17 **Copacabana** La Paz, W Bolivia 16°11′S 69°02′W
63 H14 **Copahué, Volcán** ⛰ C Chile 37°56′S 71°04′W
41 U16 **Copainalá** Chiapas, SE Mexico 17°04′N 93°13′W
32 F8 **Copalis Beach** Washington, NW USA 47°05′N 124°11′W
42 F6 **Copán** ◈ department W Honduras
Copán see Copán Ruinas
25 T14 **Copano Bay** bay NW Gulf of Mexico
42 F6 **Copán Ruinas** var. Copán. Copán, W Honduras 14°52′N 89°10′W
Copenhagen see København
107 Q19 **Copertino** Puglia, SE Italy 40°16′N 18°03′E
62 H7 **Copiapó** Atacama, N Chile 27°17′S 70°25′W
62 G8 **Copiapó, Bahía** bay N Chile
62 G7 **Copiapó, Río** ≈ N Chile
114 M12 **Çöpköy** Edirne, NW Turkey 41°14′N 26°51′E
182 I5 **Copley** South Australia 30°36′S 138°26′E
106 H9 **Copparo** Emilia-Romagna, C Italy 44°53′N 11°53′E
55 V10 **Coppename Rivier** var. Koppename. ≈ C Suriname
25 S9 **Copperas Cove** Texas, SW USA 31°07′N 97°54′W
82 J13 **Copperbelt** ◈ province C Zambia
39 S11 **Copper Center** Alaska, USA 61°57′N 145°21′W
8 K8 **Coppermine** ≈ Northwest Territories/Nunavut, N Canada
Coppermine see Kugluktuk
39 T11 **Copper River** ≈ Alaska, USA
Copper State see Arizona
116 I11 **Copșa Mică** Ger. Kleinkopisch, Hung. Kiskapus. Sibiu, C Romania 46°06′N 24°15′E
158 J14 **Coqên** Xizang Zizhiqu, W China 31°13′N 85°12′E
Coquilhatville see Mbandaka
32 E14 **Coquille** Oregon, NW USA 43°11′N 124°12′W
62 G9 **Coquimbo** Coquimbo, N Chile 30°S 71°18′W
62 G9 **Coquimbo** off. Región de Coquimbo. ◈ region C Chile
Coquimbo, Región de see Coquimbo
116 I15 **Corabia** Olt, S Romania 43°46′N 24°31′E
57 F17 **Coracora** Ayacucho, SW Peru 15°03′S 73°45′W
Cora Droma Rúisc see Carrick-on-Shannon
44 K9 **Corail** SW Haiti 18°34′N 73°53′W
183 V4 **Coraki** New South Wales, SE Australia 29°01′S 153°15′E
180 G8 **Coral Bay** Western Australia 23°02′S 113°51′E
23 Y16 **Coral Gables** Florida, SE USA 25°43′N 80°16′W
9 P8 **Coral Harbour** var. Salliq. Southampton Island, Nunavut, NE Canada 64°10′N 83°15′W
192 I9 **Coral Sea** sea SW Pacific Ocean
174 M7 **Coral Sea Basin** undersea feature N Coral Sea
192 H9 **Coral Sea Islands** ◇ Australian external territory SW Pacific Ocean
182 M12 **Corangamite, Lake** ◎ Victoria, SE Australia
Corantijn Rivier see Courantyne River
18 B14 **Coraopolis** Pennsylvania, NE USA 40°28′N 80°08′W
107 N17 **Corato** Puglia, SE Italy 41°09′N 16°25′E
103 O17 **Corbières** ▲▲ S France
103 P8 **Corbigny** Nièvre, C France 47°15′N 03°42′E
21 N7 **Corbin** Kentucky, S USA 36°57′N 84°06′W
104 L14 **Corbones** ≈ SW Spain
Corcaigh see Cork
35 R11 **Corcoran** California, W USA 36°06′N 119°33′W
47 T14 **Corcovado, Golfo** gulf S Chile
63 G18 **Corcovado, Volcán** ⛰ S Chile 43°13′S 72°45′W
104 F3 **Corcubión** Galicia, NW Spain 42°56′N 09°12′W
Corcyra Nigra see Korčula
60 Q9 **Cordeiro** Rio de Janeiro, SE Brazil 22°01′S 42°20′W
23 T6 **Cordele** Georgia, SE USA 31°59′N 83°49′W
26 L11 **Cordell** Oklahoma, C USA 35°17′N 98°59′W
103 N14 **Cordes** Tarn, S France 44°03′N 01°57′E
62 O6 **Cordillera** off. Departamento de la Cordillera. ◈ department C Paraguay
Cordillera see Cacaguatique, Cordillera
Cordillera, Departamento de la see Cordillera
182 K1 **Cordillo Downs** South Australia 26°44′S 140°37′E
62 K10 **Córdoba** Córdoba, C Argentina 31°25′S 64°11′W
41 R14 **Córdoba** Veracruz-Llave, E Mexico 18°55′N 96°55′W
104 M13 **Córdoba** var. Cordoba, Eng. Cordova; anc. Corduba. Andalucía, SW Spain 37°53′N 04°46′W
62 K11 **Córdoba** off. Provincia de Córdoba. ◈ province C Argentina
54 D7 **Córdoba** off. Departamento de Córdoba. ◈ province NW Colombia
104 L13 **Córdoba** ◈ province Andalucía, S Spain
Córdoba, Departamento de see Córdoba
Córdoba, Provincia de see Córdoba
62 K10 **Córdoba, Sierras de** ▲▲ C Argentina
23 O3 **Cordova** Alabama, S USA 33°45′N 87°10′W
39 S12 **Cordova** Alaska, USA 60°32′N 145°45′W
Cordova/Cordoba see Córdoba
Corduba see Córdoba
Corentyne River see Courantyne River
Corfu see Kérkyra
104 J9 **Coria** Extremadura, W Spain 39°59′N 06°32′W
104 J14 **Coria del Río** Andalucía, S Spain 37°18′N 06°04′W
183 S8 **Coricudgy, Mount** ▲ New South Wales, SE Australia 32°49′S 150°28′E
107 N20 **Corigliano Calabro** Calabria, SW Italy 39°36′N 16°32′E
Corinium/Corinium Dobunorum see Cirencester
23 N1 **Corinth** Mississippi, S USA 34°56′N 88°29′W
Corinth see Kórinthos
Corinth Canal see Dióryga Korinthou
Corinth, Gulf of/Corinthiacus Sinus see Korinthiakós Kólpos
Corinthus see Kórinthos
42 I9 **Corinto** Chinandega, NW Nicaragua 12°29′N 87°14′W
97 C21 **Cork** Ir. Corcaigh. S Ireland 51°54′N 07°06′W
97 C21 **Cork** Ir. Corcaigh. cultural region SW Ireland
97 C21 **Cork** ✕ Cork, SW Ireland 51°52′N 08°25′W
97 D21 **Cork Harbour** Ir. Cuan Chorcaí. inlet SW Ireland
107 I23 **Corleone** Sicilia, Italy, C Mediterranean Sea 37°49′N 13°18′E
114 N13 **Çorlu** Tekirdağ, NW Turkey 41°11′N 27°48′E
114 N12 **Çorlu Çayı** ≈ NW Turkey
Cormaiore see Courmayeur
11 V13 **Cormorant** Manitoba, C Canada 54°12′N 100°33′W
23 T2 **Cornelia** Georgia, SE USA 34°30′N 83°31′W
60 J10 **Cornélio Procópio** Paraná, S Brazil 23°07′S 50°40′W
55 V9 **Corneliskondre** Sipaliwini, N Suriname 05°21′N 56°10′W
30 J5 **Cornell** Wisconsin, N USA 45°09′N 91°10′W
13 S12 **Corner Brook** Newfoundland, Newfoundland and Labrador, E Canada 48°58′N 57°58′W
Corner Rise Seamounts see Corner Seamounts
64 I9 **Corner Seamounts** var. Corner Rise Seamounts. undersea feature NW Atlantic Ocean 35°30′N 51°30′W
116 M9 **Cornești** Rus. Korneshty. C Moldova 47°23′N 28°00′E
Corneto see Tarquinia
Cornhusker State see Nebraska
27 X8 **Corning** Arkansas, C USA 36°26′N 90°35′W
35 N5 **Corning** California, W USA 39°54′N 122°12′W
29 U15 **Corning** Iowa, C USA 40°58′N 94°46′W
18 G11 **Corning** New York, NE USA 42°08′N 77°03′W
Corn Islands see Maíz, Islas del
107 J14 **Corno Grande** ▲ C Italy 42°26′N 13°29′E
15 N13 **Cornwall** Ontario, SE Canada 45°02′N 74°45′W
97 H25 **Cornwall** cultural region SW England, United Kingdom
97 G25 **Cornwall, Cape** headland SW England, United Kingdom 50°11′N 05°39′W
54 J4 **Coro** prev. Santa Ana de Coro. Falcón, NW Venezuela 11°27′N 69°41′W
57 J18 **Corocoro** La Paz, W Bolivia 17°10′S 68°28′W
57 K17 **Coroico** La Paz, W Bolivia 16°09′S 67°45′W
184 M5 **Coromandel** Waikato, North Island, New Zealand 36°47′S 175°30′E
155 K20 **Coromandel Coast** coast E India
184 M5 **Coromandel Peninsula** peninsula North Island, New Zealand
184 M6 **Coromandel Range** ▲▲ North Island, New Zealand
171 N5 **Coron** Busuanga Island, W Philippines 12°02′N 120°10′E
35 T15 **Corona** California, W USA 33°52′N 117°34′W
37 T12 **Corona** New Mexico, SW USA 34°15′N 105°36′W
11 U17 **Coronach** Saskatchewan, S Canada 49°07′N 105°33′W
35 U17 **Coronado** California, W USA 32°41′N 117°10′W
43 N15 **Coronado, Bahía de** bay S Costa Rica
11 R15 **Coronation** Alberta, SW Canada 52°06′N 111°25′W
8 K7 **Coronation Gulf** gulf Nunavut, N Canada
194 I1 **Coronation Island** island Antarctica
39 X14 **Coronation Island** island Alexander Archipelago, Alaska, USA
61 B18 **Coronda** Santa Fe, C Argentina 31°58′S 60°56′W
63 F14 **Coronel** Bío Bío, C Chile 37°01′S 73°08′W
61 D20 **Coronel Brandsen** var. Brandsen. Buenos Aires, E Argentina 35°08′S 58°15′W
62 K4 **Coronel Cornejo** Salta, N Argentina 22°46′S 63°49′W
61 B24 **Coronel Dorrego** Buenos Aires, E Argentina 38°38′S 61°15′W
62 P6 **Coronel Oviedo** Caaguazú, SE Paraguay 25°24′S 56°30′W
61 B23 **Coronel Pringles** Buenos Aires, E Argentina 37°56′S 61°25′W
61 B23 **Coronel Suárez** Buenos Aires, E Argentina 37°30′S 61°52′W
61 E22 **Coronel Vidal** Buenos Aires, E Argentina 37°28′S 57°45′W
55 V9 **Coronie** ◈ district NW Suriname
57 G17 **Coropuna, Nevado** ▲ S Peru 15°31′S 72°31′W
Çorovoda see Çorovodë
113 L22 **Çorovodë** var. Çorovoda. Berat, S Albania 40°29′N 20°15′E
183 P11 **Corowa** New South Wales, SE Australia 36°01′S 146°22′E
42 G1 **Corozal** Corozal, N Belize 18°23′N 88°23′W
54 E6 **Corozal** Sucre, NW Colombia 09°18′N 75°19′W
42 G1 **Corozal** ◈ district N Belize
25 T14 **Corpus Christi** Texas, SW USA 27°48′N 97°24′W
25 T14 **Corpus Christi Bay** inlet Texas, SW USA
25 R14 **Corpus Christi, Lake** ⊠ Texas, SW USA
63 F16 **Corral** Los Ríos, C Chile 39°55′S 73°30′W
105 O9 **Corral de Almaguer** Castilla-La Mancha, C Spain 39°45′N 03°10′W
104 K6 **Corrales** Castilla y León, N Spain 41°22′N 05°44′W
37 R11 **Corrales** New Mexico, SW USA 35°11′N 106°37′W
Corrán Tuathail see Carrauntoohil
106 F9 **Correggio** Emilia-Romagna, C Italy 44°47′N 10°46′E
59 M16 **Corrente** Piauí, E Brazil 10°29′S 45°11′W
59 I19 **Correntes, Rio** ≈ SW Brazil
103 N12 **Corrèze** ◈ department C France
97 C17 **Corrib, Lough** Ir. Loch Coirib. ◎ W Ireland
61 C14 **Corrientes** Corrientes, NE Argentina 27°29′S 58°42′W
61 D15 **Corrientes** off. Provincia de Corrientes. ◈ province NE Argentina
44 A5 **Corrientes, Cabo** headland W Cuba 21°48′N 84°30′W
40 I13 **Corrientes, Cabo** headland SW Mexico 20°25′N 105°42′W
Corrientes, Provincia de see Corrientes
61 C16 **Corrientes, Río** ≈ NE Argentina
56 E8 **Corrientes, Río** ≈ Ecuador/Peru
25 W9 **Corrigan** Texas, SW USA 31°00′N 94°49′W
55 U9 **Corriverton** E Guyana 05°55′N 57°09′W
Corriza see Korçë
183 Q11 **Corryong** Victoria, SE Australia 36°14′S 147°54′E
103 F2 **Corse** Eng. Corsica. ◈ region France, C Mediterranean Sea
101 X13 **Corse** Eng. Corsica. island France, C Mediterranean Sea
103 Y12 **Corse, Cap** headland Corse, France, C Mediterranean Sea 43°01′N 09°25′E
103 X15 **Corse-du-Sud** ◈ department Corse, France, C Mediterranean Sea
29 P11 **Corsica** South Dakota, N USA 43°25′N 98°24′W
Corsica see Corse
25 U7 **Corsicana** Texas, SW USA 32°05′N 96°27′W
103 Y15 **Corte** Corse, France, C Mediterranean Sea 42°18′N 09°08′E
63 G16 **Corte Alto** Los Lagos, S Chile 40°58′S 73°04′W
104 I13 **Cortegana** Andalucía, S Spain 37°55′N 06°49′W
43 N15 **Cortés** var. Ciudad Cortés. Puntarenas, SE Costa Rica 08°59′N 83°32′W
42 G5 **Cortés** ◈ department NW Honduras
37 P8 **Cortez** Colorado, C USA 37°22′N 108°36′W
Cortez, Sea of see California, Golfo de
106 H6 **Cortina d'Ampezzo** Veneto, NE Italy 46°33′N 12°09′E
18 H11 **Cortland** New York, NE USA 42°34′N 76°09′W
31 V11 **Cortland** Ohio, N USA 41°19′N 80°43′W
106 H12 **Cortona** Toscana, C Italy 43°15′N 12°01′E
76 H13 **Corubal, Rio** ≈ E Guinea-Bissau
104 G10 **Coruche** Santarém, C Portugal 38°58′N 08°31′W
Çoruh see Rize
137 R11 **Çoruh Nehri** Geor. Chorokh, Rus. Chorokhi. ≈ Georgia/Turkey
136 K12 **Çorum** var. Chorum. Çorum, N Turkey 40°31′N 34°57′E
136 J12 **Çorum** var. Chorum. ◈ province N Turkey
59 H19 **Corumbá** Mato Grosso do Sul, S Brazil 19°S 57°35′W
14 D16 **Corunna** Ontario, S Canada 42°49′N 82°25′W
Corunna see A Coruña
32 F12 **Corvallis** Oregon, NW USA 44°35′N 123°16′W
64 M1 **Corvo** var. Ilha do Corvo. island Azores, Portugal, NE Atlantic Ocean
Corvo, Ilha do see Corvo
31 O16 **Corydon** Indiana, N USA 38°12′N 86°07′W
29 V16 **Corydon** Iowa, C USA 40°45′N 93°19′W
Cos see Kos
40 I9 **Cosalá** Sinaloa, C Mexico 24°25′N 106°39′W
41 R15 **Cosamaloapan** var. Cosamaloapan de Carpio. Veracruz-Llave, E Mexico 18°23′N 95°50′W
Cosamaloapan de Carpio see Cosamaloapan
107 N21 **Cosenza** anc. Consentia. Calabria, SW Italy 39°17′N 16°15′E
31 T13 **Coshocton** Ohio, N USA 40°16′N 81°51′W
42 H9 **Cosigüina, Punta** headland NW Nicaragua 12°53′N 87°42′W
29 T9 **Cosmos** Minnesota, N USA 44°56′N 94°42′W
103 O8 **Cosne-Cours-sur-Loire** Nièvre, C France 47°25′N 02°56′E
108 B9 **Cossonay** Vaud, W Switzerland 46°37′N 06°28′E
Cossyra see Pantelleria
47 R4 **Costa, Cordillera de la** var. Cordillera de Venezuela. ▲▲ N Venezuela
42 K13 **Costa Rica** off. Republic of Costa Rica. ◆ republic Central America
Costa Rica, Republic of see Costa Rica
43 N15 **Costeña, Fila** ▲▲ S Costa Rica
Costermansville see Bukavu
116 I14 **Costești** Argeș, SW Romania 44°40′N 24°53′E
37 S8 **Costilla** New Mexico, SW USA 36°58′N 105°31′W
35 O7 **Cosumnes River** ≈ California, W USA
101 O16 **Coswig** Sachsen, E Germany 51°07′N 13°36′E
101 M14 **Coswig** Sachsen-Anhalt, E Germany 51°53′N 12°26′E
Cosyra see Pantelleria
171 Q7 **Cotabato** Mindanao, S Philippines 07°13′N 124°12′E
56 C5 **Cotacachi** ▲ N Ecuador 0°29′N 78°17′W
57 L21 **Cotagaita** Potosí, S Bolivia 20°47′S 65°40′W
103 V15 **Côte d'Azur** prev. Nice. ✕ (Nice) Alpes-Maritimes, SE France 43°40′N 07°12′E
Côte d'Ivoire see Ivory Coast
Côte d'Ivoire, République de la see Ivory Coast
103 R7 **Côte d'Or** ◈ department E France
103 R8 **Côte d'Or** cultural region C France
Côte Française des Somalis see Djibouti
102 J4 **Cotentin** peninsula N France
102 G6 **Côtes d'Armor** prev. Côtes-du-Nord. ◈ department NW France
Côtes-du-Nord see Côtes d'Armor
Cöthen see Köthen
Côtière, Chaîne see Coast Mountains
40 M13 **Cotija** var. Cotija de la Paz. Michoacán, SW Mexico 19°49′N 102°39′W
Cotija de la Paz see Cotija
77 R16 **Cotonou** var. Kotonu. S Benin 06°24′N 02°26′E
77 R16 **Cotonou** ✕ S Benin 06°31′N 02°18′E
56 B6 **Cotopaxi** prev. León. ◈ province C Ecuador
56 C6 **Cotopaxi** ⛰ N Ecuador 0°42′S 78°24′W
Cotrone see Crotone
97 L21 **Cotswold Hills** var. Cotswolds. hill range S England, United Kingdom
Cotswolds see Cotswold Hills
32 F13 **Cottage Grove** Oregon, NW USA 43°48′N 123°03′W
21 S14 **Cottageville** South Carolina, SE USA 32°55′N 80°28′W
101 P14 **Cottbus** Lus. Chośebuz; prev. Kottbus. Brandenburg, E Germany 51°42′N 14°22′E
27 U9 **Cotter** Arkansas, C USA 36°16′N 92°30′W
106 A9 **Cottian Alps** Fr. Alpes Cottiennes, It. Alpi Cozie. ▲▲ France/Italy
Cottiennes, Alpes see Cottian Alps
Cotton State, The see Alabama
22 G4 **Cotton Valley** Louisiana, S USA 32°49′N 93°25′W
36 L12 **Cottonwood** Arizona, SW USA 34°43′N 112°00′W
32 M10 **Cottonwood** Idaho, NW USA 46°01′N 116°20′W
29 S9 **Cottonwood** Minnesota, N USA 44°37′N 95°41′W
25 Q7 **Cottonwood** Texas, SW USA 32°12′N 99°14′W
27 O5 **Cottonwood Falls** Kansas, C USA 38°21′N 96°33′W
36 L3 **Cottonwood Heights** Utah, W USA 40°37′N 111°48′W
29 S10 **Cottonwood River** ≈ Minnesota, N USA
45 O9 **Cotuí** C Dominican Republic 19°04′N 70°10′W
25 Q13 **Cotulla** Texas, SW USA 28°27′N 99°15′W
Cotyora see Ordu
102 I11 **Coubre, Pointe de la** headland W France 45°39′N 01°23′W
18 E12 **Coudersport** Pennsylvania, NE USA 41°45′N 78°00′W
15 S9 **Coudres, Île aux** island Québec, SE Canada
182 G11 **Couedic, Cape de** headland South Australia 36°04′S 136°43′E
Couentrey see Coventry
102 I6 **Couesnon** ≈ NW France
32 H10 **Cougar** Washington, NW USA 46°03′N 122°18′W
102 L10 **Couhé** Vienne, W France 46°18′N 00°10′E
32 K8 **Coulee City** Washington, NW USA 47°36′N 119°18′W
195 Q15 **Coulman Island** island Antarctica
103 P5 **Coulommiers** Seine-et-Marne, N France 48°49′N 03°04′E
14 K11 **Coulonge** ≈ Québec, SE Canada
14 K11 **Coulonge Est** ≈ Québec, SE Canada
35 Q9 **Coulterville** California, W USA 37°41′N 120°10′W
38 M9 **Council** Alaska, USA 64°54′N 163°40′W
32 M12 **Council** Idaho, NW USA 44°45′N 116°26′W
29 S15 **Council Bluffs** Iowa, C USA 41°16′N 95°52′W
27 O5 **Council Grove** Kansas, C USA 38°41′N 96°29′W
27 O5 **Council Grove Lake** ⊠ Kansas, C USA
32 G7 **Coupeville** Washington, NW USA 48°13′N 122°41′W
55 U12 **Courantyne River** var. Corantijn Rivier, Corentyne River. ≈ Guyana/Suriname
99 G21 **Courcelles** Hainaut, S Belgium 50°28′N 04°23′E
108 C7 **Courgenay** Jura, NW Switzerland 47°24′N 07°09′E
126 B2 **Courland Lagoon** Ger. Kurisches Haff, Rus. Kurskiy Zaliv. lagoon Lithuania/Russian Federation
118 B12 **Courland Spit** Lith. Kuršių Nerija, Rus. Kurshskaya Kosa. spit Lithuania/Russian Federation
106 A6 **Courmayeur** prev. Cormaiore. Valle d'Aosta, NW Italy 45°48′N 07°00′E
108 D7 **Courroux** Jura, NW Switzerland 47°22′N 07°23′E
10 K17 **Courtenay** Vancouver Island, British Columbia, SW Canada 49°40′N 124°58′W
21 W7 **Courtland** Virginia, NE USA 36°44′N 77°06′W
25 V10 **Courtney** Texas, SW USA 30°16′N 96°04′W
30 J4 **Court Oreilles, Lac** ◎ Wisconsin, N USA
Courtrai see Kortrijk
99 H19 **Court-Saint-Étienne** Walloon Brabant, C Belgium 50°38′N 04°34′E
22 G6 **Coushatta** Louisiana, S USA 32°00′N 93°20′W
172 I16 **Cousin** island Inner Islands, NE Seychelles
172 I16 **Cousine** island Inner Islands, NE Seychelles
102 J4 **Coutances** anc. Constantia. Manche, N France 49°04′N 01°27′W
102 K12 **Coutras** Gironde, SW France 45°01′N 00°07′W
45 U14 **Couva** Trinidad, Trinidad and Tobago 10°25′N 61°27′W
108 B8 **Couvet** Neuchâtel, W Switzerland 46°57′N 06°41′E
99 H22 **Couvin** Namur, S Belgium 50°03′N 04°30′E
116 K12 **Covasna** Ger. Kowasna, Hung. Kovászna. Covasna, E Romania 45°51′N 26°11′E
116 J11 **Covasna** ◈ county E Romania
14 E12 **Cove Island** island Ontario, S Canada
34 M5 **Covelo** California, W USA 39°46′N 123°16′W
97 M20 **Coventry** anc. Couentrey. C England, United Kingdom 52°25′N 01°30′W
Cove of Cork see Cobh
21 U5 **Covesville** Virginia, NE USA 37°52′N 78°41′W
104 I8 **Covilhã** Castelo Branco, E Portugal 40°17′N 07°30′W
23 T3 **Covington** Georgia, SE USA 33°34′N 83°52′W
31 N13 **Covington** Indiana, N USA 40°08′N 87°23′W
20 M3 **Covington** Kentucky, S USA 39°04′N 84°30′W

◆ Country
● Country Capital
◇ Dependent Territory
○ Dependent Territory Capital
◈ Administrative Regions
✕ International Airport
▲ Mountain
▲▲ Mountain Range
⛰ Volcano
≈ River
◎ Lake
⊠ Reservoir

22 K8 **Covington** Louisiana, S USA 30°28′N 90°06′W
31 Q13 **Covington** Ohio, N USA 40°07′N 84°21′W
20 F9 **Covington** Tennessee, S USA 35°32′N 89°40′W
21 S6 **Covington** Virginia, NE USA 37°48′N 80°01′W
183 Q8 **Cowal, Lake** *seasonal lake* New South Wales, SE Australia
11 W15 **Cowan** Manitoba, S Canada 51°59′N 100°36′W
18 F12 **Cowanesque River** River New York/Pennsylvania, NE USA
180 L12 **Cowan, Lake** Lake Western Australia
15 P13 **Cowansville** Québec, SE Canada 45°13′N 72°44′W
182 H8 **Cowell** South Australia 33°43′S 136°53′E
97 M23 **Cowes** S England, United Kingdom 50°45′N 01°19′W
27 Q10 **Coweta** Oklahoma, C USA 35°57′N 95°39′W
0 D6 **Cowie Seamount** *undersea feature* NE Pacific Ocean 54°15′N 149°30′W
32 G10 **Cowlitz River** River Washington, NW USA
21 Q11 **Cowpens** South Carolina, SE USA 35°01′N 81°48′W
183 R8 **Cowra** New South Wales, SE Australia 33°50′S 148°45′E
Coxen Hole *see* Roatán
59 I19 **Coxim** Mato Grosso do Sul, S Brazil 18°28′S 54°45′W
59 I19 **Coxim, Rio** River SW Brazil
Coxin Hole *see* Roatán
153 V17 **Cox's Bazar** Chittagong, S Bangladesh 21°25′N 91°59′E
76 H14 **Coyah** Conakry, W Guinea 09°45′N 13°26′W
40 K5 **Coyame** Chihuahua, N Mexico 29°29′N 105°07′W
24 L9 **Coyanosa Draw** River Texas, SW USA
Coyhaique *see* Coihaique
42 C7 **Coyolate, Río** River S Guatemala
Coyote State, The *see* South Dakota
40 I10 **Coyotitán** Sinaloa, C Mexico 23°48′N 106°37′W
41 N15 **Coyuca** *var.* Coyuca de Catalán. Guerrero, S Mexico 18°21′N 100°39′W
41 O16 **Coyuca** *var.* Coyuca de Benítez. Guerrero, S Mexico 17°01′N 100°08′W
Coyuca de Benítez/Coyuca de Catalán *see* Coyuca
29 N15 **Cozad** Nebraska, C USA 40°52′N 99°58′W
158 L14 **Cozhê** Xizang Zizhiqu, W China 31°53′N 87°51′E
Cozie, Alpi *see* Cottian Alps
Cozmeni *see* Kitsman'
40 E3 **Cozón, Cerro** ▲ NW Mexico 31°16′N 112°29′W
41 Z12 **Cozumel** Quintana Roo, E Mexico 20°29′N 86°54′W
41 Z12 **Cozumel, Isla** *island* SE Mexico
32 K8 **Crab Creek** River Washington, NW USA
44 H12 **Crab Pond Point** *headland* W Jamaica 18°07′N 78°01′W
Cracovia/Cracow *see* Kraków
83 I25 **Cradock** Eastern Cape, S South Africa 32°07′S 25°38′E
39 Y14 **Craig** Prince of Wales Island, Alaska, USA 55°29′N 133°04′W
37 Q3 **Craig** Colorado, C USA 40°31′N 107°33′W
97 F15 **Craigavon** C Northern Ireland, United Kingdom 54°28′N 06°25′W
21 T5 **Craigsville** Virginia, NE USA 38°07′N 79°21′W
101 J21 **Crailsheim** Baden-Württemberg, S Germany 49°07′N 10°04′E
116 H14 **Craiova** Dolj, SW Romania 44°19′N 23°49′E
10 K12 **Cranberry Junction** British Columbia, SW Canada 55°35′N 128°21′W
18 J8 **Cranberry Lake** Lake New York, NE USA
11 V13 **Cranberry Portage** Manitoba, C Canada 54°34′N 101°22′W
11 P17 **Cranbrook** British Columbia, SW Canada 49°29′N 115°48′W
30 M5 **Crandon** Wisconsin, N USA 45°34′N 88°54′W
32 K14 **Crane** Oregon, NW USA 43°24′N 118°35′W
24 M9 **Crane** Texas, SW USA 31°23′N 102°22′W
Crane *see* The Crane
25 S8 **Cranfills Gap** Texas, SW USA 31°46′N 97°49′W
19 O12 **Cranston** Rhode Island, NE USA 41°46′N 71°26′W
Cranz *see* Zelenogradsk
59 L15 **Craolândia** Tocantins, E Brazil 07°17′S 47°23′W
102 J7 **Craon** Mayenne, NW France 47°52′N 00°57′W
195 V16 **Crary, Cape** *headland* Antarctica
Crasna *see* Kraszna
32 G14 **Crater Lake** Lake Oregon, NW USA
33 P14 **Craters of the Moon National Monument** *national park* Idaho, NW USA
59 O14 **Crateús** Ceará, E Brazil 05°10′S 40°39′W
Crathis *see* Crati
107 N20 **Crati** *anc.* Crathis. River S Italy
11 U16 **Craven** Saskatchewan, S Canada 50°44′N 104°50′W
54 I8 **Cravo Norte** Arauca, E Colombia 06°17′N 70°15′W
28 J12 **Crawford** Nebraska, C USA 42°40′N 103°24′W
25 T8 **Crawford** Texas, SW USA 31°31′N 97°26′W
11 O17 **Crawford Bay** British Columbia, SW Canada 49°39′N 116°44′W
65 M19 **Crawford Seamount** *undersea feature* S Atlantic Ocean 40°30′S 10°00′W
31 O13 **Crawfordsville** Indiana, N USA 40°02′N 86°52′W
23 S9 **Crawfordville** Florida, SE USA 30°10′N 84°22′W
97 O23 **Crawley** SE England, United Kingdom 51°07′N 00°12′W
33 S10 **Crazy Mountains** ▲ Montana, NW USA
11 T11 **Cree** River Saskatchewan, C Canada
37 R7 **Creede** Colorado, C USA 37°51′N 106°55′W
40 I6 **Creel** Chihuahua, N Mexico 27°45′N 107°36′W
11 S11 **Cree Lake** Lake Saskatchewan, C Canada
11 V13 **Creighton** Saskatchewan, C Canada 54°46′N 101°54′W
29 Q13 **Creighton** Nebraska, C USA 42°28′N 97°54′W
103 O4 **Creil** Oise, N France 49°16′N 02°29′E
106 E8 **Crema** Lombardia, N Italy 45°22′N 09°40′E
106 E8 **Cremona** Lombardia, N Italy 45°08′N 10°02′E
Creole State *see* Louisiana
112 M10 **Crepaja** *Hung.* Cserépalja. Vojvodina, N Serbia 45°02′N 20°36′E
103 O4 **Crépy-en-Valois** Oise, N France 49°13′N 02°54′E
112 B10 **Cres** *It.* Cherso. Primorje-Gorski Kotar, NW Croatia 44°57′N 14°24′E
112 A11 **Cres** *It.* Cherso; *anc.* Crexa. *island* W Croatia
32 H14 **Crescent** Oregon, NW USA 43°27′N 121°40′W
34 K1 **Crescent City** California, W USA 41°45′N 124°14′W
23 W10 **Crescent City** Florida, SE USA 29°25′N 81°30′W
167 X10 **Crescent Group** *Chin.* Yongle Qundao, *Viet.* Nhom L Liêm. *island group* C Paracel Islands
23 W10 **Crescent Lake** Lake Florida, SE USA
29 X11 **Cresco** Iowa, C USA 43°22′N 92°06′W
61 B18 **Crespo** Entre Ríos, E Argentina 32°05′S 60°20′W
103 R13 **Crest** Drôme, E France 44°45′N 05°00′E
37 R5 **Crested Butte** Colorado, C USA 38°52′N 106°59′W
31 S12 **Crestline** Ohio, N USA 40°47′N 82°44′W
11 O17 **Creston** British Columbia, SW Canada 49°05′N 116°32′W
29 U15 **Creston** Iowa, C USA 41°03′N 94°21′W
33 V16 **Creston** Wyoming, C USA 41°40′N 107°43′W
37 S7 **Crestone Peak** ▲ Colorado, C USA 37°58′N 105°34′W
23 P8 **Crestview** Florida, SE USA 30°44′N 86°34′W
121 R10 **Cretan Trough** *undersea feature* Aegean Sea, C Mediterranean Sea
29 R16 **Crete** Nebraska, C USA 40°36′N 96°58′W
Crete *see* Kríti
103 O5 **Créteil** Val-de-Marne, N France 48°47′N 02°28′E
Crete, Sea of/Creticum, Mare *see* Kritikó Pélagos
105 X4 **Creus, Cap de** *headland* NE Spain 42°18′N 03°18′E
103 N10 **Creuse** ◆ *department* C France
102 M9 **Creuse** River C France
103 T4 **Creutzwald** Moselle, NE France 49°13′N 06°41′E
105 S12 **Crevillent** *prev.* Crevillente. Valenciana, E Spain 38°15′N 00°48′W
Crevillente *see* Crevillent
97 L18 **Crewe** C England, United Kingdom 53°05′N 02°27′W
21 V7 **Crewe** Virginia, NE USA 37°10′N 78°07′W
Crexa *see* Cres
43 Q15 **Cricamola, Río** River NW Panama
61 K14 **Criciúma** Santa Catarina, S Brazil 28°39′S 49°23′W
96 J11 **Crieff** C Scotland, United Kingdom 56°22′N 03°49′W
112 B10 **Crikvenica** *It.* Cirquenizza; *prev.* Cirkvenica, Crjkvenica. Primorje-Gorski Kotar, NW Croatia 45°12′N 14°40′E
Crimea/Crimean Oblast *see* Krym, Avtonomna Respublika
101 M16 **Crimmitschau** *var.* Krimmitschau. Sachsen, E Germany 50°48′N 12°23′E
116 G11 **Crișcior** *Hung.* Kristyor. Hunedoara, W Romania 46°09′N 22°54′E
21 Y5 **Crisfield** Maryland, NE USA 37°58′N 75°51′W
31 P3 **Crisp Point** *headland* Michigan, N USA 46°45′N 85°15′W
59 L19 **Cristalina** Goiás, C Brazil 16°43′S 47°37′W
44 J7 **Cristal, Sierra del** ▲ E Cuba
43 T14 **Cristóbal** Colón, C Panama 09°18′N 79°52′W
54 F4 **Cristóbal Colón, Pico** ▲ N Colombia 10°52′N 73°46′W
Cristur/Cristuru Săcuiesc *see* Cristuru Secuiesc
116 I11 **Cristuru Secuiesc** *prev.* Cristur, Cristuru Săcuiesc, Sitaș Cristuru, *Ger.* Kreutz, *Hung.* Székelykeresztúr, Szitás-Keresztúr. Harghita, C Romania 46°17′N 25°02′E
116 F10 **Crișul Alb** *var.* Weisse Kreisch, *Ger.* Weisse Körös, *Hung.* Fehér-Körös. River Hungary/Romania
116 F10 **Crișul Negru** *Ger.* Schwarze Körös, *Hung.* Fekete-Körös. River Hungary/Romania
116 G10 **Crișul Repede** *var.* Schnelle Kreisch, *Ger.* Schnelle Körös, *Hung.* Sebes-Körös. River Hungary/Romania
117 N10 **Criuleni** *Rus.* Kriulyany. C Moldova 47°12′N 29°09′E
Crivadia Vulcanului *see* Vulcan
Crjkvenica *see* Crikvenica
113 J17 **Crkvice** SW Montenegro 42°34′N 18°38′E
113 O17 **Crna Gora** *Alb.* Mali i Zi. ▲ FYR Macedonia/Serbia
Crna Gora *see* Montenegro
113 O20 **Crna Reka** River S FYR Macedonia
Crni Drim *see* Black Drin
109 V10 **Črni vrh** ▲ NE Slovenia 46°28′N 15°14′E
109 V13 **Črnomelj** *Ger.* Tschernembl. SE Slovenia 45°32′N 15°12′E
97 A17 **Croagh Patrick** *Ir.* Cruach Phádraig. ▲ W Ireland 53°45′N 09°39′W
112 D9 **Croatia** *off.* Republic of Croatia, *Ger.* Kroatien, *SCr.* Hrvatska. ◆ *republic* SE Europe
Croatia, Republic of *see* Croatia
Croce, Picco di *see* Wilde Kreuzspitze
15 P8 **Croche** River Québec, SE Canada
169 V7 **Crocker, Banjaran** *var.* Crocker Range. ▲ East Malaysia
Crocker Range *see* Crocker, Banjaran
25 V9 **Crockett** Texas, SW USA 31°21′N 95°30′W
67 V14 **Crocodile** *var.* Krokodil. River N South Africa
Crocodile *see* Limpopo
20 I7 **Crofton** Kentucky, S USA 37°01′N 87°25′W
29 Q12 **Crofton** Nebraska, C USA 42°43′N 97°30′W
Croia *see* Krujë
103 R16 **Croisette, Cap** *headland* SE France 43°12′N 05°21′E
102 G8 **Croisic, Pointe du** *headland* NW France 47°16′N 02°42′W
103 S13 **Croix Haute, Col de la** *pass* E France
15 U5 **Croix, Pointe à la** *headland* Québec, SE Canada 49°16′N 67°46′W
14 F13 **Croker, Cape** *headland* Ontario, S Canada 44°56′N 80°57′W
181 P1 **Croker Island** *island* Northern Territory, N Australia
96 I8 **Cromarty** N Scotland, United Kingdom 57°40′N 04°02′W
99 M21 **Crombach** Liège, E Belgium 50°14′N 06°07′E
97 Q18 **Cromer** E England, United Kingdom 52°56′N 01°06′E
185 D22 **Cromwell** Otago, South Island, New Zealand 45°03′S 169°14′E
185 H16 **Cronadun** West Coast, South Island, New Zealand 42°03′S 171°52′E
39 O11 **Crooked Creek** Alaska, USA 61°52′N 158°06′W
44 K5 **Crooked Island** *island* SE The Bahamas
44 J5 **Crooked Island Passage** *channel* SE The Bahamas
32 I13 **Crooked River** River Oregon, NW USA
29 R4 **Crookston** Minnesota, N USA 47°47′N 96°36′W
28 I10 **Crooks Tower** ▲ South Dakota, N USA 44°09′N 103°55′W
31 T14 **Crooksville** Ohio, N USA 39°46′N 82°05′W
183 R9 **Crookwell** New South Wales, SE Australia 34°28′S 149°27′E
14 L14 **Crosby** Ontario, SE Canada 44°39′N 76°13′W
97 K17 **Crosby** *var.* Great Crosby. NW England, United Kingdom 53°30′N 03°02′W
29 U6 **Crosby** Minnesota, N USA 46°30′N 93°58′W
28 K2 **Crosby** North Dakota, N USA 48°54′N 103°17′W
25 O5 **Crosbyton** Texas, SW USA 33°40′N 101°16′W
77 V16 **Cross** River Cameroon/Nigeria
23 U10 **Cross City** Florida, SE USA 29°37′N 83°08′W
Crossen *see* Krosno Odrzańskie
27 V14 **Crossett** Arkansas, C USA 33°08′N 91°58′W
97 K15 **Cross Fell** ▲ N England, United Kingdom 54°42′N 02°30′W
11 P16 **Crossfield** Alberta, SW Canada 51°24′N 114°03′W
21 Q12 **Cross Hill** South Carolina, SE USA 34°18′N 81°58′W
19 U6 **Cross Island** *island* Maine, NE USA
11 X13 **Cross Lake** Manitoba, C Canada 54°38′N 97°35′W
22 F5 **Cross Lake** Reservoir Louisiana, S USA
36 I12 **Crossman Peak** ▲ Arizona, SW USA 34°33′N 114°09′W
25 Q7 **Cross Plains** Texas, SW USA 32°07′N 99°10′W
77 V17 **Cross River** ◆ *state* SE Nigeria
20 L9 **Crossville** Tennessee, S USA 35°57′N 85°02′W
31 S8 **Croswell** Michigan, N USA 43°16′N 82°37′W
14 K13 **Crotch Lake** Lake Ontario, SE Canada
Croton/Crotona *see* Crotone
107 O21 **Crotone** *var.* Cotrone; *anc.* Croton, Crotona. Calabria, SW Italy 39°05′N 17°07′E
33 V11 **Crow Agency** Montana, NW USA 45°35′N 107°28′W
183 U7 **Crowdy Head** *headland* New South Wales, SE Australia 31°52′S 152°45′E
25 Q4 **Crowell** Texas, SW USA 33°59′N 99°45′W
183 O6 **Crowl Creek** *seasonal river* New South Wales, SE Australia
22 H9 **Crowley** Louisiana, S USA 30°11′N 92°21′W
35 S9 **Crowley, Lake** Reservoir California, W USA
27 X10 **Crowleys Ridge** *hill range* Arkansas, C USA
31 N11 **Crown Point** Indiana, N USA 41°25′N 87°22′W
37 P10 **Crownpoint** New Mexico, SW USA 35°40′N 108°09′W
33 R10 **Crow Peak** ▲ Montana, NW USA 46°17′N 111°54′W
11 P17 **Crowsnest Pass** *pass* Alberta/British Columbia, SW Canada
29 T6 **Crow Wing River** River Minnesota, N USA
97 O22 **Croydon** SE England, United Kingdom 51°21′N 00°06′W
Crozer, Mount *see* Finkol, Mount
173 P11 **Crozet Basin** *undersea feature* S Indian Ocean 39°00′S 60°00′E
173 O12 **Crozet Islands** *island group* French Southern and Antarctic Territories
173 N12 **Crozet Plateau** *var.* Crozet Plateaus. *undersea feature* SW Indian Ocean 46°00′S 51°00′E
Crozet Plateaus *see* Crozet Plateau
102 E6 **Crozon** Finistère, NW France 48°14′N 04°31′W
Cruacha Dubha, Na *see* Macgillycuddy's Reeks
Cruach Phádraig *see* Croagh Patrick
116 M14 **Crucea** Constanța, SE Romania 44°30′N 28°18′E
44 E5 **Cruces** Cienfuegos, C Cuba 22°20′N 80°17′W
107 O20 **Crucoli Torretta** Calabria, SW Italy 39°26′N 17°03′E
41 P9 **Cruillas** Tamaulipas, C Mexico 24°43′N 98°26′W
64 K9 **Cruiser Tablemount** *undersea feature* E Atlantic Ocean 32°00′N 28°00′W
61 G14 **Cruz Alta** Rio Grande do Sul, S Brazil 28°38′S 53°38′W
44 G8 **Cruz, Cabo** *headland* S Cuba 19°50′N 77°43′W
60 N9 **Cruzeiro** São Paulo, S Brazil 22°33′S 44°59′W
60 H10 **Cruzeiro do Oeste** Paraná, S Brazil 23°45′S 53°03′W
59 A15 **Cruzeiro do Sul** Acre, W Brazil 07°40′S 72°39′W
23 U11 **Crystal Bay** *bay* Florida, SE USA NE Gulf of Mexico Atlantic Ocean
182 I8 **Crystal Brook** South Australia 33°24′S 138°10′E
11 X17 **Crystal City** Manitoba, S Canada 49°07′N 98°54′W
27 X5 **Crystal City** Missouri, C USA 38°13′N 90°22′W
25 P13 **Crystal City** Texas, SW USA 28°43′N 99°51′W
30 M4 **Crystal Falls** Michigan, N USA 46°06′N 88°20′W
23 Q8 **Crystal Lake** Florida, SE USA 30°26′N 85°41′W
31 O6 **Crystal Lake** Lake Michigan, N USA
23 V11 **Crystal River** Florida, SE USA 28°54′N 82°35′W
37 Q5 **Crystal River** River Colorado, C USA
22 K6 **Crystal Springs** Mississippi, S USA 31°59′N 90°21′W
Csaca *see* Čadca
Csakathurn/Csáktornya *see* Čakovec
Csap *see* Chop
Csepén *see* Čepin
Cserépalja *see* Crepaja
Csermő *see* Cermei
Csíkszereda *see* Miercurea-Ciuc
111 L24 **Csongrád** Csongrád, SE Hungary 46°42′N 20°09′E
111 L24 **Csongrád** *off.* Csongrád Megye. ◆ *county* SE Hungary
Csongrád Megye *see* Csongrád
111 H22 **Csorna** Győr-Moson-Sopron, NW Hungary 47°37′N 17°14′E
Csucsa *see* Ciucea
111 G25 **Csurgó** Somogy, SW Hungary 46°16′N 17°09′E
Csurog *see* Čurug
54 L5 **Cúa** Miranda, N Venezuela 10°14′N 66°58′W
82 C11 **Cuale** Malanje, NW Angola 08°22′S 16°10′E
67 T12 **Cuando** *var.* Kwando. River S Africa
Cuando Cubango *see* Kuando Kubango
83 E16 **Cuangar** Kuando Kubango, S Angola 17°34′S 18°39′E
82 D11 **Cuango** Lunda Norte, NE Angola 09°10′S 17°59′E
82 C10 **Cuango** Uíge, NW Angola 06°20′S 16°42′E
82 C10 **Cuango** *var.* Kwango. River Angola/Dem. Rep. Congo *see also* Kwango
Cuango *see* Kwango
Cuan, Loch *see* Strangford Lough
82 C12 **Cuanza** *var.* Kwanza. River C Angola
Cuanza Norte *see* Kwanza Norte
Cuanza Sul *see* Kwanza Sul
61 E16 **Cuareim, Río** *var.* Rio Quaraí. River Brazil/Uruguay *see also* Quaraí, Rio
Cuareim, Río *see* Quaraí, Rio
83 D15 **Cuatir** River S Angola
40 M7 **Cuatro Ciénegas** *var.* Cuatro Ciénegas de Carranza. Coahuila, NE Mexico 27°00′N 102°03′W
Cuatro Ciénegas de Carranza *see* Cuatro Ciénegas
40 I6 **Cuauhtémoc** Chihuahua, N Mexico 28°22′N 106°52′W
41 P14 **Cuautla** Morelos, S Mexico 18°48′N 98°56′W
104 H12 **Cuba** Beja, S Portugal 38°10′N 07°54′W
27 W6 **Cuba** Missouri, C USA 38°03′N 91°24′W
37 R10 **Cuba** New Mexico, SW USA 36°01′N 107°04′W
44 E6 **Cuba** *off.* Republic of Cuba. ◆ *republic* W West Indies
47 O2 **Cuba** *island* W West Indies
82 B13 **Cubal** Benguela, W Angola 12°58′S 14°16′E
83 C15 **Cubango** *var.* Kuvango, *Port.* Vila Artur de Paiva, Vila da Ponte. Huíla, SW Angola 14°27′S 16°18′E
83 D16 **Cubango** *var.* Kavango, Kavengo, Kubango, Okavango, Okavango. River S Africa *see also* Okavango
Cubango *see* Okavango
54 H8 **Cubará** Boyacá, N Colombia 07°01′N 72°07′W
Cuba, Republic of *see* Cuba
136 I12 **Çubuk** Ankara, N Turkey 40°13′N 33°02′E
83 D14 **Cuchi** Kuando Kubango, C Angola 14°40′S 16°58′E
42 C5 **Cuchumatanes, Sierra de los** ▲ W Guatemala
Cuculaya, Rio *see* Kukalaya, Rio
82 E12 **Cucumbi** *prev.* Trás-os-Montes. Lunda Sul, NE Angola 10°13′S 19°04′E
54 G7 **Cúcuta** *var.* San José de Cúcuta. Norte de Santander, N Colombia 07°55′N 72°31′W
31 N9 **Cudahy** Wisconsin, N USA 42°54′N 87°51′W
155 J21 **Cuddalore** Tamil Nādu, SE India 11°43′N 79°46′E
155 I18 **Cuddapah** Andhra Pradesh, S India 14°30′N 78°50′E
104 M6 **Cuéllar** Castilla y León, N Spain 41°24′N 04°19′W
82 D13 **Cuemba** *var.* Coemba. Bié, C Angola 12°09′S 18°07′E
56 B8 **Cuenca** Azuay, S Ecuador 02°54′S 79°W
105 Q9 **Cuenca** *anc.* Conca. Castilla-La Mancha, C Spain 40°04′N 02°07′W
105 P9 **Cuenca** ◆ *province* Castilla-La Mancha, C Spain
40 L9 **Cuencamé** *var.* Cuencamé de Ceniceros. Durango, C Mexico 24°53′N 103°41′W
Cuencamé de Ceniceros *see* Cuencamé
105 Q8 **Cuenca, Serranía de** ▲ C Spain
Cuera *see* Chur
105 P5 **Cuerda del Pozo, Embalse de la** Reservoir N Spain
41 O14 **Cuernavaca** Morelos, S Mexico 18°57′N 99°15′W
25 T12 **Cuero** Texas, SW USA 29°06′N 97°19′W
44 I7 **Cueto** Holguín, E Cuba 20°43′N 75°54′W
41 Q13 **Cuetzalán** *var.* Cuetzalán del Progreso. Puebla, S Mexico 20°00′N 97°27′W
Cuetzalán del Progreso *see* Cuetzalán
105 Q14 **Cuevas de Almanzora** Andalucía, S Spain 37°19′N 01°52′W
Cuevas de Vinromá *see* Les Coves de Vinromá
116 H12 **Cugir** *Hung.* Kudzsir. Alba, SW Romania 45°48′N 23°25′E
59 H18 **Cuiabá** *prev.* Cuyabá. *state capital* Mato Grosso, SW Brazil 15°32′S 56°05′W
59 H19 **Cuiabá, Rio** River SW Brazil
41 R15 **Cuicatlán** *var.* San Juan Bautista Cuicatlán. Oaxaca, SE Mexico 17°49′N 96°59′W
191 W16 **Cuidado, Punta** *headland* Easter Island, Chile, E Pacific Ocean 27°08′S 109°18′W
Cúige *see* Connaught
Cúige Laighean *see* Leinster
Cúige Mumhan *see* Munster
Cuihua *see* Daguan
98 L13 **Cuijck** Noord-Brabant, SE Netherlands 51°41′N 05°56′E
Cúil an tSúdaire *see* Portarlington
42 D7 **Cuilapa** Santa Rosa, S Guatemala 14°16′N 90°18′W
42 B5 **Cuilco, Río** River W Guatemala
Cúil Mhuine *see* Collooney
Cúil Raithin *see* Coleraine
83 C14 **Cuima** Huambo, C Angola 13°16′S 15°39′E
83 E16 **Cuito** *var.* Kwito. River SE Angola
83 E15 **Cuito Cuanavale** Kuando Kubango, E Angola 15°01′S 19°07′E
41 N14 **Cuitzeo, Lago de** Lake C Mexico
27 W4 **Cuivre River** River Missouri, C USA
Çuka *see* Çukë
168 L8 **Cukai** *var.* Chukai, Kemaman. Terengganu, Peninsular Malaysia 04°15′N 103°25′E
113 L23 **Çukë** *var.* Çuka. Vlorë, S Albania 39°50′N 20°01′E
Cularo *see* Grenoble
33 Y7 **Culbertson** Montana, NW USA 48°09′N 104°30′W
28 M16 **Culbertson** Nebraska, C USA 40°13′N 100°50′W
183 P10 **Culcairn** New South Wales, SE Australia 35°41′S 147°01′E
45 W5 **Culebra** *var.* Dewey. E Puerto Rico 18°19′N 65°17′W
45 W6 **Culebra, Isla de** *island* E Puerto Rico
37 T8 **Culebra Peak** ▲ Colorado, C USA 37°07′N 105°11′W
104 J5 **Culebra, Sierra de la** ▲ NW Spain
98 J12 **Culemborg** Gelderland, C Netherlands 51°57′N 05°17′E
137 V14 **Culfa** *Rus.* Dzhul'fa. SW Azerbaijan 38°58′N 45°37′E
183 P4 **Culgoa River** River New South Wales/Queensland, SE Australia
40 I9 **Culiacán** *var.* Culiacán Rosales, Culiacán-Rosales. Sinaloa, C Mexico 24°48′N 107°25′W
Culiacán-Rosales/Culiacán Rosales *see* Culiacán
105 P14 **Cúllar-Baza** Andalucía, S Spain 37°35′N 02°34′W
105 S10 **Cullera** Valenciana, E Spain 39°10′N 00°15′W
23 P3 **Cullman** Alabama, S USA 34°10′N 86°50′W
108 B10 **Cully** Vaud, W Switzerland 46°58′N 06°46′E
Culm *see* Chełmno
Culmsee *see* Chełmża
21 V4 **Culpeper** Virginia, NE USA 38°28′N 77°59′W
185 I17 **Culverden** Canterbury, South Island, New Zealand 42°46′S 172°51′E
83 H18 **Cum** *var.* Xhumo. Central, C Botswana 21°13′S 24°38′E
55 N5 **Cumaná** Sucre, NE Venezuela 10°29′N 64°12′W
55 O5 **Cumanacoa** Sucre, NE Venezuela 10°17′N 63°58′W
54 C13 **Cumbal, Nevado de** *elevation* S Colombia
21 O7 **Cumberland** Kentucky, S USA 36°55′N 83°00′W
21 U2 **Cumberland** Maryland, NE USA 39°40′N 78°47′W
21 V6 **Cumberland** Virginia, NE USA 37°31′N 78°16′W
187 P12 **Cumberland, Cape** *var.* Cape Nahoi. *headland* Espiritu Santo, N Vanuatu 14°39′S 166°35′E
11 V14 **Cumberland House** Saskatchewan, C Canada 53°57′N 102°21′W
23 W8 **Cumberland Island** *island* Georgia, SE USA
20 L7 **Cumberland, Lake** Reservoir Kentucky, S USA
9 R5 **Cumberland Peninsula** *peninsula* Baffin Island, Nunavut, NE Canada
2 N9 **Cumberland Plateau** *plateau* E USA
30 L1 **Cumberland Point** *headland* Michigan, N USA 47°51′N 89°14′W
21 O7 **Cumberland River** River Kentucky/Tennessee, S USA
9 S6 **Cumberland Sound** *inlet* Baffin Island, Nunavut, NE Canada
96 I12 **Cumbernauld** S Scotland, United Kingdom 55°57′N 04°W
97 K15 **Cumbria** *cultural region* NW England, United Kingdom
97 K15 **Cumbrian Mountains** ▲ NW England, United Kingdom
23 S2 **Cumming** Georgia, SE USA 34°12′N 84°08′W
Cummin in Pommern *see* Kamień Pomorski
182 G9 **Cummins** South Australia 34°17′S 135°43′E
96 I13 **Cumnock** W Scotland, United Kingdom 55°32′N 04°28′W
40 G4 **Cumpas** Sonora, NW Mexico 30°N 109°48′W
136 H16 **Çumra** Konya, C Turkey 37°34′N 32°38′E
63 G15 **Cunco** Araucanía, C Chile 38°55′S 72°02′W
54 E9 **Cundinamarca** *off.* Departamento de Cundinamarca. ◆ *province* C Colombia
Cundinamarca, Departamento de *see* Cundinamarca
41 U15 **Cunduacán** Tabasco, SE Mexico 18°00′N 93°07′W
83 C16 **Cunene** ◆ *province* S Angola
83 A16 **Cunene** *var.* Kunene. River Angola/Namibia *see also* Kunene
Cunene *see* Kunene
106 A9 **Cuneo** *Fr.* Coni. Piemonte, NW Italy 44°23′N 07°32′E
83 E15 **Cunjamba** Kuando Kubango, E Angola 15°22′S 20°07′E
181 V10 **Cunnamulla** Queensland, E Australia 28°09′S 145°44′E
Čunusavvon *see* Junosuando
Cuokkarašša *see* Čohkarášša
106 B7 **Cuorgne** Piemonte, NE Italy 45°23′N 07°34′E
96 K11 **Cupar** E Scotland, United Kingdom 56°19′N 03°01′W
116 L8 **Cupcina** *Rus.* Kupchino; *prev.* Calinisc, Kalinisk. N Moldova 48°07′N 27°22′E
54 C8 **Cupica** Chocó, W Colombia 06°43′N 77°31′W
54 C8 **Cupica, Golfo de** *gulf* W Colombia
112 N13 **Ćuprija** Serbia, E Serbia 43°57′N 21°21′E
Cura *see* Villa de Cura
45 S9 **Curaçao** *prev.* Dutch West Indies. ◇ Dutch *self-governing territory* S Caribbean Sea
45 P16 **Curaçao** *island* Lesser Antilles
56 H13 **Curanja, Río** River E Peru
56 F7 **Curaray, Río** River Ecuador/Peru
116 K14 **Curcani** Călărași, SE Romania 44°11′N 26°39′E
182 H4 **Curdimurka** South Australia 29°27′S 136°56′E
103 P7 **Cure** River C France
173 Y16 **Curepipe** C Mauritius 20°19′S 57°31′E
55 R6 **Curiapo** Delta Amacuro, NE Venezuela 10°03′N 63°05′W
Curia Rhaetorum *see* Chur
62 G12 **Curicó** Maule, C Chile 35°00′S 71°15′W
Curieta *see* Krk
172 I15 **Curieuse** *island* Inner Islands, NE Seychelles
59 C16 **Curitiba** Acre, W Brazil 10°08′S 69°00′W
60 K12 **Curitiba** *prev.* Curytiba. *state capital* Paraná, S Brazil 25°25′S 49°25′W
60 J13 **Curitibanos** Santa Catarina, S Brazil 27°18′S 50°35′W
183 S6 **Curlewis** New South Wales, SE Australia 31°09′S 150°18′E
182 J6 **Curnamona** South Australia 31°39′S 139°35′E
83 A15 **Curoca** River SW Angola
183 T6 **Currabubula** New South Wales, SE Australia 31°17′S 150°43′E
59 Q14 **Currais Novos** Rio Grande do Norte, E Brazil 06°12′S 36°30′W
35 W7 **Currant** Nevada, W USA 38°43′N 115°27′W
35 W6 **Currant Mountain** ▲ Nevada, W USA 38°56′N 115°19′W
44 H2 **Current** Eleuthera Island, C The Bahamas 25°24′N 76°44′W
27 W8 **Current River** River Arkansas/Missouri, C USA
182 M14 **Currie** Tasmania, SE Australia 39°59′S 143°51′E
21 Y8 **Currituck** North Carolina, SE USA 36°29′N 76°02′W
21 Y8 **Currituck Sound** *sound* North Carolina, SE USA
39 R11 **Curry** Alaska, USA 62°36′N 150°00′W
Curtbunar *see* Tervel
116 I13 **Curtea de Argeș** *var.* Curtea-de-Arges. Argeș, S Romania 45°06′N 24°40′E
Curtea-de-Arges *see* Curtea de Argeș
116 E10 **Curtici** *Ger.* Kurtitsch, *Hung.* Kürtös. Arad, W Romania 46°21′N 21°17′E
104 H2 **Curtis** Galicia, NW Spain 43°09′N 08°10′W
28 M16 **Curtis** Nebraska, C USA 40°36′N 100°27′W
183 O6 **Curtis Group** *island group* Tasmania, SE Australia
181 Y8 **Curtis Island** *island* Queensland, SE Australia
58 K11 **Curuá, Ilha do** *island* NE Brazil
47 U7 **Curuá, Rio** River N Brazil
59 A14 **Curuçá, Rio** River NW Brazil
112 L9 **Čurug** *Hung.* Csurog. Vojvodina, N Serbia 45°30′N 20°02′E
61 D16 **Curuzú Cuatiá** Corrientes, NE Argentina 29°50′S 58°05′W
59 M19 **Curvelo** Minas Gerais, SE Brazil 18°45′S 44°27′W
18 E14 **Curwensville** Pennsylvania, NE USA 40°57′N 78°29′W
30 M3 **Curwood, Mount** ▲ Michigan, N USA 46°42′N 88°14′W
Curytiba *see* Curitiba
Curzola *see* Korčula
42 A10 **Cuscatlán** ◆ *department* C El Salvador
57 H15 **Cusco** *var.* Cuzco. Cusco, C Peru 13°35′S 72°02′W
57 H15 **Cusco** *off.* Departamento de Cusco, *var.* Cuzco. ◆ *department* C Peru
Cusco, Departamento de *see* Cusco
27 O9 **Cushing** Oklahoma, C USA 35°01′N 96°46′W
25 W8 **Cushing** Texas, SW USA 31°48′N 94°50′W
40 I6 **Cusihuiriachic** Chihuahua, N Mexico 28°16′N 106°46′W
103 P10 **Cusset** Allier, C France 46°08′N 03°27′E
23 S6 **Cusseta** Georgia, SE USA 32°18′N 84°46′W
28 J10 **Custer** South Dakota, N USA 43°46′N 103°36′W
Cüstrin *see* Kostrzyn
33 Q7 **Cut Bank** Montana, NW USA 48°38′N 112°20′W
Cutch, Gulf of *see* Kachchh, Gulf of
23 S6 **Cuthbert** Georgia, SE USA 31°46′N 84°47′W
11 S15 **Cut Knife** Saskatchewan, S Canada 52°40′N 108°54′W
23 Y16 **Cutler Ridge** Florida, SE USA 25°34′N 80°21′W
22 K10 **Cut Off** Louisiana, S USA 29°32′N 90°20′W
63 I15 **Cutral-Có** Neuquén, C Argentina 38°56′S 69°13′W
107 O21 **Cutro** Calabria, SW Italy 39°01′N 16°59′E
183 O4 **Cuttaburra Channels** *seasonal river* New South Wales, SE Australia
154 O12 **Cuttack** Odisha, E India 20°28′N 85°53′E
83 C15 **Cuvelai** Cunene, SW Angola 15°40′S 15°48′E
79 G18 **Cuvette** *var.* Région de la Cuvette. ◆ *province* C Congo
Cuvette, Région de la *see* Cuvette
173 V9 **Cuvier Basin** *undersea feature* E Indian Ocean
173 U9 **Cuvier Plateau** *undersea feature* E Indian Ocean
82 B12 **Cuvo** River W Angola
100 H9 **Cuxhaven** Niedersachsen, NW Germany 53°51′N 08°43′E
Cuyabá *see* Cuiabá
Cuyuni, Río *see* Cuyuni River
55 S8 **Cuyuni River** *var.* Río Cuyuni. River Guyana/Venezuela
Cuzco *see* Cusco
97 K22 **Cwmbran** *Wel.* Cwmbrân. SW Wales, United Kingdom 51°39′N 03°W
Cwmbrân *see* Cwmbran
28 K15 **C. W. McConaughy, Lake** Reservoir Nebraska, C USA
81 D20 **Cyangugu** SW Rwanda 02°27′S 29°00′E
110 D11 **Cybinka** *Ger.* Ziebingen. Lubuskie, W Poland 52°11′N 14°46′E
Cyclades *see* Kykládes
Cydonia *see* Chaniá
Cymru *see* Wales
20 M5 **Cynthiana** Kentucky, S USA 38°22′N 84°18′W
11 S17 **Cypress Hills** ▲ Alberta/Saskatchewan, SW Canada
Cypro-Syrian Basin *see* Cyprus Basin
121 U11 **Cyprus** *off.* Republic of Cyprus, *Gk.* Kypros, *Turk.* Kıbrıs, Kıbrıs Cumhuriyeti. ◆ *republic* E Mediterranean Sea
84 L14 **Cyprus** *Gk.* Kypros, *Turk.* Kıbrıs. *island* E Mediterranean Sea
121 W11 **Cyprus Basin** *var.* Cypro-Syrian Basin. *undersea feature* E Mediterranean Sea 34°00′N 34°00′E
Cyprus, Republic of *see* Cyprus
Cythera *see* Kýthira
Cythnos *see* Kýthnos
110 F9 **Czaplinek** *Ger.* Tempelburg. Zachodnio-pomorskie, NW Poland 53°33′N 16°14′E
Czarna Woda *see* Wda
110 G8 **Czarne** Pomorskie, N Poland 53°40′N 17°00′E
110 G10 **Czarnków** Wielkopolskie, C Poland 52°53′N 16°32′E
111 E17 **Czech Republic** *Cz.* Česká Republika. ◆ *republic* C Europe
Czegléd *see* Cegléd
110 G12 **Czempiń** Wielkopolskie, C Poland 52°10′N 16°46′E
Czenstochau *see* Częstochowa
Czerkow *see* Čerchov
Czernowitz *see* Chernivtsi
110 I8 **Czersk** Pomorskie, N Poland 53°48′N 17°58′E
111 J15 **Częstochowa** *Ger.* Czenstochau, Tschenstochau, *Rus.* Chenstokhov. Śląskie, S Poland 50°49′N 19°07′E
110 F10 **Człopa** *Ger.* Schloppe. Zachodnio-pomorskie, NW Poland 53°05′N 16°05′E
110 H8 **Człuchów** *Ger.* Schlochau. Pomorskie, NW Poland 53°41′N 17°20′E

D

163 V9 **Da'an** *var.* Dalai. Jilin, NE China 45°28′N 124°18′E
15 S10 **Daaquam** Québec, SE Canada 46°36′N 70°03′W
Daawo, Webi *see* Dawa Wenz
54 I4 **Dabajuro** Falcón, NW Venezuela 11°00′N 70°41′W
77 N15 **Dabakala** NE Ivory Coast 08°19′N 04°24′W
163 S11 **Daban** *var.* Bairin Youqi. Nei Mongol Zizhiqu, N China 43°33′N 118°40′E
111 K23 **Dabas** Pest, C Hungary 47°36′N 19°22′E
160 L8 **Daba Shan** ▲ C China
Dabba *see* Daocheng
140 J5 **Dabbāgh, Jabal** ▲ NW Saudi Arabia 27°52′N 35°48′E
54 D8 **Dabeiba** Antioquia, NW Colombia 07°01′N 76°18′W
154 E11 **Dabhoi** Gujarāt, W India 22°08′N 73°28′E
161 P8 **Dabie Shan** ▲ C China
76 J13 **Dabola** C Guinea 10°48′N 11°02′W
77 N17 **Dabou** S Ivory Coast 05°20′N 04°23′W
162 M15 **Dabqig** *prev.* Uxin Qi. Nei Mongol Zizhiqu, N China 38°29′N 108°48′E
110 P8 **Dąbrowa Białostocka** Podlaskie, NE Poland 53°38′N 23°18′E
111 M16 **Dąbrowa Tarnowska** Małopolskie, S Poland 50°10′N 21°E
119 M20 **Dabryn'** *Rus.* Dobryn'. Homyel'skaya Voblasts', SE Belarus 51°46′N 29°12′E
159 P10 **Dabsan Hu** Lake C China
161 Q13 **Dabu** *var.* Huliao. Guangdong, S China 24°19′N 116°07′E
116 H15 **Dăbuleni** Dolj, SW Romania 43°48′N 24°05′E
152 G9 **Dabwāli** Haryāna, NW India 29°56′N 74°40′E
Dacca *see* Dhaka
101 L23 **Dachau** Bayern, SE Germany 48°15′N 11°26′E
Dachuan *see* Dazhou
Dacia Bank *see* Dacia Seamount

◆ Country
● Country Capital
◇ Dependent Territory
○ Dependent Territory Capital
◆ Administrative Regions
✈ International Airport
▲ Mountain
▲ Mountain Range
Volcano
River
Lake
Reservoir

64 M10 **Dacia Seamount** *var.* Dacia Bank. *undersea feature* E Atlantic Ocean 31°10′N 13°42′W
37 T3 **Dacono** Colorado, C USA 40°04′N 104°56′W
Đắc Tô *see* Đắk Tô
Dacura *see* Dákura
23 W12 **Dade City** Florida, SE USA 28°21′N 82°12′W
152 L10 **Dadeldhurā** *var.* Dandeldhura. Far Western, W Nepal 29°12′N 80°31′E
23 Q5 **Dadeville** Alabama, S USA 32°49′N 85°45′W
Dadong *see* Donggang
103 N15 **Dadou** ~ S France
154 D12 **Dādra and Nagar Haveli** ◆ *union territory* W India
149 P14 **Dādu** Sind, SE Pakistan 26°42′N 67°48′E
167 U11 **Da Du Boloc** Kon Tum, C Vietnam 14°06′N 107°40′E
160 G9 **Dadu He** ~ C China
163 V15 **Daecheong-do** *prev.* Taechŏng-do. *island* NW South Korea
163 Y16 **Daegu**, *Jap.* Taikyū; *prev.* Taegu. SE South Korea 35°55′N 128°33′E
163 Y15 **Daejeon**, *Jap.* Taiden; *prev.* Taejŏn. C South Korea 36°20′N 127°28′E
Daerah Istimewa Aceh *see* Aceh
171 P4 **Daet** Luzon, N Philippines 14°06′N 122°57′E
160 I11 **Dafang** Guizhou, S China 27°07′N 105°40′E
Dafeng *see* Shanglin
153 W11 **Dafla Hills** ▲ NE India
11 U15 **Dafoe** Saskatchewan, S Canada 51°46′N 104°11′W
76 G10 **Dagana** N Senegal 16°28′N 15°35′W
Dagana *see* Massakory, Chad
Dagana *see* Dahana, Tajikistan
Dagcagoin *see* Zoigê
118 K11 **Dagda** SE Latvia 56°06′N 27°36′E
Dagden *see* Hiiumaa
Dagden-Sund *see* Soela Väin
127 P16 **Dagestan, Respublika** *prev.* Dagestanskaya ASSR, *Eng.* Daghestan. ◆ *autonomous republic* SW Russian Federation
Dagestanskaya ASSR *see* Dagestan, Respublika
127 R17 **Dagestanskiye Ogni** Respublika Dagestan, SW Russian Federation 42°09′N 48°08′E
Dagezhen *see* Fengning
185 A23 **Dagg Sound** *sound* South Island, New Zealand
Daghestan *see* Dagestan, Respublika
141 Y8 **Daghmar** NE Oman 23°09′N 59°01′E
Dağliq Quarabağ *see* Nagorno-Karabakh
Dagö *see* Hiiumaa
54 D11 **Dagua** Valle del Cauca, W Colombia 03°39′N 76°40′W
160 H11 **Daguan** *var.* Cuihua. Yunnan, SW China 27°42′N 103°51′E
171 N3 **Dagupan** *off.* Dagupan City. Luzon, N Philippines 16°05′N 120°21′E
Dagupan City *see* Dagupan
159 N16 **Dagzê** *var.* Dêqên. Xizang Zizhiqu, W China 29°38′N 91°15′E
147 Q13 **Dahana** *Rus.* Dagana, Dakhana. SW Tajikistan 38°03′N 69°51′E
163 V10 **Dahei Shan** ▲ N China
163 T7 **Da Hinggan Ling** *Eng.* Great Khingan Range. ▲ NE China
Dahlac Archipelago *see* Dahlak Archipelago
80 K9 **Dahlak Archipelago** *var.* Dahlac Archipelago. *island group* E Eritrea
23 T2 **Dahlonega** Georgia, SE USA 34°31′N 83°59′W
101 O14 **Dahme** Brandenburg, E Germany 52°10′N 13°47′E
100 O13 **Dahme** ~ E Germany
141 O14 **Dahm, Ramlat** *desert* NW Yemen
154 E10 **Dāhod** *prev.* Dohad. Gujarāt, W India 22°48′N 74°18′E
Dahomey *see* Benin
158 G10 **Dahongliutan** Xinjiang Uygur Zizhiqu, NW China 35°59′N 79°12′E
Dahra *see* Dara
Dahuaishu *see* Hongtong
139 R2 **Dahūk** *var.* Dohuk, *Kurd.* Dihok, Dihôk. Dahūk, N Iraq 36°52′N 43°01′E
139 R2 **Dahūk** *off.* Muḥāfa at Dahūk, *var.* Dohuk, *Kurd.* Dihok. *off. Kurd.* Parêzga-î Dihok. ◆ *governorate* N Iraq
Dahūk, Muḥāfa at *see* Dahūk
116 J15 **Daia** Giurgiu, S Romania 44°00′N 25°59′E
165 P12 **Daigo** Ibaraki, Honshū, S Japan 36°43′N 140°22′E
163 O13 **Dai Hai** ⊚ N China
Daihoku *see* Taibei
186 M8 **Dai Island** *island* N Solomon Islands
166 M8 **Daik-u** Bago, SW Myanmar (Burma) 17°46′N 96°40′E
138 H9 **Dā'īl** Dar'ā, S Syria 32°45′N 36°08′E
167 U12 **Đại Lanh** Khanh Hoa, S Vietnam 12°49′N 109°20′E
161 Q13 **Daimao Shan** ▲ SE China
105 N11 **Daimiel** Castilla-La Mancha, C Spain 39°04′N 03°37′W
115 F22 **Daimoniá** Pelopónnisos, S Greece 36°38′N 22°54′E
Dainan *see* Tainan
25 W6 **Daingerfield** Texas, SW USA 33°03′N 94°42′W
Daingin, Bá an *see* Dingle Bay
159 R13 **Dainkognubma** Xizang Zizhiqu, W China 32°26′N 97°58′E
164 K14 **Daiō-zaki** *headland* Honshū, SW Japan 34°15′N 136°50′E
Dairbhre *see* Valencia Island
61 B22 **Daireaux** Buenos Aires, E Argentina 36°34′S 61°40′W
Dairen *see* Dalian
Dairût *see* Dayrūṭ
25 X10 **Daisetta** Texas, SW USA 30°06′N 94°38′W
192 G5 **Daitō-jima** *island group* SW Japan
192 G5 **Daitō Ridge** *undersea feature* N Philippine Sea 25°30′N 133°00′E
161 N3 **Daixian** *var.* Dai Xian, Shangguan. Shanxi, C China 39°10′N 112°57′E
Dai Xian *see* Daixian
Daiyue *see* Shanyin
161 Q12 **Daiyun Shan** ▲ SE China
44 M8 **Dajabón** NW Dominican Republic 19°35′N 71°41′W
160 G8 **Dajin Chuan** ~ C China
148 J6 **Dak** ⊚ W Afghanistan
76 F11 **Dakar** ● (Senegal) W Senegal 14°44′N 17°27′W
76 F11 **Dakar** ✕ W Senegal 14°42′N 17°27′W
Đăk Glây *see* Đăk Glei
167 U10 **Đăk Glei** *prev.* Đăk Glây. Kon Tum, C Vietnam 15°05′N 107°42′E
Dakhana *see* Dahana
153 U16 **Dakhin Shahbazpur Island** *island* S Bangladesh
Dakhla *see* Ad Dakhla
76 F7 **Dakhlet Nouâdhibou** ◆ *region* NW Mauritania
Đăk Lap *see* Kiên Đưc
Đăk Nông *see* Gia Nghia
77 U11 **Dakoro** Maradi, S Niger 14°29′N 06°45′E
29 U12 **Dakota City** Iowa, C USA 42°42′N 94°13′W
29 R13 **Dakota City** Nebraska, C USA 42°25′N 96°25′W
Dakovica *see* Gjakovë
112 I10 **Đakovo** *var.* Djakovo, *Hung.* Diakovár. Osijek-Baranja, E Croatia 45°18′N 18°24′E
Dakshin *see* Deccan
167 U11 **Đăk Tô** *var.* Đăc Tô. Kon Tum, C Vietnam 14°35′N 107°55′E
43 N7 **Dákura** *var.* Dacura. Región Autónoma Atlántico Norte, NE Nicaragua 14°22′N 83°13′W
95 I14 **Dal** Akershus, S Norway 60°19′N 11°16′E
82 E12 **Dala** Lunda Sul, E Angola 11°04′S 20°15′E
108 J8 **Dalaas** Vorarlberg, W Austria 47°08′N 10°03′E
76 I13 **Dalaba** W Guinea 10°47′N 12°12′W
Dalai *see* Da'an
162 I12 **Dalain Hob** *var.* Ejin Qi. Nei Mongol Zizhiqu, N China 41°59′N 101°04′E
Dalai Nor *see* Hulun Nur
163 Q11 **Dalai Nur** *salt lake* N China
Dala-Jarna *see* Järna
95 M14 **Dalälven** ~ C Sweden
136 C16 **Dalaman** Muğla, SW Turkey 36°47′N 28°47′E
136 C16 **Dalaman** ✕ Muğla, SW Turkey 36°37′N 28°51′E
136 D16 **Dalaman Çayı** ~ SW Turkey
162 K11 **Dalandzadgad** Ömnögovĭ, S Mongolia 43°35′N 104°23′E
95 D17 **Dalane** *physical region* S Norway
189 Z2 **Dalap-Uliga-Djarrit** *var.* Delap-Uliga-Darrit, D-U-D. *island group* Ratak Chain, SE Marshall Islands
94 J12 **Dalarna** *prev.* Kopparberg. ◆ *county* C Sweden
94 L13 **Dalarna** *prev. Eng.* Dalecarlia. *cultural region* C Sweden
95 P16 **Dalarö** Stockholm, C Sweden 59°07′N 18°25′E
167 U13 **Đa Lat** Lâm Đông, S Vietnam 11°56′N 108°25′E
Dalay *see* Bayandalay
148 L12 **Dālbandīn** *var.* Dāl Bandin. Baluchistān, SW Pakistan 28°48′N 64°08′E
95 J17 **Dalbosjön** *lake bay* S Sweden
181 Y10 **Dalby** Queensland, E Australia 27°11′S 151°12′E
94 D13 **Dale** Hordaland, S Norway 60°35′N 05°48′E
94 C12 **Dale** Sogn Og Fjordane, S Norway 61°22′N 05°24′E
32 K12 **Dale** Oregon, NW USA 44°58′N 118°56′W
25 T11 **Dale** Texas, SW USA 29°56′N 97°34′W
Dalecarlia *see* Dalarna
21 W4 **Dale City** Virginia, NE USA 38°38′N 77°18′W
20 L8 **Dale Hollow Lake** ⊠ Kentucky/Tennessee, S USA
98 O8 **Dalen** Drenthe, NE Netherlands 52°42′N 06°45′E
95 E15 **Dalen** Telemark, S Norway 59°25′N 07°58′E
166 K14 **Daletme** Chin State, W Myanmar (Burma) 21°44′N 92°48′E
23 Q7 **Daleville** Alabama, S USA 31°18′N 85°42′W
98 M9 **Dalfsen** Overijssel, E Netherlands 52°31′N 06°16′E
114 M8 **Dalgopol** *var.* Dŭlgopol. Varna, E Bulgaria 43°05′N 27°24′E
24 M1 **Dalhart** Texas, SW USA 36°05′N 102°31′W
13 O13 **Dalhousie** New Brunswick, SE Canada 48°03′N 66°22′W
152 I6 **Dalhousie** Himāchal Pradesh, N India 32°32′N 76°01′E
160 F12 **Dali** *var.* Xiaguan. Yunnan, SW China 25°34′N 100°11′E
Dali *see* Idálion
163 U14 **Dalian** *var.* Dairen, Dalien, Jay Dairen, Lüda, Ta-lien, *Rus.* Dalny. Liaoning, NE China 38°53′N 121°37′E
105 O15 **Dalías** Andalucía, S Spain 36°49′N 02°50′W
Dalien *see* Dalian
Dalijan *see* Delījān
112 J9 **Dalj** *Hung.* Dalja. Osijek-Baranja, E Croatia 45°29′N 19°00′E
Dalja *see* Dalj
32 F12 **Dallas** Oregon, NW USA 44°56′N 123°20′W
25 U6 **Dallas** Texas, SW USA 32°47′N 96°48′W
25 T7 **Dallas-Fort Worth** ✕ Texas, SW USA 32°37′N 97°16′W
154 K12 **Dalli Rājhara** *var.* Dhalli Rajhara. Chhattīsgarh, C India 20°32′N 81°10′E
39 X15 **Dall Island** *island* Alexander Archipelago, Alaska, USA
38 M12 **Dall Lake** ⊚ Alaska, USA
Dállogilli *see* Korpilombolo
77 S12 **Dallol Bosso** *seasonal river* W Niger
141 U7 **Dalmā** *island* W United Arab Emirates
113 E14 **Dalmacija** *Eng.* Dalmatia, *Ger.* Dalmatien, *It.* Dalmazia. *cultural region* S Croatia
Dalmatia/Dalmatien/Dalmazia *see* Dalmacija
123 S15 **Dal'negorsk** Primorskiy Kray, SE Russian Federation 44°27′N 135°30′E
Dalny *see* Dalian
76 M16 **Daloa** C Ivory Coast 06°56′N 06°28′W
160 J11 **Dalou Shan** ▲ S China
181 X7 **Dalrymple Lake** ⊚ Queensland, E Australia
14 H14 **Dalrymple Lake** ⊚ Ontario, S Canada
181 X7 **Dalrymple, Mount** ▲ Queensland, E Australia 21°01′S 148°34′E
93 K20 **Dalsbruk** *Fin.* Taalintehdas. Varsinais-Suomi, SW Finland 60°02′N 22°31′E
95 K19 **Dalsjöfors** Västra Götaland, S Sweden 57°43′N 13°05′E
95 J17 **Dals Långed** *var.* Långed. Västra Götaland, S Sweden 58°54′N 12°20′E
153 O15 **Dāltenganj** *prev.* Daltonganj. Jhārkhand, N India 24°02′N 84°07′E
23 R2 **Dalton** Georgia, SE USA 34°46′N 84°58′W
Daltonganj *see* Dāltenganj
195 X14 **Dalton Iceberg Tongue** *ice feature* Antarctica
92 J1 **Dalvík** Norðurland Eystra, N Iceland 65°58′N 18°31′W
Dálvvadis *see* Jokkmokk
35 N8 **Daly City** California, W USA 37°44′N 122°27′W
181 P2 **Daly River** ~ Northern Territory, N Australia
181 Q3 **Daly Waters** Northern Territory, N Australia 16°21′S 133°22′E
119 F20 **Damachava** *var.* Damachova, *Pol.* Domaczewo, *Rus.* Domachëvo. Brestskaya Voblasts', SW Belarus 51°45′N 23°36′E
Damachova *see* Damachava
77 W11 **Damagaram Takaya** Zinder, S Niger 14°02′N 09°28′E
154 D12 **Damān** Damān and Diu, W India 20°25′N 72°58′E
154 B12 **Damān and Diu** ◆ *union territory* W India
75 V7 **Damanhûr** *anc.* Hermopolis Parva. N Egypt 31°03′N 30°28′E
161 O1 **Damaqun Shan** ▲ E China
79 I15 **Damara** Ombella-Mpoko, S Central African Republic 05°00′N 18°45′E
83 D18 **Damaraland** *physical region* C Namibia
171 S15 **Damar, Kepulauan** *var.* Baraf Daja Islands, Kepulauan Barat Daya. *island group* C Indonesia
168 J8 **Damar Laut** Perak, Peninsular Malaysia 04°13′N 100°36′E
171 S15 **Damar, Pulau** *island* Maluku, E Indonesia
Damas *see* Dimashq
77 Y12 **Damasak** Borno, NE Nigeria 13°10′N 12°40′E
Damasco *see* Dimashq
21 Q8 **Damascus** Virginia, NE USA 36°37′N 81°46′W
Damascus *see* Dimashq
77 X13 **Damaturu** Yobe, NE Nigeria 11°44′N 11°58′E
171 R9 **Damau** Pulau Kaburuang, N Indonesia 03°46′N 126°49′E
143 O5 **Damāvand, Qolleh-ye** ▲ N Iran 35°56′N 52°08′E
82 B10 **Damba** Uíge, NW Angola 06°44′S 15°20′E
114 M12 **Dambaslar** Tekirdağ, NW Turkey 41°13′N 27°13′E
116 J13 **Dâmbovița** *prev.* Dîmbovița. ◆ *county* SE Romania
116 J13 **Dâmbovița** *prev.* Dîmbovița. ~ S Romania
173 Y15 **D'Ambre, Île** *island* NE Mauritius
155 K24 **Dambulla** Central Province, C Sri Lanka 07°51′N 80°40′E
44 K9 **Dame-Marie** SW Haiti 18°36′N 74°26′W
44 J9 **Dame Marie, Cap** *headland* SW Haiti 18°37′N 74°24′W
143 Q4 **Dāmghān** Semnān, N Iran 36°13′N 54°22′E
Damietta *see* Dumyāṭ
138 G10 **Dāmiyā** Al Balqā', NW Jordan 32°07′N 35°33′E
146 G11 **Damla** Daşoguz Welaýaty, N Turkmenistan 40°05′N 59°15′E
Dammām *see* Ad Dammān
100 G12 **Damme** Niedersachsen, NW Germany 52°31′N 08°12′E
153 R15 **Dāmodar** ~ NE India
154 J9 **Damoh** Madhya Pradesh, C India 23°50′N 79°30′N
77 P15 **Damongo** NW Ghana 09°05′N 01°49′W
138 G7 **Damoûr** *var.* Ad Dāmūr. W Lebanon 33°36′N 35°30′E
171 N11 **Dampal, Teluk** *bay* Sulawesi, C Indonesia
180 H7 **Dampier** Western Australia 20°40′S 116°40′E
180 H6 **Dampier Archipelago** *island group* Western Australia
141 U14 **Damqawt** *var.* Damqut. E Yemen 16°35′N 52°39′E
159 O13 **Dam Qu** ~ C China
Damqut *see* Damqawt
167 R13 **Dâmrei, Chuŏr Phnum** *Fr.* Chaîne de l'Éléphant. ▲ SW Cambodia
108 C7 **Damvant** Jura, NW Switzerland 47°22′N 06°55′E
Damwâld *see* Damwoude
98 L5 **Damwoude** *Fris.* Damwâld. Fryslân, N Netherlands 53°18′N 05°59′E
159 N15 **Damxung** *var.* Gongtang. Xizang Zizhiqu, W China 30°29′N 91°02′E
80 K11 **Danakil Desert** *var.* Afar Depression, Danakil Plain. *desert* E Africa
Danakil Plain *see* Danakil Desert
35 R8 **Dana, Mount** ▲ California, W USA 37°54′N 119°13′W
76 L16 **Danané** W Ivory Coast 07°16′N 08°09′W
167 U10 **Đa Nang** *prev.* Tourane. Quang Nam-Đa Nang, C Vietnam 16°04′N 108°14′E
160 G9 **Danba** *var.* Zhanggu, *Tib.* Rongzhag. Sichuan, C China 30°54′N 101°49′E
Danborg *see* Daneborg
18 L13 **Danbury** Connecticut, NE USA 41°21′N 73°27′W
25 W12 **Danbury** Texas, SW USA 29°13′N 95°20′W
35 X15 **Danby Lake** ⊚ California, W USA
194 H4 **Danco Coast** *physical region* Antarctica
82 B11 **Dande** ~ NW Angola
Dandeldhura *see* Dadeldhurā
155 E17 **Dandeli** Karnātaka, W India 15°18′N 74°42′E
183 O12 **Dandenong** Victoria, SE Australia 38°01′S 145°13′E
163 V13 **Dandong** *var.* Tan-tung; *prev.* An-tung. Liaoning, NE China 40°10′N 124°23′E
197 Q14 **Daneborg** *var.* Danborg. ◆ N Greenland
25 V12 **Danevang** Texas, SW USA 29°03′N 96°11′W
Dänew *see* Galkynyş
Danfeng *see* Shizong
14 L12 **Danford Lake** Québec, SE Canada 45°55′N 76°12′W
19 T4 **Danforth** Maine, NE USA 45°39′N 67°54′W
37 P3 **Danforth Hills** ▲ Colorado, C USA
Dangara *see* Danghara
159 V12 **Dangchang** Gansu, C China 34°01′N 104°19′E
159 P8 **Dangchengwan** *var.* Subei, Subei Mongolzu Zizhixian. Gansu, N China 39°33′N 94°50′E
82 B10 **Dange** Uíge, NW Angola 07°55′S 15°01′E
Dangerous Archipelago *see* Tuamotu, Îles
83 E26 **Danger Point** *headland* SW South Africa 34°37′S 19°20′E
147 Q13 **Danghara** *Rus.* Dangara. SW Tajikistan 38°05′N 69°14′E
159 P8 **Danghe Nanshan** ▲ W China
80 I12 **Dangila** *var.* Dāngla. Āmara, NW Ethiopia 11°08′N 36°51′E
159 P8 **Dangjin Shankou** *pass* N China
Dangla *see* Tanggula Shan, China
Dang La *see* Tanggula Shankou, China
Dānglā *see* Dangila, Ethiopia
Dangme Chu *see* Manās
153 Y11 **Dāngori** Assam, NE India 27°40′N 95°35′E
Dang Raek, Phanom/Dangrek, Chaîne des *see* Dângrêk, Chuŏr Phnum
167 S11 **Dângrêk, Chuŏr Phnum** *var.* Phanom Dang Raek, Phanom Dong Rak, *Fr.* Chaîne des Dangrek. ▲ Cambodia/Thailand
42 G3 **Dangriga** *prev.* Stann Creek. Stann Creek, E Belize 16°59′N 88°13′W
161 P6 **Dangshan** Anhui, E China 34°22′N 116°21′E
33 T15 **Daniel** Wyoming, C USA 42°49′N 110°04′W
83 H22 **Daniëlskuil** Northern Cape, N South Africa 28°11′S 23°33′E
19 N12 **Danielson** Connecticut, NE USA 41°48′N 71°53′W
124 M15 **Danilov** Yaroslavskaya Oblast', W Russian Federation 58°11′N 40°11′E
127 O9 **Danilovka** Volgogradskaya Oblast', SW Russian Federation 50°21′N 44°03′E
Danish West Indies *see* Virgin Islands (US)
160 L7 **Dan Jiang** ~ C China
160 M7 **Danjiangkou Shuiku** ⊠ C China
141 W8 **Ḍank** *var.* Dhank. NW Oman 23°34′N 56°16′E
152 J7 **Dankhar** Himāchal Pradesh, N India 32°08′N 78°12′E
126 L6 **Dankov** Lipetskaya Oblast', W Russian Federation 53°17′N 39°07′E
42 J7 **Danlí** El Paraíso, S Honduras 14°02′N 86°34′W
Danmark *see* Denmark
Danmarksstraedet *see* Denmark Strait
95 O14 **Dannemora** Uppsala, C Sweden 60°13′N 17°49′E
18 L6 **Dannemora** New York, NE USA 44°42′N 73°42′W
100 K11 **Dannenberg** Niedersachsen, N Germany 53°05′N 11°06′E
184 N12 **Dannevirke** Manawatu-Wanganui, North Island, New Zealand 40°14′S 176°05′E
21 U8 **Dan River** ~ Virginia, NE USA
167 P8 **Dan Sai** Loei, C Thailand 17°15′N 101°04′E
18 F10 **Dansville** New York, NE USA 42°34′N 77°40′W
Dantzig *see* Gdańsk
86 E12 **Danube** *Bul.* Dunav, *Cz.* Dunaj, *Ger.* Donau, *Hung.* Duna, *Rom.* Dunărea. ~ C Europe
Danubian Plain *see* Dunavska Ravnina
166 L8 **Danubyu** Ayeyawady, SW Myanmar (Burma) 17°15′N 95°35′E
Danum *see* Doncaster
19 P11 **Danvers** Massachusetts, NE USA 42°34′N 70°54′W
27 T11 **Danville** Arkansas, C USA 35°03′N 93°22′W
31 N13 **Danville** Illinois, N USA 40°10′N 87°37′W
31 O14 **Danville** Indiana, N USA 39°45′N 86°31′W
29 Y15 **Danville** Iowa, C USA 40°52′N 91°18′W
20 M6 **Danville** Kentucky, S USA 37°40′N 84°49′W
18 G14 **Danville** Pennsylvania, NE USA 40°57′N 76°36′W
21 T6 **Danville** Virginia, NE USA 36°34′N 79°25′W
Danxian/Dan Xian *see* Danzhou
160 L17 **Danzhou** *prev.* Danxian, Dan Xian, Nada. Hainan, S China 19°31′N 109°31′E
Danzhou *see* Yichuan
Danzig *see* Gdańsk
Danziger Bucht *see* Danzig, Gulf of
110 J6 **Danzig, Gulf of** *var.* Gulf of Gdańsk, *Ger.* Danziger Bucht, *Pol.* Zakota Gdańska, *Rus.* Gdan'skaya Bukhta. *gulf* N Poland
160 F10 **Daocheng** *var.* Jinzhu, *Tib.* Dabba. Sichuan, C China 29°05′N 100°14′E
Daojiang *see* Daoxian
Daokou *see* Huaxian
104 H7 **Dão, Rio** ~ N Portugal
Daosa *see* Dausa
77 Y7 **Dao Timmi** Agadez, NE Niger 20°31′N 13°34′E
160 M13 **Daoxian** *var.* Daojiang. Hunan, S China 25°30′N 111°37′E
77 Q14 **Dapaong** N Togo 10°52′N 00°12′E
23 N8 **Daphne** Alabama, S USA 30°36′N 87°54′W
171 P7 **Dapitan** Mindanao, S Philippines 08°39′N 123°26′E
159 P9 **Da Qaidam** Qinghai, C China 37°50′N 95°18′E
163 V8 **Daqing** *var.* Sartu. Heilongjiang, NE China 46°35′N 125°00′E
163 O13 **Daqing Shan** ▲ N China
163 T11 **Daqin Tal** *var.* Naiman Qi. Nei Mongol Zizhiqu, N China 42°51′N 120°41′E
Daqm *see* Duqm
160 G8 **Da Qu** *var.* Do Qu. ~ C China
139 T5 **Dāqūq** *var.* Tāwūq. Kirkūk, N Iraq 35°08′N 44°27′E
76 G10 **Dara** *var.* Dahra. NW Senegal 15°20′N 15°28′W
138 H9 **Dar'ā** *var.* Der'a, *Fr.* Déraa. Dar'ā, SW Syria 32°37′N 36°06′E
138 H8 **Dar'ā** *off.* Muḥāfaẓat Dar'ā, *var.* Darā, Der'a, Derrá. ◆ *governorate* S Syria
143 Q12 **Dārāb** Fārs, S Iran 28°52′N 54°25′E
116 K8 **Darabani** Botoşani, NW Romania 48°10′N 26°39′E
Daraj *see* Dirj
Dar'ā, Muḥāfaẓat *see* Dar'ā
142 M8 **Dārān** Eşfahān, W Iran 33°00′N 50°27′E
167 U12 **Da Răng, Sông** *var.* Ba. ~ S Vietnam
Daraut-Korgan *see* Daroot-Korgon
77 W13 **Darazo** Bauchi, E Nigeria 11°01′N 10°27′E
139 S3 **Darband** Arbīl, N Iraq 36°15′N 44°17′E
139 V4 **Darband-i Khān, Sadd** *dam* NE Iraq
139 N1 **Darbāsīyah** *var.* Derbisīye. Al Ḥasakah, N Syria 37°06′N 40°42′E
118 C11 **Darbėnai** Klaipėda, NW Lithuania 56°02′N 21°16′E
153 Q13 **Darbhanga** Bihār, N India 26°10′N 85°54′E
38 M9 **Darby, Cape** *headland* Alaska, USA 64°19′N 162°46′W
112 I9 **Darda** *Hung.* Dárda. Osijek-Baranja, E Croatia 45°37′N 18°41′E
Dárda *see* Darda
Dardanë *see* Kamenicë
27 T11 **Dardanelle** Arkansas, C USA 35°11′N 93°09′W
27 S11 **Dardanelle, Lake** ⊠ Arkansas, C USA
Dardanelles *see* Çanakkale Boğazı
Dardanelli *see* Çanakkale
Dardo *see* Kangding
Dar-el-Beida *see* Casablanca
136 M14 **Darende** Malatya, C Turkey 38°34′N 37°29′E
81 J22 **Dar es Salaam** Dar es Salaam, E Tanzania 06°51′S 39°18′E
81 J23 **Dar es Salaam** *off.* Mkoa wa Dar es Salaam. ◆ *region* E Tanzania
81 J23 **Dar es Salaam** *off.* Mkoa wa Dar es Salaam. ◆ *region* E Tanzania
81 J22 **Dar es Salaam** ✕ Dar es Salaam, E Tanzania 06°57′S 39°17′E
Dar es Salaam, Mkoa wa *see* Dar es Salaam
Dar es Salaam, Mkoa wa *see* Dar es Salaam
185 H18 **Darfield** Canterbury, South Island, New Zealand 43°29′S 172°07′E
106 F7 **Darfo** Lombardia, N Italy 45°54′N 10°12′E
80 B10 **Darfur** *var.* Darfur Massif. *cultural region* W Sudan
Darfur Massif *see* Darfur
Darganata/Dargan-Ata *see* Birata
143 T3 **Dargaz** *var.* Darreh Gaz; *prev.* Moḥammadābād. Khorāsān-e Razavī, NE Iran 37°28′N 59°08′E
139 U4 **Dargazayn** As Sulaymānīyah, NE Iraq 35°39′N 45°00′E
183 P12 **Dargo** Victoria, SE Australia 37°29′S 147°15′E
162 L6 **Darhan** Darhan Uul, N Mongolia 49°24′N 105°57′E
163 N8 **Darhan** Hentiy, C Mongolia 46°38′N 109°25′E
Darhan *see* Büreghangay
Darhan Muminggan Lianheqi *see* Bailingmiao
162 L6 **Darhan Uul** ◆ *province* N Mongolia
23 W7 **Darien** Georgia, SE USA 31°22′N 81°25′W
43 W16 **Darién** *off.* Provincia del Darién. ◆ *province* SE Panama
Darién, Golfo del *see* Darien, Gulf of
43 X14 **Darien, Gulf of** *Sp.* Golfo del Darién. *gulf* S Caribbean Sea
Darien, Isthmus of *see* Panama, Istmo de
Darién, Provincia del *see* Darién
42 K9 **Dariense, Cordillera** ▲ C Nicaragua
43 W15 **Darién, Serranía del** ▲ Colombia/Panama
163 P10 **Dariganga** *var.* Ovoot. Sühbaatar, SE Mongolia 45°08′N 113°51′E
Dario *see* Ciudad Darío
Dariorigum *see* Vannes
Dariv *see* Darvi
Darj *see* Dirj
Darjeeling *see* Dārjiling
153 S12 **Dārjiling** *prev.* Darjeeling. West Bengal, NE India 27°00′N 88°13′E
Darkehnen *see* Ozersk
159 S12 **Darlag** *var.* Gümai. Qinghai, C China 33°43′N 99°42′E
183 T3 **Darling Downs** *hill range* Queensland, E Australia
28 M2 **Darling, Lake** ⊠ North Dakota, N USA
180 I12 **Darling Range** ▲ Western Australia
182 L8 **Darling River** ~ New South Wales, SE Australia
97 M15 **Darlington** N England, United Kingdom 54°31′N 01°34′W
21 T12 **Darlington** South Carolina, SE USA 34°19′N 79°53′W
30 K9 **Darlington** Wisconsin, N USA 42°41′N 90°08′W
110 G7 **Darłowo** Zachodnio-pomorskie, NW Poland 54°24′N 16°21′E
101 G19 **Darmstadt** Hessen, SW Germany 49°52′N 08°39′E
75 S7 **Darnah** *var.* Dérna. NE Libya 32°46′N 22°39′E
103 S6 **Darney** Vosges, NE France 48°06′N 05°58′E
182 M7 **Darnick** New South Wales, SE Australia 32°52′S 143°38′E
195 Y6 **Darnley, Cape** *cape* Antarctica
105 R7 **Daroca** Aragón, NE Spain 41°07′N 01°25′W
147 S11 **Daroot-Korgon** *var.* Daraut-Kurgan. Oshskaya Oblast', SW Kyrgyzstan 39°35′N 72°13′E
61 A23 **Darragueira** *var.* Darregueira. Buenos Aires, E Argentina 37°40′S 63°12′W
Darregueira *see* Darragueira
Darreh Gaz *see* Dargaz
142 K7 **Darreh Shahr** *var.* Darreh-ye Shahr. Īlām, W Iran 33°10′N 47°18′E
Darreh-ye Shahr *see* Darreh Shahr
32 I7 **Darrington** Washington, NW USA 48°15′N 121°36′W
25 P1 **Darrouzett** Texas, SW USA 36°27′N 100°19′W
153 S15 **Darsana** *var.* Darshana. Khulna, N Bangladesh 23°32′N 88°49′E
Darshana *see* Darsana
100 M7 **Darss** *peninsula* NE Germany
100 M7 **Darsser Ort** *headland* NE Germany 54°28′N 12°31′E
97 J24 **Dart** ~ SW England, United Kingdom
Dartang *see* Baqên
97 P22 **Dartford** SE England, United Kingdom 51°27′N 00°13′E
182 L12 **Dartmoor** Victoria, SE Australia 37°56′S 141°18′E
97 I24 **Dartmoor** *moorland* SW England, United Kingdom
13 Q15 **Dartmouth** Nova Scotia, SE Canada 44°40′N 63°35′W
97 J24 **Dartmouth** SW England, United Kingdom 50°21′N 03°34′W
15 Y6 **Dartmouth** ~ Québec, SE Canada
183 Q11 **Dartmouth Reservoir** ⊚ Victoria, SE Australia
Dartuch, Cabo *see* Artrutx, Cap d'
186 C9 **Daru** Western, SW Papua New Guinea 09°05′S 143°10′E
112 G9 **Daruvar** *Hung.* Daruvár. Bjelovar-Bilogora, NE Croatia 45°34′N 17°12′E
Daruvár *see* Daruvar
Darvaza *see* Derweze, Turkmenistan
Darvaza *see* Darvoza, Uzbekistan
Darvazskiy Khrebet *see* Darvoz, Qatorkŭhi
Darvel Bay *see* Lahad Datu, Teluk
Darvel, Teluk *see* Lahad Datu, Teluk
162 F8 **Darvi** *var.* Dariv. Govĭ-Altay, W Mongolia 46°20′N 94°11′E
162 F7 **Darvi** *var.* Bulgan. Hovd, W Mongolia 46°57′N 93°40′E
Darvīshān *see* Darwēshān
147 O10 **Darvoza** *Rus.* Darvaza. Jizzax Viloyati, C Uzbekistan 40°59′N 67°19′E
147 R13 **Darvoz, Qatorkŭhi** *Rus.* Darvazskiy Khrebet. ▲ C Tajikistan
148 L9 **Darwēshān** *var.* Garmser; *prev.* Darvīshān. Helmand, S Afghanistan 31°02′N 64°12′E
63 J15 **Darwin** Río Negro, S Argentina 39°13′S 65°41′W
181 O1 **Darwin** *prev.* Palmerston, Port Darwin. *territory capital* Northern Territory, N Australia 12°28′S 130°52′E
65 D24 **Darwin** *var.* Darwin Settlement. East Falkland, Falkland Islands 51°51′S 58°55′W
62 H8 **Darwin, Cordillera** ▲ N Chile
Darwin Settlement *see* Darwin
57 B17 **Darwin, Volcán** ▲ Galapagos Islands, Ecuador, E Pacific Ocean 0°12′S 91°17′W
149 S8 **Darya Khān** Punjab, E Pakistan 31°47′N 71°10′E
145 O15 **Dar'yalyktakyr, Ravnina** *plain* S Kazakhstan
143 T11 **Dārzīn** Kermān, S Iran 29°11′N 58°09′E
Dashhowuz *see* Daşoguz
Dashhowuz Welayaty *see* Daşoguz Welaýaty
162 K7 **Dashinchilen** *var.* Süüj. Bulgan, C Mongolia 47°49′N 104°06′E
119 O16 **Dashkawka** *Rus.* Dashkovka. Mahilyowskaya Voblasts', E Belarus 53°44′N 30°16′E
Dashkhovuz *see* Daşoguz Welaýaty
Dashkhovuzskiy Velayat *see* Daşoguz Welaýaty
Dashköpri *see* Daşköpri
Dashkovka *see* Dashkawka
148 J15 **Dasht** ~ SW Pakistan
Dasht-i *see* Bābūs, Dasht-e
Dashtidzhum *see* Dashtijum
147 R13 **Dashtijum** *Rus.* Dashtidzhum. SW Tajikistan 38°06′N 70°11′E
149 W7 **Daska** Punjab, NE Pakistan 32°15′N 74°23′E
146 J16 **Daşköpri** *var.* Dashköpri, *Rus.* Tashkepri. Mary Welaýaty, S Turkmenistan 36°15′N 62°37′E
146 H8 **Daşoguz** *Rus.* Dashkhovuz, *Turkm.* Dashhowuz; *prev.* Tashauz. Daşoguz Welaýaty, N Turkmenistan 41°51′N 59°53′E
146 E9 **Daşoguz Welaýaty** *var.* Dashhowuz Welayaty, *Rus.* Dashkhovuz, Dashkhovuzskiy Velayat. ◆ *province* N Turkmenistan
Đa, Sông *see* Black River
77 R15 **Dassa** *var.* Dassa-Zoumé. S Benin 07°46′N 02°15′E
Dassa-Zoumé *see* Dassa
29 U8 **Dassel** Minnesota, N USA 45°06′N 94°18′W
152 H3 **Dastegil Sar** ▲ N India
136 C16 **Datça** Muğla, SW Turkey 36°46′N 27°40′E
165 R4 **Date** Hokkaidō, NE Japan 42°28′N 140°51′E
154 I8 **Datia** *prev.* Duttia. Madhya Pradesh, C India 25°41′N 78°28′E
Dåtnejaevrie *see* Tunnsjøen
159 T10 **Datong** *var.* Datong Huizu Tuzu Zizhixian, Qiaotou. Qinghai, C China 37°01′N 101°33′E
161 N2 **Datong** *var.* Tatung, Ta-t'ung. Shanxi, C China 40°09′N 113°17′E
Datong *see* Tong'an
159 S8 **Datong He** ~ C China
Datong Huizu Tuzu Zizhixian *see* Datong
159 S9 **Datong Shan** ▲ C China
169 O10 **Datu, Tanjung** *headland* Indonesia/Malaysia 02°01′N 109°37′E
Datu, Teluk *see* Lahad Datu, Teluk
Daua *see* Dawa Wenz
172 H16 **Dauban, Mount** ▲ Silhouette, NE Seychelles
149 T7 **Dāūd Khel** Punjab, E Pakistan 32°52′N 71°35′E
119 G15 **Daugai** Alytus, S Lithuania 54°22′N 24°20′E
Daugava *see* Western Dvina
118 J11 **Daugavpils** *Ger.* Dünaburg; *prev. Rus.* Dvinsk. SE Latvia 55°53′N 26°34′E
Dauka *see* Dawkah
Daulatabad *see* Malāyer
101 D18 **Daun** Rheinland-Pfalz, W Germany 50°13′N 06°50′E
155 E14 **Daund** *prev.* Dhond. Mahārāshtra, W India 18°28′N 74°38′E
166 M12 **Daung Kyun** *island* S Myanmar (Burma)
11 W15 **Dauphin** Manitoba, S Canada 51°09′N 100°05′W
103 S13 **Dauphiné** *cultural region* E France
23 N9 **Dauphin Island** *island* Alabama, S USA
11 X15 **Dauphin River** Manitoba, S Canada 51°55′N 98°03′W
77 V12 **Daura** Katsina, N Nigeria 13°03′N 08°18′E
152 H12 **Dausa** *prev.* Daosa. Rājasthān, N India 26°51′N 76°21′E
Dauwa *see* Dawwah
Dāvāçi *see* Şabran
155 F18 **Dāvangere** Karnātaka, W India 14°30′N 75°52′E
171 Q8 **Davao** *off.* Davao City. Mindanao, S Philippines 07°06′N 125°36′E
Davao City *see* Davao
171 Q8 **Davao Gulf** *gulf* Mindanao, S Philippines
15 Q11 **Daveluyville** Québec, SE Canada 46°12′N 72°07′W
29 Z14 **Davenport** Iowa, C USA 41°31′N 90°35′W
32 L8 **Davenport** Washington, NW USA 47°39′N 118°09′W
43 P16 **David** Chiriquí, W Panama 08°26′N 82°26′W
15 O11 **David** ~ Québec, SE Canada
29 R15 **David City** Nebraska, C USA 41°15′N 97°07′W
David-Gorodok *see* Davyd-Haradok
11 T16 **Davidson** Saskatchewan, S Canada 51°15′N 105°59′W
21 R10 **Davidson** North Carolina, SE USA 35°29′N 80°49′W
26 K12 **Davidson** Oklahoma, C USA 34°15′N 99°06′W
39 S6 **Davidson Mountains** ▲ Alaska, USA
172 M8 **Davie Ridge** *undersea feature* W Indian Ocean 17°10′S 41°45′E
182 A1 **Davies, Mount** ▲ South Australia 26°14′S 129°14′E
35 O7 **Davis** California, W USA 38°31′N 121°46′W
27 N12 **Davis** Oklahoma, C USA 34°30′N 97°07′W
195 Y7 **Davis** *Australian research station* Antarctica 68°30′S 78°15′E
194 H3 **Davis Coast** *physical region* Antarctica
18 C16 **Davis, Mount** ▲ Pennsylvania, NE USA 39°47′N 79°10′W
24 K9 **Davis Mountains** ▲ Texas, SW USA
195 Z9 **Davis Sea** *sea* Antarctica
65 O20 **Davis Seamounts** *undersea feature* S Atlantic Ocean
196 M13 **Davis Strait** *strait* Baffin Bay/Labrador Sea
127 U5 **Davlekanovo** Respublika Bashkortostan, W Russian Federation 54°13′N 55°06′E
108 J9 **Davos** *Rmsch.* Tavau. Graubünden, E Switzerland 46°48′N 09°50′E
Davvesiida *see* Lebesby
119 J20 **Davyd-Haradok** *Pol.* Dawidgródek, *Rus.* David-Gorodok. Brestskaya Voblasts', SW Belarus 52°03′N 27°13′E
163 U12 **Dawa** Liaoning, NE China 40°55′N 122°02′E
141 O11 **Dawāsir, Wādī ad** *dry watercourse* S Saudi Arabia
81 K15 **Dawa Wenz** *var.* Daua, Webi Daawo. ~ E Africa
Dawaymah, Birkat ad *see* Umm al Baqar, Hawr
167 N10 **Dawei** *var.* Tavoy, Htawei. Tanintharyi, S Myanmar (Burma) 14°02′N 98°12′E
119 K14 **Dawhinava** *Rus.* Dolginovo. Minskaya Voblasts', N Belarus 54°39′N 27°29′E
Dawidgródek *see* Davyd-Haradok
141 V12 **Dawkah** *var.* Dauka. SW Oman 18°32′N 54°03′E
Dawlat Qaṭar *see* Qatar
24 M3 **Dawn** Texas, SW USA 34°54′N 102°10′W
Dawo *see* Maqên
140 M11 **Daws** Al Bāḥah, SW Saudi Arabia 20°19′N 41°12′E
10 H5 **Dawson** *var.* Dawson City. Yukon, NW Canada 64°04′N 139°24′W
23 S6 **Dawson** Georgia, SE USA 31°46′N 84°27′W
29 S9 **Dawson** Minnesota, N USA 44°55′N 96°03′W
Dawson City *see* Dawson

◆ Country
● Country Capital
◇ Dependent Territory
○ Dependent Territory Capital
◆ Administrative Regions
✕ International Airport
▲ Mountain
▲ Mountain Range
Volcano
~ River
⊚ Lake
⊠ Reservoir

11 N13 **Dawson Creek** British Columbia, W Canada 55°45´N 120°07´W
10 H7 **Dawson Range** ▲▲ Yukon, W Canada
181 Y9 **Dawson River** ≈ Queensland, E Australia
10 J15 **Dawsons Landing** British Columbia, SW Canada 51°33´N 127°38´W
20 I7 **Dawson Springs** Kentucky, S USA 37°10´N 87°41´W
23 S2 **Dawsonville** Georgia, SE USA 34°28´N 84°07´W
160 G8 **Dawu** *var.* Xianshui. Sichuan, C China 30°55´N 101°08´E
Dawu *see* Maqên
Dawukou *see* Huinong
141 Y10 **Dawwah** *var.* Dauwa. W Oman 20°36´N 58°52´E
102 J15 **Dax** *var.* Ax; *anc.* Aquae Augustae, Aquae Tarbelicae. Landes, SW France 43°43´N 01°03´W
Daxian *see* Dazhou
Daxiangshan *see* Gangu
Daxue *see* Wencheng
160 G9 **Daxue Shan** ▲▲ C China
Dayan *see* Lijiang
160 G12 **Dayao** *var.* Jinbi. Yunnan, SW China 25°41´N 101°23´E
Dayishan *see* Gaoyou
Dāykondī *see* Dāykundī
149 O6 **Dāykundī** *prev.* Dāykondī. ⬖ *province* C Afghanistan
183 N12 **Daylesford** Victoria, SE Australia 37°24´S 144°07´E
35 U10 **Daylight Pass** *pass* California, W USA
61 D17 **Daymán, Río** ≈ N Uruguay
Dayong *see* Zhangjiajie
Dayr *see* Ad Dayr
138 G10 **Dayr ʿAllā** *var.* Deir ʿAlla. Al Balqā', N Jordan 32°39´N 36°06´E
139 N4 **Dayr az Zawr** *var.* Deir ez Zor. Dayr az Zawr, E Syria 35°12´N 40°12´E
138 M5 **Dayr az Zawr** *off.* Muḩāfaẓat Dayr az Zawr, *var.* Dayr Az-Zor. ⬖ *governorate* E Syria
Dayr az Zawr, Muḩāfaẓat *see* Dayr az Zawr
Dayr Az-Zor *see* Dayr az Zawr
75 W9 **Dayrūṭ** *var.* Dairûṭ. C Egypt 27°34´N 30°48´E
11 Q15 **Daysland** Alberta, SW Canada 52°53´N 112°19´W
31 R14 **Dayton** Ohio, N USA 39°46´N 84°12´W
20 L10 **Dayton** Tennessee, S USA 35°30´N 85°01´W
25 W11 **Dayton** Texas, SW USA 30°03´N 94°53´W
32 L10 **Dayton** Washington, NW USA 46°19´N 117°58´W
23 X10 **Daytona Beach** Florida, SE USA 29°12´N 81°03´W
169 U12 **Dayu** Borneo, C Indonesia 01°59´S 115°04´E
161 O13 **Dayu Ling** ▲▲ S China
161 R7 **Da Yunhe** *Eng.* Grand Canal. *canal* E China
161 S11 **Dayu Shan** *island* SE China
160 K8 **Dazhou** *prev.* Dachuan, Daxian. Sichuan, C China 31°16´N 107°31´E
160 J9 **Dazhu** *var.* Zhuyang. Sichuan, C China 30°45´N 107°11´E
161 T13 **Dazhuoshui** *prev.* Tachoshui. N Taiwan 24°26´N 121°43´E
160 J9 **Dazu** *var.* Longgang. Chongqing Shi, C China 29°47´N 106°30´E
83 H24 **De Aar** Northern Cape, C South Africa 30°40´S 24°01´E
194 K5 **Deacon, Cape** *headland* Antarctica
39 R5 **Deadhorse** Alaska, USA 70°15´N 148°28´W
33 T12 **Dead Indian Peak** ▲ Wyoming, C USA 44°36´N 109°45´W
23 R9 **Dead Lake** ◎ Florida, SE USA
44 J4 **Deadman's Cay** Long Island, C The Bahamas 23°09´N 75°06´W
138 G11 **Dead Sea** *var.* Bahret Lut, Lacus Asphaltites, *Ar.* Al Baḩr al Mayyit, Baḩrat Lūt, *Heb.* Yam HaMelaḥ. *salt lake* Israel/Jordan
28 J9 **Deadwood** South Dakota, N USA 44°22´N 103°43´W
97 Q22 **Deal** SE England, United Kingdom 51°14´N 01°23´E
83 I22 **Dealesville** Free State, C South Africa 28°40´S 25°46´E
161 P10 **De'an** *var.* Puting. Jiangxi, S China 29°24´N 115°46´E
62 K9 **Deán Funes** Córdoba, C Argentina 30°25´S 64°22´W
194 L12 **Dean Island** *island* Antarctica
. **Deanuvuotna** *see* Tanafjorden
31 S10 **Dearborn** Michigan, N USA 42°16´N 83°13´W
27 R3 **Dearborn** Missouri, C USA 39°31´N 94°46´W
Deargget *see* Tärendö
32 K9 **Deary** Idaho, NW USA 46°46´N 118°33´W
32 M9 **Deary** Washington, NW USA 46°42´N 116°36´W
10 J10 **Dease** ≈ British Columbia, W Canada
10 J10 **Dease Lake** British Columbia, W Canada 58°28´N 130°04´W
35 U11 **Death Valley** California, W USA 36°25´N 116°50´W
35 U11 **Death Valley** *valley* California, W USA
92 M8 **Deatnu** *Fin.* Tenojoki, *Nor.* Tana. ≈ Finland/Norway *see also* Tana, Tenojoki
Deatnu *see* Tana
102 L4 **Deauville** Calvados, N France 49°21´N 00°06´E
117 X7 **Debal'tseve** *Rus.* Debal'tsevo. Donets'ka Oblast', SE Ukraine 48°21´N 38°26´E
Debal'tsevo *see* Debal'tseve
113 M19 **Debar** *Ger.* Dibra, *Turk.* Debre. W FYR Macedonia 41°32´N 20°33´E
39 O9 **Debauch Mountain** ▲ Alaska, USA 64°31´N 159°52´W
De Behagle *see* Laï
25 X7 **De Berry** Texas, SW USA 32°18´N 94°09´W
Debessy *see* Debesy
127 T2 **Debesy** *prev.* Debessy. Udmurtskaya Respublika, NW Russian Federation 57°41´N 53°56´E
111 N16 **Dębica** Podkarpackie, SE Poland 50°04´N 21°24´E
De Bildt *see* De Bilt
98 J11 **De Bilt** *var.* De Bildt. Utrecht, C Netherlands 52°06´N 05°11´E
123 T9 **Debin** Magadanskaya Oblast', E Russian Federation 62°18´N 150°42´E
110 N13 **Dęblin** *Rus.* Ivangorod. Lubelskie, E Poland 51°34´N 21°50´E
110 D10 **Dębno** Zachodnio-pomorskie, NW Poland 52°43´N 14°42´E
39 S10 **Deborah, Mount** ▲ Alaska, USA 63°38´N 147°13´W
33 N8 **De Borgia** Montana, NW USA 47°23´N 115°24´W
Debra Birhan *see* Debre Birhan
Debra Marcos *see* Debre Mark'os
Debra Tabor *see* Debre Tabor
Debre *see* Debar
80 J13 **Debre Birhan** *var.* Debra Birhan. Āmara, N Ethiopia 09°45´N 39°40´E
111 N22 **Debrecen** *Ger.* Debreczin, *Rom.* Debreţin; *prev.* Debreczen. Hajdú-Bihar, E Hungary 47°32´N 21°38´E
Debreczen/Debreczin *see* Debrecen
80 I12 **Debre Mark'os** *var.* Debra Marcos. Āmara, N Ethiopia 10°18´N 37°48´E
113 N19 **Debrešte** SW FYR Macedonia 41°29´N 21°20´E
80 J11 **Debre Tabor** *var.* Debra Tabor. Āmara, N Ethiopia 11°46´N 38°06´E
Debreţin *see* Debrecen
80 J13 **Debre Zeyt** Oromīya, C Ethiopia 08°41´N 39°00´E
113 L16 **Deçan** *Serb.* Dečane; *prev.* Dečani. W Kosovo 42°33´N 20°18´E
Dečane *see* Deçan
Dečani *see* Deçan
23 P2 **Decatur** Alabama, S USA 34°36´N 86°58´W
23 S3 **Decatur** Georgia, SE USA 33°46´N 84°18´W
30 L13 **Decatur** Illinois, N USA 39°50´N 88°57´W
31 Q12 **Decatur** Indiana, N USA 40°40´N 84°57´W
22 M5 **Decatur** Mississippi, S USA 32°26´N 89°06´W
29 S14 **Decatur** Nebraska, C USA 42°00´N 96°19´W
25 S6 **Decatur** Texas, SW USA 33°14´N 97°35´W
20 H9 **Decaturville** Tennessee, S USA 35°35´N 88°08´W
103 O13 **Decazeville** Aveyron, S France 44°34´N 02°18´E
155 H17 **Deccan** *Hind.* Dakshin. *plateau* C India
14 J8 **Decelles, Réservoir** ⊠ Québec, SE Canada
12 K2 **Déception** Québec, NE Canada 62°06´N 74°36´W
160 G11 **Dechang** *var.* Dezhou. Sichuan, C China 27°24´N 102°09´E
111 C15 **Děčín** *Ger.* Tetschen. Ústecký Kraj, NW Czech Republic 50°48´N 14°15´E
103 P9 **Decize** Nièvre, C France 46°51´N 03°25´E
98 I6 **De Cocksdorp** Noord-Holland, NW Netherlands 53°09´N 04°52´E
29 X11 **Decorah** Iowa, C USA 43°18´N 91°47´W
Dedeagaç/Dedeagach *see* Alexandroúpoli
188 C15 **Dededo** N Guam 13°30´N 144°51´E
98 N9 **Dedemsvaart** Overijssel, E Netherlands 52°36´N 06°28´E
19 O11 **Dedham** Massachusetts, NE USA 42°14´N 71°10´W
63 H19 **Dedo, Cerro** ▲ SW Argentina 44°46´S 71°48´W
77 O13 **Dédougou** W Burkina Faso 12°29´N 03°25´W
124 G15 **Dedovichi** Pskovskaya Oblast', W Russian Federation 57°31´N 29°53´E
Dedu *see* Wudalianchi
155 J24 **Deduru Oya** ≈ W Sri Lanka
83 N14 **Dedza** Central, S Malawi 14°20´S 34°24´E
83 N14 **Dedza Mountain** ▲ C Malawi 14°22´S 34°16´E
96 K9 **Dee** ≈ NE Scotland, United Kingdom
97 J19 **Dee** *Wel.* Afon Dyfrdwy. ≈ England/Wales, United Kingdom
Deep Bay *see* Chilumba
21 T3 **Deep Creek Lake** ⊠ Maryland, NE USA
36 J4 **Deep Creek Range** ▲▲ Utah, W USA
27 P10 **Deep Fork River** ≈ Oklahoma, C USA
14 J11 **Deep River** Ontario, SE Canada 46°04´N 77°29´W
21 T10 **Deep River** ≈ North Carolina, SE USA
183 U4 **Deepwater** New South Wales, SE Australia 29°27´S 151°52´E
31 S14 **Deer Creek Lake** ◎ Ohio, N USA
23 Z15 **Deerfield Beach** Florida, SE USA 26°19´N 80°06´W
39 N8 **Deering** Alaska, USA 66°04´N 162°43´W
38 M16 **Deer Island** *island* Alaska, USA
19 S7 **Deer Isle** *island* Maine, NE USA
13 S11 **Deer Lake** Newfoundland and Labrador, SE Canada 49°11´N 57°27´N
99 D18 **Deerlijk** West-Vlaanderen, W Belgium 50°52´N 03°21´E
33 Q10 **Deer Lodge** Montana, NW USA 46°24´N 112°43´W
32 L8 **Deer Park** Washington, NW USA 47°55´N 117°28´W
29 U5 **Deer River** Minnesota, N USA 47°19´N 93°47´W
Deés *see* Dej
Defeng *see* Liping
31 R11 **Defiance** Ohio, N USA 41°17´N 84°21´W
23 Q8 **De Funiak Springs** Florida, SE USA 30°43´N 86°07´W
95 L23 **Degeberga** Skåne, S Sweden 55°48´N 14°06´E
104 H12 **Degebe, Ribeira** ≈ S Portugal
80 M13 **Degeh Bur** Sumalē, E Ethiopia 08°08´N 43°35´E
15 U9 **Dégelis** Québec, SE Canada 47°30´N 68°38´W
77 U17 **Degema** Rivers, S Nigeria 04°46´N 06°47´E
95 L16 **Degerfors** Örebro, C Sweden 59°14´N 14°26´E
193 R14 **De Gerlache Seamounts** *undersea feature* SE Pacific Ocean
101 N21 **Deggendorf** Bayern, SE Germany 48°50´N 12°58´E
80 I11 **Degoma** Āmara, N Ethiopia 12°22´N 37°36´E
De Gordyk *see* Gorredijk
27 T12 **De Gray Lake** ⊠ Arkansas, C USA
180 J6 **De Grey River** ≈ Western Australia
126 M10 **Degtevo** Rostovskaya Oblast', SW Russian Federation 49°12´N 40°39´E
Dehbārez *see* Rūdān
Deh Bīd *see* Şafāshahr
142 M10 **Deh Dasht** Kohkīlūyeh va Būyer AḨmad, SW Iran 30°49´N 50°36´E
75 N8 **Dehibat** SE Tunisia 31°58´N 10°43´E
Dehli *see* Delhi
142 K8 **Dehlorān** Īlām, W Iran 32°41´N 47°18´E
147 N13 **Dehqonobod** *Rus.* Dekhkanabad. Qashqadaryo Viloyati, S Uzbekistan 38°24´N 66°31´E
152 J9 **Dehra Dūn** Uttararanchal, N India 30°19´N 78°04´E
153 O14 **Dehri** Bihār, N India 24°55´N 84°11´E
Deh Shū *see* Dīshū
163 W9 **Dehui** Jilin, NE China 44°23´N 125°42´E
99 D17 **Deinze** Oost-Vlaanderen, NW Belgium 50°59´N 03°32´E
Deir ʿAlla *see* Dayr ʿAllā
Deir ez Zor *see* Dayr az Zawr
Deirgeirt, Loch *see* Derg, Lough
116 H9 **Dej** *Hung.* Dés; *prev.* Deés. Cluj, NW Romania 47°08´N 23°55´E
95 K15 **Deje** Värmland, C Sweden 59°35´N 13°29´E
171 Y15 **De Jongs, Tanjung** *headland* Papua, SE Indonesia 06°56´S 138°32´E
De Jouwer *see* Joure
30 M10 **De Kalb** Illinois, N USA 41°55´N 88°45´W
22 M5 **De Kalb** Mississippi, S USA 32°46´N 88°39´W
25 W5 **De Kalb** Texas, SW USA 33°30´N 94°34´W
83 G18 **Dekar** *var.* D'Kar. Ghanzi, NW Botswana 21°31´S 21°55´E
Dekéleia *see* Dhekélia
79 K20 **Dekese** Kasai-Occidental, C Dem. Rep. Congo 03°28´S 21°24´E
Dekhkanabad *see* Dehqonobod
79 I14 **Dékoa** Kémo, C Central African Republic 06°17´N 19°07´E
98 H6 **De Koog** Noord-Holland, NW Netherlands 53°06´N 04°43´E
30 M9 **Delafield** Wisconsin, N USA 43°03´N 88°22´W
61 C23 **De La Garma** Buenos Aires, E Argentina 37°58´S 60°25´W
14 K10 **Delahey, Lac** ◎ Québec, SE Canada
80 E11 **Delami** Southern Kordofan, C Sudan 11°51´N 30°30´E
23 X11 **De Land** Florida, SE USA 29°01´N 81°18´W
35 R12 **Delano** California, W USA 35°46´N 119°15´W
29 V8 **Delano** Minnesota, N USA 45°03´N 93°46´W
36 K6 **Delano Peak** ▲ Utah, W USA 38°22´N 112°21´W
Delap-Uliga-Darrit *see* Dalap-Uliga-Djarrit
Delārām *see* Dilārām
38 F17 **Delarof Islands** *island group* Aleutian Islands, Alaska, USA
30 M9 **Delavan** Wisconsin, N USA 42°37´N 88°37´W
31 S13 **Delaware** Ohio, N USA 40°18´N 83°06´W
18 I17 **Delaware** *off.* State of Delaware, *also known as* Blue Hen State, Diamond State, First State. ⬖ *state* NE USA
18 I17 **Delaware Bay** *bay* NE USA
24 J8 **Delaware Mountains** ▲▲ Texas, SW USA
18 I12 **Delaware River** ≈ NE USA
27 Q3 **Delaware River** ≈ Kansas, C USA
18 J14 **Delaware Water Gap** *valley* New Jersey/Pennsylvania, NE USA
101 G14 **Delbrück** Nordrhein-Westfalen, W Germany 51°46´N 08°34´E
11 Q15 **Delburne** Alberta, SW Canada 52°09´N 113°11´W
172 M12 **Del Cano Rise** *undersea feature* SW Indian Ocean 45°15´S 44°15´E
113 Q18 **Delčevo** NE FYR Macedonia 41°57´N 22°45´E
Delcommune, Lac *see* Nzilo, Lac
98 O10 **Delden** Overijssel, E Netherlands 52°16´N 06°41´E
183 R12 **Delegate** New South Wales, SE Australia 37°04´S 148°57´E
De Lemmer *see* Lemmer
108 D7 **Delémont** *Ger.* Delsberg. Jura, NW Switzerland 47°22´N 07°21´E
25 R7 **De Leon** Texas, SW USA 32°06´N 98°33´W
115 F18 **Delfoí** Stereá Elláda, C Greece 38°28´N 22°31´E
98 G12 **Delft** Zuid-Holland, W Netherlands 52°01´N 04°22´E
155 J23 **Delft** *island* NW Sri Lanka
98 O5 **Delfzijl** Groningen, NE Netherlands 53°20´N 06°55´E
0 E9 **Delgada Fan** *undersea feature* NE Pacific Ocean 39°15´N 126°00´W
42 F7 **Delgado** San Salvador, SW El Salvador 13°44´N 89°07´W
82 Q12 **Delgado, Cabo** *headland* N Mozambique 10°41´S 40°40´E
162 G8 **Delger** *var.* Taygan. Govĭ-Altay, C Mongolia 46°20´N 97°22´E
163 O9 **Delgereh** *var.* Hongor. Dornogovĭ, SE Mongolia 45°49´N 111°20´E
162 J8 **Delgerhaan** *var.* Hujirt. Töv, C Mongolia 46°41´N 104°40´E
162 K9 **Delgerhangay** *var.* Hashaat. Dundgovĭ, C Mongolia 45°09´N 104°51´E
162 L9 **Delgertsogt** *var.* Amardalay. Dundgovĭ, C Mongolia 46°09´N 106°24´E
80 E6 **Delgo** Northern, N Sudan 20°08´N 30°35´E
159 R10 **Delhi** *var.* Delingha. Qinghai, C China 37°19´N 97°22´E
152 I10 **Delhi** *var.* Dehli, *Hind.* Dilli, *hist.* Shahjahanabad. *union territory capital* Delhi, N India 28°40´N 77°11´E
22 J5 **Delhi** Louisiana, S USA 32°28´N 91°29´W
18 J11 **Delhi** New York, NE USA 42°16´N 74°55´W
152 I10 **Delhi** ⬖ *union territory* NW India
136 J17 **Deli Burnu** *headland* S Turkey 36°43´N 34°55´E
136 J12 **Delice Çayı** ≈ C Turkey
55 X10 **Délices** C French Guiana 04°45´N 53°45´W
40 J6 **Delicias** *var.* Ciudad Delicias. Chihuahua, N Mexico 28°09´N 105°22´W
143 N7 **Delījān** *var.* Dalijan, Dilijan. Markazī, W Iran 34°02´N 50°39´E
112 P12 **Deli Jovan** ▲▲ E Serbia
Déli-Kárpátok *see* Carpaţii Meridionalii
8 I8 **Délįne** *prev.* Fort Franklin. Northwest Territories, NW Canada 65°10´N 123°30´W
Delingha *see* Delhi
15 Q7 **Delisle** Québec, SE Canada 48°39´N 71°42´W
11 T15 **Delisle** Saskatchewan, S Canada 51°54´N 107°01´W
101 M15 **Delitzsch** Sachsen, E Germany 51°31´N 12°19´E
33 Q12 **Dell** Montana, NW USA 44°41´N 112°42´W
24 I7 **Dell City** Texas, SW USA 31°56´N 105°12´W
103 U7 **Delle** Territoire-de-Belfort, E France 47°30´N 07°00´E
36 J9 **Dellenbaugh, Mount** ▲ Arizona, SW USA 36°06´N 113°32´W
29 R11 **Dell Rapids** South Dakota, N USA 43°50´N 96°42´W
21 Y4 **Delmar** Maryland, NE USA 38°26´N 75°32´W
18 K11 **Delmar** New York, NE USA 42°37´N 73°49´W
100 G11 **Delmenhorst** Niedersachsen, NW Germany 53°03´N 08°38´E
112 C9 **Delnice** Primorje-Gorski Kotar, NW Croatia 45°24´N 14°49´E
37 R7 **Del Norte** Colorado, C USA 37°40´N 106°21´W
39 N6 **De Long Mountains** ▲▲ Alaska, USA
183 P16 **Deloraine** Tasmania, SE Australia 41°34´S 146°43´E
11 W17 **Deloraine** Manitoba, S Canada 49°12´N 100°28´W
31 O12 **Delphi** Indiana, N USA 40°35´N 86°40´W
31 Q12 **Delphos** Ohio, N USA 40°49´N 84°20´W
23 Z15 **Delray Beach** Florida, SE USA 26°27´N 80°04´W
25 O12 **Del Rio** Texas, SW USA 29°23´N 100°56´W
Delsberg *see* Delémont
94 N11 **Delsbo** Gävleborg, C Sweden 61°49´N 16°34´E
37 P6 **Delta** Colorado, C USA 38°44´N 108°04´W
36 K5 **Delta** Utah, W USA 39°21´N 112°34´W
77 T17 **Delta** ⬖ *state* S Nigeria
55 Q6 **Delta Amacuro** *off.* Territorio Delta Amacuro. ⬖ *federal district* NE Venezuela
Delta Amacuro, Territorio *see* Delta Amacuro
39 S9 **Delta Junction** Alaska, USA 64°02´N 145°43´W
23 X11 **Deltona** Florida, SE USA 28°54´N 81°15´W
183 T5 **Delungra** New South Wales, SE Australia 29°40´S 150°49´E
162 D6 **Delüün** *var.* Rashaant. Bayan-Ölgiy, W Mongolia 47°48´N 90°45´E
154 C12 **Delvāda** Gujarāt, W India 20°46´N 71°02´E
61 B21 **Del Valle** Buenos Aires, E Argentina 35°55´S 60°42´W
Delvina *see* Delvinë
115 C15 **Delvináki** *var.* Dhelvinákion; *prev.* Pogónion. Ípeiros, W Greece 39°57´N 20°28´E
113 L23 **Delvinë** *var.* Delvina, *It.* Delvino. Vlorë, S Albania 39°56´N 20°07´E
Delvino *see* Delvinë
116 I7 **Delyatyn** Ivano-Frankivs'ka Oblast', W Ukraine 48°32´N 24°38´E
127 U5 **Dëma** ≈ W Russian Federation
105 O5 **Demanda, Sierra de la** ▲▲ W Spain
39 T5 **Demarcation Point** *headland* Alaska, USA 69°40´N 141°19´W
79 K21 **Demba** Kasai-Occidental, C Dem. Rep. Congo 05°24´S 22°16´E
172 H13 **Dembéni** Grande Comore, NW Comoros 11°50´S 43°25´E
79 M15 **Dembia** Mbomou, SE Central African Republic 05°08´N 24°25´E
Dembidollo *see* Dembī Dolo
80 H13 **Dembī Dolo** *var.* Dembidollo. Oromīya, C Ethiopia 08°33´N 34°49´E
152 K6 **Demchok** *var.* Dêmqog. China/India 32°30´N 79°42´E *see also* Dêmqog
152 L6 **Demchok** *var.* Dêmqog. *disputed region* China/India *see also* Dêmqog
98 I12 **De Meern** Utrecht, C Netherlands 52°06´N 05°00´E
99 I17 **Demer** ≈ C Belgium
64 H12 **Demerara Plain** *undersea feature* W Atlantic Ocean 10°00´N 48°00´W
64 H12 **Demerara Plateau** *undersea feature* W Atlantic Ocean
55 T9 **Demerara River** ≈ NE Guyana
126 H3 **Demidov** Smolenskaya Oblast', W Russian Federation 55°15´N 31°30´E
37 Q15 **Deming** New Mexico, SW USA 32°17´N 107°46´W
32 H6 **Deming** Washington, NW USA 48°49´N 122°13´W
58 E10 **Demini, Rio** ≈ NW Brazil
136 D13 **Demirci** Manisa, W Turkey 39°03´N 28°40´E
113 P19 **Demir Kapija** *prev.* Železna Vrata. SE FYR Macedonia 41°25´N 22°15´E
114 N11 **Demirköy** Kırklareli, NW Turkey 41°48´N 27°49´E
100 N9 **Demmin** Mecklenburg-Vorpommern, NE Germany 53°53´N 13°03´E
23 O5 **Demopolis** Alabama, S USA 32°31´N 87°50´W
31 N11 **Demotte** Indiana, N USA 41°13´N 87°07´W
158 F13 **Dêmqog** *var.* Demchok. China/India 32°36´N 79°29´E *see also* Demchok
152 L6 **Dêmqog** *var.* Demchok. *disputed region* China/India *see also* Demchok
171 Y13 **Demta** Papua, E Indonesia 02°19´S 140°06´E
121 K11 **Dem'yanka** ≈ C Russian Federation
124 H15 **Demyansk** Novgorodskaya Oblast', W Russian Federation 57°39´N 32°31´E
122 H10 **Dem'yanskoye** Tyumenskaya Oblast', C Russian Federation 59°39´N 69°15´E
103 P2 **Denain** Nord, N France 50°19´N 03°24´E
39 S10 **Denali** Alaska, USA 63°08´N 147°33´W
Denali *see* McKinley, Mount
81 M14 **Denan** Sumalē, E Ethiopia 06°40´N 43°31´E
Denau *see* Denov
97 J18 **Denbigh** *Wel.* Dinbych. NE Wales, United Kingdom 53°11´N 03°25´W
97 J18 **Denbigh** *cultural region* N Wales, United Kingdom
98 I6 **Den Burg** Noord-Holland, NW Netherlands 53°03´N 04°47´E
99 F18 **Dender** *Fr.* Dendre. ≈ W Belgium
99 F18 **Denderleeuw** Oost-Vlaanderen, NW Belgium 50°53´N 04°05´E
99 F17 **Dendermonde** *Fr.* Termonde. Oost-Vlaanderen, NW Belgium 51°02´N 04°08´E
Dendre *see* Dender
194 I9 **Dendtler Island** *island* Antarctica
98 P10 **Denekamp** Overijssel, E Netherlands 52°23´N 07°E
77 W12 **Dengas** Zinder, S Niger 13°15´N 09°43´E
Dêngka *see* Têwo
Dêngkagoin *see* Têwo
162 L13 **Dengkou** *var.* Bayan Gol. Nei Mongol Zizhiqu, N China 40°15´N 106°59´E
159 Q14 **Dêngqên** *var.* Gyamotang. Xizang Zizhiqu, W China 31°28´N 95°28´E
Deng Xian *see* Dengzhou
160 M7 **Dengzhou** *prev.* Deng Xian. Henan, C China 32°48´N 111°59´E
Dengzhou *see* Penglai
Den Haag *see* 's-Gravenhage
180 H10 **Denham** Western Australia 25°56´S 113°35´E
98 N9 **Den Ham** Overijssel, E Netherlands 52°30´N 06°31´E
44 J12 **Denham, Mount** ▲ C Jamaica 18°13´N 77°33´W
22 J8 **Denham Springs** Louisiana, S USA 30°29´N 90°57´W
98 I7 **Den Helder** Noord-Holland, NW Netherlands 52°54´N 04°45´E
105 T11 **Dénia** Valenciana, E Spain 38°51´N 00°07´E
189 Q8 **Denig** W Nauru
183 N10 **Deniliquin** New South Wales, SE Australia 35°33´S 144°58´E
29 T14 **Denison** Iowa, C USA 42°00´N 95°22´W
25 U5 **Denison** Texas, SW USA 33°45´N 96°32´W
144 L8 **Denisovka** *prev.* Ordzhonikidze. Kostanay, N Kazakhstan 52°27´N 61°42´E
136 D15 **Denizli** Denizli, SW Turkey 37°46´N 29°05´E
136 D15 **Denizli** ⬖ *province* SW Turkey
Denjong *see* Sikkim
183 S7 **Denman** New South Wales, SE Australia 32°24´S 150°43´E
195 Y10 **Denman Glacier** *glacier* Antarctica
21 R14 **Denmark** South Carolina, SE USA 33°19´N 81°08´W
95 G23 **Denmark** *off.* Kingdom of Denmark, *Dan.* Danmark; *anc.* Hafnia. ◆ *monarchy* N Europe
Denmark, Kingdom of *see* Denmark
92 H1 **Denmark Strait** *var.* Danmarksstraedet. *strait* Greenland/Iceland
45 T11 **Dennery** E Saint Lucia 13°55´N 60°53´W
98 I7 **Den Oever** Noord-Holland, NW Netherlands 52°56´N 05°01´E
147 O13 **Denov** *Rus.* Denau. Surkhondaryo Viloyati, S Uzbekistan 38°20´N 67°48´E
169 U17 **Denpasar** *prev.* Paloe. Bali, C Indonesia 08°40´S 115°14´E
116 E12 **Denta** Timiş, W Romania 45°20´N 21°15´E
21 Y3 **Denton** Maryland, NE USA 38°53´N 75°50´W
25 T6 **Denton** Texas, SW USA 33°11´N 97°08´W
186 G9 **D'Entrecasteaux Islands** *island group* SE Papua New Guinea
37 T4 **Denver** *state capital* Colorado, C USA 39°45´N 105°W
37 T4 **Denver** ✈ Colorado, C USA 39°57´N 104°38´W
24 L6 **Denver City** Texas, SW USA 32°57´N 102°49´W
152 J9 **Deoband** Uttar Pradesh, N India 29°41´N 77°40´E
Deoghar *see* Devghar
163 X15 **Deokjeok-gundo** *prev.* Tŏkchŏk-kundo. *island group* NW South Korea
154 E13 **Deolāli** Mahārāshtra, W India 19°55´N 73°49´E
154 I10 **Deori** Madhya Pradesh, C India 23°09´N 78°39´E
153 O12 **Deoria** Uttar Pradesh, N India 26°31´N 83°48´E
99 A17 **De Panne** West-Vlaanderen, W Belgium 51°06´N 02°35´E
Departamento del Quindío *see* Quindío
Departamento de Narino, *see* Nariño
54 M5 **Dependencia Federal** *off.* Territorio Dependencia Federal. ⬖ *federal dependency* N Venezuela
Dependencia Federal, Territorio *see* Dependencia Federal
30 M7 **De Pere** Wisconsin, N USA 44°26´N 88°03´W
18 D10 **Depew** New York, NE USA 42°54´N 78°41´W
99 E17 **De Pinte** Oost-Vlaanderen, NW Belgium 51°00´N 03°37´E
25 V5 **Deport** Texas, SW USA 33°31´N 95°19´W
123 Q8 **Deputatskiy** Respublika Sakha (Yakutiya), NE Russian Federation 69°18´N 139°48´E
Dêqên *see* Dagzê
27 S13 **De Queen** Arkansas, C USA 34°02´N 94°20´W
22 G8 **De Quincy** Louisiana, S USA 30°27´N 93°25´W
81 J20 **Dera** *spring/well* S Kenya 02°39´S 39°52´E
Der'a/Derá/Déraa *see* Darʿā
149 S10 **Dera Ghāzi Khān** *var.* Dera Ghāzikhān. Punjab, C Pakistan 30°01´N 70°37´E
Dera Ghāzikhān *see* Dera Ghāzi Khān
149 S8 **Dera Ismāīl Khān** Khyber Pakhtunkhwa, C Pakistan 31°51´N 70°56´E
Đeravica *see* Gjeravicë
116 L6 **Derazhnya** Khmel'nyts'ka Oblast', W Ukraine 49°16´N 27°24´E
127 R17 **Derbent** Respublika Dagestan, SW Russian Federation 42°01´N 48°16´E
147 N13 **Derbent** Surkhondaryo Viloyati, S Uzbekistan 38°15´N 66°59´E
Derbisīye *see* Darbāsīyah
79 M15 **Derbissaka** Mbomou, SE Central African Republic 05°43´N 24°48´E
180 L4 **Derby** Western Australia 17°18´S 123°37´E
97 M19 **Derby** C England, United Kingdom 52°55´N 01°30´W
27 N7 **Derby** Kansas, C USA 37°33´N 97°16´W
97 L18 **Derbyshire** *cultural region* C England, United Kingdom
112 O11 **Đerdap** *physical region* E Serbia
Derelí *see* Gónnoi
162 L9 **Deren** *var.* Tsant. Dundgovĭ, C Mongolia 46°16´N 106°55´E
171 W13 **Derew** ≈ Papua, E Indonesia
127 R8 **Dergachi** Saratovskaya Oblast', W Russian Federation 51°15´N 48°58´E
Dergachi *see* Derhachi
97 C19 **Derg, Lough** *Ir.* Loch Deirgeirt. ◎ W Ireland
117 V5 **Derhachi** *Rus.* Dergachi. Kharkivs'ka Oblast', E Ukraine 50°09´N 36°11´E
22 G8 **De Ridder** Louisiana, S USA 30°51´N 93°18´W
137 P16 **Derik** Mardin, SE Turkey 37°22´N 40°16´E
83 E20 **Derm** Hardap, C Namibia 23°38´S 18°12´E
Dermentobe *see* Diirmentobe
27 W14 **Dermott** Arkansas, C USA 33°31´N 91°26´W
Dérna *see* Darnah
Dernberg, Cape *see* Dolphin Head
22 J11 **Dernieres, Isles** *island group* Louisiana, S USA
102 I4 **Déroute, Passage de la** *strait* Channel Islands/France
Derrá *see* Darʿā
81 O16 **Derri** *prev.* Dirri. Galguduud, C Somalia 04°15´N 46°31´E
Derry *see* Londonderry
Dertona *see* Tortona
Dertosa *see* Tortosa
80 H8 **Derudeb** Red Sea, NE Sudan 17°31´N 36°07´E
112 H10 **Derventa** Republika Srpska, N Bosnia and Herzegovina 44°57´N 17°55´E
183 O16 **Derwent Bridge** Tasmania, SE Australia 42°10´S 146°13´E
183 O17 **Derwent, River** ≈ Tasmania, SE Australia
146 F10 **Derweze** *Rus.* Darvaza. Ahal Welaýaty, C Turkmenistan 40°10´N 58°27´E
Deržavinsk *see* Derzhavinsk
145 O9 **Derzhavinsk** *var.* Deržavinsk. ⬖ Akmola, C Kazakhstan
Dés *see* Dej
57 J18 **Desaguadero** Puno, S Peru 16°35´S 69°05´W
57 J18 **Desaguadero, Río** ≈ Bolivia/Peru
191 W9 **Désappointement, Îles du** *island group* Îles Tuamotu, C French Polynesia
27 W11 **Des Arc** Arkansas, C USA 34°58´N 91°30´W
14 C10 **Desbarats** Ontario, S Canada 46°20´N 83°52´W
62 H13 **Descabezado Grande, Volcán** ⛰ C Chile 35°34´S 70°40´W
102 L9 **Descartes** Indre-et-Loire, C France 46°58´N 00°40´E
11 T13 **Deschambault Lake** ◎ Saskatchewan, C Canada
Deschnaer Koppe *see* Velká Deštná
32 I11 **Deschutes River** ≈ Oregon, NW USA
80 J12 **Desē** *var.* Desse, *It.* Dessie. Āmara, N Ethiopia 11°02´N 39°39´E
63 I20 **Deseado, Río** ≈ S Argentina
106 F8 **Desenzano del Garda** Lombardia, N Italy 45°28´N 10°31´E
36 K3 **Deseret Peak** ▲ Utah, W USA 40°27´N 112°37´W
64 P6 **Deserta Grande** *island* Madeira, Portugal, NE Atlantic Ocean
64 P6 **Desertas, Ilhas** *island group* Madeira, Portugal, NE Atlantic Ocean
35 X16 **Desert Center** California, W USA 33°42´N 115°22´W
35 V15 **Desert Hot Springs** California, W USA 33°57´N 116°33´W
14 K10 **Désert, Lac** ◎ Québec, SE Canada
36 J2 **Desert Peak** ▲ Utah, W USA 41°03´N 113°22´W
31 R11 **Deshler** Ohio, N USA 41°12´N 83°55´W
Deshu *see* Dīshū
Desiderii Fanum *see* St-Dizier
106 D7 **Desio** Lombardia, N Italy 45°37´N 09°12´E
115 E15 **Deskáti** *var.* Dheskáti. Dytikí Makedonía, N Greece 39°55´N 21°49´E
28 L2 **Des Lacs River** ≈ North Dakota, N USA
27 X6 **Desloge** Missouri, C USA 37°52´N 90°31´W
11 Q12 **Desmarais** Alberta, W Canada 55°58´N 113°56´W
29 Q10 **De Smet** South Dakota, N USA 44°23´N 97°33´W
29 V14 **Des Moines** *state capital* Iowa, C USA 41°36´N 93°37´W
17 N9 **Des Moines River** ≈ C USA
117 P4 **Desna** ≈ Russian Federation/Ukraine
116 G14 **Desnăţui** ≈ S Romania
63 F24 **Desolación, Isla** *island* S Chile
29 V14 **De Soto** Iowa, C USA 41°31´N 94°00´W
23 Q4 **De Soto Falls** *waterfall* Alabama, S USA
83 I25 **Despatch** Eastern Cape, S South Africa 33°48´S 25°28´E
105 N12 **Despeñaperros, Desfiladero de** *pass* S Spain
31 N10 **Des Plaines** Illinois, N USA 42°01´N 87°52´W
115 J21 **Despotikó** *island* Kykládes, Greece, Aegean Sea
112 N12 **Despotovac** Serbia, E Serbia 44°06´N 21°25´E
101 M14 **Dessau** Sachsen-Anhalt, E Germany 51°51´N 12°15´E
Desse *see* Desē
99 J16 **Dessel** Antwerpen, N Belgium 51°15´N 05°07´E
Dessie *see* Desē
Destêrro *see* Florianópolis
23 P9 **Destin** Florida, SE USA 30°23´N 86°30´W
Deštná *see* Velká Deštná
193 T10 **Desventurados, Islas de los** *island group* W Chile
103 N1 **Desvres** Pas-de-Calais, N France 50°41´N 01°48´E
116 E12 **Deta** *Ger.* Detta. Timiş, W Romania 45°24´N 21°14´E
101 H14 **Detmold** Nordrhein-Westfalen, W Germany 51°55´N 08°52´E
31 S10 **Detroit** Michigan, N USA 42°20´N 83°03´W
25 W5 **Detroit** Texas, SW USA 33°39´N 95°16´W
31 S10 **Detroit** ≈ Canada/USA
29 S6 **Detroit Lakes** Minnesota, N USA 46°49´N 95°49´W
31 S10 **Detroit Metropolitan** ✈ Michigan, N USA 42°12´N 83°16´W
Detta *see* Deta
167 S10 **Det Udom** Ubon Ratchathani, E Thailand 14°54´N 105°03´E
111 K20 **Detva** *Hung.* Gyeva. Bankobýstricky Kraj, C Slovakia 48°35´N 19°25´E
154 G13 **Deūlgaon Rāja** Mahārāshtra, C India 20°04´N 76°08´E
99 L15 **Deurne** Noord-Brabant, SE Netherlands 51°28´N 05°47´E
99 H16 **Deurne** ✈ (Antwerpen) Antwerpen, N Belgium 51°10´N 04°28´E
Deutsch-Brod *see* Havlíčkův Brod
Deutschendorf *see* Poprad
Deutsch-Eylau *see* Iława
109 Y6 **Deutschkreutz** Burgenland, E Austria 47°37´N 16°37´E
Deutsch Krone *see* Wałcz
Deutschland/Deutschland, Bundesrepublik *see* Germany
109 V9 **Deutschlandsberg** Steiermark, SE Austria 46°52´N 15°13´E
Deutsch-Südwestafrika *see* Namibia
109 Y3 **Deutsch-Wagram** Niederösterreich, E Austria 48°19´N 16°33´E
Deux-Ponts *see* Zweibrücken
14 I11 **Deux Rivieres** Ontario, SE Canada 46°13´N 78°16´W
102 K9 **Deux-Sèvres** ⬖ *department* W France
116 G11 **Deva** *Ger.* Diemrich, *Hung.* Déva. Hunedoara, W Romania 45°55´N 22°55´E
Déva *see* Deva
Deva *see* Chester
Devana *see* Aberdeen
Devana Castra *see* Chester
Đevđelija *see* Gevgelija
136 L12 **Deveci Dağları** ▲▲ N Turkey
137 P15 **Devegeçidi Barajī** ⊠ SE Turkey
136 K15 **Develi** Kayseri, C Turkey 38°22´N 35°28´E
98 M11 **Deventer** Overijssel, E Netherlands 52°15´N 06°10´E
15 O10 **Devenyns, Lac** ◎ Québec, SE Canada
96 K8 **Deveron** ≈ NE Scotland, United Kingdom
153 R14 **Devghar** *prev.* Deoghar. Jhārkhand, NE India
27 R10 **Devil's Den** *plateau* Arkansas, C USA
35 R7 **Devils Gate** *pass* California, W USA
30 J2 **Devils Island** *island* Apostle Islands, Wisconsin, N USA
Devil's Island *see* Diable, Île du
29 P3 **Devils Lake** North Dakota, N USA 48°08´N 98°50´W
31 R10 **Devils Lake** ◎ Michigan, N USA
29 O3 **Devils Lake** ◎ North Dakota, N USA
35 W13 **Devils Playground** *desert* California, W USA
25 O11 **Devils River** ≈ Texas, SW USA
33 Y12 **Devils Tower** ▲ Wyoming, C USA 44°33´N 104°45´W
114 I11 **Devin** *prev.* Dovlen. Smolyan, S Bulgaria 41°45´N 24°24´E
25 R12 **Devine** Texas, SW USA 29°08´N 98°54´W
152 H13 **Devli** Rājasthān, N India 25°47´N 75°23´E
Devne *see* Devnya
114 N8 **Devnya** *prev.* Devne. Varna, E Bulgaria 43°15´N 27°35´E
31 U14 **Devola** Ohio, N USA 39°28´N 81°28´W
Devoll *see* Devollit, Lumi i

◆ Country
● Country Capital
◇ Dependent Territory
○ Dependent Territory Capital
⬖ Administrative Regions
✈ International Airport
▲ Mountain
▲▲ Mountain Range
⛰ Volcano
≈ River
◎ Lake
⊠ Reservoir

113 *M21* **Devollit, Lumi i** *var.* Devoll. ≈ SE Albania
11 *Q14* **Devon** Alberta, SW Canada 53°21′N 113°47′W
97 *I23* **Devon** *cultural region* SW England, United Kingdom
197 *N10* **Devon Island** *prev.* North Devon Island. *island* Parry Islands, Nunavut, NE Canada
183 *O16* **Devonport** Tasmania, SE Australia 41°14′S 146°21′E
136 *H11* **Devrek** Zonguldak, N Turkey 41°14′N 31°57′E
154 *G10* **Dewās** Madhya Pradesh, C India 22°59′N 76°03′E
De Westerein *see* Zwaagwesteinde
27 *P8* **Dewey** Oklahoma, C USA 36°48′N 95°56′W
Dewey *see* Culebra
98 *M8* **De Wijk** Drenthe, NE Netherlands 52°41′N 06°13′E
27 *W12* **De Witt** Arkansas, C USA 34°17′N 91°21′W
29 *Z14* **De Witt** Iowa, C USA 41°49′N 90°32′W
29 *R16* **De Witt** Nebraska, C USA 40°23′N 96°55′W
97 *M17* **Dewsbury** N England, United Kingdom 53°42′N 01°37′W
161 *Q10* **Dexing** Jiangxi, S China 28°51′N 117°36′E
27 *Y8* **Dexter** Missouri, C USA 36°48′N 89°57′W
37 *U14* **Dexter** New Mexico, SW USA 33°12′N 104°25′W
160 *I8* **Deyang** Sichuan, C China 31°08′N 104°23′E
182 *C4* **Dey-Dey, Lake** *salt lake* South Australia
143 *S7* **Deyhūk** Yazd, E Iran 33°18′N 57°30′E
Deynau *see* Galkynyş
Deyyer *see* Bandar-e Dayyer
142 *L8* **Dezfūl** *var.* Dizful. Khūzestān, SW Iran 32°23′N 48°28′E
129 *X4* **Dezhneva, Mys** *headland* NE Russian Federation 66°08′N 69°40′W
161 *P4* **Dezhou** Shandong, E China 37°28′N 116°18′E
Dezhou *see* Dechang
Dezh Shāhpūr *see* Marīvān
Dhaalu Atoll *see* South Nilandhe Atoll
Dhahran *see* Az̧ Z̧ahrān
Dhahran Al Khobar *see* Az̧ Z̧ahrān al Khubar
153 *U14* **Dhaka** *prev.* Dacca. ● (Bangladesh) Dhaka, C Bangladesh 23°42′N 90°22′E
153 *T15* **Dhaka** ◈ *division* C Bangladesh
Dhali *see* Idálion
Dhalli Rajhara *see* Dalli Rājhara
141 *O15* **Dhamār** W Yemen 14°31′N 44°25′E
154 *K12* **Dhamtari** Chhattīsgarh, C India 20°43′N 81°36′E
153 *Q15* **Dhanbād** Jhārkhand, NE India 23°48′N 86°27′E
152 *L10* **Dhangaḍhi** *var.* Dhangarhi. Far Western, W Nepal 28°45′N 80°38′E
Dhangarhi *see* Dhangaḍhi
Dhank *see* Ḑank
153 *R12* **Dhankuṭā** Eastern, E Nepal 27°06′N 87°21′E
152 *I6* **Dhaola Dhār** ▲▲ NE India
154 *F10* **Dhār** Madhya Pradesh, C India 22°32′N 75°24′E
153 *R12* **Dharān** *var.* Dharan Bazar. Eastern, E Nepal 26°51′N 87°18′E
Dharan Bazar *see* Dharān
155 *H21* **Dhārāpuram** Tamil Nādu, SE India 10°45′N 77°33′E
155 *H20* **Dharmapuri** Tamil Nādu, SE India 12°11′N 78°07′E
155 *H18* **Dharmavaram** Andhra Pradesh, E India 14°27′N 77°44′E
154 *M11* **Dharmjaygarh** Chhattīsgarh, C India 22°27′N 83°16′E
Dharmsāla *see* Dharmshāla
152 *I7* **Dharmshāla** *prev.* Dharmsāla. Himāchal Pradesh, N India 32°14′N 76°24′E
155 *F17* **Dhārwād** *prev.* Dharwar. Karnātaka, SW India 15°30′N 75°04′E
Dharwar *see* Dhārwād
Dhaulagiri *see* Dhawalāgiri
153 *O10* **Dhawalāgiri** *var.* Dhaulagiri. ▲ C Nepal 28°45′N 83°27′E
81 *L18* **Dheere Laaq** *var.* Lak Dera, *It.* Lach Dera. *seasonal river* Kenya/Somalia
Dhekeleia Sovereign Base Area *see* Dhekelia Sovereign Base Area
121 *Q3* **Dhekélia** *Eng.* Dhekelia, *Gk.* Dekéleia. *UK air base* SE Cyprus35°00′N 33°45′E *see* Dhekélia
121 *Q3* **Dhekelia Sovereign Base Area** *var.* Dhekeleia Sovereign Base Area. *UK military installation* E Cyprus34°59′N 33°45′E
Dhelvinákion *see* Delvináki
113 *M22* **Dhëmbelit, Maja e** ▲ S Albania 40°10′N 20°22′E
154 *O12* **Dhenkānāl** Odisha, E India 20°40′N 85°36′E
Dheskáti *see* Deskáti
138 *G11* **Dhībān** Mādabā, NW Jordan 31°30′N 35°47′E
Dhidhimótikhon *see* Didymóteicho
Dhíkti Ori *see* Díkti
139 *W11* **Dhī Qār** *off.* Muḩāfa at Dhī Qār, *var.* Al Muntafiq, An Nāşirīyah. ◈ *governorate* SE Iraq
Dhī Qār, Muḩāfa at *see* Dhī Qār
138 *I12* **Dhirwah, Wādī adh** *dry watercourse* C Jordan
Dhístomon *see* Dístomo
Dhodhekánisos *see* Dodekánisa
Dhodhóni *see* Dodóni
Dhofar *see* Z̧ufār
Dhomokós *see* Domokós
Dhond *see* Daund
155 *H17* **Dhone** Andhra Pradesh, C India 15°25′N 77°52′E
154 *B11* **Dhorāji** Gujarāt, W India 21°44′N 70°27′E
Dhráma *see* Dráma
154 *C10* **Dhrāngadhra** Gujarāt, W India 22°59′N 71°32′E
Dhrepanon, Akrotírio *see* Drépano, Akrotírio
153 *T13* **Dhuburi** Assam, NE India 26°06′N 89°55′E
154 *F12* **Dhule** *prev.* Dhulia. Mahārāshtra, C India 20°54′N 74°47′E
Dhulia *see* Dhule
Dhún Dealgan, Cuan *see* Dundalk Bay
Dhún Droma, Cuan *see* Dundrum Bay
Dhún na nGall, Bá *see* Donegal Bay
Dhū Shaykh *see* Qazānīyah
80 *Q13* **Dhuudo** Bari, NE Somalia 09°21′N 50°19′E
81 *N15* **Dhuusa Marreeb** *var.* Dusa Marreb, *It.* Dusa Mareb. Galguduud, C Somalia 05°33′N 46°24′E
115 *J24* **Día** *island* SE Greece
55 *Y9* **Diable, Île du** *var.* Devil's Island. *island* N French Guiana
15 *N12* **Diable, Rivière du** ≈ Québec, SE Canada
35 *N8* **Diablo, Mount** ▲ California, W USA 37°52′N 121°57′W
35 *O9* **Diablo Range** ▲▲ California, W USA
24 *I8* **Diablo, Sierra** ▲▲ Texas, SW USA
45 *O11* **Diablotins, Morne** ▲ N Dominica 15°30′N 61°23′W
77 *N11* **Diafarabé** Mopti, C Mali 14°09′N 05°01′W
77 *N11* **Diaka** ≈ SW Mali
Diakovár *see* Đakovo
76 *I12* **Dialakoto** S Senegal 13°21′N 13°19′W
61 *B18* **Diamante** Entre Ríos, E Argentina 32°05′S 60°40′W
62 *I12* **Diamante, Río** ≈ C Argentina
59 *M19* **Diamantina** Minas Gerais, SE Brazil 18°17′S 43°37′W
59 *N17* **Diamantina, Chapada** ▲▲ E Brazil
173 *U11* **Diamantina Fracture Zone** *tectonic feature* E Indian Ocean
181 *T8* **Diamantina River** ≈ Queensland/South Australia
38 *D9* **Diamond Head** *headland* O'ahu, Hawai'i, USA 21°15′N 157°48′W
37 *P2* **Diamond Peak** ▲ Colorado, C USA 40°56′N 108°56′W
35 *W5* **Diamond Peak** ▲ Nevada, W USA 39°34′N 115°46′W
Diamond State *see* Delaware
76 *J11* **Diamou** Kayes, SW Mali 14°04′N 11°16′W
95 *I23* **Dianalund** Sjælland, C Denmark 55°32′N 11°30′E
65 *G25* **Diana's Peak** ▲ C Saint Helena
160 *M16* **Dianbai** *var.* Shuidong. Guangdong, S China 21°30′N 111°05′E
160 *G13* **Dian Chi** ⊠ SW China
106 *B10* **Diano Marina** Liguria, NW Italy 43°55′N 08°06′E
163 *V11* **Diaobingshan** *var.* Tiefa. Liaoning, NE China 42°25′N 123°39′E
Diaoyu Dao *see* Uotsuri-shima
Diaoyutai *see* Senkaku-shotō
77 *R13* **Diapaga** E Burkina Faso 12°09′N 01°48′E
Diarbekr *see* Diyarbakır
107 *J15* **Diavolo, Passo del** *pass* C Italy
61 *B18* **Díaz** Santa Fe, C Argentina 32°22′S 61°05′W
141 *W6* **Dibā al Ḩişn** *var.* Dibāh, Dibba. Ash Shāriqah, NE United Arab Emirates 25°35′N 56°16′E
dibaga *see* Dîbege
Dibāh *see* Dibā al Ḩişn
79 *L22* **Dibaya** Kasai-Occidental, S Dem. Rep. Congo 06°31′S 22°57′E
Dibba *see* Dibā al Ḩişn
195 *W15* **Dibble Iceberg Tongue** *ice feature* Antarctica
139 *S3* **Dîbege** *Ar.* Ad Dībakah, *var.* Dībaga. Arbīl, N Iraq 35°51′N 43°49′E
113 *L19* **Dibër** ◈ *district* E Albania
83 *I20* **Dibete** Central, SE Botswana 23°45′S 26°26′E
Dibio *see* Dijon
25 *W9* **Diboll** Texas, SW USA 31°11′N 94°46′W
Dibra *see* Debar
153 *X11* **Dibrugarh** Assam, NE India 27°29′N 94°49′E
54 *G4* **Dibulla** La Guajira, N Colombia 11°14′N 73°22′W
25 *O5* **Dickens** Texas, SW USA 33°38′N 100°51′W
19 *R2* **Dickey** Maine, NE USA 47°04′N 69°05′W
30 *K9* **Dickeyville** Wisconsin, N USA 42°37′N 90°36′W
28 *K5* **Dickinson** North Dakota, N USA 46°54′N 102°48′W
0 *E6* **Dickins Seamount** *undersea feature* NE Pacific Ocean 54°30′N 137°00′W
27 *O13* **Dickson** Oklahoma, C USA 34°11′N 96°58′W
20 *I9* **Dickson** Tennessee, S USA 36°04′N 87°23′W
Dicle *see* Tigris
Dicsőszentmárton *see* Târnăveni
98 *M12* **Didam** Gelderland, E Netherlands 51°56′N 06°08′E
163 *Y8* **Didao** Heilongjiang, NE China 45°22′N 130°48′E
76 *L12* **Didiéni** Koulikoro, W Mali 13°48′N 08°01′W
Didimo *see* Dídymo
Didimotiho *see* Didymóteicho
81 *K17* **Didimtu** *spring/well* NE Kenya 02°58′N 40°07′E
67 *U9* **Didinga Hills** ▲▲ S Sudan
11 *Q16* **Didsbury** Alberta, SW Canada 51°39′N 114°09′W
152 *G11* **Dīdwāna** Rājasthān, N India 27°23′N 74°36′E
115 *G20* **Dídymo** *var.* Didimo. ▲ S Greece 37°28′N 23°12′E
114 *L12* **Didymóteicho** *var.* Dhidhimótikhon, Didimotiho. Anatolikí Makedonía kai Thráki, NE Greece 41°22′N 26°29′E
103 *S13* **Die** Drôme, E France 44°46′N 05°21′E
77 *O13* **Diébougou** SW Burkina Faso 11°00′N 03°12′W
Diedenhofen *see* Thionville
11 *S16* **Diefenbaker, Lake** ⊠ Saskatchewan, S Canada
62 *H7* **Diego de Almagro** Atacama, N Chile 26°24′S 70°10′W
63 *F23* **Diego de Almagro, Isla** *island* S Chile
61 *A20* **Diego de Alvear** Santa Fe, C Argentina 34°25′S 62°10′W
173 *Q7* **Diego Garcia** *island* S British Indian Ocean Territory
Diégo-Suarez *see* Antsirañana
99 *M23* **Diekirch** Diekirch, C Luxembourg 49°52′N 06°10′E
99 *L23* **Diekirch** ◈ *district* N Luxembourg
76 *K11* **Diéma** Kayes, W Mali 14°30′N 09°12′W
101 *H15* **Diemel** ≈ W Germany
98 *I10* **Diemen** Noord-Holland, C Netherlands 52°21′N 04°58′E
Diemrich *see* Deva
Điện Biên *see* Điện Biên Phu
167 *R6* **Điện Biên Phu** *var.* Bien Bien, Điện Biên. Lai Châu, N Vietnam 21°23′N 103°02′E
167 *S7* **Diên Châu** Nghê An, N Vietnam 18°54′N 105°35′E
99 *K18* **Diepenbeek** Limburg, NE Belgium 50°54′N 05°25′E
98 *N11* **Diepenheim** Overijssel, E Netherlands 52°10′N 06°37′E
98 *M10* **Diepenveen** Overijssel, E Netherlands 52°18′N 06°09′E
100 *G12* **Diepholz** Niedersachsen, NW Germany 52°36′N 08°23′E
102 *M3* **Dieppe** Seine-Maritime, N France 49°55′N 01°05′E
98 *M12* **Dieren** Gelderland, E Netherlands 52°03′N 06°06′E
27 *S13* **Dierks** Arkansas, C USA 34°07′N 94°01′W
99 *J17* **Diest** Vlaams Brabant, C Belgium 50°58′N 05°03′E
108 *F7* **Dietikon** Zürich, NW Switzerland 47°24′N 08°25′E
103 *R13* **Dieulefit** Drôme, E France 44°30′N 05°01′E
103 *T5* **Dieuze** Moselle, NE France 48°49′N 06°41′E
119 *H15* **Dieveniškės** Vilnius, SE Lithuania 54°12′N 25°38′E
98 *N7* **Diever** Drenthe, NE Netherlands 52°49′N 06°19′E
101 *F17* **Diez** Rheinland-Pfalz, W Germany 50°22′N 08°01′E
77 *Y12* **Diffa** Diffa, SE Niger 13°19′N 12°37′E
77 *Y10* **Diffa** ◈ *department* SE Niger
99 *L25* **Differdange** Luxembourg, SW Luxembourg 49°32′N 05°53′E
13 *O16* **Digby** Nova Scotia, SE Canada 44°37′N 65°47′W
26 *J5* **Dighton** Kansas, C USA 38°28′N 100°28′W
Dignano d'Istria *see* Vodnjan
103 *T14* **Digne** *var.* Digne-les-Bains. Alpes-de-Haute-Provence, SE France 44°05′N 06°14′E
Digne-les-Bains *see* Digne
Digoel *see* Digul, Sungai
103 *Q10* **Digoin** Saône-et-Loire, C France 46°30′N 04°E
171 *Q8* **Digos** Mindanao, S Philippines 06°46′N 125°21′E
149 *Q16* **Digri** Sind, SE Pakistan 25°11′N 69°10′E
171 *Y14* **Digul Barat, Sungai** ≈ Papua, E Indonesia
171 *Y15* **Digul, Sungai** *prev.* Digoel. ≈ Papua, E Indonesia
171 *Z14* **Digul Timur, Sungai** ≈ Papua, E Indonesia
153 *X10* **Dihāng** ≈ NE India
Dihang *see* Brahmaputra
Dihok *see* Dahūk
Dihōk *see* Dahūk
Dihok *see* Dahūk
Dihok, Parêzga-î *see* Dahūk
81 *L17* **Diinsoor** Bay, S Somalia 02°28′N 43°00′E
144 *M14* **Diirmentobe** *Kas.* Dīirmentöbe; *prev.* Dermentobe, Dyurment'yube. Kzyl-Orda, S Kazakhstan 45°46′N 63°42′E
Dīirmentöbe *see* Diirmentobe
Dijlah *see* Tigris
99 *H17* **Dijle** ≈ C Belgium
103 *R8* **Dijon** *anc.* Dibio. Côte d'Or, C France 47°21′N 05°04′E
93 *H14* **Dikanäs** Västerbotten, N Sweden 65°15′N 16°00′E
80 *L12* **Dikhil** SW Djibouti 11°08′N 42°19′E
136 *B13* **Dikili** İzmir, W Turkey 39°05′N 26°52′E
99 *B17* **Diksmuide** *var.* Dixmuide, *Fr.* Dixmude. West-Vlaanderen, W Belgium 51°02′N 02°52′E
122 *K7* **Dikson** Krasnoyarskiy Kray, N Russian Federation 73°30′N 80°35′E
115 *K25* **Díkti** *var.* Dhíkti Ori. ▲▲ Kríti, Greece, E Mediterranean Sea
77 *Z13* **Dikwa** Borno, NE Nigeria 12°00′N 13°57′E
81 *J15* **Dīla** Southern Nationalities, S Ethiopia 06°19′N 38°16′E
148 *L7* **Dilārām** *prev.* Delārām. Nīmrōz, SW Afghanistan 32°11′N 63°27′E
99 *G18* **Dilbeek** Vlaams Brabant, C Belgium 50°51′N 04°16′E
171 *Q16* **Dili** *var.* Dilli, Dilly. ● (East Timor) N East Timor 08°33′S 125°34′E
77 *Y11* **Dilia** *var.* Dillia. ≈ SE Niger
Dilijan *see* Delījān
167 *U13* **Di Linh** Lâm Đông, S Vietnam 11°38′N 108°07′E
101 *G16* **Dillenburg** Hessen, W Germany 50°45′N 08°16′E
25 *Q13* **Dilley** Texas, SW USA 28°40′N 99°10′W
Dilli *see* Dili, East Timor
Dilli *see* Delhi, India
Dillia *see* Dilia
80 *E11* **Dilling** *var.* Ad Dalanj. Southern Kordofan, C Sudan 12°02′N 29°41′E
101 *D20* **Dillingen** Saarland, SW Germany 49°20′N 06°43′E
Dillingen *see* Dillingen an der Donau
101 *J22* **Dillingen an der Donau** *var.* Dillingen. Bayern, S Germany 48°34′N 10°29′E
39 *O13* **Dillingham** Alaska, USA 59°03′N 158°30′W
33 *Q12* **Dillon** Montana, NW USA 45°14′N 112°38′W
21 *T12* **Dillon** South Carolina, SE USA 34°25′N 79°22′W
31 *T13* **Dillon Lake** ⊠ Ohio, N USA
Dilly *see* Dili
Dilman *see* Salmās
79 *K24* **Dilolo** Katanga, S Dem. Rep. Congo 10°42′S 22°21′E
115 *J20* **Dílos** *island* Kykládes, Greece, Aegean Sea
141 *Y11* **Ḑil', Ra's aḑ** *headland* E Oman 19°12′N 57°53′E
29 *R5* **Dilworth** Minnesota, N USA 46°53′N 96°38′W
138 *H7* **Dimashq** *var.* Ash Shām, Esh Sham, *Eng.* Damascus, *Fr.* Damas, *It.* Damasco. ● (Syria) Rīf Dimashq, SW Syria 33°30′N 36°19′E
138 *I7* **Dimashq** ✈ Rīf Dimashq, S Syria 33°30′N 36°19′E
Dimashq, Muḩāfaz̧at *see* Rīf Dimashq
79 *L21* **Dimbelenge** Kasai-Occidental, C Dem. Rep. Congo 05°36′S 23°04′E
77 *N16* **Dimbokro** E Ivory Coast 06°43′N 04°46′W
182 *L11* **Dimboola** Victoria, SE Australia 36°29′S 142°03′E
Dîmboviţa *see* Dâmboviţa
Dimitrov *see* Dymytrov
114 *K11* **Dimitrovgrad** Haskovo, S Bulgaria 42°03′N 25°36′E
127 *R5* **Dimitrovgrad** Ul'yanovskaya Oblast', W Russian Federation 54°14′N 49°37′E
113 *Q15* **Dimitrovgrad** *prev.* Caribrod. Serbia, SE Serbia 43°01′N 22°46′E
Dimitrovo *see* Pernik
Dimlang *see* Vogel Peak
24 *M3* **Dimmitt** Texas, SW USA 34°32′N 102°20′W
114 *F7* **Dimovo** Vidin, NW Bulgaria 43°46′N 22°47′E
59 *A16* **Dimpolis** Acre, W Brazil 09°52′S 71°51′W
115 *O23* **Dimyliá** Ródos, Dodekánisa, Greece, Aegean Sea 36°17′N 27°59′E
171 *Q6* **Dinagat Island** *island* S Philippines
153 *S13* **Dinajpur** Rajshahi, NW Bangladesh 25°38′N 88°40′E
102 *I6* **Dinan** Côtes d'Armor, NW France 48°27′N 02°02′W
99 *I21* **Dinant** Namur, S Belgium 50°16′N 04°55′E
136 *E15* **Dinar** Afyon, SW Turkey 38°05′N 30°09′E
112 *F13* **Dinara** ▲ W Croatia 43°49′N 16°42′E
Dinara *see* Dinaric Alps
102 *I5* **Dinard** Ille-et-Vilaine, NW France 48°38′N 02°04′W
112 *F13* **Dinaric Alps** *var.* Dinara. ▲▲ Bosnia and Herzegovina/Croatia
143 *N10* **Dīnār, Kūh-e** ▲ C Iran 30°51′N 51°36′E
Dinbych *see* Denbigh
155 *H22* **Dindigul** Tamil Nādu, SE India 10°23′N 78°E
83 *M19* **Dindiza** Gaza, S Mozambique 23°22′S 33°28′E
79 *H21* **Dinga** Bandundu, SW Dem. Rep. Congo 05°00′S 16°29′E
149 *V7* **Dinga** Punjab, E Pakistan 32°38′N 73°45′E
Dingchang *see* Qinxian
158 *L16* **Dinggyê** *var.* Gyangkar. Xizang Zizhiqu, W China 28°18′N 88°06′E
97 *A20* **Dingle** *Ir.* An Daingean. SW Ireland 52°09′N 10°16′W
97 *A20* **Dingle Bay** *Ir.* Bá an Daingin. *bay* SW Ireland
18 *I13* **Dingmans Ferry** Pennsylvania, NE USA 41°12′N 74°51′W
101 *N22* **Dingolfing** Bayern, SE Germany 48°37′N 12°28′E
171 *O1* **Dingras** Luzon, N Philippines 18°06′N 120°43′E
76 *J13* **Dinguiraye** N Guinea 11°19′N 10°49′W
96 *I8* **Dingwall** N Scotland, United Kingdom 57°36′N 04°26′W
159 *V10* **Dingxi** Gansu, C China 35°36′N 104°33′E
Ding Xian *see* Dingzhou
161 *Q7* **Dingyuan** Anhui, E China 32°30′N 117°40′E
161 *O3* **Dingzhou** *prev.* Ding Xian. Hebei, E China 38°31′N 114°52′E
167 *U6* **Đinh Lâp** Lang Sơn, N Vietnam 21°33′N 107°03′E
167 *T13* **Đinh Quan** *var.* Tân Phu. Đông Nai, S Vietnam 11°11′N 107°20′E
100 *E13* **Dinkel** ≈ Germany/Netherlands
101 *J21* **Dinkelsbühl** Bayern, S Germany 49°06′N 10°18′E
101 *D14* **Dinslaken** Nordrhein-Westfalen, W Germany 51°34′N 06°43′E
35 *R11* **Dinuba** California, W USA 36°32′N 119°23′W
21 *W7* **Dinwiddie** Virginia, NE USA 37°02′N 77°40′W
98 *N13* **Dinxperlo** Gelderland, E Netherlands 51°51′N 06°30′E
Dio *see* Díon
Diófás *see* Nucet
76 *M12* **Dioïla** Koulikoro, W Mali 12°28′N 06°43′W
115 *F14* **Díon** *var.* Dio; *anc.* Dium. *site of ancient city* Kentrikí Makedonía, N Greece
115 *G19* **Dióryga Korinthou** *Eng.* Corinth Canal. *canal* S Greece
76 *G12* **Diouloulou** SW Senegal 13°00′N 16°34′W
77 *N11* **Dioura** Mopti, W Mali 14°48′N 05°20′W
76 *G11* **Diourbel** W Senegal 14°39′N 16°12′W
152 *L10* **Dipāyal** Far Western, W Nepal 29°09′N 80°46′E
121 *R1* **Dipkarpaz** *Gk.* Rizokarpaso, Rizokárpason. NE Cyprus 35°36′N 34°23′E
149 *R17* **Diplo** Sind, SE Pakistan 24°29′N 69°36′E
171 *P7* **Dipolog** *var.* Dipolog City. Mindanao, S Philippines 08°31′N 123°20′E
Dipolog City *see* Dipolog
185 *C23* **Dipton** Southland, South Island, New Zealand 45°55′S 168°21′E
77 *O10* **Diré** Tombouctou, C Mali 16°12′N 03°31′W
80 *L13* **Dirē Dawa** Dirē Dawa, E Ethiopia 09°35′N 41°53′E
Dirfis *see* Dírfys
115 *H18* **Dírfys** *var.* Dirfis. ▲▲ Évvioa, C Greece
75 *N9* **Dirj** *var.* Daraj, Darj. W Libya 30°09′N 10°26′E
180 *G10* **Dirk Hartog Island** *island* Western Australia
77 *Y8* **Dirkou** Agadez, NE Niger 18°45′N 13°00′E
181 *X11* **Dirranbandi** Queensland, E Australia 28°37′S 148°13′E
Dirri *see* Derri
Dirschau *see* Tczew
37 *N6* **Dirty Devil River** ≈ Utah, W USA
32 *E10* **Disappointment, Cape** *headland* Washington, NW USA 46°16′N 124°06′W
180 *L8* **Disappointment, Lake** *salt lake* Western Australia
183 *R12* **Disaster Bay** *bay* New South Wales, SE Australia
44 *J11* **Discovery Bay** C Jamaica 18°27′N 77°24′W
182 *K13* **Discovery Bay** *inlet* SE Australia
67 *Y15* **Discovery II Fracture Zone** *tectonic feature* SW Indian Ocean
Discovery Seamount/ Discovery Seamounts *see* Discovery Tablemounts
65 *O19* **Discovery Tablemounts** *var.* Discovery Seamount, Discovery Seamounts. *undersea feature* SW Atlantic Ocean 42°00′S 00°10′E
108 *G9* **Disentis** *Rmsch.* Mustér. Graubünden, S Switzerland 46°43′N 08°52′E
39 *O10* **Dishna River** ≈ Alaska, USA
148 *K10* **Dīshū** *var.* Deshu; *prev.* Deh Shū. Helmand, S Afghanistan 30°28′N 63°21′E
Disko Bugt *see* Qeqertarsuup Tunua
195 *X4* **Dismal Mountains** ▲▲ Antarctica
28 *M14* **Dismal River** ≈ Nebraska, C USA
Disna *see* Dzisna
99 *L19* **Dison** Liège, E Belgium 50°37′N 05°52′E
153 *V12* **Dispur** *state capital* Assam, NE India 26°03′N 91°52′E
15 *R11* **Disraeli** Québec, SE Canada 45°58′N 71°21′W
115 *F18* **Dístomo** *prev.* Dhístomon. Stereá Elláda, C Greece 38°25′N 22°40′E
Dístos, Límni *see* Dýstos, Límni
59 *L18* **Distrito Federal** *Eng.* Federal District. ◈ *federal district* C Brazil
41 *P14* **Distrito Federal** ◈ *federal district* S Mexico
54 *L4* **Distrito Federal** *off.* Territorio Distrito Federal. ◈ *federal district* N Venezuela
Distrito Federal, Territorio *see* Distrito Federal
116 *J10* **Ditrău** *Hung.* Ditró. Harghita, C Romania 46°49′N 25°31′E
Ditró *see* Ditrău
154 *B12* **Diu** Damān and Diu, W India 20°42′N 70°59′E
Dium *see* Díon
109 *S13* **Divača** SW Slovenia 45°40′N 13°58′E
102 *K5* **Dives** ≈ N France
33 *Q11* **Divide** Montana, NW USA 45°44′N 112°47′W
Divin *see* Dzivin
83 *N18* **Divinhe** Sofala, E Mozambique 20°41′S 34°46′E
59 *L20* **Divinópolis** Minas Gerais, SE Brazil 20°08′S 44°55′W
127 *N13* **Divnoye** Stavropol'skiy Kray, SW Russian Federation 45°54′N 43°18′E
76 *M17* **Divo** S Ivory Coast 05°50′N 05°22′W
Divodurum Mediomatricum *see* Metz
137 *N13* **Divriği** Sivas, C Turkey 39°23′N 38°06′E
Diwaniyah *see* Ad Dīwānīyah
14 *J10* **Dix Milles, Lac** ◎ Québec, SE Canada
14 *M8* **Dix Milles, Lac des** ◎ Québec, SE Canada
Dixmude/Dixmuide *see* Diksmuide
35 *N7* **Dixon** California, W USA 38°19′N 121°49′W
30 *L10* **Dixon** Illinois, N USA 41°51′N 89°26′W
20 *I6* **Dixon** Kentucky, S USA 37°30′N 87°39′W
27 *V6* **Dixon** Missouri, C USA 37°59′N 92°05′W
37 *S9* **Dixon** New Mexico, SW USA 36°10′N 105°49′W
39 *Y15* **Dixon Entrance** *strait* Canada/USA
18 *D14* **Dixonville** Pennsylvania, NE USA 40°43′N 79°01′W
137 *T13* **Diyadin** Ağrı, E Turkey 39°33′N 43°41′E
139 *V7* **Diyālá** *off.* Muḩāfa at Diyālá, *var.* Ba'qūbah. ◈ *governorate* E Iraq
Diyālá, Muḩāfa at *see* Diyālá
Diyālá, Nahr *see* Sīrvān, Rūdkhāneh-ye
139 *V5* **Diyālá, Sirwan Nahr** *var.* Rudkhaneh-ye Sīrvān, Sīrwān. ≈ Iran/Iraq *see also* Sīrvān, Rudkhaneh-ye
137 *P15* **Diyarbakır** *var.* Diarbekr; *anc.* Amida. Diyarbakır, SE Turkey 37°55′N 40°14′E
137 *P15* **Diyarbakır** *var.* Diarbekr. ◈ *province* SE Turkey
Dizful *see* Dezfūl
79 *F16* **Dja** ≈ SE Cameroon
Djadié *see* Zadié
77 *X7* **Djado** Agadez, NE Niger 21°00′N 12°11′E
77 *X6* **Djado, Plateau du** ▲▲ NE Niger
Djailolo *see* Halmahera, Pulau
Djajapura *see* Jayapura
Djakarta *see* Jakarta
Djakovo *see* Đakovo
79 *G20* **Djambala** Plateaux, C Congo 02°32′S 14°43′E
Djambi *see* Hari, Batang
Djambi *see* Jambi
74 *M9* **Djanet** E Algeria 28°43′N 08°57′E
74 *M11* **Djanet** *prev.* Fort Charlet. SE Algeria 24°34′N 09°33′E
Djaul *see* Dyaul Island
Djawa *see* Jawa
Djéblé *see* Jablah
78 *I10* **Djédaa** Batha, C Chad 13°31′N 18°34′E
74 *J6* **Djelfa** *var.* El Djelfa. N Algeria 34°43′N 03°14′E
79 *M14* **Djéma** Haut-Mbomou, E Central African Republic 06°04′N 25°20′E
Djember *see* Jember
Djeneponto *see* Jeneponto
77 *N12* **Djenné** *var.* Jenné. Mopti, C Mali 13°55′N 04°31′W
Djérablous *see* Jarābulus
Djerba *see* Jerba, Île de
79 *F15* **Djérem** ≈ C Cameroon
Djevdjelija *see* Gevgelija
77 *P11* **Djibo** N Burkina Faso 14°09′N 01°38′W
80 *L12* **Djibouti** *var.* Jibuti. ● (Djibouti) E Djibouti 11°33′N 42°55′E
80 *L12* **Djibouti** *off.* Republic of Djibouti, *var.* Jibuti; *prev.* French Somaliland, French Territory of the Afars and Issas, *Fr.* Côte Française des Somalis, Territoire Français des Afars et des Issas. ◆ *republic* E Africa
80 *L12* **Djibouti** ✈ C Djibouti 11°29′N 42°54′E
Djibouti, Republic of *see* Djibouti
Djidjel/Djidjelli *see* Jijel
Djidji *see* Ivando
55 *W10* **Djoemoe** Sipaliwini, C Suriname 04°00′N 55°27′W
Djokjakarta *see* Yogyakarta
79 *K21* **Djoku-Punda** Kasai-Occidental, S Dem. Rep. Congo 05°27′S 20°58′E
79 *K18* **Djolu** Equateur, N Dem. Rep. Congo 0°35′N 22°30′E
Djombang *see* Jombang
Ďjorče Petrov *see* Đorče Petrov
79 *F17* **Djoua** ≈ Congo/Gabon
77 *R14* **Djougou** W Benin 09°42′N 01°38′E
79 *F16* **Djoum** Sud, S Cameroon 02°38′N 12°51′E
78 *I8* **Djourab, Erg du** *desert* N Chad
79 *P17* **Djugu** Orientale, NE Dem. Rep. Congo 01°55′N 30°31′E
Djumbir *see* Ďumbier
92 *L3* **Djúpivogur** Austurland, SE Iceland 64°40′N 14°18′W
94 *L13* **Djura** Dalarna, C Sweden 60°37′N 15°00′E
Djurdjevac *see* Đurđevac
D'Kar *see* Dekar
197 *U6* **Dmitriya Lapteva, Proliv** *strait* N Russian Federation
126 *J7* **Dmitriyev-L'govskiy** Kurskaya Oblast', W Russian Federation 52°08′N 35°09′E
Dmitriyevsk *see* Makiyivka
126 *K3* **Dmitrov** Moskovskaya Oblast', W Russian Federation 56°23′N 37°30′E
Dmitrovichi *see* Dzmitravichy
126 *J6* **Dmitrovsk-Orlovskiy** Orlovskaya Oblast', W Russian Federation 52°28′N 35°01′E
117 *R3* **Dmytrivka** Chernihivs'ka Oblast', N Ukraine 50°56′N 32°57′E
Dnepr *see* Dnieper
Dneprodzerzhinsk *see* Romaniv
Dneprodzerzhinskoye Vodokhranilishche *see* Dniprodzerzhyns'ke Vodoskhovyshche
Dnepropetrovsk *see* Dnipropetrovs'k
Dnepropetrovskaya Oblast' *see* Dnipropetrovs'ka Oblast'
Dneprorudnoye *see* Dniprorudne
Dneprovskiy Liman *see* Dniprovs'kyy Lyman
Dneprovsko-Bugskiy Kanal *see* Dnyaprowska-Buhski Kanal
Dnestr *see* Dniester
Dnestrovskiy Liman *see* Dnistrovs'kyy Lyman
86 *H11* **Dnieper** *Bel.* Dnyapro, *Rus.* Dnepr, *Ukr.* Dnipro. ≈ E Europe
117 *P3* **Dnieper Lowland** *Bel.* Prydnyaprowskaya Nizina, *Ukr.* Prydniprovs'ka Nyzovyna. *lowlands* Belarus/Ukraine
116 *M8* **Dniester** *Rom.* Nistru, *Rus.* Dnestr, *Ukr.* Dnister; *anc.* Tyras. ≈ Moldova/Ukraine
Dnipro *see* Dnieper
Dniprodzerzhyns'k *see* Romaniv
117 *T7* **Dniprodzerzhyns'ke Vodoskhovyshche** *Rus.* Dneprodzerzhinskoye Vodokhranilishche. ⊠ C Ukraine
117 *U7* **Dnipropetrovs'k** *Rus.* Dnepropetrovsk; *prev.* Yekaterinoslav. Dnipropetrovs'ka Oblast', E Ukraine 48°28′N 35°E
117 *U8* **Dnipropetrovs'k** ✈ Dnipropetrovs'ka Oblast', S Ukraine 48°20′N 35°04′E
Dnipropetrovs'k *see* Dnipropetrovs'ka Oblast'
117 *T7* **Dnipropetrovs'ka Oblast'** *var.* Dnipropetrovs'k, *Rus.* Dnepropetrovskaya Oblast'. ◈ *province* E Ukraine
117 *U9* **Dniprorudne** *Rus.* Dneprorudnoye. Zaporiz'ka Oblast', SE Ukraine 47°21′N 35°00′E
117 *Q11* **Dniprovs'kyy Lyman** *Rus.* Dneprovskiy Liman. *bay* S Ukraine
Dnister *see* Dniester
117 *O11* **Dnistrovs'kyy Lyman** *Rus.* Dnestrovskiy Liman. *inlet* S Ukraine
124 *G14* **Dno** Pskovskaya Oblast', W Russian Federation 57°48′N 29°58′E
Dnyapro *see* Dnieper
119 *H20* **Dnyaprowska-Buhski Kanal** *Rus.* Dneprovsko-Bugskiy Kanal. *canal* SW Belarus
13 *O14* **Doaktown** New Brunswick, SE Canada 46°34′N 66°06′W
78 *H13* **Doba** Logone-Oriental, S Chad 08°40′N 16°50′E
118 *E9* **Dobele** *Ger.* Doblen. W Latvia 56°36′N 23°14′E
101 *N16* **Döbeln** Sachsen, E Germany 51°07′N 13°07′E
171 *U12* **Doberai, Jazirah** *Dut.* Vogelkop. *peninsula* Papua Barat, E Indonesia
110 *F10* **Dobiegniew** *Ger.* Lubuskie, Woldenberg Neumark. Lubuskie, W Poland 52°58′N 15°43′E
Doblen *see* Dobele
81 *K18* **Dobli** *spring/well* SW Somalia 0°24′N 41°18′E
112 *H11* **Doboj** Republiks Srpska, N Bosnia and Herzegovina 44°45′N 18°03′E
143 *R12* **Doborjī** *var.* Fūrg. Fārs, S Iran 28°16′N 55°13′E
110 *L8* **Dobre Miasto** *Ger.* Guttstadt. Warmińsko-mazurskie, NE Poland 53°59′N 20°25′E
114 *N7* **Dobrich** *Rom.* Bazargic; *prev.* Tolbukhin. Dobrich, NE Bulgaria 43°35′N 27°49′E
114 *N7* **Dobrich** ◈ *province* NE Bulgaria
126 *M8* **Dobrinka** Lipetskaya Oblast', W Russian Federation 52°10′N 40°30′E
126 *M7* **Dobrinka** Volgogradskaya Oblast', SW Russian Federation 50°52′N 41°48′E
Dobrla Vas *see* Eberndorf
111 *I15* **Dobrodzień** *Ger.* Guttentag. Opolskie, S Poland 50°43′N 18°24′E
Dobrogea *see* Dobruja
117 *W7* **Dobropillya** *Rus.* Dobropol'ye. Donets'ka Oblast', SE Ukraine 48°25′N 37°02′E
Dobropol'ye *see* Dobropillya
117 *P8* **Dobrovelychkivka** Kirovohrads'ka Oblast', C Ukraine 48°22′N 31°12′E
Dobrudja/Dobrudzha *see* Dobruja
114 *O7* **Dobruja** *var.* Dobrudja, *Bul.* Dobrudzha, *Rom.* Dobrogea. *physical region* Bulgaria/Romania
119 *P19* **Dobrush** Homyel'skaya Voblasts', SE Belarus 52°25′N 31°19′E
125 *U14* **Dobryanka** Permskiy Kray, NW Russian Federation 58°28′N 56°27′E
117 *P2* **Dobryanka** Chernihivs'ka Oblast', N Ukraine 52°03′N 31°09′E
Dobryn' *see* Dabryn'
21 *R8* **Dobson** North Carolina, SE USA 36°25′N 80°45′W
59 *N20* **Doce, Rio** ≈ SE Brazil
93 *I16* **Docksta** Västernorrland, C Sweden 63°06′N 18°22′E
41 *N10* **Doctor Arroyo** Nuevo León, NE Mexico 23°40′N 100°09′W
62 *L4* **Doctor Pedro P. Peña** Boquerón, W Paraguay 22°22′S 62°23′W
171 *S11* **Dodaga** Pulau Halmahera, E Indonesia 01°06′N 128°10′E
155 *G21* **Dodda Betta** ▲ S India 11°28′N 76°44′E
Dodecanese *see* Dodekánisa
115 *M22* **Dodekánisa** *var.* Nóties Sporádes, *Eng.* Dodecanese; *prev.* Dhodhekánisos, Dodekanisos. *island group* SE Greece
Dodekanisos *see* Dodekánisa
26 *J6* **Dodge City** Kansas, C USA 37°45′N 100°01′W
30 *K9* **Dodgeville** Wisconsin, N USA 42°57′N 90°08′W
97 *H25* **Dodman Point** *headland* SW England, United Kingdom 50°13′N 04°47′W
81 *J14* **Dodola** Oromīya, C Ethiopia 07°00′N 39°15′E
81 *H22* **Dodoma** ● (Tanzania) Dodoma, C Tanzania 06°11′S 35°45′E
81 *H22* **Dodoma** ◈ *region* C Tanzania
115 *C16* **Dodóni** *var.* Dhodhóni. *site of ancient city* Ípeiros, W Greece
33 *U7* **Dodson** Montana, NW USA 48°25′N 108°18′W
25 *P3* **Dodson** Texas, SW USA 34°46′N 100°01′W
98 *M12* **Doesburg** Gelderland, E Netherlands 52°01′N 06°08′E
98 *N12* **Doetinchem** Gelderland, E Netherlands 51°58′N 06°17′E
158 *L12* **Dogai Coring** *var.* Lake Montcalm. ◎ W China
137 *N15* **Doğanşehir** Malatya, C Turkey 38°07′N 37°54′E
84 *E9* **Dogger Bank** *undersea feature* C North Sea 55°00′N 03°00′E
23 *S10* **Dog Island** *island* Florida, SE USA
14 *C7* **Dog Lake** ◎ Ontario, S Canada
106 *B9* **Dogliani** Piemonte, NE Italy 44°33′N 07°55′E
164 *H11* **Dōgo** *island* Oki-shotō, SW Japan
143 *N10* **Do Gonbadān** *var.* Dow Gonbadān, Gonbadān. Kohkīlūyeh va Būyer AḨmad, SW Iran 30°12′N 50°48′E
77 *S12* **Dogondoutchi** Dosso, SW Niger 13°36′N 04°03′E
137 *T13* **Doğubayazıt** Ağrı, E Turkey 39°33′N 44°07′E
137 *P12* **Doğu Karadeniz Dağları** *var.* Anadolu Dağları. ▲▲ NE Turkey
158 *K16* **Dogxung Zangbo** ≈ W China
Doha *see* Ad Dawḩah
Doha *see* Ad Dawḩah
Dohad *see* Dāhod
Dohuk *see* Dahūk
Dohuk *see* Dahūk
159 *N16* **Doilungdêqên** *var.* Namka. Xizang Zizhiqu, W China 29°41′N 90°58′E
114 *F12* **Doïráni, Límni** *var.* Limni Doïranis, *Bul.* Ezero Doyransko. ◎ N Greece
Doire *see* Londonderry
99 *H22* **Doische** Namur, S Belgium 50°09′N 04°43′E
59 *P17* **Dois de Julho** ✈ (Salvador) Bahia, NE Brazil 12°04′S 38°58′W
60 *H12* **Dois Vizinhos** Paraná, S Brazil 25°47′S 53°03′W
80 *H10* **Doka** Gedaref, E Sudan 13°30′N 35°47′E
Doka *see* Kéita, Bahr
Dokan *see* Dûkan
94 *H13* **Dokka** Oppland, S Norway 60°49′N 10°04′E
98 *L5* **Dokkum** Fryslân, N Netherlands 53°20′N 06°E
98 *L5* **Dokkumer Ee** ≈ N Netherlands
76 *K13* **Doko** NE Guinea 11°46′N 08°58′W
Dokshitsy *see* Dokshytsy
118 *K13* **Dokshytsy** *Rus.* Dokshitsy. Vitsyebskaya Voblasts', N Belarus 54°54′N 27°46′E
117 *X8* **Dokuchayevs'k** *var.* Dokuchayevsk. Donets'ka Oblast', SE Ukraine 47°43′N 37°41′E

◆ Country
● Country Capital
◇ Dependent Territory
○ Dependent Territory Capital
◈ Administrative Regions
✈ International Airport
▲ Mountain
▲▲ Mountain Range
✶ Volcano
≈ River
◎ Lake
⊠ Reservoir

Dokuchayevsk *see* Dokuchayevs'k
Dolak, Pulau *see* Yos Sudarso, Pulau
29 *P9* **Doland** South Dakota, N USA *44°51′N 98°06′W*
63 *J18* **Dolavón** Chaco, S Argentina *43°16′S 65°44′W*
15 *P6* **Dolbeau** Québec, SE Canada *48°52′N 72°15′W*
15 *P6* **Dolbeau-Mistassini** Québec, SE Canada *48°54′N 72°13′W*
102 *I5* **Dol-de-Bretagne** Ille-et-Vilaine, NW France *48°33′N 01°45′W*
64 *J13* **Doldrums Fracture Zone** *tectonic feature* W Atlantic Ocean
103 *S8* **Dôle** Jura, E France *47°05′N 05°30′E*
97 *J19* **Dolgellau** NW Wales, United Kingdom *52°45′N 03°54′W*
Dolginovo *see* Dawhinava
Dolgi, Ostrov *see* Dolgiy, Ostrov
125 *U2* **Dolgiy, Ostrov** *var.* Ostrov Dolgi. *island* NW Russian Federation
162 *J9* **Dölgöön** Övörhangay, C Mongolia *45°57′N 103°14′E*
107 *C20* **Dolianova** Sardegna, Italy, C Mediterranean Sea *39°23′N 09°08′E*
Dolina *see* Dolyna
123 *T13* **Dolinsk** Ostrov Sakhalin, Sakhalinskaya Oblast', SE Russian Federation *47°20′N 142°52′E*
Dolinskaya *see* Dolyns'ka
79 *F21* **Dolisie** *prev.* Loubomo. Niari, S Congo *04°12′S 12°41′E*
116 *G14* **Dolj** ◐ *county* SW Romania
98 *P5* **Dollard** *bay* NW Germany
194 *J5* **Dolleman Island** *island* Antarctica
114 *K8* **Dolna Oryahovitsa** *var.* Dolna Oryakhovits. Veliko Tarnovo, N Bulgaria *43°09′N 25°44′E*
Dolna Oryakhovits *see* Dolna Oryahovitsa
114 *N9* **Dolni Chiflik** Varna, E Bulgaria *42°59′N 27°43′E*
114 *I8* **Dolni Dabnik** *var.* Dolni Dŭbnik. Pleven, N Bulgaria *43°24′N 24°25′E*
Dolni Dŭbnik *see* Dolni Dabnik
114 *F8* **Dolni Lom** Vidin, NW Bulgaria *43°31′N 22°46′E*
Dolnja Lendava *see* Lendava
129 *F14* **Dolnośląskie** ◐ *province* SW Poland
111 *K18* **Dolný Kubín** *Hung.* Alsókubin. Zílinský Kraj, N Slovakia *49°12′N 19°17′E*
106 *H8* **Dolo** Veneto, NE Italy *45°25′N 12°06′E*
Dolomites/Dolomiti *see* Dolomitiche, Alpi
106 *H6* **Dolomitiche, Alpi** *var.* Dolomiti, *Eng.* Dolomites. ▲▲ NE Italy
Dolonnur *see* Duolun
Doloon *see* Tsogt-Ovoo
61 *E21* **Dolores** Buenos Aires, E Argentina *36°21′S 57°39′W*
42 *E3* **Dolores** Petén, N Guatemala *16°33′N 89°26′W*
171 *Q5* **Dolores** Samar, C Philippines *12°01′N 125°27′E*
105 *S12* **Dolores** Valenciana, E Spain *38°09′N 00°45′W*
61 *D19* **Dolores** Soriano, SW Uruguay *33°34′S 58°15′W*
41 *N12* **Dolores Hidalgo** *var.* Ciudad de Dolores Hidalgo. Guanajuato, C Mexico *21°10′N 100°55′W*
8 *J7* **Dolphin and Union Strait** *strait* Northwest Territories/Nunavut, N Canada
65 *D23* **Dolphin, Cape** *headland* East Falkland, Falkland Islands *51°15′S 58°57′W*
44 *H12* **Dolphin Head** *hill* W Jamaica
83 *B21* **Dolphin Head** *var.* Cape Dernberg. *headland* SW Namibia *25°33′S 14°36′E*
110 *G12* **Dolsk** *Ger.* Dolzig. Weilkopolskie, C Poland *51°59′N 17°03′E*
167 *S8* **Đô Lương** Nghê An, N Vietnam *18°51′N 105°19′E*
116 *I6* **Dolyna** *Rus.* Dolina. Ivano-Frankivs'ka Oblast', W Ukraine *48°58′N 24°01′E*
117 *R8* **Dolyns'ka** *Rus.* Dolinskaya. Kirovohrads'ka Oblast', S Ukraine *48°06′N 32°46′E*
Dolzig *see* Dolsk
Domachëvo/Domaczewo *see* Damachava
117 *P9* **Domanivka** Mykolayivs'ka Oblast', S Ukraine *47°40′N 30°56′E*
153 *S13* **Domar** Rajshahi, N Bangladesh *26°08′N 88°57′E*
108 *I9* **Domat/Ems** Graubünden, SE Switzerland *46°50′N 09°28′E*
111 *A18* **Domažlice** *Ger.* Taus. Plzeňský Kraj, W Czech Republic *49°26′N 12°54′E*
127 *X8* **Dombarovskiy** Orenburgskaya Oblast', W Russian Federation *50°53′N 59°18′E*
94 *G10* **Dombås** Oppland, S Norway *62°04′N 09°07′E*
83 *M17* **Dombe** Manica, C Mozambique *19°59′S 33°24′E*
82 *A13* **Dombe Grande** Benguela, C Angola *12°57′S 13°07′E*
103 *R10* **Dombes** *physical region* E France
111 *I25* **Dombóvár** Tolna, S Hungary *46°24′N 18°09′E*
99 *D14* **Domburg** Zeeland, SW Netherlands *51°34′N 03°30′E*
58 *L13* **Dom Eliseu** Pará, NE Brazil *04°02′S 47°31′W*
Domel Island *see* Letsôk-aw Kyun
103 *O11* **Dôme, Puy de** ▲ C France *45°46′N 03°00′E*
36 *H13* **Dome Rock Mountains** ▲▲ Arizona, SW USA
Domesnes, Cape *see* Kolkasrags
62 *G8* **Domeyko** Atacama, N Chile *28°58′S 70°54′W*
62 *H5* **Domeyko, Cordillera** ▲▲ N Chile
102 *K5* **Domfront** Orne, N France *48°35′N 00°39′W*
171 *X13* **Dom, Gunung** ▲ Papua, E Indonesia *02°41′S 137°00′E*
45 *X11* **Dominica** *off.* Commonwealth of Dominica. ◆ *republic* E West Indies
47 *S3* **Dominica** *island* Dominica
Dominica Channel *see* Martinique Passage
Dominica, Commonwealth of *see* Dominica
43 *N15* **Dominical** Puntarenas, SE Costa Rica *09°16′N 83°52′W*
45 *Q8* **Dominican Republic** ◆ *republic* C West Indies
45 *X11* **Dominica Passage** *passage* E Caribbean Sea
99 *K14* **Dommel** ≈ S Netherlands
81 *O14* **Domo** Sumalē, E Ethiopia *07°53′N 46°55′E*
126 *L4* **Domodedovo** ✕ (Moskva) Moskovskaya Oblast', W Russian Federation *55°19′N 37°55′E*
106 *C6* **Domodossola** Piemonte, NE Italy *46°07′N 08°20′E*
115 *F17* **Domokós** *var.* Dhomokós. Stereá Elláda, C Greece *39°07′N 22°18′E*
172 *I14* **Domoni** Anjouan, SE Comoros *12°15′S 44°39′E*
61 *G16* **Dom Pedrito** Rio Grande do Sul, S Brazil *31°00′S 54°40′W*
Dompoe *see* Dompu
170 *M16* **Dompu** *prev.* Dompoe. Sumbawa, C Indonesia *08°30′S 118°28′E*
Domschale *see* Domžale
62 *H13* **Domuyo, Volcán** (volcano) W Argentina *36°36′S 70°22′W*
109 *U11* **Domžale** *Ger.* Domschale. C Slovenia *46°09′N 14°33′E*
127 *O10* **Don** *var.* Duna, Tanais. ≈ SW Russian Federation
96 *K9* **Don** ≈ NE Scotland, United Kingdom
182 *M11* **Donald** Victoria, SE Australia *36°27′S 143°03′E*
22 *J9* **Donaldsonville** Louisiana, S USA *30°06′N 90°59′W*
23 *S8* **Donalsonville** Georgia, SE USA *31°02′N 84°52′W*
Donau *see* Danube
101 *G23* **Donaueschingen** Baden-Württemberg, SW Germany *47°57′N 08°30′E*
101 *K22* **Donaumoos** *wetland* S Germany
101 *K22* **Donauwörth** Bayern, S Germany *48°43′N 10°46′E*
109 *U7* **Donawitz** Steiermark, SE Austria *47°23′N 15°00′E*
117 *X7* **Donbass** *industrial region* Russian Federation/Ukraine
104 *K11* **Don Benito** Extremadura, W Spain *38°57′N 05°52′W*
97 *M17* **Doncaster** *anc.* Danum. N England, United Kingdom *53°32′N 01°07′W*
44 *K12* **Don Christophers Point** *headland* C Jamaica *18°19′N 76°48′W*
55 *V9* **Donderkamp** Sipaliwini, NW Suriname *05°18′N 56°22′W*
82 *B12* **Dondo** Kwanza Norte, NW Angola *09°40′S 14°24′E*
171 *O12* **Dondo** Sulawesi, N Indonesia *0°54′S 121°33′E*
83 *N17* **Dondo** Sofala, C Mozambique *19°41′S 34°45′E*
155 *K26* **Dondra Head** *headland* S Sri Lanka *05°57′N 80°33′E*
Dondușani *see* Dondușeni
116 *M8* **Dondușeni** *var.* Dondușani, *Rus.* Dondyushany. N Moldova *48°13′N 27°38′E*
Dondyushany *see* Dondușeni
97 *D15* **Donegal** *Ir.* Dún na nGall. Donegal, NW Ireland *54°39′N 08°06′W*
97 *D14* **Donegal** *Ir.* Dún na nGall. *cultural region* NW Ireland
97 *C15* **Donegal Bay** *Ir.* Bá Dhún na nGall. *bay* NW Ireland
84 *K10* **Donets** ≈ Russian Federation/Ukraine
117 *X8* **Donets'k** *Rus.* Donetsk; *prev.* Stalino. Donets'ka Oblast', E Ukraine *47°58′N 37°50′E*
117 *W8* **Donets'k** ✕ Donets'ka Oblast', E Ukraine *48°03′N 37°44′E*
Donets'k *see* Donets'ka Oblast'
117 *W8* **Donets'ka Oblast'** *var.* Donets'k, *Rus.* Donetskaya Oblast'; *prev. Rus.* Stalins'kaya Oblast'. ◐ *province* SE Ukraine
Donetskaya Oblast' *see* Donets'ka Oblast'
67 *P8* **Donga** ≈ Cameroon/Nigeria
157 *O13* **Dongchuan** Yunnan, SW China *26°09′N 103°10′E*
161 *Q14* **Dongchuan Dao** *prev.* Dongshan Dao. *island* SE China
Dong Dao *see* Lincoln Island
99 *I14* **Dongen** Noord-Brabant, S Netherlands *51°38′N 04°56′E*
160 *K17* **Dongfang** *var.* Basuo. Hainan, S China *19°05′N 108°40′E*
163 *Z7* **Dongfanghong** Heilongjiang, NE China *46°13′N 133°13′E*
163 *W11* **Dongfeng** Jilin, NE China *42°39′N 125°33′E*
171 *N12* **Donggala** Sulawesi, C Indonesia *0°40′S 119°44′E*
163 *V13* **Donggang** *var.* Dadong; *prev.* Donggou. Liaoning, NE China *39°52′N 124°08′E*
Donggou *see* Donggang
161 *O14* **Dongguan** Guangdong, S China *23°03′N 113°43′E*
167 *T9* **Đông Ha** Quang Tri, C Vietnam *16°45′N 107°10′E*
163 *Y14* **Donghae** *prev.* Tonghae. NE South Korea *37°26′N 129°09′E*
Dong Hai *see* East China Sea
160 *M16* **Donghai Dao** *island* S China
Dong He *see* Omon Gol
Donghe *see* Wangcang
167 *T9* **Đông Hơi** Quang Binh, C Vietnam *17°32′N 106°35′E*
Donghua *see* Huating
Donghuang *see* Xishui
108 *H10* **Dongio** Ticino, S Switzerland *46°27′N 08°58′E*
Dongkan *see* Binhai
160 *L11* **Dongkou** Hunan, S China *27°06′N 110°35′E*
167 *S8* **Đông Lê** Quang Binh, C Vietnam *17°52′N 105°49′E*
Dongliao *see* Liaoyuan
Dong-nai *see* Đông Nai, Sông
167 *U13* **Đông Nai, Sông** *var.* Dong-nai, Dong Noi, Donnai. ≈ S Vietnam
161 *N14* **Dongnan Qiuling** *plateau* SE China
163 *Y9* **Dongning** Heilongjiang, NE China *44°01′N 131°03′E*
Dong Noi *see* Đông Nai, Sông
83 *C14* **Dongo** Huíla, C Angola *14°35′S 15°51′E*
80 *E7* **Dongola** *var.* Donqola, Dunqulah. Northern, N Sudan *19°10′N 30°27′E*
79 *I17* **Dongou** Likouala, NE Congo *02°05′N 18°E*
Đông Phu *see* Đông Xoai
Dongping *see* Anhua
Dong Rak, Phanom *see* Dângrêk, Chuŏr Phnum
Dongshan Dao *see* Dongchuan Dao
Dongsha Qundao *see* Tungsha Tao
Dongsheng *see* Ordos
161 *S13* **Dongshi** *Jap.* Tōsei; *prev.* Tungshih. N Taiwan *24°13′N 120°54′E*
161 *R7* **Dongtai** Jiangsu, E China *32°50′N 120°22′E*
161 *N10* **Dongting Hu** *var.* Tung-t'ing Hu. ◎ S China
161 *P10* **Dongxiang** *var.* Xiaogang. Jiangxi, S China *28°16′N 116°32′E*
167 *T13* **Đông Xoai** *var.* Đông Phu. Sông Be, S Vietnam *11°31′N 106°55′E*
161 *Q4* **Dongying** Shandong, E China *37°27′N 118°01′E*
27 *X8* **Doniphan** Missouri, C USA *36°39′N 90°51′W*
Donja Łužica *see* Niederlausitz
10 *G7* **Donjek** ≈ Yukon, W Canada
112 *E11* **Donji Lapac** Lika-Senj, W Croatia *44°33′N 15°58′E*
112 *H8* **Donji Miholjac** Osijek-Baranja, NE Croatia *45°45′N 18°10′E*
112 *P12* **Donji Milanovac** Serbia, E Serbia *44°27′N 22°06′E*
112 *G12* **Donji Vakuf** *var.* Srbobran. ◐ Federacija Bosne I Hercegovine, C Bosnia and Herzegovina
98 *M6* **Donkerbroek** Fryslân, N Netherlands *52°58′N 05°15′E*
167 *P11* **Don Muang** ✕ (Krung Thep) Nonthaburi, C Thailand *13°51′N 100°40′E*
25 *S17* **Donna** Texas, SW USA *26°10′N 98°03′W*
15 *Q10* **Donnacona** Québec, SE Canada *46°41′N 71°46′W*
Donnai *see* Đông Nai, Sông
29 *Y16* **Donnellson** Iowa, C USA *40°38′N 91°33′W*
11 *O13* **Donnelly** Alberta, W Canada *55°42′N 117°06′W*
35 *P6* **Donner Pass** *Pass* California, W USA
101 *F19* **Donnersberg** ▲ W Germany *49°37′N 07°54′E*
Donoso *see* Miguel de la Borda
105 *P2* **Donostia-San Sebastián** País Vasco, N Spain *43°19′N 01°59′W*
115 *K21* **Donoúsa** *var.* Donoússa. *island* Kykládes, Greece, Aegean Sea
Donoússa *see* Donoúsa
35 *P8* **Don Pedro Reservoir** ⊠ California, W USA
Donqola *see* Dongola
126 *L5* **Donskoy** Tul'skaya Oblast', W Russian Federation *54°02′N 38°27′E*
81 *L16* **Doolow** Sumalē, E Ethiopia *04°10′N 42°05′E*
39 *Q7* **Doonerak, Mount** ▲ Alaska, USA *67°54′N 150°33′W*
98 *J12* **Doorn** Utrecht, C Netherlands *52°02′N 05°21′E*
Doornik *see* Tournai
31 *N6* **Door Peninsula** *peninsula* Wisconsin, N USA
80 *P13* **Dooxo Nugaaleed** *var.* Nogal Valley. *valley* E Somalia
Do Qu *see* Da Qu
106 *B7* **Dora Baltea** *anc.* Duria Major. ≈ NW Italy
180 *K7* **Dora, Lake** *salt lake* Western Australia
106 *A8* **Dora Riparia** *anc.* Duria Minor. ≈ NW Italy
Dorbiljin *see* Emin
Dorbod/Dorbod Mongolzu Zizhixian *see* Taikang
Dorbod Mongolzu Zizhixian *see* Taikang
113 *N18* **Đorče Petrov** *var.* Djorče Petrov, Gorče Petrov. N Macedonia *42°01′N 21°21′E*
14 *F16* **Dorchester** Ontario, S Canada *43°00′N 81°04′W*
97 *L24* **Dorchester** *anc.* Durnovaria. S England, United Kingdom *50°43′N 02°26′W*
9 *P7* **Dorchester, Cape** *headland* Baffin Island, Nunavut, NE Canada *65°25′N 77°25′W*
83 *D19* **Dordabis** Khomas, C Namibia *22°57′S 17°39′E*
102 *L12* **Dordogne** ◐ *department* SW France
103 *N12* **Dordogne** ≈ W France
98 *H13* **Dordrecht** *var.* Dordt, Dort. Zuid-Holland, SW Netherlands *51°48′N 04°40′E*
Dordt *see* Dordrecht
103 *P11* **Dore** ≈ C France
11 *S13* **Doré Lake** Saskatchewan, C Canada *54°37′N 107°36′W*
103 *O12* **Dore, Monts** ▲▲ C France
101 *M23* **Dorfen** Bayern, SE Germany *48°16′N 12°06′E*
107 *D18* **Dorgali** Sardegna, Italy, C Mediterranean Sea *40°18′N 09°34′E*
159 *N11* **Dorgê Co** *var.* Elsen Nur. ◎ C China
162 *E6* **Dörgön** *var.* Seer. Hovd, W Mongolia *48°18′N 92°37′E*
162 *F7* **Dörgön Nuur** ◎ NW Mongolia
77 *Q12* **Dori** N Burkina Faso *14°03′N 00°02′W*
83 *E24* **Doring** ≈ S South Africa
101 *E16* **Dormagen** Nordrhein-Westfalen, W Germany *51°06′N 06°49′E*
103 *P4* **Dormans** Marne, N France *49°03′N 03°44′E*
108 *E6* **Dornach** Solothurn, NW Switzerland *47°29′N 07°37′E*
Dorna Watra *see* Vatra Dornei
108 *J7* **Dornbirn** Vorarlberg, W Austria *47°25′N 09°46′E*
96 *I7* **Dornoch** N Scotland, United Kingdom *57°52′N 04°01′W*
96 *J7* **Dornoch Firth** *inlet* N Scotland, United Kingdom
163 *P7* **Dornod** ◐ *province* E Mongolia
163 *N10* **Dornogovĭ** ◐ *province* SE Mongolia
77 *P10* **Doro** Tombouctou, S Mali *16°07′N 00°57′W*
116 *L14* **Dorobanțu** Călărași, S Romania *44°15′N 26°55′E*
111 *J22* **Dorog** Komárom-Esztergom, N Hungary *47°43′N 18°42′E*
126 *I4* **Dorogobuzh** Smolenskaya Oblast', W Russian Federation *54°56′N 33°16′E*
116 *K8* **Dorohoi** Botoșani, NE Romania *47°57′N 26°24′E*
93 *H15* **Dorotea** Västerbotten, N Sweden *64°17′N 16°30′E*
Dorpat *see* Tartu
180 *G10* **Dorre Island** *island* Western Australia
183 *U5* **Dorrigo** New South Wales, SE Australia *30°22′S 152°43′E*
35 *N1* **Dorris** California, W USA *41°58′N 121°54′W*
14 *H13* **Dorset** Ontario, SE Canada *45°12′N 78°52′W*
97 *K23* **Dorset** *cultural region* S England, United Kingdom
101 *E14* **Dorsten** Nordrhein-Westfalen, W Germany *51°38′N 06°58′E*
Dort *see* Dordrecht
101 *F15* **Dortmund** Nordrhein-Westfalen, W Germany *51°31′N 07°28′E*
100 *F12* **Dortmund-Ems-Kanal** *canal* W Germany
136 *L17* **Dörtyol** Hatay, S Turkey *36°51′N 36°11′E*
Do Rūd *see* Dow Rūd
79 *O15* **Doruma** Orientale, N Dem. Rep. Congo *04°35′N 27°43′E*
15 *O12* **Dorval** ✕ (Montréal) Québec, SE Canada *45°27′N 73°46′W*
162 *F7* **Dörvöljin** *var.* Buga. Dzavhan, W Mongolia *47°42′N 94°53′E*
45 *T5* **Dos Bocas, Lago** ◎ C Puerto Rico
104 *K14* **Dos Hermanas** Andalucía, S Spain *37°16′N 05°55′W*
Dospad Dagh *see* Rhodope Mountains
35 *P10* **Dos Palos** California, W USA *37°00′N 120°39′W*
114 *I11* **Dospat** Smolyan, S Bulgaria *41°39′N 24°10′E*
114 *H11* **Dospat, Yazovir** ⊠ SW Bulgaria
100 *M11* **Dosse** ≈ NE Germany
77 *S12* **Dosso** Dosso, SW Niger *13°03′N 03°10′E*
77 *S12* **Dosso** ◐ *department* SW Niger
144 *G12* **Dossor** Atyrau, SW Kazakhstan *47°31′N 53°01′E*
147 *O10* **Do'stlik** Jizzax Viloyati, C Uzbekistan *40°37′N 67°59′E*
147 *V9* **Dostuk** Narynskaya Oblast', C Kyrgyzstan *41°19′N 75°40′E*
145 *X13* **Dostyk** *prev.* Druzhba. Almaty, SE Kazakhstan *45°15′N 82°29′E*
23 *R7* **Dothan** Alabama, S USA *31°13′N 85°23′W*
39 *T9* **Dot Lake** Alaska, USA *63°39′N 144°10′W*
118 *F12* **Dotnuva** Kaunas, C Lithuania *55°23′N 23°53′E*
99 *D19* **Dottignies** Hainaut, W Belgium *50°43′N 03°16′E*
103 *P2* **Douai** *prev.* Douay; *anc.* Duacum. Nord, N France *50°22′N 03°04′E*
14 *L9* **Douaire, Lac** ◎ Québec, SE Canada
79 *D16* **Douala** *var.* Duala. Littoral, W Cameroon *04°04′N 09°43′E*
79 *D16* **Douala** ✕ Littoral, W Cameroon *03°57′N 09°48′E*
102 *F6* **Douarnenez** Finistère, NW France *48°05′N 04°20′W*
102 *E6* **Douarnenez, Baie de** *bay* NW France
Douay *see* Douai
25 *O6* **Double Mountain Fork Brazos River** ≈ Texas, SW USA
23 *O3* **Double Springs** Alabama, S USA *34°09′N 87°24′W*
103 *T8* **Doubs** ◐ *department* E France
108 *C8* **Doubs** ≈ France/Switzerland
185 *A22* **Doubtful Sound** *sound* South Island, New Zealand
184 *J2* **Doubtless Bay** *bay* North Island, New Zealand
25 *X9* **Doucette** Texas, SW USA *30°48′N 94°25′W*
102 *K8* **Doué-la-Fontaine** Maine-et-Loire, NW France *47°12′N 00°16′W*
77 *O11* **Douentza** Mopti, S Mali *14°59′N 02°57′W*
65 *E24* **Douglas** East Falkland, Falkland Islands
97 *I16* **Douglas** ○ (Isle of Man) E Isle of Man *54°09′N 04°28′W*
83 *H23* **Douglas** Northern Cape, C South Africa *29°04′S 23°47′E*
39 *X13* **Douglas** Alexander Archipelago, Alaska, USA *58°12′N 134°18′W*
37 *O17* **Douglas** Arizona, SW USA *31°20′N 109°32′W*
23 *U7* **Douglas** Georgia, SE USA *31°30′N 82°51′W*
33 *Y15* **Douglas** Wyoming, C USA *42°48′N 105°23′W*
21 *O7* **Douglas Cape** *headland* Alaska, N USA *64°59′N 166°41′W*
10 *J14* **Douglas Channel** *channel* British Columbia, W Canada
182 *G3* **Douglas Creek** *seasonal river* South Australia
31 *P5* **Douglas Lake** ◎ Michigan, N USA
21 *O9* **Douglas Lake** ⊠ Tennessee, S USA
39 *Q13* **Douglas, Mount** ▲ Alaska, USA *58°51′N 153°31′W*
194 *I6* **Douglas Range** ▲▲ Alexander Island, Antarctica
161 *S14* **Douliu** *prev.* Touliu. C Taiwan *23°44′N 120°27′E*
103 *O2* **Doullens** Somme, N France *50°09′N 02°21′E*
Douma *see* Dūmā
79 *F15* **Doumé** Est, E Cameroon *04°14′N 13°27′E*
99 *E21* **Dour** Hainaut , S Belgium *50°24′N 03°47′E*
59 *K18* **Dourada, Serra** ▲▲ S Brazil
59 *I21* **Dourados** Mato Grosso do Sul, S Brazil *22°09′S 54°52′W*
103 *N5* **Dourdan** Essonne, N France *48°33′N 01°58′E*
104 *I6* **Douro** *Sp.* Duero. ≈ Portugal/Spain *see also* Duero
Douro *see* Duero
104 *G6* **Douro Litoral** *former province* N Portugal
Douvres *see* Dover
102 *K15* **Douze** ≈ SW France
183 *P17* **Dover** Tasmania, SE Australia *43°19′S 147°01′E*
97 *Q22* **Dover** *Fr.* Douvres, *Lat.* Dubris Portus. SE England, United Kingdom *51°08′N 03°E*
21 *Y3* **Dover** *state capital* Delaware, NE USA *39°09′N 75°31′W*
19 *P9* **Dover** New Hampshire, NE USA *43°10′N 70°50′W*
18 *J14* **Dover** New Jersey, NE USA *40°51′N 74°33′W*
31 *U12* **Dover** Ohio, N USA *40°31′N 81°28′W*
20 *H8* **Dover** Tennessee, S USA *36°30′N 87°50′W*
97 *Q23* **Dover, Strait of** *var.* Straits of Dover, *Fr.* Pas de Calais. *strait* England, United Kingdom/France
Dover, Straits of *see* Dover, Strait of
Dovlen *see* Devin
94 *G11* **Dovre** Oppland, S Norway *61°59′N 09°16′E*
94 *G10* **Dovrefjell** *plateau* S Norway
Dovsk *see* Dowsk
83 *M14* **Dowa** Central, C Malawi *13°40′S 33°55′E*
31 *O10* **Dowagiac** Michigan, N USA *41°58′N 86°06′W*
Dow Gonbadān *see* Do Gonbadān
148 *M2* **Dowlatābād** Fāryāb, N Afghanistan *36°30′N 64°51′E*
97 *G16* **Down** *cultural region* SE Northern Ireland, United Kingdom
33 *R16* **Downey** Idaho, NW USA *42°25′N 112°06′W*
35 *P5* **Downieville** California, W USA *39°34′N 120°49′W*
97 *G16* **Downpatrick** *Ir.* Dún Pádraig. SE Northern Ireland, United Kingdom *54°20′N 05°43′W*
26 *M3* **Downs** Kansas, C USA *39°30′N 98°33′W*
18 *J12* **Downsville** New York, NE USA *42°03′N 74°59′W*
142 *L7* **Dow Rūd** *var.* Do Rūd, Durud. Lorestān, W Iran *33°28′N 49°04′E*
29 *V12* **Dows** Iowa, C USA *42°39′N 93°30′W*
119 *O17* **Dowsk** *Rus.* Dovsk. Homyel'skaya Voblasts', SE Belarus *53°09′N 30°28′E*
35 *Q4* **Doyle** California, W USA *40°00′N 120°06′W*
18 *I15* **Doylestown** Pennsylvania, NE USA *40°18′N 75°08′W*
Doyransko, Ezero *see* Doïráni, Límni
114 *I8* **Doyrentsi** Lovech, N Bulgaria *43°13′N 24°46′E*
164 *G11* **Dōzen** *island* Oki-shotō, SW Japan
14 *K9* **Dozois, Réservoir** ⊠ Québec, SE Canada
74 *D9* **Drâa** *seasonal river* S Morocco
Drâa, Hammada du *see* Dra, Hamada du
Drabble *see* José Enrique Rodó
117 *Q5* **Drabiv** Cherkas'ka Oblast', C Ukraine *49°57′N 32°10′E*
Drable *see* José Enrique Rodó
103 *S13* **Drac** ≈ E France
Drač/Draç *see* Durrës
60 *I8* **Dracena** São Paulo, S Brazil *21°27′S 51°30′W*
98 *M6* **Drachten** Fryslân, N Netherlands *53°07′N 06°06′E*
92 *H11* **Drag** *Lapp.* Ájluokta. Nordland, C Norway *68°02′N 16°E*
116 *L14* **Dragalina** Călărași, SE Romania *44°26′N 27°19′E*
116 *I14* **Draganesti-Olt** Olt, S Romania *44°06′N 25°00′E*
116 *J14* **Drăgănești-Vlașca** Teleorman, S Romania *44°05′N 25°39′E*
116 *I13* **Drăgășani** Vâlcea, SW Romania *44°40′N 24°16′E*
114 *G9* **Dragoman** Sofia, W Bulgaria *42°32′N 22°56′E*
115 *L25* **Dragonáda** *island* SE Greece
Dragonera, Isla *see* Sa Dragonera
45 *T14* **Dragon's Mouths, The** *strait* Trinidad and Tobago/Venezuela
95 *J23* **Dragør** Sjælland, E Denmark *55°36′N 12°42′E*
114 *F10* **Dragovishtitsa** Kysutendil, W Bulgaria *42°22′N 22°39′E*
103 *U15* **Draguignan** Var, SE France *43°31′N 06°31′E*
74 *E9* **Dra, Hamada du** *var.* Hammada du Drâa, Haut Plateau du Dra. *plateau* W Algeria
Dra, Haut Plateau du *see* Dra, Hamada du
119 *H19* **Drahichyn** *Pol.* Drohiczyn Poleski, *Rus.* Drogichin. Brestskaya Voblasts', SW Belarus *52°11′N 25°10′E*
29 *N4* **Drake** North Dakota, N USA *47°54′N 100°23′W*
83 *K23* **Drakensberg** ▲▲ Lesotho/South Africa
194 *F3* **Drake Passage** *passage* Atlantic Ocean/Pacific Ocean
114 *L8* **Dralfa** Tŭrgovishte, N Bulgaria *43°17′N 26°25′E*
114 *I12* **Dráma** *var.* Dhráma. Anatolikí Makedonía kai Thráki, NE Greece *41°09′N 24°10′E*
Dramburg *see* Drawsko Pomorskie
95 *H15* **Drammen** Buskerud, S Norway *59°44′N 10°12′E*
95 *H15* **Drammensfjorden** *fjord* S Norway
92 *H1* **Drangajökull** ▲ NW Iceland *66°13′N 22°18′W*
95 *F16* **Drangedal** Telemark, S Norway *59°05′N 09°05′E*
92 *I2* **Drangsnes** Vestfirðir, NW Iceland *65°42′N 21°27′W*
Drann *see* Dravinja
109 *T10* **Drau** *var.* Drava, *Eng.* Drave, *Hung.* Dráva. ≈ C Europe *see also* Drava
Drau *see* Drava
84 *I11* **Drava** *var.* Drau, *Eng.* Drave, *Hung.* Dráva. ≈ C Europe *see also* Drau
Dráva/Drave *see* Drau/Drava
109 *W10* **Dravinja** *Ger.* Drann. ≈ NE Slovenia
109 *V9* **Dravograd** *Ger.* Unterdrauburg; *prev.* Spodnji Dravograd. N Slovenia *46°36′N 15°00′E*
110 *F10* **Drawa** ≈ NW Poland
110 *F9* **Drawno** Zachodnio-pomorskie, NW Poland *53°12′N 15°44′E*
110 *F9* **Drawsko Pomorskie** *Ger.* Dramburg. Zachodnio-pomorskie, NW Poland *53°32′N 15°48′E*
29 *R3* **Drayton** North Dakota, N USA *48°34′N 97°10′W*
11 *P14* **Drayton Valley** Alberta, SW Canada *53°15′N 115°00′W*
186 *B6* **Dreikikir** East Sepik, NW Papua New Guinea *03°42′S 142°46′E*
Dreikirchen *see* Teiuș
98 *N7* **Drenthe** ◐ *province* NE Netherlands
115 *H15* **Drépano, Akrotírio** *var.* Akrotírio Dhrepanon. *headland* N Greece *39°56′N 23°57′E*
Drepanum *see* Trapani
14 *D17* **Dresden** Ontario, S Canada *42°34′N 82°09′W*
101 *O16* **Dresden** Sachsen, E Germany *51°03′N 13°43′E*
20 *G8* **Dresden** Tennessee, S USA *36°17′N 88°42′W*
118 *M11* **Dretun'** Vitsyebskaya Voblasts', N Belarus *55°41′N 29°13′E*
102 *M5* **Dreux** *anc.* Drocae, Durocasses. Eure-et-Loir, C France *48°44′N 01°23′E*
94 *I11* **Drevsjø** Hedmark, S Norway *61°52′N 12°01′E*
22 *K3* **Drew** Mississippi, S USA *33°48′N 90°31′W*
110 *F10* **Drezdenko** *Ger.* Driesen. Lubuskie, W Poland *52°51′N 15°50′E*
98 *J12* **Driebergen** *var.* Driebergen-Rijsenburg. Utrecht, C Netherlands *52°03′N 05°17′E*
Driebergen-Rijsenburg *see* Driebergen
Driesen *see* Drezdenko
97 *N16* **Driffield** E England, United Kingdom *54°00′N 00°28′W*
65 *D25* **Driftwood Point** *headland* East Falkland, Falkland Islands *52°15′S 59°00′W*
33 *S14* **Driggs** Idaho, NW USA *43°44′N 111°06′W*
Drin *see* Drinit, Lumi i
112 *K12* **Drina** ≈ Bosnia and Herzegovina/Serbia
Drin, Gulf of *see* Drinit, Gjiri i
113 *M16* **Drini i Bardhë** *Serb.* Beli Drim. ≈ Albania/Serbia
113 *K18* **Drinit, Gjiri i** *var.* Pellg i Drinit, *Eng.* Gulf of Drin. *gulf* NW Albania
113 *L17* **Drinit, Lumi i** *var.* Drin. ≈ NW Albania
Drinit, Pellg i *see* Drinit, Gjiri i
Drinit të Zi, Lumi i *see* Black Drin
113 *L22* **Dríno** *var.* Drino, Drínos Pótamos, *Alb.* Lumi i Drinos. ≈ Albania/Greece
Drinos, Lumi i/Drínos Pótamos *see* Dríno
25 *S11* **Dripping Springs** Texas, SW USA *30°11′N 98°04′W*
25 *S15* **Driscoll** Texas, SW USA *27°40′N 97°45′W*
22 *H5* **Driskill Mountain** ▲ Louisiana, S USA *32°25′N 92°54′W*
Drissa *see* Drysa
94 *G10* **Driva** ≈ S Norway
112 *E13* **Drniš** *It.* Šibenik-Knin. Šibenik-Knin, S Croatia *43°51′N 16°10′E*
95 *H15* **Drøbak** Akershus, S Norway *59°40′N 10°40′E*
116 *G13* **Drobeta-Turnu Severin** *prev.* Turnu Severin. Mehedinți, SW Romania *44°39′N 22°40′E*
Drocae *see* Dreux
116 *M8* **Drochia** *Rus.* Drokiya. N Moldova *48°02′N 27°46′E*
97 *F17* **Drogheda** *Ir.* Droichead Átha. NE Ireland *53°43′N 06°21′W*
Drogichin *see* Drahichyn
Drogobych *see* Drohobych
Drohiczyn Poleski *see* Drahichyn
116 *H6* **Drohobych** *Pol.* Drohobycz, *Rus.* Drogobych. L'vivs'ka Oblast', NW Ukraine *49°22′N 23°33′E*
Drohobycz *see* Drohobych
Droichead Átha *see* Drogheda
Droicheadna Bandan *see* Bandon
Droichead na Banna *see* Banbridge
Droim Mór *see* Dromore
Drokiya *see* Drochia
103 *R13* **Drôme** ◐ *department* E France
103 *S13* **Drôme** ≈ E France
97 *G15* **Dromore** *Ir.* Droim Mór. SE Northern Ireland, United Kingdom *54°25′N 06°09′W*
106 *A9* **Dronero** Piemonte, NE Italy *44°28′N 07°25′E*
102 *L12* **Dronne** ≈ SW France
195 *T3* **Dronning Fabiolafjella** *var.* Mount Victor. ▲ Antarctica *72°49′S 33°01′E*
195 *Q3* **Dronning Maud Land** *physical region* Antarctica
98 *K6* **Dronrijp** *Fris.* Dronryp. Fryslân, N Netherlands *53°12′N 05°37′E*
Dronryp *see* Dronrijp
98 *L9* **Dronten** Flevoland, C Netherlands *52°31′N 05°41′E*
Drontheim *see* Trondheim
102 *L13* **Dropt** ≈ SW France
149 *T4* **Drosh** Khyber Pakhtunkhwa, NW Pakistan *35°33′N 71°48′E*
Drossen *see* Ośno Lubuskie
Drug *see* Durg
Drujba *see* Pitnak
118 *I12* **Drūkšiai** ◎ NE Lithuania
Druk-yul *see* Bhutan
11 *Q16* **Drumheller** Alberta, SW Canada *51°28′N 112°42′W*
33 *Q10* **Drummond** Montana, NW USA *46°39′N 113°12′W*
31 *R4* **Drummond Island** *island* Michigan, N USA
Drummond Island *see* Tabiteuea
21 *X7* **Drummond, Lake** ◎ Virginia, NE USA
15 *P12* **Drummondville** Québec, SE Canada *45°52′N 72°28′W*
39 *T11* **Drum, Mount** ▲ Alaska, USA *62°11′N 144°37′W*
27 *O9* **Drumright** Oklahoma, C USA *35°59′N 96°36′W*
99 *J14* **Drunen** Noord-Brabant, S Netherlands *51°41′N 05°08′E*
Druskienniki *see* Druskininkai
119 *F15* **Druskininkai** *Pol.* Druskienniki. Alytus, S Lithuania *54°00′N 24°00′E*
98 *K13* **Druten** Gelderland, SE Netherlands *51°53′N 05°37′E*
118 *K11* **Druya** Vitsyebskaya Voblasts', NW Belarus *55°47′N 27°27′E*
117 *S2* **Druzhba** Sums'ka Oblast', NE Ukraine *52°01′N 33°56′E*
Druzhba *see* Dostyk, Kazakhstan
Druzhba *see* Pitnak, Uzbekistan
123 *R7* **Druzhina** Respublika Sakha (Yakutiya), NE Russian Federation *68°01′N 144°58′E*
117 *X7* **Druzhkivka** Donets'ka Oblast', E Ukraine *48°38′N 37°31′E*
112 *E12* **Drvar** Federacija Bosne I Hercegovine, W Bosnia and Herzegovina *44°21′N 16°24′E*
113 *G15* **Drvenik** Split-Dalmacija, SE Croatia *43°10′N 17°13′E*
114 *K9* **Dryanovo** Gabrovo, N Bulgaria *42°58′N 25°28′E*
26 *G7* **Dry Cimarron River** ≈ Kansas/Oklahoma, C USA
12 *B11* **Dryden** Ontario, C Canada *49°48′N 92°48′W*
24 *M11* **Dryden** Texas, SW USA *30°01′N 102°06′W*
195 *Q14* **Drygalski Ice Tongue** *ice feature* Antarctica
118 *L11* **Drysa** *Rus.* Drissa. ≈ N Belarus
23 *V17* **Dry Tortugas** *island* Florida, SE USA
79 *D15* **Dschang** Ouest, W Cameroon *05°28′N 10°02′E*
54 *J5* **Duaca** Lara, N Venezuela *10°22′N 69°08′W*
Duacum *see* Douai
Duala *see* Douala
45 *N9* **Duarte, Pico** ▲ C Dominican Republic *19°02′N 70°57′W*
140 *J5* **Ḑubā** Tabūk, NW Saudi Arabia *27°26′N 35°42′E*
Dubai *see* Dubayy
117 *N9* **Dubăsari** *Rus.* Dubossary. NE Moldova *47°16′N 29°07′E*
117 *N9* **Dubăsari Reservoir** ⊠ NE Moldova
8 *M10* **Dubawnt** ≈ Nunavut, NW Canada
8 *L9* **Dubawnt Lake** ◎ Northwest Territories/Nunavut, N Canada
30 *L6* **Du Bay, Lake** ⊠ Wisconsin, N USA
141 *U7* **Dubayy** *Eng.* Dubai. Dubayy, NE United Arab Emirates *25°11′N 55°18′E*
141 *W7* **Dubayy** *Eng.* Dubai. ✕ Dubayy, NE United Arab Emirates *25°15′N 55°22′E*
183 *R7* **Dubbo** New South Wales, SE Australia *32°16′S 148°41′E*
108 *G7* **Dübendorf** Zürich, NW Switzerland *47°23′N 08°37′E*
97 *F18* **Dublin** *Ir.* Baile Átha Cliath; *anc.* Eblana. ● (Ireland) Dublin, E Ireland *53°20′N 06°15′W*
23 *U5* **Dublin** Georgia, SE USA *32°32′N 82°54′W*
25 *R7* **Dublin** Texas, SW USA *32°05′N 98°20′W*
97 *G18* **Dublin** *Ir.* Baile Átha Cliath; *anc.* Eblana. *cultural region* E Ireland
97 *G18* **Dublin Airport** ✕ Dublin, E Ireland *53°25′N 06°18′W*
189 *V12* **Dublon** *var.* Tonoas. *island* Chuuk Islands, C Micronesia
126 *K2* **Dubna** Moskovskaya Oblast', W Russian Federation *56°45′N 37°09′E*
111 *G19* **Dubňany** *Ger.* Dubnian. Jihomoravský Kraj, SE Czech Republic *48°54′N 17°00′E*
Dubnian *see* Dubňany
111 *I19* **Dubnica nad Váhom** *Hung.* Máriatölgyes; *prev.* Dubnicz. Trenčiansky Kraj, W Slovakia *48°58′N 18°10′E*
Dubnicz *see* Dubnica nad Váhom
116 *K4* **Dubno** Rivnens'ka Oblast', NW Ukraine *50°28′N 25°40′E*
33 *R13* **Dubois** Idaho, NW USA *44°10′N 112°13′W*
18 *D13* **Du Bois** Pennsylvania, NE USA *41°07′N 78°45′W*
33 *T14* **Dubois** Wyoming, C USA *43°31′N 109°37′W*
Dubossary *see* Dubăsari
127 *O10* **Dubovka** Volgogradskaya Oblast', SW Russian Federation *49°10′N 48°49′E*
76 *H14* **Dubréka** SW Guinea *09°48′N 13°31′W*
14 *B7* **Dubreuilville** Ontario, S Canada *48°21′N 84°31′W*
Dubris Portus *see* Dover
119 *L20* **Dubrova** Homyel'skaya Voblasts', SE Belarus *51°47′N 28°13′E*
Dubrovačko-Neretvanska Županija *see* Dubrovnik-Neretva
126 *I5* **Dubrovka** Bryanskaya Oblast', W Russian Federation *53°44′N 33°27′E*
113 *H16* **Dubrovnik** *It.* Ragusa. Dubrovnik-Neretva, SE Croatia *42°40′N 18°06′E*
113 *I16* **Dubrovnik** ✕ Dubrovnik-Neretva, SE Croatia *42°34′N 18°17′E*
113 *F16* **Dubrovnik-Neretva** *off.* Dubrovačko-Neretvanska Županija. ◐ *province* SE Croatia
Dubrovno *see* Dubrowna
116 *L2* **Dubrovytsya** Rivnens'ka Oblast', NW Ukraine *51°34′N 26°37′E*
119 *O14* **Dubrowna** *Rus.* Dubrovno. Vitsyebskaya Voblasts', N Belarus *54°35′N 30°41′E*
29 *Z13* **Dubuque** Iowa, C USA *42°30′N 90°40′W*
118 *E12* **Dubysa** ≈ C Lithuania
Đưc Cơ *see* Chư Ty
191 *V12* **Duc de Gloucester, Îles du** *Eng.* Duke of Gloucester Islands. *island group* C French Polynesia
111 *C15* **Duchcov** *Ger.* Dux. Ústecký Kraj, NW Czech Republic *50°37′N 13°45′E*

◆ Country
● Country Capital
◇ Dependent Territory
○ Dependent Territory Capital
◐ Administrative Regions
✕ International Airport
▲ Mountain
▲▲ Mountain Range
(volcano) Volcano
≈ River
◎ Lake
⊠ Reservoir

37 N3 **Duchesne** Utah, W USA 40°09′N 110°24′W
191 P17 **Ducie Island** atoll E Pitcairn Group of Islands
11 W15 **Duck Bay** Manitoba, S Canada 52°11′N 100°08′W
23 X17 **Duck Key** island Florida Keys, Florida, SE USA
11 T14 **Duck Lake** Saskatchewan, S Canada 52°52′N 106°12′W
11 V15 **Duck Mountain** ▲▲ Manitoba, S Canada
20 I9 **Duck River** ≈ Tennessee, S USA
20 M10 **Ducktown** Tennessee, S USA 35°01′N 84°24′W
167 U10 **Đức Phô** Quang Ngai, C Vietnam 14°56′N 108°55′E
Đưc Tho see Lin Camh
Đưc Trong see Liên Nghia
D-U-D see Dalap-Uliga-Djarrit
153 N15 **Dūddhinagar** var. Dūdhi. Uttar Pradesh, N India 24°09′N 83°16′E
99 M25 **Dudelange** var. Forge du Sud, Ger. Dudelingen. Luxembourg, S Luxembourg 49°28′N 06°05′E
Dudelingen see Dudelange
101 J15 **Duderstadt** Niedersachsen, C Germany 51°31′N 10°16′E
Dūdhi see Dūddhinagar
122 K8 **Dudinka** Krasnoyarskiy Kray, N Russian Federation 69°27′N 86°13′E
97 L20 **Dudley** C England, United Kingdom 52°30′N 02°05′W
154 G13 **Dudna** ≈ C India
76 L16 **Duékoué** W Ivory Coast 05°50′N 05°22′W
104 M5 **Dueñas** Castilla y León, N Spain 41°52′N 04°33′W
104 K4 **Duerna** ≈ NW Spain
105 O6 **Duero** Port. Douro. ≈ Portugal/Spain see also Douro
Duero see Douro
Duesseldorf see Düsseldorf
21 P12 **Due West** South Carolina, SE USA 34°19′N 82°23′W
195 P11 **Dufek Coast** physical region Antarctica
99 H17 **Duffel** Antwerpen, C Belgium 51°06′N 04°30′E
35 S2 **Duffer Peak** ▲ Nevada, W USA 41°40′N 118°45′W
187 Q9 **Duff Islands** island group E Solomon Islands
Dufour, Pizzo/Dufour, Punta see Dufour Spitze
108 E12 **Dufour Spitze** It. Pizzo Dufour, Punta Dufour. ▲ Italy/Switzerland 45°54′N 07°50′E
112 D9 **Duga Resa** Karlovac, C Croatia 45°25′N 15°30′E
22 H5 **Dugdemona River** ≈ Louisiana, S USA
154 J12 **Duggīpar** Mahārāshtra, C India 21°06′N 80°10′E
112 B13 **Dugi Otok** var. Isola Grossa, It. Isola Lunga. island W Croatia
113 F14 **Dugopolje** Split-Dalmacija, S Croatia 43°35′N 16°35′E
160 L8 **Du He** ≈ C China
54 M11 **Duida, Cerro** ▲ S Venezuela 03°21′N 65°45′W
Duinekerke see Dunkerque
101 E15 **Duisburg** prev. Duisburg-Hamborn. Nordrhein-Westfalen, W Germany 51°25′N 06°47′E
Duisburg-Hamborn see Duisburg
99 F14 **Duiveland** island SW Netherlands
98 M12 **Duiven** Gelderland, E Netherlands 51°57′N 06°02′E
139 W10 **Dujaylah, Hawr ad** ◎ S Iraq
160 H9 **Dujiangyan** var. Guanxian, Guan Xian. Sichuan, C China 31°01′N 103°40′E
81 L18 **Dujuuma** Shabeellaha Hoose, S Somalia 01°04′N 42°37′E
139 T3 **Dûkan** Ar. Dūkān, var. Dokan. As Sulaymānīyah, E Iraq 35°55′N 44°58′E
Dūkān see Dûkan
39 Z14 **Duke Island** island Alexander Archipelago, Alaska, USA
Dukelský Priesmy/Dukelský Průsmyk see Dukla Pass
Duke of Gloucester Islands see Duc de Gloucester, Îles du
81 F14 **Duk Faiwil** Jonglei, E South Sudan 07°30′N 31°27′E
141 T7 **Dukhān** C Qatar 25°29′N 50°48′E
Dukhan Heights see Dukhān, Jabal
143 N16 **Dukhān, Jabal** var. Dukhan Heights. hill range S Qatar
127 Q7 **Dukhovnitskoye** Saratovskaya Oblast', W Russian Federation 52°31′N 48°32′E
126 H4 **Dukhovshchina** Smolenskaya Oblast', W Russian Federation 55°15′N 32°22′E
Dukielska, Przełęcz see Dukla Pass
111 N17 **Dukla** Podkarpackie, SE Poland 49°33′N 21°40′E
Duklai Hág see Dukla Pass
111 N18 **Dukla Pass** Cz. Dukelský Průsmyk, Ger. Dukla-Pass, Hung. Duklai Hág, Pol. Przełęcz Dukielska, Slvk. Dukelský Priesmy. pass Poland/Slovakia
Dukla-Pass see Dukla Pass
Dukou see Panzhihua
118 I12 **Dūkštas** Utena, E Lithuania 55°32′N 26°21′E
Dulaan see Herlenbayan-Ulaan
159 R10 **Dulan** var. Qagan Us. Qinghai, C China 36°11′N 97°51′E
37 R8 **Dulce** New Mexico, SW USA 36°55′N 107°00′W
43 N16 **Dulce, Golfo** gulf S Costa Rica
Dulce, Golfo see Izabal, Lago de
42 K6 **Dulce Nombre de Culmí** Olancho, C Honduras 15°09′N 85°37′W
62 L9 **Dulce, Río** ≈ C Argentina
123 Q9 **Dulgalakh** ≈ NE Russian Federation
Dŭlgopol see Dalgopol
153 V14 **Dullabchara** Assam, NE India 24°25′N 92°22′E
20 D3 **Dulles** ✈ (Washington DC)Virginia, NE USA 39°00′N 77°27′W
101 E14 **Dülmen** Nordrhein-Westfalen, W Germany 51°51′N 07°17′E
114 M7 **Dulovo** Silistra, NE Bulgaria 43°51′N 27°10′E
29 W5 **Duluth** Minnesota, N USA 46°47′N 92°06′W
138 H7 **Dūmā** Fr. Douma. Rīf Dimashq, SW Syria 33°33′N 36°24′E
171 O8 **Dumagasa Point** headland Mindanao, S Philippines 07°01′N 121°54′E
171 P6 **Dumaguete** var. Dumaguete City. Negros, C Philippines 09°16′N 123°17′E
Dumaguete City see Dumaguete
168 J10 **Dumai** Sumatera, W Indonesia 01°39′N 101°28′E
183 T4 **Dumaresq River** ≈ New South Wales/Queensland, SE Australia
27 W13 **Dumas** Arkansas, C USA 33°53′N 91°29′W
25 N1 **Dumas** Texas, SW USA 35°51′N 101°57′W
138 I7 **Ḑumayr** Rīf Dimashq, W Syria 33°36′N 36°28′E
96 I12 **Dumbarton** W Scotland, United Kingdom 55°57′N 04°35′W
96 I12 **Dumbarton** cultural region C Scotland, United Kingdom
187 Q17 **Dumbéa** Province Sud, S New Caledonia 22°11′S 166°27′E
111 K19 **Ďumbier** Ger. Djumbir, Hung. Gyömbér. ▲ C Slovakia 48°54′N 19°36′E
116 I11 **Dumbrăveni** Ger. Elisabethstedt, Hung. Erzsébetváros; prev. Ebesfalva, Eppeschdorf, Ibaşfalău. Sibiu, C Romania 46°14′N 24°34′E
116 L12 **Dumbrăveni** Vrancea, E Romania 45°31′N 27°09′E
97 J14 **Dumfries** S Scotland, United Kingdom 55°04′N 03°37′W
97 J14 **Dumfries** cultural region SW Scotland, United Kingdom
153 R15 **Dumka** Jhārkhand, NE India 24°17′N 87°15′E
Dümmer see Dümmersee
100 G12 **Dümmersee** var. Dümmer. ◎ NW Germany
14 J11 **Dumoine** ≈ Québec, SE Canada
14 J10 **Dumoine, Lac** ◎ Québec, SE Canada
195 V16 **Dumont d'Urville** French research station Antarctica 66°24′S 139°38′E
195 W15 **Dumont d'Urville Sea** sea S Pacific Ocean
14 K11 **Dumont, Lac** ◎ Québec, SE Canada
75 W7 **Dumyāt** var. Dumyât, Eng. Damietta. N Egypt 31°26′N 31°48′E
Duna see Danube, C Europe
Düna see Western Dvina
Duna see Don, Russian Federation
Dünaburg see Daugavpils
111 J24 **Dunaföldvár** Tolna, C Hungary 46°48′N 18°55′E
Dunaj see Wien, Austria
Dunaj see Danube, C Europe
111 L18 **Dunajec** ≈ S Poland
111 H21 **Dunajská Streda** Hung. Dunaszerdahely. Trnavský Kraj, W Slovakia 48°N 17°28′E
Dunapentele see Dunaújváros
Dunărea see Danube
116 M13 **Dunărea Veche, Braţul** ≈ SE Romania
117 N13 **Dunării, Delta** delta SE Romania
Dunaszerdahely see Dunajská Streda
111 J23 **Dunaújváros** prev. Dunapentele, Sztálinváros. Fejér, C Hungary 47°N 18°55′E
Dunav see Danube
114 J8 **Dunavska Ravnina** Eng. Danubian Plain. lowlands N Bulgaria
114 G7 **Dunavtsi** Vidin, NW Bulgaria 43°54′N 22°49′E
123 S15 **Dunay** Primorskiy Kray, SE Russian Federation 42°53′N 132°20′E
Dunayevtsy see Dunayivtsi
116 L7 **Dunayivtsi** Rus. Dunayevtsy. Khmel'nyts'ka Oblast', NW Ukraine 48°56′N 26°50′E
185 F22 **Dunback** Otago, South Island, New Zealand 45°22′S 170°37′E
10 L17 **Duncan** Vancouver Island, British Columbia, SW Canada 48°46′N 123°10′W
37 O15 **Duncan** Arizona, SW USA 32°43′N 109°06′W
26 M12 **Duncan** Oklahoma, C USA 34°30′N 97°57′W
Duncan Island see Pinzón, Isla
151 Q20 **Duncan Passage** strait Andaman Sea/Bay of Bengal
96 K6 **Duncansby Head** headland N Scotland, United Kingdom 58°37′N 03°01′W
14 G12 **Dunchurch** Ontario, S Canada 45°36′N 79°54′W
118 D7 **Dundaga** NW Latvia 57°29′N 22°19′E
14 G14 **Dundalk** Ontario, S Canada 44°11′N 80°22′W
97 F16 **Dundalk** Ir. Dún Dealgan. Louth, NE Ireland 54°01′N 06°25′W
21 X3 **Dundalk** Maryland, NE USA 39°15′N 76°31′W
97 F16 **Dundalk Bay** Ir. Cuan Dhún Dealgan. bay NE Ireland
14 G16 **Dundas** Ontario, S Canada 43°16′N 79°55′W
180 L12 **Dundas, Lake** salt lake Western Australia
Dundbürd see Batnorov
Dún Dealgan see Dundalk
15 N13 **Dundee** Québec, SE Canada 45°01′N 74°27′W
83 K22 **Dundee** KwaZulu/Natal, E South Africa 28°09′S 30°12′E
96 K11 **Dundee** E Scotland, United Kingdom 56°28′N 03°W
31 R10 **Dundee** Michigan, N USA 41°57′N 83°39′W
25 R5 **Dundee** Texas, SW USA 33°43′N 98°52′W
194 H3 **Dundee Island** island Antarctica
162 L9 **Dundgovĭ** ◈ province C Mongolia
97 G16 **Dundrum Bay** Ir. Cuan Dhún Droma. inlet NW Irish Sea
11 T15 **Dundurn** Saskatchewan, S Canada 51°43′N 106°27′W
Dund-Us see Hovd
Dund-Us see Hovd
185 F23 **Dunedin** Otago, South Island, New Zealand 45°52′S 170°31′E
183 R7 **Dunedoo** New South Wales, SE Australia 32°04′S 149°23′E
97 D14 **Dunfanaghy** Ir. Dún Fionnachaidh. NW Ireland 55°11′N 07°59′W
96 J12 **Dunfermline** C Scotland, United Kingdom 56°04′N 03°29′W
Dún Fionnachaidh see Dunfanaghy
149 V10 **Dunga Bunga** Punjab, E Pakistan 29°51′N 73°19′E
97 F15 **Dungannon** Ir. Dún Geanainn. C Northern Ireland, United Kingdom 54°31′N 06°46′W
Dun Garbháin see Dungarvan
152 F15 **Dūngarpur** Rājasthān, N India 23°50′N 73°43′E
97 E21 **Dungarvan** Ir. Dun Garbháin. Waterford, S Ireland 52°05′N 07°37′W
101 N21 **Dungau** cultural region SE Germany
Dún Geanainn see Dungannon
97 P23 **Dungeness** headland SE England, United Kingdom 50°55′N 00°58′E
63 I23 **Dungeness, Punta** headland S Argentina 52°25′S 68°25′W
Dungloe see Dunglow
97 D14 **Dunglow** var. Dungloe, Ir. An Clochán Liath. Donegal, NW Ireland 54°57′N 08°22′W
183 T7 **Dungog** New South Wales, SE Australia 32°24′S 151°45′E
79 O16 **Dungu** Orientale, NE Dem. Rep. Congo 03°40′N 28°32′E
168 L8 **Dungun** var. Kuala Dungun. Terengganu, Peninsular Malaysia 04°47′N 103°26′E
80 I6 **Dungûnab** Red Sea, NE Sudan 21°10′N 37°09′E
15 P13 **Dunham** Québec, SE Canada 45°08′N 72°48′W
Dunheved see Launceston
Dunholme see Durham
163 X10 **Dunhua** Jilin, NE China 43°22′N 128°12′E
159 P8 **Dunhuang** Gansu, N China 40°10′N 94°40′E
182 L12 **Dunkeld** Victoria, SE Australia 37°41′S 142°19′E
103 O1 **Dunkerque** Eng. Dunkirk, Flem. Duinekerke; prev. Dunquerque. Nord, N France 51°06′N 02°34′E
97 K23 **Dunkery Beacon** ▲ SW England, United Kingdom 51°10′N 03°36′W
18 C11 **Dunkirk** New York, NE USA 42°28′N 79°19′W
Dunkirk see Dunkerque
77 P17 **Dunkwa** SW Ghana 05°59′N 01°45′W
97 G18 **Dún Laoghaire** Eng. Dunleary; prev. Kingstown. E Ireland 53°17′N 06°08′W
29 S14 **Dunlap** Iowa, C USA 41°51′N 95°36′W
20 L10 **Dunlap** Tennessee, S USA 35°22′N 85°23′W
Dunleary see Dún Laoghaire
Dún Mánmhaí see Dunmanway
97 B21 **Dunmanway** Ir. Dún Mánmhaí. Cork, SW Ireland 51°43′N 09°07′W
18 I13 **Dunmore** Pennsylvania, NE USA 41°25′N 75°37′W
21 U10 **Dunn** North Carolina, SE USA 35°18′N 78°36′W
Dún na nGall see Donegal
23 V11 **Dunnellon** Florida, SE USA 29°03′N 82°27′W
96 J6 **Dunnet Head** headland N Scotland, United Kingdom 58°40′N 03°27′W
29 N14 **Dunning** Nebraska, C USA 41°49′N 100°04′W
65 B24 **Dunnose Head Settlement** West Falkland, Falkland Islands 51°24′S 60°29′W
14 G17 **Dunnville** Ontario, S Canada 42°54′N 79°36′W
Dún Pádraig see Downpatrick
Dunquerque see Dunkerque
Dunqulah see Dongola
96 L12 **Duns** SE Scotland, United Kingdom 55°46′N 02°13′W
29 N2 **Dunseith** North Dakota, N USA 48°48′N 100°03′W
35 N2 **Dunsmuir** California, W USA 41°12′N 122°19′W
97 N21 **Dunstable** Lat. Durocobrivae. E England, United Kingdom 51°53′N 00°32′W
185 D21 **Dunstan Mountains** ▲▲ South Island, New Zealand
103 O9 **Dun-sur-Auron** Cher, C France 46°52′N 02°40′E
185 F21 **Duntroon** Canterbury, South Island, New Zealand 44°52′S 170°40′E
149 T10 **Dunyāpur** Punjab, E Pakistan 29°48′N 71°48′E
163 U5 **Duobukur He** ≈ NE China
163 R12 **Duolun** var. Dolonnur. Nei Mongol Zizhiqu, N China 42°11′N 116°30′E
167 R14 **Dương Đông** Kiên Giang, S Vietnam 10°15′N 103°58′E
114 G10 **Dupnitsa** prev. Marek, Stanke Dimitrov. Kyustendil, W Bulgaria 42°16′N 23°07′E
28 L8 **Dupree** South Dakota, N USA 45°03′N 101°36′W
33 Q7 **Dupuyer** Montana, NW USA 48°13′N 112°34′W
141 Y11 **Duqm** var. Daqm. E Oman 19°42′N 57°40′E
63 F23 **Duque de York, Isla** island S Chile
181 N4 **Durack Range** ▲▲ Western Australia
136 K10 **Durağan** Sinop, N Turkey 41°26′N 35°03′E
103 S15 **Durance** ≈ SE France
31 R9 **Durand** Michigan, N USA 42°54′N 83°58′W
30 I6 **Durand** Wisconsin, N USA 44°37′N 91°56′W
40 K10 **Durango** var. Victoria de Durango. Durango, W Mexico 24°03′N 104°38′W
105 P3 **Durango** País Vasco, N Spain 43°10′N 02°38′W
37 Q8 **Durango** Colorado, C USA 37°13′N 107°51′W
40 J9 **Durango** ◈ state C Mexico
114 O7 **Durankulak** Rom. Răcari; prev. Blatnitsa, Duranulac. Dobrich, NE Bulgaria 43°41′N 28°31′E
22 L4 **Durant** Mississippi, S USA 33°04′N 89°51′W
27 P13 **Durant** Oklahoma, C USA 33°59′N 96°24′W
Duranulac see Durankulak
105 N6 **Duratón** ≈ N Spain
61 E19 **Durazno** var. San Pedro de Durazno. Durazno, C Uruguay 33°22′S 56°31′W
61 E19 **Durazno** ◈ department C Uruguay
Durazzo see Durrës
83 K23 **Durban** var. Port Natal. KwaZulu/Natal, E South Africa 29°51′S 31°E
83 K23 **Durban** ✈ KwaZulu/Natal, E South Africa 29°55′S 31°01′E
118 C9 **Durbe** Ger. Durben. W Latvia 56°34′N 21°22′E
Durben see Durbe
99 K21 **Durbuy** Luxembourg, SE Belgium 50°21′N 05°27′E
105 N15 **Dúrcal** Andalucía, S Spain 37°00′N 03°24′W
112 F8 **Đurđevac** Ger. Sankt Georgen, Hung. Szentgyörgy; prev. Djurdjevac, Gjurgjevac. Koprivnica-Kri»zevci, N Croatia 46°02′N 17°03′E
113 K15 **Đurđevica Tara** N Montenegro 43°09′N 19°18′E
97 L24 **Durdle Door** natural arch S England, United Kingdom
158 L3 **Düre** Xinjiang Uygur Zizhiqu, W China 46°30′N 88°26′E
101 D16 **Düren** anc. Marcodurum. Nordrhein-Westfalen, W Germany 50°48′N 06°30′E
154 K12 **Durg** prev. Drug. Chhattīsgarh, C India 21°12′N 81°20′E
153 U13 **Durgapur** Dhaka, N Bangladesh 25°10′N 90°41′E
153 R15 **Durgāpur** West Bengal, NE India 23°30′N 87°20′E
14 F14 **Durham** Ontario, S Canada 44°10′N 80°48′W
97 M14 **Durham** hist. Dunholme. N England, United Kingdom 54°47′N 01°34′W
21 U9 **Durham** North Carolina, SE USA 36°N 78°54′W
97 L15 **Durham** cultural region N England, United Kingdom
168 J10 **Duri** Sumatera, W Indonesia 01°13′N 101°13′E
Duria Major see Dora Baltea
Duria Minor see Dora Riparia
Durlas see Thurles
141 P8 **Ḑurmā** Ar Riyāḑ, C Saudi Arabia 24°37′N 46°06′E
113 J15 **Durmitor** ▲ N Montenegro
96 H6 **Durness** N Scotland, United Kingdom 58°34′N 04°46′W
109 Y3 **Dürnkrut** Niederösterreich, E Austria 48°28′N 16°50′E
Durnovaria see Dorchester
Durobrivae see Rochester
Durocasses see Dreux
Durocobrivae see Dunstable
Durocortorum see Reims
Durostorum see Silistra
Durovernum see Canterbury
113 K20 **Durrës** var. Durrësi, Dursi, It. Durazzo, SCr. Drač, Turk. Draç. Durrës, W Albania 41°20′N 19°26′E
113 K19 **Durrës** ◈ district W Albania
Durrësi see Durrës
97 A21 **Dursey Island** Ir. Oileán Baoi. island SW Ireland
Dursi see Durrës
Duru see Wuchuan
Durud see Dow Rūd
114 P12 **Durusu** İstanbul, NW Turkey 41°18′N 28°41′E
114 O12 **Durusu Gölü** ◎ NW Turkey
138 I9 **Durūz, Jabal ad** ▲ SW Syria 37°00′N 32°30′E
184 K13 **D'Urville Island** island C New Zealand
171 X12 **D'Urville, Tanjung** headland Papua, E Indonesia 01°26′S 137°52′E
146 H14 **Duşak** Rus. Dushak. Ahal Welaýaty, S Turkmenistan 37°15′N 59°57′E
Dusa Mareb/Dusa Marreb see Dhuusa Marreeb
118 I11 **Dusetos** Utena, NE Lithuania 55°44′N 25°49′E
Dushak see Duşak
160 K12 **Dushan** Guizhou, S China 25°50′N 107°36′E
147 P13 **Dushanbe** var. Dyushambe; prev. Stalinabad, Taj. Stalinobod. ● (Tajikistan) W Tajikistan 38°35′N 68°44′E
147 P13 **Dushanbe** ✈ W Tajikistan 38°31′N 68°49′E
137 T9 **Dusheti** prev. Dushet'i. E Georgia 42°07′N 44°44′E
Dushet'i see Dusheti
18 H13 **Dushore** Pennsylvania, NE USA 41°30′N 76°23′W
185 A23 **Dusky Sound** sound South Island, New Zealand
101 E15 **Düsseldorf** var. Duesseldorf. Nordrhein-Westfalen, W Germany 51°14′N 06°49′E
147 P14 **Dŭstí** Rus. Dusti. SW Tajikistan 37°22′N 68°41′E
194 I9 **Dustin Island** island Antarctica
Dutch East Indies see Indonesia
Dutch Guiana see Suriname
38 L17 **Dutch Harbor** Unalaska Island, Alaska, USA 53°51′N 166°33′W
36 J3 **Dutch Mount** ▲ Utah, W USA 40°16′N 113°56′W
Dutch New Guinea see Papua
Dutch West Indies see Curaçao
83 H20 **Dutlwe** Kweneng, S Botswana 23°58′S 23°56′E
67 V16 **Du Toit Fracture Zone** tectonic feature SW Indian Ocean
125 U8 **Dutovo** Respublika Komi, NW Russian Federation 63°45′N 56°38′E
77 V13 **Dutsan Wai** var. Dutsen Wai. Kaduna, C Nigeria 10°49′N 08°15′E
77 W13 **Dutse** Jigawa, N Nigeria 11°43′N 09°25′E
Dutsen Wai see Dutsan Wai
Duttia see Datia
14 E17 **Dutton** Ontario, S Canada 42°40′N 81°28′W
36 L7 **Dutton, Mount** ▲ Utah, W USA 38°00′N 112°10′W
83 J25 **Dutywa** prev. Idutywa. Eastern Cape, SE South Africa 32°06′S 28°20′E see also Dutywa
162 E7 **Duut** Hovd, W Mongolia 47°28′N 91°52′E
14 K11 **Duval, Lac** ◎ Québec, SE Canada
127 W3 **Duvan** Respublika Bashkortostan, W Russian Federation 55°42′N 57°56′E
138 L9 **Duwaykhilat Satiḩ ar Ruwayshid** seasonal river SE Jordan
Dux see Duchcov
160 J13 **Duyang Shan** ▲▲ S China
167 T14 **Duyên Hai** Tra Vinh, S Vietnam 09°39′N 106°28′E
160 K12 **Duyun** Guizhou, S China 26°16′N 107°29′E
136 G11 **Düzce** Düzce, NW Turkey 40°51′N 31°09′E
136 K14 **Düzce** ◈ province NW Turkey
Duzdab see Zāhedān
Duzenkyr, Khrebet see Duzkyr, Khrebet
146 I16 **Duzkyr, Khrebet** prev. Khrebet Duzenkyr. ▲▲ S Turkmenistan
114 K8 **Dve Mogili** Ruse, N Bulgaria 43°35′N 25°51′E
Dvina Bay see Chëshskaya Guba
Dvinsk see Daugavpils
124 L7 **Dvinskaya Guba** bay NW Russian Federation
112 E10 **Dvor** Sisak-Moslavina, C Croatia 45°05′N 16°22′E
117 W5 **Dvorichna** Kharkivs'ka Oblast', E Ukraine 49°52′N 37°43′E
111 F16 **Dvůr Králové nad Labem** Ger. Königinhof an der Elbe. Královéhradecký Kraj, N Czech Republic 50°27′N 15°50′E
154 A10 **Dwārka** Gujarāt, W India 22°14′N 68°58′E
30 M12 **Dwight** Illinois, N USA 41°05′N 88°25′W
98 N8 **Dwingeloo** Drenthe, NE Netherlands 52°49′N 06°20′E
33 N10 **Dworshak Reservoir** ▣ Idaho, NW USA
Dyal see Dyaul Island
Dyanev see Galkynyş
Dyatlovo see Dzyatlava
186 G5 **Dyaul Island** var. Djaul, Dyal. island NE Papua New Guinea
20 G8 **Dyer** Tennessee, S USA 36°04′N 88°59′W
9 S5 **Dyer, Cape** headland Baffin Island, Nunavut, NE Canada 66°37′N 61°13′W
20 F8 **Dyersburg** Tennessee, S USA 36°02′N 89°21′W
29 Y13 **Dyersville** Iowa, C USA 42°29′N 91°07′W
97 I21 **Dyfed** cultural region SW Wales, United Kingdom
Dyfrdwy, Afon see Dee
Dyhernfurth see Brzeg Dolny
111 E19 **Dyje** var. Thaya. ≈ Austria/Czech Republic see also Thaya
Dyje see Thaya
117 T5 **Dykan'ka** Poltavs'ka Oblast', C Ukraine 49°48′N 34°33′E
127 N16 **Dykhtau** ▲ SW Russian Federation 43°01′N 42°56′E
111 A16 **Dyleń** Ger. Tillenberg. ▲ NW Czech Republic 49°58′N 12°31′E
110 K9 **Dylewska Góra** ▲ N Poland 53°33′N 19°57′E
117 O4 **Dymer** Kyyivs'ka Oblast', N Ukraine 50°50′N 30°20′E
117 W7 **Dymytrov** Rus. Dimitrov. Donets'ka Oblast', SE Ukraine 48°18′N 37°19′E
111 O17 **Dynów** Podkarpackie, SE Poland 49°49′N 22°14′E
29 X13 **Dysart** Iowa, C USA 42°10′N 92°18′W
Dysna see Dzisna
115 H18 **Dýstos, Límni** var. Límni Dístos. ◎ Évvoia, C Greece
115 D18 **Dytikí Elláda** Eng. Greece West, var. Dytikí Ellás. ◈ region C Greece
Dytikí Ellás see Dytikí Elláda
115 C14 **Dytikí Makedonía** Eng. Macedonia West. ◈ region N Greece
Dyurment'yube see Diirmentobe
127 U4 **Dyurtyuli** Respublika Bashkortostan, W Russian Federation 55°31′N 54°49′E
Dyushambe see Dushanbe
162 K7 **Dzaamar** var. Bat-Öldziyt. Töv, C Mongolia 48°10′N 104°49′E
Dzaanhushuu see Ihtamir
Dza Chu see Mekong
Dzadgay see Bömbögör
162 H8 **Dzag** Bayanhongor, C Mongolia 46°54′N 99°11′E
Dzalaa see Shinejinst
163 O11 **Dzamïn-Üüd** var. Borhoyn Tal. Dornogovĭ, SE Mongolia 43°43′N 111°53′E
172 J14 **Dzaoudzi** E Mayotte 12°48′S 45°18′E
Dzaudzhikau see Vladikavkaz
162 G7 **Dzavhan** ◈ province NW Mongolia
162 G7 **Dzavhan Gol** ≈ NW Mongolia
162 G6 **Dzavhanmandal** var. Nuga. Dzavhan, W Mongolia 48°17′N 95°07′E
Dzegstey see Ögiynuur
162 E7 **Dzereg** var. Altanteel. Hovd, W Mongolia 47°05′N 92°57′E
127 O3 **Dzerzhinsk** Nizhegorodskaya Oblast', W Russian Federation 56°20′N 43°22′E
Dzerzhinsk see Dzyarzhynsk
Dzerzhinsk see Romaniv
Dzerzhinskiy see Nar'yan-Mar
Dzerzhinskoye see Tokzhaylau
Dzerzhinskoye see Tokzhaylau
Dzerzhyns'k see Romaniv
Dzerzhyns'k see Romaniv
Džetygara see Zhitikara
Dzhailgan see Jayilgan
Dzhalagash see Zhalagash
147 T10 **Dzhalal-Abad** Kir. Jalal-Abad. Dzhalal-Abadskaya Oblast', W Kyrgyzstan 40°56′N 73°00′E
147 S9 **Dzhalal-Abadskaya Oblast'** Kir. Jalal-Abad Oblasty. ◈ province W Kyrgyzstan
Dzhalilabad see Cälilabad
Dzhambeyty see Zhympity
Dzhambul see Taraz
Dzhambulskaya Oblast' see Zhambyl
144 D9 **Dzhanibek** prev. Dzhanybek, Kaz. Zhänibek. Zapadnyy Kazakhstan, W Kazakhstan 49°27′N 46°51′E
Dzhankel'dy see Jongeldi
117 T12 **Dzhankoy** Avtonomna Respublika Krym, S Ukraine 45°40′N 34°20′E
Dzhansugurov see Zhansugirov
147 R9 **Dzhany-Bazar** var. Yangibazar. Dzhalal-Abadskaya Oblast', W Kyrgyzstan 41°40′N 70°49′E
Dzhanybek see Dzhanibek
123 P8 **Dzhardzhan** Respublika Sakha (Yakutiya), NE Russian Federation 68°47′N 123°51′E
Dzharkurgan see Jarqo'rg'on
117 S11 **Dzharylhats'ka Zatoka** gulf S Ukraine
Dzhayilgan see Jayilgan
Dzhebel see Jebel
147 T14 **Dzhelandy** SE Tajikistan 37°34′N 72°35′E
147 Y7 **Dzhergalan** Kir. Jyrgalan. Issyk-Kul'skaya Oblast', NE Kyrgyzstan 42°37′N 78°56′E
Dzhetysay see Zhetysay
Dzhezkazgan see Zhezkazgan
Dzhigirbent see Jigerbent
Dzhirgatal' see Jirgatol
Dzhizak see Jizzax
Dzhizakskaya Oblast' see Jizzax Viloyati
123 P8 **Dzhugdzhur, Khrebet** ▲▲ E Russian Federation
Dzhul'fa see Culfa
Dzhuma see Juma
Dzhungarskiy Alatau see Zhetysuskiy Alatau
Dzhusaly see Zhosaly
146 J12 **Dzhynlykum, Peski** desert E Turkmenistan
110 L9 **Działdowo** Warmińsko-Mazurskie, C Poland 53°13′N 20°12′E
111 L16 **Działoszyce** Świętokrzyskie, C Poland 50°21′N 20°19′E
41 X11 **Dzidzantún** Yucatán, E Mexico
111 G15 **Dzierżoniów** Ger. Reichenbach. Dolnośląskie, SW Poland 50°43′N 16°40′E
41 X11 **Dzilam de Bravo** Yucatán, E Mexico 21°24′N 88°52′W
118 L12 **Dzisna** Rus. Disna. Vitsyebskaya Voblasts', N Belarus 55°33′N 28°13′E
118 K12 **Dzisna** Lith. Dysna, Rus. Disna. ≈ Belarus/Lithuania
119 G20 **Dzivin** Rus. Divin. Brestskaya Voblasts', SW Belarus 51°58′N 24°33′E
119 M15 **Dzmitravichy** Rus. Dmitrovichi. Minskaya Voblasts', C Belarus 53°58′N 29°14′E
Dzogsool see Bayantsagaan
162 I5 **Dzöölön** var. Rinchinlhumbe. Hövsgöl, N Mongolia 51°06′N 99°40′E
129 S8 **Dzungaria** var. Sungaria, Zungaria. physical region W China
Dzungarian Basin see Junggar Pendi
Dzür see Tes
162 J8 **Dzüünbayan-Ulaan** var. Bayan-Ulaan. Övörhangay, C Mongolia 46°38′N 102°30′E
Dzüünbulag see Matad, Dornod, Mongolia
Dzüünbulag see Uulbayan, Sühbaatar, Mongolia
162 L8 **Dzuunmod** Töv, C Mongolia 47°45′N 107°00′E
Dzuunmod see Ider
Dzüün Soyonï Nuruu see Vostochnyy Sayan
Dzüyl see Tonhil
Dzvina see Western Dvina
119 J16 **Dzyarzhynsk** Rus. Dzerzhinsk; prev. Kaydanovo. Minskaya Voblasts', C Belarus 53°41′N 27°09′E
119 H17 **Dzyatlava** Pol. Zdzięcioł, Rus. Dyatlovo. Hrodzyenskaya Voblasts', W Belarus 53°27′N 25°23′E

E

E see Hubei
Éadan Doire see Edenderry
167 U12 **Ea Drăng** var. Ea H'leo. Đăc Lăc, S Vietnam 13°09′N 108°14′E
37 W6 **Eads** Colorado, C USA 38°28′N 102°46′W
37 O13 **Eagar** Arizona, SW USA 34°05′N 109°17′W
39 T8 **Eagle** Alaska, USA 64°47′N 141°12′W
13 S8 **Eagle** ≈ Newfoundland and Labrador, E Canada
10 I3 **Eagle** ≈ Yukon, NW Canada
29 T7 **Eagle Bend** Minnesota, N USA 46°10′N 95°02′W
28 M8 **Eagle Butte** South Dakota, N USA 44°58′N 101°13′W
29 V12 **Eagle Grove** Iowa, C USA 42°39′N 93°54′W
19 R2 **Eagle Lake** Maine, NE USA 47°01′N 68°35′W
25 U11 **Eagle Lake** Texas, SW USA 29°35′N 96°19′W
12 A11 **Eagle Lake** ◎ Ontario, S Canada
35 P3 **Eagle Lake** ◎ California, W USA
19 R3 **Eagle Lake** ◎ Maine, NE USA
29 Y3 **Eagle Mountain** ▲ Minnesota, N USA 47°54′N 90°33′W
25 T6 **Eagle Mountain Lake** ▣ Texas, SW USA
37 S9 **Eagle Nest Lake** ▣ New Mexico, SW USA
25 P13 **Eagle Pass** Texas, SW USA 28°44′N 100°31′W
65 C25 **Eagle Passage** passage SW Atlantic Ocean
35 R8 **Eagle Peak** ▲ California, W USA 38°11′N 119°22′W
35 Q2 **Eagle Peak** ▲ California, W USA 41°16′N 120°12′W
37 P13 **Eagle Peak** ▲ New Mexico, SW USA 33°39′N 109°36′W
10 I4 **Eagle Plain** Yukon, NW Canada 66°23′N 136°42′W
32 G15 **Eagle Point** Oregon, NW USA 42°28′N 122°48′W
186 P10 **Eagle Point** headland SE Papua New Guinea 10°31′S 149°53′E
39 R11 **Eagle River** Alaska, USA 61°18′N 149°38′W
30 M2 **Eagle River** Michigan, N USA 47°24′N 88°18′W
30 L4 **Eagle River** Wisconsin, N USA 45°56′N 89°15′W
21 S6 **Eagle Rock** Virginia, NE USA 37°40′N 79°46′W
36 J13 **Eagletail Mountains** ▲▲ Arizona, SW USA
Ea H'leo see Ea Drăng
167 U12 **Ea Kar** Đăc Lăc, S Vietnam 12°47′N 108°26′E
Eanjum see Anjum
Eanodat see Enontekiö
12 B10 **Ear Falls** Ontario, C Canada 50°38′N 93°13′W
27 X10 **Earle** Arkansas, C USA 35°16′N 90°28′W
35 R12 **Earlimart** California, W USA 35°52′N 119°17′W
20 I6 **Earlington** Kentucky, S USA 37°16′N 87°30′W
14 H8 **Earlton** Ontario, S Canada 47°41′N 79°46′W
29 T13 **Early** Iowa, C USA 42°27′N 95°09′W
96 J11 **Earn** ≈ N Scotland, United Kingdom
185 C21 **Earnslaw, Mount** ▲ South Island, New Zealand 44°34′S 168°26′E
24 M4 **Earth** Texas, SW USA 34°13′N 102°24′W
21 P11 **Easley** South Carolina, SE USA 34°49′N 82°36′W
East see Est
East Açores Fracture Zone see East Azores Fracture Zone
97 P19 **East Anglia** physical region E England, United Kingdom
15 Q12 **East Angus** Québec, SE Canada 45°29′N 71°39′W
195 V8 **East Antarctica** prev. Greater Antarctica. physical region Antarctica
18 E10 **East Aurora** New York, NE USA 42°44′N 78°36′W
East Australian Basin see Tasman Basin
East Azerbaijan see Āẕarbāyjān-e Sharqī
64 L9 **East Azores Fracture Zone** var. East Açores Fracture Zone. tectonic feature E Atlantic Ocean
22 M11 **East Bay** bay Louisiana, S USA
25 V11 **East Bernard** Texas, SW USA 29°32′N 96°04′W
29 V8 **East Bethel** Minnesota, N USA 45°24′N 93°14′W
East Borneo see Kalimantan Timur
97 P23 **Eastbourne** SE England, United Kingdom 50°46′N 00°16′E
15 R11 **East-Broughton** Québec, SE Canada 46°14′N 71°05′W
44 M6 **East Caicos** island E Turks and Caicos Islands
184 R7 **East Cape** headland North Island, New Zealand 37°40′S 178°31′E
174 M4 **East Caroline Basin** undersea feature SW Pacific Ocean 04°00′N 146°45′E
192 P4 **East China Sea** Chin. Dong Hai. sea W Pacific Ocean
97 P19 **East Dereham** E England, United Kingdom 52°41′N 00°55′E
30 J9 **East Dubuque** Illinois, N USA 42°29′N 90°38′W
11 S17 **Eastend** Saskatchewan, S Canada 49°29′N 108°48′W
193 S10 **Easter Fracture Zone** tectonic feature E Pacific Ocean
Easter Island see Pascua, Isla de
153 Q12 **Eastern** ◈ zone E Nepal
155 K25 **Eastern** ◈ province E Sri Lanka
82 L13 **Eastern** ◈ province E Zambia
83 H24 **Eastern Cape** off. Eastern Cape Province, Afr. Oos-Kaap. ◈ province SE South Africa
Eastern Cape Province see Eastern Cape
80 C12 **Eastern Darfur** ◈ state SW Sudan
Eastern Desert see Sahara el Sharqîya
81 F15 **Eastern Equatoria** ◈ state SE South Sudan
Eastern Euphrates see Murat Nehri
155 J17 **Eastern Ghats** ▲▲ SE India
186 E7 **Eastern Highlands** ◈ province C Papua New Guinea
Eastern Region see Ash Sharqīyah
Eastern Sayans see Vostochnyy Sayan
Eastern Scheldt see Oosterschelde
Eastern Sierra Madre see Madre Oriental, Sierra
Eastern Transvaal see Mpumalanga
11 W14 **Easterville** Manitoba, C Canada 53°06′N 99°53′W
Easterwâlde see Oosterwolde
63 M23 **East Falkland** var. Isla Soledad. island E Falkland Islands
19 P12 **East Falmouth** Massachusetts, NE USA 41°34′N 70°31′W
East Fayu see Fayu
East Flanders see Oost-Vlaanderen
39 S6 **East Fork Chandalar River** ≈ Alaska, USA
29 U12 **East Fork Des Moines River** ≈ Iowa/Minnesota, C USA
East Frisian Islands see Ostfriesische Inseln
18 K10 **East Glenville** New York, NE USA 42°53′N 73°55′W
29 R4 **East Grand Forks** Minnesota, N USA 47°54′N 97°59′W
97 O23 **East Grinstead** SE England, United Kingdom 51°08′N 00°00′W
18 M12 **East Hartford** Connecticut, NE USA 41°45′N 72°36′W
18 M13 **East Haven** Connecticut, NE USA 41°16′N 72°52′W

◆ Country
● Country Capital
◇ Dependent Territory
○ Dependent Territory Capital
◈ Administrative Regions
✈ International Airport
▲ Mountain
▲▲ Mountain Range
Volcano
≈ River
◎ Lake
▣ Reservoir

173 T9 **East Indiaman Ridge** *undersea feature* E Indian Ocean
129 V16 **East Indies** *island group* SE Asia
East Java *see* Jawa Timur
31 Q6 **East Jordan** Michigan, N USA *45°09′N 85°07′W*
East Kalimantan *see* Kalimantan Timur
East Kazakhstan *see* Vostochnyy Kazakhstan
96 I12 **East Kilbride** S Scotland, United Kingdom *55°46′N 04°10′W*
25 R7 **Eastland** Texas, SW USA *32°23′N 98°50′W*
31 Q9 **East Lansing** Michigan, N USA *42°44′N 84°28′W*
35 X11 **East Las Vegas** Nevada, W USA *36°05′N 115°02′W*
97 M23 **Eastleigh** S England, United Kingdom *50°58′N 01°22′W*
31 V12 **East Liverpool** Ohio, N USA *40°37′N 80°34′W*
83 J25 **East London** *Afr.* Oos-Londen; *prev.* Emonti, Port Rex. Eastern Cape, S South Africa *33°S 27°54′E*
96 K12 **East Lothian** *cultural region* SE Scotland, United Kingdom
12 I10 **Eastmain** Québec, E Canada *52°11′N 78°27′W*
12 J10 **Eastmain** ≈ Québec, C Canada
15 P13 **Eastman** Québec, SE Canada *45°19′N 72°18′W*
23 U6 **Eastman** Georgia, SE USA *32°12′N 83°10′W*
175 O3 **East Mariana Basin** *undersea feature* W Pacific Ocean
30 K11 **East Moline** Illinois, N USA *41°30′N 90°26′W*
186 H7 **East New Britain** ◈ *province* E Papua New Guinea
29 T15 **East Nishnabotna River** ≈ Iowa, C USA
197 V12 **East Novaya Zemlya Trough** *var.* Novaya Zemlya Trough. *undersea feature* W Kara Sea
East Nusa Tenggara *see* Nusa Tenggara Timur
21 X4 **Easton** Maryland, NE USA *38°46′N 76°04′W*
18 I14 **Easton** Pennsylvania, NE USA *40°41′N 75°13′W*
193 R16 **East Pacific Rise** *undersea feature* E Pacific Ocean *20°00′S 115°00′W*
East Pakistan *see* Bangladesh
31 V12 **East Palestine** Ohio, N USA *40°49′N 80°32′W*
30 L12 **East Peoria** Illinois, N USA *40°40′N 89°34′W*
23 S3 **East Point** Georgia, SE USA *33°40′N 84°26′W*
19 U6 **Eastport** Maine, NE USA *44°54′N 66°59′W*
27 Z8 **East Prairie** Missouri, C USA *36°46′N 89°23′W*
19 O12 **East Providence** Rhode Island, NE USA *41°48′N 71°20′W*
20 L11 **East Ridge** Tennessee, S USA *35°00′N 85°15′W*
97 N16 **East Riding** *cultural region* N England, United Kingdom
18 F9 **East Rochester** New York, NE USA *43°06′N 77°29′W*
30 K15 **East Saint Louis** Illinois, N USA *38°35′N 90°07′W*
65 K21 **East Scotia Basin** *undersea feature* SE Scotia Sea
129 Y8 **East Sea** *var.* Sea of Japan, *Rus.* Yapanskoye More. Sea NW Pacific Ocean *see also* Japan, Sea of
186 B6 **East Sepik** ◈ *province* NW Papua New Guinea
173 N4 **East Sheba Ridge** *undersea feature* W Arabian Sea *14°30′N 56°15′E*
East Siberian Sea *see* Vostochno-Sibirskoye More
18 I14 **East Stroudsburg** Pennsylvania, NE USA *41°00′N 75°10′W*
East Tasmanian Rise/East Tasmania Plateau/East Tasmania Rise *see* East Tasman Plateau
192 I12 **East Tasman Plateau** *var.* East Tasmanian Rise, East Tasmania Plateau, East Tasmania Rise. *undersea feature* SW Tasman Sea
64 L7 **East Thulean Rise** *undersea feature* N Atlantic Ocean
171 R16 **East Timor** *var.* Loro Sae; *prev.* Portuguese Timor, Timor Timur. ◆ *country* S Indonesia
21 Y6 **Eastville** Virginia, NE USA *37°22′N 75°58′W*
35 R7 **East Walker River** ≈ California/Nevada, W USA
182 D1 **Eateringinna Creek** ≈ South Australia
37 T3 **Eaton** Colorado, C USA *40°31′N 104°42′W*
15 Q12 **Eaton** ≈ Québec, SE Canada
11 S16 **Eatonia** Saskatchewan, S Canada *51°13′N 109°22′W*
31 Q10 **Eaton Rapids** Michigan, N USA *42°30′N 84°39′W*
23 U4 **Eatonton** Georgia, SE USA *33°19′N 83°23′W*
32 H9 **Eatonville** Washington, NW USA *46°51′N 122°19′W*
30 J6 **Eau Claire** Wisconsin, N USA *44°50′N 91°30′W*
12 J7 **Eau Claire, Lac à l'** ◎ Québec, SE Canada
Eau Claire, Lac à L' *see* St. Clair, Lake
30 L6 **Eau Claire River** ≈ Wisconsin, N USA
188 J16 **Eauripik Atoll** *atoll* Caroline Islands, C Micronesia
192 H7 **Eauripik Rise** *undersea feature* W Pacific Ocean *03°00′N 142°00′E*
102 K15 **Eauze** Gers, S France *43°52′N 00°06′E*
41 P11 **Ébano** San Luis Potosí, C Mexico *22°16′N 98°26′W*
97 K21 **Ebbw Vale** SE Wales, United Kingdom *51°48′N 03°13′W*
79 E17 **Ebebiyin** NE Equatorial Guinea *02°08′N 11°15′E*
95 H22 **Ebeltoft** Midtjylland, C Denmark *56°11′N 10°42′E*
109 X5 **Ebenfurth** Niederösterreich, E Austria *47°53′N 16°22′E*
18 D14 **Ebensburg** Pennsylvania, NE USA *40°28′N 78°44′W*
109 S5 **Ebensee** Oberösterreich, N Austria *47°48′N 13°46′E*
101 H20 **Eberbach** Baden-Württemberg, SW Germany *49°28′N 08°58′E*
121 U8 **Eber Gölü** *salt lake* C Turkey
109 U9 **Eberndorf** *Slvn.* Dobrla Vas. Kärnten, S Austria *46°33′N 14°35′E*
109 R4 **Eberschwang** Oberösterreich, N Austria *48°09′N 13°37′E*
100 O11 **Eberswalde-Finow** Brandenburg, E Germany *52°50′N 13°48′E*
Ebesfalva *see* Dumbrăveni
165 T4 **Ebetsu** *var.* Ebetu. Hokkaidō, NE Japan *43°08′N 141°37′E*
Ebetu *see* Ebetsu
Ebinayon *see* Evinayong
158 I4 **Ebinur Hu** ◎ NW China
138 I3 **Ebla** *Ar.* Tell Mardīkh. *site of ancient city* Idlib, NW Syria
Eblana *see* Dublin
108 H7 **Ebnat** Sankt Gallen, NE Switzerland *47°16′N 09°07′E*
107 L18 **Eboli** Campania, S Italy *40°37′N 15°03′E*
79 E16 **Ebolowa** Sud, S Cameroon *02°56′N 11°11′E*
79 N21 **Ebombo** Kasai-Oriental, C Dem. Rep. Congo *05°42′S 26°07′E*
189 T9 **Ebon Atoll** *var.* Epoon. *atoll* Ralik Chain, S Marshall Islands
Ebora *see* Évora
Eboracum *see* York
Eborodunum *see* Yverdon
101 J19 **Ebrach** Bayern, C Germany *49°49′N 10°30′E*
109 X5 **Ebreichsdorf** Niederösterreich, E Austria *47°58′N 16°24′E*
105 S6 **Ebro** ≈ NE Spain
105 N3 **Ebro, Embalse del** ⊠ N Spain
120 G7 **Ebro Fan** *undersea feature* W Mediterranean Sea
Eburacum *see* York
Ebusus *see* Ibiza
Ebusus *see* Eivissa
99 F20 **Écaussinnes-d'Enghien** Hainaut, SW Belgium *50°34′N 04°10′E*
Ecbatana *see* Hamadān
21 Q6 **Eccles** West Virginia, NE USA *37°46′N 81°16′W*
115 L14 **Eceabat** Çanakkale, NW Turkey *40°12′N 26°22′E*
171 O2 **Echague** Luzon, N Philippines *16°42′N 121°37′E*
Ech Cheliff/Ech Chleff *see* Chlef
Echeng *see* Ezhou
115 C18 **Echinádes** *island group* W Greece
114 J12 **Echínos** *var.* Ehinos, Ekhínos. Anatolikí Makedonía kai Thráki, NE Greece *41°16′N 25°00′E*
Echizen *see* Takefu
164 J12 **Echizen-misaki** *headland* Honshū, SW Japan *35°59′N 135°57′E*
Echmiadzin *see* Vagharshapat
8 J8 **Echo Bay** Northwest Territories, NW Canada *66°04′N 118°W*
35 Y11 **Echo Bay** Nevada, W USA *36°19′N 114°27′W*
36 L9 **Echo Cliffs** *cliff* Arizona, SW USA
14 C10 **Echo Lake** ◎ Ontario, S Canada
35 Q7 **Echo Summit** ▲ California, W USA *38°47′N 120°06′W*
14 L8 **Échouani, Lac** ◎ Québec, SE Canada
99 L17 **Echt** Limburg, SE Netherlands *51°07′N 05°52′E*
101 H22 **Echterdingen** ✈ (Stuttgart) Baden-Württemberg, SW Germany *48°40′N 09°13′E*
99 N24 **Echternach** Grevenmacher, E Luxembourg *49°49′N 06°25′E*
183 N11 **Echuca** Victoria, SE Australia *36°10′S 144°20′E*
104 L14 **Ecija** *anc.* Astigi. Andalucía, SW Spain *37°33′N 05°04′W*
Eckengraf *see* Viesīte
100 I7 **Eckernförde** Schleswig-Holstein, N Germany *54°28′N 09°49′E*
100 J7 **Eckernförder Bucht** *inlet* N Germany
102 L7 **Écommoy** Sarthe, NW France *47°51′N 00°15′E*
14 L10 **Écorce, Lac de l'** ◎ Québec, SE Canada
15 Q8 **Écorces, Rivière aux** ≈ Québec, SE Canada
56 C7 **Ecuador** *off.* Republic of Ecuador. ◆ *republic* NW South America
Ecuador, Republic of *see* Ecuador
95 I17 **Ed** Västra Götaland, S Sweden *58°55′N 11°55′E*
Ed *see* 'Īdi
98 I9 **Edam** Noord-Holland, C Netherlands *52°30′N 05°02′E*
96 K4 **Eday** *island* NE Scotland, United Kingdom
25 S17 **Edcouch** Texas, SW USA *26°17′N 97°57′W*
80 C11 **Ed Da'ein** Eastern Darfur, W Sudan *11°25′N 26°08′E*
80 G11 **Ed Damazin** *var.* Ad Damazīn. Blue Nile, E Sudan *11°45′N 34°20′E*
80 G8 **Ed Damer** *var.* Ad Dāmir, Ad Damar. River Nile, NE Sudan *17°37′N 33°59′E*
80 E8 **Ed Debba** Northern, N Sudan *18°02′N 30°56′E*
80 F10 **Ed Dueim** *var.* Ad Duwaym, Ad Duwēm. White Nile, C Sudan *13°38′N 32°36′E*
183 Q16 **Eddystone Point** *headland* Tasmania, SE Australia *41°01′S 148°18′E*
97 I25 **Eddystone Rocks** *rocks* SW England, United Kingdom
29 W15 **Eddyville** Iowa, C USA *41°09′N 92°37′W*
20 H7 **Eddyville** Kentucky, S USA *37°03′N 88°02′W*
98 L12 **Ede** Gelderland, C Netherlands *52°03′N 05°40′E*
77 T16 **Ede** Osun, SW Nigeria *07°40′N 04°21′E*
79 D16 **Edéa** Littoral, SW Cameroon *03°47′N 10°08′E*
111 M20 **Edelény** Borsod-Abaúj-Zemplén, NE Hungary *48°18′N 20°40′E*
183 R12 **Eden** New South Wales, SE Australia *37°04′S 149°51′E*
21 T8 **Eden** North Carolina, SE USA *36°29′N 79°46′W*
25 P9 **Eden** Texas, SW USA *31°13′N 99°51′W*
97 K14 **Eden** ≈ NW England, United Kingdom
83 I23 **Edenburg** Free State, C South Africa *29°45′S 25°57′E*
185 D24 **Edendale** Southland, South Island, New Zealand *46°18′S 168°48′E*
97 E18 **Edenderry** *Ir.* Éadan Doire. Offaly, C Ireland *53°21′N 07°03′W*
182 L11 **Edenhope** Victoria, SE Australia *37°04′S 141°15′E*
21 X8 **Edenton** North Carolina, SE USA *36°04′N 76°39′W*
101 G16 **Eder** ≈ NW Germany
101 H15 **Edersee** ⊠ W Germany
114 E13 **Édessa** *var.* Edhessa. Kentrikí Makedonía, N Greece *40°48′N 22°03′E*
Edessa *see* Şanlıurfa
Edfu *see* Idfū
29 P16 **Edgar** Nebraska, C USA *40°22′N 97°58′W*
19 P13 **Edgartown** Martha's Vineyard, Massachusetts, NE USA *41°23′N 70°30′W*
39 X13 **Edgecumbe, Mount** ▲ Baranof Island, Alaska, USA *57°03′N 135°45′W*
21 Q13 **Edgefield** South Carolina, SE USA *33°50′N 81°57′W*
29 P6 **Edgeley** North Dakota, N USA *46°19′N 98°42′W*
28 I11 **Edgemont** South Dakota, N USA *43°18′N 103°49′W*
92 O4 **Edgeøya** *island* S Svalbard
27 Q4 **Edgerton** Kansas, C USA *38°45′N 95°00′W*
29 S10 **Edgerton** Minnesota, N USA *43°52′N 96°07′W*
21 X3 **Edgewood** Maryland, NE USA *39°20′N 76°21′W*
25 V6 **Edgewood** Texas, SW USA *32°42′N 95°53′W*
Édhessa *see* Édessa
29 V9 **Edina** Minnesota, N USA *44°53′N 93°21′W*
27 U2 **Edina** Missouri, C USA *40°10′N 92°10′W*
25 S17 **Edinburg** Texas, SW USA *26°18′N 98°10′W*
65 M24 **Edinburgh** *var.* Settlement of Edinburgh. ○ (Tristan da Cunha) NW Tristan da Cunha *37°03′S 12°18′W*
96 J12 **Edinburgh** ● S Scotland, United Kingdom *55°57′N 03°13′W*
31 P14 **Edinburgh** Indiana, N USA *39°19′N 86°00′W*
96 J12 **Edinburgh** ✈ S Scotland, United Kingdom *55°57′N 03°22′W*
116 L8 **Edineţ** *var.* Edineţi, *Rus.* Yedintsy. NW Moldova *48°10′N 27°18′E*
Edineţi *see* Edineţ
Edingen *see* Enghien
136 B9 **Edirne** *Eng.* Adrianople; *anc.* Adrianopolis, Hadrianopolis. Edirne, NW Turkey *41°40′N 26°34′E*
136 B11 **Edirne** ◈ *province* NW Turkey
18 K15 **Edison** New Jersey, NE USA *40°31′N 74°24′W*
21 S15 **Edisto Island** South Carolina, SE USA *32°34′N 80°17′W*
21 R14 **Edisto River** ≈ South Carolina, SE USA
33 S10 **Edith, Mount** ▲ Montana, NW USA *46°25′N 111°10′W*
27 N10 **Edmond** Oklahoma, C USA *35°40′N 97°30′W*
32 H8 **Edmonds** Washington, NW USA *47°48′N 122°22′W*
11 Q14 **Edmonton** *province capital* Alberta, SW Canada *53°34′N 113°25′W*
20 K7 **Edmonton** Kentucky, S USA *36°59′N 85°39′W*
11 Q14 **Edmonton** ✈ Alberta, SW Canada *53°22′N 113°43′W*
29 P3 **Edmore** North Dakota, N USA *48°22′N 98°26′W*
13 N13 **Edmundston** New Brunswick, SE Canada *47°22′N 68°20′W*
25 U12 **Edna** Texas, SW USA *29°00′N 96°41′W*
39 X14 **Edna Bay** Kosciusko Island, Alaska, USA *55°54′N 133°40′W*
77 U16 **Edo** ◈ *state* S Nigeria
106 F6 **Edolo** Lombardia, N Italy *46°13′N 10°20′E*
64 L6 **Edoras Bank** *undersea feature* C Atlantic Ocean
96 G7 **Edrachillis Bay** *bay* NW Scotland, United Kingdom
136 B12 **Edremit** Balıkesir, NW Turkey *39°34′N 27°01′E*
136 B12 **Edremit Körfezi** *gulf* NW Turkey
95 P14 **Edsbro** Stockholm, C Sweden *59°54′N 18°30′E*
95 N18 **Edsbruk** Kalmar, S Sweden *58°01′N 16°30′E*
94 M12 **Edsbyn** Gävleborg, C Sweden *61°22′N 15°45′E*
11 O14 **Edson** Alberta, SW Canada *53°36′N 116°28′W*
62 K13 **Eduardo Castex** La Pampa, C Argentina *35°55′S 64°18′W*
58 F12 **Eduardo Gomes** ✈ (Manaus) Amazonas, NW Brazil *03°55′S 35°15′W*
Edwardesabad *see* Bannu
67 U9 **Edward, Lake** *var.* Albert Edward Nyanza, Edward Nyanza, Lac Idi Amin, Lake Rutanzige. ◎ Uganda/Dem. Rep. Congo
Edward Nyanza *see* Edward, Lake
22 K5 **Edwards** Mississippi, S USA *32°19′N 90°36′W*
25 O10 **Edwards Plateau** *plain* Texas, SW USA
30 J11 **Edwards River** ≈ Illinois, N USA
30 K15 **Edwardsville** Illinois, N USA *38°48′N 89°57′W*
195 X4 **Edward VIII Gulf** *bay* Antarctica
195 O13 **Edward VII Peninsula** *peninsula* Antarctica
10 J11 **Edziza, Mount** ▲ British Columbia, W Canada *57°43′N 130°39′W*
23 H16 **Edzo** *prev.* Rae-Edzo. Northwest Territories, NW Canada *62°44′N 115°55′W*
39 N12 **Eek** Alaska, USA *60°13′N 162°01′W*
99 D16 **Eeklo** *var.* Eekloo. Oost-Vlaanderen, NW Belgium *51°11′N 03°34′E*
Eekloo *see* Eeklo
39 N12 **Eek River** ≈ Alaska, USA
98 N6 **Eelde** Drenthe, NE Netherlands *53°07′N 06°30′E*
34 L5 **Eel River** ≈ California, W USA
31 P12 **Eel River** ≈ Indiana, N USA
Eems *see* Ems
98 O4 **Eemshaven** Groningen, NE Netherlands *53°28′N 06°50′E*
98 O5 **Eems Kanaal** *canal* NE Netherlands
98 M11 **Eerbeek** Gelderland, E Netherlands *52°07′N 06°04′E*
99 C17 **Eernegem** West-Vlaanderen, W Belgium *51°08′N 03°03′E*
99 J15 **Eersel** Noord-Brabant, S Netherlands *51°22′N 05°19′E*
Eesti Vabariik *see* Estonia
187 R14 **Efate** *var.* Éfaté, *Fr.* Vaté; *prev.* Sandwich Island. *island* C Vanuatu
Efate *see* Éfaté
109 S4 **Eferding** Oberösterreich, N Austria *48°18′N 14°00′E*
30 M15 **Effingham** Illinois, N USA *39°07′N 88°32′W*
117 N15 **Eforie-Nord** Constanţa, SE Romania *44°04′N 28°37′E*
117 N15 **Eforie-Sud** Constanţa, E Romania *44°00′N 28°38′E*
Efyrnwy, Afon *see* Vyrnwy
Eg *see* Hentiy
107 G23 **Egadi, Isole** *island group* S Italy
35 X6 **Egan Range** ▲ Nevada, W USA
14 K12 **Eganville** Ontario, SE Canada *45°33′N 77°03′W*
Ege Denizi *see* Aegean Sea
39 O14 **Egegik** Alaska, USA *58°13′N 157°22′W*
Egentliga Finland *see* Varsinais-Suomi
Egentliga Tavastland *see* Kanta-Häme
111 L21 **Eger** *Ger.* Erlau. Heves, NE Hungary *47°54′N 20°22′E*
Eger *see* Cheb, Czech Republic
Eger *see* Ohře, Czech Republic/Germany
173 P8 **Egeria Fracture Zone** *tectonic feature* W Indian Ocean
95 C17 **Egersund** Rogaland, S Norway *58°27′N 06°01′E*
108 J7 **Egg** Vorarlberg, NW Austria *47°27′N 09°55′E*
101 H14 **Egge-gebirge** ▲ C Germany
109 Q4 **Eggelsberg** Oberösterreich, N Austria *48°04′N 13°00′E*
109 W2 **Eggenburg** Niederösterreich, NE Austria *48°29′N 15°49′E*
101 N22 **Eggenfelden** Bayern, SE Germany *48°24′N 12°45′E*
18 J17 **Egg Harbor City** New Jersey, NE USA *39°31′N 74°39′W*
65 G25 **Egg Island** *island* W Saint Helena
183 N14 **Egg Lagoon** Tasmania, SE Australia *39°42′S 143°57′E*
99 I20 **Éghezèe** Namur, C Belgium *50°36′N 04°55′E*
92 L2 **Egilsstaðir** Austurland, E Iceland *65°14′N 14°21′W*
Egina *see* Aígina
Egindibulaq *see* Yegindybulak
Egio *see* Aígio
103 N12 **Égletons** Corrèze, C France *45°24′N 02°01′E*
98 H9 **Egmond aan Zee** Noord-Holland, NW Netherlands *52°37′N 04°37′E*
Egmont *see* Taranaki, Mount
184 J10 **Egmont, Cape** *headland* North Island, New Zealand *39°18′S 173°44′E*
Egoli *see* Johannesburg
136 I15 **Eğridir Gölü** ◎ W Turkey
Eğri Palanka *see* Kriva Palanka
95 G23 **Egtved** Syddanmark, C Denmark *55°34′N 09°18′E*
123 U5 **Egvekinot** Chukotskiy Avtonomnyy Okrug, NE Russian Federation *66°13′N 178°55′W*
75 V9 **Egypt** *off.* Arab Republic of Egypt, *Ar.* Jumhūrīyah Miṣr al 'Arabīyah, *prev.* United Arab Republic; *anc.* Aegyptus. ◆ *republic* NE Africa
30 L17 **Egypt, Lake Of** ⊠ Illinois, N USA
162 I14 **Ehen Hudag** *var.* Alx Youqi. Nei Mongol Zizhiqu, N China *39°12′N 101°40′E*
164 F14 **Ehime** *off.* Ehime-ken. ◈ *prefecture* Shikoku, SW Japan
Ehime-ken *see* Ehime
101 I23 **Ehingen** Baden-Württemberg, S Germany *48°16′N 09°43′E*
Ehinos *see* Echínos
21 R14 **Ehrhardt** South Carolina, SE USA *33°06′N 81°00′W*
108 L7 **Ehrwald** Tirol, W Austria *47°24′N 10°52′E*
191 W6 **Eiao** *island* Îles Marquises, NE French Polynesia
105 P2 **Eibar** País Vasco, N Spain *43°11′N 02°28′W*
98 O11 **Eibergen** Gelderland, E Netherlands *52°06′N 06°39′E*
109 V9 **Eibiswald** Steiermark, SE Austria *46°40′N 15°15′E*
109 P8 **Eichham** ▲ SW Austria *47°04′N 12°24′E*
101 J15 **Eichsfeld** *hill range* C Germany
101 K21 **Eichstätt** Bayern, SE Germany *48°53′N 11°11′E*
100 H8 **Eider** ≈ N Germany
94 E13 **Eidfjord** Hordaland, S Norway *60°26′N 07°05′E*
94 D13 **Eidfjorden** *fjord* S Norway
94 F9 **Eidsvåg** Møre og Romsdal, S Norway *62°46′N 08°00′E*
95 I14 **Eidsvoll** Akershus, S Norway *60°19′N 11°14′E*
92 N2 **Eidsvollfjellet** ▲ NW Svalbard *79°13′N 13°23′E*
Eier-Berg *see* Suur Munamägi
101 D18 **Eifel** *plateau* W Germany
108 E9 **Eiger** ▲ C Switzerland *46°33′N 08°02′E*
96 G10 **Eigg** *island* W Scotland, United Kingdom
155 D24 **Eight Degree Channel** *channel* India/Maldives
44 G1 **Eight Mile Rock** Grand Bahama Island, N The Bahamas *26°28′N 78°43′W*
194 J9 **Eights Coast** *physical region* Antarctica
180 K6 **Eighty Mile Beach** *beach* Western Australia
99 L18 **Eijsden** Limburg, SE Netherlands *50°47′N 05°41′E*
95 G15 **Eikeren** ◎ S Norway
Eil *see* Eyl
Eilat *see* Elat
183 O12 **Eildon** Victoria, SE Australia *37°17′S 145°57′E*
183 O12 **Eildon, Lake** ◎ Victoria, SE Australia
80 E8 **Eilei** Northern Kordofan, C Sudan *16°33′N 30°54′E*
101 N15 **Eilenburg** Sachsen, E Germany *51°28′N 12°37′E*
Eil Malk *see* Mecherchar
94 H13 **Eina** Oppland, S Norway *60°37′N 10°37′E*
138 E12 **Ein Avdat** *prev.* En 'Avedat. *well* S Israel
101 I14 **Einbeck** Niedersachsen, C Germany *51°49′N 09°52′E*
99 K15 **Eindhoven** Noord-Brabant, S Netherlands *51°26′N 05°30′E*
108 G8 **Einsiedeln** Schwyz, NE Switzerland *47°07′N 08°45′E*
Eipel *see* Ipel'
Éire *see* Ireland
Éireann, Muir *see* Irish Sea
Eirik Outer Ridge *see* Eirik Ridge
64 I6 **Eirik Ridge** *var.* Eirik Outer Ridge. *undersea feature* E Labrador Sea
92 I3 **Eiríksjökull** ▲ C Iceland *64°47′N 20°23′W*
59 B14 **Eirunepé** Amazonas, N Brazil *06°38′S 69°53′W*
99 L17 **Eisden** Limburg, NE Belgium *51°05′N 05°42′E*
83 F18 **Eiseb** ≈ Botswana/Namibia
Eisen *see* Yeongcheon
101 J16 **Eisenach** Thüringen, C Germany *50°59′N 10°19′E*
Eisenburg *see* Vasvár
109 U6 **Eisenerz** Steiermark, SE Austria *47°33′N 14°53′E*
100 Q13 **Eisenhüttenstadt** Brandenburg, E Germany *52°09′N 14°36′E*
109 U10 **Eisenkappel** *Slvn.* Železna Kapela. Kärnten, S Austria *46°27′N 14°33′E*
Eisenmarkt *see* Hunedoara
109 Y5 **Eisenstadt** Burgenland, E Austria *47°50′N 16°32′E*
Eishū *see* Yeongju
119 H15 **Eišiškes** Vilnius, SE Lithuania *54°10′N 24°59′E*
101 L15 **Eisleben** Sachsen-Anhalt, C Germany *51°32′N 11°33′E*
190 I3 **Eita** Tarawa, W Kiribati *01°21′N 173°05′E*
Eitape *see* Aitape
105 V11 **Eivissa** *var.* Iviza, *Cast.* Ibiza; *anc.* Ebusus. Ibiza, Spain, W Mediterranean Sea *38°54′N 01°26′E*
Eivissa *see* Ibiza
105 R4 **Ejea de los Caballeros** Aragón, NE Spain *42°07′N 01°09′W*
40 E8 **Ejido Insurgentes** Baja California Sur, NW Mexico *25°18′N 111°51′W*
Ejin Qi *see* Dalain Hob
Ejmiadzin/Ejmiatsin *see* Vagharshapat
77 P16 **Ejura** C Ghana *07°23′N 01°22′W*
41 R16 **Ejutla** *var.* Ejutla de Crespo. Oaxaca, SE Mexico *16°33′N 96°40′W*
Ejutla de Crespo *see* Ejutla
33 Y10 **Ekalaka** Montana, NW USA *45°52′N 104°32′W*
Ekapa *see* Cape Town
Ekaterinodar *see* Krasnodar
93 L20 **Ekenäs** *Fin.* Tammisaari. Uusimaa, SW Finland *60°00′N 23°30′E*
146 B13 **Ekerem** *Rus.* Okarem. Balkan Welaýaty, W Turkmenistan *38°06′N 53°52′E*
184 M13 **Eketahuna** Manawatu-Wanganui, North Island, New Zealand *40°41′S 175°40′E*
Ekhínos *see* Echínos
Ekiatapskiy Khrebet *see* Ekvyvatapskiy Khrebet
Ekibastuz *see* Yekibastuz
123 R13 **Ekimchan** Amurskaya Oblast', SE Russian Federation *53°04′N 132°56′E*
95 O15 **Ekoln** ◎ C Sweden
80 I7 **Ekowit** Red Sea, NE Sudan *18°46′N 37°07′E*
95 L19 **Eksjö** Jönköping, S Sweden *57°40′N 15°00′E*
93 I15 **Ekträsk** Västerbotten, N Sweden *64°28′N 19°49′E*
39 O13 **Ekuk** Alaska, USA *58°48′N 158°25′W*
123 U5 **Ekvyvatapskiy Khrebet** *prev.* Ekiatapskiy Khrebet. ▲ NE Russian Federation
12 F9 **Ekwan** ≈ Ontario, C Canada
39 O13 **Ekwok** Alaska, USA *59°21′N 157°28′W*
166 M6 **Ela** Mandalay, C Myanmar (Burma) *19°37′N 96°15′E*
El Aaiun *see* El Ayoun
81 N15 **El Ābrēd** Sumalē, E Ethiopia *05°33′N 45°12′E*
115 F22 **Elafónisos** *island* S Greece
115 F22 **Elafonísou, Porthmós** *strait* S Greece
El-Aïoun *see* El Ayoun
El 'Alamein *see* Al 'Alamayn
41 Q12 **El Alazán** Veracruz-Llave, C Mexico *21°06′N 97°43′W*
57 J18 **El Alto** *var.* La Paz. ✈ (La Paz) La Paz, W Bolivia *16°31′S 68°07′W*
Elam *see* Īlām
El Amparo *see* El Amparo de Apure
54 I8 **El Amparo de Apure** *var.* El Amparo. Apure, C Venezuela *07°07′N 70°47′W*
171 R13 **Elara** Pulau Ambelau, E Indonesia *03°49′S 127°10′E*
El Araïch/El Araïche *see* Larache
40 D6 **El Arco** Baja California Norte, NW Mexico *28°03′N 113°25′W*
El 'Arîsh *see* Al 'Arīsh
115 L25 **Elása** *island* SE Greece
El Asnam *see* Chlef
Elassón *see* Elassóna
115 E15 **Elassóna** *prev.* Elassón. Thessalía, C Greece *39°53′N 22°10′E*
105 N2 **El Astillero** Cantabria, N Spain *43°23′N 03°45′W*
138 F14 **Elat** *var.* Eilat, Elath. Southern, S Israel *29°33′N 34°57′E*
Elat, Gulf of *see* Aqaba, Gulf of
Elath *see* Elat, Israel
Elath *see* Al 'Aqabah, Jordan
115 C17 **Eláti** ▲ Lefkáda, Iónia Nisiá, Greece, C Mediterranean Sea *38°43′N 20°38′E*
188 L16 **Elato Atoll** *atoll* Caroline Islands, C Micronesia
80 C7 **El'Atrun** Northern Darfur, NW Sudan *18°11′N 26°40′E*
74 H6 **El Ayoun** *var.* El Aaiun, El-Aïoun, La Youne. NE Morocco *34°39′N 02°29′W*
137 N14 **Elâzığ** *var.* Elâzig, Elâziz. Elâzığ, E Turkey *38°41′N 39°14′E*
137 O14 **Elâzığ** *var.* Elâzig, Elâziz. ◈ *province* C Turkey
Elâzig/Elâziz *see* Elâzığ
23 Q7 **Elba** Alabama, S USA *31°24′N 86°04′W*
106 E13 **Elba, Isola d'** *island* Archipelago Toscano, C Italy
123 S13 **El'ban** Khabarovskiy Kray, E Russian Federation *50°03′N 136°34′E*
54 F6 **El Banco** Magdalena, N Colombia *09°00′N 74°01′W*
El Barco *see* O Barco
104 L8 **El Barco de Ávila** Castilla y León, N Spain *40°21′N 05°31′W*
El Barco de Valdeorras *see* O Barco
138 H7 **El Barouk, Jabal** ▲ C Lebanon
113 L20 **Elbasan** *var.* Elbasani. Elbasan, C Albania *41°07′N 20°04′E*
113 L20 **Elbasan** ◈ *district* C Albania
Elbasani *see* Elbasan
54 K6 **El Baúl** Cojedes, C Venezuela *08°59′N 68°16′W*
86 D11 **Elbe** *Cz.* Labe. ≈ Czech Republic/Germany
100 L13 **Elbe-Havel-Kanal** *canal* E Germany
100 K9 **Elbe-Lübeck-Kanal** *canal* N Germany
57 L15 **El Beni** *var.* El Beni. ◈ *department* N Bolivia
El Beni *see* El Beni
138 H7 **El Beqaa** *var.* Al Biqā', Bekaa Valley. *valley* E Lebanon
37 R5 **Elbert, Mount** ▲ Colorado, C USA *39°07′N 106°26′W*
23 U3 **Elberton** Georgia, SE USA *34°06′N 82°52′W*
100 K11 **Elbe-Seiten-Kanal** *canal* N Germany
102 M4 **Elbeuf** Seine-Maritime, N France *49°16′N 01°01′E*
Elbing *see* Elbląg
136 M15 **Elbistan** Kahramanmaraş, S Turkey *38°14′N 37°11′E*
110 K7 **Elbląg** *var.* Elblag, *Ger.* Elbing. Warmińsko-Mazurskie, NE Poland *54°10′N 19°25′E*
43 N10 **El Bluff** Región Autónoma Atlántico Sur, SE Nicaragua *12°00′N 83°40′W*
63 H17 **El Bolsón** Río Negro, W Argentina *41°59′S 71°35′W*
105 P11 **El Bonillo** Castilla-La Mancha, C Spain *38°57′N 02°32′W*
El Bordo *see* Patía
El Boulaida/El Boulaïda *see* Blida
11 T16 **Elbow** Saskatchewan, S Canada *51°07′N 106°30′W*
29 S7 **Elbow Lake** Minnesota, N USA *45°59′N 95°58′W*
127 N16 **El'brus** *var.* Gora El'brus. ▲ SW Russian Federation *42°29′E 43°21′N*
El'brus, Gora *see* El'brus
126 M15 **El'brusskiy** Karachayevo-Cherkesskaya Respublika, SW Russian Federation *43°36′N 42°06′E*
El Buhayrat *see* Lakes
El Bur *see* Ceel Buur
98 L10 **Elburg** Gelderland, E Netherlands *52°27′N 05°46′E*
105 O6 **El Burgo de Osma** Castilla y León, C Spain *41°36′N 03°04′W*
Elburz Mountains *see* Alborz, Reshteh-ye Kūhhā-ye
35 V17 **El Cajon** California, W USA *32°46′N 116°52′W*
63 H22 **El Calafate** *var.* Calafate. Santa Cruz, S Argentina *50°20′S 72°13′W*
55 Q8 **El Callao** Bolívar, E Venezuela *07°18′N 61°48′W*
25 U12 **El Campo** Texas, SW USA *29°12′N 96°16′W*
54 I7 **El Cantón** Barinas, W Venezuela *07°23′N 71°10′W*
35 Q8 **El Capitan** ▲ California, W USA *37°46′N 119°39′W*
54 H5 **El Carmelo** Zulia, NW Venezuela *10°20′N 71°48′W*
62 J5 **El Carmen** Jujuy, NW Argentina *24°24′S 65°16′W*
54 E5 **El Carmen de Bolívar** Bolívar, NW Colombia *09°43′N 75°07′W*
55 O8 **El Casabe** Bolívar, SE Venezuela *06°26′N 63°35′W*
42 M12 **El Castillo de La Concepción** Río San Juan, SE Nicaragua *11°01′N 84°24′W*
El Cayo *see* San Ignacio
35 X17 **El Centro** California, W USA *32°47′N 115°33′W*
55 N6 **El Chaparro** Anzoátegui, NE Venezuela *09°12′N 65°03′W*
105 S12 **Elche** *Cat.* Elx; *anc.* Ilici, *Lat.* Illicis. Valenciana, E Spain *38°16′N 00°41′W*
105 Q12 **Elche de la Sierra** Castilla-La Mancha, C Spain *38°27′N 02°03′W*
41 U15 **El Chichonal, Volcán** ⋀ SE Mexico *17°20′N 93°12′W*
40 C2 **El Chinero** Baja California Norte, NW Mexico
181 R1 **Elcho Island** *island* Wessel Islands, Northern Territory, N Australia
63 H18 **El Corcovado** Chubut, SW Argentina *43°31′S 71°30′W*
105 R12 **Elda** Valenciana, E Spain *38°29′N 00°47′W*
100 M10 **Elde** ≈ NE Germany
98 L12 **Elden** Gelderland, E Netherlands *51°57′N 05°53′E*
81 J16 **El Der** *spring/well* S Ethiopia *03°55′N 39°48′E*
El Dere *see* Ceel Dheere
40 B2 **El Descanso** Baja California Norte, NW Mexico *32°08′N 116°51′W*
40 E3 **El Desemboque** Sonora, NW Mexico *30°33′N 112°59′W*
54 F5 **El Difícil** *var.* Ariguaní. Magdalena, N Colombia *09°55′N 74°12′W*
123 R10 **El'dikan** Respublika Sakha (Yakutiya), NE Russian Federation *60°46′N 135°04′E*
El Djazaïr *see* Alger
El Djelfa *see* Djelfa
29 X15 **Eldon** Iowa, C USA *40°55′N 92°13′W*
27 U5 **Eldon** Missouri, C USA *38°21′N 92°34′W*
54 E13 **El Doncello** Caquetá, S Colombia *01°43′N 75°17′W*
29 W13 **Eldora** Iowa, C USA *42°21′N 93°06′W*
60 G12 **Eldorado** Misiones, NE Argentina *26°24′S 54°38′W*
40 I9 **El Dorado** Sinaloa, C Mexico *24°19′N 107°23′W*
27 U14 **El Dorado** Arkansas, C USA *33°12′N 92°40′W*
30 M17 **Eldorado** Illinois, N USA *37°48′N 88°26′W*
27 O6 **El Dorado** Kansas, C USA *37°51′N 96°52′W*
26 K12 **Eldorado** Oklahoma, C USA *34°28′N 99°39′W*
25 O9 **Eldorado** Texas, SW USA *30°53′N 100°37′W*
55 Q8 **El Dorado** Bolívar, E Venezuela *06°45′N 61°37′W*
54 F10 **El Dorado** ✈ (Bogotá) Cundinamarca, C Colombia *01°15′N 71°52′W*
El Dorado *see* California
27 O6 **El Dorado Lake** ⊠ Kansas, C USA
27 S6 **El Dorado Springs** Missouri, C USA *37°53′N 94°01′W*
81 H18 **Eldoret** Uasin Gishu, W Kenya *0°31′N 35°17′E*
29 Z14 **Eldridge** Iowa, C USA *41°39′N 90°34′W*
95 J21 **Eldsberga** Halland, S Sweden *56°36′N 13°00′E*
25 R4 **Electra** Texas, SW USA *34°01′N 98°55′W*
37 Q7 **Electra Lake** ⊠ Colorado, C USA
38 B8 **'Ele'ele** *var.* Eleele. Kaua'i, Hawaii, USA, C Pacific Ocean *21°54′N 159°35′W*
Eleele *see* 'Ele'ele
Elefantes *see* Lepelle
115 H19 **Elefsína** *prev.* Elevsís. Attikí, C Greece *38°02′N 23°33′E*
115 G19 **Eléftheres** *anc.* Eleutherae. *site of ancient city* Attikí/Stereá Elláda, C Greece
114 I13 **Eleftheroúpoli** *prev.* Elevtheroúpolis. Anatolikí Makedonía kai Thráki, NE Greece *40°55′N 24°15′E*
74 F10 **El Eglab** ▲ SW Algeria
118 F10 **Eleja** C Latvia *56°24′N 23°41′E*
Elek *see* Yelek
119 G14 **Elektrėnai** Vilnius, SE Lithuania *54°47′N 24°35′E*
126 L3 **Elektrostal'** Moskovskaya Oblast', W Russian Federation *55°47′N 38°24′E*
81 H15 **Elemi Triangle** *disputed region* Kenya/Sudan
114 K9 **Elena** Veliko Tarnovo, N Bulgaria *42°55′N 25°53′E*
54 G16 **El Encanto** Amazonas, S Colombia *01°45′S 73°12′W*
37 R14 **Elephant Butte Reservoir** ⊠ New Mexico, SW USA
Éléphant, Chaîne de l' *see* Dâmrei, Chuŏr Phnum
194 G2 **Elephant Island** *island* South Shetland Islands, Antarctica
Elephant River *see* Olifants
El Escorial *see* San Lorenzo de El Escorial
Élesd *see* Aleşd
114 F11 **Eleshnitsa** ≈ W Bulgaria
137 S13 **Eleşkirt** Ağrı, E Turkey *39°22′N 42°48′E*
42 F5 **El Estor** Izabal, E Guatemala *15°37′N 89°22′W*
Eleutherae *see* Eléftheres
44 I2 **Eleuthera Island** *island* N The Bahamas
37 S5 **Elevenmile Canyon Reservoir** ⊠ Colorado, C USA
27 W8 **Eleven Point River** ≈ Arkansas/Missouri, C USA
Elevsís *see* Elefsína
Elevtheroúpolis *see* Eleftheroúpoli
El Faiyûm *see* Al Fayyūm
80 B10 **El Fasher** *var.* Al Fāshir. Northern Darfur, W Sudan *13°37′N 25°22′E*
El Fashn *see* Al Fashn
El Ferrol/El Ferrol del Caudillo *see* Ferrol
39 W13 **Elfin Cove** Chichagof Island, Alaska, USA *58°09′N 136°16′W*
105 W4 **El Fluvià** ≈ NE Spain
40 H7 **El Fuerte** Sinaloa, W Mexico *26°28′N 108°35′W*
80 D11 **El Fula** Western Kordofan, C Sudan *11°44′N 28°20′E*
El Gedaref *see* Gedaref
80 A10 **El Geneina** *var.* Ajjinena, Al-Genain, Al Junaynah. Western Darfur, W Sudan *13°27′N 22°30′E*
81 I18 **Elgeyo/Marakwet** ◈ *county* W Kenya
96 J8 **Elgin** NE Scotland, United Kingdom *57°39′N 03°20′W*
30 M10 **Elgin** Illinois, N USA *42°02′N 88°16′W*
29 P14 **Elgin** Nebraska, C USA *41°58′N 98°04′W*
35 Y9 **Elgin** Nevada, W USA *37°19′N 114°30′W*
28 L6 **Elgin** North Dakota, N USA *46°24′N 101°51′W*
26 M12 **Elgin** Oklahoma, C USA *34°46′N 98°17′W*
25 T10 **Elgin** Texas, SW USA *30°21′N 97°22′W*
123 R9 **El'ginskiy** Respublika Sakha (Yakutiya), NE Russian Federation *64°27′N 141°57′E*
El Giza *see* Giza
74 J8 **El Goléa** *var.* Al Golea. C Algeria *30°35′N 02°59′E*
40 D2 **El Golfo de Santa Clara** Sonora, NW Mexico *31°48′N 114°40′W*
81 G18 **Elgon, Mount** ▲ E Uganda *01°07′N 34°29′E*
94 I10 **Elgpiggen** ▲ S Norway *62°13′N 11°18′E*
105 T4 **El Grado** Aragón, NE Spain *42°09′N 00°13′E*
40 L6 **El Guaje, Laguna** ⊠ NE Mexico
54 H6 **El Guayabo** Zulia, W Venezuela *08°37′N 72°20′W*
77 O6 **El Guettâra** *oasis* N Mali
76 J6 **El Ḥammâmi** *desert* N Mauritania
76 M5 **El Ḥank** *cliff* N Mauritania
El Haseke *see* Al Ḥasakah
80 H10 **El Hawata** Gedaref, E Sudan *13°25′N 34°42′E*

◆ Country
● Country Capital
◇ Dependent Territory
○ Dependent Territory Capital
◈ Administrative Regions
✈ International Airport
▲ Mountain
▲ Mountain Range
⋀ Volcano
≈ River
◎ Lake
⊠ Reservoir

◆ Country
● Country Capital
◇ Dependent Territory
○ Dependent Territory Capital
◈ Administrative Regions
✈ International Airport
▲ Mountain
▲ Mountain Range
Volcano
∽ River
◎ Lake
⊠ Reservoir

162 I8 Erdenetsogt Bayanhongor, C Mongolia 46°27′N 100°53′E
Erdenetsogt see Bayan-Ovoo
78 K7 Erdi plateau NE Chad
78 L7 Erdi Ma desert NE Chad
101 M23 Erding Bayern, SE Germany 48°18′N 11°54′E
Erdőszáda see Ardusat
Erdőszentgyörgy see Sângeorgiu de Pădure
102 I7 Erdre ≈ NW France
195 R13 Erebus, Mount ※ Ross Island, Antarctica 78°11′S 165°09′E
61 H14 Erechim Rio Grande do Sul, S Brazil 27°35′S 52°15′W
163 O7 Ereen Davaanï Nuruu ▲▲ NE Mongolia
163 Q6 Ereentsav Dornod, NE Mongolia 49°51′N 115°41′E
136 I16 Ereğli Konya, S Turkey 37°30′N 34°02′E
115 A15 Ereíkoussa island Iónia Nísiá, Greece, C Mediterranean Sea
163 O11 Erenhot var. Erlian. Nei Mongol Zizhiqu, NE China 43°35′N 112°E
104 M6 Eresma ≈ N Spain
115 K17 Eresós var. Eressós. Lésvos, E Greece 39°11′N 25°57′E
Eressós see Eresós
Ereymentaū see Yereymentau
99 K21 Érezée Luxembourg, SE Belgium 50°16′N 05°34′E
74 G7 Erfoud SE Morocco 31°29′N 04°18′W
101 D16 Erft ≈ W Germany
101 K16 Erfurt Thüringen, C Germany 50°59′N 11°02′E
137 P15 Ergani Dıyarbakır, SE Turkey 38°17′N 39°44′E
Ergel see Hatanbulag
Ergene Çayı see Ergene Irmaği
136 C10 Ergene Irmaği var. Ergene Çayı. ≈ NW Turkey
118 I9 Ērgļi C Latvia 56°55′N 25°38′E
78 H11 Erguig, Bahr ≈ SW Chad
163 S5 Ergun var. Labudalin; prev. Ergun Youqi. Nei Mongol Zizhiqu, N China 50°13′N 120°09′E
Ergun see Gegan Gol
Ergun He see Argun
Ergun Youqi see Ergun
Ergun Zuoqi see Gegen Gol
160 F12 Er Hai ▣ SW China
104 K4 Ería ≈ NW Spain
80 H8 Eriba Kassala, NE Sudan 16°37′N 36°04′E
96 I6 Eriboll, Loch inlet NW Scotland, United Kingdom
65 Q18 Erica Seamount undersea feature SW Indian Ocean 38°15′S 14°30′E
107 H23 Erice Sicilia, Italy, C Mediterranean Sea 38°02′N 12°35′E
104 E10 Ericeira Lisboa, C Portugal 38°58′N 09°25′W
96 H10 Ericht, Loch ◎ C Scotland, United Kingdom
26 J11 Erick Oklahoma, C USA 35°13′N 99°52′W
18 B11 Erie Pennsylvania, NE USA 42°07′N 80°04′W
18 E9 Erie Canal canal New York, NE USA
Érié, Lac see Erie, Lake
31 T10 Erie, Lake Fr. Lac Érié. ◎ Canada/USA
Erigabo see Ceerigaabo
77 N8 'Erîgât desert N Mali
Erigavo see Ceerigaabo
92 P2 Erik Eriksenstretet strait E Svalbard
11 X15 Eriksdale Manitoba, S Canada 50°52′N 98°07′W
189 V6 Erikub Atoll var. Ādkup. atoll Ratak Chain, C Marshall Islands
102 G4 Er, Îles d' Island group NW France
Erimanthos see Erýmanthos
165 T6 Erimo Hokkaidō, NE Japan 42°01′N 143°07′E
165 T6 Erimo-misaki headland Hokkaidō, NE Japan 41°57′N 143°12′E
20 H8 Erin Tennessee, S USA 36°19′N 87°42′W
96 E9 Eriskay island NW Scotland, United Kingdom
Erithraí see Erythrés
80 I9 Eritrea off. State of Eritrea, Ērtra. ◆ transitional government E Africa
Eritrea, State of see Eritrea
Erivan see Yerevan
101 D16 Erkelenz Nordrhein-Westfalen, W Germany 51°04′N 06°19′E
95 P15 Erken ◎ C Sweden
101 K19 Erlangen Bayern, S Germany 49°36′N 11°E
160 G9 Erlang Shan ▲ C China 29°56′N 102°24′E
Erlau see Eger
109 V5 Erlauf ≈ NE Austria
181 Q8 Erldunda Roadhouse Northern Territory, N Australia 25°13′S 133°13′E
Erlian see Erenhot
27 T15 Erling, Lake ▣ Arkansas, USA
109 O8 Erlsbach Tirol, W Austria 46°54′N 12°15′E
Ermak see Aksu
98 K10 Ermelo Gelderland, C Netherlands 52°18′N 05°38′E
83 K21 Ermelo Mpumalanga, NE South Africa 26°32′S 29°59′E
136 H17 Ermenek Karaman, S Turkey 36°38′N 32°55′E
Érmihályfalva see Valea lui Mihai
115 G20 Ermióni Pelopónnisos, S Greece 37°24′N 23°15′E
115 J20 Ermoúpoli var. Hermoupolis; prev. Ermoúpolis. Sýyros, Kykládes, Greece, Aegean Sea 37°26′N 24°55′E
Ermoúpolis see Ermoúpoli
Ernabella see Pukatja
155 G22 Ernākulam Kerala, SW India 10°04′N 76°18′E
102 J6 Ernée Mayenne, NW France 48°18′N 00°54′W
61 H14 Ernestina, Barragem ▣ S Brazil
54 E4 Ernesto Cortissoz ✕ (Barranquilla) Atlántico, N Colombia
155 H21 Erode Tamil Nādu, SE India 11°21′N 77°43′E
Eroj see Iroj
83 C19 Erongo ◈ district W Namibia
99 F21 Erquelinnes Hainaut, S Belgium 50°18′N 04°08′E
74 G7 Er-Rachidia var. Ksar al Soule. E Morocco 31°58′N 04°22′W
80 E11 Er Rahad var. Ar Rahad. Northern Kordofan, C Sudan 12°43′N 30°39′E
Er Ramle see Ramla
83 O15 Errego Zambézia, NE Mozambique 16°02′S 37°11′E
105 Q2 Errenteria Cast. Rentería. País Vasco, N Spain 43°17′N 01°54′W
Er Rif/Er Riff see Rif
97 D14 Errigal Mountain Ir. An Earagail. ▲ N Ireland 55°03′N 08°09′W
97 A15 Erris Head Ir. Ceann Iorrais. headland W Ireland 54°18′N 10°01′W
187 S15 Erromango island S Vanuatu
Error Guyot see Error Tablemount
173 O4 Error Tablemount var. Error Guyot. undersea feature W Indian Ocean 10°20′N 56°05′E
80 G11 Er Roseires Blue Nile, E Sudan 11°52′N 34°23′E
Erseka see Ersekë
113 M22 Ersekë var. Erseka, Kolonjë. Korçë, SE Albania 40°19′N 20°39′E
Érsekújvár see Nové Zámky
29 S4 Erskine Minnesota, N USA 47°42′N 96°00′W
103 V6 Erstein Bas-Rhin, NE France 48°24′N 07°39′E
108 G9 Erstfeld Uri, C Switzerland 46°49′N 08°41′E
158 M3 Ertai Xinjiang Uygur Zizhiqu, NW China 46°04′N 90°06′E
126 M7 Ertil' Voronezhskaya Oblast', W Russian Federation 51°51′N 40°46′E
Ertis see Irtysh, C Asia
Ertis see Irtyshsk, Kazakhstan
158 K2 Ertix He Rus. Chërnyy Irtysh. ≈ China/Kazakhstan
Ērtra see Eritrea
21 P9 Erwin North Carolina, SE USA 35°19′N 78°40′W
Erydropótamos see Erythropótamos
115 E19 Erýmanthos var. Erimanthos. ▲ S Greece 37°57′N 21°51′E
115 G19 Erythrés prev. Erithraí. Stereá Elláda, C Greece 38°18′N 23°20′E
114 L12 Erythropótamos Bul. Byala Reka, var. Erydropótamos. ≈ Bulgaria/Greece
160 F12 Eryuan var. Yuhu. Yunnan, SW China 26°09′N 100°01′E
109 U6 Erzbach ≈ W Austria
Erzerum see Erzurum
101 N17 Erzgebirge Cz. Krušné Hory, Eng. Ore Mountains. ▲ Czech Republic/Germany see also Krušné Hory
Erzgebirge see Krušné Hory
122 L14 Erzin Respublika Tyva, S Russian Federation 50°17′N 95°03′E
137 O13 Erzincan var. Erzinjan. Erzincan, E Turkey 39°44′N 39°30′E
137 N13 Erzincan var. Erzinjan. ◈ province NE Turkey
Erzinjan see Erzincan
Erzsébetváros see Dumbrăveni
137 Q13 Erzurum prev. Erzerum. Erzurum, NE Turkey 39°57′N 41°17′E
137 Q12 Erzurum prev. Erzerum. ◈ province NE Turkey
186 G9 Esa'ala Normanby Island, SE Papua New Guinea 09°45′S 150°47′E
165 T2 Esashi Hokkaidō, NE Japan 44°57′N 142°32′E
165 Q9 Esashi var. Esasi. Iwate, Honshū, C Japan 39°13′N 141°11′E
165 Q5 Esasho Hokkaidō, N Japan
Esasi see Esashi
95 F23 Esbjerg Syddtjylland, W Denmark 55°28′N 08°28′E
Esbo see Espoo
36 L7 Escalante Utah, W USA 37°46′N 111°36′W
36 M7 Escalante River ≈ Utah, W USA
14 L12 Escalier, Réservoir l' ▣ Québec, SE Canada
40 K7 Escalón Chihuahua, N Mexico 26°43′N 104°20′W
104 M8 Escalona Castilla-La Mancha, C Spain 40°10′N 04°24′W
23 O8 Escambia River ≈ Florida, SE USA
31 N5 Escanaba Michigan, N USA 45°45′N 87°03′W
31 N4 Escanaba River ≈ Michigan, N USA
105 R8 Escandón, Puerto de pass E Spain
41 W14 Escárcega Campeche, SE Mexico 18°33′N 90°41′W
171 O1 Escarpada Point headland Luzon, N Philippines 18°28′N 122°10′E
23 N8 Escatawpa River ≈ Alabama/Mississippi, S USA
103 P2 Escaut ≈ N France
Escaut see Scheldt
99 M25 Esch-sur-Alzette Luxembourg, S Luxembourg 49°30′N 05°59′E
101 J15 Eschwege Hessen, C Germany 51°10′N 10°03′E
101 D16 Eschweiler Nordrhein-Westfalen, W Germany 50°49′N 06°16′E
Esclaves, Grand Lac des see Great Slave Lake
45 O8 Escocesa, Bahía bay N Dominican Republic
43 W15 Escocés, Punta headland NE Panama 08°50′N 77°37′W
35 U17 Escondido California, W USA 33°07′N 117°05′W
42 M10 Escondido, Río ≈ SE Nicaragua
15 S7 Escoumins, Rivière des ≈ Québec, SE Canada
37 O13 Escudilla Mountain ▲ Arizona, SW USA 33°57′N 109°07′W
40 J11 Escuinapa var. Escuinapa de Hidalgo. Sinaloa, C Mexico 22°50′N 105°46′W
Escuinapa de Hidalgo see Escuinapa
42 C6 Escuintla Escuintla, S Guatemala 14°17′N 90°46′W
41 V17 Escuintla Chiapas, SE Mexico 15°20′N 92°40′W
42 A2 Escuintla off. Departamento de Escuintla. ◈ department S Guatemala
Escuintla, Departamento de see Escuintla
15 W7 Escuminac ≈ Québec, SE Canada
79 D16 Eséka Centre, SW Cameroon 03°40′N 10°48′E
136 I12 Esenboğa ✕ (Ankara) Ankara, C Turkey 40°05′N 33°01′E
136 D17 Eşen Çayı ≈ SW Turkey
146 B13 Esenguly Rus. Gasan-Kuli. Balkan Welaýaty, W Turkmenistan 37°29′N 53°57′E
105 T4 Ésera ≈ NE Spain
143 N8 Eşfahān Eng. Isfahan; anc. Aspadana. Eşfahān, C Iran 32°41′N 51°41′E
143 O7 Eşfahān off. Ostān-e Eşfahān. ◈ province C Iran
Eşfahān, Ostān-e see Eşfahān
105 N5 Esgueva ≈ N Spain
Eshkamesh see Ishkamish
Eshkāshem see Ishkāshim
83 L23 Eshowe KwaZulu/Natal, E South Africa 28°53′S 31°28′E
143 T5 'Eshqābād Khorāsān-e Razavī, NE Iran 36°00′N 59°01′E
Esh Sham see Rīf Dimashq
Esh Sharā see Ash Sharāh
Esik see Yesik
Esil see Yesil'
Esil see Ishim, Kazakhstan/Russian Federation
183 V2 Esk Queensland, E Australia 27°15′S 152°23′E
184 O11 Eskdale Hawke's Bay, North Island, New Zealand 39°24′S 176°51′E
Eski Dzhumaya see Targovishte
92 L2 Eskifjörður Austurland, E Iceland 65°04′N 14°01′W
139 S3 Eski Kaļak var. Askī Kalak, Kalak. Arbīl, N Iraq 36°16′N 43°40′E
95 N16 Eskilstuna Södermanland, C Sweden 59°22′N 16°31′E
8 H6 Eskimo Lakes ◎ Northwest Territories, NW Canada
9 O10 Eskimo Point headland Nunavut, E Canada 61°19′N 93°49′W
Eskimo Point see Arviat
139 Q2 Eski Mosul Nīnawá, N Iraq 36°31′N 42°45′E
Eski-Nookat see Nookat
136 F12 Eskişehir var. Eskishehr. Eskişehir, W Turkey 39°46′N 30°30′E
136 F13 Eskişehir var. Eski shehr. ◈ province NW Turkey
Eskishehr see Eskişehir
104 K5 Esla ≈ NW Spain
Eslāmābād see Eslāmābād-e Gharb
142 J6 Eslāmābād-e Gharb var. Eslāmābād; prev. Harunabad, Shāhābād. Kermānshāhān, W Iran 34°08′N 46°35′E
148 J4 Eslam Qal'eh Pash. Islam Qala. Herāt, W Afghanistan 34°41′N 61°03′E
95 K23 Eslöv Skåne, S Sweden 55°50′N 13°20′E
143 S12 Esmā'īlābād Kermān, S Iran 28°48′N 56°39′E
143 U8 Esmā'īlābād Khorāsān-e Jonūbī, E Iran 35°20′N 60°30′E
136 D14 Eşme Uşak, W Turkey 38°26′N 28°59′E
44 G6 Esmeralda Camagüey, E Cuba 21°51′N 78°10′W
63 F21 Esmeralda, Isla island S Chile
56 B5 Esmeraldas Esmeraldas, N Ecuador 0°55′N 79°40′W
56 B5 Esmeraldas ◈ province NW Ecuador
Esna see Isnā
14 B6 Esnagi Lake ◎ Ontario, S Canada
143 V14 Espakeh Sīstān va Balūchestān, SE Iran 26°54′N 60°09′E
103 O13 Espalion Aveyron, S France 44°31′N 02°45′E
España see Spain
14 E11 Espanola Ontario, S Canada 46°15′N 81°46′W
37 S10 Espanola New Mexico, SW USA 35°59′N 106°04′W
57 C18 Española, Isla var. Hood Island. island Galapagos Islands, Ecuador, E Pacific Ocean
104 M13 Espejo Andalucía, S Spain 37°40′N 04°34′W
94 C13 Espeland Hordaland, S Norway 60°22′N 05°27′E
100 G12 Espelkamp Nordrhein-Westfalen, NW Germany 52°22′N 08°37′E
38 M8 Espenberg, Cape headland Alaska, USA 66°33′N 163°36′W
180 L13 Esperance Western Australia 33°49′S 121°52′E
186 L9 Esperance, Cape headland Guadalcanal, C Solomon Islands 09°09′S 159°38′E
57 P18 Esperancita Santa Cruz, E Bolivia
61 B17 Esperanza Santa Fe, C Argentina 31°29′S 61°00′W
40 G6 Esperanza Sonora, NW Mexico 27°37′N 109°51′W
24 H9 Esperanza Texas, SW USA 31°09′N 105°40′W
194 H3 Esperanza Argentinian research station Antarctica 63°29′S 56°53′W
104 E12 Espichel, Cabo headland S Portugal 38°25′N 09°15′W
54 E10 Espinal Tolima, C Colombia 04°08′N 74°53′W
48 K10 Espinhaço, Serra do ▲ SE Brazil
104 G6 Espinho Aveiro, N Portugal 41°01′N 08°38′W
59 N18 Espinosa Minas Gerais, SE Brazil 14°58′S 42°49′W
103 O15 Espinouse ▲ S France
60 Q8 Espírito Santo off. Estado do Espírito Santo. ◈ state E Brazil
Espírito Santo, Estado do see Espírito Santo
187 P13 Espiritu Santo var. Santo. island W Vanuatu
41 Z13 Espíritu Santo, Bahía del bay SE Mexico
40 F9 Espíritu Santo, Isla del island NW Mexico
41 Y12 Espita Yucatán, SE Mexico 21°00′N 88°17′W
15 Y7 Espoir, Cap d' headland Québec, SE Canada 48°24′N 64°21′W
Esponsede/Esponsende see Esposende
93 L20 Espoo Swe. Esbo. Uusimaa, S Finland 60°10′N 24°42′E
104 G5 Esposende var. Esponsede, Esponsende. Braga, N Portugal 41°32′N 08°47′W
83 M18 Espungabera Manica, SW Mozambique 20°29′S 32°48′E
63 H17 Esquel Chubut, SW Argentina 42°55′S 71°20′W
10 L17 Esquimalt Vancouver Island, British Columbia, SW Canada 48°26′N 123°27′W
61 C16 Esquina Corrientes, NE Argentina 30°00′S 59°30′W
42 E6 Esquipulas Chiquimula, SE Guatemala 14°36′N 89°22′W
42 K9 Esquipulas Matagalpa, C Nicaragua 12°30′N 85°55′W
94 I8 Essandsjøen ◎ S Norway
74 E7 Essaouira prev. Mogador. W Morocco 31°33′N 09°40′W
Esseg see Osijek
Es Semara see Smara
99 G15 Essen Antwerpen, N Belgium 51°28′N 04°28′E
101 E15 Essen var. Essen an der Ruhr. Nordrhein-Westfalen, W Germany 51°28′N 07°01′E
Essen an der Ruhr see Essen
74 I5 Es Senia ✕ (Oran) NW Algeria 35°34′N 00°42′W
55 T8 Essequibo Islands island group N Guyana
55 T11 Essequibo River ≈ C Guyana
14 C18 Essex Ontario, S Canada 42°10′N 82°50′W
29 T16 Essex Iowa, C USA 40°49′N 95°18′W
97 P21 Essex cultural region E England, United Kingdom
31 R8 Essexville Michigan, N USA 43°37′N 83°50′W
101 H22 Esslingen var. Esslingen am Neckar. Baden-Württemberg, SW Germany 48°45′N 09°19′E
Esslingen am Neckar see Esslingen
103 N6 Essonne ◈ department N France
Es Suweida see As Suwaydā'
79 F16 Est Eng. East. ◈ province SE Cameroon
104 I1 Estaca de Bares, Punta de point NW Spain
24 M5 Estacado, Llano plain New Mexico/Texas, SW USA
41 P12 Estación Tamuín San Luis Potosí, C Mexico 22°00′N 98°44′W
63 K25 Estados, Isla de los prev. Eng. Staten Island. island S Argentina
Estado Vargas see Vargas
143 P12 Eşţahbān Fārs, S Iran
14 F11 Estaire Ontario, S Canada 46°19′N 80°47′W
59 P16 Estância Sergipe, E Brazil 11°15′S 37°28′W
37 S12 Estancia New Mexico, SW USA 34°45′N 106°03′W
104 G7 Estarreja Aveiro, N Portugal 40°45′N 08°35′W
102 M17 Estats, Pica d' Sp. Pico d'Estats. ▲ France/Spain 42°39′N 01°24′E
Estats, Pico d' see Estats, Pica d'
83 K23 Estcourt KwaZulu/Natal, E South Africa 29°00′S 29°53′E
106 H8 Este anc. Ateste. Veneto, NE Italy 45°14′N 11°40′E
42 J9 Estelí Estelí, NW Nicaragua 13°05′N 86°21′W
42 J9 Estelí ◈ department NW Nicaragua
105 Q4 Estella Bas. Lizarra. Navarra, N Spain 42°41′N 02°02′W
29 R9 Estelline South Dakota, N USA 44°34′N 96°54′W
25 P4 Estelline Texas, SW USA 34°33′N 100°26′W
104 L14 Estepa Andalucía, S Spain 37°17′N 04°52′W
104 L16 Estepona Andalucía, S Spain 36°26′N 05°09′W
39 R9 Ester Alaska, USA 64°49′N 148°03′W
11 V16 Esterhazy Saskatchewan, S Canada 50°40′N 102°02′W
37 S3 Estes Park Colorado, C USA 40°22′N 105°31′W
11 V17 Estevan Saskatchewan, S Canada 49°07′N 103°05′W
29 T11 Estherville Iowa, C USA 43°24′N 94°49′W
21 R15 Estill South Carolina, SE USA 32°45′N 81°14′W
103 Q6 Estissac Aube, N France 48°17′N 03°51′E
15 T9 Est, Lac de l' ◎ Québec, SE Canada
Estland see Estonia
11 S16 Eston Saskatchewan, S Canada 51°09′N 108°42′W
118 G5 Estonia off. Republic of Estonia, Est. Eesti Vabariik, Ger. Estland, Latv. Igaunija; prev. Estonian SSR, Rus. Estonskaya SSR. ◆ republic NE Europe
Estonian SSR see Estonia
Estonia, Republic of see Estonia
Estonskaya SSR see Estonia
104 E11 Estoril Lisboa, W Portugal 38°42′N 09°23′W
59 L14 Estreito Maranhão, E Brazil 06°34′S 47°22′W
104 I8 Estrela, Serra da ▲ C Portugal
40 D3 Estrella, Punta headland NW Mexico 30°53′N 114°45′W
104 F10 Estremadura cultural and historical region W Portugal
Estremadura see Extremadura
104 H11 Estremoz Évora, S Portugal 38°50′N 07°35′W
79 D18 Estuaire off. Province de l'Estuaire, var. L'Estuaire. ◈ province NW Gabon
Estuaire, Province de l' see Estuaire
Eszék see Osijek
111 I22 Esztergom Ger. Gran; anc. Strigonium. Komárom-Esztergom, N Hungary 47°47′N 18°44′E
152 K11 Etah Uttar Pradesh, N India 27°33′N 78°39′E
189 R17 Etal Atoll atoll Mortlock Islands, C Micronesia
99 K24 Étalle Luxembourg, SE Belgium 49°41′N 05°36′E
103 N6 Étampes Essonne, N France 48°26′N 02°10′E
182 J1 Etamunbanie, Lake salt lake South Australia
103 N1 Étaples Pas-de-Calais, N France 50°31′N 01°39′E
152 K12 Etāwah Uttar Pradesh, N India 26°46′N 79°01′E
15 R10 Etchemin ≈ Québec, SE Canada
Etchmiadzin see Vagharshapat
40 G7 Etchojoa Sonora, NW Mexico 26°54′N 109°37′W
93 L19 Etelä-Karjala Swe. Södra Karelen, Eng. South Karelia. ◈ region S Finland
93 K17 Etelä-Pohjanmaa Swe. Södra Österbotten, Eng. South Ostrobothnia. ◈ region W Finland
93 M18 Etelä-Savo Swe. Södra Savolax. ◈ region SE Finland
83 B16 Etengua Kunene, NW Namibia 17°24′S 13°05′E
99 K25 Éthe Luxembourg, SE Belgium 49°34′N 05°32′E
11 W15 Ethelbert Manitoba, S Canada 51°30′N 100°22′W
80 H12 Ethiopia off. Federal Democratic Republic of Ethiopia; prev. Abyssinia, People's Democratic Republic of Ethiopia. ◆ republic E Africa
Ethiopia, Federal Democratic Republic of see Ethiopia
80 I13 Ethiopian Highlands var. Ethiopian Plateau. plateau N Ethiopia
Ethiopian Plateau see Ethiopian Highlands
Ethiopia, People's Democratic Republic of see Ethiopia
34 M2 Etna California, W USA 41°25′N 122°53′W
18 B14 Etna Pennsylvania, NE USA 40°29′N 79°55′W
94 G12 Etna ≈ S Norway
107 L24 Etna, Monte Eng. Mount Etna. ※ Sicilia, Italy, C Mediterranean Sea 37°46′N 15°00′E
Etna, Mount see Etna, Monte
95 C15 Etne Hordaland, S Norway 59°40′N 05°55′E
Etoliko see Aitolikó
39 Y14 Etolin Island island Alexander Archipelago, Alaska, USA
38 L12 Etolin Strait strait Alaska, USA
83 C17 Etosha Pan salt lake N Namibia
79 G18 Etoumbi Cuvette Ouest, NW Congo 0°01′N 14°57′E
20 M10 Etowah Tennessee, S USA 35°19′N 84°31′W
23 S2 Etowah River ≈ Georgia, SE USA
146 B13 Etrek var. Gyzyletrek, Rus. Kizyl-Atrek. Balkan Welaýaty, W Turkmenistan 37°40′N 54°44′E
146 C13 Etrek Per. Rūd-e Atrak, Rus. Atrak, Atrek. ≈ Iran/Turkmenistan
102 L3 Étretat Seine-Maritime, N France 49°46′N 00°23′E
114 H9 Etropole Sofia, W Bulgaria 42°50′N 24°00′E
Etsch see Adige
Et Tafila see Aţ Ţafīlah
99 M23 Ettelbrück Diekirch, C Luxembourg 49°51′N 06°06′E
189 V12 Etten atoll Chuuk Islands, C Micronesia
99 H14 Etten-Leur Noord-Brabant, S Netherlands 51°34′N 04°37′E
76 G7 Et Tîdra var. Ile Tîdra. island Dakhlet Nouâdhibou, NW Mauritania
101 G21 Ettlingen Baden-Württemberg, SW Germany 48°57′N 08°25′E
102 M2 Eu Seine-Maritime, N France 50°01′N 01°24′E
193 W16 'Eua prev. Middleburg Island. island Tongatapu Group, SE Tonga
193 W15 Eua Iki island Tongatapu Group, S Tonga
Euboea see Évvoia
181 O12 Eucla Western Australia 31°41′S 128°51′E
31 U11 Euclid Ohio, N USA 41°34′N 81°33′W
27 W14 Eudora Arkansas, C USA 33°06′N 91°15′W
27 Q4 Eudora Kansas, C USA 38°56′N 95°06′W
182 J9 Eudunda South Australia 34°11′S 139°03′E
23 R6 Eufaula Alabama, S USA 31°53′N 85°09′W
27 Q11 Eufaula Oklahoma, C USA 35°16′N 95°36′W
27 Q11 Eufaula Lake var. Eufaula Reservoir. ▣ Oklahoma, C USA
Eufaula Reservoir see Eufaula Lake
32 F13 Eugene Oregon, NW USA 44°03′N 123°05′W
40 B6 Eugenia, Punta headland NW Mexico 27°48′N 115°03′W
183 Q8 Eugowra New South Wales, SE Australia 33°28′S 148°21′E
104 I2 Eume ≈ NW Spain
104 H2 Eume, Encoro do ▣ NW Spain
Eumolpias see Plovdiv
59 O18 Eunápolis Bahia, SE Brazil 16°20′S 39°36′W
22 H8 Eunice Louisiana, S USA 30°29′N 92°25′W
37 W15 Eunice New Mexico, SW USA 32°26′N 103°09′W
99 M19 Eupen Liège, E Belgium 50°38′N 06°02′E
130 B10 Euphrates Ar. Al-Furāt, Turk. Firat Nehri. ≈ SW Asia
138 L3 Euphrates Dam dam N Syria
22 M4 Eupora Mississippi, S USA 33°32′N 89°16′W
93 K19 Eura Satakunta, SW Finland 61°07′N 22°12′E
93 K19 Eurajoki Satakunta, SW Finland 61°13′N 21°45′E
0-1 Eurasian Plate tectonic feature
102 L4 Eure ◈ department N France
102 M4 Eure ≈ N France
102 M6 Eure-et-Loir ◈ department C France
34 K3 Eureka California, W USA 40°47′N 124°12′W
27 P6 Eureka Kansas, C USA 37°51′N 96°17′W
33 O6 Eureka Montana, NW USA 48°52′N 115°03′W
35 V5 Eureka Nevada, W USA 39°31′N 115°58′W
29 O7 Eureka South Dakota, N USA 45°46′N 99°37′W
36 L4 Eureka Utah, W USA 39°57′N 112°07′W
32 K10 Eureka Washington, NW USA 46°21′N 118°41′W
27 S9 Eureka Springs Arkansas, C USA 36°25′N 93°45′W
182 K6 Eurinilla Creek seasonal river South Australia
183 O11 Euroa Victoria, SE Australia 36°46′S 145°35′E
172 M9 Europa, Île island W Madagascar
104 L3 Europa, Picos de ▲ N Spain
104 L16 Europa Point headland S Gibraltar 36°07′N 05°20′W
84-85 Europe continent
98 F12 Europoort Zuid-Holland, W Netherlands 51°59′N 04°08′E
Euskadi see País Vasco
101 D17 Euskirchen Nordrhein-Westfalen, W Germany 50°40′N 06°47′E
23 W11 Eustis Florida, SE USA 28°51′N 81°41′W
182 M9 Euston New South Wales, SE Australia 34°34′S 142°45′E
23 N5 Eutaw Alabama, S USA 32°50′N 87°53′W
100 K8 Eutin Schleswig-Holstein, N Germany 54°08′N 10°38′E
10 K14 Eutsuk Lake ▣ British Columbia, SW Canada
Euxine Sea see Black Sea
83 C16 Evale Cunene, SW Angola 16°36′S 15°46′E
37 T3 Evans Colorado, C USA 40°22′N 104°41′W
11 P14 Evansburg Alberta, SW Canada 53°34′N 114°57′W
29 X13 Evansdale Iowa, C USA 42°28′N 92°16′W
183 V4 Evans Head New South Wales, SE Australia 29°07′S 153°27′E
12 J11 Evans, Lac ◎ Québec, SE Canada
37 S5 Evans, Mount ▲ Colorado, C USA 39°15′N 106°10′W
9 Q6 Evans Strait strait Nunavut, N Canada
31 N10 Evanston Illinois, N USA 42°02′N 87°41′W
33 S17 Evanston Wyoming, C USA 41°16′N 110°57′W
14 D11 Evansville Manitoulin Island, Ontario, S Canada 45°48′N 82°34′W
31 N16 Evansville Indiana, N USA 37°58′N 87°33′W
30 L9 Evansville Wisconsin, N USA 42°46′N 89°16′W
25 S8 Evant Texas, SW USA 31°28′N 98°09′W
143 P13 Evaz Fārs, S Iran 27°48′N 53°58′E
29 W4 Eveleth Minnesota, N USA 47°27′N 92°32′W
182 E3 Evelyn Creek seasonal river South Australia
181 Q2 Evelyn, Mount ▲ Northern Territory, N Australia 13°28′S 132°50′E
122 K10 Evenkiyskiy Avtonomnyy Okrug ◈ autonomous district Krasnoyarskiy Kray, N Russian Federation
183 R13 Everard, Cape headland Victoria, SE Australia 37°48′S 149°21′E
182 F6 Everard, Lake salt lake South Australia
182 C2 Everard Ranges ▲▲ South Australia
153 R11 Everest, Mount Chin. Qomolangma Feng, Nep. Sagarmāthā. ▲ China/Nepal 27°59′N 86°57′E
18 E15 Everett Pennsylvania, NE USA 40°00′N 78°22′W
32 H7 Everett Washington, NW USA 47°59′N 122°12′W
99 E17 Evergem Oost-Vlaanderen, NW Belgium 51°07′N 03°43′E
23 X16 Everglades City Florida, SE USA 25°51′N 81°22′W
23 Y16 Everglades, The wetland Florida, SE USA
23 P7 Evergreen Alabama, S USA 31°25′N 86°55′W
37 T4 Evergreen Colorado, C USA 39°37′N 105°19′W
Evergreen State see Washington
97 L21 Evesham C England, United Kingdom 52°06′N 01°57′W
103 T10 Évian-les-Bains Haute-Savoie, E France 46°22′N 06°34′E
93 K16 Evijärvi Etelä-Pohjanmaa, W Finland 63°22′N 23°30′E
79 D17 Evinayong var. Ebinayon, Evinayoung. C Equatorial Guinea 01°28′N 10°17′E
Evinayoung see Evinayong
115 E18 Évinos ≈ C Greece
95 E17 Evje Aust-Agder, S Norway 58°35′N 07°49′E
Evmolpia see Plovdiv
104 H11 Évora anc. Ebora, Lat. Liberalitas Julia. Évora, C Portugal 38°34′N 07°54′W
104 G11 Évora ◈ district S Portugal
102 M4 Évreux anc. Civitas Eburovicum. Eure, N France 49°02′N 01°10′E
102 K6 Évron Mayenne, NW France 48°10′N 00°24′W
114 L13 Évros Bul. Maritsa, Turk. Meriç; anc. Hebrus. ≈ SE Europe see also Maritsa/Meriç
Évros see Meriç
Évros see Maritsa
115 F21 Evrótas ≈ S Greece
103 O5 Évry Essonne, N France 48°38′N 02°34′E
25 O8 E. V. Spence Reservoir ▣ Texas, SW USA
115 I18 Évvoia Lat. Euboea. island C Greece
38 D9 'Ewa Beach var. Ewa Beach. O'ahu, Hawaii, USA, C Pacific Ocean 21°19′N 158°00′W
Ewa Beach see 'Ewa Beach
32 L9 Ewan Washington, NW USA 47°06′N 117°46′W
44 K12 Ewarton C Jamaica 18°11′N 77°06′W
81 J18 Ewaso Ng'iro var. Nyiro. ≈ C Kenya
29 P13 Ewing Nebraska, C USA 42°13′N 98°20′W
194 J5 Ewing Island island Antarctica
65 P17 Ewing Seamount undersea feature E Atlantic Ocean 23°20′S 08°45′E
158 L6 Ewirgol Xinjiang Uygur Zizhiqu, W China 42°56′N 87°39′E
79 G19 Ewo Cuvette, W Congo 0°55′S 14°49′E
27 S3 Excelsior Springs Missouri, C USA 39°20′N 94°13′W
97 J23 Exe ≈ SW England, United Kingdom
194 L12 Executive Committee Range ▲▲ Antarctica
14 E16 Exeter Ontario, S Canada 43°19′N 81°26′W
97 J24 Exeter anc. Isca Damnoniorum. SW England, United Kingdom 50°43′N 03°31′W
35 R11 Exeter California, W USA 36°17′N 119°08′W
19 P10 Exeter New Hampshire, NE USA 42°57′N 70°55′W
Exin see Kcynia
29 T14 Exira Iowa, C USA 41°36′N 94°55′W
97 J23 Exmoor moorland SW England, United Kingdom
21 Y6 Exmore Virginia, NE USA 37°31′N 75°48′W
180 G8 Exmouth Western Australia 22°01′S 114°06′E
97 J24 Exmouth SW England, United Kingdom 50°36′N 03°25′W
180 G8 Exmouth Gulf gulf Western Australia
173 V8 Exmouth Plateau undersea feature E Indian Ocean 16°00′S 115°00′E
83 K23 eXobho prev. Ixopo. KwaZulu/Natal, E South Africa 30°10′S 30°05′E
115 J20 Exompourgo ancient monument Tínos, Kykládes, Greece, Aegean Sea
104 I10 Extremadura var. Estremadura. ◈ autonomous community W Spain
78 F12 Extrême-Nord Eng. Extreme North. ◈ province N Cameroon
Extreme North see Extrême-Nord
44 I3 Exuma Cays islets C The Bahamas
44 I3 Exuma Sound sound C The Bahamas
81 H20 Eyasi, Lake ◎ N Tanzania
95 F17 Eydehavn Aust-Agder, S Norway 58°31′N 08°53′E
96 L12 Eyemouth SE Scotland, United Kingdom 55°52′N 02°07′W
96 G7 Eye Peninsula peninsula NW Scotland, United Kingdom
92 J4 Eyjafjallajökull ※ S Iceland 63°37′N 19°37′W
80 Q13 Eyl It. Eil. Nugaal, E Somalia 08°03′N 49°49′E
103 N11 Eymoutiers Haute-Vienne, C France 45°45′N 01°43′E
Eyo (lower course) see Uolo, Río
29 X10 Eyota Minnesota, N USA 44°00′N 92°13′W
182 H2 Eyre Basin, Lake salt lake South Australia
182 I1 Eyre Creek seasonal river Northern Territory/South Australia
174 L9 Eyre, Lake salt lake South Australia
185 C22 Eyre Mountains ▲▲ South Island, New Zealand
182 H3 Eyre North, Lake salt lake South Australia
182 G7 Eyre Peninsula peninsula South Australia
182 H4 Eyre South, Lake salt lake South Australia
95 B18 Eysturoy Dan. Østerø. island N Faroe Islands
61 D20 Ezeiza ✕ (Buenos Aires) Buenos Aires, E Argentina 34°49′S 58°30′W
Ezeres see Ezeriş
116 F12 Ezeriş Hung. Ezeres. Caraş-Severin, W Romania 45°21′N 21°55′E
161 O9 Ezhou prev. Echeng. Hubei, C China 30°23′N 114°52′E
125 R11 Ezhva Respublika Komi, NW Russian Federation 61°45′N 50°43′E
136 B12 Ezine Çanakkale, NW Turkey 39°46′N 26°20′E
Ezo see Hokkaidō
Ezra/Ezraa see Izra'

F

191 P7 Faaa Tahiti, W French Polynesia 17°32′S 149°36′W
191 P7 Faaa ✕ (Papeete) Tahiti, W French Polynesia 17°31′S 149°36′W
95 H24 Faaborg var. Fåborg. Syddtjylland, C Denmark 55°06′N 10°10′E
151 K19 Faadhippolhu Atoll var. Fadiffolu, Lhaviyani Atoll. atoll N Maldives
191 U10 Faaite atoll Îles Tuamotu, C French Polynesia
191 Q8 Faaone Tahiti, W French Polynesia 17°39′S 149°18′W
24 H8 Fabens Texas, SW USA 31°30′N 106°09′W
94 H12 Fåberg Oppland, S Norway 61°15′N 10°21′E
Fåborg see Faaborg
106 I12 Fabriano Marche, C Italy 43°20′N 12°54′E
145 U16 Fabrichnoye prev. Fabrichnyy. Almaty, SE Kazakhstan 43°12′N 76°19′E
Fabrichnyy see Fabrichnoye
54 F10 Facatativá Cundinamarca, C Colombia 04°49′N 74°22′W
77 X9 Fachi Agadez, C Niger 18°01′N 11°36′E
188 B16 Facpi Point headland W Guam
18 I13 Factoryville Pennsylvania, NE USA 41°34′N 75°45′W
78 K8 Fada Ennedi-Ouest, E Chad 17°14′N 21°32′E
77 Q13 Fada-Ngourma E Burkina Faso 12°05′N 00°26′E
123 N6 Faddeya, Zaliv bay N Russian Federation
123 Q5 Faddeyevskiy, Poluostrov island Novosibirskiye Ostrova, NE Russian Federation
141 W12 Fadhī S Oman 17°54′N 55°30′E
Fadiffolu see Faadhippolhu Atoll

◆ Country
● Country Capital
◇ Dependent Territory
○ Dependent Territory Capital
◈ Administrative Regions
✕ International Airport
▲ Mountain
▲▲ Mountain Range
※ Volcano
≈ River
◎ Lake
▣ Reservoir

106 H10 **Faenza** *anc.* Faventia. Emilia-Romagna, N Italy 44°17′N 11°53′E
Faeroe-Iceland Ridge *see* Faroe-Iceland Ridge
Faeroe Islands *see* Faroe Islands
Færøerne *see* Faroe Islands
Faeroe-Shetland Trough *see* Faroe-Shetland Trough
104 H6 **Fafe** Braga, N Portugal 41°27′N 08°11′W
80 K13 **Fafen Shet'** ≈ E Ethiopia
193 V15 **Fafo** *island* Tongatapu Group, S Tonga
192 I16 **Fagaloa Bay** *bay* Upolu, E Samoa
192 H15 **Fagamālo** Savai'i, N Samoa 13°27′S 172°22′W
116 I12 **Făgăraş** *Ger.* Fogarasch, *Hung.* Fogaras. Braşov, C Romania 45°50′N 24°59′E
191 W10 **Fagatau** *prev.* Fangatau. *atoll* Îles Tuamotu, C French Polynesia
191 X12 **Fagataufa** *prev.* Fangataufa. *island* Îles Tuamotu, SE French Polynesia
95 M20 **Fagerhult** Kalmar, S Sweden 57°07′N 15°40′E
94 G13 **Fagernes** Oppland, S Norway 60°59′N 09°14′E
92 I9 **Fagernes** Troms, N Norway 69°31′N 19°16′E
95 M14 **Fagersta** Västmanland, C Sweden 59°59′N 15°49′E
77 W13 **Faggo** *var.* Foggo. Bauchi, N Nigeria 11°22′N 09°55′E
Faghman *see* Fughmah
Fagibina, Lake *see* Faguibine, Lac
63 J25 **Fagnano, Lago** ◎ S Argentina
99 G22 **Fagne** *hill range* S Belgium
77 N10 **Faguibine, Lac** *var.* Lake Fagibina. ◎ NW Mali
Fahaheel *see* Al Fuḩayḩīl
Fahlun *see* Falun
143 U12 **Fahraj** Kermān, SE Iran 29°00′N 59°00′E
64 P5 **Faial** Madeira, Portugal, NE Atlantic Ocean 32°47′N 16°53′W
64 N2 **Faial** *var.* Ilha do Faial. *island* Azores, Portugal, NE Atlantic Ocean
Faial, Ilha do *see* Faial
108 G10 **Faido** Ticino, S Switzerland 46°30′N 08°48′E
Faifo *see* Hôi An
Failaka Island *see* Faylakah
190 G12 **Faioa, Île** *island* N Wallis and Futuna
181 W8 **Fairbairn Reservoir** ▧ Queensland, E Australia
39 R9 **Fairbanks** Alaska, USA 64°48′N 147°47′W
21 U12 **Fair Bluff** North Carolina, SE USA 34°18′N 79°02′W
31 R14 **Fairborn** Ohio, N USA 39°48′N 84°03′W
23 S3 **Fairburn** Georgia, SE USA 33°34′N 84°34′W
30 M12 **Fairbury** Illinois, N USA 40°45′N 88°30′W
29 Q17 **Fairbury** Nebraska, C USA 40°08′N 97°10′W
29 T9 **Fairfax** Minnesota, N USA 44°31′N 94°43′W
27 O8 **Fairfax** Oklahoma, C USA 36°34′N 96°42′W
21 R14 **Fairfax** South Carolina, SE USA 32°57′N 81°14′W
35 N8 **Fairfield** California, W USA 38°14′N 122°03′W
33 O14 **Fairfield** Idaho, NW USA 43°20′N 114°45′W
30 M16 **Fairfield** Illinois, N USA 38°22′N 88°23′W
29 X15 **Fairfield** Iowa, C USA 41°00′N 91°57′W
33 R8 **Fairfield** Montana, NW USA 47°36′N 111°59′W
31 Q14 **Fairfield** Ohio, N USA 39°21′N 84°34′W
25 U8 **Fairfield** Texas, SW USA 31°43′N 96°10′W
27 T7 **Fair Grove** Missouri, C USA 37°22′N 93°09′W
19 P12 **Fairhaven** Massachusetts, NE USA 41°38′N 70°51′W
23 N8 **Fairhope** Alabama, S USA 30°31′N 87°54′W
96 L4 **Fair Isle** *island* NE Scotland, United Kingdom
185 F20 **Fairlie** Canterbury, South Island, New Zealand 44°06′S 170°50′E
29 U11 **Fairmont** Minnesota, N USA 43°40′N 94°27′W
29 Q16 **Fairmont** Nebraska, C USA 40°37′N 97°36′W
21 S3 **Fairmont** West Virginia, NE USA 39°28′N 80°08′W
31 P13 **Fairmount** Indiana, N USA 40°25′N 85°39′W
18 H10 **Fairmount** New York, NE USA 43°03′N 76°14′W
29 R7 **Fairmount** North Dakota, N USA 46°02′N 96°36′W
37 S5 **Fairplay** Colorado, C USA 39°13′N 106°00′W
18 F9 **Fairport** New York, NE USA 43°06′N 77°26′W
11 O12 **Fairview** Alberta, W Canada 56°03′N 118°28′W
26 L9 **Fairview** Oklahoma, C USA 36°16′N 98°29′W
36 L4 **Fairview** Utah, W USA 39°37′N 111°26′W
35 T6 **Fairview Peak** ▲ Nevada, W USA 39°13′N 118°09′W
188 H14 **Fais** *atoll* Caroline Islands, W Micronesia
149 U8 **Faisalābād** *prev.* Lyallpur. Punjab, NE Pakistan 31°26′N 73°06′E
Faisaliya *see* Fayşalīyah
28 L8 **Faith** South Dakota, N USA 45°01′N 102°02′W
153 N12 **Faizābād** Uttar Pradesh, N India 26°46′N 82°08′E
Faizabad/Faizābād *see* Feiẕābād
45 S9 **Fajardo** E Puerto Rico 18°20′N 65°39′W
139 R9 **Fajj, Wādī al** *dry watercourse* S Iraq
140 K4 **Fajr, Bi'r** *well* NW Saudi Arabia
191 W10 **Fakahina** *atoll* Îles Tuamotu, C French Polynesia
190 L10 **Fakaofo Atoll** *island* SE Tokelau
191 U10 **Fakarava** *atoll* Îles Tuamotu, C French Polynesia
127 T2 **Fakel** Udmurtskaya Respublika, NW Russian Federation 57°35′N 53°00′E
97 P19 **Fakenham** E England, United Kingdom 52°50′N 00°51′E
171 U13 **Fakfak** Papua Barat, E Indonesia 02°55′S 132°17′E
153 T12 **Fakīragrām** Assam, NE India 26°22′N 90°15′E
114 M10 **Fakiyska Reka** ≈ SE Bulgaria
95 J24 **Fakse** Sjælland, SE Denmark 55°16′N 12°08′E
95 J24 **Fakse Bugt** *bay* SE Denmark
95 J24 **Fakse Ladeplads** Sjælland, SE Denmark 55°14′N 12°11′E
163 V11 **Faku** Liaoning, NE China 42°30′N 123°27′E
76 J14 **Falaba** N Sierra Leone 09°54′N 11°22′W
102 K5 **Falaise** Calvados, N France 48°52′N 00°12′W
114 H12 **Falakró** ▲▲ NE Greece
189 T12 **Falalu** *island* Chuuk, C Micronesia
166 L4 **Falam** Chin State, W Myanmar (Burma) 22°58′N 93°45′E
143 N8 **Falāvarjān** Eşfahān, C Iran 32°33′N 51°28′E
116 M11 **Fălciu** Vaslui, E Romania 46°19′N 28°10′E
54 I4 **Falcón** *off.* Estado Falcón. ⬖ *state* NW Venezuela
106 J12 **Falconara Marittima** Marche, C Italy 43°37′N 13°23′E
Falcone, Capo del *see* Falcone, Punta del
107 A16 **Falcone, Punta del** *var.* Capo del Falcone. *headland* Sardegna, Italy, C Mediterranean Sea 40°57′N 8°12′E
Falcón, Estado *see* Falcón
11 Y16 **Falcon Lake** Manitoba, S Canada 49°44′N 95°18′W
Falcon Lake *see* Falcón, Presa/Falcon Reservoir
41 O7 **Falcón, Presa** *var.* Falcon Lake, Falcon Reservoir. ▧ Mexico/USA *see also* Falcon Reservoir
Falcón, Presa *see* Falcon Reservoir
25 Q16 **Falcon Reservoir** *var.* Falcon Lake, Presa Falcón. ▧ Mexico/USA *see also* Falcón, Presa
Falcon Reservoir *see* Falcón, Presa
190 L10 **Fale** *island* Fakaofo Atoll, SE Tokelau
192 F15 **Faleālupo** Savai'i, NW Samoa 13°30′S 172°46′W
190 B10 **Falefatu** *island* Funafuti Atoll, C Tuvalu
192 G15 **Fālelima** Savai'i, NW Samoa 13°30′S 172°41′W
95 N18 **Falerum** Östergötland, S Sweden 58°07′N 16°15′E
Faleshty *see* Făleşti
116 M9 **Făleşti** *Rus.* Faleshty. NW Moldova 47°33′N 27°43′E
25 S15 **Falfurrias** Texas, SW USA 27°17′N 98°10′W
11 O13 **Falher** Alberta, W Canada 55°45′N 117°18′W
Falkenau an der Eger *see* Sokolov
95 J21 **Falkenberg** Halland, S Sweden 56°55′N 12°30′E
Falkenberg *see* Niemodlin
Falkenburg in Pommern *see* Złocieniec
100 N12 **Falkensee** Brandenburg, NE Germany 52°34′N 13°04′E
96 J12 **Falkirk** C Scotland, United Kingdom 56°N 03°48′W
65 I20 **Falkland Escarpment** *undersea feature* SW Atlantic Ocean 50°00′S 45°00′W
63 K24 **Falkland Islands** *var.* Falklands, Islas Malvinas. ◇ *UK dependent territory* SW Atlantic Ocean
47 W14 **Falkland Islands** *island group* SW Atlantic Ocean
65 I20 **Falkland Plateau** *var.* Argentine Rise. *undersea feature* SW Atlantic Ocean 51°00′S 50°00′W
Falklands *see* Falkland Islands
63 M23 **Falkland Sound** *var.* Estrecho de San Carlos. *strait* C Falkland Islands
Falknov nad Ohří *see* Sokolov
115 H21 **Falkonéra** *island* S Greece
95 K18 **Falköping** Västra Götaland, S Sweden 58°10′N 13°31′E
139 U8 **Fallāḩ** Wāsiţ, E Iraq 32°58′N 45°09′E
35 U16 **Fallbrook** California, W USA 33°22′N 117°15′W
189 U12 **Falleallej Pass** *passage* Chuuk Islands, C Micronesia
93 J14 **Fällfors** Västerbotten, N Sweden 65°07′N 20°46′E
194 I6 **Fallières Coast** *physical region* Antarctica
100 I11 **Fallingbostel** Niedersachsen, NW Germany 52°52′N 09°42′E
33 X9 **Fallon** Montana, NW USA 46°49′N 105°07′W
35 S5 **Fallon** Nevada, W USA 39°29′N 118°47′W
19 O12 **Fall River** Massachusetts, NE USA 41°42′N 71°09′W
27 P6 **Fall River Lake** ▧ Kansas, C USA
35 O3 **Fall River Mills** California, W USA 41°00′N 121°28′W
21 W4 **Falls Church** Virginia, NE USA 38°53′N 77°11′W
29 S17 **Falls City** Nebraska, C USA 40°03′N 95°36′W
25 S12 **Falls City** Texas, SW USA 28°58′N 98°01′W
Falluja *see* Al Fallūjah
77 S12 **Falmey** Dosso, SW Niger 12°29′N 02°58′E
45 W10 **Falmouth** Antigua, Antigua and Barbuda 17°02′N 61°47′W
44 J11 **Falmouth** W Jamaica 18°28′N 77°39′W
97 H25 **Falmouth** SW England, United Kingdom 50°08′N 05°04′W
20 M4 **Falmouth** Kentucky, S USA 38°40′N 84°20′W
19 P13 **Falmouth** Massachusetts, NE USA 41°31′N 70°36′W
21 W5 **Falmouth** Virginia, NE USA 38°19′N 77°28′W
189 U12 **Falos** *island* Chuuk, C Micronesia
83 E26 **False Bay** *Afr.* Valsbaai. *bay* SW South Africa
155 K17 **False Divi Point** *headland* E India 15°46′N 80°43′E
38 M16 **False Pass** Unimak Island, Alaska, USA 54°52′N 163°15′W
154 P12 **False Point** *headland* E India 20°23′N 86°52′E
105 U6 **Falset** Cataluña, NE Spain 41°08′N 00°49′E
95 I25 **Falster** *island* SE Denmark
116 K9 **Fălticeni** *Hung.* Falticsén. Suceava, NE Romania 47°27′N 26°20′E
Falticsén *see* Fălticeni
94 M13 **Falun** *var.* Fahlun. Kopparberg, C Sweden 60°36′N 15°36′E
Famagusta *see* Gazimağusa
Famagusta Bay *see* Gazimağusa Körfezi
62 I8 **Famatina** La Rioja, NW Argentina 28°58′S 67°46′W
99 J21 **Famenne** *physical region* SE Belgium
77 X15 **Fan** ≈ E Nigeria
76 M12 **Fana** Koulikoro, SW Mali 12°45′N 06°55′W
115 K19 **Fána** *ancient harbour* Chíos, SE Greece
189 V13 **Fanan** *island* Chuuk, C Micronesia
189 U12 **Fanapanges** *island* Chuuk, C Micronesia
115 L20 **Fanári, Akrotírio** *headland* Ikaría, Dodekánisa, Greece, Aegean Sea 37°40′N 26°21′E
45 Q13 **Fancy** Saint Vincent, Saint Vincent and the Grenadines 13°22′N 61°10′W
172 I5 **Fandriana** Fianarantsoa, SE Madagascar 20°14′S 47°21′E
167 O6 **Fang** Chiang Mai, NW Thailand 19°56′N 99°14′E
80 E13 **Fangak** Jonglei, E South Sudan 09°05′N 30°52′E
Fangatau *see* Fagatau
Fangataufa *see* Fagataufa
193 V15 **Fanga Uta** *bay* S Tonga
161 N7 **Fangcheng** Henan, C China 33°18′N 113°03′E
Fangcheng *see* Fangchenggang
160 K15 **Fangchenggang** *var.* Fangcheng Gezu Zizhixian; *prev.* Fangcheng. Guangxi Zhuangzu Zizhiqu, S China 21°49′N 108°21′E
Fangcheng Gezu Zizhixian *see* Fangchenggang
161 S15 **Fangshan** S Taiwan 22°19′N 120°41′E
163 X8 **Fangzheng** Heilongjiang, NE China 45°50′N 128°50′E
Fani *see* Fanit, Lumi i
119 K16 **Fanipal'** *Rus.* Fanipol'. Minskaya Voblasts', C Belarus 53°45′N 27°20′E
Fanipol' *see* Fanipal'
113 D22 **Fanit, Lumi i** *var.* Fani. ≈ N Albania
25 T13 **Fannin** Texas, SW USA 28°41′N 97°13′W
Fanning Island *see* Tabuaeran
94 G8 **Fannrem** Sør-Trøndelag, S Norway 63°16′N 09°48′E
106 I11 **Fano** *anc.* Colonia Julia Fanestris, Fanum Fortunae. Marche, C Italy 43°50′N 13°E
95 E23 **Fanø** *island* W Denmark
167 R5 **Fan Si Pan** ▲ N Vietnam 22°18′N 103°36′E
Fanum Fortunae *see* Fano
Fao *see* Al Fāw
141 W7 **Faq'** *var.* Al Faqa. Dubayy, E United Arab Emirates 24°42′N 55°37′E
Farab *see* Farap
185 G16 **Faraday, Mount** ▲ South Island, New Zealand 42°01′S 171°37′E
79 P16 **Faradje** Orientale, NE Dem. Rep. Congo 03°45′N 29°43′E
Faradofay *see* Tôlañaro
172 I7 **Farafangana** Fianarantsoa, SE Madagascar 22°50′S 47°50′E
148 J7 **Farāh** *var.* Farah, Fararud. Farāh, W Afghanistan 32°22′N 62°07′E
148 K7 **Farāh** ⬖ *province* W Afghanistan
148 J7 **Farāh Rūd** ≈ W Afghanistan
188 K7 **Farallon de Medinilla** *island* C Northern Mariana Islands
188 J2 **Farallon de Pajaros** *var.* Uracas. *island* N Northern Mariana Islands
76 J14 **Faranah** Haute-Guinée, S Guinea 10°02′N 10°44′W
146 K12 **Farap** *Rus.* Farab. Lebap Welaýaty, NE Turkmenistan 39°15′N 63°32′E
Fararud *see* Farāh
140 M13 **Farasān, Jazā'ir** *island group* SW Saudi Arabia
172 I5 **Faratsiho** Antananarivo, C Madagascar 19°24′S 46°57′E
188 K15 **Faraulep Atoll** *atoll* Caroline Islands, C Micronesia
99 H20 **Farciennes** Hainaut, S Belgium 50°26′N 04°33′E
105 O14 **Fardes** ≈ S Spain
191 S10 **Fare** Huahine, W French Polynesia 16°42′S 151°01′W
97 M23 **Fareham** S England, United Kingdom 50°51′N 01°10′W
39 P11 **Farewell** Alaska, USA 62°35′N 153°59′W
184 H13 **Farewell, Cape** *headland* South Island, New Zealand 40°30′S 172°39′E
Farewell, Cape *see* Nunap Isua
184 I13 **Farewell Spit** *spit* South Island, New Zealand
95 I17 **Färgelanda** Västra Götaland, S Sweden 58°34′N 11°59′E
Farghona, Wodii/ Farghona Valley *see* Fergana Valley
Farghona Wodiysi *see* Fergana Valley
23 V8 **Fargo** Georgia, SE USA 30°42′N 82°33′W
29 R5 **Fargo** North Dakota, N USA 46°53′N 96°47′W
147 S10 **Farg'ona** *Rus.* Fergana; *prev.* Novyy Margilan. Farg'ona Viloyati, E Uzbekistan 40°28′N 71°44′E
147 R10 **Farg'ona Viloyati** *Rus.* Ferganskaya Oblast'. ⬖ *province* E Uzbekistan
29 V10 **Faribault** Minnesota, N USA 44°18′N 93°16′W
152 J11 **Farīdābād** Haryāna, N India 28°26′N 77°190′E
152 H8 **Farīdkot** Punjab, NW India 30°42′N 74°47′E
153 T15 **Faridpur** Dhaka, C Bangladesh 23°29′N 89°50′E
121 P14 **Fārigh, Wādī al** ≈ N Libya
172 I4 **Farihy Alaotra** ◎ C Madagascar
94 M11 **Färila** Gävleborg, C Sweden 61°48′N 15°55′E
104 E9 **Farilhões** *island* C Portugal
76 G12 **Farim** NW Guinea-Bissau 12°30′N 15°09′W
Farish *see* Forish
141 T11 **Fāris, Qalamat** *well* SE Saudi Arabia
95 N21 **Färjestaden** Kalmar, S Sweden 56°38′N 16°30′E
149 R2 **Farkhār** Takhār, NE Afghanistan 36°39′N 69°43′E
147 Q14 **Farkhor** *Rus.* Parkhar. SW Tajikistan 37°32′N 69°22′E
116 F12 **Fârliug** *prev.* Fîrliug, *Hung.* Furluk. Caraş-Severin, SW Romania 45°21′N 21°55′E
115 M21 **Farmakonísi** *island* Dodekánisa, Greece, Aegean Sea
30 M13 **Farmer City** Illinois, N USA 40°14′N 88°38′W
31 N14 **Farmersburg** Indiana, N USA 39°14′N 87°21′W
25 U6 **Farmersville** Texas, SW USA 33°09′N 96°21′W
22 H5 **Farmerville** Louisiana, S USA 32°46′N 92°24′W
29 X16 **Farmington** Iowa, C USA 40°37′N 91°43′W
19 Q6 **Farmington** Maine, NE USA 44°40′N 70°09′W
29 V9 **Farmington** Minnesota, N USA 44°39′N 93°09′W
27 X6 **Farmington** Missouri, C USA 37°46′N 90°26′W
19 O9 **Farmington** New Hampshire, NE USA 43°23′N 71°04′W
37 P9 **Farmington** New Mexico, SW USA 36°44′N 108°13′W
36 L2 **Farmington** Utah, W USA 40°58′N 111°53′W
21 W9 **Farmville** North Carolina, SE USA 35°37′N 77°36′W
21 U6 **Farmville** Virginia, NE USA 37°17′N 78°25′W
97 N22 **Farnborough** S England, United Kingdom 51°17′N 00°46′W
97 N22 **Farnham** S England, United Kingdom 51°13′N 00°49′W
10 J7 **Faro** Yukon, W Canada 62°15′N 133°30′W
104 G14 **Faro** Faro, S Portugal 37°01′N 07°56′W
95 Q18 **Fårö** Gotland, SE Sweden 57°55′N 19°10′E
104 G14 **Faro** ⬖ *district* S Portugal
78 F13 **Faro** ≈ Cameroon/Nigeria
104 G14 **Faro** ✈ Faro, S Portugal 37°02′N 08°01′W
64 M5 **Faroe-Iceland Ridge** *var.* Faeroe-Iceland Ridge. *undersea ridge* NW Norwegian Sea
86 C8 **Faroe Islands** *var.* Faeroe Islands. *island group* N Atlantic Ocean
64 M5 **Faroe Islands** *var.* Faeroe Islands, *Dan.* Færøerne, *Faer.* Føroyar. *Self-governing territory of Denmark* N Atlantic Ocean
64 N6 **Faroe-Shetland Trough** *var.* Faeroe-Shetland Trough. *trough* NE Atlantic Ocean
Faro, Punta del *see* Peloro, Capo
95 Q18 **Fårösund** Gotland, SE Sweden 57°51′N 19°02′E
173 N7 **Farquhar Group** *island group* S Seychelles
18 B13 **Farrell** Pennsylvania, NE USA 41°12′N 80°28′W
152 K11 **Farrukhābād** Uttar Pradesh, N India 27°24′N 79°34′E
143 P11 **Fārs** *off.* Ostān-e Fārs; *anc.* Persis. ⬖ *province* S Iran
115 F16 **Fársala** Thessalía, C Greece 39°17′N 22°23′E
143 R4 **Fārsīān** Golestán, N Iran
Fars, Khalīj-e *see* Persian Gulf
95 G21 **Farsø** Nordjylland, N Denmark 56°47′N 09°21′E
Fārs, Ostān-e *see* Fārs
95 D18 **Farsund** Vest-Agder, S Norway 58°05′N 06°49′E
141 U14 **Fartak, Ra's** *headland* E Yemen 15°34′N 52°19′E
60 H13 **Fartura, Serra da** ▲▲ S Brazil
Farvel, Kap *see* Nunap Isua
24 L4 **Farwell** Texas, SW USA 34°23′N 103°03′W
194 I9 **Farwell Island** *island* Antarctica
152 L9 **Far Western** ⬖ *zone* W Nepal
148 M3 **Fāryāb** ⬖ *province* N Afghanistan
143 P12 **Fasā** Fārs, S Iran 28°55′N 53°39′E
141 U12 **Fasad, Ramlat** *desert* SW Oman
107 P17 **Fasano** Puglia, SE Italy 40°50′N 17°21′E
92 L3 **Fáskrúðsfjörður** Austurland, E Iceland 64°55′N 14°01′W
117 O5 **Fastiv** *Rus.* Fastov. Kyyivs'ka Oblast', NW Ukraine 50°08′N 29°59′E
97 B22 **Fastnet Rock** *Ir.* Carraig Aonair. *island* SW Ireland
Fastov *see* Fastiv
190 C9 **Fatato** *island* Funafuti Atoll, C Tuvalu
152 K12 **Fatehgarh** Uttar Pradesh, N India 27°22′N 79°38′E
149 U6 **Fatehjang** Punjab, E Pakistan 33°33′N 72°43′E
152 G11 **Fatehpur** Rājasthān, N India 27°59′N 74°58′E
152 L13 **Fatehpur** Uttar Pradesh, N India 25°56′N 80°55′E
126 J7 **Fatezh** Kurskaya Oblast', W Russian Federation 52°01′N 35°51′E
76 G11 **Fatick** W Senegal 14°19′N 16°27′W
104 G9 **Fátima** Santarém, W Portugal 39°37′N 08°39′W
136 M11 **Fatsa** Ordu, N Turkey 41°02′N 37°31′E
Fatshan *see* Foshan
190 D12 **Fatua, Pointe** *var.* Pointe Nord. *headland* Île Futuna, S Wallis and Futuna
191 X7 **Fatu Hiva** *island* Îles Marquises, NE French Polynesia
Fatunda *see* Fatundu
79 H21 **Fatundu** *var.* Fatunda. Bandundu, W Dem. Rep. Congo 04°08′S 17°13′E
29 O8 **Faulkton** South Dakota, N USA 45°02′N 99°07′W
116 L13 **Făurei** *prev.* Filimon Sîrbu. Brăila, SE Romania 45°05′N 27°15′E
92 G12 **Fauske** Nordland, C Norway 67°15′N 15°27′E
11 P13 **Faust** Alberta, W Canada 55°19′N 115°33′W
99 L23 **Fauvillers** Luxembourg, SE Belgium 49°52′N 05°40′E
107 J24 **Favara** Sicilia, Italy, C Mediterranean Sea 37°19′N 13°40′E
Faventia *see* Faenza
107 G23 **Favignana, Isola** *island* Isole Egadi, S Italy
12 D8 **Fawn** ≈ Ontario, SE Canada
Faxa Bay *see* Faxaflói
92 H3 **Faxaflói** *Eng.* Faxa Bay. *bay* W Iceland
78 I7 **Faya** *prev.* Faya-Largeau, Largeau. Borkou, N Chad 17°58′N 19°06′E
Faya-Largeau *see* Faya
187 Q16 **Fayaoué** Province des Îles Loyauté, C New Caledonia 20°41′S 166°31′E
138 M5 **Fayḑāt** *hill range* E Syria
23 O3 **Fayette** Alabama, S USA 33°40′N 87°49′W
29 X12 **Fayette** Iowa, C USA 42°50′N 91°48′W
22 J6 **Fayette** Mississippi, S USA 31°42′N 91°03′W
27 U4 **Fayette** Missouri, C USA 39°09′N 92°40′W
27 S9 **Fayetteville** Arkansas, C USA 36°04′N 94°10′W
21 U10 **Fayetteville** North Carolina, SE USA 35°03′N 78°53′W
20 J10 **Fayetteville** Tennessee, S USA 35°08′N 86°33′W
25 U11 **Fayetteville** Texas, SW USA 29°52′N 96°40′W
21 R5 **Fayetteville** West Virginia, NE USA 38°03′N 81°09′W
141 R4 **Faylakah** *var.* Failaka Island. *island* E Kuwait
139 T10 **Fayşalīyah** *var.* Faisaliya. Al Qādisīyah, S Iraq 31°48′N 44°36′E
189 P15 **Fayu** *var.* East Fayu. *island* Hall Islands, C Micronesia
152 G8 **Fāzilka** Punjab, NW India 30°26′N 74°04′E
Fdérick *see* Fdérik
76 I6 **Fdérik** *var.* Fdérick, *Fr.* Fort Gouraud. Tiris Zemmour, NW Mauritania 22°40′N 12°41′W
Feabhail, Loch *see* Foyle, Lough
97 B20 **Feale** ≈ SW Ireland
21 V12 **Fear, Cape** *headland* Bald Head Island, North Carolina, SE USA 33°50′N 77°57′W
35 O6 **Feather River** ≈ California, W USA
185 M14 **Featherston** Wellington, North Island, New Zealand 41°07′S 175°28′E
102 L3 **Fécamp** Seine-Maritime, N France 49°45′N 00°22′E
Fédala *see* Mohammedia
61 D17 **Federación** Entre Ríos, E Argentina 31°00′S 57°55′W
61 D17 **Federal** Entre Ríos, E Argentina 30°55′S 58°45′W
77 T15 **Federal Capital District** ⬖ *capital territory* C Nigeria
Federal Capital Territory *see* Australian Capital Territory
Federal District *see* Distrito Federal
21 Y4 **Federalsburg** Maryland, NE USA 38°41′N 75°46′W
74 M6 **Fedjaj, Chott el** *var.* Chott el Fejaj, Shaţţ al Fijāj. *salt lake* C Tunisia
94 B13 **Fedje** *island* S Norway
144 M7 **Fedorovka** Kostanay, N Kazakhstan 51°12′N 52°00′E
127 U6 **Fedorovka** Respublika Bashkortostan, W Russian Federation 53°09′N 55°07′E
Fëdory *see* Fyadory
117 U11 **Fedotova Kosa** *spit* SE Ukraine
189 V13 **Fefan** *atoll* Chuuk Islands, C Micronesia
111 O21 **Fehérgyarmat** Szabolcs-Szatmár-Bereg, E Hungary 47°59′N 22°29′E
Fehér-Körös *see* Crişul Alb
Fehértemplom *see* Bela Crkva
Fehérvölgy *see* Albac
100 L7 **Fehmarn** *island* N Germany
95 H25 **Fehmarn Belt** *Dan.* Femern Bælt, *Ger.* Fehmarnbelt. *strait* Denmark /Germany *see also* Femern Bælt
Fehmarnbelt *see* Fehmarn Belt/Femer Bælt
109 X8 **Fehring** Steiermark, SE Austria 46°56′N 16°00′E
59 B15 **Feijó** Acre, W Brazil 08°07′S 70°27′W
184 M12 **Feilding** Manawatu-Wanganui, North Island, New Zealand 40°15′S 175°34′E
Feira *see* Feira de Santana
59 O17 **Feira de Santana** *var.* Feira. Bahia, E Brazil 12°17′S 38°53′W
109 X7 **Feistritz** ≈ SE Austria
Feistritz *see* Ilirska Bistrica
161 P8 **Feixi** *var.* Shangpai; *prev.* Shangpaihe. Anhui, E China 31°40′N 117°08′E
149 S2 **Feiẕābād** *var.* Faizabad, Faizābād, Feyzābād, Fyzabad; *prev.* Feyẕābād. Badakhshān, NE Afghanistan 37°06′N 70°34′E
Fejaj, Chott el *see* Fedjaj, Chott el
111 I23 **Fejér** *off.* Fejér Megye. ⬖ *county* W Hungary
Fejér Megye *see* Fejér
95 I24 **Fejø** *island* SE Denmark
136 K15 **Feke** Adana, S Turkey 37°49′N 35°55′E
Feketehalom *see* Codlea
Fekete-Körös *see* Crişul Negru
105 Y9 **Felanitx** Mallorca, Spain, W Mediterranean Sea 39°28′N 03°08′E
109 T3 **Feldaist** ≈ N Austria
109 W8 **Feldbach** Steiermark, SE Austria 46°58′N 15°53′E
101 F24 **Feldberg** ▲ SW Germany 47°52′N 08°01′E
116 J12 **Feldioara** *Ger.* Marienburg, *Hung.* Földvár. Braşov, C Romania 45°49′N 25°36′E
108 I7 **Feldkirch** *anc.* Clunia. Vorarlberg, W Austria 47°15′N 09°38′E
109 S9 **Feldkirchen in Kärnten** *Slvn.* Trg. Kärnten, S Austria 46°42′N 14°01′E
Félegyháza *see* Kiskunfélegyháza
192 H16 **Feleolo** ✈ (Āpia) Upolu, C Samoa 13°49′S 171°59′W
104 H6 **Felgueiras** Porto, N Portugal 41°22′N 08°12′W
Felicitas Julia *see* Lisboa
172 J16 **Félicité** *island* Inner Islands, NE Seychelles
151 K20 **Felidhu Atoll** *atoll* C Maldives
41 Y13 **Felipe Carrillo Puerto** Quintana Roo, SE Mexico 19°34′N 88°02′W
97 Q21 **Felixstowe** E England, United Kingdom 51°58′N 01°20′E
103 N11 **Felletin** Creuse, C France 45°53′N 02°12′E
Fellin *see* Viljandi
Felsőbánya *see* Baia Sprie
Felsőmuzslya *see* Mužlja
Felsővisó *see* Vişeu de Sus
35 N10 **Felton** California, W USA 37°03′N 122°04′W
106 H7 **Feltre** Veneto, NE Italy 46°01′N 11°55′E
95 H25 **Femer Bælt** *Dan.* Fehmarn Belt, *Ger.* Fehmarnbelt. *strait* Denmark/Germany *see also* Fehmarn Belt
95 I24 **Femø** *island* SE Denmark
94 I10 **Femunden** ◎ S Norway
104 H2 **Fene** Galicia, NW Spain 43°28′N 08°10′W
14 I14 **Fenelon Falls** Ontario, SE Canada 44°34′N 78°43′W
189 U13 **Feneppi** *atoll* Chuuk Islands, C Micronesia
137 O11 **Fener Burnu** *headland* N Turkey 41°07′N 39°26′E
Fénérive *see* Fenoarivo Atsinanana
115 J14 **Fengári** ▲ Samothráki, E Greece 40°27′N 25°37′E
163 V13 **Fengcheng** *var.* Feng-cheng, Fenghwangcheng. Liaoning, NE China 40°28′N 124°01′E
Fengcheng *see* Lianjiang
Feng-cheng *see* Fengcheng
160 K11 **Fenggang** *var.* Longquan. Guizhou, S China 27°57′N 107°42′E
161 O7 **Feng He** ≈ C China
161 S9 **Fenghua** Zhejiang, SE China 29°40′N 121°25′E
Fenghwangcheng *see* Fengcheng
Fengjiaba *see* Wangcang
160 L9 **Fengjie** *var.* Yong'an. Sichuan, C China 31°03′N 109°31′E
160 M14 **Fengkai** *var.* Jiangkou. Guangdong, S China 23°26′N 111°28′E
161 T13 **Fenglin** *Jap.* Hōrin. C Taiwan 23°52′N 121°30′E
161 P1 **Fengning** *prev.* Dagezhen. Hebei, E China 41°12′N 116°37′E
160 E13 **Fengqing** *var.* Fengshan. Yunnan, SW China 24°38′N 99°54′E
161 O6 **Fengqiu** Henan, C China 35°02′N 114°24′E
161 Q2 **Fengrun** Hebei, E China 39°50′N 118°10′E
Fengshan *see* Luoyuan, Fujian, China
Fengshan *see* Fengqing, Yunnan, China
163 T4 **Fengshui Shan** ▲ NE China 52°20′N 123°22′E
161 P14 **Fengshun** Guangdong, S China 23°51′N 116°11′E
Fengtien *see* Liaoning, China
Fengtien *see* Shenyang, China
160 J7 **Fengxian** *var.* Feng Xian; *prev.* Shuangshipu. Shaanxi, C China 33°50′N 106°33′E
Feng Xian *see* Fengxian
Fengxiang *see* Luobei
Fengyizhen *see* Maoxian
163 P13 **Fengzhen** Nei Mongol Zizhiqu, N China 40°25′N 113°09′E
160 M6 **Fen He** ≈ C China
153 V15 **Feni** Chittagong, E Bangladesh 23°00′N 91°24′E
186 I6 **Feni Islands** *island group* NE Papua New Guinea
38 H17 **Fenimore Pass** *strait* Aleutian Islands, Alaska, USA
84 B9 **Feni Ridge** *undersea feature* N Atlantic Ocean 53°45′N 18°00′W
Fennern *see* Vändra
30 J9 **Fennimore** Wisconsin, N USA 42°58′N 90°39′W
172 J4 **Fenoarivo Atsinanana** *prev./Fr.* Fénérive. Toamasina, E Madagascar 20°52′S 46°52′E
95 I24 **Fensmark** Sjælland, SE Denmark 55°17′N 11°48′E
97 O19 **Fens, The** *wetland* E England, United Kingdom
31 R9 **Fenton** Michigan, N USA 42°48′N 83°42′W
190 K10 **Fenua Fala** *island* SE Tokelau
190 F12 **Fenuafo'ou, Île** *island* E Wallis and Futuna
190 L10 **Fenua Loa** *island* Fakaofo Atoll, E Tokelau
160 M4 **Fenyang** Shanxi, C China 37°14′N 111°40′E
117 U13 **Feodosiya** *var.* Kefe, *It.* Kaffa; *anc.* Theodosia. Avtonomna Respublika Krym, S Ukraine 45°03′N 35°24′E
94 I10 **Feragen** ◎ S Norway
74 L5 **Fer, Cap de** *headland* NE Algeria 37°05′N 07°10′E
31 O16 **Ferdinand** Indiana, N USA 38°13′N 86°51′W
Ferdinand *see* Montana, Bulgaria
Ferdinand *see* Mihail Kogălniceanu, Romania
Ferdinandsberg *see* Oţelu Roşu
143 T7 **Ferdows** *var.* Firdaus; *prev.* Tūn. Khorāsān-e Razavī, E Iran 34°00′N 58°09′E
103 Q5 **Fère-Champenoise** Marne, N France 48°45′N 03°59′E
Ferencz-József Csúcs *see* Gerlachovský štít
107 J16 **Ferentino** Lazio, C Italy 41°40′N 13°16′E
114 L13 **Féres** Anatolikí Makedonía kai Thráki, NE Greece 40°54′N 26°12′E
Fergana *see* Farg'ona
147 S10 **Fergana Valley** *var.* Farghona Valley, *Rus.* Ferganskaya Dolina, *Taj.* Wodii Farghona, *Uzb.* Farghona Wodiysi. *basin* Tajikistan/Uzbekistan
Ferganskaya Dolina *see* Fergana Valley
Ferganskaya Oblast' *see* Farg'ona Viloyati
147 U9 **Ferganskiy Khrebet** ▲▲ C Kyrgyzstan
14 F15 **Fergus** Ontario, S Canada 43°42′N 80°22′W
29 S6 **Fergus Falls** Minnesota, N USA 46°15′N 96°02′W
186 G9 **Fergusson Island** *var.* Kaluwawa. *island* SE Papua New Guinea
111 K22 **Ferihegy** ✈ (Budapest) Budapest, C Hungary 47°25′N 19°13′E
113 N17 **Ferizaj** *Serb.* Uroševac. C Kosovo 42°23′N 21°09′E
77 N14 **Ferkessédougou** N Ivory Coast 09°36′N 05°12′W
109 T10 **Ferlach** *Slvn.* Borovlje. Kärnten, S Austria 46°31′N 14°18′E
97 E16 **Fermanagh** *cultural region* SW Northern Ireland, United Kingdom
106 J13 **Fermo** *anc.* Firmum Picenum. Marche, C Italy 43°09′N 13°44′E
104 J6 **Fermoselle** Castilla y León, N Spain 41°19′N 06°24′W
97 D20 **Fermoy** *Ir.* Mainistir Fhear Maí. SW Ireland 52°08′N 08°16′W
23 W8 **Fernandina Beach** Amelia Island, Florida, SE USA 30°40′N 81°27′W
57 A17 **Fernandina, Isla** *var.* Narborough Island. *island* Galapagos Islands, Ecuador, E Pacific Ocean
47 Y5 **Fernando de Noronha** *island* E Brazil
Fernando Po/Fernando Póo *see* Bioco, Isla de
60 J7 **Fernandópolis** São Paulo, S Brazil 20°18′S 50°13′W
104 M13 **Fernán Núñez** Andalucía, S Spain 37°40′N 04°44′W
83 Q14 **Fernão Veloso, Baia de** *bay* NE Mozambique
34 K3 **Ferndale** California, W USA 40°34′N 124°16′W
32 H6 **Ferndale** Washington, NW USA 48°51′N 122°35′W
11 P17 **Fernie** British Columbia, SW Canada 49°30′N 115°00′W
35 R5 **Fernley** Nevada, W USA 39°35′N 119°15′W
Ferozepore *see* Fīrozpur
107 N18 **Ferrandina** Basilicata, S Italy 40°30′N 16°25′E
106 G9 **Ferrara** *anc.* Forum Alieni. Emilia-Romagna, N Italy 44°50′N 11°36′E
120 F9 **Ferrat, Cap** *headland* NW Algeria 35°52′N 00°24′W
107 D20 **Ferrato, Capo** *headland* Sardegna, Italy, C Mediterranean Sea 39°18′N 09°37′E
104 G12 **Ferreira do Alentejo** Beja, S Portugal 38°04′N 08°06′W
56 B11 **Ferreñafe** Lambayeque, W Peru 06°42′S 79°45′W
108 C12 **Ferret** Valais, SW Switzerland 45°57′N 07°04′E
102 I13 **Ferret, Cap** *headland* W France 44°37′N 01°15′W
22 I6 **Ferriday** Louisiana, S USA 31°37′N 91°33′W
Ferro *see* Hierro
107 D16 **Ferro, Capo** *headland* Sardegna, Italy, C Mediterranean Sea 41°09′N 09°31′E
104 H2 **Ferrol** *var.* El Ferrol; *prev.* El Ferrol del Caudillo. Galicia, NW Spain 43°29′N 08°14′W
56 B12 **Ferrol, Península de** *peninsula* W Peru
36 M5 **Ferron** Utah, W USA 39°05′N 111°07′W
21 S7 **Ferrum** Virginia, NE USA 36°54′N 80°01′W
23 O8 **Ferry Pass** Florida, SE USA 30°30′N 87°12′W
Ferryville *see* Menzel Bourguiba
29 S4 **Fertile** Minnesota, N USA 47°32′N 96°16′W
Fertő *see* Neusiedler See
Ferwerd *see* Ferwert
98 L5 **Ferwert** *Dutch.* Ferwerd. Fryslân, N Netherlands 53°21′N 05°47′E
74 G6 **Fès** *Eng.* Fez. N Morocco 34°06′N 04°57′W
79 I22 **Feshi** Bandundu, SW Dem. Rep. Congo 06°08′S 18°12′E
29 O4 **Fessenden** North Dakota, N USA 47°36′N 99°37′W
Festenberg *see* Twardogóra
27 X5 **Festus** Missouri, C USA 38°13′N 90°24′W
116 M14 **Feteşti** Ialomiţa, SE Romania 44°22′N 27°51′E
136 D17 **Fethiye** Muğla, SW Turkey 36°37′N 29°08′E
96 M1 **Fetlar** *island* NE Scotland, United Kingdom
95 I15 **Fetsund** Akershus, S Norway 59°55′N 11°03′E
12 L5 **Feuilles, Lac aux** ◎ Québec, E Canada
12 L5 **Feuilles, Rivière aux** ≈ Québec, E Canada
99 M23 **Feulen** Diekirch, C Luxembourg 49°52′N 06°03′E
103 Q11 **Feurs** Loire, E France 45°44′N 04°14′E
95 F18 **Fevik** Aust-Agder, S Norway 58°22′N 08°40′E
123 R13 **Fevral'sk** Amurskaya Oblast', SE Russian Federation 52°25′N 131°06′E
Feyzābād *see* Feiẕābād
Feyẕābād *see* Feiẕābād
Fez *see* Fès
97 J19 **Ffestiniog** NW Wales, United Kingdom 52°55′N 03°54′W
Fhóid Duibh, Cuan an *see* Blacksod Bay
62 I8 **Fiambalá** Catamarca, NW Argentina 27°45′S 67°37′W
172 I6 **Fianarantsoa** Fianarantsoa, C Madagascar 21°27′S 47°05′E
172 H6 **Fianarantsoa** ⬖ *province* SE Madagascar
78 G12 **Fianga** Mayo-Kébbi Est, SW Chad 09°57′N 15°09′E
Ficce *see* Fichē
80 J12 **Fichē** *It.* Ficce. Oromīya, C Ethiopia 09°48′N 38°43′E
101 N17 **Fichtelberg** ▲ Czech Republic/Germany 50°26′N 12°57′E
101 M18 **Fichtelgebirge** ▲▲ SE Germany
101 M19 **Fichtelnaab** ≈ SE Germany
106 E9 **Fidenza** Emilia-Romagna, N Italy 44°52′N 10°04′E
113 K21 **Fier** *var.* Fieri. Fier, SW Albania 40°44′N 19°34′E
113 K21 **Fier** ⬖ *district* W Albania
Fieri *see* Fier
Fierza *see* Fierzë

◆ Country
● Country Capital
◇ Dependent Territory
○ Dependent Territory Capital
⬖ Administrative Regions
✈ International Airport
▲ Mountain
▲▲ Mountain Range
⛰ Volcano
≈ River
◎ Lake
▧ Reservoir

113 L17 **Fierzë** *var.* Fierza. Shkodër, N Albania 42°15′N 20°02′E
113 L17 **Fierzës, Liqeni i** ◎ N Albania
108 F10 **Fiesch** Valais, SW Switzerland 46°25′N 08°09′E
106 G11 **Fiesole** Toscana, C Italy 43°50′N 11°18′E
138 G12 **Fīfah** Aṭ Ṭafīlah, W Jordan 30°55′N 35°25′E
96 K11 **Fife** *var.* Kingdom of Fife. *cultural region* E Scotland, United Kingdom
Fife, Kingdom of *see* Fife
96 K11 **Fife Ness** *headland* E Scotland, United Kingdom 56°16′N 02°35′W
Fifteen Twenty Fracture Zone *see* Barracuda Fracture Zone
103 N13 **Figeac** Lot, S France 44°37′N 02°01′E
95 N19 **Figeholm** Kalmar, SE Sweden 57°12′N 16°34′E
Figig *see* Figuig
83 J18 **Figtree** Matabeleland South, SW Zimbabwe 20°24′S 28°21′E
104 F8 **Figueira da Foz** Coimbra, W Portugal 40°09′N 08°51′W
105 X4 **Figueres** Cataluña, E Spain 42°16′N 02°57′E
74 H7 **Figuig** *var.* Figig. E Morocco 32°09′N 01°13′W
Fijājj, Shaṭṭ al *see* Fedjaj, Chott el
187 Y15 **Fiji** *off.* Republic of Fiji, *prev.* Sovereign Democratic Republic of Fiji, *prev.* Republic of the Fiji Islands, *Fij.* Viti. ◆ *republic* SW Pacific Ocean
192 K9 **Fiji** *island group* SW Pacific Ocean
Fiji Islands, Republic of the *see* Fiji
175 Q8 **Fiji Plate** *tectonic feature*
Fiji, Republic of *see* Fiji
Fiji, Sovereign Democratic Republic of *see* Fiji
105 P14 **Filabres, Sierra de los** ▲▲ SE Spain
83 K18 **Filabusi** Matabeleland South, S Zimbabwe 20°34′S 29°20′E
42 K13 **Filadelfia** Guanacaste, W Costa Rica 10°28′N 85°33′W
111 K20 **Fil'akovo** *Hung.* Fülek. Banskobýstricky Kraj, C Slovakia 48°15′N 19°53′E
195 N5 **Filchner Ice Shelf** *ice shelf* Antarctica
14 J11 **Fildegrand** ≈ Québec, SE Canada
33 O15 **Filer** Idaho, NW USA 42°34′N 114°36′W
Filevo *see* Varbitsa
116 H14 **Filiaşi** Dolj, SW Romania 44°32′N 23°31′E
115 B16 **Filiátes** Ípeiros, W Greece 39°38′N 20°16′E
115 D21 **Filiatrá** Pelopónnisos, S Greece 37°09′N 21°35′E
107 K22 **Filicudi, Isola** *island* Isole Eolie, S Italy
141 Y10 **Filim** E Oman 20°37′N 58°11′E
Filimon Sîrbu *see* Făurei
77 S11 **Filingué** Tillabéri, W Niger 14°21′N 03°22′E
Filiouri *see* Líssos
114 I13 **Fílippoi** *anc.* Philippi. *site of ancient city* Anatolikí Makedonía kai Thráki, NE Greece
95 L15 **Filipstad** Värmland, C Sweden 59°44′N 14°10′E
108 I9 **Filisur** Graubünden, S Switzerland 46°40′N 09°43′E
94 E12 **Fillefjell** ▲▲ S Norway
35 R14 **Fillmore** California, W USA 34°23′N 118°56′W
36 K5 **Fillmore** Utah, W USA 38°57′N 112°19′W
14 J10 **Fils, Lac du** ◎ Québec, SE Canada
Filyos Çayı *see* Yenice Çayı
Fimbul Ice Shelf *see* Fimbulisen
195 Q2 **Fimbutheimen** *physical region* Antarctica
106 G9 **Finale Emilia** Emilia-Romagna, C Italy 44°50′N 11°17′E
106 C10 **Finale Ligure** Liguria, NW Italy 44°11′N 08°22′E
105 P14 **Fiñana** Andalucía, S Spain 37°09′N 02°47′W
21 S6 **Fincastle** Virginia, NE USA 37°30′N 79°54′W
99 M25 **Findel** ✈ (Luxembourg) Luxembourg, C Luxembourg 49°39′N 06°16′E
96 J9 **Findhorn** ≈ N Scotland, United Kingdom
31 R12 **Findlay** Ohio, N USA 41°02′N 83°40′W
18 G11 **Finger Lakes** ◎ New York, NE USA
83 L14 **Fingoè** Tete, NW Mozambique 15°10′S 31°51′E
136 E17 **Finike** Antalya, SW Turkey 36°18′N 30°08′E
102 F6 **Finistère** ◈ *department* NW France
186 D7 **Finisterre Range** ▲▲ N Papua New Guinea
181 Q8 **Finke** Northern Territory, N Australia 25°37′S 134°35′E
109 S10 **Finkenstein** Kärnten, S Austria 46°34′N 13°53′E
189 Y15 **Finkol, Mount** *var.* Mount Crozer. ▲ Kosrae, E Micronesia 05°18′N 163°00′E
93 L17 **Finland** *off.* Republic of Finland, *Fin.* Suomen Tasavalta, Suomi. ◆ *republic* N Europe
124 F12 **Finland, Gulf of** *Est.* Soome Laht, *Fin.* Suomenlahti, *Ger.* Finnischer Meerbusen, *Rus.* Finskiy Zaliv, *Swe.* Finska Viken. *gulf* E Baltic Sea
Finland, Republic of *see* Finland
10 L11 **Finlay** ≈ British Columbia, W Canada
183 O10 **Finley** New South Wales, SE Australia 35°41′S 145°33′E
29 Q4 **Finley** North Dakota, N USA 47°30′N 97°50′W
Finnischer Meerbusen *see* Finland, Gulf of
92 K9 **Finnmark** ◈ *county* N Norway
92 K9 **Finnmarksvidda** *physical region* N Norway
92 I9 **Finnsnes** Troms, N Norway 69°16′N 18°00′E
186 E7 **Finschhafen** Morobe, C Papua New Guinea 06°35′S 147°51′E
94 E13 **Finse** Hordaland, S Norway 60°35′N 07°33′E
Finska Viken/Finskiy Zaliv *see* Finland, Gulf of
95 M17 **Finspång** Östergötland, S Sweden 58°42′N 15°45′E
108 F10 **Finsteraarhorn** ▲ Switzerland 46°33′N 08°07′E
101 O14 **Finsterwalde** Brandenburg, E Germany 51°38′N 13°43′E
185 A23 **Fiordland** *physical region* South Island, New Zealand
106 E9 **Fiorenzuola d'Arda** Emilia-Romagna, C Italy 44°57′N 09°53′E
Firat Nehri *see* Euphrates
Firdaus *see* Ferdows
18 M14 **Fire Island** *island* New York, NE USA
106 G11 **Firenze** *Eng.* Florence; *anc.* Florentia. Toscana, C Italy 43°47′N 11°15′E
106 G10 **Firenzuola** Toscana, C Italy 44°07′N 11°22′E
14 C6 **Fire River** Ontario, S Canada 48°46′N 83°34′W
Fîrliug *see* Fârliug
61 B19 **Firmat** Santa Fe, C Argentina 33°29′S 61°29′W
103 Q12 **Firminy** Loire, E France 45°22′N 04°18′E
Firmum Picenum *see* Fermo
152 J12 **Fīrozābād** Uttar Pradesh, N India 27°09′N 78°24′E
152 G8 **Fīrozpur** *var.* Ferozepore. Punjab, NW India 30°55′N 74°38′E
First State *see* Delaware
143 O12 **Fīrūzābād** Fārs, S Iran 28°51′N 52°35′E
Fischamend *see* Fischamend Markt
109 Y4 **Fischamend Markt** *var.* Fischamend. Niederösterreich, NE Austria 48°08′N 16°37′E
109 W6 **Fischbacher Alpen** ▲▲ E Austria
Fischhausen *see* Primorsk
83 D21 **Fish** *var.* Vis. ≈ S Namibia
83 F24 **Fish** *Afr.* Vis. ≈ SW South Africa
11 X15 **Fisher Branch** Manitoba, S Canada 51°09′N 97°34′W
11 X15 **Fisher River** Manitoba, S Canada 51°25′N 97°23′W
19 N13 **Fishers Island** *island* New York, NE USA
37 U8 **Fishers Peak** ▲ Colorado, C USA 37°06′N 104°27′W
9 P9 **Fisher Strait** *strait* Nunavut, N Canada
97 H21 **Fishguard** *Wel.* Abergwaun. SW Wales, United Kingdom 51°59′N 04°49′W
19 R2 **Fish River Lake** ◎ Maine, NE USA
194 K6 **Fiske, Cape** *headland* Antarctica 74°27′S 60°28′W
103 P4 **Fismes** Marne, N France 49°19′N 03°41′E
104 F3 **Fisterra, Cabo** *headland* NW Spain 42°53′N 09°16′W
19 N11 **Fitchburg** Massachusetts, NE USA 42°34′N 71°48′W
96 L3 **Fitful Head** *headland* NE Scotland, United Kingdom 59°57′N 01°24′W
95 C14 **Fitjar** Hordaland, S Norway 59°55′N 05°19′E
192 H16 **Fito, Mauga** ▲ Upolu, C Samoa 13°57′S 171°42′W
23 U6 **Fitzgerald** Georgia, SE USA 31°42′N 83°15′W
180 M5 **Fitzroy Crossing** Western Australia 18°10′S 125°40′E
63 G21 **Fitzroy, Monte** *var.* Cerro Chaltel. ▲ S Argentina 49°18′S 73°06′W
181 Y8 **Fitzroy River** ≈ Queensland, E Australia
180 L5 **Fitzroy River** ≈ Western Australia
14 E12 **Fitzwilliam Island** *island* Ontario, S Canada
107 J15 **Fiuggi** Lazio, C Italy 41°47′N 13°16′E
Fiume *see* Rijeka
107 H15 **Fiumicino** Lazio, C Italy 41°46′N 12°13′E
Fiumicino *see* Leonardo da Vinci
106 E10 **Fivizzano** Toscana, C Italy 44°13′N 10°06′E
79 O21 **Fizi** Sud-Kivu, E Dem. Rep. Congo 04°15′S 28°57′E
Fizuli *see* Füzuli
92 I11 **Fjällåsen** Norrbotten, N Sweden 67°31′N 20°08′E
95 G20 **Fjerritslev** Nordjylland, N Denmark 57°06′N 09°17′E
F.J.S. *see* Franz Josef Strauss
95 L16 **Fjugesta** Örebro, C Sweden 59°10′N 14°50′E
Fladstrand *see* Frederikshavn
37 V5 **Flagler** Colorado, C USA 39°17′N 103°04′W
23 X10 **Flagler Beach** Florida, SE USA 29°28′N 81°07′W
36 L11 **Flagstaff** Arizona, SW USA 35°12′N 111°39′W
65 H24 **Flagstaff Bay** *bay* N Saint Helena, C Atlantic Ocean
19 P5 **Flagstaff Lake** ◎ Maine, NE USA
94 E13 **Flåm** Sogn Og Fjordane, S Norway 60°51′N 07°06′E
15 O8 **Flamand** ≈ Québec, SE Canada
30 J5 **Flambeau River** ≈ Wisconsin, N USA
97 O16 **Flamborough Head** *headland* E England, United Kingdom 54°06′N 00°03′W
100 N13 **Fläming** *hill range* NE Germany
16 H8 **Flaming Gorge Reservoir** ⊞ Utah/Wyoming, NW USA
Flanders *see* Vlaanderen
Flandre *see* Vlaanderen
29 R10 **Flandreau** South Dakota, N USA 44°03′N 96°36′W
96 D6 **Flannan Isles** *island group* NW Scotland, United Kingdom
28 M6 **Flasher** North Dakota, N USA 46°25′N 101°12′W
93 G15 **Flåsjön** ◎ N Sweden
39 O11 **Flat** Alaska, USA 62°27′N 158°00′W
92 H1 **Flateyri** Vestfirðir, NW Iceland 66°03′N 23°28′W
33 P8 **Flathead Lake** ◎ Montana, NW USA
173 Y15 **Flat Island** *Fr.* Île Plate. *island* N Mauritius
25 T11 **Flatonia** Texas, SW USA 29°41′N 97°06′W
185 M14 **Flat Point** *headland* North Island, New Zealand 41°12′S 176°03′E
27 X6 **Flat River** Missouri, C USA 37°51′N 90°31′W
31 P8 **Flat River** ≈ Michigan, N USA
31 P14 **Flatrock River** ≈ Indiana, N USA
32 E6 **Flattery, Cape** *headland* Washington, NW USA 48°22′N 124°43′W
64 B12 **Flatts Village** *var.* The Flatts Village. C Bermuda 32°19′N 64°44′W
108 H7 **Flawil** Sankt Gallen, NE Switzerland 47°25′N 09°12′E
97 N22 **Fleet** S England, United Kingdom 51°16′N 00°50′W
97 K16 **Fleetwood** NW England, United Kingdom 53°55′N 03°02′W
18 H15 **Fleetwood** Pennsylvania, NE USA 40°27′N 75°49′W
95 D18 **Flekkefjord** Vest-Agder, S Norway 58°17′N 06°40′E
21 N5 **Flemingsburg** Kentucky, S USA 38°26′N 83°43′W
18 J15 **Flemington** New Jersey, NE USA 40°30′N 74°51′W
64 I7 **Flemish Cap** *undersea feature* NW Atlantic Ocean 47°00′N 45°00′W
95 N16 **Flen** Södermanland, C Sweden 59°04′N 16°39′E
100 I6 **Flensburg** Schleswig-Holstein, N Germany 54°47′N 09°26′E
100 J6 **Flensburger Förde** *inlet* Denmark/Germany
102 K5 **Flers** Orne, N France 48°45′N 00°34′W
95 C14 **Flesland** ✈ (Bergen) Hordaland, S Norway 60°18′N 05°15′E
Flessingue *see* Vlissingen
21 P10 **Fletcher** North Carolina, SE USA 35°24′N 82°29′W
31 R6 **Fletcher Pond** ◎ Michigan, N USA
102 L15 **Fleurance** Gers, S France 43°50′N 00°39′E
108 B8 **Fleurier** Neuchâtel, W Switzerland 46°55′N 06°37′E
99 H20 **Fleurus** Hainaut, S Belgium 50°28′N 04°33′E
103 N7 **Fleury-les-Aubrais** Loiret, C France 47°55′N 01°55′E
98 K10 **Flevoland** ◈ *province* C Netherlands
Flickertail State *see* North Dakota
108 H9 **Flims** Glarus, NE Switzerland 46°50′N 09°16′E
182 F8 **Flinders Island** *island* Investigator Group, South Australia
183 P14 **Flinders Island** *island* Furneaux Group, Tasmania, SE Australia
182 I6 **Flinders Ranges** ▲▲ South Australia
181 U5 **Flinders River** ≈ Queensland, NE Australia
11 V13 **Flin Flon** Manitoba, C Canada 54°47′N 101°51′W
97 K18 **Flint** NE Wales, United Kingdom 53°15′N 03°10′W
31 R9 **Flint** Michigan, N USA 43°01′N 83°41′W
97 J18 **Flint** *cultural region* NE Wales, United Kingdom
27 O7 **Flint Hills** *hill range* Kansas, C USA
191 Y6 **Flint Island** *island* Line Islands, E Kiribati
23 S4 **Flint River** ≈ Georgia, SE USA
31 R9 **Flint River** ≈ Michigan, N USA
189 X12 **Flipper Point** *headland* C Wake Island 19°18′N 166°37′E
94 I13 **Flisa** Hedmark, S Norway 60°36′N 12°02′E
94 J13 **Flisa** ≈ S Norway
122 J5 **Flissingskiy, Mys** *headland* Novaya Zemlya, NW Russian Federation 76°43′N 69°01′E
Flitsch *see* Bovec
105 U6 **Flix** Cataluña, NE Spain 41°13′N 00°32′E
95 J19 **Floda** Västra Götaland, S Sweden 57°47′N 12°20′E
101 O16 **Flöha** ≈ E Germany
25 O4 **Flomot** Texas, SW USA 34°13′N 100°58′W
29 V5 **Floodwood** Minnesota, N USA 46°55′N 92°55′W
30 M15 **Flora** Illinois, N USA 38°40′N 88°29′W
103 P14 **Florac** Lozère, S France 44°18′N 03°35′E
23 Q8 **Florala** Alabama, S USA 31°00′N 86°19′W
103 S4 **Florange** Moselle, NE France 49°21′N 06°06′E
Floreana, Isla *see* Santa María, Isla
23 O2 **Florence** Alabama, S USA 34°48′N 87°40′W
36 L14 **Florence** Arizona, SW USA 33°01′N 111°23′W
37 T6 **Florence** Colorado, C USA 38°20′N 105°06′W
27 O5 **Florence** Kansas, C USA 38°13′N 96°56′W
20 M4 **Florence** Kentucky, S USA 39°00′N 84°37′W
32 E13 **Florence** Oregon, NW USA 43°58′N 124°06′W
21 T12 **Florence** South Carolina, SE USA 34°12′N 79°44′W
25 S9 **Florence** Texas, SW USA 30°50′N 97°47′W
Florence *see* Firenze
54 E13 **Florencia** Caquetá, S Colombia 01°37′N 75°37′W
99 H21 **Florennes** Namur, S Belgium 50°15′N 04°36′E
Florentia *see* Firenze
63 J18 **Florentino Ameghino, Embalse** ◎ S Argentina
99 J24 **Florenville** Luxembourg, SE Belgium 49°42′N 05°19′E
42 E3 **Flores** Petén, N Guatemala 16°55′N 89°56′W
61 E19 **Flores** ◈ *department* S Uruguay
171 O16 **Flores** *island* Nusa Tenggara, C Indonesia
64 M1 **Flores** *island* Azores, Portugal, NE Atlantic Ocean
Floreshty *see* Floreşti
Flores, Lago de *see* Petén Itzá, Lago
Flores, Laut *see* Flores Sea
171 N15 **Flores Sea** *Ind.* Laut Flores. *sea* C Indonesia
116 M8 **Floreşti** *Rus.* Floreshty. N Moldova 47°52′N 28°19′E
25 S12 **Floresville** Texas, SW USA 29°09′N 98°10′W
59 N14 **Floriano** Piauí, E Brazil 06°45′S 43°W
61 K14 **Florianópolis** *prev.* Destêrro. *state capital* Santa Catarina, S Brazil 27°35′S 48°32′W
44 G6 **Florida** Camagüey, C Cuba 21°32′N 78°14′W
61 F19 **Florida** Florida, S Uruguay 34°04′S 56°14′W
61 F19 **Florida** ◈ *department* S Uruguay
23 U9 **Florida** *off.* State of Florida, *also known as* Peninsular State, Sunshine State. ◈ *state* SE USA
23 Y17 **Florida Bay** *bay* Florida, SE USA
54 G8 **Floridablanca** Santander, N Colombia 07°04′N 73°06′W
23 Y17 **Florida Keys** *island group* Florida, SE USA
37 Q16 **Florida Mountains** ▲▲ New Mexico, SW USA
64 D10 **Florida, Straits of** *strait* Atlantic Ocean/Gulf of Mexico
114 D13 **Flórina** *var.* Phlórina. Dytikí Makedonía, N Greece 40°48′N 21°26′E
27 X4 **Florissant** Missouri, C USA 38°48′N 90°20′W
94 C11 **Florø** Sogn Og Fjordane, S Norway 61°36′N 05°04′E
115 L22 **Floúda, Akrotírio** *headland* Astypálaia, Kykládes, Greece, Aegean Sea 36°38′N 26°23′E
21 S7 **Floyd** Virginia, NE USA 36°55′N 80°22′W
25 N4 **Floydada** Texas, SW USA 33°58′N 101°20′W
Flüela Wisshorn *see* Weisshorn
98 K7 **Fluessen** ◎ N Netherlands
105 S5 **Flúmen** ≈ NE Spain
107 C20 **Flumendosa** ≈ Sardegna, Italy, C Mediterranean Sea
31 R9 **Flushing** Michigan, N USA 43°03′N 83°51′W
Flushing *see* Vlissingen
25 O6 **Fluvanna** Texas, SW USA 32°54′N 101°06′W
186 B8 **Fly** ≈ Indonesia/Papua New Guinea
194 I10 **Flying Fish, Cape** *headland* Thurston Island, Antarctica 72°00′S 102°25′W
Flylân *see* Vlieland
Fly River *see* Western
193 Y15 **Foa** *island* Ha'apai Group, C Tonga
11 U15 **Foam Lake** Saskatchewan, S Canada 51°38′N 103°31′W
113 J14 **Foča** *var.* Srbinje. SE Bosnia and Herzegovina 43°32′N 18°46′E
116 L12 **Focşani** Vrancea, E Romania 45°45′N 27°13′E
Fogaras/Fogarasch *see* Făgăraş
107 M16 **Foggia** Puglia, SE Italy 41°28′N 15°31′E
Foggo *see* Faggo
76 D10 **Fogo** *island* Ilhas de Sotavento, SW Cape Verde
13 U11 **Fogo Island** *island* Newfoundland and Labrador, E Canada
109 U7 **Fohnsdorf** Steiermark, SE Austria 47°13′N 14°40′E
100 G7 **Föhr** *island* NW Germany
104 F14 **Fóia** ▲ S Portugal 37°19′N 08°39′W
14 I10 **Foins, Lac aux** ◎ Québec, SE Canada
103 N17 **Foix** Ariège, S France 42°58′N 01°39′E
126 I5 **Fokino** Bryanskaya Oblast', W Russian Federation 53°22′N 34°22′E
123 S15 **Fokino** Primorskiy Kray, W Russian Federation 42°58′N 132°25′E
Fola, Cnoc *see* Bloody Foreland
94 E13 **Folarskardnuten** ▲ S Norway 60°34′N 07°18′E
92 G11 **Folda** *prev.* Foldafjorden. *fjord* C Norway
Foldafjorden *see* Folda
93 F14 **Foldereid** Nord-Trøndelag, C Norway 64°58′N 12°09′E
Földvár *see* Feldioara
115 J22 **Folégandros** *island* Kykládes, Greece, Aegean Sea
23 O9 **Foley** Alabama, S USA 30°24′N 87°40′W
29 U7 **Foley** Minnesota, N USA 45°39′N 93°54′W
14 E7 **Foleyet** Ontario, S Canada 48°15′N 82°26′W
95 D14 **Folgefonna** *glacier* S Norway
106 I13 **Foligno** Umbria, C Italy 42°58′N 12°40′E
97 Q23 **Folkestone** SE England, United Kingdom 51°05′N 01°11′E
23 W8 **Folkston** Georgia, SE USA 30°49′N 82°00′W
94 H10 **Folldal** Hedmark, S Norway 62°08′N 10°00′E
25 P1 **Follett** Texas, SW USA 36°25′N 100°08′W
106 F13 **Follonica** Toscana, C Italy 42°55′N 10°45′E
21 T15 **Folly Beach** South Carolina, SE USA 32°39′N 79°56′W
35 O7 **Folsom** California, W USA 38°40′N 121°11′W
116 M12 **Folteşti** Galaţi, E Romania 45°45′N 28°00′E
172 H14 **Fomboni** Mohéli, S Comoros 12°18′S 43°46′E
18 K10 **Fonda** New York, NE USA 42°57′N 74°24′W
11 S10 **Fond-du-Lac** Saskatchewan, C Canada 59°20′N 107°09′W
30 M8 **Fond du Lac** Wisconsin, N USA 43°48′N 88°27′W
11 T10 **Fond-du-Lac** ≈ Saskatchewan, C Canada
190 G8 **Fongafale** *atoll* C Tuvalu
107 C18 **Fonni** Sardegna, Italy, C Mediterranean Sea 40°07′N 09°17′E
189 V12 **Fono** *island* Chuuk, C Micronesia
54 G4 **Fonseca** La Guajira, N Colombia 10°53′N 72°51′W
Fonseca, Golfo de *see* Fonseca, Gulf of
42 H8 **Fonseca, Gulf of** *Sp.* Golfo de Fonseca. *gulf* C Central America
103 O6 **Fontainebleau** Seine-et-Marne, N France 48°24′N 02°42′E
63 G19 **Fontana, Lago** ◎ W Argentina
21 N10 **Fontana Lake** ⊞ North Carolina, SE USA
107 L24 **Fontanarossa** ✈ (Catania) Sicilia, Italy, C Mediterranean Sea 37°28′N 15°04′E
11 N11 **Fontas** ≈ British Columbia, W Canada
58 D12 **Fonte Boa** Amazonas, N Brazil 02°32′S 66°01′W
102 J10 **Fontenay-le-Comte** Vendée, NW France 46°28′N 00°48′W
33 T16 **Fontenelle Reservoir** ⊞ Wyoming, C USA
193 Y14 **Fonualei** *island* Vava'u Group, N Tonga
111 H24 **Fonyód** Somogy, W Hungary 46°43′N 17°32′E
Foochow *see* Fuzhou
39 Q10 **Foraker, Mount** ▲ Alaska, USA 62°57′N 151°24′W
187 R14 **Forari** Éfaté, C Vanuatu 17°42′S 168°33′E
103 U4 **Forbach** Moselle, NE France 49°11′N 06°54′E
183 Q8 **Forbes** New South Wales, SE Australia 33°24′S 148°00′E
77 T17 **Forcados** Delta, S Nigeria 05°16′N 05°25′E
103 S14 **Forcalquier** Alpes-de-Haute-Provence, SE France 43°57′N 05°46′E
101 K19 **Forchheim** Bayern, SE Germany 49°43′N 11°07′E
35 R13 **Ford City** California, W USA 35°09′N 119°27′W
94 D11 **Førde** Sogn Og Fjordane, S Norway 61°27′N 05°51′E
31 N4 **Ford River** ≈ Michigan, N USA
183 O4 **Fords Bridge** New South Wales, SE Australia 29°44′S 145°25′E
20 J6 **Fordsville** Kentucky, S USA 37°36′N 86°39′W
27 U13 **Fordyce** Arkansas, C USA 33°49′N 92°23′W
76 I14 **Forécariah** SW Guinea 09°28′N 13°06′W
197 O14 **Forel, Mont** ▲ SE Greenland 66°55′N 36°45′W
11 R17 **Foremost** Alberta, SW Canada 49°30′N 111°34′W
14 D16 **Forest** Ontario, S Canada 43°05′N 82°00′W
22 L5 **Forest** Mississippi, S USA 32°22′N 89°30′W
31 S12 **Forest** Ohio, N USA 40°47′N 83°26′W
29 V11 **Forest City** Iowa, C USA 43°15′N 93°38′W
21 Q10 **Forest City** North Carolina, SE USA 35°19′N 81°52′W
32 G11 **Forest Grove** Oregon, NW USA 45°31′N 123°06′W
183 P17 **Forestier Peninsula** *peninsula* Tasmania, SE Australia
29 V8 **Forest Lake** Minnesota, N USA 45°16′N 92°59′W
23 S3 **Forest Park** Georgia, SE USA 33°37′N 84°22′W
29 Q3 **Forest River** ≈ North Dakota, N USA
15 T6 **Forestville** Québec, SE Canada 48°45′N 69°04′W
103 Q11 **Forez, Monts du** ▲▲ C France
96 K10 **Forfar** E Scotland, United Kingdom 56°38′N 02°54′W
26 J8 **Forgan** Oklahoma, C USA 36°54′N 100°32′W
Forge du Sud *see* Dudelange
101 J24 **Forggensee** ◎ S Germany
147 N10 **Forish** *Rus.* Farish. Jizzax Viloyati, C Uzbekistan 40°33′N 66°52′E
20 F9 **Forked Deer River** ≈ Tennessee, S USA
32 F7 **Forks** Washington, NW USA 47°57′N 124°22′W
92 N2 **Forlandsundet** *sound* W Svalbard
106 H10 **Forlì** *anc.* Forum Livii. Emilia-Romagna, N Italy 44°14′N 12°02′E
29 Q7 **Forman** North Dakota, N USA 46°07′N 97°39′W
97 K17 **Formby** NW England, United Kingdom 53°34′N 03°05′W
105 V11 **Formentera** *anc.* Ophiusa, *Lat.* Frumentum. *island* Islas Baleares, Spain, W Mediterranean Sea
Formentor, Cabo de *see* Formentor, Cap de
105 Y9 **Formentor, Cap de** *var.* Cabo de Formentor, Cape Formentor. *headland* Mallorca, Spain, W Mediterranean Sea 39°57′N 03°12′E
Formentor, Cape *see* Formentor, Cap de
107 J16 **Formia** Lazio, C Italy 41°16′N 13°37′E
62 O7 **Formosa** Formosa, NE Argentina 26°07′S 58°14′W
62 M6 **Formosa** *off.* Provincia de Formosa. ◈ *province* NE Argentina
Formosa/Formo'sa *see* Taiwan
Formosa, Provincia de *see* Formosa
59 I17 **Formosa, Serra** ▲▲ C Brazil
Formosa Strait *see* Taiwan Strait
95 H21 **Fornæs** *headland* C Denmark 56°26′N 10°57′E
25 U6 **Forney** Texas, SW USA 32°45′N 96°28′W
106 E9 **Fornovo di Taro** Emilia-Romagna, C Italy 44°42′N 10°07′E
117 T14 **Foros** Avtonomna Respublika Krym, S Ukraine 44°24′N 33°47′E
Føroyar *see* Faroe Islands
96 J8 **Forres** NE Scotland, United Kingdom 57°32′N 03°38′W
27 X11 **Forrest City** Arkansas, C USA 35°01′N 90°48′W
39 Y15 **Forrester Island** *island* Alexander Archipelago, Alaska, USA
25 N7 **Forsan** Texas, SW USA 32°06′N 101°22′W
181 V5 **Forsayth** Queensland, NE Australia 18°31′S 143°37′E
95 L19 **Forserum** Jönköping, S Sweden 57°42′N 14°28′E
95 K15 **Forshaga** Värmland, C Sweden 59°33′N 13°29′E
93 L19 **Forssa** Kanta-Häme, S Finland 60°49′N 23°40′E
101 Q14 **Forst** *Lus.* Baršć Łužyca. Brandenburg, E Germany 51°43′N 14°38′E
183 U7 **Forster-Tuncurry** New South Wales, SE Australia 32°11′S 152°30′E
23 T4 **Forsyth** Georgia, SE USA 33°00′N 83°57′W
27 T8 **Forsyth** Missouri, C USA 36°41′N 93°07′W
33 W10 **Forsyth** Montana, NW USA 46°16′N 106°40′W
149 U11 **Fort Abbās** Punjab, E Pakistan 29°12′N 73°00′E
12 G10 **Fort Albany** Ontario, C Canada 52°15′N 81°35′W
56 L13 **Fortaleza** Pando, N Bolivia 09°48′S 65°29′W
58 P13 **Fortaleza** *prev.* Ceará. *state capital* Ceará, NE Brazil 03°45′S 38°35′W
59 D16 **Fortaleza** Rondônia, W Brazil 08°45′S 64°06′W
56 C13 **Fortaleza, Río** ≈ W Peru
Fort-Archambault *see* Sarh
21 U3 **Fort Ashby** West Virginia, NE USA 39°30′N 78°46′W
96 I9 **Fort Augustus** N Scotland, United Kingdom 57°14′N 04°38′W
Fort-Bayard *see* Zhanjiang
33 S8 **Fort Benton** Montana, NW USA 47°49′N 110°40′W
35 Q1 **Fort Bidwell** California, W USA 41°50′N 120°07′W
34 L5 **Fort Bragg** California, W USA 39°25′N 123°48′W
31 N16 **Fort Branch** Indiana, N USA 38°15′N 87°34′W
Fort-Bretonnet *see* Bousso
33 T17 **Fort Bridger** Wyoming, C USA 41°18′N 110°19′W
Fort-Cappolani *see* Tidjikja
Fort-Carnot *see* Ikongo
Fort Charlet *see* Djanet
Fort-Chimo *see* Kuujjuaq
11 R10 **Fort Chipewyan** Alberta, C Canada 58°42′N 111°08′W
Fort Cobb Lake *see* Fort Cobb Reservoir
26 L11 **Fort Cobb Reservoir** *var.* Fort Cobb Lake. ⊞ Oklahoma, C USA
37 T3 **Fort Collins** Colorado, C USA 40°35′N 105°05′W
14 K12 **Fort-Coulonge** Québec, SE Canada 45°50′N 76°45′W
Fort-Crampel *see* Kaga Bandoro
Fort-Dauphin *see* Tôlañaro
24 K10 **Fort Davis** Texas, SW USA 30°35′N 103°54′W
37 O10 **Fort Defiance** Arizona, SW USA 35°44′N 109°04′W
45 Q12 **Fort-de-France** *prev.* Fort-Royal. ○ (Martinique) W Martinique 14°36′N 61°05′W
45 P12 **Fort-de-France, Baie de** *bay* W Martinique
Fort de Kock *see* Bukittinggi
23 P6 **Fort Deposit** Alabama, S USA 31°58′N 86°34′W
29 U13 **Fort Dodge** Iowa, C USA 42°30′N 94°10′W
13 S10 **Forteau** Québec, E Canada 51°30′N 56°55′W
106 E11 **Forte dei Marmi** Toscana, C Italy 43°59′N 10°10′E
14 H17 **Fort Erie** Ontario, S Canada 42°55′N 78°56′W
180 H7 **Fortescue River** ≈ Western Australia
19 S2 **Fort Fairfield** Maine, NE USA 46°45′N 67°51′W
Fort-Foureau *see* Kousséri
12 A11 **Fort Frances** Ontario, S Canada 48°37′N 93°23′W
Fort Franklin *see* Déline
23 R7 **Fort Gaines** Georgia, SE USA 31°36′N 85°03′W
37 T8 **Fort Garland** Colorado, C USA 37°22′N 105°24′W
21 P5 **Fort Gay** West Virginia, NE USA 38°06′N 82°35′W
Fort George *see* Chisasibi
Fort George *see* La Grande Rivière
27 Q10 **Fort Gibson** Oklahoma, C USA 35°48′N 95°15′W
27 Q9 **Fort Gibson Lake** ⊞ Oklahoma, C USA
8 H7 **Fort Good Hope** *var.* Rádeyilikóé. Northwest Territories, NW Canada 66°16′N 128°37′W
23 V4 **Fort Gordon** Georgia, SE USA 33°25′N 82°09′W
Fort Gouraud *see* Fdérik
96 I11 **Forth** ≈ C Scotland, United Kingdom
Fort Hall *see* Murang'a
24 H8 **Fort Hancock** Texas, SW USA 31°18′N 105°49′W
Fort Hertz *see* Putao
96 K12 **Forth, Firth of** *estuary* E Scotland, United Kingdom
14 L14 **Forthton** Ontario, SE Canada 44°43′N 75°31′W
14 M8 **Fortier** ≈ Québec, SE Canada
Fortín General Eugenio Garay *see* General Eugenio A. Garay
Fort Jameson *see* Chipata
Fort Johnston *see* Mangochi
19 R1 **Fort Kent** Maine, NE USA 47°15′N 68°33′W
Fort-Lamy *see* N'Djaména
23 Z15 **Fort Lauderdale** Florida, SE USA 26°07′N 80°09′W
21 R11 **Fort Lawn** South Carolina, SE USA 34°43′N 80°46′W
8 H10 **Fort Liard** *var.* Liard. Northwest Territories, W Canada 60°14′N 123°28′W
44 M8 **Fort-Liberté** NE Haiti 19°42′N 71°51′W
21 N9 **Fort Loudoun Lake** ⊞ Tennessee, S USA
37 T3 **Fort Lupton** Colorado, C USA 40°04′N 104°48′W
11 R12 **Fort MacKay** Alberta, C Canada 57°12′N 111°41′W
11 Q17 **Fort Macleod** *var.* MacLeod. Alberta, SW Canada 49°44′N 113°24′W
29 Y16 **Fort Madison** Iowa, C USA 40°37′N 91°15′W
Fort Manning *see* Mchinji
25 P9 **Fort McKavett** Texas, SW USA 30°50′N 100°07′W
11 R12 **Fort McMurray** Alberta, C Canada 56°44′N 111°23′W
8 G7 **Fort McPherson** *var.* McPherson. Northwest Territories, NW Canada 67°29′N 134°50′W
21 R11 **Fort Mill** South Carolina, SE USA 35°00′N 80°57′W
Fort-Millot *see* Ngouri
37 U3 **Fort Morgan** Colorado, C USA 40°14′N 103°48′W
23 W14 **Fort Myers** Florida, SE USA 26°39′N 81°52′W
23 W15 **Fort Myers Beach** Florida, SE USA 26°27′N 81°57′W
10 M10 **Fort Nelson** British Columbia, W Canada 58°48′N 122°44′W
10 M10 **Fort Nelson** ≈ British Columbia, W Canada
Fort Norman *see* Tulita
23 Q2 **Fort Payne** Alabama, S USA 34°23′N 85°43′W
33 W7 **Fort Peck** Montana, NW USA 48°00′N 106°28′W
33 V8 **Fort Peck Lake** ⊞ Montana, NW USA
23 Y13 **Fort Pierce** Florida, SE USA 27°28′N 80°20′W
29 N10 **Fort Pierre** South Dakota, N USA 44°21′N 100°22′W
81 E18 **Fort Portal** SW Uganda 0°39′N 30°17′E
8 J10 **Fort Providence** *var.* Providence. Northwest Territories, W Canada 61°21′N 117°39′W
11 U16 **Fort Qu'Appelle** Saskatchewan, S Canada 50°50′N 103°52′W
Fort-Repoux *see* Akjoujt
8 K10 **Fort Resolution** *var.* Resolution. Northwest Territories, W Canada 61°10′N 113°39′W
33 T13 **Fortress Mountain** ▲ Wyoming, C USA 44°20′N 109°51′W
Fort Rosebery *see* Mansa
Fort Rousset *see* Owando
Fort-Royal *see* Fort-de-France
Fort Rupert *see* Waskaganish
8 H13 **Fort St. James** British Columbia, SW Canada 54°26′N 124°15′W
11 N12 **Fort St. John** British Columbia, W Canada 56°16′N 120°52′W
Fort Sandeman *see* Zhob
11 Q14 **Fort Saskatchewan** Alberta, SW Canada 53°42′N 113°12′W
27 R6 **Fort Scott** Kansas, C USA 37°52′N 94°43′W
12 E6 **Fort Severn** Ontario, C Canada 56°N 87°40′W
31 R12 **Fort Shawnee** Ohio, N USA 40°41′N 84°08′W
144 E14 **Fort-Shevchenko** Mangistau, W Kazakhstan 44°29′N 50°16′E
Fort-Sibut *see* Sibut
8 I10 **Fort Simpson** *var.* Simpson. Northwest Territories, W Canada 61°52′N 121°23′W
8 I10 **Fort Smith** Northwest Territories, W Canada 60°01′N 111°55′W
27 R10 **Fort Smith** Arkansas, C USA 35°23′N 94°24′W
37 T13 **Fort Stanton** New Mexico, SW USA 33°28′N 105°31′W
24 L9 **Fort Stockton** Texas, SW USA 30°54′N 102°54′W
37 U12 **Fort Sumner** New Mexico, SW USA 34°28′N 104°15′W
26 K8 **Fort Supply** Oklahoma, C USA 36°34′N 99°34′W
26 K8 **Fort Supply Lake** ⊞ Oklahoma, C USA
29 O10 **Fort Thompson** South Dakota, N USA 44°01′N 99°22′W
Fort-Trinquet *see* Bîr Mogreïn
105 R12 **Fortuna** Murcia, SE Spain 38°11′N 01°07′W
34 K3 **Fortuna** California, W USA 40°35′N 124°07′W
28 J2 **Fortuna** North Dakota, N USA 48°53′N 103°46′W
23 T5 **Fort Valley** Georgia, SE USA 32°33′N 83°53′W
11 P11 **Fort Vermilion** Alberta, W Canada 58°22′N 115°59′W
Fort Victoria *see* Masvingo
31 P13 **Fortville** Indiana, N USA 39°55′N 85°51′W
23 P9 **Fort Walton Beach** Florida, SE USA 30°24′N 86°37′W
31 P12 **Fort Wayne** Indiana, N USA 41°08′N 85°08′W
96 H10 **Fort William** N Scotland, United Kingdom 56°49′N 05°07′W
25 T6 **Fort Worth** Texas, SW USA 32°44′N 97°19′W
28 M7 **Fort Yates** North Dakota, N USA 46°05′N 100°37′W
39 S7 **Fort Yukon** Alaska, USA 66°35′N 145°05′W
Forum Alieni *see* Ferrara
Forum Julii *see* Fréjus
Forum Livii *see* Forlì
143 Q15 **Forūr-e Bozorg, Jazīreh-ye** *island* S Iran
94 H7 **Fosen** *physical region* S Norway
161 N14 **Foshan** *var.* Fatshan, Fo-shan, Namhoi. Guangdong, S China 23°03′N 113°08′E
Fo-shan *see* Foshan
194 J6 **Fossil Bluff** *UK research station* Antarctica 71°30′S 68°30′W
Fossa Claudia *see* Chioggia
106 B9 **Fossano** Piemonte, NW Italy 44°33′N 07°43′E
99 H21 **Fosses-la-Ville** Namur, S Belgium 50°24′N 04°42′E
32 J12 **Fossil** Oregon, NW USA 45°01′N 120°14′W
Foss Lake *see* Foss Reservoir
106 I11 **Fossombrone** Marche, C Italy 43°42′N 12°48′E
26 K10 **Foss Reservoir** *var.* Foss Lake. ⊞ Oklahoma, C USA
29 S4 **Fosston** Minnesota, N USA 47°34′N 95°45′W
183 O13 **Foster** Victoria, SE Australia 38°40′S 146°15′E
11 T12 **Foster Lakes** ◎ Saskatchewan, C Canada
31 S12 **Fostoria** Ohio, N USA 41°09′N 83°25′W
79 D19 **Fougamou** Ngounié, C Gabon 01°16′S 10°30′E
102 J6 **Fougères** Ille-et-Vilaine, NW France 48°21′N 01°12′W
Fou-hsin *see* Fuxin
27 S14 **Fouke** Arkansas, C USA 33°15′N 93°53′W
96 K2 **Foula** *island* NE Scotland, United Kingdom
65 D24 **Foul Bay** *bay* East Falkland, Falkland Islands
97 P21 **Foulness Island** *island* SE England, United Kingdom
185 F15 **Foulwind, Cape** *headland* South Island, New Zealand 41°45′S 171°28′E
79 E15 **Foumban** Ouest, NW Cameroon 05°43′N 10°50′E
172 H13 **Foumbouni** Grande Comore, NW Comoros 11°49′S 43°30′E
195 N8 **Foundation Ice Stream** *glacier* Antarctica
37 T6 **Fountain** Colorado, C USA 38°40′N 104°42′W
36 L4 **Fountain Green** Utah, W USA 39°37′N 111°37′W
21 P11 **Fountain Inn** South Carolina, SE USA 34°41′N 82°12′W
27 S11 **Fourche LaFave River** ≈ Arkansas, C USA
33 Z13 **Four Corners** Wyoming, C USA 44°04′N 104°08′W

◆ Country
● Country Capital
◇ Dependent Territory
○ Dependent Territory Capital
◈ Administrative Regions
✈ International Airport
▲ Mountain
▲▲ Mountain Range
Volcano
≈ River
◎ Lake
⊞ Reservoir

103 Q2 **Fourmies** Nord, N France *50°01´N 04°03´E*
38 J17 **Four Mountains, Islands of** *island group* Aleutian Islands, Alaska, USA
173 P17 **Fournaise, Piton de la** ✱ SE Réunion *21°14´S 55°43´E*
14 J8 **Fournière, Lac** ◎ Québec, SE Canada
115 L20 **Foúrnoi** *island* Dodekánisa, Greece, Aegean Sea
64 K13 **Four North Fracture Zone** *tectonic feature* W Atlantic Ocean
Fouron-Saint-Martin *see* Sint-Martens-Voeren
30 L3 **Fourteen Mile Point** *headland* Michigan, N USA *46°59´N 89°07´W*
Fou-shan *see* Fushun
76 I13 **Fouta Djallon** *var.* Futa Jallon. ▲▲ W Guinea
185 C25 **Foveaux Strait** *strait* S New Zealand
35 Q11 **Fowler** California, W USA *36°35´N 119°40´W*
37 U6 **Fowler** Colorado, C USA *38°07´N 104°01´W*
31 N12 **Fowler** Indiana, N USA *40°36´N 87°20´W*
182 D7 **Fowlers Bay** *bay* South Australia
25 R13 **Fowlerton** Texas, SW USA *28°27´N 98°48´W*
142 M3 **Fowman** *var.* Fuman, Fumen. Gīlān, NW Iran *37°15´N 49°19´E*
65 C25 **Fox Bay East** West Falkland, Falkland Islands
65 C25 **Fox Bay West** West Falkland, Falkland Islands
14 J14 **Foxboro** Ontario, SE Canada *44°16´N 77°23´W*
11 O14 **Fox Creek** Alberta, W Canada *54°25´N 116°57´W*
64 G5 **Foxe Basin** *sea* Nunavut, N Canada
64 G5 **Foxe Channel** *channel* Nunavut, N Canada
95 I16 **Foxen** ◎ C Sweden
9 Q7 **Foxe Peninsula** *peninsula* Baffin Island, Nunavut, NE Canada
185 E19 **Fox Glacier** West Coast, South Island, New Zealand *43°28´S 170°00´E*
38 L17 **Fox Islands** *island* Aleutian Islands, Alaska, USA
30 M10 **Fox Lake** Illinois, N USA *42°24´N 88°10´W*
9 V12 **Fox Mine** Manitoba, C Canada *56°36´N 101°48´W*
35 R3 **Fox Mountain** ▲ Nevada, W USA *41°01´N 119°30´W*
65 E25 **Fox Point** *headland* East Falkland, Falkland Islands *51°55´S 58°24´W*
30 M11 **Fox River** 〰 Illinois/ Wisconsin, N USA
30 L7 **Fox River** 〰 Wisconsin, N USA
184 L13 **Foxton** Manawatu-Wanganui, North Island, New Zealand *40°27´S 175°18´E*
11 S16 **Fox Valley** Saskatchewan, S Canada *50°29´N 109°29´W*
11 W16 **Foxwarren** Manitoba, S Canada *50°30´N 101°09´W*
97 E14 **Foyle, Lough** *Ir.* Loch Feabhail. *inlet* N Ireland
194 H5 **Foyn Coast** *physical region* Antarctica
104 I2 **Foz** Galicia, NW Spain *43°33´N 07°16´W*
60 I12 **Foz do Areia, Represa de** ⊞ S Brazil
59 A16 **Foz do Breu** Acre, W Brazil *09°21´S 72°41´W*
83 A16 **Foz do Cunene** Namibe, SW Angola *17°11´S 11°52´E*
60 G12 **Foz do Iguaçu** Paraná, S Brazil *25°33´S 54°31´W*
58 C12 **Foz do Mamoriá** Amazonas, NW Brazil *02°28´S 66°06´W*
105 T6 **Fraga** Aragón, NE Spain *41°32´N 00°21´E*
44 F5 **Fragoso, Cayo** *island* C Cuba
61 G18 **Fraile Muerto** Cerro Largo, NE Uruguay *32°30´S 54°30´W*
99 H21 **Fraire** Namur, S Belgium *50°16´N 04°30´E*
99 L21 **Fraiture, Baraque de** *hill* SE Belgium
Frakštát *see* Hlohovec
Fram Basin *see* Amundsen Basin
99 F20 **Frameries** Hainaut, S Belgium *50°25´N 03°41´E*
19 O11 **Framingham** Massachusetts, NE USA *42°15´N 71°24´W*
60 L7 **Franca** São Paulo, S Brazil *20°33´S 47°27´W*
187 O15 **Français, Récif des** *reef* W New Caledonia
107 K14 **Francavilla al Mare** Abruzzo, C Italy *42°25´N 14°16´E*
107 P18 **Francavilla Fontana** Puglia, SE Italy *40°32´N 17°35´E*
102 M8 **France** *off.* French Republic, *It./Sp.* Francia; *prev.* Gaul, Gaule, *Lat.* Gallia. ◆ *republic* W Europe
45 O8 **Francés Viejo, Cabo** *headland* NE Dominican Republic *19°39´N 69°57´W*
79 F19 **Franceville** *var.* Massoukou, Masuku. Haut-Ogooué, E Gabon *01°40´S 13°31´E*
79 F19 **Franceville** ✈ Haut-Ogooué, E Gabon *01°38´S 13°24´E*
Francfort *see* Frankfurt am Main
103 T8 **Franche-Comté** ◈ *region* E France
Francia *see* France
29 O11 **Francis Case, Lake** ⊞ South Dakota, N USA
60 H12 **Francisco Beltrão** Paraná, S Brazil *26°05´S 53°04´W*
Francisco I. Madero *see* Villa Madero
61 A21 **Francisco Madero** Buenos Aires, E Argentina *35°52´S 62°03´W*
42 H6 **Francisco Morazán** *prev.* Tegucigalpa. ◈ *department* C Honduras
83 J18 **Francistown** North East, NE Botswana *21°08´S 27°31´E*
Franconian Forest *see* Frankenwald
Franconian Jura *see* Fränkische Alb
98 K6 **Franeker** *Fris.* Frentsjer. Fryslân, N Netherlands *53°11´N 05°33´E*
Frankenalb *see* Fränkische Alb
101 H16 **Frankenberg** Hessen, C Germany *51°04´N 08°49´E*
101 J20 **Frankenhöhe** *hill range* C Germany
31 R8 **Frankenmuth** Michigan, N USA *43°19´N 83°44´W*
101 F20 **Frankenstein** *hill* W Germany
Frankenstein/Frankenstein in Schlesien *see* Ząbkowice Śląskie
101 G20 **Frankenthal** Rheinland-Pfalz, W Germany *49°32´N 08°22´E*
101 L18 **Frankenwald** *Eng.* Franconian Forest. ▲▲ C Germany
44 J12 **Frankfield** C Jamaica *18°08´N 77°22´W*
14 J14 **Frankford** Ontario, SE Canada *44°12´N 77°36´W*
31 O13 **Frankfort** Indiana, N USA *40°16´N 86°30´W*
27 O3 **Frankfort** Kansas, C USA *39°42´N 96°25´W*
20 L5 **Frankfort** *state capital* Kentucky, S USA *38°12´N 84°52´W*
Frankfort on the Main *see* Frankfurt am Main
Frankfurt *see* Frankfurt am Main, Germany
Frankfurt *see* Słubice, Poland
101 G18 **Frankfurt am Main** *var.* Frankfurt, *Fr.* Francfort; *prev. Eng.* Frankfort on the Main. Hessen, SW Germany *50°07´N 08°41´E*
100 Q12 **Frankfurt an der Oder** Brandenburg, E Germany *52°20´N 14°32´E*
101 L21 **Fränkische Alb** *var.* Frankenalb, *Eng.* Franconian Jura. ▲▲ S Germany
101 I18 **Fränkische Saale** 〰 C Germany
101 L19 **Fränkische Schweiz** *hill range* C Germany
23 R4 **Franklin** Georgia, SE USA *33°15´N 85°06´W*
31 P14 **Franklin** Indiana, N USA *39°29´N 86°02´W*
20 J7 **Franklin** Kentucky, S USA *36°42´N 86°35´W*
22 I9 **Franklin** Louisiana, S USA *29°48´N 91°30´W*
29 O17 **Franklin** Nebraska, C USA *40°06´N 98°57´W*
21 N10 **Franklin** North Carolina, SE USA *35°12´N 83°23´W*
18 C13 **Franklin** Pennsylvania, NE USA *41°24´N 79°49´W*
20 J9 **Franklin** Tennessee, S USA *35°55´N 86°52´W*
25 U9 **Franklin** Texas, SW USA *31°02´N 96°30´W*
21 X7 **Franklin** Virginia, NE USA *36°41´N 76°58´W*
21 T4 **Franklin** West Virginia, NE USA *38°39´N 79°21´W*
30 M9 **Franklin** Wisconsin, N USA *42°53´N 88°00´W*
8 I6 **Franklin Bay** *inlet* Northwest Territories, N Canada
32 K7 **Franklin D. Roosevelt Lake** ⊞ Washington, NW USA
35 W4 **Franklin Lake** ◎ Nevada, W USA
185 B22 **Franklin Mountains** ▲▲ South Island, New Zealand
39 R5 **Franklin Mountains** ▲▲ Alaska, USA
39 N4 **Franklin, Point** *headland* Alaska, USA *70°54´N 158°48´W*
183 O17 **Franklin River** 〰 Tasmania, SE Australia
22 K8 **Franklinton** Louisiana, S USA *30°51´N 90°09´W*
21 U9 **Franklinton** North Carolina, SE USA *36°06´N 78°27´W*
Frankstadt *see* Frenštát pod Radhoštěm
25 V7 **Frankston** Texas, SW USA *32°03´N 95°30´W*
33 U12 **Frannie** Wyoming, C USA *44°57´N 108°37´W*
15 U5 **Franquelin** Québec, SE Canada *49°17´N 67°52´W*
15 U5 **Franquelin** 〰 Québec, SE Canada
83 C18 **Fransfontein** Kunene, NW Namibia *20°12´S 15°01´E*
93 H17 **Fränsta** Västernorrland, C Sweden *62°30´N 16°06´E*
122 J3 **Frantsa-Iosifa, Zemlya** *Eng.* Franz Josef Land. *island group* N Russian Federation
185 E18 **Franz Josef Glacier** West Coast, South Island, New Zealand *43°22´S 170°11´E*
Franz Josef Land *see* Frantsa-Iosifa, Zemlya
Franz-Josef Spitze *see* Gerlachovský štít
101 L23 **Franz Josef Strauss** *abbrev.* F.J.S.. ✈ (München) Bayern, SE Germany *48°07´N 11°43´E*
107 A19 **Frasca, Capo della** *headland* Sardegna, Italy, C Mediterranean Sea *39°46´N 08°27´E*
107 I15 **Frascati** Lazio, C Italy *41°48´N 12°41´E*
11 N14 **Fraser** 〰 British Columbia, SW Canada
83 G24 **Fraserburg** Western Cape, SW South Africa *31°55´S 21°31´E*
96 L8 **Fraserburgh** NE Scotland, United Kingdom *57°42´N 02°20´W*
181 Z9 **Fraser Island** *var.* Great Sandy Island. *island* Queensland, E Australia
10 L14 **Fraser Lake** British Columbia, SW Canada *54°00´N 124°45´W*
10 L15 **Fraser Plateau** *plateau* British Columbia, SW Canada
184 P10 **Frasertown** Hawke's Bay, North Island, New Zealand *38°58´S 177°25´E*
99 E19 **Frasnes-lez-Buissenal** Hainaut, SW Belgium *50°40´N 03°37´E*
108 I7 **Frastanz** Vorarlberg, NW Austria *47°13´N 09°38´E*
14 B8 **Frater** Ontario, S Canada *47°19´N 84°28´W*
Frauenbach *see* Baia Mare
Frauenburg *see* Saldus, Latvia
Frauenburg *see* Frombork, Poland
108 H6 **Frauenfeld** Thurgau, NE Switzerland *47°34´N 08°54´E*
109 Z5 **Frauenkirchen** Burgenland, E Austria *47°50´N 16°57´E*
61 D19 **Fray Bentos** Río Negro, W Uruguay *33°09´S 58°14´W*
61 F19 **Fray Marcos** Florida, S Uruguay *34°13´S 55°43´W*
29 S6 **Frazee** Minnesota, N USA *46°42´N 95°40´W*
104 M5 **Frechilla** Castilla y León, N Spain *42°08´N 04°50´W*
30 I4 **Frederic** Wisconsin, N USA *45°42´N 92°30´W*
95 G23 **Fredericia** Syddanmark, C Denmark *55°34´N 09°47´E*
21 W3 **Frederick** Maryland, NE USA *39°25´N 77°25´W*
26 L12 **Frederick** Oklahoma, C USA *34°24´N 99°03´W*
29 P7 **Frederick** South Dakota, N USA *45°49´N 98°31´W*
29 X12 **Fredericksburg** Iowa, C USA *42°58´N 92°12´W*
25 R10 **Fredericksburg** Texas, SW USA *30°17´N 98°52´W*
21 W5 **Fredericksburg** Virginia, NE USA *38°16´N 77°27´W*
39 X13 **Frederick Sound** *sound* Alaska, USA
27 X6 **Fredericktown** Missouri, C USA *37°33´N 90°19´W*
60 H13 **Frederico Westphalen** Rio Grande do Sul, S Brazil *27°22´S 53°20´W*
13 O15 **Fredericton** *province capital* New Brunswick, SE Canada *45°57´N 66°40´W*
Frederiksborgs Amt *see* Hovedstaden
Frederikshåb *see* Paamiut
95 H19 **Frederikshavn** *prev.* Fladstrand. Nordjylland, N Denmark *57°28´N 10°33´E*
95 J22 **Frederikssund** Hovedstaden, E Denmark *55°51´N 12°05´E*
45 T9 **Frederiksted** Saint Croix, S Virgin Islands (US) *17°41´N 64°51´W*
95 I22 **Frederiksværk** *var.* Frederiksværk og Hanehoved. Hovedstaden, E Denmark *55°58´N 12°02´E*
Frederiksværk og Hanehoved *see* Frederiksværk
54 E9 **Fredonia** Antioquia, W Colombia *05°57´N 75°42´W*
36 K8 **Fredonia** Arizona, SW USA *36°57´N 112°31´W*
27 P7 **Fredonia** Kansas, C USA *37°32´N 95°50´W*
18 C11 **Fredonia** New York, NE USA *42°26´N 79°19´W*
35 P4 **Fredonyer Pass** *pass* California, W USA
93 I15 **Fredrika** Västerbotten, N Sweden *64°03´N 18°25´E*
95 L14 **Fredriksberg** Dalarna, C Sweden *60°07´N 14°23´E*
Fredrikshald *see* Halden
Fredrikshamn *see* Hamina
95 H16 **Fredrikstad** Østfold, S Norway *59°12´N 10°57´E*
30 K16 **Freeburg** Illinois, N USA *38°25´N 89°54´W*
18 K15 **Freehold** New Jersey, NE USA *40°14´N 74°14´W*
18 H14 **Freeland** Pennsylvania, NE USA *41°01´N 75°54´W*
182 J5 **Freeling Heights** ▲ South Australia *30°09´S 139°24´E*
35 Q7 **Freel Peak** ▲ California, W USA *38°52´N 119°52´W*
11 Z9 **Freels, Cape** *headland* Newfoundland and Labrador, E Canada *49°16´N 53°30´W*
29 Q11 **Freeman** South Dakota, N USA *43°21´N 97°26´W*
44 G1 **Freeport** Grand Bahama Island, N The Bahamas *26°28´N 78°43´W*
30 L10 **Freeport** Illinois, N USA *42°18´N 89°37´W*
25 W12 **Freeport** Texas, SW USA *28°57´N 95°21´W*
44 G1 **Freeport** ✈ Grand Bahama Island, N The Bahamas *26°31´N 78°48´W*
25 R14 **Freer** Texas, SW USA *27°52´N 98°37´W*
83 I22 **Free State** *off.* Free State Province; *prev.* Orange Free State, *Afr.* Oranje Vrystaat. ◈ *province* C South Africa
Free State *see* Maryland
Free State Province *see* Free State
76 G15 **Freetown** ● (Sierra Leone) W Sierra Leone *08°27´N 13°16´W*
172 J16 **Frégate** *island* Inner Islands, NE Seychelles
104 J12 **Fregenal de la Sierra** Extremadura, W Spain *38°10´N 06°39´W*
182 C2 **Fregon** South Australia *26°44´S 132°03´E*
102 H5 **Fréhel, Cap** *headland* NW France *48°41´N 02°21´W*
94 F8 **Frei** Møre og Romsdal, S Norway *63°02´N 07°47´E*
101 O16 **Freiberg** Sachsen, E Germany *50°55´N 13°21´E*
101 O16 **Freiberger Mulde** 〰 E Germany
Freiburg *see* Freiburg im Breisgau, Germany
Freiburg *see* Fribourg, Switzerland
101 F23 **Freiburg im Breisgau** *var.* Freiburg, *Fr.* Fribourg-en-Brisgau. Baden-Württemberg, SW Germany *48°N 07°52´E*
Freiburg in Schlesien *see* Świebodzice
Freie Hansestadt Bremen *see* Bremen
Freie und Hansestadt Hamburg *see* Brandenburg
101 L22 **Freising** Bayern, SE Germany *48°24´N 11°45´E*
109 T3 **Freistadt** Oberösterreich, N Austria *48°31´N 14°31´E*
Freistadtl *see* Hlohovec
101 O16 **Freital** Sachsen, E Germany *51°00´N 13°40´E*
Freiwaldau *see* Jeseník
104 J6 **Freixo de Espada à Cinta** Bragança, N Portugal *41°05´N 06°49´W*
103 U15 **Fréjus** *anc.* Forum Julii. Var, SE France *43°26´N 06°44´E*
180 I13 **Fremantle** Western Australia *32°07´S 115°44´E*
35 N9 **Fremont** California, W USA *37°34´N 122°01´W*
31 Q11 **Fremont** Indiana, N USA *41°43´N 84°54´W*
29 W15 **Fremont** Iowa, C USA *41°12´N 92°26´W*
31 P8 **Fremont** Michigan, N USA *43°28´N 85°56´W*
29 R15 **Fremont** Nebraska, C USA *41°25´N 96°30´W*
31 S11 **Fremont** Ohio, N USA *41°21´N 83°08´W*
33 T14 **Fremont Peak** ▲ Wyoming, C USA *43°07´N 109°37´W*
36 M6 **Fremont River** 〰 Utah, W USA
21 O9 **French Broad River** 〰 Tennessee, S USA
21 N5 **Frenchburg** Kentucky, S USA *37°58´N 83°37´W*
18 C12 **French Creek** 〰 Pennsylvania, NE USA
32 K15 **Frenchglen** Oregon, NW USA *42°49´N 118°55´W*
55 Y10 **French Guiana** *var.* Guiana, Guyane. ◇ *French overseas department* N South America
French Guinea *see* Guinea
31 O15 **French Lick** Indiana, N USA *38°33´N 86°37´W*
185 J14 **French Pass** Marlborough, South Island, New Zealand *40°57´S 173°49´E*
191 T11 **French Polynesia** ◇ *French overseas territory* S Pacific Ocean
French Republic *see* France
14 F11 **French River** 〰 Ontario, S Canada
French Somaliland *see* Djibouti
173 P12 **French Southern and Antarctic Lands** *prev.* French Southern and Antarctic Territories, *Fr.* Terres Australes et Antarctiques Françaises. ◈ *French overseas territory* S Indian Ocean
French Southern and Antarctic Territories *see* French Southern and Antarctic Lands
French Sudan *see* Mali
French Territory of the Afars and Issas *see* Djibouti
French Togoland *see* Togo
74 J6 **Frenda** NW Algeria *35°04´N 01°03´E*
111 I18 **Frenštát pod Radhoštěm** *Ger.* Frankstadt. Moravskoslezský Kraj, E Czech Republic *49°33´N 18°10´E*
Frentsjer *see* Franeker
76 M17 **Fresco** S Ivory Coast *05°03´N 05°31´W*
195 U16 **Freshfield, Cape** *headland* Antarctica
40 L10 **Fresnillo** *var.* Fresnillo de González Echeverría. Zacatecas, C Mexico *23°11´N 102°53´W*
Fresnillo de González Echeverría *see* Fresnillo
35 Q10 **Fresno** California, W USA *36°45´N 119°48´W*
Freu, Cabo del *see* Freu, Cap des
105 Y9 **Freu, Cap des** *var.* Cabo del Freu. *cape* Mallorca, Spain, W Mediterranean Sea
101 G22 **Freudenstadt** Baden-Württemberg, SW Germany *48°28´N 08°25´E*
Freudenthal *see* Bruntál
183 Q17 **Freycinet Peninsula** *peninsula* Tasmania, SE Australia
76 H14 **Fria** W Guinea *10°27´N 13°38´W*
83 A17 **Fria, Cape** *headland* NW Namibia *18°32´S 12°00´E*
35 Q10 **Friant** California, W USA *36°56´N 119°44´W*
62 K8 **Frías** Catamarca, N Argentina *28°41´S 65°00´W*
108 D9 **Fribourg** *Ger.* Freiburg. Fribourg, W Switzerland *46°50´N 07°10´E*
108 C9 **Fribourg** *Ger.* Freiburg. ◈ *canton* W Switzerland
Fribourg-en-Brisgau *see* Freiburg im Breisgau
32 G7 **Friday Harbor** San Juan Islands, Washington, NW USA *48°31´N 123°01´W*
Friedau *see* Ormož
101 K23 **Friedberg** Bayern, S Germany *48°21´N 10°58´E*
101 H18 **Friedberg** Hessen, W Germany *50°19´N 08°46´E*
Friedeberg Neumark *see* Strzelce Krajeńskie
Friedek-Mistek *see* Frýdek-Místek
Friedland *see* Pravdinsk
101 I24 **Friedrichshafen** Baden-Württemberg, S Germany *47°39´N 09°29´E*
Friedrichstadt *see* Jaunjelgava
29 Q16 **Friend** Nebraska, C USA *40°37´N 97°16´W*
Friendly Islands *see* Tonga
55 V9 **Friendship** Coronie, N Suriname *05°50´N 56°16´W*
30 L7 **Friendship** Wisconsin, N USA *43°58´N 89°48´W*
109 T8 **Friesach** Kärnten, S Austria *46°58´N 14°24´E*
Friesche Eilanden *see* Frisian Islands
101 F22 **Friesenheim** Baden-Württemberg, SW Germany *48°27´N 07°56´E*
Friesische Inseln *see* Frisian Islands
Friesland *see* Fryslân
60 Q10 **Frio, Cabo** *headland* SE Brazil *23°01´S 41°59´W*
24 M3 **Friona** Texas, SW USA *34°38´N 102°43´W*
42 L12 **Frío, Río** 〰 N Costa Rica
25 R13 **Frio River** 〰 Texas, SW USA
99 M25 **Frisange** Luxembourg, S Luxembourg *49°31´N 06°12´E*
Frisches Haff *see* Vistula Lagoon
36 J6 **Frisco Peak** ▲ Utah, W USA *38°31´N 113°17´W*
84 F9 **Frisian Islands** *Dut.* Friesche Eilanden, *Ger.* Friesische Inseln. *island group* N Europe
18 L12 **Frissell, Mount** ▲ Connecticut, NE USA *42°01´N 73°25´W*
95 J19 **Fristad** Västra Götaland, S Sweden *57°50´N 13°01´E*
25 N2 **Fritch** Texas, SW USA *35°38´N 101°36´W*
95 J19 **Fritsla** Västra Götaland, S Sweden *57°33´N 12°47´E*
101 H16 **Fritzlar** Hessen, C Germany *51°09´N 09°16´E*
106 H6 **Friuli-Venezia Giulia** ◈ *region* NE Italy
196 L13 **Frobisher Bay** *inlet* Baffin Island, Nunavut, NE Canada
Frobisher Bay *see* Iqaluit
11 S12 **Frobisher Lake** ◎ Saskatchewan, C Canada
94 G7 **Frohavet** *sound* C Norway
Frohenbruck *see* Veselí nad Lužnicí
109 V7 **Frohnleiten** Steiermark, SE Austria *47°17´N 15°20´E*
99 G22 **Froidchapelle** Hainaut, S Belgium *50°10´N 04°18´E*
127 O9 **Frolovo** Volgogradskaya Oblast', SW Russian Federation *49°46´N 43°38´E*
110 K7 **Frombork** *Ger.* Frauenburg. Warmińsko-Mazurskie, NE Poland *54°21´N 19°40´E*
97 L22 **Frome** SW England, United Kingdom *51°15´N 02°22´W*
182 I4 **Frome Creek** *seasonal river* South Australia
182 J6 **Frome Downs** South Australia *31°17´S 139°48´E*
182 J5 **Frome, Lake** *salt lake* South Australia
Fronicken *see* Wronki
104 H10 **Fronteira** Portalegre, C Portugal *39°03´N 07°39´W*
40 M7 **Frontera** Coahuila, NE Mexico *26°55´N 101°27´W*
41 U14 **Frontera** Tabasco, SE Mexico *18°32´N 92°39´W*
40 G3 **Fronteras** Sonora, NW Mexico *30°51´N 109°33´W*
103 Q16 **Frontignan** Hérault, S France *43°27´N 03°45´E*
54 D8 **Frontino** Antioquia, NW Colombia *06°46´N 76°10´W*
21 V4 **Front Royal** Virginia, NE USA *38°56´N 78°13´W*
107 J16 **Frosinone** *anc.* Frusino. Lazio, C Italy *41°38´N 13°22´E*
107 K16 **Frosolone** Molise, C Italy *41°34´N 14°25´E*
25 U7 **Frost** Texas, SW USA *32°04´N 96°48´W*
21 U2 **Frostburg** Maryland, NE USA *39°39´N 78°55´W*
23 X13 **Frostproof** Florida, SE USA *27°45´N 81°31´W*
Frostviken *see* Kvarnbergsvattnet
95 M15 **Frövi** Örebro, C Sweden *59°28´N 15°24´E*
94 F7 **Frøya** *island* W Norway
37 P5 **Fruita** Colorado, C USA *39°10´N 108°42´W*
28 J9 **Fruitdale** South Dakota, N USA *44°39´N 103°38´W*
23 W11 **Fruitland Park** Florida, SE USA *28°51´N 81°54´W*
Frumentum *see* Formentera
147 S11 **Frunze** Batkenskaya Oblast', SW Kyrgyzstan *40°07´N 71°40´E*
Frunze *see* Bishkek
117 O9 **Frunzivka** Odes'ka Oblast', SW Ukraine *47°19´N 29°46´E*
Frusino *see* Frosinone
108 E9 **Frutigen** Bern, W Switzerland *46°35´N 07°38´E*
111 I17 **Frýdek-Místek** *Ger.* Friedek-Mistek. Moravskoslezský Kraj, E Czech Republic *49°40´N 18°22´E*
98 K6 **Fryslân** *prev.* Friesland. ◈ *province* N Netherlands
193 V16 **Fua'amotu** Tongatapu, S Tonga *21°15´S 175°08´W*
190 A9 **Fuafatu** *island* Funafuti Atoll, C Tuvalu
190 A9 **Fuagea** *island* Funafuti Atoll, C Tuvalu
190 B8 **Fualifeke** *atoll* C Tuvalu
190 A8 **Fualopa** *island* Funafuti Atoll, C Tuvalu
151 K22 **Fuammulah** *var.* Fuammulah, Gnaviyani. *atoll* S Maldives
Fuammulah *see* Fuammulah
161 R11 **Fu'an** Fujian, SE China *27°11´N 119°42´E*
Fu-chien *see* Fujian
Fu-chou *see* Fuzhou
164 G13 **Fuchū** *var.* Hutyû. Hiroshima, Honshū, SW Japan *34°35´N 133°12´E*
160 M13 **Fuchuan** *var.* Fuyang. Guangxi Zhuangzu Zizhiqu, S China *24°56´N 111°15´E*
161 R9 **Fuchun Jiang** *var.* Tsien Tang. 〰 SE China
165 R8 **Fudai** Iwate, Honshū, C Japan *39°59´N 141°50´E*
161 S11 **Fuding** *var.* Tongshan. Fujian, SE China *27°21´N 120°10´E*
81 J20 **Fudua** *spring/well* S Kenya *02°13´S 39°43´E*
104 M16 **Fuengirola** Andalucía, S Spain *36°32´N 04°38´W*
104 J12 **Fuente de Cantos** Extremadura, W Spain *38°15´N 06°18´W*
104 J11 **Fuente del Maestre** Extremadura, W Spain *38°31´N 06°26´W*
104 L12 **Fuente Obejuna** Andalucía, S Spain *38°15´N 05°25´W*
104 L6 **Fuentesaúco** Castilla y León, N Spain *41°14´N 05°30´W*
62 O3 **Fuerte Olimpo** *var.* Olimpo. Alto Paraguay, NE Paraguay *21°02´S 57°51´W*
40 H8 **Fuerte, Río** 〰 C Mexico
64 Q11 **Fuerteventura** *island* Islas Canarias, Spain, NE Atlantic Ocean
141 S14 **Fughmah** *var.* Faghman, Fugma. C Yemen *16°08´N 49°23´E*
92 M2 **Fuglehuken** *headland* W Svalbard *78°54´N 10°30´E*
Fugløy *see* Fugloy
95 B18 **Fugloy** *Dan.* Fugløy. *island* NE Faroe Islands
197 T15 **Fugløya Bank** *undersea feature* E Norwegian Sea *71°00´N 19°20´E*
Fugma *see* Fughmah
160 E11 **Fugong** Yunnan, SW China *27°00´N 98°48´E*
81 K16 **Fugugo** *spring/well* NE Kenya *03°19´N 39°39´E*
158 L2 **Fuhai** *var.* Burultokay. Xinjiang Uygur Zizhiqu, NW China *47°15´N 87°39´E*
161 P10 **Fu He** 〰 S China
Fuhkien *see* Fujian
100 J9 **Fuhlsbüttel** ✈ (Hamburg) Hamburg, N Germany *53°37´N 09°57´E*
101 L14 **Fuhne** 〰 C Germany
Fu-hsin *see* Fuxin
Fujairah *see* Al Fujayrah
164 M14 **Fuji** *var.* Huzi. Shizuoka, Honshū, S Japan *35°08´N 138°39´E*
161 Q12 **Fujian** *var.* Fu-chien, Fuhkien, Fukien, Min, Fujian Sheng. ◈ *province* SE China
160 I9 **Fu Jiang** 〰 C China
Fujian Sheng *see* Fujian
164 M14 **Fujieda** *var.* Huzieda. Shizuoka, Honshū, S Japan *34°54´N 138°15´E*
Fuji, Mount/Fujiyama *see* Fuji-san
163 Y7 **Fujin** Heilongjiang, NE China *47°12´N 132°01´E*
164 M13 **Fujinomiya** *var.* Huzinomiya. Shizuoka, Honshū, S Japan *35°16´N 138°33´E*
164 N13 **Fuji-san** *var.* Fujiyama, *Eng.* Mount Fuji. ▲ Honshū, SE Japan *35°23´N 138°44´E*
165 N14 **Fujisawa** *var.* Huzisawa. Kanagawa, Honshū, S Japan *35°22´N 139°29´E*
165 T3 **Fukagawa** *var.* Hukagawa. Hokkaidō, NE Japan *43°44´N 142°03´E*
158 L5 **Fukang** Xinjiang Uygur Zizhiqu, W China *44°07´N 87°55´E*
165 P7 **Fukaura** Aomori, Honshū, C Japan *40°38´N 139°55´E*
193 W15 **Fukave** *island* Tongatapu Group, S Tonga
Fukien *see* Fujian
164 J13 **Fukuchiyama** *var.* Hukutiyama. Kyōto, Honshū, SW Japan *35°19´N 135°08´E*
Fukue *see* Gotō
164 A13 **Fukue-jima** *island* Gotō-rettō, SW Japan
164 K12 **Fukui** *var.* Hukui. Fukui, Honshū, SW Japan *36°03´N 136°12´E*
164 K12 **Fukui** *off.* Fukui-ken, *var.* Hukui. ◈ *prefecture* Honshū, SW Japan
Fukui-ken *see* Fukui
164 D13 **Fukuoka** *var.* Hukuoka, *hist.* Najima. Fukuoka, Kyūshū, SW Japan *33°36´N 130°24´E*
164 D13 **Fukuoka** *off.* Fukuoka-ken, *var.* Hukuoka. ◈ *prefecture* Kyūshū, SW Japan
Fukuoka-ken *see* Fukuoka
165 Q6 **Fukushima** Hokkaidō, NE Japan *41°27´N 140°14´E*
165 Q12 **Fukushima** *off.* Fukushima-ken, *var.* Hukusima. ◈ *prefecture* Honshū, C Japan
Fukushima-ken *see* Fukushima
164 G13 **Fukuyama** *var.* Hukuyama. Hiroshima, Honshū, SW Japan *34°29´N 133°21´E*
76 G13 **Fulacunda** C Guinea-Bissau *11°44´N 15°03´W*
129 P8 **Fūlādī, Kūh-e** ▲ E Afghanistan *34°38´N 67°32´E*
187 Z15 **Fulaga** *island* Lau Group, E Fiji
101 I17 **Fulda** Hessen, C Germany *50°33´N 09°41´E*
29 S10 **Fulda** Minnesota, N USA *43°52´N 95°36´W*
101 I16 **Fulda** 〰 C Germany
Fülek *see* Fil'akovo
Fuli *see* Jixian
Fulin *see* Hanyuan
160 K10 **Fuling** Chongqing Shi, C China *29°45´N 107°23´E*
35 T15 **Fullerton** California, SE USA *33°53´N 117°55´W*
29 P15 **Fullerton** Nebraska, C USA *41°21´N 97°58´W*
108 M8 **Fulpmes** Tirol, W Austria *47°11´N 11°22´E*
20 G8 **Fulton** Kentucky, S USA *36°31´N 88°52´W*
23 N2 **Fulton** Mississippi, S USA *34°16´N 88°24´W*
27 V4 **Fulton** Missouri, C USA *38°50´N 91°57´W*
18 H9 **Fulton** New York, NE USA *43°18´N 76°22´W*
Fuman/Fumen *see* Fowman
103 R3 **Fumay** Ardennes, N France *49°58´N 04°42´E*
102 M13 **Fumel** Lot-et-Garonne, SW France *44°31´N 00°58´E*
190 B10 **Funafara** *atoll* C Tuvalu
190 C9 **Funafuti** ✈ Funafuti Atoll, C Tuvalu *08°30´S 179°12´E*
Funafuti *see* Vaiaku
190 C9 **Funafuti Atoll** ● (Tuvalu) C Tuvalu *08°31´N 179°11´E*
190 C9 **Funafuti Atoll** *atoll* C Tuvalu
Funan *see* Fusui
190 B9 **Funangongo** *atoll* C Tuvalu
93 F17 **Funäsdalen** Jämtland, C Sweden *62°33´N 12°33´E*
64 O6 **Funchal** Madeira, Portugal, NE Atlantic Ocean *32°40´N 16°55´W*
64 P5 **Funchal** ✈ Madeira, Portugal, NE Atlantic Ocean *32°38´N 16°53´W*
54 F5 **Fundación** Magdalena, N Colombia *10°31´N 74°09´W*
104 I8 **Fundão** *var.* Fundáo. Castelo Branco, C Portugal *40°08´N 07°30´W*
Fundáo *see* Fundão
13 O16 **Fundy, Bay of** *bay* Canada/ USA
Fünen *see* Fyn
54 C13 **Fúnes** Nariño, SW Colombia *0°59´N 77°27´W*
Fünfkirchen *see* Pécs
83 M19 **Funhalouro** Inhambane, S Mozambique *23°04´S 34°24´E*
161 R6 **Funing** Jiangsu, E China *33°43´N 119°47´E*
160 I14 **Funing** *var.* Xinhua. Yunnan, SW China *23°39´N 105°41´E*
160 M7 **Funiu Shan** ▲▲ C China
77 U13 **Funtua** Katsina, N Nigeria *11°31´N 07°19´E*
161 R12 **Fuqing** Fujian, SE China *25°40´N 119°23´E*
83 M14 **Furancungo** Tete, NW Mozambique *14°51´S 33°39´E*
116 I15 **Furculeşti** Teleorman, S Romania *43°51´N 25°07´E*
Füred *see* Balatonfüred
165 W4 **Füren-ko** ◎ Hokkaidō, NE Japan
Fürg *see* Doborjī
Furluk *see* Fârliug
Fürmanov/Furmanovka *see* Moyynkum
Furmanovo *see* Zhalpaktal
59 L20 **Furnas, Represa de** ⊞ SE Brazil
183 Q14 **Furneaux Group** *island group* Tasmania, SE Australia
Furnes *see* Veurne
160 J10 **Furong Jiang** 〰 S China
138 I5 **Furqlus** Ḥimş, W Syria *34°40´N 37°02´E*
100 F12 **Fürstenau** Niedersachsen, NW Germany *52°30´N 07°40´E*
109 X8 **Fürstenfeld** Steiermark, SE Austria *47°03´N 16°05´E*
101 L23 **Fürstenfeldbruck** Bayern, S Germany *48°10´N 11°16´E*
100 P12 **Fürstenwalde** Brandenburg, NE Germany *52°22´N 14°04´E*
101 K20 **Fürth** Bayern, S Germany *49°29´N 10°59´E*
109 W3 **Furth bei Göttweig** Niederösterreich, NW Austria *48°22´N 15°33´E*
165 R3 **Furubira** Hokkaidō, NE Japan *43°14´N 140°38´E*
94 L12 **Furudal** Dalarna, C Sweden *61°10´N 15°07´E*
164 L12 **Furukawa** *var.* Hida. Gifu, Honshū, SW Japan *36°13´N 137°11´E*
165 Q10 **Furukawa** *var.* Hurukawa, Ōsaki. Miyagi, Honshū, C Japan *38°36´N 140°57´E*
54 F10 **Fusagasugá** Cundinamarca, C Colombia *04°22´N 74°21´W*
Fusan *see* Busan
Fushë-Arëzi/Fushë-Arrësi *see* Fushë-Arrëz
113 L18 **Fushë-Arrëz** *var.* Fushë-Arëzi, Fushë-Arrësi. Shkodër, N Albania *42°05´N 20°01´E*
113 N16 **Fushë Kosovë** *Serb.* Kosovo Polje. C Kosovo *42°40´N 21°07´E*
Fushë-Kruja *see* Fushë-Krujë
113 K19 **Fushë-Krujë** *var.* Fushë-Kruja. Durrës, C Albania *41°30´N 19°43´E*
163 V12 **Fushun** *var.* Fou-shan, Fu-shun. Liaoning, NE China *41°50´N 123°54´E*
Fu-shun *see* Fushun
Fusin *see* Fuxin
108 G10 **Fusio** Ticino, S Switzerland *46°27´N 08°39´E*
163 X11 **Fusong** Jilin, NE China *42°20´N 127°17´E*
101 K24 **Füssen** Bayern, S Germany *47°34´N 10°43´E*
160 K15 **Fusui** *var.* Xinning; *prev.* Funan. Guangxi Zhuangzu Zizhiqu, S China *22°39´N 107°49´E*
Futa Jallon *see* Fouta Djallon
63 G18 **Futaleufú** Los Lagos, S Chile *43°14´S 71°50´W*
112 K10 **Futog** Vojvodina, NW Serbia *45°15´N 19°43´E*
165 O14 **Futtsu** *var.* Huttu. Chiba, Honshū, S Japan *35°11´N 139°52´E*
187 S15 **Futuna** *island* S Vanuatu
190 D12 **Futuna, Île** *island* S Wallis and Futuna
161 Q11 **Futun Xi** 〰 SE China
160 L5 **Fuxian** *var.* Fu Xian. Shaanxi, C China *36°03´N 109°19´E*
Fuxian *see* Wafangdian
Fu Xian *see* Fuxian
160 G13 **Fuxian Hu** ◎ SW China
163 U12 **Fuxin** *var.* Fou-hsin, Fu-hsin, Fusin. Liaoning, NE China *41°59´N 121°40´E*
Fuxing *see* Wangmo
161 P7 **Fuyang** Anhui, E China *32°52´N 115°51´E*
Fuyang *see* Fuchuan
161 O4 **Fuyang He** 〰 E China
163 U7 **Fuyu** Heilongjiang, NE China *47°48´N 124°26´E*
163 Z6 **Fuyuan** Heilongjiang, NE China *48°20´N 134°22´E*
Fuyu/Fu-yü *see* Songyuan
158 M3 **Fuyun** *var.* Koktokay. Xinjiang Uygur Zizhiqu, NW China *46°58´N 89°30´E*
111 L22 **Füzesabony** Heves, E Hungary *47°46´N 20°25´E*
161 R12 **Fuzhou** *var.* Foochow, Fu-chou. *province capital* Fujian, SE China *26°09´N 119°17´E*
161 P11 **Fuzhou** *prev.* Linchuan. Jiangxi, S China *27°58´N 116°20´E*
137 W13 **Füzuli** *Rus.* Fizuli. SW Azerbaijan *39°33´N 47°09´E*
119 I20 **Fyadory** *Rus.* Fëdory. Brestskaya Voblasts', SW Belarus *51°57´N 26°24´E*
95 G23 **Fyn** *Ger.* Fünen. *island* C Denmark
96 H12 **Fyne, Loch** *inlet* W Scotland, United Kingdom
95 E16 **Fyresvatnet** ◎ S Norway
FYR Macedonia/FYROM *see* Macedonia, FYR
Fyzabad *see* Feizābād

G

Gaafu Alifu Atoll *see* North Huvadhu Atoll
81 O14 **Gaalkacyo** *var.* Galka'yo, *It.* Galcaio. Mudug, C Somalia *06°42´N 47°24´E*
146 J11 **Gabakly** *Rus.* Kabakly. Lebap Welaýaty, NE Turkmenistan *39°45´N 62°30´E*
114 H8 **Gabare** Vratsa, NW Bulgaria *43°20´N 23°57´E*
102 K15 **Gabas** 〰 SW France
Gabasumdo *see* Tongde
35 T7 **Gabbs** Nevada, W USA *38°51´N 117°55´W*
82 B12 **Gabela** Kwanza Sul, W Angola *10°50´S 14°21´E*
Gaberones *see* Gaborone
189 X14 **Gabert** *island* Caroline Islands, E Micronesia
74 M7 **Gabès** *var.* Qābis. E Tunisia *33°53´N 10°03´E*
74 M6 **Gabès, Golfe de** *Ar.* Khalīj Qābis. *gulf* E Tunisia
Gablonz an der Neisse *see* Jablonec nad Nisou
Gablös *see* Cavalese
79 E18 **Gabon** *off.* Gabonese Republic. ◆ *republic* C Africa
Gabonese Republic *see* Gabon
83 I20 **Gaborone** *prev.* Gaberones. ● (Botswana) South East, SE Botswana *24°42´S 25°50´E*
83 I20 **Gaborone** ✈ South East, SE Botswana *24°45´S 25°49´E*
104 K8 **Gabriel y Galán, Embalse de** ⊞ W Spain
143 U15 **Gābrīk, Rūd-e** 〰 SE Iran
114 J9 **Gabrovo** Gabrovo, N Bulgaria *42°54´N 25°19´E*
114 J9 **Gabrovo** ◈ *province* N Bulgaria
76 H12 **Gabú** *prev.* Nova Lamego. E Guinea-Bissau *12°16´N 14°09´W*
29 O6 **Gackle** North Dakota, N USA *46°34´N 99°07´W*
113 I15 **Gacko** Republika Srpska, S Bosnia and Herzegovina *43°08´N 18°29´E*
155 F17 **Gadag** Karnātaka, W India *15°25´N 75°37´E*
93 G15 **Gäddede** Jämtland, C Sweden *64°30´N 14°15´E*

159 S12 **Gadê** var. Kequ; prev. Pagqên. Qinghai, C China 33°56′N 99°49′E
Gades/Gadier/Gadir/Gadire see Cádiz
105 P15 **Gádor, Sierra de** ▲▲ S Spain
149 S15 **Gadra** Sind, SE Pakistan 25°39′N 70°28′E
23 Q3 **Gadsden** Alabama, S USA 34°00′N 86°00′W
36 H15 **Gadsden** Arizona, SW USA 32°33′N 114°45′W
Gadyach see Hadyach
124 J3 **Gadzhiyevo** Murmanskaya Oblast', NW Russian Federation 69°16′N 33°20′E
79 H15 **Gadzi** Mambéré-Kadéï, SW Central African Republic 04°46′N 16°42′E
116 J13 **Găeşti** Dâmboviţa, S Romania 44°42′N 25°19′E
107 J17 **Gaeta** Lazio, C Italy 41°12′N 13°35′E
107 J17 **Gaeta, Golfo di** var. Gulf of Gaeta. gulf C Italy
Gaeta, Gulf of see Gaeta, Golfo di
188 L14 **Gaferut** atoll Caroline Islands, W Micronesia
21 Q10 **Gaffney** South Carolina, SE USA 35°03′N 81°40′W
Gâfle see Gävle
Gäfleborg see Gävleborg
74 M6 **Gafsa** var. Qafşah. W Tunisia 34°25′N 08°52′E
Gafurov see Ghafurov
126 J3 **Gagarin** prev. Gzhatsk. Smolenskaya Oblast', W Russian Federation 55°33′N 35°00′E
147 O10 **Gagarin** Jizzax Viloyati, C Uzbekistan 40°40′N 68°04′E
116 M12 **Găgăuzia** ◐ cultural region S Moldavia
101 G21 **Gaggenau** Baden-Württemberg, SW Germany 48°48′N 08°19′E
188 F16 **Gagil Tamil** var. Gagil-Tomil. island Caroline Islands, W Micronesia
Gagil-Tomil see Gagil Tamil
127 O4 **Gagino** Nizhegorodskaya Oblast', W Russian Federation 55°18′N 45°01′E
107 Q19 **Gagliano del Capo** Puglia, SE Italy 39°49′N 18°22′E
94 L13 **Gagnef** Dalarna, C Sweden 60°34′N 15°04′E
76 M17 **Gagnoa** C Ivory Coast 06°11′N 05°56′W
13 N10 **Gagnon** Québec, E Canada 51°56′N 68°16′W
Gago Coutinho see Lumbala N'Guimbo
137 P8 **Gagra** NW Georgia 43°17′N 40°18′E
31 S13 **Gahanna** Ohio, N USA 40°01′N 82°52′W
143 R13 **Gahkom** Hormozgān, S Iran 28°14′N 55°48′E
Gahnpa see Ganta
57 Q19 **Gaiba, Laguna** ◎ E Bolivia
153 T13 **Gaibandha** var. Gaibanda. Rajshahi, NW Bangladesh 25°21′N 89°36′E
Gaibhlte, Cnoc Mór na n see Galtymore Mountain
109 R9 **Gail** ≈ S Austria
101 I21 **Gaildorf** Baden-Württemberg, S Germany 48°41′N 10°08′E
103 N15 **Gaillac** var. Gaillac-sur-Tarn. Tarn, S France 43°54′N 01°54′E
Gaillac-sur-Tarn see Gaillac
Gaillimh see Galway
Gaillimhe, Cuan na see Galway Bay
109 Q9 **Gailtaler Alpen** ▲▲ S Austria
63 J17 **Gaimán** Chaco, S Argentina 43°15′S 65°30′W
20 K8 **Gainesboro** Tennessee, S USA 36°20′N 85°41′W
23 V10 **Gainesville** Florida, SE USA 29°39′N 82°19′W
23 T2 **Gainesville** Georgia, SE USA 34°18′N 83°49′W
27 U8 **Gainesville** Missouri, C USA 36°37′N 92°28′W
25 T5 **Gainesville** Texas, SW USA 33°39′N 97°08′W
109 X5 **Gainfarn** Niederösterreich, NE Austria 47°59′N 16°11′E
97 N18 **Gainsborough** E England, United Kingdom 53°24′N 00°48′W
182 G6 **Gairdner, Lake** salt lake South Australia
Gaissane see Gáissát
92 L8 **Gáissát** var. Gaissane. ▲▲ N Norway
43 T15 **Gaital, Cerro** ▲ C Panama 08°37′N 80°04′W
21 W3 **Gaithersburg** Maryland, NE USA 39°08′N 77°13′W
163 U13 **Gaizhou** Liaoning, NE China 40°24′N 122°17′E
Gaizina Kalns see Gaiziņkalns
118 H7 **Gaiziņkalns** var. Gaizina Kalns. ▲ E Latvia 56°51′N 25°58′E
Gajac see Villeneuve-sur-Lot
197 T10 **Gakkel Ridge** var. Arctic Mid Oceanic Ridge; prev. Nansen Cordillera. seamount range Arctic Ocean
39 S10 **Gakona** Alaska, USA 62°21′N 145°16′W
158 M16 **Gala** Xizang Zizhiqu, China 28°17′N 89°21′E
Galaassiya see Galaosiyo
114 K10 **Galabovo** , Gŭlŭbovo. Stara Zagora, C Bulgaria 42°08′N 25°51′E
Ġalâğıl see Jalājil
Galam, Pulau see Gelam, Pulau
62 J6 **Galán, Cerro** ▲ NW Argentina 25°54′S 66°45′W
111 H21 **Galanta** Hung. Galánta. Trnavský Kraj, W Slovakia 48°12′N 17°45′E
146 L11 **Galaosiyo** Rus. Galaassiya. Buxoro Viloyati, C Uzbekistan 39°53′N 64°25′E
57 B17 **Galápagos** off. Provincia de Galápagos. ◐ province W Ecuador, E Pacific Ocean
193 P8 **Galapagos Fracture Zone** tectonic feature E Pacific Ocean
Galapagos Islands see Colón, Archipiélago de
Galápagos, Islas de los see Colón, Archipiélago de
Galápagos, Provincia de see Galápagos
193 S9 **Galapagos Rise** undersea feature E Pacific Ocean 15°00′S 97°00′W
96 K13 **Galashiels** SE Scotland, United Kingdom 55°37′N 02°49′W
116 M12 **Galați** Ger. Galatz. Galați, E Romania 45°27′N 28°00′E
116 L12 **Galați** ◐ county E Romania
107 Q19 **Galatina** Puglia, SE Italy 40°10′N 18°10′E
107 Q19 **Galatone** Puglia, SE Italy 40°09′N 18°05′E
Galatz see Galați
21 R8 **Galax** Virginia, NE USA 36°40′N 80°56′W
146 J16 **Galaýmor** Rus. Kala-i-Mor. Mary Welaýaty, S Turkmenistan 35°40′N 62°28′E
Galcaio see Gaalkacyo
64 P11 **Gáldar** Gran Canaria, Islas Canarias, NE Atlantic Ocean 28°09′N 15°40′W
94 F11 **Galdhøpiggen** ▲ S Norway 61°30′N 08°08′E
40 I4 **Galeana** Chihuahua, N Mexico 30°08′N 107°38′W
41 O9 **Galeana** Nuevo León, NE Mexico 24°45′N 99°59′W
60 P9 **Galeão** ✈ (Rio de Janeiro) Rio de Janeiro, SE Brazil 22°48′S 43°16′W
171 R10 **Galela** Pulau Halmahera, E Indonesia 01°52′N 127°48′E
39 O9 **Galena** Alaska, USA 64°43′N 156°55′W
30 K10 **Galena** Illinois, N USA 42°25′N 90°25′W
27 R7 **Galena** Kansas, C USA 37°04′N 94°38′W
27 T8 **Galena** Missouri, C USA 36°45′N 93°30′W
45 V15 **Galeota Point** headland Trinidad, Trinidad and Tobago 10°07′N 60°59′W
105 P13 **Galera** Andalucía, S Spain 37°45′N 02°33′W
45 Y16 **Galera Point** headland Trinidad, Trinidad and Tobago 10°49′N 60°54′W
56 A5 **Galera, Punta** headland NW Ecuador 0°49′N 80°03′W
30 K12 **Galesburg** Illinois, N USA 40°57′N 90°22′W
30 J7 **Galesville** Wisconsin, N USA 44°04′N 91°21′W
18 F12 **Galeton** Pennsylvania, NE USA 41°43′N 77°38′W
116 H9 **Gâlgău** Hung. Galgó; prev. Gîlgău. Sălaj, NW Romania 47°17′N 23°43′E
Galgó see Gâlgău
Galgóc see Hlohovec
81 N15 **Galguduud** off. Gobolka Galguduud. ◐ region E Somalia
Galguduud, Gobolka see Galguduud
137 Q9 **Gali** W Georgia 42°40′N 41°39′E
125 N14 **Galich** Kostromskaya Oblast', NW Russian Federation 58°21′N 42°21′E
114 H7 **Galiche** Vratsa, NW Bulgaria 43°36′N 23°53′E
104 H3 **Galicia** anc. Gallaecia. ◐ autonomous community NW Spain
64 M8 **Galicia Bank** undersea feature E Atlantic Ocean 11°45′W 42°40′N
Galilee see HaGalil
181 W7 **Galilee, Lake** ◎ Queensland, NE Australia
Galilee, Sea of see Tiberias, Lake
106 E11 **Galileo Galilei** ✈ (Pisa) Toscana, C Italy 43°40′N 10°22′E
31 S12 **Galion** Ohio, N USA 40°43′N 82°47′W
Galka'yo see Gaalkacyo
146 K12 **Galkynyş** prev. Rus. Deynau, Dyanev, Turkm. Dänew. Lebap Welaýaty, NE Turkmenistan 39°16′N 63°10′E
80 H11 **Gallabat** Gedaref, E Sudan 12°57′N 36°10′E
Gallaecia see Galicia
147 O11 **G'allaorol** Jizzax Viloyati, C Uzbekistan 40°01′N 67°30′E
106 C7 **Gallarate** Lombardia, NW Italy 45°39′N 08°47′E
27 S2 **Gallatin** Missouri, C USA 39°54′N 93°57′W
20 J8 **Gallatin** Tennessee, S USA 36°22′N 86°28′W
33 R11 **Gallatin Peak** ▲ Montana, NW USA 45°22′N 111°21′W
33 R12 **Gallatin River** ≈ Montana/Wyoming, NW USA
155 J26 **Galle** prev. Point de Galle. Southern Province, SW Sri Lanka 06°04′N 80°12′E
105 S5 **Gállego** ≈ NE Spain
193 Q8 **Gallego Rise** undersea feature E Pacific Ocean 02°00′S 115°00′W
Gallegos see Río Gallegos
63 H23 **Gallegos, Río** ≈ Argentina/Chile
Gallia see France
22 K10 **Galliano** Louisiana, S USA 29°26′N 90°18′W
114 G13 **Gallikós** ≈ N Greece
37 S12 **Gallinas Peak** ▲ New Mexico, SW USA 34°14′N 105°47′W
54 H3 **Gallinas, Punta** headland NE Colombia 12°27′N 71°44′W
37 T11 **Gallinas River** ≈ New Mexico, SW USA
107 Q19 **Gallipoli** Puglia, SE Italy 40°08′N 18°E
Gallipoli see Gelibolu
Gallipoli Peninsula see Gelibolu Yarımadası
31 T15 **Gallipolis** Ohio, N USA 38°49′N 82°14′W
92 J12 **Gällivare** Lapp. Váhtjer. Norrbotten, N Sweden 67°08′N 20°39′E
109 T4 **Gallneukirchen** Oberösterreich, N Austria 48°22′N 14°25′E
93 G17 **Gällö** Jämtland, C Sweden 62°57′N 15°15′E
105 Q7 **Gallo** ≈ C Spain
107 I23 **Gallo, Capo** headland Sicilia, Italy, C Mediterranean Sea 38°13′N 13°18′E
37 P13 **Gallo Mountains** ▲▲ New Mexico, SW USA
18 G8 **Galloo Island** island New York, NE USA
97 H15 **Galloway, Mull of** headland S Scotland, United Kingdom 54°37′N 04°54′W
37 P10 **Gallup** New Mexico, SW USA 35°32′N 108°45′W
105 R5 **Gallur** Aragón, NE Spain 41°51′N 01°21′W
Gâlma see Guelma
163 N9 **Galshar** var. Buyant. Hentiy, C Mongolia 46°15′N 110°50′E
162 I6 **Galt** var. Ider. Hövsgöl, C Mongolia 48°45′N 99°52′E
35 O8 **Galt** California, W USA 38°13′N 121°19′W
74 C10 **Galtat-Zemmour** C Western Sahara 25°07′N 12°21′W
95 G22 **Galten** Midtjylland, C Denmark 56°09′N 09°54′E
Gålto see Kultsjön
97 D20 **Galtymore Mountain** Ir. Cnoc Mór na nGaibhlte. ▲ S Ireland 52°21′N 08°09′W
97 D20 **Galty Mountains** Ir. Na Gaibhlte. ▲▲ S Ireland
30 K11 **Galva** Illinois, N USA 41°10′N 90°02′W
25 X12 **Galveston** Texas, SW USA 29°17′N 94°48′W
25 W11 **Galveston Bay** inlet Texas, SW USA
25 W12 **Galveston Island** island Texas, SW USA
61 B18 **Gálvez** Santa Fe, C Argentina 32°03′S 61°14′W
97 C18 **Galway** Ir. Gaillimh. W Ireland 53°16′N 09°03′W
97 B18 **Galway** Ir. Gaillimh. cultural region W Ireland
97 B18 **Galway Bay** Ir. Cuan na Gaillimhe. bay W Ireland
83 F18 **Gam** Otjozondjupa, NE Namibia 20°10′S 20°51′E
164 L14 **Gamagōri** Aichi, Honshū, SW Japan 34°49′N 137°15′E
54 F7 **Gamarra** Cesar, N Colombia 08°21′N 73°46′W
Gámas see Kaamanen
158 L17 **Gamba** Xizang Zizhiqu, W China 28°13′N 88°32′E
Gamba see Zamtang
77 P14 **Gambaga** NE Ghana 10°32′N 00°28′W
80 G13 **Gambēla** Gambēla Hizboch, W Ethiopia 08°09′N 34°15′E
81 H14 **Gambēla Hizboch** ◐ federal region W Ethiopia
38 K10 **Gambell** Saint Lawrence Island, Alaska, USA 63°44′N 171°41′W
76 E12 **Gambia** off. Republic of The Gambia, The Gambia. ◆ republic W Africa
76 I12 **Gambia** Fr. Gambie. ≈ W Africa
64 K12 **Gambia Plain** undersea feature E Atlantic Ocean
Gambia, Republic of The see Gambia
Gambia, The see Gambia
Gambie see Gambia
31 T13 **Gambier** Ohio, N USA 40°22′N 82°24′W
191 Y13 **Gambier, Îles** island group E French Polynesia
182 G10 **Gambier Islands** island group South Australia
79 H19 **Gamboma** Plateaux, E Congo 01°53′S 15°51′E
79 G16 **Gamboula** Mambéré-Kadéï, SW Central African Republic 04°09′N 15°12′E
37 P10 **Gamerco** New Mexico, SW USA 35°34′N 108°45′W
137 V12 **Gamış Dağı** ▲ W Azerbaijan 40°18′N 46°15′E
Gamlakarleby see Kokkola
95 N18 **Gamleby** Kalmar, S Sweden 57°54′N 16°25′E
Gammelstad see Gammelstaden
93 J14 **Gammelstaden** var. Gammelstad. Norrbotten, N Sweden 65°38′N 22°05′E
Gammouda see Sidi Bouzid
155 J25 **Gampaha** Western Province, W Sri Lanka 07°05′N 80°00′E
155 K25 **Gampola** Central Province, C Sri Lanka 07°10′N 80°34′E
167 S5 **Gâm, Sông** ≈ N Vietnam
92 L7 **Gamvik** Finnmark, N Norway 71°04′N 28°08′E
150 H13 **Gan** Addu Atoll, C Maldives
Gan see Gansu, China
Gan see Jiangxi, China
Ganaane see Juba
37 O10 **Ganado** Arizona, SW USA 35°42′N 109°31′W
25 U12 **Ganado** Texas, SW USA 29°02′N 96°30′W
14 L14 **Gananoque** Ontario, SE Canada 44°21′N 76°11′W
Ganāveh see Bandar-e Gonāveh
137 V11 **Gäncä** Rus. Gyandzha; prev. Kirovabad, Yelisavetpol. W Azerbaijan 40°42′N 46°23′E
Ganchi see Ghonchí
Gand see Gent
82 B13 **Ganda** var. Mariano Machado, Port. Vila Mariano Machado. Benguela, W Angola 13°02′S 14°40′E
79 L22 **Gandajika** Kasai-Oriental, S Dem. Rep. Congo 06°42′S 24°01′E
153 O12 **Gandak** Nep. Nārāyāni. ≈ India/Nepal
13 U11 **Gander** Newfoundland and Labrador, SE Canada 48°56′N 54°33′W
13 U11 **Gander** ✈ Newfoundland and Labrador, E Canada 49°03′N 54°49′W
100 G11 **Ganderkesee** Niedersachsen, NW Germany 53°01′N 08°33′E
105 T7 **Gandesa** Cataluña, NE Spain 41°03′N 00°26′E
154 B10 **Gāndhīdhām** Gujarāt, W India 23°08′N 70°05′E
154 D10 **Gāndhīnagar** state capital Gujarāt, W India 23°12′N 72°37′E
154 F9 **Gāndhi Sāgar** ◎ C India
105 T11 **Gandia** prev. Gandía. Valenciana, E Spain 38°59′N 00°11′W
Gandía see Gandia
159 O10 **Gang** Qinghai, W China
152 G9 **Ganganagar** Rājasthān, NW India 29°54′N 73°56′E
152 I12 **Gangāpur** Rājasthān, N India 26°30′N 76°49′E
153 S17 **Ganga Sāgar** West Bengal, NE India 21°39′N 88°05′E
Gangavathi see Gangāwati
155 G17 **Gangāwati** var. Gangavathi. Karnātaka, C India 15°26′N 76°35′E
159 S9 **Gangca** var. Shaliuhe. Qinghai, C China 37°21′N 100°09′E
158 H14 **Gangdisê Shan** Eng. Kailas Range. ▲▲ W China
103 Q15 **Ganges** Hérault, S France 43°57′N 03°42′E
153 P13 **Ganges** Ben. Padma. ≈ Bangladesh/India see also Padma
Ganges see Padma
Ganges Cone see Ganges Fan
173 S3 **Ganges Fan** var. Ganges Cone. undersea feature N Bay of Bengal 12°00′N 87°00′E
153 U17 **Ganges, Mouths of the** delta Bangladesh/India
107 K23 **Gangi** anc. Engyum. Sicilia, Italy, C Mediterranean Sea 37°48′N 14°13′E
163 Y14 **Gangneung** Jap. Kōryō; prev. Kangnŭng. NE South Korea 37°47′N 128°51′E
152 K8 **Gangotri** Uttarakhand, N India 30°56′N 79°02′E
Gangra see Çankırı
153 S11 **Gangtok** state capital Sikkim, N India 27°20′N 88°39′E
159 W11 **Gangu** var. Daxiangshan. Gansu, C China 34°38′N 105°18′E
163 U5 **Gan He** ≈ NE China
171 S12 **Gani** Pulau Halmahera, E Indonesia 0°45′S 128°13′E
161 O12 **Gan Jiang** ≈ S China
163 U11 **Ganjig** var. Horqin Zuoyi Houqi. Nei Mongol Zizhiqu, N China 42°53′N 122°22′E
146 H15 **Gannaly** Ahal Welaýaty, S Turkmenistan 37°02′N 60°43′E
163 U7 **Gannan** Heilongjiang, NE China 47°58′N 123°36′E
103 P10 **Gannat** Allier, C France 46°06′N 03°12′E
33 T14 **Gannett Peak** ▲ Wyoming, C USA 43°10′N 109°39′W
29 O10 **Gannvalley** South Dakota, N USA 44°01′N 98°59′W
Ganqu see Lhünzhub
109 Y3 **Gänserndorf** Niederösterreich, NE Austria 48°21′N 16°43′E
Gansos, Lago dos see Goose Lake
159 T9 **Gansu** var. Gan, Gansu Sheng, Kansu. ◐ province N China
Gansu Sheng see Gansu
76 K16 **Ganta** var. Gahnpa. NE Liberia 07°15′N 08°59′W
182 H11 **Gantheaume, Cape** headland South Australia 36°04′S 137°28′E
Gantsevichi see Hantsavichy
161 Q6 **Ganyu** var. Qingkou. Jiangsu, E China 34°52′N 119°11′E
144 D12 **Ganyushkino** Atyrau, SW Kazakhstan 46°38′N 49°12′E
161 O12 **Ganzhou** Jiangxi, S China 25°51′N 114°59′E
Ganzhou see Zhangye
77 Q10 **Gao** Gao, E Mali 16°16′N 00°03′E
77 R10 **Gao** ◐ region SE Mali
161 O10 **Gao'an** Jiangxi, S China 28°24′N 115°27′E
Gaocheng see Litang
Gaoleshan see Xianfeng
161 R5 **Gaomi** Shandong, C China 36°23′N 119°45′E
161 N5 **Gaoping** Shanxi, C China 35°51′N 112°55′E
159 S8 **Gaotai** Gansu, N China 39°22′N 99°44′E
Gaoth Dobhair see Gweedore
77 O14 **Gaoua** SW Burkina Faso 10°18′N 03°12′W
76 I13 **Gaoual** N Guinea 11°44′N 13°14′W
161 S14 **Gaoxiong** var. Kaohsiung, Jap. Takao, Takow. S Taiwan 22°36′N 120°17′E
161 S14 **Gaoxiong** ✈ S Taiwan 22°26′N 120°32′E
161 R7 **Gaoyou** var. Dayishan. Jiangsu, E China 32°48′N 119°26′E
161 R7 **Gaoyou Hu** ◎ E China
160 M15 **Gaozhou** Guangdong, S China 21°56′N 110°49′E
103 T13 **Gap** anc. Vapincum. Hautes-Alpes, SE France 44°33′N 06°05′E
146 E9 **Gaplaňgyr Platosy** Rus. Plato Kaplangky. ridge Turkmenistan/Uzbekistan
156 G13 **Gar** var. Shiquanhe. Xizang Zizhiqu, W China 32°31′N 80°04′E
Gar see Gar Xincun
Garabekevyul see Garabekewül
146 L13 **Garabekewül** Rus. Garabekevyul, Karabekaul. Lebap Welaýaty, E Turkmenistan 38°31′N 64°04′E
146 K15 **Garabil Belentligi** Rus. Vozvyshennost' Karabil'. ▲ S Turkmenistan
146 A8 **Garabogaz** Rus. Bekdash. Balkan Welaýaty, NW Turkmenistan 41°33′N 52°33′E
146 B9 **Garabogaz Aylagy** Rus. Zaliv Kara-Bogaz-Gol. bay NW Turkmenistan
146 A9 **Garabogazköl** Rus. Kara-Bogaz-Gol. Balkan Welaýaty, NW Turkmenistan 41°03′N 52°52′E
43 V16 **Garachiné** Darién, SE Panama 08°03′N 78°22′W
43 V16 **Garachiné, Punta** headland SE Panama 08°05′N 78°23′W
146 K12 **Garagan** Rus. Karagan. Ahal Welaýaty, C Turkmenistan 38°16′N 57°34′E
54 G10 **Garagoa** Boyacá, C Colombia 05°05′N 73°20′W
146 A11 **Garagöl'** Rus. Karagel'. Balkan Welaýaty, W Turkmenistan 39°24′N 53°13′E
146 F12 **Garagum** var. Garagumy, Qara Qum, Eng. Black Sand Desert, Kara Kum; prev. Peski Karakumy. desert C Turkmenistan
146 E12 **Garagum Kanaly** var. Kara Kum Canal, Rus. Karagumskiy Kanal, Karakumskiy Kanal. canal C Turkmenistan
Garagumy see Garagum
183 S4 **Garah** New South Wales, SE Australia 29°07′S 149°37′E
64 O11 **Garajonay** ▲ Gomera, Islas Canarias, NE Atlantic Ocean 28°07′N 17°14′W
76 L13 **Garalo** Sikasso, SW Mali 10°58′N 07°26′W
Garam see Hron, Slovakia
146 L14 **Garamätnyýaz** Rus. Karamet-Niyaz. Lebap Welaýaty, E Turkmenistan 37°45′N 64°28′E
Garamszentkereszt see Žiar nad Hronom
77 Q13 **Garango** S Burkina Faso 11°45′N 00°30′W
59 Q15 **Garanhuns** Pernambuco, E Brazil 08°53′S 36°28′W
188 H5 **Garapan** Saipan, S Northern Mariana Islands 15°12′S 145°43′E
Gárasavvon see Kaaresuando
Gárassavon see Kaaresuvanto
78 J13 **Garba** Bamingui-Bangoran, N Central African Republic 09°09′N 20°24′E
Garba see Jiulong
81 L16 **Garbahaarrey** It. Garba Harre. Gedo, SW Somalia 03°14′N 42°18′E
Garba Harre see Garbahaarrey
81 J18 **Garba Tula** Isiolo, C Kenya 0°31′N 38°35′E
27 N9 **Garber** Oklahoma, C USA 36°26′N 97°35′W
34 L4 **Garberville** California, W USA 40°07′N 123°48′W
Garbo see Lhozhag
100 I12 **Garbsen** Niedersachsen, N Germany 52°25′N 09°36′E
60 K9 **Garça** São Paulo, S Brazil 22°14′S 49°36′W
104 L10 **García de Solá, Embalse de** ⊠ C Spain
103 Q14 **Gard** ◐ department S France
103 Q14 **Gard** ≈ S France
106 F7 **Garda, Lago di** var. Benaco, Eng. Lake Garda, Ger. Gardasee. ◎ NE Italy
Garda, Lake see Garda, Lago di
Gardan Divāl see Gardan Dīwāl
149 Q5 **Gardan Dīwāl** var. Gardan Divāl. Wardak, C Afghanistan 34°30′N 68°15′E
103 S15 **Gardanne** Bouches-du-Rhône, SE France 43°27′N 05°28′E
Gardasee see Garda, Lago di
100 L12 **Gardelegen** Sachsen-Anhalt, C Germany 52°31′N 11°25′E
14 B10 **Garden** ≈ Ontario, S Canada
23 X6 **Garden City** Georgia, SE USA 32°06′N 81°09′W
26 I6 **Garden City** Kansas, C USA 37°57′N 100°54′W
27 S5 **Garden City** Missouri, C USA 38°34′N 94°12′W
25 N8 **Garden City** Texas, SW USA 31°51′N 101°30′W
23 P3 **Gardendale** Alabama, S USA 33°39′N 86°48′W
31 P5 **Garden Island** island Michigan, N USA
22 M11 **Garden Island Bay** bay Louisiana, S USA
31 O5 **Garden Peninsula** peninsula Michigan, N USA
Garden State, The see New Jersey
95 I14 **Gardermoen** Akershus, S Norway 60°10′N 11°04′E
95 I14 **Gardermoen** ✈ (Oslo) Akershus, S Norway 60°12′N 11°05′E
Gardeyz/Gardez see Gardēz
149 Q6 **Gardēz** var. Gardeyz, Gordiaz; prev. Gardīz. Paktīyā, E Afghanistan 33°35′N 69°14′E
93 G14 **Gardiken** ◎ N Sweden
19 Q7 **Gardiner** Maine, NE USA 44°13′N 69°46′W
33 S12 **Gardiner** Montana, NW USA 45°02′N 110°42′W
19 N13 **Gardiners Island** island New York, NE USA
Gardner Island see Nikumaroro
19 T6 **Gardner Lake** ◎ Maine, NE USA
35 Q6 **Gardnerville** Nevada, W USA 38°55′N 119°44′W
Gardo see Qardho
106 F7 **Gardone Val Trompia** Lombardia, N Italy 45°40′N 10°11′E
Garegegasnjárga see Karigasniemi
38 F17 **Gareloi Island** island Aleutian Islands, Alaska, USA
Gares see Puente la Reina
106 B10 **Garessio** Piemonte, NE Italy 44°14′N 08°01′E
32 M9 **Garfield** Washington, NW USA 47°00′N 117°07′W
31 U11 **Garfield Heights** Ohio, N USA 41°25′N 81°36′W
Gargaliani see Gargaliánoi
115 D21 **Gargaliánoi** var. Gargaliani. Pelopónnisos, S Greece 37°04′N 21°38′E
107 N15 **Gargano, Promontorio del** headland SE Italy 41°51′N 16°11′E
108 J8 **Gargellen** Graubünden, W Switzerland 46°57′N 09°55′E
93 I14 **Gargnäs** Västerbotten, N Sweden 65°19′N 18°00′E
118 C11 **Gargždai** Klaipėda, W Lithuania 55°42′N 21°24′E
154 J13 **Garhchiroli** Mahārāshtra, C India 20°14′N 79°58′E
153 O15 **Garhwa** Jhārkhand, N India 24°07′N 83°52′E
171 V13 **Gariau** Papua Barat, E Indonesia 03°43′S 134°54′E
83 E24 **Garies** Northern Cape, W South Africa 30°30′S 18°00′E
107 K17 **Garigliano** ≈ C Italy
81 K19 **Garissa** Garissa, E Kenya 0°27′S 39°39′E
81 K19 **Garissa** ◐ county SE Kenya
21 V11 **Garland** North Carolina, SE USA 34°45′N 78°25′W
25 T6 **Garland** Texas, SW USA 32°50′N 96°37′W
36 L1 **Garland** Utah, W USA 41°43′N 112°07′W
106 D8 **Garlasco** Lombardia, N Italy 45°12′N 08°59′E
119 F14 **Garliava** Kaunas, S Lithuania 54°49′N 23°52′E
Garm see Gharm
142 M9 **Garm, Āb-e** var. Rūd-e Khersān. ≈ SW Iran
Garmik see Germik
101 K25 **Garmisch-Partenkirchen** Bayern, S Germany 47°30′N 11°05′E
143 O5 **Garmsār** prev. Qishlaq. Semnān, N Iran 35°18′N 52°22′E
Garmser see Darwēshān
29 V12 **Garner** Iowa, C USA 43°06′N 93°36′W
21 U9 **Garner** North Carolina, SE USA 35°42′N 78°36′W
27 Q5 **Garnett** Kansas, C USA 38°16′N 95°15′W
99 M25 **Garnich** Luxembourg, SW Luxembourg 49°37′N 05°57′E
182 M8 **Garnpung, Lake** salt lake New South Wales, SE Australia
Garoe see Garoowe
Garoet see Garut
153 U13 **Gāro Hills** hill range NE India
102 K13 **Garonne** anc. Garumna. ≈ S France
80 P13 **Garoowe** var. Garoe. Nugaal, N Somalia 08°24′N 48°29′E
78 F12 **Garoua** var. Garua. Nord, N Cameroon 09°17′N 13°22′E
79 G14 **Garoua Boulaï** Est, E Cameroon 05°54′N 14°33′E
77 O10 **Garou, Lac** ◎ C Mali
95 L16 **Garphyttan** Örebro, C Sweden 59°18′N 14°54′E
29 R11 **Garretson** South Dakota, N USA 43°43′N 96°30′W
31 Q11 **Garrett** Indiana, N USA 41°21′N 85°08′W
33 Q10 **Garrison** Montana, NW USA 46°32′N 112°46′W
28 M4 **Garrison** North Dakota, N USA 47°36′N 101°25′W
25 X8 **Garrison** Texas, SW USA 31°49′N 94°29′W
28 L4 **Garrison Dam** dam North Dakota, N USA
104 J9 **Garrovillas** Extremadura, W Spain 39°43′N 06°33′W
Garrygala see Magtymguly
8 L8 **Garry Lake** ◎ Nunavut, N Canada
Gars see Gars am Kamp
109 W3 **Gars am Kamp** var. Gars. Niederösterreich, NE Austria 48°35′N 15°40′E
81 K20 **Garsen** Tana River, S Kenya 02°16′S 40°07′E
Garshy see Garşy
14 F10 **Garson** Ontario, S Canada 46°33′N 80°51′W
109 T5 **Garsten** Oberösterreich, N Austria 48°00′N 14°24′E
146 A9 **Garşy** var. Garshy, Rus. Karshi. Balkan Welaýaty, NW Turkmenistan 40°45′N 52°50′E
Gartar see Qianning
102 M10 **Gartempe** ≈ C France
Gartog see Markam
Garua see Garoua
83 D21 **Garub** Karas, SW Namibia 26°33′S 16°00′E
Garumna see Garonne
169 P16 **Garut** prev. Garoet. Jawa, C Indonesia 07°15′S 107°55′E
185 C20 **Garvie Mountains** ▲▲ South Island, New Zealand
110 N12 **Garwolin** Mazowieckie, E Poland 51°54′N 21°36′E
25 U12 **Garwood** Texas, SW USA 29°25′N 96°26′W
158 G13 **Gar Xincun** prev. Gar. Xizang Zizhiqu, W China 32°04′N 80°01′E
31 N11 **Gary** Indiana, N USA 41°36′N 87°21′W
25 X7 **Gary** Texas, SW USA 32°00′N 94°21′W
158 G13 **Garyarsa** Xizang Zizhiqu, SW China 31°44′N 80°20′E
159 O13 **Gar Zangbo** ≈ W China
160 F8 **Garzê** Sichuan, C China 31°40′N 99°58′E
54 E12 **Garzón** Huila, S Colombia 02°14′N 75°37′W
Gasan-Kuli see Esenguly
31 P13 **Gas City** Indiana, N USA 40°29′N 85°36′W
102 K15 **Gascogne** Eng. Gascony. cultural region S France
26 V5 **Gasconade River** ≈ Missouri, C USA
Gascony see Gascogne
180 H9 **Gascoyne Junction** Western Australia 25°06′S 115°10′E
173 V8 **Gascoyne Plain** undersea feature E Indian Ocean
180 H9 **Gascoyne River** ≈ Western Australia
192 J11 **Gascoyne Tablemount** undersea feature N Tasman Sea 36°30′S 156°30′E
67 U6 **Gash** var. Nahr al Qāsh. ≈ W Sudan
149 X3 **Gasherbrum** ▲ NE Pakistan 35°39′N 76°34′E
Gas Hu see Gas Hure Hu
77 X12 **Gashua** Yobe, NE Nigeria 12°55′N 11°10′E
159 N9 **Gas Hure Hu** var. Gas Hu. ◎ C China
Gásluokta see Kjøpsvik
186 G7 **Gasmata** New Britain, E Papua New Guinea 06°12′S 150°25′E
23 V14 **Gasparilla Island** island Florida, SE USA
169 O13 **Gaspar, Selat** strait W Indonesia
15 Y6 **Gaspé** Québec, SE Canada 48°50′N 64°33′W
15 Z6 **Gaspé, Cap de** headland Québec, SE Canada 48°45′N 64°10′W
15 X6 **Gaspé, Péninsule de** var. Péninsule de la Gaspésie. peninsula Québec, SE Canada
Gaspésie, Péninsule de la see Gaspé, Péninsule de
77 W15 **Gassol** Taraba, E Nigeria 08°28′N 10°24′E
Gastein see Badgastein
21 R10 **Gastonia** North Carolina, SE USA 35°14′N 81°12′W
21 V8 **Gaston, Lake** ⊠ North Carolina/Virginia, SE USA
115 D19 **Gastoúni** Dytikí Elláda, S Greece 37°51′N 21°15′E
63 I17 **Gastre** Chubut, S Argentina 42°20′S 69°10′W
Gat see Ghāt
105 P15 **Gata, Cabo de** cape S Spain
Gata, Cape see Gátas, Akrotíri
105 T11 **Gata de Gorgos** Valenciana, E Spain 38°45′N 00°06′E
116 E12 **Gătaia** Ger. Gataja, Hung. Gátalja; prev. Gáttája. Timiş, W Romania 45°24′N 21°26′E
Gataja/Gátalja see Gătaia
121 P3 **Gátas, Akrotíri** var. Cape Gata. headland S Cyprus 34°34′N 33°03′E
104 J8 **Gata, Sierra de** ▲▲ W Spain
124 G13 **Gatchina** Leningradskaya Oblast', NW Russian Federation 59°34′N 30°06′E
21 P8 **Gate City** Virginia, NE USA 36°38′N 82°37′W
97 M14 **Gateshead** NE England, United Kingdom 54°57′N 01°37′W
21 X8 **Gatesville** North Carolina, SE USA 36°24′N 76°46′W
25 S8 **Gatesville** Texas, SW USA 31°26′N 97°46′W
14 L12 **Gatineau** Québec, SE Canada 45°29′N 75°40′W
14 L11 **Gatineau** ≈ Ontario/Québec, SE Canada
21 N9 **Gatlinburg** Tennessee, S USA 35°42′N 83°30′W
Gatooma see Kadoma
Gâttája see Gătaia
43 T14 **Gatún, Lago** ⊠ C Panama
59 N14 **Gaturiano** Piauí, NE Brazil 06°53′S 41°45′W
97 O22 **Gatwick** ✈ (London) SE England, United Kingdom 51°10′N 00°12′W
187 Y14 **Gau** prev. Ngau. island C Fiji
187 R12 **Gaua** var. Santa Maria. Island Banks Island, N Vanuatu
104 L16 **Gaucín** Andalucía, S Spain 36°31′N 05°19′W
Gauhāti see Guwāhāti
118 I8 **Gauja** Ger. Aa. ≈ Estonia/Latvia
118 I7 **Gaujiena** NE Latvia 57°31′N 26°24′E
94 H9 **Gauldalen** valley S Norway
21 R5 **Gauley River** ≈ West Virginia, NE USA
Gaul/Gaule see France
99 D19 **Gaurain-Ramecroix** Hainaut, SW Belgium 50°35′N 03°31′E
95 F15 **Gaustatoppen** ▲ S Norway 59°50′N 08°39′E
83 J21 **Gauteng** off. Gauteng Province; prev. Pretoria-Witwatersrand-Vereeniging. ◐ province NE South Africa
Gauteng see Johannesburg, South Africa
Gauteng see Germiston, South Africa
Gauteng Province see Gauteng
137 U11 **Gavarr** prev. Kamo. C Armenia 40°21′N 45°07′E
143 P14 **Gāvbandī** Hormozgān, S Iran 27°07′N 53°21′E
115 H25 **Gavdopoúla** island SE Greece
115 H26 **Gávdos** island SE Greece
102 K16 **Gave de Pau** var. Gave-de-Pay. ≈ SW France
Gave-de-Pay see Gave de Pau
102 J16 **Gave d'Oloron** ≈ SW France
99 E18 **Gavere** Oost-Vlaanderen, NW Belgium 50°56′N 03°41′E
94 N13 **Gävle** var. Gäfle; prev. Gefle. Gävleborg, C Sweden 60°41′N 17°09′E
94 M11 **Gävleborg** var. Gäfleborg, Gefleborg. ◐ county C Sweden
94 O13 **Gävlebukten** bay C Sweden
124 L16 **Gavrilov-Yam** Yaroslavskaya Oblast', W Russian Federation 57°19′N 39°52′E
182 I9 **Gawler** South Australia 34°38′S 138°44′E
182 G7 **Gawler Ranges** hill range South Australia
Gawso see Goaso
162 H11 **Gaxun Nur** ◎ N China
153 P14 **Gaya** Bihār, N India 24°48′N 85°E
77 S13 **Gaya** Dosso, SW Niger 11°52′N 03°28′E
Gaya see Kyjov
31 Q6 **Gaylord** Michigan, N USA 45°01′N 84°40′W
29 U9 **Gaylord** Minnesota, N USA 44°33′N 94°13′W
181 Y9 **Gayndah** Queensland, E Australia 25°37′S 151°31′E
125 T12 **Gayny** Komi-Permyatskiy Okrug, NW Russian Federation 60°19′N 54°15′E
Gaysin see Haysyn
Gayvorno see Hayvoron
138 E11 **Gaza** Ar. Ghazzah, Heb. 'Azza. NE Gaza Strip 31°30′N 34°E
83 L20 **Gaza** off. Província de Gaza. ◐ province SW Mozambique
Gaz-Achak see Gazojak
147 Q9 **G'azalkent** Rus. Gazalkent. Toshkent Viloyati, E Uzbekistan 41°30′N 69°46′E
Gazalkent see G'azalkent
Gazandzhyk/Gazanjyk see Bereket
77 V12 **Gazaoua** Maradi, S Niger 13°28′N 07°54′E
Gaza, Província de see Gaza
138 E11 **Gaza Strip** Ar. Qita Ghazzah. disputed region SW Asia
136 M16 **Gaziantep** var. Gazi Antep; prev. Aintab, Antep. Gaziantep, S Turkey 37°04′N 37°21′E
136 M17 **Gaziantep** var. Gazi Antep. ◐ province S Turkey
Gazi Antep see Gaziantep
114 M13 **Gaziköy** Tekirdağ, NW Turkey 40°45′N 27°18′E
121 Q2 **Gazimağusa** var. Famagusta, Gk. Ammóchostos. E Cyprus 35°07′N 33°57′E
121 Q2 **Gazimağusa Körfezi** var. Famagusta Bay, Gk. Kólpos Ammóchostos. bay E Cyprus
146 K11 **Gazli** Buxoro Viloyati, C Uzbekistan 40°09′N 63°28′E
146 I9 **Gazojak** Rus. Gaz-Achak. Lebap Welaýaty, NE Turkmenistan 41°12′N 61°24′E
79 K15 **Gbadolite** Equateur, NW Dem. Rep. Congo 04°14′N 20°59′E
76 K16 **Gbanga** var. Gbarnga. N Liberia 07°02′N 09°30′W
Gbarnga see Gbanga
77 S14 **Gbéroubouè** var. Béroubouay. N Benin 10°35′N 02°47′E
77 W16 **Gboko** Benue, S Nigeria 07°21′N 08°57′E
Gcuwa see Butterworth
110 J7 **Gdańsk** Fr. Dantzig, Ger. Danzig. Pomorskie, N Poland 54°22′N 18°35′E
Gdan'skaya Bukhta/Gdańsk, Gulf of see Danzig, Gulf of
Gdańska, Zakota see Danzig, Gulf of
Gdingen see Gdynia
124 F13 **Gdov** Pskovskaya Oblast', W Russian Federation 58°43′N 27°51′E
110 I6 **Gdynia** Ger. Gdingen. Pomorskie, N Poland 54°31′N 18°30′E
26 M10 **Geary** Oklahoma, C USA 35°37′N 98°19′W
Geavvú see Kevo
76 H12 **Géba, Rio** ≈ C Guinea-Bissau
136 E11 **Gebze** Kocaeli, NW Turkey 40°48′N 29°26′E
80 H10 **Gedaref** var. Al Qaḍārif, El Gedaref. Gedaref, E Sudan 14°03′N 35°24′E
80 H10 **Gedaref** ◐ state E Sudan

◆ Country ● Country Capital ◇ Dependent Territory ○ Dependent Territory Capital ◐ Administrative Regions ✈ International Airport ▲ Mountain ▲▲ Mountain Range ⊼ Volcano ≈ River ◎ Lake ⊠ Reservoir

80 B11 Gedid Ras el Fil Southern Darfur, W Sudan 12°45′N 25°45′E
99 I23 Gedinne Namur, SE Belgium 49°57′N 04°55′E
136 E13 Gediz Kütahya, W Turkey 39°04′N 29°25′E
136 C14 Gediz Nehri ≈ W Turkey
81 N14 Gedlegubē Sumalē, E Ethiopia 06°53′N 45°08′E
81 L17 Gedo off. Gobolka Gedo. ◈ region SW Somalia
Gedo, Gobolka see Gedo
95 I25 Gedser Sjælland, SE Denmark 54°34′N 11°57′E
99 I16 Geel var. Gheel. Antwerpen, N Belgium 51°10′N 04°59′E
183 N13 Geelong Victoria, SE Australia 38°10′S 144°21′E
Ge'e'mu see Golmud
99 I14 Geertruidenberg Noord-Brabant, S Netherlands 51°43′N 04°52′E
100 H10 Geeste ≈ NW Germany
100 J10 Geesthacht Schleswig-Holstein, N Germany 53°25′N 10°22′E
183 P17 Geeveston Tasmania, SE Australia 43°12′S 146°54′E
Gefle see Gävle
Gefleborg see Gävleborg
163 S5 Gegan Gol prev. Ergun, Gen He, Zuoqi. NE China
163 T5 Gegen Gol prev. Ergun Zuoqi, Genhe. Nei Mongol Zizhiqu, N China 50°48′N 121°30′E
158 G13 Gê'gyai Xizang Zizhiqu, W China 32°29′N 81°04′E
77 X12 Geidam Yobe, NE Nigeria 12°52′N 11°55′E
11 T11 Geikie ≈ Saskatchewan, C Canada
94 F13 Geilo Buskerud, S Norway 60°32′N 08°13′E
94 E10 Geiranger Møre og Romsdal, S Norway 62°07′N 07°12′E
101 I22 Geislingen var. Geislingen an der Steige. Baden-Württemberg, SW Germany 48°37′N 09°50′E
Geislingen an der Steige see Geislingen
81 F20 Geita Geita, NW Tanzania 02°52′S 32°12′E
81 F21 Geita off. Mkoa wa Geita. ◈ region N Tanzania
Geita, Mkoa wa see Geita
95 G15 Geithus Buskerud, S Norway 59°56′N 09°58′E
160 H14 Gejiu var. Kochiu. Yunnan, S China 23°22′N 103°07′E
Gëkdepe see Gökdepe
146 E9 Gekdengkui, Solonchak var. Solonchak Goklenkuy. salt marsh NW Turkmenistan
81 D14 Gel ≈ C South Sudan
107 K25 Gela prev. Terranova di Sicilia. Sicilia, Italy, C Mediterranean Sea 37°05′N 14°15′E
81 N14 Geladī SE Ethiopia 06°58′N 46°24′E
169 P13 Gelam, Pulau var. Pulau Galam. island N Indonesia
Gelaozu Miaozu Zhizhixian see Wuchuan
98 L11 Gelderland prev. Eng. Guelders. ◈ province E Netherlands
98 J13 Geldermalsen Gelderland, C Netherlands 51°53′N 05°17′E
101 D14 Geldern Nordrhein-Westfalen, W Germany 51°31′N 06°19′E
99 K15 Geldrop Noord-Brabant, SE Netherlands 51°25′N 05°34′E
99 L17 Geleen Limburg, SE Netherlands 50°57′N 05°49′E
126 K14 Gelendzhik Krasnodarskiy Kray, SW Russian Federation 44°34′N 38°06′E
Gelib see Jilib
136 B11 Gelibolu Eng. Gallipoli. Çanakkale, NW Turkey 40°25′N 26°41′E
115 L14 Gelibolu Yarımadası Eng. Gallipoli Peninsula. peninsula NW Turkey
81 O14 Gellinsor Galguduud, C Somalia 06°25′N 46°44′E
101 H18 Gelnhausen Hessen, C Germany 50°12′N 09°12′E
101 E14 Gelsenkirchen Nordrhein-Westfalen, W Germany 51°30′N 07°05′E
83 C20 Geluk Hardap, SW Namibia 24°35′S 15°48′E
99 H20 Gembloux Namur, Belgium 50°34′N 04°42′E
79 J16 Gemena Equateur, NW Dem. Rep. Congo 03°13′N 19°49′E
99 L14 Gemert Noord-Brabant, SE Netherlands 51°33′N 05°41′E
136 E11 Gemlik Bursa, NW Turkey 40°26′N 29°10′E
Gem of the Mountains see Idaho
106 J6 Gemona del Friuli Friuli-Venezia Giulia, NE Italy 46°18′N 13°12′E
Gem State see Idaho
Genalē Wenz see Juba
169 R10 Genali, Danau ◎ Borneo, N Indonesia
99 G19 Genappe Walloon Brabant, C Belgium 50°39′N 04°27′E
137 P14 Genç Bingöl, E Turkey 38°44′N 40°35′E
Genck see Genk
98 M9 Genemuiden Overijssel, E Netherlands 52°38′N 06°03′E
63 K14 General Acha La Pampa, C Argentina 37°25′S 64°38′W
61 C21 General Alvear Buenos Aires, E Argentina 36°03′S 60°01′W
62 I12 General Alvear Mendoza, W Argentina 34°59′S 67°40′W
61 B20 General Arenales Buenos Aires, E Argentina 34°21′S 61°20′W
61 D21 General Belgrano Buenos Aires, E Argentina 35°47′S 58°30′W
194 H3 General Bernardo O'Higgins Chilean research station Antarctica 63°09′S 57°13′W
41 O8 General Bravo Nuevo León, NE Mexico 25°47′N 99°04′W
62 M7 General Capdevila Chaco, N Argentina 27°25′S 61°30′W
General Carrera, Lago see Buenos Aires, Lago
41 N9 General Cepeda Coahuila, NE Mexico 25°18′N 101°24′W
63 K15 General Conesa Río Negro, E Argentina 40°06′S 64°26′W
61 G18 General Enrique Martínez Treinta y Tres, E Uruguay 33°13′S 53°47′W
62 L3 General Eugenio A. Garay var. Fortín General Eugenio Garay; prev. Yrendagüé. Nueva Asunción, NW Paraguay 20°30′S 61°56′W
61 C18 General Galarza Entre Ríos, E Argentina 32°43′S 59°24′W
61 E22 General Guido Buenos Aires, E Argentina 36°36′S 57°45′W
General José F.Uriburu see Zárate
61 E22 General Juan Madariaga Buenos Aires, E Argentina 37°00′S 57°09′W
41 O16 General Juan N Alvarez ✈ (Acapulco) Guerrero, S Mexico 16°47′N 99°47′W
61 B22 General La Madrid Buenos Aires, E Argentina 37°17′S 61°20′W
61 E21 General Lavalle Buenos Aires, E Argentina 36°25′S 56°56′W
General Machado see Camacupa
62 I8 General Manuel Belgrano, Cerro ▲ W Argentina 29°05′S 67°05′W
41 O8 General Mariano Escobero ✈ (Monterrey) Nuevo León, NE Mexico 25°47′N 100°00′W
61 B20 General O'Brien Buenos Aires, E Argentina 34°54′S 60°45′W
62 K13 General Pico La Pampa, C Argentina 35°43′S 63°45′W
62 M7 General Pinedo Chaco, N Argentina 27°17′S 61°20′W
61 B20 General Pinto Buenos Aires, E Argentina 34°45′S 61°50′W
61 E22 General Pirán Buenos Aires, E Argentina 37°16′S 57°46′W
43 N15 General, Río ≈ S Costa Rica
63 I15 General Roca Río Negro, C Argentina 39°00′S 67°35′W
171 Q8 General Santos off. General Santos City. Mindanao, S Philippines 06°10′N 125°10′E
General Santos City see General Santos
41 O9 General Terán Nuevo León, NE Mexico 25°18′N 99°40′W
114 N7 General Toshevo Rom. I.G.Duca; prev. Casim, Kasimköj. Dobrich, NE Bulgaria 43°43′N 28°04′E
61 B20 General Viamonte Buenos Aires, E Argentina 35°01′S 61°00′W
61 A20 General Villegas Buenos Aires, E Argentina 35°02′S 63°01′W
Gênes see Genova
18 E11 Genesee River ≈ New York/Pennsylvania, NE USA
30 K11 Geneseo Illinois, N USA 41°27′N 90°08′W
18 F10 Geneseo New York, NE USA 42°48′N 77°46′W
57 L14 Geneshuaya, Río ≈ N Bolivia
23 Q8 Geneva Alabama, S USA 31°01′N 85°51′W
30 M10 Geneva Illinois, N USA 41°53′N 88°18′W
29 Q16 Geneva Nebraska, C USA 40°31′N 97°36′W
18 G10 Geneva New York, NE USA 42°52′N 76°58′W
31 U10 Geneva Ohio, NE USA 41°48′N 80°53′W
Geneva see Genève
108 B10 Geneva, Lake Fr. Lac de Genève, Lac Léman, le Léman, Ger. Genfer See. ◎ France/Switzerland
108 A10 Genève Eng. Geneva, Ger. Genf, It. Ginevra. Genève, SW Switzerland 46°13′N 06°09′E
108 A11 Genève Eng. Geneva, Ger. Genf, It. Ginevra. ◈ canton SW Switzerland
108 A10 Genève var. Geneva. ✈ Vaud, SW Switzerland 46°13′N 06°06′E
Genève, Lac de see Geneva, Lake
Genf see Genève
Genfer See see Geneva, Lake
Gen He see Gegan Gol
Genhe see Gegen Gol
Genichesk see Heniches'k
104 L14 Genil ≈ S Spain
99 K18 Genk var. Genck. Limburg, NE Belgium 50°58′N 05°30′E
165 C13 Genkai-nada gulf Kyūshū, SW Japan
107 C19 Gennargentu, Monti del ▲▲ Sardegna, Italy, C Mediterranean Sea
99 M14 Gennep Limburg, SE Netherlands 51°43′N 05°58′E
30 M10 Genoa Illinois, N USA 42°06′N 88°41′W
29 Q15 Genoa Nebraska, C USA 41°27′N 97°43′W
Genoa see Genova
Genoa, Gulf of see Genova, Golfo di
106 D10 Genova Eng. Genoa, Fr. Gênes; anc. Genua. Liguria, NW Italy 44°28′N 09°E
106 D10 Genova, Golfo di Eng. Gulf of Genoa. gulf NW Italy
57 C17 Genovesa, Isla var. Tower Island. island Galapagos Islands, Ecuador, E Pacific Ocean
Genshū see Wonju
99 E17 Gent Eng. Ghent, Fr. Gand. Oost-Vlaanderen, NW Belgium 51°02′N 03°42′E
169 N16 Genteng Jawa, C Indonesia 07°21′S 106°20′E
100 M12 Genthin Sachsen-Anhalt, E Germany 52°24′N 12°10′E
27 R9 Gentry Arkansas, C USA 36°16′N 94°28′W
Genua see Genova
107 I15 Genzano di Roma Lazio, C Italy 41°42′N 12°42′E
163 Y17 Geogeum-do prev. Kŏgŭm-do. island S South Korea
163 Z16 Geogeum-do Jap. Kyōsai-tō; prev. Kŏje-do. island S South Korea
Geokchay see Göyçay
Geok-Tepe see Gökdepe
122 I3 Georga, Zemlya Eng. George Land. island Zemlya Frantsa-Iosifa, N Russian Federation
83 G26 George Western Cape, S South Africa 33°57′S 22°28′E
29 S11 George Iowa, C USA 43°20′N 96°00′W
13 O5 George ≈ Newfoundland and Labrador/Québec, E Canada
George F L Charles see Vigie
65 C25 George Island island S Falkland Islands
183 R10 George, Lake ◎ New South Wales, SE Australia
81 E18 George, Lake ◎ SW Uganda
23 W10 George, Lake ◎ Florida, SE USA
18 L8 George, Lake ⊠ New York, NE USA
George Land see Georga, Zemlya
Georgenburg see Jurbarkas
George River see Kangiqsualujjuaq
64 G8 Georges Bank undersea feature W Atlantic Ocean 41°15′N 67°30′W
185 A21 George Sound sound South Island, New Zealand
65 F15 Georgetown ○ (Ascension Island) NW Ascension Island 17°56′S 14°25′W
181 V5 Georgetown Queensland, NE Australia 18°17′S 143°37′E
183 P15 George Town Tasmania, SE Australia 41°04′S 146°48′E
44 D8 George Town var. Georgetown. ○ (Cayman Islands) Grand Cayman, SW Cayman Islands 19°16′N 81°23′W
76 H12 Georgetown E Gambia 13°33′N 14°49′W
55 T8 Georgetown ● (Guyana) N Guyana 06°46′N 58°10′W
168 I7 George Town var. Penang, Pinang. Pinang, Peninsular Malaysia 05°28′N 100°20′E
45 Y14 Georgetown Saint Vincent, Saint Vincent and the Grenadines 13°19′N 61°09′W
44 I4 George Town Great Exuma Island, C The Bahamas 23°28′N 75°47′W
21 Y4 Georgetown Delaware, NE USA 38°39′N 75°22′W
23 R6 Georgetown Georgia, SE USA 31°52′N 85°04′W
20 M5 Georgetown Kentucky, S USA 38°13′N 84°30′W
21 T13 Georgetown South Carolina, SE USA 33°23′N 79°18′W
25 S10 Georgetown Texas, SW USA 30°39′N 97°42′W
55 T8 Georgetown ✈ N Guyana 06°46′N 58°10′W
Georgetown see George Town
195 U16 George V Coast physical region Antarctica
194 J7 George VI Ice Shelf ice shelf Antarctica
194 J6 George VI Sound sound Antarctica
195 T15 George V Land physical region Antarctica
25 S14 George West Texas, SW USA 28°21′N 98°08′W
137 R9 Georgia off. Republic of Georgia, Geor. Sak'art'velo, Rus. Gruzinskaya SSR, Gruziya. ◆ republic SW Asia
23 S5 Georgia off. State of Georgia, also known as Empire State of the South, Peach State. ◈ state SE USA
14 F12 Georgian Bay lake bay Ontario, S Canada
Georgia, Republic of see Georgia
10 L17 Georgia, Strait of strait British Columbia, W Canada
Georgi Dimitrov see Kostenets
Georgi Dimitrov, Yazovir see Koprinka, Yazovir
Georgiu-Dezh see Liski
145 W10 Georgiyevka Vostochnyy Kazakhstan, E Kazakhstan 49°19′N 81°35′E
Georgiyevka see Korday
127 N15 Georgiyevsk Stavropol'skiy Kray, SW Russian Federation 44°07′N 43°22′E
100 G13 Georgsmarienhütte Niedersachsen, NW Germany 52°13′N 08°02′E
195 O1 Georg von Neumayer German research station Antarctica 70°41′S 08°18′W
101 M16 Gera Thüringen, E Germany 50°51′N 12°13′E
101 K16 Gera ≈ C Germany
99 E19 Geraardsbergen Oost-Vlaanderen, SW Belgium 50°47′N 03°53′E
115 F21 Geráki Pelopónnisos, S Greece 36°56′N 22°46′E
27 W5 Gerald Missouri, C USA 38°24′N 91°20′W
47 V8 Geral de Goiás, Serra ▲ E Brazil
185 G20 Geraldine Canterbury, South Island, New Zealand 44°06′S 171°14′E
180 H11 Geraldton Western Australia 28°48′S 114°40′E
12 E11 Geraldton Ontario, S Canada 49°44′N 86°59′W
60 J12 Geral, Serra ▲ S Brazil
103 U6 Gérardmer Vosges, NE France 48°05′N 06°54′E
Gerasa see Jarash
Gerdauen see Zheleznodorozhnyy
39 Q11 Gerdine, Mount ▲ Alaska, USA 61°40′N 152°21′W
136 H11 Gerede Bolu, N Turkey 40°48′N 32°13′E
136 H11 Gerede Çayı ≈ N Turkey
148 M8 Gereshk Helmand, SW Afghanistan 31°50′N 64°32′E
101 L24 Geretsried Bayern, S Germany 47°51′N 11°28′E
105 P14 Gérgal Andalucía, S Spain 37°07′N 02°34′W
28 I14 Gering Nebraska, C USA 41°49′N 103°39′W
35 R3 Gerlach Nevada, W USA 40°38′N 119°21′W
Gerlachfalvi Csúcs/ Gerlachovka see Gerlachovský štít
111 L18 Gerlachovský štít var. Gerlachovka, Ger. Gerlsdorfer Spitze, Hung. Gerlachfalvi Csúcs; prev. Stalinov Štít, Ger. Franz-Josef Spitze, Hung. Ferencz-József Csúcs. ▲ N Slovakia 49°12′N 20°09′E
108 E8 Gerlafingen Solothurn, NW Switzerland 47°10′N 07°35′E
Gerlsdorfer Spitze see Gerlachovský štít
Germak see Germik
German East Africa see Tanzania
Germanicopolis see Çankırı
Germanicum, Mare/ German Ocean see North Sea
Germanovichi see Hyermanavichy
German Southwest Africa see Namibia
20 E10 Germantown Tennessee, S USA 35°06′N 89°51′W
101 I15 Germany off. Federal Republic of Germany, Bundesrepublik Deutschland, Ger. Deutschland. ◆ federal republic N Europe
Germany, Federal Republic of see Germany
101 L23 Germering Bayern, SE Germany 48°07′N 11°22′E
139 V3 Germik Ar. Garmik, var. Germak. As Sulaymānīyah, E Iraq 35°49′N 46°09′E
83 J21 Germiston var. Gauteng. Gauteng, NE South Africa 26°15′S 28°10′E
Gernika see Gernika-Lumo
105 P2 Gernika-Lumo var. Gernika, Guernica, Guernica y Lumo. País Vasco, N Spain 43°19′N 02°40′W
164 L12 Gero Gifu, Honshū, SW Japan 35°48′N 137°15′E
115 F22 Geroliménas Pelopónnisos, S Greece 36°28′N 22°25′E
Gerona see Girona
99 H21 Gerpinnes Hainaut, S Belgium 50°20′N 04°38′E
102 L15 Gers ◈ department S France
102 L14 Gers ≈ S France
Gerunda see Girona
158 I13 Gêrzê var. Luring. Xizang Zizhiqu, W China 32°19′N 84°05′E
136 K10 Gerze Sinop, N Turkey 41°48′N 35°13′E
Gesoriacum see Boulogne-sur-Mer
Gessoriacum see Boulogne-sur-Mer
99 J21 Gesves Namur, SE Belgium 50°24′N 05°04′E
93 J20 Geta Åland, SW Finland 60°22′N 19°49′E
105 N8 Getafe Madrid, C Spain 40°18′N 03°44′W
95 J21 Getinge Halland, S Sweden 56°46′N 12°42′E
18 F16 Gettysburg Pennsylvania, NE USA 39°49′N 77°13′W
29 N8 Gettysburg South Dakota, N USA 45°00′N 99°57′W
194 K12 Getz Ice Shelf ice shelf Antarctica
137 S15 Gevaş Van, SE Turkey 38°16′N 43°05′E
Gevgeli see Gevgelija
113 Q20 Gevgelija var. Đevđelija, Djevdjelija, Turk. Gevgeli. SE Macedonia 41°09′N 22°30′E
103 T10 Gex Ain, E France 46°21′N 06°02′E
92 I3 Geysir physical region SW Iceland
136 F11 Geyve Sakarya, NW Turkey 40°32′N 30°18′E
80 G10 Gezira ◈ state E Sudan
109 V3 Gföhl Niederösterreich, N Austria 48°30′N 15°27′E
83 H22 Ghaap Plateau Afr. Ghaapplato. plateau C South Africa
Ghaapplato see Ghaap Plateau
Ghaba see Al Ghābah
138 I8 Ghāb, Tall ▲ SE Syria 33°09′N 37°48′E
139 Q9 Ghadaf, Wādī al dry watercourse C Iraq
Ghadamès see Ghadāmis
74 M9 Ghadāmis var. Ghadamès, Rhadames. W Libya 30°08′N 09°30′E
141 Y10 Ghadan E Oman 20°20′N 57°58′E
75 O10 Ghaddūwah C Libya 26°36′N 14°26′E
147 Q11 Ghafurov Rus. Gafurov; prev. Sovetabad. NW Tajikistan 40°13′N 69°42′E
153 N12 Ghāghara ≈ S Asia
149 P13 Ghaibi Dero Sind, SE Pakistan 27°35′N 67°42′E
141 Y10 Ghalat E Oman 21°06′N 58°51′E
139 W11 Ghamūkah, Hawr ◎ S Iraq
77 P15 Ghana off. Republic of Ghana. ◆ republic W Africa
141 X12 Ghānah spring/well S Oman 18°35′N 56°34′E
Ghanongga see Ranongga
Ghansi/Ghansiland see Ghanzi
83 F18 Ghanzi var. Khanzi. Ghanzi, W Botswana 21°39′S 21°38′E
83 G19 Ghanzi var. Ghansi, Ghansiland, Khanzi. ◈ district C Botswana
67 T14 Ghanzi var. Knanzi. ≈ Botswana/South Africa
Ghap'an see Kapan
138 F13 Gharandal Al 'Aqabah, SW Jordan 30°12′N 35°18′E
139 U14 Gharbīyah, Sha'īb al ≈ S Iraq
Gharbt, Jabal al see Liban, Jebel
74 K7 Ghardaïa N Algeria 32°30′N 03°44′E
147 R12 Gharm Rus. Garm. C Tajikistan 39°03′N 70°25′E
149 P17 Gharo Sind, SE Pakistan 24°44′N 67°35′E
139 W10 Gharrāf, Shaṭṭ al ≈ S Iraq
Gharvān see Gharyān
75 O7 Gharyān var. Gharvān. NW Libya 32°10′N 13°01′E
74 M11 Ghāt var. Gat. SW Libya 24°58′N 10°11′E
Ghawdex see Gozo
141 U8 Ghayathi Abū Ẓaby, W United Arab Emirates 23°51′N 53°01′E
Ghazāl, Baḥr al see Ghazal, Bahr el
78 H9 Ghazal, Bahr el var. Soro. seasonal river C Chad
80 E13 Ghazal, Bahr el var. Baḥr al Ghazāl. ≈ N South Sudan
74 H6 Ghazaouet NW Algeria 35°08′N 01°50′W
152 J10 Ghāziābād Uttar Pradesh, N India 28°42′N 77°28′E
153 O13 Ghāzīpur Uttar Pradesh, N India 25°36′N 83°36′E
149 Q6 Ghaznī var. Ghazni. Ghaznī, E Afghanistan 33°31′N 68°24′E
149 P7 Ghaznī ◈ province SE Afghanistan
Ghazzah see Gaza
Gheel see Geel
Ghelizâne see Relizane
Ghent see Gent
Gheorghe Brațul see Sfântu Gheorghe, Brațul
Gheorghe Gheorghiu-Dej see Onești
116 J10 Gheorgheni prev. Gheorghieni, Sin-Miclăuș, Ger. Niklasmarkt, Hung. Gyergyószentmiklós. Harghita, C Romania 46°43′N 25°36′E
Gheorghieni see Gheorgheni
116 H10 Gherla Ger. Neuschliss, Hung. Szamosújvár; prev. Armenierstadt. Cluj, NW Romania 47°02′N 23°55′E
Gheweifat see Ghuwayfāt
Ghilan see Gīlān
107 C18 Ghilarza Sardegna, Italy, C Mediterranean Sea 40°09′N 08°50′E
Ghilizane see Relizane
Ghimbi see Gīmbī
Ghiriș see Câmpia Turzii
103 Y15 Ghisonaccia Corse, France, C Mediterranean Sea 42°00′N 09°25′E
147 Q11 Ghonchī Rus. Ganchi. NW Tajikistan 39°57′N 69°10′E
153 T13 Ghoraghat Rajshahi, NW Bangladesh 25°18′N 89°20′E
148 J5 Ghōrīān prev. Ghūrīān. Herāt, W Afghanistan 34°20′N 61°26′E
149 R13 Ghotki Sind, SE Pakistan 28°00′N 69°21′E
Ghowr see Gōwr
147 T13 Ghŭdara var. Gudara, Rus. Kudara. SE Tajikistan 38°28′N 72°39′E
153 R13 Ghugri ≈ N India
147 S14 Ghund Rus. Gunt. ≈ SE Tajikistan
Ghurābīyah, Sha'īb al see Gharbīyah, Sha'īb al
Ghurdaqah see Al Ghurdaqah
Ghūrīān see Ghōrīān
141 T8 Ghuwayfāt var. Gheweifat. Abū Ẓaby, W United Arab Emirates 24°06′N 51°40′E
121 O14 Ghuzayyil, Sabkhat salt lake N Libya
126 J3 Ghzatsk Smolenskaya Oblast', W Russian Federation 55°33′N 35°00′E
115 G17 Giáltra Évvoia, C Greece 38°21′N 22°58′E
Giamame see Jamaame
167 U13 Gia Nghia var. Đak Nông. Đăc Lăc, S Vietnam 11°58′N 107°42′E
114 F13 Giannitsá var. Yiannitsá. Kentrikí Makedonía, N Greece 40°49′N 22°24′E
107 F14 Giannutri, Isola di island Archipelago Toscano, C Italy
96 F13 Giant's Causeway Ir. Clochán an Aifir. lava flow N Northern Ireland, United Kingdom
167 S15 Gia Rai Minh Hai, S Vietnam 09°14′N 105°28′E
107 L24 Giarre Sicilia, Italy, C Mediterranean Sea 37°44′N 15°12′E
44 I7 Gibara Holguín, E Cuba 21°09′N 76°11′W
29 O16 Gibbon Nebraska, C USA 40°45′N 98°50′W
32 K11 Gibbon Oregon, NW USA 45°40′N 118°22′W
33 P11 Gibbonsville Idaho, NW USA 45°33′N 113°55′W
64 A13 Gibbs Hill hill S Bermuda
92 I9 Gibostad Troms, N Norway 69°21′N 18°01′E
104 I14 Gibraleón Andalucía, S Spain 37°23′N 06°58′W
104 L16 Gibraltar ○ (Gibraltar) S Gibraltar 36°08′N 05°21′W
104 L16 Gibraltar ◇ UK dependent territory SW Europe
Gibraltar, Détroit de/ Gibraltar, Estrecho de see Gibraltar, Strait of
104 J17 Gibraltar, Strait of Fr. Détroit de Gibraltar, Sp. Estrecho de Gibraltar. strait Atlantic Ocean/ Mediterranean Sea
31 S11 Gibsonburg Ohio, N USA 41°22′N 83°19′W
30 M13 Gibson City Illinois, N USA 40°27′N 88°24′W
180 L8 Gibson Desert desert Western Australia
10 L17 Gibsons British Columbia, SW Canada 49°24′N 123°32′W
149 N12 Gīdār Baluchistān, SW Pakistan 28°16′N 66°00′E
155 I17 Giddalūr Andhra Pradesh, E India 15°24′N 78°54′E
25 U10 Giddings Texas, SW USA 30°12′N 96°59′W
27 Y8 Gideon Missouri, C USA 36°27′N 89°55′W
81 I15 Gīdolē Southern Nationalities, S Ethiopia 05°31′N 37°26′E
Giebnegáisi see Kebnekaise
118 H13 Giedraičiai Utena, E Lithuania 55°05′N 25°16′E
103 O7 Gien Loiret, C France 47°40′N 02°37′E
101 G17 Giessen Hessen, W Germany 50°35′N 08°41′E
98 O6 Gieten Drenthe, NE Netherlands 53°00′N 06°43′E
23 Y13 Gifford Florida, SE USA 27°40′N 80°24′W
9 O5 Gifford ≈ Baffin Island, Nunavut, NE Canada
100 J12 Gifhorn Niedersachsen, N Germany 52°28′N 10°33′E
11 P13 Gift Lake Alberta, W Canada 55°49′N 115°57′W
Giftokastro see Tsigansko Gradishte
164 L13 Gifu var. Gihu. Gifu, Honshū, SW Japan 35°24′N 136°46′E
164 K13 Gifu off. Gifu-ken, var. Gihu. ◈ prefecture Honshū, SW Japan
Gifu-ken see Gifu
126 M13 Gigant Rostovskaya Oblast', SW Russian Federation 46°29′N 41°18′E
40 E8 Giganta, Sierra de la ▲▲ NW Mexico
54 E12 Gigante Huila, S Colombia 02°24′N 75°34′W
114 I7 Gigen Pleven, N Bulgaria 43°40′N 24°31′E
Giggiga see Jijiga
96 G12 Gigha Island island SW Scotland, United Kingdom
107 E14 Giglio, Isola del island Archipelago Toscano, C Italy
Gihu see Gifu
146 L11 G'ijduvon Rus. Gizhduvon. Buxoro Viloyati, C Uzbekistan 40°06′N 64°38′E
104 L2 Gijón var. Xixón. Asturias, NW Spain 43°32′N 05°40′W
81 D20 Gikongoro SW Rwanda 02°30′S 29°32′E
36 K14 Gila Bend Arizona, SW USA 32°57′N 112°43′W
36 J14 Gila Bend Mountains ▲▲ Arizona, SW USA
36 I15 Gila Mountains ▲▲ Arizona, SW USA
37 N14 Gila Mountains ▲▲ Arizona, SW USA
142 M4 Gīlān off. Ostān-e Gīlān, var. Ghilan, Guilan. ◈ province NW Iran
Gīlān, Ostān-e see Gīlān
36 L14 Gila River ≈ Arizona, SW USA
29 W4 Gilbert Minnesota, N USA 47°29′N 92°27′W
Gilbert Islands see Tungaru
10 L16 Gilbert, Mount ▲ British Columbia, SW Canada 50°49′N 124°03′W
181 U4 Gilbert River ≈ Queensland, NE Australia
0 C6 Gilbert Seamounts undersea feature NE Pacific Ocean 52°50′N 150°10′W
33 S7 Gildford Montana, NW USA 48°34′N 110°21′W
83 P15 Gilé Zambézia, NE Mozambique 16°10′S 28°17′E
30 K4 Gile Flowage ◎ Wisconsin, N USA
182 G7 Giles, Lake salt lake South Australia
Gilf Kebir Plateau see Haḍabat al Jilf al Kabīr
183 R6 Gilgandra New South Wales, SE Australia 31°43′S 148°39′E
Gilgău see Gâlgău
81 I19 Gilgil Nakuru, SW Kenya 0°29′S 36°19′E
183 S4 Gil Gil Creek ≈ New South Wales, SE Australia
149 V3 Gilgit Jammu and Kashmir, NE Pakistan 35°54′N 74°20′E
149 V3 Gilgit ≈ N Pakistan
11 X11 Gillam Manitoba, C Canada 56°25′N 94°45′W
95 J22 Gilleleje Hovedstaden, E Denmark 56°05′N 12°17′E
30 K14 Gillespie Illinois, N USA 39°07′N 89°49′W
27 W13 Gillett Arkansas, C USA 34°07′N 91°22′W
33 X12 Gillette Wyoming, C USA 44°17′N 105°30′W
97 P22 Gillingham SE England, United Kingdom 51°24′N 00°33′E
195 X6 Gillock Island island Antarctica
173 O16 Gillot ✈ (St-Denis) N Réunion 20°53′S 55°31′E
65 H25 Gill Point headland E Saint Helena 15°59′S 05°38′W
30 M12 Gilman Illinois, N USA 40°44′N 87°58′W
25 W6 Gilmer Texas, SW USA 32°44′N 94°58′W
Gilolo see Halmahera, Pulau
81 G14 Gilo Wenz ≈ SW Ethiopia
35 O10 Gilroy California, W USA 37°00′N 121°34′W
123 Q12 Gilyuy ≈ SE Russian Federation
99 I14 Gilze Noord-Brabant, S Netherlands 51°33′N 04°56′E
165 R16 Gima Okinawa, Kume-jima, SW Japan
80 H13 Gīmbī It. Ghimbi. Oromīya, C Ethiopia 09°13′N 35°39′E
163 Y15 Gimcheon prev. Kimch'ŏn. C South Korea 36°08′N 128°06′E
163 Z16 Gimhae prev. Kim Hae. ✈ (Busan) SE South Korea 35°10′N 128°57′E
45 T12 Gimie, Mount ▲ C Saint Lucia 13°51′N 61°00′W
11 X16 Gimli Manitoba, S Canada 50°39′N 97°00′W
Gimma see Jīma
95 O14 Gimo Uppsala, C Sweden 60°11′N 18°12′E
102 L15 Gimone ≈ S France
Gimpoe see Gimpu
171 N12 Gimpu prev. Gimpoe. Sulawesi, C Indonesia 01°38′S 120°00′E
182 F5 Gina South Australia 29°56′S 134°33′E
Ginevra see Genève
99 J19 Gingelom Limburg, NE Belgium 50°46′N 05°09′E
180 I12 Gingin Western Australia 31°22′S 115°51′E
171 Q7 Gingoog Mindanao, S Philippines 08°47′N 125°05′E
81 K14 Gīnīr Oromīya, C Ethiopia 07°12′N 40°43′E
Giohar see Jawhar
107 O17 Gióia del Colle Puglia, SE Italy 40°47′N 16°56′E
107 M22 Gioia, Golfo di gulf S Italy
Giona see Gkióna
115 I16 Gioúra island Vóreies Sporádes, Greece, Aegean Sea
107 O17 Giovinazzo Puglia, SE Italy 41°11′N 16°40′E
Gipeswic see Ipswich
105 P3 Gipuzkoa Cast. guipuzcoa. ◈ province País Vasco, N Spain
Giran see Ilan
30 K14 Girard Illinois, N USA 39°27′N 89°46′W
27 R7 Girard Kansas, C USA 37°30′N 94°50′W
25 O6 Girard Texas, SW USA 33°18′N 100°38′W
54 E10 Girardot Cundinamarca, C Colombia 04°19′N 74°47′W
172 M7 Giraud Seamount undersea feature SW Indian Ocean 09°57′S 46°55′E
83 A15 Giraul ≈ SW Angola
96 L9 Girdle Ness headland NE Scotland, United Kingdom 57°09′N 02°04′W
137 N11 Giresun var. Kerasunt; anc. Cerasus, Pharnacia. Giresun, NE Turkey 40°55′N 38°35′E
137 N12 Giresun var. Kerasunt. ◈ province NE Turkey
137 N12 Giresun Dağları ▲▲ N Turkey
Girga see Jirjā
Girgeh see Jirjā
Girgenti see Agrigento
153 Q15 Girīdīh Jhārkhand, NE India 24°10′N 86°20′E
183 P6 Girilambone New South Wales, SE Australia 31°19′S 146°57′E
Girin see Jilin
121 W10 Girne Gk. Kerýneia, Kyrenia. N Cyprus 35°20′N 33°20′E
Giron see Kiruna
105 X5 Girona var. Gerona; anc. Gerunda. Cataluña, NE Spain 41°59′N 02°49′E
105 W5 Girona var. Gerona. ◈ province Cataluña, NE Spain
102 J12 Gironde ◈ department SW France
102 J11 Gironde estuary SW France
105 V5 Gironella Cataluña, NE Spain 42°02′N 01°53′E
103 N15 Girou ≈ S France
97 H14 Girvan W Scotland, United Kingdom 55°14′N 04°53′W
24 M9 Girvin Texas, SW USA 31°05′N 102°24′W
184 Q9 Gisborne Gisborne, North Island, New Zealand 38°41′S 178°01′E
184 P9 Gisborne off. Gisborne District. ◈ unitary authority North Island, New Zealand
Gisborne District see Gisborne
Giseifu see Uijeongbu
Gisenye see Gisenyi
81 D19 Gisenyi var. Gisenye. NW Rwanda 01°42′S 29°18′E
95 K20 Gislaved Jönköping, S Sweden 57°19′N 13°30′E
103 N4 Gisors Eure, N France 49°18′N 01°46′E
Gissar see Hisor
147 P12 Gissar Range Rus. Gissarskiy Khrebet. ▲▲ Tajikistan/Uzbekistan
Gissarskiy Khrebet see Gissar Range
99 B16 Gistel West-Vlaanderen, W Belgium 51°09′N 02°58′E
108 F9 Giswil Obwalden, C Switzerland 46°49′N 08°11′E
115 B16 Gitánes ancient monument Ípeiros, W Greece
81 E20 Gitarama C Rwanda 02°05′S 29°45′E
81 E20 Gitega C Burundi 03°20′S 29°56′E
Githio see Gýtheio
108 H11 Giubiasco Ticino, S Switzerland 46°11′N 09°01′E
106 K13 Giulianova Abruzzi, C Italy 42°45′N 13°58′E
Giulie, Alpi see Julian Alps
Giumri see Gyumri
116 M13 Giurgeni Ialomița, SE Romania 44°45′N 27°48′E
116 J15 Giurgiu Giurgiu, S Romania 43°54′N 25°58′E
116 J14 Giurgiu ◈ county SE Romania
95 F22 Give Syddanmark, C Denmark 55°51′N 09°15′E
103 R2 Givet Ardennes, N France 50°08′N 04°50′E
103 R11 Givors Rhône, E France 45°35′N 04°47′E
83 K19 Giyani Limpopo, NE South Africa 23°20′S 30°37′E
80 I13 Giyon Oromīya, C Ethiopia 08°31′N 37°56′E
75 W8 Giza var. Al Jīzah, El Gîza, Gizeh. N Egypt 30°01′N 31°13′E
75 V8 Giza, Pyramids of ancient monument N Egypt
Gizhduvon see G'ijduvon
123 U8 Gizhiga Magadanskaya Oblast', E Russian Federation 61°58′N 160°16′E
123 T9 Gizhiginskaya Guba bay E Russian Federation
186 K8 Gizo Gizo, NW Solomon Islands 08°03′S 156°49′E
110 N7 Giżycko Ger. Lötzen. Warmińsko-Mazurskie, NE Poland 54°03′N 21°48′E
Gízymałów see Hrymayliv
113 M17 Gjakovë Serb. Đakovica. W Kosovo 42°22′N 20°30′E
94 F12 Gjende ◎ S Norway
113 L16 Gjeravicë Serb. Deravica. ▲ S Serbia 42°33′N 20°08′E
95 F17 Gjerstad Aust-Agder, S Norway 58°54′N 09°03′E
113 O17 Gjilan Serb. Gnjilane. E Kosovo 42°27′N 21°28′E
Gjinokastër see Gjirokastër
113 L23 Gjirokastër var. Gjirokastra; prev. Gjinokastër, Gk. Argyrokastron, It. Argirocastro. Gjirokastër, S Albania 40°04′N 20°09′E
113 L22 Gjirokastër ◈ district S Albania
Gjirokastra see Gjirokastër
9 N7 Gjoa Haven var. Uqsuqtuuq. King William Island, Nunavut, NW Canada 68°38′N 95°57′W
94 H13 Gjøvik Oppland, S Norway 60°47′N 10°40′E
113 J22 Gjuhëzës, Kepi i headland SW Albania 40°25′N 19°19′E
Gjurgjevac see Đurđevac
115 E18 Gkióna var. Giona. ▲ C Greece
121 R3 Gkréko, Akrotíri var. Cape Greco, Pidálion. cape E Cyprus
99 I18 Glabbeek-Zuurbemde Vlaams Brabant, C Belgium 50°54′N 04°58′E
13 R14 Glace Bay Cape Breton Island, Nova Scotia, SE Canada 46°12′N 59°57′W
11 O16 Glacier British Columbia, SW Canada 51°12′N 117°33′W
39 W12 Glacier Bay inlet Alaska, USA
32 I7 Glacier Peak ▲ Washington, NW USA 48°06′N 121°06′W
159 N13 Gladaindong Feng ▲ C China 33°24′N 91°00′E
21 Q7 Glade Spring Virginia, NE USA 36°47′N 81°46′W
43 W7 Gladewater Texas, SW USA 32°32′N 94°57′W
181 Y8 Gladstone Queensland, E Australia 23°52′S 151°16′E
182 I8 Gladstone South Australia 33°16′S 138°21′E
11 X16 Gladstone Manitoba, S Canada 50°14′N 98°56′W
31 O5 Gladstone Michigan, N USA 45°51′N 87°01′W
27 R4 Gladstone Missouri, C USA 39°12′N 94°33′W
31 Q7 Gladwin Michigan, N USA 43°58′N 84°29′W
95 J15 Glafsfjorden ◎ C Sweden
92 H2 Gláma physical region NW Iceland
94 I12 Gláma var. Glommen. ≈ S Norway

◆ Country
● Country Capital
◇ Dependent Territory
○ Dependent Territory Capital
◈ Administrative Regions
✈ International Airport
▲ Mountain
▲▲ Mountain Range
Volcano
≈ River
◎ Lake
⊠ Reservoir

112 F13 **Glamoč** Federacija Bosne I Hercegovine, NE Boznia and Hercegovina 44°01′N 16°51′E
97 J22 **Glamorgan** *cultural region* S Wales, United Kingdom
95 G24 **Glamsbjerg** Syddtjylland, C Denmark 55°17′N 10°07′E
171 Q8 **Glan** Mindanao, S Philippines 05°49′N 125°11′E
109 T9 **Glan** ≈ SE Austria
101 F19 **Glan** ≈ W Germany
95 M17 **Glan** ◎ S Sweden
Glaris *see* Glarus
108 H9 **Glarner Alpen** *Eng.* Glarus Alps. ▲▲ E Switzerland
108 H8 **Glarus** Glarus, E Switzerland 47°03′N 09°04′E
108 H9 **Glarus** *Fr.* Glaris. ◆ *canton* C Switzerland
Glarus Alps *see* Glarner Alpen
27 N3 **Glasco** Kansas, C USA 39°21′N 97°50′W
96 I12 **Glasgow** S Scotland, United Kingdom 55°53′N 04°15′W
20 K7 **Glasgow** Kentucky, S USA 37°00′N 85°54′W
27 T4 **Glasgow** Missouri, C USA 39°13′N 92°51′W
33 W7 **Glasgow** Montana, NW USA 48°12′N 106°37′W
21 T6 **Glasgow** Virginia, NE USA 37°37′N 79°27′W
96 I12 **Glasgow** ✕ W Scotland, United Kingdom 55°52′N 04°27′W
11 S14 **Glaslyn** Saskatchewan, S Canada 53°20′N 108°18′W
18 I16 **Glassboro** New Jersey, NE USA 39°40′N 75°05′W
24 L10 **Glass Mountains** ▲▲ Texas, SW USA
97 K23 **Glastonbury** SW England, United Kingdom 51°09′N 02°43′W
Glatz *see* Kłodzko
101 N16 **Glauchau** Sachsen, E Germany 50°48′N 12°32′E
Glavn'a Morava *see* Velika Morava
Glavnik *see* Gllamnik
127 T1 **Glazov** Udmurtskaya Respublika, NW Russian Federation 58°06′N 52°38′E
Glda *see* Gwda
109 U8 **Gleinalpe** ▲▲ SE Austria
109 W8 **Gleisdorf** Steiermark, SE Austria 47°07′N 15°43′E
Gleiwitz *see* Gliwice
39 S11 **Glenallen** Alaska, USA 62°06′N 145°33′W
102 F7 **Glénan, Îles** *island group* NW France
185 G21 **Glenavy** Canterbury, South Island, New Zealand 44°53′S 171°04′E
10 H5 **Glenboyle** Yukon, NW Canada 63°55′N 138°43′W
21 X3 **Glen Burnie** Maryland, NE USA 39°09′N 76°37′W
36 L8 **Glen Canyon** *canyon* Utah, W USA
36 L8 **Glen Canyon Dam** *dam* Arizona, SW USA
30 K15 **Glen Carbon** Illinois, N USA 38°45′N 89°58′W
14 E17 **Glencoe** Ontario, S Canada 42°44′N 81°42′W
83 K22 **Glencoe** KwaZulu/Natal, E South Africa 28°10′S 30°15′E
29 U9 **Glencoe** Minnesota, N USA 44°46′N 94°09′W
96 H10 **Glen Coe** *valley* N Scotland, United Kingdom
36 K13 **Glendale** Arizona, SW USA 33°32′N 112°11′W
35 S15 **Glendale** California, W USA 34°09′N 118°20′W
182 G5 **Glendambo** South Australia 30°59′S 135°45′E
33 Y8 **Glendive** Montana, NW USA 47°08′N 104°42′W
33 Y15 **Glendo** Wyoming, C USA 42°27′N 105°01′W
55 S10 **Glendor Mountains** C Guyana
182 K12 **Glenelg River** ≈ South Australia/Victoria, SE Australia
29 P4 **Glenfield** North Dakota, N USA 47°25′N 98°33′W
25 V12 **Glen Flora** Texas, SW USA 29°22′N 96°12′W
181 P7 **Glen Helen** Northern Territory, N Australia 23°45′S 132°46′E
183 U5 **Glen Innes** New South Wales, SE Australia 29°42′S 151°45′E
31 P6 **Glen Lake** ◎ Michigan, N USA
10 I7 **Glenlyon Peak** ▲ Yukon, W Canada 62°32′N 134°51′W
37 N16 **Glenn, Mount** ▲ Arizona, SW USA 31°55′N 110°00′W
33 N15 **Glenns Ferry** Idaho, NW USA 42°57′N 115°18′W
23 W6 **Glennville** Georgia, SE USA 31°56′N 81°55′W
10 J10 **Glenora** British Columbia, W Canada 57°52′N 131°16′W
182 M11 **Glenorchy** Victoria, SE Australia 36°56′S 142°39′E
183 V5 **Glenreagh** New South Wales, SE Australia 30°04′S 153°00′E
33 X15 **Glenrock** Wyoming, C USA 42°51′N 105°52′W
96 K11 **Glenrothes** E Scotland, United Kingdom 56°11′N 03°09′W
18 L9 **Glens Falls** New York, NE USA 43°18′N 73°38′W
97 D14 **Glenties** *Ir.* Na Gleannta. Donegal, NW Ireland 54°47′N 08°17′W
28 L5 **Glen Ullin** North Dakota, N USA 46°49′N 101°49′W
21 R4 **Glenville** West Virginia, NE USA 38°57′N 80°51′W
27 T12 **Glenwood** Arkansas, C USA 34°19′N 93°33′W
29 S15 **Glenwood** Iowa, C USA 41°03′N 95°44′W
29 T7 **Glenwood** Minnesota, N USA 45°39′N 95°23′W
36 L5 **Glenwood** Utah, W USA 38°45′N 111°59′W
30 I5 **Glenwood City** Wisconsin, N USA 45°04′N 92°11′W
37 Q4 **Glenwood Springs** Colorado, C USA 39°33′N 107°21′W
108 F10 **Gletsch** Valais, S Switzerland 46°34′N 08°21′E
Glevum *see* Gloucester
29 U14 **Glidden** Iowa, C USA 42°03′N 94°43′W
112 E9 **Glina** *var.* Banijska Palanka. Sisak-Moslavina, NE Croatia 45°19′N 16°07′E
94 F11 **Glittertind** ▲ S Norway 61°24′N 08°19′E
111 J16 **Gliwice** *Ger.* Gleiwitz. Śląskie, S Poland 50°19′N 18°49′E
113 N16 **Gllamnik** *Serb.* Glavnik. N Kosovo 42°53′N 21°10′E
36 M14 **Globe** Arizona, SW USA 33°24′N 110°47′W
Globino *see* Hlobyne
108 L9 **Glockturm** ▲ SW Austria 46°54′N 10°42′E
116 L9 **Glodeni** *Rus.* Glodyany. N Moldova 47°47′N 27°33′E
109 S9 **Glödnitz** Kärnten, S Austria 46°57′N 14°03′E
Glodyany *see* Glodeni
Glogau *see* Głogów
109 W6 **Gloggnitz** Niederösterreich, E Austria 47°41′N 15°57′E
110 F13 **Głogów** *Ger.* Glogau, Glogow. Dolnośląskie, SW Poland 51°40′N 16°04′E
Glogow *see* Głogów
111 I16 **Głogówek** *Ger.* Oberglogau. Opolskie, S Poland 50°21′N 17°51′E
92 G12 **Glomfjord** Nordland, C Norway 66°49′N 14°00′E
Glomma *see* Gláma
Glommen *see* Gláma
93 I14 **Glommersträsk** Norrbotten, N Sweden 65°17′N 19°40′E
172 I1 **Glorieuses, Îles** *Eng.* Glorioso Islands. *island* (to France) N Madagascar
Glorioso Islands *see* Glorieuses, Îles
65 C25 **Glorious Hill** *hill* East Falkland, Falkland Islands
38 J12 **Glory of Russia Cape** *headland* Saint Matthew Island, Alaska, USA 60°36′N 172°57′W
22 J7 **Gloster** Mississippi, S USA 31°12′N 91°01′W
183 U7 **Gloucester** New South Wales, SE Australia 32°01′S 152°00′E
186 F7 **Gloucester** New Britain, E Papua New Guinea 05°30′S 148°30′E
97 L21 **Gloucester** *hist.* Caer Glou, *Lat.* Glevum. C England, United Kingdom 51°53′N 02°14′W
19 P10 **Gloucester** Massachusetts, NE USA 42°36′N 70°36′W
21 X6 **Gloucester** Virginia, NE USA 37°26′N 76°33′W
97 K21 **Gloucestershire** *cultural region* C England, United Kingdom
31 T14 **Glouster** Ohio, N USA 39°30′N 82°04′W
42 H3 **Glovers Reef** *reef* E Belize
18 K10 **Gloversville** New York, NE USA 43°03′N 74°20′W
110 K12 **Głowno** Łódź, C Poland 51°58′N 19°43′E
111 H16 **Głubczyce** *Ger.* Leobschütz. Opolskie, S Poland 50°13′N 17°50′E
126 L11 **Glubokiy** Rostovskaya Oblast', SW Russian Federation 48°34′N 40°16′E
145 W9 **Glubokoye** Vostochnyy Kazakhstan, E Kazakhstan 50°08′N 82°16′E
Glubokoye *see* Hlybokaye
111 H16 **Głuchołazy** *Ger.* Ziegenhais. Opolskie, S Poland 50°20′N 17°23′E
100 I9 **Glückstadt** Schleswig-Holstein, N Germany 53°47′N 09°26′E
Glukhov *see* Hlukhiv
Glushkevichi *see* Hlushkavichy
Glusk/Glussk *see* Hlusk
Glybokaya *see* Hlyboka
95 F21 **Glyngøre** Midtjylland, NW Denmark 56°45′N 08°55′E
127 Q9 **Gmelinka** Volgogradskaya Oblast', SW Russian Federation 50°50′N 46°51′E
109 R8 **Gmünd** Kärnten, S Austria 46°56′N 13°32′E
109 U2 **Gmünd** Niederösterreich, N Austria 48°47′N 14°59′E
Gmünd *see* Schwäbisch Gmünd
109 S5 **Gmunden** Oberösterreich, N Austria 47°56′N 13°48′E
Gmundner See *see* Traunsee
94 N10 **Gnarp** Gävleborg, C Sweden 62°03′N 17°19′E
109 W8 **Gnas** Steiermark, SE Austria 46°53′N 15°48′E
Gnaviyani *see* Fuammulah
Gnesen *see* Gniezno
95 O16 **Gnesta** Södermanland, C Sweden 59°05′N 17°20′E
110 H11 **Gniezno** *Ger.* Gnesen. Weilkopolskie, C Poland 52°33′N 17°35′E
Gnjilane *see* Gjilan
95 K20 **Gnosjö** Jönköping, S Sweden 57°22′N 13°44′E
155 E17 **Goa** *prev.* Old Goa, Vela Goa, Velha Goa. Goa, W India 15°31′N 73°56′E
155 E17 **Goa** *var.* Old Goa. ◆ *state* W India
Goabddális *see* Kåbdalis
42 H7 **Goascorán, Río** ≈ El Salvador/Honduras
77 O16 **Goaso** *var.* Gawso. W Ghana 06°49′N 02°27′W
81 K14 **Goba** Oromīya, C Ethiopia 07°02′N 39°58′E
83 C20 **Gobabed** Erongo, W Namibia 23°36′S 15°03′E
83 E19 **Gobabis** Omaheke, E Namibia 22°25′S 18°58′E
Gobannium *see* Abergavenny
64 M7 **Goban Spur** *undersea feature* NW Atlantic Ocean
63 H21 **Gobernador Gregores** Santa Cruz, S Argentina 48°43′S 70°21′W
61 F14 **Gobernador Ingeniero Virasoro** Corrientes, NE Argentina 28°06′S 56°00′W
162 L12 **Gobi** *desert* China/Mongolia
164 I14 **Gobō** Wakayama, Honshū, SW Japan 33°52′N 135°09′E
Gobolka Awdal *see* Awdal
Gobolka Sool *see* Sool
101 D14 **Goch** Nordrhein-Westfalen, W Germany 51°41′N 06°10′E
83 E20 **Gochas** Hardap, S Namibia 24°54′S 18°43′E
155 I14 **Godāvari** *var.* Godavari. ≈ C India
Godavari *see* Godāvari
155 L16 **Godāvari, Mouths of the** *delta* E India
15 V5 **Godbout** Québec, SE Canada 49°19′N 67°37′W
15 U5 **Godbout** ≈ Québec, SE Canada
15 U5 **Godbout Est** ≈ Québec, SE Canada
27 N6 **Goddard** Kansas, C USA 37°39′N 97°34′W
14 E15 **Goderich** Ontario, S Canada 43°43′N 81°43′W
Godhavn *see* Qeqertarsuaq
154 E10 **Godhra** Gujarāt, W India 22°49′N 73°40′E
Göding *see* Hodonín
111 K22 **Gödöllő** Pest, N Hungary 47°36′N 19°20′E
62 H11 **Godoy Cruz** Mendoza, W Argentina 32°59′S 68°49′W
11 Y11 **Gods** ≈ Manitoba, C Canada
11 X13 **Gods Lake** ◎ Manitoba, C Canada
11 Y13 **Gods Lake Narrows** Manitoba, C Canada 54°29′N 94°21′W
Godthaab/Godthåb *see* Nuuk
Godwin Austen, Mount *see* K2
Goede Hoop, Kaap de *see* Good Hope, Cape of
Goedgegun *see* Nhlangano
Goeie Hoop, Kaap die *see* Good Hope, Cape of
13 O7 **Goélands, Lac aux** ◎ Québec, SE Canada
98 E13 **Goeree** *island* SW Netherlands
99 F15 **Goes** Zeeland, SW Netherlands 51°30′N 03°55′E
Goettingen *see* Göttingen
19 O10 **Goffstown** New Hampshire, NE USA 43°01′N 71°34′W
14 E8 **Gogama** Ontario, S Canada 47°42′N 81°44′W
30 L3 **Gogebic, Lake** ◎ Michigan, N USA
30 K3 **Gogebic Range** *hill range* Michigan/Wisconsin, N USA
Gogi Lerr *see* Gogi, Mount
137 V13 **Gogi, Mount** *Arm.* Gogi Lerr, *Az.* Küküdağ. ▲ Armenia/Azerbaijan 39°33′N 45°35′E
124 F12 **Gogland, Ostrov** *island* NW Russian Federation
111 I15 **Gogolin** Opolskie, S Poland 50°28′N 18°04′E
Gogonou *see* Gogounou
77 S14 **Gogounou** *var.* Gogonou. N Benin 10°50′N 02°50′E
152 I10 **Gohāna** Haryāna, N India 29°06′N 76°43′E
59 K18 **Goianésia** Goiás, C Brazil 15°21′S 49°02′W
59 K18 **Goiânia** *prev.* Goyania. *state capital* Goiás, C Brazil 16°43′S 49°18′W
59 K18 **Goiás** Goiás, C Brazil 15°57′S 50°07′W
59 J18 **Goiás** *off.* Estado de Goiás; *prev.* Goiaz, Goyaz. ◆ *state* C Brazil
Goiás, Estado de *see* Goiás
Goiaz *see* Goiás
Goidhoo Atoll *see* Horsburgh Atoll
159 R14 **Goinsargoin** Xizang Zizhiqu, W China 31°56′N 98°04′E
60 H10 **Goio-Erê** Paraná, SW Brazil 24°08′S 53°07′W
99 I15 **Goirle** Noord-Brabant, S Netherlands 51°31′N 05°04′E
104 H8 **Góis** Coimbra, N Portugal 40°10′N 08°06′W
165 Q8 **Gojōme** Akita, Honshū, NW Japan 39°55′N 140°07′E
149 U9 **Gojra** Punjab, E Pakistan 31°10′N 72°43′E
136 A11 **Gökçeada** *var.* Imroz Adasī, *Gk.* Imbros. *island* NW Turkey
Gökçeada *see* İmroz
146 F13 **Gökdepe** *Rus.* Gëkdepe, Geok-Tepe. Ahal Welaýaty, C Turkmenistan 38°05′N 58°08′E
136 I10 **Gökırmak** ≈ N Turkey
Goklenkuy, Solonchak *see* Geklengkui, Solonchak
136 C16 **Gökova Körfezi** *gulf* SW Turkey
136 K15 **Göksu** ≈ S Turkey
136 L15 **Göksun** Kahramanmaraş, C Turkey 38°03′N 36°30′E
136 I17 **Göksu Nehri** ≈ S Turkey
83 J16 **Gokwe** Midlands, NW Zimbabwe 18°13′S 28°55′E
94 F13 **Gol** Buskerud, S Norway 60°42′N 08°57′E
153 X12 **Golāghāt** Assam, NE India 26°31′N 93°54′E
110 H10 **Gołańcz** Wielkopolskie, C Poland 52°57′N 17°17′E
138 G8 **Golan Heights** *Ar.* Al Jawlān, *Heb.* HaGolan. ▲▲ SW Syria
Golārā *see* Ārān-va-Bidgol
Golaya Pristan *see* Hola Prystan'
143 T11 **Golbāf** Kermān, C Iran 29°51′N 57°44′E
136 M15 **Gölbaşı** Adıyaman, S Turkey 37°46′N 37°40′E
109 P9 **Gölbner** ▲ SW Austria 46°51′N 12°31′E
30 M17 **Golconda** Illinois, N USA 37°20′N 88°29′W
35 T3 **Golconda** Nevada, W USA 40°56′N 117°29′W
136 E11 **Gölcük** Kocaeli, NW Turkey 40°44′N 29°50′E
108 I7 **Goldach** Sankt Gallen, NE Switzerland 47°28′N 09°28′E
110 N7 **Gołdap** *Ger.* Goldap. Warmińsko-Mazurskie, NE Poland 54°19′N 22°23′E
32 E15 **Gold Beach** Oregon, NW USA 42°25′N 124°27′W
Goldberg *see* Złotoryja
183 V3 **Gold Coast** *cultural region* Queensland, E Australia
68 D12 **Gold Coast** *coastal region* S Ghana
39 R10 **Gold Creek** Alaska, USA 62°48′N 149°40′W
11 O16 **Golden** British Columbia, SW Canada 51°19′N 116°58′W
37 T4 **Golden** Colorado, C USA 39°40′N 105°12′W
184 I13 **Golden Bay** *bay* South Island, New Zealand
27 R7 **Golden City** Missouri, C USA 37°23′N 94°05′W
32 I11 **Goldendale** Washington, NW USA 45°49′N 120°49′W
Goldener Tisch *see* Zlatý Stôl
44 L13 **Golden Grove** E Jamaica 17°56′N 76°17′W
14 J12 **Golden Lake** ◎ Ontario, SE Canada
22 K10 **Golden Meadow** Louisiana, S USA 29°22′N 90°15′W
Golden Sands *see* Zlatni Pyasatsi
Golden State, The *see* California
83 K16 **Golden Valley** Mashonaland West, N Zimbabwe 18°11′S 29°50′E
35 U9 **Goldfield** Nevada, W USA 37°42′N 117°15′W
Goldingen *see* Kuldīga
Goldmarkt *see* Zlatna
10 K17 **Gold River** Vancouver Island, British Columbia, SW Canada 49°41′N 126°05′W
21 V10 **Goldsboro** North Carolina, SE USA 35°23′N 78°00′W
24 M8 **Goldsmith** Texas, SW USA 31°58′N 102°36′W
25 R8 **Goldthwaite** Texas, SW USA 31°28′N 98°35′W
137 R11 **Göle** Ardahan, NE Turkey 40°47′N 42°36′E
Golema Ada *see* Ostrovo
114 H9 **Golema Planina** ▲▲ W Bulgaria
114 F9 **Golemi Vrah** *var.* Golemi Vrŭkh. ▲ W Bulgaria 42°41′N 22°38′E
Golemi Vrŭkh *see* Golemi Vrah
110 D8 **Goleniów** *Ger.* Gollnow. Zachodnio-pomorskie, NW Poland 53°34′N 14°48′E
143 R3 **Golestān** *off.* Ostān-e Golestān. ◆ *province* N Iran
Golestān, Ostān-e *see* Golestān
35 Q14 **Goleta** California, W USA 34°26′N 119°50′W
43 O16 **Golfito** Puntarenas, SE Costa Rica 08°42′N 83°10′W
25 T13 **Goliad** Texas, SW USA 28°40′N 97°26′W
113 L14 **Golija** ▲▲ SW Serbia
Golinka *see* Gongbo'gyamda
113 O16 **Goljak** ▲▲ SE Serbia
136 M12 **Gölköy** Ordu, N Turkey 40°42′N 37°37′E
Gollel *see* Lavumisa
109 X3 **Göllersbach** ≈ NE Austria
Gollnow *see* Goleniów
Golmo *see* Golmud
159 P10 **Golmud** *var.* Ge'e'mu, Golmo, *Chin.* Ko-erh-mu. Qinghai, C China 36°23′N 94°56′E
103 Y14 **Golo** ≈ Corse, France, C Mediterranean Sea
Golovanevsk *see* Holovanivs'k
Golovchin *see* Halowchyn
39 N9 **Golovin** Alaska, USA 64°33′N 162°54′W
142 M7 **Golpāyegān** *var.* Gulpaigan. Eşfahān, W Iran 33°23′N 50°18′E
Golshan *see* Ţabas
Gol'shany *see* Hal'shany
96 J7 **Golspie** N Scotland, United Kingdom 57°59′N 03°56′W
112 O11 **Golubac** Serbia, NE Serbia 44°38′N 21°36′E
110 J9 **Golub-Dobrzyń** Kujawski-pomorskie, C Poland 53°07′N 19°03′E
145 S7 **Golubovka** Pavlodar, N Kazakhstan 53°07′N 74°11′E
82 B11 **Golungo Alto** Kwanza Norte, NW Angola 09°10′S 14°45′E
114 M8 **Golyama Kamchia** *var.* Golyama Kamchiya. ≈ E Bulgaria
Golyama Kamchiya *see* Golyama Kamchia
114 L8 **Golyama Reka** ≈ N Bulgaria
Golyama Syutka *see* Golyama Syutkya
114 H11 **Golyama Syutkya** *var.* Golyama Syutka. ▲ SW Bulgaria 41°55′N 24°03′E
114 I12 **Golyam Perelik** ▲ S Bulgaria 41°37′N 24°34′E
114 I11 **Golyam Persenk** ▲ S Bulgaria 41°50′N 24°33′E
79 P19 **Goma** Nord-Kivu, NE Dem. Rep. Congo 01°36′S 29°08′E
153 N13 **Gomati** *var.* Gumti. ≈ N India
77 X14 **Gombe** Gombe, E Nigeria 10°19′N 11°02′E
67 U10 **Gombe** *var.* Igombe. ≈ E Tanzania
77 Y14 **Gombi** Adamawa, E Nigeria 10°07′N 12°45′E
Gombroon *see* Bandar-e 'Abbās
Gomel' *see* Homyel'
Gomel'skaya Oblast' *see* Homyel'skaya Voblasts'
64 N11 **Gomera** *island* Islas Canarias, Spain, NE Atlantic Ocean
40 I5 **Gómez Farías** Chihuahua, N Mexico 29°25′N 107°46′W
41 L8 **Gómez Palacio** Durango, C Mexico 25°39′N 103°30′W
158 J13 **Gomo** Xizang Zizhiqu, W China 33°37′N 86°40′E
143 T6 **Gonābād** *var.* Gunabad. Khorāsān-e Razavī, NE Iran 36°30′N 59°18′E
44 L8 **Gonaïves** *var.* Les Gonaïves. N Haiti 19°26′N 72°41′W
123 Q12 **Gonam** ≈ NE Russian Federation
44 L9 **Gonâve, Canal de la** *var.* Canal de Sud. *channel* N Caribbean Sea
44 K9 **Gonâve, Golfe de la** *gulf* N Caribbean Sea
Gonāveh *see* Bandar-e Gonāveh
44 K9 **Gonâve, Île de la** *island* C Haiti
Gonbadān *see* Do Gonbadān
143 Q3 **Gonbad-e Kāvūs** *var.* Gunbad-i-Qawus. Golestān, N Iran 37°15′N 55°11′E
152 M12 **Gonda** Uttar Pradesh, N India 27°08′N 81°58′E
Gondar *see* Gonder
80 I11 **Gonder** *var.* Gondar. Āmara, NW Ethiopia 12°36′N 37°27′E
78 J13 **Gondey** Moyen-Chari, S Chad 09°07′N 19°10′E
154 J12 **Gondia** Mahārāshtra, C India 21°27′N 80°12′E
104 G6 **Gondomar** Porto, NW Portugal 41°10′N 08°35′W
136 C12 **Gönen** Balıkesir, W Turkey 40°06′N 27°39′E
136 C12 **Gönen Çayı** ≈ NW Turkey
159 O15 **Gongbo'gyamda** *var.* Golinka. Xizang Zizhiqu, W China 30°03′N 93°10′E
Gongchang *see* Longxi
159 N16 **Gonggar** *var.* Gyixong. Xizang Zizhiqu, W China 29°18′N 90°56′E
160 G9 **Gongga Shan** ▲ C China 29°50′N 101°55′E
159 T10 **Gonghe** *var.* Qabqa. Qinghai, C China 36°20′N 100°46′E
Gongjiang *see* Yudu
158 I5 **Gongliu** *var.* Tokkuztara. Xinjiang Uygur Zizhiqu, NW China 43°29′N 82°16′E
77 W14 **Gongola** ≈ E Nigeria
Gongoleh State *see* Jonglei
183 P5 **Gongolgon** New South Wales, SE Australia 30°19′S 146°57′E
159 Q6 **Gongpoquan** Gansu, N China 41°45′N 100°27′E
Gongquan *see* Gongxian
Gongtang *see* Damxung
160 I10 **Gongxian** *var.* Gongquan, Gong Xian. Sichuan, C China 28°25′N 104°51′E
Gong Xian *see* Gongxian
157 V10 **Gongzhuling** *prev.* Huaide. Jilin, NE China 43°30′N 124°48′E
159 S14 **Gonjo** Xizang Zizhiqu, W China 30°51′N 98°16′E
107 B20 **Gonnesa** Sardegna, Italy, C Mediterranean Sea 39°15′N 08°27′E
Gonni/Gónnos *see* Gónnoi
115 F15 **Gónnoi** *var.* Gonni, Gónnos; *prev.* Dereli. Thessalía, C Greece 39°52′N 22°27′E
Gōnoura *see* Iki
35 O11 **Gonzales** California, W USA 36°30′N 121°26′W
22 J9 **Gonzales** Louisiana, S USA 30°14′N 90°55′W
25 T12 **Gonzales** Texas, SW USA 29°31′N 97°29′W
41 P11 **González** Tamaulipas, C Mexico 22°50′N 98°25′W
21 V6 **Goochland** Virginia, NE USA 37°42′N 77°54′W
195 X14 **Goodenough, Cape** *headland* Antarctica 66°15′S 126°35′E
186 F9 **Goodenough Island** *var.* Morata. *island* SE Papua New Guinea
39 N8 **Goodhope Bay** *bay* Alaska, USA
83 D26 **Good Hope, Cape of** *Afr.* Kaap de Goede Hoop, Kaap die Goeie Hoop. *headland* SW South Africa 34°19′S 18°25′E
10 K10 **Good Hope Lake** British Columbia, W Canada 59°15′N 129°18′W
83 E23 **Goodhouse** Northern Cape, W South Africa 28°54′S 18°13′E
33 O15 **Gooding** Idaho, NW USA 42°56′N 114°42′W
26 H3 **Goodland** Kansas, C USA 39°20′N 101°43′W
173 Y15 **Goodlands** NW Mauritius 20°02′S 57°39′E
20 J8 **Goodlettsville** Tennessee, S USA 36°19′N 86°42′W
39 N13 **Goodnews** Alaska, USA 59°07′N 161°35′W
25 O3 **Goodnight** Texas, SW USA 35°00′N 101°07′W
183 Q4 **Goodooga** New South Wales, SE Australia 29°09′S 147°30′E
29 N4 **Goodrich** North Dakota, N USA 47°24′N 100°07′W
25 W10 **Goodrich** Texas, SW USA 30°36′N 94°57′W
29 X10 **Goodview** Minnesota, N USA 44°04′N 91°42′W
26 H8 **Goodwell** Oklahoma, C USA 36°36′N 101°38′W
97 N17 **Goole** E England, United Kingdom 53°43′N 00°46′W
183 O8 **Goolgowi** New South Wales, SE Australia 34°00′S 145°43′E
182 I10 **Goolwa** South Australia 35°31′S 138°43′E
181 Y11 **Goondiwindi** Queensland, E Australia 28°33′S 150°22′E
98 O11 **Goor** Overijssel, E Netherlands 52°13′N 06°33′E
Goose Bay *see* Happy Valley-Goose Bay
33 V13 **Gooseberry Creek** ≈ Wyoming, C USA
21 S14 **Goose Creek** South Carolina, SE USA 32°58′N 80°01′W
63 M23 **Goose Green** *var.* Prado del Ganso. East Falkland, Falkland Islands 51°52′S 59°W
16 D8 **Goose Lake** *var.* Lago dos Gansos. ◎ California/Oregon, W USA
29 Q4 **Goose River** ≈ North Dakota, N USA
153 T16 **Gopalganj** Dhaka, S Bangladesh 23°00′N 89°48′E
153 O12 **Gopālganj** Bihār, N India 26°28′N 84°26′E
Gopher State *see* Minnesota
101 I22 **Göppingen** Baden-Württemberg, SW Germany 48°42′N 09°39′E
110 G13 **Góra** *Ger.* Guhrau. Dolnośląskie, SW Poland 51°40′N 16°03′E
110 M12 **Góra Kalwaria** Mazowieckie, C Poland 52°00′N 21°14′E
153 O12 **Gorakhpur** Uttar Pradesh, N India 26°45′N 83°23′E
Gora Kyuren *see* Kürendag
Gorany *see* Harany
113 J14 **Goražde** Federacija Bosne I Hercegovine, SE Bosnia and Herzegovina 43°39′N 18°58′E
Gorbovichi *see* Harbavichy
Gorče Petrov *see* Đorče Petrov
0 E9 **Gorda Ridges** *undersea feature* NE Pacific Ocean 41°30′N 128°00′W
Gordiaz *see* Gardēz
78 K12 **Gordil** Vakaga, N Central African Republic 09°37′N 21°42′E
23 U5 **Gordon** Georgia, SE USA 32°52′N 83°19′W
28 K12 **Gordon** Nebraska, C USA 42°48′N 102°12′W
25 R7 **Gordon** Texas, SW USA 32°32′N 98°21′W
28 L13 **Gordon Creek** ≈ Nebraska, C USA
63 I25 **Gordon, Isla** *island* S Chile
183 O17 **Gordon, Lake** ◎ Tasmania, SE Australia
183 O17 **Gordon River** ≈ Tasmania, SE Australia
21 V5 **Gordonsville** Virginia, NE USA 38°08′N 78°11′W
78 H13 **Goré** Logone-Oriental, S Chad 07°55′N 16°38′E
80 H13 **Gorē** Oromīya, C Ethiopia 08°08′N 35°33′E
185 D24 **Gore** Southland, South Island, New Zealand 46°06′S 168°58′E
14 D11 **Gore Bay** Manitoulin Island, Ontario, S Canada 45°54′N 82°28′W
25 Q5 **Goree** Texas, SW USA 33°28′N 99°31′W
137 O11 **Görele** Giresun, NE Turkey 41°00′N 39°00′E
19 N6 **Gore Mountain** ▲ Vermont, NE USA 44°55′N 71°47′W
39 R13 **Gore Point** *headland* Alaska, USA 59°12′N 150°57′W
37 R4 **Gore Range** ▲▲ Colorado, C USA
97 F19 **Gorey** *Ir.* Guaire. Wexford, SE Ireland 52°40′N 06°18′W
143 R12 **Gorgāb** Kermān, S Iran
143 Q4 **Gorgān** *var.* Astarabad, Astrabad, Gurgan, *prev.* Asterābād; *anc.* Hyrcania. Golestān, N Iran 36°53′N 54°28′E
143 Q4 **Gorgān, Rūd-e** ≈ N Iran
76 I10 **Gorgol** ◆ *region* S Mauritania
106 D12 **Gorgona, Isola di** *island* Archipelago Toscano, C Italy
19 P8 **Gorham** Maine, NE USA 43°41′N 70°27′W
137 T10 **Gori** C Georgia 42°00′N 44°07′E
98 I13 **Gorinchem** *var.* Gorkum. Zuid-Holland, C Netherlands 51°50′N 04°59′E
137 V13 **Goris** SE Armenia 39°31′N 46°20′E
124 K16 **Goritsy** Tverskaya Oblast', W Russian Federation 57°09′N 36°44′E
106 J7 **Gorizia** *Ger.* Görz. Friuli-Venezia Giulia, NE Italy 45°57′N 13°37′E
116 G13 **Gorj** ◆ *county* SW Romania
109 W12 **Gorjanci** *var.* Uskočke Planine, Žumberak, Žumberačko Gorje, *Ger.* Uskokengebirge; *prev.* Sichelburger Gerbirge. ▲▲ Croatia/Slovenia Europe *see also* Žumberačko Gorje
Görkau *see* Jirkov
Gorki *see* Horki
Gor'kiy *see* Nizhniy Novgorod
Gorkum *see* Gorinchem
95 I23 **Gørlev** Sjælland, E Denmark 55°33′N 11°14′E
111 M17 **Gorlice** Małopolskie, S Poland 49°40′N 21°09′E
101 Q15 **Görlitz** Sachsen, E Germany 51°09′N 14°58′E
Görlitz *see* Zgorzelec
Gorlovka *see* Horlivka
25 R7 **Gorman** Texas, SW USA 32°12′N 98°40′W
21 T3 **Gormania** West Virginia, NE USA 39°16′N 79°18′W
Gorna Dzhumaya *see* Blagoevgrad
114 K8 **Gorna Oryahovitsa** , Gorna Oryakhovitsa. Veliko Tŭrnovo, N Bulgaria 43°07′N 25°40′E
Gorna Oryakhovitsa *see* Gorna Oryahovitsa
114 J8 **Gorna Studena** Veliko Tŭrnovo, N Bulgaria 43°26′N 25°21′E
Gornja Mužlja *see* Mužlja
109 X9 **Gornja Radgona** *Ger.* Oberradkersburg. NE Slovenia 46°39′N 16°00′E
112 M13 **Gornji Milanovac** Serbia, C Serbia 44°01′N 20°26′E
112 G13 **Gornji Vakuf** *var.* Uskoplje. Federacija Bosne I Hercegovine, SW Bosnia and Herzegovina 43°55′N 17°34′E
122 J13 **Gorno-Altaysk** Respublika Altay, S Russian Federation 51°59′N 85°56′E
Gorno-Altayskaya Respublika *see* Altay, Respublika
123 N12 **Gorno-Chuyskiy** Irkutskaya Oblast', C Russian Federation 57°33′N 111°38′E
125 V14 **Gornozavodsk** Permskiy Kray, NW Russian Federation 58°21′N 58°24′E
125 V14 **Gornozavodsk** Ostrov Sakhalin, Sakhalinskaya Oblast', SE Russian Federation 46°34′N 141°52′E
122 I13 **Gornyak** Altayskiy Kray, S Russian Federation 50°58′N 81°24′E
123 O14 **Gornyy** Chitunskaya Oblast', S Russian Federation 51°42′N 114°16′E
127 R8 **Gornyy** Saratovskaya Oblast', W Russian Federation 51°42′N 48°26′E
Gornyy Altay *see* Altay, Respublika
127 O10 **Gornyy Balykley** Volgogradskaya Oblast', SW Russian Federation 49°37′N 45°03′E
80 I13 **Goroch'an** ▲ W Ethiopia 09°09′N 37°16′E
116 J7 **Gorodenka** *var.* Horodenka. Ivano-Frankivs'ka Oblast', W Ukraine 48°41′N 25°28′E
127 O3 **Gorodets** Nizhegorodskaya Oblast', W Russian Federation 56°36′N 43°27′E
Gorodets *see* Haradzyets
Gorodeya *see* Haradzyeya
127 P6 **Gorodishche** Penzenskaya Oblast', W Russian Federation 53°17′N 45°39′E
Gorodishche *see* Horodyshche
Gorodnya *see* Horodnya
Gorodok *see* Haradok
Gorodok/Gorodok Yagellonski *see* Horodok
126 M13 **Gorodovikovsk** Respublika Kalmykiya, SW Russian Federation 46°07′N 41°56′E
186 D7 **Goroka** Eastern Highlands, C Papua New Guinea 06°02′S 145°22′E
Gorokhov *see* Horokhiv
127 N3 **Gorokhovets** Vladimirskaya Oblast', W Russian Federation 56°12′N 42°40′E
77 Q11 **Gorom-Gorom** NE Burkina Faso 14°27′N 00°14′W
171 U13 **Gorong, Kepulauan** *island group* E Indonesia
83 M17 **Gorongosa** Sofala, C Mozambique 18°40′S 34°03′E
171 P11 **Gorontalo** Sulawesi, C Indonesia 00°33′N 123°05′E
171 O11 **Gorontalo** *off.* Propinsi Gorontalo. ◆ *province* N Indonesia
Propinsi Gorontalo *see* Gorontalo
Gorontalo, Teluk *see* Tomini, Teluk
110 L7 **Górowo Iławeckie** *Ger.* Landsberg. Warmińsko-Mazurskie, NE Poland 54°18′N 20°30′E
98 M7 **Gorredijk** *Fris.* De Gordyk. Fryslân, N Netherlands 53°00′N 06°04′E
84 C14 **Gorringe Ridge** *undersea feature* E Atlantic Ocean 36°40′N 11°35′W
98 M11 **Gorssel** Gelderland, E Netherlands 52°12′N 06°13′E
109 T8 **Görtschitz** ≈ S Austria
Goryn *see* Horyn'
145 S15 **Gory Shu-Ile** *Kaz.* Shū-Ile Taūlary; *prev.* Chu-Iliyskiye Gory. ▲▲ S Kazakhstan
Görz *see* Gorizia
110 E10 **Gorzów Wielkopolski** *Ger.* Landsberg, Landsberg an der Warthe. Lubuskie, W Poland 52°44′N 15°12′E
146 B10 **Goşabo** *var.* Goshoba, *Rus.* Koshoba. Balkan Welaýaty, NW Turkmenistan 40°28′N 54°11′E
108 G9 **Göschenen** Uri, C Switzerland 46°40′N 08°36′E
165 O11 **Gosen** Niigata, Honshū, C Japan 37°45′N 139°11′E
163 Y13 **Goseong** *prev.* Kosŏng. SE North Korea 38°41′N 128°14′E
183 T8 **Gosford** New South Wales, SE Australia 33°25′S 151°18′E
31 P11 **Goshen** Indiana, N USA 41°34′N 85°49′W
18 K13 **Goshen** New York, NE USA 41°24′N 74°17′W
Goshoba *see* Goşabo
Goshoba *see* Goşoba
165 Q7 **Goshogawara** *var.* Gosyogawara. Aomori, Honshū, C Japan 40°47′N 140°24′E
Goshquduq Qum *see* Tosquduq Qumlari
101 J14 **Goslar** Niedersachsen, C Germany 51°55′N 10°25′E
27 Y9 **Gosnell** Arkansas, C USA 35°57′N 89°58′W
146 B10 **Goşoba** *var.* Goshoba, *Rus.* Koshoba. Balkanskiy Velayat, NW Turkmenistan 40°28′N 54°11′E
112 C11 **Gospić** Lika-Senj, C Croatia 44°32′N 15°21′E
97 N23 **Gosport** S England, United Kingdom 50°48′N 01°08′W
94 D9 **Gossa** *island* S Norway
108 H7 **Gossau** Sankt Gallen, NE Switzerland 47°25′N 09°16′E
99 G20 **Gosselies** *var.* Goss'lies. Hainaut, S Belgium 50°28′N 04°26′E
77 P10 **Gossi** Tombouctou, C Mali 15°44′N 01°19′W
Goss'lies *see* Gosselies
113 N18 **Gostivar** W FYR Macedonia 41°48′N 20°55′E
Gostomel' *see* Hostomel'
110 G12 **Gostyń** *var.* Gostyn. Wielkopolskie, C Poland 51°52′N 17°00′E
Gostyn *see* Gostyń
110 K11 **Gostynin** Mazowiesckie, C Poland 52°25′N 19°27′E
Gosyogawara *see* Goshogawara
95 J18 **Göta Älv** ≈ S Sweden
95 N17 **Göta kanal** *canal* S Sweden
95 K18 **Götaland** *cultural region* S Sweden
95 H17 **Göteborg** *Eng.* Gothenburg. Västra Götaland, S Sweden 57°43′N 11°58′E
77 X16 **Gotel Mountains** ▲▲ E Nigeria
95 K17 **Götene** Västra Götaland, S Sweden 58°32′N 13°29′E
Gotera *see* San Francisco
101 K16 **Gotha** Thüringen, C Germany 50°57′N 10°43′E
29 N15 **Gothenburg** Nebraska, C USA 40°57′N 100°09′W
Gothenburg *see* Göteborg
77 R12 **Gothèye** Tillabéri, SW Niger 13°52′N 01°27′E
Gothland *see* Gotland
95 P19 **Gotland** *var.* Gothland, Gottland. ◆ *county* SE Sweden
95 O18 **Gotland** *island* SE Sweden
164 A14 **Gotō** *var.* Hukue; *prev.* Fukue. Nagasaki, Fukue-jima, SW Japan 32°41′N 128°52′E
164 B13 **Gotō-rettō** *island group* SW Japan
114 H12 **Gotse Delchev** *prev.* Nevrokop. Blagoevgrad, SW Bulgaria 41°33′N 23°42′E
95 P17 **Gotska Sandön** *island* SE Sweden
101 I15 **Göttingen** *var.* Goettingen. Niedersachsen, C Germany 51°33′N 09°55′E
Gottland *see* Gotland
93 I16 **Gottne** Västernorrland, C Sweden 63°27′N 18°25′E
Gottschee *see* Kočevje
Gottwaldov *see* Zlín
146 B11 **Goturdepe** *Rus.* Koturdepe. Balkan Welaýaty, W Turkmenistan 39°32′N 53°39′E
108 I7 **Götzis** Vorarlberg, NW Austria 47°21′N 09°40′E
98 H12 **Gouda** Zuid-Holland, C Netherlands 52°01′N 04°42′E
76 I11 **Goudiri** *var.* Goudiry. E Senegal 14°12′N 12°41′W
Goudiry *see* Goudiri
77 X12 **Goudoumaria** Diffa, S Niger 13°28′N 11°15′E
15 R9 **Gouffre, Rivière du** ≈ Québec, SE Canada
65 M19 **Gough Fracture Zone** *tectonic feature* S Atlantic Ocean
65 M19 **Gough Island** *island* Tristan da Cunha, S Atlantic Ocean
15 N8 **Gouin, Réservoir** ⊠ Québec, SE Canada
14 B10 **Goulais River** Ontario, S Canada 46°41′N 84°22′W
183 R9 **Goulburn** New South Wales, SE Australia 34°45′S 149°44′E
183 O11 **Goulburn River** ≈ Victoria, SE Australia
195 O10 **Gould Coast** *physical region* Antarctica
Goulimime *see* Guelmime
114 F13 **Gouménissa** Kentrikí Makedonía, N Greece 40°56′N 22°27′E
77 O10 **Goundam** Tombouctou, NW Mali 16°27′N 03°39′W
78 H12 **Goundi** Moundou, S Chad 09°22′N 17°21′E
78 G12 **Gounou-Gaya** Mayo-Kébbi Est, SW Chad 09°37′N 15°30′E
Gourci *see* Goursi
Gourcy *see* Goursi
102 M13 **Gourdon** Lot, S France 44°45′N 01°22′E
77 W11 **Gouré** Zinder, SE Niger 13°59′N 10°16′E
102 G6 **Gourin** Morbihan, NW France 48°07′N 03°37′W

◆ Country
● Country Capital
◇ Dependent Territory
○ Dependent Territory Capital
◆ Administrative Regions
✕ International Airport
▲ Mountain
▲▲ Mountain Range
Volcano
≈ River
◎ Lake
⊠ Reservoir

77 P10 **Gourma-Rharous** Tombouctou, C Mali 16°54′N 01°55′W
103 N4 **Gournay-en-Bray** Seine-Maritime, N France 49°29′N 01°42′E
78 J6 **Gouro** Ennedi-Ouest, N Chad 19°26′N 19°36′E
77 O12 **Goursi** var. Gourci, Gourcy. NW Burkina Faso 13°13′N 02°20′W
104 H8 **Gouveia** Guarda, N Portugal 40°29′N 07°35′W
18 I7 **Gouverneur** New York, NE USA 44°20′N 75°27′W
99 L21 **Gouvy** Luxembourg, E Belgium 50°10′N 05°55′E
45 R14 **Gouyave** var. Charlotte Town. NW Grenada 12°10′N 61°44′W
Goverla, Gora see Hoverla, Hora
59 N20 **Governador Valadares** Minas Gerais, SE Brazil 18°51′S 41°57′W
171 R8 **Governor Generoso** Mindanao, S Philippines 06°36′N 126°06′E
44 I2 **Governor's Harbour** Eleuthera Island, C The Bahamas 25°11′N 76°15′W
162 F9 **Govĭ-Altay** ◆ province SW Mongolia
162 I10 **Govĭ Altayn Nuruu** ▲ S Mongolia
154 L9 **Govind Ballabh Pant Sāgar** ⊠ C India
152 I7 **Govind Sagar** ⊠ NE India
162 M8 **Govĭ-Sumber** ◆ province C Mongolia
Govurdak see Magdanly
18 D11 **Gowanda** New York, NE USA 42°25′N 78°55′W
148 J10 **Gowd-e Zereh, Dasht-e** var. Guad-i-Zirreh. marsh SW Afghanistan
14 F8 **Gowganda** Ontario, S Canada 47°41′N 80°46′W
14 G8 **Gowganda Lake** ◎ Ontario, S Canada
148 M5 **Gōwr** prev. Ghowr. ◆ province C Afghanistan
29 U13 **Gowrie** Iowa, C USA 42°16′N 94°17′W
Gowurdak see Magdanly
61 C15 **Goya** Corrientes, NE Argentina 29°10′S 59°15′W
Goyania see Goiânia
Goyaz see Goiás
137 X11 **Göyçay** Rus. Geokchay. C Azerbaijan 40°38′N 47°44′E
137 V11 **Göygöl** ; prev. Xanlar. NW Azerbaijan 40°37′N 46°18′E
146 D10 **Goymat** Rus. Koymat. Balkan Welaýaty, NW Turkmenistan 40°23′N 55°45′E
146 D10 **Goymatdag, Gory** Rus. Gory Koymatdag. hill range Balkan Welaýaty, NW Turkmenistan
136 F12 **Göynük** Bolu, NW Turkey 40°24′N 30°45′E
165 R9 **Goyō-san** ▲ Honshū, C Japan 39°12′N 141°40′E
78 K11 **Goz Beïda** Sila, SE Chad 12°06′N 21°22′E
146 M10 **G'ozg'on** Rus. Gazgan. Navoiy Viloyati, C Uzbekistan 40°36′N 65°29′E
158 H11 **Gozha Co** ◎ W China
121 O15 **Gozo** var. Ghawdex. island N Malta
80 H9 **Goz Regeb** Kassala, NE Sudan 16°03′N 35°33′E
83 H25 **Graaff-Reinet** Eastern Cape, S South Africa 32°15′S 24°32′E
Graasten see Gråsten
76 L17 **Grabo** SW Ivory Coast 04°57′N 07°30′W
112 P11 **Grabovica** Serbia, E Serbia 44°30′N 22°29′E
110 I13 **Grabów nad Prosną** Wielkopolskie, C Poland 51°30′N 18°06′E
108 I8 **Grabs** Sankt Gallen, NE Switzerland 47°10′N 09°27′E
112 D12 **Gračac** Zadar, SW Croatia 44°18′N 15°52′E
112 I11 **Gračanica** Federacija Bosne I Hercegovine, NE Bosnia and Herzegovina 44°41′N 18°20′E
14 L11 **Gracefield** Québec, SE Canada 46°06′N 76°03′W
99 K19 **Grâce-Hollogne** Liège, E Belgium 50°38′N 05°30′E
23 R8 **Graceville** Florida, SE USA 30°57′N 85°31′W
29 R8 **Graceville** Minnesota, N USA 45°34′N 96°25′W
42 G6 **Gracias** Lempira, W Honduras 14°35′N 88°35′W
Gracias see Lempira
42 L5 **Gracias a Dios** ◆ department E Honduras
43 O6 **Gracias a Dios, Cabo de** headland Honduras/Nicaragua 15°00′N 83°10′W
64 O2 **Graciosa** var. Ilha Graciosa. island Azores, Portugal, NE Atlantic Ocean
64 Q11 **Graciosa** island Islas Canarias, Spain, NE Atlantic Ocean
Graciosa, Ilha see Graciosa
112 I11 **Gradačac** Federacija Bosne I Hercegovine, N Bosnia and Herzegovina 44°51′N 18°24′E
59 J15 **Gradaús, Serra dos** ▲ C Brazil
104 L3 **Gradefes** Castilla y León, N Spain 42°37′N 05°14′W
Gradiška see Bosanska Gradiška
Gradizhsk see Hradyz'k
106 J7 **Grado** Friuli-Venezia Giulia, NE Italy 45°41′N 13°24′E
Grado see Grau
113 P19 **Gradsko** C FYR Macedonia 41°34′N 21°56′E
37 V11 **Grady** New Mexico, SW USA 34°49′N 103°19′W
Grad Zagreb see Zagreb
29 T12 **Graettinger** Iowa, C USA 43°14′N 94°45′W
101 M23 **Grafing** Bayern, SE Germany 48°01′N 11°57′E
25 S6 **Graford** Texas, SW USA 32°56′N 98°15′W
183 V5 **Grafton** New South Wales, SE Australia 29°41′S 152°55′E
29 Q3 **Grafton** North Dakota, N USA 48°24′N 97°24′W
21 S3 **Grafton** West Virginia, NE USA 39°21′N 80°03′W
21 T9 **Graham** North Carolina, SE USA 36°05′N 79°25′W
25 R6 **Graham** Texas, SW USA 33°07′N 98°36′W
Graham Bell Island see Greem-Bell, Ostrov
10 I13 **Graham Island** island Queen Charlotte Islands, British Columbia, SW Canada
19 S6 **Graham Lake** ◎ Maine, NE USA
194 H4 **Graham Land** physical region Antarctica
37 N15 **Graham, Mount** ▲ Arizona, SW USA 32°42′N 109°52′W
Grahamstad see Grahamstown
83 I25 **Grahamstown** Afr. Grahamstad. Eastern Cape, S South Africa 33°18′S 26°32′E
Grahovo see Bosansko Grahovo
68 C12 **Grain Coast** coastal region S Liberia
169 S17 **Grajagan, Teluk** bay Jawa, S Indonesia
59 L14 **Grajaú** Maranhão, E Brazil 05°50′S 45°12′W
58 M13 **Grajaú, Rio** ≈ NE Brazil
110 O8 **Grajewo** Podlaskie, NE Poland 53°38′N 22°26′E
95 F24 **Gram** Syddanmark, SW Denmark 55°18′N 09°03′E
103 N13 **Gramat** Lot, S France 44°45′N 01°45′E
22 H5 **Grambling** Louisiana, S USA 32°31′N 92°43′W
115 C14 **Grámmos** ▲ Albania/Greece
96 I9 **Grampian Mountains** ▲ C Scotland, United Kingdom
182 L12 **Grampians, The** ▲ Victoria, SE Australia
98 O9 **Gramsbergen** Overijssel, E Netherlands 52°37′N 06°39′E
113 L21 **Gramsh** var. Gramshi. Elbasan, C Albania 40°52′N 20°12′E
Gramshi see Gramsh
Gran see Esztergom, Hungary
Gran see Hron
54 F11 **Granada** Meta, C Colombia 03°33′N 73°44′W
42 J10 **Granada** Granada, SW Nicaragua 11°55′N 85°58′W
105 N14 **Granada** Andalucía, S Spain 37°13′N 03°41′W
37 W6 **Granada** Colorado, C USA 38°00′N 102°18′W
42 J11 **Granada** ◆ department SW Nicaragua
105 N14 **Granada** ◆ province Andalucía, S Spain
63 I21 **Gran Antiplanicie Central** plain S Argentina
97 E17 **Granard** Ir. Gránard. C Ireland 53°47′N 07°30′W
Gránard see Granard
63 J20 **Gran Bajo** basin S Argentina
63 J15 **Gran Bajo del Gualicho** basin E Argentina
63 I21 **Gran Bajo de San Julián** basin SE Argentina
25 S7 **Granbury** Texas, SW USA 32°27′N 97°47′W
15 P12 **Granby** Québec, SE Canada 45°23′N 72°44′W
27 S8 **Granby** Missouri, C USA 36°55′N 94°14′W
37 S3 **Granby, Lake** ⊠ Colorado, C USA
64 O12 **Gran Canaria** var. Grand Canary. island Islas Canarias, Spain, NE Atlantic Ocean
47 T11 **Gran Chaco** var. Chaco. lowland plain South America
45 R14 **Grand Anse** SW Grenada 12°01′N 61°45′W
Grand-Anse see Portsmouth
44 G1 **Grand Bahama Island** island N The Bahamas
Grand Balé see Tui
103 U7 **Grand Ballon** Ger. Ballon de Guebwiller. ▲ NE France 47°53′N 07°06′E
13 T13 **Grand Bank** Newfoundland, Newfoundland and Labrador, SE Canada 47°06′N 55°48′W
64 I7 **Grand Banks of Newfoundland** undersea feature NW Atlantic Ocean 45°00′N 40°00′W
Grand Bassa see Buchanan
77 N17 **Grand-Bassam** var. Bassam. SE Ivory Coast 05°14′N 03°45′W
14 E16 **Grand Bend** Ontario, S Canada 43°17′N 81°46′W
76 L17 **Grand-Bérébi** var. Grand-Béréby. SW Ivory Coast 04°38′N 06°55′W
Grand-Béréby see Grand-Bérébi
45 X11 **Grand-Bourg** Marie-Galante, SE Guadeloupe 15°53′N 61°19′W
44 M6 **Grand Caicos** var. Middle Caicos. island C Turks and Caicos Islands
14 K12 **Grand Calumet, Île du** island Québec, SE Canada
97 E18 **Grand Canal** Ir. An Chanáil Mhór. canal C Ireland
Grand Canal see Da Yunhe
Grand Canary see Gran Canaria
36 K10 **Grand Canyon** Arizona, SW USA 36°01′N 112°10′W
36 J9 **Grand Canyon** canyon Arizona, SW USA
Grand Canyon State see Arizona
44 D8 **Grand Cayman** island SW Cayman Islands
11 R14 **Grand Centre** Alberta, SW Canada 54°25′N 110°13′W
76 L17 **Grand Cess** SE Liberia 04°36′N 08°12′W
108 D12 **Grand Combin** ▲ S Switzerland 45°58′N 07°27′E
32 K8 **Grand Coulee** Washington, NW USA 47°56′N 119°00′W
32 J8 **Grand Coulee** valley Washington, NW USA
45 X5 **Grand Cul-de-Sac Marin** bay N Guadeloupe
Grand Duchy of Luxembourg see Luxembourg
63 I22 **Grande, Bahía** bay S Argentina
11 N14 **Grande Cache** Alberta, W Canada 53°53′N 119°07′W
103 U12 **Grande Casse** ▲ E France 45°22′N 06°50′E
Grande Comore see Ngazidja
61 G18 **Grande, Cuchilla** hill range E Uruguay
45 S5 **Grande de Añasco, Río** ≈ W Puerto Rico
Grande de Chiloé, Isla see Chiloé, Isla de
58 J12 **Grande de Gurupá, Ilha** river island NE Brazil
57 K21 **Grande de Lipez, Río** ≈ SW Bolivia
45 U6 **Grande de Loíza, Río** ≈ E Puerto Rico
45 T5 **Grande de Manatí, Río** ≈ C Puerto Rico
42 L9 **Grande de Matagalpa, Río** ≈ C Nicaragua
40 K12 **Grande de Santiago, Río** var. Santiago. ≈ C Mexico
43 O15 **Grande de Térraba, Río** var. Río Térraba. ≈ SE Costa Rica
12 J9 **Grande Deux, Réservoir la** ⊠ Québec, E Canada
60 O10 **Grande, Ilha** island SE Brazil
11 O13 **Grande Prairie** Alberta, W Canada 55°10′N 118°52′W
74 I8 **Grand Erg Occidental** desert W Algeria
74 L9 **Grand Erg Oriental** desert Algeria/Tunisia
57 M18 **Grande, Río** ≈ C Bolivia
59 J20 **Grande, Rio** ≈ S Brazil
2 F15 **Grande, Rio** var. Río Bravo, Sp. Río Bravo del Norte, Bravo del Norte. ≈ Mexico/USA
15 Y7 **Grande-Rivière** Québec, SE Canada 48°27′N 64°37′W
15 Y6 **Grande Rivière** ≈ Québec, SE Canada
44 M8 **Grande-Rivière-du-Nord** N Haiti 19°36′N 72°10′W
62 K9 **Grande, Salina** var. Gran Salitral. salt lake C Argentina
15 S7 **Grandes-Bergeronnes** Québec, SE Canada 48°16′N 69°32′W
47 W6 **Grande, Serra** ▲ W Brazil
40 K4 **Grande, Sierra** ▲ N Mexico
103 S12 **Grandes Rousses** ▲ E France
63 K17 **Grandes, Salinas** salt lake E Argentina
45 Y5 **Grande Terre** island E West Indies
15 X5 **Grande-Vallée** Québec, SE Canada 49°14′N 65°08′W
45 Y5 **Grande Vigie, Pointe de la** headland Grande Terre, N Guadeloupe 16°31′N 61°27′W
13 N14 **Grand Falls** New Brunswick, SE Canada 47°02′N 67°46′W
13 T11 **Grand Falls** Newfoundland, Newfoundland and Labrador, SE Canada 48°57′N 55°48′W
24 L9 **Grandfalls** Texas, SW USA 31°20′N 102°51′W
21 P9 **Grandfather Mountain** ▲ North Carolina, SE USA 36°06′N 81°48′W
26 L13 **Grandfield** Oklahoma, C USA 34°15′N 98°40′W
11 N17 **Grand Forks** British Columbia, SW Canada 49°02′N 118°30′W
29 R4 **Grand Forks** North Dakota, N USA 47°54′N 97°03′W
31 O9 **Grand Haven** Michigan, N USA 43°03′N 86°13′W
Grandichi see Hrandzichy
29 P15 **Grand Island** Nebraska, C USA 40°55′N 98°20′W
31 O3 **Grand Island** island Michigan, N USA
22 K10 **Grand Isle** Louisiana, S USA 29°12′N 90°00′W
65 A23 **Grand Jason** island Jason Islands, NW Falkland Islands
37 P5 **Grand Junction** Colorado, C USA 39°03′N 108°33′W
20 F10 **Grand Junction** Tennessee, S USA 35°03′N 89°11′W
14 J9 **Grand-Lac-Victoria** Québec, SE Canada 47°33′N 77°28′W
14 J9 **Grand lac Victoria** ◎ Québec, SE Canada
77 N17 **Grand-Lahou** var. Grand Lahu. S Ivory Coast 05°09′N 05°01′W
Grand Lahu see Grand-Lahou
37 S3 **Grand Lake** Colorado, C USA 40°15′N 105°49′W
13 S11 **Grand Lake** ◎ Newfoundland and Labrador, E Canada
22 G9 **Grand Lake** ◎ Louisiana, S USA
31 R5 **Grand Lake** ◎ Michigan, N USA
31 Q13 **Grand Lake** ◎ Ohio, N USA
27 R9 **Grand Lake O' The Cherokees** var. Lake O' The Cherokees. ⊠ Oklahoma, C USA
31 Q9 **Grand Ledge** Michigan, N USA 42°45′N 84°45′W
102 I8 **Grand-Lieu, Lac de** ◎ NW France
19 U6 **Grand Manan Channel** channel Canada/USA
13 O15 **Grand Manan Island** island New Brunswick, SE Canada
29 Y4 **Grand Marais** Minnesota, N USA 47°45′N 90°19′W
15 P10 **Grand-Mère** Québec, SE Canada 46°36′N 72°41′W
37 P5 **Grand Mesa** ▲ Colorado, C USA
108 C10 **Grand Muveran** ▲ W Switzerland 46°16′N 07°12′E
104 G12 **Grândola** Setúbal, S Portugal 38°10′N 08°34′W
Grand Paradis see Gran Paradiso
187 O15 **Grand Passage** passage N New Caledonia
77 R16 **Grand-Popo** S Benin 06°19′N 01°50′E
29 Z3 **Grand Portage** Minnesota, N USA 48°00′N 89°36′W
25 T6 **Grand Prairie** Texas, SW USA 32°45′N 97°00′W
11 W14 **Grand Rapids** Manitoba, C Canada 53°12′N 99°19′W
31 P9 **Grand Rapids** Michigan, N USA 42°57′N 86°40′W
29 V5 **Grand Rapids** Minnesota, N USA 47°14′N 93°31′W
14 L10 **Grand-Remous** Québec, SE Canada 46°36′N 75°53′W
14 F15 **Grand River** ≈ Ontario, S Canada
31 P9 **Grand River** ≈ Michigan, N USA
27 T3 **Grand River** ≈ Missouri, C USA
28 M7 **Grand River** ≈ South Dakota, N USA
45 Q11 **Grand' Rivière** N Martinique 14°52′N 61°11′W
32 F11 **Grand Ronde** Oregon, NW USA 45°03′N 123°43′W
32 L11 **Grand Ronde River** ≈ Oregon/Washington, NW USA
Grand-Saint-Bernard, Col du see Great Saint Bernard Pass
25 V6 **Grand Saline** Texas, SW USA 32°40′N 95°42′W
55 X10 **Grand-Santi** W French Guiana 04°19′N 54°24′W
Grandsee see Grandson
172 J16 **Grand Sœur** island Les Sœurs, NE Seychelles
108 B9 **Grandson** prev. Grandsee. Vaud, W Switzerland 46°49′N 06°39′E
33 S14 **Grand Teton** ▲ Wyoming, C USA 43°44′N 110°48′W
31 P5 **Grand Traverse Bay** lake bay Michigan, N USA
45 N6 **Grand Turk** ○ (Turks and Caicos Islands) Grand Turk Island, S Turks and Caicos Islands 21°24′N 71°08′W
45 N6 **Grand Turk Island** island SE Turks and Caicos Islands
103 S13 **Grand Veymont** ▲ E France 44°51′N 05°32′E
11 W15 **Grandview** Manitoba, S Canada 51°11′N 100°41′W
27 R4 **Grandview** Missouri, C USA 38°53′N 94°31′W
36 I10 **Grand Wash Cliffs** cliff Arizona, SW USA
14 J8 **Granet, Lac** ◎ Québec, SE Canada
95 L14 **Grangärde** Dalarna, C Sweden 60°15′N 15°00′E
44 H12 **Grange Hill** W Jamaica 18°19′N 78°11′W
96 J12 **Grangemouth** C Scotland, United Kingdom 56°01′N 03°44′W
25 T10 **Granger** Texas, SW USA 30°43′N 97°26′W
32 J10 **Granger** Washington, NW USA 46°20′N 120°11′W
33 T17 **Granger** Wyoming, C USA 41°37′N 109°58′W
Granges see Grenchen
95 L14 **Grängesberg** Dalarna, C Sweden 60°06′N 15°00′E
33 N11 **Grangeville** Idaho, NW USA 45°55′N 116°07′W
10 K13 **Granisle** British Columbia, SW Canada 54°55′N 126°14′W
30 K15 **Granite City** Illinois, N USA 38°42′N 90°09′W
29 S9 **Granite Falls** Minnesota, N USA 44°48′N 95°33′W
21 Q9 **Granite Falls** North Carolina, SE USA 35°48′N 81°25′W
36 K12 **Granite Mountain** ▲ Arizona, SW USA 34°38′N 112°34′W
33 T12 **Granite Peak** ▲ Montana, NW USA 45°09′N 109°48′W
35 T2 **Granite Peak** ▲ Nevada, W USA 41°40′N 117°35′W
36 J3 **Granite Peak** ▲ Utah, W USA 40°09′N 113°18′W
Granite State see New Hampshire
107 H24 **Granitola, Capo** headland Sicilia, Italy, C Mediterranean Sea 37°33′N 12°39′E
185 H15 **Granity** West Coast, South Island, New Zealand 41°37′S 171°53′E
Gran Lago see Nicaragua, Lago de
63 J18 **Gran Laguna Salada** ◎ S Argentina
Gran Malvina see West Falkland
95 L18 **Gränna** Jönköping, S Sweden 58°02′N 14°30′E
105 W5 **Granollers** var. Granollérs. Cataluña, NE Spain 41°37′N 02°18′E
Granollérs see Granollers
106 A7 **Gran Paradiso** Fr. Grand Paradis. ▲ NW Italy 45°31′N 07°13′E
Gran Pilastro see Hochfeiler
Gran Salitral see Grande, Salina
Gran San Bernardo, Passo di see Great Saint Bernard Pass
Gran Santiago see Santiago
107 J14 **Gran Sasso d'Italia** ▲ C Italy
100 N11 **Gransee** Brandenburg, NE Germany 53°00′N 13°10′E
28 L15 **Grant** Nebraska, C USA 40°50′N 101°43′W
27 R1 **Grant City** Missouri, C USA 40°29′N 94°25′W
97 N19 **Grantham** E England, United Kingdom 52°55′N 00°39′W
65 D24 **Grantham Sound** sound East Falkland, Falkland Islands
194 K13 **Grant Island** island Antarctica
45 Z14 **Grantley Adams** ✈ (Bridgetown) SE Barbados 13°04′N 59°29′W
35 S7 **Grant, Mount** ▲ Nevada, W USA 38°34′N 118°47′W
96 J9 **Grantown-on-Spey** N Scotland, United Kingdom 57°11′N 03°53′W
35 W8 **Grant Range** ▲ Nevada, W USA
37 Q11 **Grants** New Mexico, SW USA 35°09′N 107°50′W
30 J4 **Grantsburg** Wisconsin, N USA 45°47′N 92°40′W
32 F15 **Grants Pass** Oregon, NW USA 42°26′N 123°20′W
36 K3 **Grantsville** Utah, W USA 40°36′N 112°27′W
21 R4 **Grantsville** West Virginia, NE USA 38°55′N 81°07′W
102 I5 **Granville** Manche, N France 48°50′N 01°35′W
11 V12 **Granville Lake** ◎ Manitoba, C Canada
25 V8 **Grapeland** Texas, SW USA 31°29′N 95°28′W
25 T6 **Grapevine** Texas, SW USA 32°55′N 97°04′W
83 K20 **Graskop** Mpumalanga, NE South Africa 24°58′S 30°49′E
95 P14 **Gräsö** Uppsala, C Sweden 60°22′N 18°30′E
93 J19 **Gräsö** island C Sweden
103 U15 **Grasse** Alpes-Maritimes, SE France 43°42′N 06°52′E
18 E14 **Grassflat** Pennsylvania, NE USA 41°00′N 78°04′W
33 U9 **Grassrange** Montana, NW USA 47°02′N 108°48′W
18 J6 **Grass River** ≈ New York, NE USA
35 P6 **Grass Valley** California, W USA 39°12′N 121°04′W
183 N14 **Grassy** Tasmania, SE Australia 40°03′S 144°04′E
28 K4 **Grassy Butte** North Dakota, N USA 47°20′N 103°13′W
21 R5 **Grassy Knob** ▲ West Virginia, NE USA 38°04′N 80°31′W
95 G24 **Gråsten** var. Graasten. Syddanmark, SW Denmark 54°55′N 09°37′E
95 J18 **Grästorp** Västra Götaland, S Sweden 58°20′N 12°40′E
Gratianopolis see Grenoble
109 V8 **Gratwein** Steiermark, SE Austria 47°08′N 15°20′E
Gratz see Graz
104 K2 **Grau** var. Grado. Asturias, N Spain 43°23′N 06°04′W
108 I9 **Graubünden** Fr. Grisons, It. Grigioni. ◆ canton SE Switzerland
Graudenz see Grudziądz
103 N15 **Graulhet** Tarn, S France 43°45′N 01°58′E
105 T4 **Graus** Aragón, NE Spain 42°11′N 00°21′E
61 I16 **Gravataí** Rio Grande do Sul, S Brazil 29°55′S 51°00′W
98 L13 **Grave** Noord-Brabant, SE Netherlands 51°45′N 05°45′E
11 T17 **Gravelbourg** Saskatchewan, S Canada 49°53′N 106°33′W
103 N1 **Gravelines** Nord, N France 51°00′N 02°07′E
Graven see Grez-Doiceau
14 H13 **Gravenhurst** Ontario, S Canada 44°55′N 79°22′W
33 O10 **Grave Peak** ▲ Idaho, NW USA 46°24′N 114°43′W
102 I11 **Grave, Pointe de** headland W France 45°33′N 01°24′W
183 S4 **Gravesend** New South Wales, SE Australia 29°37′S 150°15′E
97 P22 **Gravesend** SE England, United Kingdom 51°27′N 00°24′E
107 N17 **Gravina in Puglia** Puglia, SE Italy 40°48′N 16°25′E
103 S8 **Gray** Haute-Saône, E France 47°28′N 05°34′E
23 T4 **Gray** Georgia, SE USA 33°00′N 83°31′W
195 V16 **Gray, Cape** headland Antarctica 67°30′S 143°30′E
32 F9 **Grayland** Washington, NW USA 46°46′N 124°07′W
39 N10 **Grayling** Alaska, USA 62°55′N 160°07′W
31 Q6 **Grayling** Michigan, N USA 44°40′N 84°43′W
32 F8 **Grays Harbor** inlet Washington, NW USA
21 O5 **Grayson** Kentucky, S USA 38°21′N 82°59′W
37 S4 **Grays Peak** ▲ Colorado, C USA 39°37′N 105°49′W
30 M16 **Grayville** Illinois, N USA 38°15′N 87°59′W
109 V8 **Graz** prev. Gratz. Steiermark, SE Austria 47°05′N 15°23′E
104 L15 **Grazalema** Andalucía, S Spain 36°46′N 05°23′W
113 P15 **Grdelica** Serbia, SE Serbia 42°54′N 22°05′E
44 H1 **Great Abaco** var. Abaco Island. island N The Bahamas
Great Admiralty Island see Manus Island
Great Alfold see Great Hungarian Plain
Great Ararat see Büyükağrı Dağı
181 U8 **Great Artesian Basin** lowlands Queensland, C Australia
181 O12 **Great Australian Bight** bight S Australia
64 E11 **Great Bahama Bank** undersea feature E Gulf of Mexico 23°15′N 78°00′W
184 M4 **Great Barrier Island** island N New Zealand
181 X4 **Great Barrier Reef** reef Queensland, NE Australia
18 L11 **Great Barrington** Massachusetts, NE USA 42°11′N 73°20′W
0 F10 **Great Basin** basin W USA
8 I8 **Great Bear Lake** Fr. Grand Lac de l'Ours. ◎ Northwest Territories, NW Canada
Great Belt see Storebælt
26 L5 **Great Bend** Kansas, C USA 38°22′N 98°47′W
Great Bermuda see Bermuda
97 A20 **Great Blasket Island** Ir. An Blascaod Mór. island SW Ireland
Great Britain see Britain
151 Q23 **Great Channel** channel Andaman Sea/Indian Ocean
166 J10 **Great Coco Island** island SW Myanmar (Burma)
Great Crosby see Crosby
21 X7 **Great Dismal Swamp** wetland North Carolina/Virginia, SE USA
33 V16 **Great Divide Basin** basin Wyoming, C USA
181 W7 **Great Dividing Range** ▲ NE Australia
14 D12 **Great Duck Island** island Ontario, S Canada
Great Elder Reservoir see Waconda Lake
44 G8 **Greater Antilles** island group West Indies
129 V16 **Greater Sunda Islands** var. Sunda Islands. island group Indonesia
184 I1 **Great Exhibition Bay** inlet North Island, New Zealand
44 H4 **Great Exuma Island** island C The Bahamas
33 R8 **Great Falls** Montana, NW USA 47°30′N 111°18′W
21 R11 **Great Falls** South Carolina, SE USA 34°34′N 80°54′W
84 F9 **Great Fisher Bank** undersea feature C North Sea 57°00′N 04°00′E
Great Glen see Mor, Glen
Great Grimsby see Grimsby
44 I4 **Great Guana Cay** island C The Bahamas
64 I5 **Great Hellefiske Bank** undersea feature N Atlantic Ocean
111 L24 **Great Hungarian Plain** var. Great Alfold, Plain of Hungary, Hung. Alföld. plain SE Europe
44 L7 **Great Inagua** var. Inagua Islands. island S The Bahamas
Great Indian Desert see Thar Desert
83 G25 **Great Karoo** var. Great Karroo, High Veld, Afr. Groot Karoo, Hoë Karoo. plateau region S South Africa
Great Karroo see Great Karoo
Great Kei see Nciba
Great Khingan Range see Da Hinggan Ling
14 E11 **Great La Cloche Island** island Ontario, S Canada
183 P16 **Great Lake** ◎ Tasmania, SE Australia
Great Lake see Tônlé Sap
11 R15 **Great Lakes** lakes Ontario, Canada/USA
Great Lakes State see Michigan
97 L20 **Great Malvern** W England, United Kingdom 52°07′N 02°19′W
184 M5 **Great Mercury Island** island N New Zealand
Great Meteor Seamount see Great Meteor Tablemount
64 K10 **Great Meteor Tablemount** var. Great Meteor Seamount. undersea feature E Atlantic Ocean 30°00′N 28°30′W
31 Q14 **Great Miami River** ≈ Ohio, N USA
151 Q24 **Great Nicobar** island Nicobar Islands, India, NE Indian Ocean
97 O19 **Great Ouse** var. Ouse. ≈ E England, United Kingdom
183 Q17 **Great Oyster Bay** bay Tasmania, SE Australia
44 I13 **Great Pedro Bluff** headland W Jamaica 17°51′N 77°44′W
21 T12 **Great Pee Dee River** ≈ North Carolina/South Carolina, SE USA
129 W9 **Great Plain of China** plain E China
0 F12 **Great Plains** var. High Plains. plains Canada/USA
37 W6 **Great Plains Reservoirs** ⊠ Colorado, C USA
19 Q13 **Great Point** headland Nantucket Island, Massachusetts, NE USA 41°23′N 70°03′W
68 I13 **Great Rift Valley** var. Rift Valley. depression Asia/Africa
81 I23 **Great Ruaha** ≈ S Tanzania
18 K10 **Great Sacandaga Lake** ⊠ New York, NE USA
108 C12 **Great Saint Bernard Pass** Fr. Col du Grand-Saint-Bernard, It. Passo del Gran San Bernardo. pass Italy/Switzerland
44 F1 **Great Sale Cay** island N The Bahamas
Great Salt Desert see Kavīr, Dasht-e
36 K1 **Great Salt Lake** salt lake Utah, W USA
36 J3 **Great Salt Lake Desert** plain Utah, W USA
26 M8 **Great Salt Plains Lake** ⊠ Oklahoma, C USA
75 T9 **Great Sand Sea** desert Egypt/Libya
180 L6 **Great Sandy Desert** desert Western Australia
Great Sandy Desert see Ar Rub' al Khālī
Great Sandy Island see Fraser Island
187 Y13 **Great Sea Reef** reef Vanua Levu, N Fiji
38 H17 **Great Sitkin Island** island Aleutian Islands, Alaska, USA
8 J10 **Great Slave Lake** Fr. Grand Lac des Esclaves. ◎ Northwest Territories, NW Canada
21 O10 **Great Smoky Mountains** ▲ North Carolina/Tennessee, SE USA
10 L11 **Great Snow Mountain** ▲ British Columbia, W Canada 57°22′N 124°08′W
Great Socialist People's Libyan Arab Jamahiriya see Libya
64 A12 **Great Sound** sound Bermuda, NW Atlantic Ocean
180 M10 **Great Victoria Desert** desert South Australia/Western Australia
194 H2 **Great Wall** Chinese research station South Shetland Islands, Antarctica 61°57′S 58°23′W
19 T7 **Great Wass Island** island Maine, NE USA
97 Q19 **Great Yarmouth** var. Yarmouth. E England, United Kingdom 52°37′N 01°44′E
139 S1 **Great Zab** Ar. Az Zāb al Kabīr, Kurd. Zē-i Bādīnān, Turk. Büyükzap Suyu. ≈ Iraq/Turkey
95 I17 **Grebbestad** Västra Götaland, S Sweden 58°42′N 11°15′E
Grebenka see Hrebinka
42 M13 **Grecia** Alajuela, C Costa Rica 10°04′N 84°19′W
61 E18 **Greco** Río Negro, W Uruguay 32°49′S 57°03′W
Greco, Cape see Gkréko, Akrotíri
104 L8 **Gredos, Sierra de** ▲ W Spain
18 F9 **Greece** New York, NE USA 43°12′N 77°41′W
115 E17 **Greece** off. Hellenic Republic, Gk. Ellás; anc. Hellas. ◆ republic SE Europe
Greece Central see Stereá Elláda
Greece West see Dytikí Elláda
37 T3 **Greeley** Colorado, C USA 40°21′N 104°41′W
29 P14 **Greeley** Nebraska, C USA 41°33′N 98°31′W
122 K3 **Greem-Bell, Ostrov** Eng. Graham Bell Island. island Zemlya Frantsa-Iosifa, N Russian Federation
30 M6 **Green Bay** Wisconsin, N USA 44°32′N 88°W
31 N6 **Green Bay** lake bay Michigan/Wisconsin, N USA
21 S5 **Greenbrier River** ≈ West Virginia, NE USA
29 S2 **Greenbush** Minnesota, N USA 48°42′N 96°10′W
183 R12 **Green Cape** headland New South Wales, SE Australia 37°15′S 150°03′E
31 O14 **Greencastle** Indiana, N USA 39°38′N 86°51′W
18 F16 **Greencastle** Pennsylvania, NE USA 39°47′N 77°43′W
27 T2 **Green City** Missouri, C USA 40°16′N 92°57′W
21 O9 **Greeneville** Tennessee, S USA 36°10′N 82°50′W
35 O11 **Greenfield** California, W USA 36°19′N 121°15′W
31 P14 **Greenfield** Indiana, N USA 39°47′N 85°46′W
29 U15 **Greenfield** Iowa, C USA 41°18′N 94°27′W
18 M11 **Greenfield** Massachusetts, NE USA 42°33′N 72°34′W
27 S7 **Greenfield** Missouri, C USA 37°25′N 93°50′W
31 S14 **Greenfield** Ohio, N USA 39°21′N 83°22′W
20 G8 **Greenfield** Tennessee, S USA 36°09′N 88°48′W
30 M9 **Greenfield** Wisconsin, N USA 42°55′N 87°59′W
27 T9 **Green Forest** Arkansas, C USA 36°19′N 93°24′W
37 T7 **Greenhorn Mountain** ▲ Colorado, C USA 37°50′N 104°59′W
Green Island see Lü Dao
186 I6 **Green Islands** var. Nissan Islands. island group NE Papua New Guinea
11 S14 **Green Lake** Saskatchewan, C Canada 54°15′N 107°51′W
30 L8 **Green Lake** ◎ Wisconsin, N USA
197 O14 **Greenland** Dan. Grønland, Inuit Kalaallit Nunaat. ◇ Danish self-governing territory NE North America
84 D4 **Greenland** island NE North America
197 R13 **Greenland Plain** undersea feature N Greenland Sea
197 R14 **Greenland Sea** sea Arctic Ocean
37 R4 **Green Mountain Reservoir** ⊠ Colorado, C USA
18 M8 **Green Mountains** ▲ Vermont, NE USA
Green Mountain State see Vermont
96 H12 **Greenock** W Scotland, United Kingdom 55°57′N 04°45′W
39 T5 **Greenough, Mount** ▲ Alaska, USA 69°15′N 141°37′W
186 A6 **Green River** West Sepik, NW Papua New Guinea 03°54′S 141°08′E
37 N5 **Green River** Utah, W USA 39°00′N 110°07′W
33 U17 **Green River** Wyoming, C USA 41°33′N 109°27′W
16 H9 **Green River** ≈ W USA
30 K11 **Green River** ≈ Illinois, N USA
20 J7 **Green River** ≈ Kentucky, C USA
28 K5 **Green River** ≈ North Dakota, N USA
37 N6 **Green River** ≈ Utah, W USA
33 T16 **Green River** ≈ Wyoming, C USA
20 L7 **Green River Lake** ⊠ Kentucky, S USA
23 O5 **Greensboro** Alabama, S USA 32°42′N 87°36′W
23 U3 **Greensboro** Georgia, SE USA 33°34′N 83°10′W
21 T9 **Greensboro** North Carolina, SE USA 36°04′N 79°48′W
31 P14 **Greensburg** Indiana, N USA 39°20′N 85°28′W
26 K6 **Greensburg** Kansas, C USA 37°36′N 99°17′W
20 L7 **Greensburg** Kentucky, S USA 37°14′N 85°30′W
18 C15 **Greensburg** Pennsylvania, NE USA 40°18′N 79°33′W
37 O13 **Greens Peak** ▲ Arizona, SW USA 34°06′N 109°34′W
21 V12 **Green Swamp** wetland North Carolina, SE USA
21 O4 **Greenup** Kentucky, S USA 38°34′N 82°49′W
36 M16 **Green Valley** Arizona, SW USA 31°49′N 111°00′W
76 K17 **Greenville** var. Sino, Sinoe. SE Liberia 05°01′N 09°03′W
23 P6 **Greenville** Alabama, S USA 31°49′N 86°37′W
23 T8 **Greenville** Florida, SE USA 30°28′N 83°37′W
23 S4 **Greenville** Georgia, SE USA 33°03′N 84°42′W
30 L15 **Greenville** Illinois, N USA 38°53′N 89°24′W
20 I7 **Greenville** Kentucky, S USA 37°11′N 87°11′W
19 Q5 **Greenville** Maine, NE USA 45°26′N 69°36′W
31 P9 **Greenville** Michigan, N USA 43°10′N 85°15′W
22 J4 **Greenville** Mississippi, S USA 33°24′N 91°03′W
21 W9 **Greenville** North Carolina, SE USA 35°36′N 77°23′W
31 Q13 **Greenville** Ohio, N USA 40°06′N 84°37′W
19 O12 **Greenville** Rhode Island, NE USA 41°52′N 71°33′W
21 P11 **Greenville** South Carolina, SE USA 34°51′N 82°24′W
25 U6 **Greenville** Texas, SW USA 33°09′N 96°07′W
31 T12 **Greenwich** Ohio, N USA 41°01′N 82°31′W
27 S11 **Greenwood** Arkansas, C USA 35°13′N 94°15′W
31 O14 **Greenwood** Indiana, N USA 39°38′N 86°06′W
22 K4 **Greenwood** Mississippi, S USA 33°30′N 90°11′W
21 P12 **Greenwood** South Carolina, SE USA 34°11′N 82°10′W
21 Q12 **Greenwood, Lake** ⊠ South Carolina, SE USA
21 P11 **Greer** South Carolina, SE USA 34°56′N 82°13′W
27 V10 **Greers Ferry Lake** ⊠ Arkansas, C USA
27 S13 **Greeson, Lake** ⊠ Arkansas, C USA
45 N8 **Gregorio Luperón** ✈ N Dominican Republic 19°43′N 70°43′W
29 O12 **Gregory** South Dakota, N USA 43°11′N 99°26′W
182 J3 **Gregory, Lake** salt lake South Australia
180 J9 **Gregory Lake** ◎ Western Australia
181 V5 **Gregory Range** ▲ Queensland, E Australia
Greifenberg/Greifenberg in Pommern see Gryfice
Greifenhagen see Gryfino
100 O8 **Greifswald** Mecklenburg-Vorpommern, NE Germany 54°04′N 13°24′E
100 O8 **Greifswalder Bodden** bay NE Germany
109 U4 **Grein** Oberösterreich, N Austria 48°14′N 14°50′E
101 M17 **Greiz** Thüringen, C Germany 50°40′N 12°11′E
125 V14 **Gremyachinsk** Permskiy Kray, NW Russian Federation 58°33′N 57°52′E
Grenå see Grenaa
95 H21 **Grenaa** var. Grenå. Midtjylland, C Denmark 56°25′N 10°53′E

◆ Country
● Country Capital
◇ Dependent Territory
○ Dependent Territory Capital
◆ Administrative Regions
✈ International Airport
▲ Mountain
▲ Mountain Range
Volcano
≈ River
◎ Lake
⊠ Reservoir

22 L3 **Grenada** Mississippi, S USA 33°46′N 89°48′W
45 W15 **Grenada** ◆ commonwealth republic SE West Indies
47 S4 **Grenada** island Grenada
47 R4 **Grenada Basin** undersea feature W Atlantic Ocean 13°30′N 62°00′W
22 L3 **Grenada Lake** ⊠ Mississippi, S USA
45 Y14 **Grenadines, The** island group Grenada/St Vincent and the Grenadines
108 D7 **Grenchen** *Fr.* Granges. Solothurn, NW Switzerland 47°13′N 07°24′E
183 Q9 **Grenfell** New South Wales, SE Australia 33°54′S 148°09′E
11 V16 **Grenfell** Saskatchewan, S Canada 50°24′N 102°56′W
92 J1 **Grenivík** Norðurland Eystra, N Iceland 65°57′N 18°10′W
103 S12 **Grenoble** *anc.* Cularo, Gratianopolis. Isère, E France 45°11′N 05°42′E
28 J2 **Grenora** North Dakota, N USA 48°36′N 103°57′W
92 N8 **Grense-Jakobselv** Finnmark, N Norway 69°46′N 30°39′E
45 S14 **Grenville** E Grenada 12°07′N 61°37′W
32 G11 **Gresham** Oregon, NW USA 45°30′N 122°25′W
Gresk *see* Hresk
106 B7 **Gressoney-St-Jean** Valle d'Aosta, NW Italy 45°48′N 07°49′E
22 K9 **Gretna** Louisiana, S USA 29°54′N 90°03′W
21 T7 **Gretna** Virginia, NE USA 36°57′N 79°21′W
98 F13 **Grevelingen** inlet S North Sea
100 F13 **Greven** Nordrhein-Westfalen, NW Germany 52°07′N 07°38′E
115 D15 **Grevená** Dytikí Makedonía, N Greece 40°05′N 21°26′E
101 D16 **Grevenbroich** Nordrhein-Westfalen, W Germany 51°06′N 06°34′E
99 N24 **Grevenmacher** Grevenmacher, E Luxembourg 49°41′N 06°27′E
99 M24 **Grevenmacher** ◈ district E Luxembourg
100 K9 **Grevesmühlen** Mecklenburg-Vorpommern, N Germany 53°52′N 11°12′E
185 H16 **Grey** ≈ South Island, New Zealand
33 V12 **Greybull** Wyoming, C USA 44°29′N 108°03′W
33 U13 **Greybull River** ≈ Wyoming, C USA
65 A24 **Grey Channel** sound Falkland Islands
Greyerzer See *see* Gruyère, Lac de la
13 T10 **Grey Islands** island group Newfoundland and Labrador, E Canada
18 L10 **Greylock, Mount** ▲ Massachusetts, NE USA 42°38′N 73°09′W
185 G17 **Greymouth** West Coast, South Island, New Zealand 42°29′S 171°14′E
181 U10 **Grey Range** ▲▲ New South Wales/Queensland, E Australia
97 G18 **Greystones** *Ir.* Na Clocha Liatha. E Ireland 53°08′N 06°05′W
185 M14 **Greytown** Wellington, North Island, New Zealand 41°04′S 175°29′E
83 K23 **Greytown** KwaZulu/Natal, E South Africa 29°04′S 30°35′E
Greytown *see* San Juan del Norte
99 H19 **Grez-Doiceau** *Dut.* Graven. Walloon Brabant, C Belgium 50°43′N 04°41′E
115 J19 **Griá, Akrotírio** headland Ándros, Kykládes, Greece, Aegean Sea 37°54′N 24°57′E
127 N8 **Gribanovskiy** Voronezhskaya Oblast', W Russian Federation 51°27′N 41°53′E
78 I13 **Gribingui** ≈ N Central African Republic
35 O6 **Gridley** California, W USA 39°21′N 121°41′W
83 G23 **Griekwastad** *var.* Griquatown. Northern Cape, C South Africa 28°50′S 23°16′E
23 S4 **Griffin** Georgia, SE USA 33°15′N 84°17′W
183 O9 **Griffith** New South Wales, SE Australia 34°18′S 146°04′E
14 F13 **Griffith Island** island Ontario, S Canada
21 W10 **Grifton** North Carolina, SE USA 35°22′N 77°26′W
Grigioni *see* Graubünden
119 H14 **Grigiškes** Vilnius, SE Lithuania 54°42′N 25°00′E
117 N10 **Grigoriopol** C Moldova 47°09′N 29°18′E
147 X7 **Grigor'yevka** Issyk-Kul'skaya Oblast', E Kyrgyzstan 42°43′N 77°27′E
193 U8 **Grijalva Ridge** undersea feature E Pacific Ocean
41 U15 **Grijalva, Río** *var.* Tabasco. ≈ Guatemala/Mexico
98 N5 **Grijpskerk** Groningen, NE Netherlands 53°15′N 06°18′E
83 C22 **Grillenthal** Karas, SW Namibia 26°55′S 15°24′E
79 J15 **Grimari** Ouaka, C Central African Republic 05°44′N 20°02′E
Grimaylov *see* Hrymayliv
99 G18 **Grimbergen** Vlaams Brabant, C Belgium 50°56′N 04°22′E
183 N15 **Grim, Cape** headland Tasmania, SE Australia 40°42′S 144°42′E
100 N8 **Grimmen** Mecklenburg-Vorpommern, NE Germany 54°06′N 13°03′E
14 G16 **Grimsby** Ontario, S Canada 43°12′N 79°35′W
97 O17 **Grimsby** *prev.* Great Grimsby. E England, United Kingdom 53°35′N 00°05′W
92 J1 **Grímsey** *var.* Grimsey. island N Iceland
Grimsey *see* Grímsey
11 O12 **Grimshaw** Alberta, W Canada 56°11′N 117°37′W
95 F18 **Grimstad** Aust-Agder, S Norway 58°20′N 08°35′E
92 H4 **Grindavík** Suðurnes, W Iceland 65°57′N 18°10′W
108 F9 **Grindelwald** Bern, S Switzerland 46°37′N 08°04′E
95 F23 **Grindsted** Syddtjylland, W Denmark 55°46′N 08°56′E
29 W14 **Grinnell** Iowa, C USA 41°44′N 92°43′W
109 U10 **Grintovec** ▲ N Slovenia 46°21′N 14°31′E
Griquatown *see* Griekwastad
9 N4 **Grise Fiord** *var* Aujuittuq. Northwest Territories, Ellesmere Island, N Canada 76°10′N 83°15′W
182 H1 **Griselda, Lake** salt lake South Australia
Grisons *see* Graubünden
95 P14 **Grisslehamn** Stockholm, C Sweden 60°04′N 18°50′E
29 T15 **Griswold** Iowa, C USA 41°14′N 95°08′W
102 M1 **Griz Nez, Cap** headland N France 50°51′N 01°34′E
112 P13 **Grljan** Serbia, E Serbia 43°52′N 22°18′E
112 E11 **Grmeč** ▲▲ NW Bosnia and Herzegovina
99 H16 **Grobbendonk** Antwerpen, N Belgium 51°12′N 04°41′E
Grobin *see* Grobiņa
118 C10 **Grobiņa** *Ger.* Grobin. W Latvia 56°32′N 21°12′E
83 K20 **Groblersdal** Mpumalanga, NE South Africa 25°15′S 29°25′E
83 G23 **Groblershoop** Northern Cape, W South Africa 28°51′S 22°01′E
Gródek Jagielloński *see* Horodok
109 Q6 **Grödig** Salzburg, W Austria 47°42′N 13°06′E
111 H15 **Grodków** Opolskie, S Poland 50°42′N 17°23′E
Grodnenskaya Oblast' *see* Hrodzyenskaya Voblasts'
Grodno *see* Hrodna
110 L12 **Grodzisk Mazowiecki** Mazowieckie, C Poland 52°09′N 20°38′E
110 F12 **Grodzisk Wielkopolski** Wielkopolskie, C Poland 52°13′N 16°21′E
Grodzyanka *see* Hradzyanka
98 O12 **Groenlo** Gelderland, E Netherlands 52°02′N 06°36′E
83 E22 **Groenrivier** Karas, SE Namibia 27°27′S 18°52′E
25 U8 **Groesbeck** Texas, SW USA 31°31′N 96°35′W
98 L13 **Groesbeek** Gelderland, SE Netherlands 51°47′N 05°56′E
102 G7 **Groix, Îles de** island group NW France
110 M12 **Grójec** Mazowieckie, C Poland 51°51′N 20°52′E
65 K15 **Gröll Seamount** undersea feature C Atlantic Ocean 12°54′S 33°24′W
100 E13 **Gronau** *var.* Gronau in Westfalen. Nordrhein-Westfalen, NW Germany 52°13′N 07°02′E
Gronau in Westfalen *see* Gronau
93 F15 **Grong** Nord-Trøndelag, C Norway 64°29′N 12°19′E
95 N22 **Grönhögen** Kalmar, S Sweden 56°16′N 16°09′E
98 N5 **Groningen** Groningen, NE Netherlands 53°13′N 06°35′E
55 W9 **Groningen** Saramacca, N Suriname 05°45′N 55°31′W
98 N5 **Groningen** ◈ province NE Netherlands
Grønland *see* Greenland
108 H11 **Grono** Graubünden, S Switzerland 46°15′N 09°07′E
95 M20 **Grönskåra** Kalmar, S Sweden 57°04′N 15°45′E
25 O2 **Groom** Texas, SW USA 35°12′N 101°06′W
35 W9 **Groom Lake** ⊛ Nevada, W USA
83 H25 **Groot** ≈ S South Africa
181 S2 **Groote Eylandt** island Northern Territory, N Australia
98 M6 **Grootegast** Groningen, NE Netherlands 53°11′N 06°12′E
83 D17 **Grootfontein** Otjozondjupa, N Namibia 19°32′S 18°05′E
83 E22 **Groot Karasberge** ▲▲ S Namibia
Groot Karoo *see* Great Karoo
Groot-Kei *see* Nciba
15 V6 **Grosses-Roches** Québec, SE Canada 48°55′N 67°06′W
109 V2 **Gross-Siegharts** Niederösterreich, N Austria 48°48′N 15°25′E
45 T10 **Gros Islet** N Saint Lucia 14°04′N 60°57′W
44 L8 **Gros-Morne** NW Haiti 19°45′N 72°46′W
13 S11 **Gros Morne** ▲ Newfoundland, Newfoundland and Labrador, E Canada 49°38′N 57°45′W
103 R9 **Grosne** ≈ C France
45 S12 **Gros Piton** ▲ SW Saint Lucia 13°48′N 61°04′W
Grossa, Isola *see* Dugi Otok
Grossbetschkerek *see* Zrenjanin
Grosse Isper *see* Grosse Ysper
Grosse Kokel *see* Târnava Mare
101 M21 **Grosse Laaber** *var.* Grosse Laber. ≈ SE Germany
Grosse Laber *see* Grosse Laaber
Grosse Morava *see* Velika Morava
101 O15 **Grossenhain** Sachsen, E Germany 51°18′N 13°31′E
109 Y4 **Grossenzersdorf** Niederösterreich, NE Austria 48°12′N 16°33′E
101 O21 **Grosser Arber** ▲ SE Germany 49°07′N 13°10′E
101 K17 **Grosser Beerberg** ▲ C Germany 50°39′N 10°45′E
101 G18 **Grosser Feldberg** ▲ W Germany 50°13′N 08°28′E
109 O8 **Grosser Löffler** *It.* Monte Lovello. ▲ Austria/Italy 47°02′N 11°56′E
109 N8 **Grosser Möseler** *var.* Mesule. ▲ Austria/Italy 47°01′N 11°52′E
100 J8 **Grosser Plöner See** ⊛ N Germany
101 O21 **Grosser Rachel** ▲ SE Germany 48°59′N 13°23′E
Grosser Sund *see* Suur Väin
109 P8 **Grosses Wiesbachhorn** *var.* Wiesbachhorn. ▲ W Austria 47°09′N 12°44′E
106 F13 **Grosseto** Toscana, C Italy 42°45′N 11°07′E
101 M22 **Grosse Vils** ≈ SE Germany
109 U4 **Grosse Ysper** *var.* Grosse Isper. ≈ N Austria
101 G19 **Gross-Gerau** Hessen, W Germany 49°55′N 08°28′E
109 U3 **Gross Gerungs** Niederösterreich, N Austria 48°33′N 14°58′E
109 P8 **Grossglockner** ▲ W Austria 47°05′N 12°39′E
Grosskanizsa *see* Nagykanizsa
Gross-Karol *see* Carei
Grosskikinda *see* Kikinda
109 W9 **Grossklein** Steiermark, SE Austria 46°43′N 15°24′E
Grosskoppe *see* Velká Deštná
Grossmeseritsch *see* Velké Meziříčí
Grossmichel *see* Michalovce
101 H19 **Grossostheim** Bayern, C Germany 49°54′N 09°03′E
109 X7 **Grosspetersdorf** Burgenland, SE Austria 47°15′N 16°19′E
109 T5 **Grossraming** Oberösterreich, C Austria 47°54′N 14°34′E
101 P14 **Grossräschen** Brandenburg, E Germany 51°34′N 14°00′E
Grossrauschenbach *see* Revúca
Gross-Sankt-Johannis *see* Suure-Jaani
Gross-Schlatten *see* Abrud
Gross-Skaisgirren *see* Bol'shakovo
Gross-Steffelsdorf *see* Rimavská Sobota
Gross Strehlitz *see* Strzelce Opolskie
109 O8 **Grossvenediger** ▲ W Austria 47°07′N 12°19′E
Grosswardein *see* Oradea
Gross Wartenberg *see* Syców
109 U11 **Grosuplje** C Slovenia 46°00′N 14°36′E
99 H17 **Grote Nete** ≈ N Belgium
94 E10 **Grotli** Oppland, S Norway 62°02′N 07°36′E
19 N13 **Groton** Connecticut, NE USA 41°20′N 72°03′W
29 P8 **Groton** South Dakota, N USA 45°27′N 98°06′W
107 P18 **Grottaglie** Puglia, SE Italy 40°32′N 17°26′E
107 L17 **Grottaminarda** Campania, S Italy 41°04′N 15°02′E
106 K13 **Grottammare** Marche, C Italy 43°00′N 13°52′E
21 U5 **Grottoes** Virginia, NE USA 38°16′N 78°49′W
98 L6 **Grou** *Dutch.* Grouw. Fryslân, N Netherlands 53°07′N 05°51′E
13 N10 **Groulx, Monts** ▲▲ Québec, E Canada
14 E7 **Groundhog** ≈ Ontario, S Canada
36 J1 **Grouse Creek** Utah, W USA 41°41′N 113°52′W
36 J1 **Grouse Creek Mountains** ▲▲ Utah, W USA
Grouw *see* Grou
27 R8 **Grove** Oklahoma, C USA 36°35′N 94°46′W
31 S13 **Grove City** Ohio, N USA 39°52′N 83°05′W
18 B13 **Grove City** Pennsylvania, NE USA 41°09′N 80°02′W
23 O6 **Grove Hill** Alabama, S USA 31°42′N 87°46′W
33 S15 **Grover** Wyoming, C USA 42°48′N 110°57′W
35 P13 **Grover City** California, W USA 35°08′N 120°37′W
25 Y11 **Groves** Texas, SW USA 29°57′N 93°55′W
19 O7 **Groveton** New Hampshire, NE USA 44°35′N 71°28′W
25 W9 **Groveton** Texas, SW USA 31°04′N 95°08′W
36 J15 **Growler Mountains** ▲▲ Arizona, SW USA
Grozdovo *see* Bratya Daskalovi
127 P16 **Groznyy** Chechenskaya Respublika, SW Russian Federation 43°20′N 45°43′E
Grubeshov *see* Hrubieszów
112 G9 **Grubišno Polje** Bjelovar-Bilogora, NE Croatia 45°42′N 17°09′E
Grudovo *see* Sredets
110 J9 **Grudziądz** *Ger.* Graudenz. Kujawsko-pomorskie, C Poland 53°29′N 18°45′E
25 R17 **Grulla** *var.* La Grulla. Texas, SW USA 26°15′N 98°37′W
40 K14 **Grullo** Jalisco, SW Mexico 19°45′N 104°15′W
67 V10 **Grumeti** ≈ N Tanzania
95 K16 **Grums** Värmland, C Sweden 59°22′N 13°11′E
109 S5 **Grünau im Almtal** Oberösterreich, N Austria 47°51′N 13°56′E
101 H17 **Grünberg** Hessen, W Germany 50°36′N 08°57′E
Grünberg/Grünberg in Schlesien *see* Zielona Góra
92 H3 **Grundarfjörður** Vestfirðir, W Iceland 64°55′N 23°15′W
21 P7 **Grundy** Virginia, NE USA 37°17′N 82°06′W
29 W13 **Grundy Center** Iowa, C USA 42°21′N 92°46′W
Grüneberg *see* Zielona Góra
25 N1 **Gruver** Texas, SW USA 36°16′N 101°24′W
108 C9 **Gruyère, Lac de la** *Ger.* Greyerzer See. ⊛ SW Switzerland
108 C9 **Gruyères** Fribourg, W Switzerland 46°34′N 07°04′E
118 E11 **Gruzdžiai** Šiauliai, N Lithuania 56°06′N 23°15′E
Gruzinskaya SSR/Gruziya *see* Georgia
Gryada Akkyr *see* Akgyr Erezi
126 L7 **Gryazi** Lipetskaya Oblast', W Russian Federation 52°27′N 39°56′E
124 M14 **Gryazovets** Vologodskaya Oblast', NW Russian Federation 58°52′N 40°12′E
111 M17 **Grybów** Małopolskie, SE Poland 49°35′N 20°54′E
94 M13 **Grycksbo** Dalarna, C Sweden 60°40′N 15°30′E
110 E8 **Gryfice** *Ger.* Greifenberg, Greifenberg in Pommern. Zachodnio-pomorskie, NW Poland 53°55′N 15°11′E
110 D9 **Gryfino** *Ger.* Greifenhagen. Zachodnio-pomorskie, NW Poland 53°15′N 14°30′E
92 H9 **Gryllefjord** Troms, N Norway 69°21′N 17°07′E
95 L15 **Grythyttan** Örebro, C Sweden 59°52′N 14°31′E
108 D10 **Gstaad** Bern, W Switzerland 46°30′N 07°16′E
43 P14 **Guabito** Bocas del Toro, NW Panama 09°30′N 82°35′W
44 G7 **Guacanayabo, Golfo de** gulf S Cuba
40 I7 **Guachochi** Chihuahua, N Mexico
104 J11 **Guadajira** ≈ SW Spain
104 M13 **Guadajoz** ≈ S Spain
40 L13 **Guadalajara** Jalisco, C Mexico 20°43′N 103°24′W
105 O8 **Guadalajara** *Ar.* Wad Al-Hajarah; *anc.* Arriaca. Castilla-La Mancha, C Spain 40°37′N 03°10′W
105 O7 **Guadalajara** ◈ province Castilla-La Mancha, C Spain
104 K12 **Guadalcanal** Andalucía, S Spain 38°06′N 05°49′W
186 L10 **Guadalcanal** *off.* Guadalcanal Province. ◈ province C Solomon Islands
186 M9 **Guadalcanal** island C Solomon Islands
Guadalcanal Province *see* Guadalcanal
105 O12 **Guadalén** ≈ S Spain
105 R13 **Guadalentín** ≈ SE Spain
104 K15 **Guadalete** ≈ SW Spain
105 O13 **Guadalimar** ≈ S Spain
105 P12 **Guadalmena** ≈ S Spain
104 L11 **Guadalmez** ≈ W Spain
105 S7 **Guadalope** ≈ E Spain
104 K13 **Guadalquivir** ≈ W Spain
104 J14 **Guadalquivir, Marismas del** *var.* Las Marismas. wetland SW Spain
40 M11 **Guadalupe** Zacatecas, C Mexico 22°47′N 102°30′W
57 E16 **Guadalupe** Ica, W Peru 13°59′S 75°49′W
104 L10 **Guadalupe** Extremadura, W Spain 39°26′N 05°18′W
36 L14 **Guadalupe** Arizona, SW USA 33°20′N 111°57′W
35 P13 **Guadalupe** California, W USA 34°55′N 120°34′W
Guadalupe *see* Canelones
40 J3 **Guadalupe Bravos** Chihuahua, N Mexico 31°22′N 106°04′W
40 A4 **Guadalupe, Isla** island NW Mexico
37 U15 **Guadalupe Mountains** ▲▲ New Mexico/Texas, SW USA
24 J8 **Guadalupe Peak** ▲ Texas, SW USA 31°53′N 104°51′W
25 R11 **Guadalupe River** ≈ SW USA
104 K10 **Guadalupe, Sierra de** ▲▲ W Spain
40 K9 **Guadalupe Victoria** Durango, C Mexico 24°30′N 104°08′W
40 I8 **Guadalupe y Calvo** Chihuahua, N Mexico 26°04′N 106°58′W
105 N7 **Guadarrama** Madrid, C Spain 40°40′N 04°06′W
105 N7 **Guadarrama** ≈ C Spain
104 M7 **Guadarrama, Puerto de** pass C Spain
105 N9 **Guadarrama, Sierra de** ▲▲ C Spain
105 Q9 **Guadazaón** ≈ C Spain
45 X10 **Guadeloupe** ◇ French overseas department E West Indies
47 S3 **Guadeloupe** island group E West Indies
45 W10 **Guadeloupe Passage** passage E Caribbean Sea
104 H13 **Guadiana** ≈ Portugal/Spain
105 O13 **Guadiana Menor** ≈ S Spain
105 Q8 **Guadiela** ≈ C Spain
105 O14 **Guadix** Andalucía, S Spain 37°19′N 03°08′W
Guad-i-Zirreh *see* Gowd-e Zereh, Dasht-e
193 T12 **Guafo Fracture Zone** tectonic feature SE Pacific Ocean
63 F18 **Guafo, Isla** island S Chile
42 I6 **Guaimaca** Francisco Morazán, C Honduras 14°34′N 86°49′W
54 J12 **Guainía** *off.* Comisaría del Guainía. ◈ province E Colombia
Guainía, Comisaría del *see* Guainía
54 K12 **Guainía, Río** ≈ Colombia/Venezuela
55 O9 **Guaiquinima, Cerro** elevation SE Venezuela
60 G10 **Guaíra** Paraná, S Brazil 24°05′S 54°15′W
60 L7 **Guaíra** São Paulo, S Brazil 20°17′S 48°21′W
62 O7 **Guairá** *off.* Departamento del Guairá. ◈ department S Paraguay
Guairá, Departamento del *see* Guairá
Guaire *see* Gorey
63 F18 **Guaiteca, Isla** island S Chile
44 G6 **Guajaba, Cayo** headland C Cuba 21°50′N 77°33′W
59 D16 **Guajará-Mirim** Rondônia, W Brazil 10°50′S 65°21′W
Guajira *see* La Guajira
Guajira, Departamento de La *see* La Guajira
54 H3 **Guajira, Península de la** peninsula N Colombia
42 J6 **Gualaco** Olancho, C Honduras 15°00′N 86°03′W
34 L7 **Gualala** California, W USA 38°45′N 123°33′W
42 E5 **Gualán** Zacapa, C Guatemala 15°06′N 89°22′W
61 C19 **Gualeguay** Entre Ríos, E Argentina 33°09′S 59°20′W
61 D18 **Gualeguaychú** Entre Ríos, E Argentina 33°03′S 58°31′W
61 C18 **Gualeguay, Río** ≈ E Argentina
63 K16 **Gualicho, Salina del** salt lake E Argentina
188 B15 **Guam** ◇ US unincorporated territory W Pacific Ocean
63 F19 **Guamblin, Isla** island Archipiélago de los Chonos, S Chile
61 A22 **Guaminí** Buenos Aires, E Argentina 37°01′S 62°28′W
40 H8 **Guamúchil** Sinaloa, C Mexico 25°23′N 108°01′W
54 H4 **Guana** *var.* Misión de Guana. Zulia, NW Venezuela 11°07′N 72°17′W
44 C4 **Guanabacoa** La Habana, W Cuba 23°02′N 82°12′W
42 K13 **Guanacaste** *off.* Provincia de Guanacaste. ◈ province NW Costa Rica
42 K12 **Guanacaste, Cordillera de** ▲▲ NW Costa Rica
Guanacaste, Provincia de *see* Guanacaste
40 J8 **Guanaceví** Durango, C Mexico 25°55′N 105°51′W
44 A5 **Guanahacabibes, Golfo de** gulf W Cuba
42 K4 **Guanaja, Isla de** island Islas de la Bahía, N Honduras
44 C4 **Guanajay** La Habana, W Cuba 22°56′N 82°42′W
41 N12 **Guanajuato** Guanajuato, C Mexico 21°N 101°19′W
40 M12 **Guanajuato** ◈ state C Mexico
54 J6 **Guanare** Portuguesa, N Venezuela 09°04′N 69°45′W
54 K7 **Guanare, Río** ≈ W Venezuela
54 J6 **Guanarito** Portuguesa, NW Venezuela 08°43′N 69°12′W
160 M3 **Guancen Shan** ▲▲ C China
62 I9 **Guandacol** La Rioja, W Argentina 29°32′S 68°37′W
44 A5 **Guane** Pinar del Río, W Cuba 22°12′N 84°05′W
161 N14 **Guangdong** *var.* Guangdong Sheng, Kuang-tung, Kwangtung, Yue. ◈ province S China
Guangdong Sheng *see* Guangdong
Guanghua *see* Laohekou
Guangju *see* Gwangju
161 Q9 **Guangming Ding** ▲ Anhui, China 30°06′N 118°04′E
160 I13 **Guangnan** *var.* Liancheng. Yunnan, SW China 24°07′N 104°54′E
161 N8 **Guangshui** *prev.* Yinshan. Hubei, C China 31°41′N 113°53′E
Guangxi *see* Guangxi Zhuangzu Zizhiqu
160 K14 **Guangxi Zhuangzu Zizhiqu** *var.* Guangxi, Gui, Kuang-hsi, Kwangsi, *Eng.* Kwangsi Chuang Autonomous Region. ◈ autonomous region S China
160 J8 **Guangyuan** *var.* Kuang-yuan, Kwangyuan. Sichuan, C China 32°27′N 105°49′E
161 N14 **Guangzhou** *var.* Kuang-chou, Kwangchow, *Eng.* Canton. province capital Guangdong, S China 23°11′N 113°19′E
59 N19 **Guanhães** Minas Gerais, SE Brazil 18°46′S 42°58′W
160 I12 **Guanling** *var.* Guanling Bouyeizu Miaozu Zizhixian. Guizhou, S China 26°00′N 105°40′E
Guanling Bouyeizu Miaozu Zizhixian *see* Guanling
55 N5 **Guanta** Anzoátegui, NE Venezuela 10°15′N 64°38′W
44 J8 **Guantánamo** Guantánamo, SE Cuba 20°06′N 75°16′W
44 J8 **Guantánamo, Bahia de** *Eng.* Guantanamo Bay. US military base SE Cuba20°06′N 75°16′W
Guantanamo Bay *see* Guantánamo, Bahia de
Guanxian/Guan Xian *see* Dujiangyan
161 Q6 **Guanyun** *var.* Yishan. Jiangsu, E China 34°18′N 119°14′E
54 C12 **Guapí** Cauca, SW Colombia 02°36′N 77°54′W
43 N13 **Guápiles** Limón, NE Costa Rica 10°15′N 83°46′W
61 I15 **Guaporé** Rio Grande do Sul, S Brazil 28°55′S 51°53′W
47 S8 **Guaporé, Rio** *var.* Río Iténez. ≈ Bolivia/Brazil *see also* Río Iténez
Guaporé, Rio *see* Iténez, Río
56 B7 **Guaranda** Bolívar, C Ecuador 01°35′S 78°59′W
60 H11 **Guaraniaçu** Paraná, S Brazil 25°05′S 52°52′W
59 O20 **Guarapari** Espírito Santo, SE Brazil 20°39′S 40°31′W
60 I12 **Guarapuava** Paraná, S Brazil 25°22′S 51°28′W
60 J8 **Guararapes** São Paulo, S Brazil 21°16′S 50°37′W
105 S4 **Guara, Sierra de** ▲▲ NE Spain
60 N10 **Guaratinguetá** São Paulo, S Brazil 22°44′S 45°16′W
104 I7 **Guarda** Guarda, N Portugal 40°32′N 07°17′W
104 I7 **Guarda** ◈ district N Portugal
Guardak *see* Magdanly
104 M3 **Guardo** Castilla y León, N Spain 42°48′N 04°50′W
104 K11 **Guareña** Extremadura, W Spain 38°51′N 06°06′W
60 J11 **Guaricana, Pico** ▲ S Brazil 25°13′S 48°50′W
54 L6 **Guárico** *off.* Estado Guárico. ◈ state N Venezuela
Guárico, Estado *see* Guárico
44 J7 **Guarico, Punta** headland E Cuba 20°36′N 74°43′W
54 L7 **Guárico, Río** ≈ C Venezuela
60 M10 **Guarujá** São Paulo, SE Brazil 23°50′S 46°27′W
61 L22 **Guarulhos** ✈ (São Paulo) São Paulo, S Brazil 23°23′S 46°32′W
43 R17 **Guarumal** Veraguas, S Panama 07°48′N 81°15′W
Guasapa *see* Guasopa
40 H8 **Guasave** Sinaloa, C Mexico 25°33′N 108°29′W
54 I8 **Guasdualito** Apure, C Venezuela 07°15′N 70°40′W
55 Q7 **Guasipati** Bolívar, E Venezuela 07°28′N 61°58′W
186 I9 **Guasopa** *var.* Guasapa. Woodlark Island, SE Papua New Guinea 09°12′S 152°58′E
106 F9 **Guastalla** Emilia-Romagna, C Italy 44°54′N 10°38′E
42 D6 **Guastatoya** *var.* El Progreso. El Progreso, C Guatemala 14°51′N 90°01′W
42 D5 **Guatemala** *off.* Republic of Guatemala. ◆ republic Central America
42 A2 **Guatemala** *off.* Departamento de Guatemala. ◈ department S Guatemala
193 S7 **Guatemala Basin** undersea feature E Pacific Ocean 11°00′N 95°00′W
Guatemala City *see* Ciudad de Guatemala
Guatemala, Departamento de *see* Guatemala
Guatemala, Republic of *see* Guatemala
45 V14 **Guatuaro Point** headland Trinidad, Trinidad and Tobago 10°19′N 60°58′W
186 B8 **Guavi** ≈ SW Papua New Guinea
54 G13 **Guaviare** *off.* Comisaría Guaviare. ◈ province S Colombia
Guaviare, Comisaría *see* Guaviare
54 J11 **Guaviare, Río** ≈ E Colombia
61 E15 **Guaviravi** Corrientes, NE Argentina 29°20′S 56°50′W
54 G12 **Guayabero, Río** ≈ SW Colombia
45 U6 **Guayama** E Puerto Rico 17°59′N 66°07′W
42 J7 **Guayambre, Río** ≈ S Honduras
Guayanas, Macizo de las *see* Guiana Highlands
45 V6 **Guayanés, Punta** headland E Puerto Rico 18°03′N 65°48′W
42 J6 **Guayape, Río** ≈ C Honduras
56 B7 **Guayaquil** *var.* Santiago de Guayaquil. Guayas, SW Ecuador 02°13′S 79°54′W
Guayaquil *see* Simon Bolivar
56 A8 **Guayaquil, Golfo de** *var.* Gulf of Guayaquil. gulf SW Ecuador
Guayaquil, Gulf of *see* Guayaquil, Golfo de
56 A7 **Guayas** ◈ province W Ecuador
62 N7 **Guaycurú, Río** ≈ NE Argentina
40 F6 **Guaymas** Sonora, NW Mexico 27°56′N 110°54′W
45 U5 **Guaynabo** E Puerto Rico 18°19′N 66°05′W
80 H12 **Guba** Bīnishangul Gumuz, W Ethiopia 11°11′N 35°21′E
146 H8 **Gubadag** *Turkm.* Tel'man; *prev.* Tel'mansk. Daşoguz Welaýaty, N Turkmenistan 42°07′N 59°55′E
125 V13 **Gubakha** Permskiy Kray, NW Russian Federation 58°52′N 57°35′E
106 I12 **Gubbio** Umbria, C Italy 43°27′N 12°34′E
100 Q13 **Guben** *var.* Wilhelm-Pieck-Stadt. Brandenburg, E Germany 51°59′N 14°42′E
Guben *see* Gubin
110 D12 **Gubin** *Ger.* Guben. Lubuskie, W Poland 51°59′N 14°43′E
126 K8 **Gubkin** Belgorodskaya Oblast', W Russian Federation 51°16′N 37°32′E
162 J9 **Guchin-Us** *var.* Arguut. Övörhangay, C Mongolia 45°27′N 102°25′E
Gudara *see* Ghŭdara
105 S8 **Gúdar, Sierra de** ▲▲ E Spain
137 P8 **Gudauta** *prev.* Gudaut'a. NW Georgia 43°07′N 40°35′E
Gudaut'a *see* Gudauta
94 G12 **Gudbrandsdalen** valley S Norway
95 G21 **Gudenå** *var.* Gudenaa. ≈ C Denmark
Gudenaa *see* Gudenå
127 P16 **Gudermes** Chechenskaya Respublika, SW Russian Federation 43°23′N 46°06′E
155 J18 **Gūdūr** Andhra Pradesh, E India 14°10′N 79°51′E
146 B13 **Gudurolum** Balkan Welaýaty, W Turkmenistan 37°28′N 54°30′E
94 D13 **Gudvangen** Sogn Og Fjordane, S Norway 60°54′N 06°49′E
103 U7 **Guebwiller** Haut-Rhin, NE France 47°55′N 07°13′E
Guéckédou *see* Guékédou
14 K8 **Guéguen, Lac** ⊛ Québec, SE Canada
76 J15 **Guékédou** *var.* Guéckédou. Guinée-Forestière, S Guinea 08°33′N 10°08′W
41 R16 **Guelatao** Oaxaca, SE Mexico
Guelders *see* Gelderland
78 G11 **Guélengdeng** Mayo-Kébbi Est, W Chad 10°55′N 15°31′E
74 L5 **Guelma** *var.* Gâlma. NE Algeria 36°29′N 07°25′E
74 D8 **Guelmime** *var.* Goulimime. SW Morocco 28°59′N 10°10′W
14 G15 **Guelph** Ontario, S Canada 43°34′N 80°16′W
102 I7 **Guemené-Penfao** Loire-Atlantique, NW France 47°37′N 01°49′W
102 I7 **Guer** Morbihan, NW France 47°54′N 02°07′W
78 I11 **Guéra** *off.* Région du Guéra. ◈ region S Chad
102 H8 **Guérande** Loire-Atlantique, NW France 47°20′N 02°25′W
Guéra, Région du *see* Guéra
78 K9 **Guéréda** Wadi Fira, E Chad 14°30′N 22°05′E
103 N10 **Guéret** Creuse, C France 46°10′N 01°52′E
Guernica/Guernica y Lumo *see* Gernika-Lumo
33 Z15 **Guernsey** Wyoming, C USA 42°16′N 104°44′W
97 K25 **Guernsey** ◇ British Crown Dependency Channel Islands, NW Europe
97 K25 **Guernsey** island Channel Islands, NW Europe
76 J10 **Guérou** Assaba, S Mauritania 16°48′N 11°40′W
25 R16 **Guerra** Texas, SW USA 26°54′N 98°53′W
41 O15 **Guerrero** ◈ state S Mexico
40 D6 **Guerrero Negro** Baja California Sur, NW Mexico 27°56′N 114°04′W
103 P9 **Gueugnon** Saône-et-Loire, C France 46°36′N 04°03′E
76 M17 **Guéyo** S Ivory Coast 05°25′N 06°04′W
107 L15 **Guglionesi** Molise, C Italy 41°54′N 14°54′E
188 K5 **Guguan** island C Northern Mariana Islands
Guhrau *see* Góra
Gui *see* Guangxi Zhuangzu Zizhiqu
Guiana *see* French Guiana
47 V4 **Guiana Basin** undersea feature W Atlantic Ocean 11°00′N 52°00′W
48 G6 **Guiana Highlands** *var.* Macizo de las Guayanas. ▲▲ N South America
Guiba *see* Juba
102 I7 **Guichen** Ille-et-Vilaine, NW France 47°57′N 01°47′W
Guichi *see* Chizhou
61 E18 **Guichón** Paysandú, W Uruguay 32°30′S 57°13′W
77 U12 **Guidan-Roumji** Maradi, S Niger 13°40′N 06°41′E
Guidder *see* Guider
159 T10 **Guide** *var.* Heyin. Qinghai, C China 36°06′N 101°25′E
78 F12 **Guider** *var.* Guidder. Nord, N Cameroon 09°55′N 13°59′E
76 I11 **Guidimaka** ◈ region S Mauritania
77 W12 **Guidimouni** Zinder, S Niger 13°40′N 09°31′E
76 G10 **Guier, Lac de** *var.* Lac de Guiers. ⊛ N Senegal
Guiers, Lac de *see* Guier, Lac de
160 L14 **Guigang** *var.* Guixian, Gui Xian. Guangxi Zhuangzu Zizhiqu, S China 23°06′N 109°36′E
76 L16 **Guiglo** W Ivory Coast 06°33′N 07°29′W
54 L5 **Güigüe** Carabobo, N Venezuela 10°05′N 67°48′W
83 M20 **Guijá** Gaza, S Mozambique 24°31′S 33°02′E
42 E7 **Güija, Lago de** ⊛ El Salvador/Guatemala
160 L14 **Gui Jiang** *var.* Gui Shui. ≈ S China
104 K8 **Guijuelo** Castilla y León, N Spain 40°34′N 05°40′W
Guilan *see* Gīlān
97 N22 **Guildford** SE England, United Kingdom 51°14′N 00°35′W
19 R5 **Guildford** Maine, NE USA 45°10′N 69°22′W
19 O7 **Guildhall** Vermont, NE USA 44°34′N 71°36′W
103 R13 **Guilherand** Ardèche, E France 44°57′N 04°49′E
160 L13 **Guilin** *var.* Kuei-lin, Kweilin. Guangxi Zhuangzu Zizhiqu, S China 25°15′N 110°16′E
12 J6 **Guillaume-Delisle, Lac** ⊛ Québec, NE Canada
103 U13 **Guillestre** Hautes-Alpes, SE France 44°41′N 06°39′E
104 H6 **Guimarães** *var.* Guimaráes. Braga, N Portugal 41°26′N 08°19′W
Guimaráes *see* Guimarães
58 D11 **Guimarães Rosas, Pico** ▲ NW Brazil
23 N3 **Guin** Alabama, S USA 33°58′N 87°54′W
Güina *see* Wina
76 I14 **Guinea** *off.* Republic of Guinea, *var.* Guinée; *prev.* French Guinea, People's Revolutionary Republic of Guinea. ◆ republic W Africa
64 N13 **Guinea Basin** undersea feature E Atlantic Ocean 0°00′N 05°00′W
76 E12 **Guinea-Bissau** *off.* Republic of Guinea-Bissau, *Fr.* Guinée-Bissau, *Port.* Guiné-Bissau; *prev.* Portuguese Guinea. ◆ republic W Africa
Guinea-Bissau, Republic of *see* Guinea-Bissau
66 K7 **Guinea Fracture Zone** tectonic feature E Atlantic Ocean
64 O13 **Guinea, Gulf of** *Fr.* Golfe de Guinée. gulf E Atlantic Ocean
Guinea, People's Revolutionary Republic of *see* Guinea
Guinea, Republic of *see* Guinea
Guiné-Bissau *see* Guinea-Bissau
Guinée *see* Guinea
Guinée-Bissau *see* Guinea-Bissau
Guinée, Golfe de *see* Guinea, Gulf of
44 C4 **Güines** La Habana, W Cuba 22°50′N 82°02′W
102 G5 **Guingamp** Côtes d'Armor, NW France 48°34′N 03°09′W
Guipúzcoa *see* Gipuzkoa
44 C5 **Güira de Melena** La Habana, W Cuba 22°47′N 82°33′W
74 G8 **Guir, Hamada du** desert Algeria/Morocco
55 P5 **Güiria** Sucre, NE Venezuela 10°37′N 62°21′W
Gui Shui *see* Gui Jiang
104 H2 **Guitiriz** Galicia, NW Spain 43°10′N 07°52′W
77 N17 **Guitri** S Ivory Coast 05°31′N 05°14′W
171 Q5 **Guiuan** Samar, C Philippines 11°02′N 125°45′E
Gui Xian/Guixian *see* Guigang
160 J12 **Guiyang** *var.* Kuei-Yang, Kuei-yang, Kueyang, Kweiyang; *prev.* Kweichu. province capital Guizhou, S China 26°33′N 106°45′E
160 J12 **Guizhou** *var.* Guizhou Sheng, Kuei-chou, Kweichow, Qian. ◈ province S China
Guizhou Sheng *see* Guizhou
102 J13 **Gujan-Mestras** Gironde, SW France 44°39′N 01°04′W
154 B10 **Gujarāt** *var.* Gujerat. ◈ state W India
149 V6 **Gūjar Khān** Punjab, E Pakistan 33°19′N 73°23′E
Gujerat *see* Gujarāt
149 V7 **Gujrānwāla** Punjab, NE Pakistan 32°11′N 74°09′E
149 V7 **Gujrāt** Punjab, E Pakistan 32°34′N 74°04′E
146 B8 **Gulandag** *Rus.* Gory Kulandag. ▲▲ Balkan Welaýaty, W Turkmenistan
159 U9 **Gulang** Gansu, C China 37°31′N 102°55′E
183 R6 **Gulargambone** New South Wales, SE Australia 31°19′S 148°31′E
155 G15 **Gulbarga** Karnātaka, C India 17°22′N 76°47′E
118 J8 **Gulbene** *Ger.* Alt-Schwanenburg. NE Latvia 57°10′N 26°44′E
147 U10 **Gul'cha** *Kir.* Gülchö. Oshskaya Oblast', SW Kyrgyzstan 40°16′N 73°27′E
Gülchö *see* Gul'cha
173 T10 **Gulden Draak Seamount** undersea feature E Indian Ocean 33°45′S 101°00′E
136 J16 **Gülek Boğazı** *var.* Cilician Gates. pass S Turkey
186 D8 **Gulf** ◈ province S Papua New Guinea
23 O9 **Gulf Breeze** Florida, SE USA 30°21′N 87°09′W
Gulf of Liaotung *see* Liaodong Wan
23 V13 **Gulfport** Florida, SE USA 27°45′N 82°42′W
22 M9 **Gulfport** Mississippi, S USA 30°22′N 89°06′W
23 O9 **Gulf Shores** Alabama, S USA 30°15′N 87°40′W
183 R7 **Gulgong** New South Wales, SE Australia 32°22′S 149°31′E
160 I11 **Gulin** Sichuan, C China 28°06′N 105°47′E

◆ Country
● Country Capital
◇ Dependent Territory
○ Dependent Territory Capital
◈ Administrative Regions
✈ International Airport
▲ Mountain
▲▲ Mountain Range
Volcano
≈ River
⊛ Lake
⊠ Reservoir

171 U14 **Gulir** Pulau Kasiui, E Indonesia 04°27′S 131°41′E
Gulistan see Guliston
147 P10 **Guliston** *Rus.* Gulistan. Sirdaryo Viloyati, E Uzbekistan 40°29′N 68°46′E
163 T6 **Guliya Shan** ▲ NE China 49°42′N 122°22′E
Gulja see Yining
39 S11 **Gulkana** Alaska, USA 62°17′N 145°25′W
11 S17 **Gull Lake** Saskatchewan, S Canada 50°05′N 108°30′W
31 P10 **Gull Lake** ◎ Michigan, N USA
29 T6 **Gull Lake** ◎ Minnesota, N USA
95 L16 **Gullspång** Västra Götaland, S Sweden 58°58′N 14°04′E
136 B15 **Güllük Körfezi** *prev.* Akbük Limanı. *bay* W Turkey
152 H5 **Gulmarg** Jammu and Kashmir, NW India 34°04′N 74°25′E
Gulpaigan see Golpāyegān
99 L18 **Gulpen** Limburg, SE Netherlands 50°48′N 05°53′E
Gul'shad see Gul'shat
145 S13 **Gul'shat** *var.* Gul'shad. Karaganda, E Kazakhstan 46°37′N 74°22′E
81 F17 **Gulu** N Uganda 02°46′N 32°21′E
Gülübovo see Galabovo
114 I7 **Gulyantsi** Pleven, N Bulgaria 43°37′N 24°40′E
Gulyaypole see Hulyaypole
Guma see Pishan
Gümai see Darlag
79 K16 **Gumba** Equateur, NW Dem. Rep. Congo 02°58′N 21°23′E
Gumbinnen see Gusev
81 H24 **Gumbiro** Ruvuma, S Tanzania 10°19′S 35°40′E
146 B11 **Gumdag** *prev.* Kum-Dag. Balkan Welaýaty, W Turkmenistan 39°13′N 54°35′E
77 W12 **Gumel** Jigawa, N Nigeria 12°37′N 09°23′E
105 N5 **Gumiel de Hizán** Castilla y León, N Spain 41°46′N 03°42′W
153 P16 **Gumla** Jhārkhand, N India 23°03′N 84°36′E
Gumma see Gunma
101 F16 **Gummersbach** Nordrhein-Westfalen, W Germany 51°01′N 07°34′E
77 T13 **Gummi** Zamfara, NW Nigeria 12°07′N 05°07′E
Gumpolds see Humpolec
Gumti see Gomati
Gümülcine/Gümüljina see Komotiní
Gümüşane see Gümüşhane
137 O12 **Gümüşhane** *var.* Gümüşane, Gumushkhane. Gümüşhane, NE Turkey 40°31′N 39°27′E
137 O12 **Gümüşhane** *var.* Gümüşane, Gumushkhane. ◆ *province* NE Turkey
Gumushkhane see Gümüşhane
171 V14 **Gumzai** Pulau Kola, E Indonesia 05°27′S 134°38′E
154 H9 **Guna** Madhya Pradesh, C India 24°39′N 77°18′E
Gunabad see Gonābād
Gunan see Qijiang
Gunbad-i-Qawus see Gonbad-e Kāvūs
183 O9 **Gunbar** New South Wales, SE Australia 34°03′S 145°32′E
183 O9 **Gun Creek** *seasonal river* New South Wales, SE Australia
183 Q10 **Gundagai** New South Wales, SE Australia 35°06′S 148°03′E
79 K17 **Gundji** Equateur, N Dem. Rep. Congo 02°03′N 21°31′E
155 G20 **Gundlupet** Karnātaka, W India 11°48′N 76°42′E
136 G16 **Gündoğmuş** Antalya, S Turkey 36°50′N 32°07′E
137 O14 **Güney Doğu Toroslar** ▲▲ SE Turkey
79 J21 **Gungu** Bandundu, SW Dem. Rep. Congo 05°43′S 19°20′E
127 P17 **Gunib** Respublika Dagestan, SW Russian Federation 42°24′N 46°55′E
112 J11 **Gunja** Vukovar-Srijem, E Croatia 44°53′N 18°51′E
31 P9 **Gun Lake** ◎ Michigan, N USA
165 N12 **Gunma** *off.* Gunma-ken, *var.* Gumma. ◆ *prefecture* Honshū, S Japan
Gunma-ken see Gunma
197 P15 **Gunnbjørn Fjeld** *var.* Gunnbjørns Bjerge. ▲ C Greenland 69°03′N 29°36′W
Gunnbjørns Bjerge see Gunnbjørn Fjeld
183 S6 **Gunnedah** New South Wales, SE Australia 30°59′S 150°15′E
173 Y15 **Gunner's Quoin** *var.* Coin de Mire. *island* N Mauritius
37 R6 **Gunnison** Colorado, C USA 38°33′N 106°55′W
36 L5 **Gunnison** Utah, W USA 39°09′N 111°49′W
37 P5 **Gunnison River** ↝ Colorado, C USA
21 X2 **Gunpowder River** ↝ Maryland, NE USA
Güns see Kőszeg
163 X16 **Gunsan** *var.* Gunsan, *Jap.* Gunzan; *prev.* Kunsan. W South Korea 35°58′N 126°42′E
Gunsan see Gunsan
109 S4 **Gunskirchen** Oberösterreich, N Austria 48°07′N 13°54′E
Gunt see Ghund
155 H17 **Guntakal** Andhra Pradesh, C India 15°11′N 77°24′E
23 Q2 **Guntersville** Alabama, S USA 34°21′N 86°17′W
23 Q2 **Guntersville Lake** ⊠ Alabama, S USA
109 X4 **Guntramsdorf** Niederösterreich, E Austria 48°03′N 16°19′E
155 J16 **Guntūr** *var.* Guntur. Andhra Pradesh, SE India 16°20′N 80°27′E
168 H10 **Gunungsitoli** Pulau Nias, W Indonesia 01°11′N 97°35′E
155 M14 **Gunupur** Odisha, E India 19°04′N 83°52′E
101 J23 **Günz** ↝ S Germany
Gunzan see Gunsan
101 J22 **Günzburg** Bayern, S Germany 48°26′N 10°18′E
101 K21 **Gunzenhausen** Bayern, S Germany 49°07′N 10°45′E
Guolüezhen see Lingbao
Guovdageaidnu see Kautokeino
161 P7 **Guoyang** Anhui, E China 33°30′N 116°12′E
116 G11 **Gurahonț** *Hung.* Honctő. Arad, W Romania 46°16′N 22°21′E
Gurahumora see Gura Humorului
116 K9 **Gura Humorului** *Ger.* Gurahumora. Suceava, NE Romania 47°31′N 26°00′E
146 H8 **Gurbansoltan Eje** *prev.* Ýylanly, *Rus.* Il'yaly. Daşoguz Welaýaty, N Turkmenistan 41°57′N 59°42′E
158 K4 **Gurbantünggüt Shamo** *desert* W China
152 H7 **Gurdāspur** Punjab, N India 32°04′N 75°28′E
27 T13 **Gurdon** Arkansas, C USA 33°55′N 93°09′W
Gurdzhaani see Gurjaani
152 I10 **Gurgaon** Haryāna, N India 28°27′N 77°01′E
Gurgan see Gorgān
59 M15 **Gurguéia, Rio** ↝ NE Brazil
55 Q7 **Guri, Embalse de** ⊠ E Venezuela
137 V10 **Gurjaani** *Rus.* Gurdzhaani. E Georgia 41°42′N 45°47′E
109 T8 **Gurk** Kärnten, S Austria 46°52′N 14°17′E
109 T9 **Gurk** *Slvn.* Krka. ↝ S Austria
Gurkfeld see Krško
114 K9 **Gurkovo** *prev.* Kolupchii. Stara Zagora, C Bulgaria 42°42′N 25°46′E
109 S9 **Gurktaler Alpen** ▲▲ S Austria
146 H8 **Gurlan** *Rus.* Gurlen. Xorazm Viloyati, W Uzbekistan 41°54′N 60°18′E
Gurlen see Gurlan
83 M16 **Guro** Manica, C Mozambique 17°28′S 33°18′E
136 M14 **Gürün** Sivas, C Turkey 38°44′N 37°15′E
59 K16 **Gurupi** Tocantins, C Brazil 11°44′S 49°01′W
58 L12 **Gurupi, Rio** ↝ NE Brazil
152 E14 **Guru Sikhar** ▲ NW India 24°45′N 72°51′E
162 H8 **Gurvanbulag** *var.* Höviyn Am. Bayanhongor, C Mongolia 47°08′N 98°41′E
162 K7 **Gurvanbulag** *var.* Avdzaga. Bulgan, C Mongolia 47°43′N 103°30′E
162 I11 **Gurvantes** *var.* Urt. Ömnögovĭ, S Mongolia 43°16′N 101°00′E
Gur'yev/Gur'yevskaya Oblast' see Atyrau
77 U13 **Gusau** Zamfara, NW Nigeria 12°18′N 06°27′E
126 C3 **Gusev** *Ger.* Gumbinnen. Kaliningradskaya Oblast', W Russian Federation 54°36′N 22°14′E
146 J17 **Gushgy** *Rus.* Kushka. ↝ Mary Welaýaty, S Turkmenistan
Gushiago see Gushiegu
77 Q14 **Gushiegu** *var.* Gushiago. NE Ghana 09°54′N 00°12′W
165 S17 **Gushikawa** Okinawa, Okinawa, SW Japan 26°21′N 127°50′E
113 L16 **Gusinje** E Montenegro 42°34′N 19°51′E
126 M4 **Gus'-Khrustal'nyy** Vladimirskaya Oblast', W Russian Federation 55°39′N 40°42′E
107 B19 **Guspini** Sardegna, Italy, C Mediterranean Sea 39°33′N 08°39′E
109 X8 **Güssing** Burgenland, SE Austria 47°03′N 16°19′E
109 V6 **Gusswerk** Steiermark, E Austria 47°43′N 15°18′E
92 O2 **Gustav Adolf Land** *physical region* NE Svalbard
195 X5 **Gustav Bull Mountains** ▲▲ Antarctica
39 W13 **Gustavus** Alaska, USA 58°24′N 135°44′W
92 O1 **Gustav V Land** *physical region* NE Svalbard
35 P9 **Gustine** California, W USA 37°14′N 121°00′W
25 R8 **Gustine** Texas, SW USA 31°51′N 98°24′W
100 M9 **Güstrow** Mecklenburg-Vorpommern, NE Germany 53°48′N 12°12′E
95 N18 **Gusum** Östergötland, S Sweden 58°15′N 16°30′E
Guta/Gúta see Kolárovo
Gutenstein see Ravne na Koroškem
101 G14 **Gütersloh** Nordrhein-Westfalen, W Germany 51°54′N 08°23′E
27 N10 **Guthrie** Oklahoma, C USA 35°53′N 97°26′W
25 P5 **Guthrie** Texas, SW USA 33°38′N 100°21′W
29 U14 **Guthrie Center** Iowa, C USA 41°40′N 94°30′W
41 Q13 **Gutiérrez Zamora** Veracruz-Llave, E Mexico 20°29′N 97°07′W
Gutta see Kolárovo
29 Y12 **Guttenberg** Iowa, C USA 42°47′N 91°06′W
Guttentag see Dobrodzień
Guttstadt see Dobre Miasto
162 G8 **Guulin** Govĭ-Altay, C Mongolia 46°33′N 97°21′E
153 V12 **Guwāhāti** *prev.* Gauhāti. Assam, NE India 26°09′N 91°42′E
139 R3 **Guwēr** *var.* Al Kuwayr, Al Quwayr, Quwair. Arbīl, N Iraq 36°03′N 43°30′E
146 A10 **Guwlumaýak** *Rus.* Kuuli-Mayak. Balkan Welaýaty, NW Turkmenistan 40°14′N 52°43′E
55 R9 **Guyana** *off.* Co-operative Republic of Guyana; *prev.* British Guiana. ◆ *republic* N South America
Guyana, Co-operative Republic of see Guyana
21 P5 **Guyandotte River** ↝ West Virginia, NE USA
Guyane see French Guiana
Guyi see Sanjiang
26 H8 **Guymon** Oklahoma, C USA 36°42′N 101°30′W
146 K12 **Guynuk** Lebap Welaýaty, NE Turkmenistan 39°18′N 63°00′E
Guyong see Jiangle
21 O9 **Guyot, Mount** ▲ North Carolina/Tennessee, SE USA 35°42′N 83°15′W
183 U5 **Guyra** New South Wales, SE Australia 30°13′S 151°42′E
159 W10 **Guyuan** Ningxia, N China 35°57′N 106°13′E
Guzar see G'uzor
121 P2 **Güzelyurt** *Gk.* Kólpos Mórfu, Morphou. W Cyprus 35°12′N 33°E
121 N2 **Güzelyurt Körfezi** *var.* Morfou Bay, Morphou Bay, *Gk.* Kólpos Mórfou. *bay* W Cyprus
Guzhou see Rongjiang
40 I3 **Guzmán** Chihuahua, N Mexico 31°13′N 107°27′W
147 N13 **G'uzor** *Rus.* Guzar. Qashqadaryo Viloyati, S Uzbekistan 38°41′N 66°12′E
119 B14 **Gvardeysk** *Ger.* Tapaiu. Kaliningradskaya Oblast', W Russian Federation 54°39′N 21°02′E
Gvardeyskoye see Hvardiys'ke
183 R5 **Gwabegar** New South Wales, SE Australia 30°34′S 148°58′E
148 J16 **Gwādar** *var.* Gwadur. Baluchistān, SW Pakistan 25°09′N 62°21′E
148 J16 **Gwādar East Bay** *bay* SW Pakistan
148 J16 **Gwādar West Bay** *bay* SW Pakistan
Gwadur see Gwādar
83 J17 **Gwai** Matabeleland North, W Zimbabwe 19°17′S 27°37′E
154 I7 **Gwalior** Madhya Pradesh, C India 26°16′N 78°12′E
83 J18 **Gwanda** Matabeleland South, SW Zimbabwe 20°56′S 29°E
79 N15 **Gwane** Orientale, N Dem. Rep. Congo 04°40′N 25°51′E
163 X16 **Gwangju** *off.* Kwangju-gwangyŏksi, *var.* Guangju, Kwangchu, *Jap.* Kōshū; *prev.* Kwangju. SW South Korea 35°09′N 126°53′E
83 I17 **Gwayi** ↝ W Zimbabwe
110 G8 **Gwda** *var.* Glda, *Ger.* Küddow. ↝ NW Poland
97 C14 **Gweebarra Bay** *Ir.* Béal an Bheara. *inlet* W Ireland
97 D14 **Gweedore** *Ir.* Gaoth Dobhair. Donegal, NW Ireland 55°03′N 08°14′W
Gwelo see Gweru
97 K21 **Gwent** *cultural region* S Wales, United Kingdom
83 K17 **Gweru** *prev.* Gwelo. Midlands, C Zimbabwe 19°27′S 29°49′E
29 Q7 **Gwinner** North Dakota, N USA 46°10′N 97°42′W
77 Y13 **Gwoza** Borno, NE Nigeria 11°07′N 13°40′E
Gwy see Wye
183 R4 **Gwydir River** ↝ New South Wales, SE Australia
97 I19 **Gwynedd** *var.* Gwyneth. *cultural region* NW Wales, United Kingdom
Gwyneth see Gwynedd
159 O16 **Gyaca** *var.* Ngarrab. Xizang Zizhiqu, W China 29°06′N 92°37′E
Gya'gya see Saga
Gyaijêpozhanggê see Zhidoi
Gyaisi see Jiulong
115 M22 **Gyalí** *var.* Yialí. *island* Dodekánisa, Greece, Aegean Sea
Gyamotang see Dêngqên
Gyandzha see Gäncä
Gyangkar see Dinggyê
158 M16 **Gyangzê** Xizang Zizhiqu, W China 28°50′N 89°38′E
158 L14 **Gyaring Co** ◎ W China
159 Q12 **Gyaring Hu** ◎ C China
115 I20 **Gyáros** *var.* Yioúra. *island* Kykládes, Greece, Aegean Sea
122 J7 **Gyda** Yamalo-Nenetskiy Avtonomnyy Okrug, N Russian Federation 70°55′N 78°34′E
122 J7 **Gydanskiy Poluostrov** *Eng.* Gyda Peninsula. *peninsula* N Russian Federation
Gyda Peninsula see Gydanskiy Poluostrov
Gyêgu see Yushu
163 W15 **Gyeonggi-man** *prev.* Kyŏnggi-man. *bay* NW South Korea
163 Z16 **Gyeongju** *Jap.* Keishū; *prev.* Kyŏngju. SE South Korea 35°49′N 129°09′E
Gyéres see Câmpia Turzii
Gyergyószentmiklós see Gheorgheni
Gyergyótölgyes see Tulgheș
Gyertyámos see Cărpiniș
Gyeva see Detva
Gyigang see Zayü
Gyixong see Gonggar
95 I23 **Gyldenløveshøy** *hill range* C Denmark
181 Z10 **Gympie** Queensland, E Australia 26°05′S 152°40′E
166 L7 **Gyobingauk** Bago, SW Myanmar (Burma) 18°14′N 95°39′E
111 M23 **Gyomaendrőd** Békés, SE Hungary 46°56′N 20°50′E
Gyömbér see Ďumbier
111 L22 **Gyöngyös** Heves, NE Hungary 47°44′N 19°49′E
111 H22 **Győr** *Ger.* Raab, *Lat.* Arrabona. Győr-Moson-Sopron, NW Hungary 47°41′N 17°40′E
111 G22 **Győr-Moson-Sopron** *off.* Győr-Moson-Sopron Megye. ◆ *county* NW Hungary
Győr-Moson-Sopron Megye see Győr-Moson-Sopron
11 X15 **Gypsumville** Manitoba, S Canada 51°47′N 98°38′W
12 M4 **Gyrfalcon Islands** *island group* Northwest Territories, NE Canada
95 N14 **Gysinge** Gävleborg, C Sweden 60°16′N 16°55′E
115 F22 **Gýtheio** *var.* Githio; *prev.* Ýithion. Pelopónnisos, S Greece 36°46′N 22°34′E
146 L13 **Gyuichbirleshik** Lebap Welaýaty, E Turkmenistan 38°10′N 64°33′E
111 N24 **Gyula** *Rom.* Jula. Békés, SE Hungary 46°39′N 21°17′E
Gyulafehérvár see Alba Iulia
Gyulovo see Roza
137 T11 **Gyumri** *var.* Giumri, *Rus.* Kumayri; *prev.* Aleksandropol', Leninakan. W Armenia 40°48′N 43°51′E
146 D13 **Gyunuzyndag, Gora** ▲ Balkan Welaýaty, W Turkmenistan 38°15′N 56°25′E
146 J15 **Gyzylbaydak** *Rus.* Krasnoye Znamya. Mary Welaýaty, S Turkmenistan 36°51′N 62°24′E
Gyzyletrek see Etrek
146 D10 **Gyzylgaýa** *Rus.* Kizyl-Kaya. Balkan Welaýaty, NW Turkmenistan 40°37′N 55°15′E
146 A10 **Gyzylsuw** *Rus.* Kizyl-Su. Balkan Welaýaty, W Turkmenistan 39°49′N 53°00′E
Gyzyrlabat see Serdar
Gzhatsk see Gagarin

H

153 T12 **Ha** W Bhutan 27°17′N 89°22′E
Haabai see Ha'apai Group
99 H17 **Haacht** Vlaams Brabant, C Belgium 50°58′N 04°38′E
109 T4 **Haag** Niederösterreich, NE Austria 48°07′N 14°32′E
194 L8 **Haag Nunataks** ▲ Antarctica
92 N2 **Haakon VII Land** *physical region* NW Svalbard
98 O11 **Haaksbergen** Overijssel, E Netherlands 52°09′N 06°45′E
99 E14 **Haamstede** Zeeland, SW Netherlands 51°43′N 03°45′E
193 Y15 **Ha'ano** *island* Ha'apai Group, C Tonga
193 Y15 **Ha'apai Group** *var.* Haabai. *island group* C Tonga
93 L15 **Haapajärvi** Pohjois-Pohjanmaa, C Finland 63°45′N 25°20′E
93 L17 **Haapamäki** Pirkanmaa, C Finland 62°11′N 24°32′E
93 L15 **Haapavesi** Pohjois-Pohjanmaa, C Finland 64°09′N 25°25′E
191 N7 **Haapiti** Moorea, W French Polynesia 17°33′S 149°52′W
118 F4 **Haapsalu** *Ger.* Hapsal. Läänemaa, W Estonia 58°58′N 23°32′E
Ha'Arava see 'Arabah, Wādī al
95 G24 **Haarby** *var.* Hårby. Syddtjylland, C Denmark 55°13′N 10°07′E
98 H10 **Haarlem** *prev.* Harlem. Noord-Holland, W Netherlands 52°23′N 04°39′E
185 D19 **Haast** West Coast, South Island, New Zealand 43°53′S 169°02′E
185 C20 **Haast** ↝ South Island, New Zealand
185 D20 **Haast Pass** *pass* South Island, New Zealand
193 W16 **Ha'atua** 'Eau, E Tonga 21°23′S 174°57′W
149 P15 **Hab** ↝ SW Pakistan
141 W7 **Haba** *var.* Al Haba. Dubayy, NE United Arab Emirates 25°01′N 55°37′E
158 K2 **Habahe** *var.* Kaba. Xinjiang Uygur Zizhiqu, NW China 48°04′N 86°20′E
141 U13 **Ḥabarūt** *var.* Habrut. SW Oman 17°19′N 52°45′E
81 J18 **Habaswein** Isiolo, NE Kenya 01°01′N 39°27′E
99 L24 **Habay-la-Neuve** Luxembourg, SE Belgium 49°43′N 05°38′E
139 S8 **Ḥabbānīyah, Buḥayrat** ◎ C Iraq
Habelschwerdt see Bystrzyca Kłodzka
153 V14 **Habiganj** Sylhet, NE Bangladesh 24°23′N 91°25′E
163 Q12 **Habirag** Nei Mongol Zizhiqu, N China 42°18′N 115°40′E
95 L19 **Habo** Västra Götaland, S Sweden 57°55′N 14°05′E
123 V14 **Habomai Islands** *island group* Kuril'skiye Ostrova, SE Russian Federation
165 S2 **Haboro** Hokkaidō, NE Japan 44°19′N 141°42′E
153 S16 **Habra** West Bengal, NE India 22°39′N 88°17′E
Habrut see Ḥabarūt
143 P17 **Habshān** Abū Ẓaby, C United Arab Emirates 23°51′N 53°34′E
54 E14 **Hacha** Putumayo, S Colombia 0°02′S 75°30′W
165 X13 **Hachijō-jima** *island* Izu-shotō, SE Japan
164 L12 **Hachiman** Gifu, Honshū, SW Japan 35°46′N 136°57′E
165 P7 **Hachimori** Akita, Honshū, C Japan 40°22′N 139°59′E
165 R7 **Hachinohe** Aomori, Honshū, C Japan 40°30′N 141°29′E
165 X13 **Hachiōji** Tōkyō, Hachijō-jima, SE Japan 35°40′N 139°20′E
137 Y12 **Hacıqabal** *prev.* Qazimämmäd. SE Azerbaijan 40°03′N 48°56′E
93 G17 **Hackås** Jämtland, C Sweden 62°55′N 14°31′E
18 K14 **Hackensack** New Jersey, NE USA 40°51′N 73°57′W
75 U12 **Ḥadabat al Jilf al Kabīr** *var.* Gilf Kebir Plateau. *plateau* SW Egypt
Hadama see Nazrēt
141 W13 **Ḥadbaram** S Oman 17°27′N 55°13′E
139 U13 **Ḥaddānīyah** *well* S Iraq
96 K12 **Haddington** SE Scotland, United Kingdom 55°59′N 02°46′W
141 Z8 **Ḥadd, Ra's al** *headland* NE Oman 22°28′N 59°58′E
Haded see Xadeed
77 W12 **Hadejia** Jigawa, N Nigeria 12°22′N 10°02′E
77 W12 **Hadejia** ↝ N Nigeria
138 F9 **Hadera** *var.* Khadera; *prev.* Ḥadera. Haifa, N Israel 32°26′N 34°55′E
Ḥadera see Hadera
Hadersleben see Haderslev
95 G24 **Haderslev** *Ger.* Hadersleben. Syddanmark, SW Denmark 55°15′N 09°30′E
151 J21 **Hadhdhunmathi Atoll** *atoll* S Maldives
141 W17 **Ḥadībūh** Suquṭrā, SE Yemen 12°38′N 54°05′E
158 K9 **Hadilik** Xinjiang Uygur Zizhiqu, W China 37°51′N 86°10′E
136 H16 **Hadım** Konya, S Turkey 36°58′N 32°27′E
140 K7 **Hadīyah** Al Madīnah, W Saudi Arabia 25°36′N 38°31′E
78 H11 **Hadjer-Lamis** *off.* Région du Hadjer-Lamis. ◆ *region* SW Chad
Hadjer-Lamis, Région du see Hadjer-Lamis
8 L5 **Hadley Bay** *bay* Victoria Island, Nunavut, N Canada
167 S6 **Ha Đông** *var.* Hadong. Ha Tây, N Vietnam 20°58′N 105°46′E
Hadong see Ha Đông
Hadramaut see Ḥaḑramawt
141 R15 **Ḥaḑramawt** *Eng.* Hadramaut. ▲▲ S Yemen
Hadria see Adria
Hadrianopolis see Edirne
Hadria Picena see Apricena
95 G22 **Hadsten** Midtjylland, C Denmark 56°19′N 10°03′E
95 G21 **Hadsund** Nordjylland, N Denmark 56°43′N 10°08′E
117 S4 **Hadyach** *Rus.* Gadyach. Poltavs'ka Oblast', NE Ukraine 50°21′N 34°00′E
114 N9 **Hadzhiyska Reka** *var.* Khadzhiyska Reka. ↝ E Bulgaria
112 I13 **Hadžići** Federacija Bosne I Hercegovine, SE Bosnia and Herzegovina 43°49′N 18°12′E
163 W14 **Haeju** S North Korea 38°04′N 125°40′E
Haerbin/Haerhpin/Ha-erh-pin see Harbin
141 P5 **Ḩafar al Bāţin** Ash Sharqīyah, N Saudi Arabia 28°25′N 45°59′E
11 T15 **Hafford** Saskatchewan, S Canada 52°43′N 107°19′W
136 M13 **Hafik** Sivas, N Turkey 39°53′N 37°24′E
149 V8 **Ḥāfizābād** Punjab, E Pakistan 32°03′N 73°42′E
92 H4 **Hafnarfjörður** Höfuðborgarsvæðið, W Iceland 64°03′N 21°57′W
Hafnia see Denmark
Hafnia see København
Hafren see Severn
Hafun see Xaafuun
Hafun, Ras see Xaafuun, Raas
80 G10 **Hag 'Abdullah** Sinnar, E Sudan 13°59′N 33°35′E
81 K18 **Hagadera** Garissa, E Kenya 0°06′N 40°23′E
138 G8 **HaGalil** *Eng.* Galilee. ▲▲ N Israel
14 G10 **Hagar** Ontario, S Canada 46°27′N 80°22′W
155 G18 **Hagari** *var.* Vedāvati. ↝ W India
188 B16 **Hagåtña**, *var.* Agaña. ○ (Guam) NW Guam 13°27′N 144°45′E
100 M13 **Hagelberg** *hill* NE Germany
39 N14 **Hagemeister Island** *island* Alaska, USA
101 F15 **Hagen** Nordrhein-Westfalen, W Germany 51°22′N 07°27′E
100 K10 **Hagenow** Mecklenburg-Vorpommern, N Germany 53°27′N 11°10′E
10 K15 **Hagensborg** British Columbia, SW Canada 52°24′N 126°24′W
80 I13 **Hāgere Hiywet** *var.* Agere Hiywet, Ambo. Oromīya, C Ethiopia 09°00′N 37°55′E
33 O15 **Hagerman** Idaho, NW USA 42°48′N 114°53′W
37 U14 **Hagerman** New Mexico, SW USA 33°07′N 104°19′W
21 V2 **Hagerstown** Maryland, NE USA 39°39′N 77°44′W
14 G16 **Hagersville** Ontario, S Canada 42°58′N 80°03′W
102 J15 **Hagetmau** Landes, SW France 43°40′N 00°36′W
95 K14 **Hagfors** Värmland, C Sweden 60°03′N 13°45′E
93 G16 **Häggenås** Jämtland, C Sweden 63°24′N 14°53′E
164 E12 **Hagi** Yamaguchi, Honshū, SW Japan 34°25′N 131°22′E
167 S5 **Ha Giang** Ha Giang, N Vietnam 22°50′N 104°58′E
Hagios Evstrátios see Ágios Efstrátios
HaGolan see Golan Heights
103 T4 **Hagondange** Moselle, NE France 49°16′N 06°06′E
97 B18 **Hag's Head** *Ir.* Ceann Caillí. *headland* W Ireland 52°56′N 09°29′W
102 I3 **Hague, Cap de la** *headland* N France 49°43′N 01°56′W
103 V5 **Haguenau** Bas-Rhin, NE France 48°49′N 07°47′E
165 X16 **Hahajima-rettō** *island group* SE Japan
15 R8 **Há Há , Lac** ◎ Québec, SE Canada
172 H13 **Hahaya** ✈ (Moroni) Grande Comore, NW Comoros
22 K9 **Hahnville** Louisiana, S USA 29°58′N 90°24′W
83 E22 **Haib** Karas, S Namibia 28°12′S 18°19′E
Haibak see Aibak
149 N15 **Haibo** ↝ SW Pakistan
Haibowan see Wuhai
163 U12 **Haicheng** Liaoning, NE China 40°53′N 122°45′E
Haicheng see Haifeng
Haicheng see Haiyuan
Haida see Nový Bor
Haidarabad see Hyderābād
Haidenschaft see Ajdovščina
167 T6 **Hai Dương** Hai Hưng, N Vietnam 20°56′N 106°21′E
138 F9 **Haifa** ◆ *district* NW Israel
Haifa see Hefa
Haifa, Bay of see Mifrats Hefa
161 P14 **Haifeng** *var.* Haicheng. Guangdong, S China 22°56′N 115°19′E
Haifong see Hai Phong
161 P3 **Hai He** ↝ E China
Haikang see Leizhou
160 L17 **Haikou** *var.* Hai-k'ou, Hoihow, *Fr.* Hoï-Hao. *province capital* Hainan, S China 20°N 110°17′E
Hai-k'ou see Haikou
140 M6 **Ḩā'il** Ḩā'il, NW Saudi Arabia 27°N 42°50′E
141 N5 **Ḩā'il** *off.* Minţaqah Ḩā'il. ◆ *province* N Saudi Arabia
163 S6 **Hailar He** ↝ NE China
33 P14 **Hailey** Idaho, NW USA 43°31′N 114°18′W
14 H9 **Haileybury** Ontario, S Canada 47°27′N 79°39′W
163 X9 **Hailin** Heilongjiang, NE China 44°37′N 129°24′E
Ḩā'il, Minţaqah see Ḩā'il
Hailong see Meihekou
93 K14 **Hailuoto** *Swe.* Karlö. *island* W Finland
Haima see Haymā'
Haimen see Taizhou
160 M17 **Hainan** *var.* Hainan Sheng, Qiong. ◆ *province* S China
160 K17 **Hainan Dao** *island* S China
Hainan Sheng see Hainan
Hainan Strait see Qiongzhou Haixia
Hainasch see Ainaži
Hainau see Chojnów
99 E20 **Hainaut** ◆ *province* SW Belgium
Hainburg see Hainburg an der Donau
109 Z4 **Hainburg an der Donau** *var.* Hainburg. Niederösterreich, NE Austria 48°09′N 16°57′E
39 W12 **Haines** Alaska, USA 59°13′N 135°27′W
32 L12 **Haines** Oregon, NW USA 44°53′N 117°56′W
23 W12 **Haines City** Florida, SE USA 28°06′N 81°37′W
10 H8 **Haines Junction** Yukon, W Canada 60°45′N 137°30′W
109 W4 **Hainfeld** Niederösterreich, NE Austria 48°03′N 15°47′E
101 N16 **Hainichen** Sachsen, E Germany 50°58′N 13°08′E
Hai Ninh see Mong Cai
167 T6 **Hai Phong** *var.* Haifong, Haiphong. N Vietnam 20°50′N 106°41′E
Haiphong see Hai Phong
161 S12 **Haitan Dao** *island* SE China
44 K8 **Haiti** *off.* Republic of Haiti. ◆ *republic* C West Indies
Haiti, Republic of see Haiti
35 T11 **Haiwee Reservoir** ⊠ California, W USA
80 I7 **Haiya** Red Sea, NE Sudan 18°17′N 36°21′E
159 T10 **Haiyan** *var.* Sanjiaocheng. Qinghai, W China 36°55′N 100°54′E
160 M13 **Haiyang Shan** ▲▲ S China
159 V10 **Haiyuan** Ningxia, N China 36°32′N 105°31′E
Hajda see Nový Bor
111 M22 **Hajdú-Bihar** *off.* Hajdú-Bihar Megye. ◆ *county* E Hungary
Hajdú-Bihar Megye see Hajdú-Bihar
111 N22 **Hajdúböszörmény** Hajdú-Bihar, E Hungary 47°39′N 21°32′E
111 N22 **Hajdúhadház** Hajdú-Bihar, E Hungary 47°40′N 21°40′E
111 N21 **Hajdúnánás** Hajdú-Bihar, E Hungary 47°50′N 21°26′E
111 N22 **Hajdúszoboszló** Hajdú-Bihar, E Hungary 47°27′N 21°24′E
142 I3 **Ḥājī Ebrāhīm, Kūh-e** ▲ Iran/Iraq 36°53′N 44°56′E
165 O9 **Hajiki-zaki** *headland* Sado, C Japan 38°19′N 138°28′E
Hajine see Abū Ḩardān
153 P13 **Hājīpur** Bihār, N India 25°41′N 85°13′E
141 N14 **Ḥajjah** W Yemen 15°43′N 43°33′E
139 U11 **Ḥajjama** Al Muthannā, S Iraq 31°24′N 45°20′E
143 R12 **Ḥājjīābād** Hormozgān, C Iran
139 U14 **Ḩājj, Thaqb al** *well* S Iraq
113 L16 **Hajla** ▲ E Montenegro
110 P10 **Hajnówka** *Ger.* Hermhausen. Podlaskie, NE Poland 52°45′N 23°32′E
Haka see Hakha
Hakapehi see Punaauia
Hakāri see Hakkâri
166 K4 **Hakha** *var.* Haka. Chin State, W Myanmar (Burma) 22°42′N 93°41′E
137 T16 **Hakkâri** *var.* Çölemerik, Hakāri. Hakkâri, SE Turkey 37°36′N 43°45′E
137 T16 **Hakkâri** *var.* Hakkari. ◆ *province* SE Turkey
Hakkari see Hakkâri
92 J12 **Hakkas** Norrbotten, N Sweden 66°53′N 21°36′E
164 J14 **Hakken-zan** ▲ Honshū, SW Japan 34°11′N 135°57′E
165 R7 **Hakkōda-san** ▲ Honshū, C Japan 40°40′N 140°49′E
165 T2 **Hako-dake** ▲ Hokkaidō, NE Japan 44°40′N 142°22′E
165 R5 **Hakodate** Hokkaidō, NE Japan 41°46′N 140°43′E
164 L11 **Hakui** Ishikawa, Honshū, SW Japan 36°55′N 136°46′E
190 B16 **Hakupu** SE Niue 19°06′S 169°50′E
164 L12 **Haku-san** ▲ Honshū, SW Japan 36°07′N 136°45′E
Hakusan see Mattō
Hal see Halle
149 Q15 **Hāla** Sind, SE Pakistan 25°47′N 68°28′E
138 J3 **Ḥalab** *Eng.* Aleppo, *Fr.* Alep; *anc.* Beroea. Ḥalab, NW Syria 36°14′N 37°10′E
138 J3 **Ḥalab** *off.* Muḥāfaẓat Ḥalab, *var.* Aleppo, Halab. ◆ *governorate* NW Syria
138 J3 **Ḥalab** ✈ Ḥalab, NW Syria 36°12′N 37°10′E
Halab see Ḥalab
141 O8 **Ḥalabān** *var.* Halibān. Ar Riyāḑ, C Saudi Arabia 23°29′N 44°20′E
Ḥalabja see Helebce
Ḩalabjah see Helebce
Ḥalab, Muḥāfaẓat see Ḥalab
146 L13 **Halaç** *Rus.* Khalach. Lebap Welaýaty, E Turkmenistan 38°05′N 64°46′E
190 A16 **Halagigie Point** *headland* W Niue
75 Z11 **Halaib** SE Egypt 22°10′N 36°33′E
76 Z11 **Hala'ib Triangle** ◆ *disputed region* S Egypt / N Sudan
190 G12 **Halalo** Île Uvea, N Wallis and Futuna 13°21′S 176°11′W
167 U10 **Ha Lam** Quang Nam-Đa Nẵng, C Vietnam 15°42′N 108°24′E
Halandri see Chalándri
141 X13 **Ḥalāniyāt, Juzur al** *var.* Jazā'ir Bin Ghalfān, *Eng.* Kuria Muria Islands. *island group* S Oman
141 W13 **Ḥalāniyāt, Khalīj al** *Eng.* Kuria Muria Bay. *bay* S Oman
Halas see Kiskunhalas
38 G11 **Hālawa** *var.* Halawa. Hawaii, USA, C Pacific Ocean 20°13′N 155°46′W
Halawa see Hālawa
38 F9 **Hālawa, Cape** *var.* Cape Halawa. *headland* Moloka'i, Hawai'i, USA 21°09′N 156°43′W
Halawa, Cape see Hālawa, Cape
Halban see Tsetserleg
101 K14 **Halberstadt** Sachsen-Anhalt, C Germany 51°54′N 11°04′E
184 M12 **Halcombe** Manawatu-Wanganui, North Island, New Zealand 40°09′S 175°30′E
95 I16 **Halden** *prev.* Fredrikshald. Østfold, S Norway 59°08′N 11°20′E
100 L13 **Haldensleben** Sachsen-Anhalt, C Germany 52°18′N 11°25′E
Háldi see Halti
153 S17 **Haldia** West Bengal, NE India 22°04′N 88°02′E
152 K10 **Haldwāni** Uttarakhand, N India 29°13′N 79°31′E
163 P9 **Haldzan** *var.* Hatavch. Sühbaatar, E Mongolia 46°10′N 112°57′E
163 P9 **Haldzan** Sühbaatar, E Mongolia 46°10′N 112°57′E
38 F10 **Haleakalā** *var.* Haleakala. *crater* Maui, Hawai'i, USA
Haleakala see Haleakalā
25 N4 **Hale Center** Texas, SW USA 34°03′N 101°50′W
99 J18 **Halen** Limburg, NE Belgium 50°35′N 05°08′E
23 O2 **Haleyville** Alabama, S USA 34°13′N 87°37′W
77 O17 **Half Assini** SW Ghana 05°03′N 02°57′W
35 R8 **Half Dome** ▲ California, W USA 37°46′N 119°27′W
185 C25 **Halfmoon Bay** *var.* Oban. Stewart Island, Southland, New Zealand 46°53′S 168°08′E
182 E5 **Half Moon Lake** *salt lake* South Australia
163 R7 **Halhgol** Dornod, E Mongolia 47°57′N 118°07′E
163 S8 **Halhgol** *var.* Tsagaannuur. Dornod, E Mongolia 47°30′N 118°45′E
Haliacmon see Aliákmonas
Halibān see Ḥalabān
14 I13 **Haliburton** Ontario, SE Canada 45°03′N 78°20′W
14 I12 **Haliburton Highlands** *var.* Madawaska Highlands. *hill range* Ontario, SE Canada
13 Q15 **Halifax** *province capital* Nova Scotia, SE Canada 44°38′N 63°35′W
97 L17 **Halifax** N England, United Kingdom 53°44′N 01°52′W
21 W8 **Halifax** North Carolina, SE USA 36°19′N 77°37′W
21 U7 **Halifax** Virginia, NE USA 36°46′N 78°55′W
13 Q15 **Halifax** ✈ Nova Scotia, SE Canada 44°33′N 63°48′W
143 T13 **Halīl Rūd** *seasonal river* SE Iran
138 I6 **Ḥalīmah** ▲ Lebanon/Syria 34°12′N 36°37′E
162 G8 **Haliun** Govĭ-Altay, W Mongolia 45°55′N 96°06′E
118 I3 **Haljala** *Ger.* Halljal. Lääne-Virumaa, N Estonia 59°25′N 26°18′E
39 Q4 **Halkett, Cape** *headland* Alaska, USA 70°48′N 152°11′W
Halkida see Chalkida
96 J6 **Halkirk** N Scotland, United Kingdom 58°30′N 03°29′W
15 X7 **Hall** ↝ Québec, SE Canada
Hall see Schwäbisch Hall
93 H15 **Hälla** Västerbotten, N Sweden 63°56′N 17°20′E
96 J6 **Halladale** ↝ N Scotland, United Kingdom
95 J21 **Halland** ◆ *county* S Sweden
23 Z15 **Hallandale** Florida, SE USA 25°58′N 80°09′W
95 K22 **Hallandsås** *physical region* S Sweden
9 P6 **Hall Beach** *var* Sanirajak. Nunavut, N Canada 68°10′N 81°56′W
99 G19 **Halle** *Fr.* Hal. Vlaams Brabant, C Belgium 50°44′N 04°14′E
101 M15 **Halle** *var.* Halle an der Saale. Sachsen-Anhalt, C Germany 51°28′N 11°58′E
Halle an der Saale see Halle
35 W3 **Halleck** Nevada, W USA 40°57′N 115°27′W
95 L15 **Hällefors** Örebro, C Sweden 59°46′N 14°30′E
95 N16 **Hälleforsnäs** Södermanland, C Sweden 59°10′N 16°30′E
109 Q6 **Hallein** Salzburg, N Austria 47°41′N 13°06′E
101 L15 **Halle-Neustadt** Sachsen-Anhalt, C Germany 51°29′N 11°54′E
25 U12 **Hallettsville** Texas, SW USA 29°27′N 96°57′W
195 N4 **Halley** *UK research station* Antarctica 75°42′S 26°30′W
28 L4 **Halliday** North Dakota, N USA 47°19′N 102°19′W
37 S2 **Halligan Reservoir** ⊠ Colorado, C USA
100 G7 **Halligen** *island group* N Germany
94 G13 **Hallingdal** *valley* S Norway
38 J12 **Hall Island** *island* Alaska, USA
Hall Island see Maiana
189 P15 **Hall Islands** *island group* C Micronesia
118 H6 **Halliste** ↝ S Estonia
Halljal see Haljala
93 I15 **Hällnäs** Västerbotten, N Sweden 64°20′N 19°41′E
29 R2 **Hallock** Minnesota, N USA 48°47′N 96°56′W
9 S6 **Hall Peninsula** *peninsula* Baffin Island, Nunavut, NE Canada
20 F9 **Halls** Tennessee, S USA 35°52′N 89°24′W
95 M16 **Hallsberg** Örebro, C Sweden 59°05′N 15°07′E
181 N5 **Halls Creek** Western Australia 18°17′S 127°39′E
182 L12 **Halls Gap** Victoria, SE Australia 37°09′S 142°30′E
95 N15 **Hallstahammar** Västmanland, C Sweden 59°37′N 16°13′E
109 R6 **Hallstatt** Salzburg, W Austria 47°32′N 13°39′E
109 R6 **Hallstätter See** ◎ C Austria
95 P14 **Hallstavik** Stockholm, C Sweden 60°12′N 18°45′E
25 X7 **Hallsville** Texas, SW USA 32°31′N 94°30′W
103 P1 **Halluin** Nord, N France 50°46′N 03°07′E
171 S12 **Halmahera, Laut** *Eng.* Halmahera Sea. *sea* E Indonesia
171 R11 **Halmahera, Pulau** *prev.* Djailolo, Gilolo, Jailolo. *island* E Indonesia

◆ Country
● Country Capital
◇ Dependent Territory
○ Dependent Territory Capital
◆ Administrative Regions
✈ International Airport
▲ Mountain
▲▲ Mountain Range
Volcano
↝ River
◎ Lake
⊠ Reservoir

Halmahera Sea see Halmahera, Laut
95 J21 Halmstad Halland, S Sweden 56°41′N 12°49′E
167 T6 Ha Long prev. Hông Gai, var. Hon Gai, Hongay. Quang Ninh, N Vietnam 20°57′N 107°06′E
119 N15 Halowchyn Rus. Golovchin. Mahilyowskaya Voblasts', E Belarus 54°04′N 29°55′E
95 H20 Hals Nordjylland, N Denmark 57°00′N 10°19′E
94 F8 Halsa Møre og Romsdal, S Norway 63°04′N 08°13′E
119 I15 Hal'shany Rus. Gol'shany. Hrodzyenskaya Voblasts', W Belarus 54°15′N 26°01′E
Hälsingborg see Helsingborg
29 R5 Halstad Minnesota, N USA 47°21′N 96°49′W
27 N6 Halstead Kansas, C USA 38°00′N 97°30′W
99 G15 Halsteren Noord-Brabant, S Netherlands 51°32′N 04°16′E
93 L16 Halsua Keski-Pohjanmaa, W Finland 63°28′N 24°10′E
101 E14 Haltern Nordrhein-Westfalen, W Germany 51°45′N 07°10′E
92 J9 Halti var. Haltiatunturi, Lapp. Háldi. ▲ Finland/Norway 69°18′N 21°19′E
Haltiatunturi see Halti
116 J6 Halych Ivano-Frankivs'ka Oblast', W Ukraine 49°08′N 24°44′E
Halycus see Platani
103 P3 Ham Somme, N France 49°46′N 03°03′E
Hama see Ḩamāh
164 F12 Hamada Shimane, Honshū, SW Japan 34°54′N 132°07′E
142 L6 Hamadān anc. Ecbatana. Hamadān, W Iran 34°51′N 48°31′E
142 L6 Hamadān off. Ostān-e Hamadān. ◈ province W Iran
Hamadān, Ostān-e see Hamadān
138 I5 Ḩamāh var. Hama; anc. Epiphania, Bibl. Hamath. Ḩamāh, W Syria 35°09′N 36°44′E
138 I5 Ḩamāh off. Muḩāfaz̧at Ḩamāh, var. Hama. ◈ governorate C Syria
Ḩamāh, Muḩāfaz̧at see Ḩamāh
1665 S3 Hamamasu Hokkaidō, NE Japan 43°37′N 141°24′E
164 L14 Hamamatsu var. Hamamatu. Shizuoka, Honshū, S Japan 34°43′N 137°46′E
Hamamatu see Hamamatsu
165 W14 Hamanaka Hokkaidō, NE Japan 43°05′N 145°05′E
164 L14 Hamana-ko ◎ Honshū, S Japan
94 I13 Hamar prev. Storhammer. Hedmark, S Norway 60°57′N 10°55′E
141 U10 Ḩamārir al Kidan, Qalamat well E Saudi Arabia
164 I12 Hamasaka Hyōgo, Honshū, SW Japan 35°37′N 134°27′E
Hamath see Ḩamāh
165 T1 Hamatonbetsu Hokkaidō, NE Japan 45°07′N 142°21′E
155 K26 Hambantota Southern Province, SE Sri Lanka 06°07′N 81°07′E
Hambourg see Hamburg
100 J9 Hamburg Hamburg, N Germany 53°33′N 10°03′E
27 V14 Hamburg Arkansas, C USA 33°13′N 91°50′W
29 S16 Hamburg Iowa, C USA 40°36′N 95°39′W
18 D10 Hamburg New York, NE USA 42°40′N 78°49′W
100 I10 Hamburg Fr. Hambourg. ◈ state N Germany
148 K5 Hamdam Āb, Dasht-e Pash. Dasht-i Hamdamab. ▲ W Afghanistan
Hamdamab, Dasht-i see Hamdam Āb, Dasht-e
18 M13 Hamden Connecticut, NE USA 41°23′N 72°55′W
140 K6 Ḩamḑ, Wādī al dry watercourse W Saudi Arabia
93 K18 Hämeenkyrö Pirkanmaa, W Finland 61°39′N 23°10′E
93 L19 Hämeenlinna Swe. Tavastehus. Kanta-Häme, S Finland 61°N 24°25′E
HaMela h, Yam see Dead Sea
Hamelin see Hameln
100 I13 Hameln Eng. Hamelin. Niedersachsen, N Germany 52°07′N 09°22′E
180 I8 Hamersley Range ▲ Western Australia
163 Y12 Hamgyŏng-sanmaek ▲ N North Korea
163 X13 Hamhŭng C North Korea 39°53′N 127°31′E
159 O6 Hami var. Ha-mi, Uigh. Kumul, Qomul. Xinjiang Uygur Zizhiqu, NW China 42°48′N 93°27′E
Ha-mi see Hami
139 X10 Ḩāmid Amīn Maysān, E Iraq 32°06′N 46°53′E
141 W11 Hamīdān, Khawr oasis SE Saudi Arabia
114 L12 Hamidiye Edirne, NW Turkey 41°09′N 26°40′E
182 L12 Hamilton Victoria, SE Australia 37°45′S 142°04′E
64 B12 Hamilton ○ (Bermuda) C Bermuda 32°18′N 64°48′W
14 G16 Hamilton Ontario, S Canada 43°15′N 79°50′W
184 M7 Hamilton Waikato, North Island, New Zealand 37°49′S 175°16′E
96 I12 Hamilton S Scotland, United Kingdom 55°47′N 04°03′W
23 N3 Hamilton Alabama, S USA 34°08′N 87°59′W
38 M10 Hamilton Alaska, USA 62°54′N 163°53′W
30 J13 Hamilton Illinois, N USA 40°24′N 91°20′W
27 S3 Hamilton Missouri, C USA 39°44′N 94°00′W
33 P10 Hamilton Montana, NW USA 46°15′N 114°09′W
25 S8 Hamilton Texas, SW USA 31°42′N 98°08′W
14 G16 Hamilton ✈ Ontario, SE Canada 43°12′N 79°54′W
64 I6 Hamilton Bank undersea feature SE Labrador Sea
182 E1 Hamilton Creek seasonal river South Australia
13 R8 Hamilton Inlet inlet Newfoundland and Labrador, E Canada
27 T12 Hamilton, Lake ⊠ Arkansas, C USA
35 W6 Hamilton, Mount ▲ Nevada, W USA 39°15′N 115°30′W
75 S8 Ḩamīm, Wādī al ≈ NE Libya
93 N19 Hamina Swe. Fredrikshamn. Kymenlaakso, S Finland 60°33′N 27°15′E
11 W16 Hamiota Manitoba, S Canada 50°13′N 100°37′W
152 L13 Hamīrpur Uttar Pradesh, N India 25°57′N 80°08′E
Hamīs Musait see Khamīs Mushayt
21 T11 Hamlet North Carolina, SE USA 34°52′N 79°41′W
25 P6 Hamlin Texas, SW USA 32°52′N 100°07′W
21 P5 Hamlin West Virginia, NE USA 38°16′N 82°07′W
31 O7 Hamlin Lake ◎ Michigan, N USA
101 F14 Hamm var. Hamm in Westfalen. Nordrhein-Westfalen, W Germany 51°39′N 07°49′E
Ḩammāmāt, Khalīj al see Hammamet, Golfe de
75 N5 Hammamet, Golfe de Ar. Khalīj al Ḩammāmāt. gulf NE Tunisia
139 R3 Ḩammān al 'Alīl Nīnawé, N Iraq 36°07′N 43°15′E
139 X12 Ḩammār, Hawr al ◎ SE Iraq
93 J20 Hammarland Åland, SW Finland 60°13′N 19°45′E
93 H16 Hammarstrand Jämtland, C Sweden 63°07′N 16°27′E
93 O17 Hammaslahti Pohjois-Karjala, SE Finland 62°26′N 29°58′E
99 F17 Hamme Oost-Vlaanderen, NW Belgium 51°06′N 04°08′E
100 H10 Hamme ≈ NW Germany
95 G22 Hammel Midtjylland, C Denmark 56°15′N 09°53′E
101 I18 Hammelburg Bayern, C Germany 50°06′N 09°50′E
99 H18 Hamme-Mille Walloon Brabant, C Belgium 50°48′N 04°42′E
100 H10 Hamme-Oste-Kanal canal NW Germany
93 G16 Hammerdal Jämtland, C Sweden 63°34′N 15°19′E
92 K8 Hammerfest Finnmark, N Norway 70°40′N 23°44′E
101 D14 Hamminkeln Nordrhein-Westfalen, W Germany 51°43′N 06°36′E
Hamm in Westfalen see Hamm
26 K10 Hammon Oklahoma, C USA 35°37′N 99°22′W
31 N11 Hammond Indiana, N USA 41°35′N 87°30′W
22 K8 Hammond Louisiana, S USA 30°30′N 90°27′W
99 K20 Hamoir Liège, E Belgium 50°28′N 05°35′E
99 J21 Hamois Namur, SE Belgium 50°21′N 05°09′E
99 K16 Hamont Limburg, NE Belgium 51°15′N 05°33′E
185 F22 Hampden Otago, South Island, New Zealand 45°18′S 170°49′E
19 R6 Hampden Maine, NE USA 44°44′N 68°51′W
97 M23 Hampshire cultural region S England, United Kingdom
13 O15 Hampton New Brunswick, SE Canada 45°30′N 65°50′W
27 U14 Hampton Arkansas, C USA 33°33′N 92°28′W
29 V12 Hampton Iowa, C USA 42°44′N 93°12′W
19 P10 Hampton New Hampshire, NE USA 42°55′N 70°48′W
21 R14 Hampton South Carolina, SE USA 32°52′N 81°06′W
21 P8 Hampton Tennessee, S USA 36°16′N 82°10′W
21 X7 Hampton Virginia, NE USA 37°02′N 76°23′W
94 L11 Hamra Gävleborg, C Sweden 61°40′N 15°00′E
80 D10 Hamrat esh Sheikh Northern Kordofan, C Sudan 14°38′N 27°56′E
139 S5 Ḩamrīn, Jabal ▲ N Iraq
121 P16 Ħamrun C Malta 35°53′N 14°28′E
Ham Thuân Nam see Thuận Nam
Hāmūn, Daryācheh-ye see Şāberī, Hāmūn-e/Sīstān, Daryācheh-ye
Hamwih see Southampton
38 G10 Hāna var. Hana. Maui, Hawaii, USA, C Pacific Ocean 20°45′N 155°59′W
Hana see Hāna
21 S14 Hanahan South Carolina, SE USA 32°55′N 80°01′W
38 B8 Hanalei Kaua'i, Hawaii, USA, C Pacific Ocean 22°12′N 159°30′W
165 Q9 Hanamaki Iwate, Honshū, C Japan 39°25′N 141°04′E
38 F10 Hanamanioa, Cape headland Maui, Hawai'i, USA 20°34′N 156°22′W
190 B16 Hanan ✈ (Alofi) SW Niue
101 H18 Hanau Hessen, W Germany 50°06′N 08°56′E
162 M11 Hanbogd var. Ih Bulag. Ömnögovĭ, S Mongolia 43°04′N 107°43′E
8 L9 Hanbury ≈ Northwest Territories, NW Canada
Hâncești see Hîncești
10 M15 Hanceville British Columbia, SW Canada 51°54′N 122°56′W
23 P3 Hanceville Alabama, S USA 34°03′N 86°46′W
Hancewicze see Hantsavichy
160 L6 Hancheng Shaanxi, C China 35°22′N 110°27′E
21 V2 Hancock Maryland, NE USA 39°42′N 78°10′W
30 M3 Hancock Michigan, N USA 47°07′N 88°34′W
29 S8 Hancock Minnesota, N USA 45°30′N 95°47′W
18 I12 Hancock New York, NE USA 41°57′N 75°15′W
80 Q12 Handa Bari, NE Somalia 10°35′N 51°09′E
161 O5 Handan var. Han-tan. Hebei, E China 36°35′N 114°28′E
95 P16 Handen Stockholm, C Sweden 59°12′N 18°09′E
81 J22 Handeni Tanga, E Tanzania 05°25′S 38°04′E
37 Q7 Handies Peak ▲ Colorado, C USA 37°54′N 107°30′W
111 J19 Handlová Ger. Krickerhäu, Hung. Nyitrabánya; prev. Kriegerhaj. Trenčiansky Kraj, C Slovakia 48°45′N 18°45′E
165 O13 Haneda ✈ (Tōkyō) Tōkyō, Honshū, S Japan 35°33′N 139°45′E
138 F13 HaNegev Eng. Negev. desert S Israel
Hanfeng see Kaixian
35 Q11 Hanford California, W USA 36°19′N 119°39′W
191 V16 Hanga Roa Easter Island, Chile, E Pacific Ocean 27°09′S 109°26′W
162 I7 Hangay var. Hunt. Arhangay, C Mongolia 47°49′N 99°24′E
162 H7 Hangayn Nuruu ▲ C Mongolia
Hang-chou/Hangchow see Hangzhou
95 K20 Hånger Jönköping, S Sweden 57°06′N 13°58′E
Hangö see Hanko
161 R9 Hangzhou var. Hang-chou, Hangchow. province capital Zhejiang, SE China 30°18′N 120°07′E
162 J4 Hanh var. Turt. Hövsgöl, N Mongolia 51°30′N 100°40′E
162 F5 Hanhöhiy Uul ▲ NW Mongolia
162 K10 Hanhongor var. Ögöömör. Ömnögovĭ, S Mongolia 43°47′N 104°31′E
146 I14 Hanhowuz Rus. Khauz-Khan. Ahal Welaýaty, S Turkmenistan 37°15′N 61°12′E
146 I14 Hanhowuz Suw Howdany Rus. Khauzkhanskoye Vodoranilishche. ⊠ S Turkmenistan
137 P15 Hani Diyarbakır, SE Turkey 38°26′N 40°23′E
Hania see Chaniá
141 R11 Ḩanīsh al Kabīr, Jazīrat al island SW Yemen
Hanka, Lake see Khanka, Lake
93 M17 Hankasalmi Keski-Suomi, C Finland 62°25′N 26°27′E
29 R7 Hankinson North Dakota, N USA 46°04′N 96°54′W
93 K20 Hanko Swe. Hangö. Uusimaa, SW Finland 59°50′N 23°E
Han-kou/Han-k'ou/Hankow see Wuhan
36 M6 Hanksville Utah, W USA 38°21′N 110°43′W
152 K6 Hanle Jammu and Kashmir, NW India 32°46′N 79°01′E
185 I17 Hanmer Springs Canterbury, South Island, New Zealand 42°31′S 172°49′E
11 R16 Hanna Alberta, SW Canada 51°38′N 111°56′W
27 V3 Hannibal Missouri, C USA 39°42′N 91°23′W
180 M3 Hann, Mount ▲ Western Australia 15°53′S 125°46′E
100 I12 Hannover Eng. Hanover. Niedersachsen, NW Germany 52°23′N 09°43′E
99 J19 Hannut Liège, C Belgium 50°40′N 05°05′E
95 L22 Hanöbukten bay S Sweden
167 T6 Ha Nôi Eng. Hanoi, Fr. Hanoï. ● (Vietnam) N Vietnam 21°01′N 105°52′E
14 F14 Hanover Ontario, S Canada 44°10′N 81°03′W
31 P15 Hanover Indiana, N USA 38°42′N 85°28′W
18 G16 Hanover Pennsylvania, NE USA 39°46′N 76°57′W
21 W6 Hanover Virginia, NE USA 37°44′N 77°21′W
Hanover see Hannover
63 G23 Hanover, Isla island S Chile
Hanselbeck see Érd
195 X5 Hansen Mountains ▲ Antarctica
160 M8 Han Shui ≈ C China
152 H10 Hānsi Haryāna, NW India 29°06′N 76°01′E
95 F20 Hanstholm Midtjylland, NW Denmark 57°05′N 08°39′E
Han-tan see Handan
158 H6 Hantengri Feng var. Pik Khan-Tengri. ▲ China/Kazakhstan 42°17′N 80°11′E see also Khan-Tengri, Pik
119 I18 Hantsavichy Pol. Hancewicze, Rus. Gantsevichi. Brestskaya Voblasts', SW Belarus 52°45′N 26°27′E
9 Q6 Hantzsch ≈ Baffin Island, Nunavut, NE Canada
152 G9 Hanumāngarh Rājasthān, NW India 29°33′N 74°21′E
183 O9 Hanwood New South Wales, SE Australia 34°19′S 146°03′E
Hanyang see Wuhan
Hanyang see Caidian
160 H10 Hanyuan var. Fulin. Sichuan, C China 29°29′N 102°45′E
Hanyuan see Xihe
160 J7 Hanzhong Shaanxi, C China 33°12′N 107°06′E
191 W11 Hao atoll Îles Tuamotu, C French Polynesia
153 S16 Hāora prev. Howrah. West Bengal, NE India 22°35′N 88°20′E
78 K8 Haouach, Ouadi dry watercourse E Chad
92 K13 Haparanda Norrbotten, N Sweden 65°49′N 24°05′E
25 N3 Happy Texas, SW USA 34°44′N 101°51′W
34 M1 Happy Camp California, W USA 41°48′N 123°24′W
13 Q9 Happy Valley-Goose Bay prev. Goose Bay. Newfoundland and Labrador, E Canada 53°19′N 60°24′W
Hapsal see Haapsalu
152 J10 Hāpur Uttar Pradesh, N India 28°43′N 77°47′E
HaQatan, HaMakhtesh see HaKatan, HaMakhtesh
140 I4 Ḩaql Tabūk, NW Saudi Arabia 29°18′N 34°58′E
171 U14 Har Pulau Kai Besar, E Indonesia 05°21′S 133°09′E
Haraat see Tsagaandelger
141 R8 Ḩaraḑ var. Haradh. Ash Sharqīyah, E Saudi Arabia 24°08′N 49°02′E
Haradh see Ḩaraḑ
118 N12 Haradok Rus. Gorodok. Vitsyebskaya Voblasts', N Belarus 55°28′N 30°00′E
92 J13 Harads Norrbotten, N Sweden 66°04′N 21°04′E
119 G19 Haradzyets Rus. Gorodets. Brestskaya Voblasts', SW Belarus 52°12′N 24°40′E
119 J17 Haradzyeya Rus. Gorodeya. Minskaya Voblasts', C Belarus 53°19′N 26°32′E
191 V10 Haraiki atoll Îles Tuamotu, C French Polynesia
165 Q11 Haramachi Fukushima, Honshū, E Japan 37°40′N 140°55′E
118 M12 Harany Rus. Gorany. Vitsyebskaya Voblasts', N Belarus 55°25′N 29°03′E
83 L16 Harare prev. Salisbury. ● (Zimbabwe) Mashonaland East, NE Zimbabwe 17°47′S 31°04′E
83 L16 Harare ✈ Mashonaland East, NE Zimbabwe 17°51′S 31°06′E
78 J10 Haraz-Djombo Batha, C Chad 14°10′N 19°35′E
119 O16 Harbavichy Rus. Gorbovichi. Mahilyowskaya Voblasts', E Belarus 53°49′N 30°42′E
76 J16 Harbel W Liberia 06°19′N 10°20′W
163 W8 Harbin var. Haerbin, Ha-erh-pin, Kharbin; prev. Haerhpin, Pingkiang, Pinkiang. province capital Heilongjiang, NE China 45°45′N 126°41′E
31 S7 Harbor Beach Michigan, N USA 43°50′N 82°39′W
13 T13 Harbour Breton Newfoundland, Newfoundland and Labrador, E Canada 47°29′N 55°50′W
65 D25 Harbours, Bay of bay East Falkland, Falkland Islands
Hårby see Haarby
36 I13 Harcuvar Mountains ▲ Arizona, SW USA
108 I7 Hard Vorarlberg, NW Austria 47°29′N 09°42′E
154 H11 Harda Khas Madhya Pradesh, C India 22°22′N 77°06′E
95 D14 Hardanger physical region S Norway
95 D14 Hardangerfjorden fjord S Norway
94 E13 Hardangerjøkulen glacier S Norway
95 E14 Hardangervidda plateau S Norway
83 D20 Hardap ◈ district S Namibia
21 R15 Hardeeville South Carolina, SE USA 32°18′N 81°04′W
Hardegarijp see Hurdegaryp
98 O9 Hardenberg Overijssel, E Netherlands 52°34′N 06°38′E
183 Q9 Harden-Murrumburrah New South Wales, SE Australia 34°33′S 148°22′E
98 K10 Harderwijk Gelderland, C Netherlands 52°21′N 05°37′E
30 J14 Hardin Illinois, N USA 39°10′N 90°38′W
33 V11 Hardin Montana, NW USA 45°44′N 107°35′W
23 R5 Harding, Lake ⊠ Alabama/Georgia, SE USA
20 J6 Hardinsburg Kentucky, S USA 37°46′N 86°29′W
98 I13 Hardinxveld-Giessendam Zuid-Holland, C Netherlands 51°52′N 04°49′E
11 R15 Hardisty Alberta, SW Canada 52°42′N 111°22′W
152 L12 Hardoi Uttar Pradesh, N India 27°23′N 80°06′E
Hardwar see Haridwār
23 U4 Hardwick Georgia, SE USA 33°03′N 83°13′W
27 W9 Hardy Arkansas, C USA 36°19′N 91°29′W
94 D10 Hareid Møre og Romsdal, S Norway 62°22′N 06°02′E
8 H7 Hare Indian ≈ Northwest Territories, NW Canada
99 D18 Harelbeke var. Harlebeke. West-Vlaanderen, W Belgium 50°51′N 03°19′E
Harem see Ḩārim
100 E11 Haren Niedersachsen, NW Germany 52°47′N 07°16′E
98 N6 Haren Groningen, NE Netherlands 53°10′N 06°37′E
80 L13 Hārer E Ethiopia 09°17′N 42°19′E
95 P14 Harg Uppsala, C Sweden 60°13′N 18°25′E
Hargeisa see Hargeysa
80 M13 Hargeysa var. Hargeisa. Woqooyi Galbeed, NW Somalia 09°32′N 44°07′E
116 J10 Harghita ◈ county NE Romania
25 S17 Hargill Texas, SW USA 26°26′N 98°00′W
162 J8 Harhorin Övörhangay, C Mongolia 47°13′N 102°48′E
159 Q9 Har Hu ◎ C China
Hariana see Haryāna
141 P15 Ḩarīb W Yemen 15°08′N 45°35′E
168 M12 Hari, Batang prev. Djambi. ≈ Sumatera, W Indonesia
152 J9 Haridwār prev. Hardwar. Uttarakhand, N India 29°58′N 78°09′E
155 F18 Harihar Karnātaka, W India 14°33′N 75°44′E
185 F18 Harihari West Coast, South Island, New Zealand 43°09′S 170°35′E
138 I3 Ḩārim var. Harem. Idlib, W Syria 36°30′N 36°30′E
98 F13 Haringvliet channel SW Netherlands
98 F13 Haringvlietdam dam SW Netherlands
149 U5 Harīpur Khyber Pakhtunkhwa, NW Pakistan 34°00′N 73°01′E
148 J4 Harīrūd var. Tedzhen, Turkm. Tejen. ≈ Afghanistan/Iran see also Tejen
Harīrūd see Tejen
94 J11 Härjåhågnen Swe. Härjahågnen, var. Härjehågna. ▲ Norway/Sweden 61°43′N 12°07′E
Härjahågnen see Østrehogna
Härjåhågnen see Härjehågna
93 K18 Harjavalta Satakunta, W Finland 61°19′N 22°10′E
Härjehågna see Østrehogna
Härjehågna see Härjåhågnen
118 G4 Harjumaa var. Harju Maakond. ◈ province NW Estonia
Harju Maakond see Harjumaa
21 X11 Harkers Island North Carolina, SE USA 34°42′N 76°33′W
Harki see Hurkê
29 T14 Harlan Iowa, C USA 41°40′N 95°19′W
21 O7 Harlan Kentucky, S USA 36°50′N 83°19′W
29 N17 Harlan County Lake ⊠ Nebraska, C USA
116 L9 Hârlău var. Hîrlău. Iași, NE Romania 47°26′N 26°54′E
Harlebeke see Harelbeke
33 U7 Harlem Montana, NW USA 48°31′N 108°46′W
Harlem see Haarlem
95 G22 Harlev Midtjylland, C Denmark 56°08′N 10°00′E
98 K6 Harlingen Fris. Harns. Fryslân, N Netherlands 53°10′N 05°25′E
25 T17 Harlingen Texas, SW USA 26°12′N 97°43′W
97 O21 Harlow E England, United Kingdom 51°47′N 00°07′E
33 T10 Harlowton Montana, NW USA 46°26′N 109°49′W
94 N11 Harmånger Gävleborg, C Sweden 61°55′N 17°19′E
114 K11 Harmanli var. Kharmanli. Haskovo, S Bulgaria 41°56′N 25°54′E
114 K11 Harmanliyska Reka var. Kharmanliyska Reka. ≈ S Bulgaria
98 I11 Harmelen Utrecht, C Netherlands 52°06′N 04°58′E
29 X11 Harmony Minnesota, N USA 43°33′N 92°00′W
32 J14 Harney Basin basin Oregon, NW USA
32 J14 Harney Lake ◎ Oregon, NW USA
28 J10 Harney Peak ▲ South Dakota, N USA 43°52′N 103°31′W
93 H17 Härnösand var. Hernösand. Västernorrland, C Sweden 62°37′N 17°55′E
Harns see Harlingen
105 P4 Haro La Rioja, N Spain 42°34′N 02°52′W
40 F6 Haro, Cabo headland NW Mexico 27°50′N 110°55′W
94 D9 Harøy island S Norway
97 N21 Harpenden E England, United Kingdom 51°49′N 00°22′W
76 L18 Harper var. Cape Palmas. NE Liberia 04°25′N 07°43′W
26 M7 Harper Kansas, C USA 37°17′N 98°01′W
32 L13 Harper Oregon, NW USA 43°51′N 117°37′W
25 Q10 Harper Texas, SW USA 30°18′N 99°18′W
35 U13 Harper Lake salt flat California, W USA
39 T9 Harper, Mount ▲ Alaska, USA 64°18′N 143°54′W
95 J21 Harplinge Halland, S Sweden 56°45′N 12°45′E
36 J13 Harquahala Mountains ▲ Arizona, SW USA
141 T15 Ḩarrah SE Yemen 15°02′N 50°23′E
12 H11 Harricana ≈ Québec, SE Canada
20 M9 Harriman Tennessee, S USA 35°57′N 84°33′W
13 R11 Harrington Harbour Québec, E Canada 50°34′N 59°29′W
64 B12 Harrington Sound bay Bermuda, NW Atlantic Ocean
96 F8 Harris physical region NW Scotland, United Kingdom
27 X10 Harrisburg Arkansas, C USA 35°33′N 90°43′W
30 M17 Harrisburg Illinois, N USA 37°44′N 88°32′W
28 I14 Harrisburg Nebraska, C USA 41°33′N 103°46′W
32 F12 Harrisburg Oregon, NW USA 44°16′N 123°10′W
18 G15 Harrisburg state capital Pennsylvania, NE USA 40°16′N 76°53′W
182 F6 Harris, Lake ◎ South Australia
23 W11 Harris, Lake ◎ Florida, SE USA
83 J22 Harrismith Free State, E South Africa 28°16′S 29°08′E
27 T9 Harrison Arkansas, C USA 36°13′N 93°07′W
31 Q7 Harrison Michigan, N USA 44°02′N 84°46′W
28 I12 Harrison Nebraska, C USA 42°42′N 103°53′W
39 Q5 Harrison Bay inlet Alaska, USA
22 I6 Harrisonburg Louisiana, S USA 31°44′N 91°51′W
21 U4 Harrisonburg Virginia, NE USA 38°27′N 78°54′W
13 R7 Harrison, Cape headland Newfoundland and Labrador, E Canada 54°55′N 57°48′W
27 R5 Harrisonville Missouri, C USA 38°40′N 94°21′W
Harris Ridge see Lomonosov Ridge
192 M3 Harris Seamount undersea feature N Pacific Ocean 46°09′N 161°25′W
96 F8 Harris, Sound of strait NW Scotland, United Kingdom
31 R6 Harrisville Michigan, N USA 44°41′N 83°19′W
21 R3 Harrisville West Virginia, NE USA 39°13′N 81°04′W
20 M6 Harrodsburg Kentucky, S USA 37°45′N 84°51′W
97 M16 Harrogate N England, United Kingdom 54°N 01°33′W
25 Q4 Harrold Texas, SW USA 34°05′N 99°02′W
27 S5 Harry S. Truman Reservoir ⊠ Missouri, C USA
100 G13 Harsewinkel Nordrhein-Westfalen, W Germany 51°58′N 08°13′E
116 M14 Hârşova prev. Hîrşova. Constanța, SE Romania 44°41′N 27°56′E
92 H10 Harstad Troms, N Norway 68°48′N 16°31′E
31 O8 Hart Michigan, N USA 43°43′N 86°22′W
24 M4 Hart Texas, SW USA 34°23′N 102°07′W
10 I5 Hart ≈ Yukon, NW Canada
83 F23 Hartbees ≈ C South Africa
109 X7 Hartberg Steiermark, SE Austria 47°18′N 15°58′E
182 I10 Hart, Cape headland South Australia 35°54′S 138°01′E
95 E14 Hårteigen ▲ S Norway 60°11′N 07°01′E
23 Q7 Hartford Alabama, S USA 31°06′N 85°42′W
27 R11 Hartford Arkansas, C USA 35°01′N 94°22′W
18 M12 Hartford state capital Connecticut, NE USA 41°46′N 72°41′W
20 J6 Hartford Kentucky, S USA 37°26′N 86°57′W
31 P10 Hartford Michigan, N USA 42°12′N 85°54′W
29 R11 Hartford South Dakota, N USA 43°37′N 96°56′W
30 M8 Hartford Wisconsin, N USA 43°19′N 88°25′W
31 P13 Hartford City Indiana, N USA 40°27′N 85°22′W
29 Q13 Hartington Nebraska, C USA 42°37′N 97°15′W
13 N14 Hartland New Brunswick, SE Canada 46°18′N 67°31′W
97 H23 Hartland Point headland SW England, United Kingdom 51°01′N 04°33′W
97 M15 Hartlepool N England, United Kingdom 54°41′N 01°13′W
29 T12 Hartley Iowa, C USA 43°10′N 95°28′W
24 M1 Hartley Texas, SW USA 35°52′N 102°24′W
32 J15 Hart Mountain ▲ Oregon, NW USA 42°24′N 119°46′W
173 U10 Hartog Ridge undersea feature W Indian Ocean
93 M18 Hartola Päijät-Häme, S Finland 61°34′N 26°04′E
67 U14 Harts var. Hartz. ≈ N South Africa
23 P2 Hartselle Alabama, S USA 34°26′N 86°56′W
23 S3 Hartsfield Atlanta ✈ Georgia, SE USA 33°38′N 84°24′W
27 Q11 Hartshorne Oklahoma, C USA 34°51′N 95°33′W
21 S12 Hartsville South Carolina, SE USA 34°22′N 80°04′W
20 K8 Hartsville Tennessee, S USA 36°23′N 86°11′W
27 U7 Hartville Missouri, C USA 37°15′N 92°30′W
23 U2 Hartwell Georgia, SE USA 34°21′N 82°55′W
21 O11 Hartwell Lake ⊠ Georgia/South Carolina, SE USA
Hartz see Harts
Harunabad see Eslāmābād-e Gharb
Har-Us see Erdenebüren
162 F6 Har Us Gol ◎ Hovd, W Mongolia
162 E6 Har Us Nuur ◎ NW Mongolia
30 M10 Harvard Illinois, N USA 42°25′N 88°36′W
29 P16 Harvard Nebraska, C USA 40°37′N 98°06′W
37 R5 Harvard, Mount ▲ Colorado, C USA 38°55′N 106°19′W
31 N11 Harvey Illinois, N USA 41°36′N 87°39′W
29 N4 Harvey North Dakota, N USA 47°43′N 99°55′W
97 Q21 Harwich E England, United Kingdom 51°56′N 01°16′E
152 H10 Haryāna var. Hariana. ◈ state N India
141 Y9 Ḩaryān, Ţawī al spring/well NE Oman 21°56′N 58°33′E
101 J14 Harz ▲ C Germany
Ḩasakah, Muḩāfaz̧at al see Al Ḩasakah
165 Q9 Hasama Miyagi, Honshū, C Japan 38°42′N 141°09′E
137 Y13 Häsänabad prev. 26 Bakinskikh Komissarov. SE Azerbaijan 39°18′N 49°13′E
136 J15 Hasan Dağı ▲ C Turkey 38°09′N 34°15′E
139 T9 Ḩasan Ibn Ḩassūn An Najaf, C Iraq 32°24′N 44°13′E
149 R6 Ḩasan Khēl var. Ahmad Khel. Paktiyā, SE Afghanistan 33°46′N 69°37′E
100 F12 Hase ≈ NW Germany
Haselberg see Krasnoznamensk
100 F12 Haselünne Niedersachsen, NW Germany 52°40′N 07°28′E
Hashaat see Delgerhangay
Hashemite Kingdom of Jordan see Jordan
139 V8 Hāshimah Wāsiţ, E Iraq 33°22′N 45°56′E
142 K3 Hashtrūd var. Azaran. Āz̄arbāyjān-e Khāvarī, N Iran 37°34′N 47°10′E
141 W13 Ḩāsik S Oman 17°22′N 55°18′E
149 U10 Hāsilpur Punjab, E Pakistan 29°42′N 72°40′E
27 Q10 Haskell Oklahoma, C USA 35°49′N 95°40′W
25 Q6 Haskell Texas, SW USA 33°10′N 99°45′W
114 K11 Haskovo var. Khaskovo. Haskovo, S Bulgaria 41°56′N 25°35′E
114 K11 Haskovo var. Khaskovo. ◈ province S Bulgaria
114 M11 Hasköy Edirne, NW Turkey 41°37′N 26°51′E
95 L24 Hasle Bornholm, E Denmark 55°12′N 14°43′E
97 N23 Haslemere SE England, United Kingdom 51°06′N 00°45′W
102 I16 Hasparren Pyrénées-Atlantiques, SW France 43°23′N 01°18′W
Hassakeh see Al Ḩasakah
155 G19 Hassan Karnātaka, W India 13°01′N 76°03′E
36 J13 Hassayampa River ≈ Arizona, SW USA
101 J18 Hassberge hill range C Germany
94 N10 Hassela Gävleborg, C Sweden 62°06′N 16°45′E
99 J18 Hasselt Limburg, NE Belgium 50°56′N 05°20′E
98 M9 Hasselt Overijssel, E Netherlands 52°36′N 06°06′E
Hassetché see Al Ḩasakah
101 J18 Hassfurt Bayern, C Germany 50°02′N 10°32′E
74 L9 Hassi Bel Guebbour E Algeria 28°41′N 06°29′E
74 L8 Hassi Messaoud E Algeria 31°41′N 06°10′E
95 K22 Hässleholm Skåne, S Sweden 56°09′N 13°45′E
Hasta Colonia/Hasta Pompeia see Asti
183 O13 Hastings Victoria, SE Australia 38°18′S 145°12′E
184 O11 Hastings Hawke's Bay, North Island, New Zealand 39°39′S 176°51′E
97 P23 Hastings SE England, United Kingdom 50°51′N 00°36′E
31 P9 Hastings Michigan, N USA 42°38′N 85°17′W
29 W9 Hastings Minnesota, N USA 44°44′N 92°51′W
29 P16 Hastings Nebraska, C USA 40°35′N 98°23′W
95 K22 Hästveda Skåne, S Sweden 56°16′N 13°55′E
92 J8 Hasvik Finnmark, N Norway 70°29′N 22°08′E
37 V6 Haswell Colorado, C USA 38°27′N 103°09′W
163 N11 Hatanbulag var. Ergel. Dornogovĭ, SE Mongolia 43°10′N 109°13′E
Hatansuudal see Bayanlig
Hatavch see Haldzan
136 K17 Hatay ◈ province S Turkey
37 R15 Hatch New Mexico, SW USA 32°40′N 107°10′W
36 K7 Hatch Utah, W USA 37°39′N 112°25′W
20 F9 Hatchie River ≈ Tennessee, S USA
116 G12 Haţeg Ger. Wallenthal, Hung. Hátszeg; prev. Hatzeg, Hötzing. Hunedoara, SW Romania 45°35′N 22°57′E
165 O17 Hateruma-jima island Yaeyama-shotō, SW Japan
183 N8 Hatfield New South Wales, SE Australia 33°54′S 143°43′E
162 I5 Hatgal Hövsgöl, N Mongolia 50°24′N 100°12′E
153 V16 Hathazari Chittagong, SE Bangladesh 22°30′N 91°46′E
141 T13 Hathūt, Hiṣā' oasis NE Yemen
167 R14 Ha Tiên Kiên Giang, S Vietnam 10°24′N 104°30′E
167 T8 Ha Tinh Ha Tinh, N Vietnam 18°21′N 105°55′E
Ḩatira, Haré see Hatira, Harei
138 F12 Hatira, Harei prev. Haré Ḩatira. hill range S Israel
167 R6 Hat Lot var. Mai Son. Son La, N Vietnam 21°07′N 104°10′E
45 P16 Hato Airport ✈ (Willemstad) Curaçao 12°10′N 68°56′W
54 H9 Hato Corozal Casanare, C Colombia 06°08′N 71°45′W
Hato del Volcán see Volcán
45 P9 Hato Mayor E Dominican Republic 18°49′N 69°16′W
Hatra see Al Ḩaḑr
Hattia see Adria
Hatszeg see Haţeg
143 R16 Ḩaţţā Dubayy, NE United Arab Emirates 24°50′N 56°06′E
182 L9 Hattah Victoria, SE Australia 34°49′S 142°18′E
98 M9 Hattem Gelderland, E Netherlands 52°29′N 06°04′E
21 Z10 Hatteras Hatteras Island, North Carolina, SE USA 35°13′N 75°39′W
21 Rr10 Hatteras, Cape headland North Carolina, SE USA 35°29′N 75°33′W
21 Z9 Hatteras Island island North Carolina, SE USA
64 F10 Hatteras Plain undersea feature W Atlantic Ocean 31°00′N 71°00′W
93 G14 Hattfjelldal Troms, N Norway 65°37′N 13°58′E
22 M7 Hattiesburg Mississippi, S USA 31°20′N 89°17′W
29 Q4 Hatton North Dakota, N USA 47°38′N 97°27′W
Hatton Bank see Hatton Ridge
64 L6 Hatton Ridge var. Hatton Bank. undersea feature N Atlantic Ocean 59°00′N 17°30′W
191 W6 Hatutu island Îles Marquises, NE French Polynesia
111 K22 Hatvan Heves, NE Hungary 47°40′N 19°39′E
167 O16 Hat Yai var. Ban Hat Yai. Songkhla, SW Thailand 07°01′N 100°27′E
Hatzeg see Haţeg
Hatzfeld see Jimbolia
80 N13 Haud plateau Ethiopia/Somalia
95 D18 Hauge Rogaland, S Norway 58°20′N 06°17′E
95 C15 Haugesund Rogaland, S Norway 59°24′N 05°17′E
109 X2 Haugsdorf Niederösterreich, NE Austria 48°41′N 16°04′E
184 M9 Hauhungaroa Range ▲ North Island, New Zealand
95 E15 Haukeligrend Telemark, S Norway 59°45′N 07°33′E
93 L14 Haukipudas Pohjois-Pohjanmaa, C Finland 65°11′N 25°21′E
93 M17 Haukivesi ◎ SE Finland
93 M17 Haukivuori Etelä-Savo, E Finland 62°02′N 27°11′E
Hauptkanal see Havelländ Grosse
187 N10 Hauraha Makira-Ulawa, SE Solomon Islands 10°47′S 162°00′E
184 L5 Hauraki Gulf gulf North Island, New Zealand
185 B24 Hauroko, Lake ◎ South Island, New Zealand
167 S14 Hâu, Sông ≈ S Vietnam
92 N12 Hautajärvi Lappi, NE Finland 66°30′N 29°01′E
74 F7 Haut Atlas Eng. High Atlas. ▲ C Morocco
79 M17 Haut-Congo off. Région du Haut-Congo; prev. Haut-Zaïre. ◈ region NE Dem. Rep. Congo
103 Y14 Haute-Corse ◈ department Corse, France, C Mediterranean Sea
102 L16 Haute-Garonne ◈ department S France
79 K14 Haute-Kotto ◈ prefecture E Central African Republic
103 P12 Haute-Loire ◈ department C France
103 R6 Haute-Marne ◈ department N France
102 M3 Haute-Normandie ◈ region N France
15 U6 Hauterive Québec, SE Canada 49°11′N 68°16′W
103 T13 Hautes-Alpes ◈ department SE France
103 S7 Haute-Saône ◈ department E France
103 T10 Haute-Savoie ◈ department E France
99 M20 Hautes Fagnes Ger. Hohes Venn. ▲ E Belgium
102 K16 Hautes-Pyrénées ◈ department S France
99 L23 Haute Sûre, Lac de la ⊠ NW Luxembourg
102 M11 Haute-Vienne ◈ department C France
19 S8 Haut, Isle au island Maine, NE USA

◆ Country ● Country Capital ◇ Dependent Territory ○ Dependent Territory Capital ◈ Administrative Regions ✈ International Airport ▲ Mountain ▲ Mountain Range Volcano ≈ River ◎ Lake ⊠ Reservoir

79 M14 **Haut-Mbomou** ⬖ *prefecture* SE Central African Republic
103 Q2 **Hautmont** Nord, N France 50°15′N 03°55′E
79 F19 **Haut-Ogooué** *off.* Province du Haut-Ogooué, *var.* Le Haut-Ogooué. ⬖ *province* SE Gabon
Haut-Ogooué, Le *see* Haut-Ogooué
Haut-Ogooué, Province du *see* Haut-Ogooué
103 U7 **Haut-Rhin** ⬖ *department* NE France
74 I6 **Hauts Plateaux** *plateau* Algeria/Morocco
Haut-Zaïre *see* Haut-Congo
38 D9 **Hau'ula** *var.* Hauula. O'ahu, Hawaii, USA, C Pacific Ocean 21°36′N 157°54′W
Hauula *see* Hau'ula
101 O22 **Hauzenberg** Bayern, SE Germany 48°39′N 13°37′E
30 K13 **Havana** Illinois, N USA 40°18′N 90°03′W
Havana *see* La Habana
97 N23 **Havant** S England, United Kingdom 50°51′N 00°59′W
35 Y14 **Havasu, Lake** ⊠ Arizona/California, W USA
95 J23 **Havdrup** Sjælland, E Denmark 55°33′N 12°08′E
100 N10 **Havel** ≋ NE Germany
99 J21 **Havelange** Namur, SE Belgium 50°23′N 05°14′E
100 M11 **Havelberg** Sachsen-Anhalt, NE Germany 52°49′N 12°05′E
149 U5 **Haveliān** Khyber Pakhtunkhwa, NW Pakistan 34°05′N 73°14′E
100 N12 **Havelländ Grosse** *var.* Hauptkanal. *canal* NE Germany
14 J14 **Havelock** Ontario, SE Canada 44°22′N 77°57′W
185 J14 **Havelock** Marlborough, South Island, New Zealand 41°17′S 173°46′E
21 X11 **Havelock** North Carolina, SE USA 34°52′N 76°54′W
184 O11 **Havelock North** Hawke's Bay, North Island, New Zealand 39°40′S 176°53′E
98 M8 **Havelte** Drenthe, NE Netherlands 52°46′N 06°14′E
27 N6 **Haven** Kansas, C USA 37°54′N 97°46′W
97 H21 **Haverfordwest** SW Wales, United Kingdom 51°50′N 04°57′W
97 P20 **Haverhill** E England, United Kingdom 52°05′N 00°26′E
19 O10 **Haverhill** Massachusetts, NE USA 42°46′N 71°02′W
93 G17 **Haverö** Västernorrland, C Sweden 62°25′N 15°04′E
111 I17 **Havířov** Moravskoslezský Kraj, E Czech Republic 49°47′N 18°30′E
111 E17 **Havlíčkův Brod** *Ger.* Deutsch-Brod; *prev.* Německý Brod. Vysočina, C Czech Republic 49°38′N 15°46′E
92 K7 **Havøysund** Finnmark, N Norway 70°59′N 24°39′E
99 F20 **Havré** Hainaut, S Belgium 50°29′N 04°03′E
33 T7 **Havre** Montana, NW USA 48°33′N 109°41′W
Havre *see* le Havre
13 P11 **Havre-St-Pierre** Québec, E Canada 50°16′N 63°36′W
136 B10 **Havsa** Edirne, NW Turkey 41°32′N 26°49′E
38 D8 **Hawai'i** *off.* State of Hawai'i, *also known as* Aloha State, Paradise of the Pacific, *var.* Hawaii. ⬖ *state* USA, C Pacific Ocean
38 G12 **Hawai'i** *var.* Hawaii. *island* Hawaiian Islands, USA, C Pacific Ocean
192 M5 **Hawai'ian Islands** *prev.* Sandwich Islands. *island group* Hawaii, USA
192 L5 **Hawaiian Ridge** *undersea feature* N Pacific Ocean 24°00′N 165°00′W
193 N6 **Hawaiian Trough** *undersea feature* N Pacific Ocean
29 R12 **Hawarden** Iowa, C USA 43°00′N 96°29′W
Hawash *see* Āwash
139 P6 **Hawbayn al Gharbīyah** Al Anbār, C Iraq 34°24′N 42°06′E
185 D21 **Hawea, Lake** ◎ South Island, New Zealand
184 K11 **Hawera** Taranaki, North Island, New Zealand 39°36′S 174°16′E
20 J5 **Hawesville** Kentucky, S USA 37°53′N 86°47′W
38 G11 **Hawi** Hawaii, USA, C Pacific Ocean 20°14′N 155°50′W
38 G11 **Hāwī** *var.* Hawi. Hawaii, USA, C Pacific Ocean 20°13′N 155°49′E
Hawi *see* Hāwī
96 K13 **Hawick** SE Scotland, United Kingdom 55°24′N 02°49′W
139 S4 **Ḩawījah** Kirkūk, C Iraq 35°15′N 43°54′E
139 Y10 **Ḩawīzah, Hawr al** ◎ S Iraq
185 E21 **Hawkdun Range** ▲▲ South Island, New Zealand
184 P10 **Hawke Bay** *bay* North Island, New Zealand
182 I6 **Hawker** South Australia 31°54′S 138°25′E
184 N11 **Hawke's Bay** *off.* Hawkes Bay Region. ⬖ *region* North Island, New Zealand
149 O16 **Hawkes Bay** *bay* SE Pakistan
Hawkes Bay Region *see* Hawke's Bay
15 N12 **Hawkesbury** Ontario, SE Canada 45°36′N 74°38′W
Hawkeye State *see* Iowa
23 T5 **Hawkinsville** Georgia, SE USA 32°16′N 83°28′W
14 B7 **Hawk Junction** Ontario, S Canada 48°05′N 84°34′W
21 N10 **Haw Knob** ▲ North Carolina/Tennessee, SE USA 35°18′N 84°01′W
21 Q9 **Hawksbill Mountain** ▲ North Carolina, SE USA 35°54′N 81°53′W
33 Z16 **Hawk Springs** Wyoming, C USA 41°48′N 104°17′W
Hawlēr *see* Arbīl
29 S5 **Hawley** Minnesota, N USA 46°53′N 96°18′W
25 P7 **Hawley** Texas, SW USA 32°36′N 99°47′W
141 R14 **Ḩawrā'** C Yemen 15°39′N 48°21′E
139 P7 **Ḩawrān, Wadi** *dry watercourse* W Iraq
21 T9 **Haw River** ≋ North Carolina, SE USA
139 U5 **Hawshqūrah** Diyālá, E Iraq 34°34′N 45°33′E
35 S7 **Hawthorne** Nevada, W USA 38°30′N 118°38′W
37 W3 **Haxtun** Colorado, C USA 40°36′N 102°38′W
183 N9 **Hay** New South Wales, SE Australia 34°31′S 144°51′E
11 O10 **Hay** ≋ W Canada
171 S13 **Haya** Pulau Seram, E Indonesia 03°22′S 129°31′E
165 R9 **Hayachine-san** ▲ Honshū, C Japan 39°31′N 141°28′E
103 S4 **Hayange** Moselle, NE France 49°19′N 06°04′E
HaYarden *see* Jordan
Hayastani Hanrapetut'yun *see* Armenia
Hayasui-seto *see* Hōyo-kaikyō
39 N9 **Haycock** Alaska, USA 65°12′N 161°10′W
36 M14 **Hayden** Arizona, SW USA 33°00′N 110°46′W
37 Q3 **Hayden** Colorado, C USA 40°29′N 107°15′W
28 M10 **Hayes** South Dakota, N USA 44°20′N 101°01′W
9 X13 **Hayes** ≋ Manitoba, C Canada
11 P12 **Hayes** ≋ Nunavut, NE Canada
28 M16 **Hayes Center** Nebraska, C USA 40°30′N 101°02′W
39 S10 **Hayes, Mount** ▲ Alaska, USA 63°37′N 146°43′W
21 N11 **Hayesville** North Carolina, SE USA 35°03′N 83°49′W
35 X10 **Hayford Peak** ▲ Nevada, W USA 36°40′N 115°10′W
34 M3 **Hayfork** California, W USA 40°33′N 123°10′W
Hayir, Qasr al *see* Ḩayr al Gharbī, Qaşr al
Haylaastay *see* Sühbaatar
14 I12 **Hay Lake** ◎ Ontario, SE Canada
141 X11 **Haymā'** *var.* Haima. C Oman 19°59′N 56°20′E
136 H13 **Haymana** Ankara, C Turkey 39°26′N 32°30′E
138 J7 **Ḩaymūr, Jabal** ▲▲ W Syria
Haynau *see* Chojnów
22 G4 **Haynesville** Louisiana, S USA 32°57′N 93°08′W
23 P6 **Hayneville** Alabama, S USA 32°13′N 86°34′W
114 M12 **Hayrabolu** Tekirdağ, NW Turkey 41°14′N 27°04′E
136 C10 **Hayrabolu Deresi** ≋ NW Turkey
138 J6 **Ḩayr al Gharbī, Qaşr al** *var.* Qasr al Hayir, Qasr al Hir al Gharbi. *ruins* Ḩimş, C Syria
138 L5 **Ḩayr ash Sharqī, Qaşr al** *var.* Qasr al Hir Ash Sharqī. *ruins* Ḩimş, C Syria
162 J7 **Hayrhan** *var.* Uubulan. Arhangay, C Mongolia 48°37′N 101°58′E
162 J9 **Hayrhandulaan** *var.* Mardzad. Övörhangay, C Mongolia 45°58′N 102°06′E
8 J10 **Hay River** Northwest Territories, W Canada 60°51′N 115°42′W
26 K4 **Hays** Kansas, C USA 38°53′N 99°20′W
28 K12 **Hay Springs** Nebraska, C USA 42°40′N 102°41′W
65 H25 **Haystack, The** ▲ NE Saint Helena 15°55′S 05°40′W
27 N7 **Haysville** Kansas, C USA 37°34′N 97°21′W
117 O7 **Haysyn** *Rus.* Gaysin. Vinnyts'ka Oblast', C Ukraine 48°50′N 29°29′E
27 Y9 **Hayti** Missouri, C USA 36°13′N 89°45′W
29 Q9 **Hayti** South Dakota, N USA 44°40′N 97°22′W
117 O8 **Hayvoron** *Rus.* Gayvorno. Kirovohrads'ka Oblast', C Ukraine 48°20′S 29°52′E
35 N9 **Hayward** California, W USA 37°40′N 122°07′W
30 J4 **Hayward** Wisconsin, N USA 46°02′N 91°26′W
97 O23 **Haywards Heath** SE England, United Kingdom 51°N 00°06′W
146 A11 **Hazar** *prev. Rus.* Cheleken. Balkan Welaýaty, W Turkmenistan 39°26′N 53°07′E
143 S11 **Hazārān, Kūh-e** *var.* Kūh-e ā Hazr. ▲ SE Iran 29°26′N 57°15′E
Hazarat Imam *see* Imām Şāḩib
21 O7 **Hazard** Kentucky, S USA 37°14′N 83°11′W
137 O15 **Hazar Gölü** ◎ C Turkey
153 P15 **Hazārībāg** *var.* Hazāribāgh. Jhārkhand, N India 24°00′N 85°23′E
Hazāribāgh *see* Hazārībāg
103 O1 **Hazebrouck** Nord, N France 50°43′N 02°33′E
30 K9 **Hazel Green** Wisconsin, N USA 42°33′N 90°26′W
192 K9 **Hazel Holme Bank** *undersea feature* S Pacific Ocean 12°49′S 174°30′E
10 K13 **Hazelton** British Columbia, SW Canada 55°15′N 127°38′W
29 N6 **Hazelton** North Dakota, N USA 46°27′N 100°17′W
35 R5 **Hazen** Nevada, W USA 39°33′N 119°02′W
28 L5 **Hazen** North Dakota, N USA 47°18′N 101°37′W
38 L12 **Hazen Bay** *bay* E Bering Sea
9 N1 **Hazen, Lake** ◎ Nunavut, N Canada
139 S5 **Hazīm, Bi'r** *well* C Iraq
23 V6 **Hazlehurst** Georgia, SE USA 31°51′N 82°35′W
22 K6 **Hazlehurst** Mississippi, S USA 31°51′N 90°24′W
18 K15 **Hazlet** New Jersey, NE USA 40°24′N 74°10′W
146 I9 **Hazorasp** *Rus.* Khazarosp. Xorazm Viloyati, W Uzbekistan 41°21′N 61°01′E
147 R13 **Hazratishoh, Qatorkŭhi** *var.* Khrebet Khazretishi, *Rus.* Khrebet Khozretishi. ▲▲ S Tajikistan
Hazr, Kūh-e ā *see* Hazārān, Kūh-e
149 U6 **Hazro** Punjab, E Pakistan 33°55′N 72°33′E
23 R7 **Headland** Alabama, S USA 31°21′N 85°20′W
182 C6 **Head of Bight** *headland* South Australia 31°33′S 131°05′E
33 N10 **Headquarters** Idaho, NW USA 46°38′N 115°52′W
34 M7 **Healdsburg** California, W USA 38°36′N 122°52′W
27 N13 **Healdton** Oklahoma, C USA 34°13′N 97°29′W
183 O12 **Healesville** Victoria, SE Australia 37°41′S 145°31′E
39 R10 **Healy** Alaska, USA 63°51′N 148°58′W
173 R13 **Heard and McDonald Islands** ◇ *Australian external territory* S Indian Ocean
173 R13 **Heard Island** *island* Heard and McDonald Islands, S Indian Ocean
25 U9 **Hearne** Texas, SW USA 30°52′N 96°35′W
12 F12 **Hearst** Ontario, S Canada 49°42′N 83°40′W
194 J5 **Hearst Island** *island* Antarctica
Heart of Dixie *see* Alabama
28 L5 **Heart River** ≋ North Dakota, N USA
31 T13 **Heath** Ohio, N USA 40°01′N 82°26′W
183 N11 **Heathcote** Victoria, SE Australia 36°57′S 144°43′E
97 N22 **Heathrow** ✈ (London) SE England, United Kingdom 51°28′N 00°27′W
21 X5 **Heathsville** Virginia, NE USA 37°55′N 76°29′W
27 R11 **Heavener** Oklahoma, C USA 34°53′N 94°36′W
25 R15 **Hebbronville** Texas, SW USA 27°19′N 98°41′W
163 Q13 **Hebei** *var.* Hebei Sheng, Hopeh, Hopei, Ji; *prev.* Chihli. ⬖ *province* E China
Hebei Sheng *see* Hebei
36 M3 **Heber City** Utah, W USA 40°31′N 111°25′W
27 V10 **Heber Springs** Arkansas, C USA 35°30′N 92°01′W
161 N5 **Hebi** Henan, C China 35°57′N 114°08′E
32 F11 **Hebo** Oregon, NW USA 45°10′N 123°55′W
96 F9 **Hebrides, Sea of the** *sea* NW Scotland, United Kingdom
13 P5 **Hebron** Newfoundland and Labrador, E Canada 58°15′N 62°45′W
31 N11 **Hebron** Indiana, N USA 41°19′N 87°12′W
29 Q17 **Hebron** Nebraska, C USA 40°10′N 97°35′W
28 L5 **Hebron** North Dakota, N USA 46°54′N 102°03′W
138 F11 **Hebron** *var.* Al Khalīl, El Khalil, *Heb.* Hevron; *anc.* Kiriath-Arba. S West Bank 31°30′N 35°E
Hebrus *see* Évros/Maritsa/Meriç
95 N14 **Heby** Västmanland, C Sweden 59°56′N 16°53′E
10 I14 **Hecate Strait** *strait* British Columbia, W Canada
41 W12 **Hecelchakán** Campeche, SE Mexico 20°09′N 90°04′W
160 K13 **Hechi** *var.* Jinchengjiang. Guangxi Zhuangzu Zizhiqu, S China 24°39′N 108°02′E
101 H23 **Hechingen** Baden-Württemberg, S Germany 48°20′N 08°58′E
99 K17 **Hechtel** Limburg, NE Belgium 51°07′N 05°24′E
160 J9 **Hechuan** *var.* Heyang. Chongqing Shi, C China 30°02′N 106°15′E
29 P7 **Hecla** South Dakota, N USA 45°52′N 98°09′W
9 N1 **Hecla, Cape** *headland* Nunavut, N Canada 82°00′N 64°00′W
29 T9 **Hector** Minnesota, N USA 44°44′N 94°43′W
93 F17 **Hede** Jämtland, C Sweden 62°25′N 13°33′E
Hede *see* Sheyang
95 M14 **Hedemora** Dalarna, C Sweden 60°18′N 15°58′E
92 K13 **Hedenäset** *Finn.* Hietaniemi. Norrbotten, N Sweden 66°12′N 23°40′E
95 G23 **Hedensted** Syddanmark, C Denmark 55°47′N 09°43′E
95 N14 **Hedesunda** Gävleborg, C Sweden 60°25′N 17°00′E
95 N14 **Hedesundafjärden** ◎ C Sweden
25 O3 **Hedley** Texas, SW USA 34°52′N 100°39′W
94 I12 **Hedmark** ⬖ *county* S Norway
165 T16 **Hedo-misaki** *headland* Okinawa, SW Japan 26°55′N 128°15′E
29 X15 **Hedrick** Iowa, C USA 41°10′N 92°18′W
99 L16 **Heel** Limburg, SE Netherlands 51°12′N 06°01′E
189 Y12 **Heel Point** *point* Wake Island
98 H9 **Heemskerk** Noord-Holland, W Netherlands 52°31′N 04°40′E
98 M10 **Heerde** Gelderland, E Netherlands 52°24′N 06°02′E
98 L7 **Heerenveen** *Fris.* It Hearrenfean. Fryslân, N Netherlands 52°57′N 05°55′E
98 I8 **Heerhugowaard** Noord-Holland, NW Netherlands 52°40′N 04°50′E
92 O3 **Heer Land** *physical region* C Svalbard
99 M18 **Heerlen** Limburg, SE Netherlands 50°55′N 06°E
99 J19 **Heers** Limburg, NE Belgium 50°46′N 05°17′E
Heerwegen *see* Polkowice
98 K13 **Heesch** Noord-Brabant, S Netherlands 51°44′N 05°32′E
99 K15 **Heeze** Noord-Brabant, SE Netherlands 51°23′N 05°35′E
138 F8 **Hefa** *var.* Haifa, *hist.* Caiffa, Caiphas; *anc.* Sycaminum. Haifa, N Israel 32°49′N 34°59′E
Hefa, Mifraz *see* Mifrats Hefa
161 Q8 **Hefei** *var.* Hofei, *hist.* Luchow. *province capital* Anhui, E China 31°51′N 117°20′E
23 R3 **Heflin** Alabama, S USA 33°39′N 85°35′W
163 X7 **Hegang** Heilongjiang, NE China 47°18′N 130°16′E
164 L10 **Hegura-jima** *island* SW Japan
Heguri-jima *see* Heigun-tō
Hei *see* Heilongjiang
100 H8 **Heide** Schleswig-Holstein, N Germany 54°12′N 09°06′E
101 G20 **Heidelberg** Baden-Württemberg, SW Germany 49°24′N 08°41′E
83 J21 **Heidelberg** Gauteng, NE South Africa 26°31′S 28°21′E
22 M6 **Heidelberg** Mississippi, S USA 31°53′N 88°58′W
Heidenheim *see* Heidenheim an der Brenz
101 J22 **Heidenheim an der Brenz** *var.* Heidenheim. Baden-Württemberg, S Germany 48°41′N 10°09′E
109 U2 **Heidenreichstein** Niederösterreich, N Austria 48°53′N 15°07′E
164 F14 **Heigun-tō** *var.* Heguri-jima. *island* SW Japan
163 W5 **Heihe** *prev.* Ai-hun. Heilongjiang, NE China 50°13′N 127°29′E
162 S8 **Hei He** ≋ C China
Hei-ho *see* Nagqu
83 J22 **Heilbron** Free State, N South Africa 27°17′S 27°58′E
101 H21 **Heilbronn** Baden-Württemberg, SW Germany 49°09′N 09°13′E
Heiligenbeil *see* Mamonovo
109 Q8 **Heiligenblut** Tirol, W Austria 47°04′N 12°50′E
100 K7 **Heiligenhafen** Schleswig-Holstein, N Germany 54°22′N 10°57′E
Heiligenkreuz *see* Žiar nad Hronom
101 J15 **Heiligenstadt** Thüringen, C Germany 51°22′N 10°09′E
163 W8 **Heilongjiang** *var.* Hei, Heilongjiang Sheng, Hei-lung-chiang, Heilungkiang. ⬖ *province* NE China
Heilong Jiang *see* Amur
Heilongjiang Sheng *see* Heilongjiang
98 H9 **Heiloo** Noord-Holland, NW Netherlands 52°36′N 04°43′E
Heilsberg *see* Lidzbark Warmiński
Hei-lung-chiang/Heilungkiang *see* Heilongjiang
Heimaæy *see* Heimaey
92 I4 **Heimaey** *var.* Heimaæy. *island* S Iceland
94 H8 **Heimdal** Sør-Trøndelag, S Norway 63°21′N 10°23′E
Heinaste *see* Ainaži
93 N17 **Heinävesi** Etelä-Savo, E Finland 62°22′N 28°42′E
99 M22 **Heinerscheid** Diekirch, N Luxembourg 50°06′N 06°05′E
98 M10 **Heino** Overijssel, E Netherlands 52°26′N 06°13′E
93 M18 **Heinola** Päijät-Häme, S Finland 61°13′N 26°05′E
101 C16 **Heinsberg** Nordrhein-Westfalen, W Germany 51°02′N 06°01′E
163 U12 **Heishan** Liaoning, NE China 41°43′N 122°12′E
160 H8 **Heishui** *var.* Luhua. Sichuan, C China 32°08′N 102°54′E
99 H17 **Heist-op-den-Berg** Antwerpen, C Belgium 51°04′N 04°43′E
Heitō *see* Pingdong
171 X15 **Heitske** Papua, E Indonesia 07°02′S 138°45′E
Hejanah *see* Al Hījānah
Hejaz *see* Al Ḩijāz
160 M14 **He Jiang** ≋ S China
Hejiayan *see* Lüeyang
158 K6 **Hejing** Xinjiang Uygur Zizhiqu, NW China 42°21′N 86°19′E
Héjjasfalva *see* Vânători
Heka *see* Hoika
137 N14 **Hekimhan** Malatya, C Turkey 38°50′N 37°56′E
92 J4 **Hekla** ▲ S Iceland 63°56′N 19°42′W
Hekou *see* Yanshan, Jiangxi, China
Hekou *see* Yajiang, Sichuan, China
110 J6 **Hel** *Ger.* Hela. Pomorskie, N Poland 54°35′N 18°48′E
186 B6 **Hela** ⬖ *province* W Papua New Guinea
Hela *see* Hel
93 F17 **Helagsfjället** ▲ C Sweden 62°57′N 12°31′E
159 W8 **Helan** *var.* Xigang. Ningxia, N China 38°33′N 106°21′E
162 K14 **Helan Shan** ▲▲ N China
99 M16 **Helden** Limburg, SE Netherlands 51°20′N 06°00′E
139 V4 **Helebce** *Ar.* Ḩalabjah, *var.* Ḩalabja. As Sulaymānīyah, NE Iraq 35°11′N 45°59′E
27 X12 **Helena** Arkansas, C USA 34°32′N 90°34′W
33 R10 **Helena** *state capital* Montana, NW USA 46°36′N 112°02′W
96 H12 **Helensburgh** W Scotland, United Kingdom 56°00′N 04°45′W
184 K5 **Helensville** Auckland, North Island, New Zealand 36°42′S 174°26′E
95 L20 **Helgasjön** ◎ S Sweden
100 G8 **Helgoland** *Eng.* Heligoland. *island* NW Germany
Helgoland Bay *see* Helgoländer Bucht
100 G8 **Helgoländer Bucht** *var.* Helgoland Bay, Heligoland Bight. *bay* NW Germany
Heligoland *see* Helgoland
Heligoland Bight *see* Helgoländer Bucht
Heliopolis *see* Baalbek
92 I4 **Hella** Suðurland, SW Iceland 63°51′N 20°24′W
Hellas *see* Greece
143 N11 **Ḩelleh, Rūd-e** ≋ S Iran
98 N10 **Hellendoorn** Overijssel, E Netherlands 52°22′N 06°27′E
Hellenic Republic *see* Greece
121 Q10 **Hellenic Trough** *undersea feature* Aegean Sea, C Mediterranean Sea 22°00′E 35°30′N
94 E10 **Hellesylt** Møre og Romsdal, S Norway 62°06′N 06°51′E
98 F13 **Hellevoetsluis** Zuid-Holland, SW Netherlands 51°49′N 04°08′E
105 Q12 **Hellín** Castilla-La Mancha, C Spain 38°31′N 01°43′W
115 H19 **Hellinikon** ✈ (Athína) Attikí, C Greece 37°53′N 23°43′E
32 M12 **Hells Canyon** *valley* Idaho/Oregon, NW USA
148 L9 **Helmand** ⬖ *province* S Afghanistan
148 K10 **Helmand, Daryā-ye** *var.* Rūd-e Hīrmand. ≋ Afghanistan/Iran *see also* Hīrmand, Rūd-e
Helmand, Daryā-ye *see* Hīrmand, Rūd-e
Helmantica *see* Salamanca
101 K15 **Helme** ≋ C Germany
99 L15 **Helmond** Noord-Brabant, S Netherlands 51°29′N 05°41′E
96 J7 **Helmsdale** N Scotland, United Kingdom 58°06′N 03°36′W
100 K13 **Helmstedt** Niedersachsen, N Germany 52°14′N 11°01′E
163 Y10 **Helong** Jilin, NE China 42°38′N 129°01′E
36 M4 **Helper** Utah, W USA 39°40′N 110°52′W
100 O10 **Helpter Berge** *hill* NE Germany
95 J22 **Helsingborg** *prev.* Hälsingborg. Skåne, S Sweden 56°N 12°48′E
Helsingfors *see* Helsinki
95 J22 **Helsingør** *Eng.* Elsinore. Hovedstaden, E Denmark 56°03′N 12°38′E
93 M20 **Helsinki** *Swe.* Helsingfors. ● (Finland) Uusimaa, S Finland 60°18′N 24°58′E
97 H25 **Helston** SW England, United Kingdom 50°04′N 05°17′W
Heltau *see* Cisnădie
61 C17 **Helvecia** Santa Fe, C Argentina 31°09′S 60°09′W
97 K15 **Helvellyn** ▲ NW England, United Kingdom 54°31′N 03°00′W
Helvetia *see* Switzerland
Helwân *see* Hilwān
97 N21 **Hemel Hempstead** E England, United Kingdom 51°46′N 00°28′W
35 U16 **Hemet** California, W USA 33°45′N 116°58′W
28 J13 **Hemingford** Nebraska, C USA 42°18′N 103°02′W
21 T13 **Hemingway** South Carolina, SE USA 33°45′N 79°25′W
92 G13 **Hemnesberget** Nordland, C Norway 66°14′N 13°40′E
25 Y8 **Hemphill** Texas, SW USA 31°21′N 93°50′W
25 V11 **Hempstead** Texas, SW USA 30°06′N 96°06′W
95 P20 **Hemse** Gotland, SE Sweden 57°12′N 18°22′E
94 F13 **Hemsedal** *valley* S Norway
161 N6 **Henan** *var.* Henan Sheng, Honan, Yu. ⬖ *province* C China
184 L4 **Hen and Chickens** *island group* N New Zealand
Henan Mongolzu Zizhixian/Henan Sheng *see* Yêgainnyin
105 O7 **Henares** ≋ C Spain
165 P7 **Henashi-zaki** *headland* Honshū, C Japan 40°37′N 139°51′E
102 I16 **Hendaye** Pyrénées-Atlantiques, SW France 43°22′N 01°46′W
136 F11 **Hendek** Sakarya, NW Turkey 40°47′N 30°45′E
61 B21 **Henderson** Buenos Aires, E Argentina 36°18′S 61°43′W
20 I5 **Henderson** Kentucky, S USA 37°50′N 87°35′W
35 X11 **Henderson** Nevada, W USA 36°02′N 114°58′W
21 V8 **Henderson** North Carolina, SE USA 36°20′N 78°26′W
20 G10 **Henderson** Tennessee, S USA 35°27′N 88°40′W
25 W7 **Henderson** Texas, SW USA 32°11′N 94°48′W
30 J12 **Henderson Creek** ≋ Illinois, N USA
186 M9 **Henderson Field** ✈ (Honiara) Guadalcanal, C Solomon Islands 09°28′S 160°02′E
191 O17 **Henderson Island** *atoll* N Pitcairn Group of Islands
21 O10 **Hendersonville** North Carolina, SE USA 35°19′N 82°28′W
20 J8 **Hendersonville** Tennessee, S USA 36°18′N 86°37′W
143 O14 **Hendorābī, Jazīreh-ye** *island* S Iran
55 V10 **Hendrik Top** *var.* Hendriktop. *elevation* C Suriname
Hendriktop *see* Hendrik Top
Hendū Kosh *see* Hindu Kush
14 L12 **Heney, Lac** ◎ Québec, SE Canada
Hengchow *see* Hengyang
161 S15 **Hengch'un** S Taiwan 22°09′N 120°43′E
159 R16 **Hengduan Shan** ▲▲ SW China
98 N12 **Hengelo** Gelderland, E Netherlands 52°03′N 06°19′E
98 O10 **Hengelo** Overijssel, E Netherlands 52°16′N 06°46′E
Hengnan *see* Hengyang
161 N11 **Hengshan** Hunan, S China 27°17′N 112°51′E
160 L4 **Hengshan** Shaanxi, C China 37°57′N 109°17′E
161 O4 **Hengshui** Hebei, E China 37°42′N 115°39′E
161 N12 **Hengyang** *var.* Hengnan, Heng-yang; *prev.* Hengchow. Hunan, S China 26°55′N 112°34′E
Heng-yang *see* Hengyang
117 U11 **Heniches'k** *Rus.* Genichesk. Khersons'ka Oblast', S Ukraine 46°10′N 34°49′E
21 Z4 **Henlopen, Cape** *headland* Delaware, NE USA 38°48′N 75°06′W
Henna *see* Enna
94 M10 **Hennan** Gävleborg, C Sweden 62°01′N 15°55′E
102 G7 **Hennebont** Morbihan, NW France 47°48′N 03°17′W
30 L11 **Hennepin** Illinois, N USA 41°14′N 89°21′W
26 M9 **Hennessey** Oklahoma, C USA 36°06′N 97°54′W
100 N12 **Hennigsdorf** *var.* Hennigsdorf bei Berlin. Brandenburg, NE Germany 52°37′N 13°13′E
Hennigsdorf bei Berlin *see* Hennigsdorf
19 N9 **Henniker** New Hampshire, NE USA 43°10′N 71°47′W
25 S5 **Henrietta** Texas, SW USA 33°49′N 98°13′W
Henrique de Carvalho *see* Saurimo
30 L12 **Henry** Illinois, N USA 41°06′N 89°21′W
21 Y7 **Henry, Cape** *headland* Virginia, NE USA 36°55′N 76°01′W
27 P10 **Henryetta** Oklahoma, C USA 35°26′N 95°58′W
194 M7 **Henry Ice Rise** *ice cap* Antarctica
9 R5 **Henry Kater, Cape** *headland* Baffin Island, Nunavut, NE Canada 69°09′N 66°45′W
33 R13 **Henrys Fork** ≋ Idaho, NW USA
14 E15 **Hensall** Ontario, S Canada 43°25′N 81°28′W
100 J9 **Henstedt-Ulzburg** Schleswig-Holstein, N Germany 53°45′N 09°59′E
163 N7 **Hentiy** *var.* Batshireet, Eg. ⬖ *province* N Mongolia
162 M7 **Hentiyn Nuruu** ▲▲ N Mongolia
183 P10 **Henty** New South Wales, SE Australia 35°33′S 147°03′E
Henzada *see* Hinthada
Heping *see* Huishui
101 G19 **Heppenheim** Hessen, W Germany 49°39′N 08°38′E
32 J11 **Heppner** Oregon, NW USA 45°21′N 119°32′W
160 L15 **Hepu** *var.* Lianzhou. Guangxi Zhuangzu Zizhiqu, S China 21°40′N 109°12′E
Heracleum *see* Irákleio
92 J2 **Heraðsvötn** C Iceland
Herakleion *see* Irákleio
148 K5 **Herāt** *var.* Herat; *anc.* Aria. Herāt, W Afghanistan 34°23′N 62°11′E
148 J5 **Herāt** ⬖ *province* W Afghanistan
103 P14 **Hérault** ⬖ *department* S France
103 P15 **Hérault** ≋ S France
11 T16 **Herbert** Saskatchewan, S Canada 50°27′N 107°09′W
185 F22 **Herbert** Otago, South Island, New Zealand 45°14′S 170°48′E
38 J17 **Herbert Island** *island* Aleutian Islands, Alaska, USA
Herbertshöhe *see* Kokopo
15 Q7 **Hérbertville** Québec, SE Canada 48°23′N 71°42′W
101 G17 **Herborn** Hessen, W Germany 50°40′N 08°18′E
113 I17 **Herceg-Novi** *It.* Castelnuovo; *prev.* Ercegnovi. SW Montenegro 42°28′N 18°35′E
11 X10 **Herchmer** Manitoba, C Canada 57°25′N 94°12′W
186 E8 **Hercules Bay** *bay* E Papua New Guinea
92 K2 **Herðubreið** C Iceland 65°12′N 16°26′W
42 M13 **Heredia** Heredia, C Costa Rica 10°N 84°06′W
42 M12 **Heredia** *off.* Provincia de Heredia. ⬖ *province* N Costa Rica
Heredia, Provincia de *see* Heredia
97 K21 **Hereford** W England, United Kingdom 52°04′N 02°43′W
24 M3 **Hereford** Texas, SW USA 34°49′N 102°25′W
15 Q13 **Hereford, Mont** ▲ Québec, SE Canada 45°04′N 71°38′W
97 K21 **Herefordshire** *cultural region* W England, United Kingdom
191 U11 **Hereheretue** *atoll* Îles Tuamotu, C French Polynesia
105 N10 **Herencia** Castilla-La Mancha, C Spain 39°22′N 03°12′W
99 H18 **Herent** Vlaams Brabant, C Belgium 50°54′N 04°40′E
99 I16 **Herentals** *var.* Herenthals. Antwerpen, N Belgium 51°11′N 04°50′E
Herenthals *see* Herentals
99 H17 **Herenthout** Antwerpen, N Belgium 51°09′N 04°45′E
95 J23 **Herfølge** Sjælland, E Denmark 55°25′N 12°09′E
100 G13 **Herford** Nordrhein-Westfalen, NW Germany 52°07′N 08°41′E
27 O5 **Herington** Kansas, C USA 38°37′N 96°55′W
108 H7 **Herisau** *Fr.* Hérisau. Ausser Rhoden, NE Switzerland 47°23′N 09°17′E
Hérisau *see* Herisau
Héristal *see* Herstal
99 J18 **Herk-de-Stad** Limburg, NE Belgium 50°57′N 05°12′E
Herkulesbad/Herkulesfürdő *see* Băile Herculane
162 M8 **Herlenbayan-Ulaan** *var.* Dulaan. Hentiy, C Mongolia 47°09′N 108°48′E
Herlen Gol/Herlen He *see* Kerulen
35 Q4 **Herlong** California, W USA 40°07′N 120°06′W
97 L26 **Herm** *island* Channel Islands
109 R9 **Hermagor** *Slvn.* Šmohor. Kärnten, S Austria 46°37′N 13°24′E
29 S7 **Herman** Minnesota, N USA 45°49′N 96°08′W
96 L1 **Herma Ness** *headland* NE Scotland, United Kingdom 60°51′N 00°55′W
27 V4 **Hermann** Missouri, C USA 38°43′N 91°26′W
181 Q8 **Hermannsburg** Northern Territory, N Australia 23°59′S 132°55′E
Hermannstadt *see* Sibiu
94 E12 **Hermansverk** Sogn Og Fjordane, S Norway 61°11′N 06°52′E
138 H6 **Hermel** *var.* Hirmil. NE Lebanon 34°23′N 36°19′E
Hermhausen *see* Hajnówka
183 P6 **Hermidale** New South Wales, SE Australia 31°36′S 146°42′E
55 X9 **Herminadorp** Sipaliwini, NE Suriname 05°05′N 54°22′W
32 K11 **Hermiston** Oregon, NW USA 45°50′N 119°17′W
27 T6 **Hermitage** Missouri, C USA 37°57′N 93°21′W
186 D4 **Hermit Islands** *island group* N Papua New Guinea
25 O7 **Hermleigh** Texas, SW USA 32°37′N 100°44′W
138 G7 **Hermon, Mount** *Ar.* Jabal ash Shaykh. ▲ S Syria 35°30′N 33°30′E
Hermopolis Parva *see* Damanhūr
28 J10 **Hermosa** South Dakota, N USA 43°49′N 103°12′W
40 F5 **Hermosillo** Sonora, NW Mexico 28°59′N 110°53′W
Hermoupolis *see* Ermoúpoli
111 N20 **Hernád** *var.* Hornád, *Ger.* Kundert. ≋ Hungary/Slovakia
61 C18 **Hernández** Entre Ríos, E Argentina 32°21′S 60°02′W
23 V11 **Hernando** Florida, SE USA 28°54′N 82°22′W
22 L1 **Hernando** Mississippi, S USA 34°49′N 89°59′W
105 Q2 **Hernani** País Vasco, N Spain 43°16′N 01°59′W
99 F19 **Herne** Vlaams Brabant, C Belgium 50°45′N 04°03′E
101 E14 **Herne** Nordrhein-Westfalen, W Germany 51°32′N 07°12′E
95 F22 **Herning** Midtjylland, W Denmark 56°08′N 08°59′E
Hernösand *see* Härnösand
121 U11 **Herodotus Basin** *undersea feature* E Mediterranean Sea
121 Q12 **Herodotus Trough** *undersea feature* C Mediterranean Sea
29 T11 **Heron Lake** Minnesota, N USA 43°48′N 95°18′W
Herowābād *see* Khalkhāl
95 G16 **Herre** Telemark, S Norway 59°06′N 09°34′E
29 N7 **Herreid** South Dakota, N USA 45°49′N 100°04′W
101 H22 **Herrenberg** Baden-Württemberg, S Germany 48°36′N 08°52′E
104 L14 **Herrera** Andalucía, S Spain 37°22′N 04°50′W
43 R17 **Herrera** *off.* Provincia de Herrera. ⬖ *province* S Panama
104 L10 **Herrera del Duque** Extremadura, W Spain 39°10′N 05°03′W
104 M4 **Herrera de Pisuerga** Castilla y León, N Spain 42°35′N 04°20′W
Herrera, Provincia de *see* Herrera
41 Z13 **Herrero, Punta** *headland* SE Mexico 19°15′N 87°28′W
183 P16 **Herrick** Tasmania, SE Australia 41°07′S 147°53′E
30 L17 **Herrin** Illinois, N USA 37°48′N 89°01′W
20 M6 **Herrington Lake** ⊠ Kentucky, S USA
95 K18 **Herrljunga** Västra Götaland, S Sweden 58°05′N 13°02′E
103 N16 **Hers** ≋ S France
10 I1 **Herschel Island** *island* Yukon, NW Canada
99 I17 **Herselt** Antwerpen, C Belgium 51°04′N 04°53′E
18 G15 **Hershey** Pennsylvania, NE USA 40°17′N 76°39′W
99 K19 **Herstal** *Fr.* Héristal. Liège, E Belgium 50°40′N 05°38′E
97 O21 **Hertford** E England, United Kingdom 51°48′N 00°05′W
21 X8 **Hertford** North Carolina, SE USA 36°11′N 76°30′W
97 O21 **Hertfordshire** *cultural region* E England, United Kingdom
181 Z9 **Hervey Bay** Queensland, E Australia 25°17′S 152°48′E
101 O14 **Herzberg** Brandenburg, E Germany 51°42′N 13°15′E
99 E18 **Herzele** Oost-Vlaanderen, NW Belgium 50°52′N 03°52′E
101 K20 **Herzogenaurach** Bayern, SE Germany 49°34′N 10°52′E
109 W4 **Herzogenburg** Niederösterreich, NE Austria 48°18′N 15°43′E
Herzogenbusch *see* 's-Hertogenbosch
103 N2 **Hesdin** Pas-de-Calais, N France 50°21′N 02°00′E
160 K14 **Heshan** Guangxi Zhuangzu Zizhiqu, S China 23°45′N 108°58′E
159 X10 **Heshui** *var.* Xihuachi. Gansu, C China 35°42′N 108°06′E
99 M25 **Hespérange** Luxembourg, SE Luxembourg 49°34′N 06°10′E
35 U14 **Hesperia** California, W USA 34°25′N 117°17′W
37 P7 **Hesperus Mountain** ▲ Colorado, C USA 37°27′N 108°05′W
10 J6 **Hess** ≋ Yukon, NW Canada
Hesse *see* Hessen
101 J21 **Hesselberg** ▲ S Germany 49°04′N 10°32′E
95 I22 **Hesselo** *island* E Denmark
101 H17 **Hessen** *Eng./Fr.* Hesse. ⬖ *state* C Germany
192 L6 **Hess Tablemount** *undersea feature* C Pacific Ocean 17°49′N 174°15′W
27 N6 **Hesston** Kansas, C USA 38°08′N 97°25′W
93 G15 **Hestkjøltoppen** ▲ C Norway 64°21′N 13°57′E
97 K18 **Heswall** NW England, United Kingdom 53°20′N 03°06′W
153 P12 **Hetaudā** Central, C Nepal
Hétfalu *see* Săcele
28 K7 **Hettinger** North Dakota, N USA 46°00′N 102°38′W
101 L14 **Hettstedt** Sachsen-Anhalt, C Germany 51°39′N 11°31′E
92 P3 **Heuglin, Kapp** *headland* SE Svalbard 78°15′N 22°49′E
163 W17 **Heuksan-jedo** *var.* Hŭksan-gundo. *island group* SW South Korea
187 N10 **Heuru** Makira-Ulawa, SE Solomon Islands 10°13′S 161°25′E
99 J17 **Heusden** Limburg, NE Belgium 51°02′N 05°17′E
98 J13 **Heusden** Noord-Brabant, S Netherlands 51°43′N 05°05′E
102 K3 **Hève, Cap de la** *headland* N France 49°28′N 00°13′W
99 H18 **Heverlee** Vlaams Brabant, C Belgium 50°52′N 04°41′E
111 L22 **Heves** Heves, NE Hungary 47°36′N 20°17′E
111 L22 **Heves** *off.* Heves Megye. ⬖ *county* NE Hungary
Heves Megye *see* Heves
Hevron *see* Hebron
45 Y13 **Hewanorra** ✈ (Saint Lucia) S Saint Lucia 13°44′N 60°57′W
Hewlêr *see* Arbīl
Hewlêr *see* Arbīl
Hewlêr, Parêzga-î *see* Arbīl
Hexian *see* Hezhou
160 L6 **Heyang** Shaanxi, C China 35°14′N 110°02′E
Heyang *see* Hechuan
Heydebrech *see* Kędzierzyn-Kozle
Heydekrug *see* Šilutė
Heyin *see* Guide
97 K16 **Heysham** NW England, United Kingdom 54°02′N 02°54′W
161 O14 **Heyuan** *var.* Yuancheng. Guangdong, S China 23°41′N 114°45′E
182 L12 **Heywood** Victoria, SE Australia 38°09′S 141°38′E
180 K3 **Heywood Islands** *island group* Western Australia
161 O6 **Heze** *var.* Caozhou. Shandong, E China 35°16′N 115°27′E

◆ Country
● Country Capital
◇ Dependent Territory
○ Dependent Territory Capital
⬖ Administrative Regions
✈ International Airport
▲ Mountain
▲▲ Mountain Range
Volcano
≋ River
◎ Lake
⊠ Reservoir

159 U11 Hezheng Gansu, C China 35°29′N 103°36′E
160 M13 Hezhou var. Babu; prev. Hexian. Guangxi Zhuangzu Zizhiqu, S China 24°33′N 111°30′E
159 U11 Hezuo Gansu, C China 34°55′N 102°49′E
23 Z16 Hialeah Florida, SE USA 25°51′N 80°16′W
27 Q3 Hiawatha Kansas, C USA 39°51′N 95°34′W
36 M4 Hiawatha Utah, W USA 39°28′N 111°00′W
29 V4 Hibbing Minnesota, N USA 47°24′N 92°55′W
183 N17 Hibbs, Point headland Tasmania, SE Australia 42°37′S 145°15′E
Hibernia see Ireland
20 F8 Hickman Kentucky, S USA 36°33′N 89°11′W
21 Q9 Hickory North Carolina, SE USA 35°44′N 81°20′W
21 Q9 Hickory, Lake ⊚ North Carolina, SE USA
184 Q7 Hicks Bay Gisborne, North Island, New Zealand 37°36′S 178°18′E
25 S8 Hico Texas, SW USA 31°58′N 98°01′W
Hida see Furukawa
165 T4 Hidaka Hokkaidō, NE Japan 42°53′N 142°24′E
164 I12 Hidaka Hyōgo, Honshū, SW Japan 35°27′N 134°43′E
165 T5 Hidaka-sanmyaku ▲▲ Hokkaidō, NE Japan
41 O6 Hidalgo var. Villa Hidalgo. Coahuila, NE Mexico 27°46′N 99°54′W
41 N8 Hidalgo Nuevo León, NE Mexico 25°59′N 100°27′W
41 O10 Hidalgo Tamaulipas, C Mexico 24°16′N 99°28′W
41 O13 Hidalgo ◆ state C Mexico
40 J7 Hidalgo del Parral var. Parral. Chihuahua, N Mexico 26°58′N 105°40′W
100 N7 Hiddensee island NE Germany
80 G6 Hidiglib, Wadi ≈ NE Sudan
109 U6 Hieflau Salzburg, E Austria 47°36′N 14°34′E
187 P16 Hienghène Province Nord, C New Caledonia 20°43′S 164°54′E
Hierosolyma see Jerusalem
64 N12 Hierro var. Ferro. island Islas Canarias, Spain, NE Atlantic Ocean
Hietaniemi see Hedenäset
164 G13 Higashi-Hiroshima var. Higasihirosima. Hiroshima, Honshū, SW Japan 34°27′N 132°43′E
164 C12 Higashi-suidō strait SW Japan
Higasihirosima see Higashi-Hiroshima
25 P1 Higgins Texas, SW USA 36°06′N 100°01′W
31 P7 Higgins Lake ⊚ Michigan, N USA
27 S4 Higginsville Missouri, C USA 39°04′N 93°43′W
High Atlas see Haut Atlas
30 M5 High Falls Reservoir ⊠ Wisconsin, N USA
44 K12 Highgate C Jamaica 18°16′N 76°53′W
25 X11 High Island Texas, SW USA 29°35′N 94°24′W
31 O5 High Island island Michigan, N USA
30 K15 Highland Illinois, N USA 38°44′N 89°40′W
31 N10 Highland Park Illinois, N USA 42°10′N 87°48′W
21 O10 Highlands North Carolina, SE USA 35°04′N 83°10′W
11 O11 High Level Alberta, W Canada 58°31′N 117°08′W
29 O9 Highmore South Dakota, N USA 44°29′N 99°26′W
171 N3 High Peak ▲ Luzon, N Philippines 15°28′N 120°07′E
High Plains see Great Plains
21 S9 High Point North Carolina, SE USA 35°58′N 80°00′W
18 J13 High Point hill New Jersey, NE USA
11 P13 High Prairie Alberta, W Canada 55°27′N 116°28′W
11 Q16 High River Alberta, SW Canada 50°35′N 113°50′W
21 S9 High Rock Lake ⊠ North Carolina, SE USA
23 V9 High Springs Florida, SE USA 29°49′N 82°36′W
High Veld see Great Karoo
97 J24 High Willhays ▲ SW England, United Kingdom 50°39′N 03°58′W
97 N22 High Wycombe prev. Chepping Wycombe, Chipping Wycombe. SE England, United Kingdom 51°38′N 00°46′W
41 P12 Higos var. El Higo. Veracruz-Llave, E Mexico 21°48′N 98°25′W
102 I16 Higuer, Cap headland NE Spain 43°23′N 01°46′W
45 R5 Higüero, Punta headland W Puerto Rico 18°21′N 67°15′W
45 P9 Higüey var. Salvaleón de Higüey. E Dominican Republic 18°40′N 68°43′W
190 G11 Hihifo ✕ (Matāʻutu) Île Uvea, N Wallis and Futuna
81 N16 Hiiraan off. Gobolka Hiiraan. ◆ region C Somalia
Hiiraan, Gobolka see Hiiraan
118 E4 Hiiumaa var. Hiiumaa Maakond. ◆ province W Estonia
118 D4 Hiiumaa Ger. Dagden, Swe. Dagö. island W Estonia
Hiiumaa Maakond see Hiiumaa
Hijanah see Al Hījānah
105 S6 Híjar Aragón, NE Spain 41°10′N 00°27′W
191 V10 Hikueru atoll Îles Tuamotu, C French Polynesia
184 K3 Hikurangi Northland, North Island, New Zealand 35°37′S 174°16′E
184 Q8 Hikurangi ▲ North Island, New Zealand 37°55′S 177°59′E
192 L11 Hikurangi Trench var. Hikurangi Trough. undersea feature SW Pacific Ocean
Hikurangi Trough see Hikurangi Trench
190 B15 Hikutavake NW Niue
121 Q12 Hilāl, Ra's al headland N Libya 32°55′N 22°09′E
61 A24 Hilario Ascasubi Buenos Aires, E Argentina 39°22′S 62°39′W
101 K17 Hildburghausen Thüringen, C Germany 50°26′N 10°44′E
101 E15 Hilden Nordrhein-Westfalen, W Germany 51°12′N 06°56′E
100 I13 Hildesheim Niedersachsen, N Germany 52°09′N 09°57′E
33 T9 Hilger Montana, NW USA 47°15′N 109°18′W
153 S13 Hili var. Hilli. Rajshahi, NW Bangladesh 25°16′N 89°04′E
Hilla see Al Ḥillah
45 O14 Hillaby, Mount ▲ N Barbados 13°12′N 59°34′W
Ḥillah, Al see Bābil
95 K19 Hillared Västra Götaland, S Sweden 57°37′N 13°10′E
195 R12 Hillary Coast physical region Antarctica
42 G2 Hill Bank Orange Walk, N Belize 17°36′N 88°43′W
33 O14 Hill City Idaho, NW USA 43°18′N 115°03′W
26 K3 Hill City Kansas, C USA 39°23′N 99°51′W
29 V5 Hill City Minnesota, N USA 46°59′N 93°36′W
28 J10 Hill City South Dakota, N USA 43°54′N 103°33′W
65 C24 Hill Cove Settlement West Falkland, Falkland Islands
98 H10 Hillegom Zuid-Holland, W Netherlands 52°18′N 04°35′E
95 J22 Hillerød Hovedstaden, E Denmark 55°56′N 12°19′E
36 M7 Hillers, Mount ▲ Utah, W USA 37°53′N 110°42′W
Hilli see Hili
29 R11 Hills Minnesota, N USA 43°31′N 96°21′W
30 L14 Hillsboro Illinois, N USA 39°09′N 89°29′W
27 N5 Hillsboro Kansas, C USA 38°21′N 97°12′W
27 X5 Hillsboro Missouri, C USA 38°13′N 90°33′W
19 N10 Hillsboro New Hampshire, NE USA 43°06′N 71°52′W
37 Q14 Hillsboro New Mexico, SW USA 32°55′N 107°33′W
29 R4 Hillsboro North Dakota, N USA 47°25′N 97°03′W
31 R14 Hillsboro Ohio, N USA 39°12′N 83°36′W
32 G11 Hillsboro Oregon, NW USA 45°32′N 122°59′W
25 T8 Hillsboro Texas, SW USA 32°01′N 97°08′W
30 K8 Hillsboro Wisconsin, N USA 43°40′N 90°21′W
23 Y14 Hillsboro Canal canal Florida, SE USA
45 Y15 Hillsborough Carriacou, N Grenada 12°28′N 61°28′W
97 G15 Hillsborough E Northern Ireland, United Kingdom 54°27′N 06°06′W
21 U9 Hillsborough North Carolina, SE USA 36°04′N 79°06′W
31 Q10 Hillsdale Michigan, N USA 41°55′N 84°37′W
183 O8 Hillston New South Wales, SE Australia 33°30′S 145°33′E
21 R7 Hillsville Virginia, NE USA 36°46′N 80°44′W
96 L2 Hillswick NE Scotland, United Kingdom 60°28′N 01°37′W
Hill Tippera see Tripura
38 H11 Hilo Hawaii, USA, C Pacific Ocean 19°42′N 155°04′W
18 F9 Hilton New York, NE USA 43°17′N 77°47′W
14 C10 Hilton Beach Ontario, S Canada 46°14′N 83°51′W
21 R16 Hilton Head Island South Carolina, SE USA 32°13′N 80°45′W
21 R16 Hilton Head Island island South Carolina, SE USA
99 J15 Hilvarenbeek Noord-Brabant, S Netherlands 51°29′N 05°08′E
98 J11 Hilversum Noord-Holland, C Netherlands 52°14′N 05°10′E
75 W8 Hilwān var. Helwān, Hilwān, Hulwan, Hulwān. N Egypt 29°51′N 31°20′E
Hilwān see Hilwān
152 J7 Himāchal Pradesh ◆ state NW India
Himalaya/Himalaya Shan see Himalayas
152 M9 Himalayas var. Himalaya, Chin. Himalaya Shan. ▲▲ S Asia
171 P6 Himamaylan Negros, C Philippines 10°04′N 122°52′E
93 K15 Himanka Pohjois-Pohjanmaa, W Finland 64°04′N 23°40′E
Himara see Himarë
113 L23 Himarë var. Himara. Vlorë, S Albania 40°06′N 19°45′E
138 M2 Ḥimār, Wādī al dry watercourse N Syria
154 D9 Himatnagar Gujarāt, W India 23°38′N 73°02′E
109 Y4 Himberg Niederösterreich, E Austria 48°05′N 16°27′E
164 I13 Himeji var. Himezi. Hyōgo, Honshū, SW Japan 34°47′N 134°32′E
164 E14 Hime-jima island SW Japan
Himezi see Himeji
164 L13 Himi Toyama, Honshū, SW Japan 36°54′N 136°59′E
109 S9 Himmelberg Kärnten, S Austria 46°45′N 14°01′E
138 I5 Ḥimş var. Homs; anc. Emesa. Ḥimş, C Syria 34°44′N 36°43′E
138 K6 Ḥimş off. Muḥāfaẓat Ḥimş, var. Homs. ◆ governorate C Syria
138 I5 Ḥimş, Buḥayrat var. Buḥayrat Qaṭṭinah. ⊚ W Syria
171 R7 Hinatuan Mindanao, S Philippines 08°21′N 126°19′E
117 N10 Hînceşti var. Hânceşti; prev. Kotovsk. C Moldova 46°48′N 28°33′E
44 M9 Hinche C Haiti 19°07′N 72°00′W
181 X5 Hinchinbrook Island island Queensland, NE Australia
39 S12 Hinchinbrook Island island Alaska, USA
97 M19 Hinckley C England, United Kingdom 52°33′N 01°21′W
29 V7 Hinckley Minnesota, N USA 46°01′N 92°57′W
36 K5 Hinckley Utah, W USA 39°21′N 112°39′W
18 J9 Hinckley Reservoir ⊠ New York, NE USA
152 I12 Hindaun Rājasthān, N India 26°44′N 77°02′E
Hindenburg/Hindenburg in Oberschlesien see Zabrze
Hindiya see Al Hindīyah
21 O6 Hindman Kentucky, S USA 37°20′N 82°58′W
182 L10 Hindmarsh, Lake ⊚ Victoria, SE Australia
185 G19 Hinds Canterbury, South Island, New Zealand 44°01′S 171°33′E
185 G19 Hinds ≈ South Island, New Zealand
95 H23 Hindsholm island C Denmark
149 S4 Hindu Kush Per. Hendū Kosh. ▲▲ Afghanistan/Pakistan
155 H19 Hindupur Andhra Pradesh, E India 13°46′N 77°33′E
11 O12 Hines Creek Alberta, W Canada 56°14′N 118°36′W
23 W6 Hinesville Georgia, SE USA 31°51′N 81°36′W
154 I12 Hinganghāt Mahārāshtra, C India 20°32′N 78°52′E
149 N15 Hingol ≈ SW Pakistan
154 H13 Hingoli Mahārāshtra, C India 19°45′N 77°08′E
137 R13 Hınıs Erzurum, E Turkey 39°22′N 41°44′E
92 O2 Hinlopenstretet strait N Svalbard
92 G10 Hinnøya Lapp. Iinnasuolu. island C Norway
108 H10 Hinterrhein ≈ SW Switzerland
166 L8 Hinthada var. Henzada. Ayeyawady, SW Myanmar (Burma) 17°36′N 95°26′E
11 O14 Hinton Alberta, SW Canada 53°24′N 117°35′W
26 M10 Hinton Oklahoma, C USA 35°28′N 98°21′W
21 R6 Hinton West Virginia, NE USA 37°42′N 80°54′W
Hios see Chíos
41 N8 Hipólito Coahuila, NE Mexico 25°42′N 101°22′W
Hipponium see Vibo Valentia
164 B13 Hirado Nagasaki, Hirado-shima, SW Japan 33°22′N 129°31′E
164 B13 Hirado-shima island SW Japan
165 P16 Hirakubo-saki headland Ishigaki-jima, SW Japan 24°36′N 124°19′E
154 M11 Hīrākud Reservoir ⊠ E India
Hir al Gharbi, Qasr al see Ḥayr al Gharbī, Qaṣr al
165 Q16 Hirara Okinawa, Miyako-jima, SW Japan 24°48′N 125°17′E
Hir Ash Sharqi, Qasr al see Ḥayr ash Sharqī, Qaṣr al
164 G12 Hirata Shimane, Honshū, SW Japan 35°25′N 132°45′E
136 I13 Hirfanlı Barajı ⊠ C Turkey
155 G18 Hiriyūr Karnātaka, W India 13°58′N 76°33′E
Hîrlău see Hârlău
148 K10 Hirmand, Rūd-e var. Daryā-ye Helmand. ≈ Afghanistan/Iran see also Helmand, Daryā-ye
Hirmand, Rūd-e see Helmand, Daryā-ye
Hirmil see Hermel
165 T5 Hiroo Hokkaidō, NE Japan 42°16′N 143°16′E
165 Q7 Hirosaki Aomori, Honshū, C Japan 40°34′N 140°28′E
164 G13 Hiroshima var. Hirosima. Hiroshima, Honshū, SW Japan 34°23′N 132°26′E
164 F13 Hiroshima off. Hiroshima-ken, var. Hirosima. ◆ prefecture Honshū, SW Japan
Hiroshima-ken see Hiroshima
Hirosima see Hiroshima
Hirschberg/Hirschberg im Riesengebirge/Hirschberg in Riesengebirge see Jelenia Góra
103 Q3 Hirson Aisne, N France 49°56′N 04°05′E
Hîrşova see Hârşova
95 G19 Hirtshals Nordjylland, N Denmark 57°34′N 09°58′E
152 H10 Hisār Haryāna, NW India 29°10′N 75°45′E
114 I10 Hisarya var. Khisarya. Plovdiv, C Bulgaria 42°33′N 24°43′E
162 K7 Hishig Öndör var. Maanĭt. Bulgan, C Mongolia 48°17′N 103°29′E
186 E9 Hisiu Central, SW Papua New Guinea 09°20′S 146°48′E
147 P13 Hisor Rus. Gissar. W Tajikistan 38°34′N 68°29′E
Hispalis see Sevilla
Hispana/Hispania see Spain
44 M7 Hispaniola island Dominion Republic/Haiti
64 F11 Hispaniola Basin var. Hispaniola Trough. undersea feature SW Atlantic Ocean
Hispaniola Trough see Hispaniola Basin
Histonium see Vasto
139 R7 Hīt Al Anbār, SW Iraq 33°38′N 42°50′E
165 P14 Hita Ōita, Kyūshū, SW Japan 33°19′N 130°55′E
165 P12 Hitachi var. Hitati. Ibaraki, Honshū, S Japan 36°40′N 140°42′E
165 P12 Hitachiōta Ibaraki, Honshū, S Japan 36°32′N 140°31′E
Hitati see Hitachi
97 O21 Hitchin E England, United Kingdom 51°57′N 00°17′W
191 Q7 Hitiaa Tahiti, W French Polynesia 17°35′S 149°17′W
164 D15 Hitoyoshi var. Hitoyosi. Kumamoto, Kyūshū, SW Japan 32°12′N 130°48′E
Hitoyosi see Hitoyoshi
94 F7 Hitra prev. Hitteren. island S Norway
Hitteren see Hitra
187 Q11 Hiu island Torres Islands, N Vanuatu
165 O11 Hiuchiga-take ▲ Honshū, C Japan 36°57′N 139°18′E
191 X7 Hiva Oa island Îles Marquises, N French Polynesia
20 M10 Hiwassee Lake ⊠ North Carolina, SE USA
20 M10 Hiwassee River ≈ SE USA
95 H20 Hjallerup Nordjylland, N Denmark 57°10′N 10°10′E
95 M16 Hjälmaren Eng. Lake Hjalmar. ⊚ C Sweden
Hjalmar, Lake see Hjälmaren
95 C14 Hjellestad Hordaland, S Norway 60°15′N 05°13′E
95 D16 Hjelmeland Rogaland, S Norway 59°12′N 06°07′E
94 G10 Hjerkinn Oppland, S Norway 62°13′N 09°37′E
95 L18 Hjo Västra Götaland, S Sweden 58°18′N 14°17′E
95 G19 Hjørring Nordjylland, N Denmark 57°28′N 09°59′E
167 O1 Hkakabo Razi ▲ Myanmar (Burma)/China 28°17′N 97°28′E
166 M2 Hkamti var. Singkaling Hkamti. Sagaing, N Myanmar (Burma) 26°00′N 95°43′E
167 N1 Hkring Bum ▲ N Myanmar (Burma) 27°05′N 97°16′E
83 L21 Hlathikulu var. Hlatikulu. S Swaziland 26°58′S 31°19′E
Hlatikulu see Hlathikulu
Hliboka see Hlyboka
111 F17 Hlinsko var. Hlinsko v Čechách. Pardubický Kraj, C Czech Republic 49°46′N 15°54′E
Hlinsko v Čechách see Hlinsko
117 S6 Hlobyne Rus. Globino. Poltavs'ka Oblast', NE Ukraine 49°24′N 33°16′E
111 H20 Hlohovec Ger. Freistadtl, Hung. Galgóc; prev. Frakštát. Trnavský Kraj, W Slovakia 48°26′N 17°49′E
83 J23 Hlotse var. Leribe. NW Lesotho 28°55′S 28°01′E
111 I17 Hlučín Ger. Hultschin, Pol. Hulczyn. Moravskoslezský Kraj, E Czech Republic 49°54′N 18°11′E
117 S2 Hlukhiv Rus. Glukhov. Sums'ka Oblast', NE Ukraine 51°40′N 33°53′E
119 K21 Hlushkavichy Rus. Glushkevichi. Homyel'skaya Voblasts', SE Belarus 51°34′N 27°47′E
119 L18 Hlusk Rus. Glusk, Glussk. Mahilyowskaya Voblasts', E Belarus 52°54′N 28°41′E
116 K8 Hlyboka Ger. Hliboka, Rus. Glybokaya. Chernivets'ka Oblast', W Ukraine 48°04′N 25°56′E
118 K13 Hlybokaye Rus. Glubokoye. Vitsyebskaya Voblasts', N Belarus 55°08′N 27°41′E
77 Q16 Ho SE Ghana 06°36′N 00°28′E
167 S6 Hoa Binh Hoa Binh, N Vietnam 20°49′N 105°20′E
83 E20 Hoachanas Hardap, C Namibia 23°55′S 18°04′E
Hoai Nhon see Bông Sơn
167 T8 Hoa Lac Quang Binh, C Vietnam 17°54′N 106°24′E
167 S5 Hoang Liên Sơn ▲▲ N Vietnam
Hoang Sa, Quần Đao see Paracel Islands
83 B17 Hoanib ≈ NW Namibia
33 S15 Hoback Peak ▲ Wyoming, C USA 43°04′N 110°34′W
183 P17 Hobart prev. Hobarton, Hobart Town. state capital Tasmania, SE Australia 42°54′S 147°18′E
26 L11 Hobart Oklahoma, C USA 35°03′N 99°04′W
183 P17 Hobart ✕ Tasmania, SE Australia 42°52′S 147°28′E
Hobarton/Hobart Town see Hobart
37 W14 Hobbs New Mexico, SW USA 32°42′N 103°08′W
194 L12 Hobbs Coast physical region Antarctica
23 Z14 Hobe Sound Florida, SE USA 27°03′N 80°08′W
Hobicaurikány see Uricani
54 E12 Hobo Huila, S Colombia 02°34′N 75°28′W
99 G16 Hoboken Antwerpen, N Belgium 51°12′N 04°22′E
158 K3 Hoboksar var. Hoboksar Mongol Zizhixian. Xinjiang Uygur Zizhiqu, NW China 46°48′N 85°42′E
Hoboksar Mongol Zizhixian see Hoboksar
95 G21 Hobro Nordjylland, N Denmark 56°39′N 09°51′E
21 X10 Hobucken North Carolina, SE USA 35°15′N 76°31′W
95 O20 Hoburgen headland SE Sweden 56°54′N 18°07′E
81 P15 Hobyo It. Obbia. Mudug, E Somalia 05°16′N 48°24′E
109 R8 Hochalmspitze ▲ SW Austria 47°00′N 13°19′E
109 Q4 Hochburg Oberösterreich, N Austria 48°10′N 12°57′E
108 F8 Hochdorf Luzern, N Switzerland 47°10′N 08°16′E
109 N8 Hochfeiler It. Gran Pilastro. ▲ Austria/Italy 46°59′N 11°42′E
167 T14 Ho Chi Minh var. Ho Chi Minh City; prev. Saigon. S Vietnam 10°46′N 106°43′E
Ho Chi Minh City see Hô Chi Minh
108 I7 Höchst Vorarlberg, NW Austria 47°28′N 09°40′E
Höchstadt see Höchstadt an der Aisch
101 K19 Höchstadt an der Aisch var. Höchstadt. Bayern, C Germany 49°43′N 10°48′E
108 L9 Hochwilde It. L'Altissima. ▲ Austria/Italy 46°45′N 11°00′E
109 S7 Hochwildstelle ▲ C Austria 47°21′N 13°53′E
31 T14 Hocking River ≈ Ohio, N USA
Hoctúm see Hoctún
41 X12 Hoctún var. Hoctúm. Yucatán, E Mexico 20°48′N 89°14′W
Hodeida see Al Ḥudaydah
20 K6 Hodgenville Kentucky, S USA 37°34′N 85°45′W
11 T17 Hodgeville Saskatchewan, S Canada 50°06′N 106°55′W
76 L9 Hodh ech Chargui ◆ region E Mauritania
Hodh el Garbi see Hodh el Gharbi
76 J10 Hodh el Gharbi var. Hodh el Garbi. ◆ region S Mauritania
111 L25 Hódmezővásárhely Csongrád, SE Hungary 46°27′N 20°18′E
74 J6 Hodna, Chott El var. Chott el-Hodna, Ar. Shatt al-Hodna. salt lake N Algeria
Hodna, Chott el-/Hodna, Shatt al- see Hodna, Chott El
111 G19 Hodonín Ger. Göding. Jihomoravský Kraj, SE Czech Republic 48°52′N 17°07′E
Hödrögö see Nömrög
Hodság/Hodschag see Odžaci
39 R7 Hodzana River ≈ Alaska, USA
Hoei see Huy
99 H19 Hoeilaart Vlaams Brabant, C Belgium 50°46′N 04°28′E
Hoë Karoo see Great Karoo
98 F12 Hoek van Holland Eng. Hook of Holland. Zuid-Holland, W Netherlands 52°00′N 04°07′E
98 L11 Hoenderloo Gelderland, E Netherlands 52°08′N 05°46′E
99 L18 Hoensbroek Limburg, SE Netherlands 50°55′N 05°55′E
163 Y11 Hoeryŏng NE North Korea 42°23′N 129°46′E
99 K18 Hoeselt Limburg, NE Belgium 50°50′N 05°30′E
98 K11 Hoevelaken Gelderland, C Netherlands 52°10′N 05°27′E
Hoey see Huy
101 M18 Hof Bayern, SE Germany 50°19′N 11°55′E
Höfdhakaupstadhur see Skagaströnd
Hofei see Hefei
101 G18 Hofheim am Taunus Hessen, W Germany 50°04′N 08°27′E
Hofmarkt see Odorheiu Secuiesc
92 L3 Höfn Austurland, SE Iceland 64°14′N 15°17′W
94 N13 Hofors Gävleborg, C Sweden 60°33′N 16°21′E
92 J6 Hofsjökull glacier C Iceland
92 J1 Hofsós Norðurland Vestra, N Iceland 65°54′N 19°25′W
164 E13 Hōfu Yamaguchi, Honshū, SW Japan 34°01′N 131°34′E
92 I4 Hofuðborgarsvæðið SW Iceland
Hofuf see Al Hufūf
95 J22 Höganäs Skåne, S Sweden 56°11′N 12°39′E
183 P14 Hogan Group island group Tasmania, SE Australia
23 R4 Hogansville Georgia, SE USA 33°10′N 84°55′W
39 P8 Hogatza River ≈ Alaska, USA
28 I14 Hogback Mountain ▲ Nebraska, C USA 41°40′N 103°44′W
95 G14 Høgevarde ▲ S Norway 60°19′N 09°27′E
Högfors see Karkkila
31 P5 Hog Island island Michigan, N USA
21 Y6 Hog Island island Virginia, NE USA
Hogoley Islands see Chuuk Islands
95 N20 Högsby Kalmar, S Sweden 57°10′N 16°03′E
36 K1 Hogup Mountains ▲▲ Utah, W USA
101 E17 Hohe Acht ▲ W Germany 50°23′N 07°00′E
Hohenelbe see Vrchlabí
108 I7 Hohenems Vorarlberg, W Austria 47°23′N 09°43′E
Hohenmauth see Vysoké Mýto
Hohensalza see Inowrocław
Hohenstadt see Zábřeh
Hohenstein in Ostpreussen see Olsztynek
20 I9 Hohenwald Tennessee, S USA 35°33′N 87°31′W
101 L17 Hohenwarte-Stausee ⊠ C Germany
Hohes Venn see Hautes Fagnes
109 Q8 Hohe Tauern ▲▲ W Austria
163 O13 Hohhot var. Huhehot, Huhohaote, Mong. Kukukhoto; prev. Kweisui, Kwesui. Nei Mongol Zizhiqu, N China 40°49′N 111°37′E
162 F7 Hohmorit var. Sayn-Ust. Govĭ-Altay, W Mongolia 47°23′N 94°19′E
103 U6 Hohneck ▲ NE France 48°04′N 07°01′E
77 Q16 Hohoe E Ghana 07°08′N 00°32′E
164 E12 Hōhoku Yamaguchi, Honshū, SW Japan 34°15′N 130°56′E
159 O11 Hoh Sai Hu ⊚ C China
159 N11 Hoh Xil Hu ⊚ C China
158 L11 Hoh Xil Shan ▲▲ W China
167 U10 Hôi An prev. Faifo. Quang Nam-Đa Nǎng, C Vietnam 15°54′N 108°19′E
Hoï-Hao/Hoihow see Haikou
159 S11 Hoika prev. Heka. Qinghai, W China 35°49′N 99°50′E
81 F17 Hoima W Uganda 01°25′N 31°22′E
26 L5 Hoisington Kansas, C USA 38°31′N 98°46′W
146 D12 Hojagala Rus. Khodzhakala. Balkan Welaýaty, W Turkmenistan 38°46′N 56°14′E
146 M13 Hojambaz Rus. Khodzhambas. Lebap Welaýaty, E Turkmenistan 38°11′N 64°33′E
95 H23 Højby Syddtjylland, C Denmark 55°20′N 10°27′E
95 F24 Højer Syddanmark, SW Denmark 54°57′N 08°43′E
164 E14 Hōjō var. Hôzyô. Ehime, Shikoku, SW Japan 33°58′N 132°47′E
184 J3 Hokianga Harbour inlet SE Tasman Sea
185 F17 Hokitika West Coast, South Island, New Zealand 42°44′S 170°58′E
165 U4 Hokkai-dō ◆ territory Hokkaidō, NE Japan
165 T3 Hokkaidō prev. Ezo, Yeso, Yezo. island NE Japan
95 G15 Hokksund Buskerud, S Norway 59°45′N 09°55′E
143 S4 Hokmābād Khorāsān-e Razavī, N Iran 36°37′N 57°34′E
Hokō see Pohang
Hoko-guntō/Hoko-shotō see Penghu Liedao
Hoktemberyan see Armavir
94 F13 Hol Buskerud, S Norway 60°36′N 08°18′E
81 K19 Hola Tana River, SE Kenya 01°06′S 40°01′E
117 R11 Hola Prystan' Rus. Golaya Pristan. Khersons'ka Oblast', S Ukraine 46°31′N 32°31′E
95 I23 Holbæk Sjælland, E Denmark 55°42′N 11°42′E
Holboo see Santmargats
183 P10 Holbrook New South Wales, SE Australia 35°45′S 147°18′E
37 N11 Holbrook Arizona, SW USA 34°54′N 110°09′W
27 S5 Holden Missouri, C USA 38°42′N 93°59′W
36 K5 Holden Utah, W USA 39°06′N 112°27′W
27 O11 Holdenville Oklahoma, C USA 35°05′N 96°25′W
29 O16 Holdrege Nebraska, C USA 40°28′N 99°28′W
35 X3 Hole in the Mountain Peak ▲ Nevada, W USA 40°54′N 115°06′W
155 G20 Hole Narsipur Karnātaka, W India 12°46′N 76°14′E
111 H18 Holešov Ger. Holleschau. Zlínský Kraj, E Czech Republic 49°20′N 17°35′E
45 N14 Holetown prev. Jamestown. W Barbados 13°11′N 59°38′W
31 Q12 Holgate Ohio, N USA 41°12′N 84°06′W
44 I7 Holguín Holguín, SE Cuba 20°51′N 76°16′W
23 V12 Holiday Florida, SE USA 28°11′N 82°44′W
163 S9 Holin Gol prev. Hulingol. Nei Mongol Zizhiqu, N China 45°36′N 119°54′E
39 O12 Holitna River ≈ Alaska, USA
94 J13 Höljes Värmland, C Sweden 60°54′N 12°34′E
109 X3 Hollabrunn Niederösterreich, NE Austria 48°33′N 16°06′E
36 L3 Holladay Utah, W USA 40°39′N 111°49′W
11 X16 Holland Manitoba, S Canada 49°36′N 98°52′W
31 O9 Holland Michigan, N USA 42°47′N 86°06′W
25 T9 Holland Texas, SW USA 30°52′N 97°24′W
Holland see Netherlands
22 K4 Hollandale Mississippi, S USA 33°10′N 90°51′W
Hollandia see Jayapura
Hollandsch Diep see Hollands Diep
99 H14 Hollands Diep var. Hollandsch Diep. channel SW Netherlands
Holleschau see Holešov
25 R5 Holliday Texas, SW USA 33°49′N 98°41′W
18 E15 Hollidaysburg Pennsylvania, NE USA 40°24′N 78°22′W
21 S6 Hollins Virginia, NE USA 37°20′N 79°56′W
26 J12 Hollis Oklahoma, C USA 34°42′N 99°56′W
35 O10 Hollister California, W USA 36°51′N 121°25′W
27 T8 Hollister Missouri, C USA 36°37′N 93°13′W
93 M19 Hollola Päijät-Häme, S Finland 61°N 25°32′E
98 K4 Hollum Fryslân, N Netherlands 53°27′N 05°38′E
95 J23 Höllviken prev. Höllviksnäs. Skåne, S Sweden 55°25′N 12°57′E
Höllviksnäs see Höllviken
37 W6 Holly Colorado, C USA 38°03′N 102°07′W
31 R9 Holly Michigan, N USA 42°47′N 83°37′W
21 S14 Holly Hill South Carolina, SE USA 33°19′N 80°24′W
21 W11 Holly Ridge North Carolina, SE USA 34°31′N 77°31′W
22 L1 Holly Springs Mississippi, S USA 34°47′N 89°25′W
23 Z15 Hollywood Florida, SE USA 26°00′N 80°09′W
8 J6 Holman Victoria Island, Northwest Territories, N Canada 70°42′N 117°45′W
92 I2 Hólmavík Vestfirðir, NW Iceland 65°42′N 21°43′W
30 J7 Holmen Wisconsin, N USA 43°57′N 91°14′W
23 R8 Holmes Creek ≈ Alabama/Florida, SE USA
95 H16 Holmestrand Vestfold, S Norway 59°29′N 10°20′E
93 J16 Holmön island N Sweden
95 E22 Holmsland Klit beach W Denmark
93 J16 Holmsund Västerbotten, N Sweden 63°42′N 20°26′E
95 Q18 Holmudden headland SE Sweden 57°59′N 19°14′E
138 F10 Holon var. Kholon; prev. Holon. Tel Aviv, C Israel 32°01′N 34°46′E
Holon see Holon
163 P7 Hölönbuyr var. Bayan. Dornod, E Mongolia 47°56′N 112°58′E
117 P8 Holovanivs'k Rus. Golovanevsk. Kirovohrads'ka Oblast', C Ukraine 48°21′N 30°26′E
95 F21 Holstebro Midtjylland, W Denmark 56°22′N 08°38′E
95 F23 Holsted Syddtjylland, W Denmark 55°30′N 08°54′E
29 T13 Holstein Iowa, C USA 42°29′N 95°32′W
Holsteinborg/Holsteinsborg/Holstenborg/Holstensborg see Sisimiut
21 O8 Holston River ≈ Tennessee, S USA
31 Q9 Holt Michigan, N USA 42°38′N 84°31′W
98 N10 Holten Overijssel, E Netherlands 52°16′N 06°25′E
27 P3 Holton Kansas, C USA 39°27′N 95°44′W
27 U5 Holts Summit Missouri, C USA 38°38′N 92°07′W
35 X17 Holtville California, W USA 32°48′N 115°22′W
98 L5 Holwerd Fris. Holwert. Fryslân, N Netherlands 53°22′N 05°51′E
Holwert see Holwerd
39 O11 Holy Cross Alaska, USA 62°12′N 159°46′W
37 R4 Holy Cross, Mount Of The ▲ Colorado, C USA 39°28′N 106°28′W
97 I18 Holyhead Wel. Caer Gybi. NW Wales, United Kingdom 53°19′N 04°38′W
97 H18 Holy Island island NW Wales, United Kingdom
96 L12 Holy Island island NE England, United Kingdom
37 W3 Holyoke Colorado, C USA 40°31′N 102°18′W
18 M11 Holyoke Massachusetts, NE USA 42°12′N 72°37′W
101 I14 Holzminden Niedersachsen, C Germany 51°49′N 09°27′E
81 G19 Homa Bay Homa Bay, W Kenya 0°31′S 34°30′E
81 G19 Homa Bay ◆ county W Kenya
Homāyūnshahr see Khomeynīshahr
77 P11 Hombori Mopti, S Mali 15°13′N 01°39′W
101 E20 Homburg Saarland, SW Germany 49°20′N 07°20′E
9 R5 Home Bay bay Baffin Bay, Nunavut, NE Canada
Homenau see Humenné
39 Q13 Homer Alaska, USA 59°38′N 151°33′W
22 H4 Homer Louisiana, S USA 32°47′N 93°03′W
18 H10 Homer New York, NE USA 42°38′N 76°10′W
23 V7 Homerville Georgia, SE USA 31°02′N 82°45′W
23 Y16 Homestead Florida, SE USA 25°28′N 80°28′W
27 O9 Hominy Oklahoma, C USA 36°24′N 96°24′W
94 H8 Hommelvik Sør-Trøndelag, S Norway 63°24′N 10°48′E
95 C16 Hommersåk Rogaland, S Norway 58°55′N 05°51′E
155 H15 Homnābād Karnātaka, C India 17°46′N 77°08′E
22 J7 Homochitto River ≈ Mississippi, S USA
83 N20 Homoine Inhambane, SE Mozambique 23°51′S 35°04′E
112 O12 Homoljske Planine ▲▲ E Serbia
Homonna see Humenné
Homs see Al Khums, Libya
Homs see Ḥimş
119 P19 Homyel' Rus. Gomel'. Homyel'skaya Voblasts', SE Belarus 52°25′N 31°E
118 L12 Homyel' Vitsyebskaya Voblasts', N Belarus 55°20′N 28°52′E
119 L19 Homyel'skaya Voblasts' Rus. Gomel'skaya Oblast'. ◆ province SE Belarus
Honan see Henan, China
Honan see Luoyang, China
164 U4 Honbetsu Hokkaidō, NE Japan 43°09′N 143°46′E
Honctő see Gurahonț
54 E9 Honda Tolima, C Colombia 05°12′N 74°45′W
83 D24 Hondeklip Afr. Hondeklipbaai. Northern Cape, W South Africa 30°15′S 17°17′E
Hondeklipbaai see Hondeklip
11 Q13 Hondo Alberta, W Canada 54°43′N 113°14′W
25 Q12 Hondo Texas, SW USA 29°21′N 99°09′W
42 G1 Hondo ≈ Central America
Hondo see Honshū
Hondo see Amakusa
42 G6 Honduras off. Republic of Honduras. ◆ republic Central America
Honduras, Golfo de see Honduras, Gulf of
42 H4 Honduras, Gulf of Sp. Golfo de Honduras. gulf W Caribbean Sea
Honduras, Republic of see Honduras
11 V12 Hone Manitoba, C Canada 56°13′N 101°12′W
21 P12 Honea Path South Carolina, SE USA 34°27′N 82°23′W
95 H14 Hønefoss Buskerud, S Norway 60°10′N 10°15′E
31 S12 Honey Creek ≈ Ohio, N USA
25 V5 Honey Grove Texas, SW USA 33°34′N 95°54′W
35 Q4 Honey Lake ⊚ California, W USA
102 L4 Honfleur Calvados, N France 49°25′N 00°14′E
Hon Gai see Ha Long
161 O8 Hong'an prev. Huang'an. Hubei, C China 31°20′N 114°43′E
Hongay see Ha Long
Hong Gai see Ha Long
161 O15 Honghai Wan bay N South China Sea
Hông Hà, Sông see Red River
161 O7 Hong He ≈ C China
161 N9 Hong Hu ⊚ C China
160 L11 Hongjiang Hunan, S China 27°09′N 109°58′E
Hongjiang see Wangcang
161 O15 Hong Kong off. Hong Kong Special Administrative Region, var. Hong Kong S.A.R., Chin. Xianggang Tebie Xingzhengqu. S China 22°17′N 114°09′E
Hong Kong S.A.R. see Hong Kong
Hong Kong Special Administrative Region see Hong Kong
160 L4 Hongliu He ≈ C China
160 L4 Hongliu He ≈ C China
159 P8 Hongliuwan var. Aksay, Aksay Kazakzu Zizhixian. Gansu, N China 39°25′N 94°09′E
159 P7 Hongliuyuan Gansu, N China 41°02′N 95°24′E
Hongor see Delgereh
161 S8 Hongqiao ✕ (Shanghai) Shanghai Shi, E China 31°28′N 121°08′E
160 K14 Hongshui He ≈ S China
160 M5 Hongtong var. Dahuaishu. Shanxi, C China 36°30′N 111°42′E
164 J15 Hongū Wakayama, Honshū, SW Japan 33°50′N 135°42′E
Honguedo, Détroit d' see Honguedo Passage
15 Y5 Honguedo Passage var. Honguedo Strait, Fr. Détroit d'Honguedo. strait Québec, E Canada
Honguedo Strait see Honguedo Passage
Hongwan see Hongwansi

● Country
● Country Capital
◇ Dependent Territory
○ Dependent Territory Capital
◆ Administrative Regions
✕ International Airport
▲ Mountain
▲▲ Mountain Range
Volcano
≈ River
⊚ Lake
⊠ Reservoir

159 S8 **Hongwansi** *var.* Sunan, Sunan Yugurzu Zizhixian; *prev.* Hongwan. Gansu, N China 38°55′N 99°29′E
163 X13 **Hongwŏn** E North Korea 40°03′N 127°54′E
160 H7 **Hongyuan** *var.* Qiongxi; *prev.* Hurama. Sichuan, C China 32°49′N 102°40′E
161 Q7 **Hongze Hu** *var.* Hung-tse Hu. ◎ E China
186 L9 **Honiara** ● (Solomon Islands) Guadalcanal, C Solomon Islands 09°27′S 159°56′E
165 P8 **Honjō** *var.* Honzyô, Yurihonjō. Akita, Honshū, C Japan 39°23′N 140°03′E
93 K18 **Honkajoki** Satakunta, SW Finland 62°00′N 22°15′E
92 K7 **Honningsvåg** Finnmark, N Norway 70°58′N 25°59′E
95 I19 **Hönö** Västra Götaland, S Sweden 57°42′N 11°39′E
38 G11 **Honoka'a** Hawaii, USA, C Pacific Ocean 20°04′N 155°27′W
38 G11 **Honoka'a** *var.* Honokaa. Hawaii, USA, C Pacific Ocean 20°04′N 155°27′W
Honokaa *see* Honoka'a
38 D9 **Honolulu** *state capital* O'ahu, Hawaii, USA, C Pacific Ocean 21°18′N 157°52′W
38 H11 **Honomú** *var.* Honomu. Hawaii, USA, C Pacific Ocean 19°51′N 155°06′W
105 P10 **Honrubia** Castilla-La Mancha, C Spain 39°36′N 02°17′W
164 M12 **Honshū** *var.* Hondo, Honsyû. *island* SW Japan
Honsyû *see* Honshū
Honte *see* Westerschelde
Honzyô *see* Honjō
8 K8 **Hood** ≈ Nunavut, NW Canada
Hood Island *see* Española, Isla
32 H11 **Hood, Mount** ▲ Oregon, NW USA 45°22′N 121°41′W
32 H11 **Hood River** Oregon, NW USA 45°44′N 121°31′W
98 H10 **Hoofddorp** Noord-Holland, W Netherlands 52°18′N 04°41′E
99 G15 **Hoogerheide** Noord-Brabant, S Netherlands 51°25′N 04°19′E
98 N8 **Hoogeveen** Drenthe, NE Netherlands 52°44′N 06°30′E
98 O6 **Hoogezand-Sappemeer** Groningen, NE Netherlands 53°10′N 06°47′E
98 J8 **Hoogkarspel** Noord-Holland, NW Netherlands 52°42′N 04°59′E
98 N5 **Hoogkerk** Groningen, NE Netherlands 53°13′N 06°30′E
98 G13 **Hoogvliet** Zuid-Holland, SW Netherlands 51°51′N 04°23′E
26 I8 **Hooker** Oklahoma, C USA 36°51′N 101°12′W
97 E21 **Hook Head** *Ir.* Rinn Duáin. *headland* SE Ireland 52°07′N 06°55′W
Hook of Holland *see* Hoek van Holland
Hoolt *see* Tögrög
39 W13 **Hoonah** Chichagof Island, Alaska, USA 58°05′N 135°21′W
38 L11 **Hooper Bay** Alaska, USA 61°31′N 166°06′W
31 N13 **Hoopeston** Illinois, N USA 40°28′N 87°40′W
95 K22 **Höör** Skåne, S Sweden 55°55′N 13°33′E
98 I9 **Hoorn** Noord-Holland, NW Netherlands 52°38′N 05°04′E
18 L10 **Hoosic River** ≈ New York, NE USA
Hoosier State *see* Indiana
35 Y11 **Hoover Dam** *dam* Arizona/Nevada, W USA
Höövör *see* Baruunbayan-Ulaan
137 Q11 **Hopa** Artvin, NE Turkey 41°23′N 41°28′E
18 J14 **Hopatcong** New Jersey, NE USA 40°55′N 74°39′W
10 M17 **Hope** British Columbia, SW Canada 49°21′N 121°28′W
39 R12 **Hope** Alaska, USA 60°55′N 149°38′W
27 T14 **Hope** Arkansas, C USA 33°40′N 93°36′W
31 P14 **Hope** Indiana, N USA 39°18′N 85°46′W
29 Q5 **Hope** North Dakota, N USA 47°18′N 97°42′W
13 Q7 **Hopedale** Newfoundland and Labrador, NE Canada 55°26′N 60°14′W
Hopeh/Hopei *see* Hebei
180 K13 **Hope, Lake** *salt lake* Western Australia
41 X13 **Hopelchén** Campeche, SE Mexico 19°46′N 89°50′W
21 U11 **Hope Mills** North Carolina, SE USA 34°58′N 78°57′W
183 O7 **Hope, Mount** New South Wales, SE Australia 32°49′S 145°55′E
92 P4 **Hopen** *island* SE Svalbard
197 Q4 **Hope, Point** *headland* Alaska, USA
12 M3 **Hopes Advance, Cap** *cape* Québec, NE Canada
182 L10 **Hopetoun** Victoria, SE Australia 35°46′S 142°23′E
83 H23 **Hopetown** Northern Cape, W South Africa 29°37′S 24°05′E
21 W6 **Hopewell** Virginia, NE USA 37°16′N 77°15′W
109 O7 **Hopfgarten im Brixental** Tirol, W Austria 47°28′N 12°14′E
181 N8 **Hopkins Lake** *salt lake* Western Australia
182 M12 **Hopkins River** ≈ Victoria, SE Australia
20 I7 **Hopkinsville** Kentucky, S USA 36°50′N 87°30′W
34 M6 **Hopland** California, W USA 38°58′N 123°09′W
95 G24 **Hoptrup** Syddanmark, SW Denmark 55°09′N 09°27′E
Hoqin Zuoyi Zhongji *see* Baokang
32 F9 **Hoquiam** Washington, NW USA 46°58′N 123°53′W
29 R6 **Horace** North Dakota, N USA 46°44′N 96°54′W
117 T14 **Hora Roman-Kosh** ▲ S Ukraine 44°37′N 34°13′E
137 R12 **Horasan** Erzurum, NE Turkey 40°03′N 42°10′E
101 G22 **Horb am Neckar** Baden-Württemberg, S Germany 48°27′N 08°42′E
95 K23 **Hörby** Skåne, S Sweden 55°51′N 13°42′E
43 P16 **Horconcitos** Chiriquí, W Panama 08°20′N 82°10′W
95 C14 **Hordaland** ⬖ *county* S Norway
116 H13 **Horezu** Vâlcea, SW Romania 45°06′N 24°00′E
108 G7 **Horgen** Zürich, N Switzerland 47°16′N 08°36′E
Horgo *see* Tariat
Hōrin *see* Fenglin
163 O13 **Horinger** Nei Mongol Zizhiqu, N China 40°23′N 111°48′E
Horiult *see* Bogd
11 U17 **Horizon** Saskatchewan, S Canada 49°33′N 105°05′W
192 K9 **Horizon Bank** *undersea feature* S Pacific Ocean
192 L10 **Horizon Deep** *undersea feature* W Pacific Ocean
95 L14 **Hörken** Örebro, S Sweden 60°03′N 14°55′E
119 O15 **Horki** *Rus.* Gorki. Mahilyowskaya Voblasts', E Belarus 54°18′N 31°E
195 O10 **Horlick Mountains** ▲▲ Antarctica
117 X7 **Horlivka** *Rom.* Adâncata, *Rus.* Gorlovka. Donets'ka Oblast', E Ukraine 48°19′N 38°04′E
143 V11 **Ḩormak** Sīstān va Balūchestān, SE Iran 30°00′N 60°50′E
143 R13 **Hormozgān** *off.* Ostān-e Hormozgān. ⬖ *province* S Iran
Hormozgān, Ostān-e *see* Hormozgān
Hormoz, Tangeh-ye *see* Hormuz, Strait of
141 W6 **Hormuz, Strait of** *var.* Strait of Ormuz, *Per.* Tangeh-ye Hormoz. *strait* Iran/Oman
109 W2 **Horn** Niederösterreich, NE Austria 48°40′N 15°40′E
95 M18 **Horn** Östergötland, S Sweden 57°54′N 15°49′E
8 J9 **Horn** ≈ Northwest Territories, NW Canada
Hornád *see* Hernád
8 I6 **Hornaday** ≈ Northwest Territories, NW Canada
92 H13 **Hornavan** ◎ N Sweden
65 C24 **Hornby Mountains** *hill range* West Falkland, Falkland Islands
Horn, Cape *see* Hornos, Cabo de
97 O18 **Horncastle** E England, United Kingdom 53°12′N 00°07′W
95 N14 **Horndal** Dalarna, C Sweden 60°16′N 16°25′E
93 I16 **Hörnefors** Västerbotten, N Sweden 63°37′N 19°54′E
18 F11 **Hornell** New York, NE USA 42°19′N 77°38′W
Horné Nové Mesto *see* Kysucké Nové Mesto
12 F12 **Hornepayne** Ontario, S Canada 49°14′N 84°48′W
94 D10 **Hornindalsvatnet** ◎ S Norway
101 G22 **Hornisgrinde** ▲ SW Germany 48°37′N 08°13′E
22 M9 **Horn Island** *island* Mississippi, S USA
Hornja Łužica *see* Oberlausitz
63 J26 **Hornos, Cabo de** *Eng.* Cape Horn. *headland* S Chile 55°52′S 67°00′W
117 S10 **Hornostayivka** Khersons'ka Oblast', S Ukraine 47°00′N 33°42′E
183 T9 **Hornsby** New South Wales, SE Australia 33°44′S 151°08′E
97 O16 **Hornsea** E England, United Kingdom 53°54′N 00°10′W
94 O11 **Hornslandet** *peninsula* C Sweden
95 H22 **Hornslet** Midtjylland, C Denmark 56°19′N 10°20′E
92 O4 **Hornsundtind** ▲ S Svalbard 76°54′N 16°07′E
Horochów *see* Horokhiv
Horodenka *see* Gorodenka
117 Q2 **Horodnya** *Rus.* Gorodnya. Chernihivs'ka Oblast', NE Ukraine 51°54′N 31°30′E
116 K6 **Horodok** Khmel'nyts'ka Oblast', W Ukraine 49°10′N 26°34′E
116 H5 **Horodok** *Pol.* Gródek Jagielloński, *Rus.* Gorodok, Gorodok Yagellonski. L'vivs'ka Oblast', NW Ukraine 49°48′N 23°39′E
117 Q6 **Horodyshche** *Rus.* Gorodishche. Cherkas'ka Oblast', C Ukraine 49°19′N 31°27′E
165 T3 **Horokanai** Hokkaidō, NE Japan 44°02′N 142°08′E
116 J4 **Horokhiv** *Pol.* Horochów, *Rus.* Gorokhov. Volyns'ka Oblast', NW Ukraine 50°31′N 24°50′E
165 T4 **Horoshiri-dake** *var.* Horosiri Dake. ▲ Hokkaidō, N Japan 42°43′N 142°41′E
Horosiri Dake *see* Horoshiri-dake
111 C17 **Hořovice** *Ger.* Horowitz. Střední Čechy, W Czech Republic 49°49′N 13°53′E
Horowitz *see* Hořovice
Horqin Zuoyi Houqi *see* Ganjig
Horqin Zuoyi Zhongqi *see* Bayan Huxu
62 O5 **Horqueta** Concepción, C Paraguay 23°24′S 56°53′W
55 O12 **Horqueta Minas** Amazonas, S Venezuela 02°20′N 63°32′W
95 J20 **Horred** Västra Götaland, S Sweden 57°22′N 12°25′E
151 J19 **Horsburgh Atoll** *var.* Goidhoo Atoll. *atoll* N Maldives
20 K7 **Horse Cave** Kentucky, S USA 37°10′N 85°54′W
37 V6 **Horse Creek** ≈ Colorado, C USA
27 S6 **Horse Creek** ≈ Missouri, C USA
18 G11 **Horseheads** New York, NE USA 42°10′N 76°49′W
37 P13 **Horse Mount** ▲ New Mexico, SW USA 33°58′N 108°10′W
95 G22 **Horsens** Syddanmark, C Denmark 55°53′N 09°53′E
65 F25 **Horse Pasture Point** *headland* W Saint Helena 15°57′S 05°46′W
33 N13 **Horseshoe Bend** Idaho, NW USA 43°55′N 116°11′W
36 L13 **Horseshoe Reservoir** ⊠ Arizona, SW USA
64 M9 **Horseshoe Seamounts** *undersea feature* E Atlantic Ocean 36°30′N 15°00′W
182 L11 **Horsham** Victoria, SE Australia 36°44′S 142°13′E
97 O23 **Horsham** SE England, United Kingdom 51°01′N 00°21′W
99 M15 **Horst** Limburg, SE Netherlands 51°30′N 06°05′E
64 N2 **Horta** Faial, Azores, Portugal, NE Atlantic Ocean 38°32′N 28°39′W
105 S12 **Horta, Cap de l'** *Cast.* Cabo Huertas. *headland* SE Spain 38°21′N 00°25′W
95 H16 **Horten** Vestfold, S Norway 59°25′N 10°25′E
111 M23 **Hortobágy-Berettyó** ≈ E Hungary
27 Q3 **Horton** Kansas, C USA 39°39′N 95°31′W
8 I7 **Horton** ≈ Northwest Territories, NW Canada
95 I23 **Hørve** Sjælland, E Denmark 55°46′N 11°28′E
95 L22 **Hörvik** Blekinge, S Sweden 56°01′N 14°45′E
Horvot Haluza *see* Horvot Halutsa
14 E7 **Horwood Lake** ◎ Ontario, S Canada
116 K4 **Horyn'** *Rus.* Goryn. ≈ NW Ukraine
81 I14 **Hosa'ina** *var.* Hosseina, *It.* Hosanna. Southern Nationalities, S Ethiopia 07°38′N 37°58′E
Hosanna *see* Hosa'ina
101 H18 **Hösbach** Bayern, C Germany 50°00′N 09°12′E
Hose Mountains *see* Hose, Pegunungan
169 T9 **Hose, Pegunungan** *var.* Hose Mountains. ▲▲ East Malaysia
148 L15 **Hoshāb** Baluchistān, SW Pakistan 26°01′N 63°51′E
154 H10 **Hoshangābād** Madhya Pradesh, C India 22°44′N 77°45′E
116 L4 **Hoshcha** Rivnens'ka Oblast', NW Ukraine 50°37′N 26°38′E
152 I7 **Hoshiārpur** Punjab, NW India 31°30′N 75°59′E
Höshööt *see* Öldziyt
99 M23 **Hosingen** Diekirch, NE Luxembourg 50°01′N 06°05′E
186 G7 **Hoskins** New Britain, E Papua New Guinea 05°28′S 150°25′E
155 G17 **Hospet** Karnātaka, C India 15°16′N 76°20′E
104 K4 **Hospital de Orbigo** Castilla y León, N Spain 42°27′N 05°53′W
Hospitalet *see* L'Hospitalet de Llobregat
92 N13 **Hossa** Kainuu, E Finland 65°28′N 29°36′E
Hosseina *see* Hosa'ina
Hosszúmezjő *see* Câmpulung Moldovenesc
63 I25 **Hoste, Isla** *island* S Chile
117 O4 **Hostomel'** *Rus.* Gostomel'. Kyyivs'ka Oblast', N Ukraine 50°41′N 30°15′E
155 H20 **Hosūr** Tamil Nādu, SE India 12°45′N 77°51′E
167 N8 **Hot** Chiang Mai, NW Thailand 18°07′N 98°34′E
158 G10 **Hotan** *var.* Khotan, *Chin.* Ho-t'ien. Xinjiang Uygur Zizhiqu, NW China 37°10′N 79°51′E
158 H9 **Hotan He** ≈ NW China
83 G22 **Hotazel** Northern Cape, N South Africa 27°12′S 22°58′E
37 Q5 **Hotchkiss** Colorado, C USA 38°47′N 107°43′W
35 V7 **Hot Creek Range** ▲▲ Nevada, W USA
Hote *see* Hoti
171 T13 **Hoti** *var.* Hote. Pulau Seram, E Indonesia 02°58′S 130°19′E
Ho-t'ien *see* Hotan
Hotin *see* Khotyn
93 H15 **Hoting** Jämtland, C Sweden 64°07′N 16°14′E
162 L14 **Hotong Qagan Nur** ◎ N China
162 J8 **Hotont** Arhangay, C Mongolia 47°21′N 102°27′E
27 T12 **Hot Springs** Arkansas, C USA 34°31′N 93°03′W
28 J11 **Hot Springs** South Dakota, N USA 43°26′N 103°29′W
21 S5 **Hot Springs** Virginia, NE USA 38°00′N 79°50′W
35 Q4 **Hot Springs Peak** ▲ California, W USA 40°23′N 120°06′W
27 T12 **Hot Springs Village** Arkansas, C USA 34°39′N 93°03′W
Hotspur Bank *see* Hotspur Seamount
65 J16 **Hotspur Seamount** *var.* Hotspur Bank. *undersea feature* C Atlantic Ocean 18°00′S 35°00′W
8 J8 **Hottah Lake** ◎ Northwest Territories, NW Canada
44 K9 **Hotte, Massif de la** ▲▲ SW Haiti
99 K21 **Hotton** Luxembourg, SE Belgium 50°18′N 05°25′E
Hötzing *see* Haţeg
187 P17 **Houaïlou** Province Nord, C New Caledonia 21°17′S 165°37′E
74 K5 **Houari Boumediène** ✈ (Alger) N Algeria 36°38′N 03°15′E
167 P6 **Houayxay** *var.* Ban Houayxay. Bokèo, N Laos 20°17′N 100°27′E
103 N5 **Houdan** Yvelines, N France 48°48′N 01°36′E
99 F20 **Houdeng-Goegnies** *var.* Houdeng-Gœgnies. Hainaut, S Belgium 50°29′N 04°10′E
102 K14 **Houeillès** Lot-et-Garonne, SW France 44°15′N 00°02′E
99 L22 **Houffalize** Luxembourg, SE Belgium 50°08′N 05°47′E
30 M3 **Houghton** Michigan, N USA 47°07′N 88°34′W
31 Q7 **Houghton Lake** Michigan, N USA 44°18′N 84°45′W
31 Q7 **Houghton Lake** ◎ Michigan, N USA
19 T3 **Houlton** Maine, NE USA 46°09′N 67°50′W
160 M5 **Houma** Shanxi, C China 35°36′N 11°23′E
193 U16 **Houma** Tongatapu, S Tonga 21°18′S 174°55′W
22 J10 **Houma** Louisiana, S USA 29°35′N 90°44′W
196 V16 **Houma Taloa** *headland* Tongatapu, S Tonga 21°16′S 175°08′W
77 O13 **Houndé** SW Burkina Faso 11°34′N 03°31′W
102 J12 **Hourtin-Carcans, Lac d'** ◎ SW France
36 J5 **House Range** ▲▲ Utah, W USA
10 K13 **Houston** British Columbia, SW Canada 54°24′N 126°39′W
39 R11 **Houston** Alaska, USA 61°37′N 149°50′W
29 X10 **Houston** Minnesota, N USA 43°45′N 91°34′W
22 M3 **Houston** Mississippi, S USA 33°54′N 89°00′W
27 V7 **Houston** Missouri, C USA 37°19′N 91°59′W
25 W11 **Houston** Texas, SW USA 29°46′N 95°22′W
25 W11 **Houston** ✈ Texas, SW USA 30°03′N 95°18′W
98 J12 **Houten** Utrecht, C Netherlands 52°02′N 05°10′E
99 K17 **Houthalen** Limburg, NE Belgium 51°02′N 05°22′E
99 I22 **Houyet** Namur, SE Belgium 50°10′N 05°00′E
95 H22 **Hov** Midtjylland, C Denmark 55°54′N 10°13′E
95 L17 **Hova** Västra Götaland, S Sweden 58°52′N 14°13′E
162 E6 **Hovd** *var.* Dund-Us. Hovd, W Mongolia 48°06′N 91°22′E
162 J10 **Hovd** *var.* Dund-Us. Hovd, W Mongolia 48°06′N 91°22′E
162 E6 **Hovd** *var.* Khovd, Kobdo; *prev.* Jirgalanta. Hovd, W Mongolia 47°59′N 91°41′E
162 E7 **Hovd** ⬖ *province* W Mongolia
Hovd *see* Bogd
162 C5 **Hovd Gol** ≈ NW Mongolia
97 O23 **Hove** SE England, United Kingdom 50°49′N 00°11′W
95 I22 **Hovedstaden** *off.* Frederiksborgs Amt. ⬖ *county* E Denmark
29 N8 **Hoven** South Dakota, N USA 45°12′N 99°47′W
116 I8 **Hoverla, Hora** *Rus.* Gora Goverla. ▲ W Ukraine 48°09′N 24°30′E
Höviyn Am *see* Gurvanbulag
95 M21 **Hovmantorp** Kronoberg, S Sweden 56°47′N 15°08′E
163 N11 **Hövsgöl** Dornogovĭ, SE Mongolia 43°35′N 109°40′E
162 I5 **Hövsgöl** ⬖ *province* N Mongolia
Hovsgol, Lake *see* Hövsgöl Nuur
162 J5 **Hövsgöl Nuur** *var.* Lake Hovsgol. ◎ N Mongolia
78 L9 **Howa, Ouadi** *var.* Wādi Howar. ≈ Chad/Sudan *see also* Howar, Wādi
Howa, Ouadi *see* Howar, Wādi
27 P7 **Howard** Kansas, C USA 37°27′N 96°16′W
29 Q10 **Howard** South Dakota, N USA 43°58′N 97°31′W
25 N10 **Howard Draw** *valley* Texas, SW USA
29 U8 **Howard Lake** Minnesota, N USA 45°03′N 94°03′W
80 B8 **Howar, Wādi** *var.* Ouadi Howa. ≈ Chad/Sudan *see also* Howa, Ouadi
Howar, Wādi *see* Howa, Ouadi
25 U5 **Howe** Texas, SW USA 33°29′N 96°38′W
183 R12 **Howe, Cape** *headland* New South Wales/Victoria, SE Australia 37°30′S 149°58′E
31 R9 **Howell** Michigan, N USA 42°36′N 83°55′W
28 L9 **Howes** South Dakota, N USA 44°34′N 102°03′W
83 K23 **Howick** KwaZulu/Natal, E South Africa 29°29′S 30°13′E
Howrah *see* Hāora
167 T9 **Hô Xa** *prev.* Vinh Linh. Quang Tri, C Vietnam 17°02′N 107°03′E
27 W9 **Hoxie** Arkansas, C USA 36°03′N 90°58′W
26 J3 **Hoxie** Kansas, C USA 39°21′N 100°27′W
101 I14 **Höxter** Nordrhein-Westfalen, W Germany 51°46′N 09°22′E
158 K6 **Hoxud** *var.* Tewulike. Xinjiang Uygur Zizhiqu, NW China 42°18′N 86°51′E
96 J5 **Hoy** *island* N Scotland, United Kingdom
43 S17 **Hoya, Cerro** ▲ S Panama 07°22′N 80°38′W
94 D12 **Høyanger** Sogn Og Fjordane, S Norway 61°13′N 06°05′E
101 P15 **Hoyerswerda** *Lus.* Wojerecy. Sachsen, E Germany 51°27′N 14°18′E
164 R14 **Hōyo-kaikyō** *var.* Hayasui-seto. *strait* SW Japan
104 J8 **Hoyos** Extremadura, W Spain 40°10′N 06°43′W
29 W4 **Hoyt Lakes** Minnesota, N USA 47°31′N 92°08′W
87 V2 **Hoyvík** Streymoy, N Faroe Islands
137 O14 **Hozat** Tunceli, E Turkey 39°09′N 39°13′E
Hôzyô *see* Hōjō
167 N8 **Hpa-an** *var.* Pa-an. Kayin State, S Myanmar (Burma) 16°51′N 97°37′E
167 N8 **Hpapun** *var.* Papun. Kayin State, S Myanmar (Burma) 18°05′N 97°26′E
167 N7 **Hpasawng** *var.* Pasawng. Kayah State, C Myanmar (Burma) 18°50′N 97°16′E
Hpyu *see* Phyu
111 F16 **Hradec Králové** *Ger.* Königgrätz. Královéhradecký Kraj, N Czech Republic 50°13′N 15°50′E
Hradecký Kraj *see* Královehradecký Kraj
111 B16 **Hradiště** *Ger.* Burgstadlberg. ▲ NW Czech Republic 50°12′N 13°04′E
117 R6 **Hradyz'k** *Rus.* Gradizhsk. Poltavs'ka Oblast', NE Ukraine 49°14′N 33°07′E
119 M16 **Hradzyanka** *Rus.* Grodzyanka. Mahilyowskaya Voblasts', E Belarus 53°33′N 28°45′E
119 F16 **Hrandzichy** *Rus.* Grandichi. Hrodzyenskaya Voblasts', W Belarus 53°43′N 23°49′E
111 H18 **Hranice** *Ger.* Mährisch-Weisskirchen. Olomoucký Kraj, E Czech Republic 49°34′N 17°45′E
112 I13 **Hrasnica** Federacija Bosna I Hercegovina, SE Bosnia and Herzegovina 43°48′N 18°19′E
109 V11 **Hrastnik** C Slovenia 46°09′N 15°08′E
137 U12 **Hrazdan** *Rus.* Razdan. C Armenia 40°30′N 44°50′E
137 T12 **Hrazdan** *var.* Zanga, *Rus.* Razdan. ≈ C Armenia
117 R5 **Hrebinka** *Rus.* Grebenka. Poltavs'ka Oblast', NE Ukraine 50°08′N 32°27′E
119 K17 **Hresk** *Rus.* Gresk. Minskaya Voblasts', C Belarus 53°10′N 27°29′E
Hrisoupoli *see* Chrysoúpoli
119 F16 **Hrodna** *Rus., Pol.* Grodno. Hrodzyenskaya Voblasts', W Belarus 53°40′N 23°50′E
119 F16 **Hrodzyenskaya Voblasts'** *Rus.* Grodnenskaya Oblast'. ⬖ *province* W Belarus
111 J21 **Hron** *Ger.* Gran, *Hung.* Garam. ≈ C Slovakia
111 Q14 **Hrubieszów** *Rus.* Grubeshov. Lubelskie, E Poland 50°49′N 23°53′E
112 F13 **Hrvace** Split-Dalmacija, SE Croatia 43°46′N 16°35′E
Hrvatska *see* Croatia
112 F10 **Hrvatska Kostanjnica** *var.* Kostajnica. Sisak-Moslavina, C Croatia 45°14′N 16°35′E
Hrvatsko Grahovi *see* Bosansko Grahovo
116 K6 **Hrymayliv** *Pol.* Gízymałów, *Rus.* Grimaylov. Ternopil's'ka Oblast', W Ukraine 49°18′N 26°02′E
167 N4 **Hseni** *var.* Hsenwi. Shan State, E Myanmar (Burma) 23°20′N 97°59′E
Hsenwi *see* Hseni
Hsia-men *see* Xiamen
Hsiang-t'an *see* Xiangtan
Hsi Chiang *see* Xi Jiang
167 N6 **Hsihseng** Shan State, C Myanmar (Burma) 20°07′N 97°17′E
Hsinchu *see* Xinzhu
Hsing-K'ai Hu *see* Khanka, Lake
Hsi-ning/Hsining *see* Xining
Hsinking *see* Changchun
Hsin-yang *see* Xinyang
Hsinying *see* Xinying
167 N4 **Hsipaw** Shan State, C Myanmar (Burma) 22°32′N 97°12′E
Hsu-chou *see* Xuzhou
Hsüeh Shan *see* Xue Shan
Htawei *see* Dawei
Hu *see* Shanghai Shi
83 B18 **Huab** ≈ W Namibia
57 M21 **Huacaya** Chuquisaca, S Bolivia 20°45′S 63°42′W
57 J19 **Huachacalla** Oruro, SW Bolivia 18°47′S 68°23′W
159 X9 **Huachi** *var.* Rouyuan, Rouyuanchengzi. Gansu, C China 36°24′N 107°58′E
57 N16 **Huachi, Laguna** ◎ N Bolivia
57 D14 **Huacho** Lima, W Peru 11°05′S 77°36′W
163 Y7 **Huachuan** Heilongjiang, NE China 47°00′S 130°21′E
163 P12 **Huade** Nei Mongol Zizhiqu, N China 41°52′N 113°58′E
163 W10 **Huadian** Jilin, NE China 42°59′N 126°38′E
56 E13 **Huagaruncho, Cordillera** ▲▲ C Peru
Hua Hin *see* Ban Hua Hin
191 S10 **Huahine** *island* Îles Sous le Vent, W French Polynesia
Huahua, Rio *see* Wawa, Río
167 R8 **Huai** ≈ E Thailand
161 Q7 **Huai'an** *var.* Qingjiang. Jiangsu, E China 33°33′N 119°03′E
161 P6 **Huaibei** Anhui, E China 34°00′N 116°48′E
Huaide *see* Gongzhuling
157 T10 **Huai He** ≈ C China
160 L11 **Huaihua** Hunan, S China 27°36′N 109°57′E
161 N14 **Huaiji** Guangdong, S China 23°54′N 112°12′E
161 O2 **Huailai** *var.* Shacheng. Hebei, E China 40°22′N 115°34′E
161 P7 **Huainan** *var.* Huai-nan, Hwainan. Anhui, E China 32°37′N 116°57′E
Huai-nan *see* Huainan
161 N2 **Huairen** *var.* Yunzhong. Shanxi, C China 35°28′N 110°29′E
161 O7 **Huaiyang** Henan, C China 33°44′N 114°55′E
161 Q7 **Huaiyin** Jiangsu, E China 33°31′N 119°03′E
167 N16 **Huai Yot** Trang, SW Thailand 07°45′N 99°36′E
41 Q15 **Huajuapan** *var.* Huajuapan de León. Oaxaca, SE Mexico 17°50′N 97°48′W
Huajuapan de León *see* Huajuapan
41 O9 **Hualahuises** Nuevo León, NE Mexico 24°56′N 99°42′W
36 I11 **Hualapai Mountains** ▲▲ Arizona, SW USA
36 I11 **Hualapai Peak** ▲ Arizona, SW USA 35°04′N 113°54′W
62 J7 **Hualfin** Catamarca, N Argentina 27°15′S 66°53′W
161 T13 **Hualian** *var.* Hualien, Hwalien, *Jap.* Karen. C Taiwan 23°58′N 121°35′E
Hualien *see* Hualian
56 E10 **Huallaga, Río** ≈ N Peru
56 C11 **Huamachuco** La Libertad, C Peru 07°50′S 78°01′W
41 Q14 **Huamantla** Tlaxcala, S Mexico 19°18′N 97°57′W
82 C13 **Huambo** *Port.* Nova Lisboa. Huambo, C Angola 12°48′S 15°45′E
82 B13 **Huambo** ⬖ *province* C Angola
41 P15 **Huamuxtitlán** Guerrero, S Mexico 17°49′N 98°34′W
163 Y8 **Huanan** Heilongjiang, NE China 46°21′N 130°43′E
63 H17 **Huancache, Sierra** ▲▲ SW Argentina
57 I17 **Huancané** Puno, SE Peru 15°10′S 69°44′W
57 F16 **Huancapi** Ayacucho, C Peru 13°40′S 74°05′W
57 E15 **Huancavelica** Huancavelica, SW Peru 12°45′S 75°03′W
57 E15 **Huancavelica** *off.* Departamento de Huancavelica. ⬖ *department* W Peru
Huancavelica, Departamento de *see* Huancavelica
57 E14 **Huancayo** Junín, C Peru 12°03′S 75°14′W
57 K20 **Huanchaca, Cerro** ▲ S Bolivia 20°12′S 66°35′W
Huancheng *see* Huanxian
56 C12 **Huandoy, Nevado** ▲ W Peru 08°48′S 77°33′W
Huang'an *see* Hong'an
161 O8 **Huangchuan** Henan, C China 32°00′N 115°02′E
161 O9 **Huanggang** Hubei, C China 30°27′N 114°48′E
Huang Hai *see* Yellow Sea
157 Q8 **Huang He** *var.* Yellow River. ≈ C China
Huanghe *see* Madoi
161 Q4 **Huanghe Kou** *delta* E China
Huangheyan *see* Madoi
160 L5 **Huangling** Shaanxi, C China 35°40′N 109°14′E
161 O9 **Huangpi** Hubei, C China 30°53′N 114°22′E
163 P13 **Huangqi Hai** ◎ N China
161 Q9 **Huangshan** *var.* Tunxi. Anhui, E China 29°43′N 118°20′E
161 O9 **Huangshi** *var.* Huang-shih, Hwangshih. Hubei, C China 30°14′N 115°E
Huang-shih *see* Huangshi
160 L5 **Huangtu Gaoyuan** *plateau* C China
61 B22 **Huanguelén** Buenos Aires, E Argentina 37°02′S 61°57′W
161 S10 **Huangyan** Zhejiang, SE China 28°39′N 121°19′E
159 T10 **Huangyuan** Qinghai, C China 36°40′N 101°12′E
159 T10 **Huangzhong** *var.* Lushar. Qinghai, C China 36°31′N 101°32′E
163 W12 **Huanren** *var.* Huanren Manzu Zizhixian. Liaoning, NE China 41°16′N 125°25′E
Huanren Manzu Zizhixian *see* Huanren
57 F15 **Huanta** Ayacucho, C Peru 12°54′S 74°13′W
56 E13 **Huánuco** Huánuco, C Peru 09°58′S 76°16′W
56 D13 **Huánuco** *off.* Departamento de Huánuco. ⬖ *department* C Peru
Huánuco, Departamento de *see* Huánuco
57 K19 **Huanuni** Oruro, W Bolivia 18°15′S 66°48′W
159 X9 **Huanxian** *var.* Huancheng. Gansu, C China 36°30′N 107°20′E
161 S12 **Huaping Yu** *prev.* Huap'ing Yu. *island* N Taiwan
62 H3 **Huara** Tarapacá, N Chile 19°59′S 69°42′W
57 D14 **Huaral** Lima, W Peru 11°31′S 77°10′W
Huarás *see* Huaraz
56 D13 **Huaraz** *var.* Huarás. Ancash, W Peru 09°31′S 77°32′W
57 I16 **Huari Huari, Río** ≈ S Peru
56 C13 **Huarmey** Ancash, W Peru 10°03′S 78°08′W
40 H4 **Huásabas** Sonora, NW Mexico 29°47′N 109°18′W
56 D8 **Huasaga, Río** ≈ Ecuador/Peru
167 O15 **Hua Sai** Nakhon Si Thammarat, SW Thailand 08°02′N 100°18′E
56 D12 **Huascarán, Nevado** ▲ W Peru 09°01′S 77°27′W
62 G8 **Huasco** Atacama, N Chile 28°30′S 71°15′W
62 G8 **Huasco, Río** ≈ C Chile
159 S11 **Huashixia** Qinghai, W China
40 G7 **Huatabampo** Sonora, NW Mexico 26°49′N 109°40′W
159 W10 **Huating** *var.* Donghua. Gansu, C China 35°13′N 106°39′E
167 S7 **Huatt, Phou** ▲ N Vietnam 19°45′N 104°48′E
41 Q14 **Huatusco** *var.* Huatusco de Chicuellar. Veracruz-Llave, C Mexico 19°13′N 96°57′W
Huatusco de Chicuellar *see* Huatusco
41 P13 **Huauchinango** Puebla, S Mexico 20°11′N 98°04′W
Huaunta *see* Wounta
41 R15 **Huautla** *var.* Huautla de Jiménez. Oaxaca, SE Mexico 18°10′N 96°51′W
Huautla de Jiménez *see* Huautla
161 O5 **Huaxian** *var.* Daokou, Hua Xian. Henan, C China 35°33′N 114°30′E
Hua Xian *see* Huaxian
Huazangsi *see* Tianzhu
29 V13 **Hubbard** Iowa, C USA 42°18′N 93°18′W
25 U8 **Hubbard** Texas, SW USA 31°52′N 96°43′W
25 Q6 **Hubbard Creek Lake** ◎ Texas, SW USA
31 R6 **Hubbard Lake** ◎ Michigan, N USA
160 M9 **Hubei** *var.* E, Hubei Sheng, Hupeh, Hupei. ⬖ *province* C China
Hubei Sheng *see* Hubei
109 P8 **Huben** Tirol, W Austria 46°55′N 12°35′E
31 R13 **Huber Heights** Ohio, N USA 39°50′N 84°07′W
155 F17 **Hubli** Karnātaka, SW India 15°20′N 75°14′E
163 X12 **Huch'ang** N North Korea 41°25′N 127°04′E
97 M18 **Hucknall** C England, United Kingdom 53°02′N 01°11′W
97 L17 **Huddersfield** N England, United Kingdom 53°39′N 01°47′W
95 O16 **Huddinge** Stockholm, C Sweden 59°15′N 17°57′E
94 N11 **Hudiksvall** Gävleborg, C Sweden 61°45′N 17°12′E
29 W13 **Hudson** Iowa, C USA 42°24′N 92°27′W
19 O11 **Hudson** Massachusetts, NE USA 42°24′N 71°34′W
31 Q11 **Hudson** Michigan, N USA 41°51′N 84°21′W
30 H6 **Hudson** Wisconsin, N USA 44°59′N 92°43′W
11 V14 **Hudson Bay** Saskatchewan, S Canada 52°51′N 102°23′W
12 G6 **Hudson Bay** *bay* NE Canada
195 T16 **Hudson, Cape** *headland* Antarctica 68°15′S 154°00′E
Hudson, Détroit d' *see* Hudson Strait
27 Q9 **Hudson, Lake** ⊠ Oklahoma, C USA
18 K9 **Hudson River** ≈ New Jersey/New York, NE USA
10 M12 **Hudson's Hope** British Columbia, W Canada 56°03′N 121°59′W
12 L2 **Hudson Strait** *Fr.* Détroit d'Hudson. *strait* Northwest Territories/Québec, NE Canada
Ḩudūd ash Shamālīyah, Minţaqat al *see* Al Ḩudūd ash Shamālīyah
Hudur *see* Xuddur
167 U9 **Huê** Th,a Thiên-Huê, C Vietnam 16°28′N 107°35′E
104 J7 **Huebra** ≈ W Spain
24 H8 **Hueco Mountains** ▲▲ Texas, SW USA
116 G10 **Huedin** *Hung.* Bánffyhunyad. Cluj, NW Romania 46°52′N 23°02′E
40 J10 **Huehuento, Cerro** ▲ C Mexico 24°04′N 105°42′W
42 B5 **Huehuetenango** Huehuetenango, W Guatemala 15°19′N 91°26′W
42 B4 **Huehuetenango** *off.* Departamento de Huehuetenango. ⬖ *department* W Guatemala
Huehuetenango, Departamento de *see* Huehuetenango
40 L11 **Huejuquilla** Jalisco, SW Mexico 22°40′N 103°52′W
41 P12 **Huejutla** *var.* Huejutla de Reyes. Hidalgo, C Mexico 21°10′N 98°25′W
Huejutla de Reyes *see* Huejutla
102 G6 **Huelgoat** Finistère, NW France 48°22′N 03°45′W
105 O13 **Huelma** Andalucía, S Spain 37°39′N 03°28′W
104 I14 **Huelva** *anc.* Onuba. Andalucía, SW Spain 37°15′N 06°56′W
104 I13 **Huelva** ⬖ *province* Andalucía, SW Spain
104 J13 **Huelva** ≈ SW Spain
105 Q14 **Huercal-Overa** Andalucía, S Spain 37°23′N 01°56′W
37 Q9 **Huerfano Mountain** ▲ New Mexico, SW USA 36°25′N 107°50′W
37 T7 **Huerfano River** ≈ Colorado, C USA
Huertas, Cabo *see* Horta, Cap de l'
105 R6 **Huerva** ≈ N Spain
105 S4 **Huesca** *anc.* Osca. Aragón, NE Spain 42°08′N 00°25′W
105 T4 **Huesca** ⬖ *province* Aragón, NE Spain
105 P13 **Huéscar** Andalucía, S Spain 37°39′N 02°32′W
41 N15 **Huetamo** *var.* Huetamo de Núñez. Michoacán, SW Mexico 18°36′N 100°54′W
Huetamo de Núñez *see* Huetamo
105 P8 **Huete** Castilla-La Mancha, C Spain 40°09′N 02°42′W
23 P4 **Hueytown** Alabama, S USA 33°27′N 87°00′W
28 L16 **Hugh Butler Lake** ⊠ Nebraska, C USA
181 V6 **Hughenden** Queensland, NE Australia 20°57′S 144°16′E
182 A6 **Hughes** South Australia 30°41′S 129°31′E
39 P8 **Hughes** Alaska, USA 66°03′N 154°15′W
27 X11 **Hughes** Arkansas, C USA 34°57′N 90°28′W
25 W6 **Hughes Springs** Texas, SW USA 33°00′N 94°37′W
37 V5 **Hugo** Colorado, C USA 39°08′N 103°28′W
27 Q13 **Hugo** Oklahoma, C USA 34°01′N 95°31′W
27 Q13 **Hugo Lake** ⊠ Oklahoma, C USA
26 H7 **Hugoton** Kansas, C USA 37°11′N 101°22′W
Huhehot/Huhohaote *see* Hohhot
Huhttán *see* Kvikkjokk
161 R13 **Hui'an** *var.* Luocheng. Fujian, SE China 25°06′N 118°45′E
184 O9 **Huiarau Range** ▲▲ North Island, New Zealand
83 D22 **Huib-Hoch Plateau** *plateau* S Namibia
41 O13 **Huichapán** Hidalgo, C Mexico 20°24′N 99°40′W
Huicheng *see* Shexian
163 W13 **Hŭich'ŏn** C North Korea 40°09′N 126°17′E
83 B15 **Huíla** ⬖ *province* SW Angola
54 E12 **Huila** *off.* Departamento del Huila. ⬖ *province* S Colombia
Huila, Departamento del *see* Huila
54 D11 **Huila, Nevado del** *elevation* C Colombia
83 B15 **Huíla Plateau** *plateau* S Angola
160 G12 **Huili** Sichuan, C China 26°39′N 102°13′E
161 P4 **Huimin** Shandong, E China 37°29′N 117°30′E
163 W11 **Huinan** *var.* Chaoyang. Jilin, NE China 42°40′N 126°03′E
62 K12 **Huinca Renancó** Córdoba, C Argentina 34°51′S 64°22′W
159 V10 **Huining** *var.* Huishi. Gansu, C China 35°42′N 105°01′E
159 W8 **Huinong** *var.* Dawukou. Ningxia, N China 39°04′N 106°22′E
Huishi *see* Huining
160 J12 **Huishui** *var.* Heping. Guizhou, S China 26°07′N 106°39′E
102 L6 **Huisne** ≈ NW France
98 L12 **Huissen** Gelderland, SE Netherlands 51°57′N 05°57′E
159 N11 **Huiten Nur** ◎ C China
93 K19 **Huittinen** Satakunta, SW Finland 61°11′N 22°40′E
41 O15 **Huitzuco** *var.* Huitzuco de los Figueroa. Guerrero, S Mexico 18°18′N 99°22′W
Huitzuco de los Figueroa *see* Huitzuco
159 W11 **Huixian** *var.* Hui Xian. Gansu, C China 33°48′N 106°02′E

◆ Country
● Country Capital
◇ Dependent Territory
○ Dependent Territory Capital
⬖ Administrative Regions
✈ International Airport
▲ Mountain
▲▲ Mountain Range
Volcano
≈ River
◎ Lake
⊠ Reservoir

Hui Xian *see* Huixian
41 *V17* **Huixtla** Chiapas, SE Mexico *15°09′N 92°30′W*
160 *H12* **Huize** *var.* Zhongping. Yunnan, SW China *26°28′N 103°18′E*
98 *J10* **Huizen** Noord-Holland, C Netherlands *52°17′N 05°15′E*
161 *O14* **Huizhou** Guangdong, S China *23°02′N 114°28′E*
162 *J6* **Hujirt** Arhangay, C Mongolia *48°49′N 101°20′E*
Hujirt *see* Tsetserleg, Övörhangay, Mongolia
Hujirt *see* Delgerhaan, Töv, Mongolia
Hukagawa *see* Fukagawa
Hŭksan-gundo *see* Heuksan-jedo
Hukue *see* Gotō
Hukui *see* Fukui
83 *G20* **Hukuntsi** Kgalagadi, SW Botswana *23°59′S 21°44′E*
Hukuoka *see* Fukuoka
Hukusima *see* Fukushima
Hukutiyama *see* Fukuchiyama
Hukuyama *see* Fukuyama
163 *W8* **Hulan** Heilongjiang, NE China *45°59′N 126°37′E*
163 *W8* **Hulan He** ≈ NE China
31 *Q4* **Hulbert Lake** ◎ Michigan, N USA
Hulczyn *see* Hlučín
Huliao *see* Dabu
163 *Z8* **Hulin** Heilongjiang, NE China *45°48′N 133°06′E*
Hulingol *see* Holin Gol
14 *L12* **Hull** Québec, SE Canada *45°26′N 75°45′W*
29 *S12* **Hull** Iowa, C USA *43°11′N 96°07′W*
Hull *see* Kingston upon Hull
Hull Island *see* Orona
99 *F16* **Hulst** Zeeland, SW Netherlands *51°17′N 04°03′E*
Hulstay *see* Choybalsan
Hultschin *see* Hlučín
95 *M19* **Hultsfred** Kalmar, S Sweden *57°30′N 15°50′E*
163 *T13* **Huludao** *prev.* Jinxi, Lianshan. Liaoning, NE China *40°46′N 120°47′E*
Hulun *see* Hulun Buir
163 *S6* **Hulun Buir** *var.* Hailar; *prev.* Hulun. Nei Mongol Zizhiqu, N China *49°15′N 119°41′E*
Hu-lun Ch'ih *see* Hulun Nur
163 *Q6* **Hulun Nur** *var.* Hu-lun Ch'ih; *prev.* Dalai Nor. ◎ NE China
Hulwan/Hulwân *see* Hilwān
117 *V8* **Hulyaypole** *Rus.* Gulyaypole. Zaporiz'ka Oblast', SE Ukraine *47°41′N 36°10′E*
163 *V4* **Huma** Heilongjiang, NE China *51°40′N 126°38′E*
45 *V6* **Humacao** E Puerto Rico *18°09′N 65°50′W*
163 *U4* **Huma He** ≈ NE China
62 *J5* **Humahuaca** Jujuy, N Argentina *23°13′S 65°20′W*
59 *E14* **Humaitá** Amazonas, N Brazil *07°33′S 63°01′W*
62 *N7* **Humaitá** Ñeembucú, S Paraguay *27°02′S 58°31′W*
83 *H26* **Humansdorp** Eastern Cape, S South Africa *34°01′S 24°45′E*
27 *S6* **Humansville** Missouri, C USA *37°47′N 93°34′W*
40 *I8* **Humaya, Río** ≈ C Mexico
83 *C16* **Humbe** Cunene, SW Angola *16°37′S 14°52′E*
97 *N17* **Humber** *estuary* E England, United Kingdom
97 *N17* **Humberside** *cultural region* E England, United Kingdom
Humberto *see* Umberto
25 *W11* **Humble** Texas, SW USA *29°58′N 95°15′W*
11 *U15* **Humboldt** Saskatchewan, S Canada *52°13′N 105°09′W*
29 *U12* **Humboldt** Iowa, C USA *42°42′N 94°13′W*
27 *Q6* **Humboldt** Kansas, C USA *37°48′N 95°26′W*
29 *S17* **Humboldt** Nebraska, C USA *40°09′N 95°56′W*
35 *S3* **Humboldt** Nevada, W USA *40°36′N 118°15′W*
20 *G9* **Humboldt** Tennessee, S USA *35°49′N 88°55′W*
34 *K3* **Humboldt Bay** *bay* California, W USA
35 *S4* **Humboldt Lake** ◎ Nevada, W USA
35 *S4* **Humboldt River** ≈ Nevada, W USA
35 *T5* **Humboldt Salt Marsh** *wetland* Nevada, W USA
183 *P11* **Hume, Lake** ◎ New South Wales/Victoria, SE Australia
111 *N19* **Humenné** *Ger.* Homenau, *Hung.* Homonna. Prešovský Kraj, E Slovakia *48°57′N 21°54′E*
29 *V15* **Humeston** Iowa, C USA *40°51′N 93°30′W*
54 *J5* **Humocaro Bajo** Lara, N Venezuela *09°41′N 70°00′W*
29 *Q14* **Humphrey** Nebraska, C USA *41°38′N 97°29′W*
35 *S9* **Humphreys, Mount** ▲ California, W USA *37°11′N 118°39′W*
36 *L11* **Humphreys Peak** ▲ Arizona, SW USA *35°18′N 111°40′W*
111 *E17* **Humpolec** *Ger.* Gumpolds, Humpoletz. Vysočina, C Czech Republic *49°33′N 15°23′E*
Humpoletz *see* Humpolec
93 *K19* **Humppila** Kanta-Häme, S Finland *60°54′N 23°21′E*
32 *F8* **Humptulips** Washington, NW USA *47°13′N 123°57′W*
42 *H7* **Humuya, Río** ≈ W Honduras
75 *P9* **Hūn** N Libya *29°06′N 15°56′E*
92 *I1* **Húnaflói** *bay* NW Iceland
160 *M11* **Hunan** *var.* Hunan Sheng, Xiang. ◈ *province* S China
Hunan Sheng *see* Hunan
163 *Y10* **Hunchun** Jilin, NE China *42°51′N 130°21′E*
95 *I22* **Hundested** Hovedstaden, E Denmark *55°58′N 11°53′E*
Hundred Mile House *see* 100 Mile House
116 *G12* **Hunedoara** *Ger.* Eisenmarkt, *Hung.* Vajdahunyad. Hunedoara, SW Romania *45°45′N 22°54′E*
116 *G12* **Hunedoara** ◈ *county* W Romania
101 *I17* **Hünfeld** Hessen, C Germany *50°41′N 09°46′E*
Hungarian People's Republic *see* Hungary
111 *H23* **Hungary** *off.* Republic of Hungary, *Ger.* Ungarn, *Hung.* Magyarország, *Rom.* Ungaria, *SCr.* Mađarska, *Ukr.* Uhorshchyna; *prev.* Hungarian People's Republic. ◆ *republic* C Europe
Hungary, Plain of *see* Great Hungarian Plain
Hungary, Republic of *see* Hungary
Hungiy *see* Urgamal
163 *X13* **Hŭngnam** E North Korea *39°50′N 127°36′E*
33 *P8* **Hungry Horse Reservoir** ⊠ Montana, NW USA
Hungt'ou *see* Lan Yu
Hung-tse Hu *see* Hongze Hu
167 *T6* **Hưng Yên** Hai Hưng, N Vietnam *20°38′N 106°05′E*
Hunjiang *see* Baishan
95 *I18* **Hunnebostrand** Västra Götaland, S Sweden *58°26′N 11°19′E*
101 *E19* **Hunsrück** ▲▲ W Germany
97 *P18* **Hunstanton** E England, United Kingdom *52°57′N 00°27′E*
155 *G20* **Hunsūr** Karnātaka, E India *12°18′N 76°15′E*
Hunt *see* Hangay
100 *G12* **Hunte** ≈ NW Germany
29 *Q5* **Hunter** North Dakota, N USA *47°10′N 97°11′W*
25 *S11* **Hunter** Texas, SW USA *29°47′N 98°01′W*
185 *D20* **Hunter** ≈ South Island, New Zealand
183 *N15* **Hunter Island** *island* Tasmania, SE Australia
18 *K11* **Hunter Mountain** ▲ New York, NE USA *42°10′N 74°13′W*
185 *B23* **Hunter Mountains** ▲▲ South Island, New Zealand
183 *S7* **Hunter River** ≈ New South Wales, SE Australia
32 *L7* **Hunters** Washington, NW USA *48°07′N 118°13′W*
185 *F20* **Hunters Hills, The** *hill range* South Island, New Zealand
184 *M12* **Hunterville** Manawatu-Wanganui, North Island, New Zealand *39°55′S 175°34′E*
31 *N16* **Huntingburg** Indiana, N USA *38°18′N 86°57′W*
97 *O20* **Huntingdon** E England, United Kingdom *52°20′N 00°12′W*
18 *E15* **Huntingdon** Pennsylvania, NE USA *40°28′N 78°00′W*
20 *G9* **Huntingdon** Tennessee, S USA *36°00′N 88°25′W*
97 *O20* **Huntingdonshire** *cultural region* C England, United Kingdom
31 *P12* **Huntington** Indiana, N USA *40°52′N 85°30′W*
32 *L13* **Huntington** Oregon, NW USA *44°22′N 117°18′W*
25 *X9* **Huntington** Texas, SW USA *31°16′N 94°34′W*
36 *M5* **Huntington** Utah, W USA *39°19′N 110°57′W*
21 *P5* **Huntington** West Virginia, NE USA *38°25′N 82°27′W*
35 *T16* **Huntington Beach** California, W USA *33°39′N 118°00′W*
35 *W4* **Huntington Creek** ≈ Nevada, W USA
184 *L7* **Huntly** Waikato, North Island, New Zealand *37°34′S 175°09′E*
96 *K8* **Huntly** NE Scotland, United Kingdom *57°25′N 02°48′W*
10 *K8* **Hunt, Mount** ▲ Yukon, NW Canada *61°29′N 129°10′W*
14 *H12* **Huntsville** Ontario, S Canada *45°20′N 79°14′W*
23 *P2* **Huntsville** Alabama, S USA *34°44′N 86°35′W*
27 *S9* **Huntsville** Arkansas, C USA *36°04′N 93°46′W*
27 *U3* **Huntsville** Missouri, C USA *39°27′N 92°31′W*
20 *M8* **Huntsville** Tennessee, S USA *36°25′N 84°30′W*
25 *V10* **Huntsville** Texas, SW USA *30°43′N 95°34′W*
36 *L2* **Huntsville** Utah, W USA *41°16′N 111°47′W*
41 *W12* **Hunucmá** Yucatán, SE Mexico *20°59′N 89°55′W*
149 *W3* **Hunza** ≈ NE Pakistan
Hunza *see* Karīmābād
Hunze *see* Oostermoers Vaart
158 *H4* **Huocheng** *var.* Shuiding. Xinjiang Uygur Zizhiqu, NW China *44°03′N 80°49′E*
161 *N6* **Huojia** Henan, C China *35°14′N 113°38′E*
186 *N14* **Huon** *reef* N New Caledonia
186 *E7* **Huon Peninsula** *headland* C Papua New Guinea *06°24′S 147°50′E*
Huoshao Dao *see* Lü Dao
Huoshao Tao *see* Lan Yu
Hupeh/Hupei *see* Hubei
Hurama *see* Hongyuan
95 *H14* **Hurdalsjøen** *prev.* Hurdalssjøen. ◎ S Norway
Hurdalssjøen *see* Hurdalsjøen
14 *E13* **Hurd, Cape** *headland* Ontario, S Canada *45°12′N 81°43′W*
98 *L5* **Hurdegaryp** *Dutch.* Hardegarijp. Fryslân, N Netherlands *53°13′N 05°57′E*
29 *N4* **Hurdsfield** North Dakota, N USA *47°24′N 99°55′W*
Hüremt *see* Sayhan, Bulgan, Mongolia
Hüremt *see* Taragt, Övörhangay, Mongolia
Hurghada *see* Al Ghurdaqah
67 *V9* **Huri Hills** ▲▲ NW Kenya
Hūrkay *see* Hurkê
139 *S1* **Hurkê** *Ar.* Hūrkay, *var.* Harki. Dahūk, N Iraq *37°03′N 43°39′E*
37 *P15* **Hurley** New Mexico, SW USA *32°42′N 108°07′W*
30 *K4* **Hurley** Wisconsin, N USA *46°25′N 90°15′W*
21 *Y4* **Hurlock** Maryland, NE USA *38°37′N 75°51′W*
162 *K11* **Hürmen** *var.* Tsoohor. Ömnögovĭ, S Mongolia *43°15′N 104°04′E*
29 *P10* **Huron** South Dakota, N USA *44°19′N 98°13′W*
31 *S6* **Huron, Lake** ◎ Canada/USA
31 *N3* **Huron Mountains** *hill range* Michigan, N USA
36 *J8* **Hurricane** Utah, W USA *37°10′N 113°18′W*
21 *P5* **Hurricane** West Virginia, NE USA *38°25′N 82°01′W*
36 *J8* **Hurricane Cliffs** *cliff* Arizona, SW USA
23 *V6* **Hurricane Creek** ≈ Georgia, SE USA
94 *E12* **Hurrungane** ▲ S Norway *61°25′N 07°48′E*
101 *E16* **Hürth** Nordrhein-Westfalen, W Germany *50°52′N 06°49′E*
Hurukawa *see* Furukawa
185 *I17* **Hurunui** ≈ South Island, New Zealand
95 *F21* **Hurup** Midtjylland, NW Denmark *56°46′N 08°26′E*
117 *T14* **Hurzuf** Avtonomna Respublika Krym, S Ukraine *44°33′N 34°18′E*
Huş *see* Huşi
95 *B19* **Húsavík** *Dan.* Husevig. Sandoy, C Faroe Islands *65°24′N 06°38′W*
92 *K1* **Húsavík** Norðurland Eystra, NE Iceland *66°03′N 17°20′W*
Husevig *see* Húsavík
116 *M10* **Huşi** *var.* Huş. Vaslui, E Romania *46°40′N 28°05′E*
95 *L19* **Huskvarna** Jönköping, S Sweden *57°47′N 14°15′E*
39 *P8* **Huslia** Alaska, USA *65°42′N 156°24′W*
Husn *see* Al Ḥuṣn
95 *C15* **Husnes** Hordaland, S Norway *59°52′N 05°46′E*
94 *D8* **Hustadvika** *sea area* S Norway
Husté *see* Khust
100 *H7* **Husum** Schleswig-Holstein, N Germany *54°29′N 09°04′E*
93 *I16* **Husum** Västernorrland, C Sweden *63°21′N 19°12′E*
116 *K6* **Husyatyn** Ternopil's'ka Oblast', W Ukraine *49°04′N 26°10′E*
Huszt *see* Khust
Hutag *see* Hutag-Öndör
162 *K6* **Hutag-Öndör** *var.* Hutag. Bulgan, N Mongolia *49°22′N 102°50′E*
26 *M6* **Hutchinson** Kansas, C USA *38°03′N 97°56′W*
29 *U9* **Hutchinson** Minnesota, N USA *44°53′N 94°22′W*
23 *Y13* **Hutchinson Island** *island* Florida, SE USA
36 *L11* **Hutch Mountain** ▲ Arizona, SW USA *34°49′N 111°22′W*
141 *O14* **Ḥūth** NW Yemen *16°14′N 44°E*
186 *I7* **Hutjena** Buka Island, NE Papua New Guinea *05°19′S 154°40′E*
109 *T8* **Hüttenberg** Kärnten, S Austria *46°58′N 14°33′E*
25 *T10* **Hutto** Texas, SW USA *30°32′N 97°33′W*
Huttu *see* Futtsu
108 *E8* **Huttwil** Bern, W Switzerland *47°06′N 07°48′E*
158 *K5* **Hutubi** Xinjiang Uygur Zizhiqu, NW China *44°10′N 86°51′E*
161 *N4* **Hutuo He** ≈ C China
Hutyû *see* Fuchū
185 *E20* **Huxley, Mount** ▲ South Island, New Zealand *44°02′S 169°42′E*
99 *J20* **Huy** *Dut.* Hoei, Hoey. Liège, E Belgium *50°32′N 05°14′E*
161 *R8* **Huzhou** *var.* Wuxing. Zhejiang, SE China *30°52′N 120°06′E*
Huzi *see* Fuji
Huzieda *see* Fujieda
Huzinomiya *see* Fujinomiya
Huzisawa *see* Fujisawa
92 *I2* **Hvammstangi** Norðurland Vestra, N Iceland *65°22′N 20°54′W*
92 *K4* **Hvannadalshnjúkur** *var.* Hvannadalshnúkur. ⛰ S Iceland *64°01′N 16°39′W*
Hvannadalshnúkur *see* Hvannadalshnjúkur
113 *E15* **Hvar** *It.* Lesina. Split-Dalmacija, S Croatia *43°10′N 16°27′E*
113 *F15* **Hvar** *It.* Lesina; *anc.* Pharus. *island* S Croatia
117 *T13* **Hvardiys'ke** *Rus.* Gvardeyskoye. Avtonomna Respublika Krym, S Ukraine *45°08′N 34°01′E*
92 *I4* **Hveragerði** Suðurland, SW Iceland *64°00′N 21°13′W*
95 *E22* **Hvide Sande** Midtjylland, W Denmark *56°00′N 08°08′E*
92 *I3* **Hvitá** ≈ C Iceland
95 *G15* **Hvittingfoss** Buskerud, S Norway *59°28′N 10°00′E*
92 *I4* **Hvolsvöllur** Suðurland, SW Iceland *63°44′N 20°12′W*
Hwach'ŏn-chŏsuji *see* Paro-ho
Hwainan *see* Huainan
Hwalien *see* Hualian
83 *I16* **Hwange** *prev.* Wankie. Matabeleland North, W Zimbabwe *18°18′S 26°31′E*
Hwang-Hae *see* Yellow Sea
Hwangshih *see* Huangshi
83 *L17* **Hwedza** Mashonaland East, E Zimbabwe *18°35′S 31°35′E*
63 *G20* **Hyades, Cerro** ▲ S Chile *46°57′S 73°09′W*
162 *K6* **Hyalganat** *var.* Selenge. Bulgan, N Mongolia *49°34′N 104°18′E*
19 *Q12* **Hyannis** Massachusetts, NE USA *41°38′N 70°15′W*
28 *L13* **Hyannis** Nebraska, C USA *42°00′N 101°45′W*
162 *F6* **Hyargas Nuur** ◎ NW Mongolia
Hybla/Hybla Major *see* Paternò
39 *Y14* **Hydaburg** Prince of Wales Island, Alaska, USA *55°10′N 132°44′W*
185 *F22* **Hyde** Otago, South Island, New Zealand *45°17′S 170°17′E*
21 *O7* **Hyden** Kentucky, S USA *37°08′N 83°23′W*
18 *K12* **Hyde Park** New York, NE USA *41°46′N 73°52′W*
39 *Z14* **Hyder** Alaska, USA *55°55′N 130°01′W*
155 *I15* **Hyderābād** *var.* Haidarabad. *state capital* Telangana/Andhra Pradesh, C India *17°22′N 78°26′E*
149 *Q16* **Hyderābād** *var.* Haidarabad. Sind, SE Pakistan *25°26′N 68°22′E*
103 *T16* **Hyères** Var, SE France *43°07′N 06°08′E*
103 *T16* **Hyères, Îles d'** *island group* S France
118 *K12* **Hyermanavichy** *Rus.* Germanovichi. Vitsyebskaya Voblasts', N Belarus *55°24′N 27°48′E*
163 *X12* **Hyesan** NE North Korea *41°18′N 128°13′E*
10 *K8* **Hyland** ≈ Yukon, NW Canada
95 *K20* **Hyltebruk** Halland, S Sweden *57°N 13°14′E*
18 *D16* **Hyndman** Pennsylvania, NE USA *39°49′N 78°42′W*
33 *P14* **Hyndman Peak** ▲ Idaho, NW USA *43°45′N 114°07′W*
164 *I13* **Hyōgo** *off.* Hyōgo-ken. ◈ *prefecture* Honshū, SW Japan
Hyōgo-ken *see* Hyōgo
Hypsas *see* Belice
Hyrcania *see* Gorgān
36 *L1* **Hyrum** Utah, W USA *41°37′N 111°51′W*
93 *N14* **Hyrynsalmi** Kainuu, C Finland *64°41′N 28°30′E*
33 *V10* **Hysham** Montana, NW USA *46°16′N 107°14′W*
11 *N13* **Hythe** Alberta, W Canada *55°18′N 119°44′W*
97 *Q23* **Hythe** SE England, United Kingdom *51°05′N 01°04′E*
Hyvinge *see* Hyvinkää
93 *L19* **Hyvinkää** *Swe.* Hyvinge. Uusimaa, S Finland *60°37′N 24°50′E*

I

116 *J9* **Iacobeni** *Ger.* Jakobeny. Suceava, NE Romania *47°24′N 25°20′E*
Iader *see* Zadar
172 *I7* **Iakora** Fianarantsoa, SE Madagascar *23°04′S 46°40′E*
116 *K14* **Ialomiţa** *var.* Jalomitsa. ◈ *county* SE Romania
116 *K14* **Ialomiţa** ≈ SE Romania
117 *N10* **Ialoveni** *Rus.* Yaloveny. C Moldova *46°57′N 28°47′E*
117 *N11* **Ialpug** *var.* Ialpugul Mare, *Rus.* Yalpug. ≈ Moldova/Ukraine
Ialpugul Mare *see* Ialpug
23 *T8* **Iamonia, Lake** ◎ Florida, SE USA
116 *L13* **Ianca** Brăila, SE Romania *45°06′N 27°29′E*
116 *M10* **Iaşi** *Ger.* Jassy. Iaşi, NE Romania *47°08′N 27°38′E*
116 *L9* **Iaşi** *Ger.* Jassy, Yassy. ◈ *county* NE Romania
114 *J13* **Íasmos** Anatolikí Makedonía kai Thráki, NE Greece *41°07′N 25°12′E*
22 *H6* **Iatt, Lake** ⊠ Louisiana, S USA
58 *B11* **Iauretê** Amazonas, NW Brazil *0°37′N 69°10′W*
171 *N3* **Iba** Luzon, N Philippines *15°25′N 119°55′E*
77 *S16* **Ibadan** Oyo, SW Nigeria *07°22′N 04°01′E*
54 *E10* **Ibagué** Tolima, C Colombia *04°27′N 75°14′W*
60 *J10* **Ibaiti** Paraná, S Brazil *23°49′S 50°15′W*
36 *J4* **Ibapah Peak** ▲ Utah, W USA *39°51′N 113°55′W*
Ibar *see* Ibër
165 *P13* **Ibaraki** *off.* Ibaraki-ken. ◈ *prefecture* Honshū, S Japan
Ibaraki-ken *see* Ibaraki
56 *C5* **Ibarra** *var.* San Miguel de Ibarra. Imbabura, N Ecuador *0°23′S 78°08′W*
Ibaşfalău *see* Dumbrăveni
141 *O16* **Ibb** W Yemen *13°55′N 44°10′E*
100 *F13* **Ibbenbüren** Nordrhein-Westfalen, NW Germany *52°17′N 07°43′E*
79 *H16* **Ibenga** ≈ N Congo
113 *M15* **Ibër** *Serb.* Ibar. ≈ C Serbia
57 *I14* **Iberia** Madre de Dios, E Peru *11°21′S 69°36′W*
Iberia *see* Spain
66 *M1* **Iberian Basin** *undersea feature* E Atlantic Ocean *39°00′N 16°00′W*
Iberian Mountains *see* Ibérico, Sistema
84 *D12* **Iberian Peninsula** *physical region* Portugal/Spain
64 *M8* **Iberian Plain** *undersea feature* E Atlantic Ocean *13°30′W 43°45′N*
Ibérica, Cordillera *see* Ibérico, Sistema
105 *P6* **Ibérico, Sistema** *var.* Cordillera Ibérica, *Eng.* Iberian Mountains. ▲▲ NE Spain
12 *K7* **Iberville Lac d'** ◎ Québec, NE Canada
77 *T14* **Ibeto** Niger, W Nigeria *10°30′N 05°07′E*
77 *W15* **Ibi** Taraba, C Nigeria *08°13′N 09°46′E*
105 *S11* **Ibi** Valenciana, E Spain *38°38′N 00°35′W*
59 *L20* **Ibiá** Minas Gerais, SE Brazil *19°30′S 46°31′W*
61 *F15* **Ibicuí, Rio** ≈ S Brazil
61 *C19* **Ibicuy** Entre Ríos, E Argentina *33°44′S 59°10′W*
61 *F16* **Ibirapuitã** ≈ S Brazil
105 *V10* **Ibiza** *var.* Iviza, *Cast.* Eivissa; *anc.* Ebusus. *island* Islas Baleares, Spain, W Mediterranean Sea
Ibiza *see* Eivissa
138 *J4* **Ibn Wardān, Qaşr** *ruins* Ḥamāh, C Syria
Ibo *see* Sassandra
188 *E9* **Ibobang** Babeldaob, N Palau
171 *V13* **Ibonma** Papua Barat, E Indonesia *03°27′S 133°30′E*
59 *N17* **Ibotirama** Bahia, E Brazil *12°13′S 43°12′W*
141 *Y8* **Ibrā** NE Oman *22°45′N 58°30′E*
127 *Q4* **Ibresi** Chuvashskaya Respublika, W Russian Federation *55°22′N 47°04′E*
141 *X8* **'Ibrī** NW Oman *23°12′N 56°28′E*
164 *C16* **Ibusuki** Kagoshima, Kyūshū, SW Japan *31°15′N 130°40′E*
57 *E16* **Ica** Ica, SW Peru *14°02′S 75°48′W*
57 *E16* **Ica** *off.* Departamento de Ica. ◈ *department* SW Peru
Ica, Departamento de *see* Ica
58 *C11* **Içana** Amazonas, NW Brazil *0°22′N 67°25′W*
Icaria *see* Ikaría
58 *B13* **Içá, Rio** *var.* Río Putumayo. ≈ NW South America *see also* Putumayo, Río
Içá, Rio *see* Putumayo, Río
136 *I17* **İçel** *prev.* Ichili; *prev.* Mersin. ◈ *province* S Turkey
İçel *see* Mersin
92 *I3* **Iceland** *off.* Republic of Iceland, *Dan.* Island, *Icel.* Ísland. ◆ *republic* N Atlantic Ocean
86 *B6* **Iceland** *island* N Atlantic Ocean
64 *L5* **Iceland Basin** *undersea feature* N Atlantic Ocean *61°00′N 19°00′W*
Icelandic Plateau *see* Iceland Plateau
197 *Q15* **Iceland Plateau** *var.* Icelandic Plateau. *undersea feature* S Greenland Sea *12°00′W 69°30′N*
Iceland, Republic of *see* Iceland
155 *E16* **Ichalkaranji** Mahārāshtra, W India *16°42′N 74°28′E*
164 *D15* **Ichifusa-yama** ▲ Kyūshū, SW Japan *32°18′N 131°05′E*
Ichili *see* İçel
164 *K13* **Ichinomiya** *var.* Itinomiya. Aichi, Honshū, SW Japan *35°18′N 136°48′E*
165 *Q9* **Ichinoseki** *var.* Itinoseki. Iwate, Honshū, C Japan *38°56′N 141°08′E*
117 *R3* **Ichnya** Chernihivs'ka Oblast', NE Ukraine *50°52′N 32°24′E*
57 *L17* **Ichoa, Río** ≈ C Bolivia
Iconium *see* Konya
Iculisma *see* Angoulême
39 *U12* **Icy Bay** *inlet* Alaska, USA
39 *N5* **Icy Cape** *headland* Alaska, USA *70°19′N 161°52′W*
39 *W13* **Icy Strait** *strait* Alaska, USA
27 *R13* **Idabel** Oklahoma, C USA *33°54′N 94°50′W*
29 *T13* **Ida Grove** Iowa, C USA *42°21′N 95°28′W*
77 *U16* **Idah** Kogi, S Nigeria *07°06′N 06°45′E*
33 *N13* **Idaho** *off.* State of Idaho, *also known as* Gem of the Mountains, Gem State. ◈ *state* NW USA
33 *N14* **Idaho City** Idaho, NW USA *43°48′N 115°51′W*
33 *R14* **Idaho Falls** Idaho, NW USA *43°28′N 112°01′W*
121 *P2* **Idálion** *var.* Dali, Dhali. C Cyprus *35°00′N 33°25′E*
25 *N5* **Idalou** Texas, SW USA *33°40′N 101°40′W*
104 *I9* **Idanha-a-Nova** Castelo Branco, C Portugal *39°55′N 07°15′W*
101 *E19* **Idar-Oberstein** Rheinland-Pfalz, SW Germany *49°43′N 07°19′E*
118 *J3* **Ida-Virumaa** *var.* Ida-Viru Maakond. ◈ *province* NE Estonia
Ida-Viru Maakond *see* Ida-Virumaa
124 *J8* **Idel'** Respublika Kareliya, NW Russian Federation *64°08′N 34°12′E*
79 *C15* **Idenao** Sud-Ouest, SW Cameroon *04°04′N 09°01′E*
Idensalmi *see* Iisalmi
162 *I6* **Ider** *var.* Dzuunmod. Hövsgöl, C Mongolia *48°09′N 97°22′E*
Ider *see* Galt
75 *X10* **Idfū** *var.* Edfu. SE Egypt *24°55′N 32°52′E*
Ídhi Óros *see* Ídi
Ídhra *see* Ýdra
80 *L10* **'Īdī** *var.* Ed. SE Eritrea *13°54′N 41°39′E*
168 *H7* **Idi** Sumatera, W Indonesia *05°00′N 98°00′E*
115 *I25* **Ídi** *var.* Ídhi Óros. ▲▲ Kríti, Greece, E Mediterranean Sea
Idi Amin, Lac *see* Edward, Lake
106 *G10* **Idice** ≈ N Italy
76 *G9* **Idîni** Trarza, W Mauritania *17°58′N 15°40′W*
79 *J21* **Idiofa** Bandundu, SW Dem. Rep. Congo *05°00′S 19°38′E*
39 *O10* **Iditarod River** ≈ Alaska, USA
Idjavuotna *see* Ifjord
95 *M14* **Idkerberget** Dalarna, C Sweden *60°22′N 15°15′E*
138 *I3* **Idlib** Idlib, NW Syria *35°57′N 36°38′E*
138 *I4* **Idlib** *off.* Muḥāfaẓat Idlib. ◈ *governorate* NW Syria
Idlib, Muḥāfaẓat *see* Idlib
Idra *see* Ýdra
94 *J11* **Idre** Dalarna, C Sweden *61°52′N 12°45′E*
Idria *see* Idrija
109 *S11* **Idrija** *It.* Idria. W Slovenia *46°00′N 14°59′E*
101 *G18* **Idstein** Hessen, W Germany *50°10′N 08°16′E*
Idutywa *see* Dutywa
Idzhevan *see* Ijevan
118 *G9* **Iecava** S Latvia *56°36′N 24°10′E*
165 *T16* **Ie-jima** *var.* Ii-shima. *island* Nansei-shotō, SW Japan
99 *B18* **Ieper** *Fr.* Ypres. West-Vlaanderen, W Belgium *50°51′N 02°53′E*
115 *K25* **Ierápetra** Kríti, Greece, E Mediterranean Sea *35°00′N 25°45′E*
115 *G22* **Iérax, Akrotírio** *headland* S Greece *36°45′N 23°06′E*
Ierisós *see* Ierissós
115 *H14* **Ierissós** *var.* Ierisós. Kentrikí Makedonía, N Greece *40°24′N 23°53′E*
116 *I11* **Iernut** *Hung.* Radnót. Mureş, C Romania *46°27′N 24°15′E*
106 *J12* **Iesi** *var.* Jesi. Marche, C Italy *43°33′N 13°16′E*
92 *K9* **Iešjavri** ◎ N Norway
Iesolo *see* Jesolo
188 *K16* **Ifalik Atoll** *atoll* Caroline Islands, C Micronesia
172 *I6* **Ifanadiana** Fianarantsoa, SE Madagascar *21°19′S 47°39′E*
77 *T16* **Ife** Osun, SW Nigeria *07°25′N 04°31′E*
77 *V8* **Iferouâne** Agadez, N Niger *19°05′N 08°24′E*
Iferten *see* Yverdon
92 *L8* **Ifjord** *Lapp.* Idjavuotna. Finnmark, N Norway *70°27′N 27°06′E*
77 *R8* **Ifôghas, Adrar des** *var.* Adrar des Iforas. ▲▲ NE Mali
Iforas, Adrar des *see* Ifôghas, Adrar des
182 *D6* **Ifould lake** *salt lake* South Australia
74 *G6* **Ifrane** C Morocco *33°31′N 05°09′W*
171 *S11* **Iga** Pulau Halmahera, E Indonesia *01°23′N 128°17′E*
81 *G18* **Iganga** SE Uganda *0°34′N 33°27′E*
60 *L7* **Igarapava** São Paulo, S Brazil *20°01′S 47°46′W*
122 *K9* **Igarka** Krasnoyarskiy Kray, N Russian Federation *67°31′N 86°33′E*
Igaunija *see* Estonia
137 *T12* **Iğdir** ◈ *province* NE Turkey
I.G.Duca *see* General Toshevo
Igel *see* Jihlava
94 *N11* **Iggesund** Gävleborg, C Sweden *61°38′N 17°04′E*
39 *P7* **Igikpak, Mount** ▲ Alaska, USA *67°28′N 154°55′W*
39 *P13* **Igiugig** Alaska, USA *59°19′N 155°53′W*
Iglau/Iglawa/Igława *see* Jihlava
107 *B20* **Iglesias** Sardegna, Italy, C Mediterranean Sea *39°20′N 08°34′E*
127 *V4* **Iglino** Respublika Bashkortostan, W Russian Federation *54°51′N 56°29′E*
Igló *see* Spišská Nová Ves
9 *O6* **Igloolik** Nunavut, N Canada *69°24′N 81°55′W*
12 *B11* **Ignace** Ontario, S Canada *49°26′N 91°40′W*
118 *I12* **Ignalina** Utena, E Lithuania *55°20′N 26°10′E*
127 *Q5* **Ignatovka** Ul'yanovskaya Oblast', W Russian Federation *53°56′N 47°40′E*
124 *K12* **Ignatovo** Vologodskaya Oblast', NW Russian Federation *60°47′N 37°51′E*
114 *N11* **İğneada** Kırklareli, NW Turkey *41°54′N 27°58′E*
121 *S7* **İğneada Burnu** *headland* NW Turkey *41°54′N 28°03′E*
Igombe *see* Gombe
115 *B16* **Igoumenítsa** Ípeiros, W Greece *39°30′N 20°16′E*
127 *T2* **Igra** Udmurtskaya Respublika, NW Russian Federation *57°30′N 53°01′E*
122 *H9* **Igrim** Khanty-Mansiyskiy Avtonomnyy Okrug-Yugra, N Russian Federation *63°09′N 64°33′E*
60 *G12* **Iguaçu, Rio** *Sp.* Río Iguazú. ≈ Argentina/Brazil *see also* Iguazú, Río
Iguaçu, Río *see* Iguazú, Río
59 *I22* **Iguaçu, Salto do** *Sp.* Cataratas del Iguazú; *prev.* Victoria Falls. *waterfall* Argentina/Brazil *see also* Iguazú, Cataratas del
Iguaçu, Salto do *see* Iguazú, Cataratas del
41 *O15* **Iguala** *var.* Iguala de la Independencia. Guerrero, S Mexico *18°21′N 99°31′W*
105 *V5* **Igualada** Cataluña, NE Spain *41°35′N 01°37′E*
Iguala de la Independencia *see* Iguala
60 *G12* **Iguazú, Cataratas del** *Port.* Salto do Iguaçu; *prev.* Victoria Falls. *waterfall* Argentina/Brazil *see also* Iguaçu, Salto do
Iguazú, Cataratas del *see* Iguaçu, Salto do
62 *Q6* **Iguazú, Río** *var.* Río Iguaçu. ≈ Argentina/Brazil *see also* Iguaçu, Río
Iguazú, Río *see* Iguaçu, Rio
79 *D19* **Iguéla** *prev.* Iguéla. Ogooué-Maritime, SW Gabon *02°00′S 09°23′E*
Iguéla *see* Iguéla
Iguid, Erg *see* Iguîdi, 'Erg
67 *M5* **Iguîdi, 'Erg** *var.* Erg Iguid. *desert* Algeria/Mauritania
172 *K2* **Iharaña** *prev.* Vohémar. Antsiranana, NE Madagascar *13°22′S 50°00′E*
151 *K18* **Ihavandhippolhu Atoll** *var.* Ihavandiffulu Atoll. *atoll* N Maldives
Ihavandiffulu Atoll *see* Ihavandhippolhu Atoll
Ih Bulag *see* Hanbogd
165 *T16* **Iheya-jima** *island* Nansei-shotō, SW Japan
Ihhayrhan *see* Bayan-Önjüül
163 *N9* **Ihhet** *var.* Bayan. Dornogovĭ, SE Mongolia *46°15′N 110°16′E*
172 *I6* **Ihosy** Fianarantsoa, S Madagascar *22°23′S 46°09′E*
Ihsüüj *see* Bayanchandmanĭ
162 *I7* **Ihtamir** *var.* Dzaanhushuu. Arhangay, C Mongolia *47°36′N 101°06′E*
114 *H10* **Ihtiman** Sofia, W Bulgaria *42°26′N 23°49′E*
162 *H6* **Ih-Uul** *var.* Bayan-Uhaa. Dzavhan, C Mongolia *48°41′N 98°46′E*
162 *J6* **Ih-Uul** *var.* Selenge. Hövsgöl, N Mongolia *49°25′N 101°30′E*
93 *L14* **Ii** Pohjois-Pohjanmaa, C Finland *65°18′N 25°23′E*
164 *M13* **Iida** Nagano, Honshū, S Japan *35°32′N 137°48′E*
93 *M14* **Iijoki** ≈ C Finland
Iinnasuolu *see* Hinnøya
118 *J4* **Iisaku** *Ger.* Isaak. Ida-Virumaa, NE Estonia *59°06′N 27°19′E*
93 *M16* **Iisalmi** *var.* Idensalmi. Pohjois-Savo, C Finland *63°32′N 27°10′E*
Ii-shima *see* Ie-jima
165 *N11* **Iiyama** Nagano, Honshū, S Japan *36°52′N 138°22′E*
77 *S16* **Ijebu-Ode** Ogun, SW Nigeria *06°46′N 03°57′E*
137 *U11* **Ijevan** *Rus.* Idzhevan. N Armenia *40°53′N 45°07′E*
98 *H9* **IJmuiden** Noord-Holland, W Netherlands *52°28′N 04°38′E*
98 *M12* **IJssel** *var.* Yssel. ≈ Netherlands
98 *J8* **IJsselmeer** *prev.* Zuider Zee. ◎ N Netherlands
98 *L9* **IJsselmuiden** Overijssel, E Netherlands *52°34′N 05°55′E*
98 *I12* **IJsselstein** Utrecht, C Netherlands *52°01′N 05°02′E*
61 *G14* **Ijuí** Rio Grande do Sul, S Brazil *28°23′S 53°55′W*
61 *G14* **Ijuí, Rio** ≈ S Brazil
189 *R8* **Ijuw** NE Nauru *0°30′S 166°57′E*
99 *E16* **IJzendijke** Zeeland, SW Netherlands *51°20′N 03°36′E*
99 *A18* **Ijzer** ≈ W Belgium
Ikaahuk *see* Sachs Harbour
93 *K18* **Ikaalinen** Pirkanmaa, W Finland *61°46′N 23°05′E*
172 *I6* **Ikalamavony** Fianarantsoa, SE Madagascar *21°10′S 46°35′E*
Ikaluktutiak *see* Cambridge Bay
185 *G16* **Ikamatua** West Coast, South Island, New Zealand *42°16′S 171°42′E*
145 *P16* **Ikan** *prev.* Staroikan. Yuzhnyy Kazakhstan, S Kazakhstan *43°09′N 68°34′E*
77 *U16* **Ikare** Ondo, SW Nigeria *07°36′N 05°52′E*
115 *L20* **Ikaría** *var.* Kariot, Nicaria, Nikaria; *anc.* Icaria. *island* Dodekánisa, Greece, Aegean Sea
95 *F22* **Ikast** Midtjylland, W Denmark *56°09′N 09°10′E*
184 *O9* **Ikawhenua Range** ▲▲ North Island, New Zealand
165 *U4* **Ikeda** Hokkaidō, NE Japan *42°54′N 143°25′E*
164 *H14* **Ikeda** Tokushima, Shikoku, SW Japan *34°00′N 133°47′E*
77 *S16* **Ikeja** Lagos, SW Nigeria *06°36′N 03°16′E*
79 *L19* **Ikela** Equateur, C Dem. Rep. Congo *01°11′S 23°16′E*
164 *C13* **Iki** *prev.* Gōnoura. Nagasaki, Iki, SW Japan *33°44′N 129°41′E*
164 *C13* **Iki** *island* SW Japan
127 *O13* **Iki Burul** Respublika Kalmykiya, SW Russian Federation *45°48′N 44°44′E*
137 *P11* **İkizdere** Rize, NE Turkey *40°47′N 40°34′E*
39 *P14* **Ikolik, Cape** *headland* Kodiak Island, Alaska, USA *57°12′N 154°46′W*
77 *V17* **Ikom** Cross River, SE Nigeria *05°57′N 08°43′E*
172 *I6* **Ikongo** *prev.* Fort-Carnot. Fianarantsoa, SE Madagascar *21°52′S 47°27′E*
39 *P5* **Ikpikpuk River** ≈ Alaska, USA
190 *H1* **Iku** *prev.* Lone Tree Islet. *atoll* Tungaru, W Kiribati
164 *I12* **Ikuno** Hyōgo, Honshū, SW Japan *35°13′N 134°48′E*
190 *H16* **Ikurangi** ▲ Rarotonga, S Cook Islands *21°12′S 159°45′W*
171 *X14* **Ilaga** Papua, E Indonesia *03°54′S 137°30′E*
171 *O2* **Ilagan** Luzon, N Philippines *17°08′N 121°54′E*
142 *J7* **Īlām** *var.* Elam. Īlām, W Iran *33°37′N 46°27′E*
153 *R12* **Ilām** Eastern, E Nepal *26°52′N 87°58′E*
142 *J8* **Īlām** *off.* Ostān-e Īlām. ◈ *province* W Iran
Īlām, Ostān-e *see* Īlām
161 *T13* **Ilan** *Jap.* Giran. N Taiwan *24°45′N 121°44′E*
146 *G9* **Ilanly Obvodnitel'nyy Kanal** *canal* N Turkmenistan
122 *L12* **Ilanskiy** Krasnoyarskiy Kray, S Russian Federation *56°16′N 95°59′E*
108 *H9* **Ilanz** Graubünden, S Switzerland *46°46′N 09°10′E*
77 *S16* **Ilaro** Ogun, SW Nigeria *06°52′N 03°01′E*
57 *I17* **Ilave** Puno, S Peru *16°07′S 69°40′W*
110 *K8* **Iława** *Ger.* Deutsch-Eylau. Warmińsko-Mazurskie, NE Poland *53°36′N 19°35′E*
121 *P16* **Il-Bajja ta' Marsaxlokk** *var.* Marsaxlokk Bay. *bay* SE Malta
123 *P10* **Ilbenge** Respublika Sakha (Yakutiya), NE Russian Federation *62°52′N 124°13′E*
Ile *see* Ile/Ili He
Ile *see* Ili He
11 *S13* **Ile-à-la-Crosse** Saskatchewan, C Canada *55°29′N 108°00′W*
79 *J21* **Ilebo** *prev.* Port-Francqui. Kasai-Occidental, W Dem. Rep. Congo *04°19′S 20°32′E*
103 *N5* **Île-de-France** ◈ *region* N France
Ilek *see* Yelek
Ilerda *see* Lleida
77 *T16* **Ilesha** Osun, SW Nigeria *07°35′N 04°49′E*
187 *Q16* **Îles Loyauté, Province des** ◈ *province* E New Caledonia
11 *X12* **Ilford** Manitoba, C Canada *56°02′N 95°48′W*
116 *K14* **Ilfov** ◈ *county* S Romania
97 *I23* **Ilfracombe** SW England, United Kingdom *51°12′N 04°10′W*
136 *I11* **Ilgaz Dağları** ▲▲ N Turkey
136 *G15* **Ilgın** Konya, W Turkey *38°16′N 31°57′E*
60 *I7* **Ilha Solteira** São Paulo, S Brazil *20°28′S 51°19′W*
104 *G7* **Ílhavo** Aveiro, N Portugal *40°36′N 08°40′W*
59 *O18* **Ilhéus** Bahia, E Brazil *14°50′S 39°06′W*
129 *R7* **Ili** *var.* Ile, *Chin.* Ili He, *Rus.* Reka Ili. ≈ China/Kazakhstan *see also* Ili He
116 *G11* **Ilia** *Hung.* Marosillye. Hunedoara, SW Romania *45°57′N 22°40′E*
39 *P13* **Iliamna** Alaska, USA *59°42′N 154°49′W*
39 *P13* **Iliamna Lake** ◎ Alaska, USA
137 *N13* **İliç** Erzincan, C Turkey *39°27′N 38°34′E*
Il'ichevsk *see* Şărur, Azerbaijan
Il'ichevsk *see* Illichivs'k, Ukraine
Ilici *see* Elche
37 *V2* **Iliff** Colorado, C USA *40°46′N 103°04′W*
171 *Q7* **Iligan** *off.* Iligan City. Mindanao, S Philippines *08°12′N 124°16′E*
171 *Q7* **Iligan Bay** *bay* S Philippines
Iligan City *see* Iligan
158 *I5* **Ili He** *var.* Ili, *Kaz.* Ile, *Rus.* Reka Ili. ≈ China/Kazakhstan *see also* Ile
Ili He *see* Ili
56 *C6* **Iliniza** ▲ N Ecuador *0°37′S 78°41′W*
Ilinski *see* Il'inskiy
125 *U14* **Il'inskiy** *var.* Ilinski. Permskiy Kray, NW Russian Federation *58°33′N 55°31′E*
125 *U14* **Il'inskiy** Ostrov Sakhalin, Sakhalinskaya Oblast', SE Russian Federation *47°59′N 142°14′E*
18 *I10* **Ilion** New York, NE USA *43°01′N 75°02′W*
38 *E9* **'Īlio Point** *var.* Ilio Point. *headland* Moloka'i, Hawai'i, USA *21°13′N 157°15′W*
Ilio Point *see* 'Īlio Point

◆ Country
● Country Capital
◇ Dependent Territory
○ Dependent Territory Capital
◈ Administrative Regions
✈ International Airport
▲ Mountain
▲▲ Mountain Range
⛰ Volcano
≈ River
◎ Lake
⊠ Reservoir

109 T13 **Ilirska Bistrica** *prev.* Bistrica, *Ger.* Feistritz, Illyrisch-Feistritz, *It.* Villa del Nevoso. SW Slovenia 45°34′N 14°12′E
137 Q16 **Ilisu Barajı** ⊠ SE Turkey
155 G17 **Ilkal** Karnātaka, C India 15°59′N 76°08′E
97 M19 **Ilkeston** C England, United Kingdom 52°59′N 01°18′W
121 O16 **Il-Kullana** *headland* SW Malta 35°49′N 14°26′E
108 J8 **Ill** ≈ W Austria
103 U6 **Ill** ≈ NE France
62 G10 **Illapel** Coquimbo, C Chile 31°40′S 71°13′W
Illaue Fartak Trench *see* Alula-Fartak Trench
182 C2 **Illbillee, Mount** ▲ South Australia 27°01′S 132°13′E
102 I6 **Ille-et-Vilaine** ⬖ *department* NW France
77 T11 **Illéla** Tahoua, SW Niger 14°25′N 05°10′E
101 J24 **Iller** ≈ S Germany
101 J23 **Illertissen** Bayern, S Germany 48°13′N 10°08′E
105 X9 **Illes Baleares** ⬖ *autonomous community* E Spain
105 N8 **Illescas** Castilla-La Mancha, C Spain 40°08′N 03°51′W
Ille-sur-la-Têt *see* Ille-sur-Têt
103 O17 **Ille-sur-Têt** *var.* Ille-sur-la-Têt. Pyrénées-Orientales, S France 42°40′N 02°37′E
Illiberis *see* Elne
117 P11 **Illichivs'k** *Rus.* Il'ichevsk. Odes'ka Oblast', SW Ukraine 46°18′N 30°36′E
Illicis *see* Elche
102 M6 **Illiers-Combray** Eure-et-Loir, C France 48°18′N 01°15′E
30 K12 **Illinois** *off.* State of Illinois, *also known as* Prairie State, Sucker State. ⬖ *state* C USA
30 J13 **Illinois River** ≈ Illinois, N USA
117 N6 **Illintsi** Vinnyts'ka Oblast', C Ukraine 49°07′N 29°13′E
Illiturgis *see* Andújar
74 M10 **Illizi** SE Algeria 26°30′N 08°28′E
27 Y7 **Illmo** Missouri, C USA 37°13′N 89°30′W
Illurco *see* Lorca
Illuro *see* Mataró
Illyrisch-Feistritz *see* Ilirska Bistrica
101 K16 **Ilm** ≈ C Germany
101 K17 **Ilmenau** Thüringen, C Germany 50°40′N 10°55′E
124 H14 **Il'men', Ozero** ◎ NW Russian Federation
57 H18 **Ilo** Moquegua, SW Peru 17°42′S 71°20′W
171 O6 **Iloilo** *off.* Iloilo City. Panay Island, C Philippines 10°42′N 122°34′E
Iloilo City *see* Iloilo
112 K10 **Ilok** *Hung.* Újlak. Vojvodina, NW Serbia 45°12′N 19°22′E
93 O16 **Ilomantsi** Pohjois-Karjala, SE Finland 62°40′N 30°55′E
42 F8 **Ilopango, Lago de** *volcanic lake* C El Salvador
77 T15 **Ilorin** Kwara, W Nigeria 08°32′N 04°35′E
117 X8 **Ilovays'k** *Rus.* Ilovaysk. Donets'ka Oblast', SE Ukraine 47°55′N 38°14′E
Ilovaysk *see* Ilovays'k
127 O10 **Ilovlya** Volgogradskaya Oblast', SW Russian Federation 49°45′N 44°19′E
127 O10 **Ilovlya** ≈ SW Russian Federation
121 N15 **Il-Ponta ta' San Dimitri** *var.* Ras San Dimitri, San Dimitri Point. *headland* Gozo, NW Malta 36°04′N 14°12′E
126 K14 **Il'skiy** Krasnodarskiy Kray, SW Russian Federation 44°52′N 38°26′E
182 B2 **Iltur** South Australia 27°33′S 130°31′E
171 Y13 **Ilugwa** Papua, E Indonesia 03°42′S 139°09′E
Īluh *see* Batman
118 I11 **Ilūkste** SE Latvia 55°58′N 26°21′E
196 N13 **Ilulissat** Qaasuitsup, C Greenland 68°13′N 51°06′W
171 Y13 **Ilur** Pulau Gorong, E Indonesia 04°00′S 131°25′E
32 F10 **Ilwaco** Washington, NW USA 46°19′N 124°03′W
Il'yaly *see* Gurbansoltan Eje
İlyasbaba Burnu *see* Tekke Burnu
125 U9 **Ilych** ≈ NW Russian Federation
101 O21 **Ilz** ≈ SE Germany
111 M14 **Iłża** Radom, SE Poland 51°09′N 21°15′E
164 G13 **Imabari** *var.* Imaharu. Ehime, Shikoku, SW Japan 34°04′N 132°59′E
Imaharu *see* Imabari
165 O12 **Imaichi** *var.* Imaiti. Tochigi, Honshū, S Japan 36°43′N 139°41′E
Imaiti *see* Imaichi
164 K12 **Imajō** Fukui, Honshū, SW Japan 35°45′N 136°10′E
139 R9 **Imām Ibn Hāshim** Karbalā', C Iraq 32°46′N 43°21′E
149 Q2 **Imām Şāḩib** *var.* Emam Saheb, Hazarat Imam; *prev.* Emām Şāḩeb. Kunduz, NE Afghanistan 37°11′N 68°55′E
139 T11 **Imān 'Abd Allāh** Al Qādisīyah, S Iraq 31°36′N 44°34′E
164 F15 **Imano-yama** ▲ Shikoku, SW Japan 32°51′N 132°48′E
164 C13 **Imari** Saga, Kyūshū, SW Japan 33°18′N 129°51′E
Imarssuak Mid-Ocean Seachannel *see* Imarssuak Seachannel
64 J6 **Imarssuak Seachannel** *var.* Imarssuak Mid-Ocean Seachannel. *channel* N Atlantic Ocean
93 N18 **Imatra** Etelä-Karjala, SE Finland 61°14′N 28°50′E
164 K13 **Imazu** Shiga, Honshū, SW Japan 35°25′N 136°00′E
56 C6 **Imbabura** ⬖ *province* N Ecuador
55 R9 **Imbaimadai** W Guyana 05°44′N 60°23′W
61 K14 **Imbituba** Santa Catarina, S Brazil 28°15′S 48°44′W
27 W9 **Imboden** Arkansas, C USA 36°12′N 91°10′W
Imbros *see* Gökçeada
Imeni 26 Bakinskikh Komissarov *see* Uzboý
125 N13 **Imeni Babushkina** Vologodskaya Oblast', NW Russian Federation 59°40′N 43°04′E
126 J7 **Imeni Karla Libknekhta** Kurskaya Oblast', W Russian Federation 51°36′N 35°28′E
Imeni Mollanepesa *see* Mollanepes Adyndaky
Imeni S. A. Niyazova *see* S. A.Nyýazow Adyndaky
Imeni Sverdlova Rudnik *see* Sverdlovs'k
188 E9 **Imeong** Babeldaob, N Palau
81 L14 **Īmī** Sumalē, E Ethiopia 06°27′N 42°10′E
115 M21 **Imia** *Turk.* Kardak. *island* Dodekánisa, Greece, Aegean Sea
Imishli *see* İmişli
137 X12 **İmişli** *Rus.* Imishli. C Azerbaijan 39°54′N 48°04′E
163 X14 **Imjin-gang** ≈ North Korea/South Korea
35 S3 **Imlay** Nevada, W USA 40°39′N 118°10′W
31 S9 **Imlay City** Michigan, N USA 43°01′N 83°04′W
23 X15 **Immokalee** Florida, SE USA 26°24′N 81°25′W
77 U17 **Imo** ⬖ *state* SE Nigeria
106 G10 **Imola** Emilia-Romagna, N Italy 44°22′N 11°43′E
186 A5 **Imonda** West Sepik, NW Papua New Guinea 03°21′S 141°10′E
Imoschi *see* Imotski
113 G14 **Imotski** *It.* Imoschi. Split-Dalmacija, SE Croatia 43°28′N 17°13′E
59 L14 **Imperatriz** Maranhão, NE Brazil 05°32′S 47°28′W
106 B10 **Imperia** Liguria, NW Italy 43°53′N 08°03′E
57 E15 **Imperial** Lima, W Peru 13°04′S 76°21′W
35 X17 **Imperial** California, W USA 32°51′N 115°34′W
28 L16 **Imperial** Nebraska, C USA 40°30′N 101°37′W
24 M9 **Imperial** Texas, SW USA 31°15′N 102°40′W
35 Y17 **Imperial Dam** *dam* California, W USA
79 I17 **Impfondo** Likouala, NE Congo 01°37′N 18°04′E
153 X14 **Imphāl** *state capital* Manipur, NE India 24°47′N 93°55′E
103 P9 **Imphy** Nièvre, C France 46°55′N 03°16′E
106 G11 **Impruneta** Toscana, C Italy 43°42′N 11°16′E
115 K15 **İmroz** *var.* Gökçeada. Çanakkale, NW Turkey 40°06′N 25°50′E
Imroz Adası *see* Gökçeada
108 L7 **Imst** Tirol, W Austria 47°14′N 10°45′E
40 F3 **Imuris** Sonora, NW Mexico 30°48′N 110°52′W
164 M13 **Ina** Nagano, Honshū, S Japan 35°52′N 137°58′E
65 M18 **Inaccessible Island** *island* W Tristan da Cunha
115 F20 **Ínachos** ≈ S Greece
188 H6 **I Naftan, Puntan** *headland* Saipan, S Northern Mariana Islands
Inagua Islands *see* Little Inagua
Inagua Islands *see* Great Inagua
185 H15 **Inangahua** West Coast, South Island, New Zealand 41°51′S 171°58′E
57 I14 **Iñapari** Madre de Dios, E Peru 11°00′S 69°34′W
188 B17 **Inarajan** SE Guam 13°16′N 144°45′E
92 L10 **Inari** *Lapp.* Anár, Aanaar. Lappi, N Finland 68°54′N 27°06′E
92 L10 **Inarijärvi** *Lapp.* Aanaarjävri, *Swe.* Enareträsk. ◎ N Finland
92 L9 **Inarijoki** *Lapp.* Anárjohka. ≈ Finland/Norway
Inău *see* Ineu
165 P11 **Inawashiro-ko** *var.* Inawasiro Ko. ◎ Honshū, C Japan
Inawasiro Ko *see* Inawashiro-ko
105 X9 **Inca** Mallorca, Spain, W Mediterranean Sea 39°43′N 02°54′E
62 H7 **Inca de Oro** Atacama, N Chile 26°45′S 69°54′W
115 J15 **İnce Burnu** *cape* NW Turkey
136 K9 **İnce Burnu** *headland* N Turkey 42°06′N 34°57′E
136 I17 **İncekum Burnu** *headland* S Turkey 36°13′N 33°57′E
163 X15 **Incheon** , *Jap.* Jinsen; *prev.* Chemulpo, Inch'ŏn. NW South Korea 37°27′N 126°41′E
161 X15 **Incheon** ✈ (Seoul) NW South Korea 37°37′N 126°42′E
76 G7 **Inchiri** ⬖ *region* NW Mauritania
Inch'ŏn *see* Incheon
83 M17 **Inchope** Manica, C Mozambique 19°09′S 33°54′E
Incoronata *see* Kornat
103 Y15 **Incudine, Monte** ▲ Corse, France, C Mediterranean Sea 41°52′N 09°13′E
60 M10 **Indaiatuba** São Paulo, S Brazil 23°03′S 47°14′W
93 H17 **Indal** Västernorrland, C Sweden 62°36′N 17°06′E
93 H17 **Indalsälven** ≈ C Sweden
40 K8 **Inde** Durango, C Mexico 25°55′N 105°10′W
Indefatigable Island *see* Santa Cruz, Isla
35 S10 **Independence** California, W USA 36°48′N 118°14′W
29 X13 **Independence** Iowa, C USA 42°28′N 91°42′W
27 P7 **Independence** Kansas, C USA 37°13′N 95°43′W
20 M4 **Independence** Kentucky, S USA 38°56′N 84°32′W
27 R4 **Independence** Missouri, C USA 39°04′N 94°27′W
21 R8 **Independence** Virginia, NE USA 36°38′N 81°11′W
30 J7 **Independence** Wisconsin, N USA 44°21′N 91°25′W
197 R12 **Independence Fjord** *fjord* N Greenland
Independence Island *see* Malden Island
35 W2 **Independence Mountains** ▲▲ Nevada, W USA
57 K18 **Independencia** Cochabamba, C Bolivia 17°08′S 66°52′W
57 E16 **Independencia, Bahía de la** *bay* W Peru
Independencia, Monte *see* Adam, Mount
116 M12 **Independenţa** Galaţi, SE Romania 45°29′N 27°45′E
Inderagiri *see* Indragiri, Sungai
144 F11 **Inderbor** *prev.* Inderborskiy. Atyrau, W Kazakhstan 48°35′N 51°45′E
Inderborskiy *see* Inderbor
151 I14 **India** *off.* Republic of India, *var.* Indian Union, Union of India, *Hind.* Bhārat. ◆ *republic* S Asia
India *see* Inđija
18 D14 **Indiana** Pennsylvania, NE USA 40°37′N 79°09′W
31 N13 **Indiana** *off.* State of Indiana, *also known as* Hoosier State. ⬖ *state* N USA
31 O14 **Indianapolis** *state capital* Indiana, N USA 39°46′N 86°09′W
11 O10 **Indian Cabins** Alberta, W Canada 59°51′N 117°06′W
42 G1 **Indian Church** Orange Walk, N Belize 17°47′N 88°39′W
Indian Desert *see* Thar Desert
11 U16 **Indian Head** Saskatchewan, S Canada 50°32′N 103°41′W
31 O4 **Indian Lake** ◎ Michigan, N USA
18 K9 **Indian Lake** ◎ New York, NE USA
31 R13 **Indian Lake** ⊠ Ohio, N USA
172-173 **Indian Ocean** *ocean*
29 V15 **Indianola** Iowa, C USA 41°21′N 93°33′W
22 K4 **Indianola** Mississippi, S USA 33°27′N 90°39′W
36 J6 **Indian Peak** ▲ Utah, W USA 38°18′N 113°52′W
23 Y13 **Indian River** *lagoon* Florida, SE USA
35 W10 **Indian Springs** Nevada, W USA 36°33′N 115°40′W
23 Y14 **Indiantown** Florida, SE USA 27°01′N 80°29′W
Indian Union *see* India
59 K19 **Indiara** Goiás, S Brazil 17°12′S 50°09′W
India, Republic of *see* India
India, Union of *see* India
125 Q4 **Indiga** Nenetskiy Avtonomnyy Okrug, NW Russian Federation 67°40′N 49°01′E
123 R9 **Indigirka** ≈ NE Russian Federation
112 L10 **Inđija** *Hung.* India; *prev.* Indjija. Vojvodina, N Serbia 45°03′N 20°04′E
35 V16 **Indio** California, W USA 33°42′N 116°13′W
42 M12 **Indio, Río** ≈ SE Nicaragua
152 I10 **Indira Gandhi** ✈ (Delhi) Delhi, N India
151 Q23 **Indira Point** *headland* Andaman and Nicobar Island, India, NE Indian Ocean 6°54′N 93°54′E
Indjija *see* Inđija
129 Q13 **Indo-Australian Plate** *tectonic feature*
173 N11 **Indomed Fracture Zone** *tectonic feature* SW Indian Ocean
170 L12 **Indonesia** *off.* Republic of Indonesia, *Ind.* Republik Indonesia; *prev.* Dutch East Indies, Netherlands East Indies, United States of Indonesia. ◆ *republic* SE Asia
Indonesian Borneo *see* Kalimantan
Indonesia, Republic of *see* Indonesia
Indonesia, Republik *see* Indonesia
Indonesia, United States of *see* Indonesia
154 G10 **Indore** Madhya Pradesh, C India 22°42′N 75°51′E
168 L11 **Indragiri, Sungai** *var.* Batang Kuantan, Inderagiri. ≈ Sumatera, W Indonesia
Indramajoe/Indramaju *see* Indramayu
169 P15 **Indramayu** *prev.* Indramajoe, Indramaju. Jawa, C Indonesia 06°22′S 108°20′E
155 K14 **Indrāvati** ≈ S India
103 N9 **Indre** ⬖ *department* C France
102 M8 **Indre** ≈ C France
94 D13 **Indre Ålvik** Hordaland, S Norway 60°26′N 06°27′E
102 L8 **Indre-et-Loire** ⬖ *department* C France
Indreville *see* Châteauroux
152 G3 **Indus** *Chin.* Yindu He; *prev.* Yin-tu Ho. ≈ S Asia
Indus Cone *see* Indus Fan
173 P3 **Indus Fan** *var.* Indus Cone. *undersea feature* N Arabian Sea 16°00′N 66°30′E
149 P17 **Indus, Mouths of the** *delta* S Pakistan
83 I24 **Indwe** Eastern Cape, SE South Africa 31°28′S 27°20′E
136 I10 **İnebolu** Kastamonu, N Turkey 41°57′N 33°45′E
77 P8 **I-n-Échaï** *oasis* C Mali
114 M13 **İnecik** Tekirdağ, NW Turkey 40°55′N 27°16′E
136 E12 **İnegöl** Bursa, NW Turkey 40°06′N 29°31′E
Inessa *see* Biancavilla
116 F10 **Ineu** *Hung.* Borosjenő; *prev.* Inău. Arad, W Romania 46°26′N 21°51′E
Ineul/Ineu, Vîrful *see* Ineu, Vârful
116 J9 **Ineu, Vârful** *var.* Ineul; *prev.* Vîrful Ineu. ▲ N Romania 47°31′N 24°52′E
21 P6 **Inez** Kentucky, S USA 37°53′N 82°33′W
74 E8 **Inezgane** ✈ (Agadir) W Morocco 30°35′N 09°27′W
41 T17 **Inferior, Laguna** *lagoon* S Mexico
40 M15 **Infiernillo, Presa del** ⊠ S Mexico
Infiesto *see* L'Infiestu
93 L20 **Ingå** *Fin.* Inkoo. Uusimaa, S Finland 60°01′N 24°05′E
77 U10 **Ingal** *var.* I-n-Gall. Agadez, C Niger 16°52′N 06°57′E
I-n-Gall *see* Ingal
99 C18 **Ingelmunster** West-Vlaanderen, W Belgium 50°12′N 03°15′E
79 I18 **Ingende** Equateur, W Dem. Rep. Congo 0°15′S 18°58′E
62 L5 **Ingeniero Guillermo Nueva Juárez** Formosa, N Argentina 23°55′S 61°50′W
63 H16 **Ingeniero Jacobacci** Río Negro, C Argentina 41°18′S 69°35′W
14 F16 **Ingersoll** Ontario, S Canada 43°03′N 80°53′W
Ingettolgoy *see* Selenge
181 W5 **Ingham** Queensland, NE Australia 18°35′S 146°12′E
146 M11 **Ingichka** Samarqand Viloyati, C Uzbekistan 39°46′N 65°56′E
97 L16 **Ingleborough** ▲ N England, United Kingdom 54°07′N 02°22′W
25 T14 **Ingleside** Texas, SW USA 27°52′N 97°12′W
184 K10 **Inglewood** Taranaki, North Island, New Zealand 39°07′S 174°13′E
35 S15 **Inglewood** California, W USA 33°57′N 118°21′W
101 L21 **Ingolstadt** Bayern, S Germany 48°46′N 11°26′E
33 V9 **Ingomar** Montana, NW USA 46°34′N 107°21′W
13 R14 **Ingonish Beach** Cape Breton Island, Nova Scotia, SE Canada 46°42′N 60°22′W
153 S14 **Ingrāj Bāzār** *prev.* English Bazar. West Bengal, NE India 25°00′N 88°10′E
25 Q11 **Ingram** Texas, SW USA 30°04′N 99°14′W
195 X7 **Ingrid Christensen Coast** *physical region* Antarctica
74 K14 **I-n-Guezzam** S Algeria 19°35′N 05°49′E
Ingulets *see* Inhulets'
Inguri *see* Enguri
Ingushetia/Ingushetiya, Respublika *see* Ingushetiya, Respublika
127 O15 **Ingushetiya, Respublika** *var.* Respublika Ingushetiya, *Eng.* Ingushetia. ⬖ *autonomous republic* SW Russian Federation
83 N20 **Inhambane** Inhambane, SE Mozambique 23°52′S 35°31′E
83 M20 **Inhambane** *off.* Província de Inhambane. ⬖ *province* S Mozambique
Inhambane, Província de *see* Inhambane
83 N17 **Inhaminga** Sofala, C Mozambique 18°24′S 35°00′E
83 N20 **Inharrime** Inhambane, SE Mozambique 24°29′S 35°01′E
83 M18 **Inhassoro** Inhambane, E Mozambique 21°32′S 35°13′E
117 S9 **Inhulets'** *Rus.* Ingulets. Dnipropetrovs'ka Oblast', E Ukraine 47°43′N 33°16′E
117 R10 **Inhulets'** *Rus.* Ingulets. ≈ S Ukraine
105 Q10 **Iniesta** Castilla-La Mancha, C Spain 39°27′N 01°45′W
I-ning *see* Yining
54 K11 **Inírida, Río** ≈ E Colombia
Inis *see* Ennis
Inis Ceithleann *see* Enniskillen
Inis Córthaidh *see* Enniscorthy
Inis Díomáin *see* Ennistimon
97 A17 **Inishbofin** *Ir.* Inis Bó Finne. *island* W Ireland
97 B18 **Inisheer** *var.* Inishere, *Ir.* Inis Oírr. *island* W Ireland
Inishere *see* Inisheer
97 B18 **Inishmaan** *Ir.* Inis Meáin. *island* W Ireland
97 A18 **Inishmore** *Ir.* Árainn. *island* W Ireland
96 E13 **Inishtrahull** *Ir.* Inis Trá Tholl. *island* NW Ireland
97 A17 **Inishturk** *Ir.* Inis Toirc. *island* W Ireland
Inkoo *see* Ingå
185 J16 **Inland Kaikoura Range** ▲▲ South Island, New Zealand
Inland Sea *see* Seto-naikai
21 P11 **Inman** South Carolina, SE USA 35°03′N 82°05′W
108 L7 **Inn** ≈ C Europe
197 O11 **Innaanganeq** *var.* Kap York. *headland* NW Greenland 75°54′N 66°27′W
182 K2 **Innamincka** South Australia 27°47′S 140°45′E
92 G12 **Inndyr** Nordland, C Norway 67°01′N 14°00′E
42 G3 **Inner Channel** *inlet* SE Belize
96 F11 **Inner Hebrides** *island group* W Scotland, United Kingdom
172 H15 **Inner Islands** *var.* Central Group. *island group* NE Seychelles
Inner Mongolia/Inner Mongolian Autonomous Region *see* Nei Mongol Zizhiqu
109 I7 **Inner Rhoden** *former canton* Appenzell. ⬖ NW Switzerland
96 G8 **Inner Sound** *strait* NW Scotland, United Kingdom
100 J13 **Innerste** ≈ C Germany
181 W5 **Innisfail** Queensland, NE Australia 17°29′S 146°03′E
11 Q15 **Innisfail** Alberta, SW Canada 52°01′N 113°59′W
Inniskilling *see* Enniskillen
39 O11 **Innoko River** ≈ Alaska, USA
Innsbruch *see* Innsbruck
108 M7 **Innsbruck** *var.* Innsbruch. Tirol, W Austria 47°17′N 11°25′E
79 I19 **Inongo** Bandundu, W Dem. Rep. Congo 01°55′S 18°20′E
Inoucdjouac *see* Inukjuak
Inowrazlaw *see* Inowrocław
110 I10 **Inowrocław** *Ger.* Hohensalza; *prev.* Inowrazlaw. Kujawski-pomorskie, C Poland 52°47′N 18°15′E
57 K18 **Inquisivi** La Paz, W Bolivia 16°55′S 67°10′W
Inrin *see* Yuanlin
77 O8 **I-n-Sâkâne, 'Erg** *desert* N Mali
77 O8 **I-n-Sâkâne, 'Erg** *desert* N Mali
74 J10 **I-n-Salah** *var.* In Salah. C Algeria 27°11′N 02°31′E
127 O5 **Insar** Respublika Mordoviya, W Russian Federation 53°52′N 44°26′E
189 X15 **Insiaf** Kosrae, E Micronesia
94 L13 **Insjön** Dalarna, C Sweden 60°40′N 15°05′E
Insterburg *see* Chernyakhovsk
Insula *see* Lille
116 L13 **Însurăţei** Brăila, SE Romania 44°55′N 27°40′E
125 V6 **Inta** Respublika Komi, NW Russian Federation 66°00′N 60°10′E
77 R9 **I-n-Tebezas** Kidal, E Mali 17°58′N 01°51′E
Interamna *see* Teramo
Interamna Nahars *see* Terni
28 L11 **Interior** South Dakota, N USA 43°42′N 101°57′W
108 E9 **Interlaken** Bern, SW Switzerland 46°41′N 07°51′E
29 V2 **International Falls** Minnesota, N USA 48°38′N 93°26′W
167 O7 **Inthanon, Doi** ▲ NW Thailand 18°33′N 98°29′E
42 G7 **Intibucá** ⬖ *department* SW Honduras
42 G8 **Intipucá** La Unión, SE El Salvador 13°10′N 88°03′W
61 B15 **Intiyaco** Santa Fe, C Argentina 28°43′S 60°04′W
116 K12 **Întorsura Buzăului** *Ger.* Bozau, *Hung.* Bodzafordulő. Covasna, E Romania 45°40′N 26°02′E
22 H9 **Intracoastal Waterway** *inland waterway system* Louisiana, S USA
25 V13 **Intracoastal Waterway** *inland waterway system* Texas, SW USA
108 G11 **Intragna** Ticino, S Switzerland 46°12′N 08°42′E
165 P14 **Inubō-zaki** *headland* Honshū, S Japan 35°42′N 140°51′E
164 E14 **Inukai** Ōita, Kyūshū, SW Japan 33°05′N 131°37′E
12 I5 **Inukjuak** *var.* Inoucdjouac; *prev.* Port Harrison. Québec, NE Canada 58°28′N 77°58′W
63 I24 **Inútil, Bahía** *bay* S Chile
Inuuvik *see* Inuvik
11 R8 **Inuvik** *var.* Inuuvik. Northwest Territories, NW Canada 68°25′N 133°35′W
164 L13 **Inuyama** Aichi, Honshū, SW Japan 35°23′N 136°56′E
56 G13 **Inuya, Río** ≈ E Peru
125 U13 **In'va** ≈ NW Russian Federation
96 H11 **Inveraray** W Scotland, United Kingdom 56°13′N 05°05′W
185 C24 **Invercargill** Southland, South Island, New Zealand 46°25′S 168°22′E
183 T5 **Inverell** New South Wales, SE Australia 29°46′S 151°10′E
96 I8 **Invergordon** N Scotland, United Kingdom 57°42′N 04°02′W
11 P16 **Invermere** British Columbia, SW Canada 50°30′N 116°00′W
13 R14 **Inverness** Cape Breton Island, Nova Scotia, SE Canada 46°14′N 61°19′W
96 I8 **Inverness** N Scotland, United Kingdom 57°27′N 04°15′W
23 V11 **Inverness** Florida, SE USA 28°50′N 82°19′W
96 I9 **Inverness** *cultural region* NW Scotland, United Kingdom
96 K9 **Inverurie** NE Scotland, United Kingdom 57°14′N 02°14′W
182 F8 **Investigator Group** *island group* South Australia
173 T7 **Investigator Ridge** *undersea feature* E Indian Ocean 11°30′S 98°10′E
182 H10 **Investigator Strait** *strait* South Australia
29 R11 **Inwood** Iowa, C USA 43°16′N 96°25′W
123 S10 **Inya** ≈ E Russian Federation
Inyanga *see* Nyanga
83 M16 **Inyangani** ▲ NE Zimbabwe 18°22′S 32°57′E
83 J17 **Inyathi** Matabeleland North, SW Zimbabwe 19°39′S 28°54′E
35 T12 **Inyokern** California, W USA 35°37′N 117°48′W
35 T10 **Inyo Mountains** ▲▲ California, W USA
127 P6 **Inza** Ul'yanovskaya Oblast', W Russian Federation 53°51′N 46°21′E
127 W5 **Inzer** Respublika Bashkortostan, W Russian Federation 54°11′N 57°37′E
127 N7 **Inzhavino** Tambovskaya Oblast', W Russian Federation 52°18′N 42°28′E
115 C16 **Ioánnina** *var.* Janina, Yannina. Ípeiros, W Greece 39°39′N 20°52′E
164 B17 **Iō-jima** *var.* Iwojima. *island* Nansei-shotō, SW Japan
124 L4 **Iokan'ga** ≈ NW Russian Federation
27 Q6 **Iola** Kansas, C USA 37°55′N 95°24′W
Iolcus *see* Iolkós
115 G16 **Iolkós** *anc.* Iolcus. *site of ancient city* Thessalía, C Greece
Iolotan' *see* Ýolöten
83 A16 **Iona** Namibe, SW Angola 16°54′S 12°39′E
96 F11 **Iona** *island* W Scotland, United Kingdom
116 M15 **Ion Corvin** Constanţa, SE Romania 44°07′N 27°50′E
35 P7 **Ione** California, W USA 38°21′N 120°55′W
116 I13 **Ioneşti** Vâlcea, SW Romania 44°52′N 24°12′E
31 Q9 **Ionia** Michigan, N USA 42°59′N 85°04′W
Ionia Basin *see* Ionian Basin
121 O10 **Ionian Basin** *var.* Ionia Basin. *undersea feature* Ionian Sea, C Mediterranean Sea 36°00′N 20°00′E
115 B17 **Iónia Nisiá** *var.* Iónioi Nísoi, *Eng.* Ionian Islands. *island group* W Greece
Ionian Islands *see* Iónia Nisiá/Iónioi Nísoi
121 O10 **Ionian Sea** *Gk.* Iónio Pélagos, *It.* Mar Ionio. *sea* C Mediterranean Sea
115 B17 **Iónioi Nísoi** *Eng.* Ionian Islands. ⬖ *region* W Greece
Iónioi Nísoi *see* Iónia Nisiá
Ionio, Mar/Iónio Pélagos *see* Ionian Sea
Iordan *see* Yordon
137 U10 **Iori** *var.* Qabırrı. ≈ Azerbaijan/Georgia
Iorrais, Ceann *see* Erris Head
115 J22 **Íos** *var.* Nio. *island* Kykládes, Greece, Aegean Sea
Íos *see* Chóra
165 U15 **Io-Tori-shima** *prev.* Tori-shima. *island* Izu-shotō, SE Japan
115 I20 **Ioulís** *prev.* Kéa. Tziá, Kykládes, Greece, Aegean Sea 37°40′N 24°19′E
22 G9 **Iowa** Louisiana, S USA 30°12′N 93°00′W
29 V13 **Iowa** *off.* State of Iowa, *also known as* Hawkeye State. ⬖ *state* C USA
29 Y14 **Iowa City** Iowa, C USA 41°40′N 91°32′W
29 V13 **Iowa Falls** Iowa, C USA 42°31′N 93°15′W
25 R4 **Iowa Park** Texas, SW USA 33°57′N 98°40′W
29 Y14 **Iowa River** ≈ Iowa, C USA
119 M19 **Ipa** ≈ SE Belarus
59 N20 **Ipatinga** Minas Gerais, SE Brazil 19°32′S 42°30′W
127 N13 **Ipatovo** Stavropol'skiy Kray, SW Russian Federation 45°40′N 42°51′E
115 C16 **Ípeiros** *Eng.* Epirus. ⬖ *region* W Greece
111 J21 **Ipel'** *var.* Ipoly, *Ger.* Eipel. ≈ Hungary/Slovakia
54 C13 **Ipiales** Nariño, SW Colombia 0°52′N 77°38′W
189 V14 **Ipis** *atoll* Chuuk Islands, C Micronesia
59 A14 **Ipixuna** Amazonas, W Brazil 06°57′S 71°42′W
168 J8 **Ipoh** Perak, Peninsular Malaysia 04°36′N 101°02′E
Ipoly *see* Ipel'
187 S15 **Ipota** Erromango, S Vanuatu 18°54′S 169°19′E
79 K14 **Ippy** Ouaka, C Central African Republic 06°17′N 21°13′E
114 L13 **İpsala** Edirne, NW Turkey 40°56′N 26°23′E
Ipsario *see* Ypsário
183 V3 **Ipswich** Queensland, E Australia 27°28′S 152°40′E
97 Q20 **Ipswich** *hist.* Gipeswic. E England, United Kingdom 52°05′N 01°08′E
29 O8 **Ipswich** South Dakota, N USA 45°24′N 99°00′W
Iput' *see* Iputs'
119 P18 **Iputs'** *Rus.* Iput'. ≈ Belarus/Russian Federation
9 R7 **Iqaluit** *prev.* Frobisher Bay. *province capital* Baffin Island, Nunavut, NE Canada 63°44′N 68°28′W
Iqlid *see* Eqlīd
62 G3 **Iquique** Tarapacá, N Chile 20°15′S 70°08′W
56 G8 **Iquitos** Loreto, N Peru 03°51′S 73°13′W
25 N9 **Iraan** Texas, SW USA 30°52′N 101°52′W
79 K14 **Ira Banda** Haute-Kotto, E Central African Republic 05°57′N 22°05′E
165 P16 **Irabu-jima** *island* Miyako-shotō, SW Japan
55 Y9 **Iracoubo** N French Guiana 05°28′N 53°15′W
60 H13 **Iraí** Rio Grande do Sul, S Brazil 27°15′S 53°17′W
114 G12 **Irákleia** Kentrikí Makedonía, N Greece 41°09′N 23°16′E
115 J21 **Irákleia** *island* Kykládes, Greece, Aegean Sea
115 J25 **Irákleio** *var.* Herakleion, *Eng.* Candia; *prev.* Iráklion. Kríti, Greece, E Mediterranean Sea 35°20′N 25°08′E
115 F15 **Irákleio** *anc.* Heracleum. *castle* Kentrikí Makedonía, N Greece
115 J25 **Irákleio** ✈ Kríti, Greece, E Mediterranean Sea 35°20′N 25°10′E
Iráklion *see* Irákleio
143 O7 **Iran** *off.* Islamic Republic of Iran; *prev.* Persia. ◆ *republic* SW Asia
58 F13 **Iranduba** Amazonas, NW Brazil 03°19′S 60°09′W
85 P13 **Iranian Plate** *tectonic feature*
143 Q9 **Iranian Plateau** *var.* Plateau of Iran. *plateau* N Iran
Iran, Islamic Republic of *see* Iran
Iran Mountains *see* Iran, Pegunungan
169 U9 **Iran, Pegunungan** *var.* Iran Mountains. ▲▲ Indonesia/Malaysia
Iran, Plateau of *see* Iranian Plateau
143 W13 **Īrānshahr** Sīstān va Balūchestān, SE Iran 27°14′N 60°40′E
55 P5 **Irapa** Sucre, NE Venezuela 10°37′N 62°35′W
41 N13 **Irapuato** Guanajuato, C Mexico 20°40′N 101°23′W
139 R7 **Iraq** *off.* Republic of Iraq, *Ar.* 'Irāq. ◆ *republic* SW Asia
'Irāq *see* Iraq
Iraq, Republic of *see* Iraq
60 J12 **Irati** Paraná, S Brazil 25°25′S 50°38′W
105 R3 **Irati** ≈ N Spain
125 T8 **Irayël'** Respublika Komi, NW Russian Federation 64°28′N 55°20′E
43 N13 **Irazú, Volcán** ⛰ C Costa Rica 09°57′N 83°52′W
Irbenskiy Zaliv/Irbes Šaurums *see* Irbe Strait
118 D7 **Irbe Strait** *Est.* Kura Kurk, *Latv.* Irbes Šaurums, Irbenskiy Zaliv; *prev. Est.* Irbe Väin. *strait* Estonia/Latvia
Irbe Väin *see* Irbe Strait
138 G9 **Irbid** Irbid, N Jordan 32°33′N 35°51′E
138 G9 **Irbid** *off.* Muḩāfaz̧at Irbid. ⬖ *governorate* N Jordan
Irbid, Muḩāfaz̧at *see* Irbid
Irbil *see* Arbīl
Irbīl *see* Arbīl
109 S6 **Irdning** Steiermark, SE Austria 47°29′N 14°04′E
79 I18 **Irebu** Equateur, W Dem. Rep. Congo 0°32′S 17°44′E
97 D17 **Ireland** *off.* Republic of Ireland, *Ir.* Éire. ◆ *republic* NW Europe
84 C9 **Ireland** *Lat.* Hibernia. *island* Ireland/United Kingdom
64 A12 **Ireland Island North** *island* W Bermuda
64 A12 **Ireland Island South** *island* W Bermuda
Ireland, Republic of *see* Ireland
125 V15 **Iren'** ≈ NW Russian Federation
185 A22 **Irene, Mount** ▲ South Island, New Zealand 45°04′S 167°24′E
Irgalem *see* Yirga 'Alem
Irgiz *see* Yrghyz
Irian *see* New Guinea
Irian Barat *see* Papua
Irian Jaya *see* Papua
Irian Jaya Barat *see* Papua Barat
Irian, Teluk *see* Cenderawasih, Teluk
78 K9 **Iriba** Wadi Fira, NE Chad 15°10′N 22°11′E
127 X7 **Iriklinskoye Vodokhranilishche** ⊠ W Russian Federation
81 H23 **Iringa** Iringa, C Tanzania 07°49′S 35°39′E
81 H23 **Iringa** ⬖ *region* S Tanzania
165 O16 **Iriomote-jima** *island* Sakishima-shotō, SW Japan
42 L4 **Iriona** Colón, NE Honduras 15°55′N 85°10′W
47 U7 **Iriri** ≈ N Brazil
58 I13 **Iriri, Rio** ≈ C Brazil
Iris *see* Yeşilırmak
35 W9 **Irish, Mount** ▲ Nevada, W USA 37°39′N 115°22′W
97 H17 **Irish Sea** *Ir.* Muir Éireann. *sea* C British Isles
139 U12 **Irjal ash Shaykhīyah** Al Muthanná, S Iraq 30°49′N 44°58′E
147 U11 **Irkeshtam** Oshskaya Oblast', SW Kyrgyzstan 39°39′N 73°49′E
122 M13 **Irkutsk** Irkutskaya Oblast', S Russian Federation 52°18′N 104°15′E
122 M12 **Irkutskaya Oblast'** ⬖ *province* S Russian Federation
Irlir, Gora *see* Irlir Tog'i
146 K8 **Irlir Tog'i** *var.* Gora Irlir. ▲ N Uzbekistan 42°43′N 63°24′E
Irminger Basin *see* Reykjanes Basin
21 R12 **Irmo** South Carolina, SE USA 34°05′N 81°10′W
102 E6 **Iroise** *sea* NW France
189 X2 **Iroj** *var.* Eroj. *island* Ratak Chain, SE Marshall Islands
182 H7 **Iron Baron** South Australia 33°01′S 137°13′E
14 C10 **Iron Bridge** Ontario, S Canada 46°16′N 83°12′W
20 H10 **Iron City** Tennessee, S USA 35°01′N 87°34′W
14 I13 **Irondale** ≈ Ontario, SE Canada
182 H7 **Iron Knob** South Australia 32°46′S 137°08′E
30 M5 **Iron Mountain** Michigan, N USA 45°51′N 88°03′W
30 M4 **Iron River** Michigan, N USA 46°05′N 88°38′W
30 J3 **Iron River** Wisconsin, N USA 46°34′N 91°22′W
27 X6 **Ironton** Missouri, C USA 37°37′N 90°40′W
31 S15 **Ironton** Ohio, N USA 38°32′N 82°40′W
30 K4 **Ironwood** Michigan, N USA 46°27′N 90°10′W
12 H12 **Iroquois Falls** Ontario, S Canada 48°47′N 80°41′W
31 N12 **Iroquois River** ≈ Illinois/Indiana, N USA
164 M15 **Irō-zaki** *headland* Honshū, S Japan 34°36′N 138°49′E
Irpen' *see* Irpin'
117 O4 **Irpin'** *Rus.* Irpen'. Kyyivs'ka Oblast', N Ukraine 50°31′N 30°16′E
117 O4 **Irpin'** *Rus.* Irpen'. ≈ N Ukraine
141 Q16 **'Irqah** SW Yemen 13°42′N 47°21′E
166 L6 **Irrawaddy** *var.* Ayeyarwady. ≈ W Myanmar (Burma)
Irrawaddy *see* Ayeyawady
166 K8 **Irrawaddy, Mouths of the** *delta* SW Myanmar (Burma)
117 N4 **Irsha** ≈ N Ukraine
116 H7 **Irshava** Zakarpats'ka Oblast', W Ukraine 48°19′N 23°03′E
107 N18 **Irsina** Basilicata, S Italy 40°42′N 16°18′E
Irtish *see* Yertis
Irtysh *see* Yertis
Irtyshsk *see* Yertis
79 P17 **Irumu** Orientale, E Dem. Rep. Congo 01°27′N 29°52′E
105 Q2 **Irun** *Cast.* Irún. País Vasco, N Spain 43°20′N 01°48′W
Irún *see* Irun
Iruña *see* Pamplona
105 Q3 **Irurtzun** Navarra, N Spain 42°55′N 01°50′W
96 I13 **Irvine** W Scotland, United Kingdom 55°37′N 04°40′W
21 N6 **Irvine** Kentucky, S USA 37°42′N 83°59′W
25 T6 **Irving** Texas, SW USA 32°47′N 96°57′W
20 K5 **Irvington** Kentucky, S USA 37°52′N 86°16′W
164 C15 **Isa** *prev.* Ōkuchi, Ôkuti. Kagoshima, Kyūshū, SW Japan 32°04′N 130°36′E
Isaak *see* Iisaku
28 L8 **Isabel** South Dakota, N USA 45°21′N 101°25′W
186 L8 **Isabel** *off.* Isabel Province. ⬖ *province* N Solomon Islands
171 O8 **Isabela** Basilan Island, SW Philippines 06°41′N 122°00′E
45 S5 **Isabela** W Puerto Rico 18°30′N 67°02′W
45 N8 **Isabela, Cabo** *headland* NW Dominican Republic 19°54′N 71°03′W
57 A18 **Isabela, Isla** *var.* Albemarle Island. *island* Galapagos Islands, Ecuador, E Pacific Ocean
40 I12 **Isabela, Isla** *island* C Mexico
42 K9 **Isabella, Cordillera** ▲▲ NW Nicaragua
35 S12 **Isabella Lake** ⊠ California, W USA
31 N2 **Isabelle, Point** *headland* Michigan, N USA 47°20′N 87°56′W
Isabel Province *see* Isabel
Isabel Segunda *see* Vieques
116 M13 **Isaccea** Tulcea, E Romania 45°16′N 28°28′E
92 H1 **Ísafjarðardjúp** *inlet* NW Iceland

◆ Country
● Country Capital
◇ Dependent Territory
○ Dependent Territory Capital
⬖ Administrative Regions
✈ International Airport
▲ Mountain
▲▲ Mountain Range
⛰ Volcano
≈ River
◎ Lake
⊠ Reservoir

92 H1 **Ísafjarðardjúp** *inlet* NW Iceland
92 H1 **Ísafjörður** Vestfirðir, NW Iceland 66°04′N 23°09′W
164 C14 **Isahaya** Nagasaki, Kyūshū, SW Japan 32°51′N 130°02′E
149 S7 **Īsa Khel** Punjab, E Pakistan 32°39′N 71°20′E
172 H7 **Isalo** *var.* Massif de L'Isalo. ▲▲ SW Madagascar
79 K20 **Isandja** Kasai-Occidental, C Dem. Rep. Congo 03°03′S 21°57′E
187 R15 **Isangel** Tanna, S Vanuatu 19°34′S 169°17′E
79 M18 **Isangi** Orientale, C Dem. Rep. Congo 0°46′N 24°15′E
101 L24 **Isar** ≈ Austria/Germany
101 M23 **Isar-Kanal** *canal* SE Germany
Isbarta *see* Isparta
Isca Damnoniorum *see* Exeter
107 K18 **Ischia** *var.* Isola d'Ischia; *anc.* Aenaria. Campania, S Italy 40°44′N 13°57′E
107 J18 **Ischia, Isola d'** *island* S Italy
54 B12 **Iscuandé** *var.* Santa Bárbara. Nariño, SW Colombia 02°32′N 78°00′W
164 K14 **Ise** Mie, Honshū, SW Japan 34°29′N 136°43′E
100 J12 **Ise** ≈ N Germany
95 I23 **Isefjord** *fjord* E Denmark
Iseghem *see* Izegem
192 M14 **Iselin Seamount** *undersea feature* S Pacific Ocean 72°30′S 179°00′W
Isenhof *see* Püssi
106 E7 **Iseo** Lombardia, N Italy 45°40′N 10°03′E
103 U12 **Iseran, Col de l'** *pass* E France
103 S11 **Isère** ⬖ *department* E France
103 S12 **Isère** ≈ E France
101 F15 **Iserlohn** Nordrhein-Westfalen, W Germany 51°23′N 07°42′E
107 K16 **Isernia** *var.* Æsernia. Molise, C Italy 41°35′N 14°14′E
165 N12 **Isesaki** Gunma, Honshū, S Japan 36°19′N 139°11′E
129 Q5 **Iset'** ≈ C Russian Federation
77 S15 **Iseyin** Oyo, W Nigeria 07°56′N 03°33′E
Isfahan *see* Eşfahān
147 Q11 **Isfana** Batkenskaya Oblast', SW Kyrgyzstan 39°51′N 69°31′E
147 R11 **Isfara** N Tajikistan 40°06′N 70°34′E
149 O4 **Isfi Maidān** Gōwr, N Afghanistan 35°09′N 66°16′E
92 O3 **Isfjorden** *fjord* W Svalbard
Isgender *see* Kul'mach
Isha Baydhabo *see* Baydhabo
125 V11 **Isherim, Gora** ▲ NW Russian Federation 61°06′N 59°09′E
127 Q5 **Isheyevka** Ul'yanovskaya Oblast', W Russian Federation 54°27′N 48°18′E
165 P16 **Ishigaki** Okinawa, Ishigaki-jima, SW Japan 24°20′N 124°09′E
165 P16 **Ishigaki-jima** *island* Sakishima-shotō, SW Japan
165 R3 **Ishikari-wan** *bay* Hokkaidō, NE Japan
165 S16 **Ishikawa** *var.* Isikawa. Okinawa, Okinawa, SW Japan 26°25′N 127°47′E
164 K11 **Ishikawa** *off.* Ishikawa-ken, *var.* Isikawa. ⬖ *prefecture* Honshū, SW Japan
Ishikawa-ken *see* Ishikawa
122 H11 **Ishim** Tyumenskaya Oblast', C Russian Federation 56°13′N 69°25′E
127 V6 **Ishimbay** Respublika Bashkortostan, W Russian Federation 53°21′N 56°03′E
145 O9 **Ishimskoye** Akmola, C Kazakhstan 51°23′N 67°07′E
165 Q10 **Ishinomaki** *var.* Isinomaki. Miyagi, Honshū, C Japan 38°26′N 141°17′E
165 P13 **Ishioka** *var.* Isioka. Ibaraki, Honshū, S Japan 36°11′N 140°16′E
149 Q3 **Ishkamish** *prev.* Eshkamesh. Takhār, NE Afghanistan 36°25′N 69°11′E
149 T2 **Ishkāshim** *prev.* Eshkāshem. Badakhshān, NE Afghanistan 36°43′N 71°34′E
Ishkashim *see* Ishkoshim
Ishkashimskiy Khrebet *see* Ishkoshim, Qatorkŭhi
147 S15 **Ishkoshim** *Rus.* Ishkashim. S Tajikistan 36°46′N 71°35′E
147 S15 **Ishkoshim, Qatorkŭhi** *Rus.* Ishkashimskiy Khrebet. ▲▲ SE Tajikistan
31 N4 **Ishpeming** Michigan, N USA 46°29′N 87°40′W
147 N11 **Ishtixon** *Rus.* Ishtykhan. Samarqand Viloyati, C Uzbekistan 39°59′N 66°28′E
Ishtykhan *see* Ishtixon
Ishurdi *see* Iswardi
61 G17 **Isidoro Noblia** Cerro Largo, NE Uruguay 31°58′S 54°09′W
102 J4 **Isigny-sur-Mer** Calvados, N France 49°20′N 01°06′W
Isikawa *see* Ishikawa
136 C11 **Işıklar Dağı** ▲▲ NW Turkey
107 C19 **Isili** Sardegna, Italy, C Mediterranean Sea 39°46′N 09°06′E
122 H12 **Isil'kul'** Omskaya Oblast', C Russian Federation 54°52′N 71°07′E
Isinomaki *see* Ishinomaki
Isioka *see* Ishioka
81 I18 **Isiolo** Isiolo, C Kenya 0°20′N 37°36′E
81 J18 **Isiolo** ⬖ *county* C Kenya
79 O16 **Isiro** Orientale, NE Dem. Rep. Congo 02°51′N 27°47′E
92 P2 **Isispynten** *headland* NE Svalbard 79°51′N 26°44′E
123 P11 **Isit** Respublika Sakha (Yakutiya), NE Russian Federation 60°53′N 125°32′E
149 O2 **Iskabad Canal** *canal* N Afghanistan
147 Q9 **Iskandar** *Rus.* Iskander. Toshkent Viloyati, E Uzbekistan 41°32′N 69°46′E
Iskander *see* Iskandar
114 G10 **Iskar** *var.* Iskăr, Iskŭr. ≈ NW Bulgaria
Iskăr *see* Iskar
114 H10 **Iskar, Yazovir** *var.* Yazovir Iskŭr; *prev.* Yazovir Stalin. ⊠ W Bulgaria
121 Q2 **İskele** *var.* Trikomo, *Gk.* Tríkomon. E Cyprus 35°16′N 33°54′E
136 K17 **İskenderun** *Eng.* Alexandretta. Hatay, S Turkey 36°34′N 36°10′E
138 H2 **İskenderun Körfezi** *Eng.* Gulf of Alexandretta. *gulf* S Turkey
136 J11 **İskilip** Çorum, N Turkey 40°45′N 34°28′E
Iski-Nauket *see* Nookat
114 J11 **Iskra** *prev.* Popovo. Haskovo, S Bulgaria 41°55′N 25°12′E
Iskŭr *see* Iskar
Iskŭr, Yazovir *see* Iskar, Yazovir
41 S15 **Isla** Veracruz-Llave, SE Mexico 18°01′N 95°30′W
119 J15 **Islach** *Rus.* Isloch'. ≈ C Belarus
104 H14 **Isla Cristina** Andalucía, S Spain 37°12′N 07°20′W
Isla de León *see* San Fernando
149 U6 **Islāmābād** ● (Pakistan) Federal Capital Territory Islāmābād, NE Pakistan 33°40′N 73°08′E
149 V6 **Islāmābād** ✈ Federal Capital Territory Islāmābād, NE Pakistan 33°40′N 73°08′E
Islamabad *see* Anantnāg
149 R17 **Islāmkot** Sind, SE Pakistan 24°37′N 70°04′E
23 Y17 **Islamorada** Florida Keys, Florida, SE USA 24°55′N 80°37′W
153 P14 **Islāmpur** Bihār, N India 25°09′N 85°13′E
Islam Qala *see* Eslām Qal'eh
18 K16 **Island Beach** *spit* New Jersey, NE USA
19 S4 **Island Falls** Maine, NE USA 45°59′N 68°16′W
Island/Ísland *see* Iceland
182 H6 **Island Lagoon** ◎ South Australia
11 Y13 **Island Lake** ◎ Manitoba, C Canada
29 W5 **Island Lake Reservoir** ⊠ Minnesota, N USA
33 R13 **Island Park** Idaho, NW USA 44°27′N 111°21′W
19 N6 **Island Pond** Vermont, NE USA 44°48′N 71°51′W
184 K2 **Islands, Bay of** *inlet* North Island, New Zealand
103 R7 **Is-sur-Tille** Côte d'Or, C France 47°34′N 05°03′E
42 J3 **Islas de la Bahía** ⬖ *department* N Honduras
65 L20 **Islas Orcadas Rise** *undersea feature* S Atlantic Ocean
96 F12 **Islay** *island* SW Scotland, United Kingdom
116 I15 **Islaz** Teleorman, S Romania 43°44′N 24°45′E
29 V7 **Isle** Minnesota, N USA 46°08′N 93°28′W
102 M12 **Isle** ≈ W France
97 I16 **Isle of Man** ◇ *British Crown Dependency* NW Europe
97 I16 **Isle of Man** *island* NW Europe
21 X7 **Isle of Wight** Virginia, NE USA 36°54′N 76°41′W
97 M24 **Isle of Wight** *cultural region* S England, United Kingdom
191 Y3 **Isles Lagoon** ◎ Kiritimati, E Kiribati
37 R11 **Isleta Pueblo** New Mexico, SW USA 34°54′N 106°40′W
Isloch' *see* Islach
61 E19 **Ismael Cortinas** Flores, S Uruguay 33°57′S 57°05′W
Ismailia *see* Al Ismā'īlīya
Ismâ'ilîya *see* Al Ismā'īlīya
Ismailly *see* İsmayıllı
137 X11 **İsmayıllı** *Rus.* Ismailly. N Azerbaijan 40°47′N 48°09′E
Ismid *see* İzmit
147 S12 **Ismoili Somoní, Qullai** *prev.* Qullai Kommunizm. ▲ E Tajikistan
75 X10 **Isnā** *var.* Esna. SE Egypt 25°16′N 32°30′E
93 K18 **Isojoki** Etelä-Pohjanmaa, W Finland 62°07′N 22°00′E
82 M12 **Isoka** Muchinga, NE Zambia 10°08′S 32°43′E
Isola d'Ischia *see* Ischia
Isola d'Istria *see* Izola
Isonzo *see* Soča
15 U4 **Isoukustouc** ≈ Québec, SE Canada
136 F15 **Isparta** *var.* Isbarta. Isparta, SW Turkey 37°46′N 30°32′E
136 F15 **Isparta** *var.* Isbarta. ⬖ *province* SW Turkey
114 M7 **Isperih** *var.* Isperikh; *prev.* Kemanlar. Razgrad, N Bulgaria 43°43′N 26°49′E
Isperikh *see* Isperih
107 L26 **Ispica** Sicilia, Italy, C Mediterranean Sea 36°47′N 14°55′E
148 J14 **Ispikān** Baluchistān, SW Pakistan 26°21′N 62°15′E
137 Q12 **İspir** Erzurum, NE Turkey 40°29′N 41°02′E
138 E12 **Israel** *off.* State of Israel, *var.* Medinat Israel, *Heb.* Yisrael, Yisra'el. ◆ *republic* SW Asia
Israel, State of *see* Israel
Issa *see* Vis
55 S9 **Issano** C Guyana 05°49′N 59°28′W
76 M16 **Issia** SW Ivory Coast 06°33′N 06°33′W
Issiq Köl *see* Issyk-Kul', Ozero
103 P11 **Issoire** Puy-de-Dôme, C France 45°33′N 03°15′E
103 N9 **Issoudun** *anc.* Uxellodunum. Indre, C France 46°57′N 01°59′E
81 H22 **Issuna** Singida, C Tanzania 05°24′S 34°48′E
Issyk *see* Yesik
Issyk-Kul' *see* Balykchy
147 X7 **Issyk-Kul', Ozero** *var.* Issiq Köl, *Kir.* Ysyk-Köl. ◎ E Kyrgyzstan
147 X7 **Issyk-Kul'skaya Oblast'** *Kir.* Ysyk-Köl Oblasty. ⬖ *province* E Kyrgyzstan
149 Q7 **Īstādeh-ye Moqor, Āb-e-** *var.* Āb-i-Istāda. ◎ SE Afghanistan
136 D11 **İstanbul** *Bul.* Tsarigrad, *Eng.* Istanbul, *prev.* Constantinople; *anc.* Byzantium. İstanbul, NW Turkey 41°02′N 28°57′E
114 P12 **İstanbul** ⬖ *province* NW Turkey
114 P12 **İstanbul Boğazı** *var.* Bosporus Thracius, *Eng.* Bosphorus, Bosporus, *Turk.* Karadeniz Boğazi. *strait* NW Turkey
Istarska Županija *see* Istra
115 G19 **Isthmía** Pelopónnisos, S Greece 37°55′N 23°02′E
115 G17 **Istiaía** Évvoia, C Greece 38°57′N 23°09′E
54 D9 **Istmina** Chocó, W Colombia 05°09′N 76°42′W
23 W13 **Istokpoga, Lake** ◎ Florida, SE USA
112 A9 **Istra** *off.* Istarska Županija. ⬖ *province* NW Croatia
112 I10 **Istra** *Eng.* Istria, *Ger.* Istrien. *cultural region* NW Croatia
103 R15 **Istres** Bouches-du-Rhône, SE France 43°30′N 04°59′E
Istria/Istrien *see* Istra
153 T15 **Iswardi** *var.* Ishurdi. Rajshahi, W Bangladesh 24°10′N 89°04′E
127 V7 **Isyangulovo** Respublika Bashkortostan, W Russian Federation 52°10′N 56°38′E
62 O6 **Itá** Central, S Paraguay 25°29′S 57°21′W
59 O17 **Itaberaba** Bahia, E Brazil 12°34′S 40°21′W
59 M20 **Itabira** *prev.* Presidente Vargas. Minas Gerais, SE Brazil 19°39′S 43°14′W
59 O18 **Itabuna** Bahia, E Brazil 14°48′S 39°18′W
59 J18 **Itacaiu** Mato Grosso, S Brazil 14°49′S 51°21′W
58 G12 **Itacoatiara** Amazonas, N Brazil 03°06′S 58°22′W
54 D9 **Itagüí** Antioquia, W Colombia 06°12′N 75°40′W
60 D13 **Itá Ibaté** Corrientes, NE Argentina 27°27′S 57°24′W
60 G11 **Itaipú, Represa de** ⊠ Brazil/Paraguay
58 H13 **Itaituba** Pará, NE Brazil 04°15′S 55°56′W
60 K13 **Itajaí** Santa Catarina, S Brazil 26°50′S 48°39′W
Italia/Italiana, Republica/Italian Republic, The *see* Italy
Italian Somaliland *see* Somalia
25 T7 **Italy** Texas, SW USA 32°10′N 96°52′W
106 G12 **Italy** *off.* The Italian Republic, *It.* Italia, Repubblica Italiana. ◆ *republic* S Europe
59 O19 **Itamaraju** Bahia, E Brazil 16°58′S 39°32′W
59 C14 **Itamarati** Amazonas, W Brazil 06°13′S 68°17′W
59 M19 **Itambé, Pico de** ▲ SE Brazil 18°23′S 43°21′W
164 J13 **Itami** ✈ (Ōsaka) Ōsaka, Honshū, SW Japan 34°47′N 135°24′E
115 H15 **Ítamos** ▲ N Greece 40°06′N 23°51′E
153 W11 **Itānagar** *state capital* Arunāchal Pradesh, NE India 27°02′N 93°38′E
Itany *see* Litani
59 N19 **Itaobim** Minas Gerais, SE Brazil 16°34′S 41°27′W
59 P15 **Itaparica, Represa de** ⊠ E Brazil
58 M13 **Itapecuru-Mirim** Maranhão, E Brazil 03°24′S 44°20′W
60 Q8 **Itaperuna** Rio de Janeiro, SE Brazil 21°14′S 41°51′W
59 O18 **Itapetinga** Bahia, E Brazil 15°17′S 40°16′W
60 L10 **Itapetininga** São Paulo, S Brazil 23°36′S 48°07′W
60 K10 **Itapeva** São Paulo, S Brazil 23°58′S 48°54′W
47 W6 **Itapicuru, Rio** ≈ NE Brazil
58 O13 **Itapipoca** Ceará, E Brazil 03°29′S 39°35′W
60 M9 **Itapira** São Paulo, S Brazil 22°25′S 46°46′W
60 K8 **Itápolis** São Paulo, S Brazil 21°36′S 48°43′W
60 K10 **Itaporanga** São Paulo, S Brazil 23°43′S 49°28′W
62 P7 **Itapúa** *off.* Departamento de Itapúa. ⬖ *department* SE Paraguay
Itapúa, Departamento de *see* Itapúa
59 E15 **Itapuã do Oeste** Rondônia, W Brazil 09°21′S 63°07′W
61 E15 **Itaqui** Rio Grande do Sul, S Brazil 29°10′S 56°28′W
60 K10 **Itararé** São Paulo, S Brazil 24°07′S 49°16′W
60 K10 **Itararé, Rio** ≈ S Brazil
154 H11 **Itārsi** Madhya Pradesh, C India 22°39′N 77°48′E
25 T7 **Itasca** Texas, SW USA 32°09′N 97°09′W
Itassi *see* Vieille Case
60 D13 **Itatí** Corrientes, NE Argentina 27°16′S 58°15′W
60 K10 **Itatinga** São Paulo, S Brazil 23°08′S 48°36′W
115 F18 **Itéas, Kólpos** *gulf* C Greece
57 N15 **Iténez, Río** *var.* Rio Guaporé. ≈ Bolivia/Brazil *see also* Rio Guaporé
Iténez, Río *see* Guaporé, Rio
54 H11 **Iteviate, Río** ≈ C Colombia
100 I13 **Ith** *hill range* C Germany
31 Q8 **Ithaca** Michigan, N USA 43°17′N 84°36′W
18 H11 **Ithaca** New York, NE USA 42°26′N 76°30′W
115 C18 **Itháki** *island* Iónia Nisiá, Greece, C Mediterranean Sea
Itháki *see* Vathy
It Hearrenfean *see* Heerenveen
79 L17 **Itimbiri** ≈ N Dem. Rep. Congo
Itinomiya *see* Ichinomiya
Itinoseki *see* Ichinoseki
39 Q5 **Itkilik River** ≈ Alaska, USA
164 M11 **Itoigawa** Niigata, Honshū, C Japan 37°02′N 137°53′E
15 R6 **Itomamo, Lac** ◎ Québec, SE Canada
165 S17 **Itoman** Okinawa, SW Japan 26°05′N 127°40′E
102 M5 **Iton** ≈ N France
57 M16 **Itonamas Río** ≈ NE Bolivia
Itoupé, Mont *see* Sommet Tabulaire
Itseqqortoormiit *see* Ittoqqortoormiit
22 K4 **Itta Bena** Mississippi, S USA 33°30′N 90°19′W
107 B17 **Ittiri** Sardegna, Italy, C Mediterranean Sea 40°36′N 08°34′E
197 Q14 **Ittoqqortoormiit** *var.* Itseqqortoormiit, *Dan.* Scoresbysund, *Eng.* Scoresby Sound. Sermersooq, C Greenland 70°33′N 21°52′W
60 M10 **Itu** São Paulo, S Brazil 23°17′S 47°16′W
54 D8 **Ituango** Antioquia, NW Colombia 07°07′N 75°46′W
59 A14 **Ituí, Rio** ≈ NW Brazil
79 O20 **Itula** Sud-Kivu, E Dem. Rep. Congo 03°30′S 27°50′E
59 K19 **Itumbiara** Goiás, C Brazil 18°25′S 49°15′W
55 T9 **Ituni** E Guyana 05°24′N 58°18′W
41 X13 **Iturbide** Campeche, SE Mexico 19°41′N 89°29′W
Ituri *see* Aruwimi
123 V13 **Iturup, Ostrov** *island* Kuril'skiye Ostrova, SE Russian Federation
60 L7 **Ituverava** São Paulo, S Brazil 20°22′S 47°48′W
59 C15 **Ituxi, Rio** ≈ W Brazil
61 E14 **Ituzaingó** Corrientes, NE Argentina 27°34′S 56°44′W
101 K18 **Itz** ≈ C Germany
100 I9 **Itzehoe** Schleswig-Holstein, N Germany 53°56′N 09°31′E
23 N2 **Iuka** Mississippi, S USA 34°48′N 88°11′W
60 I11 **Ivaiporã** Paraná, S Brazil 24°16′S 51°46′W
60 I11 **Ivaí, Rio** ≈ S Brazil
92 L10 **Ivalo** *Lapp.* Avveel, Avvil. Lappi, N Finland 68°34′N 27°29′E
92 L10 **Ivalojoki** *Lapp.* Avveel. ≈ N Finland
119 H20 **Ivanava** *Pol.* Janów, Janów Poleski, *Rus.* Ivanovo. Brestskaya Voblasts', SW Belarus 52°09′N 25°32′E
79 F18 **Ivando** *var.* Djidji. ≈ Congo/Gabon
Ivangorod *see* Dęblin
Ivangrad *see* Berane
183 N7 **Ivanhoe** New South Wales, SE Australia 32°55′S 144°21′E
29 S9 **Ivanhoe** Minnesota, N USA 44°27′N 96°15′W
14 D8 **Ivanhoe** ≈ Ontario, S Canada
112 E8 **Ivanić-Grad** Sisak-Moslavina, N Croatia 45°43′N 16°23′E
117 T10 **Ivanivka** Khersons'ka Oblast', S Ukraine 46°43′N 34°28′E
117 P10 **Ivanivka** Odes'ka Oblast', SW Ukraine 46°57′N 30°26′E
113 L14 **Ivanjica** Serbia, C Serbia 43°36′N 20°14′E
112 G11 **Ivanjska** *var.* Potkozarje. Republika Srpska, NW Bosnia and Herzegovina 44°54′N 17°04′E
111 H21 **Ivanka** ✈ (Bratislava) Bratislavský Kraj, W Slovakia 48°10′N 17°13′E
117 O3 **Ivankiv** *Rus.* Ivankov. Kyyivs'ka Oblast', N Ukraine 50°55′N 29°53′E
Ivankov *see* Ivankiv
39 O15 **Ivanof Bay** Alaska, USA 55°55′N 159°28′W
116 J7 **Ivano-Frankivs'k** *Ger.* Stanislau, *Pol.* Stanisławów, *Rus.* Ivano-Frankovsk; *prev.* Stanislav. Ivano-Frankivs'ka Oblast', W Ukraine 48°55′N 24°45′E
Ivano-Frankivs'k *see* Ivano-Frankivs'ka Oblast'
116 I7 **Ivano-Frankivs'ka Oblast'** *var.* Ivano-Frankivs'k, *Rus.* Ivano-Frankovskaya Oblast'; *prev.* Stanislavskaya Oblast'. ⬖ *province* W Ukraine
Ivano-Frankovsk *see* Ivano-Frankivs'k
Ivano-Frankovskaya Oblast' *see* Ivano-Frankivs'ka Oblast'
124 M16 **Ivanovo** Ivanovskaya Oblast', W Russian Federation 57°02′N 40°58′E
Ivanovo *see* Ivanava
124 M16 **Ivanovskaya Oblast'** ⬖ *province* W Russian Federation
35 X12 **Ivanpah Lake** ◎ California, W USA
112 E7 **Ivanščica** ▲▲ NE Croatia
127 R7 **Ivanteyevka** Saratovskaya Oblast', W Russian Federation 52°13′N 49°06′E
Ivantsevichi/Ivatsevichi *see* Ivatsevichy
116 I4 **Ivanychi** Volyns'ka Oblast', NW Ukraine 50°37′N 24°22′E
119 H18 **Ivatsevichy** *Pol.* Iwacewicze, *Rus.* Ivantsevichi, Ivatsevichi. Brestskaya Voblasts', SW Belarus 52°43′N 25°21′E
114 L12 **Ivaylovgrad** Haskovo, S Bulgaria 41°32′N 26°06′E
114 K11 **Ivaylovgrad, Yazovir** ⊠ S Bulgaria
122 G9 **Ivdel'** Sverdlovskaya Oblast', C Russian Federation 60°42′N 60°07′E
Ivenets *see* Ivyanyets
116 L12 **Iveşti** Galaţi, E Romania 45°27′N 28°00′E
Ivgovuotna *see* Lyngen
Ivigtut *see* Ivittuut
59 I21 **Ivinheima** Mato Grosso do Sul, SW Brazil 22°16′S 53°52′W
196 M15 **Ivittuut** *var.* Ivigtut. Sermersooq, S Greenland 61°12′N 48°10′W
Iviza *see* Eivissa\Ibiza
172 I6 **Ivohibe** Fianarantsoa, SE Madagascar 22°28′S 46°53′E
Ivoire, Côte d' *see* Ivory Coast
76 L15 **Ivory Coast** *off.* Republic of the Ivory Coast, *Fr.* Côte d'Ivoire, République de la Côte d'Ivoire. ◆ *republic* W Africa
68 C12 **Ivory Coast** *Fr.* Côte d'Ivoire. *coastal region* S Ivory Coast
Ivory Coast, Republic of the *see* Ivory Coast
95 L22 **Ivösjön** ◎ S Sweden
106 B7 **Ivrea** *anc.* Eporedia. Piemonte, NW Italy 45°28′N 07°52′E
12 J2 **Ivujivik** Québec, NE Canada 62°26′N 77°49′W
119 J16 **Ivyanyets** *Rus.* Ivenets. Minskaya Voblasts', C Belarus 53°53′N 26°45′E
Iv'ye *see* Iwye
Iwacewicze *see* Ivatsevichy
165 R8 **Iwaizumi** Iwate, Honshū, NE Japan 39°48′N 141°46′E
165 P12 **Iwaki** Fukushima, Honshū, N Japan 37°01′N 140°52′E
164 F13 **Iwakuni** Yamaguchi, Honshū, SW Japan 34°08′N 132°06′E
165 S4 **Iwamizawa** Hokkaidō, NE Japan 43°12′N 141°47′E
165 R4 **Iwanai** Hokkaidō, NE Japan 42°51′N 140°21′E
165 Q10 **Iwanuma** Miyagi, Honshū, C Japan 38°06′N 140°51′E
164 L14 **Iwata** Shizuoka, Honshū, S Japan 34°42′N 137°51′E
165 R8 **Iwate** Iwate, Honshū, N Japan 40°03′N 141°12′E
165 R8 **Iwate** *off.* Iwate-ken. ⬖ *prefecture* Honshū, C Japan
Iwate-ken *see* Iwate
77 S16 **Iwo** Oyo, SW Nigeria 07°21′N 03°58′E
Iwojima *see* Iō-jima
119 I16 **Iwye** *Pol.* Iwje, *Rus.* Iv'ye. Hrodzyenskaya Voblasts', W Belarus 53°56′N 25°46′E
42 C4 **Ixcán, Río** ≈ Guatemala/Mexico
99 G18 **Ixelles** *Dut.* Elsene. Brussels, C Belgium 50°49′N 04°21′E
57 J16 **Ixiamas** La Paz, NW Bolivia 13°45′S 68°10′W
41 O13 **Ixmiquilpan** *var.* Ixmiquilpán. Hidalgo, C Mexico 20°30′N 99°15′W
Ixmiquilpán *see* Ixmiquilpan
Ixopo *see* eXobho
Ixtaccíhuatal, Volcán *see* Iztaccíhuati, Volcán
40 M16 **Ixtapa** Guerrero, S Mexico 17°38′N 101°29′W
41 S16 **Ixtepec** Oaxaca, SE Mexico 16°32′N 95°03′W
40 K12 **Ixtlán** *var.* Ixtlán del Río. Nayarit, C Mexico 21°02′N 104°21′W
Ixtlán del Río *see* Ixtlán
122 H11 **Iyevlevo** Tyumenskaya Oblast', C Russian Federation 57°36′N 67°20′E
164 F14 **Iyo** Ehime, Shikoku, SW Japan 33°43′N 132°42′E
164 E14 **Iyo-nada** *sea* S Japan
42 E4 **Izabal** *off.* Departamento de Izabal. ⬖ *department* E Guatemala
Izabal, Departamento de *see* Izabal
42 F5 **Izabal, Lago de** *prev.* Golfo Dulce. ◎ E Guatemala
143 O9 **Īzad Khvāst** Fārs, C Iran 31°31′N 52°09′E
41 X12 **Izamal** Yucatán, SE Mexico 20°58′N 89°00′W
127 Q16 **Izberbash** Respublika Dagestan, SW Russian Federation 42°32′N 47°51′E
99 C18 **Izegem** *prev.* Iseghem. West-Vlaanderen, W Belgium 50°55′N 03°13′E
142 M9 **Īzeh** Khūzestān, SW Iran 31°48′N 49°49′E
165 T16 **Izena-jima** *island* Nansei-shotō, SW Japan
114 N10 **Izgrev** Burgas, E Bulgaria 42°09′N 27°49′E
127 T2 **Izhevsk** *prev.* Ustinov. Udmurtskaya Respublika, NW Russian Federation 56°48′N 53°12′E
125 S7 **Izhma** Respublika Komi, NW Russian Federation 64°56′N 53°52′E
125 S7 **Izhma** ≈ NW Russian Federation
141 X8 **Izkī** NE Oman 22°45′N 57°36′E
Izmail *see* Izmayil
117 N13 **Izmayil** *Rus.* Izmail. Odes'ka Oblast', SW Ukraine 45°19′N 28°49′E
136 B14 **İzmir** *prev.* Smyrna. İzmir, W Turkey 38°25′N 27°10′E
136 C14 **İzmir** *prev.* Smyrna. ⬖ *province* W Turkey
136 E11 **İzmit** *var.* Ismid; *anc.* Astacus. Kocaeli, NW Turkey 40°47′N 29°55′E
104 M14 **Iznájar** Andalucía, S Spain 37°17′N 04°16′W
104 M14 **Iznajar, Embalse de** ⊠ S Spain
105 N14 **Iznalloz** Andalucía, S Spain 37°23′N 03°31′W
136 E11 **İznik** Bursa, NW Turkey 40°27′N 29°43′E
136 E12 **İznik Gölü** ◎ NW Turkey
126 M14 **Izobil'nyy** Stavropol'skiy Kray, SW Russian Federation 45°22′N 41°40′E
109 S13 **Izola** *It.* Isola d'Istria. SW Slovenia 45°31′N 13°40′E
138 H9 **Izra'** *var.* Ezra, Ezraa. Dar'ā, S Syria 32°52′N 36°15′E
41 P14 **Iztaccíhuati, Volcán** *var.* Volcán Ixtaccíhuatal. ⛰ S Mexico 19°07′N 98°37′W
42 C7 **Iztapa** Escuintla, SE Guatemala 13°58′N 90°42′W
Izúcar de Matamoros *see* Matamoros
165 N14 **Izu-hantō** *peninsula* Honshū, S Japan
Izuhara *see* Tsushima
164 J14 **Izumiōtsu** Ōsaka, Honshū, SW Japan 34°29′N 135°25′E
164 I14 **Izumisano** Ōsaka, Honshū, SW Japan 34°23′N 135°18′E
164 G12 **Izumo** Shimane, Honshū, SW Japan 35°22′N 132°46′E
192 H5 **Izu Trench** *undersea feature* NW Pacific Ocean
122 K6 **Izvestiy TsIK, Ostrova** *island* N Russian Federation
114 G10 **Izvor** Pernik, W Bulgaria 42°27′N 22°53′E
116 L5 **Izyaslav** Khmel'nyts'ka Oblast', W Ukraine 50°08′N 26°53′E
117 W6 **Izyum** Kharkivs'ka Oblast', E Ukraine 49°12′N 37°19′E

J

93 M18 **Jaala** Kymenlaakso, S Finland 61°04′N 26°30′E
140 J5 **Jabal ash Shifā** *desert* NW Saudi Arabia
141 U8 **Jabal az Ẓannah** *var.* Jebel Dhanna. Abū Ẓaby, W United Arab Emirates 24°10′N 52°36′E
138 E11 **Jabalīya** *var.* Jabāliyah. NE Gaza Strip 31°32′N 34°29′E
Jabāliyah *see* Jabalīya
105 N11 **Jabalón** ≈ C Spain
154 J10 **Jabalpur** *prev.* Jubbulpore. Madhya Pradesh, C India 23°10′N 79°59′E
141 N15 **Jabal Zuqar, Jazīrat** *var.* Az Zuqur. *island* SW Yemen
Jabar *see* Jawa Barat
Jabat *see* Jabwot
138 J3 **Jabbūl, Sabkhat al** *sabkha* NW Syria
181 P1 **Jabiru** Northern Territory, N Australia 12°44′S 132°48′E
138 H4 **Jablah** *var.* Jeble, *Fr.* Djéblé. Al Lādhiqīyah, W Syria 35°00′N 36°00′E
112 C11 **Jablanac** Lika-Senj, W Croatia 44°43′N 14°54′E
113 H14 **Jablanica** Federacija Bosne I Hercegovine, SW Bosnia and Herzegovina 43°39′N 17°43′E
113 M20 **Jablanica** *Alb.* Mali i Jabllanicës, *var.* Malet e Jabllanicës. ▲ Albania/FYR Macedonia *see also* Jabllanicës, Mali i
Jabllanicës, Malet e *see* Jablanica/Jabllanicës, Mali i
113 M20 **Jabllanicës, Mali i**, *Mac.* Jablanica. ▲ Albania/FYR Macedonia *see also* Jablanica
111 E15 **Jablonec nad Nisou** *Ger.* Gablonz an der Neisse. Liberecký Kraj, N Czech Republic 50°44′N 15°10′E
Jabłonków/Jablunkau *see* Jablunkov
110 J9 **Jabłonowo Pomorskie** Kujawski-pomorskie, C Poland 53°24′N 19°08′E
111 J17 **Jablunkov** *Ger.* Jablunkau, *Pol.* Jabłonków. Moravskoslezský Kraj, E Czech Republic 49°35′N 18°46′E
59 Q15 **Jaboatão** Pernambuco, E Brazil 08°05′S 35°W
60 L8 **Jaboticabal** São Paulo, S Brazil 21°15′S 48°17′W
189 U7 **Jabwot** *var.* Jabat, Jebat, Jōwat. *island* Ralik Chain, S Marshall Islands
105 S4 **Jaca** Aragón, NE Spain 42°34′N 00°33′W
42 B4 **Jacaltenango** Huehuetenango, W Guatemala 15°39′N 91°46′W
59 G14 **Jacaré-a-Canga** Pará, NE Brazil 05°59′S 57°32′W
60 N10 **Jacareí** São Paulo, S Brazil 23°18′S 45°55′W
59 I18 **Jaciara** Mato Grosso, W Brazil 15°59′S 54°57′W
59 E15 **Jaciparaná** Rondônia, W Brazil 09°20′S 64°28′W
19 P5 **Jackman** Maine, NE USA 45°35′N 70°14′W
35 X1 **Jackpot** Nevada, W USA 41°57′N 114°41′W
20 M8 **Jacksboro** Tennessee, S USA 36°19′N 84°11′W
25 S6 **Jacksboro** Texas, SW USA 33°13′N 98°11′W
23 N7 **Jackson** Alabama, S USA 31°30′N 87°53′W
35 P7 **Jackson** California, W USA 38°19′N 120°46′W
23 T4 **Jackson** Georgia, SE USA 33°17′N 83°58′W
21 O6 **Jackson** Kentucky, S USA 37°32′N 83°24′W
22 J8 **Jackson** Louisiana, S USA 30°50′N 91°13′W
31 Q10 **Jackson** Michigan, N USA 42°15′N 84°24′W
29 T11 **Jackson** Minnesota, N USA 43°38′N 95°00′W
22 K5 **Jackson** *state capital* Mississippi, S USA 32°19′N 90°12′W
27 Y7 **Jackson** Missouri, C USA 37°23′N 89°40′W
21 W8 **Jackson** North Carolina, SE USA 36°24′N 77°25′W
31 T15 **Jackson** Ohio, NE USA 39°03′N 82°40′W
20 G9 **Jackson** Tennessee, S USA 35°37′N 88°50′W
33 S14 **Jackson** Wyoming, C USA 43°28′N 110°45′W
185 C19 **Jackson Bay** *bay* South Island, New Zealand
186 E9 **Jackson Field** ✈ (Port Moresby) Central/National Capital District, S Papua New Guinea 09°28′S 147°12′E
185 C20 **Jackson Head** *headland* South Island, New Zealand 43°57′S 168°38′E
23 S8 **Jackson, Lake** ◎ Florida, SE USA
33 S13 **Jackson Lake** ◎ Wyoming, C USA
194 J6 **Jackson, Mount** ▲ Antarctica 71°43′S 63°45′W
37 U3 **Jackson Reservoir** ◎ Colorado, C USA
23 Q3 **Jacksonville** Alabama, S USA 33°48′N 85°45′W
27 V11 **Jacksonville** Arkansas, C USA 34°52′N 92°08′W
23 W8 **Jacksonville** Florida, SE USA 30°20′N 81°39′W
30 K14 **Jacksonville** Illinois, N USA 39°43′N 90°13′W
21 W11 **Jacksonville** North Carolina, SE USA 34°45′N 77°26′W
25 W7 **Jacksonville** Texas, SW USA 31°57′N 95°16′W
23 X9 **Jacksonville Beach** Florida, SE USA 30°17′N 81°23′W
44 L9 **Jacmel** *var.* Jaquemel. S Haiti 18°13′N 72°33′W
Jacob *see* Nkayi
149 Q12 **Jacobābād** Sind, SE Pakistan 28°16′N 68°30′E
55 T11 **Jacobs Ladder Falls** *waterfall* S Guyana
45 O11 **Jaco, Pointe** *headland* N Dominica 15°38′N 61°25′W
15 Q9 **Jacques-Cartier** ≈ Québec, SE Canada
13 P11 **Jacques-Cartier, Détroit de** *var.* Jacques-Cartier Passage. *strait* Gulf of St. Lawrence/St. Lawrence River, Canada
15 W6 **Jacques-Cartier, Mont** ▲ Québec, SE Canada 48°58′N 66°00′W
Jacques-Cartier Passage *see* Jacques-Cartier, Détroit de
61 H16 **Jacuí, Rio** ≈ S Brazil
60 L11 **Jacupiranga** São Paulo, S Brazil 24°42′S 48°00′W
100 G10 **Jade** ≈ NW Germany
100 G10 **Jadebusen** *bay* NW Germany
Jadotville *see* Likasi
Jadransko More/Jadransko Morje *see* Adriatic Sea
105 O7 **Jadraque** Castilla-La Mancha, C Spain 40°55′N 02°55′W
95 I22 **Jægerspris** Hovedstaden, E Denmark 55°52′N 11°59′E
56 C10 **Jaén** Cajamarca, N Peru 05°45′S 78°51′W
105 N13 **Jaén** Andalucía, SW Spain 37°46′N 03°48′W
105 N13 **Jaén** ⬖ *province* Andalucía, S Spain
95 C17 **Jæren** *physical region* S Norway
155 J23 **Jaffna** Northern Province, N Sri Lanka 09°42′N 80°03′E
155 K23 **Jaffna Lagoon** *lagoon* N Sri Lanka
19 N10 **Jaffrey** New Hampshire, NE USA 42°46′N 72°00′W
138 H13 **Jafr, Qā' al** *var.* El Jafr. *salt pan* S Jordan
152 J9 **Jagādhri** Haryāna, N India 30°11′N 77°18′E
118 H4 **Jägala** *var.* Jägala Jõgi, *Ger.* Jaggowal. ≈ NW Estonia
Jägala Jõgi *see* Jägala
Jagannath *see* Puri
155 L14 **Jagdalpur** Chhattīsgarh, C India 19°07′N 82°04′E
163 U5 **Jagdaqi** Nei Mongol Zizhiqu, N China 50°26′N 124°03′E
Jägerndorf *see* Krnov
Jaggowal *see* Jägala
139 O2 **Jaghjaghah, Nahr** ≈ N Syria
112 N13 **Jagodina** *prev.* Svetozarevo. Serbia, C Serbia 43°59′N 21°15′E
112 K12 **Jagodnja** ▲▲ W Serbia
101 I20 **Jagst** ≈ SW Germany
155 I14 **Jagtiāl** Telangana, C India 18°49′N 78°53′E
61 H18 **Jaguarão** Rio Grande do Sul, S Brazil 32°30′S 53°25′W
61 H18 **Jaguarão, Rio** *var.* Río Yaguarón. ≈ Brazil/Uruguay
60 K11 **Jaguariaíva** Paraná, S Brazil 24°15′S 49°44′W
44 D5 **Jagüey Grande** Matanzas, W Cuba 22°31′N 81°07′W
153 P14 **Jahānābād** Bihār, N India 25°13′N 84°59′E
Jahra *see* Al Jahrā'
143 P12 **Jahrom** *var.* Jahrum. Fārs, S Iran 28°35′N 53°32′E
Jahrum *see* Jahrom
Jailolo *see* Halmahera, Pulau
Jainat *see* Chai Nat
Jainti *see* Jayanti
152 H12 **Jaipur** *prev.* Jeypore. *state capital* Rājasthān, N India 26°54′N 75°47′E
153 T14 **Jaipurhat** *var.* Joypurhat. Rajshahi, NW Bangladesh 25°04′N 89°06′E
152 D11 **Jaisalmer** Rājasthān, NW India 26°55′N 70°56′E
154 O12 **Jājapur** *var.* Jajpur, Panikoilli. Odisha, E India 18°54′N 82°36′E
143 R4 **Jājarm** Khorāsān-e Shemālī, NE Iran 36°58′N 56°26′E
112 G12 **Jajce** Federacija Bosne I Hercegovine, W Bosnia and Herzegovina 44°20′N 17°16′E
Jaji *see* 'Alī Khēl
Jajpur *see* Jājapur
83 D17 **Jakalsberg** Otjozondjupa, N Namibia 19°23′S 17°28′E
169 O15 **Jakarta** *prev.* Djakarta, *Dut.* Batavia. ● (Indonesia) Jawa, C Indonesia 06°08′S 106°45′E
10 I8 **Jakes Corner** Yukon, W Canada 60°18′N 134°00′W
152 H9 **Jākhal** Haryāna, NW India 29°46′N 75°51′E
Jakobeny *see* Iacobeni
93 K16 **Jakobstad** *Fin.* Pietarsaari. Österbotten, W Finland 63°41′N 22°40′E
Jakobstadt *see* Jēkabpils
113 O18 **Jakupica** ▲▲ C FYR Macedonia
37 W15 **Jal** New Mexico, SW USA 32°07′N 103°10′W
141 P7 **Jalājil** *var.* Ğalāğıl. Ar Riyāḍ, C Saudi Arabia 25°43′N 45°22′E
149 S5 **Jalālābād** *var.* Jalalabad, Jelalabad. Nangarhār, E Afghanistan 34°26′N 70°28′E
Jalal-Abad *see* Dzhalal-Abad, Dzhalal-Abadskaya Oblast', Kyrgyzstan
Jalal-Abad Oblasty *see* Dzhalal-Abadskaya Oblast'
149 V7 **Jalālpur** Punjab, E Pakistan 32°39′N 74°11′E
149 T11 **Jalālpur Pīrwāla** Punjab, E Pakistan 29°30′N 71°20′E
152 H8 **Jalandhar** *prev.* Jullundur. Punjab, N India 31°20′N 75°37′E
42 J7 **Jalán, Río** ≈ S Honduras
42 E6 **Jalapa** Jalapa, C Guatemala 14°39′N 89°59′W
42 J7 **Jalapa** Nueva Segovia, NW Nicaragua 13°56′N 86°11′W
42 A3 **Jalapa** *off.* Departamento de Jalapa. ⬖ *department* SE Guatemala
Jalapa, Departamento de *see* Jalapa
42 E6 **Jalapa, Río** ≈ SE Guatemala
143 X13 **Jālaq** Sīstān va Balūchestān, SE Iran
93 K17 **Jalasjärvi** Etelä-Pohjanmaa, W Finland 62°30′N 22°50′E
149 O8 **Jaldak** Zābul, SE Afghanistan 32°00′N 66°45′E
60 J7 **Jales** São Paulo, S Brazil 20°15′S 50°34′W
154 P11 **Jaleshwar** *var.* Jaleswar. Odisha, NE India 21°51′N 87°15′E
Jaleswar *see* Jaleshwar
154 F12 **Jalgaon** Mahārāshtra, C India 21°01′N 75°34′E
139 W12 **Jalībah** Dhī Qār, S Iraq 30°37′N 46°31′E
139 W13 **Jalīb Shahāb** Al Muthanná, S Iraq 30°26′N 46°09′E
77 X15 **Jalingo** Taraba, E Nigeria 08°54′N 11°22′E
40 K13 **Jalisco** ⬖ *state* SW Mexico
154 G13 **Jālna** Mahārāshtra, W India 19°50′N 75°53′N
Jalomitsa *see* Ialomiţa
105 R5 **Jalón** ≈ N Spain
152 E13 **Jālor** Rājasthān, N India 25°21′N 72°43′E
112 K11 **Jalovik** Serbia, W Serbia 44°37′N 19°48′E
40 L12 **Jalpa** Zacatecas, C Mexico 21°40′N 103°W
153 S12 **Jalpāiguri** West Bengal, NE India 26°43′N 88°24′E
41 O12 **Jalpán** *var.* Jalpan. Querétaro de Arteaga, C Mexico 21°13′N 99°28′W
Jalpan *see* Jalpán
67 P2 **Jalta** *island* N Tunisia
75 S9 **Jālū** *var.* Jūlā. NE Libya 29°02′N 21°33′E
189 U8 **Jaluit Atoll** *var.* Jālwōj. *atoll* Ralik Chain, S Marshall Islands
Jālwōj *see* Jaluit Atoll
81 L18 **Jamaame** *It.* Giamame; *prev.* Margherita. Jubbada Hoose, S Somalia 0°00′N 42°43′E
77 W13 **Jamaare** ≈ NE Nigeria

◆ Country
● Country Capital
◇ Dependent Territory
○ Dependent Territory Capital
⬖ Administrative Regions
✈ International Airport
▲ Mountain
▲▲ Mountain Range
⛰ Volcano
≈ River
◎ Lake
⊠ Reservoir

44 G9 **Jamaica** ◆ *commonwealth republic* W West Indies
47 P3 **Jamaica** *island* W West Indies
44 I9 **Jamaica Channel** *channel* Haiti/Jamaica
153 T14 **Jamalpur** Dhaka, N Bangladesh 24°54′N 89°57′E
153 Q14 **Jamālpur** Bihār, NE India 25°19′N 86°30′E
168 L9 **Jamaluang** *var.* Jemaluang. Johor, Peninsular Malaysia 02°15′N 103°50′E
59 I14 **Jamanxim, Rio** ≈ C Brazil
56 B8 **Jambeli, Canal de** *channel* S Ecuador
99 I20 **Jambes** Namur, SE Belgium 50°26′N 04°51′E
168 L12 **Jambi** *var.* Telanaipura; *prev.* Djambi. Sumatera, W Indonesia 01°34′S 103°37′E
168 K12 **Jambi** *off.* Propinsi Jambi, *var.* Djambi. ◈ *province* W Indonesia
Jambi, Propinsi *see* Jambi
Jamdena *see* Yamdena, Pulau
12 H8 **James Bay** *bay* Ontario/Québec, E Canada
63 F19 **James, Isla** *island* Archipiélago de los Chonos, S Chile
181 Q8 **James Ranges** ▲▲ Northern Territory, C Australia
29 P8 **James River** ≈ North Dakota/South Dakota, N USA
21 X7 **James River** ≈ Virginia, NE USA
194 H4 **James Ross Island** *island* Antarctica
182 I8 **Jamestown** South Australia 33°13′S 138°36′E
65 G25 **Jamestown** ○ (Saint Helena) NW Saint Helena 15°56′S 05°44′W
35 P8 **Jamestown** California, W USA 37°57′N 120°25′W
20 L7 **Jamestown** Kentucky, S USA 36°58′N 85°03′W
18 D11 **Jamestown** New York, NE USA 42°05′N 79°15′W
29 P5 **Jamestown** North Dakota, N USA 46°54′N 98°42′W
20 L8 **Jamestown** Tennessee, S USA 36°24′N 84°58′W
Jamestown *see* Holetown
15 N10 **Jamet** ≈ Québec, SE Canada
41 Q17 **Jamiltepec** *var.* Santiago Jamiltepec. Oaxaca, SE Mexico 16°18′N 97°51′W
95 F20 **Jammerbugten** *bay* Skagerrak, E North Sea
152 H6 **Jammu** *prev.* Jummoo. *state capital* Jammu and Kashmir, NW India 32°43′N 74°54′E
152 I5 **Jammu and Kashmir** *var.* Jammu-Kashmir, Kashmir. ◈ *state* NW India
149 V4 **Jammu and Kashmīr** *disputed region* India/Pakistan
Jammu-Kashmir *see* Jammu and Kashmir
154 B10 **Jāmnagar** *prev.* Navanagar. Gujarāt, W India 22°28′N 70°06′E
149 S11 **Jāmpur** Punjab, E Pakistan 29°38′N 70°40′E
93 L18 **Jämsä** Keski-Suomi, C Finland 61°51′N 25°10′E
93 L18 **Jämsänkoski** Keski-Suomi, C Finland 61°54′N 25°10′E
153 Q16 **Jamshedpur** Jhārkhand, NE India 22°47′N 86°12′E
94 K9 **Jämtland** ◈ *county* C Sweden
153 Q14 **Jamūi** Bihār, NE India 24°57′N 86°14′E
Jamuna *see* Brahmaputra
153 T14 **Jamuna Nadi** ≈ N Bangladesh
Jamundá *see* Nhamundá, Rio
54 D11 **Jamundí** Valle del Cauca, SW Colombia 03°16′N 76°31′W
153 Q12 **Janakpur** Central, C Nepal 26°45′N 85°55′E
59 N18 **Janaúba** Minas Gerais, SE Brazil 15°47′S 43°16′W
58 K11 **Janaucu, Ilha** *island* NE Brazil
143 Q7 **Jandaq** Eşfahān, C Iran 34°04′N 54°26′E
64 Q11 **Jandia, Punta de** *headland* Fuerteventura, Islas Canarias, Spain, NE Atlantic Ocean 28°03′N 14°32′W
59 B14 **Jandiatuba, Rio** ≈ NW Brazil
105 N12 **Jándula** ≈ S Spain
29 V10 **Janesville** Minnesota, N USA 44°07′N 93°43′W
30 L9 **Janesville** Wisconsin, N USA 42°42′N 89°02′W
83 N20 **Jangamo** Inhambane, SE Mozambique 24°04′S 35°25′E
155 J14 **Jangaon** Telangana, C India 18°47′N 79°25′E
153 S14 **Jangīpur** West Bengal, NE India 24°31′N 88°03′E
Janina *see* Ioánnina
Janischken *see* Joniškis
112 J11 **Janja** NE Bosnia and Herzegovina 44°40′N 19°15′E
Jankovac *see* Jánoshalma
197 Q15 **Jan Mayen** ◇ *constituent part of Norway* N Atlantic Ocean
84 D5 **Jan Mayen** *island* N Atlantic Ocean
197 R15 **Jan Mayen Fracture Zone** *tectonic feature* Greenland Sea/Norwegian Sea
197 R15 **Jan Mayen Ridge** *undersea feature* Greenland Sea/Norwegian Sea 69°00′N 08°00′W
40 H3 **Janos** Chihuahua, N Mexico 30°50′N 108°10′W
111 K25 **Jánoshalma** *SCr.* Jankovac. Bács-Kiskun, S Hungary 46°19′N 19°16′E
Janów *see* Ivanava, Belarus
110 H10 **Janowiec Wielkopolski** *Ger.* Janowitz. Kujawski-pomorskie, C Poland 52°47′N 17°30′E
Janowitz *see* Janowiec Wielkopolski
Janow/Janów *see* Jonava, Lithuania
111 O15 **Janów Lubelski** Lubelski, E Poland 50°42′N 22°24′E
Janów Poleski *see* Ivanava
83 H25 **Jansenville** Eastern Cape, S South Africa 32°56′S 24°40′E
59 M18 **Januária** Minas Gerais, SE Brazil 15°28′S 44°23′W
Janūbīyah, Al Bādiyah al *see* Ash Shāmīyah
102 I7 **Janzé** Ille-et-Vilaine, NW France 47°55′N 01°28′W
154 F10 **Jaora** Madhya Pradesh, C India 23°40′N 75°10′E
131 Y9 **Japan** *var.* Nippon, *Jap.* Nihon. ◆ *monarchy* E Asia
129 Y9 **Japan** *island group* E Asia
192 H4 **Japan Basin** *undersea feature* N Sea of Japan 40°00′N 135°00′E
129 Y8 **Japan, Sea of** *var.* East Sea, *Rus.* Yapanskoye More. *sea* NW Pacific Ocean *see also* East Sea
192 H4 **Japan Trench** *undersea feature* NW Pacific Ocean 37°00′N 143°00′E
Japen *see* Yapen, Pulau
59 A15 **Japiim** *var.* Máncio Lima. Acre, W Brazil 08°00′S 73°39′W
58 D12 **Japurá** Amazonas, N Brazil 01°43′S 66°14′W
58 C12 **Japurá, Rio** *var.* Río Caquetá, Yapurá. ≈ Brazil/Colombia *see also* Caquetá, Río
Japurá, Rio *see* Caquetá, Río
43 W17 **Jaqué** Darién, SE Panama 07°31′N 78°09′W
Jaquemel *see* Jacmel
Jarablos *see* Jarābulus
138 K2 **Jarābulus** *var.* Jarablos, Jerablus, *Fr.* Djérablous. Ḩalab, N Syria 36°51′N 38°02′E
60 K13 **Jaraguá do Sul** Santa Catarina, S Brazil 26°29′S 49°07′W
104 K9 **Jaraicejo** Extremadura, W Spain 39°40′N 05°49′W
104 K9 **Jaráiz de la Vera** Extremadura, W Spain 40°04′N 05°45′W
105 O7 **Jarama** ≈ C Spain
63 J20 **Jaramillo** Santa Cruz, SE Argentina 47°10′S 67°07′W
Jarandilla de la Vega *see* Jarandilla de la Vera
104 K8 **Jarandilla de la Vera** *var.* Jarandilla de la Vega. Extremadura, W Spain 40°08′N 05°39′W
149 V9 **Jarānwāla** Punjab, E Pakistan 31°20′N 73°26′E
138 G9 **Jarash** *var.* Jerash; *anc.* Gerasa. Jarash, NW Jordan 32°17′N 35°54′E
138 G8 **Jarash** *off.* Muḩāfa at Jarash. ◈ *governorate* N Jordan
Jarash, Muḩāfa at *see* Jarash
Jarbah, Jazīrat *see* Jerba, Île de
94 N13 **Järbo** Gävleborg, C Sweden 60°43′N 16°40′E
Jardan *see* Yordon
44 F7 **Jardines de la Reina, Archipiélago de los** *island group* C Cuba
162 I8 **Jargalant** Bayanhongor, C Mongolia 47°14′N 99°43′E
162 K6 **Jargalant** Bulgan, N Mongolia 49°09′N 104°19′E
162 G7 **Jargalant** *var.* Buyanbat. Govĭ-Altay, W Mongolia 47°00′N 95°57′E
162 I6 **Jargalant** *var.* Orgil. Hövsgöl, C Mongolia 48°31′N 99°19′E
Jargalant *see* Battsengel
Jargalant *see* Bulgan, Bayan-Ölgiy, Mongolia
Jargalant *see* Biger, Govĭ-Altay, Mongolia
Jarīd, Shaţţ al *see* Jerid, Chott el
58 I11 **Jari, Rio** *var.* Jary. ≈ N Brazil
141 N7 **Jarīr, Wādī al** *dry watercourse* C Saudi Arabia
Jarja *see* Yur'ya
94 L13 **Järna** *var.* Dala-Jarna. Dalarna, C Sweden 60°31′N 14°22′E
95 O16 **Järna** Stockholm, C Sweden 59°05′N 17°35′E
102 K11 **Jarnac** Charente, W France 45°41′N 00°10′W
110 H12 **Jarocin** Wielkopolskie, C Poland 51°59′N 17°30′E
111 F16 **Jaroměř** *Ger.* Jermer. Královéhradecký Kraj, N Czech Republic 50°22′N 15°55′E
Jaroslau *see* Jarosław
111 O16 **Jarosław** *Ger.* Jaroslau, *Rus.* Yaroslav. Podkarpackie, SE Poland 50°01′N 22°41′E
93 F16 **Järpen** Jämtland, C Sweden 63°21′N 13°30′E
147 O14 **Jarqo'rg'on** *Rus.* Dzharkurgan. Surkhondaryo Viloyati, S Uzbekistan 37°31′N 67°20′E
139 P2 **Jarrāh, Wadi** *dry watercourse* NE Syria
Jars, Plain of *see* Xiangkhoang, Plateau de
162 K14 **Jartai Yanchi** ◎ N China
59 E16 **Jaru** Rondônia, W Brazil 10°24′S 62°45′W
Jarud Qi *see* Lubei
118 I4 **Järva-Jaani** *Ger.* Sankt-Johannis. Järvamaa, N Estonia 59°03′N 25°54′E
118 G5 **Järvakandi** *Ger.* Jerwakant. Raplamaa, NW Estonia 58°45′N 24°49′E
118 H4 **Järvamaa** *var.* Järva Maakond. ◈ *province* N Estonia
Järva Maakond *see* Järvamaa
93 L19 **Järvenpää** Uusimaa, S Finland 60°29′N 25°06′E
14 G17 **Jarvis** Ontario, S Canada 42°53′N 80°06′W
177 R8 **Jarvis Island** ◇ *US unincorporated territory* C Pacific Ocean
94 M11 **Järvsö** Gävleborg, C Sweden 61°43′N 16°25′E
Jary *see* Jari, Rio
112 M9 **Jaša Tomić** Vojvodina, NE Serbia 45°27′N 20°51′E
112 D12 **Jasenice** Zadar, SW Croatia 44°15′N 15°33′E
138 I11 **Jashshat al 'Adlah, Wādī al** *dry watercourse* C Jordan
77 Q16 **Jasikan** E Ghana 07°24′N 00°28′E
Jāsk *see* Bandar-e Jāsk
146 F6 **Jasliq** *Rus.* Zhaslyk. Qoraqalpog'iston Respublikasi, NW Uzbekistan 43°57′N 57°30′E
111 N17 **Jasło** Podkarpackie, SE Poland 49°45′N 21°28′E
11 U16 **Jasmin** Saskatchewan, S Canada 51°11′N 103°34′W
65 A23 **Jason Islands** *island group* NW Falkland Islands
194 I4 **Jason Peninsula** *peninsula* Antarctica
31 N15 **Jasonville** Indiana, N USA 39°09′N 87°12′W
11 O15 **Jasper** Alberta, SW Canada 52°55′N 118°05′W
14 L13 **Jasper** Ontario, SE Canada 44°50′N 75°57′W
23 O3 **Jasper** Alabama, S USA 33°49′N 87°16′W
27 T9 **Jasper** Arkansas, C USA 36°00′N 93°11′W
23 U8 **Jasper** Florida, SE USA 30°31′N 82°57′W
31 N16 **Jasper** Indiana, N USA 38°22′N 86°57′W
29 R11 **Jasper** Minnesota, N USA 43°51′N 96°24′W
27 S7 **Jasper** Missouri, C USA 37°20′N 94°18′W
20 K10 **Jasper** Tennessee, S USA 35°04′N 85°36′W
25 Y9 **Jasper** Texas, SW USA 30°55′N 94°00′W
11 O15 **Jasper National Park** *national park* Alberta/British Columbia, SW Canada
Jassy *see* Iaşi
113 N14 **Jastrebac** ▲▲ SE Serbia
112 D9 **Jastrebarsko** Zagreb, N Croatia 45°40′N 15°40′E
Jastrow *see* Jastrowie
110 G9 **Jastrowie** *Ger.* Jastrow. Wielkopolskie, C Poland 53°25′N 16°48′E
111 J17 **Jastrzębie-Zdrój** Śląskie, S Poland 49°58′N 18°34′E
111 L22 **Jászapáti** Jász-Nagykun-Szolnok, E Hungary 47°30′N 20°10′E
111 L22 **Jászberény** Jász-Nagykun-Szolnok, E Hungary 47°30′N 19°56′E
111 L23 **Jász-Nagykun-Szolnok** *off.* Jász-Nagykun-Szolnok Megye. ◈ *county* E Hungary
Jász-Nagykun-Szolnok Megye *see* Jász-Nagykun-Szolnok
59 J19 **Jataí** Goiás, C Brazil 17°58′S 51°45′W
58 G12 **Jatapu, Serra do** ▲▲ N Brazil
41 W16 **Jatate, Río** ≈ SE Mexico
Jateng *see* Jawa Tengah
149 P17 **Jāti** Sind, SE Pakistan 24°20′N 68°18′E
44 F6 **Jatibonico** Sancti Spíritus, C Cuba 21°56′N 79°11′W
169 O16 **Jatiluhur, Danau** ◎ Jawa, S Indonesia
Jatim *see* Jawa Timur
Jativa *see* Xàtiva
149 S11 **Jatoi** *prev.* Jattoi. Punjab, E Pakistan 29°29′N 70°58′E
Jattoi *see* Jatoi
60 L9 **Jaú** São Paulo, S Brazil 22°11′S 48°35′W
58 F11 **Jauaperi, Rio** ≈ N Brazil
99 I19 **Jauche** Walloon Brabant, C Belgium 50°42′N 04°55′E
Jauer *see* Jawor
Jauf *see* Al Jawf
149 U7 **Jauharābād** Punjab, E Pakistan 32°16′N 72°17′E
57 E14 **Jauja** Junín, C Peru 11°48′S 75°30′W
41 O10 **Jaumave** Tamaulipas, C Mexico 23°28′N 99°22′W
118 H10 **Jaunjelgava** *Ger.* Friedrichstadt. S Latvia 56°38′N 25°03′E
Jaunlatgale *see* Pytalovo
118 I8 **Jaunpiebalga** NE Latvia 57°10′N 26°02′E
118 E9 **Jaunpils** C Latvia 56°45′N 23°03′E
153 N13 **Jaunpur** Uttar Pradesh, N India 25°44′N 82°41′E
29 N8 **Java** South Dakota, N USA 45°29′N 99°54′W
Java *see* Jawa
105 R9 **Javalambre** ▲ E Spain 40°02′N 01°06′W
173 V7 **Java Ridge** *undersea feature* E Indian Ocean
59 A14 **Javari, Rio** *var.* Yavarí. ≈ Brazil/Peru
Javarthushuu *see* Bayan-Uul
169 Q15 **Java Sea** *Ind.* Laut Jawa. *sea* W Indonesia
173 U7 **Java Trench** *var.* Sunda Trench. *undersea feature* E Indian Ocean
143 Q10 **Javazm** *var.* Jowzam. Kermān, C Iran 30°31′N 55°01′E
105 T11 **Jávea** *Cat.* Xàbia. Valenciana, E Spain 38°48′N 00°10′E
Javhlant *see* Bayan-Ovoo
63 G20 **Javier, Isla** *island* S Chile
113 L14 **Javor** ▲▲ Bosnia and Herzegovina/Serbia
111 K20 **Javorie** *Hung.* Jávoros. ▲ S Slovakia 48°26′N 19°16′E
Jávoros *see* Javorie
93 J14 **Jävre** Norrbotten, N Sweden 65°07′N 21°31′E
192 E8 **Jawa** *Eng.* Java; *prev.* Djawa. *island* C Indonesia
169 O16 **Jawa Barat** *off.* Propinsi Jawa Barat, *var.* Jabar, *Eng.* West Java. ◈ *province* S Indonesia
Jawa Barat, Propinsi *see* Jawa Barat
Jawa, Laut *see* Java Sea
139 R3 **Jawān** Nīnawýa, NW Iraq 35°57′N 43°03′E
169 P16 **Jawa Tengah** *off.* Propinsi Jawa Tengah, *var.* Jateng, *Eng.* Central Java. ◈ *province* S Indonesia
Jawa Tengah, Propinsi *see* Jawa Tengah
169 R16 **Jawa Timur** *off.* Propinsi Jawa Timur, *var.* Jatim, *Eng.* East Java. ◈ *province* S Indonesia
Jawa Timur, Propinsi *see* Jawa Timur
81 N17 **Jawhar** *var.* Jowhar, *It.* Giohar. Shabeellaha Dhexe, S Somalia 02°37′N 45°30′E
111 F14 **Jawor** *Ger.* Jauer. Dolnośląskie, SW Poland 51°01′N 16°11′E
Jaworów *see* Yavoriv
111 J16 **Jaworzno** Śląskie, S Poland 50°13′N 19°11′E
Jaxartes *see* Syr Darya
27 R9 **Jay** Oklahoma, C USA 36°25′N 94°49′W
Jayabum *see* Chaiyaphum
Jayanath *see* Chai Nat
153 T12 **Jayanti** *prev.* Jainti. West Bengal, NE India 26°45′N 89°40′E
171 X14 **Jaya, Puncak** *prev.* Puntjak Carstensz, Puntjak Sukarno. ▲ Papua, E Indonesia 04°00′S 137°10′E
171 Z13 **Jayapura** *var.* Djajapura, *Dut.* Hollandia; *prev.* Kotabaru, Sukarnapura. Papua, E Indonesia 02°37′S 140°39′E
Jay Dairen *see* Dalian
Jayhawker State *see* Kansas
147 S12 **Jayilgan** *Rus.* Dzhailgan, Dzhayilgan. C Tajikistan 39°17′N 71°32′E
155 L14 **Jaypur** *var.* Jeypore, Jeypur. Odisha, E India 18°54′N 82°36′E
25 O6 **Jayton** Texas, SW USA 33°16′N 100°35′W
143 U13 **Jaz Mūrīān, Hāmūn-e** ◎ SE Iran
138 M4 **Jazrah** Ar Raqqah, C Syria 35°56′N 39°02′E
138 G6 **Jbaïl** *var.* Jebeil, Jubayl, Jubeil; *anc.* Biblical Gebal, Bybles. W Lebanon 34°00′N 35°45′E
25 O7 **J. B. Thomas, Lake** ⊠ Texas, SW USA
Jdaïdé *see* Judaydah
35 X12 **Jean** Nevada, W USA 35°45′N 115°20′W
22 I9 **Jeanerette** Louisiana, S USA 29°54′N 91°39′W
44 L8 **Jean-Rabel** NW Haiti 19°48′N 73°05′W
143 T12 **Jebāl Bārez, Kūh-e** ▲▲ SE Iran
Jebat *see* Jabwot
77 T15 **Jebba** Kwara, W Nigeria 09°04′N 04°50′E
Jebeil *see* Jbaïl
116 E12 **Jebel** *Hung.* Széphely; *prev. Hung.* Zsebely. Timiş, W Romania 45°33′N 21°14′E
146 B11 **Jebel** *Rus.* Dzhebel. Balkan Welaýaty, W Turkmenistan 39°42′N 54°10′E
Jebel, Bahr el *see* White Nile
Jebel Dhanna *see* Jabal az Zannah
Jeble *see* Jablah
163 Y15 **Jecheon** *Jap.* Teisen; *prev.* Chech'ŏn. N South Korea 37°06′N 128°15′E
96 K13 **Jedburgh** SE Scotland, United Kingdom 55°29′N 02°34′W
Jedda *see* Jiddah
111 L15 **Jędrzejów** *Ger.* Endersdorf. Świętokrzyskie, C Poland 50°39′N 20°18′E
100 K12 **Jeetze** *var.* Jeetzel. ≈ C Germany
Jeetzel *see* Jeetze
29 U14 **Jefferson** Iowa, C USA 42°01′N 94°22′W
21 Q8 **Jefferson** North Carolina, SE USA 36°24′N 81°33′W
25 X6 **Jefferson** Texas, SW USA 32°45′N 94°21′W
30 M9 **Jefferson** Wisconsin, N USA 43°01′N 88°48′W
27 U5 **Jefferson City** *state capital* Missouri, C USA 38°33′N 92°13′W
33 R10 **Jefferson City** Montana, NW USA 46°24′N 112°01′W
21 N9 **Jefferson City** Tennessee, S USA 36°07′N 83°29′W
35 U7 **Jefferson, Mount** ▲ Nevada, W USA 38°49′N 116°54′W
32 H12 **Jefferson, Mount** ▲ Oregon, NW USA 44°40′N 121°48′W
20 L5 **Jeffersontown** Kentucky, S USA 38°11′N 85°33′W
31 P16 **Jeffersonville** Indiana, N USA 38°16′N 85°45′W
33 V15 **Jeffrey City** Wyoming, C USA 42°29′N 107°49′W
77 T13 **Jega** Kebbi, NW Nigeria 12°15′N 04°21′E
Jehol *see* Chengde
163 X17 **Jeju** *Jap.* Saishū; *prev.* Cheju. S South Korea 33°31′N 126°34′E
163 Y17 **Jeju-do** *Jap.* Saishū; *prev.* Cheju-do, Quelpart. *island* S South Korea
163 Y17 **Jeju-haehyeop** *Eng.* Cheju Strait; *prev.* Cheju-haehyeop. *strait* S South Korea
62 P5 **Jejui-Guazú, Río** ≈ E Paraguay
118 I10 **Jēkabpils** *Ger.* Jakobstadt. S Latvia 56°30′N 25°56′E
23 W7 **Jekyll Island** *island* Georgia, SE USA
169 R13 **Jelai, Sungai** ≈ Borneo, N Indonesia
Jelalabad *see* Jalālābād
111 H14 **Jelcz-Laskowice** Dolnośląskie, SW Poland 51°01′N 17°24′E
111 E14 **Jelenia Góra** *Ger.* Hirschberg, Hirschberg im Riesengebirge, Hirschberg in Riesengebirge, Hirschberg in Schlesien. Dolnośląskie, SW Poland 50°54′N 15°48′E
153 S11 **Jelep La** *pass* N India
118 F9 **Jelgava** *Ger.* Mitau. C Latvia 56°38′N 23°47′E
112 L13 **Jelica** ▲▲ C Serbia
20 M8 **Jellico** Tennessee, S USA 36°33′N 84°06′W
95 G23 **Jelling** Syddanmark, C Denmark 55°45′N 09°24′E
169 N9 **Jemaja, Pulau** *island* W Indonesia
Jemaluang *see* Jamaluang
99 E20 **Jemappes** Hainaut, S Belgium 50°27′N 03°53′E
169 S17 **Jember** *prev.* Djember. Jawa, C Indonesia 08°07′S 113°45′E
99 I20 **Jemeppe-sur-Sambre** Namur, S Belgium 50°27′N 04°41′E
37 R10 **Jemez Pueblo** New Mexico, SW USA 35°36′N 106°43′W
158 K2 **Jeminay** *var.* Tuotiereke. Xinjiang Uygur Zizhiqu, NW China 47°28′N 85°49′E
189 U5 **Jemo Island** *atoll* Ratak Chain, C Marshall Islands
169 U11 **Jempang, Danau** ◎ Borneo, N Indonesia
101 L16 **Jena** Thüringen, C Germany 50°56′N 11°35′E
22 I6 **Jena** Louisiana, S USA 31°40′N 92°07′W
108 I8 **Jenaz** Graubünden, SE Switzerland 46°56′N 09°43′E
109 N7 **Jenbach** Tirol, W Austria 47°24′N 11°47′E
171 N15 **Jeneponto** *prev.* Djeneponto. Sulawesi, C Indonesia 05°41′S 119°42′E
138 F9 **Jenin** N West Bank 32°28′N 35°17′E
21 P7 **Jenkins** Kentucky, S USA 37°10′N 82°37′W
27 P9 **Jenks** Oklahoma, C USA 36°01′N 95°58′W
Jenné *see* Djenné
109 X8 **Jennersdorf** Burgenland, SE Austria 46°57′N 16°08′E
22 H9 **Jennings** Louisiana, S USA 30°13′N 92°39′W
11 N7 **Jenny Lind Island** *island* Nunavut, N Canada
23 Y13 **Jensen Beach** Florida, SE USA 27°15′N 80°13′W
9 P6 **Jens Munk Island** *island* Nunavut, NE Canada
163 Y15 **Jeonju** *Jap.* Zenshū; *prev.* Chŏnju. SW South Korea 35°51′N 127°08′E
59 O17 **Jequié** Bahia, E Brazil 13°52′S 40°06′W
59 O18 **Jequitinhonha, Rio** ≈ E Brazil
Jerablus *see* Jarābulus
74 H6 **Jerada** NE Morocco 34°16′N 02°07′W
Jerash *see* Jarash
75 N7 **Jerba, Île de** *var.* Djerba, Jazīrat Jarbah. *island* E Tunisia
44 K9 **Jérémie** SW Haiti 18°39′N 74°11′W
Jeréz *see* Jerez de García Salinas, Mexico
Jerez *see* Jeréz de la Frontera, Spain
40 L11 **Jerez de García Salinas** *var.* Jeréz. Zacatecas, C Mexico 22°40′N 103°00′W
104 J15 **Jerez de la Frontera** *var.* Jerez; *prev.* Xeres. Andalucía, SW Spain 36°41′N 06°08′W
104 I12 **Jerez de los Caballeros** Extremadura, W Spain 38°20′N 06°45′W
Jergucati *see* Jorgucat
138 G10 **Jericho** *Ar.* Arīḩā, *Heb.* Yeriḩo. E West Bank 31°51′N 35°27′E
74 M7 **Jerid, Chott el** *var.* Shaţţ al Jarīd. *salt lake* SW Tunisia
183 O10 **Jerilderie** New South Wales, SE Australia 35°24′S 145°43′E
Jerischmarkt *see* Câmpia Turzii
92 K11 **Jerisjärvi** ◎ NW Finland
Jermak *see* Aksu
Jermentau *see* Yereymentau
Jermer *see* Jaroměř
36 K11 **Jerome** Arizona, SW USA 34°45′N 112°06′W
33 O15 **Jerome** Idaho, NW USA 42°43′N 114°31′W
97 K26 **Jersey** ◇ *British Crown Dependency* Channel Islands, NW Europe
97 L26 **Jersey** *island* Channel Islands, NW Europe
18 K14 **Jersey City** New Jersey, NE USA 40°42′N 74°01′W
18 F13 **Jersey Shore** Pennsylvania, NE USA 41°12′N 77°13′W
30 K14 **Jerseyville** Illinois, N USA 39°07′N 90°19′W
104 K8 **Jerte** ≈ W Spain
138 F10 **Jerusalem** *Ar.* Al Quds, Al Quds ash Sharīf, *Heb.* Yerushalayim; *anc.* Hierosolyma. ● (Israel) Jerusalem, NE Israel 31°47′N 35°13′E
138 G10 **Jerusalem** ◈ *district* E Israel
183 S10 **Jervis Bay** New South Wales, SE Australia 35°09′S 150°42′E
183 S10 **Jervis Bay Territory** ◈ *territory* SE Australia
Jerwakant *see* Järvakandi
109 S10 **Jesenice** *Ger.* Assling. NW Slovenia 46°26′N 14°01′E
111 H16 **Jeseník** *Ger.* Freiwaldau. Olomoucký Kraj, E Czech Republic 50°14′N 17°12′E
Jesi *see* Iesi
106 I8 **Jesolo** *var.* Iesolo. Veneto, NE Italy 45°32′N 12°37′E
Jesselton *see* Kota Kinabalu
95 I14 **Jessheim** Akershus, S Norway 60°07′N 11°10′E
153 T15 **Jessore** Khulna, W Bangladesh 23°10′N 89°12′E
23 W6 **Jesup** Georgia, SE USA 31°36′N 81°54′W
41 S15 **Jesús Carranza** Veracruz-Llave, SE Mexico 17°30′N 95°01′W
62 K10 **Jesús María** Córdoba, C Argentina 30°59′S 64°05′W
26 K6 **Jetmore** Kansas, C USA 38°05′N 99°55′W
103 Q2 **Jeumont** Nord, N France 50°18′N 04°06′E
95 H14 **Jevnaker** Oppland, S Norway 60°15′N 10°25′E
Jewe *see* Jõhvi
25 V9 **Jewett** Texas, SW USA 31°21′N 96°08′W
19 N12 **Jewett City** Connecticut, NE USA 41°36′N 71°58′W
Jeypore *see* Jaipur, Rājasthān, India
Jeypore/Jeypur *see* Jaypur, Orissa, India
113 L17 **Jezercës, Maja e** ▲ N Albania 42°27′N 19°49′E
111 B18 **Jezerní Hora** ▲ SW Czech Republic 49°10′N 13°11′E
154 F10 **Jhābua** Madhya Pradesh, C India 22°44′N 74°37′E
152 H14 **Jhālāwār** Rājasthān, N India 24°37′N 76°12′E
Jhang/Jhang Sadar *see* Jhang Sadr
149 U9 **Jhang Sadr** *var.* Jhang, Jhang Sadar. Punjab, NE Pakistan 31°16′N 72°19′E
152 J13 **Jhānsi** Uttar Pradesh, N India 25°27′N 78°34′E
153 O16 **Jhārkhand** ◈ *state* NE India
154 M11 **Jhārsuguda** Odisha, E India 21°56′N 84°04′E
149 V7 **Jhelum** Punjab, NE Pakistan 32°55′N 73°42′E
129 P9 **Jhelum** ≈ E Pakistan
Jhenaida *see* Jhenaidaha
153 T15 **Jhenaidaha** *var.* Jhenaida, Jhenida. Khulna, Bangladesh 23°32′N 89°09′E
Jhenaidaha *see* Jhenida
153 T15 **Jhenida** *var.* Jhenaidaha. Dhaka, W Bangladesh 23°34′N 89°39′E
jhenida *see* Jhenaidaha
149 P16 **Jhimpīr** Sind, SE Pakistan 25°00′N 68°01′E
Jhind *see* Jīnd
149 R16 **Jhudo** Sind, SE Pakistan 24°58′N 69°18′E
Jhumra *see* Chak Jhumra
152 H11 **Jhunjhunūn** Rājasthān, N India 28°05′N 75°30′E
Ji *see* Hebei, China
Ji *see* Jilin, China
Jiading *see* Xinfeng
153 S14 **Jiāganj** West Bengal, NE India 24°18′N 88°07′E
Jiaji *see* Qionghai
160 J7 **Jialing Jiang** ≈ C China
163 Y7 **Jiamusi** *var.* Chia-mu-ssu, Kiamusze. Heilongjiang, NE China 46°46′N 130°19′E
161 O11 **Ji'an** Jiangxi, S China 27°08′N 115°00′E
163 W12 **Ji'an** Jilin, NE China 41°04′N 126°07′E
163 T13 **Jianchang** Liaoning, NE China 40°48′N 119°51′E
Jianchang *see* Nancheng
Jiancheng *see* Jianyang
160 F11 **Jianchuan** *var.* Jinhuan. Yunnan, SW China 26°28′N 99°49′E
158 M4 **Jiangjunmiao** Xinjiang Uygur Zizhiqu, W China 44°42′N 90°06′E
160 K11 **Jiangkou** *var.* Shuangjiang. Guizhou, S China 27°46′N 108°53′E
Jiangkou *see* Fengkai
161 Q12 **Jiangle** *var.* Guyong. Fujian, SE China 26°44′N 117°26′E
161 N15 **Jiangmen** Guangdong, S China 22°35′N 113°02′E
Jiangna *see* Yanshan
161 Q10 **Jiangshan** Zhejiang, SE China 28°41′N 118°33′E
161 Q7 **Jiangsu** *var.* Chiang-su, Jiangsu Sheng, Kiangsu, Su. ◈ *province* E China
Jiangsu *see* Nanjing
Jiangsu Sheng *see* Jiangsu
161 O11 **Jiangxi** *var.* Chiang-hsi, Gan, Jiangxi Sheng, Kiangsi. ◈ *province* S China
Jiangxi Sheng *see* Jiangxi
160 I8 **Jiangyou** *prev.* Zhongba. Sichuan, C China 31°52′N 104°52′E
161 N9 **Jianli** *var.* Rongcheng. Hubei, C China 29°51′N 112°50′E
161 Q11 **Jian'ou** Fujian, SE China 27°04′N 118°20′E
163 S12 **Jianping** *var.* Yebaishou. Liaoning, NE China 41°13′N 119°37′E
Jianshe *see* Baiyü
160 L9 **Jianshi** *var.* Yezhou. Hubei, C China 30°37′N 109°42′E
Jiantang *see* Xamgyi'nyilha
129 V11 **Jian Xi** ≈ SE China
161 Q11 **Jianyang** Fujian, SE China 27°20′N 118°01′E
160 I9 **Jianyang** *var.* Jiancheng. Sichuan, C China 30°22′N 104°31′E
163 X10 **Jiaohe** Jilin, NE China 43°41′N 127°20′E
Jiaojiang *see* Taizhou
Jiaoxian *see* Jiaozhou
161 R5 **Jiaozhou** *prev.* Jiaoxian. Shandong, E China 36°17′N 120°00′E
161 N6 **Jiaozuo** Henan, C China 35°14′N 113°13′E
Jiashan *see* Mingguang
158 F8 **Jiashi** *var.* Baren, Payzawat. Xinjiang Uygur Zizhiqu, NW China 39°27′N 76°45′E
154 L9 **Jiāwān** Madhya Pradesh, C India 24°20′N 82°17′E
161 S9 **Jiaxing** Zhejiang, SE China 30°44′N 120°46′E
161 S14 **Jiayi** *var.* Chia-i, Chiai, Chiayi, Kiayi, *Jap.* Kagi. C Taiwan 23°29′N 120°27′E
163 X6 **Jiayin** *var.* Chaoyang. Heilongjiang, NE China 48°51′N 130°24′E
159 R8 **Jiayuguan** Gansu, N China 39°47′N 98°14′E
Jibhalanta *see* Uliastay
138 M4 **Jibli** Ar Raqqah, C Syria 35°49′N 39°23′E
116 H9 **Jibou** *Hung.* Zsibó. Sălaj, NW Romania 47°15′N 23°17′E
141 Z9 **Jibsh, Ra's al** *headland* E Oman 21°20′N 59°23′E
Jibuti *see* Djibouti
111 E15 **Jičín** *Ger.* Jitschin. Královéhradecký Kraj, N Czech Republic 50°27′N 15°20′E
140 K10 **Jiddah** *Eng.* Jedda. (Saudi Arabia) Makkah, W Saudi Arabia 21°34′N 39°13′E
141 W11 **Jiddat al Ḩarāsīs** *desert* C Oman
Jiesjavrre *see* Iešjavri
160 M4 **Jiexiu** Shanxi, C China 37°00′N 111°50′E
161 P14 **Jieyang** Guangdong, S China 23°32′N 116°20′E
119 F14 **Jieznas** Kaunas, S Lithuania 54°37′N 24°10′E
Jīfa', Bi'r *see* Jif'īyah, Bi'r
141 P15 **Jif'īyah, Bi'r** *var.* Bi'r Jīfa'. *well* C Yemen
77 W13 **Jigawa** ◈ *state* N Nigeria
146 J10 **Jigerbent** *Rus.* Dzhigirbent. Lebap Welaýaty, NE Turkmenistan 40°44′N 61°56′E
44 I7 **Jiguaní** Granma, E Cuba 20°24′N 76°26′W
159 T12 **Jigzhi** *var.* Chugqênsumdo. Qinghai, C China 33°23′N 101°25′E
Jih-k'a-tse *see* Xigazê
111 E18 **Jihlava** *Ger.* Iglau, *Pol.* Igława. Vysocina, S Czech Republic 49°22′N 15°36′E
111 E18 **Jihlava** *var.* Igel, *Ger.* Iglawa. ≈ Vysočina, C Czech Republic
Jihlavský Kraj *see* Vysočina
111 C18 **Jihočeský Kraj** *prev.* Budějovický Kraj. ◈ *region* Czech Republic
111 G19 **Jihomoravský Kraj** *prev.* Brněnský Kraj. ◈ *region* SE Czech Republic
74 L5 **Jijel** *var.* Djidjel; *prev.* Djidjelli. NE Algeria 36°50′N 05°43′E
116 L9 **Jijia** ≈ N Romania
80 L13 **Jijiga** *It.* Giggiga. Sumalē, E Ethiopia 09°21′N 42°53′E
105 S12 **Jijona** *var.* Xixona. Valenciana, E Spain 38°34′N 00°29′W
81 L18 **Jilib** *It.* Gelib. Jubbada Dhexe, S Somalia 0°18′N 42°48′E
163 W10 **Jilin** *var.* Chi-lin, Girin, Kirin; *prev.* Yungki, Yunki. Jilin, NE China 43°46′N 126°32′E
163 W10 **Jilin** *var.* Chi-lin, Girin, Ji, Jilin Sheng, Kirin. ◈ *province* NE China
163 W11 **Jilin Hada Ling** ▲▲ NE China
Jilin Sheng *see* Jilin
163 S4 **Jiliu He** ≈ NE China
105 Q6 **Jiloca** ≈ N Spain
161 T12 **Jilong** *var.* Keelung, *Jap.* Kirun, Kirun', *prev.* Chilung, *prev. Sp.* Santissima Trinidad. N Taiwan 25°10′N 121°43′E
81 I14 **Jīma** *var.* Jimma, *It.* Gimma. Oromīya, C Ethiopia 07°42′N 36°51′E
44 M9 **Jimaní** W Dominican Republic 18°29′N 71°49′W
116 E11 **Jimbolia** *Ger.* Hatzfeld, *Hung.* Zsombolya. Timiş, W Romania 45°47′N 20°43′E
104 K16 **Jimena de la Frontera** Andalucía, S Spain 36°27′N 05°28′W
40 K7 **Jiménez** Chihuahua, N Mexico 27°09′N 104°54′W
41 N5 **Jiménez** Coahuila, NE Mexico 29°05′N 100°40′W
41 P9 **Jiménez** *var.* Santander Jiménez. Tamaulipas, C Mexico 24°11′N 98°29′W
40 L10 **Jiménez del Teul** Zacatecas, C Mexico 23°13′N 103°46′W
77 Y14 **Jimeta** Adamawa, E Nigeria 09°16′N 12°25′E
Jimma *see* Jīma
158 M5 **Jimsar** Xinjiang Uygur Zizhiqu, NW China 44°05′N 88°48′E
18 I14 **Jim Thorpe** Pennsylvania, NE USA 40°51′N 75°43′W
Jin *see* Shanxi
Jin *see* Tianjin Shi
161 P5 **Jinan** *var.* Chinan, Chi-nan, Tsinan. *province capital* Shandong, E China 36°43′N 116°58′E
Jin'an *see* Songpan
Jinbi *see* Dayao
159 T8 **Jinchang** Gansu, N China 38°31′N 102°07′E
161 N5 **Jincheng** Shanxi, C China 35°30′N 112°52′E
Jincheng *see* Wuding
Jinchengjiang *see* Hechi
152 I9 **Jīnd** *prev.* Jhind. Haryāna, NW India 29°29′N 76°22′E
183 Q11 **Jindabyne** New South Wales, SE Australia 36°28′S 148°36′E
163 X17 **Jin-do** *Jap.* Chin-tō; *prev.* Chin-do. *island* SW South Korea
111 O18 **Jindřichův Hradec** *Ger.* Neuhaus. Jihočeský Kraj, S Czech Republic 49°09′N 15°01′E
Jing *see* Beijing Shi
Jing *see* Jinghe, China
159 X10 **Jingchuan** Gansu, C China 35°20′N 107°45′E
161 Q10 **Jingdezhen** Jiangxi, S China 29°18′N 117°18′E
161 O12 **Jinggangshan** Jiangxi, S China 26°36′N 114°11′E
161 P3 **Jinghai** Tianjin Shi, E China 38°53′N 116°45′E
158 I4 **Jinghe** *var.* Jing. Xinjiang Uygur Zizhiqu, NW China 44°35′N 82°55′E
160 K6 **Jing He** ≈ C China
160 F15 **Jinghong** *var.* Yunjinghong. Yunnan, SW China 22°03′N 100°56′E
160 M9 **Jingmen** Hubei, C China 30°58′N 112°09′E
163 X10 **Jingpo Hu** ◎ NE China
160 M8 **Jing Shan** ▲▲ C China
159 V9 **Jingtai** *var.* Yitiaoshan. Gansu, C China 37°12′N 104°06′E
160 J14 **Jingxi** *var.* Xinjing. Guangxi Zhuangzu Zizhiqu, S China 23°10′N 106°22′E
Jing Xian *see* Jingzhou, Hunan, China
163 W11 **Jingyu** Jilin, NE China 42°23′N 126°48′E
159 V10 **Jingyuan** *var.* Wulan. Gansu, C China 36°35′N 104°40′E
160 M9 **Jingzhou** *prev.* Shashi, Sha-shih, Shasi. Hubei, C China 30°21′N 112°09′E
160 L12 **Jingzhou** *var.* Jing Xian, Jingzhou Miaouzu Dongzu Zizhixian, Quyang. Hunan, S China 26°35′N 109°40′E
Jingzhou Miaouzu Dongzu Zizhixian *see* Jingzhou
163 Z16 **Jinhae** *Jap.* Chinkai; *prev.* Chinhae. S South Korea 35°06′N 128°48′E
Jinhe *see* Jinping
161 R10 **Jinhua** Zhejiang, SE China 29°15′N 119°36′E
Jinhuan *see* Jianchuan
161 P5 **Jining** Shandong, E China 35°25′N 116°35′E
Jining *see* Ulan Qab
81 G18 **Jinja** S Uganda 0°27′N 33°14′E
161 R13 **Jinjiang** *var.* Qingyang. Fujian, SE China 24°53′N 118°36′E
161 O11 **Jin Jiang** ≈ S China
Jinjiang *see* Chengmai
163 Y16 **Jinju** *prev.* Chinju, *Jap.* Shinshū. S South Korea 35°12′N 128°06′E
171 V15 **Jin, Kepulauan** *island group* E Indonesia
161 R13 **Jinmen Dao** *var.* Chinmen Tao, Quemoy. *island* W Taiwan
42 J9 **Jinotega** Jinotega, NW Nicaragua 13°03′N 85°59′W
42 K7 **Jinotega** ◈ *department* N Nicaragua
42 J11 **Jinotepe** Carazo, SW Nicaragua 11°50′N 86°10′W
160 L13 **Jinping** *var.* Sanjiang. Guizhou, S China 26°42′N 109°13′E
160 H14 **Jinping** *var.* Jinhe. Yunnan, SW China 22°47′N 103°12′E
Jinsen *see* Incheon
160 J11 **Jinsha** Guizhou, S China 27°24′N 106°16′E
160 M10 **Jinshi** Hunan, S China 29°42′N 111°46′E
Jinshi *see* Xinning
162 I9 **Jinst** *var.* Bodĭ. Bayanhongor, C Mongolia 45°25′N 100°33′E
159 R7 **Jinta** Gansu, N China 40°01′N 98°57′E
161 Q12 **Jin Xi** ≈ SE China
Jinxi *see* Huludao
Jinxian *see* Jinzhou
161 P6 **Jinxiang** Shandong, E China 35°08′N 116°19′E
161 P8 **Jinzhai** *var.* Meishan. Anhui, E China 31°42′N 115°47′E
161 N4 **Jinzhong** *var.* Yuci. Shanxi, C China 37°34′N 112°45′E
163 T12 **Jinzhou** *var.* Chin-chou, Chinchow; *prev.* Chinhsien. Liaoning, NE China 41°07′N 121°06′E

◆ Country
● Country Capital
◇ Dependent Territory
○ Dependent Territory Capital
◈ Administrative Regions
✕ International Airport
▲ Mountain
▲▲ Mountain Range
Volcano
≈ River
◎ Lake
⊠ Reservoir

163 U14 **Jinzhou** *prev.* Jinxian. Liaoning, NE China 39°04′N 121°45′E
Jinzhu *see* Daocheng
138 H12 **Jinz, Qā' al** ◎ C Jordan
47 S8 **Jiparaná, Rio** ~ W Brazil
56 A7 **Jipijapa** Manabí, W Ecuador 01°23′S 80°35′W
42 F8 **Jiquilisco** Usulután, S El Salvador 13°19′N 88°35′W
Jirgalanta *see* Hovd
147 S12 **Jirgatol** *Rus.* Dzhirgatal'. C Tajikistan 39°13′N 71°09′E
75 X10 **Jirjā** *var.* Girga, Girgeh, Jirjā. C Egypt 26°17′N 31°58′E
Jirjā *see* Jirjā
111 B15 **Jirkov** *Ger.* Görkau. Ústecký Kraj, NW Czech Republic 50°30′N 13°27′E
143 T12 **Jiroft** *var.* Sabzawaran, Sabzvārān. Kermān, SE Iran 28°40′N 57°40′E
81 P14 **Jirriiban** Mudug, E Somalia 07°15′N 48°55′E
160 L11 **Jishou** Hunan, S China 28°20′N 109°43′E
Jisr ash Shadadi *see* Ash Shadādah
116 I14 **Jitaru** Olt, S Romania 44°27′N 24°32′E
Jitschin *see* Jičín
116 H14 **Jiu** *Ger.* Schil, Schyl, *Hung.* Zsil, Zsily. ~ S Romania
161 R11 **Jiufeng Shan** ▲▲ SE China
161 P9 **Jiujiang** Jiangxi, S China 29°45′N 115°59′E
161 O10 **Jiuling Shan** ▲▲ S China
160 G10 **Jiulong** *var.* Garba, *Tib.* Gyaisi. Sichuan, C China 29°00′N 101°30′E
161 Q13 **Jiulong Jiang** ~ SE China
161 Q12 **Jiulong Xi** ~ SE China
159 R8 **Jiuquan** *var.* Suzhou. Gansu, N China 39°47′N 98°30′E
160 K17 **Jiusuo** Hainan, S China 18°25′N 109°55′E
163 W10 **Jiutai** Jilin, NE China 44°01′N 125°51′E
160 K13 **Jiuwan Dashan** ▲▲ S China
160 I7 **Jiuzhaigou** *var.* Nongle; *prev.* Nanping. Sichuan, C China 33°25′N 104°05′E
186 C7 **Jiwaka** ◆ *province* C Papua New Guinea
148 I16 **Jiwani** Baluchistān, SW Pakistan 25°05′N 61°46′E
163 Y8 **Jixi** Heilongjiang, NE China 45°17′N 131°01′E
163 Y7 **Jixian** *var.* Fuli. Heilongjiang, NE China 46°38′N 131°04′E
160 M5 **Jixian** *var.* Ji Xian. Shanxi, C China 36°15′N 110°41′E
Ji Xian *see* Jixian
Jiza *see* Al Jīzah
141 N13 **Jīzān** *var.* Qīzān. Jīzān, SW Saudi Arabia 17°50′N 42°50′E
141 N13 **Jīzān** *var.* Minṭaqat Jīzān. ◆ *province* SW Saudi Arabia
Jīzān, Minṭaqat *see* Jīzān
140 K6 **Jizl, Wādī al** *dry watercourse* W Saudi Arabia
164 H12 **Jizō-zaki** *headland* Honshū, SW Japan 35°34′N 133°16′E
141 U14 **Jiz', Wādī al** *dry watercourse* E Yemen
147 O11 **Jizzax** *Rus.* Dzhizak. Jizzax Viloyati, C Uzbekistan 40°08′N 67°47′E
147 N10 **Jizzax Viloyati** *Rus.* Dzhizakskaya Oblast'. ◆ *province* C Uzbekistan
60 I13 **Joaçaba** Santa Catarina, S Brazil 27°08′S 51°30′W
Joal *see* Joal-Fadiout
76 F11 **Joal-Fadiout** *prev.* Joal. W Senegal 14°09′N 16°50′W
76 E10 **João Barrosa** Boa Vista, E Cape Verde 16°01′N 22°44′W
João Belo *see* Xai-Xai
João de Almeida *see* Chibia
59 Q15 **João Pessoa** *prev.* Paraíba. *state capital* Paraíba, E Brazil 07°06′S 34°53′W
25 X7 **Joaquin** Texas, SW USA 31°58′N 94°03′W
62 K6 **Joaquín V. González** Salta, N Argentina 25°06′S 64°07′W
Joazeiro *see* Juazeiro
Job'urg *see* Johannesburg
109 O7 **Jochberger Ache** ~ W Austria
Jo-ch'iang *see* Ruoqiang
92 K12 **Jock** Norrbotten, N Sweden 66°40′N 22°45′E
42 I5 **Jocón** Yoro, N Honduras 15°17′N 86°55′W
105 O13 **Jódar** Andalucía, S Spain 37°51′N 03°18′W
152 F12 **Jodhpur** Rājasthān, NW India 26°17′N 73°02′E
99 I19 **Jodoigne** Walloon Brabant, C Belgium 50°43′N 04°52′E
93 O16 **Joensuu** Pohjois-Karjala, SE Finland 62°36′N 29°45′E
37 W4 **Joes** Colorado, C USA 39°36′N 102°40′W
191 Z3 **Joe's Hill** *hill* Kiritimati, NE Kiribati
165 N11 **Jōetsu** *var.* Zyôetu. Niigata, Honshū, C Japan 37°09′N 138°13′E
83 M18 **Jofane** Inhambane, S Mozambique 21°16′S 34°21′E
153 R12 **Jogbani** Bihār, NE India 26°23′N 87°16′E
118 I5 **Jõgeva** *Ger.* Laisholm. Jõgevamaa, E Estonia 58°45′N 26°28′E
118 I4 **Jõgevamaa** *off.* Jõgeva Maakond. ◆ *province* E Estonia
Jõgeva Maakond *see* Jõgevamaa
155 E18 **Jog Falls** *Waterfall* Karnātaka, W India
143 S4 **Joghatāy** Khorāsān-e Razavī, NE Iran 36°34′N 57°00′E
153 U12 **Jogīghopa** Assam, NE India 26°14′N 90°35′E
152 I7 **Jogindarnagar** Himāchal Pradesh, N India 31°51′N 76°47′E
Jogjakarta *see* Yogyakarta
164 L11 **Jōhana** Toyama, Honshū, SW Japan 36°30′N 136°53′E
83 J21 **Johannesburg** *var.* Egoli, Erautini, Gauteng, *abbrev.* Job'urg. Gauteng, NE South Africa 26°10′S 28°02′E
35 T13 **Johannesburg** California, W USA 35°20′N 117°37′W
Johannisburg *see* Pisz
149 P14 **Johi** Sind, SE Pakistan 26°46′N 67°28′E
55 T13 **Johi Village** S Guyana 01°48′N 58°33′W
45 W10 **John A. Osborne** ✈ (Plymouth) E Montserrat 16°45′N 62°09′W
32 K13 **John Day** Oregon, NW USA 44°25′N 118°57′W
32 I11 **John Day River** ~ Oregon, NW USA
18 L14 **John F Kennedy** ✈ (New York) Long Island, New York, NE USA 40°39′N 73°45′W
21 V8 **John H. Kerr Reservoir** *var.* Buggs Island Lake, Kerr Lake. ⊠ North Carolina/Virginia, SE USA
37 V6 **John Martin Reservoir** ⊠ Colorado, C USA
96 K6 **John o'Groats** N Scotland, United Kingdom 58°38′N 03°03′W
27 P5 **John Redmond Reservoir** ⊠ Kansas, C USA
39 Q7 **John River** ~ Alaska, USA
26 H6 **Johnson** Kansas, C USA 37°33′N 101°46′W
18 M7 **Johnson** Vermont, NE USA 44°39′N 72°40′W
18 D13 **Johnsonburg** Pennsylvania, NE USA 41°28′N 78°37′W
18 H11 **Johnson City** New York, NE USA 42°06′N 75°54′W
21 P8 **Johnson City** Tennessee, S USA 36°18′N 82°21′W
25 R10 **Johnson City** Texas, SW USA 30°17′N 98°27′W
35 S12 **Johnsondale** California, W USA 35°58′N 118°32′W
10 I8 **Johnsons Crossing** Yukon, W Canada 60°30′N 133°15′W
21 T13 **Johnsonville** South Carolina, SE USA 33°50′N 79°26′W
21 Q13 **Johnston** South Carolina, SE USA 33°49′N 81°48′W
192 M6 **Johnston Atoll** ◇ *US unincorporated territory* C Pacific Ocean
175 Q3 **Johnston Atoll** *atoll* C Pacific Ocean
30 L17 **Johnston City** Illinois, N USA 37°49′N 88°55′W
180 K12 **Johnston, Lake** *salt lake* Western Australia
31 S13 **Johnstown** Ohio, N USA 40°08′N 82°39′W
18 D15 **Johnstown** Pennsylvania, NE USA 40°20′N 78°56′W
168 L10 **Johor** *var.* Johore. ◆ *state* Peninsular Malaysia
Johor Baharu *see* Johor Bahru
168 K10 **Johor Bahru** *var.* Johor Baharu, Johore Bahru. Johor, Peninsular Malaysia 01°29′N 103°44′E
Johore *see* Johor
Johore Bahru *see* Johor Bahru
118 K3 **Jõhvi** *Ger.* Jewe. Ida-Virumaa, NE Estonia 59°21′N 27°25′E
103 P7 **Joigny** Yonne, C France 47°58′N 03°24′E
Joinvile *see* Joinville
60 K12 **Joinville** *var.* Joinvile. Santa Catarina, S Brazil 26°20′S 48°55′W
103 R6 **Joinville** Haute-Marne, N France 48°26′N 05°07′E
194 H3 **Joinville Island** *island* Antarctica
41 O15 **Jojutla** *var.* Jojutla de Juárez. Morelos, S Mexico 18°38′N 99°10′W
Jojutla de Juárez *see* Jojutla
92 I12 **Jokkmokk** *Lapp.* Dálvvadis. Norrbotten, N Sweden 66°35′N 19°57′E
92 L2 **Jökuldalur** ~ E Iceland
92 K2 **Jökulsá á Fjöllum** ~ NE Iceland
Jokyakarta *see* Yogyakarta
30 M11 **Joliet** Illinois, N USA 41°33′N 88°05′W
15 O11 **Joliette** Québec, SE Canada 46°02′N 73°27′W
171 O8 **Jolo** Jolo Island, SW Philippines 06°02′N 121°00′E
94 D11 **Jølstervatnet** ◎ S Norway
169 S16 **Jombang** *prev.* Djombang. Jawa, S Indonesia 07°33′S 112°14′E
159 R14 **Jomda** Xizang Zizhiqu, W China 31°26′N 98°09′E
118 G13 **Jonava** *Ger.* Janow, *Pol.* Janów. Kaunas, C Lithuania 55°05′N 24°19′E
146 L11 **Jondor** *Rus.* Zhondor. Buxoro Viloyati, C Uzbekistan 39°46′N 64°11′E
159 V11 **Jonê** *var.* Liulin. Gansu, C China 34°36′N 103°39′E
27 X9 **Jonesboro** Arkansas, C USA 35°50′N 90°42′W
23 S4 **Jonesboro** Georgia, SE USA 33°31′N 84°21′W
30 L17 **Jonesboro** Illinois, N USA 37°25′N 89°19′W
22 H5 **Jonesboro** Louisiana, S USA 32°14′N 92°43′W
21 P8 **Jonesboro** Tennessee, S USA 36°17′N 82°28′W
19 T6 **Jonesport** Maine, NE USA 44°33′N 67°35′W
0 J4 **Jones Sound** *channel* Nunavut, N Canada
22 I6 **Jonesville** Louisiana, S USA 31°37′N 91°49′W
31 Q10 **Jonesville** Michigan, N USA 41°58′N 84°39′W
21 Q11 **Jonesville** South Carolina, SE USA 34°49′N 81°38′W
146 K10 **Jongeldi** *Rus.* Dzhankel'dy. Buxoro Viloyati, C Uzbekistan 40°50′N 63°16′E
81 F14 **Jonglei** Jonglei, E South Sudan 06°54′N 31°19′E
81 F14 **Jonglei** *var.* Gongoleh State. ◆ *state* E South Sudan
81 F14 **Jonglei Canal** *canal* E South Sudan
118 F11 **Joniškėlis** Panevėžys, N Lithuania 56°02′N 24°10′E
118 F10 **Joniškis** *Ger.* Janischken. Šiaulai, N Lithuania 56°15′N 23°36′E
95 L19 **Jönköping** Jönköping, S Sweden 57°45′N 14°10′E
95 K20 **Jönköping** ◆ *county* S Sweden
15 Q7 **Jonquière** Québec, SE Canada 48°25′N 71°16′W
41 V15 **Jonuta** Tabasco, SE Mexico 18°04′N 92°03′W
102 K12 **Jonzac** Charente-Maritime, W France 45°26′N 00°25′W
27 R7 **Joplin** Missouri, C USA 37°04′N 94°31′W
33 W8 **Jordan** Montana, NW USA 47°18′N 106°54′W
138 H12 **Jordan** *off.* Hashemite Kingdom of Jordan, *Ar.* Al Mamlaka al Urdunīya al Hashemīyah, Al Urdunn; *prev.* Transjordan. ◆ *monarchy* SW Asia
138 G9 **Jordan** *Ar.* Urdunn, *Heb.* HaYarden. ~ SW Asia
Jordan Lake *see* B. Everett Jordan Reservoir
111 K17 **Jordanów** Małopolskie, S Poland 49°39′N 19°51′E
32 M15 **Jordan Valley** Oregon, NW USA 42°59′N 117°03′W
138 G9 **Jordan Valley** *valley* N Israel
57 D15 **Jorge Chávez Internacional** *var.* Lima. ✈ (Lima) Lima, W Peru 12°07′S 77°01′W
113 L23 **Jorgucat** *var.* Jergucati, Jorgucati. Gjirokastër, S Albania 39°57′N 20°14′E
Jorgucati *see* Jorgucat
153 X12 **Jorhāt** Assam, NE India 26°45′N 94°09′E
93 J14 **Jörn** Västerbotten, N Sweden 65°03′N 20°04′E
37 R14 **Jornada Del Muerto** *valley* New Mexico, SW USA
93 N17 **Joroinen** Etelä-Savo, E Finland 62°11′N 27°50′E
95 C16 **Jørpeland** Rogaland, S Norway 59°01′N 06°04′E
77 W14 **Jos** Plateau, C Nigeria 09°59′N 08°57′E
171 Q8 **Jose Abad Santos** *var.* Trinidad. Mindanao, S Philippines 05°51′N 125°35′E
61 F19 **José Batlle y Ordóñez** *var.* Batlle y Ordóñez. Florida, C Uruguay 33°28′S 55°08′W
63 H18 **José de San Martín** Chubut, S Argentina 44°04′S 70°29′W
61 E19 **José Enrique Rodó** *var.* Rodó, José E.Rodo; *prev.* Drabble, Drable. Soriano, SW Uruguay 33°43′S 57°33′W
José E.Rodo *see* José Enrique Rodó
Josefsdorf *see* Žabalj
44 C4 **José Martí** ✈ (La Habana) Cuidad de La Habana, N Cuba 23°03′N 82°22′W
61 F19 **José Pedro Varela** *var.* José P.Varela. Lavalleja, S Uruguay 33°30′S 54°28′W
181 N2 **Joseph Bonaparte Gulf** *gulf* N Australia
37 N11 **Joseph City** Arizona, SW USA 34°56′N 110°18′W
13 O9 **Joseph, Lake** ◎ Newfoundland and Labrador, E Canada
14 G13 **Joseph, Lake** ◎ Ontario, S Canada
186 C6 **Josephstaal** Madang, N Papua New Guinea 04°42′S 144°55′E
José P.Varela *see* José Pedro Varela
59 J14 **José Rodrigues** Pará, N Brazil 05°45′S 51°20′W
152 K9 **Joshimath** Uttarakhand, N India 30°33′N 79°35′E
25 T7 **Joshua** Texas, SW USA 32°27′N 97°23′W
35 V15 **Joshua Tree** California, W USA 34°07′N 116°18′W
77 V14 **Jos Plateau** *plateau* C Nigeria
102 H6 **Josselin** Morbihan, NW France 47°57′N 02°35′W
Jos Sudarso *see* Yos Sudarso, Pulau
94 E11 **Jostedalsbreen** *glacier* S Norway
94 F12 **Jotunheimen** ▲▲ S Norway
138 G7 **Joûnié** *var.* Junīyah. W Lebanon 33°54′N 35°36′E
25 R13 **Jourdanton** Texas, SW USA 28°55′N 98°34′W
98 L7 **Joure** *Fris.* De Jouwer. Fryslân, N Netherlands 52°58′N 05°48′E
93 M18 **Joutsa** Keski-Suomi, C Finland 61°46′N 26°09′E
93 N18 **Joutseno** Etelä-Karjala, SE Finland 61°06′N 28°30′E
92 M12 **Joutsijärvi** Lappi, NE Finland 66°40′N 28°00′E
108 A9 **Joux, Lac de** ◎ W Switzerland
Jovakān *see* Jowkān
44 D5 **Jovellanos** Matanzas, W Cuba 22°49′N 81°11′W
153 V13 **Jowai** Meghālaya, NE India 25°25′N 92°21′E
Jowat *see* Jabwot
Jowhar *see* Jawhar
143 O12 **Jowkān** *var.* Jovakān. Fārs, S Iran
Jowzam *see* Javazm
149 N2 **Jowzjān** ◆ *province* N Afghanistan
Joypurhat *see* Jaipurhat
Józseffalva *see* Žabalj
J.Storm Thurmond Reservoir *see* Clark Hill Lake
45 T6 **Juana Díaz** C Puerto Rico 18°03′N 66°30′W
40 L9 **Juan Aldama** Zacatecas, C Mexico 24°20′N 103°23′W
0 E9 **Juan de Fuca Plate** *tectonic feature*
32 F7 **Juan de Fuca, Strait of** *strait* Canada/USA
Juan Fernandez Islands *see* Juan Fernández, Islas
193 S11 **Juan Fernández, Islas** *Eng.* Juan Fernández Islands. *island group* W Chile
55 O4 **Juangriego** Nueva Esparta, NE Venezuela 11°06′N 63°59′W
56 D11 **Juanjuí** *var.* Juanjuy. San Martín, N Peru 07°10′S 76°44′W
Juanjuy *see* Juanjuí
93 N16 **Juankoski** Pohjois-Savo, C Finland 63°01′N 28°24′E
Juan Lacaze *see* Juan L. Lacaze
61 E20 **Juan L. Lacaze** *var.* Juan Lacaze, Puerto Sauce; *prev.* Sauce. Colonia, SW Uruguay 34°26′S 57°25′W
62 L5 **Juan Solá** Salta, N Argentina 23°30′S 62°42′W
63 F21 **Juan Stuven, Isla** *island* S Chile
59 H16 **Juará** Mato Grosso, W Brazil 11°10′S 57°28′W
41 N7 **Juárez** *var.* Villa Juárez. Coahuila, NE Mexico 27°39′N 100°43′W
40 C2 **Juárez, Sierra de** ▲▲ NW Mexico
59 O15 **Juazeiro** *prev.* Joazeiro. Bahia, E Brazil 09°25′S 40°30′W
59 P14 **Juazeiro do Norte** Ceará, E Brazil 07°10′S 39°18′W
81 F15 **Juba** *var.* Jūbā. ● Central Equatoria, S South Sudan 04°50′N 31°35′E
81 L17 **Juba** *Amh.* Genalē Wenz, *It.* Guiba, *Som.* Ganaane, Webi Jubba. ~ Ethiopia/Somalia
194 H2 **Jubany** *Argentinian research station* Antarctica 61°57′S 58°23′W
Jubayl *see* Jbaïl
81 L18 **Jubbada Dhexe** *off.* Gobolka Jubbada Dhexe. ◆ *region* SW Somalia
Jubbada Dhexe, Gobolka *see* Jubbada Dhexe
81 K18 **Jubbada Hoose** ◆ *region* SW Somalia
Jubba, Webi *see* Juba
Jubbulpore *see* Jabalpur
Jubeil *see* Jbaïl
74 B9 **Juby, Cap** *headland* SW Morocco 27°58′N 12°56′W
105 R10 **Júcar** *var.* Jucar. ~ C Spain
40 L12 **Juchipila** Zacatecas, C Mexico 21°25′N 103°06′W
41 S16 **Juchitán** *var.* Juchitán de Zaragosa. Oaxaca, SE Mexico 16°27′N 95°W
Juchitán de Zaragosa *see* Juchitán
138 G11 **Judaea** *cultural region* Israel/West Bank
138 F11 **Judaean Hills** *Heb.* Haré Yehuda. *hill range* E Israel
138 H8 **Judaydah** *Fr.* Jdaïdé. Rīf Dimashq, W Syria 33°17′N 36°15′E
139 P11 **Judayyidat Hāmir** Al Anbār, S Iraq 31°50′N 41°50′E
109 U8 **Judenburg** Steiermark, C Austria 47°09′N 14°43′E
33 T8 **Judith River** ~ Montana, NW USA
27 V11 **Judsonia** Arkansas, C USA 35°16′N 91°38′W
141 P14 **Jufrah, Wādī al** *dry watercourse* NW Yemen
Jugar *see* Sêrxü
Jugoslavija *see* Serbia
42 K10 **Juigalpa** Chontales, S Nicaragua 12°04′N 85°21′W
100 E9 **Juist** *island* NW Germany
Juisui *see* Ruisui
59 M21 **Juiz de Fora** Minas Gerais, SE Brazil 21°47′S 43°23′W
62 J5 **Jujuy** *off.* Provincia de Jujuy. ◆ *province* N Argentina
Jujuy *see* San Salvador de Jujuy
Jujuy, Provincia de *see* Jujuy
92 J11 **Jukkasjärvi** *Lapp.* Čohkkiras. Norrbotten, N Sweden 67°52′N 20°39′E
Jula *see* Gyula, Hungary
Jūlā *see* Jālū, Libya
37 W2 **Julesburg** Colorado, C USA 40°59′N 102°15′W
Julia Beterrae *see* Béziers
57 I17 **Juliaca** Puno, SE Peru 15°32′S 70°10′W
181 U6 **Julia Creek** Queensland, C Australia 20°40′S 141°49′E
35 V17 **Julian** California, W USA 33°04′N 116°36′W
98 H7 **Julianadorp** Noord-Holland, NW Netherlands 52°53′N 04°43′E
109 S11 **Julian Alps** *Ger.* Julische Alpen, *It.* Alpi Giulie, *Slvn.* Julijske Alpe. ▲▲ Italy/Slovenia
55 V11 **Juliana Top** ▲ C Suriname 03°39′N 56°36′W
Julianehåb *see* Qaqortoq
Julijske Alpe *see* Julian Alps
40 J6 **Julimes** Chihuahua, N Mexico 28°29′N 105°21′W
Julio Briga *see* Bragança
Juliobriga *see* Logroño
61 G15 **Júlio de Castilhos** Rio Grande do Sul, S Brazil 29°14′S 53°42′W
Juliomagus *see* Angers
Julische Alpen *see* Julian Alps
Jullundur *see* Jalandhar
147 N11 **Juma** *Rus.* Dzhuma. Samarqand Viloyati, C Uzbekistan 39°43′N 66°37′E
161 O3 **Juma He** ~ E China
81 L18 **Jumboo** *prev.* Jumbo. Jubbada Hoose, S Somalia 0°12′S 42°34′E
Jumboo *see* Jumba
35 Y11 **Jumbo Peak** ▲ Nevada, W USA 36°12′N 114°09′W
105 R12 **Jumilla** Murcia, SE Spain 38°28′N 01°19′W
153 N10 **Jumla** Mid Western, NW Nepal 29°22′N 82°13′E
Jummoo *see* Jammu
Jumna *see* Yamuna
Jumporn *see* Chumphon
30 K5 **Jump River** ~ Wisconsin, N USA
154 B11 **Jūnāgadh** *var.* Junagarh. Gujarāt, W India 21°32′N 70°32′E
Junagarh *see* Jūnāgadh
161 Q6 **Junan** *var.* Shizilu. Shandong, E China 35°11′N 118°47′E
62 G11 **Juncal, Cerro** ▲ C Chile 33°03′S 70°02′W
25 Q10 **Junction** Texas, SW USA 30°31′N 99°48′W
36 K6 **Junction** Utah, W USA 38°14′N 112°13′W
27 O4 **Junction City** Kansas, C USA 39°02′N 96°51′W
32 F13 **Junction City** Oregon, NW USA 44°13′N 123°12′W
60 M10 **Jundiaí** São Paulo, S Brazil 23°10′S 46°54′W
39 X12 **Juneau** *state capital* Alaska, USA 58°13′N 134°11′W
30 M8 **Juneau** Wisconsin, N USA 43°23′N 88°42′W
105 U6 **Juneda** Cataluña, NE Spain 41°33′N 00°49′E
183 Q9 **Junee** New South Wales, SE Australia 34°51′S 147°33′E
35 R8 **June Lake** California, W USA 37°46′N 119°04′W
Jungbunzlau *see* Mladá Boleslav
158 L4 **Junggar Pendi** *Eng.* Dzungarian Basin. *basin* NW China
99 N24 **Junglinster** Grevenmacher, C Luxembourg 49°43′N 06°15′E
18 F14 **Juniata River** ~ Pennsylvania, NE USA
61 B20 **Junín** Buenos Aires, E Argentina 34°36′S 61°02′W
57 E14 **Junín** Junín, C Peru 11°11′S 76°00′W
57 F14 **Junín** *off.* Departamento de Junín. ◆ *department* C Peru
63 H15 **Junín de los Andes** Neuquén, W Argentina 39°57′S 71°05′W
Junín, Departamento de *see* Junín
57 D14 **Junín, Lago de** ◎ C Peru
Junīyah *see* Joûnié
Junkseylon *see* Phuket
160 I11 **Junlian** Sichuan, C China 28°11′N 104°31′E
25 O11 **Juno** Texas, SW USA 30°09′N 101°07′W
92 J11 **Junosuando** *Lapp.* Čunusavvon. Norrbotten, N Sweden 67°25′N 22°28′E
93 H16 **Junsele** Västernorrland, C Sweden 63°42′N 16°54′E
32 L14 **Juntura** Oregon, NW USA 43°43′N 118°05′W
93 N14 **Juntusranta** Kainuu, E Finland 65°12′N 29°30′E
118 H11 **Juodupė** Panevėžys, NE Lithuania 56°07′N 25°37′E
119 H14 **Juozapinės Kalnas** ▲ SE Lithuania 54°29′N 25°27′E
99 K19 **Juprelle** Liège, E Belgium 50°43′N 05°31′E
80 D13 **Jur** ~ W South Sudan
103 S9 **Jura** ◆ *department* E France
108 C7 **Jura** ◆ *canton* NW Switzerland
108 B8 **Jura** *var.* Jura Mountains. ▲▲ France/Switzerland
96 G12 **Jura** *island* SW Scotland, United Kingdom
Juraciszki *see* Yuratsishki
54 C8 **Juradó** Chocó, NW Colombia 07°07′N 77°45′W
Jura Mountains *see* Jura
96 G12 **Jura, Sound of** *strait* W Scotland, United Kingdom
139 V15 **Juraybīyāt, Bi'r** *well* S Iraq
118 E13 **Jurbarkas** *Ger.* Georgenburg, Jurburg. Tauragė, W Lithuania 55°04′N 22°45′E
99 F20 **Jurbise** Hainaut, SW Belgium 50°33′N 03°54′E
Jurburg *see* Jurbarkas
118 F9 **Jūrmala** C Latvia 56°57′N 23°42′E
58 D13 **Juruá** Amazonas, NW Brazil 03°08′S 65°59′W
48 F7 **Juruá, Rio** *var.* Río Yuruá. ~ Brazil/Peru
59 G16 **Juruena** Mato Grosso, W Brazil 10°32′S 58°38′W
59 G16 **Juruena, Rio** ~ W Brazil
165 Q6 **Jūsan-ko** ◎ Honshū, C Japan
25 O6 **Justiceburg** Texas, SW USA 32°57′N 101°07′W
Justinianopolis *see* Kirşehir
62 K11 **Justo Daract** San Luis, C Argentina 33°52′S 65°12′W
59 C14 **Jutaí** Amazonas, W Brazil 05°10′S 68°45′W
58 C13 **Jutaí, Rio** ~ NW Brazil
100 N13 **Jüterbog** Brandenburg, E Germany 51°58′N 13°06′E
42 E6 **Jutiapa** Jutiapa, S Guatemala 14°18′N 89°52′W
42 A3 **Jutiapa** *off.* Departamento de Jutiapa. ◆ *department* SE Guatemala
Jutiapa, Departamento de *see* Jutiapa
42 J6 **Juticalpa** Olancho, C Honduras 14°39′N 86°12′W
82 I13 **Jutila** North Western, NW Zambia 12°33′S 26°09′E
Jutland *see* Jylland
84 F8 **Jutland Bank** *undersea feature* SE North Sea 56°50′N 07°20′E
93 N16 **Juuka** Pohjois-Karjala, E Finland 63°12′N 29°17′E
93 N17 **Juva** Etelä-Savo, E Finland 61°55′N 27°54′E
Juvavum *see* Salzburg
44 A6 **Juventud, Isla de la** *var.* Isla de Pinos, *Eng.* Isle of Youth; *prev.* The Isle of the Pines. *island* W Cuba
Juwārtā *see* Chwarta
Juwārtā *see* Chemchemal
161 Q5 **Juxian** *var.* Chengyang, Ju Xian. Shandong, E China 35°33′N 118°45′E
Ju Xian *see* Juxian
161 P6 **Juye** Shandong, E China 35°26′N 116°04′E
113 O15 **Južna Morava** *Ger.* Südliche Morava. ~ SE Serbia
83 H20 **Jwaneng** Southern, S Botswana 24°35′S 24°45′E
95 I23 **Jyderup** Sjælland, E Denmark 55°40′N 11°25′E
95 F22 **Jylland** *Eng.* Jutland. *peninsula* W Denmark
Jyrgalan *see* Dzhergalan
93 M17 **Jyväskylä** Keski-Suomi, C Finland 62°08′N 25°47′E

K

38 D9 **Ka'a'awa** *var.* Kaawa. O'ahu, Hawaii, USA, C Pacific Ocean 21°33′N 157°47′W
Kaawa *see* Ka'a'awa
81 G16 **Kaabong** NE Uganda 03°30′N 34°08′E
Kaaden *see* Kadaň
Kaafu Atoll *see* Male' Atoll
55 V9 **Kaaimanston** Sipaliwini, N Suriname 05°06′N 56°04′W
Kaakhka *see* Kaka
Kaala *see* Caála
187 O16 **Kaala-Gomen** Province Nord, W New Caledonia 20°40′S 164°24′E
92 L9 **Kaamanen** *Lapp.* Gámas. Lappi, N Finland 69°05′N 27°16′E
Kaapstad *see* Cape Town
Kaarasjoki *see* Karasjok
Kaaresuanto *see* Karesuando
92 J10 **Kaaresuvanto** *Lapp.* Gárassavon. Lappi, N Finland 68°28′N 22°29′E
93 K19 **Kaarina** Varsinais-Suomi, SW Finland 60°24′N 22°25′E
99 I14 **Kaatsheuvel** Noord-Brabant, S Netherlands 51°39′N 05°02′E
93 N16 **Kaavi** Pohjois-Savo, C Finland 62°58′N 28°30′E
Kaba *see* Habahe
171 O14 **Kabaena, Pulau** *island* C Indonesia
Kabakly *see* Gabakly
76 J14 **Kabala** N Sierra Leone 09°40′N 11°36′W
81 E19 **Kabale** SW Uganda 01°15′S 29°58′E
55 U10 **Kabalebo Rivier** ~ W Suriname
79 N22 **Kabalo** Katanga, SE Dem. Rep. Congo 06°02′S 26°55′E
79 O21 **Kabambare** Maniema, E Dem. Rep. Congo 04°40′S 27°41′E
145 W13 **Kabanbay** *Kaz.* Qabanbay; *prev.* Andreyevka, *Kaz.* Andreevka. Almaty, SE Kazakhstan 45°50′N 80°34′E
145 Q9 **Kabanbay Batyr** *prev.* Rozhdestvenka. Akmola, C Kazakhstan 50°51′N 71°25′E
187 Y15 **Kabara** *prev.* Kambara. *island* Lau Group, E Fiji
Kabardino-Balkaria *see* Kabardino-Balkarskaya Respublika
126 M15 **Kabardino-Balkarskaya Respublika** *Eng.* Kabardino-Balkaria. ◆ *autonomous republic* SW Russian Federation
79 O19 **Kabare** Sud-Kivu, E Dem. Rep. Congo 02°13′S 28°40′E
171 T11 **Kabarei** Papua Barat, E Indonesia 0°01′S 130°58′E
171 P7 **Kabasalan** Mindanao, S Philippines 07°46′N 122°49′E
77 U15 **Kabba** Kogi, S Nigeria 07°48′N 06°07′E
92 I13 **Kåbdalis** *Lapp.* Goabddális. Norrbotten, N Sweden 66°08′N 20°03′E
138 M6 **Kabd aş Şārim** *hill range* E Syria
14 B7 **Kabenung Lake** ◎ Ontario, S Canada
29 W3 **Kabetogama Lake** ◎ Minnesota, N USA
Kabia, Pulau *see* Kabin, Pulau
79 M22 **Kabinda** Kasai-Oriental, SE Dem. Rep. Congo 06°09′S 24°29′E
Kabinda *see* Cabinda
171 O15 **Kabin, Pulau** *var.* Pulau Kabia. *island* W Indonesia
171 P16 **Kabir** Pulau Pantar, S Indonesia 08°15′S 124°12′E
149 T10 **Kabīrwāla** Punjab, E Pakistan 30°24′N 71°51′E
114 M9 **Kableshkovo** Burgas, E Bulgaria 42°65′N 27°34′E
78 I13 **Kabo** Ouham, NW Central African Republic 07°43′N 18°38′E
Kābol *see* Kābul
83 H14 **Kabompo** North Western, W Zambia 13°36′S 24°10′E
83 H14 **Kabompo** ~ W Zambia
79 M22 **Kabongo** Katanga, SE Dem. Rep. Congo 07°20′S 25°34′E
120 K11 **Kaboudia, Rass** *headland* E Tunisia 35°13′N 11°09′E
124 J14 **Kabozha** Novgorodskaya Oblast', W Russian Federation 58°48′N 35°00′E
Kabūd Gonbad *see* Kalāt
142 L5 **Kabūd Rāhang** Hamadān, W Iran 35°12′N 48°44′E
82 L12 **Kabuko** Muchinga, NE Zambia 11°31′S 31°16′E
149 Q5 **Kābul** *prev.* Kābol. ● (Afghanistan) Kābul, E Afghanistan 34°34′N 69°08′E
149 Q5 **Kābul** ; *prev.* Kābol. ◆ *province* E Afghanistan
149 Q5 **Kābul** ✈ Kābul, E Afghanistan 34°31′N 69°11′E
149 R5 **Kabul** *var.* Daryā-ye Kābul. ~ Afghanistan/Pakistan *see also* Kābul, Daryā-ye
Kabul *see* Kābul, Daryā-ye
149 S5 **Kābul, Daryā-ye** *var.* Kabul. ~ Afghanistan/Pakistan *see also* Kabul
Kābul, Daryā-ye *see* Kabul
79 O25 **Kabunda** Katanga, SE Dem. Rep. Congo 12°21′S 29°14′E
171 R9 **Kaburuang, Pulau** *island* Kepulauan Talaud, N Indonesia
80 G8 **Kabushiya** River Nile, NE Sudan 16°54′N 33°41′E
83 J14 **Kabwe** Central, C Zambia 14°29′S 28°25′E
186 E7 **Kabwum** Morobe, C Papua New Guinea 06°04′S 147°09′E
113 N17 **Kaçanik** *Serb.* Kačanik. S Kosovo 42°13′N 21°16′E
Kačanik *see* Kaçanik
118 F13 **Kačerginė** Kaunas, C Lithuania 54°55′N 23°40′E
117 S13 **Kacha** Avtonomna Respublika Krym, S Ukraine 44°46′N 33°33′E
154 A10 **Kachchh, Gulf of** *var.* Gulf of Cutch, Gulf of Kutch. *gulf* W India
154 I11 **Kachchhīdhāna** Madhya Pradesh, C India 21°33′N 78°54′E
149 Q11 **Kachchh, Rann of** *var.* Rann of Kachh, Rann of Kutch. *salt marsh* India/Pakistan
39 Q13 **Kachemak Bay** *bay* Alaska, USA
Kachh, Rann of *see* Kachchh, Rann of
77 V14 **Kachia** Kaduna, C Nigeria 09°52′N 08°00′E
167 N2 **Kachin State** ◆ *state* N Myanmar (Burma)
Kachiry *see* Kashyr
137 Q11 **Kaçkar Dağları** ▲▲ NE Turkey
155 C21 **Kadamatt Island** *island* Lakshadweep, India, N Indian Ocean
111 B15 **Kadaň** *Ger.* Kaaden. Ústecký Kraj, NW Czech Republic 50°24′N 13°16′E
1667 N11 **Kadan Kyun** *prev.* King Island. *island* Mergui Archipelago, S Myanmar (Burma)
187 X15 **Kadavu** *prev.* Kandavu. *island* S Fiji
187 X15 **Kadavu Passage** *channel* S Fiji
79 G16 **Kadéï** ~ Cameroon/Central African Republic
Kadhimain *see* Al Kāẓimīyah
Kadijica *see* Kadiytsa
114 M13 **Kadıköy Barajı** ⊠ NW Turkey
182 I8 **Kadina** South Australia 33°59′S 137°43′E
136 H15 **Kadınhanı** Konya, C Turkey 38°15′N 32°14′E
76 M14 **Kadiolo** Sikasso, S Mali 10°30′N 05°43′W
136 L16 **Kadirli** Osmaniye, S Turkey 37°22′N 36°05′E
114 G11 **Kadiytsa** *Mac.* Kadijica. ▲ Bulgaria/FYR Macedonia 41°48′N 22°58′E
28 L10 **Kadoka** South Dakota, N USA 43°49′N 101°30′W
127 N5 **Kadom** Ryazanskaya Oblast', W Russian Federation 54°35′N 42°27′E
83 K16 **Kadoma** *prev.* Gatooma. Mashonaland West, C Zimbabwe !8°22′S 29°55′E
80 E12 **Kadugli** Southern Kordofan, S Sudan 11°N 29°44′E
77 V14 **Kaduna** Kaduna, C Nigeria 10°32′N 07°26′E
77 V14 **Kaduna** ◆ *state* C Nigeria
77 V15 **Kaduna** ~ N Nigeria
124 K14 **Kaduy** Vologodskaya Oblast', NW Russian Federation 59°10′N 37°11′E
154 E13 **Kadwa** ~ W India
123 S9 **Kadykchan** Magadanskaya Oblast', E Russian Federation 62°54′N 146°53′E
Kadzharan *see* K'ajaran
125 T7 **Kadzherom** Respublika Komi, NW Russian Federation 64°42′N 55°51′E
Kadzhi-Say *see* Bokonbayevo
76 I10 **Kaédi** Gorgol, S Mauritania 16°12′N 13°32′W
78 G12 **Kaélé** Extrême-Nord, N Cameroon 10°05′N 14°28′E
38 C9 **Ka'ena Point** *var.* Kaena Point. *headland* O'ahu, Hawai'i, USA 21°34′N 158°16′W
184 J2 **Kaeo** Northland, North Island, New Zealand 35°03′S 173°40′E
163 X14 **Kaesŏng** *var.* Kaesŏng-si. S North Korea 37°58′N 126°31′E
Kaesŏng-si *see* Kaesŏng
Kaewieng *see* Kavieng
79 L24 **Kafakumba** Shaba, S Dem. Rep. Congo 09°39′S 23°43′E
Kafan *see* Kapan
77 V14 **Kafanchan** Kaduna, C Nigeria 09°32′N 08°18′E
Kaffa *see* Feodosiya
76 G11 **Kaffrine** C Senegal 14°07′N 15°27′W
Kafiréas, Akrotírio *see* Ntóro, Kávo
115 I19 **Kafiréos, Stenó** *strait* Évvoia/Kykládes, Greece, Aegean Sea
Kafirnigan *see* Kofarnihon
Kafo *see* Kafu
75 W7 **Kafr ash Shaykh** *var.* Kafrel Sheik, Kafr el Sheikh. N Egypt 31°07′N 30°56′E
Kafr el Sheikh *see* Kafr ash Shaykh
81 F17 **Kafu** *var.* Kafo. ~ W Uganda
83 J15 **Kafue** Lusaka, SE Zambia 15°44′S 28°10′E
83 I14 **Kafue** ~ C Zambia
67 T13 **Kafue Flats** *plain* C Zambia
164 K12 **Kaga** Ishikawa, Honshū, SW Japan 36°18′N 136°19′E
79 J14 **Kaga Bandoro** *prev.* Fort-Crampel. Nana-Grébizi, C Central African Republic 06°54′N 19°10′E
81 E18 **Kagadi** W Uganda 0°57′N 30°52′E
38 H17 **Kagalaska Island** *island* Aleutian Islands, Alaska, USA
Kagan *see* Kogon
Kaganovichabad *see* Kolkhozobod
Kagarlyk *see* Kaharlyk
164 H14 **Kagawa** *off.* Kagawa-ken. ◆ *prefecture* Shikoku, SW Japan
Kagawa-ken *see* Kagawa
154 J13 **Kagaznagar** Telangana, C India 19°25′N 79°30′E
93 J14 **Kåge** Västerbotten, N Sweden 64°49′N 21°00′E
81 E19 **Kagera** *var.* Ziwa Magharibi, *Eng.* West Lake. ◆ *region* NW Tanzania
81 E19 **Kagera** *var.* Akagera. ~ Rwanda/Tanzania *see also* Akagera
76 L5 **Kâghet** *var.* Karet. *physical region* N Mauritania
Kagi *see* Jiayi
137 S12 **Kağızman** Kars, NE Turkey 40°08′N 43°07′E
188 I6 **Kagman Point** *headland* Saipan, S Northern Mariana Islands
164 C16 **Kagoshima** *var.* Kagosima. Kagoshima, Kyūshū, SW Japan 31°37′N 130°33′E
164 C16 **Kagoshima** *off.* Kagoshima-ken, *var.* Kagosima. ◆ *prefecture* Kyūshū, SW Japan
Kagoshima-ken *see* Kagoshima
Kagosima *see* Kagoshima
Kagul *see* Cahul
Kagul, Ozero *see* Kahul, Ozero
38 B8 **Kahala Point** *headland* Kaua'i, Hawai'i, USA 22°08′N 159°17′W
81 F21 **Kahama** Shinyanga, NW Tanzania 03°48′S 32°36′E
117 P5 **Kaharlyk** *Rus.* Kagarlyk. Kyyivs'ka Oblast', N Ukraine 49°50′N 30°50′E
169 T13 **Kahayan, Sungai** ~ Borneo, C Indonesia
79 I22 **Kahemba** Bandundu, SW Dem. Rep. Congo 07°20′S 19°00′E
185 A23 **Kaherekoau Mountains** ▲▲ South Island, New Zealand
143 W14 **Kahīrī** *var.* Kūhīrī. Sīstān va Balūchestān, SE Iran 26°55′N 61°04′E
101 L16 **Kahla** Thüringen, C Germany 50°49′N 11°33′E
101 G15 **Kahler Asten** ▲ W Germany 51°11′N 08°32′E
149 Q4 **Kahmard, Daryā-ye** *prev.* Darya-i-surkhab. ~ NE Afghanistan
143 T13 **Kahnūj** Kermān, SE Iran 28°N 57°41′E
27 V1 **Kahoka** Missouri, C USA 40°24′N 91°44′W
38 E10 **Kaho'olawe** *var.* Kahoolawe. *island* Hawai'i, USA, C Pacific Ocean
Kahoolawe *see* Kaho'olawe
136 M16 **Kahramanmaraş** *var.* Kahraman Maraş, Maraş, Marash. Kahramanmaraş, S Turkey 37°34′N 36°54′E
136 L15 **Kahramanmaraş** *var.* Kahraman Maraş, Maraş, Marash. ◆ *province* C Turkey
Kahraman Maraş *see* Kahramanmaraş
Kahror/Kahror Pakka *see* Kahror Pakka
149 T11 **Kahror Pakka** *var.* Kahror, Koror Pacca. Punjab, E Pakistan 29°38′N 71°59′E
137 N15 **Kâhta** Adıyaman, S Turkey 37°48′N 38°35′E

◆ Country ● Country Capital ◇ Dependent Territory ○ Dependent Territory Capital ◆ Administrative Regions ✈ International Airport ▲ Mountain ▲▲ Mountain Range Volcano ~ River ◎ Lake ⊠ Reservoir

38 *D8* **Kahuku** O'ahu, Hawaii, USA, C Pacific Ocean *21°40′N 157°57′W*
38 *D8* **Kahuku Point** *headland* O'ahu, Hawai'i, USA *21°42′N 157°59′W*
116 *M12* **Kahul, Ozero** *var.* Lacul Cahul, *Rus.* Ozero Kagul. ◎ Moldova/Ukraine
143 *V11* **Kahūrak** Sīstān va Balūchestān, SE Iran *29°25′N 59°38′E*
184 *G13* **Kahurangi Point** *headland* South Island, New Zealand *40°41′S 171°57′E*
149 *V6* **Kahūta** Punjab, E Pakistan *33°38′N 73°27′E*
77 *S14* **Kaiama** Kwara, W Nigeria *09°37′N 03°58′E*
186 *D7* **Kaiapit** Morobe, C Papua New Guinea *06°12′S 146°09′E*
185 *I18* **Kaiapoi** Canterbury, South Island, New Zealand *43°23′S 172°40′E*
36 *K9* **Kaibab Plateau** *plain* Arizona, SW USA
171 *U14* **Kai Besar, Pulau** *island* Kepulauan Kai, E Indonesia
36 *L9* **Kaibito Plateau** *plain* Arizona, SW USA
158 *K6* **Kaidu He** *var.* Karaxahar. ≈ NW China
55 *S10* **Kaieteur Falls** *waterfall* C Guyana
161 *O6* **Kaifeng** Henan, C China *34°47′N 114°20′E*
184 *J3* **Kaihu** Northland, North Island, New Zealand *35°47′S 173°39′E*
Kaihua *see* Wenshan
171 *U14* **Kai Kecil, Pulau** *island* Kepulauan Kai, E Indonesia
169 *U16* **Kai, Kepulauan** *prev.* Kei Islands. *island group* Maluku, SE Indonesia
184 *J3* **Kaikohe** Northland, North Island, New Zealand *35°25′S 173°48′E*
185 *J16* **Kaikoura** Canterbury, South Island, New Zealand *42°22′S 173°40′E*
185 *J16* **Kaikoura Peninsula** *peninsula* South Island, New Zealand
Kailas Range *see* Gangdisê Shan
160 *K12* **Kaili** Guizhou, S China *26°34′N 107°58′E*
38 *F10* **Kailua** Maui, Hawaii, USA, C Pacific Ocean *20°53′N 156°13′W*
Kailua *see* Kalaoa
38 *G11* **Kailua-Kona** *var.* Kona. Hawaii, USA, C Pacific Ocean *19°43′N 155°58′W*
186 *B7* **Kaim** ≈ W Papua New Guinea
171 *X14* **Kaima** Papua, E Indonesia *05°36′S 138°39′E*
184 *M7* **Kaimai Range** ▲▲ North Island, New Zealand
114 *E13* **Kaïmaktsalán** *var.* Kajmakčalan. ▲ Greece/FYR Macedonia *40°57′N 21°48′E* *see also* Kajmakčalan
Kaïmaktsalán *see* Kajmakčalan
185 *C20* **Kaimanawa Mountains** ▲▲ North Island, New Zealand
118 *E4* **Käina** *Ger.* Keinis; *prev.* Keina. Hiiumaa, W Estonia *58°50′N 22°49′E*
109 *V7* **Kainach** ≈ SE Austria
164 *I14* **Kainan** Tokushima, Shikoku, SW Japan *33°36′N 134°20′E*
164 *H15* **Kainan** Wakayama, Honshū, SW Japan *34°09′N 135°12′E*
147 *U7* **Kaindy** *Kir.* Kayyngdy. Chuyskaya Oblast', N Kyrgyzstan *42°48′N 73°39′E*
77 *T14* **Kainji Dam** *dam* W Nigeria
Kainji Lake *see* Kainji Reservoir
77 *T14* **Kainji Reservoir** *var.* Kainji Lake. ⊠ W Nigeria
186 *D8* **Kaintiba** *var.* Kamina. Gulf, S Papua New Guinea *07°29′S 146°04′E*
92 *K12* **Kainulasjärvi** Norrbotten, N Sweden *67°00′N 22°31′E*
93 *M14* **Kainuu** *Swe.* Kajanaland. ◈ *region* N Finland
184 *K5* **Kaipara Harbour** *harbour* North Island, New Zealand
152 *I10* **Kairāna** Uttar Pradesh, N India *29°24′N 77°10′E*
74 *M6* **Kairouan** *var.* Al Qayrawān. E Tunisia *35°46′N 10°11′E*
Kaisaria *see* Kayseri
101 *F20* **Kaiserslautern** Rheinland-Pfalz, SW Germany *49°27′N 07°46′E*
118 *G13* **Kaišiadorys** Kaunas, S Lithuania *54°51′N 24°27′E*
184 *I2* **Kaitaia** Northland, North Island, New Zealand *35°07′S 173°13′E*
185 *E24* **Kaitangata** Otago, South Island, New Zealand *46°18′S 169°52′E*
152 *I9* **Kaithal** Haryāna, NW India *29°47′N 76°26′E*
Kaitong *see* Tongyu
169 *N13* **Kait, Tanjung** *headland* Sumatera, W Indonesia *03°13′S 106°03′E*
38 *E9* **Kaiwi Channel** *channel* Hawai'i, USA, C Pacific Ocean
160 *K9* **Kaixian** *var.* Hanfeng. Sichuan, C China *31°13′N 108°25′E*
163 *V11* **Kaiyuan** *var.* K'ai-yüan. Liaoning, NE China *42°33′N 124°04′E*
160 *H14* **Kaiyuan** Yunnan, SW China *23°42′N 103°14′E*
K'ai-yüan *see* Kaiyuan
39 *O9* **Kaiyuh Mountains** ▲▲ Alaska, USA
93 *M15* **Kajaani** *Swe.* Kajana. Kainuu, C Finland *64°17′N 27°46′E*
149 *N7* **Kajakī, Band-e** ◎ C Afghanistan
Kajan *see* Kayan, Sungai
Kajana *see* Kajaani
Kajanaland *see* Kainuu
137 *V13* **K'ajaran** *Rus.* Kadzharan. SE Armenia *39°10′N 46°09′E*
81 *I19* **Kajiado** Kajiado, S Kenya *01°51′S 36°48′E*
81 *I20* **Kajiado** ◈ *county* S Kenya
Kajisay *see* Bokonbayevo
113 *O20* **Kajmakčalan** *var.* Kaïmaktsalán. ▲ S FYR Macedonia *40°57′N 21°48′E* *see also* Kaïmaktsalán
Kajmakčalan *see* Kaïmaktsalán
Kajnar *see* Kaynar
149 *N6* **Kajrān** Dāykundī, C Afghanistan *33°12′N 65°28′E*
149 *N5* **Kaj Rūd** ≈ C Afghanistan
146 *G14* **Kaka** *Rus.* Kaakhka. Ahal Welaýaty, S Turkmenistan *37°20′N 59°37′E*
12 *C12* **Kakabeka Falls** Ontario, S Canada *48°24′N 89°40′W*
83 *F23* **Kakamas** Northern Cape, W South Africa *28°45′S 20°33′E*
81 *H18* **Kakamega** Kakamega, W Kenya *0°17′N 34°47′E*
81 *H18* **Kakamega** ◈ *county* W Kenya
112 *H13* **Kakanj** Federacija Bosne I Hercegovine, C Bosnia and Herzegovina *44°06′N 18°07′E*
185 *F22* **Kakanui Mountains** ▲▲ South Island, New Zealand
184 *K11* **Kakaramea** Taranaki, North Island, New Zealand *39°42′S 174°27′E*
76 *J16* **Kakata** C Liberia *06°35′N 10°19′W*
184 *M11* **Kakatahi** Manawatu-Wanganui, North Island, New Zealand *39°40′S 175°20′E*
113 *M23* **Kakavi** Gjirokastër, S Albania *39°55′N 20°19′E*
147 *O14* **Kakaydi** Surkhondaryo Viloyati, S Uzbekistan *37°37′N 67°30′E*
164 *F13* **Kake** Hiroshima, Honshū, SW Japan *34°37′N 132°17′E*
39 *X13* **Kake** Kupreanof Island, Alaska, USA *56°58′N 133°57′W*
171 *P14* **Kakea** Pulau Wowoni, C Indonesia *04°09′S 123°06′E*
164 *M14* **Kakegawa** Shizuoka, Honshū, S Japan *34°47′N 138°02′E*
165 *V16* **Kakeroma-jima** Kagoshima, SW Japan
143 *T6* **Kākhak** Khorāsān-e Raẕavī, E Iran
118 *L11* **Kakhanavichy** *Rus.* Kokhanovichi. Vitsyebskaya Voblasts', N Belarus *55°52′N 28°08′E*
39 *P13* **Kakhonak** Alaska, USA *59°26′N 154°48′W*
117 *S10* **Kakhovka** Khersons'ka Oblast', S Ukraine *46°40′N 33°30′E*
117 *U9* **Kakhovs'ke Vodoskhovyshche** *Rus.* Kakhovskoye Vodokhranilishche. ⊠ SE Ukraine
Kakhovskoye Vodokhranilishche *see* Kakhovs'ke Vodoskhovyshche
117 *T11* **Kakhovs'kyy Kanal** *canal* S Ukraine
Kakia *see* Khakhea
155 *L16* **Kākināda** *prev.* Cocanada. Andhra Pradesh, E India *16°56′N 82°13′E*
Käkisalmi *see* Priozersk
164 *I13* **Kakogawa** Hyōgo, Honshū, SW Japan *34°49′N 134°52′E*
81 *F18* **Kakoge** C Uganda *01°03′N 32°30′E*
145 *O7* **Kak, Ozero** ◎ N Kazakhstan
Ka-Krem *see* Malyy Yenisey
Kakshaal-Too, Khrebet *see* Kokshaal-Tau
39 *S5* **Kaktovik** Alaska, USA *70°08′N 143°37′W*
165 *Q11* **Kakuda** Miyagi, Honshū, C Japan *37°59′N 140°48′E*
165 *Q8* **Kakunodate** Akita, Honshū, C Japan *39°37′N 140°35′E*
Kalaallit Nunaat *see* Greenland
149 *T7* **Kālābāgh** Punjab, E Pakistan *33°00′N 71°35′E*
171 *Q16* **Kalabahi** Pulau Alor, S Indonesia *08°14′S 124°32′E*
188 *I5* **Kalabera** Saipan, S Northern Mariana Islands
83 *G14* **Kalabo** Western, W Zambia *15°00′S 22°37′E*
126 *M9* **Kalach** Voronezhskaya Oblast', W Russian Federation *50°24′N 41°00′E*
127 *N10* **Kalach-na-Donu** Volgogradskaya Oblast', SW Russian Federation *48°45′N 43°29′E*
166 *K5* **Kaladan** ≈ W Myanmar (Burma)
14 *K14* **Kaladar** Ontario, SE Canada *44°38′N 77°06′W*
38 *G13* **Ka Lae** *var.* South Cape, South Point. *headland* Hawai'i, USA, C Pacific Ocean *18°54′N 155°40′W*
83 *G19* **Kalahari Desert** *desert* Southern Africa
38 *B8* **Kalāheo** *var.* Kalaheo. Kaua'i, Hawaii, USA, C Pacific Ocean *21°55′N 159°31′W*
Kalaheo *see* Kalāheo
Kalaikhum *see* Qal'aikhum
Kala-i-Mor *see* Galaýmor
93 *K15* **Kalajoki** Pohjois-Pohjanmaa, W Finland *64°15′N 24°E*
Kalak *see* Eski Kaļak
Kal al Sraghna *see* El Kelâa Srarhna
32 *G10* **Kalama** Washington, NW USA *46°00′N 122°50′W*
Kalámai *see* Kalámata
115 *G14* **Kalamariá** Kentrikí Makedonía, N Greece *40°37′N 22°58′E*
115 *C15* **Kalamás** *var.* Thiamis; *prev.* Thýamis. ≈ W Greece
115 *E21* **Kalámata** *prev.* Kalámai. Pelopónnisos, S Greece *37°02′N 22°07′E*
148 *L16* **Kalamat, Khor** *Eng.* Kalmat Lagoon. *lagoon* SW Pakistan
148 *L16* **Kalamat, Khor** *Eng.* Kalmat Lagoon. *lagoon* SW Pakistan
31 *P10* **Kalamazoo** Michigan, N USA *42°17′N 85°35′W*
31 *P9* **Kalamazoo River** ≈ Michigan, N USA
Kalambaka *see* Kalampáka
117 *S13* **Kalamits'ka Zatoka** *Rus.* Kalamitskiy Zaliv. *gulf* S Ukraine
Kalamitskiy Zaliv *see* Kalamits'ka Zatoka
115 *H18* **Kálamos** Attikí, C Greece *38°16′N 23°51′E*
115 *C18* **Kálamos** *island* Iónioi Nísia, Greece, C Mediterranean Sea
115 *D15* **Kalampáka** *var.* Kalambaka. Thessalía, C Greece *39°43′N 21°36′E*
Kalan *see* Călan, Romania
Kalan *see* Tunceli, Turkey
117 *S11* **Kalanchak** Khersons'ka Oblast', S Ukraine *46°14′N 33°19′E*
38 *G11* **Kalaoa** *var.* Kailua. Hawaii, USA, C Pacific Ocean *19°43′N 155°59′W*
171 *O15* **Kalaotoa, Pulau** *island* W Indonesia
155 *J24* **Kala Oya** ≈ NW Sri Lanka
Kalarash *see* Călăraşi
93 *H17* **Kälarne** Jämtland, C Sweden *63°00′N 16°10′E*
143 *V15* **Kalar Rūd** ≈ SE Iran
169 *R9* **Kalasin** *var.* Muang Kalasin. Kalasin, E Thailand *16°29′N 103°31′E*
143 *U4* **Kalāt** *var.* Kabūd Gonbad. Khorāsān-e Raẕavī, NE Iran *37°02′N 59°46′E*
149 *O11* **Kālat** *var.* Kelat, Khelat. Baluchistān, SW Pakistan *29°01′N 66°38′E*
Kalāt *see* Qalāt
115 *J14* **Kalathriá, Ákrotirio** *headland* Samothráki, NE Greece *40°24′N 25°34′E*
193 *W147* **Kalau** *island* Tongatapu Group, SE Tonga
38 *E9* **Kalaupapa** Moloka'i, Hawaii, USA, C Pacific Ocean *21°11′N 156°59′W*
127 *N13* **Kalaus** ≈ SW Russian Federation
Kalávrita *see* Kalávryta
115 *E19* **Kalávryta** *var.* Kalávrita. Dytikí Elláda, S Greece *38°02′N 22°06′E*
141 *Y10* **Kalbān** W Oman *20°19′N 58°40′E*
Kalbar *see* Kalimantan Barat
180 *H11* **Kalbarri** Western Australia *27°43′S 114°08′E*
Kalbinskiy Khrebet *see* Khrebet Kalba
144 *G10* **Kaldygayty** ≈ W Kazakhstan
136 *I12* **Kalecik** Ankara, N Turkey *40°08′N 33°27′E*
79 *O19* **Kalehe** Sud-Kivu, E Dem. Rep. Congo *02°05′S 28°52′E*
79 *P22* **Kalemie** *prev.* Albertville. Katanga, SE Dem. Rep. Congo *05°55′S 29°09′E*
166 *L4* **Kalemyo** Sagaing, W Myanmar (Burma) *23°11′N 94°03′E*
82 *H12* **Kalene Hill** North Western, NW Zambia *11°10′S 24°12′E*
167 *T11* **Kalêng** *prev.* Phumĭ Kalêng. Stœng Trêng, NE Cambodia *13°57′N 106°17′E*
Kale Sultanie *see* Çanakkale
124 *I7* **Kalevala** Respublika Kareliya, NW Russian Federation *65°12′N 31°16′E*
166 *L4* **Kalewa** Sagaing, C Myanmar (Burma) *23°15′N 94°19′E*
Kalgan *see* Zhangjiakou
39 *Q12* **Kalgin Island** *island* Alaska, USA
180 *L12* **Kalgoorlie** Western Australia *30°51′S 121°27′E*
Kali *see* Sārda
115 *E17* **Kaliakoúda** ▲ C Greece *38°47′N 21°42′E*
114 *O8* **Kaliakra, Nos** *headland* NE Bulgaria *43°22′N 28°28′E*
115 *F19* **Kaliánoi** Pelopónnisos, S Greece *37°55′N 22°28′E*
115 *N24* **Kalí Límni** ▲ Kárpathos, SE Greece *35°34′N 27°08′E*
79 *N20* **Kalima** Maniema, E Dem. Rep. Congo *02°34′S 26°27′E*
169 *S11* **Kalimantan** *Eng.* Indonesian Borneo. ◈ *geopolitical region* Borneo, C Indonesia
169 *Q11* **Kalimantan Barat** *off.* Propinsi Kalimantan Berat, *var.* Kalbar, *Eng.* West Borneo, West Kalimantan. ◈ *province* N Indonesia
Kalimantan Barat, Propinsi *see* Kalimantan Barat
169 *T13* **Kalimantan Selatan** *off.* Propinsi Kalimantan Selatan, *var.* Kalsel, *Eng.* South Borneo, South Kalimantan. ◈ *province* N Indonesia
Kalimantan Selatan, Propinsi *see* Kalimantan Selatan
169 *R12* **Kalimantan Tengah** *off.* Propinsi Kalimantan Tengah, *var.* Kalteng, *Eng.* Central Borneo, Central Kalimantan. ◈ *province* N Indonesia
Kalimantan Tengah, Propinsi *see* Kalimantan Tengah
169 *U10* **Kalimantan Timur** *off.* Propinsi Kalimantan Timur, *var.* Kaltim, *Eng.* East Borneo, East Kalimantan. ◈ *province* N Indonesia
Kalimantan Timur, Propinsi *see* Kalimantan Timur
169 *V9* **Kalimantan Utara** *off.* Propinsi Kalimantan Utara, *var.* Kaltara, *Eng.* North Kalimantan. ◈ *province* N Indonesia
Kalimantan Utara, Propinsi *see* Kalimantan Utara
Kálimnos *see* Kálymnos
153 *S12* **Kālimpang** West Bengal, NE India *27°02′N 88°34′E*
Kalinin *see* Tver'
Kalinin *see* Boldumsaz
Kalininabad *see* Kalininobod
126 *B3* **Kaliningrad** Kaliningradskaya Oblast', W Russian Federation *54°48′N 21°33′E*
Kaliningrad *see* Kaliningradskaya Oblast'
126 *A3* **Kaliningradskaya Oblast'** *var.* Kaliningrad. ◈ *province and enclave* W Russian Federation
Kalinino *see* Tashir
147 *P14* **Kalininobod** *Rus.* Kalininabad. SW Tajikistan *37°49′N 68°55′E*
127 *O8* **Kalininsk** Saratovskaya Oblast', W Russian Federation *51°31′N 44°25′E*
Kalininsk *see* Boldumsaz
Kalinisk *see* Cupcina
119 *M19* **Kalinkavichy** *Rus.* Kalinkovichi. Homyel'skaya Voblasts', SE Belarus *52°08′N 29°19′E*
Kalinkovichi *see* Kalinkavichy
81 *G18* **Kaliro** SE Uganda *0°54′N 33°30′E*
Kalisch/Kalish *see* Kalisz
33 *O7* **Kalispell** Montana, NW USA *48°12′N 114°18′W*
110 *I13* **Kalisz** *Ger.* Kalisch, *Rus.* Kalish; *anc.* Calisia. Wielkopolskie, C Poland *51°46′N 18°04′E*
110 *F9* **Kalisz Pomorski** *Ger.* Kallies. Zachodnio-pomorskie, NW Poland *53°55′N 15°55′E*
126 *M10* **Kalitva** ≈ SW Russian Federation
81 *F21* **Kaliua** Tabora, C Tanzania *05°03′S 31°48′E*
92 *K13* **Kalix** Norrbotten, N Sweden *65°51′N 23°14′E*
92 *K12* **Kalixälven** ≈ N Sweden
92 *J11* **Kalixfors** Norrbotten, N Sweden *67°45′N 20°20′E*
145 *T8* **Kalkaman** *Kaz.* Qalqaman. Pavlodar, NE Kazakhstan *51°57′N 75°58′E*
Kalkandelen *see* Tetovo
181 *O4* **Kalkarindji** Northern Territory, N Australia *17°32′S 130°40′E*
31 *P6* **Kalkaska** Michigan, N USA *44°44′N 85°11′W*
93 *F16* **Kall** Jämtland, C Sweden *63°31′N 13°16′E*
189 *X2* **Kallalen** *var.* Calalen. *island* Ratak Chain, SE Marshall Islands
118 *J5* **Kallaste** *Ger.* Krasnogor. Tartumaa, SE Estonia *58°40′N 27°12′E*
93 *N16* **Kallavesi** ◎ SE Finland
115 *F17* **Kallídromo** ▲▲ C Greece
Kallies *see* Kalisz Pomorski
95 *M22* **Kallinge** Blekinge, S Sweden *56°14′N 15°17′E*
115 *L16* **Kalloní** Lésvos, E Greece *39°14′N 26°16′E*
93 *F16* **Kallsjön** ◎ C Sweden
95 *N21* **Kalmar** *var.* Calmar. Kalmar, S Sweden *56°40′N 16°22′E*
95 *M19* **Kalmar** *var.* Calmar. ◈ *county* S Sweden
95 *N20* **Kalmarsund** *strait* S Sweden
Kalmat Lagoon *see* Kalamat, Khor
117 *X9* **Kal'mius** ≈ E Ukraine
99 *H15* **Kalmthout** Antwerpen, N Belgium *51°24′N 04°27′E*
Kalmykia/Kalmykiya-Khal'mg Tangch, Respublika *see* Kalmykiya, Respublika
127 *O12* **Kalmykiya, Respublika** *var.* Respublika Kalmykiya-Khal'mg Tangch, *Eng.* Kalmykia; *prev.* Kalmytskaya ASSR. ◈ *autonomous republic* SW Russian Federation
Kalmytskaya ASSR *see* Kalmykiya, Respublika
118 *F9* **Kalnciems** C Latvia *56°46′N 23°37′E*
114 *L10* **Kalnitsa** ≈ SE Bulgaria
111 *J24* **Kalocsa** Bács-Kiskun, S Hungary *46°31′N 19°00′E*
114 *J9* **Kalofer** Plovdiv, C Bulgaria *42°36′N 25°00′E*
38 *E10* **Kalohi Channel** *channel* C Pacific Ocean
83 *I16* **Kalomo** Southern, S Zambia *17°02′S 26°29′E*
29 *X14* **Kalona** Iowa, C USA *41°28′N 91°42′W*
115 *K22* **Kalotási, Akrotírio** *cape* Amorgós, Kykládes, Greece, Aegean Sea
152 *J8* **Kalpa** Himāchal Pradesh, N India *31°33′N 78°16′E*
115 *C15* **Kalpáki** Ípeiros, W Greece *39°53′N 20°38′E*
155 *C22* **Kalpeni Island** *island* Lakshadweep, India, N Indian Ocean
152 *K13* **Kālpi** Uttar Pradesh, N India *26°07′N 79°44′E*
158 *G7* **Kalpin** Xinjiang Uygur Zizhiqu, NW China *40°35′N 78°52′E*
149 *P16* **Kalri Lake** ◎ SE Pakistan
Kalsel *see* Kalimantan Selatan
143 *R5* **Kāl Shūr** ≈ N Iran
39 *N11* **Kalskag** Alaska, USA *61°32′N 160°15′W*
Kalsø *see* Kalsoy
95 *B18* **Kalsoy** *Dan.* Kalsø. *island* N Faroe Islands
39 *O9* **Kaltag** Alaska, USA *64°19′N 158°43′W*
Kaltara *see* Kalimantan Utara
108 *H7* **Kaltbrunn** Sankt Gallen, NE Switzerland *47°11′N 09°00′E*
Kaltdorf *see* Pruszków
Kalteng *see* Kalimantan Tengah
Kaltim *see* Kalimantan Timur
77 *X14* **Kaltungo** Gombe, E Nigeria *09°49′N 11°22′E*
126 *K4* **Kaluga** Kaluzhskaya Oblast', W Russian Federation *54°31′N 36°16′E*
155 *J26* **Kalu Ganga** ≈ S Sri Lanka
82 *J13* **Kalulushi** Copperbelt, C Zambia *12°50′S 28°03′E*
180 *M2* **Kalumburu** Western Australia *14°11′S 126°40′E*
95 *H23* **Kalundborg** Sjælland, E Denmark *55°42′N 11°06′E*
82 *K11* **Kalungwishi** ≈ N Zambia
149 *T8* **Kalūr Kot** Punjab, E Pakistan *32°08′N 71°20′E*
116 *I6* **Kalush** *Pol.* Kałusz. Ivano-Frankivs'ka Oblast', W Ukraine *49°02′N 24°20′E*
Kałusz *see* Kalush
110 *N11* **Kałuszyn** Mazowieckie, C Poland *52°12′N 21°43′E*
155 *J26* **Kalutara** Western Province, SW Sri Lanka *06°35′N 79°59′E*
Kaluwawa *see* Fergusson Island
126 *I5* **Kaluzhskaya Oblast'** ◈ *province* W Russian Federation
119 *E14* **Kalvarija** *Pol.* Kalwaria. Marijampolė, S Lithuania *54°25′N 23°13′E*
93 *K15* **Kälviä** Keski-Pohjanmaa, W Finland *63°53′N 23°30′E*
109 *U6* **Kalwang** Steiermark, E Austria *47°25′N 14°48′E*
Kalwaria *see* Kalvarija
154 *D13* **Kalyān** Mahārāshtra, W India *19°17′N 73°11′E*
124 *K16* **Kalyazin** Tverskaya Oblast', W Russian Federation *57°15′N 37°53′E*
115 *D18* **Kalydón** *anc.* Calydon. *site of ancient city* Dytikí Elláda, C Greece
115 *M21* **Kálymnos** *var.* Kálimnos. Kálymnos, Dodekánisa, Greece, Aegean Sea *36°57′N 26°59′E*
115 *M21* **Kálymnos** *var.* Kálimnos. *island* Dodekánisa, Greece, Aegean Sea
117 *O5* **Kalynivka** Kyyivs'ka Oblast', N Ukraine *50°14′N 30°16′E*
117 *N6* **Kalynivka** Vinnyts'ka Oblast', C Ukraine *49°27′N 28°32′E*
145 *W15* **Kalzhat** *prev.* Kol'zhat. Almaty, SE Kazakhstan *43°29′N 80°37′E*
42 *M10* **Kama** *var.* Cama. Región Autónoma Atlántico Sur, SE Nicaragua *12°06′N 83°55′W*
165 *R9* **Kamaishi** *var.* Kamaisi. Iwate, Honshū, C Japan *39°18′N 141°52′E*
Kamaisi *see* Kamaishi
118 *H13* **Kamajai** Utena, E Lithuania *55°49′N 25°30′E*
Kamajai *see* Toliejai
149 *U9* **Kamālia** Punjab, NE Pakistan *30°44′N 72°39′E*
83 *I14* **Kamalondo** North Western, NW Zambia *13°42′S 25°38′E*
136 *I13* **Kaman** Kirşehir, C Turkey *39°22′N 33°43′E*
79 *O20* **Kamanyola** Sud-Kivu, E Dem. Rep. Congo *02°54′S 29°04′E*
141 *N14* **Kamarān** *island* W Yemen
55 *R9* **Kamarang** W Guyana *05°49′N 60°38′W*
Kāmāreddi/Kamareddy *see* Rāmāreddi
Kama Reservoir *see* Kamskoye Vodokhranilishche
148 *K13* **Kamarod** Baluchistān, SW Pakistan *27°34′N 63°36′E*
171 *P14* **Kamaru** Pulau Buton, C Indonesia *05°10′S 123°03′E*
77 *S13* **Kamba** Kebbi, NW Nigeria *11°50′N 03°44′E*
Kambaeng Petch *see* Kamphaeng Phet
180 *L12* **Kambalda** Western Australia *31°15′S 121°33′E*
149 *P13* **Kambar** *var.* Qambar. Sind, SE Pakistan *27°35′N 68°03′E*
Kambara *see* Kabara
76 *I14* **Kambia** W Sierra Leone *09°09′N 12°53′W*
Kambos *see* Kámpos
79 *N25* **Kambove** Katanga, SE Dem. Rep. Congo *10°50′S 26°39′E*
Kambryk *see* Cambrai
123 *V10* **Kamchatka** ≈ E Russian Federation
Kamchatka *see* Kamchatka, Poluostrov
Kamchatka Basin *see* Komandorskaya Basin
123 *U10* **Kamchatka, Poluostrov** *Eng.* Kamchatka. *peninsula* E Russian Federation
123 *V10* **Kamchatskiy Kray** ◈ *province* E Russian Federation
123 *V10* **Kamchatskiy Zaliv** *gulf* E Russian Federation
114 *N9* **Kamchia** *var.* Kamchiya. ≈ E Bulgaria
114 *L9* **Kamchia, Yazovir** *var.* Yazovir Kamchiya. ◎ E Bulgaria
Kamchiya *see* Kamchia
Kamchiya, Yazovir *see* Kamchia, Yazovir
149 *T4* **Kāmdēsh** *var.* Kamdesh; *prev.* Kāmdeysh. Nūrestān, E Afghanistan *35°25′N 71°26′E*
Kamdesh *see* Kāmdēsh
Kāmdeysh *see* Kāmdēsh
Kamen' *see* Kamyen'
Kamenets *see* Kamyanyets
Kamenets-Podol'skaya Oblast' *see* Khmel'nyts'ka Oblast'
Kamenets-Podol'skiy *see* Kam"yanets'-Podil's'kyy
113 *Q18* **Kamenica** NE Macedonia *42°03′N 22°34′E*
113 *O16* **Kamenicë** *var.* Dardanë, *Serb.* Kosovska Kamenica. E Kosovo *42°37′N 21°33′E*
112 *A11* **Kamenjak, Rt** *headland* NW Croatia
125 *O6* **Kamenka** Arkhangel'skaya Oblast', NW Russian Federation *65°55′N 44°01′E*
126 *6* **Kamenka** Penzenskaya Oblast', W Russian Federation *53°12′N 44°00′E*
127 *L8* **Kamenka** Voronezhskaya Oblast', W Russian Federation *50°44′N 39°31′E*
Kamenka *see* Taskala
Kamenka *see* Camenca
Kamenka *see* Kam"yanka
Kamenka-Bugskaya *see* Kam"yanka-Buz'ka
Kamenka Dneprovskaya *see* Kam"yanka-Dniprovs'ka
Kamen Kashirskiy *see* Kamin'-Kashyrs'kyy
Kamenka-Strumilov *see* Kam"yanka-Buz'ka
126 *L15* **Kamennomostskiy** Respublika Adygeya, SW Russian Federation *44°13′N 40°12′E*
126 *L11* **Kamenolomni** Rostovskaya Oblast', SW Russian Federation *47°36′N 40°18′E*
127 *P8* **Kamenskiy** Saratovskaya Oblast', W Russian Federation *50°56′N 45°32′E*
Kamenskoye *see* Romaniv
126 *L11* **Kamensk-Shakhtinskiy** Rostovskaya Oblast', SW Russian Federation *48°18′N 40°16′E*
101 *P15* **Kamenz** Sachsen, E Germany *51°15′N 14°06′E*
164 *J13* **Kameoka** Kyōto, Honshū, SW Japan *35°02′N 135°35′E*
126 *M3* **Kameshkovo** Vladimirskaya Oblast', W Russian Federation *56°21′N 41°01′E*
164 *C11* **Kami-Agata** Nagasaki, Tsushima, SW Japan *34°40′N 129°27′E*
33 *N10* **Kamiah** Idaho, NW USA *46°13′N 116°01′W*
Kamień Koszyrski *see* Kamin'-Kashyrs'kyy
110 *H9* **Kamień Krajeński** *Ger.* Kamin in Westpreussen. Kujawski-pomorskie, C Poland *53°31′N 17°31′E*
111 *F15* **Kamienna Góra** *Ger.* Landeshut, Landeshut in Schlesien. Dolnośląskie, SW Poland *50°48′N 16°00′E*
110 *D8* **Kamień Pomorski** *Ger.* Cummin in Pommern. Zachodnio-pomorskie, NW Poland *53°57′N 14°44′E*
165 *R5* **Kamiiso** Hokkaidō, NE Japan *41°50′N 140°38′E*
79 *L22* **Kamiji** Kasai-Oriental, S Dem. Rep. Congo *06°39′S 23°27′E*
165 *T3* **Kamikawa** Hokkaidō, NE Japan *43°51′N 142°47′E*
164 *B15* **Kami-Koshiki-jima** *island* SW Japan
79 *M23* **Kamina** Katanga, S Dem. Rep. Congo *08°42′S 25°01′E*
Kamina *see* Kaintiba
42 *C6* **Kaminaljuyú** *ruins* Guatemala, C Guatemala
Kamin in Westpreussen *see* Kamień Krajeński
116 *J2* **Kamin'-Kashyrs'kyy** *Pol.* Kamień Koszyrski, *Rus.* Kamen Kashirskiy. Volyns'ka Oblast', NW Ukraine *51°39′N 24°59′E*
Kaminka Strumiłowa *see* Kam"yanka-Buz'ka
165 *Q5* **Kaminokuni** Hokkaidō, NE Japan *41°48′N 140°05′E*
165 *P10* **Kaminoyama** Yamagata, Honshū, C Japan *38°10′N 140°16′E*
39 *Q13* **Kamishak Bay** *bay* Alaska, USA
165 *U4* **Kamishihoro** Hokkaidō, NE Japan *43°14′N 143°18′E*
Kamishli *see* Al Qāmishlī
Kamissar *see* Kamsar
164 *C11* **Kami-Tsushima** Nagasaki, Tsushima, SW Japan *34°40′N 129°27′E*
79 *O20* **Kamituga** Sud-Kivu, E Dem. Rep. Congo *03°07′S 28°10′E*
164 *B17* **Kamiyaku** Kagoshima, Yaku-shima, SW Japan *30°24′N 130°32′E*
11 *N16* **Kamloops** British Columbia, SW Canada *50°39′N 120°24′W*
107 *G25* **Kamma** Sicilia, Italy, C Mediterranean Sea *36°46′N 12°03′E*
192 *K4* **Kammu Seamount** *undersea feature* N Pacific Ocean *32°09′N 173°00′E*
109 *U11* **Kamnik** *Ger.* Stein. C Slovenia *46°13′N 14°34′E*
Kamniške Alpe *see* Kamniško-Savinjske Alpe
109 *T10* **Kamniško-Savinjske Alpe** *var.* Kamniške Alpe, Sanntaler Alpen, *Ger.* Steiner Alpen. ▲▲ N Slovenia
Kamo *see* Gavarr
165 *R3* **Kamoenai** *var.* Kamuenai. Hokkaidō, NE Japan *43°07′N 140°25′E*
165 *O14* **Kamogawa** Chiba, Honshū, S Japan *35°05′N 140°04′E*
149 *W8* **Kāmoke** Punjab, E Pakistan *31°58′N 74°15′E*
82 *L13* **Kamoto** Eastern, E Zambia *13°16′S 32°04′E*
109 *V3* **Kamp** ≈ N Austria
81 *F18* **Kampala** ● (Uganda) S Uganda *0°20′N 32°28′E*
168 *K11* **Kampar, Sungai** ≈ Sumatera, W Indonesia
98 *L9* **Kampen** Overijssel, E Netherlands *52°33′N 05°55′E*
79 *N20* **Kampene** Maniema, E Dem. Rep. Congo *03°35′S 26°40′E*
29 *Q9* **Kampeska, Lake** ◎ South Dakota, N USA
167 *O9* **Kamphaeng Phet** *var.* Kambaeng Petch. Kamphaeng Phet, W Thailand *16°28′N 99°31′E*
Kampo *see* Campo, Cameroon
Kampo *see* Ntem, Cameroon/Equatorial Guinea
167 *S12* **Kâmpóng Cham** *prev.* Kompong Cham. Kâmpóng Cham, C Cambodia *12°N 105°27′E*
167 *R12* **Kâmpóng Chhnăng** *prev.* Kompong. Kâmpóng Chhnăng, C Cambodia *12°15′N 104°40′E*
167 *R12* **Kâmpóng Khleăng** *prev.* Kompong Kleang. Siĕmréab, NW Cambodia *13°04′N 104°07′E*
Kâmpōng Saôm *see* Sihanoukville
167 *R13* **Kâmpóng Spoe** *prev.* Kompong Speu. Kâmpóng Spœ, S Cambodia *11°28′N 104°29′E*
167 *S12* **Kâmpóng Thum** *var.* Trâpeăng Vêng. Kâmpóng Thum, C Cambodia *12°37′N 104°58′E*
167 *S12* **Kâmpóng Trâbêk** *prev.* Phumĭ Kâmpóng Trâbêk, Phum Kompong Trabek. Kâmpóng Thum, C Cambodia *13°06′N 105°16′E*
121 *O2* **Kámpos** *var.* Kambos. NW Cyprus *35°03′N 32°44′E*
167 *R14* **Kâmpôt** Kâmpôt, SW Cambodia *10°37′N 104°11′E*
Kamptee *see* Kāmthi
77 *O14* **Kampti** SW Burkina Faso *10°07′N 03°22′W*
Kampuchea *see* Cambodia
Kampuchea, Democratic *see* Cambodia
Kampuchea, People's Democratic Republic of *see* Cambodia
169 *Q9* **Kampung Sirik** Sarawak, East Malaysia *02°42′N 111°28′E*
11 *V15* **Kamsack** Saskatchewan, S Canada *51°34′N 101°51′W*
76 *H13* **Kamsar** *var.* Kamissar. Guinée-Maritime, W Guinea *10°33′N 14°34′W*
127 *R4* **Kamskoye Ust'ye** Respublika Tatarstan, W Russian Federation *55°13′N 49°11′E*
125 *U14* **Kamskoye Vodokhranilishche** *var.* Kama Reservoir. ⊠ NW Russian Federation
154 *I12* **Kāmthi** *prev.* Kamptee. Mahārāshtra, C India *21°19′N 79°11′E*
Kamuela *see* Waimea
Kamuenai *see* Kamoenai
165 *T5* **Kamui-dake** ⛰ Hokkaidō, NE Japan *42°24′N 142°57′E*
165 *R3* **Kamui-misaki** *headland* Hokkaidō, NE Japan *43°20′N 140°20′E*
43 *O15* **Kámuk, Cerro** ▲ SE Costa Rica *09°15′N 83°01′W*
116 *K7* **Kam"yanets'-Podil's'kyy** *Rus.* Kamenets-Podol'skiy. Khmel'nyts'ka Oblast', W Ukraine *48°43′N 26°36′E*
117 *Q6* **Kam"yanka** *Rus.* Kamenka. Cherkas'ka Oblast', C Ukraine *49°03′N 32°06′E*
116 *I5* **Kam"yanka-Buz'ka** *prev.* Kamenka-Strumilov, *Pol.* Kaminka Strumiłowa, *Rus.* Kamenka-Bugskaya. L'vivs'ka Oblast', NW Ukraine *50°04′N 24°21′E*
117 *T9* **Kam"yanka-Dniprovs'ka** *Rus.* Kamenka Dneprovskaya. Zaporiz'ka Oblast', SE Ukraine *47°28′N 34°24′E*
119 *F19* **Kamyanyets** *Rus.* Kamenets. Brestskaya Voblasts', SW Belarus *52°24′N 23°49′E*
118 *M13* **Kamyen'** *Rus.* Kamen'. Vitsyebskaya Voblasts', N Belarus *55°01′N 28°53′E*
127 *P9* **Kamyshin** Volgogradskaya Oblast', SW Russian Federation *50°07′N 45°20′E*
127 *Q13* **Kamyzyak** Astrakhanskaya Oblast', SW Russian Federation *46°07′N 48°03′E*
12 *K8* **Kanaaupscow** ≈ Québec, C Canada
36 *K8* **Kanab** Utah, W USA *37°03′N 112°31′W*
36 *K9* **Kanab Creek** ≈ Arizona/Utah, SW USA
187 *Y14* **Kanacea** *prev.* Kanathea. Taveuni, N Fiji *16°59′S 179°54′E*
38 *G17* **Kanaga Island** *island* Aleutian Islands, Alaska, USA
38 *G17* **Kanaga Volcano** ▲ Kanaga Island, Alaska, USA *51°55′N 177°09′W*
164 *N14* **Kanagawa** *off.* Kanagawa-ken. ◈ *prefecture* Honshū, S Japan
Kanagawa-ken *see* Kanagawa
13 *Q8* **Kanairiktok** ≈ Newfoundland and Labrador, E Canada
Kanaky *see* New Caledonia
79 *K22* **Kananga** *prev.* Luluabourg. Kasai-Occidental, S Dem. Rep. Congo *05°51′S 22°22′E*
Kanara *see* Karnātaka
36 *J7* **Kanarraville** Utah, W USA *37°32′N 113°10′W*
127 *Q4* **Kanash** Chuvashskaya Respublika, W Russian Federation *55°30′N 47°27′E*
Kanathea *see* Kanacea
21 *Q4* **Kanawha River** ≈ West Virginia, NE USA
164 *L13* **Kanayama** Gifu, Honshū, SW Japan *35°46′N 137°15′E*
164 *L11* **Kanazawa** Ishikawa, Honshū, SW Japan *36°35′N 136°40′E*
166 *M4* **Kanbalu** Sagaing, C Myanmar (Burma) *23°10′N 95°31′E*
166 *L8* **Kanbe** Yangon, SW Myanmar (Burma) *16°40′N 96°01′E*
167 *O11* **Kanchanaburi** Kanchanaburi, W Thailand *14°02′N 99°32′E*
Kanchanjańghā/Kānchenjunga *see* Kangchenjunga
155 *J19* **Kānchīpuram** *prev.* Conjeeveram. Tamil Nādu, SE India *12°50′N 79°44′E*
149 *N8* **Kandahār** *Per.* Qandahār. Kandahār, S Afghanistan *31°36′N 65°48′E*
149 *N9* **Kandahār** *Per.* Qandahār. ◈ *province* SE Afghanistan
167 *S13* **Kândal** *var.* Ta Khmau. Kândal, S Cambodia *11°30′N 104°59′E*
Kandalakša *see* Kandalaksha
124 *I5* **Kandalaksha** *var.* Kandalakša, *Fin.* Kantalahti. Murmanskaya Oblast', NW Russian Federation *67°09′N 32°14′E*
Kandalaksha Gulf/Kandalakshskaya Guba *see* Kandalakshskiy Zaliv
124 *K6* **Kandalakshskiy Zaliv** *var.* Kandalakshskaya Guba, *Eng.* Kandalaksha Gulf. *bay* NW Russian Federation
83 *G17* **Kandalengoti** *var.* Kandalengoti. Ngamiland, NW Botswana *19°25′S 22°12′E*
Kandalengoti *see* Kandalengoti
169 *U13* **Kandangan** Borneo, C Indonesia *02°50′S 115°15′E*
Kandau *see* Kandava
118 *E8* **Kandava** *Ger.* Kandau. W Latvia *57°02′N 22°48′E*
Kandavu *see* Kadavu
77 *R14* **Kandé** *var.* Kanté. NE Togo *09°55′N 01°01′E*
101 *F23* **Kandel** ▲ SW Germany *48°03′N 08°00′E*
186 *C7* **Kandep** Enga, W Papua New Guinea *05°54′S 143°34′E*
149 *R12* **Kandh Kot** Sind, SE Pakistan *28°15′N 69°18′E*
77 *S13* **Kandi** N Benin *11°05′N 02°59′E*
149 *P14* **Kandiāro** Sind, SE Pakistan *27°02′N 68°16′E*
136 *F11* **Kandıra** Kocaeli, NW Turkey *41°05′N 30°08′E*
183 *S8* **Kandos** New South Wales, SE Australia *32°52′S 149°58′E*
148 *M16* **Kandrāch** *var.* Kanrach. Baluchistān, SW Pakistan *25°26′N 65°28′E*
172 *I4* **Kandreho** Mahajanga, C Madagascar *17°27′S 46°06′E*
186 *F7* **Kandrian** New Britain, E Papua New Guinea *06°14′S 149°32′E*
Kandukur *see* Kondukūr
155 *K25* **Kandy** Central Province, C Sri Lanka *07°17′N 80°40′E*
144 *I10* **Kandyagash** *Kaz.* Qandyaghash; *prev.* Oktyab'rsk. Aktyubinsk, W Kazakhstan *49°25′N 57°24′E*
18 *D12* **Kane** Pennsylvania, NE USA *41°39′N 78°47′W*
64 *I11* **Kane Fracture Zone** *tectonic feature* NW Atlantic Ocean
Kanëka *see* Kanëvka
78 *G9* **Kanem** *off.* Région du Kanem. ◈ *region* W Chad
Kanem, Région du *see* Kanem
38 *D9* **Kāne'ohe** *var.* Kaneohe. O'ahu, Hawaii, USA, C Pacific Ocean *21°25′N 157°48′W*
Kanestron, Akrotírio *see* Palioúri, Akrotírio
Kanëv *see* Kaniv
124 *M5* **Kanëvka** *var.* Kanëka. Murmanskaya Oblast', NW Russian Federation *67°07′N 39°43′E*
126 *K13* **Kanevskaya** Krasnodarskiy Kray, SW Russian Federation *46°07′N 38°57′E*

◆ Country
● Country Capital
◇ Dependent Territory
○ Dependent Territory Capital
◈ Administrative Regions
✈ International Airport
▲ Mountain
▲▲ Mountain Range
⛰ Volcano
≈ River
◎ Lake
⊠ Reservoir

Kanevskoye Vodokhranilishche *see* Kanivs'ke Vodoskhovyshche
165 *P9* **Kaneyama** Yamagata, Honshū, C Japan *38°54′N 140°20′E*
83 *G20* **Kang** Kgalagadi, C Botswana *23°41′S 22°50′E*
76 *L13* **Kangaba** Koulikoro, SW Mali *11°57′N 08°24′W*
136 *M13* **Kangal** Sivas, C Turkey *39°15′N 37°23′E*
Kangān *see* Bandar-e Kangān
168 *J6* **Kangar** Perlis, Peninsular Malaysia *06°28′N 100°10′E*
76 *L13* **Kangaré** Sikasso, S Mali *11°39′N 08°10′W*
182 *F10* **Kangaroo Island** *island* South Australia
93 *M17* **Kangasniemi** Etelä-Savo, E Finland *61°58′N 26°37′E*
142 *K6* **Kangāvar** *var.* Kangāwar. Kermānshāhān, W Iran *34°29′N 47°55′E*
Kangāwar *see* Kangāvar
153 *S11* **Kangchenjunga** *var.* Kānchenjunga, *Nep.* Kanchanjaṅghā. ▲ NE India *27°36′N 88°06′E*
160 *G9* **Kangding** *var.* Lucheng, *Tib.* Dardo. Sichuan, C China *30°03′N 101°56′E*
169 *U16* **Kangean, Kepulauan** *island group* S Indonesia
169 *T16* **Kangean, Pulau** *island* Kepulauan Kangean, S Indonesia
67 *U8* **Kangen** *var.* Kengen. ≈ E South Sudan
197 *N14* **Kangerlussuaq** *Dan.* Sondre Strømfjord. ✈ Qeqqata, W Greenland *66°59′N 50°28′E*
197 *Q15* **Kangertittivaq** *Dan.* Scoresby Sund. *fjord* E Greenland
167 *O2* **Kangfang** Kachin State, N Myanmar (Burma) *26°09′N 98°36′E*
163 *X12* **Kanggye** N North Korea *40°58′N 126°37′E*
197 *P15* **Kangikajik** *var.* Kap Brewster. *headland* E Greenland *70°10′N 22°00′W*
13 *N5* **Kangiqsualujjuaq** *prev.* George River, Port-Nouveau-Québec. Québec, E Canada *58°35′N 65°59′W*
12 *L2* **Kangiqsujuaq** *prev.* Maricourt, Wakeham Bay. Québec, NE Canada *61°35′N 72°00′W*
12 *M4* **Kangirsuk** *prev.* Bellin, Payne. Québec, E Canada *60°00′N 70°01′W*
Kangle *see* Wanzai
158 *M16* **Kangmar** Xizang Zizhiqu, W China *28°34′N 89°40′E*
Kangnŭng *see* Gangneung
79 *D18* **Kango** Estuaire, NW Gabon *0°17′N 10°00′E*
152 *I7* **Kāngra** Himāchal Pradesh, NW India *32°04′N 76°16′E*
153 *Q16* **Kangsabati Reservoir** ▣ N India
159 *O17* **Kangto** ▲ China/India *27°54′N 92°33′E*
159 *W12* **Kangxian** *var.* Kang Xian, Zuitai, Zuitaizi. Gansu, C China *33°21′N 105°40′E*
Kang Xian *see* Kangxian
76 *M15* **Kani** NW Ivory Coast *08°29′N 06°36′W*
166 *L4* **Kani** Sagaing, C Myanmar (Burma) *22°24′N 94°55′E*
79 *M23* **Kaniama** Katanga, S Dem. Rep. Congo *07°32′S 24°11′E*
Kanibadam *see* Konibodom
169 *V6* **Kanibongan** Sabah, East Malaysia *06°40′N 117°12′E*
185 *F17* **Kaniere** West Coast, South Island, New Zealand *42°45′S 171°00′E*
185 *G17* **Kaniere, Lake** ◎ South Island, New Zealand
188 *E17* **Kanifaay** Yap, W Micronesia
125 *O4* **Kanin Kamen'** ▲▲ NW Russian Federation
125 *N3* **Kanin Nos** Nenetskiy Avtonomnyy Okrug, NW Russian Federation *68°38′N 43°19′E*
125 *N3* **Kanin Nos, Mys** *cape* NW Russian Federation
125 *O5* **Kanin, Poluostrov** *peninsula* NW Russian Federation
139 *V8* **Kānī Sakht** Wāsiṭ, E Iraq *33°19′N 46°04′E*
139 *T3* **Kanî Slêman** *Ar.* Kānī Sulaymān. Arbīl, N Iraq *35°54′N 44°35′E*
Kānī Sulaymān *see* Kanî Slêman
165 *Q6* **Kanita** Aomori, Honshū, C Japan *41°04′N 140°36′E*
117 *Q5* **Kaniv** *Rus.* Kanëv. Cherkas'ka Oblast', C Ukraine *49°46′N 31°28′E*
182 *K11* **Kaniva** Victoria, SE Australia *36°25′S 141°13′E*
117 *Q5* **Kanivs'ke Vodoskhovyshche** *Rus.* Kanevskoye Vodokhranilishche. ▣ C Ukraine
112 *L8* **Kanjiža** *Ger.* Altkanischa, *Hung.* Magyarkanizsa, Ókanizsa; *prev.* Stara Kanjiža. Vojvodina, N Serbia *46°03′N 20°03′E*
93 *K18* **Kankaanpää** Satakunta, SW Finland *61°47′N 22°25′E*
30 *M12* **Kankakee** Illinois, N USA *41°07′N 87°51′W*
31 *O11* **Kankakee River** ≈ Illinois/ Indiana, N USA
76 *K14* **Kankan** E Guinea *10°25′N 09°19′W*
154 *K13* **Kānker** Chhattīsgarh, C India *20°19′N 81°29′E*
76 *J10* **Kankossa** Assaba, S Mauritania *15°54′N 11°31′W*
169 *N12* **Kanmaw Kyun** *var.* Kisseraing, Kithareng. *island* Mergui Archipelago, S Myanmar (Burma)
164 *F12* **Kanmuri-yama** ▲ Kyūshū, SW Japan *34°28′N 132°03′E*
21 *R10* **Kannapolis** North Carolina, SE USA *35°30′N 80°36′W*
93 *L16* **Kannonkoski** Keski-Suomi, C Finland *62°59′N 25°20′E*
93 *K15* **Kannus** Keski-Pohjanmaa, W Finland *63°55′N 23°55′E*
77 *V13* **Kano** Kano, N Nigeria *11°56′N 08°31′E*
77 *V13* **Kano** ◈ *state* N Nigeria
77 *V13* **Kano** ✈ Kano, N Nigeria *11°56′N 08°26′E*
164 *G14* **Kan'onji** *var.* Kanonzi. Kagawa, Shikoku, SW Japan *34°08′N 133°38′E*
Kanonzi *see* Kan'onji
26 *M5* **Kanopolis Lake** ▣ Kansas, C USA
36 *K5* **Kanosh** Utah, W USA *38°48′N 112°26′W*
169 *R9* **Kanowit** Sarawak, East Malaysia *02°03′N 112°15′E*
164 *C16* **Kanoya** Kagoshima, Kyūshū, SW Japan *31°22′N 130°50′E*
152 *L13* **Kānpur** *Eng.* Cawnpore. Uttar Pradesh, N India *26°28′N 80°21′E*
Kanrach *see* Kandrāch
164 *I14* **Kansai** ✈ (Ōsaka) Ōsaka, Honshū, SW Japan *34°25′N 135°13′E*
27 *R9* **Kansas** Oklahoma, C USA *36°14′N 94°46′W*
26 *L5* **Kansas** *off.* State of Kansas, *also known as* Jayhawker State, Sunflower State. ◈ *state* C USA
27 *R4* **Kansas City** Kansas, C USA *39°07′N 94°38′W*
27 *R4* **Kansas City** Missouri, C USA *39°06′N 94°35′W*
27 *R3* **Kansas City** ✈ Missouri, C USA *39°18′N 94°45′W*
27 *P4* **Kansas River** ≈ Kansas, C USA
122 *L14* **Kansk** Krasnoyarskiy Kray, S Russian Federation *56°11′N 95°32′E*
Kansu *see* Gansu
147 *V7* **Kant** Chuyskaya Oblast', N Kyrgyzstan *42°54′N 74°47′E*
93 *L19* **Kanta-Häme** *Swe.* Egentliga Tavastland. ◈ *region* S Finland
Kantalahti *see* Kandalaksha
167 *N16* **Kantang** *var.* Ban Kantang. Trang, SW Thailand *07°25′N 99°30′E*
115 *H25* **Kántanos** Kríti, Greece, E Mediterranean Sea *35°20′N 23°42′E*
77 *R12* **Kantchari** E Burkina Faso *12°47′N 01°37′E*
Kanté *see* Kandé
Kantemir *see* Cantemir
126 *L9* **Kantemirovka** Voronezhskaya Oblast', W Russian Federation *49°44′N 39°53′E*
167 *R11* **Kantharalak** Si Sa Ket, E Thailand *14°32′N 104°37′E*
Kantipur *see* Kathmandu
39 *Q9* **Kantishna River** ≈ Alaska, USA
191 *S3* **Kanton** *var.* Abariringa, Canton Island; *prev.* Mary Island. *atoll* Phoenix Islands, C Kiribati
97 *C20* **Kanturk** *Ir.* Ceann Toirc. Cork, SW Ireland *52°12′N 08°54′W*
55 *T11* **Kanuku Mountains** ▲▲ S Guyana
165 *O12* **Kanuma** Tochigi, Honshū, S Japan *36°34′N 139°44′E*
83 *H20* **Kanye** Southern, SE Botswana *24°55′S 25°14′E*
83 *H17* **Kanyu** North-West, C Botswana *20°04′S 24°36′E*
166 *M7* **Kanyutkwin** Bago, C Myanmar (Burma) *18°19′N 96°30′E*
79 *M24* **Kanzenze** Katanga, SE Dem. Rep. Congo *10°33′S 25°28′E*
193 *Y15* **Kao** *island* Kotu Group, W Tonga
167 *Q13* **Kaôh Kŏng** *var.* Krŏng Kaôh Kŏng. Kaôh Kông, SW Cambodia *11°37′N 102°59′E*
Kaohsiung *see* Gaoxiong
Kaokaona *see* Kirakira
83 *B17* **Kaoko Veld** ▲▲ N Namibia
76 *G11* **Kaolack** *var.* Kaolak. W Senegal *14°09′N 16°08′W*
Kaolak *see* Kaolack
Kaolan *see* Lanzhou
186 *M8* **Kaolo** San Jorge, N Solomon Islands *08°24′S 159°35′E*
83 *H14* **Kaoma** Western, W Zambia *14°50′S 24°48′E*
38 *B8* **Kapa'a** *var.* Kapaa. Kaua'i, Hawaii, USA, C Pacific Ocean *22°04′N 159°19′W*
Kapaa *see* Kapa'a
113 *J16* **Kapa Moračka** ▲ C Montenegro *42°53′N 19°01′E*
137 *V13* **Kapan** *Rus.* Kafan; *prev.* Ghap'an. SE Armenia *39°13′N 46°25′E*
82 *L13* **Kapandashila** Muchinga, NE Zambia *12°43′S 31°00′E*
79 *L23* **Kapanga** Katanga, S Dem. Rep. Congo *08°22′S 22°37′E*
Kapchagay *see* Kapshagay
Kapchagayskoye Vodokhranilishche *see* Vodokhranilishche Kapshagay
99 *F15* **Kapelle** Zeeland, SW Netherlands *51°29′N 03°58′E*
99 *G16* **Kapellen** Antwerpen, N Belgium *51°19′N 04°25′E*
95 *P15* **Kapellskär** Stockholm, C Sweden *59°43′N 19°03′E*
81 *H18* **Kapenguria** West Pokit, W Kenya *01°14′N 35°08′E*
109 *V6* **Kapfenberg** Steiermark, C Austria *47°27′N 15°18′E*
83 *J14* **Kapiri Mposhi** Central, C Zambia *13°59′S 28°40′E*
149 *R4* **Kāpīsā** ◈ *province* E Afghanistan
12 *G10* **Kapiskau** ≈ Ontario, C Canada
184 *K13* **Kapiti Island** *island* C New Zealand
78 *K9* **Kapka, Massif du** ▲▲ E Chad
Kaplamada *see* Kaubalatmada, Gunung
22 *H9* **Kaplan** Louisiana, S USA *30°00′N 92°16′W*
Kaplangky, Plato *see* Gaplaňgyr Platosy
111 *D19* **Kaplice** *Ger.* Kaplitz. Jihočeský Kraj, S Czech Republic *48°42′N 14°27′E*
Kaplitz *see* Kaplice
Kapoche *see* Capoche
171 *T12* **Kapocol** Papua Barat, E Indonesia *01°59′S 130°11′E*
167 *N14* **Kapoe** Ranong, SW Thailand *09°33′N 98°37′E*
Kapoeas *see* Kapuas, Sungai
81 *G15* **Kapoeta** Eastern Equatoria, SE South Sudan *04°50′N 33°35′E*
111 *I25* **Kapos** ≈ S Hungary
111 *H25* **Kaposvár** Somogy, SW Hungary *46°23′N 17°54′E*
94 *H13* **Kapp** Oppland, S Norway *60°42′N 10°49′E*
100 *I7* **Kappeln** Schleswig-Holstein, N Germany *54°41′N 09°56′E*
Kapronçza *see* Koprivnica
109 *P7* **Kaprun** Salzburg, C Austria *47°15′N 12°48′E*
145 *U15* **Kapshagay** *prev.* Kapchagay. Almaty, SE Kazakhstan *43°52′N 77°05′E*
Kapstad *see* Cape Town
Kapsukas *see* Marijampolė
171 *Y13* **Kaptiau** Papua, E Indonesia *02°23′S 139°51′E*
119 *L19* **Kaptsevichy** *Rus.* Koptsevichi. Homyel'skaya Voblasts', SE Belarus *52°14′N 28°19′E*
Kapuas Hulu, Banjaran/ Kapuas Hulu, Pegunungan *see* Kapuas Mountains
169 *S10* **Kapuas Mountains** *Ind.* Banjaran Kapuas Hulu, Pegunungan Kapuas Hulu. ▲▲ Indonesia/Malaysia
169 *P11* **Kapuas, Sungai** ≈ Borneo, N Indonesia
169 *T13* **Kapuas, Sungai** *prev.* Kapoeas. ≈ Borneo, C Indonesia
182 *J9* **Kapunda** South Australia *34°23′S 138°51′E*
152 *H8* **Kapūrthala** Punjab, N India *31°20′N 75°26′E*
12 *G12* **Kapuskasing** Ontario, S Canada *49°25′N 82°26′W*
14 *D6* **Kapuskasing** ≈ Ontario, S Canada
127 *P11* **Kapustin Yar** Astrakhanskaya Oblast', SW Russian Federation *48°36′N 45°49′E*
82 *K11* **Kaputa** Northern, NE Zambia *08°28′S 29°41′E*
111 *G22* **Kapuvár** Győr-Moson-Sopron, NW Hungary *47°35′N 17°01′E*
Kapydzhik, Gora *see* Qazangödağ
119 *J17* **Kapyl'** *Rus.* Kopyl'. Minskaya Voblasts', C Belarus *53°09′N 27°05′E*
43 *N9* **Kara** *var.* Cara. Región Autónoma Atlántico Sur, E Nicaragua *12°50′N 83°35′W*
77 *R14* **Kara** *var.* Lama-Kara. NE Togo *09°33′N 01°12′E*
77 *Q14* **Kara** ≈ N Togo
147 *U7* **Kara-Balta** Chuyskaya Oblast', N Kyrgyzstan *42°51′N 73°51′E*
144 *L7* **Karabalyk** *var.* Komsomolets, *Kaz.* Komsomol. Kostanay, N Kazakhstan *53°47′N 61°58′E*
144 *G11* **Karabau** *Kaz.* Qarabaū. Atyrau, W Kazakhstan *48°29′N 53°05′E*
146 *E7* **Karabaur', Uval** *Kaz.* Korabavur Pastligi, *Uzb.* Qorabowur Kirlari. *physical region* Kazakhstan/Uzbekistan
Karabekaul *see* Garabekewül
Karabil', Vozvyshennost' *see* Garabil Belentligi
Kara-Bogaz-Gol *see* Garabogazköl
Kara-Bogaz-Gol, Zaliv *see* Garabogaz Aylagy
145 *R15* **Karaboget** *Kaz.* Qaraböget. Zhambyl, S Kazakhstan *44°36′N 72°03′E*
136 *H11* **Karabük** Karabük, NW Turkey *41°12′N 32°36′E*
136 *H11* **Karabük** ◈ *province* NW Turkey
122 *L12* **Karabula** Krasnoyarskiy Kray, C Russian Federation *58°01′N 97°17′E*
145 *V14* **Karabulak** *Kaz.* Qarabulaq. Taldykorgan, SE Kazakhstan *44°53′N 78°29′E*
145 *Y11* **Karabulak** *Kaz.* Qarabulaq. Vostochnyy Kazakhstan, E Kazakhstan *47°34′N 84°40′E*
145 *Q17* **Karabulak** *Kaz.* Qarabulaq. Yuzhnyy Kazakhstan, S Kazakhstan *42°31′N 69°47′E*
Karabura *see* Yumin
136 *C17* **Kara Burnu** *headland* SW Turkey *36°34′N 28°00′E*
144 *K10* **Karabutak** *Kaz.* Qarabutaq. Aktyubinsk, W Kazakhstan *49°55′N 60°05′E*
136 *D12* **Karacabey** Bursa, NW Turkey *40°14′N 28°22′E*
114 *O12* **Karacaköy** İstanbul, NW Turkey *41°24′N 28°21′E*
114 *M12* **Karacaoğlan** Kırklareli, NW Turkey *41°30′N 27°06′E*
Karachay-Cherkessia *see* Karachayevo-Cherkesskaya Respublika
126 *L15* **Karachayevo-Cherkesskaya Respublika** *Eng.* Karachay-Cherkessia. ◈ *autonomous republic* SW Russian Federation
126 *M15* **Karachayevsk** Karachayevo-Cherkesskaya Respublika, SW Russian Federation *43°43′N 41°53′E*
126 *J6* **Karachev** Bryanskaya Oblast', W Russian Federation *53°07′N 35°56′E*
149 *O16* **Karāchi** Sind, SE Pakistan *24°51′N 67°02′E*
149 *O16* **Karāchi** ✈ Sind, S Pakistan *24°51′N 67°02′E*
Karácsonkő *see* Piatra-Neamț
155 *E15* **Karād** Mahārāshtra, W India *17°19′N 74°15′E*
136 *H16* **Karadağ** ▲ S Turkey *37°00′N 33°00′E*
147 *T10* **Karadar'ya** *Uzb.* Qoradaryo. ≈ Kyrgyzstan/Uzbekistan
Karadeniz *see* Black Sea
Karadeniz Boğazi *see* İstanbul Boğazı
146 *B13* **Karadepe** Balkan Welaýaty, W Turkmenistan *38°04′N 54°01′E*
Karadzhar *see* Qorajar
Karaferiye *see* Véroia
Karagan *see* Garagan
Karaganda *see* Karagandy
Karaganda *see* Karagandy
Karagandinskaya Oblast' *see* Karagandy
145 *R10* **Karagandy** *Kaz.* Qaraghandy; *prev.* Karaganda. Karaganda, C Kazakhstan *49°53′N 73°07′E*
145 *R10* **Karagandy** *off.* Karagandinskaya Oblast', *Kaz.* Qaraghandy Oblysy; *prev.* Karaganda. ◈ *province* C Kazakhstan
145 *T10* **Karagayly** *Kaz.* Qaraghayly. Karaganda, C Kazakhstan *49°25′N 75°31′E*
Karagel' *see* Garagöl'
123 *U9* **Karaginskiy, Ostrov** *island* E Russian Federation
197 *T1* **Karaginskiy Zaliv** *bay* E Russian Federation
137 *P13* **Karagöl Dağları** ▲▲ NE Turkey
Karagumskiy Kanal *see* Garagum Kanaly
114 *L13* **Karahisar** Edirne, NW Turkey *40°47′N 26°34′E*
127 *V3* **Karaidel'** Respublika Bashkortostan, W Russian Federation *55°50′N 56°55′E*
114 *L13* **Karaidemir Barajı** ▣ NW Turkey
155 *J21* **Kāraikāl** Puducherry, SE India *10°58′N 79°50′E*
155 *I22* **Kāraikkudi** Tamil Nādu, SE India *10°04′N 78°46′E*
Kara Irtysh *see* Kara Yertis
143 *N5* **Karaj** Alborz, N Iran *35°44′N 51°26′E*
168 *K8* **Karak** Pahang, Peninsular Malaysia *03°24′N 101°59′E*
Karak *see* Al Karak
147 *T11* **Kara-Kabak** Oshskaya Oblast', SW Kyrgyzstan *39°40′N 72°45′E*
Kara-Kala *see* Magtymguly
Karakala *see* Oqqal'a
Karakalpakstan, Respublika *see* Qoraqalpog'iston Respublikasi
Karakalpakya *see* Qoraqalpog'iston
Karakax *see* Moyu
158 *G10* **Karakax He** ≈ NW China
121 *X8* **Karakaya Barajı** ▣ C Turkey
171 *Q9* **Karakelong, Pulau** *island* N Indonesia
Karakılısse *see* Ağrı
Karak, Muḥāfaẓat al *see* Al Karak
147 *X8* **Karakol** *var.* Karakolka. Issyk-Kul'skaya Oblast', NE Kyrgyzstan *41°30′N 77°18′E*
147 *Y7* **Karakol** *prev.* Przheval'sk. Issyk-Kul'skaya Oblast', NE Kyrgyzstan *42°32′N 78°21′E*
Kara-Köl *see* Kara-Kul'
Karakolka *see* Karakol
149 *W2* **Karakoram Highway** *road* China/Pakistan
149 *Z3* **Karakoram Pass** *Chin.* Karakoram Shankou. *pass* C Asia
152 *I3* **Karakoram Range** ▲▲ C Asia
Karakoram Shankou *see* Karakoram Pass
Karaköse *see* Ağrı
145 *P14* **Karakoyyn, Ozero** *Kaz.* Qaraqoyyn. ◎ C Kazakhstan
83 *F19* **Karakubis** Ghanzi, W Botswana *22°03′S 20°36′E*
147 *T9* **Kara-Kul'** *Kir.* Kara-Köl. Dzhalal-Abadskaya Oblast', W Kyrgyzstan *40°35′N 73°36′E*
Karakul' *see* Qarokŭl, Tajikistan
Karakul' *see* Qorako'l, Uzbekistan
147 *U10* **Kara-Kul'dzha** Oshskaya Oblast', SW Kyrgyzstan *40°32′N 73°50′E*
127 *T3* **Karakulino** Udmurtskaya Respublika, NW Russian Federation *56°02′N 53°45′E*
Karakul', Ozero *see* Qarokŭl
Kara Kum *see* Garagum
Kara Kum Canal/ Karakumskiy Kanal *see* Garagum Kanaly
Karakumy, Peski *see* Garagum
83 *E17* **Karakuwisa** Okavango, NE Namibia *18°56′S 19°40′E*
122 *M13* **Karam** Irkutskaya Oblast', S Russian Federation *55°07′N 107°21′E*
Karamai *see* Karamay
169 *T14* **Karamain, Pulau** *island* N Indonesia
136 *I16* **Karaman** Karaman, S Turkey *37°11′N 33°13′E*
136 *H16* **Karaman** ◈ *province* S Turkey
114 *M8* **Karamandere** ≈ NE Bulgaria
158 *J4* **Karamay** *var.* Karamai, Kelamayi; *prev. Chin.* K'o-la-ma-i. Xinjiang Uygur Zizhiqu, NW China *45°33′N 84°45′E*
169 *U14* **Karambu** Borneo, N Indonesia *03°48′S 116°06′E*
185 *H14* **Karamea** West Coast, South Island, New Zealand *41°15′S 172°07′E*
185 *H14* **Karamea** ≈ South Island, New Zealand
185 *G15* **Karamea Bight** *gulf* South Island, New Zealand
Karamet-Niyaz *see* Garamätnyýaz
158 *K10* **Karamiran He** ≈ NW China
147 *S11* **Karamyk** Oshskaya Oblast', SW Kyrgyzstan *39°28′N 71°45′E*
169 *U17* **Karangasem** Bali, S Indonesia *08°24′S 115°40′E*
154 *H12* **Kāranja** Mahārāshtra, C India *20°30′N 77°26′E*
Karanpur *see* Karanpura
152 *F9* **Karanpura** *var.* Karanpur. Rājasthān, NW India *29°46′N 73°30′E*
Karánsebes/Karansebesch *see* Caransebeș
145 *T14* **Karaoy** *Kaz.* Qaraoy. Almaty, SE Kazakhstan *45°52′N 74°44′E*
144 *E10* **Karaozen** *Kaz.* Ülkenözen; *prev.* Bol'shoy Uzen'. ≈ Kazakhstan/Russian Federation
114 *N7* **Karapelit** *Rom.* Stejarul. Dobrich, NE Bulgaria *43°40′N 27°33′E*
136 *I15* **Karapınar** Konya, C Turkey *37°43′N 33°34′E*
83 *D22* **Karas** ◈ *district* S Namibia
147 *Y8* **Kara-Say** Issyk-Kul'skaya Oblast', NE Kyrgyzstan *41°34′N 77°56′E*
83 *E22* **Karasburg** Karas, S Namibia *27°59′S 18°46′E*
Kara Sea *see* Karskoye More
92 *K9* **Kárášjohka** *var.* Karašjokka. ≈ N Norway
Kárášjohka *see* Karasjok
92 *L9* **Karasjok** *Fin.* Kaarasjoki, *Lapp.* Kárášjohka. Finnmark, N Norway *69°27′N 25°28′E*
Karašjokka *see* Kárášjohka
Kara Strait *see* Karskiye Vorota, Proliv
145 *N8* **Karasu** *Kaz.* Qarasū. Kostanay, N Kazakhstan *52°44′N 65°29′E*
136 *F11* **Karasu** Sakarya, NW Turkey *41°07′N 30°37′E*
Kara Su *see* Mesta/Néstos
Karasubazar *see* Bilohirs'k
122 *I12* **Karasuk** Novosibirskaya Oblast', C Russian Federation *53°41′N 78°04′E*
145 *U13* **Karatal** *Kaz.* Qaratal. ≈ SE Kazakhstan
136 *K17* **Karataş** Adana, S Turkey *36°32′N 35°22′E*
145 *Q16* **Karatau** *Kaz.* Qarataū. Zhambyl, S Kazakhstan *43°09′N 70°28′E*
Karatau *see* Karatau, Khrebet
145 *P16* **Karatau, Khrebet** *var.* Karatau, *Kaz.* Qarataū. ▲▲ S Kazakhstan
144 *G13* **Karaton** *Kaz.* Qaraton. Atyrau, W Kazakhstan *46°33′N 53°31′E*
164 *C13* **Karatsu** *var.* Karatu. Saga, Kyūshū, SW Japan *33°28′N 129°48′E*
Karatu *see* Karatsu
122 *K8* **Karaul** Krasnoyarskiy Kray, N Russian Federation *70°07′N 83°12′E*
Karaulbazar *see* Qorowulbozor
Karauzyak *see* Qorao'zak
115 *D16* **Karáva** ▲ C Greece *39°19′N 21°33′E*
Karavanke *see* Karawanken
115 *F22* **Karavás** Kýthira, S Greece *36°21′N 22°57′E*
113 *J20* **Karavastasë, Laguna e** *var.* Kënet' e Karavastas, Kravasta Lagoon. *lagoon* W Albania
Karavastas, Kënet' e *see* Karavastasë, Laguna e
118 *I5* **Käravere** Tartumaa, E Estonia *58°25′N 26°29′E*
115 *L23* **Karavonísia** *island* Kykládes, Greece, Aegean Sea
169 *O15* **Karawang** *prev.* Krawang. Jawa, C Indonesia *06°13′S 107°16′E*
109 *T10* **Karawanken** *Slvn.* Karavanke. ▲▲ Austria/Serbia
Karaxahar *see* Kaidu He
137 *R13* **Karayazı** Erzurum, NE Turkey *39°40′N 42°09′E*
145 *Y11* **Kara Yertis** *prev.* Kara Irtysh *Rus.* Chërnyy Irtysh. ≈ NE Kazakhstan
145 *Q12* **Karazhal** *Kaz.* Qarazhal. Karaganda, C Kazakhstan *48°002′N 70°52′E*
139 *S9* **Karbalā'** *var.* Kerbala, Kerbela. Karbalā', S Iraq *32°37′N 44°03′E*
139 *S9* **Karbalā'** *off.* Muḥāfa at Karbalā'. ◈ *governorate* S Iraq
Karbalā', Muḥāfa at *see* Karbalā'
94 *L11* **Kårböle** Gävleborg, C Sweden *61°59′N 15°16′E*
111 *M23* **Karcag** Jász-Nagykun-Szolnok, E Hungary *47°22′N 20°51′E*
Kardak *see* Imia
114 *N7* **Kardam** Dobrich, NE Bulgaria *43°45′N 28°06′E*
Kardamila *see* Kardámyla
115 *L18* **Kardámyla** *var.* Kardamila, Kardhámila. Chíos, E Greece *38°33′N 26°04′E*
Kardeljevo *see* Ploče
Kardh *see* Qardho
Kardhámila *see* Kardámyla
Kardhítsa *see* Kardítsa
115 *E16* **Kardítsa** *var.* Kardhítsa. Thessalía, C Greece *39°22′N 21°56′E*
118 *E4* **Kärdla** *Ger.* Kertel. Hiiumaa, W Estonia *59°00′N 22°42′E*
Kărdžali *see* Kardzhali
114 *J11* **Kardzhali** *var.* Kŭrdzhali, Kărdžali, Kirdzhali. Kardzhali, S Bulgaria *41°39′N 25°23′E*
114 *K11* **Kardzhali** *var* Kŭrdzhali. ◈ *province* S Bulgaria
114 *J11* **Kardzhali, Yazovir** *var.* Yazovir Kŭrdzhali. ▣ S Bulgaria
Karelia *see* Kareliya, Respublika
119 *I16* **Karelichy** *Pol.* Korelicze, *Rus.* Korelichi. Hrodzyenskaya Voblasts', W Belarus *53°34′N 26°08′E*
124 *I10* **Kareliya, Respublika** *prev.* Karel'skaya ASSR, *Eng.* Karelia. ◈ *autonomous republic* NW Russian Federation
Karel'skaya ASSR *see* Kareliya, Respublika
81 *E22* **Karema** Katavi, W Tanzania *06°50′S 30°25′E*
Karen *see* Hualian
83 *I14* **Karenda** Central, C Zambia *14°42′S 26°52′E*
Karen State *see* Kayin State
92 *J10* **Karesuando** *Fin.* Kaaresuanto, *Lapp.* Gárasavvon. Norrbotten, N Sweden *68°25′N 22°28′E*
Karet *see* Kâgheṭ
Kareyz-e-Elyās/Kārez Iliās *see* Kārīz-e Elyās
122 *J11* **Kargasok** Tomskaya Oblast', C Russian Federation *59°01′N 80°34′E*
122 *I12* **Kargat** Novosibirskaya Oblast', C Russian Federation *55°07′N 80°19′E*
136 *J11* **Kargı** Çorum, N Turkey *41°09′N 34°32′E*
152 *I5* **Kargil** Jammu and Kashmir, NW India *34°34′N 76°06′E*
Kargilik *see* Yecheng
124 *L11* **Kargopol'** Arkhangel'skaya Oblast', NW Russian Federation *61°30′N 38°53′E*
110 *F12* **Kargowa** *Ger.* Unruhstadt. Lubuskie, W Poland *52°05′N 15°50′E*
77 *X13* **Kari** Bauchi, E Nigeria *11°13′N 10°34′E*
83 *J15* **Kariba** Mashonaland West, N Zimbabwe *16°29′S 28°48′E*
83 *J16* **Kariba, Lake** ▣ Zambia/ Zimbabwe
165 *Q4* **Kariba-yama** ▲ Hokkaidō, NE Japan *42°36′N 139°55′E*
83 *C19* **Karibib** Erongo, C Namibia *21°56′S 15°51′E*
Karies *see* Karyés
92 *L9* **Karigasniemi** *Lapp.* Garegegasnjárga. Lappi, N Finland *69°24′N 25°52′E*
184 *J2* **Karikari, Cape** *headland* North Island, New Zealand *34°47′S 173°24′E*
149 *W3* **Karīmābād** *prev.* Hunza. Jammu and Kashmir, NE Pakistan *36°23′N 74°43′E*
169 *P12* **Karimata, Kepulauan** *island group* N Indonesia
169 *P12* **Karimata, Pulau** *island* Kepulauan Karimata, N Indonesia
169 *O11* **Karimata, Selat** *strait* W Indonesia
155 *I14* **Karīmnagar** Telangana, C India *18°28′N 79°09′E*
186 *C7* **Karimui** Chimbu, C Papua New Guinea *06°19′S 144°48′E*
169 *Q15* **Karimunjawa, Pulau** *island* S Indonesia
80 *N12* **Karin** Woqooyi Galbeed, N Somalia *10°48′N 45°46′E*
Kariot *see* Ikaría
93 *L20* **Karis** *Fin.* Karjaa. Uusimaa, SW Finland *60°05′N 23°39′E*
Káristos *see* Kárystos
148 *J4* **Kārīz-e Elyās** *var.* Kareyz-e-Elyās, Kārez Iliās. Herāt, NW Afghanistan *35°26′N 61°24′E*
Karjaa *see* Karis
Karkaralinsk *see* Karkaraly
145 *T10* **Karkaraly** *prev.* Karkaralinsk. Karaganda, E Kazakhstan *49°31′N 75°53′E*
186 *D6* **Karkar Island** *island* N Papua New Guinea
143 *N7* **Karkas, Kūh-e** ▲▲ C Iran
142 *K8* **Karkheh, Rūd-e** ≈ SW Iran
115 *L20* **Karkinágri** *var.* Karkinagrio. Ikaría, Dodekánisa, Greece, Aegean Sea *37°31′N 26°01′E*
Karkinagrio *see* Karkinágri
117 *R12* **Karkinits'ka Zatoka** *Rus.* Karkinitskiy Zaliv. *gulf* S Ukraine
Karkinitskiy Zaliv *see* Karkinits'ka Zatoka
93 *L19* **Karkkila** *Swe.* Högfors. Uusimaa, S Finland *60°32′N 24°10′E*
93 *M19* **Kärkölä** Päijät-Häme, S Finland *60°52′N 25°17′E*
182 *G9* **Karkoo** South Australia *34°03′S 135°45′E*
Karkūk *see* Āltūn Kūbrî
118 *D5* **Kärla** *Ger.* Kergel. Saaremaa, W Estonia *58°20′N 22°15′E*
Karleby *see* Kokkola
110 *F7* **Karlino** *Ger.* Körlin an der Persante. Zachodinio-pomorskie, NW Poland *54°02′N 15°52′E*
137 *Q13* **Karlıova** Bingöl, E Turkey *39°16′N 41°01′E*
117 *U6* **Karlivka** Poltavs'ka Oblast', C Ukraine *49°27′N 35°08′E*
Karl-Marx-Stadt *see* Chemnitz
Karlö *see* Hailuoto
112 *C11* **Karlobag** *It.* Carlopago. Lika-Senj, W Croatia *44°31′N 15°06′E*
112 *D9* **Karlovac** *Ger.* Karlstadt, *Hung.* Károlyváros. Karlovac, C Croatia *45°29′N 15°31′E*
112 *C10* **Karlovac** *off.* Karlovačka Županija. ◈ *province* C Croatia
Karlovačka Županija *see* Karlovac
111 *A16* **Karlovarský Kraj** ◈ W Czech Republic
115 *M19* **Karlovási** *var.* Néon Karlovásion, Néon Karlovasi. Sámos, Dodekánisa, Greece, Aegean Sea *37°47′N 26°40′E*
114 *J9* **Karlovo** *prev.* Levskigrad. Plovdiv, C Bulgaria *42°38′N 24°49′E*
111 *A16* **Karlovy Vary** *Ger.* Karlsbad; *prev. Eng.* Carlsbad. Karlovarský Kraj, W Czech Republic *50°13′N 12°51′E*
Karlsbad *see* Karlovy Vary
95 *L17* **Karlsborg** Västra Götaland, S Sweden *58°32′N 14°32′E*
Karlsburg *see* Alba Iulia
95 *L22* **Karlshamn** Blekinge, S Sweden *56°10′N 14°50′E*
95 *L16* **Karlskoga** Örebro, C Sweden *59°19′N 14°33′E*
95 *M22* **Karlskrona** Blekinge, S Sweden *56°11′N 15°39′E*
101 *G21* **Karlsruhe** *var.* Carlsruhe. Baden-Württemberg, SW Germany *49°01′N 08°24′E*
95 *K16* **Karlstad** Värmland, C Sweden *59°22′N 13°36′E*
29 *R3* **Karlstad** Minnesota, N USA *48°34′N 96°31′W*
101 *I18* **Karlstadt** Bayern, C Germany *49°58′N 09°46′E*
Karlstadt *see* Karlovac
39 *Q14* **Karluk** Kodiak Island, Alaska, USA *57°34′N 154°27′W*
Karluk *see* Qarluq
119 *O17* **Karma** *Rus.* Korma. Homyel'skaya Voblasts', SE Belarus *53°07′N 30°48′E*
155 *F14* **Karmāla** Mahārāshtra, W India *18°26′N 75°08′E*
146 *M11* **Karmana** Navoiy Viloyati, C Uzbekistan *40°09′N 65°18′E*
138 *G8* **Karmi'el** *var.* Carmiel. Northern, N Israel *32°55′N 35°18′E*
95 *B16* **Karmøy** *island* S Norway
152 *I9* **Karnāl** Haryāna, N India *29°41′N 76°58′E*
153 *W15* **Karnaphuli Reservoir** ▣ NE India
155 *F17* **Karnātaka** *var.* Kanara; *prev.* Maisur, Mysore. ◈ *state* W India
25 *S13* **Karnes City** Texas, SW USA *28°54′N 97°55′W*
109 *P9* **Karnische Alpen** *It.* Alpi Carniche. ▲▲ Austria/Italy
114 *M9* **Karnobat** Burgas, E Bulgaria *42°39′N 26°59′E*
109 *Q9* **Kärnten** *off.* Land Kärten, *Eng.* Carinthi, *Slvn.* Koroška. ◈ *state* S Austria
Karnul *see* Kurnool
83 *K16* **Karoi** Mashonaland West, N Zimbabwe *16°50′S 29°40′E*
Karol *see* Carei
Károly-Fehérvár *see* Alba Iulia
Károlyváros *see* Karlovac
82 *M12* **Karonga** Northern, N Malawi *09°54′S 33°55′E*
147 *W10* **Karool-Dëbë** *Kas.* Karool-Döbö; *prev.* Karool-Tëbë. Narynskaya Oblast', C Kyrgyzstan *40°33′N 75°52′E*
Karool-Döbö *see* Karool-Dëbë
Karool-Tëbë *see* Karool-Dëbë
182 *J9* **Karoonda** South Australia *35°04′S 139°58′E*
149 *S9* **Karor** *var.* Koror Lāl Esan. Punjab, E Pakistan *31°15′N 70°58′E*
Karosa *see* Karossa
171 *N12* **Karossa** *var.* Karosa. Sulawesi, C Indonesia *01°38′S 119°21′E*
Karpaten *see* Carpathian Mountains
115 *L22* **Karpáthio Pélagos** *sea* Dodekánisa, Greece, Aegean Sea
115 *N24* **Kárpathos** Kárpathos, SE Greece *35°30′N 27°13′E*
115 *N24* **Kárpathos** *It.* Scarpanto; *anc.* Carpathos, Carpathus. *island* SE Greece
Karpathos Strait *see* Karpathou, Stenó
115 *N24* **Karpathou, Stenó** *var.* Karpathos Strait, Scarpanto Strait. *strait* Dodekánisa, Greece, Aegean Sea
Karpaty *see* Carpathian Mountains
115 *E17* **Karpenísi** *prev.* Karpenísion. Stereá Elláda, C Greece *38°55′N 21°46′E*
Karpenísion *see* Karpenísi
Karpilovka *see* Aktsyabrski
125 *O8* **Karpogory** Arkhangel'skaya Oblast', NW Russian Federation *64°01′N 44°22′E*
180 *I7* **Karratha** Western Australia *20°44′S 116°52′E*
137 *S12* **Kars** *var.* Qars. Kars, NE Turkey *40°35′N 43°05′E*
137 *S12* **Kars** *var.* Qars. ◈ *province* NE Turkey
145 *O12* **Karsakpay** *Kaz.* Qarsaqbay. Karaganda, C Kazakhstan *47°51′N 66°42′E*
93 *L15* **Kärsämäki** Pohjois-Pohjanmaa, C Finland *63°58′N 25°49′E*
Karsau *see* Kārsava
118 *K9* **Kārsava** *Ger.* Karsau; *prev. Rus.* Korsovka. E Latvia *56°46′N 27°39′E*
Karshi *see* Garşy, Turkmenistan
Karshi *see* Qarshi, Uzbekistan
Karshinskaya Step *see* Qarshi Cho'li
Karshinskiy Kanal *see* Qarshi Kanali
84 *I5* **Karskiye Vorota, Proliv** *Eng.* Kara Strait. *strait* N Russian Federation
122 *J6* **Karskoye More** *Eng.* Kara Sea. *sea* Arctic Ocean
93 *L17* **Karstula** Keski-Suomi, C Finland *62°52′N 24°48′E*
127 *Q5* **Karsun** Ul'yanovskaya Oblast', W Russian Federation *54°12′N 47°00′E*
122 *F11* **Kartaly** Chelyabinskaya Oblast', C Russian Federation *53°02′N 60°42′E*
18 *E13* **Karthaus** Pennsylvania, NE USA *41°06′N 78°03′W*
110 *I7* **Kartuzy** Pomorskie, NW Poland *54°21′N 18°11′E*
165 *R8* **Karumai** Iwate, Honshū, C Japan *40°19′N 141°27′E*
181 *U4* **Karumba** Queensland, NE Australia *17°31′S 140°51′E*
142 *L10* **Kārūn** *var.* Rūd-e Kārūn. ≈ SW Iran
92 *K13* **Karungi** Norrbotten, N Sweden *66°03′N 23°55′E*
92 *K13* **Karunki** Lappi, N Finland *66°01′N 24°06′E*
Kārūn, Rūd-e *see* Kārūn
155 *H21* **Karūr** Tamil Nādu, SE India *10°58′N 78°03′E*
93 *K17* **Karvia** Satakunta, SW Finland *62°07′N 22°34′E*
111 *J17* **Karviná** *Ger.* Karwin, *Pol.* Karwina; *prev.* Nová Karvinná. Moravskoslezský Krajj, E Czech Republic *49°50′N 18°30′E*
155 *E17* **Kārwār** Karnātaka, W India *14°50′N 74°09′E*
108 *M7* **Karwendelgebirge** ▲▲ Austria/Germany
Karwin/Karwina *see* Karviná
115 *I14* **Karyés** *var.* Karies. Ágion Óros, N Greece *40°15′N 24°15′E*
115 *I19* **Kárystos** *var.* Káristos. Évvoia, C Greece *38°01′N 24°25′E*
136 *E17* **Kaş** Antalya, SW Turkey *36°12′N 29°38′E*
39 *Y14* **Kasaan** Prince of Wales Island, Alaska, USA *55°32′N 132°24′W*
164 *I13* **Kasai** Hyōgo, Honshū, SW Japan *34°56′N 134°49′E*
79 *K21* **Kasai** *var.* Cassai, Kassai. ≈ Angola/Dem. Rep. Congo
79 *K22* **Kasai-Occidental** *off.* Région Kasai Occidental. ◈ *region* S Dem. Rep. Congo
Kasai Occidental, Région *see* Kasai-Occidental
79 *L21* **Kasai-Oriental** *off.* Région Kasai Oriental. ◈ *region* C Dem. Rep. Congo
Kasai Oriental, Région *see* Kasai-Oriental
79 *L24* **Kasaji** Katanga, S Dem. Rep. Congo *10°22′S 23°29′E*
82 *L12* **Kasama** Northern, N Zambia *10°14′S 31°12′E*
Kasan *see* Koson
83 *H16* **Kasane** North-West, NE Botswana *17°48′S 20°06′E*
81 *E23* **Kasanga** Rukwa, W Tanzania *08°27′S 31°10′E*
79 *G21* **Kasangulu** Bas-Congo, W Dem. Rep. Congo *04°33′S 15°12′E*
Kasansay *see* Kosonsoy
155 *E20* **Kāsaragod** Kerala, SW India *12°30′N 74°59′E*
Kasargen *see* Kasari
118 *P13* **Kasari** *var.* Kasari Jõgi, *Ger.* Kasargen. ≈ W Estonia
Kasari Jõgi *see* Kasari
8 *L11* **Kasba Lake** ◎ Northwest Territories, Nunavut N Canada
Kaschau *see* Košice
Kaseda *see* Minamisatsuma
83 *I14* **Kasempa** North Western, NW Zambia *13°27′S 25°49′E*
79 *O24* **Kasenga** Katanga, SE Dem. Rep. Congo *10°22′S 28°37′E*
79 *P17* **Kasenye** *var.* Kasenyi. Orientale, NE Dem. Rep. Congo *01°23′N 30°25′E*
Kasenyi *see* Kasenye
79 *O19* **Kasese** Maniema, E Dem. Rep. Congo *01°36′S 27°31′E*
81 *E18* **Kasese** SW Uganda *0°10′N 30°06′E*
152 *J11* **Kāsganj** Uttar Pradesh, N India *27°48′N 78°38′E*

◆ Country
● Country Capital
◇ Dependent Territory
○ Dependent Territory Capital
◈ Administrative Regions
✈ International Airport
▲ Mountain
▲▲ Mountain Range
⛰ Volcano
≈ River
◎ Lake
▣ Reservoir

143 U4 **Kashaf Rūd** ≈ NE Iran
143 N7 **Kāshān** Eşfahān, C Iran 33°57′N 51°31′E
126 M10 **Kashary** Rostovskaya Oblast', SW Russian Federation 49°02′N 40°58′E
39 O12 **Kashegelok** Alaska, USA 60°57′N 157°46′W
Kashgar *see* Kashi
158 E7 **Kashi** *Chin.* Kaxgar, K'o-shih, *Uigh.* Kashgar. Xinjiang Uygur Zizhiqu, NW China 39°32′N 75°58′E
164 J14 **Kashihara** *var.* Kasihara. Nara, Honshū, SW Japan 34°28′N 135°46′E
165 P13 **Kashima-nada** *gulf* S Japan
124 K15 **Kashin** Tverskaya Oblast', W Russian Federation 57°20′N 37°34′E
152 K10 **Kāshīpur** Uttarakhand, N India 29°13′N 78°58′E
126 L4 **Kashira** Moskovskaya Oblast', W Russian Federation 54°53′N 38°13′E
165 N11 **Kashiwazaki** *var.* Kasiwazaki. Niigata, Honshū, C Japan 37°22′N 138°33′E
Kashkadar'inskaya Oblast' *see* Qashqadaryo Viloyati
143 T5 **Kāshmar** *var.* Turshiz; *prev.* Solṭānābād, Torshiz. Khorāsān, NE Iran 35°13′N 58°25′E
Kashmir *see* Jammu and Kashmir
149 R12 **Kashmor** Sind, SE Pakistan 28°24′N 69°42′E
149 S5 **Kashmūnḏ Ghar** *Eng.* Kashmund Range. ▲▲ E Afghanistan
Kashmund Range *see* Kashmūnḏ Ghar
145 T7 **Kashyr** *prev.* Kachiry. Pavlodar, NE Kazakhstan 53°07′N 76°08′E
Kasi *see* Vārānasi
153 O12 **Kasia** Uttar Pradesh, N India 26°45′N 83°55′E
39 N12 **Kasigluk** Alaska, USA 60°54′N 162°31′W
Kasihara *see* Kashihara
39 R12 **Kasilof** Alaska, USA 60°20′N 151°16′W
Kasimköj *see* General Toshevo
126 M4 **Kasimov** Ryazanskaya Oblast', W Russian Federation 54°59′N 41°22′E
79 P18 **Kasindi** Nord-Kivu, E Dem. Rep. Congo 0°03′N 29°43′E
82 M12 **Kasitu** ≈ N Malawi
Kasiwazaki *see* Kashiwazaki
30 L14 **Kaskaskia River** ≈ Illinois, N USA
93 J17 **Kaskinen** *Swe.* Kaskö. Österbotten, W Finland 62°23′N 21°10′E
Kaskö *see* Kaskinen
Kas Kong *see* Kŏng, Kaôh
11 O17 **Kaslo** British Columbia, SW Canada 49°54′N 116°57′W
Käsmark *see* Kežmarok
169 T12 **Kasongan** Borneo, C Indonesia 02°01′S 113°21′E
79 N21 **Kasongo** Maniema, E Dem. Rep. Congo 04°22′S 26°42′E
79 H22 **Kasongo-Lunda** Bandundu, SW Dem. Rep. Congo 06°30′S 16°51′E
115 M24 **Kásos** *island* S Greece
Kasos Strait *see* Kásou, Stenó
115 M25 **Kásou, Stenó** *var.* Kasos Strait. *strait* Dodekánisos/Kríti, Greece, Aegean Sea
137 T10 **K'asp'i** *prev.* Kaspi. C Georgia 41°54′N 44°25′E
Kaspi *see* K'asp'i
114 M8 **Kaspichan** Shumen, NE Bulgaria 43°18′N 27°09′E
Kaspīy Mangy Oypaty *see* Caspian Depression
127 Q16 **Kaspiysk** Respublika Dagestan, SW Russian Federation 42°52′N 47°40′E
Kaspiyskiy *see* Lagan'
Kaspiyskoye More/Kaspiy Tengizi *see* Caspian Sea
Kassa *see* Košice
Kassai *see* Kasai
80 I9 **Kassala** Kassala, E Sudan 15°24′N 36°25′E
80 H9 **Kassala** ◈ *state* NE Sudan
115 G15 **Kassándra** *prev.* Pallíni; *anc.* Pallene. *peninsula* NE Greece
115 G15 **Kassándra** *headland* N Greece 39°58′N 23°22′E
115 H15 **Kassándras, Kólpos** *var.* Kólpos Toronaíos. *gulf* N Greece
139 Y11 **Kassārah** Maysān, E Iraq 31°21′N 47°25′E
101 I15 **Kassel** *prev.* Cassel. Hessen, C Germany 51°19′N 09°30′E
74 M6 **Kasserine** *var.* Al Qaşrayn. W Tunisia 35°15′N 08°52′E
14 J14 **Kasshabog Lake** ◎ Ontario, SE Canada
139 O5 **Kassīr, Sabkhat al** ◎ E Syria
29 W10 **Kasson** Minnesota, N USA 44°00′N 92°42′W
115 C17 **Kassópeia** *Var.* Kassópi. *site of ancient city* Ípeiros, W Greece
Kassópi *see* Kassópeia
136 I11 **Kastamonu** *var.* Castamoni, Kastamuni. Kastamonu, N Turkey 41°22′N 33°47′E
136 I10 **Kastamonu** *var.* Kastamuni. ◈ *province* N Turkey
Kastamuni *see* Kastamonu
Kastaneá *see* Kastaniá
115 E14 **Kastaniá** *prev.* Kastaneá. Kentrikí Makedonía, N Greece 40°25′N 22°09′E
Kastélli *see* Kíssamos
Kastellórizon *see* Megísti
115 N24 **Kástelo, Akrotírio** *prev.* Akrotírio Kastállou. *headland* Kárpathos, SE Greece 35°24′N 27°08′E
95 N21 **Kastlösa** Kalmar, S Sweden 56°25′N 16°25′E
115 D14 **Kastoría** Dytikí Makedonía, N Greece 40°33′N 21°15′E
126 K7 **Kastornoye** Kurskaya Oblast', W Russian Federation 51°49′N 38°07′E
115 I21 **Kástro** Sífnos, Kykládes, Greece, Aegean Sea 36°58′N 24°45′E
95 J23 **Kastrup** ✈ (København) København, E Denmark 55°36′N 12°39′E
119 Q17 **Kastsyukovichy** *Rus.* Kostyukovichi. Mahilyowskaya Voblasts', E Belarus 53°20′N 32°03′E
119 O18 **Kastsyukowka** *Rus.* Kostyukovka. Homyel'skaya Voblasts', SE Belarus 52°32′N 30°54′E
164 D13 **Kasuga** Fukuoka, Kyūshū, SW Japan 33°31′N 130°27′E
164 L13 **Kasugai** Aichi, Honshū, SW Japan 35°15′N 136°57′E
81 E21 **Kasulu** Kigoma, W Tanzania 04°33′S 30°06′E
164 I12 **Kasumi** Hyōgo, Honshū, SW Japan 35°36′N 134°37′E
127 R17 **Kasumkent** Respublika Dagestan, SW Russian Federation 41°39′N 48°09′E
82 M13 **Kasungu** Central, C Malawi 13°04′S 33°29′E
149 W9 **Kasūr** Punjab, E Pakistan 31°07′N 74°30′E
83 G15 **Kataba** Western, W Zambia 15°28′S 23°25′E
19 R4 **Katahdin, Mount** ▲ Maine, NE USA 45°55′N 68°52′W
79 M20 **Katako-Kombe** Kasai-Oriental, C Dem. Rep. Congo 03°24′S 24°25′E
39 T12 **Katalla** Alaska, USA 60°12′N 144°31′W
Katana *see* Qaṭanā
79 L24 **Katanga** *off.* Région du Katanga; *prev.* Shaba. ◈ *region* SE Dem. Rep. Congo
122 M11 **Katanga** ≈ C Russian Federation
Katanga, Région du *see* Katanga
154 J11 **Katāngi** Madhya Pradesh, C India 21°46′N 79°50′E
180 J13 **Katanning** Western Australia 33°45′S 117°33′E
181 P8 **Kata Tjuṯa** *var.* Mount Olga. ▲ Northern Territory, C Australia 25°20′S 130°47′E
81 F23 **Katavi** *off.* Mkoa wa Katavi. ◈ *region* SW Tanzania
Katawaz *see* Zarghūn Shahr
151 Q22 **Katchall Island** *island* Nicobar Islands, India, NE Indian Ocean
115 F14 **Kateríni** Kentrikí Makedonía, N Greece 40°15′N 22°30′E
117 P7 **Katerynopil'** Cherkas'ka Oblast', C Ukraine 49°00′N 30°59′E
166 M3 **Katha** Sagaing, N Myanmar (Burma) 24°11′N 96°20′E
181 P2 **Katherine** Northern Territory, N Australia 14°29′S 132°20′E
154 B11 **Kāthiāwār Peninsula** *peninsula* W India
153 P11 **Kathmandu** *prev.* Kantipur. ● (Nepal) Central, C Nepal 27°46′N 85°17′E
152 H7 **Kathua** Jammu and Kashmir, NW India 32°23′N 75°34′E
76 L12 **Kati** Koulikoro, SW Mali 12°41′N 08°04′W
153 R13 **Katihār** Bihār, NE India 25°33′N 87°34′E
184 N7 **Katikati** Bay of Plenty, North Island, New Zealand 37°34′S 175°55′E
83 H16 **Katima Mulilo** Caprivi, NE Namibia 17°31′S 24°20′E
77 N15 **Katiola** C Ivory Coast 08°11′N 05°04′W
191 V10 **Katiu** *atoll* Îles Tuamotu, C French Polynesia
92 J5 **Katla** 🌋 S Iceland 63°38′N 19°03′W
117 N12 **Katlabuh, Ozero** ◎ SW Ukraine
39 P14 **Katmai, Mount** ▲ Alaska, USA 58°16′N 154°57′W
154 J9 **Katni** Madhya Pradesh, C India 23°47′N 80°29′E
115 D19 **Káto Achaïa** *var.* Kato Ahaia, Káto Akhaía. Dytikí Elláda, S Greece 38°08′N 21°33′E
Kato Ahaia/Káto Akhaía *see* Káto Achaïa
121 P2 **Kato Lakatámeia** *var.* Kato Lakatamia. C Cyprus 35°07′N 33°20′E
Kato Lakatamia *see* Kato Lakatámeia
79 N22 **Katompi** Katanga, SE Dem. Rep. Congo 06°10′S 26°19′E
83 K14 **Katondwe** Lusaka, C Zambia 15°08′S 30°10′E
114 H12 **Káto Nevrokópi** *prev.* Káto Nevrokópion. Anatolikí Makedonía kai Thráki, NE Greece 41°21′N 23°51′E
Káto Nevrokópion *see* Káto Nevrokópi
81 E18 **Katonga** ≈ S Uganda
115 F15 **Káto Ólympos** ▲▲ C Greece
115 D17 **Katoúna** Dytikí Elláda, C Greece 38°47′N 21°07′E
115 E19 **Káto Vlasiá** Dytikí Makedonía, S Greece 38°02′N 21°54′E
111 J16 **Katowice** *Ger.* Kattowitz. Śląskie, S Poland 50°15′N 19°01′E
153 S15 **Kātoya** West Bengal, NE India 23°39′N 88°11′E
136 E16 **Katrançık Dağı** ▲▲ SW Turkey
95 N16 **Katrineholm** Södermanland, C Sweden 58°59′N 16°15′E
96 I11 **Katrine, Loch** ◎ C Scotland, United Kingdom
77 V12 **Katsina** Katsina, N Nigeria 12°59′N 07°33′E
77 U12 **Katsina** ◈ *state* N Nigeria
67 P8 **Katsina Ala** ≈ S Nigeria
164 C13 **Katsumoto** Nagasaki, Iki, SW Japan 33°49′N 129°42′E
165 P13 **Katsuta** *var.* Katuta. Ibaraki, Honshū, S Japan 36°24′N 140°32′E
165 O14 **Katsuura** *var.* Katuura. Chiba, Honshū, S Japan 35°09′N 140°16′E
164 K12 **Katsuyama** *var.* Katuyama. Fukui, Honshū, SW Japan 36°00′N 136°30′E
164 H12 **Katsuyama** Okayama, Honshū, SW Japan 35°06′N 133°43′E
Kattakurgan *see* Kattaqo'rg'on
147 N11 **Kattaqo'rg'on** *Rus.* Kattakurgan. Samarqand Viloyati, C Uzbekistan 39°56′N 66°11′E
115 O23 **Kattavía** Ródos, Dodekánisa, Greece, Aegean Sea 35°56′N 27°47′E
95 I21 **Kattegat** *Dan.* Kattegatt. *strait* N Europe
Kattegatt *see* Kattegat
95 P19 **Katthammarsvik** Gotland, SE Sweden 57°27′N 18°54′E
Kattowitz *see* Katowice
122 J13 **Katun'** ≈ S Russian Federation
Katuta *see* Katsuta
Katuura *see* Katsuura
Katuyama *see* Katsuyama
Katwijk *see* Katwijk aan Zee
98 G11 **Katwijk aan Zee** *var.* Katwijk. Zuid-Holland, W Netherlands 59°12′N 04°24′E
38 B8 **Kaua'i** *var.* Kauai. *island* Hawaiian Islands, Hawai'i, USA, C Pacific Ocean
Kauai *see* Kaua'i
38 C8 **Kaua'i Channel** *var.* Kauai Channel. *channel* Hawai'i, USA, C Pacific Ocean
Kauai Channel *see* Kaua'i Channel
171 R13 **Kaubalatmada, Gunung** *var.* Kaplamada. ▲ Pulau Buru, E Indonesia 03°16′S 126°17′E
191 U10 **Kauehi** *atoll* Îles Tuamotu, C French Polynesia
Kauen *see* Kaunas
101 K24 **Kaufbeuren** Bayern, S Germany 47°53′N 10°37′E
25 U7 **Kaufman** Texas, SW USA 32°35′N 96°18′W
101 I15 **Kaufungen** Hessen, C Germany 51°16′N 09°39′E
93 K17 **Kauhajoki** Etelä-Pohjanmaa, W Finland 62°26′N 22°10′E
93 K16 **Kauhava** Etelä-Pohjanmaa, W Finland 63°06′N 23°08′E
30 M7 **Kaukauna** Wisconsin, N USA 44°18′N 88°18′W
92 L11 **Kaukonen** Lappi, N Finland 67°28′N 24°49′E
38 A8 **Kaulakahi Channel** *channel* Hawai'i, USA, C Pacific Ocean
38 E9 **Kaunakakai** Moloka'i, Hawaii, USA, C Pacific Ocean 21°05′N 157°01′W
38 F12 **Kaunā Point** *var.* Kauna Point. *headland* Hawai'i, USA, C Pacific Ocean 19°02′N 155°52′W
Kauna Point *see* Kaunā Point
118 F13 **Kaunas** *Ger.* Kauen, *Pol.* Kowno; *prev. Rus.* Kovno. Kaunas, C Lithuania 54°54′N 23°57′E
118 F13 **Kaunas** ◈ *province* C Lithuania
186 C6 **Kaup** East Sepik, NW Papua New Guinea 03°50′S 144°01′E
77 U12 **Kaura Namoda** Zamfara, NW Nigeria 12°43′N 06°17′E
Kaushany *see* Căuşeni
93 K16 **Kaustinen** Keski-Pohjanmaa, W Finland 63°33′N 23°40′E
99 M23 **Kautenbach** Diekirch, NE Luxembourg 49°57′N 06°01′E
92 K10 **Kautokeino** *Lapp.* Guovdageaidnu. Finnmark, N Norway 69°N 23°01′E
Kavadar *see* Kavadarci
113 P19 **Kavadarci** *Turk.* Kavadar. C Macedonia 41°25′N 22°00′E
Kavaja *see* Kavajë
113 K20 **Kavajë** *It.* Cavaia, Kavaja. Tiranë, W Albania 41°11′N 19°33′E
114 M13 **Kavak Çayı** ≈ NW Turkey
Kavakli *see* Topolovgrad
114 I13 **Kavála** *prev.* Kaválla. Anatolikí Makedonía kai Thráki, NE Greece 40°57′N 24°26′E
114 I13 **Kaválas, Kólpos** *gulf* Aegean Sea, NE Mediterranean Sea
155 J17 **Kāvali** Andhra Pradesh, E India 15°05′N 80°02′E
Kaválla *see* Kavála
Kavango *see* Cubango/Okavango
155 C21 **Kavaratti** Lakshadweep, SW India 10°33′N 72°38′E
114 O8 **Kavarna** Dobrich, NE Bulgaria 43°27′N 28°21′E
118 G12 **Kavarskas** Utena, E Lithuania 55°27′N 24°55′E
76 I13 **Kavendou** ▲ C Guinea 10°49′N 12°14′W
Kavengo *see* Cubango/Okavango
155 F20 **Kāveri** *var.* Cauvery. ≈ S India
186 G5 **Kavieng** *var.* Kaewieng. New Ireland, NE Papua New Guinea 04°13′S 152°11′E
83 H16 **Kavimba** North-West, NE Botswana 18°02′S 24°38′E
83 I15 **Kavingu** Southern, S Zambia 15°39′S 26°03′E
143 Q6 **Kavīr, Dasht-e** *var.* Great Salt Desert. *salt pan* N Iran
Kavirondo Gulf *see* Winam Gulf
Kavkaz *see* Caucasus
95 K23 **Kävlinge** Skåne, S Sweden 55°47′N 13°05′E
82 G12 **Kavungo** Moxico, E Angola 11°31′S 22°59′E
165 Q8 **Kawabe** Akita, Honshū, C Japan 39°39′N 140°14′E
165 R9 **Kawai** Iwate, Honshū, C Japan 39°36′N 141°40′E
38 A8 **Kawaihoa Point** *headland* Ni'ihau, Hawai'i, USA, C Pacific Ocean 21°47′N 160°12′W
184 K3 **Kawakawa** Northland, North Island, New Zealand 35°23′S 174°06′E
82 I13 **Kawama** North Western, NW Zambia 13°04′S 25°59′E
82 K11 **Kawambwa** Luapula, N Zambia 09°45′S 29°10′E
154 K11 **Kawardha** Chhattīsgarh, C India 21°59′N 81°12′E
14 I14 **Kawartha Lakes** ◎ Ontario, SE Canada
165 O13 **Kawasaki** Kanagawa, Honshū, S Japan 35°32′N 139°41′E
171 R12 **Kawassi** Pulau Obi, E Indonesia 01°32′S 127°25′E
165 R6 **Kawauchi** Aomori, Honshū, C Japan 41°11′N 141°00′E
184 L5 **Kawau Island** *island* N New Zealand
184 N10 **Kaweka Range** ▲▲ North Island, New Zealand
Kawelecht *see* Puhja
184 O8 **Kawerau** Bay of Plenty, North Island, New Zealand 38°06′S 176°43′E
184 L8 **Kawhia** Waikato, North Island, New Zealand 38°04′S 174°49′E
184 K8 **Kawhia Harbour** *inlet* North Island, New Zealand
35 V8 **Kawich Peak** ▲ Nevada, W USA 38°00′N 116°27′W
35 V9 **Kawich Range** ▲▲ Nevada, W USA
14 G12 **Kawigamog Lake** ◎ Ontario, S Canada
171 P9 **Kawio, Kepulauan** *island group* N Indonesia
167 N9 **Kawkareik** Kayin State, S Myanmar (Burma) 16°33′N 98°18′E
27 O8 **Kaw Lake** ⊠ Oklahoma, C USA
166 M3 **Kawlin** Sagaing, N Myanmar (Burma) 23°48′N 95°41′E
Kawm Umbū *see* Kom Ombo
Kawthule State *see* Kayin State
Kaxgar *see* Kashi
158 D7 **Kaxgar He** ≈ NW China
158 J5 **Kax He** ≈ NW China
77 P12 **Kaya** C Burkina Faso 13°04′N 01°09′W
167 N6 **Kayah State** ◈ *state* C Myanmar (Burma)
39 T12 **Kayak Island** *island* Alaska, USA
114 M11 **Kayalıköy Barajı** ⊠ NW Turkey
166 M8 **Kayan** Yangon, SW Myanmar (Burma) 16°54′N 96°35′E
Kayangel Islands *see* Ngcheangel
155 G23 **Kāyankulam** Kerala, SW India 09°10′N 76°31′E
169 V9 **Kayan, Sungai** *prev.* Kajan. ≈ Borneo, C Indonesia
144 F14 **Kaydak, Sor** *salt flat* SW Kazakhstan
Kaydanovo *see* Dzyarzhynsk
37 N9 **Kayenta** Arizona, SW USA 36°43′N 110°15′W
76 J11 **Kayes** Kayes, W Mali 14°26′N 11°22′W
76 J11 **Kayes** ◈ *region* SW Mali
167 N8 **Kayin State** *var.* Kawthule State, Karen State. ◈ *state* S Myanmar (Burma)
145 U10 **Kaynar** *Kaz.* Qaynar, *var.* Kajnar. Vostochnyy Kazakhstan, E Kazakhstan 49°13′N 77°27′E
Kaynary *see* Căinari
83 H15 **Kayoya** Western, W Zambia 16°13′S 24°09′E
Kayrakkum *see* Qayroqqum
Kayrakkumskoye Vodokhranilishche *see* Qayroqqum, Obanbori
136 K14 **Kayseri** *var.* Kaisaria; *anc.* Caesarea Mazaca, Mazaca. Kayseri, C Turkey 38°42′N 35°28′E
136 K14 **Kayseri** *var.* Kaisaria. ◈ *province* C Turkey
36 L2 **Kaysville** Utah, W USA 41°10′N 111°55′W
Kayyngdy *see* Kaindy
14 L11 **Kazabazua** Québec, SE Canada 45°58′N 76°00′W
14 L12 **Kazabazua** ≈ Québec, SE Canada
123 Q7 **Kazach'ye** Respublika Sakha (Yakutiya), NE Russian Federation 70°38′N 135°54′E
Kazakdar'ya *see* Qozoqdaryo
146 E9 **Kazakhlyshor, Solonchak** *var.* Solonchak Shorkazakhly. *salt marsh* NW Turkmenistan
Kazakhskaya SSR/Kazakh Soviet Socialist Republic *see* Kazakhstan
144 L12 **Kazakhstan** *off.* Republic of Kazakhstan, *var.* Kazakstan, *Kaz.* Qazaqstan, Qazaqstan Respublikasy; *prev.* Kazakh Soviet Socialist Republic, *Rus.* Kazakhskaya SSR. ◆ *republic* C Asia
Kazakhstan, Republic of *see* Kazakhstan
Kazakh Uplands *see* Saryarka
Kazakstan *see* Kazakhstan
Kazalinsk *see* Kazaly
144 L14 **Kazaly** *prev.* Kazalinsk. Kzyl-Orda, S Kazakhstan 45°45′N 62°01′E
127 R4 **Kazan'** Respublika Tatarstan, W Russian Federation 55°43′N 49°07′E
8 M10 **Kazan** ≈ Nunavut, NW Canada
127 R4 **Kazan'** ✈ Respublika Tatarstan, W Russian Federation 55°46′N 49°21′E
Kazandzhik *see* Bereket
117 R8 **Kazanka** Mykolayivs'ka Oblast', S Ukraine 47°49′N 32°50′E
Kazanketken *see* Qozonketkan
114 J9 **Kazanlak** *var.* Kazanlŭk; *prev.* Kazanlik. Stara Zagora, C Bulgaria 42°38′N 25°24′E
Kazanlik *see* Kazanlak
Kazanlŭk *see* Kazanlak
165 Y16 **Kazan-rettō** *Eng.* Volcano Islands. *island group* SE Japan
Kazantip, Mys *see* Kazantyp, Mys
117 V12 **Kazantyp, Mys** *prev.* Mys Kazantip. *headland* S Ukraine 45°27′N 35°50′E
147 U9 **Kazarman** Narynskaya Oblast', C Kyrgyzstan 41°21′N 74°03′E
Kazatin *see* Kozyatyn
Kazbegi *see* Q'azbegi
Kazbegi *see* Kazbek
137 T9 **Kazbek** *var.* Kazbegi, *Geor.* Mqinvartsveri. ▲ N Georgia 42°43′N 44°28′E
82 M13 **Kazembe** Eastern, NE Zambia 12°06′S 32°45′E
143 N11 **Kāzerūn** Fārs, S Iran 29°41′N 51°38′E
125 R12 **Kazhym** Respublika Komi, NW Russian Federation 60°19′N 51°26′E
Kazi Ahmad *see* Qāzi Ahmad
136 H16 **Kazımkarabekir** Karaman, S Turkey 37°13′N 32°58′E
111 M20 **Kazincbarcika** Borsod-Abaúj-Zemplén, NE Hungary 48°15′N 20°40′E
119 H17 **Kazlowshchyna** *Pol.* Kozlowszczyzna, *Rus.* Kozlovshchina. Hrodzyenskaya Voblasts', W Belarus 53°19′N 25°18′E
119 E14 **Kazlų Rūda** Marijampolė, S Lithuania 54°45′N 23°28′E
144 E9 **Kaztalovka** Zapadnyy Kazakhstan, W Kazakhstan 49°45′N 48°42′E
79 K22 **Kazumba** Kasai-Occidental, S Dem. Rep. Congo 06°25′S 22°02′E
165 Q8 **Kazuno** Akita, Honshū, C Japan 40°14′N 140°48′E
Kazvin *see* Qazvīn
118 J12 **Kazyany** *Rus.* Koz'yany. Vitsyebskaya Voblasts', NW Belarus 55°18′N 26°52′E
122 H9 **Kazym** ≈ N Russian Federation
110 H10 **Kcynia** *Ger.* Exin. Kujawsko-pomorskie, C Poland 53°00′N 17°29′E
Kéa *see* Tziá
Kéa *see* Ioulís
38 H11 **Kea'au** *var.* Keaau. Hawaii, USA, C Pacific Ocean 19°36′N 155°01′W
Keaau *see* Kea'au
38 F11 **Keāhole Point** *var.* Keahole Point. *headland* Hawai'i, USA, C Pacific Ocean 19°43′N 156°03′W
38 G12 **Kealakekua** Hawaii, USA, C Pacific Ocean 19°31′N 155°56′W
38 H11 **Kea, Mauna** ▲ Hawai'i, USA 19°50′N 155°30′W
37 N10 **Keams** Arizona, SW USA 35°47′N 110°09′W
Kéamu *see* Aneityum
29 O16 **Kearney** Nebraska, C USA 40°42′N 99°06′W
36 L3 **Kearns** Utah, W USA 40°39′N 112°00′W
115 H20 **Kéas, Stenó** *strait* SE Greece
137 O14 **Keban Barajı** *dam* C Turkey
137 O14 **Keban Barajı** ⊠ C Turkey
77 S13 **Kebbi** ◈ *state* NW Nigeria
76 G10 **Kébémèr** NW Senegal 15°24′N 16°25′W
74 M7 **Kebili** *var.* Qibilī. C Tunisia 33°42′N 09°06′E
138 H4 **Kebir, Nahr el** ≈ NW Syria
80 A10 **Kebkabiya** Northern Darfur, W Sudan 13°39′N 24°05′E
92 I11 **Kebnekaise** *Lapp.* Giebnegáisi. ▲ N Sweden 68°01′N 18°24′E
81 M14 **K'ebrī Dehar** Sumalē, E Ethiopia 06°43′N 44°15′E
148 K15 **Kech** ≈ SW Pakistan
10 K10 **Kechika** ≈ British Columbia, W Canada
111 K23 **Kecskemét** Bács-Kiskun, C Hungary 46°54′N 19°42′E
168 J6 **Kedah** ◈ *state* Peninsular Malaysia
118 F12 **Kėdainiai** Kaunas, C Lithuania 55°19′N 24°00′E
152 K8 **Kedārnāth** Uttarakhand, N India 30°44′N 79°03′E
Kedder *see* Kehra
13 N13 **Kedgwick** New Brunswick, SE Canada 47°38′N 67°21′W
169 R16 **Kediri** Jawa, C Indonesia 07°45′S 112°01′E
171 Y13 **Kedir Sarmi** Papua, E Indonesia 02°00′S 139°01′E
163 V7 **Kedong** Heilongjiang, NE China 48°00′N 126°15′E
76 I12 **Kédougou** SE Senegal 12°35′N 12°09′W
122 I11 **Kedrovyy** Tomskaya Oblast', C Russian Federation 57°31′N 79°45′E
111 H16 **Kędzierzyn-Kozle** *Ger.* Heydebrech. Opolskie, S Poland 50°20′N 18°12′E
8 H8 **Keele** ≈ Northwest Territories, NW Canada
10 K6 **Keele Peak** ▲ Yukon, NW Canada 63°31′N 130°21′W
Keelung *see* Jilong
19 N10 **Keene** New Hampshire, NE USA 42°56′N 72°14′W
99 H17 **Keerbergen** Vlaams Brabant, C Belgium 51°01′N 04°39′E
83 E21 **Keetmanshoop** Karas, S Namibia 26°36′S 18°08′E
12 A11 **Keewatin** Ontario, S Canada 49°47′N 94°30′W
29 V4 **Keewatin** Minnesota, N USA 47°24′N 93°04′W
Kefallinía *see* Kefallonía
115 B18 **Kefallonía** *var.* Kefallinía. *island* Iónia Nisiá, Greece, C Mediterranean Sea
115 M22 **Kéfalos** Kos, Dodekánisa, Greece, Aegean Sea 36°44′N 26°58′E
171 Q17 **Kefamenanu** Timor, C Indonesia 09°31′S 124°29′E
Kefar Sava *see* Kfar Sava
Kefe *see* Feodosiya
77 V15 **Keffi** Nassarawa, C Nigeria 08°52′N 07°54′E
92 H4 **Keflavík** Suðurnes, W Iceland 64°01′N 22°35′W
92 H4 **Keflavik** ✈ (Reykjavík) Suðurnes, W Iceland 63°58′N 22°37′W
Kegalee *see* Kegalla
155 J25 **Kegalla** *var.* Kegalee, Kegalle. Sabaragamuwa Province, C Sri Lanka 07°14′N 80°21′E
Kegalle *see* Kegalla
Kegayli *see* Kegeyli
Kegel *see* Keila
145 W16 **Kegen** Almaty, SE Kazakhstan 42°58′N 79°12′E
146 H7 **Kegeyli** *prev.* Kegayli. Qoraqalpog'iston Respublikasi, W Uzbekistan 42°46′N 59°49′E
101 F22 **Kehl** Baden-Württemberg, SW Germany 48°34′N 07°49′E
118 H3 **Kehra** *Ger.* Kedder. Harjumaa, NW Estonia 59°19′N 25°21′E
117 U6 **Kehychivka** Kharkivs'ka Oblast', E Ukraine 49°18′N 35°46′E
97 L17 **Keighley** N England, United Kingdom 53°51′N 01°58′W
Kei Islands *see* Kai, Kepulauan
Keijō *see* Seoul
118 G3 **Keila** *Ger.* Kegel. Harjumaa, NW Estonia 59°18′N 24°29′E
Keilberg *see* Klínovec
83 F23 **Keimoes** Northern Cape, W South Africa 28°41′S 20°59′E
Keina/Keinis *see* Käina
Keishū *see* Gyeongju
77 T11 **Keïta** Tahoua, C Niger 14°45′N 05°40′E
78 J12 **Kéita, Bahr** *var.* Doka. ≈ S Chad
182 K10 **Keith** South Australia 36°01′S 140°22′E
96 K8 **Keith** NE Scotland, United Kingdom 57°33′N 02°57′W
26 K3 **Keith Sebelius Lake** ⊠ Kansas, C USA
32 G11 **Keizer** Oregon, NW USA 44°59′N 123°01′W
38 A8 **Kekaha** Kaua'i, Hawaii, USA, C Pacific Ocean 21°58′N 159°43′W
147 U10 **Këk-Art** *prev.* Alaykel', Alay-Kuu. Oshskaya Oblast', SW Kyrgyzstan 40°16′N 74°21′E
147 W10 **Këk-Aygyr** *var.* Keyaygyr. Narynskaya Oblast', C Kyrgyzstan 40°42′N 75°37′E
147 V9 **Këk-Dzhar** Narynskaya Oblast', C Kyrgyzstan 41°28′N 74°48′E
14 L8 **Kekek** ≈ Québec, SE Canada
185 K15 **Kekerengu** Canterbury, South Island, New Zealand 41°55′S 174°05′E
111 L21 **Kékes** ▲ N Hungary 47°53′N 19°59′E
171 P17 **Kekneno, Gunung** ▲ Timor, S Indonesia
147 S9 **Këk-Tash** *Kir.* Kök-Tash. Dzhalal-Abadskaya Oblast', W Kyrgyzstan 41°08′N 72°25′E
81 M15 **K'elafo** Sumalē, E Ethiopia 05°36′N 44°12′E
169 U10 **Kelai, Sungai** ≈ Borneo, N Indonesia
Kelamayi *see* Karamay
Kelang *see* Klang
168 K7 **Kelantan** ◈ *state* Peninsular Malaysia
Kelantan *see* Kelantan, Sungai
168 K7 **Kelantan, Sungai** *var.* Kelantan. ≈ Peninsular Malaysia
Kelat *see* Kālat
Këlcyra *see* Këlcyrë
113 L22 **Këlcyrë** *var.* Këlcyra. Gjirokastër, S Albania 40°19′N 20°10′E
Kelifskiy Uzboy *see* Kelif Uzboýy
146 L14 **Kelif Uzboýy** *Rus.* Kelifskiy Uzboy. *salt marsh* E Turkmenistan
137 O12 **Kelkit** Gümüşhane, NE Turkey 40°07′N 39°28′E
136 M12 **Kelkit Çayı** ≈ N Turkey
79 G18 **Kéllé** Cuvette-Quest, W Congo 0°04′S 14°33′E
77 W11 **Kellé** Zinder, S Niger 14°10′N 10°10′E
145 P7 **Kellerovka** Severnyy Kazakhstan, N Kazakhstan 53°51′N 69°15′E
8 I5 **Kellett, Cape** *headland* Banks Island, Northwest Territories, NW Canada 71°57′N 125°55′W
31 S11 **Kelleys Island** *island* Ohio, N USA
33 N8 **Kellogg** Idaho, NW USA 47°30′N 116°07′W
92 M12 **Kelloselkä** Lappi, N Finland 66°56′N 28°52′E
97 F17 **Kells** *Ir.* Ceanannas. Meath, E Ireland 53°44′N 06°53′W
118 E12 **Kelmė** Šiaulai, C Lithuania 55°39′N 22°57′E
99 M19 **Kelmis** *var.* La Calamine. Liège, E Belgium 50°43′N 06°01′E
78 H12 **Kélo** Tandjilé, SW Chad 09°21′N 15°50′E
83 I14 **Kelongwa** North Western, NW Zambia 13°41′S 26°19′E
11 N17 **Kelowna** British Columbia, SW Canada 49°50′N 119°29′W
11 X12 **Kelsey** Manitoba, C Canada 56°02′N 96°31′W
34 M6 **Kelseyville** California, W USA 38°58′N 122°51′W
96 K13 **Kelso** SE Scotland, United Kingdom 55°36′N 02°27′W
32 G10 **Kelso** Washington, NW USA 46°09′N 122°54′W
195 W15 **Keltie, Cape** *headland* Antarctica
Keltsy *see* Kielce
168 L9 **Keluang** *var.* Kluang. Johor, Peninsular Malaysia 02°01′N 103°18′E
168 M11 **Kelume** Pulau Lingga, W Indonesia 0°12′S 104°27′E
11 U15 **Kelvington** Saskatchewan, S Canada 52°10′N 103°30′W
124 J7 **Kem'** Respublika Kareliya, NW Russian Federation 64°55′N 34°18′E
124 I7 **Kem'** ≈ NW Russian Federation
137 O13 **Kemah** Erzincan, E Turkey 39°35′N 39°02′E
137 N13 **Kemaliye** Erzincan, C Turkey 39°16′N 38°29′E
Kemaman *see* Cukai
Kemanlar *see* Isperih
10 K14 **Kemano** British Columbia, SW Canada 53°39′N 127°58′W
Kemarat *see* Khemmarat
171 P12 **Kembani** Pulau Peleng, N Indonesia 01°32′S 122°57′E
136 F17 **Kemer** Antalya, SW Turkey 36°39′N 30°33′E
122 J12 **Kemerovo** *prev.* Shcheglovsk. Kemerovskaya Oblast', C Russian Federation 55°25′N 86°05′E
122 K12 **Kemerovskaya Oblast'** ◈ *province* S Russian Federation
92 L13 **Kemi** Lappi, NW Finland 65°46′N 24°34′E
92 M12 **Kemijärvi** *Swe.* Kemiträsk. Lappi, N Finland 66°41′N 27°24′E
92 M12 **Kemijärvi** ◎ N Finland
92 L13 **Kemijoki** ≈ NW Finland
147 V7 **Kemin** *prev.* Bystrovka. Chuyskaya Oblast', N Kyrgyzstan
92 L13 **Keminmaa** Lappi, NW Finland 65°49′N 24°34′E
Kemins Island *see* Nikumaroro
Kemiö *see* Kimito
Kemiträsk *see* Kemijärvi
127 P5 **Kemlya** Respublika Mordoviya, W Russian Federation 54°42′N 45°16′E
99 B18 **Kemmel** West-Vlaanderen, W Belgium 50°42′N 02°51′E
33 S16 **Kemmerer** Wyoming, C USA 41°47′N 110°32′W
121 O15 **Kemmuna** *var.* Comino. *island* C Malta
79 I14 **Kémo** ◈ *prefecture* S Central African Republic
25 U7 **Kemp** Texas, SW USA 32°26′N 96°13′W
93 L14 **Kempele** Pohjois-Pohjanmaa, C Finland 64°56′N 25°26′E
101 D15 **Kempen** Nordrhein-Westfalen, W Germany 51°22′N 06°25′E
25 Q5 **Kemp, Lake** ◎ Texas, SW USA
195 W5 **Kemp Land** *physical region* Antarctica
25 S9 **Kempner** Texas, SW USA 31°03′N 98°01′W
44 H3 **Kemp's Bay** Andros Island, W The Bahamas 24°02′N 77°32′W
183 U6 **Kempsey** New South Wales, SE Australia 31°05′S 152°50′E
101 J24 **Kempten** Bayern, S Germany 47°44′N 10°19′E
15 N9 **Kempt, Lac** ◎ Québec, SE Canada
183 P17 **Kempton** Tasmania, SE Australia 42°34′S 147°13′E
154 J9 **Ken** ≈ C India
39 R12 **Kenai** Alaska, USA 60°33′N 151°15′W
0 D5 **Kenai Mountains** ▲▲ Alaska, USA
39 R12 **Kenai Peninsula** *peninsula* Alaska, USA
21 V11 **Kenansville** North Carolina, SE USA 34°57′N 77°54′W
146 A10 **Kenar** *prev. Rus.* Ufra. Balkan Welaýaty, NW Turkmenistan 40°00′N 53°05′E
121 U13 **Kenâyis, Râs el-** *headland* N Egypt 31°13′N 27°53′E
97 K16 **Kendal** NW England, United Kingdom 54°20′N 02°45′W
23 Y16 **Kendall** Florida, SE USA 25°39′N 80°18′W
9 O8 **Kendall, Cape** *headland* Nunavut, E Canada 63°31′N 87°09′W
18 J15 **Kendall Park** New Jersey, NE USA 40°25′N 74°33′W
31 Q11 **Kendallville** Indiana, N USA 41°24′N 85°10′W
171 P14 **Kendari** Sulawesi, C Indonesia 03°57′S 122°36′E
169 Q13 **Kendawangan** Borneo, C Indonesia 02°32′S 110°13′E
Kendrāpāra *see* Kendrāparha
154 O12 **Kendrāparha** *var.* Kendrāpāra. Odisha, E India 20°29′N 86°25′E
154 O11 **Kendujhargarh** *prev.* Keonjihargarh. Odisha, E India 21°38′N 85°40′E
25 S13 **Kenedy** Texas, SW USA 28°49′N 97°51′W
76 J15 **Kenema** SE Sierra Leone 07°55′N 11°12′W
29 P16 **Kenesaw** Nebraska, C USA 40°37′N 98°39′W
Këneurgench *see* Köneürgenç
79 H21 **Kenge** Bandundu, SW Dem. Rep. Congo 04°52′S 16°59′E
Kengen *see* Kangen
167 O5 **Kengtung** Shan State, E Myanmar (Burma) 21°18′N 99°36′E
83 F23 **Kenhardt** Northern Cape, W South Africa 29°19′S 21°08′E
76 J12 **Kéniéba** Kayes, W Mali 12°47′N 11°16′W
Kenimekh *see* Konimex
169 U7 **Keningau** Sabah, East Malaysia 05°21′N 116°11′E
74 F6 **Kénitra** *prev.* Port-Lyautey. NW Morocco 34°20′N 06°29′W
21 V9 **Kenly** North Carolina, SE USA 35°39′N 78°16′W
97 B21 **Kenmare** *Ir.* Neidín. S Ireland 51°53′N 09°35′W
28 L2 **Kenmare** North Dakota, N USA 48°40′N 102°04′W
97 A21 **Kenmare River** *Ir.* An Ribhéar. *inlet* NE Atlantic Ocean
18 D10 **Kenmore** New York, NE USA 42°58′N 78°52′W
25 W8 **Kennard** Texas, SW USA 31°21′N 95°10′W
29 N10 **Kennebec** South Dakota, N USA 43°53′N 99°51′W
19 Q7 **Kennebec River** ≈ Maine, NE USA
19 P9 **Kennebunk** Maine, NE USA 43°22′N 70°33′W
39 R13 **Kennedy Entrance** *strait* Alaska, USA
166 L3 **Kennedy Peak** ▲ W Myanmar (Burma) 23°18′N 93°52′E
22 K9 **Kenner** Louisiana, S USA 29°57′N 90°15′W
180 I8 **Kenneth Range** ▲▲ Western Australia
27 Y9 **Kennett** Missouri, C USA 36°15′N 90°04′W
18 I16 **Kennett Square** Pennsylvania, NE USA 39°50′N 75°40′W
32 K10 **Kennewick** Washington, NW USA 46°12′N 119°08′W
12 E11 **Kenogami** ≈ Ontario, S Canada
15 Q7 **Kénogami, Lac** ◎ Québec, SE Canada
14 G8 **Kenogami Lake** Ontario, S Canada 48°04′N 80°10′W
14 F7 **Kenogamissi Lake** ◎ Ontario, S Canada
10 I6 **Keno Hill** Yukon, NW Canada 63°54′N 135°18′W
12 A11 **Kenora** Ontario, S Canada 49°47′N 94°26′W
31 N9 **Kenosha** Wisconsin, N USA 42°34′N 87°50′W
13 P14 **Kensington** Prince Edward Island, SE Canada 46°26′N 63°39′W
26 L3 **Kensington** Kansas, C USA 39°46′N 99°01′W
32 I11 **Kent** Oregon, NW USA 45°14′N 120°43′W
24 J9 **Kent** Texas, SW USA 31°03′N 104°13′W
32 H8 **Kent** Washington, NW USA 47°22′N 122°13′W
97 P22 **Kent** *cultural region* SE England, United Kingdom
145 P16 **Kentau** Yuzhnyy Kazakhstan, S Kazakhstan 43°28′N 68°41′E
183 P14 **Kent Group** *island group* Tasmania, SE Australia
31 N12 **Kentland** Indiana, N USA 40°46′N 87°26′W
31 R12 **Kenton** Ohio, N USA 40°39′N 83°36′W
8 K7 **Kent Peninsula** *peninsula* Nunavut, N Canada
115 F14 **Kentrikí Makedonía** *Eng.* Macedonia Central. ◈ *region* N Greece
20 J6 **Kentucky** *off.* Commonwealth of Kentucky, *also known as* Bluegrass State. ◈ *state* C USA
20 H8 **Kentucky Lake** ⊠ Kentucky/Tennessee, S USA
13 P15 **Kentville** Nova Scotia, SE Canada 45°04′N 64°30′W
22 K8 **Kentwood** Louisiana, S USA 30°56′N 90°30′W
31 P9 **Kentwood** Michigan, N USA 42°52′N 85°33′W
81 H17 **Kenya** *off.* Republic of Kenya. ◆ *republic* E Africa
Kenya, Mount *see* Kirinyaga

◆ Country
● Country Capital
◇ Dependent Territory
○ Dependent Territory Capital
◈ Administrative Regions
✈ International Airport
▲ Mountain
▲▲ Mountain Range
🌋 Volcano
≈ River
◎ Lake
⊠ Reservoir

- **Kenya, Republic of** *see* Kenya
- 168 L7 **Kenyir, Tasik** *var.* Tasek Kenyir. ◎ Peninsular Malaysia
- 29 W10 **Kenyon** Minnesota, N USA 44°16′N 92°59′W
- 29 Y16 **Keokuk** Iowa, C USA 40°24′N 91°22′W
- **Keonjihargarh** *see* Kendujhargarh
- **Kéos** *see* Tziá
- 29 X16 **Keosauqua** Iowa, C USA 40°43′N 91°58′W
- 29 X15 **Keota** Iowa, C USA 41°21′N 91°57′W
- 21 O11 **Keowee, Lake** ▣ South Carolina, SE USA
- 124 I7 **Kepa** *var.* Kepe. Respublika Kareliya, NW Russian Federation 65°09′N 32°15′E
- **Kepe** *see* Kepa
- 189 O13 **Kepirohi Falls** *waterfall* Pohnpei, E Micronesia
- 185 B22 **Kepler Mountains** ▲▲ South Island, New Zealand
- 111 I14 **Kępno** Wielkopolskie, C Poland 51°17′N 17°57′E
- 65 C24 **Keppel Island** *island* N Falkland Islands
- **Keppel Island** *see* Niuatoputapu
- 65 C23 **Keppel Sound** *sound* N Falkland Islands
- **Kepri** *see* Kepulauan Riau
- 136 D12 **Kepsut** Balıkesir, NW Turkey 39°41′N 28°09′E
- 168 M11 **Kepulauan Riau** *off.* Propinsi Kepulauan Riau, *var.* Kepri. ◐ *province* NW Indonesia
- **Kequ** *see* Gadê
- 171 V13 **Kerai** Papua Barat, E Indonesia 03°53′S 134°30′E
- **Kerak** *see* Al Karak
- 155 F22 **Kerala** ◐ *state* S India
- 165 R16 **Kerama-rettō** *island group* SW Japan
- 183 N10 **Kerang** Victoria, SE Australia 35°46′S 144°01′E
- **Kerasunt** *see* Giresun
- 115 H19 **Keratéa** *var.* Keratea. Attikí, C Greece 37°48′N 23°58′E
- **Keratea** *see* Keratéa
- 93 M19 **Kerava** *Swe.* Kervo. Uusimaa, S Finland 60°25′N 25°10′E
- **Kerbala/Kerbela** *see* Karbalā'
- 32 F15 **Kerby** Oregon, NW USA 42°10′N 123°39′W
- 117 W12 **Kerch** *Rus.* Kerch'. Avtonomna Respublika Krym, SE Ukraine 45°22′N 36°30′E
- **Kerch'** *see* Kerch
- **Kerchens'ka Protska/ Kerchenskiy Proliv** *see* Kerch Strait
- 117 V13 **Kerchens'kyy Pivostriv** *peninsula* S Ukraine
- 121 V4 **Kerch Strait** *var.* Bosporus Cimmerius, Enikale Strait, *Rus.* Kerchenskiy Proliv, *Ukr.* Kerchens'ka Protska. *strait* Black Sea/Sea of Azov
- **Kerdilio** *see* Kerdýlio
- 114 H13 **Kerdýlio** *var.* Kerdilio. ▲ N Greece 40°46′N 23°37′E
- 186 D8 **Kerema** Gulf, S Papua New Guinea 07°59′S 145°46′E
- **Keremitlik** *see* Lyulyakovo
- 136 I9 **Kerempe Burnu** *headland* N Turkey 42°01′N 33°20′E
- 80 J9 **Keren** *var.* Cheren. C Eritrea 15°45′N 38°22′E
- 25 U7 **Kerens** Texas, SW USA 32°07′N 96°13′W
- 184 M6 **Kerepehi** Waikato, North Island, New Zealand 37°18′S 175°33′E
- 145 P10 **Kerey, Ozero** ◎ C Kazakhstan
- **Kergel** *see* Kärla
- 173 Q12 **Kerguelen** *island* C French Southern and Antarctic Territories
- 173 Q13 **Kerguelen Plateau** *undersea feature* S Indian Ocean
- 115 C20 **Kerí** Zákynthos, Iónia Nisiá, Greece, C Mediterranean Sea 37°40′N 20°48′E
- 81 H19 **Kericho** Kericho, W Kenya 0°22′S 35°19′E
- 81 H19 **Kericho** ◐ *county* W Kenya
- 184 K2 **Kerikeri** Northland, North Island, New Zealand 35°14′S 173°58′E
- 93 O17 **Kerimäki** Etelä-Savo, E Finland 61°56′N 29°18′E
- 168 K12 **Kerinci, Gunung** ▲ Sumatera, W Indonesia 02°00′S 101°40′E
- **Keriya** *see* Yutian
- 158 H9 **Keriya He** ≈ NW China
- 98 J9 **Kerkbuurt** Noord-Holland, C Netherlands 52°29′N 05°08′E
- 98 J13 **Kerkdriel** Gelderland, C Netherlands 51°46′N 05°21′E
- 75 N6 **Kerkenah, Îles de** *var.* Kerkenna Islands, *Ar.* Juzur Qarqannah. *island group* E Tunisia
- **Kerkenna Islands** *see* Kerkenah, Îles de
- 115 M20 **Kerketévs** ▲ Sámos, Dodekánisa, Greece, Aegean Sea 37°44′N 26°39′E
- 29 T8 **Kerkhoven** Minnesota, N USA 45°12′N 95°18′W
- **Kerki** *see* Atamyrat
- **Kerkichi** *see* Kerkiçi
- 146 M14 **Kerkiçi** *Rus.* Kerkichi. Lebap Welaýaty, E Turkmenistan 37°46′N 65°18′E
- 115 F16 **Kerkíneo** *prehistoric site* Thessalía, C Greece
- 114 G12 **Kerkíni, Límni** *var.* Límni Kerkinitis. ◎ N Greece
- **Kerkinitis Límni** *see* Kerkíni, Límni
- **Kérkira** *see* Kérkyra
- 99 M18 **Kerkrade** Limburg, SE Netherlands 50°53′N 06°04′E
- **Kerkuk** *see* Āltūn Kūbrī
- 115 B16 **Kérkyra** *var.* Kérkira, *Eng.* Corfu. Kérkyra, Iónia Nisiá, Greece, C Mediterranean Sea 39°37′N 19°56′E
- 115 B16 **Kerkyra** ✕ Kérkyra, Iónia Nisiá, Greece, C Mediterranean Sea 39°36′N 19°55′E
- 115 A16 **Kérkyra** *var.* Kérkira, *Eng.* Corfu. *island* Iónia Nisiá, Greece, C Mediterranean Sea
- 192 K10 **Kermadec Islands** *island group* New Zealand, SW Pacific Ocean
- 175 R10 **Kermadec Ridge** *undersea feature* SW Pacific Ocean 30°30′S 178°30′W
- 175 R11 **Kermadec Trench** *undersea feature* SW Pacific Ocean
- 143 S10 **Kermān** *var.* Kirman; *anc.* Carmana. Kermān, C Iran 30°18′N 57°05′E
- 143 R11 **Kermān** *off.* Ostān-e Kermān, *var.* Kirman; *anc.* Carmania. ◐ *province* SE Iran
- 143 U12 **Kermān, Bīābān-e** *desert* SE Iran
- **Kermān, Ostān-e** *see* Kermān
- 142 K6 **Kermānshāh** *var.* Qahremānshahr; *prev.* Bākhtarān. Kermānshāhān, W Iran 34°19′N 47°04′E
- 143 Q9 **Kermānshāh** Yazd, C Iran 34°19′N 47°04′E
- 142 J6 **Kermānshāh** *off.* Ostān-e Kermānshāh; *prev.* Bākhtarān. ◐ *province* W Iran
- **Kermānshāh, Ostān-e** *see* Kermānshāh
- 114 L10 **Kermen** Sliven, C Bulgaria 42°30′N 26°12′E
- 24 L8 **Kermit** Texas, SW USA 31°49′N 103°07′W
- 21 P6 **Kermit** West Virginia, NE USA 37°51′N 82°24′W
- 21 S9 **Kernersville** North Carolina, SE USA 36°12′N 80°13′W
- 35 S12 **Kern River** ≈ California, W USA
- 35 S12 **Kernville** California, W USA 35°44′N 118°25′W
- 115 K21 **Kéros** *island* Kykládes, Greece, Aegean Sea
- 76 K14 **Kérouané** SE Guinea 09°16′N 09°00′W
- 101 D16 **Kerpen** Nordrhein-Westfalen, W Germany 50°51′N 06°40′E
- 146 I11 **Kerpichli** Lebap Welaýaty, NE Turkmenistan 40°12′N 61°09′E
- 24 M1 **Kerrick** Texas, SW USA 36°29′N 102°14′W
- **Kerr Lake** *see* John H. Kerr Reservoir
- 11 S15 **Kerrobert** Saskatchewan, S Canada 51°56′N 109°09′W
- 25 Q11 **Kerrville** Texas, SW USA 30°03′N 99°09′W
- 97 B20 **Kerry** *Ir.* Ciarraí. *cultural region* SW Ireland
- 21 S11 **Kershaw** South Carolina, SE USA 34°33′N 80°34′W
- **Kertel** *see* Kärdla
- 95 H23 **Kerteminde** Syddtjylland, C Denmark 55°27′N 10°40′E
- 163 Q7 **Kerulen** *Chin.* Herlen He, *Mong.* Herlen Gol. ≈ China/ Mongolia
- **Kervo** *see* Kerava
- **Kerýneia** *see* Girne
- 12 H11 **Kesagami Lake** ◎ Ontario, SE Canada
- 93 O17 **Kesälahti** Pohjois-Karjala, SE Finland 61°54′N 29°49′E
- 136 B11 **Keşan** Edirne, NW Turkey 40°52′N 26°37′E
- 165 R9 **Kesennuma** Miyagi, Honshū, C Japan 38°55′N 141°35′E
- 163 V7 **Keshan** Heilongjiang, NE China 48°00′N 125°46′E
- 30 M6 **Keshena** Wisconsin, N USA 44°54′N 88°37′W
- 136 I13 **Keskin** Kırıkkale, C Turkey 39°41′N 33°36′E
- 93 K16 **Keski-Pohjanmaa** *Swe.* Mellersta Österbotten, *Eng.* centralostrobothnia. ◐ *region* W Finland
- 93 M17 **Keski-Suomi** *Swe.* Mellersta Finland, *Eng.* Central Finland. ◐ *region* C Finland
- **Késmárk** *see* Kežmarok
- 124 I6 **Kesten'ga** *var.* Kest Enga. Respublika Kareliya, NW Russian Federation 65°53′N 31°47′E
- **Kest Enga** *see* Kesten'ga
- 98 K12 **Kesteren** Gelderland, C Netherlands 51°55′N 05°34′E
- 14 H14 **Keswick** Ontario, S Canada 44°15′N 79°26′W
- 97 K15 **Keswick** NW England, United Kingdom 54°30′N 03°04′W
- 111 H24 **Keszthely** Zala, SW Hungary 46°47′N 17°16′E
- 122 K11 **Ket'** ≈ C Russian Federation
- 77 R17 **Keta** SE Ghana 05°55′N 00°59′E
- 169 Q12 **Ketapang** Borneo, C Indonesia 01°50′S 109°59′E
- 127 O12 **Ketchenery** *prev.* Sovetskoye. Respublika Kalmykiya, SW Russian Federation 47°18′N 44°31′E
- 39 Y14 **Ketchikan** Revillagigedo Island, Alaska, USA 55°21′N 131°39′W
- 33 O14 **Ketchum** Idaho, NW USA 43°40′N 114°24′W
- **Kete/Kete Krakye** *see* Kete-Krachi
- 77 Q15 **Kete-Krachi** *var.* Kete, Kete Krakye. E Ghana 07°50′N 00°03′W
- 98 L9 **Ketelmeer** *channel* E Netherlands
- 149 P17 **Keti Bandar** Sind, SE Pakistan 23°55′N 67°31′E
- 77 S16 **Kétou** SE Benin 07°25′N 02°36′E
- 110 M7 **Kętrzyn** *Ger.* Rastenburg. Warmińsko-Mazurskie, NE Poland 54°05′N 21°24′E
- 97 N20 **Kettering** C England, United Kingdom 52°24′N 00°44′W
- 31 R14 **Kettering** Ohio, N USA 39°41′N 84°10′W
- 18 F13 **Kettle Creek** ≈ Pennsylvania, NE USA
- 32 L7 **Kettle Falls** Washington, NW USA 48°36′N 118°03′W
- 14 D16 **Kettle Point** *headland* Ontario, S Canada 43°12′N 82°01′W
- 29 V6 **Kettle River** ≈ Minnesota, N USA
- 186 B7 **Ketu** ≈ W Papua New Guinea
- 18 G10 **Keuka Lake** ◎ New York, NE USA
- **Keupriya** *see* Primorsko
- 93 L17 **Keuruu** Keski-Suomi, C Finland 62°15′N 24°34′E
- **Kevevára** *see* Kovin
- 92 L9 **Kevo** *Lapp.* Geavvú. Lappi, N Finland 69°42′N 27°08′E
- 44 M6 **Kew** North Caicos, N Turks and Caicos Islands 21°52′N 71°57′W
- 30 K11 **Kewanee** Illinois, N USA 41°15′N 89°55′W
- 31 N7 **Kewaunee** Wisconsin, N USA 44°27′N 87°31′W
- 30 M3 **Keweenaw Bay** ◎ Michigan, N USA
- 31 N2 **Keweenaw Peninsula** *peninsula* Michigan, N USA
- 31 N2 **Keweenaw Point** *peninsula* Michigan, N USA
- 29 N12 **Keya Paha River** ≈ Nebraska/South Dakota, N USA
- **Keyaygyr** *see* Këk-Aygyr
- 23 Z16 **Key Biscayne** Florida, SE USA 25°41′N 80°09′W
- 26 G8 **Keyes** Oklahoma, C USA 36°48′N 102°15′W
- 23 Y17 **Key Largo** Key Largo, Florida, SE USA 25°06′N 80°25′W
- 21 U3 **Keyser** West Virginia, NE USA 39°25′N 78°59′W
- 27 O9 **Keystone Lake** ▣ Oklahoma, C USA
- 36 L16 **Keystone Peak** ▲ Arizona, SW USA 31°52′N 111°12′W
- **Keystone State** *see* Pennsylvania
- 21 U7 **Keysville** Virginia, NE USA 37°02′N 78°28′W
- 27 T3 **Keytesville** Missouri, C USA 39°25′N 92°56′W
- 23 W17 **Key West** Florida Keys, Florida, SE USA 24°34′N 81°48′W
- 127 T1 **Kez** Udmurtskaya Respublika, NW Russian Federation 57°55′N 53°42′E
- **Kezdivásárhely** *see* Târgu Secuiesc
- 122 M12 **Kezhma** Krasnoyarskiy Kray, C Russian Federation 58°57′N 101°00′E
- 111 L18 **Kežmarok** *Ger.* Käsmark, *Hung.* Késmárk. Prešovský Kraj, E Slovakia 49°09′N 20°25′E
- **Kfar Saba** *see* Kfar Sava
- 138 F10 **Kfar Sava** *var.* Kfar Saba; *prev.* Kefar Sava. Central, C Israel 32°11′N 34°58′E
- 83 F20 **Kgalagadi** ◐ *district* SW Botswana
- 83 I20 **Kgatleng** ◐ *district* SE Botswana
- 188 F8 **Kgkeklau** Babeldaob, N Palau
- 125 R6 **Khabarikha** *var.* Chabaricha. Respublika Komi, NW Russian Federation 65°52′N 52°19′E
- 123 S14 **Khabarovsk** Khabarovskiy Kray, SE Russian Federation 48°32′N 135°08′E
- 123 R11 **Khabarovskiy Kray** ◐ *territory* E Russian Federation
- 141 W7 **Khabb** Abū Ẓaby, E United Arab Emirates 24°39′N 55°43′E
- **Khabour, Nahr al** *see* Khābūr, Nahr al
- **Khabura** *see* Al Khābūrah
- 139 N2 **Khābūr, Nahr al** *var.* Nahr al Khabour. ≈ Syria/Turkey
- **Khachmas** *see* Xaçmaz
- 80 B12 **Khadari** ≈ W Sudan
- **Khadera** *see* Hadera
- 141 X12 **Khādhil** *var.* Khudal. SE Oman 18°48′N 56°48′E
- 155 E14 **Khadki** *prev.* Kirkee. Mahārāshtra, W India 18°34′N 73°52′E
- 126 L14 **Khadyzhensk** Krasnodarskiy Kray, SW Russian Federation 44°26′N 39°31′E
- **Khadzhiyska Reka** *see* Hadzhiyska Reka
- 117 P10 **Khadzhybeys'kyy Lyman** ◎ SW Ukraine
- 138 K3 **Khafsah** Ḥalab, N Syria 36°16′N 38°03′E
- 152 M13 **Khāga** Uttar Pradesh, N India 25°47′N 81°05′E
- 153 Q13 **Khagaria** Bihār, NE India 25°31′N 86°27′E
- 149 Q13 **Khairpur** Sind, SE Pakistan 27°30′N 68°50′E
- 122 K13 **Khakasiya, Respublika** *prev.* Khakasskaya Avtonomnaya Oblast', *Eng.* Khakassia. ◐ *autonomous republic* C Russian Federation
- **Khakassia/Khakasskaya Avtonomnaya Oblast'** *see* Khakasiya, Respublika
- 167 N9 **Kha Khaeng, Khao** ▲ W Thailand 16°13′N 99°03′E
- 83 G20 **Khakhea** *var.* Kakia. Southern, S Botswana 24°41′S 23°29′E
- **Khalach** *see* Halaç
- **Khalándrion** *see* Chalándri
- 75 T7 **Khalīj as Sallūm** *Ar.* Gulf of Salûm. *gulf* Egypt/Libya
- 75 X8 **Khalīj as Suways** *var.* Suez, Gulf of. *gulf* NE Egypt
- 127 W7 **Khalilovo** Orenburgskaya Oblast', W Russian Federation 51°25′N 58°13′E
- **Khalkabad** *see* Xalqobod
- 142 L3 **Khalkhāl** *prev.* Herowābād. Ardabīl, NW Iran 37°36′N 48°36′E
- **Khalkidhikí** *see* Chalkidikí
- **Khalkís** *see* Chalkída
- 125 W3 **Khal'mer-Yu** Respublika Komi, NW Russian Federation 68°00′N 64°45′E
- 119 M14 **Khalopyenichy** *Rus.* Kholopenichi. Minskaya Voblasts', NE Belarus 54°31′N 28°58′E
- **Khalturin** *see* Orlov
- 141 Y10 **Khalūf** *var.* Al Khaluf. E Oman 20°27′N 57°59′E
- 154 K10 **Khamaria** Madhya Pradesh, C India 23°07′N 80°54′E
- 154 D11 **Khambhāt** Gujarāt, W India 22°19′N 72°39′E
- 154 C12 **Khambhāt, Gulf of** *Eng.* Gulf of Cambay. *gulf* W India
- 167 U10 **Khâm Đưc** *var.* Phươc Sơn. Quang Nam-Đa Năng, C Vietnam 15°28′N 107°49′E
- 154 G12 **Khāmgaon** Mahārāshtra, C India 20°41′N 76°34′E
- 141 O14 **Khamir** *var.* Khamr. W Yemen 16°N 43°56′E
- 141 N12 **Khamīs Mushayṭ** *var.* Hamïs Musait. 'Asīr, SW Saudi Arabia 18°19′N 42°41′E
- 155 J16 **Khammam** Telangana, India 17°16′N 80°13′E
- 123 P10 **Khampa** Respublika Sakha (Yakutiya), NE Russian Federation 63°43′N 123°02′E
- **Khamr** *see* Khamir
- 83 C19 **Khan** ≈ W Namibia
- 149 Q2 **Khānābād** Kunduz, NE Afghanistan 36°42′N 69°08′E
- **Khân Abou Châmâte/Khan Abou Ech Cham** *see* Khān Abū Shāmāt
- 138 I7 **Khān Abū Shāmāt** *var.* Khân Abou Châmâte, Khan Abou Ech Cham. Rīf Dimashq, W Syria 33°43′N 36°56′E
- **Khān al Baghdādī** *see* Al Baghdādī
- **Khān al Maḥāwīl** *see* Al Maḥāwīl
- 139 T7 **Khān al Mashāhidah** Baghdād, C Iraq 33°40′N 44°15′E
- 139 T10 **Khān al Muṣallá** An Najaf, S Iraq 32°09′N 44°20′E
- 139 U6 **Khānaqīn** Diyālá, E Iraq 34°22′N 45°22′E
- 139 T11 **Khān ar Ruḥbah** An Najaf, S Iraq 31°42′N 44°18′E
- 139 P2 **Khān as Sūr** Nīnawá, N Iraq 36°28′N 41°36′E
- 139 T8 **Khān Āzād** Baghdād, C Iraq 33°08′N 44°21′E
- 154 N13 **Khandaparha** *prev.* Khandpara. Odisha, E India 20°15′N 85°11′E
- **Khandpara** *see* Khandaparha
- 149 T2 **Khandūd** *var.* Khandud, Wakhan. Badakhshān, NE Afghanistan 36°57′N 72°19′E
- **Khandud** *see* Khandūd
- 154 G11 **Khandwa** Madhya Pradesh, C India 21°49′N 76°23′E
- 123 R10 **Khandyga** Respublika Sakha (Yakutiya), NE Russian Federation 62°39′N 135°30′E
- 149 T10 **Khānewāl** Punjab, NE Pakistan 30°18′N 71°56′E
- 149 S10 **Khāngarh** Punjab, E Pakistan 29°57′N 71°14′E
- **Khanh Hung** *see* Soc Trăng
- **Khaniá** *see* Chaniá
- **Khanka** *see* Xonqa
- 163 Z8 **Khanka, Lake** *var.* Hsing-K'ai Hu, Lake Hanka, *Chin.* Xingkai Hu, *Rus.* Ozero Khanka. ◎ China/Russian Federation
- **Khanka, Ozero** *see* Khanka, Lake
- **Khankendi** *see* Xankändi
- 123 O9 **Khannya** ≈ NE Russian Federation
- 144 D10 **Khan Ordasy** *prev.* Urda. Zapadnyy Kazakhstan, W Kazakhstan 48°52′N 47°31′E
- 149 S12 **Khānpur** Punjab, E Pakistan 28°31′N 70°30′E
- 138 I4 **Khān Shaykhūn** *var.* Khan Sheikhun. Idlib, NW Syria 35°27′N 36°38′E
- **Khan Sheikhun** *see* Khān Shaykhūn
- **Khanshyngghys** *see* Khrebet Khanshyngys
- 145 S15 **Khantau** Zhambyl, S Kazakhstan 44°13′N 73°47′E
- 145 W16 **Khan Tengri, Pik** ▲ SE Kazakhstan 42°17′N 80°11′E
- **Khan-Tengri, Pik** *see* Hantengri Feng
- **Khanthabouli** *see* Savannakhét
- 127 V8 **Khanty-Mansiysk** *prev.* Ostyako-Voguls'k. Khanty-Mansiyskiy Avtonomnyy Okrug-Yugra, C Russian Federation 61°01′N 69°E
- 125 V8 **Khanty-Mansiyskiy Avtonomnyy Okrug-Yugra** ◐ *autonomous district* C Russian Federation
- 139 R4 **Khānūqah** Nīnawýé, C Iraq 35°25′N 43°15′E
- 138 E11 **Khān Yūnis** *var.* Khān Yūnus. S Gaza Strip 31°21′N 34°18′E
- **Khān Yūnus** *see* Khān Yūnis
- **Khanzi** *see* Ghanzi
- **Khān Zūr** *see* Xan Sûr
- 167 N10 **Khao Laem Reservoir** ▣ W Thailand
- 123 O14 **Khapcheranga** Zabaykal'skiy Kray, S Russian Federation 49°46′N 112°21′E
- 127 Q12 **Kharabali** Astrakhanskaya Oblast', SW Russian Federation 47°28′N 47°14′E
- 153 R16 **Kharagpur** West Bengal, NE India 22°30′N 87°19′E
- 139 V11 **Kharā'ib 'Abd al Karīm** Al Muthanná, S Iraq 31°07′N 45°33′E
- 143 Q8 **Kharānaq** Yazd, C Iran 31°54′N 54°21′E
- **Kharbin** *see* Harbin
- **Kharchi** *see* Mārwār
- 146 H13 **Khardzhagaz** Ahal Welaýaty, C Turkmenistan 37°54′N 60°10′E
- 154 F11 **Khargon** Madhya Pradesh, C India 21°49′N 75°39′E
- 149 V7 **Khāriān** Punjab, NE Pakistan 32°52′N 73°52′E
- 117 V5 **Kharkiv** *Rus.* Khar'kov. Kharkivs'ka Oblast', NE Ukraine 50°N 36°14′E
- 117 V5 **Kharkiv** ✕ Kharkivs'ka Oblast', E Ukraine 49°54′N 36°20′E
- **Kharkiv** *see* Kharkivs'ka Oblast'
- 117 U5 **Kharkivs'ka Oblast'** *var.* Kharkiv, *Rus.* Khar'kovskaya Oblast'. ◐ *province* E Ukraine
- **Khar'kov** *see* Kharkiv
- **Khar'kovskaya Oblast'** *see* Kharkivs'ka Oblast'
- 124 L3 **Kharlovka** Murmanskaya Oblast', NW Russian Federation 68°47′N 37°09′E
- **Kharmanli** *see* Harmanli
- **Kharmanliyska Reka** *see* Harmanliyska Reka
- 124 M13 **Kharovsk** Vologodskaya Oblast', NW Russian Federation 59°57′N 40°05′E
- 80 F9 **Khartoum** *var.* El Khartûm, Khartum. ● (Sudan) Khartoum, C Sudan 15°33′N 32°32′E
- 80 F9 **Khartoum** ◐ *state* NE Sudan
- 80 F9 **Khartoum** ✕ Khartoum, C Sudan 15°36′N 32°37′E
- 80 F9 **Khartoum North** Khartoum, C Sudan 15°38′N 32°33′E
- 117 X8 **Khartsyz'k** *Rus.* Khartsyzsk. Donets'ka Oblast', SE Ukraine 48°01′N 38°10′E
- 117 X8 **Khartsyz'k** Donets'ka Oblast', E Ukraine 48°01′N 38°10′E
- **Khartsyzsk** *see* Khartsyz'k
- **Khartum** *see* Khartoum
- **Kharwazawk** *see* Xêrzok
- **Khasab** *see* Al Khaṣab
- 123 S15 **Khasan** Primorskiy Kray, SE Russian Federation 42°24′N 130°45′E
- 127 P16 **Khasavyurt** Respublika Dagestan, SW Russian Federation 43°16′N 46°33′E
- 143 W12 **Khāsh** *prev.* Vāsht. Sīstān va Balūchestān, SE Iran 28°15′N 61°11′E
- 148 K8 **Khāsh, Dasht-e** *Eng.* Khash Desert. *desert* SW Afghanistan
- **Khash Desert** *see* Khāsh, Dasht-e
- **Khashim Al Qirba/Khashm al Qirbah** *see* Khashm el Girba
- 80 H9 **Khashm el Girba** *var.* Khashim Al Qirba, Khashm al Qirbah. Kassala, E Sudan 15°00′N 35°59′E
- 138 G14 **Khashsh, Jabal al** ▲▲ S Jordan
- 137 S10 **Khashuri** C Georgia 41°59′N 43°36′E
- 153 V13 **Khāsi Hills** *hill range* NE India
- **Khaskovo** *see* Haskovo
- **Khaskovo** *see* Haskovo
- 122 M7 **Khatanga** ≈ N Russian Federation
- **Khatanga, Gulf of** *see* Khatangskiy Zaliv
- 123 N7 **Khatangskiy Zaliv** *var.* Gulf of Khatanga. *bay* N Russian Federation
- 141 W7 **Khatmat al Malāḥah** N Oman 24°58′N 56°22′E
- 143 S16 **Khaṭmat al Malāḥah** Ash Shāriqah, E United Arab Emirates
- 123 V7 **Khatyrka** Chukotskiy Avtonomnyy Okrug, NE Russian Federation 62°03′N 175°09′E
- **Khauz-Khan** *see* Hanhowuz
- **Khauzkhanskoye Vodoranilishche** *see* Hanhowuz Suw Howdany
- **Khavaling** *see* Khovaling
- **Khavast** *see* Xovos
- 139 W10 **Khawrah, Nahr al** ≈ S Iraq
- **Khawr Barakah** *see* Barka
- 141 W7 **Khawr Fakkān** *var.* Khor Fakkan. Ash Shāriqah, NE United Arab Emirates 25°22′N 56°19′E
- 140 L6 **Khaybar** Al Madīnah, NW Saudi Arabia 25°53′N 39°16′E
- **Khaybar, Kowtal-e** *see* Khyber Pass
- 147 S11 **Khaydarkan** *var.* Khaydarken. Batkenskaya Oblast', SW Kyrgyzstan 39°56′N 71°17′E
- **Khaydarken** *see* Khaydarkan
- 125 U2 **Khaypudyrskaya Guba** *bay* NW Russian Federation
- **Khayrūzuk** *see* Xêrzok
- **Khazar, Baḥr-e/Khazar, Daryā-ye** *see* Caspian Sea
- **Khazarosp** *see* Hazorasp
- **Khazretishi, Khrebet** *see* Hazratishoh, Qatorkŭhi
- 75 X11 **Khazzān Aswān** *var.* Aswan Dam. *dam* SE Egypt
- **Khelat** *see* Kālat
- 74 F6 **Khemisset** NW Morocco 33°52′N 06°04′W
- 167 R10 **Khemmarat** *var.* Kemarat. Ubon Ratchathani, E Thailand 16°03′N 105°11′E
- 74 L6 **Khenchela** *var.* Khenchla. NE Algeria 35°22′N 07°09′E
- **Khenchla** *see* Khenchela
- 74 G7 **Khénifra** C Morocco 32°59′N 05°37′W
- **Khersān, Rūd-e** *see* Garm, Āb-e
- 117 R10 **Kherson** Khersons'ka Oblast', S Ukraine 46°39′N 32°38′E
- **Kherson** *see* Khersons'ka Oblast'
- 117 S14 **Khersones, Mys** *Rus.* Mys Khersonesskiy. *headland* S Ukraine 44°34′N 33°24′E
- **Khersonesskiy, Mys** *see* Khersones, Mys
- 117 R10 **Khersons'ka Oblast'** *var.* Kherson, *Rus.* Khersonskaya Oblast'. ◐ *province* S Ukraine
- **Khersonskaya Oblast'** *see* Khersons'ka Oblast'
- 122 L8 **Kheta** ≈ N Russian Federation
- 149 U7 **Khewra** Punjab, E Pakistan 32°41′N 73°04′E
- **Khiam** *see* El Khiyam
- 124 J4 **Khibiny** ▲▲ NW Russian Federation
- 126 K3 **Khimki** Moskovskaya Oblast', W Russian Federation 55°57′N 37°48′E
- 147 S12 **Khingov** *Rus.* Obi-Khingou. ≈ C Tajikistan
- **Khíos** *see* Chíos
- 149 R15 **Khipro** Sind, SE Pakistan 25°50′N 69°24′E
- 139 S10 **Khirr, Wādī al** *dry watercourse* S Iraq
- **Khisarya** *see* Hisarya
- **Khiva/Khiwa** *see* Xiva
- 167 N9 **Khlong Khlung** Kamphaeng Phet, W Thailand 16°15′N 99°41′E
- 167 N15 **Khlong Thom** Krabi, SW Thailand 07°55′N 99°09′E
- 167 P12 **Khlung** Chantaburi, S Thailand 12°25′N 102°12′E
- **Khmel'nik** *see* Khmil'nyk
- **Khmel'nitskaya Oblast'** *see* Khmel'nyts'ka Oblast'
- **Khmel'nitskiy** *see* Khmel 'nyts'kyy
- 116 K5 **Khmel'nyts'ka Oblast'** *var.* Khmel'nyts'kyy, *Rus.* Khmel'nitskaya Oblast'; *prev.* Kamenets-Podol'skaya Oblast'. ◐ *province* NW Ukraine
- 116 L6 **Khmel 'nyts'kyy** *Rus.* Khmel'nitskiy; *prev.* Proskurov. Khmel'nyts'ka Oblast', W Ukraine 49°24′N 26°59′E
- **Khmel'nyts'kyy** *see* Khmel'nyts'ka Oblast'
- 116 M6 **Khmil'nyk** *Rus.* Khmel'nik. Vinnyts'ka Oblast', C Ukraine 49°36′N 27°59′E
- **Khobda** *see* Kobda
- 137 R9 **Khobi** W Georgia 42°20′N 41°54′E
- 119 P15 **Khodasy** *Rus.* Khodosy. Mahilyowskaya Voblasts', E Belarus 53°56′N 31°29′E
- 116 I6 **Khodoriv** *Pol.* Chodorów, *Rus.* Khodorov. L'vivs'ka Oblast', NW Ukraine 49°20′N 24°19′E
- **Khodorov** *see* Khodoriv
- **Khodosy** *see* Khodasy
- **Khodzhakala** *see* Hojagala
- **Khodzhambas** *see* Hojambaz
- **Khodzhent** *see* Khujand
- **Khodzheyli** *see* Xo'jayli
- **Khoi** *see* Khvoy
- **Khojend** *see* Khujand
- **Khokand** *see* Qo'qon
- 126 L8 **Khokhol'skiy** Voronezhskaya Oblast', W Russian Federation 51°33′N 38°43′E
- 167 P10 **Khok Samrong** Lop Buri, C Thailand 15°03′N 100°44′E
- 124 H15 **Kholm** Novgorodskaya Oblast', W Russian Federation 57°10′N 31°06′E
- **Kholm** *see* Khulm
- **Kholm** *see* Chełm
- **Kholmech'** *see* Kholmyech
- 123 T13 **Kholmsk** Ostrov Sakhalin, Sakhalinskaya Oblast', SE Russian Federation 46°57′N 142°10′E
- 119 O19 **Kholmyech** *Rus.* Kholmech'. Homyel'skaya Voblasts', SE Belarus 52°09′N 30°37′E
- **Kholon** *see* Holon
- **Kholopenichi** *see* Khalopyenichy
- 83 D19 **Khomas** ◐ *district* C Namibia
- 83 D19 **Khomas Hochland** *var.* Khomasplato. *plateau* C Namibia
- **Khomasplato** *see* Khomas Hochland
- **Khomein** *see* Khomeyn
- 142 M7 **Khomeyn** *var.* Khomein, Khumain. Markazī, W Iran 33°38′N 50°03′E
- 143 N8 **Khomeynīshahr** *prev.* Homāyūnshahr. Eşfahān, C Iran 32°42′N 51°28′E
- **Khoms** *see* Al Khums
- **Khong Sedone** *see* Muang Khôngxédôn
- 167 Q9 **Khon Kaen** *var.* Muang Khon Kaen. Khon Kaen, E Thailand 16°25′N 102°50′E
- **Khonqa** *see* Xonqa
- 167 Q9 **Khon San** Khon Kaen, E Thailand 16°40′N 101°51′E
- 123 R8 **Khonuu** Respublika Sakha (Yakutiya), NE Russian Federation 66°24′N 143°15′E
- 127 N8 **Khopër** *var.* Khoper. ≈ SW Russian Federation
- **Khoper** *see* Khopër
- 123 S14 **Khor** Khabarovskiy Kray, SE Russian Federation 47°44′N 134°48′E
- 143 U9 **Khorāsān-e Jonūbī** *off.* Ostan-e Khorāsān-e Jonūbī. ◐ *province* E Iran
- 143 U6 **Khorāsān-e Raẕavī** *off.* Ostān-e Khorāsān-e Raẕavī, *var.* Khorassan, Khurasan. ◐ *province* NE Iran
- **Khorāsān-e Raẕavī, Ostān-e** *see* Khorāsān-e Raẕavī
- 143 S3 **Khorāsān-e Shomālī** *off.* Ostan-e Khorāsān-e Shomālī. ◐ *province* NE Iran
- **Khorassan** *see* Khorāsān-e Raẕavī
- **Khorat** *see* Nakhon Ratchasima
- 154 O13 **Khordha** *prev.* Khurda. Odisha, E India 20°10′N 85°42′E
- 125 U4 **Khorey-Ver** Nenetskiy Avtonomnyy Okrug, NW Russian Federation 67°25′N 58°05′E
- **Khorezmskaya Oblast'** *see* Xorazm Viloyati
- **Khor Fakkan** *see* Khawr Fakkān
- **Khorgos** *see* Korgas
- 123 N13 **Khorinsk** Respublika Buryatiya, S Russian Federation 52°13′N 109°52′E
- 83 C18 **Khorixas** Kunene, NW Namibia 20°23′S 14°55′E
- 141 O17 **Khormaksar** *var.* Aden. ✕ ('Adan) SW Yemen 12°56′N 45°00′E
- **Khormal** *see* Xurmał
- **Khormuj** *see* Khvormūj
- **Khorog** *see* Khorugh
- 117 S5 **Khorol** Poltavs'ka Oblast', NE Ukraine 49°49′N 33°17′E
- 142 L7 **Khorramābād** *var.* Khurramabad. Lorestān, W Iran 33°29′N 48°21′E
- 143 R9 **Khorramdasht** Kermān, C Iran 31°41′N 56°10′E
- 142 K10 **Khorramshahr** *var.* Khurramshahr, Muhammerah; *prev.* Mohammerah. Khūzestān, SW Iran 30°30′N 48°09′E
- 147 S14 **Khorugh** *Rus.* Khorog. S Tajikistan 37°30′N 71°31′E
- **Khorvot Khalutsa** *see* Horvot Halutsa
- 127 Q12 **Khosheutovo** Astrakhanskaya Oblast', SW Russian Federation 47°04′N 47°49′E
- 149 R6 **Khōst** *prev.* Khowst. Khōst, E Afghanistan 33°22′N 69°57′E
- 149 S6 **Khōst** *prev.* Khowst. ◐ *province* E Afghanistan
- **Khotan** *see* Hotan
- **Khotimsk** *see* Khotsimsk
- **Khotin** *see* Khotyn
- 119 R16 **Khotsimsk** *Rus.* Khotimsk. Mahilyowskaya Voblasts', E Belarus 53°24′N 32°35′E
- 116 K7 **Khotyn** *Rom.* Hotin, *Rus.* Khotin. Chernivets'ka Oblast', W Ukraine 48°29′N 26°30′E
- 74 F7 **Khouribga** C Morocco 32°55′N 06°51′W
- 147 Q13 **Khovaling** *Rus.* Khavaling. SW Tajikistan 38°22′N 69°54′E
- **Khovd** *see* Hovd
- **Khowst** *see* Khōst
- **Khoy** *see* Khvoy
- 119 N20 **Khoyniki** Homyel'skaya Voblasts', SE Belarus 51°54′N 29°58′E
- **Khozretishi, Khrebet** *see* Hazratishoh, Qatorkŭhi
- 145 V11 **Khrebet Khanshyngys** *Kas.* Khanshyngghys; *prev.* Khrebet Kanchingiz. ▲▲ E Kazakhstan
- 145 X10 **Khrebet Kalba** *Kaz.* Qalba Zhotasy; *prev.* Kalbinskiy Khrebet. ▲▲ E Kazakhstan
- **Khrebet Kanchingiz** *see* Khrebet Khanshyngys
- **Khrebet Ketmen** *see* Khrebet Uzynkara
- 145 Y10 **Khrebet Naryn** *Kaz.* Naryn Zhotasy; *prev.* Narymskiy Khrebet. ▲▲ E Kazakhstan
- 145 W16 **Khrebet Uzynkara** *prev.* Khrebet Ketmen. ▲▲ SE Kazakhstan
- **Khrisoúpolis** *see* Chrysoúpoli
- 144 J10 **Khromtau** *Kaz.* Khromtaū. Aktyubinsk, W Kazakhstan 50°14′N 58°22′E
- **Khromtaū** *see* Khromtau
- **Khrysokhou Bay** *see* Chrysochoú, Kólpos
- 117 O7 **Khrystynivka** Cherkas'ka Oblast', C Ukraine 48°49′N 29°55′E
- 167 R10 **Khuang Nai** Ubon Ratchathani, E Thailand 15°22′N 104°33′E
- **Khudal** *see* Khādhil
- **Khudat** *see* Xudat
- 149 W9 **Khudiān** Punjab, E Pakistan 30°59′N 74°19′E
- **Khudzhand** *see* Khujand
- 83 G21 **Khuis** Kgalagadi, SW Botswana 26°37′S 21°50′E
- 147 Q11 **Khujand** *var.* Khodzhent, Khojend, *Rus.* Khudzhand; *prev.* Leninabad, *Taj.* Leninobod. N Tajikistan 40°17′N 69°37′E
- 167 R11 **Khukhan** Si Sa Ket, E Thailand 14°38′N 104°12′E
- 149 P2 **Khulm** *var.* Tashqurghan; *prev.* Kholm. Balkh, N Afghanistan 36°42′N 67°41′E
- 153 T16 **Khulna** Khulna, SW Bangladesh 22°48′N 89°32′E
- 153 T16 **Khulna** ◐ *division* SW Bangladesh
- **Khumain** *see* Khomeyn
- **Khums** *see* Al Khums
- 149 W2 **Khunjerāb Pass** *pass* China/ Pakistan
- **Khūnjerāb Pass** *see* Kunjirap Daban
- 153 P16 **Khunti** Jhārkhand, N India 23°02′N 85°19′E
- 167 N7 **Khun Yuam** Mae Hong Son, NW Thailand 18°54′N 97°54′E
- **Khurais** *see* Khurayş
- **Khurasan** *see* Khorāsān-e Raẕavī
- 141 R7 **Khurayş** *var.* Khurais. Ash Sharqīyah, C Saudi Arabia 25°06′N 48°03′E
- **Khurda** *see* Khordha
- 152 J11 **Khurja** Uttar Pradesh, N India 28°15′N 77°51′E
- **Khūrmāl** *see* Xurmał
- **Khurramabad** *see* Khorramābād
- **Khurramshahr** *see* Khorramshahr
- 149 U7 **Khushāb** Punjab, NE Pakistan 32°16′N 72°18′E
- 116 H8 **Khust** *var.* Husté, *Cz.* Chust, *Hung.* Huszt. Zakarpats'ka Oblast', W Ukraine 48°11′N 23°19′E
- 80 D11 **Khuwei** Western Kordofan, C Sudan 13°02′N 29°13′E
- 149 O13 **Khuzdār** Baluchistān, SW Pakistan 27°49′N 66°39′E
- 142 L9 **Khūzestān** *off.* Ostān-e Khūzestān, *var.* Khuzistan, *prev.* Arabistan; *anc.* Susiana. ◐ *province* SW Iran
- **Khūzestān, Ostān-e** *see* Khūzestān
- **Khuzistan** *see* Khūzestān
- **Khvājeh Ghār** *see* Khwājeh Ghār
- 127 Q7 **Khvalynsk** Saratovskaya Oblast', W Russian Federation 52°30′N 48°06′E
- 143 N12 **Khvormūj** *var.* Khormuj. Būshehr, S Iran 28°32′N 51°22′E
- 142 I2 **Khvoy** *var.* Khoi, Khoy. Āẕarbāyjān-e Bākhtarī, NW Iran 38°36′N 45°04′E
- **Khwajaghar/Khwaja-i-Ghar** *see* Khwājeh Ghār
- 149 R2 **Khwājeh Ghār** *var.* Khwajaghar, Khwaja-i-Ghar; *prev.* Khvājeh Ghār. Takhār, NE Afghanistan 37°08′N 69°24′E
- 149 U4 **Khyber Pakhtunkhwa** *prev.* North-West Frontier Province. ◐ *province* NW Pakistan
- 149 S5 **Khyber Pass** *var.* Kowtal-e Khaybar. *pass* Afghanistan/ Pakistan
- 186 L8 **Kia** Santa Isabel, N Solomon Islands 07°34′S 158°31′E
- 183 S10 **Kiama** New South Wales, SE Australia 34°41′S 150°49′E
- 79 O22 **Kiambi** Katanga, SE Dem. Rep. Congo 07°15′S 28°01′E
- 81 I19 **Kiambu** ◐ *county* C Kenya
- 27 Q12 **Kiamichi Mountains** ▲▲ Oklahoma, C USA
- 27 Q12 **Kiamichi River** ≈ Oklahoma, C USA
- 14 M10 **Kiamika, Réservoir** ▣ Québec, SE Canada
- **Kiamusze** *see* Jiamusi
- 39 N7 **Kiana** Alaska, USA 66°58′N 160°25′W
- **Kiangmai** *see* Chiang Mai
- **Kiang-ning** *see* Nanjing
- **Kiangsi** *see* Jiangxi
- **Kiangsu** *see* Jiangsu
- 93 M14 **Kiantajärvi** ◎ E Finland
- 115 F19 **Kiáto** *prev.* Kiáton. Pelopónnisos, S Greece 38°01′N 22°45′E
- **Kiáton** *see* Kiáto
- **Kiayi** *see* Jiayi
- 95 F22 **Kibæk** Midtjylland, W Denmark 56°03′N 08°52′E
- 67 T9 **Kibali** *var.* Uele (upper course). ≈ NE Dem. Rep. Congo
- 79 E20 **Kibangou** Niari, SW Congo 03°27′S 12°21′E
- **Kibarty** *see* Kybartai
- 92 M8 **Kiberg** Finnmark, N Norway 70°17′N 30°47′E
- 79 N20 **Kibombo** Maniema, E Dem. Rep. Congo 03°52′S 25°59′E
- 81 E20 **Kibondo** Kigoma, NW Tanzania 03°34′S 30°41′E
- 81 J15 **Kibre Mengist** *var.* Adola. Oromīya, C Ethiopia 05°50′N 39°06′E
- **Kıbrıs** *see* Cyprus
- **Kıbrıs/Kıbrıs Cumhuriyeti** *see* Cyprus
- 81 E20 **Kibungo** *var.* Kibungu. SE Rwanda 02°09′S 30°30′E
- **Kibungu** *see* Kibungo
- 113 N19 **Kičevo** SW FYR Macedonia 41°31′N 20°57′E
- 125 P13 **Kichmengskiy Gorodok** Vologodskaya Oblast', NW Russian Federation 60°00′N 45°52′E
- 30 J8 **Kickapoo River** ≈ Wisconsin, N USA

◆ Country · ● Country Capital · ◇ Dependent Territory · ○ Dependent Territory Capital · ◐ Administrative Regions · ✕ International Airport · ▲ Mountain · ▲▲ Mountain Range · Volcano · ≈ River · ◎ Lake · ▣ Reservoir

11 P16 **Kicking Horse Pass** pass Alberta/British Columbia, SW Canada
77 R9 **Kidal** Kidal, C Mali 18°22′N 01°21′E
77 Q8 **Kidal** ◈ region NE Mali
171 Q7 **Kidapawan** Mindanao, S Philippines 07°02′N 125°04′E
97 L20 **Kidderminster** C England, United Kingdom 52°23′N 02°14′W
76 I11 **Kidira** E Senegal 14°28′N 12°13′W
184 O11 **Kidnappers, Cape** headland North Island, New Zealand 41°13′S 175°15′E
100 J8 **Kiel** Schleswig-Holstein, N Germany 54°21′N 10°05′E
111 L15 **Kielce** Rus. Keltsy. Świętokrzyskie, C Poland 50°53′N 20°39′E
100 K7 **Kieler Bucht** bay N Germany
100 J7 **Kieler Förde** inlet N Germany
167 U13 **Kiên Đưc** var. Dak Lap. Đăc Lăc, S Vietnam 11°59′N 107°30′E
79 N24 **Kienge** Katanga, SE Dem. Rep. Congo 10°33′S 27°33′E
100 Q12 **Kietz** Brandenburg, NE Germany 52°33′N 14°36′E
Kiev see Kyyiv
Kiev Reservoir see Kyyivs'ke Vodoskhovyshche
76 J10 **Kiffa** Assaba, S Mauritania 16°38′N 11°23′W
115 H19 **Kifisiá** Attikí, C Greece 38°04′N 23°49′E
115 F18 **Kifisós** ≈ C Greece
139 U5 **Kifrī** At Ta'mím, N Iraq 34°44′N 44°58′E
81 D20 **Kigali** ● (Rwanda) C Rwanda 01°59′S 30°02′E
81 E20 **Kigali** ✈ C Rwanda 01°43′S 30°01′E
137 P13 **Kiğı** Bingöl, E Turkey 39°19′N 40°20′E
81 E21 **Kigoma** Kigoma, W Tanzania 04°52′S 29°36′E
81 E21 **Kigoma** ◈ region W Tanzania
38 F10 **Kihei** var. Kihei. Maui, Hawaii, USA, C Pacific Ocean 20°47′N 156°28′W
93 K17 **Kihniö** Pirkanmaa, W Finland 62°11′N 23°10′E
118 F6 **Kihnu** var. Kihnu Saar, Ger. Kühnö. island SW Estonia
Kihnu Saar see Kihnu
38 A8 **Kii Landing** Ni'ihau, Hawaii, USA, C Pacific Ocean 21°58′N 160°03′W
93 L14 **Kiiminki** Pohjois-Pohjanmaa, C Finland 65°05′N 25°47′E
164 J14 **Kii-Nagashima** var. Nagashima. Mie, Honshū, SW Japan 34°10′N 136°18′E
164 J14 **Kii-sanchi** ▲▲ Honshū, SW Japan
92 L11 **Kiistala** Lappi, N Finland 67°52′N 25°19′E
164 I15 **Kii-suidō** strait S Japan
165 V16 **Kikai-shima** island Nansei-shotō, SW Japan
112 M8 **Kikinda** Ger. Grosskikinda, Hung. Nagykikinda; prev. Velika Kikinda. Vojvodina, N Serbia 45°48′N 20°29′E
Kikládhes see Kykládes
165 Q5 **Kikonai** Hokkaidō, NE Japan 41°40′N 140°25′E
186 C8 **Kikori** Gulf, S Papua New Guinea 07°25′S 144°13′E
186 C8 **Kikori** ≈ W Papua New Guinea
165 O14 **Kikuchi** var. Kikuti. Kumamoto, Kyūshū, SW Japan 33°00′N 130°49′E
Kikuti see Kikuchi
127 N8 **Kikvidze** Volgogradskaya Oblast', SW Russian Federation 50°47′N 42°58′E
14 I10 **Kikwissi, Lac** ◎ Québec, SE Canada
79 I21 **Kikwit** Bandundu, W Dem. Rep. Congo 05°S 18°53′E
95 K15 **Kil** Värmland, C Sweden 59°30′N 13°20′E
94 N12 **Kilafors** Gävleborg, C Sweden 61°13′N 16°34′E
38 B8 **Kilauea** Kaua'i, Hawaii, USA, C Pacific Ocean 22°12′N 159°24′W
38 H12 **Kilauea Caldera** var. Kilauea Caldera. crater Hawai'i, USA, C Pacific Ocean
Kilauea Caldera see Kilauea Caldera
109 V4 **Kilb** Niederösterreich, C Austria 48°06′N 15°21′E
39 O12 **Kilbuck Mountains** ▲▲ Alaska, USA
163 Y12 **Kilchu** NE North Korea 40°58′N 129°22′E
97 F18 **Kilcock** Ir. Cill Choca. Kildare, E Ireland 53°25′N 06°40′W
183 V2 **Kilcoy** Queensland, E Australia 26°58′S 152°30′E
97 F18 **Kildare** Ir. Cill Dara. E Ireland 53°10′N 06°55′W
97 F18 **Kildare** Ir. Cill Dara. cultural region E Ireland
124 K2 **Kil'din, Ostrov** island NW Russian Federation
25 W7 **Kilgore** Texas, SW USA 32°23′N 94°52′W
Kilien Mountains see Qilian Shan
114 K9 **Kilifarevo** Veliko Tŭrnovo, N Bulgaria 43°00′N 25°36′E
81 K20 **Kilifi** Kilifi, SE Kenya 03°37′S 39°50′E
81 J21 **Kilifi** ◈ county SE Kenya
189 U9 **Kili Island** var. Kōle. island Ralik Chain, S Marshall Islands
149 V2 **Kilik Pass** pass Afghanistan/China
Kilimane see Quelimane
81 I21 **Kilimanjaro** ◈ region E Tanzania
81 I20 **Kilimanjaro** var. Uhuru Peak. ⛰ NE Tanzania 03°01′S 37°14′E
Kilimbangara see Kolombangara
Kilinailau Islands see Tulun Islands
81 K23 **Kilindoni** Pwani, E Tanzania 07°56′S 39°40′E
118 H6 **Kilingi-Nõmme** Ger. Kurkund. Pärnumaa, SW Estonia 58°07′N 24°00′E
136 M17 **Kilis** Kilis, S Turkey 36°43′N 37°07′E
136 M16 **Kilis** ◈ province S Turkey
117 N12 **Kiliya** Rom. Chilia-Nouă. Odes'ka Oblast', SW Ukraine 45°30′N 29°16′E
97 B19 **Kilkee** Ir. Cill Chaoi. Clare, W Ireland 52°41′N 09°38′W
97 E19 **Kilkenny** Ir. Cill Chainnigh. Kilkenny, S Ireland 52°39′N 07°15′W
97 E19 **Kilkenny** Ir. Cill Chainnigh. cultural region S Ireland
97 B18 **Kilkieran Bay** Ir. Cuan Chill Chiaráin. bay W Ireland
114 G13 **Kilkís** Kentrikí Makedonía, N Greece 40°59′N 22°55′E
97 C15 **Killala Bay** Ir. Cuan Chill Ala. inlet NW Ireland
11 R15 **Killam** Alberta, SW Canada 52°45′N 111°46′W
183 U3 **Killarney** Queensland, E Australia 28°18′S 152°15′E
11 W17 **Killarney** Manitoba, S Canada 49°12′N 99°40′W
14 E11 **Killarney** Ontario, S Canada 45°58′N 81°27′W
97 B20 **Killarney** Ir. Cill Airne. Kerry, SW Ireland 52°03′N 09°30′W
28 K4 **Killdeer** North Dakota, N USA 47°21′N 102°45′W
28 J4 **Killdeer Mountains** ▲ North Dakota, N USA
45 V15 **Killdeer River** ≈ Trinidad, Trinidad and Tobago
25 S9 **Killeen** Texas, SW USA 31°07′N 97°44′W
39 P6 **Killik River** ≈ Alaska, USA
11 T7 **Killinek Island** island Nunavut, NE Canada
Killini see Kyllíni
115 C19 **Killínis, Akrotírio** headland S Greece 37°55′N 21°07′E
97 D15 **Killybegs** Ir. Na Ceala Beaga. NW Ireland 54°38′N 08°27′W
Kilmain see Quelimane
96 I13 **Kilmarnock** W Scotland, United Kingdom 55°37′N 04°30′W
21 X6 **Kilmarnock** Virginia, NE USA 37°42′N 76°22′W
125 S16 **Kil'mez'** Kirovskaya Oblast', NW Russian Federation 56°55′N 51°03′E
127 S2 **Kil'mez'** Udmurtskaya Respublika, NW Russian Federation 57°04′N 51°22′E
125 R16 **Kil'mez'** ≈ NW Russian Federation
67 V11 **Kilombero** ≈ S Tanzania
92 J10 **Kilpisjärvi** Lappi, N Finland 69°03′N 20°50′E
97 B19 **Kilrush** Ir. Cill Rois. Clare, W Ireland 52°39′N 09°29′W
79 O24 **Kilwa** Katanga, SE Dem. Rep. Congo 09°22′S 28°19′E
Kilwa see Kilwa Kivinje
81 J24 **Kilwa Kivinje** var. Kilwa. Lindi, SE Tanzania 08°45′S 39°21′E
81 J24 **Kilwa Masoko** Lindi, SE Tanzania 08°55′S 39°31′E
171 T13 **Kilwo** Pulau Seram, E Indonesia 03°36′S 130°48′E
114 P12 **Kilyos** Istanbul, NW Turkey 41°15′N 29°01′E
37 V8 **Kim** Colorado, C USA 37°12′N 103°22′W
145 O9 **Kima** prev. Kiyma. Akmola, C Kazakhstan 51°37′N 67°31′E
169 U7 **Kimanis, Teluk** bay Sabah, East Malaysia
182 H8 **Kimba** South Australia 33°09′S 136°26′E
28 I15 **Kimball** Nebraska, C USA 41°16′N 103°40′W
29 O11 **Kimball** South Dakota, N USA 43°45′N 98°57′W
79 I21 **Kimbao** Bandundu, SW Dem. Rep. Congo 05°27′S 17°40′E
186 F7 **Kimbe** New Britain, E Papua New Guinea 05°36′S 150°10′E
186 G7 **Kimbe Bay** inlet New Britain, E Papua New Guinea
11 P17 **Kimberley** British Columbia, SW Canada 49°40′N 115°58′W
83 H23 **Kimberley** Northern Cape, C South Africa 28°45′S 24°46′E
180 M4 **Kimberley Plateau** plateau Western Australia
33 P15 **Kimberly** Idaho, NW USA 42°31′N 114°21′W
163 Y12 **Kimch'aek** prev. Sŏngjin. E North Korea 40°42′N 129°13′E
Kimch'ŏn see Gimcheon
Kim Hae see Gimhae
Kími see Kými
93 K20 **Kimito** Swe. Kemiö. Varsinais-Suomi, SW Finland 60°10′N 22°45′E
9 R7 **Kimmirut** prev. Lake Harbour. Baffin Island, Nunavut, NE Canada 62°48′N 69°49′W
165 R4 **Kimobetsu** Hokkaidō, NE Japan 42°47′N 140°55′E
115 I21 **Kímolos** island Kykládes, Greece, Aegean Sea
115 I21 **Kímolou Sífnou, Stenó** strait Kykládes, Greece, Aegean Sea
126 L5 **Kimovsk** Tul'skaya Oblast', W Russian Federation 53°59′N 38°34′E
Kimpolung see Câmpulung Moldovenesc
124 K16 **Kimry** Tverskaya Oblast', W Russian Federation 56°52′N 37°21′E
79 H21 **Kimvula** Bas-Congo, SW Dem. Rep. Congo 05°44′S 15°58′E
169 U6 **Kinabalu, Gunung** ▲ East Malaysia 05°52′N 116°08′E
Kinabatangan see Kinabatangan, Sungai
169 V7 **Kinabatangan, Sungai** var. Kinabatangan. ≈ East Malaysia
115 L21 **Kínaros** island Kykládes, Greece, Aegean Sea
11 O15 **Kinbasket Lake** ⊠ British Columbia, SW Canada
96 I7 **Kinbrace** N Scotland, United Kingdom 58°16′N 02°59′W
14 E14 **Kincardine** Ontario, S Canada 44°11′N 81°38′W
96 K10 **Kincardine** cultural region E Scotland, United Kingdom
79 K21 **Kinda** Kasai-Occidental, SE Dem. Rep. Congo 04°48′S 21°50′E
79 M24 **Kinda** Katanga, SE Dem. Rep. Congo 09°20′S 25°06′E
166 L3 **Kindat** Sagaing, N Myanmar (Burma) 23°42′N 94°29′E
109 V6 **Kindberg** Steiermark, C Austria 47°31′N 15°27′E
22 H8 **Kinder** Louisiana, S USA 30°29′N 92°51′W
98 H13 **Kinderdijk** Zuid-Holland, SW Netherlands 51°52′N 04°37′E
97 M17 **Kinder Scout** ▲ C England, United Kingdom 53°25′N 01°52′W
11 S16 **Kindersley** Saskatchewan, S Canada 51°29′N 109°08′W
76 I14 **Kindia** Guinée-Maritime, SW Guinea 10°12′N 12°26′W
64 B11 **Kindley Field** air base E Bermuda
29 R6 **Kindred** North Dakota, N USA 46°39′N 97°01′W
79 N20 **Kindu** prev. Kindu-Port-Empain. Maniema, C Dem. Rep. Congo 02°57′S 25°54′E
Kindu-Port-Empain see Kindu
127 S6 **Kinel'** Samarskaya Oblast', W Russian Federation 53°14′N 50°40′E
125 N15 **Kineshma** Ivanovskaya Oblast', W Russian Federation 57°28′N 42°08′E
King see King William's Town
140 K10 **King Abdul Aziz** ✈ (Makkah) Makkah, W Saudi Arabia 21°44′N 39°08′E
Kingait see Cape Dorset
21 X6 **King and Queen Court House** Virginia, NE USA 37°40′N 76°49′W
King Charles Islands see Kong Karls Land
King Christian IX Land see Kong Christian IX Land
King Christian X Land see Kong Christian X Land
35 O11 **King City** California, W USA 36°12′N 121°09′W
27 R2 **King City** Missouri, C USA 40°03′N 94°31′W
38 M16 **King Cove** Alaska, USA 55°03′N 162°19′W
26 M10 **Kingfisher** Oklahoma, C USA 35°53′N 97°56′W
King Frederik VI Coast see Kong Frederik VI Kyst
King Frederik VIII Land see Kong Frederik VIII Land
65 B24 **King George Bay** bay West Falkland, Falkland Islands
194 G3 **King George Island** var. King George Land. island South Shetland Islands, Antarctica
12 I6 **King George Islands** island group Northwest Territories, C Canada
King George Land see King George Island
124 G13 **Kingisepp** Leningradskaya Oblast', NW Russian Federation 59°23′N 28°37′E
183 N14 **King Island** island Tasmania, SE Australia
10 J15 **King Island** island British Columbia, SW Canada
King Island see Kadan Kyun
Kingissepp see Kuressaare
141 Q7 **King Khalid** ✈ (Ar Riyāḍ) Ar Riyāḍ, C Saudi Arabia 25°00′N 46°40′E
35 S2 **King Lear Peak** ▲ Nevada, W USA 41°13′N 118°30′W
195 Y8 **King Leopold and Queen Astrid Land** physical region Antarctica
180 M4 **King Leopold Ranges** ▲▲ Western Australia
36 I11 **Kingman** Arizona, SW USA 35°12′N 114°02′W
26 M6 **Kingman** Kansas, C USA 37°39′N 98°07′W
192 L7 **Kingman Reef** ◇ US unincorporated territory C Pacific Ocean
79 N20 **Kingombe** Maniema, E Dem. Rep. Congo 02°37′S 26°39′E
182 F5 **Kingoonya** South Australia 30°56′S 135°20′E
194 J10 **King Peninsula** peninsula Antarctica
39 P13 **King Salmon** Alaska, USA 58°41′N 156°39′W
35 Q6 **Kings Beach** California, W USA 39°13′N 120°02′W
35 R11 **Kingsburg** California, W USA 36°30′N 119°33′W
182 I10 **Kingscote** South Australia 35°41′S 137°36′E
King's County see Offaly
194 H2 **King Sejong** South Korean research station Antarctica 61°57′S 58°23′W
183 T9 **Kingsford Smith** ✈ (Sydney) New South Wales, SE Australia 33°58′S 151°09′E
11 P17 **Kingsgate** British Columbia, SW Canada 48°58′N 116°09′W
23 W8 **Kingsland** Georgia, SE USA 30°48′N 81°41′W
29 S13 **Kingsley** Iowa, C USA 42°35′N 95°58′W
97 O19 **King's Lynn** var. Bishop's Lynn, Kings Lynn, Lynn, Lynn Regis. E England, United Kingdom 52°45′N 00°24′E
Kings Lynn see King's Lynn
21 Q10 **Kings Mountain** North Carolina, SE USA 35°15′N 81°20′W
180 K4 **King Sound** sound Western Australia
37 N2 **Kings Peak** ▲ Utah, W USA 40°43′N 110°27′W
21 O8 **Kingsport** Tennessee, S USA 36°32′N 82°33′W
35 R11 **Kings River** ≈ California, W USA
183 P17 **Kingston** Tasmania, SE Australia 42°57′S 147°18′E
14 K14 **Kingston** Ontario, SE Canada 44°14′N 76°30′W
44 K13 **Kingston** ● (Jamaica) E Jamaica 17°58′N 76°48′W
185 C22 **Kingston** Otago, South Island, New Zealand 45°20′S 168°45′E
19 P12 **Kingston** Massachusetts, NE USA 41°59′N 70°43′W
27 S3 **Kingston** Missouri, C USA 39°36′N 94°02′W
18 K12 **Kingston** New York, NE USA 41°55′N 74°00′W
31 S14 **Kingston** Ohio, N USA 39°28′N 82°54′W
19 O13 **Kingston** Rhode Island, NE USA 41°28′N 71°31′W
20 M9 **Kingston** Tennessee, S USA 35°52′N 84°30′W
35 W12 **Kingston Peak** ▲ California, W USA 35°43′N 115°54′W
182 J11 **Kingston Southeast** South Australia 36°51′S 139°53′E
97 N17 **Kingston upon Hull** var. Hull. E England, United Kingdom 53°45′N 00°20′W
97 N22 **Kingston upon Thames** SE England, United Kingdom 51°26′N 00°18′W
45 P14 **Kingstown** ● (Saint Vincent and the Grenadines) Saint Vincent, Saint Vincent and the Grenadines 13°09′N 61°14′W
Kingstown see Dún Laoghaire
21 T13 **Kingstree** South Carolina, SE USA 33°40′N 79°50′W
64 L8 **Kings Trough** undersea feature E Atlantic Ocean 22°00′W 43°48′N
14 C18 **Kingsville** Ontario, S Canada 42°03′N 82°43′W
25 S15 **Kingsville** Texas, SW USA 27°32′N 97°53′W
21 W6 **King William** Virginia, NE USA 37°42′N 77°03′W
9 N7 **King William Island** island Nunavut, N Canada
83 I25 **King William's Town** var. King, Kingwilliamstown. Eastern Cape, S South Africa 32°53′S 27°24′E
Kingwilliamstown see King William's Town
21 T3 **Kingwood** West Virginia, NE USA 39°27′N 79°43′W
136 C13 **Kınık** İzmir, W Turkey 39°05′N 27°25′E
79 G21 **Kinkala** Pool, S Congo 04°18′S 14°49′E
165 R10 **Kinka-san** headland Honshū, C Japan 38°17′N 141°34′E
184 M8 **Kinleith** Waikato, North Island, New Zealand 38°16′S 175°53′E
95 J19 **Kinna** Västra Götaland, S Sweden 57°32′N 12°42′E
96 L8 **Kinnaird Head** var. Kinnairds Head. headland NE Scotland, United Kingdom 58°39′N 03°22′W
Kinnairds Head see Kinnaird Head
95 K20 **Kinnared** Halland, S Sweden 57°01′N 13°04′E
92 L7 **Kinnarodden** headland N Norway 71°07′N 27°40′E
Kinneret, Yam see Tiberias, Lake
155 K24 **Kinniyai** Eastern Province, NE Sri Lanka 08°30′N 81°11′E
93 L16 **Kinnula** Keski-Suomi, C Finland 63°24′N 25°E
14 I8 **Kinojévis** ≈ Québec, SE Canada
164 I14 **Kino-kawa** ≈ Honshū, SW Japan
11 U11 **Kinoosao** Saskatchewan, C Canada 57°06′N 101°02′W
99 L17 **Kinrooi** Limburg, NE Belgium 51°09′N 05°48′E
96 J11 **Kinross** C Scotland, United Kingdom 56°14′N 03°27′W
96 J11 **Kinross** cultural region C Scotland, United Kingdom
97 C21 **Kinsale** Ir. Cionn tSáile. Cork, SW Ireland 51°42′N 08°32′W
95 D14 **Kinsarvik** Hordaland, S Norway 60°22′N 06°43′E
79 G21 **Kinshasa** prev. Léopoldville. ● Kinshasa, W Dem. Rep. Congo 04°21′S 15°16′E
79 G21 **Kinshasa** off. Ville de Kinshasa, var. Kinshasa City. ◈ region (Dem. Rep. Congo) SW Dem. Rep. Congo
79 G21 **Kinshasa** ✈ Kinshasa, SW Dem. Rep. Congo 04°23′S 15°30′E
Kinshasa City see Kinshasa
117 U9 **Kins'ka** ≈ SE Ukraine
26 K6 **Kinsley** Kansas, C USA 37°55′N 99°26′W
21 W10 **Kinston** North Carolina, SE USA 35°16′N 77°35′W
77 P15 **Kintampo** W Ghana 08°06′N 01°40′W
182 B1 **Kintore, Mount** ▲ South Australia 26°30′S 130°24′E
96 G13 **Kintyre** peninsula W Scotland, United Kingdom
96 G13 **Kintyre, Mull of** headland W Scotland, United Kingdom 55°16′N 05°46′W
166 M4 **Kin-u** Sagaing, C Myanmar (Burma) 22°47′N 95°36′E
12 G8 **Kinushseo** ≈ Ontario, C Canada
11 P13 **Kinuso** Alberta, W Canada 55°19′N 115°23′W
154 I13 **Kinwat** Mahārāshtra, C India 19°37′N 78°12′E
81 F16 **Kinyeti** ▲ S South Sudan 03°56′N 32°52′E
101 I17 **Kinzig** ≈ SW Germany
Kioga, Lake see Kyoga, Lake
26 M8 **Kiowa** Kansas, C USA 37°01′N 98°29′W
27 P12 **Kiowa** Oklahoma, C USA 34°43′N 95°54′W
Kiparissía see Kyparissía
14 H10 **Kipawa, Lac** ◎ Québec, SE Canada
81 G24 **Kipengere Range** ▲▲ SW Tanzania
81 E23 **Kipili** Rukwa, W Tanzania 07°30′S 30°39′E
81 K20 **Kipini** Tana River, SE Kenya 02°30′S 40°30′E
11 V16 **Kipling** Saskatchewan, S Canada 50°04′N 102°45′W
38 M13 **Kipnuk** Alaska, USA 59°56′N 164°02′W
97 F18 **Kippure** Ir. Cipiúr. ▲ E Ireland 53°10′N 06°22′W
79 N25 **Kipushi** Katanga, SE Dem. Rep. Congo 11°45′S 27°20′E
187 N10 **Kirakira** var. Kaokaona. Makira-Ulawa, SE Solomon Islands 10°28′S 161°54′E
155 K14 **Kirandul** var. Bailādila. Chhattīsgarh, C India 18°46′N 81°18′E
155 I21 **Kīranūr** Tamil Nādu, SE India 11°37′N 79°10′E
119 N21 **Kiraw** Rus. Kirovo. Homyel'skaya Voblasts', SE Belarus 51°30′N 29°25′E
119 M17 **Kirawsk** Rus. Kirovsk; prev. Startsy. Mahilyowskaya Voblasts', E Belarus 53°16′N 29°29′E
118 F5 **Kirbla** Läänemaa, W Estonia 58°45′N 23°57′E
25 Y9 **Kirbyville** Texas, SW USA 30°39′N 93°53′W
114 M12 **Kircasalih** Edirne, NW Turkey 41°24′N 26°48′E
109 W8 **Kirchbach** var. Kirchbach in Steiermark. Steiermark, SE Austria 46°55′N 15°40′E
Kirchbach in Steiermark see Kirchbach
108 H7 **Kirchberg** Sankt Gallen, NE Switzerland 47°24′N 09°03′E
109 S5 **Kirchdorf an der Krems** Oberösterreich, N Austria 47°55′N 14°08′E
Kirchheim see Kirchheim unter Teck
101 I22 **Kirchheim unter Teck** var. Kirchheim. Baden-Württemberg, SW Germany 48°39′N 09°28′E
139 T1 **Kirdī Kawrāw, Qimmat** var. Sar-i Kōrāwa. ▲ NE Iraq 37°08′N 44°39′E
Kirdzhali see Kardzhali
123 N13 **Kirenga** ≈ S Russian Federation
123 N12 **Kirensk** Irkutskaya Oblast', C Russian Federation 57°37′N 107°54′E
Kirghizia see Kyrgyzstan
145 S16 **Kirghiz Range** Rus. Kirgizskiy Khrebet; prev. Alexander Range. ▲▲ Kazakhstan/Kyrgyzstan
Kirghiz SSR see Kyrgyzstan
Kirghiz Steppe see Saryarka
Kirgizskaya SSR see Kyrgyzstan
Kirgizskiy Khrebet see Kirghiz Range
79 I19 **Kiri** Bandundu, W Dem. Rep. Congo 01°29′S 19°00′E
Kiriath-Arba see Hebron
191 R3 **Kiribati** off. Republic of Kiribati. ◆ republic C Pacific Ocean
Kiribati, Republic of see Kiribati
136 L17 **Kırıkhan** Hatay, S Turkey 36°30′N 36°20′E
136 I13 **Kırıkkale** Kırıkkale, C Turkey 39°50′N 33°31′E
136 C10 **Kırıkkale** ◈ province C Turkey
124 L13 **Kirillov** Vologodskaya Oblast', NW Russian Federation 59°52′N 38°24′E
Kirin see Jilin
81 I19 **Kirinyaga** ◈ county C Kenya
81 I18 **Kirinyaga** prev. Mount Kenya. ⛰ Kirinyaga, C Kenya 0°02′S 37°19′E
124 H13 **Kirishi** var. Kirisi. Leningradskaya Oblast', NW Russian Federation 59°28′N 32°02′E
164 C16 **Kirishima-yama** ▲ Kyūshū, SW Japan 31°58′N 130°51′E
Kirisi see Kirishi
191 Y2 **Kiritimati** ✈ Kiritimati, E Kiribati 02°00′N 157°30′W
191 Y2 **Kiritimati** prev. Christmas Island. atoll Line Islands, E Kiribati
186 G9 **Kiriwina Island** Eng. Trobriand Island. island SE Papua New Guinea
186 G9 **Kiriwina Islands** var. Trobriand Islands. island group S Papua New Guinea
96 K12 **Kirkcaldy** E Scotland, United Kingdom 56°07′N 03°10′W
97 I14 **Kirkcudbright** S Scotland, United Kingdom 54°50′N 04°03′W
97 I14 **Kirkcudbright** cultural region S Scotland, United Kingdom
Kirkee see Khadki
95 I14 **Kirkenær** Hedmark, S Norway 60°27′N 12°04′E
92 M8 **Kirkenes** Fin. Kirkkoniemi. Finnmark, N Norway 69°43′N 30°02′E
92 J4 **Kirkjubæjarklaustur** Suðurland, S Iceland 63°46′N 18°03′W
Kirk-Kilissa see Kırklareli
Kirkkoniemi see Kirkenes
93 L20 **Kirkkonummi** Swe. Kyrkslätt. Uusimaa, S Finland 60°06′N 24°20′E
14 G7 **Kirkland Lake** Ontario, S Canada 48°10′N 80°02′W
136 C9 **Kırklareli** prev. Kirk-Kilissa. Kırklareli, NW Turkey 41°45′N 27°12′E
136 I13 **Kırklareli** ◈ province NW Turkey
185 F20 **Kirkliston Range** ▲▲ South Island, New Zealand
14 D10 **Kirkpatrick Lake** ◎ Ontario, S Canada
195 Q11 **Kirkpatrick, Mount** ▲ Antarctica 84°37′S 164°36′E
27 U2 **Kirksville** Missouri, C USA 40°12′N 92°35′W
139 S4 **Kirkūk** off. Muḥāfa at at Kirkūk; prev. At Ta'mīm. ◈ governorate NE Iraq
Kirkūk see Āltūn Kūbrī
Kirkūk, Muḥāfa at at see Kirkūk
139 U7 **Kir Kush** Diyālá, E Iraq 33°42′N 45°15′E
96 K5 **Kirkwall** NE Scotland, United Kingdom 58°59′N 02°58′W
83 H25 **Kirkwood** Eastern Cape, S South Africa 33°23′S 25°19′E
27 X5 **Kirkwood** Missouri, C USA 38°35′N 90°24′W
Kirman see Kermān
Kir Moab/Kir of Moab see Al Karak
126 I5 **Kirov** Kaluzhskaya Oblast', W Russian Federation 54°02′N 34°17′E
125 R14 **Kirov** prev. Vyatka. Kirovskaya Oblast', NW Russian Federation 58°35′N 49°39′E
Kirov see Balpyk Bi/Ust'yevoye
Kirovabad see Gäncä
Kirovabad see Panj, Tajikistan
Kirovakan see Vanadzor
Kirovo/Kirova see Kopbirlik, Kazakhstan
Kirovo see Kiraw, Belarus
Kirovo see Beshariq, Uzbekistan
125 R14 **Kirovo-Chepetsk** Kirovskaya Oblast', NW Russian Federation 58°33′N 50°06′E
Kirovogradskaya Oblast'/Kirovohrad see Kirovohrads'ka Oblast'
117 R7 **Kirovohrad** Rus. Kirovograd; prev. Kirovo, Yelizavetgrad, Zinov'yevsk. Kirovohrads'ka Oblast', C Ukraine 48°30′N 31°17′E
117 P7 **Kirovohrads'ka Oblast'** var. Kirovohrad, Rus. Kirovogradskaya Oblast'. ◈ province C Ukraine
Kirovo/Kirovograd see Kirovohrad
124 J4 **Kirovsk** Murmanskaya Oblast', NW Russian Federation 67°37′N 33°38′E
117 X7 **Kirovs'k** Luhans'ka Oblast', E Ukraine 48°40′N 38°39′E
Kirovsk see Kirawsk, Belarus
Kirovsk see Babadaýhan, Turkmenistan
122 E9 **Kirovskaya Oblast'** ◈ province NW Russian Federation
117 U13 **Kirovs'ke** Rus. Kirovskoye. Avtonomna Respublika Krym, S Ukraine 45°13′N 35°12′E
117 X8 **Kirovs'ke** Donets'ka Oblast', E Ukraine 48°12′N 38°20′E
Kirovskiy see Balpyk Bi
Kirovskiy see Ust'yevoye
Kirovskoye see Kyzyl-Adyr
Kirovskoye see Kirovs'ke
146 E11 **Kirpili** Ahal Welaýaty, C Turkmenistan 39°31′N 57°13′E
96 K10 **Kirriemuir** E Scotland, United Kingdom 56°38′N 03°01′W
125 S13 **Kirs** Kirovskaya Oblast', NW Russian Federation 59°22′N 52°20′E
127 N7 **Kirsanov** Tambovskaya Oblast', W Russian Federation 52°40′N 42°48′E
136 J14 **Kırşehir** anc. Justinianopolis. Kırşehir, C Turkey 39°09′N 34°08′E
136 I13 **Kırşehir** ◈ province C Turkey
149 P4 **Kirthar Range** ▲▲ S Pakistan
37 P9 **Kirtland** New Mexico, SW USA 36°43′N 108°21′W
92 J11 **Kiruna** Lapp. Giron. Norrbotten, N Sweden 67°50′N 20°16′E
79 M18 **Kirundu** Orientale, NE Dem. Rep. Congo 0°45′S 25°28′E
Kirun/Kirun' see Jilong
26 L3 **Kirwin Reservoir** ⊠ Kansas, C USA
127 Q4 **Kirya** Chuvashskaya Respublika, W Russian Federation 55°04′N 46°50′E
138 G8 **Kiryat Shmona** prev. Qiryat Shemona. Northern, N Israel 33°13′N 35°35′E
95 M18 **Kisa** Östergötland, S Sweden 58°N 15°39′E
165 P9 **Kisakata** Akita, Honshū, C Japan 39°12′N 139°55′E
Kisalföld see Little Alföld
79 L18 **Kisangani** prev. Stanleyville. Orientale, NE Dem. Rep. Congo 0°30′N 25°14′E
39 N12 **Kisaralik River** ≈ Alaska, USA
165 O14 **Kisarazu** Chiba, Honshū, S Japan 35°23′N 139°57′E
111 I22 **Kisbér** Komárom-Esztergom, NW Hungary 47°30′N 18°00′E
11 V17 **Kisbey** Saskatchewan, S Canada 49°41′N 102°39′W
122 J13 **Kiselevsk** Kemerovskaya Oblast', S Russian Federation 54°00′N 86°38′E
153 R13 **Kishanganj** Bihār, NE India 26°06′N 87°57′E
152 G12 **Kishangarh** Rājasthān, N India 26°33′N 74°52′E
Kishegyes see Mali Idoš
77 S15 **Kishi** Oyo, W Nigeria 09°01′N 03°53′E
Kishinev see Chişinău
Kishiözen see Saryozen
164 I14 **Kishiwada** var. Kisiwada. Ōsaka, Honshū, SW Japan 34°28′N 135°22′E
143 P14 **Kish, Jazireh-ye** var. Qey. island S Iran
145 R7 **Kishkenekol'** prev. Kzyltu, Kaz. Qyzyltū. Kokshetau, N Kazakhstan 53°39′N 72°22′E
138 G9 **Kishon, Nahal** prev. Nahal Qishon. ≈ N Israel
152 I6 **Kishtwār** Jammu and Kashmir, NW India 33°20′N 75°49′E
81 H19 **Kisii** Kisii, SW Kenya 0°40′S 34°47′E
81 H19 **Kisii** ◈ county W Kenya
81 J23 **Kisiju** Pwani, E Tanzania 07°25′S 39°20′E
Kisiwada see Kishiwada
Kisjenő see Chişineu-Criş
38 E17 **Kiska Island** island Aleutian Islands, Alaska, USA
Kiskapus see Copşa Mică
111 M22 **Kiskőrei-víztároló** ⊠ E Hungary
Kis-Küküllo see Târnava Mică
111 L24 **Kiskunfélegyháza** var. Félegyháza. Bács-Kiskun, C Hungary 46°42′N 19°52′E
111 K25 **Kiskunhalas** var. Halas. Bács-Kiskun, S Hungary 46°26′N 19°29′E
111 K24 **Kiskunmajsa** Bács-Kiskun, S Hungary 46°30′N 19°46′E
127 N15 **Kislovodsk** Stavropol'skiy Kray, SW Russian Federation 43°55′N 42°45′E
81 L18 **Kismaayo** var. Chisimayu, Kismayu, It. Chisimaio. Jubbada Hoose, S Somalia 0°05′S 42°35′E
Kismayu see Kismaayo
164 M13 **Kiso-sanmyaku** ▲▲ Honshū, S Japan
115 H24 **Kíssamos** prev. Kastélli. Kríti, Greece, E Mediterranean Sea 35°30′N 23°39′E
Kisseraing see Kanmaw Kyun
76 K14 **Kissidougou** Guinée-Forestière, S Guinea 09°15′N 10°08′W
23 X12 **Kissimmee** Florida, SE USA 28°17′N 81°24′W
23 X12 **Kissimmee, Lake** ◎ Florida, SE USA
23 X13 **Kissimmee River** ≈ Florida, SE USA
11 V13 **Kississing Lake** ◎ Manitoba, C Canada
111 L24 **Kistelek** Csongrád, SE Hungary 46°27′N 19°58′E
Kistna see Krishna
111 M23 **Kisújszállás** Jász-Nagykun-Szolnok, E Hungary 47°14′N 20°45′E
164 G12 **Kisuki** var. Unnan. Shimane, Honshū, SW Japan 35°25′N 133°15′E
81 H18 **Kisumu** prev. Port Florence. Kisumu, W Kenya 0°02′N 34°42′E
81 H18 **Kisumu** ◈ county W Kenya
Kisutzaneustadtl see Kysucké Nové Mesto
111 O20 **Kisvárda** Ger. Kleinwardein. Szabolcs-Szatmár-Bereg, E Hungary 48°13′N 22°03′E
81 J24 **Kiswere** Lindi, SE Tanzania 09°24′S 39°37′E
Kiszucaújhely see Kysucké Nové Mesto
76 K12 **Kita** Kayes, W Mali 13°00′N 09°28′W
Kitaa ◈ province W Greenland
Kita-Akita see Takanosu
Kitab see Kitob
165 Q4 **Kitahiyama** Hokkaidō, NE Japan 42°25′N 139°55′E
165 P12 **Kitaibaraki** Ibaraki, Honshū, S Japan 36°46′N 140°45′E
165 X16 **Kita-Iō-jima** Eng. San Alessandro. island SE Japan
165 Q9 **Kitakami** Iwate, Honshū, C Japan 39°18′N 141°05′E
165 P11 **Kitakata** Fukushima, Honshū, C Japan 37°38′N 139°52′E
164 D13 **Kitakyūshū** var. Kitakyûsyû. Fukuoka, Kyūshū, SW Japan 33°51′N 130°49′E
Kitakyûsyû see Kitakyūshū
81 H18 **Kitale** Trans Nzoia, W Kenya 01°01′N 35°01′E
165 U3 **Kitami** Hokkaidō, NE Japan 43°52′N 143°51′E
165 T2 **Kitami-sanchi** ▲▲ Hokkaidō, NE Japan
37 W5 **Kit Carson** Colorado, C USA 38°45′N 102°47′W
180 M12 **Kitchener** Western Australia 31°03′S 124°00′E
14 F16 **Kitchener** Ontario, S Canada 43°28′N 80°27′W
93 O17 **Kitee** Pohjois-Karjala, SE Finland 62°06′N 30°09′E
81 G16 **Kitgum** N Uganda 03°17′N 32°54′E
Kithareng see Kanmaw Kyun
Kíthira see Kýthira
Kíthnos see Kýthnos
8 L8 **Kitikmeot** ◈ cultural region Nunavut, N Canada
10 J13 **Kitimat** British Columbia, SW Canada 54°05′N 128°38′W
92 L11 **Kitinen** ≈ N Finland
147 N12 **Kitob** Rus. Kitab. Qashqadaryo Viloyati, S Uzbekistan 39°06′N 66°47′E
116 K7 **Kitsman'** Ger. Kotzman, Rom. Cozmeni, Rus. Kitsman. Chernivets'ka Oblast', W Ukraine 48°30′N 25°50′E
164 E14 **Kitsuki** var. Kituki. Ōita, Kyūshū, SW Japan 33°24′N 131°36′E
18 C14 **Kittanning** Pennsylvania, NE USA 40°48′N 79°28′W
19 P10 **Kittery** Maine, NE USA 43°05′N 70°44′W
92 L11 **Kittilä** Lappi, N Finland 67°39′N 24°53′E
109 Z4 **Kittsee** Burgenland, E Austria 48°06′N 17°03′E
81 J19 **Kitui** Kitui, S Kenya 01°25′S 38°00′E
81 J20 **Kitui** ◈ county S Kenya
Kituki see Kitsuki
81 G22 **Kitunda** Tabora, C Tanzania 06°47′S 33°13′E
10 K13 **Kitwanga** British Columbia, SW Canada 55°07′N 128°03′W
82 J13 **Kitwe** var. Kitwe-Nkana. Copperbelt, C Zambia 12°48′S 28°13′E
Kitwe-Nkana see Kitwe
109 O7 **Kitzbühel** Tirol, W Austria 47°27′N 12°23′E
109 O7 **Kitzbüheler Alpen** ▲▲ W Austria
101 J19 **Kitzingen** Bayern, SE Germany 49°45′N 10°11′E
153 Q14 **Kiul** Bihār, NE India 25°10′N 86°06′E
186 A7 **Kiunga** Western, SW Papua New Guinea 06°10′S 141°15′E
93 M16 **Kiuruvesi** Pohjois-Savo, C Finland 63°38′N 26°40′E
38 M7 **Kivalina** Alaska, USA 67°44′N 164°32′W
9 O10 **Kivalliq** ◈ cultural region Nunavut, N Canada
92 L13 **Kivalo** ridge C Finland
116 J3 **Kivertsi** Pol. Kiwerce, Rus. Kivertsy. Volyns'ka Oblast', NW Ukraine 50°50′N 25°31′E
Kivertsy see Kivertsi
93 L16 **Kivijärvi** Keski-Suomi, C Finland 63°09′N 25°04′E
95 L23 **Kivik** Skåne, S Sweden 55°40′N 14°15′E
118 J3 **Kiviõli** Ida-Virumaa, NE Estonia 59°20′N 27°00′E
Kivu, Lac see Kivu, Lake
67 U10 **Kivu, Lake** Fr. Lac Kivu. ◎ Rwanda/Dem. Rep. Congo
186 C9 **Kiwai Island** island SW Papua New Guinea
39 N8 **Kiwalik** Alaska, USA 66°01′N 161°50′W
Kiwerce see Kivertsi
Kiyev see Kyyiv
145 R10 **Kiyevka** Karaganda, C Kazakhstan 50°15′N 71°33′E
Kiyevskaya Oblast' see Kyyivs'ka Oblast'
Kiyevskoye Vodokhranilishche see Kyyivs'ke Vodoskhovyshche
136 D10 **Kıyıköy** Kırklareli, NW Turkey 41°37′N 28°07′E
Kiyma see Kima
125 V13 **Kizel** Permskiy Kray, NW Russian Federation 58°59′N 57°37′E
125 O12 **Kizema** var. Kizëma. Arkhangel'skaya Oblast', NW Russian Federation 61°06′N 44°51′E
Kizëma see Kizema
Kizilagach see Elhovo
136 H12 **Kızılcahamam** Ankara, N Turkey 40°28′N 32°37′E
136 J10 **Kızıl Irmak** ≈ C Turkey
Kizil Kum see Kyzyl Kum
137 P16 **Kızıltepe** Mardin, SE Turkey 37°12′N 40°36′E
Ki Zil Uzen see Qezel Owzan, Rūd-e
127 Q16 **Kizilyurt** Respublika Dagestan, SW Russian Federation 43°13′N 46°54′E
127 Q15 **Kizlyar** Respublika Dagestan, SW Russian Federation 43°51′N 46°39′E

◆ Country ● Country Capital ◇ Dependent Territory ○ Dependent Territory Capital ◈ Administrative Regions ✈ International Airport ▲ Mountain ▲▲ Mountain Range ⛰ Volcano ≈ River ◎ Lake ⊠ Reservoir

127 S3 **Kizner** Udmurtskaya Respublika, NW Russian Federation 56°19′N 51°37′E
Kizyl-Arvat see Serdar
Kizyl-Atrek see Etrek
Kizyl-Kaya see Gyzylgaýa
Kizyl-Su see Gyzylsuw
95 H16 **Kjerkøy** island S Norway
Kjølen see Kölen
92 L7 **Kjøllefjord** Finnmark, N Norway 70°55′N 27°19′E
92 H11 **Kjøpsvik** Lapp. Gásluokta. Nordland, C Norway 68°06′N 16°21′E
169 N12 **Klabat, Teluk** bay Pulau Bangka, W Indonesia
112 I12 **Kladanj** ⬖ Fedederacija Bosna I Hercegovina, E Bosnia and Herzegovina
171 X16 **Kladar** Papua, E Indonesia 08°14′S 137°46′E
111 C16 **Kladno** Středočeský, NW Czech Republic 50°10′N 14°05′E
112 P11 **Kladovo** Serbia, E Serbia 44°37′N 22°36′E
167 P12 **Klaeng** Rayong, S Thailand 12°48′N 101°41′E
109 T9 **Klagenfurt** Slvn. Celovec. Kärnten, S Austria 46°38′N 14°20′E
118 B11 **Klaipėda** Ger. Memel. Klaipeda, NW Lithuania 55°42′N 21°09′E
118 C11 **Klaipėda** ⬖ province W Lithuania
Klaksvig see Klaksvík
95 B18 **Klaksvík** Dan. Klaksvig. Faroe Islands 62°13′N 06°34′W
34 L2 **Klamath** California, W USA 41°31′N 124°02′W
32 H16 **Klamath Falls** Oregon, NW USA 42°14′N 121°47′W
34 M1 **Klamath Mountains** ▲▲ California/Oregon, W USA
34 L2 **Klamath River** ≈ California/Oregon, W USA
168 K9 **Klang** var. Kelang; prev. Port Swettenham. Selangor, Peninsular Malaysia 03°02′N 101°27′E
94 J13 **Klarälven** ≈ Norway/Sweden
111 B15 **Klášterec nad Ohří** Ger. Klösterle an der Eger. Ústecky Kraj, NW Czech Republic 50°24′N 13°10′E
111 B18 **Klatovy** Ger. Klattau. Plzeňský Kraj, W Czech Republic 49°24′N 13°16′E
Klattau see Klatovy
Klausenburg see Cluj-Napoca
39 Y14 **Klawock** Prince of Wales Island, Alaska, USA 55°33′N 133°06′W
98 P8 **Klazienaveen** Drenthe, NE Netherlands 52°43′N 07°E
Kleck see Klyetsk
110 H11 **Kłecko** Weilkopolskie, C Poland 52°37′N 17°27′E
110 I11 **Kleczew** Wielkopolskie, C Poland 52°22′N 18°12′E
10 L15 **Kleena Kleene** British Columbia, SW Canada 51°55′N 124°54′W
83 D20 **Klein Aub** Hardap, C Namibia 23°48′S 16°39′E
Kleine Donau see Mosoni-Duna
101 O14 **Kleine Elster** ≈ E Germany
Kleine Kokel see Târnava Mică
99 I16 **Kleine Nete** ≈ N Belgium
Kleines Ungarisches Tiefland see Little Alföld
83 E22 **Klein Karas** Karas, S Namibia 27°36′S 18°05′E
Kleinkopisch see Copşa Mică
Klein-Marien see Väike-Maarja
Kleinschlatten see Zlatna
83 D23 **Kleinsee** Northern Cape, W South Africa 29°43′S 17°03′E
Kleinwardein see Kisvárda
115 C16 **Kleisoúra** Ípeiros, W Greece 39°21′N 20°52′E
95 C17 **Klepp** Rogaland, S Norway 58°46′N 05°39′E
83 I22 **Klerksdorp** North-West, N South Africa 26°52′S 26°39′E
126 I5 **Kletnya** Bryanskaya Oblast', W Russian Federation 53°25′N 32°58′E
Kletsk see Klyetsk
101 D14 **Kleve** Eng. Cleves, Fr. Clèves; prev. Cleve. Nordrhein-Westfalen, W Germany 51°47′N 06°11′E
113 J16 **Kličevo** C Montenegro 42°45′N 18°58′E
119 M16 **Klichaw** Rus. Klichev. Mahilyowskaya Voblasts', E Belarus 53°29′N 29°21′E
Klichev see Klichaw
119 Q16 **Klimavichy** Rus. Klimovichi. Mahilyowskaya Voblasts', E Belarus 53°37′N 31°58′E
114 M7 **Kliment** Shumen, NE Bulgaria 43°37′N 27°00′E
Klimovichi see Klimavichy
93 G14 **Klimpfjäll** Västerbotten, N Sweden 65°05′N 14°50′E
126 K3 **Klin** Moskovskaya Oblast', W Russian Federation 56°19′N 36°45′E
Klina see Klinë
113 M16 **Klinë** Serb. Klina. W Kosovo 42°38′N 20°35′E
111 B15 **Klínovec** Ger. Keilberg. ▲ NW Czech Republic 50°23′N 12°57′E
95 P19 **Klintehamn** Gotland, SE Sweden 57°22′N 18°15′E
127 R8 **Klintsovka** Saratovskaya Oblast', W Russian Federation 51°42′N 49°17′E
126 H6 **Klintsy** Bryanskaya Oblast', W Russian Federation 52°46′N 32°21′E
95 K22 **Klippan** Skåne, S Sweden 56°08′N 13°10′E
92 G13 **Klippen** Västerbotten, N Sweden 65°50′N 15°07′E
121 P2 **Klírou** W Cyprus 35°01′N 33°11′E
114 I9 **Klisura** Plovdiv, C Bulgaria 42°40′N 24°28′E
95 F20 **Klitmøller** Midtjylland, NW Denmark 57°01′N 08°29′E
112 F11 **Ključ** Federacija Bosne I Hercegovine, NW Bosnia and Herzegovina 44°32′N 16°46′E
111 J14 **Kłobuck** Śląskie, S Poland 50°56′N 18°55′E
110 J11 **Kłodawa** Wielkopolskie, C Poland 52°14′N 18°55′E
111 G16 **Kłodzko** Ger. Glatz. Dolnośląskie, SW Poland 50°27′N 16°37′E
95 I14 **Kløfta** Akershus, S Norway 60°04′N 11°06′E
112 P12 **Klokočevac** Serbia, E Serbia 44°19′N 22°11′E
118 G3 **Klooga** Ger. Lodensee. Harjumaa, NW Estonia 59°18′N 24°10′E
99 F15 **Kloosterzande** Zeeland, SW Netherlands 51°22′N 04°01′E
113 L19 **Klos** var. Klosi. Dibër, C Albania 41°30′N 20°07′E
Klosi see Klos
Klösterle an der Eger see Klášterec nad Ohří
109 X3 **Klosterneuburg** Niederösterreich, NE Austria 48°19′N 16°20′E
108 J9 **Klosters** Graubünden, SE Switzerland 46°54′N 09°52′E
108 G7 **Kloten** Zürich, N Switzerland 47°27′N 08°35′E
108 G7 **Kloten** ✕ (Zürich) Zürich, N Switzerland 47°25′N 08°36′E
100 K12 **Klötze** Sachsen-Anhalt, C Germany 52°37′N 11°09′E
12 K3 **Klotz, Lac** ◎ Québec, NE Canada
101 O15 **Klotzsche** ✕ (Dresden) Sachsen, E Germany 51°06′N 13°44′E
10 H7 **Kluane Lake** ◎ Yukon, W Canada
Kluang see Keluang
111 I14 **Kluczbork** Ger. Kreuzburg, Kreuzburg in Oberschlesien. Opolskie, S Poland 50°59′N 18°13′E
39 W12 **Klukwan** Alaska, USA 59°24′N 135°49′W
Klyastitsy see Klyastsitsy
118 L11 **Klyastsitsy** Rus. Klyastitsy. Vitsyebskaya Voblasts', N Belarus 55°53′N 28°36′E
127 T5 **Klyavlino** Samarskaya Oblast', W Russian Federation 54°21′N 52°12′E
84 K9 **Klyaz'in** ≈ W Russian Federation
127 N3 **Klyaz'ma** ≈ W Russian Federation
119 J17 **Klyetsk** Pol. Kleck, Rus. Kletsk. Minskaya Voblasts', SW Belarus 53°04′N 26°38′E
147 S8 **Klyuchevka** Talasskaya Oblast', NW Kyrgyzstan 42°34′N 71°45′E
123 V10 **Klyuchevskaya Sopka, Vulkan** ⛰ E Russian Federation 56°03′N 160°38′E
95 D17 **Knaben** Vest-Agder, S Norway 58°46′N 07°04′E
95 K21 **Knäred** Halland, S Sweden 56°30′N 13°21′E
97 M16 **Knaresborough** N England, United Kingdom 54°01′N 01°35′W
114 H8 **Knezha** Vratsa, NW Bulgaria 43°29′N 24°04′E
25 O9 **Knickerbocker** Texas, SW USA 31°18′N 100°35′W
28 K5 **Knife River** ≈ North Dakota, N USA
10 K16 **Knight Inlet** inlet British Columbia, W Canada
39 S12 **Knight Island** island Alaska, USA
97 K20 **Knighton** E Wales, United Kingdom 52°20′N 03°01′W
35 O7 **Knights Landing** California, W USA 38°47′N 121°43′W
112 E13 **Knin** Šibenik-Knin, S Croatia 44°03′N 16°12′E
25 Q12 **Knippal** Texas, SW USA
109 U7 **Knittelfeld** Steiermark, C Austria 47°14′N 14°50′E
95 O15 **Knivsta** Uppsala, C Sweden 59°43′N 17°49′E
113 P14 **Knjaževac** Serbia, E Serbia 43°34′N 22°16′E
27 S4 **Knob Noster** Missouri, C USA 38°47′N 93°33′W
99 D15 **Knokke-Heist** West-Vlaanderen, NW Belgium 51°21′N 03°19′E
95 H20 **Knøsen** hill N Denmark
Knosós see Knossos
115 J25 **Knossos** Gk. Knosós. prehistoric site Kríti, Greece, E Mediterranean Sea
25 N7 **Knott** Texas, SW USA 32°21′N 101°35′W
194 K5 **Knowles, Cape** headland Antarctica 71°45′S 60°20′W
31 O11 **Knox** Indiana, N USA 41°17′N 86°37′W
29 O3 **Knox** North Dakota, N USA 48°19′N 99°43′W
18 C13 **Knox** Pennsylvania, NE USA 41°13′N 79°33′W
189 X8 **Knox Atoll** var. Ņadikdik, Narikrik. atoll Ratak Chain, SE Marshall Islands
10 H13 **Knox, Cape** headland Graham Island, British Columbia, SW Canada 54°05′N 133°02′W
25 P5 **Knox City** Texas, SW USA 33°25′N 99°49′W
195 Y11 **Knox Coast** physical region Antarctica
31 T12 **Knox Lake** ⊠ Ohio, N USA
23 T5 **Knoxville** Georgia, SE USA 32°44′N 83°58′W
30 K12 **Knoxville** Illinois, N USA 40°54′N 90°16′W
29 W15 **Knoxville** Iowa, C USA 41°19′N 93°06′W
21 N9 **Knoxville** Tennessee, S USA 35°58′N 83°55′W
197 P11 **Knud Rasmussen Land** physical region N Greenland
Knüll see Knüllgebirge
101 I16 **Knüllgebirge** var. Knüll. ▲▲ C Germany
124 I5 **Knyazhegubskoye Vodokhranilishche** ◎ NW Russian Federation
Knyazhevo see Sredishte
Knyazhitsy see Knyazhytsy
119 O15 **Knyazhytsy** Rus. Knyazhitsy. Mahilyowskaya Voblasts', E Belarus 54°10′N 30°28′E
83 G26 **Knysna** Western Cape, SW South Africa 34°03′S 23°03′E
Koartac see Quaqtaq
169 N13 **Koba** Pulau Bangka, W Indonesia 02°30′S 106°26′E
Kobanî see 'Ayn al 'Arab
164 D16 **Kobayashi** var. Kobayasi. Miyazaki, Kyūshū, SW Japan 32°00′N 130°58′E
Kobayasi see Kobayashi
144 I10 **Kobda** prev. Khobda, Novoalekseyevka. Aktyubinsk, W Kazakhstan 50°09′N 55°39′E
144 H9 **Kobda** Kaz. Ülkenqobda; prev. Bol'shaya Khobda. ≈ Kazakhstan/Russian Federation
Kobdo see Hovd
164 I13 **Kōbe** Hyōgo, Honshū, SW Japan 34°40′N 135°10′E
Kobelyaki see Kobelyaky
117 T6 **Kobelyaky** Rus. Kobelyaki. Poltavs'ka Oblast', NE Ukraine 49°10′N 34°13′E
95 J22 **København** Eng. Copenhagen; anc. Hafnia. ● (Denmark) Sjælland, København, E Denmark 55°43′N 12°34′E
76 K10 **Kobenni** Hodh el Gharbi, S Mauritania 15°58′N 09°24′W
171 T13 **Kobi** Pulau Seram, E Indonesia 02°56′S 129°53′E
101 F17 **Koblenz** prev. Coblenz, Fr. Coblence; anc. Confluentes. Rheinland-Pfalz, W Germany 50°21′N 07°36′E
108 F6 **Koblenz** Aargau, N Switzerland 47°34′N 08°16′E
Kobrin see Kobryn
171 V15 **Kobroor, Pulau** island Kepulauan Aru, E Indonesia
119 G19 **Kobryn** Rus. Kobrin. Brestskaya Voblasts', SW Belarus 52°13′N 24°21′E
39 O7 **Kobuk** Alaska, USA 66°54′N 156°52′W
39 O7 **Kobuk River** ≈ Alaska, USA
137 Q10 **Kobuleti** prev. K'obulet'i. W Georgia 41°47′N 41°47′E
K'obulet'i see Kobuleti
123 P10 **Kobyay** Respublika Sakha (Yakutiya), NE Russian Federation 63°36′N 126°33′E
136 E11 **Kocaeli** ⬖ province NW Turkey
113 P18 **Kočani** NE FYR Macedonia 41°55′N 22°25′E
112 K12 **Koceljevo** Serbia, W Serbia 44°28′N 19°49′E
109 U12 **Kočevje** Ger. Gottschee. S Slovenia 45°41′N 14°48′E
153 T12 **Koch Bihār** West Bengal, NE India 26°19′N 89°26′E
Kochchi see Cochin/Kochi
122 M9 **Kochechum** ≈ N Russian Federation
101 I20 **Kocher** ≈ SW Germany
125 T13 **Kochevo** Komi-Permyatskiy Okrug, NW Russian Federation 59°37′N 54°16′E
164 G14 **Kōchi** var. Kôti. Kōchi, Shikoku, SW Japan 33°31′N 133°30′E
164 G14 **Kōchi** off. Kōchi-ken, var. Kôti. ⬖ prefecture Shikoku, SW Japan
Kōchi-ken see Kōchi
Kochiu see Gejiu
Kochkor see Kochkorka
147 V8 **Kochkorka** Kir. Kochkor. Narynskaya Oblast', C Kyrgyzstan 42°09′N 75°42′E
125 V5 **Kochmes** Respublika Komi, NW Russian Federation 66°10′N 60°46′E
127 P15 **Kochubey** Respublika Dagestan, SW Russian Federation 44°25′N 46°33′E
115 I17 **Kochýlas** ▲ Skńyros, Vóreies Sporádes, Greece, Aegean Sea 38°50′N 24°35′E
110 O13 **Kock** Lubelskie, E Poland 51°39′N 22°26′E
81 J19 **Kodacho** spring/well S Kenya 01°52′S 39°22′E
155 K24 **Koddiyar Bay** bay NE Sri Lanka
39 Q14 **Kodiak** Kodiak Island, Alaska, USA 57°47′N 152°24′W
39 Q14 **Kodiak Island** island Alaska, USA
154 B12 **Kodīnār** Gujarāt, W India 20°44′N 70°46′E
124 M9 **Kodino** Arkhangel'skaya Oblast', NW Russian Federation 63°36′N 39°54′E
122 M12 **Kodinsk** Krasnoyarskiy Kray, C Russian Federation 58°37′N 99°18′E
80 F12 **Kodok** Upper Nile, NE South Sudan 09°51′N 32°07′E
117 N8 **Kodyma** Odes'ka Oblast', SW Ukraine 48°05′N 29°09′E
Koedoes see Kudus
99 B17 **Koekelare** West-Vlaanderen, W Belgium 51°07′N 02°58′E
Koeln see Köln
Koepang see Kupang
Ko-erh-mu see Golmud
99 J17 **Koersel** Limburg, NE Belgium 51°04′N 05°17′E
83 E21 **Koës** Karas, SE Namibia 25°59′S 19°08′E
Koetai see Mahakam, Sungai
Koetaradja see Banda Aceh
36 I14 **Kofa Mountains** ▲▲ Arizona, SW USA
171 Y15 **Kofarau** Papua, E Indonesia 07°29′S 140°28′E
147 P13 **Kofarnihon** Rus. Kofarnikhon; prev. Ordzhonikidzeabad, Taj. Orjonikidzeobod, Yangi-Bazar. W Tajikistan 38°32′N 68°56′E
147 P14 **Kofarnihon** Rus. Kafirnigan. ≈ SW Tajikistan
Kofarnikhon see Kofarnihon
114 M11 **Kofçaz** Kırklareli, NW Turkey 41°58′N 27°12′E
115 J25 **Kófinas** ▲ Kríti, Greece, E Mediterranean Sea 34°58′N 25°03′E
121 P3 **Kofínou** var. Kophinou. S Cyprus 34°49′N 33°24′E
109 V8 **Köflach** Steiermark, SE Austria 47°04′N 15°04′E
77 Q17 **Koforidua** SE Ghana 06°01′N 00°12′W
164 H12 **Kōfu** Tottori, Honshū, SW Japan 35°16′N 133°31′E
164 M13 **Kōfu** var. Kôhu. Yamanashi, Honshū, S Japan 35°41′N 138°33′E
81 F22 **Koga** Tabora, C Tanzania 06°08′S 32°20′E
Kogălniceanu see Mihail Kogălniceanu
13 P6 **Kogaluk** ≈ Newfoundland and Labrador, E Canada
12 J4 **Kogaluk, Riviére** ≈ Québec, NE Canada
145 V14 **Kogaly** Kaz. Qoghaly; prev. Kugaly. Almaty, SE Kazakhstan 44°30′N 78°40′E
122 I10 **Kogalym** Khanty-Mansiyskiy Avtonomnyy Okrug-Yugra, C Russian Federation 62°13′N 74°34′E
95 J23 **Køge** Sjælland, E Denmark 55°28′N 12°12′E
95 J23 **Køge Bugt** bay E Denmark
77 U16 **Kogi** ⬖ state C Nigeria
146 L11 **Kogon** Rus. Kagan. Buxoro Viloyati, C Uzbekistan 39°47′N 64°29′E
Kŏgŭm-do see Geogeum-do
Kőhalom see Rupea
149 T6 **Kohāt** Khyber Pakhtunkhwa, NW Pakistan 33°37′N 71°30′E
142 L10 **Kohgīlūyeh va Bowyer Aḩmad** off. Ostān-e Kohgīlūyeh va Bowyer Aḩmad, var. Boyer Ahmadī va Kohkīlūyeh. ⬖ province SW Iran
Kohgīlūyeh va Bowyer Aḩmad, Ostān-e see Kohgīlūyeh va Bowyer Aḩmad
118 G4 **Kohila** Ger. Koil. Raplamaa, NW Estonia 59°09′N 24°45′E
153 X13 **Kohīma** state capital Nāgāland, E India 25°40′N 94°08′E
Koh I Noh see Büyükağrı Dağı
Kohsān see Kūhestān
118 J3 **Kohtla-Järve** Ida-Virumaa, NE Estonia 59°22′N 27°21′E
Kôhu see Kōfu
Kohyl'nyk see Cogîlnic
165 N11 **Koide** Niigata, Honshū, C Japan 37°13′N 138°58′E
10 G7 **Koidern** Yukon, W Canada 61°55′N 140°22′W
76 J15 **Koidu** E Sierra Leone 08°40′N 11°01′W
118 I4 **Koigi** Järvamaa, C Estonia 58°51′N 25°45′E
Koil see Kohila
172 H13 **Koimbani** Grande Comore, NW Comoros 11°37′S 43°23′E
93 O16 **Koitere** ◎ E Finland
Koivisto see Primorsk
Kōje-do see Geogeum-do
80 J13 **K'ok'a Hāyk'** ◎ C Ethiopia
Kokand see Qo'qon
182 F6 **Kokatha** South Australia 31°17′S 135°16′E
146 M10 **Ko'kcha** Rus. Kokcha. Buxoro Viloyati, C Uzbekistan 40°30′N 64°58′E
Kokcha see Ko'kcha
Kokchetav see Kokshetau
93 K18 **Kokemäenjoki** ≈ SW Finland
171 W14 **Kokenau** var. Kokonau. Papua, E Indonesia 04°38′S 136°24′E
83 E22 **Kokerboom** Karas, SE Namibia 28°11′S 19°25′E
119 N14 **Kokhanava** Rus. Kokhanovo. Vitsyebskaya Voblasts', NE Belarus 54°28′N 29°59′E
Kokhanovichi see Kakhanavichy
Kokhanovo see Kokhanava
Kök-Janggak see Kok-Yangak
93 K16 **Kokkola** Swe. Karleby; prev. Swe. Gamlakarleby. Keski-Pohjanmaa, W Finland 63°50′N 23°10′E
158 L3 **Kok Kuduk** spring/well N China 46°03′N 87°34′E
118 H9 **Koknese** C Latvia 56°38′N 25°27′E
77 T13 **Koko** Kebbi, W Nigeria 11°25′N 04°33′E
186 E9 **Kokoda** Northern, S Papua New Guinea 08°52′S 147°44′E
76 K12 **Kokofata** Kayes, W Mali 12°48′N 09°56′W
39 N6 **Kokolik River** ≈ Alaska, USA
31 O13 **Kokomo** Indiana, N USA 40°29′N 86°07′W
Kokonau see Kokenau
Koko Nor see Qinghai, China
Koko Nor see Qinghai Hu, China
186 H6 **Kokopo** var. Kopopo; prev. Herbertshöhe. New Britain, E Papua New Guinea 04°18′S 152°17′E
Kokpekti see Kokpekty
Kokpekti see Kokpekty
145 X10 **Kokpekty** prev. Kokpekti. Vostochnyy Kazakhstan, E Kazakhstan 48°47′N 82°28′E
145 X11 **Kokpekty** prev. Kokpekti. ≈ E Kazakhstan
39 P9 **Kokrines** Alaska, USA 64°58′N 154°42′W
39 P9 **Kokrines Hills** ▲▲ Alaska, USA
145 P17 **Koksaray** Yuzhnyy Kazakhstan, S Kazakhstan 42°34′N 68°06′E
147 X9 **Kokshaal-Tau** Rus. Khrebet Kakshaal-Too. ▲▲ China/Kyrgyzstan
145 P7 **Kokshetau** Kaz. Kökshetaū; prev. Kokchetav. Kokshetau, N Kazakhstan 53°18′N 69°25′E
Kökshetaū see Kokshetau
99 A17 **Koksijde** West-Vlaanderen, W Belgium 51°07′N 02°40′E
12 M5 **Koksoak** ≈ Québec, E Canada
83 K24 **Kokstad** KwaZulu/Natal, E South Africa 30°23′S 29°23′E
145 V14 **Koksu** Kaz. Rūdnīchnyy. Almaty, SE Kazakhstan 44°39′N 78°57′E
145 W15 **Koktal** Kaz. Köktal. Almaty, SE Kazakhstan 44°05′N 79°44′E
145 Q12 **Koktas** ≈ C Kazakhstan
Kök-Tash see Këk-Tash
Koktokay see Fuyun
147 T9 **Kok-Yangak** Kir. Kök-Janggak. Dzhalal-Abadskaya Oblast', W Kyrgyzstan 41°02′N 73°11′E
158 F9 **Kokyar** Xinjiang Uygur Zizhiqu, W China 37°41′N 77°15′E
149 O13 **Kolāchi** var. Kulachi. ≈ SW Pakistan
76 J15 **Kolahun** N Liberia 08°24′N 10°02′W
171 O14 **Kolaka** Sulawesi, C Indonesia 04°04′S 121°38′E
K'o-la-ma-i see Karamay
Kola Peninsula see Kol'skiy Poluostrov
155 H19 **Kolār** Karnātaka, E India 13°10′N 78°10′E
155 H19 **Kolār Gold Fields** Karnātaka, E India 12°56′N 78°16′E
92 K11 **Kolari** Lappi, NW Finland 67°20′N 23°51′E
111 I21 **Kolárovo** Ger. Gutta; prev. Guta, Hung. Gúta. Nitriansky Kraj, SW Slovakia 47°54′N 18°01′E
113 K16 **Kolašin** E Montenegro 42°49′N 19°32′E
Kolāyat see Shir Kolāyat
95 N15 **Kolbäck** Västmanland, C Sweden 59°33′N 16°15′E
Kolbcha see Kowbcha
197 Q15 **Kolbeinsey Ridge** undersea feature Denmark Strait/Norwegian Sea 69°00′N 17°30′W
Kolberg see Kołobrzeg
95 H15 **Kolbotn** Akershus, S Norway 62°15′N 10°24′E
111 N16 **Kolbuszowa** Podkarpackie, SE Poland 50°12′N 22°07′E
126 L3 **Kol'chugino** Vladimirskaya Oblast', W Russian Federation 56°19′N 39°24′E
76 H12 **Kolda** S Senegal 12°58′N 14°58′W
95 G23 **Kolding** Syddanmark, C Denmark 55°29′N 09°30′E
79 K20 **Kole** Kasai-Oriental, SW Dem. Rep. Congo 03°30′S 22°28′E
79 M17 **Kole** Orientale, N Dem. Rep. Congo 02°08′N 25°25′E
Kōle see Kili Island
84 F6 **Kölen** Nor. Kjølen. ▲▲ Norway/Sweden
Kolepom, Pulau see Yos Sudarso, Pulau
118 H3 **Kolga Laht** Ger. Kolko-Wiek. bay N Estonia
125 Q3 **Kolguyev, Ostrov** island NW Russian Federation
155 E16 **Kolhāpur** Mahārāshtra, SW India 16°42′N 74°13′E
151 K21 **Kolhumadulu** var. Thaa Atoll. atoll S Maldives
93 O16 **Koli** var. Kolinkylä. Pohjois-Karjala, E Finland 63°06′N 29°46′E
39 O13 **Koliganek** Alaska, USA 59°43′N 157°16′W
111 E16 **Kolín** Ger. Kolin. Střední Čechy, C Czech Republic 50°02′N 15°10′E
Kolinkylä see Koli
190 E12 **Koliu** Île Futuna, W Wallis and Futuna
118 E7 **Kolka** NW Latvia 57°44′N 22°34′E
118 E7 **Kolkasrags** prev. Eng. Cape Domesnes. headland NW Latvia 57°45′N 22°35′E
153 S16 **Kolkata** prev. Calcutta. state capital West Bengal, NE India 22°30′N 88°20′E
Kolkhozabad see Kolkhozobod
147 P14 **Kolkhozobod** Rus. Kolkhozabad; prev. Kaganovichabad, Tugalan. SW Tajikistan 37°33′N 68°34′E
Kolki/Kołki see Kolky
Kolko-Wiek see Kolga Laht
116 K3 **Kolky** Pol. Kołki, Rus. Kolki. Volyns'ka Oblast', NW Ukraine 51°05′N 25°40′E
155 G20 **Kollegāl** Karnātaka, W India 12°08′N 77°06′E
98 M5 **Kollum** Fryslân, N Netherlands 53°17′N 06°09′E
Kolmar see Colmar
101 E16 **Köln** var. Koeln, Eng./Fr. Cologne, prev. Cöln; anc. Colonia Agrippina, Oppidum Ubiorum. Nordrhein-Westfalen, W Germany 50°57′N 06°57′E
110 N9 **Kolno** Podlaskie, NE Poland 53°24′N 21°57′E
110 J12 **Koło** Wielkopolskie, C Poland 52°11′N 18°39′E
38 B8 **Kōloa** var. Koloa. Kaua'i, Hawaii, USA, C Pacific Ocean 21°54′N 159°28′W
Koloa see Kōloa
110 E7 **Kołobrzeg** Ger. Kolberg. Zachodnio-pomorskie, NW Poland 54°11′N 15°34′E
126 H4 **Kolodnya** Smolenskaya Oblast', W Russian Federation 54°57′N 32°22′E
190 E13 **Kolofau, Mont** ⛰ Île Alofi, S Wallis and Futuna 14°21′S 178°02′W
125 O14 **Kologriv** Kostromskaya Oblast', NW Russian Federation 58°49′N 44°22′E
76 L12 **Kolokani** Koulikoro, W Mali 13°35′N 08°01′W
77 N13 **Koloko** W Burkina Faso 11°06′N 05°18′W
186 K8 **Kolombangara** var. Kilimbangara, Nduke. island New Georgia Islands, NW Solomon Islands
Kolomea see Kolomyya
126 L4 **Kolomna** Moskovskaya Oblast', W Russian Federation 55°03′N 38°52′E
116 J7 **Kolomyya** Ger. Kolomea. Ivano-Frankivs'ka Oblast', W Ukraine 48°31′N 25°00′E
76 M13 **Kolondiéba** Sikasso, SW Mali 11°04′N 06°55′W
193 V15 **Kolonga** Tongatapu, S Tonga 21°07′S 175°05′W
189 U16 **Kolonia** var. Colonia. Pohnpei, E Micronesia 06°57′N 158°12′E
Kolonja see Kolonjë
113 K21 **Kolonjë** var. Kolonja. Fier, C Albania 40°49′N 19°37′E
Kolonjë see Ersekë
Kolosjoki see Nikel'
Kolotambu see Avuavu
193 U15 **Kolovai** Tongatapu, S Tonga 21°05′S 175°20′W
Kolozsvár see Cluj-Napoca
112 C9 **Kolpa** Ger. Kulpa, SCr. Kupa. ≈ Croatia/Slovenia
122 J11 **Kolpashevo** Tomskaya Oblast', C Russian Federation 58°21′N 82°44′E
124 H13 **Kolpino** Leningradskaya Oblast', NW Russian Federation 59°44′N 30°39′E
100 M10 **Kölpinsee** ◎ NE Germany
Kólpos Mórfu see Güzelyurt
146 K8 **Ko'lquduq** Rus. Kulkuduk. Navoiy Viloyati, N Uzbekistan 42°36′N 63°24′E
124 K5 **Kol'skiy Poluostrov** Eng. Kola Peninsula. peninsula NW Russian Federation
127 T6 **Koltubanovskiy** Orenburgskaya Oblast', W Russian Federation 53°00′N 52°00′E
112 L11 **Kolubara** ≈ C Serbia
Kolupchii see Gurkovo
110 K13 **Koluszki** Łódźskie, C Poland 51°44′N 19°50′E
125 T6 **Kolva** ≈ NW Russian Federation
93 E14 **Kolvereid** Nord-Trøndelag, W Norway 64°47′N 11°22′E
79 M24 **Kolwezi** Katanga, S Dem. Rep. Congo 10°43′S 25°29′E
123 S7 **Kolyma** ≈ NE Russian Federation
Kolyma Lowland see Kolymskaya Nizmennost'
Kolyma Range/Kolymskiy, Khrebet see Kolymskoye Nagor'ye
123 S7 **Kolymskaya Nizmennost'** Eng. Kolyma Lowland. lowlands NE Russian Federation
123 S7 **Kolymskoye** Respublika Sakha (Yakutiya), NE Russian Federation 68°42′N 158°46′E
123 U8 **Kolymskoye Nagor'ye** var. Khrebet Kolymskiy, Eng. Kolyma Range. ▲▲ E Russian Federation
123 V5 **Kolyuchinskaya Guba** bay NE Russian Federation
Kol'zhat see Kalzhat
114 G8 **Kom** ▲ NW Bulgaria 43°10′N 23°02′E
80 I13 **Koma** Oromīya, C Ethiopia 08°19′N 36°48′E
77 X12 **Komadugu Gana** ≈ NE Nigeria
164 M13 **Komagane** Nagano, Honshū, S Japan 35°44′N 137°54′E
79 P17 **Komanda** Orientale, NE Dem. Rep. Congo 01°25′N 29°43′E
197 U1 **Komandorskaya Basin** var. Kamchatka Basin. undersea feature SW Bering Sea 57°00′N 168°00′E
125 Pp9 **Komandorskiye Ostrova** Eng. Commander Islands. island group E Russian Federation
Kománfalva see Comăneşti
111 I22 **Komárno** Ger. Komorn, Hung. Komárom. Nitriansky Kraj, SW Slovakia 47°46′N 18°07′E
111 I22 **Komárom** Komárom-Esztergom, NW Hungary 47°43′N 18°06′E
Komárom see Komárno
111 I22 **Komárom-Esztergom** off. Komárom-Esztergom Megye. ⬖ county N Hungary
Komárom-Esztergom Megye see Komárom-Esztergom
164 K11 **Komatsu** var. Komatu. Ishikawa, Honshū, SW Japan 36°25′N 136°27′E
Komatu see Komatsu
83 D17 **Kombat** Otjozondjupa, N Namibia 19°42′S 17°45′E
Kombissiguiri see Kombissiri
77 P13 **Kombissiri** var. Kombissiguiri. C Burkina Faso 12°01′N 01°27′W
188 E10 **Komebail Lagoon** lagoon N Palau
81 F20 **Kome Island** island N Tanzania
Komeyo see Wandai
117 P10 **Kominternivs'ke** Odes'ka Oblast', SW Ukraine 46°52′N 30°56′E
125 R8 **Komi, Respublika** ⬖ autonomous republic NW Russian Federation
111 I25 **Komló** Baranya, SW Hungary 46°11′N 18°15′E
Kommunarsk see Alchevs'k
Kommunizm, Qullai see Ismoili Somoní, Qullai
186 B7 **Komo** Hela, W Papua New Guinea 06°06′S 142°52′E
170 M16 **Komodo, Pulau** island Nusa Tenggara, S Indonesia
77 N15 **Komoé** var. Komoé Fleuve. ≈ E Ivory Coast
Komoé Fleuve see Komoé
75 X11 **Kom Ombo** var. Kôm Ombo, Kawm Umbū. SE Egypt 24°26′N 32°57′E
79 F20 **Komono** Lékoumou, SW Congo 03°15′S 13°14′E
171 Y16 **Komoran** Papua, E Indonesia 08°14′S 138°51′E
171 Y16 **Komoran, Pulau** island E Indonesia
Komorn see Komárno
Komornok see Comorâşte
Komosolabad see Komsomolobod
Komotau see Chomutov
114 K13 **Komotiní** var. Gümüljina, Turk. Gümülcine. Anatolikí Makedonía kai Thráki, NE Greece 41°07′N 25°27′E
113 K16 **Komovi** ▲▲ E Montenegro
117 R8 **Kompaniyivka** Kirovohrads'ka Oblast', C Ukraine 48°16′N 32°12′E
Kompong see Kâmpóng Chhnăng
Kompong Cham see Kâmpóng Cham
Kompong Kleang see Kâmpóng Khleăng
Kompong Som see Sihanoukville
Kompong Speu see Kâmpóng Spoe
Komrat see Comrat
144 G12 **Komsomol** prev. Komsomol'skiy. Atyrau, W Kazakhstan 47°18′N 53°37′E
125 W4 **Komsomol** prev. Komsomol'skiy. Respublika Komi, NW Russian Federation 67°33′N 64°00′E
122 K14 **Komsomolets, Ostrov** island Severnaya Zemlya, N Russian Federation
144 F13 **Komsomolets, Zaliv** lake gulf SW Kazakhstan
Komsomol/Komsomolets see Karabalyk, Kostanay, Kazakhstan
147 Q12 **Komsomolobod** Rus. Komosolabad. C Tajikistan 38°51′N 69°54′E
124 M16 **Komsomol'sk** Ivanovskaya Oblast', W Russian Federation 56°58′N 40°15′E
117 S6 **Komsomol's'k** Poltavs'ka Oblast', C Ukraine 49°01′N 33°37′E
146 M11 **Komsomol'sk** Navoiy Viloyati, N Uzbekistan 40°14′N 65°10′E
Komsomol'skiy see Komsomol
Komsomol'skiy see Komsomol
123 S13 **Komsomol'sk-na-Amure** Khabarovskiy Kray, SE Russian Federation 50°32′N 136°59′E
Komsomol'sk-na-Ustyurte see Kubla-Ustyurt
144 K10 **Komsomol'skoye** Aktyubinsk, NW Kazakhstan
127 Q8 **Komsomol'skoye** Saratovskaya Oblast', W Russian Federation 50°45′N 47°00′E
145 P10 **Kon** ≈ C Kazakhstan
Kona see Kailua-Kona
124 K16 **Konakovo** Tverskaya Oblast', W Russian Federation 56°42′N 36°44′E
Konar see Kunar
143 V15 **Konārak** Sīstān va Balūchestān, SE Iran 25°26′N 60°23′E
Konarhā see Kunar
27 O11 **Konawa** Oklahoma, C USA 34°57′N 96°45′W
122 H10 **Konda** ≈ C Russian Federation
154 L13 **Kondagaon** Chhattīsgarh, C India 19°38′N 81°41′E
14 K10 **Kondiaronk, Lac** ◎ Québec, SE Canada
180 J13 **Kondinin** Western Australia 32°31′S 118°15′E
81 H21 **Kondoa** Dodoma, C Tanzania 04°54′S 35°46′E
127 P6 **Kondol'** Penzenskaya Oblast', W Russian Federation 52°49′N 45°03′E
114 N10 **Kondolovo** Burgas, E Bulgaria 42°07′N 27°43′E
171 Z16 **Kondomirat** Papua, E Indonesia 08°57′S 140°55′E
124 J10 **Kondopoga** Respublika Kareliya, NW Russian Federation 62°13′N 34°17′E
Kondoz see Kunduz
155 J17 **Kondukūr** var. Kandukur. Andhra Pradesh, E India 15°17′N 79°49′E
Kondūz see Kunduz
187 P16 **Koné** Province Nord, W New Caledonia 21°04′S 164°51′E
146 E13 **Könekesir** Balkan Welaýaty, W Turkmenistan 38°16′N 56°51′E
146 G8 **Köneürgenç** var. Köneürgench, Rus. Këneurgench; prev. Kunya-Urgench. Daşoguz Welaýaty, N Turkmenistan 42°21′N 59°09′E
77 N15 **Kong** N Ivory Coast 09°10′N 04°33′W
39 S5 **Kongakut River** ≈ Alaska, USA
197 O14 **Kong Christian IX Land** Eng. King Christian IX Land. physical region SE Greenland
197 P13 **Kong Christian X Land** Eng. King Christian X Land. physical region E Greenland
197 N13 **Kong Frederik IX Land** physical region SW Greenland
197 Q12 **Kong Frederik VIII Land** Eng. King Frederik VIII Land. physical region NE Greenland
197 N15 **Kong Frederik VI Kyst** Eng. King Frederik VI Coast. physical region SE Greenland
167 P13 **Kŏng, Kaôh** prev. Kas Kong. island SW Cambodia
92 P2 **Kong Karls Land** Eng. King Charles Islands. island group SE Svalbard
81 G14 **Kong Kong** ≈ E South Sudan
Kongo see Congo (river)
83 G16 **Kongola** Caprivi, NE Namibia 17°47′S 23°24′E
79 N21 **Kongolo** Katanga, E Dem. Rep. Congo 05°20′S 26°58′E
81 F14 **Kongor** Jonglei, E South Sudan 07°09′N 31°44′E
197 Q14 **Kong Oscar Fjord** fjord E Greenland
77 P12 **Kongoussi** N Burkina Faso 13°19′N 01°31′W
95 G15 **Kongsberg** Buskerud, S Norway 59°39′N 09°38′E
92 Q2 **Kongsøya** island Kong Karls Land, E Svalbard
95 I14 **Kongsvinger** Hedmark, S Norway 60°10′N 12°00′E
Kongtong see Pingliang
167 T11 **Kông, Tônlé** var. Xê Kong. ≈ Cambodia/Laos
158 E8 **Kongur Shan** ▲ NW China 38°39′N 75°21′E
81 I22 **Kongwa** Dodoma, C Tanzania 06°13′S 36°28′E
Kong, Xê see Kông, Tônlé
Konia see Konya
147 R11 **Konibodom** Rus. Kanibadam. N Tajikistan 40°16′N 70°20′E
111 K15 **Koniecpol nad Pilicą** Śląskie, S Poland 50°47′N 19°45′E
Konieh see Konya
Königgrätz see Hradec Králové
Königinhof an der Elbe see Dvůr Králové nad Labem
101 K23 **Königsbrunn** Bayern, S Germany 48°16′N 10°52′E
101 O24 **Königssee** ◎ SE Germany
Königshütte see Chorzów
109 S8 **Königstuhl** ▲ S Austria 46°57′N 13°47′E
109 U3 **Königswiesen** Oberösterreich, N Austria 48°25′N 14°48′E
101 E17 **Königswinter** Nordrhein-Westfalen, W Germany 50°40′N 07°12′E
146 M11 **Konimex** Rus. Kenimekh. Navoiy Viloyati, N Uzbekistan 40°14′N 65°10′E
110 I12 **Konin** Ger. Kuhnau. Weilkopolskie, C Poland 52°13′N 18°17′E
Koninkrijk der Nederlanden see Netherlands
113 L24 **Konispol** var. Konispoli. Vlorë, S Albania 39°40′N 20°10′E
Konispoli see Konispol
115 C15 **Kónitsa** Ípeiros, W Greece 40°04′N 20°48′E
Konitz see Chojnice
108 D8 **Köniz** Bern, W Switzerland 46°56′N 07°25′E
113 H14 **Konjic** ⬖ Federacija Bosne I Hercegovine, S Bosnia and Herzegovina
92 J10 **Könkämäälven** ≈ Finland/Sweden
155 D14 **Konkan** plain W India
83 D22 **Konkiep** ≈ S Namibia
76 I14 **Konkouré** ≈ W Guinea

◆ Country
● Country Capital
◇ Dependent Territory
○ Dependent Territory Capital
⬖ Administrative Regions
✕ International Airport
▲ Mountain
▲▲ Mountain Range
⛰ Volcano
≈ River
◎ Lake
⊠ Reservoir

77 O11 **Konna** Mopti, S Mali 14°58′N 03°49′W
186 H6 **Konogaiang, Mount** ▲ New Ireland, NE Papua New Guinea 04°05′S 152°43′E
186 H5 **Konogogo** New Ireland, NE Papua New Guinea 03°25′S 152°09′E
108 E9 **Konolfingen** Bern, W Switzerland 46°53′N 07°36′E
77 P16 **Konongo** C Ghana 06°39′N 01°06′W
186 H5 **Konos** New Ireland, NE Papua New Guinea 03°09′S 151°47′E
124 M12 **Konosha** Arkhangel'skaya Oblast', NW Russian Federation 60°58′N 40°09′E
117 R3 **Konotop** Sums'ka Oblast', NE Ukraine 51°15′N 33°14′E
158 L7 **Konqi He** ⛆ NW China
111 L14 **Końskie** Siętokrzyskie, C Poland 51°12′N 20°23′E
Konstantinovka *see* Kostyantynivka
126 M11 **Konstantinovsk** Rostovskaya Oblast', SW Russian Federation 47°37′N 41°07′E
101 H24 **Konstanz** *var.* Constanz, *Eng.* Constance, *hist.* Kostnitz; *anc.* Constantia. Baden-Württemberg, S Germany 47°40′N 09°10′E
Konstanza *see* Constanța
77 T14 **Kontagora** Niger, W Nigeria 10°25′N 05°29′E
78 E13 **Kontcha** Nord, N Cameroon 08°00′N 12°13′E
99 G17 **Kontich** Antwerpen, N Belgium 51°08′N 04°27′E
93 O16 **Kontiolahti** Pohjois-Karjala, SE Finland 62°46′N 29°51′E
93 M15 **Kontiomäki** Kainuu, C Finland 64°20′N 28°09′E
167 U11 **Kon Tum** *var.* Kontum. Kon Tum, C Vietnam 14°23′N 108°00′E
Kontum *see* Kon Tum
Konur *see* Sulakyurt
136 H15 **Konya** *var.* Konieh, *prev.* Konia; *anc.* Iconium. Konya, C Turkey 37°51′N 32°30′E
136 H15 **Konya** *var.* Konia, Konieh. ◆ *province* C Turkey
151 E15 **Konya Reservoir** *prev.* Shivāji Sāgar. ⊠ W India
145 T13 **Konyrat** *var.* Kounradskiy, *Kaz.* Qongyrat. Karaganda, SE Kazakhstan 46°57′N 75°01′E
145 W15 **Konyrolen** Almaty, SE Kazakhstan 44°16′N 79°18′E
81 I19 **Konza** Kajiado, S Kenya 01°44′S 37°07′E
98 I9 **Koog aan den Zaan** Noord-Holland, C Netherlands 52°28′N 04°49′E
182 E7 **Koonibba** South Australia 31°55′S 133°23′E
31 O11 **Koontz Lake** Indiana, N USA 41°25′N 86°24′W
171 U12 **Koor** Papua Barat, E Indonesia 0°21′S 132°28′E
183 R9 **Koorawatha** New South Wales, SE Australia 34°03′S 148°33′E
118 J5 **Koosa** Tartumaa, E Estonia 58°31′N 27°06′E
33 N7 **Kootenai** *var.* Kootenay. ⛆ Canada/USA *see also* Kootenay
Kootenai *see* Kootenay
11 P17 **Kootenay** *var.* Kootenai. ⛆ Canada/USA *see also* Kootenai
Kootenay *see* Kootenai
83 F24 **Kootjieskolk** Northern Cape, W South Africa 31°16′S 20°21′E
113 M15 **Kopaonik** ▲ S Serbia
Kopar *see* Koper
92 K1 **Kópasker** Norðurland Eystra, N Iceland 66°15′N 16°23′W
92 H4 **Kópavogur** Höfuðborgarsvæðið, W Iceland 64°06′N 21°47′W
145 U13 **Kopbirlik** *prev.* Kirov, Kirova. Almaty, SE Kazakhstan 46°24′N 77°16′E
109 S13 **Koper** *It.* Capodistria; *prev.* Kopar. SW Slovenia 45°32′N 13°43′E
95 C16 **Kopervik** Rogaland, S Norway 59°17′N 05°20′E
Köpetdag Gershi/Kopetdag, Khrebet *see* Koppeh Dāgh
Kophinou *see* Kofinou
182 G8 **Kopi** South Australia 33°24′S 135°40′E
153 W12 **Kopili** ⛆ NE India
95 M15 **Köping** Västmanland, C Sweden 59°31′N 16°00′E
113 K17 **Koplik** *var.* Kopliku. Shkodër, NW Albania 42°12′N 19°26′E
Kopliku *see* Koplik
Kopopo *see* Kokopo
94 I11 **Koppang** Hedmark, S Norway 61°34′N 11°04′E
Kopparberg *see* Dalarna
143 S3 **Koppeh Dāgh** *Rus.* Khrebet Kopetdag, *Turkm.* Köpetdag Gershi. ▲ Iran/Turkmenistan
Koppename *see* Coppename Rivier
95 J15 **Koppom** Värmland, C Sweden 59°42′N 12°07′E
Kopreinitz *see* Koprivnica
114 K9 **Koprinka, Yazovir** *prev.* Yazovir Georgi Dimitrov. ⊠ C Bulgaria
112 F7 **Koprivnica** *Ger.* Kopreinitz, *Hung.* Kaproncza. Koprivnica-Križevci, N Croatia 46°10′N 16°49′E
112 F8 **Koprivnica-Križevci** *off.* Koprivničko-Križevačka Županija. ◆ *province* N Croatia
111 I17 **Kopřivnice** *Ger.* Nesselsdorf. Moravskoslezský Kraj, E Czech Republic 49°36′N 18°09′E
Koprivničko-Križevačka Županija *see* Koprivnica-Križevci
Köprülü *see* Veles
Koptsevichi *see* Kaptsevichy
Kopyl' *see* Kapyl'
119 O14 **Kopys'** Vitsyebskaya Voblasts', NE Belarus 54°19′N 30°18′E
113 M18 **Korab** ▲ Albania/FYR Macedonia 41°48′N 20°33′E
Korabavur Pastligi *see* Karabaur', Uval
124 M5 **Korabel'noye** Murmanskaya Oblast', NW Russian Federation 67°00′N 41°10′E
81 M14 **K'orahē** Sumalē, E Ethiopia 06°36′N 44°21′E
115 L16 **Kórakas, Akrotírio** *cape* Lésvos, E Greece
112 D9 **Korana** ⛆ C Croatia
155 L14 **Korāput** Odisha, E India 18°48′N 82°41′E
Korat *see* Nakhon Ratchasima
167 Q9 **Korat Plateau** *plateau* E Thailand
Kōrāwa, Sar-I *see* Kirdī Kawrāw, Qimmat
154 L11 **Korba** Chhattīsgarh, C India 22°25′N 82°43′E
101 H15 **Korbach** Hessen, C Germany 51°16′N 08°52′E
Korça *see* Korçë
113 M21 **Korçë** *var.* Korça, *Gk.* Korytsa, *It.* Corriza; *prev.* Koritsa. Korçë, SE Albania 40°38′N 20°47′E
113 M21 **Korçë** ◆ *district* SE Albania
113 G15 **Korčula** *It.* Curzola. Dubrovnik-Neretva, S Croatia 42°57′N 17°08′E
113 F15 **Korčula** *It.* Curzola; *anc.* Corcyra Nigra. *island* S Croatia
113 F15 **Korčulanski Kanal** *channel* S Croatia
145 T6 **Korday** *prev.* Georgiyevka. Zhambyl, SE Kazakhstan 43°03′N 74°43′E
142 J5 **Kordestān** *off.* Ostān-e Kordestān, *var.* Kurdestan. ◆ *province* W Iran
Kordestān, Ostān-e *see* Kordestān
143 P4 **Kord Küy** *var.* Kurd Kui. Golestān, N Iran 36°49′N 54°05′E
163 V13 **Korea Bay** *bay* China/North Korea
Korea, Democratic People's Republic of *see* North Korea
171 T15 **Koreare** Pulau Yamdena, E Indonesia 07°33′S 131°13′E
Korea, Republic of *see* South Korea
163 Z17 **Korea Strait** *Jap.* Chōsen-kaikyō, *Kor.* Taehan-haehyŏp. *channel* Japan/South Korea
Korelichi/Korelicze *see* Karelichy
80 J11 **Korem** Tigrai, N Ethiopia 12°32′N 39°29′E
77 U11 **Korén Adoua** ⛆ C Niger
126 I7 **Korenevo** Kurskaya Oblast', W Russian Federation 51°21′N 34°53′E
126 L13 **Korenovsk** Krasnodarskiy Kray, SW Russian Federation 45°28′N 39°25′E
116 L4 **Korets'** *Pol.* Korzec, *Rus.* Korets. Rivnens'ka Oblast', NW Ukraine 50°38′N 27°12′E
Korets *see* Korets'
194 L7 **Korff Ice Rise** *ice cap* Antarctica
145 Q10 **Korgalzhyn** *var.* Kurgal'dzhino, Kurgal'dzhinsky, *Kaz.* Qorghalzhyn. Akmola, C Kazakhstan 50°33′N 69°58′E
145 W15 **Korgas**; *prev.* Khorgos. Almaty, SE Kazakhstan 44°13′N 80°22′E
92 G13 **Korgen** Troms, N Norway 66°04′N 13°51′E
147 R9 **Korgon-Dëbë** Dzhalal-Abadskaya Oblast', W Kyrgyzstan 41°51′N 70°52′E
76 M14 **Korhogo** N Ivory Coast 09°29′N 05°39′W
115 F19 **Korinthiakós Kólpos** *Eng.* Gulf of Corinth; *anc.* Corinthiacus Sinus. *gulf* C Greece
115 F19 **Kórinthos** *anc.* Corinthus *Eng.* Corinth. Pelopónnisos, S Greece 37°56′N 22°55′E
113 M18 **Koritnik** ▲ S Serbia 42°06′N 20°34′E
Koritsa *see* Korçë
165 P11 **Kōriyama** Fukushima, Honshū, C Japan 37°25′N 140°20′E
136 E16 **Korkuteli** Antalya, SW Turkey 37°07′N 30°11′E
158 K6 **Korla** *Chin.* K'u-erh-lo. Xinjiang Uygur Zizhiqu, NW China 41°48′N 86°10′E
122 J10 **Korliki** Khanty-Mansiyskiy Avtonomnyy Okrug-Yugra, C Russian Federation 61°28′N 82°12′E
Körlin an der Persante *see* Karlino
Korma *see* Karma
14 D8 **Kormak** Ontario, S Canada 47°38′N 83°00′W
Kormakiti, Akrotíri/Kormakiti, Cape/Kormakítis *see* Koruçam Burnu
111 G23 **Körmend** Vas, W Hungary 47°02′N 16°35′E
139 S5 **Kōrmōr** Şalāḩ ad Dīn, E Iraq 35°06′N 44°47′E
112 C13 **Kornat** *It.* Incoronata. *island* W Croatia
Korneshty *see* Cornești
109 X3 **Korneuburg** Niederösterreich, NE Austria 48°22′N 16°20′E
145 P7 **Korneyevka** Severnyy Kazakhstan, N Kazakhstan 54°01′N 68°30′E
95 I17 **Kornsjø** Østfold, S Norway 58°55′N 11°40′E
77 O11 **Koro** Mopti, S Mali 14°05′N 03°06′W
187 Y14 **Koro** *island* C Fiji
186 B7 **Koroba** Hela, W Papua New Guinea 05°46′S 142°48′E
126 K8 **Korocha** Belgorodskaya Oblast', W Russian Federation 50°49′N 37°08′E
136 H12 **Köroğlu Dağları** ▲ C Turkey
183 V6 **Korogoro Point** *headland* New South Wales, SE Australia 31°03′S 153°04′E
81 J21 **Korogwe** Tanga, E Tanzania 05°10′S 38°30′E
182 L13 **Koroit** Victoria, SE Australia 38°17′S 142°22′E
187 X15 **Korolevu** Viti Levu, W Fiji 18°12′S 177°44′E
190 I17 **Koromiri** *island* S Cook Islands
171 Q8 **Koronadal** Mindanao, S Philippines 06°23′N 124°54′E
114 G13 **Korónia, Límni** *var.* Límni Korónia. ◎ N Greece
115 E22 **Koróni** Pelopónnisos, S Greece 36°47′N 21°57′E
Korónia, Límni *see* Korónia, Límni
110 I9 **Koronowo** *Ger.* Krone an der Brahe. Kujawski-pomorskie, C Poland 53°18′N 17°56′E
117 R2 **Korop** Chernihivs'ka Oblast', N Ukraine 51°35′N 32°57′E
115 H19 **Koropí** Attikí, C Greece 37°54′N 23°52′E
188 C8 **Koror** (Palau) Oreor, N Palau 07°21′N 134°28′E
Koror *see* Oreor
Koror Lāl Esan *see* Karor
Koror Pacca *see* Kahror Pakka
111 L23 **Körös** ⛆ E Hungary
Körös *see* Križevci
Kőrösbánya *see* Baia de Criș
187 Y14 **Koro Sea** *sea* C Fiji
Koroška *see* Kärnten
117 N3 **Korosten'** Zhytomyrs'ka Oblast', NW Ukraine 50°56′N 28°39′E
Korostyshev *see* Korostyshiv
117 N4 **Korostyshiv** *Rus.* Korostyshev. Zhytomyrs'ka Oblast', N Ukraine 50°18′N 29°05′E
125 V3 **Korotaikha** ⛆ NW Russian Federation
122 J9 **Korotchayevo** Yamalo-Nenetskiy Avtonomnyy Okrug, N Russian Federation 66°00′N 78°11′E
78 I8 **Koro Toro** Borkou, N Chad 16°01′N 18°27′E
39 N16 **Korovin Island** *island* Shumagin Islands, Alaska, USA
187 X14 **Korovou** Viti Levu, W Fiji 17°48′S 178°32′E
93 M17 **Korpilahti** Keski-Suomi, C Finland 62°02′N 25°34′E
92 K12 **Korpilombolo** *Lapp.* Dállogilli. Norrbotten, N Sweden 66°51′N 23°00′E
123 T13 **Korsakov** Ostrov Sakhalin, Sakhalinskaya Oblast', SE Russian Federation 46°41′N 142°45′E
93 J16 **Korsholm** *Fin.* Mustasaari. Österbotten, W Finland 63°05′N 21°43′E
95 I23 **Korsør** Sjælland, E Denmark 55°19′N 11°09′E
Korsovka *see* Kārsava
117 P6 **Korsun'-Shevchenkivs'kyy** *Rus.* Korsun'-Shevchenkovskiy. Cherkas'ka Oblast', C Ukraine 49°26′N 31°15′E
Korsun'-Shevchenkovskiy *see* Korsun'-Shevchenkivs'kyy
99 C17 **Kortemark** West-Vlaanderen, W Belgium 51°03′N 03°03′E
99 H18 **Kortenberg** Vlaams Brabant, C Belgium 50°53′N 04°33′E
99 K18 **Kortessem** Limburg, NE Belgium 50°52′N 05°22′E
99 E14 **Kortgene** Zeeland, SW Netherlands 51°34′N 03°48′E
80 F8 **Korti** Northern, N Sudan 18°06′N 31°33′E
99 C18 **Kortrijk** *Fr.* Courtrai. West-Vlaanderen, W Belgium 50°50′N 03°17′E
121 O2 **Koruçam Burnu** *var.* Cape Kormakiti, Kormakítis, *Gk.* Akrotíri Kormakíti. *headland* N Cyprus 35°24′N 32°55′E
183 O13 **Korumburra** Victoria, SE Australia 38°27′S 145°48′E
Koryak Range *see* Koryakskoye Nagor'ye
Koryakskiy Khrebet *see* Koryakskoye Nagor'ye
123 V8 **Koryakskiy Okrug** ◆ *autonomous district* E Russian Federation
123 V7 **Koryakskoye Nagor'ye** *var.* Koryakskiy Khrebet, *Eng.* Koryak Range. ▲ NE Russian Federation
125 P11 **Koryazhma** Arkhangel'skaya Oblast', NW Russian Federation 61°16′N 47°07′E
Koryŏ *see* Gangneung
Korytsa *see* Korçë
117 Q2 **Koryukivka** Chernihivs'ka Oblast', N Ukraine 51°45′N 32°16′E
Korzec *see* Korets'
115 N21 **Kos** Kos, Dodekánisa, Greece, Aegean Sea 36°53′N 27°19′E
115 M21 **Kos** *It.* Coo; *anc.* Cos. *island* Dodekánisa, Greece, Aegean Sea
125 T12 **Kosa** Komi-Permyatskiy Okrug, NW Russian Federation 59°55′N 54°54′E
125 T13 **Kosa** ⛆ NW Russian Federation
164 B12 **Kō-saki** *headland* Nagasaki, Tsushima, SW Japan 34°06′N 129°13′E
163 X13 **Kosan** SE North Korea 38°50′N 127°26′E
119 H18 **Kosava** *Rus.* Kosovo. Brestskaya Voblasts', SW Belarus 52°45′N 25°16′E
Kosch *see* Kose
Koschagyl *see* Kosshagyl
110 G12 **Kościan** *Ger.* Kosten. Wielkopolskie, C Poland 52°05′N 16°38′E
110 I7 **Kościerzyna** Pomorskie, NW Poland 54°07′N 17°55′E
22 L4 **Kosciusko** Mississippi, S USA 33°03′N 89°35′W
Kosciusko, Mount *see* Kosciuszko, Mount
183 R11 **Kosciuszko, Mount** *prev.* Mount Kosciusko. ▲ New South Wales, SE Australia 36°28′S 148°15′E
118 H4 **Kose** *Ger.* Kosch. Harjumaa, NW Estonia 59°11′N 25°10′E
25 U9 **Kosse** Texas, SW USA 31°16′N 96°38′W
114 G6 **Koshava** Vidin, NW Bulgaria 44°03′N 23°00′E
147 U9 **Kosh-Dëbë** *var.* Koshtebë. Narynskaya Oblast', C Kyrgyzstan 41°03′N 74°08′E
K'o-shih *see* Kashi
164 B12 **Koshikijima-rettō** *var.* Kosikizima Rettô. *island group* SW Japan
145 W13 **Koshkarkol', Ozero** ◎ SE Kazakhstan
30 L9 **Koshkonong, Lake** ◎ Wisconsin, N USA
Koshoba *see* Goşabo
Koshoba *see* Goşoba
164 M12 **Kōshoku** *var.* Kôsyoku. Nagano, Honshū, S Japan 36°33′N 138°09′E
Koshtebë *see* Kosh-Dëbë
Kōshū *see* Enzan
Kōshū *see* Gwangju
111 N19 **Košice** *Ger.* Kaschau, *Hung.* Kassa. Košický Kraj, E Slovakia 48°44′N 21°15′E
111 M20 **Košický Kraj** ◆ *region* E Slovakia
Kosikizima Rettô *see* Koshikijima-rettō
153 R12 **Kosi Reservoir** ⊠ E Nepal
116 J8 **Kosiv** Ivano-Frankivs'ka Oblast', W Ukraine 48°19′N 25°04′E
145 O11 **Koskol'** *Kaz.* Qosköl. Karaganda, C Kazakhstan 49°32′N 67°08′E
125 Q9 **Koslan** Respublika Komi, NW Russian Federation 63°27′N 48°52′E
Köslin *see* Koszalin
146 M12 **Koson** *Rus.* Kasan. Qashqadaryo Viloyati, S Uzbekistan 39°04′N 65°35′E
Kosŏng *see* Goseong
147 S9 **Kosonsoy** *Rus.* Kasansay. Namangan Viloyati, E Uzbekistan 41°15′N 71°28′E
113 M16 **Kosovo** *prev.* Autonomous Province of Kosovo and Metohija. ◆ *republic* SE Europe
Kosovo *see* Kosava
Kosovo and Metohija, Autonomous Province of *see* Kosovo
Kosovo Polje *see* Fushë Kosovë
Kosovska Kamenica *see* Kamenicë
Kosovska Mitrovica *see* Mitrovicë
189 X17 **Kosrae** ◆ *state* E Micronesia
189 Y14 **Kosrae** *prev.* Kusaie. *island* Caroline Islands, E Micronesia
109 P6 **Kössen** Tirol, W Austria 47°40′N 12°24′E
144 G12 **Kosshagyl** *prev.* Koschagyl, *Kaz.* Qosshaghyl. Atyrau, W Kazakhstan 46°52′N 53°46′E
76 M16 **Kossou, Lac de** ◎ C Ivory Coast
Kossukavak *see* Krumovgrad
Kostajnica *see* Hrvatska Kostanjnica
Kostamus *see* Kostomuksha
144 M7 **Kostanay** *var.* Kustanay, *Kaz.* Qostanay. Kostanay, N Kazakhstan 53°16′N 63°34′E
144 L8 **Kostanay** *var.* Kostanayskaya Oblast', *Kaz.* Qostanay Oblysy. ◆ *province* N Kazakhstan
Kostanayskaya Oblast' *see* Kostanay
Kosten *see* Kościan
114 H10 **Kostenets** *prev.* Georgi Dimitrov. Sofia, W Bulgaria 42°15′N 23°48′E
80 F10 **Kosti** White Nile, C Sudan 13°11′N 32°38′E
Kostnitz *see* Konstanz
124 H7 **Kostomuksha** *Fin.* Kostamus. Respublika Kareliya, NW Russian Federation 64°33′N 30°28′E
116 K3 **Kostopil'** *Rus.* Kostopol'. Rivnens'ka Oblast', NW Ukraine 50°20′N 26°29′E
Kostopol' *see* Kostopil'
124 M15 **Kostroma** Kostromskaya Oblast', NW Russian Federation 57°46′N 41°E
125 N14 **Kostroma** ⛆ NW Russian Federation
125 N14 **Kostromskaya Oblast'** ◆ *province* NW Russian Federation
110 D11 **Kostrzyn** *Ger.* Cüstrin, Küstrin. Lubuskie, W Poland 52°35′N 14°40′E
110 H11 **Kostrzyn** Wielkopolskie, C Poland 52°23′N 17°13′E
117 X7 **Kostyantynivka** *Rus.* Konstantinovka. Donets'ka Oblast', SE Ukraine 48°33′N 37°45′E
Kostyukovichi *see* Kastsyukovichy
Kostyukovka *see* Kastsyukowka
Kōsyoku *see* Kōshoku
125 U6 **Kos'yu** Respublika Komi, NW Russian Federation 65°39′N 59°01′E
125 U6 **Kos'yu** ⛆ NW Russian Federation
110 F7 **Koszalin** *Ger.* Köslin. Zachodnio-pomorskie, NW Poland 54°12′N 16°10′E
111 F22 **Kőszeg** *Ger.* Güns. Vas, W Hungary 47°23′N 16°33′E
152 H13 **Kota** *prev.* Kotah. Rājasthān, N India 25°14′N 75°52′E
Kota Baharu *see* Kota Bharu
Kota Bahru *see* Kota Bharu
169 U13 **Kotabaru** Pulau Laut, C Indonesia 03°15′S 116°15′E
168 K12 **Kota Baru** Sumatera, W Indonesia 01°07′S 101°43′E
Kotabaru *see* Jayapura
168 K6 **Kota Bharu** *var.* Kota Baharu, Kota Bahru. Kelantan, Peninsular Malaysia 06°07′N 102°15′E
Kotaboemi *see* Kotabumi
168 M14 **Kotabumi** *prev.* Kotaboemi. Sumatera, W Indonesia 04°50′S 104°54′E
149 S10 **Kot Addu** Punjab, E Pakistan 30°28′N 70°58′E
Kotah *see* Kota
169 U7 **Kota Kinabalu** *prev.* Jesselton. Sabah, East Malaysia 05°59′N 116°04′E
169 U7 **Kota Kinabalu** ✈ Sabah, East Malaysia 05°59′N 116°04′E
92 M12 **Kotala** Lappi, N Finland 67°01′N 29°00′E
Kotamobagoe *see* Kotamobagu
171 Q11 **Kotamobagu** *prev.* Kotamobagoe. Sulawesi, C Indonesia 0°46′N 124°21′E
155 L14 **Kotapad** *var.* Kotapārh. Odisha, E India 19°10′N 82°23′E
Kotapārh *see* Kotapad
166 N17 **Ko Ta Ru Tao** *island* SW Thailand
169 R13 **Kotawaringin, Teluk** *bay* Borneo, C Indonesia
149 Q13 **Kot Dīji** Sind, SE Pakistan 27°16′N 68°44′E
152 K9 **Kotdwāra** Uttarakhand, N India 29°44′N 78°33′E
125 Q14 **Kotel'nich** Kirovskaya Oblast', NW Russian Federation 58°19′N 48°12′E
127 N12 **Kotel'nikovo** Volgogradskaya Oblast', SW Russian Federation 47°37′N 43°07′E
123 Q6 **Kotel'nyy, Ostrov** *island* Novosibirskiye Ostrova, N Russian Federation
117 T5 **Kotel'va** Poltavs'ka Oblast', C Ukraine 50°04′N 34°46′E
101 M14 **Köthen** *var.* Cöthen. Sachsen-Anhalt, C Germany 51°46′N 11°59′E
Kôti *see* Kōchi
81 G17 **Kotido** NE Uganda 03°03′N 34°07′E
93 N19 **Kotka** Kymenlaakso, S Finland 60°28′N 26°55′E
125 P11 **Kotlas** Arkhangel'skaya Oblast', NW Russian Federation 61°14′N 46°43′E
38 M10 **Kotlik** Alaska, USA 63°01′N 163°33′W
77 Q17 **Kotoka** ✈ (Accra) S Ghana 05°41′N 00°10′W
Kotonu *see* Cotonou
113 J17 **Kotor** *It.* Cattaro. SW Montenegro 42°25′N 18°47′E
Kotor *see* Kotoriba
112 F7 **Kotoriba** *Hung.* Kotor. Medimurje, N Croatia 46°20′N 16°47′E
113 I17 **Kotorska, Boka** *It.* Bocche di Cattaro. *bay* SW Montenegro
112 H11 **Kotorsko** ◆ Republika Srpska, N Bosnia and Herzegovina
112 G11 **Kotor Varoš** ◆ Republika Srpska, N Bosnia and Herzegovina
Koto Sho/Kotosho *see* Lan Yu
126 M7 **Kotovsk** Tambovskaya Oblast', W Russian Federation 52°39′N 41°31′E
117 O9 **Kotovs'k** *Rus.* Kotovsk. Odes'ka Oblast', SW Ukraine 47°42′N 29°30′E
Kotovsk *see* Hînceşti
119 G16 **Kotra** ⛆ W Belarus
149 P16 **Kotri** Sind, SE Pakistan 25°22′N 68°18′E
109 Q9 **Kötschach** Kärnten, S Austria 46°41′N 12°57′E
155 K15 **Kottagūdem** Telangana, E India 17°36′N 80°40′E
155 F21 **Kottappadi** Kerala, SW India 11°38′N 76°03′E
155 G23 **Kottayam** Kerala, SW India 09°34′N 76°31′E
Kottbus *see* Cottbus
79 K15 **Kotto** ⛆ Central African Republic/Dem. Rep. Congo
193 X15 **Kotu Group** *island group* W Tonga
Koturdepe *see* Goturdepe
122 M9 **Kotuy** ⛆ N Russian Federation
83 M16 **Kotwa** Mashonaland East, NE Zimbabwe 16°58′S 32°46′E
39 N7 **Kotzebue** Alaska, USA 66°54′N 162°36′W
38 M7 **Kotzebue Sound** *inlet* Alaska, USA
Kotzenan *see* Chocianów
Kotzman *see* Kitsman'
77 R14 **Kouandé** NW Benin 10°20′N 01°42′E
79 J15 **Kouango** Ouaka, S Central African Republic 05°00′N 20°01′E
77 O13 **Koudougou** C Burkina Faso 12°15′N 02°23′W
98 K7 **Koudum** Fryslân, N Netherlands 52°55′N 05°26′E
115 L25 **Koufonísi** *island* SE Greece
115 K21 **Koufonísi** *island* Kykládes, Greece, Aegean Sea
38 M8 **Kougarok Mountain** ▲ Alaska, USA 65°41′N 165°29′W
79 E21 **Kouilou** ◆ *province* SW Congo
79 E20 **Kouilou** ⛆ S Congo
167 Q11 **Koŭk Kduŏch** *prev.* Phumĭ Koŭk Kduŏch. Bătdâmbâng, NW Cambodia 13°16′N 103°08′E
121 O3 **Koúklia** SW Cyprus 34°42′N 32°35′E
79 E19 **Koulamoutou** Ogooué-Lolo, C Gabon 01°07′S 12°27′E
76 L12 **Koulikoro** Koulikoro, SW Mali 12°55′N 07°31′W
76 L11 **Koulikoro** ◆ *region* SW Mali
187 P16 **Koumac** Province Nord, W New Caledonia 20°34′S 164°18′E
165 N12 **Koumi** Nagano, Honshū, S Japan 36°06′N 138°27′E
78 I13 **Koumra** Mandoul, S Chad 08°56′N 17°32′E
Kounadougou *see* Koundougou
76 M15 **Kounahiri** C Ivory Coast 07°47′N 05°51′W
76 I12 **Koundâra** Moyenne-Guinée, NW Guinea 12°28′N 13°15′W
77 N13 **Koundougou** *var.* Kounadougou. C Burkina Faso 11°43′N 04°40′W
76 H11 **Koungheul** C Senegal 14°00′N 14°48′W
Kounradskiy *see* Konyrat
25 X10 **Kountze** Texas, SW USA 30°22′N 94°20′W
77 Q13 **Koupéla** C Burkina Faso 12°07′N 00°21′W
77 N13 **Kouri** Sikasso, SW Mali 12°09′N 04°46′W
55 Y9 **Kourou** N French Guiana 05°08′N 52°37′W
114 J12 **Kouroú** ⛆ NE Greece
76 K14 **Kouroussa** C Guinea 10°40′N 09°50′W
Kousseir *see* Al Quşayr
78 G11 **Kousséri** *prev.* Fort-Foureau. Extrême-Nord, NE Cameroon 12°05′N 14°56′E
Kouteifé *see* Al Quṭayfah
76 M13 **Koutiala** Sikasso, S Mali 12°20′N 05°23′W
76 M14 **Kouto** NW Ivory Coast 09°51′N 06°25′W
93 M19 **Kouvola** Kymenlaakso, S Finland 60°50′N 26°48′E
79 G18 **Kouyou** ⛆ C Congo
112 M10 **Kovačica** *Hung.* Antafalva; *prev.* Kovacsicza. Vojvodina, N Serbia 45°08′N 20°36′E
Kovacsicza *see* Kovačica
Kővárhosszúfalu *see* Satulung
Kovászna *see* Covasna
124 I4 **Kovdor** Murmanskaya Oblast', NW Russian Federation 67°32′N 30°27′E
116 J3 **Kovel'** *Pol.* Kowel. Volyns'ka Oblast', NW Ukraine 51°14′N 24°43′E
112 M11 **Kovin** *Hung.* Kevevára; *prev.* Temes-Kubin. Vojvodina, NE Serbia 44°45′N 20°59′E
Kovno *see* Kaunas
127 N3 **Kovrov** Vladimirskaya Oblast', W Russian Federation 56°24′N 41°21′E
127 O5 **Kovylkino** Respublika Mordoviya, W Russian Federation 54°03′N 43°52′E
110 J11 **Kowal** Kujawsko-pomorskie, C Poland 52°31′N 19°09′E
110 J9 **Kowalewo Pomorskie** *Ger.* Schönsee. Kujawsko-pomorskie, N Poland 53°07′N 18°48′E
Kowasna *see* Covasna
119 M16 **Kowbcha** *Rus.* Kolbcha. Mahilyowskaya Voblasts', E Belarus 53°39′N 29°14′E
Koweit *see* Kuwait
Kowel *see* Kovel'
185 F17 **Kowhitirangi** West Coast, South Island, New Zealand 42°54′S 171°01′E
161 O15 **Kowloon** Hong Kong, S China
Kowno *see* Kaunas
159 N7 **Kox Kuduk** *well* NW China
136 D16 **Köyceğiz** Muğla, SW Turkey 36°57′N 28°40′E
125 N6 **Koyda** Arkhangel'skaya Oblast', NW Russian Federation 66°22′N 42°42′E
139 T3 **Koye** *Ar.* Küysanjaq, *var.* Koi Sanjaq. Arbīl, N Iraq 36°05′N 44°38′E
Koymat *see* Goymat
Koymatdag, Gory *see* Goymatdag, Gory
151 E15 **Koyna Reservoir** ⊠ W India
165 P9 **Koyoshi-gawa** ⛆ Honshū, C Japan
Koi Sanjaq *see* Koye
Koytash *see* Qo'ytosh
146 M14 **Köytendag** *prev. Rus.* Charshanga, Charshangy, *Turkm.* Charshangngy. Lebap Welaýaty, E Turkmenistan 37°31′N 65°58′E
39 N9 **Koyuk** Alaska, USA 64°55′N 161°09′W
39 N9 **Koyuk River** ⛆ Alaska, USA
39 O9 **Koyukuk** Alaska, USA 64°52′N 157°42′W
39 O9 **Koyukuk River** ⛆ Alaska, USA
136 J13 **Kozaklı** Nevşehir, C Turkey 39°12′N 34°48′E
136 K16 **Kozan** Adana, S Turkey 37°27′N 35°47′E
115 E14 **Kozáni** Dytikí Makedonía, N Greece 40°19′N 21°48′E
112 F10 **Kozara** ▲ NW Bosnia and Herzegovina
Kozarska Dubica *see* Bosanska Dubica
117 P3 **Kozelets'** *Rus.* Kozelets. Chernihivs'ka Oblast', NE Ukraine 50°54′N 31°09′E
Kozelets *see* Kozelets'
117 S6 **Kozel'shchyna** Poltavs'ka Oblast', C Ukraine 49°13′N 33°49′E
126 J5 **Kozel'sk** Kaluzhskaya Oblast', W Russian Federation 54°04′N 35°51′E
151 F21 **Kozhikode** *var.* Calicut. Kerala, SW India 11°17′N 75°49′E *see also* Calicut
Kozhimiz, Gora *see* Kozhymiz, Gora
124 L9 **Kozhozero, Ozero** ◎ NW Russian Federation
125 T7 **Kozhva** Respublika Komi, NW Russian Federation 65°06′N 57°00′E
125 T7 **Kozhva** ⛆ NW Russian Federation
125 U6 **Kozhym** Respublika Komi, NW Russian Federation 65°43′N 59°25′E
125 V9 **Kozhymiz, Gora** *prev.* Gora Kozhimiz. ▲ NW Russian Federation 63°13′N 58°54′E
110 N13 **Kozienice** Mazowieckie, C Poland 51°35′N 21°31′E
109 S13 **Kozina** SW Slovenia 45°36′N 13°56′E
114 H7 **Kozloduy** Vratsa, NW Bulgaria 43°48′N 23°42′E
127 Q3 **Kozlovka** Chuvashskaya Respublika, W Russian Federation 55°53′N 48°07′E
Kozlovshchina/Kozlowszczyzna *see* Kazlowshchyna
127 P3 **Koz'modem'yansk** Respublika Mariy El, W Russian Federation 56°19′N 46°33′E
116 J6 **Kozova** Ternopil's'ka Oblast', W Ukraine 49°25′N 25°09′E
113 P20 **Kožuf** ▲ Greece/Macedonia
165 N15 **Kōzu-shima** *island* E Japan
Koz'yany *see* Kazyany
117 N5 **Kozyatyn** *Rus.* Kazatin. Vinnyts'ka Oblast', C Ukraine 49°41′N 28°49′E
77 Q16 **Kpalimé** *var.* Palimé. SW Togo 06°54′N 00°38′E
77 Q16 **Kpandu** E Ghana 07°00′N 00°18′E
99 F15 **Krabbendijke** Zeeland, SW Netherlands 51°25′N 04°07′E
167 N15 **Krabi** *var.* Muang Krabi. Krabi, SW Thailand 08°04′N 98°52′E
167 N13 **Kra Buri** Ranong, SW Thailand 10°25′N 98°48′E
167 S12 **Krâchéh** *prev.* Kratie. Krâchéh, E Cambodia 12°29′N 106°01′E
95 G17 **Kragerø** Telemark, S Norway 58°54′N 09°25′E
112 M13 **Kragujevac** Serbia, C Serbia 44°01′N 20°55′E
Krainburg *see* Kranj
166 N13 **Kra, Isthmus of** *isthmus* Malaysia/Thailand
112 D12 **Krajina** *cultural region* SW Croatia
Krakatau, Pulau *see* Rakata, Pulau
Krakau *see* Kraków
111 L16 **Kraków** *Eng.* Cracow, *Ger.* Krakau; *anc.* Cracovia. Małopolskie, S Poland 50°03′N 19°58′E
100 L9 **Krakower See** ◎ NE Germany
167 Q11 **Krâlănh** Siĕmréab, NW Cambodia 13°35′N 103°27′E
45 Q16 **Kralendijk** ○ Bonaire 12°07′N 68°13′W
112 B10 **Kraljevica** *It.* Porto Re. Primorje-Gorski Kotar, NW Croatia 45°15′N 14°36′E
112 M13 **Kraljevo** *prev.* Rankovićevo. Serbia, C Serbia 43°44′N 20°40′E
111 E16 **Královehradecký Kraj** *prev.* Hradecký Kraj. ◆ *region* N Czech Republic
Kralup an der Moldau *see* Kralupy nad Vltavou
111 C16 **Kralupy nad Vltavou** *Ger.* Kralup an der Moldau. Středočeský Kraj, NW Czech Republic 50°15′N 14°20′E
117 W7 **Kramators'k** *Rus.* Kramatorsk. Donets'ka Oblast', SE Ukraine 48°43′N 37°34′E
Kramatorsk *see* Kramators'k
93 H17 **Kramfors** Västernorrland, C Sweden 62°55′N 17°50′E
Kranéa *see* Kraniá
108 M7 **Kranebitten** ✈ (Innsbruck) Tirol, W Austria 47°18′N 11°21′E
115 D15 **Kraniá** *var.* Kranéa. Dytikí Makedonía, N Greece 39°54′N 21°21′E
115 G20 **Kranídi** Pelopónnisos, S Greece 37°21′N 23°09′E
109 T11 **Kranj** *Ger.* Krainburg. NW Slovenia 46°17′N 14°16′E
115 F16 **Krannón** *battleground* Thessalía, C Greece
Kranz *see* Zelenogradsk
112 D7 **Krapina** Krapina-Zagorje, N Croatia 46°12′N 15°52′E
112 E8 **Krapina** ⛆ N Croatia
112 D8 **Krapina-Zagorje** *off.* Krapinsko-Zagorska Županija. ◆ *province* N Croatia
114 L7 **Krapinets** ⛆ NE Bulgaria
Krapinsko-Zagorska Županija *see* Krapina-Zagorje
111 I15 **Krapkowice** *Ger.* Krappitz. Opolskie, SW Poland 50°29′N 17°56′E
Krappitz *see* Krapkowice
125 O12 **Krasavino** Vologodskaya Oblast', NW Russian Federation 60°56′N 46°27′E
122 H6 **Krasino** Novaya Zemlya, Arkhangel'skaya Oblast', N Russian Federation 70°45′N 54°16′E
123 S15 **Kraskino** Primorskiy Kray, SE Russian Federation 42°40′N 130°51′E
118 J11 **Krāslava** SE Latvia 55°56′N 27°08′E
119 M14 **Krasnaluki** *Rus.* Krasnoluki. Vitsyebskaya Voblasts', N Belarus 54°37′N 28°50′E
119 P17 **Krasnapollye** *Rus.* Krasnopol'ye. Mahilyowskaya Voblasts', E Belarus 53°20′N 31°24′E
126 L15 **Krasnaya Polyana** Krasnodarskiy Kray, SW Russian Federation 43°40′N 40°13′E
Krasnaya Slabada/Krasnaya Sloboda *see* Chyrvonaya Slabada
119 J15 **Krasnaye** *Rus.* Krasnoye. Minskaya Voblasts', C Belarus 54°14′N 27°05′E
111 O14 **Kraśnik** *Ger.* Kratznick. Lubelskie, E Poland 50°56′N 22°14′E
117 O9 **Krasni Okny** Odes'ka Oblast', SW Ukraine 47°33′N 29°28′E
127 P8 **Krasnoarmeysk** Saratovskaya Oblast', W Russian Federation 51°02′N 45°42′E
Krasnoarmeysk *see* Tayynsha
Krasnoarmeysk *see* Krasnoarmiys'k/Tayynsha
123 T6 **Krasnoarmeyskiy** Chukotskiy Avtonomnyy Okrug, NE Russian Federation 69°30′N 171°44′E
117 W7 **Krasnoarmiys'k** *Rus.* Krasnoarmeysk. Donets'ka Oblast', SE Ukraine 48°17′N 37°14′E
125 P11 **Krasnoborsk** Arkhangel'skaya Oblast', NW Russian Federation 61°31′N 45°57′E
126 K14 **Krasnodar** *prev.* Ekaterinodar, Yekaterinodar. Krasnodarskiy Kray, SW Russian Federation 45°06′N 39°01′E
126 K13 **Krasnodarskiy Kray** ◆ *territory* SW Russian Federation
117 Z7 **Krasnodon** Luhans'ka Oblast', E Ukraine 48°17′N 39°44′E
Krasnogor *see* Kallaste
127 T2 **Krasnogorskoye** Udmurtskaya Respublika, NW Russian Federation 57°42′N 52°29′E
Krasnograd *see* Krasnohrad
Krasnogvardeysk *see* Bulungh'ur
126 M13 **Krasnogvardeyskoye** Stavropol'skiy Kray, SW Russian Federation 45°49′N 41°31′E
Krasnogvardeyskoye *see* Krasnohvardiys'ke
117 U6 **Krasnohrad** *Rus.* Krasnograd. Kharkivs'ka Oblast', E Ukraine 49°22′N 35°28′E
117 S12 **Krasnohvardiys'ke** *Rus.* Krasnogvardeyskoye. Avtonomna Respublika Krym, S Ukraine 45°30′N 34°19′E
123 P14 **Krasnokamensk** Zabaykal'skiy Kray, S Russian Federation 50°03′N 118°01′E
125 U14 **Krasnokamsk** Permskiy Kray, W Russian Federation 58°08′N 55°48′E
127 U8 **Krasnokholm** Orenburgskaya Oblast', W Russian Federation 51°34′N 54°11′E
117 U5 **Krasnokuts'k** *Rus.* Krasnokutsk. Kharkivs'ka Oblast', E Ukraine 50°01′N 35°03′E
Krasnokutsk *see* Krasnokuts'k
126 L7 **Krasnolesnyy** Voronezhskaya Oblast', W Russian Federation 51°53′N 39°37′E

◆ Country
● Country Capital
◇ Dependent Territory
○ Dependent Territory Capital
◆ Administrative Regions
✈ International Airport
▲ Mountain
▲ Mountain Range
Volcano
⛆ River
◎ Lake
⊠ Reservoir

Krasnoluki see Krasnaluki
Krasnoosol'skoye Vodokhranilishche see Chervonooskil's'ke Vodoskhovyshche
117 S11 **Krasnoperekops'k** Rus. Krasnoperekopsk. Avtonomna Respublika Krym, S Ukraine 45°56′N 33°47′E
Krasnoperekopsk see Krasnoperekops'k
117 U4 **Krasnopillya** Sums'ka Oblast', NE Ukraine 50°46′N 35°17′E
Krasnopol'ye see Krasnapollye
124 L5 **Krasnoshchel'ye** Murmanskaya Oblast', NW Russian Federation 67°22′N 37°03′E
127 O5 **Krasnoslobodsk** Respublika Mordoviya, W Russian Federation 54°24′N 43°51′E
127 T2 **Krasnoslobodsk** Volgogradskaya Oblast', SW Russian Federation 48°41′N 44°34′E
Krasnostav see Krasnystaw
127 V5 **Krasnousol'skiy** Respublika Bashkortostan, W Russian Federation 53°55′N 56°22′E
125 U12 **Krasnovishersk** Permskiy Kray, NW Russian Federation 60°22′N 57°04′E
Krasnovodsk see Türkmenbaşy
Krasnovodskiy Zaliv see Türkmenbasy Aylagy
146 B10 **Krasnovodskoye Plato** Turkm. Krasnowodsk Platosy. plateau NW Turkmenistan
Krasnowodsk Aylagy see Türkmenbasy Aylagy
Krasnowodsk Platosy see Krasnovodskoye Plato
122 K12 **Krasnoyarsk** Krasnoyarskiy Kray, S Russian Federation 56°05′N 92°46′E
127 X7 **Krasnoyarskiy** Orenburgskaya Oblast', W Russian Federation 51°56′N 59°54′E
122 K11 **Krasnoyarskiy Kray** ◐ territory C Russian Federation
Krasnoye see Krasnaye
Krasnoye Znamya see Gyzylbaydak
125 R11 **Krasnozatonskiy** Respublika Komi, NW Russian Federation 61°39′N 51°00′E
118 D13 **Krasnoznamensk** prev. Lasdehnen, Ger. Haselberg. Kaliningradskaya Oblast', W Russian Federation 54°57′N 22°28′E
126 K3 **Krasnoznamensk** Moskovskaya Oblast', W Russian Federation 55°40′N 37°05′E
117 R11 **Krasnoznam"yans'kyy Kanal** canal S Ukraine
111 P14 **Krasnystaw** Rus. Krasnostav. Lubelskie, SE Poland 51°N 23°10′E
126 H4 **Krasnyy** Smolenskaya Oblast', W Russian Federation 54°36′N 31°27′E
127 P2 **Krasnyye Baki** Nizhegorodskaya Oblast', W Russian Federation 57°07′N 45°12′E
127 Q13 **Krasnyye Barrikady** Astrakhanskaya Oblast', SW Russian Federation 46°14′N 47°48′E
124 K15 **Krasnyy Kholm** Tverskaya Oblast', W Russian Federation 58°04′N 37°05′E
127 Q8 **Krasnyy Kut** Saratovskaya Oblast', W Russian Federation 50°54′N 46°58′E
Krasnyy Liman see Krasnyy Lyman
117 Y7 **Krasnyy Luch** prev. Krindachevka. Luhans'ka Oblast', E Ukraine 48°09′N 38°52′E
117 X6 **Krasnyy Lyman** Rus. Krasnyy Liman. Donets'ka Oblast', SE Ukraine 49°00′N 37°50′E
127 R3 **Krasnyy Steklovar** Respublika Mariy El, W Russian Federation 56°14′N 48°49′E
127 P8 **Krasnyy Tekstil'shchik** Saratovskaya Oblast', W Russian Federation 51°35′N 45°49′E
127 R13 **Krasnyy Yar** Astrakhanskaya Oblast', SW Russian Federation 46°33′N 48°21′E
Krassóvár see Caraşova
116 L5 **Krasyliv** Khmel'nyts'ka Oblast', W Ukraine 49°38′N 26°59′E
111 O21 **Kraszna** Rom. Crasna. ≈ Hungary/Romania
Kratie see Krâchéh
113 P17 **Kratovo** NE FYR Macedonia 42°04′N 22°08′E
Kratznick see Kraśnik
171 Y13 **Krau** Papua, E Indonesia 03°15′S 140°07′E
167 Q13 **Krâvanh, Chuŏr Phnum** Eng. Cardamom Mountains, Fr. Chaîne des Cardamomes. ▲▲ W Cambodia
Kravasta Lagoon see Karavastasë, Laguna e
Krawang see Karawang
127 Q15 **Kraynovka** Respublika Dagestan, SW Russian Federation 43°58′N 47°24′E
118 D12 **Kražiai** Šiaulai, C Lithuania 55°36′N 22°41′E
27 P11 **Krebs** Oklahoma, C USA 34°55′N 95°43′W
101 D15 **Krefeld** Nordrhein-Westfalen, W Germany 51°20′N 06°34′E
Kreisstadt see Krosno Odrzańskie
115 D17 **Kremastón, Technití Límni** ▣ C Greece
Kremenchug see Kremenchuk
Kremenchugskoye Vodokhranilishche/Kremenchuk Reservoir see Kremenchuts'ke Vodoskhovyshche
117 S6 **Kremenchuk** Rus. Kremenchug. Poltavs'ka Oblast', NE Ukraine 49°04′N 33°27′E
117 R6 **Kremenchuts'ke Vodoskhovyshche** Eng. Kremenchuk Reservoir, Rus. Kremenchugskoye Vodokhranilishche. ▣ C Ukraine
116 K5 **Kremenets'** Pol. Krzemieniec, Rus. Kremenets. Ternopil's'ka Oblast', W Ukraine 50°06′N 25°43′E
Kremennaya see Kreminna
117 X6 **Kreminna** Rus. Kremennaya. Luhans'ka Oblast', E Ukraine 49°03′N 38°15′E
37 R4 **Kremmling** Colorado, C USA 40°03′N 106°23′W
109 V3 **Krems** ≈ NE Austria
Krems see Krems an der Donau
109 W3 **Krems an der Donau** var. Krems. Niederösterreich, N Austria 48°25′N 15°36′E
Kremsier see Kroměříž
109 S4 **Kremsmünster** Oberösterreich, N Austria 48°04′N 14°08′E
38 M17 **Krenitzin Islands** island Aleutian Islands, Alaska, USA
Kresena see Kresna
114 G11 **Kresna** var. Kresena. Blagoevgrad, SW Bulgaria 41°43′N 23°10′E
112 O12 **Krespoljin** Serbia, E Serbia 44°22′N 21°36′E
25 N4 **Kress** Texas, SW USA 34°21′N 101°43′W
123 V6 **Kresta, Zaliv** bay E Russian Federation
115 D20 **Kréstena** prev. Selinoús. Dytikí Elláda, S Greece 37°36′N 21°36′E
124 H14 **Kresttsy** Novgorodskaya Oblast', W Russian Federation 58°15′N 32°28′E
Kretikon Delagos see Kritikó Pélagos
118 C11 **Kretinga** Ger. Krottingen. Klaipėda, NW Lithuania 55°53′N 21°13′E
Kreutz see Cristuru Secuiesc
Kreuz see Križevci, Croatia
Kreuz see Risti, Estonia
Kreuzburg/Kreuzburg in Oberschlesien see Kluczbork
Kreuzingen see Bol'shakovo
108 H6 **Kreuzlingen** Thurgau, NE Switzerland 47°38′N 09°12′E
101 K25 **Kreuzspitze** ▲ S Germany 47°30′N 10°55′E
101 F16 **Kreuztal** Nordrhein-Westfalen, W Germany 50°58′N 08°00′E
119 I15 **Kreva** Rus. Krevo. Hrodzyenskaya Voblasts', W Belarus 54°19′N 26°17′E
Krevo see Kreva
Kría Vrísi see Krýa Vrýsi
79 D16 **Kribi** Sud, SW Cameroon 02°53′N 09°57′E
Krichëv see Krychaw
Krickerhäu/Kriegerhaj see Handlová
109 W6 **Krieglach** Steiermark, E Austria 47°33′N 15°37′E
108 F8 **Kriens** Luzern, W Switzerland 47°03′N 08°17′E
Krievija see Russian Federation
Krimmitschau see Crimmitschau
98 H12 **Krimpen aan den IJssel** Zuid-Holland, SW Netherlands 51°56′N 04°39′E
Krindachevka see Krasnyy Luch
115 G25 **Kríos, Akrotírio** headland Kríti, Greece, E Mediterranean Sea 35°17′N 23°31′E
155 J16 **Krishna** prev. Kistna. ≈ C India
155 H20 **Krishnagiri** Tamil Nādu, SE India 12°33′N 78°11′E
155 K17 **Krishna, Mouths of the** delta SE India
153 S15 **Krishnanagar** West Bengal, N India 23°22′N 88°32′E
155 G20 **Krishnarājāsāgara** var. Paradip. ▣ W India
95 N19 **Kristdala** Kalmar, S Sweden 57°24′N 16°12′E
Kristiania see Oslo
95 E18 **Kristiansand** var. Christiansand. Vest-Agder, S Norway 58°08′N 07°52′E
95 L22 **Kristianstad** Skåne, S Sweden 56°02′N 14°10′E
94 F8 **Kristiansund** var. Christiansund. Møre og Romsdal, S Norway 63°07′N 07°45′E
Kristiinankaupunki see Kristinestad
93 I14 **Kristineberg** Västerbotten, N Sweden 65°07′N 18°36′E
95 L16 **Kristinehamn** Värmland, C Sweden 59°17′N 14°09′E
93 J17 **Kristinestad** Fin. Kristiinankaupunki. Österbotten, W Finland 62°15′N 21°24′E
Kristyor see Crişcior
115 J25 **Kríti** Eng. Crete. ◐ region Greece, Aegean Sea
115 J24 **Kríti** Eng. Crete. island Greece, Aegean Sea
115 J23 **Kritikó Pélagos** var. Kretikon Delagos, Eng. Sea of Crete; anc. Mare Creticum. sea Greece, Aegean Sea
Kriulyany see Criuleni
112 I12 **Krivaja** ≈ NE Bosnia and Herzegovina
Krivaja see Mali Iđoš
113 P17 **Kriva Palanka** Turk. Eğri Palanka. NE Macedonia 42°13′N 22°19′E
Krivichi see Kryvichy
114 H8 **Krivodol** Vratsa, NW Bulgaria 43°23′N 23°30′E
126 M10 **Krivorozh'ye** Rostovskaya Oblast', SW Russian Federation 48°51′N 40°49′E
Krivoshin see Kryvoshyn
Krivoy Rog see Kryvyy Rih
112 F7 **Križevci** Ger. Kreuz, Hung. Kőrös. Varaždin, NE Croatia 46°02′N 16°32′E
112 B10 **Krk** It. Veglia. Primorje-Gorski Kotar, NW Croatia 45°01′N 14°36′E
112 B10 **Krk** It. Veglia; anc. Curieta. island NW Croatia
109 V12 **Krka** ≈ SE Slovenia
Krka see Gurk
109 R11 **Krn** ▲ NW Slovenia 46°15′N 13°37′E
111 H16 **Krnov** Ger. Jägerndorf. Moravskoslezský Kraj, E Czech Republic 50°05′N 17°40′E
Kroatien see Croatia
95 G14 **Krøderen** Buskerud, S Norway 60°06′N 09°48′E
95 G14 **Krøderen** ◎ S Norway
Kroi see Krui
95 N17 **Krokek** Östergötland, S Sweden 58°40′N 16°25′E
Krokodil see Crocodile
93 G16 **Krokom** Jämtland, C Sweden 63°20′N 14°30′E
117 S2 **Krolevets'** Rus. Krolevets. Sums'ka Oblast', NE Ukraine 51°34′N 33°24′E
Krolevets see Krolevets'
Królewska Huta see Chorzów
111 H18 **Kroměříž** Ger. Kremsier. Zlínský Kraj, E Czech Republic 49°18′N 17°24′E
98 I9 **Krommenie** Noord-Holland, C Netherlands 52°30′N 04°46′E
126 J6 **Kromy** Orlovskaya Oblast', W Russian Federation 52°41′N 35°45′E
101 L18 **Kronach** Bayern, E Germany 50°14′N 11°19′E
Krone an der Brahe see Koronowo
Krŏng Kaôh Kŏng see Kaôh Kŏng
95 K21 **Kronoberg** ◐ county S Sweden
123 V10 **Kronotskiy Zaliv** bay E Russian Federation
195 O2 **Kronprinsesse Märtha Kyst** physical region Antarctica
195 V3 **Kronprins Olav Kyst** physical region Antarctica
124 G12 **Kronshtadt** Leningradskaya Oblast', NW Russian Federation 60°01′N 29°42′E
Kronstadt see Braşov
83 I22 **Kroonstad** Free State, C South Africa 27°40′S 27°15′E
123 O12 **Kropotkin** Irkutskaya Oblast', C Russian Federation 58°30′N 115°21′E
126 L14 **Kropotkin** Krasnodarskiy Kray, SW Russian Federation 45°29′N 40°31′E
110 J11 **Krośniewice** Łódzskie, C Poland 52°14′N 19°10′E
111 N17 **Krosno** Ger. Krossen. Podkarpackie, SE Poland 49°40′N 21°46′E
110 E12 **Krosno Odrzańskie** Ger. Crossen, Kreisstadt. Lubuskie, W Poland 52°02′N 15°06′E
Krossen see Krosno
Krotoschin see Krotoszyn
110 H13 **Krotoszyn** Ger. Krotoschin. Wielkopolskie, C Poland 51°43′N 17°24′E
Krottingen see Kretinga
Krousón see Krousónas
115 J25 **Krousónas** prev. Krousón, Kroussón. Kríti, Greece, E Mediterranean Sea 35°14′N 24°59′E
Kroussón see Krousónas
Krraba see Krrabë
113 L20 **Krrabë** var. Krraba. Tiranë, C Albania 41°15′N 19°56′E
113 L17 **Krrabit, Mali i** ▲▲ N Albania
109 W12 **Krško** Ger. Gurkfeld; prev. Videm-Krško. E Slovenia 45°57′N 15°31′E
83 K19 **Kruger National Park** national park Northern, N South Africa
83 J21 **Krugersdorp** Gauteng, NE South Africa 26°06′S 27°46′E
38 D16 **Krugloi Point** headland Agattu Island, Alaska, USA 52°30′N 173°46′E
Krugloye see Kruhlaye
119 N15 **Kruhlaye** Rus. Krugloye. Mahilyowskaya Voblasts', E Belarus 54°15′N 29°48′E
168 L15 **Krui** var. Kroi. Sumatera, SW Indonesia 05°11′S 103°55′E
99 G16 **Kruibeke** Oost-Vlaanderen, N Belgium 51°10′N 04°18′E
83 G25 **Kruidfontein** Western Cape, SW South Africa 32°50′S 21°59′E
99 F15 **Kruiningen** Zeeland, SW Netherlands 51°28′N 04°01′E
Kruja see Krujë
113 L19 **Krujë** var. Kruja, It. Croia. Durrës, C Albania 41°30′N 19°48′E
Krulevshchina/Krulewshchyna see Krulyewshchyzna
118 K13 **Krulyewshchyzna** Rus. Krulevshchina, Krulewshchyna. Vitsyebskaya Voblasts', N Belarus 55°02′N 27°45′E
25 T6 **Krum** Texas, SW USA 33°15′N 97°14′W
101 J23 **Krumbach** Bayern, S Germany 48°12′N 10°21′E
113 M17 **Krumë** Kukës, NE Albania 42°11′N 20°25′E
Krummau see Český Krumlov
114 K12 **Krumovgrad** prev. Kossukavak. Yambol, E Bulgaria 41°27′N 25°40′E
114 K12 **Krumovitsa** ≈ S Bulgaria
114 L10 **Krumovo** Yambol, E Bulgaria 42°16′N 26°25′E
167 O11 **Krung Thep, Ao** var. Bight of Bangkok. bay S Thailand
Krung Thep Mahanakhon see Ao Krung Thep
Krupa/Krupa na Uni see Bosanska Krupa
119 M15 **Krupki** Minskaya Voblasts', C Belarus 54°19′N 29°08′E
95 G24 **Kruså** var. Krusaa. Syddanmark, SW Denmark 54°50′N 09°25′E
Krusaa see Kruså
113 N14 **Kruševac** Serbia, C Serbia 43°37′N 21°20′E
113 N19 **Kruševo** SW FYR Macedonia 41°22′N 21°15′E
111 A15 **Krušné Hory** Eng. Ore Mountains, Ger. Erzgebirge. ▲ Czech Republic/Germany see also Erzgebirge
Krušné Hory see Erzgebirge
39 W13 **Kruzof Island** island Alexander Archipelago, Alaska, USA
114 F13 **Krýa Vrýsi** var. Kría Vrísi. Kentrikí Makedonía, N Greece 40°41′N 22°18′E
119 P16 **Krychaw** Rus. Krichëv. Mahilyowskaya Voblasts', E Belarus 53°42′N 31°43′E
64 K11 **Krylov Seamount** undersea feature E Atlantic Ocean 17°35′N 30°07′W
Krym see Krym, Avtonomna Respublika
117 S13 **Krym, Avtonomna Respublika** var. Krym, Eng. Crimea, Crimean Oblast; prev. Rus. Krymskaya ASSR, Krymskaya Oblast'. ◐ province SE Ukraine
126 K14 **Krymsk** Krasnodarskiy Kray, SW Russian Federation 44°56′N 38°02′E
Krymskaya ASSR/Krymskaya Oblast' see Krym, Avtonomna Respublika
117 T13 **Kryms'ki Hory** ▲▲ S Ukraine
117 T13 **Kryms'kyy Pivostriv** peninsula S Ukraine
111 M18 **Krynica** Ger. Tannenhof. Małopolskie, S Poland 49°25′N 20°56′E
117 P8 **Kryve Ozero** Odes'ka Oblast', SW Ukraine 47°54′N 30°19′E
119 K14 **Kryvichy** Rus. Krivichi. Minskaya Voblasts', C Belarus 54°43′N 27°17′E
119 I18 **Kryvoshyn** Rus. Krivoshin. Brestskaya Voblasts', SW Belarus 52°52′N 26°08′E
117 S8 **Kryvyy Rih** Rus. Krivoy Rog. Dnipropetrovs'ka Oblast', SE Ukraine 47°53′N 33°24′E
117 N8 **Kryzhopil'** Vinnyts'ka Oblast', C Ukraine 48°22′N 28°51′E
Krzemieniec see Kremenets'
111 J14 **Krzepice** Śląskie, S Poland 50°58′N 18°42′E
110 F10 **Krzyż Wielkopolskie** Wielkopolskie, W Poland 52°52′N 16°03′E
Ksar al Kabir see Ksar-el-Kebir
Ksar al Soule see Er-Rachidia
74 J5 **Ksar El Boukhari** N Algeria 35°55′N 02°47′E
74 G5 **Ksar-el-Kebir** var. Alcázar, Ksar al Kabir, Ksar-el-Kébir, Ar. Al-Kasr al-Kebir, Al-Qsar al-Kbir, Sp. Alcazarquivir. NW Morocco 35°04′N 05°56′W
Ksar-el-Kébir see Ksar-el-Kebir
110 H12 **Książ Wielkopolski** Ger. Xions. Weilkopolskie, W Poland 52°03′N 17°10′E
127 O3 **Kstovo** Nizhegorodskaya Oblast', W Russian Federation 56°07′N 44°12′E
169 T8 **Kuala Belait** W Brunei 04°48′N 114°12′E
Kuala Dungun see Dungun
169 S10 **Kualakeriau** Borneo, C Indonesia
169 S12 **Kualakuayan** Borneo, C Indonesia 02°01′S 112°35′E
168 K8 **Kuala Lipis** Pahang, Peninsular Malaysia 04°11′N 102°00′E
168 K9 **Kuala Lumpur** ● (Malaysia) Kuala Lumpur, Peninsular Malaysia 03°08′N 101°42′E
168 K9 **Kuala Lumpur International** ✕ Selangor, Peninsular Malaysia 02°51′N 101°45′E
Kuala Pelabohan Kelang see Pelabuhan Klang
169 U7 **Kuala Penyu** Sabah, East Malaysia 05°37′N 115°36′E
38 E9 **Kualapu'u** var. Kualapuu. Moloka'i, Hawaii, USA, C Pacific Ocean 21°09′N 157°02′W
Kualapuu see Kualapu'u
168 L7 **Kuala Terengganu** var. Kuala Trengganu. Terengganu, Peninsular Malaysia 05°20′N 103°07′E
168 L11 **Kualatungkal** Sumatera, W Indonesia 0°49′S 103°22′E
171 P11 **Kuandang** Sulawesi, N Indonesia 0°50′N 122°55′E
163 V12 **Kuandian** var. Kuandian Manzu Zizhixian. Liaoning, NE China 40°41′N 124°46′E
Kuandian Manzu Zizhixian see Kuandian
83 E15 **Kuando Kubango** prev. Cuando Cubango. ◐ province SE Angola
Kuang-chou see Guangzhou
Kuang-hsi see Guangxi Zhuangzu Zizhiqu
Kuang-tung see Guangdong
Kuang-yuan see Guangyuan
Kuantan, Batang see Indragiri, Sungai
Kuanzhou see Qingjian
Kuba see Quba
Kubango see Cubango/Okavango
141 X8 **Kubārah** NW Oman 23°03′N 56°52′E
93 H16 **Kubbe** Västernorrland, C Sweden 63°31′N 18°04′E
80 A11 **Kubbum** Southern Darfur, W Sudan 11°47′N 23°47′E
124 L13 **Kubenskoye, Ozero** ◎ NW Russian Federation
146 G6 **Kubla-Ustyurt** Rus. Komsomol'sk-na-Ustyurte. Qoraqalpog'iston Respublikasi, NW Uzbekistan 44°06′N 58°14′E
164 G15 **Kubokawa** Kōchi, Shikoku, SW Japan 33°22′N 133°14′E
114 L7 **Kubrat** prev. Balbunar. Razgrad, N Bulgaria 43°48′N 26°31′E
112 O13 **Kučajske Planine** ▲▲ E Serbia
165 T1 **Kuccharo-ko** ◎ Hokkaidō, N Japan
112 O11 **Kučevo** Serbia, NE Serbia 44°29′N 21°42′E
Kuchan see Qūchān
169 Q10 **Kuching** prev. Sarawak. Sarawak, East Malaysia 01°32′N 110°20′E
169 Q10 **Kuching** ✕ Sarawak, East Malaysia 01°32′N 110°20′E
164 B17 **Kuchinoerabu-jima** island Nansei-shotō, SW Japan
Kuchinotsu see Minamishimabara
109 Q6 **Kuchl** Salzburg, NW Austria 47°37′N 13°12′E
148 L9 **Kūchnay Darwēshān** prev. Kūchnay Darweyshān. Helmand, S Afghanistan 31°02′N 64°10′E
Kūchnay Darweyshān see Kūchnay Darwēshān
Kuchurgan see Kuchurhan
117 O9 **Kuchurhan** Rus. Kuchurgan. ≈ NE Ukraine
Kuçova see Kuçovë
113 L21 **Kuçovë** var. Kuçova; prev. Qyteti Stalin. Berat, C Albania 40°48′N 19°55′E
136 D11 **Küçük Çekmece** İstanbul, NW Turkey 41°01′N 28°47′E
164 F14 **Kudamatsu** var. Kudamatu. Yamaguchi, Honshū, SW Japan 34°00′N 131°53′E
Kudamatu see Kudamatsu
Kudara see Ghŭdara
169 V6 **Kudat** Sabah, East Malaysia 06°54′N 116°47′E
Küddow see Gwda
155 G17 **Kūdligi** Karnātaka, W India 14°58′N 76°24′E
Kudowa see Kudowa-Zdrój
111 F16 **Kudowa-Zdrój** Ger. Kudowa. Wałbrzych, SW Poland 50°27′N 16°20′E
117 P9 **Kudryavtsivka** Mykolayivs'ka Oblast', S Ukraine 47°18′N 31°02′E
169 R16 **Kudus** prev. Koedoes. Jawa, C Indonesia 06°46′S 110°48′E
125 T13 **Kudymkar** Permskiy Kray, NW Russian Federation 59°01′N 54°40′E
Kudzsir see Cugir
Kuei-chou see Guizhou
Kuei-lin see Guilin
Kuei-Yang/Kuei-yang see Guiyang
K'u-erh-lo see Korla
Kueyang see Guiyang
Kufa see Al Kūfah
136 E14 **Küfiçayı** ≈ C Turkey
109 O6 **Kufstein** Tirol, W Austria 47°36′N 12°10′E
9 N7 **Kugaaruk** prev. Pelly Bay. Nunavut, N Canada 68°38′N 89°45′W
Kugaly see Kogaly
8 K8 **Kugluktuk** var. Qurlurtuuq; prev. Coppermine. Nunavut, NW Canada 67°49′N 115°12′W
143 Y13 **Kūhak** Sīstān va Balūchestān, SE Iran 27°10′N 63°15′E
143 R9 **Kūhbonān** Kermān, C Iran 31°23′N 56°16′E
148 J5 **Kūhestān** var. Kohsān. Herāt, W Afghanistan 34°40′N 61°11′E
Kūhīrī see Kahīrī
93 N15 **Kuhmo** Kainuu, E Finland 64°04′N 29°34′E
93 L18 **Kuhmoinen** Keski-Suomi, C Finland 61°32′N 25°09′E
Kuhnau see Konin
Kühnö see Kihnu
143 O8 **Kūhpāyeh** Eşfahān, C Iran 32°42′N 52°25′E
167 O12 **Kui Buri** var. Ban Kui Nua. Prachuap Khiri Khan, SW Thailand 12°10′N 99°49′E
Kuibyshev see Kuybyshevskoye Vodokhranilishche
82 D13 **Kuito** Port. Silva Porto. Bié, C Angola 12°21′S 16°55′E
39 X14 **Kuiu Island** island Alexander Archipelago, Alaska, USA
92 L13 **Kuivaniemi** Pohjois-Pohjanmaa, C Finland 65°34′N 25°13′E
196 M15 **Kujalleo** ◐ municipality S Greenland
Kujalleo, Kommune see Kujalleo
77 V14 **Kujama** Kaduna, C Nigeria 10°27′N 07°39′E
110 I10 **Kujawsko-pomorskie** ◐ province C Poland
165 R8 **Kuji** var. Kuzi. Iwate, Honshū, C Japan 40°12′N 141°47′E
Kujto, Ozero see Yushkozerskoye Vodokhranilishche
Kujū-renzan see Kujū-san
164 D14 **Kujū-san** var. Kujū-renzan. ▲ Kyūshū, SW Japan 33°07′N 131°13′E
43 N7 **Kukalaya, Rio** var. Rio Cuculaya, Rio Kukulaya. ≈ NE Nicaragua
113 O16 **Kukavica** var. Vlajna. ▲ SE Serbia 42°46′N 21°58′E
113 M18 **Kukës** var. Kukësi. Kukës, NE Albania 42°03′N 20°25′E
113 L18 **Kukës** ◐ district NE Albania
Kukësi see Kukës
186 D8 **Kukipi** Gulf, S Papua New Guinea 08°11′S 146°09′E
127 S3 **Kukmor** Respublika Tatarstan, W Russian Federation 56°11′N 50°56′E
Kukong see Shaoguan
39 N6 **Kukpowruk River** ≈ Alaska, USA
38 M6 **Kukpuk River** ≈ Alaska, USA
Küküdağ see Gogi, Mount
Kukukhoto see Hohhot
Kukulaya, Rio see Kukalaya, Rio
189 W12 **Kuku Point** headland NW Wake Island 19°19′N 166°36′E
146 G11 **Kukurtli** Ahal Welaýaty, C Turkmenistan 39°58′N 58°47′E
Kül see Kūl, Rūd-e
114 F7 **Kula** Vidin, NW Bulgaria 43°55′N 22°32′E
112 K9 **Kula** Vojvodina, NW Serbia 45°37′N 19°31′E
136 D14 **Kula** Manisa, W Turkey 38°33′N 28°38′E
149 S8 **Kulāchi** Khyber Pakhtunkhwa, NW Pakistan 31°58′N 70°30′E
Kulachi see Kolāchi
Kulagino see Yesbol
168 L10 **Kulai** Johor, Peninsular Malaysia 01°41′N 103°33′E
114 M7 **Kulak** ≈ NE Bulgaria
153 T11 **Kula Kangri** var. Kulhakangri. ▲ Bhutan/China 28°06′N 90°19′E
144 E13 **Kulaly, Ostrov** island SW Kazakhstan
145 S16 **Kulan** Kaz. Qulan; prev. Lugovoy, Lugovoye. Zhambyl, S Kazakhstan 42°54′N 72°45′E
147 V9 **Kulanak** Narynskaya Oblast', C Kyrgyzstan 41°18′N 75°38′E
Gory Kulandag see Gulandag
153 V14 **Kulaura** Sylhet, NE Bangladesh 24°32′N 92°02′E
118 D9 **Kuldīga** Ger. Goldingen. W Latvia 56°57′N 21°59′E
Kuldja see Yining
Kul'dzhuktau, Gory see QuljuqtovTog'lari
127 N4 **Kulebaki** Nizhegorodskaya Oblast', W Russian Federation 55°25′N 42°31′E
112 E11 **Kulen Vakuf** var. Spasovo. ◐ Federacija Bosne I Hercegovine, NW Bosnia and Herzegovina
181 Q9 **Kulgera Roadhouse** Northern Territory, N Australia 25°49′S 133°30′E
Kulhakangri see Kula Kangri
127 T1 **Kuliga** Udmurtskaya Respublika, NW Russian Federation 58°14′N 53°49′E
Kulkuduk see Ko'lquduq
118 G4 **Kullamaa** Läänemaa, W Estonia 58°52′N 24°05′E
197 O12 **Kullorsuaq** var. Kuvdlorssuak. ◐ Qaasuitsup, C Greenland
29 O6 **Kulm** North Dakota, N USA 46°18′N 98°57′W
Kulm see Chełmno
146 D12 **Kul'mach** prev. Turkm. Isgender. Balkan Welaýaty, W Turkmenistan 39°04′N 55°49′E
101 L18 **Kulmbach** Bayern, SE Germany 50°07′N 11°27′E
Kulmsee see Chełmża
147 Q14 **Kŭlob** Rus. Kulyab. SW Tajikistan 37°55′N 68°46′E
92 M13 **Kuloharju** Lappi, N Finland 65°51′N 28°10′E
125 N7 **Kuloy** Arkhangel'skaya Oblast', NW Russian Federation 64°55′N 43°35′E
125 N7 **Kuloy** ≈ NW Russian Federation
137 Q14 **Kulp** Dıyarbakır, SE Turkey 38°32′N 41°01′E
Kulpa see Kolpa
77 P14 **Kulpawn** ≈ N Ghana
143 R13 **Kūl, Rūd-e** var. Kül. ≈ S Iran
144 G12 **Kul'sary** Kaz. Qulsary. Atyrau, W Kazakhstan 46°59′N 54°02′E
153 R15 **Kulti** West Bengal, NE India 23°45′N 86°50′E
93 G14 **Kultsjön** Lapp. Gålto. ◎ N Sweden
136 I14 **Kulu** Konya, W Turkey 39°06′N 33°02′E
123 S9 **Kulu** ≈ E Russian Federation
122 I13 **Kulunda** Altayskiy Kray, S Russian Federation 52°33′N 79°04′E
Kulunda Steppe see Ravnina Kulyndy
Kulundinskaya Ravnina see Ravnina Kulyndy
182 M9 **Kulwin** Victoria, SE Australia 35°04′S 142°37′E
Kulyab see Kŭlob
117 Q3 **Kulykivka** Chernihivs'ka Oblast', N Ukraine 51°23′N 31°39′E
Kum see Qom
164 F14 **Kuma** Ehime, Shikoku, SW Japan 33°36′N 132°53′E
127 P14 **Kuma** ≈ SW Russian Federation
165 O12 **Kumagaya** Saitama, Honshū, S Japan 36°09′N 139°22′E
165 Q5 **Kumaishi** Hokkaidō, NE Japan 42°08′N 139°57′E
169 R13 **Kumai, Teluk** bay Borneo, C Indonesia
127 Y7 **Kumak** Orenburgskaya Oblast', W Russian Federation 51°16′N 60°06′E
164 C14 **Kumamoto** Kumamoto, Kyūshū, SW Japan 32°49′N 130°41′E
164 D15 **Kumamoto** off. Kumamoto-ken. ◐ prefecture Kyūshū, SW Japan
Kumamoto-ken see Kumamoto
164 J15 **Kumano** Mie, Honshū, SW Japan 33°54′N 136°08′E
Kumanova see Kumanovo
113 O17 **Kumanovo** Turk. Kumanova. N Macedonia 42°08′N 21°43′E
185 G17 **Kumara** West Coast, South Island, New Zealand 42°39′S 171°12′E
180 J8 **Kumarina Roadhouse** Western Australia 24°46′S 119°39′E
153 T15 **Kumarkhali** Khulna, W Bangladesh 23°54′N 89°16′E
77 P16 **Kumasi** prev. Coomassie. C Ghana 06°41′N 01°40′W
Kumayri see Gyumri
79 D15 **Kumba** Sud-Ouest, W Cameroon 04°39′N 09°26′E
114 N13 **Kumbağ** Tekirdağ, NW Turkey 40°51′N 27°26′E
155 J21 **Kumbakonam** Tamil Nādu, SE India 10°59′N 79°24′E
Kum-Dag see Gumdag
165 R16 **Kume-jima** island Nansei-shotō, SW Japan
127 V6 **Kumertau** Respublika Bashkortostan, W Russian Federation 52°48′N 55°48′E
Kumillā see Comilla
35 R4 **Kumiva Peak** ▲ Nevada, W USA 40°24′N 119°16′W
159 N7 **Kum Kuduk** Xinjiang Uygur Zizhiqu, W China 40°15′N 91°55′E
159 N8 **Kum Kuduk** well NW China
Kumkurgan see Qumqo'rg'on
95 M16 **Kumla** Örebro, C Sweden 59°08′N 15°09′E
136 E17 **Kumluca** Antalya, SW Turkey 36°23′N 30°17′E
100 N9 **Kummerower See** ◎ NE Germany
77 X14 **Kumo** Gombe, E Nigeria 10°03′N 11°14′E
145 O13 **Kumola** ≈ C Kazakhstan
167 N1 **Kumon Range** ▲▲ N Myanmar (Burma)
83 F22 **Kums** Karas, SE Namibia 28°07′S 19°40′E
155 E18 **Kumta** Karnātaka, W India 14°25′N 74°24′E
38 H12 **Kumukahi, Cape** headland Hawai'i, USA, C Pacific Ocean 19°31′N 154°48′W
127 Q17 **Kumukh** Respublika Dagestan, SW Russian Federation 42°10′N 47°07′E
Kumul see Hami
158 L6 **Kümüx** Xinjiang Uygur Zizhiqu, W China
127 N9 **Kumylzhenskaya** Volgogradskaya Oblast', SW Russian Federation 49°54′N 42°35′E
141 W6 **Kumzār** N Oman 26°19′N 56°26′E
43 W15 **Kuna de Wargandí** ◐ special territory NE Panama
149 S4 **Kunar** Per. Konarhā; prev. Konar. ◐ province E Afghanistan
Kunashiri see Kunashir, Ostrov
123 U14 **Kunashir, Ostrov** var. Kunashiri. island Kuril'skiye Ostrova, SE Russian Federation
43 V14 **Kuna Yala** prev. San Blas. ◐ special territory NE Panama
118 I3 **Kunda** Lääne-Virumaa, NE Estonia 59°31′N 26°33′E
152 M13 **Kunda** Uttar Pradesh, N India 25°43′N 81°31′E
155 E19 **Kundāpura** var. Coondapoor. Karnātaka, W India 13°39′N 74°41′E
79 O24 **Kundelungu, Monts** ▲▲ S Dem. Rep. Congo
Kundert see Hernád
186 D7 **Kundiawa** Chimbu, W Papua New Guinea 06°00′S 144°57′E
Kundla see Sāvarkundla
Kunduk, Ozero see Sasyk, Ozero
Kunduk, Ozero Sasyk see Sasyk, Ozero
168 L10 **Kundur, Pulau** island W Indonesia
149 Q2 **Kunduz** var. Kondūz, Qondūz; prev. Kondoz, Kundūz. Kunduz, NE Afghanistan 36°49′N 68°50′E
149 Q2 **Kunduz** prev. Kondoz. ◐ province NE Afghanistan
Kunduz/Kundūz see Kunduz
Kuneitra see Al Qunayţirah
83 B18 **Kunene** ◐ district NE Namibia
83 A16 **Kunene** var. Cunene. ≈ Angola/Namibia see also Cunene
Kunene see Cunene
Künes see Xinyuan
158 J5 **Künes He** ≈ NW China
95 I19 **Kungälv** Västra Götaland, S Sweden 57°54′N 12°00′E
147 W7 **Kungei Ala-Tau** Rus. Khrebet Kyungëy Ala-Too, Kir. Küngöy Ala-Too. ▲▲ Kazakhstan/Kyrgyzstan
Küngöy Ala-Too see Kungei Ala-Tau
Kungrad see Qo'ng'irot
95 J19 **Kungsbacka** Halland, S Sweden 57°30′N 12°05′E
95 I18 **Kungshamn** Västra Götaland, S Sweden 58°21′N 11°15′E
95 M16 **Kungsör** Västmanland, C Sweden 59°25′N 16°05′E
79 J16 **Kungu** Equateur, NW Dem. Rep. Congo 02°47′N 19°12′E
125 V15 **Kungur** Permskiy Kray, NW Russian Federation 57°24′N 56°56′E
166 L9 **Kungyangon** Yangon, SW Myanmar (Burma) 16°27′N 96°00′E
111 M22 **Kunhegyes** Jász-Nagykun-Szolnok, E Hungary 47°22′N 20°36′E
167 O5 **Kunhing** Shan State, E Myanmar (Burma) 21°17′N 98°26′E
158 D9 **Kunjirap Daban** var. Khūnjerāb Pass. pass China/Pakistan see also Khūnjerāb Pass
Kunjirap Daban see Khunjerāb Pass
Kunlun Mountains see Kunlun Shan
158 H10 **Kunlun Shan** Eng. Kunlun Mountains. ▲▲ NW China
159 P11 **Kunlun Shankou** pass C China
160 G13 **Kunming** var. K'un-ming; prev. Yunnan. province capital Yunnan, SW China 25°04′N 102°41′E
K'un-ming see Kunming
Kunø see Kunoy
95 B18 **Kunoy** Dan. Kunø. island N Faroe Islands
Kunsan see Gunsan
111 L24 **Kunszentmárton** Jász-Nagykun-Szolnok, E Hungary 46°50′N 20°19′E
111 J23 **Kunszentmiklós** Bács-Kiskun, C Hungary 47°00′N 19°07′E
181 N3 **Kununurra** Western Australia 15°50′S 128°44′E
Kunyang see Pingyang
Kunya-Urgench see Köneürgenç
Kunyé see Pins, Île des
169 T11 **Kunyi** Borneo, C Indonesia 03°23′S 119°20′E
101 I20 **Künzelsau** Baden-Württemberg, S Germany 49°22′N 09°43′E
161 S10 **Kuocang Shan** ▲▲ SE China
Kuolajärvi see Kuoloyarvi
124 H5 **Kuoloyarvi** Finn. Kuolajärvi, var. Luolajarvi. Murmanskaya Oblast', NW Russian Federation 66°58′N 29°13′E
93 N16 **Kuopio** Pohjois-Savo, C Finland 62°54′N 27°41′E
93 K17 **Kuortane** Etelä-Pohjanmaa, W Finland 62°48′N 23°30′E
93 M18 **Kuortti** Etelä-Savo, E Finland 61°25′N 26°25′E
Kupa see Kolpa
171 P17 **Kupang** prev. Koepang. Timor, C Indonesia 10°13′S 123°38′E
39 Q5 **Kuparuk River** ≈ Alaska, USA
Kupchino see Cupcina
186 E9 **Kupiano** Central, S Papua New Guinea 10°06′S 148°12′E
180 M4 **Kupingarri** Western Australia 16°46′S 125°57′E
122 I12 **Kupino** Novosibirskaya Oblast', C Russian Federation 54°22′N 77°09′E
118 H11 **Kupiškis** Panevėžys, NE Lithuania 55°51′N 24°58′E
114 L13 **Küplü** Edirne, NW Turkey 41°06′N 26°23′E
39 X13 **Kupreanof Island** island Alexander Archipelago, Alaska, USA
39 O16 **Kupreanof Point** headland Alaska, USA 55°34′N 159°36′W
112 G13 **Kupres** ◐ Federacija Bosne I Hercegovine, SW Bosnia and Herzegovina
117 W5 **Kup"yans'k** Rus. Kupyansk. Kharkivs'ka Oblast', E Ukraine 49°42′N 37°36′E
Kupyansk see Kup"yans'k

◆ Country
● Country Capital
◇ Dependent Territory
○ Dependent Territory Capital
◐ Administrative Regions
✕ International Airport
▲ Mountain
▲▲ Mountain Range
⛰ Volcano
≈ River
◎ Lake
▣ Reservoir

117 W5 **Kup"yans'k-Vuzlovyy** Kharkivs'ka Oblast', E Ukraine *49°40′N 37°41′E*
158 I6 **Kuqa** Xinjiang Uygur Zizhiqu, NW China *41°43′N 82°58′E*
Kür *see* Kura
137 W11 **Kura** *Az.* Kür, *Geor.* Mtkvari, *Turk.* Kura Nehri. ≈ SW Asia
55 R8 **Kuracki** NW Guyana *06°52′N 60°13′W*
Kura Kurk *see* Irbe Strait
147 Q10 **Kurama Range** *Rus.* Kuraminskiy Khrebet. ⛰ Tajikistan/Uzbekistan
Kuraminskiy Khrebet *see* Kurama Range
Kura Nehri *see* Kura
119 J14 **Kuranyets** *Rus.* Kurenets. Minskaya Voblasts', C Belarus *54°33′N 26°57′E*
164 H13 **Kurashiki** *var.* Kurasiki. Okayama, Honshū, SW Japan *34°35′N 133°44′E*
154 L10 **Kurasia** Chhattisgarh, C India *23°11′N 82°16′E*
Kurasiki *see* Kurashiki
164 H12 **Kurayoshi** *var.* Kurayosi. Tottori, Honshū, SW Japan *35°27′N 133°52′E*
Kurayosi *see* Kurayoshi
163 X6 **Kurbin He** ≈ NE China
Kurchum *see* Kurshim
Kurchum *see* Kurshim
137 X11 **Kürdämir** *Rus.* Kyurdamir. C Azerbaijan *40°21′N 48°08′E*
Kurdestan *see* Kordestān
139 S1 **Kurdistan** *cultural region* SW Asia
Kurd Kui *see* Kord Küy
155 F15 **Kurduvādi** Mahārāshtra, W India *18°06′N 75°31′E*
Kŭrdzhali *see* Kardzhali
Kŭrdzhali *see* Kardzhali
Kŭrdzhali, Yazovir *see* Kardzhali, Yazovir
164 F13 **Kure** Hiroshima, Honshū, SW Japan *34°15′N 132°33′E*
192 K5 **Kure Atoll** *var.* Ocean Island. *atoll* Hawaiian Islands, Hawaii, USA
136 J10 **Küre Dağları** ⛰ N Turkey
146 C11 **Kürendag** *Rus.* Gora Kyuren. ▲ W Turkmenistan *39°05′N 55°09′E*
Kurenets *see* Kuranyets
118 E6 **Kuressaare** *Ger.* Arensburg; *prev.* Kingissepp. Saaremaa, W Estonia *58°17′N 22°29′E*
122 K9 **Kureyka** Krasnoyarskiy Kray, N Russian Federation *66°22′N 87°21′E*
122 K9 **Kureyka** ≈ N Russian Federation
Kurgal'dzhino/Kurgal'dzhinsky *see* Korgalzhyn
122 G11 **Kurgan** Kurganskaya Oblast', C Russian Federation *55°30′N 65°20′E*
126 L14 **Kurganinsk** Krasnodarskiy Kray, SW Russian Federation *44°55′N 40°45′E*
122 G11 **Kurganskaya Oblast'** ◈ *province* C Russian Federation
Kurgan-Tyube *see* Qŭrghonteppa
191 O2 **Kuria** *prev.* Woodle Island. *island* Tungaru, W Kiribati
Kuria Muria Bay *see* Ḥalānīyāt, Khalīj al
Kuria Muria Islands *see* Ḥalānīyāt, Juzur al
153 T13 **Kurigram** Rajshahi, N Bangladesh *25°49′N 89°39′E*
93 K17 **Kurikka** Etelä-Pohjanmaa, W Finland *62°36′N 22°25′E*
192 I3 **Kuril Basin** *var.* Kurile Basin. *undersea basin* NW Pacific Ocean
Kurile Basin *see* Kuril Basin
Kurile Islands *see* Kuril'skiye Ostrova
Kurile-Kamchatka Depression *see* Kuril-Kamchatka Trench
Kurile Trench *see* Kuril-Kamchatka Trench
Kuril Islands *see* Kuril'skiye Ostrova
192 J3 **Kuril-Kamchatka Trench** *var.* Kurile-Kamchatka Depression, Kurile Trench. *trench* NW Pacific Ocean
127 Q9 **Kurilovka** Saratovskaya Oblast', W Russian Federation *50°39′N 48°02′E*
123 U13 **Kuril'sk** *Jap.* Shana. Kuril'skiye Ostrova, Sakhalinskaya Oblats', SE Russian Federation *45°10′N 147°51′E*
122 G11 **Kuril'skiye Ostrova** *Eng.* Kuril Islands, Kurile Islands. *island group* SE Russian Federation
42 M9 **Kurinwas, Río** ≈ E Nicaragua
Kurisches Haff *see* Courland Lagoon
Kurkund *see* Kilingi-Nõmme
126 M4 **Kurlovskiy** Vladimirskaya Oblast', W Russian Federation *55°25′N 40°39′E*
80 G12 **Kurmuk** Blue Nile, SE Sudan *10°36′N 34°16′E*
Kurna *see* Al Qurnah
155 H17 **Kurnool** *var.* Karnul. Andhra Pradesh, S India *15°51′N 78°01′E*
164 M11 **Kurobe** Toyama, Honshū, SW Japan *36°55′N 137°24′E*
165 Q7 **Kuroishi** *var.* Kuroisi. Aomori, Honshū, C Japan *40°37′N 140°34′E*
Kuroisi *see* Kuroishi
165 O12 **Kuroiso** Tochigi, Honshū, S Japan *36°58′N 140°02′E*
165 Q4 **Kuromatsunai** Hokkaidō, NE Japan *42°40′N 140°18′E*
164 B17 **Kuro-shima** *island* SW Japan
185 F21 **Kurow** Canterbury, South Island, New Zealand *44°44′S 170°29′E*
127 N15 **Kursavka** Stavropol'skiy Kray, SW Russian Federation *44°28′N 42°31′E*
118 E11 **Kuršėnai** Šiauliai, N Lithuania *56°00′N 22°56′E*
145 X10 **Kurshim** *prev.* Kurchum. Vostochnyy Kazakhstan, E Kazakhstan *48°35′N 83°37′E*
145 Y10 **Kurshim** *prev.* Kurchum. ≈ E Kazakhstan
Kurshskaya Kosa/Kuršių Nerija *see* Courland Spit
126 J7 **Kursk** Kurskaya Oblast', W Russian Federation *51°44′N 36°47′E*
126 I7 **Kurskaya Oblast'** ◈ *province* W Russian Federation
Kurskiy Zaliv *see* Courland Lagoon
113 N15 **Kuršumlija** Serbia, S Serbia *43°09′N 21°16′E*
137 R15 **Kurtalan** Siirt, SE Turkey *37°58′N 41°36′E*
Kurtbunar *see* Tervel
Kurt-Dere *see* Valchi Dol
Kurtitsch/Kürtös *see* Curtici
145 U15 **Kurtty** ≈ SE Kazakhstan
93 L18 **Kuru** Pirkanmaa, W Finland *61°51′N 23°46′E*
80 C13 **Kuru** ≈ W South Sudan
114 M13 **Kuru Dağı** ⛰ NW Turkey
158 L7 **Kuruktag** ⛰ NW China
83 G22 **Kuruman** Northern Cape, N South Africa *27°28′S 23°27′E*
67 T14 **Kuruman** ≈ W South Africa
164 D14 **Kurume** Fukuoka, Kyūshū, SW Japan *33°15′N 130°27′E*
123 N13 **Kurumkan** Respublika Buryatiya, S Russian Federation *54°13′N 110°21′E*
155 J25 **Kurunegala** North Western Province, C Sri Lanka *07°28′N 80°23′E*
55 T10 **Kurupukari** C Guyana *04°39′N 58°39′W*
125 U10 **Kur"ya** Respublika Komi, NW Russian Federation *61°38′N 57°12′E*
144 E15 **Kuryk** *var.* Yeraliyev, *Kaz.* Quryq. Mangistau, SW Kazakhstan *43°12′N 51°43′E*
136 B15 **Kuşadası** Aydın, SW Turkey *37°50′N 27°16′E*
115 M19 **Kuşadası Körfezi** *gulf* SW Turkey
164 A17 **Kusagaki-guntō** *island* SW Japan
Kusaie *see* Kosrae
145 T12 **Kusak** ≈ C Kazakhstan
Kusary *see* Qusar
167 P7 **Ku Sathan, Doi** ▲ NW Thailand *18°22′N 100°31′E*
164 J13 **Kusatsu** *var.* Kusatu. Shiga, Honshū, SW Japan *35°02′N 136°00′E*
Kusatu *see* Kusatsu
138 F11 **Kuseifa** Southern, C Israel *31°15′N 35°01′E*
136 C12 **Kuş Gölü** ◎ NW Turkey
126 L12 **Kushchevskaya** Krasnodarskiy Kray, SW Russian Federation *46°35′N 39°40′E*
164 D16 **Kushima** *var.* Kusima. Miyazaki, Kyūshū, SW Japan *31°28′N 131°14′E*
164 I15 **Kushimoto** Wakayama, Honshū, SW Japan *33°28′N 135°45′E*
165 V4 **Kushiro** *var.* Kusiro. Hokkaidō, NE Japan *42°58′N 144°24′E*
148 K4 **Kushk** *prev.* Kūshk. Herāt, W Afghanistan *34°55′N 62°20′E*
Kushka *see* Serhetabat
Kushka *see* Gushgy/Serhetabat
Kushmurun *see* Kusmuryn
Kushmurun, Ozero *see* Kusmuryn, Ozero
127 U4 **Kushnarenkovo** Respublika Bashkortostan, W Russian Federation *55°07′N 55°24′E*
Kushrabat *see* Qo'shrabot
Kushtia *see* Kustia
Kusima *see* Kushima
Kusiro *see* Kushiro
38 M13 **Kuskokwim Bay** *bay* Alaska, USA
39 P11 **Kuskokwim Mountains** ⛰ Alaska, USA
39 N12 **Kuskokwim River** ≈ Alaska, USA
145 N8 **Kusmuryn** *Kaz.* Qusmuryn; *prev.* Kushmurun. Kostanay, N Kazakhstan *52°27′N 64°31′E*
145 N8 **Kusmuryn, Ozero** *Kaz.* Qusmuryn; *prev.* Ozero Kushmurun. ◎ N Kazakhstan
108 G7 **Küsnacht** Zürich, N Switzerland *47°19′N 08°34′E*
165 V4 **Kussharo-ko** *var.* Kussyaro. ◎ Hokkaidō, NE Japan
Küssnacht *see* Küssnacht am Rigi
108 F8 **Küssnacht am Rigi** *var.* Küssnacht. Schwyz, C Switzerland *47°03′N 08°25′E*
Kussyaro *see* Kussharo-ko
Kustanay *see* Kostanay
Küstence/Küstendje *see* Constanța
100 F11 **Küstenkanal** *var.* Ems-Hunte Canal. *canal* NW Germany
153 T15 **Kustia** *var.* Kushtia. Khulna, W Bangladesh *23°54′N 89°07′E*
Küstrin *see* Kostrzyn
171 R11 **Kusu** Pulau Halmahera, E Indonesia *0°51′N 127°41′E*
170 L16 **Kuta** Pulau Lombok, S Indonesia *08°53′S 116°15′E*
139 T4 **Kutabān** Kirkūk, N Iraq *35°21′N 44°45′E*
136 E13 **Kütahya** *prev.* Kutaia. Kütahya, W Turkey *39°25′N 29°56′E*
136 E13 **Kütahya** *var.* Kutaia. ◈ *province* W Turkey
Kutai *see* Mahakam, Sungai
Kutaia *see* Kütahya
137 R9 **K'ut'aisi** W Georgia *42°16′N 42°42′E*
Kūt, Al *see* Wāsiṭ
Kut al 'Amārah *see* Al Kūt
Kut al Hai/Kūt al Ḩayy *see* Al Ḩayy
Kut al Imara *see* Al Kūt
123 Q11 **Kutana** Respublika Sakha (Yakutiya), NE Russian Federation *59°05′N 131°43′E*
Kutaradja/Kutaraja *see* Banda Aceh
165 R4 **Kutchan** Hokkaidō, NE Japan *42°54′N 140°46′E*
Kutch, Gulf of *see* Kachchh, Gulf of
Kutch, Rann of *see* Kachchh, Rann of
112 F9 **Kutina** Sisak-Moslavina, NE Croatia *45°29′N 16°45′E*
112 H9 **Kutjevo** Požega-Slavonija, NE Croatia *45°26′N 17°54′E*
111 E17 **Kutná Hora** *Ger.* Kuttenberg. Střední Čechy, C Czech Republic *49°58′N 15°18′E*
110 K12 **Kutno** Łódzkie, C Poland *52°14′N 19°23′E*
Kuttenberg *see* Kutná Hora
79 I20 **Kutu** Bandundu, W Dem. Rep. Congo *02°42′S 18°10′E*
153 V17 **Kutubdia Island** *island* SE Bangladesh
80 B10 **Kutum** Northern Darfur, W Sudan *14°10′N 24°40′E*
147 Y7 **Kuturgu** Issyk-Kul'skaya Oblast', E Kyrgyzstan *42°45′N 78°04′E*
12 M5 **Kuujjuaq** *prev.* Fort-Chimo. Québec, E Canada *58°10′N 68°15′W*
12 I7 **Kuujjuarapik** Québec, C Canada *55°07′N 78°09′W*
12 I7 **Kuujjuarapik** *prev.* Poste-de-la-Baleine. Québec, NE Canada *55°13′N 77°54′W*
Kuuli-Mayak *see* Guwlumaýak
118 I6 **Kuulsemägi** ▲ S Estonia
92 N13 **Kuusamo** Pohjois-Pohjanmaa, E Finland *65°57′N 29°15′E*
93 M19 **Kuusankoski** Kymenlaakso, S Finland *60°51′N 26°40′E*
127 W7 **Kuvandyk** Orenburgskaya Oblast', W Russian Federation *51°27′N 57°18′E*
Kuvango *see* Cubango
Kuvasay *see* Quvasoy
Kuvdlorssuak *see* Kullorsuaq
124 I16 **Kuvshinovo** Tverskaya Oblast', W Russian Federation *57°03′N 34°09′E*
141 Q4 **Kuwait** *off.* State of Kuwait, *var.* Dawlat al Kuwait, Koweit, Kuweit. ◆ *monarchy* SW Asia
Kuwait *see* Al Kuwayt
Kuwait Bay *see* Kuwayt, Jūn al
Kuwait City *see* Al Kuwayt
Kuwait, Dawlat al *see* Kuwait
Kuwait, State of *see* Kuwait
Kuwajleen *see* Kwajalein Atoll
164 K13 **Kuwana** Mie, Honshū, SW Japan *35°04′N 136°40′E*
139 X9 **Kuwayt** Maysān, E Iraq *32°26′N 47°12′E*
142 K11 **Kuwayt, Jūn al** *var.* Kuwait Bay. *bay* E Kuwait
Kuweit *see* Kuwait
117 P10 **Kuyal'nyts'kyy Lyman** ◎ SW Ukraine
122 I12 **Kuybyshev** Novosibirskaya Oblast', C Russian Federation *55°28′N 77°55′E*
Kuybyshev *see* Bolgar, Respublika Tatarstan, Russian Federation
Kuybyshev *see* Samara
117 W9 **Kuybysheve** *Rus.* Kuybyshevo. Zaporiz'ka Oblast', SE Ukraine *47°20′N 36°41′E*
Kuybyshevo *see* Kuybysheve
Kuybyshev Reservoir *see* Kuybyshevskoye Vodokhranilishche
Kuybyshevskaya Oblast' *see* Samarskaya Oblast'
Kuybyshevskiy *see* Novoishimskiy
127 R4 **Kuybyshevskoye Vodokhranilishche** *var.* Kuibyshev, *Eng.* Kuybyshev Reservoir. ⊠ W Russian Federation
123 S9 **Kuydusun** Respublika Sakha (Yakutiya), NE Russian Federation *63°15′N 143°10′E*
125 U16 **Kuyeda** Permskiy Kray, NW Russian Federation *56°23′N 55°19′E*
Küysanjaq *see* Koye
158 J4 **Kuytun** Xinjiang Uygur Zizhiqu, NW China *44°25′N 84°55′E*
122 M13 **Kuytun** Irkutskaya Oblast', S Russian Federation *54°18′N 101°28′E*
55 S12 **Kuyuwini Landing** S Guyana *02°06′N 59°14′W*
Kuzi *see* Kuji
38 M9 **Kuzitrin River** ≈ Alaska, USA
127 P6 **Kuznetsk** Penzenskaya Oblast', W Russian Federation *53°06′N 46°27′E*
116 K3 **Kuznetsovs'k** Rivnens'ka Oblast', NW Ukraine *51°21′N 25°51′E*
165 R8 **Kuzumaki** Iwate, Honshū, C Japan *40°04′N 141°26′E*
95 H24 **Kværndrup** Syddtjylland, C Denmark *55°10′N 10°31′E*
92 H9 **Kvaløya** *island* N Norway
92 K8 **Kvalsund** Finnmark, N Norway *70°30′N 23°56′E*
94 G11 **Kvam** Oppland, S Norway *61°42′N 09°43′E*
127 X7 **Kvarkeno** Orenburgskaya Oblast', W Russian Federation *52°09′N 59°44′E*
93 G15 **Kvarnbergsvattnet** *var.* Frostviken. ◎ N Sweden
112 A11 **Kvarner** *var.* Carnaro, *It.* Quarnero. *gulf* W Croatia
112 B11 **Kvarnerić** *channel* W Croatia
39 O14 **Kvichak Bay** *bay* Alaska, USA
92 H12 **Kvikkjokk** *Lapp.* Huhttán. Norrbotten, N Sweden *66°58′N 17°45′E*
95 D17 **Kvina** ≈ S Norway
95 Q1 **Kvitøya** *island* NE Svalbard
95 F16 **Kvitseid** Telemark, S Norway *59°23′N 08°31′E*
79 H20 **Kwa** ≈ W Dem. Rep. Congo
83 L23 **KwaDukuza** *prev.* Stanger. KwaZulu/Natal, E South Africa *29°20′S 31°18′E* *see also* Stanger
77 Q15 **Kwadwokurom** C Ghana *07°49′N 00°15′W*
186 M8 **Kwailibesi** Malaita, N Solomon Islands *08°25′S 160°48′E*
189 S6 **Kwajalein Atoll** *var.* Kuwajleen. *atoll* Ralik Chain, C Marshall Islands
55 W9 **Kwakoegron** Brokopondo, N Suriname *05°14′N 55°20′W*
81 J21 **Kwale** Kwale, S Kenya *04°10′S 39°27′E*
77 U17 **Kwale** Delta, S Nigeria *05°51′N 06°29′E*
81 J21 **Kwale** ◈ *county* SE Kenya
79 H20 **Kwamouth** Bandundu, W Dem. Rep. Congo *03°11′S 16°16′E*
Kwando *see* Cuando
Kwangchow *see* Guangzhou
Kwangchu *see* Gwangju
Kwangju *see* Gwangju
Kwangju-gwangyŏksi *see* Gwangju
79 H20 **Kwango** *Port.* Cuango. ≈ Angola/Dem. Rep. Congo *see also* Cuango
Kwango *see* Cuango
Kwangsi/Kwangsi Chuang Autonomous Region *see* Guangxi Zhuangzu Zizhiqu
Kwangtung *see* Guangdong
Kwangyuan *see* Guangyuan
81 F17 **Kwania, Lake** ◎ C Uganda
82 B11 **Kwanza Norte** *var.* Cuanza Norte. ◈ *province* NE Angola
82 B12 **Kwanza Sul** *prev.* Cuanza Sul. ◈ *province* NE Angola
77 S15 **Kwara** ◈ *state* SW Nigeria
83 K22 **KwaZulu/Natal** *off.* KwaZulu/Natal Province; *prev.* Natal. ◈ *province* E South Africa
KwaZulu/Natal Province *see* KwaZulu/Natal
Kweichow *see* Guizhou
Kweichu *see* Guiyang
Kweilin *see* Guilin
Kweisui *see* Hohhot
Kweiyang *see* Guiyang
83 K17 **Kwekwe** *prev.* Que Que. Midlands, C Zimbabwe *18°56′S 29°49′E*
83 G20 **Kweneng** ◈ *district* S Botswana
Kwesui *see* Hohhot
39 N12 **Kwethluk** Alaska, USA *60°48′N 161°26′W*
39 N12 **Kwethluk River** ≈ Alaska, USA
110 J8 **Kwidzyń** *Ger.* Marienwerder. Pomorskie, N Poland *53°44′N 18°55′E*
38 M13 **Kwigillingok** Alaska, USA *59°52′N 163°08′W*
186 E9 **Kwikila** Central, S Papua New Guinea *09°51′S 147°43′E*
79 I20 **Kwilu** ≈ W Dem. Rep. Congo
Kwito *see* Cuito
171 U12 **Kwoka, Gunung** ▲ Papua Barat, E Indonesia *0°34′S 132°25′E*
78 I12 **Kyabé** Moyen-Chari, S Chad *09°28′N 18°54′E*
183 O11 **Kyabram** Victoria, SE Australia *36°21′S 145°05′E*
166 M9 **Kyaikkami** *prev.* Amherst. Mon State, S Myanmar (Burma) *16°03′N 97°36′E*
166 L9 **Kyaiklat** Ayeyawady, SW Myanmar (Burma) *16°25′N 95°42′E*
166 M8 **Kyaikto** Mon State, S Myanmar (Burma) *17°16′N 97°01′E*
123 N14 **Kyakhta** Respublika Buryatiya, S Russian Federation *50°25′N 106°13′E*
182 G8 **Kyancutta** South Australia *33°10′S 135°33′E*
167 T8 **Ky Anh** Ha Tinh, N Vietnam *18°05′N 106°16′E*
166 L5 **Kyaukpadaung** Mandalay, C Myanmar (Burma) *20°50′N 95°08′E*
Kyaukpyu *see* Kyaunkpyu
166 M5 **Kyaukse** Mandalay, C Myanmar (Burma) *21°33′N 96°06′E*
166 L8 **Kyaunggon** Ayeyawady, SW Myanmar (Burma) *17°04′N 95°12′E*
166 J6 **Kyaunkpyu** *var.* Kyaukpyu. Rakhine State, W Myanmar (Burma) *19°27′N 93°33′E*
119 E14 **Kybartai** *Pol.* Kibarty. Marijampolė, S Lithuania *54°37′N 22°44′E*
152 I7 **Kyelang** Himāchal Pradesh, NW India *32°33′N 77°03′E*
111 G19 **Kyjov** *Ger.* Gaya. Jihomoravský Kraj, SE Czech Republic *49°00′N 17°07′E*
115 J21 **Kykládes** *var.* Kikládhes, *Eng.* Cyclades. *island group* SE Greece
25 S11 **Kyle** Texas, SW USA *29°59′N 97°52′W*
96 G9 **Kyle of Lochalsh** N Scotland, United Kingdom *57°18′N 05°39′W*
101 D18 **Kyll** ≈ W Germany
115 F19 **Kyllíni** *var.* Killini. ▲ S Greece
93 N20 **Kymenlaakso** *Swe.* Kymmenedalen. ◈ *region* S Finland
115 H18 **Kými** *prev.* Kími. Évvoia, C Greece *38°38′N 24°06′E*
93 M19 **Kymijoki** ≈ S Finland
115 H18 **Kýmis, Akrotírio** *headland* Évvoia, C Greece *38°39′N 24°08′E*
Kymmenedalen *see* Kymenlaakso
125 W14 **Kyn** Permskiy Kray, NW Russian Federation *57°48′N 58°38′E*
183 N12 **Kyneton** Victoria, SE Australia *37°14′S 144°28′E*
81 G17 **Kyoga, Lake** *var.* Lake Kioga. ◎ C Uganda
164 J12 **Kyōga-misaki** *headland* Honshū, SW Japan *35°46′N 135°13′E*
183 V4 **Kyogle** New South Wales, SE Australia *28°37′S 153°00′E*
Kyŏnggi-man *see* Gyeonggi-man
Kyŏngju *see* Gyeongju
Kyŏngsŏng *see* Seoul
Kyŏsai-tō *see* Geoje-do
81 F19 **Kyotera** S Uganda *0°38′S 31°34′E*
164 J13 **Kyōto** Kyōto, Honshū, SW Japan *35°01′N 135°46′E*
164 J13 **Kyōto** *off.* Kyōto-fu, *var.* Kyōto Hu. ◈ *urban prefecture* Honshū, SW Japan
Kyōto-fu/Kyōto Hu *see* Kyōto
115 D21 **Kyparissía** *var.* Kiparissia. Pelopónnisos, S Greece *37°15′N 21°40′E*
115 D20 **Kyparissiakós Kólpos** *gulf* S Greece
Kyperounda *see* Kyperoúnta
121 P3 **Kyperoúnta** *var.* Kyperounda. C Cyprus *34°57′N 33°02′E*
Kypros *see* Cyprus
115 H16 **Kyrá Panagía** *island* Vóreies Sporádes, Greece, Aegean Sea
Kyrenia *see* Girne
Kyrenia Mountains *see* Beşparmak Dağları
Kyrgyz Republic *see* Kyrgyzstan
147 U9 **Kyrgyzstan** *off.* Kyrgyz Republic, *var.* Kirghizia; *prev.* Kirgizskaya SSR, Kirghiz SSR, Republic of Kyrgyzstan. ◆ *republic* C Asia
Kyrgyzstan, Republic of *see* Kyrgyzstan
138 F11 **Kyriat Gat** *prev.* Qiryat Gat. Southern, C Israel *31°37′N 34°47′E*
100 M11 **Kyritz** Brandenburg, NE Germany *52°56′N 12°24′E*
94 G8 **Kyrkseterøra** Sør-Trøndelag, S Norway *63°17′N 09°06′E*
Kyrkslätt *see* Kirkkonummi
125 U8 **Kyrta** Respublika Komi, NW Russian Federation *64°03′N 57°41′E*
111 J18 **Kysucké Nové Mesto** *prev.* Horné Nové Mesto, *Ger.* Kisutzaneustadtl, Oberneustadtl, *Hung.* Kiszucaújhely. Zílinský Kraj, N Slovakia *49°18′N 18°48′E*
117 N12 **Kytay, Ozero** ◎ SW Ukraine
115 F23 **Kýthira** *var.* Kíthira, *It.* Cerigo, *Lat.* Cythera. Kńythira, S Greece *41°39′N 26°30′E*
115 F23 **Kýthira** *var.* Kíthira, *It.* Cerigo, *Lat.* Cythera. *island* S Greece
115 I20 **Kýthnos** Kńythnos, Kykládes, Greece, Aegean Sea *37°24′N 24°28′E*
115 I20 **Kýthnos** *var.* Kíthnos, Thermiá, *It.* Termia; *anc.* Cythnos. *island* Kykládes, Greece, Aegean Sea
115 I20 **Kýthnou, Stenó** *strait* Kykládes, Greece, Aegean Sea
Kyungëy Ala-Too, Khrebet *see* Kungei Ala-Tau
Kyurdamir *see* Kürdämir
164 D15 **Kyūshū** *var.* Kyûsyû. *island* SW Japan
192 H6 **Kyushu-Palau Ridge** *var.* Kyusyu-Palau Ridge. *undersea feature* W Pacific Ocean *20°00′N 136°00′E*
114 F10 **Kyustendil** *anc.* Pautalia. Kyustendil, W Bulgaria *42°17′N 22°42′E*
114 G11 **Kyustendil** ◈ *province* W Bulgaria
Kyûsyû *see* Kyūshū
Kyusyu-Palau Ridge *see* Kyushu-Palau Ridge
123 P8 **Kyusyur** Respublika Sakha (Yakutiya), NE Russian Federation *70°36′N 127°19′E*
183 P10 **Kywong** New South Wales, SE Australia *34°59′S 146°42′E*
117 P4 **Kyyiv** *Eng.* Kiev, *Rus.* Kiyev. ● (Ukraine) Kyyivs'ka Oblast', N Ukraine *50°26′N 30°32′E*
Kyyiv *see* Kyyivs'ka Oblast'
117 O4 **Kyyivs'ka Oblast'** *var.* Kyyiv, *Rus.* Kiyevskaya Oblast'. ◈ *province* N Ukraine
117 P3 **Kyyivs'ke Vodoskhovyshche** *Eng.* Kiev Reservoir, *Rus.* Kiyevskoye Vodokhranilishche. ⊠ N Ukraine
93 L16 **Kyyjärvi** Keski-Suomi, C Finland *63°02′N 24°34′E*
122 K14 **Kyzyl** Respublika Tyva, C Russian Federation *51°45′N 94°28′E*
147 S8 **Kyzyl-Adyr** *var.* Kirovskoye. Talasskaya Oblast', NW Kyrgyzstan *42°37′N 71°34′E*
145 V14 **Kyzylagash** *Kaz.* Qyzylaghash. Almaty, SE Kazakhstan *45°20′N 78°45′E*
146 C13 **Kyzylbair** Balkan Welaýaty, W Turkmenistan *38°13′N 55°38′E*
Kyzyl-Dzhiik, Pereval *see* Uzbel Shankou
145 S7 **Kyzylkak, Ozero** ◎ NE Kazakhstan
145 X11 **Kyzylkesek** Vostochnyy Kazakhstan, E Kazakhstan *47°56′N 82°02′E*
147 S10 **Kyzyl-Kiya** *Kir.* Kyzyl-Kyya. Batkenskaya Oblast', SW Kyrgyzstan *40°15′N 72°07′E*
144 L11 **Kyzylkol', Ozero** ◎ C Kazakhstan
122 K14 **Kyzyl Kum** *var.* Kizil Kum, Qizil Qum, *Uzb.* Qizilqum. *desert* Kazakhstan/Uzbekistan
Kyzyl-Kyya *see* Kyzyl-Kiya
145 N15 **Kyzylorda** *var.* Kzyl-Orda, Qizil Orda, Qyzylorda; *prev.* Kzylorda, Perovsk. Kyzylorda, S Kazakhstan *44°54′N 65°31′E*
144 L14 **Kyzylorda** *off.* Kyzylordinskaya Oblast', *Kaz.* Qyzylorda Oblysy. ◈ *province* S Kazakhstan
Kyzylordinskaya Oblast' *see* Kyzylorda
Kyzylrabat *see* Qizilravote
Kyzylrabot *see* Qizilrabot
Kyzylsu *see* Kyzyl-Suu
147 X7 **Kyzyl-Suu** *prev.* Pokrovka. Issyk-Kul'skaya Oblast', NE Kyrgyzstan *42°20′N 77°55′E*
147 S12 **Kyzyl-Suu** *var.* Kyzylsu. ≈ Kyrgyzstan/Tajikistan
147 X8 **Kyzyl-Tuu** Issyk-Kul'skaya Oblast', E Kyrgyzstan *42°06′N 76°54′E*
145 Q12 **Kyzylzhar** *Kaz.* Qyzylzhar. Karaganda, C Kazakhstan *48°22′N 70°00′E*
Kzyl-Orda *see* Kyzylorda
Kzylorda *see* Kyzylorda
Kzyltu *see* Kishkenekol'

L

109 X2 **Laa an der Thaya** Niederösterreich, NE Austria *48°44′N 16°23′E*
63 K15 **La Adela** La Pampa, SE Argentina *38°57′S 64°02′W*
Laagen *see* Numedalslågen
109 S5 **Laakirchen** Oberösterreich, N Austria *47°59′N 13°49′E*
Laaland *see* Lolland
104 I11 **La Albuera** Extremadura, W Spain *38°43′N 06°49′W*
105 O7 **La Alcarria** *physical region* C Spain
104 K14 **La Algaba** Andalucía, S Spain *37°27′N 06°01′W*
105 P9 **La Almarcha** Castilla-La Mancha, C Spain *39°41′N 02°23′W*
105 R6 **La Almunia de Doña Godina** Aragón, NE Spain *41°28′N 01°23′W*
41 N5 **La Amistad, Presa** ⊠ NW Mexico
118 F4 **Läänemaa** *var.* Lääne Maakond. ◈ *province* NW Estonia
Lääne Maakond *see* Läänemaa
118 I3 **Lääne-Virumaa** *off.* Lääne-Viru Maakond. ◈ *province* NE Estonia
Lääne-Viru Maakond *see* Lääne-Virumaa
62 J9 **La Antigua, Salina** *salt lake* W Argentina
99 E17 **Laarne** Oost-Vlaanderen, NW Belgium *51°03′N 03°50′E*
80 O13 **Laas Caanood** Sool, N Somalia *08°33′N 47°44′E*
41 O9 **La Ascensión** Nuevo León, NE Mexico *24°15′N 99°53′W*
80 N12 **Laas Dhaareed** Togdheer, N Somalia *10°12′N 46°09′E*
55 O4 **La Asunción** Nueva Esparta, NE Venezuela *11°06′N 63°53′W*
Laatokka *see* Ladozhskoye, Ozero
100 I13 **Laatzen** Niedersachsen, NW Germany *52°19′N 09°46′E*
38 E9 **La'au Point** *var.* Laau Point. *headland* Moloka'i, Hawai'i, USA *21°06′N 157°18′W*
Laau Point *see* La'au Point
42 D6 **La Aurora** ✈ (Ciudad de Guatemala) Guatemala, C Guatemala *14°33′N 90°30′W*
74 C9 **Laâyoune** *var.* Aaiún. ● (Western Sahara) NW Western Sahara *27°10′N 13°11′W*
126 L14 **Laba** ≈ SW Russian Federation
40 M6 **La Babia** Coahuila, NE Mexico *28°39′N 102°00′W*
15 R7 **La Baie** Québec, SE Canada *48°20′N 70°54′W*
171 P16 **Labala** Pulau Lomblen, S Indonesia *08°30′S 123°27′E*
62 K8 **La Banda** Santiago del Estero, N Argentina *27°44′S 64°14′W*
La Banda Oriental *see* Uruguay
104 K4 **La Bañeza** Castilla y León, N Spain *42°18′N 05°54′W*
167 T11 **Labăng** *prev.* Phumĭ Labăng. Rôtânôkiri, NE Cambodia *13°51′N 107°01′E*
40 M13 **La Barca** Jalisco, SW Mexico *20°20′N 102°33′W*
40 K14 **La Barra de Navidad** Jalisco, C Mexico *19°12′N 104°38′W*
187 Y13 **Labasa** *prev.* Lambasa. Vanua Levu, N Fiji *16°25′S 179°24′E*
102 H8 **la Baule-Escoublac** Loire-Atlantique, NW France *47°17′N 02°24′W*
76 I13 **Labé** NW Guinea *11°19′N 12°17′W*
Labe *see* Elbe
15 N11 **Labelle** Québec, SE Canada *46°15′N 74°43′W*
23 X14 **La Belle** Florida, SE USA *26°45′N 81°26′W*
10 H7 **Laberge, Lake** ◎ Yukon, W Canada
Labes *see* Łobez
Labiau *see* Polessk
112 A10 **Labin** *It.* Albona. Istra, NW Croatia *45°05′N 14°10′E*
126 L14 **Labinsk** Krasnodarskiy Kray, SW Russian Federation *44°39′N 40°43′E*
105 X5 **La Bisbal d'Empordà** Cataluña, NE Spain *41°58′N 03°02′E*
119 P16 **Labkovichy** *Rus.* Lobkovichi. Mahilyowskaya Voblasts', E Belarus *53°50′N 31°45′E*
15 S4 **La Blache, Lac de** ◎ Québec, SE Canada
171 P4 **Labo** Luzon, N Philippines *14°10′N 122°47′E*
Laboehanbadjo *see* Labuhanbajo
Laborca *see* Laborec
111 N18 **Laborec** *Hung.* Laborca. ≈ E Slovakia
108 D11 **La Borgne** ≈ S Switzerland
45 T12 **Laborie** SW Saint Lucia *13°45′N 61°00′W*
102 J14 **Labouheyre** Landes, SW France *44°12′N 00°55′W*
62 L12 **Laboulaye** Córdoba, C Argentina *34°05′S 63°20′W*
13 Q7 **Labrador** *cultural region* Newfoundland and Labrador, SW Canada
64 I6 **Labrador Basin** *var.* Labrador Sea Basin. *undersea feature* Labrador Sea *53°00′N 48°00′W*
13 N9 **Labrador City** Newfoundland and Labrador, E Canada *52°56′N 66°52′W*
13 Q5 **Labrador Sea** *sea* NW Atlantic Ocean
Labrador Sea Basin *see* Labrador Basin
Labrang *see* Xiahe
54 G9 **Labranzagrande** Boyacá, C Colombia *05°34′N 72°34′W*
59 D14 **Lábrea** Amazonas, N Brazil *07°20′S 64°46′W*
45 U15 **La Brea** Trinidad, Trinidad and Tobago *10°14′N 61°37′W*
15 S6 **Labrieville** Québec, SE Canada *49°15′N 69°31′W*
102 K14 **Labrit** Landes, SW France *44°03′N 00°29′W*
108 C9 **La Broye** ≈ SW Switzerland
103 N15 **Labruguière** Tarn, S France *43°32′N 02°15′E*
168 M11 **Labu** Pulau Singkep, W Indonesia *0°34′S 104°24′E*
169 T7 **Labuan** *var.* Victoria. Labuan, East Malaysia *05°20′N 115°14′E*
169 T7 **Labuan** ◈ *federal territory* East Malaysia
Labuan *see* Labuan, Pulau
169 T7 **Labuan, Pulau** *var.* Labuan. *island* East Malaysia
Labudalin *see* Ergun
171 N16 **Labuhanbajo** *prev.* Laboehanbadjo. Flores, S Indonesia *08°33′S 119°55′E*
168 J9 **Labuhanbilik** Sumatera, N Indonesia *02°30′N 100°10′E*
168 G8 **Labuhanhaji** Sumatera, W Indonesia *03°31′N 97°00′E*
Labuk *see* Labuk, Sungai
Labuk Bay *see* Labuk, Teluk
169 V7 **Labuk, Sungai** *var.* Labuk. ≈ East Malaysia
169 W6 **Labuk, Teluk** *var.* Labuk Bay, Telukan Labuk. *bay* S Sulu Sea
Labuk, Telukan *see* Labuk, Teluk
166 K9 **Labutta** Ayeyawady, SW Myanmar (Burma) *16°08′N 94°45′E*
122 I8 **Labytnangi** Yamalo-Nenetskiy Avtonomnyy Okrug, N Russian Federation *66°39′N 66°26′E*
113 K19 **Laç** *var.* Laci. Lezhë, C Albania *41°37′N 19°37′E*
78 F10 **Lac** *off.* Région du Lac. ◈ *region* W Chad
57 K19 **Lacajahuira, Rio** ≈ W Bolivia
La Calamine *see* Kelmis
62 G11 **La Calera** Valparaíso, C Chile *32°47′S 71°16′W*
13 P11 **Lac-Allard** Québec, E Canada *50°37′N 63°26′W*
104 L13 **La Campana** Andalucía, S Spain *37°35′N 05°24′W*
102 J12 **Lacanau** Gironde, SW France *44°59′N 01°04′W*
42 C2 **Lacandón, Sierra del** ⛰ Guatemala/Mexico
La Cañiza *see* A Cañiza
41 W16 **Lacantún, Río** ≈ SE Mexico
103 Q3 **la Capelle** Aisne, N France *49°58′N 03°55′E*
112 K10 **Laćarak** Vojvodina, NW Serbia *45°00′N 19°34′E*
62 L11 **La Carlota** Córdoba, C Argentina *33°30′S 63°15′W*
104 L13 **La Carlota** Andalucía, S Spain *37°40′N 04°56′W*
105 N12 **La Carolina** Andalucía, S Spain *38°15′N 03°37′W*
103 O15 **Lacaune** Tarn, S France *43°42′N 02°42′E*
15 P7 **Lac-Bouchette** Québec, SE Canada *48°14′N 72°11′W*
Laccadive Islands/Laccadive Minicoy and Amindivi Islands, the *see* Lakshadweep
11 Y16 **Lac du Bonnet** Manitoba, S Canada *50°13′N 96°04′W*
30 L4 **Lac du Flambeau** Wisconsin, N USA *45°58′N 89°51′W*
15 P8 **Lac-Édouard** Québec, SE Canada *47°39′N 72°16′W*
42 I4 **La Ceiba** Atlántida, N Honduras *15°45′N 86°29′W*
54 E9 **La Ceja** Antioquia, W Colombia *06°02′N 75°30′W*
182 J11 **Lacepede Bay** *bay* South Australia
32 G9 **Lacey** Washington, NW USA *47°01′N 122°49′W*
103 P12 **la Chaise-Dieu** Haute-Loire, C France *45°19′N 03°41′E*
114 G13 **Lachanás** Kentrikí Makedonía, N Greece *40°57′N 23°15′E*
124 L11 **Lacha, Ozero** ◎ NW Russian Federation
103 O8 **la Charité-sur-Loire** Nièvre, C France *47°10′N 02°59′E*
103 N9 **la Châtre** Indre, C France *46°35′N 01°59′E*
108 C8 **La Chaux-de-Fonds** Neuchâtel, W Switzerland *47°07′N 06°51′E*
Lach Dera *see* Dheere Laaq
108 G8 **Lachen** Schwyz, C Switzerland *47°12′N 08°51′E*
183 Q8 **Lachlan River** ≈ New South Wales, SE Australia
43 T15 **La Chorrera** Panamá, C Panama *08°51′N 79°46′W*
15 V7 **Lac-Humqui** Québec, SE Canada *48°21′N 67°32′W*
15 N12 **Lachute** Québec, SE Canada *45°39′N 74°21′W*
Lachyn *see* Laçın
Laci *see* Laç
137 W13 **Laçın** *Rus.* Lachyn. SW Azerbaijan *39°36′N 46°34′E*
103 S16 **la Ciotat** *anc.* Citharista. Bouches-du-Rhône, SE France *43°10′N 05°36′E*
18 D10 **Lackawanna** New York, NE USA *42°49′N 78°49′W*
11 Q13 **Lac La Biche** Alberta, SW Canada *54°46′N 111°59′W*
Lac la Martre *see* Wha Ti
15 R12 **Lac-Mégantic** *var.* Mégantic. Québec, SE Canada *45°35′N 70°53′W*
Lacobriga *see* Lagos
40 G5 **La Colorada** Sonora, NW Mexico *28°49′N 110°32′W*
11 Q15 **Lacombe** Alberta, SW Canada *52°30′N 113°42′W*
30 L12 **Lacon** Illinois, N USA *41°01′N 89°24′W*
43 P16 **La Concepción** *var.* Concepción. Chiriquí, W Panama *08°31′N 82°39′W*
54 H5 **La Concepción** Zulia, NW Venezuela *10°48′N 71°46′W*
107 C19 **Laconi** Sardegna, Italy, C Mediterranean Sea *39°52′N 09°02′E*
19 O9 **Laconia** New Hampshire, NE USA *43°32′N 71°29′W*
61 H19 **La Coronilla** Rocha, E Uruguay *33°44′S 53°31′W*
La Coruña *see* A Coruña
103 O11 **la Courtine** Creuse, C France *45°42′N 02°18′E*
102 J16 **Lacq** Pyrénées-Atlantiques, SW France *43°25′N 00°37′W*
Lac, Région du *see* Lac
15 P9 **La Croche** Québec, SE Canada *47°38′N 72°42′W*
29 X3 **la Croix, Lac** ◎ Canada/USA
26 K5 **La Crosse** Kansas, C USA *38°32′N 99°19′W*
21 V7 **La Crosse** Virginia, NE USA *36°41′N 78°03′W*
32 L9 **La Crosse** Washington, NW USA *46°48′N 117°51′W*
30 J7 **La Crosse** Wisconsin, N USA *43°46′N 91°12′W*
54 C13 **La Cruz** Nariño, SW Colombia *01°33′N 76°58′W*
42 K12 **La Cruz** Guanacaste, NW Costa Rica *11°05′N 85°39′W*
40 I10 **La Cruz** Sinaloa, W Mexico *23°53′N 106°53′W*
61 F19 **La Cruz** Florida, S Uruguay *33°54′S 56°11′W*
42 M9 **La Cruz de Río Grande** Región Autónoma Atlántico Sur, E Nicaragua *13°04′N 84°12′W*
54 J4 **La Cruz de Taratara** Falcón, N Venezuela *11°03′N 69°44′W*
15 Q10 **Lac-St-Charles** Québec, SE Canada *46°56′N 71°24′W*
40 M6 **La Cuesta** Coahuila, NE Mexico *28°45′N 102°26′W*

◆ Country
● Country Capital
◇ Dependent Territory
○ Dependent Territory Capital
◈ Administrative Regions
✈ International Airport
▲ Mountain
⛰ Mountain Range
Volcano
≈ River
◎ Lake
⊠ Reservoir

57 A17 **La Cumbra, Volcán** ⛰ Galapagos Islands, Ecuador, E Pacific Ocean 0°21′S 91°30′W
152 J5 **Ladākh Range** ▲▲ NE India
26 I5 **Ladder Creek** ≈ Kansas, C USA
45 X10 **la Désirade** *atoll* E Guadeloupe
Lādhiqīyah, Muḥāfaẓat al *see* Al Lādhiqīyah
Lådik *see* Lødingen
83 F25 **Ladismith** Western Cape, SW South Africa 33°30′S 21°15′E
152 G11 **Lādnūn** Rājasthān, NW India 27°36′N 74°26′E
Ladoga, Lake *see* Ladozhskoye, Ozero
115 E19 **Ládon** ≈ S Greece
54 E9 **La Dorada** Caldas, C Colombia 05°28′N 74°41′W
124 H11 **Ladozhskoye, Ozero** *Eng.* Lake Ladoga, *Fin.* Laatokka. ◎ NW Russian Federation
37 R12 **Ladron Peak** ▲ New Mexico, SW USA 34°25′N 107°04′W
124 J11 **Ladva-Vetka** Respublika Kareliya, NW Russian Federation 61°18′N 34°24′E
183 Q15 **Lady Barron** Tasmania, SE Australia 40°12′S 148°12′E
14 G9 **Lady Evelyn Lake** ◎ Ontario, S Canada
23 W11 **Lady Lake** Florida, SE USA 28°55′N 81°55′W
10 L17 **Ladysmith** Vancouver Island, British Columbia, SW Canada 48°55′N 123°45′W
83 J22 **Ladysmith** KwaZulu/Natal, E South Africa 28°34′S 29°47′E
30 J5 **Ladysmith** Wisconsin, N USA 45°27′N 91°07′W
Ladyzhenka *see* Tilekey
186 E7 **Lae** Morobe, W Papua New Guinea 06°45′S 147°00′E
189 R6 **Lae Atoll** *atoll* Ralik Chain, W Marshall Islands
40 C3 **La Encantada, Cerro de** ▲ NW Mexico 31°03′N 115°25′W
55 N11 **La Esmeralda** Amazonas, S Venezuela 03°11′N 65°33′W
42 G7 **La Esperanza** Intibucá, SW Honduras 14°19′N 88°09′W
30 K8 **La Farge** Wisconsin, N USA 43°36′N 90°39′W
23 R5 **Lafayette** Alabama, S USA 32°54′N 85°24′W
37 T4 **Lafayette** Colorado, C USA 39°59′N 105°06′W
23 R2 **La Fayette** Georgia, SE USA 34°42′N 85°16′W
31 O13 **Lafayette** Indiana, N USA 40°25′N 86°52′W
22 I9 **Lafayette** Louisiana, S USA 30°13′N 92°01′W
20 K8 **Lafayette** Tennessee, S USA 36°31′N 86°01′W
19 N7 **Lafayette, Mount** ▲ New Hampshire, NE USA 44°09′N 71°37′W
La Fe *see* Santa Fé
103 P3 **la Fère** Aisne, N France 49°41′N 03°20′E
102 L6 **la Ferté-Bernard** Sarthe, NW France 48°13′N 00°40′E
102 K5 **la Ferté-Macé** Orne, N France 48°35′N 00°21′W
103 N7 **la Ferté-St-Aubin** Loiret, C France 47°42′N 01°57′E
103 P5 **la Ferté-sous-Jouarre** Seine-et-Marne, N France 48°57′N 03°08′E
77 V15 **Lafia** Nassarawa, C Nigeria 08°29′N 08°34′E
77 T15 **Lafiagi** Kwara, W Nigeria 08°52′N 05°25′E
11 T17 **Lafleche** Saskatchewan, S Canada 49°40′N 106°28′W
102 K7 **la Flèche** Sarthe, NW France 47°42′N 00°04′W
109 X7 **Lafnitz** *Hung.* Lapines. ≈ Austria/Hungary
187 P17 **La Foa** Province Sud, S New Caledonia 21°46′S 165°49′E
20 M8 **La Follette** Tennessee, S USA 36°22′N 84°07′W
15 N12 **Lafontaine** Québec, SE Canada 45°52′N 74°01′W
22 K10 **Lafourche, Bayou** ≈ Louisiana, S USA
62 K6 **La Fragua** Santiago del Estero, N Argentina 26°06′S 64°06′W
54 H7 **La Fría** Táchira, NW Venezuela 08°13′N 72°15′W
104 J7 **La Fuente de San Esteban** Castilla y León, N Spain 40°48′N 06°14′W
186 C7 **Lagaip** ≈ W Papua New Guinea
61 B15 **La Gallareta** Santa Fe, C Argentina 29°34′S 60°23′W
127 Q14 **Lagan'** *prev.* Kaspiyskiy. Respublika Kalmykiya, SW Russian Federation 45°25′N 47°19′E
95 L20 **Lagan** Kronoberg, S Sweden 56°55′N 14°01′E
95 K21 **Lagan** ≈ S Sweden
92 L2 **Lagarfljót** *var.* Lögurinn. ◎ E Iceland
37 R7 **La Garita Mountains** ▲▲ Colorado, C USA
171 O2 **Lagawe** Luzon, N Philippines 16°46′N 121°06′E
78 F13 **Lagdo** Nord, N Cameroon 09°12′N 13°43′E
78 F13 **Lagdo, Lac de** ◎ N Cameroon
100 H13 **Lage** Nordrhein-Westfalen, W Germany 52°00′N 08°48′E
94 H12 **Lågen** ≈ S Norway
61 J14 **Lages** Santa Catarina, S Brazil 27°45′S 50°16′W
Lágesvuotna *see* Laksefjorden
149 R4 **Laghmān** ◈ *province* E Afghanistan
74 J6 **Laghouat** N Algeria 33°49′N 02°59′E
105 Q10 **La Gineta** Castilla-La Mancha, C Spain 39°08′N 02°00′W
115 E21 **Lagkáda** *var.* Langada. Pelopónnisos, S Greece 36°49′N 22°19′E
114 G13 **Lagkadás** *var.* Langades, Langadhás. Kentrikí Makedonía, N Greece 40°45′N 23°04′E
115 E20 **Lagkádia** *var.* Langádhia, *cont.* Langadia. Pelopónnisos, S Greece 37°40′N 22°01′E
54 F6 **La Gloria** Cesar, N Colombia 08°37′N 73°51′W
41 O7 **La Gloria** Nuevo León, NE Mexico
92 N3 **Lågneset** *headland* W Svalbard 77°46′N 13°44′E
104 G14 **Lagoa** Faro, S Portugal 37°07′N 08°27′W
La Goagira *see* La Guajira
Lago Agrio *see* Nueva Loja
61 I14 **Lagoa Vermelha** Rio Grande do Sul, S Brazil 28°13′S 51°32′W
137 V10 **Lagodekhi** SE Georgia 41°49′N 46°15′E
42 C7 **La Gomera** Escuintla, S Guatemala 14°05′N 91°03′W
Lagone *see* Logone
107 M19 **Lagonegro** Basilicata, S Italy 40°06′N 15°42′E
63 G16 **Lago Ranco** Los Ríos, S Chile 40°21′S 72°29′W
77 S16 **Lagos** Lagos, SW Nigeria 06°24′N 03°17′E
104 F14 **Lagos** *anc.* Lacobriga. Faro, S Portugal 37°05′N 08°40′W
77 S16 **Lagos** ◈ *state* SW Nigeria
40 M12 **Lagos de Moreno** Jalisco, SW Mexico 21°21′N 101°55′W
Lagosta *see* Lastovo
74 A12 **Lagouira** SW Western Sahara 20°55′N 17°05′W
92 O1 **Lågøya** *island* N Svalbard
32 L11 **La Grande** Oregon, NW USA 45°21′N 118°05′W
103 Q14 **la Grande-Combe** Gard, S France 44°13′N 04°01′E
12 K9 **La Grande Rivière** *var.* Fort George. ≈ Québec, SE Canada
23 R4 **La Grange** Georgia, SE USA 33°02′N 85°02′W
31 P11 **Lagrange** Indiana, N USA 41°38′N 85°25′W
20 L5 **La Grange** Kentucky, S USA 38°24′N 85°23′W
27 V2 **La Grange** Missouri, C USA 40°00′N 91°31′W
21 V10 **La Grange** North Carolina, SE USA 35°18′N 77°47′W
25 U11 **La Grange** Texas, SW USA 29°55′N 96°54′W
105 N7 **La Granja** Castilla y León, N Spain 40°53′N 04°01′W
55 Q9 **La Gran Sabana** *grassland* E Venezuela
54 H7 **La Grita** Táchira, NW Venezuela 08°09′N 71°58′W
La Grulla *see* Grulla
15 R11 **La Guadeloupe** Québec, SE Canada 45°57′N 70°56′W
65 L6 **La Guaira** Distrito Federal, N Venezuela 10°35′N 66°52′W
54 G4 **La Guajira** *off.* Departamento de La Guajira, *var.* Guajira, La Goagira. ◈ *province* NE Colombia
188 I4 **Lagua Lichan, Punta** *headland* Saipan, S Northern Mariana Islands
105 P4 **Laguardia** *Basq.* Biasteri. País Vasco, N Spain 42°32′N 02°31′W
18 K14 **La Guardia** ✈ (New York) Long Island, New York, NE USA 40°44′N 73°51′W
La Guardia/Laguardia *see* A Guarda
La Gudiña *see* A Gudiña
103 O9 **la Guerche-sur-l'Aubois** Cher, C France 46°55′N 03°00′E
103 O13 **Laguiole** Aveyron, S France 44°42′N 02°51′E
83 F26 **L'Agulhas** *var.* Agulhas. Western Cape, SW South Africa 34°49′S 20°01′E
61 K14 **Laguna** Santa Catarina, S Brazil 28°29′S 48°45′W
37 Q11 **Laguna** New Mexico, SW USA 35°03′N 107°30′W
35 T16 **Laguna Beach** California, W USA 33°32′N 117°46′W
35 Y17 **Laguna Dam** *dam* Arizona/California, W USA
40 L7 **Laguna El Rey** Coahuila, N Mexico
35 V17 **Laguna Mountains** ▲▲ California, W USA
61 B17 **Laguna Paiva** Santa Fe, C Argentina 31°21′S 60°40′W
62 H3 **Lagunas** Tarapacá, N Chile 21°01′S 69°36′W
56 E9 **Lagunas** Loreto, N Peru 05°15′S 75°24′W
57 M20 **Lagunillas** Santa Cruz, SE Bolivia 19°38′S 63°39′W
54 H6 **Lagunillas** Mérida, NW Venezuela 08°31′N 71°24′W
44 C4 **La Habana** *var.* Havana. ● (Cuba) Ciudad de La Habana, W Cuba 23°07′N 82°25′W
169 W7 **Lahad Datu** Sabah, East Malaysia 05°01′N 118°20′E
169 W7 **Lahad Datu, Teluk** *var.* Telukan Lahad Datu, Teluk Darvel, Teluk Datu; *prev.* Darvel Bay. *bay* Sabah, East Malaysia, C Pacific Ocean
Lahad Datu, Telukan *see* Lahad Datu, Teluk
38 F10 **Lahaina** Maui, Hawaii, USA, C Pacific Ocean 20°52′N 156°40′W
168 L14 **Lahat** Sumatera, W Indonesia 03°46′S 103°32′E
La Haye *see* 's-Gravenhage
Lahej *see* Laḥij
62 G9 **La Higuera** Coquimbo, N Chile 29°33′S 71°15′W
141 S13 **Lahī, Ḥisā' al** *spring/well* NE Yemen 17°28′N 50°05′E
141 O16 **Laḥij** *var.* Laḥj, *Eng.* Lahej. SW Yemen 13°04′N 44°55′E
142 M3 **Lāhījān** Gīlān, NW Iran 37°12′N 50°00′E
119 I19 **Lahishyn** *Pol.* Lohiszyn, *Rus.* Logishin. Brestskaya Voblasts', SW Belarus 52°20′N 25°59′E
Laḥj *see* Laḥij
101 F18 **Lahn** ≈ W Germany
Lähn *see* Wleń
95 J21 **Laholm** Halland, S Sweden 56°30′N 13°05′E
95 J21 **Laholmsbukten** *bay* S Sweden
35 R6 **Lahontan Reservoir** ⊠ Nevada, W USA
149 W8 **Lahore** Punjab, NE Pakistan 31°36′N 74°18′E
149 W8 **Lahore** ✈ Punjab, E Pakistan 31°34′N 74°22′E
55 Q6 **La Horqueta** Delta Amacuro, NE Venezuela 09°13′N 62°02′W
119 K15 **Lahoysk** *Rus.* Logoysk. Minskaya Voblasts', C Belarus 54°12′N 27°53′E
101 F22 **Lahr** Baden-Württemberg, S Germany 48°21′N 07°52′E
93 M19 **Lahti** *Swe.* Lahtis. Päijät-Häme, S Finland 61°N 25°40′E
Lahtis *see* Lahti
40 M14 **La Huacana** Michoacán, SW Mexico 18°56′N 101°52′W
40 K14 **La Huerta** Jalisco, SW Mexico 19°29′N 104°40′W
78 H12 **Laï** *prev.* Behagle, De Behagle. Tandjilé, S Chad 09°22′N 16°14′E
Laibach *see* Ljubljana
167 Q5 **Lai Châu** Lai Châu, N Vietnam 22°04′N 103°10′E
Laichow Bay *see* Laizhou Wan
38 D9 **Lā'ie** *var.* Laie. O'ahu, Hawaii, USA, C Pacific Ocean 21°39′N 157°55′W
Laie *see* Lā'ie
102 L5 **l'Aigle** Orne, N France 48°46′N 00°37′E
103 Q7 **Laignes** Côte d'Or, C France 47°51′N 04°24′E
93 K17 **Laihia** Österbotten, W Finland 62°58′N 22°00′E
81 I19 **Laikipia** ◈ *county* C Kenya
Laila *see* Laylā
83 F25 **Laingsburg** Western Cape, SW South Africa 33°12′S 20°51′E
109 U2 **Lainsitz** *Cz.* Lužnice. ≈ Austria/Czech Republic
96 I7 **Lairg** N Scotland, United Kingdom 58°02′N 04°23′W
81 I17 **Laisamis** Marsabit, N Kenya 01°35′N 37°49′E
Laisberg *see* Leisi
127 R4 **Laishevo** Respublika Tatarstan, W Russian Federation 55°26′N 49°27′E
Laisholm *see* Jõgeva
92 H13 **Laisvall** Norrbotten, N Sweden 66°07′N 17°10′E
93 K19 **Laitila** Varsinais-Suomi, SW Finland 60°52′N 21°40′E
161 P5 **Laiwu** Shandong, E China 36°14′N 117°40′E
161 R4 **Laixi** *var.* Shuiji. Shandong, E China 36°50′N 120°40′E
161 R4 **Laiyang** Shandong, E China 36°58′N 120°40′E
161 O3 **Laiyuan** Hebei, E China 39°19′N 114°44′E
161 R4 **Laizhou** *var.* Ye Xian. Shandong, E China 37°12′N 120°01′E
161 Q4 **Laizhou Wan** *var.* Laichow Bay. *bay* E China
37 S8 **La Jara** Colorado, C USA 37°16′N 105°57′W
61 I15 **Lajeado** Rio Grande do Sul, S Brazil 29°28′S 52°00′W
112 L12 **Lajkovac** Serbia, C Serbia 44°22′N 20°12′E
111 K23 **Lajosmizse** Bács-Kiskun, C Hungary 47°02′N 19°31′E
Lajta *see* Leitha
40 I6 **La Junta** Chihuahua, N Mexico 28°30′N 107°20′W
37 V7 **La Junta** Colorado, C USA 37°59′N 103°34′W
92 J13 **Lakaträsk** Norrbotten, N Sweden 66°16′N 21°10′E
Lak Dera *see* Dheere Laaq
Lakeamu *see* Lakekamu
29 P12 **Lake Andes** South Dakota, N USA 43°08′N 98°33′W
22 H9 **Lake Arthur** Louisiana, S USA 30°04′N 92°40′W
187 Z15 **Lakeba** *prev.* Lakemba. *island* Lau Group, E Fiji
187 Z14 **Lakeba Passage** *channel* E Fiji
29 S10 **Lake Benton** Minnesota, N USA 44°15′N 96°17′W
23 V9 **Lake Butler** Florida, SE USA 30°01′N 82°20′W
183 P8 **Lake Cargelligo** New South Wales, SE Australia 33°21′S 146°25′E
22 G9 **Lake Charles** Louisiana, S USA 30°13′N 93°13′W
27 X9 **Lake City** Arkansas, C USA 35°50′N 90°28′W
37 Q7 **Lake City** Colorado, C USA 38°01′N 107°18′W
23 V9 **Lake City** Florida, SE USA 30°12′N 82°39′W
29 U13 **Lake City** Iowa, C USA 42°16′N 94°43′W
31 P7 **Lake City** Michigan, N USA 44°22′N 85°12′W
29 W9 **Lake City** Minnesota, N USA 44°27′N 92°16′W
21 T13 **Lake City** South Carolina, SE USA 33°52′N 79°45′W
29 Q7 **Lake City** South Dakota, N USA 45°42′N 97°22′W
20 M8 **Lake City** Tennessee, S USA 36°13′N 84°09′W
10 L17 **Lake Cowichan** Vancouver Island, British Columbia, SW Canada 48°50′N 124°04′W
29 U10 **Lake Crystal** Minnesota, N USA 44°07′N 94°13′W
25 T6 **Lake Dallas** Texas, SW USA 33°07′N 97°01′W
97 K15 **Lake District** *physical region* NW England, United Kingdom
18 D10 **Lake Erie Beach** New York, NE USA 42°37′N 79°04′W
29 T11 **Lakefield** Minnesota, N USA 43°40′N 95°10′W
25 V6 **Lake Fork Reservoir** ◎ Texas, SW USA
30 M9 **Lake Geneva** Wisconsin, N USA 42°36′N 88°25′W
18 L9 **Lake George** New York, NE USA 43°25′N 73°45′W
Lake Harbour *see* Kimmirut
36 I12 **Lake Havasu City** Arizona, SW USA 34°26′N 114°20′W
25 W12 **Lake Jackson** Texas, SW USA 29°01′N 95°25′W
186 D8 **Lakekamu** *var.* Lakeamu. ≈ S Papua New Guinea
180 K13 **Lake King** Western Australia 33°09′S 119°46′E
23 V12 **Lakeland** Florida, SE USA 28°03′N 81°57′W
23 U7 **Lakeland** Georgia, SE USA 31°02′N 83°04′W
181 W4 **Lakeland Downs** Queensland, NE Australia 15°54′S 144°54′E
11 P16 **Lake Louise** Alberta, SW Canada 51°26′N 116°10′W
Lakemba *see* Lakeba
29 V11 **Lake Mills** Iowa, C USA 43°25′N 93°31′W
39 Q10 **Lake Minchumina** Alaska, USA 63°55′N 152°25′W
Lakemti *see* Nek'emtē
186 A7 **Lake Murray** Western, SW Papua New Guinea 06°35′S 141°28′E
31 R9 **Lake Orion** Michigan, N USA 42°46′N 83°14′W
190 B16 **Lakepa** NE Niue 18°59′S 169°48′E
29 T11 **Lake Park** Iowa, C USA 43°27′N 95°19′W
18 K7 **Lake Placid** New York, NE USA 44°16′N 73°57′W
18 K9 **Lake Pleasant** New York, NE USA 43°27′N 74°24′W
34 M6 **Lakeport** California, W USA 39°04′N 122°56′W
29 Q10 **Lake Preston** South Dakota, N USA 44°21′N 97°22′W
22 J5 **Lake Providence** Louisiana, S USA 32°48′N 91°10′W
185 E20 **Lake Pukaki** Canterbury, South Island, New Zealand 44°12′S 170°10′E
81 D14 **Lakes** *var.* El Buhayrat. ◈ *state* C South Sudan
183 Q12 **Lakes Entrance** Victoria, SE Australia 37°52′S 147°58′E
37 N12 **Lakeside** Arizona, SW USA 34°09′N 109°58′W
35 V17 **Lakeside** California, W USA 32°50′N 116°55′W
23 S9 **Lakeside** Florida, SE USA 30°22′N 84°18′W
28 K13 **Lakeside** Nebraska, C USA 42°01′N 102°27′W
32 E13 **Lakeside** Oregon, NW USA 43°34′N 124°10′W
21 W6 **Lakeside** Virginia, NE USA 37°36′N 77°28′W
Lake State *see* Michigan
185 F20 **Lake Tekapo** Canterbury, South Island, New Zealand 44°01′S 170°29′E
21 O10 **Lake Toxaway** North Carolina, SE USA 35°06′N 82°57′W
29 T13 **Lake View** Iowa, C USA 42°18′N 95°04′W
32 I16 **Lakeview** Oregon, NW USA 42°13′N 120°21′W
25 O3 **Lakeview** Texas, SW USA 34°38′N 100°36′W
27 W14 **Lake Village** Arkansas, C USA 33°20′N 91°19′W
23 W12 **Lake Wales** Florida, SE USA 27°54′N 81°35′W
37 T4 **Lakewood** Colorado, C USA 39°38′N 105°07′W
18 K15 **Lakewood** New Jersey, NE USA 40°04′N 74°11′W
18 C11 **Lakewood** New York, NE USA 42°03′N 79°19′W
31 T11 **Lakewood** Ohio, N USA 41°28′N 81°48′W
23 Y13 **Lakewood Park** Florida, SE USA 27°32′N 80°24′W
23 Z14 **Lake Worth** Florida, SE USA 26°37′N 80°03′W
124 H11 **Lakhdenpokh'ya** Respublika Kareliya, NW Russian Federation 61°25′N 30°05′E
152 L11 **Lakhimpur** Uttar Pradesh, N India 27°57′N 80°47′E
154 J11 **Lakhnādon** Madhya Pradesh, C India 22°34′N 79°38′E
Lakhnau *see* Lucknow
154 A9 **Lakhpat** Gujarāt, W India 23°49′N 68°54′E
119 K19 **Lakhva** Brestskaya Voblasts', SW Belarus 52°13′N 27°06′E
114 J11 **Laki** *var.* Lŭki. Plovdiv, C Bulgaria 40°50′N 24°49′E
26 I6 **Lakin** Kansas, C USA 37°57′N 101°16′W
149 S7 **Lakki** *var.* Lakki Marwat. Khyber Pakhtunkhwa, NW Pakistan 32°35′N 70°58′E
Lakki Marwat *see* Lakki
115 F21 **Lakonía** *historical region* S Greece
115 F22 **Lakonikós Kólpos** *gulf* S Greece
76 M17 **Lakota** S Ivory Coast 05°50′N 05°40′W
29 U11 **Lakota** Iowa, C USA 43°22′N 94°04′W
29 P3 **Lakota** North Dakota, N USA 48°02′N 98°20′W
Lak Sao *see* Ban Lakxao
92 L8 **Laksefjorden** *Lapp.* Lágesvuotna. *fjord* N Norway
92 K8 **Lakselv** *Lapp.* Leavdnja. Finnmark, N Norway 70°02′N 24°57′E
155 B21 **Lakshadweep** *prev.* the Laccadive Minicoy and Amindivi Islands. ◈ *union territory* India, N Indian Ocean
155 C22 **Lakshadweep** *Eng.* Laccadive Islands. *island group* India, N Indian Ocean
153 S17 **Lakshmikāntapur** West Bengal, NE India 22°05′N 88°19′E
112 G11 **Laktaši** ◈ Republika Srpska, N Bosnia and Herzegovina
149 V7 **Lāla Mūsa** Punjab, NE Pakistan 32°41′N 74°01′E
la Laon *see* Laon
114 M11 **Lalapaşa** Edirne, NW Turkey 41°52′N 26°44′E
83 P14 **Lalaua** Nampula, N Mozambique 14°21′S 38°16′E
105 S9 **L'Alcora** *var.* Alcora. Valenciana, E Spain 40°05′N 00°14′W
105 S10 **L'Alcúdia** *var.* L'Alcudia. Valenciana, E Spain 39°10′N 00°30′W
42 E8 **La Libertad** La Libertad, SW El Salvador 13°28′N 89°20′W*
42 E3 **La Libertad** Petén, N Guatemala 16°49′N 90°08′W
42 H6 **La Libertad** Comayagua, SW Honduras 14°43′N 87°36′W
40 E4 **La Libertad** *var.* Puerto Libertad. Sonora, NW Mexico 29°52′N 112°39′W
42 K10 **La Libertad** Chontales, S Nicaragua 12°12′N 85°10′W
42 A9 **La Libertad** ◈ *department* SW El Salvador
56 B11 **La Libertad** *off.* Departamento de La Libertad. ◈ *department* W Peru
La Libertad, Departamento de *see* La Libertad
62 G11 **La Ligua** Valparaíso, C Chile 31°30′S 71°16′W
139 U5 **La'lī Khān** As Sulaymānīyah, E Iraq 34°58′N 45°36′E
104 H3 **Lalín** Galicia, NW Spain 42°40′N 08°06′W
102 L13 **Lalinde** Dordogne, SW France 44°52′N 00°42′E
104 K16 **La Línea** *var.* La Línea de la Concepción. Andalucía, S Spain 36°10′N 05°21′W
La Línea de la Concepción *see* La Línea
152 J14 **Lalitpur** Uttar Pradesh, N India 24°42′N 78°24′E
153 P11 **Lalitpur** Central, C Nepal 27°45′N 85°18′E
152 K10 **Lālkua** Uttarachal, N India 29°04′N 79°31′E
153 T12 **Lalmanirhat** Rājshāhi, N Bangladesh 25°51′N 89°34′E
11 R12 **La Loche** Saskatchewan, C Canada 56°31′N 109°27′W
102 M6 **la Loupe** Eure-et-Loir, C France 48°30′N 01°04′E
99 G20 **La Louvière** Hainaut, S Belgium 50°29′N 04°15′E
L'Altissima *see* Hochwilde
104 L14 **La Luisiana** Andalucía, S Spain 37°30′N 05°14′W
37 S14 **La Luz** New Mexico, SW USA 32°58′N 105°56′W
107 D16 **la Maddalena** Sardegna, Italy, C Mediterranean Sea 41°13′N 09°25′E
62 J7 **La Madrid** Tucumán, N Argentina 27°37′S 65°16′W
Lama-Kara *see* Kara
15 S8 **La Malbaie** Québec, SE Canada 47°39′N 70°11′W
Lamam *see* Xékong
105 P10 **La Mancha** *physical region* C Spain
la Manche *see* English Channel
187 R13 **Lamap** Malekula, C Vanuatu 16°26′S 167°47′E
37 W6 **Lamar** Colorado, C USA 38°40′N 102°37′W
27 S7 **Lamar** Missouri, C USA 37°30′N 94°18′W
21 S12 **Lamar** South Carolina, SE USA 34°10′N 80°03′W
107 C19 **La Marmora, Punta** ▲ Sardegna, Italy, C Mediterranean Sea 40°00′N 09°20′E
8 I9 **La Martre, Lac** ◎ Northwest Territories, NW Canada
56 D10 **Lamas** San Martín, N Peru 06°28′S 76°31′W
42 I5 **La Masica** Atlántida, NW Honduras 15°38′N 87°08′W
103 R12 **Lamastre** Ardèche, E France 45°00′N 04°32′E
La Matepec *see* Santa Ana, Volcán de
44 I7 **La Maya** Santiago de Cuba, E Cuba 20°11′N 75°40′W
109 S5 **Lambach** Oberösterreich, N Austria 48°06′N 13°52′E
168 I11 **Lambak** Pulau Pini, W Indonesia 0°08′N 98°36′E
102 H5 **Lamballe** Côtes d'Armor, NW France 48°28′N 02°31′W
79 D18 **Lambaréné** Moyen-Ogooué, W Gabon 0°41′S 10°13′E
Lambasa *see* Labasa
56 B11 **Lambayeque** Lambayeque, W Peru 06°42′S 79°55′W
56 A10 **Lambayeque** *off.* Departamento de Lambayeque. ◈ *department* NW Peru
Lambayeque, Departamento de *see* Lambayeque
97 G17 **Lambay Island** *Ir.* Reachrainn. *island* E Ireland
186 G6 **Lambert, Cape** *headland* New Britain, E Papua New Guinea 04°15′S 151°31′E
195 W6 **Lambert Glacier** *glacier* Antarctica
29 T10 **Lamberton** Minnesota, N USA 44°14′N 95°15′W
27 X4 **Lambert-Saint Louis** ✈ Missouri, C USA 38°43′N 90°19′W
31 R11 **Lambertville** Michigan, N USA 41°46′N 83°37′W
18 J15 **Lambertville** New Jersey, NE USA 40°20′N 74°55′W
171 N12 **Lambogo** Sulawesi, N Indonesia 0°57′S 120°23′E
106 D8 **Lambro** ≈ N Italy
33 W11 **Lame Deer** Montana, NW USA 45°37′N 106°37′W
104 H6 **Lamego** Viseu, N Portugal 41°05′N 07°49′W
187 Q14 **Lamen Bay** Épi, C Vanuatu 16°36′S 168°10′E
45 X6 **Lamentin** Basse Terre, N Guadeloupe 16°16′N 61°38′W
Lamentin *see* le Lamentin
182 K10 **Lameroo** South Australia 35°22′S 140°30′E
54 F10 **La Mesa** Cundinamarca, C Colombia 04°37′N 74°30′W
35 U17 **La Mesa** California, W USA 32°44′N 117°00′W
37 R16 **La Mesa** New Mexico, SW USA 32°03′N 106°41′W
25 N6 **Lamesa** Texas, SW USA 32°43′N 101°57′W
107 N21 **Lamezia Terme** Calabria, SE Italy 38°54′N 16°13′E
115 F17 **Lamía** Stereá Elláda, C Greece 38°54′N 22°27′E
171 O8 **Lamitan** Basilan Island, SW Philippines 06°40′N 122°07′E
187 Y14 **Lamiti** Gau, C Fiji 18°00′S 179°20′E
171 T11 **Lamlam** Papua Barat, E Indonesia 0°03′S 130°46′E
188 B16 **Lamlam, Mount** ▲ SW Guam 13°20′N 144°40′E
109 Q6 **Lammer** ≈ E Austria
185 E23 **Lammerlaw Range** ▲▲ South Island, New Zealand
95 L20 **Lammhult** Kronoberg, S Sweden 57°09′N 14°35′E
93 L18 **Lammi** Kanta-Häme, S Finland 61°06′N 25°00′E
189 U11 **Lamoil** *island* Chuuk, C Micronesia
35 W3 **Lamoille** Nevada, W USA 40°47′N 115°37′W
18 M7 **Lamoille River** ≈ Vermont, NE USA
30 J13 **La Moine River** ≈ Illinois, N USA
171 P4 **Lamon Bay** *bay* Luzon, N Philippines
29 V16 **Lamoni** Iowa, C USA 40°37′N 93°56′W
35 R13 **Lamont** California, W USA 35°15′N 118°54′W
27 N8 **Lamont** Oklahoma, C USA 36°41′N 97°33′W
54 E13 **La Montañita** *var.* Montañita. Caquetá, S Colombia 01°22′N 75°25′W
43 N8 **La Mosquitia** *var.* Miskito Coast, *Eng.* Mosquito Coast. *coastal region* E Nicaragua
102 I9 **la Mothe-Achard** Vendée, NW France 46°37′N 01°37′W
188 L15 **Lamotrek Atoll** *atoll* Caroline Islands, C Micronesia
29 P6 **La Moure** North Dakota, N USA 46°21′N 98°17′W
167 O8 **Lampang** *var.* Muang Lampang. Lampang, NW Thailand 18°16′N 99°30′E
167 R9 **Lam Pao Reservoir** ⊠ E Thailand
25 S9 **Lampasas** Texas, SW USA 31°04′N 98°12′W
25 S9 **Lampasas River** ≈ Texas, SW USA
41 N7 **Lampazos** *var.* Lampazos de Naranjo. Nuevo León, NE Mexico 27°00′N 100°28′W
Lampazos de Naranjo *see* Lampazos
115 E19 **Lámpeia** Dytikí Elláda, S Greece 37°51′N 21°48′E
101 G19 **Lampertheim** Hessen, W Germany 49°36′N 08°28′E
97 I20 **Lampeter** SW Wales, United Kingdom 52°08′N 04°04′W
167 O7 **Lamphun** *var.* Lampun, Muang Lamphun. Lamphun, NW Thailand 18°36′N 99°02′E
11 X10 **Lamprey** Manitoba, C Canada 58°18′N 94°06′W
Lampun *see* Lamphun
168 M15 **Lampung** *off.* Propinsi Lampung. ◈ *province* SW Indonesia
Lampung, Propinsi *see* Lampung
126 K6 **Lamskoye** Lipetskaya Oblast', W Russian Federation 52°57′N 38°04′E
81 K20 **Lamu** Lamu, SE Kenya 02°17′S 40°54′E
81 K20 **Lamu** ◈ *county* SE Kenya
43 N14 **La Muerte, Cerro** ▲ C Costa Rica 09°33′N 83°47′W
103 S13 **la Mure** Isère, E France 44°54′N 05°48′E
37 S10 **Lamy** New Mexico, SW USA 35°27′N 105°52′W
119 J18 **Lan'** ≈ C Belarus
38 E10 **Lāna'i** *var.* Lanai. *island* Hawai'i, USA, C Pacific Ocean
38 E10 **Lāna'i City** *var.* Lanai City. Lanai, Hawaii, USA, C Pacific Ocean 20°49′N 156°55′W
Lanai City *see* Lāna'i City
99 L18 **Lanaken** Limburg, NE Belgium 50°53′N 05°39′E
171 Q7 **Lanao, Lake** *var.* Lake Sultan Alonto. ◎ Mindanao, S Philippines
96 J12 **Lanark** S Scotland, United Kingdom 55°38′N 04°25′W
96 I13 **Lanark** *cultural region* C Scotland, United Kingdom
104 L9 **La Nava de Ricomalillo** Castilla-La Mancha, C Spain 39°40′N 04°59′W
166 M13 **Lanbi Kyun** *prev.* Sullivan Island. *island* Mergui Archipelago, S Myanmar (Burma)
Lancang Jiang *see* Mekong
97 K17 **Lancashire** *cultural region* N England, United Kingdom
15 N13 **Lancaster** Ontario, SE Canada 45°10′N 74°31′W
97 K16 **Lancaster** NW England, United Kingdom 54°03′N 02°48′W
35 T14 **Lancaster** California, W USA 34°42′N 118°08′W
20 M6 **Lancaster** Kentucky, S USA 37°35′N 84°34′W
27 U1 **Lancaster** Missouri, C USA 40°32′N 92°31′W
19 O7 **Lancaster** New Hampshire, NE USA 44°29′N 71°34′W
18 D10 **Lancaster** New York, NE USA 42°54′N 78°40′W
31 T14 **Lancaster** Ohio, N USA 39°42′N 82°36′W
18 H16 **Lancaster** Pennsylvania, NE USA 40°03′N 76°18′W
21 R11 **Lancaster** South Carolina, SE USA 34°43′N 80°47′W
25 U7 **Lancaster** Texas, SW USA 32°35′N 96°45′W
21 X5 **Lancaster** Virginia, NE USA 37°48′N 76°30′W
30 J9 **Lancaster** Wisconsin, N USA 42°52′N 90°43′W
197 N10 **Lancaster Sound** *sound* Nunavut, N Canada
Lan-chou/Lan-chow/Lanchow *see* Lanzhou
107 K14 **Lanciano** Abruzzo, C Italy 42°13′N 14°23′E
111 O16 **Łańcut** Podkarpackie, SE Poland 50°04′N 22°14′E
169 Q11 **Landak, Sungai** ≈ Borneo, N Indonesia
Landao *see* Lantau Island
Landau *see* Landau an der Isar
Landau *see* Landau in der Pfalz
101 N22 **Landau an der Isar** *var.* Landau. Bayern, SE Germany 48°40′N 12°41′E
101 F20 **Landau in der Pfalz** *var.* Landau. Rheinland-Pfalz, SW Germany 49°12′N 08°07′E
Land Burgenland *see* Burgenland
108 K8 **Landeck** Tirol, W Austria 47°09′N 10°35′E
99 J19 **Landen** Vlaams Brabant, C Belgium 50°45′N 05°05′E
33 U15 **Lander** Wyoming, C USA 42°49′N 108°43′W
102 F5 **Landerneau** Finistère, NW France 48°27′N 04°16′W
95 K20 **Landeryd** Halland, S Sweden 57°04′N 13°15′E
102 J15 **Landes** ◈ *department* SW France
Landeshut/Landeshut in Schlesien *see* Kamienna Góra
105 R9 **Landete** Castilla-La Mancha, C Spain 39°54′N 01°22′W
99 M18 **Landgraaf** Limburg, SE Netherlands 50°55′N 06°04′E
102 F5 **Landivisiau** Finistère, NW France 48°31′N 04°03′W
Land Kärten *see* Kärnten
Land of Enchantment *see* New Mexico
The Land of Opportunity *see* Arkansas
Land of Steady Habits *see* Connecticut
Land of the Midnight Sun *see* Alaska
108 I8 **Landquart** Graubünden, SE Switzerland 46°58′N 09°35′E
108 J9 **Landquart** ≈ Austria/Switzerland
21 P10 **Landrum** South Carolina, SE USA 35°10′N 82°11′W
Landsberg *see* Gorzów Wielkopolski, Lubuskie, Poland
Landsberg *see* Górowo Iławeckie, Warmińsko-Mazurskie, NE Poland
101 K23 **Landsberg am Lech** Bayern, S Germany 48°03′N 10°52′E
Landsberg an der Warthe *see* Gorzów Wielkopolski
97 G25 **Land's End** *headland* SW England, United Kingdom 50°02′N 05°41′W
101 M22 **Landshut** Bayern, SE Germany 48°32′N 12°09′E
Landskron *see* Lanškroun
95 J22 **Landskrona** Skåne, S Sweden 55°52′N 12°52′E
98 I10 **Landsmeer** Noord-Holland, C Netherlands 52°26′N 04°55′E
95 J19 **Landvetter** ✈ (Göteborg) Västra Götaland, S Sweden 57°39′N 12°22′E
Landwarów *see* Lentvaris
23 R5 **Lanett** Alabama, S USA 32°52′N 85°11′W
108 C8 **La Neuveville** *var.* Neuveville, *Ger.* Neuenstadt. Neuchâtel, W Switzerland 47°05′N 07°03′E
95 G21 **Langå** *var.* Langaa. Midtjylland, C Denmark 56°23′N 09°55′E
Langaa *see* Langå
158 G14 **La'nga Co** ◎ W China
Langada *see* Lagkáda
Langades/Langadhás *see* Lagkadás
Langádhia/ *see* Lagkádia
147 T14 **Langar** *Rus.* Lyangar. SE Tajikistan 37°04′N 72°39′E
146 M10 **Langar** *Rus.* Lyangar. Navoiy Viloyati, C Uzbekistan 40°27′N 65°54′E
142 M3 **Langarūd** Gīlān, NW Iran 37°10′N 50°09′E
11 V16 **Langbank** Saskatchewan, S Canada 50°01′N 102°16′W
29 P2 **Langdon** North Dakota, N USA 48°45′N 98°22′W
103 P12 **Langeac** Haute-Loire, C France 45°06′N 03°31′E
102 L8 **Langeais** Indre-et-Loire, C France 47°22′N 00°27′E
80 I8 **Langeb, Wadi** ≈ NE Sudan
Långed *see* Dals Långed
95 G25 **Langeland** *island* S Denmark
99 B18 **Langemark** West-Vlaanderen, W Belgium 50°55′N 02°55′E
101 G18 **Langen** Hessen, W Germany 49°58′N 08°40′E
101 J22 **Langenau** Baden-Württemberg, S Germany 48°30′N 10°08′E
11 V16 **Langenburg** Saskatchewan, S Canada 50°50′N 101°43′W
108 L8 **Längenfeld** Tirol, W Austria 47°04′N 10°59′E
101 E16 **Langenfeld** Nordrhein-Westfalen, W Germany 51°06′N 06°57′E
100 I12 **Langenhagen** Niedersachsen, N Germany 52°26′N 09°45′E
100 I12 **Langenhagen** ✈ (Hannover) Niedersachsen, NW Germany 52°28′N 09°40′E
109 W3 **Langenlois** Niederösterreich, NE Austria 48°29′N 15°42′E
108 E7 **Langenthal** Bern, NW Switzerland 47°13′N 07°48′E
109 W6 **Langenwang** Steiermark, E Austria 47°34′N 15°39′E
109 X3 **Langenzersdorf** Niederösterreich, E Austria 48°20′N 16°22′E
100 F9 **Langeoog** *island* NW Germany
95 H23 **Langeskov** Syddtjylland, C Denmark 55°22′N 10°36′E
95 G16 **Langesund** Telemark, S Norway 59°00′N 09°43′E
95 G17 **Langesundsfjorden** *fjord* S Norway
94 D10 **Langevågen** Møre og Romsdal, S Norway 62°26′N 06°15′E
161 P3 **Langfang** Hebei, E China 39°30′N 116°39′E
94 E9 **Langfjorden** *fjord* S Norway
29 Q8 **Langford** South Dakota, N USA 45°36′N 97°48′W
168 I10 **Langgapayung** Sumatera, W Indonesia 01°42′N 99°57′E
106 E9 **Langhirano** Emilia-Romagna, C Italy 44°37′N 10°16′E
97 K14 **Langholm** S Scotland, United Kingdom 55°14′N 03°11′W
92 I3 **Langjökull** *glacier* C Iceland
168 I6 **Langkawi, Pulau** *island* Peninsular Malaysia
166 M14 **Langkha Tuk, Khao** ▲ SW Thailand 09°19′N 98°39′E
14 L8 **Langlade** Québec, SE Canada 48°13′N 75°58′W
10 M17 **Langley** British Columbia, SW Canada 49°07′N 122°39′W
167 S7 **Lang Mô** Thanh Hoa, N Vietnam 19°36′N 105°30′E
Langnau *see* Langnau im Emmental
108 E8 **Langnau im Emmental** *var.* Langnau. Bern, W Switzerland 46°57′N 07°47′E
103 Q13 **Langogne** Lozère, S France 44°40′N 03°52′E
102 K13 **Langon** Gironde, SW France 44°33′N 00°14′W
La Ngounié *see* Ngounié
92 G10 **Langøya** *island* C Norway
158 G14 **Langqên Zangbo** ≈ China/India
Langreo *see* Llangréu
103 S7 **Langres** Haute-Marne, N France 47°53′N 05°20′E
103 R8 **Langres, Plateau de** *plateau* C France
168 H8 **Langsa** Sumatera, W Indonesia 04°30′N 97°53′E
93 H16 **Långsele** Västernorrland, C Sweden 63°11′N 17°05′E
162 L12 **Lang Shan** ▲▲ N China
95 M14 **Långshyttan** Dalarna, C Sweden 60°26′N 16°02′E
167 T5 **Lang Sơn** *var.* Langson. Lang Sơn, N Vietnam 21°50′N 106°45′E
Langson *see* Lang Sơn
167 N14 **Lang Suan** Chumphon, SW Thailand 09°55′N 99°07′E
93 J14 **Långträsk** Norrbotten, N Sweden 65°22′N 20°19′E
25 N11 **Langtry** Texas, SW USA 29°46′N 101°25′W
103 P16 **Languedoc** *cultural region* S France
103 P15 **Languedoc-Roussillon** ◈ *region* S France
27 X10 **L'Anguille River** ≈ Arkansas, C USA
93 I16 **Långviksmon** Västernorrland, N Sweden 63°39′N 18°45′E
101 K22 **Langweid** Bayern, S Germany 48°29′N 10°50′E
160 J8 **Langzhong** Sichuan, C China 31°46′N 105°55′E
Lan Hsü *see* Lan Yu
11 U15 **Lanigan** Saskatchewan, S Canada 51°50′N 105°01′W
116 K5 **Lanivtsi** Ternopil's'ka Oblast', W Ukraine 49°52′N 26°05′E
137 Y13 **Länkäran** *Rus.* Lenkoran'. S Azerbaijan 38°46′N 48°51′E
102 L16 **Lannemezan** Hautes-Pyrénées, S France 43°08′N 00°22′E
102 G5 **Lannion** Côtes d'Armor, NW France 48°44′N 03°27′W
14 M11 **L'Annonciation** Québec, SE Canada 46°22′N 74°51′W
105 V5 **L'Anoia** ≈ NE Spain

◆ Country
● Country Capital
◇ Dependent Territory
○ Dependent Territory Capital
◈ Administrative Regions
✈ International Airport
▲ Mountain
▲▲ Mountain Range
⛰ Volcano
≈ River
◎ Lake
⊠ Reservoir

18 I15 **Lansdale** Pennsylvania, NE USA 40°14′N 75°13′W
14 L14 **Lansdowne** Ontario, SE Canada 44°25′N 76°00′W
152 K9 **Lansdowne** Uttarakhand, N India 29°50′N 78°42′E
30 M3 **L'Anse** Michigan, N USA 46°45′N 88°27′W
15 S7 **L'Anse-St-Jean** Québec, SE Canada 48°14′N 70°13′W
29 Y11 **Lansing** Iowa, C USA 43°22′N 91°11′W
27 R4 **Lansing** Kansas, C USA 39°15′N 94°54′W
31 Q9 **Lansing** *state capital* Michigan, N USA 42°44′N 84°33′W
Länsi-Suomi ◆ *province* W Finland
92 J12 **Lansjärv** Norrbotten, N Sweden 66°39′N 22°10′E
111 G17 **Lanškroun** *Ger.* Landskron. Pardubický Kraj, E Czech Republic 49°55′N 16°38′E
167 N16 **Lanta, Ko** *island* S Thailand
161 O15 **Lantau Island** *Cant.* Tai Yue Shan, *Chin.* Landao. *island* Hong Kong, S China
Lantian *see* Lianyuan
Lan-ts'ang Chiang *see* Mekong
Lantung, Gulf of *see* Liaodong Wan
171 O11 **Lanu** Sulawesi, N Indonesia 01°00′N 121°33′E
107 D19 **Lanusei** Sardegna, Italy, C Mediterranean Sea 39°55′N 09°31′E
102 H7 **Lanvaux, Landes de** *physical region* NW France
163 W8 **Lanxi** Heilongjiang, NE China 46°18′N 126°15′E
161 R10 **Lanxi** Zhejiang, SE China 29°12′N 119°27′E
La Nyanga *see* Nyanga
161 T15 **Lan Yu** *var.* Huoshao Tao, Hungt'ou, Lan Hsü, Lanyü, *Eng.* Orchid Island; *prev.* Kotosho, Koto Sho, Lan Yü. *island* SE Taiwan
Lanyü *see* Lan Yu
64 P11 **Lanzarote** *island* Islas Canarias, Spain, NE Atlantic Ocean
159 V10 **Lanzhou** *var.* Lan-chou, Lanchow, Lan-chow; *prev.* Kaolan. *province capital* Gansu, C China 36°01′N 103°52′E
106 B8 **Lanzo Torinese** Piemonte, NE Italy 45°18′N 07°26′E
171 O11 **Laoag** Luzon, N Philippines 18°11′N 120°34′E
171 Q5 **Laoang** Samar, C Philippines 12°29′N 125°01′E
167 R5 **Lao Cai** Lao Cai, N Vietnam 22°29′N 104°00′E
Laodicea/Laodicea ad Mare *see* Al Lādhiqīyah
Laoet *see* Laut, Pulau
163 T11 **Laoha He** ≈ NE China
160 M8 **Laohekou** *var.* Guanghua. Hubei, C China 32°20′N 111°42′E
Laoi, An *see* Lee
97 E19 **Laois** *prev.* Leix, Queen's County. *cultural region* C Ireland
163 W12 **Lao Ling** ▲ N China
64 Q11 **La Oliva** *var.* Oliva. Fuerteventura, Islas Canarias, Spain, NE Atlantic Ocean 28°36′N 13°53′W
Lao, Loch *see* Belfast Lough
Laolong *see* Longchuan
Lao Mangnai *see* Mangnai
103 P3 **Laon** *var.* la Laon; *anc.* Laudunum. Aisne, N France 49°34′N 03°37′E
Lao People's Democratic Republic *see* Laos
54 M3 **La Orchila, Isla** *island* N Venezuela
64 O11 **La Orotava** Tenerife, Islas Canarias, Spain, NE Atlantic Ocean 28°23′N 16°32′W
57 E14 **La Oroya** Junín, C Peru 11°36′S 75°54′W
167 Q7 **Laos** *off.* Lao People's Democratic Republic. ◆ *republic* SE Asia
161 R5 **Laoshan Wan** *bay* E China
163 Y10 **Laoye Ling** ▲ NE China
60 J12 **Lapa** Paraná, S Brazil 25°46′S 49°44′W
103 P10 **Lapalisse** Allier, C France 46°13′N 03°39′E
54 F9 **La Palma** Cundinamarca, C Colombia 05°23′N 74°24′W
42 F7 **La Palma** Chalatenango, N El Salvador 14°19′N 89°10′W
43 W16 **La Palma** Darién, SE Panama 08°24′N 78°09′W
64 N11 **La Palma** *island* Islas Canarias, Spain, NE Atlantic Ocean
104 J14 **La Palma del Condado** Andalucía, S Spain 37°23′N 06°33′W
61 F18 **La Paloma** Durazno, C Uruguay 32°54′S 55°36′W
61 G20 **La Paloma** Rocha, E Uruguay 34°37′S 54°08′W
61 A21 **La Pampa** *off.* Provincia de La Pampa. ◆ *province* C Argentina
La Pampa, Provincia de *see* La Pampa
55 P8 **La Paragua** Bolívar, E Venezuela 06°53′N 63°16′W
119 O16 **Lapatsichy** *Rus.* Lopatichi. Mahilyowskaya Voblasts', E Belarus 53°34′N 30°53′E
61 C16 **La Paz** Entre Ríos, E Argentina 30°45′S 59°36′W
62 I11 **La Paz** Mendoza, C Argentina 33°30′S 67°36′W
57 J18 **La Paz** *var.* La Paz de Ayacucho. ● (Bolivia-seat of government) La Paz, W Bolivia 16°30′S 68°13′W
42 H6 **La Paz** La Paz, SW Honduras 14°20′N 87°40′W
40 F9 **La Paz** Baja California Sur, NW Mexico 24°07′N 110°18′W
61 F20 **La Paz** Canelones, S Uruguay 34°46′S 56°13′W
57 J16 **La Paz** ◆ *department* W Bolivia
42 B9 **La Paz** ◆ *department* S El Salvador
42 G7 **La Paz** ◆ *department* SW Honduras
La Paz *see* El Alto, Bolivia
La Paz *see* Robles, Colombia
La Paz *see* La Paz Centro
40 F9 **La Paz, Bahía de** *bay* NW Mexico
42 I10 **La Paz Centro** *var.* La Paz. León, W Nicaragua 12°20′N 86°41′W
La Paz de Ayacucho *see* La Paz
54 J15 **La Pedrera** Amazonas, SE Colombia 01°19′S 69°31′W
31 S9 **Lapeer** Michigan, N USA 43°03′N 83°19′W
40 K6 **La Perla** Chihuahua, N Mexico 28°18′N 104°34′W
165 T1 **La Pérouse Strait** *Jap.* Sōya-kaikyō, *Rus.* Proliv Laperuza. *strait* Japan/Russian Federation
63 I14 **La Perra, Salitral de** *salt lake* C Argentina
Laperuza, Proliv *see* La Pérouse Strait
41 Q10 **La Pesca** Tamaulipas, C Mexico 23°49′N 97°45′W
40 M13 **La Piedad Cavadas** Michoacán, C Mexico 20°20′N 102°01′W
Lapines *see* Lafnitz
93 M16 **Lapinlahti** Pohjois-Savo, C Finland 63°21′N 27°25′E
Lápithos *see* Lapta
22 K9 **Laplace** Louisiana, S USA 30°04′N 90°28′W
45 X12 **La Plaine** SE Dominica 15°20′N 61°15′W
173 P16 **La Plaine-des-Palmistes** C Réunion
92 K11 **Lapland** *Fin.* Lappi, *Swe.* Lappland. *cultural region* N Europe
Lapland *see* Lappi
28 M8 **La Plant** South Dakota, N USA 45°06′N 100°40′W
61 D20 **La Plata** Buenos Aires, E Argentina 34°56′S 57°55′W
54 D12 **La Plata** Huila, SW Colombia 02°33′N 75°55′W
21 W4 **La Plata** Maryland, NE USA 38°32′N 76°59′W
La Plata *see* Sucre
45 U6 **la Plata, Río de** ≈ C Puerto Rico
105 W4 **La Pobla de Lillet** Cataluña, NE Spain 42°15′N 01°57′E
105 U4 **La Pobla de Segur** Cataluña, NE Spain 42°15′N 00°58′E
15 S9 **La Pocatière** Québec, SE Canada 47°21′N 70°04′W
104 K2 **La Pola** *prev.* Pola de Lena. Asturias, N Spain 43°10′N 05°49′W
104 L3 **La Pola de Gordón** Castilla y León, N Spain 42°50′N 05°38′W
104 L2 **La Pola Siero** *prev.* Pola de Siero. Asturias, N Spain 43°24′N 05°39′W
31 O11 **La Porte** Indiana, N USA 41°36′N 86°43′W
18 H13 **Laporte** Pennsylvania, NE USA 41°25′N 76°29′W
29 X13 **La Porte City** Iowa, C USA 42°19′N 92°11′W
62 J8 **La Posta** Catamarca, C Argentina 27°59′S 65°32′W
40 E8 **La Poza Grande** Baja California Sur, NW Mexico 25°50′N 112°00′W
93 K16 **Lappajärvi** Etelä-Pohjanmaa, W Finland 63°13′N 23°40′E
93 L16 **Lappajärvi** ◎ W Finland
93 N18 **Lappeenranta** *Swe.* Villmanstrand. Etelä-Karjala, SE Finland 61°04′N 28°15′E
93 J17 **Lappfjärd** *Fin.* Lapväärtti. Österbotten, W Finland 62°14′N 21°30′E
92 L12 **Lappi** *Swe.* Lappland, *Eng.* Lapland. ◆ *region* N Finland
Lappi/Lappland *see* Lapland
Lappo *see* Lapua
61 C23 **Laprida** Buenos Aires, E Argentina 37°34′S 60°45′W
25 P13 **La Pryor** Texas, SW USA 28°56′N 99°51′W
136 B11 **Lâpseki** Çanakkale, NW Turkey 40°22′N 26°42′E
121 P2 **Lapta** *Gk.* Lápithos. NW Cyprus 35°20′N 33°11′E
Laptev Sea *see* Laptevykh, More
122 N6 **Laptevykh, More** *Eng.* Laptev Sea. *sea* Arctic Ocean
93 K16 **Lapua** *Swe.* Lappo. Etelä-Pohjanmaa, W Finland 62°57′N 23°E
105 P3 **La Puebla de Arganzón** País Vasco, N Spain 42°45′N 02°49′W
104 L14 **La Puebla de Cazalla** Andalucía, S Spain 37°14′N 05°18′W
104 M9 **La Puebla de Montalbán** Castilla-La Mancha, C Spain 39°52′N 04°22′W
54 I6 **La Puerta** Trujillo, NW Venezuela 09°08′N 70°46′W
Lapurdum *see* Bayonne
40 E7 **La Purísima** Baja California Sur, NW Mexico 26°10′N 112°05′W
Lapväärtti *see* Lappfjärd
110 O10 **Łapy** Podlaskie, NE Poland 53°N 22°54′E
80 D6 **Laqiya Arba'in** Northern, NW Sudan 20°01′N 28°01′E
62 J4 **La Quiaca** Jujuy, N Argentina 22°12′S 65°36′W
107 J14 **L'Aquila** *var.* Aquila, Aquila degli Abruzzi. Abruzzo, C Italy 42°21′N 13°24′E
143 Q13 **Lār** Fārs, S Iran 27°42′N 54°19′E
55 J5 **Lara** *off.* Estado Lara. ◆ *state* NW Venezuela
104 G2 **Laracha** Galicia, NW Spain 43°14′N 08°34′W
74 G5 **Larache** *var.* al Araïch, El Araïch, *prev.* El Araïche; *anc.* Lixus. NW Morocco 35°12′N 06°10′W
Lara, Estado *see* Lara
103 T14 **Laragne-Montéglin** Hautes-Alpes, SE France 44°21′N 05°46′E
104 M13 **La Rambla** Andalucía, S Spain 37°37′N 04°44′W
33 Y17 **Laramie** Wyoming, C USA 41°18′N 105°35′W
33 X15 **Laramie Mountains** ▲ Wyoming, C USA
33 Y16 **Laramie River** ≈ Wyoming, C USA
60 H12 **Laranjeiras do Sul** Paraná, S Brazil 25°23′S 52°23′W
Larantoeka *see* Larantuka
171 P16 **Larantuka** *prev.* Larantoeka. Flores, C Indonesia 08°20′S 123°00′E
171 U15 **Larat** Pulau Larat, E Indonesia 07°07′S 131°46′E
171 U15 **Larat, Pulau** *island* Kepulauan Tanimbar, E Indonesia
95 P19 **Lärbro** Gotland, SE Sweden 57°46′N 18°49′E
106 A9 **Larche, Col de** *pass* France/Italy
14 H8 **Larder Lake** Ontario, S Canada 48°06′N 79°44′W
105 O2 **Laredo** Cantabria, N Spain 43°23′N 03°22′W
25 Q15 **Laredo** Texas, SW USA 27°30′N 99°30′W
40 H9 **La Reforma** Sinaloa, W Mexico 25°05′N 108°03′W
98 N11 **Laren** Gelderland, E Netherlands 52°12′N 06°22′E
98 J11 **Laren** Noord-Holland, C Netherlands 52°15′N 05°13′E
102 K13 **la Réole** Gironde, SW France 44°34′N 00°00′W
La Réunion *see* Réunion
Largeau *see* Faya
103 U13 **l'Argentière-la-Bessée** Hautes-Alpes, SE France 44°49′N 06°34′E
149 O4 **Lar Gerd** *var.* Largird. Balkh, N Afghanistan 35°36′N 66°48′E
Largird *see* Lar Gerd
23 V12 **Largo** Florida, SE USA 27°55′N 82°47′W
37 Q9 **Largo, Canon** *valley* New Mexico, SW USA
44 D6 **Largo, Cayo** *island* W Cuba
23 Z17 **Largo, Key** *island* Florida Keys, Florida, SE USA
96 H12 **Largs** W Scotland, United Kingdom 55°48′N 04°50′W
102 I16 **la Rhune** *var.* Larrún. ▲ France/Spain 43°19′N 01°36′W *see also* Larrún
la Rhune *see* Larrún
la Riege *see* Ariège
29 Q4 **Larimore** North Dakota, N USA 47°54′N 97°37′W
107 L15 **Larino** Molise, C Italy 41°46′N 14°50′E
Lario *see* Como, Lago di
62 J9 **La Rioja** La Rioja, NW Argentina 29°26′S 66°50′W
62 I9 **La Rioja** *off.* Provincia de La Rioja. ◆ *province* NW Argentina
105 O4 **La Rioja** ◆ *autonomous community* N Spain
La Rioja, Provincia de *see* La Rioja
115 F16 **Lárisa** *var.* Larissa. Thessalía, C Greece 39°38′N 22°27′E
Larissa *see* Lárisa
149 Q13 **Lārkāna** *var.* Larkhana. Sind, SE Pakistan 27°32′N 68°18′E
Larkhana *see* Lārkāna
Larnaca *see* Lárnaka
121 Q3 **Lárnaka** *var.* Larnaca, Larnax. SE Cyprus 34°55′N 33°39′E
121 Q3 **Lárnaka** ✈ SE Cyprus 34°52′N 33°38′E
Larnax *see* Lárnaka
97 G14 **Larne** *Ir.* Latharna. E Northern Ireland, United Kingdom 54°51′N 05°49′W
26 L5 **Larned** Kansas, C USA 38°12′N 99°05′W
104 L3 **La Robla** Castilla y León, N Spain 42°48′N 05°37′W
104 J10 **La Roca de la Sierra** Extremadura, W Spain 39°06′N 06°41′W
99 K22 **La Roche-en-Ardenne** Luxembourg, SE Belgium 50°11′N 05°35′E
102 L11 **la Rochefoucauld** Charente, W France 45°43′N 00°23′E
102 J10 **la Rochelle** *anc.* Rupella. Charente-Maritime, W France 46°09′N 01°07′W
102 I9 **la Roche-sur-Yon** *prev.* Bourbon Vendée, Napoléon-Vendée. Vendée, NW France 46°40′N 01°26′W
105 Q10 **La Roda** Castilla-La Mancha, C Spain 39°13′N 02°10′W
104 L14 **La Roda de Andalucía** Andalucía, S Spain 37°12′N 04°45′W
45 P9 **La Romana** E Dominican Republic 18°25′N 69°00′W
11 T13 **La Ronge** Saskatchewan, C Canada 55°07′N 105°18′W
11 U13 **La Ronge, Lac** ◎ Saskatchewan, C Canada
22 K10 **Larose** Louisiana, S USA 29°34′N 90°22′W
42 M7 **La Rosita** Región Autónoma Atlántico Norte, NE Nicaragua 13°55′N 84°23′W
181 Q3 **Larrimah** Northern Territory, N Australia 15°30′S 133°12′E
62 N11 **Larroque** Entre Ríos, E Argentina 33°05′S 59°06′W
105 Q2 **Larrún** *Fr.* la Rhune. ▲ France/Spain 43°18′N 01°35′W *see also* la Rhune
Larrún *see* la Rhune
195 X6 **Lars Christensen Coast** *physical region* Antarctica
39 Q14 **Larsen Bay** Kodiak Island, Alaska, USA 57°32′N 153°58′W
194 I5 **Larsen Ice Shelf** *ice shelf* Antarctica
8 M6 **Larsen Sound** *sound* Nunavut, N Canada
La Rúa *see* A Rúa de Valdeorras
102 K16 **Laruns** Pyrénées-Atlantiques, SW France 43°00′N 00°25′W
95 G16 **Larvik** Vestfold, S Norway 59°04′N 10°02′E
La-sa *see* Lhasa
171 S13 **Lasahata** Pulau Seram, E Indonesia 02°52′S 128°27′E
Lasahau *see* Lasihao
37 O6 **La Sal** Utah, W USA 38°19′N 109°14′W
14 C17 **La Salle** Ontario, S Canada 42°13′N 83°05′W
30 L11 **La Salle** Illinois, N USA 41°19′N 89°06′W
45 O9 **Las Americas** ✈ (Santo Domingo) S Dominican Republic 18°24′N 69°38′W
37 V6 **Las Animas** Colorado, C USA 38°04′N 103°13′W
108 D10 **La Sarine** *var.* Sarine. ≈ SW Switzerland
108 B9 **La Sarraz** Vaud, W Switzerland 46°40′N 06°32′E
12 H12 **La Sarre** Québec, SE Canada 48°49′N 79°12′W
54 L3 **Las Aves, Islas** *var.* Islas de Aves. *island group* N Venezuela
55 N7 **Las Bonitas** Bolívar, C Venezuela 07°50′N 65°40′W
104 K15 **Las Cabezas de San Juan** Andalucía, S Spain 36°59′N 05°56′W
61 G19 **Lascano** Rocha, E Uruguay 33°40′S 54°12′W
62 I5 **Lascar, Volcán** ⛰ N Chile 23°22′S 67°33′W
41 T15 **Las Choapas** *var.* Choapas. Veracruz-Llave, SE Mexico 17°51′N 94°00′W
37 R15 **Las Cruces** New Mexico, SW USA 32°19′N 106°49′W
Lasdehnen *see* Krasnoznamensk
La Selle *see* Selle, Pic de la
62 G9 **La Serena** Coquimbo, C Chile 29°54′S 71°18′W
104 K11 **La Serena** *physical region* W Spain
105 V4 **La Seu d'Urgell** *var.* La Seu d'Urgel; *prev.* Seo de Urgel. Cataluña, NE Spain 42°22′N 01°27′E
La Seu d'Urgel *see* La Seu d'Urgell
103 T16 **la Seyne-sur-Mer** Var, SE France 43°07′N 05°53′E
61 D21 **Las Flores** Buenos Aires, E Argentina 36°03′S 59°08′W
62 H9 **Las Flores** San Juan, W Argentina 30°14′S 69°10′W
11 S14 **Lashburn** Saskatchewan, S Canada 53°09′N 109°37′W
62 I11 **Las Heras** Mendoza, W Argentina 32°48′S 68°50′W
148 M8 **Lashkar Gāh** *var.* Lash-Kar-Gar'. Helmand, S Afghanistan 31°35′N 64°21′E
Lash-Kar-Gar' *see* Lashkar Gāh
171 P14 **Lasihao** *var.* Lasahau. Pulau Muna, C Indonesia 05°01′S 122°23′E
107 N21 **La Sila** ▲ SW Italy
63 H23 **La Silueta, Cerro** ▲ S Chile 52°22′S 72°09′W
42 L9 **La Sirena** Región Autónoma Atlántico Sur, E Nicaragua 12°59′N 84°35′W
110 J13 **Łask** Łódzkie, C Poland 51°36′N 19°06′E
109 V11 **Laško** *Ger.* Tüffer. C Slovenia 46°08′N 15°13′E
63 H14 **Las Lajas** Neuquén, W Argentina 38°31′S 70°22′W
63 H15 **Las Lajas, Cerro** ▲ W Argentina 38°49′S 70°42′W
62 M6 **Las Lomitas** Formosa, N Argentina 24°45′S 60°35′W
41 V16 **Las Margaritas** Chiapas, SE Mexico 16°15′N 91°58′W
Las Marismas *see* Guadalquivir, Marismas del
54 M6 **Las Mercedes** Guárico, N Venezuela 09°08′N 66°27′W
42 F6 **Las Minas, Cerro** ▲ W Honduras 14°33′N 88°41′W
105 O11 **La Solana** Castilla-La Mancha, C Spain 38°56′N 03°14′W
45 Q14 **La Soufrière** ⛰ Saint Vincent, Saint Vincent and the Grenadines 13°20′N 61°11′W
102 M10 **la Souterraine** Creuse, C France 46°15′N 01°28′E
62 N7 **Las Palmas** Chaco, N Argentina 27°08′S 58°45′W
43 Q16 **Las Palmas** Veraguas, W Panama 08°09′N 81°28′W
64 P12 **Las Palmas** *var.* Las Palmas de Gran Canaria. Gran Canaria, Islas Canarias, Spain, NE Atlantic Ocean 28°08′N 15°27′W
64 P12 **Las Palmas** ◆ *province* Islas Canarias, Spain, NE Atlantic Ocean
64 Q12 **Las Palmas** ✈ Gran Canaria, Islas Canarias, Spain, NE Atlantic Ocean
Las Palmas de Gran Canaria *see* Las Palmas
40 D6 **Las Palomas** Baja California Norte, W Mexico 31°44′N 107°37′W
105 P10 **Las Pedroñeras** Castilla-La Mancha, C Spain 39°27′N 02°41′W
106 E10 **La Spezia** Liguria, NW Italy 44°08′N 09°50′E
61 F20 **Las Piedras** Canelones, S Uruguay 34°42′S 56°14′W
63 J18 **Las Plumas** Chubut, S Argentina 43°46′S 67°15′W
61 B18 **Las Rosas** Santa Fe, C Argentina 32°27′S 61°30′W
Lassa *see* Lhasa
35 O4 **Lassen Peak** ⛰ California, W USA 40°27′N 121°28′W
194 K6 **Lassiter Coast** *physical region* Antarctica
109 V9 **Lassnitz** ≈ SE Austria
15 O12 **L'Assomption** Québec, SE Canada 45°48′N 73°27′W
15 N11 **L'Assomption** ≈ Québec, SE Canada
43 S17 **Las Tablas** Los Santos, S Panama 07°45′N 80°17′W
Lastarria, Volcán *see* Azufre, Volcán
37 V4 **Last Chance** Colorado, C USA 39°41′N 103°34′W
Last Frontier, The *see* Alaska
11 U16 **Last Mountain Lake** ◎ Saskatchewan, S Canada
62 H9 **Las Tórtolas, Cerro** ▲ W Argentina 29°57′S 69°49′W
61 C14 **Las Toscas** Santa Fe, C Argentina 28°22′S 59°20′W
79 F19 **Lastoursville** Ogooué-Lolo, E Gabon 0°50′S 12°43′E
113 F16 **Lastovo** *It.* Lagosta. *island* SW Croatia
113 F16 **Lastovski Kanal** *channel* SW Croatia
40 E6 **Las Tres Vírgenes, Volcán** ⛰ NW Mexico 27°27′N 112°34′W
40 F4 **Las Trincheras** Sonora, NW Mexico 30°21′N 111°27′W
55 N8 **Las Trincheras** Bolívar, E Venezuela 06°57′N 64°49′W
44 H7 **Las Tunas** *var.* Victoria de las Tunas. Las Tunas, E Cuba 20°58′N 76°59′W
La Suisse *see* Switzerland
40 I5 **Las Varas** Chihuahua, N Mexico 29°35′N 108°01′W
40 J12 **Las Varas** Nayarit, C Mexico 21°12′N 105°10′W
62 L10 **Las Varillas** Córdoba, E Argentina 31°54′S 62°45′W
35 X11 **Las Vegas** Nevada, W USA 36°09′N 115°10′W
37 T10 **Las Vegas** New Mexico, SW USA 35°35′N 105°15′W
187 P10 **Lata** Nendö, Solomon Islands 10°45′S 165°43′E
13 R10 **La Tabatière** Québec, E Canada 50°51′N 58°59′W
56 C6 **Latacunga** Cotopaxi, C Ecuador 0°58′S 78°36′W
194 I7 **Latady Island** *island* Antarctica
54 E14 **La Tagua** Putumayo, S Colombia 0°05′S 74°39′W
Latakia *see* Al Lādhiqīyah
92 J10 **Lätäseno** ≈ NW Finland
14 H9 **Latchford** Ontario, S Canada 47°20′N 79°45′W
14 J13 **Latchford Bridge** Ontario, SE Canada 45°16′N 77°29′W
193 Y14 **Late** *island* Vava'u Group, N Tonga
153 P15 **Lātehār** Jhārkhand, N India 23°48′N 84°28′E
15 R7 **Laterrière** Québec, SE Canada 48°17′N 71°10′W
102 J13 **la Teste** Gironde, SW France 44°38′N 01°04′W
25 V8 **Latexo** Texas, SW USA 31°24′N 95°28′W
18 L10 **Latham** New York, NE USA 42°45′N 73°45′W
Latharna *see* Larne
108 B9 **La Thielle** *var.* Thièle. ≈ W Switzerland
27 R3 **Lathrop** Missouri, C USA 39°33′N 94°19′W
107 I16 **Latina** *prev.* Littoria. Lazio, C Italy 41°28′N 12°53′E
41 R14 **La Tinaja** Veracruz-Llave, S Mexico
106 J7 **Latisana** Friuli-Venezia Giulia, NE Italy 45°47′N 13°01′E
Latium *see* Lazio
115 K25 **Lató** *site of ancient city* Kríti, Greece, E Mediterranean Sea
187 Q17 **La Tontouta** ✈ (Noumea) Province Sud, S New Caledonia 22°06′S 166°12′E
55 N4 **La Tortuga, Isla** *var.* Isla Tortuga. *island* N Venezuela
108 C10 **La Tour-de-Peilz** *var.* La Tour de Peilz. Vaud, SW Switzerland 46°28′N 06°52′E
La Tour de Peilz *see* La Tour-de-Peilz
103 S11 **la Tour-du-Pin** Isère, E France 45°34′N 05°25′E
102 J11 **la Tremblade** Charente-Maritime, W France 45°45′N 01°07′W
102 L10 **la Trimouille** Vienne, W France 46°27′N 01°02′E
42 J9 **La Trinidad** Estelí, NW Nicaragua 12°57′N 86°15′W
41 V16 **La Trinitaria** Chiapas, SE Mexico 16°02′N 92°00′W
45 Q11 **la Trinité** E Martinique 14°44′N 60°58′W
15 U7 **La Trinité-des-Monts** Québec, SE Canada 48°07′N 68°31′W
18 C15 **Latrobe** Pennsylvania, NE USA 40°18′N 79°19′W
183 P13 **La Trobe River** ≈ Victoria, SE Australia
Lattakia/Lattaquié *see* Al Lādhiqīyah
171 S13 **Latu** Pulau Seram, E Indonesia 03°24′S 128°37′E
15 P9 **La Tuque** Québec, SE Canada 47°26′N 72°47′W
155 G14 **Lātūr** Mahārāshtra, C India 18°24′N 76°34′E
118 G8 **Latvia** *off.* Republic of Latvia, *Ger.* Lettland, *Latv.* Latvija, Latvijas Republika; *prev.* Latvian SSR, *Rus.* Latviyskaya SSR. ◆ *republic* NE Europe
Latvian SSR/Latvija/Latvijas Republika/Latviyskaya SSR *see* Latvia
Latvia, Republic of *see* Latvia
186 H7 **Lau** New Britain, E Papua New Guinea 05°46′S 151°21′E
175 R9 **Lau Basin** *undersea feature* S Pacific Ocean
101 O15 **Lauchhammer** Brandenburg, E Germany 51°30′N 13°48′E
105 O3 **Laudio** *var.* Llodio. País Vasco, N Spain 43°08′N 02°59′W
Laudunum *see* Laon
Laudus *see* St-Lô
Lauenburg/Lauenburg in Pommern *see* Lębork
101 L20 **Lauf an der Pegnitz** Bayern, SE Germany 49°31′N 11°16′E
108 D7 **Laufen** Basel, NW Switzerland 47°26′N 07°31′E
109 P5 **Lauffen** Salzburg, NW Austria 47°54′N 12°57′E
92 I2 **Laugarbakki** Norðurland Vestra, N Iceland 65°18′N 20°51′W
92 I4 **Laugarvatn** Suðurland, SW Iceland 64°09′N 20°43′W
31 O3 **Laughing Fish Point** *headland* Michigan, N USA 46°31′N 87°01′W
187 Z14 **Lau Group** *island group* E Fiji
Lauis *see* Lugano
93 M17 **Laukaa** Keski-Suomi, C Finland 62°27′N 25°58′E
118 D12 **Laukuva** Tauragė, W Lithuania 55°37′N 22°12′E
Laun *see* Louny
183 P16 **Launceston** Tasmania, SE Australia 41°25′S 147°07′E
97 I24 **Launceston** *anc.* Dunheved. SW England, United Kingdom 50°38′N 04°21′W
54 C13 **La Unión** Nariño, SW Colombia 01°35′N 77°09′W
42 H8 **La Unión** La Unión, SE El Salvador 13°20′N 87°50′W
42 I6 **La Unión** Olancho, C Honduras 15°02′N 86°40′W
40 M15 **La Unión** Guerrero, S Mexico 17°58′N 101°49′W
41 Y14 **La Unión** Quintana Roo, E Mexico 18°00′N 101°48′W
105 S13 **La Unión** Murcia, SE Spain 37°37′N 00°54′W
54 L7 **La Unión** Barinas, C Venezuela 08°15′N 67°46′W
42 B10 **La Unión** ◆ *department* E El Salvador
38 H11 **Laupāhoehoe** *var.* Laupahoehoe. Hawaii, USA, C Pacific Ocean 20°00′N 155°15′W
Laupahoehoe *see* Laupāhoehoe
101 I23 **Laupheim** Baden-Württemberg, S Germany 48°13′N 09°54′E
181 W3 **Laura** Queensland, NE Australia 15°37′S 144°34′E
189 X2 **Laura** *atoll* Majuro Atoll, SE Marshall Islands
Laurana *see* Lovran
54 L8 **La Urbana** Bolívar, C Venezuela 07°05′N 66°58′W
21 Y4 **Laurel** Delaware, NE USA 38°33′N 75°34′W
23 V14 **Laurel** Florida, SE USA 27°07′N 82°27′W
21 W3 **Laurel** Maryland, NE USA 39°06′N 76°51′W
22 M6 **Laurel** Mississippi, S USA 31°41′N 89°10′W
33 U11 **Laurel** Montana, NW USA 45°40′N 108°46′W
29 R13 **Laurel** Nebraska, C USA 42°25′N 97°04′W
18 H15 **Laureldale** Pennsylvania, NE USA 40°24′N 75°52′W
18 C16 **Laurel Hill** *ridge* Pennsylvania, NE USA
29 T12 **Laurens** Iowa, C USA 42°51′N 94°51′W
21 P11 **Laurens** South Carolina, SE USA 34°29′N 82°01′W
Laurentian Highlands *see* Laurentian Mountains
15 P10 **Laurentian Mountains** *var.* Laurentian Highlands, *Fr.* Les Laurentides. *plateau* Newfoundland and Labrador/Québec, Canada
15 O12 **Laurentides** Québec, SE Canada 45°51′N 73°49′W
Laurentides, Les *see* Laurentian Mountains
107 M19 **Lauria** Basilicata, S Italy 40°03′N 15°50′E
194 I1 **Laurie Island** *island* Antarctica
21 T11 **Laurinburg** North Carolina, SE USA 34°46′N 79°29′W
30 M2 **Laurium** Michigan, N USA 47°14′N 88°26′W
Lauru *see* Choiseul
108 B9 **Lausanne** *It.* Losanna. Vaud, SW Switzerland 46°32′N 06°39′E
101 Q16 **Lausche** *var.* Luže. ▲ Czech Republic/Germany 50°52′N 14°39′E *see also* Luže
Lausche *see* Luže
101 Q16 **Lausitzer Bergland** *var.* Lausitzer Gebirge, *Cz.* Gory Lužycke, Lužické Hory', *Eng.* Lusatian Mountains. ▲ E Germany
Lausitzer Gebirge *see* Lausitzer Bergland
Lausitzer Neisse *see* Neisse
103 T12 **Lautaret, Col du** *pass* SE France
63 G15 **Lautaro** Araucanía, C Chile 38°30′S 71°30′W
101 F21 **Lauter** ≈ W Germany
108 I7 **Lauterach** Vorarlberg, NW Austria 47°29′N 09°44′E
101 I17 **Lauterbach** Hessen, C Germany 50°37′N 09°24′E
108 E9 **Lauterbrunnen** Bern, C Switzerland 46°36′N 07°52′E
169 U14 **Laut Kecil, Kepulauan** *island group* N Indonesia
187 X14 **Lautoka** Viti Levu, W Fiji 17°36′S 177°28′E
169 O8 **Laut, Pulau** *prev.* Laoet. *island* Borneo, C Indonesia
169 V14 **Laut, Pulau** *island* Kepulauan Natuna, W Indonesia
169 U13 **Laut, Selat** *strait* Borneo, C Indonesia
168 H8 **Laut Tawar, Danau** ◎ Sumatera, NW Indonesia
189 V14 **Lauvergne Island** *island* Chuuk, C Micronesia
98 M5 **Lauwers Meer** ◎ N Netherlands
98 M4 **Lauwersoog** Groningen, NE Netherlands 53°25′N 06°14′E
102 M14 **Lauzerte** Tarn-et-Garonne, S France 44°15′N 01°08′E
25 U13 **Lavaca Bay** *bay* Texas, SW USA
25 U12 **Lavaca River** ≈ Texas, SW USA
15 O12 **Laval** Québec, SE Canada 45°32′N 73°44′W
102 J6 **Laval** Mayenne, NW France 48°04′N 00°46′W
15 T6 **Laval** ≈ Québec, SE Canada
105 S9 **La Vall d'Uixó** *var.* Vall D'Uxó. Valenciana, E Spain 39°49′N 00°15′W
61 F19 **Lavalleja** ◆ *department* S Uruguay
15 O12 **Lavaltrie** Québec, SE Canada 45°56′N 73°14′W
186 M10 **Lavanggu** Rennell, S Solomon Islands 11°39′S 160°13′E
143 O14 **Lāvān, Jazīreh-ye** *island* S Iran
109 U8 **Lavant** ≈ S Austria
118 G5 **Lavassaare** *Ger.* Lawassaar. Pärnumaa, SW Estonia 58°29′N 24°22′E
104 L3 **La Vecilla de Curueño** Castilla y León, N Spain 42°51′N 05°24′W
45 N8 **La Vega** *var.* Concepción de la Vega. C Dominican Republic 19°15′N 70°33′W
La Vela *see* La Vela de Coro
54 J4 **La Vela de Coro** *var.* La Vela. Falcón, N Venezuela 11°30′N 69°33′W
103 N17 **Lavelanet** Ariège, S France 42°56′N 01°50′E
107 M17 **Lavello** Basilicata, S Italy 41°03′N 15°48′E
36 J8 **La Verkin** Utah, W USA 37°12′N 113°16′W
26 J8 **Laverne** Oklahoma, C USA 36°42′N 99°53′W
25 S12 **La Vernia** Texas, SW USA 29°19′N 98°07′W
93 K18 **Lavia** Satakunta, SW Finland 61°36′N 22°34′E
14 I12 **Lavieille, Lake** ◎ Ontario, SE Canada
94 C12 **Lavik** Sogn Og Fjordane, S Norway 61°06′N 05°25′E
La Vila Joíosa *see* Villajoyosa
33 U10 **Lavina** Montana, NW USA 46°18′N 108°55′W
194 H5 **Lavoisier Island** *island* Antarctica
23 U2 **Lavonia** Georgia, SE USA 34°26′N 83°06′W
103 R13 **la Voulte-sur-Rhône** Ardèche, E France 44°49′N 04°46′E
123 W5 **Lavrentiya** Chukotskiy Avtonomnyy Okrug, NE Russian Federation 65°33′N 171°12′W
115 H20 **Lávrio** *prev.* Lávrion. Attikí, C Greece 37°43′N 24°03′E
Lávrion *see* Lávrio
83 L22 **Lavumisa** *prev.* Gollel. SE Swaziland 27°18′S 31°55′E
149 T4 **Lawarai Pass** *pass* N Pakistan
Lawassaar *see* Lavassaare
141 P16 **Lawdar** SW Yemen 13°49′N 45°55′E
25 Q7 **Lawn** Texas, SW USA 32°07′N 99°45′W
195 Y4 **Law Promontory** *headland* Antarctica
77 O14 **Lawra** NW Ghana 10°40′N 02°49′W
185 E23 **Lawrence** Otago, South Island, New Zealand 45°53′S 169°43′E
31 P14 **Lawrence** Indiana, N USA 39°49′N 86°01′W
27 Q4 **Lawrence** Kansas, C USA 38°58′N 95°15′W
19 O10 **Lawrence** Massachusetts, NE USA 42°42′N 71°09′W
20 L5 **Lawrenceburg** Kentucky, S USA 38°02′N 84°53′W
20 I10 **Lawrenceburg** Tennessee, S USA 35°16′N 87°20′W
23 T3 **Lawrenceville** Georgia, SE USA 33°57′N 83°59′W
31 N15 **Lawrenceville** Illinois, N USA 38°43′N 87°40′W
21 V7 **Lawrenceville** Virginia, NE USA 36°45′N 77°50′W
27 S3 **Lawson** Missouri, C USA 39°26′N 94°12′W
26 L12 **Lawton** Oklahoma, C USA 34°35′N 98°20′W
140 I4 **Lawz, Jabal al** ▲ NW Saudi Arabia 28°45′N 35°20′E
95 L16 **Laxå** Örebro, C Sweden 59°00′N 14°37′E
125 T5 **Laya** ≈ NW Russian Federation
57 I19 **La Yarada** Tacna, SW Peru 18°14′S 70°30′W
141 S15 **Layjūn** C Yemen 15°27′N 49°16′E
141 Q9 **Laylā** *var.* Laila. Ar Riyāḍ, C Saudi Arabia 22°14′N 46°40′E
23 P4 **Lay Lake** ◎ Alabama, S USA
45 P14 **Layou** Saint Vincent, Saint Vincent and the Grenadines 13°11′N 61°16′W
La Youne *see* El Ayoun
192 L5 **Laysan Island** *island* Hawaiian Islands, Hawai'i, USA
36 L2 **Layton** Utah, W USA 41°03′N 112°00′W
34 L5 **Laytonville** California, W USA 39°39′N 123°30′W
172 H17 **Lazare, Pointe** *headland* Mahé, NE Seychelles 04°46′S 55°28′E
123 T12 **Lazarev** Khabarovskiy Kray, SE Russian Federation 52°11′N 141°18′E
112 L12 **Lazarevac** Serbia, C Serbia 44°25′N 20°17′E
65 N22 **Lazarev Sea** *sea* Antarctica
40 M15 **Lázaro Cárdenas** Michoacán, SW Mexico 17°56′N 102°13′W
119 F15 **Lazdijai** Alytus, S Lithuania 54°13′N 23°33′E
107 H15 **Lazio** *anc.* Latium. ◆ *region* C Italy
111 A16 **Lázně Kynžvart** *Ger.* Bad Königswart. Karlovarský Kraj, W Czech Republic 50°00′N 12°40′E
Lazovsk *see* Sîngerei
167 R12 **Leach** Poŭthĭsăt, W Cambodia 12°19′N 103°45′E
27 X9 **Leachville** Arkansas, C USA 35°56′N 90°15′W
28 J9 **Lead** South Dakota, N USA 44°21′N 103°45′W
11 S16 **Leader** Saskatchewan, S Canada 50°55′N 109°31′W
19 R5 **Lead Mountain** ▲ Maine, NE USA 44°53′N 68°07′W
37 R5 **Leadville** Colorado, C USA 39°15′N 106°17′W
11 V12 **Leaf Rapids** Manitoba, C Canada 56°30′N 100°02′W
22 M7 **Leaf River** ≈ Mississippi, S USA
25 W11 **League City** Texas, SW USA 29°30′N 95°05′W
92 K8 **Leaibevuotna** *Nor.* Olderfjord. Finnmark, N Norway 70°29′N 24°58′E
23 N7 **Leakesville** Mississippi, S USA 31°09′N 88°33′W
25 Q11 **Leakey** Texas, SW USA 29°44′N 99°48′W
Leal *see* Lihula
83 G15 **Lealui** Western, W Zambia 15°12′S 22°59′E
Leamhcán *see* Lucan
14 C18 **Leamington** Ontario, S Canada 42°03′N 82°35′W
Leamington/Leamington Spa *see* Royal Leamington Spa
Leammi *see* Lemmenjoki
25 S10 **Leander** Texas, SW USA 30°34′N 97°51′W
60 F13 **Leandro N. Alem** Misiones, NE Argentina 27°34′S 55°15′W
97 A20 **Leane, Lough** *Ir.* Loch Léin. ◎ SW Ireland
180 G8 **Learmonth** Western Australia 22°17′S 114°03′E
Leau *see* Zoutleeuw
L'Eau d'Heure *see* Plate Taille, Lac de la
190 D12 **Leava** Île Futuna, S Wallis and Futuna
Leavdnja *see* Lakselv
27 R3 **Leavenworth** Kansas, C USA 39°19′N 94°55′W
32 I8 **Leavenworth** Washington, NW USA 47°36′N 120°39′W
92 L8 **Leavvajohka** *var.* Levajok. Finnmark, N Norway 69°57′N 26°18′E
27 R4 **Leawood** Kansas, C USA 38°57′N 94°37′W
110 H6 **Łeba** *Ger.* Leba. Pomorskie, N Poland 54°45′N 17°32′E
110 I6 **Łeba** *Ger.* Leba. ≈ N Poland
Leba *see* Łeba
101 D20 **Lebach** Saarland, SW Germany 49°25′N 06°54′E
Łeba, Jezioro *see* Łebsko, Jezioro
171 P8 **Lebak** Mindanao, S Philippines 06°28′N 124°03′E
Lebanese Republic *see* Lebanon
31 O13 **Lebanon** Indiana, N USA 40°03′N 86°28′W
20 L6 **Lebanon** Kentucky, S USA 37°33′N 85°15′W
27 U6 **Lebanon** Missouri, C USA 37°40′N 92°40′W
19 N9 **Lebanon** New Hampshire, NE USA 43°40′N 72°15′W
32 G12 **Lebanon** Oregon, NW USA 44°32′N 122°54′W
18 H15 **Lebanon** Pennsylvania, NE USA 40°20′N 76°24′W
20 J8 **Lebanon** Tennessee, S USA 36°11′N 86°19′W
21 P7 **Lebanon** Virginia, NE USA 36°52′N 82°07′W
138 G6 **Lebanon** *off.* Lebanese Republic, *Ar.* Al Lubnān, *Fr.* Liban. ◆ *republic* SW Asia

20 K6 **Lebanon Junction** Kentucky, S USA 37°49′N 85°43′W
Lebanon, Mount *see* Liban, Jebel
146 J10 **Lebap** Lebapskiy Velayat, NE Turkmenistan 41°04′N 61°49′E
Lebapskiy Velayat *see* Lebap Welaýaty
146 J11 **Lebap Welaýaty** *Rus.* Lebapskiy Velayat; *prev. Rus.* Chardzhevskaya Oblast, *Turkm.* Chärjew Oblasty. ◆ *province* E Turkmenistan
Lebasee *see* Łebsko, Jezioro
99 F17 **Lebbeke** Oost-Vlaanderen, NW Belgium 51°00′N 04°08′E
35 S14 **Lebec** California, W USA 34°51′N 118°52′W
Lebedin *see* Lebedyn
123 Q11 **Lebedinyy** Respublika Sakha (Yakutiya), NE Russian Federation 58°23′N 125°24′E
126 L6 **Lebedyan'** Lipetskaya Oblast', W Russian Federation 53°00′N 39°11′E
117 T4 **Lebedyn** *Rus.* Lebedin. Sums'ka Oblast', NE Ukraine 50°36′N 34°30′E
12 I12 **Lebel-sur-Quévillon** Québec, SE Canada 49°01′N 76°56′W
92 L8 **Lebesby** *Lapp.* Davvesiida. Finnmark, N Norway 70°31′N 27°00′E
102 M9 **le Blanc** Indre, C France 46°38′N 01°04′E
79 L15 **Lebo** Orientale, N Dem. Rep. Congo 04°30′N 23°58′E
27 P5 **Lebo** Kansas, C USA 38°22′N 95°50′W
110 H6 **Lębork** *var.* Lębórk, *Ger.* Lauenburg, Lauenburg in Pommern. Pomorskie, N Poland 54°32′N 17°43′E
103 O17 **le Boulou** Pyrénées-Orientales, S France 42°32′N 02°50′E
108 A9 **Le Brassus** Vaud, W Switzerland 46°35′N 06°14′E
104 J15 **Lebrija** Andalucía, S Spain 36°55′N 06°04′W
110 G6 **Łebsko, Jezioro** *Ger.* Lebasee; *prev.* Jezioro Łeba. ◎ N Poland
63 F14 **Lebu** Bío Bío, C Chile 37°38′S 73°43′W
Lebyazh'ye *see* Akku
104 F6 **Leça da Palmeira** Porto, N Portugal 41°12′N 08°43′W
103 U15 **le Cannet** Alpes-Maritimes, SE France 43°35′N 07°E
Le Cap *see* Cap-Haïtien
103 P2 **le Cateau-Cambrésis** Nord, N France 50°05′N 03°32′E
107 Q18 **Lecce** Puglia, SE Italy 40°23′N 18°11′E
106 D7 **Lecco** Lombardia, N Italy 45°51′N 09°23′E
29 V10 **Le Center** Minnesota, N USA 44°23′N 93°43′W
108 J7 **Lech** Vorarlberg, W Austria 47°14′N 10°10′E
101 K22 **Lech** ≈ Austria/Germany
115 D19 **Lechainá** *var.* Lehena, Lekhainá. Dytikí Elláda, S Greece 37°57′N 21°16′E
102 J11 **le Château d'Oléron** Charente-Maritime, W France 45°53′N 01°12′W
103 R3 **le Chesne** Ardennes, N France 49°33′N 04°42′E
103 R13 **le Cheylard** Ardèche, E France 44°55′N 04°27′E
108 K7 **Lechtaler Alpen** ▲ W Austria
100 H6 **Leck** Schleswig-Holstein, N Germany 54°45′N 09°00′E
14 L9 **Lecointre, Lac** ◎ Québec, SE Canada
22 H7 **Lecompte** Louisiana, S USA 31°05′N 92°24′W
103 Q9 **le Creusot** Saône-et-Loire, C France 46°48′N 04°27′E
Lecumberri *see* Lekunberri
110 P13 **Łęczna** Lubelskie, E Poland 51°20′N 22°52′E
110 J12 **Łęczyca** *Ger.* Lentschiza, *Rus.* Lenchitsa. Łódzkie, C Poland 52°04′N 19°10′E
100 F10 **Leda** ≈ NW Germany
109 Y9 **Ledava** ≈ NE Slovenia
99 F17 **Lede** Oost-Vlaanderen, NW Belgium 50°58′N 03°59′E
104 K6 **Ledesma** Castilla y León, N Spain 41°05′N 06°00′W
45 Q12 **le Diamant** SW Martinique 14°29′N 61°02′W
172 J16 **Le Digue** *island* Inner Islands, NE Seychelles
103 Q10 **le Donjon** Allier, C France 46°19′N 03°50′E
102 M10 **le Dorat** Haute-Vienne, C France 46°14′N 01°05′E
Ledo Salinarius *see* Lons-le-Saunier
11 Q14 **Leduc** Alberta, SW Canada 53°17′N 113°30′W
123 V7 **Ledyanaya, Gora** ▲ E Russian Federation 61°51′N 171°03′E
97 C21 **Lee** *Ir.* An Laoi. ≈ SW Ireland
29 U5 **Leech Lake** ◎ Minnesota, N USA
26 K10 **Leedey** Oklahoma, C USA 35°54′N 99°21′W
97 M17 **Leeds** N England, United Kingdom 53°50′N 01°35′W
23 P4 **Leeds** Alabama, S USA 33°33′N 86°32′W
29 O3 **Leeds** North Dakota, N USA 48°19′N 99°43′W
98 N6 **Leek** Groningen, NE Netherlands 53°10′N 06°24′E
99 K15 **Leende** Noord-Brabant, SE Netherlands 51°21′N 05°34′E
100 F10 **Leer** Niedersachsen, NW Germany 53°14′N 07°26′E
98 J13 **Leerdam** Zuid-Holland, C Netherlands 51°54′N 05°06′E
98 K12 **Leersum** Utrecht, C Netherlands 52°01′N 05°26′E
23 W11 **Leesburg** Florida, SE USA 28°48′N 81°52′W
21 V3 **Leesburg** Virginia, NE USA 39°09′N 77°34′W
27 R4 **Lees Summit** Missouri, C USA 38°55′N 94°21′W
22 G7 **Leesville** Louisiana, S USA 31°08′N 93°15′W
25 S12 **Leesville** Texas, SW USA 29°22′N 97°45′W
31 U13 **Leesville Lake** ⊞ Ohio, N USA
Leesville Lake *see* Smith Mountain Lake
183 P9 **Leeton** New South Wales, SE Australia 34°33′S 146°24′E
98 L6 **Leeuwarden** *Fris.* Ljouwert. Fryslân, N Netherlands 53°15′N 05°48′E
180 I14 **Leeuwin, Cape** *headland* Western Australia 34°18′S 115°03′E
35 R8 **Lee Vining** California, W USA 37°57′N 119°07′W
45 V8 **Leeward Islands** *island group* E West Indies
Leeward Islands *see* Sotavento, Ilhas de
Leeward Islands *see* Vent, Îles Sous le
79 G20 **Léfini** ≈ SE Congo
115 C17 **Lefkáda** *prev.* Levkás. Lefkáda, Iónia Nisiá, Greece, C Mediterranean Sea 38°50′N 20°42′E
115 B17 **Lefkáda** *It.* Santa Maura, *prev.* Levkás; *anc.* Leucas. *island* Iónia Nisiá, Greece, C Mediterranean Sea
115 H25 **Lefká Óri** ▲ Kríti, Greece, E Mediterranean Sea
115 B16 **Lefkímmi** *var.* Levkímmi. Kérkyra, Iónia Nisiá, Greece, C Mediterranean Sea 39°26′N 20°05′E
Lefkosía/Lefkoşa *see* Nicosia
25 O2 **Lefors** Texas, SW USA 35°26′N 100°48′W
45 R12 **le François** E Martinique 14°36′N 60°59′W
180 L12 **Lefroy, Lake** *salt lake* Western Australia
Legaceaster *see* Chester
105 N8 **Leganés** Madrid, C Spain 40°20′N 03°46′W
Legaspi *see* Legazpi City
Leghorn *see* Livorno
110 M11 **Legionowo** Mazowieckie, C Poland 52°25′N 20°56′E
99 K24 **Léglise** Luxembourg, SE Belgium 49°48′N 05°31′E
106 G8 **Legnago** Lombardia, NE Italy 45°13′N 11°18′E
106 D7 **Legnano** Veneto, NE Italy 45°36′N 08°54′E
111 F14 **Legnica** *Ger.* Liegnitz. Dolnośląskie, SW Poland 51°12′N 16°11′E
35 Q9 **Le Grand** California, W USA 37°12′N 120°15′W
103 Q15 **le Grau-du-Roi** Gard, S France 43°32′N 04°10′E
183 U3 **Legume** New South Wales, SE Australia 28°24′S 152°20′E
102 L4 **le Havre** *Eng.* Havre; *prev.* le Havre-de-Grâce. Seine-Maritime, N France 49°30′N 00°06′E
le Havre-de-Grâce *see* le Havre
Lehena *see* Lechainá
36 L3 **Lehi** Utah, W USA 40°23′N 111°51′W
18 I14 **Lehighton** Pennsylvania, NE USA 40°49′N 75°42′W
29 O6 **Lehr** North Dakota, N USA 46°15′N 99°21′W
38 A8 **Lehua Island** *island* Hawaiian Islands, Hawai'i, USA
149 S9 **Leiāh** Punjab, NE Pakistan 30°59′N 70°58′E
109 W9 **Leibnitz** Steiermark, SE Austria 46°48′N 15°33′E
97 M19 **Leicester** *Lat.* Batae Coritanorum. C England, United Kingdom 52°38′N 01°05′W
97 M19 **Leicestershire** *cultural region* C England, United Kingdom
Leicheng *see* Leizhou
98 H11 **Leiden** *prev.* Leyden; *anc.* Lugdunum Batavorum. Zuid-Holland, W Netherlands 52°09′N 04°30′E
98 H11 **Leiderdorp** Zuid-Holland, W Netherlands 52°08′N 04°32′E
98 G11 **Leidschendam** Zuid-Holland, W Netherlands 52°05′N 04°24′E
99 D18 **Leie** *Fr.* Lys. ≈ Belgium/France
Leifear *see* Lifford
184 L4 **Leigh** Auckland, North Island, New Zealand 36°17′S 174°48′E
97 K17 **Leigh** NW England, United Kingdom 53°30′N 02°33′W
182 I5 **Leigh Creek** South Australia 30°27′S 138°23′E
23 O2 **Leighton** Alabama, S USA 34°42′N 87°31′W
97 M21 **Leighton Buzzard** E England, United Kingdom 51°55′N 00°41′W
Léim an Bhradáin *see* Leixlip
Léim An Mhadaidh *see* Limavady
Léime, Ceann *see* Loop Head, Ireland
Léime, Ceann *see* Slyne Head, Ireland
101 G20 **Leimen** Baden-Württemberg, SW Germany 49°21′N 08°40′E
100 I13 **Leine** ≈ NW Germany
101 J15 **Leinefelde** Thüringen, C Germany 51°22′N 10°19′E
Léin, Loch *see* Leane, Lough
97 D19 **Leinster** *Ir.* Cúige Laighean. *cultural region* E Ireland
97 F19 **Leinster, Mount** *Ir.* Stua Laighean. ▲ SE Ireland 52°36′N 06°45′W
119 F15 **Leipalingis** Alytus, S Lithuania 54°05′N 23°52′E
92 J12 **Leipojärvi** Norrbotten, N Sweden 67°03′N 21°15′E
31 R12 **Leipsic** Ohio, N USA 41°06′N 83°58′W
Leipsic *see* Leipzig
115 M20 **Leipsoí** *island* Dodekánisa, Greece, Aegean Sea
101 M15 **Leipzig** *Pol.* Lipsk, *hist.* Leipsic; *anc.* Lipsia. Sachsen, E Germany 51°19′N 12°24′E
101 M15 **Leipzig Halle** ✈ Sachsen, E Germany 51°26′N 12°14′E
104 G9 **Leiria** *anc.* Collipo. Leiria, C Portugal 39°45′N 08°49′W
104 F9 **Leiria** ◆ *district* C Portugal
95 C15 **Leirvik** Hordaland, S Norway 59°49′N 05°27′E
118 E5 **Leisi** *Ger.* Laisberg. Saaremaa, W Estonia 58°33′N 22°41′E
104 J3 **Leitariegos, Puerto de** *pass* NW Spain
20 J6 **Leitchfield** Kentucky, S USA 37°28′N 86°19′W
109 Y5 **Leitha** *Hung.* Lajta. ≈ Austria/Hungary
Leitir Ceanainn *see* Letterkenny
Leitmeritz *see* Litoměřice
Leitomischl *see* Litomyšl
97 D16 **Leitrim** *Ir.* Liatroim. *cultural region* NW Ireland
Leix *see* Laois
97 F18 **Leixlip** *Eng.* Salmon Leap, *Ir.* Léim an Bhradáin. Kildare, E Ireland 53°23′N 06°32′W
64 N8 **Leixões** Porto, N Portugal 41°11′N 08°41′W
161 N12 **Leiyang** Hunan, S China 26°23′N 112°49′E
160 L16 **Leizhou** *var.* Haikang, Leicheng. Guangdong, S China 20°54′N 110°05′E
160 L16 **Leizhou Bandao** *var.* Luichow Peninsula. *peninsula* S China
98 H13 **Lek** ≈ SW Netherlands
114 I13 **Lekánis** ▲ NE Greece
172 H13 **Le Kartala** ▲ Grande Comore, NW Comoros
Le Kef *see* El Kef
79 G20 **Lékéti, Monts de la** ▲ S Congo
Lekhainá *see* Lechainá
92 G11 **Leknes** Nordland, C Norway 68°07′N 13°36′E
79 F20 **Lékoumou** ◆ *province* SW Congo
94 L13 **Leksand** Dalarna, C Sweden 60°44′N 15°E
124 H8 **Leksozero, Ozero** ◎ NW Russian Federation
105 Q3 **Lekunberri** *var.* Lecumberri. Navarra, N Spain 43°00′N 01°54′W
171 S11 **Lelai, Tanjung** *headland* Pulau Halmahera, N Indonesia 01°32′N 128°43′E
45 Q12 **le Lamentin** *var.* Lamentin. C Martinique 14°37′N 61°01′W
31 P6 **Leland** Michigan, N USA 45°01′N 85°44′W
22 J4 **Leland** Mississippi, S USA 33°24′N 90°54′W
95 J16 **Lelång** *var.* Lelången. ◎ S Sweden
Lelången *see* Lelång
Lel'chitsy *see* Lyel'chytsy
le Léman *see* Geneva, Lake
Leli *see* Tianlin
25 O3 **Lelia Lake** Texas, SW USA 34°52′N 100°42′W
113 I14 **Lelija** ▲ SE Bosnia and Herzegovina 43°25′N 18°31′E
108 C8 **Le Locle** Neuchâtel, W Switzerland 47°04′N 06°45′E
189 Y14 **Lelu** Kosrae, E Micronesia
Lelu *see* Lelu Island
189 Y14 **Lelu Island** *var.* Lelu. *island* Kosrae, E Micronesia
55 W9 **Lelydorp** Wanica, N Suriname 05°36′N 55°04′W
98 K9 **Lelystad** Flevoland, C Netherlands 52°30′N 05°26′E
63 K25 **Le Maire, Estrecho de** *strait* S Argentina
168 L10 **Lemang** Pulau Rangsang, W Indonesia 01°04′N 102°44′E
186 I7 **Lemankoa** Buka Island, NE Papua New Guinea 05°06′S 154°23′E
Léman, Lac *see* Geneva, Lake
102 L6 **le Mans** Sarthe, NW France 48°N 00°12′E
29 S12 **Le Mars** Iowa, C USA 42°47′N 96°10′W
109 S3 **Lembach im Mühlkreis** Oberösterreich, N Austria 48°28′N 13°53′E
101 G23 **Lemberg** ▲ SW Germany 48°09′N 08°47′E
Lemberg *see* L'viv
Lemdiyya *see* Médéa
121 P3 **Lemesós** *var.* Limassol. SW Cyprus 34°41′N 33°02′E
100 H13 **Lemgo** Nordrhein-Westfalen, W Germany 52°02′N 08°54′E
33 P13 **Lemhi Range** ▲ Idaho, NW USA
9 S6 **Lemieux Islands** *island group* Nunavut, NE Canada
171 O11 **Lemito** Sulawesi, N Indonesia 0°34′N 121°31′E
92 L10 **Lemmenjoki** *Lapp.* Leammi. ≈ NE Finland
98 L7 **Lemmer** *Fris.* De Lemmer. Fryslân, N Netherlands 52°50′N 05°43′E
28 L7 **Lemmon** South Dakota, N USA 45°54′N 102°08′W
36 M15 **Lemmon, Mount** ▲ Arizona, SW USA 32°26′N 110°47′W
Lemnos *see* Límnos
31 O14 **Lemon, Lake** ◎ Indiana, N USA
102 J5 **le Mont St-Michel** *castle* Manche, N France
35 Q11 **Lemoore** California, W USA 36°16′N 119°48′W
189 T13 **Lemotol Bay** *bay* Chuuk Islands, C Micronesia
45 Y5 **le Moule** *var.* Moule. Grande Terre, NE Guadeloupe 16°20′N 61°21′W
Lemovices *see* Limoges
Le Moyen-Ogooué *see* Moyen-Ogooué
12 M6 **le Moyne, Lac** ◎ Québec, E Canada
93 L18 **Lempäälä** Pirkanmaa, W Finland 61°14′N 23°47′E
42 E7 **Lempa, Río** ≈ Central America
42 F7 **Lempira** *prev.* Gracias. ◆ *department* SW Honduras
Lemsalu *see* Limbaži
107 N17 **Le Murge** ▲ SE Italy
125 V6 **Lemva** ≈ NW Russian Federation
95 F21 **Lemvig** Midtjylland, W Denmark 56°31′N 08°19′E
166 K8 **Lemyethna** Ayeyawady, SW Myanmar (Burma) 17°36′N 95°08′E
30 K10 **Lena** Illinois, N USA 42°22′N 89°49′W
129 V4 **Lena** ≈ NE Russian Federation
173 N13 **Lena Tablemount** *undersea feature* S Indian Ocean 51°06′S 56°54′E
Lenchitsa *see* Łęczyca
59 N17 **Lençóis** Bahia, E Brazil 12°36′S 41°24′W
60 K9 **Lençóis Paulista** São Paulo, S Brazil 22°35′S 48°51′W
109 Y9 **Lendava** *Hung.* Lendva, *Ger.* Unterlimbach; *prev.* Dolnja Lendava. NE Slovenia 46°33′N 16°27′E
83 F20 **Lendepas** Hardap, SE Namibia 24°41′S 19°58′E
124 H9 **Lendery** *Finn.* Lentiira. Respublika Kareliya, NW Russian Federation 63°20′N 31°18′E
Lendum *see* Lens
Lendva *see* Lendava
27 R4 **Lenexa** Kansas, C USA 38°57′N 94°43′W
109 Q5 **Lengau** Oberösterreich, N Austria 48°01′N 13°17′E
145 Q17 **Lenger** Yuzhnyy Kazakhstan, S Kazakhstan 42°10′N 69°54′E
Lenghu *see* Lenghuzhen
159 O9 **Lenghuzhen** *var.* Lenghu. Qinghai, C China 38°50′N 93°25′E
159 T9 **Lenglong Ling** ▲ N China 37°40′N 102°13′E
108 D7 **Lengnau** Bern, W Switzerland 47°12′N 07°23′E
95 M20 **Lenhovda** Kronoberg, S Sweden 57°00′N 15°16′E
Lenīn *see* Uzynkol', Kazakhstan
Lenin *see* Akdepe, Turkmenistan
Leninabad *see* Khujand
Leninakan *see* Gyumri
Lenina, Pik *see* Lenin Peak
117 V12 **Lenine** *Rus.* Lenino. Avtonomna Respublika Krym, S Ukraine 45°18′N 35°47′E
Leningor *see* Leninogorsk
Leningrad *see* Sankt-Peterburg
Leningrad *see* Mu'minobod
126 L13 **Leningradskaya** Krasnodarskiy Kray, SW Russian Federation 46°19′N 39°23′E
195 S16 **Leningradskaya** *Russian research station* Antarctica 69°30′S 159°51′E
124 H12 **Leningradskaya Oblast'** ◆ *province* NW Russian Federation
Leningradskiy *see* Mu'minobod
Lenino *see* Lyenina, Belarus
Lenino *see* Lenine, Ukraine
Leninobod *see* Khujand
145 X9 **Leninogorsk** *Kaz.* Leningor. Vostochnyy Kazakhstan, E Kazakhstan 50°20′N 83°34′E
147 T12 **Lenin Peak** *Rus.* Pik Lenina, *Taj.* Qullai Lenin. ▲ Kyrgyzstan/Tajikistan 39°20′N 72°50′E
147 S8 **Leninpol'** Talasskaya Oblast', NW Kyrgyzstan 42°29′N 71°54′E
Lenin, Qullai *see* Lenin Peak
127 P11 **Leninsk** Volgogradskaya Oblast', SW Russian Federation 48°41′N 45°18′E
Leninsk *see* Baykonyr, Kazakhstan
Leninsk *see* Akdepe, Turkmenistan
Leninsk *see* Asaka, Uzbekistan
145 T8 **Leninskiy** Pavlodar, E Kazakhstan 52°13′N 76°50′E
122 I13 **Leninsk-Kuznetskiy** Kemerovskaya Oblast', S Russian Federation 54°42′N 86°16′E
125 P15 **Leninskoje** Kirovskaya Oblast', NW Russian Federation 58°19′N 47°03′E
Leninskoye *see* Uzynkol'
Lenin-Turkmenski *see* Türkmenabat
Leninváros *see* Tiszaújváros
Lenkoran' *see* Länkäran
101 F15 **Lenne** ≈ W Germany
101 G16 **Lennestadt** Nordrhein-Westfalen, W Germany 51°07′N 08°04′E
29 R11 **Lennox** South Dakota, N USA 43°21′N 96°53′W
63 J25 **Lennox, Isla** *Eng.* Lennox Island. *island* S Chile
Lennox Island *see* Lennox, Isla
21 Q9 **Lenoir** North Carolina, SE USA 35°56′N 81°31′W
20 M9 **Lenoir City** Tennessee, S USA 35°48′N 84°15′W
108 C7 **Le Noirmont** Jura, NW Switzerland 47°14′N 06°57′E
14 L9 **Lenôtre, Lac** ◎ Québec, SE Canada
29 U15 **Lenox** Iowa, C USA 40°52′N 94°33′W
103 O2 **Lens** *anc.* Lendum, Lentium. Pas-de-Calais, N France 50°26′N 02°50′E
123 O11 **Lensk** Respublika Sakha (Yakutiya), NE Russian Federation 60°43′N 115°16′E
111 F24 **Lenti** Zala, SW Hungary 46°38′N 16°30′E
Lentia *see* Linz
93 N14 **Lentiira** Kainuu, E Finland 64°22′N 29°52′E
Lentiira *see* Lendery
107 L25 **Lentini** *anc.* Leontini. Sicilia, Italy, C Mediterranean Sea 37°17′N 15°00′E
Lentium *see* Lens
Lentschiza *see* Łęczyca
93 N15 **Lentua** ◎ E Finland
119 H14 **Lentvaris** *Pol.* Landwarów. Vilnius, SE Lithuania 54°39′N 24°58′E
108 F7 **Lenzburg** Aargau, N Switzerland 47°23′N 08°11′E
109 R5 **Lenzing** Oberösterreich, N Austria 47°58′N 13°34′E
77 P13 **Léo** SW Burkina Faso 11°07′N 02°08′W
109 V7 **Leoben** Steiermark, C Austria 47°23′N 15°06′E
Leobschütz *see* Głubczyce
44 L9 **Léogâne** S Haiti 18°32′N 72°37′W
171 O11 **Leok** Sulawesi, N Indonesia 01°10′N 121°20′E
29 O7 **Leola** South Dakota, N USA 45°41′N 98°58′W
97 K20 **Leominster** W England, United Kingdom 52°09′N 02°18′W
19 N11 **Leominster** Massachusetts, NE USA 42°29′N 71°43′W
102 I15 **Léon** Landes, SW France 43°54′N 01°17′W
40 M12 **León** *var.* León de los Aldamas. Guanajuato, C Mexico 21°05′N 101°43′W
42 I10 **León** León, NW Nicaragua 12°24′N 86°52′W
104 L4 **León** Castilla y León, NW Spain 42°34′N 05°34′W
29 V16 **Leon** Iowa, C USA 40°44′N 93°45′W
42 I9 **León** ◆ *department* W Nicaragua
104 K4 **León** ◆ *province* Castilla y León, NW Spain
León *see* Cotopaxi
25 V9 **Leona** Texas, SW USA 31°09′N 95°58′W
25 U5 **Leonard** Texas, SW USA 33°22′N 96°15′W
107 H15 **Leonardo da Vinci** *prev.* Fiumicino. ✈ (Roma) Lazio, C Italy 41°48′N 12°15′E
21 X5 **Leonardtown** Maryland, NE USA 38°17′N 76°35′W
25 Q13 **Leona River** ≈ Texas, SW USA
41 Z11 **Leona Vicario** Quintana Roo, SE Mexico 20°57′N 87°06′W
101 H21 **Leonberg** Baden-Württemberg, SW Germany 48°48′N 09°01′E
62 M3 **León, Cerro** ▲ NW Paraguay 20°21′S 60°16′W
León de los Aldamas *see* León
109 T4 **Leonding** Oberösterreich, N Austria 48°17′N 14°15′E
107 I14 **Leonessa** Lazio, C Italy 42°36′N 12°56′E
107 K24 **Leonforte** Sicilia, Italy, C Mediterranean Sea 37°38′N 14°23′E
183 O13 **Leongatha** Victoria, SE Australia 38°30′S 145°56′E
Leonídi *see* Leonídio
115 F21 **Leonídio** *var.* Leonídi. Pelopónnisos, S Greece 37°11′N 22°50′E
104 J4 **León, Montes de** ▲ NW Spain
180 K11 **Leonora** Western Australia 28°52′S 121°16′E
25 S8 **Leon River** ≈ Texas, SW USA
Leontini *see* Lentini
Léopold II, Lac *see* Mai-Ndombe, Lac
99 J17 **Leopoldsburg** Limburg, NE Belgium 51°07′N 05°16′E
Léopoldville *see* Kinshasa
26 I5 **Leoti** Kansas, C USA 38°28′N 101°22′W
116 M11 **Leova** *Rus.* Leovo. SW Moldova 46°31′N 28°16′E
Leovo *see* Leova
102 G8 **le Palais** Morbihan, NW France 47°20′N 03°08′W
27 X10 **Lepanto** Arkansas, C USA 35°34′N 90°21′W
169 N13 **Lepar, Pulau** *island* W Indonesia
104 I14 **Lepe** Andalucía, S Spain 37°15′N 07°12′W
Lepel' *see* Lyepyel'
83 E25 **Lepelle** *var.* Elefantes; *prev.* Olifants. ≈ SW South Africa
83 I20 **Lephepe** *var.* Lephephe. Kweneng, SE Botswana 23°20′S 25°50′E
Lephephe *see* Lephepe
161 Q10 **Leping** Jiangxi, S China 28°57′N 117°07′E
Lépontiennes, Alpes/Lepontine, Alpi *see* Lepontine Alps
108 G10 **Lepontine Alps** *Fr.* Alpes Lépontiennes, *It.* Alpi Lepontine. ▲ Italy/Switzerland
79 G20 **Le Pool** ◆ *province* S Congo
173 O16 **Le Port** NW Réunion
103 N1 **le Portel** Pas-de-Calais, N France 50°42′N 01°35′E
93 N17 **Leppävirta** Pohjois-Savo, C Finland 62°30′N 27°50′E
45 Q11 **le Prêcheur** NW Martinique 14°48′N 61°14′W
145 V13 **Lepsi** *prev.* Lepsy. Taldykorgan, SE Kazakhstan 46°14′N 78°56′E
145 V13 **Lepsi** *prev.* Lepsy. ≈ SE Kazakhstan
Lepsy *see* Lepsi
Le Puglie *see* Puglia
103 Q12 **le Puy** *prev.* le Puy-en-Velay, *hist.* Anicium, Podium Anicensis. Haute-Loire, C France 45°03′N 03°53′E
le Puy-en-Velay *see* le Puy
45 X11 **le Raizet** *var.* Le Raizet. ✈ (Pointe-à-Pitre) Grande Terre, C Guadeloupe 16°16′N 61°31′W
107 J24 **Lercara Friddi** Sicilia, Italy, C Mediterranean Sea 37°45′N 13°37′E
78 G12 **Léré** Mayo-Kébbi Ouest, SW Chad 09°41′N 14°17′E
Leribe *see* Hlotse
106 E10 **Lerici** Liguria, NW Italy 44°06′N 09°53′E
54 I14 **Lérida** Vaupés, SE Colombia 0°01′S 70°28′W
105 U5 **Lérida** ◆ *province* Cataluña, NE Spain
Lérida *see* Lleida
105 N5 **Lerma** Castilla y León, N Spain 42°02′N 03°46′W
40 M13 **Lerma, Río** ≈ C Mexico
Lerna *see* Lérni
115 F20 **Lérni** *var.* Lerna. *prehistoric site* Pelopónnisos, S Greece
45 R11 **le Robert** E Martinique 14°41′N 60°57′W
115 M21 **Léros** *island* Dodekánisa, Greece, Aegean Sea
30 L13 **Le Roy** Illinois, N USA 40°21′N 88°45′W
27 Q6 **Le Roy** Kansas, C USA 38°04′N 95°37′W
29 W11 **Le Roy** Minnesota, N USA 43°30′N 92°30′W
18 E10 **Le Roy** New York, NE USA 42°58′N 77°58′W
Lerrnayin Gharabakh *see* Nagorno-Karabakh
95 J19 **Lerum** Västra Götaland, S Sweden 57°46′N 12°12′E
96 M2 **Lerwick** NE Scotland, United Kingdom 60°09′N 01°09′W
45 Y6 **les Abymes** *var.* Abymes. Grande Terre, C Guadeloupe 16°16′N 61°31′W
les Albères *see* Albères, Chaîne des
102 M4 **les Andelys** Eure, N France 49°15′N 01°26′E
45 Q12 **les Anses-d'Arlets** SW Martinique 14°29′N 61°05′W
104 L2 **Les Arriondes** *prev.* Arriondas. Asturias, N Spain 43°23′N 05°11′W
105 U6 **Les Borges Blanques** *var.* Borjas Blancas. Cataluña, NE Spain 41°31′N 00°52′E
Lesbos *see* Lésvos
Les Cayes *see* Cayes
31 Q4 **Les Cheneaux Islands** *island* Michigan, N USA
105 T8 **Les Coves de Vinromá** *Cast.* Cuevas de Vinromá. Valenciana, E Spain 40°18′N 00°07′E
103 T12 **les Écrins** ▲ E France 44°54′N 06°25′E
108 C10 **Le Sépey** Vaud, W Switzerland 46°21′N 07°04′E
15 T7 **Les Escoumins** Québec, SE Canada 48°21′N 69°26′W
Les Gonaïves *see* Gonaïves
160 H9 **Leshan** Sichuan, C China 29°42′N 103°43′E
108 D11 **Les Haudères** Valais, SW Switzerland 46°02′N 07°27′E
102 J9 **les Herbiers** Vendée, NW France 46°52′N 01°01′W
Lesh/Leshi *see* Lezhë
125 O8 **Leshukonskoye** Arkhangel'skaya Oblast', NW Russian Federation 64°54′N 45°48′E
Lesina *see* Hvar
107 M15 **Lesina, Lago di** ◎ SE Italy
114 K13 **Lesítse** ▲ NE Greece
94 G10 **Lesja** Oppland, S Norway 62°07′N 08°56′E
95 L15 **Lesjöfors** Värmland, C Sweden 59°57′N 14°12′E
111 O18 **Lesko** Podkarpackie, SE Poland 49°28′N 22°19′E
113 O15 **Leskovac** Serbia, SE Serbia 43°00′N 21°58′E
113 M22 **Leskovik** *var.* Leskoviku. Korçë, S Albania 40°09′N 20°39′E
Leskoviku *see* Leskovik
33 P14 **Leslie** Idaho, NW USA 43°51′N 113°28′W
31 Q10 **Leslie** Michigan, N USA 42°27′N 84°25′W
Leśna/Lesnaya *see* Lyasnaya
Lesnaya *see* Lyasnaya
102 F5 **Lesneven** Finistère, NW France 48°35′N 04°19′W
112 J11 **Lešnica** Serbia, W Serbia 44°40′N 19°18′E
125 S13 **Lesnoy** Kirovskaya Oblast', NW Russian Federation 59°49′N 52°07′E
122 G10 **Lesnoy** Sverdlovskaya Oblast', C Russian Federation 58°40′N 59°48′E
122 K12 **Lesosibirsk** Krasnoyarskiy Kray, C Russian Federation 58°13′N 92°23′E
83 J23 **Lesotho** *off.* Kingdom of Lesotho; *prev.* Basutoland. ◆ *monarchy* S Africa
Lesotho, Kingdom of *see* Lesotho
102 J12 **Lesparre-Médoc** Gironde, SW France 45°18′N 00°57′W
108 C8 **Les Ponts-de-Martel** Neuchâtel, W Switzerland 47°00′N 06°45′E
103 P1 **Lesquin** ✈ Nord, N France 50°34′N 03°07′E
102 I9 **les Sables-d'Olonne** Vendée, NW France 46°30′N 01°47′W
109 S7 **Lessach** *var.* Lessachbach. ≈ E Austria
Lessachbach *see* Lessach
45 W11 **les Saintes** *var.* Îles des Saintes. *island group* S Guadeloupe
74 L5 **Les Salines** ✈ (Annaba) NE Algeria 36°45′N 07°57′E
99 J22 **Lesse** ≈ SE Belgium
95 M21 **Lessebo** Kronoberg, S Sweden 56°45′N 15°19′E
Lesser Antarctica *see* West Antarctica
45 P15 **Lesser Antilles** *island group* E West Indies
137 T10 **Lesser Caucasus** *Rus.* Malyy Kavkaz. ▲ SW Asia
Lesser Khingan Range *see* Xiao Hinggan Ling
11 P13 **Lesser Slave Lake** ◎ Alberta, W Canada
99 E19 **Lessines** Hainaut, SW Belgium 50°43′N 03°50′E
103 R16 **les Stes-Maries-de-la-Mer** Bouches-du-Rhône, SE France 43°27′N 04°26′E
14 G15 **Lester B. Pearson** *var.* Toronto. ✈ (Toronto) Ontario, S Canada 43°59′N 81°30′W
29 U9 **Lester Prairie** Minnesota, N USA 44°52′N 94°02′W
93 L16 **Lestijärvi** Keski-Pohjanmaa, W Finland 63°29′N 24°41′E
L'Estuaire *see* Estuaire
29 U9 **Le Sueur** Minnesota, N USA 44°27′N 93°53′W
108 B8 **Les Verrières** Neuchâtel, W Switzerland 46°54′N 06°29′E
115 L17 **Lésvos** *anc.* Lesbos. *island* E Greece
110 G12 **Leszno** *Ger.* Lissa. Wielkopolskie, C Poland 51°51′N 16°35′E
83 L20 **Letaba** Northern, NE South Africa 23°44′S 31°29′E
173 P17 **Le Tampon** SW Réunion
97 O21 **Letchworth** E England, United Kingdom 51°58′N 00°14′W
111 G25 **Letenye** Zala, SW Hungary 46°25′N 16°44′E
11 Q17 **Lethbridge** Alberta, SW Canada 49°43′N 112°48′W
55 S11 **Lethem** S Guyana 03°24′N 59°45′W
83 H18 **Letiahau** ≈ W Botswana
54 J18 **Leticia** Amazonas, S Colombia 04°09′N 69°57′W
171 S16 **Leti, Kepulauan** *island group* E Indonesia
83 I18 **Letlhakane** Central, C Botswana 21°26′S 25°36′E
83 H20 **Letlhakeng** Kweneng, SE Botswana 24°05′S 25°03′E
114 J8 **Letnitsa** Lovech, N Bulgaria 43°19′N 25°02′E
103 N1 **le Touquet-Paris-Plage** Pas-de-Calais, N France 50°31′N 01°36′E
166 L8 **Letpadan** Bago, SW Myanmar (Burma) 17°46′N 94°45′E
166 K6 **Letpan** Rakhine State, W Myanmar (Burma) 19°22′N 94°11′E
102 M2 **le Tréport** Seine-Maritime, N France 50°03′N 01°21′E
166 M12 **Letsôk-aw Kyun** *var.* Letsutan Island; *prev.* Domel Island. *island* Mergui Archipelago, S Myanmar (Burma)
Letsutan Island *see* Letsôk-aw Kyun
97 E14 **Letterkenny** *Ir.* Leitir Ceanainn. Donegal, NW Ireland 54°57′N 07°44′W
Lettland *see* Latvia
116 M6 **Letychiv** Khmel'nyts'ka Oblast', W Ukraine 49°24′N 27°39′E
Lëtzebuerg *see* Luxembourg
116 H14 **Leu** Dolj, SW Romania 44°10′N 24°01′E
Leucas *see* Lefkáda
103 P17 **Leucate** Aude, S France 42°55′N 03°03′E
103 P17 **Leucate, Étang de** ◎ S France
108 E10 **Leuk** Valais, SW Switzerland 46°18′N 07°46′E
108 E10 **Leukerbad** Valais, SW Switzerland 46°22′N 07°47′E
Leusden *see* Leusden-Centrum
98 K11 **Leusden-Centrum** *var.* Leusden. Utrecht, C Netherlands 52°08′N 05°25′E
Leutensdorf *see* Litvínov
Leutschau *see* Levoča
99 H18 **Leuven** *Fr.* Louvain, *Ger.* Löwen. Vlaams Brabant, C Belgium 50°53′N 04°42′E
99 I20 **Leuze** Namur, C Belgium 50°33′N 04°55′E
Leuze *see* Leuze-en-Hainaut
99 E19 **Leuze-en-Hainaut** *var.* Leuze. Hainaut, SW Belgium 50°36′N 03°37′E
Léva *see* Levice
Levádia *see* Livádeia
Levajok *see* Leavvajohka
36 L4 **Levan** Utah, W USA 39°33′N 111°51′W
93 E16 **Levanger** Nord-Trøndelag, C Norway 63°45′N 11°18′E
106 D10 **Levanto** Liguria, W Italy 44°12′N 09°33′E
107 H23 **Levanzo, Isola di** *island* Isole Egadi, S Italy
127 Q17 **Levashi** Respublika Dagestan, SW Russian Federation 42°27′N 47°19′E
24 M5 **Levelland** Texas, SW USA 33°35′N 102°23′W
39 P13 **Levelock** Alaska, USA 59°07′N 156°51′W
101 E16 **Leverkusen** Nordrhein-Westfalen, W Germany 51°02′N 07°E
111 J21 **Levice** *Ger.* Lewentz, *Hung.* Léva, Lewenz. Nitriansky Kraj, SW Slovakia 48°14′N 18°38′E
106 G6 **Levico Terme** Trentino-Alto Adige, N Italy 46°02′N 11°19′E
115 E20 **Levídi** Pelopónnisos, S Greece 37°39′N 22°13′E
103 P14 **le Vigan** Gard, S France 43°00′N 03°36′E
184 L13 **Levin** Manawatu-Wanganui, North Island, New Zealand 40°38′S 175°17′E
15 R10 **Lévis** *var.* Levis. Québec, SE Canada 46°47′N 71°12′W
Levis *see* Lévis
21 P6 **Levisa Fork** ≈ Kentucky/Virginia, S USA
115 L21 **Levítha** *island* Kykládes, Greece, Aegean Sea
18 L14 **Levittown** Long Island, New York, NE USA 40°42′N 73°29′W
18 J15 **Levittown** Pennsylvania, NE USA 40°09′N 74°50′W
Levkás *see* Lefkáda
Levkímmi *see* Lefkímmi
111 L19 **Levoča** *Ger.* Leutschau, *Hung.* Locse. Prešovský Kraj, E Slovakia 49°01′N 20°34′E
Lévrier, Baie du *see* Nouâdhibou, Dakhlet
103 N9 **Levroux** Indre, C France 46°59′N 01°37′E
114 J8 **Levski** Pleven, N Bulgaria 43°21′N 25°11′E
Levskigrad *see* Karlovo
126 L6 **Lev Tolstoy** Lipetskaya Oblast', W Russian Federation 53°12′N 39°28′E
187 X14 **Levuka** Ovalau, C Fiji 17°42′S 178°50′E
166 L6 **Lewe** Mandalay, C Myanmar (Burma) 19°40′N 96°04′E
Lewentz/Lewenz *see* Levice
97 O23 **Lewes** SE England, United Kingdom 50°52′N 00°01′E
21 Z4 **Lewes** Delaware, NE USA 38°46′N 75°08′W
29 Q12 **Lewis And Clark Lake** ⊞ Nebraska/South Dakota, N USA
18 G14 **Lewisburg** Pennsylvania, NE USA 40°57′N 76°52′W
20 J10 **Lewisburg** Tennessee, S USA 35°29′N 86°49′W
21 S6 **Lewisburg** West Virginia, NE USA 37°49′N 80°28′W
96 F6 **Lewis, Butt of** *headland* NW Scotland, United Kingdom 58°31′N 06°18′W
96 F7 **Lewis, Isle of** *island* NW Scotland, United Kingdom
35 U4 **Lewis, Mount** ▲ Nevada, W USA 40°22′N 116°50′W
185 H16 **Lewis Pass** *pass* South Island, New Zealand
33 P7 **Lewis Range** ▲ Montana, NW USA
23 O3 **Lewis Smith Lake** ⊞ Alabama, S USA
32 M10 **Lewiston** Idaho, NW USA 46°25′N 117°01′W
19 P7 **Lewiston** Maine, NE USA 44°08′N 70°14′W
29 X10 **Lewiston** Minnesota, N USA 43°58′N 91°52′W
18 D9 **Lewiston** New York, NE USA 43°10′N 79°02′W
36 L1 **Lewiston** Utah, W USA 41°58′N 111°52′W
30 K13 **Lewistown** Illinois, N USA 40°23′N 90°09′W
33 T9 **Lewistown** Montana, NW USA 47°04′N 109°26′W
27 T14 **Lewisville** Arkansas, C USA 33°21′N 93°38′W
25 T6 **Lewisville** Texas, SW USA 33°00′N 96°57′W
25 T6 **Lewisville, Lake** ⊞ Texas, SW USA
Le Woleu-Ntem *see* Woleu-Ntem
23 U3 **Lexington** Georgia, SE USA 33°51′N 83°04′W
20 M5 **Lexington** Kentucky, S USA 38°03′N 84°30′W
22 L4 **Lexington** Mississippi, S USA 33°06′N 90°03′W
27 S4 **Lexington** Missouri, C USA 39°11′N 93°52′W
29 N16 **Lexington** Nebraska, C USA 40°46′N 99°44′W
20 S9 **Lexington** North Carolina, SE USA 35°49′N 80°15′W
27 N11 **Lexington** Oklahoma, C USA 35°00′N 97°20′W
21 R12 **Lexington** South Carolina, SE USA 33°59′N 81°15′W
20 G9 **Lexington** Tennessee, S USA 35°39′N 88°24′W
25 T10 **Lexington** Texas, SW USA 30°25′N 97°00′W

◆ Country
● Country Capital
◇ Dependent Territory
○ Dependent Territory Capital
◆ Administrative Regions
✈ International Airport
▲ Mountain
▲ Mountain Range
Volcano
≈ River
◎ Lake
⊞ Reservoir

21 T6 **Lexington** Virginia, NE USA 37°47′N 79°27′W
21 X5 **Lexington Park** Maryland, NE USA 38°16′N 76°27′W
Leyden *see* Leiden
102 J14 **Leyre** SW France
171 Q5 **Leyte** *island* C Philippines
171 Q6 **Leyte Gulf** *gulf* E Philippines
111 O16 **Leżajsk** Podkarpackie, SE Poland 50°15′N 22°25′E
Lezha *see* Lezhë
113 K18 **Lezhë** *var.* Lezha; *prev.* Lesh, Leshi. Lezhë, NW Albania 41°46′N 19°40′E
113 K18 **Lezhë** ◆ *district* NW Albania
103 O16 **Lézignan-Corbières** Aude, S France 43°12′N 02°46′E
126 J7 **L'gov** Kurskaya Oblast', W Russian Federation 51°38′N 35°17′E
159 P15 **Lhari** Xizang Zizhiqu, W China 30°34′N 93°40′E
159 N16 **Lhasa** *var.* La-sa, Lassa. Xizang Zizhiqu, W China 29°41′N 91°10′E
159 O15 **Lhasa He** W China
Lhaviyani Atoll *see* Faadhippolhu Atoll
158 K16 **Lhazê** *var.* Quxar. Xizang Zizhiqu, W China 29°07′N 87°32′E
158 K14 **Lhazhong** Xizang Zizhiqu, W China 31°58′N 86°43′E
168 H7 **Lhoksukon** Sumatera, W Indonesia 05°04′N 97°19′E
159 Q15 **Lhorong** *var.* Zito. Xizang Zizhiqu, W China 30°51′N 95°41′E
105 W6 **L'Hospitalet de Llobregat** *var.* Hospitalet. Cataluña, NE Spain 41°21′N 02°06′E
153 R11 **Lhotse** ▲ China/Nepal 28°00′N 86°55′E
159 N17 **Lhozhag** *var.* Garbo. Xizang Zizhiqu, W China 28°21′N 90°47′E
159 O16 **Lhünzê** *var.* Xingba. Xizang Zizhiqu, W China 28°25′N 92°30′E
159 N15 **Lhünzhub** *var.* Ganqu. Xizang Zizhiqu, W China 30°14′N 91°20′E
167 N8 **Li** Lamphun, NW Thailand 17°46′N 98°54′E
115 L21 **Liádi** *var.* Livádi. *island* Kykládes, Greece, Aegean Sea
161 P12 **Liancheng** *var.* Lianfeng. Fujian, SE China 25°47′N 116°42′E
Liancheng *see* Lianjiang, Guangdong, China
Liancheng *see* Qinglong, Guizhou, China
Liancheng *see* Guangnan, Yunnan, China
Lianfeng *see* Liancheng
160 K9 **Liangping** *var.* Liangshan. Sichuan, C China 30°40′N 107°46′E
Liangshan *see* Liangping
Liangzhou *see* Wuwei
161 O9 **Liangzi Hu** ◎ C China
161 R12 **Lianjiang** *var.* Fengcheng. Fujian, SE China 26°14′N 119°33′E
160 L15 **Lianjiang** *var.* Liancheng. Guangdong, S China 21°41′N 110°12′E
Lianjiang *see* Xingguo
161 O13 **Lianping** *var.* Yuanshan. Guangdong, S China 24°18′N 114°27′E
Lianshan *see* Huludao
Lian Xian *see* Lianzhou
160 M11 **Lianyuan** *prev.* Lantian. Hunan, S China 27°51′N 111°44′E
161 Q6 **Lianyungang** *var.* Xinpu. Jiangsu, E China 34°38′N 119°12′E
161 N13 **Lianzhou** *var.* Linxian; *prev.* Lian Xian. Guangdong, S China 24°48′N 112°26′E
Lianzhou *see* Hepu
Liao *see* Liaoning
161 P5 **Liaocheng** Shandong, E China 36°31′N 115°59′E
163 U13 **Liaodong Bandao** *var.* Liaotung Peninsula. *peninsula* NE China
163 T13 **Liaodong Wan** *Eng.* Gulf of Lantung, Gulf of Liaotung. *gulf* NE China
163 U11 **Liao He** NE China
163 U12 **Liaoning** *var.* Liao, Liaoning Sheng, Shengking, *hist.* Fengtien, Shenking. ◆ *province* NE China
Liaoning Sheng *see* Liaoning
Liaotung Peninsula *see* Liaodong Bandao
163 V12 **Liaoyang** *var.* Liao-yang. Liaoning, NE China 41°16′N 123°12′E
Liao-yang *see* Liaoyang
163 V11 **Liaoyuan** *var.* Dongliao, Shuang-liao, *Jap.* Chengchiatun. Jilin, NE China 42°52′N 125°09′E
163 U12 **Liaozhong** Liaoning, NE China 41°33′N 122°54′E
Liaqatabad *see* Piplān
10 M10 **Liard** W Canada
Liard *see* Fort Liard
10 L10 **Liard River** British Columbia, W Canada 59°23′N 126°05′W
149 O15 **Liāri** Baluchistān, SW Pakistan 25°43′N 66°28′E
Liatroim *see* Leitrim
189 S6 **Lib** *var.* Ellep. *island* Ralik Chain, C Marshall Islands
Liban *see* Lebanon
138 H6 **Liban, Jebel** *Ar.* Jabal al Gharbt, Jabal Lubnān, *Eng.* Mount Lebanon. ▲ C Lebanon
Libau *see* Liepāja
33 N7 **Libby** Montana, NW USA 48°25′N 115°33′W
79 I16 **Libenge** Equateur, NW Dem. Rep. Congo 03°39′N 18°39′E
26 I7 **Liberal** Kansas, C USA 37°03′N 100°56′W
27 R7 **Liberal** Missouri, C USA 37°33′N 94°31′W
Liberalitas Julia *see* Évora
111 D15 **Liberec** *Ger.* Reichenberg. Liberecký Kraj, N Czech Republic 50°45′N 15°05′E
111 D15 **Liberecký Kraj** ◆ *region* N Czech Republic
42 K12 **Liberia** Guanacaste, NW Costa Rica 10°36′N 85°26′W
76 K17 **Liberia** *off.* Republic of Liberia. ◆ *republic* W Africa
Liberia, Republic of *see* Liberia
61 D16 **Libertad** Corrientes, NE Argentina 30°01′S 57°51′W
61 E20 **Libertad** San José, S Uruguay 34°38′S 56°39′W
54 I7 **Libertad** Barinas, NW Venezuela 08°21′N 69°39′W
54 K6 **Libertad** Cojedes, N Venezuela 09°15′N 68°30′W
62 G12 **Libertador** *off.* Región del Libertador General Bernardo O'Higgins. ◆ *region* C Chile
Libertador General Bernardo O'Higgins, Región del *see* Libertador
Libertador General San Martín *see* Ciudad de Libertador General San Martín
20 L6 **Liberty** Kentucky, S USA 37°19′N 84°58′W
22 J7 **Liberty** Mississippi, S USA 31°09′N 90°49′W
27 R4 **Liberty** Missouri, C USA 39°15′N 94°22′W
18 J12 **Liberty** New York, NE USA 41°48′N 74°45′W
21 T9 **Liberty** North Carolina, SE USA 35°49′N 79°34′W
Libian Desert *see* Libyan Desert
99 J23 **Libin** Luxembourg, SE Belgium 50°01′N 05°13′E
Lībīyah, Aş Şahrā' al *see* Libyan Desert
160 K13 **Libo** *var.* Yuping. Guizhou, S China 25°28′N 107°52′E
Libohova *see* Libohovë
113 L23 **Libohovë** *var.* Libohova. Gjirokastër, S Albania 40°03′N 20°13′E
81 K18 **Liboi** Wajir, E Kenya 0°23′N 40°55′E
102 K13 **Libourne** Gironde, SW France 44°55′N 00°14′W
99 K23 **Libramont** Luxembourg, SE Belgium 49°55′N 05°21′E
113 M20 **Librazhd** *var.* Librazhdi. Elbasan, E Albania 41°10′N 20°22′E
Librazhdi *see* Librazhd
79 C18 **Libreville** ● (Gabon) Estuaire, NW Gabon 0°25′N 09°29′E
75 P10 **Libya** *off.* Great Socialist People's Libyan Arab Jamahiriya, *Ar.* Al Jamāhīrīyah al 'Arabīyah al Lībīyah ash Sha'bīyah al Ishtirākīy; *prev.* Libyan Arab Republic. ◆ *Islamic state* N Africa
Libyan Arab Republic *see* Libya
75 T11 **Libyan Desert** *var.* Libian Desert, *Ar.* Aş Şahrā' al Lībīyah. *desert* N Africa
75 T8 **Libyan Plateau** *var.* Aḍ Diffah. *plateau* Egypt/Libya
62 G12 **Licantén** Maule, C Chile 35°00′S 72°00′W
107 J25 **Licata** *anc.* Phintias. Sicilia, Italy, C Mediterranean Sea 37°07′N 13°57′E
137 P14 **Lice** Diyarbakır, SE Turkey 38°29′N 40°39′E
Licheng *see* Lipu
97 L19 **Lichfield** C England, United Kingdom 52°42′N 01°48′W
83 N14 **Lichinga** Niassa, N Mozambique 13°19′S 35°13′E
109 V3 **Lichtenau** Niederösterreich, N Austria 48°29′N 15°24′E
83 I21 **Lichtenburg** North-West, N South Africa 26°09′S 26°11′E
101 K18 **Lichtenfels** Bayern, SE Germany 50°09′N 11°04′E
98 O12 **Lichtenvoorde** Gelderland, E Netherlands 51°59′N 06°34′E
Lichtenwald *see* Sevnica
99 C17 **Lichtervelde** West-Vlaanderen, W Belgium 51°02′N 03°09′E
160 L9 **Lichuan** Hubei, C China 30°20′N 108°56′E
27 V7 **Licking** Missouri, C USA 37°30′N 91°51′W
20 M4 **Licking River** Kentucky, S USA
112 C11 **Lički Osik** Lika-Senj, C Croatia 44°36′N 15°24′E
Ličko-Senjska Županija *see* Lika-Senj
107 K19 **Licosa, Punta** *headland* S Italy 40°15′N 14°54′E
119 H16 **Lida** Hrodzyenskaya Voblasts', W Belarus 53°53′N 25°20′E
93 H17 **Liden** Västernorrland, C Sweden 62°43′N 16°49′E
29 R7 **Lidgerwood** North Dakota, N USA 46°04′N 97°09′W
Lidhorikion *see* Lidoríki
95 K21 **Lidhult** Kronoberg, S Sweden 56°49′N 13°25′E
95 P16 **Lidingö** Stockholm, C Sweden 59°22′N 18°10′E
95 K17 **Lidköping** Västra Götaland, S Sweden 58°30′N 13°10′E
Lido di Iesolo *see* Lido di Jesolo
106 I8 **Lido di Jesolo** *var.* Lido di Iesolo. Veneto, NE Italy 45°30′N 12°37′E
107 H15 **Lido di Ostia** Lazio, C Italy 41°42′N 12°19′E
Lidokhorikion *see* Lidoríki
115 E18 **Lidoríki** *prev.* Lidhorikion, Lidokhorikion. Stereá Elláda, C Greece 38°32′N 22°12′E
110 K9 **Lidzbark** Warmińsko-Mazurskie, NE Poland 53°15′N 19°49′E
110 L7 **Lidzbark Warmiński** *Ger.* Heilsberg. Olsztyn, N Poland 54°08′N 20°35′E
109 U3 **Liebenau** Oberösterreich, N Austria 48°33′N 14°48′E
181 P7 **Liebig, Mount** ▲ Northern Territory, C Australia 23°19′S 131°30′E
109 V8 **Lieboch** Steiermark, SE Austria 47°00′N 15°21′E
108 I8 **Liechtenstein** *off.* Principality of Liechtenstein. ◆ *principality* C Europe
Liechtenstein, Principality of *see* Liechtenstein
99 F18 **Liedekerke** Vlaams Brabant, C Belgium 50°51′N 04°05′E
99 K19 **Liège** *Dut.* Luik, *Ger.* Lüttich. Liège, E Belgium 50°38′N 05°35′E
99 K20 **Liège** *Dut.* Luik. ◆ *province* E Belgium
Liegnitz *see* Legnica
93 O16 **Lieksa** Pohjois-Karjala, E Finland 63°20′N 30°E
118 F10 **Lielupe** Latvia/Lithuania
118 G9 **Lielvārde** C Latvia 56°45′N 24°48′E
167 U13 **Liên Hương** *var.* Tuy Phong. Bình Thuận, S Vietnam 11°13′N 108°40′E
167 U13 **Liên Nghia** *var.* Liên Nghia *var.* Đưc Trong. Lâm Đông, S Vietnam 11°45′N 108°24′E
Liên Nghia *see* Liên Nghia
109 P9 **Lienz** Tirol, W Austria 46°50′N 12°45′E
118 B10 **Liepāja** *Ger.* Libau. W Latvia 56°32′N 21°02′E
99 H17 **Lier** *Fr.* Lierre. Antwerpen, N Belgium 51°08′N 04°35′E
95 H15 **Lierbyen** Buskerud, S Norway 59°57′N 10°14′E
99 L21 **Lierneux** Liège, E Belgium 50°12′N 05°51′E
Lierre *see* Lier
101 D18 **Lieser** W Germany
109 U7 **Liesing** E Austria
108 E6 **Liestal** Basel Landschaft, N Switzerland 47°29′N 07°43′E
Lietuva *see* Lithuania
Lievenhof *see* Līvāni
103 O2 **Liévin** Pas-de-Calais, N France 50°25′N 02°48′E
14 M9 **Lièvre, Rivière du** Québec, SE Canada
109 T6 **Liezen** Steiermark, C Austria 47°34′N 14°12′E
97 E14 **Lifford** *Ir.* Leifear. Donegal, NW Ireland 54°50′N 07°29′W
187 Q16 **Lifou** *island* Îles Loyauté, E New Caledonia
193 Y15 **Lifuka** *island* Ha'apai Group, C Tonga
171 P4 **Ligao** Luzon, N Philippines 13°16′N 123°30′E
Liger *see* Loire
42 H2 **Lighthouse Reef** *reef* E Belize
183 Q4 **Lightning Ridge** New South Wales, SE Australia 29°29′S 148°00′E
103 N9 **Lignières** Cher, C France 46°45′N 02°10′E
103 S5 **Ligny-en-Barrois** Meuse, NE France 48°42′N 05°22′E
83 P15 **Ligonha** NE Mozambique
31 P11 **Ligonier** Indiana, N USA 41°25′N 85°33′W
81 I25 **Ligunga** Ruvuma, S Tanzania 10°51′S 37°10′E
106 D9 **Ligure, Appennino** *Eng.* Ligurian Mountains. ▲ NW Italy
Ligure, Mar *see* Ligurian Sea
106 C9 **Liguria** ◆ *region* NW Italy
Ligurian Mountains *see* Ligure, Appennino
120 K6 **Ligurian Sea** *Fr.* Mer Ligurienne, *It.* Mar Ligure. *sea* N Mediterranean Sea
Ligurienne, Mer *see* Ligurian Sea
186 H5 **Lihir Group** *island group* NE Papua New Guinea
38 B8 **Līhu'e** *var.* Lihue. Kaua'i, Hawaii, USA 21°59′N 159°23′W
Lihue *see* Līhu'e
118 F5 **Lihula** *Ger.* Leal. Läänemaa, W Estonia 58°44′N 23°49′E
124 I2 **Liinakhamari** *var.* Linacmamari. Murmanskaya Oblast', NW Russian Federation 69°40′N 31°27′E
Liivi Laht *see* Riga, Gulf of
160 F11 **Lijiang** *var.* Dayan, Lijiang Naxizu Zizhixian. Yunnan, SW China 26°52′N 100°10′E
Lijiang Naxizu Zizhixian *see* Lijiang
112 C11 **Lika-Senj** *off.* Ličko-Senjska Županija. ◆ *province* W Croatia
79 N25 **Likasi** *prev.* Jadotville. Shaba, SE Dem. Rep. Congo 11°02′S 26°51′E
79 L16 **Likati** Orientale, N Dem. Rep. Congo 03°28′N 23°45′E
10 M15 **Likely** British Columbia, SW Canada 52°40′N 121°34′W
153 Y11 **Likhapāni** Assam, NE India 27°19′N 95°54′E
124 J16 **Likhoslavl'** Tverskaya Oblast', W Russian Federation 57°08′N 35°27′E
189 U5 **Likiep Atoll** *atoll* Ratak Chain, C Marshall Islands
95 D18 **Liknes** Vest-Agder, S Norway 58°20′N 06°58′E
79 H16 **Likouala** ◆ *province* NE Congo
79 H18 **Likouala** N Congo
79 H18 **Likouala aux Herbes** E Congo
190 B16 **Liku** E Niue 19°02′S 169°47′E
Likupang, Selat *see* Bangka, Selat
27 Y8 **Lilbourn** Missouri, C USA 36°35′N 89°37′W
103 X14 **L'Île-Rousse** Corse, France, C Mediterranean Sea 42°39′N 08°59′E
109 W5 **Lilienfeld** Niederösterreich, NE Austria 48°01′N 15°36′E
161 N11 **Liling** Hunan, S China 27°42′N 113°49′E
95 J18 **Lilla Edet** Västra Götaland, S Sweden 58°08′N 12°08′E
103 P1 **Lille** *var.* l'Isle, *Dut.* Rijssel, *Flem.* Ryssel, *prev.* Lisle; *anc.* Insula. Nord, N France 50°38′N 03°04′E
95 G24 **Lillebælt** *var.* Lille Bælt, *Eng.* Little Belt. *strait* S Denmark
Lille Bælt *see* Lillebælt
102 L3 **Lillebonne** Seine-Maritime, N France 49°30′N 00°34′E
94 H12 **Lillehammer** Oppland, S Norway 61°07′N 10°28′E
103 O1 **Lillers** Pas-de-Calais, N France 50°34′N 02°26′E
95 F18 **Lillesand** Aust-Agder, S Norway 58°15′N 08°23′E
95 I15 **Lillestrøm** Akershus, S Norway 59°58′N 11°05′E
93 F18 **Lillhärdal** Jämtland, C Sweden 61°51′N 14°04′E
21 U10 **Lillington** North Carolina, SE USA 35°25′N 78°50′W
105 O9 **Lillo** Castilla-La Mancha, C Spain 39°43′N 03°19′W
10 M16 **Lillooet** British Columbia, SW Canada 50°41′N 121°59′W
83 M14 **Lilongwe** ● (Malawi) Central, W Malawi 13°58′S 33°48′E
83 M14 **Lilongwe** ✈ Central, W Malawi 13°46′S 33°44′E
83 M14 **Lilongwe** W Malawi
171 P7 **Liloy** Mindanao, S Philippines 08°04′N 122°42′E
Lilybaeum *see* Marsala
182 J7 **Lilydale** South Australia 32°57′S 140°00′E
183 P16 **Lilydale** Tasmania, SE Australia 41°17′S 147°13′E
113 J14 **Lim** SE Europe
57 D15 **Lima** ● (Peru) Lima, W Peru 12°06′S 78°W
94 K13 **Lima** Dalarna, C Sweden 60°55′N 13°19′E
31 R12 **Lima** Ohio, NE USA 40°43′N 84°06′W
57 D14 **Lima** ◆ *department* W Peru
Lima *see* Jorge Chávez Internacional
137 Y13 **Liman** *prev.* Port-Iliç. SE Azerbaijan 38°54′N 48°49′E
111 L17 **Limanowa** Małopolskie, S Poland 49°43′N 20°25′E
104 G5 **Lima, Rio** *Sp.* Limia. Portugal/Spain *see also* Limia
Lima, Rio *see* Limia
168 M11 **Limas** Pulau Sebangka, W Indonesia 0°09′N 104°31′E
Limassol *see* Lemesós
97 F14 **Limavady** *Ir.* Léim An Mhadaidh. NW Northern Ireland, United Kingdom 55°03′N 06°57′W
63 J14 **Limay Mahuida** La Pampa, C Argentina 37°09′S 66°40′W
63 H15 **Limay, Río** W Argentina
101 N16 **Limbach-Oberfrohna** Sachsen, E Germany 50°52′N 12°46′E
81 F22 **Limba Limba** C Tanzania
118 G7 **Limbaži** *Est.* Lemsalu. N Latvia 57°33′N 24°46′E
44 M8 **Limbé** N Haiti 19°44′N 72°25′W
99 L19 **Limbourg** Liège, E Belgium 50°37′N 05°56′E
99 K17 **Limburg** ◆ *province* NE Belgium
99 L16 **Limburg** ◆ *province* SE Netherlands
101 F17 **Limburg an der Lahn** Hessen, W Germany 50°22′N 08°04′E
94 K13 **Limedsforsen** Dalarna, C Sweden 60°52′N 13°25′E
60 L9 **Limeira** São Paulo, S Brazil 22°34′S 47°25′W
97 C19 **Limerick** *Ir.* Luimneach. Limerick, SW Ireland 52°40′N 08°38′W
97 C20 **Limerick** *Ir.* Luimneach. *cultural region* SW Ireland
19 S2 **Limestone** Maine, NE USA 46°52′N 67°49′W
25 U9 **Limestone, Lake** ◎ Texas, SW USA
39 P12 **Lime Village** Alaska, USA 61°21′N 155°26′W
95 F20 **Limfjorden** *fjord* N Denmark
95 J23 **Limhamn** Skåne, S Sweden 55°34′N 12°57′E
104 H5 **Limia** *Port.* Rio Lima. Portugal/Spain *see also* Lima, Rio
Limia *see* Lima, Rio
93 L14 **Liminka** Pohjois-Pohjanmaa, C Finland 64°48′N 25°19′E
Limín Vathéos *see* Sámos
115 G17 **Límni** Évvoia, C Greece 38°46′N 23°20′E
115 J15 **Límnos** *anc.* Lemnos. *island* E Greece
102 M11 **Limoges** *anc.* Augustoritum Lemovicensium, Lemovices. Haute-Vienne, C France 45°51′N 01°16′E
43 O13 **Limón** *var.* Puerto Limón. Limón, E Costa Rica 09°59′N 83°02′W
42 K4 **Limón** Colón, NE Honduras 15°50′N 85°31′W
37 U5 **Limon** Colorado, C USA 39°15′N 103°41′W
43 N13 **Limón** *off.* Provincia de Limón. ◆ *province* E Costa Rica
106 A10 **Limone Piemonte** Piemonte, NE Italy 44°12′N 07°37′E
Limones *see* Valdéz
Limón, Provincia de *see* Limón
Limonum *see* Poitiers
103 N11 **Limousin** ◆ *region* C France
103 N16 **Limoux** Aude, S France 43°03′N 02°13′E
83 J20 **Limpopo** *off.* Limpopo Province; *prev.* Northern, Northern Transvaal. ◆ *province* NE South Africa
83 L19 **Limpopo** *var.* Crocodile. S Africa
Limpopo Province *see* Limpopo
160 K17 **Limu Ling** S China
113 M20 **Lin** *var.* Lini. Elbasan, E Albania 41°03′N 20°37′E
Linacmamari *see* Liinakhamari
62 G13 **Linares** Maule, C Chile 35°50′S 71°37′W
54 C13 **Linares** Nariño, SW Colombia 01°24′N 77°30′W
41 O9 **Linares** Nuevo León, NE Mexico 24°54′N 99°38′W
105 N12 **Linares** Andalucía, S Spain 38°05′N 03°38′W
107 G15 **Linaro, Capo** *headland* C Italy 42°01′N 11°49′E
106 D8 **Linate** ✈ (Milano) Lombardia, N Italy 45°27′N 09°18′E
167 T8 **Lin Camh** *prev.* Đưc Tho. Ha Tinh, N Vietnam 18°30′N 105°36′E
160 F13 **Lincang** Yunnan, SW China 23°55′N 100°03′E
Lincheng *see* Lingao
Linchuan *see* Fuzhou
61 B20 **Lincoln** Buenos Aires, E Argentina 34°55′S 61°30′W
185 H19 **Lincoln** Canterbury, South Island, New Zealand 43°37′S 172°30′E
97 N18 **Lincoln** *anc.* Lindum, Lindum Colonia. E England, United Kingdom 53°14′N 00°33′W
35 O6 **Lincoln** California, W USA 38°52′N 121°18′W
30 L13 **Lincoln** Illinois, N USA 40°09′N 89°21′W
27 M4 **Lincoln** Kansas, C USA 39°03′N 98°09′W
19 S5 **Lincoln** Maine, NE USA 45°22′N 68°30′W
27 T5 **Lincoln** Missouri, C USA 38°23′N 93°19′W
29 R16 **Lincoln** *state capital* Nebraska, C USA 40°46′N 96°43′W
32 F11 **Lincoln City** Oregon, NW USA 44°57′N 124°01′W
167 X10 **Lincoln Island** *Chin.* Dong Dao, *Viet.* Đao Linh Côn. *island* E Paracel Islands
197 Q11 **Lincoln Sea** *sea* Arctic Ocean
97 N18 **Lincolnshire** *cultural region* E England, United Kingdom
21 R10 **Lincolnton** North Carolina, SE USA 35°27′N 81°16′W
25 V7 **Lindale** Texas, SW USA 32°31′N 95°24′W
101 I25 **Lindau** *var.* Lindau am Bodensee. Bayern, S Germany 47°33′N 09°41′E
Lindau am Bodensee *see* Lindau
123 P9 **Linde** NE Russian Federation
55 T9 **Linden** E Guyana 05°58′N 58°12′W
23 O6 **Linden** Alabama, S USA 32°18′N 87°48′W
20 H9 **Linden** Tennessee, S USA 35°38′N 87°50′W
25 X6 **Linden** Texas, SW USA 33°01′N 94°22′W
44 H2 **Linden Pindling** ✈ New Providence, C The Bahamas 25°00′N 77°26′W
18 J16 **Lindenwold** New Jersey, NE USA 39°47′N 74°58′W
95 M15 **Lindesberg** Örebro, C Sweden 59°36′N 15°15′E
95 D18 **Lindesnes** *headland* S Norway 57°58′N 07°03′E
Líndhos *see* Líndos
81 K24 **Lindi** Lindi, SE Tanzania 10°S 39°41′E
81 J24 **Lindi** ◆ *region* SE Tanzania
79 N17 **Lindi** NE Dem. Rep. Congo
163 V7 **Lindian** Heilongjiang, NE China 47°15′N 124°51′E
185 E21 **Lindis Pass** *pass* South Island, New Zealand
83 J22 **Lindley** Free State, C South Africa 27°52′S 27°55′E
95 J19 **Lindome** Västra Götaland, S Sweden 57°34′N 12°05′E
163 S10 **Lindong** *var.* Bairin Zuoqi. Nei Mongol Zizhiqu, N China 43°59′N 119°24′E
115 O23 **Líndos** *var.* Líndhos. Ródos, Dodekánisa, Greece, Aegean Sea 36°05′N 28°05′E
14 I14 **Lindsay** Ontario, SE Canada 44°21′N 78°44′W
35 R11 **Lindsay** California, W USA 36°11′N 119°06′W
33 X8 **Lindsay** Montana, NW USA 47°13′N 105°10′W
27 N11 **Lindsay** Oklahoma, C USA 34°50′N 97°37′W
27 N5 **Lindsborg** Kansas, C USA 38°34′N 97°39′W
95 N21 **Lindsdal** Kalmar, S Sweden 56°44′N 16°18′E
Lindum/Lindum Colonia *see* Lincoln
191 W3 **Line Islands** *island group* E Kiribati
Linëvo *see* Linova
160 M5 **Linfen** *var.* Lin-fen. Shanxi, C China 36°08′N 111°34′E
Lin-fen *see* Linfen
104 L2 **L'Infiestu** *prev.* Infiesto. Asturias, N Spain 43°21′N 05°21′W
155 F18 **Linganamakki Reservoir** ⊠ SW India
160 L17 **Lingao** *var.* Lincheng. Hainan, S China 19°44′N 109°23′E
171 N3 **Lingayen** Luzon, N Philippines 16°00′N 120°12′E
160 M6 **Lingbao** *var.* Guolüezhen. Henan, C China 34°34′N 110°50′E
94 N12 **Lingbo** Gävleborg, C Sweden 61°04′N 16°45′E
Lingcheng *see* Lingshan, Guangxi, China
Lingcheng *see* Beiliu, Guangxi, China
Lingeh *see* Bandar-e Lengeh
100 E12 **Lingen** *var.* Lingen an der Ems. Niedersachsen, NW Germany 52°31′N 07°19′E
Lingen an der Ems *see* Lingen
168 M11 **Lingga, Kepulauan** *island group* W Indonesia
168 L11 **Lingga, Pulau** *island* Kepulauan Lingga, W Indonesia
14 J14 **Lingham Lake** ◎ Ontario, SE Canada
94 M13 **Linghed** Dalarna, C Sweden 60°48′N 15°55′E
33 Z15 **Lingle** Wyoming, C USA 42°07′N 104°21′W
18 G15 **Linglestown** Pennsylvania, NE USA 40°20′N 76°46′W
160 M12 **Lingling** *prev.* Yongzhou, Zhishan. Hunan, S China 26°13′N 111°36′E
79 K18 **Lingomo 11** Equateur, NW Dem. Rep. Congo 0°42′N 21°59′E
160 L15 **Lingshan** *var.* Lingcheng. Guangxi Zhuangzu Zizhiqu, S China 22°28′N 109°19′E
160 L17 **Lingshui** *var.* Lingshui Lizu Zizhixian. Hainan, S China 18°35′N 110°03′E
Lingshui Lizu Zizhixian *see* Lingshui
155 G16 **Lingsugūr** Karnātaka, C India 16°13′N 76°33′E
107 L23 **Linguaglossa** Sicilia, Italy, C Mediterranean Sea 37°51′N 15°06′E
76 H10 **Linguère** N Senegal 15°24′N 15°06′W
159 W8 **Lingwu** Ningxia, N China 38°04′N 106°21′E
Lingxi *see* Yongshun, Hunan, China
Lingxi *see* Cangnan, Zhejiang, China
Lingxian/Ling Xian *see* Yanling
163 S12 **Lingyuan** Liaoning, NE China 41°09′N 119°24′E
163 U4 **Linhai** Heilongjiang, NE China 51°30′N 124°18′E
161 S10 **Linhai** *var.* Taizhou. Zhejiang, SE China 28°54′N 121°08′E
59 O20 **Linhares** Espírito Santo, SE Brazil 19°22′S 40°04′W
Linh Côn, Đao *see* Lincoln Island
Linhe *see* Bayannur
Lini *see* Lin
Linik, Chiyā-ē *see* Linkì, Chiya-i
Linjiang *see* Shanghang
139 S1 **Linkì, Chiya-i** *Ar.* Jabal Linkī, *var.* Chiyā-ē Linik, Chiyā-ī Linkī Kurezūr. ▲ N Iraq
Linkī, Jabal *see* Linkì, Chiya-i
95 M18 **Linköping** Östergötland, S Sweden 58°25′N 15°37′E
163 Y8 **Linkou** Heilongjiang, NE China 45°18′N 130°17′E
118 F11 **Linkuva** Šiaulai, N Lithuania 56°06′N 23°58′E
27 V5 **Linn** Missouri, C USA 38°29′N 91°51′W
25 S16 **Linn** Texas, SW USA 26°32′N 98°06′W
27 T2 **Linneus** Missouri, C USA 39°53′N 93°10′W
96 H10 **Linnhe, Loch** *inlet* W Scotland, United Kingdom
119 G19 **Linova** *Rus.* Linëvo. Brestskaya Voblasts', SW Belarus 52°29′N 24°30′E
161 O5 **Linqing** Shandong, E China 36°51′N 115°42′E
60 K8 **Lins** São Paulo, S Brazil 21°40′S 49°44′W
93 F17 **Linsell** Jämtland, C Sweden 62°10′N 14°E
160 J9 **Linshui** Sichuan, C China 30°24′N 106°54′E
44 K12 **Linstead** C Jamaica 18°08′N 77°02′W
159 U11 **Lintan** Gansu, N China 34°43′N 101°27′E
159 V11 **Lintao** *var.* Taoyang. Gansu, C China 35°23′N 103°54′E
15 S12 **Lintère** Québec, SE Canada
108 H8 **Linth** NW Switzerland
108 H8 **Linthal** Glarus, NE Switzerland 46°59′N 08°57′E
31 N15 **Linton** Indiana, N USA 39°01′N 87°10′W
29 N6 **Linton** North Dakota, N USA 46°16′N 100°13′W
163 R11 **Linxi** Nei Mongol Zizhiqu, N China 43°29′N 117°59′E
159 U11 **Linxia** *var.* Linxia Huizu Zizhizhou. Gansu, C China 35°34′N 103°08′E
Linxia Huizu Zizhizhou *see* Linxia
Linxian *see* Lianzhou
161 P4 **Linyi** *var.* Yishi. Shandong, E China 37°12′N 116°54′E
161 Q6 **Linyi** Shandong, S China 37°06′N 118°18′E
160 M6 **Linyi** Shanxi, C China 35°10′N 118°18′E
109 T4 **Linz** *anc.* Lentia. Oberösterreich, N Austria 48°19′N 14°18′E
159 S8 **Linze** *var.* Shahe; *prev.* Shahepu. Gansu, N China 39°06′N 100°03′E
44 J13 **Lionel Town** C Jamaica 17°49′N 77°14′W
103 Q16 **Lion, Golfe du** *Eng.* Gulf of Lion, Gulf of Lions; *anc.* Sinus Gallicus. *gulf* S France
Lion, Gulf of/Lions, Gulf of *see* Lion, Golfe du
83 K16 **Lions Den** Mashonaland West, N Zimbabwe 17°16′S 30°00′E
14 F13 **Lion's Head** Ontario, S Canada 44°59′N 81°16′W
Lios Ceannúir, Bá *see* Liscannor Bay
Lios Mór *see* Lismore
Lios na gCearrbhach *see* Lisburn
Lios Tuathail *see* Listowel
79 G17 **Liouesso** Sangha, N Congo 01°02′N 15°43′E
Liozno *see* Lyozna
171 O4 **Lipa** *off.* Lipa City. Luzon, N Philippines 13°57′N 121°10′E
Lipa City *see* Lipa
25 S7 **Lipan** Texas, SW USA 32°31′N 98°03′W
Lipari Islands/Lipari, Isole *see* Eolie, Isole
107 L22 **Lipari, Isola** *island* Isole Eolie, S Italy
116 L8 **Lipcani** *Rus.* Lipkany. N Moldova 48°16′N 26°47′E
93 N17 **Liperi** Pohjois-Karjala, SE Finland 62°33′N 29°29′E
126 L7 **Lipetsk** Lipetskaya Oblast', W Russian Federation 52°37′N 39°38′E
126 K6 **Lipetskaya Oblast'** ◆ *province* W Russian Federation
57 K22 **Lípez, Cordillera de** ▲ SW Bolivia
110 E10 **Lipiany** *Ger.* Lippehne. Zachodnio-pomorskie, W Poland 53°00′N 14°58′E
112 G9 **Lipik** Požega-Slavonija, NE Croatia 45°24′N 17°08′E
124 L12 **Lipin Bor** Vologodskaya Oblast', NW Russian Federation 60°12′N 38°04′E
160 L12 **Liping** *var.* Defeng. Guizhou, S China 26°16′N 109°08′E
Lipkany *see* Lipcani
119 H15 **Lipnishki** Hrodzyenskaya Voblasts', W Belarus 54°00′N 25°37′E
110 J10 **Lipno** Kujawsko-pomorskie, C Poland 52°52′N 19°11′E
116 F11 **Lipova** *Hung.* Lippa. Arad, W Romania 46°05′N 21°42′E
Lipovets *see* Lypovets'
Lippa *see* Lipova
101 E14 **Lippe** W Germany
Lippehne *see* Lipiany
101 G14 **Lippstadt** Nordrhein-Westfalen, W Germany 51°41′N 08°20′E
25 P1 **Lipscomb** Texas, SW USA 36°14′N 100°16′W
Lipsia/Lipsk *see* Leipzig
Liptau-Sankt-Nikolaus/Liptószentmiklós *see* Liptovský Mikuláš
111 K19 **Liptovský Mikuláš** *Ger.* Liptau-Sankt-Nikolaus, *Hung.* Liptószentmiklós. Žilinský Kraj, N Slovakia 49°06′N 19°36′E
183 O13 **Liptrap, Cape** *headland* Victoria, SE Australia 38°55′S 145°58′E
160 L13 **Lipu** *var.* Licheng. Guangxi Zhuangzu Zizhiqu, S China 24°25′N 110°15′E
81 G17 **Lira** N Uganda 02°15′N 32°55′E
57 F15 **Lircay** Huancavelica, C Peru 12°59′S 74°44′W
107 K17 **Liri** C Italy
144 M8 **Lisakovsk** Kostanay, NW Kazakhstan 52°32′N 62°32′E
79 K17 **Lisala** Equateur, N Dem. Rep. Congo 02°10′N 21°29′E
104 F11 **Lisboa** *Eng.* Lisbon; *anc.* Felicitas Julia, Olisipo. ● (Portugal) Lisboa, W Portugal 38°44′N 09°08′W
104 F10 **Lisboa** *Eng.* Lisbon. ◆ *district* C Portugal
19 N7 **Lisbon** New Hampshire, NE USA 44°11′N 71°52′W
29 Q6 **Lisbon** North Dakota, N USA 46°27′N 97°42′W
Lisbon *see* Lisboa
19 Q8 **Lisbon Falls** Maine, NE USA 44°00′N 70°03′W
97 G15 **Lisburn** *Ir.* Lios na gCearrbhach. E Northern Ireland, United Kingdom 54°31′N 06°03′W
38 L6 **Lisburne, Cape** *headland* Alaska, USA 68°52′N 166°13′W
97 B19 **Liscannor Bay** *Ir.* Bá Lios Ceannúir. *inlet* W Ireland
113 Q18 **Lisec** ▲ E FYR Macedonia 41°46′N 22°30′E
160 F13 **Lishe Jiang** SW China
Lishi *see* Lüliang
163 V10 **Lishu** Jilin, NE China 43°25′N 124°19′E
161 R10 **Lishui** Zhejiang, SE China 28°27′N 119°25′E
192 L5 **Lisianski Island** *island* Hawaiian Islands, Hawai'i, USA
Lisichansk *see* Lysychans'k
102 L4 **Lisieux** *anc.* Noviomagus. Calvados, N France 49°09′N 00°13′E
126 L8 **Liski** *prev.* Georgiu-Dezh. Voronezhskaya Oblast', W Russian Federation 51°00′N 39°36′E
103 N4 **l'Isle-Adam** Val-d'Oise, N France 49°07′N 02°13′E
Lisle/l'Isle *see* Lille
103 R15 **l'Isle-sur-la-Sorgue** Vaucluse, SE France 43°55′N 05°03′E
15 S9 **L'Islet** Québec, SE Canada 47°07′N 70°18′W
183 V4 **Lismore** New South Wales, SE Australia 28°48′S 153°12′E
182 M12 **Lismore** Victoria, SE Australia 37°59′S 143°18′E
97 D20 **Lismore** *Ir.* Lios Mór. S Ireland 52°10′N 07°10′W
Lissa *see* Vis, Croatia
Lissa *see* Leszno, Poland
98 H11 **Lisse** Zuid-Holland, W Netherlands 52°15′N 04°33′E
114 K13 **Lissos** *var.* Filiouri. NE Greece
95 D18 **Lista** *peninsula* S Norway
95 D18 **Listafjorden** *fjord* S Norway
195 R13 **Lister, Mount** ▲ Antarctica 78°12′S 161°46′E
126 M8 **Listopadovka** Voronezhskaya Oblast', W Russian Federation 51°54′N 41°08′E
14 F15 **Listowel** Ontario, S Canada 43°44′N 80°57′W
97 B20 **Listowel** *Ir.* Lios Tuathail. Kerry, SW Ireland 52°27′N 09°29′W
160 L14 **Litang** Guangxi Zhuangzu Zizhiqu, S China 23°09′N 109°08′E
160 F9 **Litang** *var.* Gaocheng. Sichuan, C China 30°03′N 100°12′E
160 F10 **Litang Qu** C China
55 X12 **Litani** *var.* Itany. French Guiana/Suriname
138 G8 **Lītani, Nahr el** *var.* Nahr al Litant. C Lebanon
Litant, Nahr al *see* Lītani, Nahr el
Litauen *see* Lithuania
30 K14 **Litchfield** Illinois, N USA 39°17′N 89°52′W
29 U8 **Litchfield** Minnesota, N USA 45°09′N 94°31′W
36 K13 **Litchfield Park** Arizona, SW USA 33°29′N 112°21′W
183 S8 **Lithgow** New South Wales, SE Australia 33°30′S 150°09′E
115 I26 **Líthino, Akrotírio** *headland* Kríti, Greece, E Mediterranean Sea 34°55′N 24°43′E
118 D12 **Lithuania** *off.* Republic of Lithuania, *Ger.* Litauen, *Lith.* Lietuva, *Pol.* Litwa, *Rus.* Litva; *prev.* Lithuanian SSR, *Rus.* Litovskaya SSR. ◆ *republic* NE Europe
Lithuanian SSR *see* Lithuania
Lithuania, Republic of *see* Lithuania
109 U11 **Litija** *Ger.* Littai. C Slovenia 46°03′N 14°50′E
18 H15 **Lititz** Pennsylvania, NE USA 40°09′N 76°18′W
115 F15 **Litóchoro** *var.* Litohoro, Litókhoron. Kentrikí Makedonía, N Greece 40°06′N 22°30′E
Litohoro/Litókhoron *see* Litóchoro
111 C15 **Litoměřice** *Ger.* Leitmeritz. Ústecký Kraj, NW Czech Republic 50°33′N 14°10′E
111 F17 **Litomyšl** *Ger.* Leitomischl. Pardubický Kraj, C Czech Republic 49°54′N 16°18′E
111 G17 **Litovel** *Ger.* Littau. Olomoucký Kraj, E Czech Republic 49°42′N 17°05′E
123 S13 **Litovko** Khabarovskiy Kray, SE Russian Federation 49°22′N 135°10′E
Litovskaya SSR *see* Lithuania
Littai *see* Litija
Littau *see* Litovel
44 G1 **Little Abaco** *var.* Abaco Island. *island* N The Bahamas
111 I21 **Little Alföld** *Ger.* Kleines Ungarisches Tiefland, *Hung.* Kisalföld, *Slvk.* Podunajská Rovina. *plain* Hungary/Slovakia
151 Q20 **Little Andaman** *island* Andaman Islands, India, NE Indian Ocean
26 M5 **Little Arkansas River** Kansas, C USA
184 L4 **Little Barrier Island** *island* N New Zealand
Little Belt *see* Lillebælt
38 M11 **Little Black River** Alaska, USA
27 O2 **Little Blue River** Kansas/Nebraska, C USA
44 D8 **Little Cayman** *island* E Cayman Islands
11 X11 **Little Churchill** Manitoba, C Canada
166 J10 **Little Coco Island** *island* SW Myanmar (Burma)
36 L10 **Little Colorado River** Arizona, SW USA
14 E11 **Little Current** Manitoulin Island, Ontario, S Canada 45°57′N 81°56′W

◆ Country
● Country Capital
◇ Dependent Territory
○ Dependent Territory Capital
◆ Administrative Regions
✈ International Airport
▲ Mountain
▲ Mountain Range
Volcano
River
◎ Lake
⊠ Reservoir

12 E11 **Little Current** ≈ Ontario, S Canada
38 L8 **Little Diomede Island** *island* Alaska, USA
44 I4 **Little Exuma** *island* C The Bahamas
29 U7 **Little Falls** Minnesota, N USA *45°59′N 94°21′W*
18 J10 **Little Falls** New York, NE USA *43°02′N 74°51′W*
24 M5 **Littlefield** Texas, SW USA *33°56′N 102°20′W*
29 V3 **Littlefork** Minnesota, N USA *48°24′N 93°33′W*
29 V3 **Little Fork River** ≈ Minnesota, N USA
11 N16 **Little Fort** British Columbia, SW Canada *51°27′N 120°15′W*
11 Y14 **Little Grand Rapids** Manitoba, C Canada *52°06′N 95°29′W*
97 N23 **Littlehampton** SE England, United Kingdom *50°48′N 00°33′W*
35 T2 **Little Humboldt River** ≈ Nevada, W USA
44 K6 **Little Inagua** *var.* Inagua Islands. *island* S The Bahamas
21 Q4 **Little Kanawha River** ≈ West Virginia, NE USA
83 F25 **Little Karoo** *plateau* S South Africa
39 O16 **Little Koniuji Island** *island* Shumagin Islands, Alaska, USA
44 H12 **Little London** W Jamaica *18°15′N 78°13′W*
13 R10 **Little Mecatina** *Fr.* Rivière du Petit Mécatina. ≈ Newfoundland and Labrador/Québec, E Canada
96 F8 **Little Minch, The** *strait* NW Scotland, United Kingdom
27 T13 **Little Missouri River** ≈ Arkansas, USA
28 J7 **Little Missouri River** ≈ NW USA
28 J3 **Little Muddy River** ≈ North Dakota, N USA
151 Q22 **Little Nicobar** *island* Nicobar Islands, India, NE Indian Ocean
27 R6 **Little Osage River** ≈ Missouri, C USA
97 P20 **Little Ouse** ≈ E England, United Kingdom
149 V2 **Little Pamir** *Pash.* Pāmīr-e Khord, *Rus.* Malyy Pamir. ▲▲ Afghanistan/Tajikistan
21 U12 **Little Pee Dee River** ≈ North Carolina/South Carolina, SE USA
27 V10 **Little Red River** ≈ Arkansas, C USA
Little Rhody *see* Rhode Island
185 I19 **Little River** Canterbury, South Island, New Zealand *43°45′S 172°49′E*
21 U12 **Little River** South Carolina, SE USA *33°52′N 78°36′W*
27 Y9 **Little River** ≈ Arkansas/ Missouri, C USA
27 R13 **Little River** ≈ Arkansas/ Oklahoma, USA
23 T7 **Little River** ≈ Georgia, SE USA
22 H6 **Little River** ≈ Louisiana, S USA
25 T10 **Little River** ≈ Texas, SW USA
27 V12 **Little Rock** *state capital* Arkansas, C USA *34°45′N 92°17′W*
31 N8 **Little Sable Point** *headland* Michigan, N USA *43°38′N 86°32′W*
103 U11 **Little Saint Bernard Pass** *Fr.* Col du Petit St-Bernard, *It.* Colle del Piccolo San Bernardo. *pass* France/Italy
36 K7 **Little Salt Lake** ◎ Utah, W USA
180 K8 **Little Sandy Desert** *desert* Western Australia
29 S13 **Little Sioux River** ≈ Iowa, C USA
38 E17 **Little Sitkin Island** *island* Aleutian Islands, Alaska, USA
11 O13 **Little Smoky** Alberta, W Canada *54°35′N 117°06′W*
11 O14 **Little Smoky** ≈ Alberta, W Canada
37 P3 **Little Snake River** ≈ Colorado, C USA
64 A12 **Little Sound** *bay* Bermuda, NW Atlantic Ocean
37 T4 **Littleton** Colorado, C USA *39°36′N 105°01′W*
19 N7 **Littleton** New Hampshire, NE USA *44°18′N 71°46′W*
18 D11 **Little Valley** New York, NE USA *42°15′N 78°47′W*
30 M15 **Little Wabash River** ≈ Illinois, N USA
14 D10 **Little White River** ≈ Ontario, S Canada
28 M12 **Little White River** ≈ South Dakota, N USA
25 R5 **Little Wichita River** ≈ Texas, SW USA
142 I4 **Little Zab** *Ar.* Nahraz Zāb aş Şaghīr, *Kurd.* Zē-i Kōya, *Per.* Rūdkhāneh-ye Zāb-e Kūchek. ≈ Iran/Iraq
79 D15 **Littoral** ◈ *province* W Cameroon
Littoria *see* Latina
Litva/Litwa *see* Lithuania
111 B15 **Litvínov** *Ger.* Leutensdorf. Ústecký Kraj, NW Czech Republic *50°38′N 13°30′E*
116 M6 **Lityn** Vinnyts'ka Oblast', C Ukraine *49°19′N 28°06′E*
Liu-chou/Liuchow *see* Liuzhou
163 W11 **Liuhe** Jilin, NE China *42°15′N 125°49′E*
Liujiaxia *see* Yongjing
Liulin *see* Jonê
Liupanshui *see* Lupanshui
83 Q15 **Liúpo** Nampula, NE Mozambique *15°36′S 39°57′E*
83 G14 **Liuwa Plain** *plain* W Zambia
160 L13 **Liuzhou** *var.* Liu-chou, Liuchow. Guangxi Zhuangzu Zizhiqu, S China *24°09′N 108°55′E*
116 H8 **Livada** *Hung.* Sárköz. Satu Mare, NW Romania *47°52′N 23°04′E*
115 J20 **Liváda, Akrotírio** *headland* Tínos, Kykládes, Greece, Aegean Sea *37°36′N 25°15′E*
115 F18 **Livádeia** *prev.* Levádia. Stereá Elláda, C Greece *38°24′N 22°51′E*
Livádi *see* Liádi
Livanátai *see* Livanátes
115 G18 **Livanátes** *prev.* Livanátai. Stereá Elláda, C Greece *38°43′N 23°03′E*
118 I10 **Līvāni** *Ger.* Lievenhof. SE Latvia *56°22′N 26°12′E*
65 E25 **Lively Island** *island* SE Falkland Islands
65 D25 **Lively Sound** *sound* SE Falkland Islands
39 R8 **Livengood** Alaska, USA *65°31′N 148°32′W*
106 I7 **Livenza** ≈ NE Italy
35 O6 **Live Oak** California, W USA *39°17′N 121°41′W*
23 U9 **Live Oak** Florida, SE USA *30°18′N 82°59′W*
35 O9 **Livermore** California, W USA *37°40′N 121°46′W*
20 I6 **Livermore** Kentucky, S USA *37°31′N 87°08′W*
19 Q7 **Livermore Falls** Maine, NE USA *44°30′N 70°09′W*
24 J10 **Livermore, Mount** ▲ Texas, SW USA *30°37′N 104°10′W*
13 P16 **Liverpool** Nova Scotia, SE Canada *44°03′N 64°43′W*
97 K17 **Liverpool** NW England, United Kingdom *53°25′N 02°55′W*
183 S7 **Liverpool Range** ▲▲ New South Wales, SE Australia
42 F4 **Livingston** Izabal, E Guatemala *15°50′N 88°44′W*
96 J12 **Livingston** C Scotland, United Kingdom *55°51′N 03°31′W*
23 N5 **Livingston** Alabama, S USA *32°35′N 88°12′W*
35 P9 **Livingston** California, W USA *37°22′N 120°45′W*
22 J8 **Livingston** Louisiana, S USA *30°30′N 90°45′W*
33 S11 **Livingston** Montana, NW USA *45°40′N 110°33′W*
20 L8 **Livingston** Tennessee, S USA *36°22′N 85°20′W*
25 W9 **Livingston** Texas, SW USA *30°42′N 94°58′W*
83 I16 **Livingstone** *var.* Maramba. Southern, S Zambia *17°51′S 25°48′E*
185 B22 **Livingstone Mountains** ▲▲ South Island, New Zealand
80 K13 **Livingstone Mountains** ▲▲ S Tanzania
82 N12 **Livingstonia** Northern, N Malawi *10°29′S 34°06′E*
194 G4 **Livingston Island** *island* Antarctica
25 W9 **Livingston, Lake** ⊡ Texas, SW USA
112 F13 **Livno** ◈ Federicija Bosna I Hercegovina, SW Bosnia and Herzegovina
126 K7 **Livny** Orlovskaya Oblast', W Russian Federation *52°25′N 37°42′E*
93 M14 **Livojoki** ≈ C Finland
31 R10 **Livonia** Michigan, N USA *42°22′N 83°22′W*
106 E11 **Livorno** *Eng.* Leghorn. Toscana, C Italy *43°32′N 10°18′E*
Livramento *see* Santana do Livramento
141 U8 **Līwā** *var.* Al Liwā'. *oasis region* S United Arab Emirates
81 I24 **Liwale** Lindi, SE Tanzania *09°46′S 37°56′E*
159 W9 **Liwang** Ningxia, N China *36°42′N 106°05′E*
83 N15 **Liwonde** Southern, S Malawi *15°01′S 35°15′E*
159 V11 **Lixian** *var.* Li Xian. Gansu, C China *34°15′N 105°07′E*
160 H8 **Lixian** *var.* Li Xian, Zagunao. Sichuan, C China *31°27′N 103°06′E*
Li Xian *see* Lixian
Lixian Jiang *see* Black River
115 B18 **Lixoúri** *prev.* Lixoúrion. Kefallinía, Iónia Nisiá, Greece, C Mediterranean Sea *38°14′N 20°24′E*
Lixoúrion *see* Lixoúri
Lixus *see* Larache
33 U15 **Lizard Head Peak** ▲ Wyoming, C USA *42°47′N 109°12′W*
97 H25 **Lizard Point** *headland* SW England, United Kingdom *49°57′N 05°12′W*
Lizarra *see* Estella
112 L12 **Ljig** Serbia, C Serbia *44°14′N 20°16′E*
Ljouwert *see* Leeuwarden
Ljubelj *see* Loibl Pass
109 U11 **Ljubljana** *Ger.* Laibach, *It.* Lubiana; *anc.* Aemona, Emona. ● (Slovenia) C Slovenia *46°03′N 14°29′E*
109 T11 **Ljubljana** ✈ C Slovenia *46°14′N 14°26′E*
113 N17 **Ljuboten** *Alb.* Luboten. ▲ S Serbia *42°12′N 21°06′E*
95 P19 **Ljugarn** Gotland, SE Sweden *57°23′N 18°45′E*
84 G7 **Ljungan** ≈ N Sweden
93 F17 **Ljungan** ≈ C Sweden
95 K21 **Ljungby** Kronoberg, S Sweden *56°49′N 13°55′E*
95 M17 **Ljungsbro** Östergötland, S Sweden *58°31′N 15°30′E*
95 I18 **Ljungskile** Västra Götaland, S Sweden *58°14′N 11°55′E*
94 M11 **Ljusdal** Gävleborg, C Sweden *61°50′N 16°10′E*
94 M11 **Ljusnan** ≈ C Sweden
94 N12 **Ljusne** Gävleborg, C Sweden *61°11′N 17°07′E*
95 P15 **Ljusterö** Stockholm, C Sweden *59°31′N 18°40′E*
109 X9 **Ljutomer** *Ger.* Luttenberg. NE Slovenia *46°31′N 16°12′E*
63 G15 **Llaima, Volcán** ⋏ S Chile *39°01′S 71°38′W*
105 X4 **Llançà** *var.* Llansá. Cataluña, NE Spain *42°23′N 03°08′E*
97 J21 **Llandovery** C Wales, United Kingdom *52°01′N 03°47′W*
97 J20 **Llandrindod Wells** E Wales, United Kingdom *52°15′N 03°23′W*
97 J18 **Llandudno** N Wales, United Kingdom *53°19′N 03°49′W*
97 I21 **Llanelli** *prev.* Llanelly. SW Wales, United Kingdom *51°41′N 04°12′W*
Llanelly *see* Llanelli
104 M2 **Llanes** Asturias, N Spain *43°25′N 04°46′W*
97 K19 **Llangollen** NE Wales, United Kingdom *52°58′N 03°10′W*
104 K2 **Llangréu** *var.* Langreo, Sama de Langreo. Asturias, N Spain *43°18′N 05°40′W*
25 R10 **Llano** Texas, SW USA *30°49′N 98°42′W*
25 Q10 **Llano River** ≈ Texas, SW USA
54 I9 **Llanos** *physical region* Colombia/Venezuela
63 G16 **Llanquihue, Lago** ◎ S Chile
Llansá *see* Llançà
105 U5 **Lleida** *Cast.* Lérida; *anc.* Ilerda. Cataluña, NE Spain *41°38′N 00°35′E*
104 K12 **Llerena** Extremadura, W Spain *38°13′N 06°00′W*
Linki Kurezûr, Chiyā-î *see* Linkî, Chiyā-î
105 S9 **Llíria** Valenciana, E Spain *39°38′N 00°36′W*
105 W4 **Llívia** Cataluña, NE Spain *42°27′N 02°00′E*
Llodio *see* Laudio
105 X5 **Lloret de Mar** Cataluña, NE Spain *41°42′N 02°51′E*
Llorri *see* Toretta de l'Orrí
10 L11 **Lloyd George, Mount** ▲ British Columbia, W Canada *57°46′N 124°57′W*
11 R14 **Lloydminster** Alberta/ Saskatchewan, SW Canada *53°18′N 110°00′W*
104 K2 **Lluanco** *var.* Luanco. Asturias, N Spain *43°36′N 05°48′W*
105 X9 **Llucmajor** Mallorca, Spain, W Mediterranean Sea *39°29′N 02°53′E*
36 L6 **Loa** Utah, W USA *38°24′N 111°38′W*
169 S8 **Loagan Bunut** ◎ East Malaysia
38 G12 **Loa, Mauna** ▲ Hawai'i, USA *19°28′N 155°39′W*
Loanda *see* Luanda
79 J22 **Loange** ≈ S Dem. Rep. Congo
79 E21 **Loango** Kouilou, S Congo *04°38′S 11°50′E*
106 B10 **Loano** Liguria, NW Italy *44°07′N 08°15′E*
62 H4 **Loa, Río** ≈ N Chile
83 I20 **Lobatse** *var.* Lobatsi. Kgatleng, SE Botswana *25°11′S 25°40′E*
Lobatsi *see* Lobatse
101 Q15 **Löbau** Sachsen, E Germany *51°07′N 14°40′E*
79 H16 **Lobaye** ◈ *prefecture* SW Central African Republic
79 I16 **Lobaye** ≈ SW Central African Republic
99 G21 **Lobbes** Hainaut, S Belgium *50°21′N 04°16′E*
61 D23 **Lobería** Buenos Aires, E Argentina *38°08′S 58°48′W*
110 F8 **Łobez** *Ger.* Labes. Zacodnio-pomorskie, NW Poland *53°38′N 15°39′E*
82 A13 **Lobito** Benguela, W Angola *12°20′S 13°34′E*
Lobkovichi *see* Labkovichy
Lob Nor *see* Lop Nur
171 V13 **Lobo** Papua Barat, E Indonesia *03°41′S 134°06′E*
104 J11 **Lobón** Extremadura, W Spain *38°51′N 06°38′W*
61 D20 **Lobos** Buenos Aires, E Argentina *35°11′S 59°08′W*
40 E4 **Lobos, Cabo** *headland* NW Mexico *29°53′N 112°43′W*
40 F6 **Lobos, Isla** *island* NW Mexico
Lobositz *see* Lovosice
Lobsens *see* Łobżenica
Loburi *see* Lop Buri
110 H9 **Łobżenica** *Ger.* Lobsens. Wielkopolskie, C Poland *53°19′N 17°11′E*
108 G11 **Locarno** *Ger.* Luggarus. Ticino, S Switzerland *46°11′N 08°48′E*
96 E9 **Lochboisdale** NW Scotland, United Kingdom *57°08′N 07°17′W*
98 N11 **Lochem** Gelderland, E Netherlands *52°10′N 06°25′E*
102 M8 **Loches** Indre-et-Loire, C France *47°08′N 01°00′E*
Loch Garman *see* Wexford
96 H12 **Lochgilphead** W Scotland, United Kingdom *56°02′N 05°27′W*
96 H7 **Lochinver** N Scotland, United Kingdom *58°10′N 05°15′W*
96 F8 **Lochmaddy** NW Scotland, United Kingdom *57°35′N 07°10′W*
96 J10 **Lochnagar** ▲ C Scotland, United Kingdom *56°58′N 03°09′W*
99 E17 **Lochristi** Oost-Vlaanderen, NW Belgium *51°07′N 03°49′E*
96 I9 **Lochy, Loch** ◎ N Scotland, United Kingdom
182 G8 **Lock** South Australia *33°37′S 135°45′E*
97 J14 **Lockerbie** S Scotland, United Kingdom *55°11′N 03°27′W*
27 S13 **Lockesburg** Arkansas, C USA *33°58′N 94°10′W*
183 P10 **Lockhart** New South Wales, SE Australia *35°15′S 146°43′E*
25 S11 **Lockhart** Texas, SW USA *29°54′N 97°41′W*
18 F13 **Lock Haven** Pennsylvania, NE USA *41°08′N 77°27′W*
25 N4 **Lockney** Texas, SW USA *34°06′N 101°27′W*
100 O12 **Löcknitz** ≈ NE Germany
18 E9 **Lockport** New York, NE USA *43°09′N 78°40′W*
167 T13 **Lôc Ninh** Sông Be, S Vietnam *11°51′N 106°35′E*
107 N23 **Locri** Calabria, SW Italy *38°16′N 16°16′E*
Locse *see* Levoča
27 T2 **Locust Creek** ≈ Missouri, C USA
23 P3 **Locust Fork** ≈ Alabama, S USA
27 Q9 **Locust Grove** Oklahoma, C USA *36°12′N 95°10′W*
94 E11 **Lodalskåpa** ▲ S Norway *61°47′N 07°10′E*
183 N10 **Loddon River** ≈ Victoria, SE Australia
Lodensee *see* Klooga
103 P15 **Lodève** *anc.* Luteva. Hérault, S France *43°44′N 03°19′E*
124 I12 **Lodeynoye Pole** Leningradskaya Oblast', NW Russian Federation *60°41′N 33°29′E*
33 V11 **Lodge Grass** Montana, NW USA *45°19′N 107°20′W*
28 J15 **Lodgepole Creek** ≈ Nebraska/Wyoming, C USA
149 T11 **Lodhrān** Punjab, E Pakistan *29°32′N 71°40′E*
106 D8 **Lodi** Lombardia, NW Italy *45°19′N 09°30′E*
35 O8 **Lodi** California, W USA *38°07′N 121°17′W*
31 T12 **Lodi** Ohio, N USA *41°00′N 82°01′W*
92 H10 **Lødingen** *Lapp.* Lådik. Nordland, C Norway *68°25′N 16°00′E*
79 L20 **Lodja** Kasai-Oriental, C Dem. Rep. Congo *03°29′S 23°25′E*
37 O3 **Lodore, Canyon of** *canyon* Colorado, C USA
105 Q4 **Lodosa** Navarra, N Spain *42°26′N 02°05′W*
81 H16 **Lodwar** Turkana, NW Kenya *03°06′N 35°38′E*
110 K13 **Łódź** *Rus.* Lodz. Łódź, C Poland *51°51′N 19°26′E*
110 E12 **Łódzkie** ◈ *province* C Poland
167 P8 **Loei** *var.* Loey, Muang Loei. Loei, C Thailand *17°32′N 101°40′E*
98 I11 **Loenen** Utrecht, C Netherlands *52°13′N 05°01′E*
167 R9 **Loeng Nok Tha** Yasothon, E Thailand *16°12′N 104°31′E*
83 F24 **Loeriesfontein** Northern Cape, W South Africa *30°59′S 19°29′E*
Loewoek *see* Luwuk
Loey *see* Loei
76 J16 **Lofa** ≈ N Liberia
109 P6 **Lofer** Salzburg, C Austria *47°37′N 12°42′E*
92 F11 **Lofoten** *var.* Lofoten Islands. *island group* C Norway
Lofoten Islands *see* Lofoten
95 N18 **Loftahammar** Kalmar, S Sweden *57°55′N 16°45′E*
127 O10 **Log** Volgogradskaya Oblast', SW Russian Federation *49°32′N 43°52′E*
77 S12 **Loga** Dosso, SW Niger *13°40′N 03°15′E*
29 S14 **Logan** Iowa, C USA *41°38′N 95°47′W*
26 K3 **Logan** Kansas, C USA *39°39′N 99°34′W*
31 T14 **Logan** Ohio, N USA *39°32′N 82°24′W*
36 L1 **Logan** Utah, W USA *41°45′N 111°50′W*
21 P6 **Logan** West Virginia, NE USA *37°52′N 82°00′W*
35 Y10 **Logandale** Nevada, W USA *36°36′N 114°28′W*
19 O11 **Logan International** ✈ (Boston) Massachusetts, NE USA *42°22′N 71°00′W*
11 N16 **Logan Lake** British Columbia, SW Canada *50°28′N 120°42′W*
23 Q4 **Logan Martin Lake** ⊡ Alabama, S USA
10 G8 **Logan, Mount** ▲ Yukon, W Canada *60°32′N 140°34′W*
32 I7 **Logan, Mount** ▲ Washington, NW USA *48°32′N 120°57′W*
33 P7 **Logan Pass** *pass* Montana, NW USA
31 O12 **Logansport** Indiana, N USA *40°44′N 86°25′W*
22 F6 **Logansport** Louisiana, S USA *31°58′N 94°00′W*
149 Q5 **Lōgar** *prev.* Lowgar. ◈ *province* E Afghanistan
67 R11 **Loge** ≈ NW Angola
Logishin *see* Lahishyn
Log na Coille *see* Lugnaquillia Mountain
78 G11 **Logone** *var.* Lagone. ≈ Cameroon/Chad
78 G13 **Logone-Occidental** *off.* Région du Logone-Occidental. ◈ *region* SW Chad
78 H13 **Logone Occidental** ≈ SW Chad
Logone-Occidental, Région du *see* Logone-Occidental
78 G13 **Logone-Oriental** *off.* Région du Logone-Oriental. ◈ *region* SW Chad
78 H13 **Logone Oriental** ≈ SW Chad
Logone Oriental *see* Pendé
Logone-Oriental, Région du *see* Logone-Oriental
L'Ogooué-Ivindo *see* Ogooué-Ivindo
L'Ogooué-Lolo *see* Ogooué-Lolo
L'Ogooué-Maritime *see* Ogooué-Maritime
Logoysk *see* Lahoysk
105 P4 **Logroño** *anc.* Vareia, *Lat.* Juliobriga. La Rioja, N Spain *42°28′N 02°26′W*
104 L10 **Logrosán** Extremadura, W Spain *39°21′N 05°29′W*
95 G20 **Løgstør** Nordjylland, N Denmark *56°57′N 09°19′E*
95 H22 **Løgten** Midtjylland, C Denmark *56°17′N 10°20′E*
95 F24 **Løgumkloster** Syddanmark, SW Denmark *55°04′N 08°58′E*
Lögurinn *see* Lagarfljót
153 P15 **Lohārdaga** Jhārkhand, N India *23°27′N 84°42′E*
152 H10 **Lohāru** Haryāna, N India *28°28′N 75°50′E*
101 D15 **Lohausen** ✈ (Düsseldorf) Nordrhein-Westfalen, W Germany *51°18′N 06°51′E*
189 O14 **Lohd** Pohnpei, E Micronesia
92 L12 **Lohiniva** Lappi, N Finland *67°09′N 25°04′E*
Lohiszyn *see* Lahishyn
93 L20 **Lohja** *var.* Lojo. Uusimaa, S Finland *60°14′N 24°07′E*
169 V11 **Lohjanan** Borneo, C Indonesia
25 Q9 **Lohn** Texas, SW USA *31°15′N 99°22′W*
100 G12 **Löhne** Niedersachsen, NW Germany *52°40′N 08°13′E*
Lohr *see* Lohr am Main
101 I18 **Lohr am Main** *var.* Lohr. Bayern, C Germany *50°00′N 09°30′E*
109 T10 **Loibl Pass** *Ger.* Loiblpass, *Slvn.* Ljubelj. *pass* Austria/ Slovenia
Loiblpass *see* Loibl Pass
167 N6 **Loikaw** Kayah State, C Myanmar (Burma) *19°40′N 97°17′E*
93 K19 **Loimaa** Varsinais-Suomi, SW Finland *60°51′N 23°03′E*
103 O6 **Loing** ≈ C France
167 R6 **Loi, Phou** ▲ N Laos *20°18′N 103°14′E*
102 L7 **Loir** ≈ C France
103 Q11 **Loire** ◈ *department* E France
102 M7 **Loire** *var.* Liger. ≈ C France
102 I7 **Loire-Atlantique** ◈ *department* NW France
103 O7 **Loiret** ◈ *department* C France
102 M8 **Loir-et-Cher** ◈ *department* C France
101 L24 **Loisach** ≈ SE Germany
56 B9 **Loja** Loja, S Ecuador *03°59′S 79°16′W*
104 M14 **Loja** Andalucía, S Spain *37°10′N 04°09′W*
56 B9 **Loja** ◈ *province* S Ecuador
Lojo *see* Lohja
116 J4 **Lokachi** Volyns'ka Oblast', NW Ukraine *50°44′N 24°39′E*
79 M20 **Lokandu** Maniema, C Dem. Rep. Congo *02°34′S 25°44′E*
92 M11 **Lokan Tekojärvi** ◎ NE Finland
137 Z11 **Lökbatan** *Rus.* Lokbatan. E Azerbaijan *40°21′N 49°43′E*
Lokbatan *see* Lökbatan
99 F17 **Lokeren** Oost-Vlaanderen, NW Belgium *51°06′N 03°59′E*
Lokhvitsa *see* Lokhvytsya
117 S4 **Lokhvytsya** *Rus.* Lokhvitsa. Poltavs'ka Oblast', NE Ukraine *50°22′N 33°16′E*
81 H17 **Lokichar** Turkana, NW Kenya *02°23′N 35°40′E*
81 G16 **Lokichokio** Turkana, NW Kenya *04°16′N 34°22′E*
81 H16 **Lokitaung** Turkana, NW Kenya *04°15′N 35°45′E*
92 M11 **Lokka** Lappi, N Finland *67°48′N 27°41′E*
94 G8 **Løkken Verk** Sør-Trøndelag, S Norway *63°06′N 09°43′E*
124 G16 **Loknya** Pskovskaya Oblast', W Russian Federation *56°48′N 30°08′E*
77 V15 **Loko** Nassarawa, C Nigeria *08°00′N 07°48′E*
77 U15 **Lokoja** Kogi, C Nigeria *07°48′N 06°45′E*
81 H17 **Lokori** Turkana, W Kenya *01°56′N 36°03′E*
77 R16 **Lokossa** S Benin *06°38′N 01°43′E*
118 I3 **Loksa** *Ger.* Loxa. Harjumaa, NW Estonia *59°32′N 25°42′E*
9 T7 **Loks Land** *island* Nunavut, NE Canada
80 C13 **Lol** ≈ NW South Sudan
76 K15 **Lola** SE Guinea *07°52′N 08°29′W*
35 Q5 **Lola, Mount** ▲ California, W USA *39°27′N 120°20′W*
81 H20 **Loliondo** Arusha, NE Tanzania *02°03′S 35°46′E*
95 H25 **Lolland** *prev.* Laaland. *island* S Denmark
186 G6 **Lolobau Island** *island* E Papua New Guinea
79 E16 **Lolodorf** Sud, SW Cameroon *03°17′N 10°50′E*
114 G7 **Lom** *prev.* Lom-Palanka. Montana, NW Bulgaria *43°49′N 23°16′E*
114 G7 **Lom** ≈ NW Bulgaria
79 M19 **Lomami** ≈ C Dem. Rep. Congo
57 I17 **Lomas** Arequipa, SW Peru *15°29′S 74°54′W*
63 I23 **Lomas, Bahía** *bay* S Chile
61 D20 **Lomas de Zamora** Buenos Aires, E Argentina *34°53′S 58°26′W*
61 D20 **Loma Verde** Buenos Aires, E Argentina *35°16′S 58°24′W*
180 K4 **Lombadina** Western Australia *16°39′S 122°54′E*
106 E6 **Lombardia** *Eng.* Lombardy. ◈ *region* N Italy
Lombardy *see* Lombardia
102 M15 **Lombez** Gers, S France *43°29′N 00°54′E*
171 Q16 **Lomblen, Pulau** *island* Nusa Tenggara, S Indonesia
173 W7 **Lombok Basin** *undersea feature* E Indian Ocean *09°50′S 116°00′E*
170 L16 **Lombok, Pulau** *island* Nusa Tenggara, C Indonesia
77 Q16 **Lomé** ● (Togo) S Togo *06°08′N 01°13′E*
77 Q16 **Lomé** ✈ S Togo *06°08′N 01°13′E*
79 L19 **Lomela** Kasai-Oriental, C Dem. Rep. Congo *02°19′S 23°15′E*
25 R9 **Lometa** Texas, SW USA *31°13′N 98°23′W*
79 F16 **Lomié** Est, SE Cameroon *03°09′N 13°35′E*
30 M8 **Lomira** Wisconsin, N USA *43°36′N 88°26′W*
95 K23 **Lomma** Skåne, S Sweden *55°41′N 13°05′E*
99 J16 **Lommel** Limburg, N Belgium *51°14′N 05°19′E*
96 I11 **Lomond, Loch** ◎ C Scotland, United Kingdom
197 R9 **Lomonosov Ridge** *var.* Harris Ridge, *Rus.* Khrebet Homonsova. *undersea feature* Arctic Ocean *88°00′N 140°00′E*
Lomonsova, Khrebet *see* Lomonosov Ridge
Lom-Palanka *see* Lom
Lomphat *see* Lumphăt
35 P14 **Lompoc** California, W USA *34°39′N 120°29′W*
167 P9 **Lom Sak** *var.* Muang Lom Sak. Phetchabun, C Thailand *16°45′N 101°12′E*
110 N9 **Łomża** *Rus.* Lomzha. Podlaskie, NE Poland *53°11′N 22°04′E*
Lomzha *see* Łomża
Lonaula *see* Lonāvale
155 D14 **Lonāvale** *prev.* Lonaula. Mahārāshtra, W India *18°45′N 73°27′E*
63 G15 **Loncoche** Araucanía, C Chile *39°22′S 72°34′W*
63 H14 **Loncopue** Neuquén, W Argentina *38°04′S 70°43′W*
99 G17 **Londerzeel** Vlaams Brabant, C Belgium *51°00′N 04°19′E*
Londinium *see* London
14 E16 **London** Ontario, S Canada *42°59′N 81°13′W*
191 Y2 **London** Kiritimati, E Kiribati *02°00′N 157°28′W*
97 O22 **London** *anc.* Augusta, *Lat.* Londinium. ● (United Kingdom) SE England, United Kingdom *51°30′N 00°10′W*
21 N7 **London** Kentucky, S USA *37°07′N 84°05′W*
31 S13 **London** Ohio, NE USA *39°52′N 83°27′W*
25 Q10 **London** Texas, SW USA *30°40′N 99°33′W*
97 O22 **London City** ✈ SE England, United Kingdom *51°31′N 00°07′E*
97 E14 **Londonderry** *var.* Derry, *Ir.* Doire. NW Northern Ireland, United Kingdom *55°00′N 07°19′W*
97 F14 **Londonderry** *cultural region* NW Northern Ireland, United Kingdom
180 M2 **Londonderry, Cape** *cape* Western Australia
63 H25 **Londonderry, Isla** *island* S Chile
43 O7 **Londres, Cayos** *reef* NE Nicaragua
60 I10 **Londrina** Paraná, S Brazil *23°18′S 51°13′W*
27 N13 **Lone Grove** Oklahoma, C USA *34°11′N 97°15′W*
14 E12 **Lonely Island** *island* Ontario, S Canada
35 T8 **Lone Mountain** ▲ Nevada, W USA *38°01′N 117°28′W*
25 V6 **Lone Oak** Texas, SW USA *33°02′N 95°58′W*
35 T11 **Lone Pine** California, W USA *36°36′N 118°04′W*
Lone Star State *see* Texas
Lone Tree Islet *see* Iku
83 D14 **Longa** Kuando Kubango, C Angola *14°44′S 18°36′E*
82 B12 **Longa** ≈ W Angola
83 E15 **Longa** ≈ SE Angola
Long'an *see* Pingwu
163 W11 **Longan Shan** ▲▲ NE China
197 S4 **Longa, Proliv** *Eng.* Long Strait. *strait* NE Russian Federation
44 J13 **Long Bay** *bay* W Jamaica
21 V13 **Long Bay** *bay* North Carolina/South Carolina, E USA
35 T16 **Long Beach** California, W USA *33°46′N 118°11′W*
22 M9 **Long Beach** Mississippi, S USA *30°21′N 89°09′W*
18 L14 **Long Beach** Long Island, New York, NE USA *40°34′N 73°38′W*
32 F9 **Long Beach** Washington, NW USA *46°21′N 124°03′W*
18 K16 **Long Beach Island** *island* New Jersey, NE USA
65 M25 **Longbluff** *headland* SW Tristan da Cunha
23 U13 **Longboat Key** *island* Florida, SE USA
18 K15 **Long Branch** New Jersey, NE USA *40°18′N 73°59′W*
44 J5 **Long Cay** *island* SE The Bahamas
Longcheng *see* Xiaoxian
161 P14 **Longchuan** *var.* Laolong. Guangdong, S China *24°07′N 115°10′E*
Longchuan *see* Nanhua
Longchuan Jiang *see* Shweli
32 K12 **Long Creek** Oregon, NW USA *44°40′N 119°07′W*
159 W10 **Longde** Ningxia, N China *35°37′N 106°07′E*
183 P16 **Longford** Tasmania, SE Australia *41°41′S 147°03′E*
97 D17 **Longford** *Ir.* An Longfort. Longford, C Ireland *53°45′N 07°50′W*
97 E17 **Longford** *Ir.* An Longfort. *cultural region* C Ireland
Longgang *see* Dazu
161 P1 **Longhua** Hebei, E China *41°18′N 117°44′E*
169 U11 **Longiram** Borneo, C Indonesia *0°02′S 115°36′E*
12 H8 **Long Island** *island* Nunavut, C Canada
186 D7 **Long Island** *var.* Arop Island. *island* N Papua New Guinea
44 J4 **Long Island** *island* C The Bahamas
18 L14 **Long Island** *island* New York, NE USA
Long Island *see* Bermuda
18 M14 **Long Island Sound** *sound* NE USA
163 U7 **Longjiang** Heilongjiang, NE China *47°20′N 123°09′E*
161 K13 **Long Jiang** ≈ S China
163 Y10 **Longjing** *var.* Yanji. Jilin, NE China *42°48′N 129°26′E*
161 R4 **Longkou** Shandong, E China *37°40′N 120°21′E*
12 E11 **Longlac** Ontario, S Canada *49°47′N 86°34′W*
19 S1 **Long Lake** ◎ Maine, NE USA
31 O6 **Long Lake** ◎ Michigan, N USA
31 R5 **Long Lake** ◎ Michigan, N USA
29 N6 **Long Lake** ◎ North Dakota, N USA
30 J4 **Long Lake** ⊡ Wisconsin, N USA
99 K23 **Longlier** Luxembourg, SE Belgium *49°51′N 05°27′E*
160 I13 **Longlin** *var.* Longlin Gezu Zizhixian, Xinzhou. Guangxi Zhuangzu Zizhiqu, S China *24°46′N 105°19′E*
Longlin Gezu Zizhixian *see* Longlin
37 T3 **Longmont** Colorado, C USA *40°09′N 105°07′W*
157 P10 **Longnan** *var.* Wudu. Gansu, C China *33°27′N 104°57′E*
29 N13 **Long Pine** Nebraska, C USA *42°32′N 99°42′W*
Longping *see* Luodian
14 F17 **Long Point** *headland* Ontario, S Canada *42°33′N 80°15′W*
14 K15 **Long Point** *headland* Ontario, SE Canada *43°56′N 76°53′W*
184 P10 **Long Point** *headland* North Island, New Zealand *39°07′S 177°41′E*
30 L2 **Long Point** *headland* Michigan, N USA *47°50′N 89°09′W*
14 G17 **Long Point Bay** *lake bay* Ontario, S Canada
29 T7 **Long Prairie** Minnesota, N USA *45°58′N 94°52′W*
Longquan *see* Fenggang
Longquan *see* Yanggao
13 S11 **Long Range Mountains** *hill range* Newfoundland and Labrador, E Canada
65 H25 **Long Range Point** *headland* SE Saint Helena *16°00′S 05°41′W*
181 V8 **Longreach** Queensland, E Australia *23°31′S 144°18′E*
160 H7 **Longriba** Sichuan, C China *32°32′N 102°20′E*
160 L10 **Longshan** *var.* Min'an. Hunan, S China *29°25′N 109°28′E*
37 S3 **Longs Peak** ▲ Colorado, C USA *40°15′N 105°37′W*
Long Strait *see* Longa, Proliv
102 K8 **Longué** Maine-et-Loire, NW France *47°23′N 00°07′W*
13 P11 **Longue-Pointe** Québec, E Canada *50°20′N 64°13′W*
103 S4 **Longuyon** Meurthe-et-Moselle, NE France *49°25′N 05°37′E*
25 W7 **Longview** Texas, SW USA *32°30′N 94°45′W*
32 G10 **Longview** Washington, NW USA *46°08′N 122°56′W*
65 H25 **Longwood** C Saint Helena
25 P7 **Longworth** Texas, SW USA *32°37′N 100°20′W*
103 S3 **Longwy** Meurthe-et-Moselle, NE France *49°31′N 05°46′E*
159 V11 **Longxi** *var.* Gongchang. Gansu, C China *35°00′N 104°34′E*
Longxian *see* Wengyuan
167 S14 **Long Xuyên** *var.* Longxuyen. An Giang, S Vietnam *10°23′N 105°25′E*
Longxuyen *see* Long Xuyên
161 Q13 **Longyan** Fujian, SE China *25°06′N 117°02′E*
92 O3 **Longyearbyen** ○ (Svalbard) Spitsbergen, W Svalbard *78°12′N 15°39′E*
160 J15 **Longzhou** Guangxi Zhuangzu Zizhiqu, S China *22°22′N 106°46′E*
Longzhouping *see* Changyang
100 F12 **Löningen** Niedersachsen, NW Germany *52°43′N 07°42′E*
27 V11 **Lonoke** Arkansas, C USA *34°46′N 91°56′W*
95 L21 **Lönsboda** Skåne, S Sweden *56°24′N 14°19′E*
103 S9 **Lons-le-Saunier** *anc.* Ledo Salinarius. Jura, E France *46°41′N 05°32′E*
31 O15 **Loogootee** Indiana, N USA *38°40′N 86°54′W*
31 Q9 **Looking Glass River** ≈ Michigan, N USA
21 X11 **Lookout, Cape** *headland* North Carolina, SE USA *34°36′N 76°31′W*
39 O6 **Lookout Ridge** *ridge* Alaska, USA
Lookransar *see* Lūnkaransar
181 N11 **Loongana** Western Australia *30°53′S 127°15′E*
99 I14 **Loon op Zand** Noord-Brabant, S Netherlands *51°38′N 05°05′E*
97 A19 **Loop Head** *Ir.* Ceann Léime. *promontory* W Ireland
109 V4 **Loosdorf** Niederösterreich, NE Austria *48°13′N 15°25′E*
158 G10 **Lop** Xinjiang Uygur Zizhiqu, NW China *37°06′N 80°12′E*
112 J11 **Lopare** ◈ Republika Srpska, NE Bosnia and Herzegovina
Lopatichi *see* Lapatsichy
127 P7 **Lopatino** Penzenskaya Oblast', W Russian Federation *52°38′N 45°46′E*
167 P10 **Lop Buri** *var.* Loburi. Lop Buri, C Thailand *14°49′N 100°37′E*
25 R16 **Lopeno** Texas, SW USA *26°42′N 99°06′W*
79 C18 **Lopez, Cap** *headland* W Gabon *0°39′S 08°44′E*
98 I12 **Lopik** Utrecht, C Netherlands *51°58′N 04°57′E*
Lop Nor *see* Lop Nur
158 M7 **Lop Nur** *var.* Lob Nor, Lop Nor, Lo-pu Po. *seasonal lake* NW China
Lopnur *see* Yuli
79 K17 **Lopori** ≈ NW Dem. Rep. Congo
98 O5 **Loppersum** Groningen, NE Netherlands *53°20′N 06°45′E*
92 I8 **Lopphavet** *sound* N Norway
Lo-pu Po *see* Lop Nur
Lora *see* Lowrah
182 F3 **Lora Creek** *seasonal river* South Australia
104 K13 **Lora del Río** Andalucía, S Spain *37°39′N 05°32′W*
148 M11 **Lora, Hāmūn-i** *wetland* SW Pakistan
31 T11 **Lorain** Ohio, N USA *41°27′N 82°10′W*
25 O7 **Loraine** Texas, SW USA *32°24′N 100°42′W*
31 R13 **Loramie, Lake** ⊡ Ohio, N USA
105 Q13 **Lorca** *Ar.* Lurka; *anc.* Eliocroca, *Lat.* Illurco. Murcia, S Spain *37°40′N 01°41′W*
192 I10 **Lord Howe Island** *island* E Australia
Lord Howe Island *see* Ontong Java Atoll
175 O10 **Lord Howe Rise** *undersea feature* SW Pacific Ocean
192 J10 **Lord Howe Seamounts** *undersea feature* W Pacific Ocean
37 P15 **Lordsburg** New Mexico, SW USA *32°19′N 108°42′W*
186 E5 **Lorengau** *var.* Lorungau. Manus Island, N Papua New Guinea *02°01′S 147°15′E*
25 N5 **Lorenzo** Texas, SW USA *33°40′N 101°31′W*
142 K7 **Lorestān** *off.* Ostān-e Lorestān, *var.* Luristan. ◈ *province* W Iran
Lorestān, Ostān-e *see* Lorestān
57 M17 **Loreto** El Beni, N Bolivia *15°13′S 64°44′W*
106 J12 **Loreto** Marche, C Italy *43°25′N 13°37′E*
40 F8 **Loreto** Baja California Sur, NW Mexico *25°59′N 111°22′W*
40 M11 **Loreto** Zacatecas, C Mexico *22°15′N 102°00′W*
56 E9 **Loreto** *off.* Departamento de Loreto. ◈ *department* NE Peru
Loreto, Departamento de *see* Loreto
81 K18 **Lorian Swamp** *swamp* E Kenya
54 E6 **Lorica** Córdoba, NW Colombia *09°14′N 75°50′W*
102 G7 **Lorient** *prev.* l'Orient. Morbihan, NW France *47°45′N 03°22′W*
l'Orient *see* Lorient
111 K22 **Lőrinci** Heves, NE Hungary *47°46′N 19°40′E*
14 G11 **Loring** Ontario, S Canada *45°55′N 79°59′W*
33 V6 **Loring** Montana, NW USA *48°49′N 107°48′W*
103 R13 **Loriol-sur-Drôme** Drôme, E France *44°46′N 04°51′E*
21 U12 **Loris** South Carolina, SE USA *34°03′N 78°53′W*
57 I18 **Loriscota, Laguna** ◎ S Peru
183 N13 **Lorne** Victoria, SE Australia *38°33′S 143°57′E*
96 G11 **Lorn, Firth of** *inlet* W Scotland, United Kingdom
Loro Sae *see* East Timor
101 F24 **Lörrach** Baden-Württemberg, S Germany *47°38′N 07°40′E*

◆ Country
● Country Capital
◇ Dependent Territory
○ Dependent Territory Capital
◈ Administrative Regions
✈ International Airport
▲ Mountain
▲▲ Mountain Range
⋏ Volcano
≈ River
◎ Lake
⊡ Reservoir

103 *T5* **Lorraine** ⬖ *region* NE France
Lorungau *see* Lorengau
94 *L11* **Los** Gävleborg, C Sweden *61°43´N 15°15´E*
35 *P14* **Los Alamos** California, W USA *34°44´N 120°16´W*
37 *S10* **Los Alamos** New Mexico, SW USA *35°52´N 106°17´W*
42 *F5* **Los Amates** Izabal, E Guatemala *15°14´N 89°06´W*
63 *G14* **Los Ángeles** Bío Bío, C Chile *37°30´S 72°18´W*
35 *S15* **Los Angeles** California, W USA *34°03´N 118°15´W*
35 *S15* **Los Angeles** ✈ California, W USA *33°54´N 118°24´W*
35 *T13* **Los Angeles Aqueduct** *aqueduct* California, W USA
Losanna *see* Lausanne
63 *H20* **Los Antiguos** Santa Cruz, SW Argentina *46°36´S 71°31´W*
189 *Q16* **Losap Atoll** *atoll* C Micronesia
35 *P10* **Los Banos** California, W USA *37°00´N 120°39´W*
104 *K16* **Los Barrios** Andalucía, S Spain *36°11´N 05°30´W*
62 *L5* **Los Blancos** Salta, N Argentina *23°36´S 62°35´W*
42 *L12* **Los Chiles** Alajuela, NW Costa Rica *11°00´N 84°42´W*
105 *O2* **Los Corrales de Buelna** Cantabria, N Spain *43°15´N 04°04´W*
25 *T17* **Los Fresnos** Texas, SW USA *26°03´N 97°28´W*
35 *N9* **Los Gatos** California, W USA *37°13´N 121°58´W*
127 *P10* **Loshchina** Volgogradskaya Oblast', SW Russian Federation *48°58´N 46°14´E*
110 *O11* **Łosice** Mazowieckie, C Poland *52°13´N 22°42´E*
112 *B11* **Lošinj** *Ger.* Lussin, *It.* Lussino. *island* W Croatia
Los Jardines *see* Ngetik Atoll
63 *G15* **Los Lagos** Los Ríos, C Chile *39°50´S 72°50´W*
63 *F17* **Los Lagos** *off.* Regíon de los Lagos. ⬖ *region* C Chile
los Lagos, Regíon de *see* Los Lagos
Loslau *see* Wodzisław Śląski
64 *N11* **Los Llanos de Aridane** *var.* Los Llanos de Aridane. La Palma, Islas Canarias, Spain, NE Atlantic Ocean *28°39´N 17°54´W*
Los Llanos de Aridane *see* Los Llanos de Aridane
37 *R11* **Los Lunas** New Mexico, SW USA *34°48´N 106°43´W*
63 *I16* **Los Menucos** Río Negro, C Argentina *40°52´S 68°07´W*
40 *H8* **Los Mochis** Sinaloa, C Mexico *25°48´N 108°58´W*
35 *N4* **Los Molinos** California, W USA *40°00´N 122°05´W*
104 *M9* **Los Navalmorales** Castilla-La Mancha, C Spain *39°43´N 04°38´W*
25 *S15* **Los Olmos Creek** ≈ Texas, SW USA
Losonc/Losontz *see* Lučenec
167 *S5* **Lô, Sông** *var.* Panlong Jiang. ≈ China/Vietnam
44 *B5* **Los Palacios** Pinar del Río, W Cuba *22°35´N 83°16´W*
104 *K14* **Los Palacios y Villafranca** Andalucía, S Spain *37°10´N 05°55´W*
37 *R12* **Los Pinos Mountains** ▲▲ New Mexico, SW USA
37 *R11* **Los Ranchos de Albuquerque** New Mexico, SW USA *35°09´N 106°37´W*
40 *M14* **Los Reyes** Michoacán, SW Mexico *19°36´N 102°29´W*
63 *G15* **Los Ríos** ⬖ *region* C Chile
56 *B7* **Los Ríos** ⬖ *province* C Ecuador
64 *O11* **Los Rodeos** ✈ (Santa Cruz de Ternerife) Tenerife, Islas Canarias, Spain, NE Atlantic Ocean *28°27´N 16°20´W*
54 *L4* **Los Roques, Islas** *island group* N Venezuela
43 *S17* **Los Santos** Los Santos, S Panama *07°56´N 80°23´W*
43 *S17* **Los Santos** *off.* Provincia de Los Santos. ⬖ *province* S Panama
Los Santos *see* Los Santos de Maimona
104 *J12* **Los Santos de Maimona** *var.* Los Santos. Extremadura, W Spain *38°27´N 06°22´W*
Los Santos, Provincia de *see* Los Santos
98 *P10* **Losser** Overijssel, E Netherlands *52°16´N 06°25´E*
96 *J8* **Lossiemouth** NE Scotland, United Kingdom *57°43´N 03°18´W*
61 *B14* **Los Tábanos** Santa Fe, C Argentina *28°27´S 59°57´W*
54 *J4* **Los Taques** Falcón, N Venezuela *11°50´N 70°16´W*
14 *G11* **Lost Channel** Ontario, S Canada *45°54´N 80°20´W*
54 *L5* **Los Teques** Miranda, N Venezuela *10°25´N 67°01´W*
35 *Q12* **Lost Hills** California, W USA *35°35´N 119°40´W*
36 *I7* **Lost Peak** ▲ Utah, W USA *37°30´N 113°57´W*
33 *P11* **Lost Trail Pass** *pass* Montana, NW USA
186 *G9* **Losuia** Kiriwina Island, SE Papua New Guinea *08°29´S 151°03´E*
62 *G10* **Los Vilos** Coquimbo, C Chile *31°56´S 71°35´W*
105 *N10* **Los Yébenes** Castilla-La Mancha, C Spain *39°35´N 03°52´W*
103 *N13* **Lot** ⬖ *department* S France
103 *N13* **Lot** ≈ S France
63 *F14* **Lota** Bío Bío, C Chile *37°07´S 73°10´W*
81 *G15* **Lotagipi Swamp** *wetland* Kenya/Sudan
102 *K14* **Lot-et-Garonne** ⬖ *department* SW France
83 *K21* **Lothair** Mpumalanga, NE South Africa *26°23´S 30°26´E*
33 *R7* **Lothair** Montana, NW USA *48°28´N 111°15´W*
79 *L20* **Loto** Kasai-Oriental, C Dem. Rep. Congo *02°48´S 22°30´E*
108 *E10* **Lötschbergtunnel** *tunnel* Valais, SW Switzerland
25 *T9* **Lott** Texas, SW USA *31°12´N 97°02´W*
124 *H3* **Lotta** *var.* Lutto. ≈ Finland/Russian Federation
184 *Q7* **Lottin Point** *headland* North Island, New Zealand *37°26´S 178°07´E*
Lötzen *see* Giżycko
Loualaba *see* Lualaba
167 *P6* **Louangnamtha** *var.* Luong Nam Tha. Louang Namtha, N Laos *20°55´N 101°24´E*
167 *Q7* **Louangphabang** *var.* Louangphrabang, Luang Prabang. Louangphabang, N Laos *19°51´N 102°08´E*
Louangphrabang *see* Louangphabang
194 *H5* **Loubet Coast** *physical region* Antarctica
Loubomo *see* Dolisie
Louch *see* Loukhi
102 *H6* **Loudéac** Côtes d'Armor, NW France *48°11´N 02°45´W*
160 *M11* **Loudi** Hunan, S China *27°51´N 111°59´E*
79 *F21* **Loudima** Bouenza, S Congo *04°06´S 13°05´E*
20 *M9* **Loudon** Tennessee, S USA *35°43´N 84°19´W*
31 *T12* **Loudonville** Ohio, N USA *40°38´N 82°13´W*
102 *L8* **Loudun** Vienne, W France *47°01´N 00°05´E*
102 *K7* **Loué** Sarthe, NW France *48°00´N 00°14´W*
76 *G10* **Louga** NW Senegal *15°36´N 16°15´W*
97 *M19* **Loughborough** C England, United Kingdom *52°47´N 01°11´W*
97 *C18* **Loughrea** *Ir.* Baile Locha Riach. Galway, W Ireland *53°12´N 08°34´W*
103 *S9* **Louhans** Saône-et-Loire, C France *46°38´N 05°12´E*
21 *P5* **Louisa** Kentucky, S USA *38°06´N 82°37´W*
21 *V5* **Louisa** Virginia, NE USA *38°02´N 78°00´W*
21 *V9* **Louisburg** North Carolina, SE USA *36°05´N 78°18´W*
25 *U12* **Louise** Texas, SW USA *29°07´N 96°22´W*
15 *P11* **Louiseville** Québec, SE Canada *46°15´N 72°54´W*
27 *W3* **Louisiana** Missouri, C USA *39°25´N 91°03´W*
22 *G8* **Louisiana** *off.* State of Louisiana, *also known as* Creole State, Pelican State. ⬖ *state* S USA
83 *K19* **Louis Trichardt** *prev.* Makhado. Northern, NE South Africa *23°01´S 29°43´E*
23 *V4* **Louisville** Georgia, SE USA *33°00´N 82°24´W*
30 *M15* **Louisville** Illinois, N USA *38°46´N 88°32´W*
20 *K5* **Louisville** Kentucky, S USA *38°15´N 85°46´W*
22 *M4* **Louisville** Mississippi, S USA *33°07´N 89°03´W*
29 *S15* **Louisville** Nebraska, C USA *41°00´N 96°09´W*
192 *L11* **Louisville Ridge** *undersea feature* S Pacific Ocean
124 *J6* **Loukhi** *var.* Louch. Respublika Kareliya, NW Russian Federation *66°05´N 33°04´E*
79 *H19* **Loukoléla** Cuvette, E Congo *01°04´S 17°10´E*
104 *G14* **Loulé** Faro, S Portugal *37°08´N 08°02´W*
111 *C16* **Louny** *Ger.* Laun. Ústecký Kraj, NW Czech Republic *50°22´N 13°50´E*
29 *O15* **Loup City** Nebraska, C USA *41°16´N 98°58´W*
29 *P15* **Loup River** ≈ Nebraska, C USA
15 *S9* **Loup, Rivière du** ≈ Québec, SE Canada
12 *K7* **Loups Marins, Lacs des** ◎ Québec, NE Canada
102 *K16* **Lourdes** Hautes-Pyrénées, S France *43°06´N 00°03´W*
Lourenço Marques *see* Maputo
104 *F11* **Loures** Lisboa, C Portugal *38°50´N 09°10´W*
104 *F10* **Lourinhã** Lisboa, C Portugal *39°14´N 09°19´W*
115 *C16* **Loúros** ≈ W Greece
104 *G8* **Lousã** Coimbra, N Portugal *40°07´N 08°15´W*
Loushanguan *see* Tongzi
160 *M10* **Lou Shui** ≈ C China
183 *O5* **Louth** New South Wales, SE Australia *30°34´S 145°07´E*
97 *O18* **Louth** E England, United Kingdom *53°19´N 00°00´E*
97 *F17* **Louth** *Ir.* Lú. *cultural region* NE Ireland
115 *H15* **Loutrá** Kentrikí Makedonía, N Greece *39°55´N 23°37´E*
115 *G19* **Loutráki** Pelopónnisos, S Greece *37°55´N 22°55´E*
Louvain *see* Leuven
99 *H19* **Louvain-la Neuve** Walloon Brabant, C Belgium *50°39´N 04°36´E*
14 *J8* **Louvicourt** Québec, SE Canada *48°04´N 77°22´W*
102 *M4* **Louviers** Eure, N France *49°13´N 01°11´E*
30 *K14* **Lou Yaeger, Lake** ⊠ Illinois, N USA
93 *J15* **Lövånger** Västerbotten, N Sweden *64°22´N 21°19´E*
124 *J14* **Lovat'** ≈ NW Russian Federation
113 *J17* **Lovćen** ▲ SW Montenegro *42°22´N 18°49´E*
114 *I8* **Lovech** Lovech, N Bulgaria *43°08´N 24°45´E*
114 *I9* **Lovech** ⬖ *province* N Bulgaria
25 *V9* **Lovelady** Texas, SW USA *31°07´N 95°27´W*
37 *T3* **Loveland** Colorado, C USA *40°24´N 105°04´W*
33 *U12* **Lovell** Wyoming, C USA *44°50´N 108°23´W*
Lovello, Monte *see* Grosser Löffler
35 *S4* **Lovelock** Nevada, W USA *40°11´N 118°30´W*
106 *E7* **Lovere** Lombardia, N Italy *45°51´N 10°06´E*
30 *L10* **Loves Park** Illinois, N USA *42°19´N 89°03´W*
26 *M2* **Lovewell Reservoir** ⊠ Kansas, C USA
93 *M19* **Loviisa** *Swe.* Lovisa. Uusimaa, S Finland *60°27´N 26°15´E*
37 *V15* **Loving** New Mexico, SW USA *32°17´N 104°06´W*
21 *U6* **Lovingston** Virginia, NE USA *37°46´N 78°54´W*
37 *V14* **Lovington** New Mexico, SW USA *32°56´N 103°21´W*
Lovisa *see* Loviisa
111 *C15* **Lovosice** *Ger.* Lobositz. Ústecký Kraj, NW Czech Republic *50°30´N 14°02´E*
124 *K4* **Lovozero** Murmanskaya Oblast', NW Russian Federation *68°00´N 35°03´E*
124 *K4* **Lovozero, Ozero** ◎ NW Russian Federation
112 *B9* **Lovran** *It.* Laurana. Primorje-Gorski Kotar, NW Croatia *45°16´N 14°15´E*
116 *E11* **Lovrin** *Ger.* Lowrin. Timiş, W Romania *45°58´N 20°49´E*
82 *E10* **Lóvua** Lunda Norte, NE Angola *07°21´S 20°09´E*
82 *G12* **Lóvua** Moxico, E Angola *11°33´S 23°35´E*
65 *D25* **Low Bay** *bay* East Falkland, Falkland Islands
9 *P9* **Low, Cape** *headland* Nunavut, E Canada *63°05´N 85°27´W*
33 *N10* **Lowell** Idaho, NW USA *46°07´N 115°36´W*
19 *O10* **Lowell** Massachusetts, NE USA *42°38´N 71°19´W*
Löwen *see* Leuven
Löwenberg in Schlesien *see* Lwówek Śląski
Lower Austria *see* Niederösterreich
Lower Bann *see* Bann
Lower California *see* Baja California
Lower Danube *see* Niederösterreich
185 *L14* **Lower Hutt** Wellington, North Island, New Zealand *41°13´S 174°51´E*
39 *N11* **Lower Kalskag** Alaska, USA *61°30´N 160°28´W*
35 *O1* **Lower Klamath Lake** ◎ California, W USA
35 *Q2* **Lower Lake** ◎ California/Nevada, W USA
97 *E15* **Lower Lough Erne** ◎ SW Northern Ireland, United Kingdom
Lower Lusatia *see* Niederlausitz
Lower Normandy *see* Basse-Normandie
10 *K9* **Lower Post** British Columbia, W Canada *59°53´N 128°19´W*
29 *T4* **Lower Red Lake** ◎ Minnesota, N USA
Lower Rhine *see* Neder Rijn
Lower Saxony *see* Niedersachsen
97 *Q19* **Lowestoft** E England, United Kingdom *52°29´N 01°45´E*
Lowgar *see* Lōgar
182 *H7* **Low Hill** South Australia *32°17´S 136°46´E*
110 *K12* **Łowicz** Łódzkie, C Poland *52°06´N 19°55´E*
33 *N13* **Lowman** Idaho, NW USA *44°04´N 115°37´W*
149 *P8* **Lowrah** *var.* Lora. ≈ SE Afghanistan
Lowrin *see* Lovrin
183 *N17* **Low Rocky Point** *headland* Tasmania, SE Australia *42°59´S 145°28´E*
18 *I8* **Lowville** New York, NE USA *43°47´N 75°29´W*
Loxa *see* Loksa
182 *K9* **Loxton** South Australia *34°30´S 140°36´E*
81 *G21* **Loya** Tabora, C Tanzania *04°57´S 33°53´E*
30 *K6* **Loyal** Wisconsin, N USA *44°45´N 90°30´W*
18 *G13* **Loyalsock Creek** ≈ Pennsylvania, NE USA
35 *Q5* **Loyalton** California, W USA *39°39´N 120°16´W*
Lo-yang *see* Luoyang
187 *Q16* **Loyauté, Îles** *island group* S New Caledonia
Loyev *see* Loyew
119 *O20* **Loyew** *Rus.* Loyev. Homyel'skaya Voblasts', SE Belarus *51°56´N 30°48´E*
125 *S13* **Loyno** Kirovskaya Oblast', NW Russian Federation *59°44´N 52°42´E*
103 *P13* **Lozère** ⬖ *department* S France
103 *Q14* **Lozère, Mont** ▲ S France *44°27´N 03°44´E*
112 *J11* **Loznica** Serbia, W Serbia *44°32´N 19°13´E*
114 *L8* **Loznitsa** Razgrad, N Bulgaria *43°22´N 26°36´E*
117 *V7* **Lozova** *Rus.* Lozovaya. Kharkivs'ka Oblast', E Ukraine *48°54´N 36°23´E*
Lozovaya *see* Lozova
105 *N7* **Lozoyuela** Madrid, C Spain *40°55´N 03°36´W*
Lu *see* Shandong, China
Lú *see* Louth, Ireland
82 *F12* **Luacano** Moxico, E Angola *11°19´S 21°30´E*
79 *N21* **Lualaba** *Fr.* Loualaba. ≈ SE Dem. Rep. Congo
83 *H14* **Luampa** Western, NW Zambia *15°02´S 24°27´E*
83 *H15* **Luampa Kuta** Western, W Zambia *15°22´S 24°40´E*
161 *P8* **Lu'an** Anhui, E China *31°46´N 116°31´E*
Luanco *see* Lluanco
82 *A11* **Luanda** *var.* Loanda, *Port.* São Paulo de Loanda. ● Luanda, NW Angola *08°48´S 13°17´E*
82 *A11* **Luanda** ⬖ *province* (Angola) NW Angola
82 *A11* **Luanda** ✈ Luanda, NW Angola *08°49´S 13°16´E*
82 *D12* **Luando** ≈ C Angola
Luang *see* Tapi, Mae Nam
83 *G14* **Luanginga** *var.* Luanguinga. ≈ Angola/Zambia
167 *N15* **Luang, Khao** ▲ SW Thailand *08°21´N 99°46´E*
Luang Prabang *see* Louangphabang
167 *P8* **Luang Prabang Range** *Th.* Thiukhaoluang Phrahang. ▲▲ Laos/Thailand
167 *N16* **Luang, Thale** *lagoon* S Thailand
Luangua, Rio *see* Luangwa
82 *E11* **Luangue** ≈ NE Angola
Luanguinga *see* Luanginga
83 *K15* **Luangwa** *var.* Aruângua. Lusaka, C Zambia *15°36´S 30°27´E*
83 *K14* **Luangwa** *var.* Aruângua, Rio Luangua. ≈ Mozambique/Zambia
161 *Q2* **Luan He** ≈ E China
190 *G11* **Luaniva, Île** *island* E Wallis and Futuna
161 *P2* **Luanping** *var.* Anjiangying. Hebei, E China *40°55´N 117°19´E*
82 *J13* **Luanshya** Copperbelt, C Zambia *13°09´S 28°24´E*
62 *K13* **Luan Toro** La Pampa, C Argentina *36°14´S 64°15´W*
161 *Q2* **Luanxian** *var.* Luan Xian. Hebei, E China *39°46´N 118°46´E*
Luan Xian *see* Luanxian
82 *J12* **Luapula** ⬖ *province* N Zambia
79 *O25* **Luapula** ≈ Dem. Rep. Congo/Zambia
104 *J2* **Luarca** Asturias, N Spain *43°33´N 06°31´W*
169 *R10* **Luar, Danau** ◎ Borneo, N Indonesia
79 *L25* **Luashi** Katanga, S Dem. Rep. Congo *10°54´S 23°55´E*
82 *G12* **Luau** *Port.* Vila Teixeira de Sousa. Moxico, NE Angola *10°42´S 22°12´E*
79 *C16* **Luba** *prev.* San Carlos. Isla de Bioco, NW Equatorial Guinea *03°26´N 08°36´E*
42 *F4* **Lubaantun** *ruins* Toledo, S Belize
111 *P16* **Lubaczów** *var.* Lúbaczów. Podkarpackie, SE Poland *50°10´N 23°08´E*
Lubale *see* Lubalo
82 *E11* **Lubalo** Lunda Norte, NE Angola *09°02´S 19°11´E*
82 *E11* **Lubalo** *var.* Lubale. ≈ Angola/Dem. Rep. Congo
118 *J9* **Lubāna** E Latvia *56°55´N 29°43´E*
Lubānas Ezers *see* Lubāns
171 *N4* **Lubang Island** *island* N Philippines
83 *B15* **Lubango** *Port.* Sá da Bandeira. Huíla, SW Angola *14°55´S 13°33´E*
118 *J9* **Lubāns** *var.* Lubānas Ezers. ◎ E Latvia
79 *M21* **Lubao** Kasai-Oriental, C Dem. Rep. Congo *05°21´S 25°42´E*
110 *O13* **Lubartów** *Ger.* Qumälisch. Lublin, E Poland *51°29´N 22°38´E*
100 *G13* **Lübbecke** Nordrhein-Westfalen, NW Germany *52°18´N 08°37´E*
100 *O13* **Lübben** Brandenburg, E Germany *51°56´N 13°52´E*
101 *P14* **Lübbenau** Brandenburg, E Germany *51°52´N 13°57´E*
25 *N5* **Lubbock** Texas, SW USA *33°35´N 101°51´W*
Lubcz *see* Lyubcha
19 *U6* **Lubec** Maine, NE USA *44°49´N 67°00´W*
100 *K9* **Lübeck** Schleswig-Holstein, N Germany *53°52´N 10°41´E*
100 *K8* **Lübecker Bucht** *bay* N Germany
79 *M21* **Lubefu** Kasai-Oriental, C Dem. Rep. Congo *04°43´S 24°25´E*
163 *T10* **Lubei** *var.* Jarud Qi. Nei Mongol Zizhiqu, N China *44°25´N 121°12´E*
111 *O14* **Lubelska, Wyżyna** *plateau* SE Poland
110 *O13* **Lubelskie** ⬖ *province* E Poland
Lubembe *see* Luembe
Lüben *see* Lubin
144 *H9* **Lubenka** Zapadnyy Kazakhstan, W Kazakhstan *50°27´N 54°07´E*
79 *P18* **Lubero** Nord-Kivu, E Dem. Rep. Congo *0°10´S 29°12´E*
79 *L22* **Lubi** ≈ S Dem. Rep. Congo
Lubiana *see* Ljubljana
110 *J11* **Lubień Kujawski** Kujawsko-pomorskie, C Poland *52°25´N 19°10´E*
67 *T11* **Lubilandji** ≈ S Dem. Rep. Congo
110 *F13* **Lubin** *Ger.* Lüben. Dolnośląskie, SW Poland *51°23´N 16°12´E*
111 *O14* **Lublin** *Rus.* Lyublin. Lubelskie, E Poland *51°15´N 22°33´E*
111 *J15* **Lubliniec** Śląskie, S Poland *50°41´N 18°41´E*
Lubnān, Jabal *see* Liban, Jebel
117 *R5* **Lubny** Poltavs'ka Oblast', NE Ukraine *50°00´N 33°00´E*
Luboml *see* Lyuboml'
110 *G11* **Luboń** *Ger.* Peterhof. Wielkolpolskie, C Poland *52°23´N 16°54´E*
Luboten *see* Ljuboten
110 *D12* **Lubsko** *Ger.* Sommerfeld. Lubuskie, W Poland *51°47´N 14°57´E*
79 *N24* **Lubudi** Katanga, SE Dem. Rep. Congo *09°57´S 25°59´E*
168 *L13* **Lubuklinggau** Sumatera, W Indonesia *03°10´S 102°52´E*
79 *N25* **Lubumbashi** *prev.* Élisabethville. Shaba, SE Dem. Rep. Congo *11°40´S 27°31´E*
83 *I14* **Lubungu** Central, C Zambia *14°28´S 26°30´E*
110 *E12* **Lubuskie** ⬖ *province* W Poland
Lubuskie *see* Dobiegniew
79 *N18* **Lubutu** Maniema, E Dem. Rep. Congo *0°48´S 26°39´E*
Luca *see* Lucca
82 *C11* **Lucala** ≈ W Angola
14 *E16* **Lucan** Ontario, S Canada *43°10´N 81°22´W*
97 *F18* **Lucan** *Ir.* Leamhcán. Dublin, E Ireland *53°22´N 06°27´W*
Lucanian Mountains *see* Lucano, Appennino
107 *M18* **Lucano, Appennino** *Eng.* Lucanian Mountains. ▲▲ S Italy
82 *F11* **Lucapa** *var.* Lukapa. Lunda Norte, NE Angola *08°24´S 20°42´E*
29 *V15* **Lucas** Iowa, C USA *41°01´N 93°26´W*
61 *C18* **Lucas González** Entre Ríos, E Argentina *32°25´S 59°33´W*
65 *C25* **Lucas Point** *headland* West Falkland, Falkland Islands *52°10´S 60°22´W*
31 *S15* **Lucasville** Ohio, N USA *38°52´N 83°00´W*
106 *F11* **Lucca** *anc.* Luca. Toscana, C Italy *43°50´N 10°30´E*
44 *H12* **Lucea** W Jamaica *18°26´N 78°11´W*
97 *H15* **Luce Bay** *inlet* SW Scotland, United Kingdom
22 *M8* **Lucedale** Mississippi, S USA *30°55´N 88°35´W*
171 *O4* **Lucena** *off.* Lucena City. Luzon, N Philippines *13°57´N 121°38´E*
104 *M14* **Lucena** Andalucía, S Spain *37°25´N 04°29´W*
Lucena City *see* Lucena
105 *S8* **Lucena del Cid** Valenciana, E Spain *40°07´N 00°15´W*
111 *D15* **Lučenec** *Ger.* Losontz, *Hung.* Losonc. Banskobystrický Kraj, C Slovakia *48°21´N 19°37´E*
Lucentum *see* Alicante
107 *M16* **Lucera** Puglia, SE Italy *41°30´N 15°19´E*
Lucerna/Lucerne *see* Luzern
Lucerne, Lake of *see* Vierwaldstätter See
40 *J4* **Lucero** Chihuahua, N Mexico *30°50´N 106°30´W*
123 *S14* **Luchegorsk** Primorskiy Kray, SE Russian Federation *46°26´N 134°10´E*
105 *Q13* **Luchena** ≈ SE Spain
Lucheng *see* Kangding
82 *N13* **Lucheringo** *var.* Luchulingo. ≈ N Mozambique
Luchesa *see* Luchosa
Luchin *see* Luchyn
118 *N13* **Luchosa** *Rus.* Luchesa. ≈ N Belarus
100 *K11* **Lüchow** Mecklenburg-Vorpommern, N Germany *52°57´N 11°10´E*
Luchow *see* Hefei
Luchulingo *see* Lucheringo
119 *N17* **Luchyn** *Rus.* Luchin. Homyel'skaya Voblasts', SE Belarus *53°01´N 30°01´E*
55 *U11* **Lucie Rivier** ≈ W Suriname
182 *K11* **Lucindale** South Australia *36°57´S 140°20´E*
83 *A14* **Lucira** Namibe, SW Angola *13°51´S 12°35´E*
Łuck *see* Luts'k
101 *O14* **Luckau** Brandenburg, E Germany *51°50´N 13°42´E*
100 *N13* **Luckenwalde** Brandenburg, E Germany *52°05´N 13°11´E*
14 *E15* **Lucknow** Ontario, S Canada *43°58´N 81°30´W*
152 *L12* **Lucknow** *var.* Lakhnau. *state capital* Uttar Pradesh, N India *26°50´N 80°54´E*
102 *J10* **Luçon** Vendée, NW France *46°27´N 01°10´W*
44 *I7* **Lucrecia, Cabo** *headland* E Cuba *21°00´N 75°34´W*
82 *F13* **Lucusse** Moxico, E Angola *12°32´S 20°46´E*
Lüda *see* Dalian
114 *M9* **Luda Kamchia** *var.* Luda Kamchiya. ≈ E Bulgaria
Luda Kamchiya *see* Luda Kamchia
161 *T14* **Lü Dao** *var.* Huoshao Dao, Lütao, *Eng.* Green Island; *prev.* Lü Tao. *island* SE Taiwan
Ludasch *see* Luduş
114 *I10* **Luda Yana** ≈ C Bulgaria
112 *F7* **Ludbreg** Varaždin, N Croatia *46°15´N 16°36´E*
29 *P7* **Ludden** North Dakota, N USA *45°58´N 98°07´W*
101 *F15* **Lüdenscheid** Nordrhein-Westfalen, W Germany *51°13´N 07°38´E*
83 *C21* **Lüderitz** *prev.* Angra Pequena. Karas, SW Namibia *26°38´S 15°10´E*
152 *H8* **Ludhiāna** Punjab, N India *30°56´N 75°52´E*
31 *O7* **Ludington** Michigan, N USA *43°58´N 86°27´W*
97 *K20* **Ludlow** W England, United Kingdom *52°20´N 02°28´W*
35 *W14* **Ludlow** California, W USA *34°43´N 116°07´W*
28 *J7* **Ludlow** South Dakota, N USA *45°48´N 103°21´W*
18 *M9* **Ludlow** Vermont, NE USA *43°24´N 72°39´W*
114 *L7* **Ludogorie** *physical region* NE Bulgaria
23 *W6* **Ludowici** Georgia, SE USA *31°42´N 81°44´W*
Ludsan *see* Ludza
116 *I10* **Luduş** *Ger.* Ludasch, *Hung.* Marosludas. Mureş, C Romania *46°28´N 24°05´E*
95 *M14* **Ludvika** Dalarna, C Sweden *60°08´N 15°14´E*
101 *H21* **Ludwigsburg** Baden-Württemberg, SW Germany *48°54´N 09°12´E*
100 *O13* **Ludwigsfelde** Brandenburg, NE Germany *52°17´N 13°15´E*
101 *G20* **Ludwigshafen** *var.* Ludwigshafen am Rhein. Rheinland-Pfalz, W Germany *49°29´N 08°24´E*
Ludwigshafen am Rhein *see* Ludwigshafen
101 *L20* **Ludwigskanal** *canal* SE Germany
100 *L10* **Ludwigslust** Mecklenburg-Vorpommern, N Germany *53°19´N 11°29´E*
118 *K10* **Ludza** *Ger.* Ludsan. E Latvia *56°32´N 27°41´E*
79 *K21* **Luebo** Kasai-Occidental, SW Dem. Rep. Congo *05°19´S 21°21´E*
25 *Q6* **Lueders** Texas, SW USA *32°46´N 99°38´W*
79 *N20* **Lueki** Maniema, C Dem. Rep. Congo *03°25´S 25°50´E*
82 *F10* **Luembe** *var.* Lubembe. ≈ Angola/Dem. Rep. Congo
82 *E13* **Luena** *var.* Lwena, *Port.* Luso. Moxico, E Angola *11°47´S 19°52´E*
79 *M24* **Luena** Katanga, SE Dem. Rep. Congo *09°28´S 25°45´E*
82 *K12* **Luena** Northern, NE Zambia *10°40´S 30°21´E*
82 *F13* **Luena** ≈ E Angola
83 *F16* **Luengue** ≈ SE Angola
67 *V13* **Luenha** ≈ W Mozambique
83 *G15* **Lueti** ≈ Angola/Zambia
160 *J7* **Lüeyang** *var.* Hejiayan. Shaanxi, C China *33°12´N 106°31´E*
161 *P14* **Lufeng** Guangdong, S China *22°59´N 115°40´E*
79 *N24* **Lufira** ≈ SE Dem. Rep. Congo
79 *N25* **Lufira, Lac de Retenue de la** *var.* Lac Tshangalele. ◎ SE Dem. Rep. Congo
25 *W8* **Lufkin** Texas, SW USA *31°21´N 94°47´W*
82 *L11* **Lufubu** ≈ N Zambia
124 *G14* **Luga** Leningradskaya Oblast', NW Russian Federation *58°43´N 29°46´E*
124 *G13* **Luga** ≈ NW Russian Federation
Luganer See *see* Lugano, Lago di
108 *H11* **Lugano** *Ger.* Lauis. Ticino, S Switzerland *46°01´N 08°57´E*
108 *H12* **Lugano, Lago di** *var.* Ceresio, *Ger.* Luganer See. ◎ S Switzerland
Lugansk *see* Luhans'k
187 *Q13* **Luganville** Espiritu Santo, C Vanuatu *15°31´S 167°12´E*
Lugdunum *see* Lyon
Lugdunum Batavorum *see* Leiden
83 *O15* **Lugela** Zambézia, NE Mozambique *16°27´S 36°47´E*
83 *O16* **Lugela** ≈ C Mozambique
82 *P13* **Lugenda, Rio** ≈ N Mozambique
Luggarus *see* Locarno
Lugh Ganana *see* Luuq
97 *G19* **Lugnaquillia Mountain** *Ir.* Log na Coille. ▲ E Ireland *52°58´N 06°27´W*
106 *H10* **Lugo** Emilia-Romagna, N Italy *44°25´N 11°53´E*
104 *I3* **Lugo** *anc.* Lugus Augusti. Galicia, NW Spain *43°N 07°33´W*
104 *I3* **Lugo** ⬖ *province* Galicia, NW Spain
21 *R12* **Lugoff** South Carolina, SE USA *34°13´N 80°41´W*
116 *F12* **Lugoj** *Ger.* Lugosch, *Hung.* Lugos. Timiş, W Romania *45°41´N 21°56´E*
Lugos/Lugosch *see* Lugoj
Lugovoy/Lugovoye *see* Kulan
158 *I13* **Lugu** Xizang Zizhiqu, W China *33°26´N 84°10´E*
Lugus Augusti *see* Lugo
Luguvallium/Luguvallum *see* Carlisle
117 *Y7* **Luhans'k** *Rus.* Lugansk; *prev.* Voroshilovgrad. Luhans'ka Oblast', E Ukraine *48°32´N 39°21´E*
117 *Y7* **Luhans'k** ✈ Luhans'ka Oblast', E Ukraine *48°25´N 39°24´E*
Luhans'k *see* Luhans'ka Oblast'
117 *X6* **Luhans'ka Oblast'** *var.* Luhans'k; *prev.* Voroshilovgrad, *Rus.* Voroshilovgradskaya Oblast'. ⬖ *province* E Ukraine
161 *Q7* **Luhe** Jiangsu, E China *32°20´N 118°52´E*
171 *S13* **Luhu** Pulau Seram, E Indonesia *03°20´S 127°58´E*
Luhua *see* Heishui
160 *G8* **Luhuo** *var.* Xindu, *Tib.* Zhaggo. Sichuan, C China *31°18´N 100°39´E*
116 *M3* **Luhyny** Zhytomyrs'ka Oblast', N Ukraine *51°06´N 28°24´E*
83 *G15* **Lui** ≈ W Zambia
83 *G16* **Luiana** ≈ SE Angola
83 *L15* **Luia, Rio** *var.* Ruya. ≈ Mozambique/Zimbabwe
Luichow Peninsula *see* Leizhou Bandao
Luik *see* Liège
82 *C13* **Luimbale** Huambo, C Angola *12°15´S 15°19´E*
Luimneach *see* Limerick
106 *D6* **Luino** Lombardia, N Italy *46°00´N 08°45´E*
82 *E13* **Luio** ≈ E Angola
92 *L11* **Luiro** ≈ NE Finland
79 *N25* **Luishia** Katanga, SE Dem. Rep. Congo *11°18´S 27°08´E*
59 *M19* **Luislândia do Oeste** Minas Gerais, SE Brazil *17°59´S 45°35´W*
40 *K5* **Luis L. León, Presa** ⊠ N Mexico
45 *U5* **Luis Muñoz Marín** *var.* Luis Muñoz Marin. ✈ NE Puerto Rico *18°27´N 66°05´W*
Luis Muñoz Marin *see* Luis Muñoz Marín
195 *N5* **Luitpold Coast** *physical region* Antarctica
79 *K22* **Luiza** Kasai-Occidental, S Dem. Rep. Congo *07°11´S 22°27´E*
61 *D20* **Luján** Buenos Aires, E Argentina *34°34´S 59°07´W*
79 *N24* **Lukafu** Katanga, SE Dem. Rep. Congo *10°28´S 27°32´E*
Lukapa *see* Lucapa
112 *I11* **Lukavac** ⬖ Federacija Bosne I Hercegovine, NE Bosnia and Herzegovina
79 *I20* **Lukenie** ≈ C Dem. Rep. Congo
Lŭki *see* Laki
79 *H19* **Lukolela** Equateur, W Dem. Rep. Congo *01°10´S 17°11´E*
Lukoml'skaye, Vozyera *see* Lukomskaye, Vozyera
Lukoml'skoye, Ozero *see* Lukomskaye, Vozyera
119 *M14* **Lukomskaye, Vozyera** *Rus.* Ozero Lukoml'skoye; *prev.* Vozyera Lukoml'skaye. ◎ N Belarus
114 *I8* **Lukovit** Lovech, N Bulgaria *43°11´N 24°10´E*
110 *O12* **Łuków** *Ger.* Bogendorf. Lubelskie, E Poland *51°57´N 22°22´E*
127 *O4* **Lukoyanov** Nizhegorodskaya Oblast', W Russian Federation *55°02´N 44°26´E*
Lukransar *see* Lūnkaransar
79 *N22* **Lukuga** ≈ SE Dem. Rep. Congo
79 *F21* **Lukula** Bas-Congo, SW Dem. Rep. Congo *05°23´S 12°57´E*
83 *G14* **Lukulu** Western, NW Zambia *14°24´S 23°12´E*
189 *R17* **Lukunor Atoll** *atoll* Mortlock Islands, C Micronesia
82 *J12* **Lukwesa** Luapula, NE Zambia *10°03´S 28°42´E*
93 *K14* **Luleå** Norrbotten, N Sweden *65°35´N 22°10´E*
92 *J13* **Luleälven** ≈ N Sweden
136 *C10* **Lüleburgaz** Kırklareli, NW Turkey *41°25´N 27°22´E*
160 *M4* **Lüliang** *var.* Lishi. Shanxi, C China *37°27´N 111°05´E*
160 *M4* **Lüliang Shan** ▲▲ C China
79 *O21* **Lulimba** Maniema, E Dem. Rep. Congo *04°42´S 28°38´E*
22 *K9* **Luling** Louisiana, S USA *29°55´N 90°22´W*
25 *T11* **Luling** Texas, SW USA *29°40´N 97°39´W*
79 *I18* **Lulonga** ≈ NW Dem. Rep. Congo
79 *K22* **Lulua** ≈ S Dem. Rep. Congo
Luluabourg *see* Kananga
192 *L17* **Luma** Ta'ū, E American Samoa *14°15´S 169°30´W*
169 *S17* **Lumajang** Jawa, C Indonesia *08°06´S 113°13´E*
158 *G12* **Lumajangdong Co** ◎ W China
82 *G13* **Lumbala Kaquengue** Moxico, E Angola *12°40´S 22°34´E*
83 *F14* **Lumbala N'Guimbo** *var.* Nguimbo, Gago Coutinho, *Port.* Vila Gago Coutinho. Moxico, E Angola *14°08´S 21°25´E*
21 *T11* **Lumber River** ≈ North Carolina/South Carolina, SE USA
Lumber State *see* Maine
22 *L8* **Lumberton** Mississippi, S USA *31°00´N 89°27´W*
21 *U11* **Lumberton** North Carolina, SE USA *34°37´N 79°00´W*
105 *R4* **Lumbier** Navarra, N Spain *42°39´N 01°19´W*
83 *Q15* **Lumbo** Nampula, NE Mozambique *15°S 40°40´E*
124 *M4* **Lumbovka** Murmanskaya Oblast', NW Russian Federation *67°41´N 40°31´E*
104 *J7* **Lumbrales** Castilla y León, N Spain *40°57´N 06°43´W*
153 *W13* **Lumding** Assam, NE India *25°46´N 93°10´E*
82 *F12* **Lumege** *var.* Lumeje. Moxico, E Angola *11°30´S 20°57´E*
Lumeje *see* Lumege
99 *J17* **Lummen** Limburg, NE Belgium *50°58´N 05°12´E*
93 *J20* **Lumparland** Åland, SW Finland *60°06´N 20°15´E*
167 *T11* **Lumphăt** *prev.* Lomphat. Rôtânôkiri, NE Cambodia *13°30´N*
11 *U16* **Lumsden** Saskatchewan, S Canada *50°39´N 104°52´W*
185 *C23* **Lumsden** Southland, South Island, New Zealand *45°43´S 168°26´E*
169 *N14* **Lumut, Tanjung** *headland* Sumatera, W Indonesia *03°47´S 105°55´E*
157 *P4* **Lün** Töv, C Mongolia *47°51´N 105°11´E*
116 *I13* **Lunca Corbului** Argeş, S Romania *44°41´N 24°46´E*
95 *K23* **Lund** Skåne, S Sweden *55°42´N 13°10´E*
35 *X6* **Lund** Nevada, W USA *38°50´N 115°00´W*
82 *D11* **Lunda Norte** ⬖ *province* NE Angola
82 *E12* **Lunda Sul** ⬖ *province* NE Angola
82 *M13* **Lundazi** Eastern, NE Zambia *12°19´S 33°11´E*
95 *G16* **Lunde** Telemark, S Norway *61°31´N 06°38´E*
Lundenburg *see* Břeclav
95 *C17* **Lundevatnet** ◎ S Norway
Lundi *see* Runde
97 *I23* **Lundy** *island* SW England, United Kingdom
100 *J10* **Lüneburg** Niedersachsen, N Germany *53°15´N 10°25´E*
100 *J11* **Lüneburger Heide** *heathland* NW Germany
103 *Q15* **Lunel** Hérault, S France *43°40´N 04°08´E*
101 *F14* **Lünen** Nordrhein-Westfalen, W Germany *51°37´N 07°31´E*
13 *P16* **Lunenburg** Nova Scotia, SE Canada *44°23´N 64°21´W*
21 *V7* **Lunenburg** Virginia, NE USA *36°56´N 78°15´W*
103 *T5* **Lunéville** Meurthe-et-Moselle, NE France *48°35´N 06°30´E*
83 *I14* **Lunga** ≈ C Zambia
Lunga, Isola *see* Dugi Otok
158 *H12* **Lungdo** Xizang Zizhiqu, W China *33°45´N 82°09´E*
158 *I14* **Lunggar** Xizang Zizhiqu, W China *31°10´N 84°01´E*
76 *I15* **Lungi** ✈ (Freetown) W Sierra Leone *08°36´N 13°10´W*
Lungkiang *see* Qiqihar
Lungleh *see* Lunglei
153 *W15* **Lunglei** *prev.* Lungleh. Mizoram, NE India *22°55´N 92°49´E*
158 *L15* **Lungsang** Xizang Zizhiqu, W China *29°50´N 88°27´E*
82 *E13* **Lungué-Bungo** *var.* Lungwebungu. ≈ Angola/Zambia *see also* Lungwebungu
Lungué-Bungo *see* Lungwebungu
83 *G14* **Lungwebungu** *var.* Lungué-Bungo. ≈ Angola/Zambia *see also* Lungué-Bungo
Lungwebungu *see* Lungué-Bungo
152 *F12* **Lūni** Rājasthān, N India *26°03´N 73°00´E*
152 *F12* **Lūni** ≈ N India
Luninets *see* Luninyets
35 *S7* **Luning** Nevada, W USA *38°29´N 118°10´W*
Łuniniec *see* Luninyets
127 *P6* **Lunino** Penzenskaya Oblast', W Russian Federation *53°35´N 45°12´E*
119 *J19* **Luninyets** *Pol.* Łuniniec, *Rus.* Luninets. Brestskaya Voblasts', SW Belarus *52°15´N 26°48´E*
152 *F10* **Lūnkaransar** *var.* Lookransar, Lukransar. Rājasthān, NW India *28°32´N 73°50´E*
119 *G17* **Lunna** *Pol.* Łunna, *Rus.* Lunno. Hrodzyenskaya Voblasts', W Belarus *53°27´N 24°16´E*
Łunna *see* Lunna
Lunno *see* Lunna
76 *I15* **Lunsar** W Sierra Leone *08°41´N 12°32´W*
83 *K14* **Lunsemfwa** ≈ C Zambia
158 *J6* **Luntai** *var.* Bügür. Xinjiang Uygur Zizhiqu, NW China *41°48´N 84°14´E*
98 *K11* **Lunteren** Gelderland, C Netherlands *52°05´N 05°38´E*
109 *U5* **Lunz am See** Niederösterreich, C Austria *47°54´N 15°01´E*
163 *Y7* **Luobei** *var.* Fengxiang. Heilongjiang, NE China *47°35´N 130°50´E*
Luocheng *see* Hui'an, Fujian, China
Luocheng *see* Luoding, Guangdong, China
160 *J13* **Luodian** *var.* Longping. Guizhou, S China *25°25´N 106°49´E*
160 *M15* **Luoding** *var.* Luocheng. Guangdong, S China *22°44´N 111°28´E*

◆ Country
● Country Capital
◇ Dependent Territory
○ Dependent Territory Capital
⬖ Administrative Regions
✈ International Airport
▲ Mountain
▲▲ Mountain Range
Volcano
≈ River
◎ Lake
⊠ Reservoir

161 N7 **Luohe** Henan, C China 33°37′N 114°00′E
160 M6 **Luo He** ≈ C China
160 L5 **Luo He** ≈ C China
L i Liêm, Nhom *see* Crescent Group
Luolajarvi *see* Kuoloyarvi
Luong Nam Tha *see* Louangnamtha
160 L13 **Luoqing Jiang** ≈ S China
161 O8 **Luoshan** Henan, C China 32°12′N 114°30′E
161 O12 **Luoxiao Shan** ▲ S China
161 N6 **Luoyang** *var.* Honan, Lo-yang. Henan, C China 34°41′N 112°25′E
161 R12 **Luoyuan** *var.* Fengshan. Fujian, SE China 26°29′N 119°32′E
79 F21 **Luozi** Bas-Congo, W Dem. Rep. Congo 04°57′S 14°08′E
83 J17 **Lupane** Matabeleland North, W Zimbabwe 18°54′S 27°44′E
160 I12 **Lupanshui** *var.* Liupanshui; *prev.* Shuicheng. Guizhou, S China 26°38′N 104°49′E
169 R10 **Lupar, Batang** ≈ East Malaysia
Lupatia *see* Altamura
116 G12 **Lupeni** *Hung.* Lupény. Hunedoara, SW Romania 45°20′N 23°10′E
Lupény *see* Lupeni
82 N13 **Lupiliche** Niassa, N Mozambique 11°36′S 35°15′E
83 E14 **Lupire** Kuando Kubango, E Angola 14°39′S 19°39′E
79 L22 **Luputa** Kasai-Oriental, S Dem. Rep. Congo 07°07′S 23°43′E
121 P16 **Luqa** ✕ (Valletta) S Malta 35°53′N 14°27′E
159 U11 **Luqu** *var.* Ma'ai. Gansu, C China 34°34′N 102°27′E
45 U5 **Luquillo, Sierra de** ▲ E Puerto Rico
26 L4 **Luray** Kansas, C USA 39°06′N 98°41′W
21 U4 **Luray** Virginia, NE USA 38°40′N 78°28′W
103 T7 **Lure** Haute-Saône, E France 47°42′N 06°30′E
82 D11 **Luremo** Lunda Norte, NE Angola 08°32′S 17°55′E
97 F15 **Lurgan** *Ir.* An Lorgain. S Northern Ireland, United Kingdom 54°28′N 06°20′W
57 K18 **Luribay** La Paz, W Bolivia 17°05′S 67°37′W
Luring *see* Gêrzê
83 Q14 **Lúrio** Nampula, NE Mozambique 13°32′S 40°34′E
83 P14 **Lúrio, Rio** ≈ NE Mozambique
Luristan *see* Lorestān
Lurka *see* Lorca
83 J15 **Lusaka** ● (Zambia) Lusaka, SE Zambia 15°24′S 28°17′E
83 J15 **Lusaka** ◆ *province* C Zambia
83 J15 **Lusaka** ✕ Lusaka, C Zambia 15°10′S 28°22′E
79 L21 **Lusambo** Kasai-Oriental, C Dem. Rep. Congo 04°59′S 23°26′E
186 F8 **Lusancay Islands and Reefs** *island group* SE Papua New Guinea
79 I21 **Lusanga** Bandundu, SW Dem. Rep. Congo 04°55′S 18°40′E
79 N21 **Lusangi** Maniema, E Dem. Rep. Congo 04°39′S 27°10′E
Lusatian Mountains *see* Lausitzer Bergland
Lushar *see* Huangzhong
Lushnja *see* Lushnjë
113 K21 **Lushnjë** *var.* Lushnja. Fier, C Albania 40°54′N 19°43′E
81 J21 **Lushoto** Tanga, E Tanzania 04°48′S 38°20′E
102 L10 **Lusignan** Vienne, W France 46°25′N 00°06′E
33 Z15 **Lusk** Wyoming, C USA 42°45′N 104°27′W
Luso *see* Luena
102 L10 **Lussac-les-Châteaux** Vienne, W France 46°23′N 00°44′E
Lussin/Lussino *see* Lošinj
Lussinpiccolo *see* Mali Lošinj
108 I7 **Lustenau** Vorarlberg, W Austria 47°26′N 09°42′E
Lütao *see* Lü Dao
Lü Tao *see* Lü Dao
Lūt, Baḩrat/Lut, Bahret *see* Dead Sea
22 K9 **Lutcher** Louisiana, S USA 30°02′N 90°42′W
143 T9 **Lūt, Dasht-e** *var.* Kavīr-e Lūt. *desert* E Iran
83 F14 **Lutembo** Moxico, E Angola 13°30′S 21°21′E
Lutetia/Lutetia Parisiorum *see* Paris
Luteva *see* Lodève
14 G15 **Luther Lake** ◎ Ontario, S Canada
186 K8 **Luti** Choiseul, NW Solomon Islands 07°13′S 157°01′E
Lūt, Kavīr-e *see* Lūt, Dasht-e
97 N21 **Luton** E England, United Kingdom 51°53′N 00°25′W
97 N21 **Luton** ✕ (London) SE England, United Kingdom 51°54′N 00°24′W
108 B10 **Lutry** Vaud, SW Switzerland 46°31′N 06°32′E
8 K10 **Łutselk'e** *prev.* Snowdrift. Northwest Territories, W Canada 62°24′N 110°42′W
8 K10 **Łutselk'e** *var.* Snowdrift. ≈ Northwest Territories, NW Canada
29 Y4 **Lutsen** Minnesota, N USA 47°39′N 90°37′W
116 J4 **Luts'k** *Pol.* Łuck, *Rus.* Lutsk. Volyns'ka Oblast', NW Ukraine 50°45′N 25°23′E
Lutsk *see* Luts'k
Luttenberg *see* Ljutomer
Lüttich *see* Liège
83 G25 **Luttig** Western Cape, SW South Africa 32°33′S 22°13′E
Lutto *see* Lotta
82 E13 **Lutuai** Moxico, E Angola 12°38′S 20°06′E
117 Y7 **Lutuhyne** Luhans'ka Oblast', E Ukraine 48°24′N 39°12′E
171 V14 **Lutur, Pulau** *island* Kepulauan Aru, E Indonesia
23 V12 **Lutz** Florida, SE USA 28°09′N 82°27′W
Lutzow-Holm Bay *see* Lützow Holmbukta
195 V2 **Lützow Holmbukta** *var.* Lutzow-Holm Bay. *bay* Antarctica
81 L16 **Luuq** *It.* Lugh Ganana. Gedo, SW Somalia 03°42′N 42°34′E
92 M12 **Luusua** Lappi, NE Finland 66°28′N 27°16′E
23 Q6 **Luverne** Alabama, S USA 31°43′N 86°15′W
29 S11 **Luverne** Minnesota, N USA 43°39′N 96°12′W
79 O22 **Luvua** ≈ SE Dem. Rep. Congo
82 F13 **Luvuei** Moxico, E Angola 13°08′S 21°09′E
81 H24 **Luwego** ≈ S Tanzania
82 K12 **Luwingu** Northern, NE Zambia 10°13′S 29°58′E
171 P12 **Luwuk** *prev.* Loewoek. Sulawesi, C Indonesia 0°56′S 122°47′E
23 N3 **Luxapallila Creek** ≈ Alabama/Mississippi, S USA
99 M25 **Luxembourg** ● (Luxembourg) Luxembourg, S Luxembourg 49°37′N 06°08′E
99 M25 **Luxembourg** *off.* Grand Duchy of Luxembourg, *var.* Lëtzebuerg, Luxemburg. ◆ *monarchy* NW Europe
99 J23 **Luxembourg** ◆ *province* SE Belgium
99 L24 **Luxembourg** ◆ *district* S Luxembourg
31 N6 **Luxemburg** Wisconsin, N USA 44°32′N 87°42′W
Luxemburg *see* Luxembourg
103 U7 **Luxeuil-les-Bains** Haute-Saône, E France 47°49′N 06°22′E
160 E13 **Luxi** *prev.* Mangshi. Yunnan, SW China 24°27′N 98°31′E
82 E10 **Luxico** ≈ Angola/Dem. Rep. Congo
75 X10 **Luxor** *Ar.* Al Uqşur. E Egypt 25°39′N 32°39′E
75 X10 **Luxor** ✕ C Egypt 25°39′N 32°48′E
160 M4 **Luya Shan** ▲ C China
102 J15 **Luy de Béarn** ≈ SW France
102 J15 **Luy de France** ≈ SW France
125 P12 **Luza** Kirovskaya Oblast', NW Russian Federation 60°38′N 47°13′E
125 Q12 **Luza** ≈ NW Russian Federation
104 I16 **Luz, Costa de la** *coastal region* SW Spain
111 K20 **Luže** *var.* Lausche. ▲ Czech Republic/Germany 50°51′N 14°40′E *see also* Lausche
Luže *see* Lausche
108 F8 **Luzern** *Fr.* Lucerne, *It.* Lucerna. Luzern, C Switzerland 47°03′N 08°17′E
108 E8 **Luzern** *Fr.* Lucerne. ◆ *canton* C Switzerland
160 L13 **Luzhai** Guangxi Zhuangzu Zizhiqu, S China 24°31′N 109°46′E
118 K12 **Luzhki** Vitsyebskaya Voblasts', N Belarus 55°21′N 27°52′E
160 I10 **Luzhou** Sichuan, C China 28°55′N 105°25′E
Lužická Nisa *see* Neisse
Lužické Hory' *see* Lausitzer Bergland
Lužnice *see* Lainsitz
171 O2 **Luzon** *island* N Philippines
171 N1 **Luzon Strait** *strait* Philippines/Taiwan
Lużyckie, Gory *see* Lausitzer Bergland
116 I5 **L'viv** *Ger.* Lemberg, *Pol.* Lwów, *Rus.* L'vov. L'vivs'ka Oblast', W Ukraine 49°49′N 24°05′E
L'viv *see* L'vivs'ka Oblast'
116 I4 **L'vivs'ka Oblast'** *var.* L'viv, *Rus.* L'vovskaya Oblast'. ◆ *province* NW Ukraine
L'vov *see* L'viv
L'vovskaya Oblast' *see* L'vivs'ka Oblast'
Lwena *see* Luena
Lwów *see* L'viv
110 F11 **Lwówek** *Ger.* Neustadt bei Pinne. Wielkopolskie, C Poland 52°27′N 16°10′E
111 E14 **Lwówek Śląski** *Ger.* Löwenberg in Schlesien. Jelenia Góra, SW Poland 51°06′N 15°35′E
119 I18 **Lyakhavichy** *Rus.* Lyakhovichi. Brestskaya Voblasts', SW Belarus 53°02′N 26°16′E
Lyakhovichi *see* Lyakhavichy
185 B22 **Lyall, Mount** ▲ South Island, New Zealand 45°14′S 167°31′E
Lyallpur *see* Faisalābād
Lyangar *see* Langar
124 H11 **Lyaskelya** Respublika Kareliya, NW Russian Federation 61°42′N 31°06′E
119 I18 **Lyasnaya** *Rus.* Lesnaya. Brestskaya Voblasts', SW Belarus 52°59′N 25°46′E
119 F19 **Lyasnaya** *Pol.* Leśna, *Rus.* Lesnaya. ≈ SW Belarus
124 H15 **Lychkovo** Novgorodskaya Oblast', W Russian Federation 57°55′N 32°24′E
93 I15 **Lycksele** Västerbotten, N Sweden 64°34′N 18°40′E
18 G13 **Lycoming Creek** ≈ Pennsylvania, NE USA
Lycopolis *see* Asyūţ
195 N3 **Lyddan Island** *island* Antarctica
Lydenburg *see* Mashishing
119 L20 **Lyel'chytsy** *Rus.* Lel'chitsy. Homyel'skaya Voblasts', SE Belarus 51°47′N 28°20′E
119 P14 **Lyenina** *Rus.* Lenino. Mahilyowskaya Voblasts', E Belarus 54°25′N 31°08′E
118 L13 **Lyepyel'** *Rus.* Lepel'. Vitsyebskaya Voblasts', N Belarus 54°54′N 28°44′E
25 S17 **Lyford** Texas, SW USA 26°24′N 97°47′W
95 E17 **Lygna** ≈ S Norway
18 G14 **Lykens** Pennsylvania, NE USA 40°33′N 76°42′W
115 E21 **Lykódimo** ▲ S Greece 36°56′N 21°49′E
97 K24 **Lyme Bay** *bay* S England, United Kingdom
97 K24 **Lyme Regis** S England, United Kingdom 50°44′N 02°56′W
110 L7 **Łyna** *Ger.* Alle. ≈ N Poland
29 P12 **Lynch** Nebraska, C USA 42°49′N 98°27′W
20 J10 **Lynchburg** Tennessee, S USA 35°17′N 86°22′W
21 T6 **Lynchburg** Virginia, NE USA 37°24′N 79°09′W
21 T12 **Lynches River** ≈ South Carolina, SE USA
32 H6 **Lynden** Washington, NW USA 48°57′N 122°27′W
182 I5 **Lyndhurst** South Australia 30°19′S 138°20′E
27 Q5 **Lyndon** Kansas, C USA 38°37′N 95°40′W
19 N7 **Lyndonville** Vermont, NE USA 44°31′N 71°58′W
95 D18 **Lyngdal** Vest-Agder, S Norway 58°10′N 07°08′E
92 I9 **Lyngen** *Lapp.* Ivgovuotna. *inlet* Arctic Ocean
95 G17 **Lyngør** Aust-Agder, S Norway 58°38′N 09°05′E
92 I9 **Lyngseidet** Troms, N Norway 69°36′N 20°07′E
19 P11 **Lynn** Massachusetts, NE USA 42°28′N 70°57′W
Lynn *see* King's Lynn
23 R9 **Lynn Haven** Florida, SE USA 30°15′N 85°39′W
11 V11 **Lynn Lake** Manitoba, C Canada 56°51′N 101°01′W
Lynn Regis *see* King's Lynn
118 I13 **Lyntupy** Vitsyebskaya Voblasts', NW Belarus 55°03′N 26°19′E
103 R11 **Lyon** *Eng.* Lyons; *anc.* Lugdunum. Rhône, E France 45°46′N 04°50′E
8 I6 **Lyon, Cape** *headland* Northwest Territories, NW Canada 69°47′N 123°10′W
18 K6 **Lyon Mountain** ▲ New York, NE USA 44°42′N 73°52′W
103 Q11 **Lyonnais, Monts du** ▲ C France
65 N25 **Lyon Point** *headland* SE Tristan da Cunha 37°06′S 12°13′W
182 E5 **Lyons** South Australia 30°40′S 133°50′E
37 T3 **Lyons** Colorado, C USA 40°13′N 105°16′W
23 V6 **Lyons** Georgia, SE USA 32°12′N 82°19′W
26 M5 **Lyons** Kansas, C USA 38°22′N 98°13′W
29 R14 **Lyons** Nebraska, C USA 41°56′N 96°28′W
18 G10 **Lyons** New York, NE USA 43°03′N 76°58′W
Lyons *see* Lyon
118 O13 **Lyozna** *Rus.* Liozno. Vitsyebskaya Voblasts', NE Belarus 55°02′N 30°48′E
117 S4 **Lypova Dolyna** Sums'ka Oblast', NE Ukraine 50°36′N 33°50′E
117 N6 **Lypovets'** *Rus.* Lipovets. Vinnyts'ka Oblast', C Ukraine 49°13′N 29°06′E
Lys *see* Leie
111 I18 **Lysá Hora** ▲ E Czech Republic 49°31′N 18°27′E
95 D16 **Lysefjorden** *fjord* S Norway
95 I18 **Lysekil** Västra Götaland, S Sweden 58°16′N 11°26′E
Lyši *see* Akdoğan
33 V14 **Lysite** Wyoming, C USA 43°16′N 107°42′W
127 P3 **Lyskovo** Nizhegorodskaya Oblast', W Russian Federation 56°04′N 45°01′E
108 D8 **Lyss** Bern, W Switzerland 47°04′N 07°19′E
95 H22 **Lystrup** Midtjylland, C Denmark 56°14′N 10°14′E
125 V14 **Lys'va** Permskiy Kray, NW Russian Federation 58°04′N 57°48′E
117 P6 **Lysyanka** Cherkas'ka Oblast', C Ukraine 49°15′N 30°50′E
117 X6 **Lysychans'k** *Rus.* Lisichansk. Luhans'ka Oblast', E Ukraine 48°52′N 38°27′E
97 K17 **Lytham St Anne's** NW England, United Kingdom 53°45′N 03°01′W
185 I19 **Lyttelton** South Island, New Zealand 43°35′S 172°44′E
10 M17 **Lytton** British Columbia, SW Canada 50°12′N 121°34′W
119 L18 **Lyuban'** Minskaya Voblasts', S Belarus 52°48′N 28°00′E
119 L18 **Lyubanskoye Vodaskhovishcha** *Rus.* Lyubanskoye Vodokhranilishche. ▣ C Belarus
Lyubanskoye Vodokhranilishche *see* Lyubanskoye Vodaskhovishcha
116 M5 **Lyubar** Zhytomyrs'ka Oblast', N Ukraine 49°54′N 27°48′E
Lyubashëvka *see* Lyubashivka
117 O8 **Lyubashivka** *Rus.* Lyubashëvka. Odes'ka Oblast', SW Ukraine 47°49′N 30°18′E
119 I16 **Lyubcha** *Pol.* Lubcz. Hrodzyenskaya Voblasts', W Belarus 53°45′N 26°04′E
126 L4 **Lyubertsy** Moskovskaya Oblast', W Russian Federation 55°37′N 38°02′E
116 K2 **Lyubeshiv** Volyns'ka Oblast', NW Ukraine 51°46′N 25°33′E
124 M14 **Lyubim** Yaroslavskaya Oblast', NW Russian Federation 58°21′N 40°46′E
114 K11 **Lyubimets** Haskovo, S Bulgaria 41°51′N 26°03′E
Lyublin *see* Lublin
116 I3 **Lyuboml'** *Pol.* Luboml. Volyns'ka Oblast', NW Ukraine 51°12′N 24°01′E
Lyubotin *see* Lyubotyn
117 U5 **Lyubotyn** *Rus.* Lyubotin. Kharkivs'ka Oblast', E Ukraine 49°57′N 35°57′E
126 I5 **Lyudinovo** Kaluzhskaya Oblast', W Russian Federation 53°52′N 34°28′E
127 T2 **Lyuk** Udmurtskaya Respublika, NW Russian Federation 56°55′N 52°45′E
114 M9 **Lyulyakovo** *prev.* Keremitlik. Burgas, E Bulgaria 42°53′N 27°05′E
119 I18 **Lyusina** *Rus.* Lyusino. Brestskaya Voblasts', SW Belarus 52°38′N 26°31′E
Lyusino *see* Lyusina

M

138 G9 **Ma'ād** Irbid, N Jordan 32°37′N 35°36′E
Ma'ai *see* Luqu
Maalahti *see* Malax
Maale *see* Male'
138 G13 **Ma'ān** Ma'ān, SW Jordan 30°11′N 35°45′E
138 H13 **Ma'ān** *off.* Muḩāfaẓat Ma'ān, *var.* Ma'an, Ma'ān. ◆ *governorate* S Jordan
93 M16 **Maaninka** Pohjois-Savo, C Finland 63°10′N 27°19′E
Maanīt *see* Bayan, Töv, Mongolia
Maanīt *see* Hishig Öndör, Bulgan, Mongolia
Ma'ān, Muḩāfaẓat *see* Ma'ān
93 N15 **Maanselkä** Kainuu, C Finland 63°54′N 28°28′E
161 Q8 **Ma'anshan** Anhui, E China 31°45′N 118°32′E
188 F16 **Maap** *island* Caroline Islands, W Micronesia
118 H3 **Maardu** *Ger.* Maart. Harjumaa, NW Estonia 59°28′N 24°56′E
Ma'aret-en-Nu'man *see* Ma'arrat an Nu'mān
99 K16 **Maarheeze** Noord-Brabant, SE Netherlands 51°19′N 05°37′E
Maarianhamina *see* Mariehamn
138 I4 **Ma'arrat an Nu'mān** *var.* Ma'aret-en-Nu'man, *Fr.* Maarret enn Naamâne. Idlib, NW Syria 35°40′N 36°40′E
Maarret enn Naamâne *see* Ma'arrat an Nu'mān
98 I11 **Maarssen** Utrecht, C Netherlands 52°08′N 05°03′E
Maart *see* Maardu
99 L17 **Maas** *Fr.* Meuse. ≈ W Europe *see also* Meuse
Maas *see* Meuse
99 M15 **Maasbree** Limburg, SE Netherlands 51°22′N 06°03′E
99 L17 **Maaseik** *prev.* Maeseyck. Limburg, NE Belgium 51°05′N 05°48′E
171 Q6 **Maasin** Leyte, C Philippines 10°10′N 124°55′E
99 L17 **Maasmechelen** Limburg, NE Belgium 50°58′N 05°42′E
98 G12 **Maassluis** Zuid-Holland, SW Netherlands 51°55′N 04°15′E
99 L18 **Maastricht** *var.* Maestricht; *anc.* Traiectum ad Mosam, Traiectum Tungorum. Limburg, SE Netherlands 50°51′N 05°42′E
183 N18 **Maatsuyker Group** *island group* Tasmania, SE Australia
Maba *see* Qujiang
83 L20 **Mabalane** Gaza, S Mozambique 23°43′S 32°37′E
25 V7 **Mabank** Texas, SW USA 32°22′N 96°06′W
97 O18 **Mablethorpe** E England, United Kingdom 53°21′N 00°14′E
171 V12 **Maboi** Papua Barat, E Indonesia 01°00′S 134°02′E
83 M19 **Mabote** Inhambane, S Mozambique 22°03′S 34°09′E
32 J10 **Mabton** Washington, NW USA 46°13′N 120°00′W
83 H20 **Mabutsane** Southern, S Botswana 24°24′S 23°34′E
63 G19 **Macá, Cerro** ▲ S Chile 45°07′S 73°11′W
60 Q9 **Macaé** Rio de Janeiro, SE Brazil 22°21′S 41°48′W
82 N13 **Macaloge** Niassa, N Mozambique 12°27′S 35°25′E
Macan *see* Bonerate, Kepulauan
161 N15 **Macao** *off.* Macao Special Administrative Region, *var.* Macao S.A.R., *Chin.* Aomen Tebie Xingzhengqu, *Port.* Região Administrativa Especial de Macau. Guangdong, SE China 22°06′N 113°30′E
104 H9 **Mação** Santarém, C Portugal 39°33′N 08°00′W
Macao S.A.R. *see* Macao
Macao Special Administrative Region *see* Macao
58 J11 **Macapá** *state capital* Amapá, N Brazil 0°04′N 51°04′W
43 S17 **Macaracas** Los Santos, S Panama 07°46′N 80°31′W
55 P6 **Macare, Caño** ≈ NE Venezuela
55 Q6 **Macareo, Caño** ≈ NE Venezuela
Macarsca *see* Makarska
182 L12 **Macarthur** Victoria, SE Australia 38°04′S 142°02′E
MacArthur *see* Ormoc
56 C7 **Macas** Morona Santiago, SE Ecuador 02°22′S 78°08′W
Macassar *see* Makassar
59 Q14 **Macau** Rio Grande do Norte, E Brazil 05°05′S 36°37′W
Macău *see* Makó, Hungary
Macau, Região Administrativa Especial de *see* Macao
65 E24 **Macbride Head** *headland* East Falkland, Falkland Islands 51°25′S 57°55′W
23 V9 **Macclenny** Florida, SE USA 30°16′N 82°07′W
97 L18 **Macclesfield** C England, United Kingdom 53°16′N 02°07′W
192 F6 **Macclesfield Bank** *undersea feature* N South China Sea 15°50′N 114°20′E
MacCluer Gulf *see* Berau, Teluk
181 N7 **Macdonald, Lake** *salt lake* Western Australia
181 Q7 **Macdonnell Ranges** ▲ Northern Territory, C Australia
96 K8 **Macduff** NE Scotland, United Kingdom 57°40′N 02°29′W
104 I6 **Macedo de Cavaleiros** Bragança, N Portugal 41°31′N 06°57′W
Macedonia *see* Macedonia, FYR
Macedonia Central *see* Kentrikí Makedonía
Macedonia East and Thrace *see* Anatolikí Makedonía kai Thráki
113 O19 **Macedonia, FYR** *off.* the Former Yugoslav Republic of Macedonia, *var.* Macedonia, *Mac.* Makedonija, *abbrev.* FYR Macedonia, FYROM. ◆ *republic* SE Europe
Macedonia, the Former Yugoslav Republic of *see* Macedonia, FYR
Macedonia West *see* Dytikí Makedonía
59 Q16 **Maceió** *state capital* Alagoas, E Brazil 09°40′S 35°44′W
76 K15 **Macenta** SE Guinea 08°31′N 09°32′W
106 J12 **Macerata** Marche, C Italy 43°18′N 13°27′E
11 S11 **MacFarlane** ≈ Saskatchewan, C Canada
182 H7 **Macfarlane, Lake** *var.* Lake Mcfarlane. ◎ South Australia
Macgillicuddy's Reeks Mountains *see* Macgillycuddy's Reeks
97 B21 **Macgillycuddy's Reeks** *var.* Macgillicuddy's Reeks Mountains, *Ir.* Na Cruacha Dubha. ▲ SW Ireland
11 X16 **MacGregor** Manitoba, S Canada 49°58′N 98°49′W
149 O10 **Mach** Baluchistān, SW Pakistan 29°52′N 67°20′E
56 C6 **Machachi** Pichincha, C Ecuador 0°33′S 78°34′W
83 M19 **Machaila** Gaza, S Mozambique 22°15′S 32°57′E
Machaire Fíolta *see* Magherafelt
Machaire Rátha *see* Maghera
81 I19 **Machakos** Machakos, S Kenya 01°31′S 37°16′E
81 I20 **Machakos** ◆ *county* C Kenya
56 B8 **Machala** El Oro, SW Ecuador 03°20′S 79°57′W
83 J19 **Machaneng** Central, SE Botswana 23°12′S 27°30′E
83 M18 **Machanga** Sofala, E Mozambique 20°56′S 35°04′E
80 G13 **Machar Marshes** *wetland* SE Sudan
102 I8 **Machecoul** Loire-Atlantique, NW France 46°59′N 01°51′W
161 O8 **Macheng** Hubei, C China 31°10′N 115°00′E
155 J16 **Mācherla** Andhra Pradesh, C India 16°29′N 79°25′E
153 O11 **Māchhāpuchhre** ▲ C Nepal 28°30′N 83°57′E
19 T6 **Machias** Maine, NE USA 44°44′N 67°28′W
19 T6 **Machias River** ≈ Maine, NE USA
19 R3 **Machias River** ≈ Maine, NE USA
64 P5 **Machico** Madeira, Portugal, NE Atlantic Ocean 32°43′N 16°47′W
155 K16 **Machilīpatnam** *var.* Bandar Masulipatnam. Andhra Pradesh, E India 16°12′N 81°11′E
54 G5 **Machiques** Zulia, NW Venezuela 10°04′N 72°37′W
57 G15 **Machu Picchu** Cusco, C Peru 13°08′S 72°30′W
83 M20 **Macia** *var.* Vila de Macia. Gaza, S Mozambique 25°02′S 33°08′E
Macías Nguema Biyogo *see* Bioco, Isla de
116 M13 **Măcin** Tulcea, SE Romania 45°15′N 28°09′E
183 T4 **Macintyre River** ≈ New South Wales/Queensland, SE Australia
181 Y7 **Mackay** Queensland, NE Australia 21°10′S 149°10′E
181 O7 **Mackay, Lake** *salt lake* Northern Territory/Western Australia
10 M13 **Mackenzie** British Columbia, W Canada 55°18′N 123°09′W
8 I9 **Mackenzie** ≈ Northwest Territories, NW Canada
195 Y6 **Mackenzie Bay** *bay* Antarctica
10 J1 **Mackenzie Bay** *bay* NW Canada
2 D9 **Mackenzie Delta** *delta* Northwest Territories, NW Canada
197 P8 **Mackenzie King Island** *island* Queen Elizabeth Islands, Northwest Territories, N Canada
8 H8 **Mackenzie Mountains** ▲ Northwest Territories, NW Canada
31 Q5 **Mackinac, Straits of** ◎ Michigan, N USA
194 K5 **Mackintosh, Cape** *headland* Antarctica 72°52′S 60°00′W
11 R15 **Macklin** Saskatchewan, S Canada 52°19′N 109°51′W
183 V6 **Macksville** New South Wales, SE Australia 30°39′S 152°54′E
183 V5 **Maclean** New South Wales, SE Australia 29°30′S 153°15′E
83 J24 **Maclear** Eastern Cape, SE South Africa 31°05′S 28°22′E
183 U6 **Macleay River** ≈ New South Wales, SE Australia
MacLeod *see* Fort Macleod
180 G9 **Macleod, Lake** ◎ Western Australia
10 I6 **Macmillan** ≈ Yukon, NW Canada
30 J12 **Macomb** Illinois, N USA 40°27′N 90°40′W
107 B18 **Macomer** Sardegna, Italy, C Mediterranean Sea 40°15′N 08°47′E
82 Q13 **Macomia** Cabo Delgado, NE Mozambique 12°15′S 40°06′E
103 R10 **Mâcon** *anc.* Matisco, Matisco Ædourum. Saône-et-Loire, C France 46°19′N 04°49′E
23 T5 **Macon** Georgia, SE USA 32°49′N 83°41′W
23 N4 **Macon** Mississippi, S USA 33°06′N 88°33′W
27 U3 **Macon** Missouri, C USA 39°44′N 92°27′W
22 J6 **Macon, Bayou** ≈ Arkansas/Louisiana, S USA
82 G13 **Macondo** Moxico, E Angola 12°31′S 23°45′E
83 M16 **Macossa** Manica, C Mozambique 17°51′S 33°54′E
11 T12 **Macoun Lake** ◎ Saskatchewan, C Canada
30 K14 **Macoupin Creek** ≈ Illinois, N USA
Macouria *see* Tonate
83 N18 **Macovane** Inhambane, SE Mozambique 21°30′S 35°07′E
183 N17 **Macquarie Harbour** *inlet* Tasmania, SE Australia
192 J13 **Macquarie Island** *island* New Zealand, SW Pacific Ocean
183 T8 **Macquarie, Lake** *lagoon* New South Wales, SE Australia
183 Q6 **Macquarie Marshes** *wetland* New South Wales, SE Australia
175 O13 **Macquarie Ridge** *undersea feature* SW Pacific Ocean 57°00′S 159°00′E
183 Q6 **Macquarie River** ≈ New South Wales, SE Australia
183 P17 **Macquarie River** ≈ Tasmania, SE Australia
195 V5 **Mac. Robertson Land** *physical region* Antarctica
97 C21 **Macroom** *Ir.* Maigh Chromtha. Cork, SW Ireland 51°54′N 08°57′W
42 G5 **Macuelizo** Santa Bárbara, NW Honduras 15°21′N 88°31′W
182 G2 **Macumba River** ≈ South Australia
57 I16 **Macusani** Puno, S Peru 14°05′S 70°24′W
56 E8 **Macusari, Río** ≈ N Peru
41 U15 **Macuspana** Tabasco, SE Mexico 17°43′N 92°36′W
138 G10 **Mādabā** *var.* Ma'dabā, Madeba; *anc.* Medeba. Mādabā, NW Jordan 31°44′N 35°48′E
138 G11 **Mādabā** *off.* Muḩāfaẓat Mādabā. ◆ *governorate* C Jordan
Ma'dabā *see* Mādabā
Mādabā, Muḩāfaẓat *see* Mādabā
172 G2 **Madagascar** *off.* Democratic Republic of Madagascar, *Malg.* Madagasikara; *prev.* Malagasy Republic. ◆ *republic* W Indian Ocean
172 I5 **Madagascar** *island* W Indian Ocean
128 L17 **Madagascar Basin** *undersea feature* W Indian Ocean 27°00′S 53°00′E
Madagascar, Democratic Republic of *see* Madagascar
128 L16 **Madagascar Plain** *undersea feature* W Indian Ocean 19°00′S 52°00′E
67 Y14 **Madagascar Plateau** *var.* Madagascar Ridge, Madagascar Rise, *Rus.* Madagaskarskiy Khrebet. *undersea feature* W Indian Ocean 30°00′S 45°00′E
Madagascar Rise/Madagascar Ridge *see* Madagascar Plateau
Madagasikara *see* Madagascar
Madagaskarskiy Khrebet *see* Madagascar Plateau
64 N2 **Madalena** Pico, Azores, Portugal, NE Atlantic Ocean 38°32′N 28°15′W
77 Y6 **Madama** Agadez, NE Niger 21°54′N 13°43′E
114 J12 **Madan** Smolyan, S Bulgaria 41°29′N 24°56′E
155 I19 **Madanapalle** Andhra Pradesh, E India 13°33′N 78°31′E
186 D7 **Madang** Madang, N Papua New Guinea 05°14′S 145°45′E
186 C6 **Madang** ◆ *province* N Papua New Guinea
146 G7 **Madaniyat** *Rus.* Madeniyet. Qoraqalpog'iston Respublikasi, W Uzbekistan 42°48′N 59°00′E
Madanīyīn *see* Médenine
77 U11 **Madaoua** Tahoua, SW Niger 14°06′N 06°01′E
Madaras *see* Vtáčnik
153 U15 **Madaripur** Dhaka, C Bangladesh 23°09′N 90°11′E
77 U12 **Madarounfa** Maradi, S Niger 13°16′N 07°07′E
Mad'arské Hory *see* Hungary
146 B13 **Madau** Balkan Welaýaty, W Turkmenistan 38°11′N 54°46′E
186 H9 **Madau Island** *island* SE Papua New Guinea
19 S1 **Madawaska** Maine, NE USA 47°19′N 68°19′W
14 J13 **Madawaska** ≈ Ontario, SE Canada
Madawaska Highlands *see* Haliburton Highlands
166 M4 **Madaya** Mandalay, C Myanmar (Burma) 22°12′N 96°05′E
107 K17 **Maddaloni** Campania, S Italy 41°03′N 14°23′E
29 O3 **Maddock** North Dakota, N USA 47°57′N 99°31′W
99 I14 **Made** Noord-Brabant, S Netherlands 51°41′N 04°48′E
Madeba *see* Mādabā
64 L9 **Madeira** *var.* Ilha de Madeira. *island* Madeira, Portugal, NE Atlantic Ocean
Madeira, Ilha de *see* Madeira
64 O5 **Madeira Islands** *Port.* Região Autónoma da Madeira. ◆ *autonomous region* Madeira, Portugal, NE Atlantic Ocean
64 L9 **Madeira Plain** *undersea feature* E Atlantic Ocean
Madeira, Região Autónoma da *see* Madeira Islands
64 L9 **Madeira Ridge** *undersea feature* E Atlantic Ocean 35°30′N 15°45′W
59 F14 **Madeira, Rio** *var.* Río Madera. ≈ Bolivia/Brazil *see also* Madera, Río
Madeira, Rio *see* Madera, Río
101 J25 **Mädelegabel** ▲ Austria/Germany 47°18′N 10°19′E
15 X6 **Madeleine** ≈ Québec, SE Canada
15 X5 **Madeleine, Cap de la** *headland* Québec, SE Canada 49°13′N 65°20′W
13 Q13 **Madeleine, Îles de la** *Eng.* Magdalen Islands. *island group* Québec, E Canada
29 U10 **Madelia** Minnesota, N USA 44°03′N 94°26′W
35 P3 **Madeline** California, W USA 41°02′N 120°28′W
30 K3 **Madeline Island** *island* Apostle Islands, Wisconsin, N USA
137 O15 **Maden** Elazığ, SE Turkey 38°24′N 39°42′E
145 V12 **Madeniyet** Vostochnyy Kazakhstan, E Kazakhstan 47°51′N 78°37′E
Madeniyet *see* Madaniyat
40 H5 **Madera** Chihuahua, N Mexico 29°10′N 108°10′W
35 Q10 **Madera** California, W USA 36°57′N 120°02′W
56 L13 **Madera, Río** *Port.* Rio Madeira. ≈ Bolivia/Brazil *see also* Madeira, Rio
Madera, Río *see* Madeira, Rio
106 D6 **Madesimo** Lombardia, N Italy 46°20′N 09°26′E
141 O14 **Madhāb, Wādī** *dry watercourse* NW Yemen
153 R13 **Madhepura** *prev.* Madhipure. Bihār, NE India 25°56′N 86°48′E
Madhipure *see* Madhepura
153 Q13 **Madhubani** Bihār, N India 26°21′N 86°05′E
153 Q15 **Madhupur** Jhārkhand, NE India 24°17′N 86°38′E
154 I10 **Madhya Pradesh** *prev.* Central Provinces and Berar. ◆ *state* C India
57 K15 **Madidi, Río** ≈ W Bolivia
155 F20 **Madikeri** *prev.* Mercara. Karnātaka, W India 12°29′N 75°40′E
27 O13 **Madill** Oklahoma, C USA 34°06′N 96°46′W
79 G21 **Madimba** Bas-Congo, SW Dem. Rep. Congo 04°58′S 15°08′E
138 M4 **Ma'dīn** Ar Raqqah, C Syria 35°45′N 39°36′E
Madīnah, Minṭaqat al *see* Al Madīnah
76 M14 **Madinani** NW Ivory Coast 09°37′N 06°57′W
141 O17 **Madinat ash Sha'b** *prev.* Al Ittiḩād. SW Yemen 12°52′N 44°55′E
138 K3 **Madinat ath Thawrah** *var.* Ath Thawrah. Ar Raqqah, N Syria 35°36′N 39°00′E
173 O6 **Madingley Rise** *undersea feature* W Indian Ocean
79 E21 **Madingo-Kayes** Kouilou, S Congo 04°27′S 11°43′E
79 F21 **Madingou** Bouenza, S Congo 04°10′S 13°33′E
Madioen *see* Madiun
23 U8 **Madison** Florida, SE USA 30°27′N 83°24′W
23 T3 **Madison** Georgia, SE USA 33°37′N 83°28′W
31 P15 **Madison** Indiana, N USA 38°44′N 85°22′W
27 P6 **Madison** Kansas, C USA 45′01′N 96′11′W
19 Q6 **Madison** Maine, NE USA 44°48′N 69°52′W
29 S9 **Madison** Minnesota, N USA 45°00′N 96°12′W
22 K5 **Madison** Mississippi, S USA 32°27′N 90°07′W
29 Q14 **Madison** Nebraska, C USA 41°49′N 97°27′W
29 R10 **Madison** South Dakota, N USA 44°00′N 97°06′W
21 V5 **Madison** Virginia, NE USA 38°23′N 78°16′W
21 Q5 **Madison** West Virginia, NE USA 38°03′N 81°50′W
30 L9 **Madison** *state capital* Wisconsin, N USA 43°04′N 89°22′W
21 T6 **Madison Heights** Virginia, NE USA 37°25′N 79°07′W
20 I6 **Madisonville** Kentucky, S USA 37°20′N 87°30′W
20 M10 **Madisonville** Tennessee, S USA 35°31′N 84°21′W
25 V9 **Madisonville** Texas, SW USA 30°58′N 95°56′W
Madisonville *see* Taiohae
169 R16 **Madiun** *prev.* Madioen. Jawa, C Indonesia 07°37′S 111°33′E
Madjene *see* Majene
14 J14 **Madoc** Ontario, SE Canada 44°31′N 77°27′W
Madoera *see* Madura, Pulau
81 J18 **Mado Gashi** Garissa, E Kenya 0°40′N 39°09′E
159 R11 **Madoi** *var.* Huanghe; *prev.* Huangheyan. Qinghai, C China 34°53′N 98°07′E
189 O13 **Madolenihmw** Pohnpei, E Micronesia
118 I9 **Madona** *Ger.* Modohn. E Latvia 56°51′N 26°10′E
107 J23 **Madonie** ▲ Sicilia, Italy, C Mediterranean Sea
141 Y11 **Madrakah, Ra's** *headland* E Oman 18°56′N 57°54′E
32 I12 **Madras** Oregon, NW USA 44°39′N 121°08′W
Madras *see* Chennai
Madras *see* Tamil Nādu
57 H14 **Madre de Dios** *off.* Departamento de Madre de Dios. ◆ *department* E Peru
Madre de Dios, Departamento de *see* Madre de Dios
63 F22 **Madre de Dios, Isla** *island* S Chile
57 J14 **Madre de Dios, Río** ≈ Bolivia/Peru
0 H15 **Madre del Sur, Sierra** ▲ S Mexico
41 Q9 **Madre, Laguna** *lagoon* NE Mexico
25 T16 **Madre, Laguna** *lagoon* Texas, SW USA
37 Q12 **Madre Mount** ▲ New Mexico, SW USA 34°18′N 107°54′W
0 H13 **Madre Occidental, Sierra** *var.* Western Sierra Madre. ▲ C Mexico
0 H13 **Madre Oriental, Sierra** *var.* Eastern Sierra Madre. ▲ C Mexico
41 U17 **Madre, Sierra** *var.* Sierra de Soconusco. ▲ Guatemala/Mexico
37 R2 **Madre, Sierra** ▲ Colorado/Wyoming, C USA
105 N8 **Madrid** ● (Spain) Madrid, C Spain 40°25′N 03°43′W
29 V14 **Madrid** Iowa, C USA 41°52′N 93°49′W
105 N7 **Madrid** ◆ *autonomous community* C Spain
105 N10 **Madridejos** Castilla-La Mancha, C Spain 39°29′N 03°32′W
104 L7 **Madrigal de las Altas Torres** Castilla y León, N Spain 41°05′N 05°00′W
104 K10 **Madrigalejo** Extremadura, W Spain 39°08′N 05°36′W
34 L3 **Mad River** ≈ California, W USA
42 J8 **Madriz** ◆ *department* NW Nicaragua
104 K10 **Madroñera** Extremadura, W Spain 39°25′N 05°46′W
181 N12 **Madura** Western Australia 31°52′S 127°01′E
Madura *see* Madurai
155 H22 **Madurai** *prev.* Madura, Mathurai. Tamil Nādu, S India 09°55′N 78°07′E
169 S16 **Madura, Pulau** *prev.* Madoera. *island* C Indonesia
169 S16 **Madura, Selat** *strait* C Indonesia

◆ Country
● Country Capital
◇ Dependent Territory
○ Dependent Territory Capital
◆ Administrative Regions
✕ International Airport
▲ Mountain
▲ Mountain Range
Volcano
≈ River
◎ Lake
▣ Reservoir

127 Q17 **Madzhalis** Respublika Dagestan, SW Russian Federation 42°12′N 47°46′E
114 K12 **Madzharovo** Haskovo, S Bulgaria 41°36′N 25°52′E
83 M14 **Madzimoyo** Eastern, E Zambia 13°42′S 32°34′E
165 O12 **Maebashi** *var.* Maebasi, Mayebashi. Gunma, Honshū, S Japan 36°24′N 139°02′E
Maebasi *see* Maebashi
167 O6 **Mae Chan** Chiang Rai, NW Thailand 20°13′N 99°52′E
167 N7 **Mae Hong Son** *var.* Maehongson, Muai To. Mae Hong Son, NW Thailand 19°16′N 97°56′E
Maehongson *see* Mae Hong Son
Mae Nam Khong *see* Mekong
167 Q7 **Mae Nam Nan** ≈ NW Thailand
167 O10 **Mae Nam Tha Chin** ≈ W Thailand
167 P7 **Mae Nam Yom** ≈ W Thailand
37 O3 **Maeser** Utah, W USA 40°28′N 109°35′W
Maeseyck *see* Maaseik
167 N9 **Mae Sot** *var.* Ban Mae Sot. Tak, W Thailand 16°44′N 98°32′E
44 H8 **Maestra, Sierra** ▲▲ E Cuba
Maestricht *see* Maastricht
167 O7 **Mae Suai** *var.* Ban Mae Suai. Chiang Rai, NW Thailand 19°43′N 99°30′E
167 O7 **Mae Tho, Doi** ▲ NW Thailand 18°56′N 99°20′E
172 I4 **Maevatanana** Mahajanga, C Madagascar 16°57′S 46°50′E
187 R13 **Maéwo** *prev.* Aurora. *island* C Vanuatu
171 S11 **Mafa** Pulau Halmahera, E Indonesia 0°01′N 127°50′E
83 I23 **Mafeteng** W Lesotho 29°48′S 27°15′E
99 J21 **Maffe** Namur, SE Belgium 50°21′N 05°19′E
183 P12 **Maffra** Victoria, SE Australia 37°59′S 147°03′E
81 K23 **Mafia** *island* E Tanzania
81 J23 **Mafia Channel** *sea waterway* E Tanzania
83 I21 **Mafikeng** *off.* Mahikeng. North-West, N South Africa 25°53′S 25°39′E
60 J12 **Mafra** Santa Catarina, S Brazil 26°08′S 49°47′W
104 F10 **Mafra** Lisboa, C Portugal 38°57′N 09°19′W
143 Q17 **Mafraq** Abū Ẓaby, C United Arab Emirates 24°21′N 54°33′E
Mafraq/Muḥāfaẓat al Mafraq *see* Al Mafraq
123 T10 **Magadan** Magadanskaya Oblast', E Russian Federation 59°38′N 150°50′E
123 T9 **Magadanskaya Oblast'** ◈ *province* E Russian Federation
108 G11 **Magadino** Ticino, S Switzerland 46°09′N 08°50′E
63 G23 **Magallanes** *var.* Región de Magallanes y de la Antártica Chilena. ◈ *region* S Chile
Magallanes *see* Punta Arenas
Magallanes, Estrecho de *see* Magellan, Strait of
Magallanes y de la Antártica Chilena, Región de *see* Magallanes
14 I10 **Maganasipi, Lac** ◎ Québec, SE Canada
54 F6 **Magangué** Bolívar, N Colombia 09°14′N 74°46′W
191 Y13 **Magareva** *var.* Mangareva. *island* Îles Tuamotu, SE French Polynesia
77 V12 **Magaria** Zinder, S Niger 13°00′N 08°55′E
186 F10 **Magarida** Central, SW Papua New Guinea 10°10′S 149°21′E
171 O2 **Magat** ≈ Luzon, N Philippines
27 T11 **Magazine Mountain** ▲ Arkansas, C USA 35°10′N 93°38′W
76 I15 **Magburaka** C Sierra Leone 08°44′N 11°57′W
123 Q13 **Magdagachi** Amurskaya Oblast', SE Russian Federation 53°25′N 125°41′E
62 O12 **Magdalena** Buenos Aires, E Argentina 35°05′S 57°30′W
57 M15 **Magdalena** El Beni, N Bolivia 13°22′S 64°07′W
40 F4 **Magdalena** Sonora, NW Mexico 30°38′N 110°59′W
37 Q13 **Magdalena** New Mexico, SW USA 34°07′N 107°14′W
54 F5 **Magdalena** *off.* Departamento del Magdalena. ◈ *province* N Colombia
40 E9 **Magdalena, Bahía** *bay* W Mexico
Magdalena, Departamento del *see* Magdalena
63 G19 **Magdalena, Isla** *island* Archipiélago de los Chonos, S Chile
40 D8 **Magdalena, Isla** *island* NW Mexico
47 P6 **Magdalena, Río** ≈ C Colombia
40 F4 **Magdalena, Río** ≈ NW Mexico
Magdalen Islands *see* Madeleine, Îles de la
147 N14 **Magdanly** *Rus.* Govurdak; *prev.* gowurdak, Guardak. Lebap Welaýaty, E Turkmenistan 37°50′N 66°06′E
100 L13 **Magdeburg** Sachsen-Anhalt, C Germany 52°08′N 11°39′E
22 L6 **Magee** Mississippi, S USA 31°52′N 89°43′W
169 Q16 **Magelang** Jawa, C Indonesia 07°28′S 110°11′E
192 K7 **Magellan Rise** *undersea feature* C Pacific Ocean
63 H24 **Magellan, Strait of** *Sp.* Estrecho de Magallanes. *strait* Argentina/Chile
106 D7 **Magenta** Lombardia, NW Italy 45°28′N 08°52′E
Magerøy *see* Magerøya
92 K7 **Magerøya** *var.* Magerøy, *Lapp.* Máhkarávju. *island* N Norway
164 C17 **Mage-shima** *island* Nansei-shotō, SW Japan
108 G11 **Maggia** Ticino, S Switzerland 46°15′N 08°42′E
108 G10 **Maggia** ≈ SW Switzerland
Maggiore, Lago *see* Maggiore, Lake
106 C6 **Maggiore, Lake** *It.* Lago Maggiore. ◎ Italy/Switzerland
44 I12 **Maggotty** W Jamaica 18°09′N 77°46′W
76 I10 **Maghama** Gorgol, S Mauritania 15°31′N 12°50′W
97 F14 **Maghera** *Ir.* Machaire Rátha. C Northern Ireland, United Kingdom 54°51′N 06°40′W
97 F15 **Magherafelt** *Ir.* Machaire Fíolta. C Northern Ireland, United Kingdom 54°45′N 06°36′W
188 H6 **Magicienne Bay** *bay* Saipan, S Northern Mariana Islands
105 O13 **Magina** ▲ S Spain 37°43′N 03°24′W
81 H24 **Magingo** Ruvuma, S Tanzania 09°57′S 35°23′E
112 H11 **Maglaj** ◈ Federacija Bosne I Hercegovine, N Bosnia and Herzegovina
107 Q19 **Maglie** Puglia, SE Italy 40°07′N 18°18′E
114 K10 **Maglizh** *var.* Mŭglizh. Stara Zagora, C Bulgaria 42°36′N 25°32′E
36 L2 **Magna** Utah, W USA 40°42′N 112°06′W
Magnesia *see* Manisa
14 G12 **Magnetawan** ≈ Ontario, S Canada
27 T14 **Magnolia** Arkansas, C USA 33°17′N 93°16′W
22 K7 **Magnolia** Mississippi, S USA 31°08′N 90°27′W
25 V10 **Magnolia** Texas, SW USA 30°12′N 95°46′W
Magnolia State *see* Mississippi
95 J15 **Magnor** Hedmark, S Norway 59°57′N 12°14′E
187 Y14 **Mago** *prev.* Mango. *island* Lau Group, E Fiji
83 L15 **Màgoé** Tete, NW Mozambique 15°50′S 31°42′E
15 Q13 **Magog** Québec, SE Canada 45°16′N 72°09′W
83 J15 **Magoye** Southern, S Zambia 16°00′S 27°34′E
41 Q12 **Magozal** Veracruz-Llave, C Mexico 21°33′N 97°57′W
14 B7 **Magpie** ≈ Ontario, S Canada
11 Q17 **Magrath** Alberta, SW Canada 49°27′N 112°52′W
105 R10 **Magre** ≈ Valenciana, E Spain
76 I9 **Magta' Lahjar** *var.* Magta Lahjar, Magta' Lahjar, Magtá Lahjar. Brakna, SW Mauritania 17°27′N 13°07′W
146 D12 **Magtymguly** *prev.* Garrygala, *Rus.* Kara-Kala. Balkan Welaýaty, W Turkmenistan 38°27′N 56°15′E
83 L20 **Magude** Maputo, S Mozambique 25°02′S 32°40′E
77 Y12 **Magumeri** Borno, NE Nigeria 12°07′N 12°48′E
189 O14 **Magur Islands** *island group* Caroline Islands, C Micronesia
166 L6 **Magway** *var.* Magwe. Magway, W Myanmar (Burma) 20°08′N 94°55′E
166 L6 **Magway** *var.* Magwe. ◈ *region* C Myanmar (Burma)
Magwe *see* Magway
Magyar-Becse *see* Bečej
Magyarkanizsa *see* Kanjiža
Magyarország *see* Hungary
Magyarzsombor *see* Zimbor
142 J4 **Mahābād** *var.* Mehabad; *prev.* Sāūjbulāgh. Āžarbāyjān-e Gharbī, NW Iran 36°44′N 45°44′E
172 H5 **Mahabo** Toliara, W Madagascar 20°22′S 44°39′E
Maha Chai *see* Samut Sakhon
155 D14 **Mahād** Mahārāshtra, W India 18°04′N 73°21′E
81 N17 **Mahadday Weyne** Shabeellaha Dhexe, C Somalia 02°55′N 45°30′E
79 Q17 **Mahagi** Orientale, NE Dem. Rep. Congo 02°16′N 30°59′E
Mahāil *see* Muhāyil
172 I4 **Mahajamba** *seasonal river* NW Madagascar
152 G10 **Mahājan** Rājasthān, NW India 28°47′N 73°50′E
172 I3 **Mahajanga** *var.* Majunga. Mahajanga, NW Madagascar 15°40′S 46°20′E
172 I3 **Mahajanga** ◈ *province* W Madagascar
172 I3 **Mahajanga** ✈ Mahajanga, NW Madagascar 15°40′S 46°20′E
169 U10 **Mahakam, Sungai** *var.* Koetai, Kutai. ≈ Borneo, C Indonesia
83 I19 **Mahalapye** *var.* Mahalatswe. Central, SE Botswana 23°02′S 26°53′E
Mahalatswe *see* Mahalapye
Mahalla el Kubra *see* El Mahalla el Kubra
171 O13 **Mahalona** Sulawesi, C Indonesia 02°37′S 121°26′E
Mahameru *see* Semeru, Gunung
143 S11 **Māhān** Kermān, E Iran 30°00′N 57°00′E
154 N12 **Mahanādi** ≈ E India
172 J5 **Mahanoro** Toamasina, E Madagascar 19°53′S 48°48′E
153 P13 **Mahārājganj** Bihār, N India 26°07′N 84°31′E
154 G13 **Mahārāshtra** ◈ *state* W India
172 I4 **Mahavavy** *seasonal river* N Madagascar
155 K24 **Mahaweli Ganga** ≈ C Sri Lanka
Mahbés *see* El Mahbas
155 J15 **Mahbūbābād** Telangana, E India 17°35′N 80°00′E
155 H16 **Mahbūbnagar** Telangana, C India 16°46′N 78°01′E
140 M8 **Mahd adh Dhahab** Al Madīnah, W Saudi Arabia 23°33′N 40°56′E
55 S9 **Mahdia** C Guyana 05°16′N 59°08′W
75 N6 **Mahdia** *var.* Al Mahdīyah, Mehdia. NE Tunisia 35°14′N 11°06′E
155 F20 **Mahe** *Fr.* Mahé; *prev.* Mayyali. Puducherry, SW India 11°41′N 75°31′E
172 I16 **Mahé** ✈ Mahé, NE Seychelles 04°37′S 55°27′E
172 H16 **Mahé** *island* Inner Islands, NE Seychelles
Mahé *see* Mahe
173 Y17 **Mahebourg** SE Mauritius 20°24′S 57°42′E
152 I11 **Mahendragarh** *prev.* Mohendergarh. Haryāna, N India 28°17′N 76°14′E
152 L10 **Mahendranagar** Far Western, W Nepal 28°58′N 80°13′E
81 I23 **Mahenge** Morogoro, SE Tanzania 08°41′S 36°41′E
185 F22 **Maheno** Otago, South Island, New Zealand 45°10′S 170°51′E
154 D9 **Mahesāna** Gujarāt, W India 23°37′N 72°28′E
154 F11 **Maheshwar** Madhya Pradesh, C India 22°11′N 75°40′E
153 V17 **Maheskhali Island** *var.* Maiskhal Island. *island* SE Bangladesh
151 F14 **Mahi** ≈ N India
184 Q10 **Mahia Peninsula** *peninsula* North Island, New Zealand
Mahikeng *see* Mafikeng
119 O16 **Mahilyow** *Rus.* Mogilëv. Mahilyowskaya Voblasts', E Belarus 53°55′N 30°23′E
119 M16 **Mahilyowskaya Voblasts'** *Rus.* Mogilëvskaya Oblast'. ◈ *province* E Belarus
191 P7 **Mahina** Tahiti, W French Polynesia 17°29′S 149°27′W
185 E23 **Mahinerangi, Lake** ◎ South Island, New Zealand
Máhkarávju *see* Magerøya
83 L22 **Mahlabatini** KwaZulu/Natal, E South Africa 28°15′S 31°28′E
166 L5 **Mahlaing** Mandalay, C Myanmar (Burma) 21°03′N 95°44′E
109 X8 **Mahldorf** Steiermark, SE Austria 46°54′N 15°55′E
Maḥmūd-e 'Erāqī *see* Maḥmūd-e Rāqī
149 R4 **Maḥmūd-e Rāqī** *var.* Mahmūd-e 'Erāqī. Kāpīsā, NE Afghanistan 35°01′N 69°20′E
Mahmudiya *see* Al Maḥmūdīyah
29 S5 **Mahnomen** Minnesota, N USA 47°19′N 95°58′W
152 K14 **Mahoba** Uttar Pradesh, N India 25°18′N 79°53′E
Mahón *see* Maó
18 D14 **Mahoning Creek Lake** ⊠ Pennsylvania, NE USA
105 Q10 **Mahora** Castilla-La Mancha, C Spain 39°13′N 01°44′W
Mähren *see* Moravia
Mährisch-Budwitz *see* Moravské Budějovice
Mährisch-Kromau *see* Moravský Krumlov
Mährisch-Neustadt *see* Uničov
Mährisch-Schönberg *see* Šumperk
Mährisch-Trübau *see* Moravská Třebová
Mährisch-Weisskirchen *see* Hranice
Māh-Shahr *see* Bandar-e Māhshahr
79 N19 **Mahulu** Maniema, E Dem. Rep. Congo 01°04′S 27°10′E
154 C12 **Mahuva** Gujarāt, W India 21°06′N 71°46′E
114 N11 **Mahya Dağı** ▲ NW Turkey 41°47′N 27°34′E
105 T6 **Maials** *var.* Mayals. Cataluña, NE Spain 41°22′N 00°30′E
191 O2 **Maiana** *prev.* Hall Island. *atoll* Tungaru, W Kiribati
191 S11 **Maiao** *var.* Tapuaemanu, Tubuai-Manu. *island* Îles du Vent, W French Polynesia
54 H4 **Maicao** La Guajira, N Colombia 11°23′N 72°16′W
Mai Ceu/Mai Chio *see* Maych'ew
103 U8 **Maîche** Doubs, E France 47°15′N 06°43′E
149 Q5 **Maīdān Shahr** *var.* Maydān Shahr; *prev.* Meydān Shahr. Wardak, E Afghanistan 34°27′N 68°48′E
97 N22 **Maidenhead** S England, United Kingdom 51°32′N 00°44′W
11 S15 **Maidstone** Saskatchewan, S Canada 53°06′N 109°21′W
97 P22 **Maidstone** SE England, United Kingdom 51°17′N 00°31′E
77 Y13 **Maiduguri** Borno, NE Nigeria 11°51′N 13°10′E
108 I8 **Maienfeld** Sankt Gallen, NE Switzerland 47°01′N 09°30′E
116 J12 **Măieruş** *Hung.* Szászmagyarós. Braşov, C Romania 45°55′N 25°30′E
Maigh Chromtha *see* Macroom
Maigh Eo *see* Mayo
55 N9 **Maigualida, Sierra** ▲▲ S Venezuela
154 K9 **Maihar** Madhya Pradesh, C India 24°18′N 80°46′E
154 K11 **Maikala Range** ▲▲ C India
67 T10 **Maiko** ≈ W Dem. Rep. Congo
Mailand *see* Milano
152 L11 **Mailāni** Uttar Pradesh, N India 28°17′N 80°20′E
149 U10 **Māilsi** Punjab, E Pakistan 29°46′N 72°15′E
147 R8 **Maimak** Talasskaya Oblast', NW Kyrgyzstan 42°40′N 71°12′E
Maimāna *see* Maimanah
148 M3 **Maimanah** *var.* Maimāna, Maymana; *prev.* Meymaneh. Fāryāb, NW Afghanistan 35°57′N 64°48′E
171 V13 **Maimawa** Papua Barat, E Indonesia 03°21′S 133°36′E
Maimuna *see* Al Maymūnah
101 G18 **Main** ≈ C Germany
115 F22 **Maïna** *ancient monument* Pelopónnisos, S Greece
115 E20 **Maïnalo** ▲▲ S Greece
101 L22 **Mainburg** Bayern, SE Germany 48°40′N 11°48′E
Main Camp *see* Banana
14 E12 **Main Channel** *lake channel* Ontario, S Canada
79 I20 **Mai-Ndombe, Lac** *prev.* Lac Léopold II. ◎ W Dem. Rep. Congo
101 K20 **Main-Donau-Kanal** *canal* SE Germany
19 R6 **Maine** *off.* State of Maine, *also known as* Lumber State, Pine Tree State. ◈ *state* NE USA
102 K6 **Maine** *cultural region* NW France
102 J7 **Maine-et-Loire** ◈ *department* NW France
19 Q9 **Maine, Gulf of** *gulf* NE USA
77 X12 **Maïné-Soroa** Diffa, SE Niger 13°14′N 12°00′E
167 N2 **Maingkwan** *var.* Mungkawn. Kachin State, N Myanmar (Burma) 26°20′N 96°37′E
Main Island *see* Bermuda
Mainistir Fhear Maí *see* Fermoy
Mainistirna Búille *see* Boyle
Mainistir na Corann *see* Midleton
Mainistir na Féile *see* Abbeyfeale
96 J5 **Mainland** *island* N Scotland, United Kingdom
96 L2 **Mainland** *island* NE Scotland, United Kingdom
159 P16 **Mainling** *var.* Tungdor. Xizang Zizhiqu, W China 29°12′N 94°06′E
152 K12 **Mainpuri** Uttar Pradesh, N India 27°14′N 79°01′E
103 N5 **Maintenon** Eure-et-Loir, C France 48°35′N 01°34′E
172 H4 **Maintirano** Mahajanga, W Madagascar 18°01′S 44°03′E
93 M15 **Mainua** Kainuu, C Finland 64°05′N 27°28′E
101 G18 **Mainz** *Fr.* Mayence. Rheinland-Pfalz, SW Germany 50°00′N 08°16′E
76 I9 **Maio** *var.* Vila do Maio. Maio, S Cape Verde 15°07′N 23°12′W
76 E10 **Maio** *var.* Mayo. *island* Ilhas de Sotavento, SE Cape Verde
62 G12 **Maipo, Río** ≈ C Chile
62 H12 **Maipo, Volcán** ⛰ W Argentina 34°09′S 69°51′W
61 E22 **Maipú** Buenos Aires, E Argentina 36°52′S 57°52′W
62 I11 **Maipú** Mendoza, E Argentina 33°00′S 68°46′W
62 H11 **Maipú** Santiago, C Chile 33°30′S 70°52′W
106 A9 **Maira** ≈ NW Italy
108 I10 **Maira** *It.* Mera. ≈ Italy/Switzerland
153 V12 **Mairābari** Assam, NE India 26°28′N 92°22′E
44 K7 **Maisí** Guantánamo, E Cuba 20°13′N 74°08′W
118 H13 **Maišiagala** Vilnius, SE Lithuania 54°52′N 25°03′E
Maiskhal Island *see* Maheskhali Island
167 N13 **Mai Sombun** Chumphon, SW Thailand 10°49′N 99°13′E
Mai Son *see* Hat Lot
Maisur *see* Karnātaka, India
Maisur *see* Mysore, India
183 T8 **Maitland** New South Wales, SE Australia 32°33′S 151°33′E
182 I9 **Maitland** South Australia 34°21′S 137°42′E
14 F15 **Maitland** ≈ Ontario, S Canada
195 R1 **Maitri** *Indian research station* Antarctica 70°03′S 08°59′E
159 N15 **Maizhokunggar** Xizang Zizhiqu, W China 29°50′N 91°40′E
43 O10 **Maíz, Islas del** *var.* Corn Islands. *island group* SE Nicaragua
164 J12 **Maizuru** Kyōto, Honshū, SW Japan 35°30′N 135°20′E
54 F6 **Majagual** Sucre, N Colombia 08°36′N 74°39′W
41 Z13 **Majahual** Quintana Roo, E Mexico 18°43′N 87°43′W
Mājeej *see* Mejit Island
171 N13 **Majene** *prev.* Madjene. Sulawesi, C Indonesia 03°33′S 118°59′E
43 V15 **Majé, Serranía de** ▲▲ E Panama
112 I11 **Majevica** ▲▲ NE Bosnia and Herzegovina
81 H15 **Maji** Southern Nationalities, S Ethiopia 06°11′N 35°32′E
141 X7 **Majīs** NW Oman 24°25′N 56°34′E
Majorca *see* Mallorca
105 X9 **Major, Puig** ▲ Mallorca, Spain, W Mediterranean Sea 39°50′N 2°50′E
Mājro *see* Majuro Atoll
Majunga *see* Mahajanga
189 Y3 **Majuro** ✈ Majuro Atoll, SE Marshall Islands 07°05′N 171°08′E
189 Y2 **Majuro Atoll** *var.* Mājro. *atoll* Ratak Chain, SE Marshall Islands
189 X2 **Majuro Lagoon** *lagoon* Majuro Atoll, SE Marshall Islands
76 H11 **Maka** C Senegal 13°40′N 14°12′W
79 F20 **Makabana** Niari, SW Congo 03°28′S 12°36′E
38 D9 **Mākaha** *var.* Makaha. O'ahu, Hawaii, USA, C Pacific Ocean 21°28′N 158°13′W
38 B8 **Makahu'ena Point** *var.* Makahuena Point. *headland* Kaua'i, Hawai'i, USA 21°52′N 159°28′W
38 D9 **Makakilo City** O'ahu, Hawaii, USA, C Pacific Ocean 21°21′N 158°05′W
83 H18 **Makalamabedi** Central, C Botswana 20°19′S 23°51′E
Makale *see* Mek'elē
158 K17 **Makalu** *Chin.* Makaru Shan. ▲ China/Nepal 27°53′N 87°09′E
81 G23 **Makampi** Mbeya, S Tanzania 08°00′S 33°17′E
Makanchi *see* Makanshy
145 X12 **Makanshy** *prev.* Makanchi. Vostochnyy Kazakhstan, E Kazakhstan 46°47′N 82°00′E
42 M8 **Makantaka** Región Autónoma Atlántico Norte, NE Nicaragua 13°13′N 84°04′W
190 B16 **Makapu Point** *headland* W Niue 18°59′S 169°56′E
185 C24 **Makarewa** Southland, South Island, New Zealand 46°17′S 168°16′E
117 O4 **Makariv** Kyyivs'ka Oblast', N Ukraine 50°28′N 29°49′E
185 D20 **Makarora** ≈ South Island, New Zealand
123 T13 **Makarov** Ostrov Sakhalin, Sakhalinskaya Oblast', SE Russian Federation 48°24′N 142°37′E
197 R9 **Makarov Basin** *undersea feature* Arctic Ocean
192 I5 **Makarov Seamount** *undersea feature* W Pacific Ocean 29°30′N 153°30′E
113 F15 **Makarska** *It.* Macarsca. Split-Dalmacija, SE Croatia 43°18′N 17°00′E
Makaru Shan *see* Makalu
125 O15 **Makar'yev** Kostromskaya Oblast', NW Russian Federation 57°52′N 43°46′E
82 L11 **Makasa** Northern, NE Zambia 09°42′S 31°54′E
Makasar *see* Makassar
Makasar, Selat *see* Makassar Straits
170 M14 **Makassar** *var.* Macassar, Makasar; *prev.* Ujungpandang. Sulawesi, C Indonesia 05°09′S 119°28′E
192 F7 **Makassar Straits** *Ind.* Makasar Selat. *strait* C Indonesia
144 G12 **Makat** *Kaz.* Maqat. Atyrau, SW Kazakhstan 47°40′N 53°28′E
191 T10 **Makatea** *island* Îles Tuamotu, C French Polynesia
139 U7 **Makātū** Diyālá, E Iraq 33°55′N 45°25′E
172 H6 **Makay** *var.* Massif du Makay. ▲▲ SW Madagascar
Makay, Massif du *see* Makay
114 J12 **Makaza** *pass* Bulgaria/Greece
Makedonija *see* Macedonia, FYR
190 B16 **Makefu** W Niue 18°59′S 169°55′W
191 V10 **Makemo** *atoll* Îles Tuamotu, C French Polynesia
76 I15 **Makeni** C Sierra Leone 08°57′N 12°02′W
Makenzen *see* Orlyak
Makeyevka *see* Makiyivka
127 Q16 **Makhachkala** *prev.* Petrovsk-Port. Respublika Dagestan, SW Russian Federation 42°58′N 47°30′E
Makhado *see* Louis Trichardt
144 F11 **Makhambet** Atyrau, W Kazakhstan 47°35′N 51°35′E
Makharadze *see* Ozurgeti
139 W13 **Makhfar al Buşayyah** Al Muthannā, S Iraq 30°09′N 46°09′E
Makhmūr *see* Mexmûr
138 I11 **Makhrūq, Wadi al** *dry watercourse* E Jordan
139 R4 **Makḥūl, Jabal** ▲▲ C Iraq
141 R13 **Makhyah, Wādī** *dry watercourse* N Yemen
171 V13 **Maki** Papua Barat, E Indonesia 03°00′S 134°10′E
185 G21 **Makikihi** Canterbury, South Island, New Zealand 44°36′S 171°09′E
191 O2 **Makin** *prev.* Pitt Island. *atoll* Tungaru, W Kiribati
81 I20 **Makindu** Makueni, S Kenya 02°15′S 37°49′E
145 Q8 **Makinsk** Akmola, N Kazakhstan 52°40′N 70°28′E
Makira *see* Makira-Ulawa
Makira *see* San Cristobal
187 N10 **Makira-Ulawa** *prev.* Makira. ◈ *province* SE Solomon Islands
117 X8 **Makiyivka** *Rus.* Makeyevka; *prev.* Dmitriyevsk. Donets'ka Oblast', E Ukraine 47°57′N 37°47′E
140 L10 **Makkah** *Eng.* Mecca. Makkah, W Saudi Arabia 21°28′N 39°50′E
140 M10 **Makkah** *var.* Minṭaqat Makkah. ◈ *province* W Saudi Arabia
Makkah, Minṭaqat *see* Makkah
13 R7 **Makkovik** Newfoundland and Labrador, NE Canada 55°06′N 59°07′W
98 K6 **Makkum** Fryslân, N Netherlands 53°03′N 05°25′E
111 M25 **Makó** *Rom.* Macău. Csongrád, SE Hungary 46°14′N 20°28′E
Mako *see* Makung
14 G9 **Makobe Lake** ◎ Ontario, S Canada
79 F18 **Makokou** Ogooué-Ivindo, NE Gabon 0°38′N 12°47′E
81 G23 **Makongolosi** Mbeya, S Tanzania 08°24′S 33°09′E
81 E19 **Makota** SW Uganda 0°37′S 30°12′E
79 G18 **Makoua** Cuvette, C Congo 0°01′S 15°40′E
110 M10 **Maków Mazowiecki** Mazowieckie, C Poland 52°51′N 21°06′E
111 K17 **Maków Podhalański** Małopolskie, S Poland 49°43′N 19°40′E
143 V14 **Makran** *cultural region* Iran/Pakistan
152 G12 **Makrāna** Rājasthān, N India 27°02′N 74°44′E
143 U15 **Makran Coast** *coastal region* SE Iran
119 F20 **Makrany** *Rus.* Mokrany. Brestskaya Voblasts', SW Belarus 51°50′N 24°15′E
Makrinoros *see* Makrynóros
115 H20 **Makrónisosi** *island* Kykládes, Greece, Aegean Sea
115 D17 **Makrynóros** *var.* Makrinoros. ▲▲ C Greece
115 G19 **Makryplági** ▲ C Greece 38°00′N 23°06′E
Maksamaa *see* Maxmo
Maksatha *see* Maksatikha
Maksaticha *see* Maksatikha
124 J15 **Maksatikha** *var.* Maksatha, Maksaticha. Tverskaya Oblast', W Russian Federation 57°49′N 35°46′E
154 G10 **Maksi** Madhya Pradesh, C India 23°20′N 76°36′E
142 I1 **Mākū** Āžarbāyjān-e Gharbī, NW Iran 39°20′N 44°50′E
81 I20 **Makueni** ◈ *county* S Kenya
153 Y11 **Mākum** Assam, NE India 27°28′N 95°28′E
Makum *see* Makung
161 R14 **Makung** *prev.* Mako, Makun. W Taiwan 23°35′N 119°35′E
164 B16 **Makurazaki** Kagoshima, Kyūshū, SW Japan 31°16′N 130°18′E
77 V15 **Makurdi** Benue, C Nigeria 07°42′N 08°36′E
38 L17 **Makushin Volcano** ⛰ Unalaska Island, Alaska, USA 53°53′N 166°55′W
83 K16 **Makwiro** Mashonaland West, N Zimbabwe 17°58′S 30°25′E
57 D15 **Mala** Lima, W Peru 12°40′S 76°36′W
93 I14 **Malå** Västerbotten, N Sweden 65°12′N 18°45′E
Mala *see* Mallow, Ireland
Mala *see* Malaita, Solomon Islands
171 P8 **Malabang** Mindanao, S Philippines
155 E21 **Malabār Coast** *coast* SW India
79 C16 **Malabo** *prev.* Santa Isabel. ● (Equatorial Guinea) Isla de Bioco, NW Equatorial Guinea 03°43′N 08°52′E
79 C16 **Malabo** ✈ Isla de Bioco, N Equatorial Guinea 03°43′N 08°51′E
Malaca *see* Málaga
Malacca *see* Melaka
168 I7 **Malacca, Strait of** *Ind.* Selat Malaka. *strait* Indonesia/Malaysia
Malacka *see* Malacky
111 G20 **Malacky** *Hung.* Malacka. Bratislavský Kraj, W Slovakia 48°26′N 17°01′E
33 R16 **Malad City** Idaho, NW USA 42°10′N 112°16′W
117 Q4 **Mala Divytsya** Chernihivs'ka Oblast', N Ukraine 50°40′N 32°13′E
119 J15 **Maladzyechna** *Pol.* Molodeczno, *Rus.* Molodechno. Minskaya Voblasts', C Belarus 54°19′N 26°51′E
190 D12 **Malaee** Île Futuna, N Wallis and Futuna
190 G12 **Mala'etoli** Île Uvea, E Wallis and Futuna
54 G8 **Málaga** Santander, C Colombia 06°44′N 72°45′W
104 M15 **Málaga** *anc.* Malaca. Andalucía, S Spain 36°43′N 04°25′W
37 V15 **Malaga** New Mexico, SW USA 32°10′N 104°04′W
104 L15 **Málaga** ◈ *province* Andalucía, S Spain
104 M15 **Málaga** ✈ Andalucía, S Spain 36°38′N 04°36′W
Malagasy Republic *see* Madagascar
105 N10 **Malagón** Castilla-La Mancha, C Spain 39°10′N 03°51′W
97 G18 **Malahide** *Ir.* Mullach Íde. Dublin, E Ireland 53°27′N 06°09′W
187 N9 **Malaita** *off.* Malaita Province. ◈ *province* N Solomon Islands
187 N8 **Malaita** *var.* Mala. *island* N Solomon Islands
Malaita Province *see* Malaita
80 F13 **Malakal** Upper Nile, NE South Sudan 09°31′N 31°40′E
112 C10 **Mala Kapela** ▲▲ NW Croatia
25 V7 **Malakoff** Texas, SW USA 32°10′N 96°00′W
Malakula *see* Malekula
149 V7 **Malakwāl** *var.* Mālikwāla. Punjab, E Pakistan 32°32′N 73°18′E
186 E7 **Malalamai** Madang, W Papua New Guinea 05°49′S 146°44′E
171 O13 **Malamala** Sulawesi, C Indonesia 03°21′S 120°58′E
169 S17 **Malang** Jawa, C Indonesia 07°59′S 112°45′E
83 O14 **Malanga** Niassa, N Mozambique 13°27′S 36°05′E
Malange *see* Malanje
92 I9 **Malangen** *sound* N Norway
82 C11 **Malanje** *var.* Malange. Malanje, NW Angola 09°34′S 16°25′E
82 C11 **Malanje** *var.* Malange. ◈ *province* N Angola
148 M16 **Malān, Rās** *cape* SW Pakistan
77 S13 **Malanville** NE Benin 11°50′N 03°23′E
Malapane *see* Ozimek
155 F21 **Malappuram** Kerala, SW India 11°00′N 76°02′E
43 T17 **Mala, Punta** *headland* S Panama 07°28′N 79°58′W
148 L15 **Malar** Baluchistān, SW Pakistan 26°19′N 64°55′E
95 N16 **Mälaren** ◎ C Sweden
62 H13 **Malargüe** Mendoza, W Argentina 35°32′S 69°35′W
14 J8 **Malartic** Québec, SE Canada 48°09′N 78°09′W
119 F20 **Malaryta** *Pol.* Maloryta, *Rus.* Malorita. Brestskaya Voblasts', SW Belarus 51°47′N 24°05′E
63 J19 **Malaspina** Chubut, SE Argentina 44°56′S 66°52′W
10 G8 **Malaspina Glacier** *glacier* Yukon, W Canada
39 U12 **Malaspina Glacier** *glacier* Alaska, USA
137 N15 **Malatya** *anc.* Melitene. Malatya, SE Turkey 38°22′N 38°18′E
136 M14 **Malatya** ◈ *province* C Turkey
Malavate *see* Pédima
117 Q7 **Mala Vyska** *Rus.* Malaya Viska. Kirovohrads'ka Oblast', S Ukraine 48°37′N 31°36′E
83 M14 **Malawi** *off.* Republic of Malaŵi; *prev.* Nyasaland, Nyasaland Protectorate. ◆ *republic* S Africa
Malawi, Lake *see* Nyasa, Lake
Malawi, Republic of *see* Malawi
93 J17 **Malax** *Fin.* Maalahti. Österbotten, W Finland 62°55′N 21°30′E
124 H14 **Malaya Vishera** Novgorodskaya Oblast', W Russian Federation 58°52′N 32°12′E
Malaya Viska *see* Mala Vyska
171 Q7 **Malaybalay** Mindanao, S Philippines 08°10′N 125°08′E
142 L6 **Malāyer** *prev.* Daulatabad. Hamadān, W Iran 34°20′N 48°47′E
168 J7 **Malay Peninsula** *peninsula* Malaysia/Thailand
168 L7 **Malaysia** *off.* Malaysia, *var.* Federation of Malaysia; *prev.* the separate territories of Federation of Malaya, Sarawak and Sabah (North Borneo) and Singapore. ◆ *monarchy* SE Asia
Malaysia, Federation of *see* Malaysia
137 R14 **Malazgirt** Muş, E Turkey 39°09′N 42°30′E
15 R8 **Malbaie** ≈ Québec, SE Canada
77 T12 **Malbaza** Tahoua, S Niger 13°57′N 05°32′E
110 J7 **Malbork** *Ger.* Marienburg, Marienburg in Westpreussen. Pomorskie, N Poland 54°01′N 19°03′E
100 N9 **Malchin** Mecklenburg-Vorpommern, N Germany 53°43′N 12°46′E
100 M9 **Malchiner See** ◎ NE Germany
99 D16 **Maldegem** Oost-Vlaanderen, NW Belgium 51°12′N 03°27′E
98 L13 **Malden** Gelderland, SE Netherlands 51°47′N 05°51′E
19 O11 **Malden** Massachusetts, NE USA 42°25′N 71°04′W
27 Y8 **Malden** Missouri, C USA 36°33′N 89°58′W
191 X4 **Malden Island** *prev.* Independence Island. *atoll* E Kiribati
173 Q6 **Maldives** *off.* Maldivian Divehi, Republic of Maldives. ◆ *republic* N Indian Ocean
Maldives, Republic of *see* Maldives
Maldivian Divehi *see* Maldives
97 P21 **Maldon** E England, United Kingdom 51°44′N 00°40′E
61 G20 **Maldonado** Maldonado, S Uruguay 34°57′S 54°59′W
61 G20 **Maldonado** ◈ *department* S Uruguay
41 P17 **Maldonado, Punta** *headland* S Mexico 16°18′N 98°31′W
106 G6 **Malè** Trentino-Alto Adige, N Italy 46°21′N 10°51′E
151 K19 **Male'** *Div.* Maale. ● (Maldives) Male' Atoll, C Maldives 04°10′N 73°29′E
76 K13 **Maléa** *var.* Maléya. NE Guinea 11°46′N 09°43′W
Maléas, Ákra *see* Agriliá, Akrotírio
115 G22 **Maléas, Akrotírio** *headland* S Greece 36°25′N 23°11′E
151 K19 **Male' Atoll** *var.* Kaafu Atoll. *atoll* C Maldives
Malebo, Pool *see* Stanley Pool
154 E12 **Mālegaon** Mahārāshtra, W India 20°33′N 74°32′E
81 F15 **Malek** Jonglei, E South Sudan 06°04′N 31°36′E
187 Q13 **Malekula** *var.* Malakula; *prev.* Mallicolo. *island* W Vanuatu
189 Y15 **Malem** Kosrae, E Micronesia 05°16′N 163°01′E
83 O15 **Malema** Nampula, N Mozambique 14°57′S 37°28′E
79 N23 **Malemba-Nkulu** Katanga, SE Dem. Rep. Congo 08°01′S 26°48′E
139 T1 **Malḥe Mela** *Ar.* Marī Mīlā, *var.* Mārī Mīlah. Arbīl, E Iraq 36°58′N 44°42′E
124 K9 **Malen'ga** Respublika Kareliya, NW Russian Federation 63°50′N 36°21′E
95 M20 **Målerås** Kalmar, S Sweden 56°55′N 15°34′E
103 O6 **Malesherbes** Loiret, C France 48°18′N 02°25′E
115 G18 **Malesína** Stereá Elláda, E Greece 38°37′N 23°15′E
Maléya *see* Maléa
127 O15 **Malgobek** Respublika Ingushetiya, SW Russian Federation 43°34′N 44°34′E
105 X5 **Malgrat de Mar** Cataluña, NE Spain 41°39′N 02°45′E
80 C9 **Malha** Northern Darfur, W Sudan 15°07′N 26°00′E
139 Q5 **Malḥah** *var.* Malḥāt. Ṣalāḥ ad Dīn, C Iraq 34°44′N 42°41′E
Malḥāt *see* Malḥah
32 K14 **Malheur Lake** ◎ Oregon, NW USA
32 L14 **Malheur River** ≈ Oregon, NW USA
76 I13 **Mali** NW Guinea 12°08′N 12°29′W
77 O9 **Mali** *off.* Republic of Mali, *Fr.* République du Mali; *prev.* French Sudan, Sudanese Republic. ◆ *republic* W Africa
171 Q16 **Maliana** W East Timor 08°57′S 125°25′E
167 O2 **Mali Hka** ≈ N Myanmar (Burma)
Mali Idjoš *see* Mali Idoš
112 K8 **Mali Idoš** *var.* Mali Idjoš, *Hung.* Kishegyes; *prev.* Krivaja. Vojvodina, N Serbia 45°43′N 19°40′E
113 M18 **Mali i Sharrit** *Serb.* Šar Planina. ▲▲ FYR Macedonia/Serbia
Mali i Zi *see* Crna Gora
112 K9 **Mali Kanal** *canal* N Serbia
171 P12 **Maliku** Sulawesi, N Indonesia 0°36′S 123°13′E
Malik, Wadi al *see* Milk, Wadi el
Malikwāla *see* Malakwāl
167 N11 **Mali Kyun** *var.* Tavoy Island. *island* Mergui Archipelago, S Myanmar (Burma)
95 M19 **Målilla** Kalmar, S Sweden 57°24′N 15°49′E
112 B11 **Mali Lošinj** *It.* Lussinpiccolo. Primorje-Gorski Kotar, W Croatia 44°31′N 14°28′E
Malin *see* Malyn
171 P7 **Malindang, Mount** ▲ Mindanao, S Philippines 08°12′N 123°37′E
81 K20 **Malindi** Kilifi, SE Kenya 03°14′S 40°05′E
Malines *see* Mechelen
96 E13 **Malin Head** *Ir.* Cionn Mhálanna. *headland* NW Ireland 55°37′N 07°37′W
171 O11 **Malino, Gunung** ▲ Sulawesi, N Indonesia 0°44′N 120°45′E
113 M21 **Maliq** *var.* Maliqi. Korçë, SE Albania 40°45′N 20°45′E
Maliqi *see* Maliq
Mali, Republic of *see* Mali
Mali, République du *see* Mali
171 Q8 **Malita** Mindanao, S Philippines 06°13′N 125°39′E
Maljovica *see* Malyovitsa
154 G12 **Malkāpur** Mahārāshtra, C India 20°52′N 76°18′E
136 B10 **Malkara** Tekirdağ, NW Turkey 40°54′N 26°54′E
119 J19 **Mal'kavichy** *Rus.* Mal'kovichi. Brestskaya Voblasts', SW Belarus 52°31′N 26°36′E
Malkiye *see* Al Mālikīyah
114 L11 **Malko Sharkovo, Yazovir** ⊠ SE Bulgaria

◆ Country
● Country Capital
◇ Dependent Territory
○ Dependent Territory Capital
◈ Administrative Regions
✈ International Airport
▲ Mountain
▲▲ Mountain Range
⛰ Volcano
≈ River
◎ Lake
⊠ Reservoir

114 N11 **Malko Tarnovo** var. Malko Tŭrnovo. Burgas, E Bulgaria 42°00′N 27°33′E
Malko Tŭrnovo see Malko Tarnovo
Mal'kovichi see Mal'kavichy
183 R12 **Mallacoota** Victoria, SE Australia 37°34′S 149°45′E
96 G10 **Mallaig** N Scotland, United Kingdom 57°04′N 05°48′W
182 I9 **Mallala** South Australia 34°29′S 138°30′E
75 W9 **Mallawi** var. Mallawī. C Egypt 27°44′N 30°50′E
Mallawī see Mallawi
105 R5 **Mallén** Aragón, NE Spain 41°53′N 01°25′W
106 F5 **Malles Venosta** Ger. Mals im Vinschgau. Trentino-Alto Adige, N Italy 46°40′N 10°37′E
Mallicolo see Malekula
109 Q8 **Mallnitz** Salzburg, S Austria 46°58′N 13°09′E
105 W9 **Mallorca** Eng. Majorca; anc. Baleares Major. island Islas Baleares, Spain, W Mediterranean Sea
97 C20 **Mallow** Ir. Mala. SW Ireland 52°08′N 08°39′W
93 E15 **Malm** Nord-Trøndelag, C Norway 64°04′N 11°12′E
95 L19 **Malmbäck** Jönköping, S Sweden 57°34′N 14°30′E
92 J12 **Malmberget** Lapp. Malmivaara. Norrbotten, N Sweden 67°09′N 20°39′E
99 M20 **Malmédy** Liège, E Belgium 50°26′N 06°02′E
83 E25 **Malmesbury** Western Cape, SW South Africa 33°28′S 18°43′E
Malmivaara see Malmberget
95 N16 **Malmköping** Södermanland, C Sweden 59°08′N 16°49′E
95 K23 **Malmö** Skåne, S Sweden 55°36′N 13°E
95 K23 **Malmo** ✈ Skåne, S Sweden 55°33′N 13°23′E
45 Q16 **Malmok** headland N Bonaire 12°16′N 68°21′W
95 M18 **Malmslätt** Östergötland, S Sweden 58°25′N 15°30′E
125 R16 **Malmyzh** Kirovskaya Oblast', NW Russian Federation 56°30′N 50°37′E
187 Q13 **Malo** island W Vanuatu
126 J7 **Maloarkhangel'sk** Orlovskaya Oblast', W Russian Federation 52°25′N 36°37′E
Maloelap see Maloelap Atoll
189 V6 **Maloelap Atoll** var. Maloelap. atoll E Marshall Islands
Maloenda see Malunda
108 I10 **Maloja** Graubünden, S Switzerland 46°25′N 09°42′E
82 L12 **Malole** Northern, NE Zambia 10°05′S 31°37′E
171 O3 **Malolos** Luzon, N Philippines 14°51′N 120°49′E
18 K6 **Malone** New York, NE USA 44°51′N 74°18′W
79 K25 **Malonga** Katanga, S Dem. Rep. Congo 10°26′S 23°10′E
111 L17 **Małopolskie** ◈ province SE Poland
Malorita/Maloryta see Malaryta
124 K9 **Maloshuyka** Arkhangel'skaya Oblast', NW Russian Federation 63°43′N 37°20′E
Mal'ovitsa see Malyovitsa
145 V15 **Malovodnoye** Almaty, SE Kazakhstan 43°31′N 77°42′E
94 C10 **Måløy** Sogn Og Fjordane, S Norway 61°57′N 05°06′E
126 K4 **Maloyaroslavets** Kaluzhskaya Oblast', W Russian Federation 55°03′N 36°31′E
122 G7 **Malozemel'skaya Tundra** physical region NW Russian Federation
104 J10 **Malpartida de Cáceres** Extremadura, W Spain 39°26′N 06°30′W
104 K9 **Malpartida de Plasencia** Extremadura, W Spain 39°59′N 06°03′W
106 C7 **Malpensa** ✈ (Milano) Lombardia, N Italy 45°41′N 08°40′E
76 J6 **Malqteïr** desert N Mauritania
Mals im Vinschgau see Malles Venosta
118 J10 **Malta** SE Latvia 56°19′N 27°11′E
33 V7 **Malta** Montana, NW USA 48°21′N 107°52′W
120 M11 **Malta** off. Republic of Malta. ◆ republic C Mediterranean Sea
109 R8 **Malta** var. Maltabach. ∽ S Austria
120 M11 **Malta** island Malta, C Mediterranean Sea
Maltabach see Malta
Malta, Canale di see Malta Channel
120 M11 **Malta Channel** It. Canale di Malta. strait Italy/Malta
83 D20 **Maltahöhe** Hardap, SW Namibia 24°50′S 17°00′E
Malta, Republic of see Malta
97 N16 **Malton** N England, United Kingdom 54°07′N 00°50′W
171 R13 **Maluku** off. Propinsi Maluku, Dut. Molukken, Eng. Moluccas. ◈ province E Indonesia
171 R13 **Maluku** Dut. Molukken, Eng. Moluccas; prev. Spice Islands. island group E Indonesia
Maluku, Laut see Molucca Sea
Maluku, Propinsi see Maluku
171 R11 **Maluku Utara** off. Propinsi Maluku Utara. ◈ province E Indonesia
Maluku Utara, Propinsi see Maluku Utara
77 V13 **Malumfashi** Katsina, N Nigeria 11°51′N 07°39′E
171 N13 **Malunda** prev. Maloenda. Sulawesi, C Indonesia 02°58′S 118°52′E
94 K13 **Malung** Dalarna, C Sweden 60°40′N 13°45′E
94 K13 **Malungsfors** Dalarna, C Sweden 60°43′N 13°34′E
186 M8 **Maluu** var. Malu'u. Malaita, N Solomon Islands 08°22′S 160°39′E
Malu'u see Maluu
155 D16 **Mālvan** Mahārāshtra, W India 16°05′N 73°28′E
Malventum see Benevento
27 U12 **Malvern** Arkansas, C USA 34°21′N 92°50′W
29 S15 **Malvern** Iowa, C USA 40°59′N 95°36′W
44 I13 **Malvern** ▲ W Jamaica 17°59′N 77°42′W
Malvina, Isla Gran see West Falkland
Malvinas, Islas see Falkland Islands
117 N4 **Malyn** Rus. Malin. Zhytomyrs'ka Oblast', N Ukraine 50°46′N 29°14′E
114 G10 **Malyovitsa** var. Maljovica, Mal'ovitsa. ▲ W Bulgaria 42°12′N 23°19′E
127 O11 **Malyye Derbety** Respublika Kalmykiya, SW Russian Federation 47°57′N 44°39′E
Malyy Kavkaz see Lesser Caucasus
123 Q6 **Malyy Lyakhovskiy, Ostrov** island NE Russian Federation
Malyy Pamir see Little Pamir
122 N5 **Malyy Taymyr, Ostrov** island Severnaya Zemlya, N Russian Federation
Malyy Uzen' see Saryozen
122 L14 **Malyy Yenisey** var. Ka-Krem. ∽ S Russian Federation
127 S3 **Mamadysh** Respublika Tatarstan, W Russian Federation 55°46′N 51°22′E
117 N14 **Mamaia** Constanţa, E Romania 44°13′N 28°37′E
187 W14 **Mamanuca Group** island group Yasawa Group, W Fiji
146 L13 **Mamash** Lebap Welaýaty, E Turkmenistan 38°24′N 64°12′E
79 O17 **Mambasa** Orientale, NE Dem. Rep. Congo 01°20′N 29°05′E
171 X13 **Mamberamo, Sungai** ∽ Papua, E Indonesia
79 G15 **Mambéré** ∽ SW Central African Republic
79 G15 **Mambéré-Kadéï** ◈ prefecture SW Central African Republic
Mambij see Manbij
79 H18 **Mambili** ∽ W Congo
83 M18 **Mambone** var. Nova Mambone. Inhambane, E Mozambique 20°59′S 35°04′E
171 O4 **Mamburao** Mindoro, N Philippines 13°16′N 120°36′E
172 I16 **Mamelles** island Inner Islands, NE Seychelles
99 M25 **Mamer** Luxembourg, SW Luxembourg 49°37′N 06°01′E
102 L6 **Mamers** Sarthe, NW France 48°21′N 00°22′E
79 D15 **Mamfe** Sud-Ouest, W Cameroon 05°46′N 09°18′E
145 P6 **Mamlyutka** Severnyy Kazakhstan, N Kazakhstan 54°54′N 68°36′E
36 M15 **Mammoth** Arizona, SW USA 32°43′N 110°38′W
33 S12 **Mammoth Hot Springs** Wyoming, C USA 44°57′N 110°40′W
Mamoedjoe see Mamuju
119 A14 **Mamonovo** Ger. Heiligenbeil. Kaliningradskaya Oblast', W Russian Federation 54°28′N 19°57′E
57 L14 **Mamoré, Rio** ∽ Bolivia/Brazil
76 I14 **Mamou** W Guinea 10°24′N 12°05′W
22 H8 **Mamou** Louisiana, S USA 30°37′N 92°25′W
172 I14 **Mamoudzou** ○ (Mayotte) C Mayotte 12°48′S 45°E
172 I3 **Mampikony** Mahajanga, N Madagascar 16°03′S 47°39′E
77 P16 **Mampong** C Ghana 07°06′N 01°20′W
110 M7 **Mamry, Jezioro** Ger. Mauersee. ◎ NE Poland
171 N13 **Mamuju** prev. Mamoedjoe. Sulawesi, S Indonesia 02°41′S 118°55′E
83 F19 **Mamuno** Ghanzi, W Botswana 22°15′S 20°02′E
113 K19 **Mamuras** var. Mamurasi, Mamurras. Lezhë, C Albania 41°34′N 19°42′E
Mamurasi/Mamurras see Mamuras
76 L16 **Man** W Ivory Coast 07°24′N 07°33′W
55 X9 **Mana** NW French Guiana 05°40′N 53°49′W
56 A6 **Manabí** ◈ province W Ecuador
42 G4 **Manabique, Punta** var. Cabo Tres Puntas. headland E Guatemala 15°57′N 88°37′W
54 G11 **Manacacías, Río** ∽ C Colombia
58 F13 **Manacapuru** Amazonas, N Brazil 03°16′S 60°37′W
105 Y9 **Manacor** Mallorca, Spain, W Mediterranean Sea 39°35′N 03°12′E
171 Q11 **Manado** prev. Menado. Sulawesi, C Indonesia 01°32′N 124°55′E
188 H5 **Managaha** island S Northern Mariana Islands
99 G20 **Manage** Hainaut, S Belgium 50°30′N 04°14′E
42 J10 **Managua** ● (Nicaragua) Managua, W Nicaragua 12°08′N 86°15′W
42 J10 **Managua** ◈ department W Nicaragua
42 J10 **Managua** ✈ Managua, W Nicaragua 12°07′N 86°11′W
42 J10 **Managua, Lago de** var. Xolotlán. ◎ W Nicaragua
Manah see Bilād Manah
18 K16 **Manahawkin** New Jersey, NE USA 39°39′N 74°12′W
184 K11 **Manaia** Taranaki, North Island, New Zealand 39°33′S 174°07′E
172 J6 **Manakara** Fianarantsoa, SE Madagascar 22°09′S 48°E
152 J7 **Manāli** Himāchal Pradesh, NW India 32°12′N 77°06′E
Ma, Nam see Sông Ma
Manama see Al Manāmah
186 D6 **Manam Island** island N Papua New Guinea
67 Y13 **Mananara Avaratra** ∽ SE Madagascar
182 M9 **Manangatang** Victoria, SE Australia 35°04′S 142°53′E
172 J6 **Mananjary** Fianarantsoa, SE Madagascar 21°13′S 48°20′E
76 L14 **Manankoro** Sikasso, SW Mali 10°33′N 07°25′W
76 J12 **Manantali, Lac de** ◎ W Mali
Manāos see Manaus
185 B23 **Manapouri** Southland, South Island, New Zealand 45°33′S 167°38′E
185 B23 **Manapouri, Lake** ◎ South Island, New Zealand
58 F13 **Manaquiri** Amazonas, NW Brazil 03°27′S 60°37′W
Manar see Mannar
158 K5 **Manas** Xinjiang Uygur Zizhiqu, NW China 44°16′N 86°12′E
153 U12 **Manās** var. Dangme Chu. ∽ Bhutan/India
153 P10 **Manāslu** var. Manaslu. ▲ C Nepal 28°33′N 84°33′E
147 R8 **Manas, Gora** ▲ Kyrgyzstan/Uzbekistan 42°17′N 71°04′E
158 K3 **Manas Hu** ◎ NW China
Manaslu see Manāslu
37 S8 **Manassa** Colorado, C USA 37°10′N 105°56′W
21 W4 **Manassas** Virginia, NE USA 38°45′N 77°28′W
45 T5 **Manatí** C Puerto Rico 18°26′N 66°29′W
186 E8 **Manau** Northern, S Papua New Guinea 08°02′S 148°00′E
54 H4 **Manaure** La Guajira, N Colombia 11°46′N 72°28′W
58 F12 **Manaus** prev. Manáos. state capital Amazonas, NW Brazil 03°06′S 60°W
136 G17 **Manavgat** Antalya, SW Turkey 36°47′N 31°28′E
184 M13 **Manawatu** ∽ North Island, New Zealand
184 L11 **Manawatu-Wanganui** off. Manawatu-Wanganui Region. ◈ region North Island, New Zealand
Manawatu-Wanganui Region see Manawatu-Wanganui
171 R7 **Manay** Mindanao, S Philippines 07°12′N 126°29′E
138 K2 **Manbij** var. Mambij, Fr. Membidj. Ḩalab, N Syria 36°32′N 37°55′E
105 N13 **Mancha Real** Andalucía, S Spain 37°47′N 03°37′W
102 I4 **Manche** ◈ department N France
97 L17 **Manchester** Lat. Mancunium. NW England, United Kingdom 53°30′N 02°15′W
23 S5 **Manchester** Georgia, SE USA 32°51′N 84°37′W
29 Y13 **Manchester** Iowa, C USA 42°28′N 91°27′W
21 N7 **Manchester** Kentucky, S USA 37°09′N 83°46′W
19 O10 **Manchester** New Hampshire, NE USA 42°59′N 71°26′W
20 K10 **Manchester** Tennessee, S USA 35°28′N 86°05′W
18 M9 **Manchester** Vermont, NE USA 43°09′N 73°03′W
97 L18 **Manchester** ✈ NW England, United Kingdom 53°21′N 02°16′W
149 P15 **Manchhar Lake** ◎ SE Pakistan
Man-chou-li see Manzhouli
129 X7 **Manchurian Plain** plain NE China
Máncio Lima see Japiim
Mancunium see Manchester
148 J15 **Mand** Baluchistān, SW Pakistan 26°06′N 61°58′E
Mand see Mand, Rūd-e
81 H25 **Manda** Njombe, SW Tanzania 10°30′S 34°37′E
172 H6 **Mandabe** Toliara, W Madagascar 21°02′S 44°56′E
162 M10 **Mandah** var. Töhöm. Dornogovĭ, SE Mongolia 44°25′N 108°18′E
95 E18 **Mandal** Vest-Agder, S Norway 58°02′N 07°30′E
Mandal see Arbulag, Hövsgöl, Mongolia
Mandal see Batsümber, Töv, Mongolia
166 L5 **Mandalay** Mandalay, C Myanmar (Burma) 21°57′N 96°04′E
166 M6 **Mandalay** ◈ region C Myanmar (Burma)
162 L9 **Mandalgovĭ** Dundgovĭ, C Mongolia 45°47′N 106°18′E
139 V7 **Mandalī** Diyālá, E Iraq 33°43′N 45°33′E
162 K10 **Mandal-Ovoo** var. Sharhulsan. Ömnögovĭ, S Mongolia 44°43′N 104°06′E
95 E18 **Mandalselva** ∽ S Norway
163 P11 **Mandalt** var. Sonid Zuoqi. Nei Mongol Zizhiqu, N China 43°49′N 113°36′E
28 M5 **Mandan** North Dakota, N USA 46°49′N 100°53′W
Mandargiri Hill see Mandār Hill
153 R14 **Mandār Hill** prev. Mandargiri Hill. Bihār, NE India 24°51′N 87°03′E
170 M13 **Mandar, Teluk** bay Sulawesi, C Indonesia
107 C19 **Mandas** Sardegna, Italy, C Mediterranean Sea 39°40′N 09°07′E
Mandasor see Mandsaur
81 L16 **Mandera** Mandera, NE Kenya 03°56′N 41°53′E
81 K17 **Mandera** ◈ county NE Kenya
33 V13 **Manderson** Wyoming, C USA 44°13′N 107°55′W
44 J12 **Mandeville** C Jamaica 18°02′N 77°31′W
22 K9 **Mandeville** Louisiana, S USA 30°21′N 90°04′W
152 I7 **Mandi** Himāchal Pradesh, NW India 31°40′N 76°59′E
76 K14 **Mandiana** E Guinea 10°37′N 08°39′W
Mandi Būrewāla see Būrewāla
Mandidzudzure see Chimanimani
83 M15 **Mandié** Manica, NW Mozambique 16°27′S 33°28′E
83 N14 **Mandimba** Niassa, N Mozambique 14°21′S 35°40′E
57 Q19 **Mandioré, Laguna** ◎ E Bolivia
154 J10 **Mandla** Madhya Pradesh, C India 22°36′N 80°23′E
83 M20 **Mandlakazi** var. Manjacaze. Gaza, S Mozambique 24°47′S 33°50′E
95 E24 **Mandø** var. Manø. island W Denmark
Mandoúdhion/Mandoudi see Mantoúdi
78 I13 **Mandoul** off. Région du Mandoul. ◈ region S Chad
Mandoul, Région du see Mandoul
115 G19 **Mándra** Attikí, C Greece 38°04′N 23°30′E
172 I7 **Mandrare** ∽ S Madagascar
114 M10 **Mandra, Yazovir** salt lake SE Bulgaria
107 L23 **Mandrazzi, Portella** pass Sicilia, Italy, C Mediterranean Sea
172 J3 **Mandritsara** Mahajanga, N Madagascar 15°49′S 48°50′E
143 O13 **Mand, Rūd-e** var. Mand. ∽ S Iran
154 F9 **Mandsaur** prev. Mandasor. Madhya Pradesh, C India 24°03′N 75°10′E
154 F11 **Māndu** Madhya Pradesh, C India 22°22′N 75°24′E
169 W8 **Mandul, Pulau** island N Indonesia
83 G15 **Mandundu** Western, W Zambia 16°34′S 22°18′E
180 I13 **Mandurah** Western Australia 32°31′S 115°41′E
107 P18 **Manduria** Puglia, SE Italy 40°24′N 17°38′E
155 G20 **Mandya** Karnātaka, C India 12°34′N 76°55′E
77 P12 **Mané** C Burkina Faso 12°59′N 01°21′W
106 E8 **Manerbio** Lombardia, NW Italy 45°22′N 10°09′E
Manevichi see Manevychi
116 K3 **Manevychi** Pol. Maniewicze, Rus. Manevichi. Volyns'ka Oblast', NW Ukraine 51°18′N 25°29′E
107 N16 **Manfredonia** Puglia, SE Italy 41°38′N 15°54′E
107 N16 **Manfredonia, Golfo di** gulf Adriatic Sea, N Mediterranean Sea
77 P13 **Manga** C Burkina Faso 11°41′N 01°04′W
59 L16 **Mangabeiras, Chapada das** ▲ E Brazil
79 J20 **Mangai** Bandundu, W Dem. Rep. Congo 03°58′S 19°32′E
190 L17 **Mangaia** island group S Cook Islands
184 M9 **Mangakino** Waikato, North Island, New Zealand 38°23′S 175°47′E
116 M15 **Mangalia** anc. Callatis. Constanţa, SE Romania 43°48′N 28°35′E
78 J11 **Mangalmé** Guéra, SE Chad 12°26′N 19°37′E
155 E19 **Mangalore** Karnātaka, W India 12°54′N 74°51′E
Mangareva see Magareva
83 I23 **Mangaung** Free State, C South Africa 29°10′S 26°19′E
Mangaung see Bloemfontein
154 K9 **Mangawān** Madhya Pradesh, C India 24°39′N 81°33′E
184 M11 **Mangaweka** Manawatu-Wanganui, North Island, New Zealand 39°49′S 175°47′E
184 N11 **Mangaweka** ▲ North Island, New Zealand 39°51′S 176°06′E
79 P17 **Mangbwalu** Orientale, NE Dem. Rep. Congo 02°06′N 30°04′E
139 R1 **Mangêsh** Ar. Māngish, var. Manguāsh. Dahūk, N Iraq 37°03′N 43°04′E
101 L24 **Mangfall** ∽ SE Germany
169 P13 **Manggar** Pulau Belitung, W Indonesia 02°52′S 108°13′E
Mangghystaū Ūstirti' see Mangystau, Plato
166 M2 **Mangin Range** ▲▲ N Myanmar (Burma)
Māngīsh see Mangêsh
Mangish see Mangêsh
Mangistau see Mangystau
146 H8 **Mang'it** Rus. Mangit. Qoraqalpog'iston Respublikasi, W Uzbekistan 42°06′N 60°02′E
Mangit see Mang'it
54 A13 **Manglares, Cabo** headland SW Colombia 01°36′N 79°02′W
149 V6 **Mangla Reservoir** ⊠ NE Pakistan
159 N9 **Mangnai** var. Lao Mangnai. Qinghai, C China 37°52′N 91°45′E
Mango see Mago, Fiji
Mango see Sansanné-Mango, Togo
Mangoche see Mangochi
83 N14 **Mangochi** var. Mangoche; prev. Fort Johnston. Southern, SE Malawi 14°30′S 35°15′E
77 N14 **Mangodara** SW Burkina Faso 09°49′N 04°22′W
172 H6 **Mangoky** ∽ W Madagascar
171 Q12 **Mangole, Pulau** island Kepulauan Sula, E Indonesia
184 J2 **Mangonui** Northland, North Island, New Zealand 35°00′S 173°30′E
Mangqystaū Oblysy see Mangystau
Mangqystaū Shyghanaghy see Mangystau Zaliv
Mangshi see Luxi
104 H7 **Mangualde** Viseu, N Portugal 40°36′N 07°46′W
61 H18 **Mangueira, Lagoa** ◎ S Brazil
77 X6 **Manguéni, Plateau du** ▲ NE Niger
163 T4 **Mangui** Nei Mongol Zizhiqu, N China 52°02′N 122°13′E
26 K11 **Mangum** Oklahoma, C USA 34°52′N 99°30′W
79 O18 **Manguredjipa** Nord-Kivu, E Dem. Rep. Congo 0°28′N 28°33′E
83 L16 **Mangwendi** Mashonaland East, E Zimbabwe 18°22′S 31°24′E
Mangyshlak, Plato see Mangystau, Plato
Mangyshlakskiy Zaliv see Mangystau Zaliv
Mangyshlaskaya see Mangystau
144 F15 **Mangystau** Kaz. Mangqystaū Oblysy; prev. Mangistau; prev. Mangyshlaskaya. ◈ province SW Kazakhstan
144 F15 **Mangystau, Plato** plateau SW Kazakhstan
144 E14 **Mangystau Zaliv** Kaz. Mangqystaū Shyghanaghy; prev. Mangyshlakskiy Zaliv. gulf SW Kazakhstan
162 E7 **Manhan** var. Tögrög. Hovd, W Mongolia 47°24′N 92°06′E
Manhan see Alag-Erdene
27 O4 **Manhattan** Kansas, C USA 39°11′N 96°35′W
99 L21 **Manhay** Luxembourg, SE Belgium 50°13′N 05°43′E
83 L21 **Manhiça** prev. Vila de Manhiça. Maputo, S Mozambique 25°25′S 32°49′E
83 L21 **Manhoca** Maputo, S Mozambique 26°49′S 32°36′E
59 N20 **Manhuaçu** Minas Gerais, SE Brazil 20°16′S 42°01′W
117 W9 **Manhush** prev. Pershotravneve. Donets'ka Oblast', E Ukraine 47°03′N 37°20′E
54 H10 **Maní** Casanare, C Colombia 04°50′N 72°15′W
143 R11 **Mānī** Kermān, C Iran
83 M17 **Manica** var. Vila de Manica. Manica, W Mozambique 18°56′S 32°52′E
83 M17 **Manica** off. Província de Manica. ◈ province W Mozambique
83 L17 **Manicaland** ◈ province E Zimbabwe
Manica, Província de see Manica
15 U5 **Manic Deux, Réservoir** ⊠ Québec, SE Canada
Manich see Manych
59 F14 **Manicoré** Amazonas, N Brazil 05°48′S 61°16′W
13 N11 **Manicouagan** Québec, SE Canada 50°40′N 68°46′W
13 N11 **Manicouagan** ∽ Québec, SE Canada
15 U6 **Manicouagan, Péninsule de** peninsula Québec, SE Canada
13 N11 **Manicouagan, Réservoir** ⊠ Québec, E Canada
15 T4 **Manic Trois, Réservoir** ⊠ Québec, SE Canada
79 M20 **Maniema** off. Région du Maniema. ◈ region E Dem. Rep. Congo
Maniema, Région du see Maniema
Maniewicze see Manevychi
160 F8 **Maniganggo** Sichuan, C China 32°01′N 99°04′E
11 Y15 **Manigotagan** Manitoba, S Canada 51°06′N 96°18′W
153 R13 **Manihāri** Bihār, N India 25°21′N 87°37′E
191 U9 **Manihi** island Îles Tuamotu, C French Polynesia
190 L13 **Manihiki** atoll N Cook Islands
175 U8 **Manihiki Plateau** undersea feature C Pacific Ocean
196 M14 **Maniitsoq** var. Manîtsoq, Dan. Sukkertoppen. ◈ Qeqqata, S Greenland
153 T15 **Manikganj** Dhaka, C Bangladesh 23°52′N 90°00′E
152 M14 **Mānikpur** Uttar Pradesh, N India 25°04′N 81°06′E
171 N4 **Manila** off. City of Manila. ● (Philippines) Luzon, N Philippines 14°34′N 120°59′E
27 Y9 **Manila** Arkansas, C USA 35°52′N 90°10′W
Manila, City of see Manila
189 N16 **Manila Reef** reef W Micronesia
183 T6 **Manilla** New South Wales, SE Australia 30°44′S 150°43′E
192 P6 **Maniloa** island Tongatapu Group, S Tonga
123 U8 **Manily** Krasnoyarskiy Kray, E Russian Federation 62°33′N 165°03′E
171 V12 **Manim, Pulau** island E Indonesia
168 I11 **Maninjau, Danau** ◎ Sumatera, W Indonesia
153 W13 **Manipur** ◈ state NE India
153 X14 **Manipur Hills** hill range E India
136 C14 **Manisa** var. Manissa, prev. Saruhan; anc. Magnesia. Manisa, W Turkey 38°36′N 27°29′E
136 C13 **Manisa** var. Manissa. ◈ province W Turkey
Manissa see Manisa
31 O7 **Manistee** Michigan, N USA 44°14′N 86°19′W
31 P7 **Manistee River** ∽ Michigan, N USA
31 O4 **Manistique** Michigan, N USA 45°57′N 86°15′W
31 P4 **Manistique Lake** ◎ Michigan, N USA
11 W13 **Manitoba** ◈ province S Canada
11 X16 **Manitoba, Lake** ◎ Manitoba, S Canada
11 X17 **Manitou** Manitoba, S Canada 49°12′N 98°28′W
31 N2 **Manitou Island** island Michigan, N USA
14 H11 **Manitou Lake** ◎ Ontario, SE Canada
12 G15 **Manitoulin Island** island Ontario, S Canada
37 T5 **Manitou Springs** Colorado, C USA 38°51′N 104°56′W
14 G12 **Manitouwabing Lake** ◎ Ontario, S Canada
12 E12 **Manitouwadge** Ontario, S Canada 49°14′N 85°51′W
12 G15 **Manitowaning** Manitoulin Island, Ontario, S Canada 45°44′N 81°50′W
14 B7 **Manitowik Lake** ◎ Ontario, S Canada
31 N7 **Manitowoc** Wisconsin, N USA 44°04′N 87°40′W
Manîtsoq see Maniitsoq
139 O7 **Mānī', Wādī al** dry watercourse W Iraq
12 J14 **Maniwaki** Québec, SE Canada 46°22′N 75°58′W
171 W13 **Maniwori** Papua, E Indonesia 02°49′S 136°00′E
54 E10 **Manizales** Caldas, W Colombia 05°03′N 73°52′W
112 F11 **Manjača** ▲ NW Bosnia and Herzegovina
Manjacaze see Mandlakazi
180 J14 **Manjimup** Western Australia 34°18′S 116°14′E
109 V4 **Mank** Niederösterreich, C Austria 48°06′N 15°13′E
79 I17 **Mankanza** Equateur, NW Dem. Rep. Congo 01°40′N 19°08′E
153 N12 **Mankāpur** Uttar Pradesh, N India 27°03′N 82°12′E
26 M3 **Mankato** Kansas, C USA 39°48′N 98°13′W
29 U10 **Mankato** Minnesota, N USA 44°10′N 94°00′W
117 O7 **Man'kivka** Cherkas'ka Oblast', C Ukraine 48°58′N 30°10′E
76 M15 **Mankono** C Ivory Coast 08°01′N 06°09′W
11 T17 **Mankota** Saskatchewan, S Canada 49°25′N 107°05′W
155 K23 **Mankulam** Northern Province, N Sri Lanka 09°07′N 80°27′E
162 L10 **Manlay** var. Üydzen. Ömnögovĭ, S Mongolia 44°08′N 106°48′E
39 Q9 **Manley Hot Springs** Alaska, USA 65°00′N 150°37′W
18 H10 **Manlius** New York, NE USA 43°00′N 75°58′W
105 W5 **Manlleu** Cataluña, NE Spain 41°59′N 02°17′E
29 V11 **Manly** Iowa, C USA 43°17′N 93°12′W
154 E13 **Manmād** Mahārāshtra, W India 20°15′N 74°29′E
182 J7 **Mannahill** South Australia 32°29′S 139°58′E
155 J23 **Mannar** var. Manar. Northern Province, NW Sri Lanka 09°01′N 79°53′E
155 I24 **Mannar, Gulf of** gulf India/Sri Lanka
155 J23 **Mannar Island** island N Sri Lanka
Mannersdorf see Mannersdorf am Leithagebirge
109 Y5 **Mannersdorf am Leithagebirge** var. Mannersdorf. Niederösterreich, E Austria 47°59′N 16°36′E
109 Y6 **Mannersdorf an der Rabnitz** Burgenland, E Austria 47°25′N 16°32′E
101 G20 **Mannheim** Baden-Württemberg, SW Germany 49°29′N 08°29′E
11 O12 **Manning** Alberta, W Canada 56°53′N 117°39′W
29 T14 **Manning** Iowa, C USA 41°54′N 95°03′W
28 K5 **Manning** North Dakota, N USA 47°15′N 102°48′W
21 S13 **Manning** South Carolina, SE USA 33°42′N 80°12′W
191 Y2 **Manning, Cape** headland Kiritimati, NE Kiribati 02°02′N 157°26′W
21 S3 **Mannington** West Virginia, NE USA 39°31′N 80°20′W
182 A1 **Mann Ranges** ▲▲ South Australia
107 C19 **Mannu** ∽ Sardegna, Italy, C Mediterranean Sea
11 R14 **Mannville** Alberta, SW Canada 53°19′N 111°08′W
76 J15 **Mano** ∽ Liberia/Sierra Leone
Manø see Mandø
61 F15 **Manoel Viana** Rio Grande do Sul, S Brazil 29°33′S 55°28′W
39 O13 **Manokotak** Alaska, USA 59°00′N 158°58′W
171 V12 **Manokwari** Papua Barat, E Indonesia 0°53′S 134°05′E
79 N22 **Manono** Shaba, SE Dem. Rep. Congo 07°18′S 27°25′E
25 T10 **Manor** Texas, SW USA 30°20′N 97°33′W
97 D16 **Manorhamilton** Ir. Cluainín. Leitrim, NW Ireland 54°18′N 08°10′W
103 S15 **Manosque** Alpes-de-Haute-Provence, SE France 43°50′N 05°47′E
12 L11 **Manouane, Lac** ◎ Québec, SE Canada
163 W12 **Manp'o** var. Manp'ojin. NW North Korea 41°10′N 126°24′E
Manp'ojin see Manp'o
191 T4 **Manra** prev. Sydney Island. atoll Phoenix Islands, C Kiribati
105 V5 **Manresa** Cataluña, NE Spain 41°43′N 01°50′E
152 H9 **Mānsa** Punjab, NW India 30°00′N 75°25′E
82 J12 **Mansa** prev. Fort Rosebery. Luapula, N Zambia 11°14′S 28°55′E
76 G12 **Mansa Konko** C Gambia 13°26′N 15°29′W
15 Q11 **Manseau** Québec, SE Canada 46°23′N 71°59′W
149 U5 **Mānsehra** Khyber Pakhtunkhwa, NW Pakistan 34°23′N 73°18′E
9 Q9 **Mansel Island** island Nunavut, NE Canada
183 O12 **Mansfield** Victoria, SE Australia 37°04′S 146°06′E
97 M18 **Mansfield** C England, United Kingdom 53°09′N 01°11′W
27 S11 **Mansfield** Arkansas, C USA 35°03′N 94°15′W
22 G6 **Mansfield** Louisiana, S USA 32°02′N 93°42′W
19 O12 **Mansfield** Massachusetts, NE USA 42°00′N 71°11′W
31 T12 **Mansfield** Ohio, N USA 40°45′N 82°31′W
18 G12 **Mansfield** Pennsylvania, NE USA 41°46′N 77°02′W
18 M7 **Mansfield, Mount** ▲ Vermont, NE USA 44°31′N 72°49′W
59 M16 **Mansidão** Bahia, E Brazil 10°46′S 44°04′W
102 L11 **Mansle** Charente, W France 45°52′N 00°11′E
76 G12 **Mansôa** C Guinea-Bissau 12°08′N 15°18′W
47 V8 **Manso, Rio** ∽ C Brazil
Manşūra see Al Manşūrah
Mansurabad see Mehrān, Rūd-e
56 A6 **Manta** Manabí, W Ecuador 0°59′S 80°44′W
56 A6 **Manta, Bahía de** bay W Ecuador
57 F14 **Mantaro, Río** ∽ C Peru
35 O8 **Manteca** California, W USA 37°48′N 121°13′W
54 J7 **Mantecal** Apure, C Venezuela 07°34′N 69°07′W
31 N11 **Manteno** Illinois, N USA 41°15′N 87°49′W
21 Y9 **Manteo** Roanoke Island, North Carolina, SE USA 35°54′N 75°42′W
Mantes-Gassicourt see Mantes-la-Jolie
103 N5 **Mantes-la-Jolie** prev. Mantes-Gassicourt, Mantes-sur-Seine; anc. Medunta. Yvelines, N France 48°59′N 01°43′E
Mantes-sur-Seine see Mantes-la-Jolie
36 L5 **Manti** Utah, W USA 39°16′N 111°38′W
Mantinea see Mantíneia
115 F20 **Mantíneia** anc. Mantinea. site of ancient city Pelopónnisos, S Greece
59 M21 **Mantiqueira, Serra da** ▲▲ S Brazil
29 W10 **Mantorville** Minnesota, N USA 44°04′N 92°45′W
115 G17 **Mantoúdi** var. Mandoudi; prev. Mandoúdhion. Évvoia, C Greece 38°47′N 23°29′E
Mantoue see Mantova
106 F8 **Mantova** Eng. Mantua, Fr. Mantoue. Lombardia, NW Italy 45°10′N 10°47′E
93 M19 **Mäntsälä** Uusimaa, S Finland 60°38′N 25°21′E
93 L17 **Mänttä** Pirkanmaa, W Finland 62°00′N 24°36′E
Mantua see Mantova
125 O14 **Manturovo** Kostromskaya Oblast', NW Russian Federation 58°19′N 44°42′E
93 M18 **Mäntyharju** Etelä-Savo, SE Finland 61°25′N 26°53′E
92 M13 **Mäntyjärvi** Lappi, N Finland 66°00′N 27°35′E
190 L16 **Manuae** island S Cook Islands
191 Q10 **Manuae** atoll Îles Sous le Vent, W French Polynesia
192 L16 **Manu'a Islands** island group E American Samoa
40 L5 **Manuel Benavides** Chihuahua, N Mexico 29°07′N 103°52′W
61 D21 **Manuel J. Cobo** Buenos Aires, E Argentina 35°49′S 57°54′W
58 M12 **Manuel Luís, Recife** reef E Brazil
59 I14 **Manuel Zinho** Pará, N Brazil 07°21′S 54°47′W
191 V11 **Manuhangi** prev. Manuhangi. atoll Îles Tuamotu, C French Polynesia
Manuhangi see Manuhagi
185 E22 **Manuherikia** ∽ South Island, New Zealand
171 P13 **Manui, Pulau** island N Indonesia
Manukau see Manurewa
184 L6 **Manukau Harbour** harbour North Island, New Zealand
191 Z2 **Manulu Lagoon** ◎ Kiritimati, E Kiribati
182 J7 **Manunda Creek** seasonal river South Australia
57 K15 **Manupari, Río** ∽ N Bolivia
184 L6 **Manurewa** var. Manukau. Auckland, North Island, New Zealand 37°01′S 174°55′E
57 K15 **Manurimi, Río** ∽ NW Bolivia
186 D5 **Manus** ◈ province N Papua New Guinea
186 D5 **Manus Island** var. Great Admiralty Island. island N Papua New Guinea
171 T16 **Manuwui** Pulau Babar, E Indonesia 07°47′S 129°39′E
29 Q3 **Manvel** North Dakota, N USA 48°07′N 97°15′W
33 Z14 **Manville** Wyoming, C USA 42°45′N 104°38′W
22 G6 **Many** Louisiana, S USA 31°34′N 93°28′W
81 H21 **Manyara, Lake** ◎ NE Tanzania
126 L12 **Manych** var. Manich. ∽ SW Russian Federation
83 H14 **Manyinga** North Western, NW Zambia 13°28′S 24°18′E
105 O11 **Manzanares** Castilla-La Mancha, C Spain 39°N 03°23′W
44 H7 **Manzanillo** Granma, E Cuba 20°21′N 77°07′W
40 K14 **Manzanillo** Colima, SW Mexico 19°00′N 104°19′W
40 K14 **Manzanillo, Bahía** bay SW Mexico
37 S11 **Manzano Mountains** ▲▲ New Mexico, SW USA
37 R12 **Manzano Peak** ▲ New Mexico, SW USA 34°35′N 106°27′W
163 R6 **Manzhouli** var. Man-chou-li. Nei Mongol Zizhiqu, N China 49°36′N 117°28′E
Manzil Bū Ruqaybah see Menzel Bourguiba
139 X9 **Manzilīyah** Maysān, E Iraq 32°26′N 47°01′E
83 L21 **Manzini** prev. Bremersdorp. C Swaziland 26°30′S 31°22′E
83 L21 **Manzini** ✈ (Mbabane) C Swaziland 26°36′S 31°25′E
78 G10 **Mao** Kanem, W Chad 14°06′N 15°17′E
45 N8 **Mao** NW Dominican Republic 19°37′N 71°04′W
105 Z9 **Maó** Cast. Mahón, Eng. Port Mahon; anc. Portus Magonis. Menorca, Spain, W Mediterranean Sea 39°54′N 04°15′E
Maoemere see Maumere
159 W9 **Maojing** Gansu, N China 36°26′N 106°36′E
171 Y14 **Maoke, Pegunungan** Dut. Sneeuw-gebergte, Eng. Snow Mountains. ▲▲ Papua, E Indonesia
Maol Réidh, Caoc see Mweelrea
160 M15 **Maoming** Guangdong, S China 21°46′N 110°51′E
160 H8 **Maoxian** var. Mao Xian; prev. Fengyizhen. Sichuan, C China 31°42′N 103°48′E
Mao Xian see Maoxian
83 L19 **Mapai** Gaza, SW Mozambique 22°52′S 32°00′E
158 H15 **Mapam Yumco** ◎ W China
83 I15 **Mapanza** Southern, S Zambia 16°16′S 26°54′E
54 J4 **Mapararí** Falcón, N Venezuela 10°52′N 69°27′W
41 U17 **Mapastepec** Chiapas, SE Mexico 15°28′N 93°00′W
169 V9 **Mapat, Pulau** island N Indonesia
171 Y15 **Mapi** Papua, E Indonesia 07°02′S 139°24′E
171 V11 **Mapia, Kepulauan** island group E Indonesia
40 L8 **Mapimí** Durango, C Mexico 25°50′N 103°50′W
83 N19 **Mapinhane** Inhambane, SE Mozambique 22°14′S 35°07′E
55 N7 **Mapire** Monagas, NE Venezuela 07°48′N 64°40′W
11 S17 **Maple Creek** Saskatchewan, S Canada 49°55′N 109°28′W
31 Q9 **Maple River** ∽ Michigan, N USA
29 P7 **Maple River** ∽ North Dakota/South Dakota, N USA
29 S13 **Mapleton** Iowa, C USA 42°10′N 95°47′W
29 U10 **Mapleton** Minnesota, N USA 43°55′N 93°57′W
29 R5 **Mapleton** North Dakota, N USA 46°51′N 97°04′W
32 F13 **Mapleton** Oregon, NW USA 44°01′N 123°56′W

◆ Country
● Country Capital
◇ Dependent Territory
○ Dependent Territory Capital
◈ Administrative Regions
✈ International Airport
▲ Mountain
▲▲ Mountain Range
Volcano
∽ River
◎ Lake
⊠ Reservoir

36 L3 **Mapleton** Utah, W USA 40°07′N 111°37′W
192 K5 **Mapmaker Seamounts** *undersea feature* N Pacific Ocean 25°00′N 165°00′E
186 B6 **Maprik** East Sepik, NW Papua New Guinea 03°38′S 143°02′E
83 L21 **Maputo** *prev.* Lourenço Marques. ● (Mozambique) Maputo, S Mozambique 25°58′S 32°35′E
83 L21 **Maputo** ◆ *province* S Mozambique
67 V14 **Maputo** ⇝ S Mozambique
83 L21 **Maputo** ✕ Maputo, S Mozambique 25°47′S 32°36′E
Maqat *see* Makat
113 M19 **Maqellarë** Dibër, C Albania 41°36′N 20°29′E
159 S12 **Maqên** *var.* Dawo; *prev.* Dawu. Qinghai, C China 34°32′N 100°17′E
159 S11 **Maqên Kangri** ▲ C China 34°44′N 99°25′E
141 X7 **Maqiz al Kurbá** N Oman 24°13′N 56°48′E
159 U12 **Maqu** *var.* Nyinma. Gansu, C China 34°02′N 102°00′E
104 M9 **Maqueda** Castilla-La Mancha, C Spain 40°04′N 04°22′W
82 B9 **Maquela do Zombo** Uíge, NW Angola 06°06′S 15°12′E
63 I16 **Maquinchao** Río Negro, C Argentina 41°19′S 68°47′W
29 Z13 **Maquoketa** Iowa, C USA 42°03′N 90°42′W
29 Y13 **Maquoketa River** ⇝ Iowa, C USA
14 F13 **Mar** Ontario, S Canada 44°48′N 81°12′W
95 F14 **Mår** ◎ S Norway
81 G19 **Mara** ◆ *region* N Tanzania
58 D12 **Maraã** Amazonas, NW Brazil 01°48′S 65°21′W
191 P8 **Maraa** Tahiti, W French Polynesia 17°44′S 149°34′W
191 O8 **Maraa, Pointe** *headland* Tahiti, W French Polynesia 17°44′S 149°34′W
59 K14 **Marabá** Pará, NE Brazil 05°23′S 49°10′W
54 H5 **Maracaibo** Zulia, NW Venezuela 10°40′N 71°39′W
Maracaibo, Gulf of *see* Venezuela, Golfo de
54 H5 **Maracaibo, Lago de** *var.* Lake Maracaibo. *inlet* NW Venezuela
Maracaibo, Lake *see* Maracaibo, Lago de
58 K10 **Maracá, Ilha de** *island* NE Brazil
59 H20 **Maracaju, Serra de** ▲ S Brazil
58 I11 **Maracanaquará, Planalto** ▲ NE Brazil
54 L5 **Maracay** Aragua, N Venezuela 10°15′N 67°36′W
Marada *see* Marādah
75 R9 **Marādah** *var.* Marada. N Libya 29°16′N 19°29′E
77 U12 **Maradi** Maradi, S Niger 13°30′N 07°05′E
77 U11 **Maradi** ◆ *department* S Niger
81 E21 **Maragarazi** *var.* Muragarazi. ⇝ Burundi/Tanzania
Maragha *see* Marāgheh
142 J3 **Marāgheh** *var.* Maragha. Āžarbāyjān-e Khāvarī, NW Iran 37°21′N 46°14′E
141 P7 **Marāh** *var.* Marrāt. Ar Riyāḍ, C Saudi Arabia 25°04′N 45°30′E
55 N11 **Marahuaca, Cerro** ▲ S Venezuela 03°37′N 65°25′W
27 R5 **Marais des Cygnes River** ⇝ Kansas/Missouri, C USA
58 L11 **Marajó, Baía de** *bay* N Brazil
59 K12 **Marajó, Ilha de** *island* N Brazil
191 O2 **Marakei** *atoll* Tungaru, W Kiribati
Marakesh *see* Marrakech
81 I18 **Maralal** Samburu, C Kenya 01°05′N 36°42′E
83 G21 **Maralaleng** Kgalagadi, S Botswana 25°42′S 22°39′E
145 U8 **Maraldy, Ozero** ◎ NE Kazakhstan
182 C5 **Maralinga** South Australia 30°16′S 131°35′E
Máramarossziget *see* Sighetu Marmației
187 N9 **Maramasike** *var.* Small Malaita. *island* N Solomon Islands
Maramba *see* Livingstone
194 H3 **Marambio** *Argentinian research station* Antarctica 64°22′S 57°18′W
116 H9 **Maramureș** ◆ *county* NW Romania
36 L15 **Marana** Arizona, SW USA 32°24′N 111°12′W
105 P7 **Maranchón** Castilla-La Mancha, C Spain 41°02′N 02°11′W
142 J2 **Marand** *var.* Merend. Āžarbāyjān-e Sharqī, NW Iran 38°25′N 45°40′E
Marandellas *see* Marondera
58 L13 **Maranhão** *off.* Estado do Maranhão. ◆ *state* E Brazil
104 H10 **Maranhão, Barragem do** ⊠ C Portugal
Maranhão, Estado do *see* Maranhão
149 O11 **Mārān, Koh-i** ▲ SW Pakistan 29°24′N 66°50′E
106 J7 **Marano, Laguna di** *lagoon* NE Italy
56 E9 **Marañón, Río** ⇝ N Peru
102 J10 **Marans** Charente-Maritime, W France 46°19′N 00°58′W
83 M20 **Marão** Inhambane, S Mozambique 24°15′S 34°09′E
185 B23 **Mararoa** ⇝ South Island, New Zealand
Maraş/Marash *see* Kahramanmaraş
107 M19 **Maratea** Basilicata, S Italy 39°57′N 15°44′E
104 G11 **Marateca** Setúbal, S Portugal 38°34′N 08°40′W
115 B20 **Marathiá, Akrotírio** *headland* Zákynthos, Iónia Nisiá, Greece, C Mediterranean Sea 37°39′N 20°49′E
12 E12 **Marathon** Ontario, S Canada 48°44′N 86°23′W
23 Y17 **Marathon** Florida Keys, Florida, SE USA 24°42′N 81°05′W
24 L10 **Marathon** Texas, SW USA 30°10′N 103°14′W
Marathón *see* Marathónas
115 H19 **Marathónas** *prev.* Marathón. Attikí, C Greece 38°09′N 23°57′E
169 W9 **Maratua, Pulau** *island* N Indonesia
59 O18 **Maraú** Bahia, SE Brazil 14°07′S 39°02′W
143 R3 **Marāveh Tappeh** Golestán, N Iran 37°53′N 55°57′E
24 L11 **Maravillas Creek** ⇝ Texas, SW USA
186 D8 **Marawaka** Eastern Highlands, C Papua New Guinea 06°56′S 145°54′E
171 Q7 **Marawi** Mindanao, S Philippines 07°59′N 124°16′E
Märäzä *see* Qobustan
Marbat *see* Mirbāṭ
104 L16 **Marbella** Andalucía, S Spain 36°31′N 04°50′W
180 J7 **Marble Bar** Western Australia 21°13′S 119°48′E
36 L9 **Marble Canyon** *canyon* Arizona, SW USA
25 S10 **Marble Falls** Texas, SW USA 30°34′N 98°16′W
27 Y7 **Marble Hill** Missouri, C USA 37°18′N 89°58′W
33 T15 **Marbleton** Wyoming, C USA 42°31′N 110°06′W
Marburg *see* Marburg an der Lahn, Germany
Marburg *see* Maribor, Slovenia
101 H16 **Marburg an der Lahn** *hist.* Marburg. Hessen, W Germany 50°49′N 08°46′E
111 H23 **Marcal** ⇝ W Hungary
42 G7 **Marcala** La Paz, SW Honduras 14°11′N 88°00′W
111 H24 **Marcali** Somogy, SW Hungary 46°33′N 17°29′E
83 A16 **Marca, Ponta da** *headland* SW Angola 16°31′S 11°42′E
59 I16 **Marcelândia** Mato Grosso, W Brazil 11°18′S 54°49′W
27 T3 **Marceline** Missouri, C USA 39°42′N 92°57′W
60 I13 **Marcelino Ramos** Rio Grande do Sul, S Brazil 27°31′S 51°57′W
55 Y12 **Marcel, Mont** ▲ S French Guiana 02°32′N 53°00′W
97 O19 **March** E England, United Kingdom 52°37′N 00°13′E
109 Z3 **March** *var.* Morava. ⇝ C Europe *see also* Morava
March *see* Morava
106 I12 **Marche** *Eng.* Marches. ◆ *region* C Italy
103 N11 **Marche** *cultural region* C France
99 J21 **Marche-en-Famenne** Luxembourg, SE Belgium 50°13′N 05°21′E
104 K14 **Marchena** Andalucía, S Spain 37°20′N 05°24′W
57 B17 **Marchena, Isla** *var.* Bindloe Island. *island* Galapagos Islands, Ecuador, E Pacific Ocean
Marches *see* Marche
99 J20 **Marchin** Liège, E Belgium 50°30′N 05°17′E
181 S1 **Marchinbar Island** *island* Wessel Islands, Northern Territory, N Australia
62 L9 **Mar Chiquita, Laguna** ◎ C Argentina
103 Q10 **Marcigny** Saône-et-Loire, C France 46°16′N 04°04′E
23 W16 **Marco** Florida, SE USA 25°56′N 81°43′W
Marcodurum *see* Düren
59 O15 **Marcolândia** Pernambuco, E Brazil 07°21′S 40°40′W
106 I8 **Marco Polo** ✕ (Venezia) Veneto, NE Italy 45°30′N 12°21′E
Marcounda *see* Markounda
Marcq *see* Mark
116 M8 **Mărculești** *Rus.* Markuleshty. N Moldova 47°54′N 28°14′E
29 S12 **Marcus** Iowa, C USA 42°49′N 95°48′W
39 S11 **Marcus Baker, Mount** ▲ Alaska, USA 61°26′N 147°45′W
192 I5 **Marcus Island** *var.* Minami Tori Shima. *island* E Japan
18 K8 **Marcy, Mount** ▲ New York, NE USA 44°06′N 73°55′W
149 T5 **Mardān** Khyber Pakhtunkhwa, N Pakistan 34°14′N 71°59′E
63 N14 **Mar del Plata** Buenos Aires, E Argentina 38°57′S 57°32′W
137 Q16 **Mardin** Mardin, SE Turkey 37°19′N 40°43′E
137 Q16 **Mardin** ◆ *province* SE Turkey
137 Q16 **Mardin Dağları** ▲ SE Turkey
Mardzad *see* Hayrhandulaan
187 R17 **Maré** *island* Îles Loyauté, E New Caledonia
Marea Neagră *see* Black Sea
105 Z8 **Mare de Déu del Toro** *var.* El Toro. ▲ Menorca, Spain, W Mediterranean Sea 39°59′N 04°06′E
181 W4 **Mareeba** Queensland, NE Australia 17°03′S 145°30′E
96 G8 **Maree, Loch** ◎ N Scotland, United Kingdom
Mareeq *see* Mereeg
Marek *see* Dupnitsa
76 J11 **Maréna** Kayes, W Mali 14°36′N 10°57′W
190 I2 **Marenanuka** *atoll* Tungaru, W Kiribati
29 X14 **Marengo** Iowa, C USA 41°48′N 92°04′W
102 J11 **Marennes** Charente-Maritime, W France 45°47′N 01°04′W
107 G23 **Marettimo, Isola** *island* Isole Egadi, S Italy
24 K10 **Marfa** Texas, SW USA 30°19′N 104°03′W
57 P17 **Marfil, Laguna** ◎ E Bolivia
Margarets *see* Marhanets'
25 Q4 **Margaret** Texas, SW USA 34°00′N 99°38′W
180 I14 **Margaret River** Western Australia 33°58′S 115°10′E
186 C7 **Margarima** Hela, W Papua New Guinea 06°00′S 143°23′E
55 N4 **Margarita, Isla de** *island* N Venezuela
115 I25 **Margarites** Kríti, Greece, E Mediterranean Sea 35°19′N 24°40′E
97 Q22 **Margate** *prev.* Mergate. SE England, United Kingdom 51°24′N 01°24′E
23 Z15 **Margate** Florida, SE USA 26°14′N 80°12′W
Margelan *see* Marg'ilon
103 P13 **Margeride, Montagnes de la** ▲ C France
Margherita *see* Jamaame
107 N16 **Margherita di Savoia** Puglia, SE Italy 41°23′N 16°09′E
Margherita, Lake *see* Ābaya Hāyk'
81 E18 **Margherita Peak** *Fr.* Pic Marguerite. ▲ Uganda/Dem. Rep. Congo 0°28′N 29°58′E
149 O4 **Marghi** Bāmyān, N Afghanistan 35°10′N 66°26′E
116 G9 **Marghita** *Hung.* Margitta. Bihor, NW Romania 47°20′N 22°20′E
Margilan *see* Marg'ilon
147 S10 **Marg'ilon** *var.* Margelan, *Rus.* Margilan. Farg'ona Viloyati, E Uzbekistan 40°29′N 71°43′E
116 K8 **Marginea** Suceava, NE Romania 47°49′N 25°47′E
Margitta *see* Marghita
148 K9 **Mārgow, Dasht-e** *desert* SW Afghanistan
99 L18 **Margraten** Limburg, SE Netherlands 50°49′N 05°49′E
10 M15 **Marguerite** British Columbia, SW Canada 52°17′N 122°10′W
15 V3 **Marguerite** ⇝ Quebec, SE Canada
194 I6 **Marguerite Bay** *bay* Antarctica
Marguerite, Pic *see* Margherita Peak
117 T9 **Marhanets'** *Rus.* Marganets. Dnipropetrovs'ka Oblast', E Ukraine 47°35′N 34°37′E
186 B9 **Mari** Western, SW Papua New Guinea 09°10′S 141°39′E
191 V12 **Maria** *atoll* Groupe Actéon, SE French Polynesia
191 R12 **Maria** *island* Îles Australes, SW French Polynesia
40 I12 **María Cleofas, Isla** *island* C Mexico
62 H4 **María Elena** *var.* Oficina María Elena. Antofagasta, N Chile 22°18′S 69°40′W
95 G21 **Mariager** Midtjylland, C Denmark 56°39′N 09°59′E
61 C22 **María Ignacia** Buenos Aires, E Argentina 37°24′S 59°30′W
183 P17 **Maria Island** *island* Tasmania, SE Australia
40 H12 **María Madre, Isla** *island* C Mexico
40 I12 **María Magdalena, Isla** *island* C Mexico
192 H6 **Mariana Islands** *island group* Guam/Northern Mariana Islands
175 N3 **Mariana Trench** *var.* Challenger Deep. *undersea feature* W Pacific Ocean 15°00′N 147°30′E
153 X12 **Mariāni** Assam, NE India 26°39′N 94°18′E
27 X11 **Marianna** Arkansas, C USA 34°46′N 90°49′W
23 R8 **Marianna** Florida, SE USA 30°46′N 85°13′W
172 J16 **Marianne** *island* Inner Islands, NE Seychelles
95 M19 **Mariannelund** Jönköping, S Sweden 57°37′N 15°33′E
61 D15 **Mariano I. Loza** Corrientes, NE Argentina 29°22′S 58°12′W
Mariano Machado *see* Ganda
111 A16 **Mariánské Lázně** *Ger.* Marienbad. Karlovarský Kraj, W Czech Republic 49°57′N 12°43′E
Máriaradna *see* Radna
33 S7 **Marias River** ⇝ Montana, NW USA
Maria-Theresiopel *see* Subotica
Máriatölgyes *see* Dubnica nad Váhom
184 H1 **Maria van Diemen, Cape** *headland* North Island, New Zealand 34°27′S 172°38′E
109 V5 **Mariazell** Steiermark, E Austria 47°47′N 15°20′E
141 P15 **Ma'rib** W Yemen 15°28′N 45°25′E
95 I25 **Maribo** Sjælland, S Denmark 54°47′N 11°30′E
109 W9 **Maribor** *Ger.* Marburg. NE Slovenia 46°34′N 15°40′E
Marica *see* Maritsa
35 R13 **Maricopa** California, W USA 35°03′N 119°24′W
Maricourt *see* Kangiqsujuaq
81 D15 **Maridi** Western Equatoria, SW South Sudan 04°55′N 29°30′E
194 M11 **Marie Byrd Land** *physical region* Antarctica
193 P14 **Marie Byrd Seamount** *undersea feature* N Amundsen Sea 70°00′S 118°00′W
45 X11 **Marie-Galante** *var.* Ceyre to the Caribs. *island* SE Guadeloupe
45 Y6 **Marie-Galante, Canal de** *channel* S Guadeloupe
93 J20 **Mariehamn** *Fin.* Maarianhamina. Åland, SW Finland 60°05′N 19°55′E
44 C4 **Mariel** La Habana, W Cuba 23°02′N 82°44′W
99 H22 **Mariembourg** Namur, S Belgium 50°07′N 04°30′E
Marienbad *see* Mariánské Lázně
Marienburg *see* Alūksne, Latvia
Marienburg *see* Malbork, Poland
Marienburg *see* Feldioara, Romania
Marienburg in Westpreussen *see* Malbork
Marienhausen *see* Viļaka
83 D20 **Mariental** Hardap, SW Namibia 24°35′S 17°56′E
18 D13 **Marienville** Pennsylvania, NE USA 41°27′N 79°07′W
Marienwerder *see* Kwidzyń
58 C12 **Marié, Rio** ⇝ NW Brazil
95 K17 **Mariestad** Västra Götaland, S Sweden 58°42′N 13°50′E
23 S3 **Marietta** Georgia, SE USA 33°57′N 84°34′W
31 U14 **Marietta** Ohio, N USA 39°25′N 81°27′W
27 N13 **Marietta** Oklahoma, C USA 33°57′N 97°08′W
81 H18 **Marigat** Baringo, W Kenya 0°29′S 35°59′E
103 S16 **Marignane** Bouches-du-Rhône, SE France 43°25′N 05°12′E
Marignano *see* Melegnano
45 O11 **Marigot** NE Dominica 15°32′N 61°18′W
122 K12 **Mariinsk** Kemerovskaya Oblast', S Russian Federation 56°13′N 87°27′E
127 Q3 **Mariinskiy Posad** Respublika Mariy El, W Russian Federation 56°07′N 47°44′E
119 E14 **Marijampolė** *prev.* Kapsukas. Marijampolė, S Lithuania 54°33′N 23°21′E
Marikostenovo *see* Marikostinovo
114 G12 **Marikostinovo** *prev.* Marikostenovo. Blagoevgrad, SW Bulgaria 41°25′N 23°21′E
60 J9 **Marília** São Paulo, S Brazil 22°13′S 49°58′W
82 D11 **Marimba** Malanje, NW Angola 08°18′S 16°58′E
Mari Mila *see* Małe Mela
Māri Milah *see* Małe Mela
104 G4 **Marín** Galicia, NW Spain 42°23′N 08°43′W
35 N10 **Marina** California, W USA 36°40′N 121°48′W
Mar"ina Gorka *see* Mar"ina Horka
119 L17 **Mar"ina Horka** *Rus.* Mar'ina Gorka. Minskaya Voblasts', C Belarus 53°31′N 28°09′E
171 O4 **Marinduque** *island* C Philippines
31 S9 **Marine City** Michigan, N USA 42°43′N 82°29′W
31 N6 **Marinette** Wisconsin, N USA 45°06′N 87°38′W
60 I10 **Maringá** Paraná, S Brazil 23°26′S 51°55′W
83 N16 **Maringuè** Sofala, C Mozambique 17°57′S 34°23′E
104 F9 **Marinha Grande** Leiria, C Portugal 39°45′N 08°55′W
107 I15 **Marino** Lazio, C Italy 41°46′N 12°40′E
59 A15 **Mário Lobão** Acre, W Brazil 08°21′S 72°58′W
23 O5 **Marion** Alabama, S USA 32°37′N 87°19′W
27 Y11 **Marion** Arkansas, C USA 35°12′N 90°12′W
30 L17 **Marion** Illinois, N USA 37°43′N 88°55′W
31 P13 **Marion** Indiana, N USA 40°32′N 85°40′W
29 X13 **Marion** Iowa, C USA 42°01′N 91°36′W
27 O5 **Marion** Kansas, C USA 38°22′N 97°02′W
20 H6 **Marion** Kentucky, S USA 37°19′N 88°06′W
21 P9 **Marion** North Carolina, SE USA 35°43′N 82°00′W
31 S12 **Marion** Ohio, N USA 40°35′N 83°08′W
21 T12 **Marion** South Carolina, SE USA 34°11′N 79°23′W
21 Q7 **Marion** Virginia, NE USA 36°51′N 81°30′W
27 O5 **Marion Lake** ⊠ Kansas, C USA
21 S13 **Marion, Lake** ⊠ South Carolina, SE USA
27 S8 **Marionville** Missouri, C USA 37°00′N 93°38′W
55 N7 **Maripa** Bolívar, E Venezuela 07°27′N 65°10′W
55 X11 **Maripasoula** W French Guiana 03°43′N 54°04′W
35 Q9 **Mariposa** California, W USA 37°28′N 119°59′W
61 G19 **Mariscala** Lavalleja, S Uruguay 34°03′S 54°47′W
62 M4 **Mariscal Estigarribia** Boquerón, NW Paraguay 22°03′S 60°39′W
56 C6 **Mariscal Sucre** *var.* Quito. ✕ (Quito) Pichincha, C Ecuador 0°21′S 78°37′W
30 K16 **Marissa** Illinois, N USA 38°15′N 89°45′W
103 U14 **Maritime Alps** *Fr.* Alpes Maritimes, *It.* Alpi Marittime. ▲ France/Italy
Maritimes, Alpes *see* Maritime Alps
Maritime Territory *see* Primorskiy Kray
114 K11 **Maritsa** *var.* Marica, *Gk.* Évros, *Turk.* Meriç; *anc.* Hebrus. ⇝ SW Europe *see also* Évros/Meriç
Maritsa *see* Simeonovgrad, Bulgaria
Marittime, Alpi *see* Maritime Alps
Maritzburg *see* Pietermaritzburg
117 X9 **Mariupol'** *prev.* Zhdanov. Donets'ka Oblast', SE Ukraine 47°06′N 37°34′E
55 Q6 **Mariusa, Caño** ⇝ NE Venezuela
142 J5 **Marīvān** *prev.* Dezh Shāhpūr. Kordestān, W Iran 35°30′N 46°09′E
127 R3 **Mariyets** Respublika Mariy El, W Russian Federation 56°31′N 49°48′E
118 G4 **Märjamaa** *Ger.* Merjama. Raplamaa, NW Estonia 58°54′N 24°21′E
99 I15 **Mark** *Fr.* Marcq. ⇝ Belgium/Netherlands
81 N17 **Marka** *var.* Merca. Shabeellaha Hoose, S Somalia 01°43′N 44°45′E
145 Z10 **Markakol', Ozero** *Kaz.* Marqaköl. ◎ E Kazakhstan
76 M12 **Markala** Ségou, W Mali 13°38′N 06°07′W
159 S15 **Markam** *var.* Gartog. Xizang Zizhiqu, W China 29°40′N 98°33′E
95 K21 **Markaryd** Kronoberg, S Sweden 56°26′N 13°35′E
142 L7 **Markazī** *off.* Ostān-e Markazī. ◆ *province* W Iran
Markazī, Ostān-e *see* Markazī
14 F14 **Markdale** Ontario, S Canada 44°19′N 80°37′W
27 X10 **Marked Tree** Arkansas, C USA 35°31′N 90°25′W
98 N11 **Markelo** Overijssel, E Netherlands 52°15′N 06°30′E
98 J9 **Markermeer** ◎ C Netherlands
97 N20 **Market Harborough** C England, United Kingdom 52°30′N 00°57′W
97 N18 **Market Rasen** E England, United Kingdom 53°23′N 00°21′W
123 O10 **Markha** ⇝ NE Russian Federation
12 H16 **Markham** Ontario, S Canada 43°54′N 79°16′W
25 V12 **Markham** Texas, SW USA 28°57′N 96°04′W
186 E7 **Markham** ⇝ C Papua New Guinea
195 Q11 **Markham, Mount** ▲ Antarctica 82°58′S 163°30′E
110 M11 **Marki** Mazowieckie, C Poland 52°20′N 21°07′E
158 F8 **Markit** Xinjiang Uygur Zizhiqu, NW China 38°55′N 77°40′E
117 Y5 **Markivka** *Rus.* Markovka. Luhans'ka Oblast', E Ukraine 49°34′N 39°35′E
35 Q7 **Markleeville** California, W USA 38°41′N 119°46′W
98 L8 **Marknesse** Flevoland, N Netherlands 52°44′N 05°54′E
79 H14 **Markounda** *var.* Marcounda. Ouham, NW Central African Republic 07°38′N 17°00′E
Markovka *see* Markivka
123 U7 **Markovo** Chukotskiy Avtonomnyy Okrug, NE Russian Federation 64°43′N 170°13′E
127 P8 **Marks** Saratovskaya Oblast', W Russian Federation 51°40′N 46°44′E
22 K2 **Marks** Mississippi, S USA 34°15′N 90°16′W
22 I7 **Marksville** Louisiana, S USA 31°07′N 92°04′W
101 I19 **Marktheidenfeld** Bayern, C Germany 49°50′N 09°36′E
101 J24 **Marktoberdorf** Bayern, S Germany 47°45′N 10°36′E
101 M18 **Marktredwitz** Bayern, E Germany 50°N 12°04′E
Markt-Übelbach *see* Übelbach
27 V3 **Mark Twain Lake** ⊠ Missouri, C USA
Markuleshty *see* Mărculești
101 E14 **Marl** Nordrhein-Westfalen, W Germany 51°38′N 07°06′E
182 E2 **Marla** South Australia 27°19′S 133°35′E
181 Y8 **Marlborough** Queensland, E Australia 22°55′S 150°07′E
97 M22 **Marlborough** S England, United Kingdom 51°25′N 01°45′W
185 I15 **Marlborough** *off.* Marlborough District. ◆ *unitary authority* South Island, New Zealand
Marlborough District *see* Marlborough
103 P3 **Marle** Aisne, N France 49°44′N 03°47′E
31 S8 **Marlette** Michigan, N USA 43°20′N 83°05′W
25 T9 **Marlin** Texas, SW USA 31°20′N 96°55′W
21 S5 **Marlinton** West Virginia, NE USA 38°14′N 80°06′W
26 M12 **Marlow** Oklahoma, C USA 34°39′N 97°57′W
155 E17 **Marmagao** Goa, W India 15°26′N 73°50′E
Marmanda *see* Marmande
102 L13 **Marmande** *anc.* Marmanda. Lot-et-Garonne, SW France 44°30′N 00°10′E
136 C11 **Marmara** Balıkesir, NW Turkey 40°36′N 27°34′E
136 D11 **Marmara Denizi** *Eng.* Sea of Marmara. *sea* NW Turkey
114 N13 **Marmaraereğlisi** Tekirdağ, NW Turkey 40°59′N 27°57′E
Marmara, Sea of *see* Marmara Denizi
136 C16 **Marmaris** Muğla, SW Turkey 36°52′N 28°17′E
28 J6 **Marmarth** North Dakota, N USA 46°17′N 103°55′W
21 Q5 **Marmet** West Virginia, NE USA 38°12′N 81°31′W
106 H5 **Marmolada, Monte** ▲ N Italy 46°36′N 11°58′E
104 M13 **Marmolejo** Andalucía, S Spain 38°03′N 04°10′W
14 J14 **Marmora** Ontario, SE Canada 44°29′N 77°40′W
39 Q14 **Marmot Bay** *bay* Alaska, USA
103 Q4 **Marne** ◆ *department* N France
103 Q4 **Marne** ⇝ N France
137 U10 **Marneuli** *prev.* Borchalo, Sarvani. S Georgia 41°28′N 44°45′E
78 I13 **Maro** Moyen-Chari, S Chad 08°25′N 18°46′E
54 L12 **Maroa** Amazonas, S Venezuela 02°40′N 67°33′W
172 J3 **Maroantsetra** Toamasina, NE Madagascar 15°23′S 49°44′E
191 W11 **Marokau** *atoll* Îles Tuamotu, C French Polynesia
172 J5 **Marolambo** Toamasina, E Madagascar 20°03′S 48°08′E
172 J2 **Maromokotro** ▲ N Madagascar
83 L16 **Marondera** *prev.* Marandellas. Mashonaland East, NE Zimbabwe 18°11′S 31°33′E
55 X9 **Maroni** *Dut.* Marowijne. ⇝ French Guiana/Suriname
183 V2 **Maroochydore-Mooloolaba** Queensland, E Australia 26°36′S 153°04′E
171 N14 **Maros** Sulawesi, C Indonesia 04°59′S 119°35′E
116 H11 **Maros** *var.* Mureș, Mureșul, *Ger.* Marosch, Mieresch. ⇝ Hungary/Romania *see also* Mureș
Marosch *see* Maros/Mureș
Marosheviz *see* Toplița
Marosillye *see* Ilia
Marosludas *see* Luduș
Marosújvár/ Marosújvárakna *see* Ocna Mureș
Marosvásárhely *see* Târgu Mureș
191 V14 **Marotiri** *var.* Îlots de Bass, Morotiri. *island group* Îles Australes, SW French Polynesia
78 G12 **Maroua** Extrême-Nord, N Cameroon 10°35′N 14°20′E
55 X12 **Marouini Rivier** ⇝ SE Suriname
172 I3 **Marovoay** Mahajanga, NW Madagascar 16°05′S 46°40′E
55 W9 **Marowijne** ◆ *district* NE Suriname
Marowijne *see* Maroni
Marqaköl *see* Markakol', Ozero
193 P8 **Marquesas Fracture Zone** *tectonic feature* E Pacific Ocean
Marquesas Islands *see* Marquises, Îles
23 W17 **Marquesas Keys** *island group* Florida, SE USA
29 Y12 **Marquette** Iowa, C USA 43°02′N 91°10′W
31 N3 **Marquette** Michigan, N USA 46°32′N 87°24′W
103 N1 **Marquise** Pas-de-Calais, N France 50°49′N 01°42′E
191 X7 **Marquises, Îles** *Eng.* Marquesas Islands. *island group* N French Polynesia
183 Q6 **Marra Creek** ⇝ New South Wales, SE Australia
80 B10 **Marra Hills** *plateau* W Sudan
80 B11 **Marra, Jebel** ▲ W Sudan 12°59′N 24°16′E
74 E7 **Marrakech** *var.* Marakesh, *Eng.* Marrakesh; *prev.* Morocco. W Morocco 31°39′N 07°58′W
Marrakesh *see* Marrakech
Marrāt *see* Marāh
183 N15 **Marrawah** Tasmania, SE Australia 40°56′S 144°41′E
182 I4 **Marree** South Australia 29°40′S 138°06′E
81 L17 **Marrehan** ▲ SW Somalia
83 N17 **Marromeu** Sofala, C Mozambique 18°18′S 35°58′E
104 J17 **Marroquí, Punta** *headland* SW Spain 36°01′N 05°39′W
183 N8 **Marrowie Creek** *seasonal river* New South Wales, SE Australia
83 O14 **Marrupa** Niassa, N Mozambique 13°10′S 37°30′E
182 D1 **Marryat** South Australia 26°22′S 133°22′E
75 Y10 **Marsā al 'Alam** *var.* Marsa 'Alam. SE Egypt 25°03′N 33°44′E
Marsa 'Alam *see* Marsā al 'Alam
75 R8 **Marsá al Burayqah** *var.* Al Burayqah. N Libya 30°21′N 19°37′E
81 J17 **Marsabit** Marsabit, N Kenya 02°20′N 37°59′E
81 I17 **Marsabit** ◆ *county* N Kenya
107 H23 **Marsala** *anc.* Lilybaeum. Sicilia, Italy, C Mediterranean Sea 37°48′N 12°26′E
75 U7 **Marsá Maṭrūḥ** *var.* Maṭrūḥ; *anc.* Paraetonium. NW Egypt 31°21′N 27°15′E
Marsaxlokk Bay *see* Il-Bajja ta' Marsaxlokk
65 G15 **Mars Bay** *bay* Ascension Island, C Atlantic Ocean
101 H15 **Marsberg** Nordrhein-Westfalen, W Germany 51°28′N 08°51′E
11 R15 **Marsden** Saskatchewan, S Canada 52°50′N 109°45′W
98 H7 **Marsdiep** *strait* NW Netherlands
103 R16 **Marseille** *Eng.* Marseilles; *anc.* Massilia. Bouches-du-Rhône, SE France 43°19′N 05°22′E
Marseille-Marigrane *see* Provence
30 M11 **Marseilles** Illinois, N USA 41°19′N 88°42′W
Marseilles *see* Marseille
76 J16 **Marshall** W Liberia 06°10′N 10°23′W
39 N11 **Marshall** Alaska, USA 61°52′N 162°04′W
27 U9 **Marshall** Arkansas, C USA 35°54′N 92°40′W
31 N14 **Marshall** Illinois, N USA 39°23′N 87°41′W
31 Q10 **Marshall** Michigan, N USA 42°16′N 84°57′W
29 S9 **Marshall** Minnesota, N USA 44°26′N 95°48′W
27 T4 **Marshall** Missouri, C USA 39°07′N 93°12′W
21 O9 **Marshall** North Carolina, SE USA 35°48′N 82°43′W
25 X6 **Marshall** Texas, SW USA 32°33′N 94°22′W
189 S4 **Marshall Islands** *off.* Republic of the Marshall Islands. ◆ *republic* W Pacific Ocean
175 Q3 **Marshall Islands** *island group* W Pacific Ocean
Marshall Islands, Republic of the *see* Marshall Islands
192 K6 **Marshall Seamounts** *undersea feature* SW Pacific Ocean 10°00′N 165°00′E
29 W13 **Marshalltown** Iowa, C USA 42°01′N 92°54′W
19 P12 **Marshfield** Massachusetts, NE USA 42°04′N 70°40′W
27 T7 **Marshfield** Missouri, C USA 37°20′N 92°55′W
30 K6 **Marshfield** Wisconsin, N USA 44°41′N 90°12′W
44 H1 **Marsh Harbour** Great Abaco, W The Bahamas 26°31′N 77°03′W
19 S3 **Mars Hill** Maine, NE USA 46°31′N 67°51′W
21 P9 **Mars Hill** North Carolina, SE USA 35°49′N 82°33′W
22 H10 **Marsh Island** *island* Louisiana, S USA
21 S11 **Marshville** North Carolina, SE USA 34°59′N 80°22′W
15 W5 **Marsoui** Québec, SE Canada 49°12′N 65°58′W
15 R8 **Mars, Rivière à** ⇝ Québec, SE Canada
95 O15 **Märsta** Stockholm, C Sweden 59°37′N 17°52′E
95 H24 **Marstal** Syddtjylland, C Denmark 54°52′N 10°32′E
95 I19 **Marstrand** Västra Götaland, S Sweden 57°54′N 11°31′E
25 U8 **Mart** Texas, SW USA 31°32′N 96°49′W
Martaban *see* Mottama
Martaban, Gulf of *see* Mottama, Gulf of
107 Q19 **Martano** Puglia, SE Italy 40°12′N 18°19′E
Martapoera *see* Martapura
169 T13 **Martapura** *prev.* Martapoera. Borneo, C Indonesia 03°25′S 114°51′E
99 L23 **Martelange** Luxembourg, SE Belgium 49°50′N 05°43′E
114 L7 **Marten** Ruse, N Bulgaria 43°57′N 26°08′E
14 H10 **Marten River** Ontario, S Canada 46°43′N 79°45′W
11 T15 **Martensville** Saskatchewan, S Canada 52°15′N 106°42′W
Marteskirch *see* Târnăveni
Martes Tolosane *see* Martres-Tolosane
115 K25 **Mártha** Kríti, Greece, E Mediterranean Sea 35°03′N 25°22′E
183 Q6 **Marthaguy Creek** ⇝ New South Wales, SE Australia
19 P13 **Martha's Vineyard** *island* Massachusetts, NE USA
108 C11 **Martigny** Valais, SW Switzerland 46°06′N 07°04′E
103 R16 **Martigues** Bouches-du-Rhône, SE France 43°24′N 05°03′E
111 J19 **Martin** *Ger.* Sankt Martin, *Hung.* Turócszentmárton; *prev.* Turčianský Svätý Martin. Žilinský Kraj, N Slovakia 49°03′N 18°54′E
28 L11 **Martin** South Dakota, N USA 43°10′N 101°43′W
20 G8 **Martin** Tennessee, S USA 36°20′N 88°51′W
105 S7 **Martín** ⇝ E Spain
107 P18 **Martina Franca** Puglia, SE Italy 40°42′N 17°20′E
185 M14 **Martinborough** Wellington, North Island, New Zealand 41°12′S 175°28′E
25 S11 **Martindale** Texas, SW USA 29°49′N 97°49′W
35 N8 **Martinez** California, W USA 38°00′N 122°12′W
23 V3 **Martinez** Georgia, SE USA 33°31′N 82°04′W
41 Q13 **Martínez de La Torre** Veracruz-Llave, E Mexico 20°05′N 97°02′W
45 Y12 **Martinique** ◇ *French overseas department* E West Indies
1 O15 **Martinique** *island* E West Indies
Martinique Channel *see* Martinique Passage
45 X12 **Martinique Passage** *var.* Dominica Channel, Martinique Channel. *channel* Dominica/Martinique
23 Q5 **Martin Lake** ⊠ Alabama, S USA
115 G18 **Martíno** *prev.* Martínon. Stereá Elláda, C Greece 38°34′N 23°13′E
Martínon *see* Martíno
194 J11 **Martin Peninsula** *peninsula* Antarctica
39 S5 **Martin Point** *headland* Alaska, USA 70°06′N 143°04′W
109 V3 **Martinsberg** Niederösterreich, NE Austria 48°23′N 15°09′E
21 V3 **Martinsburg** West Virginia, NE USA 39°28′N 77°59′W
31 V13 **Martins Ferry** Ohio, N USA 40°06′N 80°43′W
Martinskirch *see* Târnăveni
31 O14 **Martinsville** Indiana, N USA 39°25′N 86°25′W
21 S8 **Martinsville** Virginia, NE USA 36°43′N 79°53′W
65 K16 **Martin Vaz, Ilhas** *island group* E Brazil
144 I9 **Martok** *prev.* Martuk. Aktyubinsk, NW Kazakhstan 50°45′N 56°30′E
184 M12 **Marton** Manawatu-Wanganui, North Island, New Zealand 40°05′S 175°22′E
105 N13 **Martos** Andalucía, S Spain 37°44′N 03°58′W
102 M16 **Martres-Tolosane** *var.* Martes Tolosane. Haute-Garonne, S France 43°13′N 01°00′E
92 M11 **Martti** Lappi, NE Finland 67°28′N 28°20′E
Martuk *see* Martok
137 U12 **Martuni** E Armenia 40°07′N 45°20′E
58 L11 **Marudá** Pará, E Brazil 05°25′S 49°04′W
169 V6 **Marudu, Teluk** *bay* East Malaysia
149 O8 **Ma'rūf** Kandahār, SE Afghanistan 31°34′N 67°06′E
164 H13 **Marugame** Kagawa, Shikoku, SW Japan 34°17′N 133°46′E
185 H16 **Maruia** ⇝ South Island, New Zealand
98 M6 **Marum** Groningen, NE Netherlands 53°07′N 06°16′E
187 R13 **Marum, Mount** ▲ Ambrym, C Vanuatu 16°15′S 168°07′E
79 P23 **Marungu** ▲ SE Dem. Rep. Congo
191 Y12 **Marutea** *atoll* Groupe Actéon, C French Polynesia
143 O11 **Marv Dasht** *var.* Mervdasht. Fārs, S Iran 29°50′N 52°40′E
103 P13 **Marvejols** Lozère, S France 44°35′N 03°16′E
27 X12 **Marvell** Arkansas, C USA 34°33′N 90°52′W
36 L6 **Marvine, Mount** ▲ Utah, W USA 38°40′N 111°38′W
139 Q7 **Marwānīyah** Al Anbār, C Iraq 33°58′N 42°31′E
152 F13 **Mārwār** *var.* Kharchi, Marwar Junction. Rājasthān, N India 25°41′N 73°42′E
Marwar Junction *see* Mārwār
11 R14 **Marwayne** Alberta, SW Canada 53°30′N 110°25′W
146 I14 **Mary** *prev.* Merv. Mary Welaýaty, S Turkmenistan 37°25′N 61°48′E
Mary *see* Mary Welaýaty
181 Z9 **Maryborough** Queensland, E Australia 25°32′S 152°36′E
182 M11 **Maryborough** Victoria, SE Australia 37°05′S 143°47′E
Maryborough *see* Port Laoise
83 G23 **Marydale** Northern Cape, W South Africa 29°25′S 22°06′E
117 W8 **Mar"yinka** Donets'ka Oblast', E Ukraine 47°57′N 37°27′E
Mary Island *see* Kanton
21 W4 **Maryland** *off.* State of Maryland, *also known as* America in Miniature, Cockade State, Free State, Old Line State. ◆ *state* NE USA
Maryland, State of *see* Maryland
25 P7 **Maryneal** Texas, SW USA 32°12′N 100°25′W
97 J15 **Maryport** NW England, United Kingdom 54°45′N 03°28′W
13 U13 **Marystown** Newfoundland, Newfoundland and Labrador, SE Canada 47°10′N 55°10′W
36 K6 **Marysvale** Utah, W USA 38°26′N 112°14′W

◆ Country
● Country Capital
◇ Dependent Territory
○ Dependent Territory Capital
◆ Administrative Regions
✕ International Airport
▲ Mountain
▲ Mountain Range
⛰ Volcano
⇝ River
◎ Lake
⊠ Reservoir

35 *O6* **Marysville** California, W USA *39°07′N 121°35′W*
27 *O3* **Marysville** Kansas, C USA *39°48′N 96°37′W*
31 *S13* **Marysville** Michigan, N USA *42°54′N 82°29′W*
31 *S9* **Marysville** Ohio, NE USA *40°13′N 83°22′W*
32 *H7* **Marysville** Washington, NW USA *48°03′N 122°10′W*
27 *R2* **Maryville** Missouri, C USA *40°20′N 94°53′W*
21 *N9* **Maryville** Tennessee, S USA *35°45′N 83°59′W*
146 *I15* **Mary Welaýaty** *var.* Mary, *Rus.* Maryyskiy Velayat. ◈ *province* S Turkmenistan
Maryyskiy Velayat *see* Mary Welaýaty
Marzūq *see* Murzuq
42 *J11* **Masachapa** *var.* Puerto Masachapa. Managua, W Nicaragua *11°47′N 86°31′W*
81 *G19* **Masai Mara National Reserve** *reserve* SW Kenya
81 *I21* **Masai Steppe** *grassland* NW Tanzania
81 *F19* **Masaka** SW Uganda *0°20′S 31°46′E*
169 *T15* **Masalembo Besar, Pulau** *island* S Indonesia
137 *Y13* **Masallı** *Rus.* Masally. S Azerbaijan *39°03′N 48°39′E*
Masally *see* Masallı
171 *N13* **Masamba** Sulawesi, C Indonesia *02°33′S 120°20′E*
Masampo *see* Masan
163 *Y16* **Masan** *prev.* Masampo. S South Korea *35°11′N 128°36′E*
Masandam Peninsula *see* Musandam Peninsula
81 *J25* **Masasi** Mtwara, SE Tanzania *10°43′S 38°48′E*
Masawa/Massawa *see* Mits'iwa
42 *J10* **Masaya** Masaya, W Nicaragua *11°59′N 86°06′W*
42 *J10* **Masaya** ◈ *department* W Nicaragua
171 *P5* **Masbate** Masbate, N Philippines *12°21′N 123°34′E*
171 *P5* **Masbate** *island* C Philippines
74 *I6* **Mascara** *var.* Mouaskar. NW Algeria *35°20′N 00°09′E*
173 *O7* **Mascarene Basin** *undersea feature* W Indian Ocean *15°00′S 56°00′E*
173 *O9* **Mascarene Islands** *island group* W Indian Ocean
173 *N9* **Mascarene Plain** *undersea feature* W Indian Ocean *19°00′S 52°00′E*
173 *O7* **Mascarene Plateau** *undersea feature* W Indian Ocean *10°00′S 60°00′E*
194 *H5* **Mascart, Cape** *headland* Adelaide Island, Antarctica
62 *J10* **Mascasín, Salinas de** *salt lake* C Argentina
40 *K13* **Mascota** Jalisco, C Mexico *20°31′N 104°46′W*
15 *O12* **Mascouche** Québec, SE Canada *45°46′N 73°37′W*
124 *J9* **Masel'gskaya** Respublika Kareliya, NW Russian Federation *63°09′N 34°22′E*
83 *J23* **Maseru** ● (Lesotho) W Lesotho *29°21′S 27°35′E*
83 *J23* **Maseru** ✈ W Lesotho *29°27′S 27°37′E*
Mashaba *see* Mashava
160 *K14* **Mashan** *var.* Baishan. Guangxi Zhuangzu Zizhiqu, S China *23°40′N 108°10′E*
83 *K17* **Mashava** *prev.* Mashaba. Masvingo, SE Zimbabwe *20°03′S 30°29′E*
143 *U4* **Mashhad** *var.* Meshed. Khorāsān-e Raẕavī, NE Iran *36°16′N 59°34′E*
165 *S3* **Mashike** Hokkaidō, NE Japan *43°51′N 141°30′E*
83 *K20* **Mashishing** *prev.* Lydenburg. Mpumalanga, NE South Africa *25°10′S 30°29′E*
Mashīz *see* Bardsīr
149 *N14* **Mashkai** ≈ SW Pakistan
143 *X13* **Māshkel** *var.* Rūd-i Māshkel, Rūd-e Māshkīd. ≈ Iran/Pakistan
148 *K12* **Māshkel, Hāmūn-i** *salt marsh* SW Pakistan
Māshkel, Rūd-i/Māshkīd, Rūd-e *see* Māshkel
83 *K15* **Mashonaland Central** ◈ *province* N Zimbabwe
83 *K16* **Mashonaland East** ◈ *province* NE Zimbabwe
83 *J16* **Mashonaland West** ◈ *province* NW Zimbabwe
Mashtagi *see* Maştağa
141 *S14* **Masīlah, Wādī al** *dry watercourse* SE Yemen
79 *I21* **Masi-Manimba** Bandundu, SW Dem. Rep. Congo *04°47′S 17°54′E*
81 *F17* **Masindi** W Uganda *01°41′N 31°45′E*
81 *I19* **Masinga Reservoir** ⊠ S Kenya
Masīra *see* Maşīrah, Jazīrat
Masira, Gulf of *see* Maşīrah, Khalīj
141 *Y10* **Maşīrah, Jazīrat** *var.* Masīra. *island* E Oman
141 *Y10* **Maşīrah, Khalīj** *var.* Gulf of Masira. *bay* E Oman
Masis *see* Büyükağrı Dağı
79 *O19* **Masisi** Nord-Kivu, E Dem. Rep. Congo *01°25′S 28°50′E*
Masjed-e Soleymān *see* Masjed Soleymān
142 *L9* **Masjed Soleymān** *var.* Masjed-e Soleymān, Masjid-i Sulaiman. Khūzestān, SW Iran *31°59′N 49°18′E*
Masjid-i Sulaiman *see* Masjed Soleymān
Maskat *see* Masqaţ
139 *Q7* **Maskhān** Al Anbār, C Iraq *33°41′N 42°46′E*
141 *X8* **Maskin** *var.* Miskin. NW Oman *23°28′N 56°46′E*
97 *B17* **Mask, Lough** *Ir.* Loch Measca. ◎ W Ireland
114 *N10* **Maslen Nos** *headland* E Bulgaria *42°19′N 27°47′E*
172 *K3* **Masoala, Tanjona** *headland* NE Madagascar *15°59′N 50°13′E*
Masohi *see* Amahai
31 *Q9* **Mason** Michigan, N USA *42°33′N 84°25′W*
31 *R14* **Mason** Ohio, N USA *39°21′N 84°18′W*
25 *Q10* **Mason** Texas, SW USA *30°45′N 99°15′W*
21 *P4* **Mason** West Virginia, NE USA *39°01′N 82°01′W*
185 *B25* **Mason Bay** *bay* Stewart Island, New Zealand
30 *K13* **Mason City** Illinois, N USA *40°12′N 89°42′W*
29 *V12* **Mason City** Iowa, C USA *43°09′N 93°12′W*
18 *B16* **Masontown** Pennsylvania, NE USA *39°49′N 79°53′W*
141 *Y8* **Masqaţ** *var.* Maskat, *Eng.* Muscat. ● (Oman) NE Oman *23°35′N 58°36′E*
106 *E10* **Massa** Toscana, C Italy *44°02′N 10°07′E*
18 *M11* **Massachusetts** *off.* Commonwealth of Massachusetts, *also known as* Bay State, Old Bay State, Old Colony State. ◈ *state* NE USA
19 *P11* **Massachusetts Bay** *bay* Massachusetts, NE USA
35 *R2* **Massacre Lake** ◎ Nevada, W USA
107 *O18* **Massafra** Puglia, SE Italy *40°35′N 17°08′E*
108 *G11* **Massagno** Ticino, S Switzerland *46°01′N 08°55′E*
78 *G11* **Massaguet** Hadjer-Lamis, W Chad *12°28′N 15°26′E*
Massakori *see* Massakory
78 *G10* **Massakory** *var.* Massakori; *prev.* Dagana. Hadjer-Lamis, W Chad *13°02′N 15°43′E*
78 *H11* **Massalassef** Hadjer-Lamis, SW Chad *11°37′N 17°09′E*
106 *F13* **Massa Marittima** Toscana, C Italy *43°03′N 10°55′E*
82 *B11* **Massangano** Kwanza Norte, NW Angola *09°40′S 14°13′E*
83 *M18* **Massangena** Gaza, S Mozambique *21°34′S 32°57′E*
80 *K9* **Massawa Channel** *channel* E Eritrea
18 *J6* **Massena** New York, NE USA *44°55′N 74°53′W*
78 *H11* **Massenya** Chari-Baguirmi, SW Chad *11°21′N 16°09′E*
10 *I13* **Masset** Graham Island, British Columbia, SW Canada *54°00′N 132°09′W*
102 *L16* **Masseube** Gers, S France *43°26′N 00°33′E*
14 *E11* **Massey** Ontario, S Canada *46°13′N 82°06′W*
103 *P12* **Massiac** Cantal, C France *45°16′N 03°13′E*
103 *P12* **Massif Central** *plateau* C France
Massif de L'Isalo *see* Isalo
Massilia *see* Marseille
31 *U12* **Massillon** Ohio, N USA *40°48′N 81°31′W*
77 *N12* **Massina** Ségou, W Mali *13°58′N 05°24′W*
83 *N19* **Massinga** Inhambane, SE Mozambique *23°20′S 35°25′E*
83 *L20* **Massingir** Gaza, SW Mozambique *23°51′S 31°58′E*
195 *Z10* **Masson Island** *island* Antarctica
Massoukou *see* Franceville
137 *Z11* **Maştağa** *Rus.* Mashtagi, Mastaga. E Azerbaijan *40°31′N 50°01′E*
Mastanli *see* Momchilgrad
184 *M13* **Masterton** Wellington, North Island, New Zealand *40°56′S 175°40′E*
18 *M14* **Mastic** Long Island, New York, NE USA *40°48′N 72°50′W*
149 *O10* **Mastung** Baluchistān, SW Pakistan *29°44′N 66°56′E*
119 *J20* **Mastva** *Rus.* Mostva. ≈ SW Belarus
119 *G17* **Masty** *Rus.* Mosty. Hrodzyenskaya Voblasts', W Belarus *53°25′N 24°32′E*
164 *F12* **Masuda** Shimane, Honshū, SW Japan *34°40′N 131°50′E*
92 *J11* **Masugnsbyn** Norrbotten, N Sweden *67°28′N 22°01′E*
Masuku *see* Franceville
83 *K17* **Masvingo** *prev.* Fort Victoria, Nyanda, Victoria. Masvingo, SE Zimbabwe *20°05′S 30°50′E*
83 *K18* **Masvingo** *prev.* Victoria. ◈ *province* SE Zimbabwe
138 *H5* **Maşyāf** *Fr.* Misiaf. Ḩamāh, C Syria *35°04′N 36°21′E*
110 *E9* **Maszewo** Zachodniopomorskie, NW Poland *53°29′N 15°01′E*
83 *I17* **Matabeleland North** ◈ *province* W Zimbabwe
83 *J18* **Matabeleland South** ◈ *province* S Zimbabwe
82 *O13* **Mataca** Niassa, N Mozambique *12°27′S 36°13′E*
14 *G8* **Matachewan** Ontario, S Canada *47°58′N 80°37′W*
163 *Q8* **Matad** *var.* Dzüünbulag. Dornod, E Mongolia *46°48′N 115°21′E*
79 *F22* **Matadi** Bas-Congo, W Dem. Rep. Congo *05°49′S 13°31′E*
25 *O4* **Matador** Texas, SW USA *34°01′N 100°50′W*
42 *J9* **Matagalpa** Matagalpa, C Nicaragua *12°53′N 85°56′W*
42 *K9* **Matagalpa** ◈ *department* W Nicaragua
12 *I12* **Matagami** Québec, S Canada *49°47′N 77°38′W*
25 *U13* **Matagorda** Texas, SW USA *28°40′N 96°57′W*
25 *U13* **Matagorda Bay** *inlet* Texas, SW USA
25 *U14* **Matagorda Island** *island* Texas, SW USA
25 *V13* **Matagorda Peninsula** *headland* Texas, SW USA *28°34′N 96°01′W*
191 *Q8* **Mataiea** Tahiti, W French Polynesia *17°46′S 149°25′W*
191 *T9* **Mataiva** *atoll* Îles Tuamotu, C French Polynesia
183 *O7* **Matakana** New South Wales, SE Australia *32°59′S 145°53′E*
184 *N7* **Matakana Island** *island* NE New Zealand
83 *C15* **Matala** Huíla, SW Angola *14°45′S 15°02′E*
190 *G12* **Matala'a Pointe** *headland* Île Uvea, N Wallis and Futuna *13°20′S 176°08′W*
155 *K25* **Matale** Central Province, C Sri Lanka *07°29′N 80°38′E*
190 *E12* **Matalesina, Pointe** *headland* Île Alofi, W Wallis and Futuna
76 *I10* **Matam** NE Senegal *15°40′N 13°18′W*
184 *M8* **Matamata** Waikato, North Island, New Zealand *37°49′S 175°45′E*
77 *V12* **Matamey** Zinder, S Niger *13°27′N 08°27′E*
40 *L8* **Matamoros** Coahuila, NE Mexico *25°34′N 103°13′W*
41 *P15* **Matamoros** *var.* Izúcar de Matamoros. Puebla, S Mexico *18°38′N 98°30′W*
41 *Q8* **Matamoros** Tamaulipas, C Mexico *25°50′N 97°31′W*
75 *S13* **Ma'ţan as Sārah** SE Libya *21°45′N 21°55′E*
82 *J12* **Matanda** Luapula, N Zambia *11°24′S 28°25′E*
81 *J24* **Matandu** ≈ S Tanzania
15 *V6* **Matane** Québec, SE Canada *48°50′N 67°31′W*
15 *V6* **Matane** ≈ Québec, SE Canada
77 *S12* **Matankari** Dosso, SW Niger *13°39′N 04°03′E*
39 *R11* **Matanuska River** ≈ Alaska, USA
54 *G7* **Matanza** Santander, N Colombia *07°22′N 73°02′W*
44 *D4* **Matanzas** Matanzas, NW Cuba *23°N 81°32′W*
15 *V7* **Matapédia** ≈ Québec, SE Canada
15 *V6* **Matapédia, Lac** ◎ Québec, SE Canada
190 *B17* **Mata Point** *headland* SE Niue *19°07′S 169°51′E*
190 *D12* **Matapu, Pointe** *headland* Île Futuna, W Wallis and Futuna
62 *G12* **Mataquito, Río** ≈ C Chile
155 *K26* **Matara** Southern Province, S Sri Lanka *05°58′N 80°33′E*
115 *D18* **Matarágka** *var.* Mataránga. Dytikí Elláda, C Greece *38°32′N 21°28′E*
171 *K16* **Mataram** Pulau Lombok, C Indonesia *08°36′S 116°07′E*
Mataránga *see* Matarágka
181 *Q3* **Mataranka** Northern Territory, N Australia *14°55′S 133°03′E*
105 *W6* **Mataró** *anc.* Illuro. Cataluña, E Spain *41°32′N 02°27′E*
184 *O8* **Matata** Bay of Plenty, North Island, New Zealand *37°54′S 176°45′E*
192 *K16* **Matātula, Cape** *headland* Tutuila, W American Samoa *14°15′S 170°35′W*
185 *D24* **Mataura** Southland, South Island, New Zealand *46°12′S 168°53′E*
185 *D24* **Mataura** ≈ South Island, New Zealand
Mata Uta *see* Matā'utu
192 *H16* **Matāutu** Upolu, C Samoa *13°57′S 171°55′W*
190 *G11* **Matā'utu** *var.* Mata Uta. ○ (Wallis and Futuna) Île Uvea, Wallis and Futuna *13°22′S 176°12′W*
190 *G12* **Matā'utu, Baie de** *bay* Île Uvea, Wallis and Futuna
191 *P7* **Matavai, Baie de** *bay* Tahiti, W French Polynesia
190 *I16* **Matavera** Rarotonga, S Cook Islands *21°13′S 159°44′W*
191 *V16* **Mataveri** Easter Island, Chile, E Pacific Ocean *27°10′S 109°27′W*
191 *V17* **Mataveri** ✈ (Easter Island) Easter Island, Chile, E Pacific Ocean *27°10′S 109°27′W*
184 *P9* **Matawai** Gisborne, North Island, New Zealand *38°23′S 177°31′E*
15 *O10* **Matawin** ≈ Québec, SE Canada
145 *V13* **Matay** Almaty, SE Kazakhstan *45°53′N 78°45′E*
14 *K8* **Matchi-Manitou, Lac** ◎ Québec, SE Canada
41 *O10* **Matehuala** San Luis Potosí, C Mexico *23°40′N 100°40′W*
45 *V13* **Matelot** Trinidad, Trinidad and Tobago *10°48′N 61°06′W*
83 *M15* **Matenge** Tete, NW Mozambique *15°22′S 33°47′E*
107 *O18* **Matera** Basilicata, S Italy *40°39′N 16°35′E*
111 *O21* **Mátészalka** Szabolcs-Szatmár-Bereg, E Hungary *47°58′N 22°17′E*
93 *H17* **Matfors** Västernorrland, C Sweden *62°21′N 17°02′E*
102 *K11* **Matha** Charente-Maritime, W France *45°50′N 00°13′W*
0 *F15* **Mathematicians Seamounts** *undersea feature* E Pacific Ocean *15°00′N 111°00′W*
21 *X6* **Mathews** Virginia, NE USA *37°26′N 76°20′W*
25 *S14* **Mathis** Texas, SW USA *28°05′N 97°49′W*
152 *J11* **Mathura** *prev.* Muttra. Uttar Pradesh, N India *27°30′N 77°42′E*
Mathurai *see* Madurai
171 *R7* **Mati** Mindanao, S Philippines *06°58′N 126°11′E*
Matianus *see* Orūmīyeh, Daryācheh-ye
Matiara *see* Matiāri
149 *Q15* **Matiāri** *var.* Matiara. Sind, SE Pakistan *25°38′N 68°29′E*
41 *S16* **Matías Romero** Oaxaca, SE Mexico *16°53′N 95°02′W*
43 *O13* **Matina** Limón, E Costa Rica *10°06′N 83°18′W*
14 *D10* **Matinenda Lake** ◎ Ontario, S Canada
19 *R8* **Matinicus Island** *island* Maine, NE USA
Matisco/Matisco Ædourum *see* Mâcon
113 *K19* **Matit, Lumi i** ≈ NW Albania
149 *Q16* **Mātli** Sind, SE Pakistan *25°06′N 68°37′E*
97 *M18* **Matlock** C England, United Kingdom *53°08′N 01°32′W*
59 *F18* **Mato Grosso** *prev.* Vila Bela da Santissima Trindade. Mato Grosso, W Brazil *14°53′S 59°58′W*
59 *G17* **Mato Grosso** *off.* Estado de Mato Grosso; *prev.* Matto Grosso. ◈ *state* W Brazil
60 *H8* **Mato Grosso do Sul** *off.* Estado de Mato Grosso do Sul. ◈ *state* S Brazil
Mato Grosso do Sul, Estado de *see* Mato Grosso do Sul
Mato Grosso, Estado de *see* Mato Grosso
59 *I18* **Mato Grosso, Planalto de** *plateau* C Brazil
83 *L21* **Matola** Maputo, S Mozambique *25°57′S 32°27′E*
104 *G6* **Matosinhos** *prev.* Matozinhos. Porto, NW Portugal *41°11′N 08°42′W*
Matou *see* Pingguo
55 *Z10* **Matoury** NE French Guiana *04°49′N 52°17′W*
Matozinhos *see* Matosinhos
111 *L21* **Mátra** ▲▲ N Hungary
141 *Y8* **Maţraḩ** *var.* Mutrah. NE Oman *23°35′N 58°31′E*
116 *L12* **Mătrașești** Vrancea, E Romania *45°53′N 27°14′E*
108 *M8* **Matrei am Brenner** Tirol, W Austria *47°09′N 11°28′E*
109 *P8* **Matrei in Osttirol** Tirol, W Austria *47°04′N 12°31′E*
76 *I15* **Matru** SW Sierra Leone *07°37′N 12°08′W*
Maţrûḩ *see* Marsā Maţrūḩ
165 *U16* **Matsubara** *var.* Matubara. Kagoshima, Tokuno-shima, SW Japan *32°58′N 129°56′E*
161 *S12* **Matsu Dao** *var.* Mazu Tao; *prev.* Matsu Tao. *island* NW Taiwan
164 *G12* **Matsue** *var.* Matsuye, Matue. Shimane, Honshū, SW Japan *35°27′N 133°04′E*
165 *Q6* **Matsumae** Hokkaidō, NE Japan *41°27′N 140°04′E*
164 *M12* **Matsumoto** *var.* Matumoto. Nagano, Honshū, S Japan *36°18′N 137°58′E*
164 *K14* **Matsusaka** *var.* Matsuzaka, Matusaka. Mie, Honshū, SW Japan *34°33′N 136°31′E*
Matsu Tao *see* Matsu Dao
Matsutō *see* Mattō
164 *F14* **Matsuyama** *var.* Matuyama. Ehime, Shikoku, SW Japan *33°50′N 132°47′E*
Matsuye *see* Matsue
Matsuzaka *see* Matsusaka
164 *M14* **Matsuzaki** Shizuoka, Honshū, S Japan *34°43′N 138°45′E*
14 *F8* **Mattagami** ≈ Ontario, S Canada
14 *F8* **Mattagami Lake** ◎ Ontario, S Canada
62 *K12* **Mattaldi** Córdoba, C Argentina *34°26′S 64°14′W*
21 *Y9* **Mattamuskeet, Lake** ◎ North Carolina, SE USA
21 *W6* **Mattaponi River** ≈ Virginia, NE USA
14 *I11* **Mattawa** Ontario, SE Canada *46°19′N 78°42′W*
14 *I11* **Mattawa** ≈ Ontario, SE Canada
19 *S5* **Mattawamkeag** Maine, NE USA *45°30′N 68°20′W*
19 *S4* **Mattawamkeag Lake** ◎ Maine, NE USA
108 *D11* **Matterhorn** *It.* Monte Cervino. ▲ Italy/Switzerland *45°58′N 07°36′E* *see also* Cervino, Monte
32 *L12* **Matterhorn** *var.* Sacajawea Peak. ▲ Oregon, NW USA *45°12′N 117°18′W*
35 *W1* **Matterhorn** ▲ Nevada, W USA *41°48′N 115°22′W*
Matterhorn *see* Cervino, Monte
35 *R8* **Matterhorn Peak** ▲ California, W USA *38°06′N 119°19′W*
109 *Y5* **Mattersburg** Burgenland, E Austria *47°45′N 16°24′E*
108 *E11* **Matter Vispa** ≈ S Switzerland
55 *R7* **Matthews Ridge** N Guyana *07°30′N 60°07′W*
44 *K7* **Matthew Town** Great Inagua, S The Bahamas *20°56′N 73°41′W*
109 *Q4* **Mattighofen** Oberösterreich, NW Austria *48°07′N 13°09′E*
107 *N16* **Mattinata** Puglia, SE Italy *41°41′N 16°01′E*
141 *T9* **Maţţī, Sabkhat** *salt flat* Saudi Arabia/United Arab Emirates
18 *M14* **Mattituck** Long Island, New York, NE USA *40°59′N 72°31′W*
164 *L11* **Mattō** *var.* Hakusan, Matsutō. Ishikawa, Honshū, SW Japan *36°31′N 136°34′E*
Matto Grosso *see* Mato Grosso
30 *M14* **Mattoon** Illinois, N USA *39°28′N 88°22′W*
57 *L16* **Mattos, Río** ≈ C Bolivia
Mattu *see* Metu
169 *R9* **Matu** Sarawak, East Malaysia *02°39′N 111°31′E*
Matubara *see* Matsubara
57 *E14* **Matucana** Lima, W Peru *11°54′S 76°25′W*
Matue *see* Matsue
187 *Y15* **Matuku** *island* S Fiji
112 *B9* **Matulji** Primorje-Gorski Kotar, NW Croatia *45°21′N 14°18′E*
Matumoto *see* Matsumoto
55 *P5* **Maturín** Monagas, NE Venezuela *09°45′N 63°10′W*
Matusaka *see* Matsusaka
Matuyama *see* Matsuyama
126 *K11* **Matveyev Kurgan** Rostovskaya Oblast', SW Russian Federation *47°31′N 38°55′E*
127 *O8* **Matyshevo** Volgogradskaya Oblast', SW Russian Federation *50°53′N 44°09′E*
153 *O13* **Mau** *var.* Maunāth Bhanjan. Uttar Pradesh, N India *25°57′N 83°33′E*
83 *O14* **Maúa** Niassa, N Mozambique *13°53′S 37°10′E*
102 *M17* **Maubermé, Pic de** *var.* Tuc de Moubermé, *Sp.* Pico Mauberme; *prev.* Tuc de Maubermé. ▲ France/Spain *42°48′N 00°54′E* *see also* Moubermé, Tuc de
Maubermé, Pic de *see* Moubermé, Tuc de
Mauberme, Pico *see* Maubermé, Pic de/Moubermé, Tuc de
Maubermé, Tuc de *see* Maubermé, Pic de/Moubermé, Tuc de
103 *Q2* **Maubeuge** Nord, N France *50°16′N 04°00′E*
166 *L8* **Maubin** Ayeyawady, SW Myanmar (Burma) *16°44′N 95°37′E*
152 *L13* **Maudaha** Uttar Pradesh, N India *25°41′N 80°07′E*
183 *N9* **Maude** New South Wales, SE Australia *34°30′S 144°20′E*
195 *P3* **Maudheimvidda** *physical region* Antarctica
65 *N22* **Maud Rise** *undersea feature* S Atlantic Ocean
109 *Q4* **Mauerkirchen** Oberösterreich, NW Austria *48°11′N 13°08′E*
Mauersee *see* Mamry, Jezioro
188 *K2* **Maug Islands** *island group* N Northern Mariana Islands
103 *Q15* **Mauguio** Hérault, S France *43°37′N 04°01′E*
193 *N5* **Maui** *island* Hawai'i, USA, C Pacific Ocean
190 *M16* **Mauke** *atoll* S Cook Islands
62 *G13* **Maule** *var.* Región del Maule. ◈ *region* C Chile
102 *J9* **Mauléon** Deux-Sèvres, W France *46°55′N 00°45′W*
102 *J16* **Mauléon-Licharre** Pyrénées-Atlantiques, SW France *43°14′N 00°51′W*
Maule, Región del *see* Maule
62 *G13* **Maule, Río** ≈ C Chile
63 *G17* **Maullín** Los Lagos, S Chile *41°38′S 73°35′W*
31 *R11* **Maumee** Ohio, N USA *41°34′N 83°40′W*
31 *Q12* **Maumee River** ≈ Indiana/Ohio, N USA
27 *U11* **Maumelle** Arkansas, C USA *34°51′N 92°24′W*
27 *T11* **Maumelle, Lake** ⊠ Arkansas, C USA
171 *O16* **Maumere** *prev.* Maoemere. Flores, S Indonesia *08°35′S 122°13′E*
83 *G17* **Maun** North-West, C Botswana *20°01′S 23°28′E*
Maunāth Bhanjan *see* Mau
Maunawai *see* Waimea
190 *H16* **Maungaroa** ▲ Rarotonga, S Cook Islands *21°13′S 159°48′W*
184 *K3* **Maungatapere** Northland, North Island, New Zealand *35°46′S 174°10′E*
184 *K4* **Maungaturoto** Northland, North Island, New Zealand *36°06′S 174°21′E*
166 *J5* **Maungdaw** *var.* Zullapara. Rakhine State, W Myanmar (Burma) *20°51′N 92°23′E*
191 *R10* **Maupiti** *var.* Maurua. *island* Îles Sous le Vent, W French Polynesia
152 *K14* **Mau Rānīpur** Uttar Pradesh, N India *25°14′N 79°07′E*
22 *K9* **Maurepas, Lake** ◎ Louisiana, S USA
103 *T16* **Maures** ▲▲ SE France
103 *O12* **Mauriac** Cantal, C France *45°13′N 02°21′E*
Maurice *see* Mauritius
65 *J20* **Maurice Ewing Bank** *undersea feature* SW Atlantic Ocean *51°00′S 43°00′W*
182 *C4* **Maurice, Lake** *salt lake* South Australia
18 *I17* **Maurice River** ≈ New Jersey, NE USA
25 *Y10* **Mauriceville** Texas, SW USA *30°13′N 93°52′W*
98 *K12* **Maurik** Gelderland, C Netherlands *51°57′N 05°25′E*
76 *H8* **Mauritania** *off.* Islamic Republic of Mauritania, *Ar.* Mūrītānīyah. ◆ *republic* W Africa
Mauritania, Islamic Republic of *see* Mauritania
173 *W15* **Mauritius** *off.* Republic of Mauritius, *Fr.* Maurice. ◆ *republic* W Indian Ocean
128 *M17* **Mauritius** *island* W Indian Ocean
Mauritius, Republic of *see* Mauritius
173 *N9* **Mauritius Trench** *undersea feature* W Indian Ocean
102 *H6* **Mauron** Morbihan, NW France *48°06′N 02°16′W*
103 *N13* **Maurs** Cantal, C France *44°45′N 02°12′E*
Maurua *see* Maupiti
Maury Mid-Ocean Channel *see* Maury Seachannel
64 *L6* **Maury Seachannel** *var.* Maury Mid-Ocean Channel. *undersea feature* N Atlantic Ocean *56°33′N 24°00′W*
30 *K8* **Mauston** Wisconsin, N USA *43°46′N 90°06′W*
109 *R8* **Mauterndorf** Salzburg, NW Austria *47°09′N 13°39′E*
109 *T4* **Mauthausen** Oberösterreich, N Austria *48°13′N 14°30′E*
109 *Q9* **Mauthen** Kärnten, S Austria *46°39′N 12°58′E*
83 *F15* **Mavinga** Kuando Kubango, SE Angola *15°44′S 20°21′E*
83 *M17* **Mavita** Manica, W Mozambique *19°31′S 33°09′E*
115 *K22* **Mavrópetra, Akrotírio** *headland* Santoríni, Kykládes, Greece, Aegean Sea *36°28′N 25°22′E*
115 *F16* **Mavrovoúni** ▲ C Greece *39°37′N 22°45′E*
184 *Q8* **Mawhai Point** *headland* North Island, New Zealand *38°08′S 178°24′E*
166 *L3* **Mawlaik** Sagaing, C Myanmar (Burma) *23°40′N 94°26′E*
Mawlamyaing *see* Mawlamyine
166 *M9* **Mawlamyine** *var.* Mawlamyaing, Moulmein. Mon State, S Myanmar (Burma) *16°30′N 97°39′E*
166 *L8* **Mawlamyinegyunn** *var.* Moulmeingyun. Ayeyawady, SW Myanmar (Burma) *16°24′N 95°15′E*
141 *N14* **Mawr, Wādī** *dry watercourse* NW Yemen
Mawşil, Al *see* Nīnawá
195 *X5* **Mawson** *Australian research station* Antarctica *67°24′S 63°16′E*
195 *X5* **Mawson Coast** *physical region* Antarctica
28 *M4* **Max** North Dakota, N USA *47°48′N 101°18′W*
41 *W12* **Maxcanú** Yucatán, SE Mexico *20°35′N 90°00′W*
Maxesibebi *see* Mount Ayliff
109 *Q5* **Maxglan** ✈ (Salzburg) Salzburg, W Austria *47°46′N 13°00′E*
93 *K16* **Maxmo** *Fin.* Maksamaa. Österbotten, W Finland *63°13′N 22°04′E*
21 *T11* **Maxton** North Carolina, SE USA *34°47′N 79°34′W*
25 *R8* **May** Texas, SW USA *31°58′N 98°54′W*
186 *B6* **May** ≈ NW Papua New Guinea
123 *R10* **Maya** ≈ E Russian Federation
151 *Q19* **Māyābandar** Andaman and Nicobar Islands, India, E Indian Ocean *12°43′N 92°52′E*
Mayadin *see* Al Mayādīn
44 *L5* **Mayaguana** *island* SE The Bahamas
44 *L5* **Mayaguana Passage** *passage* SE The Bahamas
45 *S6* **Mayagüez** W Puerto Rico *18°12′N 67°08′W*
45 *R6* **Mayagüez, Bahía de** *bay* W Puerto Rico
Mayals *see* Maials
79 *G20* **Mayama** Pool, SE Congo *03°50′S 14°52′E*
37 *V8* **Maya, Mesa De** ▲ Colorado, C USA *37°06′N 103°30′W*
143 *R4* **Mayamey** Semnān, N Iran *36°50′N 55°50′E*
42 *F3* **Maya Mountains** *Sp.* Montañas Mayas. ▲▲ Belize/Guatemala
44 *I7* **Mayarí** Holguín, E Cuba *20°41′N 75°42′W*
Mayas, Montañas *see* Maya Mountains
18 *I17* **May, Cape** *headland* New Jersey, NE USA *38°55′N 74°57′W*
80 *J11* **Maych'ew** *var.* Mai Chio, *It.* Mai Ceu. Tigray, N Ethiopia *12°55′N 39°30′E*
138 *I2* **Maydān Ikbiz** Ḩalab, N Syria *36°51′N 36°40′E*
Maydān Shahr *see* Maīdān Shahr
80 *O12* **Maydh** Sanaag, N Somalia *10°57′N 47°07′E*
Maydī *see* Mīdī
Mayebashi *see* Maebashi
Mayence *see* Mainz
102 *K6* **Mayenne** Mayenne, NW France *48°18′N 00°37′W*
102 *J6* **Mayenne** ◈ *department* NW France
102 *J7* **Mayenne** ≈ N France
36 *K12* **Mayer** Arizona, SW USA *34°25′N 112°15′W*
22 *J4* **Mayersville** Mississippi, S USA *32°54′N 91°04′W*
11 *P14* **Mayerthorpe** Alberta, SW Canada *53°59′N 115°06′W*
21 *S12* **Mayesville** South Carolina, SE USA *34°00′N 80°10′W*
185 *G19* **Mayfield** Canterbury, South Island, New Zealand *43°50′S 171°24′E*
33 *N14* **Mayfield** Idaho, NW USA *43°24′N 115°56′W*
20 *G7* **Mayfield** Kentucky, S USA *36°45′N 88°40′W*
36 *L5* **Mayfield** Utah, W USA *39°06′N 111°42′W*
Mayhan *see* Sant
37 *T14* **Mayhill** New Mexico, SW USA *32°52′N 105°28′W*
Maykain *see* Maykayyn
145 *T9* **Maykayyn** *prev.* Maykain *Kaz.* Mayqayyng. Pavlodar, NE Kazakhstan *51°27′N 75°52′E*
126 *L14* **Maykop** Respublika Adygeya, SW Russian Federation *44°36′N 40°07′E*
Maylibash *see* Maylybas
Mayli-Say *see* Mayluu-Suu
147 *T9* **Mayluu-Suu** *prev.* Mayli-Say, *Kir.* Mayly-Say. Dzhalal-Abadskaya Oblast', W Kyrgyzstan *41°16′N 72°27′E*
144 *L14* **Maylybas** *prev.* Maylibash. Kzylorda, S Kazakhstan *45°51′N 62°37′E*
Mayly-Say *see* Mayluu-Suu
Maymana *see* Maīmanah
Maymyo *see* Pyin-Oo-Lwin
123 *V7* **Mayn** ≈ NE Russian Federation
127 *Q5* **Mayna** Ul'yanovskaya Oblast', W Russian Federation *54°04′N 47°20′E*
21 *N8* **Maynardville** Tennessee, S USA *36°15′N 83°48′W*
14 *J13* **Maynooth** Ontario, SE Canada *45°14′N 77°54′W*
10 *I6* **Mayo** Yukon, NW Canada *63°37′N 135°48′W*
23 *U9* **Mayo** Florida, SE USA *30°03′N 83°10′W*
97 *B16* **Mayo** *Ir.* Maigh Eo. *cultural region* W Ireland
Mayo *see* Maio
78 *G12* **Mayo-Kébbi Est** *off.* Région du Mayo-Kébbi Est. ◈ *region* SW Chad
Mayo-Kébbi Est, Région du *see* Mayo-Kébbi Est
78 *G13* **Mayo-Kébbi Ouest** *off.* Région du mayo-Kébbi Ouest. ◈ *region* SW Chad
Mayo-Kébbi Ouest, Région du *see* Mayo-Kébbi Ouest
79 *F19* **Mayoko** Niari, SW Congo *02°19′S 12°47′E*
171 *P4* **Mayon Volcano** ⛰ Luzon, N Philippines *13°15′N 123°41′E*
61 *A24* **Mayor Buratovich** Buenos Aires, E Argentina *39°15′S 62°35′W*
104 *L4* **Mayorga** Castilla y León, N Spain *42°10′N 05°16′W*
184 *N6* **Mayor Island** *island* NE New Zealand
Mayor Pablo Lagerenza *see* Capitán Pablo Lagerenza
173 *I14* **Mayotte** ◇ *French overseas department* E Africa
Mayoumba *see* Mayumba
44 *J13* **May Pen** C Jamaica *17°58′N 77°15′W*
Mayqayyng *see* Maykayyn
171 *O1* **Mayraira Point** *headland* Luzon, N Philippines *18°36′N 120°47′E*
109 *N8* **Mayrhofen** Tirol, W Austria *47°09′N 11°52′E*
186 *A6* **May River** East Sepik, NW Papua New Guinea *04°24′S 141°52′E*
139 *Y10* **Maysān** *off.* Muḩāfa at Maysān, *var.* Al 'Amārah, Mīsān. ◈ *governorate* SE Iraq
Maysān, Muḩāfa at *see* Maysān
123 *R13* **Mayskiy** Amurskaya Oblast', SE Russian Federation *52°13′N 129°30′E*
127 *O15* **Mayskiy** Kabardino-Balkarskaya Respublika, SW Russian Federation *43°37′N 44°04′E*
145 *U9* **Mayskoye** Pavlodar, NE Kazakhstan *50°55′N 78°11′E*
18 *J17* **Mays Landing** New Jersey, NE USA *39°27′N 74°44′W*
21 *N4* **Maysville** Kentucky, S USA *38°38′N 83°46′W*
27 *R2* **Maysville** Missouri, C USA *39°53′N 94°21′W*
79 *D20* **Mayumba** *var.* Mayoumba. Nyanga, S Gabon *03°23′S 10°38′E*
31 *S8* **Mayville** Michigan, N USA *43°18′N 83°16′W*
18 *C11* **Mayville** New York, NE USA *42°15′N 79°32′W*
29 *Q4* **Mayville** North Dakota, N USA *47°27′N 97°17′W*
Mayyali *see* Mahe
Mayyit, Al Baḩr al *see* Dead Sea
83 *J15* **Mazabuka** Southern, S Zambia *15°52′S 27°46′E*
Mazaca *see* Kayseri
Mazagan *see* El-Jadida
32 *J7* **Mazama** Washington, NW USA *48°34′N 120°26′W*
103 *O15* **Mazamet** Tarn, S France *43°30′N 02°21′E*
143 *O4* **Māzandarān** *off.* Ostān-e Māzandarān. ◈ *province* N Iran
Māzandarān, Ostān-e *see* Māzandarān
156 *F7* **Mazar** Xinjiang Uygur Zizhiqu, NW China *36°28′N 77°oo′E*
107 *H24* **Mazara del Vallo** Sicilia, Italy, C Mediterranean Sea *37°39′N 12°36′E*
149 *O2* **Mazār-e Sharīf** *var.* Mazār-i Sharif. Balkh, N Afghanistan *36°44′N 67°06′E*
Mazār-i Sharif *see* Mazār-e Sharīf
105 *R13* **Mazarrón** Murcia, SE Spain *37°36′N 01°19′W*
105 *R14* **Mazarrón, Golfo de** *gulf* SE Spain
55 *S9* **Mazaruni River** ≈ N Guyana
42 *B6* **Mazatenango** Suchitepéquez, SW Guatemala *14°31′N 91°30′W*
40 *I10* **Mazatlán** Sinaloa, C Mexico *23°15′N 106°24′W*
36 *L12* **Mazatzal Mountains** ▲▲ Arizona, SW USA
118 *D10* **Mažeikiai** Telšiai, NW Lithuania *56°19′N 22°22′E*
118 *D7* **Mazirbe** NW Latvia *57°39′N 22°16′E*
40 *G5* **Mazocahui** Sonora, NW Mexico *29°32′N 110°09′W*
57 *I18* **Mazocruz** Puno, S Peru *16°41′S 69°42′W*
Mazoe, Rio *see* Mazowe
79 *N21* **Mazomeno** Maniema, E Dem. Rep. Congo *04°54′S 27°13′E*
159 *Q6* **Mazong Shan** ▲ N China *41°40′N 97°10′E*
83 *L16* **Mazowe** *var.* Rio Mazoe. ≈ Mozambique/Zimbabwe
110 *M11* **Mazowieckie** ◈ *province* C Poland
Mazra'a *see* Al Mazra'ah
138 *G6* **Mazraat Kfar Debiâne** C Lebanon *34°00′N 35°51′E*
118 *H7* **Mazsalaca** *Est.* Väike-Salatsi, *Ger.* Salisburg. N Latvia *57°52′N 25°03′E*
110 *L9* **Mazury** *physical region* NE Poland
Mazu Tao *see* Matsu Dao
119 *M20* **Mazyr** *Rus.* Mozyr'. Homyel'skaya Voblasts', SE Belarus *52°04′N 29°15′E*
107 *K25* **Mazzarino** Sicilia, Italy, C Mediterranean Sea *37°18′N 14°13′E*
Mba *see* Ba
83 *L21* **Mbabane** ● (Swaziland) NW Swaziland *26°24′S 31°13′E*
Mbacké *see* Mbaké
77 *N16* **Mbahiakro** E Ivory Coast *07°33′N 04°19′W*
79 *I16* **Mbaïki** *var.* M'Baiki. Lobaye, SW Central African Republic *03°52′N 17°58′E*
M'Baiki *see* Mbaïki
79 *F14* **Mbakaou, Lac de** ◎ C Cameroon
76 *G11* **Mbaké** *var.* Mbacké. W Senegal *14°47′N 15°54′W*
82 *L11* **Mbala** *prev.* Abercorn. Northern, NE Zambia *08°50′S 31°23′E*
83 *J18* **Mbalabala** *prev.* Balla Balla. Matabeleland South, SW Zimbabwe *20°27′S 29°03′E*
81 *G18* **Mbale** E Uganda *01°04′N 34°12′E*
79 *E16* **Mbalmayo** *var.* M'Balmayo. Centre, S Cameroon *03°30′N 11°31′E*
M'Balmayo *see* Mbalmayo
81 *H25* **Mbamba Bay** Ruvuma, S Tanzania *11°15′S 34°44′E*
79 *I18* **Mbandaka** *prev.* Coquilhatville. Equateur, NW Dem. Rep. Congo *0°07′N 18°12′E*
82 *B9* **M'banza Kongo** Zaire Province, NW Angola *06°11′S 14°16′E*
79 *G21* **Mbanza-Ngungu** Bas-Congo, W Dem. Rep. Congo *05°19′S 14°45′E*
67 *V11* **Mbarangandu** ≈ E Tanzania
81 *E19* **Mbarara** SW Uganda *0°36′S 30°40′E*
79 *L15* **Mbari** ≈ SE Central African Republic
81 *I24* **Mbarika Mountains** ▲▲ S Tanzania
78 *F13* **Mbé** Nord, N Cameroon *07°51′N 13°36′E*
81 *J24* **Mbemkuru** *var.* Mbwemkuru. ≈ S Tanzania
Mbengga *see* Beqa
172 *H13* **Mbéni** Grande Comore, NW Comoros
83 *K18* **Mberengwa** Midlands, S Zimbabwe *20°29′S 29°55′E*
81 *G24* **Mbeya** Mbeya, SW Tanzania *08°54′S 33°29′E*
81 *G23* **Mbeya** ◈ *region* S Tanzania
83 *J24* **Mbhashe** *prev.* Mbashe. ≈ S South Africa
79 *E19* **Mbigou** Ngounié, C Gabon *01°54′S 12°00′E*
Mbilua *see* Vella Lavella
79 *F19* **Mbinda** Niari, SW Congo *02°07′S 12°52′E*
79 *D17* **Mbini** W Equatorial Guinea *01°34′N 09°39′E*
Mbini *see* Uolo, Río
83 *L18* **Mbizi** Masvingo, SE Zimbabwe *21°23′S 30°54′E*
81 *G23* **Mbogo** Mbeya, W Tanzania *07°24′S 33°26′E*
79 *N15* **Mboki** Haut-Mbomou, SE Central African Republic *05°18′N 25°52′E*
83 *K21* **Mbombela** *prev.* Nelspruit. Mpumalanga, NE South Africa *25°28′S 30°58′E* *see also* Nelspruit
79 *G18* **Mbomo** Cuvette, NW Congo *0°25′N 14°42′E*
79 *L15* **Mbomou** ◈ *prefecture* SE Central African Republic
Mbomou/M'Bomu/Mbomu *see* Bomu
76 *F11* **Mbour** W Senegal *14°22′N 16°54′W*
76 *I10* **Mbout** Gorgol, S Mauritania *16°02′N 12°38′W*

◆ Country
● Country Capital
◇ Dependent Territory
○ Dependent Territory Capital
◈ Administrative Regions
✈ International Airport
▲ Mountain
▲▲ Mountain Range
⛰ Volcano
≈ River
◎ Lake
⊠ Reservoir

79 J14 **Mbrès** *var.* Mbrés. Nana-Grébizi, C Central African Republic 06°40′N 19°46′E
Mbrés *see* Mbrès
79 L22 **Mbuji-Mayi** *prev.* Bakwanga. Kasai-Oriental, S Dem. Rep. Congo 06°05′S 23°30′E
81 H21 **Mbulu** Manyara, N Tanzania 03°45′S 35°33′E
186 E5 **M'bunai** *var.* Bunai. Manus Island, N Papua New Guinea 02°08′S 147°13′E
62 N8 **Mburucuyá** Corrientes, NE Argentina 28°03′S 58°15′W
Mbutha *see* Buca
Mbwemkuru *see* Mbemkuru
81 G21 **Mbwikwe** Singida, C Tanzania 05°19′S 34°09′E
13 O15 **McAdam** New Brunswick, SE Canada 45°34′N 67°20′W
25 O5 **McAdoo** Texas, SW USA 33°41′N 100°58′W
35 V2 **McAfee Peak** ▲ Nevada, W USA 41°31′N 115°57′W
27 P11 **McAlester** Oklahoma, C USA 34°56′N 95°46′W
25 S17 **McAllen** Texas, SW USA 26°12′N 98°14′W
21 S11 **McBee** South Carolina, SE USA 34°30′N 80°12′W
11 N14 **McBride** British Columbia, SW Canada 53°21′N 120°19′W
24 M9 **McCamey** Texas, SW USA 31°08′N 102°13′W
33 R15 **McCammon** Idaho, NW USA 42°38′N 112°10′W
35 X11 **McCarran** ✈ (Las Vegas) Nevada, W USA 36°04′N 115°07′W
39 T11 **McCarthy** Alaska, USA 61°25′N 142°55′W
30 M5 **McCaslin Mountain** *hill* Wisconsin, N USA
25 O2 **McClellan Creek** ≋ Texas, SW USA
21 T14 **McClellanville** South Carolina, SE USA 33°07′N 79°27′W
195 R12 **McClintock, Mount** ▲ Antarctica 80°09′S 156°42′E
35 N2 **McCloud** California, W USA 41°15′N 122°09′W
35 N3 **McCloud River** ≋ California, W USA
35 Q9 **McClure, Lake** ⊠ California, W USA
197 O8 **McClure Strait** *strait* Northwest Territories, N Canada
29 N4 **McClusky** North Dakota, N USA 47°27′N 100°25′W
21 T11 **McColl** South Carolina, SE USA 34°40′N 79°33′W
22 K7 **McComb** Mississippi, S USA 31°14′N 90°27′W
18 E16 **McConnellsburg** Pennsylvania, NE USA 39°56′N 78°00′W
31 T14 **McConnelsville** Ohio, N USA 39°39′N 81°51′W
28 M17 **McCook** Nebraska, C USA 40°12′N 100°38′W
21 P13 **McCormick** South Carolina, SE USA 33°55′N 82°19′W
11 W16 **McCreary** Manitoba, S Canada 50°48′N 99°34′W
27 W11 **McCrory** Arkansas, C USA 35°15′N 91°12′W
25 T10 **McDade** Texas, SW USA 30°15′N 97°15′W
23 O8 **McDavid** Florida, SE USA 30°51′N 87°18′W
35 T1 **McDermitt** Nevada, W USA 41°57′N 117°43′W
23 S4 **McDonough** Georgia, SE USA 33°27′N 84°09′W
36 L12 **McDowell Mountains** ▲▲ Arizona, SW USA
20 H8 **McEwen** Tennessee, S USA 36°06′N 87°37′W
35 R12 **McFarland** California, W USA 35°41′N 119°14′W
Mcfarlane, Lake *see* Macfarlane, Lake
27 P12 **McGee Creek Lake** ⊠ Oklahoma, C USA
27 W13 **McGehee** Arkansas, C USA 33°37′N 91°24′W
35 X5 **Mcgill** Nevada, W USA 39°24′N 114°46′W
14 K11 **McGillivray, Lac** ◎ Québec, SE Canada
39 P10 **McGrath** Alaska, USA 62°57′N 155°36′W
25 T8 **McGregor** Texas, SW USA 31°26′N 97°24′W
33 O12 **McGuire, Mount** ▲ Idaho, NW USA 45°10′N 114°36′W
83 M14 **Mchinji** *prev.* Fort Manning. Central, W Malawi 13°48′S 32°55′E
28 M7 **McIntosh** South Dakota, N USA 45°56′N 101°21′W
9 S7 **McKeand** ≋ Baffin Island, Nunavut, NE Canada
191 R4 **McKean Island** *island* Phoenix Islands, C Kiribati
30 J13 **McKee Creek** ≋ Illinois, N USA
18 C15 **Mckeesport** Pennsylvania, NE USA 40°18′N 79°48′W
21 V7 **McKenney** Virginia, NE USA 36°57′N 77°42′W
20 G8 **McKenzie** Tennessee, S USA 36°07′N 88°31′W
185 B20 **McKerrow, Lake** ◎ South Island, New Zealand
39 Q10 **McKinley, Mount** *var.* Denali. ▲ Alaska, USA 63°04′N 151°00′W
39 R10 **McKinley Park** Alaska, USA 63°42′N 149°01′W
34 K3 **McKinleyville** California, W USA 40°56′N 124°06′W
25 U6 **McKinney** Texas, SW USA 33°14′N 96°37′W
26 I5 **McKinney, Lake** ◎ Kansas, C USA
28 M7 **McLaughlin** South Dakota, N USA 45°48′N 100°48′W
25 O2 **McLean** Texas, SW USA 35°13′N 100°36′W
30 M16 **Mcleansboro** Illinois, N USA 38°05′N 88°32′W
11 O13 **McLennan** Alberta, W Canada 55°42′N 116°50′W
14 L9 **McLennan, Lac** ◎ Québec, SE Canada
10 M13 **McLeod Lake** British Columbia, W Canada 55°03′N 123°02′W
8 L6 **M'Clintock Channel** *channel* Nunavut, N Canada
27 N10 **McLoud** Oklahoma, C USA 35°26′N 97°05′W
32 G15 **McLoughlin, Mount** ▲ Oregon, NW USA 42°27′N 122°18′W
37 U15 **McMillan, Lake** ⊠ New Mexico, SW USA
32 G11 **McMinnville** Oregon, NW USA 45°14′N 123°12′W
20 K9 **McMinnville** Tennessee, S USA 35°40′N 85°49′W
195 R13 **McMurdo** *US research station* Antarctica 77°40′S 167°16′E
37 N13 **Mcnary** Arizona, SW USA 34°04′N 109°51′W
24 H9 **McNary** Texas, SW USA 31°15′N 105°46′W
27 N5 **McPherson** Kansas, C USA 38°22′N 97°41′W
McPherson *see* Fort McPherson
23 U6 **McRae** Georgia, SE USA 32°04′N 82°54′W
29 P4 **McVille** North Dakota, N USA 47°45′N 98°10′W
83 J25 **Mdantsane** Eastern Cape, SE South Africa 32°55′S 27°39′E
167 T6 **Me** Ninh Binh, N Vietnam 20°21′N 105°49′E
26 J7 **Meade** Kansas, C USA 37°17′N 100°21′W
39 O5 **Meade River** ≋ Alaska, USA
35 Y11 **Mead, Lake** ⊠ Arizona/Nevada, W USA
24 M5 **Meadow** Texas, SW USA 33°20′N 102°12′W
11 S14 **Meadow Lake** Saskatchewan, C Canada 54°90′N 108°30′W
35 Y10 **Meadow Valley Wash** ≋ Nevada, W USA
22 J7 **Meadville** Mississippi, S USA 31°28′N 90°51′W
18 B12 **Meadville** Pennsylvania, NE USA 41°38′N 80°09′W
14 F14 **Meaford** Ontario, S Canada 44°35′N 80°35′W
Meáin, Inis *see* Inishmaan
104 G8 **Mealhada** Aveiro, N Portugal 40°22′N 08°27′W
13 R8 **Mealy Mountains** ▲▲ Newfoundland and Labrador, E Canada
11 O10 **Meander River** Alberta, W Canada 59°02′N 117°42′W
32 E11 **Meares, Cape** *headland* Oregon, NW USA 45°29′N 123°59′W
47 V6 **Mearim, Rio** ≋ NE Brazil
Measca, Loch *see* Mask, Lough
97 F17 **Meath** *Ir.* An Mhí. *cultural region* E Ireland
11 T14 **Meath Park** Saskatchewan, S Canada 53°25′N 105°18′W
103 O5 **Meaux** Seine-et-Marne, N France 48°47′N 02°54′E
21 T9 **Mebane** North Carolina, SE USA 36°06′N 79°16′W
171 U12 **Mebo, Gunung** ▲ Papua Barat, E Indonesia 01°10′S 133°53′E
94 I8 **Mebonden** Sør-Trøndelag, S Norway 63°13′N 11°00′E
82 A10 **Mebridege** ≋ NW Angola
35 W16 **Mecca** California, W USA 33°34′N 116°04′W
Mecca *see* Makkah
29 Y14 **Mechanicsville** Iowa, C USA 41°54′N 91°15′W
18 L10 **Mechanicville** New York, NE USA 42°54′N 73°41′W
99 H17 **Mechelen** *Eng.* Mechlin, *Fr.* Malines. Antwerpen, C Belgium 51°02′N 04°29′E
188 C8 **Mecherchar** *var.* Eil Malk. *island* Palau Islands, Palau
101 D17 **Mechernich** Nordrhein-Westfalen, W Germany 50°36′N 06°39′E
126 L12 **Mechetinskaya** Rostovskaya Oblast', SW Russian Federation 46°46′N 40°30′E
114 J11 **Mechka** ≋ S Bulgaria
Mechlin *see* Mechelen
61 D23 **Mechongué** Buenos Aires, E Argentina 38°09′S 58°13′W
115 L14 **Mecidiye** Edirne, NW Turkey 40°39′N 26°33′E
101 I24 **Meckenbeuren** Baden-Württemberg, S Germany 47°42′N 09°34′E
100 L8 **Mecklenburger Bucht** *bay* N Germany
100 M10 **Mecklenburgische Seenplatte** *wetland* NE Germany
100 L9 **Mecklenburg-Vorpommern** ◈ *state* NE Germany
83 Q15 **Meconta** Nampula, NE Mozambique 15°01′S 39°52′E
111 I25 **Mecsek** ▲▲ SW Hungary
83 P14 **Mecubúri** ≋ N Mozambique
83 Q14 **Mecúfi** Cabo Delgado, NE Mozambique 13°20′S 40°32′E
82 O13 **Mecula** Niassa, N Mozambique 12°03′S 37°37′E
168 I8 **Medan** Sumatera, E Indonesia 03°35′N 98°39′E
61 A24 **Médanos** *var.* Medanos. Buenos Aires, E Argentina 38°52′S 62°45′W
61 C19 **Médanos** Entre Ríos, E Argentina 33°28′S 59°07′W
155 K24 **Medawachchiya** North Central Province, N Sri Lanka 08°32′N 80°30′E
106 C8 **Mede** Lombardia, N Italy 45°06′N 08°43′E
74 J5 **Médéa** *var.* El Mediyya, Lemdiyya. N Algeria 36°15′N 02°48′E
Medeba *see* Mādabā
54 E8 **Medellín** Antioquia, NW Colombia 06°15′N 75°36′W
100 H9 **Medem** ≋ NW Germany
98 J8 **Medemblik** Noord-Holland, NW Netherlands 52°47′N 05°06′E
75 N7 **Médenine** *var.* Madanīyīn. SE Tunisia 33°23′N 10°30′E
76 G9 **Mederdra** Trarza, SW Mauritania 16°56′N 15°40′W
Medeshamstede *see* Peterborough
42 F4 **Medesto Mendez** Izabal, NE Guatemala 15°54′N 89°13′W
19 O11 **Medford** Massachusetts, NE USA 42°25′N 71°08′W
27 N8 **Medford** Oklahoma, C USA 36°48′N 97°45′W
32 G15 **Medford** Oregon, NW USA 42°20′N 122°52′W
30 K6 **Medford** Wisconsin, N USA 45°08′N 90°22′W
39 P10 **Medfra** Alaska, USA 63°06′N 154°42′W
116 M14 **Medgidia** Constanţa, SE Romania 44°15′N 28°16′E
Medgyes *see* Mediaş
43 O5 **Media Luna, Arrecifes de la** *reef* E Honduras
60 G11 **Medianeira** Paraná, S Brazil 25°15′S 54°07′W
29 Y15 **Mediapolis** Iowa, C USA 41°00′N 91°09′W
116 I11 **Mediaş** *Ger.* Mediasch, *Hung.* Medgyes. Sibiu, C Romania 46°10′N 24°20′E
41 S15 **Medias Aguas** Veracruz-Llave, SE Mexico 17°40′N 95°02′W
Mediasch *see* Mediaş
106 G10 **Medicina** Emilia-Romagna, C Italy 44°29′N 11°41′E
33 X16 **Medicine Bow** Wyoming, C USA 41°52′N 106°11′W
37 S2 **Medicine Bow Mountains** ▲▲ Colorado/Wyoming, C USA
33 X16 **Medicine Bow River** ≋ Wyoming, C USA
11 R17 **Medicine Hat** Alberta, SW Canada 50°03′N 110°41′W
26 L7 **Medicine Lodge** Kansas, C USA 37°18′N 98°35′W
26 L7 **Medicine Lodge River** ≋ Kansas/Oklahoma, C USA
112 E7 **Međimurje** *off.* Međimurska Županija. ◈ *province* N Croatia
Međimurska Županija *see* Međimurje
54 G10 **Medina** Cundinamarca, C Colombia 04°31′N 73°21′W
18 E9 **Medina** New York, NE USA 43°13′N 78°23′W
29 O5 **Medina** North Dakota, N USA 46°53′N 99°18′W
31 T11 **Medina** Ohio, N USA 41°08′N 81°51′W
25 Q11 **Medina** Texas, SW USA 29°46′N 99°14′W
Medina *see* Al Madīnah
105 P6 **Medinaceli** Castilla y León, N Spain 41°10′N 02°26′W
104 L6 **Medina del Campo** Castilla y León, N Spain 41°18′N 04°55′W
104 L5 **Medina de Ríoseco** Castilla y León, N Spain 41°53′N 05°03′W
Médina Gonassé *see* Médina Gounas
76 H12 **Médina Gounas** *var.* Médina Gonassé. S Senegal 13°06′N 13°49′W
25 S12 **Medina River** ≋ Texas, SW USA
104 K16 **Medina Sidonia** Andalucía, S Spain 36°28′N 05°55′W
Medinat Israel *see* Israel
119 H14 **Medininkai** Vilnius, SE Lithuania 54°32′N 25°40′E
153 R16 **Medinīpur** West Bengal, NE India 22°25′N 87°24′E
Mediolanum *see* Saintes, France
Mediolanum *see* Milano, Italy
Mediomatrica *see* Metz
121 Q11 **Mediterranean Ridge** *undersea feature* C Mediterranean Sea 34°00′N 23°00′E
121 O16 **Mediterranean Sea** *Fr.* Mer Méditerranée. *sea* Africa/Asia/Europe
Méditerranée, Mer *see* Mediterranean Sea
79 N17 **Medje** Orientale, NE Dem. Rep. Congo 02°27′N 27°14′E
Medjerda, Oued *see* Mejerda
114 G7 **Medkovets** Montana, NW Bulgaria 43°39′N 23°22′E
93 J15 **Medle** Västerbotten, N Sweden 64°45′N 20°45′E
127 W7 **Mednogorsk** Orenburgskaya Oblast', W Russian Federation 51°24′N 57°37′E
123 W9 **Mednyy, Ostrov** *island* E Russian Federation
102 J12 **Médoc** *cultural region* SW France
159 Q16 **Mêdog** Xizang Zizhiqu, W China 29°26′N 95°26′E
28 J5 **Medora** North Dakota, N USA 46°56′N 103°40′W
79 E17 **Médouneu** Woleu-Ntem, N Gabon 0°58′N 10°50′E
106 I7 **Meduna** ≋ NE Italy
Medunta *see* Mantes-la-Jolie
Medvedica *see* Medveditsa
124 J16 **Medveditsa** *var.* Medvedica. ≋ W Russian Federation
127 O9 **Medveditsa** ≋ SW Russian Federation
112 E8 **Medvednica** ▲▲ NE Croatia
125 R15 **Medvedok** Kirovskaya Oblast', NW Russian Federation 57°23′N 50°01′E
123 S6 **Medvezh'i, Ostrova** *island group* NE Russian Federation
124 J9 **Medvezh'yegorsk** Respublika Kareliya, NW Russian Federation 62°56′N 34°26′E
109 T11 **Medvode** *Ger.* Zwischenwässern. NW Slovenia 46°09′N 14°21′E
126 J4 **Medyn'** Kaluzhskaya Oblast', W Russian Federation 54°59′N 35°52′E
180 J10 **Meekatharra** Western Australia 26°37′S 118°35′E
37 Q4 **Meeker** Colorado, C USA 40°02′N 107°54′W
13 T12 **Meelpaeg Lake** ◎ Newfoundland, Newfoundland and Labrador, E Canada
Meemu Atoll *see* Mulakatholhu
Meenen *see* Menen
101 M16 **Meerane** Sachsen, E Germany 50°50′N 12°28′E
101 D15 **Meerbusch** Nordrhein-Westfalen, W Germany 51°19′N 06°43′E
98 I12 **Meerkerk** Zuid-Holland, C Netherlands 51°55′N 05°00′E
99 L18 **Meerssen** *var.* Mersen. Limburg, SE Netherlands 50°53′N 05°45′E
152 J10 **Meerut** Uttar Pradesh, N India 29°01′N 77°41′E
33 U13 **Meeteetse** Wyoming, C USA 44°10′N 108°53′W
99 K17 **Meeuwen** Limburg, NE Belgium 51°04′N 05°36′E
81 J16 **Mēga** Oromīya, C Ethiopia 04°03′N 38°15′E
81 J16 **Mēga Escarpment** *escarpment* S Ethiopia
Megála Kalívia *see* Megála Kalývia
115 E16 **Megála Kalývia** *var.* Megála Kalívia. Thessalía, C Greece 39°30′N 21°48′E
115 H14 **Megáli Panagiá** *var.* Megáli Panayía. Kentrikí Makedonía, N Greece 40°24′N 23°42′E
Megáli Panayía *see* Megáli Panagiá
Megáli Préspa, Límni *see* Prespa, Lake
114 K12 **Megálo Livádi** ▲ Bulgaria/Greece 41°18′N 25°51′E
115 E20 **Megalópoli** *prev.* Megalópolis. Pelopónnisos, S Greece 37°24′N 22°08′E
Megalópolis *see* Megalópoli
171 U12 **Megamo** Papua Barat, E Indonesia 0°55′S 131°46′E
115 C18 **Meganísi** *island* Iónia Nisiá, Greece, C Mediterranean Sea
Meganom, Mys *see* Mehanom, Mys
Mégantic *see* Lac-Mégantic
15 R12 **Mégantic, Mont** ▲ Québec, SE Canada 45°27′N 71°09′W
115 G19 **Mégara** Attikí, C Greece 38°00′N 23°20′E
25 R5 **Megargel** Texas, SW USA 33°27′N 98°55′W
98 K13 **Megen** Noord-Brabant, S Netherlands 51°49′N 05°34′E
153 U13 **Meghālaya** ◈ *state* NE India
153 U16 **Meghna Nadi** ≋ S Bangladesh
137 V14 **Meghri** *Rus.* Megri. SE Armenia 38°57′N 46°15′E
115 Q23 **Megísti** *var.* Kastellórizon. *island* SE Greece
Megri *see* Meghri
Mehabad *see* Mahābād
116 F13 **Mehadia** *Hung.* Mehádia. Caraş-Severin, SW Romania 44°53′N 22°20′E
Mehádia *see* Mehadia
92 L7 **Mehamn** Finnmark, N Norway 71°01′N 27°46′E
117 U13 **Mehanom, Mys** *Rus.* Mys Meganom. *headland* S Ukraine 44°48′N 35°04′E
149 P14 **Mehar** Sind, SE Pakistan 27°12′N 67°51′E
180 J8 **Meharry, Mount** ▲ Western Australia 23°17′S 118°48′E
Mehdia *see* Mahdia
116 G14 **Mehedinţi** ◈ *county* SW Romania
153 S15 **Meherpur** Khulna, W Bangladesh 23°47′N 88°40′E
21 W8 **Meherrin River** ≋ North Carolina/Virginia, SE USA
Meheso *see* Mī'ēso
191 T11 **Mehetia** *island* Îles du Vent, W French Polynesia
118 K6 **Mehikoorma** Tartumaa, E Estonia 58°14′N 27°29′E
Me Hka *see* Nmai Hka
143 N5 **Mehrabad** ✈ (Tehrān) Tehrān, N Iran 35°46′N 51°07′E
142 J7 **Mehrān** Īlām, W Iran 33°07′N 46°10′E
143 Q14 **Mehrān, Rūd-e** *prev.* Mansurabad. ≋ W Iran
143 Q9 **Mehrīz** Yazd, C Iran 31°32′N 54°28′E
149 R5 **Mehtar Lām** *var.* Mehtarlām, Meterlam, Methariam, Metharlam. Laghmān, E Afghanistan 34°39′N 70°10′E
Mehtarlām *see* Mehtar Lām
103 N8 **Mehun-sur-Yèvre** Cher, C France 47°09′N 02°15′E
79 G14 **Meiganga** Adamaoua, NE Cameroon 06°31′N 14°07′E
160 H10 **Meigu** *var.* Bapu. Sichuan, C China 28°16′N 103°20′E
163 W11 **Meihekou** *var.* Hailong. Jilin, NE China 42°31′N 125°40′E
99 L15 **Meijel** Limburg, SE Netherlands 51°22′N 05°52′E
Meijiang *see* Ningdu
166 M5 **Meiktila** Mandalay, C Myanmar (Burma) 20°53′N 95°54′E
Meilbhe, Loch *see* Melvin, Lough
108 G7 **Meilen** Zürich, N Switzerland 47°17′N 08°39′E
Meilu *see* Wuchuan
101 J17 **Meiningen** Thüringen, C Germany 50°34′N 10°25′E
108 F9 **Meiringen** Bern, S Switzerland 46°42′N 08°13′E
Meishan *see* Jinzhai
101 O15 **Meissen** *Ger.* Meißen. Sachsen, E Germany 51°10′N 13°28′E
Meißen *see* Meissen
101 I15 **Meissner** ▲ C Germany 51°13′N 09°52′E
99 K25 **Meix-devant-Virton** Luxembourg, SE Belgium 49°36′N 05°27′E
Meixian *see* Meizhou
Meixing *see* Xinjin
161 P13 **Meizhou** *var.* Meixian, Mei Xian. Guangdong, S China 24°21′N 116°05′E
67 P2 **Mejerda** *var.* Oued Medjerda, Wādī Majardah. ≋ Algeria/Tunisia *see also* Medjerda, Oued
42 F7 **Mejicanos** San Salvador, C El Salvador 13°50′N 89°13′W
Méjico *see* Mexico
62 G5 **Mejillones** Antofagasta, N Chile 23°03′S 70°25′W
189 V5 **Mejit Island** *var.* Mājeej. *island* Ratak Chain, NE Marshall Islands
79 F17 **Mékambo** Ogooué-Ivindo, NE Gabon 01°03′N 13°50′E
80 J10 **Mek'elē** *var.* Makale. Tigray, N Ethiopia 13°36′N 39°29′E
74 I10 **Mekerrhane, Sebkha** *var.* Sebkha Meqerghane, Sebkra Mekerrhane. *salt flat* C Algeria
Mekerrhane, Sebkra *see* Mekerrhane, Sebkha
76 G10 **Mékhé** NW Senegal 15°02′N 16°40′W
146 G14 **Mekhinli** Ahal Welaýaty, C Turkmenistan 37°28′N 59°20′E
15 P9 **Mékinac, Lac** ◎ Québec, SE Canada
Meklong *see* Samut Songkhram
74 G6 **Meknès** N Morocco 33°54′N 05°27′W
129 U12 **Mekong** *var.* Lan-ts'ang Chiang, *Cam.* Mékôngk, *Chin.* Lancang Jiang, *Lao.* Mènam Khong, Nam Khong, *Th.* Mae Nam Khong, *Tib.* Dza Chu, *Vtn.* Sông Tiên Giang. ≋ SE Asia
Mékôngk *see* Mekong
167 T15 **Mekong, Mouths of the** *delta* S Vietnam
38 L12 **Mekoryuk** Nunivak Island, Alaska, USA 60°23′N 166°11′W
77 R14 **Mékrou** ≋ N Benin
168 K9 **Melaka** *var.* Malacca. Melaka, Peninsular Malaysia 02°14′N 102°14′E
168 L9 **Melaka** *var.* Malacca. ◈ *state* Peninsular Malaysia
Melaka, Selat *see* Malacca, Strait of
175 O6 **Melanesia** *island group* W Pacific Ocean
175 P5 **Melanesian Basin** *undersea feature* W Pacific Ocean 0°05′N 160°35′E
171 R9 **Melanguane** Pulau Karakelang, N Indonesia 04°02′N 126°43′E
169 R11 **Melawi, Sungai** ≋ Borneo, N Indonesia
183 N12 **Melbourne** *state capital* Victoria, SE Australia 37°51′S 144°56′E
27 V9 **Melbourne** Arkansas, C USA 36°04′N 91°54′W
23 Y12 **Melbourne** Florida, SE USA 28°04′N 80°36′W
29 W14 **Melbourne** Iowa, C USA 41°57′N 93°07′W
92 G10 **Melbu** Nordland, C Norway 68°31′N 14°50′E
Melchor de Mencos *see* Ciudad Melchor de Mencos
63 F19 **Melchor, Isla** *island* Archipiélago de los Chonos, S Chile
40 M9 **Melchor Ocampo** Zacatecas, C Mexico 24°45′N 101°38′W
14 C11 **Meldrum Bay** Manitoulin Island, Ontario, S Canada 45°55′N 83°06′W
Meleda *see* Mljet
106 D8 **Melegnano** *prev.* Marignano. Lombardia, N Italy 45°22′N 09°19′E
188 F9 **Melekeok** ● Babeldaob, N Palau 07°30′N 134°37′E
112 L9 **Melenci** *Hung.* Melencze. Vojvodina, N Serbia 45°32′N 20°18′E
Melencze *see* Melenci
127 N4 **Melenki** Vladimirskaya Oblast', W Russian Federation 55°21′N 41°37′E
127 V6 **Meleuz** Respublika Bashkortostan, W Russian Federation 52°55′N 55°54′E
12 L6 **Mélèzes, Rivière aux** ≋ Québec, C Canada
78 I11 **Melfi** Guéra, S Chad 11°05′N 17°57′E
107 M17 **Melfi** Basilicata, S Italy 41°00′N 15°33′E
11 U14 **Melfort** Saskatchewan, S Canada 52°52′N 104°38′W
104 H4 **Melgaço** Viana do Castelo, N Portugal 42°07′N 08°15′W
105 N4 **Melgar de Fernamental** Castilla y León, N Spain 42°24′N 04°15′W
74 L6 **Melghir, Chott** *var.* Chott Melrhir. *salt lake* E Algeria
94 H8 **Melhus** Sør-Trøndelag, S Norway 63°17′N 10°18′E
104 H3 **Melide** Galicia, NW Spain 42°54′N 08°01′W
Meligalá *see* Meligalás
115 E21 **Meligalás** *prev.* Meligalá. Pelopónnisos, S Greece 37°13′N 21°58′E
60 L12 **Mel, Ilha do** *island* S Brazil
120 E10 **Melilla** *anc.* Rusaddir, Russadir. Melilla, Spain, N Africa 35°18′N 02°56′W
71 N1 **Melilla** *enclave* Spain, N Africa
63 G18 **Melimoyu, Monte** ▲ S Chile 44°06′S 72°49′W
169 V11 **Melintang, Danau** ◎ Borneo, N Indonesia
117 U7 **Melioratyvne** Dnipropetrovs'ka Oblast', E Ukraine 48°35′N 35°18′E
62 G11 **Melipilla** Santiago, C Chile 33°42′S 71°15′W
115 I25 **Mélissa, Akrotírio** *headland* Kríti, Greece, E Mediterranean Sea 35°06′N 24°33′E
11 W17 **Melita** Manitoba, S Canada 49°16′N 100°59′W
Melita *see* Mljet
Melitene *see* Malatya
107 M23 **Melito di Porto Salvo** Calabria, SW Italy 37°55′N 15°48′E
117 U10 **Melitopol'** Zaporiz'ka Oblast', SE Ukraine 46°49′N 35°23′E
109 V4 **Melk** Niederösterreich, NE Austria 48°14′N 15°21′E
95 K15 **Mellan-Fryken** ◎ C Sweden
99 E17 **Melle** Oost-Vlaanderen, NW Belgium 51°N 03°48′E
100 G13 **Melle** Niedersachsen, NW Germany 52°12′N 08°19′E
Mellersta Finland *see* Keski-Suomi
Mellersta Österbotten *see* Keski-Pohjanmaa
95 J17 **Mellerud** Västra Götaland, S Sweden 58°42′N 12°27′E
102 K10 **Melle-sur-Bretonne** Deux-Sèvres, W France 46°13′N 00°07′W
29 P8 **Mellette** South Dakota, N USA 45°07′N 98°29′W
121 O15 **Mellieħa** E Malta 35°58′N 14°21′E
80 B10 **Mellit** Northern Darfur, W Sudan 14°07′N 25°34′E
75 N7 **Mellita** ✈ SE Tunisia 33°47′N 10°51′E
63 G21 **Mellizo Sur, Cerro** ▲ S Chile 48°27′S 73°10′W
100 G9 **Mellum** *island* NW Germany
83 L22 **Melmoth** KwaZulu/Natal, E South Africa 28°35′S 31°25′E
111 D16 **Mělník** *Ger.* Melnik. Středočeský Kraj, NW Czech Republic 50°21′N 14°30′E
122 J12 **Mel'nikovo** Tomskaya Oblast', C Russian Federation 56°35′N 84°11′E
61 G18 **Melo** Cerro Largo, NE Uruguay 32°22′S 54°10′W
Melodunum *see* Melun
Melrhir, Chott *see* Melghir, Chott
183 P7 **Melrose** New South Wales, SE Australia 32°41′S 146°58′E
182 I7 **Melrose** South Australia 32°52′S 138°16′E
29 T7 **Melrose** Minnesota, N USA 45°40′N 94°46′W
33 Q11 **Melrose** Montana, NW USA 45°33′N 112°41′W
37 V12 **Melrose** New Mexico, SW USA 34°25′N 103°37′W
108 I8 **Mels** Sankt Gallen, NE Switzerland 47°03′N 09°26′E
Melsetter *see* Chimanimani
33 V9 **Melstone** Montana, NW USA 46°37′N 107°49′W
101 I16 **Melsungen** Hessen, C Germany 51°08′N 09°33′E
92 L12 **Meltaus** Lappi, NW Finland 66°54′N 25°18′E
97 N19 **Melton Mowbray** C England, United Kingdom 52°46′N 01°04′W
82 Q13 **Meluco** Cabo Delgado, NE Mozambique 12°31′S 39°39′E
103 O5 **Melun** *anc.* Melodunum. Seine-et-Marne, N France 48°32′N 02°40′E
80 F12 **Melut** Upper Nile, NE South Sudan 10°27′N 32°13′E
27 P5 **Melvern Lake** ⊠ Kansas, C USA
11 V16 **Melville** Saskatchewan, S Canada 50°57′N 102°49′W
Melville Bay/Melville Bugt *see* Qimusseriarsuaq
45 O11 **Melville Hall** ✈ (Dominica) NE Dominica 15°33′N 61°19′W
181 O1 **Melville Island** *island* Northern Territory, N Australia
197 O8 **Melville Island** *island* Parry Islands, Northwest Territories, NW Canada
11 W9 **Melville, Lake** ◎ Newfoundland and Labrador, E Canada
9 O7 **Melville Peninsula** *peninsula* Nunavut, NE Canada
Melville Sound *see* Viscount Melville Sound
25 Q9 **Melvin** Texas, SW USA 31°12′N 99°34′W
97 D15 **Melvin, Lough** *Ir.* Loch Meilbhe. ◎ S Northern Ireland, United Kingdom/Ireland
169 S12 **Memala** Borneo, C Indonesia 01°44′S 112°36′E
113 L22 **Memaliaj** Gjirokastër, S Albania 40°21′N 19°56′E
83 Q14 **Memba** Nampula, NE Mozambique 14°07′S 40°33′E
83 Q14 **Memba, Baia de** *inlet* NE Mozambique
Membidj *see* Manbij
Memel *see* Neman, NE Europe
Memel *see* Klaipeda, Lithuania
101 J23 **Memmingen** Bayern, S Germany 47°59′N 10°11′E
27 U1 **Memphis** Missouri, C USA 40°28′N 92°11′W
20 E10 **Memphis** Tennessee, S USA 35°09′N 90°03′W
25 P3 **Memphis** Texas, SW USA 34°43′N 100°34′W
20 E10 **Memphis** ✈ Tennessee, S USA 35°02′N 89°57′W
15 Q13 **Memphrémagog, Lac** *var.* Lake Memphremagog. ◎ Canada/USA *see also* Lake Memphremagog
19 N6 **Memphremagog, Lake** *var.* Lac Memphrémagog. ◎ Canada/USA *see also* Memphrémagog, Lac
117 Q2 **Mena** Chernihivs'ka Oblast', NE Ukraine 51°30′N 32°15′E
27 S12 **Mena** Arkansas, C USA 34°40′N 94°15′W
Menaam *see* Menaldum
Menado *see* Manado
106 D6 **Menaggio** Lombardia, N Italy 46°03′N 09°14′E
29 T6 **Menahga** Minnesota, N USA 46°45′N 95°06′W
77 R10 **Ménaka** Goa, E Mali 15°55′N 02°25′E
98 K5 **Menaldum** *Fris.* Menaam. Fryslân, N Netherlands 53°14′N 05°38′E
Mènam Khong *see* Mekong
74 E7 **Menara** ✈ (Marrakech) C Morocco 31°36′N 08°00′W
25 Q9 **Menard** Texas, SW USA 30°56′N 99°48′W
193 Q12 **Menard Fracture Zone** *tectonic feature* E Pacific Ocean
30 M7 **Menasha** Wisconsin, N USA 44°13′N 88°25′W
Mencezi Garagum *see* Merkezi Garagumy
193 U9 **Mendaña Fracture Zone** *tectonic feature* E Pacific Ocean
169 S13 **Mendawai, Sungai** ≋ Borneo, C Indonesia
103 P13 **Mende** *anc.* Mimatum. Lozère, S France 44°32′N 03°30′E
81 J14 **Mendebo** ▲▲ C Ethiopia
80 J9 **Mendefera** *prev.* Adi Ugri. S Eritrea 14°53′N 38°51′E
197 S7 **Mendeleyev Ridge** *undersea feature* Arctic Ocean
127 T3 **Mendeleyevsk** Respublika Tatarstan, W Russian Federation 55°54′N 52°19′E
101 F15 **Menden** Nordrhein-Westfalen, W Germany 51°26′N 07°47′E
22 L6 **Mendenhall** Mississippi, S USA 31°57′N 89°52′W
38 L13 **Mendenhall, Cape** *headland* Nunivak Island, Alaska, USA 59°45′N 166°10′W
41 P9 **Méndez** *var.* Villa de Méndez. Tamaulipas, C Mexico 25°06′N 98°32′W
80 H13 **Mendī** Oromīya C Ethiopia 09°43′N 35°07′E
186 C7 **Mendi** Southern Highlands, W Papua New Guinea 06°13′S 143°39′E
97 K22 **Mendip Hills** *var.* Mendips. *hill range* S England, United Kingdom
Mendips *see* Mendip Hills
34 L6 **Mendocino** California, W USA 39°18′N 123°48′W
34 J3 **Mendocino, Cape** *headland* California, W USA 40°26′N 124°24′W
0 B8 **Mendocino Fracture Zone** *tectonic feature* NE Pacific Ocean
35 P10 **Mendota** California, W USA 36°44′N 120°24′W
30 L11 **Mendota** Illinois, N USA 41°32′N 89°04′W
30 K8 **Mendota, Lake** ◎ Wisconsin, N USA
62 I11 **Mendoza** Mendoza, W Argentina 33°00′S 68°47′W
62 I12 **Mendoza** *off.* Provincia de Mendoza. ◈ *province* W Argentina
Mendoza, Provincia de *see* Mendoza
108 H12 **Mendrisio** Ticino, S Switzerland 45°53′N 08°59′E
168 L10 **Mendung** Pulau Mendol, W Indonesia 0°33′N 103°09′E
54 I5 **Mene de Mauroa** Falcón, NW Venezuela 10°39′N 71°04′W
54 I5 **Mene Grande** Zulia, NW Venezuela 09°51′N 70°57′W
136 B14 **Menemen** İzmir, W Turkey 38°34′N 27°03′E
99 C18 **Menen** *var.* Meenen, *Fr.* Menin. West-Vlaanderen, W Belgium 50°48′N 03°07′E
163 Q8 **Menengiyn Tal** *plain* E Mongolia
189 R9 **Meneng Point** *headland* SW Nauru 0°33′S 166°57′E
92 L10 **Menesjärvi** *Lapp.* Menešjávri. Lappi, N Finland 68°39′N 26°22′E
Menešjávri *see* Menesjärvi
107 I24 **Menfi** Sicilia, Italy, C Mediterranean Sea 37°36′N 12°59′E
161 P7 **Mengcheng** Anhui, E China 33°15′N 116°33′E
160 F15 **Menghai** Yunnan, SW China 22°02′N 100°18′E
160 F15 **Mengla** Yunnan, SW China 21°30′N 101°33′E
65 F24 **Menguera Point** *headland* East Falkland, Falkland Islands
160 M13 **Mengzhu Ling** ▲▲ S China
160 H14 **Mengzi** Yunnan, SW China 23°20′N 103°32′E
114 H13 **Meníkio** *var.* Menoíkio. ▲ NE Greece 40°50′N 12°40′E
Menin *see* Menen
182 L7 **Menindee** New South Wales, SE Australia 32°24′S 142°25′E
182 L7 **Menindee Lake** ◎ New South Wales, SE Australia
182 J10 **Meningie** South Australia 35°43′S 139°20′E
103 O5 **Mennecy** Essonne, N France 48°34′N 02°25′E
29 Q12 **Menno** South Dakota, N USA 43°14′N 97°34′W
114 H13 **Menoíkio** ▲▲ NE Greece
Menoíkio *see* Meníkio
31 N5 **Menominee** Michigan, N USA 45°06′N 87°36′W
30 M5 **Menominee River** ≋ Michigan/Wisconsin, N USA
30 M8 **Menomonee Falls** Wisconsin, N USA 43°11′N 88°09′W
30 I6 **Menomonie** Wisconsin, N USA 44°52′N 91°55′W
83 D14 **Menongue** *var.* Vila Serpa Pinto, *Port.* Serpa Pinto. Kuando Kubango, C Angola 14°38′S 17°39′E
120 H8 **Menorca** *Eng.* Minorca; *anc.* Balearis Minor. *island* Islas Baleares, Spain, W Mediterranean Sea
105 S13 **Menor, Mar** *lagoon* SE Spain
39 S10 **Mentasta Lake** ◎ Alaska, USA
39 S10 **Mentasta Mountains** ▲▲ Alaska, USA
168 I13 **Mentawai, Kepulauan** *island group* W Indonesia
168 I12 **Mentawai, Selat** *strait* W Indonesia
168 M12 **Mentok** Pulau Bangka, W Indonesia 02°01′S 105°10′E
103 V15 **Menton** *It.* Mentone. Alpes-Maritimes, SE France 43°47′N 07°30′E
24 K8 **Mentone** Texas, SW USA 31°42′N 103°36′W
Mentone *see* Menton
31 U11 **Mentor** Ohio, N USA 41°40′N 81°20′W
169 U10 **Menyapa, Gunung** ▲ Borneo, N Indonesia 01°04′N 116°01′E
159 T9 **Menyuan** *var.* Menyuan Huizu Zizhixian. Qinghai, C China 37°27′N 101°33′E
Menyuan Huizu Zizhixian *see* Menyuan
74 M5 **Menzel Bourguiba** *var.* Manzil Bū Ruqaybah; *prev.* Ferryville. N Tunisia 37°09′N 09°51′E
136 M15 **Menzelet Barajı** ⊠ C Turkey
127 T4 **Menzelinsk** Respublika Tatarstan, W Russian Federation 55°44′N 53°00′E
180 K11 **Menzies** Western Australia 29°42′S 121°04′E
195 V6 **Menzies, Mount** ▲ Antarctica 73°32′S 61°02′E
40 J6 **Meoqui** Chihuahua, N Mexico 28°18′N 105°30′W
83 N14 **Meponda** Niassa, NE Mozambique 13°20′S 34°53′E
98 M8 **Meppel** Drenthe, NE Netherlands 52°42′N 06°12′E
100 E12 **Meppen** Niedersachsen, NW Germany 52°42′N 07°18′E
Meqerghane, Sebkha *see* Mekerrhane, Sebkha
105 T6 **Mequinenza, Embalse de** ⊠ NE Spain
30 M8 **Mequon** Wisconsin, N USA 43°13′N 87°57′W
Mera *see* Maira
182 D3 **Meramangye, Lake** *salt lake* South Australia
27 W5 **Meramec River** ≋ Missouri, C USA
Meran *see* Merano
168 K13 **Merangin** ≋ Sumatera, W Indonesia
106 G5 **Merano** *Ger.* Meran. Trentino-Alto Adige, N Italy 46°40′N 11°10′E
168 K8 **Merapuh Lama** Pahang, Peninsular Malaysia 04°37′N 101°58′E
106 D7 **Merate** Lombardia, N Italy 45°42′N 09°26′E
169 U13 **Meratus, Pegunungan** ▲▲ Borneo, N Indonesia
171 Y16 **Merauke, Sungai** ≋ Papua, E Indonesia
182 L9 **Merbein** Victoria, SE Australia 34°11′S 142°03′E
99 F21 **Merbes-le-Château** Hainaut, S Belgium 50°19′N 04°09′E
Merca *see* Marka
54 C13 **Mercaderes** Cauca, SW Colombia 01°46′N 77°09′W
Mercara *see* Madikeri

◆ Country
● Country Capital
◇ Dependent Territory
○ Dependent Territory Capital
◈ Administrative Regions
✈ International Airport
▲ Mountain
▲▲ Mountain Range
Volcano
≋ River
◎ Lake
⊠ Reservoir

35 P9 **Merced** California, W USA 37°17′N 120°30′W
61 C20 **Mercedes** Buenos Aires, E Argentina 34°42′S 59°30′W
61 D15 **Mercedes** Corrientes, NE Argentina 29°09′S 58°05′W
61 D19 **Mercedes** Soriano, SW Uruguay 33°16′S 58°01′W
25 S17 **Mercedes** Texas, SW USA 26°09′N 97°54′W
Mercedes *see* Villa Mercedes
35 R9 **Merced Peak** ▲ California, W USA 37°34′N 119°30′W
35 P9 **Merced River** ≈ California, W USA
18 B13 **Mercer** Pennsylvania, NE USA 41°14′N 80°14′W
99 G18 **Merchtem** Vlaams Brabant, C Belgium 50°57′N 04°14′E
15 O13 **Mercier** Québec, SE Canada 45°15′N 73°45′W
25 Q9 **Mercury** Texas, SW USA 31°23′N 99°09′W
184 M5 **Mercury Islands** *island group* N New Zealand
19 O9 **Meredith** New Hampshire, NE USA 43°36′N 71°28′W
65 B25 **Meredith, Cape** *var.* Cabo Belgrano. *headland* West Falkland, Falkland Islands 52°15′S 60°40′W
37 V6 **Meredith, Lake** ◎ Colorado, C USA
25 N2 **Meredith, Lake** ⊡ Texas, SW USA
81 O16 **Mereeg** *var.* Mareeq, *It.* Meregh. Galguduud, E Somalia 03°47′N 47°19′E
117 V5 **Merefa** Kharkivs'ka Oblast', E Ukraine 49°49′N 36°05′E
Meregh *see* Mereeg
99 E17 **Merelbeke** Oost-Vlaanderen, NW Belgium 51°00′N 03°45′E
Merend *see* Marand
167 T12 **Méreuch** Môndól Kiri, E Cambodia 13°01′N 107°26′E
Mergate *see* Margate
Mergui *see* Myeik
Mergui Archipelago *see* Myeik Archipelago
114 L12 **Meriç** Edirne, NW Turkey 41°12′N 26°24′E
114 L12 **Meriç** *Bul.* Maritsa, *Gk.* Évros; *anc.* Hebrus. ≈ SE Europe *see also* Évros/Maritsa
41 X12 **Mérida** Yucatán, SW Mexico 20°58′N 89°35′W
104 J11 **Mérida** *anc.* Augusta Emerita. Extremadura, W Spain 38°55′N 06°20′W
54 I6 **Mérida** Mérida, W Venezuela 08°36′N 71°08′W
54 H7 **Mérida** *off.* Estado Mérida. ◈ *state* W Venezuela
Mérida, Estado *see* Mérida
18 M13 **Meriden** Connecticut, NE USA 41°32′N 72°48′W
22 M5 **Meridian** Mississippi, S USA 32°24′N 88°43′W
25 S8 **Meridian** Texas, SW USA 31°56′N 97°40′W
102 J13 **Mérignac** Gironde, SW France 44°50′N 00°40′W
102 J13 **Mérignac** ✈ (Bordeaux) Gironde, SW France 44°51′N 00°44′W
93 J18 **Merikarvia** Satakunta, SW Finland 61°51′N 21°30′E
183 R12 **Merimbula** New South Wales, SE Australia 36°52′S 149°51′E
182 L9 **Meringur** Victoria, SE Australia 34°26′S 141°19′E
Merín, Laguna *see* Mirim Lagoon
97 I19 **Merioneth** *cultural region* W Wales, United Kingdom
188 A11 **Merir** *island* Palau Islands, N Palau
188 B17 **Merizo** SW Guam 13°15′N 144°40′E
Merjama *see* Märjamaa
Merke *see* Merki
25 P7 **Merkel** Texas, SW USA 32°28′N 100°00′W
146 E12 **Merkezi Garagumy** *var.* Mencezi Garagum, *Rus.* Tsentral'nyye Nizmennyye Garagumy. *desert* C Turkmenistan
145 S16 **Merki** *prev.* Merke. Zhambyl, S Kazakhstan 42°48′N 73°10′E
119 F15 **Merkinė** Alytus, S Lithuania 54°09′N 24°11′E
99 G16 **Merksem** Antwerpen, N Belgium 51°17′N 04°26′E
99 I15 **Merksplas** Antwerpen, N Belgium 51°22′N 04°54′E
Merkulovichi *see* Myerkulavichy
119 G15 **Merkys** ≈ S Lithuania
32 F15 **Merlin** Oregon, NW USA 42°34′N 123°23′W
61 C20 **Merlo** Buenos Aires, E Argentina 34°39′S 58°45′W
138 G8 **Meron, Harei** *prev.* Haré Meron. ▲ N Israel 35°06′N 33°00′E
74 K6 **Merouane, Chott** *salt lake* NE Algeria
80 F7 **Merowe** Northern, N Sudan 18°29′N 31°49′E
180 J12 **Merredin** Western Australia 31°31′S 118°18′E
97 I14 **Merrick** ▲ S Scotland, United Kingdom 55°09′N 04°28′W
32 H16 **Merrill** Oregon, NW USA 42°00′N 121°37′W
30 L5 **Merrill** Wisconsin, N USA 45°12′N 89°43′W
31 N11 **Merrillville** Indiana, N USA 41°28′N 87°19′W
19 O10 **Merrimack River** ≈ Massachusetts/New Hampshire, NE USA
28 L12 **Merriman** Nebraska, C USA 42°54′N 101°42′W
11 N17 **Merritt** British Columbia, SW Canada 50°09′N 120°49′W
23 Y12 **Merritt Island** Florida, SE USA 28°21′N 80°42′W
23 Y11 **Merritt Island** *island* Florida, SE USA
28 M12 **Merritt Reservoir** ⊡ Nebraska, C USA
183 S7 **Merriwa** New South Wales, SE Australia 32°09′S 150°24′E
183 O8 **Merriwagga** New South Wales, SE Australia 33°51′S 145°38′E
22 G8 **Merryville** Louisiana, S USA 30°45′N 93°32′W
80 K9 **Mersa Fat'ma** E Eritrea 14°52′N 40°16′E
102 M7 **Mer St-Aubin** Loir-et-Cher, C France 47°42′N 01°31′E
99 M24 **Mersch** Luxembourg, C Luxembourg 49°45′N 06°06′E
101 M15 **Merseburg** Sachsen-Anhalt, C Germany 51°22′N 12°00′E
Mersen *see* Meerssen
97 K18 **Mersey** ≈ NW England, United Kingdom
136 J17 **Mersin** *var.* İçel. İçel, S Turkey 36°50′N 34°39′E
Mersin *see* İçel
168 L9 **Mersing** Johor, Peninsular Malaysia 02°25′N 103°50′E
118 E8 **Mērsrags** NW Latvia 57°21′N 23°05′E
Merta *see* Merta City
152 G12 **Merta City** *var.* Merta. Rājasthān, N India 26°40′N 74°04′E
152 F12 **Merta Road** Rājasthān, N India 26°42′N 73°54′E
97 J21 **Merthyr Tydfil** S Wales, United Kingdom 51°46′N 03°23′W
104 H13 **Mértola** Beja, S Portugal 37°38′N 07°40′W
144 G14 **Mertvyy Kultuk, Sor** *salt flat* SW Kazakhstan
195 V16 **Mertz Glacier** *glacier* Antarctica
99 M24 **Mertzig** Diekirch, C Luxembourg 49°50′N 06°00′E
25 O9 **Mertzon** Texas, SW USA 31°16′N 100°50′W
103 N4 **Méru** Oise, N France 49°15′N 02°07′E
81 I18 **Meru** Meru, C Kenya 0°03′N 37°38′E
81 I19 **Meru** ◈ *county* C Kenya
81 I20 **Meru, Mount** ▲ NE Tanzania 03°12′S 36°45′E
Merv *see* Mary
Mervdasht *see* Marv Dasht
136 K11 **Merzifon** Amasya, N Turkey 40°52′N 35°28′E
101 D20 **Merzig** Saarland, SW Germany 49°27′N 06°39′E
36 L14 **Mesa** Arizona, SW USA 33°25′N 111°49′W
29 V4 **Mesabi Range** ▲▲ Minnesota, N USA
54 H6 **Mesa Bolívar** Mérida, NW Venezuela 08°30′N 71°38′W
107 Q18 **Mesagne** Puglia, SE Italy 40°33′N 17°49′E
39 P12 **Mesa Mountain** ▲ Alaska, USA 60°26′N 155°14′W
115 J25 **Mesará** *lowland* Kríti, Greece, E Mediterranean Sea
37 S14 **Mescalero** New Mexico, SW USA 33°09′N 105°46′W
101 G15 **Meschede** Nordrhein-Westfalen, W Germany 51°21′N 08°16′E
137 Q12 **Mescit Dağları** ▲▲ NE Turkey
189 V13 **Mesegon** *island* Chuuk, C Micronesia
Meseritz *see* Międzyrzecz
54 F11 **Mesetas** Meta, C Colombia 03°14′N 74°09′W
Meshchera Lowland *see* Meshcherskaya Nizmennost'
Meshcherskaya Nizina *see* Meshcherskaya Nizmennost'
126 M4 **Meshcherskaya Nizmennost'** *var.* Meshcherskaya Nizina, *Eng.* Meshchera Lowland. *basin* W Russian Federation
126 J5 **Meshchovsk** Kaluzhskaya Oblast', W Russian Federation 54°21′N 35°23′E
125 R9 **Meshchura** Respublika Komi, NW Russian Federation 63°18′N 50°56′E
Meshed *see* Mashhad
Meshed-i-Sar *see* Bābolsar
80 E13 **Meshra'er Req** Warap, W South Sudan 08°30′N 29°27′E
37 R15 **Mesilla** New Mexico, SW USA 32°15′N 106°49′W
108 H10 **Mesocco** *Ger.* Misox. Ticino, S Switzerland 46°18′N 09°13′E
115 D18 **Mesolóngi** *prev.* Mesolóngion. Dytikí Elláda, W Greece 38°21′N 21°26′E
Mesolóngion *see* Mesolóngi
14 E8 **Mesomikenda Lake** ◎ Ontario, S Canada
61 D15 **Mesopotamia** *var.* Mesopotamia Argentina. *physical region* NE Argentina
Mesopotamia Argentina *see* Mesopotamia
35 Y10 **Mesquite** Nevada, W USA 36°47′N 114°04′W
82 Q13 **Messalo, Rio** *var.* Mualo. ≈ NE Mozambique
Messana/Messene *see* Messina
99 L25 **Messancy** Luxembourg, SE Belgium 49°36′N 05°49′E
107 M23 **Messina** *var.* Messana, Messene; *anc.* Zancle. Sicilia, Italy, C Mediterranean Sea 38°12′N 15°33′E
Messina *see* Musina
Messina, Strait of *see* Messina, Stretto di
107 M23 **Messina, Stretto di** *Eng.* Strait of Messina. *strait* SW Italy
115 E21 **Messíni** Pelopónnisos, S Greece 37°03′N 22°00′E
115 E21 **Messinía** *peninsula* S Greece
115 E22 **Messiniakós Kólpos** *gulf* S Greece
122 J8 **Messoyakha** ≈ N Russian Federation
114 H11 **Mesta** *Gk.* Néstos, *Turk.* Kara Su. ≈ Bulgaria/Greece *see also* Néstos
Mesta *see* Néstos
Mestghanem *see* Mostaganem
137 R8 **Mest'ia** *prev.* Mestia, *var.* Mestiya. N Georgia 43°03′N 42°50′E
Mestia *see* Mest'ia
Mestiya *see* Mest'ia
115 K18 **Mestón, Akrotírio** *cape* Chíos, E Greece
106 H8 **Mestre** Veneto, NE Italy 45°30′N 12°14′E
59 M16 **Mestre, Espigão** ▲▲ E Brazil
169 N14 **Mesuji** ≈ Sumatera, W Indonesia
Mesule *see* Grosser Möseler
10 J10 **Meszah Peak** ▲ British Columbia, W Canada 58°31′N 131°28′W
54 G11 **Meta** *off.* Departamento del Meta. ◈ *province* C Colombia
15 Q8 **Metabetchouane** ≈ Québec, SE Canada
Meta, Departamento del *see* Meta
9 S7 **Meta Incognita Peninsula** *peninsula* Baffin Island, Nunavut, NE Canada
22 K9 **Metairie** Louisiana, S USA 29°58′N 90°09′W
32 M6 **Metaline Falls** Washington, NW USA 48°51′N 117°21′W
62 K6 **Metán** Salta, N Argentina 25°29′S 64°57′W
82 N13 **Metangula** Niassa, N Mozambique 12°41′S 34°50′E
42 E7 **Metapán** Santa Ana, NW El Salvador 14°20′N 89°28′W
54 K9 **Meta, Río** ≈ Colombia/Venezuela
106 I11 **Metauro** ≈ C Italy
80 H11 **Metema** Āmara, N Ethiopia 12°53′N 36°10′E
115 D15 **Metéora** *religious building* Thessalía, C Greece
65 O20 **Meteor Rise** *undersea feature* SW Indian Ocean 46°00′S 05°30′E
186 G5 **Meteran** New Hanover, NE Papua New Guinea 02°40′S 150°12′E
Meterlam *see* Mehtar Lām
115 G20 **Methanon** *peninsula* S Greece
Methariam/Metharlam *see* Mehtar Lām
32 J6 **Methow River** ≈ Washington, NW USA
19 O10 **Methuen** Massachusetts, NE USA 42°43′N 71°10′W
185 G19 **Methven** Canterbury, South Island, New Zealand 43°37′S 171°38′E
Metis *see* Metz
113 G15 **Metković** Dubrovnik-Neretva, SE Croatia 43°02′N 17°37′E
39 Y14 **Metlakatla** Annette Island, Alaska, USA 55°07′N 131°34′W
109 V13 **Metlika** *Ger.* Möttling. SE Slovenia 45°38′N 15°18′E
109 T8 **Metnitz** Kärnten, S Austria 46°58′N 14°09′E
27 W12 **Meto, Bayou** ≈ Arkansas, C USA
168 M15 **Metro** Sumatera, W Indonesia 05°05′S 105°20′E
30 M17 **Metropolis** Illinois, N USA 37°09′N 88°43′W
Metropolitan *see* Santiago
35 N8 **Metropolitan Oakland** ✈ California, W USA 37°42′N 122°13′W
115 D15 **Métsovo** *prev.* Métsovon. Ípeiros, C Greece 39°47′N 21°12′E
Métsovon *see* Métsovo
23 V5 **Metter** Georgia, SE USA 32°24′N 82°03′W
99 H21 **Mettet** Namur, S Belgium 50°19′N 04°43′E
101 D20 **Mettlach** Saarland, SW Germany 49°28′N 06°37′E
Mettu *see* Metu
80 H13 **Metu** *var.* Mattu, Mettu. Oromīya, C Ethiopia 08°18′N 35°39′E
138 G8 **Metula** *prev.* Metulla. Northern, N Israel 33°16′N 35°35′E
169 T10 **Metulang** Borneo, N Indonesia 01°28′N 114°40′E
Metulla *see* Metula
103 T4 **Metz** *anc.* Divodurum Mediomatricum, Mediomatrica, Metis. Moselle, NE France 49°07′N 06°09′E
101 H22 **Metzingen** Baden-Württemberg, S Germany 48°31′N 09°16′E
168 G8 **Meulaboh** Sumatera, W Indonesia 04°10′N 96°09′E
99 D18 **Meulebeke** West-Vlaanderen, W Belgium 50°57′N 03°18′E
103 U6 **Meurthe** ≈ NE France
103 S5 **Meurthe-et-Moselle** ◈ *department* NE France
103 S4 **Meuse** ◈ *department* NE France
84 F10 **Meuse** *Dut.* Maas. ≈ W Europe *see also* Maas
Meuse *see* Maas
Mexcala, Río *see* Balsas, Río
25 U8 **Mexia** Texas, SW USA 31°40′N 96°28′W
58 K11 **Mexiana, Ilha** *island* NE Brazil
40 C1 **Mexicali** Baja California Norte, NW Mexico 32°34′N 115°26′W
Mexicanos, Estados Unidos *see* Mexico
41 O14 **México** *var.* Ciudad de México, *Eng.* Mexico City. ● (Mexico) México, C Mexico 19°26′N 99°08′W
27 V4 **Mexico** Missouri, C USA 19°10′N 99°04′W
18 H9 **Mexico** New York, NE USA 43°27′N 76°14′W
40 L7 **Mexico** *off.* United Mexican States, *var.* Méjico, México, *Sp.* Estados Unidos Mexicanos. ◆ *federal republic* N Central America
41 O13 **México** ◈ *state* S Mexico
México *see* Mexico
0 J13 **Mexico Basin** *var.* Sigsbee Deep. *undersea feature* C Gulf of Mexico 25°00′N 92°00′W
Mexico City *see* México
México, Golfo de *see* Mexico, Gulf of
44 B4 **Mexico, Gulf of** *Sp.* Golfo de México. *gulf* W Atlantic Ocean
139 R4 **Mexmûr** *Ar.* Makhmūr. Arbīl, N Iraq 35°47′N 43°32′E
Meyadine *see* Al Mayādīn
Meydān Shahr *see* Maīdān Shahr
39 Y14 **Meyers Chuck** Etolin Island, Alaska, USA 55°44′N 132°15′W
Meymaneh *see* Maīmanah
143 N7 **Meymeh** Eşfahān, C Iran 33°29′N 51°09′E
123 V7 **Meynypil'gyno** Chukotskiy Avtonomnyy Okrug, NE Russian Federation 62°33′N 177°00′E
108 A10 **Meyrin** Genève, SW Switzerland 46°14′N 06°05′E
166 L7 **Mezaligon** Ayeyawady, SW Myanmar (Burma) 17°53′N 95°12′E
41 O15 **Mezcala** Guerrero, S Mexico 17°55′N 99°34′W
114 H8 **Mezdra** Vratsa, NW Bulgaria 43°08′N 23°42′E
103 P16 **Mèze** Hérault, S France 43°26′N 03°37′E
125 O6 **Mezen'** Arkhangel'skaya Oblast', NW Russian Federation 65°54′N 44°10′E
125 P8 **Mezen'** ≈ NW Russian Federation
Mezen, Bay of *see* Mezenskaya Guba
103 Q13 **Mézenc, Mont** ▲ C France 44°57′N 04°15′E
125 O8 **Mezenskaya Guba** *var.* Bay of Mezen. *bay* NW Russian Federation
Mezha *see* Myazha
122 H6 **Mezhdusharskiy, Ostrov** *island* Novaya Zemlya, N Russian Federation
Mezhëvo *see* Myezhava
Mezhgor'ye *see* Mizhhir"ya
117 V8 **Mezhova** Dnipropetrovs'ka Oblast', E Ukraine 48°15′N 36°44′E
10 J12 **Meziadin Junction** British Columbia, W Canada 56°06′N 129°15′W
111 G16 **Mezileské Sedlo** *var.* Przełęcz Międzyleska. *pass* Czech Republic/Poland
102 L14 **Mézin** Lot-et-Garonne, SW France 44°03′N 00°16′E
111 M24 **Mezőberény** Békés, SE Hungary 46°49′N 21°00′E
111 M25 **Mezőhegyes** Békés, SE Hungary 46°20′N 20°48′E
111 M25 **Mezőkovácsháza** Békés, SE Hungary 46°24′N 20°52′E
111 M21 **Mezőkövesd** Borsod-Abaúj-Zemplén, NE Hungary 47°49′N 20°32′E
Mezőtelegd *see* Tileagd
111 M23 **Mezőtúr** Jász-Nagykun-Szolnok, E Hungary 47°N 20°37′E
40 K10 **Mezquital** Durango, C Mexico 23°31′N 104°19′W
106 G6 **Mezzolombardo** Trentino-Alto Adige, N Italy 46°13′N 11°08′E
82 L13 **Mfuwe** Muchinga, N Zambia 13°00′S 31°45′E
121 O15 **Mġarr** Gozo, N Malta 36°01′N 14°18′E
126 H6 **Mglin** Bryanskaya Oblast', W Russian Federation 53°01′N 32°54′E
Mhálanna, Cionn *see* Malin Head
154 G10 **Mhow** Madhya Pradesh, C India 22°32′N 75°49′E
Miadzioł Nowy *see* Myadzyel
171 O6 **Miagao** Panay Island, C Philippines 10°40′N 122°15′E
41 R17 **Miahuatlán** *var.* Miahuatlán de Porfirio Díaz. Oaxaca, SE Mexico 16°21′N 96°36′W
Miahuatlán de Porfirio Díaz *see* Miahuatlán
104 K10 **Miajadas** Extremadura, W Spain 39°10′N 05°54′W
Miajlar *see* Myājlār
36 M14 **Miami** Arizona, SW USA 33°23′N 110°53′W
23 Z16 **Miami** Florida, SE USA 25°46′N 80°12′W
27 R8 **Miami** Oklahoma, C USA 36°53′N 94°54′W
25 O2 **Miami** Texas, SW USA 35°42′N 100°37′W
23 Z16 **Miami** ✈ Florida, SE USA 25°47′N 80°16′W
23 Z16 **Miami Beach** Florida, SE USA 25°47′N 80°08′W
23 Y15 **Miami Canal** *canal* Florida, SE USA
31 R14 **Miamisburg** Ohio, N USA 39°38′N 84°17′W
149 U10 **Miān Channūn** Punjab, E Pakistan 30°24′N 72°27′E
142 J4 **Mīāndowāb** *var.* Mianduab, Miyāndoāb. Āzarbāyjān-e Gharbī, NW Iran 36°57′N 46°06′E
172 H5 **Miandrivazo** Toliara, C Madagascar 19°31′S 45°29′E
Mianduab *see* Mīāndowāb
142 K3 **Mīāneh** *var.* Miyāneh. Āzarbāyjān-e Sharqī, NW Iran 37°23′N 47°45′E
161 T12 **Mianhua Yu** *prev.* Mienhua Yü. *island* N Taiwan
149 O16 **Miāni Hōr** *lagoon* S Pakistan
160 G10 **Mianning** Sichuan, C China 28°34′N 102°12′E
149 T7 **Miānwāli** Punjab, NE Pakistan 32°32′N 71°33′E
160 J7 **Mianxian** *var.* Mian Xian. Shaanxi, C China 33°12′N 106°36′E
Mian Xian *see* Mianxian
160 I8 **Mianyang** Sichuan, C China 31°29′N 104°43′E
Mianyang *see* Xiantao
161 R3 **Miaodao Qundao** *island group* E China
161 S13 **Miaoli** N Taiwan 24°33′N 120°48′E
122 F11 **Miass** Chelyabinskaya Oblast', C Russian Federation 55°00′N 59°55′E
110 G8 **Miastko** *Ger.* Rummelsburg in Pommern. Pomorskie, N Poland 54°N 16°58′E
Miava *see* Myjava
11 O15 **Mica Creek** British Columbia, SW Canada 51°58′N 118°29′W
160 J7 **Micang Shan** ▲▲ C China
Mi Chai *see* Nong Khai
111 O19 **Michalovce** *Ger.* Grossmichel, *Hung.* Nagymihály. Košický Kraj, E Slovakia 48°46′N 21°55′E
99 M20 **Michel, Baraque** *hill* E Belgium
39 S5 **Michelson, Mount** ▲ Alaska, USA 69°19′N 144°16′W
45 P9 **Miches** E Dominican Republic 18°59′N 69°03′W
30 M4 **Michigamme, Lake** ◎ Michigan, N USA
30 M4 **Michigamme Reservoir** ⊡ Michigan, N USA
31 N4 **Michigamme River** ≈ Michigan, N USA
31 O7 **Michigan** *off.* State of Michigan, *also known as* Great Lakes State, Lake State, Wolverine State. ◈ *state* N USA
31 O11 **Michigan City** Indiana, N USA 41°43′N 86°52′W
31 O8 **Michigan, Lake** ◎ N USA
31 P2 **Michipicoten Bay** *lake bay* Ontario, N Canada
14 A8 **Michipicoten Island** *island* Ontario, S Canada
14 B7 **Michipicoten River** Ontario, S Canada 47°56′N 84°48′W
Michurin *see* Tsarevo
126 M6 **Michurinsk** Tambovskaya Oblast', W Russian Federation 52°56′N 40°31′E
Mico, Punta/Mico, Punto *see* Monkey Point
42 L10 **Mico, Río** ≈ SE Nicaragua
45 T12 **Micoud** SE Saint Lucia 13°49′N 60°54′W
189 N16 **Micronesia** *off.* Federated States of Micronesia. ◆ *federation* W Pacific Ocean
175 P4 **Micronesia** *island group* W Pacific Ocean
Micronesia, Federated States of *see* Micronesia
169 O9 **Midai, Pulau** *island* Kepulauan Natuna, W Indonesia
Mid-Atlantic Cordillera *see* Mid-Atlantic Ridge
65 M17 **Mid-Atlantic Ridge** *var.* Mid-Atlantic Cordillera, Mid-Atlantic Rise, Mid-Atlantic Swell. *undersea feature* Atlantic Ocean 0°00′N 20°00′W
Mid-Atlantic Rise/Mid-Atlantic Swell *see* Mid-Atlantic Ridge
99 E15 **Middelburg** Zeeland, SW Netherlands 51°30′N 03°36′E
83 H24 **Middelburg** Eastern Cape, S South Africa 31°28′S 25°01′E
83 K21 **Middelburg** Mpumalanga, NE South Africa 25°47′S 29°28′E
95 G23 **Middelfart** Syddtjylland, C Denmark 55°30′N 09°44′E
98 G13 **Middelharnis** Zuid-Holland, SW Netherlands 51°45′N 04°10′E
99 B16 **Middelkerke** West-Vlaanderen, W Belgium 51°12′N 02°51′E
98 I9 **Middenbeemster** Noord-Holland, C Netherlands 52°33′N 04°55′E
98 I8 **Middenmeer** Noord-Holland, NW Netherlands 52°48′N 04°58′E
35 Q2 **Middle Alkali Lake** ◎ California, W USA
193 S6 **Middle America Trench** *undersea feature* E Pacific Ocean 15°00′N 95°00′W
151 P19 **Middle Andaman** *island* Andaman Islands, India, NE Indian Ocean
Middle Atlas *see* Moyen Atlas
21 R3 **Middlebourne** West Virginia, NE USA 39°30′N 80°53′W
23 W9 **Middleburg** Florida, SE USA 30°03′N 81°55′W
Middleburg Island *see* 'Eua
Middle Caicos *see* Grand Caicos
25 N8 **Middle Concho River** ≈ Texas, SW USA
Middle Congo *see* Congo (Republic of)
39 R6 **Middle Fork Chandalar River** ≈ Alaska, USA
39 Q7 **Middle Fork Koyukuk River** ≈ Alaska, USA
33 O12 **Middle Fork Salmon River** ≈ Idaho, NW USA
11 T15 **Middle Lake** Saskatchewan, S Canada 52°31′N 105°16′W
28 L13 **Middle Loup River** ≈ Nebraska, C USA
185 E22 **Middlemarch** Otago, South Island, New Zealand 45°30′S 170°07′E
31 T15 **Middleport** Ohio, N USA 39°00′N 82°03′W
29 U14 **Middle Raccoon River** ≈ Iowa, C USA
29 R3 **Middle River** ≈ Minnesota, N USA
21 N8 **Middlesboro** Kentucky, S USA 36°37′N 83°42′W
97 M15 **Middlesbrough** N England, United Kingdom 54°35′N 01°14′W
42 G3 **Middlesex** Stann Creek, C Belize 17°00′N 88°31′W
97 N22 **Middlesex** *cultural region* SE England, United Kingdom
13 P15 **Middleton** Nova Scotia, SE Canada 44°56′N 65°04′W
20 F10 **Middleton** Tennessee, S USA 35°05′N 88°57′W
30 L9 **Middleton** Wisconsin, N USA 43°06′N 89°30′W
39 S13 **Middleton Island** *island* Alaska, USA
34 M7 **Middletown** California, W USA 38°44′N 122°39′W
21 Y2 **Middletown** Delaware, NE USA 39°25′N 75°39′W
18 K15 **Middletown** New Jersey, NE USA 40°23′N 74°08′W
18 K13 **Middletown** New York, NE USA 41°27′N 74°25′W
31 R14 **Middletown** Ohio, N USA 39°33′N 84°19′W
18 G15 **Middletown** Pennsylvania, NE USA 40°11′N 76°42′W
141 N14 **Mīdī** *var.* Maydī. NW Yemen 16°18′N 42°51′E
103 O16 **Midi, Canal du** *canal* S France
102 K17 **Midi de Bigorre, Pic du** ▲ S France 42°57′N 00°08′E
102 K17 **Midi d'Ossau, Pic du** ▲ SW France 42°51′N 00°27′W
173 R7 **Mid-Indian Basin** *undersea feature* N Indian Ocean 10°00′S 80°00′E
173 P7 **Mid-Indian Ridge** *var.* Central Indian Ridge. *undersea feature* C Indian Ocean 12°00′S 66°00′E
103 N14 **Midi-Pyrénées** ◈ *region* S France
25 N8 **Midkiff** Texas, SW USA 31°35′N 101°51′W
14 G13 **Midland** Ontario, S Canada 44°45′N 79°53′W
31 R8 **Midland** Michigan, N USA 43°37′N 84°15′W
28 M10 **Midland** South Dakota, N USA 44°04′N 101°07′W
24 M8 **Midland** Texas, SW USA 32°N 102°05′W
83 K17 **Midlands** ◈ *province* C Zimbabwe
97 D21 **Midleton** *Ir.* Mainistir na Corann. SW Ireland 51°55′N 08°10′W
25 T7 **Midlothian** Texas, SW USA 32°28′N 96°59′W
96 K12 **Midlothian** *cultural region* S Scotland, United Kingdom
172 I7 **Midongy Atsimo** Fianarantsoa, S Madagascar 21°58′S 47°47′E
102 K15 **Midou** ≈ SW France
192 J6 **Mid-Pacific Mountains** *var.* Mid-Pacific Seamounts. *undersea feature* NW Pacific Ocean 20°00′N 178°00′W
Mid-Pacific Seamounts *see* Mid-Pacific Mountains
171 Q7 **Midsayap** Mindanao, S Philippines 07°12′N 124°31′E
95 F21 **Midtjylland** ◈ *county* NW Denmark
36 L3 **Midway** Utah, W USA 40°30′N 111°28′W
192 L5 **Midway Islands** ◇ *US unincorporated territory* C Pacific Ocean
33 X14 **Midwest** Wyoming, C USA 43°24′N 106°15′W
27 N10 **Midwest City** Oklahoma, C USA 35°28′N 98°24′W
152 M10 **Mid Western** ◈ *zone* W Nepal
98 P5 **Midwolda** Groningen, NE Netherlands 53°12′N 07°00′E
137 Q16 **Midyat** Mardin, SE Turkey 37°25′N 41°20′E
114 F8 **Midzhur** *SCr.* Midžor. ▲ Bulgaria/Serbia 43°24′N 22°41′E *see also* Midžor
Midzhur *see* Midžor
113 Q14 **Midžor** *Bul.* Midzhur. ▲ Bulgaria/Serbia 43°24′N 22°40′E *see also* Midzhur
Midžor *see* Midzhur
164 K14 **Mie** *off.* Mie-ken. ◈ *prefecture* Honshū, SW Japan
111 L16 **Miechów** Małopolskie, S Poland 50°21′N 20°01′E
110 F11 **Międzychód** *Ger.* Mitteldorf. Wielkopolskie, C Poland 52°36′N 15°53′E
Międzyleska, Przełęcz *see* Mezileské Sedlo
110 O12 **Międzyrzec Podlaski** Lubelskie, E Poland 52°N 22°47′E
110 E11 **Międzyrzecz** *Ger.* Meseritz. Lubuskie, W Poland 52°26′N 15°33′E
Mie-ken *see* Mie
102 L16 **Miélan** Gers, S France 43°26′N 00°19′E
111 N16 **Mielec** Podkarpackie, SE Poland 50°18′N 21°27′E
95 L21 **Mien** ◎ S Sweden
Mienhua Yü *see* Mianhua Yu
41 O8 **Mier** Tamaulipas, C Mexico 26°28′N 99°10′W
116 J11 **Miercurea-Ciuc** *Ger.* Szeklerburg, *Hung.* Csíkszereda. Harghita, C Romania 46°24′N 25°48′E
Mieresch *see* Maros/Mureş
104 K2 **Mieres del Camín** *var.* Mieres del Camino. Asturias, NW Spain 43°15′N 05°46′W
Mieres del Camino *see* Mieres del Camín
99 K15 **Mierlo** Noord-Brabant, SE Netherlands 51°27′N 05°37′E
41 O10 **Mier y Noriega** Nuevo León, NE Mexico 23°24′N 100°06′W
Mies *see* Stříbro
80 K13 **Mī'ēso** *var.* Meheso, Miesso. Oromīya, C Ethiopia 09°13′N 40°47′E
Miesso *see* Mī'ēso
110 D10 **Mieszkowice** *Ger.* Bärwalde Neumark. Zachodnio-pomorskie, W Poland 52°45′N 14°24′E
18 G14 **Mifflinburg** Pennsylvania, NE USA 40°55′N 77°03′W
18 F14 **Mifflintown** Pennsylvania, NE USA 40°34′N 77°24′W
138 F8 **Mifrats Hefa** *Eng.* Bay of Haifa; *prev.* MifrazHefa. *bay* N Israel
81 G19 **Migori** ◈ *county* W Kenya
41 R15 **Miguel Alemán, Presa** ⊡ SE Mexico
40 L9 **Miguel Asua** *var.* Miguel Auza. Zacatecas, C Mexico 24°17′N 103°29′W
Miguel Auza *see* Miguel Asua
43 S15 **Miguel de la Borda** *var.* Donoso. Colón, C Panama 09°09′N 80°20′W
41 N13 **Miguel Hidalgo** ✈ (Guadalajara) Jalisco, SW Mexico 20°52′N 101°09′W
40 H7 **Miguel Hidalgo, Presa** ⊡ W Mexico
116 J14 **Mihăileşti** Giurgiu, S Romania 44°20′N 25°54′E
116 M14 **Mihail Kogălniceanu** *var.* Kogălniceanu; *prev.* Caramurat, Ferdinand. Constanţa, SE Romania 44°22′N 28°27′E
117 N14 **Mihai Viteazu** Constanţa, SE Romania 44°37′N 28°41′E
136 G12 **Mihalıççık** Eskişehir, NW Turkey 39°52′N 31°30′E
164 G13 **Mihara** Hiroshima, Honshū, SW Japan 34°24′N 133°04′E
165 N14 **Mihara-yama** ⛰ Miyako-jima, SE Japan 34°43′N 139°23′E
105 S8 **Mijares** ≈ E Spain
98 I11 **Mijdrecht** Utrecht, C Netherlands 52°12′N 04°52′E
165 S4 **Mikasa** Hokkaidō, NE Japan 43°15′N 141°57′E
Mikashevichi *see* Mikashevichy
119 K19 **Mikashevichy** *Pol.* Mikaszewicze, *Rus.* Mikashevichi. Brestskaya Voblasts', SW Belarus 52°13′N 27°28′E
Mikaszewicze *see* Mikashevichy
126 L5 **Mikhaylov** Ryazanskaya Oblast', W Russian Federation 54°12′N 39°03′E
Mikhaylovgrad *see* Montana
195 Z8 **Mikhaylov Island** *island* Antarctica
145 T6 **Mikhaylovka** Pavlodar, N Kazakhstan 53°49′N 76°31′E
127 N9 **Mikhaylovka** Volgogradskaya Oblast', SW Russian Federation 50°06′N 43°17′E
Mikhaylovka *see* Mykhaylivka
81 K24 **Mikindani** Mtwara, SE Tanzania 10°16′S 40°05′E
93 N18 **Mikkeli** *Swe.* Sankt Michel. Etelä-Savo, SE Finland 61°41′N 27°14′E
110 M8 **Mikołajki** *Ger.* Nikolaiken. Warmińsko-Mazurskie, NE Poland 53°49′N 21°31′E
Míkonos *see* Mýkonos
114 I9 **Mikre** Lovech, N Bulgaria 43°01′N 24°31′E
114 C13 **Mikrí Préspa, Límni** ◎ N Greece
125 P4 **Mikulkin, Mys** *headland* NW Russian Federation 67°50′N 46°36′E
81 I23 **Mikumi** Morogoro, SE Tanzania 07°22′S 37°00′E
125 R10 **Mikun'** Respublika Komi, NW Russian Federation 62°20′N 50°02′E
164 K13 **Mikuni** Fukui, Honshū, SW Japan 36°12′N 136°09′E
165 X13 **Mikura-jima** *island* E Japan
29 V7 **Milaca** Minnesota, N USA 45°45′N 93°40′W
62 J10 **Milagro** La Rioja, C Argentina 31°00′S 66°01′W
56 B7 **Milagro** Guayas, SW Ecuador 02°11′S 79°36′W
31 P4 **Milakokia Lake** ◎ Michigan, N USA
30 J1 **Milan** Illinois, N USA 41°27′N 90°33′W
31 R10 **Milan** Michigan, N USA 42°05′N 83°40′W
27 T2 **Milan** Missouri, C USA 40°12′N 93°08′W
37 Q11 **Milan** New Mexico, SW USA 35°10′N 107°53′W
20 G9 **Milan** Tennessee, S USA 35°55′N 88°45′W
Milan *see* Milano
95 F15 **Miland** Telemark, S Norway 59°57′N 08°48′E
83 N15 **Milange** Zambézia, NE Mozambique 16°09′S 35°44′E
106 D8 **Milano** *Eng.* Milan, *Ger.* Mailand; *anc.* Mediolanum. Lombardia, N Italy 45°28′N 09°10′E
25 U10 **Milano** Texas, SW USA 30°42′N 96°51′W
136 C15 **Milas** Muğla, SW Turkey 37°17′N 27°46′E
119 K21 **Milashavichy** *Rus.* Milashevichi. Homyel'skaya Voblasts', SE Belarus 51°39′N 27°56′E
Milashevichi *see* Milashavichy
119 I18 **Milavidy** *Rus.* Milovidy. Brestskaya Voblasts', SW Belarus 52°54′N 25°51′E
107 L23 **Milazzo** *anc.* Mylae. Sicilia, Italy, C Mediterranean Sea 38°13′N 15°15′E
29 R8 **Milbank** South Dakota, N USA 45°12′N 96°36′W
19 T7 **Milbridge** Maine, NE USA 44°31′N 67°55′W
100 L11 **Milde** ≈ C Germany
14 F14 **Mildmay** Ontario, S Canada 44°03′N 81°07′W
182 L9 **Mildura** Victoria, SE Australia 34°13′S 142°09′E
137 X12 **Mil Düzü** *Rus.* Mil'skaya Ravnina, Mil'skaya Step'. *physical region* C Azerbaijan
160 H13 **Mile** *var.* Miyang. Yunnan, SW China 24°28′N 103°26′E
Mile *see* Mili Atoll
181 Y10 **Miles** Queensland, E Australia 26°41′S 150°15′E
25 P8 **Miles** Texas, SW USA 31°36′N 100°10′W
33 X9 **Miles City** Montana, NW USA 46°24′N 105°48′W
11 U17 **Milestone** Saskatchewan, S Canada 50°00′N 104°24′W
107 N22 **Mileto** Calabria, SW Italy 38°35′N 16°03′E
107 K16 **Miletto, Monte** ▲ C Italy 41°28′N 14°21′E
18 M13 **Milford** Connecticut, NE USA 41°12′N 73°01′W
21 Y3 **Milford** *var.* Milford City. Delaware, NE USA 38°54′N 75°25′W
29 T11 **Milford** Iowa, C USA 43°19′N 95°09′W
19 S6 **Milford** Maine, NE USA 44°57′N 68°37′W
29 R16 **Milford** Nebraska, C USA 40°46′N 97°03′W
19 O10 **Milford** New Hampshire, NE USA 42°49′N 71°38′W
18 J13 **Milford** Pennsylvania, NE USA 41°20′N 74°48′W
25 T7 **Milford** Texas, SW USA 32°07′N 96°57′W
36 K6 **Milford** Utah, W USA 38°22′N 112°57′W
Milford *see* Milford Haven
Milford City *see* Milford
97 H21 **Milford Haven** *prev.* Milford. SW Wales, United Kingdom 51°44′N 05°02′W
27 O4 **Milford Lake** ⊡ Kansas, C USA
185 B21 **Milford Sound** Southland, South Island, New Zealand 44°41′S 167°57′E
185 B21 **Milford Sound** *inlet* South Island, New Zealand
Milhau *see* Millau
Milḩ, Baḩr al *see* Razāzah, Buḩayrat ar
139 T10 **Milḩ, Wādī al** *dry watercourse* S Iraq
189 W8 **Mili Atoll** *var.* Mile. *atoll* Ratak Chain, SE Marshall Islands
110 H13 **Milicz** Dolnośląskie, SW Poland 51°32′N 17°15′E
107 L25 **Militello in Val di Catania** Sicilia, Italy, C Mediterranean Sea 37°17′N 14°47′E
11 R17 **Milk River** Alberta, SW Canada 49°10′N 112°06′W
44 J13 **Milk River** ≈ C Jamaica
33 W7 **Milk River** ≈ Montana, NW USA
80 D9 **Milk, Wadi el** *var.* Wadi al Malik. ≈ C Sudan
99 L14 **Mill** Noord-Brabant, SE Netherlands 51°42′N 05°46′E
103 P14 **Millau** *var.* Milhau; *anc.* Æmilianum. Aveyron, S France 44°06′N 03°05′E
14 I14 **Millbrook** Ontario, SE Canada 44°09′N 78°26′W
23 U4 **Milledgeville** Georgia, SE USA 33°04′N 83°13′W
12 C12 **Mille Lacs, Lac des** ◎ Ontario, S Canada
29 V6 **Mille Lacs Lake** ◎ Minnesota, N USA
23 V4 **Millen** Georgia, SE USA 32°50′N 81°56′W
191 Y5 **Millennium Island** *prev.* Caroline Island, Thornton Island. *atoll* Line Islands, E Kiribati
29 O9 **Miller** South Dakota, N USA 44°31′N 98°59′W
30 K5 **Miller Dam Flowage** ◎ Wisconsin, N USA
39 U12 **Miller, Mount** ▲ Alaska, USA 60°29′N 142°16′W

◆ Country
● Country Capital
◇ Dependent Territory
○ Dependent Territory Capital
◈ Administrative Regions
✈ International Airport
▲ Mountain
▲▲ Mountain Range
⛰ Volcano
≈ River
◎ Lake
⊡ Reservoir

126 *L10* **Millerovo** Rostovskaya Oblast’, SW Russian Federation *48°57′N 40°26′E*

37 *N17* **Miller Peak** ▲ Arizona, SW USA *31°23′N 110°17′W*

31 *T12* **Millersburg** Ohio, N USA *40°33′N 81°55′W*

18 *G15* **Millersburg** Pennsylvania, NE USA *40°31′N 76°56′W*

185 *D23* **Millers Flat** Otago, South Island, New Zealand *45°42′S 169°25′E*

25 *Q8* **Millersview** Texas, SW USA *31°26′N 99°44′W*

106 *B10* **Millesimo** Piemonte, NE Italy *44°24′N 08°09′E*

12 *C12* **Milles Lacs, Lac des** ⊛ Ontario, SW Canada

25 *Q13* **Millett** Texas, SW USA *28°33′N 99°10′W*

103 *N11* **Millevaches, Plateau de** *plateau* C France

182 *K12* **Millicent** South Australia *37°29′S 140°01′E*

98 *M13* **Millingen aan den Rijn** Gelderland, SE Netherlands *51°52′N 06°02′E*

20 *E10* **Millington** Tennessee, S USA *35°20′N 89°54′W*

19 *R4* **Millinocket** Maine, NE USA *45°38′N 68°45′W*

19 *R4* **Millinocket Lake** ⊛ Maine, NE USA

195 *Z11* **Mill Island** *island* Antarctica

183 *T3* **Millmerran** Queensland, E Australia *27°53′S 151°15′E*

109 *R9* **Millstatt** Kärnten, S Austria *46°45′N 13°36′E*

97 *B19* **Milltown Malbay** *Ir.* Sráid na Cathrach. W Ireland *52°51′N 09°23′W*

18 *J17* **Millville** New Jersey, NE USA *39°24′N 75°01′W*

27 *S13* **Millwood Lake** ⊠ Arkansas, C USA

Milne Bank *see* Milne Seamounts

186 *G10* **Milne Bay** ⬖ *province* SE Papua New Guinea

64 *J8* **Milne Seamounts** *var.* Milne Bank. *undersea feature* N Atlantic Ocean

29 *Q6* **Milnor** North Dakota, N USA *46°15′N 97°27′W*

19 *R5* **Milo** Maine, NE USA *45°15′N 69°01′W*

115 *I22* **Mílos** *island* Kykládes, Greece, Aegean Sea

Mílos *see* Pláka

110 *H11* **Miłosław** Wielkopolskie, C Poland *52°13′N 17°28′E*

113 *K19* **Milot** *var.* Miloti. Lezhë, C Albania *41°42′N 19°43′E*

Miloti *see* Milot

117 *Z5* **Milove** Luhans’ka Oblast’, E Ukraine *49°22′N 40°09′E*

Milovidy *see* Milavidy

182 *L4* **Milparinka** New South Wales, SE Australia *29°48′S 141°57′E*

35 *N9* **Milpitas** California, W USA *37°25′N 121°54′W*

Mil’skaya Ravnina/ Mil’skaya Step’ *see* Mil Düzü

14 *G15* **Milton** Ontario, S Canada *43°31′N 79°53′W*

185 *E24* **Milton** Otago, South Island, New Zealand *46°08′S 169°59′E*

21 *Y4* **Milton** Delaware, NE USA *38°48′N 75°21′W*

23 *P8* **Milton** Florida, SE USA *30°37′N 87°02′W*

18 *G14* **Milton** Pennsylvania, NE USA *41°01′N 76°49′W*

18 *L7* **Milton** Vermont, NE USA *44°37′N 73°04′W*

32 *K11* **Milton-Freewater** Oregon, NW USA *45°54′N 118°24′W*

97 *N21* **Milton Keynes** SE England, United Kingdom *52°N 00°43′W*

27 *N3* **Miltonvale** Kansas, C USA *39°21′N 97°27′W*

161 *N10* **Miluo** Hunan, S China *28°52′N 113°00′E*

30 *M9* **Milwaukee** Wisconsin, N USA *43°03′N 87°56′W*

Milyang *see* Miryang

Mimatum *see* Mende

37 *Q15* **Mimbres Mountains** ▲▲ New Mexico, SW USA

182 *D2* **Mimili** South Australia *27°01′S 132°33′E*

102 *J14* **Mimizan** Landes, SW France *44°12′N 01°12′W*

Mimmaya *see* Minmaya

79 *E19* **Mimongo** Ngounié, C Gabon *01°36′S 11°44′E*

Min *see* Fujian

35 *T7* **Mina** Nevada, W USA *38°23′N 118°07′W*

143 *S14* **Mīnāb** Hormozgān, SE Iran *27°08′N 57°02′E*

Mînâ Baranîs *see* Baranīs

149 *R9* **Mīna Bāzār** Baluchistān, SW Pakistan *30°58′N 69°11′E*

Minami-Awaji *see* Nandan

165 *X17* **Minami-Iō-jima** *Eng.* San Augustine. *island* SE Japan

165 *R5* **Minami-Kayabe** Hokkaidō, NE Japan *41°54′N 140°58′E*

164 *B16* **Minamisatsuma** *var.* Kaseda. Kagoshima, Kyūshū, SW Japan *31°25′N 130°17′E*

164 *C14* **Minamishimabara** *var.* Kuchinotsu. Nagasaki, Kyūshū, SW Japan *32°36′N 130°11′E*

164 *C17* **Minamitane** Kagoshima, Tanega-shima, SW Japan *30°23′N 130°54′E*

Minami Tori Shima *see* Marcus Island

Min’an *see* Longshan

62 *J4* **Mina Pirquitas** Jujuy, NW Argentina *22°48′S 66°24′W*

173 *O3* **Mīnā’ Qābūs** NE Oman

61 *F19* **Minas** Lavalleja, S Uruguay *34°20′S 55°15′W*

13 *P15* **Minas Basin** *bay* Nova Scotia, SE Canada

61 *F17* **Minas de Corrales** Rivera, NE Uruguay *31°35′S 55°20′W*

44 *A5* **Minas de Matahambre** Pinar del Río, W Cuba *22°34′N 83°57′W*

104 *J13* **Minas de Ríotinto** Andalucía, S Spain *37°40′N 06°36′W*

60 *K7* **Minas Gerais** *off.* Estado de Minas Gerais. ⬖ *state* E Brazil

Minas Gerais, Estado de *see* Minas Gerais

42 *E5* **Minas, Sierra de las** ▲▲ E Guatemala

41 *T15* **Minatitlán** Veracruz-Llave, E Mexico *17°59′N 94°32′W*

166 *L6* **Minbu** Magway, W Myanmar (Burma) *20°09′N 94°52′E*

149 *V10* **Minchinābād** Punjab, E Pakistan *30°10′N 73°40′E*

63 *G17* **Minchinmávida, Volcán** ♨ S Chile *42°51′S 72°23′W*

96 *G7* **Minch, The** *var.* North Minch. *strait* NW Scotland, United Kingdom

106 *F8* **Mincio** *anc.* Mincius. ≈ N Italy

Mincius *see* Mincio

26 *M11* **Minco** Oklahoma, C USA *35°18′N 97°56′W*

171 *Q7* **Mindanao** *island* S Philippines

Mindanao Sea *see* Bohol Sea

101 *J23* **Mindel** ≈ S Germany

101 *J23* **Mindelheim** Bayern, S Germany *48°03′N 10°30′E*

Mindello *see* Mindelo

76 *C9* **Mindelo** *var.* Mindello; *prev.* Porto Grande. São Vicente, N Cape Verde *16°54′N 25°01′W*

14 *I13* **Minden** Ontario, SE Canada *44°54′N 78°41′W*

100 *H13* **Minden** *anc.* Minthun. Nordrhein-Westfalen, NW Germany *52°18′N 08°55′E*

22 *G5* **Minden** Louisiana, S USA *32°37′N 93°17′W*

29 *O16* **Minden** Nebraska, C USA *40°30′N 98°57′W*

35 *Q6* **Minden** Nevada, W USA *38°58′N 119°47′W*

182 *L8* **Mindona Lake** *seasonal lake* New South Wales, SE Australia

171 *O4* **Mindoro** *island* N Philippines

171 *N5* **Mindoro Strait** *strait* W Philippines

97 *J23* **Minehead** SW England, United Kingdom *51°13′N 03°29′W*

97 *E21* **Mine Head** *Ir.* Mionn Ard. *headland* S Ireland *51°58′N 07°36′W*

59 *J19* **Mineiros** Goiás, C Brazil *17°34′S 52°33′W*

25 *V6* **Mineola** Texas, SW USA *32°39′N 95°29′W*

25 *S13* **Mineral** Texas, SW USA *28°32′N 97°54′W*

127 *N15* **Mineral’nyye Vody** Stavropol’skiy Kray, SW Russian Federation *44°13′N 43°06′E*

30 *K9* **Mineral Point** Wisconsin, N USA *42°54′N 90°09′W*

25 *S6* **Mineral Wells** Texas, SW USA *32°48′N 98°06′W*

36 *K6* **Minersville** Utah, W USA *38°12′N 112°56′W*

31 *U12* **Minerva** Ohio, N USA *40°43′N 81°06′W*

107 *N17* **Minervino Murge** Puglia, SE Italy *41°06′N 16°05′E*

103 *O16* **Minervois** *physical region* S France

158 *I10* **Minfeng** *var.* Niya. Xinjiang Uygur Zizhiqu, NW China *37°07′N 82°43′E*

79 *O25* **Minga** Katanga, SE Dem. Rep. Congo *11°06′S 27°57′E*

137 *W11* **Mingäçevir** *Rus.* Mingechaur, Mingechevir. C Azerbaijan *40°46′N 47°02′E*

137 *W11* **Mingäçevir Su Anbarı** *Rus.* Mingechaurskoye Vodokhranilishche, Mingechevirskoye Vodokhranilishche. ⊠ NW Azerbaijan

166 *L8* **Mingaladon** ✈ (Yangon) Yangon, SW Myanmar (Burma) *16°55′N 96°11′E*

13 *P11* **Mingan** Québec, E Canada *50°19′N 64°02′W*

Mingāora *see* Saīdu

146 *K8* **Mingbuloq** *Rus.* Mynbulak. Navoiy Viloyati, N Uzbekistan *42°18′N 62°53′E*

146 *K9* **Mingbuloq Botig‘I** *Rus.* Vpadina Mynbulak. *depression* N Uzbekistan

Mingechaur/Mingechevir *see* Mingäçevir

Mingechaurskoye Vodokhranilishche/ Mingechevirskoye Vodokhranilishche *see* Mingäçevir Su Anbarı

161 *Q7* **Mingguang** *prev.* Jiashan. Anhui, SE China *32°45′N 117°59′E*

166 *L4* **Mingin** Sagaing, C Myanmar (Burma) *22°51′N 94°30′E*

105 *Q10* **Minglanilla** Castilla-La Mancha, C Spain *39°32′N 01°36′W*

31 *V13* **Mingo Junction** Ohio, N USA *40°19′N 80°36′W*

Mingora *see* Saīdu

163 *V7* **Mingshui** Heilongjiang, NE China *47°10′N 125°53′E*

Mingtekl Daban *see* Mintaka Pass

Mingu *see* Zhenfeng

83 *Q14* **Minguri** Nampula, NE Mozambique *14°30′S 40°37′E*

Mingzhou *see* Suide

159 *U10* **Minhe** *var.* Chuankou; *prev.* Minhe Huizu Tuzu Zizhixian, Shangchuankou. Qinghai, C China *36°21′N 102°40′E*

Minhe Huizu Tuzu Zizhixian *see* Minhe

166 *L6* **Minhla** Magway, W Myanmar (Burma) *19°58′N 95°03′E*

167 *S14* **Minh Lương** Kiên Giang, S Vietnam *09°52′N 105°10′E*

104 *G5* **Minho** *former province* N Portugal

104 *G5* **Minho, Rio** *Sp.* Miño. ≈ Portugal/Spain *see also* Miño

Minho, Rio *see* Miño

155 *C24* **Minicoy Island** *island* SW India

33 *P15* **Minidoka** Idaho, NW USA *42°45′N 113°29′W*

118 *C11* **Minija** ≈ W Lithuania

180 *G9* **Minilya** Western Australia *23°45′S 114°03′E*

14 *E8* **Minisinakwa Lake** ⊛ Ontario, S Canada

45 *T12* **Ministre Point** *headland* S Saint Lucia *13°42′N 60°57′W*

11 *V15* **Minitonas** Manitoba, S Canada *52°07′N 101°02′W*

Minius *see* Miño

186 *D7* **Minj** Jiwaka, Papua New Guinea *05°55′S 144°37′E*

161 *R12* **Min Jiang** ≈ SE China

160 *H10* **Min Jiang** ≈ C China

182 *H9* **Minlaton** South Australia *34°52′S 137°33′E*

159 *S9* **Minle** Gansu, N China

165 *Q6* **Minmaya** *var.* Mimmaya. Aomori, Honshū, C Japan *41°10′N 140°24′E*

77 *U14* **Minna** Niger, C Nigeria *09°33′N 06°33′E*

165 *P16* **Minna-jima** *island* Sakishima-shotō, SW Japan

27 *N4* **Minneapolis** Kansas, C USA *39°08′N 97°43′W*

29 *U9* **Minneapolis** Minnesota, N USA *44°59′N 93°16′W*

29 *V8* **Minneapolis-Saint Paul** ✈ Minnesota, N USA *44°53′N 93°13′W*

11 *W16* **Minnedosa** Manitoba, S Canada *50°14′N 99°50′W*

26 *J7* **Minneola** Kansas, C USA *37°26′N 100°00′W*

29 *S7* **Minnesota** *off.* State of Minnesota, *also known as* Gopher State, New England of the West, North Star State. ⬖ *state* N USA

29 *S9* **Minnesota River** ≈ Minnesota/South Dakota, N USA

29 *V9* **Minnetonka** Minnesota, N USA *44°55′N 93°28′W*

29 *O3* **Minnewaukan** North Dakota, N USA *48°04′N 99°14′W*

182 *F7* **Minnipa** South Australia *32°52′S 135°07′E*

104 *G5* **Miño** *var.* Mino, Minius, *Port.* Rio Minho. ≈ Portugal/Spain *see also* Minho, Rio

Miño *see* Minho, Rio

30 *L4* **Minocqua** Wisconsin, N USA *45°53′N 89°42′W*

30 *L12* **Minonk** Illinois, N USA *40°54′N 89°01′W*

Minorca *see* Menorca

28 *M3* **Minot** North Dakota, N USA *48°15′N 101°19′W*

159 *U8* **Minqin** Gansu, N China *38°35′N 103°07′E*

119 *J16* **Minsk** ● (Belarus) Horad Minsk, C Belarus *53°52′N 27°34′E*

Minsk-2 *see* Minsk National

Minskaya Oblast’ *see* Minskaya Voblasts’

119 *K16* **Minskaya Voblasts’** *prev. Rus.* Minskaya Oblast’. ⬖ *province* C Belarus

Minskaya Vozvyshennost’ *see* Minskaye Wzvyshsha

119 *J16* **Minskaye Wzvyshsha** *Rus.* Minskaya Vozvyshennost’. ▲▲ C Belarus

Minsk, Gorod *see* Minsk, Horad

119 *J16* **Minsk, Horad** *Russ.* Gorod Minsk. ⬖ *province* C Belarus

110 *N12* **Mińsk Mazowiecki** *var.* Nowo-Minsk. Mazowieckie, C Poland *52°10′N 21°31′E*

119 *L16* **Minsk National** *prev.* Minsk-2. ✈ Minskaya Voblasts’, C Belarus *53°52′N 27°58′E*

31 *Q13* **Minster** Ohio, N USA *40°23′N 84°22′W*

79 *F15* **Minta** Centre, C Cameroon *04°34′N 12°54′E*

149 *W2* **Mintaka Pass** *Chin.* Mingtekl Daban. *pass* China/ Pakistan

115 *D20* **Mínthi** ▲▲ S Greece

Minthun *see* Minden

13 *O14* **Minto** New Brunswick, SE Canada *46°05′N 66°05′W*

10 *H6* **Minto** Yukon, W Canada *62°33′N 136°45′W*

39 *R9* **Minto** Alaska, USA *65°07′N 149°22′W*

29 *Q3* **Minto** North Dakota, N USA *48°17′N 97°22′W*

12 *K6* **Minto, Lac** ⊛ Québec, C Canada

195 *R16* **Minto, Mount** ▲ Antarctica *71°38′S 169°11′E*

11 *U17* **Minton** Saskatchewan, S Canada *49°12′N 104°33′W*

189 *R15* **Minto Reef** *atoll* Caroline Islands, C Micronesia

37 *R4* **Minturn** Colorado, C USA *39°34′N 106°21′W*

107 *J16* **Minturno** Lazio, C Italy *41°15′N 13°47′E*

122 *K13* **Minusinsk** Krasnoyarskiy Kray, S Russian Federation *53°37′N 91°49′E*

108 *G11* **Minusio** Ticino, S Switzerland *46°11′N 08°47′E*

79 *E17* **Minvoul** Woleu-Ntem, N Gabon *02°08′N 12°12′E*

141 *R13* **Minwakh** N Yemen *16°55′N 48°04′E*

159 *V11* **Minxian** *var.* Min Xian, Minyang. Gansu, C China *34°20′N 104°09′E*

Min Xian *see* Minxian

Minya *see* Al Minyā

Minyang *see* Minxian

31 *R6* **Mio** Michigan, N USA *44°40′N 84°09′W*

Mionn Ard *see* Mine Head

Miory *see* Myory

158 *L5* **Miquan** Xinjiang Uygur Zizhiqu, NW China *44°04′N 87°40′E*

119 *I17* **Mir** Hrodzyenskaya Voblasts’, W Belarus *53°25′N 26°28′E*

106 *H8* **Mira** Veneto, NE Italy *45°25′N 12°07′E*

12 *K15* **Mirabel** *var.* Montreal. ✈ (Montréal) Québec, SE Canada *45°27′N 73°47′W*

60 *Q8* **Miracema** Rio de Janeiro, SE Brazil *21°24′S 42°10′W*

54 *G9* **Miraflores** Boyacá, C Colombia *05°07′N 73°09′W*

40 *G10* **Miraflores** Baja California Sur, NW Mexico *23°24′N 109°45′W*

44 *L9* **Miragoâne** S Haiti *18°25′N 73°07′W*

155 *E16* **Miraj** Mahārāshtra, W India *16°51′N 74°42′E*

61 *E23* **Miramar** Buenos Aires, E Argentina *38°15′S 57°50′W*

103 *R15* **Miramas** Bouches-du-Rhône, SE France *43°33′N 05°02′E*

102 *K12* **Mirambeau** Charente-Maritime, W France *45°23′N 00°33′W*

102 *L13* **Miramont-de-Guyenne** Lot-et-Garonne, SW France *44°34′N 00°20′E*

115 *L25* **Mirampéllou Kólpos** *gulf* Kríti, Greece, E Mediterranean Sea

158 *L8* **Miran** Xinjiang Uygur Zizhiqu, NW China *39°13′N 88°58′E*

54 *M5* **Miranda** *off.* Estado Miranda. ⬖ *state* N Venezuela

Miranda de Corvo *see* Miranda do Corvo

105 *O3* **Miranda de Ebro** La Rioja, N Spain *42°41′N 02°57′W*

104 *G8* **Miranda do Corvo** *var.* Miranda de Corvo. Coimbra, N Portugal *40°05′N 08°20′W*

104 *J6* **Miranda do Douro** Bragança, N Portugal *41°30′N 06°16′W*

Miranda, Estado *see* Miranda

102 *L15* **Mirande** Gers, S France *43°31′N 00°25′E*

104 *I6* **Mirandela** Bragança, N Portugal *41°28′N 07°10′W*

25 *R15* **Mirando City** Texas, SW USA *27°24′N 99°00′W*

106 *G9* **Mirandola** Emilia-Romagna, N Italy *44°52′N 11°04′E*

60 *I8* **Mirandópolis** São Paulo, S Brazil *21°10′S 51°03′W*

104 *G13* **Mira, Rio** ≈ S Portugal

60 *K8* **Mirassol** São Paulo, S Brazil *20°50′S 49°30′W*

104 *J3* **Miravalles** ▲ NW Spain *42°52′N 06°45′W*

42 *L12* **Miravalles, Volcán** ♨ NW Costa Rica *10°43′N 85°07′W*

141 *W13* **Mirbāṭ** *var.* Marbat. S Oman *17°03′N 54°44′E*

44 *M9* **Mirebalais** C Haiti *18°51′N 72°08′W*

103 *T6* **Mirecourt** Vosges, NE France *48°19′N 06°04′E*

103 *N16* **Mirepoix** Ariège, S France *43°05′N 01°51′E*

Mirgorod *see* Myrhorod

139 *W10* **Mīr Ḩājī Khalīl** Wāsiţ, E Iraq *32°11′N 46°19′E*

169 *T8* **Miri** Sarawak, East Malaysia *04°23′N 113°59′E*

77 *W12* **Miria** Zinder, S Niger *13°39′N 09°15′E*

182 *F5* **Mirikata** South Australia *29°56′S 135°13′E*

54 *K4* **Mirimire** Falcón, N Venezuela *11°14′N 68°39′W*

61 *H18* **Mirim Lagoon** *var.* Lake Mirim, *Sp.* Laguna Merín. *lagoon* Brazil/Uruguay

Mirim, Lake *see* Mirim Lagoon

Mírina *see* Mýrina

172 *H14* **Miringoni** Mohéli, S Comoros *12°17′S 93°39′E*

143 *W11* **Mīrjāveh** Sīstān va Balūchestān, SE Iran *29°04′N 61°24′E*

195 *Z9* **Mirny** *Russian research station* Antarctica *66°25′S 93°09′E*

124 *M10* **Mirnyy** Arkhangel’skaya Oblast’, NW Russian Federation *62°50′N 40°20′E*

123 *O10* **Mirnyy** Respublika Sakha (Yakutiya), NE Russian Federation *62°30′N 113°58′E*

Mironovka *see* Myronivka

110 *F9* **Mirosławiec** Zachodnio-pomorskie, NW Poland *53°21′N 16°04′E*

Mirovo *see* Vrattsa

100 *N10* **Mirow** Mecklenburg-Vorpommern, N Germany *53°16′N 12°48′E*

152 *G6* **Mīrpur** Jammu and Kashmir, NW India *33°06′N 73°49′E*

Mīrpur *see* New Mīrpur

149 *P17* **Mīrpur Batoro** Sind, SE Pakistan *24°40′N 68°15′E*

149 *Q16* **Mīrpur Khās** Sind, SE Pakistan *25°31′N 69°01′E*

149 *P17* **Mīrpur Sakro** Sind, SE Pakistan *24°32′N 67°38′E*

143 *T14* **Mīr Shahdād** Hormozgān, S Iran *26°15′N 58°29′E*

Mirtoan Sea *see* Mirtóo Pélagos

115 *G21* **Mirtóo Pélagos** *Eng.* Mirtoan Sea; *anc.* Myrtoum Mare. *sea* S Greece

163 *Z16* **Miryang** *var.* Milyang, *Jap.* Mitsuō. SE South Korea *35°30′N 128°46′E*

Mirzachirla *see* Murzechirla

164 *E14* **Misaki** Ehime, Shikoku, SW Japan *33°22′N 132°04′E*

Mīsān *see* Maysān

41 *Q13* **Misantla** Veracruz-Llave, E Mexico *19°54′N 96°51′W*

165 *R7* **Misawa** Aomori, Honshū, C Japan *40°42′N 141°26′E*

57 *G14* **Mishagua, Río** ≈ C Peru

163 *Z8* **Mishan** Heilongjiang, NE China *45°30′N 131°53′E*

31 *O11* **Mishawaka** Indiana, N USA *41°40′N 86°10′W*

39 *N6* **Misheguk Mountain** ▲ Alaska, USA *68°13′N 161°11′W*

165 *N14* **Mishima** *var.* Misima. Shizuoka, Honshū, S Japan *35°08′N 138°54′E*

164 *E12* **Mi-shima** *island* SW Japan

127 *V4* **Mishkino** Respublika Bashkortostan, W Russian Federation *55°31′N 55°57′E*

153 *Y10* **Mishmi Hills** *hill range* NE India

161 *N11* **Mi Shui** ≈ S China

Misiaf *see* Maşyāf

107 *J23* **Misilmeri** Sicilia, Italy, C Mediterranean Sea *38°03′N 13°27′E*

Misima *see* Mishima

14 *C7* **Missinaibi Lake** ⊛ Ontario, S Canada

Misión de Guana *see* Guana

60 *F13* **Misiones** *off.* Provincia de Misiones. ⬖ *province* NE Argentina

62 *P8* **Misiones** *off.* Departamento de las Misiones. ⬖ *department* S Paraguay

Misiones, Departamento de las *see* Misiones

Misiones, Provincia de *see* Misiones

Misión San Fernando *see* San Fernando

Miskin *see* Maskin

Miskito Coast *see* La Mosquitia

43 *O7* **Miskitos, Cayos** *island group* NE Nicaragua

111 *M21* **Miskolc** Borsod-Abaúj-Zemplén, NE Hungary *48°05′N 20°46′E*

171 *T12* **Misool, Pulau** *island* Papua Barat, E Indonesia

Misox *see* Mesocco

29 *Y3* **Misquah Hills** *hill range* Minnesota, N USA

75 *P7* **Mişrātah** *var.* Misurata. NW Libya *32°23′N 15°06′E*

75 *P7* **Mişrātah, Râs** *headland* N Libya *32°22′N 15°16′E*

14 *C7* **Missanabie** Ontario, S Canada *48°18′N 84°04′W*

58 *E10* **Missão Catrimani** Roraima, N Brazil *01°26′N 62°05′W*

14 *D6* **Missinaibi** ≈ Ontario, S Canada

11 *T13* **Missinipe** Saskatchewan, C Canada *55°36′N 104°45′W*

28 *M11* **Mission** South Dakota, N USA *43°16′N 100°38′W*

25 *S17* **Mission** Texas, SW USA *26°13′N 98°19′W*

12 *F10* **Missisa Lake** ⊛ Ontario, C Canada

18 *M6* **Missisquoi Bay** *lake bay* Canada/USA

14 *C10* **Mississagi** ≈ Ontario, S Canada

14 *G15* **Mississauga** Ontario, S Canada *43°38′N 79°36′W*

31 *P12* **Mississinewa Lake** ⊠ Indiana, N USA

31 *P12* **Mississinewa River** ≈ Indiana/Ohio, N USA

22 *K4* **Mississippi** *off.* State of Mississippi, *also known as* Bayou State, Magnolia State. ⬖ *state* SE USA

14 *K13* **Mississippi** ≈ Ontario, SE Canada

47 *N1* **Mississippi Fan** *undersea feature* N Gulf of Mexico *26°45′N 88°30′W*

14 *L13* **Mississippi Lake** ⊛ Ontario, SE Canada

22 *M10* **Mississippi Delta** *delta* Louisiana, S USA

0 *J11* **Mississippi River** ≈ C USA

22 *M9* **Mississippi Sound** *sound* Alabama/Mississippi, S USA

33 *P9* **Missoula** Montana, NW USA *46°54′N 114°03′W*

27 *T5* **Missouri** *off.* State of Missouri, *also known as* Bullion State, Show Me State. ⬖ *state* C USA

25 *V11* **Missouri City** Texas, SW USA *29°37′N 95°32′W*

0 *J10* **Missouri River** ≈ C USA

15 *Q6* **Mistassibi** ≈ Québec, SE Canada

15 *P6* **Mistassini** ≈ Québec, SE Canada

12 *J11* **Mistassini, Lac** ⊛ Québec, SE Canada

109 *Y3* **Mistelbach an der Zaya** Niederösterreich, NE Austria *48°34′N 16°33′E*

107 *L24* **Misterbianco** Sicilia, Italy, C Mediterranean Sea *37°31′N 15°01′E*

95 *N19* **Misterhult** Kalmar, S Sweden *57°28′N 16°34′E*

12 *K11* **Mistissini** *var.* Baie-du-Poste. Québec, SE Canada *50°20′N 73°50′W*

57 *H17* **Misti, Volcán** ♨ S Peru *16°20′S 71°22′W*

Mistras *see* Mystrás

107 *K23* **Mistretta** *anc.* Amestratus. Sicilia, Italy, C Mediterranean Sea *37°56′N 14°22′E*

164 *F12* **Misumi** Shimane, Honshū, SW Japan *34°47′N 132°00′E*

Misurata *see* Mişrātah

83 *O14* **Mitande** Niassa, N Mozambique *14°06′S 36°03′E*

40 *J13* **Mita, Punta de** *headland* C Mexico *20°46′N 105°31′W*

55 *W12* **Mitaraka, Massif du** ▲ NE South America *02°18′N 54°31′W*

Mitau *see* Jelgava

181 *X9* **Mitchell** Queensland, E Australia *26°29′S 148°00′E*

14 *E15* **Mitchell** Ontario, S Canada *43°28′N 81°11′W*

28 *I13* **Mitchell** Nebraska, C USA *41°56′N 103°48′W*

32 *J12* **Mitchell** Oregon, NW USA *44°34′N 120°09′W*

29 *P11* **Mitchell** South Dakota, N USA *43°42′N 98°01′W*

23 *P5* **Mitchell Lake** ⊠ Alabama, S USA

31 *P7* **Mitchell, Lake** ⊛ Michigan, N USA

21 *P9* **Mitchell, Mount** ▲ North Carolina, SE USA *35°46′N 82°16′W*

181 *V3* **Mitchell River** ≈ Queensland, NE Australia

97 *D20* **Mitchelstown** *Ir.* Baile Mhistéala. SW Ireland *52°20′N 08°16′W*

14 *M9* **Mitchinamécus, Lac** ⊛ Québec, SE Canada

Mitèmboni *see* Mitemele, Río

79 *D17* **Mitemele, Río** *var.* Mitèmboni, Temboni, Utamboni. ≈ S Equatorial Guinea

149 *S12* **Mithān Kot** Punjab, E Pakistan *28°53′N 70°25′E*

149 *T7* **Mitha Tiwāna** Punjab, E Pakistan *32°18′N 72°08′E*

149 *R17* **Mithi** Sind, SE Pakistan *24°43′N 69°53′E*

Míthimna *see* Míthymna

Mi Tho *see* My Tho

115 *L16* **Míthymna** *var.* Míthimna. Lésvos, E Greece *39°20′N 26°12′E*

190 *L16* **Mitiaro** *island* S Cook Islands

Mitilíni *see* Mytilíni

15 *U7* **Mitis** ≈ Québec, SE Canada

41 *R16* **Mitla** Oaxaca, SE Mexico *16°56′N 96°19′W*

165 *P13* **Mito** Ibaraki, Honshū, S Japan *36°21′N 140°26′E*

92 *N2* **Mitra, Kapp** *headland* W Svalbard *79°07′N 11°11′E*

184 *M13* **Mitre** ▲ North Island, New Zealand *40°46′S 175°27′E*

185 *B21* **Mitre Peak** ▲ South Island, New Zealand *44°37′S 167°45′E*

39 *O15* **Mitrofania Island** *island* Alaska, USA

Mitrovica/Mitrovicë *see* Kosovska Mitrovica, Serbia

Mitrovica/Mitrowitz *see* Sremska Mitrovica, Serbia

113 *M16* **Mitrovicë** *Serb.* Mitrovica, Kosovska Mitrovica, Titova Mitrovica. N Kosovo *42°54′N 20°52′E*

172 *H12* **Mitsamiouli** Grande Comore, NW Comoros *11°22′S 43°19′E*

172 *I3* **Mitsinjo** Mahajanga, NW Madagascar *16°00′S 45°52′E*

80 *J9* **Mits’iwa** *var.* Masawa, Massawa. E Eritrea *15°37′N 39°27′E*

172 *H13* **Mitsoudjé** Grande Comore, NW Comoros

138 *F12* **Mitspe Ramon** *prev.* Mizpe Ramon. Southern, S Israel *30°38′N 34°47′E*

165 *T5* **Mitsuishi** Hokkaidō, NE Japan *42°12′N 142°40′E*

165 *O11* **Mitsuke** *var.* Mituke. Niigata, Honshū, C Japan *37°30′N 138°54′E*

Mitsuō *see* Miryang

164 *C12* **Mitsushima** Nagasaki, Tsushima, SW Japan *34°16′N 129°18′E*

100 *G12* **Mittelandkanal** *canal* NW Germany

108 *J7* **Mittelberg** Vorarlberg, NW Austria *47°19′N 10°09′E*

Mitteldorf *see* Międzychód

Mittelstadt *see* Baia Sprie

Mitterburg *see* Pazin

109 *P7* **Mittersill** Salzburg, NW Austria *47°16′N 12°27′E*

Mittimatalik *see* Pond Inlet

101 *N16* **Mittweida** Sachsen, E Germany *50°59′N 12°57′E*

54 *J13* **Mitú** Vaupés, SE Colombia *01°07′N 70°05′W*

Mituke *see* Mitsuke

Mitumba, Chaîne des/ Mitumba Range *see* Mitumba, Monts

79 *O22* **Mitumba, Monts** *var.* Chaîne des Mitumba, Mitumba Range. ▲▲ E Dem. Rep. Congo

79 *N23* **Mitwaba** Katanga, SE Dem. Rep. Congo *08°37′S 27°20′E*

79 *E18* **Mitzic** Woleu-Ntem, N Gabon *0°48′N 11°30′E*

82 *K11* **Miueru Wantipa, Lake** ⊛ N Zambia

165 *N14* **Miura** Kanagawa, Honshū, S Japan *35°08′N 139°37′E*

165 *Q10* **Miyagi** *off.* Miyagi-ken. ⬖ *prefecture* Honshū, C Japan

Miyagi-ken *see* Miyagi

138 *M7* **Miyāh, Wādī al** *dry watercourse* E Syria

165 *X13* **Miyake** Tōkyō, Miyako-jima, SE Japan *34°35′N 135°33′E*

165 *Q16* **Miyake-jima** *island* Sakishima-shotō, SW Japan

165 *R8* **Miyako** Iwate, Honshū, C Japan *39°39′N 141°57′E*

164 *D16* **Miyakonojō** *var.* Miyakonzyô. Miyazaki, Kyūshū, SW Japan *31°42′N 131°04′E*

Miyakonzyô *see* Miyakonojō

165 *Q16* **Miyako-shotō** *island group* SW Japan

144 *G11* **Miyaly** Atyrau, W Kazakhstan *48°52′N 53°55′E*

Miyāndoāb *see* Mīāndowāb

Miyāneh *see* Mīāneh

Miyang *see* Mile

164 *D16* **Miyazaki** Miyazaki, Kyūshū, SW Japan *31°55′N 131°24′E*

164 *D16* **Miyazaki** *off.* Miyazaki-ken. ⬖ *prefecture* Kyūshū, SW Japan

Miyazaki-ken *see* Miyazaki

164 *J12* **Miyazu** Kyōto, Honshū, SW Japan *35°33′N 135°12′E*

Miyory *see* Myory

164 *G12* **Miyoshi** *var.* Miyosi. Hiroshima, Honshū, SW Japan *34°48′N 132°51′E*

Miyosi *see* Miyoshi

Miza *see* Mizë

81 *H14* **Mīzan Teferī** Southern Nationalities, S Ethiopia *06°57′N 35°30′E*

Mizda *see* Mizdah

75 *O8* **Mizdah** *var.* Mizda. NW Libya *31°26′N 12°59′E*

113 *K20* **Mizë** *var.* Miza. Fier, W Albania *40°58′N 19°32′E*

97 *A22* **Mizen Head** *Ir.* Carn Uí Néid. *headland* SW Ireland *51°26′N 09°50′W*

116 *H7* **Mizhhir”ya** *Rus.* Mezhgor’ye. Zakarpats’ka Oblast’, W Ukraine *48°30′N 23°30′E*

160 *L4* **Mizhi** Shaanxi, C China *37°50′N 110°03′E*

114 *H7* **Mizia** *var.* Miziya. Vratsa, NW Bulgaria *43°42′N 23°52′E*

116 *K13* **Mizil** Prahova, SE Romania *45°00′N 26°29′E*

Miziya *see* Mizia

153 *W15* **Mizo Hills** *hill range* E India

153 *W15* **Mizoram** ⬖ *state* NE India

Mizpe Ramon *see* Mitspe Ramon

57 *L19* **Mizque** Cochabamba, C Bolivia *17°57′S 65°10′W*

57 *M19* **Mizque, Río** ≈ C Bolivia

165 *Q9* **Mizusawa** *var.* Ōshū. Iwate, Honshū, C Japan *39°10′N 141°07′E*

95 *M18* **Mjölby** Östergötland, S Sweden *58°19′N 15°10′E*

95 *G15* **Mjøndalen** Buskerud, S Norway *59°45′N 09°58′E*

95 *J19* **Mjörn** ⊛ S Sweden

94 *I13* **Mjøsa** *var.* Mjøsen. ⊛ S Norway

Mjøsen *see* Mjøsa

81 *G21* **Mkalama** Singida, C Tanzania *04°09′S 34°35′E*

80 *K13* **Mkata** ≈ C Tanzania

83 *K14* **Mkushi** Central, C Zambia *13°40′S 29°26′E*

83 *L22* **Mkuze** KwaZulu/Natal, E South Africa *27°37′S 32°03′E*

81 *J22* **Mkwaja** Tanga, E Tanzania *05°42′S 38°48′E*

111 *D16* **Mladá Boleslav** *Ger.* Jungbunzlau. Středočeský Kraj, N Czech Republic *50°26′N 14°55′E*

112 *M12* **Mladenovac** Serbia, C Serbia *44°27′N 20°42′E*

114 *L11* **Mladinovo** Haskovo, S Bulgaria *41°57′N 26°13′E*

113 *O17* **Mlado Nagoričane** N FYR Macedonia *42°11′N 21°49′E*

Mlanje *see* Mulanje

112 *N12* **Mlava** ≈ E Serbia

110 *L9* **Mława** Mazowieckie, C Poland *53°07′N 20°23′E*

113 *G16* **Mljet** *It.* Meleda; *anc.* Melita. *island* S Croatia

167 *S11* **Mlu Prey** *prev.* Phumĭ Mlu Prey. Preăh Vihéar, N Cambodia *13°48′N 105°16′E*

116 *K4* **Mlyniv** Rivnens’ka Oblast’, NW Ukraine *50°31′N 25°36′E*

83 *I21* **Mmabatho** North-West, N South Africa *25°51′S 25°37′E*

83 *I19* **Mmashoro** Central, E Botswana *21°56′S 26°39′E*

44 *J7* **Moa** Holguín, E Cuba *20°42′N 74°57′W*

76 *J15* **Moa** ≈ Guinea/Sierra Leone

37 *O6* **Moab** Utah, W USA *38°35′N 109°34′W*

181 *V1* **Moa Island** *island* Queensland, NE Australia

187 *Y15* **Moala** *island* S Fiji

83 *L21* **Moamba** Maputo, SW Mozambique *25°35′S 32°13′E*

79 *F19* **Moanda** *var.* Mouanda. Haut-Ogooué, SE Gabon *01°31′S 13°07′E*

83 *M15* **Moatize** Tete, NW Mozambique *16°04′S 33°43′E*

79 *P22* **Moba** Katanga, E Dem. Rep. Congo *07°03′S 29°52′E*

Mobay *see* Montego Bay

79 *K15* **Mobaye** Basse-Kotto, S Central African Republic *04°19′N 21°17′E*

79 *K15* **Mobayi-Mbongo** Equateur, NW Dem. Rep. Congo *04°21′N 21°10′E*

25 *P2* **Mobeetie** Texas, SW USA *35°33′N 100°25′W*

27 *U3* **Moberly** Missouri, C USA *39°25′N 92°26′W*

23 *N8* **Mobile** Alabama, S USA *30°42′N 88°03′W*

23 *N9* **Mobile Bay** *bay* Alabama, S USA

23 *N8* **Mobile River** ≈ Alabama, S USA

29 *N8* **Mobridge** South Dakota, N USA *45°32′N 100°25′W*

Mobutu Sese Seko, Lac *see* Albert, Lake

45 *N8* **Moca** N Dominican Republic *19°26′N 70°33′W*

83 *Q15* **Moçambique** Nampula, NE Mozambique *15°00′S 40°44′E*

Moçâmedes *see* Namibe

167 *S6* **Môc Châu** Sơn La, N Vietnam *20°49′N 104°38′E*

187 *Z15* **Moce** *island* Lau Group, E Fiji

Mocha *see* Al Mukhā

193 *T11* **Mocha Fracture Zone** *tectonic feature* SE Pacific Ocean

63 *F14* **Mocha, Isla** *island* C Chile

56 *C12* **Moche, Río** ≈ W Peru

167 *S14* **Môc Hoa** Long An, S Vietnam *10°46′N 105°56′E*

83 *I20* **Mochudi** Kgatleng, SE Botswana *24°25′S 26°07′E*

82 *Q13* **Mocímboa da Praia** *var.* Vila de Mocímboa da Praia. Cabo Delgado, N Mozambique *11°17′S 40°21′E*

94 *L13* **Mockfjärd** Dalarna, C Sweden *60°30′N 14°57′E*

21 *R9* **Mocksville** North Carolina, SE USA *35°53′N 80°33′W*

32 *F8* **Moclips** Washington, NW USA *47°11′N 124°13′W*

82 *C13* **Môco** *var.* Morro de Môco. ▲ W Angola *12°36′S 15°09′E*

54 *D13* **Mocoa** Putumayo, SW Colombia *01°07′N 76°38′W*

60 *M8* **Mococa** São Paulo, S Brazil *21°30′S 47°00′W*

Môco, Morro de *see* Môco

40 *H8* **Mocorito** Sinaloa, C Mexico *25°24′N 107°55′W*

40 *J4* **Moctezuma** Chihuahua, N Mexico *30°10′N 106°28′W*

41 *N11* **Moctezuma** San Luis Potosí, C Mexico *22°46′N 101°06′W*

40 *G4* **Moctezuma** Sonora, NW Mexico *29°50′N 109°40′W*

41 *P12* **Moctezuma, Río** ≈ C Mexico

Mó, Cuan *see* Clew Bay

83 *O16* **Mocuba** Zambézia, NE Mozambique *16°50′S 37°02′E*

103 *U12* **Modane** Savoie, E France *45°14′N 06°41′E*

106 *F9* **Modena** *anc.* Mutina. Emilia-Romagna, N Italy *44°39′N 10°55′E*

36 *I7* **Modena** Utah, W USA *37°46′N 113°54′W*

35 *O9* **Modesto** California, W USA *37°38′N 121°02′W*

107 *L25* **Modica** *anc.* Motyca. Sicilia, Italy, C Mediterranean Sea *36°52′N 14°45′E*

83 *J20* **Modimolle** *prev.* Nylstroom. Limpopo, NE South Africa *24°39′N 28°23′E*

79 *K17* **Modjamboli** Equateur, N Dem. Rep. Congo *02°27′N 22°03′E*

109 *X4* **Mödling** Niederösterreich, NE Austria *48°06′N 16°18′E*

Modohn *see* Madona

Modot *see* Tsenhermandal

171 *V14* **Modowi** Papua Barat, E Indonesia *04°05′S 134°39′E*

112 *I12* **Modračko Jezero** ⊠ NE Bosnia and Herzegovina

112 *I10* **Modriča** Republika Srpska, N Bosnia and Herzegovina *44°57′N 18°17′E*

183 *O13* **Moe** Victoria, SE Australia *38°11′S 146°18′E*

Moearatewe *see* Muaratewe

Moei, Mae Nam *see* Thaungyin

94 *H13* **Moelv** Hedmark, S Norway *60°55′N 10°47′E*

92 *I10* **Moen** Troms, N Norway *69°08′N 18°35′E*

Möen *see* Møn, Denmark

Moen *see* Weno, Micronesia

Moena *see* Muna, Pulau

36 *M10* **Moenkopi Wash** ≈ Arizona, SW USA

185 *F22* **Moeraki Point** *headland* South Island, New Zealand *45°23′S 170°52′E*

99 *F16* **Moerbeke** Oost-Vlaanderen, NW Belgium *51°11′N 03°57′E*

99 *H14* **Moerdijk** Noord-Brabant, S Netherlands *51°42′N 04°37′E*

Moero, Lac *see* Mweru, Lake

101 *D15* **Moers** *var.* Mörs. Nordrhein-Westfalen, W Germany *51°27′N 06°36′E*

Moesi *see* Musi, Air

Moeskroen *see* Mouscron

96 *J13* **Moffat** S Scotland, United Kingdom *55°29′N 03°36′W*

185 *C22* **Moffat Peak** ▲ South Island, New Zealand *44°57′S 168°10′E*

79 *N19* **Moga** Sud-Kivu, E Dem. Rep. Congo *02°16′S 26°54′E*

152 *H8* **Moga** Punjab, N India *30°49′N 75°13′E*

Mogadiscio/Mogadishu *see* Muqdisho

Mogador *see* Essaouira

104 *J6* **Mogadouro** Bragança, N Portugal *41°20′N 06°43′W*

167 *N2* **Mogaung** Kachin State, N Myanmar (Burma) *25°20′N 96°54′E*

110 *L13* **Mogielnica** Mazowieckie, C Poland *51°40′N 20°42′E*

Mogilëv *see* Mahilyow

Mogilev-Podol’skiy *see* Mohyliv-Podil’s’kyy

Mogilëvskaya Oblast’ *see* Mahilyowskaya Voblasts’

◆ Country
● Country Capital
◇ Dependent Territory
○ Dependent Territory Capital
⬖ Administrative Regions
✈ International Airport
▲ Mountain
▲▲ Mountain Range
♨ Volcano
≈ River
⊛ Lake
⊠ Reservoir

110 I11 Mogilno Kujawsko-pomorskie, C Poland 52°39´N 17°58´E
83 Q15 Mogincual Nampula, NE Mozambique 15°33´S 40°28´E
114 E13 Moglenítsas ≈ N Greece
106 H8 Mogliano Veneto Veneto, NE Italy 45°34´N 12°14´E
113 M21 Moglicë Korçë, SE Albania 40°43´N 20°22´E
123 O13 Mogocha Zabaykal'skiy Kray, S Russian Federation 53°39´N 119°47´E
122 J11 Mogochin Tomskaya Oblast', C Russian Federation 57°42´N 83°24´E
80 F13 Mogogh Jonglei, E South Sudan 08°26´N 31°19´E
171 U12 Mogoi Papua Barat, E Indonesia 01°44´S 133°13´E
166 M4 Mogok Mandalay, C Myanmar (Burma) 22°55´N 96°29´E
37 P14 Mogollon Mountains ▲▲ New Mexico, SW USA
36 M12 Mogollon Rim cliff Arizona, SW USA
61 E23 Mogotes, Punta headland E Argentina 38°03´S 57°31´W
42 J8 Mogotón ▲ NW Nicaragua 13°45´N 86°22´W
104 I14 Moguer Andalucía, S Spain 37°15´N 06°52´W
111 J26 Mohács Baranya, SW Hungary 46°N 18°40´E
185 C20 Mohaka ≈ North Island, New Zealand
28 M2 Mohall North Dakota, N USA 48°45´N 101°30´W
Moḥammadābād see Dargaz
143 U12 Moḥammadābād-e Rigān Kermān, SE Iran 28°39´N 59°01´E
74 F6 Mohammedia prev. Fédala. NW Morocco 33°46´N 07°16´W
74 F6 Mohammed V ✈ (Casablanca) W Morocco 33°07´N 08°28´W
Mohammerah see Khorramshahr
36 H10 Mohave, Lake ⊠ Arizona/Nevada, W USA
36 I12 Mohave Mountains ▲▲ Arizona, SW USA
36 I15 Mohawk Mountains ▲▲ Arizona, SW USA
18 J10 Mohawk River ≈ New York, NE USA
163 T3 Mohe var. Xilinji. Heilongjiang, NE China 53°01´N 122°26´E
95 L20 Moheda Kronoberg, S Sweden 57°00´N 14°34´E
Mohéli see Mwali
Mohendergarh see Mahendragarh
38 K12 Mohican, Cape headland Nunivak Island, Alaska, USA 60°12´N 167°25´W
Mohn see Muhu
101 G15 Möhne ≈ W Germany
101 G15 Möhne-Stausee ⊠ W Germany
92 P2 Mohn, Kapp headland NW Svalbard 79°26´N 25°44´E
197 S14 Mohns Ridge undersea feature Greenland Sea/Norwegian Sea 72°30´N 05°00´E
57 I17 Moho Puno, SE Peru 15°21´S 69°32´W
Mohokare see Caledon
95 L17 Moholm Västra Götaland, S Sweden 58°37´N 14°04´E
36 J11 Mohon Peak ▲ Arizona, SW USA 34°55´N 113°07´W
81 J23 Mohoro Pwani, E Tanzania 08°09´S 39°10´E
Mohra see Moravice
Mohrungen see Morąg
116 M7 Mohyliv-Podil's'kyy Rus. Mogilev-Podol'skiy. Vinnyts'ka Oblast', C Ukraine 48°29´N 27°49´E
95 D17 Moi Rogaland, S Norway 58°27´N 06°32´E
Moili see Mwali
116 K11 Moinești Hung. Mojnest. Bacău, E Romania 46°27´N 26°31´E
Móinteach Mílic see Mountmellick
14 J14 Moira ≈ Ontario, SE Canada
92 G13 Mo i Rana Nordland, C Norway 66°19´N 14°10´E
153 X14 Moirāng Manipur, NE India 24°29´N 93°45´E
115 J25 Moíres Kríti, Greece, E Mediterranean Sea 35°03´N 24°51´E
118 H6 Mõisaküla Ger. Moiseküll. Viljandimaa, S Estonia 58°05´N 25°12´E
Moiseküll see Mõisaküla
15 W4 Moisie Québec, E Canada 50°12´N 66°06´W
15 W3 Moisie ≈ Québec, SE Canada
102 M14 Moissac Tarn-et-Garonne, S France 44°07´N 01°05´E
78 I13 Moïssala Mandoul, S Chad 08°21´N 17°46´E
55 O7 Moitaco Bolívar, E Venezuela 08°00´N 64°22´W
95 P15 Möja Stockholm, C Sweden 59°25´N 18°55´E
105 Q14 Mojácar Andalucía, S Spain 37°09´N 01°50´W
35 T13 Mojave California, W USA 35°03´N 118°10´W
35 V13 Mojave Desert plain California, W USA
35 V13 Mojave River ≈ California, W USA
60 L9 Moji-Mirim var. Moji-Mirim. São Paulo, S Brazil 22°26´S 46°55´W
Moji-Mirim see Moji-Mirim
113 K15 Mojkovac E Montenegro 42°57´N 19°34´E
Mojnest see Moinești
Mōka see Mooka
153 Q13 Mokāma prev. Mokameh, Mukama. Bihār, N India 25°24´N 85°55´E
Mokameh see Mokāma
79 O25 Mokambo Katanga, SE Dem. Rep. Congo 12°23´S 28°21´E
38 D9 Mōkapu Point var. Mokapu Point. headland O'ahu, Hawai'i, USA 21°27´N 157°43´W
184 L9 Mokau Waikato, North Island, New Zealand 38°42´S 174°37´E
184 L9 Mokau ≈ North Island, New Zealand
35 P7 Mokelumne River ≈ California, W USA
83 J23 Mokhotlong NE Lesotho 29°19´S 29°06´E
Mokil Atoll see Mwokil Atoll
95 N14 Möklinta Västmanland, C Sweden 60°04´N 16°34´E
Mokna see Mokra Gora
184 L4 Mokohinau Islands island group N New Zealand
153 X12 Mokokchūng Nāgāland, NE India 26°20´N 94°30´E
78 F12 Mokolo Extrême-Nord, N Cameroon 10°49´N 13°54´E
83 J20 Mokopane prev. Potgietersrus. Limpopo, NE South Africa 24°09´S 28°58´E
185 D24 Mokoreta ≈ South Island, New Zealand
163 X17 Mokpo Jap. Moppo; prev. Mokp'o. SW South Korea 34°50´N 126°26´E
Mokp'o see Mokpo
113 L16 Mokra Gora Alb. Mokna. ▲▲ S Serbia
Mokrany see Makrany
127 O5 Moksha ≈ W Russian Federation
143 X12 Mok Sukhteh-ye Pāyīn Sīstān va Balūchestān, SE Iran
Moktama see Mottama
77 T14 Mokwa Niger, W Nigeria 09°19´N 05°01´E
99 J16 Mol prev. Moll. Antwerpen, N Belgium 51°11´N 05°07´E
107 O17 Mola di Bari Puglia, SE Italy 41°03´N 17°05´E
Molai see Moláoi
41 P13 Molango Hidalgo, C Mexico 20°48´N 98°44´W
115 F22 Moláoi var. Molai. Pelopónnisos, S Greece 36°48´N 22°51´E
41 Z12 Molas del Norte, Punta var. Punta Molas. headland SE Mexico 20°34´N 86°43´W
Molas, Punta see Molas del Norte, Punta
105 R11 Molatón ▲ C Spain 38°58´N 01°19´W
97 K18 Mold NE Wales, United Kingdom 53°10´N 03°08´W
Moldau see Vltava, Czech Republic
Moldau see Moldova
Moldavia see Moldova
Moldavian SSR/Moldavskaya SSR see Moldova
94 E9 Molde Møre og Romsdal, S Norway 62°44´N 07°08´E
Moldotau, Khrebet see Moldo-Too, Khrebet
147 V9 Moldo-Too, Khrebet prev. Khrebet Moldotau. ▲▲ C Kyrgyzstan
116 L9 Moldova off. Republic of Moldova, var. Moldavia; prev. Moldavian SSR, Rus. Moldavskaya SSR. ◆ republic SE Europe
116 K9 Moldova Eng. Moldavia, Ger. Moldau. former province NE Romania
116 K9 Moldova ≈ N Romania
116 F13 Moldova Nouă Ger. Neumoldowa, Hung. Újmoldova. Caraş-Severin, SW Romania 44°45´N 21°39´E
Moldova, Republic of see Moldova
116 F13 Moldova Veche Ger. Altmoldowa, Hung. Ómoldova. Caraş-Severin, SW Romania 44°45´N 21°13´E
Moldoveanul see Vârful Moldoveanu
83 I20 Molepolole Kweneng, SE Botswana 24°25´S 25°30´E
44 L8 Môle-St-Nicolas NW Haiti 19°46´N 73°19´W
118 H13 Molėtai Utena, E Lithuania 55°14´N 25°25´E
107 O17 Molfetta Puglia, SE Italy 41°12´N 16°35´E
171 P11 Molibagu Sulawesi, N Indonesia 0°25´N 123°57´E
62 G12 Molina Maule, C Chile 35°06´S 71°18´W
105 Q7 Molina de Aragón Castilla-La Mancha, C Spain 40°50´N 01°54´W
105 R13 Molina de Segura Murcia, SE Spain 38°03´N 01°11´W
30 J11 Moline Illinois, N USA 41°30´N 90°31´W
27 P7 Moline Kansas, C USA 37°21´N 96°18´W
79 P23 Moliro Katanga, SE Dem. Rep. Congo 08°11´S 30°31´E
107 K16 Molise ◈ region S Italy
95 K15 Molkom Värmland, C Sweden 59°36´N 13°43´E
109 Q9 Möll ≈ S Austria
Moll see Mol
146 I14 Mollanepes Adyndaky Rus. Imeni Mollanepesa. Mary Welaýaty, S Turkmenistan 37°36´N 61°54´E
95 J22 Mölle Skåne, S Sweden 56°15´N 12°19´E
57 H18 Mollendo Arequipa, SW Peru 17°02´S 72°01´W
105 U5 Mollerussa Cataluña, NE Spain 41°37´N 00°53´E
108 H8 Mollis Glarus, NE Switzerland 47°05´N 09°03´E
95 J19 Mölndal Västra Götaland, S Sweden 57°39´N 12°05´E
95 J19 Mölnlycke Västra Götaland, S Sweden 57°39´N 12°09´E
117 U9 Molochans'k Rus. Molochansk. Zaporiz'ka Oblast', SE Ukraine 47°10´N 35°38´E
117 U10 Molochna Rus. Molochnaya. ≈ S Ukraine
Molochnaya see Molochna
117 U10 Molochnyy Lyman bay N Black Sea
Molodechno/Molodeczno see Maladzyechna
195 V3 Molodezhnaya Russian research station Antarctica 67°33´S 46°12´E
124 J14 Mologa ≈ NW Russian Federation
38 E9 Moloka'i var. Molokai. island Hawaiian Islands, Hawai'i, USA
175 X3 Molokai Fracture Zone tectonic feature NE Pacific Ocean
124 K15 Molokovo Tverskaya Oblast', W Russian Federation 58°10´N 36°43´E
125 Q14 Moloma ≈ NW Russian Federation
183 R8 Molong New South Wales, SE Australia 33°07´S 148°52´E
83 H21 Molopo seasonal river Botswana/South Africa
115 F17 Mólos Stereá Elláda, C Greece 38°48´N 22°39´E
171 O11 Molosipat Sulawesi, N Indonesia 0°28´N 121°08´E
Molotov see Severodvinsk, Arkhangel'skaya Oblast', Russian Federation
Molotov see Perm', Permskaya Oblast', Russian Federation
79 G17 Moloundou Est, SE Cameroon 02°03´N 15°14´E
103 U5 Molsheim Bas-Rhin, NE France 48°33´N 07°30´E
11 X13 Molson Lake ◎ Manitoba, C Canada
Moluccas see Maluku
171 Q12 Molucca Sea Ind. Laut Maluku. sea E Indonesia
Molukken see Maluku
83 O15 Molumbo Zambézia, N Mozambique 15°33´S 36°19´E
171 T15 Molu, Pulau island Maluku, E Indonesia
83 P16 Moma Nampula, NE Mozambique 16°42´S 39°12´E
171 X14 Momats ≈ Papua, E Indonesia
42 J11 Mombacho, Volcán 🌋 SW Nicaragua 11°49´N 85°58´W
81 K21 Mombasa Mombasa, SE Kenya 04°04´N 39°40´E
81 K21 Mombasa ◈ county SE Kenya
81 J21 Mombasa ✈ Mombasa, SE Kenya 04°01´S 39°31´E
Mombetsu see Monbetsu
114 J12 Momchilgrad prev. Mastanli. Kardzhali, S Bulgaria 41°33´N 25°25´E
99 F23 Momignies Hainaut, S Belgium 50°02´N 04°10´E
54 E6 Momil Córdoba, NW Colombia 09°15´N 75°40´W
42 I10 Momotombo, Volcán 🌋 W Nicaragua 12°25´N 86°33´W
56 B5 Mompiche, Ensenada de bay NW Ecuador
79 K18 Mompono Equateur, NW Dem. Rep. Congo 0°11´N 21°31´E
54 F6 Mompós Bolívar, NW Colombia 09°15´N 74°29´W
95 J24 Møn prev. Møen. island SE Denmark
36 L4 Mona Utah, W USA 39°49´N 111°52´W
Mona, Canal de la see Mona Passage
96 E8 Monach Islands island group NW Scotland, United Kingdom
103 V14 Monaco var. Monaco-Ville; anc. Monoecus. ● (Monaco) S Monaco 43°46´N 07°23´E
103 V14 Monaco off. Principality of Monaco. ◆ monarchy W Europe
Monaco see München
Monaco Basin see Canary Basin
Monaco, Principality of see Monaco
Monaco-Ville see Monaco
96 I9 Monadhliath Mountains ▲▲ N Scotland, United Kingdom
55 O6 Monagas off. Estado Monagas. ◈ state NE Venezuela
Monagas, Estado see Monagas
97 F16 Monaghan Ir. Muineachán. Monaghan, N Ireland 54°15´N 06°58´W
97 E16 Monaghan Ir. Muineachán. cultural region N Ireland
43 S16 Monagrillo Herrera, S Panama 08°00´N 80°28´W
24 L8 Monahans Texas, SW USA 31°35´N 102°54´W
45 Q9 Mona, Isla island W Puerto Rico
45 Q9 Mona Passage Sp. Canal de la Mona. channel Dominican Republic/Puerto Rico
43 O14 Mona, Punta headland E Costa Rica 09°44´N 82°48´W
155 K25 Monaragala Uva Province, SE Sri Lanka 06°52´N 81°22´E
33 S9 Monarch Montana, NW USA 47°04´N 110°51´W
10 H14 Monarch Mountain ▲ British Columbia, SW Canada 51°59´N 125°56´W
Monasterio see Monesterio
Monasterzyska see Monastyrys'ka
Monastir see Bitola
Monastyriska see Monastyrys'ka
117 O7 Monastyryshche Cherkas'ka Oblast', C Ukraine 48°59´N 29°47´E
116 J6 Monastyrys'ka Pol. Monasterzyska, Rus. Monastyriska. Ternopil's'ka Oblast', W Ukraine 49°05´N 25°10´E
79 E15 Monatélé Centre, SW Cameroon 04°16´N 11°12´E
165 U2 Monbetsu var. Mombetsu, Monbetu. Hokkaidō, NE Japan 44°23´N 143°22´E
Monbetu see Monbetsu
106 B8 Moncalieri Piemonte, NW Italy 45°00´N 07°41´E
104 G4 Monção Viana do Castelo, N Portugal 42°03´N 08°29´W
105 Q5 Moncayo ▲ N Spain 41°43´N 01°51´W
105 Q5 Moncayo, Sierra del ▲▲ N Spain
124 J4 Monchegorsk Murmanskaya Oblast', NW Russian Federation 67°56´N 32°47´E
101 D15 Mönchengladbach prev. München-Gladbach. Nordrhein-Westfalen, W Germany 51°12´N 06°35´E
104 F14 Monchique Faro, S Portugal 37°19´N 08°33´W
104 G14 Monchique, Serra de ▲▲ S Portugal
21 S14 Moncks Corner South Carolina, SE USA 33°12´N 80°00´W
41 N7 Monclova Coahuila, NE Mexico 26°55´N 101°25´W
Moncorvo see Torre de Moncorvo
13 P14 Moncton New Brunswick, SE Canada 46°04´N 64°50´W
104 F8 Mondego, Cabo headland N Portugal 40°10´N 08°58´W
104 G8 Mondego, Rio ≈ N Portugal
104 I2 Mondoñedo Galicia, NW Spain 43°25´N 07°22´W
99 N25 Mondorf-les-Bains Grevenmacher, SE Luxembourg 49°30´N 06°16´E
102 M7 Mondoubleau Loir-et-Cher, C France 48°00´N 00°49´E
106 B9 Mondovì Piemonte, NW Italy 44°23´N 07°56´E
30 J6 Mondovi Wisconsin, N USA 44°34´N 91°40´W
Mondragón see Arrasate
107 J17 Mondragone Campania, S Italy 41°07´N 13°53´E
109 R5 Mondsee ◎ N Austria
115 G22 Monemvasiá var. Monemvasía. Pelopónnisos, S Greece 36°22´N 23°03´E
18 B15 Monessen Pennsylvania, NE USA 40°07´N 79°51´W
104 J12 Monesterio var. Monasterio. Extremadura, W Spain 38°05´N 06°16´W
14 L8 Monet Québec, SE Canada 48°09´N 75°37´W
27 S8 Monett Missouri, C USA 36°55´N 93°55´W
27 X9 Monette Arkansas, C USA 35°53´N 90°20´W
14 G11 Monetville Ontario, S Canada 46°08´N 80°24´W
106 J7 Monfalcone Friuli-Venezia Giulia, NE Italy 45°49´N 13°32´E
104 H10 Monforte Portalegre, C Portugal 39°03´N 07°26´W
104 I4 Monforte de Lemos Galicia, NW Spain 42°32´N 07°30´W
79 L16 Monga Orientale, N Dem. Rep. Congo 04°12´N 22°49´E
81 I24 Monga Lindi, SE Tanzania 09°05´S 37°51´E
81 F15 Mongalla Central Equatoria, S South Sudan 05°12´N 31°42´E
153 U11 Mongar E Bhutan 27°16´N 91°07´E
167 U6 Mong Cai var. Hai Ninh. Quang Ninh, N Vietnam 21°33´N 107°56´E
180 I11 Mongers Lake salt lake Western Australia
186 K8 Mongga Kolombangara, NW Solomon Islands 07°51´S 157°00´E
167 O6 Möng Hpayak Shan State, E Myanmar (Burma) 20°56´N 100°00´E
Monghyr see Munger
106 B10 Mongioie ▲ NW Italy 44°13´N 07°46´E
153 T16 Mongla var. Mungla. Khulna, S Bangladesh 22°18´N 89°34´E
188 C15 Mongmong C Guam
167 N6 Möng Nai Shan State, E Myanmar (Burma) 20°28´N 97°51´E
78 I11 Mongo Guéra, C Chad 12°12´N 18°40´E
76 I14 Mongo ≈ N Sierra Leone
163 I8 Mongolia Mong. Mongol Uls. ◆ republic E Asia
129 V8 Mongolia, Plateau of plateau E Mongolia
Mongolküre see Zhaosu
Mongol Uls see Mongolia
79 E17 Mongomo E Equatorial Guinea 01°39´N 11°18´E
162 M7 Möngönmorit var. Bulag. Töv, C Mongolia 48°09´N 108°33´E
77 Y12 Mongonu var. Monguno. Borno, NE Nigeria 12°42´N 13°37´E
Mongora see Saidu
78 K11 Mongororo Sila, SE Chad 12°03´N 22°26´E
Mongos, Chaîne des see Bongo, Massif des
79 I16 Mongoumba Lobaye, SW Central African Republic 03°39´N 18°30´E
Mongrove, Punta see Cayacal, Punta
83 G15 Mongu Western, W Zambia 15°13´S 23°09´E
76 I10 Mônguel Gorgol, SW Mauritania 16°25´N 13°08´W
Monguno see Mongonu
167 N4 Möng Yai Shan State, E Myanmar (Burma) 22°25´N 98°02´E
167 O5 Möng Yang Shan State, E Myanmar (Burma) 21°52´N 99°31´E
167 N3 Möng Yu Shan State, E Myanmar (Burma) 24°00´N 97°57´E
Mönhbulag see Yösöndzüyl
163 O8 Mönhhaan var. Bayasgalant. Sühbaatar, E Mongolia 46°55´N 112°11´E
162 E7 Mönhhayrhan var. Tsenher. Hovd, W Mongolia 47°07´N 92°04´E
Mönh Saridag see Munku-Sardyk, Gora
186 P9 Moni ≈ S Papau New Guinea
115 I15 Moní Megístis Lávras monastery Kentrikí Makedonía, N Greece
115 F18 Moní Osíou Loúkas monastery Stereá Elláda, C Greece
54 F9 Moniquirá Boyacá, C Colombia 05°54´N 73°35´W
103 Q12 Monistrol-sur-Loire Haute-Loire, C France 45°19´N 04°12´E
35 V7 Monitor Range ▲▲ Nevada, W USA
115 I14 Moní Vatopedíou monastery Kentrikí Makedonía, N Greece
Monkchester see Newcastle upon Tyne
83 N14 Monkey Bay Southern, SE Malawi 14°09´S 34°53´E
43 N11 Monkey Point var. Punta Mico, Punte Mono, Punto Mico. headland SE Nicaragua 11°37´N 83°39´W
Monkey River see Monkey River Town
42 G3 Monkey River Town var. Monkey River. Toledo, SE Belize 16°22´N 88°29´W
14 M13 Monkland Ontario, SE Canada 45°11´N 74°51´W
79 J19 Monkoto Equateur, NW Dem. Rep. Congo 01°39´S 20°41´E
97 K21 Monmouth Wel. Trefynwy. SE Wales, United Kingdom 51°50´N 02°43´W
30 J12 Monmouth Illinois, N USA 40°54´N 90°39´W
32 F12 Monmouth Oregon, NW USA 44°51´N 123°13´W
97 K21 Monmouth cultural region SE Wales, United Kingdom
98 I10 Monnickendam Noord-Holland, C Netherlands 52°28´N 05°02´E
77 R15 Mono ≈ C Togo
Monoecus see Monaco
35 R8 Mono Lake ◎ California, W USA
115 O23 Monólithos Ródos, Dodekánisa, Greece, Aegean Sea 36°08´N 27°45´E
19 Q12 Monomoy Island island Massachusetts, NE USA
31 O12 Monon Indiana, N USA 40°52´N 86°54´W
29 Y12 Monona Iowa, C USA 43°03´N 91°23´W
30 L9 Monona Wisconsin, N USA 43°03´N 89°18´W
18 B15 Monongahela Pennsylvania, NE USA 40°10´N 79°54´W
18 B16 Monongahela River ≈ NE USA
107 P17 Monopoli Puglia, SE Italy 40°57´N 17°18´E
Mono, Punte see Monkey Point
111 K23 Monor Pest, C Hungary 47°21´N 19°27´E
Monostor see Beli Manastir
78 K8 Monou Ennedi-Ouest, NE Chad 16°22´N 22°15´E
105 S12 Monóvar Cat. Monòver. Valenciana, E Spain 38°26´N 00°50´W
Monòver see Monóvar
105 R7 Monreal del Campo Aragón, NE Spain 40°47´N 01°20´W
107 I23 Monreale Sicilia, Italy, C Mediterranean Sea 38°05´N 13°17´E
23 T3 Monroe Georgia, SE USA 33°47´N 83°42´W
29 W14 Monroe Iowa, C USA 41°31´N 93°06´W
22 I5 Monroe Louisiana, S USA 32°32´N 92°06´W
31 S10 Monroe Michigan, N USA 41°55´N 83°24´W
18 K13 Monroe New York, NE USA 41°18´N 74°09´W
21 S11 Monroe North Carolina, SE USA 35°00´N 80°35´W
36 L6 Monroe Utah, W USA 38°37´N 112°07´W
32 H7 Monroe Washington, NW USA 47°51´N 121°58´W
30 L9 Monroe Wisconsin, N USA 42°35´N 89°39´W
27 V3 Monroe City Missouri, C USA 39°39´N 91°43´W
31 O15 Monroe Lake ⊠ Indiana, N USA
23 O7 Monroeville Alabama, S USA 31°31´N 87°19´W
18 C15 Monroeville Pennsylvania, NE USA 40°24´N 79°44´W
76 J16 Monrovia ● (Liberia) W Liberia 06°18´N 10°48´W
76 J16 Monrovia ✈ W Liberia 06°22´N 10°50´W
105 T7 Monroyo Aragón, NE Spain 40°47´N 00°03´W
99 F20 Mons Dut. Bergen. Hainaut, S Belgium 50°28´N 03°58´E
104 I8 Monsanto Castelo Branco, C Portugal 40°02´N 07°07´W
106 H8 Monselice Veneto, NE Italy 45°15´N 11°47´E
166 M9 Mon State ◈ state S Myanmar (Burma)
98 G12 Monster Zuid-Holland, W Netherlands 52°01´N 04°10´E
95 N20 Mönsterås Kalmar, S Sweden 57°03´N 16°27´E
101 F17 Montabaur Rheinland-Pfalz, W Germany 50°25´N 07°48´E
106 G8 Montagnana Veneto, NE Italy 45°14´N 11°31´E
35 N1 Montague California, W USA 41°43´N 122°31´W
25 S5 Montague Texas, SW USA 33°40´N 97°44´W
183 S11 Montague Island island New South Wales, SE Australia
39 S12 Montague Island island Alaska, USA
39 S13 Montague Strait strait N Gulf of Alaska
102 J8 Montaigu Vendée, NW France 46°58´N 01°18´W
Montaigu see Scherpenheuvel
105 S7 Montalbán Aragón, NE Spain 40°49´N 00°48´W
106 G13 Montalcino Toscana, C Italy 43°01´N 11°34´E
104 H5 Montalegre Vila Real, N Portugal 41°49´N 07°48´W
114 G8 Montana prev. Ferdinand, Mikhaylovgrad. Montana, NW Bulgaria 43°25´N 23°14´E
108 D10 Montana Valais, SW Switzerland 46°23´N 07°29´E
39 R11 Montana Alaska, USA 62°06´N 150°03´W
114 G8 Montana ◈ province NW Bulgaria
33 T9 Montana off. State of Montana, also known as Mountain State, Treasure State. ◈ state NW USA
104 J10 Montánchez Extremadura, W Spain 39°15´N 06°07´W
Montañita see La Montañita
15 Q8 Mont-Apica Québec, SE Canada 47°57´N 71°24´W
104 G10 Montargil Portalegre, C Portugal 39°05´N 08°10´W
104 G10 Montargil, Barragem de ⊠ C Portugal
103 O7 Montargis Loiret, C France 48°N 02°44´E
103 O4 Montataire Oise, N France 49°16´N 02°24´E
102 M14 Montauban Tarn-et-Garonne, S France 44°01´N 01°20´E
19 N14 Montauk Long Island, New York, NE USA 41°01´N 71°58´W
19 N14 Montauk Point headland Long Island, New York, NE USA 41°04´N 71°51´W
103 Q7 Montbard Côte d'Or, C France 47°35´N 04°25´E
103 U7 Montbéliard Doubs, E France 47°31´N 06°49´E
25 W11 Mont Belvieu Texas, SW USA 29°51´N 94°53´W
105 U6 Montblanc prev. Montblanch. Cataluña, NE Spain 41°23´N 01°10´E
Montblanch see Montblanc
103 Q11 Montbrison Loire, E France 45°37´N 04°04´E
Montcalm, Lake see Dogai Coring
103 Q9 Montceau-les-Mines Saône-et-Loire, C France 46°40´N 04°19´E
103 U12 Mont Cenis, Col du pass E France
102 K15 Mont-de-Marsan Landes, SW France 43°54´N 00°30´W
103 O3 Montdidier Somme, N France 49°39´N 02°35´E
187 Q17 Mont-Dore Province Sud, S New Caledonia 22°18´S 166°34´E
20 K10 Monteagle Tennessee, S USA 35°15´N 85°47´W
57 M20 Monteagudo Chuquisaca, S Bolivia 19°48´S 63°57´W
41 R16 Monte Albán ruins Oaxaca, S Mexico
105 R11 Montealegre del Castillo Castilla-La Mancha, C Spain 38°48´N 01°18´W
59 N18 Monte Azul Minas Gerais, SE Brazil 15°13´S 42°53´W
14 M12 Montebello Québec, SE Canada 45°40´N 74°56´W
106 H7 Montebelluna Veneto, NE Italy 45°46´N 12°03´E
60 G13 Montecarlo Misiones, NE Argentina 26°38´S 54°45´W
61 D16 Monte Caseros Corrientes, NE Argentina 30°15´S 57°39´W
60 J13 Monte Castelo Santa Catarina, S Brazil 26°34´S 50°12´W
106 F11 Montecatini Terme Toscana, C Italy 43°53´N 10°46´E
42 H7 Montecillos, Cordillera de ▲▲ W Honduras
62 I12 Monte Comán Mendoza, W Argentina 34°35´S 67°53´W
44 M8 Monte Cristi var. San Fernando de Monte Cristi. NW Dominican Republic 19°52´N 71°39´W
58 C13 Monte Cristo Amazonas, W Brazil 03°14´S 68°00´W
107 E14 Montecristo, Isola di island Archipelago Toscano, C Italy
Monte Croce Carnico, Passo di see Plöcken Pass
58 J12 Monte Dourado Pará, NE Brazil 0°48´S 52°32´W
40 L11 Monte Escobedo Zacatecas, C Mexico 22°19´N 103°30´W
106 I13 Montefalco Umbria, C Italy 42°54´N 12°40´E
107 H14 Montefiascone Lazio, C Italy 42°33´N 12°01´E
105 N14 Montefrío Andalucía, S Spain 37°19´N 04°00´W
44 I11 Montego Bay var. Mobay. W Jamaica 18°28´N 77°55´W
Montego Bay see Sangster
104 J8 Montehermoso Extremadura, W Spain 40°05´N 06°20´W
104 F10 Montejunto, Serra de ▲ C Portugal 39°10´N 09°01´W
Monteleone di Calabria see Vibo Valentia
54 E7 Montelíbano Córdoba, NW Colombia 08°02´N 75°29´W
103 R13 Montélimar anc. Acunum Acusio, Montilium Adhemari. Drôme, E France 44°33´N 04°45´E
104 K15 Montellano Andalucía, S Spain 37°00´N 05°34´W
35 Y2 Montello Nevada, W USA 41°18´N 114°10´W
30 L8 Montello Wisconsin, N USA 43°47´N 89°20´W
63 J18 Montemayor, Meseta de plain SE Argentina
41 O9 Montemorelos Nuevo León, NE Mexico 25°10´N 99°52´W
104 G11 Montemor-o-Novo Évora, S Portugal 38°38´N 08°13´W
104 G8 Montemor-o-Velho var. Montemor-o-Vélho. Coimbra, N Portugal 40°11´N 08°41´W
Montemor-o-Vélho see Montemor-o-Velho
104 H7 Montemuro, Serra de ▲ N Portugal 40°59´N 07°59´W
102 K12 Montendre Charente-Maritime, W France 45°17´N 00°24´W
61 I15 Montenegro Rio Grande do Sul, S Brazil 29°40´S 51°32´W
113 J16 Montenegro Serb. Crna Gora. ◆ republic SW Europe
62 G10 Monte Patria Coquimbo, N Chile 30°40´S 71°00´W
45 O9 Monte Plata E Dominican Republic 18°50´N 69°47´W
83 P14 Montepuez Cabo Delgado, N Mozambique 13°09´S 39°00´E
83 P14 Montepuez ≈ N Mozambique
106 G13 Montepulciano Toscana, C Italy 43°02´N 11°51´E
62 L6 Monte Quemado Santiago del Estero, N Argentina 25°48´S 62°52´W
103 O6 Montereau-Faut-Yonne anc. Condate. Seine-St-Denis, N France 48°23´N 02°57´E
35 N11 Monterey California, W USA 36°36´N 121°53´W
20 L9 Monterey Tennessee, S USA 36°09´N 85°16´W
21 T5 Monterey Virginia, NE USA 38°24´N 79°36´W
Monterey see Monterrey
35 N10 Monterey Bay bay California, W USA
54 D6 Montería Córdoba, NW Colombia 08°45´N 75°54´W
57 N18 Montero Santa Cruz, C Bolivia 17°20´S 63°15´W
62 J7 Monteros Tucumán, C Argentina 27°12´S 65°30´W
104 I5 Monterrei Galicia, NW Spain 41°56´N 07°27´W
41 O8 Monterrey var. Monterey. Nuevo León, NE Mexico 25°41´N 100°16´W
32 F9 Montesano Washington, NW USA 46°58´N 123°37´W
107 M19 Montesano sulla Marcellana Campania, S Italy 40°15´N 15°41´E
107 N16 Monte Sant' Angelo Puglia, SE Italy 41°43´N 15°58´E
59 O16 Monte Santo Bahia, E Brazil 10°25´S 39°18´W
107 D18 Monte Santu, Capo di headland Sardegna, Italy, C Mediterranean Sea 40°05´N 09°43´E
59 M19 Montes Claros Minas Gerais, SE Brazil 16°45´S 43°52´W
107 K14 Montesilvano Marina Abruzzo, C Italy 42°28´N 14°07´E
23 P4 Montevallo Alabama, S USA 33°06´N 86°51´W
106 G12 Montevarchi Toscana, C Italy 43°32´N 11°34´E
61 F20 Montevideo ● (Uruguay) Montevideo, S Uruguay 34°55´S 56°10´W
29 S9 Montevideo Minnesota, N USA 44°56´N 95°43´W
37 S7 Monte Vista Colorado, C USA 37°33´N 106°08´W
23 T5 Montezuma Georgia, SE USA 32°18´N 84°01´W
29 W14 Montezuma Iowa, C USA 41°35´N 92°31´W
26 J6 Montezuma Kansas, C USA 37°33´N 100°25´W
103 U12 Montgenèvre, Col de pass France/Italy
97 K20 Montgomery E Wales, United Kingdom 52°38´N 03°05´W
23 Q5 Montgomery state capital Alabama, S USA 32°22´N 86°18´W
29 V9 Montgomery Minnesota, N USA 44°26´N 93°34´W
18 G13 Montgomery Pennsylvania, NE USA 41°08´N 76°52´W
21 Q5 Montgomery West Virginia, NE USA 38°07´N 81°19´W
97 K19 Montgomery cultural region E Wales, United Kingdom
Montgomery see Sāhīwāl
27 V4 Montgomery City Missouri, C USA 38°57´N 91°27´W
35 S8 Montgomery Pass pass Nevada, W USA
102 K12 Montguyon Charente-Maritime, W France 45°12´N 00°13´W
108 C10 Monthey Valais, SW Switzerland 46°15´N 06°56´E
27 V13 Monticello Arkansas, C USA 33°38´N 91°49´W
23 T4 Monticello Florida, SE USA 30°33´N 83°52´W
23 T8 Monticello Georgia, SE USA 33°18´N 83°40´W
30 M13 Monticello Illinois, N USA 40°01´N 88°34´W
31 O12 Monticello Indiana, N USA 40°45´N 86°46´W
29 Y13 Monticello Iowa, C USA 42°14´N 91°11´W
20 L7 Monticello Kentucky, S USA 36°50´N 84°50´W
29 V8 Monticello Minnesota, N USA 45°19´N 93°45´W
22 K7 Monticello Mississippi, S USA 31°33´N 90°06´W
27 V2 Monticello Missouri, C USA 40°07´N 91°42´W
18 J12 Monticello New York, NE USA 41°39´N 74°41´W
37 O7 Monticello Utah, W USA 37°52´N 109°20´W
106 F8 Montichiari Lombardia, N Italy 45°24´N 10°27´E
102 M12 Montignac Dordogne, SW France 45°24´N 00°54´E
99 G21 Montignies-le-Tilleul var. Montigny-le-Tilleul. Hainaut, S Belgium 50°22´N 04°23´E
14 J8 Montigny, Lac de ◎ Québec, SE Canada
103 S6 Montigny-le-Roi Haute-Marne, N France 48°02´N 05°28´E
Montigny-le-Tilleul see Montignies-le-Tilleul
43 R16 Montijo Veraguas, S Panama 07°59´N 80°58´W
104 F11 Montijo Setúbal, W Portugal 38°42´N 08°59´W
104 J11 Montijo Extremadura, W Spain 38°55´N 06°38´W
Montilium Adhemari see Montélimar
104 M13 Montilla Andalucía, S Spain 37°36´N 04°39´W
102 L3 Montivilliers Seine-Maritime, N France 49°31´N 00°10´E
15 U7 Mont-Joli Québec, SE Canada 48°36´N 68°14´W
14 M10 Mont-Laurier Québec, SE Canada 46°33´N 75°31´W
15 X5 Mont-Louis Québec, SE Canada 49°15´N 65°46´W
103 N17 Mont-Louis var. Mont Louis. Pyrénées-Orientales, S France 42°30´N 02°08´E
103 O10 Montluçon Allier, C France 46°21´N 02°37´E
15 R10 Montmagny Québec, SE Canada 47°00´N 70°31´W
103 S3 Montmédy Meuse, NE France 49°31´N 05°21´E
103 P5 Montmirail Marne, N France 48°53´N 03°31´E
15 R9 Montmorency ≈ Québec, SE Canada
102 M10 Montmorillon Vienne, W France 46°26´N 00°52´E
107 J14 Montorio al Vomano Abruzzo, C Italy 42°31´N 13°39´E
104 M13 Montoro Andalucía, S Spain 38°00´N 04°21´W
33 S16 Montpelier Idaho, NW USA 42°19´N 111°18´W
29 P6 Montpelier North Dakota, N USA 46°40´N 98°34´W
18 M7 Montpelier state capital Vermont, NE USA 44°16´N 72°32´W
103 Q15 Montpellier Hérault, S France 43°37´N 03°52´E
102 L12 Montpon-Ménestérol Dordogne, SW France 45°01´N 00°10´E
12 K15 Montréal Eng. Montreal. Québec, SE Canada 45°30´N 73°36´W
14 G8 Montreal ≈ Ontario, S Canada
Montreal see Mirabel
11 T14 Montreal Lake ◎ Saskatchewan, C Canada
14 B9 Montreal River Ontario, S Canada 47°13´N 84°36´W
103 N2 Montreuil Pas-de-Calais, N France 50°28´N 01°46´E
102 K8 Montreuil-Bellay Maine-et-Loire, NW France 47°07´N 00°10´W
108 C10 Montreux Vaud, SW Switzerland 46°27´N 06°55´E
108 B9 Montricher Vaud, W Switzerland 46°37´N 06°24´E
96 K10 Montrose E Scotland, United Kingdom 56°43´N 02°29´W
27 W14 Montrose Arkansas, C USA 33°18´N 91°29´W
37 Q6 Montrose Colorado, C USA 38°29´N 107°53´W

◆ Country
● Country Capital
◇ Dependent Territory
○ Dependent Territory Capital
◈ Administrative Regions
✈ International Airport
▲ Mountain
▲▲ Mountain Range
🌋 Volcano
≈ River
◎ Lake
⊠ Reservoir

29 Y16 **Montrose** Iowa, C USA 40°31′N 91°24′W
18 H12 **Montrose** Pennsylvania, NE USA 41°49′N 75°53′W
21 X5 **Montross** Virginia, NE USA 38°04′N 76°51′W
15 O12 **Mont-St-Hilaire** Québec, SE Canada 45°34′N 73°10′W
103 S3 **Mont-St-Martin** Meurthe-et-Moselle, NE France 49°31′N 05°51′E
45 V10 **Montserrat** *var.* Emerald Isle. ◇ *UK dependent territory* E West Indies
105 V5 **Montserrat** ▲ NE Spain 41°39′N 01°44′E
104 M7 **Montuenga** Castilla y León, N Spain 41°04′N 04°38′W
99 M19 **Montzen** Liège, E Belgium 50°42′N 05°59′E
37 N8 **Monument Valley** *valley* Arizona/Utah, SW USA
166 L4 **Monywa** Sagaing, C Myanmar (Burma) 22°05′N 95°12′E
106 D7 **Monza** Lombardia, N Italy 45°35′N 09°16′E
83 J15 **Monze** Southern, S Zambia 16°20′S 27°29′E
105 T5 **Monzón** Aragón, NE Spain 41°54′N 00°12′E
25 T9 **Moody** Texas, SW USA 31°18′N 97°21′W
98 L13 **Mook** Limburg, SE Netherlands 51°45′N 05°52′E
165 O12 **Mooka** *var.* Môka. Tochigi, Honshū, S Japan 36°27′N 139°59′E
182 K3 **Moomba** South Australia 28°07′S 140°12′E
14 G13 **Moon** ≈ Ontario, S Canada
Moon *see* Muhu
181 Y10 **Moonie** Queensland, E Australia 27°46′S 150°22′E
193 O5 **Moonless Mountains** *undersea feature* E Pacific Ocean 30°40′N 140°00′W
182 L13 **Moonlight Head** *headland* Victoria, SE Australia 38°47′S 143°12′E
Moon-Sund *see* Väinameri
182 H8 **Moonta** South Australia 34°03′S 137°36′E
Moor *see* Mór
180 I12 **Moora** Western Australia 30°23′S 116°05′E
98 H12 **Moordrecht** Zuid-Holland, C Netherlands 51°59′N 04°40′E
33 T9 **Moore** Montana, NW USA 47°00′N 109°40′W
27 N11 **Moore** Oklahoma, C USA 35°21′N 97°30′W
25 R12 **Moore** Texas, SW USA 29°03′N 99°01′W
191 S10 **Moorea** *island* Îles du Vent, W French Polynesia
21 U3 **Moorefield** West Virginia, NE USA 39°04′N 78°59′W
23 X14 **Moore Haven** Florida, SE USA 26°49′N 81°05′W
180 J11 **Moore, Lake** ◎ Western Australia
19 N7 **Moore Reservoir** ⊠ New Hampshire/Vermont, NE USA
44 G1 **Moores Island** *island* N The Bahamas
21 R10 **Mooresville** North Carolina, SE USA 35°34′N 80°48′W
29 R5 **Moorhead** Minnesota, N USA 46°51′N 96°44′W
22 K4 **Moorhead** Mississippi, S USA 33°27′N 90°30′W
99 F18 **Moorsel** Oost-Vlaanderen, C Belgium 50°58′N 04°06′E
99 C18 **Moorslede** West-Vlaanderen, W Belgium 50°53′N 03°03′E
18 L8 **Moosalamoo, Mount** ▲ Vermont, NE USA 43°55′N 73°03′W
101 M22 **Moosburg in der Isar** Bayern, SE Germany 48°28′N 11°55′E
33 S14 **Moose** Wyoming, C USA 43°38′N 110°42′W
12 H11 **Moose** ≈ Ontario, S Canada
12 H10 **Moose Factory** Ontario, S Canada 51°16′N 80°32′W
19 Q4 **Moosehead Lake** ◎ Maine, NE USA
11 U16 **Moose Jaw** Saskatchewan, S Canada 50°23′N 105°35′W
11 V14 **Moose Lake** Manitoba, C Canada 53°42′N 100°22′W
29 W6 **Moose Lake** Minnesota, N USA 46°28′N 92°46′W
19 P6 **Mooselookmeguntic Lake** ◎ Maine, NE USA
39 R12 **Moose Pass** Alaska, USA 60°28′N 149°21′W
19 P5 **Moose River** ≈ Maine, NE USA
18 J9 **Moose River** ≈ New York, NE USA
11 V16 **Moosomin** Saskatchewan, S Canada 50°09′N 101°41′W
12 H10 **Moosonee** Ontario, SE Canada 51°18′N 80°40′W
19 N12 **Moosup** Connecticut, NE USA 41°42′N 71°51′W
83 N16 **Mopeia** Zambézia, NE Mozambique 17°59′S 35°43′E
83 H18 **Mopipi** Central, C Botswana 21°07′S 24°55′E
Moppo *see* Mokpo
77 N11 **Mopti** Mopti, C Mali 14°30′N 04°15′W
77 O11 **Mopti** ◈ *region* S Mali
57 H18 **Moquegua** Moquegua, SE Peru 17°07′S 70°55′W
57 H18 **Moquegua** *off.* Departamento de Moquegua. ◈ *department* S Peru
Moquegua, Departamento de *see* Moquegua
111 I23 **Mór** *Ger.* Moor. Fejér, C Hungary 47°21′N 18°12′E
78 G11 **Mora** Extrême-Nord, N Cameroon 11°02′N 14°07′E
104 G11 **Mora** Évora, S Portugal 38°56′N 08°10′W
105 N9 **Mora** Castilla-La Mancha, C Spain 39°40′N 03°46′W
94 L12 **Mora** Dalarna, C Sweden 61°N 14°30′E
29 V7 **Mora** Minnesota, N USA 45°52′N 93°18′W
37 T10 **Mora** New Mexico, SW USA 35°56′N 105°16′W
113 J17 **Morača** ≈ S Montenegro
152 K10 **Morādābād** Uttar Pradesh, N India 28°50′N 78°45′E
105 U6 **Móra d'Ebre** *var.* Mora de Ebro. Cataluña, NE Spain 41°05′N 00°38′E
Mora de Ebro *see* Móra d'Ebre
105 S8 **Mora de Rubielos** Aragón, NE Spain 40°15′N 00°45′W
172 H4 **Morafenobe** Mahajanga, W Madagascar 17°49′S 44°54′E
110 K8 **Morąg** *Ger.* Mohrungen. Warmińsko-Mazurskie, N Poland 53°55′N 19°56′E
111 L25 **Mórahalom** Csongrád, S Hungary 46°14′N 19°52′E
105 N11 **Moral de Calatrava** Castilla-La Mancha, C Spain 38°50′N 03°34′W
63 G19 **Moraleda, Canal** *strait* SE Pacific Ocean
54 J3 **Morales** Bolívar, N Colombia 08°17′N 73°52′W
54 D12 **Morales** Cauca, SW Colombia 02°46′N 76°44′W
42 F5 **Morales** Izabal, E Guatemala 15°28′N 88°46′W
172 J5 **Moramanga** Toamasina, E Madagascar 18°57′S 48°13′E
27 Q6 **Moran** Kansas, C USA 37°55′N 95°10′W
25 Q7 **Moran** Texas, SW USA 32°33′N 99°10′W
181 X7 **Moranbah** Queensland, NE Australia 22°01′S 148°08′E
44 L13 **Morant Bay** E Jamaica 17°53′N 76°25′W
96 G10 **Morar, Loch** ◎ N Scotland, United Kingdom
Morata *see* Goodenough Island
105 Q12 **Moratalla** Murcia, SE Spain 38°11′N 01°53′W
108 C8 **Morat, Lac de** *Ger.* Murtensee. ◎ W Switzerland
84 I11 **Morava** *var.* March. ≈ C Europe *see also* March
Morava *see* March
Morava *see* Moravia, Czech Republic
Morava *see* Velika Morava, Serbia
29 W15 **Moravia** Iowa, C USA 40°53′N 92°49′W
111 F18 **Moravia** *Cz.* Morava, *Ger.* Mähren. *cultural region* E Czech Republic
111 H17 **Moravice** *Ger.* Mohra. ≈ NE Czech Republic
116 E12 **Moravița** Timiș, SW Romania 45°15′N 21°17′E
111 G17 **Moravská Třebová** *Ger.* Mährisch-Trübau. Pardubický Kraj, C Czech Republic 49°47′N 16°40′E
111 E19 **Moravské Budějovice** *Ger.* Mährisch-Budwitz. Vysočina, C Czech Republic 49°03′N 15°48′E
111 H17 **Moravskoslezský Kraj** *prev.* Ostravský Kraj. ◈ *region* E Czech Republic
111 F19 **Moravský Krumlov** *Ger.* Mährisch-Kromau. Jihomoravský Kraj, SE Czech Republic 48°58′N 16°30′E
96 J8 **Moray** *cultural region* N Scotland, United Kingdom
96 J8 **Moray Firth** *inlet* N Scotland, United Kingdom
42 B10 **Morazán** ◈ *department* NE El Salvador
154 C10 **Morbi** Gujarāt, W India 22°51′N 70°49′E
102 G7 **Morbihan** ◈ *department* NW France
Mörbisch *see* Mörbisch am See
109 Y5 **Mörbisch am See** *var.* Mörbisch. Burgenland, E Austria 47°43′N 16°40′E
95 N21 **Mörbylånga** Kalmar, S Sweden 56°31′N 16°25′E
102 J14 **Morcenx** Landes, SW France 44°04′N 00°55′W
Morcheh Khort *see* Mūrcheh Khvort
163 T5 **Mordaga** Nei Mongol Zizhiqu, N China 51°15′N 120°47′E
11 X17 **Morden** Manitoba, S Canada 49°12′N 98°05′W
Mordovia *see* Mordoviya, Respublika
127 N5 **Mordoviya, Respublika** *prev.* Mordovskaya ASSR, *Eng.* Mordovia, Mordvinia. ◈ *autonomous republic* W Russian Federation
126 M7 **Mordovo** Tambovskaya Oblast', W Russian Federation 52°05′N 40°49′E
Mordovskaya ASSR/Mordvinia *see* Mordoviya, Respublika
Morea *see* Pelopónnisos
28 K8 **Moreau River** ≈ South Dakota, N USA
97 K16 **Morecambe** NW England, United Kingdom 54°04′N 02°53′W
97 K16 **Morecambe Bay** *inlet* NW England, United Kingdom
183 S4 **Moree** New South Wales, SE Australia 29°29′S 149°53′E
21 N5 **Morehead** Kentucky, S USA 38°11′N 83°27′W
21 X11 **Morehead City** North Carolina, SE USA 34°43′N 76°43′W
27 Y8 **Morehouse** Missouri, C USA 36°51′N 89°41′W
108 E10 **Mörel** Valais, SW Switzerland 46°22′N 08°03′E
54 D13 **Morelia** Caquetá, S Colombia 01°30′N 75°43′W
41 N14 **Morelia** Michoacán, S Mexico 19°40′N 101°11′W
105 T7 **Morella** Valenciana, E Spain 40°37′N 00°06′W
40 I7 **Morelos** Chihuahua, N Mexico 26°37′N 107°37′W
41 O15 **Morelos** ◈ *state* S Mexico
154 H7 **Morena** Madhya Pradesh, C India 26°30′N 78°04′E
104 L12 **Morena, Sierra** ▲▲ S Spain
37 O14 **Morenci** Arizona, SW USA 33°05′N 109°21′W
31 R11 **Morenci** Michigan, N USA 41°43′N 84°13′W
116 J13 **Moreni** Dâmbovița, S Romania 44°59′N 25°39′E
94 D9 **Møre og Romsdal** ◈ *county* S Norway
10 I14 **Moresby Island** *island* Queen Charlotte Islands, British Columbia, SW Canada
183 W2 **Moreton Island** *island* Queensland, E Australia
103 O3 **Moreuil** Somme, N France 49°47′N 02°28′E
35 V7 **Morey Peak** ▲ Nevada, W USA 38°40′N 116°16′W
125 U4 **More-Yu** ≈ NW Russian Federation
103 T9 **Morez** Jura, E France 46°33′N 06°01′E
Morfou Bay/Mórfou, Kólpos *see* Güzelyurt Körfezi
182 J8 **Morgan** South Australia 34°02′S 139°39′E
23 S7 **Morgan** Georgia, SE USA 31°31′N 84°34′W
25 S8 **Morgan** Texas, SW USA 32°01′N 97°36′W
22 J10 **Morgan City** Louisiana, S USA 29°42′N 91°12′W
20 H6 **Morganfield** Kentucky, S USA 37°41′N 87°55′W
35 O10 **Morgan Hill** California, W USA 37°05′N 121°38′W
21 Q9 **Morganton** North Carolina, SE USA 35°44′N 81°43′W
20 J7 **Morgantown** Kentucky, S USA 37°12′N 86°42′W
21 S2 **Morgantown** West Virginia, NE USA 39°38′N 79°57′W
108 B10 **Morges** Vaud, SW Switzerland 46°31′N 06°30′E
Morghāb, Daryā-ye *see* Murgap
Morghāb, Daryā-ye *see* Murghāb, Daryā-ye
96 I9 **Mor, Glen** *var.* Glen Albyn, Great Glen. *valley* N Scotland, United Kingdom
103 T5 **Morhange** Moselle, NE France 48°56′N 06°37′E
158 M5 **Mori** *var.* Mori Kazak Zizhixian. Xinjiang Uygur Zizhiqu, NW China 43°48′N 90°21′E
165 R5 **Mori** Hokkaidō, NE Japan 42°04′N 140°36′E
35 Y6 **Moriah, Mount** ▲ Nevada, W USA 39°16′N 114°10′W
37 S11 **Moriarty** New Mexico, SW USA 34°59′N 106°03′W
54 J12 **Morichal** Guaviare, E Colombia 02°09′N 70°35′W
Mori Kazak Zizhixian *see* Mori
Morin Dawa Daurzu Zizhiqi *see* Nirji
11 Q14 **Morinville** Alberta, SW Canada 53°48′N 113°38′W
165 R8 **Morioka** Iwate, Honshū, C Japan 39°42′N 141°08′E
183 T8 **Morisset** New South Wales, SE Australia 33°07′S 151°32′E
165 Q8 **Moriyoshi-zan** ▲ Honshū, C Japan 39°58′N 140°32′E
92 K13 **Morjärv** Norrbotten, N Sweden 66°03′N 22°45′E
127 R3 **Morki** Respublika Mariy El, W Russian Federation 56°27′N 49°01′E
123 N10 **Morkoka** ≈ NE Russian Federation
102 F5 **Morlaix** Finistère, NW France 48°35′N 03°50′W
95 M20 **Mörlunda** Kalmar, S Sweden 57°19′N 15°52′E
107 N19 **Mormanno** Calabria, SW Italy 39°54′N 15°58′E
36 L11 **Mormon Lake** ◎ Arizona, SW USA
35 Y10 **Mormon Peak** ▲ Nevada, W USA 36°59′N 114°25′W
Mormon State *see* Utah
45 Y5 **Morne-à-l'Eau** Grande Terre, N Guadeloupe 16°20′N 61°31′W
29 Y15 **Morning Sun** Iowa, C USA 41°06′N 91°15′W
193 S12 **Mornington Abyssal Plain** *undersea feature* SE Pacific Ocean 50°00′S 90°00′W
63 F22 **Mornington, Isla** *island* S Chile
181 T4 **Mornington Island** *island* Wellesley Islands, Queensland, N Australia
115 E18 **Mórnos** ≈ C Greece
149 P14 **Moro** Sind, SE Pakistan 26°36′N 67°59′E
32 I11 **Moro** Oregon, NW USA 45°30′N 120°46′W
186 E8 **Morobe** Morobe, C Papua New Guinea 07°46′S 147°35′E
186 E8 **Morobe** ◈ *province* C Papua New Guinea
31 N12 **Morocco** Indiana, N USA 40°57′N 87°27′W
74 E8 **Morocco** *off.* Kingdom of Morocco, *Ar.* Al Mamlakah. ◆ *monarchy* N Africa
Morocco *see* Marrakech
Morocco, Kingdom of *see* Morocco
81 I22 **Morogoro** Morogoro, E Tanzania 06°49′S 37°40′E
81 H24 **Morogoro** ◈ *region* SE Tanzania
171 Q7 **Moro Gulf** *gulf* S Philippines
41 N13 **Moroleón** Guanajuato, C Mexico 20°00′N 101°13′W
172 H6 **Morombe** Toliara, W Madagascar 21°47′S 43°21′E
44 G5 **Morón** Ciego de Ávila, C Cuba 22°08′N 78°39′W
163 N8 **Mörön** Hentiy, C Mongolia 47°21′N 110°21′E
162 I6 **Mörön** Hövsgöl, N Mongolia 49°39′N 100°08′E
54 K5 **Morón** Carabobo, N Venezuela 10°29′N 68°11′W
Morón *see* Morón de la Frontera
56 D8 **Morona, Río** ≈ N Peru
56 C8 **Morona Santiago** ◈ *province* E Ecuador
172 H5 **Morondava** Toliara, W Madagascar 20°19′S 44°17′E
104 K14 **Morón de la Frontera** *var.* Morón. Andalucía, S Spain 37°07′N 05°27′W
172 G13 **Moroni** ● (Comoros) Grande Comore, NW Comoros 11°41′S 43°16′E
171 S10 **Morotai, Pulau** *island* Maluku, E Indonesia
Morotiri *see* Marotiri
81 H17 **Moroto** NE Uganda 02°32′N 34°41′E
Morozov *see* Bratan
126 M11 **Morozovsk** Rostovskaya Oblast', SW Russian Federation 48°21′N 41°54′E
97 L14 **Morpeth** N England, United Kingdom 55°10′N 01°41′W
Morphou *see* Güzelyurt
Morphou Bay *see* Güzelyurt Körfezi
28 I13 **Morrill** Nebraska, C USA 41°57′N 103°55′W
27 U11 **Morrilton** Arkansas, C USA 35°09′N 92°45′W
11 Q16 **Morrin** Alberta, SW Canada 51°40′N 112°45′W
184 M7 **Morrinsville** Waikato, North Island, New Zealand 37°41′S 175°32′E
11 X16 **Morris** Manitoba, S Canada 49°22′N 97°21′W
30 M11 **Morris** Illinois, N USA 41°21′N 88°25′W
29 S8 **Morris** Minnesota, N USA 45°32′N 95°53′W
14 M13 **Morrisburg** Ontario, SE Canada 44°55′N 75°07′W
197 R11 **Morris Jesup, Kap** *headland* N Greenland 83°33′N 32°40′W
182 B1 **Morris, Mount** ▲ South Australia 26°04′S 131°03′E
30 K10 **Morrison** Illinois, N USA 41°48′N 89°58′W
36 K13 **Morristown** Arizona, SW USA 33°48′N 112°34′W
18 J14 **Morristown** New Jersey, NE USA 40°48′N 74°29′W
21 O8 **Morristown** Tennessee, S USA 36°13′N 83°18′W
42 L11 **Morrito** Río San Juan, SW Nicaragua 11°37′N 85°05′W
35 P13 **Morro Bay** California, W USA 35°21′N 120°51′W
95 L22 **Mörrum** Blekinge, S Sweden 56°11′N 14°45′E
83 N16 **Morrumbala** Zambézia, NE Mozambique 17°17′S 35°35′E
83 N20 **Morrumbene** Inhambane, SE Mozambique 23°41′S 35°25′E
95 F21 **Mors** *island* NW Denmark
Mörs *see* Moers
25 N1 **Morse** Texas, SW USA 36°03′N 101°28′W
127 N6 **Morshansk** Tambovskaya Oblast', W Russian Federation 53°27′N 41°46′E
102 L5 **Mortagne-au-Perche** Orne, N France 48°32′N 00°31′E
102 J8 **Mortagne-sur-Sèvre** Vendée, NW France 47°00′N 00°57′W
104 G8 **Mortágua** Viseu, N Portugal 40°24′N 08°14′W
102 J5 **Mortain** Manche, N France 48°39′N 00°51′W
106 C8 **Mortara** Lombardia, N Italy 45°15′N 08°44′E
59 J17 **Mortes, Rio das** ≈ C Brazil
182 M12 **Mortlake** Victoria, SE Australia 38°06′S 142°48′E
Mortlock Group *see* Takuu Islands
189 Q17 **Mortlock Islands** *prev.* Nomoi Islands. *island group* C Micronesia
29 T9 **Morton** Minnesota, N USA 44°33′N 94°58′W
22 L5 **Morton** Mississippi, S USA 32°21′N 89°39′W
24 M5 **Morton** Texas, SW USA 33°40′N 102°45′W
32 H9 **Morton** Washington, NW USA 46°33′N 122°16′W
0 D7 **Morton Seamount** *undersea feature* NE Pacific Ocean 50°15′N 142°45′W
45 U15 **Moruga** Trinidad, Trinidad and Tobago 10°04′N 61°16′W
183 P9 **Morundah** New South Wales, SE Australia 34°57′S 146°18′E
191 X12 **Moruroa** *var.* Mururoa. *atoll* Îles Tuamotu, SE French Polynesia
183 S11 **Moruya** New South Wales, SE Australia 35°55′S 150°04′E
103 Q8 **Morvan** *physical region* C France
185 G21 **Morven** Canterbury, South Island, New Zealand 44°51′S 171°07′E
183 O13 **Morwell** Victoria, SE Australia 38°14′S 146°25′E
125 N6 **Morzhovets, Ostrov** *island* NW Russian Federation
126 J4 **Mosal'sk** Kaluzhskaya Oblast', W Russian Federation 54°30′N 34°55′E
101 H20 **Mosbach** Baden-Württemberg, SW Germany 49°21′N 09°06′E
95 E18 **Mosby** Vest-Agder, S Norway 58°12′N 07°55′E
33 V9 **Mosby** Montana, NW USA 46°58′N 107°53′W
32 M9 **Moscow** Idaho, NW USA 46°43′N 117°00′W
20 F10 **Moscow** Tennessee, S USA 35°04′N 89°27′W
Moscow *see* Moskva
101 D19 **Mosel** *Fr.* Moselle. ≈ W Europe *see also* Moselle
Mosel *see* Moselle
103 T4 **Moselle** ◈ *department* NE France
103 T6 **Moselle** *Ger.* Mosel. ≈ W Europe *see also* Mosel
Moselle *see* Mosel
32 K9 **Moses Lake** ◎ Washington, NW USA
83 I18 **Mosetse** Central, E Botswana 20°40′S 26°38′E
92 H4 **Mosfellsbær** Höfuðborgarsvæðið, SW Iceland 65°09′N 21°43′W
183 N8 **Mossgiel** New South Wales, SE Australia 33°16′S 144°34′E
185 F23 **Mosgiel** Otago, South Island, New Zealand 45°51′S 170°22′E
124 M11 **Mosha** ≈ NW Russian Federation
81 I20 **Moshi** Kilimanjaro, NE Tanzania 03°21′S 37°19′E
110 G12 **Mosina** Wielkolpolskie, C Poland 52°15′N 16°50′E
30 L6 **Mosinee** Wisconsin, N USA 44°45′N 89°39′W
92 F13 **Mosjøen** Nordland, C Norway 65°49′N 13°12′E
123 S12 **Moskal'vo** Ostrov Sakhalin, Sakhalinskaya Oblast', SE Russian Federation 53°36′N 142°31′E
92 I13 **Moskosel** Norrbotten, N Sweden 65°52′N 19°30′E
126 K4 **Moskovskaya Oblast'** ◈ *province* W Russian Federation
Moskovskiy *see* Moskva
126 J3 **Moskva** *Eng.* Moscow. ● (Russian Federation) Gorod Moskva, W Russian Federation 55°45′N 37°42′E
147 Q14 **Moskva** *Rus.* Moskovskiy; *prev.* Chubek. SW Tajikistan 37°41′N 69°33′E
126 L4 **Moskva** ≈ W Russian Federation
83 I20 **Mosomane** Kgatleng, SE Botswana 24°04′S 26°15′E
Moson and Magyaróvár *see* Mosonmagyaróvár
111 H21 **Mosoni-Duna** *Ger.* Kleine Donau. ≈ NW Hungary
111 H21 **Mosonmagyaróvár** *Ger.* Wieselburg-Ungarisch-Altenburg; *prev.* Moson and Magyaróvár, *Ger.* Wieselburg and Ungarisch-Altenburg. Győr-Moson-Sopron, NW Hungary 47°52′N 17°15′E
Mospino *see* Mospyne
117 X8 **Mospyne** *Rus.* Mospino. Donets'ka Oblast', E Ukraine 47°53′N 38°03′E
54 B12 **Mosquera** Nariño, SW Colombia 02°32′N 78°24′W
37 U10 **Mosquero** New Mexico, SW USA 35°46′N 103°57′W
Mosquito Coast *see* La Mosquitia
31 U11 **Mosquito Creek Lake** ⊠ Ohio, N USA
Mosquito Gulf *see* Mosquitos, Golfo de los
23 X11 **Mosquito Lagoon** *wetland* Florida, SE USA
43 N10 **Mosquito, Punta** *headland* E Nicaragua 12°18′N 83°38′W
43 W14 **Mosquito, Punta** *headland* NE Panama 09°06′N 77°52′W
43 Q15 **Mosquitos, Golfo de los** *Eng.* Mosquito Gulf. *gulf* N Panama
95 H16 **Moss** Østfold, S Norway 59°25′N 10°40′E
Mossâmedes *see* Namibe
22 G8 **Moss Bluff** Louisiana, S USA 30°18′N 93°11′W
185 C23 **Mossburn** Southland, South Island, New Zealand 45°40′S 168°15′E
83 G26 **Mosselbaai** *var.* Mosselbai, *Eng.* Mossel Bay. Western Cape, SW South Africa 34°11′S 22°08′E
Mosselbai/Mossel Bay *see* Mosselbaai
79 F20 **Mossendjo** Niari, SW Congo 02°57′S 12°40′E
101 H22 **Mössingen** Baden-Württemberg, S Germany 48°22′N 09°01′E
181 W4 **Mossman** Queensland, NE Australia 16°34′S 145°27′E
59 P14 **Mossoró** Rio Grande do Norte, NE Brazil 05°11′S 37°20′W
23 N9 **Moss Point** Mississippi, S USA 30°24′N 88°31′W
183 S9 **Moss Vale** New South Wales, SE Australia 34°33′S 150°20′E
32 G9 **Mossyrock** Washington, NW USA 46°32′N 122°30′W
111 B15 **Most** *Ger.* Brüx. Ústecký Kraj, NW Czech Republic 50°30′N 13°37′E
162 E7 **Möst** *var.* Ulaantolgoy. Hovd, W Mongolia 46°39′N 92°50′E
121 P16 **Mosta** *var.* Musta. C Malta 35°54′N 14°25′E
74 I5 **Mostaganem** *var.* Mestghanem. NW Algeria 35°54′N 00°05′E
113 H14 **Mostar** Federacija Bosne I Hercegovine, S Bosnia and Herzegovina 43°21′N 17°47′E
61 J17 **Mostardas** Rio Grande do Sul, S Brazil 31°02′S 50°51′W
116 K14 **Mostiștea** ≈ S Romania
Mostva *see* Mastva
Mosty *see* Masty
116 H5 **Mostys'ka** L'vivs'ka Oblast', W Ukraine 49°47′N 23°09′E
Mosul *see* Al Mawṣil
95 F15 **Møsvatnet** ◎ S Norway
80 J12 **Mot'a** Āmara, N Ethiopia 11°03′N 38°03′E
79 H16 **Motaba** ≈ N Congo
105 O10 **Mota del Cuervo** Castilla-La Mancha, C Spain 39°30′N 02°52′W
104 L5 **Mota del Marqués** Castilla y León, N Spain 41°38′N 05°11′W
42 F5 **Motagua, Río** ≈ Guatemala/Honduras
119 H19 **Motal'** Brestskaya Voblasts', SW Belarus 52°13′N 25°34′E
95 L17 **Motala** Östergötland, S Sweden 58°34′N 15°05′E
191 X7 **Motane** *island* Îles Marquises, NE French Polynesia
152 K13 **Moth** Uttar Pradesh, N India 25°44′N 78°56′E
Mother of Presidents/Mother of States *see* Virginia
96 I12 **Motherwell** C Scotland, United Kingdom 55°48′N 04°W
153 P12 **Motīhāri** Bihār, N India 26°40′N 84°55′E
105 Q10 **Motilla del Palancar** Castilla-La Mancha, C Spain 39°34′N 01°55′W
184 N7 **Motiti Island** *island* NE New Zealand
65 E25 **Motley Island** *island* SE Falkland Islands
83 J19 **Motloutse** ≈ E Botswana
41 V17 **Motozintla de Mendoza** Chiapas, SE Mexico 15°21′N 92°14′W
105 N15 **Motril** Andalucía, S Spain 36°45′N 03°30′W
116 G13 **Motru** Gorj, SW Romania 44°49′N 22°56′E
165 Q4 **Motsuta-misaki** *headland* Hokkaidō, NE Japan 42°36′N 139°48′E
28 L6 **Mott** North Dakota, N USA 46°21′N 102°17′W
166 M9 **Mottama** *var.* Martaban, Moktama. Mon State, S Myanmar (Burma) 16°32′N 97°35′E
166 L9 **Mottama, Gulf of** *var.* Gulf of Martaban. *gulf* S Myanmar (Burma)
Möttling *see* Metlika
107 O18 **Mottola** Puglia, SE Italy 40°38′N 17°02′E
184 P8 **Motu** ≈ North Island, New Zealand
185 I14 **Motueka** Tasman, South Island, New Zealand 41°08′S 173°00′E
185 I14 **Motueka** ≈ South Island, New Zealand
Motu Iti *see* Tupai
41 X12 **Motul** *var.* Motul de Felipe Carrillo Puerto. Yucatán, SE Mexico 21°06′N 89°17′W
Motul de Felipe Carrillo Puerto *see* Motul
191 U17 **Motu Nui** *island* Easter Island, Chile, E Pacific Ocean
191 Q10 **Motu One** *var.* Bellingshausen. *atoll* Îles Sous le Vent, W French Polynesia
190 I16 **Motutapu** *island* E Cook Islands
193 V15 **Motu Tapu** *island* Tongatapu Group, S Tonga
184 L5 **Motutapu Island** *island* N New Zealand
Motyca *see* Modica
Mouanda *see* Moanda
Mouaskar *see* Mascara
105 U3 **Moubermé, Tuc de** *Fr.* Pic de Maubermé, *Sp.* Pico Mauberme; *prev.* Tuc de Maubermé. ▲ France/Spain 42°48′N 00°57′E *see also* Maubermé, Pic de
Moubermé, Tuc de *see* Maubermé, Pic de
45 N7 **Mouchoir Passage** *passage* SE Turks and Caicos Islands
76 I9 **Moudjéria** Tagant, SW Mauritania 17°52′N 12°20′W
108 C9 **Moudon** Vaud, W Switzerland 46°41′N 06°49′E
Mouhoun *see* Black Volta
79 E19 **Mouila** Ngounié, C Gabon 01°50′S 11°02′E
79 K14 **Mouka** Haute-Kotto, C Central African Republic 07°12′N 21°52′E
Moukden *see* Shenyang
183 N10 **Moulamein** New South Wales, SE Australia 35°06′S 144°03′E
Moulamein Creek *see* Billabong Creek
74 F6 **Moulay-Bousselham** NW Morocco 35°00′N 06°22′W
Moule *see* le Moule
80 M11 **Moulhoulé** N Djibouti 12°34′N 43°06′E
103 P9 **Moulins** Allier, C France 46°34′N 03°20′E
Moulmein *see* Mawlamyine
Moulmeingyun *see* Mawlamyinegyunn
74 G6 **Moulouya** *var.* Mulucha, Muluya, Mulwiya. *seasonal river* NE Morocco
23 O2 **Moulton** Alabama, S USA 34°28′N 87°18′W
29 W16 **Moulton** Iowa, C USA 40°41′N 92°40′W
25 T11 **Moulton** Texas, SW USA 29°34′N 97°08′W
23 T7 **Moultrie** Georgia, SE USA 31°10′N 83°47′W
21 S14 **Moultrie, Lake** ⊠ South Carolina, SE USA
22 K3 **Mound Bayou** Mississippi, S USA 33°52′N 90°43′W
30 L17 **Mound City** Illinois, N USA 37°06′N 89°09′W
27 R6 **Mound City** Kansas, C USA 38°07′N 94°49′W
27 Q2 **Mound City** Missouri, C USA 40°07′N 95°13′W
29 N7 **Mound City** South Dakota, N USA 45°44′N 100°03′W
78 H13 **Moundou** Logone-Occidental, SW Chad 08°35′N 16°01′E
27 P10 **Mounds** Oklahoma, C USA 35°52′N 96°03′W
21 R2 **Moundsville** West Virginia, NE USA 39°54′N 80°44′W
167 R11 **Moŭng** *prev.* Phumĭ Moŭng. Siĕmréab, NW Cambodia 13°45′N 103°35′E
167 Q12 **Moŭng Roessei** Bătdâmbâng, W Cambodia 12°47′N 103°28′E
Moun Hou *see* Black Volta
8 H8 **Mountain** ≈ Northwest Territories, NW Canada
37 S12 **Mountainair** New Mexico, SW USA 34°31′N 106°14′W
35 V1 **Mountain City** Nevada, W USA 41°48′N 115°58′W
21 Q8 **Mountain City** Tennessee, S USA 36°28′N 81°48′W
27 U7 **Mountain Grove** Missouri, C USA 37°07′N 92°15′W
27 U9 **Mountain Home** Arkansas, C USA 36°19′N 92°24′W
33 N15 **Mountain Home** Idaho, NW USA 43°07′N 115°42′W
25 Q11 **Mountain Home** Texas, SW USA 30°11′N 99°19′W
29 W4 **Mountain Iron** Minnesota, N USA 47°31′N 92°37′W
29 T10 **Mountain Lake** Minnesota, N USA 43°57′N 94°54′W
23 S3 **Mountain Park** Georgia, SE USA 34°04′N 84°24′W
35 W12 **Mountain Pass** *pass* California, W USA
27 T12 **Mountain Pine** Arkansas, C USA 34°34′N 93°10′W
39 Y14 **Mountain Point** Annette Island, Alaska, USA 55°17′N 131°31′W
Mountain State *see* Montana
Mountain State *see* West Virginia
27 V7 **Mountain View** Arkansas, C USA 35°52′N 92°07′W
38 H12 **Mountain View** Hawaii, USA, C Pacific Ocean 19°32′N 155°03′W
27 V10 **Mountain View** Missouri, C USA 37°00′N 91°42′W
38 M11 **Mountain Village** Alaska, USA 62°06′N 163°42′W
21 R8 **Mount Airy** North Carolina, SE USA 36°30′N 80°36′W
83 K24 **Mount Ayliff** *Xh.* Maxesibebi. Eastern Cape, SE South Africa 30°48′S 29°23′E
29 U16 **Mount Ayr** Iowa, C USA 40°42′N 94°14′W
182 J9 **Mount Barker** South Australia 35°06′S 138°52′E
180 J14 **Mount Barker** Western Australia 34°42′S 117°40′E
183 P11 **Mount Beauty** Victoria, SE Australia 36°47′S 147°12′E
14 E16 **Mount Brydges** Ontario, S Canada 42°54′N 81°29′W
31 N16 **Mount Carmel** Illinois, N USA 38°23′N 87°46′W
30 K10 **Mount Carroll** Illinois, N USA 42°05′N 89°59′W
31 S9 **Mount Clemens** Michigan, N USA 42°36′N 82°52′W
185 E19 **Mount Cook** Canterbury, South Island, New Zealand 43°47′S 170°06′E
83 L16 **Mount Darwin** Mashonaland Central, NE Zimbabwe 16°45′S 31°39′E
19 S7 **Mount Desert Island** *island* Maine, NE USA
23 W11 **Mount Dora** Florida, SE USA 28°48′N 81°38′W
182 G5 **Mount Eba** South Australia 30°11′S 135°40′E
25 W8 **Mount Enterprise** Texas, SW USA 31°53′N 94°40′W
182 J4 **Mount Fitton** South Australia 29°55′S 139°26′E
83 J24 **Mount Fletcher** Eastern Cape, SE South Africa 30°41′S 28°30′E
14 F15 **Mount Forest** Ontario, S Canada 43°58′N 80°44′W
182 K12 **Mount Gambier** South Australia 37°47′S 140°49′E
181 W5 **Mount Garnet** Queensland, NE Australia 17°41′S 145°07′E
21 P6 **Mount Gay** West Virginia, NE USA 37°49′N 82°00′W
31 S12 **Mount Gilead** Ohio, N USA 40°33′N 82°49′W
186 C7 **Mount Hagen** Western Highlands, C Papua New Guinea 05°54′S 144°13′E
18 J16 **Mount Holly** New Jersey, NE USA 39°59′N 74°46′W
21 R10 **Mount Holly** North Carolina, SE USA 35°18′N 81°01′W
27 T12 **Mount Ida** Arkansas, C USA 34°32′N 93°38′W
181 T6 **Mount Isa** Queensland, C Australia 20°48′S 139°32′E
21 U4 **Mount Jackson** Virginia, NE USA 38°45′N 78°38′W
18 D12 **Mount Jewett** Pennsylvania, NE USA 41°43′N 78°37′W
18 L13 **Mount Kisco** New York, NE USA 41°12′N 73°42′W
18 B15 **Mount Lebanon** Pennsylvania, NE USA 40°21′N 80°03′W
182 J8 **Mount Lofty Ranges** ▲▲ South Australia
180 J10 **Mount Magnet** Western Australia 28°09′S 117°52′E
184 N7 **Mount Maunganui** Bay of Plenty, North Island, New Zealand 37°39′S 176°11′E
97 E18 **Mountmellick** *Ir.* Móinteach Mílic. Laois, C Ireland 53°07′N 07°20′W
30 L10 **Mount Morris** Illinois, N USA 42°03′N 89°25′W
31 R9 **Mount Morris** Michigan, N USA 43°07′N 83°42′W
18 F10 **Mount Morris** New York, NE USA 42°43′N 77°51′W
18 B16 **Mount Morris** Pennsylvania, NE USA 39°43′N 80°06′W
30 K15 **Mount Olive** Illinois, N USA 39°04′N 89°43′W
21 V10 **Mount Olive** North Carolina, SE USA 35°12′N 78°03′W
21 N4 **Mount Olivet** Kentucky, S USA 38°32′N 84°01′W
29 Y15 **Mount Pleasant** Iowa, C USA 40°57′N 91°33′W
31 Q8 **Mount Pleasant** Michigan, N USA 43°36′N 84°46′W
18 C15 **Mount Pleasant** Pennsylvania, NE USA 40°07′N 79°33′W
21 T14 **Mount Pleasant** South Carolina, SE USA 32°47′N 79°51′W
20 I9 **Mount Pleasant** Tennessee, S USA 35°32′N 87°11′W
25 W6 **Mount Pleasant** Texas, SW USA 33°10′N 94°49′W
36 L4 **Mount Pleasant** Utah, W USA 39°33′N 111°27′W
63 N23 **Mount Pleasant** ✈ (Stanley) East Falkland, Falkland Islands
97 G25 **Mount's Bay** *inlet* SW England, United Kingdom
35 N2 **Mount Shasta** California, W USA 41°18′N 122°19′W
30 J13 **Mount Sterling** Illinois, N USA 39°59′N 90°44′W
21 N5 **Mount Sterling** Kentucky, S USA 38°03′N 83°56′W
18 E15 **Mount Union** Pennsylvania, NE USA 40°21′N 77°51′W
23 V6 **Mount Vernon** Georgia, SE USA 32°10′N 82°35′W
30 L16 **Mount Vernon** Illinois, N USA 38°19′N 88°54′W
20 M6 **Mount Vernon** Kentucky, S USA 37°20′N 84°20′W
27 S7 **Mount Vernon** Missouri, C USA 37°05′N 93°49′W
31 T13 **Mount Vernon** Ohio, N USA 40°23′N 82°29′W
32 K13 **Mount Vernon** Oregon, NW USA 44°22′N 119°07′W
25 W6 **Mount Vernon** Texas, SW USA 33°11′N 95°13′W
32 H7 **Mount Vernon** Washington, NW USA 48°25′N 122°19′W
20 L5 **Mount Washington** Kentucky, S USA 38°03′N 85°33′W
182 F8 **Mount Wedge** South Australia 33°29′S 135°08′E
30 L14 **Mount Zion** Illinois, N USA 39°46′N 88°52′W
181 Y9 **Moura** Queensland, NE Australia 24°34′S 149°57′E
58 F12 **Moura** Amazonas, NW Brazil 01°32′S 61°43′W
104 H12 **Moura** Beja, S Portugal 38°08′N 07°27′W
104 I12 **Mourão** Évora, S Portugal 38°22′N 07°20′W
76 L11 **Mourdiah** Koulikoro, W Mali 14°28′N 07°31′W
78 K7 **Mourdi, Dépression du** *desert lowland* Chad/Sudan
102 J16 **Mourenx** Pyrénées-Atlantiques, SW France 43°24′N 00°37′W
Mourgana *see* Mourgkána
115 C15 **Mourgkána** *var.* Mourgana. ▲ Albania/Greece 39°48′N 20°24′E
97 G16 **Mourne Mountains** *Ir.* Beanna Boirche. ▲▲ SE Northern Ireland, United Kingdom
115 I15 **Moúrtzeflos, Akrotírio** *headland* Límnos, E Greece 40°00′N 25°02′E
99 C19 **Mouscron** *Dut.* Moeskroen. Hainaut, W Belgium 50°44′N 03°14′E
Mouse River *see* Souris River
78 H10 **Moussoro** Bahr el Gazel, W Chad 13°41′N 16°31′E
103 T11 **Moûtiers** Savoie, E France 45°28′N 06°31′E
172 J14 **Moutsamoudou** *var.* Moutsamudou. Anjouan, SE Comoros 12°10′S 44°25′E
Moutsamudou *see* Moutsamoudou
74 K11 **Mouydir, Monts de** ▲▲ S Algeria
79 F20 **Mouyondzi** Bouenza, S Congo 03°58′S 13°57′E
115 E16 **Mouzáki** *prev.* Mouzákion. Thessalía, C Greece 39°25′N 21°40′E
Mouzákion *see* Mouzáki
29 S13 **Moville** Iowa, C USA 42°30′N 96°04′W
82 E13 **Moxico** ◈ *province* E Angola
172 I14 **Moya** Anjouan, SE Comoros 12°18′S 44°27′E
40 L12 **Moyahua** Zacatecas, C Mexico 21°18′N 103°09′W
81 J16 **Moyalē** Oromīya, C Ethiopia 03°34′N 38°58′E

◆ Country
● Country Capital
◇ Dependent Territory
○ Dependent Territory Capital
◈ Administrative Regions
✈ International Airport
▲ Mountain
▲▲ Mountain Range
Volcano
≈ River
◎ Lake
⊠ Reservoir

76 I15 **Moyamba** W Sierra Leone 08°04′N 12°30′W
74 G7 **Moyen Atlas** *Eng.* Middle Atlas. ▲ N Morocco
78 H13 **Moyen-Chari** *off.* Région du Moyen-Chari. ◆ *region* S Chad
Moyen-Chari, Région du *see* Moyen-Chari
Moyen-Congo *see* Congo (Republic of)
83 J24 **Moyeni** *var.* Quthing. SW Lesotho 30°25′S 27°43′E
79 D18 **Moyen-Ogooué** *off.* Province du Moyen-Ogooué, *var.* Le Moyen-Ogooué. ◆ *province* C Gabon
Moyen-Ogooué, Province du *see* Moyen-Ogooué
103 S4 **Moyeuvre-Grande** Moselle, NE France 49°15′N 06°03′E
33 N7 **Moyie Springs** Idaho, NW USA 48°43′N 116°15′W
146 G6 **Mo'ynoq** *Rus.* Muynak. Qoraqalpog'iston Respublikasi, NW Uzbekistan 43°45′N 59°03′E
81 F16 **Moyo** NW Uganda 03°38′N 31°43′E
56 D10 **Moyobamba** San Martín, NW Peru 06°04′S 76°56′W
78 H10 **Moyto** Hadjer-Lamis, W Chad 12°35′N 16°33′E
158 G9 **Moyu** *var.* Karakax. Xinjiang Uygur Zizhiqu, NW China 37°16′N 79°39′E
122 M9 **Moyyero** ≈ N Russian Federation
145 S15 **Moyynkum** *var.* Furmanovka, *Kaz.* Fūrmanov. Zhambyl, S Kazakhstan 44°15′N 72°55′E
145 Q15 **Moyynkum, Peski** *Kaz.* Moyynqum. *desert* S Kazakhstan
Moyynqum *see* Moyynkum, Peski
145 S12 **Moyynty** Karaganda, C Kazakhstan 47°10′N 73°24′E
145 S12 **Moyynty** ≈ Karaganda, C Kazakhstan
Mozambika, Lakandranon' i *see* Mozambique Channel
83 M18 **Mozambique** *off.* Republic of Mozambique; *prev.* People's Republic of Mozambique, Portuguese East Africa. ◆ *republic* S Africa
Mozambique Basin *see* Natal Basin
Mozambique, Canal de *see* Mozambique Channel
83 P17 **Mozambique Channel** *Fr.* Canal de Mozambique, *Mal.* Lakandranon' i Mozambika. *strait* W Indian Ocean
172 L11 **Mozambique Escarpment** *var.* Mozambique Scarp. *undersea feature* SW Indian Ocean 33°00′S 36°30′E
Mozambique, People's Republic of *see* Mozambique
172 L10 **Mozambique Plateau** *var.* Mozambique Rise. *undersea feature* SW Indian Ocean 32°00′S 35°00′E
Mozambique, Republic of *see* Mozambique
Mozambique Rise *see* Mozambique Plateau
Mozambique Scarp *see* Mozambique Escarpment
127 O15 **Mozdok** Respublika Severnaya Osetiya, SW Russian Federation 43°48′N 44°42′E
57 K17 **Mozetenes, Serranías de** ▲ C Bolivia
126 J4 **Mozhaysk** Moskovskaya Oblast', W Russian Federation 55°31′N 36°01′E
127 T3 **Mozhga** Udmurtskaya Respublika, NW Russian Federation 56°24′N 52°13′E
Mozyr' *see* Mazyr
79 P22 **Mpala** Katanga, E Dem. Rep. Congo 06°43′S 29°28′E
79 G19 **Mpama** ≈ C Congo
81 E22 **Mpanda** Katavi, W Tanzania 06°21′S 31°01′E
82 L11 **Mpande** Northern, NE Zambia 09°13′S 31°42′E
83 J18 **Mphoengs** Matabeleland South, SW Zimbabwe 21°04′S 27°56′E
81 F18 **Mpigi** S Uganda 0°14′N 32°19′E
82 L13 **Mpika** Muchinga, NE Zambia 11°50′S 31°30′E
83 J14 **Mpima** Central, C Zambia 14°25′S 28°34′E
82 J13 **Mpongwe** Copperbelt, C Zambia 13°25′S 28°13′E
82 K11 **Mporokoso** Northern, N Zambia 09°22′S 30°06′E
79 H20 **Mpouya** Plateaux, SE Congo 02°38′S 16°13′E
77 P16 **Mpraeso** C Ghana 06°36′N 00°43′W
82 L11 **Mpulungu** Northern, N Zambia 08°50′S 31°06′E
83 K21 **Mpumalanga** *prev.* Eastern Transvaal, *Afr.* Oos-Transvaal. ◆ *province* NE South Africa
83 D16 **Mpungu** Okavango, N Namibia 17°36′S 18°16′E
81 I22 **Mpwapwa** Dodoma, C Tanzania 06°21′S 36°29′E
Mqinvartsveri *see* Kazbek
110 M8 **Mrągowo** *Ger.* Sensburg. Warmińsko-Mazurskie, NE Poland 53°53′N 21°19′E
127 V6 **Mrakovo** Respublika Bashkortostan, W Russian Federation 52°43′N 56°36′E
172 I13 **Mramani** Anjouan, E Comoros 12°18′N 44°39′E
166 K5 **Mrauk-oo** *var.* Mrauk U, Myohaung. Rakhine State, W Myanmar (Burma) 20°35′N 93°12′E
Mrauk U *see* Mrauk-oo
112 F12 **Mrkonjić Grad** ◆ Republika Srpska, W Bosnia and Herzegovina
110 H9 **Mrocza** Kujawsko-pomorskie, C Poland 53°15′N 17°38′E
124 I14 **Msta** ≈ NW Russian Federation
Mstislavl' *see* Mstsislaw
119 P15 **Mstsislaw** *Rus.* Mstislavl'. Mahilyowskaya Voblasts', E Belarus 54°01′N 31°43′E
83 J24 **Mthatha** *prev.* Umtata. Eastern Cape, SE South Africa 31°33′S 28°47′E *see also* Umtata
Mtkvari *see* Kura
Mtoko *see* Mutoko
126 K6 **Mtsensk** Orlovskaya Oblast', W Russian Federation 53°17′N 36°34′E
81 K24 **Mtwara** Mtwara, SE Tanzania 10°17′S 40°11′E
81 J25 **Mtwara** ◆ *region* SE Tanzania
104 G14 **Mu** ▲ S Portugal 37°24′N 08°04′W
193 V15 **Mu'a** Tongatapu, S Tonga 21°11′S 175°07′W
Muai To *see* Mae Hong Son
83 P16 **Mualama** Zambézia, NE Mozambique 16°51′S 38°21′E
Mualo *see* Messalo, Rio
79 E22 **Muanda** Bas-Congo, SW Dem. Rep. Congo 05°53′S 12°17′E
Muang Chiang Rai *see* Chiang Rai
167 R6 **Muang Ham** Houaphan, N Laos 20°19′N 104°00′E
167 S8 **Muang Hinboun** Khammouan, C Laos 17°37′N 104°37′E
Muang Kalasin *see* Kalasin
Muang Khammouan *see* Thakhèk
167 S11 **Muang Không** Champasak, S Laos 14°08′N 105°48′E
167 S10 **Muang Khôngxédôn** *var.* Khong Sedone. Salavan, S Laos 15°34′N 105°46′E
Muang Khon Kaen *see* Khon Kaen
167 Q6 **Muang Khoua** Phôngsali, N Laos 21°07′N 102°31′E
Muang Krabi *see* Krabi
Muang Lampang *see* Lampang
Muang Lamphun *see* Lamphun
Muang Loei *see* Loei
Muang Lom Sak *see* Lom Sak
Muang Nakhon Sawan *see* Nakhon Sawan
167 Q6 **Muang Namo** Oudômxai, N Laos 20°58′N 101°46′E
Muang Nan *see* Nan
167 Q6 **Muang Ngoy** Louangphabang, N Laos 20°43′N 102°42′E
167 Q5 **Muang Ou Tai** Phôngsali, N Laos 22°06′N 101°59′E
Muang Pak Lay *see* Pak Lay
Muang Pakxan *see* Pakxan
167 T10 **Muang Pakxong** Champasak, S Laos 15°10′N 106°17′E
167 S9 **Muang Phalan** *var.* Muang Phalane. Savannakhét, S Laos 16°40′N 105°33′E
Muang Phalane *see* Muang Phalan
Muang Phan *see* Phan
Muang Phayao *see* Phayao
Muang Phichit *see* Phichit
167 T9 **Muang Phin** Savannakhét, S Laos 16°31′N 106°01′E
Muang Phitsanulok *see* Phitsanulok
Muang Phrae *see* Phrae
Muang Roi Et *see* Roi Et
Muang Sakon Nakhon *see* Sakon Nakhon
Muang Samut Prakan *see* Samut Prakan
167 P6 **Muang Sing** Louang Namtha, N Laos 21°12′N 101°09′E
Muang Ubon *see* Ubon Ratchathani
Muang Uthai Thani *see* Uthai Thani
167 P7 **Muang Vangviang** Viangchan, C Laos 18°53′N 102°27′E
Muang Xaignabouri *see* Xaignabouli
Muang Xay *see* Oudômxai
167 S9 **Muang Xépôn** *var.* Sepone. Savannakhét, S Laos 16°40′N 106°15′E
168 K10 **Muar** *var.* Bandar Maharani. Johor, Peninsular Malaysia 02°01′N 102°35′E
168 J9 **Muara** Sumatera, W Indonesia 02°18′N 98°54′E
168 L13 **Muarabeliti** Sumatera, W Indonesia 03°13′S 103°00′E
168 K12 **Muarabungo** Sumatera, W Indonesia 01°28′S 102°06′E
168 L13 **Muaraenim** Sumatera, W Indonesia 03°40′S 103°48′E
169 T11 **Muarajuloi** Borneo, C Indonesia 0°12′S 114°03′E
169 U12 **Muarakaman** Borneo, C Indonesia 0°09′S 116°43′E
168 H12 **Muarasigep** Pulau Siberut, W Indonesia 01°01′S 98°48′E
168 L12 **Muaratembesi** Sumatera, W Indonesia 01°40′S 103°08′E
169 T12 **Muaratewe** *var.* Muaratewe; *prev.* Moearatewe. Borneo, C Indonesia 0°58′S 114°52′E
Muaratewe *see* Muaratewe
169 U10 **Muarawahau** Borneo, N Indonesia 01°03′N 116°48′E
138 G13 **Mubārak, Jabal** ▲ S Jordan 29°19′N 35°13′E
153 N13 **Mubārakpur** Uttar Pradesh, N India 26°05′N 83°19′E
Mubarek *see* Muborak
81 F18 **Mubende** SW Uganda 0°35′N 31°24′E
77 Y14 **Mubi** Adamawa, NE Nigeria 10°15′N 13°18′E
146 M12 **Muborak** *Rus.* Mubarek. Qashqadaryo Viloyati, S Uzbekistan 39°17′N 65°10′E
171 U12 **Mubrani** Papua Barat, E Indonesia 0°42′S 133°25′E
82 L12 **Muchinga** ◆ *province* NE Zambia
67 U12 **Muchinga Escarpment** *escarpment* NE Zambia
127 N7 **Muchkapskiy** Tambovskaya Oblast', W Russian Federation 51°51′N 42°25′E
96 G10 **Muck** *island* W Scotland, United Kingdom
82 Q13 **Mucojo** Cabo Delgado, N Mozambique 12°05′S 40°30′E
82 F12 **Muconda** Lunda Sul, NE Angola 10°37′S 21°19′E
54 I10 **Muco, Río** ≈ E Colombia
83 O16 **Mucubela** Zambézia, NE Mozambique 16°51′S 37°48′E
42 J5 **Mucupina, Monte** ▲ N Honduras 15°07′N 86°36′W
136 J14 **Mucur** Kırşehir, C Turkey 39°05′N 34°25′E
143 U8 **Mūd** Khorāsān-e Janūbī, E Iran 32°41′N 59°30′E
163 Y9 **Mudanjiang** *var.* Mu-tan-chiang. Heilongjiang, NE China 44°33′N 129°40′E
163 Y9 **Mudan Jiang** ≈ NE China
136 D11 **Mudanya** Bursa, NW Turkey 40°23′N 28°53′E
28 K8 **Mud Butte** South Dakota, N USA 45°00′N 102°51′W
155 G16 **Muddebihāl** Karnātaka, C India 16°26′N 76°07′E
27 P12 **Muddy Boggy Creek** ≈ Oklahoma, C USA
36 M6 **Muddy Creek** ≈ Utah, W USA
37 V7 **Muddy Creek Reservoir** ⊠ Colorado, C USA
33 W15 **Muddy Gap** Wyoming, C USA 42°21′N 107°27′W
35 Y11 **Muddy Peak** ▲ Nevada, W USA 36°17′N 114°40′W
183 R7 **Mudgee** New South Wales, SE Australia 32°37′S 149°36′E
29 S3 **Mud Lake** ◎ Minnesota, N USA
29 P7 **Mud Lake Reservoir** ⊠ South Dakota, N USA
167 N9 **Mudon** Mon State, S Myanmar (Burma) 16°17′N 97°40′E
81 O14 **Mudug** *off.* Gobolka Mudug. ◆ *region* NE Somalia
81 O14 **Mudug** *var.* Mudugh. *plain* N Somalia
Mudug, Gobolka *see* Mudug
Mudugh *see* Mudug
83 Q15 **Muecate** Nampula, NE Mozambique 14°56′S 39°38′E
82 Q13 **Mueda** Cabo Delgado, NE Mozambique 11°40′S 39°31′E
42 L10 **Muelle de los Bueyes** Región Autónoma Atlántico Sur, SE Nicaragua 12°03′N 84°34′W
Muenchen *see* München
83 M14 **Muende** Tete, NW Mozambique 14°22′S 33°00′E
25 T5 **Muenster** Texas, SW USA 33°39′N 97°22′W
Muenster *see* Münster
43 O6 **Muerto, Cayo** *reef* NE Nicaragua
41 T17 **Muerto, Mar** *lagoon* SE Mexico
64 F11 **Muertos Trough** *undersea feature* N Caribbean Sea
83 H14 **Mufaya Kuta** Western, NW Zambia 14°30′S 24°18′E
82 J13 **Mufulira** Copperbelt, C Zambia 12°33′S 28°16′E
161 O10 **Mufu Shan** ▲ C China
Mugalla *see* Yutian
Mugalzhar Taūlary *see* Mugodzhary, Gory
137 Y12 **Muğan Düzü** *Rus.* Muganskaya Ravnina, Muganskaya Step'. *physical region* S Azerbaijan
Muganskaya Ravnina/ Muganskaya Step' *see* Muğan Düzü
106 K8 **Múggia** Friuli-Venezia Giulia, NE Italy 45°36′N 13°48′E
153 N14 **Mughal Sarāi** Uttar Pradesh, N India 25°18′N 83°07′E
Mughla *see* Muğla
141 W11 **Mughshin** *var.* Muqshin. S Oman 19°26′N 54°38′E
147 S12 **Mughsu** *Rus.* Muksu. ≈ C Tajikistan
164 H14 **Mugi** Tokushima, Shikoku, SW Japan 33°39′N 134°24′E
136 C16 **Muğla** *var.* Mughla. Muğla, SW Turkey 37°13′N 28°22′E
136 C16 **Muğla** *var.* Mughla. ◆ *province* SW Turkey
Müglizh *see* Maglizh
144 J11 **Mugodzhary, Gory** *Kaz.* Mugalzhar Taūlary. ▲ W Kazakhstan
83 O15 **Mugulama** Zambézia, NE Mozambique 16°01′S 37°33′E
Muḥāfazat Hims *see* Ḥimṣ
139 U9 **Muḥammad** Wāsiṭ, E Iraq 32°46′N 45°14′E
139 R8 **Muḥammadīyah** Al Anbār, C Iraq 33°22′N 42°48′E
80 I6 **Muhammad Qol** Red Sea, NE Sudan 20°53′N 37°09′E
75 Y9 **Muḥammad, Rās** *headland* E Egypt 27°45′N 34°18′E
Muhammerah *see* Khorramshahr
140 M12 **Muḥāyil** *var.* Mahāil. 'Asīr, SW Saudi Arabia 18°34′N 42°01′E
139 O7 **Muḥaywīr** Al Anbār, W Iraq 33°35′N 41°06′E
101 H21 **Mühlacker** Baden-Württemberg, SW Germany 48°57′N 08°51′E
Mühlbach *see* Sebeş
Mühldorf *see* Mühldorf am Inn
101 N23 **Mühldorf am Inn** *var.* Mühldorf. Bayern, SE Germany 48°14′N 12°32′E
101 J15 **Mühlhausen** *var.* Mühlhausen in Thüringen. Thüringen, C Germany 51°13′N 10°28′E
Mühlhausen in Thüringen *see* Mühlhausen
195 Q2 **Mühlig-Hofmanfjella** *Eng.* Mühlig-Hofmann Mountains. ▲ Antarctica
93 L14 **Muhos** Pohjois-Pohjanmaa, C Finland 64°48′N 26°00′E
138 K6 **Mūḥ, Sabkhat al** ◎ C Syria
118 E5 **Muhu** *Ger.* Mohn, Moon. *island* W Estonia
81 F19 **Muhutwe** Kagera, NW Tanzania 01°31′S 31°41′E
Muhu Väin *see* Väinameri
98 J10 **Muiden** Noord-Holland, C Netherlands 52°19′N 05°04′E
193 W15 **Mui Hopohoponga** *headland* Tongatapu, S Tonga 21°09′S 175°02′W
Muinchille *see* Cootehill
Muineachán *see* Monaghan
97 F19 **Muine Bheag** *Eng.* Bagenalstown. Carlow, SE Ireland 52°42′N 06°57′W
56 B5 **Muisne** Esmeraldas, NW Ecuador 0°35′N 79°58′W
83 P14 **Muite** Nampula, NE Mozambique 14°02′S 39°06′E
41 Z11 **Mujeres, Isla** *island* E Mexico
116 G7 **Mukacheve** *Hung.* Munkács, *Rus.* Mukachevo. Zakarpats'ka Oblast', W Ukraine 48°27′N 22°45′E
Mukachevo *see* Mukacheve
169 R9 **Mukah** Sarawak, East Malaysia 02°56′N 112°02′E
Mukalla *see* Al Mukallā
Mukama *see* Mokama
Mukāshafa/Mukashshafah *see* Mukayshīfah
139 S6 **Mukayshīfah** *var.* Mukāshafa, Mukashshafah. Şalāḩ ad Dīn, N Iraq 34°24′N 43°44′E
167 R9 **Mukdahan** Mukdahan, E Thailand 16°31′N 104°43′E
Mukden *see* Shenyang
165 Y15 **Mukojima-rettō** *Eng.* Parry group. *island group* SE Japan
146 M14 **Mukry** Lebap Welaýaty, E Turkmenistan 37°39′N 65°37′E
Muksu *see* Mughsu
Muktagacha *see* Muktagachha
153 U14 **Muktagachha** *var.* Muktagacha. N Bangladesh 24°46′N 90°16′E
82 K13 **Mukuku** Central, C Zambia 12°10′S 29°50′E
82 K11 **Mukupa Kaoma** Northern, NE Zambia 09°55′S 30°19′E
81 I18 **Mukutan** Baringo, W Kenya 01°06′N 36°16′E
83 F16 **Mukwe** Caprivi, NE Namibia 18°01′S 21°24′E
105 R13 **Mula** Murcia, SE Spain 38°02′N 01°29′W
151 K20 **Mulakatholhu** *var.* Meemu Atoll, Mulaku Atoll. *atoll* C Maldives
Mulaku Atoll *see* Mulakatholhu
83 J15 **Mulalika** Lusaka, C Zambia 15°37′S 28°48′E
163 X8 **Mulan** Heilongjiang, NE China 45°57′N 128°00′E
83 N15 **Mulanje** *var.* Mlanje. Southern, S Malawi 16°05′S 35°29′E
40 H5 **Mulatos** Sonora, NW Mexico 28°42′N 108°44′W
23 P3 **Mulberry Fork** ≈ Alabama, S USA
39 P12 **Mulchatna River** ≈ Alaska, USA
125 W4 **Mul'da** Respublika Komi, NW Russian Federation 67°29′N 63°55′E
101 M14 **Mulde** ≈ E Germany
27 R10 **Muldrow** Oklahoma, C USA 35°25′N 94°34′W
40 E7 **Mulegé** Baja California Sur, NW Mexico 26°54′N 112°00′W
108 I10 **Mulegns** Graubünden, S Switzerland 46°30′N 09°36′E
79 M21 **Mulenda** Kasai-Oriental, C Dem. Rep. Congo 04°19′S 24°55′E
24 M4 **Muleshoe** Texas, SW USA 34°13′N 102°43′W
83 O15 **Mulevala** Zambézia, NE Mozambique 16°25′S 37°35′E
183 P5 **Mulgoa Creek** *seasonal river* New South Wales, SE Australia
105 O15 **Mulhacén** *var.* Cerro de Mulhacén. ▲ S Spain 37°07′N 03°11′W
Mulhacén, Cerro de *see* Mulhacén
Mülhausen *see* Mulhouse
101 E24 **Mülheim** Baden-Württemberg, SW Germany 47°50′N 07°37′E
101 E15 **Mülheim** *var.* Mulheim an der Ruhr. Nordrhein-Westfalen, W Germany 51°25′N 06°50′E
Mulheim an der Ruhr *see* Mülheim
103 U7 **Mulhouse** *Ger.* Mülhausen. Haut-Rhin, NE France 47°45′N 07°20′E
160 G11 **Muli** *var.* Qiaowa, Muli Zangzu Zizhixian. Sichuan, C China 27°49′N 101°10′E
171 X15 **Muli** *channel* Papua, E Indonesia
Mülig-Hofmann Mountains *see* Mühlig-Hofmanfjella
163 Y9 **Muling** Heilongjiang, NE China 44°54′N 130°35′E
Muli Zangzu Zizhixian *see* Muli
Mullach Íde *see* Malahide
Mullaitivu *see* Mullaittivu
155 K23 **Mullaittivu** *var.* Mullaitivu. Northern Province, N Sri Lanka 09°15′N 80°48′E
33 N8 **Mullan** Idaho, NW USA 47°28′N 115°48′W
28 M13 **Mullen** Nebraska, C USA 42°02′N 101°01′W
183 Q6 **Mullengudgery** New South Wales, SE Australia 31°42′S 147°24′E
21 Q6 **Mullens** West Virginia, NE USA 37°34′N 81°22′W
Müller-gerbergte *see* Muller, Pegunungan
169 T10 **Muller, Pegunungan** *Dut.* Müller-gerbergte. ▲ Borneo, C Indonesia
31 Q5 **Mullett Lake** ◎ Michigan, N USA
18 J16 **Mullica River** ≈ New Jersey, NE USA
25 R8 **Mullin** Texas, SW USA 31°33′N 98°40′W
97 E17 **Mullingar** *Ir.* An Muileann gCearr. C Ireland 53°32′N 07°20′W
21 T12 **Mullins** South Carolina, SE USA 34°12′N 79°15′W
96 G11 **Mull, Isle of** *island* W Scotland, United Kingdom
127 R5 **Mullovka** Ul'yanovskaya Oblast', W Russian Federation 54°13′N 49°19′E
95 K19 **Mullsjö** Västra Götaland, S Sweden 57°56′N 13°55′E
183 V4 **Mullumbimby** New South Wales, SE Australia 28°34′S 153°28′E
83 H15 **Mulobezi** Western, SW Zambia 16°48′S 25°11′E
83 C15 **Mulondo** Huíla, SW Angola 15°41′S 15°09′E
83 G15 **Mulonga Plain** *plain* W Zambia
79 N23 **Mulongo** Katanga, SE Dem. Rep. Congo 07°44′S 26°57′E
149 T10 **Multān** Punjab, E Pakistan 30°12′N 71°30′E
93 L17 **Multia** Keski-Suomi, C Finland 62°27′N 24°49′E
Mulucha *see* Moulouya
83 J14 **Mulungushi** Central, C Zambia 14°15′S 28°27′E
83 K14 **Mulungwe** Central, C Zambia 13°57′S 29°51′E
Muluya *see* Moulouya
27 N7 **Mulvane** Kansas, C USA 37°28′N 97°14′W
183 O10 **Mulwala** New South Wales, SE Australia 35°59′S 146°00′E
Mulwiya *see* Moulouya
182 K6 **Mulyungarie** South Australia 31°29′S 140°45′E
154 D13 **Mumbai** *prev.* Bombay. *state capital* Mahārāshtra, W India 18°56′N 72°51′E
154 D13 **Mumbai** ✕ Mahārāshtra, W India 19°10′N 72°51′E
83 D14 **Mumbué** Bié, C Angola 13°52′S 17°15′E
186 E8 **Mumeng** Morobe, C Papua New Guinea 06°57′S 146°34′E
171 V12 **Mumi** Papua Barat, E Indonesia 01°33′S 134°09′E
Muminabad/Mū'minobod *see* Mu'minobod
147 Q13 **Mu'minobod** *Rus.* Leningradskiy, Muminabad; *prev.* Leningrad. SW Tajikistan 38°03′N 69°50′E
127 Q13 **Mumra** Astrakhanskaya Oblast', SW Russian Federation 45°46′N 47°46′E
41 X12 **Muna** Yucatán, SE Mexico 20°29′N 89°41′W
123 O9 **Muna** ≈ NE Russian Federation
152 C12 **Munābāo** Rājasthān, NW India 25°46′N 70°19′E
Munamägi *see* Suur Munamägi
171 O14 **Muna, Pulau** *prev.* Moena. *island* C Indonesia
101 L18 **Münchberg** Bayern, E Germany 50°10′N 11°50′E
101 L23 **München** *var.* Muenchen, *Eng.* Munich, *It.* Monaco. Bayern, SE Germany 48°09′N 11°34′E
München-Gladbach *see* Mönchengladbach
108 E6 **Münchenstein** Basel Landschaft, NW Switzerland 47°31′N 07°38′E
10 L10 **Muncho Lake** British Columbia, W Canada 58°52′N 125°40′W
31 P13 **Muncie** Indiana, N USA 40°11′N 85°22′W
18 G13 **Muncy** Pennsylvania, NE USA 41°10′N 76°46′W
11 Q14 **Mundare** Alberta, SW Canada 53°34′N 112°20′W
25 Q5 **Munday** Texas, SW USA 33°27′N 99°37′W
31 N10 **Mundelein** Illinois, N USA 42°15′N 88°00′W
101 I15 **Münden** Niedersachsen, C Germany 52°16′N 08°54′E
105 Q12 **Mundo** ≈ S Spain
82 B12 **Munenga** Kwanza Sul, NW Angola 10°03′S 14°40′E
105 P11 **Munera** Castilla-La Mancha, C Spain 39°03′N 02°29′W
20 E9 **Munford** Tennessee, S USA 35°27′N 89°49′W
20 K7 **Munfordville** Kentucky, S USA 37°17′N 85°55′W
182 D5 **Mungala** South Australia 30°36′S 132°57′E
83 M16 **Mungári** Manica, C Mozambique 17°09′S 33°33′E
79 O16 **Mungbere** Orientale, NE Dem. Rep. Congo 02°38′N 28°30′E
153 Q13 **Munger** *prev.* Monghyr. Bihār, NE India 25°23′N 86°28′E
182 I2 **Mungeranie** South Australia 28°02′S 138°42′E
Mu Nggava *see* Rennell
169 O10 **Mungguresak, Tanjung** *headland* Borneo, N Indonesia 01°57′N 109°19′E
Mungiki *see* Bellona
183 R4 **Mungindi** New South Wales, SE Australia 28°59′S 149°00′E
Mungkawn *see* Maingkwan
Mungla *see* Mongla
82 C13 **Mungo** Huambo, W Angola 11°49′S 16°16′E
188 F16 **Munguuy Bay** *bay* Yap, W Micronesia
82 E13 **Munhango** Bié, C Angola 12°12′S 18°34′E
Munich *see* München
105 S7 **Muniesa** Aragón, NE Spain 41°02′N 00°49′W
31 O4 **Munising** Michigan, N USA 46°24′N 86°39′W
Munkács *see* Mukacheve
95 I17 **Munkedal** Västra Götaland, S Sweden 58°28′N 11°38′E
95 K15 **Munkfors** Värmland, C Sweden 59°50′N 13°35′E
122 M14 **Munku-Sardyk, Gora** *var.* Mönh Saridag. ▲ Mongolia/Russian Federation 51°45′N 100°22′E
99 E18 **Munkzwalm** Oost-Vlaanderen, NW Belgium 50°53′N 03°44′E
167 R10 **Mun, Mae Nam** ≈ E Thailand
153 U15 **Munshiganj** Dhaka, C Bangladesh 23°32′N 90°32′E
108 D8 **Münsingen** Bern, W Switzerland 46°53′N 07°34′E
103 U6 **Munster** Haut-Rhin, NE France 48°03′N 07°09′E
100 J11 **Munster** Niedersachsen, NW Germany 52°59′N 10°07′E
100 F13 **Münster** *var.* Muenster, Münster in Westfalen. Nordrhein-Westfalen, W Germany 51°58′N 07°38′E
108 F10 **Münster** Valais, S Switzerland 46°31′N 08°18′E
97 B20 **Munster** *Ir.* Cúige Mumhan. *cultural region* S Ireland
Münsterberg in Schlesien *see* Ziębice
Münster in Westfalen *see* Münster
100 E13 **Münsterland** *cultural region* NW Germany
100 F13 **Münster-Osnabruck** ✕ Nordrhein-Westfalen, NW Germany 52°08′N 07°41′E
31 R4 **Munuscong Lake** ◎ Michigan, N USA
83 K17 **Munyati** ≈ C Zimbabwe
109 R3 **Münzkirchen** Oberösterreich, N Austria 48°29′N 13°37′E
92 K11 **Muodoslompolo** Norrbotten, N Sweden 67°57′N 23°31′E
92 M13 **Muojärvi** ◎ NE Finland
167 S6 **Mường Khên** Hoa Binh, N Vietnam 20°34′N 105°18′E
Muong Sai *see* Oudômxai
167 Q7 **Muong Xiang Ngeun** *var.* Xieng Ngeun. Louangphabang, N Laos 19°43′N 102°09′E
92 K11 **Muonio** Lappi, N Finland 67°58′N 23°40′E
Muonioälv/Muoniojoki *see* Muonionjoki
92 K11 **Muonionjoki** *var.* Muoniojoki, *Swe.* Muonioälv. ≈ Finland/Sweden
Muorjek *see* Murjek
83 N17 **Mupa** ≈ C Mozambique
83 E16 **Mupini** Okavango, NE Namibia 17°55′S 19°34′E
80 F8 **Muqaddam, Wadi** ≈ N Sudan
138 K9 **Muqāṭ** Al Mafraq, E Jordan 32°28′N 38°04′E
81 N17 **Muqdisho** *Eng.* Mogadishu, *It.* Mogadiscio. ● (Somalia) Banaadir, S Somalia 02°06′N 45°27′E
81 N17 **Muqdisho** ✕ Banaadir, E Somalia 01°58′N 45°18′E
Muqshin *see* Mughshin
109 T8 **Mur** *SCr.* Mura. ≈ C Europe
109 X9 **Mura** ≈ NE Slovenia
Mura *see* Mur
137 T14 **Muradiye** Van, E Turkey 39°N 43°44′E
Muragarazi *see* Maragarazi
165 O10 **Murakami** Niigata, Honshū, C Japan 38°13′N 139°28′E
63 G22 **Murallón, Cerro** ▲ S Argentina 49°49′S 73°25′W
81 E20 **Muramvya** C Burundi 03°18′S 29°41′E
81 I19 **Murang'a** *prev.* Fort Hall. Murang'a, SW Kenya 0°43′S 37°10′E
81 I19 **Murang'a** ◆ *county* C Kenya
81 H16 **Murangering** Turkana, NW Kenya 03°48′N 35°29′E
Murapara *see* Murupara
140 M5 **Murār, Bi'r al** *well* NW Saudi Arabia
125 Q13 **Murashi** Kirovskaya Oblast', NW Russian Federation 59°27′N 48°02′E
103 O12 **Murat** Cantal, C France 45°07′N 02°52′E
114 N12 **Muratlı** Tekirdağ, NW Turkey 41°12′N 27°30′E
137 R14 **Murat Nehri** *var.* Eastern Euphrates; *anc.* Arsanias. ≈ NE Turkey
107 D20 **Muravera** Sardegna, Italy, C Mediterranean Sea 39°24′N 09°34′E
165 P10 **Murayama** Yamagata, Honshū, C Japan 38°29′N 140°21′E
121 R13 **Murayşah, Ra's al** *headland* N Libya 31°58′N 25°00′E
104 I6 **Murça** Vila Real, N Portugal 41°24′N 07°28′W
80 Q11 **Murcanyo** Bari, NE Somalia 11°39′N 50°27′E
143 N8 **Mūrcheh Khvort** *var.* Morcheh Khort. Eşfahān, C Iran 33°07′N 51°26′E
185 H15 **Murchison** Tasman, South Island, New Zealand 41°48′S 172°19′E
185 B22 **Murchison Mountains** ▲ South Island, New Zealand
180 I10 **Murchison River** ≈ Western Australia
105 R13 **Murcia** Murcia, SE Spain 37°59′N 01°08′W
105 Q13 **Murcia** ◆ *autonomous community* SE Spain
103 O13 **Mur-de-Barrez** Aveyron, S France 44°48′N 02°39′E
182 G8 **Murdinga** South Australia 33°46′S 135°46′E
28 M10 **Murdo** South Dakota, N USA 43°53′N 100°42′W
15 X6 **Murdochville** Québec, SE Canada 48°57′N 65°30′W
109 W9 **Mureck** Steiermark, SE Austria 46°42′N 15°46′E
114 M13 **Mürefte** Tekirdağ, NW Turkey 40°40′N 27°15′E
116 I10 **Mureş** ◆ *county* N Romania
84 J11 **Mureş** ≈ Hungary/Romania
Mureşul *see* Maros/Mureş
102 M16 **Muret** Haute-Garonne, S France 43°28′N 01°19′E
27 T13 **Murfreesboro** Arkansas, C USA 34°04′N 93°42′W
21 W8 **Murfreesboro** North Carolina, SE USA 36°26′N 77°06′W
20 J9 **Murfreesboro** Tennessee, S USA 35°50′N 86°25′W
Murgab *see* Morghāb, Daryā-ye/Murgap
Murgab *see* Murghob
146 I14 **Murgap** Mary Welaýaty, S Turkmenistan 37°19′N 61°48′E
146 J16 **Murgap** *var.* Deryasy Murgap, Murghab, *Pash.* Daryā-ye Morghāb, *Rus.* Murgab. ≈ Afghanistan/Turkmenistan *see also* Morghāb, Daryā-ye
Murgap *see* Morghāb, Daryā-ye
Murgap, Deryasy *see* Morghāb, Daryā-ye/Murgap
114 H9 **Murgash** ▲ W Bulgaria 42°51′N 23°58′E
Murghab *see* Morghāb, Daryā-ye/Murgap
148 M4 **Murghāb, Daryā-ye** *Rus.* Murgab, Murghab, *Turk.* Murgap, Deryasy Murgap; *prev.* Morghāb, Daryā-ye. ≈ Afghanistan/Turkmenistan *see also* Murgap
147 U13 **Murghob** *Rus.* Murgab. SE Tajikistan 38°11′N 74°E
147 U13 **Murghob** *Rus.* Murgab. ≈ SE Tajikistan
181 Z10 **Murgon** Queensland, E Australia 26°08′S 152°04′E
190 I16 **Muri** Rarotonga, S Cook Islands 21°15′S 159°44′W
108 F7 **Muri** Aargau, W Switzerland 47°16′N 08°21′E
108 D8 **Muri** *var.* Muri bei Bern. Bern, W Switzerland 46°55′N 07°30′E
104 K3 **Murias de Paredes** Castilla y León, N Spain 42°51′N 06°11′W
Muri bei Bern *see* Muri
82 F11 **Muriege** Lunda Sul, NE Angola 09°55′S 21°12′E
189 P14 **Murilo Atoll** *atoll* Hall Islands, C Micronesia
Mūrītānīyah *see* Mauritania
100 N10 **Müritz** *var.* Müritzee. ◎ NE Germany
Müritzee *see* Müritz
100 L10 **Müritz-Elde-Wasserstrasse** *canal* N Germany
184 K6 **Muriwai Beach** Auckland, North Island, New Zealand 36°56′S 174°28′E
92 J13 **Murjek** *Lapp.* Muorjek. Norrbotten, N Sweden 66°27′N 20°54′E
124 J3 **Murmansk** Murmanskaya Oblast', NW Russian Federation 68°59′N 33°08′E
124 I4 **Murmanskaya Oblast'** ◆ *province* NW Russian Federation
197 V14 **Murmansk Rise** *undersea feature* SW Barents Sea 71°00′N 37°00′E
124 J3 **Murmashi** Murmanskaya Oblast', NW Russian Federation 68°49′N 32°43′E
126 M5 **Murmino** Ryazanskaya Oblast', W Russian Federation 54°31′N 40°01′E
101 K24 **Murnau** Bayern, SE Germany 47°41′N 11°12′E
103 X16 **Muro, Capo di** *headland* Corse, France, C Mediterranean Sea 41°45′N 08°40′E
107 M18 **Muro Lucano** Basilicata, S Italy 40°48′N 15°33′E
127 N4 **Murom** Vladimirskaya Oblast', W Russian Federation 55°33′N 42°03′E
122 I11 **Muromtsevo** Omskaya Oblast', C Russian Federation 56°18′N 75°15′E
165 R5 **Muroran** Hokkaidō, NE Japan 42°20′N 140°58′E
104 G3 **Muros** Galicia, NW Spain 42°47′N 09°04′W
104 F3 **Muros e Noia, Ría de** *estuary* NW Spain
164 H15 **Muroto** Kōchi, Shikoku, SW Japan 33°16′N 134°10′E
164 H15 **Muroto-zaki** Shikoku, SW Japan
116 L7 **Murovani Kurylivtsi** Vinnyts'ka Oblast', C Ukraine 48°43′N 27°31′E
110 G11 **Murowana Goślina** Wielkopolskie, W Poland 52°33′N 16°59′E
32 M14 **Murphy** Idaho, NW USA 43°14′N 116°36′W
21 N10 **Murphy** North Carolina, SE USA 35°05′N 84°02′W
35 P8 **Murphys** California, W USA 38°07′N 120°27′W
30 L17 **Murphysboro** Illinois, N USA 37°45′N 89°20′W
29 V15 **Murray** Iowa, C USA 41°03′N 93°56′W
20 H8 **Murray** Kentucky, S USA 36°35′N 88°20′W
182 J10 **Murray Bridge** South Australia 35°10′S 139°17′E
175 X2 **Murray Fracture Zone** *tectonic feature* NE Pacific Ocean
192 H11 **Murray, Lake** ◎ SW Papua New Guinea
21 P12 **Murray, Lake** ⊠ South Carolina, SE USA
10 K8 **Murray, Mount** ▲ Yukon, NW Canada 60°49′N 128°57′W
Murray Range *see* Murray Ridge
173 O3 **Murray Ridge** *var.* Murray Range. *undersea feature* N Arabian Sea 21°45′N 61°50′E
183 N10 **Murray River** ≈ SE Australia
182 K10 **Murrayville** Victoria, SE Australia 35°17′S 141°12′E
149 U5 **Murree** Punjab, E Pakistan 33°55′N 73°26′E
101 I21 **Murrhardt** Baden-Württemberg, S Germany 49°00′N 09°34′E
183 O9 **Murrumbidgee River** ≈ New South Wales, SE Australia
83 P15 **Murrupula** Nampula, NE Mozambique 15°26′S 38°46′E
183 T7 **Murrurundi** New South Wales, SE Australia 31°47′S 150°51′E
109 X9 **Murska Sobota** *Ger.* Olsnitz. NE Slovenia 46°41′N 16°09′E
154 G12 **Murtajāpur** *prev.* Murtazapur. Mahārāshtra, C India 20°43′N 77°28′E
77 S16 **Murtala Muhammed** ✕ (Lagos) Ogun, SW Nigeria 06°31′N 03°12′E
Murtazapur *see* Murtajāpur
108 C8 **Murten** Neuchâtel, W Switzerland 46°55′N 07°06′E
Murtensee *see* Morat, Lac de
182 L11 **Murtoa** Victoria, SE Australia 36°39′S 142°27′E
92 N13 **Murtovaara** Pohjois-Pohjanmaa, E Finland 65°40′N 29°25′E
Murua Island *see* Woodlark Island
155 D14 **Murud** Mahārāshtra, W India 18°27′N 72°56′E
184 O9 **Murupara** *var.* Murapara. Bay of Plenty, North Island, New Zealand 38°27′S 176°41′E
Mururoa *see* Moruroa
Murviedro *see* Sagunto
154 J9 **Murwāra** Madhya Pradesh, N India 23°50′N 80°23′E
183 V4 **Murwillumbah** New South Wales, SE Australia 28°20′S 153°24′E
146 H11 **Murzechirla** *prev.* Mirzachirla. Ahal Welaýaty, C Turkmenistan 39°33′N 60°02′E
Murzuk *see* Murzuq
75 O11 **Murzuq** *var.* Marzūq, Murzuk. SW Libya 25°55′N 13°55′E
Murzuq, Edeyin *see* Murzuq, Idhān
75 N11 **Murzuq, Ḥammādat** *plateau* W Libya
75 O11 **Murzuq, Idhān** *var.* Edeyin Murzuq. *desert* SW Libya
109 W6 **Mürzzuschlag** Steiermark, E Austria 47°35′N 15°41′E
137 Q14 **Muş** *var.* Mush. Muş, E Turkey 38°45′N 41°30′E
137 Q14 **Muş** *var.* Mush. ◆ *province* E Turkey
118 G11 **Mūša** ≈ Latvia/Lithuania
186 F9 **Musa** ≈ S Papua New Guinea
Mūsa, Gebel *see* Mūsá, Jabal
Musaiyib *see* Al Musayyib
75 X8 **Mūsá, Jabal** *var.* Gebel Mûsa. ▲ NE Egypt 28°33′N 33°51′E
149 R9 **Musā Khel** *var.* Mūsa Khel Bāzār. Baluchistān, SW Pakistan 30°53′N 69°52′E
Mūsa Khel Bāzār *see* Musā Khel
139 Z13 **Mūsá, Khowr-e** *bay* Iraq/Kuwait
114 H10 **Musala** ▲ W Bulgaria 42°12′N 23°36′E
168 H10 **Musala, Pulau** *island* W Indonesia

83 I15 **Musale** Southern, S Zambia *15°27´S 26°50´E*
141 Y9 **Muṣalla** NE Oman *22°20´N 58°03´E*
141 W6 **Musandam Peninsula** *Ar.* Masandam Peninsula. *peninsula* N Oman
Musay'id *see* Umm Sa'id
Muscat *see* Masqaṭ
Muscat and Oman *see* Oman
29 Y14 **Muscatine** Iowa, C USA *41°25´N 91°03´W*
Muscat Sīb Airport *see* Seeb
31 O15 **Muscatuck River** ≈ Indiana, N USA
30 K8 **Muscoda** Wisconsin, N USA *43°11´N 90°27´W*
185 F19 **Musgrave, Mount** ▲ South Island, New Zealand *43°48´S 170°43´E*
181 P9 **Musgrave Ranges** ▲▲ South Australia
Mush *see* Muş
138 H12 **Mushayyish, Qaṣr al** *castle* Ma'ān, C Jordan
79 H20 **Mushie** Bandundu, W Dem. Rep. Congo *03°00´S 16°55´E*
168 M13 **Musi, Air** *prev.* Moesi. ≈ Sumatera, W Indonesia
192 M4 **Musicians Seamounts** *undersea feature* N Pacific Ocean
83 K19 **Musina** *prev.* Messina. Limpopo, NE South Africa *22°18´S 30°02´E*
54 D8 **Musinga, Alto** ▲ NW Colombia *06°49´N 76°24´W*
29 T2 **Muskeg Bay** *lake bay* Minnesota, N USA
31 O8 **Muskegon** Michigan, N USA *43°13´N 86°15´W*
31 O8 **Muskegon Heights** Michigan, N USA *43°12´N 86°14´W*
31 P8 **Muskegon River** ≈ Michigan, N USA
31 T14 **Muskingum River** ≈ Ohio, N USA
95 P16 **Muskö** Stockholm, C Sweden *58°58´N 18°10´E*
Muskogean *see* Tallahassee
27 Q10 **Muskogee** Oklahoma, C USA *35°45´N 95°21´W*
14 H13 **Muskoka, Lake** ◎ Ontario, S Canada
80 H8 **Musmar** Red Sea, NE Sudan *18°13´N 35°40´E*
83 K14 **Musofu** Central, C Zambia *13°31´S 29°02´E*
81 G19 **Musoma** Mara, N Tanzania *01°31´S 33°49´E*
82 L13 **Musoro** Central, C Zambia *13°21´S 31°04´E*
186 F4 **Mussau Island** *island* NE Papua New Guinea
98 P7 **Musselkanaal** Groningen, NE Netherlands *52°55´N 07°01´E*
33 V9 **Musselshell River** ≈ Montana, NW USA
82 C12 **Mussende** Kwanza Sul, NW Angola *10°33´S 16°02´E*
102 L12 **Mussidan** Dordogne, SW France *45°03´N 00°22´E*
99 L25 **Musson** Luxembourg, SE Belgium *49°33´N 05°42´E*
152 J9 **Mussoorie** Uttarakhand, N India *30°26´N 78°04´E*
Musta *see* Mosta
152 M13 **Mustafābād** Uttar Pradesh, N India *25°54´N 81°17´E*
136 D12 **Mustafakemalpaşa** Bursa, NW Turkey *40°03´N 28°25´E*
Mustafa-Pasha *see* Svilengrad
81 M15 **Mustahīl** Sumalē, E Ethiopia *05°18´N 44°34´E*
24 M7 **Mustang Draw** *valley* Texas, SW USA
25 T14 **Mustang Island** *island* Texas, SW USA
Mustasaari *see* Korsholm
Mustér *see* Disentis
63 I19 **Musters, Lago** ◎ S Argentina
45 Y14 **Mustique** *island* C Saint Vincent and the Grenadines
118 I6 **Mustla** Viljandimaa, S Estonia *58°12´N 25°50´E*
118 J4 **Mustvee** *Ger.* Tschorna. Jõgevamaa, E Estonia *58°51´N 26°59´E*
42 L9 **Musún, Cerro** ▲ NE Nicaragua *13°01´N 85°02´W*
183 T7 **Muswellbrook** New South Wales, SE Australia *32°17´S 150°55´E*
111 M18 **Muszyna** Małopolskie, SE Poland *49°21´N 20°54´E*
75 V10 **Mūṭ** *var.* Mut. C Egypt *25°28´N 28°58´E*
136 I17 **Mut** İçel, S Turkey *36°38´N 33°27´E*
109 V9 **Muta** N Slovenia *46°37´N 15°09´E*
190 B15 **Mutalau** N Niue *18°56´S 169°50´E*
Mu-tan-chiang *see* Mudanjiang
82 I13 **Mutanda** North Western, NW Zambia *12°24´S 26°13´E*
59 O17 **Mutá, Ponta do** *headland* E Brazil *13°54´S 38°54´W*
83 L17 **Mutare** *var.* Mutari; *prev.* Umtali. Manicaland, E Zimbabwe *18°55´S 32°36´E*
Mutari *see* Mutare
54 D8 **Mutatá** Antioquia, NW Colombia *07°16´N 76°32´W*
Muthannār, Muḥāfa at al *see* Al Muthannā
Mutina *see* Modena
83 L16 **Mutoko** *prev.* Mtoko. Mashonaland East, NE Zimbabwe *17°24´S 32°13´E*
81 J20 **Mutomo** Kitui, S Kenya *01°50´S 38°13´E*
Mutrah *see* Maṭraḥ
79 M24 **Mutshatsha** Katanga, S Dem. Rep. Congo *10°40´S 24°26´E*
165 R6 **Mutsu** *var.* Mutu. Aomori, Honshū, N Japan *41°18´N 141°11´E*
165 R6 **Mutsu-wan** *bay* N Japan
108 E6 **Muttenz** Basel Landschaft, NW Switzerland *47°31´N 07°39´E*
185 A26 **Muttonbird Islands** *island group* SW New Zealand
Muttra *see* Mathura
Mutu *see* Mutsu
83 O15 **Mutuáli** Nampula, N Mozambique *14°51´S 37°01´E*
82 D13 **Mutumbo** Bié, C Angola *13°10´S 17°22´E*
189 Y14 **Mutunte, Mount** *var.* Mount Buache. ▲ Kosrae, E Micronesia *05°21´N 163°00´E*
155 K24 **Mutur** Eastern Province, E Sri Lanka *08°27´N 81°15´E*
92 L13 **Muurola** Lappi, NW Finland *66°22´N 25°20´E*
162 M14 **Mu Us Shadi** *var.* Ordos Desert; *prev.* Mu Us Shamo. *desert* N China
Mu Us Shamo *see* Mu Us Shadi
82 B11 **Muxima** Bengo, NW Angola *09°33´S 13°58´E*
124 I8 **Muyezerskiy** Respublika Kareliya, NW Russian Federation *63°54´N 32°00´E*
81 E20 **Muyinga** NE Burundi *02°54´S 30°19´E*
42 K9 **Muy Muy** Matagalpa, C Nicaragua *12°43´N 85°35´W*
Muynak *see* Mo'ynoq
79 N22 **Muyumba** Katanga, SE Dem. Rep. Congo *07°13´S 27°02´E*
149 V5 **Muzaffarābād** Jammu and Kashmir, NE Pakistan *34°23´N 73°34´E*
149 S10 **Muzaffargarh** Punjab, E Pakistan *30°04´N 71°15´E*
152 J9 **Muzaffarnagar** Uttar Pradesh, N India *29°28´N 77°42´E*
153 P13 **Muzaffarpur** Bihār, N India *26°07´N 85°23´E*
158 H6 **Muzat He** ≈ W China
83 L15 **Muze** Tete, NW Mozambique *15°05´S 31°16´E*
122 H8 **Muzhi** Yamalo-Nenetskiy Avtonomnyy Okrug, N Russian Federation *65°25´N 64°28´E*
102 H7 **Muzillac** Morbihan, NW France *47°34´N 02°30´W*
Muzkol, Khrebet *see* Muzqŭl, Qatorkŭhi
112 L9 **Mužlja** *Hung.* Felsőmuzslya; *prev.* Gornja Mužlja. Vojvodina, N Serbia *45°21´N 20°25´E*
54 F9 **Muzo** Boyacá, C Colombia *05°34´N 74°07´W*
83 J15 **Muzoka** Southern, S Zambia *16°39´S 27°18´E*
39 Y15 **Muzon, Cape** *headland* Dall Island, Alaska, USA *54°39´N 132°41´W*
40 M6 **Múzquiz** Coahuila, NE Mexico *27°54´N 101°30´W*
147 U13 **Muzqŭl, Qatorkŭhi** *Rus.* Khrebet Muzkol. ▲▲ SE Tajikistan
Muztag *see* Muztag Feng
158 D8 **Muztagata** ▲ NW China *38°16´N 75°03´E*
158 K10 **Muztag Feng** *var.* Muztag. ▲ W China *36°26´N 87°15´E*
83 K17 **Mvuma** *prev.* Umvuma. Midlands, C Zimbabwe *19°17´S 30°32´E*
172 H13 **Mwali** *var.* Moili, *Fr.* Mohéli. *island* S Comoros
82 L13 **Mwanya** Eastern, E Zambia *12°40´S 32°15´E*
79 N23 **Mwanza** Katanga, SE Dem. Rep. Congo *07°49´S 26°49´E*
81 F20 **Mwanza** ◐ *region* N Tanzania
81 G20 **Mwanza** ◐ *state capital* Mwanza, NW Tanzania
82 M13 **Mwase Lundazi** Eastern, E Zambia *12°26´S 33°20´E*
97 B17 **Mweelrea** *Ir.* Caoc Maol Réidh. ▲ W Ireland *53°37´N 09°47´W*
79 K21 **Mweka** Kasai-Occidental, C Dem. Rep. Congo *04°52´S 21°38´E*
82 K12 **Mwenda** Luapula, N Zambia *10°30´S 30°21´E*
79 L22 **Mwene-Ditu** Kasai-Oriental, S Dem. Rep. Congo *07°06´S 23°34´E*
83 L18 **Mwenezi** ≈ S Zimbabwe
79 O20 **Mwenga** Sud-Kivu, E Dem. Rep. Congo *03°00´S 28°28´E*
82 K11 **Mweru, Lake** *var.* Lac Moero. ◎ Dem. Rep. Congo/Zambia
82 H13 **Mwinilunga** North Western, NW Zambia *11°44´S 24°24´E*
189 V16 **Mwokil Atoll** *prev.* Mokil Atoll. *atoll* Caroline Islands, E Micronesia
Myadel' *see* Myadzyel
118 J13 **Myadzyel** *Pol.* Miadzioł Nowy, *Rus.* Myadel'. Minskaya Voblasts', N Belarus *54°51´N 26°51´E*
152 C12 **Myājlār** *var.* Miajlar. Rājasthān, NW India *26°16´N 70°21´E*
123 T9 **Myakit** Magadanskaya Oblast', E Russian Federation *61°23´N 151°58´E*
23 W13 **Myakka River** ≈ Florida, SE USA
124 L14 **Myaksa** Vologodskaya Oblast', NW Russian Federation *58°54´N 38°15´E*
183 U8 **Myall Lake** ◎ New South Wales, SE Australia
166 L7 **Myanaung** Ayeyawady, SW Myanmar (Burma) *18°17´N 95°19´E*
166 M4 **Myanmar (Burma)** *off.* Republic of the Union of Myanmar; *prev.* Union of Myanmar, *var.* Burma. ◆ *transitional democracy* SE Asia
Myanmar, Republic of the Union of *see* Myanmar (Burma)
Myanmar, Union of *see* Myanmar (Burma)
166 K8 **Myaungmya** Ayeyawady, SW Myanmar (Burma) *16°33´N 94°55´E*
Myaydo *see* Aunglan
118 N11 **Myazha** *Rus.* Mezha. Vitsyebskaya Voblasts', NE Belarus *55°41´N 30°25´E*
167 N12 **Myeik** *var.* Mergui. Taninthayi, S Myanmar (Burma) *12°26´N 98°34´E*
166 M12 **Myeik Archipelago** *var.* Mergui Archipelago. *island group* S Myanmar (Burma)
119 O18 **Myerkulavichy** *Rus.* Merkulovichi. Homyel'skaya Voblasts', SE Belarus *52°58´N 30°36´E*
119 N14 **Myezhava** *Rus.* Mezhëvo. Vitsyebskaya Voblasts', NE Belarus *54°38´N 30°20´E*
Myggenaes *see* Mykines
166 L5 **Myingyan** Mandalay, C Myanmar (Burma) *21°25´N 95°20´E*
167 N12 **Myitkyina** Kachin State, N Myanmar (Burma) *25°24´N 97°25´E*
166 M5 **Myittha** Mandalay, C Myanmar (Burma) *21°21´N 96°06´E*
111 H19 **Myjava** *Hung.* Miava. Trenčiansky Kraj, W Slovakia *48°45´N 17°35´E*
Myjeldino *see* Myyëldino
117 U9 **Mykhaylivka** *Rus.* Mikhaylovka. Zaporiz'ka Oblast', SE Ukraine *47°16´N 35°14´E*
95 A18 **Mykines** *Dan.* Myggenaes. *island* W Faroe Islands
116 I5 **Mykolayiv** L'vivs'ka Oblast', W Ukraine *49°34´N 23°58´E*
117 Q10 **Mykolayiv** *Rus.* Nikolayev. Mykolayivs'ka Oblast', S Ukraine *46°58´N 31°59´E*
117 Q10 **Mykolayiv** ✈ Mykolayivs'ka Oblast', S Ukraine *47°02´N 31°54´E*
Mykolayiv *see* Mykolayivs'ka Oblast'
117 S13 **Mykolayivka** Avtonomna Respublika Krym, S Ukraine *44°58´N 33°37´E*
117 P9 **Mykolayivka** Odes'ka Oblast', SW Ukraine *47°34´N 30°48´E*
117 P9 **Mykolayivs'ka Oblast'** *var.* Mykolayiv, *Rus.* Nikolayevskaya Oblast'. ◐ *province* S Ukraine
115 J20 **Mýkonos** Mńykonos, Kykládes, Greece, Aegean Sea *37°27´N 25°20´E*
115 K20 **Mýkonos** *var.* Míkonos. *island* Kykládes, Greece, Aegean Sea
125 R7 **Myla** Respublika Komi, NW Russian Federation *65°24´N 50°51´E*
Mylae *see* Milazzo
93 M19 **Myllykoski** Kymenlaakso, S Finland *60°45´N 26°52´E*
153 U14 **Mymensing** *var.* Mymensingh. Dhaka, N Bangladesh *24°45´N 90°23´E*
Mymensingh *see* Mymensing
93 K19 **Mynämäki** Varsinais-Suomi, SW Finland *60°41´N 22°00´E*
145 S14 **Mynaral** *Kaz.* Myngaral. Zhambyl, S Kazakhstan *45°25´N 73°37´E*
Mynbulak *see* Mingbuloq
Mynbulak, Vpadina *see* Mingbuloq Botig'I
Myngaral *see* Mynaral
Myohaung *see* Mrauk-oo
163 W13 **Myohyang-sanmaek** ▲▲ C North Korea
164 M11 **Myōkō-san** ▲ Honshū, S Japan *36°54´N 138°05´E*
83 J15 **Myooye** Central, C Zambia *15°11´S 27°10´E*
118 K12 **Myory** *prev.* Miyory, *Rus.* Miory. Vitsyebskaya Voblasts', N Belarus *55°39´N 27°39´E*
92 J4 **Mýrdalsjökull** *glacier* S Iceland
92 G10 **Myre** Nordland, C Norway *68°54´N 15°04´E*
117 S5 **Myrhorod** *Rus.* Mirgorod. Poltavs'ka Oblast', NE Ukraine *49°58´N 33°37´E*
115 J15 **Mýrina** *var.* Mírina. Límnos, SE Greece *39°52´N 25°04´E*
117 P5 **Myronivka** *Rus.* Mironovka. Kyyivs'ka Oblast', N Ukraine *49°40´N 30°59´E*
21 U13 **Myrtle Beach** South Carolina, SE USA *33°41´N 78°53´W*
32 F14 **Myrtle Creek** Oregon, NW USA *43°01´N 123°19´W*
183 P11 **Myrtleford** Victoria, SE Australia *36°34´S 146°45´E*
32 E14 **Myrtle Point** Oregon, NW USA *43°04´N 124°08´W*
115 K25 **Mýrtos** Kríti, Greece, E Mediterranean Sea *35°00´N 25°34´E*
Myrtoum Mare *see* Mirtóo Pélagos
93 G17 **Myrviken** Jämtland, C Sweden *62°59´N 14°19´E*
95 I15 **Mysen** Østfold, S Norway *59°33´N 11°20´E*
124 L15 **Myshkin** Yaroslavskaya Oblast', NW Russian Federation *57°47´N 38°28´E*
111 K17 **Myślenice** Malopolskie, S Poland *49°50´N 19°55´E*
110 D10 **Myślibórz** Zachodnio-pomorskie, NW Poland *52°55´N 14°51´E*
155 G20 **Mysore** *var.* Maisur. Karnātaka, W India *12°18´N 76°37´E*
Mysore *see* Karnātaka
115 F21 **Mystrás** *var.* Mistras. Pelopónnisos, S Greece *37°03´N 22°22´E*
111 K15 **Myszków** Śląskie, S Poland *50°36´N 19°20´E*
167 T14 **My Tho** *var.* Mi Tho. Tiên Giang, S Vietnam *10°21´N 106°21´E*
Mytilene *see* Mytilíni
115 L17 **Mytilíni** *var.* Mitilíni; *anc.* Mytilene. Lésvos, E Greece *39°06´N 26°33´E*
126 K3 **Mytishchi** Moskovskaya Oblast', W Russian Federation *56°00´N 37°51´E*
37 N3 **Myton** Utah, W USA *40°11´N 110°03´W*
92 K2 **Mývatn** ◎ C Iceland
125 T11 **Myyëldino** *var.* Myjeldino. Respublika Komi, NW Russian Federation *61°46´N 54°48´E*
82 M13 **Mzimba** Northern, NW Malawi *11°56´S 33°36´E*
82 M12 **Mzuzu** Northern, N Malawi *11°23´S 34°03´E*

N

101 M19 **Naab** ≈ SE Germany
98 G12 **Naaldwijk** Zuid-Holland, W Netherlands *52°00´N 04°13´E*
38 G12 **Nā'ālehu** *var.* Naalehu. Hawaii, USA, C Pacific Ocean *19°04´N 155°36´W*
93 K19 **Naantali** *Swe.* Nådendal. Varsinais-Suomi, SW Finland *60°28´N 22°05´E*
98 J10 **Naarden** Noord-Holland, C Netherlands *52°18´N 05°10´E*
109 U4 **Naarn** ≈ N Austria
97 F18 **Naas** *Ir.* An Nás, Nás na Ríogh. Kildare, C Ireland *53°13´N 06°39´W*
92 M9 **Näätämöjoki** *Lapp.* Njávdám. ≈ NE Finland
83 E23 **Nababeep** *var.* Nababiep. Northern Cape, W South Africa *29°36´S 17°46´E*
Nababiep *see* Nababeep
Nabadwip *see* Navadwīp
164 J14 **Nabari** Mie, Honshū, SW Japan *34°37´N 136°05´E*
Nabatié *see* Nabatîyé
138 G8 **Nabatîyé** *var.* An Nabatīyah at Taḥtā, Nabatié, Nabatiyet et Tahta. SW Lebanon *33°18´N 35°36´E*
Nabatiyet et Tahta *see* Nabatîyé
187 X14 **Nabavatu** Vanua Levu, N Fiji *16°35´S 178°55´E*
190 I2 **Nabeina** *island* Tungaru, W Kiribati
127 T4 **Naberezhnyye Chelny** *prev.* Brezhnev. Respublika Tatarstan, W Russian Federation *55°43´N 52°21´E*
39 T10 **Nabesna** Alaska, USA *62°22´N 143°00´W*
39 T10 **Nabesna River** ≈ Alaska, USA
75 N5 **Nabeul** *var.* Nābul. NE Tunisia *36°32´N 10°45´E*
152 I9 **Nābha** Punjab, NW India *30°22´N 76°12´E*
171 W13 **Nabire** Papua, E Indonesia *03°23´S 135°31´E*
141 O15 **Nabī Shu'ayb, Jabal an** ▲ W Yemen *15°24´N 44°04´E*
138 F10 **Nablus** *var.* Năbulus, *Heb.* Shekhem; *anc.* Neapolis, *Bibl.* Shechem. N West Bank *32°13´N 35°16´E*
187 X14 **Nabouwalu** Vanua Levu, N Fiji *17°00´S 178°43´E*
Nābul *see* Nabeul
Năbulus *see* Nablus
187 Y13 **Nabuna** Vanua Levu, N Fiji *16°13´S 179°46´E*
83 Q14 **Nacala** Nampula, NE Mozambique *14°30´S 40°37´E*
42 H8 **Nacaome** Valle, S Honduras *13°30´N 87°31´W*
Na Cealla Beaga *see* Killybegs
Na-Ch'ii *see* Nagqu
164 J15 **Nachikatsuura** *var.* Nachi-Katsuura. Wakayama, Honshū, SE Japan *33°37´N 135°54´E*
Nachi-Katsuura *see* Nachikatsuura
81 J24 **Nachingwea** Lindi, SE Tanzania *10°21´S 38°46´E*
111 F16 **Náchod** Královéhradecký Kraj, N Czech Republic *50°26´N 16°10´E*
Na Clocha Liatha *see* Greystones
40 G3 **Naco** Sonora, NW Mexico *31°16´N 109°56´W*
25 X8 **Nacogdoches** Texas, SW USA *31°36´N 94°40´W*
40 G4 **Nacozari de García** Sonora, NW Mexico *30°27´N 109°43´W*
Nada *see* Danzhou
77 O14 **Nadawli** NW Ghana *10°30´N 02°40´W*
104 I3 **Nadela** Galicia, NW Spain *42°58´N 07°33´W*
Nådendal *see* Naantali
144 M7 **Nadezhdinka** *prev.* Nadezhdinskiy. Kostanay, N Kazakhstan *53°46´N 63°44´E*
Nadezhdinskiy *see* Nadezhdinka
Nadgan *see* Nadqān, Qalamat
187 W14 **Nadi** *prev.* Nandi. Viti Levu, W Fiji *17°47´S 177°32´E*
187 X14 **Nadi** *prev.* Nandi. ✈ Viti Levu, W Fiji *17°46´S 177°28´E*
154 D10 **Nadiād** Gujarāt, W India *22°42´N 72°55´E*
Nadikdik *see* Knox Atoll
116 E11 **Nădlac** *Ger.* Nadlak, *Hung.* Nagylak. Arad, W Romania *46°10´N 20°47´E*
Nadlak *see* Nădlac
74 H6 **Nador** *prev.* Villa Nador. NE Morocco *35°10´N 05°22´W*
141 S9 **Nadqān, Qalamat** *var.* Nadgan. *well* E Saudi Arabia
111 N22 **Nádudvar** Hajdú-Bihar, E Hungary *47°36´N 21°09´E*
121 O15 **Nadur** Gozo, N Malta *36°03´N 14°18´E*
187 X13 **Naduri** *prev.* Nanduri. Vanua Levu, N Fiji *16°26´S 179°08´E*
116 I7 **Nadvirna** *Pol.* Nadwórna, *Rus.* Nadvornaya. Ivano-Frankivs'ka Oblast', W Ukraine *48°27´N 24°30´E*
124 J8 **Nadvoitsy** Respublika Kareliya, NW Russian Federation *63°53´N 34°17´E*
Nadvornaya/Nadwórna *see* Nadvirna
122 I9 **Nadym** Yamalo-Nenetskiy Avtonomnyy Okrug, N Russian Federation *65°25´N 72°40´E*
122 I9 **Nadym** ≈ C Russian Federation
186 E7 **Nadzab** Morobe, C Papua New Guinea *06°36´S 146°46´E*
95 C17 **Nærbø** Rogaland, S Norway *58°40´N 05°39´E*
95 I24 **Næstved** Sjælland, SE Denmark *55°12´N 11°47´E*
77 X13 **Nafada** Gombe, E Nigeria *11°02´N 11°18´E*
108 H8 **Näfels** Glarus, NE Switzerland *47°06´N 09°04´E*
115 E18 **Náfpaktos** *var.* Návpaktos. Dytikí Elláda, C Greece *38°23´N 21°50´E*
115 F20 **Náfplio** *prev.* Návplion. Pelopónnisos, S Greece *37°34´N 22°50´E*
139 U6 **Naft Khāneh** Diyālá, E Iraq *34°01´N 45°26´E*
149 N13 **Nāg** Baluchistān, SW Pakistan *27°43´N 65°31´E*
171 P4 **Naga** *off.* Naga City; *prev.* Nueva Caceres. Luzon, N Philippines *13°36´N 123°10´E*
Nagaarzê *see* Nagarzê
Naga City *see* Naga
12 F11 **Nagagami** ≈ Ontario, S Canada
164 F14 **Nagahama** Ehime, Shikoku, SW Japan *33°36´N 132°29´E*
153 X12 **Nāga Hills** ▲▲ NE India
165 P10 **Nagai** Yamagata, Honshū, C Japan *38°08´N 140°00´E*
Na Gaibhlte *see* Galty Mountains
39 N16 **Nagai Island** *island* Shumagin Islands, Alaska, USA
153 X12 **Nāgāland** ◐ *state* NE India
164 M11 **Nagano** Nagano, Honshū, S Japan *36°39´N 138°11´E*
164 M12 **Nagano** *off.* Nagano-ken. ◐ *prefecture* Honshū, S Japan
Nagano-ken *see* Nagano
165 N11 **Nagaoka** Niigata, Honshū, C Japan *37°26´N 138°48´E*
153 W12 **Nagaon** *prev.* Nowgong. Assam, NE India *26°21´N 92°41´E*
155 J21 **Nāgappattinam** *var.* Negapatam, Negapattinam. Tamil Nādu, SE India *10°45´N 79°50´E*
Nagara Nayok *see* Nakhon Nayok
Nagara Panom *see* Nakhon Phanom
Nagara Pathom *see* Nakhon Pathom
Nagara Sridharmaraj *see* Nakhon Si Thammarat
Nagara Svarga *see* Nakhon Sawan
155 H16 **Nāgārjuna Sāgar** ⊠ E India
42 I10 **Nagarote** León, SW Nicaragua *12°15´N 86°35´W*
158 M16 **Nagarzê** *var.* Nagaarzê. Xizang Zizhiqu, W China *28°57´N 90°26´E*
164 C14 **Nagasaki** Nagasaki, Kyūshū, SW Japan *32°45´N 129°52´E*
164 C14 **Nagasaki** *off.* Nagasaki-ken. ◐ *prefecture* Kyūshū, SW Japan
Nagasaki-ken *see* Nagasaki
Nagashima *see* Kii-Nagashima
164 E12 **Nagato** Yamaguchi, Honshū, SW Japan *34°22´N 131°10´E*
152 F11 **Nāgaur** Rājasthān, NW India *27°12´N 73°48´E*
154 F10 **Nāgda** Madhya Pradesh, C India *23°30´N 75°29´E*
98 L8 **Nagele** Flevoland, N Netherlands *52°39´N 05°43´E*
155 H24 **Nāgercoil** Tamil Nādu, SE India *08°11´N 77°30´E*
153 X12 **Nāginimāra** Nāgāland, NE India *26°44´N 94°51´E*
Na Gleannta *see* Glenties
165 T16 **Nago** Okinawa, Okinawa, SW Japan *26°36´N 127°59´E*
154 K9 **Nāgod** Madhya Pradesh, C India *24°34´N 80°34´E*
155 J26 **Nagoda** Southern Province, S Sri Lanka *06°13´N 80°13´E*
101 G22 **Nagold** Baden-Württemberg, SW Germany *48°33´N 08°43´E*
137 V12 **Nagorno-Karabakh** *var.* Nagorno- Karabakhskaya Avtonomnaya Oblast, *Arm.* Lerrnayin Gharabakh, *Az.* Dağliq Quarabağ, *Rus.* Nagornyy Karabakh. *former autonomous region* SW Azerbaijan
Nagorno- Karabakhskaya Avtonomnaya Oblast *see* Nagorno-Karabakh
123 Q12 **Nagornyy** Respublika Sakha (Yakutiya), NE Russian Federation *55°53´N 124°58´E*
Nagornyy Karabakh *see* Nagorno-Karabakh
125 R13 **Nagorsk** Kirovskaya Oblast', NW Russian Federation *59°18´N 50°49´E*
164 K13 **Nagoya** Aichi, Honshū, SW Japan *35°10´N 136°53´E*
154 I12 **Nāgpur** Mahārāshtra, C India *21°09´N 79°06´E*
156 K10 **Nagqu** *Chin.* Na-Ch'ii; *prev.* Hei-ho. Xizang Zizhiqu, W China *31°30´N 91°57´E*
152 J8 **Nāg Tibba Range** ▲▲ N India
45 O8 **Nagua** NE Dominican Republic *19°25´N 69°49´W*
111 H25 **Nagyatád** Somogy, SW Hungary *46°15´N 17°25´E*
Nagybánya *see* Baia Mare
Nagybecskerek *see* Zrenjanin
Nagydisznód *see* Cisnădie
Nagyenyed *see* Aiud
111 N21 **Nagykálló** Szabolcs-Szatmár-Bereg, E Hungary *47°50´N 21°47´E*
111 G25 **Nagykanizsa** *Ger.* Grosskanizsa. Zala, SW Hungary *46°27´N 17°E*
Nagykároly *see* Carei
111 K22 **Nagykáta** Pest, C Hungary *47°25´N 19°45´E*
Nagykikinda *see* Kikinda
111 K23 **Nagykőrös** Pest, C Hungary *47°01´N 19°46´E*
Nagy-Küküllő *see* Târnava Mare
Nagylak *see* Nădlac
Nagymihály *see* Michalovce
Nagyrőce *see* Revúca
Nagysomkút *see* Şomcuta Mare
Nagysurány *see* Šurany
Nagyszalonta *see* Salonta
Nagyszeben *see* Sibiu
Nagyszentmiklós *see* Sânnicolau Mare
Nagyszöllös *see* Vynohradiv
Nagyszombat *see* Trnava
Nagytapolcsány *see* Topoľčany
Nagyvárad *see* Oradea
165 S17 **Naha** Okinawa, Okinawa, SW Japan *26°10´N 127°40´E*
152 J8 **Nāhan** Himāchal Pradesh, NW India *30°33´N 77°18´E*
Nahang, Rūd-e *see* Nihing
138 F8 **Nahariya** *prev.* Nahariyya. Northern, N Israel *33°01´N 35°05´E*
Nahariyya *see* Nahariya
142 L6 **Nahāvand** *var.* Nehavend. Hamadān, W Iran *34°13´N 48°21´E*
101 F19 **Nahe** ≈ SW Germany
Na H-Iarmhidhe *see* Westmeath
189 O13 **Nahnalaud** ▲ Pohnpei, E Micronesia
Nahoi, Cape *see* Cumberland, Cape
Nahtavárr *see* Nattavaara
63 H16 **Nahuel Huapí, Lago** ◎ W Argentina
23 W7 **Nahunta** Georgia, SE USA *31°11´N 81°58´W*
40 J6 **Naica** Chihuahua, N Mexico *27°53´N 105°30´W*
11 U15 **Naicam** Saskatchewan, S Canada *52°26´N 104°30´W*
Naiman Qi *see* Daqin Tal
158 M4 **Naimin Bulak** *spring* NW China
13 P6 **Nain** Newfoundland and Labrador, NE Canada *56°33´N 61°46´W*
143 P8 **Nā'īn** Eşfahān, C Iran *32°52´N 53°05´E*
152 K10 **Naini Tāl** Uttarakhand, N India *29°22´N 79°26´E*
154 J11 **Nainpur** Madhya Pradesh, C India *22°26´N 80°10´E*
96 J8 **Nairn** N Scotland, United Kingdom *57°36´N 03°51´W*
96 I8 **Nairn** *cultural region* NE Scotland, United Kingdom
81 I19 **Nairobi** ● (Kenya) Nairobi City, S Kenya *01°17´S 36°50´E*
81 I19 **Nairobi** ✈ Nairobi City, S Kenya *01°21´S 37°01´E*
81 I19 **Nairobi City** ◐ *county* Nairobi City, Kenya, Africa
82 P13 **Nairoto** Cabo Delgado, NE Mozambique *12°22´S 39°05´E*
118 G3 **Naissaar** *island* N Estonia
Naissus *see* Niš
187 Z14 **Naitaba** *var.* Naitauba; *prev.* Naitamba. *island* Lau Group, E Fiji
Naitamba/Naitauba *see* Naitaba
81 I19 **Naivasha** Nakuru, SW Kenya *0°44´S 36°26´E*
81 H19 **Naivasha, Lake** ◎ SW Kenya
Najaf *see* An Najaf
143 N8 **Najafābād** *var.* Nejafabad. Eşfahān, C Iran *32°38´N 51°23´E*
Najaf, Muḥāfa at an *see* An Najaf
141 N7 **Najd** *var.* Nejd. *cultural region* C Saudi Arabia
105 O4 **Nájera** La Rioja, N Spain *42°25´N 02°45´W*
105 P4 **Najerilla** ≈ N Spain
163 U7 **Naji** *var.* Arun Qi. Nei Mongol Zizhiqu, N China *48°05´N 123°28´E*
152 J9 **Najībābād** Uttar Pradesh, N India *29°37´N 78°19´E*
Najima *see* Fukuoka
163 Y11 **Najin** NE North Korea *42°13´N 130°16´E*
139 T9 **Najm al Ḥassūn** Bābil, C Iraq *32°24´N 44°13´E*
141 O13 **Najrān** *var.* Abā as Su'ūd. Najrān, S Saudi Arabia *17°31´N 44°09´E*
141 P12 **Najrān** *var.* Minṭaqat al Najrān. ◐ *province* S Saudi Arabia
Najrān, Minṭaqat al *see* Najrān
165 T2 **Nakagawa** Hokkaidō, NE Japan *44°49´N 142°04´E*
38 F9 **Nākālele Point** *var.* Nakalele Point. *headland* Maui, Hawai'i, USA *21°01´N 156°35´W*
164 D13 **Nakama** Fukuoka, Kyūshū, SW Japan *33°53´N 130°48´E*
Nakambé *see* White Volta
Nakamti *see* Nek'emtē
164 F15 **Nakamura** *var.* Shimanto. Kōchi, Shikoku, SW Japan *33°00´N 132°55´E*
186 H7 **Nakanai Mountains** ▲▲ New Britain, E Papua New Guinea
164 H11 **Nakano-shima** *island* Oki-shotō, SW Japan
165 Q6 **Nakasato** Aomori, Honshū, C Japan *40°58´N 140°26´E*
165 T5 **Nakasatsunai** Hokkaidō, NE Japan *42°42´N 143°09´E*
165 W4 **Nakashibetsu** Hokkaidō, NE Japan *43°31´N 144°58´E*
81 F18 **Nakasongola** C Uganda *01°19´N 32°28´E*
165 T1 **Nakatonbetsu** Hokkaidō, NE Japan *44°58´N 142°18´E*
164 L13 **Nakatsugawa** *var.* Nakatugawa. Gifu, Honshū, SW Japan *35°30´N 137°29´E*
Nakatugawa *see* Nakatsugawa
Nakdong *see* Nakdong-gang
163 Y15 **Nakdong-gang** *var.* Nakdong, *Jap.* Rakutō-kō; *prev.* Naktong-gang. ≈ C South Korea
Nakel *see* Nakło nad Notecią
Nakfa *see* Nakh'fa
80 J8 **Nakh'fa** *var.* Nakfa. N Eritrea *16°38´N 38°26´E*
Nakhichevan' *see* Naxçıvan
123 S15 **Nakhodka** Primorskiy Kray, SE Russian Federation *42°46´N 132°48´E*
122 J8 **Nakhodka** Yamalo-Nenetskiy Avtonomnyy Okrug, N Russian Federation *67°48´N 77°48´E*
Nakhon Navok *see* Nakhon Nayok
167 P11 **Nakhon Nayok** *var.* Nagara Nayok, Nakhon Navok. Nakhon Nayok, C Thailand *14°15´N 101°12´E*
167 O11 **Nakhon Pathom** *var.* Nagara Pathom, Nakorn Pathom. Nakhon Pathom, W Thailand *13°49´N 100°06´E*
167 R8 **Nakhon Phanom** *var.* Nagara Panom. Nakhon Phanom, E Thailand *17°22´N 104°46´E*
167 Q10 **Nakhon Ratchasima** *var.* Khorat, Korat. Nakhon Ratchasima, E Thailand *15°N 102°06´E*
167 O9 **Nakhon Sawan** *var.* Muang Nakhon Sawan, Nagara Svarga. Nakhon Sawan, W Thailand *15°42´N 100°06´E*
167 N15 **Nakhon Si Thammarat** *var.* Nagara Sridharmaraj, Nakhon Sithamnaraj. Nakhon Si Thammarat, SW Thailand *08°24´N 99°58´E*
Nakhon Sithamnaraj *see* Nakhon Si Thammarat
139 Y11 **Nakhrash** Al Başrah, SE Iraq *31°13´N 47°24´E*
10 I9 **Nakina** British Columbia, W Canada *59°12´N 132°48´W*
110 H9 **Nakło nad Notecią** *Ger.* Nakel. Kujawsko-pomorskie, C Poland *53°08´N 17°35´E*
39 P13 **Naknek** Alaska, USA *58°45´N 157°01´W*
152 H8 **Nakodar** Punjab, NW India *31°06´N 75°31´E*
82 M11 **Nakonde** Muchinga, NE Zambia *09°22´S 42°47´E*
Nakorn Pathom *see* Nakhon Pathom
95 H24 **Nakskov** Sjælland, SE Denmark *54°50´N 11°10´E*
Naktong-gang *see* Nakdong-gang
81 H18 **Nakuru** Nakuru, SW Kenya *0°16´S 36°04´E*
81 I19 **Nakuru** ◐ *county* W Kenya
81 H19 **Nakuru, Lake** ◎ Nakuru, C Kenya
11 O17 **Nakusp** British Columbia, SW Canada *50°14´N 117°48´W*
149 N15 **Nāl** ≈ W Pakistan
162 M7 **Nalayh** Töv, C Mongolia *47°48´N 107°17´E*
153 V12 **Nalbāri** Assam, NE India *26°36´N 91°49´E*
63 G19 **Nalcayec, Isla** *island* Archipiélago de los Chonos, S Chile
127 N15 **Nal'chik** Kabardino-Balkarskaya Respublika, SW Russian Federation *43°30´N 43°39´E*
155 I16 **Nalgonda** Telangana, C India *17°04´N 79°15´E*
153 S14 **Nalhāti** West Bengal, NE India *24°19´N 87°53´E*
153 U14 **Nalitabari** Dhaka, N Bangladesh *25°06´N 90°11´E*
155 I17 **Nallamala Hills** ▲▲ E India
136 G12 **Nallıhan** Ankara, NW Turkey *40°12´N 31°22´E*
104 K2 **Nalón** ≈ NW Spain
167 N3 **Nalong** Kachin State, N Myanmar (Burma) *24°42´N 97°27´E*
75 N8 **Nālūt** NW Libya *31°52´N 10°59´E*
171 T14 **Nama** Pulau Manawoka, E Indonesia *04°07´S 131°22´E*
189 Q16 **Nama** *island* C Micronesia
83 O16 **Namacurra** Zambézia, NE Mozambique *17°31´S 37°03´E*
188 F9 **Namai Bay** *bay* Babeldaob, N Palau
29 W2 **Namakan Lake** ◎ Canada/USA
143 O6 **Namak, Daryācheh-ye** *marsh* N Iran
143 T6 **Namak, Kavīr-e** *salt pan* NE Iran
167 O6 **Namaklwe** Shan State, E Myanmar (Burma) *19°45´N 99°01´E*
Namaksār, Kowl-e/ Namakzār, Daryācheh-ye *see* Namakzar
148 I5 **Namakzar** *Pash.* Daryācheh-ye Namakzār, Kowl-e Namaksār. *marsh* Afghanistan/Iran
171 V15 **Namalau** Pulau Jursian, E Indonesia *05°50´S 134°43´E*
81 I20 **Namanga** Kajiado, S Kenya *02°33´S 36°48´E*
147 S10 **Namangan** Namangan Viloyati, E Uzbekistan *40°59´N 71°34´E*
Namanganskaya Oblast' *see* Namangan Viloyati
147 R10 **Namangan Viloyati** *Rus.* Namanganskaya Oblast'. ◐ *province* E Uzbekistan
83 Q14 **Namapa** Nampula, NE Mozambique *13°43´S 39°48´E*
83 C21 **Namaqualand** *physical region* S Namibia
81 G18 **Namasagali** C Uganda *01°02´N 32°58´E*
186 H6 **Namatanai** New Ireland, NE Papua New Guinea *03°40´S 152°26´E*
83 I14 **Nambala** Central, C Zambia *15°04´S 26°56´E*
81 J23 **Nambanje** Lindi, SE Tanzania *08°37´S 38°21´E*
183 V2 **Nambour** Queensland, E Australia *26°40´S 152°52´E*
183 V6 **Nambucca Heads** New South Wales, SE Australia *30°37´S 153°00´E*
159 N15 **Nam Co** ◎ W China
167 R5 **Nâm Cum** Lai Châu, N Vietnam *22°37´N 103°12´E*
Namdik *see* Namorik Atoll
167 T6 **Nam Đinh** Nam Ha, N Vietnam *20°25´N 106°12´E*
99 I20 **Namêche** Namur, SE Belgium *50°29´N 05°02´E*
30 J4 **Namekagon Lake** ◎ Wisconsin, N USA
188 F10 **Namekakl Passage** *passage* Babeldaob, N Palau
Namen *see* Namur
83 P15 **Nametil** Nampula, NE Mozambique *15°46´S 39°21´E*
163 X14 **Nam-gang** ≈ C North Korea
163 Y16 **Nam-gang** ≈ S South Korea
163 Y17 **Namhae-do** *Jap.* Nankai-tō. *island* S South Korea
Namhoi *see* Foshan
83 C19 **Namib Desert** *desert* W Namibia
83 A15 **Namibe** *Port.* Moçâmedes, Mossâmedes. Namibe, SW Angola *15°10´S 12°09´E*
83 A15 **Namibe** ◐ *province* SE Angola
83 C18 **Namibia** *off.* Republic of Namibia, *var.* South West Africa, *Afr.* Suidwes-Afrika, *Ger.* Deutsch-Südwestafrika; *prev.* German Southwest Africa, South-West Africa. ◆ *republic* S Africa
65 O17 **Namibia Plain** *undersea feature* S Atlantic Ocean
Namibia, Republic of *see* Namibia
165 Q11 **Namie** Fukushima, Honshū, C Japan *37°29´N 140°58´E*
165 Q7 **Namioka** Aomori, Honshū, C Japan *40°43´N 140°34´E*
40 I5 **Namiquipa** Chihuahua, N Mexico *29°15´N 107°25´W*
159 P15 **Namjagbarwa Feng** ▲ W China *29°39´N 95°00´E*
Namka *see* Doilungdêqên
Nam Khong *see* Mekong
171 R13 **Namlea** Pulau Buru, E Indonesia *03°12´S 127°06´E*
158 L16 **Namling** Xizang Zizhiqu, W China *29°40´N 88°58´E*
Namnetes *see* Nantes
167 R8 **Nam Ngum** ≈ C Laos
Namo *see* Namu Atoll
183 R5 **Namoi River** ≈ New South Wales, SE Australia
189 Q17 **Namoluk Atoll** *atoll* Mortlock Islands, C Micronesia
189 O15 **Namonuito Atoll** *atoll* Caroline Islands, C Micronesia
189 T9 **Namorik Atoll** *var.* Namdik. *atoll* Ralik Chain, S Marshall Islands
167 Q6 **Nam Ou** ≈ N Laos
32 M14 **Nampa** Idaho, NW USA *43°32´N 116°33´W*
76 M11 **Nampala** Ségou, W Mali *15°21´N 05°32´W*
163 W14 **Namp'o** SW North Korea *38°46´N 125°25´E*
83 P15 **Nampula** Nampula, NE Mozambique *15°09´S 39°14´E*

◆ Country
● Country Capital
◇ Dependent Territory
○ Dependent Territory Capital
◐ Administrative Regions
✈ International Airport
▲ Mountain
▲▲ Mountain Range
Volcano
≈ River
◎ Lake
⊠ Reservoir

83 P15 **Nampula** *off.* Província de Nampula. ◈ *province* NE Mozambique
Nampula, Província de *see* Nampula
163 W13 **Namsan-ni** NW North Korea 40°25′N 125°01′E
Namslau *see* Namysłów
93 E15 **Namsos** Nord-Trøndelag, C Norway 64°28′N 11°31′E
93 F14 **Namsskogan** Nord-Trøndelag, C Norway 64°57′N 13°04′E
167 O6 **Nam Teng** ≈ E Myanmar (Burma)
167 P6 **Nam Tha** ≈ N Laos
123 Q10 **Namtsy** Respublika Sakha (Yakutiya), NE Russian Federation 62°42′N 129°30′E
167 N4 **Namtu** Shan State, E Myanmar (Burma) 23°04′N 97°26′E
10 J15 **Namu** British Columbia, SW Canada 51°46′N 127°49′W
189 T7 **Namu Atoll** *var.* Namo. *atoll* Ralik Chain, C Marshall Islands
187 Y15 **Namuka-i-lau** *island* Lau Group, E Fiji
83 O15 **Namuli, Mont** ▲ NE Mozambique 15°15′S 37°33′E
83 P14 **Namuno** Cabo Delgado, N Mozambique 13°39′S 38°50′E
99 I20 **Namur** *Dut.* Namen. Namur, SE Belgium 50°28′N 04°52′E
99 H21 **Namur** *Dut.* Namen. ◈ *province* S Belgium
83 D17 **Namutoni** Kunene, N Namibia 18°49′S 16°55′E
163 Y16 **Namwon** *Jap.* Nangen; *prev.* Namwŏn. S South Korea 35°24′N 127°20′E
Namwŏn *see* Namwon
111 H14 **Namysłów** *Ger.* Namslau. Opole, SW Poland 51°03′N 17°41′E
167 P7 **Nan** *var.* Muang Nan. Nan, NW Thailand 18°47′N 100°50′E
79 G15 **Nana** ≈ W Central African Republic
165 R5 **Nanae** Hokkaidō, NE Japan 41°55′N 140°40′E
79 I14 **Nana-Grébizi** ◈ *prefecture* N Central African Republic
10 L17 **Nanaimo** Vancouver Island, British Columbia, SW Canada 49°08′N 123°58′W
38 C9 **Nānākuli** *var.* Nanakuli. Oʻahu, Hawaii, USA, C Pacific Ocean 21°23′N 158°09′W
79 G15 **Nana-Mambéré** ◈ *prefecture* W Central African Republic
161 R13 **Nan'an** Fujian, SE China 24°57′N 118°22′E
183 U2 **Nanango** Queensland, E Australia 26°42′S 151°58′E
164 L11 **Nanao** Ishikawa, Honshū, SW Japan 37°03′N 136°58′E
161 Q14 **Nan'ao Dao** *island* S China
164 L10 **Nanatsu-shima** *island* SW Japan
56 F8 **Nanay, Río** ≈ NE Peru
160 J8 **Nanbu** Sichuan, C China 31°19′N 106°02′E
163 X7 **Nancha** Heilongjiang, NE China 47°09′N 129°17′E
161 P10 **Nanchang** *var.* Nan-ch'ang, Nanch'ang-hsien. *province capital* Jiangxi, S China 28°38′N 115°58′E
Nan-ch'ang *see* Nanchang
Nanch'ang-hsien *see* Nanchang
161 P11 **Nancheng** *var.* Jianchang. Jiangxi, S China 27°37′N 116°37′E
Nan-ching *see* Nanjing
160 J9 **Nanchong** Sichuan, C China 30°47′N 106°03′E
160 J10 **Nanchuan** Chongqing Shi, C China 29°06′N 107°13′E
103 T5 **Nancy** Meurthe-et-Moselle, NE France 48°40′N 06°11′E
185 A22 **Nancy Sound** *sound* South Island, New Zealand
152 L9 **Nanda Devi** ▲ NW India 30°27′N 80°00′E
42 J11 **Nandaime** Granada, SW Nicaragua 11°45′N 86°02′W
160 K13 **Nandan** *var.* Minami-Awaji. Guangxi Zhuangzu Zizhiqu, S China 25°03′N 107°31′E
155 H14 **Nānded** Mahārāshtra, C India 19°11′N 77°21′E
183 S5 **Nandewar Range** ▲▲ New South Wales, SE Australia
81 H18 **Nandi** ◈ *county* W Kenya
Nandi *see* Nadi
160 E13 **Nanding He** ≈ China/Vietnam
Nándorhgy *see* Oţelu Roşu
154 E11 **Nandurbār** Mahārāshtra, W India 21°22′N 74°18′E
Nanduri *see* Naduri
155 I17 **Nandyāl** Andhra Pradesh, E India 15°30′N 78°28′E
161 P11 **Nanfeng** *var.* Qincheng. Jiangxi, S China 27°15′N 116°16′E
Nang *see* Nangxian
79 E15 **Nanga Eboko** Centre, C Cameroon 04°38′N 12°21′E
Nangah Serawai *see* Nangaserawai
149 W4 **Nanga Parbat** ▲ India/Pakistan 35°15′N 74°36′E
169 R11 **Nangapinoh** Borneo, C Indonesia 0°21′S 111°44′E
149 R5 **Nangarhār** ◈ *province* E Afghanistan
169 S11 **Nangaserawai** *var.* Nangah Serawai. Borneo, C Indonesia 0°20′S 112°26′E
169 Q12 **Nangatayap** Borneo, C Indonesia 01°30′S 110°33′E
Nangen *see* Namwon
103 P5 **Nangis** Seine-et-Marne, N France 48°36′N 03°02′E
163 X13 **Nangnim-sanmaek** ▲▲ C North Korea
161 O4 **Nangong** Hebei, E China 37°22′N 115°20′E
159 Q14 **Nangqên** *var.* Xangda. Qinghai, C China 32°05′N 96°28′E
167 Q10 **Nang Rong** Buri Ram, E Thailand 14°37′N 102°48′E
159 O16 **Nangxian** *var.* Nang. Xizang Zizhiqu, W China 29°04′N 93°03′E
Nan Hai *see* South China Sea
160 L8 **Nan He** ≈ C China
160 F12 **Nanhua** *var.* Longchuan. Yunnan, SW China 25°15′N 101°15′E
Naniwa *see* Ōsaka
155 G20 **Nanjangūd** Karnātaka, W India 12°07′N 76°40′E
161 Q8 **Nanjing** *var.* Nan-ching, Nanking; *prev.* Chianning, Chian-ning, Kiang-ning, Jiangsu. *province capital* Jiangsu, E China 32°03′N 118°47′E
Nankai-tō *see* Namhae-do
161 O12 **Nankang** *var.* Rongjiang. Jiangxi, S China 25°42′N 114°45′E
Nanking *see* Nanjing
161 N13 **Nan Ling** ▲▲ S China
160 L15 **Nanliu Jiang** ≈ S China
189 P13 **Nan Madol** *ruins* Temwen Island, E Micronesia
160 K15 **Nanning** *var.* Nan-ning; *prev.* Yung-ning. Guangxi Zhuangzu Zizhiqu, S China 22°50′N 108°19′E
Nan-ning *see* Nanning
196 M15 **Nanortalik** Kujalleo, S Greenland 60°08′M 45°14′W
Nanouki *see* Aranuka
160 H13 **Nanpan Jiang** ≈ S China
152 M11 **Nānpāra** Uttar Pradesh, N India 27°51′N 81°30′E
161 Q12 **Nanping** *var.* Nan-p'ing; *prev.* Yenping. Fujian, SE China 26°40′N 118°07′E
Nan-p'ing *see* Nanping
Nanping *see* Jiuzhaigou
Nanpu *see* Pucheng
161 R12 **Nanri Dao** *island* SE China
165 S16 **Nansei-shotō** *Eng.* Ryukyu Islands. *island group* SW Japan
Nansei Syotō Trench *see* Ryukyu Trench
197 T10 **Nansen Basin** *undersea feature* Arctic Ocean
Nansen Cordillera *see* Gakkel Ridge
129 T9 **Nan Shan** ▲▲ C China
Nansha Qundao *see* Spratly Islands
12 K3 **Nantais, Lac** ◎ Québec, NE Canada
103 N5 **Nanterre** Hauts-de-Seine, N France 48°53′N 02°13′E
102 I8 **Nantes** *Bret.* Naoned; *anc.* Condivincum, Namnetes. Loire-Atlantique, NW France 47°12′N 01°32′W
14 G17 **Nanticoke** Ontario, S Canada 42°49′N 80°04′W
18 H13 **Nanticoke** Pennsylvania, NE USA 41°12′N 76°00′W
21 Y4 **Nanticoke River** ≈ Delaware/Maryland, NE USA
11 Q17 **Nanton** Alberta, SW Canada 50°21′N 113°47′W
161 S8 **Nantong** Jiangsu, E China 32°00′N 120°52′E
161 S13 **Nantou** *prev.* Nant'ou. W Taiwan 23°54′N 120°51′E
Nant'ou *see* Nantou
103 S10 **Nantua** Ain, E France 46°10′N 05°34′E
19 Q13 **Nantucket** Nantucket Island, Massachusetts, NE USA 41°15′N 70°05′W
19 Q13 **Nantucket Island** *island* Massachusetts, NE USA
19 Q13 **Nantucket Sound** *sound* Massachusetts, NE USA
82 P13 **Nantulo** Cabo Delgado, N Mozambique 12°30′S 39°03′E
189 O12 **Nanuh** Pohnpei, E Micronesia
190 D6 **Nanumaga** *var.* Nanumanga. *atoll* NW Tuvalu
Nanumanga *see* Nanumaga
190 D5 **Nanumea Atoll** *atoll* NW Tuvalu
59 O19 **Nanuque** Minas Gerais, SE Brazil 17°49′S 40°21′W
171 R10 **Nanusa, Kepulauan** *island group* N Indonesia
Nanwei Dao *see* Spratly Island
163 U4 **Nanweng He** ≈ NE China
160 I10 **Nanxi** Sichuan, C China 28°54′N 104°59′E
161 N10 **Nanxian** *var.* Nan Xian, Nanzhou. Hunan, S China 29°23′N 112°18′E
Nan Xian *see* Nanxian
161 N7 **Nanyang** *var.* Nan-yang. Henan, C China 32°59′N 112°29′E
Nan-yang *see* Nanyang
161 P6 **Nanyang Hu** ◎ E China
165 P10 **Nan'yō** Yamagata, Honshū, C Japan 38°04′N 140°06′E
81 I18 **Nanyuki** Laikipia, C Kenya 0°01′N 37°05′E
160 M8 **Nanzhang** Hubei, C China 31°47′N 111°48′E
Nanzhou *see* Nanxian
105 T11 **Nao, Cabo de La** *headland* E Spain 38°43′N 00°13′E
12 M9 **Naococane, Lac** ◎ Québec, E Canada
153 S14 **Naogaon** Rajshahi, NW Bangladesh 24°49′N 88°59′E
Naokot *see* Naukot
187 R13 **Naone** Maewo, C Vanuatu 15°03′S 168°06′E
Naoned *see* Nantes
115 E14 **Náousa** Kentrikí Makedonía, N Greece 40°38′N 22°04′E
35 N8 **Napa** California, W USA 38°15′N 122°17′W
39 O11 **Napaimiut** Alaska, USA 61°32′N 158°46′W
39 N12 **Napakiak** Alaska, USA 60°42′N 161°57′W
122 J7 **Napalkovo** Yamalo-Nenetskiy Avtonomnyy Okrug, N Russian Federation 70°06′N 73°43′E
12 I16 **Napanee** Ontario, SE Canada 44°13′N 76°57′W
39 N12 **Napaskiak** Alaska, USA 60°42′N 161°46′W
167 S5 **Na Phăc** Cao Băng, N Vietnam 22°24′N 105°54′E
184 O11 **Napier** Hawke's Bay, North Island, New Zealand 39°30′S 176°55′E
195 X3 **Napier Mountains** ▲▲ Antarctica
15 O13 **Napierville** Québec, SE Canada 45°12′N 73°25′W
23 W15 **Naples** Florida, SE USA 26°08′N 81°48′W
25 W5 **Naples** Texas, SW USA 33°12′N 94°40′W
Naples *see* Napoli
160 I14 **Napo** Guangxi Zhuangzu Zizhiqu, S China 23°21′N 105°47′E
56 C6 **Napo** ◈ *province* NE Ecuador
29 O6 **Napoleon** North Dakota, N USA 46°30′N 99°46′W
31 R11 **Napoleon** Ohio, N USA 41°23′N 84°07′W
Napoléon-Vendée *see* la Roche-sur-Yon
22 J9 **Napoleonville** Louisiana, S USA 29°55′N 91°01′W
107 K17 **Napoli** *Eng.* Naples, *Ger.* Neapel; *anc.* Neapolis. Campania, S Italy 40°52′N 14°15′E
107 J18 **Napoli, Golfo di** *gulf* S Italy
57 F7 **Napo, Río** ≈ Ecuador/Peru
191 W9 **Napuka** *island* Îles Tuamotu, C French Polynesia
142 J3 **Naqadeh** Āžarbāyjān-e Bākhtarī, NW Iran 36°57′N 45°24′E
139 U6 **Naqnah** Diyālá, E Iraq 34°13′N 45°33′E
Nar *see* Nera
164 J14 **Nara** Nara, Honshū, SW Japan 34°41′N 135°49′E
76 L11 **Nara** Koulikoro, W Mali 15°04′N 07°19′W
149 R14 **Nāra Canal** *irrigation canal* S Pakistan
182 K11 **Naracoorte** South Australia 37°02′S 140°45′E
183 P8 **Naradhan** New South Wales, SE Australia 33°37′S 146°19′E
Naradhivas *see* Narathiwat
56 B8 **Naranjal** Guayas, W Ecuador 02°43′S 79°38′W
57 Q19 **Naranjos** Santa Cruz, E Bolivia
41 Q12 **Naranjos** Veracruz-Llave, E Mexico 21°21′N 97°41′W
159 Q6 **Naran Sebstein Bulag** *spring* NW China
164 B14 **Narao** Nagasaki, Nakadōri-jima, SW Japan 32°40′N 129°03′E
155 J16 **Narasaraopet** Andhra Pradesh, E India 16°16′N 80°06′E
158 J5 **Narat** Xinjiang Uygur Zizhiqu, W China 43°20′N 84°02′E
167 P17 **Narathiwat** *var.* Naradhivas. Narathiwat, SW Thailand 06°25′N 101°48′E
37 V10 **Nara Visa** New Mexico, SW USA 35°35′N 103°06′W
Nārāyāni *see* Gandak
Narbada *see* Narmada
Narbo Martius *see* Narbonne
103 P16 **Narbonne** *anc.* Narbo Martius. Aude, S France 43°11′N 03°E
Narborough Island *see* Fernandina, Isla
104 J2 **Narcea** ≈ NW Spain
152 J9 **Narendranagar** Uttarakhand, N India 30°10′N 78°21′E
Nares Abyssal Plain *see* Nares Plain
64 G11 **Nares Plain** *var.* Nares Abyssal Plain. *undersea feature* NW Atlantic Ocean 23°30′N 63°00′W
Nares Stræde *see* Nares Strait
197 P10 **Nares Strait** *Dan.* Nares Stræde. *strait* Canada/Greenland
110 O9 **Narew** ≈ E Poland
155 F17 **Nargund** Karnātaka, W India 15°43′N 75°23′E
83 D20 **Narib** Hardap, S Namibia 24°11′S 17°46′E
Narikrik *see* Knox Atoll
Narin Gol *see* Omon Gol
54 B13 **Nariño** *off.* Departamento de Narino. ◈ *province* SW Colombia
165 P13 **Narita** Chiba, Honshū, S Japan 35°46′N 140°20′E
165 P13 **Narita** ✈ (Tōkyō) Chiba, Honshū, S Japan 35°45′N 140°23′E
Narıya *see* An Nuʻayrīyah
162 F5 **Nariyn Gol** ≈ Mongolia/Russian Federation
162 J8 **Nariynteel** *var.* Tsagaan-Ovoo. Övörhangay, C Mongolia 45°57′N 101°25′E
152 J8 **Nārkanda** Himāchal Pradesh, NW India 31°14′N 77°27′E
92 L13 **Narkaus** Lappi, NW Finland 66°13′N 26°09′E
154 E11 **Narmada** *var.* Narbada. ≈ C India
152 H11 **Narnaul** *var.* Nārnaul. Haryāna, N India 28°04′N 76°10′E
107 I14 **Narni** Umbria, C Italy 42°31′N 12°31′E
107 J24 **Naro** Sicilia, Italy, C Mediterranean Sea 37°18′N 13°48′E
Narodichi *see* Narodychi
125 V7 **Narodnaya, Gora** ▲ NW Russian Federation 65°04′N 60°12′E
117 N3 **Narodychi** *Rus.* Narodichi. Zhytomyrs'ka Oblast', N Ukraine 51°11′N 29°01′E
126 J4 **Naro-Fominsk** Moskovskaya Oblast', W Russian Federation 55°25′N 36°41′E
81 H19 **Narok** Narok, SW Kenya 01°04′S 35°54′E
81 H20 **Narok** ◈ *county* SW Kenya
104 H2 **Narón** Galicia, NW Spain 43°31′N 08°08′W
183 S11 **Narooma** New South Wales, SE Australia 36°16′S 150°08′E
Narova *see* Narva
Narovlya *see* Narowlya
149 W8 **Nārowāl** Punjab, E Pakistan 32°04′N 74°54′E
119 N20 **Narowlya** *Rus.* Narovlya. Homyel'skaya Voblasts', SE Belarus 51°48′N 29°30′E
93 J17 **Närpes** *Fin.* Närpiö. Österbotten, W Finland 62°28′N 21°19′E
Närpiö *see* Närpes
183 S5 **Narrabri** New South Wales, SE Australia 30°21′S 149°48′E
183 P9 **Narrandera** New South Wales, SE Australia 34°46′S 146°32′E
183 Q4 **Narran Lake** ◎ New South Wales, SE Australia
183 Q4 **Narran River** ≈ New South Wales/Queensland, SE Australia
180 J13 **Narrogin** Western Australia 32°53′S 117°17′E
183 Q7 **Narromine** New South Wales, SE Australia 32°16′S 148°15′E
21 R6 **Narrows** Virginia, NE USA 37°19′N 80°48′W
196 M15 **Narsarsuaq** ✈ Kujalleo, S Greenland 61°07′N 45°03′W
154 I10 **Narsimhapur** Madhya Pradesh, C India 22°58′N 79°15′E
153 U15 **Narsingdi** *var.* Narsinghdi. Dhaka, C Bangladesh 23°56′N 90°40′E
Narsinghdi *see* Narsingdi
154 H9 **Narsinghgarh** Madhya Pradesh, C India 23°42′N 77°08′E
163 Q11 **Nart** Nei Mongol Zizhiqu, N China 42°54′N 115°55′E
Nartës, Gjol i/Nartës, Laguna e *see* Nartës, Liqeni i
113 J22 **Nartës, Liqeni i** *var.* Gjol i Nartës, Laguna e Nartës. ◎ SW Albania
115 F17 **Nartháki** ▲ C Greece 39°12′N 22°24′E
127 O15 **Nartkala** Kabardino-Balkarskaya Respublika, SW Russian Federation 43°34′N 43°55′E
118 K3 **Narva** Ida-Virumaa, NE Estonia 59°23′N 28°12′E
118 K4 **Narva** *prev.* Narova. ≈ Estonia/Russian Federation
118 J3 **Narva Bay** *Est.* Narva Laht, *Ger.* Narwa-Bucht, *Rus.* Narvskiy Zaliv. *bay* Estonia/Russian Federation
Narva Laht *see* Narva Bay
124 F13 **Narva Reservoir** *Est.* Narva Veehoidla, *Rus.* Narvskoye Vodokhranilishche. ⊠ Estonia/Russian Federation
Narva Veehoidla *see* Narva Reservoir
92 H10 **Narvik** Nordland, C Norway 68°26′N 17°24′E
Narvskiy Zaliv *see* Narva Bay
Narvskoye Vodokhranilishche *see* Narva Reservoir
Narwa-Bucht *see* Narva Bay
152 I9 **Narwāna** Haryāna, NW India 29°36′N 76°11′E
125 R4 **Nar'yan-Mar** *prev.* Beloshchel'ye, Dzerzhinskiy. Nenetskiy Avtonomnyy Okrug, NW Russian Federation 67°38′N 53°E
122 J12 **Narym** Tomskaya Oblast', C Russian Federation 58°59′N 81°20′E
Narymskiy Khrebet *see* Khrebet Naryn
147 W9 **Naryn** Narynskaya Oblast', C Kyrgyzstan 41°24′N 76°E
147 U8 **Naryn** ≈ Kyrgyzstan/Uzbekistan
145 W16 **Narynkol** *Kaz.* Narynqol. Almaty, SE Kazakhstan 42°45′N 80°12′E
Naryn Oblasty *see* Narynskaya Oblast'
Narynqol *see* Narynkol
147 V9 **Narynskaya Oblast'** *Kir.* Naryn Oblasty. ◈ *province* C Kyrgyzstan
Naryn Zhotasy *see* Khrebet Naryn
126 J6 **Naryshkino** Orlovskaya Oblast', W Russian Federation 53°00′N 35°41′E
95 L14 **Nås** Dalarna, C Sweden 60°28′N 14°30′E
10 J11 **Nass** ≈ British Columbia, SW Canada
92 G13 **Nasafjellet** *Lapp.* Násávárre. ▲ C Norway 66°29′N 15°23′E
93 H16 **Näsåker** Västernorrland, C Sweden 63°27′N 16°55′E
187 Y14 **Nasau** Koro, C Fiji 17°20′S 179°26′E
190 J13 **Nassau** *island* N Cook Islands
116 I9 **Năsăud** *Ger.* Nussdorf, *Hung.* Naszód. Bistriţa-Năsăud, N Romania 47°16′N 24°24′E
Násávárre *see* Nasafjellet
103 P13 **Nasbinals** Lozère, S France 44°40′N 03°03′E
Na Sceirí *see* Skerries
Nase *see* Naze
185 E22 **Naseby** Otago, South Island, New Zealand 45°02′S 170°09′E
143 R10 **Nāşerīyeh** Kermān, C Iran
25 X5 **Nash** Texas, SW USA 33°26′N 94°04′W
154 E13 **Nāshik** *prev.* Nāsik. Mahārāshtra, W India 20°05′N 73°48′E
56 E7 **Nashiño, Río** ≈ Ecuador/Peru
29 W12 **Nashua** Iowa, C USA 42°57′N 92°32′W
33 W7 **Nashua** Montana, NW USA 48°06′N 106°16′W
19 O10 **Nashua** New Hampshire, NE USA 42°45′N 71°26′W
27 S13 **Nashville** Arkansas, C USA 33°57′N 93°50′W
23 U7 **Nashville** Georgia, SE USA 31°12′N 83°15′W
30 L16 **Nashville** Illinois, N USA 38°20′N 89°22′W
31 O14 **Nashville** Indiana, N USA 39°13′N 86°15′W
21 V9 **Nashville** North Carolina, SE USA 35°58′N 78°00′W
20 J8 **Nashville** *state capital* Tennessee, S USA 36°11′N 86°48′W
20 J9 **Nashville** ✈ Tennessee, S USA 36°06′N 86°44′W
64 H10 **Nashville Seamount** *undersea feature* NW Atlantic Ocean 30°00′N 57°20′W
112 H9 **Našice** Osijek-Baranja, E Croatia 45°29′N 18°05′E
110 M11 **Nasielsk** Mazowieckie, C Poland 52°33′N 20°46′E
93 K18 **Näsijärvi** ◎ SW Finland
Nāsik *see* Nāshik
80 G13 **Nasir** Upper Nile, NE South Sudan 08°37′N 33°06′E
148 K15 **Nasīrābād** Baluchistān, SW Pakistan 26°15′N 62°32′E
Nasir, Buhayrat/Nâşir, Buḩeiret *see* Nasser, Lake
Nāsiri *see* Ahvāz
Nasiriya *see* An Nāşirīyah
Nāşirīyah, An *see* Dhī Qār
Nás na Ríogh *see* Naas
107 L23 **Naso** Sicilia, Italy, C Mediterranean Sea 38°07′N 14°46′E
Nasratabad *see* Zābol
77 V15 **Nassarawa** Nassarawa, C Nigeria 08°33′N 07°42′E
44 H2 **Nassau** ● (The Bahamas) New Providence, N The Bahamas 25°03′N 77°21′W
23 W8 **Nassau Sound** *sound* Florida, SE USA
108 L7 **Nassereith** Tirol, W Austria 47°19′N 10°51′E
80 F5 **Nasser, Lake** *var.* Buhayrat Nasir, Buḩayrat Nāşir, Buḩeiret Nâşir. ◎ Egypt/Sudan
95 L19 **Nässjö** Jönköping, S Sweden 57°39′N 14°40′E
99 K22 **Nassogne** Luxembourg, SE Belgium 50°08′N 05°19′E
12 J6 **Nastapoka Islands** *island group* Northwest Territories, C Canada
93 M19 **Nastola** Päijät-Häme, S Finland 60°57′N 25°56′E
171 O4 **Nasugbu** Luzon, N Philippines 14°03′N 120°39′E
94 N11 **Näsviken** Gävleborg, C Sweden 61°46′N 16°55′E
Naszód *see* Năsăud
83 I17 **Nata** Central, NE Botswana 20°11′S 26°10′E
54 E11 **Natagaima** Tolima, C Colombia 03°38′N 75°07′W
59 Q14 **Natal** *state capital* Rio Grande do Norte, E Brazil 05°46′S 35°15′W
168 I11 **Natal** Sumatera, N Indonesia 0°32′N 99°07′E
Natal *see* KwaZulu/Natal
173 L10 **Natal Basin** *var.* Mozambique Basin. *undersea feature* W Indian Ocean 30°00′S 40°00′E
25 R12 **Natalia** Texas, SW USA 29°11′N 98°51′W
67 W15 **Natal Valley** *undersea feature* SW Indian Ocean 31°00′S 33°15′E
Natanya *see* Netanya
143 O7 **Naţanz** Eşfahān, C Iran 33°31′N 51°55′E
13 Q11 **Natashquan** Québec, E Canada 50°10′N 61°50′W
13 Q10 **Natashquan** ≈ Newfoundland and Labrador/Québec, E Canada
22 J7 **Natchez** Mississippi, S USA 31°34′N 91°24′W
22 G6 **Natchitoches** Louisiana, S USA 31°45′N 93°05′W
108 E10 **Naters** Valais, S Switzerland 46°22′N 08°00′E
Nathanya *see* Netanya
92 O3 **Nathorst Land** *physical region* W Svalbard
186 E9 **National Capital District** ◈ *province* S Papua New Guinea
35 U17 **National City** California, W USA 32°40′N 117°06′W
184 M10 **National Park** Manawatu-Wanganui, North Island, New Zealand 39°11′S 175°22′E
77 R14 **Natitingou** NW Benin 10°21′N 01°26′E
40 B5 **Natividad, Isla** *island* NW Mexico
165 Q10 **Natori** Miyagi, Honshū, C Japan 38°12′N 140°51′E
18 C14 **Natrona Heights** Pennsylvania, NE USA 40°37′N 79°42′W
81 H20 **Natron, Lake** ◎ Kenya/Tanzania
Natsrat *see* Natzrat
166 L7 **Nattalin** Bago, C Myanmar (Burma) 18°25′N 95°34′E
92 J12 **Nattavaara** *Lapp.* Nahtavárr. Norrbotten, N Sweden 66°45′N 20°58′E
109 S3 **Natternbach** Oberösterreich, N Austria 48°26′N 13°44′E
95 M22 **Nättraby** Blekinge, S Sweden 56°12′N 15°30′E
169 P10 **Natuna Besar, Pulau** *island* Kepulauan Natuna, W Indonesia
Natuna Islands *see* Natuna, Kepulauan
169 O9 **Natuna, Kepulauan** *var.* Natuna Islands. *island group* W Indonesia
169 N9 **Natuna, Laut** *Eng.* Natuna Sea. *sea* W Indonesia
Natuna Sea *see* Natuna, Laut
21 N6 **Natural Bridge** *tourist site* Kentucky, C USA
173 V11 **Naturaliste Fracture Zone** *tectonic feature* E Indian Ocean
174 J10 **Naturaliste Plateau** *undersea feature* E Indian Ocean
138 G9 **Natzrat** *var.* Natsrat, *Ar.* En Nazira, *Eng.* Nazareth; *prev.* Nazerat. Northern, N Israel 32°42′N 35°18′E
Nau *see* Nov
103 O14 **Naucelle** Aveyron, S France 44°10′N 02°19′E
83 D20 **Nauchas** Hardap, C Namibia 23°40′S 16°19′E
108 K9 **Nauders** Tirol, W Austria 46°52′N 10°31′E
Naugard *see* Nowogard
118 F12 **Naujamiestis** Panevėžys, C Lithuania 55°42′N 24°10′E
118 E10 **Naujoji Akmenė** Šiaulai, NW Lithuania 56°20′N 22°57′E
149 R16 **Naukot** *var.* Naokot. Sind, SE Pakistan 24°52′N 69°27′E
101 L16 **Naumburg** *var.* Naumburg an der Saale. Sachsen-Anhalt, C Germany 51°09′N 11°48′E
Naumburg am Queis *see* Nowogrodziec
Naumburg an der Saale *see* Naumburg
191 W15 **Naunau** *ancient monument* Easter Island, Chile, E Pacific Ocean
138 G10 **Nāʻūr** ʻAmmān, W Jordan 31°52′N 35°50′E
189 Q8 **Nauru** *off.* Republic of Nauru; *prev.* Pleasant Island. ◆ *republic* W Pacific Ocean
175 P5 **Nauru** *island* W Pacific Ocean
189 Q9 **Nauru International** ✈ S Nauru
Nauru, Republic of *see* Nauru
Nausari *see* Navsāri
19 Q12 **Nauset Beach** *beach* Massachusetts, NE USA
Naushahra *see* Nowshera
149 P14 **Naushahro Fīroz** Sind, SE Pakistan 26°51′N 68°11′E
Naushara *see* Nowshera
187 X14 **Nausori** Viti Levu, W Fiji 18°01′S 178°31′E
56 F9 **Nauta** Loreto, N Peru 04°31′S 73°36′W
153 O12 **Nautanwa** Uttar Pradesh, N India 27°26′N 83°25′E
41 R13 **Nautla** Veracruz-Llave, E Mexico 20°15′N 96°45′W
Nauzad *see* Now Zād
41 N6 **Nava** Coahuila, NE Mexico 28°28′N 100°45′W
Navabad *see* Navobod
104 L6 **Nava del Rey** Castilla y León, N Spain 41°19′N 05°04′W
153 S15 **Navadwīp** *prev.* Nabadwip. West Bengal, NE India 23°24′N 88°23′E
104 M9 **Navahermosa** Castilla-La Mancha, C Spain 39°39′N 04°25′W
119 I16 **Navahrudak** *Pol.* Nowogródek, *Rus.* Novogrudok. Hrodzyenskaya Voblasts', W Belarus 53°36′N 25°50′E
119 I16 **Navahrudskaye Wzvyshsha** *Rus.* Novogrudskaya Vozvyshennost'. ▲▲ W Belarus
36 M8 **Navajo Mount** ▲ Utah, W USA 37°00′N 110°52′W
37 Q9 **Navajo Reservoir** ⊠ New Mexico, SW USA
104 K9 **Navalmoral de la Mata** Extremadura, W Spain 39°54′N 05°33′W
104 K10 **Navalvillar de Pelea** Extremadura, W Spain 39°05′N 05°27′W
97 F17 **Navan** *Ir.* An Uaimh. E Ireland 53°39′N 06°41′W
Navanagar *see* Jāmnagar
118 L12 **Navapolatsk** *Rus.* Novopolotsk. Vitsyebskaya Voblasts', N Belarus 55°34′N 28°35′E
149 P6 **Nāvar, Dasht-e** *Pash.* Dasht-i-Nawar. *desert* C Afghanistan
123 W6 **Navarin, Mys** *headland* NE Russian Federation 62°18′N 179°06′E
63 I25 **Navarino, Isla** *island* S Chile
105 Q4 **Navarra** *Eng./Fr.* Navarre. ◈ *autonomous community* N Spain
Navarre *see* Navarra
105 P4 **Navarrete** La Rioja, N Spain 42°26′N 02°34′W
61 C20 **Navarro** Buenos Aires, E Argentina 35°00′S 59°15′W
105 O12 **Navas de San Juan** Andalucía, S Spain 38°11′N 03°19′W
25 V10 **Navasota** Texas, SW USA 30°23′N 96°05′W
25 U9 **Navasota River** ≈ Texas, SW USA
44 I9 **Navassa Island** ◇ *US unincorporated territory* C West Indies
119 L19 **Navasyolki** *Rus.* Novosëlki. Homyel'skaya Voblasts', SE Belarus 52°24′N 28°33′E
119 H17 **Navayel'nya** *Pol.* Nowojelnia, *Rus.* Novoyel'nya. Hrodzyenskaya Voblasts', W Belarus 53°28′N 25°35′E
171 Y13 **Naver** Papua, E Indonesia 03°27′S 139°45′E
118 H5 **Navesti** ≈ C Estonia
104 J2 **Navia** Asturias, N Spain 43°33′N 06°43′W
104 J2 **Navia** ≈ NW Spain
59 I21 **Naviraí** Mato Grosso do Sul, SW Brazil 23°01′S 54°09′W
126 I6 **Navlya** Bryanskaya Oblast', W Russian Federation 52°47′N 34°28′E
187 X13 **Navoalevu** Vanua Levu, N Fiji 16°22′S 179°28′E
147 P13 **Navobod** *Rus.* Navabad. W Tajikistan 38°37′N 68°42′E
147 R12 **Navobod** *Rus.* Navabad, Novabad. C Tajikistan 39°00′N 70°06′E
Navoi *see* Navoiy
146 M11 **Navoiy** *Rus.* Navoi. Navoiy Viloyati, C Uzbekistan 40°05′N 65°23′E
Navoiyskaya Oblast' *see* Navoiy Viloyati
146 K8 **Navoiy Viloyati** *Rus.* Navoiyskaya Oblast'. ◈ *province* N Uzbekistan
40 G7 **Navojoa** Sonora, NW Mexico 27°04′N 109°28′W
Navolat *see* Navolato
40 H9 **Navolato** *var.* Navolat. Sinaloa, C Mexico 24°46′N 107°42′W
187 Q13 **Navonda** Ambae, C Vanuatu 15°21′S 167°58′E
Návpaktos *see* Náfpaktos
Návplion *see* Náfplio
77 P14 **Navrongo** N Ghana 10°51′N 01°03′W
154 D12 **Navsāri** *var.* Nausari. Gujarāt, W India 20°55′N 72°55′E
187 X15 **Navua** Viti Levu, W Fiji 18°15′S 178°10′E
138 H8 **Nawá** Darʻā, S Syria 32°53′N 36°03′E
Nawabashah *see* Nawābshāh
153 S14 **Nawabganj** Rajshahi, NW Bangladesh 24°35′N 88°21′E
153 S14 **Nawābganj** Uttar Pradesh, N India 26°52′N 82°09′E
149 Q15 **Nawābshāh** *var.* Nawabashah. Sind, S Pakistan 26°15′N 68°26′E
153 P14 **Nawāda** Bihār, N India 24°54′N 85°33′E
152 H11 **Nawalgarh** Rājasthān, N India 27°48′N 75°21′E
Nawāl, Sabkhat an *see* Noual, Sebkhet en
Nawar, Dasht-i- *see* Nāvar, Dasht-e
167 N4 **Nawnghkio** *var.* Nawngkio. Shan State, C Myanmar (Burma) 22°17′N 96°50′E
Nawngkio *see* Nawnghkio
137 U13 **Naxçıvan** *Rus.* Nakhichevan'. SW Azerbaijan 39°14′N 45°24′E
160 I10 **Naxi** Sichuan, C China 28°50′N 105°20′E
115 K21 **Náxos** *var.* Naxos. Náxos, Kykládes, Greece, Aegean Sea 36°07′N 25°24′E
115 K21 **Náxos** *island* Kykládes, Greece, Aegean Sea
40 J11 **Nayarit** ◈ *state* C Mexico
187 Y14 **Nayau** *island* Lau Group, E Fiji
143 S8 **Nāy Band** Yazd, E Iran 32°26′N 57°30′E
165 T2 **Nayoro** Hokkaidō, NE Japan 44°22′N 142°27′E
166 M7 **Nay Pyi Taw** ◈ *union territory* C Myanmar (Burma)
104 F9 **Nazaré** *var.* Nazare. Leiria, C Portugal 39°36′N 09°04′W
Nazare *see* Nazaré
24 M4 **Nazareth** Texas, SW USA 34°32′N 102°06′W
Nazareth *see* Natzrat
173 O8 **Nazareth Bank** *undersea feature* W Indian Ocean
40 K9 **Nazas** Durango, C Mexico 25°15′N 104°06′W
57 F16 **Nazca** Ica, S Peru 14°53′S 74°54′W
0 L17 **Nazca Plate** *tectonic feature*
193 U9 **Nazca Ridge** *undersea feature* E Pacific Ocean 22°00′S 82°00′W
165 V15 **Naze** *var.* Nase. Kagoshima, Amami-ōshima, SW Japan 28°21′N 129°30′E
Nazerat *see* Natzrat
137 R14 **Nazik Gölü** ◎ E Turkey
136 C15 **Nazilli** Aydın, SW Turkey 37°55′N 28°20′E
137 P14 **Nazımiye** Tunceli, E Turkey 39°12′N 39°51′E
Nazinon *see* Red Volta
10 L15 **Nazko** British Columbia, SW Canada 52°57′N 123°44′W
127 O16 **Nazran'** Respublika Ingushetiya, SW Russian Federation 43°14′N 44°47′E
80 J13 **Nazrēt** *var.* Adama, Hadama. Oromīya, C Ethiopia 08°31′N 39°20′E
Nazwāh *see* Nizwa
82 J13 **Nchanga** Copperbelt, C Zambia 12°30′S 27°53′E
82 J11 **Nchelenge** Luapula, N Zambia 09°20′S 28°50′E
Ncheu *see* Ntcheu
83 J25 **Nciba** *Eng.* Great Kei; *prev.* Groot-Kei. ≈ S South Africa
Ndaghamcha, Sebkra de *see* Te-n-Dghâmcha, Sebkhet
81 G21 **Ndala** Tabora, C Tanzania 04°45′S 33°15′E
82 B11 **N'Dalatando** *Port.* Salazar, Vila Salazar. Kwanza Norte, NW Angola 09°19′S 14°48′E
77 S14 **Ndali** C Benin 09°50′N 02°46′E
81 E18 **Ndeke** SW Uganda 0°11′S 30°04′E
78 J13 **Ndélé** Bamingui-Bangoran, N Central African Republic 08°24′N 20°41′E
79 E19 **Ndendé** Ngounié, S Gabon 02°21′S 11°20′E
79 E20 **Ndindi** Nyanga, S Gabon 03°47′S 11°06′E
78 G11 **N'Djaména** *var.* Ndjamena; *prev.* Fort-Lamy. ● (Chad) Ville de N'Djaména, W Chad 12°08′N 15°02′E
78 G11 **N'Djaména** ✈ Ville de N'Djaména, W Chad 12°09′N 15°00′E
Ndjamena *see* N'Djaména
N'Djaména, Région de la Ville de *see* N'Djaména, Ville de
78 G11 **N'Djaména, Ville de** ◈ *region* SW Chad
79 D18 **Ndjolé** Moyen-Ogooué, W Gabon 0°07′S 10°45′E
82 J13 **Ndola** Copperbelt, C Zambia 12°59′S 28°35′E
Ndrhamcha, Sebkha de *see* Te-n-Dghâmcha, Sebkhet
79 L15 **Ndu** Orientale, N Dem. Rep. Congo 04°36′N 22°49′E
81 H21 **Nduguti** Singida, C Tanzania 04°19′S 34°40′E
186 M9 **Nduindui** Guadalcanal, C Solomon Islands 09°46′S 159°54′E
Nduke *see* Kolombangara
Ndzouani *see* Nzwani
115 F16 **Néa Anchiálos** *var.* Nea Anhialos, Néa Ankhíalos. Thessalía, C Greece 39°16′N 22°49′E
Nea Anhialos/Néa Ankhíalos *see* Néa Anchiálos
115 H18 **Néa Artáki** Évvoia, C Greece 38°31′N 23°39′E
97 F15 **Neagh, Lough** ◎ E Northern Ireland, United Kingdom
32 F7 **Neah Bay** Washington, NW USA 48°21′N 124°39′W
115 J22 **Nea Kaméni** *island* Kykládes, Greece, Aegean Sea
181 O8 **Neale, Lake** ◎ Northern Territory, C Australia
182 G2 **Neales River** *seasonal river* South Australia
115 G14 **Néa Moudanía** *var.* Néa Moudhaniá. Kentrikí Makedonía, N Greece 40°14′N 23°17′E
Néa Moudhaniá *see* Néa Moudanía
116 K10 **Neamţ** ◈ *county* NE Romania
Neapel *see* Napoli
115 D14 **Neápoli** *prev.* Neápolis. Dytikí Makedonía, N Greece 40°19′N 21°23′E
115 K25 **Neápoli** Kríti, Greece, E Mediterranean Sea 35°15′N 25°37′E
115 G22 **Neápoli** Pelopónnisos, S Greece 36°29′N 23°05′E
Neápolis *see* Neápoli, Greece
Neapolis *see* Napoli, Italy
Neapolis *see* Nablus, West Bank
38 D16 **Near Islands** *island group* Aleutian Islands, Alaska, USA
97 J21 **Neath** S Wales, United Kingdom 51°40′N 03°48′W
114 H13 **Néa Zíchni** *var.* Néa Zíkhni; *prev.* Néa Zíkhna. Kentrikí Makedonía, NE Greece 41°02′N 23°50′E
Néa Zíkhna/Néa Zíkhni *see* Néa Zíchni
42 C5 **Nebaj** Quiché, W Guatemala 15°25′N 91°05′W
77 P13 **Nebbou** S Burkina Faso 11°22′N 01°49′W
Nebitdag *see* Balkanabat
54 M13 **Neblina, Pico da** ▲ NW Brazil 0°49′N 66°31′W
124 I13 **Nebolchi** Novgorodskaya Oblast', W Russian Federation 59°08′N 33°19′E
36 L4 **Nebo, Mount** ▲ Utah, W USA 39°47′N 111°46′W
28 L14 **Nebraska** *off.* State of Nebraska, *also known as* Blackwater State, Cornhusker State, Tree Planters State. ◈ *state* C USA
29 S16 **Nebraska City** Nebraska, C USA 40°38′N 95°52′W
107 K23 **Nebrodi, Monti** *var.* Monti Caronie. ▲▲ Sicilia, Italy, C Mediterranean Sea
10 L14 **Nechako** ≈ British Columbia, SW Canada
29 Q2 **Neche** North Dakota, N USA 48°57′N 97°33′W
25 V8 **Neches** Texas, SW USA 31°51′N 95°28′W
25 W8 **Neches River** ≈ Texas, SW USA
101 H20 **Neckar** ≈ SW Germany
101 H20 **Neckarsulm** Baden-Württemberg, SW Germany 49°12′N 09°13′E
192 L5 **Necker Island** *island* C British Virgin Islands
175 U3 **Necker Ridge** *undersea feature* N Pacific Ocean
61 D23 **Necochea** Buenos Aires, E Argentina 38°34′S 58°42′W
104 H2 **Neda** Galicia, NW Spain 43°29′N 08°09′W
115 E20 **Néda** *var.* Nédas. ≈ S Greece
Nédas *see* Néda
114 J12 **Nedelino** Smolyan, S Bulgaria 41°27′N 25°05′E
25 Y11 **Nederland** Texas, SW USA 29°58′N 93°59′W
Nederland *see* Netherlands
98 K12 **Neder Rijn** *Eng.* Lower Rhine. ≈ C Netherlands

◆ Country
● Country Capital
◇ Dependent Territory
○ Dependent Territory Capital
◈ Administrative Regions
✈ International Airport
▲ Mountain
▲▲ Mountain Range
Volcano
≈ River
◎ Lake
⊠ Reservoir

99 L16 **Nederweert** Limburg, SE Netherlands *51°17′N 05°45′E*
95 G16 **Nedre Tokke** ◎ S Norway
Nedrigaylov *see* Nedryhayliv
117 S3 **Nedryhayliv** *Rus.* Nedrigaylov. Sums'ka Oblast', NE Ukraine *50°51′N 33°54′E*
98 O11 **Neede** Gelderland, E Netherlands *52°08′N 06°36′E*
33 T13 **Needle Mountain** ▲ Wyoming, C USA *44°03′N 109°33′W*
35 Y14 **Needles** California, W USA *34°50′N 114°37′W*
97 M24 **Needles, The** *rocks* S England, United Kingdom
62 O7 **Neembucú** *off.* Departamento de Ñeembucú. ◆ *department* SW Paraguay
Neembucú, Departamento de *see* Ñeembucú
30 M7 **Neenah** Wisconsin, N USA *44°09′N 88°26′W*
11 W16 **Neepawa** Manitoba, S Canada *50°14′N 99°29′W*
99 K16 **Neerpelt** Limburg, NE Belgium *51°13′N 05°26′E*
74 M6 **Nefta** ✈ W Tunisia *34°03′N 08°05′E*
126 L15 **Neftegorsk** Krasnodarskiy Kray, SW Russian Federation *44°21′N 39°40′E*
127 U3 **Neftekamsk** Respublika Bashkortostan, W Russian Federation *56°07′N 54°13′E*
127 O14 **Neftekumsk** Stavropol'skiy Kray, SW Russian Federation *44°45′N 45°00′E*
Neftezavodsk *see* Seýdi
82 C10 **Negage** *var.* N'Gage. Uíge, NW Angola *07°47′S 15°27′E*
Negapatam/Negapattinam *see* Nāgappattinam
169 T17 **Negara** Bali, Indonesia *08°21′S 114°35′E*
169 T13 **Negara** Borneo, C Indonesia *02°40′S 115°05′E*
Negara Brunei Darussalam *see* Brunei
31 N4 **Negaunee** Michigan, N USA *46°30′N 87°36′W*
81 J15 **Negēlē** *var.* Negelli, *It.* Neghelli. Oromīya, C Ethiopia *05°13′N 39°43′E*
Negelli *see* Negēlē
Negeri Pahang Darul Makmur *see* Pahang
Negeri Selangor Darul Ehsan *see* Selangor
168 K9 **Negeri Sembilan** *var.* Negri Sembilan. ◆ *state* Peninsular Malaysia
92 P3 **Negerpynten** *headland* S Svalbard *77°15′N 22°40′E*
Negev *see* HaNegev
Neghelli *see* Negēlē
116 I12 **Negoiu** *var.* Negoiul. ▲ S Romania *45°34′N 24°34′E*
Negoiul *see* Negoiu
82 P13 **Negomane** *var.* Negomano. Cabo Delgado, N Mozambique *11°22′S 38°32′E*
Negomano *see* Negomane
155 J25 **Negombo** Western Province, SW Sri Lanka *07°13′N 79°51′E*
191 W11 **Negonego** *prev.* Nengonengo. *atoll* Îles Tuamotu, C French Polynesia
Negoreloye *see* Nyeharelaye
112 P12 **Negotin** Serbia, E Serbia *44°14′N 22°31′E*
113 P19 **Negotino** C Macedonia *41°29′N 22°04′E*
56 A10 **Negra, Punta** *headland* NW Peru *06°03′S 81°08′W*
104 G3 **Negreira** Galicia, NW Spain *42°54′N 08°46′W*
116 L10 **Negreşti** Vaslui, E Romania *46°50′N 27°28′E*
Negreşti *see* Negreşti-Oaş
116 H8 **Negreşti-Oaş** *Hung.* Avasfelsőfalu; *prev.* Negreşti. Satu Mare, NE Romania *47°56′N 23°22′E*
44 H12 **Negril** W Jamaica *18°16′N 78°21′W*
Negri Sembilan *see* Negeri Sembilan
63 K15 **Negro, Río** ≈ E Argentina
62 N7 **Negro, Río** ≈ NE Argentina
57 N17 **Negro, Río** ≈ E Bolivia
48 F6 **Negro, Río** ≈ N South America
61 E18 **Negro, Río** ≈ Brazil/Uruguay
62 O5 **Negro, Río** ≈ C Paraguay
Negro, Río *see* Chixoy, Río, Guatemala/Mexico
Negro, Río *see* Sico Tinto, Río, Honduras
171 P6 **Negros** *island* C Philippines
116 M15 **Negru Vodă** Constanţa, SE Romania *43°49′N 28°12′E*
13 P13 **Neguac** New Brunswick, SE Canada *47°16′N 65°04′W*
14 B7 **Negwazu, Lake** ◎ Ontario, S Canada
Négyfalu *see* Săcele
32 F10 **Nehalem** Oregon, NW USA *45°42′N 123°55′W*
32 F10 **Nehalem River** ≈ Oregon, NW USA
Nehavend *see* Nahāvand
143 V9 **Nehbandān** Khorāsān-e Jonūbī, E Iran *31°00′N 60°00′E*
163 V6 **Nehe** Heilongjiang, NE China *48°28′N 124°52′E*
193 Y14 **Neiafu** 'Uta Vava'u, N Tonga *18°36′S 173°58′W*
45 N9 **Neiba** *var.* Neyba. SW Dominican Republic *18°31′N 71°25′W*
Néid, Carn Uí *see* Mizen Head
92 M9 **Neiden** Finnmark, N Norway *69°41′N 29°23′E*
Neidín *see* Kenmare
Néifinn *see* Nephin
103 S10 **Neige, Crêt de la** ▲ E France *46°18′N 05°58′E*
173 O16 **Neiges, Piton des** ▲ C Réunion *21°05′S 55°28′E*
15 R9 **Neiges, Rivière des** ≈ Québec, SE Canada
160 I10 **Neijiang** Sichuan, C China *29°32′N 105°03′E*
30 K6 **Neillsville** Wisconsin, N USA *44°34′N 90°36′W*
Nei Monggol Zizhiqu/Nei Mongol *see* Nei Mongol Zizhiqu
163 Q10 **Nei Mongol Gaoyuan** *plateau* NE China
163 O12 **Nei Mongol Zizhiqu** *var.* Nei Mongol, *Eng.* Inner Mongolia, Inner Mongolian Autonomous Region; *prev.* Nei Monggol Zizhiqu. ◆ *autonomous region* N China
161 O4 **Neiqiu** Hebei, E China *37°22′N 114°34′E*
Neiriz *see* Neyrīz
101 Q16 **Neisse** *Pol.* Nisa *Cz.* Lužická Nisa, *Ger.* Lausitzer Neisse, Nysa Łużycka. ≈ C Europe
Neisse *see* Nysa
54 E11 **Neiva** Huila, S Colombia *02°58′N 75°15′W*
160 M7 **Neixiang** Henan, C China *33°08′N 111°50′E*
Nejafabad *see* Najafābād
11 V9 **Nejanilini Lake** ◎ Manitoba, C Canada
Nejd *see* Najd
80 I13 **Nek'emtē** *var.* Lakemti, Nakamti. Oromīya, C Ethiopia *09°06′N 36°31′E*
126 M9 **Nekhayevskaya** Volgogradskaya Oblast', SW Russian Federation *50°25′N 41°44′E*
30 K7 **Nekoosa** Wisconsin, N USA *44°19′N 89°54′W*
Neksø Bornholm *see* Nexø
104 H7 **Nelas** Viseu, N Portugal *40°32′N 07°52′W*
124 H16 **Nelidovo** Tverskaya Oblast', W Russian Federation *56°13′N 32°45′E*
29 P13 **Neligh** Nebraska, C USA *42°07′N 98°01′W*
123 R11 **Nel'kan** Khabarovskiy Kray, E Russian Federation *57°44′N 136°09′E*
92 M10 **Nellim** *var.* Nellimö, *Lapp.* Njellim. Lappi, N Finland *68°49′N 28°18′E*
Nellimö *see* Nellim
155 J18 **Nellore** Andhra Pradesh, E India *14°29′N 80°00′E*
61 B17 **Nelson** Santa Fe, C Argentina *31°16′S 60°45′W*
11 O17 **Nelson** British Columbia, SW Canada *49°29′N 117°17′W*
185 I14 **Nelson** Nelson, South Island, New Zealand *41°17′S 173°17′E*
97 L17 **Nelson** NW England, United Kingdom *53°51′N 02°13′W*
29 P17 **Nelson** Nebraska, C USA *40°12′N 98°04′W*
185 J14 **Nelson** ◆ *unitary authority* South Island, New Zealand
11 X12 **Nelson** ≈ Manitoba, C Canada
183 U8 **Nelson Bay** New South Wales, SE Australia *32°48′S 152°10′E*
182 K13 **Nelson, Cape** *headland* Victoria, SE Australia *38°25′S 141°33′E*
63 G23 **Nelson, Estrecho** *strait* SE Pacific Ocean
11 W12 **Nelson House** Manitoba, C Canada *55°49′N 98°51′W*
30 J4 **Nelson Lake** ◎ Wisconsin, N USA
31 T14 **Nelsonville** Ohio, N USA *39°27′N 82°13′W*
27 S2 **Nelsoon River** ≈ Iowa/Missouri, C USA
83 K21 **Nelspruit** Mpumalanga, NE South Africa *25°28′S 30°58′E*
Nelspruit *see* Mbombela
76 L10 **Néma** Hodh ech Chargui, SE Mauritania *16°32′N 07°12′W*
118 D13 **Neman** *Ger.* Ragnit. Kaliningradskaya Oblast', W Russian Federation *55°01′N 22°00′E*
84 I9 **Neman** *Bel.* Nyoman, *Ger.* Memel, *Lith.* Nemunas, *Pol.* Niemen. ≈ NE Europe
Nemausus *see* Nîmes
115 F19 **Neméa** Pelopónnisos, S Greece *37°49′N 22°40′E*
Německý Brod *see* Havlíčkův Brod
14 D7 **Nemegosenda** ≈ Ontario, S Canada
14 D8 **Nemegosenda Lake** ◎ Ontario, S Canada
119 H14 **Nemencinė** Vilnius, SE Lithuania *54°50′N 25°29′E*
Nemetocenna *see* Arras
Nemirov *see* Nemyriv
103 O6 **Nemours** Seine-et-Marne, N France *48°16′N 02°41′E*
Nemunas *see* Neman
165 W4 **Nemuro** Hokkaidō, NE Japan *43°20′N 145°35′E*
165 W4 **Nemuro-hantō** *peninsula* Hokkaidō, NE Japan
165 W3 **Nemuro-kaikyō** *strait* Japan/Russian Federation
165 W4 **Nemuro-wan** *bay* N Japan
116 H5 **Nemyriv** *Rus.* Nemirov. L'vivs'ka Oblast', NW Ukraine *50°08′N 23°28′E*
117 N7 **Nemyriv** *Rus.* Nemirov. Vinnyts'ka Oblast', C Ukraine *48°58′N 28°50′E*
97 D19 **Nenagh** *Ir.* An tAonach. Tipperary, C Ireland *52°52′N 08°12′W*
39 R9 **Nenana** Alaska, USA *64°33′N 149°05′W*
39 R9 **Nenana River** ≈ Alaska, USA
187 P10 **Nendö** *var.* Swallow Island. *island* Santa Cruz Islands, E Solomon Islands
97 O19 **Nene** ≈ E England, United Kingdom
125 R4 **Nenetskiy Avtonomnyy Okrug** ◆ *autonomous district* Arkhangel'skaya Oblast', NW Russian Federation
Nengonengo *see* Negonego
163 V6 **Nenjiang** Heilongjiang, NE China *49°11′N 125°18′E*
163 U6 **Nen Jiang** *var.* Nonni. ≈ NE China
189 P16 **Neoch** *atoll* Caroline Islands, C Micronesia
115 D18 **Neochóri** Dytikí Elláda, C Greece *38°23′N 21°14′E*
27 Q7 **Neodesha** Kansas, C USA *37°25′N 95°40′W*
29 S14 **Neola** Iowa, C USA *41°27′N 95°40′W*
115 E16 **Néo Monastíri** *var.* Néon Monastiri. Thessalía, C Greece *39°22′N 21°55′E*
Néon Karlovasi/Néon Karlovásion *see* Karlovási
Néon Monastiri *see* Néo Monastíri
27 R8 **Neosho** Missouri, C USA *36°53′N 94°24′W*
27 Q7 **Neosho River** ≈ Kansas/Oklahoma, C USA
123 N12 **Nepa** ≈ C Russian Federation
153 N10 **Nepal** *off.* Nepal. ◆ *monarchy* S Asia
Nepal *see* Nepal
152 M11 **Nepālganj** Mid Western, SW Nepal *28°04′N 81°37′E*
14 L13 **Nepean** Ontario, SE Canada *45°19′N 75°54′W*
36 L4 **Nephi** Utah, W USA *39°43′N 111°50′W*
97 B16 **Nephin** *Ir.* Néifinn. ▲ W Ireland *54°00′N 09°21′W*
67 T9 **Nepoko** ≈ NE Dem. Rep. Congo
18 K15 **Neptune** New Jersey, NE USA *40°10′N 74°03′W*
182 G10 **Neptune Islands** *island group* South Australia
107 I14 **Nera** *anc.* Nar. ≈ C Italy
102 L14 **Nérac** Lot-et-Garonne, SW France *44°08′N 00°21′E*
111 D16 **Neratovice** *Ger.* Neratowitz. Středočeský Kraj, C Czech Republic *50°16′N 14°31′E*
Neratowitz *see* Neratovice
123 O13 **Nercha** ≈ S Russian Federation
123 O13 **Nerchinsk** Zabaykal'skiy Kray, S Russian Federation *52°01′N 116°25′E*
123 P14 **Nerchinskiy Zavod** Zabaykal'skiy Kray, S Russian Federation *51°13′N 119°25′E*
124 M15 **Nerekhta** Kostromskaya Oblast', NW Russian Federation *57°27′N 40°33′E*
118 H10 **Nereta** S Latvia *56°12′N 25°18′E*
106 K13 **Nereto** Abruzzo, C Italy *42°49′N 13°50′E*
113 H15 **Neretva** ≈ Bosnia and Herzegovina/Croatia
115 C17 **Nerikós** *ruins* Lefkáda, Iónia Nisiá, Greece, C Mediterranean Sea
83 F15 **Neriquinha** Kuando Kubango, SE Angola *15°44′S 21°34′E*
118 I13 **Neris** *Bel.* Viliya, *Pol.* Wilia; *prev. Pol.* Wilja. ≈ Belarus/Lithuania
Neris *see* Viliya
105 N15 **Nerja** Andalucía, S Spain *36°45′N 03°35′W*
124 L16 **Nerl'** ≈ W Russian Federation
105 P12 **Nerpio** Castilla-La Mancha, C Spain *38°08′N 02°18′W*
104 J13 **Nerva** Andalucía, S Spain *37°40′N 06°31′W*
98 L4 **Nes** Fryslân, N Netherlands *53°28′N 05°46′E*
94 G13 **Nesbyen** Buskerud, S Norway *60°36′N 09°35′E*
114 M9 **Nesebar** *var.* Nesebûr. Burgas, E Bulgaria *42°40′N 27°43′E*
Nesebûr *see* Nesebar
Neshcherdo, Ozero *see* Nyeshcharda, Vozyera
92 L2 **Neskaupstaður** Austurland, E Iceland *65°08′N 13°45′W*
92 F13 **Nesna** Nordland, C Norway *66°11′N 12°54′E*
26 K5 **Ness City** Kansas, C USA *38°27′N 99°54′W*
Nesselsdorf *see* Kopřivnice
108 H7 **Nesslau** Sankt Gallen, NE Switzerland *47°13′N 09°12′E*
96 I9 **Ness, Loch** ◎ N Scotland, United Kingdom
Nesterov *see* Zhovkva
114 I12 **Néstos** *Bul.* Mesta, *Turk.* Kara Su. ≈ Bulgaria/Greece *see also* Mesta
Néstos *see* Mesta
95 C14 **Nesttun** Hordaland, S Norway *60°19′N 05°16′E*
Nesvizh *see* Nyasvizh
138 F9 **Netanya** *var.* Natanya, Nathanya. Central, C Israel *32°20′N 34°51′E*
98 I9 **Netherlands** *off.* Kingdom of the Netherlands, *var.* Holland, *Dut.* Koninkrijk der Nederlanden, Nederland. ◆ *monarchy* NW Europe
Netherlands East Indies *see* Indonesia
Netherlands Guiana *see* Suriname
Netherlands, Kingdom of the *see* Netherlands
Netherlands New Guinea *see* Papua
116 L4 **Netishyn** Khmel'nyts'ka Oblast', W Ukraine *50°20′N 26°38′E*
138 E11 **Netivot** Southern, S Israel *31°26′N 34°36′E*
107 O21 **Neto** ≈ S Italy
9 Q6 **Nettilling Lake** ◎ Baffin Island, Nunavut, N Canada
29 V3 **Nett Lake** ◎ Minnesota, N USA
107 I16 **Nettuno** Lazio, C Italy *41°27′N 12°40′E*
Netum *see* Noto
41 U16 **Netzahualcóyotl, Presa** SE Mexico
Netze *see* Noteć
Neu Amerika *see* Puławy
Neubetsche *see* Novi Bečej
Neubidschow *see* Nový Bydžov
Neubistritz *see* Nová Bystřice
100 N9 **Neubrandenburg** Mecklenburg-Vorpommern, NE Germany *53°33′N 13°16′E*
101 K22 **Neuburg an der Donau** Bayern, S Germany *48°43′N 11°10′E*
108 C8 **Neuchâtel** *Ger.* Neuenburg. Neuchâtel, W Switzerland *46°59′N 06°55′E*
108 C8 **Neuchâtel** *Ger.* Neuenburg. ◆ *canton* W Switzerland
108 C8 **Neuchâtel, Lac de** *Ger.* Neuenburger See. ◎ W Switzerland
Neudorf *see* Spišská Nová Ves
100 L10 **Neue Elde** *canal* N Germany
Neuenburg *see* Neuchâtel
Neuenburg an der Elbe *see* Nymburk
Neuenburger See *see* Neuchâtel, Lac de
108 F7 **Neuenhof** Aargau, N Switzerland *47°27′N 08°17′E*
100 H11 **Neuenland** ✈ (Bremen) Bremen, NW Germany *53°03′N 08°46′E*
Neuenstadt *see* La Neuveville
101 C18 **Neuerburg** Rheinland-Pfalz, W Germany *50°01′N 06°13′E*
99 K24 **Neufchâteau** Luxembourg, SE Belgium *49°50′N 05°26′E*
103 S6 **Neufchâteau** Vosges, NE France *48°21′N 05°42′E*
102 M3 **Neufchâtel-en-Bray** Seine-Maritime, N France *49°44′N 01°26′E*
109 S3 **Neufelden** Oberösterreich, N Austria *48°27′N 14°01′E*
Neugradisk *see* Nova Gradiška
Neuhaus *see* Jindřichův Hradec
Neuhäusel *see* Nové Zámky
108 G6 **Neuhausen** *var.* Neuhausen am Rheinfall. Schaffhausen, N Switzerland *47°24′N 08°37′E*
Neuhausen am Rheinfall *see* Neuhausen
101 I17 **Neuhof** Hessen, C Germany *50°26′N 09°34′E*
Neuhof *see* Zgierz
Neukuhren *see* Pionerskiy
Neu-Langenburg *see* Tukuyu
109 W4 **Neulengbach** Niederösterreich, NE Austria *48°10′N 15°53′E*
113 G15 **Neum** Federacija Bosne I Hercegovine, S Bosnia and Herzegovina *42°57′N 17°33′E*
Neumark *see* Nowy Targ, Małopolskie, Poland
Neumark *see* Nowe Miasto Lubawskie, Warmińsko-Mazurskie, Poland
Neumarkt *see* Neumarkt im Hausruckkreis, Oberösterreich, Austria
Neumarkt *see* Neumarkt am Wallersee, Salzburg, Austria
Neumarkt *see* Środa Śląska, Dolnośląskie, Poland
Neumarkt *see* Târgu Secuiesc, Covasna, Romania
Neumarkt *see* Târgu Mureş
109 Q5 **Neumarkt am Wallersee** *var.* Neumarkt. Salzburg, NW Austria *47°55′N 13°16′E*
109 R4 **Neumarkt im Hausruckkreis** *var.* Neumarkt. Oberösterreich, N Austria *48°16′N 13°40′E*
101 L20 **Neumarkt in der Oberpfalz** Bayern, SE Germany *49°16′N 11°28′E*
Neumarktl *see* Tržič
Neumoldowa *see* Moldova Nouă
100 J8 **Neumünster** Schleswig-Holstein, N Germany *54°04′N 09°59′E*
109 X5 **Neunkirchen** *var.* Neunkirchen am Steinfeld. Niederösterreich, E Austria *47°44′N 16°05′E*
101 E20 **Neunkirchen** Saarland, SW Germany *49°21′N 07°11′E*
Neunkirchen am Steinfeld *see* Neunkirchen
Neuoderberg *see* Bohumín
63 I15 **Neuquén** Neuquén, SE Argentina *39°03′S 68°36′W*
63 H14 **Neuquén** *off.* Provincia de Neuquén. ◆ *province* W Argentina
Neuquén, Provincia de *see* Neuquén
63 H14 **Neuquén, Río** ≈ W Argentina
Neurode *see* Nowa Ruda
100 N11 **Neuruppin** Brandenburg, NE Germany *52°56′N 12°49′E*
Neusalz an der Oder *see* Nowa Sól
Neu Sandec *see* Nowy Sącz
101 K22 **Neusäss** Bayern, S Germany *48°24′N 10°49′E*
Neusatz *see* Novi Sad
Neuschliss *see* Gherla
21 N8 **Neuse River** ≈ North Carolina, SE USA
109 Z5 **Neusiedl am See** Burgenland, E Austria *47°58′N 16°51′E*
111 G22 **Neusiedler See** *Hung.* Fertő. ◎ Austria/Hungary
Neusohl *see* Banská Bystrica
101 D15 **Neuss** *anc.* Novaesium, Novesium. Nordrhein-Westfalen, W Germany *51°12′N 06°42′E*
Neuss *see* Nyon
Neustadt *see* Neustadt bei Coburg, Bayern, Germany
Neustadt *see* Neustadt an der Aisch, Bayern, Germany
Neustadt *see* Prudnik, Opole, Poland
Neustadt *see* Baia Mare, Maramureş, Romania
100 I12 **Neustadt am Rübenberge** Niedersachsen, N Germany *52°30′N 09°28′E*
101 J19 **Neustadt an der Aisch** *var.* Neustadt. Bayern, C Germany *49°34′N 10°36′E*
Neustadt an der Haardt *see* Neustadt an der Weinstrasse
101 F20 **Neustadt an der Weinstrasse** *prev.* Neustadt an der Haardt, *hist.* Niewenstat; *anc.* Nova Civitas. Rheinland-Pfalz, SW Germany *49°21′N 08°09′E*
101 K18 **Neustadt bei Coburg** *var.* Neustadt. Bayern, C Germany *50°19′N 11°06′E*
Neustadt bei Pinne *see* Lwówek
Neustadt in Oberschlesien *see* Prudnik
Neustadtl *see* Novo mesto
Neustadtl in Mähren *see* Nové Město na Moravě
Neustettin *see* Szczecinek
108 M8 **Neustift im Stubaital** *var.* Stubaital. Tirol, W Austria *47°07′N 11°26′E*
100 N10 **Neustrelitz** Mecklenburg-Vorpommern, NE Germany *53°22′N 13°05′E*
Neutitschein *see* Nový Jičín
Neutra *see* Nitra
101 J22 **Neu-Ulm** Bayern, S Germany *48°23′N 10°02′E*
Neuveville *see* La Neuveville
103 N12 **Neuvic** Corrèze, C France *45°23′N 02°16′E*
Neuwarp *see* Nowe Warpno
100 G9 **Neuwerk** *island* NW Germany
101 E17 **Neuwied** Rheinland-Pfalz, W Germany *50°26′N 07°28′E*
Neuzen *see* Terneuzen
124 H12 **Neva** ≈ NW Russian Federation
29 V14 **Nevada** Iowa, C USA *42°01′N 93°27′W*
27 R6 **Nevada** Missouri, C USA *37°51′N 94°22′W*
35 R5 **Nevada** *off.* State of Nevada, *also known as* Battle Born State, Sagebrush State, Silver State. ◆ *state* W USA
35 P6 **Nevada City** California, W USA *39°15′N 121°02′W*
105 O14 **Nevada, Sierra** ▲ S Spain
35 P6 **Nevada, Sierra** ▲ W USA
62 I13 **Nevado, Sierra del** ▲ W Argentina
124 G16 **Nevel'** Pskovskaya Oblast', W Russian Federation *56°01′N 29°54′E*
123 T14 **Nevel'sk** Ostrov Sakhalin, Sakhalinskaya Oblast', SE Russian Federation *46°41′N 141°54′E*
123 Q13 **Never** Amurskaya Oblast', SE Russian Federation *53°58′N 124°04′E*
127 Q6 **Neverkino** Penzenskaya Oblast', W Russian Federation *52°53′N 46°46′E*
103 P9 **Nevers** *anc.* Noviodunum. Nièvre, C France *47°00′N 03°09′E*
18 J12 **Neversink River** ≈ New York, NE USA
183 Q6 **Nevertire** New South Wales, SE Australia *31°52′S 147°42′E*
113 H15 **Nevesinje** ◆ Republika Srpska, S Bosnia and Herzegovina
118 G12 **Nevežis** ≈ C Lithuania
138 F11 **Neve Zohar** *prev.* Newé Zohar. Southern, E Israel *31°07′N 35°23′E*
126 M14 **Nevinnomyssk** Stavropol'skiy Kray, SW Russian Federation *44°39′N 41°57′E*
45 W10 **Nevis** *island* Saint Kitts and Nevis
Nevoso, Monte *see* Veliki Snežnik
Nevrokop *see* Gotse Delchev
136 J14 **Nevşehir** *var.* Nevshehr. Nevşehir, C Turkey *38°38′N 34°43′E*
136 J14 **Nevşehir** *var.* Nevshehr. ◆ *province* C Turkey
Nevshehr *see* Nevşehir
122 G10 **Nev'yansk** Sverdlovskaya Oblast', C Russian Federation *57°26′N 60°15′E*
81 J25 **Newala** Mtwara, SE Tanzania *10°59′S 39°18′E*
31 P16 **New Albany** Indiana, N USA *38°17′N 85°50′W*
22 M2 **New Albany** Mississippi, S USA *34°29′N 89°00′W*
29 Y11 **New Albin** Iowa, C USA *43°30′N 91°17′W*
55 U8 **New Amsterdam** E Guyana *06°17′N 57°31′W*
183 Q4 **New Angledool** New South Wales, SE Australia *29°06′S 147°54′E*
21 Y2 **Newark** Delaware, NE USA *39°42′N 75°45′W*
18 K14 **Newark** New Jersey, NE USA *40°42′N 74°12′W*
18 G10 **Newark** New York, NE USA *43°01′N 77°04′W*
31 T13 **Newark** Ohio, N USA *40°03′N 82°24′W*
Newark *see* Newark-on-Trent
35 W5 **Newark Lake** ◎ Nevada, W USA
97 N18 **Newark-on-Trent** *var.* Newark. C England, United Kingdom *53°05′N 00°49′W*
22 M7 **New Augusta** Mississippi, S USA *31°12′N 89°03′W*
19 P12 **New Bedford** Massachusetts, NE USA *41°38′N 70°55′W*
32 G11 **Newberg** Oregon, NW USA *45°18′N 122°58′W*
21 X10 **New Bern** North Carolina, SE USA *35°05′N 77°04′W*
20 F8 **Newbern** Tennessee, S USA *36°06′N 89°15′W*
31 P4 **Newberry** Michigan, N USA *46°21′N 85°30′W*
21 Q12 **Newberry** South Carolina, SE USA *34°17′N 81°39′W*
18 F15 **New Bloomfield** Pennsylvania, NE USA *40°24′N 77°08′W*
25 X5 **New Boston** Texas, SW USA *33°27′N 94°25′W*
25 S11 **New Braunfels** Texas, SW USA *29°43′N 98°09′W*
31 Q13 **New Bremen** Ohio, N USA *40°26′N 84°22′W*
97 F18 **Newbridge** *Ir.* An Droichead Nua. Kildare, C Ireland *53°11′N 06°48′W*
18 B14 **New Brighton** Pennsylvania, NE USA *40°44′N 80°18′W*
18 M12 **New Britain** Connecticut, NE USA *41°37′N 72°45′W*
186 G7 **New Britain** *island* E Papua New Guinea
192 I8 **New Britain Trench** *undersea feature* W Pacific Ocean
18 J15 **New Brunswick** New Jersey, NE USA *40°29′N 74°27′W*
15 V8 **New Brunswick** *Fr.* Nouveau-Brunswick. ◆ *province* SE Canada
18 K13 **Newburgh** New York, NE USA *41°30′N 74°00′W*
97 M22 **Newbury** S England, United Kingdom *51°25′N 01°20′W*
19 P10 **Newburyport** Massachusetts, NE USA *42°49′N 70°53′W*
77 T14 **New Bussa** Niger, W Nigeria *09°50′N 04°32′E*
187 O17 **New Caledonia** *var.* Kanaky, *Fr.* Nouvelle-Calédonie. ◇ *French self-governing territory of special status* SW Pacific Ocean
187 O15 **New Caledonia** *island* SW Pacific Ocean
175 O10 **New Caledonia Basin** *undersea feature* W Pacific Ocean
183 T8 **Newcastle** New South Wales, SE Australia *32°55′S 151°46′E*
13 O14 **Newcastle** New Brunswick, SE Canada *47°01′N 65°36′W*
14 I15 **Newcastle** Ontario, SE Canada *43°55′N 78°35′W*
83 K22 **Newcastle** KwaZulu/Natal, E South Africa *27°45′S 29°55′E*
97 G16 **Newcastle** *Ir.* An Caisleán Nua. SE Northern Ireland, United Kingdom *54°12′N 05°54′W*
31 P13 **New Castle** Indiana, N USA *39°56′N 85°21′W*
20 L5 **New Castle** Kentucky, S USA *38°28′N 85°10′W*
27 N11 **Newcastle** Oklahoma, C USA *35°15′N 97°36′W*
18 B13 **New Castle** Pennsylvania, NE USA *41°00′N 80°22′W*
25 R6 **Newcastle** Texas, SW USA *33°11′N 98°44′W*
36 J7 **Newcastle** Utah, W USA *37°40′N 113°31′W*
21 S6 **New Castle** Virginia, NE USA *37°31′N 80°09′W*
33 Z13 **Newcastle** Wyoming, C USA *43°52′N 104°14′W*
97 L14 **Newcastle** ✈ NE England, United Kingdom *55°03′N 01°42′W*
Newcastle *see* Newcastle upon Tyne
97 L18 **Newcastle-under-Lyme** C England, United Kingdom *53°02′N 02°14′W*
97 M14 **Newcastle upon Tyne** *var.* Newcastle, *hist.* Monkchester, *Lat.* Pons Aelii. NE England, United Kingdom *54°59′N 01°35′W*
181 Q4 **Newcastle Waters** Northern Territory, N Australia *17°20′S 133°26′E*
Newchwang *see* Yingkou
18 K13 **New City** New York, NE USA *41°08′N 73°57′W*
31 U13 **Newcomerstown** Ohio, N USA *40°16′N 81°36′W*
18 G15 **New Cumberland** Pennsylvania, NE USA *40°13′N 76°52′W*
21 R1 **New Cumberland** West Virginia, NE USA *40°30′N 80°35′W*
152 I10 **New Delhi** ● (India) Delhi, N India *28°35′N 77°15′E*
11 O17 **New Denver** British Columbia, SW Canada *49°58′N 117°21′W*
28 J9 **Newell** South Dakota, N USA *44°42′N 103°25′W*
21 Q13 **New Ellenton** South Carolina, SE USA *33°25′N 81°41′W*
22 J6 **Newellton** Louisiana, S USA *32°04′N 91°14′W*
28 K6 **New England** North Dakota, N USA *46°32′N 102°52′W*
19 P8 **New England** *cultural region* NE USA
New England of the West *see* Minnesota
183 U5 **New England Range** ▲▲ New South Wales, SE Australia
64 G9 **New England Seamounts** *var.* Bermuda-New England Seamount Arc. *undersea feature* W Atlantic Ocean *38°00′N 61°00′W*
38 M14 **Newenham, Cape** *headland* Alaska, USA *58°39′N 162°10′W*
Newé Zohar *see* Neve Zohar
18 D9 **Newfane** New York, NE USA *43°16′N 78°40′W*
97 M23 **New Forest** *physical region* S England, United Kingdom
13 T12 **Newfoundland** *Fr.* Terre-Neuve. *island* Newfoundland and Labrador, SE Canada
13 R9 **Newfoundland and Labrador** *Fr.* Terre Neuve. ◆ *province* E Canada
65 J8 **Newfoundland Basin** *undersea feature* NW Atlantic Ocean *45°00′N 40°00′W*
64 I8 **Newfoundland Ridge** *undersea feature* NW Atlantic Ocean
64 J8 **Newfoundland Seamounts** *undersea feature* N Sargasso Sea
18 G16 **New Freedom** Pennsylvania, NE USA *39°43′N 76°41′W*
186 K9 **New Georgia** *island* New Georgia Islands, NW Solomon Islands
186 K8 **New Georgia Islands** *island group* NW Solomon Islands
186 L8 **New Georgia Sound** *var.* The Slot. *sound* E Solomon Sea
30 L9 **New Glarus** Wisconsin, N USA *42°50′N 89°38′W*
13 Q15 **New Glasgow** Nova Scotia, SE Canada *45°36′N 62°38′W*
New Goa *see* Panaji
186 A6 **New Guinea** *Dut.* Nieuw Guinea, *Ind.* Irian. *island* Indonesia/Papua New Guinea
192 H8 **New Guinea Trench** *undersea feature* SW Pacific Ocean
32 I6 **Newhalem** Washington, NW USA *48°40′N 121°18′W*
39 P13 **Newhalen** Alaska, USA *59°43′N 154°54′W*
29 X13 **Newhall** Iowa, C USA *42°00′N 91°58′W*
14 F16 **New Hamburg** Ontario, S Canada *43°24′N 80°37′W*
19 N9 **New Hampshire** *off.* State of New Hampshire, *also known as* Granite State. ◆ *state* NE USA
29 W12 **New Hampton** Iowa, C USA *43°03′N 92°19′W*
186 G5 **New Hanover** *island* NE Papua New Guinea
97 P23 **Newhaven** SE England, United Kingdom *50°48′N 00°00′E*
18 M13 **New Haven** Connecticut, NE USA *41°18′N 72°55′W*
31 Q12 **New Haven** Indiana, N USA *41°02′N 84°59′W*
27 W5 **New Haven** Missouri, C USA *38°34′N 91°15′W*
10 K13 **New Hazelton** British Columbia, SW Canada *55°15′N 127°30′W*
New Hebrides *see* Vanuatu
175 P9 **New Hebrides Trench** *undersea feature* N Coral Sea
18 H15 **New Holland** Pennsylvania, NE USA *40°06′N 76°05′W*
22 I9 **New Iberia** Louisiana, S USA *30°00′N 91°51′W*
186 G5 **New Ireland** ◆ *province* NE Papua New Guinea
186 G5 **New Ireland** *island* NE Papua New Guinea
65 A24 **New Island** *island* W Falkland Islands
18 J15 **New Jersey** *off.* State of New Jersey, *also known as* The Garden State. ◆ *state* NE USA
18 C14 **New Kensington** Pennsylvania, NE USA *40°33′N 79°45′W*
21 W6 **New Kent** Virginia, NE USA *37°32′N 76°59′W*
27 O8 **Newkirk** Oklahoma, C USA *36°54′N 97°03′W*
21 Q9 **Newland** North Carolina, SE USA *36°04′N 81°50′W*
28 L6 **New Leipzig** North Dakota, N USA *46°21′N 101°54′W*
14 H9 **New Liskeard** Ontario, S Canada *47°31′N 79°41′W*
22 G7 **Newllano** Louisiana, S USA *31°06′N 93°16′W*
19 N13 **New London** Connecticut, NE USA *41°21′N 72°04′W*
29 Y15 **New London** Iowa, C USA *40°55′N 91°24′W*
29 T8 **New London** Minnesota, N USA *45°18′N 94°56′W*
27 V3 **New London** Missouri, C USA *39°34′N 91°24′W*
30 M7 **New London** Wisconsin, N USA *44°25′N 88°44′W*
27 Y8 **New Madrid** Missouri, C USA *36°34′N 89°32′W*
180 J8 **Newman** Western Australia *23°18′S 119°45′E*
194 M13 **Newman Island** *island* Antarctica
14 H15 **Newmarket** Ontario, S Canada *44°03′N 79°27′W*
97 P20 **Newmarket** E England, United Kingdom *52°18′N 00°28′E*
19 P10 **Newmarket** New Hampshire, NE USA *43°04′N 70°53′W*
21 U4 **New Market** Virginia, NE USA *38°39′N 78°40′W*
21 R2 **New Martinsville** West Virginia, NE USA *39°39′N 80°52′W*
31 U14 **New Matamoras** Ohio, N USA *39°32′N 81°04′W*
32 M12 **New Meadows** Idaho, NW USA *44°57′N 116°16′W*
26 R12 **New Mexico** *off.* State of New Mexico, *also known as* Land of Enchantment, Sunshine State. ◆ *state* SW USA
149 V6 **New Mīrpur** *var.* Mīrpur. Sind, SE Pakistan *33°11′N 73°46′E*
151 N15 **New Moore Island** *Island* E India
23 S4 **Newnan** Georgia, SE USA *33°22′N 84°48′W*
183 P17 **New Norfolk** Tasmania, SE Australia *42°46′S 147°02′E*
22 K9 **New Orleans** Louisiana, S USA *30°00′N 90°01′W*
22 K9 **New Orleans** ✈ Louisiana, S USA *29°59′N 90°17′W*
18 K12 **New Paltz** New York, NE USA *41°44′N 74°04′W*
31 U12 **New Philadelphia** Ohio, N USA *40°29′N 81°27′W*
184 K10 **New Plymouth** Taranaki, North Island, New Zealand *39°04′S 174°06′E*
97 M24 **Newport** S England, United Kingdom *50°42′N 01°18′W*
97 K22 **Newport** SE Wales, United Kingdom *51°35′N 03°W*
27 W10 **Newport** Arkansas, C USA *35°36′N 91°16′W*
31 N13 **Newport** Indiana, N USA *39°52′N 87°24′W*
20 M3 **Newport** Kentucky, S USA *39°05′N 84°27′W*
29 W9 **Newport** Minnesota, N USA *44°52′N 93°00′W*
32 F12 **Newport** Oregon, NW USA *44°39′N 124°04′W*
19 O13 **Newport** Rhode Island, NE USA *41°29′N 71°17′W*
21 O9 **Newport** Tennessee, S USA *35°58′N 83°13′W*
19 N6 **Newport** Vermont, NE USA *44°56′N 72°13′W*
32 M7 **Newport** Washington, NW USA *48°08′N 117°05′W*
21 X7 **Newport News** Virginia, NE USA *36°59′N 76°26′W*
97 N20 **Newport Pagnell** SE England, United Kingdom *52°05′N 00°44′W*
23 U12 **New Port Richey** Florida, SE USA *28°14′N 82°42′W*
29 V9 **New Prague** Minnesota, N USA *44°32′N 93°34′W*
44 H3 **New Providence** *island* N The Bahamas
97 I20 **New Quay** SW Wales, United Kingdom *52°13′N 04°22′W*
97 H24 **Newquay** SW England, United Kingdom *50°27′N 05°03′W*
29 V10 **New Richland** Minnesota, N USA *43°53′N 93°29′W*
15 X7 **New-Richmond** Québec, SE Canada *48°12′N 65°52′W*
31 R15 **New Richmond** Ohio, N USA *38°57′N 84°16′W*
30 I5 **New Richmond** Wisconsin, N USA *45°09′N 92°31′W*
42 G1 **New River** ≈ N Belize
55 T12 **New River** ≈ SE Guyana
21 R6 **New River** ≈ West Virginia, NE USA
42 G1 **New River Lagoon** ◎ N Belize
22 J8 **New Roads** Louisiana, S USA *30°42′N 91°26′W*
18 L14 **New Rochelle** New York, NE USA *40°55′N 73°44′W*
29 O4 **New Rockford** North Dakota, N USA *47°40′N 99°08′W*
97 P23 **New Romney** SE England, United Kingdom *50°58′N 00°56′E*
97 F20 **New Ross** *Ir.* Ros Mhic Thriúin. Wexford, SE Ireland *52°24′N 06°56′W*
97 F16 **Newry** *Ir.* An tIúr. SE Northern Ireland, United Kingdom *54°11′N 06°20′W*
28 M5 **New Salem** North Dakota, N USA *46°51′N 101°24′W*
New Sarum *see* Salisbury
29 W14 **New Sharon** Iowa, C USA *41°28′N 92°39′W*
New Siberian Islands *see* Novosibirskiye Ostrova
23 X11 **New Smyrna Beach** Florida, SE USA *29°01′N 80°55′W*
183 O7 **New South Wales** ◆ *state* SE Australia
39 O13 **New Stuyahok** Alaska, USA *59°27′N 157°18′W*
21 N8 **New Tazewell** Tennessee, S USA *36°26′N 83°36′W*
152 K9 **New Tehri** *prev.* Tehri. Uttarakhand, N India *30°12′N 78°29′E*
38 M12 **Newtok** Alaska, USA *60°56′N 164°37′W*
23 S7 **Newton** Georgia, SE USA *31°18′N 84°20′W*
29 W14 **Newton** Iowa, C USA *41°42′N 93°03′W*
27 N6 **Newton** Kansas, C USA *38°02′N 97°22′W*
19 O11 **Newton** Massachusetts, NE USA *42°19′N 71°10′W*
22 M5 **Newton** Mississippi, S USA *32°19′N 89°09′W*
18 J14 **Newton** New Jersey, NE USA *41°03′N 74°45′W*
21 R9 **Newton** North Carolina, SE USA *35°42′N 81°14′W*
25 Y9 **Newton** Texas, SW USA *30°51′N 93°45′W*
97 J24 **Newton Abbot** SW England, United Kingdom *50°33′N 03°34′W*
96 K13 **Newton St Boswells** SE Scotland, United Kingdom *55°34′N 02°40′W*
97 I14 **Newton Stewart** S Scotland, United Kingdom *54°58′N 04°30′W*
92 O2 **Newtontoppen** ▲ C Svalbard *78°57′N 17°34′E*
97 J20 **Newtown** E Wales, United Kingdom *52°32′N 03°19′W*
28 K3 **New Town** North Dakota, N USA *47°58′N 102°30′W*

◆ Country
● Country Capital
◇ Dependent Territory
○ Dependent Territory Capital
◆ Administrative Regions
✈ International Airport
▲ Mountain
▲▲ Mountain Range
Volcano
≈ River
◎ Lake
Reservoir

97 G15 **Newtownabbey** *Ir.* Baile na Mainistreach. E Northern Ireland, United Kingdom *54°40′N 05°57′W*
97 G15 **Newtownards** *Ir.* Baile Nua na hArda. SE Northern Ireland, United Kingdom *54°36′N 05°41′W*
29 U10 **New Ulm** Minnesota, N USA *44°20′N 94°28′W*
28 K10 **New Underwood** South Dakota, N USA *44°05′N 102°46′W*
25 V10 **New Waverly** Texas, SW USA *30°32′N 95°28′W*
18 K14 **New York** New York, NE USA *40°45′N 73°57′W*
18 G10 **New York** ◆ *state* NE USA
35 X13 **New York Mountains** ▲ California, W USA
184 K12 **New Zealand** ◆ *commonwealth republic* SW Pacific Ocean
95 M24 **Nexø** *var.* Neksø Bornholm. E Denmark *55°04′N 15°09′E*
125 O15 **Neya** Kostromskaya Oblast', NW Russian Federation *58°19′N 43°51′E*
Neyba *see* Neiba
143 Q12 **Neyrīz** *var.* Neiriz, Niriz. Fārs, S Iran *29°14′N 54°18′E*
143 T4 **Neyshābūr** *var.* Nishapur. Khorāsān-Razavi, NE Iran *36°15′N 58°47′E*
155 J21 **Neyveli** Tamil Nādu, SE India *11°36′N 79°26′E*
Nezhin *see* Nizhyn
33 N10 **Nezperce** Idaho, NW USA *46°14′N 116°15′W*
22 H8 **Nezpique, Bayou** ↝ Louisiana, S USA
77 Y13 **Ngadda** ↝ NE Nigeria
N'Gage *see* Negage
185 G16 **Ngahere** West Coast, South Island, New Zealand *42°22′S 171°29′E*
77 Z12 **Ngala** Borno, NE Nigeria *12°19′N 14°11′E*
158 K16 **Ngamring** Xizang Zizhiqu, W China *29°16′N 87°10′E*
81 K19 **Ngangerabeli Plain** *plain* SE Kenya
158 I14 **Ngangla Ringco** ◎ W China
158 H13 **Nganglong Kangri** ▲ W China *32°55′N 81°00′E*
158 K15 **Ngangzê Co** ◎ W China
79 F14 **Ngaoundéré** *var.* N'Gaoundéré. Adamaoua, N Cameroon *07°20′N 13°35′E*
N'Gaoundéré *see* Ngaoundéré
81 E20 **Ngara** Kagera, NW Tanzania *02°30′S 30°40′E*
188 F8 **Ngardmau Bay** *bay* Babeldaob, N Palau
188 F7 **Ngaregur** *island* Palau Islands, N Palau
Ngarrab *see* Gyaca
184 L7 **Ngaruawahia** Waikato, North Island, New Zealand *37°41′S 175°10′E*
184 N11 **Ngaruroro** ↝ North Island, New Zealand
190 I16 **Ngatangiia** Rarotonga, S Cook Islands *21°14′S 159°44′W*
184 M6 **Ngatea** Waikato, North Island, New Zealand *37°16′S 175°29′E*
166 L8 **Ngathainggyaung** Ayeyawady, SW Myanmar (Burma) *17°22′N 95°04′E*
Ngatik *see* Ngetik Atoll
Ngau *see* Gau
Ngawa *see* Aba
172 G12 **Ngazidja** *Fr.* Grande Comore, *var.* Njazidja. *island* NW Comoros
188 C7 **Ngcheangel** *var.* Kayangel Islands. *island* Palau Islands, N Palau
188 E10 **Ngchemiangel** Babeldaob, N Palau
188 C8 **Ngeaur** *var.* Angaur. *island* Palau Islands, S Palau
188 E10 **Ngerkeai** Babeldaob, N Palau
188 F9 **Ngermechau** Babeldaob, N Palau *07°35′N 134°39′E*
188 C8 **Ngeruktabel** *prev.* Urukthapel. *island* Palau Islands, S Palau
188 F8 **Ngetbong** Babeldaob, N Palau *07°37′N 134°35′E*
189 T17 **Ngetik Atoll** *var.* Ngatik; *prev.* Los Jardines. *atoll* Caroline Islands, E Micronesia
188 E10 **Ngetkip** Babeldaob, N Palau
Nghia Dan *see* Thai Hoa
N'Giva *see* Ondjiva
79 G20 **Ngo** Plateaux, SE Congo *02°28′S 15°43′E*
167 S7 **Ngoc Lac** Thanh Hoa, N Vietnam *20°06′N 105°21′E*
79 G17 **Ngoko** ↝ Cameroon/Congo
81 H19 **Ngorengore** Narok, SW Kenya *01°01′S 35°26′E*
159 Q11 **Ngoring Hu** ◎ C China
Ngorolaka *see* Banifing
81 H20 **Ngorongoro Crater** *crater* N Tanzania
79 D19 **Ngounié** *off.* Province de la Ngounié, *var.* La Ngounié. ◆ *province* S Gabon
79 D19 **Ngounié** ↝ Congo/Gabon
Ngounié, Province de la *see* Ngounié
78 H10 **Ngoura** *var.* NGoura. Hadjer-Lamis, W Chad *12°52′N 16°27′E*
NGoura *see* Ngoura
78 G10 **Ngouri** *var.* NGouri; *prev.* Fort-Millot. Lac, W Chad *13°42′N 15°19′E*
NGouri *see* Ngouri
77 Y10 **Ngourti** Diffa, E Niger *15°22′N 13°13′E*
77 Y11 **Nguigmi** *var.* N'Guigmi. Diffa, SE Niger *14°17′N 13°07′E*
N'Guigmi *see* Nguigmi
Nguimbo *see* Lumbala N'Guimbo
188 F15 **Ngulu Atoll** *atoll* Caroline Islands, W Micronesia
187 R14 **Nguna** *island* C Vanuatu
N'Gunza *see* Sumbe
169 U17 **Ngurah Rai** ✕ (Bali) Bali, S Indonesia *8°40′S 115°14′E*
77 W12 **Nguru** Yobe, NE Nigeria *12°55′N 10°31′E*
Ngwaketze *see* Southern
83 I16 **Ngweze** ↝ S Zambia
83 M17 **Nhamatanda** Sofala, C Mozambique *19°16′S 34°10′E*
58 G12 **Nhamundá, Rio** *var.* Jamundá, Yamundá. ↝ N Brazil
60 J7 **Nhandeara** São Paulo, S Brazil *20°40′S 50°03′W*
82 D12 **Nharêa** *var.* N'Harea, Nhareia. Bié, W Angola *11°38′S 16°58′E*
N'Harea *see* Nharêa
Nhareia *see* Nharêa
167 V12 **Nha Trang** Khanh Hoa, S Vietnam *12°15′N 109°10′E*
182 L11 **Nhill** Victoria, SE Australia *36°21′S 141°38′E*
83 L22 **Nhlangano** *prev.* Goedgegun. SW Swaziland *27°06′S 31°12′E*
181 S1 **Nhulunbuy** Northern Territory, N Australia *12°16′S 136°46′E*
77 N10 **Niafounké** Tombouctou, W Mali *15°54′N 03°58′W*
31 N5 **Niagara** Wisconsin, N USA *45°45′N 87°57′W*
14 H16 **Niagara** ↝ Ontario, S Canada
14 G15 **Niagara Escarpment** *hill range* Ontario, S Canada
14 H16 **Niagara Falls** Ontario, S Canada *43°05′N 79°06′W*
18 D9 **Niagara Falls** New York, NE USA *43°06′N 79°04′W*
14 H16 **Niagara Falls** *waterfall* Canada/USA
76 K12 **Niagassola** *var.* Nyagassola. Haute-Guinée, NE Guinea *12°24′N 09°03′W*
77 R12 **Niamey** ● (Niger) Niamey, SW Niger *13°28′N 02°03′E*
77 R12 **Niamey** ✕ Niamey, SW Niger *13°28′N 02°14′E*
77 R14 **Niamtougou** N Togo *09°50′N 01°08′E*
79 O16 **Niangara** Orientale, NE Dem. Rep. Congo *03°45′N 27°54′E*
77 O10 **Niangay, Lac** ◎ E Mali
77 N14 **Niangoloko** SW Burkina Faso *10°15′N 04°53′W*
27 U6 **Niangua River** ↝ Missouri, C USA
79 O17 **Nia-Nia** Orientale, NE Dem. Rep. Congo *01°26′N 27°38′E*
19 N13 **Niantic** Connecticut, NE USA *41°19′N 72°11′W*
163 U7 **Nianzishan** Heilongjiang, NE China *47°31′N 122°53′E*
79 E20 **Niari** ◆ *province* SW Congo
168 H10 **Nias, Pulau** *island* W Indonesia
82 O13 **Niassa** *off.* Província do Niassa. ◆ *province* N Mozambique
Niassa, Província do *see* Niassa
191 U10 **Niau** *island* Îles Tuamotu, C French Polynesia
95 G20 **Nibe** Nordjylland, N Denmark *56°59′N 09°39′E*
189 Q8 **Nibok** N Nauru *0°31′S 166°55′E*
118 C10 **Nīca** W Latvia *56°21′N 21°03′E*
Nicaea *see* Nice
42 J9 **Nicaragua** *off.* Republic of Nicaragua. ◆ *republic* Central America
42 K11 **Nicaragua, Lago de** *var.* Cocibolca, Gran Lago, *Eng.* Lake Nicaragua. ◎ S Nicaragua
Nicaragua, Lake *see* Nicaragua, Lago de
64 D11 **Nicaraguan Rise** *undersea feature* NW Caribbean Sea *16°00′N 80°00′W*
Nicaragua, Republic of *see* Nicaragua
Nicaria *see* Ikaría
107 N21 **Nicastro** Calabria, SW Italy *38°59′N 16°20′E*
103 V15 **Nice** *It.* Nizza; *anc.* Nicaea. Alpes-Maritimes, SE France *43°43′N 07°13′E*
Nice *see* Côte d'Azur
Nicephorium *see* Ar Raqqah
12 M9 **Nichicun, Lac** ◎ Québec, E Canada
164 D16 **Nichinan** *var.* Nitinan. Miyazaki, Kyūshū, SW Japan *31°36′N 131°23′E*
44 E4 **Nicholas Channel** *channel* N Cuba
Nicholas II Land *see* Severnaya Zemlya
149 U2 **Nicholas Range** *Pash.* Selselehye Kuhe Vākhān, *Taj.* Qatorkŭhi Vakhon. ▲ Afghanistan/Tajikistan
20 M6 **Nicholasville** Kentucky, S USA *37°52′N 84°34′W*
44 G2 **Nicholls Town** Andros Island, NW The Bahamas *25°07′N 78°01′W*
21 U12 **Nichols** South Carolina, SE USA *34°13′N 79°09′W*
55 U9 **Nickerie** ◆ *district* NW Suriname
55 V9 **Nickerie Rivier** ↝ NW Suriname
151 P22 **Nicobar Islands** *island group* India, E Indian Ocean
116 L9 **Nicolae Bălcescu** Botoşani, NE Romania *47°33′N 26°52′E*
15 P11 **Nicolet** Québec, SE Canada *46°13′N 72°37′W*
15 Q12 **Nicolet** ↝ Québec, SE Canada
31 Q4 **Nicolet, Lake** ◎ Michigan, N USA
29 U10 **Nicollet** Minnesota, N USA *44°16′N 94°11′W*
61 F19 **Nico Pérez** Florida, S Uruguay *33°30′S 55°10′W*
Nicopolis *see* Nikopol, Bulgaria
Nicopolis *see* Nikópoli, Greece
121 P2 **Nicosia** *Gk.* Lefkosía, *Turk.* Lefkoşa. ● (Cyprus) C Cyprus *35°10′N 33°23′E*
107 K24 **Nicosia** Sicilia, Italy, C Mediterranean Sea *37°45′N 14°24′E*
107 N22 **Nicotera** Calabria, SW Italy *38°33′N 15°55′E*
42 K13 **Nicoya** Guanacaste, W Costa Rica *10°09′N 85°26′W*
42 L14 **Nicoya, Golfo de** *gulf* W Costa Rica
42 L14 **Nicoya, Península de** *peninsula* NW Costa Rica
Nictheroy *see* Niterói
118 B12 **Nida** *Ger.* Nidden. Klaipėda, SW Lithuania *55°20′N 21°00′E*
111 L15 **Nida** ↝ S Poland
Nidaros *see* Trondheim
108 D8 **Nidau** Bern, W Switzerland *47°07′N 07°15′E*
101 H17 **Nidda** ↝ W Germany
Nidden *see* Nida
95 F17 **Nidelva** ↝ S Norway
108 F9 **Nidwalden** ◆ *canton* C Switzerland
110 L9 **Nidzica** *Ger.* Niedenburg. Warmińsko-Mazurskie, NE Poland *53°22′N 20°27′E*
100 H6 **Niebüll** Schleswig-Holstein, N Germany *54°47′N 08°51′E*
Niedenburg *see* Nidzica
99 N25 **Niederanven** Luxembourg, C Luxembourg *49°39′N 06°15′E*
103 V4 **Niederbronn-les-Bains** Bas-Rhin, NE France *48°57′N 07°37′E*
Niederdonau *see* Niederösterreich
109 S7 **Niedere Tauern** ▲ C Austria
101 P14 **Niederlausitz** *Eng.* Lower Lusatia, *Lus.* Donja Łužica. *physical region* E Germany
109 U5 **Niederösterreich** *off.* Land Niederösterreich, *Eng.* Lower Austria, *Ger.* Niederdonau; *prev.* Lower Danube. ◆ *state* NE Austria
Niederösterreich, Land *see* Niederösterreich
100 G12 **Niedersachsen** *Eng.* Lower Saxony, *Fr.* Basse-Saxe. ◆ *state* NW Germany
79 D17 **Niefang** *var.* Sevilla de Niefang. NW Equatorial Guinea *01°52′N 10°12′E*
83 G23 **Niekerkshoop** Northern Cape, W South Africa *29°21′S 22°49′E*
99 G17 **Niel** Antwerpen, N Belgium *51°07′N 04°20′E*
Niélé *see* Niellé
76 M14 **Niellé** *var.* Niélé. N Ivory Coast *10°12′N 05°38′W*
79 O22 **Niemba** Katanga, E Dem. Rep. Congo *05°58′S 28°24′E*
111 G15 **Niemcza** *Ger.* Nimptsch. Dolnośląskie, SW Poland *50°45′N 16°52′E*
Niemen *see* Neman
92 J13 **Niemisel** Norrbotten, N Sweden *66°00′N 22°00′E*
111 H15 **Niemodlin** *Ger.* Falkenberg. Opolskie, SW Poland *50°37′N 17°45′E*
76 M13 **Niéna** Sikasso, SW Mali *11°24′N 06°20′W*
100 H12 **Nienburg** Niedersachsen, N Germany *52°37′N 09°12′E*
100 N13 **Nieplitz** ↝ NE Germany
111 L16 **Niepołomice** Małopolskie, S Poland *50°02′N 20°12′E*
101 D14 **Niers** ↝ Germany/ Netherlands
101 Q15 **Niesky** *Lus.* Niska. Sachsen, E Germany *51°16′N 14°49′E*
Nieśwież *see* Nyasvizh
Nieuport *see* Nieuwpoort
98 O8 **Nieuw-Amsterdam** Drenthe, NE Netherlands *52°43′N 06°52′E*
55 W9 **Nieuw Amsterdam** Commewijne, NE Suriname *05°53′N 55°05′W*
99 M14 **Nieuw-Bergen** Limburg, SE Netherlands *51°36′N 06°04′E*
98 O7 **Nieuw-Buinen** Drenthe, NE Netherlands *52°57′N 06°55′E*
98 J12 **Nieuwegein** Utrecht, C Netherlands *52°03′N 05°06′E*
98 P6 **Nieuwe Pekela** Groningen, NE Netherlands *53°04′N 06°58′E*
98 P5 **Nieuweschans** Groningen, NE Netherlands *53°10′N 07°10′E*
Nieuw Guinea *see* New Guinea
98 I11 **Nieuwkoop** Zuid-Holland, C Netherlands *52°09′N 04°46′E*
98 M9 **Nieuwleusen** Overijssel, S Netherlands *52°34′N 06°16′E*
98 J11 **Nieuw-Loosdrecht** Noord-Holland, C Netherlands *52°12′N 05°08′E*
55 U9 **Nieuw Nickerie** Nickerie, NW Suriname *05°56′N 57°W*
98 P5 **Nieuwolda** Groningen, NE Netherlands *53°15′N 06°58′E*
99 B17 **Nieuwpoort** *var.* Nieuport. West-Vlaanderen, W Belgium *51°08′N 02°45′E*
99 G14 **Nieuw-Vossemeer** Noord-Brabant, S Netherlands *51°34′N 04°13′E*
98 P7 **Nieuw-Weerdinge** Drenthe, NE Netherlands *52°51′N 07°00′E*
40 L10 **Nieves** Zacatecas, C Mexico *24°00′N 102°57′W*
64 O11 **Nieves, Pico de las** ▲ Gran Canaria, Islas Canarias, Spain, NE Atlantic Ocean *27°58′N 15°34′W*
103 P8 **Nièvre** ◆ *department* C France
Niewenstat *see* Neustadt an der Weinstrasse
136 J15 **Niğde** Niğde, C Turkey *37°58′N 34°42′E*
136 J15 **Niğde** ◆ *province* C Turkey
83 J21 **Nigel** Gauteng, NE South Africa *26°25′S 28°28′E*
77 V10 **Niger** *off.* Republic of Niger. ◆ *republic* W Africa
77 T14 **Niger** ◆ *state* C Nigeria
67 P8 **Niger** ↝ W Africa
Niger Cone *see* Niger Fan
67 P9 **Niger Delta** *delta* S Nigeria
67 P9 **Niger Fan** *var.* Niger Cone. *undersea feature* E Atlantic Ocean *04°15′N 05°00′E*
77 T13 **Nigeria** *off.* Federal Republic of Nigeria. ◆ *federal republic* W Africa
Nigeria, Federal Republic of *see* Nigeria
77 T17 **Niger, Mouths of the** *delta* S Nigeria
Niger, Republic of *see* Niger
185 C24 **Nightcaps** Southland, South Island, New Zealand *45°58′S 168°03′E*
14 F7 **Night Hawk Lake** ◎ Ontario, S Canada
65 M19 **Nightingale Island** *island* S Tristan da Cunha, S Atlantic Ocean
38 M12 **Nightmute** Alaska, USA *60°28′N 164°43′W*
114 G13 **Nigrita** Kentrikí Makedonía, NE Greece *40°54′N 23°29′E*
148 J15 **Nihing** *Per.* Rūd-e Nahang. ↝ Iran/Pakistan
191 V10 **Nihiru** *atoll* Îles Tuamotu, C French Polynesia
Nihommatsu *see* Nihonmatsu
Nihon *see* Japan
165 P11 **Nihonmatsu** *var.* Nihommatsu, Nihonmatu. Fukushima, Honshū, C Japan *37°34′N 140°25′E*
Nihonmatu *see* Nihonmatsu
62 I12 **Nihuil, Embalse del** ◎ W Argentina
165 O10 **Niigata** Niigata, Honshū, C Japan *37°55′N 139°01′E*
165 O11 **Niigata** *off.* Niigata-ken. ◆ *prefecture* Honshū, C Japan
Niigata-ken *see* Niigata
165 G14 **Niihama** Ehime, Shikoku, SW Japan *33°57′N 133°15′E*
38 A8 **Ni'ihau** *var.* Niihau. *island* Hawai'i, USA, C Pacific Ocean
165 X12 **Nii-jima** *island* E Japan
165 H12 **Niimi** Okayama, Honshū, SW Japan *35°00′N 133°27′E*
165 O10 **Niitsu** *var.* Niitu. Niigata, Honshū, C Japan *37°48′N 139°09′E*
Niitu *see* Niitsu
105 P15 **Níjar** Andalucía, S Spain *36°57′N 02°13′W*
98 K11 **Nijkerk** Gelderland, C Netherlands *52°13′N 05°30′E*
99 H16 **Nijlen** Antwerpen, N Belgium *51°10′N 04°40′E*
98 L13 **Nijmegen** *Ger.* Nimwegen; *anc.* Noviomagus. Gelderland, SE Netherlands *51°50′N 05°52′E*
98 N10 **Nijverdal** Overijssel, E Netherlands *52°22′N 06°28′E*
190 G16 **Nikao** Rarotonga, S Cook Islands
Nikaria *see* Ikaría
124 I2 **Nikel'** *Finn.* Kolosjoki. Murmanskaya Oblast', NW Russian Federation *69°25′N 30°12′E*
171 Q17 **Nikiniki** Timor, S Indonesia *10°S 124°30′E*
129 Q15 **Nikitin Seamount** *undersea feature* E Indian Ocean *05°48′S 84°48′E*
77 S14 **Nikki** E Benin *09°55′N 03°12′E*
Niklasmarkt *see* Gheorgheni
39 P10 **Nikolai** Alaska, USA *63°00′N 154°22′W*
Nikolaiken *see* Mikołajki
Nikolainkaupunki *see* Vaasa
Nikolayev *see* Mykolayiv
145 O6 **Nikolayevka** Severnyy Kazakhstan, N Kazakhstan
Nikolayevka *see* Zhetigen
127 P9 **Nikolayevsk** Volgogradskaya Oblast', SW Russian Federation *50°03′N 45°30′E*
Nikolayevskaya Oblast' *see* Mykolayivs'ka Oblast'
123 S12 **Nikolayevsk-na-Amure** Khabarovskiy Kray, SE Russian Federation *53°04′N 140°39′E*
127 P6 **Nikol'sk** Penzenskaya Oblast', W Russian Federation *53°46′N 46°03′E*
125 O13 **Nikol'sk** Vologodskaya Oblast', NW Russian Federation *59°35′N 45°31′E*
Nikol'sk *see* Ussuriysk
38 K17 **Nikolski** Umnak Island, Alaska, USA *52°56′N 168°52′W*
Nikol'skiy *see* Satpayev
127 V7 **Nikol'skoye** Orenburgskaya Oblast', W Russian Federation *52°01′N 55°48′E*
Nikol'sk-Ussuriyskiy *see* Ussuriysk
114 J7 **Nikopol** *anc.* Nicopolis. Pleven, N Bulgaria *43°43′N 24°55′E*
117 S9 **Nikopol'** Dnipropetrovs'ka Oblast', SE Ukraine *47°34′N 34°23′E*
115 C17 **Nikópoli** *anc.* Nicopolis. *site of ancient city* Ípeiros, W Greece
136 M12 **Niksar** Tokat, N Turkey *40°35′N 36°59′E*
143 V14 **Nīkshahr** Sīstān va Balūchestān, SE Iran *26°15′N 60°10′E*
113 J16 **Nikšić** C Montenegro *42°47′N 18°56′E*
191 R4 **Nikumaroro** ; *prev.* Gardner Island. *atoll* Phoenix Islands, C Kiribati
191 P3 **Nikunau** *var.* Nukunau; *prev.* Byron Island. *atoll* Tungaru, W Kiribati
155 G21 **Nilambūr** Kerala, SW India *11°17′N 76°15′E*
35 X16 **Niland** California, W USA *33°14′N 115°31′W*
80 G8 **Nile** *former province* NW Uganda
67 T3 **Nile** *Ar.* Nahr an Nīl. ↝ N Africa
75 W7 **Nile Delta** *delta* N Egypt
67 T3 **Nile Fan** *undersea feature* E Mediterranean Sea *33°00′N 31°00′E*
31 O11 **Niles** Michigan, N USA *41°49′N 86°15′W*
31 V11 **Niles** Ohio, N USA *41°10′N 80°46′W*
155 F20 **Nīleswaram** Kerala, SW India *12°18′N 75°07′E*
14 K10 **Nilgaut, Lac** ◎ Québec, SE Canada
149 O6 **Nīlī** Dāykundī, C Afghanistan *33°43′N 66°07′E*
158 I5 **Nilka** Xinjiang Uygur Zizhiqu, NW China *43°46′N 82°33′E*
Nīl, Nahr an *see* Nile
93 N16 **Nilsiä** Pohjois-Savo, C Finland *63°13′N 28°00′E*
154 F9 **Nīmach** Madhya Pradesh, C India *24°27′N 74°56′E*
152 G14 **Nimbāhera** Rājasthān, N India *24°38′N 74°45′E*
76 L15 **Nimba, Monts** *var.* Nimba Mountains. ▲ W Africa
Nimba Mountains *see* Nimba, Monts
Nimburg *see* Nymburk
103 Q15 **Nîmes** *anc.* Nemausus, Nismes. Gard, S France *43°49′N 04°20′E*
152 H11 **Nīm ka Thāna** Rājasthān, N India *27°42′N 75°50′E*
183 R11 **Nimmitabel** New South Wales, SE Australia *36°34′S 149°18′E*
Nimptsch *see* Niemcza
195 R11 **Nimrod Glacier** *glacier* Antarctica
148 K8 **Nīmrōz** *var.* Nimroze; *prev.* Chakhānsūr, Nīmrūz. ◆ *province* SW Afghanistan
Nimroze *see* Nīmrōz
Nīmrūz *see* Nīmrōz
81 F16 **Nimule** Eastern Equatoria, S South Sudan *03°35′N 32°03′E*
Nimwegen *see* Nijmegen
139 Q3 **Nīnawá** *off.* Muḥāfa at Ninawá, *var.* Al Mawşil, Nineveh. ◆ *governorate* NW Iraq
Ninawá, Muḥāfa at *see* Nīnawá
155 C23 **Nine Degree Channel** *channel* India/Maldives
18 G9 **Ninemile Point** *headland* New York, NE USA *43°31′N 76°22′W*
173 S8 **Ninetyeast Ridge** *undersea feature* E Indian Ocean *04°00′S 90°00′E*
183 P13 **Ninety Mile Beach** *beach* Victoria, SE Australia
184 I2 **Ninety Mile Beach** *beach* North Island, New Zealand
21 P12 **Ninety Six** South Carolina, SE USA *34°10′N 82°01′W*
Nineveh *see* Nīnawá
163 Y9 **Ning'an** Heilongjiang, NE China *44°20′N 129°28′E*
161 S9 **Ningbo** *var.* Ning-po, Yin-hsien; *prev.* Ninghsien. Zhejiang, SE China *29°54′N 121°33′E*
161 U12 **Ningde** Fujian, SE China *26°48′N 119°33′E*
161 P12 **Ningdu** *var.* Meijiang. Jiangxi, S China *26°28′N 115°53′E*
Ning'er *see* Pu'er
186 A7 **Ningerum** Western, SW Papua New Guinea *05°40′S 141°10′E*
161 R9 **Ningguo** Anhui, E China *30°33′N 118°58′E*
161 S9 **Ninghai** Zhejiang, SE China *29°18′N 121°26′E*
Ning-hsia *see* Ningxia
Ninghsien *see* Ningbo
160 J15 **Ningming** *var.* Chengzhong. Guangxi Zhuangzu Zizhiqu, S China *22°07′N 106°43′E*
160 H11 **Ningnan** *var.* Pisha. Sichuan, C China *26°59′N 102°49′E*
Ning-po *see* Ningbo
Ningsia/Ningsia Hui/ Ningsia Hui Autonomous Region *see* Ningxia
160 J5 **Ningxia** *off.* Ningxia Huizu Zizhiqu, *var.* Ning-hsia, Ningsia, *Eng.* Ningsia Hui, Ningsia Hui Autonomous Region. ◆ *autonomous region* N China
Ningxia Huizu Zizhiqu *see* Ningxia
159 X10 **Ningxian** *var.* Xinning. Gansu, N China *35°30′N 108°05′E*
167 T7 **Ninh Binh** Ninh Binh, N Vietnam *20°14′N 106°00′E*
167 V12 **Ninh Hoa** Khanh Hoa, S Vietnam *12°28′N 109°07′E*
186 C4 **Ninigo Group** *island group* N Papua New Guinea
39 Q12 **Ninilchik** Alaska, USA *60°03′N 151°40′W*
27 N7 **Ninnescah River** ↝ Kansas, C USA
195 U16 **Ninnis Glacier** *glacier* Antarctica
165 R8 **Ninohe** Iwate, Honshū, C Japan *40°16′N 141°18′E*
99 F18 **Ninove** Oost-Vlaanderen, C Belgium *50°50′N 04°02′E*
171 O4 **Ninoy Aquino** ✕ (Manila) Luzon, N Philippines *14°26′N 121°00′E*
Nio *see* Íos
29 P12 **Niobrara** Nebraska, C USA *42°43′N 97°59′W*
28 M12 **Niobrara River** ↝ Nebraska/Wyoming, C USA
79 I20 **Nioki** Bandundu, W Dem. Rep. Congo *02°44′S 17°42′E*
76 M11 **Niono** Ségou, C Mali *14°18′N 05°59′W*
76 K11 **Nioro** *var.* Nioro du Sahel. Kayes, W Mali *15°13′N 09°39′W*
76 G11 **Nioro du Rip** SW Senegal *13°44′N 15°48′W*
Nioro du Sahel *see* Nioro
102 K10 **Niort** Deux-Sèvres, W France *46°21′N 00°25′W*
172 H14 **Nioumachoua** Mohéli, S Comoros *12°21′S 43°43′E*
186 C7 **Nipa** Southern Highlands, W Papua New Guinea *06°11′S 143°27′E*
11 U14 **Nipawin** Saskatchewan, S Canada *52°23′N 104°01′W*
12 D12 **Nipigon** Ontario, S Canada *49°02′N 88°15′W*
12 D11 **Nipigon, Lake** ◎ Ontario, S Canada
11 S13 **Nipin** ↝ Saskatchewan, C Canada
14 G11 **Nipissing, Lake** ◎ Ontario, S Canada
35 P13 **Nipomo** California, W USA *35°02′N 120°28′W*
Nippon *see* Japan
138 K6 **Niqniqīyah, Jabal an** ▲ C Syria
62 I9 **Niquivil** San Juan, W Argentina *30°25′S 68°42′W*
171 Y13 **Nirabotong** Papua, E Indonesia *02°35′S 140°08′E*
Niriz *see* Neyrīz
163 U7 **Nirji** *var.* Morin Dawa Daurzu Zizhiqi. Nei Mongol Zizhiqu, N China *48°21′N 124°32′E*
155 I14 **Nirmal** Telangana, C India *19°04′N 78°21′E*
153 Q13 **Nirmāli** Bihār, NE India *26°18′N 86°35′E*
113 O14 **Niš** *Eng.* Nish, *Ger.* Nisch; *anc.* Naissus. Serbia, SE Serbia *43°21′N 21°53′E*
104 H9 **Nisa** Portalegre, C Portugal *39°31′N 07°39′W*
Nisa *see* Neisse
141 P4 **Niṣab** Al Ḥudūd ash Shamālīyah, N Saudi Arabia *29°11′N 44°43′E*
141 Q15 **Niṣāb** *var.* Anṣāb. SW Yemen *14°24′N 46°47′E*
113 P14 **Nišava** *Bul.* Nishava. ↝ Bulgaria/Serbia *see also* Nishava
Nišava *see* Nishava
107 K25 **Niscemi** Sicilia, Italy, C Mediterranean Sea *37°09′N 14°23′E*
Nisch/Nish *see* Niš
165 R4 **Niseko** Hokkaidō, NE Japan *42°50′N 140°43′E*
Nishapur *see* Neyshābūr
114 G9 **Nishava** *var.* Nišava. ↝ Bulgaria/Serbia *see also* Nišava
Nishava *see* Nišava
118 L11 **Nishcha** ↝ N Belarus
165 C17 **Nishinoomote** Kagoshima, Tanega-shima, SW Japan *30°42′N 130°59′E*
165 X15 **Nishino-shima** *Eng.* Rosario. *island* Ogasawara-shotō, SE Japan
165 I13 **Nishiwaki** *var.* Nisiwaki. Hyōgo, Honshū, SW Japan *34°59′N 134°58′E*
141 U14 **Nishtūn** SE Yemen *15°47′N 52°08′E*
Nisibin *see* Nusaybin
Nisiros *see* Nísyros
Nisiwaki *see* Nishiwaki
Niska *see* Niesky
113 O14 **Niška Banja** Serbia, SE Serbia *43°18′N 22°01′E*
12 D6 **Niskibi** ↝ Ontario, C Canada
111 O15 **Nisko** Podkrapackie, SE Poland *50°31′N 22°09′E*
10 H7 **Nisling** ↝ Yukon, W Canada
99 H22 **Nismes** Namur, S Belgium *50°04′N 04°31′E*
Nismes *see* Nîmes
116 M10 **Nisporeni** *Rus.* Nisporeny. W Moldova *47°04′N 28°10′E*
Nisporeny *see* Nisporeni
95 K20 **Nissan** ↝ S Sweden
Nissan Islands *see* Green Islands
95 F16 **Nisser** ◎ S Norway
95 E21 **Nissum Bredning** *inlet* NW Denmark
29 U6 **Nisswa** Minnesota, N USA *46°31′N 94°17′W*
Nistru *see* Dniester
115 M22 **Nísyros** *var.* Nisiros. *island* Dodekánisa, Greece, Aegean Sea
118 H8 **Nītaure** C Latvia *57°05′N 25°12′E*
60 P10 **Niterói** *prev.* Nictheroy. Rio de Janeiro, SE Brazil *22°54′S 43°06′W*
14 F16 **Nith** ↝ Ontario, S Canada
96 J13 **Nith** ↝ S Scotland, United Kingdom
Nitinan *see* Nichinan
111 I21 **Nitra** *Ger.* Neutra, *Hung.* Nyitra. Nitriansky Kraj, SW Slovakia *48°20′N 18°05′E*
111 I20 **Nitra** *Ger.* Neutra, *Hung.* Nyitra. ↝ W Slovakia
111 I21 **Nitriansky Kraj** ◆ *region* SW Slovakia
21 Q5 **Nitro** West Virginia, NE USA *38°24′N 81°50′W*
Niuatobutabu *see* Niuatoputapu
193 X13 **Niuatoputapu** *var.* Niuatobutabu; *prev.* Keppel Island. *island* N Tonga
193 U15 **Niu'Aunofa** *headland* Tongatapu, S Tonga *21°03′S 175°19′W*
Niuchwang *see* Yingkou
190 B16 **Niue** ◇ *self-governing territory in free association with New Zealand* S Pacific Ocean
190 F10 **Niulakita** *var.* Nurakita. *atoll* S Tuvalu
190 E6 **Niutao** *atoll* NW Tuvalu
93 L15 **Nivala** Pohjois-Pohjanmaa, C Finland *63°56′N 25°00′E*
102 I15 **Nive** ↝ SW France
99 G19 **Nivelles** Walloon Brabant, C Belgium *50°36′N 04°04′E*
103 P8 **Nivernais** *cultural region* C France
15 N8 **Niverville, Lac** ◎ Québec, SE Canada
27 T7 **Nixa** Missouri, C USA *37°02′N 93°17′W*
35 R5 **Nixon** Nevada, W USA *39°48′N 119°24′W*
25 S12 **Nixon** Texas, SW USA *29°16′N 97°45′W*
Niya *see* Minfeng
Niyazov *see* Nyýazow
155 H14 **Nizāmābād** Telangana, C India *18°40′N 78°05′E*
155 H15 **Nizām Sāgar** ◎ C India
125 N16 **Nizhegorodskaya Oblast'** ◇ *province* W Russian Federation
Nizhnegorskiy *see* Nyzhn'ohirs'kyy
127 S4 **Nizhnekamsk** Respublika Tatarstan, W Russian Federation *55°36′N 51°45′E*
127 U3 **Nizhnekamskoye Vodokhranilishche** ⊠ W Russian Federation
123 S14 **Nizhneleninskoye** Yevreyskaya Avtonomnaya Oblast', SE Russian Federation *47°50′N 132°30′E*
122 L13 **Nizhneudinsk** Irkutskaya Oblast', S Russian Federation *54°48′N 98°51′E*
122 I10 **Nizhnevartovsk** Khanty-Mansiyskiy Avtonomnyy Okrug-Yugra, C Russian Federation *60°57′N 76°40′E*
123 Q7 **Nizhneyansk** Respublika Sakha (Yakutiya), NE Russian Federation *71°25′N 135°59′E*
127 Q11 **Nizhniy Baskunchak** Astrakhanskaya Oblast', SW Russian Federation *48°15′N 46°49′E*
127 O6 **Nizhniy Lomov** Penzenskaya Oblast', W Russian Federation *53°32′N 43°39′E*
127 P3 **Nizhniy Novgorod** *prev.* Gor'kiy. Nizhegorodskaya Oblast', W Russian Federation *56°17′N 44°E*
125 T8 **Nizhniy Odes** Respublika Komi, NW Russian Federation *63°42′N 54°59′E*
Nizhniy Pyandzh *see* Panji Poyon
122 G10 **Nizhniy Tagil** Sverdlovskaya Oblast', C Russian Federation *57°57′N 59°51′E*
125 T9 **Nizhnyaya-Omra** Respublika Komi, NW Russian Federation *62°46′N 55°54′E*
125 P5 **Nizhnyaya Pësha** Nenetskiy Avtonomnyy Okrug, NW Russian Federation *66°54′N 47°37′E*
117 Q3 **Nizhyn** *Rus.* Nezhin. Chernihivs'ka Oblast', NE Ukraine *51°03′N 31°54′E*
136 M17 **Nizip** Gaziantep, S Turkey *37°02′N 37°47′E*
141 X8 **Nizwa** *var.* Nazwāh. NE Oman *23°00′N 57°50′E*
Nizza *see* Nice
106 C9 **Nizza Monferrato** Piemonte, NE Italy *44°47′N 08°22′E*
Njávdám *see* Näätämöjoki
Njazidja *see* Ngazidja
Njellim *see* Nellim
81 H24 **Njombe** Njombe, S Tanzania *09°20′S 34°47′E*
81 H25 **Njombe** *off.* Mkoa wa Njombe. ◆ *region* S Tanzania
81 G23 **Njombe** ↝ C Tanzania
Njombe, Mkoa wa *see* Njombe
Njuk, Ozero *see* Nyuk, Ozero
Njuksenica *see* Nyuksenitsa
92 I10 **Njunis** ▲ N Norway *68°47′N 19°24′E*
Njurunda *see* Njurundabommen
93 H17 **Njurundabommen** *prev.* Njurunda. Västernorrland, C Sweden *62°15′N 17°24′E*
94 N11 **Njutånger** Gävleborg, C Sweden *61°37′N 17°04′E*
79 D14 **Nkambe** Nord-Ouest, NW Cameroon *06°35′N 10°44′E*
Nkata Bay *see* Nkhata Bay
79 F21 **Nkayi** *prev.* Jacob. Bouenza, S Congo *04°11′S 13°17′E*
83 J17 **Nkayi** Matabeleland North, W Zimbabwe *19°00′S 28°54′E*
82 N13 **Nkhata Bay** *var.* Nkata Bay. Northern, N Malawi *11°37′S 34°20′E*
81 E22 **Nkonde** Kigoma, N Tanzania *06°16′S 30°17′E*
79 D15 **Nkongsamba** *var.* N'Kongsamba. Littoral, W Cameroon *04°59′N 09°53′E*
N'Kongsamba *see* Nkongsamba
83 E16 **Nkurenkuru** Okavango, N Namibia *17°38′S 18°39′E*
77 Q15 **Nkwanta** E Ghana *08°18′N 00°27′E*
167 O2 **Nmai Hka** *var.* Me Hka. ↝ N Myanmar (Burma)
Noardwâlde *see* Noordwolde
39 N7 **Noatak** Alaska, USA *67°34′N 162°58′W*
39 N7 **Noatak River** ↝ Alaska, USA
164 E15 **Nobeoka** Miyazaki, Kyūshū, SW Japan *32°34′N 131°37′E*
27 N11 **Noble** Oklahoma, C USA *35°08′N 97°23′W*
31 P13 **Noblesville** Indiana, N USA *40°03′N 86°00′W*
165 R5 **Noboribetsu** *var.* Noboribetu. Hokkaidō, NE Japan *42°27′N 141°08′E*
Noboribetu *see* Noboribetsu
59 H18 **Nobres** Mato Grosso, W Brazil *14°44′S 56°15′W*
107 N21 **Nocera Terinese** Calabria, S Italy *39°03′N 16°10′E*
41 Q16 **Nochixtlán** *var.* Asunción Nochixtlán. Oaxaca, SE Mexico *17°29′N 97°17′W*
25 S5 **Nocona** Texas, SW USA *33°47′N 97°43′W*
63 K21 **Nodales, Bahía de los** *bay* S Argentina
27 Q2 **Nodaway River** ↝ Iowa/ Missouri, C USA
27 R8 **Noel** Missouri, C USA *36°33′N 94°29′W*
40 H3 **Nogales** Chihuahua, NW Mexico *18°50′N 97°12′W*
40 F3 **Nogales** Sonora, NW Mexico *31°17′N 110°53′W*
36 M17 **Nogales** Arizona, SW USA *31°20′N 110°55′W*
Nogal Valley *see* Dooxo Nugaaleed
102 K15 **Nogaro** Gers, S France *43°46′N 00°01′W*
110 J7 **Nogat** ↝ N Poland
164 D12 **Nōgata** Fukuoka, Kyūshū, SW Japan *33°46′N 130°42′E*
127 P15 **Nogayskaya Step'** *steppe* SW Russian Federation
102 M6 **Nogent-le-Rotrou** Eure-et-Loir, C France *48°19′N 00°50′E*
103 O4 **Nogent-sur-Oise** Oise, N France *49°16′N 02°28′E*
103 P6 **Nogent-sur-Seine** Aube, N France *48°30′N 03°31′E*
122 L10 **Noginsk** Krasnoyarskiy Kray, N Russian Federation *64°28′N 91°09′E*
126 L3 **Noginsk** Moskovskaya Oblast', W Russian Federation *55°51′N 38°23′E*
123 T12 **Nogliki** Ostrov Sakhalin, Sakhalinskaya Oblast', SE Russian Federation *51°44′N 143°14′E*
164 K12 **Nōgōhaku-san** ▲ Honshū, SW Japan *35°46′N 136°30′E*
162 D5 **Nogoonnuur** Bayan-Ölgiy, NW Mongolia *49°31′N 89°48′E*
61 C18 **Nogoyá** Entre Ríos, E Argentina *32°25′S 59°50′W*
111 K21 **Nógrád** *off.* Nógrád Megye. ◆ *county* N Hungary
Nógrád Megye *see* Nógrád
105 U5 **Noguera Pallaresa** ↝ NE Spain
105 U4 **Noguera Ribagorçana** ↝ NE Spain
101 E19 **Nohfelden** Saarland, SW Germany *49°35′N 07°08′E*
38 A8 **Nohili Point** *headland* Kaua'i, Hawai'i, USA *22°03′N 159°48′W*
104 G3 **Noia** Galicia, NW Spain *42°48′N 08°52′W*
103 N16 **Noire, Montagne** ▲ S France
14 J10 **Noire, Rivière** ↝ Québec, SE Canada
15 P12 **Noire, Rivière** ↝ Québec, SE Canada
Noire, Rivi`ere *see* Black River
102 G6 **Noires, Montagnes** ▲ NW France
102 H8 **Noirmoutier-en-l'Île** Vendée, NW France *47°00′N 02°15′W*
102 H8 **Noirmoutier, Île de** *island* NW France
187 Q10 **Noka** Nendö, E Solomon Islands *10°42′S 165°57′E*
83 G17 **Nokaneng** North West, NW Botswana *19°40′S 22°12′E*
93 L18 **Nokia** Pirkanmaa, W Finland *61°29′N 23°30′E*
148 K11 **Nok Kundi** Baluchistān, SW Pakistan *28°49′N 62°39′E*
30 L14 **Nokomis** Illinois, N USA *39°18′N 89°17′W*
30 K5 **Nokomis, Lake** ⊠ Wisconsin, N USA
78 G9 **Nokou** Kanem, W Chad *14°36′N 14°45′E*
187 Q12 **Nokuku** Espiritu Santo, W Vanuatu *14°56′S 166°34′E*
95 J18 **Nol** Västra Götaland, S Sweden *57°55′N 12°03′E*
79 H16 **Nola** Sangha-Mbaéré, SW Central African Republic *03°29′N 16°05′E*
25 P7 **Nolan** Texas, SW USA *32°15′N 100°15′W*
125 R15 **Nolinsk** Kirovskaya Oblast', NW Russian Federation *57°35′N 49°54′E*
Nolsø *see* Nólsoy
95 B19 **Nólsoy** *Dan.* Nolsø. *island* E Faroe Islands
186 B7 **Nomad** Western, SW Papua New Guinea *06°11′S 142°13′E*

◆ Country
● Country Capital
◇ Dependent Territory
○ Dependent Territory Capital
◆ Administrative Regions
✕ International Airport
▲ Mountain
▲ Mountain Range
Volcano
↝ River
◎ Lake
⊠ Reservoir

164 *B16* **Noma-zaki** Kyūshū, SW Japan
40 *K10* **Nombre de Dios** Durango, C Mexico *23°51′N 104°14′W*
42 *I5* **Nombre de Dios, Cordillera** ▲▲ N Honduras
38 *M9* **Nome** Alaska, USA *64°30′N 165°24′W*
29 *Q6* **Nome** North Dakota, N USA *46°39′N 97°49′W*
38 *M9* **Nome, Cape** *headland* Alaska, USA *64°25′N 165°00′W*
162 *K11* **Nomgon** ***var.*** Sangiyn Dalay. Ömnögovĭ, S Mongolia *42°50′N 105°04′E*
14 *M11* **Nominingue, Lac** ⊚ Québec, SE Canada
Nomoi Islands ***see*** Mortlock Islands
164 *B16* **Nomo-zaki** *headland* Kyūshū, SW Japan *32°34′N 129°45′E*
162 *G6* **Nömrög** ***var.*** Hödrögö. Dzavhan, N Mongolia *48°51′N 96°48′E*
193 *X15* **Nomuka** *island* Nomuka Group, C Tonga
193 *X15* **Nomuka Group** *island group* W Tonga
189 *Q15* **Nomwin Atoll** *atoll* Hall Islands, C Micronesia
8 *L10* **Nonacho Lake** ⊚ Northwest Territories, NW Canada
Nondaburi ***see*** Nonthaburi
39 *P12* **Nondalton** Alaska, USA *59°58′N 154°51′W*
163 *V10* **Nong'an** Jilin, NE China *44°25′N 125°10′E*
169 *P10* **Nong Bua Khok** Nakhon Ratchasima, C Thailand *15°23′N 101°51′E*
167 *Q9* **Nong Bua Lamphu** Udon Thani, E Thailand *17°11′N 102°27′E*
167 *R7* **Nông Hèt** Xiangkhoang, N Laos *19°27′N 104°02′E*
Nongkaya ***see*** Nong Khai
167 *Q8* **Nong Khai** ***var.*** Mi Chai, Nongkaya. Nong Khai, E Thailand *17°52′N 102°44′E*
Nongle ***see*** Jiuzhaigou
167 *N14* **Nong Met** Surat Thani, SW Thailand *09°27′N 99°09′E*
83 *L22* **Nongoma** KwaZulu/Natal, E South Africa *27°54′S 31°40′E*
167 *P9* **Nong Phai** Phetchabun, C Thailand *15°58′N 101°02′E*
153 *U13* **Nongstoin** Meghālaya, NE India *25°24′N 91°19′E*
83 *C19* **Nonidas** Erongo, N Namibia *22°36′S 14°40′E*
Nonni ***see*** Nen Jiang
40 *I7* **Nonoava** Chihuahua, N Mexico *27°24′N 106°18′W*
191 *O3* **Nonouti** ***prev.*** Sydenham Island. *atoll* Tungaru, W Kiribati
167 *O11* **Nonthaburi** ***var.*** Nondaburi, Nontha Buri. Nonthaburi, C Thailand *13°48′N 100°11′E*
Nontha Buri ***see*** Nonthaburi
102 *L11* **Nontron** Dordogne, SW France *45°34′N 00°41′E*
147 *T10* **Nookat** ***var.*** Iski-Nauket; ***prev.*** Eski-Nookat. Oshskaya Oblast', SW Kyrgyzstan *40°18′N 72°29′E*
181 *P1* **Noonamah** Northern Territory, N Australia *12°46′S 131°08′E*
28 *K2* **Noonan** North Dakota, N USA *48°51′N 102°57′W*
Noonu ***see*** South Miladhunmadulu Atoll
99 *E14* **Noord-Beveland** ***var.*** North Beveland. *island* SW Netherlands
99 *J14* **Noord-Brabant** ***Eng.*** North Brabant. ⬖ *province* S Netherlands
98 *H7* **Noorder Haaks** *spit* NW Netherlands
98 *H9* **Noord-Holland** ***Eng.*** North Holland. ⬖ *province* NW Netherlands
Noordhollandsch Kanaal ***see*** Noordhollands Kanaal
98 *H8* **Noordhollands Kanaal** ***var.*** Noordhollandsch Kanaal. *canal* NW Netherlands
Noord-Kaap ***see*** Northern Cape
98 *L8* **Noordoostpolder** *island* N Netherlands
45 *P16* **Noordpunt** *headland* N Curaçao *12°21′N 69°08′W*
98 *I8* **Noord-Scharwoude** Noord-Holland, NW Netherlands *52°42′N 04°48′E*
Noordwes ***see*** North-West
98 *G11* **Noordwijk aan Zee** Zuid-Holland, W Netherlands *52°15′N 04°25′E*
98 *H11* **Noordwijkerhout** Zuid-Holland, W Netherlands *52°16′N 04°30′E*
98 *M7* **Noordwolde** ***Fris.*** Noardwâlde. Fryslân, N Netherlands *52°54′N 06°10′E*
Noordzee ***see*** North Sea
98 *H10* **Noordzee-Kanaal** *canal* NW Netherlands
93 *K18* **Noormarkku** ***Swe.*** Norrmark. Satakunta, SW Finland *61°35′N 21°54′E*
39 *N8* **Noorvik** Alaska, USA *66°50′N 161°01′W*
10 *J17* **Nootka Sound** *inlet* British Columbia, W Canada
82 *A9* **Nóqui** Zaire Province, NW Angola *05°54′S 13°30′E*
95 *L15* **Nora** Örebro, C Sweden *59°31′N 15°02′E*
147 *Q13* **Norak** ***Rus.*** Nurek. W Tajikistan *38°23′N 69°14′E*
13 *I13* **Noranda** Québec, SE Canada *48°16′N 79°03′W*
29 *W12* **Nora Springs** Iowa, C USA *43°08′N 93°00′W*
95 *M14* **Norberg** Västmanland, C Sweden *60°04′N 15°56′E*
14 *K13* **Norcan Lake** ⊚ Ontario, SE Canada
197 *R12* **Nord** N Greenland *81°38′N 12°51′W*
78 *F13* **Nord** ***Eng.*** North. ⬖ *province* N Cameroon
103 *P2* **Nord** ⬖ *department* N France
92 *P1* **Nordaustlandet** *island* NE Svalbard
95 *G24* **Nordborg** ***Ger.*** Nordburg. Syddanmark, SW Denmark *55°04′N 09°41′E*
Nordburg ***see*** Nordborg
95 *F23* **Nordby** Syddtjylland, W Denmark *55°27′N 08°25′E*
11 *P15* **Nordegg** Alberta, SW Canada *52°27′N 116°06′W*
100 *E9* **Norden** Niedersachsen, NW Germany *53°36′N 07°12′E*
100 *G10* **Nordenham** Niedersachsen, NW Germany *53°30′N 08°29′E*
122 *M6* **Nordenshel'da, Arkhipelag** *island group* N Russian Federation
92 *O3* **Nordenskiold Land** *physical region* W Svalbard
100 *E9* **Norderney** *island* NW Germany
100 *J9* **Norderstedt** Schleswig-Holstein, N Germany *53°42′N 09°59′E*
94 *D11* **Nordfjord** *fjord* S Norway
94 *C11* **Nordfjord** *physical region* S Norway
94 *D11* **Nordfjordeid** Sogn og Fjordane, S Norway *61°54′N 06°E*
92 *G11* **Nordfold** Nordland, C Norway *67°48′N 15°16′E*
Nordfriesische Inseln ***see*** North Frisian Islands
100 *H7* **Nordfriesland** *cultural region* N Germany
Nordgrønland ***see*** Avannaarsua
101 *K15* **Nordhausen** Thüringen, C Germany *51°31′N 10°48′E*
25 *T13* **Nordheim** Texas, SW USA *28°55′N 97°36′W*
94 *C13* **Nordhordland** *physical region* S Norway
100 *E12* **Nordhorn** Niedersachsen, NW Germany *52°26′N 07°04′E*
172 *H16* **Nord, Île du** *island* Inner Islands, NE Seychelles
95 *F20* **Nordjylland** ⬖ *county* N Denmark
92 *K7* **Nordkapp** ***Eng.*** North Cape. *headland* N Norway *25°47′E 71°10′N*
92 *O1* **Nordkapp** *headland* N Svalbard *80°31′N 19°58′E*
79 *N19* **Nord-Kivu** ***off.*** Région du Nord Kivu. ⬖ *region* E Dem. Rep. Congo
Nord Kivu, Région du ***see*** Nord-Kivu
92 *G12* **Nordland** ⬖ *county* C Norway
101 *J21* **Nördlingen** Bayern, S Germany *48°49′N 10°28′E*
93 *I16* **Nordmaling** Västerbotten, N Sweden *63°35′N 19°30′E*
95 *K15* **Nordmark** Värmland, C Sweden *59°52′N 14°04′E*
Nord, Mer du ***see*** North Sea
94 *F8* **Nordmøre** *physical region* S Norway
100 *I8* **Nord-Ostee-Kanal** *canal* N Germany
0 *J3* **Nordøstrundingen** *cape* NE Greenland
79 *D14* **Nord-Ouest** ***Eng.*** North-West. ⬖ *province* NW Cameroon
Nord-Ouest, Territoires du ***see*** Northwest Territories
103 *N2* **Nord-Pas-de-Calais** ⬖ *region* N France
101 *F19* **Nordpfälzer Bergland** ▲▲ W Germany
Nord, Pointe ***see*** Fatua, Pointe
187 *P16* **Nord, Province** ⬖ *province* C New Caledonia
101 *D14* **Nordrhein-Westfalen** ***Eng.*** North Rhine-Westphalia, ***Fr.*** Rhénanie du Nord-Westphalie. ⬖ *state* W Germany
Nordsee/Nordsjøen/Nordsøen ***see*** North Sea
100 *H7* **Nordstrand** *island* N Germany
93 *E15* **Nord-Trøndelag** ⬖ *county* C Norway
92 *I1* **Norðurfjörður** Vestfirðir, NW Iceland *66°01′N 21°33′W*
92 *J1* **Norðurland Eystra** ⬖ *region* N Iceland
92 *I2* **Norðurland Vestra** ⬖ *region* N Iceland
97 *E19* **Nore** ***Ir.*** An Fheoir. ≈ S Ireland
29 *Q14* **Norfolk** Nebraska, C USA *42°01′N 97°25′W*
21 *X7* **Norfolk** Virginia, NE USA *36°51′N 76°17′W*
97 *P19* **Norfolk** *cultural region* E England, United Kingdom
192 *K10* **Norfolk Island** ⬖ *Australian self-governing territory* SW Pacific Ocean
175 *P9* **Norfolk Ridge** *undersea feature* W Pacific Ocean
27 *U8* **Norfork Lake** ⊞ Arkansas/Missouri, C USA
98 *N6* **Norg** Drenthe, NE Netherlands *53°04′N 06°28′E*
Norge ***see*** Norway
95 *D14* **Norheimsund** Hordaland, S Norway *60°22′N 06°09′E*
25 *S16* **Norias** Texas, SW USA *26°47′N 97°45′W*
164 *L12* **Norikura-dake** ▲ Honshū, S Japan *36°06′N 137°33′E*
122 *K8* **Noril'sk** Krasnoyarskiy Kray, N Russian Federation *69°21′N 88°02′E*
14 *I13* **Norland** Ontario, SE Canada *44°46′N 78°48′W*
21 *V8* **Norlina** North Carolina, SE USA *36°26′N 78°11′W*
30 *L13* **Normal** Illinois, N USA *40°30′N 88°59′W*
27 *N11* **Norman** Oklahoma, C USA *35°13′N 97°27′W*
Norman ***see*** Tulita
186 *G9* **Normanby Island** *island* SE Papua New Guinea
Normandes, Îles ***see*** Channel Islands
58 *G9* **Normandia** Roraima, N Brazil *03°57′N 59°39′W*
102 *L5* **Normandie** ***Eng.*** Normandy. *cultural region* N France
102 *J5* **Normandie, Collines de** *hill range* NW France
Normandy ***see*** Normandie
25 *V9* **Normangee** Texas, SW USA *31°01′N 96°06′W*
21 *Q10* **Norman, Lake** ⊞ North Carolina, SE USA
44 *K13* **Norman Manley** ✕ (Kingston) E Jamaica *17°55′N 76°46′W*
181 *U5* **Norman River** ≈ Queensland, NE Australia
181 *U4* **Normanton** Queensland, NE Australia *17°49′S 141°08′E*
8 *I8* **Norman Wells** Northwest Territories, NW Canada *65°18′N 126°42′W*
12 *H12* **Normétal** Québec, S Canada *48°59′N 79°23′W*
163 *O7* **Norovlin** ***var.*** Uldz. Hentiy, NE Mongolia *48°47′N 112°01′E*
11 *V15* **Norquay** Saskatchewan, S Canada *51°51′N 102°04′W*
Norra Karelen ***see*** Pohjois-Karjala
93 *G15* **Norråker** Jämtland, C Sweden *64°25′N 15°40′E*
94 *N12* **Norrala** Gävleborg, C Sweden *61°22′N 17°04′E*
Norra Ny ***see*** Stöllet
Norra Österbotten ***see*** Pohjois-Pohjanmaa
Norra Savolax ***see*** Pohjois-Savo
92 *G13* **Norra Storjfjället** ▲ N Sweden *65°57′N 15°15′E*
92 *I13* **Norrbotten** ⬖ *county* N Sweden
94 *N11* **Norrdellen** ⊚ C Sweden
95 *G23* **Nørre Aaby** ***var.*** Nørre Åby. Syddtjylland, C Denmark *55°28′N 09°53′E*
Nørre Åby ***see*** Nørre Aaby
95 *I24* **Nørre Alslev** Sjælland, SE Denmark *54°54′N 11°53′E*
95 *E23* **Nørre Nebel** Syddtjylland, W Denmark *55°45′N 08°16′E*
95 *G20* **Nørresundby** Nordjylland, N Denmark *57°05′N 09°55′E*
21 *N8* **Norris Lake** ⊞ Tennessee, S USA
18 *I15* **Norristown** Pennsylvania, NE USA *40°07′N 75°20′W*
95 *N17* **Norrköping** Östergötland, S Sweden *58°35′N 16°10′E*
Norrmark ***see*** Noormarkku
94 *N13* **Norrsundet** Gävleborg, C Sweden *60°55′N 17°09′E*
95 *P15* **Norrtälje** Stockholm, C Sweden *59°46′N 18°42′E*
180 *L12* **Norseman** Western Australia *32°16′S 121°46′E*
93 *I14* **Norsjö** Västerbotten, N Sweden *64°55′N 19°30′E*
95 *G16* **Norsjø** ⊚ S Norway
123 *R13* **Norsk** Amurskaya Oblast', SE Russian Federation *52°20′N 129°57′E*
Norske Havet ***see*** Norwegian Sea
187 *Q13* **Norsup** Malekula, C Vanuatu *16°05′S 167°24′E*
191 *V15* **Norte, Cabo** *headland* Easter Island, Chile, E Pacific Ocean *27°03′S 109°24′W*
54 *F7* **Norte de Santander** ***off.*** Departamento de Norte de Santander. ⬖ *province* N Colombia
Norte de Santander, Departamento de ***see*** Norte de Santander
61 *E21* **Norte, Punta** *headland* E Argentina *36°17′S 56°46′W*
21 *R13* **North** South Carolina, SE USA *33°37′N 81°06′W*
North ***see*** Nord
18 *L10* **North Adams** Massachusetts, NE USA *42°40′N 73°06′W*
113 *L17* **North Albanian Alps** ***Alb.*** Bjeshkët e Namuna, ***SCr.*** Prokletije. ▲▲ SE Europe
97 *M15* **Northallerton** N England, United Kingdom *54°20′N 01°26′W*
180 *J12* **Northam** Western Australia *31°40′S 116°40′E*
83 *J20* **Northam** Northern, N South Africa *24°56′S 27°18′E*
1 **North America** *continent*
1 *N12* **North American Basin** *undersea feature* W Sargasso Sea *30°00′N 60°00′W*
0 *C5* **North American Plate** *tectonic feature*
18 *M11* **North Amherst** Massachusetts, NE USA *42°24′N 72°31′W*
97 *N20* **Northampton** C England, United Kingdom *52°14′N 00°54′W*
97 *M20* **Northamptonshire** *cultural region* C England, United Kingdom
151 *P18* **North Andaman** *island* Andaman Islands, India, NE Indian Ocean
65 *D25* **North Arm** East Falkland, Falkland Islands *52°06′S 59°21′W*
21 *Q13* **North Augusta** South Carolina, SE USA *33°30′N 81°58′W*
173 *W8* **North Australian Basin** ***Fr.*** Bassin Nord de l' Australie. *undersea feature* E Indian Ocean
31 *R11* **North Baltimore** Ohio, N USA *41°10′N 83°40′W*
11 *T15* **North Battleford** Saskatchewan, S Canada *52°47′N 108°19′W*
14 *H11* **North Bay** Ontario, S Canada *46°20′N 79°28′W*
12 *H6* **North Belcher Islands** *island group* Belcher Islands, Nunavut, C Canada
29 *R15* **North Bend** Nebraska, C USA *41°27′N 96°46′W*
32 *E14* **North Bend** Oregon, NW USA *43°24′N 124°13′W*
96 *K12* **North Berwick** SE Scotland, United Kingdom *56°04′N 02°44′W*
North Beveland ***see*** Noord-Beveland
North Borneo ***see*** Sabah
183 *P5* **North Bourke** New South Wales, SE Australia *30°03′S 145°56′E*
North Brabant ***see*** Noord-Brabant
182 *F2* **North Branch Neales** *seasonal river* South Australia
44 *M6* **North Caicos** *island* NW Turks and Caicos Islands
26 *L10* **North Canadian River** ≈ Oklahoma, C USA
31 *U12* **North Canton** Ohio, N USA *40°52′N 81°24′W*
13 *R13* **North, Cape** *headland* Cape Breton Island, Nova Scotia, SE Canada *47°06′N 60°24′W*
184 *I1* **North Cape** *headland* North Island, New Zealand *34°23′S 173°02′E*
186 *G5* **North Cape** *headland* New Ireland, NE Papua New Guinea *02°33′S 150°48′E*
North Cape ***see*** Nordkapp
18 *J17* **North Cape May** New Jersey, NE USA *38°59′N 74°55′W*
12 *C9* **North Caribou Lake** ⊚ Ontario, C Canada
21 *U10* **North Carolina** ***off.*** State of North Carolina, ***also known as*** Old North State, Tar Heel State, Turpentine State. ⬖ *state* SE USA
North Celebes ***see*** Sulawesi Utara
155 *J24* **North Central** ⬖ *province* N Sri Lanka
31 *S4* **North Channel** *lake channel* Canada/USA
97 *G14* **North Channel** *strait* Northern Ireland/Scotland, United Kingdom
21 *S14* **North Charleston** South Carolina, SE USA *32°53′N 79°59′W*
31 *N10* **North Chicago** Illinois, N USA *42°19′N 87°50′W*
195 *Y10* **Northcliffe Glacier** *glacier* Antarctica
31 *Q14* **North College Hill** Ohio, N USA *39°13′N 84°33′W*
25 *O8* **North Concho River** ≈ Texas, SW USA
19 *O8* **North Conway** New Hampshire, NE USA *44°03′N 71°06′W*
27 *V14* **North Crossett** Arkansas, C USA *33°10′N 91°56′W*
28 *L4* **North Dakota** ***off.*** State of North Dakota, ***also known as*** Flickertail State, Peace Garden State, Sioux State. ⬖ *state* N USA
North Devon Island ***see*** Devon Island
97 *O22* **North Downs** *hill range* SE England, United Kingdom
18 *C11* **North East** Pennsylvania, NE USA *42°13′N 79°49′W*
83 *I18* **North East** ⬖ *district* NE Botswana
65 *G15* **North East Bay** *bay* Ascension Island, C Atlantic Ocean
38 *L10* **Northeast Cape** *headland* Saint Lawrence Island, Alaska, USA *63°16′N 168°50′W*
North East Frontier Agency/North East Frontier Agency of Assam ***see*** Arunāchal Pradesh
65 *E25* **North East Island** *island* E Falkland Islands
189 *V11* **Northeast Island** *island* Chuuk, C Micronesia
44 *L12* **North East Point** *headland* E Jamaica *18°09′N 76°19′W*
191 *Z2* **Northeast Point** *headland* Kiritimati, E Kiribati *10°23′S 105°45′E*
44 *L6* **Northeast Point** *headland* Great Inagua, S The Bahamas *21°18′N 73°01′W*
44 *K5* **Northeast Point** *headland* Acklins Island, SE The Bahamas *22°43′N 73°50′W*
44 *H2* **Northeast Providence Channel** *channel* N The Bahamas
101 *J14* **Northeim** Niedersachsen, C Germany *51°42′N 10°E*
29 *X14* **North English** Iowa, C USA *41°30′N 92°04′W*
138 *G8* **Northern** ⬖ *district* N Israel
82 *M12* **Northern** ⬖ *region* N Malawi
186 *F8* **Northern** ***var.*** Oro. ⬖ *province* S Papua New Guinea
155 *J23* **Northern** ⬖ *province* N Sri Lanka
80 *D7* **Northern** ⬖ *state* N Sudan
82 *K12* **Northern** ⬖ *province* NE Zambia
Northern ***see*** Limpopo
80 *B13* **Northern Bahr el Ghazal** ⬖ *state* NW South Sudan
Northern Border Region ***see*** Al Ḩudūd ash Shamālīyah
83 *F24* **Northern Cape** ***off.*** Northern Cape Province, ***Afr.*** Noord-Kaap. ⬖ *province* W South Africa
Northern Cape Province ***see*** Northern Cape
190 *K14* **Northern Cook Islands** *island group* N Cook Islands
80 *B8* **Northern Darfur** ⬖ *state* NW Sudan
Northern Dvina ***see*** Severnaya Dvina
97 *F14* **Northern Ireland** ***var.*** The Six Counties. *cultural region* Northern Ireland, United Kingdom
97 *F14* **Northern Ireland** ***var.*** The Six Counties. ⬖ *political division* Northern Ireland, United Kingdom
80 *D9* **Northern Kordofan** ⬖ *state* C Sudan
187 *Z14* **Northern Lau Group** *island group* Lau Group, NE Fiji
188 *K3* **Northern Mariana Islands** ◇ *US commonwealth territory* W Pacific Ocean
Northern Rhodesia ***see*** Zambia
Northern Sporades ***see*** Vóreies Sporádes
182 *D1* **Northern Territory** ⬖ *territory* N Australia
Northern Transvaal ***see*** Limpopo
Northern Ural Hills ***see*** Severnyye Uvaly
84 *I9* **North European Plain** *plain* N Europe
27 *V2* **North Fabius River** ≈ Missouri, C USA
65 *D24* **North Falkland Sound** *sound* N Falkland Islands
29 *V9* **Northfield** Minnesota, N USA *44°27′N 93°10′W*
19 *O9* **Northfield** New Hampshire, NE USA *43°26′N 71°34′W*
175 *Q8* **North Fiji Basin** *undersea feature* N Coral Sea
97 *Q22* **North Foreland** *headland* SE England, United Kingdom *51°22′N 01°26′E*
35 *P6* **North Fork American River** ≈ California, W USA
39 *R7* **North Fork Chandalar River** ≈ Alaska, USA
28 *K7* **North Fork Grand River** ≈ North Dakota/South Dakota, N USA
21 *O6* **North Fork Kentucky River** ≈ Kentucky, S USA
39 *Q7* **North Fork Koyukuk River** ≈ Alaska, USA
39 *Q10* **North Fork Kuskokwim River** ≈ Alaska, USA
26 *K11* **North Fork Red River** ≈ Oklahoma/Texas, SW USA
26 *K3* **North Fork Solomon River** ≈ Kansas, C USA
23 *W14* **North Fort Myers** Florida, SE USA *26°40′N 81°52′W*
31 *P5* **North Fox Island** *island* Michigan, N USA
100 *G6* **North Frisian Islands** ***var.*** Nordfriesische Inseln. *island group* N Germany
197 *N9* **North Geomagnetic Pole** *pole* Arctic Ocean
18 *M13* **North Haven** Connecticut, NE USA *41°25′N 72°51′W*
184 *J5* **North Head** *headland* North Island, New Zealand *36°23′S 174°01′E*
18 *L6* **North Hero** Vermont, NE USA *44°49′N 73°14′W*
35 *O7* **North Highlands** California, W USA *38°40′N 121°25′W*
North Holland ***see*** Noord-Holland
81 *I16* **North Horr** Marsabit, N Kenya *03°17′N 37°08′E*
151 *K21* **North Huvadhu Atoll** ***var.*** Gaafu Alifu Atoll. *atoll* S Maldives
65 *A24* **North Island** *island* W Falkland Islands
184 *N9* **North Island** *island* N New Zealand
21 *U14* **North Island** *island* South Carolina, SE USA
31 *O11* **North Judson** Indiana, N USA *41°12′N 86°44′W*
North Kalimantan ***see*** Kalimantan Utara
North Karelia ***see*** Pohjois-Karjala
North Kazakhstan ***see*** Severnyy Kazakhstan
31 *V10* **North Kingsville** Ohio, N USA *41°54′N 80°41′W*
163 *Y13* **North Korea** ***off.*** Democratic People's Republic of Korea, ***Kor.*** Chosŏn-minjujuŭi-inmin-kanghwaguk. ◆ *republic* E Asia
153 *X11* **North Lakhimpur** Assam, NE India *27°10′N 94°00′E*
184 *J3* **Northland** ***off.*** Northland Region. ⬖ *region* North Island, New Zealand
192 *K11* **Northland Plateau** *undersea feature* S Pacific Ocean
Northland Region ***see*** Northland
35 *X11* **North Las Vegas** Nevada, W USA *36°12′N 115°07′W*
31 *O11* **North Liberty** Indiana, N USA *41°36′N 86°22′W*
29 *X14* **North Liberty** Iowa, C USA *41°45′N 91°36′W*
27 *V12* **North Little Rock** Arkansas, C USA *34°46′N 92°15′W*
28 *M13* **North Loup River** ≈ Nebraska, C USA
151 *K18* **North Maalhosmadulu Atoll** ***var.*** North Malosmadulu Atoll, Raa Atoll. *atoll* N Maldives
31 *U10* **North Madison** Ohio, N USA *41°48′N 81°03′W*
North Malosmadulu Atoll ***see*** North Maalhosmadulu Atoll
31 *P12* **North Manchester** Indiana, N USA *41°00′N 85°45′W*
31 *P6* **North Manitou Island** *island* Michigan, N USA
29 *U10* **North Mankato** Minnesota, N USA *44°11′N 94°03′W*
23 *Z15* **North Miami** Florida, SE USA *25°54′N 80°11′W*
151 *K18* **North Miladhunmadulu Atoll** ***var.*** Shaviyani Atoll. *atoll* N Maldives
North Minch ***see*** Minch, The
23 *W15* **North Naples** Florida, SE USA *26°13′N 81°47′W*
175 *P8* **North New Hebrides Trench** *undersea feature* N Coral Sea
23 *Y15* **North New River Canal** ≈ Florida, SE USA
151 *K20* **North Nilandhe Atoll** *atoll* C Maldives
36 *L2* **North Ogden** Utah, W USA *41°18′N 111°57′W*
North Ossetia ***see*** Severnaya Osetiya-Alaniya, Respublika
North Ostrobothnia ***see*** Pohjois-Pohjanmaa
35 *S10* **North Palisade** ▲ California, W USA *37°06′N 118°31′W*
189 *U11* **North Pass** *passage* Chuuk Islands, C Micronesia
28 *M15* **North Platte** Nebraska, C USA *41°07′N 100°46′W*
33 *X17* **North Platte River** ≈ C USA
65 *G14* **North Point** *headland* Ascension Island, C Atlantic Ocean
172 *I16* **North Point** *headland* Mahé, NE Seychelles *04°23′S 55°28′E*
31 *R5* **North Point** *headland* Michigan, N USA *45°21′N 83°30′W*
31 *S6* **North Point** *headland* Michigan, N USA *45°01′N 83°16′W*
39 *S9* **North Pole** Alaska, USA *64°42′N 147°09′W*
197 *R9* **North Pole** *pole* Arctic Ocean
23 *O4* **Northport** Alabama, S USA *33°13′N 87°34′W*
23 *W14* **North Port** Florida, SE USA *27°03′N 82°15′W*
32 *L6* **Northport** Washington, NW USA *48°54′N 117°48′W*
32 *L12* **North Powder** Oregon, NW USA *45°00′N 117°56′W*
29 *U13* **North Raccoon River** ≈ Iowa, C USA
North Rhine-Westphalia ***see*** Nordrhein-Westfalen
97 *M16* **North Riding** *cultural region* N England, United Kingdom
96 *G5* **North Rona** *island* NW Scotland, United Kingdom
96 *K4* **North Ronaldsay** *island* NE Scotland, United Kingdom
36 *L2* **North Salt Lake** Utah, W USA *40°51′N 111°54′W*
11 *P15* **North Saskatchewan** ≈ Alberta/Saskatchewan, S Canada
35 *X5* **North Schell Peak** ▲ Nevada, W USA *39°25′N 114°34′W*
North Scotia Ridge ***see*** South Georgia Ridge
86 *D10* **North Sea** ***Dan.*** Nordsøen, ***Dut.*** Noordzee, ***Fr.*** Mer du Nord, ***Ger.*** Nordsee, ***Nor.*** Nordsjøen; ***prev.*** German Ocean, ***Lat.*** Mare Germanicum. *sea* NW Europe
35 *T6* **North Shoshone Peak** ▲ Nevada, W USA *39°08′N 117°28′W*
North Siberian Lowland/North Siberian Plain ***see*** Severo-Sibirskaya Nizmennost'
29 *R13* **North Sioux City** South Dakota, N USA *42°31′N 96°28′W*
North Solomons ***see*** Bougainville
96 *K4* **North Sound, The** *sound* N Scotland, United Kingdom
183 *T4* **North Star** New South Wales, SE Australia *28°55′S 150°25′E*
North Star State ***see*** Minnesota
183 *V3* **North Stradbroke Island** *island* Queensland, E Australia
North Sulawesi ***see*** Sulawesi Utara
North Sumatra ***see*** Sumatera Utara
14 *D17* **North Sydenham** ≈ Ontario, S Canada
18 *H9* **North Syracuse** New York, NE USA *43°07′N 76°07′W*
184 *K9* **North Taranaki Bight** *gulf* North Island, New Zealand
12 *H9* **North Twin Island** *island* Nunavut, C Canada
96 *E8* **North Uist** *island* NW Scotland, United Kingdom
97 *L14* **Northumberland** *cultural region* N England, United Kingdom
181 *Y7* **Northumberland Isles** *island group* Queensland, NE Australia
13 *Q14* **Northumberland Strait** *strait* SE Canada
32 *G14* **North Umpqua River** ≈ Oregon, NW USA
45 *Q13* **North Union** Saint Vincent , Saint Vincent and the Grenadines *13°15′N 61°07′W*
10 *L17* **North Vancouver** British Columbia, SW Canada *49°21′N 123°05′W*
18 *K9* **Northville** New York, NE USA *43°13′N 74°08′W*
97 *Q19* **North Walsham** E England, United Kingdom *52°49′N 01°22′E*
39 *T10* **Northway** Alaska, USA *62°57′N 141°56′W*
83 *G17* **North-West** ⬖ *district* NW Botswana
83 *G21* **North-West** ***off.*** North-West Province, ***Afr.*** Noordwes. ⬖ *province* N South Africa
North-West ***see*** Nord-Ouest
64 *I6* **Northwest Atlantic Mid-Ocean Canyon** *undersea feature* N Atlantic Ocean
180 *G8* **North West Cape** *cape* Western Australia
38 *J9* **Northwest Cape** *headland* Saint Lawrence Island, Alaska, USA *63°46′N 171°45′W*
155 *J24* **North Western** ⬖ *province* W Sri Lanka
82 *H13* **North Western** ⬖ *province* W Zambia
North-West Frontier Province ***see*** Khyber Pakhtunkhwa
96 *H8* **North West Highlands** ▲▲ N Scotland, United Kingdom
192 *J4* **Northwest Pacific Basin** *undersea feature* NW Pacific Ocean *40°00′N 150°00′E*
191 *Y2* **Northwest Point** *headland* Kiritimati, E Kiribati *10°25′S 105°35′E*
44 *G1* **Northwest Providence Channel** *channel* N The Bahamas
North-West Province ***see*** North-West
13 *Q8* **North West River** Newfoundland and Labrador, E Canada *53°30′N 60°10′W*
8 *J9* **Northwest Territories** ***Fr.*** Territoires du Nord-Ouest. ⬖ *territory* NW Canada
97 *K18* **Northwich** C England, United Kingdom *53°16′N 02°32′W*
25 *Q5* **North Wichita River** ≈ Texas, SW USA
18 *J17* **North Wildwood** New Jersey, NE USA *39°00′N 74°45′W*
21 *R9* **North Wilkesboro** North Carolina, SE USA *36°09′N 81°09′W*
19 *P8* **North Windham** Maine, NE USA *43°51′N 70°25′W*
197 *Q6* **Northwind Plain** *undersea feature* Arctic Ocean
29 *V11* **Northwood** Iowa, C USA *43°26′N 93°13′W*
29 *Q4* **Northwood** North Dakota, N USA *47°43′N 97°34′W*
97 *M15* **North York Moors** *moorland* N England, United Kingdom
25 *V9* **North Zulch** Texas, SW USA *30°54′N 96°06′W*
26 *K2* **Norton** Kansas, C USA *39°51′N 99°53′W*
31 *S13* **Norton** Ohio, N USA *40°25′N 83°04′W*
21 *P7* **Norton** Virginia, NE USA *36°56′N 82°37′W*
39 *N9* **Norton Bay** *bay* Alaska, USA
Norton de Matos ***see*** Balombo
31 *O9* **Norton Shores** Michigan, N USA *43°10′N 86°15′W*
38 *M10* **Norton Sound** *inlet* Alaska, USA
27 *Q3* **Nortonville** Kansas, C USA *39°25′N 95°19′W*
102 *I8* **Nort-sur-Erdre** Loire-Atlantique, NW France *47°27′N 01°30′W*
195 *N2* **Norvegia, Cape** *headland* Antarctica *71°16′S 12°25′W*
18 *L13* **Norwalk** Connecticut, NE USA *41°08′N 73°28′W*
29 *V14* **Norwalk** Iowa, C USA *41°28′N 93°40′W*
31 *S11* **Norwalk** Ohio, N USA *41°14′N 82°37′W*
19 *P7* **Norway** Maine, NE USA *44°13′N 70°30′W*
31 *N5* **Norway** Michigan, N USA *45°47′N 87°54′W*
93 *E17* **Norway** ***off.*** Kingdom of Norway, ***Nor.*** Norge. ◆ *monarchy* N Europe
11 *X13* **Norway House** Manitoba, C Canada *53°59′N 97°50′W*
Norway, Kingdom of ***see*** Norway
197 *R16* **Norwegian Basin** *undersea feature* NW Norwegian Sea *68°00′N 02°00′W*
84 *D6* **Norwegian Sea** ***var.*** Norske Havet. *sea* NE Atlantic Ocean
197 *S17* **Norwegian Trench** *undersea feature* NE North Sea *59°00′N 04°30′E*
14 *F16* **Norwich** Ontario, S Canada *42°57′N 80°37′W*
97 *Q19* **Norwich** E England, United Kingdom *52°38′N 01°18′E*
19 *N13* **Norwich** Connecticut, NE USA *41°30′N 72°02′W*
18 *I11* **Norwich** New York, NE USA *42°31′N 75°31′W*
29 *U9* **Norwood** Minnesota, N USA *44°46′N 93°55′W*
31 *Q15* **Norwood** Ohio, N USA *39°07′N 84°27′W*
14 *H11* **Nosbonsing, Lake** ⊚ Ontario, S Canada
Nösen ***see*** Bistriţa
165 *T1* **Noshappu-misaki** *headland* Hokkaidō, NE Japan *45°26′N 141°38′E*
165 *P7* **Noshiro** ***var.*** Nosiro; ***prev.*** Noshirominato. Akita, Honshū, C Japan *40°11′N 140°02′E*
Noshirominato/Nosiro ***see*** Noshiro
117 *Q3* **Nosivka** ***Rus.*** Nosovka. Chernihivs'ka Oblast', NE Ukraine *50°55′N 31°37′E*
67 *T14* **Nosop** ***var.*** Nossob, Nossop. ≈ Botswana/Namibia
83 *E20* **Nossop** ≈ E Namibia
125 *S4* **Nosovaya** Nenetskiy Avtonomnyy Okrug, NW Russian Federation *68°12′N 54°33′E*
Nosovka ***see*** Nosivka
143 *V11* **Noşratābād** Sīstān va Balūchestān, E Iran *29°53′N 59°57′E*
95 *J18* **Nossebro** Västra Götaland, S Sweden *58°12′N 12°42′E*
96 *K6* **Noss Head** *headland* N Scotland, United Kingdom *58°29′N 03°03′W*
Nossi-Bé ***see*** Be, Nosy
Nossob/Nossop ***see*** Nosop
172 *J2* **Nosy Be** ✕ Antsirānana, N Madagascar *23°36′S 47°36′E*
172 *J6* **Nosy Varika** Fianarantsoa, SE Madagascar *20°36′S 48°31′E*
14 *L10* **Notawassi** ≈ Québec, SE Canada
14 *M9* **Notawassi, Lac** ⊚ Québec, SE Canada
36 *J5* **Notch Peak** ▲ Utah, W USA *39°08′N 113°24′W*
110 *G10* **Noteć** ***Ger.*** Netze. ≈ NW Poland
Nóties Sporádes ***see*** Dodekánisa
115 *J22* **Nótion Aigaíon** ***Eng.*** Aegean South. ⬖ *region* E Greece
115 *H18* **Nótios Evvoïkós Kólpos** *gulf* E Greece
115 *B16* **Nótio Stenó Kérkyras** *strait* W Greece
107 *L25* **Noto** ***anc.*** Netum. Sicilia, Italy, C Mediterranean Sea *36°53′N 15°05′E*
164 *M10* **Noto** Ishikawa, Honshū, SW Japan *37°18′N 137°11′E*
95 *G15* **Notodden** Telemark, S Norway *59°35′N 09°18′E*
107 *L25* **Noto, Golfo di** *gulf* Sicilia, Italy, C Mediterranean Sea
164 *L10* **Noto-hantō** *peninsula* Honshū, SW Japan
164 *L11* **Noto-jima** *island* SW Japan
13 *T11* **Notre Dame Bay** *bay* Newfoundland, Newfoundland and Labrador, E Canada
15 *P6* **Notre-Dame-de-Lorette** Québec, SE Canada *49°05′N 72°24′W*
14 *L11* **Notre-Dame-de-Pontmain** Québec, SE Canada *46°18′N 75°37′W*
15 *T8* **Notre-Dame-du-Lac** Québec, SE Canada *47°36′N 68°48′W*
15 *Q6* **Notre-Dame-du-Rosaire** Québec, SE Canada *48°48′N 71°27′W*
15 *U8* **Notre-Dame, Monts** ▲▲ Québec, S Canada
77 *R16* **Notsé** S Togo *06°59′N 01°12′E*
14 *G14* **Nottawasaga** ≈ Ontario, S Canada
14 *G14* **Nottawasaga Bay** *lake bay* Ontario, S Canada
12 *I11* **Nottaway** ≈ Québec, SE Canada
23 *S1* **Nottely Lake** ⊞ Georgia, SE USA
95 *H16* **Nøtterøy** *island* S Norway
97 *M19* **Nottingham** C England, United Kingdom *52°58′N 01°10′W*
9 *E14* **Nottingham Island** *island* Nunavut, NE Canada
97 *N18* **Nottinghamshire** *cultural region* C England, United Kingdom
21 *V7* **Nottoway** Virginia, NE USA *37°07′N 78°03′W*
21 *V7* **Nottoway River** ≈ Virginia, NE USA
76 *G7* **Nouâdhibou** ***prev.*** Port-Étienne. Dakhlet Nouâdhibou, W Mauritania *20°54′N 17°01′W*
76 *G7* **Nouâdhibou** ✕ Dakhlet Nouâdhibou, W Mauritania *20°59′N 17°02′W*
76 *F7* **Nouâdhibou, Dakhlet** ***prev.*** Baie du Lévrier. *bay* W Mauritania
76 *F7* **Nouâdhibou, Râs** ***prev.*** Cap Blanc. *headland* NW Mauritania *20°48′N 17°03′W*
76 *G9* **Nouakchott** ● (Mauritania) Nouakchott District, SW Mauritania *18°09′N 15°58′W*
76 *G9* **Nouakchott** ✕ Trarza, SW Mauritania *18°18′N 15°54′W*
120 *J11* **Noual, Sebkhet en** ***var.*** Sabkhat an Nawāl. *salt flat* C Tunisia
76 *G8* **Nouâmghâr** ***var.*** Nouamrhar. Dakhlet Nouâdhibou, W Mauritania *19°22′N 16°31′W*
Nouamrhar ***see*** Nouâmghâr
Nouă Sulița ***see*** Novoselytsya
187 *Q17* **Nouméa** ○ (New Caledonia) Province Sud, S New Caledonia *22°13′S 166°29′E*
79 *E15* **Noun** ≈ C Cameroon
77 *N12* **Nouna** W Burkina Faso *12°44′N 03°54′W*
83 *H24* **Noupoort** Northern Cape, C South Africa *31°11′S 24°57′E*

◆ Country
● Country Capital
◇ Dependent Territory
○ Dependent Territory Capital
⬖ Administrative Regions
✕ International Airport
▲ Mountain
▲▲ Mountain Range
🌋 Volcano
≈ River
⊚ Lake
⊞ Reservoir

Nouveau-Brunswick *see* New Brunswick
Nouveau-Comptoir *see* Wemindji
15 *T4* **Nouvel, Lacs** ◎ Québec, SE Canada
15 *W7* **Nouvelle** Québec, SE Canada *48°07′N 66°16′W*
15 *W7* **Nouvelle** ≈ Québec, SE Canada
Nouvelle-Calédonie *see* New Caledonia
Nouvelle Écosse *see* Nova Scotia
103 *R3* **Nouzonville** Ardennes, N France *49°49′N 04°45′E*
147 *Q11* **Nov** ***Rus.*** Nau. NW Tajikistan *40°10′N 69°16′E*
59 *I21* **Nova Alvorada** Mato Grosso do Sul, SW Brazil *21°25′S 54°19′W*
Novabad *see* Navobod
111 *D19* **Nová Bystřice** ***Ger.*** Neubistritz. Jihočeský Kraj, S Czech Republic *49°N 15°05′E*
116 *H13* **Novaci** Gorj, SW Romania *45°07′N 23°37′E*
Nova Civitas *see* Neustadt an der Weinstrasse
Novaesium *see* Neuss
60 *H10* **Nova Esperança** Paraná, S Brazil *23°09′S 52°13′W*
106 *H11* **Novafeltria** Marche, C Italy *43°54′N 12°18′E*
60 *Q9* **Nova Friburgo** Rio de Janeiro, SE Brazil *22°16′S 42°34′W*
82 *D12* **Nova Gaia** ***var.*** Cambundi-Catembo. Malanje, NE Angola *10°09′S 17°31′E*
109 *S12* **Nova Gorica** W Slovenia *45°57′N 13°40′E*
112 *G10* **Nova Gradiška** ***Ger.*** Neugradisk, ***Hung.*** Újgradiska. Brod-Posavina, NE Croatia *45°15′N 17°23′E*
60 *K7* **Nova Granada** São Paulo, S Brazil *20°33′S 49°19′W*
60 *O10* **Nova Iguaçu** Rio de Janeiro, SE Brazil *22°31′S 44°05′W*
117 *S10* **Nova Kakhovka** ***Rus.*** Novaya Kakhovka. Khersons'ka Oblast', SE Ukraine *46°45′N 33°20′E*
Nová Karvinná *see* Karviná
Nova Lamego *see* Gabú
Nova Lisboa *see* Huambo
112 *C11* **Novalja** Lika-Senj, W Croatia *44°33′N 14°53′E*
119 *M14* **Novalukoml'** ***Rus.*** Novolukoml'. Vitsyebskaya Voblasts', N Belarus *54°40′N 29°09′E*
Nova Mambone *see* Mambone
83 *P16* **Nova Nabúri** Zambézia, NE Mozambique *16°47′S 38°55′E*
117 *Q9* **Nova Odesa** ***var.*** Novaya Odessa. Mykolayivs'ka Oblast', S Ukraine *47°19′N 31°45′E*
60 *H10* **Nova Olímpia** Paraná, S Brazil *23°28′S 53°12′W*
61 *I15* **Nova Prata** Rio Grande do Sul, S Brazil *28°45′S 51°37′W*
14 *H12* **Novar** Ontario, S Canada *45°26′N 79°14′W*
106 *C7* **Novara** ***anc.*** Novaria. Piemonte, NW Italy *45°27′N 08°36′E*
Novaria *see* Novara
13 *P15* **Nova Scotia** ***Fr.*** Nouvelle Écosse. ⬖ *province* SE Canada
0 *M9* **Nova Scotia** *physical region* SE Canada
34 *M8* **Novato** California, W USA *38°06′N 122°35′W*
192 *M7* **Nova Trough** *undersea feature* W Pacific Ocean
116 *L7* **Nova Ushytsya** Khmel'nyts'ka Oblast', W Ukraine *48°50′N 27°16′E*
83 *M17* **Nova Vanduzi** Manica, C Mozambique *18°54′S 33°18′E*
117 *U5* **Nova Vodolaha** ***Rus.*** Novaya Vodolaga. Kharkivs'ka Oblast', E Ukraine *49°43′N 35°49′E*
123 *O12* **Novaya Chara** Zabaykal'skiy Kray, S Russian Federation *56°45′N 117°58′E*
122 *M12* **Novaya Igirma** Irkutskaya Oblast', C Russian Federation *57°08′N 103°52′E*
Novaya Kakhovka *see* Nova Kakhovka
Novaya Kazanka *see* Zhanakazan
124 *I12* **Novaya Ladoga** Leningradskaya Oblast', NW Russian Federation *60°03′N 32°15′E*
127 *R5* **Novaya Malykla** Ul'yanovskaya Oblast', W Russian Federation *54°13′N 49°55′E*
Novaya Odessa *see* Nova Odesa
123 *Q5* **Novaya Sibir', Ostrov** *island* Novosibirskiye Ostrova, NE Russian Federation
Novaya Vodolaga *see* Nova Vodolaha
122 *I6* **Novaya Zemlya** *island group* N Russian Federation
Novaya Zemlya Trough *see* East Novaya Zemlya Trough
114 *K10* **Nova Zagora** Sliven, C Bulgaria *42°29′N 26°00′E*
105 *S12* **Novelda** Valenciana, E Spain *38°24′N 00°45′W*
111 *H19* **Nové Mesto nad Váhom** ***Ger.*** Waagneustadtl, ***Hung.*** Vágújhely. Trenčiansky Kraj, W Slovakia *48°46′N 17°50′E*
111 *F17* **Nové Město na Moravě** ***Ger.*** Neustadtl in Mähren. Vysočina, C Czech Republic *49°34′N 16°05′E*
Novesium *see* Neuss
111 *I21* **Nové Zámky** ***Ger.*** Neuhäusel, ***Hung.*** Érsekújvár. Nitriansky Kraj, SW Slovakia *49°00′N 18°10′E*
Novgorod *see* Velikiy Novgorod
Novgorod-Severskiy *see* Novhorod-Sivers'kyy
122 *C7* **Novgorodskaya Oblast'** ⬖ *province* W Russian Federation
117 *R8* **Novhorodka** Kirovohrads'ka Oblast', C Ukraine *48°21′N 32°38′E*
117 *R2* **Novhorod-Sivers'kyy** ***Rus.*** Novgorod-Severskiy. Chernihivs'ka Oblast', NE Ukraine *52°00′N 33°15′E*
31 *R10* **Novi** Michigan, N USA *42°28′N 83°28′W*
Novi *see* Novi Vinodolski
112 *L9* **Novi Bečej** ***prev.*** Új-Becse, Vološinovo, ***Ger.*** Neubetsche, ***Hung.*** Törökbecse. Vojvodina, N Serbia *45°36′N 20°09′E*
116 *M3* **Novi Bilokorovychi** ***Rus.*** Belokorovichi; ***prev.*** Bilokorovychi. Zhytomyrs'ka Oblast', N Ukraine *51°07′N 28°02′E*
25 *Q8* **Novice** Texas, SW USA *32°00′N 99°38′W*
112 *A9* **Novigrad** Istra, NW Croatia *45°19′N 13°33′E*
Novi Grad *see* Bosanski Novi
114 *G9* **Novi Iskar** Sofia Grad, W Bulgaria *42°46′N 23°19′E*
106 *C9* **Novi Ligure** Piemonte, NW Italy *44°46′N 08°47′E*
99 *L22* **Noville** Luxembourg, SE Belgium *50°04′N 05°46′E*
194 *I10* **Noville Peninsula** *peninsula* Thurston Island, Antarctica
Noviodunum *see* Soissons, Aisne, France
Noviodunum *see* Nevers, Nièvre, France
Noviodunum *see* Nyon, Vaud, Switzerland
Noviomagus *see* Lisieux, Calvados, France
Noviomagus *see* Nijmegen, Netherlands
114 *M8* **Novi Pazar** Shumen, NE Bulgaria *43°20′N 27°12′E*
113 *M15* **Novi Pazar** ***Turk.*** Yenipazar. Serbia, S Serbia *43°09′N 20°31′E*
112 *K10* **Novi Sad** ***Ger.*** Neusatz, ***Hung.*** Újvidék. Vojvodina, N Serbia *45°16′N 19°49′E*
117 *T6* **Novi Sanzhary** Poltavs'ka Oblast', C Ukraine *49°21′N 34°18′E*
112 *H12* **Novi Travnik** ***prev.*** Pučarevo. Federacija Bosne I Hercegovine, C Bosnia and Herzegovina *44°12′N 17°39′E*
112 *B10* **Novi Vinodolski** ***var.*** Novi. Primorje-Gorski Kotar, NW Croatia *45°08′N 14°46′E*
58 *F12* **Novo Airão** Amazonas, N Brazil *02°06′S 61°20′W*
Novoalekseyevka *see* Kobda
127 *N9* **Novoanninskiy** Volgogradskaya Oblast', SW Russian Federation *50°31′N 42°43′E*
58 *F13* **Novo Aripuanã** Amazonas, N Brazil *05°05′S 60°20′W*
117 *P7* **Novoarkhangel's'k** Kirovohrads'ka Oblast', C Ukraine *48°39′N 30°48′E*
117 *Y6* **Novoaydar** Luhans'ka Oblast', E Ukraine *49°00′N 39°00′E*
117 *X9* **Novoazovs'k** ***Rus.*** Novoazovsk. Donets'ka Oblast', E Ukraine *47°07′N 38°06′E*
123 *R14* **Novobureyskiy** Amurskaya Oblast', SE Russian Federation *49°42′N 129°46′E*
127 *Q3* **Novocheboksarsk** Chuvashskaya Respublika, W Russian Federation *56°07′N 47°33′E*
127 *R5* **Novocheremshansk** Ul'yanovskaya Oblast', W Russian Federation *54°23′N 50°08′E*
126 *L12* **Novocherkassk** Rostovskaya Oblast', SW Russian Federation *47°23′N 40°E*
127 *R6* **Novodevich'ye** Samarskaya Oblast', W Russian Federation *53°33′N 48°51′E*
124 *M8* **Novodvinsk** Arkhangel'skaya Oblast', NW Russian Federation *64°22′N 40°49′E*
Novograd-Volynskiy *see* Novohrad-Volyns'kyy
Novogrudok *see* Navahrudak
Novogrudskaya Vozvyshennost' *see* Navahrudskaye Wzvyshsha
61 *I15* **Novo Hamburgo** Rio Grande do Sul, S Brazil *29°42′S 51°07′W*
59 *H16* **Novo Horizonte** Mato Grosso, W Brazil *11°19′S 57°11′W*
60 *K8* **Novo Horizonte** São Paulo, S Brazil *21°27′S 49°14′W*
116 *M4* **Novohrad-Volyns'kyy** ***Rus.*** Novograd-Volynskiy. Zhytomyrs'ka Oblast', N Ukraine *50°34′N 27°32′E*
145 *O7* **Novoishimskiy** ***prev.*** Kuybyshevskiy. Severnyy Kazakhstan, N Kazakhstan *53°15′N 66°51′E*
Novokazalinsk *see* Ayteke Bi
126 *M8* **Novokhopersk** Voronezhskaya Oblast', W Russian Federation *51°09′N 41°34′E*
127 *R6* **Novokuybyshevsk** Samarskaya Oblast', W Russian Federation *53°06′N 49°56′E*
122 *J13* **Novokuznetsk** ***prev.*** Stalinsk. Kemerovskaya Oblast', S Russian Federation *53°45′N 87°12′E*
195 *R1* **Novolazarevskaya** *Russian research station* Antarctica *70°42′S 11°31′E*
Novolukoml' *see* Novalukoml'
109 *V12* **Novo mesto** ***Ger.*** Rudolfswert; ***prev. Ger.*** Neustadtl. SE Slovenia *45°49′N 15°09′E*
126 *K15* **Novomikhaylovskiy** Krasnodarskiy Kray, SW Russian Federation *44°18′N 38°49′E*
112 *L8* **Novo Miloševo** Vojvodina, N Serbia *45°43′N 20°20′E*
Novomirgorod *see* Novomyrhorod
126 *L5* **Novomoskovsk** Tul'skaya Oblast', W Russian Federation *54°05′N 38°23′E*
117 *U7* **Novomoskovs'k** ***Rus.*** Novomoskovsk. Dnipropetrovs'ka Oblast', E Ukraine *48°38′N 35°15′E*
117 *V8* **Novomykolayivka** Zaporiz'ka Oblast', SE Ukraine *47°58′N 35°54′E*
117 *Q7* **Novomyrhorod** ***Rus.*** Novomirgorod. Kirovohrads'ka Oblast', S Ukraine *48°46′N 31°39′E*
127 *N8* **Novonikolayevskiy** Volgogradskaya Oblast', SW Russian Federation *50°55′N 42°24′E*
127 *P10* **Novonikol'skoye** Volgogradskaya Oblast', SW Russian Federation *49°23′N 45°06′E*
127 *X7* **Novoorsk** Orenburgskaya Oblast', W Russian Federation *51°21′N 59°03′E*
126 *M13* **Novopokrovskaya** Krasnodarskiy Kray, SW Russian Federation *45°58′N 40°43′E*
Novopolotsk *see* Navapolatsk
117 *Y5* **Novopskov** Luhans'ka Oblast', E Ukraine *49°33′N 39°07′E*
Novoradomsk *see* Radomsko
Novo Redondo *see* Sumbe
127 *R8* **Novorepnoye** Saratovskaya Oblast', W Russian Federation *51°04′N 48°34′E*
126 *K14* **Novorossiysk** Krasnodarskiy Kray, SW Russian Federation *44°50′N 37°38′E*
Novorossiyskiy/Novorossiyskoye *see* Akzhar
124 *F15* **Novorzhev** Pskovskaya Oblast', W Russian Federation *57°01′N 29°19′E*
Novoselitsa *see* Novoselytsya
117 *S12* **Novoselivs'ke** Avtonomna Respublika Krym, S Ukraine *45°26′N 33°37′E*
Novosëlki *see* Navasyolki
114 *G6* **Novo Selo** Vidin, NW Bulgaria *44°08′N 22°48′E*
113 *M14* **Novo Selo** Serbia, C Serbia *43°39′N 20°54′E*
116 *K8* **Novoselytsya** ***Rom.*** Nouă Sulița, ***Rus.*** Novoselitsa. Chernivets'ka Oblast', W Ukraine *48°14′N 26°18′E*
127 *U7* **Novosergiyevka** Orenburgskaya Oblast', W Russian Federation *52°04′N 53°40′E*
126 *L11* **Novoshakhtinsk** Rostovskaya Oblast', SW Russian Federation *47°48′N 39°51′E*
122 *J12* **Novosibirsk** Novosibirskaya Oblast', C Russian Federation *55°04′N 83°05′E*
122 *J12* **Novosibirskaya Oblast'** ⬖ *province* C Russian Federation
122 *M4* **Novosibirskiye Ostrova** ***Eng.*** New Siberian Islands. *island group* N Russian Federation
126 *K6* **Novosil'** Orlovskaya Oblast', W Russian Federation *53°00′N 37°59′E*
124 *G16* **Novosokol'niki** Pskovskaya Oblast', W Russian Federation *56°21′N 30°07′E*
127 *Q6* **Novospasskoye** Ul'yanovskaya Oblast', W Russian Federation *53°08′N 47°48′E*
Novotroickoje *see* Birlik
127 *X8* **Novotroitsk** Orenburgskaya Oblast', W Russian Federation *51°10′N 58°18′E*
Novotroitskoye *see* Brlik, Kazakhstan
Novotroitskoye *see* Novotroyits'ke, Ukraine
117 *T11* **Novotroyits'ke** ***Rus.*** Novotroitskoye. Khersons'ka Oblast', S Ukraine *46°21′N 34°21′E*
Novoukrainka *see* Novoukrayinka
117 *Q8* **Novoukrayinka** ***Rus.*** Novoukrainka. Kirovohrads'ka Oblast', C Ukraine *48°19′N 31°33′E*
127 *Q5* **Novoul'yanovsk** Ul'yanovskaya Oblast', W Russian Federation *54°10′N 48°19′E*
127 *W8* **Novoural'sk** Orenburgskaya Oblast', W Russian Federation *51°19′N 56°57′E*
Novo-Urgench *see* Urganch
116 *I4* **Novovolyns'k** ***Rus.*** Novovolynsk. Volyns'ka Oblast', NW Ukraine *50°46′N 24°09′E*
117 *S9* **Novovorontsovka** Khersons'ka Oblast', S Ukraine *47°28′N 33°55′E*
147 *Y7* **Novovoznesenovka** Issyk-Kul'skaya Oblast', E Kyrgyzstan *42°36′N 78°44′E*
125 *R14* **Novovyatsk** Kirovskaya Oblast', NW Russian Federation *58°30′N 49°42′E*
Novoyel'nya *see* Navayel'nya
117 *O6* **Novozhyvotiv** Vinnyts'ka Oblast', C Ukraine *49°16′N 29°31′E*
126 *H6* **Novozybkov** Bryanskaya Oblast', W Russian Federation *52°36′N 31°58′E*
112 *F9* **Novska** Sisak-Moslavina, NE Croatia *45°20′N 16°58′E*
Nový Bohumín *see* Bohumín
111 *D15* **Nový Bor** ***Ger.*** Haida; ***prev.*** Bor u České Lípy, Hajda. Liberecký Kraj, N Czech Republic *50°46′N 14°32′E*
111 *E16* **Nový Bydžov** ***Ger.*** Neubidschow. Královéhradecký Kraj, N Czech Republic *50°15′N 15°27′E*
119 *G18* **Novy Dvor** ***Rus.*** Novyy Dvor. Hrodzyenskaya Voblasts', W Belarus *53°48′N 24°34′E*
111 *I17* **Nový Jičín** ***Ger.*** Neutitschein. Moravskoslezský Kraj, E Czech Republic *49°36′N 18°00′E*
118 *K12* **Novy Pahost** ***Rus.*** Novyy Pogost. Vitsyebskaya Voblasts', NW Belarus *55°30′N 27°29′E*
Novyy Bug *see* Novyy Buh
117 *R9* **Novyy Buh** ***Rus.*** Novyy Bug. Mykolayivs'ka Oblast', S Ukraine *47°40′N 32°30′E*
117 *Q4* **Novyy Bykiv** Chernihivs'ka Oblast', N Ukraine *50°36′N 31°39′E*
Novyy Dvor *see* Novy Dvor
Novyye Aneny *see* Anenii Noi
127 *P7* **Novyye Burasy** Saratovskaya Oblast', W Russian Federation *52°10′N 46°00′E*
Novyy Margilan *see* Farg'ona
126 *K8* **Novyy Oskol** Belgorodskaya Oblast', W Russian Federation *50°43′N 37°55′E*
Novyy Pogost *see* Novy Pahost
127 *R2* **Novyy Tor"yal** Respublika Mariy El, W Russian Federation *56°59′N 48°53′E*
123 *N12* **Novyy Uoyan** Respublika Buryatiya, S Russian Federation *56°06′N 111°27′E*
122 *J9* **Novyy Urengoy** Yamalo-Nenetskiy Avtonomnyy Okrug, N Russian Federation *66°06′N 76°25′E*
Novyy Uzen' *see* Zhanaozen
111 *N16* **Nowa Dęba** Podkarpackie, SE Poland *50°31′N 21°53′E*
111 *G15* **Nowa Ruda** ***Ger.*** Neurode. Dolnośląskie, SW Poland *50°34′N 16°30′E*
110 *F12* **Nowa Sól** ***var.*** Nowasól, ***Ger.*** Neusalz an der Oder. Lubuskie, W Poland *51°47′N 15°43′E*
Nowasól *see* Nowa Sól
27 *Q8* **Nowata** Oklahoma, C USA *36°42′N 95°36′W*
142 *M6* **Nowbarān** Markazī, W Iran *35°07′N 49°51′E*
110 *J8* **Nowe** Kujawski-pomorskie, N Poland *53°40′N 18°44′E*
110 *K9* **Nowe Miasto Lubawskie** ***Ger.*** Neumark. Warmińsko-Mazurskie, NE Poland *53°24′N 19°36′E*
110 *L13* **Nowe Miasto nad Pilicą** Mazowieckie, C Poland *51°37′N 20°34′E*
110 *D8* **Nowe Warpno** ***Ger.*** Neuwarp. Zachodnio-pomorskie, NW Poland *53°52′N 14°12′E*
Nowgong *see* Nagaon
110 *E8* **Nowogard** ***var.*** Nowógard, ***Ger.*** Naugard. Zachodnio-pomorskie, NW Poland *53°41′N 15°09′E*
110 *N9* **Nowogród** Podlaskie, NE Poland *53°14′N 21°52′E*
Nowogródek *see* Navahrudak
111 *E14* **Nowogrodziec** ***Ger.*** Naumburg am Queis. Dolnośląskie, SW Poland *51°12′N 15°24′E*
Nowojelnia *see* Navayel'nya
Nowo-Minsk *see* Mińsk Mazowiecki
33 *V13* **Nowood River** ≈ Wyoming, C USA
Nowo-Święciany *see* Švenčionėliai
183 *S10* **Nowra-Bomaderry** New South Wales, SE Australia *34°51′S 150°41′E*
149 *T5* **Nowshera** ***var.*** Naushahra, Naushara. Khyber Pakhtunkhwa, NE Pakistan *34°00′N 72°00′E*
110 *J7* **Nowy Dwór Gdański** ***Ger.*** Tiegenhof. Pomorskie, N Poland *54°12′N 19°03′E*
110 *L11* **Nowy Dwór Mazowiecki** Mazowieckie, C Poland *52°26′N 20°43′E*
111 *M17* **Nowy Sącz** ***Ger.*** Neu Sandec. Małopolskie, S Poland *49°36′N 20°42′E*
111 *L18* **Nowy Targ** ***Ger.*** Neumark. Małopolskie, S Poland *49°28′N 20°00′E*
110 *F11* **Nowy Tomyśl** ***var.*** Nowy Tomysl. Wielkopolskie, C Poland *52°18′N 16°07′E*
Nowy Tomysl *see* Nowy Tomyśl
148 *M7* **Now Zād** ***var.*** Nauzad. Helmand, S Afghanistan *32°22′N 64°32′E*
23 *N4* **Noxubee River** ≈ Alabama/Mississippi, S USA
122 *I10* **Noyabr'sk** Yamalo-Nenetskiy Avtonomnyy Okrug, N Russian Federation *63°08′N 75°19′E*
102 *L8* **Noyant** Maine-et-Loire, NW France *47°28′N 00°08′W*
39 *X14* **Noyes Island** *island* Alexander Archipelago, Alaska, USA
103 *O3* **Noyon** Oise, N France *49°35′N 03°E*
102 *I7* **Nozay** Loire-Atlantique, NW France *47°34′N 01°36′W*
82 *L12* **Nsando** Northern, NE Zambia *10°22′S 31°14′E*
83 *N16* **Nsanje** Southern, S Malawi *16°57′S 35°10′E*
77 *Q17* **Nsawam** SE Ghana *05°47′N 00°19′W*
79 *E16* **Nsimalen** ✈ Centre, C Cameroon *19°15′N 81°22′E*
82 *K12* **Nsombo** Northern, NE Zambia *10°35′S 29°58′E*
82 *H13* **Ntambu** North Western, NW Zambia *12°21′S 25°03′E*
83 *N14* **Ntcheu** ***var.*** Ncheu. Central, S Malawi *14°49′S 34°37′E*
79 *D17* **Ntem** ***prev.*** Campo, Kampo. ≈ Cameroon/Equatorial Guinea
83 *I14* **Ntemwa** North Western, NW Zambia *14°03′S 26°13′E*
Ntlenyana, Mount *see* Thabana Ntlenyana
79 *I19* **Ntomba, Lac** ***var.*** Lac Tumba. ◎ NW Dem. Rep. Congo
115 *I19* **Ntóro, Kávo** ***prev.*** Akrotírio Kafiréas. *cape* Évvoia, C Greece
81 *E19* **Ntungamo** SW Uganda *0°54′S 30°16′E*
81 *E18* **Ntusi** SW Uganda *0°05′N 31°13′E*
83 *H18* **Ntwetwe Pan** *salt lake* NE Botswana
93 *M15* **Nuasjärvi** ◎ C Finland
80 *F11* **Nuba Mountains** ▲▲ C Sudan
68 *J9* **Nubian Desert** *desert* NE Sudan
116 *G10* **Nucet** ***Hung.*** Diófás. Bihor, W Romania *46°28′N 22°35′E*
Nu Chiang *see* Salween
145 *U9* **Nuclear Testing Ground** *nuclear site* Pavlodar, E Kazakhstan
56 *E9* **Nucuray, Río** ≈ N Peru
25 *R14* **Nueces River** ≈ Texas, SW USA
9 *V9* **Nueltin Lake** ◎ Manitoba/Northwest Territories, C Canada
99 *K15* **Nuenen** Noord-Brabant, S Netherlands *51°29′N 05°36′E*
62 *G6* **Nuestra Señora, Bahía** *bay* N Chile
61 *D14* **Nuestra Señora Rosario de Caa Catí** Corrientes, NE Argentina *27°48′S 57°42′W*
54 *J9* **Nueva Antioquia** Vichada, E Colombia *06°04′N 69°30′W*
Nueva Caceres *see* Naga
41 *O7* **Nueva Ciudad Guerrera** Tamaulipas, C Mexico *26°32′N 99°13′W*
55 *N4* **Nueva Esparta** ***off.*** Estado Nueva Esparta. ⬖ *state* NE Venezuela
Nueva Esparta, Estado *see* Nueva Esparta
44 *C5* **Nueva Gerona** Isla de la Juventud, S Cuba *21°49′N 82°49′W*
42 *H8* **Nueva Guadalupe** San Miguel, E El Salvador *13°30′N 88°21′W*
42 *M11* **Nueva Guinea** Región Autónoma Atlántico Sur, SE Nicaragua *11°40′N 84°22′W*
61 *D19* **Nueva Helvecia** Colonia, SW Uruguay *34°16′S 57°53′W*
63 *J25* **Nueva, Isla** *island* S Chile
40 *M14* **Nueva Italia** Michoacán, SW Mexico *19°01′N 102°06′W*
56 *D6* **Nueva Loja** ***var.*** Lago Agrio. Sucumbíos, NE Ecuador
42 *F6* **Nueva Ocotepeque** ***prev.*** Ocotepeque. Ocotepeque, W Honduras *14°25′N 89°10′W*
61 *D19* **Nueva Palmira** Colonia, SW Uruguay *33°53′S 58°25′W*
41 *N6* **Nueva Rosita** Coahuila, NE Mexico *27°58′N 101°11′W*
42 *E7* **Nueva San Salvador** ***prev.*** Santa Tecla. La Libertad, SW El Salvador *13°40′N 89°18′W*
42 *J8* **Nueva Segovia** ⬖ *department* NW Nicaragua
Nueva Tabarca *see* Plana, Isla
Nueva Villa de Padilla *see* Nuevo Padilla
61 *B21* **Nueve de Julio** Buenos Aires, E Argentina *35°29′S 60°52′W*
44 *H6* **Nuevitas** Camagüey, E Cuba *21°34′N 77°18′W*
61 *D18* **Nuevo Berlín** Río Negro, W Uruguay *32°59′S 58°03′W*
40 *I4* **Nuevo Casas Grandes** Chihuahua, N Mexico *30°23′N 107°54′W*
43 *T14* **Nuevo Chagres** Colón, C Panama *09°14′N 80°05′W*
41 *W15* **Nuevo Coahuila** Campeche, E Mexico *17°53′N 90°46′W*
63 *K17* **Nuevo, Golfo** *gulf* S Argentina
41 *O7* **Nuevo Laredo** Tamaulipas, NE Mexico *27°28′N 99°32′W*
41 *N8* **Nuevo León** ⬖ *state* NE Mexico
41 *P10* **Nuevo Padilla** ***var.*** Nueva Villa de Padilla. Tamaulipas, C Mexico *24°01′N 98°48′W*
56 *E6* **Nuevo Rocafuerte** Orellana, E Ecuador *0°59′S 75°27′W*
Nuga *see* Dzavhanmandal
80 *O13* **Nugaal** ***off.*** Gobolka Nugaal. ⬖ *region* N Somalia
Nugaal, Gobolka *see* Nugaal
185 *E24* **Nugget Point** *headland* South Island, New Zealand *46°26′S 169°49′E*
186 *J5* **Nuguria Islands** *island group* E Papua New Guinea
184 *P10* **Nuhaka** Hawke's Bay, North Island, New Zealand *39°03′S 177°43′E*
138 *M10* **Nuhaydayn, Wādī an** *dry watercourse* W Iraq
190 *E7* **Nui Atoll** *atoll* W Tuvalu
Nu Jiang *see* Salween
Nûk *see* Nuuk
182 *G7* **Nukey Bluff** *hill* South Australia
Nukha *see* Şäki
123 *T9* **Nukh Yablonevyy, Gora** ▲ E Russian Federation *60°26′N 151°45′E*
186 *K7* **Nukiki** Choiseul, NW Solomon Islands *06°45′S 156°30′E*
186 *B6* **Nuku** West Sepik, NW Papua New Guinea *03°48′S 142°23′E*
193 *W15* **Nuku** *island* Tongatapu Group, NE Tonga
193 *Y16* **Nuku'alofa** ● (Tonga) Tongatapu, S Tonga *21°08′S 175°13′W*
193 *U15* **Nuku'alofa** Tongatapu, S Tonga *21°09′S 175°14′W*
190 *G12* **Nukuatea** *island* N Wallis and Futuna
190 *F7* **Nukufetau Atoll** *atoll* C Tuvalu
190 *G12* **Nukuhifala** *island* E Wallis and Futuna
191 *W7* **Nuku Hiva** *island* Îles Marquises, NE French Polynesia
191 *W7* **Nuku Hiva Island** *island* Îles Marquises, N French Polynesia
190 *F9* **Nukulaelae Atoll** ***var.*** Nukulailai. *atoll* E Tuvalu
Nukulailai *see* Nukulaelae Atoll
190 *G11* **Nukuloa** *island* N Wallis and Futuna
186 *L6* **Nukumanu Islands** ***prev.*** Tasman Group. *island group* NE Papua New Guinea
Nukunau *see* Nikunau
190 *J9* **Nukunonu Atoll** *island* C Tokelau
190 *J9* **Nukunonu Village** Nukunonu Atoll, C Tokelau
189 *S18* **Nukuoro Atoll** *atoll* Caroline Islands, S Micronesia
146 *H8* **Nukus** Qoraqalpog'iston Respublikasi, W Uzbekistan *42°29′N 59°32′E*
190 *G11* **Nukutapu** *island* N Wallis and Futuna
39 *O9* **Nulato** Alaska, USA *64°43′N 158°06′W*
39 *O10* **Nulato Hills** ▲▲ Alaska, USA
105 *T9* **Nules** Valenciana, E Spain *39°52′N 00°10′W*
Nuling *see* Sultan Kudarat
182 *C6* **Nullarbor** South Australia *31°28′S 130°57′E*
180 *M11* **Nullarbor Plain** *plateau* South Australia/Western Australia
163 *S12* **Nulu'erhu Shan** ▲▲ N China
77 *X14* **Numan** Adamawa, E Nigeria *09°26′N 11°58′E*
165 *S3* **Numata** Hokkaidō, NE Japan *43°48′N 141°55′E*
81 *C15* **Numatinna** ≈ W South Sudan
95 *F14* **Numedalen** *valley* S Norway
95 *G14* **Numedalslågen** ***var.*** Laagen. ≈ S Norway
93 *L19* **Nummela** Uusimaa, S Finland *60°21′N 24°20′E*
183 *O11* **Numurkah** Victoria, SE Australia *36°04′S 145°28′E*
196 *L16* **Nunap Isua** ***var.*** Uummannarsuaq, ***Dan.*** Kap Farvel, ***Eng.*** Cape Farewell. *cape* S Greenland
9 *N8* **Nunavut** ⬖ *territory* N Canada
54 *H9* **Nunchia** Casanare, C Colombia *05°37′N 72°13′W*
97 *M20* **Nuneaton** C England, United Kingdom *52°32′N 01°28′W*
153 *W14* **Nungba** Manipur, NE India *24°46′N 93°25′E*
38 *L12* **Nunivak Island** *island* Alaska, USA
152 *I5* **Nun Kun** ▲ NW India *34°01′N 76°04′E*
98 *L10* **Nunspeet** Gelderland, E Netherlands *52°21′N 05°45′E*
107 *C18* **Nuoro** Sardegna, Italy, C Mediterranean Sea *40°20′N 09°20′E*
75 *R12* **Nuqayy, Jabal** *hill range* S Libya
54 *C9* **Nuquí** Chocó, W Colombia *05°44′N 77°16′W*
143 *O4* **Nūr** Māzandarān, N Iran *36°32′N 52°00′E*
145 *Q9* **Nura** ≈ N Kazakhstan
143 *N11* **Nūrābād** Fārs, C Iran *30°08′N 51°30′E*
Nurakita *see* Niulakita
Nurata *see* Nurota
Nuratau, Khrebet *see* Nurota Tizmasi
136 *L17* **Nur Dağları** ▲▲ S Turkey
Nurek *see* Norak
Nuremberg *see* Nürnberg
Nūrestān *see* Nūristān
136 *M15* **Nurhak** Kahramanmaraş, S Turkey *37°57′N 37°21′E*
182 *J9* **Nuriootpa** South Australia *34°28′S 139°00′E*
149 *S4* **Nūristān** ***prev.*** Nūrestān. ⬖ *province* C Afghanistan
127 *S5* **Nurlat** Respublika Tatarstan, W Russian Federation *54°26′N 50°48′E*
93 *N15* **Nurmes** Pohjois-Karjala, E Finland *63°31′N 29°10′E*
101 *K20* **Nürnberg** ***Eng.*** Nuremberg. Bayern, S Germany *49°27′N 11°05′E*
101 *K20* **Nürnberg** ✈ Bayern, SE Germany *49°29′N 11°04′E*
146 *M10* **Nurota** ***Rus.*** Nurata. Navoiy Viloyati, C Uzbekistan *40°41′N 65°43′E*
147 *N10* **Nurota Tizmasi** ***Rus.*** Khrebet Nuratau. ▲▲ C Uzbekistan
149 *T8* **Nūrpur** Punjab, E Pakistan *31°54′N 71°55′E*
183 *P6* **Nurri, Mount** *hill* New South Wales, SE Australia
25 *T13* **Nursery** Texas, SW USA *28°55′N 97°05′W*
169 *V17* **Nusa Tenggara Barat** ***off.*** Propinsi Nusa Tenggara Barat, ***Eng.*** West Nusa Tenggara. ⬖ *province* S Indonesia
Nusa Tenggara Barat, Propinsi *see* Nusa Tenggara Barat
171 *O16* **Nusa Tenggara Timur** ***off.*** Propinsi Nusa Tenggara Timur, ***Eng.*** East Nusa Tenggara. ⬖ *province* S Indonesia
Nusa Tenggara Timur, Propinsi *see* Nusa Tenggara Timur
171 *U14* **Nusawulan** Papua Barat, E Indonesia *04°03′S 132°56′E*
137 *Q16* **Nusaybin** ***var.*** Nisibin. Manisa, SE Turkey *37°08′N 41°11′E*
39 *O14* **Nushagak Bay** *bay* Alaska, USA
39 *O13* **Nushagak Peninsula** *headland* Alaska, USA *58°39′N 159°03′W*
39 *O13* **Nushagak River** ≈ Alaska, USA
160 *E11* **Nu Shan** ▲▲ SW China
149 *N11* **Nushki** Baluchistān, SW Pakistan *29°33′N 66°01′E*
Nussdorf *see* Năsăud
112 *J9* **Nuštar** Vukovar-Srijem, E Croatia *45°20′N 18°48′E*
99 *L18* **Nuth** Limburg, SE Netherlands *50°55′N 05°52′E*
100 *N13* **Nuthe** ≈ NE Germany
Nutmeg State *see* Connecticut
39 *T10* **Nutzotin Mountains** ▲▲ Alaska, USA
64 *I5* **Nuuk** ***var.*** Nûk, ***Dan.*** Godthaab, Godthåb. ○ (Greenland) Sermersooq, SW Greenland *64°15′N 51°35′W*
92 *L13* **Nuupas** Lappi, NW Finland *66°01′N 26°19′E*
191 *O7* **Nuupere, Pointe** *headland* Moorea, W French Polynesia *17°35′S 149°47′W*
191 *O7* **Nuuroa, Pointe** *headland* Tahiti, W French Polynesia
Nüürst *see* Baganuur
Nuwara *see* Nuwara Eliya
155 *K25* **Nuwara Eliya** ***var.*** Nuwara. Central Province, S Sri Lanka *06°58′N 80°46′E*
182 *E7* **Nuyts Archipelago** *island group* South Australia
83 *F17* **Nxaunxau** North West, NW Botswana *18°57′S 21°18′E*
39 *N12* **Nyac** Alaska, USA *61°00′N 159°56′W*
122 *H9* **Nyagan'** Khanty-Mansiyskiy Avtonomnyy Okrug-Yugra, N Russian Federation *62°10′N 65°32′E*
Nyagassola *see* Niagassola
Nyagquka *see* Yajiang
81 *I18* **Nyahururu** Nyandarua, C Kenya *0°04′N 36°22′E*
182 *M10* **Nyah West** Victoria, SE Australia *35°14′S 143°18′E*
158 *M15* **Nyainqêntanglha Feng** ▲ W China *30°20′N 90°28′E*
159 *N15* **Nyainqêntanglha Shan** ▲▲ W China
80 *B11* **Nyala** Southern Darfur, W Sudan *12°01′N 24°50′E*
83 *M16* **Nyamapanda** Mashonaland East, NE Zimbabwe *16°59′S 32°52′E*
81 *H19* **Nyamira** ⬖ *county* W Kenya
81 *H25* **Nyamtumbo** Ruvuma, S Tanzania *10°33′S 36°08′E*
Nyanda *see* Masvingo
81 *I19* **Nyandarua** ⬖ *county* C Kenya
124 *M11* **Nyandoma** Arkhangel'skaya Oblast', NW Russian Federation *61°39′N 40°10′E*
83 *M16* **Nyanga** ***prev.*** Inyanga. Manicaland, E Zimbabwe *18°13′S 32°46′E*
79 *D20* **Nyanga** ***off.*** Province de la Nyanga, ***var.*** La Nyanga. ⬖ *province* SW Gabon
79 *E20* **Nyanga** ≈ Congo/Gabon
Nyanga, Province de la *see* Nyanga
81 *F20* **Nyantakara** Kagera, NW Tanzania *03°05′S 31°23′E*
81 *E21* **Nyanza-Lac** S Burundi *04°16′S 29°38′E*
68 *J14* **Nyasa, Lake** ***var.*** Lake Malawi; ***prev.*** Lago Nyassa. ◎ E Africa
Nyasaland/Nyasaland Protectorate *see* Malawi
Nyassa, Lago *see* Nyasa, Lake
119 *J17* **Nyasvizh** ***Pol.*** Nieśwież, ***Rus.*** Nesvizh. Minskaya Voblasts', C Belarus *53°13′N 26°40′E*
166 *M8* **Nyaunglebin** Bago, SW Myanmar (Burma) *17°59′N 94°44′E*
166 *M5* **Nyaung-u** Magway, C Myanmar (Burma) *21°03′N 95°44′E*
95 *H24* **Nyborg** Syddtjylland, C Denmark *55°19′N 10°48′E*
95 *N21* **Nybro** Kalmar, S Sweden *56°45′N 15°54′E*
119 *J16* **Nyeharelaye** ***Rus.*** Negoreloye. Minskaya Voblasts', C Belarus *53°36′N 27°04′E*
195 *W3* **Nye Mountains** ▲▲ Antarctica
81 *I19* **Nyeri** Nyeri, C Kenya *0°25′S 36°56′E*
81 *I19* **Nyeri** ⬖ *county* C Kenya
118 *M11* **Nyeshcharda, Vozyera** ***Rus.*** Ozero Neshcherdo. ◎ N Belarus
92 *O2* **Ny-Friesland** *physical region* N Svalbard
95 *L14* **Nyhammar** Dalarna, C Sweden *60°19′N 14°55′E*
160 *F7* **Nyikog Qu** ≈ C China
83 *L14* **Nyimba** Eastern, E Zambia *14°33′S 30°49′E*
159 *P15* **Nyingchi** ***var.*** Bayizhen. Xizang Zizhiqu, W China *29°27′N 94°43′E*
159 *P15* **Nyingchi** ***var.*** Pula. Xizang Zizhiqu, W China *29°34′N 94°33′E*
Nyinma *see* Maqu
111 *O21* **Nyírbátor** Szabolcs-Szatmár-Bereg, E Hungary *47°50′N 22°09′E*
111 *N21* **Nyíregyháza** Szabolcs-Szatmár-Bereg, NE Hungary *47°57′N 21°43′E*
Nyiro *see* Ewaso Ng'iro
Nyitra *see* Nitra
Nyitrabánya *see* Handlová
93 *K16* **Nykarleby** ***Fin.*** Uusikaarlepyy. Österbotten, W Finland *63°22′N 22°30′E*
95 *F21* **Nykøbing** Midtjylland, NW Denmark *56°48′N 08°52′E*
95 *I25* **Nykøbing** Sjælland, SE Denmark *54°47′N 11°53′E*
95 *I22* **Nykøbing** Sjælland, C Denmark *55°56′N 11°41′E*
95 *N17* **Nyköping** Södermanland, S Sweden *58°45′N 17°03′E*
95 *L15* **Nykroppa** Värmland, C Sweden *59°37′N 14°18′E*
Nyland *see* Uusimaa
Nylstroom *see* Modimolle
183 *P7* **Nymagee** New South Wales, SE Australia *32°06′S 146°19′E*
183 *V5* **Nymboida** New South Wales, SE Australia *29°57′S 152°45′E*
183 *U5* **Nymboida River** ≈ New South Wales, SE Australia
111 *D16* **Nymburk** ***var.*** Neuenburg an der Elbe, ***Ger.*** Nimburg. Středočesky Kraj, C Czech Republic *50°12′N 15°00′E*
95 *O16* **Nynäshamn** Stockholm, C Sweden *58°54′N 17°55′E*
183 *Q6* **Nyngan** New South Wales, SE Australia *31°36′S 147°07′E*
Nyoman *see* Neman
108 *A10* **Nyon** ***Ger.*** Neuss; ***anc.*** Noviodunum. Vaud, SW Switzerland *46°23′N 06°15′E*
79 *D16* **Nyong** ≈ SW Cameroon
103 *S14* **Nyons** Drôme, E France *44°22′N 05°06′E*
79 *D14* **Nyos, Lac** ***Eng.*** Lake Nyos. ◎ NW Cameroon
Nyos, Lake *see* Nyos, Lac
125 *U11* **Nyrob** ***var.*** Nyrov. Permskiy Kray, NW Russian Federation *60°41′N 56°42′E*
Nyrov *see* Nyrob
111 *H15* **Nysa** ***Ger.*** Neisse. Opolskie, S Poland *50°28′N 17°20′E*
32 *M13* **Nyssa** Oregon, NW USA *43°52′N 116°59′W*
Nysa Łużycka *see* Neisse
Nyslott *see* Savonlinna
Nystad *see* Uusikaupunki
95 *I25* **Nysted** Sjælland, SE Denmark *54°40′N 11°41′E*
125 *U14* **Nytva** Permskiy Kray, NW Russian Federation *57°56′N 55°22′E*
165 *P8* **Nyūdō-zaki** *headland* Honshū, C Japan *39°59′N 139°40′E*
125 *P9* **Nyukhcha** Arkhangel'skaya Oblast', NW Russian Federation *63°24′N 46°34′E*
124 *H8* **Nyuk, Ozero** ***var.*** Ozero Njuk. ◎ NW Russian Federation
125 *O12* **Nyuksenitsa** ***var.*** Njuksenica. Vologodskaya Oblast', NW Russian Federation *60°25′N 44°12′E*
79 *O22* **Nyunzu** Katanga, SE Dem. Rep. Congo *05°55′S 28°00′E*
123 *O10* **Nyurba** Respublika Sakha (Yakutiya), NE Russian Federation *63°17′N 118°15′E*
123 *O11* **Nyuya** Respublika Sakha (Yakutiya), NE Russian Federation *60°33′N 116°10′E*
146 *K12* **Nyýazow** ***Rus.*** Niyazov. Lebap Welaýaty, NE Turkmenistan *39°13′N 63°16′E*
117 *T10* **Nyzhni Sirohozy** Khersons'ka Oblast', S Ukraine *46°49′N 34°21′E*
117 *U12* **Nyzhn'ohirs'kyy** ***Rus.*** Nizhnegorskiy. Avtonomna Respublika Krym, S Ukraine *45°26′N 34°42′E*
NZ *see* New Zealand
81 *G21* **Nzega** Tabora, C Tanzania *04°13′S 33°11′E*
76 *K15* **Nzérékoré** SE Guinea *07°45′N 08°49′W*

◆ Country
● Country Capital
◇ Dependent Territory
○ Dependent Territory Capital
⬖ Administrative Regions
✈ International Airport
▲ Mountain
▲▲ Mountain Range
Volcano
≈ River
◎ Lake
⊞ Reservoir

82 A10 **N'Zeto** prev. Ambrizete. Zaire Province, NW Angola 07°14′S 12°52′E
79 M24 **Nzilo, Lac** prev. Lac Delcommune. ◎ SE Dem. Rep. Congo
172 I13 **Nzwani** Fr. Anjouan, var. Ndzouani. island SE Comoros

O

29 O11 **Oacoma** South Dakota, N USA 43°49′N 99°25′W
29 N9 **Oahe Dam** dam South Dakota, N USA
28 M9 **Oahe, Lake** ⊠ North Dakota/South Dakota, N USA
38 C9 **Oa'hu** var. Oahu. island Hawai'ian Islands, Hawai'i, USA
165 V4 **O-Akan-dake** ▲ Hokkaidō, NE Japan 43°26′N 144°09′E
182 K8 **Oakbank** South Australia 33°07′S 140°36′E
19 P13 **Oak Bluffs** Martha's Vineyard, New York, NE USA 41°25′N 70°32′W
36 K4 **Oak City** Utah, W USA 39°22′N 112°19′W
37 R3 **Oak Creek** Colorado, C USA 40°16′N 106°57′W
35 P8 **Oakdale** California, W USA 37°46′N 120°51′W
22 H8 **Oakdale** Louisiana, S USA 30°49′N 92°39′W
29 P7 **Oakes** North Dakota, N USA 46°08′N 98°05′W
22 J4 **Oak Grove** Louisiana, S USA 32°51′N 91°25′W
97 N19 **Oakham** C England, United Kingdom 52°41′N 00°45′W
32 H7 **Oak Harbor** Washington, NW USA 48°17′N 122°38′W
21 R5 **Oak Hill** West Virginia, NE USA 37°59′N 81°09′W
35 N8 **Oakland** California, W USA 37°48′N 122°16′W
29 T15 **Oakland** Iowa, C USA 41°18′N 95°22′W
19 Q7 **Oakland** Maine, NE USA 44°32′N 69°43′W
21 T3 **Oakland** Maryland, NE USA 39°24′N 79°25′W
29 R14 **Oakland** Nebraska, C USA 41°50′N 96°28′W
31 N11 **Oak Lawn** Illinois, N USA 41°43′N 87°45′W
33 P16 **Oakley** Idaho, NW USA 42°13′N 113°54′W
26 I4 **Oakley** Kansas, C USA 39°08′N 100°53′W
31 N10 **Oak Park** Illinois, N USA 41°53′N 87°46′W
11 X16 **Oak Point** Manitoba, S Canada 50°23′N 97°00′W
32 G13 **Oakridge** Oregon, NW USA 43°45′N 122°27′W
20 M9 **Oak Ridge** Tennessee, S USA 36°02′N 84°12′W
184 K10 **Oakura** Taranaki, North Island, New Zealand 39°07′S 173°58′E
22 L7 **Oak Vale** Mississippi, S USA 31°26′N 89°57′W
14 G16 **Oakville** Ontario, S Canada 43°27′N 79°41′W
25 V8 **Oakwood** Texas, SW USA 31°34′N 95°51′W
185 F22 **Oamaru** Otago, South Island, New Zealand 45°10′S 170°51′E
96 F13 **Oa, Mull of** headland W Scotland, United Kingdom 55°35′N 06°20′W
171 O11 **Oan** Sulawesi, N Indonesia 01°16′N 121°25′E
185 J17 **Oaro** Canterbury, South Island, New Zealand 42°29′S 173°30′E
35 X2 **Oasis** Nevada, W USA 41°01′N 114°29′W
195 S15 **Oates Land** physical region Antarctica
183 P17 **Oatlands** Tasmania, SE Australia 42°21′S 147°23′E
36 I11 **Oatman** Arizona, SW USA 35°03′N 114°19′W
41 R16 **Oaxaca** var. Oaxaca de Juárez; prev. Antequera. Oaxaca, SE Mexico 17°04′N 96°41′W
41 Q16 **Oaxaca** ⬖ state SE Mexico
Oaxaca de Juárez see Oaxaca
122 I19 **Ob'** ≈ C Russian Federation
145 X9 **Oba** prev. Uba. ≈ E Kazakhstan
14 G9 **Obabika Lake** ◎ Ontario, S Canada
Obagan see Ubagan
118 M12 **Obal'** Rus. Obol'. Vitsyebskaya Voblasts', N Belarus 55°22′N 29°17′E
79 E16 **Obala** Centre, SW Cameroon 04°09′N 11°32′E
14 C6 **Oba Lake** ◎ Ontario, S Canada
164 J12 **Obama** Fukui, Honshū, SW Japan 35°32′N 135°45′E
96 H11 **Oban** W Scotland, United Kingdom 56°25′N 05°29′W
Oban see Halfmoon Bay
Obando see Puerto Inírida
104 I4 **O Barco** var. El Barco, El Barco de Valdeorras, O Barco de Valdeorras. Galicia, NW Spain 42°24′N 07°00′W
O Barco de Valdeorras see O Barco
Obbia see Hobyo
93 J16 **Obbola** Västerbotten, N Sweden 63°41′N 20°16′E
Obbrovazzo see Obrovac
Obchuga see Abchuha
Obdorsk see Salekhard
Óbecse see Bečej
118 I11 **Obeliai** Panevėžys, NE Lithuania 55°57′N 25°47′E
60 F13 **Oberá** Misiones, NE Argentina 27°29′S 55°08′W
108 E8 **Oberburg** Bern, W Switzerland 47°00′N 07°37′E
109 Q9 **Oberdrauburg** Salzburg, S Austria 46°45′N 12°59′E
Oberglogau see Głogówek
109 W4 **Ober Grafendorf** Niederösterreich, NE Austria 48°09′N 15°33′E
101 E15 **Oberhausen** Nordrhein-Westfalen, W Germany 51°27′N 06°50′E
Oberhollabrunn see Tulln
Oberlaibach see Vrhnika
101 Q15 **Oberlausitz** var. Hornja Łužica. physical region E Germany
26 J2 **Oberlin** Kansas, C USA 39°49′N 100°33′W
22 H8 **Oberlin** Louisiana, S USA 30°37′N 92°45′W
31 T11 **Oberlin** Ohio, N USA 41°17′N 82°13′W
103 U5 **Obernai** Bas-Rhin, NE France 48°28′N 07°30′E
109 R4 **Obernberg am Inn** Oberösterreich, N Austria 48°19′N 13°20′E
Oberndorf see Oberndorf am Neckar
101 G23 **Oberndorf am Neckar** var. Oberndorf. Baden-Württemberg, SW Germany 48°18′N 08°32′E
109 Q5 **Oberndorf bei Salzburg** Salzburg, W Austria 47°57′N 12°57′E
Oberneustadtl see Kysucké Nové Mesto
183 S8 **Oberon** New South Wales, SE Australia 33°42′S 149°50′E
109 Q4 **Oberösterreich** off. Land Oberösterreich, Eng. Upper Austria. ⬖ state NW Austria
Oberösterreich, Land see Oberösterreich
Oberpahlen see Põltsamaa
101 M19 **Oberpfälzer Wald** ▲▲ SE Germany
109 Y6 **Oberpullendorf** Burgenland, E Austria 47°32′N 16°30′E
Oberradkersburg see Gornja Radgona
101 G18 **Oberursel** Hessen, W Germany 50°12′N 08°34′E
109 Q8 **Obervellach** Salzburg, S Austria 46°56′N 13°10′E
109 X7 **Oberwart** Burgenland, SE Austria 47°18′N 16°12′E
Oberwischau see Vişeu de Sus
109 T7 **Oberwölz** var. Oberwölz-Stadt. Steiermark, SE Austria 47°12′N 14°20′E
Oberwölz-Stadt see Oberwölz
31 S13 **Obetz** Ohio, N USA 39°52′N 82°57′W
Ob', Gulf of see Obskaya Guba
58 H12 **Óbidos** Pará, NE Brazil 01°52′S 55°30′W
104 F10 **Óbidos** Leiria, C Portugal 39°21′N 09°09′W
Obidovichi see Abidavichy
147 Q13 **Obigarm** W Tajikistan 38°42′N 69°34′E
165 T2 **Obihiro** Hokkaidō, NE Japan 42°56′N 143°10′E
Obi-Khingou see Khingov
147 P13 **Obikiik** SW Tajikistan 38°07′N 68°36′E
Obilić see Obiliq
113 N16 **Obiliq** Serb. Obilić. N Kosovo 42°50′N 20°57′E
127 O12 **Obil'noye** Respublika Kalmykiya, SW Russian Federation 47°31′N 44°24′E
20 F8 **Obion** Tennessee, S USA 36°15′N 89°11′W
20 F8 **Obion River** ≈ Tennessee, S USA
171 S12 **Obi, Pulau** island Maluku, E Indonesia
165 S2 **Obira** Hokkaidō, NE Japan 44°01′N 141°39′E
127 N11 **Oblivskaya** Rostovskaya Oblast', SW Russian Federation 48°34′N 42°31′E
123 R14 **Obluch'ye** Yevreyskaya Avtonomnaya Oblast', SE Russian Federation 48°59′N 131°18′E
126 K4 **Obninsk** Kaluzhskaya Oblast', W Russian Federation 55°06′N 36°40′E
114 J8 **Obnova** Pleven, N Bulgaria 43°26′N 25°04′E
79 N15 **Obo** Haut-Mbomou, E Central African Republic 05°20′N 26°29′E
159 T9 **Obo** Qinghai, C China 37°57′N 101°03′E
80 M11 **Obock** E Djibouti 11°57′N 43°09′E
Obol' see Obal'
Obolyanka see Abalyanka
171 V13 **Obome** Papua Barat, E Indonesia 03°42′S 133°21′E
110 G11 **Oborniki** Wielkolpolskie, W Poland 52°38′N 16°48′E
79 G19 **Obouya** Cuvette, C Congo 0°56′S 15°41′E
126 J8 **Oboyan'** Kurskaya Oblast', W Russian Federation 51°12′N 36°15′E
124 M9 **Obozerskiy** Arkhangel'skaya Oblast', NW Russian Federation 63°26′N 40°20′E
112 L11 **Obrenovac** Serbia, N Serbia 44°39′N 20°12′E
112 D12 **Obrovac** It. Obbrovazzo. Zadar, SW Croatia 44°12′N 15°40′E
Obrovo see Abrova
35 Q3 **Observation Peak** ▲ California, W USA 40°48′N 120°07′W
122 J8 **Obskaya Guba** Eng. Gulf of Ob. gulf N Russian Federation
173 N13 **Ob' Tablemount** undersea feature S Indian Ocean 50°16′S 51°59′E
173 T10 **Ob' Trench** undersea feature E Indian Ocean
77 P16 **Obuasi** S Ghana 06°15′N 01°36′W
117 P5 **Obukhiv** Rus. Obukhov. Kyyivs'ka Oblast', N Ukraine 50°05′N 30°37′E
Obukhov see Obukhiv
125 U14 **Obva** ≈ NW Russian Federation
108 F8 **Obwalden** ⬖ canton C Switzerland
117 V10 **Obytichna Kosa** spit SE Ukraine
117 V10 **Obytichna Zatoka** gulf SE Ukraine
114 N9 **Obzor** Burgas, E Bulgaria 42°52′N 27°53′E
105 O3 **Oca** ≈ N Spain
23 W10 **Ocala** Florida, SE USA 29°11′N 82°08′W
40 M7 **Ocampo** Coahuila, NE Mexico 27°18′N 102°24′W
54 G7 **Ocaña** Norte de Santander, N Colombia 08°16′N 73°21′W
105 N9 **Ocaña** Castilla-La Mancha, C Spain 39°57′N 03°30′W
104 H4 **O Carballiño** Cast. Carballiño. Galicia, NW Spain 42°26′N 08°05′W
37 T9 **Ocate** New Mexico, SW USA 36°09′N 105°03′W
Ocavango see Okavango
57 D14 **Occidental, Cordillera** ▲▲ W South America
21 Q6 **Oceana** West Virginia, NE USA 37°41′N 81°37′W
21 Z4 **Ocean City** Maryland, NE USA 38°20′N 75°05′W
18 J17 **Ocean City** New Jersey, NE USA 39°15′N 74°33′W
10 K15 **Ocean Falls** British Columbia, SW Canada 52°24′N 127°42′W
Ocean Island see Banaba
Ocean Island see Kure Atoll
64 J9 **Oceanographer Fracture Zone** tectonic feature NW Atlantic Ocean
35 U17 **Oceanside** California, W USA 33°12′N 117°23′W
22 M9 **Ocean Springs** Mississippi, S USA 30°24′N 88°49′W
Ocean State see Rhode Island
25 O9 **O C Fisher Lake** ⊠ Texas, SW USA
117 Q10 **Ochakiv** Rus. Ochakov. Mykolayivs'ka Oblast', S Ukraine 46°36′N 31°33′E
Ochakov see Ochakiv
Ochamchira see Ochamchire
137 Q9 **Ochamchire** Rus. Ochamchira; prev. Och'amch'ire. W Georgia 42°45′N 41°30′E
Och'amch'ire see Ochamchire
Ochansk see Okhansk
125 T15 **Ochër** Permskiy Kray, NW Russian Federation 57°54′N 54°40′E
115 I19 **Óchi** ▲ Évvoia, C Greece 38°03′N 24°27′E
165 W4 **Ochiishi-misaki** headland Hokkaidō, NE Japan 43°10′N 145°29′E
23 S9 **Ochlockonee River** ≈ Florida/Georgia, SE USA
44 K12 **Ocho Rios** C Jamaica 18°24′N 77°06′W
Ochrida see Ohrid
Ochrida, Lake see Ohrid, Lake
101 J19 **Ochsenfurt** Bayern, C Germany 49°39′N 10°03′E
23 U7 **Ocilla** Georgia, SE USA 31°35′N 83°15′W
94 N13 **Ockelbo** Gävleborg, C Sweden 60°51′N 16°46′E
Ocker see Oker
95 I19 **Öckerö** Västra Götaland, S Sweden 57°43′N 11°39′E
23 U6 **Ocmulgee River** ≈ Georgia, SE USA
116 H11 **Ocna Mureş** Hung. Marosújvár; prev. Ocna Mureşului, prev. Hung. Marosújvárakna. Alba, C Romania 46°25′N 23°53′E
Ocna Mureşului see Ocna Mureş
116 H11 **Ocna Sibiului** Ger. Salzburg, Hung. Vizakna. Sibiu, C Romania 45°52′N 23°59′E
116 H13 **Ocnele Mari** prev. Vioara. Vâlcea, S Romania 45°03′N 24°18′E
116 L7 **Ocniţa** Rus. Oknitsa. N Moldova 48°25′N 27°30′E
23 U4 **Oconee, Lake** ⊠ Georgia, SE USA
23 U5 **Oconee River** ≈ Georgia, SE USA
30 M9 **Oconomowoc** Wisconsin, N USA 43°06′N 88°29′W
30 M6 **Oconto** Wisconsin, N USA 44°55′N 87°52′W
30 M6 **Oconto Falls** Wisconsin, N USA 44°52′N 88°06′W
30 M6 **Oconto River** ≈ Wisconsin, N USA
104 I3 **O Corgo** Galicia, NW Spain 42°56′N 07°25′W
41 V16 **Ocosingo** Chiapas, SE Mexico 17°04′N 92°15′W
42 J8 **Ocotal** Nueva Segovia, NW Nicaragua 13°38′N 86°28′W
42 F6 **Ocotepeque** ⬖ department W Honduras
Ocotepeque see Nueva Ocotepeque
40 L13 **Ocotlán** Jalisco, SW Mexico 20°21′N 102°42′W
41 R16 **Ocotlán** var. Ocotlán de Morelos. Oaxaca, SE Mexico 16°49′N 96°49′W
Ocotlán de Morelos see Ocotlán
41 U16 **Ocozocuautla** Chiapas, SE Mexico 16°46′N 93°22′W
21 Y10 **Ocracoke Island** island North Carolina, SE USA
102 I3 **Octeville** Manche, N France 49°37′N 01°39′W
October Revolution Island see Oktyabr'skoy Revolyutsii, Ostrov
43 R17 **Ocú** Herrera, S Panama 07°55′N 80°43′W
83 Q14 **Ocua** Cabo Delgado, NE Mozambique 13°37′S 39°44′E
Ocumare see Ocumare del Tuy
54 M5 **Ocumare del Tuy** var. Ocumare. Miranda, N Venezuela 10°07′N 66°47′W
77 P17 **Oda** SE Ghana 05°55′N 00°56′W
165 G12 **Ōda** var. Oda. Shimane, Honshū, SW Japan 35°10′N 132°29′E
92 K3 **Ódaðahraun** lava flow C Iceland
165 Q7 **Ōdate** Akita, Honshū, C Japan 40°18′N 140°34′E
165 N14 **Odawara** Kanagawa, Honshū, S Japan 35°15′N 139°08′E
95 D14 **Odda** Hordaland, S Norway 60°03′N 06°34′E
95 G22 **Odder** Midtjylland, C Denmark 55°59′N 10°10′E
Oddur see Xuddur
29 T13 **Odebolt** Iowa, C USA 42°18′N 95°15′W
104 H14 **Odeleite** Faro, S Portugal 37°02′N 07°29′W
25 Q4 **Odell** Texas, SW USA 34°19′N 99°24′W
25 T14 **Odem** Texas, SW USA 27°57′N 97°34′W
104 F13 **Odemira** Beja, S Portugal 37°35′N 08°38′W
136 C14 **Ödemiş** İzmir, SW Turkey 38°11′N 27°58′E
Ödenburg see Sopron
83 I22 **Odendaalsrus** Free State, C South Africa 27°52′S 26°42′E
Odenpäh see Otepää
95 H23 **Odense** Syddtjylland, C Denmark 55°24′N 10°23′E
101 H19 **Odenwald** ▲▲ W Germany
84 H10 **Oder** Cz./Pol. Odra. ≈ C Europe
Oderberg see Bohumín
100 P11 **Oderbruch** wetland Germany/Poland
Oderhaff see Szczeciński, Zalew
100 O11 **Oder-Havel-Kanal** canal NE Germany
Oderhellen see Odorheiu Secuiesc
100 P13 **Oder-Spree-Kanal** canal NE Germany
Odertal see Zdzieszowice
106 I7 **Oderzo** Veneto, NE Italy 45°48′N 12°33′E
177 P10 **Odesa** Rus. Odessa. Odes'ka Oblast', SW Ukraine 46°29′N 30°44′E
24 M8 **Odessa** Texas, SW USA 31°51′N 102°22′W
32 K8 **Odessa** Washington, NW USA 47°19′N 118°41′W
Odesa see Odes'ka Oblast'
95 L18 **Ödeshög** Östergötland, S Sweden 58°13′N 14°40′E
117 O9 **Odes'ka Oblast'** var. Odesa, Rus. Odesskaya Oblast'. ⬖ province SW Ukraine
Odessa see Odesa
Odesskaya Oblast' see Odes'ka Oblast'
122 H12 **Odesskoye** Omskaya Oblast', C Russian Federation 54°15′N 72°45′E
Odessus see Varna
102 F6 **Odet** ≈ NW France
104 I14 **Odiel** ≈ SW Spain
76 L14 **Odienné** NW Ivory Coast 09°32′N 07°35′W
171 O4 **Odiongan** Tablas Island, C Philippines 12°23′N 122°01′E
153 P17 **Odisha** prev. Orissa. ⬖ state NE India
116 L12 **Odobeşti** Vrancea, E Romania 45°46′N 27°06′E
110 H13 **Odolanów** Ger. Adelnau. Wielkopolskie, C Poland 51°35′N 17°42′E
167 R13 **Ôdôngk** Kâmpóng Spoe, S Cambodia 11°48′N 104°45′E
25 N6 **O'donnell** Texas, SW USA 32°57′N 101°49′W
98 O7 **Odoorn** Drenthe, NE Netherlands 52°52′N 06°49′E
Odorhei see Odorheiu Secuiesc
116 J11 **Odorheiu Secuiesc** Ger. Oderhellen, Hung. Vámosudvarhely; prev. Odorhei, Ger. Hofmarkt. Harghita, C Romania 46°18′N 25°19′E
Odra see Oder
112 J9 **Odžaci** Ger. Hodschag, Hung. Hodság. Vojvodina, NW Serbia 45°31′N 19°15′E
59 N14 **Oeiras** Piauí, E Brazil 07°00′S 42°07′W
104 F11 **Oeiras** Lisboa, C Portugal 38°41′N 09°18′W
101 G14 **Oelde** Nordrhein-Westfalen, W Germany 51°49′N 08°09′E
28 J11 **Oelrichs** South Dakota, N USA 43°08′N 103°13′W
101 M17 **Oelsnitz** Sachsen, E Germany 50°22′N 12°12′E
Oels/Oels in Schlesien see Oleśnica
29 X12 **Oelwein** Iowa, C USA 42°40′N 91°54′W
Oeniadae see Oiniádes
191 N17 **Oeno Island** atoll Pitcairn Group of Islands, C Pacific Ocean
Oesel see Saaremaa
108 L7 **Oetz** var. Ötz. Tirol, W Austria 47°15′N 10°56′E
137 P11 **Of** Trabzon, NE Turkey 40°57′N 40°17′E
30 K15 **O'Fallon** Illinois, N USA 38°35′N 89°54′W
27 W4 **O'Fallon** Missouri, C USA 38°54′N 90°31′W
107 N16 **Ofanto** ≈ S Italy
97 D18 **Offaly** Ir. Ua Uíbh Fhailí; prev. King's County. cultural region C Ireland
101 H18 **Offenbach** var. Offenbach am Main. Hessen, W Germany 50°06′N 08°46′E
Offenbach am Main see Offenbach
101 F22 **Offenburg** Baden-Württemberg, SW Germany 48°28′N 07°57′E
182 C2 **Officer Creek** seasonal river South Australia
Oficina María Elena see María Elena
Oficina Pedro de Valdivia see Pedro de Valdivia
115 K22 **Ofidoússa** island Kykládes, Greece, Aegean Sea
Ofiral see Sharm ash Shaykh
92 H10 **Ofotfjorden** fjord N Norway
192 L16 **Ofu** island Manua Islands, E American Samoa
165 R9 **Ōfunato** Iwate, Honshū, C Japan 39°04′N 141°43′E
165 P8 **Oga** Akita, Honshū, C Japan 39°56′N 139°47′E
Ogaadeen see Ogaden
165 Q9 **Ogachi** Akita, Honshū, C Japan 39°03′N 140°26′E
165 P9 **Ogachi-tōge** pass Honshū, C Japan
81 N14 **Ogaden** Som. Ogaadeen. plateau Ethiopia/Somalia
165 P8 **Oga-hantō** peninsula Honshū, C Japan
165 K13 **Ōgaki** Gifu, Honshū, SW Japan 35°22′N 136°35′E
28 L15 **Ogallala** Nebraska, C USA 41°09′N 101°44′W
168 M14 **Ogan, Air** ≈ Sumatera, W Indonesia
165 Y15 **Ogasawara-shotō** Eng. Bonin Islands. island group SE Japan
14 I9 **Ogascanane, Lac** ◎ Québec, SE Canada
165 R7 **Ogawara-ko** ◎ Honshū, C Japan
77 T15 **Ogbomosho** var. Ogmoboso. Oyo, W Nigeria 08°10′N 04°16′E
29 U13 **Ogden** Iowa, C USA 42°03′N 94°01′W
36 L2 **Ogden** Utah, W USA 41°09′N 111°58′W
18 I6 **Ogdensburg** New York, NE USA 44°42′N 75°25′W
23 W5 **Ogeechee River** ≈ Georgia, SE USA
Oger see Ogre
165 N10 **Ogi** Niigata, Sado, C Japan 37°49′N 138°16′E
10 H5 **Ogilvie** Yukon, NW Canada 63°34′N 139°44′W
10 H4 **Ogilvie** ≈ Yukon, NW Canada
10 H5 **Ogilvie Mountains** ▲▲ Yukon, NW Canada
Oginskiy Kanal see Ahinski Kanal
162 J7 **Ögiynuur** var. Dzegstey. Arhangay, C Mongolia 47°38′N 102°31′E
146 F6 **Og'iyon Sho'rxogi** wetland NW Uzbekistan
146 B10 **Oglanly** Balkan Welaýaty, W Turkmenistan 39°56′N 54°25′E
23 T5 **Oglethorpe** Georgia, SE USA 32°17′N 84°03′W
23 T2 **Oglethorpe, Mount** ▲ Georgia, SE USA 34°29′N 84°19′W
106 F7 **Oglio** anc. Ollius. ≈ N Italy
Ogmoboso see Ogbomosho
103 T8 **Ognon** ≈ E France
123 R13 **Ogodzha** Amurskaya Oblast', S Russian Federation 52°51′N 132°49′E
77 W16 **Ogoja** Cross River, S Nigeria 06°37′N 08°48′E
12 C10 **Ogoki** ≈ Ontario, S Canada
12 D11 **Ogoki Lake** ◎ Ontario, C Canada
Ögöömör see Hanhongor
79 F19 **Ogooué** ≈ Congo/Gabon
79 E18 **Ogooué-Ivindo** off. Province de l'Ogooué-Ivindo, var. L'Ogooué-Ivindo. ⬖ province N Gabon
Ogooué-Ivindo, Province de l' see Ogooué-Ivindo
79 E19 **Ogooué-Lolo** off. Province de l'Ogooué-Lolo, var. L'Ogooué-Lolo. ⬖ province C Gabon
Ogooué-Lolo, Province de l' see Ogooué-Lolo
79 C19 **Ogooué-Maritime** off. Province de l'Ogooué-Maritime, var. L'Ogooué-Maritime. ⬖ province W Gabon
Ogooué-Maritime, Province de l' see Ogooué-Maritime
165 D14 **Ogōri** Fukuoka, Kyūshū, SW Japan 33°24′N 130°34′E
114 H7 **Ogosta** ≈ NW Bulgaria
112 Q9 **Ogražden** Bul. Ograzhden. ▲ Bulgaria/FYR Macedonia see also Ograzhden
114 G12 **Ograzhden** Mac. Orgražden. ▲ Bulgaria/FYR Macedonia see also Orgražden
Ograzhden see Ogražden
118 G9 **Ogre** Ger. Oger. C Latvia 56°49′N 24°36′E
118 H9 **Ogre** ≈ C Latvia
112 C10 **Ogulin** Karlovac, NW Croatia 45°15′N 15°13′E
77 S16 **Ogun** ⬖ state SW Nigeria
Ogurdzhaly, Ostrov see OgurjalyAdasy
146 A12 **OgurjalyAdasy** Rus. Ogurdzhaly, Ostrov. island W Turkmenistan
77 U16 **Ogwashi-Uku** Delta, S Nigeria 06°08′N 06°38′E
185 B23 **Ohai** Southland, South Island, New Zealand 45°56′S 167°59′E
147 Q10 **Ohangaron** Rus. Akhangaran. Toshkent Viloyati, E Uzbekistan 40°56′N 69°37′E
147 Q10 **Ohangaron** Rus. Akhangaran. ≈ E Uzbekistan
83 C16 **Ohangwena** ⬖ district N Namibia
30 M10 **O'Hare** ✈ (Chicago) Illinois, N USA 41°59′N 87°56′W
165 R6 **Ōhata** Aomori, Honshū, C Japan 41°23′N 141°09′E
184 L13 **Ohau** Manawatu-Wanganui, North Island, New Zealand 40°40′S 175°15′E
185 E20 **Ohau, Lake** ◎ South Island, New Zealand
Ohcejohka see Utsjoki
99 J20 **Ohey** Namur, SE Belgium 50°26′N 05°07′E
191 X15 **O'Higgins, Cabo** headland Easter Island, Chile, E Pacific Ocean 27°05′S 109°15′W
O'Higgins, Lago see San Martín, Lago
31 S12 **Ohio** off. State of Ohio, also known as Buckeye State. ⬖ state N USA
0 L10 **Ohio River** ≈ N USA
Ohlau see Oława
101 H16 **Ohm** ≈ C Germany
193 W16 **Ohonua** 'Eua, E Tonga 21°20′S 174°57′W
23 V5 **Ohoopee River** ≈ Georgia, SE USA
100 L12 **Ohre** Ger. Eger. ≈ Czech Republic/Germany
Ohri see Ohrid
113 M20 **Ohrid** Turk. Ochrida, Ohri. SW FYR Macedonia 41°07′N 20°48′E
113 M20 **Ohrid, Lake** var. Lake Ochrida, Alb. Liqeni i Ohrit, Mac. Ohridsko Ezero. ◎ Albania/FYR Macedonia
Ohridsko Ezero/Ohrit, Liqeni i see Ohrid, Lake
184 L9 **Ohura** Manawatu-Wanganui, North Island, New Zealand 38°51′S 174°58′E
58 J9 **Oiapoque** Amapá, E Brazil 03°54′N 51°46′W
58 J10 **Oiapoque, Rio** var. Fleuve l'Oyapok, Oyapock. ≈ Brazil/French Guiana see also Oyapok, Fleuve l'
Oiapoque, Rio see Oyapok, Fleuve l'
54 G8 **Oiba** Santander, C Colombia 06°16′N 73°18′W
15 O9 **Oies, Île aux** island Québec, SE Canada
92 L13 **Oijärvi** Pohjois-Pohjanmaa, C Finland 65°38′N 26°05′E
92 L12 **Oikarainen** Lappi, N Finland 66°30′N 25°46′E
188 F10 **Oikuul** Babeldaob, N Palau
18 C13 **Oil City** Pennsylvania, NE USA 41°25′N 79°42′W
18 C12 **Oil Creek** ≈ Pennsylvania, NE USA
35 R13 **Oildale** California, W USA 35°25′N 119°01′W
Oileán Ciarral see Castleisland
Oil Islands see Chagos Archipelago
115 D18 **Oiniádes** anc. Oeniadae. site of ancient city Dytikí Elláda, W Greece
115 L18 **Oinoússes** island E Greece
Oírr, Inis see Inisheer
99 J15 **Oirschot** Noord-Brabant, S Netherlands 51°30′N 05°18′E
103 N4 **Oise** ⬖ department N France
103 P3 **Oise** ≈ N France
99 J14 **Oisterwijk** Noord-Brabant, S Netherlands 51°35′N 05°12′E
45 O14 **Oistins** S Barbados 13°04′N 59°33′W
165 E4 **Ōita** Ōita, Kyūshū, SW Japan 33°15′N 131°35′E
165 D14 **Ōita** off. Ōita-ken. ⬖ prefecture Kyūshū, SW Japan
Ōita-ken see Ōita
115 E17 **Oíti** ▲ C Greece 38°48′N 22°12′E
165 S4 **Oiwake** Hokkaidō, NE Japan 42°54′N 141°49′E
35 R14 **Ojai** California, W USA 34°25′N 119°15′W
94 K13 **Öje** Dalarna, C Sweden 60°49′N 13°54′E
93 J14 **Öjebyn** Norrbotten, N Sweden 65°20′N 21°26′E
165 B13 **Ojika-jima** island SW Japan
40 K5 **Ojinaga** Chihuahua, N Mexico 29°31′N 104°26′W
40 M11 **Ojo Caliente** var. Ojocaliente. Zacatecas, C Mexico 22°35′N 102°18′W
Ojocaliente see Ojo Caliente
40 D6 **Ojo de Liebre, Laguna** var. Laguna Scammon, Scammon Lagoon. lagoon NW Mexico
62 I7 **Ojos del Salado, Cerro** ▲ W Argentina 27°04′S 68°34′W
105 R7 **Ojos Negros** Aragón, NE Spain 40°43′N 01°30′W
40 M12 **Ojuelos de Jalisco** Aguascalientes, C Mexico 21°52′N 101°40′W
127 N4 **Oka** ≈ W Russian Federation
83 D19 **Okahandja** Otjozondjupa, C Namibia 21°58′S 16°55′E
184 L9 **Okahukura** Manawatu-Wanganui, North Island, New Zealand 38°48′S 175°13′E
184 J3 **Okaihau** Northland, North Island, New Zealand 35°19′S 173°45′E
83 D18 **Okakarara** Otjozondjupa, N Namibia 20°33′S 17°20′E
13 P5 **Okak Islands** island group Newfoundland and Labrador, E Canada
10 M17 **Okanagan** ≈ British Columbia, SW Canada
11 N17 **Okanagan Lake** ◎ British Columbia, SW Canada
Ókanizsa see Kanjiža
83 C16 **Okankolo** Oshikoto, N Namibia 17°57′S 16°28′E
32 K6 **Okanogan River** ≈ Washington, NW USA
83 D18 **Okaputa** Otjozondjupa, N Namibia 20°09′S 16°56′E
149 V9 **Okāra** Punjab, E Pakistan 30°49′N 73°31′E
26 M10 **Okarche** Oklahoma, C USA 35°43′N 97°58′W
Okarem see Ekerem
189 X14 **Okat Harbor** harbor Kosrae, E Micronesia
22 M5 **Okatibbee Creek** ≈ Mississippi, S USA
83 C17 **Okaukuejo** Kunene, N Namibia 19°10′S 15°23′E
Okavanggo see Cubango/Okavango
83 E17 **Okavango** ⬖ district NW Namibia
83 G17 **Okavango** var. Cubango, Kavango, Kavengo, Kubango, Okavanggo, Port. Ocavango. ≈ S Africa see also Cubango
Okavango see Cubango
83 G17 **Okavango Delta** wetland N Botswana
164 M12 **Okaya** Nagano, Honshū, S Japan 36°03′N 138°00′E
164 H13 **Okayama** Okayama, Honshū, SW Japan 34°40′N 133°54′E
164 H13 **Okayama** off. Okayama-ken. ⬖ prefecture Honshū, SW Japan
Okayama-ken see Okayama
164 L14 **Okazaki** Aichi, Honshū, C Japan 34°58′N 137°10′E
110 M12 **Okęcie** ✈ (Warszawa) Mazowieckie, C Poland 52°08′N 20°57′E
23 Y13 **Okeechobee** Florida, SE USA 27°14′N 80°49′W
23 Y14 **Okeechobee, Lake** ◎ Florida, SE USA
26 M9 **Okeene** Oklahoma, C USA 36°07′N 98°19′W
23 V8 **Okefenokee Swamp** wetland Georgia, SE USA
97 J24 **Okehampton** SW England, United Kingdom 50°44′N 04°W
27 P10 **Okemah** Oklahoma, C USA 35°25′N 96°20′W
77 U16 **Okene** Kogi, S Nigeria 07°31′N 06°15′E
100 K13 **Oker** var. Ocker. ≈ NW Germany
101 J14 **Oker-Stausee** ◎ C Germany
123 T12 **Okha** Ostrov Sakhalin, Sakhalinskaya Oblast', SE Russian Federation 53°33′N 142°55′E
125 U15 **Okhansk** var. Ochansk. Permskiy Kray, NW Russian Federation 57°44′N 55°20′E
123 S10 **Okhotsk** Khabarovskiy Kray, E Russian Federation 59°21′N 143°15′E
192 J2 **Okhotsk, Sea of** sea NW Pacific Ocean
117 T4 **Okhtyrka** Rus. Akhtyrka. Sums'ka Oblast', NE Ukraine 50°19′N 34°54′E
83 E23 **Okiep** Northern Cape, W South Africa 29°39′S 17°53′E
Oki-guntō see Oki-shotō
164 H11 **Oki-kaikyō** strait SW Japan
165 P16 **Okinawa** Okinawa, SW Japan 26°20′N 127°47′E
165 S16 **Okinawa** off. Okinawa-ken. ⬖ prefecture Okinawa, SW Japan
165 S16 **Okinawa** island SW Japan
Okinawa-ken see Okinawa
165 U16 **Okinoerabu-jima** island Nansei-shotō, SW Japan
164 F15 **Okino-shima** island SW Japan
164 H11 **Oki-shotō** var. Oki-guntō. island group SW Japan
77 T16 **Okitipupa** Ondo, SW Nigeria 06°33′N 04°43′E
166 L8 **Okkan** Bago, SW Myanmar (Burma) 17°30′N 95°52′E
27 N10 **Oklahoma** off. State of Oklahoma, also known as The Sooner State. ⬖ state C USA
27 N11 **Oklahoma City** state capital Oklahoma, C USA 35°28′N 97°31′W
25 Q4 **Oklaunion** Texas, SW USA 34°07′N 99°07′W
23 W10 **Oklawaha River** ≈ Florida, SE USA
27 P10 **Okmulgee** Oklahoma, C USA 35°38′N 95°59′W
Oknitsa see Ocniţa
22 M3 **Okolona** Mississippi, S USA 34°00′N 88°45′W
165 U2 **Okoppe** Hokkaidō, NE Japan 44°27′N 143°06′E
11 Q16 **Okotoks** Alberta, SW Canada 50°46′N 113°57′W
80 H6 **Oko, Wadi** ≈ NE Sudan
79 G19 **Okoyo** Cuvette, W Congo 01°28′S 15°04′E
77 S15 **Okpara** ≈ Benin/Nigeria
92 J8 **Øksfjord** Finnmark, N Norway 70°13′N 22°22′E
125 R4 **Oksino** Nenetskiy Avtonomnyy Okrug, NW Russian Federation 67°33′N 52°15′E
92 G13 **Oksskolten** ▲ C Norway 66°00′N 14°18′E
Oksu see Oqsu
144 M8 **Oktaybr'skiy** Kostanay, N Kazakhstan
186 B7 **Ok Tedi** Western, W Papua New Guinea
Oktemberyan see Armavir
166 M7 **Oktwin** Bago, C Myanmar (Burma) 18°47′N 96°21′E
127 R6 **Oktyabr'sk** Samarskaya Oblast', W Russian Federation 53°13′N 48°36′E
Oktyab'rsk see Kandyagash
125 N12 **Oktyabr'skiy** Arkhangel'skaya Oblast', NW Russian Federation 61°03′N 43°16′E
122 E10 **Oktyabr'skiy** Kamchatskiy Kray, E Russian Federation 52°35′N 156°18′E
127 T5 **Oktyabr'skiy** Respublika Bashkortostan, W Russian Federation 54°28′N 53°29′E
127 O11 **Oktyabr'skiy** Volgogradskaya Oblast', SW Russian Federation 48°00′N 43°35′E
Oktyabr'skiy see Aktsyabrski
127 V7 **Oktyabr'skoye** Orenburgskaya Oblast', W Russian Federation 52°22′N 55°39′E
122 M5 **Oktyabr'skoy Revolyutsii, Ostrov** Eng. October Revolution Island. island Severnaya Zemlya, N Russian Federation
Ōkuchi see Isa
124 I14 **Okulovka** var. Okulovka. Novgorodskaya Oblast', W Russian Federation 58°24′N 33°16′E
Okulovka see Okulovka
165 Q4 **Okushiri-tō** var. Okusiri Tô. island NE Japan
Okusiri Tô see Okushiri-tō
77 S15 **Okuta** Kwara, W Nigeria 09°18′N 03°09′E
Ôkuti see Isa
83 F19 **Okwa** var. Chapman's. ≈ Botswana/Namibia
123 T10 **Ola** Magadanskaya Oblast', E Russian Federation 59°36′N 151°18′E
27 T11 **Ola** Arkansas, C USA 35°01′N 93°13′W
Ola see Ala
35 T11 **Olacha Peak** ▲ California, W USA 36°15′N 118°07′W
92 J1 **Ólafsfjörður** Norðurland Eystra, N Iceland 66°04′N 18°36′W
92 H3 **Ólafsvík** Vesturland, W Iceland 64°52′N 23°45′W
Oláhbrettye see Bretea-Română
Oláhszentgyőrgy see Sângeorz-Băi
Oláh-Toplicza see Toplița
118 F9 **Olaine** C Latvia 56°47′N 23°56′E
35 T11 **Olancha** California, W USA 36°16′N 118°00′W
42 J5 **Olanchito** Yoro, C Honduras 15°30′N 86°34′W
42 J6 **Olancho** ⬖ department E Honduras
95 O20 **Öland** island S Sweden
95 O19 **Ölands norra udde** headland S Sweden 57°21′N 17°06′E
95 N22 **Ölands södra udde** headland S Sweden 56°12′N 16°26′E
182 K7 **Olary** South Australia 32°18′S 140°16′E
27 R4 **Olathe** Kansas, C USA 38°52′N 94°50′W
61 C22 **Olavarría** Buenos Aires, E Argentina 36°57′S 60°20′W
92 O2 **Olav V Land** physical region C Svalbard
111 H14 **Oława** Ger. Ohlau. Dolnośląskie, SW Poland 50°57′N 17°18′E
107 D17 **Olbia** prev. Terranova Pausania. Sardegna, Italy, C Mediterranean Sea 40°55′N 09°30′E
44 G5 **Old Bahama Channel** channel The Bahamas/Cuba
Old Bay State/Old Colony State see Massachusetts
10 H2 **Old Crow** Yukon, NW Canada 67°34′N 139°55′W
Old Dominion see Virginia
Oldeberkeap see Oldeberkoop
98 M7 **Oldeberkoop** Fris. Oldeberkeap. Fryslân, N Netherlands 52°55′N 06°07′E
98 L10 **Oldebroek** Gelderland, E Netherlands 52°27′N 05°54′E
98 L8 **Oldemarkt** Overijssel, N Netherlands 52°49′N 05°58′E
94 E11 **Olden** Sogn Og Fjordane, C Norway 61°52′N 06°44′E
100 G10 **Oldenburg** Niedersachsen, NW Germany 53°09′N 08°13′E
100 K8 **Oldenburg** var. Oldenburg in Holstein. Schleswig-Holstein, N Germany 54°17′N 10°55′E
Oldenburg in Holstein see Oldenburg
98 P10 **Oldenzaal** Overijssel, E Netherlands 52°19′N 06°53′E
Olderfjord see Leaibevuotna
18 J8 **Old Forge** New York, NE USA 43°42′N 74°59′W
Old Goa see Goa
97 L17 **Oldham** NW England, United Kingdom 53°36′N 02°W
39 Q14 **Old Harbor** Kodiak Island, Alaska, USA 57°12′N 153°18′W

◆ Country
● Country Capital
◇ Dependent Territory
○ Dependent Territory Capital
⬖ Administrative Regions
✈ International Airport
▲ Mountain
▲▲ Mountain Range
Ѫ Volcano
≈ River
◎ Lake
⊠ Reservoir

44 J13 **Old Harbour** C Jamaica 17°56′N 77°06′W
97 C22 **Old Head of Kinsale** Ir. An Seancheann. headland SW Ireland 51°37′N 08°33′W
20 J8 **Old Hickory Lake** ⊠ Tennessee, S USA
Old Line State see Maryland
Old North State see North Carolina
81 I17 **Ol Doinyo Lengeyo** ▲ C Kenya
11 Q16 **Olds** Alberta, SW Canada 51°50′N 114°06′W
19 O7 **Old Speck Mountain** ▲ Maine, NE USA 44°34′N 70°55′W
19 S6 **Old Town** Maine, NE USA 44°55′N 68°39′W
11 T17 **Old Wives Lake** ◎ Saskatchewan, S Canada
162 J7 **Öldziyt** var. Höshööt. Arhangay, C Mongolia 48°06′N 102°34′E
162 J8 **Öldziyt** var. Ulaan-Uul. Bayanhongor, C Mongolia 46°03′N 100°52′E
162 L10 **Öldziyt** var. Rashaant. Dundgovĭ, C Mongolia 44°54′N 106°32′E
162 K8 **Öldziyt** var. Sangiyn Dalay. Övörhangay, C Mongolia 46°35′N 103°18′E
Öldziyt see Erdenemandal, Arhangay, Mongolia
Öldziyt see Sayhandulaan, Dornogovĭ, Mongolia
188 H6 **Oleai** var. San Jose. Saipan, S Northern Mariana Islands
18 E11 **Olean** New York, NE USA 42°04′N 78°24′W
110 O7 **Olecko** Ger. Treuburg. Warmińsko-Mazurskie, NE Poland 54°02′N 22°29′E
106 C7 **Oleggio** Piemonte, NE Italy 45°36′N 08°37′E
123 P11 **Olëkma** Amurskaya Oblast', SE Russian Federation 57°00′N 120°27′E
123 P12 **Olëkma** ≋ C Russian Federation
123 P11 **Olëkminsk** Respublika Sakha (Yakutiya), NE Russian Federation 60°25′N 120°25′E
117 W7 **Oleksandrivka** Donets'ka Oblast', E Ukraine 48°42′N 36°56′E
117 R7 **Oleksandrivka** Rus. Aleksandrovka. Kirovohrads'ka Oblast', C Ukraine 48°59′N 32°14′E
117 Q9 **Oleksandrivka** Mykolayivs'ka Oblast', S Ukraine 47°42′N 31°17′E
117 S7 **Oleksandriya** Rus. Aleksandriya. Kirovohrads'ka Oblast', C Ukraine 48°42′N 33°07′E
93 B20 **Ølen** Hordaland, S Norway 59°36′N 05°48′E
124 J4 **Olenegorsk** Murmanskaya Oblast', NW Russian Federation 68°06′N 33°15′E
123 N9 **Olenëk** Respublika Sakha (Yakutiya), NE Russian Federation 68°28′N 112°18′E
123 N9 **Olenëk** ≋ NE Russian Federation
123 O7 **Olenëkskiy Zaliv** bay N Russian Federation
124 K6 **Olenitsa** Murmanskaya Oblast', NW Russian Federation 66°27′N 35°21′E
102 I11 **Oléron, Île d'** island W France
111 H14 **Oleśnica** Ger. Oels, Oels in Schlesien. Dolnośląskie, SW Poland 51°13′N 17°20′E
111 I15 **Olesno** Ger. Rosenberg. Opolskie, S Poland 50°53′N 18°23′E
116 M3 **Olevs'k** Rus. Olevsk. Zhytomyrs'ka Oblast', N Ukraine 51°12′N 27°38′E
Olevsk see Olevs'k
123 S15 **Ol'ga** Primorskiy Kray, SE Russian Federation 43°41′N 135°06′E
Olga, Mount see Kata Tjuta
92 P2 **Olgastretet** strait E Svalbard
162 D5 **Ölgiy** Bayan-Ölgiy, W Mongolia 48°57′N 89°59′E
95 F23 **Ølgod** Syddtjylland, W Denmark 55°44′N 08°37′E
104 H14 **Olhão** Faro, S Portugal 37°01′N 07°50′W
93 L14 **Olhava** Pohjois-Pohjanmaa, C Finland 65°28′N 25°25′E
105 V5 **Oliana** Cataluña, NE Spain 42°04′N 01°19′E
112 B12 **Olib** It. Ulbo. island W Croatia
83 B16 **Olifa** Kunene, NW Namibia 17°25′S 14°27′E
83 E20 **Olifants** var. Elephant River. ≋ E Namibia
Olifants see Lepelle
83 I20 **Olifants Drift** var. Oliphants Drift. Kgatleng, SE Botswana 24°13′S 26°52′E
83 G22 **Olifantshoek** Northern Cape, N South Africa 27°56′S 22°45′E
188 L15 **Olimarao Atoll** atoll Caroline Islands, C Micronesia
Ólimbos see Ólympos
Olimpo see Fuerte Olimpo
59 Q15 **Olinda** Pernambuco, E Brazil 08°S 34°51′W
Olinthos see Ólynthos
Oliphants Drift see Olifants Drift
Olisipo see Lisboa
Olita see Alytus
105 Q4 **Olite** Navarra, N Spain 42°29′N 01°40′W
62 K10 **Oliva** Córdoba, C Argentina 32°03′S 63°34′W
105 T11 **Oliva** Valenciana, E Spain 38°55′N 00°09′W
Oliva see La Oliva
104 I12 **Oliva de la Frontera** Extremadura, W Spain 38°17′N 06°54′W
Olivares see Olivares de Júcar
62 H9 **Olivares, Cerro de** ▲ N Chile 30°25′S 69°52′W
105 P9 **Olivares de Júcar** var. Olivares. Castilla-La Mancha, C Spain 39°45′N 02°21′W
22 L1 **Olive Branch** Mississippi, S USA 34°58′N 89°49′W
21 O5 **Olive Hill** Kentucky, S USA 38°18′N 83°10′W
35 O6 **Olivehurst** California, W USA 39°05′N 121°33′W
104 G7 **Oliveira de Azeméis** Aveiro, N Portugal 40°49′N 08°29′W
104 I11 **Olivenza** Extremadura, W Spain 38°41′N 07°06′W
11 N17 **Oliver** British Columbia, SW Canada 49°10′N 119°37′W
103 N7 **Olivet** Loiret, C France 47°53′N 01°53′E
29 Q12 **Olivet** South Dakota, N USA 43°14′N 97°40′W
29 T9 **Olivia** Minnesota, N USA 44°46′N 94°59′W
185 C20 **Olivine Range** ▲ South Island, New Zealand
108 H10 **Olivone** Ticino, S Switzerland 46°32′N 08°55′E
Ölkeyek see Ol'keyyek
144 L11 **Ol'keyyek** Kaz. Ölkeyek; prev. Ul'kayak. ≋ C Kazakhstan
127 O9 **Ol'khovka** Volgogradskaya Oblast', SW Russian Federation 49°54′N 44°36′E
111 K16 **Olkusz** Małopolskie, S Poland 50°18′N 19°33′E
22 I6 **Olla** Louisiana, S USA 31°54′N 92°14′W
62 I4 **Ollagüe, Volcán** var. Oyahue, Volcán Oyahue. ⛰ N Chile 21°25′S 68°10′W
189 U13 **Ollan** island Chuuk, C Micronesia
188 F7 **Ollei** Babeldaob, N Palau 07°43′N 134°37′E
Ollius see Oglio
108 C10 **Ollon** Vaud, W Switzerland 46°19′N 07°00′E
147 Q10 **Olmaliq** Rus. Almalyk. Toshkent Viloyati, E Uzbekistan 40°51′N 69°39′E
104 M6 **Olmedo** Castilla y León, N Spain 41°17′N 04°41′W
56 B10 **Olmos** Lambayeque, W Peru 06°00′S 79°43′W
Olmütz see Olomouc
30 M15 **Olney** Illinois, N USA 38°43′N 88°05′W
25 R5 **Olney** Texas, SW USA 33°22′N 98°45′W
95 L22 **Olofström** Blekinge, S Sweden 56°16′N 14°33′E
187 N9 **Olomburi** Malaita, N Solomon Islands 09°00′S 161°09′E
111 H17 **Olomouc** Ger. Olmütz, Pol. Ołomuniec. Olomoucký Kraj, E Czech Republic 49°36′N 17°13′E
111 H18 **Olomoucký Kraj** ◈ region E Czech Republic
Olomuniec see Olomouc
122 D7 **Olonets** Respublika Kareliya, NW Russian Federation 60°58′N 33°01′E
171 N3 **Olongapo** off. Olongapo City. Luzon, N Philippines 14°52′N 120°16′E
Olongapo City see Olongapo
102 J16 **Oloron-Ste-Marie** Pyrénées-Atlantiques, SW France 43°12′N 00°35′W
192 L16 **Olosega** island Manua Islands, E American Samoa
105 W4 **Olot** Cataluña, NE Spain 42°11′N 02°30′E
146 K12 **Olot** Rus. Alat. Buxoro Viloyati, C Uzbekistan 39°22′N 63°42′E
112 I12 **Olovo** Federacija Bosne I Hercegovine, E Bosnia and Herzegovina 44°08′N 18°35′E
123 O14 **Olovyannaya** Zabaykal'skiy Kray, S Russian Federation 50°59′N 115°24′E
123 T7 **Oloy** ≋ NE Russian Federation
101 F16 **Olpe** Nordrhein-Westfalen, W Germany 51°02′N 07°51′E
109 N8 **Olperer** ▲ SW Austria 47°03′N 11°36′E
Olshanka see Vil'shanka
Ol'shany see Al'shany
Olsnitz see Murska Sobota
98 M10 **Olst** Overijssel, E Netherlands 52°19′N 06°06′E
110 L8 **Olsztyn** Ger. Allenstein. Warmińsko-Mazurskie, N Poland 53°46′N 20°28′E
110 L8 **Olsztynek** Ger. Hohenstein in Ostpreussen. Warmińsko-Mazurskie, N Poland 53°35′N 20°17′E
116 I14 **Olt** ◈ county SW Romania
116 I14 **Olt** var. Oltul, Ger. Alt. ≋ S Romania
108 E7 **Olten** Solothurn, NW Switzerland 47°22′N 07°55′E
116 K14 **Oltenița** prev. Eng. Oltenitsa; anc. Constantiola. Călărași, SE Romania 44°05′N 26°40′E
Oltenitsa see Oltenița
116 H14 **Olteț** ≋ S Romania
24 M4 **Olton** Texas, SW USA 34°10′N 102°07′W
137 R12 **Oltu** Erzurum, NE Turkey 40°34′N 41°59′E
Oltul see Olt
146 G7 **Oltynko'l** Qoraqalpog'iston Respublikasi, NW Uzbekistan 43°04′N 58°51′E
Oluan Pi see Eluan Bi
Olubló see Stará L'ubovňa
137 R11 **Olur** Erzurum, NE Turkey 40°49′N 42°08′E
104 L15 **Olvera** Andalucía, S Spain 36°56′N 05°15′W
Ol'viopol' see Pervomays'k
Olwanpi, Cape see Eluan Bi
115 D20 **Olympía** Dytikí Elláda, S Greece 37°39′N 21°36′E
32 G9 **Olympia** state capital Washington, NW USA 47°02′N 122°54′W
182 H5 **Olympic Dam** South Australia 30°25′S 136°56′E
32 F7 **Olympic Mountains** ▲ Washington, NW USA
121 O3 **Ólympos** var. Troodos, Eng. Mount Olympus. ▲ C Cyprus 34°55′N 32°49′E
115 F15 **Ólympos** var. Ólimbos, Eng. Mount Olympus. ▲ N Greece 40°04′N 22°24′E
115 L17 **Ólympos** ▲ Lésvos, E Greece 39°03′N 26°20′E
16 C5 **Olympus, Mount** ▲ Washington, NW USA 47°48′N 123°42′W
Olympus, Mount see Ólympos
115 G14 **Ólynthos** var. Olinthos; anc. Olynthus. site of ancient city Kentrikí Makedonía, N Greece
Olynthus see Ólynthos
117 Q3 **Olyshivka** Chernihivs'ka Oblast', N Ukraine 51°13′N 31°19′E
123 W8 **Olyutorskiy, Mys** headland E Russian Federation 59°56′N 170°22′E
123 V8 **Olyutorskiy Zaliv** bay E Russian Federation
186 M10 **Om** ≋ W Papua New Guinea
129 S6 **Om'** ≋ N Russian Federation
158 I13 **Oma** Xizang Zizhiqu, W China 32°30′N 83°14′E
165 R6 **Ōma** Aomori, Honshū, C Japan 41°31′N 140°54′E
125 P6 **Oma** ≋ NW Russian Federation
164 M12 **Ōmachi** var. Ōmati. Nagano, Honshū, S Japan 36°30′N 137°51′E
165 Q8 **Ōmagari** Akita, Honshū, C Japan 39°29′N 140°29′E
97 E15 **Omagh** Ir. An Ómaigh. W Northern Ireland, United Kingdom 54°36′N 07°18′W
29 S15 **Omaha** Nebraska, C USA 41°14′N 95°57′W
83 E19 **Omaheke** ◈ district W Namibia
141 W10 **Oman** off. Sultanate of Oman, Ar. Salṭanat 'Umān; prev. Muscat and Oman. ◆ monarchy SW Asia
129 O10 **Oman Basin** var. Bassin d'Oman. undersea feature N Indian Ocean 23°20′N 63°00′E
Oman, Bassin d' see Oman Basin
129 N10 **Oman, Gulf of** Ar. Khalīj 'Umān. gulf N Arabian Sea
Oman, Sultanate of see Oman
184 J3 **Omapere** Northland, North Island, New Zealand 35°32′S 173°24′E
185 E20 **Omarama** Canterbury, South Island, New Zealand 44°29′S 169°57′E
112 F11 **Omarska** ◈ Republika Srpska, NW Bosnia and Herzegovina
83 C18 **Omaruru** Erongo, NW Namibia 21°28′S 15°56′E
83 C19 **Omaruru** ≋ W Namibia
83 E17 **Omatako** ≋ NE Namibia
Ōmati see Ōmachi
83 E18 **Omawewozonyanda** Omaheke, E Namibia 21°30′S 19°34′E
165 R6 **Ōma-zaki** headland Honshū, C Japan 41°32′N 140°53′E
Omba see Ambae
Ombai see Alor, Pulau
83 C16 **Ombalantu** Omusati, N Namibia 17°33′S 14°58′E
79 H15 **Ombella-Mpoko** ◈ prefecture S Central African Republic
Ombetsu see Onbetsu
83 B17 **Ombombo** Kunene, NW Namibia 18°43′S 13°53′E
79 D19 **Omboué** Ogooué-Maritime, W Gabon 01°38′S 09°20′E
106 G13 **Ombrone** ≋ C Italy
80 F9 **Omdurman** var. Umm Durmān. Khartoum, C Sudan 15°37′N 32°29′E
165 N13 **Ōme** Tōkyō, Honshū, S Japan 35°48′N 139°17′E
106 C6 **Omegna** Piemonte, NE Italy 45°54′N 08°25′E
183 P12 **Omeo** Victoria, SE Australia 37°09′S 147°36′E
138 F11 **Omer** Southern, C Israel 31°16′N 34°51′E
41 P16 **Ometepec** Guerrero, S Mexico 16°39′N 98°23′W
42 K11 **Ometepe, Isla de** island S Nicaragua
Om Hager see Om Hajer
80 I10 **Om Ḥajer** var. Om Hager. SW Eritrea 14°19′N 36°46′E
165 J13 **Ōmihachiman** Shiga, Honshū, SW Japan 35°08′N 136°04′E
10 L12 **Omineca Mountains** ▲ British Columbia, W Canada
113 F14 **Omiš** It. Almissa. Split-Dalmacija, S Croatia 43°27′N 16°41′E
112 B10 **Omišalj** Primorje-Gorski Kotar, NW Croatia 45°10′N 14°33′E
83 D19 **Omitara** Khomas, C Namibia 22°18′S 18°01′E
41 O16 **Omitlán, Río** ≋ S Mexico
39 X14 **Ommaney, Cape** headland Baranof Island, Alaska, USA 56°10′N 134°40′W
98 N9 **Ommen** Overijssel, E Netherlands 52°31′N 06°25′E
163 N7 **Ömnödelger** var. Bayanbulag. Hentiy, C Mongolia 47°54′N 109°51′E
162 K11 **Ömnögovĭ** ◈ province S Mongolia
191 X7 **Omoa** Fatu Hira, NE French Polynesia 10°30′S 138°41′E
Omo Botego see Omo Wenz
Ömoldova see Moldova Veche
123 T7 **Omolon** Chukotskiy Avtonomnyy Okrug, NE Russian Federation 65°11′N 160°33′E
123 T7 **Omolon** ≋ NE Russian Federation
123 Q8 **Omoloy** ≋ NE Russian Federation
162 I12 **Omon Gol** Chin. Dong He, var. Narin Gol. ≋ N China
165 P8 **Omono-gawa** ≋ Honshū, C Japan
81 I14 **Omo Wenz** var. Omo Botego. ≋ Ethiopia/Kenya
122 H12 **Omsk** Omskaya Oblast', C Russian Federation 55°N 73°22′E
122 H11 **Omskaya Oblast'** ◈ province C Russian Federation
165 U2 **Ōmu** Hokkaidō, NE Japan 44°36′N 142°55′E
110 M9 **Omulew** ≋ NE Poland
116 J12 **Omul, Vârful** prev. Virful Omu. ▲ C Romania 45°24′N 25°26′E
83 D16 **Omundaungilo** Ohangwena, N Namibia 17°28′S 16°39′E
164 C14 **Ōmura** Nagasaki, Kyūshū, SW Japan 32°56′N 129°58′E
83 B17 **Omusati** ◈ district N Namibia
164 C14 **Ōmuta** Fukuoka, Kyūshū, SW Japan 33°01′N 130°27′E
125 S14 **Omutninsk** Kirovskaya Oblast', NW Russian Federation 58°37′N 52°08′E
Omu, Vîrful see Omul, Vârful
29 V7 **Onamia** Minnesota, N USA 46°04′N 93°40′W
21 Y5 **Onancock** Virginia, NE USA 37°42′N 75°45′W
14 E10 **Onaping Lake** ◎ Ontario, S Canada
30 M12 **Onarga** Illinois, N USA 40°39′N 88°00′W
15 R6 **Onatchiway, Lac** ◎ Québec, SE Canada
29 S14 **Onawa** Iowa, C USA 42°01′N 96°06′W
165 U5 **Onbetsu** var. Ombetsu. Hokkaidō, NE Japan 42°52′N 143°54′E
83 B16 **Oncócua** Cunene, SW Angola 16°37′S 13°23′E
105 S9 **Onda** Valenciana, E Spain 39°58′N 00°17′W
111 N18 **Ondava** ≋ NE Slovakia
83 C16 **Ondjiva** Cunene, S Angola 17°02′S 15°42′E
77 T16 **Ondo** Ondo, SW Nigeria 07°07′N 04°50′E
77 T16 **Ondo** ◈ state SW Nigeria
163 N8 **Öndörhaan** var. Undur Khan; prev. Tsetsen Khan. Hentiy, E Mongolia 47°21′N 110°42′E
162 M9 **Öndörshil** var. Böhöt. Dundgovĭ, C Mongolia 45°13′N 108°12′E
162 L8 **Öndörshireet** var. Bayshint. Töv, C Mongolia 47°22′N 105°04′E
162 I7 **Öndör-Ulaan** var. Teel. Arhangay, C Mongolia 48°01′N 100°30′E
83 D18 **Ondundazongonda** Otjozondjupa, N Namibia 20°28′S 18°00′E
151 K21 **One and Half Degree Channel** channel S Maldives
187 Z15 **Oneata** island Lau Group, E Fiji
124 L9 **Onega** Arkhangel'skaya Oblast', NW Russian Federation 63°54′N 37°59′E
124 L9 **Onega** ≋ NW Russian Federation
Onega Bay see Onezhskaya Guba
Onega, Lake see Onezhskoye Ozero
18 I10 **Oneida** New York, NE USA 43°05′N 75°39′W
20 M8 **Oneida** Tennessee, S USA 36°30′N 84°30′W
18 H9 **Oneida Lake** ◎ New York, NE USA
29 P13 **O'Neill** Nebraska, C USA 42°28′N 98°38′W
123 V12 **Onekotan, Ostrov** island Kuril'skiye Ostrova, SE Russian Federation
23 P3 **Oneonta** Alabama, S USA 33°57′N 86°28′W
18 J11 **Oneonta** New York, NE USA 42°27′N 75°03′W
190 I16 **Oneroa** island S Cook Islands
116 K11 **Onești** Hung. Onyest; prev. Gheorghe Gheorghiu-Dej. Bacău, E Romania 46°14′N 26°46′E
193 V15 **Onevai** island Tongatapu Group, S Tonga
108 A11 **Onex** Genève, SW Switzerland 46°12′N 06°04′E
124 K8 **Onezhskaya Guba** Eng. Onega Bay. bay NW Russian Federation
122 D7 **Onezhskoye Ozero** Eng. Lake Onega. ◎ NW Russian Federation
83 C16 **Ongandjera** Omusati, N Namibia 17°49′S 15°06′E
184 N12 **Ongaonga** Hawke's Bay, North Island, New Zealand 39°57′S 176°21′E
Ongi see Sayhan-Ovoo, Dundgovĭ, Mongolia
Ongi see Uyanga
163 W14 **Ongjin** SW North Korea 37°56′N 125°22′E
155 J17 **Ongole** Andhra Pradesh, E India 15°33′N 80°03′E
Ongon see Bürd
Ongtüstik Qazaqstan Oblysy see Yuzhnyy Kazakhstan
99 I21 **Onhaye** Namur, S Belgium 50°15′N 04°51′E
166 M8 **Onhne** Bago, SW Myanmar (Burma) 17°02′N 96°28′E
137 S9 **Oni** N Georgia 42°36′N 43°13′E
29 N9 **Onida** South Dakota, N USA 44°42′N 100°03′W
164 F15 **Onigajō-yama** ▲ Shikoku, SW Japan 33°10′N 132°37′E
172 H7 **Onilahy** ≋ S Madagascar
77 U16 **Onitsha** Anambra, S Nigeria 06°09′N 06°48′E
164 K12 **Ōno** Fukui, Honshū, SW Japan 35°59′N 136°30′E
164 I13 **Ono** Hyōgo, Honshū, SW Japan 34°52′N 134°55′E
187 K15 **Ono** island SW Fiji
164 E13 **Onoda** Yamaguchi, Honshū, SW Japan 34°00′N 131°11′E
187 Z16 **Ono-i-lau** island SE Fiji
164 D13 **Ōnojō** var. Ōnoyō. Fukuoka, Kyūshū, SW Japan 33°34′N 130°29′E
163 O7 **Onon Gol** ≋ N Mongolia
Ononte see Orontes
55 N6 **Onoto** Anzoátegui, NE Venezuela 09°36′N 65°12′W
191 O3 **Onotoa** prev. Clerk Island. atoll Tungaru, W Kiribati
Ōnoyō see Ōnojō
83 E23 **Onseepkans** Northern Cape, W South Africa 28°44′S 19°18′E
104 F4 **Ons, Illa de** island NW Spain
180 H7 **Onslow** Western Australia 21°42′S 115°08′E
21 W11 **Onslow Bay** bay North Carolina, E USA
98 P6 **Onstwedde** Groningen, NE Netherlands 53°01′N 07°04′E
164 C16 **On-take** ▲ Kyūshū, SW Japan 31°35′N 130°39′E
35 T15 **Ontario** California, W USA 34°03′N 117°39′W
32 M13 **Ontario** Oregon, NW USA 44°01′N 116°57′W
12 D10 **Ontario** ◈ province S Canada
11 P14 **Ontario, Lake** ◎ Canada/USA
0 L9 **Ontario Peninsula** peninsula Canada/USA
Onteniente see Ontinyent
105 S11 **Ontinyent** var. Onteniente. Valenciana, E Spain 38°49′N 00°37′W
93 N15 **Ontojärvi** ◎ E Finland
30 L3 **Ontonagon** Michigan, N USA 46°52′N 89°18′W
30 L3 **Ontonagon River** ≋ Michigan, N USA
186 M7 **Ontong Java Atoll** prev. Lord Howe Island. atoll N Solomon Islands
175 N5 **Ontong Java Rise** undersea feature W Pacific Ocean 01°00′N 157°00′E
Onuba see Huelva
55 W9 **Onverwacht** Para, N Suriname 05°36′N 55°12′W
Onyest see Onești
Oodeypore see Udaipur
182 J7 **Oodla Wirra** South Australia 32°52′S 139°05′E
182 F2 **Oodnadatta** South Australia 27°34′S 135°27′E
182 C5 **Ooldea** South Australia 30°29′S 131°50′E
27 Q8 **Oologah Lake** ⊠ Oklahoma, C USA
Oos-Kaap see Eastern Cape
Oos-Londen see East London
99 E17 **Oostakker** Oost-Vlaanderen, NW Belgium 51°06′N 03°46′E
99 D15 **Oostburg** Zeeland, SW Netherlands 51°20′N 03°30′E
98 K9 **Oostelijk-Flevoland** polder C Netherlands
99 B16 **Oostende** Eng. Ostend, Fr. Ostende. West-Vlaanderen, NW Belgium 51°13′N 02°55′E
99 B16 **Oostende** ✕ West-Vlaanderen, NW Belgium 51°12′N 02°55′E
98 L12 **Oosterbeek** Gelderland, SE Netherlands 51°59′N 05°51′E
99 I14 **Oosterhout** Noord-Brabant, S Netherlands 51°38′N 04°51′E
98 O6 **Oostermoers Vaart** var. Hunze. ≋ NE Netherlands
99 F14 **Oosterschelde** Eng. Eastern Scheldt. inlet SW Netherlands
99 E14 **Oosterscheldedam** dam SW Netherlands
98 M7 **Oosterwolde** Fris. Easterwâlde. Fryslân, N Netherlands 53°01′N 06°15′E
98 I9 **Oosthuizen** Noord-Holland, NW Netherlands 52°34′N 05°00′E
99 H16 **Oostmalle** Antwerpen, N Belgium 51°18′N 04°44′E
Oos-Transvaal see Mpumalanga
99 E15 **Oost-Souburg** Zeeland, SW Netherlands 51°28′N 03°36′E
99 E17 **Oost-Vlaanderen** Eng. East Flanders. ◈ province NW Belgium
98 J5 **Oost-Vlieland** Fryslân, N Netherlands 53°19′N 05°02′E
98 F12 **Oostvoorne** Zuid-Holland, SW Netherlands 51°55′N 04°06′E
Ootacamund see Udagamandalam
98 O10 **Ootmarsum** Overijssel, E Netherlands 52°25′N 06°55′E
10 K14 **Ootsa Lake** ⊠ British Columbia, SW Canada
Ooty see Udagamandalam
114 L8 **Opaka** Targovishte, N Bulgaria 43°26′N 26°12′E
79 M18 **Opala** Orientale, C Dem. Rep. Congo 0°40′S 24°20′E
125 Q13 **Oparino** Kirovskaya Oblast', NW Russian Federation 59°52′N 48°14′E
14 H8 **Opasatica, Lac** ◎ Québec, SE Canada
112 B9 **Opatija** It. Abbazia. Primorje-Gorski Kotar, NW Croatia 45°18′N 14°15′E
111 N15 **Opatów** Świętokrzyskie, C Poland 50°45′N 21°27′E
111 I17 **Opava** Ger. Troppau. Moravskoslezský Kraj, E Czech Republic 49°56′N 17°53′E
111 H16 **Opava** Ger. Oppa. ≋ NE Czech Republic
Ópazova see Stara Pazova
Ópécska see Pecica
14 E8 **Opeepeesway Lake** ◎ Ontario, S Canada
23 R5 **Opelika** Alabama, S USA 32°39′N 85°22′W
22 I8 **Opelousas** Louisiana, S USA 30°31′N 92°04′W
186 G6 **Open Bay** bay New Britain, E Papua New Guinea
14 I12 **Opeongo Lake** ◎ Ontario, SE Canada
99 K17 **Opglabbeek** Limburg, NE Belgium 51°04′N 05°39′E
33 W6 **Opheim** Montana, NW USA 48°50′N 106°24′W
39 P10 **Ophir** Alaska, USA 63°08′N 156°31′W
Ophiusa see Formentera
79 N18 **Opienge** Orientale, E Dem. Rep. Congo 0°15′N 27°25′E
185 G20 **Opihi** ≋ South Island, New Zealand
12 J9 **Opinaca** ≋ Québec, C Canada
12 J10 **Opinaca, Réservoir** ⊠ Québec, E Canada
117 T5 **Opishnya** Rus. Oposhnya. Poltavs'ka Oblast', NE Ukraine 49°58′N 34°37′E
98 I8 **Opmeer** Noord-Holland, NW Netherlands 52°43′N 04°56′E
77 V17 **Opobo** Akwa Ibom, S Nigeria 04°36′N 07°37′E
124 F16 **Opochka** Pskovskaya Oblast', W Russian Federation 56°42′N 28°40′E
110 L13 **Opoczno** Łodzkie, C Poland 51°24′N 20°18′E
111 I15 **Opole** Ger. Oppeln. Opolskie, S Poland 50°40′N 17°56′E
111 H15 **Opolskie** ◈ province S Poland
Opornyy see Borankul
104 G4 **O Porriño** var. Porriño. Galicia, NW Spain 42°10′N 08°38′W
Oporto see Porto
Oposhnya see Opishnya
184 P8 **Opotiki** Bay of Plenty, North Island, New Zealand 38°02′S 177°18′E
23 Q7 **Opp** Alabama, S USA 31°16′N 86°14′W
Oppa see Opava
94 G9 **Oppdal** Sør-Trøndelag, S Norway 62°36′N 09°41′E
Oppeln see Opole
107 N23 **Oppido Mamertina** Calabria, SW Italy 38°17′N 15°58′E
Oppidum Ubiorum see Köln
94 F12 **Oppland** ◈ county S Norway
118 J12 **Opsa** Vitsyebskaya Voblasts', NW Belarus 55°32′N 26°50′E
26 I8 **Optima Lake** ⊠ Oklahoma, C USA
184 J11 **Opunake** Taranaki, North Island, New Zealand 39°27′S 173°52′E
191 N6 **Opunohu, Baie d'** bay Moorea, W French Polynesia
83 B17 **Opuwo** Kunene, NW Namibia 18°03′S 13°54′E
146 H6 **Oqqal'a** var. Akkala, Rus. Karakala. Qoraqalpog'iston Respublikasi, NW Uzbekistan 43°43′N 59°25′E
147 V13 **Oqsu** Rus. Oksu. ≋ SE Tajikistan
147 P14 **Oqtogh, Qatorkŭhi** Rus. Khrebet Aktau. ▲ SW Tajikistan
146 M11 **Oqtosh** Rus. Aktash. Samarqand Viloyati, C Uzbekistan 39°22′N 65°46′E
147 N11 **Oqtov Tizmasi** var. Khrebet Aktau. ▲ C Uzbekistan
30 J12 **Oquawka** Illinois, N USA 40°56′N 90°56′W
144 J10 **Or'** Kaz. Or. ≋ Kazakhstan/Russian Federation
36 M15 **Oracle** Arizona, SW USA 32°36′N 110°46′W
147 N13 **O'radaryo** Rus. Uradar'ya. ≋ S Uzbekistan
116 F9 **Oradea** prev. Oradea Mare, Ger. Grosswardein, Hung. Nagyvárad. Bihor, NW Romania 47°03′N 21°56′E
Oradea Mare see Oradea
Orahovac see Rahovec
112 H9 **Orahovica** Virovitica-Podravina, NE Croatia 45°33′N 17°54′E
152 K13 **Orai** Uttar Pradesh, N India 26°00′N 79°26′E
92 K12 **Orajärvi** Lappi, NW Finland 66°54′N 24°04′E
138 F9 **Or'Akiva** prev. Or 'Aqiva. Haifa, N Israel 32°40′N 34°58′E
Oral see Ural'sk
74 I5 **Oran** var. Ouahran, Wahran. NW Algeria 35°42′N 00°37′W
183 R8 **Orange** New South Wales, SE Australia 33°16′S 149°06′E
103 R14 **Orange** anc. Arausio. Vaucluse, SE France 44°06′N 04°52′E
25 Y10 **Orange** Texas, SW USA 30°05′N 93°43′W
21 V5 **Orange** Virginia, NE USA 38°14′N 78°07′W
21 R13 **Orangeburg** South Carolina, SE USA 33°28′N 80°53′W
58 J9 **Orange, Cabo** headland NE Brazil 04°24′N 51°33′W
29 S12 **Orange City** Iowa, C USA 43°00′N 96°03′W
Orange Cone see Orange Fan
172 J10 **Orange Fan** var. Orange Cone. undersea feature SW Indian Ocean 32°00′S 12°00′E
Orange Free State see Free State
25 S14 **Orange Grove** Texas, SW USA 27°57′N 97°56′W
18 K13 **Orange Lake** New York, NE USA 41°32′N 74°06′W
23 V10 **Orange Lake** ◎ Florida, SE USA
Orange Mouth/Orangemund see Oranjemund
23 W9 **Orange Park** Florida, SE USA 30°10′N 81°42′W
83 E23 **Orange River** Afr. Oranjerivier. ≋ S Africa
14 G15 **Orangeville** Ontario, S Canada 43°55′N 80°06′W
36 M5 **Orangeville** Utah, W USA 39°14′N 111°03′W
42 G1 **Orange Walk** Orange Walk, N Belize 18°06′N 88°30′W
42 F1 **Orange Walk** ◈ district NW Belize
100 N11 **Oranienburg** Brandenburg, NE Germany 52°46′N 13°15′E
98 O7 **Oranjekanaal** canal NE Netherlands
83 D23 **Oranjemund** var. Orangemund; prev. Orange Mouth. Karas, SW Namibia 28°33′S 16°28′E
Oranjerivier see Orange River
45 N16 **Oranjestad** ○ (Aruba) W Aruba 12°31′N 70°W
Oranje Vrystaat see Free State
Orany see Varėna
83 H18 **Orapa** Central, C Botswana 21°16′S 25°22′E
Or 'Aqiva see Or'Akiva
112 I10 **Orašje** ◈ Federacija Bosna I Herecegovina, N Bosnia and Herzegovina
116 G11 **Orăștie** Ger. Broos, Hung. Szászváros. Hunedoara, W Romania 45°50′N 23°11′E
Orașul Stalin see Brașov
111 K18 **Orava** Hung. Árva, Pol. Orawa. ≋ N Slovakia
Oravainen see Oravais
93 K16 **Oravais** Fin. Oravainen. Österbotten, W Finland 63°18′N 22°25′E
Oravicabánya see Oravița
116 F13 **Oravița** Ger. Orawitza, Hung. Oravicabánya. Caraș-Severin, SW Romania 45°02′N 21°43′E
Orawa see Orava
185 B24 **Orawia** Southland, South Island, New Zealand 46°01′S 167°49′E
Orawitza see Oravița
103 P16 **Orb** ≋ S France
106 C9 **Orba** ≋ NW Italy
158 H12 **Orba Co** ◎ W China
108 B9 **Orbe** Vaud, W Switzerland 46°42′N 06°28′E
107 G14 **Orbetello** Toscana, C Italy 42°28′N 11°15′E
104 K3 **Orbigo** ≋ NW Spain
183 Q12 **Orbost** Victoria, SE Australia 37°44′S 148°28′E
95 O14 **Örbyhus** Uppsala, C Sweden 60°15′N 17°43′E
194 I1 **Orcadas** Argentinian research station South Orkney Islands, Antarctica 60°37′S 44°48′W
105 P12 **Orcera** Andalucía, S Spain 38°20′N 02°36′W
33 P9 **Orchard Homes** Montana, NW USA 46°52′N 114°01′W
37 P5 **Orchard Mesa** Colorado, C USA 39°02′N 108°33′W
18 D10 **Orchard Park** New York, NE USA 42°46′N 78°44′W
Orchid Island see Lan Yu
115 G18 **Orchómenos** var. Orhomenos, Orkhómenos, prev. Skripón; anc. Orchomenus. Stereá Elláda, C Greece 38°29′N 22°58′E
Orchomenus see Orchómenos
106 B7 **Orco** ≋ NW Italy
103 R8 **Or, Côte d'** physical region C France
29 O14 **Ord** Nebraska, C USA 41°36′N 98°55′W
119 O15 **Ordats'** Rus. Ordat'. Mahilyowskaya Voblasts', E Belarus 54°10′N 30°42′E
36 K8 **Orderville** Utah, W USA 37°16′N 112°38′W
104 H2 **Ordes** Galicia, NW Spain 43°04′N 08°25′W
35 V14 **Ord Mountain** ▲ California, W USA 34°41′N 116°46′W
163 N14 **Ordos** prev. Dongsheng. Nei Mongol Zizhiqu, N China 39°51′N 110°00′E
Ordos Desert see Mu Us Shadi
188 B16 **Ordot** C Guam
137 N11 **Ordu** anc. Cotyora. Ordu, N Turkey 41°N 37°52′E
136 M11 **Ordu** ◈ province N Turkey
137 V14 **Ordubad** SW Azerbaijan 38°55′N 46°00′E
Orduña see Urduña
37 U6 **Ordway** Colorado, C USA 38°13′N 103°45′W
117 T9 **Ordzhonikidze** Dnipropetrovs'ka Oblast', E Ukraine 47°39′N 34°08′E
Ordzhonikidze see Denisovka, Kazakhstan
Ordzhonikidze see Vladikavkaz, Russian Federation
Ordzhonikidze see Yenakiyeve, Ukraine
Ordzhonikidzeabad see Kofarnihon
55 U9 **Orealla** E Guyana 05°13′N 57°17′W
113 G15 **Orebić** It. Sabbioncello. Dubrovnik-Neretva, S Croatia 42°58′N 17°12′E
95 M16 **Örebro** Örebro, C Sweden 59°18′N 15°12′E
95 L16 **Örebro** ◈ county C Sweden
25 W6 **Ore City** Texas, SW USA 32°48′N 94°43′W
30 L10 **Oregon** Illinois, N USA 42°00′N 89°19′W
27 Q2 **Oregon** Missouri, C USA 39°59′N 95°08′W
31 R11 **Oregon** Ohio, N USA 41°38′N 83°29′W
32 H13 **Oregon** off. State of Oregon, also known as Beaver State, Sunset State, Valentine State, Webfoot State. ◈ state NW USA
32 G11 **Oregon City** Oregon, NW USA 45°21′N 122°36′W
Oregon, State of see Oregon
95 P14 **Öregrund** Uppsala, C Sweden 60°19′N 18°30′E
Orekhov see Orikhiv
126 L3 **Orekhovo-Zuyevo** Moskovskaya Oblast', W Russian Federation 55°46′N 39°01′E
Orekhovsk see Arekhawsk
126 J6 **Orël** Orlovskaya Oblast', W Russian Federation 52°57′N 36°06′E
Orel see Oril'
56 E11 **Orellana** Loreto, N Peru 06°53′S 75°10′W
56 E6 **Orellana** ◈ province NE Ecuador
104 L11 **Orellana, Embalse de** ⊠ W Spain
36 L3 **Orem** Utah, W USA 40°18′N 111°42′W
Ore Mountains see Erzgebirge/Krušné Hory
127 V7 **Orenburg** prev. Chkalov. Orenburgskaya Oblast', W Russian Federation 51°46′N 55°12′E
127 V7 **Orenburg** ✕ Orenburgskaya Oblast', W Russian Federation 51°54′N 55°15′E
127 T7 **Orenburgskaya Oblast'** ◈ province W Russian Federation
Orense see Ourense
188 C8 **Oreor** var. Koror. island N Palau
185 B24 **Orepuki** Southland, South Island, New Zealand 46°17′S 167°45′E
114 L12 **Orestiáda** prev. Orestiás. Anatolikí Makedonía kai Thráki, NE Greece 41°30′N 26°31′E
Orestiás see Orestiáda
Øresund/Öresund see Sound, The
185 C23 **Oreti** ≋ South Island, New Zealand
184 L5 **Orewa** Auckland, North Island, New Zealand 36°34′S 174°43′E
65 A25 **Orford, Cape** headland West Falkland, Falkland Islands 52°00′S 61°04′W
44 B5 **Órganos, Sierra de los** ▲ W Cuba
37 R15 **Organ Peak** ▲ New Mexico, SW USA 32°17′N 106°35′W
105 N9 **Orgaz** Castilla-La Mancha, C Spain 39°39′N 03°52′W
Orgeyev see Orhei
Orgil see Jargalant
105 O15 **Orgiva** var. Órjiva. Andalucía, S Spain 36°54′N 03°25′W
163 O10 **Örgön** var. Senj. Dornogovĭ, SE Mongolia 44°34′N 110°58′E
Örgön see Bayangovĭ
Orgražden see Ograzhden
117 N9 **Orhei** var. Orheiu, Rus. Orgeyev. N Moldova 47°25′N 28°48′E
Orheiu see Orhei
105 R3 **Orhi** var. Orhy, Pico de Orhy, Pic d'Orhy. ▲ France/Spain 42°55′N 01°01′W see also Orhy
Orhi see Orhy
Orhomenos see Orchómenos
162 K6 **Orhon** ◈ province N Mongolia
162 L6 **Orhon Gol** ≋ N Mongolia
102 J16 **Orhy** var. Orhi, Pico de Orhy, Pic d'Orhy. ▲ France/Spain 43°00′N 01°00′W see also Orhi
Orhy see Orhi
Orhy, Pic d'/Orhy, Pico de see Orhi/Orhy
34 L2 **Orick** California, W USA 41°16′N 124°03′W
32 L6 **Orient** Washington, NW USA 48°51′N 118°14′W
48 D6 **Oriental, Cordillera** ▲ Bolivia/Peru
48 D6 **Oriental, Cordillera** ▲ C Colombia
57 H16 **Oriental, Cordillera** ▲ C Peru
63 M15 **Oriente** Buenos Aires, E Argentina 38°45′S 60°37′W

◆ Country
● Country Capital
◇ Dependent Territory
○ Dependent Territory Capital
◈ Administrative Regions
✕ International Airport
▲ Mountain
▲ Mountain Range
⛰ Volcano
≋ River
◎ Lake
⊠ Reservoir

105 R12 **Orihuela** Valenciana, E Spain 38°05′N 00°56′W
117 V9 **Orikhiv** *Rus.* Orekhov. Zaporiz'ka Oblast', SE Ukraine 47°32′N 35°48′E
113 K22 **Orikum** *var.* Orikumi. Vlorë, SW Albania 40°20′N 19°28′E
Orikumi *see* Orikum
117 V6 **Oril'** *Rus.* Orel. ~ E Ukraine
14 H14 **Orillia** Ontario, S Canada 44°36′N 79°26′W
93 M19 **Orimattila** Päijät-Häme, S Finland 60°48′N 25°40′E
33 Y15 **Orin** Wyoming, C USA 42°39′N 105°10′W
47 R4 **Orinoco, Río** ~ Colombia/Venezuela
186 C9 **Oriomo** Western, SW Papua New Guinea 08°53′S 143°13′E
30 K11 **Orion** Illinois, N USA 41°21′N 90°22′W
29 Q5 **Oriska** North Dakota, N USA 46°54′N 97°46′W
Orissa *see* Odisha
Orissaar *see* Orissaare
118 E5 **Orissaare** *Ger.* Orissaar. Saaremaa, W Estonia 58°34′N 23°05′E
107 B19 **Oristano** Sardegna, Italy, C Mediterranean Sea 39°54′N 08°35′E
107 A19 **Oristano, Golfo di** *gulf* Sardegna, Italy, C Mediterranean Sea
54 D13 **Orito** Putumayo, SW Colombia 0°49′N 76°57′W
93 L18 **Orivesi** Häme, W Finland 61°39′N 24°21′E
93 N17 **Orivesi** ◎ Etelä-Savo, SE Finland
58 H12 **Oriximiná** Pará, NE Brazil 01°45′S 55°50′W
41 Q14 **Orizaba** Veracruz-Llave, E Mexico 18°51′N 97°08′W
41 Q14 **Orizaba, Volcán Pico de** *var.* Citlaltépetl. ▲ S Mexico 19°00′N 97°15′W
95 I16 **Ørje** Østfold, S Norway 59°28′N 11°40′E
113 I16 **Orjen** ⛰ Bosnia and Herzegovina/Montenegro
Orjiva *see* Orgiva
Orjonikidzeobod *see* Kofarnihon
94 G8 **Orkanger** Sør-Trøndelag, S Norway 63°17′N 09°52′E
94 G8 **Orkdalen** *valley* S Norway
95 K22 **Örkelljunga** Skåne, S Sweden 56°17′N 13°20′E
Orkhaniye *see* Botevgrad
Orkhómenos *see* Orchómenos
94 H9 **Orkla** ~ S Norway
Orkney *see* Orkney Islands
65 J22 **Orkney Deep** *undersea feature* Scotia Sea/Weddell Sea
96 J4 **Orkney Islands** *var.* Orkney, Orkneys. *island group* N Scotland, United Kingdom
Orkneys *see* Orkney Islands
24 K8 **Orla** Texas, SW USA 31°48′N 103°55′W
35 N5 **Orland** California, W USA 39°43′N 122°12′W
23 X11 **Orlando** Florida, SE USA 28°32′N 81°23′W
23 X12 **Orlando** ✈ Florida, SE USA 28°24′N 81°16′W
107 K23 **Orlando, Capo d'** *headland* Sicilia, Italy, C Mediterranean Sea 38°10′N 14°44′E
Orlau *see* Orlová
103 N6 **Orléanais** *cultural region* C France
103 N7 **Orléans** *anc.* Aurelianum. Loiret, C France 47°54′N 01°53′E
34 L2 **Orleans** California, W USA 41°16′N 123°36′W
19 Q12 **Orleans** Massachusetts, NE USA 41°48′N 69°57′W
15 R10 **Orléans, Île d'** *island* Québec, SE Canada
Orléansville *see* Chlef
111 F16 **Orlice** *Ger.* Adler. ~ NE Czech Republic
122 L13 **Orlik** Respublika Buryatiya, S Russian Federation 52°32′N 99°36′E
125 Q14 **Orlov** *prev.* Khalturin. Kirovskaya Oblast', NW Russian Federation 58°34′N 48°57′E
111 I17 **Orlová** *Ger.* Orlau, *Pol.* Orłowa. Moravskoslezský Kraj, E Czech Republic 49°50′N 18°21′E
Orlov, Mys *see* Orlovskiy, Mys
126 I6 **Orlovskaya Oblast'** ◈ *province* W Russian Federation
124 M5 **Orlovskiy, Mys** *var.* Mys Orlov. *headland* NW Russian Federation 67°14′N 41°17′E
Orłowa *see* Orlová
103 O5 **Orly** ✈ (Paris) Essonne, N France 48°43′N 02°24′E
119 G16 **Orlya** Hrodzyenskaya Voblasts', W Belarus 53°30′N 24°59′E
114 M7 **Orlyak** *prev.* Makenzen, Trubchular, *Rom.* Trupcilar. Dobrich, NE Bulgaria 43°39′N 27°21′E
148 L16 **Ormāra** Baluchistān, SW Pakistan 25°14′N 64°36′E
171 P5 **Ormoc** *off.* Ormoc City, *var.* MacArthur. Leyte, C Philippines 11°02′N 124°35′E
Ormoc City *see* Ormoc
23 X10 **Ormond Beach** Florida, SE USA 29°16′N 81°04′W
109 X10 **Ormož** *Ger.* Friedau. NE Slovenia 46°24′N 16°09′E
14 J13 **Ormsby** Ontario, SE Canada 44°52′N 77°45′W
97 K17 **Ormskirk** NW England, United Kingdom 53°35′N 02°54′W
Ormsö *see* Vormsi
15 N13 **Ormstown** Québec, SE Canada 45°08′N 73°57′W
Ormuz, Strait of *see* Hormuz, Strait of
103 T8 **Ornans** Doubs, E France 47°06′N 06°06′E
102 K5 **Orne** ◈ *department* N France
102 K5 **Orne** ~ N France
92 G12 **Ørnes** Nordland, C Norway 66°51′N 13°43′E
110 L7 **Orneta** Warmińsko-Mazurskie, NE Poland 54°07′N 20°10′E
95 P16 **Ornö** Stockholm, C Sweden 59°03′N 18°28′E
37 Q3 **Orno Peak** ▲ Colorado, C USA 40°06′N 107°06′W
93 I16 **Örnsköldsvik** Västernorrland, C Sweden 63°16′N 18°45′E
163 X13 **Oro** E North Korea 39°59′N 127°27′E
Oro *see* Northern
45 T6 **Orocovis** C Puerto Rico 18°13′N 66°22′W
54 H10 **Orocué** Casanare, E Colombia 04°51′N 71°21′W
77 N13 **Orodara** SW Burkina Faso 11°00′N 04°54′W
105 S4 **Oroel, Peña de** ▲ N Spain 42°30′N 00°31′W
33 N10 **Orofino** Idaho, NW USA 46°28′N 116°15′W
162 I9 **Orog Nuur** ◎ S Mongolia
35 U14 **Oro Grande** California, W USA 34°36′N 117°19′W
37 S15 **Orogrande** New Mexico, SW USA 32°24′N 106°04′W
191 Q7 **Orohena, Mont** 🌋 Tahiti, W French Polynesia 17°37′S 149°27′W
Orolaunum *see* Arlon
Orol Dengizi *see* Aral Sea
189 S15 **Oroluk Atoll** *atoll* Caroline Islands, C Micronesia
80 J13 **Oromīya** *var.* Oromo. ◈ C Ethiopia
Oromo *see* Oromīya
13 O15 **Oromocto** New Brunswick, SE Canada 45°50′N 66°28′W
191 S4 **Orona** *prev.* Hull Island. *atoll* Phoenix Islands, C Kiribati
191 V17 **Orongo** *ancient monument* Easter Island, Chile, E Pacific Ocean
138 I3 **Orontes** *var.* Ononte, Nahr el Aassi, *Ar.* Nahr al 'Āṣī. ~ SW Asia
104 L9 **Oropesa** Castilla-La Mancha, C Spain 39°55′N 05°10′W
Oropesa *see* Oropesa del Mar
105 T8 **Oropesa del Mar** *var.* Oropesa, Orpesa, *Cat.* Orpes. Valenciana, E Spain 40°06′N 00°07′E
Oropeza *see* Cochabamba
Oroqen Zizhiqi *see* Alihe
171 P7 **Oroquieta** *var.* Oroquieta City. Mindanao, S Philippines 08°27′N 123°46′E
Oroquieta City *see* Oroquieta
40 J8 **Oro, Río del** ~ C Mexico
59 O14 **Orós, Açude** ⊠ E Brazil
107 D18 **Orosei, Golfo di** *gulf* Tyrrhenian Sea, C Mediterranean Sea
111 M24 **Orosháza** Békés, SE Hungary 46°33′N 20°40′E
Orosirá Rodhópis *see* Rhodope Mountains
111 I22 **Oroszlány** Komárom-Esztergom, W Hungary 47°28′N 18°16′E
188 B16 **Orote Peninsula** *peninsula* W Guam
123 T9 **Orotukan** Magadanskaya Oblast', E Russian Federation 62°18′N 150°46′E
35 O5 **Oroville** California, W USA 39°29′N 121°35′W
32 K6 **Oroville** Washington, NW USA 48°56′N 119°25′W
35 O5 **Oroville, Lake** ⊠ California, W USA
0 G15 **Orozco Fracture Zone** *tectonic feature* E Pacific Ocean
Orpes *see* Oropesa del Mar
Orpesa *see* Oropesa del Mar
64 I7 **Orphan Knoll** *undersea feature* NW Atlantic Ocean 51°00′N 47°00′W
29 V3 **Orr** Minnesota, N USA 48°03′N 92°48′W
95 M21 **Orrefors** Kalmar, S Sweden 56°48′N 15°45′E
182 I7 **Orroroo** South Australia 32°46′S 138°38′E
31 T12 **Orrville** Ohio, N USA 40°50′N 81°45′W
94 L12 **Orsa** Dalarna, C Sweden 61°07′N 14°40′E
Orschowa *see* Orşova
Orschütz *see* Orzyc
119 O14 **Orsha** Vitsyebskaya Voblasts', NE Belarus 54°30′N 30°26′E
127 Q2 **Orshanka** Respublika Mariy El, W Russian Federation 56°54′N 47°54′E
108 C11 **Orsières** Valais, SW Switzerland 46°00′N 07°09′E
127 X8 **Orsk** Orenburgskaya Oblast', W Russian Federation 51°13′N 58°35′E
116 F13 **Orşova** *Ger.* Orschowa, *Hung.* Orsova. Mehedinţi, SW Romania 44°42′N 22°22′E
94 D10 **Ørsta** Møre og Romsdal, S Norway 62°12′N 06°09′E
95 O15 **Örsundsbro** Uppsala, C Sweden 59°45′N 17°19′E
136 D16 **Ortaca** Muğla, SW Turkey 36°49′N 28°43′E
83 I21 **O.R. Tambo** ✈ (Johannesburg) Gauteng, NE South Africa 26°08′S 28°01′E
107 M16 **Orta Nova** Puglia, SE Italy 41°20′N 15°43′E
136 I17 **Orta Toroslar** ⛰ S Turkey
54 E11 **Ortega** Tolima, W Colombia 03°57′N 75°11′W
104 H1 **Ortegal, Cabo** *headland* NW Spain 43°46′N 07°54′W
Ortelsburg *see* Szczytno
102 J15 **Orthez** Pyrénées-Atlantiques, SW France 43°29′N 00°46′W
60 J10 **Ortigueira** Paraná, S Brazil 24°10′S 50°55′W
104 H1 **Ortigueira** Galicia, NW Spain 43°40′N 07°50′W
106 H5 **Ortisei** *Ger.* Sankt-Ulrich. Trentino-Alto Adige, N Italy 46°35′N 11°42′E
40 F6 **Ortiz** Sonora, NW Mexico 28°18′N 110°40′W
54 L5 **Ortiz** Guárico, N Venezuela 09°37′N 67°20′W
Ortler *see* Ortles
106 F5 **Ortles** *Ger.* Ortler. ▲ N Italy 46°29′N 10°33′E
107 K14 **Ortona** Abruzzo, C Italy 42°21′N 14°24′E
57 K14 **Ortón, Río** ~ N Bolivia
29 R8 **Ortonville** Minnesota, N USA 45°18′N 96°26′W
147 W8 **Orto-Tokoy** Issyk-Kul'skaya Oblast', NE Kyrgyzstan 42°20′N 76°03′E
93 J15 **Örträsk** Västerbotten, N Sweden 64°10′N 19°00′E
100 J12 **Örtze** ~ NW Germany
Oruba *see* Aruba
142 I3 **Orūmīyeh** *var.* Rizaiyeh, Urmia, Urmiyeh; *prev.* Reza'iyeh. Āẕarbāyjān-e Gharbī, NW Iran 37°33′N 45°06′E
142 J3 **Orūmīyeh, Daryācheh-ye** *var.* Matianus, Sha Hi, Urumi Yeh, *Eng.* Lake Urmia; *prev.* Daryācheh-ye Reẕā'īyeh. ◎ NW Iran
57 K19 **Oruro** Oruro, W Bolivia 17°58′S 67°06′W
57 J19 **Oruro** ◈ *department* W Bolivia
95 I18 **Orust** *island* S Sweden
Orūzgān *see* Urūzgān
106 H13 **Orvieto** *anc.* Velsuna. Umbria, C Italy 42°43′N 12°06′E
194 K7 **Orville Coast** *physical region* Antarctica
114 H7 **Oryahovo** *var.* Oryakhovo. Vratsa, NW Bulgaria 43°44′N 23°58′E
Oryakhovo *see* Oryahovo
Oryokko *see* Yalu
117 R5 **Orzhytsya** Poltavs'ka Oblast', C Ukraine 49°48′N 32°40′E
110 M9 **Orzyc** *Ger.* Orschütz. ~ NE Poland
110 N8 **Orzysz** *Ger.* Arys. Warmińsko-Mazurskie, NE Poland 53°49′N 21°54′E
98 K13 **Oss** Noord-Brabant, S Netherlands 51°46′N 05°32′E
94 I10 **Os** Hedmark, S Norway 62°29′N 11°14′E
125 U15 **Osa** Permskiy Kray, NW Russian Federation 57°16′N 55°22′E
115 F15 **Óssa** ▲ C Greece
104 H11 **Ossa** ▲ S Portugal 38°43′N 07°33′W
29 W11 **Osage** Iowa, C USA 43°16′N 92°48′W
27 U5 **Osage Beach** Missouri, C USA 38°09′N 92°37′W
27 P5 **Osage City** Kansas, C USA 38°37′N 95°49′W
27 U7 **Osage Fork River** ~ Missouri, C USA
27 U5 **Osage River** ~ Missouri, C USA
164 J13 **Ōsaka** *hist.* Naniwa. Ōsaka, Honshū, SW Japan 34°38′N 135°28′E
164 I13 **Ōsaka** *off.* Ōsaka-fu, *var.* Ōsaka Hu. ◈ *urban prefecture* Honshū, SW Japan
Ōsaka-fu/Ōsaka Hu *see* Ōsaka
145 R10 **Osakarovka** Karaganda, C Kazakhstan 50°32′N 72°39′E
Ōsaki *see* Furukawa
29 T7 **Osakis** Minnesota, N USA 45°51′N 95°08′W
114 I9 **Osam** *var.* Osŭm. ~ N Bulgaria
43 N16 **Osa, Península de** *peninsula* S Costa Rica
60 M10 **Osasco** São Paulo, S Brazil 23°32′S 46°46′W
27 R5 **Osawatomie** Kansas, C USA 38°30′N 94°57′W
26 L3 **Osborne** Kansas, C USA 39°26′N 98°42′W
173 S8 **Osborn Plateau** *undersea feature* E Indian Ocean
95 L21 **Osby** Skåne, S Sweden 56°24′N 14°00′E
Osca *see* Huesca
92 N2 **Oscar II Land** *physical region* W Svalbard
27 Y10 **Osceola** Arkansas, C USA 35°43′N 89°58′W
29 V15 **Osceola** Iowa, C USA 41°01′N 93°45′W
27 S6 **Osceola** Missouri, C USA 38°01′N 93°41′W
29 Q15 **Osceola** Nebraska, C USA 41°09′N 97°28′W
101 N15 **Oschatz** Sachsen, E Germany 51°17′N 13°10′E
100 K13 **Oschersleben** Sachsen-Anhalt, C Germany 52°02′N 11°14′E
31 R7 **Oscoda** Michigan, N USA 44°25′N 83°19′W
Ösel *see* Saaremaa
94 H6 **Osen** Sør-Trøndelag, S Norway 64°17′N 10°29′E
93 I12 **Osensjøen** ◎ S Norway
164 A14 **Ōse-zaki** Fukue-jima, SW Japan
147 T10 **Osh** Oshskaya Oblast', SW Kyrgyzstan 40°34′N 72°46′E
83 C16 **Oshakati** Oshana, N Namibia 17°46′S 15°43′E
83 C16 **Oshana** ◈ *district* N Namibia
14 H15 **Oshawa** Ontario, SE Canada 43°54′N 78°50′W
165 R10 **Oshika-hantō** *peninsula* Honshū, C Japan
83 C16 **Oshikango** Ohangwena, N Namibia 17°29′S 15°54′E
83 C17 **Oshikoto** *var.* Otjikoto. ◈ *district* N Namibia
165 P5 **Ō-shima** *island* NE Japan
165 N14 **Ō-shima** *island* S Japan
165 Q5 **Oshima-hantō** ⛰ Hokkaidō, NE Japan
83 D17 **Oshivelo** Oshikoto, N Namibia 18°37′S 17°10′E
28 K14 **Oshkosh** Nebraska, C USA 41°25′N 102°21′W
30 M7 **Oshkosh** Wisconsin, N USA 44°01′N 88°32′W
Oshmyany *see* Ashmyany
Osh Oblasty *see* Oshskaya Oblast'
77 T16 **Oshogbo** *var.* Osogbo. Osun, W Nigeria 07°42′N 04°31′E
147 S11 **Oshskaya Oblast'** *Kir.* Osh Oblasty. ◈ *province* SW Kyrgyzstan
Ōshū *see* Mizusawa
Oshun *see* Osun
79 J20 **Oshwe** Bandundu, C Dem. Rep. Congo 03°24′S 19°32′E
Osiek *see* Osijek
112 I9 **Osijek** *prev.* Osiek, Osjek, *Ger.* Esseg, *Hung.* Eszék. Osijek-Baranja, E Croatia 45°33′N 18°41′E
112 I9 **Osijek-Baranja** *off.* Osječko-Baranjska Županija. ◈ *province* E Croatia
106 J12 **Osimo** Marche, C Italy 43°28′N 13°29′E
122 M12 **Osinovka** Irkutskaya Oblast', C Russian Federation 56°19′N 101°55′E
Osintorf *see* Asintorf
112 N11 **Osipaonica** Serbia, NE Serbia 44°34′N 21°00′E
Osipenko *see* Berdyans'k
Osipovichi *see* Asipovichy
Osječko-Baranjska Županija *see* Osijek-Baranja
Osjek *see* Osijek
29 W15 **Oskaloosa** Iowa, C USA 41°17′N 92°38′W
27 Q4 **Oskaloosa** Kansas, C USA 39°14′N 95°21′W
95 N20 **Oskarshamn** Kalmar, S Sweden 57°16′N 16°25′E
95 J21 **Oskarström** Halland, S Sweden 56°48′N 13°00′E
14 M8 **Oskélanéo** Québec, SE Canada 48°06′N 75°12′W
Öskemen *see* Ust'-Kamenogorsk
117 W5 **Oskil** *Rus.* Oskil. ~ Russian Federation/Ukraine
Oskil *see* Oskil
93 D20 **Oslo** *prev.* Christiania, Kristiania. ● (Norway) Oslo, S Norway 59°54′N 10°44′E
93 D20 **Oslo** ◈ *county* S Norway
93 D21 **Oslofjorden** *fjord* S Norway
155 G15 **Osmānābād** Mahārāshtra, C India 18°09′N 76°06′E
136 J11 **Osmancık** Çorum, N Turkey 40°58′N 34°50′E
136 L16 **Osmaniye** Osmaniye, S Turkey 37°04′N 36°15′E
136 L16 **Osmaniye** ◈ *province* S Turkey
95 O16 **Ösmo** Stockholm, C Sweden 58°58′N 17°55′E
118 E3 **Osmussaar** *island* W Estonia
100 G13 **Osnabrück** Niedersachsen, NW Germany 52°09′N 07°42′E
110 D11 **Ośno Lubuskie** *Ger.* Drossen. Lubuskie, W Poland 52°28′N 14°51′E
Osogbo *see* Oshogbo
113 P19 **Osogov Mountains** *var.* Osogovske Planine, Osogovska Planina, *Mac.* Osogovski Planini. ⛰ Bulgaria/FYR Macedonia
Osogovske Planine/ Osogovska Planina/ Osogovski Planini *see* Osogov Mountains
165 R6 **Osore-zan** ▲ Honshū, C Japan 41°18′N 141°06′E
Oşorhei *see* Târgu Mureş
61 J16 **Osório** Rio Grande do Sul, S Brazil 29°53′S 50°17′W
63 G16 **Osorno** Los Lagos, C Chile 40°39′S 73°05′W
104 M4 **Osorno** Castilla y León, N Spain 42°24′N 04°22′W
11 N17 **Osoyoos** British Columbia, SW Canada 49°02′N 119°31′W
95 C14 **Osøyro** Hordaland, S Norway 60°11′N 05°30′E
54 J6 **Ospino** Portuguesa, N Venezuela 09°17′N 69°26′W
23 X6 **Ossabaw Island** *island* Georgia, SE USA
23 X6 **Ossabaw Sound** *sound* Georgia, SE USA
183 O16 **Ossa, Mount** ▲ Tasmania, SE Australia 41°55′S 146°03′E
104 H11 **Ossa, Serra d'** ⛰ SE Portugal
77 U16 **Osse** ~ S Nigeria
30 J6 **Osseo** Wisconsin, N USA 44°33′N 91°13′W
109 S9 **Ossiacher See** ◎ S Austria
18 K13 **Ossining** New York, NE USA 41°10′N 73°50′W
123 V9 **Ossora** Krasnoyarskiy Kray, E Russian Federation 59°16′N 163°02′E
124 I15 **Ostashkov** Tverskaya Oblast', W Russian Federation 57°08′N 33°10′E
100 H9 **Oste** ~ NW Germany
Ostee *see* Baltic Sea
Ostend/Ostende *see* Oostende
117 P3 **Oster** Chernihivs'ka Oblast', N Ukraine 50°57′N 30°55′E
93 J16 **Österbotten** *Fin.* Pohjanmaa, *Eng.* Ostrobothnia. ◈ *region* W Finland
95 O14 **Österbybruk** Uppsala, C Sweden 60°13′N 17°55′E
95 M19 **Österbymo** Östergötland, S Sweden 57°49′N 15°15′E
94 K12 **Österdalälven** ~ C Sweden
94 I12 **Østerdalen** *valley* S Norway
95 L18 **Östergötland** ◈ *county* S Sweden
100 H10 **Osterholz-Scharmbeck** Niedersachsen, NW Germany 53°13′N 08°46′E
Östermark *see* Teuva
Østerø *see* Eysturoy
101 J14 **Osterode am Harz** Niedersachsen, C Germany 51°43′N 10°15′E
Osterode/Osterode in Ostpreussen *see* Ostróda
Østerøy *see* Osterøyni
94 C13 **Osterøyni** *prev.* Osterøy. *island* S Norway
Österreich *see* Austria
93 G16 **Östersund** Jämtland, C Sweden 63°10′N 14°44′E
95 N14 **Östervåla** Västmanland, C Sweden 60°10′N 17°13′E
101 H22 **Ostfildern** Baden-Württemberg, SW Germany 48°43′N 09°16′E
95 H16 **Østfold** ◈ *county* S Norway
100 E9 **Ostfriesische Inseln** *Eng.* East Frisian Islands. *island group* NW Germany
100 F10 **Ostfriesland** *historical region* NW Germany
95 P14 **Östhammar** Uppsala, C Sweden 60°15′N 18°25′E
Ostia Aterni *see* Pescara
106 G8 **Ostiglia** Lombardia, N Italy 45°04′N 11°09′E
94 J14 **Östmark** Värmland, C Sweden 60°15′N 12°45′E
95 K22 **Östra Ringsjön** ◎ S Sweden
111 I17 **Ostrava** Moravskoslezský Kraj, E Czech Republic 49°50′N 18°15′E
Ostravský Kraj *see* Moravskoslezský Kraj
94 J11 **Østrehogna** *Swe.* Härjahågnen, Härjehågna. ▲ Norway/Sweden 61°43′N 12°07′E
Ostrobothnia *see* Österbotten
110 K8 **Ostróda** *Ger.* Osterode, Osterode in Ostpreussen. Warmińsko-Mazurskie, NE Poland 53°42′N 19°59′E
Ostrog/Ostróg *see* Ostroh
126 L8 **Ostrogozhsk** Voronezhskaya Oblast', W Russian Federation 50°52′N 39°00′E
116 L4 **Ostroh** *Pol.* Ostróg, *Rus.* Ostrog. Rivnens'ka Oblast', NW Ukraine 50°20′N 26°29′E
110 N9 **Ostrołęka** *Ger.* Wiesenhof, *Rus.* Ostrolenka. Mazowieckie, C Poland 53°06′N 21°34′E
Ostrolenka *see* Ostrołęka
111 A16 **Ostrov** *Ger.* Schlackenwerth. Karlovarský Kraj, W Czech Republic 50°18′N 12°56′E
124 F15 **Ostrov** *Latv.* Austrava. Pskovskaya Oblast', W Russian Federation 57°21′N 28°18′E
Ostrovets *see* Ostrowiec Świętokrzyski
113 M21 **Ostrovicës, Mali i** ▲ SE Albania 40°36′N 20°25′E
165 Z2 **Ostrov Iturup** *island* NE Russian Federation
124 M4 **Ostrovnoy** Murmanskaya Oblast', NW Russian Federation 68°00′N 39°40′E
114 L7 **Ostrovo** *prev.* Golema Ada. Razgrad, N Bulgaria 43°40′N 26°37′E
125 N15 **Ostrovskoye** Kostromskaya Oblast', NW Russian Federation 57°46′N 42°18′E
Ostrów *see* Ostrów Wielkopolski
Ostrowiec *see* Ostrowiec Świętokrzyski
111 M14 **Ostrowiec Świętokrzyski** *var.* Ostrowiec, *Rus.* Ostrovets. Świętokrzyskie, C Poland 50°55′N 21°23′E
110 P13 **Ostrów Lubelski** Lubelskie, E Poland 51°29′N 22°57′E
110 N10 **Ostrów Mazowiecka** *var.* Ostrów Mazowiecki. Mazowieckie, NE Poland 52°49′N 21°53′E
Ostrów Mazowiecki *see* Ostrów Mazowiecka
Ostrowo *see* Ostrów Wielkopolski
110 H13 **Ostrów Wielkopolski** *var.* Ostrów, *Ger.* Ostrowo. Wielkopolskie, C Poland 51°40′N 17°47′E
Ostryna *see* Astryna
110 I13 **Ostrzeszów** Wielkopolskie, C Poland 51°25′N 17°55′E
107 P18 **Ostuni** Puglia, SE Italy 40°44′N 17°35′E
Ostyako-Voguls'k *see* Khanty-Mansiysk
Osum *see* Osumit, Lumi i
Osŭm *see* Osam
164 C17 **Ōsumi-hantō** ⛰ Kyūshū, SW Japan
164 C17 **Ōsumi-kaikyō** *strait* SW Japan
113 L22 **Osumit, Lumi i** *var.* Osum. ~ SE Albania
77 T16 **Osun** *var.* Oshun. ◈ *state* SW Nigeria
104 L14 **Osuna** Andalucía, S Spain 37°14′N 05°06′W
60 J8 **Osvaldo Cruz** São Paulo, S Brazil 21°49′S 50°52′W
Osveya *see* Asvyeya
18 J7 **Oswegatchie River** ~ New York, NE USA
27 Q7 **Oswego** Kansas, C USA 37°11′N 95°10′W
18 H9 **Oswego** New York, NE USA 43°27′N 76°13′W
97 K19 **Oswestry** W England, United Kingdom 52°51′N 03°06′W
111 J16 **Oświęcim** *Ger.* Auschwitz. Małopolskie, S Poland 50°02′N 19°11′E
185 E22 **Otago** *off.* Otago Region. ◈ *region* South Island, New Zealand
185 F23 **Otago Peninsula** *peninsula* South Island, New Zealand
Otago Region *see* Otago
165 F13 **Ōtake** Hiroshima, Honshū, SW Japan 34°13′N 132°12′E
184 L13 **Otaki** Wellington, North Island, New Zealand 40°46′S 175°08′E
93 M15 **Otanmäki** Kainuu, C Finland 64°07′N 27°04′E
145 T15 **Otar** Zhambyl, SE Kazakhstan 43°30′N 75°13′E
165 R4 **Otaru** Hokkaidō, NE Japan 43°14′N 140°59′E
185 C24 **Otatara** Southland, South Island, New Zealand 46°26′S 168°18′E
185 C24 **Otautau** Southland, South Island, New Zealand 46°10′S 168°01′E
93 M18 **Otava** Etelä-Savo, E Finland 61°37′N 27°07′E
111 B18 **Otava** *Ger.* Wottawa. ~ SW Czech Republic
56 C6 **Otavalo** Imbabura, N Ecuador 0°13′N 78°15′W
83 D17 **Otavi** Otjozondjupa, N Namibia 19°35′S 17°25′E
165 P12 **Ōtawara** Tochigi, Honshū, S Japan 36°52′N 140°01′E
83 B16 **Otchinjau** Cunene, SW Angola 16°31′S 13°54′E
116 F12 **Oţelu Roşu** *Ger.* Ferdinandsberg, *Hung.* Nándorhgy. Caras-Severin, SW Romania 45°30′N 22°22′E
185 E21 **Otematata** Canterbury, South Island, New Zealand 44°37′S 170°12′E
118 I6 **Otepää** *Ger.* Odenpäh. Valgamaa, SE Estonia 58°01′N 26°30′E
144 G14 **Otes** *Kaz.* Say-Ötesh; *prev.* Say-Utës. Mangistau, SW Kazakhstan 44°20′N 53°32′E
162 H7 **Otgon** *var.* Buyant. Dzavhan, C Mongolia 47°14′N 97°14′E
32 K9 **Othello** Washington, NW USA 46°49′N 119°10′W
83 K23 **oThongathi** *prev.* Tongaat, *var.* uThongathi. KwaZulu/Natal, E South Africa 29°35′S 31°07′E *see also* Tongaat
115 A15 **Othonoí** *island* Iónia Nisiá, Greece, C Mediterranean Sea
Othris *see* Óthrys
115 F17 **Óthrys** *var.* Othris. ⛰ C Greece
77 Q14 **Oti** ~ N Togo
40 K10 **Otinapa** Durango, C Mexico 24°01′N 104°58′W
185 G17 **Otira** West Coast, South Island, New Zealand 42°52′S 171°33′E
37 V3 **Otis** Colorado, C USA 40°09′N 102°57′W
12 L10 **Otish, Monts** ⛰ Québec, E Canada
83 C17 **Otjikondo** Kunene, N Namibia 19°50′S 15°23′E
Otjikoto *see* Oshikoto
83 E18 **Otjinene** Omaheke, NE Namibia 21°10′S 18°43′E
83 D18 **Otjiwarongo** Otjozondjupa, N Namibia 20°29′S 16°36′E
83 D18 **Otjosondu** *var.* Otjosundu. Otjozondjupa, C Namibia 21°19′S 17°51′E
Otjosundu *see* Otjosondu
83 D18 **Otjozondjupa** ◈ *district* C Namibia
112 C11 **Otočac** Lika-Senj, W Croatia 44°52′N 15°13′E
Otog Qi *see* Ulan
112 J10 **Otok** Vukovar-Srijem, E Croatia 45°10′N 18°52′E
116 K14 **Otopeni** ✈ (Bucureşti) Ilfov, S Romania 44°34′N 26°09′E
184 L8 **Otorohanga** Waikato, North Island, New Zealand 38°10′S 175°14′E
12 D9 **Otoskwin** ~ Ontario, C Canada
165 G14 **Ōtoyo** Kōchi, Shikoku, SW Japan 33°45′N 133°42′E
95 E16 **Otra** ~ S Norway
107 R19 **Otranto** Puglia, SE Italy 40°08′N 18°28′E
Otranto, Canale d' *see* Otranto, Strait of
107 Q18 **Otranto, Strait of** *It.* Canale d'Otranto. *strait* Albania/Italy
111 H18 **Otrokovice** *Ger.* Otrokowitz. Zlínský Kraj, E Czech Republic 49°13′N 17°33′E
Otrokowitz *see* Otrokovice
31 P10 **Otsego** Michigan, N USA 42°27′N 85°42′W
31 Q6 **Otsego Lake** ◎ Michigan, N USA
18 I11 **Otselic River** ~ New York, NE USA
164 J14 **Ōtsu** *var.* Ôtu. Shiga, Honshū, SW Japan 35°03′N 135°49′E
94 G11 **Otta** Oppland, S Norway 61°46′N 09°33′E
189 U13 **Otta** *island* Chuuk, C Micronesia
94 F11 **Otta** ~ S Norway
189 U13 **Otta Pass** *passage* Chuuk Islands, C Micronesia
95 J22 **Ottarp** Skåne, S Sweden 55°55′N 12°55′E
14 L12 **Ottawa** ● (Canada) Ontario, SE Canada 45°24′N 75°41′W
30 L11 **Ottawa** Illinois, N USA 41°21′N 88°50′W
27 Q5 **Ottawa** Kansas, C USA 38°35′N 95°16′W
31 R12 **Ottawa** Ohio, N USA 41°01′N 84°03′W
14 L12 **Ottawa** *var.* Uplands. ✈ Ontario, SE Canada 45°19′N 75°39′W
14 M12 **Ottawa** *Fr.* Outaouais. ~ Ontario/Québec, SE Canada
9 R10 **Ottawa Islands** *island group* Nunavut, C Canada
18 L8 **Otter Creek** ~ Vermont, NE USA
36 L6 **Otter Creek Reservoir** ⊠ Utah, W USA
98 L11 **Otterlo** Gelderland, E Netherlands 52°06′N 05°46′E
94 D9 **Otterøya** *island* S Norway
29 S6 **Otter Tail Lake** ◎ Minnesota, N USA
29 R7 **Otter Tail River** ~ Minnesota, C USA
95 H23 **Otterup** Syddtjylland, C Denmark 55°31′N 10°25′E
99 H19 **Ottignies** Wallon Brabant, C Belgium 50°40′N 04°34′E
101 L23 **Ottobrunn** Bayern, SE Germany 48°02′N 11°40′E
29 X15 **Ottumwa** Iowa, C USA 41°00′N 92°24′W
Ôtu *see* Ōtsu
83 B16 **Otuazuma** Kunene, NW Namibia 17°52′S 13°16′E
77 V16 **Oturkpo** Benue, S Nigeria 07°12′N 08°06′E
193 Y15 **Otu Tolu Group** *island group* SE Tonga
182 M13 **Otway, Cape** *headland* Victoria, SE Australia 38°52′S 143°31′E
63 H24 **Otway, Seno** *inlet* S Chile
Ötz *see* Oetz
108 L8 **Ötztaler Ache** ~ W Austria
108 L9 **Ötztaler Alpen** *It.* Alpi Venoste. ⛰ SW Austria
27 T12 **Ouachita, Lake** ⊠ Arkansas, C USA
27 R11 **Ouachita Mountains** ⛰ Arkansas/Oklahoma, C USA
27 U13 **Ouachita River** ~ Arkansas/Louisiana, C USA
Ouadaï *see* Ouaddaï
76 J7 **Ouadâne** *var.* Ouadane. Adrar, C Mauritania 20°57′N 11°35′W
78 K13 **Ouadda** Haute-Kotto, N Central African Republic 08°02′N 22°22′E
78 J10 **Ouaddaï** *off.* Région du Ouaddaï, *var.* Ouadaï, Wadai. ◈ *region* SE Chad
Ouaddaï, Région du *see* Ouaddaï
77 P13 **Ouagadougou** *var.* Wagadugu. ● (Burkina Faso) C Burkina 12°20′N 01°32′W
77 P13 **Ouagadougou** ✈ C Burkina Faso 12°21′N 01°27′W
77 O12 **Ouahigouya** NW Burkina Faso 13°31′N 02°20′W
Ouahran *see* Oran
79 J14 **Ouaka** ◈ *prefecture* C Central African Republic
79 J15 **Ouaka** ~ S Central African Republic
Oualam *see* Ouallam
76 M9 **Oualâta** *var.* Oualata. Hodh ech Chargui, SE Mauritania 17°18′N 07°00′W
77 R11 **Ouallam** *var.* Oualam. Tillabéri, W Niger 14°23′N 02°09′E
172 H14 **Ouanani** Mohéli, S Comoros 12°19′S 93°98′E
55 Z10 **Ouanary** E French Guiana 04°11′N 51°40′W
78 L13 **Ouanda Djallé** Vakaga, NE Central African Republic 08°53′N 22°47′E
79 N14 **Ouando** Haut-Mbomou, SE Central African Republic 05°57′N 25°57′E
79 L15 **Ouango** Mbomou, S Central African Republic 04°19′N 22°30′E
77 N14 **Ouangolodougou** *var.* Wangolodougou. N Ivory Coast 09°58′N 05°09′W
172 I13 **Ouani** Anjouan, SE Comoros
79 M15 **Ouara** ~ E Central African Republic
76 K7 **Ouarâne** *desert* C Mauritania
15 O11 **Ouareau** ~ Québec, SE Canada
74 K7 **Ouargla** *var.* Wargla. NE Algeria 32°00′N 05°16′E
74 F8 **Ouarzazate** S Morocco 30°54′N 06°55′W
77 Q11 **Ouatagouna** Gao, E Mali 15°06′N 00°41′E
74 G6 **Ouazzane** *var.* Ouezzane, *Ar.* Wazan, Wazzan. N Morocco 34°52′N 05°35′W
Oubangui *see* Ubangi
Oubangui-Chari *see* Central African Republic
Oubangui-Chari, Territoire de l' *see* Central African Republic
Oubari, Edeyen d' *see* Awbāri, Idhān
98 G13 **Oud-Beijerland** Zuid-Holland, SW Netherlands 51°50′N 04°25′E
98 F13 **Ouddorp** Zuid-Holland, SW Netherlands 51°49′N 03°55′E
77 P9 **Oudeïka** *oasis* C Mali
98 G13 **Oude Maas** ~ SW Netherlands
99 E18 **Oudenaarde** *Fr.* Audenarde. Oost-Vlaanderen, SW Belgium 50°50′N 03°37′E
99 H14 **Oudenbosch** Noord-Brabant, S Netherlands 51°35′N 04°32′E
98 P6 **Oude Pekela** Groningen, NE Netherlands 53°06′N 07°00′E
Ouderkerk *see* Ouderkerk aan den Amstel
98 I10 **Ouderkerk aan den Amstel** *var.* Ouderkerk. Noord-Holland, C Netherlands 52°18′N 04°54′E
98 I6 **Oudeschild** Noord-Holland, NW Netherlands 53°01′N 04°51′E
99 G14 **Oude-Tonge** Zuid-Holland, SW Netherlands 51°40′N 04°13′E
98 I12 **Oudewater** Utrecht, C Netherlands 52°02′N 04°54′E
Oudjda *see* Oujda
Oudkerk *see* Aldtsjerk
167 Q6 **Oudômxai** *var.* Muang Xay, Muong Sai, Xai. Oudômxai, N Laos 20°41′N 102°00′E
102 J7 **Oudon** ~ NW France
98 I9 **Oudorp** Noord-Holland, NW Netherlands 52°39′N 04°47′E
83 G25 **Oudtshoorn** Western Cape, SW South Africa 33°35′S 22°14′E
99 I16 **Oud-Turnhout** Antwerpen, N Belgium 51°19′N 05°01′E
74 F7 **Oued-Zem** C Morocco 32°53′N 06°30′W
187 P16 **Ouégoa** Province Nord, C New Caledonia 20°22′S 164°25′E
76 L13 **Ouéléssébougou** *var.* Ouolossébougou. Koulikoro, SW Mali 11°58′N 07°51′W
77 N16 **Ouéllé** E Ivory Coast 07°18′N 04°01′W
77 R16 **Ouémé** ~ C Benin
77 O13 **Ouessa** S Burkina Faso 11°02′N 02°44′W
102 D5 **Ouessant, Île d'** *Eng.* Ushant. *island* NW France
79 H17 **Ouésso** Sangha, NW Congo 01°38′N 16°03′E
79 D15 **Ouest** *Eng.* West. ◈ *province* W Cameroon
190 G11 **Ouest, Baie del'** *bay* Îles Wallis, E Wallis and Futuna
15 Y7 **Ouest, Pointe de l'** *headland* Québec, SE Canada 48°08′N 64°57′W
Ouezzane *see* Ouazzane
99 K20 **Ouffet** Liège, E Belgium 50°30′N 05°31′E
79 H14 **Ouham** ◈ *prefecture* NW Central African Republic
78 I13 **Ouham** ~ Central African Republic/Chad
79 G14 **Ouham-Pendé** ◈ *prefecture* W Central African Republic
77 R16 **Ouidah** *Eng.* Whydah, Wida. S Benin 06°23′N 02°08′E
74 H6 **Oujda** *Ar.* Oudjda, Ujda. NE Morocco 34°45′N 01°53′W
76 I7 **Oujeft** Adrar, C Mauritania 20°05′N 13°00′W
93 L15 **Oulainen** Pohjois-Pohjanmaa, C Finland 64°14′N 24°50′E
Ould Yanja *see* Ould Yenjé
76 J10 **Ould Yenjé** *var.* Ould Yanja. Guidimaka, S Mauritania 15°33′N 11°43′W
93 L14 **Oulu** *Swe.* Uleåborg. Pohjois-Pohjanmaa, C Finland 65°01′N 25°28′E
Oulu *Swe.* Uleåborg. ◈ *province* N Finland
93 L15 **Oulujärvi** *Swe.* Uleträsk. ◎ C Finland
93 M14 **Oulujoki** *Swe.* Uleälv. ~ C Finland
93 L14 **Oulunsalo** Pohjois-Pohjanmaa, C Finland 64°55′N 25°19′E
106 A8 **Oulx** Piemonte, NE Italy 45°05′N 06°41′E
78 J9 **Oum-Chalouba** Ennedi-Ouest, NE Chad 15°48′N 20°46′E
76 M16 **Oumé** C Ivory Coast 06°25′N 05°23′W
74 F7 **Oum er Rbia** ~ C Morocco
78 J10 **Oum-Hadjer** Batha, E Chad 13°18′N 19°41′E
92 K10 **Ounasjoki** ~ N Finland
78 J7 **Ounianga Kébir** Ennedi-Ouest, N Chad 19°06′N 20°29′E
Ouolossébougou *see* Ouéléssébougou
Oup *see* Auob
99 K19 **Oupeye** Liège, E Belgium 50°42′N 05°38′E
99 N21 **Our** ~ NW Europe
37 Q7 **Ouray** Colorado, C USA 38°01′N 107°40′W
103 R7 **Ource** ~ C France
104 G9 **Ourém** Santarém, C Portugal 39°40′N 08°32′W
104 H4 **Ourense** *Cast.* Orense, *Lat.* Aurium. Galicia, NW Spain 42°20′N 07°52′W
104 I4 **Ourense** *Cast.* Orense. ◈ *province* Galicia, NW Spain
59 O15 **Ouricuri** Pernambuco, E Brazil 07°51′S 40°05′W
60 J9 **Ourinhos** São Paulo, S Brazil 22°59′S 49°52′W

◆ Country ● Country Capital ◇ Dependent Territory ○ Dependent Territory Capital ◈ Administrative Regions ✈ International Airport ▲ Mountain ⛰ Mountain Range 🌋 Volcano ~ River ◎ Lake ⊠ Reservoir

104 G13 **Ourique** Beja, S Portugal 37°38′N 08°13′W
59 M20 **Ouro Preto** Minas Gerais, NE Brazil 20°25′S 43°30′W
Ours, Grand Lac de l' *see* Great Bear Lake
99 K20 **Ourthe** ≈ E Belgium
165 Q9 **Ōu-sanmyaku** ▲▲ Honshū, C Japan
97 M17 **Ouse** ≈ N England, United Kingdom
Ouse *see* Great Ouse
102 H7 **Oust** ≈ NW France
Outaouais *see* Ottawa
15 T4 **Outardes Quatre, Réservoir** ⊠ Québec, SE Canada
15 T5 **Outardes, Rivière aux** ≈ Québec, SE Canada
96 E8 **Outer Hebrides** *var.* Western Isles. *island group* NW Scotland, United Kingdom
30 K3 **Outer Island** *island* Apostle Islands, Wisconsin, N USA
35 S16 **Outer Santa Barbara Passage** *passage* California, SW USA
83 C18 **Outjo** Kunene, N Namibia 20°08′S 16°08′E
11 T16 **Outlook** Saskatchewan, S Canada 51°30′N 107°03′W
93 N16 **Outokumpu** Pohjois-Karjala, E Finland 62°43′N 29°05′E
96 M2 **Out Skerries** *island group* NE Scotland, United Kingdom
187 Q16 **Ouvéa** *island* Îles Loyauté, NE New Caledonia
103 S14 **Ouvèze** ≈ SE France
182 L9 **Ouyen** Victoria, SE Australia 35°07′S 142°19′E
39 Q14 **Ouzinkie** Kodiak Island, Alaska, USA 57°54′N 152°27′W
137 O13 **Ovacık** Tunceli, E Turkey 39°23′N 39°13′E
106 C9 **Ovada** Piemonte, NE Italy 44°41′N 08°39′E
187 X14 **Ovalau** *island* C Fiji
62 G9 **Ovalle** Coquimbo, N Chile 30°33′S 71°16′W
83 C17 **Ovamboland** *physical region* N Namibia
54 L10 **Ovana, Cerro** ▲ S Venezuela 04°41′N 66°54′W
104 G7 **Ovar** Aveiro, N Portugal 40°52′N 08°38′W
114 L10 **Ovcharitsa, Yazovir** ⊠ SE Bulgaria
54 E6 **Ovejas** Sucre, NW Colombia 09°32′N 75°14′W
101 E16 **Overath** Nordrhein-Westfalen, W Germany 50°55′N 07°16′E
98 F13 **Overflakkee** *island* SW Netherlands
99 H19 **Overijse** Vlaams Brabant, C Belgium 50°46′N 04°32′E
98 N10 **Overijssel** ◐ *province* E Netherlands
98 M9 **Overijssels Kanaal** *canal* E Netherlands
92 K13 **Överkalix** Norrbotten, N Sweden 66°19′N 22°49′E
27 R4 **Overland Park** Kansas, C USA 38°57′N 94°41′W
99 L14 **Overloon** Noord-Brabant, SE Netherlands 51°35′N 05°54′E
99 K16 **Overpelt** Limburg, NE Belgium 51°13′N 05°24′E
35 Y10 **Overton** Nevada, W USA 36°32′N 114°25′W
25 W7 **Overton** Texas, SW USA 32°16′N 94°58′W
92 K13 **Övertorneå** Norrbotten, N Sweden 66°22′N 23°40′E
95 N18 **Överum** Kalmar, S Sweden 57°58′N 16°20′E
92 G13 **Överuman** ⊚ N Sweden
117 P11 **Ovidiopol'** Odes'ka Oblast', SW Ukraine 46°15′N 30°27′E
116 M14 **Ovidiu** Constanța, SE Romania 44°16′N 28°34′E
45 N10 **Oviedo** SW Dominican Republic 17°47′N 71°22′W
104 K2 **Oviedo** *anc.* Asturias. Asturias, NW Spain 43°21′N 05°50′W
104 K2 **Oviedo** ✈ Asturias, N Spain 43°21′N 05°50′W
Ovilava *see* Wels
118 D7 **Oviši** W Latvia 57°34′N 21°43′E
146 K10 **Ovminzatovo Tog'lari** *Rus.* Gory Auminzatau. ▲▲ N Uzbekistan
Övögdiy *see* Telmen
Ovoot *see* Darīganga
157 O4 **Övörhangay** ◐ *province* C Mongolia
94 E12 **Øvre Årdal** Sogn Og Fjordane, S Norway 61°18′N 07°48′E
95 J14 **Övre Fryken** ⊚ C Sweden
92 J11 **Övre Soppero** *Lapp.* Badje-Sohppar. Norrbotten, N Sweden 68°07′N 21°40′E
117 N3 **Ovruch** Zhytomyrs'ka Oblast', N Ukraine 51°20′N 58°50′E
Övt *see* Bat-Öldziy
185 E24 **Owaka** Otago, South Island, New Zealand 46°27′S 169°42′E
79 H18 **Owando** *prev.* Fort Rousset. Cuvette, C Congo 0°29′S 15°55′E
164 J14 **Owase** Mie, Honshū, SW Japan 34°04′N 136°11′E
27 P9 **Owasso** Oklahoma, C USA 36°16′N 95°51′W
29 V10 **Owatonna** Minnesota, N USA 44°04′N 93°13′W
173 O4 **Owen Fracture Zone** *tectonic feature* W Arabian Sea
185 H15 **Owen, Mount** ▲ South Island, New Zealand 41°32′S 172°33′E
185 H15 **Owen River** Tasman, South Island, New Zealand 41°40′S 172°28′E
44 D8 **Owen Roberts** ✈ Grand Cayman, Cayman Islands 19°15′N 81°22′W
20 I6 **Owensboro** Kentucky, S USA 37°46′N 87°07′W
35 T11 **Owens Lake** *salt flat* California, W USA
14 F14 **Owen Sound** Ontario, S Canada 44°34′N 80°56′W
14 F13 **Owen Sound** ⊚ Ontario, S Canada
35 T10 **Owens River** ≈ California, W USA
186 F9 **Owen Stanley Range** ▲▲ S Papua New Guinea
27 V5 **Owensville** Missouri, C USA 38°21′N 91°30′W
20 M4 **Owenton** Kentucky, S USA 38°33′N 84°50′W
77 U17 **Owerri** Imo, S Nigeria 05°19′N 07°07′E
184 M10 **Owhango** Manawatu-Wanganui, North Island, New Zealand 39°01′S 175°22′E
21 N5 **Owingsville** Kentucky, S USA 38°09′N 83°46′W
77 T16 **Owo** Ondo, SW Nigeria 07°10′N 05°31′E
31 R9 **Owosso** Michigan, N USA 43°00′N 84°10′W
35 V1 **Owyhee** Nevada, W USA 41°57′N 116°07′W
32 L14 **Owyhee, Lake** ⊚ Oregon, NW USA
32 L15 **Owyhee River** ≈ Idaho/Oregon, NW USA
92 K1 **Öxarfjörður** *var.* Axarfjördhur. *fjord* N Iceland
94 K12 **Oxberg** Dalarna, C Sweden 61°07′N 14°10′E
11 V17 **Oxbow** Saskatchewan, S Canada 49°16′N 102°12′W
95 O17 **Oxelösund** Södermanland, S Sweden 58°40′N 17°10′E
185 H18 **Oxford** Canterbury, South Island, New Zealand 43°18′S 172°10′E
97 M21 **Oxford** *Lat.* Oxonia. S England, United Kingdom 51°46′N 01°15′W
23 Q3 **Oxford** Alabama, S USA 33°36′N 85°50′W
22 L2 **Oxford** Mississippi, S USA 34°23′N 89°30′W
29 N16 **Oxford** Nebraska, C USA 40°15′N 99°37′W
18 I11 **Oxford** New York, NE USA 42°21′N 75°39′W
21 U8 **Oxford** North Carolina, SE USA 36°22′N 78°37′W
31 Q14 **Oxford** Ohio, N USA 39°30′N 84°45′W
18 H16 **Oxford** Pennsylvania, NE USA 39°46′N 75°57′W
11 X12 **Oxford House** Manitoba, C Canada 54°55′N 95°13′W
29 Y13 **Oxford Junction** Iowa, C USA 41°58′N 90°57′W
11 X12 **Oxford Lake** ⊚ Manitoba, C Canada
97 M21 **Oxfordshire** *cultural region* S England, United Kingdom
Oxia *see* Oxyá
41 X12 **Oxkutzcab** Yucatán, SE Mexico 20°18′N 89°26′W
35 R15 **Oxnard** California, W USA 34°12′N 119°10′W
Oxonia *see* Oxford
14 I12 **Oxtongue** ≈ Ontario, SE Canada
Oxus *see* Amu Darya
115 E15 **Oxyá** *var.* Oxia. ▲ C Greece 39°46′N 21°56′E
164 L11 **Oyabe** Toyama, Honshū, SW Japan 36°42′N 136°52′E
Oyahue/Oyahue, Volcán *see* Ollagüe, Volcán
165 O12 **Oyama** Tochigi, Honshū, S Japan 36°19′N 139°46′E
47 U5 **Oyapock** ≈ E French Guiana
Oyapock *see* Oiapoque, Rio/Oyapok, Fleuve l'
55 Z10 **Oyapok, Baie de L'** *bay* Brazil/French Guiana South America W Atlantic Ocean
55 Z11 **Oyapok, Fleuve l'** *var.* Rio Oiapoque, Oyapock. ≈ Brazil/French Guiana *see also* Oiapoque, Rio
Oyapok, Fleuve l' *see* Oiapoque, Rio
79 E17 **Oyem** Woleu-Ntem, N Gabon 01°34′N 11°31′E
11 R16 **Oyen** Alberta, SW Canada 51°20′N 110°28′W
95 I15 **Øyeren** ⊚ S Norway
Oygon *see* Tüdevtey
96 I7 **Oykel** ≈ N Scotland, United Kingdom
123 R9 **Oymyakon** Respublika Sakha (Yakutiya), NE Russian Federation 63°28′N 142°22′E
79 H19 **Oyo** Cuvette, C Congo 01°17′S 16°00′E
77 S15 **Oyo** Oyo, W Nigeria 07°51′N 03°57′E
77 S15 **Oyo** ◐ *state* SW Nigeria
56 D13 **Oyón** Lima, C Peru 10°39′S 76°44′W
103 S10 **Oyonnax** Ain, E France 46°16′N 05°39′E
146 L10 **Oyoqog'itma** *Rus.* Ayakagytma. Buxoro Viloyati, C Uzbekistan 40°37′N 64°26′E
146 M9 **Oyoqquduq** *Rus.* Ayakkuduk. Navoiy Viloyati, N Uzbekistan 41°16′N 65°12′E
32 F9 **Oysterville** Washington, NW USA 46°33′N 124°03′W
95 D14 **Øystese** Hordaland, S Norway 60°23′N 06°13′E
145 S16 **Oytal** Zhambyl, S Kazakhstan 42°54′N 73°21′E
147 U10 **Oy-Tal** Oshskaya Oblast', SW Kyrgyzstan 40°23′N 74°04′E
147 T10 **Oy-Tal** ≈ SW Kyrgyzstan
145 Q15 **Oyyk** *prev.* Uyuk. Zhambyl, S Kazakhstan 43°46′N 70°55′E
144 H10 **Oyyl** *prev.* Uil. Aktyubinsk, W Kazakhstan 49°06′N 54°41′E
144 H10 **Oyyl** *prev.* Uil. ≈ W Kazakhstan
Ozarichi *see* Azarychy
23 R7 **Ozark** Alabama, S USA 31°27′N 85°38′W
27 S10 **Ozark** Arkansas, C USA 35°30′N 93°50′W
27 T8 **Ozark** Missouri, C USA 37°01′N 93°12′W
27 T8 **Ozark Plateau** *plain* Arkansas/Missouri, C USA
27 T6 **Ozarks, Lake of the** ⊠ Missouri, C USA
192 L10 **Ozbourn Seamount** *undersea feature* W Pacific Ocean 26°00′S 174°49′W
111 L20 **Ózd** Borsod-Abaúj-Zemplén, NE Hungary 48°15′N 20°18′E
112 D11 **Ozeblin** ▲ C Croatia 44°37′N 15°52′E
123 V11 **Ozernovskiy** Kamchatskiy Kray, E Russian Federation 51°28′N 156°32′E
144 M7 **Ozërnoye** *var.* Ozërnyy. Kostanay, N Kazakhstan 53°29′N 63°14′E
124 J15 **Ozërnyy** Tverskaya Oblast', W Russian Federation 57°55′N 33°45′E
Ozërnyy *see* Ozërnoye
Ozero Azhbulat *see* Ozero Ul'ken Azhibolat
Ozero Segozero *see* Segozerskoye Vodokhranilishche
115 D18 **Ozerós, Límni** ⊚ W Greece
145 T7 **Ozero Ul'ken Azhibolat** *prev.* Ozero Azhbulat. ⊚ NE Kazakhstan
122 G11 **Ozërsk** Chelyabinskaya Oblast', C Russian Federation 55°44′N 60°59′E
119 D14 **Ozersk** *prev.* Darkehnen, *Ger.* Angerapp. Kaliningradskaya Oblast', W Russian Federation 54°23′N 21°59′E
126 L4 **Ozery** Moskovskaya Oblast', W Russian Federation 54°51′N 38°37′E
Özgön *see* Uzgen
107 C17 **Ozieri** Sardegna, Italy, C Mediterranean Sea 40°35′N 09°01′E
111 I15 **Ozimek** *Ger.* Malapane. Opolskie, SW Poland 50°41′N 18°16′E
127 R8 **Ozinki** Saratovskaya Oblast', W Russian Federation 51°16′N 49°45′E
25 O10 **Ozona** Texas, SW USA 30°43′N 101°13′W
Ozorkov *see* Ozorków
110 J12 **Ozorków** *Rus.* Ozorkov. Łódýz, C Poland 52°00′N 19°17′E
164 F14 **Ōzu** Ehime, Shikoku, SW Japan 33°30′N 132°33′E
137 R10 **Ozurgeti** *prev.* Makharadze, Ozurget'i. W Georgia 41°57′N 42°01′E
Ozurget'i *see* Ozurgeti

P

99 J17 **Paal** Limburg, NE Belgium 51°03′N 05°08′E
196 M14 **Paamiut** *var.* Pâmiut, *Dan.* Frederikshåb. Sermersooq, S Greenland 61°59′N 49°40′W
Pa-an *see* Hpa-an
101 L22 **Paar** ≈ SE Germany
83 E26 **Paarl** Western Cape, SW South Africa 33°45′S 18°58′E
93 L15 **Paavola** Pohjois-Pohjanmaa, C Finland 64°34′N 25°15′E
96 E8 **Pabbay** *island* NW Scotland, United Kingdom
153 T15 **Pabna** Rajshahi, W Bangladesh 24°02′N 89°15′E
109 U4 **Pabneukirchen** Oberösterreich, N Austria 48°19′N 14°49′E
118 H13 **Pabradė** *Pol.* Podbrodzie. Vilnius, SE Lithuania 54°58′N 25°43′E
56 L13 **Pacahuaras, Río** ≈ N Bolivia
Pacaraima, Sierra/Pacaraim, Serra *see* Pakaraima Mountains
56 B11 **Pacasmayo** La Libertad, W Peru 07°27′S 79°33′W
42 D6 **Pacaya, Volcán de** ⛰ S Guatemala 14°19′N 90°36′W
115 K23 **Pacheía** *var.* Pachía. *island* Kykládes, Greece, Aegean Sea
Pachía *see* Pacheía
107 L26 **Pachino** Sicilia, Italy, C Mediterranean Sea 36°43′N 15°06′E
56 F12 **Pachitea, Río** ≈ C Peru
154 I11 **Pachmarhi** Madhya Pradesh, C India 22°36′N 78°18′E
121 P3 **Páchna** *var.* Pakhna. SW Cyprus 34°47′N 32°48′E
115 H25 **Páchnes** ▲ Kríti, Greece, E Mediterranean Sea 35°19′N 24°00′E
54 F9 **Pacho** Cundinamarca, C Colombia 05°09′N 74°08′W
154 F12 **Pāchora** Mahārāshtra, C India 20°52′N 75°28′E
41 P13 **Pachuca** *var.* Pachuca de Soto. Hidalgo, C Mexico 20°05′N 98°46′W
Pachuca de Soto *see* Pachuca
27 W5 **Pacific** Missouri, C USA 38°28′N 90°44′W
192 L14 **Pacific-Antarctic Ridge** *undersea feature* S Pacific Ocean 62°00′S 157°00′W
32 F8 **Pacific Beach** Washington, NW USA 47°09′N 124°12′W
35 N10 **Pacific Grove** California, W USA 36°35′N 121°54′W
29 S15 **Pacific Junction** Iowa, C USA 41°01′N 95°48′W
192-193 **Pacific Ocean** *ocean*
129 Z10 **Pacific Plate** *tectonic feature*
113 J15 **Pačir** ▲ N Montenegro 43°19′N 19°07′E
182 L5 **Packsaddle** New South Wales, SE Australia 30°42′S 141°55′E
32 H9 **Packwood** Washington, NW USA 46°37′N 121°38′W
Padalung *see* Phatthalung
168 J12 **Padang** Sumatera, W Indonesia 01°S 100°21′E
168 L9 **Padang Endau** Pahang, Peninsular Malaysia 02°38′N 103°37′E
Padangpandjang *see* Padangpanjang
168 I11 **Padangpanjang** *prev.* Padangpandjang. Sumatera, W Indonesia 0°30′S 100°26′E
168 I10 **Padangsidempuan** *prev.* Padangsidimpoean. Sumatera, W Indonesia 01°23′N 99°15′E
Padangsidimpoean *see* Padangsidempuan
124 I9 **Padany** Respublika Kareliya, NW Russian Federation 63°18′N 33°20′E
93 M18 **Padasjoki** Päijät-Häme, S Finland 61°20′N 25°21′E
57 M22 **Padcaya** Tarija, S Bolivia 21°52′S 64°46′W
101 H14 **Paderborn** Nordrhein-Westfalen, NW Germany 51°43′N 08°45′E
Padeșul/Padeș, Vîrful *see* Padeș, Vârful
116 F12 **Padeș, Vârful** *var.* Padeșul; *prev.* Vîrful Padeș. ▲ W Romania 45°39′N 22°19′E
112 L10 **Padinska Skela** Serbia, N Serbia 44°58′N 20°25′E
153 S14 **Padma** *var.* Ganges. ≈ Bangladesh/India *see also* Ganges
Padma *see* Brahmaputra
Padma *see* Ganges
106 H8 **Padova** *Eng.* Padua; *anc.* Patavium. Veneto, NE Italy 45°24′N 11°52′E
82 A10 **Padrão, Ponta do** *headland* NW Angola 06°06′S 12°18′E
25 T16 **Padre Island** *island* Texas, SW USA
104 G3 **Padrón** Galicia, NW Spain 42°44′N 08°40′W
118 K13 **Padsvillye** *Rus.* Podsvil'ye. Vitsyebskaya Voblasts', N Belarus 55°09′N 27°58′E
182 K11 **Padthaway** South Australia 36°39′S 140°30′E
Padua *see* Padova
20 G7 **Paducah** Kentucky, S USA 37°03′N 88°36′W
25 P4 **Paducah** Texas, SW USA 34°01′N 100°18′W
105 N15 **Padul** Andalucía, S Spain 37°02′N 03°37′W
191 P8 **Paea** Tahiti, W French Polynesia 17°41′S 149°35′W
185 L14 **Paekakariki** Wellington, North Island, New Zealand 41°00′S 174°58′E
163 X11 **Paektu-san** *var.* Baitou Shan. ▲ China/North Korea 42°00′N 128°03′W
Paengnyŏng *see* Baengnyong-do
184 M7 **Paeroa** Waikato, North Island, New Zealand 37°23′S 175°39′E
54 D12 **Páez** Cauca, SW Colombia 02°37′N 76°00′W
121 O3 **Páfos** *var.* Paphos. W Cyprus 34°46′N 32°26′E
121 O3 **Páfos** ✈ SW Cyprus 34°46′N 32°25′E
83 L19 **Pafúri** Gaza, SW Mozambique 22°27′S 31°21′E
112 C12 **Pag** *It.* Pago. Lika-Senj, SW Croatia 44°26′N 15°01′E
112 B11 **Pag** *It.* Pago. *island* Zadar, C Croatia
171 P7 **Pagadian** Mindanao, S Philippines 07°47′N 123°22′E
168 J13 **Pagai Selatan, Pulau** *island* Kepulauan Mentawai, W Indonesia
168 J13 **Pagai Utara, Pulau** *island* Kepulauan Mentawai, W Indonesia
188 K4 **Pagan** *island* C Northern Mariana Islands
115 G16 **Pagasitikós Kólpos** *gulf* E Greece
36 L8 **Page** Arizona, SW USA 36°54′N 111°28′W
29 Q5 **Page** North Dakota, N USA 47°09′N 97°33′W
118 D13 **Pagėgiai** *Ger.* Pogegen. Tauragė, SW Lithuania 55°08′N 21°54′E
21 S11 **Pageland** South Carolina, SE USA 34°46′N 80°23′W
81 G16 **Pager** ≈ NE Uganda
149 Q5 **Paghmān** Kābul, E Afghanistan 34°33′N 68°55′E
Pago *see* Pag
188 C16 **Pago Bay** *bay* E Guam, W Pacific Ocean
115 M20 **Pagóndas** *var.* Pagóndhas. Sámos, Dodekánisa, Greece, Aegean Sea 37°41′N 26°50′E
Pagóndhas *see* Pagóndas
192 J16 **Pago Pago** ○ (American Samoa) Tutuila, W American Samoa 14°16′S 170°43′W
37 R8 **Pagosa Springs** Colorado, C USA 37°13′N 107°01′W
Pagqên *see* Gadê
38 H12 **Pāhala** *var.* Pahala. Hawaii, USA, C Pacific Ocean 19°12′N 155°28′W
168 K8 **Pahang** *var.* Negeri Pahang Darul Makmur. ◐ *state* Peninsular Malaysia
Pahang *see* Pahang, Sungai
168 L8 **Pahang, Sungai** *var.* Pahang, Sungei Pahang. ≈ Peninsular Malaysia
149 S8 **Pahārpur** Khyber Pakhtunkhwa, NW Pakistan 32°06′N 71°00′E
185 B24 **Pahia Point** *headland* South Island, New Zealand 46°19′S 167°42′E
184 M13 **Pahiatua** Manawatu-Wanganui, North Island, New Zealand 40°30′S 175°49′E
38 H12 **Pāhoa** *var.* Pahoa. Hawaii, USA, C Pacific Ocean 19°29′N 154°56′W
23 Y14 **Pahokee** Florida, SE USA 26°49′N 80°40′W
35 X9 **Pahranagat Range** ▲▲ Nevada, W USA
35 W11 **Pahrump** Nevada, W USA 36°11′N 115°58′W
Pahsien *see* Chongqing
35 V9 **Pahute Mesa** ▲▲ Nevada, W USA
167 N7 **Pai** Mae Hong Son, NW Thailand 19°24′N 98°26′E
38 F10 **Pa'ia** *var.* Paia. Maui, Hawaii, USA, C Pacific Ocean 20°54′N 156°22′W
Paia *see* Pa'ia
Pai-ch'eng *see* Baicheng
118 H4 **Paide** *Ger.* Weissenstein. Järvamaa, N Estonia 58°55′N 25°36′E
97 J24 **Paignton** SW England, United Kingdom 50°26′N 03°34′W
184 K3 **Paihia** Northland, North Island, New Zealand 35°18′S 174°06′E
93 M18 **Päijänne** ⊚ C Finland
Päijänne-Tavastland *see* Päijät-Häme
93 M19 **Päijät-Häme** *Swe.* Päijänne-Tavastland. ◐ *region* S Finland
114 F13 **Páïko** ▲▲ N Greece
57 M17 **Paila, Río** ≈ C Bolivia
167 Q12 **Pailĭn** Bătdâmbâng, W Cambodia 12°51′N 102°34′E
Pailing *see* Chun'an
54 F6 **Pailitas** Cesar, N Colombia 08°58′N 73°38′W
38 F9 **Pailolo Channel** *channel* Hawai'i, USA, C Pacific Ocean
93 K19 **Paimio** *Swe.* Pemar. Varsinais-Suomi, SW Finland 60°27′N 22°42′E
165 O16 **Paimi-saki** *var.* Yaeme-saki. *headland* Iriomote-jima, SW Japan 24°18′N 123°40′E
102 G5 **Paimpol** Côtes d'Armor, NW France 48°47′N 03°03′W
168 J12 **Painan** Sumatera, W Indonesia 01°22′S 100°33′E
63 G23 **Paine, Cerro** ▲ S Chile 51°01′S 72°57′W
31 U11 **Painesville** Ohio, N USA 41°43′N 81°15′W
31 S14 **Paint Creek** ≈ Ohio, N USA
36 L10 **Painted Desert** *desert* Arizona, SW USA
Paint Hills *see* Wemindji
30 M4 **Paint River** ≈ Michigan, N USA
25 P8 **Paint Rock** Texas, SW USA 31°32′N 99°56′W
21 O6 **Paintsville** Kentucky, S USA 37°48′N 82°48′W
Paisance *see* Piacenza
96 I12 **Paisley** W Scotland, United Kingdom 55°50′N 04°26′W
32 I15 **Paisley** Oregon, NW USA 42°40′N 120°31′W
105 O3 **País Vasco** *Basq.* Euskadi, *Eng.* The Basque Country, *Sp.* Provincias Vascongadas. ◐ *autonomous community* N Spain
56 A9 **Paita** Piura, NW Peru 05°11′S 81°09′W
169 V6 **Paitan, Teluk** *bay* Sabah, East Malaysia
104 H7 **Paiva, Rio** ≈ N Portugal
92 K12 **Pajala** Norrbotten, N Sweden 67°12′N 23°19′E
104 K3 **Pajares, Puerto de** *pass* NW Spain
54 G9 **Pajarito** Boyacá, C Colombia 05°18′N 72°43′W
54 G4 **Pajaro** La Guajira, S Colombia 11°41′N 72°37′W
Pakanbaru *see* Pekanbaru
55 Q10 **Pakaraima Mountains** *var.* Serra Pacaraim, Sierra Pacaraima. ▲▲ N South America
Pākaur *see* Pākur
167 P10 **Pak Chong** Nakhon Ratchasima, C Thailand 14°38′N 101°22′E
123 V8 **Pakhachi** Krasnoyarskiy Kray, E Russian Federation 60°36′N 168°59′E
Pakhna *see* Páchna
189 U16 **Pakin Atoll** *atoll* Caroline Islands, E Micronesia
149 Q12 **Pakistan** *off.* Islamic Republic of Pakistan, *var.* Islami Jamhuriya e Pakistan. ◆ *republic* S Asia
Pakistan, Islamic Republic of *see* Pakistan
Pakistan, Islami Jamhuriya e *see* Pakistan
167 P8 **Pak Lay** *var.* Muang Pak Lay. Xaignabouli, C Laos 18°06′N 101°21′E
Paknam *see* Samut Prakan
166 L5 **Pakokku** Magway, C Myanmar (Burma) 21°20′N 95°05′E
110 I10 **Pakość** *Ger.* Pakosch. Kujawski-pomorskie, C Poland 52°47′N 18°03′E
Pakosch *see* Pakość
149 V10 **Pākpattan** Punjab, E Pakistan 30°20′N 73°27′E
167 O15 **Pak Phanang** *var.* Ban Pak Phanang. Nakhon Si Thammarat, SW Thailand 08°20′N 100°10′E
112 G9 **Pakrac** *Hung.* Pakrácz. Požega-Slavonija, NE Croatia 45°26′N 17°09′E
Pakrácz *see* Pakrac
118 F11 **Pakruojis** Šiaulai, N Lithuania 56°N 23°51′E
111 J24 **Paks** Tolna, S Hungary 46°38′N 18°51′E
Pak Sane *see* Pakxan
Paksé *see* Pakxé
167 Q10 **Pak Thong Chai** Nakhon Ratchasima, C Thailand 14°43′N 102°01′E
Paktīā *see* Paktiyā
149 Q7 **Paktīkā** ◐ *province* SE Afghanistan
149 R6 **Paktiyā** *prev.* Paktīā. ◐ *province* SE Afghanistan
171 N12 **Pakuli** Sulawesi, C Indonesia 01°14′S 119°55′E
153 S14 **Pākur** *var.* Pākaur. Jharkhand, N India 24°48′N 87°14′E
81 F17 **Pakwach** NW Uganda 02°28′N 31°28′E
167 R8 **Pakxan** *var.* Muang Pakxan, Pak Sane. Bolikhamxai, C Laos 18°27′N 103°38′E
167 S10 **Pakxé** *var.* Paksé. Champasak, S Laos 15°09′N 105°49′E
78 G12 **Pala** Mayo-Kébbi Ouest, SW Chad 09°22′N 14°54′E
61 A17 **Palacios** Santa Fe, C Argentina 30°43′S 61°37′W
25 V13 **Palacios** Texas, SW USA 28°42′N 96°13′W
105 X5 **Palafrugell** Cataluña, NE Spain 41°55′N 03°10′E
107 L24 **Palagonia** Sicilia, Italy, C Mediterranean Sea 37°20′N 14°45′E
113 E17 **Palagruža** *It.* Pelagosa. *island* SW Croatia
115 G20 **Palaiá Epídavros** Pelopónnisos, S Greece 37°38′N 23°09′E
121 P3 **Palaichóri** *var.* Palekhori. C Cyprus 34°55′N 33°06′E
115 H25 **Palaiochóra** Kríti, Greece, E Mediterranean Sea 35°14′N 23°37′E
115 A15 **Palaiolastrítsa** *religious building* Kérkyra, Iónia Nisiá, Greece, C Mediterranean Sea
115 J19 **Palaiópoli** Ándros, Kykládes, Greece, Aegean Sea 37°49′N 24°49′E
103 N5 **Palaiseau** Essonne, N France 48°41′N 02°14′E
154 N11 **Pāla Laharha** Odisha, E India 21°25′N 85°18′E
83 G19 **Palamakoloi** Ghanzi, C Botswana 23°10′S 22°22′E
115 E16 **Palamás** Thessalía, C Greece 39°28′N 22°05′E
105 X5 **Palamós** Cataluña, NE Spain 41°51′N 03°06′E
118 J5 **Palamuse** *Ger.* Sankt-Bartholomäi. Jõgevamaa, E Estonia 58°41′N 26°35′E
183 Q14 **Palana** Tasmania, SE Australia 39°48′S 147°54′E
123 U9 **Palana** Krasnoyarskiy Kray, E Russian Federation 59°05′N 159°59′E
118 C11 **Palanga** *Ger.* Polangen. Klaipėda, NW Lithuania 05°54′N 21°05′E
143 V10 **Palangān, Kūh-e** ▲▲ E Iran
Palangkaraja *see* Palangkaraya
169 T12 **Palangkaraya** *prev.* Palangkaraja. Borneo, C Indonesia 02°16′S 113°55′E
155 H22 **Palani** Tamil Nādu, SE India 10°30′N 77°24′E
Palanka *see* Bačka Palanka
154 D9 **Pālanpur** Gujarāt, W India 24°12′N 72°29′E
Palantia *see* Palencia
83 I19 **Palapye** Central, SE Botswana 22°37′S 27°06′E
155 J19 **Pālār** ≈ SE India
104 H3 **Palas de Rei** Galicia, NW Spain 42°52′N 07°51′W
123 T9 **Palatka** Magadanskaya Oblast', E Russian Federation 60°09′N 150°33′E
23 W10 **Palatka** Florida, SE USA 29°39′N 81°38′W
188 B9 **Palau** *var.* Belau. ◆ *republic* W Pacific Ocean
Palau *see* Palau Islands
129 Y14 **Palau Islands** *var.* Palau. *island group* N Palau
192 G16 **Palauli Bay** *bay* Savai'i, C Samoa, C Pacific Ocean
167 N11 **Palaw** Taninthayi, S Myanmar (Burma) 12°57′N 98°39′E
170 M6 **Palawan** *island* W Philippines
171 N6 **Palawan Passage** *passage* W Philippines
192 E7 **Palawan Trough** *undersea feature* S South China Sea 07°00′N 115°00′E
155 H23 **Pālayankottai** Tamil Nādu, SE India 08°42′N 77°46′E
107 L25 **Palazzola Acreide** *anc.* Acrae. Sicilia, Italy, C Mediterranean Sea 37°04′N 14°54′E
118 G3 **Paldiski** *prev.* Baltiski, *Eng.* Baltic Port, *Ger.* Baltischport. Harjumaa, NW Estonia 59°22′N 24°08′E
112 I13 **Pale** Republika Srpska, SE Bosnia and Herzegovina 43°49′N 18°35′E
Palekhori *see* Palaichóri
168 L13 **Palembang** Sumatera, W Indonesia 02°59′S 104°45′E
63 G18 **Palena** Los Lagos, S Chile 43°40′S 71°50′W
63 G18 **Palena, Río** ≈ S Chile
104 M5 **Palencia** *anc.* Palantia, Pallantia. Castilla y León, NW Spain 42°23′N 04°32′W
104 M3 **Palencia** ◐ *province* Castilla y León, N Spain
35 X15 **Palen Dry Lake** ⊚ California, W USA
41 V15 **Palenque** Chiapas, SE Mexico 17°32′N 91°59′W
41 V15 **Palenque** *var.* Ruinas de Palenque. *ruins* Chiapas, SE Mexico
45 O9 **Palenque, Punta** *headland* S Dominican Republic 18°13′N 70°08′W
Palenque, Ruinas de *see* Palenque
Palerme *see* Palermo
107 I23 **Palermo** *Fr.* Palerme; *anc.* Panhormus, Panormus. Sicilia, Italy, C Mediterranean Sea 38°08′N 13°23′E
25 V8 **Palestine** Texas, SW USA 31°45′N 93°39′W
25 V7 **Palestine, Lake** ⊠ Texas, SW USA
107 I15 **Palestrina** Lazio, C Italy 41°49′N 12°53′E
166 K5 **Paletwa** Chin State, W Myanmar (Burma) 21°25′N 92°49′E
155 G21 **Pālghāt** *var.* Palakkad. Kerala, SW India 10°46′N 76°42′E *see also* Palakkad
152 F13 **Pāli** Rājasthān, N India 25°48′N 73°21′E
167 N16 **Palian** Trang, SW Thailand
189 O12 **Palikir** ● (Micronesia) Pohnpei, E Micronesia 06°58′N 158°13′E
Palimé *see* Kpalimé
107 L19 **Palinuro, Capo** *headland* S Italy 40°02′N 15°16′E
115 H15 **Palioúri, Akrotírio** *var.* Akrotírio Kanestron. *headland* N Greece 39°55′N 23°45′E
33 R14 **Palisades Reservoir** ⊠ Idaho, NW USA
99 J23 **Paliseul** Luxembourg, SE Belgium 49°55′N 05°09′E
154 C11 **Pālitāna** Gujarāt, W India 21°30′N 71°50′E
118 F4 **Palivere** Läänemaa, W Estonia 58°59′N 23°58′E
41 V14 **Palizada** Campeche, SE Mexico 18°15′N 92°03′W
93 L18 **Pälkäne** Pirkanmaa, W Finland 61°22′N 24°16′E
155 J22 **Palk Strait** *strait* India/Sri Lanka
155 J23 **Pallai** Northern Province, NW Sri Lanka 09°35′N 80°20′E
Pallantia *see* Palencia
106 C6 **Pallanza** Piemonte, NE Italy 45°57′N 08°32′E
127 Q9 **Pallasovka** Volgogradskaya Oblast', SW Russian Federation 50°06′N 46°52′E
Pallene/Pallíni *see* Kassándra
185 L15 **Palliser Bay** *bay* North Island, New Zealand
185 L15 **Palliser, Cape** *headland* North Island, New Zealand 41°37′S 175°16′E
191 U9 **Palliser, Îles** *island group* Îles Tuamotu, C French Polynesia
82 Q12 **Palma** Cabo Delgado, N Mozambique 10°46′S 40°30′E
105 X9 **Palma** *var.* Palma de Mallorca. Mallorca, Spain, W Mediterranean Sea 39°35′N 02°39′E
105 X9 **Palma** ✈ Mallorca, Spain, W Mediterranean Sea
105 X10 **Palma, Badia de** *bay* Mallorca, Spain, W Mediterranean Sea
104 L13 **Palma del Río** Andalucía, S Spain 37°42′N 05°16′W
Palma de Mallorca *see* Palma
107 J25 **Palma di Montechiaro** Sicilia, Italy, C Mediterranean Sea 37°12′N 13°46′E
106 J7 **Palmanova** Friuli-Venezia Giulia, NE Italy 45°54′N 13°20′E
54 J7 **Palmarito** Apure, C Venezuela 07°36′N 70°08′W
43 N15 **Palmar Sur** Puntarenas, SE Costa Rica 08°54′N 83°27′W
60 I12 **Palmas** Paraná, S Brazil 26°29′S 52°00′W
59 K16 **Palmas** *var.* Palmas do Tocantins. Tocantins, C Brazil 10°24′S 48°19′W
76 L18 **Palmas, Cape** *Fr.* Cap des Palmès. *headland* SW Ivory Coast 04°18′N 07°31′W
Palmas do Tocantins *see* Palmas
54 D11 **Palmaseca** ✈ (Cali) Valle del Cauca, SW Colombia 03°31′N 76°27′W
107 B21 **Palmas, Golfo di** *gulf* Sardegna, Italy, C Mediterranean Sea
44 I7 **Palma Soriano** Santiago de Cuba, E Cuba 20°10′N 76°00′W
23 Y12 **Palm Bay** Florida, SE USA 28°01′N 80°35′W
35 T14 **Palmdale** California, W USA 34°34′N 118°07′W
61 H14 **Palmeira das Missões** Rio Grande do Sul, S Brazil 27°54′S 53°20′W
82 A11 **Palmeirinhas, Ponta das** *headland* NW Angola 09°04′S 13°02′E
39 R11 **Palmer** Alaska, USA 61°36′N 149°06′W
19 N11 **Palmer** Massachusetts, NE USA 42°09′N 72°19′W
25 U7 **Palmer** Texas, SW USA 32°25′N 96°40′W
194 H4 **Palmer** *US research station* Antarctica 64°37′S 64°01′W
15 R11 **Palmer** ≈ Québec, SE Canada
37 T5 **Palmer Lake** Colorado, C USA 39°07′N 104°55′W
194 J6 **Palmer Land** *physical region* Antarctica
14 F15 **Palmerston** Ontario, SE Canada 43°51′N 80°49′W
185 F22 **Palmerston** Otago, South Island, New Zealand 45°27′S 170°42′E
190 K15 **Palmerston** *island* S Cook Islands
Palmerston *see* Darwin
184 M12 **Palmerston North** Manawatu-Wanganui, North Island, New Zealand 40°20′S 175°52′E
23 V13 **Palmetto** Florida, SE USA 27°31′N 82°34′W
The Palmetto State *see* South Carolina
107 M22 **Palmi** Calabria, SW Italy 38°21′N 15°51′E
54 D11 **Palmira** Valle del Cauca, W Colombia 03°33′N 76°17′W
56 F8 **Palmira, Río** ≈ N Peru
61 D19 **Palmitas** Soriano, SW Uruguay 33°27′S 57°48′W
Palmnicken *see* Yantarnyy
35 V15 **Palm Springs** California, W USA 33°48′N 116°33′W
27 V2 **Palmyra** Missouri, C USA 39°48′N 91°31′W
18 G10 **Palmyra** New York, NE USA 43°02′N 77°13′W
18 G15 **Palmyra** Pennsylvania, NE USA 40°18′N 76°35′W
21 V5 **Palmyra** Virginia, NE USA 37°53′N 78°17′W
Palmyra *see* Tudmur
192 L7 **Palmyra Atoll** ◇ *US incorporated territory* C Pacific Ocean
154 P12 **Palmyras Point** *headland* E India 20°46′N 87°00′E
35 N9 **Palo Alto** California, W USA 37°26′N 122°08′W
25 O1 **Palo Duro Creek** ≈ Texas, SW USA
Paloe *see* Denpasar, Bali, C Indonesia
Paloe *see* Palu
168 L9 **Paloh** Johor, Peninsular Malaysia 02°10′N 103°11′E
80 F12 **Paloich** Upper Nile, NE South Sudan 10°29′N 32°31′E
40 I3 **Palomas** Chihuahua, N Mexico 31°45′N 107°38′W
107 I15 **Palombara Sabina** Lazio, C Italy 42°04′N 12°45′E
105 S13 **Palos, Cabo de** *headland* SE Spain 37°38′N 00°42′W
104 I14 **Palos de la Frontera** Andalucía, S Spain 37°14′N 06°53′W
60 G11 **Palotina** Paraná, S Brazil 24°16′S 53°49′W
32 M9 **Palouse** Washington, NW USA 46°54′N 117°04′W
32 L9 **Palouse River** ≈ Washington, NW USA
35 Y16 **Palo Verde** California, W USA 33°25′N 114°43′W
57 E16 **Palpa** Ica, W Peru 14°35′S 75°09′W
95 M16 **Pålsboda** Örebro, C Sweden 59°04′N 15°21′E
93 M15 **Paltamo** Kainuu, C Finland 64°25′N 27°50′E
171 N12 **Palu** *prev.* Paloe. Sulawesi, C Indonesia 0°54′S 119°52′E
137 P14 **Palu** Elazığ, E Turkey 38°43′N 39°56′E
152 I11 **Palwal** Haryāna, N India 28°15′N 77°18′E
123 U6 **Palyavaam** ≈ NE Russian Federation
77 Q13 **Pama** SE Burkina Faso 11°13′N 00°46′E
172 J14 **Pamandzi** ✈ (Mamoudzou) Petite-Terre, E Mayotte
Pamangkat *see* Pemangkat
143 R11 **Pā Mazār** Kermān, C Iran
83 N19 **Pambarra** Inhambane, SE Mozambique 21°57′S 35°06′E
171 X12 **Pamdai** Papua, E Indonesia 01°58′S 137°19′E
103 N16 **Pamiers** Ariège, S France 43°07′N 01°37′E
147 T14 **Pamir** *var.* Daryā-ye Pāmīr, *Taj.* Dar"yoi Pomir. ≈ Afghanistan/Tajikistan *see also* Pāmīr, Daryā-ye
Pamir *see* Pāmīr, Daryā-ye
149 U1 **Pāmīr, Daryā-ye** *var.* Pamir, *Taj.* Dar"yoi Pomir. ≈ Afghanistan/Tajikistan *see also* Pamir
Pāmīr, Daryā-ye *see* Pamir
Pāmīr-e Khord *see* Little Pamir
Pamir/Pāmīr, Daryā-ye *see* Pamirs
129 Q8 **Pamirs** *Pash.* Daryā-ye Pāmīr, *Rus.* Pamir. ▲▲ C Asia
Pâmiut *see* Paamiut
21 X10 **Pamlico River** ≈ North Carolina, SE USA
21 Y10 **Pamlico Sound** *sound* North Carolina, SE USA
25 O2 **Pampa** Texas, SW USA 35°32′N 100°58′W
Pampa Aullagas, Lago *see* Poopó, Lago
61 B21 **Pampa Húmeda** *grassland* E Argentina
56 A10 **Pampa las Salinas** *salt lake* NW Peru
57 F15 **Pampas** Huancavelica, C Peru 12°22′S 74°53′W
62 K13 **Pampas** *plain* C Argentina
55 O4 **Pampatar** Nueva Esparta, NE Venezuela 11°03′N 63°51′W
Pampeluna *see* Pamplona

◆ Country
● Country Capital
◇ Dependent Territory
○ Dependent Territory Capital
◐ Administrative Regions
✈ International Airport
▲ Mountain
▲▲ Mountain Range
⛰ Volcano
≈ River
⊚ Lake
⊠ Reservoir

104 H8 **Pampilhosa da Serra** var. Pampilhosa de Serra. Coimbra, N Portugal 40°03′N 07°58′W
173 Y15 **Pamplemousses** N Mauritius 20°06′S 57°34′E
54 G7 **Pamplona** Norte de Santander, N Colombia 07°24′N 72°38′W
105 Q3 **Pamplona** Basq. Iruña, prev. Pampeluna; anc. Pompaelo. Navarra, N Spain 42°49′N 01°39′W
114 I11 **Pamporovo** prev. Vasil Kolarov. Smolyan, S Bulgaria 41°39′N 24°45′E
136 D15 **Pamukkale** Denizli, W Turkey 37°51′N 29°13′E
21 W5 **Pamunkey River** ≈ Virginia, NE USA
152 K5 **Pamzal** Jammu and Kashmir, NW India 34°17′N 78°50′E
30 L14 **Pana** Illinois, N USA 39°23′N 89°04′W
41 Y11 **Panabá** Yucatán, SE Mexico 21°20′N 88°16′W
35 Y8 **Panaca** Nevada, W USA 37°47′N 114°24′W
115 E19 **Panachaikó** ▲▲ S Greece
14 F11 **Panache Lake** ◎ Ontario, S Canada
114 I10 **Panagyurishte** Pazardzhik, C Bulgaria 42°30′N 24°11′E
168 M16 **Panaitan, Pulau** island S Indonesia
115 D18 **Panaitolikó** ▲▲ C Greece
155 E17 **Panaji** var. Pangim, Panjim, New Goa. state capital Goa, W India 15°31′N 73°52′E
43 T15 **Panamá** var. Ciudad de Panamá, Eng. Panama City. ● (Panama) Panamá, C Panama 08°57′N 79°33′W
43 T14 **Panama** off. Republic of Panama. ◆ republic Central America
43 U14 **Panamá** off. Provincia de Panamá. ◈ province E Panama
43 U15 **Panamá, Bahía de** bay N Gulf of Panama
193 T7 **Panama Basin** undersea feature E Pacific Ocean 05°00′N 83°30′W
43 T15 **Panama Canal** canal E Panama
23 R9 **Panama City** Florida, SE USA 30°09′N 85°39′W
43 T14 **Panama City** ✈ Panamá, C Panama 09°02′N 79°24′W
Panama City see Panamá
23 Q9 **Panama City Beach** Florida, SE USA 30°10′N 85°48′W
43 T17 **Panamá, Golfo de** var. Gulf of Panama. gulf S Panama
Panama, Gulf of see Panamá, Golfo de
Panama, Isthmus of see Panama, Istmo de
43 T15 **Panama, Istmo de** Eng. Isthmus of Panama; prev. Isthmus of Darien. isthmus E Panama
Panamá, Provincia de see Panamá
Panama, Republic of see Panama
35 U11 **Panamint Range** ▲▲ California, W USA
107 L22 **Panarea, Isola** island Isole Eolie, S Italy
106 G9 **Panaro** ≈ N Italy
171 P5 **Panay Island** island C Philippines
35 W7 **Pancake Range** ▲▲ Nevada, W USA
112 M11 **Pančevo** Ger. Pantschowa, Hung. Pancsova. Vojvodina, N Serbia 44°53′N 20°40′E
113 M15 **Pančićev Vrh** ▲ SW Serbia 43°16′N 20°49′E
116 L12 **Panciu** Vrancea, E Romania 45°54′N 27°08′E
116 F10 **Pâncota** Hung. Pankota; prev. Pîncota. Arad, W Romania 46°20′N 21°45′E
Pancsova see Pančevo
83 N20 **Panda** Inhambane, SE Mozambique 24°02′S 34°45′E
171 X12 **Pandaidori, Kepulauan** island group E Indonesia
25 N11 **Pandale** Texas, SW USA 30°09′N 101°34′W
169 P12 **Pandang Tikar, Pulau** island N Indonesia
61 F20 **Pan de Azúcar** Maldonado, S Uruguay 34°45′S 55°14′W
118 H11 **Pandėlys** Panevėžys, NE Lithuania 56°04′N 25°18′E
155 F15 **Pandharpur** Mahārāshtra, W India 17°42′N 75°24′E
182 J1 **Pandie Pandie** South Australia 26°06′S 139°26′E
171 O12 **Pandiri** Sulawesi, C Indonesia 01°32′S 120°47′E
61 F20 **Pando** Canelones, S Uruguay 34°44′S 55°58′W
57 J14 **Pando** ◈ department N Bolivia
192 K9 **Pandora Bank** undersea feature W Pacific Ocean
95 G20 **Pandrup** Nordjylland, N Denmark 57°14′N 09°42′E
79 J15 **Pandu** Equateur, NW Dem. Rep. Congo 05°03′N 19°14′E
153 V12 **Pandu** Assam, NE India 26°08′N 91°37′E
Paneas see Bāniyās
59 F15 **Panelas** Mato Grosso, W Brazil 09°06′S 60°41′W
118 G12 **Panevėžys** Panevėžys, C Lithuania 55°44′N 24°21′E
118 G11 **Panevėžys** ◈ province NW Lithuania
Panfilov see Zharkent
127 N9 **Panfilovo** Volgogradskaya Oblast', SW Russian Federation 50°25′N 42°55′E
79 N17 **Panga** Orientale, N Dem. Rep. Congo 01°52′N 26°18′E
193 Y15 **Pangai** Lifuka, C Tonga 19°50′S 174°23′W
114 H13 **Pangaío** ▲▲ N Greece
79 G20 **Pangala** Pool, S Congo 03°26′S 14°38′E
81 J22 **Pangani** Tanga, E Tanzania 05°27′S 39°00′E
81 I21 **Pangani** ≈ NE Tanzania
186 K8 **Panggoe** Choiseul, NW Solomon Islands 07°00′S 157°05′E
79 N20 **Pangi** Maniema, E Dem. Rep. Congo 03°12′S 26°39′E
Pangim see Panaji
168 H8 **Pangkalanbrandan** Sumatera, W Indonesia 04°00′N 98°15′E
Pangkalanbun see Pangkalanbuun
169 R13 **Pangkalanbuun** var. Pangkalanbun. Borneo, C Indonesia 02°43′S 111°38′E
169 N12 **Pangkalpinang** Pulau Bangka, W Indonesia 02°05′S 106°09′E
11 U17 **Pangman** Saskatchewan, S Canada 49°37′N 104°33′W
Pang-Nga see Phang-Nga
9 S6 **Pangnirtung** Baffin Island, Nunavut, NE Canada 66°05′N 65°45′W
152 K6 **Pangong Tso** var. Bangong Co. ◎ China/India see also Bangong Co
Pangong Tso see Banggong Co
36 K7 **Panguitch** Utah, W USA 37°49′N 112°26′W
186 J7 **Panguna** Bougainville Island, NE Papua New Guinea 06°22′S 155°20′E
171 N8 **Pangutaran Group** island group Sulu Archipelago, SW Philippines
25 N2 **Panhandle** Texas, SW USA 35°21′N 101°24′W
Panhormus see Palermo
171 W14 **Paniai, Danau** ◎ Papua, E Indonesia
79 L21 **Pania-Mutombo** Kasai-Oriental, C Dem. Rep. Congo 05°09′S 23°49′E
187 P16 **Panié, Mont** ▲ C New Caledonia 20°33′S 164°41′E
Panikoilli see Jājapur
152 I10 **Pānīpat** Haryāna, N India 29°18′N 77°00′E
147 Q14 **Panj** Rus. Pyandzh; prev. Kirovabad. SW Tajikistan 37°39′N 69°55′E
147 P15 **Panj** Rus. Pyandzh. ≈ Afghanistan/Tajikistan
149 O5 **Panjāb** Bāmyān, C Afghanistan 34°21′N 67°00′E
147 O12 **Panjakent** Rus. Pendzhikent. W Tajikistan 39°28′N 67°33′E
148 L14 **Panjgūr** Baluchistān, SW Pakistan 26°58′N 64°05′E
Panjim see Panaji
163 U12 **Panjin** Liaoning, NE China 41°11′N 122°05′E
147 P14 **Panji Poyon** Rus. Nizhniy Pyandzh. SW Tajikistan 37°14′N 68°32′E
149 Q4 **Panjshayr** prev. Panjshīr. ≈ E Afghanistan
149 S4 **Panjshīr** ◈ province NE Afghanistan
Panjshīr see Panjshayr
Pankota see Pâncota
77 W14 **Pankshin** Plateau, C Nigeria 09°21′N 09°27′E
163 Y10 **Pan Ling** ▲▲ N China
Panlong Jiang see Lô, Sông
154 J9 **Panna** Madhya Pradesh, C India 24°43′N 80°11′E
99 M16 **Panningen** Limburg, SE Netherlands 51°20′N 05°59′E
149 R13 **Pāno Āqil** Sind, SE Pakistan 27°55′N 69°18′E
121 P3 **Páno Léfkara** S Cyprus 34°52′N 33°18′E
121 O3 **Páno Panagiá** var. Pano Panayia. W Cyprus 34°55′N 32°38′E
Pano Panayia see Páno Panagiá
Panopolis see Akhmīm
29 U14 **Panora** Iowa, C USA 41°41′N 94°21′W
60 I8 **Panorama** São Paulo, S Brazil 21°22′S 51°51′W
115 I24 **Pánormos** Kríti, Greece, E Mediterranean Sea 35°24′N 24°42′E
Panormus see Palermo
163 W11 **Panshi** Jilin, NE China 42°56′N 126°02′E
Panshi Yu see Passu Keah
59 H19 **Pantanal** var. Pantanalmato-Grossense. swamp SW Brazil
Pantanalmato-Grossense see Pantanal
61 H16 **Pântano Grande** Rio Grande do Sul, S Brazil 30°12′S 52°24′W
171 Q16 **Pantar, Pulau** island Kepulauan Alor, S Indonesia
21 X9 **Pantego** North Carolina, SE USA 35°34′N 76°39′W
107 G25 **Pantelleria** anc. Cossyra, Cosyra. Sicilia, Italy, C Mediterranean Sea 36°47′N 12°00′E
107 G25 **Pantelleria, Isola di** island SW Italy
Pante Makasar/Pante Macassar/Pante Makassar see Ponte Macassar
152 K10 **Pantnagar** Uttarakhand, N India 29°00′N 79°28′E
115 A15 **Pantokrátoras** ▲ Kérkyra, Iónia Nisiá, Greece, C Mediterranean Sea 39°45′N 19°51′E
Pantschowa see Pančevo
41 P11 **Pánuco** Veracruz-Llave, E Mexico 22°01′N 98°13′W
41 P11 **Pánuco, Río** ≈ C Mexico
160 I12 **Panxian** Guizhou, S China 25°45′N 104°39′E
168 I10 **Panyabungan** Sumatera, N Indonesia 0°55′N 99°30′E
77 W14 **Panyam** Plateau, C Nigeria 09°28′N 09°13′E
157 N13 **Panzhihua** prev. Dukou, Tu-k'ou. Sichuan, C China 26°35′N 101°41′E
79 I22 **Panzi** Bandundu, SW Dem. Rep. Congo 07°10′S 17°55′E
42 E5 **Panzós** Alta Verapaz, E Guatemala 15°21′N 89°40′W
Pao-chi/Paoki see Baoji
Pao-king see Shaoyang
107 N20 **Paola** Calabria, SW Italy 39°21′N 16°03′E
121 P16 **Paola** E Malta 35°52′N 14°30′E
27 R5 **Paola** Kansas, C USA 38°34′N 94°54′W
31 O15 **Paoli** Indiana, N USA 38°35′N 86°25′W
187 R14 **Paonangisu** Éfaté, C Vanuatu 17°33′S 168°23′E
171 S13 **Paoni** var. Pauni. Pulau Seram, E Indonesia 02°48′S 129°03′E
37 Q5 **Paonia** Colorado, C USA 38°52′N 107°35′W
191 O7 **Paopao** Moorea, W French Polynesia 17°28′S 149°48′W
Pao-shan see Baoshan
Pao-ting see Baoding
Pao-t'ou/Paotow see Baotou
79 H14 **Paoua** Ouham-Pendé, W Central African Republic 07°22′N 16°25′E
Pap see Pop
111 H23 **Pápa** Veszprém, W Hungary 47°20′N 17°29′E
42 J12 **Papagayo, Golfo de** gulf NW Costa Rica
38 H11 **Pāpa'ikou** var. Papaikou. Hawaii, USA, C Pacific Ocean 19°45′N 155°06′W
41 R15 **Papaloapan, Río** ≈ S Mexico
184 L6 **Papakura** Auckland, North Island, New Zealand 37°03′S 174°57′E
41 Q13 **Papantla** var. Papantla de Olarte. Veracruz-Llave, E Mexico 20°30′N 97°21′W
Papantla de Olarte see Papantla
191 P8 **Papara** Tahiti, W French Polynesia 17°45′S 149°33′W
184 K4 **Paparoa** Northland, North Island, New Zealand 36°06′S 174°12′E
185 G16 **Paparoa Range** ▲▲ South Island, New Zealand
115 K20 **Pápas, Akrotírio** headland Ikaría, Dodekánisa, Greece, Aegean Sea 37°31′N 25°58′E
96 L2 **Papa Stour** island NE Scotland, United Kingdom
184 L6 **Papatoetoe** Auckland, North Island, New Zealand 36°58′S 174°52′E
185 E25 **Papatowai** Otago, South Island, New Zealand 46°33′S 169°33′E
96 K4 **Papa Westray** island NE Scotland, United Kingdom
191 T10 **Papeete** ○ (French Polynesia) Tahiti, W French Polynesia 17°32′S 149°34′W
100 F11 **Papenburg** Niedersachsen, NW Germany 53°04′N 07°24′E
98 H13 **Papendrecht** Zuid-Holland, SW Netherlands 51°50′N 04°42′E
191 Q7 **Papenoo** Tahiti, W French Polynesia 17°29′S 149°25′W
191 Q7 **Papenoo Rivière** ≈ Tahiti, W French Polynesia
191 N7 **Papetoai** Moorea, W French Polynesia 17°29′S 149°52′W
92 L3 **Papey** island E Iceland
Paphos see Páfos
40 H5 **Papigochic, Río** ≈ NW Mexico
118 E10 **Papilė** Šiaulai, NW Lithuania 56°08′N 22°51′E
29 S15 **Papillion** Nebraska, C USA 41°09′N 96°02′W
15 T5 **Papinachois** ≈ Québec, SE Canada
171 X13 **Papua** var. Irian Barat, West Irian, West New Guinea, West Papua; prev. Dutch New Guinea, Irian Jaya, Netherlands New Guinea. ◈ province E Indonesia
Papua and New Guinea, Territory of see Papua New Guinea
171 V10 **Papua Barat** off. Propinsi Papua Barat; prev. Irian Jaya Barat, Eng. West Papua. ◈ province E Indonesia
186 C9 **Papua, Gulf of** gulf S Papua New Guinea
186 C8 **Papua New Guinea** off. Independent State of Papua New Guinea; prev. Territory of Papua and New Guinea, abbrev. P.N.G. ◆ commonwealth republic NW Melanesia
Papua New Guinea, Independent State of see Papua New Guinea
192 H8 **Papua Plateau** undersea feature N Coral Sea
112 G9 **Papuk** ▲▲ NE Croatia
Papun see Hpapun
42 L14 **Paquera** Puntarenas, W Costa Rica 09°52′N 84°56′W
58 I13 **Pará** off. Estado do Pará. ◈ state NE Brazil
55 V9 **Para** ◈ district N Suriname
Pará see Belém
180 I8 **Paraburdoo** Western Australia 23°07′S 117°40′E
57 E16 **Paracas, Península de** peninsula W Peru
59 L19 **Paracatu** Minas Gerais, NE Brazil 17°14′S 46°52′W
192 E6 **Paracel Islands** Chin. Xisha Qundao, Viet. Quân Đao Hoang Sa. ◇ disputed territory SE Asia
182 I6 **Parachilna** South Australia 31°09′S 138°23′E
149 R6 **Pārachinār** Khyber Pakhtunkhwa, NW Pakistan 33°56′N 70°04′E
112 N13 **Paraćin** Serbia, C Serbia 43°51′N 21°25′E
Paradip see Krishnarājāsāgara
14 K8 **Paradis** Québec, SE Canada 48°13′N 76°36′W
39 N11 **Paradise** var. Paradise Hill. Alaska, USA 62°28′N 160°09′W
35 O5 **Paradise** California, W USA 39°42′N 121°39′W
35 X11 **Paradise** Nevada, W USA 36°05′N 115°10′W
Paradise Hill see Paradise
37 R11 **Paradise Hills** New Mexico, SW USA 35°12′N 106°42′W
Paradise of the Pacific see Hawai'i
36 L13 **Paradise Valley** Arizona, SW USA 33°31′N 111°56′W
35 T2 **Paradise Valley** Nevada, W USA 41°30′N 117°30′W
115 O22 **Paradisi** ✈ (Ródos) Ródos, Dodekánisa, Greece, Aegean Sea 36°24′N 28°08′E
154 P12 **Pārādwīp** Odisha, E India 20°17′N 86°42′E
Pará, Estado do see Pará
Paraetonium see Marsā Maṭrūḥ
117 R4 **Parafiyivka** Chernihivs'ka Oblast', N Ukraine 50°53′N 32°40′E
36 K7 **Paragonah** Utah, W USA 37°53′N 112°46′W
27 X9 **Paragould** Arkansas, C USA 36°02′N 90°30′W
47 X8 **Paraguaçu** var. Paraguassú. ≈ E Brazil
60 J9 **Paraguaçu Paulista** São Paulo, S Brazil 22°22′S 50°35′W
54 H4 **Paraguaipoa** Zulia, NW Venezuela 11°21′N 71°58′W
62 O6 **Paraguarí** Paraguarí, S Paraguay 25°36′S 57°06′W
62 O7 **Paraguarí** off. Departamento de Paraguarí. ◈ department S Paraguay
Paraguarí, Departamento de see Paraguarí
57 O16 **Paraguá, Río** ≈ NE Bolivia
55 O8 **Paragua, Río** ≈ SE Venezuela
Paraguassú see Paraguaçu
62 N5 **Paraguay** ◆ republic C South America
47 U10 **Paraguay** var. Río Paraguay. ≈ C South America
Paraguay, Río see Paraguay
Parahiba/Parahyba see Paraíba
59 P15 **Paraíba** off. Estado da Paraíba; prev. Parahiba, Parahyba. ◈ state E Brazil
Paraíba see João Pessoa
60 P9 **Paraíba do Sul, Rio** ≈ SE Brazil
Paraíba, Estado da see Paraíba
Parainen see Pargas
43 N14 **Paraíso** Cartago, C Costa Rica 09°51′N 83°50′W
41 U14 **Paraíso** Tabasco, SE Mexico 18°26′N 93°10′W
57 O17 **Paraíso, Río** ≈ E Bolivia
Parajd see Praid
77 S14 **Parakou** C Benin 09°23′N 02°40′E
115 F20 **Paralía Tyrou** Pelopónnisos, S Greece 37°17′N 22°50′E
121 Q2 **Paralímni** E Cyprus 35°02′N 34°00′E
115 G18 **Paralímni, Límni** ◎ C Greece
55 W8 **Paramaribo** ● (Suriname) Paramaribo, N Suriname 05°52′N 55°14′W
55 W9 **Paramaribo** ◈ district N Suriname
55 W9 **Paramaribo** ✈ Paramaribo, N Suriname 05°52′N 55°14′W
Paramithiá see Paramythiá
56 C13 **Paramonga** Lima, W Peru 10°42′S 77°50′W
123 V12 **Paramushir, Ostrov** island SE Russian Federation
115 C16 **Paramythiá** var. Paramithiá. Ípeiros, W Greece 39°28′N 20°31′E
62 M10 **Paraná** Entre Ríos, E Argentina 31°48′S 60°29′W
60 H11 **Paraná** off. Estado do Paraná. ◈ state S Brazil
47 U11 **Paraná** var. Alto Paraná. ≈ C South America
Paraná, Estado do see Paraná
60 K12 **Paranaguá** Paraná, S Brazil 25°32′S 48°36′W
59 J20 **Paranaíba, Rio** ≈ E Brazil
61 C19 **Paraná Ibicuy, Río** ≈ E Argentina
59 H15 **Paranaíta** Mato Grosso, W Brazil 09°35′S 57°01′W
60 H9 **Paranapanema, Rio** ≈ S Brazil
60 K11 **Paranapiacaba, Serra do** ▲▲ S Brazil
60 H9 **Paranavaí** Paraná, S Brazil 23°02′S 52°36′W
143 N5 **Parandak** Markazī, W Iran 35°19′N 50°40′E
114 I12 **Paranésti** var. Paranestio. Anatolikí Makedonía kai Thráki, NE Greece 41°16′N 24°31′E
Paranestio see Paranésti
191 W11 **Paraoa** atoll Îles Tuamotu, C French Polynesia
184 L13 **Paraparaumu** Wellington, North Island, New Zealand 40°55′S 175°01′E
57 N20 **Parapetí, Río** ≈ SE Bolivia
54 L10 **Paraque, Cerro** ▲ W Venezuela 06°00′S 67°00′W
154 I11 **Parāsiya** Madhya Pradesh, C India 22°11′N 78°50′E
115 M23 **Paraspóri, Akrotírio** headland Kárpathos, SE Greece 35°54′N 27°15′E
60 O10 **Parati** Rio de Janeiro, SE Brazil 23°15′S 44°42′W
59 K14 **Parauapebas** Pará, N Brazil 06°03′S 49°48′W
103 Q10 **Paray-le-Monial** Saône-et-Loire, C France 46°27′N 04°07′E
Parbatsar see Parvatsar
154 G13 **Parbhani** Mahārāshtra, C India 19°16′N 76°51′E
100 L10 **Parchim** Mecklenburg-Vorpommern, N Germany 53°26′N 11°51′E
Parchwitz see Prochowice
110 P13 **Parczew** Lubelskie, E Poland 51°40′N 23°E
60 L8 **Pardo, Rio** ≈ S Brazil
111 E16 **Pardubice** Ger. Pardubitz. Pardubický Kraj, C Czech Republic 50°01′N 15°47′E
111 E17 **Pardubický Kraj** ◈ region N Czech Republic
Pardubitz see Pardubice
119 F16 **Parechcha** Pol. Porzecze, Rus. Porech'ye. Hrodzyenskaya Voblasts', W Belarus 53°53′N 24°08′E
59 F17 **Parecis, Chapada dos** var. Serra dos Parecis. ▲▲ W Brazil
Parecis, Serra dos see Parecis, Chapada dos
104 M4 **Paredes de Nava** Castilla y León, N Spain 42°09′N 04°42′W
189 U12 **Parem** island Chuuk, C Micronesia
189 O12 **Parem Island** island E Micronesia
184 I1 **Parengarenga Harbour** inlet North Island, New Zealand
15 N8 **Parent** Québec, SE Canada 47°55′N 74°36′W
102 J14 **Parentis-en-Born** Landes, SW France 44°22′N 01°04′W
Parenzo see Poreč
185 G20 **Pareora** Canterbury, South Island, New Zealand 44°28′S 171°12′E
171 N14 **Parepare** Sulawesi, C Indonesia 04°S 119°40′E
115 B16 **Párga** Ípeiros, W Greece 39°18′N 20°19′E
93 K20 **Pargas** Swe. Parainen. Varsinais-Suomi, SW Finland 60°18′N 22°20′E
64 O5 **Pargo, Ponta do** headland Madeira, Portugal, NE Atlantic Ocean 32°48′N 17°17′W
Paria, Golfo de see Paria, Gulf of
55 N6 **Pariaguán** Anzoátegui, NE Venezuela 08°51′N 64°43′W
45 X17 **Paria, Gulf of** var. Golfo de Paria. gulf Trinidad and Tobago/Venezuela
57 I15 **Pariamanu, Río** ≈ E Peru
36 L8 **Paria River** ≈ Utah, W USA
Parichi see Parychy
40 M14 **Paricutín, Volcán** ⩕ C Mexico 19°25′N 102°20′W
43 P16 **Parida, Isla** island SW Panama
55 T8 **Parika** NE Guyana 06°51′N 58°25′W
93 O18 **Parikkala** Etelä-Karjala, SE Finland 61°33′N 29°34′E
58 E10 **Parima, Serra** var. Sierra Parima. ▲▲ Brazil/Venezuela see also Parima, Sierra
55 N11 **Parima, Sierra** var. Serra Parima. ▲▲ Brazil/Venezuela see also Parima, Serra
57 F17 **Parinacochas, Laguna** ◎ SW Peru
56 A9 **Pariñas, Punta** headland NW Peru 04°45′S 81°22′W
58 H12 **Parintins** Amazonas, N Brazil 02°38′S 56°45′W
103 O5 **Paris** anc. Lutetia, Lutetia Parisiorum, Parisii. ● (France) Paris, N France 48°52′N 02°19′E
191 Y2 **Paris** Kiritimati, E Kiribati 01°55′N 157°30′W
27 S11 **Paris** Arkansas, C USA 35°17′N 93°46′W
33 S16 **Paris** Idaho, NW USA 42°14′N 111°24′W
31 N14 **Paris** Illinois, N USA 39°36′N 87°42′W
20 M5 **Paris** Kentucky, S USA 38°13′N 84°15′W
27 V3 **Paris** Missouri, C USA 39°28′N 92°00′W
20 H8 **Paris** Tennessee, S USA 36°19′N 88°20′W
25 V5 **Paris** Texas, SW USA 33°41′N 95°33′W
Parisii see Paris
43 S16 **Parita** Herrera, S Panama 08°01′N 80°30′W
43 S16 **Parita, Bahía de** bay S Panama
93 K18 **Parkano** Pirkanmaa, W Finland 62°03′N 23°E
Parkan/Párkány see Štúrovo
27 N6 **Park City** Kansas, C USA 37°48′N 97°19′W
36 L3 **Park City** Utah, W USA 40°39′N 111°30′W
36 I12 **Parker** Arizona, SW USA 34°07′N 114°16′W
23 R9 **Parker** Florida, SE USA 30°07′N 85°36′W
29 R11 **Parker** South Dakota, N USA 43°24′N 97°08′W
35 Z14 **Parker Dam** California, W USA 34°17′N 114°08′W
29 W13 **Parkersburg** Iowa, C USA 42°34′N 92°47′W
21 Q3 **Parkersburg** West Virginia, NE USA 39°17′N 81°33′W
29 T7 **Parkers Prairie** Minnesota, N USA 46°09′N 95°19′W
171 P8 **Parker Volcano** ⩕ Mindanao, S Philippines 06°09′N 124°52′E
181 W13 **Parkes** New South Wales, SE Australia 33°10′S 148°10′E
30 K4 **Park Falls** Wisconsin, N USA 45°57′N 90°25′W
Parkhar see Farkhor
14 E16 **Parkhill** Ontario, S Canada 43°11′N 81°39′W
29 T5 **Park Rapids** Minnesota, N USA 46°55′N 95°03′W
29 Q3 **Park River** North Dakota, N USA 48°24′N 97°44′W
29 Q11 **Parkston** South Dakota, N USA 43°24′N 97°58′W
10 L17 **Parksville** Vancouver Island, British Columbia, SW Canada 49°13′N 124°13′W
37 S3 **Parkview Mountain** ▲ Colorado, C USA 40°19′N 106°08′W
105 N8 **Parla** Madrid, C Spain 40°13′N 03°48′W
29 S8 **Parle, Lac qui** ▣ Minnesota, N USA
155 G14 **Parli Vaijnāth** Mahārāshtra, C India 18°53′N 76°36′E
106 F9 **Parma** Emilia-Romagna, N Italy 44°50′N 10°20′E
31 T11 **Parma** Ohio, N USA 41°24′N 81°43′W
Parnahyba see Parnaíba
58 N13 **Parnaíba** var. Parnahyba. Piauí, E Brazil 02°58′S 41°46′W
65 J14 **Parnaíba Ridge** undersea feature C Atlantic Ocean
58 N13 **Parnaíba, Rio** ≈ NE Brazil
115 F18 **Parnassós** ▲▲ C Greece
185 J17 **Parnassus** Canterbury, South Island, New Zealand 42°41′S 173°18′E
182 H10 **Parndana** South Australia 35°48′S 137°13′E
115 H19 **Párnitha** ▲▲ C Greece
Parnon see Párnonas
115 F21 **Párnonas** var. Parnon. ▲▲ S Greece
118 G5 **Pärnu** Ger. Pernau, Latv. Pērnava; prev. Rus. Pernov. Pärnumaa, SW Estonia 58°24′N 24°32′E
118 G6 **Pärnu** var. Parnu Jõgi, Ger. Pernau. ≈ SW Estonia
118 G5 **Pärnu-Jaagupi** Ger. Sankt-Jakobi. Pärnumaa, SW Estonia 58°36′N 24°30′E
Parnu Jõgi see Pärnu
118 G5 **Pärnu Laht** Ger. Pernauer Bucht. bay SW Estonia
118 F5 **Pärnumaa** var. Pärnu Maakond. ◈ province SW Estonia
Pärnu Maakond see Pärnumaa
153 T11 **Paro** W Bhutan 27°23′N 89°31′E
153 T11 **Paro** ✈ (Thimphu) W Bhutan 27°23′N 89°31′E
185 G17 **Paroa** West Coast, South Island, New Zealand 42°31′S 171°10′E
163 X14 **Paro-ho** var. Hwach'ŏn-chŏsuji; prev. P'aro-ho. ▣ N South Korea
P'aro-ho see Paro-ho
115 J21 **Pároikiá** prev. Páros. Páros, Kykládes, Greece, Aegean Sea 37°04′N 25°06′E
183 N6 **Paroo River** seasonal river New South Wales/Queensland, SE Australia
Paropamisus Range see Sefīd Kūh, Selseleh-ye
Paropamisus Range see Sefīd Kūh, Selseleh-ye
115 J21 **Páros** island Kykládes, Greece, Aegean Sea
Páros see Pároikiá
36 K7 **Parowan** Utah, W USA 37°50′N 112°49′W
103 U13 **Parpaillon** ▲▲ SE France
108 I9 **Parpan** Graubünden, S Switzerland 46°46′N 09°32′E
62 G13 **Parral** Maule, C Chile 36°08′S 71°52′W
Parral see Hidalgo del Parral
183 T9 **Parramatta** New South Wales, SE Australia 33°49′S 150°59′E
21 Y6 **Parramore Island** island Virginia, NE USA
40 M8 **Parras** var. Parras de la Fuente. Coahuila, NE Mexico 25°26′N 102°07′W
Parras de la Fuente see Parras
42 M14 **Parrita** Puntarenas, S Costa Rica 09°30′N 84°20′W
Parry group see Mukojima-rettō
14 G13 **Parry Island** island Ontario, S Canada
197 O9 **Parry Islands** island group Nunavut, NW Canada
14 G12 **Parry Sound** Ontario, S Canada 45°21′N 80°03′W
110 F7 **Parsęta** Ger. Persante. ≈ NW Poland
28 L3 **Parshall** North Dakota, N USA 47°57′N 102°07′W
27 Q7 **Parsons** Kansas, C USA 37°20′N 95°15′W
20 H9 **Parsons** Tennessee, S USA 35°39′N 88°07′W
21 T3 **Parsons** West Virginia, NE USA 39°06′N 79°43′W
Parsonstown see Birr
100 P11 **Parsteiner See** ◎ NE Germany
107 I24 **Partanna** Sicilia, Italy, C Mediterranean Sea 37°43′N 12°54′E
108 J8 **Partenen** Graubünden, E Switzerland 46°58′N 10°01′E
102 K9 **Parthenay** Deux-Sèvres, W France 46°39′N 00°13′W
95 J19 **Partille** Västra Götaland, S Sweden 57°43′N 12°12′E
107 I23 **Partinico** Sicilia, Italy, C Mediterranean Sea 38°03′N 13°07′E
111 I20 **Partizánske** prev. Šimonovany, Hung. Simony. Trenčiansky Kraj, W Slovakia 48°35′N 18°23′E
58 H11 **Paru de Oeste, Rio** ≈ N Brazil
182 K9 **Paruna** South Australia 34°45′S 140°43′E
58 I11 **Paru, Rio** ≈ N Brazil
Parvān see Parwān
155 M14 **Pārvatipuram** Andhra Pradesh, E India 17°01′N 81°47′E
152 G12 **Parvatsar** prev. Parbatsar. Rājasthān, N India 26°52′N 74°49′E
114 J11 **Parvomay** , Pŭrvomay; prev. Borisovgrad. Plovdiv, C Bulgaria 42°06′N 25°13′E
149 Q5 **Parwān** prev. Parvān. E Afghanistan
158 I15 **Paryang** Xizang Zizhiqu, W China 30°04′N 83°28′E
119 M18 **Parychy** Rus. Parichi. Homyel'skaya Voblasts', SE Belarus 52°48′N 29°25′E
83 J21 **Parys** Free State, C South Africa 26°55′S 27°28′E
35 T15 **Pasadena** California, W USA 34°09′N 118°09′W
25 W11 **Pasadena** Texas, SW USA 29°41′N 95°13′W
56 B8 **Pasaje** El Oro, SW Ecuador 03°23′S 79°50′W
137 T9 **Pasanauri** prev. P'asanauri. N Georgia 42°21′N 44°40′E
P'asanauri see Pasanauri
168 I13 **Pasapuat** Pulau Pagai Utara, W Indonesia 02°36′S 99°58′E
Pasawng see Hpasawng
114 L13 **Paşayiğit** Edirne, NW Turkey 40°58′N 26°38′E
23 N9 **Pascagoula** Mississippi, S USA 30°21′N 88°32′W
22 M8 **Pascagoula River** ≈ Mississippi, S USA
116 F12 **Paşcani** Hung. Páskán. Iaşi, NE Romania 47°14′N 26°46′E
109 T4 **Pasching** Oberösterreich, N Austria 48°16′N 14°10′E
32 K10 **Pasco** Washington, NW USA 46°13′N 119°06′W
56 E13 **Pasco** off. Departamento de Pasco. ◈ department C Peru
Pasco, Departamento de see Pasco
191 N11 **Pascua, Isla de** var. Rapa Nui, Easter Island. island E Pacific Ocean
63 G21 **Pascua, Río** ≈ S Chile
103 N1 **Pas-de-Calais** ◈ department N France
100 P10 **Pasewalk** Mecklenburg-Vorpommern, NE Germany 53°31′N 13°59′E
11 T10 **Pasfield Lake** ◎ Saskatchewan, C Canada
Pa-shih Hai-hsia see Bashi Channel
Pashkeni see Bolyarovo
Pashmakli see Smolyan
153 X10 **Pāsighāt** Arunāchel Pradesh, NE India 28°08′N 95°13′E
137 Q12 **Pasinler** Erzurum, NE Turkey 39°59′N 41°41′E
Pasi Oloy, Qatorkŭhi see Zaalayskiy Khrebet
42 E3 **Pasión, Río de la** ≈ N Guatemala
168 J12 **Pasirganting** Sumatera, W Indonesia 02°04′S 100°51′E
Pasirpangarayan see Bagansiapiapi
168 K6 **Pasir Puteh** var. Pasir Putih. Kelantan, Peninsular Malaysia 05°50′N 102°24′E
Pasir Putih see Pasir Puteh
169 R9 **Pasir, Tanjung** headland East Malaysia 02°24′N 111°12′E
95 N20 **Påskallavik** Kalmar, S Sweden 57°10′N 16°25′E
Páskán see Paşcani
Paskevicha, Zaliv see Tushchybas, Zaliv
110 K7 **Pasłęk** Ger. Preußisch Holland. Warmińsko-Mazurskie, NE Poland 54°03′N 19°40′E
110 K7 **Pasłęka** Ger. Passarge. ≈ N Poland
148 K16 **Pasni** Baluchistān, SW Pakistan 25°13′N 63°30′E
63 I18 **Paso de Indios** Chubut, S Argentina 43°52′S 69°06′W
54 L7 **Paso del Caballo** Guárico, N Venezuela 08°19′N 67°08′W
61 E15 **Paso de los Libres** Corrientes, NE Argentina 29°43′S 57°09′W
61 E18 **Paso de los Toros** Tacuarembó, C Uruguay 32°45′S 56°30′W
Pasoeroean see Pasuruan
35 P12 **Paso Robles** California, W USA 35°37′N 120°42′W
15 Y7 **Paspébiac** Québec, SE Canada 48°03′N 65°10′W
11 U14 **Pasquia Hills** ▲▲ Saskatchewan, S Canada
149 W7 **Pasrūr** Punjab, E Pakistan 32°12′N 74°42′E
30 M1 **Passage Island** island Michigan, N USA
65 B24 **Passage Islands** island group W Falkland Islands
8 K5 **Passage Point** headland Banks Island, Northwest Territories, NW Canada 73°31′N 115°12′W
Passarge see Pasłęka
115 C15 **Passarón** ancient monument Ípeiros, W Greece
Passarowitz see Požarevac
101 O22 **Passau** Bayern, SE Germany 48°34′N 13°28′E
22 M9 **Pass Christian** Mississippi, S USA 30°19′N 89°15′W
107 L26 **Passero, Capo** headland Sicilia, Italy, C Mediterranean Sea 36°40′N 15°09′E
171 P5 **Passi** Panay Island, C Philippines 11°05′N 122°37′E
61 H14 **Passo Fundo** Rio Grande do Sul, S Brazil 28°16′S 52°20′W
60 H13 **Passo Fundo, Barragem de** ▣ S Brazil
61 H15 **Passo Real, Barragem de** ▣ S Brazil
59 L20 **Passos** Minas Gerais, NE Brazil 20°45′S 46°38′W
167 X10 **Passu Keah** Chin. Panshi Yu, Viet. Đao Bach Quy. island S Paracel Islands
118 J13 **Pastavy** Pol. Postawy, Rus. Postavy. Vitsyebskaya Voblasts', NW Belarus 55°07′N 26°50′E
56 D7 **Pastaza** ◈ province E Ecuador
56 D9 **Pastaza, Río** ≈ Ecuador/Peru
61 A21 **Pasteur** Buenos Aires, E Argentina 35°10′S 62°14′W
15 V3 **Pasteur** ≈ Québec, SE Canada
147 Q12 **Pastigav** Rus. Pastigov. W Tajikistan 39°27′N 69°16′E
Pastigov see Pastigav
54 C13 **Pasto** Nariño, SW Colombia 01°12′N 77°17′W
38 M10 **Pastol Bay** bay Alaska, USA
37 O8 **Pastora Peak** ▲ Arizona, SW USA 36°48′N 109°10′W
105 O8 **Pastrana** Castilla-La Mancha, C Spain 40°24′N 02°55′W
169 S16 **Pasuruan** prev. Pasoeroean. Jawa, C Indonesia 07°38′S 112°44′E
118 F11 **Pasvalys** Panevėžys, N Lithuania 56°03′N 24°24′E
111 K21 **Pásztó** Nógrád, N Hungary 47°57′N 19°41′E
189 U12 **Pata** var. Patta. atoll Chuuk Islands, C Micronesia
36 M16 **Patagonia** Arizona, SW USA 31°32′N 110°45′W
63 H20 **Patagonia** physical region Argentina/Chile
Patalung see Phatthalung
154 D9 **Pātan** Gujarāt, W India 23°51′N 72°11′E
154 J10 **Pātan** Madhya Pradesh, C India 23°19′N 79°41′E
171 S11 **Patani** Pulau Halmahera, E Indonesia 0°19′N 128°46′E
Patani see Pattani
15 V7 **Patapédia Est** ≈ Québec, SE Canada
116 K13 **Pătârlagele** prev. Pătîriagele. Buzău, SE Romania 45°19′N 26°21′E
Patavium see Padova
182 L5 **Patawarta Hill** ▲ South Australia 30°57′S 138°42′E
182 L10 **Patchewollock** Victoria, SE Australia 35°24′S 142°11′E
184 K11 **Patea** Taranaki, North Island, New Zealand 39°48′S 174°35′E
184 K11 **Patea** ≈ North Island, New Zealand
77 U15 **Pategi** Kwara, C Nigeria 08°39′N 05°46′E
81 K20 **Pate Island** var. Patta Island. island SE Kenya
105 S10 **Paterna** Valenciana, E Spain 39°30′N 00°24′W
109 R9 **Paternion** Slvn. Špatrjan. Kärnten, S Austria 46°40′N 13°43′E
107 L24 **Paternò** anc. Hybla, Hybla Major. Sicilia, Italy, C Mediterranean Sea 37°34′N 14°55′E
32 J7 **Pateros** Washington, NW USA 48°01′N 119°55′W
18 J14 **Paterson** New Jersey, NE USA 40°55′N 74°12′W
32 J10 **Paterson** Washington, NW USA 45°56′N 119°37′W
185 C25 **Paterson Inlet** inlet Stewart Island, New Zealand
98 N6 **Paterswolde** Drenthe, NE Netherlands 53°07′N 06°32′E
152 H7 **Pathānkot** Himāchal Pradesh, N India 32°16′N 75°43′E
166 K8 **Pathein** var. Bassein. Ayeyawady, SW Myanmar (Burma) 16°46′N 94°45′E
33 W15 **Pathfinder Reservoir** ▣ Wyoming, C USA
167 O11 **Pathum Thani** var. Patumdhani, Prathum Thani. Pathum Thani, C Thailand 14°03′N 100°29′E
54 C12 **Patía** var. El Bordo. Cauca, SW Colombia 02°07′N 76°57′W
152 I9 **Patiāla** var. Puttiala. Punjab, NW India 30°21′N 76°27′E
54 B12 **Patía, Río** ≈ SW Colombia
188 D15 **Pati Point** headland NE Guam 13°36′N 144°39′E
Pătîriagele see Pătârlagele
56 C13 **Pativilca** Lima, W Peru 10°44′S 77°45′W
166 M1 **Pātkai Bum** var. Patkai Range. ▲▲ Myanmar (Burma)/India
Patkai Range see Pātkai Bum
115 L20 **Pátmos** Pátmos, Dodekánisa, Greece, Aegean Sea 37°18′N 26°32′E
115 L20 **Pátmos** island Dodekánisa, Greece, Aegean Sea
153 P13 **Patna** var. Azimabad. state capital Bihār, N India 25°36′N 85°11′E
154 M12 **Patnāgarh** Odisha, E India 20°42′N 83°12′E
171 O5 **Patnongon** Panay Island, C Philippines 10°56′N 122°03′E

◆ Country
● Country Capital
◇ Dependent Territory
○ Dependent Territory Capital
◈ Administrative Regions
✈ International Airport
▲ Mountain
▲▲ Mountain Range
⩕ Volcano
≈ River
◎ Lake
▣ Reservoir

137 S13 Patnos Ağrı, E Turkey 39°14′N 42°52′E
60 H12 Pato Branco Paraná, S Brazil 26°20′S 52°40′W
31 O16 Patoka Lake ◎ Indiana, N USA
92 L9 Patoniva Lapp. Buoddobohki. Lappi, N Finland 69°44′N 27°01′E
113 K21 Patos var. Patosi. Fier, SW Albania 40°40′N 19°37′E
Patos see Patos de Minas
59 K19 Patos de Minas var. Patos. Minas Gerais, NE Brazil 18°35′S 46°32′W
Patosi see Patos
61 I17 Patos, Lagoa dos lagoon S Brazil
62 J9 Patquía La Rioja, C Argentina 30°02′S 66°54′W
115 E19 Pátra Eng. Patras; prev. Pátrai. Dytikí Elláda, S Greece 38°14′N 21°45′E
115 D18 Patraïkós Kólpos gulf S Greece
Pátrai/Patras see Pátra
92 G2 Patreksfjörður Vestfirðir, W Iceland 65°33′N 23°54′W
24 M7 Patricia Texas, SW USA 32°34′N 102°00′W
63 F21 Patricio Lynch, Isla island S Chile
Patta see Pata
Patta Island see Pate Island
167 O16 Pattani var. Patani. Pattani, SW Thailand 06°50′N 101°20′E
167 P12 Pattaya Chon Buri, S Thailand 12°57′N 100°53′E
19 S4 Patten Maine, NE USA 45°58′N 68°27′W
35 O9 Patterson California, W USA 37°27′N 121°07′W
22 J10 Patterson Louisiana, S USA 29°41′N 91°18′W
35 R7 Patterson, Mount ▲ California, W USA 38°27′N 119°16′W
31 P4 Patterson, Point headland Michigan, N USA 45°58′N 85°39′W
107 L23 Patti Sicilia, Italy, C Mediterranean Sea 38°08′N 14°58′E
107 L23 Patti, Golfo di gulf Sicilia, Italy
93 L14 Pattijoki Pohjois-Pohjanmaa, W Finland 64°41′N 24°40′E
193 Q4 Patton Escarpment undersea feature E Pacific Ocean
27 S2 Pattonsburg Missouri, C USA 40°03′N 94°08′W
0 D6 Patton Seamount undersea feature NE Pacific Ocean 54°40′N 150°30′W
10 J12 Pattullo, Mount ▲ British Columbia, W Canada 56°18′N 129°43′W
153 U16 Patuakhali var. Patukhali. Barisal, S Bangladesh 22°20′N 90°20′E
42 M5 Patuca, Río ~ E Honduras
Patukhali see Patuakhali
Patumdhani see Pathum Thani
40 M14 Pátzcuaro Michoacán, SW Mexico 19°30′N 101°38′W
42 C6 Patzicía Chimaltenango, S Guatemala 14°38′N 90°52′W
102 K16 Pau Pyrénées-Atlantiques, SW France 43°18′N 00°22′W
102 J12 Pauillac Gironde, SW France 45°12′N 00°44′W
166 L5 Pauk Magway, W Myanmar (Burma) 21°25′N 94°30′E
8 I6 Paulatuk Northwest Territories, NW Canada 69°23′N 124°W
42 K5 Paulayá, Río ~ NE Honduras
22 M6 Paulding Mississippi, S USA 32°01′N 89°01′W
31 Q12 Paulding Ohio, N USA 41°08′N 84°34′W
29 S12 Paullina Iowa, C USA 42°58′N 95°41′W
59 P15 Paulo Afonso Bahia, E Brazil 09°21′S 38°14′W
38 M16 Pauloff Harbor var. Pavlor Harbour. Sanak Island, Alaska, USA 54°26′N 162°43′W
27 N12 Pauls Valley Oklahoma, C USA 34°46′N 97°14′W
166 L7 Paungde Bago, C Myanmar (Burma) 18°30′N 95°30′E
Pauni see Paoni
152 K9 Pauri Uttaranchal, N India 30°08′N 78°48′E
Pautalia see Kyustendil
142 J5 Pāveh Kermānshāhān, NW Iran 35°02′N 46°15′E
114 I9 Pavel Banya Stara Zagora, C Bulgaria 42°35′N 25°19′E
126 L5 Pavelets Ryazanskaya Oblast′, W Russian Federation 53°47′N 39°22′E
106 D8 Pavia anc. Ticinum. Lombardia, N Italy 45°10′N 09°10′E
118 C9 Pāvilosta W Latvia 56°52′N 21°12′E
125 P14 Pavino Kostromskaya Oblast′, NW Russian Federation 59°10′N 46°09′E
114 J8 Pavlikeni Veliko Tarnovo, N Bulgaria 43°14′N 25°20′E
145 T8 Pavlodar Pavlodar, NE Kazakhstan 52°21′N 76°59′E
145 S9 Pavlodar off. Pavlodarskaya Oblast′, Kaz. Pavlodar Oblysy. ◆ province NE Kazakhstan
Pavlodar Oblysy/Pavlodarskaya Oblast′ see Pavlodar
Pavlograd see Pavlohrad
117 U7 Pavlohrad Rus. Pavlograd. Dnipropetrovs′ka Oblast′, E Ukraine 48°32′N 35°50′E
Pavlor Harbour see Pauloff Harbor
145 R9 Pavlovka Akmola, C Kazakhstan 51°22′N 72°35′E
127 V4 Pavlovka Respublika Bashkortostan, W Russian Federation 55°28′N 56°36′E
127 Q7 Pavlovka Ul′yanovskaya Oblast′, W Russian Federation 52°40′N 47°08′E
127 N3 Pavlovo Nizhegorodskaya Oblast′, W Russian Federation 55°59′N 43°03′E
126 L9 Pavlovsk Voronezhskaya Oblast′, W Russian Federation 50°26′N 40°08′E
126 L13 Pavlovskaya Krasnodarskiy Kray, SW Russian Federation 46°06′N 39°52′E
117 S7 Pavlysh Kirovohrads′ka Oblast′, C Ukraine 48°54′N 33°20′E
106 F10 Pavullo nel Frignano Emilia-Romagna, C Italy 44°19′N 10°52′E
27 P8 Pawhuska Oklahoma, C USA 36°42′N 96°21′W
21 U13 Pawleys Island South Carolina, SE USA 33°27′N 79°07′W
30 K14 Pawnee Illinois, N USA 39°35′N 89°34′W
27 O9 Pawnee Oklahoma, C USA 36°21′N 96°50′W
37 U2 Pawnee Buttes ▲ Colorado, C USA 40°49′N 103°58′W
29 S17 Pawnee City Nebraska, C USA 40°06′N 96°09′W
26 K5 Pawnee River ~ Kansas, C USA
167 N6 Pawn, Nam ~ C Myanmar (Burma)
31 O10 Paw Paw Michigan, N USA 42°12′N 86°09′W
31 O10 Paw Paw Lake Michigan, N USA 42°12′N 86°16′W
19 O12 Pawtucket Rhode Island, NE USA 41°52′N 71°22′W
Pax Augusta see Badajoz
115 I25 Paximádia island SE Greece
Pax Julia see Beja
115 B16 Paxoí island Iónia Nisiá, Greece, C Mediterranean Sea
39 S10 Paxson Alaska, USA 62°58′N 145°27′W
147 O11 Paxtakor Jizzax Viloyati, C Uzbekistan 40°21′N 67°54′E
30 M13 Paxton Illinois, N USA 40°27′N 88°06′W
124 J11 Pay Respublika Kareliya, NW Russian Federation 61°10′N 34°24′E
166 M8 Payagyi Bago, SW Myanmar (Burma) 17°28′N 96°32′E
108 C9 Payerne Ger. Peterlingen. Vaud, W Switzerland 46°49′N 06°57′E
32 M13 Payette Idaho, NW USA 44°04′N 116°55′W
32 M13 Payette River ~ Idaho, NW USA
125 V2 Pay-Khoy, Khrebet ▲▲ NW Russian Federation
Payne see Kangirsuk
12 K4 Payne, Lac ◎ Québec, NE Canada
29 T8 Paynesville Minnesota, N USA 45°22′N 94°42′W
169 S8 Payong, Tanjung cape East Malaysia
Payo Obispo see Chetumal
61 D18 Paysandú Paysandú, W Uruguay 32°21′S 58°05′W
61 D17 Paysandú ◆ department W Uruguay
102 I7 Pays de la Loire ◆ region NW France
36 L12 Payson Arizona, SW USA 34°13′N 111°19′W
36 L4 Payson Utah, W USA 40°02′N 111°43′W
125 W4 Payyer, Gora ▲ NW Russian Federation 66°49′N 64°33′E
Payzawat see Jiashi
137 Q11 Pazar Rize, NE Turkey 41°10′N 40°53′E
136 F10 Pazarbaşı Burnu headland NW Turkey 41°12′N 30°18′E
136 M16 Pazarcık Kahramanmaraş, S Turkey 37°31′N 37°19′E
114 I10 Pazardzhik prev. Tatar Pazardzhik. Pazardzhik, SW Bulgaria 42°11′N 24°21′E
64 H11 Pazardzhik ◆ province C Bulgaria
54 H9 Paz de Ariporo Casanare, E Colombia 05°54′N 71°52′W
112 A10 Pazin Ger. Mitterburg, It. Pisino. Istra, NW Croatia 45°14′N 13°56′E
42 D7 Paz, Río ~ El Salvador/Guatemala
113 O18 Pčinja ~ N Macedonia
193 V15 Pea Tongatapu, S Tonga 21°10′S 175°14′W
27 O6 Peabody Kansas, C USA 38°10′N 97°06′W
11 O12 Peace ~ Alberta/British Columbia, W Canada
Peace Garden State see North Dakota
11 Q10 Peace Point Alberta, C Canada 59°11′N 112°12′W
11 O12 Peace River Alberta, W Canada 56°15′N 117°18′W
23 W13 Peace River ~ Florida, SE USA
11 N17 Peachland British Columbia, SW Canada 49°49′N 119°48′W
36 J10 Peach Springs Arizona, SW USA 35°33′N 113°27′W
Peach State see Georgia
23 S4 Peachtree City Georgia, SE USA 33°24′N 84°36′W
189 Y13 Peacock Point point SE Wake Island
97 M18 Peak District physical region C England, United Kingdom
183 Q7 Peak Hill New South Wales, SE Australia 32°39′S 148°12′E
65 G15 Peak, The ▲ C Ascension Island
105 O13 Peal de Becerro Andalucía, S Spain 37°55′N 03°08′W
189 X11 Peale Island island N Wake Island
37 O6 Peale, Mount ▲ Utah, W USA 38°26′N 109°13′W
39 O4 Peard Bay bay Alaska, USA
23 Q7 Pea River ~ Alabama/Florida, S USA
25 W11 Pearland Texas, SW USA 29°33′N 95°17′W
38 D9 Pearl City O′ahu, Hawaii, USA, C Pacific Ocean 21°24′N 157°58′W
38 D9 Pearl Harbor inlet O′ahu, Hawai′i, USA, C Pacific Ocean
Pearl Islands see Perlas, Archipiélago de las
Pearl Lagoon see Perlas, Laguna de
22 M5 Pearl River ~ Louisiana/Mississippi, S USA
25 Q13 Pearsall Texas, SW USA 28°54′N 99°07′W
23 U7 Pearson Georgia, SE USA 31°18′N 82°51′W
25 P4 Pease River ~ Texas, SW USA
12 F7 Peawanuck Ontario, C Canada 54°55′N 85°31′W
12 E8 Peawanuk ~ Ontario, S Canada
83 P16 Pebane Zambézia, NE Mozambique 17°14′S 38°10′E
65 C23 Pebble Island island N Falkland Islands
65 C23 Pebble Island Settlement Pebble Island, N Falkland Islands 51°20′S 59°40′W
Peć see Pejë
25 R8 Pecan Bayou ~ Texas, SW USA
22 H10 Pecan Island Louisiana, S USA 29°39′N 92°26′W
60 L12 Peças, Ilha das island S Brazil
30 L10 Pecatonica River ~ Illinois/Wisconsin, N USA
108 G10 Peccia Ticino, S Switzerland 46°24′N 08°39′E
Pechenegi see Pechenihy
Pechenezhskoye Vodokhranilishche see Pecheniz′ke Vodoskhovyshche
124 I2 Pechenga Fin. Petsamo. Murmanskaya Oblast′, NW Russian Federation 69°34′N 31°14′E
117 V5 Pechenihy Rus. Pechenegi. Kharkivs′ka Oblast′, E Ukraine 49°49′N 36°57′E
117 V5 Pecheniz′ke Vodoskhovyshche Rus. Pechenezhskoye Vodokhranilishche. ⊠ E Ukraine
125 U7 Pechora Respublika Komi, NW Russian Federation 65°09′N 57°09′E
125 R6 Pechora ~ NW Russian Federation
Pechora Bay see Pechorskaya Guba
Pechora Sea see Pechorskoye More
125 S3 Pechorskaya Guba Eng. Pechora Bay. bay NW Russian Federation
122 H7 Pechorskoye More Eng. Pechora Sea. sea NW Russian Federation
116 E11 Pecica Ger. Petschka, Hung. Ópécska. Arad, W Romania 46°09′N 21°06′E
24 K8 Pecos Texas, SW USA 31°25′N 103°30′W
25 N11 Pecos River ~ New Mexico/Texas, SW USA
111 I25 Pécs Ger. Fünfkirchen, Lat. Sopianae. Baranya, SW Hungary 46°05′N 18°11′E
43 T17 Pedasí Los Santos, S Panama 07°36′N 80°04′W
Pedde see Pedja
183 O17 Pedder, Lake ◎ Tasmania, SE Australia
44 M10 Pedernales SW Dominican Republic 18°02′N 71°41′W
55 Q5 Pedernales Delta Amacuro, NE Venezuela 09°58′N 62°15′W
25 R10 Pedernales River ~ Texas, SW USA
62 H6 Pedernales, Salar de salt lake N Chile
Pedhoulas see Pedoulás
55 X11 Pédima var. Malavate. SW French Guiana 03°15′N 54°08′W
182 F1 Pedirka South Australia 26°41′S 135°11′E
171 S11 Pediwang Pulau Halmahera, E Indonesia 01°29′N 127°57′E
118 I5 Pedja var. Pedja Jõgi, Ger. Pedde. ~ E Estonia
Pedja Jõgi see Pedja
121 O3 Pedoulás var. Pedhoulas. W Cyprus 34°58′N 32°51′E
59 N18 Pedra Azul Minas Gerais, NE Brazil 16°02′S 41°17′W
104 I3 Pedrafita, Porto de var. Puerto de Piedrafita. pass NW Spain
76 E9 Pedra Lume Sal, NE Cape Verde 16°47′N 22°54′W
43 P16 Pedregal Chiriquí, W Panama 09°04′N 79°25′W
54 J4 Pedregal Falcón, N Venezuela 11°04′N 70°08′W
40 L9 Pedriceña Durango, C Mexico 25°08′N 103°46′W
60 L11 Pedro Barros São Paulo, S Brazil 24°12′S 47°22′W
39 Q13 Pedro Bay Alaska, USA 59°47′N 154°06′W
62 H4 Pedro de Valdivia var. Oficina Pedro de Valdivia. Antofagasta, N Chile 22°33′S 69°38′W
62 P4 Pedro Juan Caballero Amambay, E Paraguay 22°34′S 55°41′W
63 L15 Pedro Luro Buenos Aires, E Argentina 39°30′S 62°38′W
105 O10 Pedro Muñoz Castilla-La Mancha, C Spain 39°25′N 02°56′W
155 J22 Pedro, Point headland NW Sri Lanka 09°54′N 80°08′E
182 K9 Peebinga South Australia 34°56′S 140°56′E
96 J13 Peebles SE Scotland, United Kingdom 55°40′N 03°15′W
31 S15 Peebles Ohio, N USA 38°57′N 83°23′W
96 J12 Peebles cultural region SE Scotland, United Kingdom
18 K13 Peekskill New York, NE USA 41°17′N 73°54′W
97 I16 Peel W Isle of Man 54°13′N 04°04′W
8 G7 Peel ~ Northwest Territories/Yukon, NW Canada
8 K5 Peel Point headland Victoria Island, Northwest Territories, NW Canada 73°22′N 114°33′W
8 M5 Peel Sound passage Nunavut, N Canada
100 N9 Peene ~ NE Germany
99 K17 Peer Limburg, NE Belgium 51°08′N 05°29′E
14 H14 Pefferlaw Ontario, S Canada 44°18′N 79°11′W
185 I18 Pegasus Bay bay South Island, New Zealand
121 O3 Pégeia var. Peyia. SW Cyprus 34°52′N 32°24′E
109 V7 Peggau Steiermark, SE Austria 47°10′N 15°20′E
101 L19 Pegnitz Bayern, SE Germany 49°45′N 11°33′E
101 L19 Pegnitz ~ SE Germany
105 T11 Pego Valenciana, E Spain 38°51′N 00°08′W
Pegu see Bago
Pegu see Bago
189 N13 Pehleng Pohnpei, E Micronesia
114 M12 Pehlivanköy Kırklareli, NW Turkey 41°21′N 26°55′E
77 R14 Péhonko C Benin 10°14′N 01°57′E
61 B21 Pehuajó Buenos Aires, E Argentina 35°48′S 61°53′W
Pei-ching see Beijing/Beijing Shi
100 J13 Peine Niedersachsen, C Germany 52°19′N 10°14′E
Pei-p′ing see Beijing/Beijing Shi
Peipsi Järv/Peipus-See see Peipus, Lake
118 J5 Peipus, Lake Est. Peipsi Järv, Ger. Peipus-See, Rus. Chudskoye Ozero. ◎ Estonia/Russian Federation
115 H19 Peiraías prev. Piraiévs, Eng. Piraeus. Attikí, C Greece 37°57′N 23°42′E
Peisern see Pyzdry
60 I8 Peixe, Rio do ~ S Brazil
59 I16 Peixoto de Azevedo Mato Grosso, W Brazil 10°18′S 55°03′W
168 O11 Pejantan, Pulau island W Indonesia
113 L16 Pejë Serb. Peć. W Kosovo 42°40′N 20°19′E
112 N11 Pek ~ E Serbia
Pèk see Phônsavan
169 Q16 Pekalongan Jawa, C Indonesia 06°54′S 109°37′E
168 K11 Pekanbaru var. Pakanbaru. Sumatera, W Indonesia 0°31′N 101°27′E
30 L12 Pekin Illinois, N USA 40°34′N 89°38′W
Peking see Beijing/Beijing Shi
Pelabohan Kelang/Pelabuhan Kelang see Pelabuhan Klang
168 J9 Pelabuhan Klang var. Kuala Pelabohan Kelang, Pelabohan Kelang, Pelabuhan Kelang, Port Klang, Port Swettenham. Selangor, Peninsular Malaysia 02°57′N 101°24′E
120 L11 Pelagie, Isole island group SW Italy
Pelagosa see Palagruža
22 L5 Pelahatchie Mississippi, S USA 32°19′N 89°48′W
169 T14 Pelaihari var. Pleihari. Borneo, C Indonesia 03°48′S 114°45′E
103 U14 Pelat, Mont ▲ SE France 44°16′N 06°46′E
116 F12 Peleaga, Vârful prev. Virful Peleaga. ▲ W Romania 45°23′N 22°52′E
Peleaga, Vîrful see Peleaga, Vârful
123 O11 Peleduy Respublika Sakha (Yakutiya), NE Russian Federation 59°39′N 112°36′E
14 C18 Pelee Island island Ontario, S Canada
45 Q11 Pelée, Montagne ▲ N Martinique 14°47′N 61°10′W
14 D18 Pelee, Point headland Ontario, S Canada 41°56′N 82°30′W
171 P12 Pelei Pulau Peleng, N Indonesia 01°26′S 123°27′E
Peleliu see Beliliou
171 P12 Peleng, Pulau island Kepulauan Banggai, N Indonesia
23 T7 Pelham Georgia, SE USA 31°07′N 84°09′W
111 E18 Pelhřimov Ger. Pilgram. Vysočina, C Czech Republic 49°26′N 15°14′E
39 W13 Pelican Chichagof Island, Alaska, USA 57°52′N 136°05′W
191 Z3 Pelican Lagoon ◎ Kiritimati, E Kiribati
29 V3 Pelican Lake ◎ Minnesota, N USA
29 U6 Pelican Lake ◎ Minnesota, N USA
30 L5 Pelican Lake ◎ Wisconsin, N USA
44 G1 Pelican Point Grand Bahama Island, N The Bahamas 26°39′N 78°09′W
83 B19 Pelican Point headland W Namibia 22°55′S 14°25′E
29 S6 Pelican Rapids Minnesota, N USA 46°34′N 96°04′W
Pelican State see Louisiana
11 U13 Pelikan Narrows Saskatchewan, C Canada 55°11′N 102°51′W
115 L18 Pelinaío ▲ Chíos, E Greece 38°31′N 26°01′E
Pelinnaeum see Pelinnaío
115 E16 Pelinnaío anc. Pelinnaeum. ruins Thessalía, C Greece
113 N20 Pelister ▲ SW FYR Macedonia 41°00′N 21°12′E
113 G15 Pelješac peninsula S Croatia
92 M12 Pelkosenniemi Lappi, NE Finland 67°06′N 27°30′E
29 W15 Pella Iowa, C USA 41°24′N 92°55′W
114 F13 Pélla site of ancient city Kentrikí Makedonía, N Greece
23 Q3 Pell City Alabama, S USA 33°35′N 86°17′W
61 A22 Pellegrini Buenos Aires, E Argentina 36°16′S 63°07′W
92 K12 Pello Lappi, NW Finland 66°47′N 24°E
100 G7 Pellworm island N Germany
10 H6 Pelly ~ Yukon, NW Canada
Pelly Bay see Kugaaruk
10 I8 Pelly Mountains ▲▲ Yukon, W Canada
Pélmonostor see Beli Manastir
37 P13 Pelona Mountain ▲ New Mexico, SW USA 33°40′N 108°06′W
Peloponnese/Peloponnesus see Pelopónnisos
115 E20 Pelopónnisos Eng. Peloponnese. ◆ region S Greece
115 E20 Pelopónnisos var. Morea, Eng. Peloponnese; anc. Peloponnesus. peninsula S Greece
107 L23 Peloritani, Monti anc. Pelorus and Neptunius. ▲▲ Sicilia, Italy, C Mediterranean Sea
107 M22 Peloro, Capo var. Punta del Faro. headland S Italy 38°15′N 15°39′E
Pelorus and Neptunius see Peloritani, Monti
61 H17 Pelotas Rio Grande do Sul, S Brazil 31°45′S 52°20′W
61 I14 Pelotas, Rio ~ S Brazil
92 K10 Peltovuoma Lapp. Bealdovuopmi. Lappi, N Finland 68°23′N 24°12′E
19 R4 Pemadumcook Lake ◎ Maine, NE USA
169 Q16 Pemalang Jawa, C Indonesia 06°53′S 109°07′E
169 P10 Pemangkat var. Pamangkat. Borneo, C Indonesia 01°11′N 109°00′E
Pemar see Paimio
168 I9 Pematangsiantar Sumatera, W Indonesia 02°59′N 99°01′E
83 Q14 Pemba prev. Port Amelia, Porto Amélia. Cabo Delgado, NE Mozambique 13°S 40°35′E
81 J22 Pemba ◆ region E Tanzania
81 K21 Pemba island E Tanzania
83 Q14 Pemba, Baia de inlet NE Mozambique
81 J21 Pemba Channel channel E Tanzania
180 J14 Pemberton Western Australia 34°27′S 116°09′E
10 M16 Pemberton British Columbia, SW Canada 50°19′N 122°49′W
29 Q2 Pembina North Dakota, N USA 48°58′N 97°14′W
11 P15 Pembina ~ Alberta, SW Canada
29 Q2 Pembina ~ Canada/USA
171 X16 Pembre Papua, E Indonesia 07°49′S 138°01′E
14 K12 Pembroke Ontario, SE Canada 45°49′N 77°08′W
97 H21 Pembroke SW Wales, United Kingdom 51°41′N 04°55′W
23 W6 Pembroke Georgia, SE USA 32°09′N 81°35′W
21 U11 Pembroke North Carolina, SE USA 34°40′N 79°12′W
21 R7 Pembroke Virginia, NE USA 37°19′N 80°38′W
97 H21 Pembroke cultural region SW Wales, United Kingdom
Pembuang, Sungai see Seruyan, Sungai
43 S15 Peña Blanca, Cerro ▲ C Panama 08°39′N 80°39′W
104 K8 Peña de Francia, Sierra de la ▲▲ W Spain
104 G6 Penafiel var. Peñafiel. Porto, N Portugal 41°12′N 08°17′W
105 N6 Peñafiel Castilla y León, N Spain 41°36′N 04°07′W
Peñafiel see Penafiel
Peñagolosa see Penyagolosa
105 N7 Peñalara, Pico de ▲ C Spain 40°52′N 03°55′W
171 X16 Penambo, Banjaran var. Banjaran Tama Abu, Penambo Range. ▲▲ Indonesia/Malaysia
Penambo Range see Penambo, Banjaran
41 O10 Peña Nevada, Cerro ▲ C Mexico 23°46′N 99°52′W
Penang see Pinang
Penang see George Town
Penang see Pinang, Pulau, Peninsular Malaysia
60 J8 Penápolis São Paulo, S Brazil 21°23′S 50°02′W
104 L7 Peñaranda de Bracamonte Castilla y León, N Spain 40°54′N 05°13′W
105 S8 Peñarroya ▲ E Spain 40°24′N 00°42′W
104 L12 Peñarroya-Pueblonuevo Andalucía, S Spain 38°21′N 05°18′W
97 K22 Penarth S Wales, United Kingdom 51°27′N 03°11′W
104 K1 Peñas, Cabo de headland N Spain 43°39′N 05°52′W
63 F20 Penas, Golfo de gulf S Chile
Pen-ch′i see Benxi
79 H14 Pendé var. Logone Oriental. ~ Central African Republic/Chad
76 I14 Pendembu E Sierra Leone 09°06′N 12°12′W
29 R13 Pender Nebraska, C USA 42°06′N 96°42′W
Penderma see Bandırma
32 K11 Pendleton Oregon, NW USA 45°40′N 118°47′W
32 M7 Pend Oreille, Lake ◎ Idaho, NW USA
32 M7 Pend Oreille River ~ Idaho/Washington, NW USA
Pendzhikent see Panjakent
Peneius see Pineiós
104 G8 Penela Coimbra, N Portugal 40°02′N 08°23′W
14 G13 Penetanguishene Ontario, S Canada 44°45′N 79°55′W
151 H15 Penganga ~ C India
P′engchia Yu see Pengjia Yu
79 M21 Penge Kasai-Oriental, C Dem. Rep. Congo 05°29′S 24°38′E
Penghu Archipelago/P′enghu Ch′üntao/Penghu Islands see Penghu Liedao
161 R14 Penghu Liedao var. P′enghu Ch′üntao, Penghu Islands, P′enghu Liehtao, Eng. Penghu Archipelago, Pescadores, Jap. Hoko-guntō, Hoko-shotō. island group W Taiwan
P′enghu Liehtao see Penghu Liedao
161 S14 Penghu Shuidao var. Pescadores Channel, P′enghu Shuitao. channel W Taiwan
P′enghu Shuitao see Penghu Shuidao
161 T12 Pengjia Yu prev. P′engchia Yu. island N Taiwan
161 R4 Penglai var. Dengzhou. Shandong, E China 37°50′N 120°45′E
Peng-pu see Bengbu
Penhsihu see Benxi
Penibético, Sistema see Béticos, Sistemas
104 F10 Peniche Leiria, W Portugal 39°21′N 09°23′W
169 U17 Penida, Nusa island S Indonesia
Peninsular State see Florida
105 T8 Peñíscola var. Peñíscola. Valenciana, E Spain 40°22′N 00°24′E
40 M13 Pénjamo Guanajuato, C Mexico 20°26′N 101°44′W
Penki see Benxi
102 F7 Penmarch, Pointe de headland NW France 47°46′N 04°34′W
107 L15 Penna, Punta della headland C Italy 42°10′N 14°43′E
107 K14 Penne Abruzzo, C Italy 42°28′N 13°57′E
Penner see Penneru
155 J18 Penneru var. Penner. ~ C India
182 I10 Penneshaw South Australia 35°45′S 137°57′E
18 C14 Penn Hills Pennsylvania, NE USA 40°28′N 79°53′W
Penninae, Alpes/Pennine, Alpi see Pennine Alps
108 D11 Pennine Alps Fr. Alpes Pennines, It. Alpi Pennine, Lat. Alpes Penninae. ▲▲ Italy/Switzerland
Pennine Chain see Pennines
97 L15 Pennines var. Pennine Chain. ▲▲ N England, United Kingdom
Pennines, Alpes see Pennine Alps
21 O8 Pennington Gap Virginia, NE USA 36°45′N 83°01′W
18 I16 Penns Grove New Jersey, NE USA 39°42′N 75°27′W
18 I16 Pennsville New Jersey, NE USA 39°37′N 75°29′W
18 E14 Pennsylvania off. Commonwealth of Pennsylvania, also known as Keystone State. ◆ state NE USA
18 G10 Penn Yan New York, NE USA 42°39′N 77°02′W
124 H16 Peno Tverskaya Oblast′, W Russian Federation 56°55′N 32°44′E
19 R7 Penobscot Bay bay Maine, NE USA
19 S5 Penobscot River ~ Maine, NE USA
182 K12 Penola South Australia 37°24′S 140°50′E
40 K9 Peñón Blanco Durango, C Mexico 25°12′N 100°50′W
182 E7 Penong South Australia 31°57′S 133°01′E
43 S16 Penonomé Coclé, C Panama 08°29′N 80°22′W
190 L13 Penrhyn atoll N Cook Islands
192 M9 Penrhyn Basin undersea feature C Pacific Ocean
183 S9 Penrith New South Wales, SE Australia 33°45′S 150°48′E
97 K15 Penrith NW England, United Kingdom 54°40′N 02°44′W
23 O9 Pensacola Florida, SE USA 30°25′N 87°13′W
23 O9 Pensacola Bay bay Florida, SE USA
195 N7 Pensacola Mountains ▲▲ Antarctica
182 L12 Penshurst Victoria, SE Australia 37°54′S 142°19′E
187 R13 Pentecost Fr. Pentecôte. island C Vanuatu
15 V4 Pentecôte ~ Québec, SE Canada
Pentecôte see Pentecost
15 V4 Pentecôte, Lac ◎ Québec, SE Canada
8 H15 Penticton British Columbia, SW Canada 49°29′N 119°38′W
96 J6 Pentland Firth strait N Scotland, United Kingdom
96 J12 Pentland Hills hill range S Scotland, United Kingdom
171 Q12 Penu Pulau Taliabu, E Indonesia 01°43′S 125°09′E
155 H18 Penukonda Andhra Pradesh, E India 14°04′N 77°38′E
166 L7 Penwegon Bago, C Myanmar (Burma) 18°14′N 96°34′E
24 M8 Penwell Texas, SW USA 31°45′N 102°32′W
105 S8 Penyagolosa var. Peñagolosa. ▲ E Spain 40°10′N 00°15′W
97 J21 Pen y Fan ▲ SE Wales, United Kingdom 51°52′N 03°25′W
97 L16 Pen-y-ghent ▲ N England, United Kingdom 54°11′N 02°15′W
127 O6 Penza Penzenskaya Oblast′, W Russian Federation 53°11′N 45°E
97 G25 Penzance SW England, United Kingdom 50°08′N 05°33′W
127 N6 Penzenskaya Oblast′ ◆ province W Russian Federation
123 U7 Penzhina ~ E Russian Federation
123 U9 Penzhinskaya Guba bay E Russian Federation
Penzig see Pieńsk
36 K13 Peoria Arizona, SW USA 33°34′N 112°14′W
30 L12 Peoria Illinois, N USA 40°42′N 89°35′W
30 L12 Peoria Heights Illinois, N USA 40°45′N 89°34′W
31 N11 Peotone Illinois, N USA 41°19′N 87°47′W
18 J11 Pepacton Reservoir ⊠ New York, NE USA
76 I15 Pepel W Sierra Leone 08°39′N 13°04′W
30 I6 Pepin, Lake ◎ Minnesota/Wisconsin, N USA
99 L20 Pepinster Liège, E Belgium 50°34′N 05°49′E
113 L20 Peqin var. Peqini. Elbasan, C Albania 41°03′N 19°46′E
Peqini see Peqin
40 D7 Pequeña, Punta headland NW Mexico 26°13′N 112°34′W
168 J8 Perak ◆ state Peninsular Malaysia
105 R7 Perales del Alfambra Aragón, NE Spain 40°38′N 01°00′W
115 C15 Pérama var. Perama. Ípeiros, W Greece 39°42′N 20°51′E
Perama see Pérama
92 M13 Perä-Posio Lappi, NE Finland 66°10′N 27°56′E
15 Z6 Percé Québec, SE Canada 48°32′N 64°14′W
15 Z6 Percé, Rocher island Québec, S Canada
102 L5 Perche, Collines de ▲ N France
109 X4 Perchtoldsdorf Niederösterreich, NE Austria 48°06′N 16°16′E
180 L6 Percival Lakes lakes Western Australia
105 T3 Perdido, Monte ▲ NE Spain 42°41′N 00°01′E
23 O8 Perdido River ~ Alabama/Florida, S USA
Perece Vela Basin see West Mariana Basin
116 G7 Perechyn Zakarpats′ka Oblast′, W Ukraine 48°45′N 22°28′E
54 E10 Pereira Risaralda, W Colombia 04°47′N 75°46′W
60 I7 Pereira Barreto São Paulo, S Brazil 20°37′S 51°07′W
59 G15 Pereirinha Pará, N Brazil 08°18′S 57°30′W
127 N10 Perelazovskiy Volgogradskaya Oblast′, SW Russian Federation 49°10′N 42°30′E
127 S7 Perelyub Saratovskaya Oblast′, W Russian Federation 51°52′N 50°19′E
31 P7 Pere Marquette River ~ Michigan, N USA
Peremyshl see Przemyśl
116 I5 Peremyshlyany L′vivs′ka Oblast′, W Ukraine 49°42′N 24°33′E
Pereshchepino see Pereshchepyne
116 L9 Pereshchepyne Rus. Pereshchepino. Dnipropetrovs′ka Oblast′, E Ukraine 48°59′N 35°22′E
124 L16 Pereslavl′-Zalesskiy Yaroslavskaya Oblast′, W Russian Federation 56°42′N 38°45′E
117 Y7 Pereval′s′k Luhans′ka Oblast′, E Ukraine 48°28′N 38°54′E
127 U7 Perevolotskiy Orenburgskaya Oblast′, W Russian Federation 51°54′N 54°05′E
Pereyaslav-Khmel′nitskiy see Pereyaslav-Khmel′nyts′kyy
117 Q5 Pereyaslav-Khmel′nyts′kyy Rus. Pereyaslav-Khmel′nitskiy. Kyyivs′ka Oblast′, N Ukraine 50°05′N 31°28′E
109 U4 Perg Oberösterreich, N Austria 48°15′N 14°38′E
61 B19 Pergamino Buenos Aires, E Argentina 33°56′S 60°38′W
106 G6 Pergine Valsugana Ger. Persen. Trentino-Alto Adige, N Italy 46°04′N 11°13′E
29 S6 Perham Minnesota, N USA 46°35′N 95°34′W
93 L16 Perho Keski-Pohjanmaa, W Finland 63°15′N 24°25′E
116 E11 Periam Ger. Perjamosch, Hung. Perjámos. Timiș, W Romania 46°02′N 20°54′E
15 Q6 Péribonca ~ Québec, SE Canada
12 L11 Péribonca, Lac ◎ Québec, SE Canada
15 Q6 Péribonca, Petite Rivière ~ Québec, SE Canada
15 Q7 Péribonka Québec, SE Canada 48°45′N 72°01′W
40 I9 Pericos Sinaloa, C Mexico 25°03′N 107°42′W
169 O10 Perigi Borneo, C Indonesia
102 L12 Périgueux anc. Vesuna. Dordogne, SW France 45°12′N 00°41′E
54 G5 Perijá, Serranía de ▲▲ Columbia/Venezuela
115 H17 Peristéra island Vóreies Sporádes, Greece, Aegean Sea
63 H20 Perito Moreno Santa Cruz, S Argentina 46°35′S 71°W
155 G22 Periyāl var. Periyār. ~ SW India
Periyār see Periyāl
155 G23 Periyār Lake ◎ S India
Perjámos/Perjamosch see Periam
27 O9 Perkins Oklahoma, C USA 35°58′N 97°01′W
116 L7 Perkivtsi Chernivets′ka Oblast′, W Ukraine 48°28′N 26°48′E
43 U15 Perlas, Archipiélago de las Eng. Pearl Islands. island group SE Panama
43 O10 Perlas, Cayos de reef SE Nicaragua
43 N9 Perlas, Laguna de Eng. Pearl Lagoon. lagoon E Nicaragua
43 N10 Perlas, Punta de headland E Nicaragua 12°22′N 83°30′W
100 L11 Perleberg Brandenburg, N Germany 53°04′N 11°52′E
Perlepe see Prilep
168 I6 Perlis ◆ state Peninsular Malaysia
125 U14 Perm′ prev. Molotov. Permskiy Kray, NW Russian Federation 58°01′N 56°10′E
113 M22 Përmet var. Përmeti, Prëmet. Gjirokastër, S Albania 40°12′N 20°24′E
Përmeti see Përmet
125 U15 Permskiy Kray ◆ province NW Russian Federation
59 P15 Pernambuco off. Estado de Pernambuco. ◆ state E Brazil
Pernambuco see Recife
Pernambuco Abyssal Plain see Pernambuco Plain
Pernambuco, Estado de see Pernambuco
47 Y6 Pernambuco Plain var. Pernambuco Abyssal Plain. undersea feature E Atlantic Ocean 07°30′S 27°00′W
65 K15 Pernambuco Seamounts undersea feature C Atlantic Ocean
182 H6 Pernatty Lagoon salt lake South Australia
Pernau see Pärnu
Pernauer Bucht see Pärnu Laht
Përnava see Pärnu
114 G9 Pernik prev. Dimitrovo. Pernik, W Bulgaria 42°36′N 23°02′E
114 G10 Pernik ◆ province W Bulgaria
93 K20 Perniö Swe. Bjärnå. Varsinais-Suomi, SW Finland 60°13′N 23°10′E
109 X5 Pernitz Niederösterreich, E Austria 47°54′N 15°58′E
Pernov see Pärnu
103 O3 Péronne Somme, N France 49°56′N 02°57′E
14 L8 Péronne, Lac ◎ Québec, SE Canada
106 A8 Perosa Argentina Piemonte, NE Italy 45°02′N 07°10′E
41 Q14 Perote Veracruz-Llave, E Mexico 19°32′N 97°16′W
Pérouse see Perugia
191 W15 Pérouse, Bahía de la bay Easter Island, Chile, E Pacific Ocean
Perovsk see Kyzylorda
103 O17 Perpignan Pyrénées-Orientales, S France 42°41′N 02°53′E
113 M20 Përrenjas var. Përrenjasi, Prenjas, Prenjasi. Elbasan, E Albania 41°04′N 20°34′E
Përrenjasi see Përrenjas
92 O2 Perriertoppen ▲ C Svalbard 79°10′N 17°01′E

◆ Country
● Country Capital
◇ Dependent Territory
○ Dependent Territory Capital
◆ Administrative Regions
✈ International Airport
▲ Mountain
▲▲ Mountain Range
Volcano
~ River
◎ Lake
⊠ Reservoir

25 *S6* **Perrin** Texas, SW USA *32°59′N 98°03′W*

23 *Y16* **Perrine** Florida, SE USA *25°36′N 80°21′W*

37 *S12* **Perro, Laguna del** ◎ New Mexico, SW USA

102 *G5* **Perros-Guirec** Côtes d'Armor, NW France *48°49′N 03°28′W*

23 *T9* **Perry** Florida, SE USA *30°07′N 83°34′W*

23 *T5* **Perry** Georgia, SE USA *32°27′N 83°43′W*

29 *U14* **Perry** Iowa, C USA *41°50′N 94°06′W*

18 *E10* **Perry** New York, NE USA *42°43′N 78°00′W*

27 *N9* **Perry** Oklahoma, C USA *36°17′N 97°18′W*

27 *Q3* **Perry Lake** ⊠ Kansas, C USA

31 *R11* **Perrysburg** Ohio, N USA *41°33′N 83°37′W*

25 *O1* **Perryton** Texas, SW USA *36°23′N 100°48′W*

39 *O15* **Perryville** Alaska, USA *55°55′N 159°08′W*

27 *U11* **Perryville** Arkansas, C USA *35°00′N 92°48′W*

27 *Y6* **Perryville** Missouri, C USA *37°43′N 89°51′W*

Persante *see* Parsęta

Persen *see* Pergine Valsugana

Pershay *see* Pyarshai

117 *V7* **Pershotravens'k** Dnipropetrovs'ka Oblast', E Ukraine *48°19′N 36°22′E*

Pershotravneve *see* Manhush

Persia *see* Iran

141 *T5* **Persian Gulf** *var.* The Gulf, *Ar.* Khalīj al 'Arabī, *Per.* Khalīj-e Fars. *Gulf* SW Asia *see also* Gulf, The

141 *T5* **Persian Gulf** *var.* Gulf, The, *Ar.* Khalīj al 'Arabī, *Per.* Khalīj-e Fars. *gulf* SW Asia *see also* Persian Gulf

Persis *see* Fārs

95 *K22* **Perstorp** Skåne, S Sweden *56°08′N 13°23′E*

137 *O14* **Pertek** Tunceli, C Turkey *38°53′N 39°19′E*

183 *P16* **Perth** Tasmania, SE Australia *41°39′S 147°11′E*

180 *I13* **Perth** *state capital* Western Australia *31°58′S 115°49′E*

14 *L13* **Perth** Ontario, SE Canada *44°54′N 76°15′W*

96 *J11* **Perth** C Scotland, United Kingdom *56°24′N 03°28′W*

96 *J10* **Perth** *cultural region* C Scotland, United Kingdom

180 *I12* **Perth** ✈ Western Australia *31°51′S 116°06′E*

173 *V10* **Perth Basin** *undersea feature* SE Indian Ocean *28°30′S 110°00′E*

103 *S15* **Pertuis** Vaucluse, SE France *43°42′N 05°30′E*

103 *Y16* **Pertusato, Capo** *headland* Corse, France, C Mediterranean Sea *41°22′N 09°10′E*

30 *L11* **Peru** Illinois, N USA *41°18′N 89°09′W*

31 *P12* **Peru** Indiana, N USA *40°45′N 86°04′W*

57 *E13* **Peru** *off.* Republic of Peru. ◆ *republic* W South America

Peru *see* Beru

193 *T9* **Peru Basin** *undersea feature* E Pacific Ocean *15°00′S 85°00′W*

193 *U8* **Peru-Chile Trench** *undersea feature* E Pacific Ocean *20°00′S 73°00′W*

112 *F13* **Perućko Jezero** ◎ S Croatia

106 *H13* **Perugia** *Fr.* Pérouse; *anc.* Perusia. Umbria, C Italy *43°06′N 12°24′E*

Perugia, Lake of *see* Trasimeno, Lago

61 *D15* **Perugorría** Corrientes, NE Argentina *29°21′S 58°35′W*

60 *M11* **Peruíbe** São Paulo, S Brazil *24°18′S 47°01′W*

155 *B21* **Perumalpār** *reef* India, N Indian Ocean

Peru, Republic of *see* Peru

Perusia *see* Perugia

99 *D20* **Péruwelz** Hainaut, SW Belgium *50°30′N 03°35′E*

137 *R15* **Pervari** Siirt, SE Turkey *37°55′N 42°32′E*

127 *O4* **Pervomaysk** Nizhegorodskaya Oblast', W Russian Federation *54°52′N 43°49′E*

117 *X7* **Pervomays'k** Luhans'ka Oblast', E Ukraine *48°38′N 38°36′E*

117 *P8* **Pervomays'k** *prev.* Ol'viopol'. Mykolayivs'ka Oblast', S Ukraine *48°02′N 30°51′E*

117 *S12* **Pervomays'ke** Avtonomna Respublika Krym, S Ukraine *45°43′N 33°49′E*

127 *V7* **Pervomayskiy** Orenburgskaya Oblast', W Russian Federation *51°32′N 54°58′E*

126 *M6* **Pervomayskiy** Tambovskaya Oblast', W Russian Federation *53°15′N 40°20′E*

117 *V6* **Pervomays'kyy** Kharkivs'ka Oblast', E Ukraine *49°24′N 36°12′E*

122 *F10* **Pervoural'sk** Sverdlovskaya Oblast', C Russian Federation *56°58′N 59°50′E*

123 *V11* **Pervyy Kuril'skiy Proliv** *strait* E Russian Federation

99 *I19* **Perwez** Walloon Brabant, C Belgium *50°39′N 04°49′E*

106 *I11* **Pesaro** *anc.* Pisaurum. Marche, C Italy *43°55′N 12°53′E*

35 *N9* **Pescadero** California, W USA *37°15′N 122°23′W*

Pescadores *see* Penghu Liedao

Pescadores Channel *see* Penghu Shuidao

107 *K14* **Pescara** *anc.* Aternum, Ostia Aterni. Abruzzo, C Italy *42°28′N 14°13′E*

107 *K15* **Pescara** ≈ C Italy

106 *F11* **Pescia** Toscana, C Italy *43°54′N 10°41′E*

108 *C8* **Peseux** Neuchâtel, W Switzerland *46°59′N 06°53′E*

125 *P6* **Pësha** ≈ NW Russian Federation

149 *T5* **Peshāwar** Khyber Pakhtunkhwa, N Pakistan *34°01′N 71°33′E*

149 *T6* **Peshāwar** ✈ Khyber Pakhtunkhwa, N Pakistan *34°01′N 71°40′E*

113 *M19* **Peshkopi** *var.* Peshkopia, Peshkopija. Dibër, NE Albania *41°40′N 20°25′E*

Peshkopia/Peshkopija *see* Peshkopi

114 *I11* **Peshtera** Pazardzhik, C Bulgaria *42°02′N 24°18′E*

31 *N6* **Peshtigo** Wisconsin, N USA *45°04′N 87°43′W*

31 *N6* **Peshtigo River** ≈ Wisconsin, N USA

Peski *see* Pyaski

125 *S13* **Peskovka** Kirovskaya Oblast', NW Russian Federation *59°04′N 52°17′E*

103 *S8* **Pesmes** Haute-Saône, E France *47°17′N 05°33′E*

104 *H6* **Peso da Régua** *var.* Pêso da Regua. Vila Real, N Portugal *41°10′N 07°47′W*

40 *F5* **Pesqueira** Sonora, NW Mexico *29°22′N 110°58′W*

102 *J13* **Pessac** Gironde, SW France *44°46′N 00°42′W*

111 *J23* **Pest** *off.* Pest Megye. ◈ *county* C Hungary

Pest Megye *see* Pest

124 *J14* **Pestovo** Novgorodskaya Oblast', W Russian Federation *58°37′N 35°48′E*

40 *M15* **Petacalco, Bahía** *bay* W Mexico

Petach-Tikva *see* Petah Tikva

138 *F10* **Petah Tikva** *var.* Petach-Tikva, Petah Tiqva, Petakh Tikva; *prev.* Petah Tiqwa. Tel Aviv, C Israel *32°05′N 34°53′E*

Petah Tiqwa *see* Petah Tikva

93 *L17* **Petäjävesi** Keski-Suomi, C Finland *62°17′N 25°10′E*

Petakh Tikva/Petah Tiqva *see* Petah Tikva

22 *M7* **Petal** Mississippi, S USA *31°21′N 89°15′W*

115 *I19* **Petalioí** *island* C Greece

115 *H19* **Petalión, Kólpos** *gulf* E Greece

115 *J19* **Pétalo** ▲ Ándros, Kykládes, Greece, Aegean Sea *37°51′N 24°50′E*

34 *M8* **Petaluma** California, W USA *38°15′N 122°37′W*

99 *L25* **Pétange** Luxembourg, SW Luxembourg *49°33′N 05°53′E*

54 *M5* **Petare** Miranda, N Venezuela *10°31′N 66°50′W*

41 *N16* **Petatlán** Guerrero, S Mexico *17°31′N 101°16′W*

83 *L14* **Petauke** Eastern, E Zambia *14°12′S 31°16′E*

14 *J12* **Petawawa** Ontario, SE Canada *45°54′N 77°18′W*

14 *J11* **Petawawa** ≈ Ontario, SE Canada

Petchaburi *see* Phetchaburi

42 *D2* **Petén** *off.* Departamento del Petén. ◈ *department* N Guatemala

Petén, Departamento del *see* Petén

42 *D2* **Petén Itzá, Lago** *var.* Lago de Flores. ◎ N Guatemala

30 *K7* **Petenwell Lake** ⊠ Wisconsin, N USA

14 *D6* **Peterbell** Ontario, S Canada *48°34′N 83°19′W*

182 *I7* **Peterborough** South Australia *32°59′S 138°51′E*

14 *I14* **Peterborough** Ontario, SE Canada *44°19′N 78°20′W*

97 *N20* **Peterborough** *prev.* Medeshamstede. E England, United Kingdom *52°35′N 00°15′W*

19 *N10* **Peterborough** New Hampshire, NE USA *42°51′N 71°54′W*

96 *L8* **Peterhead** NE Scotland, United Kingdom *57°30′N 01°46′W*

Peterhof *see* Luboń

Peter I øy *see* Peter I Øy

193 *Q14* **Peter I Øy** ◈ *Norwegian dependency* Antarctica

194 *H9* **Peter I Øy** *var.* Peter I øy. *island* Antarctica

97 *M14* **Peterlee** N England, United Kingdom *54°45′N 01°18′W*

Peterlingen *see* Payerne

197 *P14* **Petermann Bjerg** ▲ C Greenland *73°16′N 27°59′W*

11 *S12* **Peter Pond Lake** ◎ Saskatchewan, C Canada

39 *X13* **Petersburg** Mytkof Island, Alaska, USA *56°43′N 132°51′W*

30 *K13* **Petersburg** Illinois, N USA *40°01′N 89°52′W*

31 *N16* **Petersburg** Indiana, N USA *38°30′N 87°16′W*

29 *Q3* **Petersburg** North Dakota, N USA *47°59′N 97°59′W*

25 *N5* **Petersburg** Texas, SW USA *33°52′N 101°36′W*

21 *V7* **Petersburg** Virginia, NE USA *37°14′N 77°24′W*

21 *T4* **Petersburg** West Virginia, NE USA *39°01′N 79°09′W*

100 *H12* **Petershagen** Nordrhein-Westfalen, NW Germany *52°22′N 08°58′E*

55 *S9* **Peters Mine** *var.* Peter's Mine. N Guyana *06°13′N 59°18′W*

107 *O21* **Petilia Policastro** Calabria, SW Italy *39°07′N 16°48′E*

44 *M9* **Pétionville** S Haiti *18°29′N 72°16′W*

45 *X6* **Petit-Bourg** Basse Terre, C Guadeloupe *16°12′N 61°36′W*

15 *Y5* **Petit-Cap** Québec, SE Canada *49°00′N 64°26′W*

45 *Y6* **Petit Cul-de-Sac Marin** *bay* C Guadeloupe

44 *M9* **Petite-Rivière-de-l'Artibonite** C Haiti *19°10′N 72°30′W*

173 *X16* **Petite Rivière Noire, Piton de la** ▲ C Mauritius

15 *R9* **Petite-Rivière-St-François** Québec, SE Canada *47°18′N 70°34′W*

44 *L9* **Petit-Goâve** S Haiti *18°27′N 72°51′W*

Petitjean *see* Sidi-Kacem

13 *N10* **Petit Lac Manicouagan** ◎ Québec, E Canada

19 *T7* **Petit Manan Point** *headland* Maine, NE USA *44°23′N 67°54′W*

Petit Mécatina, Rivière du *see* Little Mecatina

11 *N10* **Petitot** ≈ Alberta/British Columbia, W Canada

45 *S12* **Petit Piton** ▲ SW Saint Lucia *13°49′N 61°03′W*

Petit-Popo *see* Aného

Petit St-Bernard, Col du *see* Little Saint Bernard Pass

13 *O8* **Petitsikapau Lake** ◎ Newfoundland and Labrador, E Canada

92 *L11* **Petkula** Lappi, N Finland *67°41′N 26°44′E*

41 *X12* **Peto** Yucatán, SE Mexico *20°09′N 88°55′W*

62 *G10* **Petorca** Valparaíso, C Chile *32°18′S 70°49′W*

31 *Q5* **Petoskey** Michigan, N USA *45°51′N 88°03′W*

138 *G14* **Petra** *archaeological site* Ma'ān, W Jordan

Petra *see* Wādī Mūsā

115 *F14* **Pétras, Sténa** *pass* N Greece

123 *S16* **Petra Velikogo, Zaliv** *bay* SE Russian Federation

Petrel *see* Petrer

14 *K15* **Petre, Point** *headland* Ontario, SE Canada *43°49′N 77°07′W*

105 *S12* **Petrer** *var.* Petrel. Valenciana, E Spain *38°28′N 00°46′W*

125 *U11* **Petretsovo** Permskiy Kray, NW Russian Federation *61°22′N 57°21′E*

114 *G12* **Petrich** Blagoevgrad, SW Bulgaria *41°25′N 23°12′E*

187 *P15* **Petrie, Récif** *reef* N New Caledonia

37 *N11* **Petrified Forest** *prehistoric site* Arizona, SW USA

Petrikau *see* Piotrków Trybunalski

Petrikov *see* Pyetrykaw

116 *H12* **Petrila** *Hung.* Petrilla. Hunedoara, W Romania *45°27′N 23°25′E*

Petrilla *see* Petrila

112 *E9* **Petrinja** Sisak-Moslavina, C Croatia *45°27′N 16°14′E*

Petroaleksandrovsk *see* To'rtkok'l

Petrőcz *see* Bački Petrovac

124 *G12* **Petrodvorets** *Fin.* Pietarhovi. Leningradskaya Oblast', NW Russian Federation *59°53′N 29°52′E*

Petrograd *see* Sankt-Peterburg

Petrokov *see* Piotrków Trybunalski

54 *G6* **Petrólea** Norte de Santander, NE Colombia *08°30′N 72°35′W*

14 *D16* **Petrolia** Ontario, S Canada *42°54′N 82°07′W*

25 *S4* **Petrolia** Texas, SW USA *34°00′N 98°13′W*

59 *O15* **Petrolina** Pernambuco, E Brazil *09°22′S 40°30′W*

45 *T6* **Petrona, Punta** *headland* C Puerto Rico *17°57′N 66°23′W*

Petropavl *see* Petropavlovsk

117 *V7* **Petropavlivka** Dnipropetrovs'ka Oblast', E Ukraine *48°28′N 36°28′E*

145 *P6* **Petropavlovsk** *Kaz.* Petropavl. Severnyy Kazakhstan, N Kazakhstan *54°47′N 69°06′E*

123 *V11* **Petropavlovsk-Kamchatskiy** Kamchatskiy Kray, E Russian Federation *53°03′N 158°43′E*

60 *P9* **Petrópolis** Rio de Janeiro, SE Brazil *22°30′S 43°28′W*

116 *H12* **Petroşani** *var.* Petroşeni, *Ger.* Petroschen, *Hung.* Petrozsény. Hunedoara, W Romania *45°25′N 23°22′E*

Petroschen/Petroşeni *see* Petroşani

Petroskoi *see* Petrozavodsk

112 *N12* **Petrovac** Serbia, E Serbia *42°22′N 21°25′E*

Petrovac *see* Bosanski Petrovac

113 *J17* **Petrovac na Moru** S Montenegro *42°11′N 19°00′E*

Petrovac/Petrovácz *see* Bački Petrovac

117 *S8* **Petrove** Kirovohrads'ka Oblast', C Ukraine *48°22′N 33°12′E*

113 *O18* **Petrovec** C FYR Macedonia *41°57′N 21°37′E*

Petrovgrad *see* Zrenjanin

127 *P7* **Petrovsk** Saratovskaya Oblast', W Russian Federation *52°20′N 45°23′E*

Petrovsk-Port *see* Makhachkala

127 *P9* **Petrov Val** Volgogradskaya Oblast', SW Russian Federation *50°10′N 45°16′E*

124 *J11* **Petrozavodsk** *Fin.* Petroskoi. Respublika Kareliya, NW Russian Federation *61°46′N 34°19′E*

Petrozsény *see* Petroşani

83 *D20* **Petrusdal** Hardap, C Namibia *23°42′S 17°23′E*

117 *T7* **Petrykivka** Dnipropetrovs'ka Oblast', E Ukraine *48°44′N 34°42′E*

Petsamo *see* Pechenga

Petschka *see* Pecica

Pettau *see* Ptuj

109 *S5* **Pettenbach** Oberösterreich, C Austria *47°58′N 14°03′E*

25 *S13* **Pettus** Texas, SW USA *28°34′N 97°49′W*

122 *G12* **Petukhovo** Kurganskaya Oblast', C Russian Federation *55°04′N 67°49′E*

Petuna *see* Songyuan

109 *R4* **Peuerbach** Oberösterreich, N Austria *48°19′N 13°45′E*

62 *G12* **Peumo** Libertador, C Chile *34°20′S 71°12′W*

123 *T6* **Pevek** Chukotskiy Avtonomnyy Okrug, NE Russian Federation *69°41′N 170°19′E*

27 *X5* **Pevely** Missouri, C USA *38°16′N 90°24′W*

Peyia *see* Pégeia

102 *J15* **Peyrehorade** Landes, SW France *43°33′N 01°05′W*

124 *J14* **Peza** ≈ NW Russian Federation

103 *P16* **Pézenas** Hérault, S France *43°28′N 03°25′E*

111 *H20* **Pezinok** *Ger.* Bösing, *Hung.* Bazin. Bratislavský Kraj, W Slovakia *48°17′N 17°15′E*

101 *L22* **Pfaffenhofen an der Ilm** Bayern, SE Germany *48°31′N 11°30′E*

108 *G7* **Pfäffikon** Schwyz, C Switzerland *47°11′N 08°46′E*

101 *F20* **Pfälzer Wald** *hill range* W Germany

101 *N22* **Pfarrkirchen** Bayern, SE Germany *48°25′N 12°56′E*

101 *G21* **Pforzheim** Baden-Württemberg, SW Germany *48°53′N 08°42′E*

101 *H24* **Pfullendorf** Baden-Württemberg, S Germany *47°55′N 09°16′E*

108 *K8* **Pfunds** Tirol, W Austria *46°56′N 10°30′E*

101 *G19* **Pfungstadt** Hessen, W Germany *49°48′N 08°36′E*

83 *L20* **Phalaborwa** *var.* Ba-Pahalaborwa. Limpopo, NE South Africa *23°59′S 31°04′E*

152 *E11* **Phalodi** Rājasthān, NW India *27°06′N 72°22′E*

152 *E12* **Phalsund** Rājasthān, NW India *26°22′N 71°56′E*

155 *E15* **Phaltan** Mahārāshtra, W India *18°01′N 74°31′E*

167 *O7* **Phan** *var.* Muang Phan. Chiang Rai, NW Thailand *19°34′N 99°44′E*

167 *O14* **Phangan, Ko** *island* SW Thailand

166 *M15* **Phang-Nga** *var.* Pang-Nga, Phangnga. Phangnga, SW Thailand *08°29′N 98°31′E*

Phangnga *see* Phang-Nga

Phan Rang/Phanrang *see* Phan Rang-Thap Cham

167 *V13* **Phan Rang-Thap Cham** *var.* Phanrang, Phan Rang, Phan Rang Thap Cham. Ninh Thuân, S Vietnam *11°34′N 109°00′E*

167 *U13* **Phan Ri** Binh Thuân, S Vietnam *11°11′N 108°31′E*

167 *U13* **Phan Thiêt** Binh Thuân, S Vietnam *10°56′N 108°06′E*

Pharnacia *see* Giresun

25 *S17* **Pharr** Texas, SW USA *26°11′N 98°10′W*

Pharus *see* Hvar

167 *N16* **Phatthalung** *var.* Padalung, Patalung. Phatthalung, SW Thailand *07°38′N 100°04′E*

167 *O7* **Phayao** *var.* Muang Phayao. Phayao, NW Thailand *19°10′N 99°55′E*

11 *U10* **Phelps Lake** ◎ Saskatchewan, C Canada

21 *X9* **Phelps Lake** ◎ North Carolina, SE USA

23 *R5* **Phenix City** Alabama, S USA *32°28′N 85°00′W*

Phet Buri *see* Phetchaburi

167 *O11* **Phetchaburi** *var.* Bejraburi, Petchaburi, Phet Buri. Phetchaburi, SW Thailand *13°05′N 99°58′E*

167 *O9* **Phichit** *var.* Bichitra, Muang Phichit, Pichit. Phichit, C Thailand *16°29′N 100°21′E*

22 *M5* **Philadelphia** Mississippi, S USA *32°45′N 89°06′W*

18 *I7* **Philadelphia** New York, NE USA *44°10′N 75°40′W*

18 *I16* **Philadelphia** Pennsylvania, NE USA *40°N 75°10′W*

18 *I16* **Philadelphia** ✈ Pennsylvania, NE USA *39°51′N 75°13′W*

Philadelphia *see* 'Amman

28 *L10* **Philip** South Dakota, N USA *44°02′N 101°39′W*

99 *H22* **Philippeville** Namur, S Belgium *50°12′N 04°33′E*

Philippeville *see* Skikda

21 *S3* **Philippi** West Virginia, NE USA *39°08′N 80°03′W*

Philippi *see* Fílippoi

195 *Y9* **Philippi Glacier** *glacier* Antarctica

192 *G6* **Philippine Basin** *undersea feature* W Pacific Ocean *17°00′N 132°00′E*

129 *X12* **Philippine Plate** *tectonic feature*

171 *O5* **Philippines** *off.* Republic of the Philippines. ◆ *republic* SE Asia

129 *X13* **Philippines** *island group* W Pacific Ocean

171 *P3* **Philippine Sea** *sea* W Pacific Ocean

Philippines, Republic of the *see* Philippines

192 *F6* **Philippine Trench** *undersea feature* W Philippine Sea

83 *H23* **Philippolis** Free State, C South Africa *30°16′S 25°16′E*

Philippopolis *see* Plovdiv

Philippopolis *see* Shahbā', Syria

45 *V9* **Philipsburg** ○ Sint Maarten *17°58′N 63°02′W*

33 *P10* **Philipsburg** Montana, NW USA *46°19′N 113°17′W*

39 *R6* **Philip Smith Mountains** ▲▲ Alaska, USA

152 *H8* **Phillaur** Punjab, N India *31°02′N 75°50′E*

183 *N13* **Phillip Island** *island* Victoria, SE Australia

25 *N2* **Phillips** Texas, SW USA *35°39′N 101°21′W*

30 *K5* **Phillips** Wisconsin, N USA *45°42′N 90°23′W*

26 *K3* **Phillipsburg** Kansas, C USA *39°45′N 99°19′W*

18 *I14* **Phillipsburg** New Jersey, NE USA *40°39′N 75°09′W*

21 *S7* **Philpott Lake** ⊠ Virginia, NE USA

Phintias *see* Licata

167 *P9* **Phitsanulok** *var.* Bisnulok, Muang Phitsanulok, Pitsanulok. Phitsanulok, C Thailand *16°49′N 100°15′E*

Phlórina *see* Flórina

Phnom Penh *see* Phnum Penh

167 *S13* **Phnum Penh** *var.* Phnom Penh. ● (Cambodia) Phnum Penh, S Cambodia *11°35′N 104°55′E*

167 *S11* **Phnum Tbêng Meanchey** Preăh Vihéar, N Cambodia *13°45′N 104°58′E*

36 *K13* **Phoenix** *state capital* Arizona, SW USA *33°27′N 112°04′W*

Phoenix Island *see* Rawaki

191 *R3* **Phoenix Islands** *island group* C Kiribati

18 *I15* **Phoenixville** Pennsylvania, NE USA *40°07′N 75°31′W*

83 *K22* **Phofung** *var.* Mont-aux-Sources. ▲ N Lesotho *28°47′S 28°52′E*

167 *Q10* **Phon** Khon Kaen, E Thailand *15°47′N 102°35′E*

167 *Q5* **Phôngsali** *var.* Phong Saly. Phôngsali, N Laos *21°40′N 102°04′E*

Phong Saly *see* Phôngsali

Phônhông *see* Ban Phônhông

167 *R7* **Phônsaven** *var.* Pèk, Xieng Khouang; *prev.* Xiangkhoang. Xiangkhoang, N Laos *19°19′N 103°23′E*

167 *R5* **Phô Rang** *var.* Bao Yèn. Lao Cai, N Vietnam *22°12′N 104°27′E*

Phort Láirge, Cuan *see* Waterford Harbour

Phou Louang *see* Annamite Mountains

167 *N10* **Phra Chedi Sam Ong** Kanchanaburi, W Thailand *15°18′N 98°26′E*

167 *O8* **Phrae** *var.* Muang Phrae, Prae. Phrae, NW Thailand *18°07′N 100°09′E*

Phra Nakhon Si Ayutthaya *see* Ayutthaya

167 *M14* **Phra Thong, Ko** *island* SW Thailand

Phu Cŭơng *see* Thu Dâu Môt

166 *M15* **Phuket** *var.* Bhuket, Puket, *Mal.* Ujung Salang; *prev.* Junkseylon, Salang. Phuket, SW Thailand *07°52′N 98°22′E*

166 *M15* **Phuket** ✈ Phuket, SW Thailand *08°03′N 98°16′E*

166 *M15* **Phuket, Ko** *island* SW Thailand

154 *N12* **Phulabāni** *prev.* Phulbani. Odisha, E India *20°30′N 84°18′E*

Phulbani *see* Phulabāni

167 *U9* **Phu Lôc** Th,a Thiên-Huê, C Vietnam *16°13′N 107°53′E*

167 *R13* **Phumĭ Chŏâm** Kâmpóng Spœ, SW Cambodia *11°42′N 103°58′E*

Phumĭ Kalêng *see* Kalêng

Phumĭ Kâmpóng Trâbêk *see* Kâmpóng Trâbêk

Phumĭ Koŭk Kduŏch *see* Koŭk Kduŏch

Phumĭ Labăng *see* Labăng

Phumĭ Mlu Prey *see* Mlu Prey

Phumĭ Moŭng *see* Moŭng

Phumĭ Prâmaôy *see* Prâmaôy

Phumĭ Samĭt *see* Samĭt

Phumĭ Sâmraông *see* Sâmraông

Phumĭ Siĕmbok *see* Siĕmbok

Phumĭ Thalabârĭvăt *see* Thalabârĭvăt

Phumĭ Veal Renh *see* Veal Renh

Phumĭ Yeay Sên *see* Yeay Sên

Phum Kompong Trabek *see* Kâmpóng Trâbêk

Phum Samrong *see* Sâmraông

167 *V11* **Phu My** Binh Đinh, C Vietnam *14°07′N 109°05′E*

Phung Hiêp *see* Tân Hiêp

153 *T12* **Phuntsholing** SW Bhutan *26°52′N 89°26′E*

167 *V13* **Phư σ c Dân** Ninh Thuân, S Vietnam *11°28′N 108°53′E*

167 *R15* **Phước Long** Minh Hai, S Vietnam *09°27′N 105°25′E*

Phước Sơn *see* Khâm Đưc

167 *R14* **Phu Quôc, Đao** *var.* Phu Quoc Island. *island* S Vietnam

Phu Quoc Island *see* Phu Quôc, Đao

167 *S6* **Phu Tho** Vinh Phu, N Vietnam *21°23′N 105°13′E*

Phu Vinh *see* Tra Vinh

166 *M7* **Phyu** *var.* Hpyu, Pyu. Bago, C Myanmar (Burma) *18°29′N 96°28′E*

189 *T13* **Piaanu Pass** *passage* Chuuk Islands, C Micronesia

106 *E8* **Piacenza** *Fr.* Paisance; *anc.* Placentia. Emilia-Romagna, N Italy *45°02′N 09°42′E*

107 *K14* **Pianella** Abruzzo, C Italy *42°23′N 14°04′E*

107 *M15* **Pianosa, Isola** *island* Archipelago Toscano, C Italy

171 *U13* **Piar** Papua Barat, E Indonesia *02°49′S 132°46′E*

45 *U14* **Piarco** *var.* Port of Spain. ✈ (Port-of-Spain) Trinidad, Trinidad and Tobago *10°36′N 61°21′W*

110 *M12* **Piaseczno** Mazowieckie, C Poland *52°03′N 21°00′E*

116 *I15* **Piatra** Teleorman, S Romania *43°49′N 25°10′E*

116 *L10* **Piatra-Neamţ** *Hung.* Karácsonkő. Neamţ, NE Romania *46°54′N 26°23′E*

Piauhy *see* Piauí

59 *N15* **Piauí** *off.* Estado do Piauí; *prev.* Piauhy. ◈ *state* E Brazil

Piauí, Estado do *see* Piauí

106 *I7* **Piave** ≈ NE Italy

107 *K24* **Piazza Armerina** *var.* Chiazza. Sicilia, Italy, C Mediterranean Sea *37°23′N 14°22′E*

81 *G14* **Pibor** *Amh.* Pībor Wenz. ≈ Ethiopia/South Sudan

81 *G14* **Pibor Post** Jonglei, E South Sudan *06°50′N 33°06′E*

Pībor Wenz *see* Pibor

Pibrans *see* Příbram

36 *K11* **Picacho Butte** ▲ Arizona, SW USA *35°12′N 112°44′W*

40 *D4* **Picachos, Cerro** ▲ NW Mexico *29°15′N 114°04′W*

103 *O4* **Picardie** *Eng.* Picardy. ◈ *region* N France

Picardy *see* Picardie

22 *L8* **Picayune** Mississippi, S USA *30°31′N 89°40′W*

Piccolo San Bernardo, Colle di *see* Little Saint Bernard Pass

PicdeBälaitous *see* Balaïtous

62 *K5* **Pichanal** Salta, N Argentina *23°18′S 64°10′W*

147 *P12* **Pichandar** W Tajikistan *38°44′N 68°51′E*

27 *R8* **Picher** Oklahoma, C USA *36°59′N 94°49′W*

62 *G12* **Pichilemu** Libertador, C Chile *34°25′S 72°00′W*

40 *F9* **Pichilingue** Baja California Sur, NW Mexico *24°20′N 110°17′W*

56 *B6* **Pichincha** ◈ *province* N Ecuador

56 *C6* **Pichincha** ▲ N Ecuador *0°12′S 78°39′W*

Pichit *see* Phichit

41 *U15* **Pichucalco** Chiapas, SE Mexico *17°32′N 93°07′W*

22 *L5* **Pickens** Mississippi, S USA *32°52′N 89°58′W*

21 *O11* **Pickens** South Carolina, SE USA *34°53′N 82°42′W*

14 *G11* **Pickerel** ≈ Ontario, S Canada

14 *H15* **Pickering** Ontario, S Canada *43°50′N 79°03′W*

97 *N16* **Pickering** N England, United Kingdom *54°14′N 00°47′W*

31 *S13* **Pickerington** Ohio, N USA *39°52′N 82°45′W*

12 *C10* **Pickle Lake** Ontario, C Canada *51°30′N 90°10′W*

29 *P12* **Pickstown** South Dakota, N USA *43°02′N 98°31′W*

25 *V6* **Pickton** Texas, SW USA *33°01′N 95°19′W*

23 *N1* **Pickwick Lake** ⊠ S USA

64 *N2* **Pico** *var.* Ilha do Pico. *island* Azores, Portugal, NE Atlantic Ocean

63 *J19* **Pico de Salamanca** Chubut, SE Argentina *45°26′S 67°26′W*

1 *P9* **Pico Fracture Zone** *tectonic feature* NW Atlantic Ocean

Pico, Ilha do *see* Pico

59 *O14* **Picos** Piauí, E Brazil *07°05′S 41°24′W*

63 *I20* **Pico Truncado** Santa Cruz, SE Argentina *46°45′S 68°00′W*

183 *S9* **Picton** New South Wales, SE Australia *34°12′S 150°36′E*

14 *K15* **Picton** Ontario, SE Canada *43°59′N 77°09′W*

185 *K14* **Picton** Marlborough, South Island, New Zealand *41°18′S 174°00′E*

63 *H15* **Picún Leufú, Arroyo** ≈ SW Argentina

Pidálion *see* Gkréko, Akrotíri

155 *K25* **Pidurutalagala** ▲ S Sri Lanka *07°03′N 80°47′E*

116 *K6* **Pidvolochys'k** Ternopil's'ka Oblast', W Ukraine *49°31′N 26°09′E*

107 *K16* **Piedimonte Matese** Campania, S Italy *41°20′N 14°30′E*

27 *X7* **Piedmont** Missouri, C USA *37°09′N 90°42′W*

21 *P11* **Piedmont** South Carolina, SE USA *34°42′N 82°27′W*

17 *S12* **Piedmont** *escarpment* E USA

Piedmont *see* Piemonte

31 *U13* **Piedmont Lake** ⊠ Ohio, N USA

104 *M11* **Piedrabuena** Castilla-La Mancha, C Spain *39°02′N 04°10′W*

Piedrafita, Puerto de *see* Pedrafita, Porto de

104 *L8* **Piedrahita** Castilla y León, N Spain *40°27′N 05°20′W*

41 *N6* **Piedras Negras** *var.* Ciudad Porfirio Díaz. Coahuila, NE Mexico *28°40′N 100°32′W*

61 *E21* **Piedras, Punta** *headland* E Argentina *35°27′S 57°04′W*

57 *I14* **Piedras, Río de las** ≈ E Peru

111 *J16* **Piekary Śląskie** Śląskie, S Poland *50°24′N 18°58′E*

93 *M17* **Pieksämäki** Etelä-Savo, E Finland *62°18′N 27°10′E*

109 *V5* **Pielach** ≈ NE Austria

93 *M16* **Pielavesi** Pohjois-Savo, C Finland *63°14′N 26°45′E*

93 *M16* **Pielavesi** ◎ C Finland

93 *N16* **Pielinen** *var.* Pielisjärvi. ◎ E Finland

Pielisjärvi *see* Pielinen

106 *A8* **Piemonte** *Eng.* Piedmont. ◈ *region* NW Italy

111 *L18* **Pieniny** ▲▲ S Poland

111 *E14* **Pieńsk** *Ger.* Penzig. Dolnośląskie, SW Poland *51°14′N 15°03′E*

29 *Q13* **Pierce** Nebraska, C USA *42°12′N 97°31′W*

11 *R14* **Pierceland** Saskatchewan, C Canada

115 *E14* **Piéria** ▲▲ N Greece

29 *N10* **Pierre** *state capital* South Dakota, N USA *44°22′N 100°21′W*

102 *K16* **Pierrefitte-Nestalas** Hautes-Pyrénées, S France *42°57′N 00°04′W*

103 *R14* **Pierrelatte** Drôme, E France *44°22′N 04°40′E*

15 *P11* **Pierreville** Québec, SE Canada *46°05′N 72°48′W*

15 *O7* **Pierriche** ≈ Québec, SE Canada

111 *H20* **Piešt'any** *Ger.* Pistyan, *Hung.* Pöstyén. Tranavský Kraj, W Slovakia *48°37′N 17°48′E*

109 *X5* **Piesting** ≈ E Austria

Pietarhovi *see* Petrodvorets

Pietari *see* Sankt-Peterburg

Pietarsaari *see* Jakobstad

83 *K23* **Pietermaritzburg** *var.* Maritzburg. KwaZulu/Natal, E South Africa *29°35′S 30°23′E*

Pietersburg *see* Polokwane

107 *K24* **Pietraperzia** Sicilia, Italy, C Mediterranean Sea *37°25′N 14°08′E*

107 *N22* **Pietra Spada, Passo della** *pass* SW Italy

Piet Retief *see* eMkhondo

116 *J10* **Pietrosul, Vârful** *prev.* Vîrful Pietrosu. ▲ N Romania *47°06′N 25°09′E*

106 *I6* **Pieve di Cadore** Veneto, NE Italy *46°27′N 12°22′E*

14 *C18* **Pigeon Bay** *lake bay* Ontario, S Canada

27 *X8* **Piggott** Arkansas, C USA *36°22′N 90°11′W*

83 *L21* **Piggs Peak** NW Swaziland *25°58′S 31°17′E*

Pigs, Bay of *see* Cochinos, Bahía de

61 *A23* **Pigüé** Buenos Aires, E Argentina *37°38′S 62°27′W*

41 *O12* **Piguícas** ▲ C Mexico *21°08′N 99°37′W*

193 *W15* **Piha Passage** *passage* S Tonga

Pihkva Järv *see* Pskov, Lake

93 *N18* **Pihlajavesi** ◎ SE Finland

93 *J18* **Pihlava** Satakunta, SW Finland *61°33′N 21°36′E*

93 *L16* **Pihtipudas** Keski-Suomi, C Finland *63°20′N 25°37′E*

40 *L14* **Pihuamo** Jalisco, SW Mexico *19°20′N 103°21′W*

189 *U11* **Piis Moen** *var.* Pis. *atoll* Chuuk Islands, C Micronesia

41 *U17* **Pijijiapán** Chiapas, SE Mexico *15°42′N 93°12′W*

98 *G12* **Pijnacker** Zuid-Holland, W Netherlands *52°01′N 04°26′E*

42 *H5* **Pijol, Pico** ▲ NW Honduras *15°07′N 87°39′W*

Pikaar *see* Bikar Atoll

124 *I13* **Pikalevo** Leningradskaya Oblast', NW Russian Federation *59°33′N 34°04′E*

188 *M15* **Pikelot** *island* Caroline Islands, C Micronesia

30 *M5* **Pike River** ≈ Wisconsin, N USA

37 *T5* **Pikes Peak** ▲ Colorado, C USA *38°51′N 105°06′W*

21 *P6* **Pikeville** Kentucky, S USA *37°29′N 82°33′W*

20 *L9* **Pikeville** Tennessee, S USA *35°35′N 85°11′W*

Pikinni *see* Bikini Atoll

79 *H18* **Pikounda** Sangha, C Congo *0°30′N 16°44′E*

110 *G9* **Piła** *Ger.* Schneidemühl. Wielkopolskie, C Poland *53°09′N 16°44′E*

62 *N6* **Pilagá, Riacho** ≈ NE Argentina

61 *D20* **Pilar** Buenos Aires, E Argentina *34°28′S 58°55′W*

62 *N7* **Pilar** *var.* Villa del Pilar. Ñeembucú, S Paraguay *26°55′S 58°20′W*

62 *N6* **Pilcomayo, Rio** ≈ C South America

147 *R12* **Pildon** *Rus.* Pil'don. C Tajikistan *39°10′N 71°00′E*

Piles *see* Pylés

Pilgram *see* Pelhřimov

152 *L10* **Pīlībhīt** Uttar Pradesh, N India *28°37′N 79°48′E*

110 *M13* **Pilica** ≈ C Poland

115 *G16* **Pílio** ▲▲ C Greece

111 *J22* **Pilisvörösvár** Pest, N Hungary *47°38′N 18°55′E*

65 *G15* **Pillar Bay** *bay* Ascension Island, C Atlantic Ocean

183 *P17* **Pillar, Cape** *headland* Tasmania, SE Australia *43°13′S 147°58′E*

Pillau *see* Baltiysk

183 *R5* **Pilliga** New South Wales, SE Australia *30°22′S 148°53′E*

44 *H8* **Pilón** Granma, E Cuba *19°54′N 77°20′W*

Pilos *see* Pýlos

11 *W17* **Pilot Mound** Manitoba, S Canada *49°12′N 98°49′W*

21 *S8* **Pilot Mountain** North Carolina, SE USA *36°23′N 80°28′W*

39 *O14* **Pilot Point** Alaska, USA *57°33′N 157°34′W*

25 *T5* **Pilot Point** Texas, SW USA *33°24′N 96°57′W*

32 *K11* **Pilot Rock** Oregon, NW USA *45°28′N 118°49′W*

38 *M11* **Pilot Station** Alaska, USA *61°56′N 162°52′W*

Pilsen *see* Plzeň

111 *K18* **Pilsko** ▲ N Slovakia *49°31′N 19°21′E*

Pilten *see* Piltene

118 *D8* **Piltene** *Ger.* Pilten. W Latvia *57°14′N 21°41′E*

111 *M16* **Pilzno** Podkarpackie, SE Poland *49°58′N 21°18′E*

Pilzno *see* Plzeň

37 *N14* **Pima** Arizona, SW USA *32°49′N 109°50′W*

58 *H13* **Pimenta** Pará, N Brazil *04°32′S 56°17′W*

59 *F16* **Pimenta Bueno** Rondônia, W Brazil *11°40′S 61°14′W*

56 *B11* **Pimentel** Lambayeque, W Peru *06°51′S 79°53′W*

105 *S6* **Pina** Aragón, NE Spain *41°28′N 00°31′W*

119 *I20* **Pina** ≈ SW Belarus

40 *E2* **Pinacate, Sierra del** ▲ NW Mexico *31°49′N 113°30′W*

63 *H22* **Pináculo, Cerro** ▲ S Argentina *50°46′S 72°07′W*

191 *X11* **Pinaki** *atoll* Îles Tuamotu, E French Polynesia

37 *N15* **Pinaleno Mountains** ▲▲ Arizona, SW USA

171 *P4* **Pinamalayan** Mindoro, N Philippines *13°00′N 121°30′E*

169 *Q10* **Pinang** Borneo, C Indonesia *0°36′N 109°11′W*

168 *J7* **Pinang** *var.* Penang. ◈ *state* Peninsular Malaysia

Pinang *see* George Town

Pinang *see* Pinang, Pulau, Peninsular Malaysia

168 *J7* **Pinang, Pulau** *var.* Penang, Pinang; *prev.* Prince of Wales Island. *island* Peninsular Malaysia

44 *B5* **Pinar del Río** Pinar del Río, W Cuba *22°24′N 83°42′W*

114 *N11* **Pınarhisar** Kırklareli, NW Turkey *41°37′N 27°32′E*

171 *O3* **Pinatubo, Mount** ⛰ Luzon, N Philippines *15°08′N 120°21′E*

11 *Y16* **Pinawa** Manitoba, S Canada *50°09′N 95°52′W*

11 *Q17* **Pincher Creek** Alberta, SW Canada *49°31′N 113°53′W*

30 *L16* **Pinckneyville** Illinois, N USA *38°04′N 89°22′W*

Pîncota *see* Pâncota

111 *L15* **Pińczów** Świętokrzyskie, C Poland *50°30′N 20°30′E*

149 *U7* **Pind Dādan Khān** Punjab, E Pakistan *32°36′N 73°07′E*

Píndhos/Píndhos Óros *see* Píndos

149 *V8* **Pindi Bhattiān** Punjab, E Pakistan *31°54′N 73°20′E*

149 *U6* **Pindi Gheb** Punjab, E Pakistan *33°16′N 72°21′E*

115 *D15* **Píndos** *var.* Píndhos Óros, *Eng.* Pindus Mountains; *prev.* Píndhos. ▲▲ C Greece

Pindus Mountains *see* Píndos

18 *J16* **Pine Barrens** *physical region* New Jersey, NE USA

27 *V12* **Pine Bluff** Arkansas, C USA *34°15′N 92°00′W*

23 *X11* **Pine Castle** Florida, SE USA *28°28′N 81°22′W*

29 *V7* **Pine City** Minnesota, N USA *45°49′N 92°55′W*

181 *P2* **Pine Creek** Northern Territory, N Australia *13°51′S 131°51′E*

35 *V4* **Pine Creek** ≈ Nevada, W USA

18 *F13* **Pine Creek** ≈ Pennsylvania, NE USA

27 *Q13* **Pine Creek Lake** ⊠ Oklahoma, C USA

33 *T15* **Pinedale** Wyoming, C USA *42°52′N 109°51′W*

11 *X15* **Pine Dock** Manitoba, S Canada *51°34′N 96°47′W*

11 *Y16* **Pine Falls** Manitoba, S Canada *50°29′N 96°12′W*

◆ Country
● Country Capital
◇ Dependent Territory
○ Dependent Territory Capital
◈ Administrative Regions
✈ International Airport
▲ Mountain
▲▲ Mountain Range
⛰ Volcano
≈ River
◎ Lake
⊠ Reservoir

35 R10 **Pine Flat Lake** ⊠ California, W USA
125 N8 **Pinega** Arkhangel'skaya Oblast', NW Russian Federation 64°40′N 43°24′E
125 N8 **Pinega** ≈ NW Russian Federation
15 N12 **Pine Hill** Québec, SE Canada 45°44′N 74°30′W
11 T12 **Pinehouse Lake** ◎ Saskatchewan, C Canada
21 T10 **Pinehurst** North Carolina, SE USA 35°12′N 79°28′W
115 D19 **Pineiós** ≈ S Greece
115 E16 **Pineiós** *var.* Piniós; *anc.* Peneius. ≈ C Greece
29 W10 **Pine Island** Minnesota, N USA 44°12′N 92°39′W
23 V15 **Pine Island** *island* Florida, SE USA
194 K10 **Pine Island Glacier** *glacier* Antarctica
25 X9 **Pineland** Texas, SW USA 31°15′N 93°58′W
23 V13 **Pinellas Park** Florida, SE USA 27°50′N 82°42′W
10 M13 **Pine Pass** *pass* British Columbia, W Canada
8 J10 **Pine Point** Northwest Territories, W Canada 60°52′N 114°30′W
28 K12 **Pine Ridge** South Dakota, N USA 43°01′N 102°33′W
29 U6 **Pine River** Minnesota, N USA 46°43′N 94°24′W
31 Q8 **Pine River** ≈ Michigan, N USA
30 M4 **Pine River** ≈ Wisconsin, N USA
106 A8 **Pinerolo** Piemonte, NE Italy 44°56′N 07°21′E
115 I15 **Píneś, Akrotírio** *var.* Akrotírio Pínnes. *headland* N Greece 40°06′N 24°19′E
25 W6 **Pines, Lake O' the** ⊠ Texas, SW USA
Pines, The Isle of the *see* Juventud, Isla de la
Pine Tree State *see* Maine
21 N7 **Pineville** Kentucky, S USA 36°47′N 83°43′W
22 H7 **Pineville** Louisiana, S USA 31°19′N 92°25′W
27 R8 **Pineville** Missouri, C USA 36°36′N 94°23′W
21 R10 **Pineville** North Carolina, SE USA 35°04′N 80°53′W
21 Q6 **Pineville** West Virginia, NE USA 37°35′N 81°34′W
33 V8 **Piney Buttes** *physical region* Montana, NW USA
163 W9 **Ping'an** Jilin, NE China 44°36′N 127°13′E
160 H14 **Pingbian** *var.* Pingbian Miaozu Zizhixian, Yuping. Yunnan, SW China 22°51′N 103°28′E
Pingbian Miaozu Zizhixian *see* Pingbian
157 S9 **Pingdingshan** Henan, C China 33°52′N 113°20′E
161 S14 **Pingdong** *Jap.* Heitō; *prev.* P'ingtung. S Taiwan 22°40′N 120°30′E
161 R4 **Pingdu** Shandong, E China 36°50′N 119°55′E
189 W16 **Pingelap Atoll** *atoll* Caroline Islands, E Micronesia
160 K14 **Pingguo** *var.* Matou. Guangxi Zhuangzu Zizhiqu, S China 23°24′N 107°30′E
161 Q13 **Pinghe** *var.* Xiaoxi. Fujian, SE China 24°30′N 117°19′E
P'ing-hsiang *see* Pingxiang
161 N10 **Pingjiang** Hunan, S China 28°44′N 113°33′E
Pingkiang *see* Harbin
160 L8 **Pingli** Shaanxi, C China 32°27′N 109°21′E
159 W10 **Pingliang** *var.* Kongtong, P'ing-liang. Gansu, C China 35°27′N 106°38′E
P'ing-liang *see* Pingliang
159 W8 **Pingluo** Ningxia, N China 38°55′N 106°31′E
Pingma *see* Tiandong
167 O7 **Ping, Mae Nam** ≈ W Thailand
161 Q1 **Pingquan** Hebei, E China 41°02′N 118°35′E
29 P5 **Pingree** North Dakota, N USA 47°07′N 98°54′W
Pingsiang *see* Pingxiang
P'ingtung *see* Pingdong
160 I8 **Pingwu** *var.* Long'an. Sichuan, C China 32°33′N 104°32′E
160 J15 **Pingxiang** Guangxi Zhuangzu Zizhiqu, S China 22°03′N 106°44′E
161 O11 **Pingxiang** *var.* P'ing-hsiang; *prev.* Pingsiang. Jiangxi, S China 27°42′N 113°50′E
Pingxiang *see* Tongwei
161 S11 **Pingyang** *var.* Kunyang. Zhejiang, SE China 27°46′N 120°37′E
161 P5 **Pingyi** Shandong, E China 35°30′N 117°38′E
161 P5 **Pingyin** Shandong, E China 36°18′N 116°24′E
60 H13 **Pinhalzinho** Santa Catarina, S Brazil 26°53′S 52°57′W
60 I12 **Pinhão** Paraná, S Brazil 25°46′S 51°32′W
61 H17 **Pinheiro Machado** Rio Grande do Sul, S Brazil 31°34′S 53°22′W
104 I7 **Pinhel** Guarda, N Portugal 40°47′N 07°03′W
Piniós *see* Pineiós
168 I11 **Pini, Pulau** *island* Kepulauan Batu, W Indonesia
109 Y7 **Pinka** ≈ SE Austria
109 X7 **Pinkafeld** Burgenland, SE Austria 47°23′N 16°08′E
Pinkiang *see* Harbin
10 M12 **Pink Mountain** British Columbia, W Canada 57°10′N 122°36′W
166 M3 **Pinlebu** Sagaing, N Myanmar (Burma) 24°02′N 95°21′E
38 J12 **Pinnacle Island** *island* Alaska, USA
180 I12 **Pinnacles, The** *tourist site* Western Australia
182 K10 **Pinnaroo** South Australia 35°17′S 140°54′E
Pinne *see* Pniewy
100 I9 **Pinneberg** Schleswig-Holstein, N Germany 53°40′N 09°49′E
Pínnes, Akrotírio *see* Píneś, Akrotírio
Pinos, Isla de *see* Juventud, Isla de la
35 R14 **Pinos, Mount** ▲ California, W USA 34°48′N 119°09′W
105 R12 **Pinoso** Valenciana, E Spain 38°25′N 01°02′W
105 N14 **Pinos-Puente** Andalucía, S Spain 37°16′N 03°46′W
41 Q17 **Pinotepa Nacional** *var.* Santiago Pinotepa Nacional. Oaxaca, SE Mexico 16°20′N 98°02′W
114 F13 **Pínovo** ▲ N Greece 41°06′N 22°19′E
187 R17 **Pins, Île des** *var.* Kunyé. *island* E New Caledonia
119 I20 **Pinsk** *Pol.* Pińsk. Brestskaya Voblasts', SW Belarus 52°07′N 26°07′E
14 D18 **Pins, Pointe aux** *headland* Ontario, S Canada 42°14′N 81°53′W
57 B16 **Pinta, Isla** *var.* Abingdon. *island* Galapagos Islands, Ecuador, E Pacific Ocean
125 Q12 **Pinyug** Kirovskaya Oblast', NW Russian Federation 60°12′N 47°45′E
57 B17 **Pinzón, Isla** *var.* Duncan Island. *island* Galapagos Islands, Ecuador, E Pacific Ocean
35 Y8 **Pioche** Nevada, W USA 37°57′N 114°30′W
106 F13 **Piombino** Toscana, C Italy 42°54′N 10°30′E
0 C9 **Pioneer Fracture Zone** *tectonic feature* NE Pacific Ocean
122 L5 **Pioner, Ostrov** *island* Severnaya Zemlya, N Russian Federation
118 A13 **Pionerskiy** *Ger.* Neukuhren. Kaliningradskaya Oblast', W Russian Federation 54°57′N 20°16′E
110 N13 **Pionki** Mazowieckie, C Poland 51°30′N 21°27′E
184 L9 **Piopio** Waikato, North Island, New Zealand 38°27′S 175°00′E
110 K13 **Piotrków Trybunalski** *Ger.* Petrikau, *Rus.* Petrokov. Łodzkie, C Poland 51°25′N 19°42′E
152 F12 **Pīpār Road** Rājasthān, N India 26°25′N 73°29′E
115 I16 **Pipéri** *island* Vóreies Sporádes, Greece, Aegean Sea
29 S10 **Pipestone** Minnesota, N USA 44°00′N 96°19′W
12 C9 **Pipestone** ≈ Ontario, C Canada
61 E21 **Pipinas** Buenos Aires, E Argentina 35°32′S 57°20′W
149 T7 **Piplān** *prev.* Liaqatabad. Punjab, E Pakistan 32°17′N 71°24′E
15 R5 **Pipmuacan, Réservoir** ⊠ Québec, SE Canada
Piqan *see* Shanshan
31 R13 **Piqua** Ohio, N USA 40°08′N 84°14′W
105 P5 **Piqueras, Puerto de** *pass* N Spain
60 H11 **Piquiri, Rio** ≈ S Brazil
60 L9 **Piracicaba** São Paulo, S Brazil 22°45′S 47°40′W
Piraeus/Piraiévs *see* Peiraías
60 K10 **Piraju** São Paulo, S Brazil 23°12′S 49°24′W
60 K9 **Pirajuí** São Paulo, S Brazil 21°58′S 49°27′W
63 G21 **Pirámide, Cerro** ▲ S Chile 49°06′S 73°32′W
Piramiva *see* Pyramíva
109 R13 **Piran** *It.* Pirano. SW Slovenia 45°35′N 13°35′E
62 N6 **Pirané** Formosa, N Argentina 25°42′S 59°06′W
59 J18 **Piranhas** Goiás, S Brazil 16°24′S 51°51′W
Pirano *see* Piran
142 I4 **Pīrānshahr** Āžarbāyjān-e Gharbī, NW Iran 36°41′N 45°08′E
59 M19 **Pirapora** Minas Gerais, NE Brazil 17°20′S 44°54′W
60 I9 **Pirapózinho** São Paulo, S Brazil 22°17′S 51°31′W
61 G19 **Piraraja** Lavalleja, S Uruguay 33°44′S 54°45′W
60 L9 **Pirassununga** São Paulo, S Brazil 21°58′S 47°23′W
45 V6 **Pirata, Monte** ▲ E Puerto Rico 18°06′N 65°33′W
60 I13 **Piratuba** Santa Catarina, S Brazil 27°26′S 51°47′W
114 I9 **Pirdop** *prev.* Strednogorie. Sofia, W Bulgaria 42°44′N 24°09′E
191 P7 **Pirea** Tahiti, W French Polynesia
59 K18 **Pirenópolis** Goiás, S Brazil 15°48′S 49°00′W
153 S13 **Pirganj** Rajshahi, NW Bangladesh 25°51′N 88°25′E
Pirgi *see* Pyrgí
Pírgos *see* Pýrgos
61 F20 **Piriápolis** Maldonado, S Uruguay 34°51′S 55°15′W
114 G11 **Pirin** ▲▲ SW Bulgaria
Pirineos *see* Pyrenees
58 N13 **Piripiri** Piauí, E Brazil 04°15′S 41°46′W
118 H4 **Pirita** *var.* Pirita Jõgi. ≈ NW Estonia
Pirita Jõgi *see* Pirita
54 J6 **Píritu** Portuguesa, N Venezuela 09°21′N 69°16′W
93 L18 **Pirkanmaa** *Swe.* Birkaland. ◆ *region* W Finland
93 L18 **Pirkkala** Pirkanmaa, W Finland 61°27′N 23°47′E
101 F20 **Pirmasens** Rheinland-Pfalz, SW Germany 49°12′N 07°37′E
101 P16 **Pirna** Sachsen, E Germany 50°57′N 13°56′E
Piroe *see* Piru
113 Q15 **Pirot** Serbia, SE Serbia 43°12′N 22°34′E
152 H6 **Pir Panjāl Range** ▲▲ NE India
43 W16 **Pirre, Cerro** ▲ SE Panama 07°54′N 77°42′W
137 Y11 **Pirsaat** *Rus.* Pirsagat. ≈ E Azerbaijan
Pirsagat *see* Pirsaat
143 V11 **Pīr Shūrān, Selseleh-ye** ▲▲ SE Iran
92 M12 **Pirttikoski** Lappi, N Finland 66°20′N 27°08′E
Pirttikylä *see* Pörtom
171 R13 **Piru** *prev.* Piroe. Pulau Seram, E Indonesia 03°01′S 128°10′E
Piryatin *see* Pyryatyn
Pis *see* Piis Moen
106 F11 **Pisa** *var.* Pisae. Toscana, C Italy 43°43′N 10°23′E
Pisae *see* Pisa
189 V12 **Pisar** *atoll* Chuuk Islands, C Micronesia
Pisaurum *see* Pesaro
14 M10 **Piscatosine, Lac** ◎ Québec, SE Canada
109 W7 **Pischeldorf** Steiermark, SE Austria 47°11′N 15°48′E
Pischk *see* Simeria
107 L19 **Pisciotta** Campania, S Italy 40°07′N 15°13′E
57 E16 **Pisco** Ica, SW Peru 13°46′S 76°12′W
116 G9 **Pişcolt** *Hung.* Piskolt. Satu Mare, NW Romania 47°35′N 22°18′E
57 E16 **Pisco, Río** ≈ E Peru
111 C18 **Písek** Budějovický Kraj, S Czech Republic 49°19′N 14°07′E
31 R14 **Pisgah** Ohio, N USA 39°19′N 84°22′W
Pisha *see* Ningnan
158 F9 **Pishan** *var.* Guma. Xinjiang Uygur Zizhiqu, NW China 37°36′N 78°45′E
117 N8 **Pishchanka** Vinnyts'ka Oblast', C Ukraine 48°12′N 28°52′E
113 K21 **Pishë** Fier, SW Albania 40°40′N 19°22′E
143 X14 **Pīshīn** Sīstān va Balūchestān, SE Iran 26°05′N 61°46′E
149 O9 **Pishin** Khyber Pakhtunkhwa, NW Pakistan 30°33′N 67°01′E
149 N11 **Pishin Lora** *var.* Psein Lora, *Pash.* Pseyn Bowr. ≈ SW Pakistan
Pishma *see* Pizhma
Pishpek *see* Bishkek
171 O14 **Pising** Pulau Kabaena, C Indonesia 05°07′S 121°50′E
Pisino *see* Pazin
Piski *see* Simeria
Piskolt *see* Pişcolt
147 Q9 **Piskom** *Rus.* Pskem. ≈ E Uzbekistan
Piskom Tizmasi *see* Pskemskiy Khrebet
35 P13 **Pismo Beach** California, W USA 35°08′N 120°38′W
77 P12 **Pissila** C Burkina Faso 13°07′N 00°51′W
62 H8 **Pissis, Monte** ▲ N Argentina 27°45′S 68°43′W
41 X12 **Piste** Yucatán, E Mexico 20°40′N 88°34′W
107 O18 **Pisticci** Basilicata, S Italy 40°23′N 16°33′E
106 F11 **Pistoia** *anc.* Pistoria, Pistoriæ. Toscana, C Italy 43°57′N 10°53′E
32 E15 **Pistol River** Oregon, NW USA 42°13′N 124°23′W
Pistoria/Pistoriæ *see* Pistoia
15 U5 **Pistuacanis** ≈ Québec, SE Canada
Pistyan *see* Piešťany
104 M5 **Pisuerga** ≈ N Spain
110 N8 **Pisz** *Ger.* Johannisburg. Warmińsko-Mazurskie, NE Poland 53°37′N 21°49′E
76 I13 **Pita** NW Guinea 11°05′N 12°15′W
54 D12 **Pitalito** Huila, S Colombia 01°51′N 76°01′W
60 I11 **Pitanga** Paraná, S Brazil 24°45′S 51°43′W
182 M9 **Pitarpunga Lake** *salt lake* New South Wales, SE Australia
Pitcairn Group of Islands *see* Pitcairn, Henderson, Ducie and Oeno Islands
193 P10 **Pitcairn, Henderson, Ducie and Oeno Islands** *var.* Pitcairn Group of Islands. ◇ *UK overseas territory* C Pacific Ocean
191 O14 **Pitcairn Island** *island* S Pitcairn Group of Islands
93 J14 **Piteå** Norrbotten, N Sweden 65°19′N 21°30′E
92 I13 **Piteälven** ≈ N Sweden
116 I13 **Pitești** Argeș, S Romania 44°53′N 24°49′E
Pithagorio *see* Pythagóreio
180 I12 **Pithara** Western Australia 30°31′S 116°38′E
103 N6 **Pithiviers** Loiret, C France 48°10′N 02°15′E
152 L9 **Pithorāgarh** Uttarakhand, N India 29°35′N 80°12′E
188 B16 **Piti** W Guam 13°28′N 144°42′E
106 G13 **Pitigliano** Toscana, C Italy 42°38′N 11°40′E
40 F3 **Pitiquito** Sonora, NW Mexico 30°39′N 112°00′W
Pitkäranta *see* Pitkyaranta
38 M11 **Pitkas Point** Alaska, USA 62°01′N 163°17′W
124 H11 **Pitkyaranta** *Fin.* Pitkäranta. Respublika Kareliya, NW Russian Federation 61°34′N 31°27′E
96 J10 **Pitlochry** C Scotland, United Kingdom 56°47′N 03°48′W
18 I16 **Pitman** New Jersey, NE USA 39°43′N 75°06′W
146 I9 **Pitnak** *var.* Drujba, *Rus.* Druzhba. Xorazm Viloyati, W Uzbekistan 41°14′N 61°13′E
112 G8 **Pitomača** Virovitica-Podravina, NE Croatia 45°57′N 17°14′E
35 O2 **Pit River** ≈ California, W USA
63 G15 **Pitrufquén** Araucanía, S Chile 38°59′S 72°40′W
Pitsanulok *see* Phitsanulok
Pitschen *see* Byczyna
Pitsunda *see* Bich'vinta
109 X6 **Pitten** ≈ E Austria
10 J14 **Pitt Island** *island* British Columbia, W Canada
Pitt Island *see* Makin
22 M3 **Pittsboro** Mississippi, S USA 33°55′N 89°20′W
21 T9 **Pittsboro** North Carolina, SE USA 35°46′N 79°12′W
27 R7 **Pittsburg** Kansas, C USA 37°24′N 94°42′W
25 W6 **Pittsburg** Texas, SW USA 33°00′N 94°58′W
18 B14 **Pittsburgh** Pennsylvania, NE USA 40°26′N 80°W
30 J14 **Pittsfield** Illinois, N USA 39°36′N 90°48′W
19 R6 **Pittsfield** Maine, NE USA 44°46′N 69°22′W
18 L11 **Pittsfield** Massachusetts, NE USA 42°27′N 73°15′W
183 U3 **Pittsworth** Queensland, E Australia 27°43′S 151°36′E
62 I8 **Pituil** La Rioja, NW Argentina 28°33′S 67°24′W
56 A10 **Piura** Piura, NW Peru 05°11′S 80°41′W
56 A9 **Piura** *off.* Departamento de Piura. ◆ *department* NW Peru
Piura, Departamento de *see* Piura
35 S13 **Piute Peak** ▲ California, W USA 35°27′N 118°24′W
113 J15 **Piva** ≈ NW Montenegro
117 V5 **Pivdenne** Kharkivs'ka Oblast', E Ukraine 49°52′N 36°04′E
117 P8 **Pivdennyy Buh** *Rus.* Yuzhnyy Bug. ≈ S Ukraine
54 F5 **Pivijay** Magdalena, N Colombia 10°31′N 74°36′W
109 T13 **Pivka** *prev.* Sent Peter, *Ger.* Sankt Peter, *It.* San Pietro del Carso. SW Slovenia 45°41′N 14°12′E
117 U13 **Pivnichno-Kryms'kyy Kanal** *canal* S Ukraine
113 J15 **Pivsko Jezero** ◎ NW Montenegro
111 M18 **Piwniczna** Małopolskie, S Poland 49°26′N 20°43′E
35 R12 **Pixley** California, W USA 35°58′N 119°18′W
125 Q15 **Pizhma** *var.* Pishma. ≈ NW Russian Federation
13 U13 **Placentia** Newfoundland, Newfoundland and Labrador, SE Canada 47°12′N 53°58′W
Placentia *see* Piacenza
13 U13 **Placentia Bay** *inlet* Newfoundland, Newfoundland and Labrador, SE Canada
171 P5 **Placer** Masbate, N Philippines 11°54′N 123°54′E
35 P7 **Placerville** California, W USA 38°42′N 120°48′W
44 F5 **Placetas** Villa Clara, C Cuba 22°18′N 79°40′W
113 Q18 **Plačkovica** ▲▲ E Macedonia
36 L2 **Plain City** Utah, W USA 41°18′N 112°05′W
22 G4 **Plain Dealing** Louisiana, S USA 32°54′N 93°42′W
31 O14 **Plainfield** Indiana, N USA 39°42′N 86°18′W
18 K14 **Plainfield** New Jersey, NE USA 40°37′N 74°25′W
33 O8 **Plains** Montana, NW USA 47°27′N 114°52′W
24 L6 **Plains** Texas, SW USA 33°12′N 102°50′W
29 X10 **Plainview** Minnesota, N USA 44°10′N 92°10′W
29 Q13 **Plainview** Nebraska, C USA 42°21′N 97°47′W
25 N4 **Plainview** Texas, SW USA 34°12′N 101°43′W
26 K4 **Plainville** Kansas, C USA 39°13′N 99°18′W
115 I22 **Pláka** *var.* Mílos. Mílos, Kykládes, Greece, Aegean Sea 36°44′N 24°25′E
115 J15 **Pláka, Akrotírio** *headland* Límnos, E Greece 40°02′N 25°25′E
113 N19 **Plakenska Planina** ▲▲ SW Macedonia
44 K5 **Plana Cays** *islets* SE The Bahamas
105 S12 **Plana, Isla** *var.* Nueva Tabarca. *island* E Spain
59 L18 **Planaltina** Goiás, S Brazil 15°35′S 47°28′W
83 O14 **Planalto Moçambicano** *plateau* N Mozambique
112 N10 **Plandište** Vojvodina, NE Serbia 45°13′N 21°07′E
100 N13 **Plane** ≈ NE Germany
54 E6 **Planeta Rica** Córdoba, NW Colombia 08°24′N 75°39′W
29 P11 **Plankinton** South Dakota, N USA 43°43′N 98°28′W
30 M11 **Plano** Illinois, N USA 41°39′N 88°32′W
25 U6 **Plano** Texas, SW USA 33°01′N 96°42′W
23 W12 **Plant City** Florida, SE USA 28°01′N 82°06′W
22 J9 **Plaquemine** Louisiana, S USA 30°17′N 91°13′W
104 K9 **Plasencia** Extremadura, W Spain 40°02′N 06°05′W
110 P7 **Płaska** Podlaskie, NE Poland 53°55′N 23°18′E
112 C10 **Plaški** Karlovac, C Croatia 45°04′N 15°21′E
113 N19 **Plasnica** SW FYR Macedonia 41°28′N 21°07′E
13 N14 **Plaster Rock** New Brunswick, SE Canada 46°55′N 67°24′W
107 J24 **Platani** *anc.* Halycus. ≈ Sicilia, Italy, C Mediterranean Sea
115 G17 **Platanía** Thessalía, C Greece 39°09′N 23°15′E
115 G24 **Plátanos** Kríti, Greece, E Mediterranean Sea 35°27′N 23°34′E
65 H18 **Plata, Río de la** *var.* River Plate. *estuary* Argentina/Uruguay
77 V15 **Plateau** ◆ *state* C Nigeria
79 G19 **Plateaux** *var.* Région des Plateaux. ◆ *province* C Congo
Plateaux, Région des *see* Plateaux
Plate, Île *see* Flat Island
92 P1 **Platen, Kapp** *headland* NE Svalbard 80°30′N 22°46′E
Plate, River *see* Plata, Río de la
99 G22 **Plate Taille, Lac de la** *var.* L'Eau d'Heure. ⊠ SE Belgium
Plathe *see* Płoty
39 N13 **Platinum** Alaska, USA 59°01′N 161°49′W
54 F5 **Plato** Magdalena, N Colombia 09°47′N 74°47′W
29 O11 **Platte** South Dakota, N USA 43°20′N 98°51′W
27 R3 **Platte City** Missouri, C USA 39°22′N 94°47′W
Plattensee *see* Balaton
27 R3 **Platte River** ≈ Iowa/Missouri, USA
29 Q15 **Platte River** ≈ Nebraska, C USA
37 T3 **Platteville** Colorado, C USA 40°13′N 104°49′W
30 K9 **Platteville** Wisconsin, N USA 42°44′N 90°27′W
101 N21 **Plattling** Bayern, SE Germany 48°45′N 12°52′E
27 R3 **Plattsburg** Missouri, C USA 39°33′N 94°27′W
18 L6 **Plattsburgh** New York, NE USA 44°42′N 73°29′W
29 S15 **Plattsmouth** Nebraska, C USA 41°00′N 95°52′W
101 M17 **Plauen** *var.* Plauen im Vogtland. Sachsen, E Germany 50°31′N 12°08′E
Plauen im Vogtland *see* Plauen
100 M10 **Plauer See** ◎ NE Germany
113 L16 **Plav** E Montenegro 42°36′N 19°57′E
118 I10 **Pļaviņas** *Ger.* Stockmannshof. S Latvia 56°37′N 25°40′E
126 K5 **Plavsk** Tul'skaya Oblast', W Russian Federation 53°42′N 37°21′E
41 Z12 **Playa del Carmen** Quintana Roo, E Mexico 20°37′N 87°04′W
40 J12 **Playa Los Corchos** Nayarit, SW Mexico 21°91′N 105°28′W
37 P16 **Playas Lake** ◎ New Mexico, SW USA
41 S15 **Playa Vicenté** Veracruz-Llave, SE Mexico 17°42′N 95°01′W
Plây Cu *see* Plei Ku
28 L3 **Plaza** North Dakota, N USA 48°00′N 102°00′W
63 I15 **Plaza Huincul** Neuquén, C Argentina 38°55′S 69°14′W
36 L3 **Pleasant Grove** Utah, W USA 40°21′N 111°44′W
29 V14 **Pleasant Hill** Iowa, C USA 41°34′N 93°31′W
27 R4 **Pleasant Hill** Missouri, C USA 38°47′N 94°16′W
Pleasant Island *see* Nauru
36 K13 **Pleasant, Lake** ⊠ Arizona, SW USA
19 P8 **Pleasant Mountain** ▲ Maine, NE USA 44°01′N 70°47′W
27 R5 **Pleasanton** Kansas, C USA 38°09′N 94°43′W
25 R12 **Pleasanton** Texas, SW USA 28°58′N 98°28′W
185 G20 **Pleasant Point** Canterbury, South Island, New Zealand 44°16′S 171°09′E
19 R5 **Pleasant River** ≈ Maine, NE USA
18 J17 **Pleasantville** New Jersey, NE USA 39°22′N 74°31′W
103 N12 **Pléaux** Cantal, C France 45°08′N 02°10′E
111 B19 **Plechý** *var.* Plöckenstein. ▲ Austria/Czech Republic 48°45′N 13°50′E
Pleebo *see* Plibo
Pleihari *see* Pelaihari
167 U11 **Plei Ku** *prev.* Plây Cu. Gia Lai, C Vietnam 13°57′N 108°01′E
101 M16 **Pleisse** ≈ E Germany
Plencia *see* Plentzia
184 O7 **Plenty, Bay of** *bay* North Island, New Zealand
33 Y6 **Plentywood** Montana, NW USA 48°46′N 104°33′W
105 O2 **Plentzia** *var.* Plencia. País Vasco, N Spain 43°25′N 02°56′W
102 H5 **Plérin** Côtes d'Armor, NW France 48°33′N 02°46′W
124 M10 **Plesetsk** Arkhangel'skaya Oblast', NW Russian Federation 62°41′N 40°14′E
Pleshchenitsy *see* Plyeshchanitsy
Pleskau *see* Pskov
Pleskauer See *see* Pskov, Lake
Pleskava *see* Pskov
112 E8 **Pleso International** ✈ (Zagreb) Zagreb, NW Croatia 45°45′N 16°00′E
Pless *see* Pszczyna
15 Q11 **Plessisville** Québec, SE Canada 46°14′N 71°46′W
110 H12 **Pleszew** Wielkopolskie, C Poland 51°54′N 17°47′E
12 L10 **Plétipi, Lac** ◎ Québec, SE Canada
101 F15 **Plettenberg** Nordrhein-Westfalen, W Germany 51°13′N 07°52′E
114 I8 **Pleven** *prev.* Plevna. Pleven, N Bulgaria 43°25′N 24°36′E
114 I8 **Pleven** ◆ *province* N Bulgaria
Plevlja/Plevlje *see* Pljevlja
Plevna *see* Pleven
Plezzo *see* Bovec
Pliberk *see* Bleiburg
76 L17 **Plibo** *var.* Pleebo. SE Liberia 04°38′N 07°41′W
121 R11 **Pliny Trench** *undersea feature* C Mediterranean Sea
118 K13 **Plisa** *Rus.* Plissa. Vitsyebskaya Voblasts', N Belarus 55°13′N 27°57′E
Plissa *see* Plisa
112 D11 **Plitvica Selo** Lika-Senj, W Croatia 44°53′N 15°36′E
112 D11 **Plješevica** ▲▲ C Croatia
113 K14 **Pljevlja** *prev.* Plevlja, Plevlje. N Montenegro 43°21′N 19°21′E
Ploça *see* Ploçë
Plocce *see* Ploče
113 K22 **Ploçë** *var.* Ploça. Vlorë, SW Albania 40°24′N 19°41′E
113 G15 **Ploče** *It.* Plocce; *prev.* Kardeljevo. Dubrovnik-Neretva, SE Croatia 43°02′N 17°25′E
110 K11 **Płock** *Ger.* Plozk. Mazowieckie, C Poland 52°32′N 19°40′E
109 Q10 **Plöcken Pass** *Ger.* Plöckenpass, *It.* Passo di Monte Croce Carnico. *pass* SW Austria
Plöckenpass *see* Plöcken Pass
Plöckenstein *see* Plechý
99 B19 **Ploegsteert** Hainaut, W Belgium 50°45′N 02°52′E
102 H6 **Ploërmel** Morbihan, NW France 47°57′N 02°24′W
Ploești *see* Ploiești
116 K13 **Ploiești** *prev.* Ploești. Prahova, SE Romania 44°56′N 26°03′E
115 L17 **Plomári** *prev.* Plomárion. Lésvos, E Greece 38°58′N 26°24′E
Plomárion *see* Plomári
103 O12 **Plomb du Cantal** ▲ C France 45°03′N 02°48′E
183 V6 **Plomer, Point** *headland* New South Wales, SE Australia 31°19′S 153°00′E
100 J8 **Plön** Schleswig-Holstein, N Germany 54°10′N 10°25′E
110 L11 **Płońsk** Mazowieckie, C Poland 52°38′N 20°23′E
119 J20 **Plotnitsa** Brestskaya Voblasts', SW Belarus 52°03′N 26°39′E
110 E8 **Płoty** *Ger.* Plathe. Zachodnio-pomorskie, NW Poland 53°48′N 15°16′E
102 G7 **Plouay** Morbihan, NW France 47°54′N 03°14′W
111 D15 **Ploučnice** *Ger.* Polzen. ≈ NE Czech Republic
114 I10 **Plovdiv** *prev.* Eumolpias; *anc.* Evmolpia, Philippopolis, *Lat.* Trimontium. Plovdiv, C Bulgaria 42°09′N 24°47′E
114 I11 **Plovdiv** ◆ *province* C Bulgaria
30 L6 **Plover** Wisconsin, N USA 44°30′N 89°33′W
Plozk *see* Płock
27 U11 **Plumerville** Arkansas, C USA 35°09′N 92°38′W
19 P10 **Plum Island** *island* Massachusetts, NE USA
32 M9 **Plummer** Idaho, NW USA 47°19′N 116°54′W
83 J18 **Plumtree** Matabeleland South, SW Zimbabwe 20°30′S 27°50′E
118 D11 **Plungė** Telšiai, W Lithuania 55°55′N 21°53′E
113 J15 **Plužine** NW Montenegro 43°08′N 18°49′E
119 K14 **Plyeshchanitsy** *Rus.* Pleshchenitsy. Minskaya Voblasts', N Belarus 54°25′N 27°50′E
97 I24 **Plymouth** SW England, United Kingdom 50°23′N 04°10′W
31 O11 **Plymouth** Indiana, N USA 41°20′N 86°19′W
19 P12 **Plymouth** Massachusetts, NE USA 41°57′N 70°40′W
19 N8 **Plymouth** New Hampshire, NE USA 43°43′N 71°39′W
21 X9 **Plymouth** North Carolina, SE USA 35°53′N 76°46′W
30 M8 **Plymouth** Wisconsin, N USA 43°48′N 87°58′W
Plymouth *see* Brades
97 J20 **Plynlimon** ▲ C Wales, United Kingdom 52°27′N 03°48′W
124 G14 **Plyussa** Pskovskaya Oblast', W Russian Federation 58°27′N 29°21′E
111 B17 **Plzeň** *Ger.* Pilsen, *Pol.* Pilzno. Plzeňský Kraj, W Czech Republic 49°45′N 13°23′E
111 B17 **Plzeňský Kraj** ◆ *region* W Czech Republic
110 F11 **Pniewy** *Ger.* Pinne. Wielkolpolskie, W Poland 52°31′N 16°14′E
77 P13 **Pô** S Burkina Faso 11°11′N 01°10′W
106 D8 **Po** ≈ N Italy
42 M13 **Poás, Volcán** ⛰ NW Costa Rica 10°12′N 84°12′W
77 S16 **Pobè** S Benin 07°00′N 02°41′E
123 S8 **Pobeda, Gora** ▲ NE Russian Federation 65°28′N 145°44′E
Pobeda Peak *see* Pobedy, Pik/Tomur Feng
147 Z7 **Pobedy, Pik** *Chin.* Tomür Feng. ▲ China/Kyrgyzstan 42°02′N 80°02′E *see also* Tomür Feng
Pobedy, Pik *see* Tomür Feng
110 H11 **Pobiedziska** *Ger.* Pudewitz. Wielkolpolskie, C Poland 52°30′N 17°19′E
Po, Bocche del *see* Po, Foci del
27 W9 **Pocahontas** Arkansas, C USA 36°15′N 91°00′W
29 U12 **Pocahontas** Iowa, C USA 42°44′N 94°40′W
33 Q15 **Pocatello** Idaho, NW USA 42°52′N 112°27′W
167 S13 **Pochentong** ✈ (Phnum Penh) Phnum Penh, S Cambodia 11°24′N 104°52′E
126 I6 **Pochep** Bryanskaya Oblast', W Russian Federation 52°56′N 33°20′E
126 H4 **Pochinok** Smolenskaya Oblast', W Russian Federation 54°21′N 32°29′E
41 R17 **Pochutla** *var.* San Pedro Pochutla. Oaxaca, SE Mexico 15°45′N 96°30′W
62 I6 **Pocitos, Salar** *var.* Salar Quiróm. *salt lake* NW Argentina
101 O22 **Pocking** Bayern, SE Germany 48°22′N 13°17′E
186 I10 **Pocklington Reef** *reef* SE Papua New Guinea
59 P15 **Poço da Cruz, Açude** ⊠ E Brazil
27 R11 **Pocola** Oklahoma, C USA 35°13′N 94°28′W
21 Y5 **Pocomoke City** Maryland, NE USA 38°04′N 75°34′W
59 L21 **Poços de Caldas** Minas Gerais, NE Brazil 21°48′S 46°33′W
124 H14 **Podberez'ye** Novgorodskaya Oblast', NW Russian Federation 58°42′N 31°22′E
Podbrodzie *see* Pabradė
125 U8 **Podcher'ye** Respublika Komi, NW Russian Federation 63°55′N 57°34′E
111 E16 **Poděbrady** *Ger.* Podiebrad. Středočeský Kraj, C Czech Republic 50°10′N 15°06′E
126 L9 **Podgorenskiy** Voronezhskaya Oblast', W Russian Federation 50°22′N 39°43′E
113 J17 **Podgorica** *prev.* Titograd. ● S Montenegro 42°25′N 19°16′E
113 K17 **Podgorica** ✈ S Montenegro 42°22′N 19°16′E
109 T13 **Podgrad** SW Slovenia 45°31′N 14°09′E
Podiebrad *see* Poděbrady
116 M5 **Podil's'ka Vysochina** *plateau* W Ukraine
Podium Anicensis *see* le Puy
122 L11 **Podkamennaya Tunguska** *Eng.* Stony Tunguska. ≈ C Russian Federation
111 N17 **Podkarpackie** ◆ *province* SW Poland
Pod Klošter *see* Arnoldstein
119 P4 **Podlaskie** ◆ *province* Mazowieckie, C Poland
127 Q8 **Podlesnoye** Saratovskaya Oblast', W Russian Federation 51°51′N 47°03′E
126 K4 **Podol'sk** Moskovskaya Oblast', W Russian Federation 55°24′N 37°30′E
76 H10 **Podor** N Senegal 16°40′N 14°57′W
125 P12 **Podosinovets** Kirovskaya Oblast', NW Russian Federation 60°15′N 47°06′E
124 I12 **Podporozh'ye** Leningradskaya Oblast', NW Russian Federation 60°52′N 34°00′E
Podravska Slatina *see* Slatina
112 J13 **Podromanlja** Republika Srpska, SE Bosnia and Herzegovina 43°55′N 18°46′E
Podsvil'ye *see* Padsvillye
116 L9 **Podu Iloaiei** *prev.* Podul Iloaiei. Iași, NE Romania 47°13′N 27°16′E
113 N15 **Podujevë** *Serb.* Podujevo. N Kosovo 42°56′N 21°13′E
Podujevo *see* Podujevë
Podul Iloaiei *see* Podu Iloaiei
Podunajská Rovina *see* Little Alföld
124 M12 **Podyuga** Arkhangel'skaya Oblast', NW Russian Federation 61°04′N 40°46′E
56 A9 **Poechos, Embalse** ◎ NW Peru
55 W10 **Poeketi** Sipaliwini, E Suriname
100 L8 **Poel** *island* N Germany
83 M20 **Poelela, Lagoa** ◎ S Mozambique
Poerwodadi *see* Purwodadi
Poerwokerto *see* Purwokerto
Poerworedjo *see* Purworejo
Poetovio *see* Ptuj
83 E23 **Pofadder** Northern Cape, W South Africa 29°09′S 19°25′E
106 I9 **Po, Foci del** *var.* Bocche del Po. ≈ NE Italy
116 E12 **Pogăniș** ≈ W Romania
Pogegen *see* Pagėgiai
106 G12 **Poggibonsi** Toscana, C Italy 43°28′N 11°09′E
107 I14 **Poggio Mirteto** Lazio, C Italy 42°17′N 12°42′E
109 V4 **Pöggstall** Niederösterreich, N Austria 48°19′N 15°10′E
116 L13 **Pogoanele** Buzău, SE Romania 44°55′N 27°00′E
Pogónion *see* Delvinákí
113 M21 **Pogradec** *var.* Pogradeci. Korçë, SE Albania 40°54′N 20°40′E
Pogradeci *see* Pogradec
123 S15 **Pogranichnyy** Primorskiy Kray, SE Russian Federation 44°18′N 131°33′E
38 M16 **Pogromni Volcano** ⛰ Unimak Island, Alaska, USA 54°34′N 164°41′W
163 Z15 **Pohang** *Jap.* Hokō; *prev.* P'ohang. E South Korea 36°02′N 129°20′E
P'ohang *see* Pohang
15 T9 **Pohénégamook, Lac** ◎ Québec, SE Canada
93 L20 **Pohja** *Swe.* Pojo. Uusimaa, SW Finland 60°07′N 23°30′E
Pohjanlahti *see* Bothnia, Gulf of
Pohjanmaa *see* Österbotten
93 O16 **Pohjois-Karjala** *Swe.* Norra Karelen, *Eng.* North Karelia. ◆ *region* E Finland
93 L14 **Pohjois-Pohjanmaa** *Swe.* Norra Österbotten, *Eng.* North Ostrobothnia. ◆ *region* N Finland
93 M17 **Pohjois-Savo** *Swe.* Norra Savolax. ◆ *region* C Finland
189 U16 **Pohnpei** ◆ *state* E Micronesia
189 O12 **Pohnpei** ✈ Pohnpei, E Micronesia
189 O12 **Pohnpei** *prev.* Ponape Ascension Island. *island* E Micronesia
111 F19 **Pohořelice** *Ger.* Pohrlitz. Jihomoravský Kraj, SE Czech Republic 48°58′N 16°30′E
109 V10 **Pohorje** *Ger.* Bacher. ▲▲ N Slovenia
117 N6 **Pohrebyshche** Vinnyts'ka Oblast', C Ukraine 49°31′N 29°16′E
Pohrlitz *see* Pohořelice
161 P9 **Po Hu** ⊠ E China
116 G15 **Poiana Mare** Dolj, S Romania 43°55′N 23°02′E
Poictiers *see* Poitiers
127 N6 **Poim** Penzenskaya Oblast', W Russian Federation 53°03′N 43°11′E
159 N15 **Poindo** Xigzang Zizhiqu, W China 29°58′N 91°20′E
195 Y13 **Poinsett, Cape** *headland* Antarctica 65°35′S 113°00′E
29 R9 **Poinsett, Lake** ◎ South Dakota, N USA
22 I10 **Point Au Fer Island** *island* Louisiana, S USA
39 X14 **Point Baker** Prince of Wales Island, Alaska, USA 56°19′N 133°31′W
25 U13 **Point Comfort** Texas, SW USA 28°40′N 96°33′W
Point de Galle *see* Galle
44 K10 **Pointe à Gravois** *headland* SW Haiti 18°00′N 73°53′W
22 L10 **Pointe a la Hache** Louisiana, S USA 29°34′N 89°48′W
45 Y6 **Pointe-à-Pitre** Grande Terre, C Guadeloupe 16°14′N 61°32′W
15 U7 **Pointe-au-Père** Québec, SE Canada 48°31′N 68°27′W
15 V5 **Pointe-aux-Anglais** Québec, SE Canada 49°40′N 67°09′W
45 T10 **Pointe Du Cap** *headland* N Saint Lucia 14°06′N 60°56′W
79 E21 **Pointe-Noire** Kouilou, S Congo 04°46′S 11°53′E
45 X6 **Pointe Noire** Basse Terre, W Guadeloupe 16°14′N 61°47′W
79 E21 **Pointe-Noire** ✈ Kouilou, S Congo 04°45′S 11°55′E
45 U15 **Point Fortin** Trinidad, Trinidad and Tobago 10°12′N 61°41′W
38 M6 **Point Hope** Alaska, USA 68°21′N 166°48′W
39 N5 **Point Lay** Alaska, USA 69°42′N 162°57′W
18 B16 **Point Marion** Pennsylvania, NE USA 39°44′N 79°53′W
18 K16 **Point Pleasant** New Jersey, NE USA 40°04′N 74°00′W
21 P4 **Point Pleasant** West Virginia, NE USA 38°53′N 82°07′W
45 R14 **Point Salines** ✈ (St. George's) SW Grenada 12°00′N 61°47′W
102 L9 **Poitiers** *prev.* Poictiers; *anc.* Limonum. Vienne, W France 46°35′N 00°19′E
102 K9 **Poitou** *cultural region* W France
102 K10 **Poitou-Charentes** ◆ *region* W France
103 N3 **Poix-de-Picardie** Somme, N France 49°47′N 01°58′E
Pojo *see* Pohja
37 S10 **Pojoaque** New Mexico, SW USA 35°52′N 106°01′W

◆ Country
● Country Capital
◇ Dependent Territory
○ Dependent Territory Capital
◆ Administrative Regions
✈ International Airport
▲ Mountain
▲▲ Mountain Range
⛰ Volcano
≈ River
◎ Lake
⊠ Reservoir

152 E11 Pokaran Rājasthān, NW India 26°55′N 71°55′E
183 R4 Pokataroo New South Wales, SE Australia 29°37′S 148°43′E
119 P18 Pokats' Rus. Pokot'. ≈ SE Belarus
29 V5 Pokegama Lake ◎ Minnesota, N USA
184 L6 Pokeno Waikato, North Island, New Zealand 37°15′S 175°01′E
153 O11 Pokharā Western, C Nepal 28°14′N 84°E
127 T6 Pokhvistnevo Samarskaya Oblast', W Russian Federation 53°38′N 52°07′E
55 W10 Pokigron Sipaliwini, C Suriname 04°31′N 55°23′W
92 L10 Pokka Lapp. Bohkká. Lappi, N Finland 68°11′N 25°45′E
79 N16 Poko Orientale, NE Dem. Rep. Congo 03°08′N 26°52′E
Pokot' see Pokats'
Po-ko-to Shan see Bogda Shan
147 S7 Pokrovka Talasskaya Oblast', NW Kyrgyzstan 42°45′N 71°33′E
Pokrovka see Kyzyl-Suu
117 V8 Pokrovs'ke Rus. Pokrovskoye. Dnipropetrovs'ka Oblast', E Ukraine 47°58′N 36°15′E
Pokrovskoye see Pokrovs'ke
Pola see Pula
37 N10 Polacca Arizona, SW USA 35°49′N 110°21′W
Pola de Laviana see Pola de Llaviana
Pola de Lena see La Pola
104 L2 Pola de Llaviana var. Pola de Laviana. Asturias, N Spain 43°15′N 05°33′W
Pola de Siero see La Pola Siero
191 Y3 Poland Kiritimati, E Kiribati 01°52′N 157°33′W
110 H12 Poland off. Republic of Poland, var. Polish Republic, Pol. Polska, Rzeczpospolita Polska; prev. Pol. Polska Rzeczpospolita Ludowa, The Polish People's Republic. ◆ republic C Europe
Poland, Republic of see Poland
Polangen see Palanga
110 G7 Polanów Ger. Pollnow. Zachodnio-pomorskie, NW Poland 54°07′N 16°38′E
136 H13 Polatlı Ankara, C Turkey 39°34′N 32°08′E
118 L12 Polatsk Rus. Polotsk. Vitsyebskaya Voblasts', N Belarus 55°29′N 28°47′E
110 F8 Połczyn-Zdrój Ger. Bad Polzin. Zachodnio-pomorskie, NW Poland 53°44′N 16°02′E
Pol-e-'Alam see Pul-e-'Alam
Polekhatum see Pulhatyn
Pol-e Khomrī see Pul-e Khumrī
197 S10 Pole Plain undersea feature Arctic Ocean
Pol-e-Safid see Pol-e Sefid
143 P5 Pol-e Sefid var. Pol-e-Safid, Pul-i-Sefid. Māzandarān, N Iran 36°05′N 53°01′E
118 B13 Polessk Ger. Labiau. Kaliningradskaya Oblast', W Russian Federation 54°52′N 21°06′E
Polesskoye see Polis'ke
171 N13 Polewali Sulawesi, C Indonesia 03°26′S 119°23′E
114 G11 Polezhan ▲ SW Bulgaria 41°42′N 23°28′E
78 F13 Poli Nord, N Cameroon 09°31′N 13°10′E
Poli see Pólis
107 M19 Policastro, Golfo di gulf S Italy
110 D8 Police Ger. Politz. Zachodnio-pomorskie, NW Poland 53°34′N 14°34′E
172 I17 Police, Pointe headland Mahé, NE Seychelles 04°48′S 55°31′E
115 L17 Polichnítos var. Polihnitos, Políkhnitos. Lésvos, E Greece 39°04′N 26°10′E
Poligiros see Polýgyros
107 P17 Polignano a Mare Puglia, SE Italy 40°59′N 17°13′E
103 S9 Poligny Jura, E France 46°51′N 05°42′E
Polihnitos see Polichnítos
Polikastro/Políkastron see Polýkastro
Políkhnitos see Polichnítos
171 O3 Polillo Islands island group N Philippines
109 Q9 Polinik ▲ SW Austria 46°54′N 13°10′E
115 J15 Políochni var. Polýochni. site of ancient city Límnos, E Greece
121 O2 Pólis var. Poli. W Cyprus 35°02′N 32°27′E
Polish People's Republic, The see Poland
Polish Republic see Poland
117 O3 Polis'ke Rus. Polesskoye. Kyyivs'ka Oblast', N Ukraine 51°16′N 29°27′E
107 N22 Polistena Calabria, SW Italy 38°25′N 16°05′E
Politz see Police
Políyiros see Polýgyros
29 V14 Polk City Iowa, C USA 41°46′N 93°42′W
110 F13 Polkowice Ger. Heerwegen. Dolnośląskie, W Poland 51°32′N 16°06′E
155 G22 Pollāchi Tamil Nādu, SE India 10°38′N 77°00′E
109 W7 Pöllau Steiermark, SE Austria 47°18′N 15°46′E
189 T13 Polle atoll Chuuk Islands, C Micronesia
105 X9 Pollença Mallorca, Spain, W Mediterranean Sea 39°52′N 03°01′E
Pollnow see Polanów
29 N7 Pollock South Dakota, N USA 45°53′N 100°15′W
92 L8 Polmak Finnmark, N Norway 70°01′N 28°04′E
30 L10 Polo Illinois, N USA 41°59′N 89°34′W
193 V15 Poloa island Tongatapu Group, N Tonga
42 E5 Polochic, Río ≈ C Guatemala
Pologi see Polohy
117 V9 Polohy Rus. Pologi. Zaporiz'ka Oblast', SE Ukraine 47°30′N 36°18′E
83 K20 Polokwane prev. Pietersburg. Limpopo, NE South Africa 23°54′S 29°23′E
14 M10 Polonais, Lac des ◎ Québec, SE Canada
61 G20 Polonio, Cabo headland E Uruguay 34°22′S 53°46′W
155 K24 Polonnaruwa North Central Province, C Sri Lanka 07°56′N 81°02′E
116 L5 Polonne Rus. Polonnoye. Khmel'nyts'ka Oblast', NW Ukraine 50°10′N 27°30′E
Polonnoye see Polonne
Polotsk see Polatsk
Polovinka see Ugleural'skiy
109 T7 Pöls var. Pölsbach. ≈ E Austria
Pölsbach see Pöls
Polska/Polska, Rzeczpospolita/Polska Rzeczpospolita Ludowa see Poland
114 L10 Polski Gradets Stara Zagora, C Bulgaria 42°12′N 26°06′E
114 K8 Polski Trambesh var. Polski Trŭmbesh. Veliko Tarnovo, N Bulgaria 43°22′N 25°38′E
Polski Trŭmbesh see Polski Trambesh
33 P8 Polson Montana, NW USA 47°41′N 114°09′W
117 T6 Poltava Poltavs'ka Oblast', NE Ukraine 49°33′N 34°32′E
Poltava see Poltavs'ka Oblast'
117 R5 Poltavs'ka Oblast' var. Poltava, Rus. Poltavskaya Oblast'. ◈ province NE Ukraine
Poltavskaya Oblast' see Poltavs'ka Oblast'
Poltoratsk see Aşgabat
118 I5 Põltsamaa Ger. Oberpahlen. Jõgevamaa, E Estonia 58°40′N 26°00′E
118 I4 Põltsamaa var. Põltsamaa Jõgi. ≈ C Estonia
Põltsamaa Jõgi see Põltsamaa
122 I8 Poluy ≈ N Russian Federation
118 J6 Põlva Ger. Pölwe. Põlvamaa, SE Estonia 58°04′N 27°06′E
93 N16 Polvijärvi Pohjois-Karjala, SE Finland 62°53′N 29°20′E
Pölwe see Põlva
115 I22 Polýaigos island Kykládes, Greece, Aegean Sea
115 I22 Polyaigou Folégandrou, Stenó strait Kykládes, Greece, Aegean Sea
124 J3 Polyarnyy Murmanskaya Oblast', NW Russian Federation 69°10′N 33°21′E
125 W5 Polyarnyy Ural ▲▲ NW Russian Federation
115 G14 Polýgyros var. Poligiros, Políyiros. Kentrikí Makedonía, N Greece 40°21′N 23°27′E
114 F13 Polýkastro var. Polikastro; prev. Políkastron. Kentrikí Makedonía, N Greece 41°01′N 22°33′E
193 O9 Polynesia island group C Pacific Ocean
Polýochni see Políochni
41 Y13 Polyuc Quintana Roo, E Mexico
109 V10 Polzela C Slovenia 46°18′N 15°04′E
Polzen see Ploučnice
56 D12 Pomabamba Ancash, C Peru 08°48′S 77°30′W
185 D23 Pomahaka ≈ South Island, New Zealand
106 F12 Pomarance Toscana, C Italy 43°19′N 10°53′E
104 G9 Pombal Leiria, C Portugal 39°55′N 08°38′W
76 D9 Pombas Santo Antão, NW Cape Verde 17°09′N 25°02′W
83 N19 Pomene Inhambane, SE Mozambique 22°57′S 35°34′E
110 G8 Pomerania cultural region Germany/Poland
110 D7 Pomeranian Bay Ger. Pommersche Bucht, Pol. Zatoka Pomorska. bay Germany/Poland
31 T15 Pomeroy Ohio, N USA 39°01′N 82°01′W
32 L10 Pomeroy Washington, NW USA 46°28′N 117°36′W
117 Q8 Pomichna Kirovohrads'ka Oblast', C Ukraine 48°07′N 31°25′E
186 H7 Pomio New Britain, E Papua New Guinea 05°31′S 151°30′E
Pomir, Dar"yoi see Pamir/Pāmīr, Daryā-ye
27 T6 Pomme de Terre Lake ⊠ Missouri, C USA
29 S8 Pomme de Terre River ≈ Minnesota, N USA
Pommersche Bucht see Pomeranian Bay
35 T15 Pomona California, W USA 34°03′N 117°45′W
114 N9 Pomorie Burgas, E Bulgaria 42°32′N 27°39′E
Pomorska, Zatoka see Pomeranian Bay
110 H8 Pomorskie ◈ province N Poland
125 Q4 Pomorskiy Proliv strait NW Russian Federation
125 T10 Pomozdino Respublika Komi, NW Russian Federation 62°11′N 54°13′E
Pompaelo see Pamplona
23 Z15 Pompano Beach Florida, SE USA 26°14′N 80°06′W
107 K18 Pompei Campania, S Italy 40°45′N 14°27′E
33 V10 Pompeys Pillar Montana, NW USA 45°58′N 107°55′W
Ponape Ascension Island see Pohnpei
29 R13 Ponca Nebraska, C USA 42°35′N 96°42′W
27 O8 Ponca City Oklahoma, C USA 36°41′N 97°04′W
45 T6 Ponce C Puerto Rico 18°01′N 66°36′W
23 X10 Ponce de Leon Inlet inlet Florida, SE USA
22 K8 Ponchatoula Louisiana, S USA 30°26′N 90°26′W
26 M8 Pond Creek Oklahoma, C USA 36°40′N 97°48′W
155 J20 Pondicherry var. Puducheri, Fr. Pondichéry. Puducherry, SE India 11°59′N 79°50′E
Pondicherry see Puducherry
Pondichéry see Puducherry
197 N11 Pond Inlet var. Mittimatalik. Baffin Island, Nunavut, NE Canada 72°37′N 77°56′W
187 P16 Ponérihouen Province Nord, C New Caledonia 21°04′S 165°24′E
104 J4 Ponferrada Castilla y León, NW Spain 42°33′N 06°35′W
184 N13 Pongaroa Manawatu-Wanganui, North Island, New Zealand 40°36′S 176°08′E
167 Q12 Pong Nam Ron Chantaburi, S Thailand 12°55′N 102°15′E
81 C14 Pongo ≈ W South Sudan
152 I7 Pong Reservoir ⊠ N India
111 N14 Poniatowa Lubelskie, E Poland 51°11′N 22°05′E
167 R12 Pônley Kâmpóng Chhnăng, C Cambodia 12°26′N 104°25′E
155 I20 Ponnaiyār ≈ SE India
11 Q15 Ponoka Alberta, SW Canada 52°42′N 113°33′W
127 U6 Ponomarevka Orenburgskaya Oblast', W Russian Federation 53°16′N 54°10′E
169 Q17 Ponorogo Jawa, C Indonesia 07°51′S 111°30′E
122 F6 Ponoy ≈ NW Russian Federation
102 K11 Pons Charente-Maritime, W France 45°31′N 00°31′W
Pons see Ponts
Pons Aelii see Newcastle upon Tyne
Pons Vetus see Pontevedra
99 G20 Pont-à-Celles Hainaut, S Belgium 50°31′N 04°21′E
102 K16 Pontacq Pyrénées-Atlantiques, SW France 43°11′N 00°06′W
64 P3 Ponta Delgada São Miguel, Azores, Portugal, NE Atlantic Ocean 37°29′N 25°40′W
64 P3 Ponta Delgada ✈ São Miguel, Azores, Portugal, NE Atlantic Ocean 37°28′N 25°40′W
64 N2 Ponta do Pico ▲ Pico, Azores, Portugal, NE Atlantic Ocean 38°28′N 28°25′W
60 J11 Ponta Grossa Paraná, S Brazil 25°07′S 50°09′W
103 S5 Pont-à-Mousson Meurthe-et-Moselle, NE France 48°55′N 06°03′E
103 T9 Pontarlier Doubs, E France 46°54′N 06°20′E
106 G11 Pontassieve Toscana, C Italy 43°46′N 11°28′E
102 L4 Pont-Audemer Eure, N France 49°22′N 00°31′E
22 K9 Pontchartrain, Lake ◎ Louisiana, S USA
102 I8 Pontchâteau Loire-Atlantique, NW France 47°26′N 02°04′W
103 R10 Pont-de-Vaux Ain, E France 46°25′N 04°57′E
104 G4 Ponteareas Galicia, NW Spain 42°11′N 08°29′W
106 J6 Pontebba Friuli-Venezia Giulia, NE Italy 46°32′N 13°18′E
104 G4 Ponte Caldelas Galicia, NW Spain 42°23′N 08°30′W
107 J16 Pontecorvo Lazio, C Italy 41°27′N 13°40′E
104 G5 Ponte da Barca Viana do Castelo, N Portugal 41°48′N 08°25′W
104 G5 Ponte de Lima Viana do Castelo, N Portugal 41°46′N 08°35′W
106 F11 Pontedera Toscana, C Italy 43°40′N 10°38′E
104 H10 Ponte de Sor Portalegre, C Portugal 39°15′N 08°01′W
104 H2 Pontedeume Galicia, NW Spain 43°22′N 08°09′W
106 F6 Ponte di Legno Lombardia, N Italy 46°16′N 10°31′E
11 T17 Ponteix Saskatchewan, S Canada 49°43′N 107°22′W
171 Q16 Ponte Macassar var. Pante Macassar, Pante Makasar, Pante Makassar. W East Timor 09°11′S 124°27′E
59 N20 Ponte Nova Minas Gerais, NE Brazil 20°25′S 42°54′W
59 G18 Pontes e Lacerda Mato Grosso, W Brazil 15°14′S 59°21′W
104 G4 Pontevedra anc. Pons Vetus. Galicia, NW Spain 42°25′N 08°39′W
104 G3 Pontevedra ◈ province Galicia, NW Spain
104 G4 Pontevedra, Ría de estuary NW Spain
30 M12 Pontiac Illinois, N USA 40°54′N 88°36′W
31 R9 Pontiac Michigan, N USA 42°38′N 83°17′W
169 P11 Pontianak Borneo, C Indonesia 0°05′S 109°16′E
107 I16 Pontino, Agro plain C Italy
Pontisarae see Pontoise
102 H6 Pontivy Morbihan, NW France 48°04′N 02°58′W
102 F6 Pont-l'Abbé Finistère, NW France 47°52′N 04°14′W
103 N4 Pontoise anc. Briva Isarae, Cergy-Pontoise, Pontisarae. Val-d'Oise, N France 49°03′N 02°05′E
11 W13 Ponton Manitoba, C Canada 54°36′N 99°02′W
102 J5 Pontorson Manche, N France 48°33′N 01°31′W
22 M2 Pontotoc Mississippi, S USA 34°15′N 89°00′W
25 R9 Pontotoc Texas, SW USA 30°52′N 98°57′W
106 E10 Pontremoli Toscana, C Italy 44°24′N 09°55′E
108 J10 Pontresina Graubünden, S Switzerland 46°29′N 09°52′E
105 U5 Ponts var. Pons. Cataluña, NE Spain 41°55′N 01°12′E
103 R14 Pont-St-Esprit Gard, S France 44°15′N 04°37′E
97 K21 Pontypool Wel. Pontypŵl. SE Wales, United Kingdom 51°43′N 03°02′W
97 J22 Pontypridd S Wales, United Kingdom 51°37′N 03°22′W
Pontypŵl see Pontypool
43 R17 Ponuga Veraguas, SE Panama 07°50′N 80°58′W
184 L6 Ponui Island island N New Zealand
119 K14 Ponya ≈ N Belarus
107 I17 Ponza, Isola di Island Isole Ponziane, S Italy
Ponziane, Isole island C Italy
182 F7 Poochera South Australia 32°45′S 134°51′E
97 L24 Poole S England, United Kingdom 50°43′N 01°59′W
25 S6 Poolville Texas, SW USA 33°00′N 97°55′W
Poona see Pune
182 M8 Pooncarie New South Wales, SE Australia 33°26′S 142°37′E
183 N6 Poopelloe Lake seasonal lake New South Wales, SE Australia
57 K19 Poopó Oruro, C Bolivia 18°23′S 66°58′W
57 K19 Poopó, Lago var. Lago Pampa Aullagas. ◎ W Bolivia
184 L3 Poor Knights Islands island N New Zealand
39 P10 Poorman Alaska, USA 64°05′N 155°34′W
182 E3 Pootnoura South Australia 28°31′S 134°09′E
147 R10 Pop Rus. Pap. Namangan Viloyati, E Uzbekistan 40°49′N 71°06′E
117 X7 Popasna Rus. Popasnaya. Luhans'ka Oblast', E Ukraine 48°38′N 38°24′E
Popasnaya see Popasna
54 D12 Popayán Cauca, SW Colombia 02°27′N 76°32′W
99 B18 Poperinge West-Vlaanderen, W Belgium 50°52′N 02°44′E
123 N7 Popigay Krasnoyarskiy Kray, N Russian Federation 71°54′N 110°45′E
123 N7 Popigay ≈ N Russian Federation
117 O5 Popil'nya Zhytomyrs'ka Oblast', N Ukraine 49°57′N 29°24′E
182 K8 Popiltah Lake seasonal lake New South Wales, SE Australia
33 X7 Poplar Montana, NW USA 48°06′N 105°12′W
11 Y14 Poplar ≈ Manitoba, C Canada
27 X8 Poplar Bluff Missouri, C USA 36°45′N 90°23′W
33 X6 Poplar River ≈ Montana, NW USA
41 P14 Popocatépetl ⌅ S Mexico 18°59′N 98°37′W
79 H21 Popokabaka Bandundu, SW Dem. Rep. Congo 05°42′S 16°35′E
107 J15 Popoli Abruzzo, C Italy 42°09′N 13°51′E
186 F9 Popondetta Northern, S Papua New Guinea 08°45′S 148°15′E
112 F9 Popovača Sisak-Moslavina, NE Croatia 45°35′N 16°37′E
114 L8 Popovo Targovishte, N Bulgaria 43°20′N 26°14′E
Popovo see Iskra
Popper see Poprad
30 M5 Popple River ≈ Wisconsin, N USA
111 L19 Poprad Ger. Deutschendorf, Hung. Poprád. Prešovský Kraj, E Slovakia 49°04′N 20°16′E
111 L18 Poprad Ger. Popper, Hung. Poprád. ≈ Poland/Slovakia
111 L19 Poprad-Tatry ✈ (Poprad) Prešovský Kraj, E Slovakia 49°04′N 20°21′E
21 X7 Poquoson Virginia, NE USA 37°08′N 76°21′W
149 O15 Porāli ≈ SW Pakistan
184 N12 Porangahau Hawke's Bay, North Island, New Zealand 40°19′S 176°36′E
59 K17 Porangatu Goiás, C Brazil 13°28′S 49°14′W
119 G18 Porazava Pol. Porozow, Rus. Porozovo. Hrodzyenskaya Voblasts', W Belarus 52°56′N 24°22′E
154 A11 Porbandar Gujarāt, W India 21°40′N 69°40′E
10 I13 Porcher Island island British Columbia, SW Canada
104 M13 Porcuna Andalucía, S Spain 37°52′N 04°12′W
14 F7 Porcupine Ontario, S Canada 48°31′N 81°07′W
64 M6 Porcupine Bank undersea feature N Atlantic Ocean
11 V15 Porcupine Hills ▲▲ Manitoba/Saskatchewan, S Canada
30 L3 Porcupine Mountains hill range Michigan, N USA
64 M7 Porcupine Plain undersea feature E Atlantic Ocean 16°00′W 49°00′N
8 G7 Porcupine River ≈ Canada/USA
106 I7 Pordenone anc. Portenau. Friuli-Venezia Giulia, NE Italy 45°58′N 12°39′E
54 H9 Pore Casanare, E Colombia 05°42′N 71°59′W
112 A9 Poreč It. Parenzo. Istra, NW Croatia 45°16′N 13°36′E
60 I9 Porecatu Paraná, S Brazil 22°46′S 51°22′W
Porech'ye see Parechcha
127 P4 Poretskoye Chuvashskaya Respublika, W Russian Federation 55°12′N 46°20′E
77 Q13 Porga N Benin 11°04′N 00°58′E
186 B7 Porgera Enga, W Papua New Guinea 05°32′S 143°08′E
93 K18 Pori Swe. Björneborg. Satakunta, SW Finland 61°28′N 21°50′E
185 L14 Porirua Wellington, North Island, New Zealand 41°08′S 174°51′E
92 I12 Porjus Lapp. Bárjås. Norrbotten, N Sweden 66°55′N 19°55′E
124 G14 Porkhov Pskovskaya Oblast', W Russian Federation 57°46′N 29°27′E
55 O4 Porlamar Nueva Esparta, NE Venezuela 10°57′N 63°51′W
102 I8 Pornic Loire-Atlantique, NW France 47°07′N 02°07′W
186 B7 Poroma Southern Highlands, W Papua New Guinea 06°15′S 143°34′E
123 T13 Poronaysk Ostrov Sakhalin, Sakhalinskaya Oblast', SE Russian Federation 49°15′N 143°00′E
115 G20 Póros Póros, S Greece 37°30′N 23°27′E
115 C19 Póros Kefallinía, Iónia Nisiá, Greece, C Mediterranean Sea 38°09′N 20°46′E
115 G20 Póros island S Greece
81 G24 Poroto Mountains ▲▲ SW Tanzania
112 B10 Porozina Primorje-Gorski Kotar, NW Croatia 45°07′N 14°17′E
Porozovo/Porozow see Porazava
195 X15 Porpoise Bay bay Antarctica
65 G15 Porpoise Point headland NE Ascension Island 07°54′S 14°22′W
65 C25 Porpoise Point headland East Falkland, Falkland Islands 52°20′S 59°18′W
108 C6 Porrentruy Jura, NW Switzerland 47°25′N 07°06′E
106 F10 Porretta Terme Emilia-Romagna, C Italy 44°10′N 11°01′E
Porriño see O Porriño
92 L7 Porsangenfjorden Lapp. Porsáŋgguvuotna. fjord N Norway
92 K8 Porsangerhalvøya peninsula N Norway
Porsáŋgguvuotna see Porsangenfjorden
95 G16 Porsgrunn Telemark, S Norway 59°08′N 09°38′E
136 E13 Porsuk Çayı ≈ C Turkey
Porsy see Boldumsaz
57 N18 Portachuelo Santa Cruz, C Bolivia 17°21′S 63°24′W
182 I9 Port Adelaide South Australia 34°49′S 138°31′E
97 F15 Portadown Ir. Port An Dúnáin. S Northern Ireland, United Kingdom 54°26′N 06°27′W
31 P10 Portage Michigan, N USA 42°12′N 85°34′W
18 D15 Portage Pennsylvania, NE USA 40°23′N 78°40′W
30 K8 Portage Wisconsin, N USA 43°33′N 89°29′W
30 M3 Portage Lake ◎ Michigan, N USA
11 X16 Portage la Prairie Manitoba, S Canada 49°58′N 98°20′W
31 R11 Portage River ≈ Ohio, N USA
27 Y8 Portageville Missouri, C USA 36°25′N 89°42′W
28 L2 Portal North Dakota, N USA 48°57′N 102°33′W
10 L17 Port Alberni Vancouver Island, British Columbia, SW Canada 49°11′N 124°49′W
14 E15 Port Albert Ontario, S Canada 43°51′N 81°42′W
104 I10 Portalegre anc. Ammaia, Amoea. Portalegre, E Portugal 39°17′N 07°25′W
104 H10 Portalegre ◈ district C Portugal
37 V12 Portales New Mexico, SW USA 34°11′N 103°19′W
39 X14 Port Alexander Baranof Island, Alaska, USA 56°15′N 134°39′W
83 I25 Port Alfred Eastern Cape, S South Africa 33°31′S 26°55′E
10 J16 Port Alice Vancouver Island, British Columbia, SW Canada 50°23′N 127°24′W
22 J8 Port Allen Louisiana, S USA 30°27′N 91°12′W
Port Amelia see Pemba
Port An Dúnáin see Portadown
32 G7 Port Angeles Washington, NW USA 48°06′N 123°26′W
44 L12 Port Antonio NE Jamaica 18°10′N 76°27′W
115 D16 Pórta Panagiá religious building Thessalía, C Greece
25 T14 Port Aransas Texas, SW USA 27°49′N 97°03′W
97 E18 Portarlington Ir. Cúil an tSúdaire. Laois/Offaly, C Ireland 53°10′N 07°11′W
183 P17 Port Arthur Tasmania, SE Australia 43°09′S 147°51′E
25 Y11 Port Arthur Texas, SW USA 29°55′N 93°56′W
96 G12 Port Askaig W Scotland, United Kingdom 55°51′N 06°06′W
182 I7 Port Augusta South Australia 32°29′S 137°44′E
44 M9 Port-au-Prince hait. Pòtoprens. ● (Haiti) C Haiti 18°33′N 72°20′W
22 I8 Port Barre Louisiana, S USA 30°33′N 91°57′W
Port-Bergé see Boriziny
151 Q19 Port Blair Andaman and Nicobar Islands, SE India 11°40′N 92°44′E
25 X12 Port Bolivar Texas, SW USA 29°21′N 94°45′W
105 X4 Portbou Cataluña, NE Spain 42°26′N 03°10′E
77 N17 Port Bouet ✈ (Abidjan) SE Ivory Coast 05°17′N 03°55′W
182 I8 Port Broughton South Australia 33°39′S 137°55′E
14 F17 Port Burwell Ontario, S Canada 42°39′N 80°47′W
12 G17 Port Burwell Québec, NE Canada 60°25′N 64°49′W
182 M13 Port Campbell Victoria, SE Australia 38°37′S 143°00′E
15 V4 Port-Cartier Québec, SE Canada 50°00′N 66°55′W
185 F23 Port Chalmers Otago, South Island, New Zealand 45°46′S 170°37′E
23 W14 Port Charlotte Florida, SE USA 27°00′N 82°07′W
38 L9 Port Clarence Alaska, USA 65°15′N 166°51′W
10 I13 Port Clements Graham Island, British Columbia, SW Canada 53°37′N 132°12′W
31 S11 Port Clinton Ohio, N USA 41°30′N 82°56′W
14 H17 Port Colborne Ontario, S Canada 42°51′N 79°16′W
15 Y7 Port-Daniel Québec, SE Canada 48°10′N 64°58′W
Port Darwin see Darwin
183 O17 Port Davey headland Tasmania, SE Australia 43°19′S 145°54′E
44 K8 Port-de-Paix NW Haiti 19°56′N 72°52′W
181 W4 Port Douglas Queensland, NE Australia 16°33′S 145°27′E
10 J13 Port Edward British Columbia, SW Canada 54°11′N 130°16′W
83 K24 Port Edward KwaZulu/Natal, SE South Africa 31°03′S 30°14′E
58 J12 Portel Pará, NE Brazil 01°58′S 50°45′W
104 H12 Portel Évora, S Portugal 38°18′N 07°42′W
14 E14 Port Elgin Ontario, S Canada 44°26′N 81°22′W
45 Y14 Port Elizabeth Bequia, Saint Vincent and the Grenadines 13°01′N 61°15′W
83 I26 Port Elizabeth Eastern Cape, S South Africa 33°58′S 25°36′E
96 G13 Port Ellen W Scotland, United Kingdom 55°37′N 06°12′W
Portenau see Pordenone
97 H16 Port Erin SW Isle of Man 54°05′N 04°47′W
45 O13 Porter Point headland Saint Vincent, Saint Vincent and the Grenadines 13°22′N 61°10′W
185 G18 Porters Pass pass South Island, New Zealand
83 E25 Porterville Western Cape, SW South Africa 33°03′S 19°00′E
35 R12 Porterville California, W USA 36°03′N 119°03′W
Port-Étienne see Nouâdhibou
182 L13 Port Fairy Victoria, SE Australia 38°24′S 142°13′E
184 M4 Port Fitzroy Great Barrier Island, Auckland, NE New Zealand 36°10′S 175°21′E
Port Florence see Kisumu
Port-Francqui see Ilebo
79 C18 Port-Gentil Ogooué-Maritime, W Gabon 0°40′S 08°50′E
182 I7 Port Germein South Australia 33°02′S 138°01′E
22 J6 Port Gibson Mississippi, S USA 31°57′N 90°58′W
39 Q13 Port Graham Alaska, USA 59°21′N 151°49′W
77 U17 Port Harcourt Rivers, S Nigeria 04°43′N 07°02′E
10 J16 Port Hardy Vancouver Island, British Columbia, SW Canada 50°41′N 127°30′W
Port Harrison see Inukjuak
13 R14 Port Hawkesbury Cape Breton Island, Nova Scotia, SE Canada 45°36′N 61°22′W
180 I6 Port Hedland Western Australia 20°23′S 118°40′E
39 O15 Port Heiden Alaska, USA 56°54′N 158°40′W
97 I19 Porthmadog var. Porthmadoc. NW Wales, United Kingdom 52°55′N 04°08′W
14 I15 Port Hope Ontario, SE Canada 43°58′N 78°18′W
13 S9 Port Hope Simpson Newfoundland and Labrador, E Canada 52°30′N 56°18′W
65 C24 Port Howard Settlement West Falkland, Falkland Islands
31 T9 Port Huron Michigan, N USA 42°58′N 82°25′W
107 K17 Portici Campania, S Italy 40°48′N 14°20′E
Port-Iliç see Liman
104 G14 Portimão var. Vila Nova de Portimão. Faro, S Portugal 37°08′N 08°32′W
25 T17 Port Isabel Texas, SW USA 26°04′N 97°13′W
18 J13 Port Jervis New York, NE USA 41°22′N 74°39′W
55 S7 Port Kaituma NW Guyana 07°42′N 59°52′W
126 K12 Port Katon Rostovskaya Oblast', SW Russian Federation 46°52′N 38°46′E
183 S9 Port Kembla New South Wales, SE Australia 34°30′S 150°54′E
182 F8 Port Kenny South Australia 33°09′S 134°38′E
Port Klang see Pelabuhan Klang
Port Láirge see Waterford
183 S8 Portland New South Wales, SE Australia 33°24′S 150°00′E
182 L13 Portland Victoria, SE Australia 38°21′S 141°38′E
184 K4 Portland Northland, North Island, New Zealand 35°48′S 174°19′E
31 Q13 Portland Indiana, N USA 40°25′N 84°58′W
19 P8 Portland Maine, NE USA 43°41′N 70°16′W
31 Q9 Portland Michigan, N USA 42°51′N 84°52′W
29 Q4 Portland North Dakota, N USA 47°28′N 97°22′W
32 G11 Portland Oregon, NW USA 45°31′N 122°41′W
20 J8 Portland Tennessee, S USA 36°34′N 86°31′W
25 T14 Portland Texas, SW USA 27°52′N 97°19′W
32 G11 Portland ✈ Oregon, NW USA 45°36′N 122°34′W
182 L13 Portland Bay bay Victoria, SE Australia
44 K13 Portland Bight bay S Jamaica
97 L24 Portland Bill var. Bill of Portland. headland S England, United Kingdom 50°31′N 02°28′W
Portland, Bill of see Portland Bill
183 P15 Portland, Cape headland Tasmania, SE Australia 40°46′S 147°58′E
10 J12 Portland Inlet inlet British Columbia, W Canada
184 P11 Portland Island island E New Zealand
65 F15 Portland Point headland SW Ascension Island
44 J13 Portland Point headland C Jamaica 17°42′N 77°10′W
103 P16 Port-la-Nouvelle Aude, S France 43°01′N 03°04′E
Portlaoighise see Port Laoise
97 E18 Port Laoise var. Portlaoise, Ir. Portlaoighise; prev. Maryborough. C Ireland 53°02′N 07°17′W
Portlaoise see Port Laoise
25 U13 Port Lavaca Texas, SW USA 28°36′N 96°39′W
182 G9 Port Lincoln South Australia 34°43′S 135°49′E
39 Q14 Port Lions Kodiak Island, Alaska, USA 57°55′N 152°48′W
76 I15 Port Loko W Sierra Leone 08°50′N 12°50′W
65 E24 Port Louis East Falkland, Falkland Islands 51°31′S 58°07′W
45 Y5 Port-Louis Grande Terre, N Guadeloupe 16°25′N 61°32′W
173 X16 Port Louis ● (Mauritius) NW Mauritius 20°10′S 57°30′E
Port Louis see Scarborough
Port-Lyautey see Kénitra
182 K12 Port MacDonnell South Australia 38°04′S 140°40′E
183 U7 Port Macquarie New South Wales, SE Australia 31°26′S 152°55′E
Portmadoc see Porthmadog
Port Mahon see Maó
44 K12 Port Maria C Jamaica 18°22′N 76°54′W
10 K16 Port McNeill Vancouver Island, British Columbia, SW Canada 50°34′N 127°06′W
13 P11 Port-Menier Île d'Anticosti, Québec, E Canada 49°49′N 64°19′W
39 N15 Port Moller Alaska, USA 56°00′N 160°31′W
44 L13 Port Morant E Jamaica 17°53′N 76°20′W
44 K13 Portmore C Jamaica 17°58′N 76°52′W
186 D9 Port Moresby ● (Papua New Guinea) Central/National Capital District, SW Papua New Guinea 09°28′S 147°12′E
Port Natal see Durban
25 Y11 Port Neches Texas, SW USA 29°59′N 93°57′W
182 G9 Port Neill South Australia 34°06′S 136°19′E
15 S6 Portneuf ≈ Québec, SE Canada
15 R6 Portneuf, Lac ◎ Québec, SE Canada
83 D23 Port Nolloth Northern Cape, W South Africa 29°17′S 16°51′E
18 J17 Port Norris New Jersey, NE USA 39°13′N 75°00′W
Port-Nouveau-Québec see Kangiqsualujjuaq
104 G6 Porto Eng. Oporto; anc. Portus Cale. Porto, NW Portugal 41°09′N 08°37′W
104 G6 Porto var. Pôrto. ◈ district N Portugal
104 G6 Porto ✈ Porto, W Portugal 41°09′N 08°37′W
Pôrto see Porto
61 I16 Porto Alegre var. Pôrto Alegre. state capital Rio Grande do Sul, S Brazil 30°03′S 51°10′W
Porto Alexandre see Tombua
82 B12 Porto Amboim Kwanza Sul, NW Angola 10°47′S 13°43′E
Porto Amélia see Pemba
Porto Bello see Portobelo
43 T14 Portobelo var. Porto Bello, Puerto Bello. Colón, N Panama 09°33′N 79°37′W
60 G10 Pôrto Camargo Paraná, S Brazil 23°23′S 53°47′W
25 U13 Port O'Connor Texas, SW USA 28°26′N 96°26′W
Pôrto de Mós see Porto de Moz
58 J12 Porto de Moz var. Pôrto de Mós. Pará, NE Brazil 01°45′S 52°15′W
64 O5 Porto do Moniz Madeira, Portugal, NE Atlantic Ocean
59 H16 Porto dos Gaúchos Mato Grosso, W Brazil 11°32′S 57°16′W
Porto Edda see Sarandë
107 J24 Porto Empedocle Sicilia, Italy, C Mediterranean Sea 37°18′N 13°32′E
59 H20 Porto Esperança Mato Grosso do Sul, SW Brazil 19°36′S 57°24′W
106 E13 Portoferraio Toscana, C Italy 42°49′N 10°18′E
96 G6 Port of Ness NW Scotland, United Kingdom 58°29′N 06°15′W
45 U14 Port-of-Spain ● (Trinidad and Tobago) Trinidad, Trinidad and Tobago 10°39′N 61°30′W
Port of Spain see Piarco
103 X15 Porto, Golfe de gulf Corse, France, C Mediterranean Sea
Porto Grande see Mindelo
106 I7 Portogruaro Veneto, NE Italy 45°46′N 12°50′E
35 P5 Portola California, W USA 39°48′N 120°28′W
187 Q13 Port-Olry Espiritu Santo, C Vanuatu 15°03′S 167°04′E
93 J17 Pörtom Fin. Pirttikylä. Österbotten, W Finland 62°42′N 21°40′E
Port Omna see Portumna
59 G21 Porto Murtinho Mato Grosso do Sul, SW Brazil 21°42′S 57°52′W
59 K16 Porto Nacional Tocantins, C Brazil 10°41′S 48°19′W
77 S16 Porto-Novo ● (Benin) S Benin 06°29′N 02°37′E
23 X10 Port Orange Florida, SE USA 29°06′N 80°59′W
32 G8 Port Orchard Washington, NW USA 47°32′N 122°38′W
Porto Re see Kraljevica
32 E15 Port Orford Oregon, NW USA 42°45′N 124°30′W
Porto Rico see Puerto Rico
106 J13 Porto San Giorgio Marche, C Italy 43°10′N 13°47′E
107 F14 Porto San Stefano Toscana, C Italy 42°26′N 11°09′E
64 P5 Porto Santo var. Vila Baleira. Porto Santo, Madeira, Portugal, NE Atlantic Ocean 33°04′N 16°20′W
64 Q5 Porto Santo ✈ Porto Santo, Madeira, Portugal, NE Atlantic Ocean 33°04′N 16°20′W
64 P5 Porto Santo var. Ilha do Porto Santo. island Madeira, Portugal, NE Atlantic Ocean
Porto Santo, Ilha do see Porto Santo
60 H9 Porto São José Paraná, S Brazil 22°43′S 53°10′W
59 O19 Porto Seguro Bahia, E Brazil 16°25′S 39°07′W
107 B17 Porto Torres Sardegna, Italy, C Mediterranean Sea 40°50′N 08°23′E
59 J23 Porto União Santa Catarina, S Brazil 26°15′S 51°04′W
103 Y16 Porto-Vecchio Corse, France, C Mediterranean Sea 41°35′N 09°17′E

◆ Country
● Country Capital
◇ Dependent Territory
○ Dependent Territory Capital
◈ Administrative Regions
✈ International Airport
▲ Mountain
▲▲ Mountain Range
⌅ Volcano
≈ River
◎ Lake
⊠ Reservoir

59 E15 **Porto Velho** *var.* Velho. *state capital* Rondônia, W Brazil *08°45′S 63°54′W*
56 A6 **Portoviejo** *var.* Puertoviejo. Manabí, W Ecuador *01°03′S 80°31′W*
185 B26 **Port Pegasus** *bay* Stewart Island, New Zealand
14 H15 **Port Perry** Ontario, SE Canada *44°08′N 78°57′W*
183 N12 **Port Phillip Bay** *harbour* Victoria, SE Australia
182 I8 **Port Pirie** South Australia *33°11′S 138°01′E*
96 G9 **Portree** N Scotland, United Kingdom *57°26′N 06°12′W*
Port Rex *see* East London
Port Rois *see* Portrush
44 K13 **Port Royal** E Jamaica *17°55′N 76°52′W*
21 R15 **Port Royal** South Carolina, SE USA *32°22′N 80°41′W*
21 R15 **Port Royal Sound** *inlet* South Carolina, SE USA
97 F14 **Portrush** *Ir.* Port Rois. N Northern Ireland, United Kingdom *55°12′N 06°40′W*
Port Said *see* Būr Sa'īd
23 R9 **Port Saint Joe** Florida, SE USA *29°49′N 85°18′W*
23 Y11 **Port Saint John** Florida, SE USA *28°28′N 80°46′W*
103 R16 **Port-St-Louis-du-Rhône** Bouches-du-Rhône, SE France *43°22′N 04°48′E*
44 K10 **Port Salut** SW Haiti *18°04′N 73°55′W*
65 E24 **Port Salvador** *inlet* East Falkland, Falkland Islands
65 D24 **Port San Carlos** East Falkland, Falkland Islands *51°30′S 58°59′W*
13 S10 **Port Saunders** Newfoundland, Newfoundland and Labrador, SE Canada *50°40′N 57°17′W*
83 K24 **Port Shepstone** KwaZulu/Natal, E South Africa *30°44′S 30°28′E*
45 O11 **Portsmouth** *var.* Grand-Anse. NW Dominica *15°34′N 61°27′W*
97 N24 **Portsmouth** S England, United Kingdom *50°48′N 01°05′W*
19 P10 **Portsmouth** New Hampshire, NE USA *43°04′N 70°47′W*
31 S15 **Portsmouth** Ohio, N USA *38°43′N 83°00′W*
21 X7 **Portsmouth** Virginia, NE USA *36°50′N 76°18′W*
14 E17 **Port Stanley** Ontario, S Canada *42°39′N 81°12′W*
Port Stanley *see* Stanley
65 B25 **Port Stephens** *inlet* West Falkland, Falkland Islands
65 B25 **Port Stephens Settlement** West Falkland, Falkland Islands
97 F14 **Portstewart** *Ir.* Port Stíobhaird. N Northern Ireland, United Kingdom *55°11′N 06°43′W*
Port Stíobhaird *see* Portstewart
83 K24 **Port St. Johns** Eastern Cape, SE South Africa *31°37′S 29°32′E*
80 I7 **Port Sudan** Red Sea, NE Sudan *19°37′N 37°14′E*
22 L10 **Port Sulphur** Louisiana, S USA *29°28′N 89°41′W*
Port Swettenham *see* Klang/Pelabuhan Klang
97 J22 **Port Talbot** S Wales, United Kingdom *51°36′N 03°47′W*
92 L11 **Porttipahdan Tekojärvi** ◎ N Finland
32 G7 **Port Townsend** Washington, NW USA *48°07′N 122°45′W*
104 H9 **Portugal** *off.* Portuguese Republic. ◆ *republic* SW Europe
105 O2 **Portugalete** País Vasco, N Spain *43°19′N 03°01′W*
54 J6 **Portuguesa** *off.* Estado Portuguesa. ⬖ *state* N Venezuela
Portuguesa, Estado *see* Portuguesa
Portuguese East Africa *see* Mozambique
Portuguese Guinea *see* Guinea-Bissau
Portuguese Republic *see* Portugal
Portuguese Timor *see* East Timor
Portuguese West Africa *see* Angola
97 D18 **Portumna** *Ir.* Port Omna. Galway, W Ireland *53°06′N 08°13′W*
Portus Cale *see* Porto
Portus Magnus *see* Almería
Portus Magonis *see* Maó
103 P17 **Port-Vendres** *var.* Port Vendres. Pyrénées-Orientales, S France *42°31′N 03°06′E*
182 H9 **Port Victoria** South Australia *34°34′S 137°31′E*
187 Q14 **Port-Vila** *var.* Vila. ● (Vanuatu) Éfaté, C Vanuatu *17°45′S 168°21′E*
Port Vila *see* Bauer Field
182 I9 **Port Wakefield** South Australia *34°13′S 138°10′E*
31 N8 **Port Washington** Wisconsin, N USA *43°23′N 87°54′W*
57 J14 **Porvenir** Pando, NW Bolivia *11°15′S 68°43′W*
63 I24 **Porvenir** Magallanes, S Chile *53°18′S 70°22′W*
61 D18 **Porvenir** Paysandú, W Uruguay *32°23′S 57°59′W*
93 M19 **Porvoo** *Swe.* Borgå. Uusimaa, S Finland *60°25′N 25°40′E*
Porzecze *see* Parechcha
104 M10 **Porzuna** Castilla-La Mancha, C Spain *39°10′N 04°10′W*
61 E14 **Posadas** Misiones, NE Argentina *27°27′S 55°52′W*
104 L13 **Posadas** Andalucía, S Spain *37°48′N 05°06′W*
Poschega *see* Požega
108 J11 **Poschiavino** ≈ Italy/Switzerland
108 J10 **Poschiavo** *Ger.* Puschlav. Graubünden, S Switzerland *46°19′N 10°02′E*
112 D12 **Posedarje** Zadar, SW Croatia *44°12′N 15°27′E*
Posen *see* Poznań
124 L14 **Poshekhon'ye** Yaroslavskaya Oblast', W Russian Federation *58°31′N 39°07′E*
92 M13 **Posio** Lappi, NE Finland *66°06′N 28°16′E*
Poskam *see* Zepu
Posnania *see* Poznań
1713 O12 **Poso** Sulawesi, C Indonesia *01°23′S 120°45′E*
171 O12 **Poso, Danau** ◎ Sulawesi, C Indonesia
137 R10 **Posof** Ardahan, NE Turkey *41°30′N 42°33′E*
25 R6 **Possum Kingdom Lake** ⊠ Texas, SW USA
25 N6 **Post** Texas, SW USA *33°14′N 101°24′W*
Postavy/Postawy *see* Pastavy
Poste-de-la-Baleine *see* Kuujjuarapik
99 M17 **Posterholt** Limburg, SE Netherlands *51°07′N 06°02′E*
83 G22 **Postmasburg** Northern Cape, N South Africa *28°20′S 23°05′E*
Pôsto Diuarum *see* Campo de Diauarum
59 I16 **Pôsto Jacaré** Mato Grosso, W Brazil *12°S 53°27′W*
109 T12 **Postojna** *Ger.* Adelsberg, *It.* Postumia. SW Slovenia *45°48′N 14°12′E*
Postumia *see* Postojna
29 X12 **Postville** Iowa, C USA *43°04′N 91°34′W*
Pöstyén *see* Piešt'any
113 G14 **Posušje** Federacija Bosne I Hercegovine, SW Bosnia and Herzegovina *43°28′N 17°20′E*
171 O16 **Pota** Flores, C Indonesia *08°21′S 120°50′E*
115 G23 **Potamós** Antikýthira, S Greece *35°53′N 23°17′E*
55 S9 **Potaru River** ≈ C Guyana
83 I21 **Potchefstroom** North-West, N South Africa *26°42′S 27°06′E*
27 R11 **Poteau** Oklahoma, C USA *35°03′N 94°36′W*
25 R12 **Poteet** Texas, SW USA *29°02′N 98°34′W*
115 G14 **Poteídaia** *site of ancient city* Kentrikí Makedonía, N Greece
Potentia *see* Potenza
107 M18 **Potenza** *anc.* Potentia. Basilicata, S Italy *40°40′N 15°50′E*
185 A24 **Poteriteri, Lake** ◎ South Island, New Zealand
104 M2 **Potes** Cantabria, N Spain *43°10′N 04°41′W*
Potgietersrus *see* Mokopane
25 S12 **Poth** Texas, SW USA *29°04′N 98°04′W*
32 J9 **Potholes Reservoir** ⊠ Washington, NW USA
137 Q9 **Poti** *prev.* P'ot'i. W Georgia *42°10′N 41°42′E*
P'ot'i *see* Poti
77 X13 **Potiskum** Yobe, NE Nigeria *11°38′N 11°07′E*
Potkozarje *see* Ivanjska
32 M9 **Potlatch** Idaho, NW USA *46°55′N 116°51′W*
33 N9 **Pot Mountain** ▲ Idaho, NW USA *46°44′N 115°24′W*
113 H14 **Potoci** Federacija Bosne I Hercegovine, S Bosnia and Herzegovina *43°24′N 17°52′E*
21 V3 **Potomac River** ≈ NE USA
Pòtoprens *see* Port-au-Prince
57 L20 **Potosí** Potosí, S Bolivia *19°35′S 65°51′W*
42 H9 **Potosí** Chinandega, NW Nicaragua *12°58′N 87°30′W*
27 W6 **Potosi** Missouri, C USA *37°57′N 90°49′W*
57 K21 **Potosí** ⬖ *department* SW Bolivia
62 H7 **Potrerillos** Atacama, N Chile *26°30′S 69°25′W*
42 H5 **Potrerillos** Cortés, NW Honduras *15°10′N 87°58′W*
62 H8 **Potro, Cerro del** ▲ N Chile *28°22′S 69°34′W*
100 N12 **Potsdam** Brandenburg, NE Germany *52°24′N 13°04′E*
18 J7 **Potsdam** New York, NE USA *44°40′N 74°58′W*
109 X5 **Pottendorf** Niederösterreich, E Austria *47°55′N 16°23′E*
109 X5 **Pottenstein** Niederösterreich, E Austria *47°58′N 16°07′E*
18 I15 **Pottstown** Pennsylvania, NE USA *40°15′N 75°39′W*
18 H14 **Pottsville** Pennsylvania, NE USA *40°40′N 76°10′W*
155 L25 **Pottuvil** Eastern Province, SE Sri Lanka *06°53′N 81°49′E*
149 U6 **Potwar Plateau** *plateau* NE Pakistan
102 J7 **Pouancé** Maine-et-Loire, W France *47°46′N 01°11′W*
15 R6 **Poulin de Courval, Lac** ◎ Québec, SE Canada
18 L9 **Poultney** Vermont, NE USA *43°31′N 73°12′W*
187 O16 **Poum** Province Nord, W New Caledonia *20°15′S 164°03′E*
59 L21 **Pouso Alegre** Minas Gerais, NE Brazil *22°13′S 45°56′W*
192 I16 **Poutasi** Upolu, SE Samoa *14°00′S 171°43′W*
167 R12 **Poŭthĭsăt** *prev.* Pursat. Poŭthĭsăt, W Cambodia *12°32′N 103°55′E*
167 R12 **Poŭthĭsăt, Stœng** *prev.* Pursat. ≈ W Cambodia
102 J9 **Pouzauges** Vendée, NW France *46°47′N 00°54′W*
Po, Valle del *see* Po Valley
106 F8 **Po Valley** *It.* Valle del Po. *valley* N Italy
111 I19 **Považská Bystrica** *Ger.* Waagbistritz, *Hung.* Vágbeszterce. Trenčiansky Kraj, W Slovakia *49°07′N 18°26′E*
124 J10 **Povenets** Respublika Kareliya, NW Russian Federation *62°50′N 34°47′E*
184 Q9 **Poverty Bay** *inlet* North Island, New Zealand
112 K12 **Povlen** ▲▲ W Serbia
104 G6 **Póvoa de Varzim** Porto, NW Portugal *41°22′N 08°46′W*
127 N8 **Povorino** Voronezhskaya Oblast', W Russian Federation *51°10′N 42°16′E*
Povungnituk *see* Puvirnituq
Rivière de Povungnituk *see* Puvirnituq, Riviere de
14 H11 **Powassan** Ontario, S Canada *46°05′N 79°21′W*
35 U17 **Poway** California, W USA *32°57′N 117°02′W*
33 W14 **Powder River** Wyoming, C USA *43°01′N 106°57′W*
33 Y10 **Powder River** ≈ Montana/Wyoming, NW USA
32 L12 **Powder River** ≈ Oregon, NW USA
33 W13 **Powder River Pass** *pass* Wyoming, C USA
33 U12 **Powell** Wyoming, C USA *44°45′N 108°45′W*
65 I22 **Powell Basin** *undersea feature* NW Weddell Sea
36 M8 **Powell, Lake** ◎ Utah, W USA
37 R4 **Powell, Mount** ▲ Colorado, C USA *39°25′N 106°20′W*
10 L17 **Powell River** British Columbia, SW Canada *49°54′N 124°34′W*
31 N5 **Powers** Michigan, N USA *45°40′N 87°29′W*
28 K2 **Powers Lake** North Dakota, N USA *48°33′N 102°37′W*
21 V6 **Powhatan** Virginia, NE USA *37°33′N 77°56′W*
31 V13 **Powhatan Point** Ohio, N USA *39°49′N 80°49′W*
97 J20 **Powys** *cultural region* E Wales, United Kingdom
187 P17 **Poya** Province Nord, C New Caledonia *21°19′S 165°07′E*
161 P10 **Poyang Hu** ◎ S China
30 L7 **Poygan, Lake** ◎ Wisconsin, N USA
109 Y2 **Poysdorf** Niederösterreich, NE Austria *48°40′N 16°38′E*
112 N11 **Požarevac** *Ger.* Passarowitz. Serbia, NE Serbia *44°37′N 21°11′E*
41 Q13 **Poza Rica** *var.* Poza Rica de Hidalgo. Veracruz-Llave, E Mexico *20°34′N 97°26′W*
Poza Rica de Hidalgo *see* Poza Rica
112 L13 **Požega** *prev.* Slavonska Požega, *Ger.* Poschega, *Hung.* Pozsega. Požega-Slavonija, NE Croatia *45°19′N 17°42′E*
112 H9 **Požega-Slavonija** *off.* Požeško-Slavonska Županija. ⬖ *province* NE Croatia
Požeško-Slavonska Županija *see* Požega-Slavonija
125 U13 **Pozhva** Komi-Permyatskiy Okrug, NW Russian Federation *59°07′N 56°04′E*
110 G11 **Poznań** *Ger.* Posen, Posnania. Wielkolpolskie, C Poland *52°24′N 16°56′E*
105 O13 **Pozo Alcón** Andalucía, S Spain *37°43′N 02°55′W*
62 H3 **Pozo Almonte** Tarapacá, N Chile *20°16′S 69°50′W*
104 L12 **Pozoblanco** Andalucía, S Spain *38°23′N 04°48′W*
105 Q11 **Pozo Cañada** Castilla-La Mancha, C Spain *38°49′N 01°45′W*
62 N5 **Pozo Colorado** Presidente Hayes, C Paraguay *23°26′S 58°51′W*
63 J20 **Pozos, Punta** *headland* S Argentina *47°55′S 65°46′W*
Pozsega *see* Požega
Pozsony *see* Bratislava
55 N5 **Pozuelos** Anzoátegui, NE Venezuela *10°11′N 64°39′W*
107 L26 **Pozzallo** Sicilia, Italy, C Mediterranean Sea *36°44′N 14°51′E*
107 K17 **Pozzuoli** *anc.* Puteoli. Campania, S Italy *40°49′N 14°07′E*
77 P17 **Pra** ≈ S Ghana
111 C19 **Prachatice** *Ger.* Prachatitz. Jihočeský Kraj, S Czech Republic *49°01′N 14°02′E*
Prachatitz *see* Prachatice
167 P11 **Prachin Buri** *var.* Prachinburi. Prachin Buri, C Thailand *14°05′N 101°23′E*
Prachinburi *see* Prachin Buri
Prachuab Girikhand *see* Prachuap Khiri Khan
167 O12 **Prachuap Khiri Khan** *var.* Prachuab Girikhand. Prachuap Khiri Khan, SW Thailand *11°50′N 99°49′E*
111 H16 **Praděd** *Ger.* Altvater. ▲ NE Czech Republic *50°06′N 17°14′E*
54 D11 **Pradera** Valle del Cauca, SW Colombia *03°23′N 76°11′W*
103 O17 **Prades** Pyrénées-Orientales, S France *42°36′N 02°22′E*
59 O19 **Prado** Bahia, SE Brazil *17°13′S 39°15′W*
54 E11 **Prado** Tolima, C Colombia *03°45′N 74°55′W*
Prado del Ganso *see* Goose Green
Prae *see* Phrae
95 I24 **Præstø** Sjælland, SE Denmark *55°08′N 12°03′E*
Prag/Praga/Prague *see* Praha
27 O10 **Prague** Oklahoma, C USA *35°29′N 96°40′W*
111 D16 **Praha** *Eng.* Prague, *Ger.* Prag, *Pol.* Praga. ● (Czech Republic) Středočeský Kraj, NW Czech Republic *50°06′N 14°26′E*
116 J13 **Prahova** ⬖ *county* SE Romania
116 J13 **Prahova** ≈ S Romania
76 E10 **Praia** ● (Cape Verde) Santiago, S Cape Verde *14°55′N 23°31′W*
83 M21 **Praia do Bilene** Gaza, S Mozambique *25°18′S 33°10′E*
83 M20 **Praia do Xai-Xai** Gaza, S Mozambique *25°04′S 33°43′E*
116 J10 **Praid** *Hung.* Parajd. Harghita, C Romania *46°33′N 25°06′E*
26 J3 **Prairie Dog Creek** ≈ Kansas/Nebraska, C USA
30 J9 **Prairie du Chien** Wisconsin, N USA *43°02′N 91°08′W*
27 S9 **Prairie Grove** Arkansas, C USA *35°58′N 94°19′W*
31 P10 **Prairie River** ≈ Michigan, N USA
Prairie State *see* Illinois
25 V11 **Prairie View** Texas, SW USA *30°05′N 95°59′W*
167 Q10 **Prakhon Chai** Buri Ram, E Thailand *14°36′N 103°04′E*
109 R4 **Pram** ≈ N Austria
167 Q12 **Prâmaôy** *prev.* Phumĭ Prâmaôy. Poŭthĭsăt, W Cambodia *12°13′N 103°05′E*
109 S4 **Prambachkirchen** Oberösterreich, N Austria *48°18′N 13°50′E*
118 H2 **Prangli** *island* N Estonia
154 J13 **Prānhita** ≈ C India
172 I15 **Praslin** *island* Inner Islands, NE Seychelles
115 O23 **Prasonísi, Akrotírio** *cape* Ródos, Dodekánisa, Greece, Aegean Sea
111 I14 **Praszka** Opolskie, S Poland *51°05′N 18°29′E*
Pratas Island *see* Tungsha Tao
119 M18 **Pratasy** *Rus.* Protasy. Homyel'skaya Voblasts', SE Belarus *52°47′N 29°05′E*
167 Q10 **Prathai** Nakhon Ratchasima, E Thailand *15°31′N 102°42′E*
Prathet Thai *see* Thailand
Prathum Thani *see* Pathum Thani
63 F21 **Prat, Isla** *island* S Chile
106 G11 **Prato** Toscana, C Italy *43°53′N 11°05′E*
103 O17 **Prats-de-Mollo-la-Preste** Pyrénées-Orientales, S France *42°25′N 02°28′E*
26 L6 **Pratt** Kansas, C USA *37°40′N 98°45′W*
108 E6 **Pratteln** Basel Landschaft, NW Switzerland *47°32′N 07°42′E*
193 O2 **Pratt Seamount** *undersea feature* N Pacific Ocean *56°09′N 142°30′W*
23 P5 **Prattville** Alabama, S USA *32°27′N 86°27′W*
Praust *see* Pruszcz Gdański
119 B14 **Pravdinsk** *Ger.* Friedland. Kaliningradskaya Oblast', W Russian Federation *54°26′N 21°01′E*
104 K2 **Pravia** Asturias, N Spain *43°30′N 06°06′W*
118 L12 **Prazaroki** *Rus.* Prozoroki. Vitsyebskaya Voblasts', N Belarus *55°18′N 28°13′E*
Prázsmár *see* Prejmer
167 S11 **Preăh Vihéar** Preăh Vihéar, N Cambodia *13°57′N 104°48′E*
116 J12 **Predeal** *Hung.* Predeál. Brașov, C Romania *45°30′N 25°31′E*
109 S8 **Predlitz** Steiermark, SE Austria *47°04′N 13°54′E*
11 V15 **Preeceville** Saskatchewan, S Canada *51°58′N 102°40′W*
Preenkuln *see* Priekule
102 K6 **Pré-en-Pail** Mayenne, NW France *48°27′N 00°15′W*
109 T4 **Pregarten** Oberösterreich, N Austria *48°21′N 14°31′E*
54 H7 **Pregonero** Táchira, NW Venezuela *08°02′N 71°35′W*
118 J10 **Preiļi** *Ger.* Preli. SE Latvia *56°17′N 26°52′E*
116 J12 **Prejmer** *Ger.* Tartlau, *Hung.* Prázsmár. Brașov, S Romania *45°42′N 25°49′E*
113 J16 **Prekornica** ▲▲ C Montenegro
Preli *see* Preiļi
Prëmet *see* Përmet
100 M12 **Premnitz** Brandenburg, NE Germany *52°33′N 12°22′E*
25 S15 **Premont** Texas, SW USA *27°21′N 98°07′W*
113 H14 **Prenj** ▲▲ S Bosnia and Herzegovina
Prenjas/Prenjasi *see* Përrenjas
22 L7 **Prentiss** Mississippi, S USA *31°36′N 89°52′W*
Preny *see* Prienai
100 O10 **Prenzlau** Brandenburg, NE Germany *53°19′N 13°52′E*
123 N11 **Preobrazhenka** Irkutskaya Oblast', C Russian Federation *60°01′N 108°00′E*
166 J9 **Preparis Island** *island* SW Myanmar (Burma)
Prerau *see* Přerov
111 H18 **Přerov** *Ger.* Prerau. Olomoucký Kraj, E Czech Republic *49°27′N 17°27′E*
Preschau *see* Prešov
14 M14 **Prescott** Ontario, SE Canada *44°43′N 75°33′W*
36 K12 **Prescott** Arizona, SW USA *34°33′N 112°26′W*
27 T13 **Prescott** Arkansas, C USA *33°49′N 93°25′W*
32 L10 **Prescott** Washington, NW USA *46°17′N 118°21′W*
30 H6 **Prescott** Wisconsin, N USA *44°46′N 92°45′W*
185 A24 **Preservation Inlet** *inlet* South Island, New Zealand
112 O7 **Preševo** Serbia, SE Serbia *42°20′N 21°38′E*
29 N10 **Presho** South Dakota, N USA *43°54′N 100°03′W*
58 M13 **Presidente Dutra** Maranhão, E Brazil *05°17′S 44°30′W*
60 I8 **Presidente Epitácio** São Paulo, S Brazil *21°45′S 52°07′W*
62 N5 **Presidente Hayes** *off.* Departamento de Presidente Hayes. ⬖ *department* C Paraguay
Presidente Hayes, Departamento de *see* Presidente Hayes
60 I9 **Presidente Prudente** São Paulo, S Brazil *22°09′S 51°24′W*
Presidente Stroessner *see* Ciudad del Este
Presidente Vargas *see* Itabira
60 I8 **Presidente Venceslau** São Paulo, S Brazil *21°52′S 51°51′W*
193 O10 **President Thiers Seamount** *undersea feature* C Pacific Ocean *24°39′S 145°50′W*
24 J11 **Presidio** Texas, SW USA *29°33′N 104°22′W*
Preslav *see* Veliki Preslav
111 M19 **Prešov** *var.* Preschau, *Ger.* Eperies, *Hung.* Eperjes. Prešovský Kraj, E Slovakia *49°N 21°14′E*
111 M19 **Prešovský Kraj** ⬖ *region* E Slovakia
113 N20 **Prespa, Lake** *Alb.* Liqen i Prespës, *Gk.* Límni Megáli Préspa, Limni Prespa, *Mac.* Prespansko Ezero, *Serb.* Prespansko Jezero. ◎ SE Europe
Prespa, Limni/Prespansko Ezero/Prespansko Jezero/Prespës, Liqen i *see* Prespa, Lake
19 S2 **Presque Isle** Maine, NE USA *46°40′N 68°01′W*
18 B11 **Presque Isle** *headland* Pennsylvania, NE USA *42°09′N 80°06′W*
Pressburg *see* Bratislava
77 P17 **Prestea** SW Ghana *05°22′N 02°07′W*
111 B17 **Přeštice** *Ger.* Pschestitz. Plzeňský Kraj, W Czech Republic *49°36′N 13°19′E*
97 K17 **Preston** NW England, United Kingdom *53°46′N 02°42′W*
23 S6 **Preston** Georgia, SE USA *32°03′N 84°32′W*
33 R16 **Preston** Idaho, NW USA *42°06′N 111°52′W*
29 Z13 **Preston** Iowa, C USA *42°03′N 90°24′W*
29 X11 **Preston** Minnesota, N USA *43°41′N 92°06′W*
21 O6 **Prestonsburg** Kentucky, S USA *37°40′N 82°46′W*
96 I13 **Prestwick** W Scotland, United Kingdom *55°31′N 04°39′W*
83 145 **Pretoria** *var.* Epitoli. ● Gauteng, NE South Africa *25°41′S 28°12′E*
Pretoria-Witwatersrand-Vereeniging *see* Gauteng
Pretusha *see* Pretushë
113 M21 **Pretushë** *var.* Pretusha. Korçë, SE Albania *40°50′N 20°45′E*
Preussisch Eylau *see* Bagrationovsk
Preußisch Holland *see* Pasłęk
Preussisch-Stargard *see* Starogard Gdański
115 C17 **Préveza** Ípeiros, W Greece *38°59′N 20°44′E*
37 V3 **Prewitt Reservoir** ⊠ Colorado, C USA
167 S13 **Prey Vêng** Prey Vêng, S Cambodia *11°30′N 105°20′E*
Priaral'skiye Karakumy, Peski *see* Priaral'skiy Karakum
144 M12 **Priaral'skiy Karakum** *prev.* Priaral'skiye Karakumy, Peski. *desert* SW Kazakhstan
123 P14 **Priargunsk** Zabaykal'skiy Kray, S Russian Federation *50°25′N 119°12′E*
38 K14 **Pribilof Islands** *island group* Alaska, USA
113 K14 **Priboj** Serbia, W Serbia *43°34′N 19°33′E*
111 C17 **Příbram** *Ger.* Pibrans. Středočeský Kraj, W Czech Republic *49°41′N 14°02′E*
36 M4 **Price** Utah, W USA *39°35′N 110°49′W*
37 N5 **Price River** ≈ Utah, W USA
23 N8 **Prichard** Alabama, S USA *30°44′N 88°04′W*
25 R8 **Priddy** Texas, SW USA *31°39′N 98°30′W*
105 P8 **Priego** Castilla-La Mancha, C Spain *40°26′N 02°19′W*
104 M14 **Priego de Córdoba** Andalucía, S Spain *37°27′N 04°12′W*
118 C10 **Priekule** *Ger.* Preenkuln. SW Latvia *56°26′N 21°36′E*
118 C12 **Priekulė** *Ger.* Prökuls. Klaipėda, W Lithuania *55°36′N 21°16′E*
119 F14 **Prienai** *Pol.* Preny. Kaunas, S Lithuania *54°37′N 23°56′E*
83 G23 **Prieska** Northern Cape, C South Africa *29°40′S 22°45′E*
32 M7 **Priest Lake** ⊠ Idaho, NW USA
32 M7 **Priest River** Idaho, NW USA *48°10′N 117°02′W*
104 M3 **Prieta, Peña** ▲ N Spain *43°01′N 04°42′W*
40 J10 **Prieto, Cerro** ▲ C Mexico *24°10′N 105°21′W*
111 J19 **Prievidza** *var.* Priewitz, *Ger.* Priwitz, *Hung.* Privigye. Trenčiansky Kraj, W Slovakia *48°47′N 18°35′E*
Priewitz *see* Prievidza
112 F10 **Prijedor** ⬖ Republika Srpska, NW Bosnia and Herzegovina
113 K14 **Prijepolje** Serbia, W Serbia *43°24′N 19°39′E*
Prikaspiyskaya Nizmennost' *see* Caspian Depression
113 O19 **Prilep** *Turk.* Perlepe. S FYR Macedonia *41°21′N 21°34′E*
108 B9 **Prilly** Vaud, SW Switzerland *46°32′N 06°38′E*
Priluki *see* Pryluky
62 L10 **Primero, Río** ≈ C Argentina
29 S12 **Primghar** Iowa, C USA *43°05′N 95°37′W*
112 B9 **Primorje-Gorski Kotar** *off.* Primorsko-Goranska Županija. ⬖ *province* NW Croatia
118 A13 **Primorsk** *Ger.* Fischhausen. Kaliningradskaya Oblast', W Russian Federation *54°45′N 20°00′E*
124 G12 **Primorsk** *Fin.* Koivisto. Leningradskaya Oblast', NW Russian Federation *60°20′N 28°39′E*
123 S14 **Primorskiy Kray** *prev.* *Eng.* Maritime Territory. ⬖ *territory* SE Russian Federation
114 N10 **Primorsko** *prev.* Keupriya. Burgas, E Bulgaria *42°15′N 27°45′E*
126 K13 **Primorsko-Akhtarsk** Krasnodarskiy Kray, SW Russian Federation *46°03′N 38°44′E*
Primorsko-Goranska Županija *see* Primorje-Gorski Kotar
Primorsk/Primorskoye *see* Prymors'k
113 D14 **Primošten** Šibenik-Knin, S Croatia *43°34′N 15°57′E*
11 R13 **Primrose Lake** ◎ Saskatchewan, C Canada
11 T14 **Prince Albert** Saskatchewan, S Canada *53°09′N 105°43′W*
83 G25 **Prince Albert** Western Cape, SW South Africa *33°13′S 22°03′E*
8 J5 **Prince Albert Peninsula** *peninsula* Victoria Island, Northwest Territories, NW Canada
8 J6 **Prince Albert Sound** *inlet* Northwest Territories, N Canada
J5 8 **Prince Alfred, Cape** *headland* Northwest Territories, NW Canada
9 P6 **Prince Charles Island** *island* Nunavut, NE Canada
195 W6 **Prince Charles Mountains** ▲▲ Antarctica
Prince-Édouard, Île-du *see* Prince Edward Island
172 M13 **Prince Edward Fracture Zone** *tectonic feature* SW Indian Ocean
13 P14 **Prince Edward Island** *Fr.* Île-du Prince-Édouard. ⬖ *province* SE Canada
13 Q14 **Prince Edward Island** *Fr.* Île-du Prince-Édouard. *island* SE Canada
173 M12 **Prince Edward Islands** *island group* S South Africa
21 X4 **Prince Frederick** Maryland, NE USA *38°32′N 76°35′W*
10 M14 **Prince George** British Columbia, SW Canada *53°55′N 122°49′W*
21 W6 **Prince George** Virginia, NE USA *37°13′N 77°13′W*
197 O8 **Prince Gustaf Adolf Sea** *sea* Nunavut, N Canada
197 Q3 **Prince of Wales, Cape** *headland* Alaska, USA *65°39′N 168°12′W*
181 V1 **Prince of Wales Island** *island* Queensland, E Australia
8 L5 **Prince of Wales Island** *island* Queen Elizabeth Islands, Nunavut, NW Canada
39 Y14 **Prince of Wales Island** *island* Alexander Archipelago, Alaska, USA
Prince of Wales Island *see* Pinang, Pulau
8 J5 **Prince of Wales Strait** *strait* Northwest Territories, N Canada
197 O8 **Prince Patrick Island** *island* Parry Islands, Northwest Territories, NW Canada
9 N5 **Prince Regent Inlet** *channel* Nunavut, N Canada
10 J13 **Prince Rupert** British Columbia, SW Canada *54°18′N 130°17′W*
Prince's Island *see* Príncipe
21 Y5 **Princess Anne** Maryland, NE USA *38°12′N 75°42′W*
Princess Astrid Coast *see* Prinsesse Astrid Kyst
181 W2 **Princess Charlotte Bay** *bay* Queensland, NE Australia
195 W7 **Princess Elizabeth Land** *physical region* Antarctica
10 J14 **Princess Royal Island** *island* British Columbia, SW Canada
45 U15 **Princes Town** Trinidad, Trinidad and Tobago *10°16′N 61°23′W*
11 N17 **Princeton** British Columbia, SW Canada *49°25′N 120°35′W*
30 L11 **Princeton** Illinois, N USA *41°22′N 89°27′W*
31 N16 **Princeton** Indiana, N USA *38°21′N 87°33′W*
29 Z14 **Princeton** Iowa, C USA *41°40′N 90°21′W*
20 H7 **Princeton** Kentucky, S USA *37°06′N 87°52′W*
29 V8 **Princeton** Minnesota, N USA *45°34′N 93°34′W*
27 S1 **Princeton** Missouri, C USA *40°22′N 93°37′W*
18 J15 **Princeton** New Jersey, NE USA *40°21′N 74°39′W*
21 R6 **Princeton** West Virginia, NE USA *37°23′N 81°06′W*
39 S12 **Prince William Sound** *inlet* Alaska, USA
67 P9 **Príncipe** *var.* Príncipe Island, *Eng.* Prince's Island. *island* N Sao Tome and Principe
Príncipe Island *see* Príncipe
32 I13 **Prineville** Oregon, NW USA *44°19′N 120°50′W*
28 J11 **Pringle** South Dakota, N USA *43°34′N 103°34′W*
25 N1 **Pringle** Texas, SW USA *35°55′N 101°28′W*
99 H14 **Prinsenbeek** Noord-Brabant, S Netherlands *51°36′N 04°42′E*
98 L6 **Prinses Margriet Kanaal** *canal* N Netherlands
195 R1 **Prinsesse Astrid Kyst** *Eng.* Princess Astrid Coast. *physical region* Antarctica
195 T2 **Prinsesse Ragnhild Kyst** *physical region* Antarctica
195 U2 **Prins Harald Kyst** *physical region* Antarctica
92 N2 **Prins Karls Forland** *island* W Svalbard
43 N8 **Prinzapolka** Región Autónoma Atlántico Norte, NE Nicaragua *13°19′N 83°35′W*
42 L8 **Prinzapolka, Río** ≈ NE Nicaragua
122 H9 **Priob'ye** Khanty-Mansiyskiy Avtonomnyy Okrug-Yugra, N Russian Federation *62°25′N 65°36′E*
104 H1 **Prior, Cabo** *headland* NW Spain *43°33′N 08°21′W*
29 V9 **Prior Lake** Minnesota, N USA *44°42′N 93°25′W*
124 H11 **Priozersk** *Fin.* Käkisalmi. Leningradskaya Oblast', NW Russian Federation *61°02′N 30°07′E*
119 J20 **Pripet** *Bel.* Prypyats', *Ukr.* Pryp"yat'. ≈ Belarus/Ukraine
119 J20 **Pripet Marshes** *wetland* Belarus/Ukraine
113 N16 **Prishtinë** , *Serb.* Priština. C Kosovo *42°40′N 21°10′E*
126 J8 **Pristen'** Kurskaya Oblast', W Russian Federation *51°15′N 36°47′E*
Priština *see* Prishtinë
100 M10 **Pritzwalk** Brandenburg, NE Germany *53°10′N 12°11′E*
103 R13 **Privas** Ardèche, E France *44°45′N 04°35′E*
107 I16 **Priverno** Lazio, C Italy *41°28′N 13°11′E*
Privigye *see* Prievidza
112 C12 **Privlaka** Zadar, SW Croatia *44°15′N 15°07′E*
124 M15 **Privolzhsk** Ivanovskaya Oblast', NW Russian Federation *57°24′N 41°16′E*
127 P7 **Privolzhskaya Vozvyshennost'** *var.* Volga Uplands. ▲▲ W Russian Federation
127 P8 **Privolzhskoye** Saratovskaya Oblast', W Russian Federation *51°08′N 45°57′E*
Priwitz *see* Prievidza
127 N13 **Priyutnoye** Respublika Kalmykiya, SW Russian Federation *46°08′N 43°33′E*
113 M17 **Prizren** S Kosovo *42°14′N 20°46′E*
107 I24 **Prizzi** Sicilia, Italy, C Mediterranean Sea *37°44′N 13°26′E*
113 P18 **Probištip** NE FYR Macedonia *42°00′N 22°06′E*
169 S16 **Probolinggo** Jawa, C Indonesia *07°45′S 113°12′E*
Probstberg *see* Wyszków
111 F14 **Prochowice** *Ger.* Parchwitz. Dolnośląskie, SW Poland *51°15′N 16°22′E*
29 W5 **Proctor** Minnesota, N USA *46°46′N 92°13′W*
25 R8 **Proctor** Texas, SW USA *31°57′N 98°25′W*
25 R8 **Proctor Lake** ⊠ Texas, SW USA
155 I18 **Proddatūr** Andhra Pradesh, E India *14°45′N 78°34′E*
104 H9 **Proença-a-Nova** *var.* Proença a Nova. Castelo Branco, C Portugal *39°45′N 07°56′W*
Proença a Nova *see* Proença-a-Nova
99 I21 **Profondeville** Namur, SE Belgium *50°22′N 04°52′E*
41 W11 **Progreso** Yucatán, SE Mexico *21°14′N 89°41′W*
123 R14 **Progress** Amurskaya Oblast', SE Russian Federation *49°40′N 129°30′E*
127 O15 **Prokhladnyy** Kabardino-Balkarskaya Respublika, SW Russian Federation *43°48′N 44°02′E*
Prokletije *see* North Albanian Alps
Prökuls *see* Priekulė
113 O15 **Prokuplje** Serbia, SE Serbia *43°15′N 21°35′E*
124 H14 **Proletariy** Novgorodskaya Oblast', W Russian Federation *58°24′N 31°40′E*
126 M12 **Proletarsk** Rostovskaya Oblast', SW Russian Federation *46°42′N 41°48′E*
127 N13 **Proletarskoye Vodokhranilishche** *salt lake* SW Russian Federation
Prome *see* Pyay
60 J8 **Promissão** São Paulo, S Brazil *21°33′S 49°51′W*
60 J8 **Promissão, Represa de** ⊠ S Brazil
125 V4 **Promyshlennyy** Respublika Komi, NW Russian Federation *67°36′N 64°E*
119 O16 **Pronya** ≈ E Belarus
10 M11 **Prophet River** British Columbia, W Canada *58°07′N 122°39′W*
30 K11 **Prophetstown** Illinois, N USA *41°40′N 89°56′W*
Propinsi Kepulauan Riau *see* Kepulauan Riau
Propinsi Papua Barat *see* Papua Barat
59 P16 **Propriá** Sergipe, E Brazil *10°15′S 36°51′W*
103 X16 **Propriano** Corse, France, C Mediterranean Sea *41°41′N 08°54′E*
Prościejów *see* Prostějov
Proskurov *see* Khmel 'nyts'kyy
114 H12 **Prosotsáni** Anatolikí Makedonía kai Thráki, NE Greece *41°11′N 23°59′E*
171 Q7 **Prosperidad** Mindanao, S Philippines *08°36′N 125°54′E*
32 J10 **Prosser** Washington, NW USA *46°12′N 119°46′W*
Prossnitz *see* Prostějov
111 G18 **Prostějov** *Ger.* Prossnitz, *Pol.* Prościejów. Olomoucký Kraj, E Czech Republic *49°29′N 17°08′E*
117 V8 **Prosyana** Dnipropetrovs'ka Oblast', E Ukraine *48°07′N 36°22′E*
111 L16 **Proszowice** Malopolskie, S Poland *50°12′N 20°15′E*
Protasy *see* Pratasy
172 J11 **Protea Seamount** *undersea feature* SW Indian Ocean *36°50′S 18°05′E*
115 D21 **Próti** *island* S Greece
114 N8 **Provadia** , Provadiya. Varna, E Bulgaria *43°10′N 27°29′E*
Provadiya *see* Provadia
103 T14 **Provence** *cultural region* SE France
103 S15 **Provence** *prev.* Marseille-Marigrane. ✈ (Marseille) Bouches-du-Rhône, SE France *43°25′N 05°15′E*
103 T14 **Provence-Alpes-Côte d'Azur** ⬖ *region* SE France
20 H6 **Providence** Kentucky, S USA *37°23′N 87°47′W*
19 N12 **Providence** *state capital* Rhode Island, NE USA *41°50′N 71°26′W*
36 L1 **Providence** Utah, W USA *41°42′N 111°49′W*
Providence *see* Fort Providence
Providence *see* Providence Atoll
67 X10 **Providence Atoll** *var.* Providence. *atoll* S Seychelles
14 D12 **Providence Bay** Manitoulin Island, Ontario, S Canada *45°39′N 82°16′W*
23 R6 **Providence Canyon** *valley* Alabama/Georgia, S USA
22 I5 **Providence, Lake** ◎ Louisiana, S USA
35 X13 **Providence Mountains** ▲▲ California, W USA
44 L6 **Providenciales** *island* W Turks and Caicos Islands
19 Q12 **Provincetown** Massachusetts, NE USA *42°01′N 70°10′W*
103 P5 **Provins** Seine-et-Marne, N France *48°34′N 03°18′E*
36 L3 **Provo** Utah, W USA *40°14′N 111°39′W*
11 R15 **Provost** Alberta, SW Canada *52°24′N 110°16′W*
112 G13 **Prozor** Federacija Bosne I Hercegovine, SW Bosnia and Herzegovina *43°46′N 17°38′E*
Prozoroki *see* Prazaroki
60 I11 **Prudentópolis** Paraná, S Brazil *25°12′S 50°58′W*
39 R5 **Prudhoe Bay** Alaska, USA *70°16′N 148°18′W*
39 R4 **Prudhoe Bay** *bay* Alaska, USA
111 H16 **Prudnik** *Ger.* Neustadt, Neustadt in Oberschlesien. Opole, SW Poland *50°20′N 17°34′E*
119 J16 **Prudy** Minskaya Voblasts', C Belarus *53°47′N 26°32′E*
101 D18 **Prüm** Rheinland-Pfalz, W Germany *50°15′N 06°27′E*
101 D18 **Prüm** ≈ W Germany

◆ Country
● Country Capital
◇ Dependent Territory
○ Dependent Territory Capital
⬖ Administrative Regions
✈ International Airport
▲ Mountain
▲▲ Mountain Range
Volcano
≈ River
◎ Lake
⊠ Reservoir

Prusa see Bursa
110 J7 **Pruszcz Gdański** *Ger.* Praust. Pomorskie, N Poland *54°16′N 18°36′E*
110 M12 **Pruszków** *Ger.* Kaltdorf. Mazowieckie, C Poland *52°09′N 20°49′E*
116 K8 **Prut** *Ger.* Pruth. ≈ E Europe
Pruth see Prut
108 L8 **Prutz** Tirol, W Austria *47°07′N 10°42′E*
Pružana see Pruzhany
119 G19 **Pruzhany** *Pol.* Prużana. Brestskaya Voblasts', SW Belarus *52°33′N 24°28′E*
124 I11 **Pryazha** Respublika Kareliya, NW Russian Federation *61°42′N 33°39′E*
117 U10 **Pryazovs'ke** Zaporiz'ka Oblast', SE Ukraine *46°43′N 35°39′E*
Prychornomor'ska Nyzovyna see Black Sea Lowland
Prydniprovs'ka Nyzovyna/Prydnyaprowskaya Nizina see Dnieper Lowland
195 Y7 **Prydz Bay** *bay* Antarctica
117 R4 **Pryluky** *Rus.* Priluki. Chernihivs'ka Oblast', NE Ukraine *50°35′N 32°23′E*
117 V10 **Prymors'k** *Rus.* Primorsk; *prev.* Primorskoye. Zaporiz'ka Oblast', SE Ukraine *46°44′N 36°19′E*
117 U13 **Prymors'kyy** Avtonomna Respublika Krym, S Ukraine *45°09′N 35°33′E*
27 Q9 **Pryor** Oklahoma, C USA *36°19′N 95°19′W*
33 U11 **Pryor Creek** ≈ Montana, NW USA
Pryp"yat'/Prypyats' see Pripet
110 M10 **Przasnysz** Mazowieckie, C Poland *53°01′N 20°51′E*
111 K14 **Przedbórz** Łodzkie, S Poland *51°04′N 19°51′E*
111 P17 **Przemyśl** *Rus.* Peremyshl. Podkarpackie, C Poland *49°47′N 22°47′E*
111 O16 **Przeworsk** Podkarpackie, SE Poland *50°04′N 22°30′E*
Przheval'sk see Karakol
110 L13 **Przysucha** Mazowieckie, SE Poland *51°22′N 20°36′E*
115 H18 **Psachná** *var.* Psahna, Psakhná. Évvoia, C Greece *38°35′N 23°39′E*
Psahna/Psakhná see Psachná
115 K18 **Psará** *island* E Greece
115 I16 **Psathoúra** *island* Vóreies Sporádes, Greece, Aegean Sea
Pschestitz see Přeštice
Psein Lora see Pishin Lora
117 S5 **Psel** *Rus.* Psël. ≈ Russian Federation/Ukraine
Psël see Psel
115 M21 **Psérimos** *island* Dodekánisa, Greece, Aegean Sea
Pseyn Bowr see Pishin Lora
Pskem see Piskom
147 R8 **Pskemskiy Khrebet** *Uzb.* Piskom Tizmasi. ▲▲ Kyrgyzstan/Uzbekistan
124 F14 **Pskov** *Ger.* Pleskau, *Latv.* Pleskava. Pskovskaya Oblast', W Russian Federation *58°32′N 31°15′E*
118 K6 **Pskov, Lake** *Est.* Pihkva Järv, *Ger.* Pleskauer See, *Rus.* Pskovskoye Ozero. ◎ Estonia/Russian Federation
124 F15 **Pskovskaya Oblast'** ◐ *province* W Russian Federation
Pskovskoye Ozero see Pskov, Lake
112 G9 **Psunj** ▲▲ NE Croatia
111 J17 **Pszczyna** *Ger.* Pless. Śląskie, S Poland *49°59′N 18°54′E*
Ptačnik/Ptacsnik see Vtáčnik
115 D17 **Ptéri** ▲ C Greece *39°08′N 21°32′E*
Ptich' see Ptsich
115 E14 **Ptolemaïda** *prev.* Ptolemaïs. Dytikí Makedonía, N Greece *40°34′N 21°42′E*
Ptolemaïs see Ptolemaïda, Greece
Ptolemaïs see 'Akko, Israel
119 M19 **Ptsich** *Rus.* Ptich'. Homyel'skaya Voblasts', SE Belarus *52°11′N 28°49′E*
119 M18 **Ptsich** *Rus.* Ptich'. ≈ SE Belarus
109 X10 **Ptuj** *Ger.* Pettau; *anc.* Poetovio. NE Slovenia *46°26′N 15°54′E*
61 A23 **Puán** Buenos Aires, E Argentina *37°35′S 62°45′W*
192 H15 **Pu'apu'a** Savai'i, C Samoa *13°32′S 172°09′W*
192 G15 **Puava, Cape** *headland* Savai'i, NW Samoa
Pubao see Baingoin
56 F12 **Pucallpa** Ucayali, C Peru *08°21′S 74°33′W*
57 J17 **Pucarani** La Paz, NW Bolivia *16°25′S 68°29′W*
Pučarevo see Novi Travnik
157 U12 **Pucheng** Shaanxi, SE China *35°00′N 109°34′E*
160 L6 **Pucheng** *var.* Nanpu. Fujian, C China *27°59′N 118°31′E*
125 N16 **Puchezh** Ivanovskaya Oblast', W Russian Federation *56°58′N 41°08′E*
111 I19 **Púchov** *Hung.* Puhó. Trenčiansky Kraj, W Slovakia *49°08′N 18°15′E*
116 J13 **Pucioasa** Dâmbovița, S Romania *45°04′N 25°23′E*
110 I6 **Puck** Pomorskie, N Poland *54°43′N 18°24′E*
30 L8 **Puckaway Lake** ◎ Wisconsin, N USA
63 G15 **Pucón** Araucanía, S Chile *39°18′S 71°52′W*
93 M14 **Pudasjärvi** Pohjois-Pohjanmaa, C Finland *65°20′N 27°02′E*
148 L8 **Pūdeh Tal, Shelleh-ye** ≈ SW Afghanistan
127 S1 **Pudem** Udmurtskaya Respublika, NW Russian Federation *58°18′N 52°08′E*
Pudewitz see Pobiedziska
124 K11 **Pudozh** Respublika Kareliya, NW Russian Federation *61°48′N 36°30′E*
97 M17 **Pudsey** N England, United Kingdom *53°48′N 01°40′W*
Puduchcheri see Puducherry
151 I20 **Puducherry** *prev.* Pondicherry, *var.* Puduchcheri, *Fr.* Pondichéry. ◐ *union territory* India
151 H21 **Pudukkottai** Tamil Nādu, SE India *10°23′N 78°47′E*
171 Z13 **Pue** Papua, E Indonesia *02°42′S 140°36′E*
41 P14 **Puebla** *var.* Puebla de Zaragoza. Puebla, S Mexico *19°02′N 98°13′W*
41 P15 **Puebla** ◐ *state* S Mexico
104 L11 **Puebla de Alcocer** Extremadura, W Spain *38°59′N 05°14′W*
Puebla de Don Fabrique see Puebla de Don Fadrique
105 P13 **Puebla de Don Fadrique** *var.* Puebla de Don Fabrique. Andalucía, S Spain *37°58′N 02°25′W*
104 J11 **Puebla de la Calzada** Extremadura, W Spain *38°54′N 06°38′W*
104 J5 **Puebla de Sanabria** Castilla y León, N Spain *42°04′N 06°38′W*
Puebla de Trives see A Pobla de Trives
Puebla de Zaragoza see Puebla
37 T6 **Pueblo** Colorado, C USA *38°15′N 104°37′W*
37 N10 **Pueblo Colorado Wash** *valley* Arizona, SW USA
61 C16 **Pueblo Libertador** Corrientes, NE Argentina *30°13′S 59°23′W*
40 J10 **Pueblo Nuevo** Durango, C Mexico *23°24′N 105°21′W*
42 J8 **Pueblo Nuevo** Estelí, NW Nicaragua *13°21′N 86°30′W*
54 J3 **Pueblo Nuevo** Falcón, N Venezuela *11°59′N 69°57′W*
42 B6 **Pueblo Nuevo Tiquisate** *var.* Tiquisate. Escuintla, SW Guatemala *14°16′N 91°21′W*
41 Q11 **Pueblo Viejo, Laguna de** *lagoon* E Mexico
63 J14 **Puelches** La Pampa, C Argentina *38°08′S 65°56′W*
104 L14 **Puente-Genil** Andalucía, S Spain *37°23′N 04°45′W*
105 Q3 **Puente la Reina** *Bas.* Gares. Navarra, N Spain *42°40′N 01°49′W*
104 L12 **Puente Nuevo, Embalse de** ⊠ S Spain
57 D14 **Puente Piedra** Lima, W Peru *11°49′S 77°01′W*
160 F14 **Pu'er** *var.* Ning'er. Yunnan, SW China *23°09′N 100°58′E*
45 V6 **Puerca, Punta** *headland* E Puerto Rico *18°13′N 65°36′W*
37 R12 **Puerco, Rio** ≈ New Mexico, SW USA
57 J17 **Puerto Acosta** La Paz, W Bolivia *15°33′S 69°15′W*
63 G19 **Puerto Aisén** Aisén, S Chile *45°24′S 72°42′W*
41 R17 **Puerto Ángel** Oaxaca, SE Mexico *15°39′N 96°29′W*
Puerto Argentino see Stanley
41 T17 **Puerto Arista** Chiapas, SE Mexico *15°55′N 93°47′W*
43 O16 **Puerto Armuelles** Chiriquí, SW Panama *08°19′N 82°51′W*
Puerto Arrecife see Arrecife
54 D14 **Puerto Asís** Putumayo, SW Colombia *0°31′N 76°31′W*
54 L9 **Puerto Ayacucho** Amazonas, SW Venezuela *05°45′N 67°37′W*
57 C18 **Puerto Ayora** Galapagos Islands, Ecuador, E Pacific Ocean *0°45′S 90°19′W*
57 C18 **Puerto Baquerizo Moreno** *var.* Baquerizo Moreno. Galapagos Islands, Ecuador, E Pacific Ocean *0°54′S 89°37′W*
42 G4 **Puerto Barrios** Izabal, E Guatemala *15°42′N 88°34′W*
Puerto Bello see Portobelo
54 F8 **Puerto Berrío** Antioquia, C Colombia *06°28′N 74°28′W*
54 F9 **Puerto Boyacá** Boyacá, C Colombia *05°58′N 74°36′W*
54 K4 **Puerto Cabello** Carabobo, N Venezuela *10°29′N 68°02′W*
43 N7 **Puerto Cabezas** *var.* Bilwi. Región Autónoma Atlántico Norte, NE Nicaragua *14°05′N 83°22′W*
54 L9 **Puerto Carreño** Vichada, E Colombia *06°08′N 67°30′W*
54 E4 **Puerto Colombia** Atlántico, N Colombia *10°59′N 74°57′W*
42 H4 **Puerto Cortés** Cortés, NW Honduras *15°50′N 87°55′W*
54 J4 **Puerto Cumarebo** Falcón, N Venezuela *11°29′N 69°21′W*
Puerto de Cabras see Puerto del Rosario
55 Q5 **Puerto de Hierro** Sucre, NE Venezuela *10°40′N 62°03′W*
64 O11 **Puerto de la Cruz** Tenerife, Islas Canarias, Spain, NE Atlantic Ocean *28°24′N 16°33′W*
64 Q11 **Puerto del Rosario** *var.* Puerto de Cabras. Fuerteventura, Islas Canarias, Spain, NE Atlantic Ocean *28°29′N 13°52′W*
63 J20 **Puerto Deseado** Santa Cruz, SE Argentina *47°46′S 65°53′W*
40 F8 **Puerto Escondido** Baja California Sur, NW Mexico *25°48′N 111°20′W*
41 R17 **Puerto Escondido** Oaxaca, SE Mexico *15°48′N 96°57′W*
60 G12 **Puerto Esperanza** Misiones, NE Argentina *26°01′S 54°39′W*
54 H10 **Puerto Gaitán** Meta, C Colombia *04°20′N 72°10′W*
Puerto Gallegos see Río Gallegos
60 G12 **Puerto Iguazú** Misiones, NE Argentina *25°39′S 54°35′W*
56 F12 **Puerto Inca** Huánuco, N Peru *09°22′S 74°54′W*
54 L11 **Puerto Inírida** *var.* Obando. Guainía, E Colombia *03°48′N 67°54′W*
42 K13 **Puerto Jesús** Guanacaste, NW Costa Rica *10°08′N 85°26′W*
41 Z11 **Puerto Juárez** Quintana Roo, SE Mexico *21°06′N 86°46′W*
55 N5 **Puerto La Cruz** Anzoátegui, NE Venezuela *10°14′N 64°40′W*
54 E14 **Puerto Leguízamo** Putumayo, S Colombia *0°14′S 74°45′W*
43 N5 **Puerto Lempira** Gracias a Dios, E Honduras *15°14′N 83°48′W*
Puerto Libertad see La Libertad
54 I11 **Puerto Limón** Meta, E Colombia *04°00′N 71°09′W*
54 D13 **Puerto Limón** Putumayo, SW Colombia *01°02′N 76°30′W*
Puerto Limón see Limón
105 N11 **Puertollano** Castilla-La Mancha, C Spain *38°41′N 04°07′W*
63 K17 **Puerto Lobos** Chubut, SE Argentina *42°00′S 64°58′W*
54 I3 **Puerto López** La Guajira, N Colombia *11°54′N 71°21′W*
105 Q14 **Puerto Lumbreras** Murcia, SE Spain *37°35′N 01°49′W*
41 V17 **Puerto Madero** Chiapas, SE Mexico *14°44′N 92°25′W*
63 K17 **Puerto Madryn** Chubut, S Argentina *42°45′S 65°02′W*
Puerto Magdalena see Bahía Magdalena
57 J15 **Puerto Maldonado** Madre de Dios, E Peru *12°37′S 69°11′W*
Puerto Masachapa see Masachapa
Puerto México see Coatzacoalcos
63 G17 **Puerto Montt** Los Lagos, C Chile *41°28′S 72°57′W*
41 Z12 **Puerto Morelos** Quintana Roo, SE Mexico *20°47′N 86°54′W*
54 L10 **Puerto Nariño** Vichada, E Colombia *04°57′N 67°51′W*
63 H23 **Puerto Natales** Magallanes, S Chile *51°42′S 72°28′W*
43 X15 **Puerto Obaldía** Kuna Yala, NE Panama *08°38′N 77°26′W*
44 H6 **Puerto Padre** Las Tunas, E Cuba *21°13′N 76°35′W*
54 L9 **Puerto Páez** Apure, C Venezuela *06°10′N 67°30′W*
40 E3 **Puerto Peñasco** Sonora, NW Mexico *31°20′N 113°35′W*
55 N5 **Puerto Píritu** Anzoátegui, NE Venezuela *10°04′N 65°00′W*
45 N8 **Puerto Plata** *var.* San Felipe de Puerto Plata. N Dominican Republic *19°46′N 70°42′W*
Puerto Presidente Stroessner see Ciudad del Este
171 N6 **Puerto Princesa** *off.* Puerto Princesa City. Palawan, W Philippines *09°48′N 118°43′E*
Puerto Princesa City see Puerto Princesa
Puerto Príncipe see Camagüey
Puerto Quellón see Quellón
60 F13 **Puerto Rico** Misiones, NE Argentina *26°48′S 54°59′W*
57 K14 **Puerto Rico** Pando, N Bolivia *11°07′S 67°32′W*
54 E12 **Puerto Rico** Caquetá, S Colombia *01°54′N 75°13′W*
45 U5 **Puerto Rico** *off.* Commonwealth of Puerto Rico; *prev.* Porto Rico. ◇ *US commonwealth territory* C West Indies
64 F11 **Puerto Rico** *island* C West Indies
Puerto Rico, Commonwealth of see Puerto Rico
64 G11 **Puerto Rico Trench** *undersea feature* NE Caribbean Sea
54 I8 **Puerto Rondón** Arauca, E Colombia *06°16′N 71°05′W*
Puerto San José see San José
63 J21 **Puerto San Julián** *var.* San Julián. Santa Cruz, SE Argentina *49°14′S 67°41′W*
63 I22 **Puerto Santa Cruz** *var.* Santa Cruz. Santa Cruz, SE Argentina *50°05′S 68°31′W*
Puerto Sauce see Juan L. Lacaze
57 Q20 **Puerto Suárez** Santa Cruz, E Bolivia *18°59′S 57°47′W*
54 D13 **Puerto Umbría** Putumayo, SW Colombia *0°52′N 76°36′W*
40 J13 **Puerto Vallarta** Jalisco, SW Mexico *20°36′N 105°15′W*
63 G16 **Puerto Varas** Los Lagos, C Chile *41°20′S 73°00′W*
42 M13 **Puerto Viejo** Heredia, NE Costa Rica *10°27′N 84°00′W*
Puertoviejo see Portoviejo
57 B18 **Puerto Villamil** *var.* Villamil. Galapagos Islands, Ecuador, E Pacific Ocean *0°57′S 91°00′W*
54 F8 **Puerto Wilches** Santander, N Colombia *07°22′N 73°53′W*
63 H20 **Pueyrredón, Lago** *var.* Lago Cochrane. ◎ S Argentina
127 R7 **Pugachëv** Saratovskaya Oblast', W Russian Federation *52°06′N 48°50′E*
127 T3 **Pugachëvo** Udmurtskaya Respublika, NW Russian Federation *56°38′N 53°03′E*
32 H8 **Puget Sound** *sound* Washington, NW USA
107 O17 **Puglia** *var.* Le Puglie, *Eng.* Apulia. ◐ *region* SE Italy
107 N17 **Puglia, Canosa di** *anc.* Canusium. Puglia, SE Italy *41°13′N 16°04′E*
118 I6 **Puhja** *Ger.* Kawelecht. Tartumaa, SE Estonia *58°20′N 26°19′E*
Puhó see Púchov
105 V4 **Puigcerdà** Cataluña, NE Spain *42°25′N 01°53′E*
Puigmal see Puigmal d'Err
103 N17 **Puigmal d'Err** *var.* Puigmal. ▲ S France *42°24′N 02°07′E*
76 I16 **Pujehun** S Sierra Leone *07°23′N 11°44′W*
Puka see Pukë
185 E20 **Pukaki, Lake** ◎ South Island, New Zealand
38 F10 **Pukalani** Maui, Hawaii, USA, C Pacific Ocean *20°50′N 156°20′W*
190 J13 **Pukapuka** *atoll* N Cook Islands
191 X9 **Pukapuka** *atoll* Îles Tuamotu, E French Polynesia
Pukari Neem see Purekkari Neem
191 X11 **Pukarua** *var.* Pukaruha. *atoll* Îles Tuamotu, E French Polynesia
Pukaruha see Pukarua
14 A7 **Pukaskwa** ≈ Ontario, S Canada
11 V12 **Pukatawagan** Manitoba, C Canada *55°46′N 101°14′W*
191 X16 **Pukatikei, Maunga** ✲ Easter Island, Chile, E Pacific Ocean
182 C1 **Pukatja** *var.* Ernabella. South Australia *26°18′S 132°13′E*
163 Y12 **Pukch'ŏng** E North Korea *40°13′N 128°20′E*
113 L18 **Pukë** *var.* Puka. Shkodër, N Albania *42°03′N 19°53′E*
184 L6 **Pukekohe** Auckland, North Island, New Zealand *37°12′S 174°54′E*
184 L7 **Pukemiro** Waikato, North Island, New Zealand *37°37′S 175°02′E*
190 D12 **Puke, Mont** ▲ Île Futuna, W Wallis and Futuna
Puket see Phuket
185 C20 **Puketeraki Range** ▲▲ South Island, New Zealand
184 N13 **Puketoi Range** ▲▲ North Island, New Zealand
185 F21 **Pukeuri Junction** Otago, South Island, New Zealand *45°01′S 171°01′E*
119 L16 **Pukhavichy** *Rus.* Pukhovichi. Minskaya Voblasts', C Belarus *53°32′N 28°15′E*
Pukhovichi see Pukhavichy
124 M10 **Puksoozero** Arkhangel'skaya Oblast', NW Russian Federation *62°37′N 40°29′E*
112 A10 **Pula** *It.* Pola; *prev.* Pulj. Istra, NW Croatia *44°53′N 13°51′E*
Pula see Nyingchi
163 U14 **Pulandian** *var.* Xinjin. Liaoning, NE China *39°25′N 121°58′E*
163 T14 **Pulandian Wan** *bay* NE China
189 O15 **Pulap Atoll** *atoll* Caroline Islands, C Micronesia
18 H9 **Pulaski** New York, NE USA *43°34′N 76°06′W*
20 I10 **Pulaski** Tennessee, S USA *35°11′N 87°00′W*
21 R7 **Pulaski** Virginia, NE USA *37°03′N 80°47′W*
171 Y14 **Pulau, Sungai** ≈ Papua, E Indonesia
110 N13 **Puławy** *Ger.* Neu Amerika. Lubelskie, E Poland *51°25′N 21°57′E*
149 R5 **Pul-e-'Alam** *prev.* Pol-e-'Alam. Lōgar, E Afghanistan *33°59′N 69°02′E*
149 Q3 **Pul-e Khumrī** *prev.* Pol-e Khomrī. Baghlān, NE Afghanistan *35°55′N 68°45′E*
146 I16 **Pulhatyn** *Rus.* Polekhatum; *prev.* Pul'-I-Khatum. Ahal Welaýaty, S Turkmenistan *36°01′N 61°08′E*
101 E16 **Pulheim** Nordrhein-Westfalen, W Germany *51°00′N 06°48′E*
155 J19 **Pulicat Lake** *lagoon* SE India
Pul'-I-Khatum see Pulhatyn
Pul-i-Sefid see Pol-e Sefid
Pulj see Pula
109 W2 **Pulkau** ≈ NE Austria
93 L15 **Pulkkila** Pohjois-Pohjanmaa, C Finland *64°15′N 25°53′E*
122 C7 **Pulkovo** ✈ (Sankt-Peterburg) Leningradskaya Oblast', NW Russian Federation *60°06′N 30°23′E*
32 M9 **Pullman** Washington, NW USA *46°43′N 117°10′W*
108 B10 **Pully** Vaud, SW Switzerland *46°31′N 06°40′E*
40 F7 **Púlpita, Punta** *headland* NW Mexico *26°30′N 111°28′W*
110 M10 **Pułtusk** Mazowieckie, C Poland *52°41′N 21°04′E*
158 H10 **Pulu** Xinjiang Uygur Zizhiqu, W China *36°10′N 81°29′E*
137 P13 **Pülümür** Tunceli, E Turkey *39°30′N 39°54′E*
189 N16 **Pulusuk** *island* Caroline Islands, C Micronesia
189 N16 **Puluwat Atoll** *atoll* Caroline Islands, C Micronesia
25 N11 **Pumpville** Texas, SW USA *29°55′N 101°43′W*
191 P7 **Punaauia** *var.* Hakapehi. Tahiti, W French Polynesia *17°38′S 149°37′W*
56 B8 **Puná, Isla** *island* SW Ecuador
185 G16 **Punakaiki** West Coast, South Island, New Zealand *42°07′S 171°21′E*
153 T11 **Punakha** C Bhutan *27°38′N 89°50′E*
57 L18 **Punata** Cochabamba, C Bolivia *17°32′S 65°50′W*
155 E14 **Pune** *prev.* Poona. Mahārāshtra, W India *18°32′N 73°52′E*
83 M17 **Pungoè, Rio** *var.* Púnguè, Pungwe. ≈ C Mozambique
21 X10 **Pungo River** ≈ North Carolina, SE USA
Púnguè/Pungwe see Pungoè, Rio
79 N19 **Punia** Maniema, E Dem. Rep. Congo *01°28′S 26°25′E*
62 H8 **Punilla, Sierra de la** ▲▲ W Argentina
161 P14 **Puning** Guangdong, S China *23°24′N 116°14′E*
62 G10 **Punitaqui** Coquimbo, C Chile *30°50′S 71°29′W*
149 T9 **Punjab** *prev.* West Punjab, Western Punjab. ◐ *province* E Pakistan
152 H8 **Punjab** *state* NW India
129 Q9 **Punjab Plains** *plain* N India
93 O17 **Punkaharju** *var.* Punkasalmi. Etelä-Savo, E Finland *61°45′N 29°21′E*
Punkasalmi see Punkaharju
57 I17 **Puno** Puno, SE Peru *15°53′S 70°03′W*
57 H17 **Puno** *off.* Departamento de Puno. ◐ *department* S Peru
Puno, Departamento de see Puno
61 B24 **Punta Alta** Buenos Aires, E Argentina *38°54′S 62°01′W*
63 H24 **Punta Arenas** *prev.* Magallanes. Magallanes, S Chile *53°10′S 70°56′W*
45 T6 **Punta, Cerro de** ▲ C Puerto Rico *18°10′N 66°36′W*
43 T15 **Punta Chame** Panamá, C Panama *08°39′N 79°42′W*
57 G17 **Punta Colorada** Arequipa, SW Peru *16°17′S 72°31′W*
40 F9 **Punta Coyote** Baja California Sur, NW Mexico
62 G8 **Punta de Díaz** Atacama, N Chile *28°03′S 70°36′W*
61 G20 **Punta del Este** Maldonado, S Uruguay *34°59′S 54°58′W*
63 K17 **Punta Delgada** Chubut, SE Argentina *42°46′S 63°40′W*
55 O5 **Punta de Mata** Monagas, NE Venezuela *09°48′N 63°38′W*
55 O4 **Punta de Piedras** Nueva Esparta, NE Venezuela *10°57′N 64°06′W*
42 F4 **Punta Gorda** Toledo, SE Belize *16°07′N 88°47′W*
43 N11 **Punta Gorda** Región Autónoma Atlántico Sur, SE Nicaragua *11°31′N 83°46′W*
23 W14 **Punta Gorda** Florida, SE USA *26°55′N 82°03′W*
42 M11 **Punta Gorda, Río** ≈ SE Nicaragua
62 H6 **Punta Negra, Salar de** *salt lake* N Chile
40 D5 **Punta Prieta** Baja California Norte, NW Mexico *28°56′N 114°11′W*
42 L13 **Puntarenas** Puntarenas, W Costa Rica *09°58′N 84°50′W*
42 L13 **Puntarenas** *off.* Provincia de Puntarenas. ◐ *province* W Costa Rica
Puntarenas, Provincia de see Puntarenas
80 P13 **Puntland** *cultural region* NE Somalia
54 J4 **Punto Fijo** Falcón, N Venezuela *11°42′N 70°13′W*
105 S4 **Puntón de Guara** ▲ N Spain *42°18′N 00°13′W*
18 D14 **Punxsutawney** Pennsylvania, NE USA *40°55′N 78°57′W*
93 M14 **Puolanka** Kainuu, C Finland *64°51′N 27°42′E*
57 J17 **Pupuya, Nevado** ▲ W Bolivia *15°04′S 69°01′W*
Puqi see Chibi
57 F16 **Puquio** Ayacucho, S Peru *14°44′S 74°07′W*
122 J9 **Pur** ≈ N Russian Federation
186 D7 **Purari** ≈ S Papua New Guinea
27 N11 **Purcell** Oklahoma, C USA *35°00′N 97°21′W*
11 O16 **Purcell Mountains** ▲▲ British Columbia, SW Canada
105 P14 **Purchena** Andalucía, S Spain *37°21′N 02°21′W*
27 S8 **Purdy** Missouri, C USA *36°49′N 93°55′W*
118 I2 **Purekkari Neem** *prev.* Pukari Neem. *headland* N Estonia *59°33′N 24°49′E*
37 U7 **Purgatoire River** ≈ Colorado, C USA
Purgstall see Purgstall an der Erlauf
109 V5 **Purgstall an der Erlauf** *var.* Purgstall. Niederösterreich, NE Austria *48°01′N 15°08′E*
154 O13 **Puri** *var.* Jagannath. Odisha, E India *19°52′N 85°49′E*
Puriramya see Buriram
109 X4 **Purkersdorf** Niederösterreich, NE Austria *48°13′N 16°12′E*
98 I9 **Purmerend** Noord-Holland, C Netherlands *52°30′N 04°56′E*
151 G16 **Pūrna** ≈ C India
Purnea see Pūrnia
153 R13 **Pūrnia** *prev.* Purnea. Bihār, NE India *25°47′N 87°28′E*
Pursat see Poŭthĭsăt, Poŭthĭsăt, W Cambodia
Pursat see Poŭthĭsăt, Stœng, W Cambodia
Purulia see Puruliya
150 L13 **Puruliya** *prev.* Purulia. West Bengal, NE India *23°20′N 86°24′E*
47 G7 **Purus, Rio** *var.* Río Purús. ≈ Brazil/Peru
186 C9 **Purutu Island** *island* SW Papua New Guinea
93 N17 **Puruvesi** ◎ SE Finland
22 L7 **Purvis** Mississippi, S USA *31°08′N 89°24′W*
Pŭrvomay see Parvomay
169 R16 **Purwodadi** *prev.* Poerwodadi. Jawa, C Indonesia *07°05′S 110°53′E*
169 P16 **Purwokerto** *prev.* Poerwokerto. Jawa, C Indonesia *07°25′S 109°14′E*
169 P16 **Purworejo** *prev.* Poerworedjo. Jawa, C Indonesia *07°45′S 110°04′E*
20 H8 **Puryear** Tennessee, S USA *36°25′N 88°21′W*
154 H13 **Pusad** Mahārāshtra, C India *19°56′N 77°40′E*
Pusan see Busan
Pusan-gwangyŏksi see Busan
168 H7 **Pusatgajo, Pegunungan** ▲▲ Sumatera, NW Indonesia
Puschlav see Poschiavo
124 G13 **Pushkin** *prev.* Tsarskoye Selo. Leningradskaya Oblast', NW Russian Federation *59°42′N 30°24′E*
126 L3 **Pushkino** Moskovskaya Oblast', W Russian Federation *55°57′N 37°45′E*
127 Q8 **Pushkino** Saratovskaya Oblast', W Russian Federation *51°09′N 47°00′E*
Pushkino see Biläsuvar
111 M22 **Püspökladány** Hajdú-Bihar, E Hungary *47°20′N 21°05′E*
118 J3 **Püssi** *Ger.* Isenhof. Ida-Virumaa, NE Estonia *59°22′N 27°04′E*
116 I5 **Pustomyty** L'vivs'ka Oblast', W Ukraine *49°43′N 23°55′E*
124 F16 **Pustoshka** Pskovskaya Oblast', W Russian Federation *56°21′N 29°16′E*
Pusztakalán see Călan
167 N1 **Putao** *prev.* Fort Hertz. Kachin State, N Myanmar (Burma) *27°22′N 97°24′E*
184 M8 **Putaruru** Waikato, North Island, New Zealand *38°03′S 175°48′E*
Puteoli see Pozzuoli
161 R12 **Putian** Fujian, SE China *25°32′N 119°02′E*
107 O17 **Putignano** Puglia, SE Italy *40°51′N 17°08′E*
Puting see De'an
Putivl' see Putyvl'
41 Q16 **Putla** *var.* Putla de Guerrero. Oaxaca, SE Mexico *17°01′N 97°56′W*
Putla de Guerrero see Putla
19 N12 **Putnam** Connecticut, NE USA *41°56′N 71°52′W*
25 Q7 **Putnam** Texas, SW USA *32°22′N 99°11′W*
18 M10 **Putney** Vermont, NE USA *42°59′N 72°30′W*
111 L20 **Putnok** Borsod-Abaúj-Zemplén, NE Hungary *48°18′N 20°25′E*
Putorana, Gory/Putorana Mountains see Putorana, Plato
122 L8 **Putorana, Plato** *var.* Gory Putorana, *Eng.* Putorana Mountains. ▲▲ N Russian Federation
168 K9 **Putrajaya** ● (Malaysia) Kuala Lumpur, Peninsular Malaysia *02°57′N 101°42′E*
62 H2 **Putre** Arica y Parinacota, N Chile *18°11′S 69°30′W*
155 J24 **Puttalam** North Western Province, W Sri Lanka *08°02′N 79°55′E*
155 J24 **Puttalam Lagoon** *lagoon* W Sri Lanka
99 H17 **Putte** Antwerpen, C Belgium *51°04′N 04°39′E*
94 E10 **Puttegga** ▲ S Norway *62°13′N 07°40′E*
98 K11 **Putten** Gelderland, C Netherlands *52°15′N 05°36′E*
100 K7 **Puttgarden** Schleswig-Holstein, N Germany *54°30′N 11°13′E*
Puttiala see Patiāla
101 D20 **Püttlingen** Saarland, SW Germany *49°16′N 06°52′E*
54 D14 **Putumayo** *off.* Intendencia del Putumayo. ◐ *province* S Colombia
Putumayo, Intendencia del see Putumayo
48 E7 **Putumayo, Río** *var.* Içá, Rio. ≈ NW South America *see also* Içá, Rio
Putumayo, Río see Içá, Rio
169 P11 **Putus, Tanjung** *headland* Borneo, N Indonesia *0°27′S 109°04′E*
116 J8 **Putyla** Chernivets'ka Oblast', W Ukraine *47°59′N 25°04′E*
117 S3 **Putyvl'** *Rus.* Putivl'. Sums'ka Oblast', NE Ukraine *51°21′N 33°53′E*
93 M18 **Puula** ◎ SE Finland
93 N18 **Puumala** Etelä-Savo, E Finland *61°31′N 28°12′E*
118 I5 **Puurmani** *Ger.* Talkhof. Jõgevamaa, E Estonia *58°36′N 26°17′E*
99 G17 **Puurs** Antwerpen, N Belgium *51°05′N 04°17′E*
38 F10 **Pu'u 'Ula'ula** *var.* Red Hill. ▲ Maui, Hawai'i, USA *20°42′N 156°16′W*
38 A8 **Pu'uwai** *var.* Puuwai. Ni'ihau, Hawaii, USA, C Pacific Ocean *21°54′N 160°11′W*
12 J4 **Puvirnituq** *prev.* Povungnituk. Québec, NE Canada *60°10′N 77°20′W*
12 J3 **Puvirnituq, Rivière de** *prev.* Rivière de Povungnituk. ≈ Québec, NE Canada
32 H8 **Puyallup** Washington, NW USA *47°11′N 122°17′W*
161 O5 **Puyang** Henan, C China *35°40′N 115°00′E*
103 O11 **Puy-de-Dôme** ◐ *department* C France
103 N15 **Puylaurens** Tarn, S France *43°33′N 02°01′E*
102 M13 **Puy-l'Evêque** Lot, S France *44°31′N 01°10′E*
103 N17 **Puymorens, Col de** *pass* S France
56 C7 **Puyo** Pastaza, C Ecuador *01°30′S 77°58′W*
185 A24 **Puysegur Point** *headland* South Island, New Zealand *46°09′S 166°38′E*
148 J8 **Pūzak, Hāmūn-e** *Pash.* Hāmūn-i-Puzak. ◎ SW Afghanistan
Puzak, Hāmūn-i- see Pūzak, Hāmūn-e
81 J23 **Pwani** *Eng.* Coast. ◐ *region* E Tanzania
79 O23 **Pweto** Katanga, SE Dem. Rep. Congo *08°28′S 28°52′E*
97 I19 **Pwllheli** NW Wales, United Kingdom *52°54′N 04°23′W*
189 O14 **Pwok** Pohnpei, E Micronesia
122 I9 **Pyakupur** ≈ N Russian Federation
124 M6 **Pyalitsa** Murmanskaya Oblast', NW Russian Federation *66°17′N 39°56′E*
124 K10 **Pyal'ma** Respublika Kareliya, NW Russian Federation *62°24′N 35°56′E*
Pyandzh see Panj
166 L9 **Pyapon** Ayeyawady, SW Myanmar (Burma) *16°15′N 95°40′E*
119 J15 **Pyarshai** *Rus.* Pershay. Minskaya Voblasts', C Belarus *54°02′N 26°41′E*
114 I10 **Pyasachnik, Yazovir** *var.* Yazovir Pyasŭchnik. ⊠ C Bulgaria
122 K8 **Pyasina** ≈ N Russian Federation
119 G17 **Pyaski** *Rus.* Peski; *prev.* Pyeski. Hrodzyenskaya Voblasts', W Belarus *53°21′N 24°38′E*
Pyasŭchnik, Yazovir see Pyasachnik, Yazovir
Pyatikhatki see P"yatykhatky
117 S7 **P"yatykhatky** *Rus.* Pyatikhatki. Dnipropetrovs'ka Oblast', E Ukraine *48°23′N 33°43′E*
166 M6 **Pyawbwe** Mandalay, C Myanmar (Burma) *20°39′N 96°04′E*
166 L7 **Pyay** *var.* Prome, Pye. Bago, C Myanmar (Burma) *18°50′N 95°14′E*
127 T3 **Pychas** Udmurtskaya Respublika, NW Russian Federation *56°30′N 52°33′E*
Pye see Pyay
166 K6 **Pyechin** Chin State, W Myanmar (Burma) *20°01′N 93°36′E*
163 X15 **Pyeongtaek** *prev.* P'yŏngt'aek. NW South Korea *37°00′N 127°04′E*
Pyeski see Pyaski
119 L19 **Pyetrykaw** *Rus.* Petrikov. Homyel'skaya Voblasts', SE Belarus *52°08′N 28°30′E*
93 O17 **Pyhäjärvi** ◎ SE Finland
93 M16 **Pyhäjärvi** ◎ C Finland
93 L15 **Pyhäjoki** Pohjois-Pohjanmaa, W Finland *64°28′N 24°15′E*
93 L15 **Pyhäjoki** ≈ W Finland
93 M15 **Pyhäntä** Pohjois-Pohjanmaa, C Finland *64°07′N 26°19′E*
93 M16 **Pyhäsalmi** Pohjois-Pohjanmaa, C Finland *63°38′N 26°E*
93 O17 **Pyhäselkä** ◎ SE Finland
93 M19 **Pyhtää** *Swe.* Pyttis. Kymenlaakso, S Finland *60°29′N 26°40′E*
166 M5 **Pyin-Oo-Lwin** *var.* Maymyo. Mandalay, C Myanmar (Burma) *22°03′N 96°30′E*
115 N24 **Pylés** *var.* Piles. Kárpathos, SE Greece *35°31′N 27°08′E*
115 D21 **Pýlos** *var.* Pilos. Pelopónnisos, S Greece *36°55′N 21°42′E*
18 B12 **Pymatuning Reservoir** ⊠ Ohio/Pennsylvania, NE USA
P'yŏngt'aek see Pyeongtaek
163 V14 **P'yŏngyang** *var.* P'yŏngyang-si, *Eng.* Pyongyang. ● (North Korea) SW North Korea *39°04′N 125°46′E*
P'yŏngyang-si see P'yŏngyang
35 Q4 **Pyramid Lake** ◎ Nevada, W USA
37 P15 **Pyramid Mountains** ▲▲ New Mexico, SW USA
37 R5 **Pyramid Peak** ▲ Colorado, C USA *39°04′N 106°57′W*
115 D17 **Pyramíva** *var.* Piramiva. ▲ C Greece *39°08′N 21°18′E*
Pyrenaei Montes see Pyrenees
86 B12 **Pyrenees** *Fr.* Pyrénées, *Sp.* Pirineos; *anc.* Pyrenaei Montes. ▲▲ SW Europe
102 J16 **Pyrénées-Atlantiques** ◐ *department* SW France
103 N17 **Pyrénées-Orientales** ◐ *department* S France
115 L19 **Pyrgí** *var.* Pirgi. Chíos, E Greece *38°13′N 26°01′E*
115 D20 **Pýrgos** *var.* Pírgos. Dytikí Elláda, S Greece *37°40′N 21°27′E*
Pyritz see Pyrzyce
115 E19 **Pýrros** ≈ S Greece
117 R4 **Pyryatyn** *Rus.* Piryatin. Poltavs'ka Oblast', NE Ukraine *50°14′N 32°31′E*
110 D9 **Pyrzyce** *Ger.* Pyritz. Zachodnio-pomorskie, NW Poland *53°09′N 14°53′E*
124 F15 **Pytalovo** *Latv.* Abrene; *prev.* Jaunlatgale. Pskovskaya Oblast', W Russian Federation *57°06′N 27°55′E*
115 M20 **Pythagóreio** *var.* Pithagorio. Sámos, Dodekánisa, Greece, Aegean Sea *37°42′N 26°57′E*
14 L11 **Pythonga, Lac** ◎ Québec, SE Canada
Pyttis see Pyhtää
Pyu see Phyu
166 M8 **Pyuntaza** Bago, SW Myanmar (Burma) *17°51′N 96°44′E*
153 N11 **Pyuthān** Mid Western, W Nepal *28°09′N 82°50′E*
110 H12 **Pyzdry** *Ger.* Peisern. Wielkopolskie, C Poland *52°10′N 17°42′E*

Q

138 H13 **Qā' al Jafr** ◎ S Jordan
197 O11 **Qaanaaq** *var.* Qânâq, *Dan.* Thule. ◐ Qaasuitsup, N Greenland
197 P12 **Qaasuitsup** *off.* Qaasuitsup Kommunia. ◐ *municipality* W Greenland
Qaasuitsup Kommunia see Qaasuitsup
Qabanbay see Kabanbay
138 G7 **Qabb Eliâs** E Lebanon *33°46′N 35°49′E*
Qabil see Al Qābil
Qabirri see Iori
Qābis see Gabès
Qābis, Khalīj see Gabès, Golfe de
Qabqa see Gonghe
141 S14 **Qabr Hūd** C Yemen *16°02′N 49°36′E*
Qacentina see Constantine
Qādes see Qādis
148 L4 **Qādis** *prev.* Qādes. Bādghīs, NW Afghanistan *34°48′N 63°26′E*
139 T11 **Qādisīyah** Al Qādisīyah, S Iraq *31°43′N 44°28′E*
Qādisīyah, Muḥāfa at al see Al Qādisīyah
143 O4 **Qā'emshahr** *prev.* 'Alīābad, Shāhī. Māzandarān, N Iran *36°31′N 52°49′E*
143 U7 **Qā'en** *var.* Qain, Qāyen. Khorāsān-e Jonūbī, E Iran *33°43′N 59°07′E*
141 U13 **Qafa** *spring/well* SW Oman *17°46′N 52°55′E*
Qafṣah see Gafsa
163 Q12 **Qagan Nur** *var.* Xulun Hobot Qagan, Zhengxiangbai Qi. Nei Mongol Zizhiqu, N China *42°10′N 114°57′E*
163 V9 **Qagan Nur** ◎ NE China
163 Q11 **Qagan Nur** ◎ N China
Qagan Us see Dulan
158 H13 **Qagcaka** Xizang Zizhiqu, W China *32°32′N 81°52′E*
Qagchêng see Xiangcheng
Qahremānshahr see Kermānshāh
159 Q10 **Qaidam He** ≈ C China
156 L8 **Qaidam Pendi** *basin* C China
Qain see Qā'en
Qala Āhangarān see Chaghcharān
Qalā Diza see Qeladize
Qal'at Sālih see Qal'at Şāliḩ
149 N4 **Qal'ah Shahr** *Pash.* Qala Shāhar; *prev.* Qal'eh Shahr. Sar-e Pul, N Afghanistan *35°34′N 65°38′E*
148 L4 **Qal'ah-ye Now** *var.* Qala Nau; *prev.* Qal'eh-ye Now. Bādghīs, NW Afghanistan *35°N 63°08′E*
147 R13 **Qal'aikhum** *Rus.* Kalaikhum. S Tajikistan *38°28′N 70°49′E*
Qala Nau see Qal'ah-ye Now
141 V17 **Qalansīyah** Suquṭrā, W Yemen *12°40′N 53°30′E*
Qala Panja see Qal'eh-ye Panjeh
Qala Shāhar see Qal'ah Shahr
149 O8 **Qalāt** *Per.* Kalāt. Zābul, S Afghanistan *32°10′N 66°54′E*
139 W9 **Qal'at Aḥmad** Maysān, E Iraq *32°24′N 46°46′E*

◆ Country
● Country Capital
◇ Dependent Territory
○ Dependent Territory Capital
◐ Administrative Regions
✈ International Airport
▲ Mountain
▲▲ Mountain Range
✲ Volcano
≈ River
◎ Lake
⊠ Reservoir

141 N11 **Qal'at Bīshah** 'Asīr, SW Saudi Arabia *19°59′N 42°38′E*
138 H4 **Qal'at Burzay** Ḩamāh, W Syria *35°37′N 36°16′E*
Qal 'at Dīzah *see* Qeładize
139 W9 **Qal'at Huşayn** Maysān, E Iraq *32°19′N 46°46′E*
139 V10 **Qal'at Majnūnah** Al Qādisīyah, S Iraq *31°39′N 45°44′E*
139 X11 **Qal'at Şāliḩ** *var.* Qal'ah Sālih. Maysān, E Iraq *31°30′N 47°24′E*
139 V10 **Qal'at Sukkar** Dhī Qār, SE Iraq *31°52′N 46°05′E*
Qalba Zhotasy *see* Khrebet Kalba
143 Q12 **Qal'eh Bīābān** Fārs, S Iran
Qal'eh Shahr *see* Qal'ah Shahr
Qal'eh-ye Now *see* Qal'ah-ye Now
149 T2 **Qal'eh-ye Panjeh** *var.* Qala Panja. Badakhshān, NE Afghanistan *36°56′N 72°15′E*
Qalqaman *see* Kalkaman
Qamanittuaq *see* Baker Lake
Qamar Bay *see* Qamar, Ghubbat al
141 U14 **Qamar, Ghubbat al** *Eng.* Qamar Bay. *bay* Oman/Yemen
141 V13 **Qamar, Jabal al** ⛰ SW Oman
147 N12 **Qamashi** Qashqadaryo Viloyati, S Uzbekistan *38°52′N 66°30′E*
Qambar *see* Kambar
159 R14 **Qamdo** Xizang Zizhiqu, W China *31°09′N 97°09′E*
75 R7 **Qamīnis** NE Libya *31°48′N 20°04′E*
Qamishly *see* Al Qāmishlī
Qânâq *see* Qaanaaq
Qandahār *see* Kandahār
80 Q11 **Qandala** Bari, NE Somalia *11°30′N 50°00′E*
Qandyaghash *see* Kandyagash
138 L2 **Qanṭārī** Ar Raqqah, N Syria *36°24′N 39°16′E*
Qapiciğ Dağı *see* Qazangödağ
158 H5 **Qapqal** *var.* Qapqal Xibe Zizhixian. Xinjiang Uygur Zizhiqu, NW China *43°46′N 81°09′E*
Qapqal Xibe Zizhixian *see* Qapqal
Qapshagay Böyeni *see* Vodokhranilishche Kapshagay
Qapugtang *see* Zadoi
196 M15 **Qaqortoq** *Dan.* Julianehåb. ◐ Kujalleo, S Greenland
139 T4 **Qara Anjīr** Kirkūk, N Iraq *35°30′N 44°37′E*
Qarabāgh *see* Qarah Bāgh
Qarabaū *see* Karabau
Qaraböget *see* Karaboget
Qarabulaq *see* Karabulak
Qarabutaq *see* Karabutak
Qaraghandy/Qaraghandy Oblysy *see* Karagandy
Qaraghayly *see* Karagayly
Qara Gol *see* Qere Gol
75 U8 **Qārah** *var.* Qâra. NW Egypt *29°34′N 26°28′E*
Qārah *see* Qārah
148 J4 **Qarah Bāgh** *var.* Qarabāgh. Herāt, NW Afghanistan *35°06′N 61°33′E*
Qarah Gawl *see* Qere Gol
138 G7 **Qaraoun, Lac de** *var.* Buḩayrat al Qir'awn. ◎ S Lebanon
Qaraoy *see* Karaoy
Qaraqoyyn *see* Karakoyyn, Ozero
Qara Qum *see* Garagum
Qarasū *see* Karasu
Qaratal *see* Karatal
Qarataū *see* Karatau, Khrebet, Kazakhstan
Qarataū *see* Karatau, Zhambyl, Kazakhstan
Qaraton *see* Karaton
Qarazhal *see* Karazhal
80 P13 **Qardho** *var.* Kardh, *It.* Gardo. Bari, N Somalia *09°34′N 49°30′E*
142 M6 **Qareh Chāy** ≈ N Iran
142 K2 **Qareh Sū** ≈ NW Iran
Qariateïne *see* Al Qaryatayn
Qarkilik *see* Ruoqiang
147 O13 **Qarluq** *Rus.* Karluk. Surkhondaryo Viloyati, S Uzbekistan *38°17′N 67°39′E*
147 U12 **Qarokŭl** *Rus.* Karakul'. E Tajikistan *39°07′N 73°33′E*
147 T12 **Qarokŭl** *Rus.* Ozero Karakul'. ◎ E Tajikistan
Qarqan *see* Qiemo
158 K9 **Qarqan He** ≈ NW China
Qarqannah, Juzur *see* Kerkenah, Îles de
149 O1 **Qarqīn** Jowzjān, N Afghanistan *37°25′N 66°03′E*
Qars *see* Kars
Qarsaqbay *see* Karsakpay
146 M12 **Qarshi** *Rus.* Karshi; *prev.* Bek-Budi. Qashqadaryo Viloyati, S Uzbekistan *38°54′N 65°48′E*
146 L12 **Qarshi Cho'li** *Rus.* Karshinskaya Step. *grassland* S Uzbekistan
146 M13 **Qarshi Kanali** *Rus.* Karshinskiy Kanal. *canal* Turkmenistan/Uzbekistan
Qaryatayn *see* Al Qaryatayn
Qāsh, Nahr al *see* Gash
146 M12 **Qashqadaryo Viloyati** *Rus.* Kashkadar'inskaya Oblast'. ◐ *province* S Uzbekistan
Qasigianguit *see* Qasigiannguit
197 N13 **Qasigiannguit** *var.* Qasigianguit, *Dan.* Christianshåb. ◐ Qaasuitsup, C Greenland
Qaşīm, Minṭaqat *see* Al Qaşīm
75 V10 **Qaşr al Farāfirah** *var.* Qasr Farâfra. W Egypt *27°00′N 27°59′E*
139 P8 **Qaşr 'Amīj** Al Anbār, C Iraq *33°30′N 41°52′E*
139 R9 **Qaşr Darwīshāh** Karbalā', C Iraq *32°36′N 43°27′E*
142 J6 **Qaşr-e Shīrīn** Kermānshāhān, W Iran *34°32′N 45°36′E*
Qasr Farâfra *see* Qaşr al Farāfirah
Qassim *see* Al Qaşīm
141 O16 **Qa'ṭabah** SW Yemen *13°51′N 44°42′E*
138 H7 **Qaṭanā** *var.* Katana. Rīf Dimashq, S Syria *33°27′N 36°04′E*
143 N15 **Qatar** *off.* State of Qatar, *Ar.* Dawlat Qaṭar. ◆ *monarchy* SW Asia
Qatar, State of *see* Qatar
Qatrana *see* Al Qaṭrānah
143 Q12 **Qaṭrūyeh** Fārs, S Iran *29°08′N 54°42′E*
Qattara Depression/Qaṭṭārah, Munkhafaḍ al *see* Qaṭṭārah, Munkhafaḍ al
75 U8 **Qaṭṭārah, Munkhafaḍ al** *var.* Munkhafaḍ al Qaṭṭārah *var.* Monkhafad el Qaṭṭâra, *Eng.* Qattara Depression. *desert* NW Egypt
Qaṭṭâra, Monkhafad el *see* Qaṭṭārah, Munkhafaḍ al
Qattinah, Buhayrat *see* Ḩimş, Buḩayrat
Qausuittuq *see* Resolute
Qaydār *see* Qeydār
Qāyen *see* Qā'en
Qaynar *see* Kaynar
147 Q11 **Qayroqqum** *Rus.* Kayrakkum. NW Tajikistan *40°16′N 69°46′E*
147 Q10 **Qayroqqum, Obanbori** *Rus.* Kayrakkumskoye Vodokhranilishche. ▣ NW Tajikistan
137 V13 **Qazangödağ** *Rus.* Gora Kapydzhik, *Turk.* Qapiciğ Dağı. ▲ SW Azerbaijan *39°18′N 46°00′E*
139 U7 **Qazānīyah** *var.* Dhū Shaykh. Diyālá, E Iraq *33°39′N 45°33′E*
Qazaqstan/Qazaqstan Respublikasy *see* Kazakhstan
137 T9 **Q'azbegi** *Rus.* Kazbegi; *prev.* Qazbegi. NE Georgia *42°39′N 44°36′E*
Qazbegi *see* Q'azbegi
149 P15 **Qāzi Ahmad** *var.* Kazi Ahmad. Sind, SE Pakistan *26°19′N 68°08′E*
Qazimämmäd *see* Hacıqabal
Qazris *see* Cáceres
142 M4 **Qazvīn** *var.* Kazvin. Qazvīn, N Iran *36°16′N 50°E*
142 M5 **Qazvīn** *off.* Ostān-e Qazvīn. ◐ *province* N Iran
Qazvīn, Ostān-e *see* Qazvīn
139 U3 **Qeładize** *Ar.* Qal 'at Dīzah, *var.* Qaļā Diza. As Sulaymānīyah, NE Iraq *36°11′N 45°07′E*
187 Z13 **Qelelevu Lagoon** *lagoon* NE Fiji
Qena *see* Qinā
113 L23 **Qeparo** Vlorë, S Albania *40°04′N 19°49′E*
Qeqertarssuaq *see* Qeqertarsuaq
197 N13 **Qeqertarsuaq** *var.* Qeqertarssuaq, *Dan.* Godhavn. ◐ Qaasuitsup, S Greenland
196 M13 **Qeqertarsuaq** *island* W Greenland
197 N13 **Qeqertarsuup Tunua** *Dan.* Disko Bugt. *inlet* W Greenland
197 N14 **Qeqqata** *off.* Qeqqata Kommunia. ◐ *municipality* W Greenland
Qeqqata Kommunia *see* Qeqqata
139 U4 **Qere Gol** *Ar.* Qarah Gawl, *var.* Qara Gol. As Sulaymānīyah, NE Iraq *35°21′N 45°38′E*
Qerveh *see* Qorveh
143 S14 **Qeshm** Hormozgān, S Iran *26°58′N 56°17′E*
143 R14 **Qeshm** *var.* Jazīreh-ye Qeshm, Qeshm Island. *island* S Iran
Qeshm Island/Qeshm, Jazīreh-ye *see* Qeshm
Qey *see* Kīsh, Jazīreh-ye
142 L4 **Qeydār** *var.* Qaydār. Zanjān, NW Iran *36°50′N 47°40′E*
142 K5 **Qezel Owzan, Rūd-e** *var.* Ki Zil Uzen, Qi Zil Uzun. ≈ NW Iran
Qian *see* Guizhou
161 Q2 **Qian'an** Heilongjiang, E China *45°00′N 124°00′E*
161 R10 **Qiandao Hu** *prev.* Xin'anjiang Shuiku. ▣ SE China
Qiandaohu *see* Chun'an
Qian Gorlo/Qian Gorlos/Qian Gorlos Mongolzu Zizhixian/Quianguozhen *see* Qianguo
163 V9 **Qianguo** *var.* Qian Gorlo, Qian Gorlos, Qian Gorlos Mongolzu Zizhixian, Quianguozhen. Jilin, NE China *45°08′N 124°48′E*
161 N9 **Qianjiang** Hubei, C China *30°23′N 112°58′E*
160 K10 **Qianjiang** Sichuan, C China *29°30′N 108°45′E*
160 L14 **Qian Jiang** ≈ S China
160 G9 **Qianning** *var.* Gartar. Sichuan, C China *30°27′N 101°24′E*
163 U13 **Qian Shan** ⛰ NE China
160 H10 **Qianwei** *var.* Yujin. Sichuan, C China *29°15′N 103°52′E*
160 J11 **Qianxi** Guizhou, S China *27°00′N 106°01′E*
Qiaotou *see* Datong
Qiaowa *see* Muli
159 Q7 **Qiaowan** Gansu, N China *40°37′N 96°40′E*
Qibilī *see* Kebili
158 K9 **Qiemo** *var.* Qarqan. Xinjiang Uygur Zizhiqu, NW China *38°09′N 85°30′E*
160 J10 **Qijiang** *var.* Gunan. Chongqing Shi, C China *29°01′N 106°40′E*
159 N5 **Qijiaojing** Xinjiang Uygur Zizhiqu, NW China *43°29′N 91°35′E*
Qike *see* Xunke
9 N5 **Qikiqtaaluk** ◐ *cultural region* Nunavut, N Canada
9 R5 **Qikiqtarjuaq** *prev.* Broughton Island. Nunavut, NE Canada *67°35′N 63°55′W*
149 P9 **Qila Saifullāh** Baluchistān, SW Pakistan *30°45′N 68°08′E*
159 S9 **Qilian** *var.* Babao. Qinghai, C China *38°09′N 100°08′E*
159 N8 **Qilian Shan** *var.* Kilien Mountains. ⛰ N China
197 O11 **Qimusseriarsuaq** *Dan.* Melville Bugt, *Eng.* Melville Bay. *bay* NW Greenland
75 X10 **Qinā** *var.* Qena; *anc.* Caene, Caenepolis. E Egypt *26°12′N 32°49′E*
159 W11 **Qin'an** Gansu, C China *34°49′N 105°50′E*
Qincheng *see* Nanfeng
Qing *see* Qinghai
163 W7 **Qing'an** Heilongjiang, NE China *46°53′N 127°29′E*
159 X10 **Qingcheng** *var.* Xifeng. Gansu, C China *35°46′N 107°35′E*
161 R5 **Qingdao** *var.* Ching-Tao, Ch'ing-tao, Tsingtao, Tsintao, *Ger.* Tsingtau. Shandong, E China *36°31′N 120°55′E*
163 V8 **Qinggang** Heilongjiang, NE China *46°41′N 126°05′E*
Qinggil *see* Qinghe
159 P11 **Qinghai** *var.* Chinghai, Koko Nor, Qing, Qinghai Sheng, Tsinghai. ◐ *province* C China
159 S10 **Qinghai Hu** *var.* Ch'ing Hai, Tsing Hai, *Mong.* Koko Nor. ◎ C China
Qinghai Sheng *see* Qinghai
158 M3 **Qinghe** *var.* Qinggil. Xinjiang Uygur Zizhiqu, NW China *46°42′N 90°19′E*
160 L4 **Qingjian** *var.* Kuanzhou; *prev.* Xiuyan. Shaanxi, C China *37°10′N 110°09′E*
160 L9 **Qing Jiang** ≈ C China
Qingjiang *see* Huai'an
Qingkou *see* Ganyu
160 I12 **Qinglong** *var.* Liancheng. Guizhou, S China *25°49′N 105°10′E*
161 Q2 **Qinglong** Hebei, E China *40°24′N 118°57′E*
Qingshan *see* Wudalianchi
159 R12 **Qingshuihe** Qinghai, C China *33°47′N 97°10′E*
Qingyang *see* Jinjiang
161 N14 **Qingyuan** Guangdong, S China *23°42′N 113°02′E*
163 V11 **Qingyuan** *var.* Qingyuan Manzu Zizhixian. Liaoning, NE China *42°08′N 124°55′E*
Qingyuan *see* Shandan
Qingyuan *see* Weiyuan
Qingyuan Manzu Zizhixian *see* Qingyuan
158 L13 **Qingzang Gaoyuan** *var.* Xizang Gaoyuan, *Eng.* Plateau of Tibet. *plateau* W China
161 Q4 **Qingzhou** *prev.* Yidu. Shandong, E China *36°41′N 118°29′E*
157 R9 **Qin He** ≈ C China
161 Q2 **Qinhuangdao** Hebei, E China *39°57′N 119°31′E*
160 K7 **Qin Ling** ⛰ C China
161 N5 **Qinxian** *var.* Dingchang, Qin Xian. Shanxi, C China *36°46′N 112°42′E*
Qin Xian *see* Qinxian
161 N6 **Qinyang** Henan, C China *35°05′N 112°56′E*
160 K15 **Qinzhou** Guangxi Zhuangzu Zizhiqu, S China *22°09′N 108°36′E*
Qiong *see* Hainan
160 L17 **Qionghai** *prev.* Jiaji. Hainan, S China *19°12′N 110°26′E*
160 H9 **Qionglai** Sichuan, C China *30°24′N 103°28′E*
160 H8 **Qionglai Shan** ⛰ C China
Qiongxi *see* Hongyuan
160 L17 **Qiongzhou Haixia** *var.* Hainan Strait. *strait* S China
163 U7 **Qiqihar** *var.* Ch'i-ch'i-ha-erh, Tsitsihar; *prev.* Lungkiang. Heilongjiang, NE China *47°23′N 124°E*
Qir *see* Qīr-va-Kārzīn
158 H10 **Qira** Xinjiang Uygur Zizhiqu, NW China *37°05′N 80°45′E*
Qir'awn, Buḩayrat al *see* Qaraoun, Lac de
143 P12 **Qīr-va-Kārzīn** *var.* Qir. Fārs, S Iran *28°27′N 53°04′E*
Qiryat Gat *see* Kyriat Gat
Qiryat Shemona *see* Kiryat Shmona
Qishlaq *see* Garmsār
141 U14 **Qishn** SE Yemen *15°29′N 51°44′E*
Qishon, Naḥal *see* Kishon, Nahal
Qita Ghazzah *see* Gaza Strip
156 K5 **Qitai** Xinjiang Uygur Zizhiqu, NW China *44°N 89°34′E*
163 Y8 **Qitaihe** Heilongjiang, NE China *45°45′N 130°53′E*
141 W12 **Qitbīt, Wādī** *dry watercourse* S Oman
161 O5 **Qixian** *var.* Qi Xian, Zhaoge. Henan, C China *35°35′N 114°10′E*
Qi Xian *see* Qixian
Qīzān *see* Jīzān
Qizil Orda *see* Kyzylorda
Qizil Qum/Qizilqum *see* Kyzyl Kum
147 V14 **Qizilrabot** *Rus.* Kyzylrabot. SE Tajikistan *37°29′N 74°44′E*
146 J10 **Qizilravote** *Rus.* Kyzylrabat. Buxoro Viloyati, C Uzbekistan *40°35′N 62°09′E*
Qi Zil Uzun *see* Qezel Owzan, Rūd-e
139 S4 **Qizil Yār** Kirkūk, N Iraq *35°26′N 44°12′E*
164 J12 **Qkutango-hantō** *peninsula* Honshū, SW Japan
137 Y11 **Qobustan** *prev.* Märäzä. E Azerbaijan *40°32′N 48°56′E*
Qoghaly *see* Kogaly
Qogir Feng *see* K2
143 N6 **Qom** *var.* Kum, Qum. Qom, N Iran *34°43′N 50°54′E*
143 N6 **Qom** *off.* Osān-e Qom. ◐ *province* N Iran
Qomisheh *see* Shahreẕā
Qomolangma Feng *see* Everest, Mount
Qom, Osān-e *see* Qom
142 M7 **Qom, Rūd-e** ≈ C Iran
Qomsheh *see* Shahreẕā
Qomul *see* Hami
Qondūz *see* Kunduz
146 G7 **Qo'ng'irot** *Rus.* Kungrad. Qoraqalpog'iston Respublikasi, NW Uzbekistan *43°01′N 58°49′E*
Qongyrat *see* Konyrat
Qoqek *see* Tacheng
147 R10 **Qo'qon** *var.* Khokand, *Rus.* Kokand. Farg'ona Viloyati, E Uzbekistan *40°34′N 70°55′E*
Qorabowur Kirlari *see* Karabaur', Uval
Qoradaryo *see* Karadar'ya
146 G6 **Qorajar** *Rus.* Karadzhar. Qoraqalpog'iston Respublikasi, NW Uzbekistan *43°34′N 58°35′E*
146 K12 **Qorako'l** *Rus.* Karakul'. Buxoro Viloyati, C Uzbekistan *39°27′N 63°45′E*
146 H7 **Qorao'zak** *Rus.* Karauzyak. Qoraqalpog'iston Respublikasi, NW Uzbekistan *43°07′N 60°03′E*
146 E5 **Qoraqalpog'iston** *Rus.* Karakalpakya. Qoraqalpog'iston Respublikasi, NW Uzbekistan *44°45′N 56°06′E*
146 G7 **Qoraqalpog'iston Respublikasi** *Rus.* Respublika Karakalpakstan. ◐ *autonomous republic* NW Uzbekistan
Qorghalzhyn *see* Korgalzhyn
138 H6 **Qornet es Saouda** ▲ NE Lebanon *36°06′N 34°06′E*
146 L12 **Qorowulbozor** *Rus.* Karaulbazar. Buxoro Viloyati, C Uzbekistan *39°28′N 64°49′E*
142 K5 **Qorveh** *var.* Qerveh, Qurveh. Kordestān, W Iran *35°09′N 47°48′E*
147 N11 **Qo'shrabot** *Rus.* Kushrabat. Samarqand Viloyati, C Uzbekistan *40°15′N 66°40′E*
Qosköl *see* Koskol'
Qosshaghyl *see* Kosshagyl
Qostanay/Qostanay Oblysy *see* Kostanay
143 P12 **Qoṭbābād** Fārs, S Iran *28°52′N 53°40′E*
143 R13 **Qoṭbābād** Hormozgān, S Iran *27°49′N 56°00′E*
138 H6 **Qoubaïyât** *var.* Al Qubayyāt. N Lebanon *37°00′N 34°30′E*
Qoussantîna *see* Constantine
Qowowuyag *see* Cho Oyu
147 O11 **Qo'ytosh** *Rus.* Koytash. Jizzax Viloyati, C Uzbekistan *40°13′N 67°19′E*
146 G7 **Qozonketkan** *Rus.* Kazanketken. Qoraqalpog'iston Respublikasi, W Uzbekistan *42°59′N 59°21′E*
146 H6 **Qozoqdaryo** *Rus.* Kazakdar'ya. Qoraqalpog'iston Respublikasi, NW Uzbekistan *43°26′N 59°47′E*
19 N11 **Quabbin Reservoir** ▣ Massachusetts, NE USA
100 F12 **Quakenbrück** Niedersachsen, NW Germany *52°41′N 07°57′E*
18 I15 **Quakertown** Pennsylvania, NE USA *40°26′N 75°17′W*
182 M10 **Quambatook** Victoria, SE Australia *35°52′S 143°28′E*
25 Q4 **Quanah** Texas, SW USA *34°17′N 99°46′W*
167 V10 **Quang Ngai** *var.* Quangngai, Quang Nghia. Quang Ngai, C Vietnam *15°09′N 108°50′E*
Quangngai *see* Quang Ngai
Quang Nghia *see* Quang Ngai
167 T9 **Quang Tri** *var.* Triêu Hai. Quang Tri, C Vietnam *16°46′N 107°11′E*
Quanjiang *see* Suichuan
Quan Long *see* Ca Mau
152 L4 **Quanshuigou** China/India *35°40′N 79°28′E*
161 R13 **Quanzhou** *var.* Ch'uan-chou, Tsinkiang; *prev.* Chin-chiang. Fujian, SE China *24°56′N 118°31′E*
160 M12 **Quanzhou** Guangxi Zhuangzu Zizhiqu, S China *25°59′N 111°02′E*
11 V16 **Qu'Appelle** ≈ Saskatchewan, S Canada
12 M3 **Quaqtaq** *prev.* Koartac. Québec, NE Canada *60°50′N 69°30′W*
61 E16 **Quaraí** Rio Grande do Sul, S Brazil *30°38′S 56°25′W*
59 H24 **Quaraí, Rio** *Sp.* Río Cuareim. ≈ Brazil/Uruguay *see also* Cuareim, Río
Quaraí, Rio *see* Cuareim, Río
171 N13 **Quarles, Pegunungan** ⛰ Sulawesi, C Indonesia
Quarnero *see* Kvarner
107 C20 **Quartu Sant' Elena** Sardegna, Italy, C Mediterranean Sea *39°15′N 09°12′E*
29 X13 **Quasqueton** Iowa, C USA *42°23′N 91°45′W*
173 X16 **Quatre Bornes** W Mauritius *20°15′S 57°28′E*
172 I17 **Quatre Bornes** Mahé, NE Seychelles
137 X10 **Quba** *Rus.* Kuba. N Azerbaijan *41°22′N 48°30′E*
Qubba *see* Ba'qūbah
143 T3 **Qūchān** *var.* Kuchan. Khorāsān-e Razavī, NE Iran *37°12′N 58°28′E*
183 R10 **Queanbeyan** New South Wales, SE Australia *35°24′S 149°17′E*
15 Q10 **Québec** *var.* Quebec. *province capital* Québec, SE Canada *46°50′N 71°15′W*
14 K10 **Québec** *var.* Quebec. ◐ *province* SE Canada
61 D17 **Quebracho** Paysandú, W Uruguay *31°58′S 57°53′W*
101 K14 **Quedlinburg** Sachsen-Anhalt, C Germany *51°48′N 11°09′E*
138 H10 **Queen Alia** ✈ ('Ammān) 'Ammān, C Jordan
10 L16 **Queen Bess, Mount** ▲ British Columbia, SW Canada *51°15′N 124°29′W*
10 I14 **Queen Charlotte** British Columbia, SW Canada *53°18′N 132°04′W*
65 B24 **Queen Charlotte Bay** *bay* West Falkland, W Falkland Islands
10 H14 **Queen Charlotte Islands** *Fr.* Îles de la Reine-Charlotte. *island group* British Columbia, SW Canada
10 I15 **Queen Charlotte Sound** *sea area* British Columbia, W Canada
10 J16 **Queen Charlotte Strait** *strait* British Columbia, W Canada
27 U1 **Queen City** Missouri, C USA *40°24′N 92°34′W*
25 X5 **Queen City** Texas, SW USA *33°09′N 94°09′W*
197 O9 **Queen Elizabeth Islands** *Fr.* Îles de la Reine-Élisabeth. *island group* Nunavut, N Canada
195 Y10 **Queen Mary Coast** *physical region* Antarctica
65 N24 **Queen Mary's Peak** ▲ C Tristan da Cunha
196 M8 **Queen Maud Gulf** *gulf* Arctic Ocean
195 P11 **Queen Maud Mountains** ⛰ Antarctica
Queen's County *see* Laois
181 U7 **Queensland** ◐ *state* N Australia
192 I9 **Queensland Plateau** *undersea feature* N Coral Sea
183 O16 **Queenstown** Tasmania, SE Australia *42°06′S 145°33′E*
185 C22 **Queenstown** Otago, South Island, New Zealand *45°01′S 168°44′E*
83 I24 **Queenstown** Eastern Cape, S South Africa *31°52′S 26°50′E*
Queenstown *see* Cobh
32 F8 **Queets** Washington, NW USA *47°31′N 124°19′W*
61 D18 **Queguay Grande, Río** ≈ W Uruguay
59 O16 **Queimadas** Bahia, E Brazil *10°59′S 39°38′W*
82 D11 **Quela** Malanje, NW Angola *09°18′S 17°07′E*
83 O16 **Quelimane** *var.* Kilimane, Kilmain, Quilimane. Zambézia, NE Mozambique *17°53′S 36°51′E*
63 G18 **Quellón** *var.* Puerto Quellón. Los Lagos, S Chile *43°05′S 73°38′W*
Quelpart *see* Jeju-do
37 P12 **Quemado** New Mexico, SW USA *34°18′N 108°29′W*
25 O12 **Quemado** Texas, SW USA *28°58′N 100°36′W*
44 K7 **Quemado, Punta de** *headland* E Cuba *20°13′N 74°07′W*
Quemoy *see* Jinmen Dao
62 K13 **Quemú Quemú** La Pampa, E Argentina *36°03′S 63°36′W*
155 E17 **Quepem** Goa, W India *15°13′N 74°03′E*
42 M14 **Quepos** Puntarenas, S Costa Rica *09°28′N 84°10′W*
Que Que *see* Kwekwe
61 D23 **Quequén** Buenos Aires, E Argentina *38°30′S 58°44′W*
61 D23 **Quequén Grande, Río** ≈ E Argentina
61 C23 **Quequén Salado, Río** ≈ E Argentina
Quera *see* Chur
41 N13 **Querétaro** Querétaro de Arteaga, C Mexico *20°36′N 100°24′W*
40 F4 **Querobabi** Sonora, NW Mexico *30°02′N 111°02′W*
42 M13 **Quesada** *var.* Ciudad Quesada, San Carlos. Alajuela, N Costa Rica *10°19′N 84°26′W*
105 O13 **Quesada** Andalucía, S Spain *37°52′N 03°05′W*
161 O7 **Queshan** Henan, C China *32°48′N 114°03′E*
10 M15 **Quesnel** British Columbia, SW Canada *52°59′N 122°30′W*
37 S9 **Questa** New Mexico, SW USA *36°41′N 105°37′W*
102 H7 **Questembert** Morbihan, NW France *47°39′N 02°24′W*
57 K22 **Quetena, Río** ≈ SW Bolivia
149 O10 **Quetta** Baluchistān, SW Pakistan *30°15′N 67°E*
Quetzalcoalco *see* Coatzacoalcos
Quetzaltenango *see* Quezaltenango
56 B6 **Quevedo** Los Ríos, C Ecuador *01°02′S 79°27′W*
42 B6 **Quezaltenango** *var.* Quetzaltenango. Quezaltenango, W Guatemala *14°50′N 91°30′W*
42 A2 **Quezaltenango** *off.* Departamento de Quezaltenango, *var.* Quetzaltenango. ◐ *department* SW Guatemala
Quezaltenango, Departamento de *see* Quezaltenango
42 E6 **Quezaltepeque** Chiquimula, SE Guatemala *14°38′N 89°25′W*
170 M6 **Quezon** Palawan, W Philippines *09°13′N 118°01′E*
161 P5 **Qufu** Shandong, E China *35°37′N 117°05′E*
82 B12 **Quibala** Kwanza Sul, NW Angola *10°44′S 14°58′E*
82 B11 **Quibaxe** *var.* Quibaxi. Kwanza Norte, NW Angola *08°30′S 14°36′E*
Quibaxi *see* Quibaxe
54 D9 **Quibdó** Chocó, W Colombia *05°40′N 76°38′W*
102 G7 **Quiberon** Morbihan, NW France *47°30′N 03°07′W*
102 G7 **Quiberon, Baie de** *bay* NW France
54 J5 **Quíbor** Lara, N Venezuela *09°55′N 69°35′W*
42 C4 **Quiché** *off.* Departamento del Quiché. ◐ *department* W Guatemala
Quiché, Departamento del *see* Quiché
99 E21 **Quiévrain** Hainaut, S Belgium *50°25′N 03°41′E*
40 I9 **Quila** Sinaloa, C Mexico *24°24′N 107°11′W*
83 B14 **Quilengues** Huíla, SW Angola *14°09′S 14°04′E*
Quilimane *see* Quelimane
57 G15 **Quillabamba** Cusco, C Peru *12°49′S 72°41′W*
57 L18 **Quillacollo** Cochabamba, C Bolivia *17°26′S 66°16′W*
62 H4 **Quillagua** Antofagasta, N Chile *21°33′S 69°32′W*
103 N17 **Quillan** Aude, S France *42°52′N 02°11′E*
11 U15 **Quill Lakes** ◎ Saskatchewan, S Canada
62 G11 **Quillota** Valparaíso, C Chile *32°54′S 71°16′W*
155 G23 **Quilon** *var.* Kollam. Kerala, SW India *08°57′N 76°37′E* *see also* Kollam
181 V9 **Quilpie** Queensland, C Australia *26°39′S 144°15′E*
149 O4 **Quil-Qala** Bāmyān, N Afghanistan *35°13′N 67°02′E*
62 L7 **Quimilí** Santiago del Estero, C Argentina *27°35′S 62°25′W*
57 O19 **Quimome** Santa Cruz, E Bolivia *17°45′S 61°15′W*
102 F6 **Quimper** *anc.* Quimper Corentin. Finistère, NW France *48°00′N 04°05′W*
Quimper Corentin *see* Quimper
102 G7 **Quimperlé** Finistère, NW France *47°52′N 03°33′W*
32 F8 **Quinault** Washington, NW USA *47°27′N 123°53′W*
32 F8 **Quinault River** ≈ Washington, NW USA
35 P5 **Quincy** California, W USA *39°56′N 120°56′W*
23 S8 **Quincy** Florida, SE USA *30°35′N 84°34′W*
30 I13 **Quincy** Illinois, N USA *39°56′N 91°24′W*
19 O11 **Quincy** Massachusetts, NE USA *42°15′N 71°00′W*
32 J9 **Quincy** Washington, NW USA *47°13′N 119°51′W*
54 E10 **Quindío** *off.* Departamento del Quindío. ◐ *province* C Colombia
54 E10 **Quindío, Nevado del** ▲ C Colombia *04°42′N 75°25′W*
62 J10 **Quines** San Luis, C Argentina *32°15′S 65°46′W*
39 N13 **Quinhagak** Alaska, USA *59°45′N 161°55′W*
76 G13 **Quinhámel** W Guinea-Bissau *11°52′N 15°52′W*
Qui Nhon/Quinhon *see* Quy Nhơn
Quinindé *see* Rosa Zárate
25 U6 **Quinlan** Texas, SW USA *32°54′N 96°08′W*
61 H17 **Quinta** Rio Grande do Sul, S Brazil *32°05′S 52°18′W*
105 O10 **Quintanar de la Orden** Castilla-La Mancha, C Spain *39°36′N 03°03′W*
41 X13 **Quintana Roo** ◐ *state* SE Mexico
105 S6 **Quinto** Aragón, NE Spain *41°25′N 00°31′W*
108 G10 **Quinto** Ticino, S Switzerland *46°32′N 08°44′E*
27 Q11 **Quinton** Oklahoma, C USA *35°07′N 95°22′W*
62 K12 **Quinto, Río** ≈ C Argentina
82 A10 **Quinzau** Zaire Province, NW Angola *06°50′S 12°48′E*
14 H8 **Quinze, Lac des** ◎ Québec, SE Canada
83 B15 **Quipungo** Huíla, C Angola *14°49′S 14°29′E*
62 G13 **Quirihue** Bío Bío, C Chile *36°15′S 72°35′W*
82 D12 **Quirima** Malanje, NW Angola *10°51′S 18°06′E*
183 T6 **Quirindi** New South Wales, SE Australia *31°29′S 150°40′E*
55 P5 **Quiriquire** Monagas, NE Venezuela *09°59′N 63°14′W*
14 D10 **Quirke Lake** ◎ Ontario, S Canada
61 B21 **Quiroga** Buenos Aires, E Argentina *35°18′S 61°22′W*
104 I4 **Quiroga** Galicia, NW Spain *42°28′N 07°15′W*
Quiróm, Salar *see* Pocitos, Salar
56 B9 **Quiroz, Río** ≈ NW Peru
82 Q13 **Quissanga** Cabo Delgado, NE Mozambique *12°24′S 40°33′E*
83 M20 **Quissico** Inhambane, S Mozambique *24°42′S 34°44′E*
25 O4 **Quitaque** Texas, SW USA *34°22′N 101°03′W*
82 Q13 **Quiterajo** Cabo Delgado, NE Mozambique *11°37′S 40°22′E*
23 T6 **Quitman** Georgia, SE USA *30°46′N 83°33′W*
22 M6 **Quitman** Mississippi, S USA *32°02′N 88°43′W*
25 V6 **Quitman** Texas, SW USA *32°37′N 95°26′W*
56 C6 **Quito** ● (Ecuador) Pichincha, N Ecuador *0°14′S 78°30′W*
Quito *see* Mariscal Sucre
58 P13 **Quixadá** Ceará, E Brazil *04°57′S 39°04′W*
83 Q15 **Quixaxe** Nampula, NE Mozambique *15°15′S 40°07′E*
161 N13 **Qujiang** *var.* Maba. Guangdong, S China *24°47′N 113°34′E*
160 J9 **Qu Jiang** ≈ C China
161 R10 **Qu Jiang** ≈ SE China
160 H12 **Qujing** Yunnan, SW China *25°39′N 103°52′E*
Qulan *see* Kulan
Qulin Gol *see* Chaor He
146 L10 **QuljuqtovTog'lari** *Rus.* Gory Kul'dzhuktau. ⛰ C Uzbekistan
Qulsary *see* Kul'sary
Qulyndy Zhazyghy *see* Ravnina Kulyndy
Qum *see* Qom
Qumälisch *see* Lubartów
159 P11 **Qumar He** ≈ C China
159 Q12 **Qumarlêb** *var.* Yuegai; *prev.* Yuegaitan. Qinghai, C China *34°06′N 95°54′E*
Qumisheh *see* Shahreẕā
147 O14 **Qumqo'rg'on** *Rus.* Kumkurgan. Surkhondaryo Viloyati, S Uzbekistan *37°54′N 67°31′E*
Qunaytirah/Qunayṭirah, Muḩāfaẓat al *see* Al Qunayṭirah
189 V12 **Quoi** *island* Chuuk, C Micronesia
9 N8 **Quoich** ≈ Nunavut, NE Canada
83 E26 **Quoin Point** *headland* SW South Africa *34°48′S 19°39′E*
182 I7 **Quorn** South Australia *32°22′S 138°03′E*
Qurein *see* Al Kuwayt
147 P14 **Qŭrghonteppa** *Rus.* Kurgan-Tyube. SW Tajikistan *37°51′N 68°42′E*
Qurlurtuuq *see* Kugluktuk
Qurveh *see* Qorveh
Quryq *see* Kuryk
Qusair *see* Al Quşayr
137 X10 **Qusar** *Rus.* Kusary. NE Azerbaijan *41°26′N 48°27′E*
Quşayr *see* Al Quşayr
Quseir *see* Al Quşayr
142 I2 **Qūshchī** Āžarbāyjān-e Gharbi, N Iran *37°59′N 45°05′E*
Qusmuryn *see* Kushmurun, Kostanay, Kazakhstan
Qusmuryn *see* Kusmuryn, Ozero
Quṭayfah/Qutayfe/Quteife *see* Al Quṭayfah
Quthing *see* Moyeni
147 S10 **Quvasoy** *Rus.* Kuvasay. Farg'ona Viloyati, E Uzbekistan *40°17′N 71°53′E*
Quwair *see* Guwēr
Quxar *see* Lhazê
Qu Xian *see* Quzhou
159 N16 **Qüxü** *var.* Xoi. Xizang Zizhiqu, W China *29°25′N 90°48′E*
Quyang *see* Jingzhou
167 V11 **Quy Nhơn** *var.* Quinhon, Qui Nhon. Binh Đinh, C Vietnam *13°47′N 109°11′E*
161 R10 **Quzhou** *var.* Qu Xian. Zhejiang, SE China *28°55′N 118°54′E*
Qyteti Stalin *see* Kuçovë
Qyzylaghash *see* Kyzylagash
Qyzylorda *see* Kyzylorda
Qyzyltū *see* Kishkenekol'
Qyzylzhar *see* Kyzylzhar

R

Raa Atoll *see* North Maalhosmadulu Atoll
109 R4 **Raab** Oberösterreich, N Austria *48°19′N 13°40′E*
109 X8 **Raab** *Hung.* Rába. ≈ Austria/Hungary *see also* Rába
Raab *see* Rába
Raab *see* Győr
109 V2 **Raabs an der Thaya** Niederösterreich, E Austria *48°51′N 15°28′E*
93 L14 **Raahe** *Swe.* Brahestad. Pohjois-Pohjanmaa, W Finland *64°42′N 24°31′E*
98 M10 **Raalte** Overijssel, E Netherlands *52°23′N 06°16′E*
99 I14 **Raamsdonksveer** Noord-Brabant, S Netherlands *51°42′N 04°54′E*
92 L12 **Raanujärvi** Lappi, NW Finland *66°39′N 24°40′E*
96 G9 **Raasay** *island* NW Scotland, United Kingdom
118 H3 **Raasiku** *Ger.* Rasik. Harjumaa, NW Estonia *59°22′N 25°11′E*
112 B11 **Rab** *It.* Arbe. Primorje-Gorski Kotar, NW Croatia *44°46′N 14°46′E*
112 B11 **Rab** *It.* Arbe. *island* NW Croatia
171 N16 **Raba** Sumbawa, S Indonesia *08°27′S 118°45′E*
111 G22 **Rába** *Ger.* Raab. ≈ Austria/Hungary *see also* Raab
Rába *see* Raab
112 A10 **Rabac** Istra, NW Croatia *45°03′N 14°09′E*
104 I2 **Rábade** Galicia, NW Spain *43°07′N 07°37′W*
80 F10 **Rabak** White Nile, C Sudan *13°12′N 32°44′E*
186 G9 **Rabaraba** Milne Bay, SE Papua New Guinea *10°00′S 149°50′E*
102 K16 **Rabastens-de-Bigorre** Hautes-Pyrénées, S France *43°23′N 00°10′E*
121 O16 **Rabat** W Malta *35°51′N 14°25′E*
74 F6 **Rabat** *var.* al Dar al Baida. ● (Morocco) NW Morocco *34°02′N 06°51′W*
Rabat *see* Victoria
186 H6 **Rabaul** New Britain, E Papua New Guinea *04°13′S 152°11′E*
Rabbah Ammon/Rabbath Ammon *see* 'Amman
28 K8 **Rabbit Creek** ≈ South Dakota, N USA
14 H10 **Rabbit Lake** ◎ Ontario, S Canada
187 Y14 **Rabi** *prev.* Rambi. *island* N Fiji
140 K9 **Rābigh** Makkah, W Saudi Arabia *22°51′N 39°E*
42 D5 **Rabinal** Baja Verapaz, C Guatemala *15°05′N 90°26′W*
168 G9 **Rabi, Pulau** *island* NW Indonesia, East Indies
111 L17 **Rabka** Małopolskie, S Poland *49°38′N 20°E*
155 F16 **Rabkavi** Karnātaka, W India *16°40′N 75°03′E*
Râbniţa *see* Rîbniţa
109 Y6 **Rabnitz** ≈ E Austria
124 J7 **Rabocheostrovsk** Respublika Kareliya, NW Russian Federation *64°58′N 34°46′E*
23 U1 **Rabun Bald** ▲ Georgia, SE USA *34°58′N 83°18′W*
75 S11 **Rabyānah** SE Libya *24°07′N 21°58′E*
75 S11 **Rabyānah, Ramlat** *var.* Rebiana Sand Sea, Şaḩrā' Rabyānah. *desert* SE Libya
Rabyānah, Şaḩrā' *see* Rabyānah, Ramlat
116 L11 **Răcăciuni** Bacău, E Romania *46°20′N 27°00′E*
Racaka *see* Riwoqê
107 J24 **Racalmuto** Sicilia, Italy, C Mediterranean Sea *37°25′N 13°44′E*
116 J14 **Răcari** Dâmboviţa, SE Romania *44°37′N 25°43′E*
Răcari *see* Durankulak
116 F13 **Răcăşdia** *Hung.* Rakasd. Caraş-Severin, SW Romania *44°58′N 21°36′E*
106 B9 **Racconigi** Piemonte, NE Italy *44°45′N 07°41′E*
31 T15 **Raccoon Creek** ≈ Ohio, N USA
13 V13 **Race, Cape** *headland* Newfoundland, Newfoundland and Labrador, E Canada *46°40′N 53°05′W*
22 K10 **Raceland** Louisiana, S USA *29°43′N 90°36′W*
19 Q12 **Race Point** *headland* Massachusetts, NE USA *42°03′N 70°14′W*
167 S14 **Rach Gia** Kiên Giang, S Vietnam *10°01′N 105°05′E*
167 S14 **Rach Gia, Vinh** *bay* S Vietnam
76 J8 **Rachid** Tagant, C Mauritania *18°48′N 11°41′W*
110 L10 **Raciąż** Mazowieckie, C Poland *52°46′N 20°04′E*
111 I16 **Racibórz** *Ger.* Ratibor. Śląskie, S Poland *50°05′N 18°10′E*
31 N9 **Racine** Wisconsin, N USA *42°42′N 87°50′W*
14 D7 **Racine Lake** ◎ Ontario, S Canada
111 J23 **Ráckeve** Pest, C Hungary *47°10′N 18°58′E*
Rácz-Becse *see* Bečej
141 O15 **Radā'** *var.* Ridā'. W Yemen *14°24′N 44°49′E*

◆ Country ● Country Capital ◇ Dependent Territory ○ Dependent Territory Capital ◐ Administrative Regions ✈ International Airport ▲ Mountain ⛰ Mountain Range Volcano ≈ River ◎ Lake ▣ Reservoir

113 O15 **Radan** ▲ SE Serbia *42°59′N 21°31′E*
63 J19 **Rada Tilly** Chubut, SE Argentina *45°54′S 67°33′W*
116 K8 **Rădăuţi** *Ger.* Radautz, *Hung.* Rádóc. Suceava, N Romania *47°49′N 25°58′E*
116 L8 **Rădăuţi-Prut** Botoşani, NE Romania *48°14′N 26°47′E*
Radautz *see* Rădăuţi
Radbusa *see* Radbuza
111 A17 **Radbuza** *Ger.* Radbusa. ⚓ SE Czech Republic
20 K6 **Radcliff** Kentucky, S USA *37°50′N 85°57′W*
139 O2 **Radd, Wādī ar** *dry watercourse* N Syria
95 H16 **Råde** Østfold, S Norway *59°21′N 10°53′E*
109 V11 **Radeče** *Ger.* Ratschach. C Slovenia *46°01′N 15°10′E*
Radein *see* Radenci
116 J4 **Radekhiv** *Pol.* Radziechów, *Rus.* Radekhov. L'vivs'ka Oblast', W Ukraine *50°17′N 24°39′E*
Radekhov *see* Radekhiv
109 X9 **Radenci** *Ger.* Radein; *prev.* Radinci. NE Slovenia *46°36′N 16°02′E*
109 S9 **Radenthein** Kärnten, S Austria *46°48′N 13°42′E*
Rádeyilikóé *see* Fort Good Hope
21 R7 **Radford** Virginia, NE USA *37°07′N 80°34′W*
154 C9 **Rādhanpur** Gujarāt, W India *23°52′N 71°49′E*
Radinci *see* Radenci
127 Q6 **Radishchevo** Ul'yanovskaya Oblast', W Russian Federation *52°49′N 47°54′E*
12 I9 **Radisson** Québec, E Canada *53°47′N 77°35′W*
11 P16 **Radium Hot Springs** British Columbia, SW Canada *50°39′N 116°09′W*
116 F11 **Radna** *Hung.* Máriaradna. Arad, W Romania *46°05′N 21°41′E*
Rádnávrre *see* Randijaure
114 K10 **Radnevo** Stara Zagora, C Bulgaria *42°17′N 25°58′E*
97 J20 **Radnor** *cultural region* E Wales, United Kingdom
Radnót *see* Iernut
Rádóc *see* Rădăuţi
101 H24 **Radolfzell am Bodensee** Baden-Württemberg, S Germany *47°43′N 08°58′E*
110 M13 **Radom** Mazowieckie, C Poland *51°23′N 21°08′E*
116 I14 **Radomireşti** Olt, S Romania *44°06′N 25°00′E*
111 K14 **Radomsko** *Rus.* Novoradomsk. Łódzkie, C Poland *51°04′N 19°25′E*
117 N4 **Radomyshl'** Zhytomyrs'ka Oblast', N Ukraine *50°30′N 29°16′E*
113 P19 **Radoviš** *prev.* Radovište. E Macedonia *41°39′N 22°26′E*
Radovište *see* Radoviš
Radøy *see* Radøyni
94 B13 **Radøyni** *prev.* Radøy. *island* S Norway
109 R7 **Radstadt** Salzburg, NW Austria *47°24′N 13°31′E*
182 E8 **Radstock, Cape** *headland* South Australia *33°11′S 134°18′E*
119 G15 **Radun'** Hrodzyenskaya Voblasts', W Belarus *54°03′N 25°00′E*
109 U10 **Raduha** ▲ N Slovenia *46°24′N 14°46′E*
126 M3 **Raduzhnyy** Vladimirskaya Oblast', W Russian Federation *55°59′N 40°15′E*
118 F11 **Radviliškis** Šiaulai, N Lithuania *55°48′N 23°32′E*
11 U17 **Radville** Saskatchewan, S Canada *49°28′N 104°19′W*
140 K7 **Radwā, Jabal** ▲ W Saudi Arabia *24°31′N 38°21′E*
111 P16 **Radymno** Podkarpackie, SE Poland *49°57′N 22°49′E*
116 J5 **Radyvyliv** Rivnens'ka Oblast', NW Ukraine *50°07′N 25°12′E*
Radziechów *see* Radekhiv
110 I11 **Radziejów** Kujawsko-pomorskie, C Poland *52°36′N 18°33′E*
110 O12 **Radzyń Podlaski** Lubelskie, E Poland *51°48′N 22°37′E*
8 J7 **Rae** ⚓ Nunavut, NW Canada
152 M13 **Rāe Bareli** Uttar Pradesh, N India *26°14′N 81°14′E*
Rae-Edzo *see* Edzo
21 T11 **Raeford** North Carolina, SE USA *34°59′N 79°15′W*
99 M19 **Raeren** Liège, E Belgium *50°42′N 06°06′E*
9 N7 **Rae Strait** *strait* Nunavut, N Canada
184 L11 **Raetihi** Manawatu-Wanganui, North Island, New Zealand *39°29′S 175°16′E*
Raevavae *see* Raivavae
Rafa *see* Rafah
62 M10 **Rafaela** Santa Fe, E Argentina *31°16′S 61°25′W*
54 E5 **Rafael Núñez** ✈ (Cartagena) Bolívar, NW Colombia *10°27′N 75°31′W*
138 E11 **Rafah** *var.* Rafa, Rafaḥ, *Heb.* Rafiah, Raphiah. SW Gaza Strip *31°18′N 34°15′E*
79 L15 **Rafaï** Mbomou, SE Central African Republic *05°01′N 23°51′E*
141 O4 **Rafḥah** Al Ḥudūd ash Shamāliyah, N Saudi Arabia *29°41′N 43°29′E*
Rafiah *see* Rafah
143 R10 **Rafsanjān** Kermān, C Iran *30°25′N 56°E*
80 B13 **Raga** Western Bahr el Ghazal, W South Sudan *08°28′N 25°41′E*
19 S8 **Ragged Island** *island* Maine, NE USA
44 I5 **Ragged Island Range** *island group* S The Bahamas
184 L7 **Raglan** Waikato, North Island, New Zealand *37°48′S 174°54′E*
22 G8 **Ragley** Louisiana, S USA *30°31′N 93°13′W*
Ragnit *see* Neman
107 K25 **Ragusa** Sicilia, Italy, C Mediterranean Sea *36°56′N 14°42′E*
Ragusa *see* Dubrovnik
Ragusavecchia *see* Cavtat
171 P14 **Raha** Pulau Muna, C Indonesia *04°50′S 122°43′E*
119 N17 **Rahachow** *Rus.* Rogachëv. Homyel'skaya Voblasts', SE Belarus *53°03′N 30°03′E*
67 U6 **Rahad** *var.* Nahr ar Rahad. ⚓ W Sudan
Rahad, Nahr ar *see* Rahad
Rahaeng *see* Tak
138 F11 **Rahat** Southern, C Israel *31°20′N 34°43′E*
140 L8 **Rahaṭ, Ḥarrat** *lava flow* W Saudi Arabia
149 S12 **Rahīmyār Khān** Punjab, SE Pakistan *28°27′N 70°21′E*
95 I14 **Råholt** Akershus, S Norway *60°16′N 11°10′E*
113 M17 **Rahovec** *Serb.* Orahovac. W Kosovo *42°24′N 20°40′E*
191 S10 **Raiatea** *island* Îles Sous le Vent, W French Polynesia
155 H16 **Rāichūr** Karnātaka, C India *16°15′N 77°20′E*
Raidestos *see* Tekirdağ
153 S13 **Rāiganj** West Bengal, NE India *25°38′N 88°11′E*
154 M11 **Raigarh** Chhattīsgarh, C India *21°53′N 83°28′E*
183 O16 **Railton** Tasmania, SE Australia *41°24′S 146°28′E*
36 L8 **Rainbow Bridge** *natural arch* Utah, W USA
23 Q3 **Rainbow City** Alabama, S USA *33°57′N 86°02′W*
11 N11 **Rainbow Lake** Alberta, W Canada *58°30′N 119°24′W*
21 R5 **Rainelle** West Virginia, NE USA *37°57′N 80°46′W*
32 G10 **Rainier** Oregon, NW USA *46°05′N 122°55′W*
32 H9 **Rainier, Mount** ⚠ Washington, NW USA *46°51′N 121°45′W*
23 Q2 **Rainsville** Alabama, S USA *34°29′N 85°51′W*
12 B11 **Rainy Lake** ◎ Canada/USA
12 A11 **Rainy River** Ontario, C Canada *48°44′N 94°33′W*
Raippaluoto *see* Replot
154 K12 **Raipur** Chhattīsgarh, C India *21°16′N 81°42′E*
154 H10 **Raisen** Madhya Pradesh, C India *23°21′N 77°49′E*
15 N13 **Raisin** ⚓ Ontario, SE Canada
31 R11 **Raisin, River** ⚓ Michigan, N USA
191 U13 **Raivavae** *var.* Raevavae. Îles Australes, SW French Polynesia
149 W9 **Rāiwind** Punjab, E Pakistan *31°14′N 74°10′E*
171 T12 **Raja Ampat, Kepulauan** *island group* E Indonesia
155 L16 **Rājahmundry** Andhra Pradesh, E India *17°05′N 81°42′E*
155 I18 **Rājampet** Andhra Pradesh, E India *14°09′N 79°10′E*
Rajang *see* Rajang, Batang
169 S9 **Rajang, Batang** *var.* Rajang. ⚓ East Malaysia
149 S11 **Rājanpur** Punjab, E Pakistan *29°05′N 70°25′E*
155 H23 **Rājapālaiyam** Tamil Nādu, SE India *09°26′N 77°36′E*
152 E12 **Rājasthān** ◆ *state* NW India
153 T15 **Rajbari** Dhaka, C Bangladesh *23°47′N 89°39′E*
153 R12 **Rajbiraj** Eastern, E Nepal *26°34′N 86°52′E*
154 G9 **Rājgarh** Madhya Pradesh, C India *24°01′N 76°42′E*
152 H10 **Rājgarh** Rājasthān, NW India *28°38′N 75°21′E*
153 P14 **Rājgīr** Bihār, N India *25°01′N 85°26′E*
110 O8 **Rajgród** Podlaskie, NE Poland *53°43′N 22°40′E*
154 L12 **Rājim** Chhattīsgarh, C India *20°57′N 81°58′E*
112 C11 **Rajinac, Mali** ▲ W Croatia *44°47′N 15°04′E*
154 B10 **Rājkot** Gujarāt, W India *22°18′N 70°47′E*
153 R14 **Rājmahal** Jhārkhand, NE India *25°03′N 87°49′E*
153 Q14 **Rājmahal Hills** *hill range* N India
154 K12 **Rāj Nāndgaon** Chhattīsgarh, C India *21°06′N 81°02′E*
152 I8 **Rājpura** Punjab, NW India *30°29′N 76°40′E*
153 S14 **Rajshahi** *prev.* Rampur Boalia. Rajshahi, W Bangladesh *24°24′N 88°40′E*
153 S13 **Rajshahi** ◆ *division* NW Bangladesh
190 K13 **Rakahanga** *atoll* N Cook Islands
185 H19 **Rakaia** Canterbury, South Island, New Zealand *43°45′S 172°02′E*
185 G19 **Rakaia** ⚓ South Island, New Zealand
152 H3 **Rakaposhi** ▲ N India *36°06′N 74°31′E*
Rakasd *see* Răcăşdia
169 N15 **Rakata, Pulau** *var.* Pulau Krakatau. *island* S Indonesia
141 U10 **Rakbah, Qalamat ar** *well* SE Saudi Arabia
166 K6 **Rakhine State** *var.* Arakan State. ◆ *state* W Myanmar (Burma)
116 I8 **Rakhiv** Zakarpats'ka Oblast', W Ukraine *48°05′N 24°15′E*
141 V13 **Rakhyūt** SW Oman *16°41′N 53°09′E*
192 K9 **Rakiraki** Viti Levu, W Fiji *17°22′S 178°10′E*
126 J8 **Rakitnoye** Belgorodskaya Oblast', W Russian Federation *50°50′N 35°51′E*
Rakka *see* Ar Raqqah
118 I4 **Rakke** Lääne-Virumaa, NE Estonia *58°58′N 26°14′E*
95 I16 **Rakkestad** Østfold, S Norway *59°25′N 11°17′E*
110 F12 **Rakoniewice** *Ger.* Rakwitz. Wielkopolskie, C Poland *52°09′N 16°10′E*
Rakonitz *see* Rakovník
83 H18 **Rakops** Central, C Botswana *21°01′S 24°20′E*
111 C16 **Rakovník** *Ger.* Rakonitz. Středočeský Kraj, W Czech Republic *50°07′N 13°44′E*
114 J10 **Rakovski** Plovdiv, C Bulgaria *42°16′N 24°58′E*
Rakutō-kō *see* Nakdong-gang
118 I3 **Rakvere** *Ger.* Wesenberg. Lääne-Virumaa, N Estonia *59°21′N 26°20′E*
Rakwitz *see* Rakoniewice
22 L6 **Raleigh** Mississippi, S USA *32°01′N 89°30′W*
21 U9 **Raleigh** *state capital* North Carolina, SE USA *35°46′N 78°38′W*
21 Y11 **Raleigh Bay** *bay* North Carolina, SE USA
21 U9 **Raleigh-Durham** ✈ North Carolina, SE USA *35°54′N 78°45′W*
189 S6 **Ralik Chain** *island group* Ralik Chain, W Marshall Islands
25 N5 **Ralls** Texas, SW USA *33°40′N 101°23′W*
18 G13 **Ralston** Pennsylvania, NE USA *41°29′N 76°57′W*
141 O16 **Ramādah** W Yemen *13°35′N 43°50′E*
Ramadi *see* Ar Ramādī
105 N2 **Ramales de la Victoria** Cantabria, N Spain *43°15′N 03°28′W*
138 F10 **Ramallah** C West Bank *31°55′N 35°12′E*
61 C19 **Ramallo** Buenos Aires, E Argentina *33°30′S 60°01′W*
155 H20 **Rāmanagaram** Karnātaka, E India *12°45′N 77°16′E*
155 I23 **Rāmanāthapuram** Tamil Nādu, SE India *09°23′N 78°53′E*
154 N12 **Rāmapur** Odisha, E India *21°48′N 84°00′E*
155 J14 **Rāmāreddi** *var.* Kāmāreddi, Kamareddy. Telangana, C India *18°19′N 78°23′E*
138 F10 **Ramat Gan** Tel Aviv, W Israel *32°04′N 34°48′E*
103 T6 **Rambervillers** Vosges, NE France *48°15′N 06°50′E*
Rambi *see* Rabi
103 N5 **Rambouillet** Yvelines, N France *48°39′N 01°50′E*
186 E5 **Rambutyo Island** *island* N Papua New Guinea
153 Q12 **Ramechhāp** Central, C Nepal *27°20′N 86°05′E*
183 R12 **Rame Head** *headland* Victoria, SE Australia *37°48′S 149°30′E*
126 L4 **Ramenskoye** Moskovskaya Oblast', W Russian Federation *55°31′N 38°24′E*
124 J15 **Rameshki** Tverskaya Oblast', W Russian Federation *57°21′N 36°05′E*
153 P14 **Rāmgarh** Jhārkhand, N India *23°37′N 85°32′E*
152 D11 **Rāmgarh** Rājasthān, NW India *27°29′N 70°38′E*
142 M9 **Rāmhormoz** *var.* Ram Hormuz, Ramuz. Khūzestān, SW Iran *31°15′N 49°38′E*
Ram Hormuz *see* Rāmhormoz
Ram, Jebel *see* Ramm, Jabal
138 F10 **Ramla** *var.* Ramle, Ramleh, *Ar.* Er Ramle. Central, C Israel *31°56′N 34°52′E*
Ramle/Ramleh *see* Ramla
138 F14 **Ramm, Jabal** *var.* Jebel Ram. ▲ SW Jordan *29°34′N 35°24′E*
152 K10 **Rāmnagar** Uttarakhand, N India *29°23′N 79°07′E*
95 N15 **Ramnäs** Västmanland, C Sweden *59°46′N 16°16′E*
Râmnicul-Sărat *see* Râmnicu Sărat
116 L12 **Râmnicu Sărat** *prev.* Râmnicul-Sărat, Rîmnicu-Sărat. Buzău, E Romania *45°24′N 27°06′E*
116 I13 **Râmnicu Vâlcea** *prev.* Rîmnicu Vîlcea. Vâlcea, C Romania *45°04′N 24°32′E*
Ramokgwebana *see* Ramokgwebane
83 J18 **Ramokgwebane** *var.* Ramokgwebana. Central, NE Botswana *20°38′S 27°40′E*
126 L7 **Ramon'** Voronezhskaya Oblast', W Russian Federation *51°51′N 39°18′E*
35 V17 **Ramona** California, W USA *33°02′N 116°52′W*
56 A10 **Ramón, Laguna** ◎ NW Peru
14 G7 **Ramore** Ontario, S Canada *48°26′N 80°19′W*
40 M11 **Ramos** San Luis Potosí, C Mexico *22°48′N 101°55′W*
41 N8 **Ramos Arizpe** Coahuila, NE Mexico *25°35′N 100°59′W*
40 J9 **Ramos, Río de** ⚓ C Mexico
83 I21 **Ramotswa** South East, S Botswana *24°56′S 25°50′E*
39 R8 **Rampart** Alaska, USA *65°30′N 150°10′W*
8 H8 **Ramparts** ⚓ Northwest Territories, NW Canada
152 K10 **Rāmpur** Uttar Pradesh, N India *28°48′N 79°03′E*
154 F9 **Rāmpura** Madhya Pradesh, C India *24°30′N 75°32′E*
Rampur Boalia *see* Rajshahi
166 K6 **Ramree Island** *island* W Myanmar (Burma)
141 W6 **Rams** *var.* Ar Rams. Ra's al Khaymah, NE United Arab Emirates *25°52′N 56°01′E*
143 N4 **Rāmsar** *prev.* Sakhtsar. Māzandarān, N Iran *36°55′N 50°39′E*
93 H16 **Ramsele** Västernorrland, N Sweden *63°33′N 16°35′E*
21 T9 **Ramseur** North Carolina, SE USA *35°43′N 79°39′W*
97 I16 **Ramsey** NE Isle of Man *54°19′N 04°24′W*
97 I16 **Ramsey Bay** *bay* NE Isle of Man
14 E9 **Ramsey Lake** ◎ Ontario, S Canada
97 Q22 **Ramsgate** SE England, United Kingdom *51°20′N 01°25′E*
94 M10 **Ramsjö** Gävleborg, C Sweden *62°10′N 15°40′E*
154 I12 **Rāmtek** Mahārāshtra, C India *21°28′N 79°28′E*
Ramtha *see* Ar Ramthā
Ramuz *see* Rāmhormoz
118 G12 **Ramygala** Panevėžys, C Lithuania *55°30′N 24°18′E*
152 H14 **Rāna Pratāp Sāgar** ◎ N India
169 V7 **Ranau** Sabah, East Malaysia *05°56′N 116°43′E*
168 L14 **Ranau, Danau** ◎ Sumatera, W Indonesia
62 H12 **Rancagua** Libertador, C Chile *34°10′S 70°45′W*
99 G22 **Rance** Hainaut, S Belgium *50°09′N 04°16′E*
102 H6 **Rance** ⚓ NW France
60 J9 **Rancharia** São Paulo, S Brazil *22°13′S 50°53′W*
153 P15 **Rānchi** Jhārkhand, N India *23°22′N 85°20′E*
61 D21 **Ranchos** Buenos Aires, E Argentina *35°32′S 58°22′W*
37 S9 **Ranchos De Taos** New Mexico, SW USA *36°21′N 105°36′W*
63 G16 **Ranco, Lago** ◎ C Chile
95 C16 **Randaberg** Rogaland, S Norway *59°00′N 05°38′E*
29 U7 **Randall** Minnesota, N USA *46°05′N 94°30′W*
107 L23 **Randazzo** Sicilia, Italy, C Mediterranean Sea *37°52′N 14°57′E*
95 G21 **Randers** Midtjylland, C Denmark *56°28′N 10°03′E*
92 I12 **Randijaure** *Lapp.* Rádnávrre. ◎ N Sweden
21 T9 **Randleman** North Carolina, SE USA *35°49′N 79°48′W*
19 O11 **Randolph** Massachusetts, NE USA *42°09′N 71°02′W*
29 Q13 **Randolph** Nebraska, C USA *42°25′N 97°05′W*
36 M1 **Randolph** Utah, W USA *41°40′N 111°10′W*
100 P9 **Randow** ⚓ NE Germany
95 H14 **Randsfjorden** ◎ S Norway
92 K13 **Råneå** Norrbotten, N Sweden *65°52′N 22°17′E*
92 G12 **Ranelva** ⚓ C Norway
93 F15 **Ranemsletta** Nord-Trøndelag, C Norway *64°36′N 11°55′E*
76 H10 **Ranérou** C Senegal *15°17′N 14°00′W*
Ranes *see* Ringvassøya
185 E22 **Ranfurly** Otago, South Island, New Zealand *45°07′S 170°06′E*
167 P17 **Rangae** Narathiwat, SW Thailand *06°15′N 101°45′E*
153 V16 **Rangamati** Chittagong, SE Bangladesh *22°40′N 92°10′E*
184 I2 **Rangaunu Bay** *bay* North Island, New Zealand
19 P6 **Rangeley** Maine, NE USA *44°58′N 70°37′W*
37 O4 **Rangely** Colorado, C USA *40°05′N 108°48′W*
25 R7 **Ranger** Texas, SW USA *32°28′N 98°40′W*
14 C9 **Ranger Lake** Ontario, S Canada *46°51′N 83°34′W*
14 C9 **Ranger Lake** ◎ Ontario, S Canada
153 V12 **Rangia** Assam, NE India *26°26′N 91°38′E*
185 I18 **Rangiora** Canterbury, South Island, New Zealand *43°19′S 172°34′E*
191 T9 **Rangiroa** *atoll* Îles Tuamotu, W French Polynesia
184 N9 **Rangitaiki** ⚓ North Island, New Zealand
185 F19 **Rangitata** ⚓ South Island, New Zealand
184 M12 **Rangitikei** ⚓ North Island, New Zealand
184 L6 **Rangitoto Island** *island* N New Zealand
Rangkasbitoeng *see* Rangkasbitung
169 N16 **Rangkasbitung** *prev.* Rangkasbitoeng. Jawa, SW Indonesia *06°21′S 106°12′E*
167 P9 **Rang, Khao** ▲ C Thailand *16°13′N 99°03′E*
147 V13 **Rangkŭl** *Rus.* Rangkul'. SE Tajikistan *38°30′N 74°24′E*
Rangkul' *see* Rangkŭl
Rangoon *see* Yangon
153 T13 **Rangpur** Rajshahi, N Bangladesh *25°46′N 89°20′E*
155 F18 **Rānibennur** Karnātaka, W India *14°36′N 75°39′E*
153 R15 **Rāniganj** West Bengal, NE India *23°34′N 87°13′E*
149 Q13 **Rānīpur** Sind, SE Pakistan *27°17′N 68°34′E*
Rāniyah *see* Ranye
25 N9 **Rankin** Texas, SW USA *31°14′N 101°56′W*
9 O9 **Rankin Inlet** Nunavut, C Canada *62°52′N 92°14′W*
183 P8 **Rankins Springs** New South Wales, SE Australia *33°51′S 146°16′E*
Rankovićevo *see* Kraljevo
108 I7 **Rankweil** Vorarlberg, W Austria *47°17′N 09°40′E*
Rann *see* Brežice
127 T8 **Ranneye** Orenburgskaya Oblast', W Russian Federation *51°28′N 52°29′E*
96 I10 **Rannoch, Loch** ◎ C Scotland, United Kingdom
191 U17 **Rano Kau** *var.* Rano Kao. *crater* Easter Island, Chile, E Pacific Ocean
167 N14 **Ranong** Ranong, SW Thailand *09°59′N 98°40′E*
186 J8 **Ranongga** *var.* Ghanongga. *island* NW Solomon Islands
191 W16 **Rano Raraku** *ancient monument* Easter Island, Chile, E Pacific Ocean
171 V12 **Ransiki** Papua Barat, E Indonesia *01°27′S 134°12′E*
92 K12 **Rantajärvi** Norrbotten, N Sweden *66°45′N 23°39′E*
93 N17 **Rantasalmi** Etelä-Savo, SE Finland *62°02′N 28°22′E*
169 U13 **Rantau** Borneo, C Indonesia *02°56′S 115°09′E*
168 L10 **Rantau, Pulau** *var.* Pulau Tebingtinggi. *island* W Indonesia
171 N13 **Rantepao** Sulawesi, C Indonesia *02°58′S 119°58′E*
30 M13 **Rantoul** Illinois, N USA *40°19′N 88°08′W*
93 L15 **Rantsila** Pohjois-Pohjanmaa, C Finland *64°31′N 25°40′E*
92 L13 **Ranua** Lappi, NW Finland *65°55′N 26°34′E*
Rānya *see* Ranye
139 T3 **Ranye** *Ar.* Rāniyah, *var.* Rānya. As Sulaymānīyah, NE Iraq *36°15′N 44°53′E*
157 X3 **Raohe** Heilongjiang, NE China *46°49′N 134°00′E*
74 H9 **Raoui, Erg er** *desert* W Algeria
193 O10 **Rapa** *island* Îles Australes, S French Polynesia
191 V14 **Rapa Iti** *island* Îles Australes, SW French Polynesia
106 D10 **Rapallo** Liguria, NW Italy *44°21′N 09°13′E*
Rapa Nui *see* Pascua, Isla de
Raphiah *see* Rafah
21 V5 **Rapidan River** ⚓ Virginia, NE USA
28 J10 **Rapid City** South Dakota, N USA *44°05′N 103°14′W*
15 P8 **Rapide-Blanc** Québec, SE Canada *47°48′N 72°57′W*
14 I8 **Rapide-Deux** Québec, SE Canada *47°56′N 78°33′W*
118 K6 **Räpina** *Ger.* Rappin. Põlvamaa, SE Estonia *58°06′N 27°27′E*
118 G4 **Rapla** *Ger.* Rappel. Raplamaa, NW Estonia *59°00′N 24°46′E*
118 G4 **Raplamaa** *var.* Rapla Maakond. ◆ *province* NW Estonia
Rapla Maakond *see* Raplamaa
21 X6 **Rappahannock River** ⚓ Virginia, NE USA
Rappel *see* Rapla
108 G7 **Rapperswil** Sankt Gallen, NE Switzerland *47°14′N 08°50′E*
Rappin *see* Räpina
153 N12 **Rāpti** ⚓ N India
57 K16 **Rápulo, Río** ⚓ E Bolivia
Raqqah/Raqqah, Muḥāfaẓat al *see* Ar Raqqah
18 J8 **Raquette Lake** ◎ New York, NE USA
18 J6 **Raquette River** ⚓ New York, NE USA
191 V10 **Raraka** *atoll* Îles Tuamotu, C French Polynesia
191 V10 **Raroia** *atoll* Îles Tuamotu, C French Polynesia
190 H15 **Rarotonga** ✈ Rarotonga, S Cook Islands, C Pacific Ocean *21°15′S 159°45′W*
190 H16 **Rarotonga** *island* S Cook Islands, C Pacific Ocean
147 P12 **Rarz** W Tajikistan *39°23′N 68°43′E*
Ras al'Ain *see* Ra's al 'Ayn
139 N2 **Ra's al 'Ayn** *var.* Ras al'Ain. Al Ḥasakah, N Syria *36°52′N 40°05′E*
138 H3 **Ra's al Basīṭ** Al Lādhiqīyah, W Syria *35°57′N 35°55′E*
Ra's al-Hafǧī *see* Ra's al Khafjī
141 R5 **Ra's al Khafjī** *var.* Ra's al-Hafǧī. Ash Sharqīyah, NE Saudi Arabia *28°22′N 48°30′E*
Ras al-Khaimah/Ras al Khaimah *see* Ra's al Khaymah
143 R15 **Ra's al Khaymah** *var.* Ras al Khaimah. Ra's al Khaymah, NE United Arab Emirates *25°44′N 55°55′E*
143 R15 **Ra's al Khaymah** *var.* Ras al-Khaimah. ✈ Ra's al Khaymah, NE United Arab Emirates *25°37′N 55°51′E*
138 G13 **Ra's an Naqb** Ma'ān, S Jordan *30°00′N 35°29′E*
61 B26 **Rasa, Punta** *headland* E Argentina *40°50′S 62°15′W*
171 V12 **Rasawi** Papua Barat, E Indonesia *02°04′S 134°02′E*
Râşcani *see* Rişcani
80 J10 **Ras Dashen Terara** ▲ N Ethiopia *13°12′N 38°09′E*
Rasdhoo Atoll *see* Rasdu Atoll
151 K19 **Rasdu Atoll** *var.* Rasdhoo Atoll. *atoll* C Maldives
118 E12 **Raseiniai** Kaunas, C Lithuania *55°23′N 23°06′E*
75 X8 **Ra's Ghārib** *var.* Ras Ghârib. E Egypt *28°16′N 33°01′E*
Ras Ghârib *see* Ra's Ghārib
162 J6 **Rashaant** Hövsgöl, N Mongolia *49°08′N 101°27′E*
Rashaant *see* Delüün, Bayan-Ölgiy, Mongolia
Rashaant *see* Öldziyt, Dundgovĭ, Mongolia
75 V7 **Rashīd** *Eng.* Rosetta. N Egypt *31°25′N 30°25′E*
139 Y11 **Rashīd** Al Baṣrah, E Iraq *31°15′N 47°31′E*
142 M3 **Rasht** *var.* Resht. Gīlān, NW Iran *37°18′N 49°38′E*
Rashwan *see* Reshwan
Rashwān *see* Reshwan
Rasik *see* Raahe
113 M15 **Raška** Serbia, C Serbia *43°18′N 20°37′E*
119 P15 **Rasna** *Rus.* Ryasna. Mahilyowskaya Voblasts', E Belarus *54°01′N 31°12′E*
116 J12 **Râşnov** *prev.* Rîşno, Rozsnyó, *Hung.* Barcarozsnyó. Braşov, C Romania *45°35′N 25°27′E*
118 L11 **Rasony** *Rus.* Rossony. Vitsyebskaya Voblasts', N Belarus *55°53′N 28°50′E*
Ra's Shamrah *see* Ugarit
127 N7 **Rasskazovo** Tambovskaya Oblast', W Russian Federation *52°42′N 41°45′E*
119 O16 **Rasta** *Rus.* Resta. ⚓ E Belarus
Rastadt *see* Rastatt
Rastâne *see* Ar Rastān
141 S6 **Ra's Tannūrah** *Eng.* Ras Tanura. Ash Sharqīyah, NE Saudi Arabia *26°44′N 50°04′E*
Ras Tanura *see* Ra's Tannūrah
101 G21 **Rastatt** *var.* Rastadt. Baden-Württemberg, SW Germany *48°51′N 08°13′E*
Rastenburg *see* Kętrzyn
149 V7 **Rasūlnagar** Punjab, E Pakistan *32°20′N 73°51′E*
189 U6 **Ratak Chain** *island group* Ratak Chain, E Marshall Islands
119 K15 **Ratamka** *Rus.* Ratomka. Minskaya Voblasts', C Belarus *53°56′N 27°21′E*
93 G17 **Ratān** Jämtland, C Sweden *62°28′N 14°35′E*
152 G11 **Ratangarh** Rājasthān, NW India *28°02′N 74°39′E*
Rat Buri *see* Ratchaburi
167 O11 **Ratchaburi** *var.* Rat Buri. Ratchaburi, W Thailand *13°30′N 99°50′E*
29 W15 **Rathbun Lake** ▣ Iowa, C USA
Ráth Caola *see* Rathkeale
166 K5 **Rathedaung** Rakhine State, W Myanmar (Burma) *20°30′N 92°48′E*
100 M12 **Rathenow** Brandenburg, NE Germany *52°37′N 12°21′E*
97 C19 **Rathkeale** *Ir.* Ráth Caola. Limerick, SW Ireland *52°32′N 08°56′W*
96 F13 **Rathlin Island** *Ir.* Reachlainn. *island* N Northern Ireland, United Kingdom
97 C20 **Ráth Luirc** *Ir.* An Ráth. Cork, SW Ireland *52°22′N 08°44′W*
Ratibor *see* Racibórz
Ratisbon/Ratisbona/Ratisbonne *see* Regensburg
Rātische Alpen *see* Rhaetian Alps
38 E17 **Rat Island** *island* Aleutian Islands, Alaska, USA
38 E17 **Rat Islands** *island group* Aleutian Islands, Alaska, USA
154 F10 **Ratlām** *prev.* Rutlam. Madhya Pradesh, C India *23°23′N 75°04′E*
155 D15 **Ratnāgiri** Mahārāshtra, W India *17°00′N 73°20′E*
155 K26 **Ratnapura** Sabaragamuwa Province, S Sri Lanka *06°41′N 80°25′E*
Ratne *see* Ratno
116 J2 **Ratno** *var.* Ratne. Volyns'ka Oblast', NW Ukraine *51°40′N 24°33′E*
Ratomka *see* Ratamka
37 U8 **Raton** New Mexico, SW USA *36°54′N 104°27′W*
139 O7 **Ratqah, Wādī ar** *dry watercourse* W Iraq
Ratschach *see* Radeče
167 O16 **Rattaphum** Songkhla, SW Thailand *07°07′N 100°16′E*
26 L6 **Rattlesnake Creek** ⚓ Kansas, C USA
94 L13 **Rättvik** Dalarna, C Sweden *60°53′N 15°12′E*
100 K9 **Ratzeburg** Mecklenburg-Vorpommern, N Germany *53°41′N 10°48′E*
100 K9 **Ratzeburger See** ◎ N Germany
10 J10 **Ratz, Mount** ▲ British Columbia, SW Canada *57°22′N 132°17′W*
61 D22 **Rauch** Buenos Aires, E Argentina *36°45′S 59°05′W*
41 U16 **Raudales** Chiapas, SE Mexico
Raudhatain *see* Ar Rawḍatayn
Raudnitz an der Elbe *see* Roudnice nad Labem
92 K1 **Raufarhöfn** Norðurland Eystra, NE Iceland *66°27′N 15°58′W*
94 H13 **Raufoss** Oppland, S Norway *60°44′N 10°39′E*
Raukawa *see* Cook Strait
184 Q8 **Raukumara** ▲ North Island, New Zealand *37°46′S 178°07′E*
192 K11 **Raukumara Plain** *undersea feature* N Coral Sea
184 P8 **Raukumara Range** ▲▲ North Island, New Zealand
Raulakela *see* Rāurkela
95 F15 **Rauland** Telemark, S Norway *59°41′N 07°57′E*
93 J19 **Rauma** *Swe.* Raumo. Satakunta, SW Finland *61°09′N 21°30′E*
94 F10 **Rauma** ⚓ S Norway
Raumo *see* Rauma
118 H8 **Rauna** C Latvia *57°19′N 25°34′E*
169 T17 **Raung, Gunung** ▲ Jawa, S Indonesia *08°00′S 114°07′E*
154 N11 **Rāurkela** *var.* Rāulakela, Rourkela. Odisha, E India *22°13′N 84°53′E*
95 J22 **Raus** Skåne, S Sweden *56°01′N 12°48′E*
165 W3 **Rausu** Hokkaidō, NE Japan *44°00′N 145°06′E*
165 W3 **Rausu-dake** ▲ Hokkaidō, NE Japan *44°06′N 145°04′E*
116 M9 **Răut** *var.* Răuţel. ⚓ C Moldova
93 M17 **Rautalampi** Pohjois-Savo, C Finland *62°37′N 26°48′E*
93 N16 **Rautavaara** Pohjois-Savo, C Finland *63°30′N 28°19′E*
Răuţel *see* Răut
93 O18 **Rautjärvi** Etelä-Karjala, SE Finland *61°21′N 29°20′E*
Rautu *see* Sosnovo
191 V11 **Ravahere** *atoll* Îles Tuamotu, C French Polynesia
107 J25 **Ravanusa** Sicilia, Italy, C Mediterranean Sea *37°16′N 13°59′E*
143 S9 **Rāvar** Kermān, C Iran *31°15′N 56°51′E*
147 Q11 **Ravat** Batkenskaya Oblast', SW Kyrgyzstan *39°54′N 70°06′E*
18 K11 **Ravena** New York, NE USA *42°28′N 73°49′W*
106 H10 **Ravenna** Emilia-Romagna, N Italy *44°28′N 12°15′E*
29 O15 **Ravenna** Nebraska, C USA *41°01′N 98°54′W*
31 U11 **Ravenna** Ohio, N USA *41°09′N 81°14′W*
101 I24 **Ravensburg** Baden-Württemberg, S Germany *47°47′N 09°37′E*
181 W4 **Ravenshoe** Queensland, NE Australia *17°29′S 145°28′E*
180 K13 **Ravensthorpe** Western Australia *33°37′S 120°03′E*
21 Q4 **Ravenswood** West Virginia, NE USA *38°57′N 81°45′W*
149 U9 **Rāvi** ⚓ India/Pakistan
112 C9 **Ravna Gora** Primorje-Gorski Kotar, NW Croatia *45°20′N 14°54′E*
109 U10 **Ravne na Koroškem** *Ger.* Gutenstein. N Slovenia *46°33′N 14°57′E*
145 T7 **Ravnina Kulyndy** *prev.* Kulunda Steppe, *Kaz.* Qulyndy Zhazyghy, *Rus.* Kulundinskaya Ravnina. *grassland* Kazakhstan/Russian Federation
139 P6 **Rāwah** Al Anbār, W Iraq *34°32′N 41°54′E*
191 T4 **Rawaki** *prev.* Phoenix Island. *atoll* Phoenix Islands, C Kiribati
149 U6 **Rāwalpindi** Punjab, NE Pakistan *33°38′N 73°06′E*
110 L13 **Rawa Mazowiecka** Łódzkie, C Poland *51°47′N 20°16′E*
Rawāndiz *see* Rewanduz
171 U12 **Rawas** Papua Barat, E Indonesia *01°07′S 132°12′E*
139 O4 **Rawḍah** ◎ E Syria
110 G13 **Rawicz** *Ger.* Rawitsch. Wielkopolskie, C Poland *51°37′N 16°51′E*
Rawitsch *see* Rawicz
180 M11 **Rawlinna** Western Australia *31°01′S 125°36′E*
33 W16 **Rawlins** Wyoming, C USA *41°47′N 107°14′W*
63 K17 **Rawson** Chubut, SE Argentina *43°22′S 65°01′W*
159 R16 **Rawu** Xizang Zizhiqu, W China *29°30′N 96°42′E*
153 P12 **Raxaul** Bihār, N India *26°58′N 84°51′E*
28 K3 **Ray** North Dakota, N USA *48°19′N 103°11′W*
169 S11 **Raya, Bukit** ▲ Borneo, C Indonesia *0°40′S 112°40′E*
155 I18 **Rāyachoti** Andhra Pradesh, E India *14°03′N 78°43′E*
Rāyadrug *see* Rāyagarha
Rāyagada *see* Rāyagarha
155 M14 **Rāyagarha** *prev.* Rāyadrug, *var.* Rāyagada. Odisha, E India *19°10′N 83°28′E*
138 H7 **Rayak** *var.* Rayaq, Riyāq. E Lebanon *33°51′N 36°03′E*
Rayaq *see* Rayak
139 T2 **Rayat** *Ar.* Rāyāt, *var.* Rāyat. Arbīl, E Iraq *36°39′N 44°56′E*
Rāyāt *see* Rayat
Rāyat *see* Rayat
169 N12 **Raya, Tanjung** *cape* Pulau Bangka, W Indonesia
13 R13 **Ray, Cape** *headland* Newfoundland, Newfoundland and Labrador, E Canada *47°38′N 59°15′W*
123 Q13 **Raychikhinsk** Amurskaya Oblast', SE Russian Federation *49°47′N 129°19′E*
127 U5 **Rayevskiy** Respublika Bashkortostan, W Russian Federation *54°04′N 54°58′E*
11 Q17 **Raymond** Alberta, SW Canada *49°30′N 112°41′W*
22 K6 **Raymond** Mississippi, S USA *32°15′N 90°25′W*
32 F9 **Raymond** Washington, NW USA *46°41′N 123°43′W*
183 T8 **Raymond Terrace** New South Wales, SE Australia *32°47′S 151°45′E*
25 T17 **Raymondville** Texas, SW USA *26°30′N 97°48′W*
11 U16 **Raymore** Saskatchewan, S Canada *51°24′N 104°34′W*
39 Q8 **Ray Mountains** ▲▲ Alaska, USA
22 H9 **Rayne** Louisiana, S USA *30°13′N 92°15′W*
41 O12 **Rayón** San Luis Potosí, C Mexico *21°54′N 99°33′W*
40 G4 **Rayón** Sonora, NW Mexico *29°45′N 110°33′W*
167 P12 **Rayong** Rayong, S Thailand *12°42′N 101°17′E*
25 T5 **Ray Roberts, Lake** ◎ Texas, SW USA
18 E15 **Raystown Lake** ▣ Pennsylvania, NE USA
141 V13 **Raysūt** SW Oman *16°58′N 54°02′E*
27 R4 **Raytown** Missouri, C USA *39°00′N 94°27′W*
22 I5 **Rayville** Louisiana, S USA *32°29′N 91°45′W*
142 L5 **Razan** Hamadān, W Iran *35°22′N 48°58′E*
139 S9 **Razāzah, Buḥayrat ar** *var.* Baḥr al Milḥ. ◎ C Iraq
114 L9 **Razboyna** ▲ E Bulgaria *42°54′N 26°31′E*
Razdan *see* Hrazdan
Razdolnoye *see* Rozdol'ne
Razelm, Lacul *see* Razim, Lacul
Razga *see* Rezge
114 L8 **Razgrad** Razgrad, N Bulgaria *43°33′N 26°31′E*
114 L8 **Razgrad** ◆ *province* NE Bulgaria
114 I10 **Razhevo Konare** *var.* Rŭzhevo Konare. Plovdiv, C Bulgaria *42°16′N 24°58′E*
117 N13 **Razim, Lacul** *prev.* Lacul Razelm. *lagoon* NW Black Sea
Razkah *see* Rezge
114 G11 **Razlog** Blagoevgrad, SW Bulgaria *41°53′N 23°28′E*
118 K10 **Rāznas Ezers** ◎ SE Latvia
102 E6 **Raz, Pointe du** *headland* NW France *48°06′N 04°52′W*
Reachlainn *see* Rathlin Island
Reachrainn *see* Lambay Island
97 N22 **Reading** S England, United Kingdom *51°28′N 00°59′W*
18 H15 **Reading** Pennsylvania, NE USA *40°20′N 75°55′W*
48 C7 **Real, Cordillera** ▲▲ C Ecuador
62 K12 **Realicó** La Pampa, C Argentina *35°02′S 64°14′W*
25 R15 **Realitos** Texas, SW USA *27°26′N 98°31′W*
108 G9 **Realp** Uri, C Switzerland *46°36′N 08°32′E*
167 Q12 **Reăng Kesei** Bătdâmbâng, W Cambodia *12°57′N 103°15′E*
191 Y11 **Reao** *atoll* Îles Tuamotu, E French Polynesia
Reate *see* Rieti
Greater Antarctica *see* East Antarctica
180 L11 **Rebecca, Lake** ◎ Western Australia
Rebiana Sand Sea *see* Rabyānah, Ramlat
124 H8 **Reboly** *Finn.* Repola. Respublika Kareliya, NW Russian Federation *63°51′N 30°49′E*
165 S1 **Rebun** Rebun-tō, NE Japan *45°19′N 141°02′E*
165 S1 **Rebun-tō** *island* NE Japan
106 J12 **Recanati** Marche, C Italy *43°23′N 13°34′E*
Rechitsa *see* Rechytsa
109 Y7 **Rechnitz** Burgenland, SE Austria *47°19′N 16°26′E*
119 J20 **Rechytsa** *Rus.* Rechitsa. Brestskaya Voblasts', SW Belarus *51°51′N 26°48′E*
119 O19 **Rechytsa** *Rus.* Rechitsa. Homyel'skaya Voblasts', SE Belarus *52°22′N 30°23′E*
59 Q15 **Recife** *prev.* Pernambuco. *state capital* Pernambuco, E Brazil *08°06′S 34°53′W*
83 I26 **Recife, Cape** *Afr.* Kaap Recife. *headland* S South Africa *34°03′S 25°37′E*
Recife, Kaap *see* Recife, Cape
172 I16 **Récifs, Îles aux** *island* Inner Islands, NE Seychelles
101 E14 **Recklinghausen** Nordrhein-Westfalen, W Germany *51°37′N 07°12′E*
100 M8 **Recknitz** ⚓ NE Germany
99 K23 **Recogne** Luxembourg, SE Belgium *49°56′N 05°20′E*
61 C15 **Reconquista** Santa Fe, C Argentina *29°08′S 59°38′W*
195 O6 **Recovery Glacier** *glacier* Antarctica
59 G15 **Recreio** Mato Grosso, W Brazil *08°13′S 58°15′W*
27 X9 **Rector** Arkansas, C USA *36°15′N 90°17′W*
110 E9 **Recz** *Ger.* Reetz Neumark. Zachodnio-pomorskie, NW Poland *53°16′N 15°32′E*

◆ Country
● Country Capital
◇ Dependent Territory
○ Dependent Territory Capital
◆ Administrative Regions
✈ International Airport
▲ Mountain
▲▲ Mountain Range
⚠ Volcano
⚓ River
◎ Lake
▣ Reservoir

99 L24 **Redange** *var.* Redange-sur-Attert. Diekirch, W Luxembourg 49°46′N 05°53′E
Redange-sur-Attert *see* Redange
18 C13 **Redbank Creek** ≈ Pennsylvania, NE USA
13 S9 **Red Bay** Québec, E Canada 51°40′N 56°37′W
23 N2 **Red Bay** Alabama, S USA 34°26′N 88°08′W
35 N4 **Red Bluff** California, W USA 40°09′N 122°14′W
24 J8 **Red Bluff Reservoir** ◎ New Mexico/Texas, SW USA
30 K16 **Red Bud** Illinois, N USA 38°12′N 89°59′W
30 J5 **Red Cedar River** ≈ Wisconsin, N USA
11 R17 **Redcliff** Alberta, SW Canada 50°06′N 110°48′W
83 K17 **Redcliff** Midlands, C Zimbabwe 19°00′S 29°49′E
182 L9 **Red Cliffs** Victoria, SE Australia 34°21′S 142°12′E
29 P17 **Red Cloud** Nebraska, C USA 40°05′N 98°31′W
22 L8 **Red Creek** ≈ Mississippi, S USA
11 P15 **Red Deer** Alberta, SW Canada 52°15′N 113°48′W
11 Q16 **Red Deer** ≈ Alberta, SW Canada
39 O11 **Red Devil** Alaska, USA 61°45′N 157°18′W
35 N3 **Redding** California, W USA 40°33′N 122°26′W
97 L20 **Redditch** W England, United Kingdom 52°19′N 01°56′W
29 P9 **Redfield** South Dakota, N USA 44°51′N 98°31′W
24 J12 **Redford** Texas, SW USA 29°31′N 104°19′W
45 V13 **Redhead** Trinidad, Trinidad and Tobago 10°44′N 60°58′W
182 I8 **Red Hill** South Australia 33°34′S 138°13′E
Red Hill *see* Pu‘u ‘Ula‘ula
26 K7 **Red Hills** *hill range* Kansas, C USA
13 T12 **Red Indian Lake** ◎ Newfoundland, Newfoundland and Labrador, E Canada
124 J16 **Redkino** Tverskaya Oblast’, W Russian Federation 56°41′N 36°07′E
12 A10 **Red Lake** Ontario, C Canada 51°00′N 93°55′W
36 I10 **Red Lake** *salt flat* Arizona, SW USA
29 S4 **Red Lake Falls** Minnesota, N USA 47°52′N 96°16′W
29 R4 **Red Lake River** ≈ Minnesota, N USA
35 U15 **Redlands** California, W USA 34°03′N 117°10′W
18 G16 **Red Lion** Pennsylvania, NE USA 39°53′N 76°36′W
33 U11 **Red Lodge** Montana, NW USA 45°11′N 109°15′W
32 H13 **Redmond** Oregon, NW USA 44°16′N 121°10′W
36 L5 **Redmond** Utah, W USA 39°00′N 111°51′W
32 H8 **Redmond** Washington, NW USA 47°40′N 122°07′W
Rednitz *see* Regnitz
29 T15 **Red Oak** Iowa, C USA 41°00′N 95°10′W
18 K12 **Red Oaks Mill** New York, NE USA 41°39′N 73°52′W
102 I7 **Redon** Ille-et-Vilaine, NW France 47°39′N 02°05′W
45 W10 **Redonda** *island* SW Antigua and Barbuda
104 G4 **Redondela** Galicia, NW Spain 42°17′N 08°36′W
104 H11 **Redondo** Évora, S Portugal 38°38′N 07°32′W
39 Q12 **Redoubt Volcano** ▲ Alaska, USA 60°29′N 152°44′W
11 Y16 **Red River** ≈ Canada/USA
129 U12 **Red River** *var.* Yuan, *Chin.* Yuan Jiang, *Vtn.* Sông Hông Hà. ≈ China/Vietnam
25 W4 **Red River** ≈ S USA
22 H7 **Red River** ≈ Louisiana, S USA
30 M6 **Red River** ≈ Wisconsin, N USA
Red Rock, Lake *see* Red Rock Reservoir
29 W14 **Red Rock Reservoir** *var.* Lake Red Rock. ⊠ Iowa, C USA
80 H7 **Red Sea** ◈ *state* NE Sudan
75 Y9 **Red Sea** *var.* Sinus Arabicus. *sea* Africa/Asia
21 T11 **Red Springs** North Carolina, SE USA 34°49′N 79°10′W
8 I9 **Redstone** ≈ Northwest Territories, NW Canada
11 V17 **Redvers** Saskatchewan, S Canada 49°31′N 101°33′W
77 P13 **Red Volta** *var.* Nazinon, *Fr.* Volta Rouge. ≈ Burkina Faso/Ghana
11 Q14 **Redwater** Alberta, SW Canada 53°57′N 113°06′W
28 M16 **Red Willow Creek** ≈ Nebraska, C USA
29 W9 **Red Wing** Minnesota, N USA 44°33′N 92°31′W
35 N9 **Redwood City** California, W USA 37°29′N 122°13′W
29 T9 **Redwood Falls** Minnesota, N USA 44°33′N 95°07′W
31 P7 **Reed City** Michigan, N USA 43°52′N 85°30′W
28 K6 **Reeder** North Dakota, N USA 46°03′N 102°55′W
35 R11 **Reedley** California, W USA 36°35′N 119°27′W
33 T11 **Reedpoint** Montana, NW USA 45°41′N 109°33′W
30 K8 **Reedsburg** Wisconsin, N USA 43°33′N 90°03′W
32 E13 **Reedsport** Oregon, NW USA 43°42′N 124°06′W
187 Q9 **Reef Islands** *island group* Santa Cruz Islands, E Solomon Islands
185 H16 **Reefton** West Coast, South Island, New Zealand 42°07′S 171°53′E
20 F8 **Reelfoot Lake** ◎ Tennessee, S USA
97 D17 **Ree, Lough** *Ir.* Loch Rí. ◎ C Ireland
Reengus *see* Rīngas
35 U4 **Reese River** ≈ Nevada, W USA
98 M8 **Reest** ≈ E Netherlands
Reetz Neumark *see* Recz
Reevhtse *see* Røssvatnet
137 N13 **Refahiye** Erzincan, C Turkey 39°54′N 38°45′E
23 N4 **Reform** Alabama, S USA 33°22′N 88°01′W
95 K20 **Reftele** Jönköping, S Sweden 57°10′N 13°34′E
25 T14 **Refugio** Texas, SW USA 28°19′N 97°18′W
110 E8 **Rega** ≈ NW Poland
Regar *see* Tursunzoda
101 O21 **Regen** Bayern, SE Germany 48°57′N 13°10′E
101 M20 **Regen** ≈ SE Germany
101 M21 **Regensburg** *Eng.* Ratisbon, *Fr.* Ratisbonne, *hist.* Ratisbona; *anc.* Castra Regina, Reginum. Bayern, SE Germany 49°01′N 12°06′E
101 M21 **Regenstauf** Bayern, SE Germany 49°06′N 12°07′E
148 M10 **Rēgestān** *var.* Registan *prev.* Rīgestān. S Afghanistan
74 I10 **Reggane** C Algeria 26°46′N 00°09′E
98 N9 **Regge** ≈ E Netherlands
Reggio *see* Reggio nell’Emilia
Reggio Calabria *see* Reggio di Calabria
107 M23 **Reggio di Calabria** *var.* Reggio Calabria, *Gk.* Rhegion; *anc.* Regium, Rhegium. Calabria, SW Italy 38°06′N 15°39′E
Reggio Emilia *see* Reggio nell’Emilia
106 F9 **Reggio nell’Emilia** *var.* Reggio Emilia, *abbrev.* Reggio; *anc.* Regium Lepidum. Emilia-Romagna, N Italy 44°42′N 10°37′E
116 I10 **Reghin** *Ger.* Sächsisch-Reen, *Hung.* Szászrégen; *prev.* Reghinul Săsesc, *Ger.* Sächsisch-Regen. Mureş, C Romania 46°46′N 24°41′E
Reghinul Săsesc *see* Reghin
11 U16 **Regina** *province capital* Saskatchewan, S Canada 50°25′N 104°39′W
55 Z10 **Régina** E French Guiana 04°20′N 52°07′W
11 U16 **Regina** ✈ Saskatchewan, S Canada 50°21′N 104°43′W
11 U16 **Regina Beach** Saskatchewan, S Canada 50°44′N 105°03′W
Reginum *see* Regensburg
Région du Haut-Congo *see* Haut-Congo
Registan *see* Rēgestān
60 L11 **Registro** São Paulo, S Brazil 24°30′S 47°50′W
Regium *see* Reggio di Calabria
Regium Lepidum *see* Reggio nell’Emilia
101 K19 **Regnitz** *var.* Rednitz. ≈ SE Germany
40 K10 **Regocijo** Durango, W Mexico 23°35′N 105°11′W
104 H12 **Reguengos de Monsaraz** Évora, S Portugal 38°25′N 07°32′W
101 M18 **Rehau** Bayern, E Germany 50°15′N 12°03′E
83 D19 **Rehoboth** Hardap, C Namibia 23°18′S 17°03′E
21 Z4 **Rehoboth Beach** Delaware, NE USA 38°42′N 75°03′W
138 F10 **Rehovot** ; *prev.* Reḥovot. Central, C Israel 31°54′N 34°49′E
Reḥovot *see* Rehovot
81 J20 **Rei** *spring/well* S Kenya 03°24′S 39°18′E
Reichenau *see* Rychnov nad Kněžnou
Reichenau *see* Bogatynia, Poland
101 M17 **Reichenbach** *var.* Reichenbach im Vogtland. Sachsen, E Germany 50°36′N 12°18′E
Reichenbach *see* Dzierżoniów
Reichenbach im Vogtland *see* Reichenbach
Reichenberg *see* Liberec
181 O11 **Reid** Western Australia 30°49′S 128°24′E
23 V6 **Reidsville** Georgia, SE USA 32°05′N 82°07′W
21 T8 **Reidsville** North Carolina, SE USA 36°21′N 79°39′W
Reifnitz *see* Ribnica
97 O22 **Reigate** SE England, United Kingdom 51°14′N 00°13′W
Reikjavik *see* Reykjavík
102 I10 **Ré, Île de** *island* W France
37 N15 **Reiley Peak** ▲ Arizona, SW USA 32°24′N 110°09′W
103 Q4 **Reims** *Eng.* Rheims; *anc.* Durocortorum, Remi. Marne, N France 49°16′N 04°01′E
63 G23 **Reina Adelaida, Archipiélago** *island group* S Chile
45 O16 **Reina Beatrix** ✈ (Oranjestad) C Aruba 12°30′N 69°57′W
108 F7 **Reinach** Aargau, W Switzerland 47°16′N 08°12′E
108 E6 **Reinach** Basel Landschaft, NW Switzerland 47°30′N 07°36′E
64 O11 **Reina Sofía** ✈ (Tenerife) Tenerife, Islas Canarias, Spain, NE Atlantic Ocean
29 W13 **Reinbeck** Iowa, C USA 42°19′N 92°36′W
100 J10 **Reinbek** Schleswig-Holstein, N Germany 53°31′N 10°15′E
11 U12 **Reindeer** ≈ Saskatchewan, C Canada
11 U11 **Reindeer Lake** ◎ Manitoba/Saskatchewan, C Canada
Reine-Charlotte, Îles de la *see* Queen Charlotte Islands
Reine-Élisabeth, Îles de la *see* Queen Elizabeth Islands
94 F13 **Reineskarvet** ▲ S Norway 60°38′N 07°48′E
184 H1 **Reinga, Cape** *headland* North Island, New Zealand 34°24′S 172°40′E
105 N3 **Reinosa** Cantabria, N Spain 43°01′N 04°09′W
109 R8 **Reisseck** ▲ S Austria 46°57′N 13°21′E
21 W3 **Reisterstown** Maryland, NE USA 39°27′N 76°46′W
Reisui *see* Yeosu
98 N5 **Reitdiep** ≈ NE Netherlands
191 V10 **Reitoru** *atoll* Îles Tuamotu, C French Polynesia
95 M17 **Rejmyre** Östergötland, S Sweden 58°49′N 15°55′E
Reka *see* Rijeka
Reka Ili *see* Ile/Ili He
Rekarne *see* Tumbo
Rekhovot *see* Rehovot
8 K9 **Reliance** Northwest Territories, C Canada 62°45′N 109°08′W
33 U16 **Reliance** Wyoming, C USA 41°42′N 109°13′W
74 I5 **Relizane** *var.* Ghelîzâne, Ghilizane. NW Algeria 35°45′N 00°33′E
182 I7 **Remarkable, Mount** ▲ South Australia 32°46′S 138°08′E
54 E8 **Remedios** Antioquia, N Colombia 07°02′N 74°42′W
43 Q16 **Remedios** Veraguas, W Panama 08°13′N 81°48′W
42 D8 **Remedios, Punta** *headland* SW El Salvador 13°31′N 89°48′W
Remi *see* Reims
99 N25 **Remich** Grevenmacher, SE Luxembourg 49°33′N 06°23′E
99 J19 **Remicourt** Liège, E Belgium 50°40′N 05°19′E
14 H8 **Rémigny, Lac** ◎ Québec, SE Canada
55 Z10 **Rémire** NE French Guiana 04°52′N 52°16′W
127 N13 **Remontnoye** Rostovskaya Oblast’, SW Russian Federation 46°35′N 43°38′E
171 U14 **Remoon** Pulau Kur, E Indonesia 05°18′S 131°59′E
99 L20 **Remouchamps** Liège, E Belgium 50°29′N 05°43′E
103 R15 **Remoulins** Gard, S France 43°56′N 04°34′E
173 X16 **Rempart, Mont du** *hill* W Mauritius
101 E15 **Remscheid** Nordrhein-Westfalen, W Germany 51°10′N 07°11′E
29 S12 **Remsen** Iowa, C USA 42°48′N 95°58′W
94 I12 **Rena** Hedmark, S Norway 61°08′N 11°21′E
94 I11 **Renåa** ≈ S Norway
Renaix *see* Ronse
118 H7 **Rencēni** N Latvia 57°43′N 25°25′E
118 D9 **Renda** W Latvia 57°04′N 22°18′E
107 N20 **Rende** Calabria, SW Italy 39°19′N 16°10′E
99 K21 **Rendeux** Luxembourg, SE Belgium 50°15′N 05°28′E
Rendina *see* Rentína
30 L16 **Rend Lake** ⊠ Illinois, N USA
186 K9 **Rendova** *island* New Georgia Islands, NW Solomon Islands
100 I8 **Rendsburg** Schleswig-Holstein, N Germany 54°18′N 09°40′E
108 B9 **Renens** Vaud, SW Switzerland 46°32′N 06°36′E
14 K12 **Renfrew** Ontario, SE Canada 45°28′N 76°44′W
96 I12 **Renfrew** *cultural region* SW Scotland, United Kingdom
168 L11 **Rengat** Sumatera, W Indonesia 0°26′S 102°38′E
153 W12 **Rengma Hills** ▲▲ NE India
62 H12 **Rengo** Libertador, C Chile 34°24′S 70°50′W
116 M12 **Reni** Odes’ka Oblast’, SW Ukraine 45°30′N 28°18′E
80 F11 **Renk** Upper Nile, NE South Sudan 11°48′N 32°49′E
93 L19 **Renko** Kanta-Häme, S Finland 60°52′N 24°16′E
98 L12 **Renkum** Gelderland, SE Netherlands 51°58′N 05°43′E
182 K9 **Renmark** South Australia 34°12′S 140°43′E
186 L10 **Rennell** *var.* Mu Nggava. *island* S Solomon Islands
186 M9 **Rennell and Bellona** *prev.* Central. ◈ *province* S Solomon Islands
181 Q4 **Renner Springs Roadhouse** Northern Territory, N Australia 18°12′S 133°48′E
102 I6 **Rennes** *Bret.* Roazon; *anc.* Condate. Ille-et-Vilaine, NW France 48°08′N 01°40′W
195 S16 **Rennick Glacier** *glacier* Antarctica
11 Y16 **Rennie** Manitoba, S Canada 49°51′N 95°28′W
35 Q5 **Reno** Nevada, W USA 39°32′N 119°49′W
106 H10 **Reno** ≈ N Italy
35 Q5 **Reno-Cannon** ✈ Nevada, W USA 39°26′N 119°42′W
83 F24 **Renoster** ≈ SW South Africa
15 T5 **Renouard, Lac** ◎ Québec, SE Canada
18 F13 **Renovo** Pennsylvania, NE USA 41°19′N 77°42′W
161 O3 **Renqiu** Hebei, E China 38°49′N 116°02′E
160 I9 **Renshou** Sichuan, C China 30°02′N 104°09′E
31 N12 **Rensselaer** Indiana, N USA 40°57′N 87°09′W
18 L11 **Rensselaer** New York, NE USA 42°38′N 73°44′W
Rentería *see* Errenteria
115 E17 **Rentína** *var.* Rendina. Thessalía, C Greece 39°04′N 21°58′E
29 T9 **Renville** Minnesota, N USA 44°48′N 95°13′W
77 O13 **Réo** W Burkina Faso 12°20′N 02°28′W
15 O12 **Repentigny** Québec, SE Canada 45°42′N 73°28′W
146 K13 **Repetek** Lebap Welaýaty, E Turkmenistan 38°40′N 63°12′E
93 J16 **Replot** *Fin.* Raippaluoto. *island* W Finland
Repola *see* Reboly
Reppen *see* Rzepin
Reps *see* Rupea
27 T7 **Republic** Missouri, C USA 37°07′N 93°28′W
32 K7 **Republic** Washington, NW USA 48°39′N 118°44′W
27 N3 **Republican River** ≈ Kansas/Nebraska, C USA
9 O7 **Repulse Bay** Northwest Terretories, N Canada 66°35′N 86°20′W
56 F9 **Requena** Loreto, NE Peru 05°05′S 73°52′W
105 R10 **Requena** Valenciana, E Spain 39°29′N 01°08′W
103 O14 **Réquista** Aveyron, S France 44°00′N 02°31′E
136 M12 **Reşadiye** Tokat, N Turkey 40°24′N 37°19′E
Reschenpass *see* Resia, Passo di
Reschitza *see* Reşiţa
113 N20 **Resen** *Turk.* Resne. SW FYR Macedonia 41°07′N 21°00′E
60 J11 **Reserva** Paraná, S Brazil 24°40′S 50°52′W
11 V15 **Reserve** Saskatchewan, S Canada 52°24′N 102°37′W
37 P13 **Reserve** New Mexico, SW USA 33°42′N 108°45′W
Reshetilovka *see* Reshetylivka
117 S6 **Reshetylivka** *Rus.* Reshetilovka. Poltavs’ka Oblast’, NE Ukraine 49°34′N 34°05′E
Resht *see* Rasht
139 S2 **Reshwan** *Ar.* Rashwān, *var.* Ŗashwan. Arbīl, N Iraq 36°28′N 43°54′E
106 F5 **Resia, Passo di** *Ger.* Reschenpass. *pass* Austria/Italy
Resicabánya *see* Reşiţa
62 N7 **Resistencia** Chaco, NE Argentina 27°27′S 58°56′W
116 F12 **Reşiţa** *Ger.* Reschitza, *Hung.* Resicabánya. Caraş-Severin, W Romania 45°14′N 21°58′E
Resne *see* Resen
197 N9 **Resolute** *Inuit* Qausuittuq. Cornwallis Island, Nunavut, N Canada 74°41′N 94°54′W
Resolution *see* Fort Resolution
9 T7 **Resolution Island** *island* Nunavut, NE Canada
185 A23 **Resolution Island** *island* SW New Zealand
Resta *see* Rasta
15 W7 **Restigouche** Québec, SE Canada 48°02′N 66°42′W
11 W17 **Reston** Manitoba, S Canada 49°33′N 101°03′W
14 H11 **Restoule Lake** ◎ Ontario, S Canada
54 F10 **Restrepo** Meta, C Colombia 04°20′N 73°29′W
42 B6 **Retalhuleu** Retalhuleu, SW Guatemala 14°31′N 91°40′W
42 A1 **Retalhuleu** *off.* Departamento de Retalhuleu. ◈ *department* SW Guatemala
Retalhuleu, Departamento de *see* Retalhuleu
97 N18 **Retford** C England, United Kingdom 53°18′N 00°52′W
103 Q3 **Rethel** Ardennes, N France 49°31′N 04°22′E
Rethimno/Réthimnon *see* Réthymno
115 I25 **Réthymno** *prev.* Rethimno, Réthimnon. Kríti, Greece, E Mediterranean Sea 35°21′N 24°29′E
Retiche, Alpi *see* Rhaetian Alps
99 J16 **Retie** Antwerpen, N Belgium 51°18′N 05°05′E
111 J21 **Rétság** Nógrád, N Hungary 47°57′N 19°08′E
109 W2 **Retz** Niederösterreich, NE Austria 48°46′N 15°58′E
173 N15 **Réunion** *off.* La Réunion. ◇ *French overseas department* W Indian Ocean
128 L17 **Réunion** *island* W Indian Ocean
105 U6 **Reus** Cataluña, E Spain 41°10′N 01°06′E
108 F7 **Reuss** ≈ NW Switzerland
99 J15 **Reusel** Noord-Brabant, S Netherlands 51°21′N 05°10′E
Reuţel *see* Ciuhuru
101 H22 **Reutlingen** Baden-Württemberg, S Germany 48°30′N 09°13′E
108 L7 **Reutte** Tirol, W Austria 47°30′N 10°44′E
99 M16 **Reuver** Limburg, SE Netherlands 51°17′N 06°05′E
28 K7 **Reva** South Dakota, N USA 45°30′N 103°03′W
Reval/Revel *see* Tallinn
124 J4 **Revda** Murmanskaya Oblast’, NW Russian Federation 67°57′N 34°29′E
122 F6 **Revda** Sverdlovskaya Oblast’, C Russian Federation 56°48′N 59°42′E
103 N16 **Revel** Haute-Garonne, S France 43°27′N 01°59′E
11 O16 **Revelstoke** British Columbia, SW Canada 51°02′N 118°12′W
43 N13 **Reventazón, Río** ≈ E Costa Rica
106 G9 **Revere** Lombardia, N Italy 45°03′N 11°07′E
39 Y14 **Revillagigedo Island** *island* Alexander Archipelago, Alaska, USA
193 R7 **Revillagigedo Islands** Mexico
103 R3 **Revin** Ardennes, N France 49°57′N 04°39′E
92 O3 **Revnosa** *headland* C Svalbard 78°03′N 18°52′E
Revolyutsii, Pik *see* Revolyutsiya, Qullai
147 T13 **Revolyutsiya, Qullai** *Rus.* Pik Revolyutsii. ▲ SE Tajikistan 38°40′N 72°26′E
111 L19 **Revúca** *Ger.* Grossrauschenbach, *Hung.* Nagyrőce. Banskobystrický Kraj, C Slovakia 48°40′N 20°10′E
154 K9 **Rewa** Madhya Pradesh, C India 24°32′N 81°18′E
139 T2 **Rewanduz** *Ar.* Rawāndūz, *var.* Ŗawāndiz. Arbīl, N Iraq 36°38′N 44°32′E
152 I11 **Rewāri** Haryāna, N India 28°14′N 76°38′E
33 R14 **Rexburg** Idaho, NW USA 43°49′N 111°47′W
78 G13 **Rey Bouba** Nord, NE Cameroon 08°40′N 14°11′E
92 L3 **Reyðarfjörður** Austurland, E Iceland 65°02′N 14°12′W
57 K16 **Reyes** El Beni, NW Bolivia 14°17′S 67°18′W
34 L8 **Reyes, Point** *headland* California, W USA 37°59′N 123°01′W
54 B12 **Reyes, Punta** *headland* SW Colombia 02°44′N 78°08′W
136 L17 **Reyhanlı** Hatay, S Turkey 36°16′N 36°35′E
43 U16 **Rey, Isla del** *island* Archipiélago de las Perlas, SE Panama
92 H2 **Reykhólar** Vestfirðir, W Iceland 65°28′N 22°12′W
92 K2 **Reykjahlíð** Norðurland Eystra, NE Iceland 65°37′N 16°54′W
197 O16 **Reykjanes Basin** *var.* Irminger Basin. *undersea feature* N Atlantic Ocean 62°30′N 33°30′W
197 N17 **Reykjanes Ridge** *undersea feature* N Atlantic Ocean 62°00′N 27°00′W
92 H4 **Reykjavík** *var.* Reikjavik. ● (Iceland) Höfudhborgarsvaedhi, W Iceland 64°08′N 21°54′W
18 D13 **Reynoldsville** Pennsylvania, NE USA 41°04′N 78°51′W
41 P8 **Reynosa** Tamaulipas, C Mexico 26°03′N 98°19′W
Reẓa’īyeh *see* Orūmīyeh
Reẓā’īyeh, Daryācheh-ye *see* Orūmīyeh, Daryācheh-ye
102 I8 **Rezé** Loire-Atlantique, NW France 47°10′N 01°36′W
118 K10 **Rēzekne** *Ger.* Rositten; *prev. Rus.* Rezhitsa. SE Latvia 56°31′N 27°22′E
139 U2 **Rezge** *Ar.* Razkah, *var.* Ŗazga. As Sulaymānīyah, E Iraq 36°25′N 45°06′E
Rezhitsa *see* Rēzekne
117 N9 **Rezina** NE Moldova 47°44′N 28°58′E
114 N11 **Rezovo** *Turk.* Rezve. Burgas, E Bulgaria 42°00′N 28°00′E
114 N11 **Rezovska Reka** *Turk.* Rezve Deresi. ≈ Bulgaria/Turkey *see also* Rezve Deresi
Rezovska Reka *see* Rezve Deresi
Rezve *see* Rezovo
114 N11 **Rezve Deresi** *Bul.* Rezovska Reka. ≈ Bulgaria/Turkey *see also* Rezovska Reka
Rezve Deresi *see* Rezovska Reka
Rhadames *see* Ghadāmis
Rhaedestus *see* Tekirdağ
108 J10 **Rhaetian Alps** *Fr.* Alpes Rhétiques, *Ger.* Rätische Alpen, *It.* Alpi Retiche. ▲▲ C Europe
108 I8 **Rhätikon** ▲▲ C Europe
101 G14 **Rheda-Wiedenbrück** Nordrhein-Westfalen, W Germany 51°51′N 08°17′E
98 M12 **Rheden** Gelderland, E Netherlands 52°01′N 06°03′E
Rhegion/Rhegium *see* Reggio di Calabria
Rheims *see* Reims
Rhein *see* Rhine
101 E17 **Rheinbach** Nordrhein-Westfalen, W Germany 50°37′N 06°57′E
100 F13 **Rheine** *var.* Rheine in Westfalen. Nordrhein-Westfalen, NW Germany 52°17′N 07°27′E
Rheine in Westfalen *see* Rheine
Rheinfeld *see* Rheinfelden
101 F24 **Rheinfelden** Baden-Württemberg, S Germany 47°34′N 07°46′E
108 E6 **Rheinfelden** *var.* Rheinfeld. Aargau, N Switzerland 47°33′N 07°47′E
101 E17 **Rheinisches Schiefergebirge** *var.* Rhine State Uplands, *Eng.* Rhenish Slate Mountains. ▲▲ W Germany
101 D18 **Rheinland-Pfalz** *Eng.* Rhineland-Palatinate, *Fr.* Rhénanie-Palatinat. ◈ *state* W Germany
101 G18 **Rhein Main** ✈ (Frankfurt am Main) Hessen, W Germany 50°03′N 08°33′E
Rhénanie du Nord-Westphalie *see* Nordrhein-Westfalen
Rhénanie-Palatinat *see* Rheinland-Pfalz
98 K12 **Rhenen** Utrecht, C Netherlands 52°01′N 06°02′E
Rhenish Slate Mountains *see* Rheinisches Schiefergebirge
Rhétiques, Alpes *see* Rhaetian Alps
100 N10 **Rhin** ≈ NE Germany
Rhin *see* Rhine
84 F10 **Rhine** *Dut.* Rijn, *Fr.* Rhin, *Ger.* Rhein. ≈ W Europe
30 L5 **Rhinelander** Wisconsin, N USA 45°39′N 89°23′W
Rhineland-Palatinate *see* Rheinland-Pfalz
Rhine State Uplands *see* Rheinisches Schiefergebirge
100 N11 **Rhinkanal** *canal* NE Germany
81 F17 **Rhino Camp** NW Uganda 02°58′N 31°24′E
74 D7 **Rhir, Cap** *headland* W Morocco 30°40′N 09°54′W
106 D7 **Rho** Lombardia, N Italy 45°32′N 09°02′E
19 N12 **Rhode Island** *off.* State of Rhode Island and Providence Plantations, *also known as* Little Rhody, Ocean State. ◈ *state* NE USA
19 O13 **Rhode Island** *island* Rhode Island, NE USA
19 O13 **Rhode Island Sound** *sound* Maine/Rhode Island, NE USA
Rhodes *see* Ródos
Rhode-Saint-Genèse *see* Sint-Genesius-Rode
84 L14 **Rhodes Basin** *undersea feature* E Mediterranean Sea 35°55′N 28°30′E
Rhodesia *see* Zimbabwe
114 I12 **Rhodope Mountains** *var.* Rodhópi Óri, *Bul.* Rhodope Planina, Rodopi, *Gk.* Orosirá Rodhópis, *Turk.* Dospad Dagh. ▲▲ Bulgaria/Greece
Rhodope Planina *see* Rhodope Mountains
101 I18 **Rhön** ▲▲ C Germany
103 Q10 **Rhône** ◈ *department* E France
86 C12 **Rhône** ≈ France/Switzerland
103 R12 **Rhône-Alpes** ◈ *region* E France
98 G13 **Rhoon** Zuid-Holland, SW Netherlands 51°52′N 04°25′E
96 G9 **Rhum** *var.* Rum. *island* W Scotland, United Kingdom
Rhuthun *see* Ruthin
97 J18 **Rhyl** NE Wales, United Kingdom 53°19′N 03°28′W
59 K18 **Rialma** Goiás, S Brazil 15°22′S 49°35′W
104 L3 **Riaño** Castilla y León, N Spain 42°59′N 05°00′W
105 O9 **Riansáres** ≈ C Spain
152 H6 **Riāsi** Jammu and Kashmir, NW India 33°03′N 74°51′E
168 K10 **Riau** *off.* Propinsi Riau. ◈ *province* W Indonesia
Riau Archipelago *see* Riau, Kepulauan
168 M11 **Riau, Kepulauan** *var.* Riau Archipelago, *Dut.* Riouw-Archipel. *island group* W Indonesia
Riau, Propinsi *see* Riau
105 O6 **Riaza** Castilla y León, N Spain 41°17′N 03°29′W
105 N6 **Riaza** ≈ N Spain
81 K17 **Riba** *spring/well* NE Kenya 01°56′N 40°38′E
104 H4 **Ribadavia** Galicia, NW Spain 42°17′N 08°08′W
104 J2 **Ribadeo** Galicia, NW Spain 43°32′N 07°04′W
104 L2 **Ribadesella** *var.* Ribeseya. Asturias, N Spain 43°27′N 05°04′W
104 G10 **Ribatejo** *former province* C Portugal
83 P15 **Ribáuè** Nampula, N Mozambique 14°56′S 38°19′E
97 K17 **Ribble** ≈ NW England, United Kingdom
95 F23 **Ribe** Syddtjylland, W Denmark 55°20′N 08°47′E
Ribeira *see* Sanata Uxía de Ribeira
64 O5 **Ribeira Brava** Madeira, Portugal, NE Atlantic Ocean 32°39′N 17°04′W
64 P3 **Ribeira Grande** São Miguel, Azores, Portugal, NE Atlantic Ocean 37°34′N 25°32′W
60 L8 **Ribeirão Preto** São Paulo, S Brazil 21°09′S 47°48′W
60 L11 **Ribeira, Rio** ≈ S Brazil
107 I24 **Ribera** Sicilia, Italy, C Mediterranean Sea 37°31′N 13°16′E
57 L14 **Riberalta** El Beni, N Bolivia 11°01′S 66°04′W
105 W4 **Ribes de Freser** Cataluña, NE Spain 42°18′N 02°11′E
Ribeseya *see* Ribadesella
30 L6 **Rib Mountain** ▲ Wisconsin, N USA 44°55′N 89°41′W
109 U12 **Ribnica** *Ger.* Reifnitz. S Slovenia 45°46′N 14°40′E
117 N9 **Rîbniţa** *var.* Râbniţa, *Rus.* Rybnitsa. NE Moldova 47°46′N 29°01′E
100 M8 **Ribnitz-Damgarten** Mecklenburg-Vorpommern, NE Germany 54°14′N 12°25′E
111 D16 **Říčany** *Ger.* Ritschan. Středočesky Kraj, W Czech Republic 49°59′N 14°40′E
29 U7 **Rice** Minnesota, N USA 45°42′N 94°10′W
30 J5 **Rice Lake** Wisconsin, N USA 45°33′N 91°43′W
14 E8 **Rice Lake** ◎ Ontario, S Canada
14 I15 **Rice Lake** ◎ Ontario, SE Canada
23 V3 **Richard B. Russell Lake** ⊠ Georgia, SE USA
25 U6 **Richardson** Texas, SW USA 32°55′N 96°44′W
11 R11 **Richardson** ≈ Alberta, C Canada
10 I3 **Richardson Mountains** ▲▲ Yukon, NW Canada
185 C21 **Richardson Mountains** ▲▲ South Island, New Zealand
42 F3 **Richardson Peak** ▲ SE Belize 16°34′N 88°46′W
76 G10 **Richard Toll** N Senegal 16°28′N 15°44′W
28 L5 **Richardton** North Dakota, N USA 46°52′N 102°19′W
14 F13 **Rich, Cape** *headland* Ontario, S Canada 44°42′N 80°37′W
102 L8 **Richelieu** Indre-et-Loire, C France 47°01′N 00°18′E
33 P15 **Richfield** Idaho, NW USA 43°03′N 114°11′W
36 K5 **Richfield** Utah, W USA 38°45′N 112°05′W
18 J10 **Richfield Springs** New York, NE USA 42°52′N 74°57′W
18 M6 **Richford** Vermont, NE USA 44°59′N 72°37′W
27 R6 **Rich Hill** Missouri, C USA 38°06′N 94°22′W
13 P14 **Richibucto** New Brunswick, SE Canada 46°42′N 64°54′W
108 G8 **Richisau** Glarus, NE Switzerland 47°00′N 08°54′E
23 S6 **Richland** Georgia, SE USA 32°05′N 84°40′W
27 U6 **Richland** Missouri, C USA 37°51′N 92°24′W
25 U8 **Richland** Texas, SW USA 31°55′N 96°25′W
32 K10 **Richland** Washington, NW USA 46°17′N 119°16′W
30 K8 **Richland Center** Wisconsin, N USA 43°22′N 90°24′W
21 W11 **Richlands** North Carolina, SE USA 34°52′N 77°33′W
21 Q7 **Richlands** Virginia, NE USA 37°05′N 81°47′W
25 R9 **Richland Springs** Texas, SW USA 31°16′N 98°56′W
183 S8 **Richmond** New South Wales, SE Australia 33°36′S 150°44′E
10 L17 **Richmond** British Columbia, SW Canada 49°07′N 123°09′W
14 L13 **Richmond** Ontario, SE Canada 45°12′N 75°49′W
15 Q12 **Richmond** Québec, SE Canada 45°39′N 72°07′W
185 I14 **Richmond** Tasman, South Island, New Zealand 41°25′S 173°04′E
35 N8 **Richmond** California, W USA 37°57′N 122°22′W
31 Q14 **Richmond** Indiana, N USA 39°50′N 84°51′W
20 M6 **Richmond** Kentucky, S USA 37°45′N 84°19′W
27 S4 **Richmond** Missouri, C USA 39°15′N 93°59′W
25 V11 **Richmond** Texas, SW USA 29°36′N 95°48′W
36 L1 **Richmond** Utah, W USA 41°55′N 111°51′W
21 W6 **Richmond** *state capital* Virginia, NE USA 37°33′N 77°28′W
14 H15 **Richmond Hill** Ontario, S Canada 43°51′N 79°24′W
185 J15 **Richmond Range** ▲▲ South Island, New Zealand
27 S12 **Rich Mountain** ▲ Arkansas, C USA 34°37′N 94°17′W
31 S13 **Richwood** Ohio, N USA 40°25′N 83°18′W
21 R5 **Richwood** West Virginia, NE USA 38°13′N 80°31′W
104 K5 **Ricobayo, Embalse de** ⊠ NW Spain
Ricomagus *see* Riom
Ridā’ *see* Radā’
127 T5 **Ridder** Respublika Tatarstan, W Russian Federation 54°34′N 52°27′E
98 H13 **Ridderkerk** Zuid-Holland, SW Netherlands 51°52′N 04°35′E
33 N16 **Riddle** Idaho, NW USA 42°07′N 116°09′W
32 F14 **Riddle** Oregon, NW USA 42°57′N 123°21′W
14 L13 **Rideau** ≈ Ontario, SE Canada
35 T12 **Ridgecrest** California, W USA 35°37′N 117°40′W
18 L13 **Ridgefield** Connecticut, NE USA 41°16′N 73°30′W
22 K5 **Ridgeland** Mississippi, S USA 32°25′N 90°07′W
21 R15 **Ridgeland** South Carolina, SE USA 32°30′N 80°59′W
20 F8 **Ridgely** Tennessee, S USA 36°15′N 89°29′W
14 D17 **Ridgetown** Ontario, S Canada 42°27′N 81°52′W
21 R12 **Ridgeway** South Carolina, SE USA 34°17′N 80°56′W
Ridgeway *see* Ridgway
18 D13 **Ridgway** *var.* Ridgeway. Pennsylvania, NE USA 41°24′N 78°40′W
11 W16 **Riding Mountain** ▲▲ Manitoba, S Canada
Ried *see* Ried im Innkreis
109 R4 **Ried im Innkreis** *var.* Ried. Oberösterreich, NW Austria 48°13′N 13°29′E
109 X8 **Riegersburg** Steiermark, SE Austria 47°03′N 15°52′E
108 E6 **Riehen** Basel-Stadt, NW Switzerland 47°35′N 07°39′E
92 J9 **Riehppegáisá** *var.* Rieppe. ▲ N Norway 69°38′N 21°31′E
99 K18 **Riemst** Limburg, NE Belgium 50°49′N 05°36′E
Rieppe *see* Riehppegáisá
101 O15 **Riesa** Sachsen, E Germany 51°18′N 13°18′E
63 H24 **Riesco, Isla** *island* S Chile
107 K25 **Riesi** Sicilia, Italy, C Mediterranean Sea 37°17′N 14°05′E
83 I23 **Riet** ≈ SW South Africa
83 F25 **Riet** ≈ SW South Africa
118 D11 **Rietavas** Telšiai, W Lithuania 55°43′N 21°56′E
83 F19 **Rietfontein** Omaheke, E Namibia 21°58′S 20°58′E
107 I14 **Rieti** *anc.* Reate. Lazio, C Italy 42°24′N 12°51′E
84 D14 **Rif** *var.* Riff, Er Rif, Er Riff. ▲▲ N Morocco
138 I8 **Rīf Dimashq** *off.* Muḥāfaẓat Dimashq, *var.* Damascus, *Ar.* Ash Sham, Ash Shām, Damasco, Esh Sham, *Fr.* Damas. ◈ *governorate* S Syria
Riff *see* Rif
37 Q4 **Rifle** Colorado, C USA 39°30′N 107°46′W
31 R7 **Rifle River** ≈ Michigan, N USA
Rift Valley *see* Great Rift Valley
118 F9 **Rīga** *Eng.* Riga. ● C Latvia 56°57′N 24°08′E
Rigaer Bucht *see* Riga, Gulf of
118 F6 **Riga, Gulf of** *Est.* Liivi Laht, *Ger.* Rigaer Bucht, *Latv.* Rīgas Jūras Līcis, *Rus.* Rizhskiy Zaliv; *prev. Est.* Riia Laht. *gulf* Estonia/Latvia
Rīgas Jūras Līcis *see* Riga, Gulf of
15 N12 **Rigaud** ≈ Ontario/Québec, SE Canada
33 R14 **Rigby** Idaho, NW USA 43°40′N 111°54′W
Rīgestān *see* Rēgestān
32 M11 **Riggins** Idaho, NW USA 45°24′N 116°18′W
13 R8 **Rigolet** Newfoundland and Labrador, NE Canada 51°10′N 58°25′W
78 G9 **Rig-Rig** Kanem, W Chad 14°16′N 14°21′E
118 F4 **Riguldi** Läänemaa, W Estonia 59°07′N 23°34′E
Riia Laht *see* Riga, Gulf of
93 L19 **Riihimäki** Kanta-Häme, S Finland 60°45′N 24°45′E
195 U2 **Riiser-Larsen Peninsula** *peninsula* Antarctica
65 P22 **Riiser-Larsen Sea** *sea* Antarctica
Riisser Larsen Ice Shelf *see* Riiser-Larsenisen
40 D2 **Riíto** Sonora, NW Mexico 32°06′N 114°57′W
112 B9 **Rijeka** *Ger.* Sankt Veit am Flaum, *It.* Fiume, *Slvn.* Reka; *anc.* Tarsatica. Primorje-Gorski Kotar, NW Croatia 45°20′N 14°26′E
99 I14 **Rijen** Noord-Brabant, S Netherlands 51°35′N 04°55′E
99 H15 **Rijkevorsel** Antwerpen, N Belgium 51°23′N 04°43′E
Rijn *see* Rhine
98 G11 **Rijnsburg** Zuid-Holland, W Netherlands 52°12′N 04°27′E
Rijssel *see* Lille
98 N10 **Rijssen** Overijssel, E Netherlands 52°19′N 06°30′E
98 G12 **Rijswijk** *Eng.* Ryswick. Zuid-Holland, W Netherlands 52°03′N 04°20′E
92 I10 **Riksgränsen** Norrbotten, N Sweden 68°24′N 18°15′E
165 U4 **Rikubetsu** Hokkaidō, NE Japan 43°30′N 143°43′E
165 R9 **Rikuzen-Takata** Iwate, Honshū, C Japan 39°03′N 141°38′E
27 O4 **Riley** Kansas, C USA 39°18′N 96°49′W
99 I17 **Rillaar** Vlaams Brabant, C Belgium 50°58′N 04°58′E
Rí, Loch *see* Ree, Lough
114 G11 **Rilska Reka** ≈ W Bulgaria
77 T12 **Rima** ≈ N Nigeria
141 N7 **Rimah, Wādī ar** *var.* Wādī ar Rummah. *dry watercourse* C Saudi Arabia
Rimaszombat *see* Rimavská Sobota
191 R12 **Rimatara** *island* Îles Australes, SW French Polynesia
111 L20 **Rimavská Sobota** *Ger.* Gross-Steffelsdorf, *Hung.* Rimaszombat. Bankobystrický Kraj, C Slovakia 48°24′N 20°01′E
11 Q15 **Rimbey** Alberta, SW Canada 52°39′N 114°10′W
95 P15 **Rimbo** Stockholm, C Sweden 59°44′N 18°21′E
95 M18 **Rimforsa** Östergötland, S Sweden 58°06′N 15°40′E
106 I11 **Rimini** *anc.* Ariminum. Emilia-Romagna, N Italy 44°03′N 12°33′E
Rîmnicu-Sărat *see* Râmnicu Sărat
Rîmnicu Vîlcea *see* Râmnicu Vâlcea
149 Y3 **Rimo Muztāgh** ▲▲ India/Pakistan
15 U7 **Rimouski** Québec, SE Canada 48°26′N 68°32′W
158 M16 **Rinbung** Xizang Zizhiqu, W China 29°15′N 89°40′E
Rinchinlhumbe *see* Dzöölön
62 I5 **Rincón, Cerro** ▲ N Chile 24°01′S 67°19′W
104 M15 **Rincón de la Victoria** Andalucía, S Spain 36°43′N 04°18′W

◆ Country
● Country Capital
◇ Dependent Territory
○ Dependent Territory Capital
◈ Administrative Regions
✈ International Airport
▲ Mountain
▲▲ Mountain Range
Volcano
≈ River
◎ Lake
⊠ Reservoir

Rincón del Bonete, Lago Artificial de *see* Río Negro, Embalse del
105 *Q4* **Rincón de Soto** La Rioja, N Spain *42°15´N 01°50´W*
94 *G8* **Rindal** Møre og Romsdal, S Norway *63°02´N 09°09´E*
115 *J20* **Ríneia** *island* Kykládes, Greece, Aegean Sea
152 *H11* **Rīngas** *prev.* Reengus, Rīngus. Rājasthān, N India *27°18´N 75°27´E*
95 *H24* **Ringe** Syddtjylland, C Denmark *55°14´N 10°30´E*
94 *H11* **Ringebu** Oppland, S Norway *61°31´N 10°09´E*
Ringen *see* Rõngu
186 *K8* **Ringgi** Kolombangara, NW Solomon Islands *08°03´S 157°08´E*
23 *R1* **Ringgold** Georgia, SE USA *34°55´N 85°06´W*
22 *G5* **Ringgold** Louisiana, S USA *32°19´N 93°16´W*
25 *S5* **Ringgold** Texas, SW USA *33°47´N 97°56´W*
95 *E22* **Ringkøbing** Midtjylland, W Denmark *56°04´N 08°22´E*
95 *E22* **Ringkøbing Fjord** *fjord* W Denmark
33 *S10* **Ringling** Montana, NW USA *46°15´N 110°48´W*
27 *N13* **Ringling** Oklahoma, C USA *34°12´N 97°35´W*
94 *H13* **Ringsaker** Hedmark, S Norway *60°54´N 10°45´E*
95 *I23* **Ringsted** Sjælland, E Denmark *55°28´N 11°48´E*
Rīngus *see* Rīngas
92 *I9* **Ringvassøya** *Lapp.* Ráneš. *island* N Norway
18 *K13* **Ringwood** New Jersey, NE USA *41°06´N 74°15´W*
Rinn Duáin *see* Hook Head
100 *H13* **Rinteln** Niedersachsen, NW Germany *52°10´N 09°04´E*
115 *E18* **Río** Dytikí Elláda, S Greece *38°18´N 21°48´E*
Rio *see* Rio de Janeiro
56 *C7* **Riobamba** Chimborazo, C Ecuador *01°44´S 78°40´W*
60 *P9* **Rio Bonito** Rio de Janeiro, SE Brazil *22°42´S 42°38´W*
59 *C16* **Rio Branco** *state capital* Acre, W Brazil *09°59´S 67°49´W*
61 *H18* **Río Branco** Cerro Largo, NE Uruguay *32°32´S 53°28´W*
Rio Branco, Território de *see* Roraima
41 *P8* **Río Bravo** Tamaulipas, C Mexico *25°57´N 98°03´W*
63 *G16* **Río Bueno** Los Ríos, C Chile *40°20´S 72°55´W*
55 *P5* **Río Caribe** Sucre, NE Venezuela *10°43´N 63°06´W*
54 *M5* **Río Chico** Miranda, N Venezuela *10°18´N 66°00´W*
63 *H18* **Río Cisnes** Aisén, S Chile *44°29´S 71°15´W*
60 *L9* **Rio Claro** São Paulo, S Brazil *22°19´S 47°35´W*
45 *V14* **Rio Claro** Trinidad, Trinidad and Tobago *10°18´N 61°11´W*
54 *J5* **Río Claro** Lara, N Venezuela *09°54´N 69°23´W*
63 *K15* **Río Colorado** Río Negro, E Argentina *39°01´S 64°05´W*
62 *K11* **Río Cuarto** Córdoba, C Argentina *33°06´S 64°20´W*
60 *P10* **Rio de Janeiro** *var.* Rio. *state capital* Rio de Janeiro, SE Brazil *22°53´S 43°17´W*
60 *P9* **Rio de Janeiro** *off.* Estado do Rio de Janeiro. ◈ *state* SE Brazil
Rio de Janeiro, Estado do *see* Rio de Janeiro
43 *R17* **Río de Jesús** Veraguas, S Panama *07°58´N 81°01´W*
34 *K3* **Rio Dell** California, W USA *40°30´N 124°07´W*
60 *K13* **Rio do Sul** Santa Catarina, S Brazil *27°15´S 49°37´W*
63 *I23* **Río Gallegos** *var.* Gallegos, Puerto Gallegos. Santa Cruz, S Argentina *51°40´S 69°21´W*
63 *J24* **Río Grande** Tierra del Fuego, S Argentina *53°45´S 67°46´W*
61 *I18* **Rio Grande** *var.* São Pedro do Rio Grande do Sul. Rio Grande do Sul, S Brazil *32°03´S 52°08´W*
40 *L10* **Río Grande** Zacatecas, C Mexico *23°50´N 103°20´W*
42 *J9* **Río Grande** León, NW Nicaragua *12°59´N 86°34´W*
45 *V5* **Río Grande** E Puerto Rico *18°23´N 65°51´W*
24 *I9* **Rio Grande** ≈ Texas, SW USA
25 *R17* **Rio Grande City** Texas, SW USA *26°24´N 98°50´W*
59 *P14* **Rio Grande do Norte** *off.* Estado do Rio Grande do Norte. ◈ *state* E Brazil
Rio Grande do Norte, Estado do *see* Rio Grande do Norte
61 *G15* **Rio Grande do Sul** *off.* Estado do Rio Grande do Sul. ◈ *state* S Brazil
Rio Grande do Sul, Estado do *see* Rio Grande do Sul
65 *M17* **Rio Grande Fracture Zone** *tectonic feature* C Atlantic Ocean
65 *J18* **Rio Grande Gap** *undersea feature* S Atlantic Ocean
Rio Grande Plateau *see* Rio Grande Rise
65 *J18* **Rio Grande Rise** *var.* Rio Grande Plateau. *undersea feature* SW Atlantic Ocean *31°00´S 35°00´W*
54 *G4* **Ríohacha** La Guajira, N Colombia *11°23´N 72°47´W*
43 *S16* **Río Hato** Coclé, C Panama *08°21´N 80°10´W*
25 *T17* **Rio Hondo** Texas, SW USA *26°14´N 97°34´W*
56 *D10* **Rioja** San Martín, N Peru *06°02´S 77°10´W*
41 *Y11* **Río Lagartos** Yucatán, SE Mexico *21°35´N 88°08´W*
103 *P11* **Riom** *anc.* Ricomagus. Puy-de-Dôme, C France *45°54´N 03°06´E*
104 *F10* **Rio Maior** Santarém, C Portugal *39°20´N 08°55´W*
103 *O12* **Riom-ès-Montagnes** Cantal, C France *45°15´N 02°39´E*
60 *J12* **Rio Negro** Paraná, S Brazil *26°06´S 49°46´W*
63 *I15* **Río Negro** *off.* Provincia de Río Negro. ◈ *province* C Argentina
61 *D18* **Río Negro** ◈ *department* W Uruguay
47 *V12* **Río Negro, Embalse del** *var.* Lago Artificial de Rincón del Bonete. ⊠ C Uruguay
Río Negro, Provincia de *see* Río Negro
107 *M17* **Rionero in Vulture** Basilicata, S Italy *40°55´N 15°40´E*
137 *S9* **Rioni** ≈ W Georgia
105 *P12* **Riópar** Castilla-La Mancha, C Spain *38°31´N 02°27´W*
61 *H16* **Rio Pardo** Rio Grande do Sul, S Brazil *29°41´S 52°25´W*
37 *R11* **Rio Rancho Estates** New Mexico, SW USA *35°14´N 106°40´W*
42 *L11* **Río San Juan** ◈ *department* S Nicaragua
54 *E9* **Ríosucio** Caldas, W Colombia *05°26´N 75°44´W*
54 *C7* **Ríosucio** Chocó, NW Colombia *07°25´N 77°05´W*
62 *K10* **Río Tercero** Córdoba, C Argentina *32°15´S 64°08´W*
42 *K5* **Río Tinto, Sierra** ▲▲ NE Honduras
54 *J5* **Río Tocuyo** Lara, N Venezuela *10°18´N 70°00´W*
Riouw-Archipel *see* Riau, Kepulauan
59 *J19* **Rio Verde** Goiás, C Brazil *17°50´S 50°55´W*
41 *O12* **Río Verde** *var.* Rioverde. San Luis Potosí, C Mexico *21°58´N 100°W*
Rioverde *see* Río Verde
35 *O8* **Rio Vista** California, W USA *38°09´N 121°42´W*
112 *M11* **Ripanj** Serbia, N Serbia *44°37´N 20°30´E*
106 *J13* **Ripatransone** Marche, C Italy *43°00´N 13°45´E*
22 *M2* **Ripley** Mississippi, S USA *34°43´N 88°57´W*
31 *R15* **Ripley** Ohio, N USA *38°45´N 83°51´W*
20 *F9* **Ripley** Tennessee, S USA *35°43´N 89°30´W*
21 *Q4* **Ripley** West Virginia, NE USA *38°49´N 81°44´W*
105 *W4* **Ripoll** Cataluña, NE Spain *42°12´N 02°12´E*
97 *M16* **Ripon** N England, United Kingdom *54°07´N 01°31´W*
30 *M7* **Ripon** Wisconsin, N USA *43°52´N 88°48´W*
107 *L24* **Riposto** Sicilia, Italy, C Mediterranean Sea *37°44´N 15°13´E*
99 *L14* **Rips** Noord-Brabant, SE Netherlands *51°31´N 05°49´E*
54 *D9* **Risaralda** *off.* Departamento de Risaralda. ◈ *province* C Colombia
Risaralda, Departamento de *see* Risaralda
116 *L8* **Rîşcani** *var.* Râşcani, *Rus.* Ryshkany. NW Moldova *47°55´N 27°31´E*
152 *J9* **Rishikesh** Uttarakhand, N India *30°06´N 78°16´E*
165 *S1* **Rishiri-tō** *var.* Risiri Tô. *island* NE Japan
165 *S1* **Rishiri-yama** ⛰ Rishiri-tō, NE Japan *45°11´N 141°11´E*
25 *R7* **Rising Star** Texas, SW USA *32°06´N 98°57´W*
31 *Q15* **Rising Sun** Indiana, N USA *38°58´N 84°53´W*
Risiri Tô *see* Rishiri-tō
102 *L4* **Risle** ≈ N France
Rîşno *see* Râşnov
27 *V13* **Rison** Arkansas, C USA *33°58´N 92°11´W*
95 *G17* **Risør** Aust-Agder, S Norway *58°44´N 09°15´E*
92 *H10* **Risøyhamn** Nordland, C Norway *69°00´N 15°37´E*
101 *I23* **Riss** ≈ S Germany
118 *G4* **Risti** *Ger.* Kreuz. Läänemaa, W Estonia *59°01´N 24°01´E*
15 *V8* **Ristigouche** ≈ Québec, SE Canada
93 *N18* **Ristiina** Etelä-Savo, E Finland *61°32´N 27°15´E*
93 *N14* **Ristijärvi** Kainuu, C Finland *64°30´N 28°15´E*
188 *C14* **Ritidian Point** *headland* N Guam *13°39´N 144°51´E*
Ritschan *see* Říčany
35 *R9* **Ritter, Mount** ▲ California, W USA *37°40´N 119°10´W*
31 *T12* **Rittman** Ohio, N USA *40°58´N 81°46´W*
32 *L9* **Ritzville** Washington, NW USA *47°07´N 118°22´W*
Riva *see* Riva del Garda
61 *A21* **Rivadavia** Buenos Aires, E Argentina *35°29´S 62°59´W*
106 *F7* **Riva del Garda** *var.* Riva. Trentino-Alto Adige, N Italy *45°54´N 10°50´E*
106 *B8* **Rivarolo Canavese** Piemonte, W Italy *45°21´N 07°42´E*
42 *K11* **Rivas** Rivas, SW Nicaragua *11°26´N 85°50´W*
42 *J11* **Rivas** ◈ *department* SW Nicaragua
103 *R11* **Rive-de-Gier** Loire, E France *45°31´N 04°36´E*
61 *A22* **Rivera** Buenos Aires, E Argentina *37°13´S 63°14´W*
61 *F16* **Rivera** Rivera, NE Uruguay *30°54´S 55°31´W*
61 *F17* **Rivera** ◈ *department* NE Uruguay
35 *P9* **Riverbank** California, W USA *37°43´N 120°59´W*
76 *K17* **River Cess** SW Liberia *05°28´N 09°32´W*
28 *M4* **Riverdale** North Dakota, N USA *47°29´N 101°22´W*
30 *I6* **River Falls** Wisconsin, N USA *44°52´N 92°38´W*
11 *T16* **Riverhurst** Saskatchewan, S Canada *50°52´N 106°49´W*
183 *O10* **Riverina** *physical region* New South Wales, SE Australia
80 *G8* **River Nile** ◈ *state* NE Sudan
63 *F19* **Rivero, Isla** *island* Archipiélago de los Chonos, S Chile
11 *W16* **Rivers** Manitoba, S Canada *50°02´N 100°14´W*
77 *U17* **Rivers** ◈ *state* S Nigeria
185 *D23* **Riversdale** Southland, South Island, New Zealand *45°54´S 168°44´E*
83 *F26* **Riversdale** Western Cape, SW South Africa *34°05´S 21°15´E*
35 *U15* **Riverside** California, W USA *33°58´N 117°25´W*
25 *W9* **Riverside** Texas, SW USA *30°51´N 95°24´W*
37 *U3* **Riverside Reservoir** ⊠ Colorado, C USA
10 *K15* **Rivers Inlet** British Columbia, SW Canada *51°43´N 127°19´W*
10 *K15* **Rivers Inlet** *inlet* British Columbia, SW Canada
11 *X15* **Riverton** Manitoba, S Canada *51°00´N 97°00´W*
185 *C24* **Riverton** Southland, South Island, New Zealand *46°20´S 168°02´E*
30 *L13* **Riverton** Illinois, N USA *39°50´N 89°31´W*
36 *L3* **Riverton** Utah, W USA *40°32´N 111°57´W*
33 *V15* **Riverton** Wyoming, C USA *43°01´N 108°22´W*
14 *G10* **River Valley** Ontario, S Canada *46°36´N 80°09´W*
13 *P14* **Riverview** New Brunswick, SE Canada *46°03´N 64°47´W*
103 *O17* **Rivesaltes** Pyrénées-Orientales, S France *42°46´N 02°48´E*
36 *H11* **Riviera** Arizona, SW USA *35°06´N 114°36´W*
25 *S15* **Riviera** Texas, SW USA *27°15´N 97°48´W*
23 *Z14* **Riviera Beach** Florida, SE USA *26°46´N 80°03´W*
15 *Q10* **Rivière-à-Pierre** Québec, SE Canada *46°59´N 72°12´W*
15 *T9* **Rivière-Bleue** Québec, SE Canada *47°26´N 69°02´W*
15 *T8* **Rivière-du-Loup** Québec, SE Canada *47°49´N 69°32´W*
173 *Y15* **Rivière du Rempart** NE Mauritius *20°06´S 57°41´E*
45 *R12* **Rivière-Pilote** S Martinique *14°29´N 60°54´W*
173 *O17* **Rivière St-Etienne, Pointe de la** *headland* SW Réunion
13 *S10* **Rivière-St-Paul** Québec, E Canada *51°26´N 57°52´W*
Rivière Sèche *see* Bel Air
116 *K4* **Rivne** *Pol.* Równe, *Rus.* Rovno. Rivnens'ka Oblast', NW Ukraine *50°37´N 26°16´E*
Rivne *see* Rivnens'ka Oblast'
116 *K3* **Rivnens'ka Oblast'** *var.* Rivne, *Rus.* Rovenskaya Oblast'. ◈ *province* NW Ukraine
106 *B8* **Rivoli** Piemonte, NW Italy *45°04´N 07°31´E*
159 *Q14* **Riwoqê** *var.* Racaka. Xizang Zizhiqu, W China *31°10´N 96°25´E*
99 *H19* **Rixensart** Walloon Brabant, C Belgium *50°43´N 04°32´E*
Riyadh/Riyāḍ, Minṭaqat ar *see* Ar Riyāḍ
Riyāq *see* Rayak
Rizaiyeh *see* Orūmīyeh
137 *P11* **Rize** Rize, NE Turkey *41°03´N 40°33´E*
137 *P11* **Rize** *prev.* Çoruh. ◈ *province* NE Turkey
161 *R5* **Rizhao** Shandong, E China *35°23´N 119°32´E*
Rizhskiy Zaliv *see* Riga, Gulf of
Rizokarpaso/Rizokárpason *see* Dipkarpaz
107 *O21* **Rizzuto, Capo** *headland* S Italy *38°54´N 17°05´E*
95 *F15* **Rjukan** Telemark, S Norway *59°54´N 08°33´E*
95 *D16* **Rjuven** ▲▲ S Norway
76 *H9* **Rkîz** Trarza, W Mauritania *16°50´N 15°20´W*
115 *Q23* **Ro** *prev.* Ágios Geórgios. *island* SE Greece
95 *H14* **Roa** Oppland, S Norway *60°16´N 10°38´E*
105 *N5* **Roa** Castilla y León, N Spain *41°42´N 03°55´W*
45 *T9* **Road Town** ○ (British Virgin Islands) Tortola, C British Virgin Islands *18°28´N 64°39´W*
96 *F6* **Roag, Loch** *inlet* NW Scotland, United Kingdom
37 *O5* **Roan Cliffs** *cliff* Colorado/Utah, W USA
21 *P9* **Roan High Knob** *var.* Roan Mountain. ▲ North Carolina/Tennessee, SE USA *36°09´N 82°07´W*
Roan Mountain *see* Roan High Knob
103 *Q10* **Roanne** *anc.* Rodunma. Loire, E France *46°03´N 04°04´E*
23 *R4* **Roanoke** Alabama, S USA *33°09´N 85°22´W*
21 *S7* **Roanoke** Virginia, NE USA *37°16´N 79°57´W*
21 *Z9* **Roanoke Island** *island* North Carolina, SE USA
21 *W8* **Roanoke Rapids** North Carolina, SE USA *36°27´N 77°39´W*
21 *X9* **Roanoke River** ≈ North Carolina/Virginia, SE USA
37 *O4* **Roan Plateau** *plain* Utah, W USA
37 *R5* **Roaring Fork River** ≈ Colorado, C USA
25 *O5* **Roaring Springs** Texas, SW USA *33°54´N 100°51´W*
42 *J4* **Roatán** *var.* Coxen Hole, Coxin Hole. Islas de la Bahía, N Honduras *16°19´N 86°33´W*
42 *I4* **Roatán, Isla de** *island* Islas de la Bahía, N Honduras
Roat Kampuchea *see* Cambodia
Roazon *see* Rennes
143 *T7* **Robāṭ-e Chāh Gonbad** Yazd, E Iran *33°24´N 57°43´E*
143 *R7* **Robāṭ-e Khān** Yazd, C Iran *33°24´N 56°04´E*
143 *T7* **Robāṭ-e Khvosh Āb** Khorāsān-e Razavī, E Iran
143 *R8* **Robāṭ-e Posht-e Bādām** Yazd, NE Iran *33°01´N 55°34´E*
143 *Q8* **Robaṭ-e Rīzāb** Yazd, C Iran
175 *S8* **Robbie Ridge** *undersea feature* W Pacific Ocean
21 *T10* **Robbins** North Carolina, SE USA *35°25´N 79°35´W*
183 *N15* **Robbins Island** *island* Tasmania, SE Australia
21 *N10* **Robbinsville** North Carolina, SE USA *35°18´N 83°49´W*
182 *J12* **Robe** South Australia *37°11´S 139°48´E*
21 *W9* **Robersonville** North Carolina, SE USA *35°49´N 77°15´W*
45 *V10* **Robert L. Bradshaw** ✕ (Basseterre) Saint Kitts, Saint Kitts and Nevis *17°16´N 62°43´W*
25 *P8* **Robert Lee** Texas, SW USA *31°50´N 100°30´W*
35 *V5* **Roberts Creek Mountain** ▲ Nevada, W USA *39°52´N 116°16´W*
93 *J15* **Robertsfors** Västerbotten, N Sweden *64°12´N 20°50´E*
27 *R11* **Robert S. Kerr Reservoir** ⊠ Oklahoma, C USA
38 *L12* **Roberts Mountain** ▲ Nunivak Island, Alaska, USA *60°01´N 166°15´W*
83 *F26* **Robertson** Western Cape, SW South Africa *33°48´S 19°53´E*
194 *H4* **Robertson Island** *island* Antarctica
76 *J16* **Robertsport** W Liberia *06°45´N 11°15´W*
182 *J8* **Robertstown** South Australia *34°00´S 139°04´E*
Robert Williams *see* Caála
15 *P7* **Roberval** Québec, SE Canada *48°31´N 72°16´W*
31 *N15* **Robinson** Illinois, N USA *39°00´N 87°44´W*
193 *U11* **Róbinson Crusoe, Isla** *island* Islas Juan Fernández, Chile, E Pacific Ocean
180 *J9* **Robinson Range** ▲▲ Western Australia
182 *M9* **Robinvale** Victoria, SE Australia *34°37´S 142°45´E*
105 *P11* **Robledo** Castilla-La Mancha, C Spain *38°45´N 02°27´W*
54 *G5* **Robles** *var.* La Paz, Robles La Paz. Cesar, N Colombia *10°24´N 73°11´W*
Robles La Paz *see* Robles
11 *V15* **Roblin** Manitoba, S Canada *51°15´N 101°20´W*
11 *S17* **Robsart** Saskatchewan, S Canada *49°22´N 109°15´W*
11 *N15* **Robson, Mount** ▲ British Columbia, SW Canada *53°09´N 119°16´W*
25 *T14* **Robstown** Texas, SW USA *27°47´N 97°40´W*
25 *P6* **Roby** Texas, SW USA *32°42´N 100°23´W*
104 *E11* **Roca, Cabo da** *cape* C Portugal
Rocadas *see* Xangongo
41 *S14* **Roca Partida, Punta** *headland* C Mexico *18°43´N 95°11´W*
47 *X6* **Rocas, Atol das** *island* E Brazil
107 *L18* **Roccadaspide** *var.* Rocca d'Aspide. Campania, S Italy *40°25´N 15°12´E*
Rocca d'Aspide *see* Roccadaspide
107 *K15* **Roccaraso** Abruzzo, C Italy *41°49´N 14°01´E*
106 *H10* **Rocca San Casciano** Emilia-Romagna, C Italy *44°06´N 11°51´E*
106 *G13* **Roccastrada** Toscana, C Italy *43°00´N 11°09´E*
61 *G20* **Rocha** Rocha, E Uruguay *34°30´S 54°22´W*
61 *G19* **Rocha** ◈ *department* E Uruguay
97 *L17* **Rochdale** NW England, United Kingdom *53°38´N 02°09´W*
102 *L11* **Rochechouart** Haute-Vienne, C France *45°49´N 00°49´E*
99 *J22* **Rochefort** Namur, SE Belgium *50°10´N 05°13´E*
102 *J11* **Rochefort** *var.* Rochefort sur Mer. Charente-Maritime, W France *45°57´N 00°58´W*
Rochefort sur Mer *see* Rochefort
125 *N10* **Rochegda** Arkhangel'skaya Oblast', NW Russian Federation *62°37´N 43°21´E*
30 *L10* **Rochelle** Illinois, N USA *41°54´N 89°03´W*
25 *Q9* **Rochelle** Texas, SW USA *31°13´N 99°10´W*
15 *V3* **Rochers Ouest, Rivière aux** ≈ Québec, SE Canada
97 *O22* **Rochester** *anc.* Durobrivae. SE England, United Kingdom *51°24´N 00°30´E*
31 *O12* **Rochester** Indiana, N USA *41°03´N 86°13´W*
29 *W10* **Rochester** Minnesota, N USA *44°01´N 92°28´W*
19 *O9* **Rochester** New Hampshire, NE USA *43°18´N 70°58´W*
18 *F9* **Rochester** New York, NE USA *43°09´N 77°37´W*
25 *P5* **Rochester** Texas, SW USA *33°19´N 99°51´W*
31 *S9* **Rochester Hills** Michigan, N USA *42°39´N 83°04´W*
Rocheuses, Montagnes/Rockies *see* Rocky Mountains
64 *M6* **Rockall** *island* N Atlantic Ocean, United Kingdom
64 *L6* **Rockall Bank** *undersea feature* N Atlantic Ocean
84 *B8* **Rockall Rise** *undersea feature* N Atlantic Ocean *59°00´N 14°00´W*
84 *C9* **Rockall Trough** *undersea feature* N Atlantic Ocean *57°00´N 12°00´W*
35 *U2* **Rock Creek** ≈ Nevada, W USA
25 *T10* **Rockdale** Texas, SW USA *30°39´N 96°58´W*
195 *N12* **Rockefeller Plateau** *plateau* Antarctica
30 *K11* **Rock Falls** Illinois, N USA *41°46´N 89°41´W*
23 *Q5* **Rockford** Alabama, S USA *32°53´N 86°11´W*
30 *L10* **Rockford** Illinois, N USA *42°16´N 89°06´W*
15 *Q12* **Rock Forest** Québec, SE Canada *45°21´N 71°58´W*
11 *T17* **Rockglen** Saskatchewan, S Canada *49°11´N 105°57´W*
181 *Y8* **Rockhampton** Queensland, E Australia *23°31´S 150°31´E*
21 *R11* **Rock Hill** South Carolina, SE USA *34°55´N 81°01´W*
180 *I13* **Rockingham** Western Australia *32°16´S 115°21´E*
21 *T11* **Rockingham** North Carolina, SE USA *34°56´N 79°47´W*
30 *J11* **Rock Island** Illinois, N USA *41°30´N 90°34´W*
25 *U12* **Rock Island** Texas, SW USA *29°31´N 96°33´W*
14 *C10* **Rock Lake** Ontario, S Canada *46°25´N 83°49´W*
29 *O2* **Rock Lake** North Dakota, N USA *48°45´N 99°12´W*
14 *I12* **Rock Lake** ◎ Ontario, SE Canada
14 *M12* **Rockland** Ontario, SE Canada *45°33´N 75°16´W*
19 *R7* **Rockland** Maine, NE USA *44°08´N 69°06´W*
182 *L11* **Rocklands Reservoir** ⊠ Victoria, SE Australia
35 *O7* **Rocklin** California, W USA *38°48´N 121°13´W*
23 *R3* **Rockmart** Georgia, SE USA *34°00´N 85°02´W*
31 *N16* **Rockport** Indiana, N USA *37°53´N 87°04´W*
27 *Q1* **Rock Port** Missouri, C USA *40°26´N 95°30´W*
25 *T14* **Rockport** Texas, SW USA *28°02´N 99°04´W*
32 *I7* **Rockport** Washington, NW USA *48°28´N 121°36´W*
29 *S11* **Rock Rapids** Iowa, C USA *43°25´N 96°10´W*
30 *K11* **Rock River** ≈ Illinois/Wisconsin, N USA
44 *I3* **Rock Sound** Eleuthera Island, C The Bahamas *24°52´N 76°10´W*
25 *P11* **Rocksprings** Texas, SW USA *30°02´N 100°14´W*
33 *U17* **Rock Springs** Wyoming, C USA *41°35´N 109°12´W*
55 *T9* **Rockstone** C Guyana *05°58´N 58°33´W*
29 *S12* **Rock Valley** Iowa, C USA *43°12´N 96°17´W*
31 *N14* **Rockville** Indiana, N USA *39°45´N 87°15´W*
21 *W3* **Rockville** Maryland, NE USA *39°05´N 77°10´W*
25 *U6* **Rockwall** Texas, SW USA *32°56´N 96°27´W*
29 *U13* **Rockwell City** Iowa, C USA *42°24´N 94°37´W*
31 *S10* **Rockwood** Michigan, N USA *42°04´N 83°15´W*
20 *M9* **Rockwood** Tennessee, S USA *35°52´N 84°41´W*
25 *Q8* **Rockwood** Texas, SW USA *31°29´N 99°23´W*
37 *U6* **Rocky Ford** Colorado, C USA *38°03´N 103°45´W*
14 *D9* **Rocky Island Lake** ◎ Ontario, S Canada
21 *V9* **Rocky Mount** North Carolina, SE USA *35°56´N 77°48´W*
21 *S7* **Rocky Mount** Virginia, NE USA *37°00´N 79°53´W*
33 *Q8* **Rocky Mountain** ▲ Montana, NW USA *47°45´N 112°46´W*
11 *P15* **Rocky Mountain House** Alberta, SW Canada *52°24´N 114°52´W*
37 *T3* **Rocky Mountain National Park** *national park* Colorado, C USA
2 *E12* **Rocky Mountains** *var.* Rockies, *Fr.* Montagnes Rocheuses. ▲▲ Canada/USA
42 *H1* **Rocky Point** *headland* NE Belize *18°21´N 88°04´W*
83 *A17* **Rocky Point** *headland* NW Namibia *19°01´S 12°27´E*
95 *F14* **Rødberg** Buskerud, S Norway *60°16´N 09°00´E*
95 *I25* **Rødby** Sjælland, SE Denmark *54°42´N 11°24´E*
95 *I25* **Rødbyhavn** Sjælland, SE Denmark *54°39´N 11°24´E*
13 *T10* **Roddickton** Newfoundland, Newfoundland and Labrador, SE Canada *50°51´N 56°03´W*
95 *F23* **Rødding** Syddanmark, SW Denmark *55°22´N 09°04´E*
95 *M22* **Rödeby** Blekinge, S Sweden *56°16´N 15°35´E*
98 *N6* **Roden** Drenthe, NE Netherlands *53°08´N 06°26´E*
62 *H9* **Rodeo** San Juan, W Argentina *30°12´S 69°06´W*
103 *O14* **Rodez** *anc.* Segodunum. Aveyron, S France *44°21´N 02°34´E*
Rodholívos *see* Rodolívos
Rodhópi Óri *see* Rhodope Mountains
Ródhos/Rodi *see* Ródos
107 *N15* **Rodi Gargancio** Puglia, SE Italy *41°54´N 15°51´E*
101 *N20* **Roding** Bayern, SE Germany *49°12´N 12°30´E*
113 *J19* **Rodinit, Kepi i** *headland* W Albania *41°35´N 19°27´E*
116 *I9* **Rodnei, Munţii** ▲▲ N Romania
184 *L4* **Rodney, Cape** *headland* North Island, New Zealand *36°16´S 174°48´E*
38 *L9* **Rodney, Cape** *headland* Alaska, USA *64°39´N 166°24´W*
124 *M16* **Rodniki** Ivanovskaya Oblast', W Russian Federation *57°04´N 41°45´E*
119 *Q16* **Rodnya** Mahilyowskaya Voblasts', E Belarus *53°31´N 32°07´E*
Rodó *see* José Enrique Rodó
114 *H13* **Rodolívos** *var.* Rodholívos. Kentrikí Makedonía, NE Greece *40°55´N 24°00´E*
Rodopi *see* Rhodope Mountains
115 *O22* **Ródos** *var.* Ródhos, *Eng.* Rhodes, *It.* Rodi. Ródos, Dodekánisa, Greece, Aegean Sea *36°26´N 28°14´E*
115 *O22* **Ródos** *var.* Ródhos, *Eng.* Rhodes, *It.* Rodi; *anc.* Rhodos. *island* Dodekánisa, Greece, Aegean Sea
Rodosto *see* Tekirdağ
59 *A14* **Rodrigues** Amazonas, W Brazil *06°50´S 73°45´W*
173 *P8* **Rodrigues** *var.* Rodriquez. *island* E Mauritius
Rodriquez *see* Rodrigues
Rodunma *see* Roanne
180 *I7* **Roebourne** Western Australia *20°49´S 117°04´E*
83 *J20* **Roedtan** Limpopo, NE South Africa *24°37´S 29°05´E*
98 *H11* **Roelofarendsveen** Zuid-Holland, W Netherlands *52°12´N 04°37´E*
Roepat *see* Rupat, Pulau
Roer *see* Rur
99 *M16* **Roermond** Limburg, SE Netherlands *51°12´N 06°E*
99 *C18* **Roeselare** *Fr.* Roulers; *prev.* Rousselaere. West-Vlaanderen, W Belgium *50°57´N 03°08´E*
9 *P8* **Roes Welcome Sound** *strait* Nunavut, N Canada
Roeteng *see* Ruteng
Rofreit *see* Rovereto
Rogachëv *see* Rahachow
57 *L15* **Rogagua, Laguna** ◎ NW Bolivia
95 *C16* **Rogaland** ◈ *county* S Norway
25 *Y9* **Roganville** Texas, SW USA *30°49´N 93°54´W*
109 *W11* **Rogaška Slatina** *Ger.* Rohitsch-Sauerbrunn; *prev.* Rogatec-Slatina. E Slovenia *46°13´N 15°38´E*
Rogatec-Slatina *see* Rogaška Slatina
112 *J13* **Rogatica** Republika Srpska, SE Bosnia and Herzegovina *43°50´N 18°55´E*
Rogatin *see* Rohatyn
93 *F17* **Rogen** ◎ C Sweden
27 *S9* **Rogers** Arkansas, C USA *36°19´N 94°07´W*
29 *P5* **Rogers** North Dakota, N USA *47°03´N 98°12´W*
25 *T9* **Rogers** Texas, SW USA *30°53´N 97°10´W*
31 *R5* **Rogers City** Michigan, N USA *45°25´N 83°49´W*
Roger Simpson Island *see* Abemama
35 *T14* **Rogers Lake** *salt flat* California, W USA
21 *Q8* **Rogers, Mount** ▲ Virginia, NE USA *36°39´N 81°32´W*
33 *O16* **Rogerson** Idaho, NW USA *42°11´N 114°36´W*
11 *O16* **Rogers Pass** *pass* British Columbia, SW Canada
21 *O8* **Rogersville** Tennessee, S USA *36°26´N 83°01´W*
99 *L16* **Roggel** Limburg, SE Netherlands *51°16´N 05°55´E*
Roggeveen *see* Roggewein, Cabo
193 *R10* **Roggeveen Basin** *undersea feature* E Pacific Ocean *31°30´S 95°30´W*
191 *X16* **Roggewein, Cabo** *var.* Roggeveen. *headland* Easter Island, Chile, E Pacific Ocean *27°07´S 109°15´W*
103 *Y13* **Rogliano** Corse, France, C Mediterranean Sea *42°58´N 09°25´E*
107 *N21* **Rogliano** Calabria, SW Italy *39°09´N 16°18´E*
92 *G12* **Rognan** Nordland, C Norway *67°04´N 15°21´E*
100 *K10* **Rögnitz** ≈ N Germany
Rogozhina/Rogozhinë *see* Rrogozhinë
110 *G10* **Rogoźno** Wielkopolskie, C Poland *52°46´N 16°58´E*
32 *E15* **Rogue River** ≈ Oregon, NW USA
116 *I6* **Rohatyn** *Rus.* Rogatin. Ivano-Frankivs'ka Oblast', W Ukraine *49°25´N 24°35´E*
189 *O14* **Rohi** Pohnpei, E Micronesia
Rohitsch-Sauerbrunn *see* Rogaška Slatina
149 *Q13* **Rohri** Sind, SE Pakistan *27°39´N 68°57´E*
152 *I10* **Rohtak** Haryāna, N India *28°57´N 76°38´E*
Roi Ed *see* Roi Et
167 *R9* **Roi Et** *var.* Muang Roi Et, Roi Ed. Roi Et, E Thailand *16°05´N 103°38´E*
191 *U9* **Roi Georges, Îles du** *island group* Îles Tuamotu, C French Polynesia
153 *Y10* **Roing** Arunāchal Pradesh, NE India *28°06´N 95°46´E*
118 *E7* **Roja** NW Latvia *57°31´N 22°44´E*
61 *B20* **Rojas** Buenos Aires, E Argentina *34°10´S 60°45´W*
149 *R12* **Rojhān** Punjab, E Pakistan *28°39´N 70°00´E*
41 *Q12* **Rojo, Cabo** *headland* C Mexico *21°33´N 97°19´W*
45 *Q10* **Rojo, Cabo** *headland* W Puerto Rico *17°57´N 67°10´W*
168 *K10* **Rokan Kiri, Sungai** ≈ Sumatera, W Indonesia
118 *I11* **Rokiškis** Panevėžys, NE Lithuania *55°58´N 25°35´E*
165 *R7* **Rokkasho** Aomori, Honshū, C Japan *40°59´N 141°22´E*
111 *B17* **Rokycany** *Ger.* Rokytzan. Plzeňský Kraj, W Czech Republic *49°45´N 13°36´E*
117 *P6* **Rokytne** Kyyivs'ka Oblast', N Ukraine *49°40´N 30°29´E*
116 *L3* **Rokytne** Rivnens'ka Oblast', NW Ukraine *51°19´N 27°09´E*
Rokytzan *see* Rokycany
158 *L11* **Rola Co** ◎ W China
29 *V13* **Roland** Iowa, C USA *42°10´N 93°30´W*
95 *D15* **Røldal** Hordaland, S Norway *59°52´N 06°49´E*
98 *O7* **Rolde** Drenthe, NE Netherlands *52°58´N 06°39´E*
29 *O2* **Rolette** North Dakota, N USA *48°39´N 99°50´W*
27 *V6* **Rolla** Missouri, C USA *37°56´N 91°47´W*
29 *O2* **Rolla** North Dakota, N USA *48°51´N 99°37´W*
108 *A10* **Rolle** Vaud, W Switzerland *46°27´N 06°19´E*
181 *X8* **Rolleston** Queensland, E Australia *24°30´S 148°36´E*
185 *H19* **Rolleston** Canterbury, South Island, New Zealand *43°34´S 172°24´E*
185 *G18* **Rolleston Range** ▲▲ South Island, New Zealand
14 *H8* **Rollet** Québec, SE Canada *47°56´N 79°14´W*
22 *J4* **Rolling Fork** Mississippi, S USA *32°54´N 90°52´W*
20 *L6* **Rolling Fork** ≈ Kentucky, S USA
14 *J11* **Rolphton** Ontario, SE Canada *46°09´N 77°43´W*
Röm *see* Rømø
181 *X10* **Roma** Queensland, E Australia *26°37´S 148°54´E*
107 *I15* **Roma** *Eng.* Rome. ● (Italy) Lazio, C Italy *41°53´N 12°30´E*
95 *P19* **Roma** Gotland, SE Sweden *57°31´N 18°28´E*
21 *T14* **Romain, Cape** *headland* South Carolina, SE USA *33°00´N 79°21´W*
13 *P11* **Romaine** ≈ Newfoundland and Labrador/Québec, E Canada
25 *R17* **Roma Los Saenz** Texas, SW USA *26°24´N 99°01´W*
114 *H8* **Roman** Vratsa, NW Bulgaria *43°09´N 23°56´E*
116 *L10* **Roman** *Hung.* Románvásár. Neamţ, NE Romania *46°46´N 26°56´E*
64 *M13* **Romanche Fracture Zone** *tectonic feature* E Atlantic Ocean
61 *C15* **Romang** Santa Fe, C Argentina *29°30´S 59°46´W*
171 *R15* **Romang, Pulau** *var.* Pulau Roma. *island* Kepulauan Damar, E Indonesia
171 *R15* **Romang, Selat** *strait* Nusa Tenggara, S Indonesia
116 *J11* **Romania** *Bul.* Rumŭniya, *Ger.* Rumänien, *Hung.* Románia, *Rom.* România, *SCr.* Rumunjska, *Ukr.* Rumuniya, *prev.* Republica Socialistă România, Roumania, Rumania, Socialist Republic of Romania, *prev.Rom.* Romînia. ◆ *republic* SE Europe
România, Republica Socialistă *see* Romania
Romania, Socialist Republic of *see* Romania
117 *T7* **Romaniv** *Rus.* Dneprodzerzhinsk, *prev.* Dniprodzerzhyns'k, *prev.* Kamenskoye. Dnipropetrovs'ka Oblast', E Ukraine *48°30´N 34°35´E*
117 *X7* **Romaniv** *Rus.* Dzerzhinsk; *prev.* Dzerzhyns'k. Donets'ka Oblast', SE Ukraine *48°21´N 37°50´E*
116 *M5* **Romaniv** *prev.* Dzerzhyns'k. Zhytomyrs'ka Oblast', N Ukraine *50°07´N 27°56´E*
23 *W16* **Romano, Cape** *headland* Florida, SE USA *25°51´N 81°40´W*
44 *G5* **Romano, Cayo** *island* C Cuba
123 *O13* **Romanovka** Respublika Buryatiya, S Russian Federation *53°10´N 112°34´E*
127 *N8* **Romanovka** Saratovskaya Oblast', W Russian Federation *51°45´N 42°45´E*
108 *I6* **Romanshorn** Thurgau, NE Switzerland *47°34´N 09°23´E*
103 *R12* **Romans-sur-Isère** Drôme, E France *45°03´N 05°03´E*
189 *U12* **Romanum** *island* Chuuk, C Micronesia
Románvásár *see* Roman
39 *S5* **Romanzof Mountains** ▲▲ Alaska, USA
Roma, Pulau *see* Romang, Pulau
103 *S4* **Rombas** Moselle, NE France *49°15´N 06°04´E*
23 *R2* **Rome** Georgia, SE USA *34°01´N 85°02´W*
18 *I9* **Rome** New York, NE USA *43°13´N 75°28´W*
Rome *see* Roma
31 *S9* **Romeo** Michigan, N USA *42°48´N 83°00´W*
Römerstadt *see* Rýmařov
Rometan *see* Romiton
103 *P5* **Romilly-sur-Seine** Aube, N France *48°31´N 03°44´E*
Romînia *see* Romania
146 *L11* **Romiton** *Rus.* Rometan. Buxoro Viloyati, C Uzbekistan *39°56´N 64°21´E*
21 *U3* **Romney** West Virginia, NE USA *39°21´N 78°44´W*
117 *S4* **Romny** Sums'ka Oblast', NE Ukraine *50°45´N 33°30´E*
95 *E24* **Rømø** *Ger.* Röm. *island* SW Denmark
117 *S5* **Romodan** Poltavs'ka Oblast', NE Ukraine *50°00´N 33°20´E*
127 *P5* **Romodanovo** Respublika Mordoviya, W Russian Federation *54°25´N 45°24´E*
Romorantin *see* Romorantin-Lanthenay
103 *N8* **Romorantin-Lanthenay** *var.* Romorantin. Loir-et-Cher, C France *47°22´N 01°44´E*
94 *F9* **Romsdal** *physical region* S Norway
94 *F10* **Romsdalen** *valley* S Norway
94 *E9* **Romsdalsfjorden** *fjord* S Norway
33 *P8* **Ronan** Montana, NW USA *47°31´N 114°06´W*
59 *M14* **Roncador** Maranhão, E Brazil *05°48´S 45°08´W*
186 *M7* **Roncador Reef** *reef* N Solomon Islands
59 *J17* **Roncador, Serra do** ▲▲ C Brazil
21 *S6* **Ronceverte** West Virginia, NE USA *37°45´N 80°27´W*
107 *H14* **Ronciglione** Lazio, C Italy *42°16´N 12°15´E*
104 *L15* **Ronda** Andalucía, S Spain *36°45´N 05°10´W*
94 *G11* **Rondane** ▲▲ S Norway
104 *L15* **Ronda, Serranía de** ▲▲ S Spain
95 *H22* **Rønde** Midtjylland, C Denmark *56°18´N 10°28´E*
Ronde, Île *see* Round Island
Rōṇdik *see* Rongrik Atoll
59 *E16* **Rondônia** *off.* Estado de Rondônia; *prev.* Território de Rondônia. ◈ *state* W Brazil
Rondônia, Estado de *see* Rondônia
Rondônia, Território de *see* Rondônia
59 *I18* **Rondonópolis** Mato Grosso, W Brazil *16°29´S 54°37´W*
94 *G11* **Rondslottet** ▲ S Norway *61°54´N 09°48´E*
95 *P20* **Ronehamn** Gotland, SE Sweden *57°10´N 18°30´E*
160 *L13* **Rong'an** *var.* Chang'an, Rongan. Guangxi Zhuangzu Zizhiqu, S China *25°14´N 109°20´E*
Rongan *see* Rong'an
Rongcheng *see* Rongxian, Guangxi, China
Rongcheng *see* Jianli, Hubei, China
189 *R4* **Rongelap Atoll** *var.* Rōnlap. *atoll* Ralik Chain, NW Marshall Islands
Rongerik *see* Rongrik Atoll
160 *K12* **Rongjiang** *var.* Guzhou. Guizhou, S China *25°59´N 108°27´E*
160 *L13* **Rong Jiang** ≈ S China
Rongjiang *see* Nankang
Rong, Kas *see* Rŭng, Kaôh
167 *P8* **Rong Kwang** Phrae, NW Thailand *18°19´N 100°18´E*
189 *T4* **Rongrik Atoll** *var.* Rōṇdik, Rongerik. *atoll* Ralik Chain, N Marshall Islands
189 *X2* **Rongrong** *island* SE Marshall Islands
160 *L13* **Rongshui** *var.* Rongshui Miaozu Zizhixian. Guangxi Zhuangzu Zizhiqu, S China *25°05´N 109°09´E*
Rongshui Miaozu Zizhixian *see* Rongshui
118 *I6* **Rõngu** *Ger.* Ringen. Tartumaa, SE Estonia *58°10´N 26°17´E*
Rongwo *see* Tongren
160 *L15* **Rongxian** *var.* Rongzhou; *prev.* Rongcheng. Guangxi Zhuangzu Zizhiqu, S China *22°52´N 110°33´E*
Rongzhag *see* Danba
Rongzhou *see* Rongxian
Roniu, Mont *see* Rooniu, Mont
189 *N13* **Ronkiti** Pohnpei, E Micronesia *06°48´N 158°10´E*
Rōnlap *see* Rongelap Atoll
95 *L24* **Rønne** Bornholm, E Denmark *55°07´N 14°43´E*
95 *M22* **Ronneby** Blekinge, S Sweden *56°12´N 15°18´E*
194 *J7* **Ronne Entrance** *inlet* Antarctica

◆ Country
● Country Capital
◇ Dependent Territory
○ Dependent Territory Capital
◈ Administrative Regions
✕ International Airport
▲ Mountain
▲▲ Mountain Range
⛰ Volcano
≈ River
◎ Lake
⊠ Reservoir

194 L6 Ronne Ice Shelf ice shelf Antarctica
99 E19 Ronse Fr. Renaix. Oost-Vlaanderen, SW Belgium 50°45′N 03°36′E
30 K14 Roodhouse Illinois, N USA 39°28′N 90°22′W
83 C19 Rooibank Erongo, W Namibia 23°04′S 14°34′E
Rooke Island see Umboi Island
65 N24 Rookery Point headland NE Tristan da Cunha 37°03′S 12°15′W
191 R8 Rooniu, Mont prev. Mont Roniu. Tahiti, W French Polynesia 17°49′S 149°12′W
171 V13 Roon, Pulau island E Indonesia
173 V7 Roo Rise undersea feature E Indian Ocean
152 J9 Roorkee Uttarakhand, N India 29°51′N 77°54′E
99 H15 Roosendaal Noord-Brabant, S Netherlands 51°32′N 04°29′E
25 P10 Roosevelt Texas, SW USA 30°28′N 100°06′W
37 N3 Roosevelt Utah, W USA 40°18′N 109°59′W
47 T8 Roosevelt W Brazil
195 O13 Roosevelt Island island Antarctica
10 L10 Roosevelt, Mount ▲ British Columbia, W Canada 58°28′N 125°22′W
11 P17 Roosville British Columbia, SW Canada 48°59′N 115°03′W
29 X10 Root River Minnesota, N USA
Ropar see Rūpnagar
111 N16 Ropczyce Podkarpackie, SE Poland 50°04′N 21°31′E
181 Q3 Roper Bar Northern Territory, N Australia 14°45′S 134°30′E
24 M5 Ropesville Texas, SW USA 33°24′N 102°09′W
102 K14 Roquefort Landes, SW France 44°01′N 00°18′W
61 C21 Roque Pérez Buenos Aires, E Argentina 35°25′S 59°24′W
58 E10 Roraima off. Estado de Roraima; prev. Território de Rio Branco, Território de Roraima. ◆ state N Brazil
Roraima, Estado de see Roraima
58 F9 Roraima, Mount ▲ N South America 05°10′N 60°36′W
Roraima, Território de see Roraima
94 I9 Røros Sør-Trøndelag, S Norway 62°37′N 11°25′E
108 I7 Rorschach Sankt Gallen, NE Switzerland 47°28′N 09°30′E
93 E14 Rørvik Nord-Trøndelag, C Norway 64°54′N 11°15′E
119 G17 Ros' Rus. Ross'. Hrodzyenskaya Voblasts', W Belarus 53°25′N 24°24′E
185 F17 Ross West Coast, South Island, New Zealand 42°54′S 170°52′E
119 G17 Ros' Rus. Ross'. W Belarus
10 J7 Ross Yukon, W Canada
117 O6 Ros' N Ukraine
44 K7 Rosa, Lake ◎ Great Inagua, S The Bahamas
32 M9 Rosalia Washington, NW USA 47°14′N 117°22′W
191 W15 Rosalia, Punta headland Easter Island, Chile, E Pacific Ocean 27°04′S 109°19′W
45 P12 Rosalie E Dominica 15°22′N 61°15′W
35 T14 Rosamond California, W USA 34°51′N 118°09′W
35 S14 Rosamond Lake salt flat California, W USA
96 H8 Ross and Cromarty cultural region N Scotland, United Kingdom
61 B18 Rosario Santa Fe, C Argentina 32°56′S 60°39′W
40 J11 Rosario Sinaloa, C Mexico 23°00′N 105°51′W
40 G6 Rosario Sonora, NW Mexico 27°53′N 109°18′W
62 O6 Rosario San Pedro, C Paraguay 24°26′S 57°06′W
61 E20 Rosario Colonia, SW Uruguay 34°20′S 57°26′W
54 H5 Rosario Zulia, NW Venezuela 10°18′N 72°19′W
Rosario see Nishino-shima
Rosario see Rosarito
40 B4 Rosario, Bahía del bay NW Mexico
62 K6 Rosario de la Frontera Salta, N Argentina 25°50′S 65°00′W
61 C18 Rosario del Tala Entre Ríos, E Argentina 32°20′S 59°10′W
61 F16 Rosário do Sul Rio Grande do Sul, S Brazil 30°15′S 54°55′W
59 H18 Rosário Oeste Mato Grosso, W Brazil 14°50′S 56°25′W
40 B1 Rosarito var. Rosario. Baja California Norte, NW Mexico 32°25′N 117°04′W
40 D5 Rosarito Baja California Norte, NW Mexico 28°27′N 113°58′W
40 F7 Rosarito Baja California Sur, NW Mexico 26°28′N 111°41′W
104 L9 Rosarito, Embalse del ⊠ W Spain
107 N22 Rosarno Calabria, SW Italy 38°29′N 15°59′E
56 B5 Rosa Zárate var. Quinindé. Esmeraldas, SW Ecuador 0°14′N 79°28′W
Roscianum see Rossano
29 O8 Roscoe South Dakota, N USA 45°24′N 99°19′W
25 P7 Roscoe Texas, SW USA 32°27′N 100°32′W
102 F5 Roscoff Finistère, NW France 48°43′N 04°00′W
Ros Comáin see Roscommon
97 C17 Roscommon Ir. Ros Comáin. C Ireland 53°38′N 08°11′W
31 Q7 Roscommon Michigan, N USA 44°30′N 84°35′W
97 C17 Roscommon Ir. Ros Comáin. cultural region C Ireland
Ros. Cré see Roscrea
97 D19 Roscrea Ir. Ros. Cré. C Ireland 52°57′N 07°47′W
14 H13 Rosseau Ontario, S Canada 45°15′N 79°38′W
45 X12 Roseau prev. Charlotte Town. ● (Dominica) SW Dominica 15°17′N 61°23′W
29 S2 Roseau Minnesota, N USA 48°51′N 95°45′W
173 Y16 Rose Belle SE Mauritius 20°24′S 57°36′E
183 O16 Rosebery Tasmania, SE Australia 41°51′S 145°33′E
21 U11 Roseboro North Carolina, SE USA 34°58′N 78°31′W
25 T9 Rosebud Texas, SW USA 31°04′N 96°58′W
33 W10 Rosebud Creek Montana, NW USA
32 F14 Roseburg Oregon, NW USA 43°13′N 123°21′W
22 J3 Rosedale Mississippi, S USA 33°51′N 91°01′W
99 H21 Rosée Namur, S Belgium 50°15′N 04°43′E
55 U8 Rose Hall E Guyana 06°14′N 57°30′W
173 X16 Rose Hill W Mauritius 20°14′S 57°29′E
80 H12 Roseires, Reservoir var. Lake Rusayris. ⊠ E Sudan
Rosenau see Rožnov pod Radhoštěm
Rosenau see Rožňava
25 V11 Rosenberg Texas, SW USA 29°33′N 95°48′W
Rosenberg see Olesno, Poland
Rosenberg see Ružomberok, Slovakia
100 I10 Rosengarten Niedersachsen, N Germany 53°24′N 09°54′E
101 M24 Rosenheim Bayern, S Germany 47°51′N 12°08′E
Rosenhof see Zilupe
105 X4 Roses Cataluña, NE Spain 42°15′N 03°11′E
105 X4 Roses, Golf de gulf NE Spain
107 K14 Roseto degli Abruzzi Abruzzo, C Italy 42°39′N 14°01′E
11 S16 Rosetown Saskatchewan, S Canada 51°34′N 107°59′W
Rosetta see Rashîd
35 O7 Roseville California, W USA 38°44′N 121°16′W
30 J12 Roseville Illinois, N USA 40°42′N 90°40′W
29 V8 Roseville Minnesota, N USA 45°00′N 93°09′W
29 R7 Rosholt South Dakota, N USA 45°51′N 96°42′W
106 F12 Rosignano Marittimo Toscana, C Italy 43°24′N 10°28′E
116 I14 Roşiori de Vede Teleorman, S Romania 44°06′N 25°00′E
114 K8 Rositsa N Bulgaria
Rositten see Rēzekne
95 J23 Roskilde Sjælland, E Denmark 55°39′N 12°07′E
Ros Láir see Rosslare
126 H5 Roslavl' Smolenskaya Oblast', W Russian Federation 54°N 32°57′E
32 I8 Roslyn Washington, NW USA 47°13′N 120°52′W
99 K14 Rosmalen Noord-Brabant, S Netherlands 51°43′N 05°21′E
Ros Mhic Thriúin see New Ross
113 P19 Rosoman C FYR Macedonia 41°31′N 21°55′E
102 F6 Rosporden Finistère, NW France 47°58′N 03°54′W
Ross' see Ros'
107 O20 Rossano anc. Roscianum. Calabria, SW Italy 39°35′N 16°38′E
22 L5 Ross Barnett Reservoir ⊠ Mississippi, S USA
11 W16 Rossburn Manitoba, S Canada 50°42′N 100°49′W
14 H13 Rosseau, Lake ◎ Ontario, S Canada
186 I10 Rossel Island prev. Yela Island. island SE Papua New Guinea
195 P12 Ross Ice Shelf ice shelf Antarctica
13 P16 Rossignol, Lake ◎ Nova Scotia, SE Canada
83 C19 Rössing Erongo, W Namibia 22°31′S 14°52′E
195 Q14 Ross Island island Antarctica
Rossitten see Rybachiy
Rossiyskaya Federatsiya see Russian Federation
11 N17 Rossland British Columbia, SW Canada 49°03′N 117°49′W
97 F20 Rosslare Ir. Ros Láir. Wexford, SE Ireland 52°16′N 06°23′W
97 F20 Rosslare Harbour Wexford, SE Ireland 52°15′N 06°20′W
101 M14 Rosslau Sachsen-Anhalt, E Germany 51°52′N 12°15′E
76 G10 Rosso Trarza, SW Mauritania 16°36′N 15°50′W
103 X14 Rosso, Cap headland Corse, France, C Mediterranean Sea 42°25′N 08°22′E
93 H16 Rossön Jämtland, C Sweden 63°54′N 16°21′E
97 K21 Ross-on-Wye W England, United Kingdom 51°55′N 02°34′W
Rossony see Rasony
126 L9 Rossosh' Voronezhskaya Oblast', W Russian Federation 50°10′N 39°34′E
181 Q7 Ross River Northern Territory, N Australia 23°36′S 134°30′E
10 J7 Ross River Yukon, W Canada 61°57′N 132°26′W
195 O15 Ross Sea sea Antarctica
92 G13 Røssvatnet Lapp. Reevhtse. ◎ C Norway
23 R1 Rossville Georgia, SE USA 34°59′N 85°22′W
Rostak see Ar Rustāq
143 P14 Rostāq Hormozgān, S Iran 26°48′N 53°50′E
117 N5 Rostavytsya N Ukraine
11 T15 Rosthern Saskatchewan, S Canada 52°40′N 106°20′W
100 M8 Rostock Mecklenburg-Vorpommern, NE Germany 54°05′N 12°08′E
124 L16 Rostov Yaroslavskaya Oblast', W Russian Federation 57°11′N 39°19′E
Rostov see Rostov-na-Donu
126 L12 Rostov-na-Donu var. Rostov, Eng. Rostov-on-Don. Rostovskaya Oblast', SW Russian Federation 47°16′N 39°45′E
Rostov-on-Don see Rostov-na-Donu
126 L10 Rostovskaya Oblast' ◆ province SW Russian Federation
93 J14 Rosvik Norrbotten, N Sweden 65°26′N 21°48′E
23 S3 Roswell Georgia, SE USA 34°01′N 84°21′W
37 U14 Roswell New Mexico, SW USA 33°23′N 104°31′W
94 K12 Rot Dalarna, C Sweden 61°16′N 14°04′E
101 I23 Rot S Germany
104 J15 Rota Andalucía, S Spain 36°39′N 06°20′W
188 K9 Rota island S Northern Mariana Islands
25 P6 Rotan Texas, SW USA 32°51′N 100°28′W
Rotcher Island see Tamana
100 I11 Rotenburg Niedersachsen, NW Germany 53°06′N 09°25′E
Rotenburg see Rotenburg an der Fulda
101 I16 Rotenburg an der Fulda var. Rotenburg. Thüringen, C Germany 51°00′N 09°43′E
101 L18 Roter Main E Germany
101 K20 Roth Bayern, SE Germany 49°15′N 11°06′E
101 G16 Rothaargebirge ▲ W Germany
Rothenburg see Rothenburg ob der Tauber
101 J20 Rothenburg ob der Tauber var. Rothenburg. Bayern, S Germany 49°23′N 10°10′E
194 H6 Rothera UK research station Antarctica 67°28′S 68°31′W
185 I17 Rotherham Canterbury, South Island, New Zealand 42°42′S 172°56′E
97 M17 Rotherham N England, United Kingdom 53°26′N 01°20′W
96 H12 Rothesay W Scotland, United Kingdom 55°49′N 05°03′W
108 E7 Rothrist Aargau, N Switzerland 47°18′N 07°54′E
194 H6 Rothschild Island island Antarctica
171 P17 Roti, Pulau island S Indonesia
95 H14 Rotnes Akershus, S Norway 60°08′N 10°45′E
183 O8 Roto New South Wales, SE Australia 33°04′S 145°27′E
184 N8 Rotoiti, Lake ◎ North Island, New Zealand
Rotomagus see Rouen
107 N19 Rotondella Basilicata, S Italy 40°12′N 16°30′E
103 X15 Rotondo, Monte ▲ Corse, France, C Mediterranean Sea 42°15′N 09°03′E
185 I15 Rotoroa, Lake ◎ South Island, New Zealand
184 N8 Rotorua Bay of Plenty, North Island, New Zealand 38°10′S 176°14′E
184 N8 Rotorua, Lake ◎ North Island, New Zealand
101 N22 Rott SE Germany
108 F10 Rotten S Switzerland
109 T6 Rottenmann Steiermark, E Austria 47°31′N 14°18′E
98 H12 Rotterdam Zuid-Holland, SW Netherlands 51°55′N 04°30′E
18 K10 Rotterdam New York, NE USA 42°46′N 73°57′W
95 M21 Rottnen ◎ S Sweden
98 N4 Rottumeroog island Waddeneilanden, NE Netherlands
98 N4 Rottumerplaat island Waddeneilanden, NE Netherlands
101 G23 Rottweil Baden-Württemberg, S Germany 48°10′N 08°38′E
191 O7 Rotui, Mont Moorea, W French Polynesia 17°30′S 149°50′W
103 P1 Roubaix Nord, N France 50°42′N 03°10′E
111 C15 Roudnice nad Labem Ger. Raudnitz an der Elbe. Ústecký Kraj, NW Czech Republic 50°25′N 14°14′E
102 M4 Rouen anc. Rotomagus. Seine-Maritime, N France 49°26′N 01°05′E
171 X13 Rouffaer Reserves reserve Papua, E Indonesia
15 N10 Rouge, Rivière Québec, SE Canada
20 J6 Rough River Kentucky, S USA
20 J6 Rough River Lake ⊠ Kentucky, S USA
Rouhaïbé see Ar Ruḩaybah
102 K11 Rouillac Charente, W France 45°46′N 00°04′W
Roulers see Roeselare
Roumania see Romania
173 Y15 Round Island var. Île Ronde. island NE Mauritius
14 J12 Round Lake ◎ Ontario, SE Canada
35 U7 Round Mountain Nevada, W USA 38°42′N 117°04′W
25 R10 Round Mountain Texas, SW USA 30°25′N 98°20′W
183 U5 Round Mountain ▲ New South Wales, SE Australia 30°22′S 152°13′E
25 S10 Round Rock Texas, SW USA 30°30′N 97°40′W
33 U10 Roundup Montana, NW USA 46°27′N 108°32′W
55 Y10 Roura NE French Guiana 04°44′N 52°16′W
Rourkela see Rāurkela
96 J4 Rousay island N Scotland, United Kingdom
Rousselaere see Roeselare
103 O17 Roussillon cultural region S France
15 V7 Routhierville Québec, SE Canada 48°09′N 67°07′W
99 K25 Rouvroy Luxembourg, SE Belgium 49°33′N 05°28′E
14 I7 Rouyn-Noranda Québec, SE Canada 48°16′N 79°03′W
Rouyuan see Huachi
Rouyuanchengzi see Huachi
92 L12 Rovaniemi Lappi, N Finland 66°29′N 25°40′E
106 E7 Rovato Lombardia, N Italy 45°34′N 10°03′E
125 N11 Rovdino Arkhangel'skaya Oblast', NW Russian Federation 61°36′N 42°28′E
Roven'ki see Roven'ky
117 Y8 Roven'ky var. Roven'ki. Luhans'ka Oblast', E Ukraine 48°05′N 39°20′E
Rovenskaya Oblast' see Rivnens'ka Oblast'
Rovenskaya Sloboda see Rovyenskaya Slabada
106 G7 Rovereto Ger. Rofreit. Trentino-Alto Adige, N Italy 45°53′N 11°03′E
167 S12 Rôviĕng Tbong Preăh Vihéar, N Cambodia 13°18′N 105°06′E
Rovigno see Rovinj
106 H8 Rovigo Veneto, NE Italy 45°04′N 11°48′E
112 A10 Rovinj It. Rovigno. Istra, NW Croatia 45°06′N 13°39′E
54 E10 Rovira Tolima, C Colombia 04°15′N 75°15′W
Rovno see Rivne
127 P9 Rovnoye Saratovskaya Oblast', W Russian Federation 50°43′N 46°03′E
82 Q12 Rovuma, Rio var. Ruvuma. Mozambique/Tanzania see also Ruvuma
Rovuma, Rio see Ruvuma
119 O19 Rovyenskaya Slabada Rus. Rovenskaya Sloboda. Homyel'skaya Voblasts', SE Belarus 52°13′N 30°19′E
183 R5 Rowena New South Wales, SE Australia 29°51′S 148°55′E
21 T11 Rowland North Carolina, SE USA 34°32′N 79°17′W
9 P5 Rowley Baffin Island, Nunavut, NE Canada
9 P6 Rowley Island island Nunavut, NE Canada
173 W8 Rowley Shoals reef NW Australia
Równe see Rivne
171 O4 Roxas Mindoro, N Philippines 12°36′N 121°29′E
171 P5 Roxas City Panay Island, C Philippines 11°33′N 122°43′E
21 U8 Roxboro North Carolina, SE USA 36°24′N 79°00′W
185 D23 Roxburgh Otago, South Island, New Zealand 45°32′S 169°18′E
96 K13 Roxburgh cultural region SE Scotland, United Kingdom
182 H5 Roxby Downs South Australia 30°29′S 136°56′E
95 M17 Roxen ◎ S Sweden
25 V5 Roxton Texas, SW USA 33°33′N 95°43′W
15 P12 Roxton-Sud Québec, SE Canada 45°30′N 72°35′W
33 U8 Roy Montana, NW USA 47°19′N 108°55′W
37 U10 Roy New Mexico, SW USA 35°56′N 104°12′W
97 E17 Royal Canal Ir. An Chanáil Ríoga. canal C Ireland
30 L1 Royale, Isle island Michigan, N USA
37 S6 Royal Gorge valley Colorado, C USA
97 M20 Royal Leamington Spa var. Leamington, Leamington Spa. C England, United Kingdom 52°18′N 01°31′W
97 O23 Royal Tunbridge Wells var. Tunbridge Wells. SE England, United Kingdom 51°08′N 00°16′E
24 L9 Royalty Texas, SW USA 31°21′N 102°51′W
102 J11 Royan Charente-Maritime, W France 45°37′N 01°01′W
65 B24 Roy Cove Settlement West Falkland, Falkland Islands 51°32′S 60°23′W
103 O3 Roye Somme, N France 49°42′N 02°46′E
95 H15 Røyken Buskerud, S Norway 59°47′N 10°21′E
93 F14 Røyrvik Nord-Trøndelag, C Norway 64°53′N 13°30′E
25 U6 Royse City Texas, SW USA 32°58′N 96°19′W
97 O21 Royston E England, United Kingdom 52°05′N 00°01′W
23 U2 Royston Georgia, SE USA 34°17′N 83°06′W
114 L10 Roza prev. Gyulovo. Yambol, E Bulgaria 42°29′N 26°30′E
113 L16 Rožaje E Montenegro 42°51′N 20°11′E
110 M10 Różan Mazowieckie, C Poland 52°36′N 21°27′E
117 O10 Rozdil'na Odes'ka Oblast', SW Ukraine 46°51′N 30°03′E
117 S12 Rozdol'ne Rus. Razdolnoye. Avtonomna Respublika Krym, S Ukraine 45°45′N 33°27′E
Rozhdestvenka see Kabanbay Batyr
116 I6 Rozhnyativ Ivano-Frankivs'ka Oblast', W Ukraine 48°58′N 24°07′E
116 J3 Rozhyshche Volyns'ka Oblast', NW Ukraine 50°54′N 25°16′E
Roznau am Radhost see Rožnov pod Radhoštěm
111 L19 Rožňava Ger. Rosenau, Hung. Rozsnyó. Košický Kraj, E Slovakia 48°41′N 20°32′E
116 K10 Roznov Neamţ, NE Romania 46°47′N 26°33′E
111 I18 Rožnov pod Radhoštěm Ger. Rosenau, Roznau am Radhost. Zlínský Kraj, E Czech Republic 49°28′N 18°09′E
Rózsahegy see Ružomberok
Rozsnyó see Râşnov, Romania
Rozsnyó see Rožňava, Slovakia
113 K18 Rranxë Shkodër, NW Albania 41°58′N 19°27′E
113 L18 Rrëshen var. Rresheni, Rrshen. Lezhë, C Albania 41°46′N 19°54′E
Rresheni see Rrëshen
Rrogozhina see Rrogozhinë
113 K20 Rrogozhinë var. Rogozhina, Rogozhinë, Rrogozhina. Tiranë, W Albania 41°04′N 19°40′E
Rrshen see Rrëshen
112 O13 Rtanj ▲ E Serbia 43°45′N 21°54′E
127 O7 Rtishchevo Saratovskaya Oblast', W Russian Federation 52°16′N 43°46′E
184 N12 Ruahine Range var. Ruarine. ▲▲ North Island, New Zealand
185 L14 Ruamahanga North Island, New Zealand
Ruanda see Rwanda
184 M10 Ruapehu, Mount North Island, New Zealand 39°15′S 175°33′E
185 C25 Ruapuke Island island SW New Zealand
Ruarine see Ruahine Range
184 O9 Ruatahuna Bay of Plenty, North Island, New Zealand 38°38′S 176°56′E
184 Q8 Ruatoria Gisborne, North Island, New Zealand 37°54′S 178°18′E
184 K4 Ruawai Northland, North Island, New Zealand 36°08′S 174°04′E
15 N8 Ruban Québec, SE Canada
81 I22 Rubeho Mountains ▲▲ C Tanzania
165 U3 Rubeshibe Hokkaidō, NE Japan 43°49′N 143°37′E
Rubezhnoye see Rubizhne
113 L18 Rubik Lezhë, C Albania 41°46′N 19°48′E
54 H7 Rubio Táchira, W Venezuela 07°42′N 72°23′W
117 X6 Rubizhne Rus. Rubezhnoye. Luhans'ka Oblast', E Ukraine 49°01′N 38°22′E
81 F20 Rubondo Island island N Tanzania
122 I13 Rubtsovsk Altayskiy Kray, S Russian Federation 51°34′N 81°11′E
39 P9 Ruby Alaska, USA 64°44′N 155°29′W
35 W3 Ruby Dome ▲ Nevada, W USA 40°35′N 115°25′W
35 W4 Ruby Lake ◎ Nevada, W USA
35 W4 Ruby Mountains ▲▲ Nevada, W USA
33 Q12 Ruby Range ▲▲ Montana, NW USA
118 C10 Rucava SW Latvia 56°09′N 21°10′E
143 S13 Rūdān var. Dehbārez. Hormozgān, S Iran 27°30′N 57°10′E
Rudelstadt see Ciechanowiec
Rudensk see Rudzyensk
119 G14 Rūdiškės Vilnius, S Lithuania 54°31′N 24°49′E
95 H24 Rudkøbing Syddtjylland, C Denmark 54°57′N 10°43′E
125 S13 Rudnichnyy Kirovskaya Oblast', NW Russian Federation 59°37′N 52°28′E
Rūdnīchnyy see Koksu
Rudny see Rudnyy
126 H4 Rudnya Smolenskaya Oblast', W Russian Federation 54°55′N 31°10′E
127 O8 Rudnya Volgogradskaya Oblast', SW Russian Federation 50°54′N 44°27′E
144 M7 Rudnyy var. Rudny. Kostanay, N Kazakhstan 53°N 63°05′E
122 K3 Rudol'fa, Ostrov island Zemlya Frantsa-Iosifa, NW Russian Federation
Rudolf, Lake see Turkana, Lake
Rudolfswert see Novo mesto
101 L17 Rudolstadt Thüringen, C Germany 50°44′N 11°20′E
31 Q4 Rudyard Michigan, N USA 46°15′N 84°36′W
33 S7 Rudyard Montana, NW USA 48°33′N 110°37′W
119 K16 Rudzyensk Rus. Rudensk. Minskaya Voblasts', C Belarus 53°36′N 27°52′E
104 L6 Rueda Castilla y León, N Spain 41°24′N 04°58′W
114 F10 Ruen ▲ Bulgaria/FYR Macedonia 42°10′N 22°31′E
80 G10 Rufa'a Gezira, C Sudan 14°49′N 33°21′E
102 L10 Ruffec Charente, W France 46°01′N 00°10′E
21 R14 Ruffin South Carolina, SE USA 33°00′N 80°48′W
81 J23 Rufiji E Tanzania
61 A20 Rufino Santa Fe, C Argentina 34°16′S 62°45′W
76 F11 Rufisque W Senegal 14°44′N 17°18′W
83 K14 Rufunsa Lusaka, C Zambia 15°02′S 29°35′E
118 J9 Rugāji E Latvia 57°01′N 27°07′E
161 R7 Rugao Jiangsu, E China 32°27′N 120°35′E
97 M20 Rugby C England, United Kingdom 52°22′N 01°18′W
29 N3 Rugby North Dakota, N USA 48°24′N 100°00′W
100 N7 Rügen headland NE Germany 54°25′N 13°21′E
Ruhaybeh see Ar Ruḩaybah
81 E19 Ruhengeri NW Rwanda 01°39′S 29°16′E
Ruhja see Rūjiena
100 M10 Ruhner Berg hill N Germany
118 F7 Ruhnu var. Ruhnu Saar, Swe. Runö. island SW Estonia
Ruhnu Saar see Ruhnu
101 G15 Ruhr W Germany
91 W6 Ruhr Valley industrial region W Germany
161 S11 Rui'an var. Rui an. Zhejiang, SE China 27°51′N 120°39′E
Rui an see Rui'an
161 P10 Ruichang Jiangxi, S China 29°46′N 115°37′E
24 J11 Ruidosa Texas, SW USA 30°00′N 104°40′W
37 S14 Ruidoso New Mexico, SW USA 33°19′N 105°40′W
161 P12 Ruijin Jiangxi, S China 25°52′N 116°01′E
160 D13 Ruili Yunnan, SW China 24°04′N 97°49′E
98 N8 Ruinen Drenthe, NE Netherlands 52°46′N 06°19′E
99 D17 Ruiselede West-Vlaanderen, W Belgium 51°03′N 03°21′E
161 T13 Ruisui prev. Juisui. C Taiwan 23°43′N 121°28′E
64 P5 Ruivo de Santana, Pico ▲ Madeira, Portugal, NE Atlantic Ocean 32°46′N 16°57′W
40 J12 Ruiz Nayarit, SW Mexico 22°00′N 105°09′W
54 E10 Ruiz, Nevado del W Colombia 04°53′N 75°22′W
138 J9 Rujaylah, Ḩarrat ar salt lake N Jordan
Rujen see Rūjiena
118 H7 Rūjiena Est. Ruhja, Ger. Rujen. N Latvia 57°54′N 25°22′E
79 I18 Ruki W Dem. Rep. Congo
81 E22 Rukwa ◆ region SW Tanzania
81 F23 Rukwa, Lake ◎ SE Tanzania
25 P6 Rule Texas, SW USA 33°10′N 99°53′W
22 K3 Ruleville Mississippi, S USA 33°43′N 90°33′W
Rum see Rhum
112 K10 Ruma Vojvodina, N Serbia 45°02′N 19°51′E
Rumadiya see Ar Ramādī
141 Q7 Rumāḩ Ar Riyāḑ, C Saudi Arabia 25°35′N 47°09′E
Rumaitha see Ar Rumaythah
Rumania/Rumänien see Romania
Rumänisch-Sankt-Georgen see Sângeorz-Băi
139 Y13 Rumaylah Al Başrah, SE Iraq 30°16′N 47°22′E
139 P2 Rumaylah, Wādī dry watercourse NE Syria
171 U13 Rumbati Papua Barat, E Indonesia 02°44′S 132°04′E
81 E14 Rumbek Lakes, C South Sudan 06°50′N 29°42′E
Rumburg see Rumburk
111 D14 Rumburk Ger. Rumburg. Ústecký Kraj, NW Czech Republic 50°58′N 14°35′E
44 J4 Rum Cay island C The Bahamas
99 M26 Rumelange Luxembourg, S Luxembourg 49°28′N 06°02′E
99 D20 Rumes Hainaut, SW Belgium 50°33′N 03°19′E
19 P7 Rumford Maine, NE USA 44°31′N 70°31′W
110 I6 Rumia Pomorskie, N Poland 54°36′N 18°21′E
113 J17 Rumija ▲ S Montenegro
103 T11 Rumilly Haute-Savoie, E France 45°52′N 05°57′E
139 O6 Rūmiyah Al Anbār, W Iraq 34°28′N 41°17′E
Rummah, Wādī ar see Rimah, Wādī ar
Rummelsburg in Pommern see Miastko
165 S3 Rumoi Hokkaidō, NE Japan 43°57′N 141°40′E
82 M12 Rumphi var. Rumpi. Northern, N Malawi 11°00′S 33°51′E
Rumpi see Rumphi
29 V7 Rum River Minnesota, N USA
188 F16 Rumung island Caroline Islands, W Micronesia
Rumuniya/Rumŭniya/Rumunjska see Romania
185 G16 Runanga West Coast, South Island, New Zealand 42°25′S 171°15′E
184 P7 Runaway, Cape headland North Island, New Zealand 37°33′S 177°59′E
97 K18 Runcorn C England, United Kingdom 53°20′N 02°44′W
118 K10 Rundāni var. Rundāni. E Latvia 56°19′N 27°51′E
Rundāni see Rundāni
83 L18 Runde var. Lundi. SE Zimbabwe
83 E16 Rundu var. Runtu. Okavango, NE Namibia 17°55′S 19°45′E
93 I16 Rundvik Västerbotten, N Sweden 63°31′N 19°22′E
81 G20 Runere Mwanza, N Tanzania 03°06′S 33°18′E
25 S13 Runge Texas, SW USA 28°52′N 97°42′W
167 Q13 Rŭng, Kaôh prev. Kas Rong. island SW Cambodia
79 O16 Rungu Orientale, NE Dem. Rep. Congo 03°11′N 27°52′E
81 F23 Rungwa Katavi, W Tanzania 07°18′S 31°40′E
81 G22 Rungwa Singida, C Tanzania 06°54′S 33°33′E
94 M14 Runn ◎ C Sweden
24 M4 Running Water Draw valley New Mexico/Texas, SW USA
Runö see Ruhnu
Runtu see Rundu
189 V12 Ruo island Caroline Islands, C Micronesia
158 L9 Ruoqiang var. Jo-ch'iang, Uigh. Charkhlik, Charkhliq, Qarkilik. Xinjiang Uygur Zizhiqu, NW China 38°59′N 88°08′E
159 S7 Ruo Shui N China
92 L8 Ruostekfielbmá var. Rustefjelbma Finnmark. Finnmark, N Norway 70°25′N 28°10′E
93 L18 Ruovesi Pirkanmaa, W Finland 61°59′N 24°05′E
112 B9 Rupa Primorje-Gorski Kotar, NW Croatia 45°29′N 14°15′E
182 M11 Rupanyup Victoria, SE Australia 36°38′S 142°37′E
168 K9 Rupat, Pulau prev. Roepat. island W Indonesia
168 K10 Rupat, Selat strait Sumatera, W Indonesia
116 J11 Rupea Ger. Reps, Hung. Kőhalom; prev. Cohalm. Braşov, C Romania 46°02′N 25°13′E
99 G17 Rupel N Belgium
Rupella see la Rochelle
33 P15 Rupert Idaho, NW USA 42°37′N 113°40′W
21 R5 Rupert West Virginia, NE USA 37°57′N 80°40′W
Rupert House see Waskaganish
12 J10 Rupert, Rivière de Québec, C Canada
152 I8 Rūpnagar var. Ropar. Punjab, India
194 M13 Ruppert Coast physical region Antarctica
100 N11 Ruppiner Kanal canal NE Germany
55 S11 Rupununi River S Guyana
101 D16 Rur Dut. Roer. Germany/Netherlands
58 H13 Rurópolis Presidente Medici Pará, N Brazil 04°05′S 55°26′W
191 S12 Rurutu island Îles Australes, SW French Polynesia
Rusaddir see Melilla
83 L17 Rusape Manicaland, E Zimbabwe 18°32′S 32°07′E
Rusayris, Lake see Roseires, Reservoir
Ruschuk/Rusçuk see Ruse
114 K7 Ruse var. Ruschuk, Rustchuk, Turk. Rusçuk. Ruse, N Bulgaria 43°50′N 25°59′E
109 W10 Ruše NE Slovenia 46°31′N 15°30′E
114 L7 Ruse ◆ province N Bulgaria
114 K7 Rusenski Lom N Bulgaria
97 G17 Rush Ir. An Ros. Dublin, E Ireland 53°32′N 06°06′W
161 S4 Rushan var. Xiacun. Shandong, E China 36°55′N 121°26′E
Rushan see Rŭshon
Rushanskiy Khrebet see Rushon, Qatorkŭhi
29 V7 Rush City Minnesota, N USA 45°41′N 92°56′W
37 V5 Rush Creek Colorado, C USA
29 X10 Rushford Minnesota, N USA 43°48′N 91°45′W
154 N13 Rushikulya E India
14 D8 Rush Lake ◎ Ontario, S Canada
30 M7 Rush Lake ◎ Wisconsin, N USA
28 J10 Rushmore, Mount ▲ South Dakota, N USA 43°52′N 103°27′W
147 S13 Rŭshon Rus. Rushan. S Tajikistan 37°58′N 71°31′E
147 S14 Rushon, Qatorkŭhi Rus. Rushanskiy Khrebet. ▲▲ SE Tajikistan
26 M12 Rush Springs Oklahoma, C USA 34°46′N 97°57′W
45 V15 Rushville Trinidad, Trinidad and Tobago 10°07′N 61°03′W
30 J13 Rushville Illinois, N USA 40°07′N 90°33′W
28 K12 Rushville Nebraska, C USA 42°41′N 102°28′W
183 O11 Rushworth Victoria, SE Australia 36°36′S 145°03′E
25 W8 Rusk Texas, SW USA 31°49′N 95°11′W
93 I14 Ruskele Västerbotten, N Sweden 64°49′N 18°55′E
118 C12 Rusnė Klaipėda, W Lithuania 55°18′N 21°19′E
114 M10 Rusokastrenska Reka E Bulgaria
Russadir see Melilla
109 X3 Russbach NE Austria
11 V16 Russell Manitoba, S Canada 50°47′N 101°17′W
184 K2 Russell Northland, North Island, New Zealand 35°17′S 174°07′E
26 L4 Russell Kansas, C USA 38°54′N 98°51′W
21 O4 Russell Kentucky, S USA 38°30′N 82°43′W
20 L7 Russell Springs Kentucky, S USA 37°02′N 85°03′W
23 O2 Russellville Alabama, S USA 34°30′N 87°43′W
27 T11 Russellville Arkansas, C USA 35°17′N 93°06′W
20 J7 Russellville Kentucky, S USA 36°50′N 86°54′W
101 G18 Rüsselsheim Hessen, W Germany 50°00′N 08°25′E
Russia see Russian Federation
Russian America see Alaska
122 J11 Russian Federation off. Russian Federation, var. Russia, Latv. Krievija, Rus. Rossiyskaya Federatsiya. ◆ republic Asia/Europe
Russian Federation see Russian Federation
39 N11 Russian Mission Alaska, USA 61°48′N 161°23′W
34 M7 Russian River California, W USA
122 J5 Russkaya Gavan' Novaya Zemlya, Arkhangel'skaya Oblast', N Russian Federation 76°13′N 62°48′E
122 J5 Russkiy, Ostrov island N Russian Federation
109 Y5 Rust Burgenland, E Austria 47°48′N 16°42′E
Rustaq see Ar Rustāq
137 U10 Rustavi prev. Rust'avi. SE Georgia 41°36′N 45°00′E
Rust'avi see Rustavi
21 T7 Rustburg Virginia, NE USA 37°17′N 79°07′W
Rustchuk see Ruse
Rustefjelbma Finnmark see Ruostekfielbmá
83 I21 Rustenburg North-West, N South Africa 25°40′S 27°15′E
22 H5 Ruston Louisiana, S USA 32°31′N 92°38′W
81 E21 Rutana SE Burundi 04°01′S 30°01′E
62 I4 Rutana, Volcán N Chile 22°43′S 67°52′W
Rutanzige, Lake see Edward, Lake
Rutba see Ar Ruţbah
104 M14 Rute Andalucía, S Spain 37°20′N 04°23′W
171 N16 Ruteng prev. Roeteng. Flores, C Indonesia 08°35′S 120°28′E
194 L8 Rutford Ice Stream ice feature Antarctica
35 X6 Ruth Nevada, W USA 39°15′N 115°00′W
101 G15 Rüthen Nordrhein-Westfalen, W Germany 51°30′N 08°28′E
14 D17 Rutherford Ontario, S Canada 42°39′N 82°06′W
21 Q10 Rutherfordton North Carolina, SE USA 35°23′N 81°57′W
97 J18 Ruthin Wel. Rhuthun. NE Wales, United Kingdom 53°05′N 03°18′W
108 G7 Rüti Zürich, N Switzerland 47°16′N 08°51′E
Rutlam see Ratlām
18 M9 Rutland Vermont, NE USA 43°37′N 72°59′W
97 N19 Rutland cultural region C England, United Kingdom
21 N8 Rutledge Tennessee, S USA 36°16′N 83°31′W
158 G12 Rutog var. Rutög, Rutok. Xizang Zizhiqu, W China 33°27′N 79°43′E
Rutok see Rutog
79 P19 Rutshuru Nord-Kivu, E Dem. Rep. Congo 01°11′S 29°28′E
98 L8 Rutten Flevoland, N Netherlands 52°49′N 05°44′E
127 Q17 Rutul Respublika Dagestan, SW Russian Federation 41°35′N 47°30′E
93 L14 Ruukki Pohjois-Pohjanmaa, C Finland 64°40′N 25°35′E
98 N11 Ruurlo Gelderland, E Netherlands 52°04′N 06°27′E
143 S15 Ru'ūs al Jibāl cape Oman/United Arab Emirates
138 I7 Ru'ūs aţ Ţiwāl, Jabal ▲ W Syria
81 H23 Ruvuma ◆ region SE Tanzania

◆ Country
● Country Capital
◇ Dependent Territory
○ Dependent Territory Capital
Administrative Regions
✕ International Airport
▲ Mountain
▲▲ Mountain Range
Volcano
River
◎ Lake
⊠ Reservoir

81 I25 Ruvuma var. Rio Rovuma. ≈ Mozambique/Tanzania see also Rovuma, Rio
Ruvuma see Rovuma, Rio
Ruwais see Ar Ruways
138 L9 Ruwayshid, Wadi ar dry watercourse NE Jordan
141 Z10 Ruways, Ra's ar headland E Oman 20°58′N 59°00′E
79 P18 Ruwenzori ▲▲ Dem. Rep. Congo/Uganda
141 Y8 Ruwī NE Oman 23°33′N 58°31′E
114 F9 Ruy ▲ Bulgaria/Serbia 42°52′N 22°35′E
Ruya see Luia, Rio
81 E20 Ruyigi E Burundi 03°28′S 30°19′E
127 P5 Ruzayevka Respublika Mordoviya, W Russian Federation 54°04′N 44°56′E
119 G18 Ruzhany Brestskaya Voblasts', SW Belarus 52°52′N 24°53′E
Rŭzhevo Konare see Razhevo Konare
Ruzhin see Ruzhyn
114 G7 Ruzhintsi Vidin, NW Bulgaria 43°38′N 22°50′E
161 N6 Ruzhou Henan, C China 34°10′N 112°51′E
117 N5 Ruzhyn Rus. Ruzhin. Zhytomyrs'ka Oblast', N Ukraine 49°42′N 29°01′E
111 K19 Ružomberok Ger. Rosenberg, Hung. Rózsahegy. Zilinský Kraj, N Slovakia 49°04′N 19°19′E
111 C16 Ruzyně ✈ (Praha) Praha, C Czech Republic
81 D19 Rwanda off. Rwandese Republic; prev. Ruanda. ◆ republic C Africa
Rwandese Republic see Rwanda
95 G22 Ry Midtjylland, C Denmark 56°06′N 09°46′E
Ryasna see Rasna
126 L5 Ryazan' Ryazanskaya Oblast', W Russian Federation 54°37′N 39°37′E
126 L5 Ryazanskaya Oblast' ◐ province W Russian Federation
126 M6 Ryazhsk Ryazanskaya Oblast', W Russian Federation 53°42′N 40°09′E
118 B13 Rybachiy Ger. Rossitten. Kaliningradskaya Oblast', W Russian Federation 55°09′N 20°49′E
124 J2 Rybachiy, Poluostrov peninsula NW Russian Federation
Rybach'ye see Balykchy
124 L15 Rybinsk prev. Andropov. Yaroslavskaya Oblast', W Russian Federation 58°03′N 38°53′E
124 K14 Rybinskoye Vodokhranilishche Eng. Rybinsk Reservoir, Rybinsk Sea. ⊠ W Russian Federation
Rybinsk Reservoir/Rybinsk Sea see Rybinskoye Vodokhranilishche
111 I16 Rybnik Śląskie, S Poland 50°05′N 18°31′E
Rybnitsa see Rîbnița
111 F16 Rychnov nad Kněžnou Ger. Reichenau. Královéhradecký Kraj, N Czech Republic 50°10′N 16°17′E
110 I12 Rychwał Wielkopolskie, C Poland 52°04′N 18°10′E
11 O13 Rycroft Alberta, W Canada 55°45′N 118°42′W
95 L21 Ryd Kronoberg, S Sweden 56°27′N 14°44′E
95 L20 Rydaholm Jönköping, S Sweden 56°57′N 14°19′E
194 I8 Rydberg Peninsula peninsula Antarctica
97 P23 Rye SE England, United Kingdom 50°57′N 00°42′E
33 T10 Ryegate Montana, NW USA 46°21′N 109°12′W
35 S3 Rye Patch Reservoir ⊠ Nevada, W USA
95 D15 Ryfylke physical region S Norway
95 H16 Rygge Østfold, S Norway 59°22′N 10°45′E
110 N13 Ryki Lubelskie, E Poland 51°38′N 21°57′E
Rykovo see Yenakiyeve
126 I7 Ryl'sk Kurskaya Oblast', W Russian Federation 51°34′N 34°41′E
183 S8 Rylstone New South Wales, SE Australia 32°48′S 149°58′E
111 H17 Rýmařov Ger. Römerstadt. Moravskoslezský Kraj, E Czech Republic 49°56′N 17°15′E
144 E11 Ryn-Peski desert W Kazakhstan
165 N10 Ryōtsu var. Ryôtu. Niigata, Sado, C Japan 38°06′N 138°28′E
Ryôtu see Ryōtsu
110 K10 Rypin Kujawsko-pomorskie, C Poland 53°03′N 19°25′E
Ryshkany see Rîșcani
Ryssel see Lille
Ryswick see Rijswijk
95 M24 Rytterknægten hill E Denmark
Ryukyu Islands see Nansei-shotō
192 G5 Ryukyu Trench var. Nansei Syotō Trench. undersea feature S East China Sea 24°45′N 128°00′E
110 D11 Rzepin Ger. Reppen. Lubuskie, W Poland 52°20′N 14°48′E
111 N16 Rzeszów Podkarpackie, SE Poland 50°03′N 22°01′E
124 I16 Rzhev Tverskaya Oblast', W Russian Federation 56°17′N 34°22′E
Rzhishchev see Rzhyshchiv
117 P5 Rzhyshchiv Rus. Rzhishchev. Kyyivs'ka Oblast', N Ukraine 49°58′N 31°02′E

S

138 E11 Sa'ad Southern, W Israel 31°27′N 34°31′E
109 P7 Saalach ≈ W Austria
101 L14 Saale ≈ C Germany
101 L17 Saalfeld var. Saalfeld an der Saale. Thüringen, C Germany 50°39′N 11°22′E
Saalfeld see Zalewo
Saalfeld an der Saale see Saalfeld
108 C8 Saane ≈ W Switzerland
101 D19 Saar Fr. Sarre. ≈ France/Germany
101 E20 Saarbrücken Fr. Sarrebruck. Saarland, SW Germany 49°13′N 07°01′E
Saarburg see Sarrebourg
118 D6 Sääre var. Sjar. Saaremaa, W Estonia 57°57′N 21°53′E
Saare see Saaremaa
118 D5 Saaremaa off. Saare Maakond. ◐ province W Estonia
118 E6 Saaremaa Ger. Oesel, Ösel; prev. Saare. island W Estonia
Saare Maakond see Saaremaa
92 L12 Saarenkylä Lappi, N Finland 66°31′N 25°51′E
Saargemund see Sarreguemines
93 L17 Saarijärvi Keski-Suomi, C Finland 62°42′N 25°16′E
Saar in Mähren see Žďár nad Sázavou
92 M10 Saariselkä Lapp. Suoločielgi. Lappi, N Finland 68°27′N 27°29′E
92 L10 Saariselkä hill range NE Finland
101 D20 Saarland Fr. Sarre. ◐ state SW Germany
Saarlautern see Saarlouis
101 D20 Saarlouis prev. Saarlautern. Saarland, SW Germany 49°19′N 06°45′E
108 E11 Saaser Vispa ≈ S Switzerland
137 X12 Saatlı Rus. Saatly. C Azerbaijan 39°57′N 48°24′E
Saatly see Saatlı
Saaz see Žatec
45 V9 Saba ◇ Dutch special municipality Sint Maarten
138 J7 Sab' Ābār var. Sab'a Biyar, Sa'b Bi'ār. Ḥimṣ, C Syria 33°46′N 37°41′E
Sab'a Biyar see Sab' Ābār
112 K11 Šabac Serbia, W Serbia 44°45′N 19°42′E
105 W5 Sabadell Cataluña, E Spain 41°33′N 02°07′E
164 K12 Sabae Fukui, Honshū, SW Japan 36°00′N 136°12′E
169 V7 Sabah prev. British North Borneo, North Borneo. ◐ state East Malaysia
168 J8 Sabak var. Sabak Bernam. Selangor, Peninsular Malaysia 03°45′N 100°59′E
Sabak Bernam see Sabak
38 D16 Sabak, Cape headland Agattu Island, Alaska, USA 52°21′N 173°43′E
81 J20 Sabaki ≈ S Kenya
142 L2 Sabalān, Kuhhā-ye ▲ NW Iran 38°21′N 47°47′E
154 H7 Sabalgarh Madhya Pradesh, C India 26°18′N 77°28′E
44 E4 Sabana, Archipiélago de island group C Cuba
42 H7 Sabanagrande var. Sabana Grande. Francisco Morazán, S Honduras 13°48′N 87°15′W
Sabana Grande see Sabanagrande
54 E5 Sabanalarga Atlántico, N Colombia 10°38′N 74°55′W
41 W14 Sabancuy Campeche, SE Mexico 18°58′N 91°11′W
45 N8 Sabaneta NW Dominican Republic 19°30′N 71°21′W
54 J4 Sabaneta Falcón, N Venezuela 11°17′N 70°00′W
188 H4 Sabaneta, Puntan prev. Ushi Point. headland Saipan, S Northern Mariana Islands 15°17′N 145°49′E
171 X14 Sabang Papua, E Indonesia 04°33′S 138°42′E
116 L10 Săbăoani Neamț, NE Romania 47°01′N 26°51′E
155 J26 Sabaragamuwa ◐ province C Sri Lanka
Sabaria see Szombathely
154 D10 Sābarmati ≈ NW India
171 S10 Sabatai Pulau Morotai, E Indonesia 02°04′N 128°23′E
141 Q15 Sab'atayn, Ramlat as desert C Yemen
107 I16 Sabaudia Lazio, C Italy 41°17′N 13°02′E
57 J19 Sabaya Oruro, S Bolivia 19°09′S 68°21′W
Sa'b Bi'ār see Sab' Ābār
Sabbioncello see Orebić
148 J8 Šāberī, Hāmūn-e var. Daryācheh-ye Hāmun, Daryācheh-ye Sīstān. ◎ Afghanistan/Iran see also Sīstān, Daryācheh-ye
Šāberī, Hāmūn-e see Sīstān, Daryācheh-ye
27 P2 Sabetha Kansas, C USA 39°54′N 95°48′W
75 P10 Sabhā C Libya 27°02′N 14°26′E
67 V13 Sabi var. Save. ≈ Mozambique/Zimbabwe see also Save
Sabi see Save
118 E8 Sabile Ger. Zabeln. NW Latvia 57°03′N 22°33′E
31 R14 Sabina Ohio, N USA 39°29′N 83°38′W
40 I3 Sabinal Chihuahua, N Mexico 30°59′N 107°29′W
25 Q12 Sabinal Texas, SW USA 29°19′N 99°28′W
25 Q11 Sabinal River ≈ Texas, SW USA
105 S4 Sabiñánigo Aragón, NE Spain 42°31′N 00°22′W
41 N6 Sabinas Coahuila, NE Mexico 27°52′N 101°04′W
41 O8 Sabinas Hidalgo Nuevo León, NE Mexico 26°29′N 100°09′W
41 N6 Sabinas, Río ≈ NE Mexico
22 F9 Sabine Lake ◎ Louisiana/Texas, S USA
92 O3 Sabine Land physical region C Svalbard
25 W7 Sabine River ≈ Louisiana/Texas, SW USA
137 X12 Sabirabad C Azerbaijan 40°00′N 48°27′E
Sabkha see As Sabkhah
171 O4 Sablayan Mindoro, N Philippines 12°48′N 120°48′E
13 P16 Sable, Cape headland Newfoundland and Labrador, SE Canada 43°21′N 65°40′W
23 X17 Sable, Cape headland Florida, SE USA 25°12′N 81°06′W
13 R16 Sable Island island Nova Scotia, SE Canada
14 L11 Sables, Lac des ◎ Québec, SE Canada
14 E10 Sables, Rivière aux ≈ Ontario, S Canada
102 K7 Sablé-sur-Sarthe Sarthe, NW France 47°49′N 00°19′W
125 U7 Sablya, Gora ▲ NW Russian Federation 64°46′N 58°52′E
77 U14 Sabon Birnin Gwari Kaduna, C Nigeria 10°43′N 06°39′E
77 V11 Sabon Kafi Zinder, C Niger 14°37′N 08°46′E
104 I6 Sabor, Rio ≈ N Portugal
14 J8 Sabourin, Lac ◎ Québec, SE Canada
137 Y10 Şabran prev. Dävägi. NE Azerbaijan 41°15′N 48°58′E
102 J14 Sabres Landes, SW France 44°07′N 00°46′W
195 X13 Sabrina Coast physical region Antarctica
140 M11 Sabt al Ulāyā 'Asīr, SW Saudi Arabia 19°33′N 41°58′E
104 I8 Sabugal Guarda, N Portugal 40°20′N 07°05′W
29 Z13 Sabula Iowa, C USA 42°04′N 90°10′W
141 N13 Şabyā Jīzān, SW Saudi Arabia 17°50′N 42°50′E
Sabzawar see Sabzevār
Sabzawaran see Jīroft
143 S4 Sabzevār var. Sabzawar. Khorāsān-e Razavī, NE Iran 36°13′N 57°38′E
Sabzvārān see Jīroft
Sacajawea Peak see Matterhorn
82 C9 Sacandica Uíge, NW Angola 06°01′S 15°57′E
42 A2 Sacatepéquez off. Departmento de Sacatepéquez. ◐ department S Guatemala
Sacatepéquez, Departamento de see Sacatepéquez
104 F11 Sacavém Lisboa, W Portugal 38°47′N 09°06′W
29 T13 Sac City Iowa, C USA 42°25′N 94°59′W
105 P8 Sacedón Castilla-La Mancha, C Spain 40°29′N 02°44′W
116 J12 Săcele Ger. Vierdörfer, Hung. Négyfalu; prev. Ger. Sieben Dörfer, Hung. Hétfalu. Brașov, C Romania 45°36′N 25°40′E
163 Y16 Sacheon Jap. Sansenhō; prev. Sach'ŏn, Samch'ŏnpŏ. S South Korea 34°55′N 128°07′E
12 C7 Sachigo ≈ Ontario, C Canada
12 C8 Sachigo Lake Ontario, C Canada 53°52′N 92°16′W
12 C8 Sachigo Lake ◎ Ontario, C Canada
Sach'ŏn see Sacheon
101 O15 Sachsen Eng. Saxony, Fr. Saxe. ◐ state E Germany
101 K14 Sachsen-Anhalt Eng. Saxony-Anhalt. ◐ state C Germany
109 R9 Sachsenburg Salzburg, S Austria 46°49′N 13°23′E
Sachsenfeld see Žalec
8 I5 Sachs Harbour var. Ikaahuk. Banks Island, Northwest Territories, N Canada 72°N 125°14′W
Sächsisch-Reen/Sächsisch-Regen see Reghin
18 H8 Sackets Harbor New York, NE USA 43°57′N 76°06′W
13 P14 Sackville New Brunswick, SE Canada 45°54′N 64°23′W
19 P9 Saco Maine, NE USA 43°32′N 70°25′W
19 P8 Saco River ≈ Maine/New Hampshire, NE USA
35 O7 Sacramento state capital California, W USA 38°35′N 121°30′W
37 T14 Sacramento Mountains ▲▲ New Mexico, SW USA
35 N6 Sacramento River ≈ California, W USA
35 N5 Sacramento Valley valley California, W USA
36 I10 Sacramento Wash valley Arizona, SW USA
105 N15 Sacratif, Cabo headland S Spain 36°41′N 03°30′W
116 F9 Săcueni prev. Săcuieni, Hung. Székelyhíd. Bihor, W Romania 47°20′N 22°05′E
Săcuieni see Săcueni
105 R4 Sádaba Aragón, NE Spain 42°15′N 01°16′W
Sá da Bandeira see Lubango
138 I6 Şadad Ḥimş, W Syria 34°19′N 36°52′E
141 O13 Şa'dah NW Yemen 16°59′N 43°45′E
167 O16 Sadao Songkhla, SW Thailand 06°39′N 100°30′E
142 L8 Sadd-e Dez, Daryācheh-ye ◎ W Iran
19 S3 Saddleback Mountain hill Maine, NE USA
19 P6 Saddleback Mountain ▲ Maine, NE USA 44°57′N 70°27′W
141 W13 Sadh S Oman 17°11′N 55°08′E
76 J11 Sadiola Kayes, W Mali 13°48′N 11°47′W
149 R12 Sādiqābād Punjab, E Pakistan 28°16′N 70°10′E
153 Y10 Sadiya Assam, NE India 27°49′N 95°38′E
139 W9 Sa'dīyah, Hawr as ◎ E Iraq
Sado see Sadoga-shima
165 N9 Sadoga-shima var. Sado. island C Japan
104 F12 Sado, Rio ≈ S Portugal
114 I8 Sadovets Pleven, N Bulgaria 43°19′N 24°21′E
114 J11 Sadovo Plovdiv, C Bulgaria 42°07′N 24°56′E
127 O11 Sadovoye Respublika Kalmykiya, SW Russian Federation 47°51′N 44°34′E
105 Y9 Sa Dragonera var. Isla Dragonera. island Islas Baleares, Spain, W Mediterranean Sea
114 I10 Saedinenie var. Săedinenie, Sŭedinenie. Plovdiv, C Bulgaria 42°14′N 24°36′E
Săedinenie see Saedinenie
105 P9 Saelices Castilla-La Mancha, C Spain 39°55′N 02°49′W
Saena Julia see Siena
Saetabicula see Alzira
114 O12 Safaalan Tekirdağ, NW Turkey 41°26′N 28°07′E
Safad see Tsefat
Şafāqis see Sfax
143 P10 Şafāshahr var. Deh Bīd. Fārs, C Iran 30°50′N 53°50′E
192 I16 Safata Bay bay Upolu, Samoa, C Pacific Ocean
Safed see Tsefat
139 X11 Şaffāf, Ḥawr aş marshy lake S Iraq
95 J16 Säffle Värmland, C Sweden 59°08′N 12°55′E
37 N15 Safford Arizona, SW USA 32°46′N 109°41′W
74 E7 Safi W Morocco 32°19′N 09°14′W
126 I4 Safonovo Smolenskaya Oblast', W Russian Federation 55°05′N 33°12′E
136 H11 Safranbolu Karabük, NW Turkey 41°16′N 32°41′E
139 Y13 Safwān Al Başrah, SE Iraq 30°06′N 47°44′E
158 J16 Saga var. Gya'gya. Xizang Zizhiqu, W China 29°22′N 85°19′E
164 C14 Saga Saga, Kyūshū, SW Japan 33°14′N 130°16′E
164 C13 Saga off. Saga-ken. ◐ prefecture Kyūshū, SW Japan
165 P10 Sagae Yamagata, Honshū, C Japan 38°22′N 140°12′E
166 L3 Sagaing Sagaing, C Myanmar (Burma) 21°55′N 95°56′E
166 L5 Sagaing ◐ region N Myanmar (Burma)
Saga-ken see Saga
165 N13 Sagamihara Kanagawa, Honshū, S Japan 35°34′N 139°22′E
165 N14 Sagami-nada inlet SW Japan
Sagan see Żagań
29 Y3 Saganaga Lake ◎ Minnesota, N USA
155 F18 Sāgar Karnātaka, W India 14°09′N 75°02′E
154 I9 Sāgar prev. Saugor. Madhya Pradesh, C India 23°53′N 78°46′E
15 S8 Sagard Québec, SE Canada 48°01′N 70°03′W
Sagarmāthā see Everest, Mount
Sagebrush State see Nevada
143 V11 Sāghand Yazd, C Iran 32°33′N 55°12′E
19 N14 Sag Harbor Long Island, New York, NE USA 40°59′N 72°15′W
Saghez see Saqqez
31 R8 Saginaw Michigan, N USA 43°25′N 83°57′W
31 R8 Saginaw Bay lake bay Michigan, N USA
Sagiz see Sagyz
64 H6 Saglek Bank undersea feature W Labrador Sea
13 P5 Saglek Bay bay SW Labrador Sea
Saglouc/Sagluk see Salluit
103 X15 Sagonne, Golfe de gulf Corse, France, C Mediterranean Sea
105 P13 Sagra ▲ S Spain 37°59′N 02°33′W
104 F14 Sagres Faro, S Portugal 37°01′N 08°56′W
37 S7 Saguache Colorado, C USA 38°05′N 106°05′W
44 J7 Sagua de Tánamo Holguín, E Cuba 20°38′N 75°14′W
44 E5 Sagua la Grande Villa Clara, C Cuba 22°48′N 80°06′W
15 R7 Saguenay ≈ Québec, SE Canada
74 C9 Saguia al Hamra var. As Saqia al Hamra. ≈ N Western Sahara
105 S9 Sagunto Cat. Sagunt, Ar. Murviedro; anc. Saguntum. Valenciana, E Spain 39°40′N 00°17′W
Sagunt/Saguntum see Sagunto
144 H11 Sagyz prev. Sagiz. Atyrau, W Kazakhstan 48°12′N 54°56′E
138 H10 Saḥāb 'Ammān, NW Jordan 31°52′N 36°00′E
54 E6 Sahagún Córdoba, NW Colombia 08°58′N 75°30′W
104 L4 Sahagún Castilla y León, N Spain 42°23′N 05°02′W
141 X8 Şaḥam N Oman 24°06′N 56°52′E
68 F9 Sahara desert Libya/Algeria
Sahara el Gharbîya see Şaḥrā' al Gharbīyah
75 X9 Sahara el Sharqîya var. Aş Şaḥrā' ash Sharqīyah, Eng. Arabian Desert, Eastern Desert. desert E Egypt
Saharan Atlas see Atlas Saharien
152 J9 Sahāranpur Uttar Pradesh, N India 29°58′N 77°33′E
64 L10 Saharan Seamounts var. Saharian Seamounts. undersea feature E Atlantic Ocean 25°00′N 20°00′W
Saharian Seamounts see Saharan Seamounts
153 Q13 Saharsa Bihār, NE India 25°54′N 86°36′E
67 O7 Sahel physical region C Africa
153 R14 Sāhibganj Jhārkhand, NE India 25°15′N 87°40′E
139 Q7 Sahīliyah Al Anbār, C Iraq 33°43′N 42°42′E
138 H4 Sāḥilīyah, Jibāl as ▲▲ NW Syria
114 M13 Şahin Tekirdağ, NW Turkey 41°01′N 26°51′E
149 U9 Sāhīwāl prev. Montgomery. Punjab, E Pakistan 30°40′N 73°05′E
149 U8 Sāhīwāl Punjab, E Pakistan 31°57′N 72°22′E
141 W11 Şaḥmah, Ramlat as desert C Oman
75 U9 Şaḥrā' al Gharbīyah var. Sahara el Gharbîya, Eng. Western Desert. desert C Egypt
139 T13 Şaḥrā' al Ḥijārah desert S Iraq
40 H5 Sahuaripa Sonora, NW Mexico 29°02′N 109°14′W
36 M16 Sahuarita Arizona, SW USA 31°24′N 110°55′W
40 L13 Sahuayo var. Sahuayo de José Mariá Morelos; prev. Sahuayo de Díaz, Sahuayo de Porfirio Díaz. Michoacán, SW Mexico 20°05′N 102°42′W
Sahuayo de Díaz/Sahuayo de José Mariá Morelos/Sahuayo de Porfirio Díaz see Sahuayo
173 W8 Sahul Shelf undersea feature N Timor Sea
167 P17 Sai Buri Pattani, SW Thailand 06°42′N 101°37′E
74 I6 Saïda NW Algeria 34°50′N 00°10′E
138 G7 Saïda var. Şaydā, Sayida; anc. Sidon. W Lebanon 33°20′N 35°24′E
Sa'īdābād see Sīrjān
80 B13 Sa'id Bundas Western Bahr el Ghazal, W South Sudan 08°24′N 24°53′E
186 E7 Saidor Madang, N Papua New Guinea 05°38′S 146°28′E
153 S13 Saidpur var. Syedpur. Rajshahi, NW Bangladesh 25°48′N 89°E
149 U5 Saīdu var. Mingora, Mongora; prev. Mingāora. Khyber Pakhtunkhwa, N Pakistan 34°45′N 72°21′E
108 C7 Saignelégier Jura, NW Switzerland 47°18′N 07°03′E
164 H11 Saigō Shimane, Dōgo, SW Japan 36°12′N 133°18′E
Saigon see Hô Chi Minh
163 P11 SaihanTal var. Sonid Youqi. Nei Mongol Zizhiqu, N China 42°45′N 112°36′E
162 I12 Saihan Toroi Nei Mongol Zizhiqu, N China 41°44′N 100°29′E
Sai Hun see Syr Darya
92 M11 Saija Lappi, NE Finland 67°07′N 28°46′E
164 G14 Saijō Ehime, Shikoku, SW Japan 33°55′N 133°10′E
164 E15 Saiki Ōita, Kyūshū, SW Japan 32°57′N 131°52′E
93 N18 Saimaa ◎ SE Finland
93 N18 Saimaa Canal Fin. Saimaan Kanava, Rus. Saymenskiy Kanal. canal Finland/Russian Federation
Saimaan Kanava see Saimaa Canal
40 L10 Saín Alto Zacatecas, C Mexico 23°36′N 103°14′W
96 L12 St Abb's Head headland SE Scotland, United Kingdom 55°54′N 02°07′W
11 Y16 St. Adolphe Manitoba, S Canada 49°39′N 96°55′W
103 O15 St-Affrique Aveyron, S France 43°57′N 02°52′E
15 Q10 St-Agapit Québec, SE Canada 46°22′N 71°37′W
97 O21 St Albans anc. Verulamium. E England, United Kingdom 51°46′N 00°21′W
18 L6 Saint Albans Vermont, NE USA 44°49′N 73°07′W
21 Q5 Saint Albans West Virginia, NE USA 38°21′N 81°47′W
St. Alban's Head see St Aldhelm's Head
11 Q14 St. Albert Alberta, SW Canada 53°38′N 113°38′W
97 M24 St Aldhelm's Head var. St. Alban's Head. headland S England, United Kingdom 50°34′N 02°04′W
15 O11 St-Alexis-des-Monts Québec, SE Canada 46°30′N 73°08′W
103 P2 St-Amand-les-Eaux Nord, N France 50°27′N 03°25′E
103 O9 St-Amand-Montrond var. St-Amand-Mont-Rond. Cher, C France 46°43′N 02°29′E
173 P16 St-André NE Réunion
14 M12 St-André-Avellin Québec, SE Canada 45°45′N 75°04′W
Saint-André, Cap see Vilanandro, Tanjona
102 K12 St-André-de-Cubzac Gironde, SW France 45°01′N 00°26′W
96 K11 St Andrews E Scotland, United Kingdom 56°20′N 02°49′W
23 Q9 Saint Andrews Bay bay Florida, SE USA
23 W7 Saint Andrew Sound sound Georgia, SE USA
Saint Anna Trough see Svyataya Anna Trough
44 J11 St. Ann's Bay C Jamaica 18°26′N 77°12′W
13 T10 St. Anthony Newfoundland and Labrador, SE Canada 51°22′N 55°34′W
33 R13 Saint Anthony Idaho, NW USA 43°56′N 111°38′W
182 M11 Saint Arnaud Victoria, SE Australia 36°39′S 143°15′E
185 I15 St.Arnaud Range ▲▲ South Island, New Zealand
13 R10 St-Augustin Québec, E Canada 51°13′N 58°39′W
23 X9 Saint Augustine Florida, SE USA 29°54′N 81°19′W
97 H24 St Austell SW England, United Kingdom 50°21′N 04°47′W
103 T4 St-Avold Moselle, NE France 49°06′N 06°43′E
103 N17 St-Barthélemy ▲▲ S France
102 L17 St-Béat Haute-Garonne, S France 42°55′N 00°39′E
97 I15 St Bees Head headland NW England, United Kingdom 54°30′N 03°39′W
173 P16 St-Benoit E Réunion
103 T13 St-Bonnet Hautes-Alpes, SE France 44°41′N 06°04′E
St.Botolph's Town see Boston
97 G21 St Brides Bay inlet SW Wales, United Kingdom
102 H5 St-Brieuc Côtes d'Armor, NW France 48°31′N 02°45′W
102 H5 St-Brieuc, Baie de bay NW France
102 L7 St-Calais Sarthe, NW France 47°55′N 00°48′E
15 Q10 St-Casimir Québec, SE Canada 46°40′N 72°05′W
14 H16 St. Catharines Ontario, S Canada 43°10′N 79°15′W
45 S14 St. Catherine, Mount ▲ N Grenada 12°10′N 61°41′W
64 C11 St Catherine Point headland E Bermuda
23 X6 Saint Catherines Island island Georgia, SE USA
97 M24 St Catherine's Point headland S England, United Kingdom 50°34′N 01°17′W
103 N13 St-Céré Lot, S France 44°52′N 01°53′E
108 A10 St. Cergue Vaud, W Switzerland 46°25′N 06°10′E
103 R11 St-Chamond Loire, E France 45°29′N 04°32′E
33 S16 Saint Charles Idaho, NW USA 42°05′N 111°23′W
27 X4 Saint Charles Missouri, C USA 38°48′N 90°29′W
103 P13 St-Chély-d'Apcher Lozère, S France 44°51′N 03°16′E
Saint Christopher and Nevis, Federation of see Saint Kitts and Nevis
Saint Christopher-Nevis see Saint Kitts and Nevis
31 S9 Saint Clair Michigan, N USA 42°49′N 82°29′W
183 O17 St. Clair, Lake ◎ Tasmania, SE Australia
14 C17 St. Clair, Lake var. Lac à L'Eau Claire. ◎ Canada/USA
31 S10 Saint Clair Shores Michigan, N USA 42°30′N 82°53′W
103 S10 St-Claude anc. Condate. Jura, E France 46°23′N 05°52′E
45 X6 St-Claude Basse Terre, SW Guadeloupe 16°02′N 61°42′W
23 X12 Saint Cloud Florida, SE USA 28°15′N 81°15′W
29 U8 Saint Cloud Minnesota, N USA 45°34′N 94°10′W
45 T9 Saint Croix island S Virgin Islands (US)
30 J4 Saint Croix Flowage ⊠ Wisconsin, N USA
19 T5 Saint Croix River ≈ Canada/USA
29 W7 Saint Croix River ≈ Minnesota/Wisconsin, N USA
45 S14 St. David's SE Grenada 12°01′N 61°40′W
97 H21 St David's SW Wales, United Kingdom 51°53′N 05°16′W
97 G21 St David's Head headland SW Wales, United Kingdom 51°54′N 05°19′W
64 C12 St David's Island island E Bermuda
173 O16 St-Denis ○ (Réunion) NW Réunion 20°55′S 14°34′E
103 U6 St-Dié Vosges, NE France 48°17′N 06°57′E
103 R5 St-Dizier anc. Desiderii Fanum. Haute-Marne, N France 48°39′N 05°00′E
15 N11 St-Donat Québec, SE Canada 46°19′N 74°15′W
15 N11 Ste-Adèle Québec, SE Canada 45°58′N 74°10′W
15 N11 Ste-Agathe-des-Monts Québec, SE Canada 46°03′N 74°19′W
11 Y16 Ste. Anne Manitoba, S Canada 49°40′N 96°40′W
45 R12 Ste-Anne Grande Terre, E Guadeloupe 16°13′N 61°23′W
45 Y6 Ste-Anne SE Martinique 14°26′N 60°53′W
15 Q10 Ste-Anne ≈ Québec, SE Canada
172 I16 Sainte Anne island Inner Islands, NE Seychelles
15 W6 Ste-Anne-des-Monts Québec, SE Canada 49°07′N 66°29′W
14 M10 Ste-Anne-du-Lac Québec, SE Canada 46°51′N 75°20′W
15 S10 Ste-Apolline Québec, SE Canada 46°47′N 70°15′W
15 R10 Ste-Claire Québec, SE Canada 46°30′N 70°40′W
15 Q10 Ste-Croix Québec, SE Canada 46°36′N 71°42′W
108 B8 Ste. Croix Vaud, SW Switzerland 46°50′N 06°31′E
103 P14 Ste-Énimie Lozère, S France 44°21′N 03°26′E
27 Y6 Sainte Genevieve Missouri, C USA 37°57′N 90°01′W
103 S12 St-Egrève Isère, E France 45°15′N 05°41′E
39 T12 Saint Elias, Cape headland Kayak Island, Alaska, USA 59°48′N 144°36′W
39 U11 St Elias, Mount ▲ Alaska, USA 60°18′N 140°57′W
10 G8 Saint Elias Mountains ▲▲ Canada/USA
55 Y10 St-Élie N French Guiana 04°50′N 53°21′W
103 O10 St-Eloy-les-Mines Puy-de-Dôme, C France 46°07′N 02°50′E
15 R10 Ste-Marie Québec, SE Canada 46°28′N 71°00′W
45 Q11 Ste-Marie NE Martinique 14°47′N 61°00′W
173 P16 Ste-Marie NE Réunion
103 U6 Ste-Marie-aux-Mines Haut-Rhin, NE France 48°16′N 07°12′E
Sainte Marie, Cap see Vohimena, Tanjona
102 L8 Ste-Maure-de-Touraine Indre-et-Loire, C France 47°06′N 00°38′E
103 R4 Ste-Menehould Marne, NE France 49°06′N 04°54′E
Ste-Perpétue see Ste-Perpétue-de-l'Islet
15 S9 Ste-Perpétue-de-l'Islet var. Ste-Perpétue. Québec, SE Canada 47°02′N 69°58′W
45 X11 Ste-Rose Basse Terre, N Guadeloupe 16°20′N 61°42′W
173 P16 Ste-Rose E Réunion
11 W15 Ste. Rose du Lac Manitoba, S Canada 51°04′N 99°31′W
102 J11 Saintes anc. Mediolanum. Charente-Maritime, W France 45°45′N 00°37′W
45 X7 Saintes, Canal des channel SW Guadeloupe
Saintes, Îles des see les Saintes
173 P16 Ste-Suzanne N Réunion
15 P10 Ste-Thècle Québec, SE Canada 46°48′N 72°31′W
103 Q12 St-Étienne Loire, E France 45°26′N 04°23′E
102 M4 St-Étienne-du-Rouvray Seine-Maritime, N France 49°22′N 01°07′E
Saint Eustatius see Sint Eustatius
14 M11 Ste-Véronique Québec, SE Canada 46°30′N 74°58′W
15 P7 St-Félicien Québec, SE Canada 48°38′N 72°29′W
15 O11 St-Félix-de-Valois Québec, SE Canada 46°10′N 73°26′W
103 Y14 St-Florent Corse, France, C Mediterranean Sea 42°41′N 09°19′E
103 Y14 St-Florent, Golfe de gulf Corse, France, C Mediterranean Sea
103 P6 St-Florentin Yonne, C France 48°00′N 03°46′E
103 N9 St-Florent-sur-Cher Cher, C France 47°00′N 02°13′E
103 P12 St-Flour Cantal, C France 45°02′N 03°05′E
26 H2 Saint Francis Kansas, C USA 39°45′N 101°31′W
83 H26 St. Francis, Cape headland S South Africa 34°11′S 24°45′E
27 X10 Saint Francis River ≈ Arkansas/Missouri, C USA
22 J8 Saint Francisville Louisiana, S USA 30°46′N 91°22′W
45 Y6 St-François Grande Terre, E Guadeloupe 16°15′N 61°17′W
15 Q12 St-François ≈ Québec, SE Canada
27 X7 Saint Francois Mountains ▲ Missouri, C USA
St-Gall/Saint Gall/St. Gallen see Sankt Gallen
St-Gall see St-Gall/Saint Gall/St.Gallen
102 L16 St-Gaudens Haute-Garonne, S France 43°07′N 00°43′E
15 R12 St-Gédéon Québec, SE Canada 45°51′N 70°36′W
181 X10 Saint George Queensland, E Australia 28°05′S 148°40′E
64 B12 St George N Bermuda 32°24′N 64°42′W
38 K15 Saint George Saint George Island, Alaska, USA 56°34′N 169°30′W
21 S14 Saint George South Carolina, SE USA 33°12′N 80°34′W
36 J8 Saint George Utah, W USA 37°06′N 113°35′W
13 R12 St. George, Cape headland Newfoundland and Labrador, E Canada 48°26′N 59°17′W
186 I6 St. George, Cape headland New Ireland, NE Papua New Guinea 04°49′S 152°52′E
38 J15 Saint George Island island Pribilof Islands, Alaska, USA
23 S10 Saint George Island island Florida, SE USA
99 J19 Saint-Georges Liège, E Belgium 50°36′N 05°20′E
15 R11 St-Georges Québec, SE Canada 46°08′N 70°40′W
55 Z11 St-Georges E French Guiana 03°55′N 51°49′W
45 R14 St. George's ● (Grenada) SW Grenada 12°04′N 61°45′W
13 R12 St. George's Bay inlet Newfoundland and Labrador, E Canada
97 G21 Saint George's Channel channel Ireland/Wales, United Kingdom
186 H6 St. George's Channel channel NE Papua New Guinea
64 B11 St George's Island island E Bermuda
99 I21 Saint-Gérard Namur, S Belgium 50°20′N 04°47′E
St-Germain see St-Germain-en-Laye
15 P12 St-Germain-de-Grantham Québec, SE Canada 45°49′N 72°32′W
103 N5 St-Germain-en-Laye var. St-Germain. Yvelines, N France 48°53′N 02°04′E
102 H8 St-Gildas, Pointe du headland NW France 47°09′N 02°25′W
103 R15 St-Gilles Gard, S France 43°41′N 04°24′E
102 I9 St-Gilles-Croix-de-Vie Vendée, NW France 46°41′N 01°55′W
173 O16 St-Gilles-les-Bains W Réunion 21°02′S 55°14′E
102 M16 St-Girons Ariège, S France 42°58′N 01°07′E
Saint Gotthard see Szentgotthárd
108 G9 St. Gotthard Tunnel tunnel Ticino, S Switzerland
97 H22 St Govan's Head headland SW Wales, United Kingdom 51°35′N 04°55′W
34 M7 Saint Helena California, W USA 38°29′N 122°30′W
67 O12 Saint Helena island C Atlantic Ocean
Saint Helena see Saint Helena, Ascension and Tristan da Cunha
65 F24 Saint Helena, Ascension and Tristan da Cunha terr. Saint Helena, Ascension, Tristan da Cunha. ◇ UK overseas territory C Atlantic Ocean
65 M16 Saint Helena Fracture Zone tectonic feature C Atlantic Ocean
34 M7 Saint Helena, Mount ▲ California, W USA 38°40′N 122°39′W
21 S15 Saint Helena Sound inlet South Carolina, SE USA
31 Q7 Saint Helen, Lake ◎ Michigan, N USA
183 Q16 Saint Helens Tasmania, SE Australia 41°21′S 148°15′E
97 K18 St Helens NW England, United Kingdom 53°28′N 02°44′W
32 G10 Saint Helens Oregon, NW USA 45°54′N 122°50′W
32 H10 Saint Helens, Mount ☄ Washington, NW USA 46°24′N 121°49′W
97 L26 St Helier ○ (Jersey) S Jersey, Channel Islands 49°12′N 02°07′W
15 S9 St-Hilarion Québec, SE Canada 47°34′N 70°24′W
99 K22 Saint-Hubert Luxembourg, SE Belgium 50°02′N 05°23′E
15 P12 St-Hyacinthe Québec, SE Canada 45°38′N 72°57′W
St.Iago de la Vega see Spanish Town
31 Q4 Saint Ignace Michigan, N USA 45°53′N 84°44′W
15 O10 St-Ignace-du-Lac Québec, SE Canada 46°43′N 73°49′W
12 D12 St. Ignace Island island Ontario, S Canada
108 C7 St. Imier Bern, W Switzerland 47°09′N 06°55′E
97 G25 St Ives SW England, United Kingdom 50°12′N 05°29′W
29 U10 Saint James Minnesota, N USA 44°00′N 94°36′W
10 I15 St. James, Cape headland Graham Island, British Columbia, SW Canada 51°57′N 131°04′W

◆ Country
● Country Capital
◇ Dependent Territory
○ Dependent Territory Capital
◐ Administrative Regions
✈ International Airport
▲ Mountain
▲▲ Mountain Range
☄ Volcano
≈ River
◎ Lake
⊠ Reservoir

15 O13 **St-Jean** *var.* St-Jean-sur-Richelieu. Québec, SE Canada 45°15′N 73°16′W
55 X9 **St-Jean** NW French Guiana 05°25′N 54°05′W
Saint-Jean-d'Acre *see* Akko
102 K11 **St-Jean-d'Angély** Charente-Maritime, W France 45°57′N 00°31′W
103 N7 **St-Jean-de-Braye** Loiret, C France 47°54′N 01°58′E
102 I16 **St-Jean-de-Luz** Pyrénées-Atlantiques, SW France 43°24′N 01°40′W
103 T12 **St-Jean-de-Maurienne** Savoie, E France 45°17′N 06°21′E
102 I9 **St-Jean-de-Monts** Vendée, NW France 46°45′N 02°00′W
103 Q14 **St-Jean-du-Gard** Gard, S France 44°06′N 03°49′E
15 Q7 **St-Jean, Lac** ◎ Québec, SE Canada
102 I16 **St-Jean-Pied-de-Port** Pyrénées-Atlantiques, SW France 43°10′N 01°14′W
15 S9 **St-Jean-Port-Joli** Québec, SE Canada 47°13′N 70°16′W
St-Jean-sur-Richelieu *see* St-Jean
15 N12 **St-Jérôme** Québec, SE Canada 45°47′N 74°01′W
25 T5 **Saint Jo** Texas, SW USA 33°42′N 97°33′W
13 O15 **St. John** New Brunswick, SE Canada 45°16′N 66°03′W
26 L6 **Saint John** Kansas, C USA 37°59′N 98°44′W
19 Q2 **Saint John** *Fr.* Saint-John. ≈ Canada/USA
76 K16 **Saint John** ≈ C Liberia
45 T9 **Saint John** *island* C Virgin Islands (US)
Saint-John *see* Saint John
22 I6 **Saint John, Lake** ◎ Louisiana, S USA
45 W10 **St John's** ● (Antigua and Barbuda) Antigua, Antigua and Barbuda 17°06′N 61°50′W
13 V12 **St. John's** *province capital* Newfoundland and Labrador, E Canada 47°34′N 52°41′W
37 O12 **Saint Johns** Arizona, SW USA 34°28′N 109°22′W
31 Q9 **Saint Johns** Michigan, N USA 43°01′N 84°31′W
13 V12 **St. John's** ✈ Newfoundland and Labrador, E Canada 47°22′N 52°45′W
23 X11 **Saint Johns River** ≈ Florida, SE USA
103 Q11 **St-Jost-St-Rambert** Loire, E France 45°30′N 04°13′E
45 N12 **St. Joseph** W Dominica 15°24′N 61°26′W
173 P17 **St-Joseph** S Réunion
22 J6 **Saint Joseph** Louisiana, S USA 31°56′N 91°14′W
31 O10 **Saint Joseph** Michigan, N USA 42′ 05′N 86′ 30′W
27 R3 **Saint Joseph** Missouri, C USA 39°46′N 94°49′W
20 I10 **Saint Joseph** Tennessee, S USA 35°02′N 87°29′W
22 R9 **Saint Joseph Bay** *bay* Florida, SE USA
15 R11 **St-Joseph-de-Beauce** Québec, SE Canada 46°20′N 70°52′W
12 C10 **St. Joseph, Lake** ◎ Ontario, C Canada
31 Q11 **Saint Joseph River** ≈ N USA
14 C11 **Saint Joseph's Island** *island* Ontario, S Canada
15 N11 **St-Jovite** Québec, SE Canada 46°07′N 74°35′W
St Julian's *see* San Ġiljan
St-Julien *see* St-Julien-en-Genevois
103 T10 **St-Julien-en-Genevois** *var.* St-Julien. Haute-Savoie, E France 46°07′N 06°06′E
102 M11 **St-Junien** Haute-Vienne, C France 45°52′N 00°54′E
96 D8 **St Kilda** *island* NW Scotland, United Kingdom
45 V10 **Saint Kitts** *island* Saint Kitts and Nevis
45 U10 **Saint Kitts and Nevis** *off.* Federation of Saint Christopher and Nevis, *var.* Saint Christopher-Nevis. ◆ *commonwealth republic* E West Indies
11 X16 **St. Laurent** Manitoba, S Canada 50°20′N 97°55′W
St-Laurent *see* St-Laurent-du-Maroni
55 X9 **St-Laurent-du-Maroni** *var.* St-Laurent. NW French Guiana 05°29′N 54°03′W
St-Laurent, Fleuve *see* St. Lawrence
102 J12 **St-Laurent-Médoc** Gironde, SW France 45°11′N 00°50′W
13 N12 **St. Lawrence** *Fr.* Fleuve St-Laurent. ≈ Canada/USA
13 Q12 **St. Lawrence, Gulf of** *gulf* NW Atlantic Ocean
38 K10 **Saint Lawrence Island** *island* Alaska, USA
14 M14 **Saint Lawrence River** ≈ Canada/USA
99 L25 **Saint-Léger** Luxembourg, SE Belgium 49°36′N 05°39′E
13 N14 **St. Léonard** New Brunswick, SE Canada 47°10′N 67°55′W
15 P11 **St-Léonard** Québec, SE Canada 46°06′N 72°18′W
173 O17 **St-Leu** W Réunion 21°09′S 55°17′E
102 J4 **St-Lô** *anc.* Briovera, Laudus. Manche, N France 49°07′N 01°08′W
11 T15 **St. Louis** Saskatchewan, S Canada 52°50′N 105°43′W
103 V7 **St-Louis** Haut-Rhin, NE France 47°35′N 07°34′E
173 O17 **St-Louis** S Réunion
76 G10 **Saint Louis** NW Senegal 15°59′N 16°30′W
27 X4 **Saint Louis** Missouri, C USA 38°38′N 90°15′W
29 W5 **Saint Louis River** ≈ Minnesota, N USA
103 T7 **St-Loup-sur-Semouse** Haute-Saône, E France 47°53′N 06°15′E
15 O12 **St-Luc** Québec, SE Canada 45°19′N 73°18′W
45 X13 **Saint Lucia** ◆ *commonwealth republic* SE West Indies
47 S3 **Saint Lucia** *island* SE West Indies
83 L22 **St. Lucia, Cape** *headland* E South Africa 28°29′S 32°26′E
45 Y13 **Saint Lucia Channel** *channel* Martinique/Saint Lucia
23 Y14 **Saint Lucie Canal** *canal* Florida, SE USA
23 Z13 **Saint Lucie Inlet** *inlet* Florida, SE USA
96 L2 **St Magnus Bay** *bay* N Scotland, United Kingdom
102 K10 **St-Maixent-l'École** Deux-Sèvres, W France 46°24′N 00°13′W
11 Y16 **St. Malo** Manitoba, S Canada 49°16′N 96°58′W
102 I5 **St-Malo** Ille-et-Vilaine, NW France 48°39′N 02°W
102 H4 **St-Malo, Golfe de** *gulf* NW France
44 L9 **St-Marc** C Haiti 19°08′N 72°41′W
44 L9 **St-Marc, Canal de** *channel* W Haiti
103 S12 **St-Marcellin-le-Mollard** Isère, E France 45°12′N 05°18′E
55 Y12 **Saint-Marcel, Mont** ▲ S French Guiana 2°32′N 53°00′E
96 K5 **St Margaret's Hope** NE Scotland, United Kingdom 58°50′N 02°57′W
32 M9 **Saint Maries** Idaho, NW USA 47°19′N 116°37′W
23 T9 **Saint Marks** Florida, SE USA 30°09′N 84°12′W
108 D11 **St. Martin** Valais, SW Switzerland 46°09′N 07°27′E
Saint Martin *see* Sint Maarten
31 O5 **Saint Martin Island** *island* Michigan, N USA
22 I9 **Saint Martinville** Louisiana, S USA 30°09′N 91°51′W
185 E20 **St. Mary, Mount** ▲ South Island, New Zealand 44°16′S 169°42′E
186 E8 **St. Mary, Mount** ▲ S Papua New Guinea 08°06′S 147°00′E
182 I6 **Saint Mary Peak** ▲ South Australia 31°25′S 138°39′E
183 Q16 **Saint Marys** Tasmania, SE Australia 41°34′S 148°13′E
14 E16 **St. Marys** Ontario, S Canada 43°15′N 81°08′W
38 M11 **Saint Marys** Alaska, USA 62°03′N 163°10′W
23 W8 **Saint Marys** Georgia, SE USA 30°44′N 81°30′W
27 P4 **Saint Marys** Kansas, C USA 39°09′N 96°00′W
31 Q4 **Saint Marys** Ohio, N USA 40°31′N 84°22′W
21 R3 **Saint Marys** West Virginia, NE USA 39°24′N 81°13′W
23 W8 **Saint Marys River** ≈ Florida/Georgia, SE USA
31 Q4 **Saint Marys River** ≈ Michigan, N USA
102 D6 **St-Mathieu, Pointe** *headland* NW France 48°17′N 04°56′W
38 J12 **Saint Matthew Island** *island* Alaska, USA
21 R13 **Saint Matthews** South Carolina, SE USA 33°40′N 80°44′W
St.Matthew's Island *see* Zadetkyi Kyun
186 G4 **St.Matthias Group** *island group* NE Papua New Guinea
108 C11 **St. Maurice** Valais, SW Switzerland 46°09′N 07°28′E
15 P9 **St-Maurice** ≈ Québec, SE Canada
102 J13 **St-Médard-en-Jalles** Gironde, SW France 44°54′N 00°43′W
39 N10 **Saint Michael** Alaska, USA 63°28′N 162°02′W
15 N10 **St-Michel-des-Saints** Québec, SE Canada 46°39′N 73°54′W
103 S5 **St-Mihiel** Meuse, NE France 48°57′N 05°33′E
108 J10 **St. Moritz** *Ger.* Sankt Moritz, *Rmsch.* San Murezzan. Graubünden, SE Switzerland 46°30′N 09°51′E
102 H8 **St-Nazaire** Loire-Atlantique, NW France 47°17′N 02°12′W
Saint Nicholas *see* São Nicolau
Saint-Nicolas *see* Sint-Niklaas
103 N1 **St-Omer** Pas-de-Calais, N France 50°45′N 02°15′E
102 J11 **Saintonge** *cultural region* W France
15 S9 **St-Pacôme** Québec, SE Canada 47°22′N 69°56′W
15 S10 **St-Pamphile** Québec, SE Canada 46°57′N 69°46′W
15 S9 **St-Pascal** Québec, SE Canada 47°32B 69°48′W
14 J11 **St-Patrice, Lac** ◎ Québec, SE Canada
11 R14 **St. Paul** Alberta, SW Canada 54°00′N 111°18′W
173 O16 **St-Paul** NW Réunion
38 K14 **Saint Paul** Saint Paul Island, Alaska, USA 57°08′N 170°13′W
29 V8 **Saint Paul** *state capital* Minnesota, N USA 45°N 93°10′W
29 P15 **Saint Paul** Nebraska, C USA 41°13′N 98°26′W
21 P7 **Saint Paul** Virginia, NE USA 36°53′N 82°18′W
77 Q17 **Saint Paul, Cape** *headland* S Ghana 05°44′N 00°55′E
103 O17 **St-Paul-de-Fenouillet** Pyrénées-Orientales, S France 42°49′N 02°29′E
65 K14 **Saint Paul Fracture Zone** *tectonic feature* E Atlantic Ocean
38 J14 **Saint Paul Island** *island* Pribilof Islands, Alaska, USA
102 J15 **St-Paul-les-Dax** Landes, SW France 43°45′N 01°01′W
21 U11 **Saint Pauls** North Carolina, SE USA 34°45′N 78°56′W
Saint Paul's Bay *see* San Pawl il Baħar
191 R16 **St Paul's Point** *headland* Pitcairn Island, Pitcairn Islands
29 U10 **Saint Peter** Minnesota, N USA 44°21′N 93°58′W
97 L26 **St Peter Port** ○ (Guernsey) C Guernsey, Channel Islands 49°28′N 02°33′W
23 V13 **Saint Petersburg** Florida, SE USA 27°47′N 82°37′W
Saint Petersburg *see* Sankt-Peterburg
23 V13 **Saint Petersburg Beach** Florida, SE USA 27°43′N 82°43′W
173 P17 **St-Philippe** SE Réunion 21°21′S 55°46′E
45 Q11 **St-Pierre** NW Martinique 14°44′N 61°11′W
173 O17 **St-Pierre** SW Réunion
13 S13 **St-Pierre and Miquelon** *Fr.* Îles St-Pierre et Miquelon. ◇ *French overseas collectivity* NE North America
15 P11 **St-Pierre, Lac** ◎ Québec, SE Canada
102 F5 **St-Pol-de-Léon** Finistère, NW France 48°42′N 04°00′W
103 O2 **St-Pol-sur-Ternoise** Pas-de-Calais, N France 50°22′N 02°21′E
St. Pons *see* St-Pons-de-Thomières
103 O16 **St-Pons-de-Thomières** *var.* St. Pons. Hérault, S France 43°28′N 02°48′E
103 P10 **St-Pourçain-sur-Sioule** Allier, C France 46°19′N 03°16′E
15 S11 **St-Prosper** Québec, SE Canada 46°14′N 70°28′W
103 P3 **St-Quentin** Aisne, N France 49°51′N 03°17′E
15 R10 **St-Raphaël** Québec, SE Canada 46°47′N 70°46′W
103 U15 **St-Raphaël** Var, SE France 43°26′N 06°46′E
15 Q10 **St-Raymond** Québec, SE Canada 46°53′N 71°49′W
33 O9 **Saint Regis** Montana, NW USA 47°18′N 115°06′W
18 J7 **Saint Regis River** ≈ New York, NE USA
103 R15 **St-Rémy-de-Provence** Bouches-du-Rhône, SE France 43°48′N 04°49′E
102 M9 **St-Savin** Vienne, W France 46°34′N 00°53′E
Saint-Sébastien,Cap *see* Anorontany, Tanjona
23 X7 **Saint Simons Island** *island* Georgia, SE USA
191 Y2 **Saint Stanislas Bay** *bay* Kiritimati, E Kiribati
13 O15 **St. Stephen** New Brunswick, SE Canada 45°12′N 67°18′W
45 V9 **St Barthélemy** ◈ *French overseas collectivity* E Caribbean Sea
39 X12 **Saint Terese** Alaska, USA 58°28′N 134°46′W
14 E17 **St. Thomas** Ontario, S Canada 42°46′N 81°12′W
29 Q2 **Saint Thomas** North Dakota, N USA 48°37′N 97°28′W
45 T9 **Saint Thomas** *island* W Virgin Islands (US)
Saint Thomas *see* São Tomé, Sao Tome and Principe
Saint Thomas *see* Charlotte Amalie, Virgin Islands (US)
15 P10 **St-Tite** Québec, SE Canada 46°42′N 72°32′W
45 V9 **St Martin** ◈ *French overseas collectivity* E Caribbean Sea
Saint-Trond *see* Sint-Truiden
103 U16 **St-Tropez** Var, SE France 43°16′N 06°39′E
Saint Ubes *see* Setúbal
102 L3 **St-Valéry-en-Caux** Seine-Maritime, N France 49°53′N 00°42′E
103 Q9 **St-Vallier** Saône-et-Loire, C France 46°39′N 04°19′E
106 B7 **St-Vincent** Valle d'Aosta, NW Italy 45°47′N 07°42′E
45 Q14 **Saint Vincent** *island* N Saint Vincent and the Grenadines
Saint Vincent *see* São Vicente
45 W14 **Saint Vincent and the Grenadines** ◆ *commonwealth republic* SE West Indies
Saint-Vincent, Cap *see* Ankaboa, Tanjona
Saint Vincent, Cape *see* São Vicente, Cabo de
102 I15 **St-Vincent-de-Tyrosse** Landes, SW France 43°39′N 01°16′W
182 I9 **Saint Vincent, Gulf** *gulf* South Australia
23 R10 **Saint Vincent Island** *island* Florida, SE USA
45 T12 **Saint Vincent Passage** *passage* Saint Lucia/Saint Vincent and the Grenadines
183 N18 **Saint Vincent, Point** *headland* Tasmania, SE Australia 43°19′S 145°50′E
Saint-Vith *see* Sankt-Vith
11 S14 **St. Walburg** Saskatchewan, S Canada 53°38′N 109°12′W
St Wolfgangsee *see* Wolfgangsee
102 M11 **St-Yrieix-la-Perche** Haute-Vienne, C France 45°31′N 01°12′E
Saint Yves *see* Setúbal
188 H5 **Saipan** *island* ● (Northern Mariana Islands) S Northern Mariana Islands
188 H6 **Saipan Channel** *channel* S Northern Mariana Islands
188 H6 **Saipan International** ✈ Saipan, S Northern Mariana Islands
74 G6 **Saïs** ✈ (Fès) C Morocco 33°58′N 04°48′W
Saishū *see* Jeju-do
Saishū *see* Jeju
102 J16 **Saison** ≈ SW France
169 R10 **Sai, Sungai** ≈ Borneo, N Indonesia
165 N13 **Saitama** *off.* Saitama-ken. ◈ *prefecture* Honshū, S Japan
Saitama *see* Urawa
Saitama-ken *see* Saitama
Saiyid Abid *see* Sayyid 'Abīd
57 J19 **Sajama, Nevado** ▲ W Bolivia 17°57′S 68°51′W
141 V13 **Sājir, Ras** *headland* S Oman 16°42′N 53°40′E
111 M20 **Sajószentpéter** Borsod-Abaúj-Zemplén, NE Hungary 48°13′N 20°43′E
83 F24 **Sak** ≈ SW South Africa
81 J18 **Saka** Tana River, E Kenya 0°11′S 39°27′E
167 P11 **Sa Kaeo** Prachin Buri, C Thailand 13°47′N 102°03′E
164 J14 **Sakai** Ōsaka, Honshū, SW Japan 34°35′N 135°28′E
164 H14 **Sakaide** Kagawa, Shikoku, SW Japan 34°19′N 133°50′E
164 H12 **Sakaiminato** Tottori, Honshū, SW Japan 35°34′N 133°12′E
140 M3 **Sakākah** Al Jawf, NW Saudi Arabia 29°56′N 40°10′E
28 L4 **Sakakawea, Lake** ⊠ North Dakota, N USA
12 J9 **Sakami, Lac** ◎ Québec, C Canada
79 O26 **Sakania** Katanga, SE Dem. Rep. Congo 12°44′S 28°34′E
146 K12 **Sakar** Lebap Welaýaty, E Turkmenistan 38°57′N 63°46′E
172 H7 **Sakaraha** Toliara, SW Madagascar 22°54′S 44°31′E
146 I14 **Sakarçäge** *var.* Sakarchäge, *Rus.* Sakar-Chaga. Mary Welaýaty, C Turkmenistan 37°40′N 61°33′E
Sakar-Chaga/Sakarchäge *see* Sakarçäge
Sak'art'velo *see* Georgia
136 F11 **Sakarya** ◈ *province* NW Turkey
136 F12 **Sakarya Nehri** ≈ NW Turkey
165 P9 **Sakata** Yamagata, Honshū, C Japan 38°54′N 139°51′E
123 P9 **Sakha (Yakutiya), Respublika** *var.* Respublika Yakutiya, *Eng.* Yakutia. ◈ *autonomous republic* NE Russian Federation
Sakhalin *see* Sakhalin, Ostrov
192 I3 **Sakhalin, Ostrov** *var.* Sakhalin. *island* SE Russian Federation
123 U12 **Sakhalinskaya Oblast'** ◈ *province* SE Russian Federation
123 T12 **Sakhalinskiy Zaliv** *gulf* E Russian Federation
Sakhnovshchina *see* Sakhnovshchyna
117 U6 **Sakhnovshchyna** *Rus.* Sakhnovshchina. Kharkivs'ka Oblast', E Ukraine 49°08′N 35°52′E
Sakhon Nakhon *see* Sakon Nakhon
Sakhtsar *see* Rāmsar
137 W10 **Şäki** *Rus.* Sheki; *prev.* Nukha. NW Azerbaijan 41°09′N 47°10′E
Saki *see* Saky
118 E13 **Šakiai** *Ger.* Schaken. Marijampolė, S Lithuania 54°57′N 23°04′E
165 O16 **Sakishima-shotō** *var.* Sakisima Syotô. *island group* SW Japan
Sakisima Syotô *see* Sakishima-shotō
Sakiz *see* Saqqez
Sakiz-Adasi *see* Chíos
155 F19 **Sakleshpur** Karnātaka, E India 12°58′N 75°45′E
167 S9 **Sakon Nakhon** *var.* Muang Sakon Nakhon, Sakhon Nakhon. Sakon Nakhon, E Thailand 17°10′N 104°08′E
149 P15 **Sakrand** Sind, SE Pakistan 26°06′N 68°20′E
83 F24 **Sak River** *Afr.* Sakrivier. Northern Cape, W South Africa 30°49′S 20°24′E
Sakrivier *see* Sak River
Saksaul'skiy *see* Saksaul'skoye
144 K13 **Saksaul'skoye** *var.* Saksaul'skiy, *Kaz.* Sekseüil. Kzylorda, S Kazakhstan 47°07′N 61°06′E
95 I25 **Sakskøbing** Sjælland, SE Denmark 54°48′N 11°39′E
165 N12 **Saku** Nagano, Honshū, S Japan 36°17′N 138°29′E
117 S13 **Saky** *Rus.* Saki. Avtonomna Respublika Krym, S Ukraine 45°09′N 33°36′E
76 E9 **Sal** *island* Ilhas de Barlavento, NE Cape Verde
127 N12 **Sal** ≈ SW Russian Federation
111 I21 **Sal'a** *Hung.* Sellye, Vágsellye. Nitriansky Kraj, SW Slovakia 48°09′N 17°51′E
95 N15 **Sala** Västmanland, C Sweden 59°55′N 16°38′E
15 N13 **Salaberry-de-Valleyfield** *var.* Valleyfield. Québec, SE Canada 45°15′N 74°07′W
118 G7 **Salacgrīva** *Est.* Salatsi. N Latvia 57°45′N 24°21′E
107 M18 **Sala Consilina** Campania, S Italy 40°23′N 15°35′E
40 C2 **Salada, Laguna** ◎ NW Mexico
61 D14 **Saladas** Corrientes, NE Argentina 28°15′S 58°40′W
61 C21 **Saladillo** Buenos Aires, E Argentina 35°40′S 59°50′W
61 B16 **Saladillo, Río** ≈ C Argentina
25 T9 **Salado** Texas, SW USA 30°57′N 97°32′W
63 J16 **Salado, Arroyo** ≈ SE Argentina
61 D21 **Salado, Río** ≈ E Argentina
62 J12 **Salado, Río** ≈ C Argentina
41 N7 **Salado, Río** ≈ NE Mexico
37 Q12 **Salado, Rio** ≈ New Mexico, SW USA
143 N6 **Salafchegān** *var.* Sarafjagān. Quom, N Iran 34°28′N 50°28′E
77 Q15 **Salaga** C Ghana 08°31′N 00°37′W
139 S6 **Şalāḩ ad Dīn** *off.* Muḩāfa at Şalāḩ ad Dīn, *var.* Salāhuddīn. ◈ *governorate* C Iraq
Şalāḩ ad Dīn, Muḩāfa at *see* Şalāḩ ad Dīn
Salāhuddīn *see* Şalāḩ ad Dīn
192 G5 **Sala'ilua** Savai'i, W Samoa 13°39′S 172°33′W
116 G9 **Sălaj** ◈ *county* NW Romania
83 H20 **Salajwe** Kweneng, SE Botswana 23°40′S 24°46′E
78 H9 **Salal** Bahr el Gazel, W Chad 14°48′N 17°12′E
80 I6 **Salala** Red Sea, NE Sudan 21°17′N 36°16′E
141 V13 **Şalālah** SW Oman 17°01′N 54°04′E
42 D5 **Salamá** Baja Verapaz, C Guatemala 15°06′N 90°18′W
42 J6 **Salamá** Olancho, C Honduras 14°48′N 86°34′W
62 G10 **Salamanca** Coquimbo, C Chile 31°47′S 70°58′W
41 N13 **Salamanca** Guanajuato, C Mexico 20°34′N 101°12′W
104 K7 **Salamanca** *anc.* Helmantica, Salmantica. Castilla y León, NW Spain 40°58′N 05°40′W
18 D11 **Salamanca** New York, NE USA 42°09′N 78°43′W
104 J7 **Salamanca** ◈ *province* Castilla y León, W Spain
63 J19 **Salamanca, Pampa de** *plain* S Argentina
78 J12 **Salamat** *off.* Région du Salamat. ◈ *region* SE Chad
78 I12 **Salamat, Bahr** ≈ S Chad
Salamat, Région du *see* Salamat
54 F5 **Salamina** Magdalena, N Colombia 10°30′N 74°48′W
115 G19 **Salamína** *var.* Salamís. Salamína, C Greece 37°58′N 23°29′E
115 G19 **Salamína** *island* C Greece
Salamís *see* Salamína
138 I5 **Salamīyah** *var.* As Salamīyah. Ḩamāh, W Syria 35°01′N 37°02′E
31 P12 **Salamonie Lake** ⊠ Indiana, N USA
31 P12 **Salamonie River** ≈ Indiana, N USA
Salang *see* Phuket
192 I16 **Salani** Upolu, SE Samoa 14°00′S 171°35′W
118 C11 **Salantai** Klaipėda, NW Lithuania 56°05′N 21°36′E
104 K2 **Salas** Asturias, N Spain 43°25′N 06°15′W
105 O5 **Salas de los Infantes** Castilla y León, N Spain 42°02′N 03°17′W
102 M16 **Salat** ≈ S France
189 V13 **Salat** *island* Chuuk, C Micronesia
169 Q16 **Salatiga** Jawa, C Indonesia 07°15′S 110°34′E
189 V13 **Salat Pass** *passage* W Pacific Ocean
Salatsi *see* Salacgrīva
167 T10 **Salavan** *var.* Saravan, Saravane. Salavan, S Laos 15°43′N 106°26′E
127 V6 **Salavat** Respublika Bashkortostan, W Russian Federation 53°20′N 55°54′E
56 C12 **Salaverry** La Libertad, N Peru 08°14′S 78°55′W
171 T12 **Salawati, Pulau** *island* E Indonesia
193 R10 **Sala y Gomez** *island* Chile, E Pacific Ocean
Sala y Gomez Fracture Zone *see* Sala y Gomez Ridge
193 S10 **Sala y Gomez Ridge** *var.* Sala y Gomez Fracture Zone. *tectonic feature* SE Pacific Ocean
61 A22 **Salazar** Buenos Aires, E Argentina 36°20′S 62°11′W
54 G7 **Salazar** Norte de Santander, N Colombia 07°46′N 72°46′W
Salazar *see* N'Dalatando
173 P16 **Salazie** C Réunion 21°02′S 55°33′E
103 N8 **Salbris** Loir-et-Cher, C France 47°25′N 02°02′E
57 G15 **Salcantay, Nevado** ▲ C Peru 13°21′S 72°31′W
45 O8 **Salcedo** N Dominican Republic 19°26′N 70°25′W
39 S9 **Salcha River** ≈ Alaska, USA
119 H15 **Šalčininkai** Vilnius, SE Lithuania 54°20′N 25°26′E
Saldae *see* Béjaïa
54 E11 **Saldaña** Tolima, C Colombia 03°57′N 75°01′W
104 M4 **Saldaña** Castilla y León, N Spain 42°31′N 04°44′W
83 E25 **Saldanha** Western Cape, SW South Africa 33°00′S 17°56′E
Salduba *see* Zaragoza
61 B23 **Saldungaray** Buenos Aires, E Argentina 38°15′S 61°45′W
118 D9 **Saldus** *Ger.* Frauenburg. W Latvia 56°40′N 22°29′E
183 P13 **Sale** Victoria, SE Australia 38°06′S 147°06′E
74 F6 **Salé** NW Morocco 34°07′N 06°40′W
74 F6 **Salé** ✈ (Rabat) W Morocco 34°09′N 06°30′W
Salehābād *see* Andīmeshk
170 M16 **Saleh, Teluk** *bay* Nusa Tenggara, S Indonesia
122 H8 **Salekhard** *prev.* Obdorsk. Yamalo-Nenetskiy Avtonomnyy Okrug, N Russian Federation 66°33′N 66°35′E
192 H16 **Sālelologa** Savai'i, C Samoa 13°42′S 172°10′W
155 H21 **Salem** Tamil Nādu, SE India 11°38′N 78°08′E
27 V9 **Salem** Arkansas, C USA 36°21′N 91°49′W
30 L15 **Salem** Illinois, N USA 38°37′N 88°57′W
31 P15 **Salem** Indiana, N USA 38°38′N 86°06′W
19 P11 **Salem** Massachusetts, NE USA 42°30′N 70°51′W
27 V6 **Salem** Missouri, C USA 37°39′N 91°32′W
18 I16 **Salem** New Jersey, NE USA 39°33′N 75°26′W
31 U12 **Salem** Ohio, N USA 40°52′N 80°51′W
32 G12 **Salem** *state capital* Oregon, NW USA 44°57′N 123°01′W
29 Q11 **Salem** South Dakota, N USA 43°43′N 97°23′W
36 L4 **Salem** Utah, W USA 40°03′N 111°40′W
21 S7 **Salem** Virginia, NE USA 37°16′N 80°00′W
21 R3 **Salem** West Virginia, NE USA 39°15′N 80°32′W
107 H23 **Salemi** Sicilia, Italy, C Mediterranean Sea 37°48′N 12°48′E
Salemy *see* As Sālimī
94 K12 **Sälen** Dalarna, C Sweden 61°11′N 13°14′E
107 Q18 **Salentina, Campi** Puglia, SE Italy 40°23′N 18°01′E
107 Q18 **Salentina, Penisola** *peninsula* SE Italy
107 L18 **Salerno** *anc.* Salernum. Campania, S Italy 40°40′N 14°44′E
107 L18 **Salerno, Golfo di** *Eng.* Gulf of Salerno. *gulf* S Italy
Salerno, Gulf of *see* Salerno, Golfo di
Salernum *see* Salerno
97 K17 **Salford** NW England, United Kingdom 53°30′N 02°16′W
Salgir *see* Salhyr
111 K21 **Salgótarján** Nógrád, N Hungary 48°07′N 19°47′E
59 O15 **Salgueiro** Pernambuco, E Brazil 08°04′S 39°05′W
94 C13 **Salhus** Hordaland, S Norway 60°30′N 05°15′E
117 T12 **Salhyr** *Rus.* Salgir. ≈ S Ukraine
171 Q9 **Salibabu, Pulau** *island* N Indonesia
37 S6 **Salida** Colorado, C USA 38°29′N 105°57′W
102 J15 **Salies-de-Béarn** Pyrénées-Atlantiques, SW France 43°28′N 00°55′W
136 C14 **Salihli** Manisa, W Turkey 38°29′N 28°08′E
119 K18 **Salihorsk** *Rus.* Soligorsk. Minskaya Voblasts', S Belarus 52°48′N 27°32′E
119 K18 **Salihorskaye Vodaskhovishcha** *Rus.* Soligorskoye Vodokhranilishche. ⊠ C Belarus
83 N14 **Salima** Central, C Malawi 13°44′S 34°21′E
166 L5 **Salin** Magway, W Myanmar (Burma) 20°30′N 94°40′E
27 N4 **Salina** Kansas, C USA 38°53′N 97°36′W
36 L5 **Salina** Utah, W USA 38°57′N 111°54′W
41 S17 **Salina Cruz** Oaxaca, SE Mexico 16°11′N 95°12′W
107 L22 **Salina, Isola** *island* Isole Eolie, S Italy
44 J5 **Salina Point** *headland* Acklins Island, SE The Bahamas 22°10′N 74°16′W
56 A7 **Salinas** Guayas, W Ecuador 02°15′S 80°58′W
40 M11 **Salinas** *var.* Salinas de Hidalgo. San Luis Potosí, C Mexico 22°36′N 101°41′W
45 T6 **Salinas** C Puerto Rico 17°57′N 66°18′W
35 O10 **Salinas** California, W USA 36°41′N 121°40′W
Salinas, Cabo de *see* Salines, Cap de ses
Salinas de Hidalgo *see* Salinas
82 A13 **Salinas, Ponta das** *headland* W Angola 12°50′S 12°57′E
45 O10 **Salinas, Punta** *headland* S Dominican Republic 18°11′N 70°32′W
Salinas, Río *see* Chixoy, Río
35 O11 **Salinas River** ≈ California, W USA
22 H6 **Saline Lake** ◎ Louisiana, S USA
25 R17 **Salineno** Texas, SW USA 26°29′N 99°06′W
27 V14 **Saline River** ≈ Arkansas, C USA
30 M17 **Saline River** ≈ Illinois, N USA
105 X10 **Salines, Cap de ses** *var.* Cabo de Salinas. *headland* Mallorca, Spain, W Mediterranean Sea 39°15′N 03°03′E
Salisburg *see* Mazsalaca
45 O12 **Salisbury** *var.* Baroui. W Dominica 15°26′N 61°27′W
97 M23 **Salisbury** *var.* New Sarum. S England, United Kingdom 51°05′N 01°48′W
21 Y4 **Salisbury** Maryland, NE USA 38°22′N 75°37′W
27 T3 **Salisbury** Missouri, C USA 39°25′N 92°48′W
21 S9 **Salisbury** North Carolina, SE USA 35°40′N 80°28′W
Salisbury *see* Harare
9 Q7 **Salisbury Island** *island* Nunavut, NE Canada
Salisbury, Lake *see* Bisina, Lake
97 L23 **Salisbury Plain** *plain* S England, United Kingdom
21 R14 **Salkehatchie River** ≈ South Carolina, SE USA
138 I9 **Şalkhad** As Suwaydā', SW Syria 32°29′N 36°42′E
92 M12 **Salla** Lappi, NE Finland 66°50′N 28°40′E
Sállan *see* Sørøya
103 U11 **Sallanches** Haute-Savoie, E France 45°55′N 06°37′E
105 V5 **Sallent** Cataluña, NE Spain 41°48′N 01°52′E
Salliq *see* Coral Harbour
61 A22 **Salliqueló** Buenos Aires, E Argentina 36°45′S 62°55′W
27 R10 **Sallisaw** Oklahoma, C USA 35°27′N 94°49′W
80 I7 **Sallom** Red Sea, NE Sudan 19°17′N 37°02′E
12 J2 **Salluit** *prev.* Saglouc, Sagluk. Québec, NE Canada 62°10′N 75°40′W
Sallyana *see* Şalyān
13 S11 **Sally's Cove** Newfoundland and Labrador, E Canada 49°43′N 58°00′W
139 W9 **Salmān Bin 'Arāzah** Maysān, E Iraq 32°33′N 46°36′E
Salmantica *see* Salamanca
142 I2 **Salmās** *prev.* Dilman, Shāpūr. Āžarbāyjān-e Gharbī, NW Iran 38°13′N 44°50′E
124 I11 **Salmi** Respublika Kareliya, NW Russian Federation 61°21′N 31°55′E
33 P12 **Salmon** Idaho, NW USA 45°10′N 113°54′W
11 N16 **Salmon Arm** British Columbia, SW Canada 50°41′N 119°18′W
192 L5 **Salmon Bank** *undersea feature* N Pacific Ocean 26°55′N 176°28′W
Salmon Leap *see* Leixlip
34 L2 **Salmon Mountains** ⛰ California, W USA
14 J15 **Salmon Point** *headland* Ontario, SE Canada 43°51′N 77°15′W
33 N11 **Salmon River** ≈ Idaho, NW USA
18 K6 **Salmon River** ≈ New York, NE USA
33 N12 **Salmon River Mountains** ⛰ Idaho, NW USA
18 I9 **Salmon River Reservoir** ⊠ New York, NE USA
93 K19 **Salo** Varsinais-Suomi, SW Finland 60°23′N 23°10′E
106 F7 **Salò** Lombardia, N Italy 45°37′N 10°30′E
Salona/Salonae *see* Solin
103 S15 **Salon-de-Provence** Bouches-du-Rhône, SE France 43°39′N 05°05′E
Salonica/Salonika *see* Thessaloníki
115 I14 **Salonikiós, Akrotírio** *var.* Akrotírio Salonikós. *headland* Thásos, E Greece 40°34′N 24°39′E
Salonikós, Akrotírio *see* Salonikiós, Akrotírio
116 F10 **Salonta** *Hung.* Nagyszalonta. Bihor, W Romania 46°49′N 21°40′E
104 I9 **Salor** ≈ W Spain
105 U6 **Salou** Cataluña, NE Spain 41°05′N 01°08′E
76 H11 **Saloum** ≈ C Senegal
42 H4 **Sal, Punta** *headland* NW Honduras 15°55′N 87°36′W
92 N3 **Salpynten** *headland* W Svalbard 78°12′N 12°11′E
138 I3 **Salqīn** Idlib, W Syria 36°09′N 36°27′E
93 F14 **Salsbruket** Nord-Trøndelag, C Norway 64°49′N 11°48′E
126 M13 **Sal'sk** Rostovskaya Oblast', SW Russian Federation 46°30′N 41°31′E
107 J25 **Salso** ≈ Sicilia, Italy, C Mediterranean Sea
107 K25 **Salso** ≈ Sicilia, Italy, C Mediterranean Sea
106 E9 **Salsomaggiore Terme** Emilia-Romagna, N Italy 44°49′N 09°58′E
Salt *see* As Salţ
62 J6 **Salta** Salta, NW Argentina 24°47′S 65°23′W
62 K6 **Salta** *off.* Provincia de Salta. ◈ *province* N Argentina
Salta, Provincia de *see* Salta
97 I24 **Saltash** SW England, United Kingdom 50°26′N 04°14′W
24 I8 **Salt Basin** *basin* Texas, SW USA
11 V16 **Saltcoats** Saskatchewan, S Canada 51°06′N 102°12′W
30 L13 **Salt Creek** ≈ Illinois, N USA
24 J9 **Salt Draw** ≈ Texas, SW USA
97 F21 **Saltee Islands** *island group* SE Ireland
92 G12 **Saltfjorden** *inlet* C Norway
24 I8 **Salt Flat** Texas, SW USA 31°43′N 105°05′W
27 N8 **Salt Fork Arkansas River** ≈ Oklahoma, C USA
31 T13 **Salt Fork Lake** ⊠ Ohio, N USA
26 J11 **Salt Fork Red River** ≈ Oklahoma/Texas, C USA
95 J23 **Saltholm** *island* E Denmark
41 N8 **Saltillo** Coahuila, NE Mexico 25°30′N 101°W
182 L5 **Salt Lake** *salt lake* New South Wales, SE Australia
37 V15 **Salt Lake** ◎ New Mexico, SW USA
36 K2 **Salt Lake City** *state capital* Utah, W USA 40°45′N 111°54′W
61 C20 **Salto** Buenos Aires, E Argentina 34°18′S 60°17′W
61 D17 **Salto** Salto, N Uruguay 31°23′S 57°58′W
61 E17 **Salto** ◈ *department* N Uruguay
107 I14 **Salto** ≈ C Italy
62 Q6 **Salto del Guairá** Canindeyú, E Paraguay 24°06′S 54°22′W
61 D17 **Salto Grande, Embalse de** *var.* Lago de Salto Grande. ⊠ Argentina/Uruguay
Salto Grande, Lago de *see* Salto Grande, Embalse de
35 W16 **Salton Sea** ◎ California, W USA
60 I12 **Salto Santiago, Represa de** ⊠ S Brazil
149 U7 **Salt Range** ⛰ E Pakistan
36 M13 **Salt River** ≈ Arizona, SW USA
20 L5 **Salt River** ≈ Kentucky, S USA
27 V3 **Salt River** ≈ Missouri, C USA
95 F17 **Saltrød** Aust-Agder, S Norway 58°28′N 08°49′E
95 P16 **Saltsjöbaden** Stockholm, C Sweden 59°15′N 18°20′E
92 G12 **Saltstraumen** Nordland, C Norway 67°16′N 14°42′E
21 Q7 **Saltville** Virginia, NE USA 36°52′N 81°48′W
Saluces/Saluciae *see* Saluzzo
21 Q12 **Saluda** South Carolina, SE USA 34°00′N 81°47′W
21 X6 **Saluda** Virginia, NE USA 37°36′N 76°36′W
21 Q12 **Saluda River** ≈ South Carolina, SE USA
Salûm *see* As Sallūm
152 F14 **Sālūmbar** Rājasthān, N India 24°16′N 74°04′E
Salûm, Gulf of *see* Khalīj as Sallūm
171 O11 **Salumpaga** Sulawesi, N Indonesia 01°18′N 120°58′E
155 M14 **Sālūr** Andhra Pradesh, E India 18°31′N 83°16′E
55 Y9 **Salut, Îles du** *island group* N French Guiana
106 A9 **Saluzzo** *Fr.* Saluces; *anc.* Saluciae. Piemonte, NW Italy 44°39′N 07°29′E
63 F23 **Salvación, Bahía** *bay* S Chile
59 P17 **Salvador** *prev.* São Salvador. *state capital* Bahia, E Brazil 12°58′S 38°29′W
65 E24 **Salvador** East Falkland, Falkland Islands 51°28′S 58°22′W
22 K10 **Salvador, Lake** ◎ Louisiana, S USA
Salvaleón de Higüey *see* Higüey
104 F10 **Salvaterra de Magos** Santarém, C Portugal 39°01′N 08°47′W
41 N13 **Salvatierra** Guanajuato, C Mexico 20°14′N 100°52′W
105 P3 **Salvatierra** *Basq.* Agurain. País Vasco, N Spain 42°52′N 02°23′W
Salwa/Salwah *see* As Salwā
166 M7 **Salween** *Bur.* Thanlwin, *Chin.* Nu Chiang, Nu Jiang. ≈ SE Asia
137 Y12 **Salyan** *Rus.* Sal'yany. SE Azerbaijan 39°36′N 48°57′E
153 N11 **Şalyān** *var.* Sallyana. Mid Western, W Nepal 28°22′N 82°10′E
Sal'yany *see* Salyan
21 O6 **Salyersville** Kentucky, S USA 37°43′N 83°06′W
109 V6 **Salza** ≈ E Austria
109 Q7 **Salzach** ≈ Austria/Germany
109 Q6 **Salzburg** *anc.* Juvavum. Salzburg, N Austria 47°48′N 13°03′E
109 O8 **Salzburg** *off.* Land Salzburg. ◈ *state* C Austria
Salzburg *see* Ocna Sibiului
Salzburg Alps *see* Salzburger Kalkalpen

◆ Country
● Country Capital
◇ Dependent Territory
○ Dependent Territory Capital
◈ Administrative Regions
✈ International Airport
▲ Mountain
⛰ Mountain Range
Volcano
≈ River
◎ Lake
⊠ Reservoir

109 Q7 **Salzburger Kalkalpen** *Eng.* Salzburg Alps. ⛰ C Austria
Salzburg, Land *see* Salzburg
100 J13 **Salzgitter** *prev.* Watenstedt-Salzgitter. Niedersachsen, C Germany 52°07′N 10°24′E
101 G14 **Salzkotten** Nordrhein-Westfalen, W Germany 51°40′N 08°36′E
100 K11 **Salzwedel** Sachsen-Anhalt, N Germany 52°51′N 11°10′E
152 D11 **Sām** Rājasthān, NW India 26°50′N 70°30′E
Šamac *see* Bosanski Šamac
54 G9 **Samacá** Boyacá, C Colombia 05°28′N 73°33′W
40 I7 **Samachique** Chihuahua, N Mexico 27°17′N 107°28′W
141 Y8 **Şamad** NE Oman 22°47′N 58°12′E
Sama de Langreo *see* Sama, Spain
Samaden *see* Samedan
57 M19 **Samaipata** Santa Cruz, C Bolivia 18°08′S 63°53′W
Samakhixai *see* Attapu
Samakov *see* Samokov
42 B6 **Samalá, Río** ≈ SW Guatemala
40 J3 **Samalayuca** Chihuahua, N Mexico 31°25′N 106°30′W
155 L16 **Sāmalkot** Andhra Pradesh, E India 17°03′N 82°15′E
45 P8 **Samaná** *var.* Santa Bárbara de Samaná. E Dominican Republic 19°14′N 69°20′W
45 P8 **Samaná, Bahía de** *bay* E Dominican Republic
44 K4 **Samana Cay** *island* SE The Bahamas
136 K17 **Samandağı** Hatay, S Turkey 36°07′N 35°55′E
149 P3 **Samangān** ◈ *province* N Afghanistan
Samangān *see* Aibak
165 T5 **Samani** Hokkaidō, NE Japan 42°07′N 142°57′E
54 C13 **Samaniego** Nariño, SW Colombia 01°22′N 77°35′W
171 Q5 **Samar** *island* C Philippines
127 S6 **Samara** *prev.* Kuybyshev. Samarskaya Oblast', W Russian Federation 53°15′N 50°15′E
127 T7 **Samara** ≈ W Russian Federation
127 S6 **Samara** ✈ Samarskaya Oblast', W Russian Federation 53°11′N 50°27′E
117 V7 **Samara** ≈ E Ukraine
186 G10 **Samarai** Milne Bay, SE Papua New Guinea 10°36′S 150°39′E
Samarang *see* Semarang
123 T14 **Samarga** Khabarovskiy Kray, SE Russian Federation 47°43′N 139°08′E
138 G9 **Samarian Hills** *hill range* N Israel
54 L9 **Samariapo** Amazonas, C Venezuela 05°16′N 67°43′W
169 V11 **Samarinda** Borneo, C Indonesia 00°30′S 117°09′E
Samarkand *see* Samarqand
Samarkandskaya Oblast' *see* Samarqand Viloyati
Samarkandski/Samarkandskoye *see* Temirtau
Samarobriva *see* Amiens
147 N11 **Samarqand** *Rus.* Samarkand. Samarqand Viloyati, C Uzbekistan 39°40′N 66°56′E
146 M11 **Samarqand Viloyati** *Rus.* Samarkandskaya Oblast'. ◈ *province* C Uzbekistan
139 S6 **Sāmarrā'** Şalāḩ ad Dīn, C Iraq 34°13′N 43°52′E
127 R7 **Samarskaya Oblast'** *prev.* Kuybyshevskaya Oblast'. ◈ *province* W Russian Federation
153 Q13 **Samastīpur** Bihār, N India 25°52′N 85°47′E
76 L14 **Samatiguila** NW Ivory Coast 09°51′N 07°36′W
Samawa *see* As Samāwah
137 Y11 **Şamaxı** *Rus.* Shemakha. E Azerbaijan 40°38′N 48°34′E
79 K18 **Samba** Equateur, NW Dem. Rep. Congo 00°13′N 21°17′E
79 N21 **Samba** Maniema, E Dem. Rep. Congo 04°41′S 26°23′E
152 H6 **Samba** Jammu and Kashmir, NW India 32°32′N 75°08′E
169 W10 **Sambaliung, Pegunungan** ⛰ Borneo, N Indonesia
154 M11 **Sambalpur** Odisha, E India 21°28′N 84°04′E
67 X12 **Sambao** ≈ W Madagascar
169 Q10 **Sambas, Sungai** ≈ Borneo, N Indonesia
172 K2 **Sambava** Antsiranana, NE Madagascar 14°16′S 50°10′E
152 J10 **Sambhal** Uttar Pradesh, N India 28°35′N 78°34′E
152 H12 **Sāmbhar Salt Lake** ◎ N India
107 N21 **Sambiase** Calabria, SW Italy 38°58′N 16°16′E
116 H5 **Sambir** *Rus.* Sambor. L'vivs'ka Oblast', NW Ukraine 49°31′N 23°10′E
82 C13 **Sambo** Huambo, C Angola 13°07′S 16°06′E
Sambor *see* Sambir
61 E21 **Samborombón, Bahía** *bay* NE Argentina
99 H20 **Sambre** ≈ Belgium/France
43 V16 **Sambú, Río** ≈ SE Panama
81 I18 **Samburu** ◈ *county* N Kenya
163 Z14 **Samcheok** *Jap.* Sanchoku; *prev.* Samch'ŏk. NE South Korea 37°21′N 129°12′E
Samch'ŏk *see* Samcheok
Samch'ŏnpŏ *see* Sacheon
81 I21 **Same** Kilimanjaro, NE Tanzania 04°04′S 37°41′E
108 J10 **Samedan** *Ger.* Samaden. Graubünden, S Switzerland 46°31′N 09°51′E
82 K12 **Samfya** Luapula, N Zambia 11°22′S 29°34′E
141 W13 **Samḩān, Jabal** ⛰ SW Oman
115 C18 **Sámi** Kefallonía, Iónia Nisiá, Greece, C Mediterranean Sea 38°15′N 20°39′E
56 F10 **Samiria, Río** ≈ N Peru
Samirum *see* Semīrom
167 Q13 **Samĭt** *prev.* Phumĭ Samĭt. Kaôh Kŏng, SW Cambodia 10°54′N 103°09′E
137 V11 **Şämkir** *Rus.* Shamkhor. NW Azerbaijan 40°51′N 46°03′E
167 S7 **Sam, Nam** *Vtn.* Sông Chu. ≈ Laos/Vietnam
Samnān *see* Semnān
Sam Neua *see* Xam Nua
75 P10 **Samnū** C Libya 27°19′N 15°01′E
192 H15 **Samoa** *off.* Independent State of Samoa, *var.* Sāmoa; *prev.* Western Samoa. ◆ *monarchy* W Polynesia
192 L9 **Sāmoa** *island group* C Pacific Ocean
Sāmoa *see* Samoa
175 T9 **Samoa Basin** *undersea feature* W Pacific Ocean
Samoa, Independent State of *see* Samoa
112 D8 **Samobor** Zagreb, N Croatia 45°48′N 15°38′E
114 H10 **Samokov** *var.* Samakov. Sofia, W Bulgaria 42°19′N 23°34′E
111 H21 **Šamorín** *Ger.* Sommerein, *Hung.* Somorja. Trnavský Kraj, W Slovakia 48°01′N 17°18′E
115 M19 **Sámos** *prev.* Limín Vathéos. Sámos, Dodekánisa, Greece, Aegean Sea 37°45′N 26°58′E
115 M20 **Sámos** *island* Dodekánisa, Greece, Aegean Sea
Samosch *see* Szamos
168 I9 **Samosir, Pulau** *island* W Indonesia
Samothrace *see* Samothráki
115 K14 **Samothráki** Samothráki, NE Greece 40°28′N 25°31′E
115 J14 **Samothráki** *anc.* Samothrace. *island* NE Greece
115 A15 **Samothráki** *island* Iónia Nisiá, Greece, C Mediterranean Sea
Samotschin *see* Szamocin
Sampê *see* Xiangcheng
169 S13 **Sampit** Borneo, C Indonesia 02°30′S 112°30′E
169 S12 **Sampit, Sungai** ≈ Borneo, N Indonesia
Sampoku *see* Sanpoku
186 H7 **Sampun** New Britain, E Papua New Guinea 05°19′S 152°06′E
79 N24 **Sampwe** Katanga, SE Dem. Rep. Congo 09°17′S 27°22′E
167 R11 **Sâmraông** *prev.* Phumĭ Sâmraông, Phum Samrong. Siĕmréab, NW Cambodia 14°11′N 103°31′E
25 X8 **Sam Rayburn Reservoir** ⊠ Texas, SW USA
167 Q6 **Sam Sao, Phou** ⛰ Laos/Thailand
95 H22 **Samsø** *island* E Denmark
95 H23 **Samsø Bælt** *channel* E Denmark
167 T7 **Sâm Sơn** Thanh Hoa, N Vietnam 19°44′N 105°53′E
136 L11 **Samsun** *anc.* Amisus. Samsun, N Turkey 41°17′N 36°22′E
136 K11 **Samsun** ◈ *province* N Turkey
137 R9 **Samt'redia** *prev.* Samtredia. W Georgia 42°09′N 42°20′E
Samtredia *see* Samt'redia
59 E15 **Samuel, Represa de** ⊠ W Brazil
167 O14 **Samui, Ko** *island* SW Thailand
Samundari *see* Samundri
149 U9 **Samundri** *var.* Samundari. Punjab, E Pakistan 31°04′N 72°58′E
137 X10 **Samur** ≈ Azerbaijan/Russian Federation
137 Y11 **Samur-Abşeron Kanalı** *Rus.* Sam ur-Apsheronskiy Kanal. *canal* E Azerbaijan
Sam ur-Apsheronskiy Kanal *see* Samur-Abşeron Kanalı
167 O11 **Samut Prakan** *var.* Muang Samut Prakan, Paknam. Samut Prakan, C Thailand 13°36′N 100°36′E
167 O11 **Samut Sakhon** *var.* Maha Chai, Samut Sakorn, Tha Chin. Samut Sakhon, C Thailand 13°31′N 100°15′E
Samut Sakorn *see* Samut Sakhon
167 O11 **Samut Songkhram** *prev.* Meklong. Samut Songkhram, SW Thailand 13°25′N 100°01′E
77 N12 **San** Ségou, C Mali 13°21′N 04°57′W
111 O15 **San** ≈ SE Poland
141 O15 **Şan'ā'** *Eng.* Sana. ● (Yemen) W Yemen 15°24′N 44°14′E
112 F11 **Sana** ≈ NW Bosnia and Herzegovina
80 O12 **Sanaag** *off.* Gobolka Sanaag. ◈ *region* N Somalia
Sanaag, Gobolka *see* Sanaag
114 J8 **Sanadinovo** Pleven, N Bulgaria 43°33′N 25°00′E
195 P1 **Sanae** *South African research station* Antarctica 70°19′S 01°31′W
139 Y10 **Sanāf, Hawr as** ◎ S Iraq
79 E15 **Sanaga** ≈ C Cameroon
54 D12 **San Agustín** Huila, SW Colombia 01°53′N 76°14′W
171 R8 **San Agustin, Cape** *headland* Mindanao, S Philippines 06°17′N 126°12′E
37 Q13 **San Agustin, Plains of** *plain* New Mexico, SW USA
38 M16 **Sanak Islands** *island* Aleutian Islands, Alaska, USA
San Alessandro *see* Kita-Iō-jima
193 U10 **San Ambrosio, Isla** *Eng.* San Ambrosio Island. *island* W Chile
San Ambrosio Island *see* San Ambrosio, Isla
171 Q12 **Sanana** Pulau Sanana, E Indonesia 02°04′S 125°58′E
171 Q12 **Sanana, Pulau** *island* Maluku, E Indonesia
142 K5 **Sanandaj** *prev.* Sinneh. Kordestān, W Iran 35°18′N 47°01′E
35 P8 **San Andreas** California, W USA 38°10′N 120°40′W
2 C13 **San Andreas Fault** *fault* W USA
54 G8 **San Andrés** Santander, C Colombia 06°52′N 72°53′W
61 C20 **San Andrés de Giles** Buenos Aires, E Argentina 34°27′S 59°27′W
37 R14 **San Andres Mountains** ⛰ New Mexico, SW USA
41 S15 **San Andrés Tuxtla** *var.* Tuxtla. Veracruz-Llave, E Mexico 18°28′N 95°15′W
25 P8 **San Angelo** Texas, SW USA 31°28′N 100°26′W
107 A20 **San Antioco, Isola di** *island* W Italy
42 F4 **San Antonio** Toledo, S Belize 16°13′N 89°02′W
62 G11 **San Antonio** Valparaíso, C Chile 33°35′S 71°38′W
188 H6 **San Antonio** Saipan, S Northern Mariana Islands
37 R13 **San Antonio** New Mexico, SW USA 33°53′N 106°52′W
25 R12 **San Antonio** Texas, SW USA 29°25′N 98°30′W
54 M11 **San Antonio** Amazonas, S Venezuela 03°31′N 66°47′W
54 I7 **San Antonio** Barinas, C Venezuela 07°24′N 71°28′W
55 O5 **San Antonio** Monagas, NE Venezuela 10°03′N 63°45′W
25 S12 **San Antonio** ✈ Texas, SW USA 29°31′N 98°11′W
San Antonio *see* San Antonio del Táchira
San Antonio Abad *see* Sant Antoni de Portmany
25 U13 **San Antonio Bay** *inlet* Texas, SW USA
61 E22 **San Antonio, Cabo** *headland* E Argentina 36°45′S 56°40′W
44 A5 **San Antonio, Cabo de** *headland* W Cuba 21°51′N 84°58′W
105 T11 **San Antonio, Cabo de** *headland* E Spain 38°50′N 00°09′E
54 H7 **San Antonio de Caparo** Táchira, W Venezuela 07°34′N 71°28′W
62 J5 **San Antonio de los Cobres** Salta, NE Argentina 24°10′S 66°17′W
54 H7 **San Antonio del Táchira** *var.* San Antonio. Táchira, W Venezuela 07°48′N 72°28′W
35 T15 **San Antonio, Mount** ▲ California, W USA 34°18′N 117°37′W
63 K16 **San Antonio Oeste** Río Negro, E Argentina 40°45′S 64°58′W
25 T13 **San Antonio River** ≈ Texas, SW USA
54 J5 **Sanare** Lara, N Venezuela 09°45′N 69°39′W
103 T16 **Sanary-sur-Mer** Var, SE France 43°07′N 05°48′E
104 G3 **Sanata Uxía de Ribeira** *var.* Ribeira. Galicia, NW Spain 42°33′N 09°01′W
25 X8 **San Augustine** Texas, SW USA 31°32′N 94°09′W
San Augustine *see* Minami-Iō-jima
141 T13 **Sanāw** *var.* Sanaw. NE Yemen 18°N 51°E
41 O11 **San Bartolo** San Luis Potosí, C Mexico 22°20′N 100°05′W
107 L16 **San Bartolomeo in Galdo** Campania, S Italy 41°24′N 15°01′E
106 K13 **San Benedetto del Tronto** Marche, C Italy 42°57′N 13°53′E
42 E3 **San Benito** Petén, N Guatemala 16°56′N 89°53′W
25 T17 **San Benito** Texas, SW USA 26°07′N 97°37′W
54 E6 **San Benito Abad** Sucre, N Colombia 08°56′N 75°02′W
35 P11 **San Benito Mountain** ▲ California, W USA 36°21′N 120°37′W
35 O10 **San Benito River** ≈ California, W USA
108 H10 **San Bernardino** Graubünden, S Switzerland 46°21′N 09°13′E
35 U15 **San Bernardino** California, W USA 34°06′N 117°15′W
35 U15 **San Bernardino Mountains** ⛰ California, W USA
62 H11 **San Bernardo** Santiago, C Chile 33°37′S 70°45′W
40 J8 **San Bernardo** Durango, C Mexico 25°58′N 105°27′W
164 G12 **Sanbe-san** ▲ Kyūshū, SW Japan 35°09′N 132°36′E
San Bizenti-Barakaldo *see* San Vicente de Barakaldo
40 J12 **San Blas** Nayarit, C Mexico 21°35′N 105°20′W
40 H8 **San Blas** Sinaloa, C Mexico 26°05′N 108°44′W
San Blas *see* Kuna Yala
43 U14 **San Blas, Archipiélago de** *island group* NE Panama
23 Q10 **San Blas, Cape** *headland* Florida, SE USA 29°39′N 85°21′W
43 V14 **San Blas, Cordillera de** ⛰ NE Panama
62 J8 **San Blas de los Sauces** Catamarca, NW Argentina 28°18′S 67°12′W
106 G8 **San Bonifacio** Veneto, NE Italy 45°22′N 11°14′E
29 S12 **Sanborn** Iowa, C USA 43°10′N 95°39′W
40 M7 **San Buenaventura** Coahuila, NE Mexico 27°04′N 101°32′W
105 S5 **San Caprasio** ▲ N Spain 41°45′N 00°26′W
62 G13 **San Carlos** Bío Bío, C Chile 36°25′S 71°58′W
40 E9 **San Carlos** Baja California Sur, NW Mexico 24°52′N 112°15′W
41 N5 **San Carlos** Coahuila, NE Mexico 29°00′N 100°51′W
41 P9 **San Carlos** Tamaulipas, C Mexico 24°36′N 98°42′W
42 L12 **San Carlos** Río San Juan, S Nicaragua 11°06′N 84°46′W
43 T16 **San Carlos** Panamá, C Panama 08°29′N 79°58′W
171 N3 **San Carlos** *off.* San Carlos City. Luzon, N Philippines 15°57′N 120°18′E
61 G20 **San Carlos** Maldonado, S Uruguay 34°46′S 54°58′W
36 M14 **San Carlos** Arizona, SW USA 33°21′N 110°27′W
54 K5 **San Carlos** Cojedes, N Venezuela 09°39′N 68°35′W
San Carlos *see* Quesada, Costa Rica
San Carlos *see* Luba, Equatorial Guinea
61 B17 **San Carlos Centro** Santa Fe, C Argentina 31°45′S 61°05′W
171 P6 **San Carlos City** Negros, C Philippines 10°34′N 123°24′E
San Carlos City *see* San Carlos
San Carlos de Ancud *see* Ancud
63 H16 **San Carlos de Bariloche** Río Negro, SW Argentina 41°08′S 71°15′W
61 B21 **San Carlos de Bolívar** Buenos Aires, E Argentina 36°15′S 61°06′W
54 H6 **San Carlos del Zulia** Zulia, W Venezuela 09°01′N 71°58′W
54 L12 **San Carlos de Río Negro** Amazonas, S Venezuela 01°54′N 67°54′W
San Carlos, Estrecho de *see* Falkland Sound
36 M14 **San Carlos Reservoir** ⊠ Arizona, SW USA
42 M12 **San Carlos, Río** ≈ N Costa Rica
65 D24 **San Carlos Settlement** East Falkland, Falkland Islands
61 C23 **San Cayetano** Buenos Aires, E Argentina 38°20′S 59°37′W
103 O8 **Sancerre** Cher, C France 47°19′N 02°53′E
158 G7 **Sanchakou** Xinjiang Uygur Zizhiqu, NW China 39°56′N 78°28′E
Sanchoku *see* Samcheok
41 O12 **San Ciro** San Luis Potosí, C Mexico 21°40′N 99°50′W
105 P10 **San Clemente** Castilla-La Mancha, C Spain 39°24′N 02°25′W
35 T16 **San Clemente** California, W USA 33°25′N 117°36′W
61 E21 **San Clemente del Tuyú** Buenos Aires, E Argentina 36°22′S 56°43′W
35 S17 **San Clemente Island** *island* Channel Islands, California, W USA
103 O9 **Sancoins** Cher, C France 46°49′N 03°00′E
61 B16 **San Cristóbal** Santa Fe, C Argentina 30°20′S 61°14′W
44 B4 **San Cristóbal** Pinar del Río, W Cuba 22°43′N 83°03′W
45 O9 **San Cristóbal** *var.* Benemérita de San Cristóbal. S Dominican Republic 18°27′N 70°07′W
54 H7 **San Cristóbal** Táchira, W Venezuela 07°46′N 72°15′W
187 N10 **San Cristobal** *var.* Makira. *island* SE Solomon Islands
San Cristóbal *see* San Cristóbal de Las Casas
41 U16 **San Cristóbal de Las Casas** *var.* San Cristóbal. Chiapas, SE Mexico 16°44′N 92°40′W
187 N10 **San Cristóbal, Isla** *var.* Chatham Island. *island* Galapagos Islands, Ecuador, E Pacific Ocean
42 D5 **San Cristóbal Verapaz** Alta Verapaz, C Guatemala 15°21′N 90°22′W
44 F6 **Sancti Spíritus** Sancti Spíritus, C Cuba 21°54′N 79°27′W
103 O11 **Sancy, Puy de** ▲ C France 45°33′N 02°48′E
95 D15 **Sand** Rogaland, S Norway 59°28′N 06°16′E
169 W7 **Sandakan** Sabah, East Malaysia 05°52′N 118°04′E
182 K9 **Sandalwood** South Australia 34°51′S 140°13′E
Sandalwood Island *see* Sumba, Pulau
94 D11 **Sandane** Sogn Og Fjordane, S Norway 61°47′N 06°14′E
114 G12 **Sandanski** *prev.* Sveti Vrach. Blagoevgrad, SW Bulgaria 41°36′N 23°19′E
Sandaohezi *see* Shawan
76 J11 **Sandaré** Kayes, W Mali 14°36′N 10°22′W
95 J19 **Sandared** Västra Götaland, S Sweden 57°43′N 12°47′E
94 N12 **Sandarne** Gävleborg, C Sweden 61°15′N 17°10′E
Sandaun *see* West Sepik
96 K4 **Sanday** *island* NE Scotland, United Kingdom
31 P15 **Sand Creek** ≈ Indiana, N USA
95 H15 **Sande** Vestfold, S Norway 59°34′N 10°13′E
95 H16 **Sandefjord** Vestfold, S Norway 59°10′N 10°15′E
77 O15 **Sandégué** E Ivory Coast 07°59′N 03°33′W
77 P14 **Sandema** N Ghana 10°42′N 01°17′W
37 O11 **Sanders** Arizona, SW USA 35°13′N 109°21′W
24 M11 **Sanderson** Texas, SW USA 30°08′N 102°25′W
23 U4 **Sandersville** Georgia, SE USA 32°58′N 82°48′W
92 H4 **Sandgerði** Suðurnes, SW Iceland 64°01′N 22°42′W
28 K14 **Sand Hills** ⛰ Nebraska, C USA
25 S14 **Sandia** Texas, SW USA 27°59′N 97°52′W
35 T17 **San Diego** California, W USA 32°43′N 117°09′W
25 S14 **San Diego** Texas, SW USA 27°47′N 98°15′W
136 F14 **Sandıklı** Afyon, W Turkey 38°28′N 30°17′E
152 L12 **Sandila** Uttar Pradesh, N India 27°05′N 80°37′E
San Dimitri Point *see* Il-Ponta ta' San Dimitri
San Dimitri, Ras *see* Il-Ponta ta' San Dimitri
168 J13 **Sanding, Selat** *strait* W Indonesia
30 J3 **Sand Island** *island* Apostle Islands, Wisconsin, N USA
95 C16 **Sandnes** Rogaland, S Norway 58°51′N 05°45′E
92 F13 **Sandnessjøen** Nordland, C Norway 66°00′N 12°37′E
Sandø *see* Sandoy
79 L24 **Sandoa** Katanga, S Dem. Rep. Congo 09°41′S 22°56′E
111 N15 **Sandomierz** *Rus.* Sandomir. Świętokrzyskie, C Poland 50°42′N 21°45′E
Sandomir *see* Sandomierz
54 C13 **Sandoná** Nariño, SW Colombia 01°18′N 77°28′W
106 I7 **San Donà di Piave** Veneto, NE Italy 45°38′N 12°34′E
124 K14 **Sandovo** Tverskaya Oblast', W Russian Federation 58°26′N 36°30′E
Sandoway *see* Thandwe
97 M24 **Sandown** S England, United Kingdom 50°40′N 01°11′W
95 B19 **Sandoy** *Dan.* Sandø. *island* C Faroe Islands
39 N16 **Sand Point** Popof Island, Alaska, USA 55°20′N 160°30′W
32 M7 **Sandpoint** Idaho, NW USA 48°16′N 116°33′W
65 N24 **Sand Point** *headland* E Tristan da Cunha
31 R7 **Sand Point** *headland* Michigan, N USA 43°54′N 83°24′W
93 H14 **Sandsele** Västerbotten, N Sweden 65°16′N 17°40′E
10 I14 **Sandspit** Moresby Island, British Columbia, SW Canada 53°14′N 131°50′W
27 P9 **Sand Springs** Oklahoma, C USA 36°08′N 96°06′W
29 W7 **Sandstone** Minnesota, N USA 46°07′N 92°51′W
36 K15 **Sand Tank Mountains** ⛰ Arizona, SW USA
31 S8 **Sandusky** Michigan, N USA 43°26′N 82°50′W
31 S11 **Sandusky** Ohio, N USA 41°27′N 82°42′W
31 S12 **Sandusky River** ≈ Ohio, N USA
83 D22 **Sandverhaar** Karas, S Namibia 26°50′S 17°25′E
95 L24 **Sandvig** Bornholm, E Denmark 55°15′N 14°45′E
95 H15 **Sandvika** Akershus, S Norway 59°54′N 10°29′E
94 N13 **Sandviken** Gävleborg, C Sweden 60°38′N 16°50′E
30 M11 **Sandwich** Illinois, N USA 41°39′N 88°37′W
Sandwich Island *see* Efate
Sandwich Islands *see* Hawai'ian Islands
153 V16 **Sandwip** *island* SE Bangladesh
11 U12 **Sandy Bay** Saskatchewan, C Canada 55°31′N 102°14′W
183 N16 **Sandy Cape** *headland* Tasmania, SE Australia 41°27′S 144°43′E
36 L3 **Sandy City** Utah, W USA 40°36′N 111°53′W
31 U12 **Sandy Creek** ≈ Ohio, N USA
21 O5 **Sandy Hook** Kentucky, S USA 38°05′N 83°09′W
18 K15 **Sandy Hook** *headland* New Jersey, NE USA 40°27′N 73°59′W
Sandykachi/Sandykgachy *see* Sandykgaçy
146 J15 **Sandykaçy** *var.* Sandykgachy. Maryyskiy Velayat, S Turkmenistan 36°34′N 62°28′E
Sandykgachy *see* Sandykaçy
146 J15 **Sandykgaçy** *var.* Sandykgachy, *Rus.* Sandykachi. Mary Welaýaty, S Turkmenistan 36°34′N 62°28′E
146 L13 **Sandykly Gumy** *Rus.* Peski Sandykly. *desert* E Turkmenistan
Sandykly, Peski *see* Sandykly Gumy
11 Q13 **Sandy Lake** Alberta, W Canada 55°50′N 113°30′W
12 B8 **Sandy Lake** Ontario, C Canada 53°00′N 93°25′W
12 B8 **Sandy Lake** ◎ Ontario, C Canada
23 S3 **Sandy Springs** Georgia, SE USA 33°57′N 84°23′W
24 H8 **San Elizario** Texas, SW USA 31°35′N 106°16′W
99 L25 **Sanem** Luxembourg, SW Luxembourg 49°33′N 05°56′E
42 K5 **San Esteban** Olancho, C Honduras 15°19′N 85°52′W
105 O6 **San Esteban de Gormaz** Castilla y León, N Spain 41°34′N 03°13′W
40 E5 **San Esteban, Isla** *island* NW Mexico
San Eugenio/San Eugenio del Cuareim *see* Artigas
62 H11 **San Felipe** *var.* San Felipe de Aconcagua. Valparaíso, C Chile 32°45′S 70°42′W
40 D3 **San Felipe** Baja California Norte, NW Mexico 31°03′N 114°52′W
40 N12 **San Felipe** Guanajuato, C Mexico 21°30′N 101°15′W
54 K5 **San Felipe** Yaracuy, NW Venezuela 10°25′N 68°40′W
44 B5 **San Felipe, Cayos de** *island group* W Cuba
San Felipe de Aconcagua *see* San Felipe
San Felipe de Puerto Plata *see* Puerto Plata
37 R11 **San Felipe Pueblo** New Mexico, SW USA 35°25′N 106°27′W
San Feliú de Guixols *see* Sant Feliu de Guíxols
193 T10 **San Félix, Isla** *Eng.* San Felix Island. *island* W Chile
San Felix Island *see* San Félix, Isla
40 C4 **San Fernando** *var.* Misión San Fernando. Baja California Norte, NW Mexico 29°58′N 115°14′W
41 P9 **San Fernando** Tamaulipas, C Mexico 24°50′N 98°10′W
171 N2 **San Fernando** Luzon, N Philippines 16°45′N 120°21′E
171 O3 **San Fernando** Luzon, N Philippines 15°01′N 120°41′E
104 J16 **San Fernando** *prev.* Isla de León. Andalucía, S Spain 36°28′N 06°12′W
45 U14 **San Fernando** Trinidad, Trinidad and Tobago 10°17′N 61°27′W
35 S15 **San Fernando** California, W USA 34°16′N 118°26′W
54 L7 **San Fernando** *var.* San Fernando de Apure. Apure, C Venezuela 07°54′N 67°28′W
San Fernando de Apure *see* San Fernando
54 L11 **San Fernando de Atabapo** Amazonas, S Venezuela 04°00′N 67°42′W
62 L8 **San Fernando del Valle de Catamarca** *var.* Catamarca. Catamarca, NW Argentina 28°28′S 65°46′W
San Fernando de Monte Cristi *see* Monte Cristi
41 P9 **San Fernando, Río** ≈ C Mexico
23 X11 **Sanford** Florida, SE USA 28°48′N 81°16′W
19 P9 **Sanford** Maine, NE USA 43°26′N 70°46′W
21 T10 **Sanford** North Carolina, SE USA 35°29′N 79°10′W
25 N2 **Sanford** Texas, SW USA 35°42′N 101°31′W
39 T10 **Sanford, Mount** ▲ Alaska, USA 62°21′N 144°12′W
42 G8 **San Francisco** *var.* Gotera, San Francisco Gotera. Morazán, E El Salvador 13°41′N 88°06′W
43 R16 **San Francisco** Veraguas, C Panama 08°19′N 80°59′W
171 N2 **San Francisco** *var.* Aurora. Luzon, N Philippines 13°22′N 122°31′E
35 L8 **San Francisco** California, W USA 37°47′N 122°25′W
54 H5 **San Francisco** Zulia, NW Venezuela 10°36′N 71°39′W
34 M8 **San Francisco** ✈ California, W USA 37°37′N 122°23′W
35 N9 **San Francisco Bay** *bay* California, W USA
61 C24 **San Francisco de Bellocq** Buenos Aires, E Argentina 38°42′S 60°01′W
40 I6 **San Francisco de Borja** Chihuahua, N Mexico 27°57′N 106°42′W
42 J6 **San Francisco de la Paz** Olancho, C Honduras 14°55′N 86°14′W
40 J7 **San Francisco del Oro** Chihuahua, N Mexico 26°52′N 105°50′W
40 M12 **San Francisco del Rincón** Jalisco, SW Mexico 21°00′N 101°51′W
45 O8 **San Francisco de Macorís** C Dominican Republic 19°19′N 70°15′W
San Francisco de Satipo *see* Satipo
San Francisco Gotera *see* San Francisco
San Francisco Telixtlahuaca *see* Telixtlahuaca
107 K23 **San Fratello** Sicilia, Italy, C Mediterranean Sea 38°00′N 14°35′E
San Fructuoso *see* Tacuarembó
82 C12 **Sanga** Kwanza Sul, NW Angola 11°10′S 15°27′E
56 C5 **San Gabriel** Carchi, N Ecuador 0°35′N 77°48′W
159 S15 **Sa'ngain** Xizang Zizhiqu, W China 30°47′N 98°45′E
154 E13 **Sangamner** Mahārāshtra, W India 19°37′N 74°18′E
152 H12 **Sāngāner** Rājasthān, N India 26°48′N 75°48′E
Sangan, Koh-i- *see* Sangān, Kūh-e
149 N6 **Sangān, Kūh-e** *Pash.* Koh-i-Sangan. ▲ C Afghanistan
123 P10 **Sangar** Respublika Sakha (Yakutiya), NE Russian Federation 63°48′N 127°37′E
155 H15 **Sangāreddi** Telangana, India 17°37′N 78°04′E
169 V11 **Sangasanga** Borneo, C Indonesia 0°36′S 117°12′E
103 N1 **Sangatte** Pas-de-Calais, N France 50°56′N 01°41′E
107 B19 **San Gavino Monreale** Sardegna, Italy, C Mediterranean Sea 39°33′N 08°47′E
57 D16 **Sangayan, Isla** *island* W Peru
30 L14 **Sangchris Lake** ⊠ Illinois, N USA
171 N16 **Sangeang, Pulau** *island* S Indonesia
116 I10 **Sângeorgiu de Pădure** *prev.* Erdăt-Sângeorz, Sîngeorgiu de Pădure, *Hung.* Erdőszentgyörgy. Mureș, C Romania 46°27′N 24°50′E
116 I9 **Sângeorz-Băi** *var.* Singeroz Băi, *Ger.* Rumänisch-Sankt-Georgen, *Hung.* Oláhszentgyőrgy; *prev.* Sîngeorz-Băi. Bistriţa-Năsăud, N Romania 47°24′N 24°40′E
35 R10 **Sanger** California, W USA 36°42′N 119°33′W
25 T5 **Sanger** Texas, SW USA 33°21′N 97°10′W
Sângerei *see* Sîngerei
101 L15 **Sangerhausen** Sachsen-Anhalt, C Germany 51°29′N 11°18′E
45 S6 **San Germán** W Puerto Rico 18°05′N 67°02′W
San Germano *see* Cassino
161 N2 **Sanggan He** ≈ E China
169 Q11 **Sanggau** Borneo, C Indonesia 0°08′N 110°35′E
79 G17 **Sangha** ◈ *province* N Congo
79 H16 **Sangha** ≈ Central African Republic/Congo
79 G16 **Sangha-Mbaéré** ◈ *prefecture* SW Central African Republic
149 Q15 **Sānghar** Sind, SE Pakistan 26°10′N 68°59′E
115 F22 **Sangiás** ▲ S Greece 36°39′N 22°24′E
Sangihe, Kepulauan *see* Sangir, Kepulauan
171 Q9 **Sangihe, Pulau** *var.* Sangir. *island* N Indonesia
54 G8 **San Gil** Santander, C Colombia 06°35′N 73°08′W
121 P16 **San Ġiljan** *var.* St Julian's. N Malta 35°55′N 14°29′E
106 F12 **San Gimignano** Toscana, C Italy 43°30′N 11°00′E
148 M8 **Sangin** *var.* Sangin. Helmand, S Afghanistan 32°03′N 64°50′E
Sangin *see* Sangin
107 O21 **San Giovanni in Fiore** Calabria, SW Italy 39°15′N 16°42′E
107 M16 **San Giovanni Rotondo** Puglia, SE Italy 41°43′N 15°44′E
106 G12 **San Giovanni Valdarno** Toscana, C Italy 43°34′N 11°31′E
Sangir *see* Sangihe, Pulau
171 Q10 **Sangir, Kepulauan** *var.* Kepulauan Sangihe. *island group* N Indonesia
Sangiyn Dalay *see* Erdenedalay, Dundgovĭ, Mongolia
Sangiyn Dalay *see* Erdene, Govĭ-Altay, Mongolia
Sangiyn Dalay *see* Nomgon, Ömnögovĭ, Mongolia
Sangiyn Dalay *see* Öldziyt, Övörhangay, Mongolia
163 Y15 **Sangju** *Jap.* Shōshū. C South Korea 36°26′N 128°09′E
167 R11 **Sangkha** Surin, E Thailand 14°36′N 103°43′E
169 W10 **Sangkulirang** Borneo, N Indonesia 01°00′N 117°56′E
169 W10 **Sangkulirang, Teluk** *bay* Borneo, N Indonesia
155 E16 **Sāngli** Mahārāshtra, W India 16°55′N 74°37′E
79 E16 **Sangmélima** Sud, S Cameroon 02°57′N 11°56′E
35 V15 **San Gorgonio Mountain** ▲ California, W USA 34°06′N 116°50′W
37 T8 **Sangre de Cristo Mountains** ⛰ Colorado/New Mexico, C USA
61 A20 **San Gregorio** Santa Fe, C Argentina 34°18′S 62°02′W
61 F18 **San Gregorio de Polanco** Tacuarembó, C Uruguay 32°37′S 55°50′W
45 V14 **Sangre Grande** Trinidad, Trinidad and Tobago 10°35′N 61°08′W
159 N16 **Sangri** Xizang Zizhiqu, W China 29°17′N 92°01′E
152 H9 **Sangrūr** Punjab, NW India 30°16′N 75°52′E
44 I11 **Sangster** *off.* Sir Donald Sangster International Airport, *var.* Montego Bay. ✈ (Montego Bay) W Jamaica 18°30′N 77°54′W
59 G17 **Sangue, Rio do** ≈ W Brazil
105 R4 **Sangüesa** *Bas.* Zangoza. Navarra, N Spain 42°34′N 01°17′W
61 C16 **San Gustavo** Entre Ríos, E Argentina 30°41′S 59°23′W
Sangyuan *see* Wuqiao
40 C6 **San Hipólito, Punta** *headland* NW Mexico 26°57′N 114°00′W
23 W15 **Sanibel** Sanibel Island, Florida, SE USA 26°27′N 82°01′W
23 V15 **Sanibel Island** *island* Florida, SE USA
60 F13 **San Ignacio** Misiones, NE Argentina 27°15′S 55°32′W
42 F2 **San Ignacio** *prev.* Cayo, El Cayo. Cayo, W Belize 17°09′N 89°02′W
57 L16 **San Ignacio** El Beni, N Bolivia 14°54′S 65°35′W
57 O18 **San Ignacio** Santa Cruz, E Bolivia 16°23′S 60°59′W
42 M14 **San Ignacio** *var.* San Ignacio de Acosta. San José, W Costa Rica 09°46′N 84°10′W
40 E6 **San Ignacio** Baja California Sur, NW Mexico 27°18′N 112°51′W
40 J10 **San Ignacio** Sinaloa, W Mexico 23°55′N 106°25′W
56 B9 **San Ignacio** Cajamarca, N Peru 05°09′S 79°00′W
San Ignacio de Acosta *see* San Ignacio
40 D7 **San Ignacio, Laguna** *lagoon* W Mexico
12 I6 **Sanikiluaq** Belcher Islands, Nunavut, C Canada 55°20′N 77°50′W
171 O3 **San Ildefonso Peninsula** *peninsula* Luzon, N Philippines
Saniquillie *see* Sanniquellie
Sanirajak *see* Hall Beach
61 D20 **San Isidro** Buenos Aires, E Argentina 34°28′S 58°31′W
43 N14 **San Isidro** *var.* San Isidro de El General. San José, SE Costa Rica 09°28′N 83°42′W
San Isidro de El General *see* San Isidro
54 E5 **San Jacinto** Bolívar, N Colombia 09°53′N 75°06′W
35 U16 **San Jacinto** California, W USA 33°47′N 116°58′W
35 V15 **San Jacinto Peak** ▲ California, W USA 33°48′N 116°40′W
61 F14 **San Javier** Misiones, NE Argentina 27°50′S 55°06′W
61 C16 **San Javier** Santa Fe, C Argentina 30°35′S 59°59′W
105 S13 **San Javier** Murcia, SE Spain 37°49′N 00°50′W
61 D18 **San Javier** Río Negro, W Uruguay 32°41′S 58°08′W
61 C16 **San Javier, Río** ≈ C Argentina
160 L12 **Sanjiang** *var.* Guyi, Sanjiang Dongzu Zizhixian. Guangxi Zhuangzu Zizhiqu, S China 25°46′N 109°26′E
Sanjiang *see* Jinping, Guizhou
Sanjiang Dongzu Zizhixian *see* Sanjiang
Sanjiaocheng *see* Haiyan
165 N11 **Sanjō** *var.* Sanzyô. Niigata, Honshū, C Japan 37°39′N 139°00′E
57 M15 **San Joaquín** El Beni, N Bolivia 13°06′S 64°46′W
55 O6 **San Joaquín** Anzoátegui, NE Venezuela 09°21′N 64°30′W
35 O9 **San Joaquin River** ≈ California, W USA
35 P10 **San Joaquin Valley** *valley* California, W USA
61 A18 **San Jorge** Santa Fe, C Argentina 31°50′S 61°50′W
40 D3 **San Jorge, Bahía de** *bay* NW Mexico
63 J19 **San Jorge, Golfo** *var.* Gulf of San Jorge. *gulf* S Argentina
San Jorge, Gulf of *see* San Jorge, Golfo
San Jorge, Isla de *see* Weddell Island
61 F14 **San José** Misiones, NE Argentina 27°46′S 55°47′W
57 P19 **San José** *var.* San José de Chiquitos. Santa Cruz, E Bolivia 14°13′S 68°05′W
42 M14 **San José** ● (Costa Rica) San José, C Costa Rica 09°55′N 84°05′W
42 C7 **San José** *var.* Puerto San José. Escuintla, S Guatemala 14°00′N 90°50′W
40 G6 **San José** Sonora, NW Mexico 27°32′N 110°09′W
188 K8 **San Jose** Tinian, S Northern Mariana Islands 15°00′S 145°38′E
105 U11 **San José** Eivissa, Spain, W Mediterranean Sea 38°55′N 01°18′E
35 N9 **San Jose** California, W USA 37°18′N 121°53′W
54 H5 **San José** Zulia, NW Venezuela 10°02′N 72°24′W
42 M14 **San José** *off.* Provincia de San José. ◈ *province* W Costa Rica
61 E19 **San José** ◈ *department* S Uruguay
42 M13 **San José** ✈ Alajuela, C Costa Rica 10°03′N 84°12′W

◆ Country
● Country Capital
◇ Dependent Territory
○ Dependent Territory Capital
◈ Administrative Regions
✈ International Airport
▲ Mountain
⛰ Mountain Range
Volcano
≈ River
◎ Lake
⊠ Reservoir

San José see San José del Guaviare, Colombia
San Jose see Oleai
San José see Sant Josep de sa Talaia, Ibiza, Spain
San José see San José de Mayo, Uruguay
171 O3 **San Jose City** Luzon, N Philippines 15°49′N 120°57′E
San José de Chiquitos see San José
San José de Cúcuta see Cúcuta
61 D16 **San José de Feliciano** Entre Ríos, E Argentina 30°26′S 58°46′W
55 O6 **San José de Guanipa** var. El Tigrito. Anzoátegui, NE Venezuela 08°54′N 64°10′W
62 I9 **San José de Jáchal** San Juan, W Argentina 30°15′S 68°46′W
40 G10 **San José del Cabo** Baja California Sur, NW Mexico 23°01′N 109°40′W
54 G12 **San José del Guaviare** var. San José. Guaviare, S Colombia 02°34′N 72°38′W
61 E20 **San José de Mayo** var. San José. San José, S Uruguay 34°20′S 56°42′W
54 I10 **San José de Ocuné** Vichada, E Colombia 04°10′N 70°21′W
41 O9 **San José de Raíces** Nuevo León, NE Mexico 24°32′N 100°15′W
63 K17 **San José, Golfo** gulf E Argentina
40 F9 **San José, Isla** island NW Mexico
43 U16 **San José, Isla** island SE Panama
25 U14 **San Jose Island** island Texas, SW USA
San José, Provincia de see San José
62 I10 **San Juan** San Juan, W Argentina 31°37′S 68°27′W
45 N9 **San Juan** var. San Juan de la Maguana. C Dominican Republic 18°49′N 71°12′W
57 E17 **San Juan** Ica, S Peru 15°22′S 75°07′W
45 U5 **San Juan** ○ (Puerto Rico) NE Puerto Rico 18°28′N 66°06′W
62 H10 **San Juan** off. Provincia de San Juan. ◈ province W Argentina
San Juan see San Juan de los Morros
62 O7 **San Juan Bautista** Misiones, S Paraguay 26°40′S 57°08′W
35 O10 **San Juan Bautista** California, W USA 36°50′N 121°34′W
San Juan Bautista see Villahermosa
San Juan Bautista Cuicatlán see Cuicatlán
San Juan Bautista Tuxtepec see Tuxtepec
79 C17 **San Juan, Cabo** headland S Equatorial Guinea 01°09′N 09°25′E
San Juan de Alicante see Sant Joan d'Alacant
54 H7 **San Juan de Colón** Táchira, NW Venezuela 08°02′N 72°17′W
40 L9 **San Juan de Guadalupe** Durango, C Mexico 25°12′N 100°50′W
San Juan de la Maguana see San Juan
54 G4 **San Juan del Cesar** La Guajira, N Colombia 10°45′N 73°00′W
40 L15 **San Juan de Lima, Punta** headland SW Mexico 18°34′N 103°40′W
42 I8 **San Juan de Limay** Estelí, NW Nicaragua 13°10′N 86°36′W
43 N12 **San Juan del Norte** var. Greytown. Río San Juan, SE Nicaragua 10°58′N 83°40′W
54 K4 **San Juan de los Cayos** Falcón, N Venezuela 11°11′N 68°27′W
40 M12 **San Juan de los Lagos** Jalisco, C Mexico 21°15′N 102°15′W
54 L5 **San Juan de los Morros** var. San Juan. Guárico, N Venezuela 09°53′N 67°23′W
40 K9 **San Juan del Río** Durango, C Mexico 25°12′N 100°50′W
41 O13 **San Juan del Río** Querétaro de Arteaga, C Mexico 20°24′N 100°00′W
42 J11 **San Juan del Sur** Rivas, SW Nicaragua 11°16′N 85°51′W
54 M9 **San Juan de Manapiare** Amazonas, S Venezuela 05°15′N 66°05′W
40 E7 **San Juanico** Baja California Sur, NW Mexico
40 D7 **San Juanico, Punta** headland NW Mexico 26°01′N 112°17′W
32 G6 **San Juan Islands** island group Washington, NW USA
40 I6 **San Juanito** Chihuahua, N Mexico
40 I12 **San Juanito, Isla** island C Mexico
37 R8 **San Juan Mountains** ▲▲ Colorado, C USA
54 E5 **San Juan Nepomuceno** Bolívar, NW Colombia 09°57′N 75°06′W
44 E5 **San Juan, Pico** ▲ C Cuba 21°58′N 80°10′W
San Juan, Provincia de see San Juan
191 W15 **San Juan, Punta** headland Easter Island, Chile, E Pacific Ocean 27°03′S 109°22′W
42 M12 **San Juan, Río** ≈ Costa Rica/Nicaragua
41 S15 **San Juan, Río** ≈ SE Mexico
37 O8 **San Juan River** ≈ Colorado/Utah, W USA
San Julián see Puerto San Julián
61 B17 **San Justo** Santa Fe, C Argentina 30°47′S 60°32′W
109 W5 **Sankt Aegyd am Neuwalde** Niederösterreich, E Austria 47°51′N 15°34′E
109 U9 **Sankt Andrä** Slvn. Šent Andraž. Kärnten, S Austria 46°46′N 14°49′E
Sankt Andrä see Szentendre
Sankt Anna see Săntana
108 K8 **Sankt Anton-am-Arlberg** Vorarlberg, W Austria 47°08′N 10°11′E
101 E16 **Sankt Augustin** Nordrhein-Westfalen, W Germany 50°46′N 07°10′E
Sankt-Bartholomäi see Palamuse
101 F24 **Sankt Blasien** Baden-Württemberg, SW Germany 47°43′N 08°09′E
109 R3 **Sankt Florian am Inn** Oberösterreich, N Austria 48°24′N 13°27′E
108 I7 **Sankt Gallen** var. St. Gallen, Eng. Saint Gall, Fr. St-Gall. Sankt Gallen, NE Switzerland 47°25′N 09°23′E
108 H8 **Sankt Gallen** var. St.Gallen, Eng. Saint Gall, Fr. St-Gall. ◆ canton NE Switzerland
108 J8 **Sankt Gallenkirch** Vorarlberg, W Austria 47°00′N 10°59′E
109 Q5 **Sankt Georgen** Salzburg, N Austria 47°59′N 12°57′E
Sankt Georgen see Đurđevac
Sankt-Georgen see Sfântu Gheorghe
109 R6 **Sankt Gilgen** Salzburg, NW Austria 47°46′N 13°21′E
Sankt Gotthard see Szentgotthárd
101 E20 **Sankt Ingbert** Saarland, SW Germany 49°17′N 07°07′E
Sankt-Jakobi see Viru-Jaagupi, Lääne-Virumaa, Estonia
Sankt-Jakobi see Pärnu-Jaagupi, Pärnumaa, Estonia
Sankt Johann see Sankt Johann in Tirol
109 T7 **Sankt Johann am Tauern** Steiermark, E Austria 47°20′N 14°27′E
109 Q7 **Sankt Johann im Pongau** Salzburg, NW Austria 47°22′N 13°13′E
109 P6 **Sankt Johann in Tirol** var. Sankt Johann. Tirol, W Austria 47°32′N 12°26′E
Sankt-Johannis see Järva-Jaani
108 L8 **Sankt Leonhard** Tirol, W Austria 47°05′N 10°53′E
Sankt Margarethen see Sankt Margarethen im Burgenland
109 Y5 **Sankt Margarethen im Burgenland** var. Sankt Margarethen. Burgenland, E Austria 47°49′N 16°38′E
Sankt Martin see Martin
109 X8 **Sankt Martin an der Raab** Burgenland, SE Austria 46°59′N 16°12′E
109 U7 **Sankt Michael in Obersteiermark** Steiermark, SE Austria 47°21′N 14°59′E
Sankt Michel see Mikkeli
Sankt Moritz see St. Moritz
108 E11 **Sankt Niklaus** Valais, S Switzerland 46°09′N 07°48′E
109 S7 **Sankt Nikolai** var. Sankt Nikolai im Sölktal. Steiermark, SE Austria 47°18′N 14°04′E
Sankt Nikolai im Sölktal see Sankt Nikolai
109 U9 **Sankt Paul** var. Sankt Paul im Lavanttal. Kärnten, S Austria 46°42′N 14°53′E
Sankt Paul im Lavanttal see Sankt Paul
Sankt Peter see Pivka
109 W9 **Sankt Peter am Ottersbach** Steiermark, SE Austria 46°49′N 15°48′E
124 J13 **Sankt-Peterburg** prev. Leningrad, Petrograd, Eng. Saint Petersburg, Fin. Pietari. Leningradskaya Oblast', NW Russian Federation 59°55′N 30°25′E
100 H8 **Sankt Peter-Ording** Schleswig-Holstein, N Germany 54°18′N 08°37′E
109 V4 **Sankt Pölten** Niederösterreich, N Austria 48°14′N 15°38′E
109 W7 **Sankt Ruprecht** var. Sankt Ruprecht an der Raab. Steiermark, SE Austria 47°10′N 15°41′E
Sankt Ruprecht an der Raab see Sankt Ruprecht
Sankt-Ulrich see Ortisei
109 T4 **Sankt Valentin** Niederösterreich, C Austria 48°11′N 14°33′E
Sankt Veit am Flaum see Rijeka
109 T9 **Sankt Veit an der Glan** Slvn. Št. Vid. Kärnten, S Austria 46°47′N 14°22′E
99 M21 **Sankt-Vith** var. Saint-Vith. Liège, E Belgium 50°17′N 06°07′E
101 E20 **Sankt Wendel** Saarland, SW Germany 49°28′N 07°10′E
109 R6 **Sankt Wolfgang** Salzburg, NW Austria 47°43′N 13°30′E
79 K21 **Sankuru** ≈ C Dem. Rep. Congo
40 D8 **San Lázaro, Cabo** headland NW Mexico 24°46′N 112°15′W
137 O16 **Sanlıurfa** prev. Sanli Urfa, Urfa; anc. Edessa. şanlıurfa, S Turkey 37°08′N 38°45′E
137 O16 **Şanlıurfa** prev. Urfa. ◈ province SE Turkey
Sanli Urfa see Şanlıurfa
137 O16 **Şanlıurfa Yaylası** plateau SE Turkey
61 B18 **San Lorenzo** Santa Fe, C Argentina 32°45′S 60°45′W
57 M21 **San Lorenzo** Tarija, S Bolivia 21°25′S 64°45′W
56 C5 **San Lorenzo** Esmeraldas, N Ecuador 01°15′N 78°51′W
42 H8 **San Lorenzo** Valle, S Honduras 13°24′N 87°27′W
56 A6 **San Lorenzo, Cabo** headland W Ecuador 0°57′S 80°49′W
105 N8 **San Lorenzo de El Escorial** var. El Escorial. Madrid, C Spain 40°36′N 04°07′W
40 E5 **San Lorenzo, Isla** island NW Mexico
57 C14 **San Lorenzo, Isla** island W Peru
63 G20 **San Lorenzo, Monte** ▲ S Argentina 47°40′S 72°12′W
40 I9 **San Lorenzo, Río** ≈ C Mexico
104 J15 **Sanlúcar de Barrameda** Andalucía, S Spain 36°46′N 06°21′W
104 J14 **Sanlúcar la Mayor** Andalucía, S Spain 37°24′N 06°13′W
40 E6 **San Lucas** var. Cabo San Lucas. Baja California Sur, NW Mexico 27°14′N 112°15′W
40 F11 **San Lucas** Baja California Sur, NW Mexico 22°50′N 109°52′W
40 G11 **San Lucas, Cabo** var. San Lucas Cape. headland NW Mexico 22°52′N 109°53′W
San Lucas Cape see San Lucas, Cabo
62 J11 **San Luis** San Luis, C Argentina 33°18′S 66°18′W
42 E4 **San Luis** Petén, NE Guatemala 16°16′N 89°27′W
42 M7 **San Luis** Región Autónoma Atlántico Norte, NE Nicaragua 13°59′N 84°10′W
36 H15 **San Luis** Arizona, SW USA 32°27′N 114°45′W
37 T8 **San Luis** Colorado, C USA 37°09′N 105°24′W
54 J4 **San Luis** Falcón, N Venezuela 11°09′N 69°39′W
62 J11 **San Luis** off. Provincia de San Luis. ◈ province C Argentina
41 N12 **San Luis de la Paz** Guanajuato, C Mexico 21°15′N 100°33′W
40 K8 **San Luis del Cordero** Durango, C Mexico 25°25′N 104°09′W
40 D4 **San Luis, Isla** island NW Mexico
42 E6 **San Luis Jilotepeque** Jalapa, SE Guatemala 14°40′N 89°42′W
57 M16 **San Luis, Laguna de** ◎ NW Bolivia
35 P13 **San Luis Obispo** California, W USA 35°17′N 120°40′W
37 R7 **San Luis Peak** ▲ Colorado, C USA 37°59′N 106°55′W
41 N11 **San Luis Potosí** San Luis Potosí, C Mexico 22°10′N 100°57′W
41 N11 **San Luis Potosí** ◈ state C Mexico
San Luis, Provincia de see San Luis
35 O10 **San Luis Reservoir** ⊠ California, W USA
40 D2 **San Luis Río Colorado** var. San Luis Río Colorado. Sonora, NW Mexico 32°26′N 114°48′W
San Luis Río Colorado see San Luis Río Colorado
37 S8 **San Luis Valley** basin Colorado, C USA
107 C19 **Sanluri** Sardegna, Italy, C Mediterranean Sea 39°34′N 08°54′E
61 D23 **San Manuel** Buenos Aires, E Argentina 37°47′S 58°50′W
36 M15 **San Manuel** Arizona, SW USA 32°36′N 110°37′W
106 F11 **San Marcello Pistoiese** Toscana, C Italy 44°03′N 10°46′E
107 N20 **San Marco Argentano** Calabria, SW Italy 39°31′N 16°07′E
54 E6 **San Marcos** Sucre, N Colombia 08°38′N 75°10′W
42 M14 **San Marcos** San José, C Costa Rica 09°39′N 84°00′W
42 B5 **San Marcos** San Marcos, W Guatemala 14°58′N 91°48′W
42 F6 **San Marcos** Ocotepeque, SW Honduras 14°23′N 88°57′W
41 O16 **San Marcos** Guerrero, S Mexico 16°45′N 99°22′W
25 S11 **San Marcos** Texas, SW USA 29°54′N 97°57′W
42 A5 **San Marcos** off. Departamento de San Marcos. ◈ department W Guatemala
San Marcos de Arica see Arica
San Marcos, Departamento de see San Marcos
40 E6 **San Marcos, Isla** island NW Mexico
106 H11 **San Marino** ● (San Marino) C San Marino 43°54′N 12°27′E
106 I11 **San Marino** off. Republic of San Marino. ◆ republic S Europe
San Marino, Republic of see San Marino
62 I11 **San Martín** Mendoza, C Argentina 33°05′S 68°28′W
54 F11 **San Martín** Meta, C Colombia 03°43′N 73°42′W
56 D11 **San Martín** off. Departamento de San Martín. ◈ department C Peru
194 I5 **San Martín** Argentinian research station Antarctica 68°18′S 67°03′W
63 H16 **San Martín de los Andes** Neuquén, W Argentina 40°11′S 71°22′W
San Martín, Departamento de see San Martín
104 M8 **San Martín de Valdeiglesias** Madrid, C Spain 40°21′N 04°24′W
63 G21 **San Martín, Lago** var. Lago O'Higgins. ◎ S Argentina
106 H6 **San Martino di Castrozza** Trentino-Alto Adige, N Italy 46°16′N 11°50′E
57 N16 **San Martín, Río** ≈ N Bolivia
San Martín Texmelucan see Texmelucan
35 N9 **San Mateo** California, W USA 37°33′N 122°19′W
55 O6 **San Mateo** Anzoátegui, NE Venezuela 09°48′N 64°36′W
42 B4 **San Mateo Ixtatán** Huehuetenango, W Guatemala 15°50′N 91°30′W
57 Q18 **San Matías** Santa Cruz, E Bolivia 16°20′S 58°24′W
63 K16 **San Matías, Golfo** var. Gulf of San Matías. gulf E Argentina
San Matías, Gulf of see San Matías, Golfo
15 O8 **Sanmaur** Québec, SE Canada 47°52′N 73°47′W
161 T10 **Sanmen Wan** bay E China
160 M6 **Sanmenxia** var. Shan Xian. Henan, C China 34°46′N 111°17′E
Sânmiclăuş Mare see Sânnicolau Mare
61 D14 **San Miguel** Corrientes, NE Argentina 28°02′S 57°41′W
57 L16 **San Miguel** El Beni, N Bolivia 13°52′S 61°06′W
42 G8 **San Miguel** San Miguel, SE El Salvador 13°27′N 88°11′W
40 L6 **San Miguel** Coahuila, N Mexico 29°10′N 101°28′W
40 J9 **San Miguel** var. San Miguel de Cruces. Durango, C Mexico 24°25′N 105°55′W
43 U16 **San Miguel** Panamá, SE Panama 08°27′N 78°51′W
35 P12 **San Miguel** California, W USA 35°45′N 120°42′W
42 B9 **San Miguel** ◈ department E El Salvador
41 N13 **San Miguel de Allende** Guanajuato, C Mexico 20°56′N 100°48′W
San Miguel de Cruces see San Miguel
San Miguel de Ibarra see Ibarra
61 D21 **San Miguel del Monte** Buenos Aires, E Argentina 35°26′S 58°50′W
62 J7 **San Miguel de Tucumán** var. Tucumán. Tucumán, N Argentina 26°47′S 65°15′W
43 V16 **San Miguel, Golfo de** gulf S Panama
35 P15 **San Miguel Island** island California, W USA
42 L11 **San Miguelito** Río San Juan, S Nicaragua 11°22′N 84°54′W
43 T15 **San Miguelito** Panamá, C Panama 08°58′N 79°31′W
57 N18 **San Miguel, Río** ≈ E Bolivia
56 D6 **San Miguel, Río** ≈ Colombia/Ecuador
40 I7 **San Miguel, Río** ≈ N Mexico
42 G8 **San Miguel, Volcán de** ⛰ SE El Salvador 13°27′N 88°18′W
161 Q12 **Sanming** Fujian, SE China 26°11′N 117°37′E
106 F11 **San Miniato** Toscana, C Italy 43°40′N 10°53′E
San Murezzan see St. Moritz
Sannār see Sennar
107 M15 **Sannicandro Garganico** Puglia, SE Italy 41°50′N 15°32′E
40 H6 **San Nicolás** Sonora, NW Mexico 28°31′N 109°24′W
61 C19 **San Nicolás de los Arroyos** Buenos Aires, E Argentina 33°20′S 60°13′W
35 R16 **San Nicolas Island** island Channel Islands, California, W USA
Sânnicolaul-Mare see Sânnicolau Mare
116 E11 **Sânnicolau Mare** var. Sânnicolaul-Mare, Hung. Nagyszentmiklós; prev. Sânmiclăuş Mare, Sînnicolau Mare. Timiş, W Romania 46°05′N 20°38′E
123 Q6 **Sannikova, Proliv** strait NE Russian Federation
76 K16 **Sanniquellie** var. Saniquillie. NE Liberia 07°24′N 08°45′W
165 R7 **Sannohe** Aomori, Honshū, C Japan 40°23′N 141°16′E
Sanntaler Alpen see Kamniško-Savinjske Alpe
111 O17 **Sanok** Podkarpackie, SE Poland 49°31′N 22°14′E
54 E5 **San Onofre** Sucre, NW Colombia 09°45′N 75°33′W
57 K21 **San Pablo** Potosí, S Bolivia 21°43′S 66°38′W
171 O4 **San Pablo** off. San Pablo City. Luzon, N Philippines 14°04′N 121°16′E
San Pablo Balleza see Balleza
35 N8 **San Pablo Bay** bay California, W USA
San Pablo City see San Pablo
40 C6 **San Pablo, Punta** headland NW Mexico 27°12′N 114°30′W
43 R16 **San Pablo, Río** ≈ C Panama
171 P4 **San Pascual** Burias Island, C Philippines 13°06′N 122°59′E
121 Q16 **San Pawl il Bahar** Eng. Saint Paul's Bay. E Malta 35°57′N 14°24′E
61 C19 **San Pedro** Buenos Aires, E Argentina 33°43′S 59°45′W
62 K5 **San Pedro** Jujuy, N Argentina 24°12′S 64°55′W
60 G13 **San Pedro** Misiones, NE Argentina 26°38′S 54°12′W
42 H1 **San Pedro** Corozal, NE Belize 17°58′N 87°55′W
76 M17 **San-Pédro** S Ivory Coast 04°45′N 06°37′W
40 L8 **San Pedro** var. San Pedro de las Colonias. Coahuila, NE Mexico 25°47′N 102°57′W
62 O5 **San Pedro** San Pedro, SE Paraguay 24°08′S 57°08′W
62 O6 **San Pedro** off. Departamento de San Pedro. ◈ department C Paraguay
44 G6 **San Pedro** ≈ C Cuba
77 N16 **San Pedro** ✕ (Yamoussoukro) C Ivory Coast 06°49′N 05°14′W
San Pedro see San Pedro del Pinatar
42 D5 **San Pedro Carchá** Alta Verapaz, C Guatemala 15°30′N 90°12′W
35 S16 **San Pedro Channel** channel California, W USA
62 I5 **San Pedro de Atacama** Antofagasta, N Chile 22°52′S 68°10′W
San Pedro de Durazno see Durazno
40 G5 **San Pedro de la Cueva** Sonora, NW Mexico 29°17′N 109°47′W
San Pedro de las Colonias see San Pedro
56 B11 **San Pedro de Lloc** La Libertad, NW Peru 07°26′S 79°31′W
105 S13 **San Pedro del Pinatar** var. San Pedro. Murcia, SE Spain 37°50′N 00°47′W
45 P9 **San Pedro de Macorís** SE Dominican Republic 18°30′N 69°18′W
San Pedro, Departamento de see San Pedro
40 C3 **San Pedro Mártir, Sierra** ▲▲ NW Mexico
San Pedro Pochutla see Pochutla
42 D2 **San Pedro, Río** ≈ Guatemala/Mexico
40 K10 **San Pedro, Río** ≈ C Mexico
104 J10 **San Pedro, Sierra de** ▲▲ W Spain
42 G5 **San Pedro Sula** Cortés, NW Honduras 15°26′N 88°01′W
San Pedro Tapanatepec see Tapanatepec
62 I4 **San Pedro, Volcán** ⛰ N Chile 21°46′S 68°13′W
106 E7 **San Pellegrino Terme** Lombardia, N Italy 45°53′N 09°42′E
25 T16 **San Perlita** Texas, SW USA 26°30′N 97°38′W
San Pietro see Supetar
San Pietro del Carso see Pivka
107 A20 **San Pietro, Isola di** island W Italy
32 K7 **Sanpoil River** ≈ Washington, NW USA
165 O9 **Sanpoku** var. Sampoku. Niigata, Honshū, C Japan 38°32′N 139°33′E
40 C3 **San Quintín** Baja California Norte, NW Mexico 30°28′N 115°58′W
40 B3 **San Quintín, Bahía de** bay NW Mexico
40 B3 **San Quintín, Cabo** headland NW Mexico 30°22′N 116°01′W
62 I12 **San Rafael** Mendoza, W Argentina 34°44′S 68°15′W
41 N9 **San Rafael** Nuevo León, NE Mexico 25°01′N 100°33′W
34 M8 **San Rafael** California, W USA 37°58′N 122°31′W
37 Q11 **San Rafael** New Mexico, SW USA 35°03′N 107°52′W
54 H4 **San Rafael** var. El Mojan. Zulia, NW Venezuela 10°58′N 71°45′W
42 J8 **San Rafael del Norte** Jinotega, NW Nicaragua 13°12′N 86°06′W
42 J10 **San Rafael del Sur** Managua, SW Nicaragua 11°51′N 86°24′W
36 M5 **San Rafael Knob** ▲ Utah, W USA 38°46′N 110°45′W
35 Q14 **San Rafael Mountains** ▲▲ California, W USA
42 M13 **San Ramón** Alajuela, C Costa Rica 10°04′N 84°31′W
57 E14 **San Ramón** Junín, C Peru 11°08′S 75°18′W
61 F19 **San Ramón** Canelones, S Uruguay 34°18′S 55°55′W
62 K5 **San Ramón de la Nueva Orán** Salta, N Argentina 23°08′S 64°20′W
57 O16 **San Ramón, Río** ≈ E Bolivia
106 B11 **San Remo** Liguria, NW Italy 43°48′N 07°47′E
54 J3 **San Román, Cabo** headland NW Venezuela 12°10′N 70°01′W
61 C15 **San Roque** Corrientes, NE Argentina 28°35′S 58°45′W
188 I4 **San Roque** Saipan, S Northern Mariana Islands 15°15′S 145°47′E
104 K16 **San Roque** Andalucía, S Spain 36°13′N 05°23′W
25 R9 **San Saba** Texas, SW USA 31°13′N 98°44′W
25 Q9 **San Saba River** ≈ Texas, SW USA
61 D17 **San Salvador** Entre Ríos, E Argentina 31°38′S 58°30′W
42 F7 **San Salvador** ● (El Salvador) San Salvador, SW El Salvador 13°42′N 89°12′W
42 A10 **San Salvador** ◈ department C El Salvador
42 F8 **San Salvador** ✕ La Paz, S El Salvador 13°27′N 89°04′W
44 K4 **San Salvador** prev. Watlings Island. island E The Bahamas
62 J5 **San Salvador de Jujuy** var. Jujuy. Jujuy, N Argentina 24°10′S 65°20′W
42 F7 **San Salvador, Volcán de** ⛰ C El Salvador 13°58′N 89°14′W
77 Q14 **Sansanné-Mango** var. Mango. N Togo 10°21′N 00°28′E
45 S5 **San Sebastián** W Puerto Rico 18°21′N 67°00′W
63 J24 **San Sebastián, Bahía** bay S Argentina
Sansenhō see Sacheon
106 H12 **Sansepolcro** Toscana, C Italy 43°35′N 12°12′E
107 M16 **San Severo** Puglia, SE Italy 41°41′N 15°23′E
112 F11 **Sanski Most** ◈ Federacija Bosne I Hercegovine, NW Bosnia and Herzegovina
171 W12 **Sansundi** Papua, E Indonesia 0°42′S 135°48′E
162 K9 **Sant** var. Mayhan. Övörhangay, C Mongolia 46°02′N 104°00′E
104 K11 **Santa Amalia** Extremadura, W Spain 39°00′N 06°01′W
60 F13 **Santa Ana** Misiones, NE Argentina 27°22′S 55°34′W
57 L16 **Santa Ana** El Beni, N Bolivia 13°43′S 65°37′W
42 E7 **Santa Ana** Santa Ana, NW El Salvador 13°59′N 89°34′W
40 F4 **Santa Ana** Sonora, NW Mexico 30°31′N 111°08′W
35 T16 **Santa Ana** California, W USA 33°45′N 117°52′W
55 N6 **Santa Ana** Nueva Esparta, NE Venezuela 09°15′N 64°39′W
42 A9 **Santa Ana** ◈ department NW El Salvador
Santa Ana de Coro see Coro
35 U16 **Santa Ana Mountains** ▲▲ California, W USA
42 E7 **Santa Ana, Volcán de** var. La Matepec. ⛰ W El Salvador 13°49′N 89°36′W
42 G6 **Santa Bárbara** Santa Bárbara, W Honduras 14°56′N 88°11′W
40 J7 **Santa Barbara** Chihuahua, N Mexico 26°46′N 105°46′W
35 Q14 **Santa Barbara** California, W USA 34°24′N 119°40′W
54 L11 **Santa Bárbara** Amazonas, S Venezuela 03°55′N 67°06′W
54 I7 **Santa Bárbara** Barinas, W Venezuela 07°48′N 71°10′W
42 F5 **Santa Bárbara** ◈ department NW Honduras
Santa Bárbara see Iscuandé
35 Q15 **Santa Barbara Channel** channel California, W USA
Santa Bárbara de Samaná see Samaná
35 R16 **Santa Barbara Island** island Channel Islands, California, W USA
54 E5 **Santa Catalina** Bolívar, N Colombia 10°36′N 75°17′W
43 R15 **Santa Catalina** Ngöbe Bugle, W Panama 08°46′N 81°18′W
104 G2 **Santa Cataliña de Armada** Galicia, NW Spain 43°02′N 08°49′W
35 T17 **Santa Catalina, Gulf of** gulf California, W USA
40 F8 **Santa Catalina, Isla** island NW Mexico
35 S16 **Santa Catalina Island** island Channel Islands, California, W USA
41 N8 **Santa Catarina** Nuevo León, NE Mexico 25°39′N 100°30′W
60 H13 **Santa Catarina** off. Estado de Santa Catarina. ◈ state S Brazil
Santa Catarina de Tepehuanes see Tepehuanes
Santa Catarina, Estado de see Santa Catarina
60 L13 **Santa Catarina, Ilha de** island S Brazil
45 Q16 **Santa Catherina** Curaçao 12°07′N 68°46′W
44 E5 **Santa Clara** Villa Clara, C Cuba 22°25′N 78°01′W
35 N9 **Santa Clara** California, W USA 37°20′N 121°57′W
36 J8 **Santa Clara** Utah, W USA 37°07′N 113°39′W
Santa Clara see Santa Clara de Olimar
61 F18 **Santa Clara de Olimar** var. Santa Clara. Cerro Largo, NE Uruguay 32°50′S 54°54′W
61 A17 **Santa Clara de Saguier** Santa Fe, C Argentina 31°21′S 61°50′W
Santa Coloma see Santa Coloma de Gramenet
105 X5 **Santa Coloma de Farners** var. Santa Coloma de Farnés. Cataluña, NE Spain 41°52′N 02°39′E
Santa Coloma de Farnés see Santa Coloma de Farners
Santa Coloma de Gramanet see Santa Coloma de Gramenet
105 W6 **Santa Coloma de Gramenet** var. Santa Coloma; prev. Santa Coloma de Gramanet. Cataluña, NE Spain 41°28′N 02°14′E
Santa Comba see Uaco Cungo
104 H8 **Santa Comba Dão** Viseu, N Portugal 40°23′N 08°07′W
82 C10 **Santa Cruz** Uíge, NW Angola 06°56′S 16°25′E
57 N19 **Santa Cruz** var. Santa Cruz de la Sierra. Santa Cruz, C Bolivia 17°49′S 63°11′W
62 G12 **Santa Cruz** Libertador, C Chile 34°38′S 71°27′W
42 K13 **Santa Cruz** Guanacaste, W Costa Rica 10°15′N 85°35′W
44 I12 **Santa Cruz** W Jamaica 18°03′N 77°43′W
64 P6 **Santa Cruz** Madeira, Portugal, NE Atlantic Ocean 32°43′N 16°47′W
35 N10 **Santa Cruz** California, W USA 36°58′N 122°01′W
63 H20 **Santa Cruz** off. Provincia de Santa Cruz. ◈ province S Argentina
57 O18 **Santa Cruz** ◈ department E Bolivia
Santa Cruz see Puerto Santa Cruz
Santa Cruz see Viru-Viru
Santa Cruz Barillas see Barillas
59 O18 **Santa Cruz Cabrália** Bahia, E Brazil 16°17′S 39°03′W
Santa Cruz de El Seibo see El Seibo
64 N11 **Santa Cruz de la Palma** La Palma, Islas Canarias, Spain, NE Atlantic Ocean 28°41′N 17°46′W
Santa Cruz de la Sierra see Santa Cruz
105 O9 **Santa Cruz de la Zarza** Castilla-La Mancha, C Spain 39°59′N 03°10′W
42 C5 **Santa Cruz del Quiché** Quiché, W Guatemala 15°02′N 91°06′W
105 N8 **Santa Cruz del Retamar** Castilla-La Mancha, C Spain 40°08′N 04°14′W
Santa Cruz del Seibo see El Seibo
44 G7 **Santa Cruz del Sur** Camagüey, C Cuba 20°44′N 78°00′W
105 O11 **Santa Cruz de Mudela** Castilla-La Mancha, C Spain 38°37′N 03°27′W
64 Q11 **Santa Cruz de Tenerife** Tenerife, Islas Canarias, Spain, NE Atlantic Ocean 28°28′N 16°15′W
64 P11 **Santa Cruz de Tenerife** ◈ province Islas Canarias, Spain, NE Atlantic Ocean
60 K9 **Santa Cruz do Rio Pardo** São Paulo, S Brazil 22°52′S 49°37′W
61 H15 **Santa Cruz do Sul** Rio Grande do Sul, S Brazil 29°42′S 52°25′W
57 C17 **Santa Cruz, Isla** var. Indefatigable Island, Isla Chávez. island Galapagos Islands, Ecuador, E Pacific Ocean
40 F8 **Santa Cruz, Isla** island NW Mexico
35 Q15 **Santa Cruz Island** island California, W USA
187 Q10 **Santa Cruz Islands** island group E Solomon Islands
Santa Cruz, Provincia de see Santa Cruz
63 I22 **Santa Cruz, Río** ≈ S Argentina
36 L15 **Santa Cruz River** ≈ Arizona, SW USA
61 C17 **Santa Elena** Entre Ríos, E Argentina 30°58′S 59°47′W
42 F2 **Santa Elena** Cayo, W Belize 17°08′N 89°04′W
25 R16 **Santa Elena** Texas, SW USA 26°43′N 98°30′W
56 A7 **Santa Elena, Bahía de** bay W Ecuador
55 R10 **Santa Elena de Uairén** Bolívar, E Venezuela 04°40′N 61°03′W
42 K12 **Santa Elena, Península** peninsula NW Costa Rica
56 A7 **Santa Elena, Punta** headland W Ecuador 02°11′S 81°00′W
104 L11 **Santa Eufemia** Andalucía, S Spain 38°36′N 04°54′W
107 N21 **Santa Eufemia, Golfo di** gulf S Italy
105 S4 **Santa Eulalia de Gállego** Aragón, NE Spain 42°16′N 00°46′W
105 V11 **Santa Eulalia del Río** Ibiza, Spain, W Mediterranean Sea 39°00′N 01°33′E
61 B17 **Santa Fe** Santa Fe, C Argentina 31°36′S 60°47′W
44 C6 **Santa Fé** var. La Fe. Isla de la Juventud, W Cuba 21°45′N 82°45′W
43 R16 **Santa Fé** Veraguas, C Panama 08°29′N 80°50′W
105 N14 **Santa Fe** Andalucía, S Spain 37°11′N 03°43′W
37 S10 **Santa Fe** state capital New Mexico, SW USA 35°41′N 105°56′W
61 B15 **Santa Fe** off. Provincia de Santa Fe. ◈ province C Argentina
Santa Fe see Bogotá
Santa Fe de Bogotá see Bogotá
60 J7 **Santa Fé do Sul** São Paulo, S Brazil 20°13′S 50°56′W
57 B18 **Santa Fe, Isla** var. Barrington Island. island Galapagos Islands, Ecuador, E Pacific Ocean
Santa Fe, Provincia de see Santa Fe
23 V9 **Santa Fe River** ≈ Florida, SE USA
59 M15 **Santa Filomena** Piauí, E Brazil 09°06′S 45°52′W
40 G10 **Santa Genoveva** ▲ NW Mexico 23°07′N 109°56′W
153 S14 **Santahar** Rajshahi, NW Bangladesh 24°45′N 89°03′E
60 G11 **Santa Helena** Paraná, S Brazil 24°53′S 54°19′W
54 J5 **Santa Inés** Lara, N Venezuela 10°37′N 69°18′W
63 G24 **Santa Inés, Isla** island S Chile
62 J13 **Santa Isabel** La Pampa, C Argentina 36°11′S 66°59′W
43 U14 **Santa Isabel** Colón, N Panama 09°31′N 79°12′W
186 L8 **Santa Isabel** var. Bughotu. island N Solomon Islands
Santa Isabel see Malabo
58 D11 **Santa Isabel do Rio Negro** Amazonas, NW Brazil 0°40′S 64°56′W
61 C15 **Santa Lucia** Corrientes, NE Argentina 28°58′S 59°05′W
57 I17 **Santa Lucía** Puno, S Peru 15°45′S 70°34′W
61 F20 **Santa Lucía** var. Santa Lucia. Canelones, S Uruguay 34°26′S 56°25′W
42 B6 **Santa Lucía Cotzumalguapa** Escuintla, SW Guatemala 14°20′N 91°00′W
107 L23 **Santa Lucia del Mela** Sicilia, Italy, C Mediterranean Sea 38°08′N 15°17′E
35 O11 **Santa Lucia Range** ▲▲ California, W USA
40 D9 **Santa Margarita, Isla** island NW Mexico
62 J7 **Santa María** Catamarca, N Argentina 26°51′S 66°02′W
61 G15 **Santa Maria** Rio Grande do Sul, S Brazil 29°41′S 53°48′W
35 P13 **Santa Maria** California, W USA 34°56′N 120°25′W
64 Q4 **Santa Maria** ✕ Santa Maria, Azores, Portugal, NE Atlantic Ocean
64 P3 **Santa Maria** island Azores, Portugal, NE Atlantic Ocean
Santa Maria see Gaua
Santa María Asunción Tlaxiaco see Tlaxiaco
40 G9 **Santa María, Bahía** bay W Mexico
83 L21 **Santa Maria, Cabo de** headland S Mozambique 26°05′S 32°58′E
104 G15 **Santa Maria, Cabo de** headland S Portugal 36°57′N 07°55′W
44 J4 **Santa Maria, Cape** headland Long Island, C The Bahamas 23°40′N 75°20′W
107 J17 **Santa Maria Capua Vetere** Campania, S Italy 41°05′N 14°15′E
104 G7 **Santa Maria da Feira** Aveiro, N Portugal 40°55′N 08°32′W
59 M17 **Santa Maria da Vitória** Bahia, E Brazil 13°26′S 44°09′W
55 N9 **Santa María de Erebato** Bolívar, SE Venezuela 05°09′N 64°50′W
55 N6 **Santa María de Ipire** Guárico, C Venezuela 08°51′N 65°21′W
Santa Maria del Buen Aire see Buenos Aires
40 J8 **Santa María del Oro** Durango, C Mexico 25°57′N 105°22′W
41 N12 **Santa María del Río** San Luis Potosí, C Mexico 21°48′N 100°42′W
Santa Maria di Castellabate see Castellabate
107 Q20 **Santa Maria di Leuca, Capo** headland SE Italy 39°48′N 18°21′E
108 K10 **Santa Maria-im-Munstertal** Graubünden, SE Switzerland 46°36′N 10°25′E
57 B18 **Santa María, Isla** var. Isla Floreana, Charles Island. island Galapagos Islands, Ecuador, E Pacific Ocean
40 J3 **Santa María, Laguna de** ◎ N Mexico
61 G16 **Santa Maria, Rio** ≈ S Brazil
43 R16 **Santa María, Río** ≈ C Panama
36 J12 **Santa Maria River** ≈ Arizona, SW USA
107 G15 **Santa Marinella** Lazio, C Italy 42°01′N 11°51′E
54 F4 **Santa Marta** Magdalena, N Colombia 11°14′N 74°13′W
104 J11 **Santa Marta** Extremadura, W Spain 38°37′N 06°39′W
54 F4 **Santa Marta, Sierra Nevada de** ▲▲ NE Colombia
Santa Maura see Lefkáda
35 S15 **Santa Monica** California, W USA 34°01′N 118°29′W
116 F10 **Sântana** Ger. Sankt Anna, Hung. Újszentanna; prev. Sîntana. Arad, W Romania 46°20′N 21°30′E
61 F16 **Santana, Coxilha de** hill range S Brazil

◆ Country ● Country Capital ◇ Dependent Territory ○ Dependent Territory Capital ◈ Administrative Regions ✕ International Airport ▲ Mountain ▲▲ Mountain Range ⛰ Volcano ≈ River ◎ Lake ⊠ Reservoir

61 *H16* **Santana da Boa Vista** Rio Grande do Sul, S Brazil *30°52′S 53°03′W*

61 *F16* **Santana do Livramento** *prev.* Livramento. Rio Grande do Sul, S Brazil *30°52′S 55°30′W*

105 *N2* **Santander** Cantabria, N Spain *43°28′N 03°48′W*

54 *F8* **Santander** *off.* Departamento de Santander. ◈ *province* C Colombia

Santander, Departamento de *see* Santander

Santander Jiménez *see* Jiménez

Sant'Andrea *see* Svetac

107 *B20* **Sant'Antioco** Sardegna, Italy, C Mediterranean Sea *39°03′N 08°28′E*

105 *V11* **Sant Antoni de Portmany** *Cas.* San Antonio Abad. Ibiza, Spain, W Mediterranean Sea *38°58′N 01°18′E*

105 *Y10* **Santanyí** Mallorca, Spain, W Mediterranean Sea *39°22′N 03°07′E*

104 *J13* **Santa Olalla del Cala** Andalucía, S Spain *37°54′N 06°13′W*

35 *R15* **Santa Paula** California, W USA *34°21′N 119°03′W*

36 *L4* **Santaquin** Utah, W USA *39°58′N 111°46′W*

58 *I12* **Santarém** Pará, N Brazil *02°26′S 54°41′W*

104 *G10* **Santarém** *anc.* Scalabis. Santarém, W Portugal *39°14′N 08°40′W*

104 *G10* **Santarém** ◈ *district* C Portugal

44 *F4* **Santaren Channel** *channel* W The Bahamas

54 *K10* **Santa Rita** Vichada, E Colombia *04°51′N 68°27′W*

188 *B16* **Santa Rita** SW Guam

42 *H5* **Santa Rita** Cortés, NW Honduras *15°10′N 87°54′W*

40 *E9* **Santa Rita** Baja California Sur, NW Mexico *27°29′N 100°33′W*

54 *H5* **Santa Rita** Zulia, NW Venezuela *10°35′N 71°30′W*

59 *I19* **Santa Rita de Araguaia** Goiás, S Brazil *17°17′S 53°13′W*

59 *M16* **Santa Rita de Cassia** *var.* Cássia. Bahia, E Brazil *11°03′S 44°16′W*

61 *D14* **Santa Rosa** Corrientes, NE Argentina *28°18′S 58°04′W*

62 *K13* **Santa Rosa** La Pampa, C Argentina *36°38′S 64°15′W*

61 *G14* **Santa Rosa** Rio Grande do Sul, S Brazil *27°50′S 54°29′W*

58 *E10* **Santa Rosa** Roraima, N Brazil *03°41′N 62°29′W*

56 *B8* **Santa Rosa** El Oro, SW Ecuador *03°29′S 79°57′W*

57 *I16* **Santa Rosa** Puno, S Peru *14°38′S 70°45′W*

34 *M7* **Santa Rosa** California, W USA *38°27′N 122°42′W*

37 *U11* **Santa Rosa** New Mexico, SW USA *34°54′N 104°43′W*

55 *O6* **Santa Rosa** Anzoátegui, NE Venezuela *09°37′N 64°20′W*

42 *A3* **Santa Rosa** *off.* Departamento de Santa Rosa. ◈ *department* SE Guatemala

Santa Rosa *see* Santa Rosa de Copán

63 *J15* **Santa Rosa, Bajo de** *basin* E Argentina

42 *F6* **Santa Rosa de Copán** *var.* Santa Rosa. Copán, W Honduras *14°48′N 88°43′W*

54 *E8* **Santa Rosa de Osos** Antioquia, C Colombia *06°40′N 75°27′W*

Santa Rosa, Departamento de *see* Santa Rosa

35 *Q15* **Santa Rosa Island** *island* California, W USA

23 *O9* **Santa Rosa Island** *island* Florida, SE USA

40 *E6* **Santa Rosalía** Baja California Sur, NW Mexico *27°20′N 112°20′W*

54 *K6* **Santa Rosalía** Portuguesa, NW Venezuela *09°02′N 69°01′W*

188 *C15* **Santa Rosa, Mount** ▲ NE Guam

35 *V16* **Santa Rosa Mountains** ▲▲ California, W USA

35 *T2* **Santa Rosa Range** ▲▲ Nevada, W USA

62 *M8* **Santa Sylvina** Chaco, N Argentina *27°49′S 61°09′W*

Santa Tecla *see* Nueva San Salvador

61 *B19* **Santa Teresa** Santa Fe, C Argentina *33°30′S 60°45′W*

59 *O20* **Santa Teresa** Espírito Santo, SE Brazil *19°51′S 40°49′W*

61 *E21* **Santa Teresita** Buenos Aires, E Argentina *36°32′S 56°41′W*

61 *H19* **Santa Vitória do Palmar** Rio Grande do Sul, S Brazil *33°32′S 53°25′W*

35 *Q14* **Santa Ynez River** ≋ California, W USA

Sant Carles de la Rápida *see* Sant Carles de la Ràpita

105 *U7* **Sant Carles de la Ràpita** *var.* Sant Carles de la Rápida. Cataluña, NE Spain *40°37′N 00°36′E*

105 *W5* **Sant Celoni** Cataluña, NE Spain *41°39′N 02°25′E*

35 *U17* **Santee** California, W USA *32°50′N 116°58′W*

21 *T13* **Santee River** ≋ South Carolina, SE USA

40 *K15* **San Telmo, Punta** *headland* SW Mexico *18°19′N 103°30′W*

107 *O17* **Santeramo in Colle** Puglia, SE Italy *40°47′N 16°45′E*

107 *M23* **San Teresa di Riva** Sicilia, Italy, C Mediterranean Sea *38°00′N 15°25′E*

105 *X5* **Sant Feliu de Guíxols** *var.* San Feliú de Guixols. Cataluña, NE Spain *41°47′N 03°02′E*

105 *W6* **Sant Feliu de Llobregat** Cataluña, NE Spain *41°22′N 02°00′E*

106 *C7* **Santhià** Piemonte, NE Italy *45°21′N 08°11′E*

61 *F15* **Santiago** Rio Grande do Sul, S Brazil *29°11′S 54°52′W*

62 *H11* **Santiago** *var.* Gran Santiago. ● (Chile) Santiago, C Chile *33°30′S 70°40′W*

45 *N8* **Santiago** *var.* Santiago de los Caballeros. N Dominican Republic *19°27′N 70°42′W*

40 *G10* **Santiago** Baja California Sur, NW Mexico *23°32′N 109°47′W*

41 *O8* **Santiago** Nuevo León, NE Mexico *25°22′N 100°09′W*

43 *R16* **Santiago** Veraguas, S Panama *08°06′N 80°59′W*

57 *E16* **Santiago** Ica, SW Peru *14°14′S 75°44′W*

62 *H11* **Santiago** *off.* Región Metropolitana de Santiago, *var.* Metropolitan. ◈ *region* C Chile

76 *D10* **Santiago** *var.* São Tiago. *island* Ilhas de Sotavento, S Cape Verde

62 *H11* **Santiago** ✈ Santiago, C Chile *33°27′S 70°40′W*

104 *G3* **Santiago** ✈ Galicia, NW Spain

Santiago *see* Santiago de Cuba, Cuba

Santiago *see* Grande de Santiago, Río, Mexico

Santiago *see* Santiago de Compostela

42 *B6* **Santiago Atitlán** Sololá, SW Guatemala *14°39′N 91°12′W*

43 *Q16* **Santiago, Cerro** ▲ W Panama *08°27′N 81°42′W*

104 *G3* **Santiago de Compostela** *var.* Santiago, *Eng.* Compostella; *anc.* Campus Stellae. Galicia, NW Spain *42°52′N 08°33′W*

44 *I8* **Santiago de Cuba** *var.* Santiago. Santiago de Cuba, E Cuba *20°01′N 75°51′W*

Santiago de Guayaquil *see* Guayaquil

62 *K8* **Santiago del Estero** Santiago del Estero, C Argentina *27°51′S 64°16′W*

61 *A15* **Santiago del Estero** *off.* Provincia de Santiago del Estero. ◈ *province* N Argentina

Santiago del Estero, Provincia de *see* Santiago del Estero

40 *I8* **Santiago de los Caballeros** Sinaloa, W Mexico *25°33′N 107°22′W*

Santiago de los Caballeros *see* Santiago, Dominican Republic

Santiago de los Caballeros *see* Ciudad de Guatemala, Guatemala

42 *F8* **Santiago de María** Usulután, SE El Salvador *13°28′N 88°28′W*

104 *F12* **Santiago do Cacém** Setúbal, S Portugal *38°01′N 08°42′W*

40 *J12* **Santiago Ixcuintla** Nayarit, C Mexico *21°50′N 105°11′W*

Santiago Jamiltepec *see* Jamiltepec

24 *L11* **Santiago Mountains** ▲▲ Texas, SW USA

40 *J9* **Santiago Papasquiaro** Durango, C Mexico *25°00′N 105°27′W*

Santiago Pinotepa Nacional *see* Pinotepa Nacional

Santiago, Región Metropolitana de *see* Santiago

56 *C8* **Santiago, Río** ≋ N Peru

40 *M10* **San Tiburcio** Zacatecas, C Mexico *24°08′N 101°29′W*

105 *N2* **Santillana** Cantabria, N Spain *43°24′N 04°06′W*

54 *I5* **San Timoteo** Zulia, NW Venezuela *09°50′N 71°05′W*

Santi Quaranta *see* Sarandë

Santissima Trinidad *see* Jilong

105 *O12* **Santisteban del Puerto** Andalucía, S Spain *38°15′N 03°11′W*

105 *S12* **Sant Joan d'Alacant** *Cast.* San Juan de Alicante. Valenciana, E Spain *38°26′N 00°27′W*

105 *U7* **Sant Jordi, Golf de** *gulf* NE Spain

105 *U11* **Sant Josep de sa Talaia** *var.* San Jose. Ibiza, Spain, W Mediterranean Sea *38°55′N 1°18′E*

162 *G6* **Santmargats** *var.* Holboo. Dzavhan, W Mongolia *48°35′N 95°25′E*

105 *T8* **Sant Mateu** Valenciana, E Spain *40°28′N 00°10′E*

25 *S7* **Santo** Texas, SW USA *32°35′N 98°06′W*

Santo *see* Espiritu Santo

60 *M10* **Santo Amaro, Ilha de** *island* SE Brazil

61 *G14* **Santo Ângelo** Rio Grande do Sul, S Brazil *28°17′S 54°15′W*

76 *C9* **Santo Antão** *island* Ilhas de Barlavento, N Cape Verde

60 *J10* **Santo Antônio da Platina** Paraná, S Brazil *23°20′S 50°05′W*

58 *C13* **Santo Antônio do Içá** Amazonas, N Brazil *03°05′S 67°56′W*

57 *Q18* **Santo Corazón, Río** ≋ E Bolivia

44 *E5* **Santo Domingo** Villa Clara, C Cuba *22°35′N 80°15′W*

45 *O9* **Santo Domingo** *prev.* Ciudad Trujillo. ● (Dominican Republic) SE Dominican Republic *18°30′N 69°57′W*

40 *E8* **Santo Domingo** Baja California Sur, NW Mexico *25°34′N 112°00′W*

40 *M10* **Santo Domingo** San Luis Potosí, C Mexico *23°18′N 101°42′W*

42 *L10* **Santo Domingo** Chontales, S Nicaragua *12°15′N 84°59′W*

105 *P4* **Santo Domingo de la Calzada** La Rioja, N Spain *42°26′N 02°57′W*

56 *B6* **Santo Domingo de los Colorados** Pichincha, NW Ecuador *0°13′S 79°09′W*

Santo Domingo Tehuantepec *see* Tehuantepec

55 *O6* **San Tomé** Anzoátegui, NE Venezuela *08°58′N 64°08′W*

San Tomé de Guayana *see* Ciudad Guayana

105 *R13* **Santomera** Murcia, SE Spain *38°03′N 01°05′W*

105 *O2* **Santoña** Cantabria, N Spain *43°27′N 03°28′W*

Santorin *see* Santoríni

115 *K22* **Santoríni** *var.* Santorin, *prev.* Thíra; *anc.* Thera. *island* Kykládes, Greece, Aegean Sea

60 *M10* **Santos** São Paulo, S Brazil *23°56′S 46°22′W*

65 *J17* **Santos Plateau** *undersea feature* SW Atlantic Ocean *25°00′S 43°00′W*

104 *G6* **Santo Tirso** Porto, N Portugal *41°20′N 08°25′W*

40 *B2* **Santo Tomás** Baja California Norte, NW Mexico *31°32′N 116°26′W*

42 *L10* **Santo Tomás** Chontales, S Nicaragua *12°04′N 85°02′W*

42 *G5* **Santo Tomás de Castilla** Izabal, E Guatemala *15°40′N 88°36′W*

40 *B2* **Santo Tomás, Punta** *headland* NW Mexico *31°30′N 116°40′W*

57 *H16* **Santo Tomás, Río** ≋ C Peru

57 *B18* **Santo Tomás, Volcán** 🌋 Galapagos Islands, Ecuador, E Pacific Ocean *0°46′S 91°01′W*

61 *F14* **Santo Tomé** Corrientes, NE Argentina *28°31′S 56°03′W*

Santo Tomé de Guayana *see* Ciudad Guayana

98 *H10* **Santpoort** Noord-Holland, W Netherlands *52°26′N 04°38′E*

Santurce *see* Santurtzi

105 *O2* **Santurtzi** *var.* Santurce, Santurzi. País Vasco, N Spain *43°20′N 03°03′W*

Santurzi *see* Santurtzi

63 *G20* **San Valentín, Cerro** ▲ S Chile *46°36′S 73°17′W*

42 *F8* **San Vicente** San Vicente, C El Salvador *13°38′N 88°42′W**

40 *C2* **San Vicente** Baja California Norte, NW Mexico *31°20′N 116°15′W*

188 *H6* **San Vicente** Saipan, S Northern Mariana Islands

42 *B9* **San Vicente** ◈ *department* E El Salvador

104 *I10* **San Vicente de Alcántara** Extremadura, W Spain *39°21′N 07°07′W*

105 *N2* **San Vicente de Barakaldo** *var.* Baracaldo, *Basq.* San Bizenti-Barakaldo. País Vasco, N Spain *43°17′N 02°59′W*

57 *E15* **San Vicente de Cañete** *var.* Cañete. Lima, W Peru *13°06′S 76°23′W*

104 *M2* **San Vicente de la Barquera** Cantabria, N Spain *43°23′N 04°24′W*

54 *E12* **San Vicente del Caguán** Caquetá, S Colombia *02°07′N 74°47′W*

42 *F8* **San Vincente, Volcán de** 🌋 C El Salvador *13°34′N 88°50′W*

43 *O15* **San Vito** Puntarenas, SE Costa Rica *08°49′N 82°58′W*

106 *I7* **San Vito al Tagliamento** Friuli-Venezia Giulia, NE Italy *45°54′N 12°55′E*

107 *H23* **San Vito, Capo** *headland* Sicilia, Italy, C Mediterranean Sea *38°11′N 12°41′E*

107 *P18* **San Vito dei Normanni** Puglia, SE Italy *40°40′N 17°42′E*

160 *L17* **Sanya** *var.* Ya Xian. Hainan, S China *18°25′N 109°27′E*

83 *J16* **Sanyati** ≋ N Zimbabwe

25 *Q16* **San Ygnacio** Texas, SW USA *27°04′N 99°26′W*

160 *L6* **Sanyuan** Shaanxi, C China *34°40′N 108°56′E*

123 *P11* **Sanyyakhtakh** Respublika Sakha (Yakutiya), NE Russian Federation *60°34′N 124°09′E*

146 *J15* **S. A.Nyýazow Adyndaky** *Rus.* Imeni S. A. Niyazova. Maryyskiy Velayat, S Turkmenistan *36°44′N 62°23′E*

82 *C10* **Sanza Pombo** Uíge, NW Angola *07°20′S 16°00′E*

Sanzyô *see* Sanjō

104 *G14* **São Bartolomeu de Messines** Faro, S Portugal *37°12′N 08°16′W*

60 *M10* **São Bernardo do Campo** São Paulo, S Brazil *23°45′S 46°34′W*

61 *F15* **São Borja** Rio Grande do Sul, S Brazil *28°35′S 56°01′W*

104 *H14* **São Brás de Alportel** Faro, S Portugal *37°09′N 07°55′W*

60 *M10* **São Caetano do Sul** São Paulo, S Brazil *23°37′S 46°34′W*

60 *L9* **São Carlos** São Paulo, S Brazil *22°02′S 47°53′W*

59 *P16* **São Cristóvão** Sergipe, E Brazil *10°59′S 37°10′W*

61 *F15* **São Fancisco de Assis** Rio Grande do Sul, S Brazil *29°32′S 55°07′W*

58 *K13* **São Félix** Pará, NE Brazil *06°43′S 51°56′W*

São Félix *see* São Félix do Araguaia

59 *J16* **São Félix do Araguaia** *var.* São Félix. Mato Grosso, W Brazil *11°36′S 50°40′W*

59 *J14* **São Félix do Xingu** Pará, NE Brazil *06°38′S 51°59′W*

60 *Q9* **São Fidélis** Rio de Janeiro, SE Brazil *21°37′S 41°40′W*

76 *D10* **São Filipe** Fogo, S Cape Verde *14°52′N 24°29′W*

60 *K12* **São Francisco do Sul** Santa Catarina, S Brazil *26°17′S 48°39′W*

60 *K12* **São Francisco, Ilha de** *island* S Brazil

59 *P16* **São Francisco, Rio** ≋ E Brazil

61 *G16* **São Gabriel** Rio Grande do Sul, S Brazil *30°17′S 54°17′W*

60 *P10* **São Gonçalo** Rio de Janeiro, SE Brazil *22°48′S 43°03′W*

81 *H23* **Sao Hill** Iringa, S Tanzania *08°19′S 35°11′E*

60 *R9* **São João da Barra** Rio de Janeiro, SE Brazil *21°39′S 41°04′W*

104 *G7* **São João da Madeira** Aveiro, N Portugal *40°52′N 08°28′W*

58 *M12* **São João de Cortes** Maranhão, E Brazil *02°30′S 44°27′W*

59 *M21* **São João del Rei** Minas Gerais, NE Brazil *21°08′S 44°15′W*

59 *N15* **São João do Piauí** Piauí, E Brazil *08°21′S 42°14′W*

59 *N14* **São João dos Patos** Maranhão, E Brazil *06°29′S 43°44′W*

58 *C11* **São Joaquim** Amazonas, NW Brazil *0°08′S 67°10′W*

61 *J14* **São Joaquim** Santa Catarina, S Brazil *28°20′S 49°55′W*

60 *L7* **São Joaquim da Barra** São Paulo, S Brazil *20°36′S 47°50′W*

64 *N2* **São Jorge** *island* Azores, Portugal, NE Atlantic Ocean

61 *K14* **São José** Santa Catarina, S Brazil *27°34′S 48°39′W*

60 *M8* **São José do Rio Pardo** São Paulo, S Brazil *21°37′S 46°52′W*

60 *K8* **São José do Rio Preto** São Paulo, S Brazil *20°50′S 49°20′W*

60 *N10* **São Jose dos Campos** São Paulo, S Brazil *23°07′S 45°52′W*

61 *I17* **São Lourenço do Sul** Rio Grande do Sul, S Brazil *31°25′S 52°00′W*

58 *M12* **São Luís** *state capital* Maranhão, NE Brazil *02°34′S 44°16′W*

58 *F11* **São Luís** Roraima, N Brazil *01°11′N 60°15′W*

58 *M12* **São Luís, Ilha de** *island* NE Brazil

61 *F14* **São Luiz Gonzaga** Rio Grande do Sul, S Brazil *28°24′S 54°58′W*

São Mandol *see* São Manuel, Rio

47 *U8* **Sao Manuel** ≋ C Brazil

59 *H15* **São Manuel, Rio** *var.* São Mandol, Teles Pirés. ≋ C Brazil

58 *C11* **São Marcelino** Amazonas, NW Brazil *0°53′N 67°16′W*

58 *N12* **São Marcos, Baía de** *bay* N Brazil

59 *O20* **São Mateus** Espírito Santo, SE Brazil *18°44′S 39°53′W*

60 *J12* **São Mateus do Sul** Paraná, S Brazil *25°58′S 50°29′W*

64 *P3* **São Miguel** *island* Azores, Portugal, NE Atlantic Ocean

60 *G13* **São Miguel d'Oeste** Santa Catarina, S Brazil *26°45′S 53°34′W*

45 *P9* **Saona, Isla** *island* SE Dominican Republic

172 *H12* **Saondzou** ▲ Grande Comore, NW Comoros

103 *R10* **Saône** ≋ E France

103 *Q9* **Saône-et-Loire** ◈ *department* C France

76 *D9* **São Nicolau** *Eng.* Saint Nicholas. *island* Ilhas de Barlavento, N Cape Verde

60 *M10* **São Paulo** *state capital* São Paulo, S Brazil *23°33′S 46°39′W*

60 *K9* **São Paulo** *off.* Estado de São Paulo. ◈ *state* S Brazil

São Paulo de Loanda *see* Luanda

São Paulo, Estado de *see* São Paulo

São Pedro do Rio Grande do Sul *see* Rio Grande

104 *H7* **São Pedro do Sul** Viseu, N Portugal *40°46′N 07°58′W*

64 *K13* **São Pedro e São Paulo** *undersea feature* C Atlantic Ocean *01°25′N 28°54′W*

59 *M14* **São Raimundo das Mangabeiras** Maranhão, E Brazil *07°00′S 45°30′W*

59 *Q14* **São Roque, Cabo de** *headland* E Brazil *05°29′S 35°16′W*

São Salvador *see* Salvador, Brazil

São Salvador/São Salvador do Congo *see* M'Banza Congo, Angola

60 *N10* **São Sebastião, Ilha de** *island* S Brazil

83 *N19* **São Sebastião, Ponta** *headland* C Mozambique *22°09′S 35°33′E*

104 *F13* **São Teotónio** Beja, S Portugal *37°30′N 08°41′W*

São Tiago *see* Santiago

79 *B18* **São Tomé** ● (Sao Tome and Principe) São Tomé, S Sao Tome and Principe *0°22′N 06°41′E*

79 *B18* **São Tomé** ✈ São Tomé, S Sao Tome and Principe *0°24′N 06°39′E*

79 *B18* **São Tomé** *Eng.* Saint Thomas. *island* S Sao Tome and Principe

79 *B17* **Sao Tome and Principe** *off.* Democratic Republic of Sao Tome and Principe. ◆ *republic* E Atlantic Ocean

Sao Tome and Principe, Democratic Republic of *see* Sao Tome and Principe

74 *H9* **Saoura, Oued** ≋ NW Algeria

60 *M10* **São Vicente** *Eng.* Saint Vincent. São Paulo, S Brazil *23°55′S 46°25′W*

64 *O5* **São Vicente** Madeira, Portugal, NE Atlantic Ocean *32°48′N 17°03′W*

76 *C9* **São Vicente** *Eng.* Saint Vincent. *island* Ilhas de Barlavento, N Cape Verde

104 *F14* **São Vicente, Cabo de** *Eng.* Cape Saint Vincent, *Port.* Cabode São Vicente. *cape* S Portugal

São Vicente, Cabo de *see* São Vicente, Cabo de

Sápai *see* Sápes

Sapaleri, Cerro *see* Zapaleri, Cerro

Saparoea *see* Saparua

171 *S13* **Saparua** *prev.* Saparoea. Pulau Saparau, C Indonesia *03°35′S 128°37′E*

168 *L11* **Sapat** Sumatera, W Indonesia *0°18′S 103°18′E*

77 *U17* **Sapele** Delta, S Nigeria *05°54′N 05°43′E*

23 *X7* **Sapelo Island** *island* Georgia, SE USA

23 *X7* **Sapelo Sound** *sound* Georgia, SE USA

114 *K13* **Sápes** *var.* Sápai. Anatolikí Makedonía kai Thráki, NE Greece *41°02′N 25°44′E*

Sapiéntza *see* Sapiénza

115 *D22* **Sapiénza** *var.* Sapiéntza. *island* S Greece

138 *F12* **Sapir** *prev.* Sappir. Southern, S Israel *30°43′N 35°11′E*

61 *I15* **Sapiranga** Rio Grande do Sul, S Brazil *29°39′S 50°58′W*

114 *K13* **Sápka** ▲▲ NE Greece

105 *X9* **Sa Pobla** Mallorca, Spain, W Mediterranean Sea *39°46′N 03°03′E*

56 *D11* **Saposoa** San Martín, N Peru *06°53′S 76°45′W*

119 *F16* **Sapotskin** *Pol.* Sopockinie, *Rus.* Sapotskino, Sopotskin. Hrodzyenskaya Voblasts', W Belarus *53°50′N 23°39′E*

77 *P13* **Sapoui** *var.* Sapouy. S Burkina Faso *11°34′N 01°44′W*

Sapouy *see* Sapoui

Sappir *see* Sapir

165 *S4* **Sapporo** Hokkaidō, NE Japan *43°05′N 141°21′E*

107 *M19* **Sapri** Campania, S Italy *40°05′N 15°36′E*

169 *T16* **Sapudi, Pulau** *island* S Indonesia

27 *P9* **Sapulpa** Oklahoma, C USA *36°00′N 96°06′W*

142 *J4* **Saqqez** *var.* Saghez, Sakiz, Saqqiz. Kordestān, NW Iran *36°31′N 46°16′E*

Saqqiz *see* Saqqez

139 *U8* **Sarābādī** Wāsiţ, E Iraq *33°00′N 44°52′E*

167 *P10* **Sara Buri** *var.* Saraburi. Saraburi, C Thailand *14°32′N 100°53′E*

Saraburi *see* Sara Buri

Sarafjagān *see* Salafchegān

24 *K9* **Saragosa** Texas, SW USA *31°03′N 103°39′W*

Saragossa *see* Zaragoza

Saragt *see* Sarahs

56 *B8* **Saraguro** Loja, S Ecuador *03°42′S 79°18′W*

146 *I15* **Sarahs** *var.* Saragt, *Rus.* Serakhs. Ahal Welaýaty, S Turkmenistan *36°33′N 61°10′E*

126 *M6* **Sarai** Ryazanskaya Oblast', W Russian Federation *53°43′N 39°59′E*

Sarāi *see* Sarāy

154 *M12* **Saraipāli** Chhattīsgarh, C India *21°21′N 83°01′E*

149 *T9* **Sarāi Sidhu** Punjab, E Pakistan *30°35′N 72°02′E*

93 *M15* **Säräisniemi** Kainuu, C Finland *64°25′N 26°50′E*

113 *I14* **Sarajevo** ● (Bosnia and Herzegovina) Federacija Bosne I Hercegovine, SE Bosnia and Herzegovina *43°53′N 18°24′E*

112 *I13* **Sarajevo** ✈ Federacija Bosne I Hercegovine, C Bosnia and Herzegovina *43°49′N 18°21′E*

143 *V4* **Sarakhs** Khorāsān-e Razavī, NE Iran *36°50′N 61°00′E*

115 *H17* **Sarakíniko, Akrotírio** *headland* Évvoia, C Greece *38°46′N 23°43′E*

115 *I18* **Sarakinó** *island* Vóreies Sporádes, Greece, Aegean Sea

127 *V7* **Saraktash** Orenburgskaya Oblast', W Russian Federation *51°46′N 56°23′E*

30 *L15* **Sara, Lake** ⊞ Illinois, N USA

23 *N8* **Saraland** Alabama, S USA *30°49′N 88°04′W*

55 *V9* **Saramacca** ◈ *district* N Suriname

55 *V10* **Saramacca Rivier** ≋ C Suriname

166 *M2* **Saramati** ▲ N Myanmar (Burma) *25°46′N 95°01′E*

145 *R10* **Saran'** *Kaz.* Saran. Karaganda, C Kazakhstan *49°47′N 73°02′E*

18 *K7* **Saranac Lake** New York, NE USA *44°18′N 74°06′W*

18 *K7* **Saranac River** ≋ New York, NE USA

Saranda *see* Sarandë

113 *L23* **Sarandë** *var.* Saranda, *It.* Porto Edda; *prev.* Santi Quaranta. Vlorë, S Albania *39°53′N 20°E*

61 *H14* **Sarandi** Rio Grande do Sul, S Brazil *27°57′S 52°58′W*

61 *F19* **Sarandí del Yí** Durazno, C Uruguay *33°18′S 55°38′W*

61 *F19* **Sarandí Grande** Florida, S Uruguay *33°43′S 56°19′W*

171 *Q8* **Sarangani Islands** *island group* S Philippines

127 *P5* **Saransk** Respublika Mordoviya, W Russian Federation *54°11′N 45°10′E*

115 *C14* **Sarantáporos** ≋ N Greece

114 *H9* **Sarantsi** Sofia, W Bulgaria *42°43′N 23°46′E*

127 *T3* **Sarapul** Udmurtskaya Respublika, NW Russian Federation *56°26′N 53°52′E*

Sarâqeb *see* Sarāqib

138 *I3* **Sarāqib** *Fr.* Sarâqeb. Idlib, N Syria *35°52′N 36°48′E*

54 *J5* **Sarare** Lara, N Venezuela *09°47′N 69°10′W*

55 *O10* **Sarariña** Amazonas, S Venezuela *04°10′N 64°31′W*

143 *S10* **Sar Ashk** Kermān, C Iran

23 *V13* **Sarasota** Florida, SE USA *27°20′N 82°31′W*

117 *O11* **Sarata** Odes'ka Oblast', SW Ukraine *46°01′N 29°40′E*

116 *I10* **Sărăţel** *Hung.* Szeretfalva. Bistriţa-Năsăud, N Romania *47°02′N 24°24′E*

25 *X10* **Saratoga** Texas, SW USA *30°15′N 94°31′W*

18 *K10* **Saratoga Springs** New York, NE USA *43°04′N 73°47′W*

127 *P8* **Saratov** Saratovskaya Oblast', W Russian Federation *51°33′N 45°58′E*

127 *P8* **Saratovskaya Oblast'** ◈ *province* W Russian Federation

127 *Q7* **Saratovskoye Vodokhranilishche** ⊞ W Russian Federation

143 *X13* **Sarāvān** Sīstān va Balūchestān, SE Iran *27°11′N 62°35′E*

Saravan/Saravane *see* Salavan

169 *S9* **Sarawak** ◈ *state* East Malaysia

Sarawak *see* Kuching

139 *U6* **Sarāy** *var.* Sarāi. Diyālá, E Iraq *34°06′N 45°06′E*

136 *D10* **Saray** Tekirdağ, NW Turkey *41°27′N 27°56′E*

76 *J12* **Saraya** SE Senegal *12°50′N 11°45′W*

143 *W14* **Sarbāz** Sīstān va Balūchestān, SE Iran *26°38′N 61°13′E*

143 *U8* **Sarbīsheh** Khorāsān-e Janūbī, E Iran *32°35′N 59°50′E*

111 *J24* **Sárbogárd** Fejér, C Hungary *46°54′N 18°36′E*

Şărcad *see* Sarkad

27 *S7* **Sarcoxie** Missouri, C USA *37°04′N 94°07′W*

152 *L11* **Sārda** *Nep.* Kali. ≋ India/Nepal

152 *G10* **Sardārshahr** Rājasthān, NW India *28°30′N 74°30′E*

107 *C18* **Sardegna** *Eng.* Sardinia. ◈ *region* Italy, C Mediterranean Sea

107 *A18* **Sardegna** *Eng.* Sardinia. *island* Italy, C Mediterranean Sea

42 *K13* **Sardinal** Guanacaste, NW Costa Rica *10°30′N 85°38′W*

54 *G7* **Sardinata** Norte de Santander, N Colombia *08°07′N 72°47′W*

Sardinia *see* Sardegna

120 *K8* **Sardinia-Corsica Trough** *undersea feature* Tyrrhenian Sea, C Mediterranean Sea

22 *L2* **Sardis** Mississippi, S USA *34°25′N 89°55′W*

22 *L2* **Sardis Lake** ⊞ Mississippi, S USA

27 *P12* **Sardis Lake** ⊞ Oklahoma, C USA

92 *H12* **Sarek** ▲▲ N Sweden

92 *H11* **Sarektjåhkkå** ▲ N Sweden *67°28′N 17°56′E*

Sar-e Pol *see* Sar-e Pul

Sar-e Pol *see* Sar-e Pol-e Żahāb

142 *J6* **Sar-e Pol-e Żahāb** *var.* Sar-e Pol, Sar-i Pul. Kermānshāhān, W Iran *34°28′N 45°52′E*

149 *N3* **Sar-e Pul** *var.* Sar-i-Pul; *prev.* Sar-e Pol. Sar-e Pul, N Afghanistan *36°16′N 65°55′E*

149 *O3* **Sar-e Pul** ◈ *province* N Afghanistan

Sarera, Teluk *see* Cenderawasih, Teluk

147 *T13* **Sarez, Kŭli** *Rus.* Sarezskoye Ozero. ◎ SE Tajikistan

Sarezskoye Ozero *see* Sarez, Kŭli

64 *G10* **Sargasso Sea** *sea* W Atlantic Ocean

149 *U8* **Sargodha** Punjab, NE Pakistan *32°06′N 72°48′E*

78 *I13* **Sarh** *prev.* Fort-Archambault. Moyen-Chari, S Chad *09°08′N 18°22′E*

143 *P4* **Sārī** *var.* Sari, Sāri. Māzandarān, N Iran *36°37′N 53°05′E*

115 *N23* **Saría** *island* SE Greece

Sariasiya *see* Sariosiyo

40 *F3* **Saric** Sonora, NW Mexico *31°08′N 111°22′W*

188 *K6* **Sarigan** *island* C Northern Mariana Islands

136 *D14* **Sarīgöl** Manisa, SW Turkey *38°16′N 28°41′E*

139 *T6* **Sārihah** At Ta'mím, E Iraq *34°34′N 44°38′E*

137 *R12* **Sarıkamış** Kars, NE Turkey *40°18′N 42°36′E*

169 *R9* **Sarikei** Sarawak, East Malaysia *02°07′N 111°30′E*

147 *U12* **Sarikol Range** *Rus.* Sarykol'skiy Khrebet. ▲▲ China/Tajikistan

181 *Y7* **Sarina** Queensland, NE Australia *21°34′S 149°12′E*

Sarine *see* La Sarine

105 *S5* **Sariñena** Aragón, NE Spain *41°47′N 00°10′W*

147 *O13* **Sariosiyo** *Rus.* Sariasiya. Surkhondaryo Viloyati, S Uzbekistan *38°25′N 67°51′E*

Sar-i-Pul *see* Sar-e Pol, Afghanistan

Sar-i Pul *see* Sar-e Pol-e Żahāb, Iran

Sariqamish Kŭli *see* Sarygamyş Köli

149 *V1* **Sarī Qūl** *Rus.* Ozero Zurkul', *Taj.* Zŭrkŭl. ◎ Afghanistan/Tajikistan *see also* Zŭrkŭl

Sarī Qūl *see* Zŭrkŭl

75 *Q12* **Sarīr Tībistī** *var.* Serir Tibesti. *desert* S Libya

25 *S15* **Sarita** Texas, SW USA *27°14′N 97°48′W*

163 *W14* **Sariwŏn** SW North Korea *38°30′N 125°52′E*

114 *P12* **Sarıyer** İstanbul, NW Turkey *41°11′N 29°03′E*

97 *L26* **Sark** *Fr.* Sercq. *island* Channel Islands

111 *N24* **Sarkad** *Rom.* Şărcad. Békés, SE Hungary *46°44′N 21°25′E*

145 *W14* **Sarkand** *Kaz.* Sarqan. Almaty, SW Kazakhstan *45°24′N 79°55′E*

152 *D11* **Sarkāri Tala** Rājasthān, NW India *27°39′N 70°52′E*

136 *G15* **Şarkîkaraağaç** *var.* Şarki Karaağaç. Isparta, SW Turkey *38°04′N 31°22′E*

Şarki Karaağaç *see* Şarkîkaraağaç

136 *L13* **Şarkışla** Sivas, C Turkey *39°21′N 36°27′E*

136 *C11* **Şarköy** Tekirdağ, NW Turkey *40°37′N 27°07′E*

Sárköz *see* Livada

Sarlat *see* Sarlat-la-Canéda

102 *M13* **Sarlat-la-Canéda** *var.* Sarlat. Dordogne, SW France *44°54′N 01°12′E*

109 *S3* **Sarleinsbach** Oberösterreich, N Austria *48°33′N 13°55′E*

Šarma *see* Ash Sharmah

171 *Y12* **Sarmi** Papua, E Indonesia *01°51′S 138°45′E*

63 *I19* **Sarmiento** Chubut, S Argentina *45°38′S 69°07′W*

63 *H25* **Sarmiento, Monte** ▲ S Chile *54°28′S 70°49′W*

94 *J11* **Särna** Dalarna, C Sweden *61°40′N 13°10′E*

108 *F8* **Sarnen** Obwalden, C Switzerland *46°54′N 08°15′E*

108 *F9* **Sarner See** ◎ C Switzerland

14 *D16* **Sarnia** Ontario, S Canada *42°58′N 82°23′W*

116 *L3* **Sarny** Rivnens'ka Oblast', NW Ukraine *51°20′N 26°35′E*

171 *O13* **Saroako** Sulawesi, C Indonesia *02°31′S 121°18′E*

118 *L13* **Sarochyna** *Rus.* Sorochino. Vitsyebskaya Voblasts', N Belarus *55°12′N 28°45′E*

168 *L12* **Sarolangun** Sumatera, W Indonesia *02°17′S 102°39′E*

165 *U3* **Saroma** Hokkaidō, NE Japan *44°01′N 143°43′E*

165 *V3* **Saroma-ko** ◎ Hokkaidō, NE Japan

Saronic Gulf *see* Saronikós Kólpos

115 *H20* **Saronikós Kólpos** *Eng.* Saronic Gulf. *gulf* S Greece

106 *D7* **Saronno** Lombardia, N Italy *45°38′N 09°02′E*

136 *B11* **Saros Körfezi** *gulf* NW Turkey

111 *N20* **Sárospatak** Borsod-Abaúj-Zemplén, NE Hungary *48°18′N 21°30′E*

127 *O4* **Sarov** *prev.* Sarova. Respublika Mordoviya, SW Russian Federation *54°39′N 43°09′E*

Sarova *see* Sarov

127 *P12* **Sarpa** Respublika Kalmykiya, SW Russian Federation *47°00′N 45°42′E*

127 *P12* **Sarpa, Ozero** ◎ SW Russian Federation

Šar Planina *see* Mali i Sharrit

95 *I16* **Sarpsborg** Østfold, S Norway *59°16′N 11°07′E*

139 *U5* **Sarqalā** At Ta'mím, N Iraq

Sarqan *see* Sarkand

103 *U4* **Sarralbe** Moselle, NE France *49°02′N 07°01′E*

Sarre *see* Saar, France/Germany

Sarre *see* Saarland, Germany

103 *U5* **Sarrebourg** *Ger.* Saarburg. Moselle, NE France *48°43′N 07°03′E*

Sarrebruck *see* Saarbrücken

103 *U4* **Sarreguemines** *prev.* Saargemund. Moselle, NE France *49°06′N 07°04′E*

104 *I3* **Sarria** Galicia, NW Spain *42°47′N 07°25′W*

105 *S8* **Sarrión** Aragón, NE Spain *40°09′N 00°49′W*

42 *F4* **Sarstoon** *Sp.* Río Sarstún. ≋ Belize/Guatemala

Sarstún, Río *see* Sarstoon

123 *Q9* **Sartang** ≋ NE Russian Federation

103 *X16* **Sartène** Corse, France, C Mediterranean Sea *41°38′N 08°58′E*

102 *K7* **Sarthe** ◈ *department* NW France

102 *K7* **Sarthe** ≋ N France

115 *H15* **Sárti** Kentrikí Makedonía, N Greece *40°05′N 24°E*

Sartu *see* Daqing

165 *T1* **Sarufutsu** Hokkaidō, NE Japan *45°20′N 142°03′E*

Saruhan *see* Manisa

152 *G9* **Sarūpsar** Rājasthān, NW India *29°25′N 73°50′E*

137 *U13* **Şärur** *prev.* Il'ichevsk. SW Azerbaijan *39°30′N 44°59′E*

Sarvani *see* Marneuli

111 *G23* **Sárvár** Vas, W Hungary *47°15′N 16°59′E*

143 *P11* **Sarvestān** Fārs, S Iran *29°16′N 53°13′E*

171 *W12* **Sarwon** Papua, E Indonesia *0°58′S 136°08′E*

145 *P17* **Saryagash** *Kaz.* Saryaghash. Yuzhnyy Kazakhstan, S Kazakhstan *41°29′N 69°10′E*

Saryaghash *see* Saryagash

145 *R9* **Saryarka** *Eng.* Kazakh Uplands, Kirghiz Steppe. *uplands* C Kazakhstan

147 *W8* **Sary-Bulak** Narynskaya Oblast', C Kyrgyzstan *41°56′N 75°44′E*

147 *U10* **Sary-Bulak** Oshskaya Oblast', SW Kyrgyzstan *40°49′N 73°44′E*

117 *S14* **Sarych, Mys** *headland* S Ukraine *44°23′N 33°44′E*

147 *Z7* **Sary-Dzhaz** *var.* Aksu He. ≋ China/Kyrgyzstan *see also* Aksu He

Sary-Dzhaz *see* Aksu He

146 *F8* **Sarygamyş Köli** *Rus.* Sarykamyshkoye Ozero, *Uzb.* Sariqamish Kŭli. *salt lake* Kazakhstan/Uzbekistan

144 *G13* **Sarykamys** *Kaz.* Saryqamys. Mangistau, SW Kazakhstan *45°58′N 53°30′E*

Sarykamyshkoye Ozero *see* Sarygamyş Köli

145 *N7* **Sarykol'** *prev.* Uritskiy. Kustanay, N Kazakhstan *53°19′N 65°34′E*

Sarykol'skiy Khrebet *see* Sarikol Range

144 *M10* **Sarykopa, Ozero** ◎ C Kazakhstan

145 *V15* **Saryozek** *Kaz.* Saryözek. Almaty, SE Kazakhstan *44°22′N 77°57′E*

144 *E10* **Saryozen** *Kaz.* Kishiözen; *prev.* Malyy Uzen'. ≋ Kazakhstan/Russian Federation

Saryqamys *see* Sarykamys

145 *S13* **Saryshagan** *Kaz.* Saryshaghan. Karaganda, SE Kazakhstan *46°05′N 73°38′E*

Saryshaghan *see* Saryshagan

145 *O13* **Sarysu** ≋ S Kazakhstan

147 *T11* **Sary-Tash** Oshskaya Oblast', SW Kyrgyzstan *39°44′N 73°14′E*

145 *T12* **Saryterek** Karaganda, C Kazakhstan *47°46′N 74°06′E*

Saryyazynskoye Vodokhranilishche *see* Saryýazy Suw Howdany

146 *J15* **Saryýazy Suw Howdany** *Rus.* Saryyazynskoye Vodokhranilishche. ⊞ S Turkmenistan

145 *T14* **Saryyesik-Atyrau, Peski** *desert* E Kazakhstan

106 *E10* **Sarzana** Liguria, NW Italy *44°07′N 09°59′E*

188 *B17* **Sasalaguan, Mount** ▲ S Guam

153 *O14* **Sasarām** Bihār, N India *24°58′N 84°01′E*

186 *M8* **Sasari, Mount** ▲ Santa Isabel, N Solomon Islands *08°09′S 159°32′E*

164 *C13* **Sasebo** Nagasaki, Kyūshū, SW Japan *33°10′N 129°42′E*

14 *I9* **Saseginaga, Lac** ◎ Québec, SE Canada

Saseno *see* Sazan

11 *R13* **Saskatchewan** ◈ *province* SW Canada

11 *U14* **Saskatchewan** ≋ Manitoba/Saskatchewan, C Canada

11 *T15* **Saskatoon** Saskatchewan, S Canada *52°10′N 106°40′W*

11 *T15* **Saskatoon** ✈ Saskatchewan, S Canada *52°15′N 107°05′W*

123 *N7* **Saskylakh** Respublika Sakha (Yakutiya), NE Russian Federation *71°56′N 114°07′E*

42 *L7* **Saslaya, Cerro** ▲ N Nicaragua *13°52′N 85°06′W*

38 *G17* **Sasmik, Cape** *headland* Tanaga Island, Alaska, USA *51°36′N 177°55′W*

◆ Country
● Country Capital
◇ Dependent Territory
○ Dependent Territory Capital
◈ Administrative Regions
✈ International Airport
▲ Mountain
▲▲ Mountain Range
🌋 Volcano
≋ River
◎ Lake
⊞ Reservoir

119 N19 **Sasnovy Bor** *Rus.* Sosnovyy Bor. Homyel'skaya Voblasts', SE Belarus 52°32′N 29°35′E
127 N5 **Sasovo** Ryazanskaya Oblast', W Russian Federation 54°19′N 41°54′E
25 S12 **Saspamco** Texas, SW USA 29°13′N 98°18′W
109 W9 **Sass** *var.* Sassbach. ≈ SE Austria
76 M17 **Sassandra** S Ivory Coast 04°58′N 06°08′W
76 M17 **Sassandra** *var.* Ibo, Sassandra Fleuve. ≈ S Ivory Coast
Sassandra Fleuve *see* Sassandra
107 B17 **Sassari** Sardegna, Italy, C Mediterranean Sea 40°44′N 08°33′E
Sassbach *see* Sass
98 H11 **Sassenheim** Zuid-Holland, W Netherlands 52°14′N 04°31′E
Sassmacken *see* Valdemārpils
100 O7 **Sassnitz** Mecklenburg-Vorpommern, NE Germany 54°32′N 13°39′E
99 E16 **Sas van Gent** Zeeland, SW Netherlands 51°13′N 03°48′E
145 W12 **Sasykkol', Ozero** ◎ E Kazakhstan
117 O12 **Sasyk, Ozero** *Rus.* Ozero Sasyk Kunduk, *var.* Ozero Kunduk. ◎ SW Ukraine
76 J12 **Satadougou** Kayes, SW Mali 12°40′N 11°25′W
93 K18 **Satakunta** ◈ *region* W Finland
164 C17 **Sata-misaki** Kyūshū, SW Japan
26 I7 **Satanta** Kansas, C USA 37°23′N 102°00′W
155 E15 **Sātāra** Mahārāshtra, W India 17°41′N 73°59′E
192 G15 **Sātaua** Savai'i, NW Samoa 13°26′S 172°40′W
188 M16 **Satawal** *island* Caroline Islands, C Micronesia
189 R17 **Satawan Atoll** *atoll* Mortlock Islands, C Micronesia
Sätbaev *see* Satpayev
23 Y12 **Satellite Beach** Florida, SE USA 28°10′N 80°35′W
95 M14 **Säter** Dalarna, C Sweden 60°21′N 15°45′E
Sathmar *see* Satu Mare
23 V7 **Satilla River** ≈ Georgia, SE USA
57 F14 **Satipo** *var.* San Francisco de Satipo. Junín, C Peru 11°19′S 74°37′W
122 F11 **Satka** Chelyabinskaya Oblast', C Russian Federation 55°08′N 58°54′E
153 T16 **Satkhira** Khulna, SW Bangladesh 22°43′N 89°06′E
146 J13 **Şatlyk** *Rus.* Shatlyk. Mary Welaýaty, C Turkmenistan 37°55′N 61°00′E
154 K9 **Satna** *prev.* Sutna. Madhya Pradesh, C India 24°33′N 80°50′E
103 R11 **Satolas** ✈ (Lyon) Rhône, E France 45°44′N 05°01′E
111 N20 **Sátoraljaújhely** Borsod-Abaúj-Zemplén, NE Hungary 48°24′N 21°39′E
145 O12 **Satpayev** *Kaz.* Sätbaev; *prev.* Nikol'skiy. Karaganda, C Kazakhstan 47°59′N 67°27′E
154 G11 **Sātpura Range** ▲▲ C India
165 Q10 **Satsuma-Sendai** Miyagi, Honshū, C Japan 38°16′N 140°52′E
Satsuma-Sendai *see* Sendai
167 P12 **Sattahip** *var.* Ban Sattahip, Ban Sattahipp. Chon Buri, S Thailand 12°36′N 100°56′E
92 L11 **Sattanen** Lappi, NE Finland 67°31′N 26°35′E
Satul *see* Satun
116 H9 **Satulung** *Hung.* Kővárhosszúfalu. Maramureş, N Romania 47°34′N 23°26′E
Satul-Vechi *see* Staro Selo
116 G8 **Satu Mare** *Ger.* Sathmar, *Hung.* Szatmárrnémeti. Satu Mare, NW Romania 47°46′N 22°55′E
116 G8 **Satu Mare** ◈ *county* NW Romania
167 N16 **Satun** *var.* Satul, Setul. Satun, SW Thailand 06°40′N 100°01′E
192 G16 **Satupa'iteau** Savai'i, W Samoa 13°46′S 172°26′W
Sau *see* Sava
14 F14 **Sauble** ≈ Ontario, S Canada
14 F13 **Sauble Beach** Ontario, S Canada 44°36′N 81°15′W
61 C16 **Sauce** Corrientes, NE Argentina 30°05′S 58°46′W
Sauce *see* Juan L. Lacaze
36 K15 **Sauceda Mountains** ▲▲ Arizona, SW USA
61 C17 **Sauce de Luna** Entre Ríos, E Argentina 31°15′S 59°09′W
63 L15 **Sauce Grande, Río** ≈ E Argentina
40 K6 **Saucillo** Chihuahua, N Mexico 28°01′N 105°17′W
95 D15 **Sauda** Rogaland, S Norway 59°38′N 06°23′E
145 Q16 **Saudakent** *Kaz.* Saūdakent; *prev.* Baykadam, *Kaz.* Bayqadam. Zhambyl, S Kazakhstan 43°49′N 69°56′E
92 J2 **Sauðárkrókur** Norðurland Vestra, N Iceland 65°45′N 19°39′W
141 P9 **Saudi Arabia** *off.* Kingdom of Saudi Arabia, Al 'Arabīyah as Su'ūdīyah, *Ar.* Al Mamlakah al 'Arabīyah as Su'ūdīyah. ◆ *monarchy* SW Asia
Saudi Arabia, Kingdom of *see* Saudi Arabia
101 D19 **Sauer** *var.* Sûre. ≈ NW Europe *see also* Sûre
Sauer *see* Sûre
101 F15 **Sauerland** *forest* W Germany
14 F14 **Saugeen** ≈ Ontario, S Canada
18 K12 **Saugerties** New York, NE USA 42°04′N 73°55′W
Saugor *see* Sāgar
10 K15 **Saugstad, Mount** ▲ British Columbia, SW Canada 52°12′N 126°35′W
Sāūjbulāgh *see* Mahābād
102 J11 **Saujon** Charente-Maritime, W France 45°40′N 00°54′W
29 T7 **Sauk Centre** Minnesota, N USA 45°44′N 94°57′W
30 L8 **Sauk City** Wisconsin, N USA 43°16′N 89°43′W
29 U7 **Sauk Rapids** Minnesota, N USA 45°35′N 94°09′W
55 Y11 **Saül** C French Guiana 03°37′N 53°12′W
103 O7 **Sauldre** ≈ C France
101 I23 **Saulgau** Baden-Württemberg, SW Germany 48°03′N 09°28′E
103 Q8 **Saulieu** Côte d'Or, C France 47°15′N 04°15′E
118 G8 **Saulkrasti** C Latvia 57°14′N 24°25′E
15 S6 **Sault-aux-Cochons, Rivière du** ≈ Québec, SE Canada
31 Q4 **Sault Sainte Marie** Michigan, N USA 46°29′N 84°22′W
12 F14 **Sault Ste. Marie** Ontario, S Canada 46°30′N 84°17′W
145 P7 **Saumalkol'** *prev.* Volodarskoye. Severnyy Kazakhstan, N Kazakhstan 53°19′N 68°05′E
190 E13 **Sauma, Pointe** *headland* Île Alofi, W Wallis and Futuna 14°21′S 177°58′W
171 T16 **Saumlaki** *var.* Saumlakki. Pulau Yamdena, E Indonesia 07°53′S 131°18′E
Saumlakki *see* Saumlaki
15 R12 **Saumon, Rivière au** ≈ Québec, SE Canada
102 K8 **Saumur** Maine-et-Loire, NW France 47°16′N 00°04′W
185 F23 **Saunders, Cape** *headland* South Island, New Zealand 45°53′S 170°40′E
195 N13 **Saunders Coast** *physical region* Antarctica
65 B23 **Saunders Island** *island* NW Falkland Islands
65 C24 **Saunders Island Settlement** Saunders Island, NW Falkland Islands 51°22′S 60°05′W
82 F11 **Saurimo** *Port.* Henrique de Carvalho, Vila Henrique de Carvalho. Lunda Sul, NE Angola 09°39′S 20°24′E
55 S11 **Sauriwaunawa** S Guyana 03°10′N 59°51′W
82 D12 **Sautar** Malanje, NW Angola 11°10′S 18°26′E
45 S13 **Sauteurs** N Grenada 12°14′N 61°38′W
102 K13 **Sauveterre-de-Guyenne** Gironde, SW France 44°43′N 00°02′W
119 O14 **Sava** Mahilyowskaya Voblasts', E Belarus 54°22′N 30°49′E
42 J5 **Savá** Colón, N Honduras 15°30′N 86°16′W
84 H11 **Sava** *Eng.* Save, *Ger.* Sau, *Hung.* Száva. ≈ SE Europe
33 Y8 **Savage** Montana, NW USA 47°28′N 104°17′W
183 N16 **Savage River** Tasmania, SE Australia 41°34′S 145°15′E
77 R15 **Savalou** S Benin 07°59′N 01°58′E
30 K10 **Savanna** Illinois, N USA 42°05′N 90°09′W
23 X6 **Savannah** Georgia, SE USA 32°02′N 81°01′W
27 R2 **Savannah** Missouri, C USA 39°57′N 94°49′W
20 H10 **Savannah** Tennessee, S USA 35°12′N 88°15′W
21 O12 **Savannah River** ≈ Georgia/South Carolina, SE USA
167 S9 **Savannakhét** *var.* Khanthabouli. Savannakhét, S Laos 16°38′N 104°49′E
44 H12 **Savanna-La-Mar** W Jamaica 18°13′N 78°08′W
12 B10 **Savant Lake** ◎ Ontario, S Canada
155 F17 **Savanūr** Karnātaka, W India 14°58′N 75°19′E
93 J16 **Sävar** Västerbotten, N Sweden 63°52′N 20°33′E
Savaria *see* Szombathely
154 C11 **Sāvarkundla** *var.* Kundla. Gujarāt, W India 21°21′N 71°20′E
116 F11 **Săvârşin** *Hung.* Soborsin; *prev.* Săvirşin. Arad, W Romania 46°00′N 22°15′E
136 C13 **Savaştepe** Balıkesir, W Turkey 39°20′N 27°38′E
147 P11 **Savat** *Rus.* Savat. Sirdaryo Viloyati, E Uzbekistan 40°03′N 68°35′E
Savat *see* Savat
Sávdijári *see* Skaulo
77 R15 **Savè** SE Benin 08°04′N 02°29′E
83 N18 **Save** Inhambane, E Mozambique 21°07′S 34°35′E
102 L16 **Save** ≈ S France
83 L17 **Save** *var.* Sabi. ≈ Mozambique/Zimbabwe *see also* Sabi
Save *see* Sava
Save *see* Sabi
142 M6 **Sāveh** Markazī, W Iran 35°00′N 50°22′E
116 L8 **Săveni** Botoşani, NE Romania 47°57′N 26°52′E
103 N16 **Saverdun** Ariège, S France 43°15′N 01°34′E
103 U5 **Saverne** *var.* Zabern; *anc.* Tres Tabernae. Bas-Rhin, NE France 48°45′N 07°22′E
106 B9 **Savigliano** Piemonte, NW Italy 44°39′N 07°39′E
Savigsivik *see* Savissivik
109 U10 **Savinja** ≈ N Slovenia
106 H11 **Savio** ≈ C Italy
Săvirşin *see* Săvârşin
197 O11 **Savissivik** *var.* Savigsivik. ◈ Qaasuitsup, N Greenland
93 N18 **Savitaipale** Etelä-Karjala, SE Finland 61°12′N 27°43′E
113 J15 **Šavnik** C Montenegro 42°57′N 19°04′E
108 I9 **Savognin** Graubünden, S Switzerland 46°34′N 09°35′E
103 T12 **Savoie** ◈ *department* E France
106 C10 **Savona** Liguria, NW Italy 44°18′N 08°29′E
93 N17 **Savonlinna** *Swe.* Nyslott. Etelä-Savo, E Finland 61°51′N 28°56′E
93 N17 **Savonranta** Etelä-Savo, E Finland 62°10′N 29°10′E
38 K10 **Savoonga** Saint Lawrence Island, Alaska, USA 63°40′N 170°29′W
30 M13 **Savoy** Illinois, N USA 40°03′N 88°15′W
117 O8 **Savran'** Odes'ka Oblast', SW Ukraine 48°07′N 30°00′E
137 R11 **Şavşat** Artvin, NE Turkey 41°15′N 42°30′E
95 L19 **Sävsjö** Jönköping, S Sweden 57°25′N 14°40′E
Savu, Kepulauan *see* Sawu, Kepulauan
92 M11 **Savukoski** Lappi, NE Finland 67°17′N 28°14′E
Savu, Pulau *see* Sawu, Pulau
187 Y14 **Savusavu** Vanua Levu, N Fiji 16°48′S 179°20′E
171 O17 **Savu Sea** *Ind.* Laut Sawu. *sea* S Indonesia
83 H17 **Savute** North-West, N Botswana 18°33′S 24°06′E
139 N7 **Şawāb Uqlat** *well* W Iraq
138 M7 **Sawāb, Wādī as** *dry watercourse* W Iraq
152 H13 **Sawāi Mādhopur** Rājasthān, N India 26°00′N 76°22′E
Sawakin *see* Suakin
167 R8 **Sawang Daen Din** Sakon Nakhon, E Thailand 17°28′N 103°27′E
167 O8 **Sawankhalok** *var.* Swankalok. Sukhothai, NW Thailand 17°19′N 99°50′E
165 P13 **Sawara** Chiba, Honshū, S Japan 35°52′N 140°31′E
37 R5 **Sawatch Range** ▲▲ Colorado, C USA
141 N12 **Sawdā', Jabal** ▲ SW Saudi Arabia 18°15′N 42°26′E
75 P9 **Sawdā', Jabal as** ▲▲ C Libya
Sawdirī *see* Sodiri
97 F14 **Sawel Mountain** ▲ C Northern Ireland, United Kingdom 54°49′N 07°04′W
75 X10 **Sawhāj** *var.* Sawhāj, *var.* Sohâg, Suliag. C Egypt 26°28′N 31°44′E
Sawhāj *see* Sawhāj
77 O15 **Sawla** N Ghana 09°14′N 02°26′W
141 X12 **Şawqirah** *var.* Suqrah. S Oman 18°16′N 56°34′E
141 X12 **Şawqirah, Dawḩat** *var.* Ghubbat Sawqirah, Sukra Bay, Suqrah Bay. *bay* S Oman
Sawqirah, Ghubbat *see* Şawqirah, Dawḩat
183 V5 **Sawtell** New South Wales, SE Australia 30°22′S 153°04′E
138 K7 **Şawt, Wādī aş** *dry watercourse* S Syria
171 O17 **Sawu, Kepulauan** *var.* Kepulauan Savu. *island group* S Indonesia
Sawu, Laut *see* Savu Sea
171 O17 **Sawu, Pulau** *var.* Pulau Sawu. *island* Kepulauan Sawu, S Indonesia
105 S12 **Sax** Valenciana, E Spain 38°33′N 00°49′W
Saxe *see* Sachsen
108 C11 **Saxon** Valais, SW Switzerland 46°07′N 07°09′E
Saxony *see* Sachsen
Saxony-Anhalt *see* Sachsen-Anhalt
77 R12 **Say** Niamey, SW Niger 13°08′N 02°20′E
15 V7 **Sayabec** Québec, SE Canada 48°33′N 67°42′W
Sayaboury *see* Xaignabouli
145 U12 **Sayak** *Kaz.* Sayaq. Karaganda, E Kazakhstan 46°54′N 77°17′E
57 D14 **Sayán** Lima, W Peru 11°10′S 77°08′W
129 T6 **Sayanskiy Khrebet** ▲▲ S Russian Federation
Sayaq *see* Sayak
146 K13 **Saýat** *Rus.* Sayat. Lebap Welaýaty, E Turkmenistan 38°44′N 63°51′E
42 D3 **Sayaxché** Petén, N Guatemala 16°34′N 90°14′W
Şaydā/Sayida *see* Saïda
162 J7 **Sayhan** *var.* Hüremt. Bulgan, C Mongolia 48°40′N 102°33′E
163 N10 **Sayhandulaan** *var.* Öldziyt. Dornogovĭ, SE Mongolia 44°42′N 109°10′E
162 K9 **Sayhan-Ovoo** *var.* Ongi. Dundgovĭ, C Mongolia 45°27′N 103°58′E
141 T15 **Sayḩūt** E Yemen 15°18′N 51°16′E
29 U14 **Saylorville Lake** ⊞ Iowa, C USA
Saymenskiy Kanal *see* Saimaa Canal
163 N10 **Saynshand** Dornogovĭ, SE Mongolia 44°51′N 110°07′E
Saynshand *see* Sevrey
Sayn-Ust *see* Hohmorĭt
Say-Ötesh *see* Otes
138 J7 **Şayqal, Baḩr** ◎ S Syria
Sayrab *see* Sayrob
158 H4 **Sayram Hu** ◎ NW China
26 K11 **Sayre** Oklahoma, C USA 35°18′N 99°38′W
18 H12 **Sayre** Pennsylvania, NE USA 41°57′N 76°30′W
18 K15 **Sayreville** New Jersey, NE USA 40°27′N 74°19′W
147 N13 **Sayrob** *Rus.* Sayrab. Surkhondaryo Viloyati, S Uzbekistan 38°03′N 66°54′E
40 L13 **Sayula** Jalisco, SW Mexico 19°52′N 103°36′W
141 R14 **Say'ūn** *var.* Saywūn. C Yemen 15°53′N 48°32′E
Say-Utës *see* Otes
10 K16 **Sayward** Vancouver Island, British Columbia, SW Canada 50°20′N 126°01′W
Saywūn *see* Say'ūn
Sayyāl *see* As Sayyāl
139 U8 **Sayyid 'Abīd** *var.* Saiyid Abid. Wāsiţ, E Iraq 32°51′N 45°07′E
113 J22 **Sazan** *var.* Ishulli i Sazanit, *It.* Saseno. *island* SW Albania
Sazanit, Ishulli i *see* Sazan
Sazau/Sazawa *see* Sázava
111 E17 **Sázava** *var.* Sazau, *Ger.* Sazawa. ≈ C Czech Republic
114 K10 **Sazliyka** ≈ C Bulgaria
124 J14 **Sazonovo** Vologodskaya Oblast', NW Russian Federation 59°04′N 35°10′E
102 G6 **Scaër** Finistère, NW France 48°00′N 03°40′W
97 J15 **Scafell Pike** ▲ NW England, United Kingdom 54°26′N 03°10′W
Scalabis *see* Santarém
96 M2 **Scalloway** N Scotland, United Kingdom 60°10′N 01°17′W
38 M11 **Scammon Bay** Alaska, USA 61°50′N 165°34′W
Scammon Lagoon/Scammon, Laguna *see* Ojo de Liebre, Laguna
84 F7 **Scandinavia** *geophysical region* NW Europe
Scania *see* Skåne
96 K5 **Scapa Flow** *sea basin* N Scotland, United Kingdom
107 K26 **Scaramia, Capo** *headland* Sicilia, Italy, C Mediterranean Sea 36°46′N 14°29′E
14 H15 **Scarborough** Ontario, SE Canada 43°46′N 79°14′W
45 Z16 **Scarborough** *prev.* Port Louis. Tobago, Trinidad and Tobago 11°11′N 60°45′W
97 N16 **Scarborough** N England, United Kingdom 54°17′N 00°24′W
185 I17 **Scargill** Canterbury, South Island, New Zealand 42°57′S 172°57′E
96 E7 **Scarp** *island* NW Scotland, United Kingdom
Scarpanto *see* Kárpathos
Scarpanto Strait *see* Karpathou, Stenó
107 G25 **Scauri** Sicilia, Italy, C Mediterranean Sea 36°45′N 12°06′E
Scealg, Bá na *see* Ballinskelligs Bay
Scebeli *see* Shebeli
100 K10 **Schaale** ≈ N Germany
100 K9 **Schaalsee** ◎ N Germany
99 G18 **Schaerbeek** Brussels, C Belgium 50°52′N 04°21′E
108 G6 **Schaffhausen** *Fr.* Schaffhouse. Schaffhausen, N Switzerland 47°42′N 08°38′E
108 G6 **Schaffhausen** *Fr.* Schaffhouse. ◈ *canton* N Switzerland
Schaffhouse *see* Schaffhausen
98 I8 **Schagen** Noord-Holland, NW Netherlands 52°47′N 04°47′E
Schaken *see* Šakiai
98 M10 **Schalkhaar** Overijssel, E Netherlands 52°16′N 06°10′E
109 R3 **Schärding** Oberösterreich, N Austria 48°27′N 13°26′E
100 G9 **Scharhörn** *island* NW Germany
Schässburg *see* Sighişoara
Schaulen *see* Šiauliai
30 M10 **Schaumburg** Illinois, N USA 42°01′N 88°04′W
Schebschi Mountains *see* Shebshi Mountains
98 P6 **Scheemda** Groningen, NE Netherlands 53°10′N 06°58′E
100 I10 **Scheessel** Niedersachsen, NW Germany 53°11′N 09°33′E
13 N8 **Schefferville** Québec, E Canada 54°50′N 67°W
Schelde *see* Scheldt
99 D18 **Scheldt** *Dut.* Schelde, *Fr.* Escaut. ≈ W Europe
35 X5 **Schell Creek Range** ▲▲ Nevada, W USA
18 K10 **Schenectady** New York, NE USA 42°48′N 73°57′W
99 I17 **Scherpenheuvel** *Fr.* Montaigu. Vlaams Brabant, C Belgium 51°00′N 04°57′E
98 K11 **Scherpenzeel** Gelderland, C Netherlands 52°07′N 05°30′E
25 S12 **Schertz** Texas, SW USA 29°33′N 98°16′W
98 G11 **Scheveningen** Zuid-Holland, W Netherlands 52°07′N 04°18′E
98 G12 **Schiedam** Zuid-Holland, SW Netherlands 51°55′N 04°25′E
99 M24 **Schieren** Diekirch, NE Luxembourg 49°50′N 06°06′E
98 M4 **Schiermonnikoog** *Fris.* Skiermûntseach. Fryslân, N Netherlands 53°28′N 06°09′E
98 M4 **Schiermonnikoog** *Fris.* Skiermûntseach. *island* Waddeneilanden, N Netherlands
99 K14 **Schijndel** Noord-Brabant, S Netherlands 51°37′N 05°27′E
Schil *see* Jiu
99 H16 **Schilde** Antwerpen, N Belgium 51°14′N 04°35′E
Schillen *see* Zhilino
103 V5 **Schiltigheim** Bas-Rhin, NE France 48°38′N 07°47′E
106 G7 **Schio** Veneto, NE Italy 45°42′N 11°21′E
98 H10 **Schiphol** ✈ (Amsterdam) Noord-Holland, C Netherlands 52°18′N 04°48′E
Schippenbeil *see* Sępopol
Schiria *see* Şiria
Schivelbein *see* Świdwin
115 D22 **Schíza** *island* S Greece
175 U3 **Schjetman Reef** *reef* Antarctica
Schlackenwerth *see* Ostrov
109 R7 **Schladming** Steiermark, SE Austria 47°24′N 13°42′E
Schlan *see* Slaný
Schlanders *see* Silandro
100 I7 **Schlei** *inlet* N Germany
101 D17 **Schleiden** Nordrhein-Westfalen, W Germany 50°31′N 06°30′E
Schlelau *see* Szydłowiec
100 I7 **Schleswig** Schleswig-Holstein, N Germany 54°31′N 09°34′E
29 T13 **Schleswig** Iowa, C USA 42°10′N 95°27′W
100 H8 **Schleswig-Holstein** ◈ *state* N Germany
Schlettstadt *see* Sélestat
108 F7 **Schlieren** Zürich, N Switzerland 47°23′N 08°27′E
Schlochau *see* Człuchów
Schloppe *see* Człopa
101 I18 **Schlüchtern** Hessen, C Germany 50°19′N 09°27′E
101 J17 **Schmalkalden** Thüringen, C Germany 50°42′N 10°26′E
109 W2 **Schmida** ≈ NE Austria
65 P19 **Schmidt-Ott Seamount** *var.* Schmitt-Ott Seamount, Schmitt-Ott Tablemount. *undersea feature* SW Indian Ocean 39°37′S 13°00′E
Schmiegel *see* Śmigiel
Schmitt-Ott Seamount/Schmitt-Ott Tablemount *see* Schmidt-Ott Seamount
15 V3 **Schmon** ≈ Québec, SE Canada
101 M18 **Schneeberg** ▲ W Germany 50°03′N 11°51′E
Schneeberg *see* Veliki Snežnik
Schnee-Eifel *see* Schneifel
Schneekoppe *see* Sněžka
Schneidemühl *see* Piła
101 D18 **Schneifel** *var.* Schnee-Eifel. *plateau* W Germany
Schnelle Körös/Schnelle Kreisch *see* Crişul Repede
100 I11 **Schneverdingen** *var.* Schneverdingen (Wümme). Niedersachsen, NW Germany 53°07′N 09°48′E
Schneverdingen (Wümme) *see* Schneverdingen
Schoden *see* Skuodas
45 Q12 **Schœlcher** W Martinique 14°37′N 61°08′W
18 K10 **Schoharie** New York, NE USA 42°40′N 74°20′W
18 K11 **Schoharie Creek** ≈ New York, NE USA
115 J21 **Schoinoússa** *island* Kykládes, Greece, Aegean Sea
100 L13 **Schönebeck** Sachsen-Anhalt, C Germany 52°01′N 11°45′E
Schöneck *see* Skarszewy
100 O12 **Schönefeld** ✈ (Berlin) Berlin, NE Germany 52°23′N 13°29′E
101 K24 **Schongau** Bayern, S Germany 47°49′N 10°54′E
100 K13 **Schöningen** Niedersachsen, C Germany 52°07′N 10°58′E
Schönlanke *see* Trzcianka
Schönsee *see* Kowalewo Pomorskie
31 P10 **Schoolcraft** Michigan, N USA 42°05′N 85°39′W
98 O8 **Schoonebeek** Drenthe, NE Netherlands 52°39′N 06°57′E
98 I12 **Schoonhoven** Zuid-Holland, C Netherlands 51°57′N 04°51′E
98 H8 **Schoorl** Noord-Holland, NW Netherlands 52°42′N 04°40′E
Schooten *see* Schoten
101 F24 **Schopfheim** Baden-Württemberg, SW Germany 47°39′N 07°49′E
101 I21 **Schorndorf** Baden-Württemberg, S Germany 48°48′N 09°31′E
100 F10 **Schortens** Niedersachsen, NW Germany 53°31′N 07°57′E
99 H16 **Schoten** *var.* Schooten. Antwerpen, N Belgium 51°15′N 04°30′E
183 Q17 **Schouten Island** *island* Tasmania, SE Australia
186 C5 **Schouten Islands** *island group* NW Papua New Guinea
98 E13 **Schouwen** *island* SW Netherlands
Schreiberhau *see* Szklarska Poręba
109 U2 **Schrems** Niederösterreich, E Austria 48°48′N 15°05′E
101 L22 **Schrobenhausen** Bayern, SE Germany 48°33′N 11°14′E
18 L8 **Schroon Lake** ◎ New York, NE USA
108 J8 **Schruns** Vorarlberg, W Austria 47°04′N 09°54′E
Schubin *see* Szubin
25 U11 **Schulenburg** Texas, SW USA 29°40′N 96°54′W
Schuls *see* Scuol
108 E8 **Schüpfheim** Luzern, C Switzerland 47°02′N 07°23′E
35 S6 **Schurz** Nevada, W USA 38°55′N 118°48′W
101 I24 **Schussen** ≈ S Germany
Schüttenhofen *see* Sušice
29 R15 **Schuyler** Nebraska, C USA 41°25′N 97°04′W
18 L10 **Schuylerville** New York, NE USA 43°05′N 73°34′W
101 K20 **Schwabach** Bayern, SE Germany 49°20′N 11°02′E
Schwabenalb *see* Schwäbische Alb
101 I23 **Schwäbische Alb** *var.* Schwabenalb, *Eng.* Swabian Jura. ▲▲ S Germany
101 I22 **Schwäbisch Gmünd** *var.* Gmünd. Baden-Württemberg, SW Germany 48°49′N 09°48′E
101 I21 **Schwäbisch Hall** *var.* Hall. Baden-Württemberg, SW Germany 49°07′N 09°45′E
101 H16 **Schwalm** ≈ C Germany
109 V9 **Schwanberg** Steiermark, SE Austria 46°46′N 15°12′E
108 H8 **Schwanden** Glarus, E Switzerland 47°02′N 09°04′E
101 M20 **Schwandorf** Bayern, SE Germany 49°20′N 12°07′E
109 S5 **Schwanenstadt** Oberösterreich, NW Austria 48°03′N 13°47′E
169 S11 **Schwaner, Pegunungan** ▲▲ Borneo, N Indonesia
109 W5 **Schwarza** ≈ E Austria
109 P9 **Schwarzach** ≈ S Austria
101 M20 **Schwarzach** *Cz.* Černice. ≈ Czech Republic/Germany
Schwarzach *see* Schwarzach im Pongau
109 Q7 **Schwarzach im Pongau** *var.* Schwarzach. Salzburg, NW Austria 47°19′N 13°09′E
Schwarzawa *see* Svratka
101 N14 **Schwarze Elster** ≈ E Germany
Schwarze Körös *see* Crişul Negru
108 D9 **Schwarzenburg** Bern, W Switzerland 46°51′N 07°28′E
83 D21 **Schwarzrand** ▲▲ S Namibia
101 G23 **Schwarzwald** *Eng.* Black Forest. ▲▲ SW Germany
Schwarzwasser *see* Wda
39 P7 **Schwatka Mountains** ▲▲ Alaska, USA
109 N7 **Schwaz** Tirol, W Austria 47°21′N 11°44′E
109 Y4 **Schwechat** Niederösterreich, NE Austria 48°09′N 16°29′E
109 Y4 **Schwechat** ✈ (Wien) Wien, E Austria 48°04′N 16°31′E
100 P11 **Schwedt** Brandenburg, NE Germany 53°04′N 14°16′E
101 D19 **Schweich** Rheinland-Pfalz, SW Germany 49°49′N 06°44′E
Schweidnitz *see* Świdnica
101 J18 **Schweinfurt** Bayern, SE Germany 50°03′N 10°13′E
Schweiz *see* Switzerland
100 L9 **Schwerin** Mecklenburg-Vorpommern, N Germany 53°38′N 11°25′E
Schwerin *see* Skwierzyna
100 L9 **Schweriner See** ◎ N Germany
Schwertberg *see* Świecie
101 F15 **Schwerte** Nordrhein-Westfalen, W Germany 51°27′N 07°34′E
Schwiebus *see* Świebodzin
100 P13 **Schwielochsee** ◎ NE Germany
Schwihau *see* Švihov
Schwiz *see* Schwyz
108 G8 **Schwyz** *var.* Schwiz. Schwyz, C Switzerland 47°02′N 08°39′E
108 G8 **Schwyz** *var.* Schwiz. ◈ *canton* C Switzerland
14 J11 **Schyan** ≈ Québec, SE Canada
Schyl *see* Jiu
107 I24 **Sciacca** Sicilia, Italy, C Mediterranean Sea 37°31′N 13°05′E
Sciasciamana *see* Shashemenē
107 L26 **Scicli** Sicilia, Italy, C Mediterranean Sea 36°48′N 14°43′E
97 F25 **Scilly, Isles of** *island group* SW England, United Kingdom
111 H17 **Ščinawa** *Ger.* Steinau an der Elbe. Dolnośląskie, SW Poland 51°22′N 16°27′E
Scio *see* Chíos
31 S14 **Scioto River** ≈ Ohio, N USA
36 L5 **Scipio** Utah, W USA 39°15′N 112°06′W
33 X6 **Scobey** Montana, NW USA 48°47′N 105°25′W
183 T7 **Scone** New South Wales, SE Australia 32°02′S 150°51′E
Scoresby Sound/Scoresbysund *see* Ittoqqortoormiit
Scoresby Sund *see* Kangertittivaq
Scorno, Punta dello *see* Caprara, Punta
34 K3 **Scotia** California, W USA 40°34′N 124°07′W
47 Y14 **Scotia Plate** *tectonic feature*
47 V15 **Scotia Ridge** *undersea feature* S Atlantic Ocean
194 H2 **Scotia Sea** *sea* SW Atlantic Ocean
29 Q12 **Scotland** South Dakota, N USA 43°09′N 97°43′W
25 R5 **Scotland** Texas, SW USA 33°37′N 98°27′W
96 H11 **Scotland** ◈ *national region* Scotland, U K
21 W8 **Scotland Neck** North Carolina, SE USA 36°07′N 77°25′W
195 R13 **Scott Base** *NZ research station* Antarctica 77°52′S 167°18′E
10 J16 **Scott, Cape** *headland* Vancouver Island, British Columbia, SW Canada 50°43′N 128°24′W
26 I5 **Scott City** Kansas, C USA 38°28′N 100°55′W
27 Y7 **Scott City** Missouri, C USA 37°13′N 89°31′W
195 R14 **Scott Coast** *physical region* Antarctica
18 C15 **Scottdale** Pennsylvania, NE USA 40°05′N 79°35′W
195 Y11 **Scott Glacier** *glacier* Antarctica
195 Q17 **Scott Island** *island* Antarctica
26 L11 **Scott, Mount** ▲ Oklahoma, USA 34°52′N 98°34′W
32 G15 **Scott, Mount** ▲ Oregon, NW USA 42°53′N 122°06′W
34 M1 **Scott River** ≈ California, W USA
28 I13 **Scottsbluff** Nebraska, C USA 41°52′N 103°40′W
23 Q2 **Scottsboro** Alabama, S USA 34°40′N 86°01′W
31 P15 **Scottsburg** Indiana, N USA 38°42′N 85°47′W
183 P16 **Scottsdale** Tasmania, SE Australia 41°13′S 147°30′E
36 L13 **Scottsdale** Arizona, SW USA 33°31′N 111°54′W
45 O12 **Scotts Head Village** *var.* Cachacrou. S Dominica 15°12′N 61°22′W
192 L14 **Scott Shoal** *undersea feature* S Pacific Ocean
20 K7 **Scottsville** Kentucky, S USA 36°45′N 86°11′W
29 U14 **Scranton** Iowa, C USA 42°01′N 94°33′W
18 I13 **Scranton** Pennsylvania, NE USA 41°25′N 75°40′W
186 B6 **Screw** ≈ NW Papua New Guinea
29 R14 **Scribner** Nebraska, C USA 41°40′N 96°40′W
Scrobesbyrig' *see* Shrewsbury
14 I14 **Scugog** ≈ Ontario, SE Canada
14 I14 **Scugog, Lake** ◎ Ontario, SE Canada
97 N17 **Scunthorpe** E England, United Kingdom 53°35′N 00°39′W
108 K9 **Scuol** *Ger.* Schuls. Graubünden, E Switzerland 46°51′N 10°21′E
Scupi *see* Skopje
Scutari *see* Shkodër
113 K17 **Scutari, Lake** *Alb.* Liqeni i Shkodrës, *SCr.* Skadarsko Jezero. ◎ Albania/Montenegro
Scyros *see* Skýros
Scythopolis *see* Beit She'an
138 E11 **Sderot** *prev.* Sederot. Southern, S Israel 31°31′N 34°35′E
25 U13 **Seadrift** Texas, SW USA 28°25′N 96°42′W
21 Y4 **Seaford** *var.* Seaford City. Delaware, NE USA 38°39′N 75°35′W
Seaford City *see* Seaford
14 E15 **Seaforth** Ontario, S Canada 43°33′N 81°25′W
24 M6 **Seagraves** Texas, SW USA 32°56′N 102°33′W
11 X9 **Seal** ≈ Manitoba, C Canada
182 M10 **Sea Lake** Victoria, SE Australia 35°34′S 142°51′E
83 G26 **Seal, Cape** *headland* S South Africa 34°06′S 23°18′E
65 D26 **Sea Lion Islands** *island group* SE Falkland Islands
19 S8 **Seal Island** *island* Maine, NE USA
25 V11 **Sealy** Texas, SW USA 29°46′N 96°09′W
35 X12 **Searchlight** Nevada, W USA 35°27′N 114°54′W
27 V11 **Searcy** Arkansas, C USA 35°14′N 91°43′W
19 R7 **Searsport** Maine, NE USA 44°28′N 68°54′W
35 N10 **Seaside** California, W USA 36°36′N 121°51′W
32 F10 **Seaside** Oregon, NW USA 45°57′N 123°55′W
18 K16 **Seaside Heights** New Jersey, NE USA 39°56′N 74°03′W
32 H8 **Seattle** Washington, NW USA 47°35′N 122°20′W
32 H9 **Seattle-Tacoma** ✈ Washington, NW USA 47°04′N 122°27′W
185 J16 **Seaward Kaikoura Range** ▲▲ South Island, New Zealand
42 J9 **Sébaco** Matagalpa, W Nicaragua 12°51′N 86°08′W
19 P8 **Sebago Lake** ◎ Maine, NE USA
169 S13 **Sebangan, Teluk** *bay* Borneo, C Indonesia
Sebaste/Sebastia *see* Sivas
23 Y12 **Sebastian** Florida, SE USA 27°55′N 80°31′W
40 C5 **Sebastían Vizcaíno, Bahía** *bay* NW Mexico
19 R6 **Sebasticook Lake** ◎ Maine, NE USA
34 M7 **Sebastopol** California, W USA 38°22′N 122°50′W
Sebastopol *see* Sevastopol'
169 W8 **Sebatik, Pulau** *island* N Indonesia
19 R5 **Sebec Lake** ◎ Maine, NE USA
76 K12 **Sébékoro** Kayes, W Mali 13°00′N 09°03′W
Sebenico *see* Šibenik
40 G6 **Seberi, Cerro** ▲ NW Mexico 27°49′N 110°18′W
116 H11 **Sebeş** *Ger.* Mühlbach, *Hung.* Szászsebes; *prev.* Sebeşu Săsesc. Alba, W Romania 45°58′N 23°34′E
Sebes-Körös *see* Crişul Repede
Sebeşu Săsesc *see* Sebeş
31 R8 **Sebewaing** Michigan, N USA 43°43′N 83°27′W
124 F16 **Sebezh** Pskovskaya Oblast', W Russian Federation 56°19′N 28°31′E
137 N12 **Şebinkarahisar** Giresun, N Turkey 40°19′N 38°25′E
116 F11 **Sebiş** *Hung.* Borossebes. Arad, W Romania 46°21′N 22°09′E
Sebkra Azz el Matti *see* Azzel Matti, Sebkha
19 Q4 **Seboomook Lake** ◎ Maine, NE USA
74 G6 **Sebou** *var.* Sebu. ≈ N Morocco
20 I6 **Sebree** Kentucky, S USA 37°34′N 87°30′W
23 X13 **Sebring** Florida, SE USA 27°30′N 81°26′W
Sebta *see* Ceuta
Sebu *see* Sebou
169 U13 **Sebuku, Pulau** *island* N Indonesia
169 W8 **Sebuku, Teluk** *bay* Borneo, N Indonesia
106 F10 **Secchia** ≈ N Italy
10 L17 **Sechelt** British Columbia, SW Canada 49°25′N 123°37′W
56 C12 **Sechin, Río** ≈ W Peru
56 A10 **Sechura, Bahía de** *bay* NW Peru
185 A22 **Secretary Island** *island* SW New Zealand
155 I15 **Secunderābād** *var.* Sikandarabad. Telangana, C India 17°30′N 78°33′E
57 L17 **Sécure, Río** ≈ C Bolivia
118 D10 **Seda** Telšiai, NW Lithuania 56°10′N 22°04′E
27 T5 **Sedalia** Missouri, C USA 38°42′N 93°15′W
103 R3 **Sedan** Ardennes, N France 49°42′N 04°56′E
27 P7 **Sedan** Kansas, C USA 37°07′N 96°11′W
105 N3 **Sedano** Castilla y León, N Spain 42°43′N 03°43′W
104 H10 **Seda, Ribeira de** *stream* C Portugal
185 K15 **Seddon** Marlborough, South Island, New Zealand 41°42′S 174°05′E
185 H15 **Seddonville** West Coast, South Island, New Zealand 41°35′S 171°59′E
143 U7 **Sedeh** Khorāsān-e Janūbī, E Iran 33°18′N 59°12′E
Sederot *see* Sderot
65 B23 **Sedge Island** *island* NW Falkland Islands
76 G12 **Sédhiou** SW Senegal 12°39′N 15°33′W
11 U16 **Sedley** Saskatchewan, S Canada 50°06′N 103°51′W
Sedlez *see* Siedlce
117 Q2 **Sedniv** Chernihivs'ka Oblast', N Ukraine 51°39′N 31°34′E
36 L11 **Sedona** Arizona, SW USA 34°52′N 111°45′W
Sedunum *see* Sion
118 F12 **Šeduva** Šiaulai, N Lithuania 55°45′N 23°46′E
141 Y8 **Seeb** *var.* Muscat Sīb Airport. ✈ (Masqaţ) NE Oman 23°36′N 58°27′E
Seeb *see* As Sīb
108 M7 **Seefeld-in-Tirol** Tirol, W Austria 47°19′N 11°16′E
83 E22 **Seeheim Noord** Karas, S Namibia 26°50′S 17°45′E
Seeland *see* Sjælland
195 N9 **Seelig, Mount** ▲ Antarctica 81°45′S 102°15′W
Seenu Atoll *see* Addu Atoll
Seeonee *see* Seoni
Seer *see* Dörgön
102 L5 **Sées** Orne, N France 48°36′N 00°11′E
101 J14 **Seesen** Niedersachsen, C Germany 51°54′N 10°11′E
Seesker Höhe *see* Szeska Góra
100 J10 **Seevetal** Niedersachsen, N Germany 53°24′N 10°01′E
109 V6 **Seewiesen** Steiermark, E Austria 47°37′N 15°16′E
136 J13 **Şefaatli** *var.* Kızılkoca. Yozgat, C Turkey 39°32′N 34°45′E
143 V9 **Sefīdābeh** Khorāsān-e Janūbī, E Iran 31°05′N 60°30′E
149 N3 **Sefīd, Darya-ye** *Pash.* Āb-i-safed. ≈ N Afghanistan
148 K5 **Sefīd Kūh, Selseleh-ye** *Eng.* Paropamisus Range. ▲ W Afghanistan 34°00′N 61°40′E
148 K5 **Sefīd Kūh, Selseleh-ye** *Eng.* Paropamisus Range. ▲▲ W Afghanistan
142 M4 **Sefīd, Rūd-e** ≈ NW Iran
74 G6 **Sefrou** N Morocco 33°51′N 04°49′W
185 E19 **Sefton, Mount** ▲ South Island, New Zealand 43°43′S 169°58′E
171 S13 **Segaf, Kepulauan** *island group* E Indonesia

◆ Country
● Country Capital
◇ Dependent Territory
○ Dependent Territory Capital
◈ Administrative Regions
✈ International Airport
▲ Mountain
▲▲ Mountain Range
Volcano
≈ River
◎ Lake
⊞ Reservoir

169 W7 **Segama, Sungai** ≈ East Malaysia
168 L9 **Segamat** Johor, Peninsular Malaysia 02°30′N 102°48′E
77 S13 **Ségbana** NE Benin 10°56′N 03°42′E
Segestica see Sisak
Segesvár see Sighişoara
171 T12 **Seget** Papua Barat, E Indonesia 01°21′S 131°04′E
Segewold see Sigulda
124 J9 **Segezha** Respublika Kareliya, NW Russian Federation 63°39′N 34°24′E
Seghedin see Szeged
Segna see Senj
107 I16 **Segni** Lazio, C Italy 41°41′N 13°02′E
Segodunum see Rodez
105 S9 **Segorbe** Valenciana, E Spain 39°51′N 00°30′W
76 M12 **Ségou** var. Segu. Ségou, C Mali 13°26′N 06°12′W
76 M12 **Ségou** ◈ region SW Mali
54 E8 **Segovia** Antioquia, N Colombia 07°08′N 74°39′W
105 N7 **Segovia** Castilla y León, C Spain 40°57′N 04°07′W
104 M6 **Segovia** ◈ province Castilla y León, N Spain
Segoviao Wangkí see Coco, Río
124 J9 **Segozerskoye Vodokhranilishche** prev. Ozero Segozero. ⊠ NW Russian Federation
102 J7 **Segré** Maine-et-Loire, NW France 47°41′N 00°51′W
105 U5 **Segre** ≈ NE Spain
Segu see Ségou
38 I17 **Seguam Island** island Aleutian Islands, Alaska, USA
38 I17 **Seguam Pass** strait Aleutian Islands, Alaska, USA
77 Y7 **Séguédine** Agadez, NE Niger 20°12′N 13°03′E
76 M15 **Séguéla** W Ivory Coast 07°58′N 06°44′W
25 S11 **Seguin** Texas, SW USA 29°34′N 97°58′W
38 E17 **Segula Island** island Aleutian Islands, Alaska, USA
62 K10 **Segundo, Río** ≈ C Argentina
105 Q12 **Segura** ≈ S Spain
105 P13 **Segura, Sierra de** ▲▲ S Spain
83 G18 **Sehithwa** North-West, N Botswana 20°28′S 22°43′E
154 H10 **Sehore** Madhya Pradesh, C India 23°12′N 77°08′E
186 G9 **Sehulea** Normanby Island, S Papua New Guinea 09°55′S 151°10′E
149 P15 **Sehwān** Sindh, SE Pakistan 26°26′N 67°52′E
109 V8 **Seiersberg** Steiermark, SE Austria 47°01′N 15°22′E
26 L9 **Seiling** Oklahoma, C USA 36°09′N 98°55′W
103 S9 **Seille** ≈ E France
99 J20 **Seilles** Namur, SE Belgium 50°31′N 05°12′E
93 K17 **Seinäjoki** Swe. Östermyra. Etelä-Pohjanmaa, W Finland 62°45′N 22°55′E
12 B12 **Seine** ≈ Ontario, S Canada
102 M4 **Seine** ≈ N France
102 K4 **Seine, Baie de la** bay N France
Seine, Banc de la see Seine Seamount
103 O5 **Seine-et-Marne** ◈ department N France
102 L3 **Seine-Maritime** ◈ department N France
84 B14 **Seine Plain** undersea feature E Atlantic Ocean 34°00′N 12°15′W
84 B15 **Seine Seamount** var. Banc de la Seine. undersea feature E Atlantic Ocean 33°45′N 14°25′W
102 E6 **Sein, Île de** island NW France
171 Y14 **Seinma** Papua, E Indonesia 04°10′S 138°54′E
Seisbierrum see Sexbierum
109 U5 **Seitenstetten Markt** Niederösterreich, C Austria 48°03′N 14°41′E
Seiyo see Uwa
Seiyu see Chŏnju
95 H22 **Sejerø** island E Denmark
110 P7 **Sejny** Podlaskie, NE Poland 54°09′N 23°21′E
163 X15 **Sejong City** ● (South Korea) E South Korea 36°29′N 127°16′E
81 G20 **Seke** Simiyu, N Tanzania 03°16′S 33°31′E
164 L13 **Seki** Gifu, Honshū, SW Japan 35°30′N 136°54′E
161 U12 **Sekibi-sho** Chin. Chiwei Yu. island (disputed) China/ Japan/Taiwan
165 U3 **Sekihoku-tōge** pass Hokkaidō, NE Japan
Sekondi see Sekondi-Takoradi
77 P17 **Sekondi-Takoradi** var. Sekondi. S Ghana 04°55′N 01°45′W
80 J11 **Sek'ot'a** Āmara, N Ethiopia 12°41′N 39°05′E
Sekseŭil see Saksaul'skoye
32 I9 **Selah** Washington, NW USA 46°39′N 120°31′W
168 J8 **Selangor** var. Negeri Selangor Darul Ehsan. ◈ state Peninsular Malaysia
Selânik see Thessaloníki
167 R10 **Selaphum** Roi Et, E Thailand 16°00′N 103°54′E
171 T16 **Selaru, Pulau** island Kepulauan Tanimbar, E Indonesia
171 U13 **Selassi** Papua Barat, E Indonesia 03°16′S 132°50′E
168 J7 **Selatan, Selat** strait Peninsular Malaysia
168 K10 **Selatpanjang** Pulau Rantau, W Indonesia 01°00′N 102°44′E
39 N8 **Selawik** Alaska, USA 66°36′N 160°00′W
39 N8 **Selawik Lake** ◎ Alaska, USA
171 N14 **Selayar, Selat** strait Sulawesi, C Indonesia
95 C14 **Selbjørnsfjorden** fjord S Norway
94 H8 **Selbusjøen** ◎ S Norway
97 M17 **Selby** N England, United Kingdom 53°49′N 01°06′W
29 N8 **Selby** South Dakota, N USA 45°30′N 100°01′W
21 Z4 **Selbyville** Delaware, NE USA 38°28′N 75°12′W
136 B15 **Selçuk** var. Akıncılar. İzmir, SW Turkey 37°56′N 27°25′E
39 Q13 **Seldovia** Alaska, USA 59°26′N 151°42′W
107 M18 **Sele** anc. Silarius. ≈ S Italy
83 J19 **Selebi-Phikwe** Central, E Botswana 21°58′S 27°48′E
42 B5 **Selegua, Río** ≈ W Guatemala
129 X7 **Selemdzha** ≈ SE Russian Federation
129 U7 **Selenga** Mong. Selenge Mörön. ≈ Mongolia/ Russian Federation
79 I19 **Selenge** Bandundu, W Dem. Rep. Congo 01°58′S 18°11′E
162 K6 **Selenge** var. Ingettolgoy. Bulgan, N Mongolia 49°27′N 103°59′E
162 L6 **Selenge** ◈ province N Mongolia
Selenge see Hyalganat, Bulgan, Mongolia
Selenge see Ih-Uul, Hövsgöl, Mongolia
Selenge Mörön see Selenga
123 N14 **Selenginsk** Respublika Buryatiya, S Russian Federation 52°00′N 106°40′E
Selenica see Selenicë
113 K22 **Selenicë** var. Selenica. Vlorë, SW Albania 40°32′N 19°38′E
123 Q8 **Selennyakh** ≈ NE Russian Federation
100 J8 **Selenter See** ◎ N Germany
Sele Sound see Soela Väin
103 U6 **Sélestat** Ger. Schlettstadt. Bas-Rhin, NE France 48°16′N 07°28′E
Seleucia see Silifke
92 I4 **Selfoss** Suðurland, SW Iceland 63°56′N 20°59′W
28 M7 **Selfridge** North Dakota, N USA 46°01′N 100°52′W
76 I15 **Seli** ≈ N Sierra Leone
76 J11 **Sélibabi** var. Sélibaby. Guidimaka, S Mauritania 15°14′N 12°11′W
Sélibaby see Sélibabi
Selidovka/Selidovo see Selydove
124 I15 **Seliger, Ozero** ◎ W Russian Federation
36 J11 **Seligman** Arizona, SW USA 35°20′N 112°56′W
27 S8 **Seligman** Missouri, C USA 36°31′N 93°56′W
80 E6 **Selima Oasis** oasis N Sudan
76 L13 **Sélingué, Lac de** ⊠ S Mali
Selinoús see Kréstena
18 G14 **Selinsgrove** Pennsylvania, NE USA 40°47′N 76°51′W
Selishche see Syelishcha
124 I16 **Selizharovo** Tverskaya Oblast', W Russian Federation 56°50′N 33°24′E
94 C10 **Selje** Sogn Og Fjordane, S Norway 62°02′N 05°22′E
11 X16 **Selkirk** Manitoba, S Canada 50°10′N 96°52′W
96 K13 **Selkirk** SE Scotland, United Kingdom 55°36′N 02°48′W
96 K13 **Selkirk** cultural region SE Scotland, United Kingdom
11 O16 **Selkirk Mountains** ▲▲ British Columbia, SW Canada
193 T11 **Selkirk Rise** undersea feature SE Pacific Ocean
115 F21 **Sellasía** Pelopónnisos, S Greece 37°14′N 22°24′E
44 M9 **Selle, Pic de la** var. La Selle. ▲ SE Haiti 18°18′N 71°55′W
102 M8 **Selles-sur-Cher** Loir-et-Cher, C France 47°16′N 01°31′E
36 K16 **Sells** Arizona, SW USA 31°54′N 111°52′W
Sellye see Sal'a
23 P5 **Selma** Alabama, S USA 32°24′N 87°01′W
35 Q11 **Selma** California, W USA 36°33′N 119°37′W
20 G10 **Selmer** Tennessee, S USA 35°10′N 88°34′W
173 N17 **Sel, Pointe au** headland W Réunion
Selselehye Kuhe Vākhān see Nicholas Range
127 S2 **Selty** Udmurtskaya Respublika, NW Russian Federation 57°19′N 52°09′E
Selukwe see Shurugwi
62 L9 **Selva** Santiago del Estero, N Argentina 29°46′S 62°02′W
11 T9 **Selwyn Lake** ◎ Northwest Territories/Saskatchewan, C Canada
10 K6 **Selwyn Mountains** ▲▲ Yukon, NW Canada
181 T6 **Selwyn Range** ▲▲ Queensland, C Australia
117 W8 **Selydove** var. Selidovka, Rus. Selidovo. Donets'ka Oblast', SE Ukraine 48°06′N 37°16′E
Selzaete see Zelzate
Seman see Semanit, Lumi i
168 M15 **Semangka, Teluk** bay Sumatera, SW Indonesia
113 D22 **Semanit, Lumi i** var. Seman. ≈ W Albania
169 Q16 **Semarang** var. Samarang. Jawa, C Indonesia 06°58′S 110°29′E
169 Q10 **Sematan** Sarawak, East Malaysia 01°50′N 109°44′E
171 P17 **Semau, Pulau** island S Indonesia
169 V8 **Sembakung, Sungai** ≈ Borneo, N Indonesia
79 G17 **Sembé** Sangha, NW Congo 01°38′N 14°35′E
169 S13 **Sembulu, Danau** ◎ Borneo, N Indonesia
139 Q2 **Sēmēl** Ar. Sumayl, var. Summēl. Dahūk, N Iraq 36°52′N 42°51′E
Semendria see Smederevo
117 R1 **Semenivka** Chernihivs'ka Oblast', N Ukraine 52°10′N 32°37′E
117 S6 **Semenivka** Rus. Semenovka. Poltavs'ka Oblast', NE Ukraine 49°36′N 33°10′E
127 O3 **Semenov** Nizhegorodskaya Oblast', W Russian Federation 56°47′N 44°27′E
Semenovka see Semenivka
169 S17 **Semeru, Gunung** var. Mahameru. ▲ Jawa, S Indonesia 08°01′S 112°53′E
145 V9 **Semey** prev. Semipalatinsk. Vostochnyy Kazakhstan, E Kazakhstan 50°26′N 80°16′E
Semezhevo see Syemyezhava
126 L7 **Semiluki** Voronezhskaya Oblast', W Russian Federation 51°46′N 39°00′E
33 W16 **Seminoe Reservoir** ⊠ Wyoming, C USA
27 O11 **Seminole** Oklahoma, C USA 35°13′N 96°40′W
24 M6 **Seminole** Texas, SW USA 32°43′N 102°39′W
23 S8 **Seminole, Lake** ⊠ Florida/ Georgia, SE USA
Semiozernoye see Auliyekol'
Semipalatinsk see Semey
143 O9 **Semīrom** var. Samirum. Eşfahān, C Iran 31°20′N 51°50′E
38 F17 **Semisopochnoi Island** island Aleutian Islands, Alaska, USA
169 R11 **Semitau** Borneo, C Indonesia 0°30′N 111°59′E
81 E18 **Semliki** ≈ Uganda/Dem. Rep. Congo
143 P5 **Semnān** var. Samnān. Semnān, N Iran 35°37′N 53°21′E
143 Q5 **Semnān** off. Ostān-e Semnān. ◈ province N Iran
Semnān, Ostān-e see Semnān
99 K24 **Semois** ≈ SE Belgium
108 E8 **Sempacher See** ◎ C Switzerland
Sena see Vila de Sena
30 L12 **Senachwine Lake** ◎ Illinois, N USA
59 O14 **Senador Pompeu** Ceará, E Brazil 05°30′S 39°25′W
Sena Gallica see Senigallia
59 C15 **Sena Madureira** Acre, W Brazil 09°05′S 68°41′W
155 L25 **Senanayake Samudra** ⊠ E Sri Lanka
83 G15 **Senanga** Western, SW Zambia 16°09′S 23°16′E
27 Y9 **Senath** Missouri, C USA 36°07′N 90°09′W
22 L2 **Senatobia** Mississippi, S USA 34°37′N 89°58′W
164 C16 **Sendai** var. Satsuma-Sendai. Kagoshima, Kyūshū, SW Japan 31°49′N 130°17′E
165 Q11 **Sendai-wan** bay E Japan
101 J23 **Senden** Bayern, S Germany 48°18′N 10°04′E
154 F11 **Sendhwa** Madhya Pradesh, C India 21°38′N 75°04′E
111 H21 **Senec** Ger. Wartberg, Hung. Szenc; prev. Szempcz. Bratislavský Kraj, W Slovakia 48°14′N 17°24′E
27 P3 **Seneca** Kansas, C USA 39°50′N 96°04′W
27 R8 **Seneca** Missouri, C USA 36°50′N 94°36′W
32 K13 **Seneca** Oregon, NW USA 44°06′N 118°57′W
21 O11 **Seneca** South Carolina, SE USA 34°41′N 82°57′W
18 G11 **Seneca Lake** ◎ New York, NE USA
31 U13 **Senecaville Lake** ⊠ Ohio, N USA
76 G11 **Senegal** off. Republic of Senegal, Fr. Sénégal. ◆ republic W Africa
76 H9 **Senegal** Fr. Sénégal. ≈ W Africa
Senegal, Republic of see Senegal
31 O4 **Seney Marsh** wetland Michigan, N USA
101 P14 **Senftenberg** Brandenburg, E Germany 51°31′N 14°01′E
82 L11 **Senga Hill** Northern, NE Zambia 09°26′S 31°12′E
158 G13 **Sênggê Zangbo** ≈ W China
171 Z13 **Senggi** Papua, E Indonesia 03°26′S 140°46′E
127 R5 **Sengiley** Ul'yanovskaya Oblast', W Russian Federation 53°54′N 48°51′E
63 I19 **Senguerr, Río** ≈ S Argentina
83 J16 **Sengwa** ≈ C Zimbabwe
Senia see Senj
111 H19 **Senica** Ger. Senitz, Hung. Szenice. Trnavský Kraj, W Slovakia 48°40′N 17°22′E
Senica see Sjenica
106 J11 **Senigallia** anc. Sena Gallica. Marche, C Italy 43°43′N 13°13′E
136 F15 **Senirkent** Isparta, SW Turkey 38°07′N 30°34′E
Senitz see Senica
112 C10 **Senj** Ger. Zengg, It. Segna; anc. Senia. Lika-Senj, NW Croatia 44°58′N 14°55′E
Senj see Örgön
92 H9 **Senja** prev. Senjen. island N Norway
Senjen see Senja
161 U12 **Senkaku-shotō** Chin. Diaoyutai. island group (disputed) SW Japan
137 R12 **Şenkaya** Erzurum, NE Turkey 40°33′N 42°17′E
83 I16 **Senkobo** Southern, S Zambia 17°38′S 25°58′E
103 O4 **Senlis** Oise, N France 49°13′N 02°33′E
Senmonorom see Sênmônoŭrôm
167 T12 **Sênmônoŭrôm** var. Senmonorom. Môndól Kiri, E Cambodia 12°27′N 107°12′E
80 G10 **Sennar** var. Sannār. Sinnar, C Sudan 13°31′N 33°38′E
Senno see Syanno
Senones see Sens
109 W11 **Senovo** E Slovenia 46°01′N 15°24′E
103 P6 **Sens** anc. Agendicum, Senones. Yonne, C France 48°12′N 03°17′E
Sensburg see Mrągowo
167 S11 **Sên, Stœng** ≈ C Cambodia
42 F7 **Sensuntepeque** Cabañas, NE El Salvador 13°52′N 88°38′W
112 L8 **Senta** Hung. Zenta. Vojvodina, N Serbia 45°57′N 20°04′E
Šent Andraž see Sankt Andrä
171 Y13 **Sentani, Danau** ◎ Papua, E Indonesia
28 J5 **Sentinel Butte** ▲ North Dakota, N USA 46°52′N 103°50′W
10 M13 **Sentinel Peak** ▲ British Columbia, W Canada 54°51′N 122°02′W
59 N16 **Sento Sé** Bahia, E Brazil 09°51′S 41°56′W
Šent Peter see Pivka
Št. Vid see Sankt Veit an der Glan
Seo de Urgel see La Seu d'Urgell
163 X17 **Seogwipo** prev. Sŏgwip'o. S South Korea 33°14′N 126°33′E
154 I7 **Seondha** Madhya Pradesh, C India 26°09′N 78°47′E
163 Y17 **Seongsan** prev. Sŏngsan. S South Korea
154 J11 **Seoni** prev. Seonee. Madhya Pradesh, C India 22°06′N 79°36′E
163 X14 **Seoul**, Jap. Keijō; prev. Kyŏngsŏng, Sŏul. ● (South Korea) NW South Korea 37°30′N 126°58′E
83 I17 **Sepako** Central, NE Botswana 19°50′S 26°29′E
184 I13 **Separation Point** headland South Island, New Zealand 40°46′S 172°58′E
169 V10 **Sepasu** Borneo, N Indonesia 0°44′N 117°38′E
186 B6 **Sepik** ≈ Indonesia/Papua New Guinea
Sepone see Muang Xépôn
110 M7 **Sępopol** Ger. Schippenbeil. Warmińsko-Mazurskie, N Poland 54°16′N 21°09′E
116 F10 **Şepreuş** Hung. Seprős. Arad, W Romania 46°34′N 21°44′E
Seprős see Şepreuş
Şepşi-Sângeorz/ Sepsiszentgyörgy see Sfântu Gheorghe
15 W4 **Sept-Îles** Québec, SE Canada 50°11′N 66°19′W
105 N6 **Sepúlveda** Castilla y León, N Spain 41°18′N 03°45′W
104 K8 **Sequeros** Castilla y León, N Spain 40°31′N 06°04′W
104 L5 **Sequillo** ≈ NW Spain
32 G7 **Sequim** Washington, NW USA 48°04′N 123°06′W
35 S11 **Sequoia National Park** national park California, W USA
137 Q14 **Şerafettin Dağları** ▲▲ E Turkey
127 N10 **Serafimovich** Volgogradskaya Oblast', SW Russian Federation 49°34′N 42°43′E
171 Q10 **Serai** Sulawesi, N Indonesia 01°45′N 124°58′E
99 K19 **Seraing** Liège, E Belgium 50°37′N 05°31′E
Sêraitang see Baima
Serakhs see Sarahs
171 W13 **Serami** Papua, E Indonesia 02°11′S 136°46′E
171 R13 **Seram, Laut** Eng. Ceram Sea. sea E Indonesia
Serampore/Serampur see Shrīrāmpur
171 S13 **Seram, Pulau** var. Serang, Eng. Ceram. island Maluku, E Indonesia
169 N15 **Serang** Jawa, C Indonesia 06°07′S 106°09′E
Serang see Seram, Pulau
169 P9 **Serasan, Pulau** island Kepulauan Natuna, W Indonesia
169 P9 **Serasan, Selat** strait Indonesia/Malaysia
112 M13 **Serbia** off. Federal Republic of Serbia; prev. Yugoslavia, SCr. Jugoslavija. ◆ federal republic SE Europe
112 M12 **Serbia** Ger. Serbien, Serb. Srbija. ◈ republic Serbia
Serbia, Federal Republic of see Serbia
Serbien see Serbia
Sercq see Sark
146 D12 **Serdar** prev. Rus. Gyzyrlabat, Kizyl-Arvat. Balkan Welaýaty, W Turkmenistan 39°02′N 56°15′E
Serdica see Sofia
Serdobol' see Sortavala
127 O7 **Serdobsk** Penzenskaya Oblast', W Russian Federation 52°30′N 44°16′E
145 X9 **Serebryansk** Vostochnyy Kazakhstan, E Kazakhstan 49°44′N 83°16′E
123 Q12 **Serebryanyy Bor** Respublika Sakha (Yakutiya), NE Russian Federation 56°40′N 124°46′E
111 H20 **Sered'** Hung. Szered. Trnavský Kraj, W Slovakia 48°19′N 17°45′E
117 S1 **Seredyna-Buda** Sums'ka Oblast', NE Ukraine 52°09′N 34°00′E
118 E13 **Seredžius** Tauragė, C Lithuania 55°04′N 23°24′E
136 I14 **Şereflikoçhisar** Ankara, C Turkey 38°56′N 33°31′E
106 D7 **Seregno** Lombardia, N Italy 45°39′N 09°12′E
103 P7 **Serein** ≈ C France
168 K9 **Seremban** Negeri Sembilan, Peninsular Malaysia 02°42′N 101°54′E
81 H20 **Serengeti Plain** plain N Tanzania
82 K13 **Serenje** Central, E Zambia 13°12′S 30°15′E
Seres see Sérres
116 J5 **Seret** ≈ W Ukraine
Seret/Sereth see Siret
115 I21 **Serifopoúla** island Kykládes, Greece, Aegean Sea
127 P4 **Sergach** Nizhegorodskaya Oblast', W Russian Federation 55°31′N 45°29′E
29 S13 **Sergeant Bluff** Iowa, C USA 42°24′N 96°19′W
163 P7 **Sergelen** Dornod, NE Mongolia 48°31′N 114°01′E
Sergelen see Tuvshinshiree
168 H8 **Sergeulangit, Pegunungan** ▲▲ Sumatera, NW Indonesia
122 L5 **Sergeya Kirova, Ostrova** island N Russian Federation
Sergeyevichi see Syarhyeyevichy
145 O7 **Sergeyevka** Severnyy Kazakhstan, N Kazakhstan 53°53′N 67°25′E
Sergiopol see Ayagoz
59 P16 **Sergipe** off. Estado de Sergipe. ◈ state E Brazil
Sergipe, Estado de see Sergipe
126 L3 **Sergiyev Posad** Moskovskaya Oblast', W Russian Federation 56°21′N 38°10′E
124 K5 **Sergozero, Ozero** ◎ NW Russian Federation
146 J17 **Serhetabat** prev. Rus. Gushgy, Kushka. Mary Welaýaty, S Turkmenistan 35°19′N 62°17′E
169 Q10 **Serian** Sarawak, East Malaysia 01°10′N 110°35′E
115 I21 **Sérifos** anc. Seriphos. island Kykládes, Greece, Aegean Sea
115 I21 **Sérifou, Stenó** strait SE Greece
136 F16 **Serik** Antalya, SW Turkey 36°55′N 31°06′E
106 E7 **Serio** ≈ N Italy
Seriphos see Sérifos
Serir Tibesti see Sarīr Tībistī
Sêrkog see Sêrtar
197 O14 **Sermersooq** off. Kommuneqarfik Sermersooq. ◈ municipality S Greenland
Sermersooq, Kommuneqarfik see Sermersooq
127 S5 **Sernovodsk** Samarskaya Oblast', W Russian Federation 53°56′N 51°16′E
127 R2 **Sernur** Respublika Mariy El, W Russian Federation 56°55′N 49°09′E
110 M11 **Serock** Mazowieckie, C Poland 52°30′N 21°03′E
61 B18 **Serodino** Santa Fe, C Argentina 32°33′S 60°52′W
Seroei see Serui
105 P14 **Serón** Andalucía, S Spain 37°20′N 02°28′W
99 E14 **Serooskerke** Zeeland, SW Netherlands 51°42′N 03°52′E
105 T6 **Serós** Cataluña, NE Spain 41°27′N 00°24′E
122 G10 **Serov** Sverdlovskaya Oblast', C Russian Federation 59°42′N 60°32′E
83 I19 **Serowe** Central, SE Botswana 22°26′S 26°44′E
104 H13 **Serpa** Beja, S Portugal 37°56′N 07°36′W
Serpa Pinto see Menongue
182 A4 **Serpentine Lakes** salt lake South Australia
45 T15 **Serpent's Mouth, The** Sp. Boca de la Serpiente. strait Trinidad and Tobago/ Venezuela
Serpiente, Boca de la see Serpent's Mouth, The
126 K4 **Serpukhov** Moskovskaya Oblast', W Russian Federation 54°54′N 37°26′E
104 I10 **Serra de São Mamede** ▲ C Portugal 39°18′N 07°19′W
60 K13 **Serra do Mar** ▲▲ S Brazil
Sérrai see Sérres
107 N22 **Serra San Bruno** Calabria, SW Italy 38°32′N 16°18′E
103 S14 **Serres** Hautes-Alpes, SE France 44°26′N 05°42′E
114 H13 **Sérres** var. Seres; prev. Sérrai. Kentrikí Makedonía, NE Greece 41°03′N 23°33′E
62 J9 **Serrezuela** Córdoba, C Argentina 30°38′S 65°26′W
59 O16 **Serrinha** Bahia, E Brazil 11°38′S 38°56′W
59 M19 **Serro** var. Sêrro. Minas Gerais, NE Brazil 18°38′S 43°22′W
Sêrro see Serro
Sert see Siirt
104 H9 **Sertã** var. Sertá. Castelo Branco, C Portugal 39°48′N 08°05′W
Sertá see Sertã
60 L8 **Sertãozinho** São Paulo, S Brazil 21°04′S 47°55′W
160 F7 **Sêrtar** var. Sêrkog. Sichuan, C China 32°18′N 100°18′E
124 G12 **Sertolovo** Leningradskaya Oblast', NW Russian Federation 60°08′N 30°06′E
171 W13 **Serui** prev. Seroei. Papua, E Indonesia 01°53′S 136°15′E
83 J19 **Serule** Central, E Botswana 21°58′S 27°20′E
169 S12 **Seruyan, Sungai** var. Sungai Pembuang. ≈ Borneo, N Indonesia
115 E14 **Sérvia** Dytikí Makedonía, N Greece 40°12′N 22°01′E
160 E7 **Sêrxü** var. Jugar. Sichuan, C China 32°54′N 98°06′E
123 R13 **Seryshevo** Amurskaya Oblast', SE Russian Federation 51°03′N 128°16′E
Sesana see Sežana
169 V8 **Sesayap, Sungai** ≈ Borneo, N Indonesia
Sesdlets see Siedlce
79 N17 **Sese** Orientale, N Dem. Rep. Congo 02°13′N 25°52′E
81 F18 **Sese Islands** island group S Uganda
83 H16 **Sesheke** var. Sesheko. Western, SE Zambia 17°28′S 24°20′E
Sesheko see Sesheke
106 C8 **Sesia** anc. Sessites. ≈ NW Italy
104 F11 **Sesimbra** Setúbal, S Portugal 38°26′N 09°06′W
115 N22 **Seskliό** island Dodekánisa, Greece, Aegean Sea
30 L16 **Sesser** Illinois, N USA 38°05′N 89°03′W
Sessites see Sesia
106 G11 **Sesto Fiorentino** Toscana, C Italy 43°50′N 11°12′E
106 E7 **Sesto San Giovanni** Lombardia, N Italy 45°32′N 09°14′E
106 A8 **Sestriere** Piemonte, NE Italy 45°00′N 06°54′E
106 D10 **Sestri Levante** Liguria, NW Italy 44°16′N 09°22′E
107 C20 **Sestu** Sardegna, Italy, C Mediterranean Sea 39°15′N 09°06′E
112 E8 **Sesvete** Zagreb, N Croatia 45°50′N 16°03′E
118 G12 **Šeta** Kaunas, C Lithuania 55°17′N 24°16′E
Setabis see Xàtiva
165 Q4 **Setana** Hokkaidō, NE Japan 42°27′N 139°52′E
103 Q16 **Sète** prev. Cette. Hérault, S France 43°24′N 03°42′E
58 J11 **Sete Ilhas** Amapá, NE Brazil 01°06′N 52°06′W
59 L20 **Sete Lagoas** Minas Gerais, NE Brazil 19°29′S 44°15′W
60 G10 **Sete Quedas, Ilha das** island S Brazil
92 I10 **Setermoen** Troms, N Norway 68°51′N 18°20′E
95 E17 **Setesdal** valley S Norway
43 W16 **Setetule, Cerro** ▲ SE Panama 07°51′N 77°37′W
21 Q5 **Seth** West Virginia, NE USA 38°06′N 81°40′W
Setia see Sezze
74 K5 **Sétif** var. Stif. N Algeria 36°11′N 05°24′E
164 L5 **Seto** Aichi, Honshū, SW Japan 35°14′N 137°06′E
164 G13 **Seto-naikai** Eng. Inland Sea. sea S Japan
165 V16 **Setouchi** var. Setoushi. Kagoshima, Amami-Ō-shima, SW Japan 44°19′N 142°58′E
Setoushi see Setouchi
74 F6 **Settat** W Morocco 33°03′N 07°37′W
79 D20 **Setté Cama** Ogooué-Maritime, SW Gabon 02°32′S 09°46′E
11 W13 **Setting Lake** ◎ Manitoba, C Canada
97 L16 **Settle** N England, United Kingdom 54°04′N 02°17′W
189 Y12 **Settlement** E Wake Island 19°17′N 166°38′E
104 F11 **Setúbal** Eng. Saint Ubes, Saint Yves. Setúbal, W Portugal 38°31′N 08°54′W
104 F11 **Setúbal** ◈ district S Portugal
104 F12 **Setúbal, Baía de** bay W Portugal
Setul see Satun
12 B10 **Seul, Lac** ◎ Ontario, S Canada
103 R8 **Seurre** Côte d'Or, C France 47°00′N 05°09′E
137 U11 **Sevan** C Armenia 40°32′N 44°56′E
137 V12 **Sevana Lich** Eng. Lake Sevan, Rus. Ozero Sevan. ◎ E Armenia
Sevan, Lake/Sevan, Ozero see Sevana Lich
77 N11 **Sévaré** Mopti, C Mali 14°32′N 04°06′W
117 S14 **Sevastopol'** Eng. Sebastopol. Avtonomna Respublika Krym, S Ukraine 44°36′N 33°33′E
25 R14 **Seven Sisters** Texas, SW USA 27°57′N 98°34′W
10 K13 **Seven Sisters Peaks** ▲ British Columbia, SW Canada 54°57′N 128°10′W
99 M15 **Sevenum** Limburg, SE Netherlands 51°25′N 06°01′E
103 P14 **Séverac-le-Château** Aveyron, S France 44°18′N 03°03′E
14 H13 **Severn** ≈ Ontario, S Canada
97 L21 **Severn** Wel. Hafren. ≈ England/Wales, United Kingdom
125 O11 **Severnaya Dvina** var. Northern Dvina. ≈ NW Russian Federation
127 N16 **Severnaya Osetiya-Alaniya, Respublika** Eng. North Ossetia; prev. Respublika Severnaya Osetiya, Severo-Osetinskaya SSR. ◈ autonomous republic SW Russian Federation
Severnaya Osetiya, Respublika see Severnaya Osetiya-Alaniya, Respublika
122 M5 **Severnaya Zemlya** var. Nicholas II Land. island group N Russian Federation
127 T5 **Severnoye** Orenburgskaya Oblast', W Russian Federation 52°03′N 52°31′E
35 S3 **Severn Troughs Range** ▲▲ Nevada, W USA
125 W3 **Severnyy** Respublika Komi, NW Russian Federation 67°38′N 64°13′E
144 I13 **Severnyy Chink Ustyurta** ▲ W Kazakhstan
125 Q13 **Severnyy Uvaly** var. Northern Ural Hills. hill range NW Russian Federation
145 O6 **Severnyy Kazakhstan** off. Severo-Kazakhstanskaya Oblast', var. North Kazakhstan, Kaz. Soltüstik Qazaqstan Oblysy. ◈ province N Kazakhstan
122 I6 **Severnyy, Ostrov** NW Russian Federation
125 V9 **Severnyy Ural** ▲▲ NW Russian Federation
Severo-Alichurskiy Khrebet see Alichuri Shimolí, Qatorkŭhi
123 N12 **Severobaykal'sk** Respublika Buryatiya, S Russian Federation 55°39′N 109°17′E
Severodonetsk see Syeverodonets'k
124 M8 **Severodvinsk** prev. Molotov, Sudostroy. Arkhangel'skaya Oblast', NW Russian Federation 64°32′N 39°50′E
Severo-Kazakhstanskaya Oblast' see Severnyy Kazakhstan
123 U11 **Severo-Kuril'sk** Sakhalinskaya Oblast', SE Russian Federation 50°38′N 155°57′E
124 J3 **Severomorsk** Murmanskaya Oblast', NW Russian Federation 69°00′N 33°16′E
Severo-Osetinskaya SSR see Severnaya Osetiya-Alaniya, Respublika
122 M7 **Severo-Sibirskaya Nizmennost'** var. North Siberian Plain, Eng. North Siberian Lowland. lowlands N Russian Federation
122 G10 **Severoural'sk** Sverdlovskaya Oblast', C Russian Federation 60°09′N 59°58′E
122 L11 **Severo-Yeniseyskiy** Krasnoyarskiy Kray, C Russian Federation 60°29′N 93°13′E
122 J12 **Seversk** Tomskaya Oblast', C Russian Federation 53°37′N 84°47′E
126 M11 **Severskiy Donets** Ukr. Sivers'kyy Donets'. ≈ Russian Federation/ Ukraine see also Sivers'kyy Donets'
Severskiy Donets see Sivers'kyy Donets'
92 M9 **Sevettijärvi** Lappi, N Finland 69°31′N 28°40′E
36 M5 **Sevier Bridge Reservoir** ⊠ Utah, W USA
36 J4 **Sevier Desert** plain Utah, W USA
36 J5 **Sevier Lake** ◎ Utah, W USA
21 N9 **Sevierville** Tennessee, S USA 35°53′N 83°34′W
104 J14 **Sevilla** Eng. Seville; anc. Hispalis. Andalucía, SW Spain 37°24′N 05°59′W
104 J13 **Sevilla** ◈ province Andalucía, SW Spain
Sevilla de Niefang see Niefang
43 O16 **Sevilla, Isla** island SW Panama
Seville see Sevilla
114 J9 **Sevlievo** Gabrovo, N Bulgaria 43°01′N 25°06′E
Sevluš/Sevlyush see Vynohradiv
109 V11 **Sevnica** Ger. Lichtenwald. E Slovenia 46°00′N 15°20′E
162 J11 **Sevrey** var. Saynshand. Ömnögovĭ, S Mongolia 43°30′N 102°08′E
126 I7 **Sevsk** Bryanskaya Oblast', W Russian Federation 52°03′N 34°31′E
76 J15 **Sewa** ≈ E Sierra Leone
39 R12 **Seward** Alaska, USA 60°06′N 149°26′W
29 R15 **Seward** Nebraska, C USA 40°52′N 97°06′W
197 Q3 **Seward Peninsula** peninsula Alaska, USA
Seward's Folly see Alaska
62 H12 **Sewell** Libertador, C Chile 34°05′S 70°25′W
98 K5 **Sexbierum** Fris. Seisbierrum. Fryslân, N Netherlands 53°13′N 05°28′E
11 O13 **Sexsmith** Alberta, W Canada 55°18′N 118°45′W
41 W13 **Seybaplaya** Campeche, SE Mexico 19°39′N 90°36′W
173 N6 **Seychelles** off. Republic of Seychelles. ◆ republic W Indian Ocean
67 Z9 **Seychelles** island group NE Seychelles
173 N6 **Seychelles Bank** var. Le Banc des Seychelles. undersea feature W Indian Ocean 04°45′S 55°30′E
Seychelles, Le Banc des see Seychelles Bank
Seychelles, Republic of see Seychelles
172 H17 **Seychellois, Morne** ▲ Mahé, NE Seychelles
146 J12 **Seýdi** Rus. Seydi; prev. Neftezavodsk. Lebap Welaýaty, E Turkmenistan 39°31′N 62°53′E
136 G16 **Seydişehir** Konya, SW Turkey 37°25′N 31°51′E
92 L2 **Seyðisfjörður** Austurland, E Iceland 65°15′N 14°00′W
136 J13 **Seyfe Gölü** ◎ C Turkey
Seyhan see Adana
136 K16 **Seyhan Barajı** ⊠ S Turkey
136 K17 **Seyhan Nehri** ≈ S Turkey
136 F13 **Seyitgazi** Eskişehir, W Turkey 39°27′N 30°42′E
126 J7 **Seym** ≈ W Russian Federation
117 S3 **Seym** ≈ N Ukraine
123 T9 **Seymchan** Magadanskaya Oblast', E Russian Federation 62°54′N 152°27′E
114 N12 **Seymen** Tekirdağ, NW Turkey 41°06′N 27°56′E
183 O11 **Seymour** Victoria, SE Australia 37°01′S 145°10′E
83 I25 **Seymour** Eastern Cape, S South Africa 32°33′S 26°46′E
29 W16 **Seymour** Iowa, C USA 40°40′N 93°07′W
27 U7 **Seymour** Missouri, C USA 37°09′N 92°46′W
25 Q5 **Seymour** Texas, SW USA 33°36′N 99°16′W
114 M12 **Şeytan Deresi** ≈ NW Turkey
109 S12 **Sežana** It. Sesana. SW Slovenia 45°42′N 13°52′E
103 P5 **Sézanne** Marne, N France 48°43′N 03°41′E
107 I16 **Sezze** anc. Setia. Lazio, C Italy 41°29′N 13°04′E
Sfákia see Chóra Sfakíon
115 D21 **Sfaktiría** island S Greece
116 J11 **Sfântu Gheorghe** Ger. Sankt-Georgen, Hung. Sepsiszentgyörgy; prev. Şepşi-Sângeorz, Sfintu Gheorghe. Covasna, C Romania 45°52′N 25°49′E
117 N13 **Sfântu Gheorghe, Braţul** var. Gheorghe Braţul. ≈ E Romania
75 N6 **Sfax** Ar. Şafāqis. E Tunisia 34°45′N 10°45′E
75 N6 **Sfax** ✕ E Tunisia 34°43′N 10°37′E
Sfintu Gheorghe see Sfântu Gheorghe
98 H13 **'s-Gravendeel** Zuid-Holland, SW Netherlands 51°48′N 04°36′E
98 F11 **'s-Gravenhage** var. Den Haag, Eng. The Hague, Fr. La Haye. ● (Netherlands-seat of government) Zuid-Holland, W Netherlands 52°07′N 04°17′E
98 G12 **'s-Gravenzande** Zuid-Holland, W Netherlands 52°00′N 04°10′E
Shaan/Shaanxi Sheng see Shaanxi
159 X11 **Shaanxi** var. Shaan, Shaanxi Sheng, Shan-hsi, Shenshi, Shensi. ◈ province C China
Shaartuz see Shahrtuz
Shaba see Katanga
Shabani see Zvishavane
81 N17 **Shabeellaha Dhexe** off. Gobolka Shabeellaha Dhexe. ◈ region E Somalia
Shabeellaha Dhexe, Gobolka see Shabeellaha Dhexe
81 L17 **Shabeellaha Hoose** off. Gobolka Shabeellaha Hoose. ◈ region S Somalia
Shabeellaha Hoose, Gobolka see Shabeellaha Hoose
Shabeelle, Webi see Shebeli
114 O7 **Shabla** Dobrich, NE Bulgaria 43°33′N 28°31′E
114 O7 **Shabla, Nos** headland NE Bulgaria 43°30′N 28°36′E
13 N9 **Shabogama Lake** ◎ Newfoundland and Labrador, E Canada
79 N20 **Shabunda** Sud-Kivu, E Dem. Rep. Congo 02°42′S 27°20′E
141 Q15 **Shabwah** C Yemen 15°09′N 46°46′E
158 F8 **Shache** var. Yarkant. Xinjiang Uygur Zizhiqu, NW China 38°27′N 77°16′E
Shacheng see Huailai
195 R12 **Shackleton Coast** physical region Antarctica
195 Z10 **Shackleton Ice Shelf** ice shelf Antarctica
Shaddādī see Ash Shadādah
28 K7 **Shadehill Reservoir** ⊠ South Dakota, N USA

◆ Country
● Country Capital
◇ Dependent Territory
○ Dependent Territory Capital
◈ Administrative Regions
✕ International Airport
▲ Mountain
▲▲ Mountain Range
Volcano
≈ River
◎ Lake
⊠ Reservoir

122 G11 **Shadrinsk** Kurganskaya Oblast', C Russian Federation 56°08′N 63°18′E
31 O12 **Shafer, Lake** Indiana, N USA
35 R13 **Shafter** California, W USA 35°27′N 119°15′W
24 J11 **Shafter** Texas, SW USA 29°49′N 104°18′W
97 L23 **Shaftesbury** S England, United Kingdom 51°01′N 02°12′W
185 F22 **Shag** South Island, New Zealand
145 V9 **Shagan** E Kazakhstan
39 O11 **Shageluk** Alaska, USA 62°40′N 159°33′W
122 K14 **Shagonar** Respublika Tyva, S Russian Federation 51°31′N 93°06′E
185 F22 **Shag Point** *headland* South Island, New Zealand 45°28′S 170°50′E
144 J12 **Shagyray, Plato** *plain* SW Kazakhstan
Shāhābād *see* Eslāmābād-e Gharb
168 K9 **Shah Alam** Selangor, Peninsular Malaysia 03°02′N 101°31′E
117 O12 **Shahany, Ozero** SW Ukraine
138 H9 **Shahbā'** *anc.* Philippopolis. As Suwaydā', S Syria 32°50′N 36°38′E
Shahbān *see* Ad Dayr
149 P17 **Shāh Bandar** Sind, SE Pakistan 23°59′N 67°54′E
149 P13 **Shahdād Kot** Sind, SW Pakistan 27°49′N 67°49′E
143 T10 **Shahdād, Namakzār-e** *salt pan* E Iran
149 Q15 **Shāhdādpur** Sind, SE Pakistan 25°56′N 68°40′E
154 K10 **Shahdol** Madhya Pradesh, C India 23°19′N 81°26′E
161 N7 **Sha He** C China
Shahe *see* Linze
Shahepu *see* Linze
153 N13 **Shāhganj** Uttar Pradesh, N India 26°03′N 82°41′E
152 C11 **Shāhgarh** Rājasthān, NW India 27°08′N 69°56′E
Sha Hi *see* Orūmīyeh, Daryācheh-ye
Shāhī *see* Qā'emshahr
Shahjahanabad *see* Delhi
152 L11 **Shāhjahānpur** Uttar Pradesh, N India 27°53′N 79°55′E
149 U7 **Shāhpur** Punjab, E Pakistan 32°15′N 72°32′E
Shāhpur *see* Shāhpur Chākar
152 G13 **Shāhpura** Rājasthān, N India 25°38′N 75°01′E
149 Q15 **Shāhpur Chākar** *var.* Shāhpur. Sind, SE Pakistan 26°11′N 68°44′E
148 M5 **Shahrak** Gōwr, C Afghanistan 34°09′N 64°16′E
143 Q11 **Shahr-e Bābak** Kermān, C Iran 30°08′N 55°04′E
143 N8 **Shahr-e Kord** *var.* Shahr Kord. Chahār Maḩall va Bakhtīārī, C Iran 32°20′N 50°52′E
143 O9 **Shahreẕā** *var.* Qomisheh, Qumisheh, Shahriza; *prev.* Qomsheh. Eşfahān, C Iran 32°01′N 51°51′E
147 S10 **Shahrikhon** *Rus.* Shakhrikhan. Andijon Viloyati, E Uzbekistan 40°42′N 72°03′E
147 P11 **Shahriston** *Rus.* Shakhristan. NW Tajikistan 39°45′N 68°47′E
Shahriza *see* Shahreẕā
Shahr-i-Zabul *see* Zābol
Shahr Kord *see* Shahr-e Kord
147 P14 **Shahrtuz** *Rus.* Shaartuz. SW Tajikistan 37°13′N 68°05′E
143 Q4 **Shāhrūd** *prev.* Emāmrūd, Emāmshahr. Semnān, N Iran 36°30′N 55°E
Shahsavār/Shahsawar *see* Tonekābon
Shaikh Ābid *see* Shaykh 'Ābid
Shaikh Fāris *see* Shaykh Fāris
Shaikh Najm *see* Shaykh Najm
138 K5 **Shā'ir, Jabal** ▲ C Syria 34°51′N 37°49′E
154 G10 **Shājāpur** Madhya Pradesh, C India 23°27′N 76°21′E
80 J8 **Shakal, Ras** *headland* NE Sudan 18°04′N 38°34′E
83 G17 **Shakawe** North West, NW Botswana 18°25′S 21°53′E
Shakhdarinskiy Khrebet *see* Shokhdara, Qatorkŭhi
Shakhrikhan *see* Shahrikhon
Shakhrisabz *see* Sharixon
Shakhristan *see* Shahriston
Shakhtërsk *see* Zuhres
145 R10 **Shakhtinsk** Karaganda, C Kazakhstan 49°40′N 72°37′E
126 L11 **Shakhty** Rostovskaya Oblast', SW Russian Federation 47°45′N 40°14′E
127 P2 **Shakhun'ya** Nizhegorodskaya Oblast', W Russian Federation 57°42′N 46°36′E
77 S15 **Shaki** Oyo, W Nigeria 08°37′N 03°25′E
81 J15 **Shakiso** Oromīya, C Ethiopia 05°33′N 38°48′E
29 V9 **Shakopee** Minnesota, N USA 44°48′N 93°31′W
165 R3 **Shakotan-misaki** *headland* Hokkaidō, NE Japan 43°22′N 140°28′E
39 N9 **Shaktoolik** Alaska, USA 64°19′N 161°05′W
81 J14 **Shala Hāyk'** C Ethiopia
124 M10 **Shalakusha** Arkhangel'skaya Oblast', NW Russian Federation 62°16′N 40°16′E
145 U8 **Shalday** Pavlodar, NE Kazakhstan 51°57′N 78°51′E
127 P16 **Shali** Chechenskaya Respublika, SW Russian Federation 43°03′N 45°55′E
141 W12 **Shalīm** *var.* Shelim. S Oman 18°07′N 55°39′E
Shalīr, Āveh-ye *see* Shilayr, Wādī
Shaliuhe *see* Gangca
144 K12 **Shalkar** *var.* Chelkar. Aktyubinsk, W Kazakhstan 47°50′N 59°29′E
144 F9 **Shalkar, Ozero** *prev.* Chelkar Ozero. W Kazakhstan
21 V12 **Shallotte** North Carolina, SE USA 33°58′N 78°21′W
25 N5 **Shallowater** Texas, SW USA 33°41′N 102°00′W
124 K11 **Shal'skiy** Respublika Kareliya, NW Russian Federation 61°45′N 36°02′E
160 F9 **Shaluli Shan** ▲ C China
81 F22 **Shama** C Tanzania
11 Z11 **Shamattawa** Manitoba, C Canada 55°52′N 92°05′W
12 F8 **Shamattawa** Ontario, C Canada
Shām, Bādiyat ash *see* Syrian Desert
Shamiya *see* Ash Shāmīyah
141 X8 **Shām, Jabal ash** *var.* Jebel Sham. ▲ NW Oman 23°21′N 57°08′E
Sham, Jebel *see* Shām, Jabal ash
Shamkhor *see* Şämkir
18 G14 **Shamokin** Pennsylvania, NE USA 40°47′N 76°33′W
25 P2 **Shamrock** Texas, SW USA 35°12′N 100°15′W
Shana *see* Kuril'sk
Sha'nabī, Jabal ash *see* Chambi, Jebel
139 Y12 **Shanāwah** Al Başrah, E Iraq 30°57′N 47°25′E
Shancheng *see* Taining
159 T8 **Shandan** *var.* Qingyuan. Gansu, N China 38°50′N 101°08′E
Shandī *see* Shendi
161 Q5 **Shandong** *var.* Lu, Shandong Sheng, Shantung. ◆ *province* E China
161 R4 **Shandong Bandao** *var.* Shantung Peninsula. *peninsula* E China
Shandong Sheng *see* Shandong
139 U8 **Shandrūkh** Diyālá, E Iraq 33°20′N 45°19′E
83 J17 **Shangani** W Zimbabwe
161 O15 **Shangchuan Dao** *island* S China
Shangchuankou *see* Minhe
163 P12 **Shangdu** Nei Mongol Zizhiqu, N China 41°32′N 113°33′E
161 O11 **Shanggao** *var.* Aoyang. Jiangxi, S China 28°16′N 114°55′E
Shangguan *see* Daixian
161 S8 **Shanghai** *var.* Shang-hai. Shanghai Shi, E China 31°14′N 121°28′E
161 S8 **Shanghai Shi** *var.* Hu, Shanghai. ◆ *municipality* E China
161 P13 **Shanghang** *var.* Linjiang. Fujian, SE China 25°03′N 116°25′E
160 K14 **Shanglin** *var.* Dafeng. Guangxi Zhuangzu Zizhiqu, S China 23°26′N 108°32′E
160 L7 **Shangluo** *prev.* Shangxian, Shangzhou. Shaanxi, C China 33°51′N 109°55′E
83 G15 **Shangombo** Western, W Zambia 16°28′S 22°10′E
Shangpai/Shangpaihe *see* Feixi
161 O6 **Shangqiu** *var.* Zhuji. Henan, C China 34°24′N 115°37′E
161 Q10 **Shangrao** Jiangxi, S China 28°27′N 117°57′E
Shangxian *see* Shangluo
161 S9 **Shangyu** *var.* Baiguan. Zhejiang, SE China 30°03′N 120°52′E
163 X9 **Shangzhi** Heilongjiang, NE China 45°13′N 127°59′E
Shangzhou *see* Shangluo
Shanhe *see* Zhengning
163 W9 **Shanhetun** Heilongjiang, NE China 44°42′N 127°12′E
Shan-hsi *see* Shaanxi, China
Shan-hsi *see* Shanxi, China
159 O6 **Shankou** Xinjiang Uygur Zizhiqu, W China 42°02′N 94°08′E
184 M13 **Shannon** Manawatu-Wanganui, North Island, New Zealand 40°32′S 175°24′E
97 C17 **Shannon** *Ir.* An tSionainn. W Ireland
97 B19 **Shannon** ✈ W Ireland 52°42′N 08°57′W
167 N6 **Shan Plateau** *plateau* E Myanmar (Burma)
158 M6 **Shanshan** *var.* Piqan. Xinjiang Uygur Zizhiqu, NW China 42°53′N 90°18′E
Shansi *see* Shanxi
167 N5 **Shan State** ◆ *state* E Myanmar (Burma)
Shantar Islands *see* Shantarskiye Ostrova
123 S12 **Shantarskiye Ostrova** *Eng.* Shantar Islands. *island group* E Russian Federation
161 Q14 **Shantou** *var.* Shan-t'ou, Swatow. Guangdong, S China 23°23′N 116°39′E
Shan-t'ou *see* Shantou
Shantung *see* Shandong
Shantung Peninsula *see* Shandong Bandao
163 O14 **Shanxi** *var.* Jin, Shan-hsi, Shansi, Shanxi Sheng. ◆ *province* C China
161 P6 **Shanxian** *var.* Shan Xian. Shandong, E China 34°51′N 116°09′E
Shan Xian *see* Sanmenxia
Shan Xian *see* Shanxian
Shanxi Sheng *see* Shanxi
160 L7 **Shanyang** Shaanxi, C China 33°35′N 109°48′E
161 N13 **Shanyin** *var.* Daiyue. Shanxi, C China E Asia 39°30′N 112°56′E
161 O13 **Shaoguan** *var.* Shao-kuan, *Cant.* Kukong; *prev.* Ch'u-chiang. Guangdong, S China 24°57′N 113°38′E
Shao-kuan *see* Shaoguan
161 Q11 **Shaowu** Fujian, SE China 27°24′N 117°26′E
161 S9 **Shaoxing** Zhejiang, SE China 30°02′N 120°35′E
160 M12 **Shaoyang** *var.* Tangdukou. Hunan, S China 26°54′N 111°14′E
160 M11 **Shaoyang** *var.* Baoqing, Shao-yang; *prev.* Pao-king. Hunan, S China 27°13′N 111°31′E
Shao-yang *see* Shaoyang
96 K5 **Shapinsay** *island* NE Scotland, United Kingdom
125 S4 **Shapkina** NW Russian Federation
Shāpūr *see* Salmās
158 M4 **Shaqiuhe** Xinjiang Uygur Zizhiqu, W China 45°00′N 88°52′E
Shaqlāwa *see* Sheqlawe
Shaqlāwah *see* Sheqlawe
138 I8 **Shaqqā** As Suwaydā', S Syria 32°53′N 36°42′E
141 P7 **Shaqrā'** Ar Riyāḑ, C Saudi Arabia 25°11′N 45°08′E
Shaqrā *see* Shuqrah
145 W10 **Shar** *var.* Charsk. Vostochnyy Kazakhstan, E Kazakhstan 49°33′N 81°03′E
149 O6 **Sharan** Dāykundī, SE Afghanistan 33°28′N 66°19′E
149 Q7 **Sharan** *var.* Zareh Sharan. Paktīkā, E Afghanistan 33°08′N 68°47′E
Sharaqpur *see* Sharqpur
145 U8 **Sharbakty** *Kaz.* Sharbaqty; *prev.* Shcherbakty. Pavlodar, E Kazakhstan 52°28′N 78°00′E
Sharbaqty *see* Sharbakty
141 X12 **Sharbatāt** S Oman 17°57′N 56°14′E
Sharbatāt, Ra's *see* Sharbithāt, Ras
141 X12 **Sharbithāt, Ras** *var.* Ra's Sharbatāt. *headland* S Oman 17°55′N 56°30′E
14 K14 **Sharbot Lake** Ontario, SE Canada 44°45′N 76°46′W
145 P17 **Shardara** *var.* Chardara. Yuzhnyy Kazakhstan, S Kazakhstan 41°15′N 68°01′E
Shardara Dalasy *see* Step' Shardara
145 P17 **Shardarinskoye Vodokhranilishche** *prev.* Chardarinskoye Vodokhranilishche. S Kazakhstan
162 F8 **Sharga** Govĭ-Altay, W Mongolia 46°16′N 95°32′E
Sharga *see* Tsagaan-Uul
116 M7 **Sharhorod** Vinnyts'ka Oblast', C Ukraine 48°46′N 28°05′E
Sharhulsan *see* Mandal-Ovoo
165 V3 **Shari** Hokkaidō, NE Japan 43°54′N 144°42′E
Shari *see* Chari
139 T6 **Shārī, Buḩayrat** C Iraq
147 N12 **Sharixon** *Rus.* Shakhrisabz. Qashqadaryo Viloyati, S Uzbekistan 39°01′N 66°45′E
Sharjah *see* Ash Shāriqah
118 K12 **Sharkawshchyna** *var.* Sharkowshchyna, *Pol.* Szarkowszczyzna, *Rus.* Sharkovshchina. Vitsyebskaya Voblasts', NW Belarus 55°27′N 27°28′E
180 G9 **Shark Bay** *bay* Western Australia
141 Y9 **Sharkh** E Oman 21°20′N 59°04′E
Sharkovshchina/ Sharkowshchyna *see* Sharkawshchyna
127 U6 **Sharlyk** Orenburgskaya Oblast', W Russian Federation 52°52′N 54°45′E
75 Y9 **Sharm ash Shaykh** *var.* Ofiral, Sharm el Sheikh. E Egypt 27°51′N 34°16′E
Sharm el Sheikh *see* Sharm ash Shaykh
18 B13 **Sharon** Pennsylvania, NE USA 41°12′N 80°28′W
26 H4 **Sharon Springs** Kansas, C USA 38°54′N 101°46′W
31 Q14 **Sharonville** Ohio, N USA 39°16′N 84°24′W
Sharourah *see* Sharūrah
29 O10 **Sharpe, Lake** South Dakota, N USA
Sharqī, Al Jabal ash/Sharqi, Jebel esh *see* Anti-Lebanon
Sharqīyah, Al Minṭaqah ash *see* Ash Sharqīyah
138 I6 **Sharqīyat an Nabk, Jabal** ▲ W Syria
149 W8 **Sharqpur** *var.* Sharaqpur. Punjab, E Pakistan 31°29′N 74°08′E
141 Q13 **Sharūrah** *var.* Sharourah. Najrān, S Saudi Arabia 17°29′N 47°05′E
125 O14 **Shar'ya** Kostromskaya Oblast', NW Russian Federation 58°22′N 45°30′E
145 W15 **Sharyn** *prev.* Charyn. Almaty, SE Kazakhstan 43°48′N 79°22′E
145 V15 **Sharyn** *var.* Charyn. SE Kazakhstan
122 K13 **Sharypovo** Krasnoyarskiy Kray, C Russian Federation 55°33′N 89°12′E
83 J18 **Shashe** Central, NE Botswana 21°25′S 27°28′E
83 J18 **Shashe** *var.* Shashi. Botswana/Zimbabwe
81 J14 **Shashemenē** *var.* Shashemene, Shashhamana, *It.* Sciasciamana. Oromīya, C Ethiopia 07°16′N 38°38′E
Shashemenne/ Shashhamana *see* Shashemenē
Shashi *see* Shashe
Shashi/Sha-shih/Shasi *see* Jingzhou, Hubei
35 N3 **Shasta Lake** California, W USA
35 N2 **Shasta, Mount** ▲ California, W USA 41°24′N 122°11′W
127 O4 **Shatki** Nizhegorodskaya Oblast', W Russian Federation 55°09′N 44°04′E
Shatlyk *see* Şatlyk
Shatra *see* Ash Shaţrah
119 K17 **Shatsk** Minskaya Voblasts', C Belarus 53°25′N 27°41′E
127 N5 **Shatsk** Ryazanskaya Oblast', W Russian Federation 54°02′N 41°38′E
26 J9 **Shattuck** Oklahoma, C USA 36°16′N 99°52′W
145 P16 **Shaul'dir** *prev.* Shaul'der. Yuzhnyy Kazakhstan, S Kazakhstan 42°45′N 68°21′E
Shaul'der *see* Shaul'dir
11 S17 **Shaunavon** Saskatchewan, S Canada 49°49′N 108°25′W
Shavat *see* Shovot
Shaviyani Atoll *see* North Miladhunmadulu Atoll
158 K4 **Shawan** *var.* Sandaohezi. Xinjiang Uygur Zizhiqu, NW China 44°21′N 85°37′E
14 G12 **Shawanaga** Ontario, S Canada 45°29′N 80°16′W
30 M6 **Shawano** Wisconsin, N USA 44°46′N 88°38′W
30 M6 **Shawano Lake** Wisconsin, N USA
15 P10 **Shawinigan** *prev.* Shawinigan Falls. Québec, SE Canada 46°33′N 72°45′W
Shawinigan Falls *see* Shawinigan
15 P10 **Shawinigan-Sud** Québec, SE Canada 46°30′N 72°43′W
138 J5 **Shawmarīyah, Jabal ash** ▲ C Syria
27 O11 **Shawnee** Oklahoma, C USA 35°20′N 96°55′W
14 K12 **Shawville** Québec, SE Canada 45°37′N 76°31′W
145 Q16 **Shayan** *var.* Chayan. Yuzhnyy Kazakhstan, S Kazakhstan 42°59′N 69°22′E
Shaykh *see* Ash Shakk
139 W9 **Shaykh 'Ābid** *var.* Shaikh Ābid. Wāsiţ, E Iraq 32°40′N 46°09′E
139 Y10 **Shaykh Fāris** *var.* Shaikh Fāris. Maysān, E Iraq 32°06′N 47°39′E
139 T7 **Shaykh Ḩātim** Baghdād, E Iraq 33°29′N 44°15′E
Shaykh, Jabal ash *see* Hermon, Mount
139 X10 **Shaykh Najm** *var.* Shaikh Najm. Maysān, E Iraq 32°04′N 46°54′E
139 W9 **Shaykh Sa'd** Maysān, E Iraq 32°35′N 46°16′E
147 T14 **Shazud** SE Tajikistan 37°45′N 72°22′E
119 N18 **Shchadryn** *Rus.* Shchedrin. Homyel'skaya Voblasts', SE Belarus 52°53′N 29°33′E
119 H18 **Shchara** SW Belarus
Shchedrin *see* Shchadryn
Shcheglovsk *see* Kemerovo
126 K5 **Shchëkino** Tul'skaya Oblast', W Russian Federation 53°57′N 37°33′E
125 S7 **Shchel'yayur** Respublika Komi, NW Russian Federation 65°19′N 53°27′E
Shcherbakty *see* Sharbakty
126 K7 **Shchigry** Kurskaya Oblast', W Russian Federation 51°53′N 36°49′E
Shchitkovichi *see* Shchytkavichy
117 Q2 **Shchors** Chernihivs'ka Oblast', N Ukraine 51°49′N 31°58′E
117 T8 **Shchors'k** Dnipropetrovs'ka Oblast', E Ukraine 48°20′N 34°07′E
Shchuchin *see* Shchuchyn
145 Q7 **Shchuchinsk** *prev.* Shchuchye. Akmola, N Kazakhstan 52°57′N 70°10′E
Shchuchye *see* Shchuchinsk
119 G16 **Shchuchyn** *Pol.* Szczuczyn Nowogródzki, *Rus.* Shchuchin. Hrodzyenskaya Voblasts', W Belarus 53°36′N 24°45′E
119 K17 **Shchytkavichy** *Rus.* Shchitkovichi. Minskaya Voblasts', C Belarus 53°13′N 27°59′E
122 J13 **Shebalino** Respublika Altay, S Russian Federation 51°16′N 85°41′E
126 J9 **Shebekino** Belgorodskaya Oblast', W Russian Federation 50°25′N 36°55′E
Shebelē Wenz, Wabē *see* Shebeli
81 L14 **Shebeli** *Amh.* Wabē Shebelē Wenz, *It.* Scebeli, *Som.* Webi Shabeelle. Ethiopia/Somalia
113 M20 **Shebenikut, Maja e** ▲ E Albania 41°13′N 20°27′E
Sheberghān *see* Shibirghān
144 F14 **Shebir** Mangistau, SW Kazakhstan 44°52′N 52°01′E
31 N8 **Sheboygan** Wisconsin, N USA 43°46′N 87°44′W
77 X15 **Shebshi Mountains** *var.* Schebschi Mountains. ▲ E Nigeria
Shechem *see* Nablus
Shedadi *see* Ash Shadādah
13 P14 **Shediac** New Brunswick, SE Canada 46°13′N 64°35′W
126 L15 **Shedok** Krasnodarskiy Kray, SW Russian Federation 44°12′N 40°49′E
Sheekh *see* Shiikh
38 M11 **Sheenjek River** Alaska, USA
96 D13 **Sheep Haven** *Ir.* Cuan na gCaorach. *inlet* N Ireland
35 X10 **Sheep Range** ▲ Nevada, W USA
98 M13 **'s-Heerenberg** Gelderland, E Netherlands 51°52′N 06°15′E
97 P22 **Sheerness** SE England, United Kingdom 51°27′N 00°45′E
13 Q15 **Sheet Harbour** Nova Scotia, SE Canada 44°56′N 62°31′W
185 H18 **Sheffield** Canterbury, South Island, New Zealand 43°22′S 172°01′E
97 M18 **Sheffield** N England, United Kingdom 53°23′N 01°30′W
23 O2 **Sheffield** Alabama, S USA 34°46′N 87°42′W
29 V12 **Sheffield** Iowa, C USA 42°53′N 93°13′W
25 N10 **Sheffield** Texas, SW USA 30°42′N 101°49′W
63 H22 **Shehuen, Río** S Argentina
Shekhem *see* Nablus
149 V8 **Shekhūpura** Punjab, NE Pakistan 31°42′N 74°08′E
Sheki *see* Şäki
124 L14 **Sheksna** Vologodskaya Oblast', NW Russian Federation 59°11′N 38°32′E
123 T5 **Shelagskiy, Mys** *headland* NE Russian Federation 70°04′N 170°39′E
27 V3 **Shelbina** Missouri, C USA 39°41′N 92°02′W
13 P16 **Shelburne** Nova Scotia, SE Canada 43°47′N 65°20′W
14 G14 **Shelburne** Ontario, S Canada 44°04′N 80°12′W
33 R7 **Shelby** Montana, NW USA 48°30′N 111°52′W
21 Q10 **Shelby** North Carolina, SE USA 35°15′N 81°34′W
31 S12 **Shelby** Ohio, N USA 40°52′N 82°39′W
30 L14 **Shelbyville** Illinois, N USA 39°24′N 88°47′W
31 P14 **Shelbyville** Indiana, N USA 39°31′N 85°46′W
20 L5 **Shelbyville** Kentucky, S USA 38°13′N 85°12′W
27 V2 **Shelbyville** Missouri, C USA 39°48′N 92°01′W
20 J10 **Shelbyville** Tennessee, S USA 35°29′N 86°30′W
25 X8 **Shelbyville** Texas, SW USA 31°42′N 94°03′W
30 L14 **Shelbyville, Lake** Illinois, N USA
29 S12 **Sheldon** Iowa, C USA 43°10′N 95°51′W
38 M11 **Sheldons Point** Alaska, USA 62°31′N 165°03′W
145 V15 **Shelek** *prev.* Chilik. Almaty, SE Kazakhstan 43°35′N 78°12′E
145 V15 **Shelek** *prev.* Chilik. SE Kazakhstan
Shelekhov Gulf *see* Shelikhova, Zaliv
123 U9 **Shelikhova, Zaliv** *Eng.* Shelekhov Gulf. *gulf* E Russian Federation
39 P14 **Shelikof Strait** *strait* Alaska, USA
Shelim *see* Shalīm
11 T14 **Shellbrook** Saskatchewan, S Canada 53°14′N 106°24′W
28 L3 **Shell Creek** North Dakota, N USA
Shellif *see* Chelif, Oued
22 I10 **Shell Keys** *island group* Louisiana, S USA
30 I4 **Shell Lake** Wisconsin, N USA 45°44′N 91°56′W
29 W12 **Shell Rock** Iowa, C USA 42°42′N 92°34′W
185 C26 **Shelter Point** *headland* Stewart Island, New Zealand 47°04′S 168°13′E
18 L13 **Shelton** Connecticut, NE USA 41°19′N 73°06′W
32 G8 **Shelton** Washington, NW USA 47°13′N 123°06′W
Shemakha *see* Şamaxı
145 W9 **Shemonaikha** Vostochnyy Kazakhstan, E Kazakhstan 50°38′N 81°55′E
127 Q4 **Shemursha** Chuvashskaya Respublika, W Russian Federation 54°57′N 47°27′E
38 D16 **Shemya Island** *island* Aleutian Islands, Alaska, USA
29 T16 **Shenandoah** Iowa, C USA 40°46′N 95°23′W
21 U4 **Shenandoah** Virginia, NE USA 38°26′N 78°34′W
21 U4 **Shenandoah Mountains** *ridge* West Virginia, NE USA
21 V3 **Shenandoah River** West Virginia, NE USA
77 W15 **Shendam** Plateau, C Nigeria 08°52′N 09°30′E
80 G8 **Shendi** *var.* Shandī. River Nile, NE Sudan 16°41′N 33°22′E
76 I15 **Shenge** SW Sierra Leone 07°54′N 12°54′W
146 L10 **Shengeldi** *Rus.* Chingildi. Navoiy Viloyati, N Uzbekistan 40°59′N 64°13′E
145 U15 **Shengel'dy** Almaty, SE Kazakhstan 44°04′N 77°31′E
113 K18 **Shëngjin** *var.* Shëngjini. Lezhë, NW Albania 41°49′N 19°34′E
Shëngjini *see* Shëngjin
Shengking *see* Liaoning
Sheng Xian/Shengxian *see* Shengzhou
161 S9 **Shengzhou** *var.* Shengxian, Sheng Xian. Zhejiang, SE China 29°36′N 120°42′E
Shenking *see* Liaoning
125 N11 **Shenkursk** Arkhangel'skaya Oblast', NW Russian Federation 62°10′N 42°58′E
160 L3 **Shenmu** Shaanxi, C China 38°49′N 110°27′E
113 L19 **Shën Noj i Madh** ▲ C Albania 41°23′N 20°07′E
160 L8 **Shennong Ding** ▲ C China 31°24′N 110°16′E
Shenshi/Shensi *see* Shaanxi
163 V12 **Shenyang** *Chin.* Shen-yang, *Eng.* Moukden, Mukden; *prev.* Fengtien. *province capital* Liaoning, NE China 41°50′N 123°26′E
Shen-yang *see* Shenyang
161 O15 **Shenzhen** Guangdong, S China 22°39′N 114°02′E
154 G8 **Sheopur** Madhya Pradesh, C India 25°41′N 76°42′E
116 L5 **Shepetivka** *Rus.* Shepetovka. Khmel'nyts'ka Oblast', NW Ukraine 50°12′N 27°01′E
Shepetovka *see* Shepetivka
25 W10 **Shepherd** Texas, SW USA 30°30′N 95°00′W
187 R14 **Shepherd Islands** *island group* C Vanuatu
20 K5 **Shepherdsville** Kentucky, S USA 38°00′N 85°42′W
183 O11 **Shepparton** Victoria, SE Australia 36°25′S 145°26′E
97 P22 **Sheppey, Isle of** *island* SE England, United Kingdom
139 T2 **Sheqlawe** *Ar.* Shaqlāwah, *var.* Shaqlāwa. Arbīl, E Iraq 36°24′N 44°21′E
Sherabad *see* Sherobod
9 O4 **Sherard, Cape** *headland* Nunavut, N Canada 74°36′N 80°10′W
97 L23 **Sherborne** S England, United Kingdom 50°58′N 02°30′W
76 H16 **Sherbro Island** *island* SW Sierra Leone
15 Q12 **Sherbrooke** Québec, SE Canada 45°23′N 71°55′W
29 T11 **Sherburn** Minnesota, N USA 43°39′N 94°43′W
78 H6 **Sherda** Tibesti, N Chad 20°04′N 16°48′E
80 G7 **Shereik** River Nile, N Sudan 18°44′N 33°37′E
126 K3 **Sheremet'yevo** ✈ (Moskva) Moskovskaya Oblast', W Russian Federation 56°05′N 37°10′E
153 P14 **Sherghāti** Bihār, N India 24°35′N 84°51′E
27 U12 **Sheridan** Arkansas, C USA 34°18′N 92°22′W
33 W12 **Sheridan** Wyoming, C USA 44°47′N 106°59′W
182 G8 **Sheringa** South Australia 33°51′S 135°13′E
25 U5 **Sherman** Texas, SW USA 33°39′N 96°35′W
194 J10 **Sherman Island** *island* Antarctica
19 S4 **Sherman Mills** Maine, NE USA 45°51′N 68°23′W
29 O15 **Sherman Reservoir** Nebraska, C USA
147 N14 **Sherobod** *Rus.* Sherabad. Surkhondaryo Viloyati, S Uzbekistan 37°43′N 66°59′E
147 O13 **Sherobod** *Rus.* Sherabad. S Uzbekistan
153 T14 **Sherpur** Dhaka, N Bangladesh 25°00′N 90°01′E
37 T4 **Sherrelwood** Colorado, C USA 39°49′N 105°00′W
99 J14 **'s-Hertogenbosch** *Fr.* Bois-le-Duc, *Ger.* Herzogenbusch. Noord-Brabant, S Netherlands 51°41′N 05°19′E
28 M2 **Sherwood** North Dakota, N USA 48°55′N 101°36′W
11 Q14 **Sherwood Park** Alberta, SW Canada 53°34′N 113°04′W
56 F13 **Sheshea, Río** E Peru
143 T5 **Sheshtamad** Khorāsān-e Razavī, NE Iran 36°03′N 57°45′E
29 S10 **Shetek, Lake** Minnesota, N USA
96 M2 **Shetland Islands** *island group* NE Scotland, United Kingdom
144 F14 **Shetpe** Mangistau, SW Kazakhstan 44°06′N 52°03′E
154 C11 **Shetrunji** W India
Shevchenko *see* Aktau
117 W5 **Shevchenkove** Kharkivs'ka Oblast', E Ukraine 49°40′N 37°13′E
81 H14 **Shewa Gīmīra** Southern Nationalities, S Ethiopia 07°12′N 35°49′E
161 Q9 **Shexian** *var.* Huicheng, She Xian. Anhui, E China 29°53′N 118°27′E
She Xian *see* Shexian
161 R6 **Sheyang** *prev.* Hede. Jiangsu, E China 33°49′N 120°13′E
29 O4 **Sheyenne** North Dakota, N USA 47°49′N 99°08′W
29 P4 **Sheyenne River** North Dakota, N USA
96 G7 **Shiant Islands** *island group* NW Scotland, United Kingdom
123 U12 **Shiashkotan, Ostrov** *island* Kuril'skiye Ostrova, SE Russian Federation
31 R9 **Shiawassee River** Michigan, N USA
141 R14 **Shibām** C Yemen 15°49′N 48°24′E
Shibarghān *see* Shibirghān
165 O10 **Shibata** *var.* Sibata. Niigata, Honshū, C Japan 37°57′N 139°20′E
Shiberghan/Shiberghān *see* Shibirghān
Shibh Jazīrat Sīnā' *see* Sinai
75 W8 **Shibīn al Kawm** *var.* Shibîn el Kôm. N Egypt 30°33′N 31°00′E
Shibîn el Kôm *see* Shibīn al Kawm
149 N2 **Shibirghān** *var.* Shibarghān, Shiberghan, Shiberghān; *prev.* Sheberghān. Jowzjān, N Afghanistan 36°41′N 65°45′E
143 O13 **Shīb, Kūh-e** ▲ S Iran
12 D8 **Shibogama Lake** Ontario, C Canada
Shibotsu-jima *see* Zelënyy, Ostrov
164 B16 **Shibushi** Kagoshima, Kyūshū, SW Japan 31°27′N 131°05′E
189 U13 **Shichiyo Islands** *island group* Chuuk, C Micronesia
Shickshock Mountains *see* Chic-Chocs, Monts
shiderti *see* Shiderty
Shiderti *see* Shiderty
145 S8 **Shiderty** *prev.* Shiderti. Pavlodar, NE Kazakhstan 51°40′N 74°50′E
145 S9 **Shiderty** *prev.* Shiderti. N Kazakhstan
96 G10 **Shiel, Loch** N Scotland, United Kingdom
164 J13 **Shiga** *off.* Shiga-ken, *var.* Siga. ◆ *prefecture* Honshū, SW Japan
Shiga-ken *see* Shiga
Shigatse *see* Xigazê
141 U13 **Shiḩan** *oasis* NE Yemen
Shih-chia-chuang/Shihmen *see* Shijiazhuang
158 K4 **Shihezi** Xinjiang Uygur Zizhiqu, NW China 44°21′N 85°59′E
Shiichi *see* Shyichy
80 N12 **Shiikh** *prev.* Sheekh. Togdheer, N Somalia 10°01′N 45°21′E
113 K19 **Shijak** *var.* Shijaku. Durrës, W Albania 41°21′N 19°34′E
Shijaku *see* Shijak
161 O4 **Shijiazhuang** *var.* Shih-chia-chuang; *prev.* Shihmen. *province capital* Hebei, E China 38°04′N 114°28′E
165 R5 **Shikabe** Hokkaidō, NE Japan 42°03′N 140°45′E
149 Q13 **Shikārpur** Sind, S Pakistan 27°59′N 68°39′E
127 Q7 **Shikhany** Saratovskaya Oblast', W Russian Federation 52°07′N 47°13′E
189 V12 **Shiki Islands** *island group* Chuuk, C Micronesia
164 G14 **Shikoku** *var.* Sikoku. *island* SW Japan
192 H5 **Shikoku Basin** *var.* Sikoku Basin. *undersea feature* N Philippine Sea 28°00′N 137°00′E
164 G14 **Shikoku-sanchi** ▲ Shikoku, SW Japan
165 X4 **Shikotan, Ostrov** *Jap.* Shikotan-tō. *island* NE Russian Federation
Shikotan-tō *see* Shikotan, Ostrov
165 R4 **Shikotsu-ko** *var.* Sikotu Ko. Hokkaidō, NE Japan
81 N15 **Shilabo** Sumalē, E Ethiopia 06°05′N 44°48′E
139 V3 **Shilayr, Wādī** *var.* Āw-e Shilēr, Āveh-ye Shalīr. E Iraq
127 X7 **Shil'da** Orenburgskaya Oblast', W Russian Federation 51°46′N 59°48′E
Shilēr, Āw-e *see* Shilayr, Wādī
153 S12 **Shiliguri** *var.* Siliguri. West Bengal, NE India 26°46′N 88°24′E
Shiliu *see* Changjiang
129 V7 **Shilka** S Russian Federation
18 H15 **Shillington** Pennsylvania, NE USA 40°18′N 75°57′W
153 V13 **Shillong** *state capital* Meghālaya, NE India 25°37′N 91°54′E
126 M5 **Shilovo** Ryazanskaya Oblast', W Russian Federation 54°18′N 40°53′E
164 C14 **Shimabara** *var.* Simabara. Nagasaki, Kyūshū, SW Japan 32°48′N 130°20′E
164 C14 **Shimabara-wan** *bay* SW Japan
164 F12 **Shimane** *off.* Shimane-ken, *var.* Simane. ◆ *prefecture* Honshū, SW Japan
164 G11 **Shimane-hantō** *peninsula* Honshū, SW Japan
Shimane-ken *see* Shimane
123 Q13 **Shimanovsk** Amurskaya Oblast', SE Russian Federation 52°00′N 127°36′E
Shimanto *see* Nakamura
Shimbir Berris *see* Shimbiris
80 O12 **Shimbiris** *var.* Shimbir Berris. ▲ N Somalia 10°43′N 47°10′E
165 T4 **Shimizu** Hokkaidō, NE Japan 42°58′N 142°54′E
164 M14 **Shimizu** *var.* Simizu. Shizuoka, Honshū, S Japan 35°01′N 138°29′E
152 I8 **Shimla** *prev.* Simla. *state capital* Himāchal Pradesh, N India 31°07′N 77°09′E
165 N14 **Shimoda** *var.* Simoda. Shizuoka, Honshū, S Japan 34°40′N 138°55′E
165 O13 **Shimodate** *var.* Simodate. Ibaraki, Honshū, S Japan 36°20′N 140°00′E
155 F18 **Shimoga** Karnātaka, W India 13°56′N 75°31′E
164 C15 **Shimo-jima** *island* SW Japan
164 B15 **Shimo-Koshiki-jima** *island* SW Japan
81 J21 **Shimoni** Kwale, S Kenya 04°40′S 39°22′E
164 D13 **Shimonoseki** *var.* Simonoseki, *hist.* Akamagaseki, Bakan. Yamaguchi, Honshū, SW Japan 33°57′N 130°54′E
124 G14 **Shimsk** Novgorodskaya Oblast', NW Russian Federation 58°12′N 30°43′E
141 W7 **Shināş** N Oman 24°45′N 56°24′E
148 J6 **Shīndand** *prev.* Shīndand. Herāt, W Afghanistan 33°19′N 62°09′E
Shīndand *see* Shīndand
Shinei *see* Xinying
162 H10 **Shinejinst** *var.* Dzalaa. Bayanhongor, C Mongolia 44°29′N 99°19′E
25 T12 **Shiner** Texas, SW USA 29°25′N 97°10′W
167 N1 **Shingbwiyang** Kachin State, N Myanmar (Burma) 26°40′N 96°14′E
Shingozha *see* Shynkozha
164 J15 **Shingū** *var.* Singū. Wakayama, Honshū, SW Japan 33°41′N 135°57′E
14 F8 **Shining Tree** Ontario, S Canada 47°36′N 81°12′W
165 P9 **Shinjō** *var.* Sinzyô. Yamagata, Honshū, C Japan 38°47′N 140°17′E
96 I7 **Shin, Loch** N Scotland, United Kingdom
21 S3 **Shinnston** West Virginia, NE USA 39°22′N 80°19′W
138 I6 **Shinshār** *Fr.* Chinnchâr. Ḩimş, W Syria 34°36′N 36°45′E
Shinshū *see* Jinju
165 T4 **Shintoku** Hokkaidō, NE Japan 43°03′N 142°50′E
81 G20 **Shinyanga** Shinyanga, NW Tanzania 03°40′S 33°25′E
81 G20 **Shinyanga** ◆ *region* N Tanzania
165 Q10 **Shiogama** *var.* Siogama. Miyagi, Honshū, C Japan 38°19′N 141°00′E
164 M12 **Shiojiri** *var.* Sioziri. Nagano, Honshū, S Japan 36°08′N 137°58′E
164 I15 **Shiono-misaki** *headland* Honshū, SW Japan 33°25′N 135°45′E
165 Q12 **Shioya-zaki** *headland* Honshū, C Japan 37°00′N 140°57′E
114 J9 **Shipchenski Prohod** , Shipchenski Prokhod. *pass* C Bulgaria
Shipchenski Prokhod *see* Shipchenski Prohod
160 G14 **Shiping** Yunnan, SW China 23°45′N 102°23′E
13 P13 **Shippagan** *var.* Shippegan. New Brunswick, SE Canada 47°45′N 64°44′W
Shippegan *see* Shippagan
18 F15 **Shippensburg** Pennsylvania, NE USA 40°03′N 77°31′W
37 P9 **Shiprock** New Mexico, SW USA 36°47′N 108°41′W
37 O9 **Ship Rock** ▲ New Mexico, SW USA 36°41′N 108°50′W
15 R6 **Shipshaw** Québec, SE Canada
123 V10 **Shipunskiy, Mys** *headland* E Russian Federation 53°04′N 159°57′E
160 K7 **Shiquan** Shaanxi, C China 33°05′N 108°15′E
Shiquanhe *see* Gar
122 K13 **Shira** Respublika Khakasiya, S Russian Federation 54°35′N 89°58′E
Shirajganj Ghat *see* Sirajganj
165 P12 **Shirakawa** *var.* Sirakawa. Fukushima, Honshū, C Japan 37°07′N 140°11′E
164 M13 **Shirane-san** ▲ Honshū, S Japan 35°39′N 138°13′E
165 U14 **Shiranuka** Hokkaidō, NE Japan 42°57′N 144°01′E
195 N12 **Shirase Coast** *physical region* Antarctica
165 U3 **Shiratаki** Hokkaidō, NE Japan 43°55′N 143°14′E
143 O11 **Shīrāz** *var.* Shīrāz. Fārs, S Iran 29°38′N 52°34′E
83 N15 **Shire** *var.* Chire. Malawi/Mozambique
Shiree *see* Tsagaanhayrhan
Shireet *see* Bayandelger
165 W3 **Shiretoko-hantō** *headland* Hokkaidō, NE Japan 44°06′N 145°07′E
165 W3 **Shiretoko-misaki** *headland* Hokkaidō, NE Japan 44°20′N 145°19′E
127 N5 **Shiringushi** Respublika Mordoviya, W Russian Federation 53°50′N 42°49′E
149 M3 **Shīrīn Tagāb** Fāryāb, N Afghanistan 36°49′N 65°01′E
149 N2 **Shīrīn Tagāb** N Afghanistan

◆ Country
● Country Capital
◇ Dependent Territory
○ Dependent Territory Capital
Administrative Regions
✈ International Airport
▲ Mountain
Mountain Range
Volcano
River
Lake
Reservoir

165 *R6* **Shiriya-zaki** *headland* Honshū, C Japan 41°24′N 141°27′E
144 *I12* **Shirkala, Gryada** *plain* W Kazakhstan
152 *F11* **Shir Kolāyat** *var.* Kolāyat. Rājasthān, NW India 27°56′N 73°02′E
165 *P10* **Shiroishi** *var.* Siroisi. Miyagi, Honshū, C Japan 38°00′N 140°38′E
Shirokoye *see* Shyroke
165 *O10* **Shirone** *var.* Sirone. Niigata, Honshū, C Japan 37°46′N 139°00′E
164 *L12* **Shirotori** Gifu, Honshū, SW Japan 35°53′N 136°52′E
197 *T1* **Shirshov Ridge** *undersea feature* W Bering Sea
Shirshütür/Shirshyutyur, Peski *see* Şirşütür Gumy
143 *T3* **Shīrvān** *var.* Shirwān. Khorāsān-e Shomālī, NE Iran 37°25′N 57°55′E
Shirwa, Lake *see* Chilwa, Lake
Shirwān *see* Shīrvān
159 *N5* **Shisanjianfang** Xinjiang Uygur Zizhiqu, W China 43°10′N 91°15′E
38 *M16* **Shishaldin Volcano** ▲ Unimak Island, Alaska, USA 54°45′N 163°58′W
Shishchitsy *see* Shyshchytsy
83 *G16* **Shishikola** North West, N Botswana 18°09′S 23°08′E
38 *M8* **Shishmaref** Alaska, USA 66°15′N 166°04′W
Shisur *see* Ash Shişar
164 *L13* **Shitara** Aichi, Honshū, SW Japan 35°06′N 137°33′E
152 *D12* **Shiv** Rājasthān, NW India 26°11′N 71°14′E
Shivāji Sāgar *see* Konya Reservoir
154 *H8* **Shivpuri** Madhya Pradesh, C India 25°28′N 77°41′E
36 *J9* **Shivwits Plateau** *plain* Arizona, SW USA
Shiwalik Range *see* Siwalik Range
160 *M8* **Shiyan** Hubei, C China 32°31′N 110°45′E
145 *O15* **Shiyeli** *prev.* Chiili. Kzylorda, S Kazakhstan 44°13′N 66°46′E
Shizilu *see* Junan
160 *H13* **Shizong** *var.* Danfeng. Yunnan, SW China 24°53′N 104′E
165 *R10* **Shizugawa** Miyagi, Honshū, NE Japan 38°40′N 141°26′E
165 *T5* **Shizunai** Hokkaidō, NE Japan 42°20′N 142°24′E
165 *M14* **Shizuoka** *var.* Sizuoka. Shizuoka, Honshū, S Japan 34°59′N 138°20′E
164 *M13* **Shizuoka** *off.* Shizuoka-ken, *var.* Sizuoka. ◈ *prefecture* Honshū, S Japan
Shizuoka-ken *see* Shizuoka
Shklov *see* Shklow
119 *N15* **Shklow** *Rus.* Shklov. Mahilyowskaya Voblasts', E Belarus 54°13′N 30°18′E
113 *K18* **Shkodër** *var.* Shkodra, *It.* Scutari, *SCr.* Skadar. Shkodër, NW Albania 42°03′N 19°31′E
113 *K17* **Shkodër** ◈ *district* NW Albania
Shkodra *see* Shkodër
Shkodrës, Liqeni i *see* Scutari, Lake
113 *L20* **Shkumbinit, Lumi i** *var.* Shkumbî, Shkumbin. ≈ C Albania
Shkumbî/Shkumbin *see* Shkumbinit, Lumi i
Shligigh, Cuan *see* Sligo Bay
122 *L4* **Shmidta, Ostrov** *island* Severnaya Zemlya, N Russian Federation
183 *S10* **Shoalhaven River** ≈ New South Wales, SE Australia
11 *W16* **Shoal Lake** Manitoba, S Canada 50°28′N 100°36′W
31 *O15* **Shoals** Indiana, N USA 38°40′N 86°47′W
164 *I13* **Shōdo-shima** *island* SW Japan
Shōka *see* Zhanghua
122 *M5* **Shokal'skogo, Proliv** *strait* N Russian Federation
147 *T14* **Shokhdara, Qatorkŭhi** *Rus.* Shakhdarinskiy Khrebet. ▲▲ SE Tajikistan
145 *T15* **Shokpar** *Kaz.* Shoqpar; *prev.* Chokpar. Zhambyl, S Kazakhstan 43°49′N 74°25′E
145 *P15* **Sholakkorgan** *var.* Chulakkurgan. Yuzhnyy Kazakhstan, S Kazakhstan 43°45′N 69°10′E
145 *N9* **Sholaksay** Kostanay, N Kazakhstan 51°45′N 64°45′E
Sholāpur *see* Solāpur
Sholdaneshty *see* Şoldăneşti
145 *W15* **Shonzhy** *prev.* Chundzha. Almaty, SE Kazakhstan 43°32′N 79°28′E
Shoqpar *see* Shokpar
155 *G21* **Shoranūr** Kerala, SW India 10°53′N 76°06′E
155 *G16* **Shorāpur** Karnātaka, C India 16°34′N 76°48′E
147 *O14* **Sho'rchi** *Rus.* Shurchi. Surkhondaryo Viloyati, S Uzbekistan 37°58′N 67°40′E
30 *M11* **Shorewood** Illinois, N USA 41°31′N 88°12′W
Shorkazakhly, Solonchak *see* Kazakhlyshor, Solonchak
145 *Q9* **Shortandy** Akmola, C Kazakhstan 51°45′N 71°01′E
149 *O2* **Shōr Tappeh** *var.* Shortepa, Shor Tepe; *prev.* Shūr Tappeh. Balkh, N Afghanistan 37°22′N 66°49′E
Shortepa/Shor Tepe *see* Shōr Tappeh
186 *J7* **Shortland Island** *var.* Alu. *island* Shortland Islands, NW Solomon Islands
Shosambetsu *see* Shosanbetsu
165 *S2* **Shosanbetsu** *var.* Shosambetsu. Hokkaidō, NE Japan 44°31′N 141°47′E
33 *O15* **Shoshone** Idaho, NW USA 42°56′N 114°24′W
35 *T6* **Shoshone Mountains** ▲▲ Nevada, W USA
33 *U12* **Shoshone River** ≈ Wyoming, C USA
83 *I19* **Shoshong** Central, SE Botswana 23°02′S 26°31′E
33 *V14* **Shoshoni** Wyoming, C USA 43°13′N 108°06′W
Shōshū *see* Sangju
117 *S2* **Shostka** Sums'ka Oblast', NE Ukraine 51°52′N 33°30′E
185 *C21* **Shotover** ≈ South Island, New Zealand
146 *H9* **Shovot** *Rus.* Shavat. Xorazm Viloyati, W Uzbekistan 41°41′N 60°13′E
37 *N12* **Show Low** Arizona, SW USA 34°15′N 110°01′W
Show Me State *see* Missouri
125 *O4* **Shoyna** Nenetskiy Avtonomnyy Okrug, NW Russian Federation 67°50′N 44°09′E
124 *M11* **Shozhma** Arkhangel'skaya Oblast', NW Russian Federation 61°57′N 40°10′E
117 *Q7* **Shpola** Cherkas'ka Oblast', N Ukraine 49°00′N 31°27′E
Shqipëria/Shqipërisë, Republika e *see* Albania
22 *G5* **Shreveport** Louisiana, S USA 32°32′N 93°45′W
97 *K19* **Shrewsbury** *hist.* Scrobesbyrig'. W England, United Kingdom 52°43′N 02°45′W
152 *D11* **Shri Mohangarh** *prev.* Sri Mohangorh. Rājasthān, NW India 27°17′N 71°18′E
153 *S16* **Shrīrāmpur** *prev.* Serampore, Serampur. West Bengal, NE India 22°44′N 88°20′E
97 *K19* **Shropshire** *cultural region* W England, United Kingdom
113 *N17* **Shtime** *Serb.* Štimlje. C Kosovo 42°27′N 21°03′E
145 *S16* **Shu** *Kaz.* Shū. Zhambyl, SE Kazakhstan 43°34′N 73°41′E
129 *Q7* **Shu** *Kaz.* Shū; *prev.* Chu. ≈ Kazakhstan/Kyrgyzstan
160 *G13* **Shuangbai** *var.* Tuodian. Yunnan, SW China 24°45′N 101°38′E
163 *W9* **Shuangcheng** Heilongjiang, NE China 45°20′N 126°21′E
Shuangcheng *see* Zherong
160 *E14* **Shuangjiang** *var.* Weiyuan. Yunnan, SW China 23°28′N 99°43′E
Shuangjiang *see* Jiangkou
Shuangjiang *see* Tongdao
163 *U10* **Shuangliao** *var.* Zhengjiatun. Jilin, NE China 43°31′N 123°32′E
Shuang-liao *see* Liaoyuan
Shuangshipu *see* Fengxian
163 *Y7* **Shuangyashan** *var.* Shuang-ya-shan. Heilongjiang, NE China 46°37′N 131°10′E
Shuang-ya-shan *see* Shuangyashan
Shu'aymiah *see* Shu'aymīyah
141 *W12* **Shu'aymīyah** *var.* Shu'aymiah. S Oman 17°55′N 55°39′E
Shubarkuduk *see* Shubarkudyk
144 *I10* **Shubarkudyk** *prev.* Shubarkuduk, *Kaz.* Shubarqudyq. Aktyubinsk, W Kazakhstan 49°09′N 56°31′E
Shubarqudyq *see* Shubarkudyk
145 *N12* **Shubar-Tengiz, Ozero** ◎ C Kazakhstan
39 *S5* **Shublik Mountains** ▲▲ Alaska, USA
Shubrā al Khaymah *see* Shubrâ el Kheima
121 *U13* **Shubrâ el Kheima** *var.* Shubrā al Khaymah. N Egypt 30°06′N 31°15′E
158 *E8* **Shufu** *var.* Tuokezhake. Xinjiang Uygur Zizhiqu, NW China 39°18′N 75°43′E
147 *S14* **Shughnon, Qatorkŭhi** *Rus.* Shugnanskiy Khrebet. ▲▲ SE Tajikistan
Shugnanskiy Khrebet *see* Shughnon, Qatorkŭhi
161 *Q6* **Shu He** ≈ E China
Shuicheng *see* Lupanshui
Shuiding *see* Huocheng
Shuidong *see* Dianbai
Shuiji *see* Laixi
Shū-Ile Taūlary *see* Gory Shu-Ile
Shuilocheng *see* Zhuanglang
Shuiluo *see* Zhuanglang
149 *T10* **Shujāābād** Punjab, E Pakistan 29°53′N 71°23′E
Shū, Kazakhstan *see* Shu
Shū, Kazakhstan/Kyrgyzstan *see* Shu
163 *W9* **Shulan** Jilin, NE China 44°28′N 126°57′E
158 *E8* **Shule** Xinjiang Uygur Zizhiqu, NW China 39°19′N 76°06′E
Shuleh *see* Shule He
159 *Q8* **Shule He** *var.* Shuleh, Sulo. ≈ C China
30 *K9* **Shullsburg** Wisconsin, N USA 42°37′N 90°12′W
Shulu *see* Xinji
39 *N16* **Shumagin Islands** *island group* Alaska, USA
146 *G7* **Shumanay** Qoraqalpog'iston Respublikasi, W Uzbekistan 42°42′N 58°56′E
114 *M8* **Shumen** Shumen, NE Bulgaria 43°17′N 26°57′E
114 *M8* **Shumen** ◈ *province* NE Bulgaria
127 *P4* **Shumerlya** Chuvashskaya Respublika, W Russian Federation 55°31′N 46°24′E
122 *G11* **Shumikha** Kurganskaya Oblast', C Russian Federation 55°12′N 63°09′E
118 *M12* **Shumilina** *Rus.* Shumilino. Vitsyebskaya Voblasts', NE Belarus 55°18′N 29°37′E
Shumilino *see* Shumilina
123 *V11* **Shumshu, Ostrov** *island* SE Russian Federation
116 *K5* **Shums'k** Ternopil's'ka Oblast', W Ukraine 50°06′N 26°04′E
Shūnan *see* Tokuyama
39 *O7* **Shungnak** Alaska, USA 66°53′N 157°08′W
Shunsen *see* Chuncheon
Shuoxian *see* Shuozhou
161 *N3* **Shuozhou** *var.* Shuoxian. Shanxi, C China 39°20′N 112°25′E
141 *P16* **Shuqrah** *var.* Shaqrā. SW Yemen 13°26′N 45°44′E
Shurab *see* Shŭrob
Shurchi *see* Sho'rchi
147 *R11* **Shŭrob** *Rus.* Shurab. NW Tajikistan 40°02′N 70°31′E
143 *T10* **Shūr, Rūd-e** ≈ E Iran
Shūr Tappeh *see* Shōr Tappeh
83 *K17* **Shurugwi** *prev.* Selukwe. Midlands, C Zimbabwe 19°40′S 30°00′E
142 *L8* **Shūsh** *anc.* Susa, *Bibl.* Shushan. Khūzestān, SW Iran 32°12′N 48°20′E
Shushan *see* Shūsh
142 *L9* **Shūshtar** *var.* Shustar, Shushter. Khūzestān, SW Iran 32°03′N 48°51′E
Shushter/Shustar *see* Shūshtar
141 *T9* **Shuṭfah, Qalamat** *well* E Saudi Arabia
139 *V9* **Shuwayjah, Hawr ash** *var.* Hawr as Suwayqiyah. ◎ E Iraq
124 *M16* **Shuya** Ivanovskaya Oblast', W Russian Federation 56°51′N 41°24′E
39 *Q14* **Shuyak Island** *island* Alaska, USA
166 *M4* **Shwebo** Sagaing, C Myanmar (Burma) 22°35′N 95°42′E
166 *L7* **Shwedaung** Bago, W Myanmar (Burma) 18°44′N 95°12′E
166 *M7* **Shwegyin** Bago, SW Myanmar (Burma) 17°56′N 96°59′E
167 *N4* **Shweli** *Chin.* Longchuan Jiang. ≈ Myanmar (Burma)/China
166 *M6* **Shwemyo** Mandalay, C Myanmar (Burma) 20°04′N 96°13′E
145 *S14* **Shyganak** *var.* Čiganak, Chiganak, *Kaz.* Shyghanaq. Zhambyl, SE Kazakhstan 45°10′N 73°55′E
Shyghanaq *see* Shyganak
Shyghys Qazagastan Oblysy *see* Vostochnyy Kazakhstan
Shyghys Qongyrat *see* Shygys Konyrat
145 *T12* **Shygys Konyrat** , *Kaz.* Shyghys Qongyrat. Karaganda, C Kazakhstan 47°01′N 75°05′E
119 *M19* **Shyichy** *Rus.* Shiichi. Homyel'skaya Voblasts', SE Belarus 52°15′N 29°14′E
145 *Q17* **Shymkent** *prev.* Chimkent. Yuzhnyy Kazakhstan, S Kazakhstan 42°19′N 69°36′E
144 *H9* **Shyngghyrlau** *prev.* Chingirlau. Zapadnyy Kazakhstan, W Kazakhstan 51°10′N 53°44′E
144 *G9* **Shyngyrlau** *prev.* Utva. ≈ NW Kazakhstan
145 *W11* **Shynkozha** *prev.* Shingozha. Vostochnyy Kazakhstan, E Kazakhstan 47°46′N 80°38′E
152 *J5* **Shyok** Jammu and Kashmir, NW India 34°13′N 78°12′E
117 *S9* **Shyroke** *Rus.* Shirokoye. Dnipropetrovs'ka Oblast', E Ukraine 47°41′N 33°16′E
117 *O9* **Shyryayeve** Odes'ka Oblast', SW Ukraine 47°21′N 30°11′E
117 *S5* **Shyshaky** Poltavs'ka Oblast', C Ukraine 49°54′N 34°00′E
119 *K17* **Shyshchytsy** *Rus.* Shishchitsy. Minskaya Voblasts', C Belarus 53°13′N 27°33′E
149 *Y3* **Siachen Muztāgh** ▲▲ NE Pakistan
Siadehan *see* Tākestān
148 *M13* **Siāhān Range** ▲▲ W Pakistan
142 *I1* **Sīāh Chashmeh** *var.* Chālderān. Āẕarbāyjān-e Gharbī, N Iran 39°02′N 44°22′E
149 *W7* **Siālkot** Punjab, NE Pakistan 32°29′N 74°35′E
186 *E7* **Sialum** Morobe, C Papua New Guinea 06°02′S 147°37′E
Siam *see* Thailand
Siam, Gulf of *see* Thailand, Gulf of
Sian *see* Xi'an
Siang *see* Brahmaputra
Siangtan *see* Xiangtan
169 *N8* **Siantan, Pulau** *island* Kepulauan Anambas, W Indonesia
54 *H11* **Siare, Río** ≈ C Colombia
171 *R6* **Siargao Island** *island* S Philippines
186 *F72* **Siassi** Umboi Island, C Papua New Guinea 05°34′S 147°50′E
115 *D14* **Siátista** Dytikí Makedonía, N Greece 40°16′N 21°34′E
166 *K4* **Siatlai** Chin State, W Myanmar (Burma) 22°05′N 93°36′E
171 *P6* **Siaton** Negros, C Philippines 09°03′N 123°03′E
171 *P6* **Siaton Point** *headland* Negros, C Philippines 09°03′N 123°00′E
118 *F11* **Šiauliai** *Ger.* Schaulen. Šiauliai, N Lithuania 55°55′N 23°21′E
118 *E11* **Šiauliai** ◈ *province* N Lithuania
171 *Q10* **Siau, Pulau** *island* N Indonesia
83 *J15* **Siavonga** Southern, SE Zambia 16°33′S 28°42′E
81 *G18* **Siaya** ◈ *county* W Kenya
Siazan' *see* Siyäzän
Sibah *see* As Sībah
107 *N20* **Sibari** Calabria, S Italy 39°45′N 16°26′E
Sibata *see* Shibata
127 *X6* **Sibay** Respublika Bashkortostan, W Russian Federation 52°40′N 58°39′E
93 *M19* **Sibbo** *Fin.* Sipoo. Uusimaa, S Finland 60°22′N 25°20′E
112 *D13* **Šibenik** *It.* Sebenico. Šibenik-Knin, S Croatia 43°43′N 15°54′E
Sibenik *see* Šibenik-Knin
112 *E13* **Šibenik-Knin** *off.* Šibenska Županija, *var.* Sibenik. ◈ *province* S Croatia
Šibenik-Knin *see* Drniš
Šibenska Županija *see* Šibenik-Knin
Siberia *see* Sibir'
Siberoet *see* Siberut, Pulau
168 *H12* **Siberut, Pulau** *prev.* Siberoet. *island* Kepulauan Mentawai, W Indonesia
168 *I12* **Siberut, Selat** *strait* W Indonesia
149 *P11* **Sibi** Baluchistān, SW Pakistan 29°31′N 67°54′E
186 *B9* **Sibidiri** Western, SW Papua New Guinea 08°58′S 142°14′E
123 *N10* **Sibir'** *var.* Siberia. *physical region* NE Russian Federation
79 *F20* **Sibiti** Lékoumou, S Congo 03°41′S 13°20′E
81 *G21* **Sibiti** ≈ C Tanzania
116 *I12* **Sibiu** *Ger.* Hermannstadt, *Hung.* Nagyszeben. Sibiu, C Romania 45°48′N 24°09′E
116 *I11* **Sibiu** ◈ *county* C Romania
29 *S11* **Sibley** Iowa, C USA 43°24′N 95°45′W
153 *Y11* **Sibsāgar** *var.* Sivasagar. Assam, NE India 26°59′N 94°38′E
169 *R9* **Sibu** Sarawak, East Malaysia 02°18′N 111°49′E
42 *G2* **Sibun** ≈ E Belize
79 *I15* **Sibut** *prev.* Fort-Sibut. Kémo, S Central African Republic 05°44′N 19°07′E
171 *P4* **Sibuyan Island** *island* C Philippines
189 *U1* **Sibylla Island** *island* N Marshall Islands
11 *N16* **Sicamous** British Columbia, SW Canada 50°49′N 118°52′W
Sichelburger Gerbirge *see* Gorjanci
167 *N14* **Sichon** *var.* Ban Sichon, Si Chon. Nakhon Si Thammarat, SW Thailand 09°03′N 99°51′E
Si Chon *see* Sichon
160 *H9* **Sichuan** *var.* Chuan, Sichuan Sheng, Ssu-ch'uan, Szechuan, Szechwan. ◈ *province* C China
160 *I9* **Sichuan Pendi** *basin* C China
Sichuan Sheng *see* Sichuan
103 *S16* **Sicie, Cap** *headland* SE France 43°03′N 05°50′E
107 *J24* **Sicilia** *Eng.* Sicily; *anc.* Trinacria. ◈ *region* Italy, C Mediterranean Sea
107 *M24* **Sicilia** *Eng.* Sicily; *anc.* Trinacria. *island* Italy, C Mediterranean Sea
Sicilian Channel *see* Sicily, Strait of
107 *H24* **Sicily, Strait of** *var.* Sicilian Channel. *strait* C Mediterranean Sea
42 *K5* **Sico Tinto, Río** *var.* Río Negro. ≈ NE Honduras
57 *H16* **Sicuani** Cusco, S Peru 14°21′S 71°13′W
112 *J10* **Šid** Vojvodina, NW Serbia 45°07′N 19°13′E
115 *A15* **Sidári** Kérkyra, Iónia Nisiá, Greece, C Mediterranean Sea 39°47′N 19°43′E
169 *Q11* **Sidas** Borneo, C Indonesia 0°24′N 109°46′E
98 *O5* **Siddeburen** Groningen, NE Netherlands 53°15′N 06°52′E
154 *D9* **Siddhapur** *prev.* Siddhpur, Sidhpur. Gujarāt, W India 23°57′N 72°28′E
Siddhpur *see* Siddhapur
155 *I15* **Siddipet** Telangana, C India 18°10′N 78°54′E
77 *N14* **Sidéradougou** SW Burkina Faso 10°39′N 04°16′W
107 *N23* **Siderno** Calabria, SW Italy 38°18′N 16°19′E
Siders *see* Sierre
154 *L9* **Sidhi** Madhya Pradesh, C India 24°24′N 81°54′E
Sidhirókastron *see* Sidirókastro
Sidhpur *see* Siddhapur
75 *U7* **Sîdi Barrâni** NW Egypt 31°38′N 25°58′E
74 *I6* **Sidi Bel Abbès** *var.* Sidi bel Abbès, Sidi-Bel-Abbès. NW Algeria 35°12′N 00°43′W
74 *E7* **Sidi-Bennour** W Morocco 32°39′N 08°28′W
74 *M6* **Sidi Bouzid** *var.* Gammouda, Sīdī Bu Zayd. C Tunisia 35°05′N 09°20′E
Sīdī Bu Zayd *see* Sidi Bouzid
74 *D8* **Sidi-Ifni** SW Morocco 29°33′N 10°04′W
74 *G6* **Sidi-Kacem** *prev.* Petitjean. N Morocco 34°21′N 05°46′W
114 *G12* **Sidirókastro** *prev.* Sidhirókastron. Kentrikí Makedonía, NE Greece 41°14′N 23°23′E
194 *L12* **Sidley, Mount** ▲ Antarctica 76°39′S 124°48′W
29 *S16* **Sidney** Iowa, C USA 40°45′N 95°39′W
33 *Y7* **Sidney** Montana, NW USA 47°42′N 104°10′W
28 *J15* **Sidney** Nebraska, C USA 41°09′N 102°57′W
18 *I11* **Sidney** New York, NE USA 42°18′N 75°21′W
31 *R13* **Sidney** Ohio, N USA 40°16′N 84°09′W
23 *T2* **Sidney Lanier, Lake** ⊠ Georgia, SE USA
Sidon *see* Saïda
122 *J9* **Sidorovsk** Yamalo-Nenetskiy Avtonomnyy Okrug, N Russian Federation 66°34′N 82°12′E
Sidra *see* Surt
Sidra/Sidra, Gulf of *see* Surt, Khalīj, N Libya
Siebenbürgen *see* Transylvania
Sieben Dörfer *see* Săcele
110 *O12* **Siedlce** *Ger.* Sedlez, *Rus.* Sesdlets. Mazowieckie, C Poland 52°10′N 22°18′E
101 *E16* **Sieg** ≈ W Germany
101 *F16* **Siegen** Nordrhein-Westfalen, W Germany 50°53′N 08°02′E
109 *X4* **Sieghartskirchen** Niederösterreich, E Austria 48°13′N 16°01′E
167 *S12* **Siĕmbok** *prev.* Phumĭ Siĕmbok. Stœng Trêng, N Cambodia 13°28′N 105°59′E
110 *O11* **Siemiatycze** Podlaskie, NE Poland 52°27′N 22°52′E
167 *T11* **Siĕmpang** Stœng Trêng, NE Cambodia 14°07′N 106°24′E
167 *R11* **Siĕmréab** *prev.* Siemreap. Siĕmréab, NW Cambodia 13°21′N 103°50′E
Siemreap *see* Siĕmréab
106 *G12* **Siena** *Fr.* Sienne; *anc.* Saena Julia. Toscana, C Italy 43°20′N 11°20′E
Sienne *see* Siena
92 *K12* **Sieppijärvi** Lappi, NW Finland 67°09′N 23°58′E
110 *J13* **Sieradz** Sieradz, C Poland 51°36′N 18°42′E
110 *K10* **Sierpc** Mazowieckie, C Poland 52°51′N 19°44′E
24 *I9* **Sierra Blanca** Texas, SW USA 31°10′N 105°22′W
37 *S14* **Sierra Blanca Peak** ▲ New Mexico, SW USA 33°22′N 105°48′W
35 *P5* **Sierra City** California, W USA 39°34′N 120°35′W
63 *I16* **Sierra Colorada** Río Negro, S Argentina 40°37′S 67°48′W
63 *J16* **Sierra Grande** Río Negro, E Argentina 41°34′S 65°21′W
76 *G15* **Sierra Leone** *off.* Republic of Sierra Leone. ◆ *republic* W Africa
64 *M13* **Sierra Leone Basin** *undersea feature* E Atlantic Ocean 05°00′N 17°00′W
66 *K8* **Sierra Leone Fracture Zone** *tectonic feature* E Atlantic Ocean
Sierra Leone, Republic of *see* Sierra Leone
Sierra Leone Ridge *see* Sierra Leone Rise
64 *L13* **Sierra Leone Rise** *var.* Sierra Leone Ridge, Sierra Leone Schwelle. *undersea feature* E Atlantic Ocean 05°30′N 21°00′W
Sierra Leone Schwelle *see* Sierra Leone Rise
40 *L7* **Sierra Mojada** Coahuila, NE Mexico 27°13′N 103°42′W
37 *N16* **Sierra Vista** Arizona, SW USA 31°33′N 110°18′W
108 *D10* **Sierre** *Ger.* Siders. Valais, SW Switzerland 46°18′N 07°33′E
36 *L16* **Sierrita Mountains** ▲▲ Arizona, SW USA
Siete Moai *see* Ahu Akivi
76 *M15* **Sifié** W Ivory Coast 07°59′N 06°55′W
115 *I21* **Sífnos** *anc.* Siphnos. *island* Kykládes, Greece, Aegean Sea
115 *I21* **Sífnou, Stenó** *strait* SE Greece
Siga *see* Shiga
103 *P16* **Sigean** Aude, S France 43°02′N 02°58′E
Sighet *see* Sighetu Marmaţiei
Sighetul Marmaţiei *see* Sighetu Marmaţiei
116 *I8* **Sighetu Marmaţiei** *var.* Sighet, Sighetul Marmaţiei, *Hung.* Máramarossziget. Maramureş, N Romania 47°56′N 23°53′E
116 *I11* **Sighişoara** *Ger.* Schässburg, *Hung.* Segesvár. Mureş, C Romania 46°12′N 24°48′E
168 *G7* **Sigli** Sumatera, W Indonesia 05°21′N 95°56′E
92 *J1* **Siglufjörður** Norðurland Vestra, N Iceland 66°09′N 18°56′W
101 *H23* **Sigmaringen** Baden-Württemberg, S Germany 48°04′N 09°12′E
101 *N20* **Signalberg** ▲ SE Germany 49°30′N 12°34′E
36 *I13* **Signal Peak** ▲ Arizona, SW USA 33°20′N 114°03′W
Signan *see* Xi'an
194 *H1* **Signy** *UK research station* South Orkney Islands, Antarctica 60°27′S 45°35′W
29 *X15* **Sigourney** Iowa, C USA 41°19′N 92°12′W
115 *K17* **Sígri, Akrotírio** *headland* Lésvos, E Greece 39°12′N 25°49′E
Sigsbee Deep *see* Mexico Basin
47 *N2* **Sigsbee Escarpment** *undersea feature* N Gulf of Mexico 26°00′N 92°30′W
56 *C8* **Sigsig** Azuay, S Ecuador 03°04′S 78°50′W
95 *O15* **Sigtuna** Stockholm, C Sweden 59°36′N 17°44′E
42 *H6* **Siguatepeque** Comayagua, W Honduras 14°39′N 87°48′W
105 *P7* **Sigüenza** Castilla-La Mancha, C Spain 41°04′N 02°38′W
105 *R4* **Sigües** Aragón, NE Spain 42°39′N 01°00′W
76 *K13* **Siguiri** NE Guinea 11°26′N 09°08′W
118 *G8* **Sigulda** *Ger.* Segewold. C Latvia 57°08′N 24°51′E
167 *Q14* **Sihanoukville** *var.* Kâmpōng Saôm; *prev.* Kompong Som. Kâmpóng Saôm, SW Cambodia 10°38′N 103°30′E
108 *G8* **Sihlsee** ◎ NW Switzerland
93 *K18* **Siikainen** Satakunta, SW Finland 61°52′N 21°49′E
93 *M16* **Siilinjärvi** Pohjois-Savo, C Finland 63°05′N 27°40′E
137 *R15* **Siirt** *var.* Sert; *anc.* Tigranocerta. Siirt, SE Turkey 37°56′N 41°56′E
137 *R15* **Siirt** *var.* Sert. ◈ *province* SE Turkey
187 *N8* **Sikaiana** *var.* Stewart Islands. *island group* W Solomon Islands
Sikandarabad *see* Secunderābād
152 *J11* **Sikandra Rao** Uttar Pradesh, N India 27°42′N 78°21′E
10 *M11* **Sikanni Chief** British Columbia, W Canada 57°16′N 122°44′W
10 *M11* **Sikanni Chief** ≈ British Columbia, W Canada
152 *H11* **Sīkar** Rājasthān, N India 27°33′N 75°12′E
76 *M13* **Sikasso** Sikasso, S Mali 11°21′N 05°43′W
76 *L13* **Sikasso** ◈ *region* SW Mali
167 *N3* **Sikaw** Kachin State, C Myanmar (Burma) 23°50′N 97°04′E
83 *H14* **Sikelenge** Western, W Zambia 14°51′S 24°13′E
27 *Y7* **Sikeston** Missouri, C USA 36°52′N 89°35′W
93 *J14* **Sikfors** Norrbotten, N Sweden 65°29′N 21°17′E
123 *T14* **Sikhote-Alin', Khrebet** ▲▲ SE Russian Federation
Siking *see* Xi'an
115 *J22* **Síkinos** *island* Kykládes, Greece, Aegean Sea
153 *S11* **Sikkim** *Tib.* Denjong. ◈ *state* N India
111 *I26* **Siklós** Baranya, SW Hungary 45°51′N 18°18′E
Sikoku *see* Shikoku
Sikoku Basin *see* Shikoku Basin
83 *G14* **Sikongo** Western, W Zambia 15°03′S 22°07′E
Sikotu Ko *see* Shikotsu-ko
Sikouri/Sikoúrion *see* Sykoúrio
123 *P8* **Siktyakh** Respublika Sakha (Yakutiya), NE Russian Federation 69°45′N 124°42′E
78 *K11* **Sila** *off.* Région du Sila. ◈ *region* E Chad
118 *D12* **Šilalė** Tauragė, W Lithuania 55°29′N 22°10′E
106 *G5* **Silandro** *Ger.* Schlanders. Trentino-Alto Adige, N Italy 46°39′N 10°55′E
41 *N12* **Silao** Guanajuato, C Mexico 20°56′N 101°28′W
Sila, Région du *see* Sila
Silarius *see* Sele
153 *W14* **Silchar** Assam, NE India 24°49′N 92°48′E
108 *G9* **Silenen** Uri, C Switzerland 46°49′N 08°39′E
21 *T9* **Siler City** North Carolina, SE USA 35°43′N 79°27′W
33 *U11* **Silesia** Montana, NW USA 45°32′N 108°52′W
110 *F13* **Silesia** *physical region* SW Poland
74 *K12* **Silet** S Algeria 22°45′N 04°51′E
Sileti *see* Silety
Siletitengiz *see* Siletyteniz, Ozero
145 *R8* **Silety** *prev.* Sileti. ≈ N Kazakhstan
145 *R7* **Siletyteniz, Ozero** *Kaz.* Siletitengiz. ◎ N Kazakhstan
172 *H16* **Silhouette** *island* Inner Islands, SE Seychelles
136 *I17* **Silifke** *anc.* Seleucia. İçel, S Turkey 36°22′N 33°57′E
Siliguri *see* Shiliguri
156 *J10* **Siling Co** ◎ W China
Silinhot *see* Xilinhot
192 *G14* **Silisili, Mauga** ▲ Savai'i, C Samoa 13°37′S 172°26′W
114 *M6* **Silistra** *var.* Silistria; *anc.* Durostorum. Silistra, NE Bulgaria 44°06′N 27°17′E
114 *M7* **Silistra** ◈ *province* NE Bulgaria
Silistria *see* Silistra
136 *D10* **Silivri** İstanbul, NW Turkey 41°05′N 28°15′E
94 *L13* **Siljan** ◎ C Sweden
95 *G22* **Silkeborg** Midtjylland, C Denmark 56°10′N 09°34′E
108 *M8* **Sill** ≈ W Austria
105 *S10* **Silla** Valenciana, E Spain 39°22′N 00°25′W
62 *H3* **Sillajguay, Cordillera** ▲ N Chile 19°45′S 68°39′W
118 *K3* **Sillamäe** *Ger.* Sillamäggi. Ida-Virumaa, NE Estonia 59°23′N 27°45′E
Sillamäggi *see* Sillamäe
Sillein *see* Žilina
109 *P9* **Sillian** Tirol, W Austria 46°45′N 12°25′E
112 *B10* **Šilo** Primorje-Gorski Kotar, NW Croatia 45°09′N 14°39′E
27 *R9* **Siloam Springs** Arkansas, C USA 36°11′N 94°32′W
25 *X10* **Silsbee** Texas, SW USA 30°21′N 94°10′W
143 *W15* **Sīlūp, Rūd-e** ≈ SE Iran
118 *C12* **Šilutė** *Ger.* Heydekrug. Klaipėda, W Lithuania 55°20′N 21°30′E
137 *Q15* **Silvan** Dıyarbakır, SE Turkey 38°08′N 41°E
108 *J10* **Silvaplana** Graubünden, S Switzerland 46°27′N 09°45′E
Silva Porto *see* Kuito
58 *M12* **Silva, Recife do** *reef* E Brazil
154 *D12* **Silvassa** Dādra and Nagar Haveli, W India 20°13′N 73°03′E
29 *X4* **Silver Bay** Minnesota, N USA 47°17′N 91°15′W
37 *P15* **Silver City** New Mexico, SW USA 32°47′N 108°16′W
18 *D10* **Silver Creek** New York, NE USA 42°32′N 79°10′W
37 *N12* **Silver Creek** ≈ Arizona, SW USA
27 *P4* **Silver Lake** Kansas, C USA 39°06′N 95°51′W
32 *I14* **Silver Lake** Oregon, NW USA 43°07′N 121°04′W
35 *T9* **Silver Peak Range** ▲▲ Nevada, W USA
21 *W3* **Silver Spring** Maryland, NE USA 39°00′N 77°01′W
Silver State *see* Colorado
Silver State *see* Nevada
37 *Q7* **Silverton** Colorado, C USA 37°48′N 107°39′W
18 *K16* **Silverton** New Jersey, NE USA 40°00′N 74°09′W
32 *G11* **Silverton** Oregon, NW USA 45°00′N 122°46′W
25 *N4* **Silverton** Texas, SW USA 34°28′N 101°18′W
104 *G14* **Silves** Faro, S Portugal 37°11′N 08°26′W
54 *D12* **Silvia** Cauca, SW Colombia 02°37′N 76°24′W
108 *J9* **Silvrettagruppe** ▲▲ Austria/Switzerland
Sily-Vajdej *see* Vulcan
108 *L7* **Silz** Tirol, W Austria 47°17′N 11°00′E
172 *I13* **Sima** Anjouan, SE Comoros 12°11′S 44°18′E
Simabara *see* Shimabara
83 *H15* **Simakando** Western, W Zambia 16°43′S 24°46′E
Simane *see* Shimane
119 *L20* **Simanichy** *Rus.* Simonichi. Homyel'skaya Voblasts', SE Belarus 51°53′N 28°05′E
160 *F14* **Simao** Yunnan, SW China 22°30′N 101°06′E
153 *P12* **Simarā** Central, C Nepal 27°14′N 85°00′E
14 *I8* **Simard, Lac** ◎ Québec, SE Canada
136 *D13* **Simav** Kütahya, W Turkey 39°05′N 28°59′E
136 *D13* **Simav Çayı** ≈ NW Turkey
79 *L18* **Simba** Orientale, N Dem. Rep. Congo 0°46′N 22°54′E
186 *C7* **Simbai** Madang, N Papua New Guinea 05°12′S 144°33′E
Simbirsk *see* Ul'yanovsk
Simbu *see* Chimbu
14 *F17* **Simcoe** Ontario, S Canada 42°50′N 80°19′W
14 *H14* **Simcoe, Lake** ◎ Ontario, S Canada
80 *J11* **Sīmēn** ▲▲ N Ethiopia
114 *K11* **Simeonovgrad** *prev.* Maritsa. Haskovo, S Bulgaria 42°03′N 25°36′E
116 *G11* **Simeria** *Ger.* Pischk, *Hung.* Piski. Hunedoara, W Romania 45°51′N 23°00′E
107 *L24* **Simeto** ≈ Sicilia, Italy, C Mediterranean Sea
168 *G9* **Simeulue, Pulau** *island* NW Indonesia
117 *T13* **Simferopol'** Avtonomna Respublika Krym, S Ukraine 44°55′N 33°06′E
117 *T13* **Simferopol'** ✈ Avtonomna Respublika Krym, S Ukraine 44°55′N 34°04′E
Simi *see* Sými
152 *M9* **Simikot** Far Western, NW Nepal 30°02′N 81°49′E
54 *F7* **Simití** Bolívar, N Colombia 07°57′N 73°57′W
114 *G11* **Simitla** Blagoevgrad, SW Bulgaria 41°57′N 23°06′E
35 *S15* **Simi Valley** California, W USA 34°16′N 118°47′W
81 *G21* **Simiyu** *off.* Mkoa wa Simiyu. ◈ *region* N Tanzania
Simiyu, Mkoa wa *see* Simiyu
Simizu *see* Shimizu
Simla *see* Shimla
Şimlăul Silvaniei/Şimleul Silvaniei *see* Şimleu Silvaniei
116 *G9* **Şimleu Silvaniei** *Hung.* Szilágysomlyó; *prev.* Şimlăul Silvaniei, Şimleul Silvaniei. Sălaj, NW Romania 47°12′N 22°49′E
Simmer *see* Simmerbach
101 *E19* **Simmerbach** *var.* Simmer. ≈ W Germany
101 *F18* **Simmern** Rheinland-Pfalz, W Germany 50°00′N 07°30′E
22 *I7* **Simmesport** Louisiana, S USA 30°58′N 91°48′W
119 *F14* **Simnas** Alytus, S Lithuania 54°23′N 23°40′E
92 *L13* **Simo** Lappi, NW Finland 65°40′N 25°04′E
Simoda *see* Shimoda
Simodate *see* Shimodate
92 *M13* **Simojärvi** ◎ N Finland
92 *L13* **Simojoki** ≈ NW Finland
41 *U15* **Simojovel** *var.* Simojovel de Allende. Chiapas, SE Mexico 17°14′N 92°40′W
Simojovel de Allende *see* Simojovel
56 *B7* **Simon Bolivar** *var.* Guayaquil. ✈ (Quayaquil) Guayas, W Ecuador 02°16′S 79°54′W
54 *L5* **Simon Bolívar** ✈ (Caracas) Vargas, N Venezuela 10°33′N 66°54′W
Simonichi *see* Simanichy
14 *M12* **Simon, Lac** ◎ Québec, SE Canada
Simonoseki *see* Shimonoseki
Šimonovany *see* Partizánske
Simonstad *see* Simon's Town
83 *E26* **Simon's Town** *var.* Simonstad. Western Cape, SW South Africa 34°12′S 18°26′E
Simony *see* Partizánske
Simpeln *see* Simplon
99 *M18* **Simpelveld** Limburg, SE Netherlands 50°50′N 05°59′E
108 *E11* **Simplon** *var.* Simpeln. Valais, SW Switzerland 46°13′N 08°01′E
108 *E11* **Simplon Pass** *pass* S Switzerland
106 *C6* **Simplon Tunnel** *tunnel* Italy/Switzerland
Simpson *see* Fort Simpson
182 *G1* **Simpson Desert** *desert* Northern Territory/South Australia
10 *J9* **Simpson Peak** ▲ British Columbia, W Canada 59°43′N 131°29′W
9 *N7* **Simpson Peninsula** *peninsula* Nunavut, NE Canada
21 *P11* **Simpsonville** South Carolina, SE USA 34°44′N 82°15′W
95 *L23* **Simrishamn** Skåne, S Sweden 55°34′N 14°20′E
123 *U13* **Simushir, Ostrov** *island* Kuril'skiye Ostrova, SE Russian Federation
168 *G9* **Sinabang** Sumatera, W Indonesia 02°27′N 96°24′E
81 *N15* **Sina Dhaqa** Galguduud, C Somalia 05°21′N 46°21′E
75 *X8* **Sinai** *var.* Sinai Peninsula, *Ar.* Shibh Jazirat Sīnā', Sīnā. *physical region* NE Egypt
116 *J12* **Sinaia** Prahova, SE Romania 45°20′N 25°33′E
188 *B16* **Sinajana** C Guam 13°28′N 144°45′E
40 *H8* **Sinaloa** ◈ *state* C Mexico
54 *H4* **Sinamaica** Zulia, NW Venezuela 11°06′N 71°52′W
163 *X14* **Sinan-ni** SE North Korea 38°37′N 127°43′E
Sīnā/Sinai Peninsula *see* Sinai
Sīnāwan *see* Sīnāwin
75 *N8* **Sīnāwin** *var.* Sīnāwan. NW Libya 31°00′N 10°37′E
83 *J16* **Sinazongwe** Southern, S Zambia 17°14′S 27°27′E
166 *L6* **Sinbaungwe** Magway, W Myanmar (Burma) 19°44′N 95°10′E
166 *L5* **Sinbyugyun** Magway, W Myanmar (Burma) 20°38′N 94°40′E
54 *E6* **Since** Sucre, NW Colombia 09°14′N 75°08′W
54 *E6* **Sincelejo** Sucre, NW Colombia 09°17′N 75°23′W
23 *U4* **Sinclair, Lake** ⊠ Georgia, SE USA
10 *M14* **Sinclair Mills** British Columbia, SW Canada 54°03′N 121°37′W
154 *I8* **Sind** ≈ N India
Sind *see* Sindh
95 *H19* **Sindal** Nordjylland, N Denmark 57°29′N 10°13′E
171 *P7* **Sindañgan** Mindanao, S Philippines 08°09′N 122°59′E
79 *D19* **Sindara** Ngounié, W Gabon 01°07′S 10°41′E
152 *E13* **Sindari** *prev.* Sindri. Rājasthān, N India 25°32′N 71°58′E
114 *N8* **Sindel** Varna, E Bulgaria 43°07′N 27°35′E
101 *H22* **Sindelfingen** Baden-Württemberg, SW Germany 48°43′N 09°E
155 *G16* **Sindgi** Karnātaka, C India 17°01′N 76°22′E
149 *Q14* **Sindh** *prev.* Sind. ◈ *province* SE Pakistan
118 *G5* **Sindi** *Ger.* Zintenhof. Pärnumaa, SW Estonia 58°28′N 24°41′E
136 *C13* **Sındırgı** Balıkesir, W Turkey 39°13′N 28°10′E
77 *N14* **Sindou** SW Burkina Faso 10°35′N 05°04′W
Sindri *see* Sindari

◆ Country
● Country Capital
◇ Dependent Territory
○ Dependent Territory Capital
◈ Administrative Regions
✈ International Airport
▲ Mountain
▲▲ Mountain Range
Volcano
≈ River
◎ Lake
⊠ Reservoir

149 T9 **Sind Sāgar Doāb** *desert* E Pakistan
126 M11 **Sinegorskiy** Rostovskaya Oblast', SW Russian Federation 48°01′N 40°52′E
123 S9 **Sinegor'ye** Magadanskaya Oblast', E Russian Federation 62°04′N 150°33′E
114 O12 **Sinekli** İstanbul, NW Turkey 41°13′N 28°13′E
104 F12 **Sines** Setúbal, S Portugal 37°58′N 08°52′W
104 F12 **Sines, Cabo de** *headland* S Portugal 37°57′N 08°55′W
92 L12 **Sinettä** Lappi, NW Finland 66°39′N 25°25′E
186 H6 **Sinewit, Mount** ▲ New Britain, C Papua New Guinea 04°42′S 151°58′E
80 G11 **Singa** *var.* Sinja, Sinjah. Sinnar, E Sudan 13°11′N 33°55′E
78 J12 **Singako** Moyen-Chari, S Chad 09°52′N 19°31′E
Singan *see* Xi'an
168 K10 **Singapore** ● (Singapore) S Singapore 01°17′N 103°48′E
168 L10 **Singapore** *off.* Republic of Singapore. ◆ *republic* SE Asia
Singapore, Republic of *see* Singapore
169 U17 **Singaraja** Bali, C Indonesia 08°06′S 115°04′E
167 O10 **Sing Buri** *var.* Singhaburi. Sing Buri, C Thailand 14°56′N 100°21′E
101 H24 **Singen** Baden-Württemberg, S Germany 47°46′N 08°50′E
Singeorgiu de Pădure *see* Sângeorgiu de Pădure
Singeorz-Băi/Singeroz Băi *see* Sângeorz-Băi
116 M9 **Singerei** *var.* Sângerei; *prev.* Lazovsk. N Moldova 47°38′N 28°08′E
Singhaburi *see* Sing Buri
81 H21 **Singida** Singida, C Tanzania 04°45′S 34°48′E
81 G22 **Singida** ◆ *region* C Tanzania
Singidunum *see* Beograd
Singkaling Hkamti *see* Hkamti
171 N14 **Singkang** Sulawesi, C Indonesia 04°09′S 119°58′E
168 J11 **Singkarak, Danau** ⊚ Sumatera, W Indonesia
169 N10 **Singkawang** Borneo, C Indonesia 0°57′N 108°57′E
168 M11 **Singkep, Pulau** *island* Kepulauan Lingga, W Indonesia
168 H9 **Singkilbaru** Sumatera, W Indonesia 02°18′N 97°47′E
183 T7 **Singleton** New South Wales, SE Australia 32°38′S 151°00′E
Singora *see* Songkhla
Singû *see* Shingū
Sining *see* Xining
107 D17 **Siniscola** Sardegna, Italy, C Mediterranean Sea 40°34′N 09°42′E
113 F14 **Sinj** Split-Dalmacija, SE Croatia 43°41′N 16°37′E
Sinjajevina *see* Sinjavina
139 P3 **Sinjār** Nīnawýa, NW Iraq 36°20′N 41°51′E
139 P2 **Sinjār, Jabal** ▲▲ N Iraq
Sinja/Sinjah *see* Singa
113 K15 **Sinjavina** *var.* Sinjajevina. ▲▲ C Montenegro
80 I7 **Sinkat** Red Sea, NE Sudan 18°52′N 36°51′E
Sinkiang/Sinkiang Uighur Autonomous Region *see* Xinjiang Uygur Zizhiqu
Sinmartin *see* Târnăveni
163 V13 **Sinmi-do** *island* NW North Korea
101 I18 **Sinn** ⛐ C Germany
Sinnamarie *see* Sinnamary
55 Y9 **Sinnamary** *var.* Sinnamarie. N French Guiana 05°23′N 53°00′W
80 G11 **Sinnar** ◆ *state* E Sudan
Sinneh *see* Sanandaj
18 E13 **Sinnemahoning Creek** ⛐ Pennsylvania, NE USA
Sînnicolau Mare *see* Sânnicolau Mare
Sinoe, Lacul *see* Sinoie, Lacul
Sinoia *see* Chinhoyi
117 N14 **Sinoie, Lacul** *prev.* Lacul Sinoe. *lagoon* SE Romania
59 H16 **Sinop** Mato Grosso, W Brazil 11°38′S 55°27′W
136 K10 **Sinop** *anc.* Sinope. Sinop, N Turkey 42°02′N 35°09′E
136 J10 **Sinop** ◆ *province* N Turkey
136 K10 **Sinop Burnu** *headland* N Turkey 42°02′N 35°12′E
Sinope *see* Sinop
Sino/Sinoe *see* Greenville
163 Y12 **Sinp'o** E North Korea 40°01′N 128°10′E
101 H20 **Sinsheim** Baden-Württemberg, SW Germany 49°15′N 08°53′E
Sîntana *see* Sântana
169 R11 **Sintang** Borneo, C Indonesia 0°03′N 111°31′E
99 F14 **Sint Annaland** Zeeland, SW Netherlands 51°36′N 04°07′E
98 L5 **Sint Annaparochie** *Fris.* Sint Anne. Fryslân, N Netherlands 53°20′N 05°46′E
Sint Anne *see* Sint Annaparochie
45 V9 **Sint Eustatius** *var.* Statia, *Eng.* Saint Eustatius. ◇ *Dutch special municipality* Sint Maarten
99 G19 **Sint-Genesius-Rode** *Fr.* Rhode-Saint-Genèse. Vlaams Brabant, C Belgium 50°45′N 04°21′E
99 F16 **Sint-Gillis-Waas** Oost-Vlaanderen, N Belgium 51°14′N 04°08′E
99 H17 **Sint-Katelijne-Waver** Antwerpen, C Belgium 51°05′N 04°31′E
99 E18 **Sint-Lievens-Houtem** Oost-Vlaanderen, NW Belgium 50°55′N 03°52′E
45 V9 **Sint Maarten** *Eng.* Saint Martin. ◇ *Dutch self-governing territory* NE Caribbean Sea
99 F14 **Sint Maartensdijk** Zeeland, SW Netherlands 51°33′N 04°05′E
99 L19 **Sint-Martens-Voeren** *Fr.* Fouron-Saint-Martin. Limburg, NE Belgium 50°46′N 05°49′E
99 J14 **Sint-Michielsgestel** Noord-Brabant, S Netherlands 51°38′N 05°21′E
Sîn-Miclăuş *see* Gheorgheni
45 O16 **Sint Nicholaas** S Aruba 12°25′N 69°52′W
99 F16 **Sint-Niklaas** *Fr.* Saint-Nicolas. Oost-Vlaanderen, N Belgium 51°10′N 04°09′E
99 K14 **Sint-Oedenrode** Noord-Brabant, S Netherlands 51°34′N 05°28′E
25 T14 **Sinton** Texas, SW USA 28°02′N 97°33′W
99 G14 **Sint Philipsland** Zeeland, SW Netherlands 51°37′N 04°11′E
99 G19 **Sint-Pieters-Leeuw** Vlaams Brabant, C Belgium 50°47′N 04°16′E
104 E11 **Sintra** *prev.* Cintra. Lisboa, W Portugal 38°48′N 09°22′W
99 J18 **Sint-Truiden** *Fr.* Saint-Trond. Limburg, NE Belgium 50°48′N 05°13′E
99 H14 **Sint Willebrord** Noord-Brabant, S Netherlands 51°33′N 04°35′E
163 V13 **Sinŭiju** W North Korea 40°08′N 124°33′E
80 P13 **Sinujiif** Nugaal, NE Somalia 08°33′N 49°05′E
Sinus Aelaniticus *see* Aqaba, Gulf of
Sinus Gallicus *see* Lion, Golfe du
Sinyang *see* Xinyang
Sinyavka *see* Sinyawka
119 I18 **Sinyawka** *Rus.* Sinyavka. Minskaya Voblasts', SW Belarus 52°57′N 26°29′E
Sinying *see* Xinying
Sinyukha *see* Synyukha
Sinzyô *see* Shinjō
111 I24 **Sió** ⛐ W Hungary
171 O7 **Siocon** Mindanao, S Philippines 07°37′N 122°09′E
111 I24 **Siófok** Somogy, Hungary 46°54′N 18°03′E
Siogama *see* Shiogama
83 G15 **Sioma** Western, SW Zambia 16°39′S 23°36′E
108 D11 **Sion** *Ger.* Sitten; *anc.* Sedunum. Valais, SW Switzerland 46°15′N 07°23′E
103 O11 **Sioule** ⛐ C France
29 S12 **Sioux Center** Iowa, C USA 43°04′N 96°10′W
29 R13 **Sioux City** Iowa, C USA 42°30′N 96°24′W
29 R11 **Sioux Falls** South Dakota, N USA 43°33′N 96°45′W
12 B11 **Sioux Lookout** Ontario, S Canada 49°27′N 94°06′W
29 T12 **Sioux Rapids** Iowa, C USA 42°53′N 95°09′W
Sioux State *see* North Dakota
Siozeri *see* Shiojiri
171 P6 **Sipalay** Negros, C Philippines 09°46′N 122°25′E
55 V11 **Sipaliwini** ◆ *district* S Suriname
45 U15 **Siparia** Trinidad, Trinidad and Tobago 10°08′N 61°31′W
Siphnos *see* Sífnos
163 V11 **Siping** *var.* Ssu-p'ing, Szeping; *prev.* Ssu-p'ing-chieh. Jilin, NE China 43°09′N 124°22′E
11 X12 **Sipiwesk** Manitoba, C Canada 55°28′N 97°16′W
11 W13 **Sipiwesk Lake** ⊚ Manitoba, C Canada
195 O11 **Siple Coast** *physical region* Antarctica
194 K13 **Siple Island** *island* Antarctica
194 K13 **Siple, Mount** ▲ Siple Island, Antarctica 73°25′S 126°24′W
Sipoo *see* Sibbo
112 G12 **Šipovo** Republika Srpska, W Bosnia and Herzegovina 44°16′N 17°05′E
23 O4 **Sipsey River** ⛐ Alabama, S USA
168 I13 **Sipura, Pulau** *island* W Indonesia
0 G16 **Siqueiros Fracture Zone** *tectonic feature* E Pacific Ocean
42 L10 **Siquia, Río** ⛐ SE Nicaragua
43 N13 **Siquirres** Limón, E Costa Rica 10°05′N 83°30′W
54 J5 **Siquisique** Lara, N Venezuela 10°36′N 69°45′W
155 G19 **Sīra** Karnātaka, W India 13°46′N 76°54′E
95 D16 **Sira** ⛐ S Norway
167 P12 **Si Racha** *var.* Ban Si Racha, Si Racha. Chon Buri, S Thailand 13°10′N 100°57′E
Si Racha *see* Si Racha
107 L25 **Siracusa** *Eng.* Syracuse. Sicilia, Italy, C Mediterranean Sea 37°04′N 15°17′E
153 T14 **Sirajganj** *var.* Shirajganj Ghat. Rajshahi, C Bangladesh 24°27′N 89°42′E
Sirakawa *see* Shirakawa
11 N14 **Sir Alexander, Mount** ▲ British Columbia, W Canada 54°00′N 120°33′W
137 O12 **Şiran** Gümüşhane, NE Turkey 40°12′N 39°07′E
77 Q12 **Sirba** ⛐ E Burkina Faso
143 O17 **Şīr Banī Yās** *island* W United Arab Emirates
95 D17 **Sirdalsvatnet** ⊚ S Norway
Sir Darya/Sirdaryo *see* Syr Darya
147 P10 **Sirdaryo** Sirdaryo Viloyati, E Uzbekistan 40°46′N 68°34′E
147 O11 **Sirdaryo Viloyati** *Rus.* Syrdar'inskaya Oblast'. ◆ *province* E Uzbekistan
Sir Donald Sangster International Airport *see* Sangster
181 S3 **Sir Edward Pellew Group** *island group* Northern Territory, NE Australia
116 K8 **Siret** *Ger.* Sereth, *Hung.* Szeret. Suceava, N Romania 47°55′N 26°05′E
116 K8 **Siret** *var.* Siretul, *Ger.* Sereth, *Rus.* Seret. ⛐ Romania/Ukraine
Siretul *see* Siret
140 K3 **Sirḥān, Wādī as** *dry watercourse* Jordan/Saudi Arabia
152 I8 **Sirhind** Punjab, N India 30°39′N 76°28′E
116 F11 **Şiria** *Ger.* Schiria. Arad, W Romania 46°16′N 21°38′E
Siria *see* Syria
143 S14 **Sīrīk** Hormozgān, SE Iran 26°32′N 57°07′E
167 P8 **Sirikit Reservoir** ⊚ N Thailand
58 K12 **Sirituba, Ilha** *island* NE Brazil
143 R11 **Sīrjān** *prev.* Sa'īdābād. Kermān, S Iran 29°29′N 55°39′E
182 H9 **Sir Joseph Banks Group** *island group* South Australia
92 K11 **Sirkka** Lappi, N Finland 67°49′N 24°48′E
Sirna *see* Sýrna
137 R16 **Şırnak** Şırnak, SE Turkey 37°33′N 42°27′E
137 S16 **Şırnak** ◆ *province* SE Turkey
Siroisi *see* Shiroishi
155 J14 **Sironcha** Mahārāshtra, C India 18°51′N 80°03′E
Sirone *see* Shirone
Síros *see* Sýros
Sirotino *see* Sirotsina
118 M12 **Sirotsina** *Rus.* Sirotino. Vitsyebskaya Voblasts', N Belarus 55°23′N 29°37′E
152 H9 **Sirsa** Haryāna, NW India 29°32′N 75°04′E
173 Y17 **Sir Seewoosagur Ramgoolam** ✈ (Port Louis) SE Mauritius
155 E18 **Sirsi** Karnātaka, W India 14°46′N 74°49′E
146 K12 **Şirşütür Gumy** *var.* Shirshütür, *Rus.* Peski Shirshyutyur. *desert* E Turkmenistan
Sirte *see* Surt
182 A2 **Sir Thomas, Mount** ▲ South Australia 27°09′S 129°49′E
Sirti, Gulf of *see* Surt, Khalīj
137 Y12 **Şirvan** *prev.* Äli-Bayramli. SE Azerbaijan 39°57′N 48°54′E
142 J5 **Sīrvān, Rūdkhāneh-ye** *var.* Nahr Diyālá, Sirwan. ⛐ Iran/Iraq *see also* Diyālá, Nahr
Sīrvān, Rudkhaneh-ye *see* Diyālá, Sirwan Nahr
118 H13 **Širvintos** Vilnius, SE Lithuania 55°01′N 24°58′E
Sirwan *see* Diyālá, Nahr/Sīrvān, Rudkhaneh-ye
11 N15 **Sir Wilfrid Laurier, Mount** ▲ British Columbia, SW Canada 52°45′N 119°51′W
14 M10 **Sir-Wilfrid, Mont** ▲ Québec, SE Canada 46°57′N 75°33′W
Sisačko-Moslavačka Županija *see* Sisak-Moslavina
112 E9 **Sisak** *var.* Siscia, *Ger.* Sissek, *Hung.* Sziszek; *anc.* Segestica. Sisak-Moslavina, C Croatia 45°28′N 16°21′E
167 R10 **Si Sa Ket** *var.* Sisaket, Sri Saket. Si Sa Ket, E Thailand 15°08′N 104°18′E
Sisaket *see* Si Sa Ket
112 E9 **Sisak-Moslavina** *off.* Sisačko-Moslavačka Županija. ◆ *province* C Croatia
167 O8 **Si Satchanalai** Sukhothai, NW Thailand
Siscia *see* Sisak
83 G22 **Sishen** Northern Cape, NW South Africa 27°47′S 22°59′E
137 V13 **Sisian** SE Armenia 39°31′N 46°03′E
197 N13 **Sisimiut** *var.* Holsteinborg, Holsteinsborg, Holstenborg, Holstensborg. Qeqqata, S Greenland 67°07′N 53°42′W
30 M1 **Siskiwit Bay** *lake bay* Michigan, N USA
34 L1 **Siskiyou Mountains** ▲▲ California/Oregon, W USA
Sisŏphŏn *see* Bântéay Méan Choăy
108 E7 **Sissach** Basel Landschaft, NW Switzerland 47°28′N 07°48′E
186 B5 **Sissano** West Sepik, NW Papua New Guinea 03°02′S 142°01′E
Sissek *see* Sisak
29 R7 **Sisseton** South Dakota, N USA 45°39′N 97°03′W
143 W9 **Sīstān, Daryācheh-ye** *var.* Daryācheh-ye Hāmūn, Hāmūn-e Şāberī. ⊚ Afghanistan/Iran *see also* Şāberī, Hāmūn-e
Sīstān, Daryācheh-ye *see* Şāberī, Hāmūn-e
143 V12 **Sīstān va Balūchestān** *off.* Ostān-e Sīstān va Balūchestān, *var.* Balūchestān va Sīstān. ◆ *province* SE Iran
Sīstān va Balūchestān, Ostān-e *see* Sīstān va Balūchestān
103 T14 **Sisteron** Alpes-de-Haute-Provence, SE France 44°12′N 05°55′E
32 H13 **Sisters** Oregon, NW USA 44°17′N 121°33′W
65 G15 **Sisters Peak** ▲ N Ascension Island 07°56′S 14°23′W
21 R3 **Sistersville** West Virginia, NE USA 39°33′N 81°00′W
Sistova *see* Svishtov
Sitakund *see* Sitakunda
153 V16 **Sitakunda** *var.* Sitakund. Chittagong, SE Bangladesh 22°35′N 91°40′E
153 P12 **Sītāmarhi** Bihār, N India 26°36′N 85°30′E
152 L11 **Sītāpur** Uttar Pradesh, N India 27°33′N 80°40′E
Sitaş Cristuru *see* Cristuru Secuiesc
115 L25 **Siteía** *var.* Sitía. Kríti, Greece, E Mediterranean Sea 35°13′N 26°06′E
105 V6 **Sitges** Cataluña, NE Spain 41°14′N 01°49′E
115 H15 **Sithonía' Atavyros** *peninsula* NE Greece
Sitía *see* Siteía
54 F4 **Sitionuevo** Magdalena, N Colombia 10°46′N 74°43′W
39 X13 **Sitka** Baranof Island, Alaska, USA 57°03′N 135°19′W
39 Q15 **Sitkinak Island** *island* Trinity Islands, Alaska, USA
Sittang *see* Sittoung
99 L17 **Sittard** Limburg, SE Netherlands 51°00′N 05°52′E
Sitten *see* Sion
108 H7 **Sitter** ⛐ NW Switzerland
109 U10 **Sittersdorf** Kärnten, S Austria 46°31′N 14°34′E
166 M7 **Sittoung** *var.* Sittang. ⛐ S Myanmar (Burma)
166 K6 **Sittwe** *var.* Akyab. Rakhine State, W Myanmar (Burma) 22°09′N 92°51′E
42 L8 **Siuna** Región Autónoma Atlántico Norte, NE Nicaragua 13°44′N 84°46′W
153 R15 **Siuri** West Bengal, NE India 23°54′N 87°32′E
Siut *see* Asyūţ
123 Q13 **Sivaki** Amurskaya Oblast', SE Russian Federation 52°39′N 126°43′E
136 M13 **Sivas** *anc.* Sebastia, Sebaste. Sivas, C Turkey 39°44′N 37°01′E
136 M13 **Sivas** ◆ *province* C Turkey
Sivasagar *see* Sibsāgar
137 O15 **Siverek** Şanlıurfa, S Turkey 37°46′N 39°19′E
117 X6 **Sivers'k** Donets'ka Oblast', E Ukraine 48°52′N 38°07′E
124 G13 **Siverskiy** Leningradskaya Oblast', NW Russian Federation 59°21′N 30°01′E
117 X6 **Sivers'kyy Donets'** *Rus.* Severskiy Donets. ⛐ Russian Federation/Ukraine *see also* Severskiy Donets
Sivers'kyy Donets' *see* Severskiy Donets
125 W5 **Sivomaskinskiy** Respublika Komi, NW Russian Federation 66°42′N 62°33′E
136 G13 **Sivrihisar** Eskişehir, W Turkey 39°29′N 31°32′E
99 F22 **Sivry** Hainaut, S Belgium 50°10′N 04°11′E
123 V9 **Sivuchiy, Mys** *headland* E Russian Federation 56°45′N 163°13′E
Siwa *see* Sīwah
75 U9 **Sīwah** *var.* Siwa. NW Egypt 29°11′N 25°32′E
152 J9 **Siwalik Range** *var.* Shiwalik Range. ▲▲ India/Nepal
153 O13 **Siwān** Bihār, N India 26°14′N 84°21′E
43 O14 **Sixaola, Río** ⛐ Costa Rica/Panama
Six Counties, The *see* Northern Ireland
103 T16 **Six-Fours-les-Plages** Var, SE France 43°05′N 05°50′E
161 Q7 **Sixian** *var.* Si Xian. Anhui, E China 33°29′N 117°53′E
Si Xian *see* Sixian
22 J9 **Six Mile Lake** ⊚ Louisiana, S USA
Siyāh Gūz *see* Syagwêz
155 L25 **Siyambalanduwa** Uva Province, SE Sri Lanka 06°54′N 81°32′E
137 Y10 **Siyäzän** *Rus.* Siazan'. NE Azerbaijan 41°05′N 49°05′E
Sizebolu *see* Sozopol
Sizuoka *see* Shizuoka
95 I24 **Sjælland** ◆ *county* SE Denmark
95 I24 **Sjælland** *Eng.* Zealand, *Ger.* Seeland. *island* E Denmark
Sjar *see* Sääre
113 L15 **Sjenica** *Turk.* Seniça. Serbia, SW Serbia 43°16′N 20°01′E
94 G11 **Sjoa** ⛐ S Norway
95 K23 **Sjöbo** Skåne, S Sweden 55°37′N 13°45′E
94 E9 **Sjøholt** Møre og Romsdal, S Norway 62°29′N 06°50′E
92 O1 **Sjuøyane** *island group* N Svalbard
Skadar *see* Shkodër
Skadarsko Jezero *see* Scutari, Lake
117 R11 **Skadovs'k** Khersons'ka Oblast', S Ukraine 46°07′N 32°53′E
95 I24 **Skælskør** Sjælland, E Denmark 55°16′N 11°18′E
92 I2 **Skagaströnd** *prev.* Höfdhakaupstadhur. Nordurland Vestra, N Iceland 65°49′N 20°18′W
95 H19 **Skagen** Nordjylland, N Denmark 57°44′N 10°37′E
Skagerak *see* Skagerrak
95 L16 **Skagern** ⊚ C Sweden
197 T17 **Skagerrak** *var.* Skagerak. *channel* N Europe
94 G12 **Skaget** ▲ S Norway 61°19′N 09°07′E
32 H7 **Skagit River** ⛐ Washington, NW USA
39 W12 **Skagway** Alaska, USA 59°27′N 135°18′W
92 K8 **Skaidi** Finnmark, N Norway 70°26′N 24°31′E
115 F21 **Skála** Pelopónnisos, S Greece 36°51′N 22°39′E
116 K6 **Skalat** *Pol.* Skałat. Ternopil's'ka Oblast', W Ukraine 49°27′N 25°59′E
95 J22 **Skälderviken** *inlet* Denmark/Sweden
92 I12 **Skalka** *Lapp.* Skalkká. ⊚ N Sweden
Skalkká *see* Skalka
114 I12 **Skalotí** Anatolikí Makedonía kai Thrákí, NE Greece 41°24′N 24°16′E
95 G22 **Skanderborg** Midtjylland, C Denmark 56°02′N 09°57′E
95 K22 **Skåne** *prev. Eng.* Scania. ◆ *county* S Sweden
75 N6 **Skanès** ✈ (Sousse) E Tunisia 35°36′N 10°56′E
95 C15 **Skånevik** Hordaland, S Norway 59°43′N 06°35′E
95 M18 **Skänninge** Östergötland, S Sweden 58°24′N 15°05′E
95 J23 **Skanör med Falsterbo** Skåne, S Sweden 55°24′N 12°48′E
115 H17 **Skantzoúra** *island* Vóreies Sporádes, Greece, Aegean Sea
95 K18 **Skara** Västra Götaland, S Sweden 58°23′N 13°25′E
95 M17 **Skärblacka** Östergötland, S Sweden 58°34′N 15°54′E
95 I18 **Skärhamn** Västra Götaland, S Sweden 57°59′N 11°33′E
95 I14 **Skarnes** Hedmark, S Norway 60°14′N 11°41′E
119 M21 **Skarodnaye** *Rus.* Skorodnoye. Homyel'skaya Voblasts', SE Belarus 51°38′N 28°50′E
110 I8 **Skarszewy** *Ger.* Schöneck. Pomorskie, NW Poland 54°04′N 18°25′E
111 M14 **Skarżysko-Kamienna** Świętokrzyskie, C Poland 51°07′N 20°52′E
114 H8 **Skat** *var.* Skŭt. NW Bulgaria
95 K16 **Skattkärr** Värmland, C Sweden 59°25′N 13°42′E
118 D12 **Skaudvilė** Tauragė, SW Lithuania 55°25′N 22°33′E
92 J12 **Skaulo** *Lapp.* Sávdijári. Norrbotten, N Sweden 67°21′N 21°03′E
111 K17 **Skawina** Małopolskie, S Poland 49°59′N 19°49′E
10 K12 **Skeena** ⛐ British Columbia, SW Canada
10 J11 **Skeena Mountains** ▲▲ British Columbia, W Canada
97 O18 **Skegness** E England, United Kingdom 53°10′N 00°21′E
92 J4 **Skeidarársandur** *coast* S Iceland
93 J15 **Skellefteå** Västerbotten, N Sweden 64°45′N 20°58′E
93 I14 **Skellefteälven** ⛐ N Sweden
93 J15 **Skelleftehamn** Västerbotten, N Sweden 64°41′N 21°13′E
25 O2 **Skellytown** Texas, SW USA 35°34′N 101°10′W
95 J19 **Skene** Västra Götaland, S Sweden 57°30′N 12°34′E
97 G17 **Skerries** *Ir.* Na Sceirí. Dublin, E Ireland 53°35′N 06°07′W
95 H15 **Ski** Akershus, S Norway 59°43′N 10°50′E
115 G17 **Skiáthos** Skiáthos, Vóreies Sporádes, Greece, Aegean Sea 39°10′N 23°30′E
115 G17 **Skiáthos** *island* Vóreies Sporádes, Greece, Aegean Sea
27 P9 **Skiatook** Oklahoma, C USA 36°22′N 96°00′W
27 P9 **Skiatook Lake** ▣ Oklahoma, C USA
97 B22 **Skibbereen** *Ir.* An Sciobairín. Cork, SW Ireland 51°33′N 09°15′W
92 I9 **Skibotn** Troms, N Norway 69°22′N 20°18′E
119 F16 **Skidal'** *Rus.* Skidel'. Hrodzyenskaya Voblasts', W Belarus 53°35′N 24°15′E
97 K15 **Skiddaw** ▲ NW England, United Kingdom 54°37′N 03°07′W
Skidel' *see* Skidal'
25 T14 **Skidmore** Texas, SW USA 28°13′N 97°40′W
95 G16 **Skien** Telemark, S Norway 59°14′N 09°37′E
Skiermûntseach *see* Schiermonnikoog
110 L12 **Skierniewice** Łódzkie, C Poland 51°58′N 20°10′E
74 L5 **Skikda** *prev.* Philippeville. NE Algeria 36°51′N 07°E
30 M16 **Skillet Fork** ⛐ Illinois, N USA
95 L19 **Skillingaryd** Jönköping, S Sweden 57°27′N 14°05′E
115 B19 **Skinári, Akrotírio** *headland* Iónia Nisiá, Greece 37°55′N 20°57′E
95 M15 **Skinnskatteberg** Västmanland, C Sweden 59°50′N 15°41′E
182 M12 **Skipton** Victoria, SE Australia 37°44′S 143°21′E
97 L16 **Skipton** N England, United Kingdom 53°57′N 02°W
Skiropoula *see* Skyropoúla
Skíros *see* Skýros
95 F21 **Skive** Midtjylland, NW Denmark 56°34′N 09°02′E
94 F11 **Skjåk** Oppland, S Norway 61°52′N 08°21′E
95 F22 **Skjern** Midtjylland, W Denmark 55°57′N 08°30′E
95 F22 **Skjern Å** *var.* Skjern Aa. ⛐ W Denmark
Skjern Aa *see* Skjern Å
92 G12 **Skjerstad** Nordland, C Norway 67°14′N 15°00′E
92 J8 **Skjervøy** Troms, N Norway 70°03′N 20°56′E
92 I10 **Skjold** Troms, N Norway 69°03′N 19°18′E
111 I17 **Skoczów** Śląskie, S Poland 49°49′N 18°45′E
109 T11 **Škofja Loka** *Ger.* Bischoflack. NW Slovenia 46°12′N 14°16′E
94 N12 **Skog** Gävleborg, C Sweden 61°10′N 16°49′E
95 K16 **Skoghall** Värmland, C Sweden 59°20′N 13°30′E
31 N10 **Skokie** Illinois, N USA 42°01′N 87°43′W
116 H6 **Skole** L'vivs'ka Oblast', W Ukraine 49°04′N 23°29′E
115 D19 **Skóllis** ▲ S Greece 37°58′N 21°33′E
167 S13 **Skon** Kâmpóng Cham, C Cambodia 12°56′N 104°36′E
115 H17 **Skópelos** Skópelos, Vóreies Sporádes, Greece, Aegean Sea 39°07′N 23°43′E
115 H17 **Skópelos** *island* Vóreies Sporádes, Greece, Aegean Sea
126 L5 **Skopin** Ryazanskaya Oblast', W Russian Federation 53°46′N 39°37′E
113 N18 **Skopje** *var.* Üsküb, *Turk.* Üsküp, *prev.* Skoplje; *anc.* Scupi. ● (FYR Macedonia) N FYR Macedonia 42°N 21°28′E
113 O18 **Skopje** ✈ N FYR Macedonia 41°58′N 21°35′E
Skoplje *see* Skopje
110 I8 **Skórcz** *Ger.* Skurz. Pomorskie, N Poland 53°46′N 18°43′E
Skorodnoye *see* Skarodnaye
93 H16 **Skorped** Västernorrland, C Sweden 63°23′N 17°55′E
95 G21 **Skørping** Nordjylland, N Denmark 56°50′N 09°55′E
95 K18 **Skövde** Västra Götaland, S Sweden 58°24′N 13°52′E
123 Q13 **Skovorodino** Amurskaya Oblast', SE Russian Federation 54°03′N 123°47′E
19 Q6 **Skowhegan** Maine, NE USA 44°46′N 69°41′W
11 W15 **Skownan** Manitoba, S Canada 51°55′N 99°34′W
94 H13 **Skreia** Oppland, S Norway 60°37′N 11°00′E
Skripón *see* Orchómenos
118 H9 **Skrīveri** S Latvia 56°39′N 25°08′E
118 J11 **Skrudaliena** SE Latvia 55°50′N 26°42′E
118 D9 **Skrunda** W Latvia 56°39′N 22°01′E
95 C16 **Skudeneshavn** Rogaland, S Norway 59°10′N 05°16′E
95 C16 **Skudnesfjorden** S Norway
83 L20 **Skukuza** Mpumalanga, NE South Africa 25°01′S 31°35′E
97 B22 **Skull** *Ir.* An Scoil. SW Ireland 51°32′N 09°34′W
22 L3 **Skuna River** ⛐ Mississippi, S USA
29 X15 **Skunk River** ⛐ Iowa, C USA
Skuø *see* Skúvoy
118 C10 **Skuodas** *Ger.* Schoden, *Pol.* Szkudy. Klaipėda, NW Lithuania 56°16′N 21°30′E
95 K23 **Skurup** Skåne, S Sweden 55°28′N 13°30′E
Skurz *see* Skórcz
Skŭt *see* Skat
94 O13 **Skutskär** Uppsala, C Sweden 60°39′N 17°25′E
95 B19 **Skúvoy** *Dan.* Skuø. *island* C Faroe Islands
Skvira *see* Skvyra
117 O5 **Skvyra** *Rus.* Skvira. Kyyivs'ka Oblast', N Ukraine 49°44′N 29°42′E
39 Q11 **Skwentna** Alaska, USA 61°56′N 151°03′W
110 E11 **Skwierzyna** *Ger.* Schwerin. Lubuskie, W Poland 52°37′N 15°27′E
96 G9 **Skye, Isle of** *island* NW Scotland, United Kingdom
36 K13 **Sky Harbor** ✈ (Phoenix) Arizona, SW USA 33°26′N 112°00′W
32 I8 **Skykomish** Washington, NW USA 47°40′N 121°20′W
Skylge *see* Terschelling
63 F19 **Skyring, Peninsula** *peninsula* S Chile
63 H24 **Skyring, Seno** *inlet* S Chile
115 H17 **Skyropoúla** *var.* Skiropoula. *island* Vóreies Sporádes, Greece, Aegean Sea
115 I17 **Skýros** *var.* Skíros. Skýros, Vóreies Sporádes, Greece, Aegean Sea 38°55′N 24°34′E
115 I17 **Skýros** *var.* Skíros; *anc.* Scyros. *island* Vóreies Sporádes, Greece, Aegean Sea
118 J12 **Slabodka** *Rus.* Slobodka. Vitsyebskaya Voblasts', NW Belarus 55°41′N 27°11′E
95 I23 **Slagelse** Sjælland, E Denmark 55°25′N 11°22′E
93 I14 **Slagnäs** Norrbotten, N Sweden 65°36′N 18°10′E
39 T10 **Slana** Alaska, USA 62°46′N 144°00′W
97 F20 **Slaney** *Ir.* An tSláine. ⛐ SE Ireland
116 J13 **Slănic** Prahova, SE Romania 45°14′N 25°58′E
116 K11 **Slănic Moldova** Bacău, E Romania 46°12′N 26°23′E
113 H16 **Slano** Dubrovnik-Neretva, SE Croatia 42°47′N 17°54′E
124 F13 **Slantsy** Leningradskaya Oblast', NW Russian Federation 59°06′N 28°00′E
111 C16 **Slaný** *Ger.* Schlan. Střední Čechy, NW Czech Republic 50°14′N 14°05′E
111 K16 **Śląskie** ◆ *province* S Poland
12 C10 **Slate Falls** Ontario, S Canada 51°11′N 91°32′W
27 T4 **Slater** Missouri, C USA 39°13′N 93°04′W
112 H9 **Slatina** *Hung.* Szlatina; *prev.* Podravska Slatina. Virovitica-Podravina, NE Croatia 45°40′N 17°46′E
116 I14 **Slatina** Olt, S Romania 44°27′N 24°21′E
25 N5 **Slaton** Texas, SW USA 33°26′N 101°38′W
95 H14 **Slattum** Akershus, S Norway 60°00′N 10°55′E
11 R10 **Slave** ⛐ Alberta/Northwest Territories, C Canada
68 E12 **Slave Coast** *coastal region* W Africa
11 P13 **Slave Lake** Alberta, SW Canada 55°17′N 114°46′W
122 I13 **Slavgorod** Altayskiy Kray, S Russian Federation 52°55′N 78°46′E
Slavgorod *see* Slawharad
Slavonia *see* Slavonija
112 G9 **Slavonija** *Eng.* Slavonia, *Ger.* Slawonien, *Hung.* Szlavonia, Szlavonország. *cultural region* NE Croatia
112 H10 **Slavonski Brod** *Ger.* Brod, *Hung.* Bród; *prev.* Brod, Brod na Savi. Brod-Posavina, NE Croatia 45°09′N 18°00′E
112 G10 **Slavonski Brod-Posavina** *off.* Brodsko-Posavska Županija, *var.* Brod-Posavina. ◆ *province* NE Croatia
116 L4 **Slavuta** Khmel'nyts'ka Oblast', NW Ukraine 50°18′N 26°52′E
117 P2 **Slavutych** Chernihivs'ka Oblast', N Ukraine 51°31′N 30°47′E
123 R15 **Slavyanka** Primorskiy Kray, SE Russian Federation 42°46′N 131°19′E
114 J8 **Slavyanovo** Pleven, N Bulgaria 43°28′N 24°52′E
Slavyansk *see* Slov"yans'k
126 K14 **Slavyansk-na-Kubani** Krasnodarskiy Kray, SW Russian Federation 45°16′N 38°09′E
119 N20 **Slavyechna** *Rus.* Slovechna. ⛐ Belarus/Ukraine
119 O16 **Slawharad** *Rus.* Slavgorod. Mahilyowskaya Voblasts', E Belarus 53°27′N 31°00′E
110 G7 **Sławno** Zachodnio-pomorskie, NW Poland 54°23′N 16°43′E
Slawonien *see* Slavonija
29 S10 **Slayton** Minnesota, N USA 43°59′N 95°45′W
97 N18 **Sleaford** E England, United Kingdom 52°59′N 00°28′W
97 A20 **Slea Head** *Ir.* Ceann Sléibhe. *headland* SW Ireland 52°05′N 10°25′W
96 G9 **Sleat, Sound of** *strait* NW Scotland, United Kingdom
Sledyuki *see* Slyedzyuki
12 I5 **Sleeper Islands** *island group* Nunavut, C Canada
31 O6 **Sleeping Bear Point** *headland* Michigan, N USA 44°54′N 86°02′W
29 T10 **Sleepy Eye** Minnesota, N USA 44°18′N 94°43′W
39 O11 **Sleetmute** Alaska, USA 61°42′N 157°10′W
Sléibhe, Ceann *see* Slea Head
Slēmānī *see* As Sulaymānīyah
Slēmānī, Parêzga-î *see* As Sulaymānīyah
195 O5 **Slessor Glacier** *glacier* Antarctica
22 L9 **Slidell** Louisiana, S USA 30°16′N 89°46′W
18 K12 **Slide Mountain** ▲ New York, NE USA 42°00′N 74°23′W
98 I13 **Sliedrecht** Zuid-Holland, C Netherlands 51°50′N 04°46′E
121 P16 **Sliema** N Malta 35°54′N 14°31′E
97 G16 **Slieve Donard** ▲ SE Northern Ireland, United Kingdom 54°10′N 05°57′W
Sligeach *see* Sligo
97 D16 **Sligo** *Ir.* Sligeach. Sligo, NW Ireland 54°17′N 08°28′W
97 C16 **Sligo** *Ir.* Sligeach. *cultural region* NW Ireland
97 C15 **Sligo Bay** *Ir.* Cuan Shligigh. *inlet* NW Ireland
18 B13 **Slippery Rock** Pennsylvania, NE USA 41°02′N 80°02′W
95 P19 **Slite** Gotland, SE Sweden 57°37′N 18°46′E
114 M8 **Slivak** Shumen, NE Bulgaria 43°27′N 26°56′E
114 L9 **Sliven** *var.* Slivno. Sliven, C Bulgaria 42°42′N 26°21′E
114 L10 **Sliven** ◆ *province* C Bulgaria
114 G9 **Slivnitsa** Sofia, W Bulgaria 42°51′N 23°01′E
Slivno *see* Sliven
114 L7 **Slivo Pole** Ruse, N Bulgaria 43°57′N 26°15′E
29 S13 **Sloan** Iowa, C USA 42°13′N 96°13′W
35 X12 **Sloan** Nevada, W USA 35°56′N 115°13′W
Slobodka *see* Slabodka
125 R14 **Slobodskoy** Kirovskaya Oblast', NW Russian Federation 58°43′N 50°12′E
Slobodzeya *see* Slobozia
117 O10 **Slobozia** *Rus.* Slobodzeya. E Moldova 46°45′N 29°42′E
116 L14 **Slobozia** Ialomiţa, SE Romania 44°34′N 27°23′E
98 O5 **Slochteren** Groningen, NE Netherlands 53°13′N 06°48′E
119 H17 **Slonim** *Pol.* Słonim. Hrodzyenskaya Voblasts', W Belarus 53°06′N 25°19′E
Słonim *see* Slonim
98 K7 **Sloter Meer** ⊚ N Netherlands
Slot, The *see* New Georgia Sound
97 N22 **Slough** S England, United Kingdom 51°31′N 00°36′W
111 J20 **Slovakia** *off.* Slovenská Republika, *Ger.* Slowakei, *Hung.* Szlovákia, *Slvk.* Slovensko. ◆ *republic* C Europe
Slovak Ore Mountains *see* Slovenské rudohorie
Slovechna *see* Slavyechna
109 S12 **Slovenia** *off.* Republic of Slovenia, *Ger.* Slowenien, *Slvn.* Slovenija. ◆ *republic* SE Europe
Slovenia, Republic of *see* Slovenia
Slovenija *see* Slovenia
109 V10 **Slovenj Gradec** *Ger.* Windischgraz. N Slovenia 46°29′N 15°05′E
109 W10 **Slovenska Bistrica** *Ger.* Windischfeistritz. NE Slovenia 46°21′N 15°27′E
Slovenská Republika *see* Slovakia
109 W10 **Slovenske Konjice** E Slovenia 46°21′N 15°28′E
111 K20 **Slovenské rudohorie** *Eng.* Slovak Ore Mountains, *Ger.* Slowakisches Erzgebirge, Ungarisches Erzgebirge. ▲▲ C Slovakia
Slovensko *see* Slovakia
117 Y7 **Slov"yanoserbs'k** Luhans'ka Oblast', E Ukraine 48°41′N 39°00′E
117 W6 **Slov"yans'k** *Rus.* Slavyansk. Donets'ka Oblast', E Ukraine 48°51′N 37°38′E
Slowakei *see* Slovakia
Slowakisches Erzgebirge *see* Slovenské rudohorie
Slowenien *see* Slovenia
110 D11 **Słubice** *Ger.* Frankfurt. Lubuskie, W Poland 52°20′N 14°35′E
119 K19 **Sluch** *Rus.* Sluch'. ⛐ C Belarus
116 L4 **Sluch** ⛐ NW Ukraine
99 D16 **Sluis** Zeeland, SW Netherlands 51°18′N 03°22′E
112 D10 **Slunj** *Hung.* Szluin. Karlovac, C Croatia 45°06′N 15°35′E
110 I11 **Słupca** Wielkopolskie, C Poland 52°17′N 17°52′E
110 G6 **Słupia** *Ger.* Stolpe. ⛐ NW Poland
110 G6 **Słupsk** *Ger.* Stolp. Pomorskie, N Poland 54°28′N 17°01′E
119 K18 **Slutsk** Minskaya Voblasts', S Belarus 53°02′N 27°32′E
119 O16 **Slyedzyuki** *Rus.* Sledyuki. Mahilyowskaya Voblasts', E Belarus 53°35′N 30°22′E
97 A17 **Slyne Head** *Ir.* Ceann Léime. *headland* W Ireland 53°25′N 10°11′W
27 U14 **Smackover** Arkansas, C USA 33°21′N 92°43′W
95 L20 **Småland** *cultural region* S Sweden
95 K20 **Smålandsstenar** Jönköping, S Sweden 57°10′N 13°24′E
Small Malaita *see* Maramasike
13 O8 **Smallwood Reservoir** ⊚ Newfoundland and Labrador, S Canada
119 N14 **Smalyany** *Rus.* Smolyany. Vitsyebskaya Voblasts', NE Belarus 54°36′N 30°04′E
119 L15 **Smalyavichy** *Rus.* Smolevichi. Minskaya Voblasts', C Belarus 54°02′N 28°05′E
74 C9 **Smara** *var.* Es Semara. N Western Sahara 26°45′N 11°44′W
119 I14 **Smarhon'** *Pol.* Smorgonie, *Rus.* Smorgon'. Hrodzyenskaya Voblasts', W Belarus 54°29′N 26°24′E
112 M11 **Smederevo** *Ger.* Semendria. Serbia, N Serbia 44°41′N 20°56′E
112 M12 **Smederevska Palanka** Serbia, C Serbia 44°24′N 20°56′E
95 M14 **Smedjebacken** Dalarna, C Sweden 60°08′N 15°25′E
116 L13 **Smeeni** Buzău, SE Romania 45°00′N 26°52′E
Smela *see* Smila
107 D16 **Smeralda, Costa** *cultural region* Sardegna, Italy, C Mediterranean Sea
111 J22 **Śmigiel** *Ger.* Schmiegel. Wielkopolskie, C Poland 52°02′N 16°33′E
117 Q6 **Smila** *Rus.* Smela. Cherkas'ka Oblast', C Ukraine 49°15′N 31°54′E
98 N7 **Smilde** Drenthe, NE Netherlands 52°57′N 06°28′E
11 S16 **Smiley** Saskatchewan, S Canada 51°40′N 109°24′W
25 T12 **Smiley** Texas, SW USA 29°16′N 97°38′W
Smilten *see* Smiltene
118 I8 **Smiltene** *Ger.* Smilten. N Latvia 57°25′N 25°53′E
123 T13 **Smirnykh** Ostrov Sakhalin, Sakhalinskaya Oblast', SE Russian Federation 49°43′N 142°48′E
11 Q13 **Smith** Alberta, W Canada 55°06′N 113°57′W
39 P4 **Smith Bay** *bay* Alaska, NW USA

◆ Country
● Country Capital
◇ Dependent Territory
○ Dependent Territory Capital
◆ Administrative Regions
✈ International Airport
▲ Mountain
▲▲ Mountain Range
Volcano
⛐ River
⊚ Lake
▣ Reservoir

12 I3 **Smith, Cape** *headland* Québec, NE Canada 60°50′N 78°06′W
26 L3 **Smith Center** Kansas, C USA 39°46′N 98°46′W
10 K13 **Smithers** British Columbia, SW Canada 54°45′N 127°10′W
21 V10 **Smithfield** North Carolina, SE USA 35°30′N 78°21′W
36 L1 **Smithfield** Utah, W USA 41°50′N 111°49′W
21 X7 **Smithfield** Virginia, NE USA 36°41′N 76°38′W
12 I3 **Smith Island** *island* Nunavut, C Canada
Smith Island *see* Sumisu-jima
20 H7 **Smithland** Kentucky, S USA 37°06′N 88°24′W
21 T7 **Smith Mountain Lake** *var.* Leesville Lake. ⊠ Virginia, NE USA
34 L1 **Smith River** California, W USA 41°54′N 124°09′W
33 R9 **Smith River** ≈ Montana, NW USA
14 L13 **Smiths Falls** Ontario, SE Canada 44°54′N 76°01′W
33 N13 **Smiths Ferry** Idaho, NW USA 44°19′N 116°04′W
20 K7 **Smiths Grove** Kentucky, S USA 37°01′N 86°14′W
183 N15 **Smithton** Tasmania, SE Australia 40°54′S 145°06′E
18 L14 **Smithtown** Long Island, New York, NE USA 40°52′N 73°13′W
20 K9 **Smithville** Tennessee, S USA 35°59′N 85°49′W
25 T11 **Smithville** Texas, SW USA 30°04′N 97°32′W
Šmohor *see* Hermagor
35 Q4 **Smoke Creek Desert** *desert* Nevada, W USA
11 O14 **Smoky** ≈ Alberta, W Canada
182 E7 **Smoky Bay** South Australia 32°22′S 133°57′E
183 V6 **Smoky Cape** *headland* New South Wales, SE Australia 30°54′S 153°06′E
26 L4 **Smoky Hill River** ≈ Kansas, C USA
26 L4 **Smoky Hills** *hill range* Kansas, C USA
11 Q14 **Smoky Lake** Alberta, SW Canada 54°08′N 112°26′W
94 E8 **Smøla** *island* W Norway
126 H4 **Smolensk** Smolenskaya Oblast′, W Russian Federation 54°48′N 32°08′E
126 H4 **Smolenskaya Oblast′** ⬖ *province* W Russian Federation
Smolensk-Moscow Upland *see* Smolensko-Moskovskaya Vozvyshennost′
126 J3 **Smolensko-Moskovskaya Vozvyshennost′** *var.* Smolensk-Moscow Upland. ▲▲ W Russian Federation
Smolevichi *see* Smalyavichy
115 C15 **Smólikas** *var.* Smolikás. ▲ W Greece 40°06′N 20°54′E
114 I12 **Smolyan** *prev.* Pashmakli. Smolyan, S Bulgaria 41°34′N 24°42′E
114 I12 **Smolyan** ⬖ *province* S Bulgaria
Smolyany *see* Smalyany
33 S15 **Smoot** Wyoming, C USA 42°37′N 110°55′W
12 G12 **Smooth Rock Falls** Ontario, S Canada 49°17′N 81°37′W
Smorgon′/Smorgonie *see* Smarhon′
95 K23 **Smygehamn** Skåne, S Sweden 55°19′N 13°25′E
194 I7 **Smyley Island** *island* Antarctica
21 Y3 **Smyrna** Delaware, NE USA 39°18′N 75°36′W
23 S3 **Smyrna** Georgia, SE USA 33°52′N 84°30′W
20 J9 **Smyrna** Tennessee, S USA 36°00′N 86°30′W
Smyrna *see* İzmir
97 I16 **Snaefell** ▲ C Isle of Man 54°15′N 04°29′W
92 H3 **Snæfellsjökull** ▲ W Iceland 64°51′N 23°51′W
92 J3 **Snækollur** ▲ C Iceland 64°38′N 19°18′W
10 J4 **Snake** ≈ Yukon, NW Canada
29 O8 **Snake Creek** ≈ South Dakota, N USA
183 P13 **Snake Island** *island* Victoria, SE Australia
35 Y6 **Snake Range** ▲▲ Nevada, W USA
32 K10 **Snake River** ≈ NW USA
29 V6 **Snake River** ≈ Minnesota, N USA
28 L12 **Snake River** ≈ Nebraska, C USA
33 Q14 **Snake River Plain** *plain* Idaho, NW USA
93 F15 **Snåsa** Nord-Trøndelag, C Norway 64°16′N 12°25′E
21 O8 **Sneedville** Tennessee, S USA 36°31′N 83°13′W
98 K6 **Sneek** *Fris.* Snits. Fryslân, N Netherlands 53°02′N 05°40′E
Sneeuw-gebergte *see* Maoke, Pegunungan
95 F22 **Snejbjerg** Midtjylland, C Denmark 56°08′N 08°55′E
122 K9 **Snezhnogorsk** Krasnoyarskiy Kray, N Russian Federation 68°06′N 87°37′E
124 J3 **Snezhnogorsk** Murmanskaya Oblast′, NW Russian Federation 69°12′N 33°20′E
Snezhnoye *see* Snizhne
111 G15 **Sněžka** *Ger.* Schneekoppe, *Pol.* Śnieżka. ▲ N Czech Republic/Poland 50°42′N 15°55′E
110 N8 **Śniardwy, Jezioro** *Ger.* Spirdingsee. ◎ NE Poland
Sniečkus *see* Visaginas
Śnieżka *see* Sněžka
117 R10 **Snihurivka** Mykolayivs′ka Oblast′, S Ukraine 47°05′N 32°48′E
116 I5 **Snilov** ✈ (L′viv) L′vivs′ka Oblast′, W Ukraine 49°45′N 23°59′E
111 O19 **Snina** *Hung.* Szinna. Prešovský Kraj, E Slovakia 49°N 22°10′E
Snits *see* Sneek
117 Y8 **Snizhne** *Rus.* Snezhnoye. Donets′ka Oblast′, SE Ukraine 48°01′N 38°46′E
94 G10 **Snøhetta** *var.* Snohetta. ▲ S Norway 62°22′N 09°08′E
92 G12 **Snøtinden** ▲ C Norway 66°39′N 13°50′E
97 I18 **Snowdon** ▲ NW Wales, United Kingdom 53°04′N 04°04′W
97 I18 **Snowdonia** ▲▲ NW Wales, United Kingdom
Snowdrift *see* Łutselk′e
Snowdrift *see* Lutselk′e
37 N12 **Snowflake** Arizona, SW USA 34°30′N 110°04′W
21 Y5 **Snow Hill** Maryland, NE USA 38°11′N 75°23′W
21 W10 **Snow Hill** North Carolina, SE USA 35°26′N 77°39′W
194 H3 **Snowhill Island** *island* Antarctica
11 V13 **Snow Lake** Manitoba, C Canada 54°56′N 100°02′W
37 R5 **Snowmass Mountain** ▲ Colorado, C USA 39°07′N 107°04′W
18 M10 **Snow, Mount** ▲ Vermont, NE USA 42°56′N 72°52′W
34 M5 **Snow Mountain** ▲ California, W USA 39°44′N 123°01′W
Snow Mountains *see* Maoke, Pegunungan
33 N7 **Snowshoe Peak** ▲ Montana, NW USA 48°15′N 115°44′W
182 I8 **Snowtown** South Australia 33°49′S 138°13′E
36 K1 **Snowville** Utah, W USA 41°59′N 112°42′W
35 X3 **Snow Water Lake** ◎ Nevada, W USA
183 Q11 **Snowy Mountains** ▲▲ New South Wales/Victoria, SE Australia
183 Q12 **Snowy River** ≈ New South Wales/Victoria, SE Australia
44 K5 **Snug Corner** Acklins Island, SE The Bahamas 22°31′N 73°51′W
167 T13 **Snuŏl** Krâchéh, E Cambodia 12°04′N 106°26′E
116 J7 **Snyatyn** Ivano-Frankivs′ka Oblast′, W Ukraine 48°30′N 25°50′E
26 L12 **Snyder** Oklahoma, C USA 34°37′N 98°56′W
25 O6 **Snyder** Texas, SW USA 32°43′N 100°54′W
172 H3 **Soalala** Mahajanga, W Madagascar 16°05′S 45°21′E
172 J4 **Soanierana-Ivongo** Toamasina, E Madagascar 16°53′S 49°35′E
171 R11 **Soasiu** *var.* Tidore. Pulau Tidore, E Indonesia 0°40′N 127°25′E
54 G8 **Soatá** Boyacá, C Colombia 06°23′N 72°40′W
172 I5 **Soavinandriana** Antananarivo, C Madagascar 19°09′S 46°43′E
77 V13 **Soba** Kaduna, C Nigeria 10°58′N 08°06′E
163 Y16 **Sobaek-sanmaek** ▲▲ S South Korea
80 F13 **Sobat** ≈ NE South Sudan
171 Z14 **Sobger, Sungai** ≈ Papua, E Indonesia
171 V13 **Sobiei** Papua Barat, E Indonesia 02°31′S 134°30′E
126 M3 **Sobinka** Vladimirskaya Oblast′, W Russian Federation 56°00′N 39°55′E
127 S7 **Sobolevo** Orenburgskaya Oblast′, W Russian Federation 51°57′N 51°42′E
Soborsin *see* Săvârșin
164 D15 **Sobo-san** ▲ Kyūshū, SW Japan 32°50′N 131°16′E
111 G14 **Sobótka** Dolnośląskie, SW Poland 50°53′N 16°48′E
59 O15 **Sobradinho** Bahia, E Brazil 09°33′S 40°56′W
Sobradinho, Barragem de *see* Sobradinho, Represa de
59 O16 **Sobradinho, Represa de** *var.* Barragem de Sobradinho. ⊠ E Brazil
58 O13 **Sobral** Ceará, E Brazil 03°45′S 40°20′W
105 T4 **Sobrarbe** *physical region* NE Spain
109 R10 **Soča** *It.* Isonzo. ≈ Italy/Slovenia
110 L11 **Sochaczew** Mazowieckie, C Poland 52°15′N 20°15′E
126 L15 **Sochi** Krasnodarskiy Kray, SW Russian Federation 43°35′N 39°46′E
114 G13 **Sochós** *var.* Sohos, Sokhós. Kentrikí Makedonía, N Greece 40°49′N 23°23′E
191 R11 **Société, Archipel de la** *var.* Archipel de Tahiti, Îles de la Société, *Eng.* Society Islands. *island group* W French Polynesia
Société, Îles de la/Society Islands *see* Société, Archipel de la
21 T11 **Society Hill** South Carolina, SE USA 34°28′N 79°54′W
175 W9 **Society Ridge** *undersea feature* C Pacific Ocean
62 I5 **Socompa, Volcán** ⛰ N Chile 24°18′S 68°03′W
Soconusco, Sierra de *see* Madre, Sierra
54 G8 **Socorro** Santander, C Colombia 06°30′N 73°16′W
37 R13 **Socorro** New Mexico, SW USA 33°58′N 106°55′W
Socotra *see* Suquṭrā
167 S14 **Soc Trăng** *var.* Khanh Hung. Soc Trăng, S Vietnam 09°36′N 105°58′E
105 P10 **Socuéllamos** Castilla-La Mancha, C Spain 39°18′N 02°48′W
35 W13 **Soda Lake** *salt flat* California, W USA
92 L11 **Sodankylä** Lappi, N Finland 67°26′N 26°35′E
Sodari *see* Sodiri
33 R15 **Soda Springs** Idaho, NW USA 42°39′N 111°36′W
Soddo/Soddu *see* Sodo
20 L10 **Soddy Daisy** Tennessee, S USA 35°14′N 85°11′W
95 N14 **Söderfors** Uppsala, C Sweden 60°23′N 17°14′E
94 N12 **Söderhamn** Gävleborg, C Sweden 61°19′N 17°10′E
95 N17 **Söderköping** Östergötland, S Sweden 58°28′N 16°20′E
95 N17 **Södermanland** ⬖ *county* C Sweden
95 O16 **Södertälje** Stockholm, C Sweden 59°11′N 17°39′E
80 D10 **Sodiri** *var.* Sawdirī, Sodari. Northern Kordofan, C Sudan 14°23′N 29°06′E
81 I14 **Sodo** *var.* Soddo, Soddu. Southern Nationalities, S Ethiopia 06°49′N 37°43′E
Södra Karelen *see* Etelä-Karjala
Södra Österbotten *see* Etelä-Pohjanmaa
Södra Savolax *see* Etelä-Savo
95 M19 **Södra Vi** Kalmar, S Sweden 57°45′N 15°45′E
18 G9 **Sodus Point** *headland* New York, NE USA 43°16′N 76°59′W
171 Q17 **Soe** *prev.* Soë. Timor, C Indonesia 09°51′S 124°29′E
Soebang *see* Subang
Soekaboemi *see* Sukabumi
169 N15 **Soekarno-Hatta** ✈ (Jakarta) Jawa, S Indonesia
Soëla-Sund *see* Soela Väin
118 E5 **Soela Väin** *prev. Eng.* Sele Sound, *Ger.* Dagden-Sund, Soëla-Sund. *strait* W Estonia
Soemba *see* Sumba, Pulau
Soembawa *see* Sumbawa
Soemenep *see* Sumenep
Soengaipenoeh *see* Sungaipenuh
Soerabaja *see* Surabaya
Soerakarta *see* Surakarta
101 G14 **Soest** Nordrhein-Westfalen, W Germany 51°34′N 08°06′E
98 J11 **Soest** Utrecht, C Netherlands 52°10′N 05°20′E
100 F11 **Soeste** ≈ NW Germany
98 J11 **Soesterberg** Utrecht, C Netherlands 52°07′N 05°17′E
115 E16 **Sofádes** *var.* Sofádhes. Thessalía, C Greece 39°20′N 22°06′E
Sofádhes *see* Sofádes
83 N18 **Sofala** Sofala, C Mozambique 20°04′S 34°43′E
83 N17 **Sofala** ⬖ *province* C Mozambique
83 N18 **Sofala, Baia de** *bay* C Mozambique
114 G10 **Sofia** *var.* Sophia, Sofiya, *Eng.* Sofia, *Lat.* Serdica. ● (Bulgaria) Sofia Grad, W Bulgaria 42°42′N 23°20′E
114 H9 **Sofia** ⬖ *province* W Bulgaria
114 G9 **Sofia** ✈ Sofia Grad, W Bulgaria 42°42′N 23°26′E
172 J3 **Sofia** *seasonal river* NW Madagascar
Sofia *see* Sofia
114 G9 **Sofia Grad** ⬖ *municipality* W Bulgaria
115 G19 **Sofikó** Pelopónnisos, S Greece 37°46′N 23°04′E
Sofi-Kurgan *see* Sopu-Korgon
Sofiya *see* Sofia
Sofiyevka *see* Sofiyivka
117 S8 **Sofiyivka** *Rus.* Sofiyevka. Dnipropetrovs′ka Oblast′, E Ukraine 48°04′N 33°55′E
123 R12 **Sofiysk** Khabarovskiy Kray, SE Russian Federation 51°32′N 139°46′E
123 R13 **Sofiysk** Khabarovskiy Kray, SE Russian Federation 52°20′N 133°37′E
124 I6 **Sofporog** Respublika Kareliya, NW Russian Federation 65°48′N 31°30′E
115 L23 **Sofraná** *prev.* Záfora. *island* Kykládes, Greece, Aegean Sea
165 Y14 **Sōfu-gan** *island* Izu-shotō, SE Japan
156 K10 **Sog** Xizang Zizhiqu, W China 31°52′N 93°40′E
54 G9 **Sogamoso** Boyacá, C Colombia 05°43′N 72°56′W
136 I11 **Soğanlı Çayı** ≈ N Turkey
94 E12 **Sogn** *physical region* S Norway
Sogndal *see* Sogndalsfjøra
94 E12 **Sogndalsfjøra** *var.* Sogndal. Sogn Og Fjordane, S Norway 61°13′N 07°05′E
95 E18 **Søgne** Vest-Agder, S Norway 58°05′N 07°49′E
94 D12 **Sognefjorden** *fjord* NE North Sea
94 C12 **Sogn Og Fjordane** ⬖ *county* S Norway
162 I11 **Sogo Nur** ◎ N China
159 T12 **Sogruma** Qinghai, W China 32°32′N 100°52′E
Sŏgwip′o *see* Seogwipo
Sohâg *see* Sawhāj
Sohar *see* Şuḩār
64 H9 **Sohm Plain** *undersea feature* NW Atlantic Ocean
100 H7 **Soholmer Au** ≈ N Germany
Sohos *see* Sochós
Sohrau *see* Żory
99 F20 **Soignies** Hainaut, SW Belgium 50°35′N 04°04′E
159 R15 **Soila** Xizang Zizhiqu, W China 30°40′N 97°07′E
103 P4 **Soissons** *anc.* Augusta Suessionum, Noviodunum. Aisne, N France 49°23′N 03°20′E
164 H13 **Sōja** Okayama, Honshū, SW Japan 34°40′N 133°42′E
152 F13 **Sojat** Rājasthān, N India 25°53′N 73°45′E
163 W13 **Sŏjosŏn-man** *inlet* W North Korea
116 I4 **Sokal′** *Rus.* Sokal. L′vivs′ka Oblast′, NW Ukraine 50°29′N 24°17′E
163 Y14 **Sokcho** *prev.* Sokch′o. N South Korea 38°07′N 128°34′E
Sokch′o *see* Sokcho
136 B15 **Söke** Aydın, SW Turkey 37°46′N 27°24′E
189 N12 **Sokehs Island** *island* E Micronesia
79 M24 **Sokele** Katanga, SE Dem. Rep. Congo 09°54′S 24°38′E
147 R11 **Sokh** *Uzb.* Sŭkh. ≈ Kyrgyzstan/Uzbekistan
Sokh *see* So′x
Sokhós *see* Sochós
137 Q8 **Sokhumi** *Rus.* Sukhumi. NW Georgia 43°02′N 41°01′E
113 O14 **Sokobanja** Serbia, E Serbia 43°39′N 21°51′E
77 R15 **Sokodé** C Togo 08°58′N 01°10′E
123 T10 **Sokol** Magadanskaya Oblast′, E Russian Federation 59°51′N 150°56′E
124 M13 **Sokol** Vologodskaya Oblast′, NW Russian Federation 59°26′N 40°09′E
110 P9 **Sokółka** Podlaskie, NE Poland 53°24′N 23°31′E
76 M11 **Sokolo** Ségou, W Mali 14°43′N 06°02′W
111 A16 **Sokolov** *Ger.* Falkenau an der Eger; *prev.* Falknov nad Ohří. Karlovarský Kraj, W Czech Republic 50°10′N 12°38′E
111 O16 **Sokołów Małopolski** Podkarpackie, SE Poland 50°12′N 22°07′E
110 O11 **Sokołów Podlaski** Mazowieckie, C Poland 52°26′N 22°14′E
76 G11 **Sokone** W Senegal 13°53′N 16°22′W
77 T12 **Sokoto** Sokoto, NW Nigeria 13°05′N 05°16′E
77 T12 **Sokoto** ⬖ *state* NW Nigeria
77 S12 **Sokoto** ≈ NW Nigeria
Sokotra *see* Suquṭrā
147 U7 **Sokuluk** Chuyskaya Oblast′, N Kyrgyzstan 42°53′N 74°19′E
116 L7 **Sokyryany** Chernivets′ka Oblast′, W Ukraine 48°28′N 27°25′E
95 C16 **Sola** Rogaland, S Norway 58°53′N 05°36′E
187 R12 **Sola** Vanua Lava, N Vanuatu 13°51′S 167°34′E
95 C17 **Sola** ✈ (Stavanger) Rogaland, S Norway 58°54′N 05°36′E
81 H18 **Solai** Nakuru, W Kenya 0°02′N 36°03′E
152 I8 **Solan** Himāchal Pradesh, N India 30°54′N 77°06′E
185 A25 **Solander Island** *island* SW New Zealand
Solano *see* Bahía Solano
155 F15 **Solāpur** *var.* Sholāpur. Mahārāshtra, W India 17°43′N 75°54′E
93 H16 **Solberg** Västernorrland, C Sweden 63°48′N 17°40′E
116 K9 **Solca** *Ger.* Solka. Suceava, N Romania 47°40′N 25°50′E
105 O16 **Sol, Costa del** *coastal region* S Spain
106 F5 **Solda** *Ger.* Sulden. Trentino-Alto Adige, N Italy 46°33′N 10°35′E
117 N9 **Şoldăneşti** *Rus.* Sholdaneshty. N Moldova 47°49′N 28°45′E
Soldau *see* Wkra
108 L8 **Sölden** Tirol, W Austria 46°58′N 11°01′E
27 P3 **Soldier Creek** ≈ Kansas, C USA
39 R12 **Soldotna** Alaska, USA 60°29′N 151°03′W
110 I10 **Solec Kujawski** Kujawsko-pomorskie, C Poland 53°04′N 18°09′E
61 B16 **Soledad** Santa Fe, C Argentina 30°38′S 60°52′W
55 E4 **Soledad** Atlántico, N Colombia 10°54′N 74°48′W
35 O11 **Soledad** California, W USA 36°25′N 121°19′W
55 O7 **Soledad** Anzoátegui, NE Venezuela 08°10′N 63°36′W
61 H15 **Soledade** Rio Grande do Sul, S Brazil 28°50′S 52°30′W
Isla Soledad *see* East Falkland
103 Y15 **Solenzara** Corse, France, C Mediterranean Sea 41°55′N 09°24′E
Soleure *see* Solothurn
94 C12 **Solheim** Hordaland, S Norway 60°54′N 05°30′E
125 N14 **Soligalich** Kostromskaya Oblast′, NW Russian Federation 59°05′N 42°15′E
Soligorsk *see* Salihorsk
Soligorskoye Vodokhranilishche *see* Salihorskaye Vodaskhovishcha
97 L20 **Solihull** C England, United Kingdom 52°25′N 01°45′W
125 U13 **Solikamsk** Permskiy Kray, NW Russian Federation 59°37′N 56°46′E
127 V8 **Sol′-Iletsk** Orenburgskaya Oblast′, W Russian Federation 51°09′N 55°05′E
57 G17 **Solimana, Nevado** ▲ S Peru 15°24′S 72°49′W
58 E13 **Solimões, Rio** ≈ C Brazil
113 E14 **Solin** *It.* Salona; *anc.* Salonae. Split-Dalmacija, S Croatia 43°33′N 16°29′E
101 E15 **Solingen** Nordrhein-Westfalen, W Germany 51°10′N 07°05′E
Solka *see* Solca
93 H16 **Sollefteå** Västernorrland, C Sweden 63°09′N 17°15′E
95 O15 **Sollentuna** Stockholm, C Sweden 59°26′N 17°56′E
105 X9 **Sóller** Mallorca, Spain, W Mediterranean Sea 39°46′N 02°42′E
94 L13 **Sollerön** Dalarna, C Sweden 60°55′N 14°34′E
101 I14 **Solling** *hill range* C Germany
95 O16 **Solna** Stockholm, C Sweden 59°22′N 17°58′E
126 K3 **Solnechnogorsk** Moskovskaya Oblast′, W Russian Federation 56°07′N 37°04′E
123 R10 **Solnechnyy** Khabarovskiy Kray, SE Russian Federation 50°41′N 136°42′E
123 S13 **Solnechnyy** Respublika Sakha (Yakutiya), NE Russian Federation 60°13′N 137°42′E
Solo *see* Surakarta
107 L17 **Solofra** Campania, S Italy 40°49′N 14°48′E
168 J11 **Solok** Sumatera, W Indonesia 0°45′S 100°42′E
42 C6 **Sololá** Sololá, W Guatemala 14°46′N 91°09′W
42 A2 **Sololá** *off.* Departamento de Sololá. ⬖ *department* SW Guatemala
Sololá, Departamento de *see* Sololá
81 J16 **Sololo** Marsabit, N Kenya 03°31′N 38°39′E
42 C4 **Soloma** Huehuetenango, W Guatemala 15°38′N 91°25′W
38 M9 **Solomon** Alaska, USA 64°33′N 164°26′W
27 N4 **Solomon** Kansas, C USA 38°55′N 97°22′W
187 N9 **Solomon Islands** *prev.* British Solomon Islands Protectorate. ◆ *commonwealth republic* W Solomon Islands N Melanesia W Pacific Ocean
186 L7 **Solomon Islands** *island group* Papua New Guinea/Solomon Islands
26 M3 **Solomon River** ≈ Kansas, C USA
186 H8 **Solomon Sea** *sea* W Pacific Ocean
31 U11 **Solon** Ohio, N USA 41°23′N 81°26′W
117 T8 **Solone** Dnipropetrovs′ka Oblast′, E Ukraine 48°12′N 34°49′E
171 P16 **Solor, Kepulauan** *island group* S Indonesia
126 M4 **Solotcha** Ryazanskaya Oblast′, W Russian Federation 54°43′N 39°50′E
108 D7 **Solothurn** *Fr.* Soleure. Solothurn, NW Switzerland 47°13′N 07°32′E
108 D7 **Solothurn** *Fr.* Soleure. ⬖ *canton* NW Switzerland
124 J7 **Solovetskiye Ostrova** *island group* NW Russian Federation
105 V5 **Solsona** Cataluña, NE Spain 42°00′N 01°31′E
113 E14 **Šolta** *It.* Solta. *island* S Croatia
Solṭānābād *see* Kāshmar
142 L4 **Solṭānīyeh** Zanjān, NW Iran 36°24′N 48°50′E
100 I11 **Soltau** Niedersachsen, NW Germany 52°59′N 09°50′E
124 G14 **Sol′tsy** Novgorodskaya Oblast′, W Russian Federation 58°09′N 30°23′E
Soltūstik Qazaqstan Oblysy *see* Severnyy Kazakhstan
Solun *see* Thessaloníki
113 O19 **Solunska Glava** ▲ C FYR Macedonia 41°43′N 21°24′E
95 L22 **Sölvesborg** Blekinge, S Sweden 56°04′N 14°35′E
97 J15 **Solway Firth** *inlet* England/Scotland, United Kingdom
82 I13 **Solwezi** North Western, NW Zambia 12°11′S 26°23′E
165 Q11 **Sōma** Fukushima, Honshū, C Japan 37°49′N 140°52′E
136 C13 **Soma** Manisa, W Turkey 39°10′N 27°36′E
Somali *see* Sumalē
81 O15 **Somalia** *off.* Somali Democratic Republic, *Som.* Jamuuriyada Demuqraadiga Soomaaliyeed, Soomaaliya; *prev.* Italian Somaliland, Somaliland Protectorate. ◆ *republic* E Africa
173 N6 **Somali Basin** *undersea feature* W Indian Ocean 0°00′N 52°00′E
Somali Democratic Republic *see* Somalia
80 N12 **Somaliland** ⬖ *disputed territory* N Somalia
Somaliland Protectorate *see* Somalia
67 Y8 **Somali Plain** *undersea feature* W Indian Ocean 01°00′N 51°30′E
112 J8 **Sombor** *Hung.* Zombor. Vojvodina, NW Serbia 45°46′N 19°07′E
99 H20 **Sombreffe** Namur, S Belgium 50°32′N 04°39′E
40 L10 **Sombrerete** Zacatecas, C Mexico 23°38′N 103°40′W
45 V8 **Sombrero** *island* N Anguilla
151 Q21 **Sombrero Channel** *channel* Nicobar Islands, India
116 H9 **Şomcuta Mare** *Hung.* Nagysomkút; *prev.* Somcuța Mare. Maramureș, N Romania 47°29′N 23°30′E
Somcuța Mare *see* Şomcuta Mare
167 R9 **Somdet** Kalasin, E Thailand 16°41′N 103°44′E
99 L15 **Someren** Noord-Brabant, SE Netherlands 51°23′N 05°42′E
93 L19 **Somero** Varsinais-Suomi, SW Finland 60°37′N 23°30′E
33 P7 **Somers** Montana, NW USA 48°04′N 114°16′W
64 A12 **Somerset** *var.* Somerset Village. W Bermuda 32°18′N 64°53′W
37 Q5 **Somerset** Colorado, C USA 38°55′N 107°27′W
20 M7 **Somerset** Kentucky, S USA 37°05′N 84°36′W
19 O12 **Somerset** Massachusetts, NE USA 41°46′N 71°07′W
97 K23 **Somerset** *cultural region* SW England, United Kingdom
Somerset East *see* Somerset-Oos
64 A12 **Somerset Island** *island* W Bermuda
197 N9 **Somerset Island** *island* Queen Elizabeth Islands, Nunavut, NW Canada
Somerset Nile *see* Victoria Nile
83 I25 **Somerset-Oos** *var.* Somerset East. Eastern Cape, S South Africa 32°44′S 25°35′E
Somerset Village *see* Somerset
83 E26 **Somerset-Wes** *var.* Somerset West. Western Cape, SW South Africa 34°05′S 18°51′E
Somerset West *see* Somerset-Wes
Somers Islands *see* Bermuda
18 J17 **Somers Point** New Jersey, NE USA 39°18′N 74°34′W
19 P9 **Somersworth** New Hampshire, NE USA 43°15′N 70°52′W
36 H15 **Somerton** Arizona, SW USA 32°36′N 114°42′W
18 J14 **Somerville** New Jersey, NE USA 40°34′N 74°36′W
20 F10 **Somerville** Tennessee, S USA 35°14′N 89°24′W
25 U10 **Somerville** Texas, SW USA 30°21′N 96°31′W
25 T10 **Somerville Lake** ⊠ Texas, SW USA
Someș/Somesch/Someșul *see* Szamos
103 N2 **Somme** ⬖ *department* N France
103 N2 **Somme** ≈ N France
95 L18 **Sommen** Jönköping, S Sweden 58°07′N 14°58′E
95 M18 **Sommen** ◎ S Sweden
101 K16 **Sömmerda** Thüringen, C Germany 51°10′N 11°07′E
Sommerein *see* Šamorín
Sommerfeld *see* Lubsko
55 Y11 **Sommet Tabulaire** *var.* Mont Itoupé. ▲▲ S French Guiana
111 H25 **Somogy** *off.* Somogy Megye. ⬖ *county* SW Hungary
Somogy Megye *see* Somogy
Somorja *see* Šamorín
105 N7 **Somosierra, Puerto de** *pass* N Spain
187 Y14 **Somosomo** Taveuni, N Fiji 16°46′S 179°57′W
42 I9 **Somotillo** Chinandega, NW Nicaragua 13°01′N 86°53′W
42 I8 **Somoto** Madriz, NW Nicaragua 13°29′N 86°36′W
110 I11 **Sompolno** Wielkopolskie, C Poland 52°24′N 18°30′E
102 J17 **Somport, Col du** *var.* Puerto de Somport, *Sp.* Somport; *anc.* Summus Portus. *pass* France/Spain *see also* Somport
99 K15 **Son** Noord-Brabant, S Netherlands 51°32′N 05°34′E
95 H15 **Son** Akershus, S Norway 59°32′N 10°42′E
154 L9 **Son** *var.* Sone. ≈ C India
43 R16 **Soná** Veraguas, W Panama 08°00′N 81°20′W
Sonag *see* Zêkog
Sonapur *see* Subarnapur
95 G24 **Sønderborg** *Ger.* Sonderburg. Syddanmark, SW Denmark 54°55′N 09°48′E
Sonderburg *see* Sønderborg
101 K15 **Sondershausen** Thüringen, C Germany 51°22′N 10°52′E
Sondre Strømfjord *see* Kangerlussuaq
106 E6 **Sondrio** Lombardia, N Italy 46°11′N 09°52′E
Sone *see* Son
Sonepur *see* Subarnapur
57 K22 **Sonequera** ▲ S Bolivia 22°06′S 67°10′W
167 V12 **Sông Câu** Phu Yên, C Vietnam 13°26′N 109°12′E
167 R15 **Sông Đốc** Minh Hai, S Vietnam 09°03′N 104°51′E
81 H25 **Songea** Ruvuma, S Tanzania 10°42′S 35°39′E
163 X10 **Songhua Hu** ◎ NE China
163 Y7 **Songhua Jiang** *var.* Sungari. ≈ NE China
161 S8 **Songjiang** Shanghai Shi, E China 31°01′N 121°14′E
Sŏngjin *see* Kimch′aek
167 O16 **Songkhla** *var.* Songkla, *Mal.* Singora. Songkhla, SW Thailand 07°12′N 100°35′E
Songkla *see* Songkhla
163 T13 **Song Ling** ▲▲ NE China
129 U12 **Sông Ma** *Laos* Ma, Nam. ≈ Laos/Vietnam
163 W14 **Songnim** SW North Korea 38°43′N 125°40′E
82 B10 **Songo** Uíge, NW Angola 07°30′S 14°56′E
83 M15 **Songo** Tete, NW Mozambique 15°36′S 32°45′E
79 F21 **Songololo** Bas-Congo, SW Dem. Rep. Congo 05°40′S 14°05′E
160 H7 **Songpan** *var.* Jin′an, *Tib.* Sungpu. Sichuan, C China 32°49′N 103°39′E
Sŏngsan *see* Seongsan
161 R11 **Songxi** Fujian, SE China 27°33′N 118°46′E
160 M6 **Songxian** *var.* Song Xian. Henan, C China 34°11′N 112°04′E
Song Xian *see* Songxian
161 R10 **Songyang** *var.* Xiping; *prev.* Songyin. Zhejiang, SE China 28°29′N 119°27′E
Songyin *see* Songyang
163 V9 **Songyuan** *var.* Fu-yü, Petuna; *prev.* Fuyu. Jilin, NE China 45°10′N 124°52′E
Sonid Youqi *see* SaihanTal
Sonid Zuoqi *see* Mandalt
152 I10 **Sonīpat** Haryāna, N India 29°00′N 77°01′E
93 M15 **Sonkajärvi** Pohjois-Savo, C Finland 63°40′N 27°30′E
167 R6 **Sơn La** Sơn La, N Vietnam 21°20′N 103°55′E
149 O16 **Sonmiāni** Baluchistān, S Pakistan 25°24′N 66°37′E
149 O16 **Sonmiāni Bay** *bay* S Pakistan
101 K18 **Sonneberg** Thüringen, C Germany 50°22′N 11°10′E
101 N24 **Sonntagshorn** ▲ Austria/Germany 47°40′N 12°42′E
Sonoita *see* Sonoyta
40 E3 **Sonoita, Rio** *var.* Río Sonoyta. ≈ Mexico/USA
35 N7 **Sonoma** California, W USA 38°16′N 122°28′W
35 T3 **Sonoma Peak** ▲ Nevada, W USA 40°50′N 117°34′W
35 P8 **Sonora** California, W USA 37°58′N 120°22′W
25 O10 **Sonora** Texas, SW USA 30°34′N 100°39′W
40 F5 **Sonora** ⬖ *state* NW Mexico
35 X17 **Sonoran Desert** *var.* Desierto de Altar. *desert* Mexico/USA *see also* Altar, Desierto de
40 G5 **Sonora, Río** ≈ NW Mexico
40 E2 **Sonoyta** *var.* Sonoita. Sonora, NW Mexico 31°49′N 112°50′W
Sonoyta, Río *see* Sonoita, Rio
142 K6 **Sonqor** *var.* Sunqur. Kermānshāhān, W Iran 34°45′N 47°39′E
105 N9 **Sonseca** *var.* Sonseca con Casalgordo. Castilla-La Mancha, C Spain 39°40′N 03°59′W
Sonseca con Casalgordo *see* Sonseca
54 E9 **Sonsón** Antioquia, W Colombia 05°45′N 75°18′W
42 E7 **Sonsonate** Sonsonate, W El Salvador 13°44′N 89°43′W
42 A9 **Sonsonate** ⬖ *department* SW El Salvador
188 A10 **Sonsorol Islands** *island group* S Palau
112 J9 **Sonta** *Hung.* Szond; *prev.* Szonta. Vojvodina, NW Serbia 45°34′N 19°06′E
167 S6 **Sơn Tây** *var.* Sontay. Ha Tây, N Vietnam 21°06′N 105°32′E
Sontay *see* Sơn Tây
101 J25 **Sonthofen** Bayern, S Germany 47°31′N 10°16′E
Soochow *see* Suzhou
80 O13 **Sool** *off.* Gobolka Sool. ⬖ *region* N Somalia
Soomaaliya/Soomaaliyeed, Jamuuriyada Demuqraadiga *see* Somalia
Soome Laht *see* Finland, Gulf of
23 V5 **Soperton** Georgia, SE USA 32°22′N 82°35′W
167 S6 **Sop Hao** Houaphan, N Laos 20°33′N 104°25′E
Sophia *see* Sofia
171 S10 **Sopi** Pulau Morotai, E Indonesia 02°36′N 128°32′E
Sopianae *see* Pécs
171 U13 **Sopinusa** Papua Barat, E Indonesia 03°31′S 132°55′E
81 B14 **Sopo** ≈ W South Sudan
Sopockinie/Sopotskin/Sopotskino *see* Sapotskin
114 I9 **Sopot** Plovdiv, C Bulgaria 42°39′N 24°43′E
110 I7 **Sopot** *Ger.* Zoppot. Pomorskie, N Poland 54°26′N 18°33′E
167 O8 **Sop Prap** *var.* Ban Sop Prap. Lampang, NW Thailand 17°55′N 99°20′E
111 G22 **Sopron** *Ger.* Ödenburg. Győr-Moson-Sopron, NW Hungary 47°40′N 16°35′E
147 U11 **Sopu-Korgon** *var.* Sofi-Kurgan. Oshskaya Oblast′, SW Kyrgyzstan 40°03′N 73°30′E
152 H5 **Sopur** Jammu and Kashmir, NW India 34°19′N 74°30′E
107 J15 **Sora** Lazio, C Italy 41°43′N 13°37′E
154 N13 **Sorada** Odisha, E India 19°46′N 84°29′E
93 H17 **Söråker** Västernorrland, C Sweden 62°32′N 17°32′E
57 J17 **Sorata** La Paz, W Bolivia 15°47′S 68°38′W
Sorau/Sorau in der Niederlausitz *see* Żary
105 Q14 **Sorbas** Andalucía, S Spain 37°06′N 02°06′W
94 N11 **Sördellen** ◎ C Sweden
Sord/Sórd Choluim Chille *see* Swords
15 O11 **Sorel** Québec, SE Canada 46°03′N 73°06′W
183 P17 **Sorell** Tasmania, SE Australia 42°49′S 147°34′E
183 O17 **Sorell, Lake** ◎ Tasmania, SE Australia
106 E8 **Soresina** Lombardia, N Italy 45°17′N 09°51′E
95 D14 **Sørfjorden** *fjord* S Norway
94 N11 **Sörforsa** Gävleborg, C Sweden 61°45′N 17°00′E
103 R14 **Sorgues** Vaucluse, SE France 44°N 04°52′E
136 K13 **Sorgun** Yozgat, C Turkey 39°49′N 35°10′E
105 P5 **Soria** Castilla y León, N Spain 41°47′N 02°26′W
105 P6 **Soria** ⬖ *province* Castilla y León, N Spain
61 D19 **Soriano** Soriano, SW Uruguay 33°25′S 58°21′W
61 D19 **Soriano** ⬖ *department* SW Uruguay
92 O4 **Sørkapp** *headland* SW Svalbard 76°34′N 16°33′E
143 T5 **Sorkh, Kūh-e** ▲▲ NE Iran
95 I23 **Sorø** Sjælland, E Denmark 55°26′N 11°34′E
Soro *see* Ghazal, Bahr el
116 M8 **Soroca** *Rus.* Soroki. N Moldova 48°10′N 28°18′E
60 L10 **Sorocaba** São Paulo, S Brazil 23°29′S 47°27′W
Sorochino *see* Sarochyna
127 T7 **Sorochinsk** Orenburgskaya Oblast′, W Russian Federation 52°26′N 53°10′E
Soroki *see* Soroca
188 H15 **Sorol** *atoll* Caroline Islands, W Micronesia
171 T12 **Sorong** Papua Barat, E Indonesia 0°49′S 131°16′E
81 G17 **Soroti** C Uganda 01°42′N 33°37′E
Sørøy *see* Sørøya
92 J8 **Sørøya** *var.* Sørøy, *Lapp.* Sállan. *island* N Norway
104 G11 **Sorraia, Rio** ≈ C Portugal
92 I10 **Sørreisa** Troms, N Norway 69°08′N 18°09′E
107 K18 **Sorrento** *anc.* Surrentum. Campania, S Italy 40°37′N 14°23′E
104 H10 **Sor, Ribeira de** *stream* C Portugal
195 T3 **Sør Rondane** *Eng.* Sor Rondane Mountains. ▲▲ Antarctica
Sor Rondane Mountains *see* Sør Rondane
93 H14 **Sorsele** Västerbotten, N Sweden 65°31′N 17°34′E
107 B17 **Sorso** Sardegna, Italy, C Mediterranean Sea 40°46′N 08°33′E
171 P4 **Sorsogon** Luzon, N Philippines 12°57′N 124°04′E
105 U4 **Sort** Cataluña, NE Spain 42°25′N 01°07′E
124 H11 **Sortavala** *prev.* Serdobol′. Respublika Kareliya, NW Russian Federation 61°45′N 30°37′E
107 L25 **Sortino** Sicilia, Italy, C Mediterranean Sea 37°10′N 15°02′E
92 G10 **Sortland** Nordland, C Norway 68°44′N 15°25′E
94 G9 **Sør-Trøndelag** ⬖ *county* S Norway
95 I15 **Sørumsand** Akershus, S Norway 59°58′N 11°13′E
118 D6 **Sõrve Säär** *headland* SW Estonia 57°54′N 22°02′E
95 K22 **Sösdala** Skåne, S Sweden 56°00′N 13°36′E
105 R4 **Sos del Rey Católico** Aragón, NE Spain 42°30′N 01°13′W
93 F15 **Sösjöfjällen** ▲ C Sweden 63°51′N 13°15′E
126 K7 **Sosna** ≈ W Russian Federation
62 H12 **Sosneado, Cerro** ▲ W Argentina 34°44′S 69°52′W
125 S9 **Sosnogorsk** Respublika Komi, NW Russian Federation 63°33′N 53°55′E
124 J8 **Sosnovets** Respublika Kareliya, NW Russian Federation 64°25′N 34°23′E
Sosnovets *see* Sosnowiec
127 Q3 **Sosnovka** Chuvashskaya Respublika, W Russian Federation 56°18′N 47°14′E
125 S16 **Sosnovka** Kirovskaya Oblast′, NW Russian Federation 56°15′N 51°20′E
124 M6 **Sosnovka** Murmanskaya Oblast′, NW Russian Federation 66°28′N 40°31′E
126 M6 **Sosnovka** Tambovskaya Oblast′, W Russian Federation 53°14′N 41°19′E
124 H12 **Sosnovo** *Fin.* Rautu. Leningradskaya Oblast′, NW Russian Federation 60°30′N 30°13′E
127 V3 **Sosnovyy Bor** Respublika Bashkortostan, W Russian Federation 55°51′N 57°09′E
Sosnovyy Bor *see* Sasnovy Bor
111 J16 **Sosnowiec** *Ger.* Sosnowitz, *Rus.* Sosnovets. Śląskie, S Poland 50°16′N 19°07′E
Sosnowitz *see* Sosnowiec
117 R2 **Sosnytsya** Chernihivs′ka Oblast′, N Ukraine 51°31′N 32°30′E
109 V10 **Šoštanj** N Slovenia 46°23′N 15°03′E
122 G10 **Sos′va** Sverdlovskaya Oblast′, C Russian Federation 59°13′N 61°58′E
54 D12 **Sotará, Volcán** ⛰ S Colombia 02°04′N 76°40′W
76 D10 **Sotavento, Ilhas de** *var.* Leeward Islands. *island group* S Cape Verde

◆ Country
● Country Capital
◇ Dependent Territory
○ Dependent Territory Capital
⬖ Administrative Regions
✈ International Airport
▲ Mountain
▲▲ Mountain Range
⛰ Volcano
≈ River
◎ Lake
⊠ Reservoir

93 N15 Sotkamo Kainuu, C Finland 64°06′N 28°30′E
109 W11 Sotla ≈ E Slovenia
41 P10 Soto la Marina Tamaulipas, C Mexico 23°44′N 98°10′W
41 P10 Soto la Marina, Río ≈ C Mexico
95 B14 Sotra island S Norway
41 X12 Sotuta Yucatán, SE Mexico 20°34′N 89°00′W
79 F17 Souanké Sangha, NW Congo 02°03′N 14°02′E
76 M17 Soubré S Ivory Coast 05°50′N 06°35′W
115 H24 Soúda var. Soúdha, Eng. Suda. Kríti, Greece, E Mediterranean Sea 35°29′N 24°04′E
Soúdha see Soúda
Soueida see As Suwaydā'
114 L12 Souflí prev. Souflíon. Anatolikí Makedonía kai Thráki, NE Greece 41°12′N 26°18′E
Souflíon see Souflí
45 S11 Soufrière W Saint Lucia 13°51′N 61°03′W
45 X6 Soufrière 🌋 Basse Terre, S Guadeloupe 16°03′N 61°39′W
102 M13 Souillac Lot, S France 44°53′N 01°29′E
173 Y17 Souillac S Mauritius 20°31′S 57°31′E
74 M5 Souk Ahras NE Algeria 36°14′N 08°00′E
Souk el Arba du Rharb/ Souk-el-Arba-du-Rharb/ Souk-el-Arba-el-Rhab see Souk-el-Arba-Rharb
74 E6 Souk-el-Arba-Rharb var. Souk el Arba du Rharb, Souk-el-Arba-du-Rharb, Souk-el-Arba-el-Rhab. NW Morocco 34°38′N 06°00′W
Soukhné see As Sukhnah
Sŏul see Seoul
102 J11 Soulac-sur-Mer Gironde, SW France 45°31′N 01°06′W
99 L19 Soumagne Liège, E Belgium 50°36′N 05°48′E
18 M14 Sound Beach Long Island, New York, NE USA 40°56′N 72°58′W
95 J22 Sound, The Dan. Øresund, Swe. Öresund. strait Denmark/Sweden
115 H20 Soúnio, Akrotírio headland C Greece 37°39′N 24°01′E
138 F8 Soûr var. Şūr; anc. Tyre. SW Lebanon 33°18′N 35°30′E
Sources, Mont-aux- see Phofung
104 G8 Soure Coimbra, N Portugal 40°04′N 08°38′W
11 W17 Souris Manitoba, S Canada 49°38′N 100°17′W
13 Q14 Souris Prince Edward Island, SE Canada 46°22′N 62°16′W
28 L2 Souris River var. Mouse River. ≈ Canada/USA
25 X10 Sour Lake Texas, SW USA 30°08′N 94°24′W
115 F17 Soúrpi Thessalía, C Greece 39°07′N 22°55′E
104 H11 Sousel Portalegre, C Portugal 38°57′N 07°40′W
75 N6 Sousse var. Sūsah. NE Tunisia 35°46′N 10°38′E
14 H11 South ≈ Ontario, S Canada
South see Sud
83 G23 South Africa off. Republic of South Africa, Afr. Suid-Afrika. ◆ republic S Africa
South Africa, Republic of see South Africa
46-47 South America continent
2 J17 South American Plate tectonic feature
97 M23 Southampton hist. Hamwih, Lat. Clausentum. S England, United Kingdom 50°54′N 01°23′W
19 N14 Southampton Long Island, New York, NE USA 40°52′N 72°22′W
9 P8 Southampton Island island Nunavut, NE Canada
151 P20 South Andaman island Andaman Islands, India, NE Indian Ocean
13 Q6 South Aulatsivik Island island Newfoundland and Labrador, E Canada
182 E4 South Australia ◈ state S Australia
South Australian Abyssal Plain see South Australian Plain
192 G11 South Australian Basin undersea feature SW Indian Ocean 38°00′S 126°00′E
173 X12 South Australian Plain var. South Australian Abyssal Plain. undersea feature SE Indian Ocean
37 R13 South Baldy ▲ New Mexico, SW USA 33°59′N 107°11′W
23 Y14 South Bay Florida, SE USA 26°39′N 80°43′W
14 E12 South Baymouth Manitoulin Island, Ontario, S Canada 45°33′N 82°01′W
30 L10 South Beloit Illinois, N USA 42°29′N 89°02′W
31 O11 South Bend Indiana, N USA 41°40′N 86°15′W
25 R6 South Bend Texas, SW USA 32°58′N 98°39′W
32 F9 South Bend Washington, NW USA 46°38′N 123°48′W
South Beveland see Zuid-Beveland
South Borneo see Kalimantan Selatan
21 U7 South Boston Virginia, NE USA 36°42′N 78°58′W
182 F2 South Branch Neales seasonal river South Australia
21 U3 South Branch Potomac River ≈ West Virginia, NE USA
185 H19 Southbridge Canterbury, South Island, New Zealand 43°49′S 172°17′E
19 N12 Southbridge Massachusetts, NE USA 42°03′N 72°00′W
183 P17 South Bruny Island island Tasmania, SE Australia
18 L7 South Burlington Vermont, NE USA 44°27′N 73°08′W
44 M6 South Caicos island S Turks and Caicos Islands
South Cape see Ka Lae
23 V3 South Carolina off. State of South Carolina, also known as The Palmetto State. ◈ state SE USA
South Carpathians see Carpații Meridionali
South Celebes see Sulawesi Selatan
21 Q5 South Charleston West Virginia, NE USA 38°22′N 81°42′W
192 D7 South China Basin undersea feature SE South China Sea 15°00′N 115°00′E
169 R8 South China Sea Chin. Nan Hai, Ind. Laut Cina Selatan, Vtn. Biên Đông. sea SE Asia
33 Z10 South Dakota off. State of South Dakota, also known as The Coyote State, Sunshine State. ◈ state N USA
23 X10 South Daytona Florida, SE USA 29°09′N 81°01′W
37 R10 South Domingo Pueblo New Mexico, SW USA 35°28′N 106°24′W
97 N23 South Downs hill range SE England, United Kingdom
83 I21 South East ◈ district SE Botswana
65 H15 South East Bay bay Ascension Island, C Atlantic Ocean
183 O17 South East Cape headland Tasmania, SE Australia 43°36′S 146°52′E
38 K10 Southeast Cape headland Saint Lawrence Island, Alaska, USA 62°56′N 169°39′W
South-East Celebes see Sulawesi Tenggara
192 G12 Southeast Indian Ridge undersea feature Indian Ocean/Pacific Ocean 50°00′S 110°00′E
Southeast Island see Tagula Island
193 P13 Southeast Pacific Basin var. Belling Hausen Mulde. undersea feature SE Pacific Ocean 60°00′S 115°00′W
65 H15 South East Point headland SE Ascension Island
183 O14 South East Point headland Victoria, S Australia 39°10′S 146°21′E
191 Z3 South East Point headland Kiritimati, NE Kiribati 01°42′N 157°10′W
44 L5 Southeast Point headland Mayaguana, SE The Bahamas 22°15′N 72°44′W
South-East Sulawesi see Sulawesi Tenggara
11 U12 Southend Saskatchewan, C Canada 56°20′N 103°14′W
97 P22 Southend-on-Sea E England, United Kingdom 51°33′N 00°43′E
83 H20 Southern var. Bangwaketse, Ngwaketze. ◈ district SE Botswana
138 E13 Southern ◈ district S Israel
83 N15 Southern ◈ region S Malawi
155 J26 Southern ◈ province S Sri Lanka
83 I15 Southern ◈ province S Zambia
185 E19 Southern Alps ⛰ South Island, New Zealand
190 K15 Southern Cook Islands island group S Cook Islands
180 K12 Southern Cross Western Australia 31°17′S 119°15′E
80 A12 Southern Darfur ◈ state W Sudan
186 B7 Southern Highlands ◈ province W Papua New Guinea
11 V11 Southern Indian Lake ◎ Manitoba, C Canada
80 E11 Southern Kordofan ◈ state C Sudan
187 Z15 Southern Lau Group island group Lau Group, SE Fiji
81 I15 Southern Nationalities ◈ region S Ethiopia
173 S13 Southern Ocean ocean
21 T10 Southern Pines North Carolina, SE USA 35°10′N 79°23′W
96 I13 Southern Uplands ⛰ S Scotland, United Kingdom
Southern Urals see Yuzhnyy Ural
183 P16 South Esk River ≈ Tasmania, SE Australia
11 U16 Southey Saskatchewan, S Canada 50°53′N 104°27′W
27 V2 South Fabius River ≈ Missouri, C USA
31 S10 Southfield Michigan, N USA 42°28′N 83°12′W
192 K10 South Fiji Basin undersea feature S Pacific Ocean 26°00′S 175°00′E
97 Q22 South Foreland headland SE England, United Kingdom 51°08′N 01°22′E
35 P7 South Fork American River ≈ California, W USA
28 K7 South Fork Grand River ≈ South Dakota, N USA
35 T12 South Fork Kern River ≈ California, W USA
39 Q7 South Fork Koyukuk River ≈ Alaska, USA
39 Q11 South Fork Kuskokwim River ≈ Alaska, USA
26 H2 South Fork Republican River ≈ C USA
26 L3 South Fork Solomon River ≈ Kansas, C USA
31 P5 South Fox Island island Michigan, N USA
20 G8 South Fulton Tennessee, S USA 36°28′N 88°53′W
195 U10 South Geomagnetic Pole pole Antarctica
65 J20 South Georgia island South Georgia and the South Sandwich Islands, SW Atlantic Ocean
65 K21 South Georgia and the South Sandwich Islands ◇ UK Dependent Territory SW Atlantic Ocean
47 Y14 South Georgia Ridge var. North Scotia Ridge. undersea feature SW Atlantic Ocean 54°00′S 40°00′W
181 Q1 South Goulburn Island island Northern Territory, N Australia
153 U16 South Hatia Island island SE Bangladesh
31 O10 South Haven Michigan, N USA 42°24′N 86°16′W
21 V7 South Hill Virginia, NE USA 36°43′N 78°07′W
South Holland see Zuid-Holland
21 P8 South Holston Lake ⊠ Tennessee/Virginia, S USA
175 N1 South Honshu Ridge undersea feature W Pacific Ocean
26 M6 South Hutchinson Kansas, C USA 38°01′N 97°56′W
151 K21 South Huvadhu Atoll atoll S Maldives
173 U14 South Indian Basin undersea feature Indian Ocean/Pacific Ocean 60°00′S 120°00′E
11 W11 South Indian Lake Manitoba, C Canada 56°48′N 98°56′W
81 I17 South Island island NW Kenya
185 C20 South Island island S New Zealand
65 B23 South Jason island Jason Islands, NW Falkland Islands
South Kalimantan see Kalimantan Selatan
South Karelia see Etelä-Karjala
South Kazakhstan see Yuzhnyy Kazakhstan
163 X15 South Korea off. Republic of Korea, Kor. Taehan Min'guk. ◆ republic E Asia
35 Q6 South Lake Tahoe California, W USA 38°56′N 119°57′W
25 N6 Southland Texas, SW USA 33°16′N 101°31′W
185 B23 Southland off. Southland Region. ◈ region South Island, New Zealand
Southland Region see Southland
29 N15 South Loup River ≈ Nebraska, C USA
151 K19 South Maalhosmadulu Atoll atoll N Maldives
14 E15 South Maitland ≈ Ontario, S Canada
192 E8 South Makassar Basin undersea feature E Java Sea
31 O6 South Manitou Island island Michigan, N USA
151 K18 South Miladhunmadulu Atoll var. Noonu. atoll N Maldives
21 X8 South Mills North Carolina, SE USA 36°28′N 76°18′W
8 H9 South Nahanni ≈ Northwest Territories, NW Canada
39 P13 South Naknek Alaska, USA 58°39′N 157°01′W
14 M13 South Nation ≈ Ontario, SE Canada
44 F9 South Negril Point headland W Jamaica 18°14′N 78°21′W
151 K20 South Nilandhe Atoll var. Dhaalu Atoll. atoll C Maldives
36 L2 South Ogden Utah, W USA 41°09′N 111°58′W
18 M14 Southold Long Island, New York, NE USA 41°03′N 72°24′W
194 H1 South Orkney Islands island group Antarctica
137 S9 South Ossetia former autonomous region SW Georgia
South Ostrobothnia see Etelä-Pohjanmaa
South Pacific Basin see Southwest Pacific Basin
19 P7 South Paris Maine, NE USA 44°14′N 70°33′W
189 U13 South Pass passage Chuuk Islands, C Micronesia
33 U15 South Pass pass Wyoming, C USA
20 K10 South Pittsburg Tennessee, S USA 35°00′N 85°42′W
28 K15 South Platte River ≈ Colorado/Nebraska, C USA
31 T16 South Point Ohio, N USA 38°25′N 82°35′W
65 G15 South Point headland S Ascension Island
31 R6 South Point headland Michigan, N USA 44°51′N 83°17′W
South Point see Ka Lae
195 Q9 South Pole pole Antarctica
183 P17 Southport Tasmania, SE Australia 43°26′S 146°57′E
97 K17 Southport NW England, United Kingdom 53°39′N 03°01′W
21 V12 Southport North Carolina, SE USA 33°55′N 78°00′W
19 P8 South Portland Maine, NE USA 43°38′N 70°14′W
14 H12 South River Ontario, S Canada 45°50′N 79°23′W
21 U11 South River ≈ North Carolina, SE USA
96 K5 South Ronaldsay island NE Scotland, United Kingdom
36 L2 South Salt Lake Utah, W USA 40°42′N 111°52′W
65 L21 South Sandwich Islands island group SW Atlantic Ocean
65 K21 South Sandwich Trench undersea feature SW Atlantic Ocean 56°30′S 25°00′W
11 S16 South Saskatchewan ≈ Alberta/Saskatchewan, S Canada
65 I21 South Scotia Ridge undersea feature S Scotia Sea
11 V10 South Seal ≈ Manitoba, C Canada
194 G4 South Shetland Islands island group Antarctica
65 H22 South Shetland Trough undersea feature Atlantic Ocean/Pacific Ocean 61°00′S 59°30′W
97 M14 South Shields NE England, United Kingdom 55°N 01°25′W
29 R13 South Sioux City Nebraska, C USA 42°28′N 96°24′W
192 J9 South Solomon Trench undersea feature W Pacific Ocean
183 V3 South Stradbroke Island island Queensland, E Australia
81 E15 South Sudan off. Republic of South Sudan. ◆ E Africa
South Sulawesi see Sulawesi Selatan
South Sumatra see Sumatera Selatan
184 K11 South Taranaki Bight bight SE Tasman Sea
South Tasmania Plateau see Tasman Plateau
36 M15 South Tucson Arizona, SW USA 32°11′N 110°56′W
12 H9 South Twin Island island Nunavut, C Canada
South Tyrol see Trentino-Alto Adige
96 E9 South Uist island NW Scotland, United Kingdom
South-West see Sud-Ouest
South-West Africa/South West Africa see Namibia
65 F15 South West Bay bay Ascension Island, C Atlantic Ocean
183 N18 South West Cape headland Tasmania, SE Australia 43°34′S 146°01′E
185 B26 South West Cape headland Stewart Island, New Zealand 47°15′S 167°28′E
38 J10 Southwest Cape headland Saint Lawrence Island, Alaska, USA 63°19′N 171°27′W
Southwest Indian Ocean Ridge see Southwest Indian Ridge
173 N11 Southwest Indian Ridge var. Southwest Indian Ocean Ridge. undersea feature SW Indian Ocean 43°00′S 40°00′E
192 L10 Southwest Pacific Basin var. South Pacific Basin. undersea feature SE Pacific Ocean 40°00′S 150°00′W
191 X3 South West Point headland Kiritimati, NE Kiribati 01°53′N 157°34′E
65 G25 South West Point headland SW Saint Helena 16°00′S 05°48′W
44 H2 Southwest Point headland Great Abaco, N The Bahamas 25°50′N 77°12′W
25 P5 South Wichita River ≈ Texas, SW USA
97 Q20 Southwold E England, United Kingdom 52°15′N 01°36′E
19 Q12 South Yarmouth Massachusetts, NE USA 41°38′N 70°09′W
116 J10 Sovata Hung. Szováta. Mureș, C Romania 46°36′N 25°04′E
107 N22 Soverato Calabria, SW Italy 38°40′N 16°31′E
121 O4 Sovereign Base Area uk military installation S Cyprus
Sovetabad see Ghafurov
126 C2 Sovetsk Ger. Tilsit. Kaliningradskaya Oblast', W Russian Federation 53°04′N 21°52′E
125 Q15 Sovetsk Kirovskaya Oblast', NW Russian Federation 57°37′N 49°02′E
127 N10 Sovetskaya Rostovskaya Oblast', SW Russian Federation 49°00′N 42°09′E
Sovetskoye see Ketchenery
146 I15 Sovet"yab prev. Sovet"yap. Ahal Welaýaty, S Turkmenistan 36°29′N 61°13′E
Sovet"yap see Sovet"yab
117 U12 Sovyets'kyy Avtonomna Respublika Krym, S Ukraine 45°20′N 34°54′E
83 I18 Sowa var. Sua. Central, NE Botswana 20°33′S 26°18′E
83 I18 Sowa Pan var. Sua Pan. salt lake NE Botswana
83 J21 Soweto Gauteng, NE South Africa 26°08′S 27°54′E
147 R11 So'x Rus. Sokh. Farg'ona Viloyati, E Uzbekistan 39°56′N 71°10′E
Sōya-kaikyō see La Pérouse Strait
165 T1 Sōya-misaki headland Hokkaidō, NE Japan 45°31′N 141°55′E
125 N7 Soyana ≈ NW Russian Federation
146 A8 Soye, Mys var. Mys Suz. headland NW Turkmenistan 41°47′N 52°27′E
82 A10 Soyo Zaire Province, NW Angola 06°07′S 12°18′E
80 J10 Soyra ▲ C Eritrea 14°46′N 39°29′E
145 P15 Sozak Kaz. Sozaq; prev. Suzak. Yuzhnyy Kazakhstan, S Kazakhstan 44°09′N 68°28′E
Sozaq see Sozak
119 P16 Sozh ≈ NE Europe
114 N10 Sozopol prev. Sizebolu; anc. Apollonia. Burgas, E Bulgaria 42°25′N 27°42′E
99 L20 Spa Liège, E Belgium 50°29′N 05°52′E
194 I7 Spaatz Island island Antarctica
144 M14 Space Launching Centre space station Kzylorda, S Kazakhstan
105 O7 Spain off. Kingdom of Spain, Sp. España; anc. Hispania, Iberia, Lat. Hispana. ◆ monarchy SW Europe
Spain, Kingdom of see Spain
Spalato see Split
97 O19 Spalding E England, United Kingdom 52°49′N 00°06′W
14 D11 Spanish Ontario, S Canada 46°12′N 82°21′W
36 L3 Spanish Fork Utah, W USA 40°09′N 111°40′W
64 B12 Spanish Point headland C Bermuda 32°18′N 64°49′W
14 E9 Spanish River ≈ Ontario, S Canada
44 K13 Spanish Town hist. St.Iago de la Vega. C Jamaica 18°N 76°57′W
Spánta, Akrotírio see Spátha, Akrotírio
35 Q5 Sparks Nevada, W USA 39°32′N 119°45′W
Sparnacum see Épernay
95 N16 Sparreholm Södermanland, C Sweden 59°04′N 16°51′E
23 U4 Sparta Georgia, SE USA 33°16′N 82°58′W
30 K16 Sparta Illinois, N USA 38°07′N 89°42′W
31 P9 Sparta Michigan, N USA 43°09′N 85°42′W
21 R8 Sparta North Carolina, SE USA 36°30′N 81°07′W
20 L9 Sparta Tennessee, S USA 35°55′N 85°30′W
30 J7 Sparta Wisconsin, N USA 43°57′N 90°50′W
Sparta see Spárti
21 Q11 Spartanburg South Carolina, SE USA 34°56′N 81°57′W
115 F21 Spárti Eng. Sparta. Pelopónnisos, S Greece 37°05′N 22°25′E
107 B21 Spartivento, Capo headland Sardegna, Italy, C Mediterranean Sea 38°52′N 08°50′E
11 P17 Sparwood British Columbia, SW Canada 49°45′N 114°45′W
126 J4 Spas-Demensk Kaluzhskaya Oblast', W Russian Federation 54°22′N 34°16′E
126 M4 Spas-Klepiki Ryazanskaya Oblast', W Russian Federation 55°08′N 40°15′E
Spasovo see Kulen Vakuf
123 R15 Spassk-Dal'niy Primorskiy Kray, SE Russian Federation 44°34′N 132°52′E
126 M5 Spassk-Ryazanskiy Ryazanskaya Oblast', W Russian Federation 54°25′N 40°21′E
115 H19 Spáta Attikí, C Greece 37°58′N 23°55′E
121 Q11 Spátha, Akrotírio var. Akrotírio Spánta. headland Kríti, Greece, E Mediterranean Sea 35°42′N 23°44′E
Špatrjan see Paternion
28 I9 Spearfish South Dakota, N USA 44°29′N 103°51′W
25 O1 Spearman Texas, SW USA 36°12′N 101°13′W
65 C25 Speedwell Island island S Falkland Islands
65 C25 Speedwell Island Settlement S Falkland Islands 52°13′S 59°41′W
65 G25 Speery Island island S Saint Helena
45 N14 Speightstown NW Barbados 13°15′N 59°39′W
106 I13 Spello Umbria, C Italy 43°00′N 12°41′E
39 R12 Spenard Alaska, USA 61°09′N 150°03′W
Spence Bay see Taloyoak
31 O14 Spencer Indiana, N USA 39°18′N 86°46′W
29 T12 Spencer Iowa, C USA 43°09′N 95°07′W
29 P12 Spencer Nebraska, C USA 42°52′N 98°42′W
21 S9 Spencer North Carolina, SE USA 35°41′N 80°26′W
20 L9 Spencer Tennessee, S USA 35°46′N 85°27′W
21 Q4 Spencer West Virginia, NE USA 38°48′N 81°22′W
30 K6 Spencer Wisconsin, N USA 44°46′N 90°17′W
182 G10 Spencer, Cape headland South Australia 35°17′S 136°52′E
39 V13 Spencer, Cape headland Alaska, USA 58°12′N 136°39′W
182 H9 Spencer Gulf gulf South Australia
18 F9 Spencerport New York, NE USA 43°11′N 77°48′W
31 Q12 Spencerville Ohio, N USA 40°42′N 84°21′W
115 E17 Spercheiáda var. Sperhiada, Sperkhiás. Stereá Elláda, C Greece 38°54′N 22°07′E
115 E17 Spercheiós ≈ C Greece
Sperhiada see Spercheiáda
95 G14 Sperillen ◎ S Norway
Sperkhiás see Spercheiáda
101 I18 Spessart hill range C Germany
Spétsai see Spétses
115 G21 Spétses prev. Spétsai. Spétses, S Greece 37°16′N 23°09′E
115 G21 Spétses island S Greece
96 J8 Spey ≈ NE Scotland, United Kingdom
101 G20 Speyer Eng. Spires; anc. Civitas Nemetum, Spira. Rheinland-Pfalz, SW Germany 49°18′N 08°26′E
101 G20 Speyerbach ≈ W Germany
107 N20 Spezzano Albanese Calabria, SW Italy 39°40′N 16°17′E
Spice Islands see Maluku
100 F9 Spiekeroog island NW Germany
109 W9 Spielfeld Steiermark, SE Austria 46°43′N 15°36′E
65 N21 Spiess Seamount undersea feature S Atlantic Ocean 53°00′S 02°00′W
108 E9 Spiez Bern, W Switzerland 46°42′N 07°41′E
98 G13 Spijkenisse Zuid-Holland, SW Netherlands 51°52′N 04°19′E
39 T6 Spike Mountain ▲ Alaska, USA 67°42′N 141°39′W
115 I25 Spíli Kríti, Greece, E Mediterranean Sea 35°12′N 24°33′E
108 D10 Spillgerten ▲ W Switzerland 46°34′N 07°25′E
118 F9 Spilve ✈ (Rīga) C Latvia 56°55′N 24°03′E
107 N17 Spinazzola Puglia, SE Italy 40°58′N 16°06′E
149 O9 Spīn Bōldak prev. Spin Buldak. Kandahār, S Afghanistan 31°01′N 66°23′E
Spin Buldak see Spīn Bōldak
Spira see Speyer
Spirdingsee see Śniardwy, Jezioro
Spires see Speyer
29 T11 Spirit Lake Iowa, C USA 43°25′N 95°06′W
29 T11 Spirit Lake ◎ Iowa, C USA
11 N13 Spirit River Alberta, W Canada 55°46′N 118°51′W
11 S14 Spiritwood Saskatchewan, S Canada 53°18′N 107°33′W
27 R11 Spiro Oklahoma, C USA 35°14′N 94°37′W
111 L19 Spišská Nová Ves Ger. Neudorf, Zipser Neudorf, Hung. Igló. Košický Kraj, E Slovakia 48°58′N 20°35′E
137 T11 Spitak NW Armenia 40°51′N 44°17′E
92 O2 Spitsbergen island NW Svalbard
Spittal see Spittal an der Drau
109 R9 Spittal an der Drau var. Spittal. Kärnten, S Austria 46°48′N 13°30′E
109 V3 Spitz Niederösterreich, NE Austria 48°24′N 15°22′E
94 D9 Spjelkavik Møre og Romsdal, S Norway 62°28′N 06°22′E
25 W10 Splendora Texas, SW USA 30°13′N 95°09′W
113 E14 Split It. Spalato. Split-Dalmacija, S Croatia 43°31′N 16°27′E
113 E14 Split ✈ Split-Dalmacija, S Croatia 43°33′N 16°19′E
113 E14 Split-Dalmacija off. Splitsko-Dalmatinska Županija. ◈ province S Croatia
11 X12 Split Lake ◎ Manitoba, C Canada
Splitsko-Dalmatinska Županija see Split-Dalmacija
108 H10 Splügen Graubünden, S Switzerland 46°33′N 09°18′E
Spodnji Dravograd see Dravograd
25 P12 Spofford Texas, SW USA 29°10′N 100°24′W
118 J11 Špoģi SE Latvia 56°03′N 26°47′E
32 L8 Spokane Washington, NW USA 47°40′N 117°26′W
32 L8 Spokane River ≈ Washington, NW USA
106 I13 Spoleto Umbria, C Italy 42°44′N 12°44′E
30 I4 Spooner Wisconsin, N USA 45°51′N 91°49′W
30 K12 Spoon River ≈ Illinois, N USA
21 W5 Spotsylvania Virginia, NE USA 38°12′N 77°35′W
32 L8 Sprague Washington, NW USA 47°19′N 117°55′W
170 J5 Spratly Island Chin. Nanwei Dao, Viet. Dao Tr ng Sa L n. island SW Spratly Islands
192 E6 Spratly Islands Chin. Nansha Qundao, Viet. Quân Dao Tr ng Sa. ◇ disputed territory SE Asia
32 J12 Spray Oregon, NW USA 44°30′N 119°38′W
112 I11 Spreča ≈ N Bosnia and Herzegovina
100 P13 Spree ≈ E Germany
100 P13 Spreewald wetland NE Germany
101 P14 Spremberg Brandenburg, E Germany 51°34′N 14°22′E
25 W11 Spring Texas, SW USA 30°03′N 95°24′W
31 Q10 Spring Arbor Michigan, N USA 42°12′N 84°33′W
83 E23 Springbok Northern Cape, W South Africa 29°44′S 17°56′E
18 I15 Spring City Pennsylvania, NE USA 40°10′N 75°33′W
20 L9 Spring City Tennessee, S USA 35°41′N 84°51′W
36 L4 Spring City Utah, W USA 39°28′N 111°30′W
35 W3 Spring Creek Nevada, W USA 40°45′N 115°40′W
27 S9 Springdale Arkansas, C USA 36°11′N 94°07′W
31 Q14 Springdale Ohio, N USA 39°17′N 84°29′W
100 I13 Springe Niedersachsen, N Germany 52°13′N 09°33′E
37 U9 Springer New Mexico, SW USA 36°21′N 104°35′W
37 W7 Springfield Colorado, C USA 37°24′N 102°36′W
23 W5 Springfield Georgia, SE USA 32°21′N 81°20′W
30 K14 Springfield state capital Illinois, N USA 39°48′N 89°39′W
20 L6 Springfield Kentucky, S USA 37°42′N 85°18′W
18 M12 Springfield Massachusetts, NE USA 42°06′N 72°32′W
29 T10 Springfield Minnesota, N USA 44°15′N 94°58′W
27 T7 Springfield Missouri, C USA 37°13′N 93°18′W
31 R13 Springfield Ohio, N USA 39°55′N 83°49′W
32 G13 Springfield Oregon, NW USA 44°03′N 123°01′W
29 Q12 Springfield South Dakota, N USA 42°51′N 97°54′W
20 J8 Springfield Tennessee, S USA 36°30′N 86°54′W
18 M9 Springfield Vermont, NE USA 43°18′N 72°27′W
30 K14 Springfield, Lake ⊠ Illinois, N USA
55 T8 Spring Garden NE Guyana 06°58′N 58°34′W
30 K8 Spring Green Wisconsin, N USA 43°10′N 90°02′W
29 X11 Spring Grove Minnesota, N USA 43°33′N 91°38′W
13 P15 Springhill Nova Scotia, SE Canada 45°40′N 64°04′W
23 V12 Spring Hill Florida, SE USA 28°28′N 82°36′W
27 R4 Spring Hill Kansas, C USA 38°44′N 94°49′W
22 G4 Springhill Louisiana, S USA 33°01′N 93°27′W
20 I9 Spring Hill Tennessee, S USA 35°46′N 86°55′W
21 U10 Spring Lake North Carolina, SE USA 35°10′N 78°58′W
24 M4 Springlake Texas, SW USA 34°13′N 102°18′W
35 W11 Spring Mountains ⛰ Nevada, W USA
65 B24 Spring Point West Falkland, Falkland Islands 51°49′S 60°27′W
27 W9 Spring River ≈ Arkansas/Missouri, C USA
27 S7 Spring River ≈ Missouri/Oklahoma, C USA
83 J21 Springs Gauteng, NE South Africa 26°16′S 28°26′E
185 H16 Springs Junction West Coast, South Island, New Zealand 42°21′S 172°11′E
181 X8 Springsure Queensland, E Australia 24°09′S 148°06′E
29 W11 Spring Valley Minnesota, N USA 43°41′N 92°23′W
18 K13 Spring Valley New York, NE USA 41°10′N 73°58′W
29 N12 Springview Nebraska, C USA 42°49′N 99°45′W
18 D11 Springville New York, NE USA 42°27′N 78°52′W
36 L3 Springville Utah, W USA 40°10′N 111°36′W
Sprottau see Szprotawa
15 V4 Sproule, Pointe headland Québec, SE Canada 49°47′N 67°02′W
11 Q14 Spruce Grove Alberta, SW Canada 53°36′N 113°55′W
21 T4 Spruce Knob ▲ West Virginia, NE USA 38°40′N 79°37′W
35 X3 Spruce Mountain ▲ Nevada, W USA 40°33′N 114°46′W
21 P9 Spruce Pine North Carolina, SE USA 35°55′N 82°03′W
98 G13 Spui ≈ SW Netherlands
107 O19 Spulico, Capo headland S Italy 39°57′N 16°38′E
25 O5 Spur Texas, SW USA 33°28′N 100°51′W
97 O17 Spurn Head headland E England, United Kingdom 53°34′N 00°06′E
99 H20 Spy Namur, S Belgium 50°29′N 04°43′E
95 I15 Spydeberg Østfold, S Norway 59°36′N 11°04′E
185 J17 Spy Glass Point headland South Island, New Zealand 42°33′S 173°31′E
10 L17 Squamish British Columbia, SW Canada 49°41′N 123°11′W
19 O8 Squam Lake ◎ New Hampshire, NE USA
19 S2 Squa Pan Mountain ▲ Maine, NE USA 46°36′N 68°09′W
39 N16 Squaw Harbor Unga Island, Alaska, USA 55°12′N 160°41′W
14 E11 Squaw Island island Ontario, S Canada
107 O22 Squillace, Golfo di gulf S Italy
107 Q18 Squinzano Puglia, SE Italy 40°26′N 18°03′E
Sráid na Cathrach see Milltown Malbay
167 S11 Srâlau Stœng Trêng, N Cambodia 14°03′N 105°46′E
Srath an Urláir see Stranorlar
112 G10 Srbac ◈ Republika Srpska, N Bosnia and Herzegovina
Srbija see Serbia
Srbinje see Foča
112 K9 Srbobran var. Bácsszenttamás, Hung. Szenttamás. Vojvodina, N Serbia 45°33′N 19°46′E
Srbobran see Donji Vakuf
167 R13 Srê Âmběl Kaôh Kông, SW Cambodia 11°07′N 103°46′E
112 K13 Srebrenica Republika Srpska, E Bosnia and Herzegovina 44°04′N 19°18′E
112 I11 Srebrenik Federacija Bosne I Hercegovine, NE Bosnia and Herzegovina 44°42′N 18°30′E
114 K10 Sredets prev. Syulemeshlii. Stara Zagora, C Bulgaria 42°16′N 25°40′E
114 M10 Sredets prev. Grudovo. ◈ Burgas, E Bulgaria
114 M10 Sredetska Reka ≈ SE Bulgaria
123 U9 Sredinnyy Khrebet ⛰ E Russian Federation
114 N7 Sredishte Rom. Beibunar; prev. Knyazhevo. Dobrich, NE Bulgaria 43°51′N 27°30′E
114 I10 Sredna Gora ⛰ C Bulgaria
123 R7 Srednekolymsk Respublika Sakha (Yakutiya), NE Russian Federation 67°28′N 153°52′E
126 K7 Srednerusskaya Vozvyshennost' Eng. Central Russian Upland. ⛰ W Russian Federation
122 L9 Srednesibirskoye Ploskogor'ye var. Central Siberian Uplands, Eng. Central Siberian Plateau. ⛰ N Russian Federation
125 V13 Sredniy Ural ⛰ NW Russian Federation
167 T12 Srê Khtŭm Môndól Kiri, E Cambodia 12°10′N 106°52′E
110 G12 Śrem Wielkopolskie, C Poland 52°07′N 17°00′E
112 K10 Sremska Mitrovica prev. Mitrovica, Ger. Mitrowitz. Vojvodina, NW Serbia 44°58′N 19°37′E
167 R11 Srêng, Stœng ≈ NW Cambodia
167 R11 Srê Noy Siĕmréab, NW Cambodia 13°47′N 104°03′E
Srepok, Sông see Srêpŏk, Tônle
167 T12 Srêpŏk, Tônle var. Sông Srepok. ≈ Cambodia/Vietnam
123 P13 Sretensk Zabaykal'skiy Kray, S Russian Federation 52°14′N 117°33′E
169 R10 Sri Aman Sarawak, East Malaysia 01°13′N 111°25′E
117 R4 Sribne Chernihivs'ka Oblast', N Ukraine 50°40′N 32°55′E
Sri Jayawardanapura see Sri Jayewardenapura Kotte
155 J25 Sri Jayewardenapura Kotte var. Sri Jayawardanapura. ● (legislative) Western Province, W Sri Lanka 06°54′N 79°58′E
155 M14 Srīkākulam Andhra Pradesh, E India 18°18′N 83°54′E
155 J25 Sri Lanka off. Democratic Socialist Republic of Sri Lanka; prev. Ceylon. ◆ republic S Asia
130 F14 Sri Lanka island S Asia
Sri Lanka, Democratic Socialist Republic of see Sri Lanka
153 V14 Srimangal Sylhet, E Bangladesh 24°19′N 91°40′E
Sri Mohangorh see Shri Mohangarh
152 H5 Srinagar state capital Jammu and Kashmir, N India 34°07′N 74°50′E
167 N10 Srinagarind Reservoir ◎ W Thailand
155 F19 Sringeri Karnātaka, W India 13°26′N 75°13′E
155 K25 Sri Pada Eng. Adam's Peak. ▲ S Sri Lanka 06°49′N 80°25′E
Sri Saket see Si Sa Ket
111 G14 Środa Śląska Ger. Neumarkt. Dolnośląskie, SW Poland 51°10′N 16°30′E
110 H12 Środa Wielkopolska Wielkopolskie, C Poland 52°13′N 17°17′E
Srpska Kostajnica see Bosanska Kostajnica
113 G14 Srpska, Republika ◈ republic Bosnia and Herzegovina
Srpski Brod see Bosanski Brod
Ssu-ch'uan see Sichuan
Ssu-p'ing/Ssu-p'ing-chieh see Siping
Stablo see Stavelot
99 G15 Stabroek Antwerpen, N Belgium 51°21′N 04°22′E
Stackeln see Strenči
96 I5 Stack Skerry island N Scotland, United Kingdom
100 I9 Stade Niedersachsen, NW Germany 53°36′N 09°29′E
94 C10 Stadlandet peninsula S Norway
109 R5 Stadl-Paura Oberösterreich, NW Austria 48°05′N 13°52′E
119 L20 Stadolichy Rus. Stodolichi. Homyel'skaya Voblasts', SE Belarus 51°44′N 28°30′E
98 P7 Stadskanaal Groningen, NE Netherlands 53°N 06°55′E
101 H16 Stadtallendorf Hessen, C Germany 50°49′N 09°01′E
101 K23 Stadtbergen Bayern, S Germany 48°21′N 10°50′E
108 G7 Stäfa Zürich, NE Switzerland 47°14′N 08°45′E
95 K23 Staffanstorp Skåne, S Sweden 55°38′N 13°13′E
101 K18 Staffelstein Bayern, C Germany 50°05′N 11°00′E
97 L19 Stafford C England, United Kingdom 52°48′N 02°07′W
26 L6 Stafford Kansas, C USA 37°57′N 98°36′W
21 W4 Stafford Virginia, NE USA 38°26′N 77°27′W

◆ Country ● Country Capital ◇ Dependent Territory ○ Dependent Territory Capital ◈ Administrative Regions ✈ International Airport ▲ Mountain ⛰ Mountain Range 🌋 Volcano ≈ River ◎ Lake ⊠ Reservoir

97 L19 **Staffordshire** *cultural region* C England, United Kingdom
19 N12 **Stafford Springs** Connecticut, NE USA 41°57′N 72°18′W
115 H14 **Stágira** Kentrikí Makedonía, N Greece 40°31′N 23°46′E
118 G7 **Staicele** N Latvia 57°52′N 24°48′E
Staierdorf-Anina *see* Anina
109 V8 **Stainz** Steiermark, SE Austria 46°55′N 15°18′E
Stájerlakanina *see* Anina
117 Y7 **Stakhanov** Luhans'ka Oblast', E Ukraine 48°30′N 38°42′E
108 E11 **Stalden** Valais, SW Switzerland 46°12′N 07°55′E
15 S8 **St-Alexandre** Québec, SE Canada 47°39′N 69°36′W
Stalin *see* Varna
Stalinabad *see* Dushanbe
Stalingrad *see* Volgograd
Staliniri *see* Tskhinvali
Stalino *see* Donets'k
Stalinobod *see* Dushanbe
Stalinov Štít *see* Gerlachovský štít
Stalinsk *see* Novokuznetsk
Stalins'kaya Oblast' *see* Donets'ka Oblast'
Stalinski Zaliv *see* Varnenski Zaliv
Stalin, Yazovir *see* Iskar, Yazovir
111 N15 **Stalowa Wola** Podkarpackie, SE Poland 50°35′N 22°02′E
114 I11 **Stamboliyski** Plovdiv, C Bulgaria 42°09′N 24°32′E
15 Q7 **St-Ambroise** Québec, SE Canada 48°35′N 71°19′W
97 N19 **Stamford** E England, United Kingdom 52°39′N 00°32′W
18 L14 **Stamford** Connecticut, NE USA 41°03′N 73°32′W
25 P6 **Stamford** Texas, SW USA 32°55′N 99°49′W
25 Q6 **Stamford, Lake** ⊡ Texas, SW USA
108 I10 **Stampa** Graubünden, SE Switzerland 46°21′N 09°35′E
Stampalia *see* Astypálaia
27 T14 **Stamps** Arkansas, C USA 33°22′N 93°30′W
92 G11 **Stamsund** Nordland, C Norway 68°07′N 13°50′E
27 R2 **Stanberry** Missouri, C USA 40°12′N 94°33′W
195 O3 **Stancomb-Wills Glacier** *glacier* Antarctica
83 K21 **Standerton** Mpumalanga, E South Africa 26°57′S 29°14′E
31 R7 **Standish** Michigan, N USA 43°59′N 83°58′W
20 M6 **Stanford** Kentucky, S USA 37°30′N 84°40′W
33 S9 **Stanford** Montana, NW USA 47°08′N 110°15′W
95 P19 **Stånga** Gotland, SE Sweden 57°16′N 18°30′E
94 I13 **Stange** Hedmark, S Norway 60°40′N 11°05′E
83 L23 **Stanger** KwaZulu/Natal, E South Africa 29°20′S 31°18′E
Stanger *see* KwaDukuza
Stanimaka *see* Asenovgrad
Stanislau *see* Ivano-Frankivs'k
35 P8 **Stanislaus River** ≈ California, W USA
Stanislav *see* Ivano-Frankivs'k
Stanislavskaya Oblast' *see* Ivano-Frankivs'ka Oblast'
Stanisławów *see* Ivano-Frankivs'k
Stanke Dimitrov *see* Dupnitsa
183 O15 **Stanley** Tasmania, SE Australia 40°48′S 145°18′E
65 E24 **Stanley** *var.* Port Stanley, Puerto Argentino. ○ (Falkland Islands) East Falkland, Falkland Islands 51°45′S 57°56′W
33 O13 **Stanley** Idaho, NW USA 44°12′N 114°58′W
28 L3 **Stanley** North Dakota, N USA 48°19′N 102°23′W
21 U4 **Stanley** Virginia, NE USA 38°34′N 78°30′W
30 J6 **Stanley** Wisconsin, N USA 44°58′N 90°54′W
79 G21 **Stanley Pool** *var.* Pool Malebo. ◎ Congo/Dem. Rep. Congo
155 H20 **Stanley Reservoir** ⊡ S India
Stanleyville *see* Kisangani
42 G3 **Stann Creek** ◈ *district* SE Belize
Stann Creek *see* Dangriga
123 Q12 **Stanovoy Khrebet** ▲▲ SE Russian Federation
108 F8 **Stans** Nidwalden, C Switzerland 46°57′N 08°23′E
97 O21 **Stansted** ✕ (London) Essex, E England, United Kingdom 51°53′N 00°16′E
183 U4 **Stanthorpe** Queensland, E Australia 28°35′S 151°52′E
21 N6 **Stanton** Kentucky, S USA 37°51′N 83°51′W
31 Q8 **Stanton** Michigan, N USA 43°19′N 85°04′W
29 Q14 **Stanton** Nebraska, C USA 41°57′N 97°13′W
28 L5 **Stanton** North Dakota, N USA 47°19′N 101°22′W
25 N7 **Stanton** Texas, SW USA 32°07′N 101°47′W
32 H7 **Stanwood** Washington, NW USA 48°14′N 122°22′W
117 Y7 **Stanychno-Luhans'ke** Luhans'ka Oblast', E Ukraine 48°39′N 39°30′E
108 K7 **Stanzach** Tirol, W Austria 47°24′N 10°36′E
98 M9 **Staphorst** Overijssel, E Netherlands 52°38′N 06°12′E
14 D18 **Staples** Ontario, S Canada 42°09′N 82°34′W
29 T6 **Staples** Minnesota, N USA 46°21′N 94°47′W
28 M14 **Stapleton** Nebraska, C USA 41°29′N 100°40′W
25 S8 **Star** Texas, SW USA 31°27′N 98°16′W
111 M14 **Starachowice** Świętokrzyskie, C Poland 51°04′N 21°02′E
Stara Kanjiža *see* Kanjiža
111 M18 **Stará L'ubovňa** *Ger.* Altlublau, *Hung.* Ólubló. Prešovský Kraj, E Slovakia 49°19′N 20°40′E
112 L10 **Stara Pazova** *Ger.* Altpasua, *Hung.* Ópazova. Vojvodina, N Serbia 44°59′N 20°10′E
Stara Planina *see* Balkan Mountains
114 L9 **Stara Reka** ≈ C Bulgaria
116 M5 **Stara Synyava** Khmel'nyts'ka Oblast', W Ukraine 49°39′N 27°39′E
116 I2 **Stara Vyzhivka** Volyns'ka Oblast', NW Ukraine 51°27′N 24°25′E
Staraya Belitsa *see* Staraya Byelitsa
119 M14 **Staraya Byelitsa** *Rus.* Staraya Belitsa. Vitsyebskaya Voblasts', NE Belarus 54°42′N 29°38′E
127 R5 **Staraya Mayna** Ul'yanovskaya Oblast', W Russian Federation 54°36′N 48°57′E
119 O18 **Staraya Rudnya** Homyel'skaya Voblasts', SE Belarus 52°50′N 30°17′E
124 H14 **Staraya Russa** Novgorodskaya Oblast', W Russian Federation 57°59′N 31°18′E
114 K10 **Stara Zagora** *Lat.* Augusta Trajana. Stara Zagora, C Bulgaria 42°26′N 25°39′E
114 K10 **Stara Zagora** ◈ *province* C Bulgaria
29 S8 **Starbuck** Minnesota, N USA 45°36′N 95°31′W
191 W4 **Starbuck Island** *prev.* Volunteer Island. *island* E Kiribati
27 V13 **Star City** Arkansas, C USA 33°56′N 91°52′W
112 F13 **Staretina** ▲▲ W Bosnia and Herzegovina
Stargard in Pommern *see* Stargard Szczeciński
110 E9 **Stargard Szczeciński** *Ger.* Stargard in Pommern. Zachodnio-pomorskie, NW Poland 53°20′N 15°02′E
187 N10 **Star Harbour** *harbour* San Cristobal, SE Solomon Islands
Stari Bečej *see* Bečej
113 F15 **Stari Grad** *It.* Cittavecchia. Split-Dalmacija, S Croatia 43°11′N 16°36′E
124 J16 **Staritsa** Tverskaya Oblast', W Russian Federation 56°28′N 34°51′E
23 V9 **Starke** Florida, SE USA 29°56′N 82°07′W
22 M4 **Starkville** Mississippi, S USA 33°27′N 88°49′W
186 B7 **Star Mountains** *Ind.* Pegunungan Sterren. ▲▲ Indonesia/Papua New Guinea
101 L23 **Starnberg** Bayern, SE Germany 48°00′N 11°19′E
101 L24 **Starnberger See** ◎ SE Germany
Starobel'sk *see* Starobil's'k
117 X8 **Starobesheve** Donets'ka Oblast', E Ukraine 47°45′N 38°01′E
117 Y6 **Starobil's'k** *Rus.* Starobel'sk. Luhans'ka Oblast', E Ukraine 49°16′N 38°56′E
119 K18 **Starobin** *var.* Starobyn. Minskaya Voblasts', S Belarus 52°44′N 27°28′E
Starobyn *see* Starobin
126 H6 **Starodub** Bryanskaya Oblast', W Russian Federation 52°30′N 32°56′E
110 I8 **Starogard Gdański** *Ger.* Preussisch-Stargard. Pomorskie, N Poland 53°57′N 18°29′E
Staroikan *see* Ikan
Starokonstantinov *see* Starokostyantyniv
116 L5 **Starokostyantyniv** *Rus.* Starokonstantinov. Khmel'nyts'ka Oblast', NW Ukraine 49°43′N 27°13′E
126 K12 **Starominskaya** Krasnodarskiy Kray, SW Russian Federation 46°31′N 39°03′E
114 L7 **Staro Selo** *Rom.* Satul-Vechi; *prev.* Star-Smil. Silistra, NE Bulgaria 43°58′N 26°32′E
126 K12 **Staroshcherbinovskaya** Krasnodarskiy Kray, SW Russian Federation 46°36′N 38°42′E
127 V6 **Starosubkhangulovo** Respublika Bashkortostan, W Russian Federation 53°05′N 57°22′E
35 S4 **Star Peak** ▲ Nevada, W USA 40°31′N 118°09′W
15 T8 **St-Arsène** Québec, SE Canada 47°55′N 69°21′W
Star-Smil *see* Staro Selo
97 J25 **Start Point** *headland* SW England, United Kingdom 50°13′N 03°38′W
Startsy *see* Kirawsk
Starum *see* Stavoren
119 L18 **Staryya Darohi** *Rus.* Staryye Dorogi. Minskaya Voblasts', S Belarus 53°02′N 28°16′E
Staryye Dorogi *see* Staryya Darohi
127 T2 **Staryye Zyattsy** Udmurtskaya Respublika, NW Russian Federation 57°22′N 52°42′E
117 U13 **Staryy Krym** Avtonomna Respublika Krym, S Ukraine 45°03′N 35°06′E
126 K8 **Staryy Oskol** Belgorodskaya Oblast', W Russian Federation 51°21′N 37°52′E
116 H6 **Staryy Sambir** L'vivs'ka Oblast', W Ukraine 49°27′N 23°00′E
101 L14 **Stassfurt** *var.* Staßfurt. Sachsen-Anhalt, C Germany 51°51′N 11°35′E
Staßfurt *see* Stassfurt
111 M15 **Staszów** Świętokrzyskie, C Poland 50°33′N 21°07′E
29 W13 **State Center** Iowa, C USA 42°01′N 93°09′W
18 E14 **State College** Pennsylvania, NE USA 40°48′N 77°52′W
18 K15 **Staten Island** *island* New York, NE USA
Staten Island *see* Estados, Isla de los
23 U8 **Statenville** Georgia, SE USA 30°42′N 83°00′W
23 W5 **Statesboro** Georgia, SE USA 32°28′N 81°47′W
States, The *see* United States of America
21 R9 **Statesville** North Carolina, SE USA 35°46′N 80°54′W
95 G16 **Stathelle** Telemark, S Norway 59°01′N 09°40′E
Statia *see* Sint Eustatius
30 K15 **Staunton** Illinois, N USA 39°00′N 89°47′W
21 T5 **Staunton** Virginia, NE USA 38°10′N 79°05′W
95 C16 **Stavanger** Rogaland, S Norway 58°58′N 05°43′E
99 L21 **Stavelot** *Dut.* Stablo. Liège, E Belgium 50°24′N 05°56′E
95 G16 **Stavern** Vestfold, S Norway 59°00′N 10°01′E
Stavers Island *see* Vostok Island
98 J7 **Stavoren** *Fris.* Starum. Fryslân, N Netherlands 52°52′N 05°22′E
115 K21 **Stavrí, Akrotírio** *var.* Akrotírio Stavrós. *headland* Naxos, Kykládes, Greece, Aegean Sea 37°12′N 25°32′E
126 M14 **Stavropol'** *prev.* Voroshilovsk. Stavropol'skiy Kray, SW Russian Federation 45°02′N 41°58′E
Stavropol' *see* Tol'yatti
126 M14 **Stavropol'skaya Vozvyshennost'** ▲ SW Russian Federation
126 M14 **Stavropol'skiy Kray** ◈ *territory* SW Russian Federation
115 H14 **Stavrós** Kentrikí Makedonía, N Greece 40°39′N 23°43′E
115 J24 **Stavrós, Akrotírio** *headland* Kríti, Greece, E Mediterranean Sea 35°25′N 24°57′E
Stavrós, Akrotírio *see* Stavrí, Akrotírio
114 I12 **Stavroúpoli** *prev.* Stavroúpolis. Anatolikí Makedonía kai Thráki, NE Greece 41°12′N 24°45′E
Stavroúpolis *see* Stavroúpoli
117 O6 **Stavyshche** Kyyivs'ka Oblast', N Ukraine 49°23′N 30°10′E
182 M11 **Stawell** Victoria, SE Australia 37°06′S 142°52′E
110 N9 **Stawiski** Podlaskie, NE Poland 53°22′N 22°08′E
14 G14 **Stayner** Ontario, S Canada 44°25′N 80°05′W
14 D17 **St. Clair** ≈ Canada/USA
37 R3 **Steamboat Springs** Colorado, C USA 40°28′N 106°51′W
15 U4 **Ste-Anne, Lac** ◎ Québec, SE Canada
20 M8 **Stearns** Kentucky, S USA 36°39′N 84°27′W
39 N10 **Stebbins** Alaska, USA 63°30′N 162°15′W
15 U7 **Ste-Blandine** Québec, SE Canada 48°22′N 68°27′W
108 K7 **Steeg** Tirol, W Austria 47°15′N 10°18′E
27 Y9 **Steele** Missouri, C USA 36°04′N 89°49′W
29 N5 **Steele** North Dakota, N USA 46°51′N 99°55′W
194 J5 **Steele Island** *island* Antarctica
30 K16 **Steeleville** Illinois, N USA 38°00′N 89°39′W
27 W6 **Steelville** Missouri, C USA 37°57′N 91°21′W
99 G14 **Steenbergen** Noord-Brabant, S Netherlands 51°35′N 04°19′E
Steenkool *see* Bintuni
11 O10 **Steen River** Alberta, W Canada 59°37′N 117°17′W
98 M8 **Steenwijk** Overijssel, N Netherlands 52°47′N 06°07′E
65 A23 **Steeple Jason** *island* Jason Islands, NW Falkland Islands
174 J8 **Steep Point** *headland* Western Australia 26°09′S 113°11′E
116 L9 **Ştefăneşti** Botoşani, NE Romania 47°44′N 27°15′E
Stefanie, Lake *see* Ch'ew Bahir
8 L5 **Stefansson Island** *island* Nunavut, N Canada
117 O10 **Ştefan Vodă** *Rus.* Suvorovo. SE Moldova 46°33′N 29°39′E
63 H18 **Steffen, Cerro** ▲ S Chile 44°27′S 71°42′W
108 D9 **Steffisburg** Bern, C Switzerland 46°47′N 07°38′E
95 J24 **Stege** Sjælland, SE Denmark 54°59′N 12°18′E
116 G10 **Ştei** *Hung.* Vaskohsziklás. Bihor, W Romania 46°34′N 22°28′E
Steier *see* Steyr
Steierdorf/Steierdorf-Anina *see* Anina
109 T7 **Steiermark** *off.* Land Steiermark, *Eng.* Styria. ◈ *state* C Austria
Steiermark, Land *see* Steiermark
101 J19 **Steigerwald** *hill range* C Germany
99 L17 **Stein** Limburg, SE Netherlands 50°58′N 05°45′E
Stein *see* Stein an der Donau
Stein *see* Kamnik, Slovenia
108 M8 **Steinach** Tirol, W Austria 47°07′N 11°30′E
Steinamanger *see* Szombathely
109 W3 **Stein an der Donau** *var.* Stein. Niederösterreich, NE Austria 48°25′N 15°35′E
Steinau an der Elbe *see* Ścinawa
11 Y16 **Steinbach** Manitoba, S Canada 49°32′N 96°40′W
Steiner Alpen *see* Kamniško-Savinjske Alpe
99 L24 **Steinfort** Luxembourg, W Luxembourg 49°39′N 05°55′E
100 H12 **Steinhuder Meer** ◎ NW Germany
93 E15 **Steinkjer** Nord-Trøndelag, C Norway 64°01′N 11°29′E
Stejarul *see* Karapelit
99 F16 **Stekene** Oost-Vlaanderen, NW Belgium 51°13′N 04°04′E
83 E26 **Stellenbosch** Western Cape, SW South Africa 33°56′S 18°51′E
98 F13 **Stellendam** Zuid-Holland, SW Netherlands 51°48′N 04°01′E
39 T12 **Steller, Mount** ▲ Alaska, USA 60°36′N 142°49′W
103 Y14 **Stello, Monte** ▲ Corse, France, C Mediterranean Sea 42°49′N 09°24′E
106 F5 **Stelvio, Passo dello** *pass* Italy/Switzerland
15 S7 **Ste-Marguerite Nord-Est** ≈ Québec, SE Canada
15 V4 **Ste-Marguerite, Pointe** *headland* Québec, SE Canada 50°01′N 66°43′W
12 J14 **Ste-Marie, Lac** ⊡ Québec, S Canada
103 R3 **Stenay** Meuse, NE France 49°29′N 05°12′E
100 L12 **Stendal** Sachsen-Anhalt, C Germany 52°36′N 11°52′E
118 E8 **Stende** NW Latvia 57°09′N 22°33′E
182 H10 **Stenhouse Bay** South Australia 35°15′S 136°58′E
95 J23 **Stenløse** Hovedstaden, E Denmark 55°47′N 12°13′E
95 L19 **Stensjön** Jönköping, S Sweden 57°36′N 14°42′E
95 K18 **Stenstorp** Västra Götaland, S Sweden 58°15′N 13°45′E
95 I18 **Stenungsund** Västra Götaland, S Sweden 58°05′N 11°49′E
Stepanakert *see* Xankändi
137 T11 **Step'anavan** N Armenia 41°00′N 44°27′E
100 K9 **Stepenitz** ≈ N Germany
29 O10 **Stephan** South Dakota, N USA 44°12′N 99°25′W
29 R3 **Stephen** Minnesota, N USA 48°27′N 96°54′W
27 T14 **Stephens** Arkansas, C USA 33°25′N 93°04′W
184 J13 **Stephens, Cape** *headland* D'Urville Island, Marlborough, SW New Zealand 40°42′S 173°56′E
21 V3 **Stephens City** Virginia, NE USA 39°03′N 78°10′W
182 L6 **Stephens Creek** New South Wales, SE Australia 31°51′S 141°30′E
184 K13 **Stephens Island** *island* C New Zealand
31 N5 **Stephenson** Michigan, N USA 45°27′N 87°36′W
13 S12 **Stephenville** Newfoundland, Newfoundland and Labrador, SE Canada 48°33′N 58°34′W
25 S7 **Stephenville** Texas, SW USA 32°12′N 98°13′W
Step' Nardara *see* Shardara, Step'
145 R8 **Stepnogorsk** Akmola, C Kazakhstan 52°04′N 72°18′E
127 O15 **Stepnoye** Stavropol'skiy Kray, SW Russian Federation 44°18′N 44°34′E
145 Q8 **Stepnyak** Akmola, N Kazakhstan 52°52′N 70°49′E
145 P17 **Step' Shardara** *Kaz.* Shardara Dalasy; *prev.* Step' Nardara. *grassland* S Kazakhstan
192 J17 **Steps Point** *headland* Tutuila, W American Samoa 14°23′S 170°46′W
115 F17 **Stereá Elláda** *Eng.* Greece Central *var.* Stereá Ellás. ◈ *region* C Greece
Stereá Ellás *see* Stereá Elláda
83 J24 **Sterkspruit** Eastern Cape, SE South Africa 30°31′S 27°22′E
127 U6 **Sterlibashevo** Respublika Bashkortostan, W Russian Federation 53°19′N 55°12′E
39 R12 **Sterling** Alaska, USA 60°32′N 150°51′W
37 V3 **Sterling** Colorado, C USA 40°37′N 103°12′W
30 K11 **Sterling** Illinois, N USA 41°47′N 89°42′W
26 M5 **Sterling** Kansas, C USA 38°12′N 98°12′W
25 O8 **Sterling City** Texas, SW USA 31°50′N 101°00′W
31 S9 **Sterling Heights** Michigan, N USA 42°34′N 83°01′W
21 W3 **Sterling Park** Virginia, NE USA 39°00′N 77°24′W
37 V2 **Sterling Reservoir** ⊡ Colorado, C USA
22 I5 **Sterlington** Louisiana, S USA 32°42′N 92°05′W
127 U6 **Sterlitamak** Respublika Bashkortostan, W Russian Federation 53°39′N 56°00′E
Sternberg *see* Šternberk
111 H17 **Šternberk** *Ger.* Sternberg. Olomoucký Kraj, E Czech Republic 49°45′N 17°20′E
141 V17 **Sterōh** Suquṭrā, S Yemen 12°21′N 53°50′E
Sterren, Pegunungan *see* Star Mountains
110 G11 **Stęszew** Wielkopolskie, C Poland 52°16′N 16°41′E
Stettin *see* Szczecin
Stettiner Haff *see* Szczeciński, Zalew
11 Q15 **Stettler** Alberta, SW Canada 52°21′N 112°40′W
31 V13 **Steubenville** Ohio, N USA 40°21′N 80°37′W
97 O21 **Stevenage** E England, United Kingdom 51°55′N 00°14′W
23 Q1 **Stevenson** Alabama, S USA 34°52′N 85°50′W
32 H11 **Stevenson** Washington, NW USA 45°43′N 121°54′W
182 E1 **Stevenson Creek** *seasonal river* South Australia
39 Q13 **Stevenson Entrance** *strait* Alaska, USA
30 L6 **Stevens Point** Wisconsin, N USA 44°32′N 89°33′W
39 R8 **Stevens Village** Alaska, USA 66°01′N 149°02′W
33 P10 **Stevensville** Montana, NW USA 46°30′N 114°05′W
93 E25 **Stevns Klint** *headland* E Denmark 55°15′N 12°25′E
10 J12 **Stewart** British Columbia, W Canada 55°58′N 129°52′W
10 I6 **Stewart** ≈ Yukon, NW Canada
10 I6 **Stewart Crossing** Yukon, NW Canada 63°22′N 136°37′W
63 H25 **Stewart, Isla** *island* S Chile
185 B25 **Stewart Island** *island* S New Zealand
Stewart Islands *see* Sikaiana
181 W6 **Stewart, Mount** ▲ Queensland, E Australia 20°11′S 145°29′E
10 H6 **Stewart River** Yukon, NW Canada 63°17′N 139°24′W
27 R3 **Stewartsville** Missouri, C USA 39°45′N 94°30′W
11 S16 **Stewart Valley** Saskatchewan, S Canada 50°34′N 107°47′W
29 W10 **Stewartville** Minnesota, N USA 43°51′N 92°29′W
Steyerlak-Anina *see* Anina
109 T5 **Steyr** *var.* Steier. Oberösterreich, N Austria 48°02′N 14°26′E
109 T5 **Steyr** ≈ NW Austria
15 T7 **St-Fabien** Québec, SE Canada 48°19′N 68°51′W
15 R11 **St-François, Lac** ◎ Québec, SE Canada
83 E25 **St. Helena Bay** *bay* SW South Africa
15 T8 **St-Hubert** Québec, SE Canada 47°46′N 69°15′W
29 P11 **Stickney** South Dakota, N USA 43°24′N 98°23′W
98 L5 **Stiens** Fryslân, N Netherlands 53°15′N 05°45′E
Stif *see* Sétif
27 Q11 **Stigler** Oklahoma, C USA 35°16′N 95°08′W
107 N18 **Stigliano** Basilicata, S Italy 40°24′N 16°13′E
95 N17 **Stigtomta** Södermanland, C Sweden 58°48′N 16°47′E
10 I11 **Stikine** ≈ British Columbia, W Canada
Stilida/Stilís *see* Stylída
95 G22 **Stilling** Midtjylland, C Denmark 56°04′N 10°00′E
29 W8 **Stillwater** Minnesota, N USA 45°03′N 92°48′W
27 O9 **Stillwater** Oklahoma, C USA 36°07′N 97°03′W
35 S5 **Stillwater Range** ▲▲ Nevada, W USA
18 I8 **Stillwater Reservoir** ⊡ New York, NE USA
107 O22 **Stilo, Punta** *headland* S Italy 38°27′N 16°36′E
27 R10 **Stilwell** Oklahoma, C USA 35°48′N 94°37′W
Štimlje *see* Shtime
25 N1 **Stinnett** Texas, SW USA 35°49′N 101°27′W
113 P18 **Štip** E FYR Macedonia 41°45′N 22°10′E
Stira *see* Stýra
96 J12 **Stirling** C Scotland, United Kingdom 56°07′N 03°57′W
96 I12 **Stirling** *cultural region* C Scotland, United Kingdom
180 J14 **Stirling Range** ▲▲ Western Australia
15 R8 **St-Jean** ≈ Québec, SE Canada
93 E16 **Stjørdalshalsen** Nord-Trøndelag, C Norway 63°27′N 10°57′E
83 L22 **St. Lucia** KwaZulu/Natal, E South Africa 28°22′S 32°25′E
Stochid *see* Stokhid
101 H24 **Stockach** Baden-Württemberg, S Germany 47°51′N 09°01′E
25 S12 **Stockdale** Texas, SW USA 29°14′N 97°57′W
109 X3 **Stockerau** Niederösterreich, NE Austria 48°24′N 16°13′E
93 H20 **Stockholm** ● (Sweden) Stockholm, C Sweden 59°17′N 18°03′E
95 O15 **Stockholm** ◈ *county* C Sweden
Stockmannshof *see* Pļaviņas
97 L18 **Stockport** NW England, United Kingdom 53°25′N 02°10′W
65 K15 **Stocks Seamount** *undersea feature* C Atlantic Ocean 11°42′S 33°48′W
35 O8 **Stockton** California, W USA 37°56′N 121°19′W
26 L3 **Stockton** Kansas, C USA 39°27′N 99°17′W
27 S6 **Stockton** Missouri, C USA 37°43′N 93°49′W
30 K3 **Stockton Island** *island* Apostle Islands, Wisconsin, N USA
27 S7 **Stockton Lake** ⊡ Missouri, C USA
97 M15 **Stockton-on-Tees** *var.* Stockton on Tees. N England, United Kingdom 54°34′N 01°19′W
Stockton on Tees *see* Stockton-on-Tees
24 M10 **Stockton Plateau** *plain* Texas, SW USA
28 M16 **Stockville** Nebraska, C USA 40°33′N 100°20′W
93 H17 **Stöde** Västernorrland, C Sweden 62°27′N 16°34′E
Stodolichi *see* Stadolichy
113 M19 **Stogovo Karaorman** ▲ W FYR Macedonia
Stoke *see* Stoke-on-Trent
97 L19 **Stoke-on-Trent** *var.* Stoke. C England, United Kingdom 53°N 02°10′W
182 M15 **Stokes Point** *headland* Tasmania, SE Australia 40°09′S 143°55′E
116 J2 **Stokhid** *Pol.* Stochód, *Rus.* Stokhod. ≈ NW Ukraine
Stokhod *see* Stokhid
92 I4 **Stokkseyri** Suðurland, SW Iceland 63°49′N 21°00′W
92 G10 **Stokmarknes** Nordland, C Norway 68°34′N 14°55′E
Stol *see* Veliki Krš
113 H15 **Stolac** Federacija Bosne I Hercegovine, S Bosnia and Herzegovina 43°04′N 17°58′E
Stolbce *see* Stowbtsy
101 D16 **Stolberg** *var.* Stolberg im Rheinland. Nordrhein-Westfalen, W Germany 50°45′N 06°15′E
Stolberg im Rheinland *see* Stolberg
123 P6 **Stolbovoy, Ostrov** *island* NE Russian Federation
Stolbtsy *see* Stowbtsy
119 J20 **Stolin** Brestskaya Voblasts', SW Belarus 51°53′N 26°51′E
95 K14 **Stöllet** *var.* Norra Ny. Värmland, C Sweden 60°24′N 13°15′E
Stolp *see* Słupsk
Stolpe *see* Słupia
Stolpmünde *see* Ustka
115 F15 **Stómio** Thessalía, C Greece 39°51′N 22°45′E
14 J11 **Stonecliffe** Ontario, SE Canada 46°12′N 77°58′W
96 L10 **Stonehaven** NE Scotland, United Kingdom 56°59′N 02°14′W
97 M23 **Stonehenge** *ancient monument* Wiltshire, S England, United Kingdom
23 T3 **Stone Mountain** ▲ Georgia, SE USA 33°48′N 84°10′W
11 X16 **Stonewall** Manitoba, S Canada 50°08′N 97°20′W
21 S3 **Stonewood** West Virginia, NE USA 39°15′N 80°18′W
14 D17 **Stoney Point** Ontario, S Canada 42°18′N 82°32′W
92 H10 **Stonglandseidet** Troms, N Norway 69°03′N 17°06′E
65 N25 **Stonybeach Bay** *bay* Tristan da Cunha, SE Atlantic Ocean
35 N5 **Stony Creek** ≈ California, W USA
65 N25 **Stonyhill Point** *headland* S Tristan da Cunha
14 I14 **Stony Lake** ◎ Ontario, SE Canada
11 Q14 **Stony Plain** Alberta, SW Canada 53°31′N 114°04′W
21 R9 **Stony Point** North Carolina, SE USA 35°51′N 81°04′W
18 G8 **Stony Point** *headland* New York, NE USA 43°50′N 76°18′W
11 T10 **Stony Rapids** Saskatchewan, C Canada 59°14′N 105°48′W
39 P11 **Stony River** Alaska, USA 61°48′N 156°37′W
Stony Tunguska *see* Podkamennaya Tunguska
12 G10 **Stooping** ≈ Ontario, C Canada
100 I9 **Stör** ≈ N Germany
95 M15 **Storå** Örebro, S Sweden 59°44′N 15°10′E
95 J16 **Stora Gla** ◎ C Sweden
95 I16 **Stora Le** *Nor.* Store Le. ◎ Norway/Sweden
92 I12 **Stora Lulevatten** ◎ N Sweden
92 H13 **Storavan** ◎ N Sweden
93 I20 **Storby** Åland, SW Finland 60°12′N 19°33′E
94 E10 **Stordalen** Møre og Romsdal, S Norway 62°22′N 07°00′E
95 H23 **Storebælt** *var.* Store Bælt, *Eng.* Great Belt, Storebelt. *channel* Baltic Sea/Kattegat
Store Bælt *see* Storebælt
Storebelt *see* Storebælt
95 M19 **Storebro** Kalmar, S Sweden 57°36′N 15°30′E
95 J24 **Store Heddinge** Sjælland, SE Denmark 55°19′N 12°24′E
Store Le *see* Stora Le
93 E16 **Støren** Sør-Trøndelag, S Norway 63°02′N 10°16′E
92 O4 **Storfjorden** *fjord* S Norway
95 L15 **Storfors** Värmland, C Sweden 59°33′N 14°16′E
92 G13 **Storforshei** Nordland, C Norway 66°25′N 14°25′E
Storhammer *see* Hamar
93 F16 **Storlien** Jämtland, C Sweden 63°18′N 12°10′E
183 P17 **Storm Bay** *inlet* Tasmania, SE Australia
29 T12 **Storm Lake** Iowa, C USA 42°38′N 95°12′W
29 S13 **Storm Lake** ◎ Iowa, C USA
96 G7 **Stornoway** NW Scotland, United Kingdom 58°13′N 06°23′W
Storojineţ *see* Storozhynets'
92 P1 **Storøya** *island* NE Svalbard
125 S10 **Storozhevsk** Respublika Komi, NW Russian Federation 61°56′N 52°18′E
Storozhinets *see* Storozhynets'
116 K8 **Storozhynets'** *Ger.* Storozynetz, *Rom.* Storojineţ, *Rus.* Storozhinets. Chernivets'ka Oblast', W Ukraine 48°11′N 25°42′E
Storozynetz *see* Storozhynets'
92 H11 **Storriten** *Lapp.* Stuorrarijjda. ▲ C Norway 68°09′N 17°12′E
19 N12 **Storrs** Connecticut, NE USA 41°48′N 72°15′W
94 I11 **Storsjøen** ◎ S Norway
94 N13 **Storsjön** ◎ C Sweden
93 F16 **Storsjön** ◎ C Sweden
92 J9 **Storslett** Troms, N Norway 69°45′N 21°03′E
94 H11 **Storsølnkletten** ▲ S Norway 61°52′N 11°32′E
92 I9 **Storsteinnes** Troms, N Norway 69°13′N 19°14′E
93 J14 **Storsund** Norrbotten, N Sweden 65°36′N 20°40′E
94 I9 **Storsylen** *Swe.* Sylarna. ▲ Norway/Sweden 63°00′N 12°14′E
92 H11 **Stortoppen** ▲ N Sweden 67°33′N 17°27′E
93 H14 **Storuman** Västerbotten, N Sweden 65°05′N 17°10′E
93 H14 **Storuman** ◎ N Sweden
94 N13 **Storvik** Gävleborg, C Sweden 60°37′N 16°30′E
95 O14 **Storvreta** Uppsala, C Sweden 59°58′N 17°42′E
100 L10 **Störwasserstrasse** *canal* N Germany
29 V13 **Story City** Iowa, C USA 42°10′N 93°36′W
11 V17 **Stoughton** Saskatchewan, S Canada 49°40′N 103°01′W
19 O11 **Stoughton** Massachusetts, NE USA 42°07′N 71°06′W
30 L9 **Stoughton** Wisconsin, N USA 42°56′N 89°12′W
97 L23 **Stour** ≈ E England, United Kingdom
97 P21 **Stour** ≈ S England, United Kingdom
27 T5 **Stover** Missouri, C USA 38°26′N 92°59′W
95 G21 **Støvring** Nordjylland, N Denmark 56°53′N 09°52′E
119 J17 **Stowbtsy** *Pol.* Stołbce, *Rus.* Stolbtsy. Minskaya Voblasts', C Belarus 53°29′N 26°44′E
25 X11 **Stowell** Texas, SW USA 29°47′N 94°22′W
97 P20 **Stowmarket** E England, United Kingdom 52°05′N 00°54′E
114 N8 **Stozher** Dobrich, NE Bulgaria 43°27′N 27°49′E
97 E14 **Strabane** *Ir.* An Srath Bán. W Northern Ireland, United Kingdom 54°49′N 07°27′W
121 S11 **Strabo Trench** *undersea feature* C Mediterranean Sea
27 T7 **Strafford** Missouri, C USA 37°16′N 93°07′W
183 N17 **Strahan** Tasmania, SE Australia 42°10′S 145°18′E
111 C18 **Strakonice** *Ger.* Strakonitz. Jihočeský Kraj, S Czech Republic 49°14′N 13°55′E
Strakonitz *see* Strakonice
100 N8 **Stralsund** Mecklenburg-Vorpommern, NE Germany 54°18′N 13°06′E
99 L16 **Stramproy** Limburg, SE Netherlands 51°12′N 05°43′E
83 E26 **Strand** Western Cape, SW South Africa 34°06′S 18°50′E
94 E10 **Stranda** Møre og Romsdal, S Norway 62°18′N 06°56′E
97 G15 **Strangford Lough** *Ir.* Loch Cuan. *inlet* E Northern Ireland, United Kingdom
95 N16 **Strängnäs** Södermanland, C Sweden 59°22′N 17°02′E
97 E14 **Stranorlar** *Ir.* Srath an Urláir. NW Ireland 54°48′N 07°46′W
97 H14 **Stranraer** S Scotland, United Kingdom 54°54′N 05°02′W
11 U16 **Strasbourg** Saskatchewan, S Canada 51°05′N 104°58′W
103 V5 **Strasbourg** *Ger.* Strassburg; *anc.* Argentoratum. Bas-Rhin, NE France 48°35′N 07°45′E
109 T8 **Strassburg** Kärnten, S Austria 46°54′N 14°21′E
37 U4 **Strasburg** Colorado, C USA 39°42′N 104°13′W
29 N7 **Strasburg** North Dakota, N USA 46°07′N 100°10′W
31 U12 **Strasburg** Ohio, N USA 40°35′N 81°31′W
21 U3 **Strasburg** Virginia, NE USA 38°59′N 78°21′W
117 N10 **Strășeni** *var.* Strasheny. C Moldova 47°07′N 28°37′E
Strasheny *see* Strășeni
Strassburg *see* Strasbourg, France
Strassburg *see* Aiud, Romania
99 M25 **Strassen** Luxembourg, S Luxembourg 49°37′N 06°05′E
109 R5 **Strasswalchen** Salzburg, C Austria 47°59′N 13°19′E
14 F16 **Stratford** Ontario, S Canada 43°22′N 81°00′W
184 K10 **Stratford** Taranaki, North Island, New Zealand 39°20′S 174°16′E
35 Q11 **Stratford** California, W USA 36°10′N 119°47′W
29 V13 **Stratford** Iowa, C USA 42°16′N 93°55′W
27 O12 **Stratford** Oklahoma, C USA 34°48′N 96°57′W
25 N1 **Stratford** Texas, SW USA 36°21′N 102°05′W
30 K6 **Stratford** Wisconsin, N USA 44°53′N 90°13′W
Stratford *see* Stratford-upon-Avon
97 M20 **Stratford-upon-Avon** *var.* Stratford. C England, United Kingdom 52°12′N 01°41′W
183 O17 **Strathgordon** Tasmania, SE Australia 42°49′S 146°04′E
11 Q16 **Strathmore** Alberta, SW Canada 51°05′N 113°20′W
35 R11 **Strathmore** California, W USA 36°07′N 119°04′W
14 E16 **Strathroy** Ontario, S Canada 42°57′N 81°40′W
96 I6 **Strathy Point** *headland* N Scotland, United Kingdom 58°36′N 04°04′W
37 W4 **Stratton** Colorado, C USA 39°16′N 102°34′W
19 P6 **Stratton** Maine, NE USA 45°08′N 70°25′W
18 M10 **Stratton Mountain** ▲ Vermont, NE USA 43°05′N 72°55′W
101 N21 **Straubing** Bayern, SE Germany 48°53′N 12°35′E
100 O12 **Strausberg** Brandenburg, E Germany 52°34′N 13°52′E
32 K13 **Strawberry Mountain** ▲ Oregon, NW USA 44°18′N 118°43′W
29 X12 **Strawberry Point** Iowa, C USA 42°40′N 91°31′W
36 M3 **Strawberry Reservoir** ⊡ Utah, W USA
36 M4 **Strawberry River** ≈ Utah, W USA
25 R7 **Strawn** Texas, SW USA 32°33′N 98°30′W
113 P17 **Straža** ▲ Bulgaria/FYR Macedonia 42°16′N 22°13′E
111 I19 **Strážov** *Hung.* Sztrazsó. ▲ NW Slovakia 48°59′N 18°29′E
182 F7 **Streaky Bay** South Australia 32°49′S 134°13′E
182 E7 **Streaky Bay** *bay* South Australia
30 L12 **Streator** Illinois, N USA 41°07′N 88°50′W
Streckenbach *see* Świdnik
Strednogorie *see* Pirdop
111 C17 **Středočeský Kraj** ◈ *region* C Czech Republic
29 O6 **Streeter** North Dakota, N USA 46°37′N 99°23′W
25 U8 **Streetman** Texas, SW USA 31°52′N 96°19′W
116 G13 **Strehaia** Mehedinți, SW Romania 44°37′N 23°10′E
Strehlen *see* Strzelin
114 I10 **Strelcha** Pazardzhik, C Bulgaria 42°28′N 24°21′E
122 L12 **Strelka** Krasnoyarskiy Kray, C Russian Federation 58°05′N 92°54′E
124 L6 **Strel'na** ≈ NW Russian Federation
118 H7 **Strenči** *Ger.* Stackeln. N Latvia 57°38′N 25°42′E
15 V6 **St-René-de-Matane** Québec, SE Canada 48°42′N 67°22′W
108 K8 **Strengen** Tirol, W Austria 47°07′N 10°25′E
106 C6 **Stresa** Piemonte, NE Italy 45°52′N 08°32′E
Streshin *see* Streshyn
119 N18 **Streshyn** *Rus.* Streshin. Homyel'skaya Voblasts', SE Belarus 52°43′N 30°07′E
95 B18 **Streymoy** *Dan.* Strømø. *island* N Faroe Islands
95 G23 **Strib** Syddtjylland, C Denmark 55°33′N 09°47′E
111 A17 **Stříbro** *Ger.* Mies. Plzeňský Kraj, W Czech Republic 49°44′N 12°55′E
186 B7 **Strickland** ≈ SW Papua New Guinea
Striegau *see* Strzegom
Strigonium *see* Esztergom
98 H13 **Strijen** Zuid-Holland, SW Netherlands 51°45′N 04°34′E
63 H21 **Strobel, Lago** ◎ S Argentina
61 B25 **Stroeder** Buenos Aires, E Argentina 40°11′S 62°35′W
115 C20 **Strofádes** *island* Iónia Nisiá, Greece, C Mediterranean Sea
Strofilia *see* Strofyliá
115 G17 **Strofyliá** *var.* Strofilia. Évvoia, C Greece 38°49′N 23°25′E
100 O10 **Strom** ≈ NE Germany
107 L22 **Stromboli** ⋀ Isola Stromboli, SW Italy 38°48′N 15°13′E
107 L22 **Stromboli, Isola** *island* Isole Eolie, S Italy
96 H9 **Stromeferry** N Scotland, United Kingdom 57°20′N 05°35′W
96 J5 **Stromness** N Scotland, United Kingdom 58°57′N 03°18′W
Strømø *see* Streymoy
94 N11 **Strömsbruk** Gävleborg, C Sweden 61°52′N 17°19′E
29 Q15 **Stromsburg** Nebraska, C USA 41°06′N 97°36′W
95 K21 **Strömsnäsbruk** Kronoberg, S Sweden 56°35′N 13°45′E
95 I17 **Strömstad** Västra Götaland, S Sweden 58°56′N 11°11′E
93 G16 **Strömsund** Jämtland, C Sweden 63°51′N 15°35′E
93 G15 **Ströms Vattudal** *valley* N Sweden
27 V14 **Strong** Arkansas, C USA 33°06′N 92°19′W
Strongilí *see* Strongylí

◆ Country
● Country Capital
◇ Dependent Territory
○ Dependent Territory Capital
◈ Administrative Regions
✕ International Airport
▲ Mountain
▲▲ Mountain Range
⋀ Volcano
≈ River
◎ Lake
⊡ Reservoir

107 *O21* **Strongoli** Calabria, SW Italy *39°17′N 17°03′E*
31 *T11* **Strongsville** Ohio, N USA *41°18′N 81°50′W*
115 *Q23* **Strongylí** ***var.*** Strongilí. *island* SE Greece
96 *K5* **Stronsay** *island* NE Scotland, United Kingdom
97 *L21* **Stroud** C England, United Kingdom *51°46′N 02°15′W*
27 *O10* **Stroud** Oklahoma, C USA *35°45′N 96°39′W*
18 *I14* **Stroudsburg** Pennsylvania, NE USA *40°59′N 75°12′W*
95 *F21* **Struer** Midtjylland, W Denmark *56°29′N 08°37′E*
113 *M20* **Struga** SW FYR Macedonia *41°11′N 20°40′E*
Strugi-Kranyse ***see*** Strugi-Krasnyye
124 *G14* **Strugi-Krasnyye** ***var.*** Strugi-Kranyse. Pskovskaya Oblast', W Russian Federation *58°19′N 29°09′E*
114 *G11* **Struma** ***Gk.*** Strymónas. ≈ Bulgaria/Greece ***see also*** Strymónas
Struma ***see*** Strymónas
97 *G21* **Strumble Head** *headland* SW Wales, United Kingdom *52°01′N 05°05′W*
Strumeshnitsa ***see*** Strumica
113 *Q19* **Strumica** E FYR Macedonia *41°27′N 22°39′E*
113 *Q19* **Strumica** ***Bulg.*** Strumeshnitsa. ≈ Bulgaria/FYR Macedonia
114 *G11* **Strumyani** Blagoevgrad, SW Bulgaria *41°41′N 23°13′E*
31 *V12* **Struthers** Ohio, N USA *41°03′N 80°36′W*
114 *I10* **Stryama** ≈ C Bulgaria
114 *G13* **Strymónas** ***Bul.*** Struma. ≈ Bulgaria/Greece ***see also*** Struma
Strymónas ***see*** Struma
115 *H14* **Strymonikós Kólpos** *gulf* N Greece
116 *I6* **Stryy** L'vivs'ka Oblast', NW Ukraine *49°16′N 23°51′E*
116 *H6* **Stryy** ≈ W Ukraine
111 *F14* **Strzegom** ***Ger.*** Striegau. Wałbrzych, SW Poland *50°59′N 16°20′E*
110 *E10* **Strzelce Krajeńskie** ***Ger.*** Friedeberg Neumark. Lubuskie, W Poland *52°52′N 15°30′E*
111 *I15* **Strzelce Opolskie** ***Ger.*** Gross Strehlitz. Opolskie, SW Poland *50°31′N 18°19′E*
182 *K3* **Strzelecki Creek** *seasonal river* South Australia
182 *J3* **Strzelecki Desert** *desert* South Australia
111 *G15* **Strzelin** ***Ger.*** Strehlen. Dolnośląskie, SW Poland *50°48′N 17°03′E*
110 *I11* **Strzelno** Kujawsko-pomorski, C Poland *52°38′N 18°11′E*
111 *N17* **Strzyżów** Podkarpackie, SE Poland *49°52′N 21°46′E*
15 *S8* **St-Siméon** Québec, SE Canada *47°50′N 69°55′W*
Stua Laighean ***see*** Leinster, Mount
23 *Y13* **Stuart** Florida, SE USA *27°12′N 80°15′W*
29 *U14* **Stuart** Iowa, C USA *41°30′N 94°19′W*
29 *O13* **Stuart** Nebraska, C USA *42°36′N 99°08′W*
21 *S8* **Stuart** Virginia, NE USA *36°38′N 80°19′W*
10 *L13* **Stuart** ≈ British Columbia, SW Canada
39 *N10* **Stuart Island** *island* Alaska, USA
10 *L13* **Stuart Lake** ◎ British Columbia, SW Canada
185 *B22* **Stuart Mountains** ▲▲ South Island, New Zealand
182 *F3* **Stuart Range** *hill range* South Australia
Stubaital ***see*** Neustift im Stubaital
95 *I24* **Stubbekøbing** Sjælland, SE Denmark *54°53′N 12°04′E*
45 *P14* **Stubbs** Saint Vincent, Saint Vincent and the Grenadines *13°08′N 61°09′W*
109 *V6* **Stübming** ≈ E Austria
114 *J11* **Studen Kladenets, Yazovir** ⊠ S Bulgaria
185 *G21* **Studholme** Canterbury, South Island, New Zealand *44°44′S 171°08′E*
Stuhlweissenberg ***see*** Székesfehérvár
Stuhm ***see*** Sztum
12 *C7* **Stull Lake** ◎ Ontario, C Canada
Stuorrarijjda ***see*** Storriten
126 *L4* **Stupino** Moskovskaya Oblast', W Russian Federation *54°54′N 38°06′E*
27 *U4* **Sturgeon** Missouri, C USA *39°13′N 92°16′W*
14 *G10* **Sturgeon** ≈ Ontario, S Canada
31 *N6* **Sturgeon Bay** Wisconsin, N USA *44°51′N 87°21′W*
14 *G11* **Sturgeon Falls** Ontario, S Canada *46°22′N 79°57′W*
12 *C11* **Sturgeon Lake** ◎ Ontario, S Canada
30 *M3* **Sturgeon River** ≈ Michigan, N USA
20 *H6* **Sturgis** Kentucky, S USA *37°33′N 87°58′W*
31 *P11* **Sturgis** Michigan, N USA *41°48′N 85°25′W*
28 *J9* **Sturgis** South Dakota, N USA *44°24′N 103°30′W*
112 *D10* **Šturlić** ◈ Federacija Bosne I Hercegovine, NW Bosnia and Herzegovina
111 *J22* **Štúrovo** ***Hung.*** Párkány; ***prev.*** Parkan. Nitriansky Kraj, SW Slovakia *47°49′N 18°40′E*
182 *L4* **Sturt, Mount** *hill* New South Wales, SE Australia
181 *P4* **Sturt Plain** *plain* Northern Territory, N Australia
181 *T9* **Sturt Stony Desert** *desert* South Australia
83 *J25* **Stutterheim** Eastern Cape, S South Africa *32°35′S 27°26′E*
101 *H21* **Stuttgart** Baden-Württemberg, SW Germany *48°47′N 09°12′E*
27 *W12* **Stuttgart** Arkansas, C USA *34°30′N 91°32′W*
92 *H2* **Stykkishólmur** Vesturland, W Iceland *65°04′N 22°43′W*
115 *F17* **Stylída** ***var.*** Stilida, Stilís. Stereá Elláda, C Greece *38°55′N 22°37′E*
116 *K2* **Styr** ***Rus.*** Styr'. ≈ Belarus/Ukraine
115 *I19* **Stýra** ***var.*** Stira. Évvoia, C Greece *38°10′N 24°13′E*
Styria ***see*** Steiermark
15 *Y5* **St-Yvon** Québec, SE Canada *49°09′N 64°51′W*
Su ***see*** Jiangsu
Sua ***see*** Sowa
171 *Q17* **Suai** W East Timor *09°19′S 125°16′E*
54 *G9* **Suaita** Santander, C Colombia *06°07′N 73°30′W*
80 *I7* **Suakin** ***var.*** Sawakin. Red Sea, NE Sudan *19°06′N 37°17′E*
161 *T13* **Su'ao** ***Jap.*** Suō. N Taiwan *24°33′N 121°48′E*
Suao ***see*** Suau
Sua Pan ***see*** Sowa Pan
40 *G6* **Suaqui Grande** Sonora, NW Mexico *28°22′N 109°52′W*
61 *A16* **Suardi** Santa Fe, C Argentina *30°32′S 61°58′W*
54 *D11* **Suárez** Cauca, SW Colombia *02°55′N 76°41′W*
186 *G10* **Suau** ***var.*** Suao. Suaul Island, SE Papua New Guinea *10°39′S 150°03′E*
118 *G12* **Subačius** Panevėžys, NE Lithuania *55°46′N 24°45′E*
168 *K9* **Subang** ***prev.*** Soebang. Jawa, C Indonesia *06°32′S 107°45′E*
169 *O16* **Subang** ✈ (Kuala Lumpur) Pahang, Peninsular Malaysia
129 *S10* **Subansiri** ≈ NE India
154 *M12* **Subarnapur** ***prev.*** Sonapur, Sonepur. Odisha, E India *20°50′N 83°58′E*
118 *I11* **Subate** SE Latvia *56°00′N 25°54′E*
139 *N5* **Subaykhān** Dayr az Zawr, E Syria *34°52′N 40°35′E*
Subei/Subei Mongolzu Zizhixian ***see*** Dangchengwan
169 *P9* **Subi Besar, Pulau** *island* Kepulauan Natuna, W Indonesia
Subiyah ***see*** Aş Şubayḩīyah
26 *I7* **Sublette** Kansas, C USA *37°28′N 100°52′W*
112 *K8* **Subotica** ***Ger.*** Maria-Theresiopel, ***Hung.*** Szabadka. Vojvodina, N Serbia *46°06′N 19°41′E*
116 *K9* **Suceava** ***Ger.*** Suczawa, ***Hung.*** Szucsava. Suceava, NE Romania *47°41′N 26°16′E*
116 *J9* **Suceava** ◈ *county* NE Romania
116 *K9* **Suceava** ***Ger.*** Suczawa. ≈ N Romania
112 *E12* **Sučevići** Zadar, SW Croatia *44°13′N 16°04′E*
111 *K17* **Sucha Beskidzka** Małopolskie, S Poland *49°44′N 19°36′E*
111 *M14* **Suchedniów** Świętokrzyskie, C Poland *51°01′N 20°49′E*
42 *A2* **Suchitepéquez** ***off.*** Departamento de Suchitepéquez. ◈ *department* SW Guatemala
Suchitepéquez, Departamento de ***see*** Suchitepéquez
Su-chou ***see*** Suzhou
Suchow ***see*** Suzhou, Jiangsu, China
Suchow ***see*** Xuzhou, Jiangsu, China
97 *D17* **Suck** ≈ C Ireland
Sucker State ***see*** Illinois
186 *F9* **Suckling, Mount** ▲ S Papua New Guinea *09°36′S 149°00′E*
57 *L19* **Sucre** ***hist.*** Chuquisaca, La Plata. ● (Bolivia-legal capital) Chuquisaca, S Bolivia *18°53′S 65°25′W*
54 *E6* **Sucre** Santander, N Colombia *08°50′N 74°22′W*
56 *A7* **Sucre** Manabí, W Ecuador *01°21′S 80°27′W*
54 *E6* **Sucre** ***off.*** Departamento de Sucre. ◈ *province* N Colombia
55 *O5* **Sucre** ***off.*** Estado Sucre. ◈ *state* NE Venezuela
Sucre, Departamento de ***see*** Sucre
Sucre, Estado ***see*** Sucre
56 *D6* **Sucumbíos** ◈ *province* NE Ecuador
113 *G15* **Sućuraj** Split-Dalmacija, S Croatia *43°07′N 17°10′E*
58 *K10* **Sucuriju** Amapá, NE Brazil *01°31′N 50°W*
Suczawa ***see*** Suceava
79 *E16* **Sud** ***Eng.*** South. ◈ *province* S Cameroon
124 *K13* **Suda** ≈ NW Russian Federation
Suda ***see*** Soúda
117 *U13* **Sudak** Avtonomna Respublika Krym, S Ukraine *44°52′N 34°57′E*
24 *M4* **Sudan** Texas, SW USA *34°04′N 102°32′W*
80 *C10* **Sudan** ***off.*** Republic of Sudan, ***Ar.*** Jumhuriyat as-Sudan; ***prev.*** Anglo-Egyptian Sudan. ◆ *republic* N Africa
Sudanese Republic ***see*** Mali
Sudan, Jumhuriyat as- ***see*** Sudan
Sudan, Republic of ***see*** Sudan
14 *F10* **Sudbury** Ontario, S Canada *46°29′N 81°W*
97 *P20* **Sudbury** E England, United Kingdom *52°04′N 00°43′E*
Sud, Canal de ***see*** Gonâve, Canal de la
80 *E13* **Sudd** *swamp region* C South Sudan
100 *K10* **Sude** ≈ N Germany
Suderø ***see*** Sudhuroy
Sudest Island ***see*** Tagula Island
111 *E15* **Sudeten** ***var.*** Sudetes, Sudetic Mountains, ***Cz./Pol.*** Sudety. ▲▲ Czech Republic/Poland
Sudetes/Sudetic Mountains/Sudety ***see*** Sudeten
95 *B19* **Sudhuroy** ***Dan.*** Suderø. *island* S Faroe Islands
124 *M15* **Sudislavl'** Kostromskaya Oblast', NW Russian Federation *57°55′N 41°45′E*
Südkarpaten ***see*** Carpaţii Meridionalii
79 *N20* **Sud-Kivu** ***off.*** Région Sud Kivu. ◈ *region* E Dem. Rep. Congo
Sud Kivu, Région ***see*** Sud-Kivu
Südliche Morava ***see*** Južna Morava
100 *E12* **Süd-Nord-Kanal** *canal* NW Germany
126 *M3* **Sudogda** Vladimirskaya Oblast', W Russian Federation *55°58′N 40°57′E*
Sudostroy ***see*** Severodvinsk
79 *C15* **Sud-Ouest** ***Eng.*** South-West. ◈ *province* W Cameroon
173 *X17* **Sud Ouest, Pointe** *headland* SW Mauritius *20°27′S 57°18′E*
187 *P17* **Sud, Province** ◈ *province* S New Caledonia
92 *G1* **Suðureyri** Vestfirðir, NW Iceland *66°08′N 23°31′W*
92 *J4* **Suðurland** ◈ *region* S Iceland
92 *H4* **Suðurnes** ◈ *region* SW Iceland
126 *J8* **Sudzha** Kurskaya Oblast', W Russian Federation *51°12′N 35°19′E*
81 *D15* **Sue** ≈ W South Sudan
105 *S10* **Sueca** Valenciana, E Spain *39°13′N 00°19′W*
Sŭedinenie ***see*** Saedinenie
Suero ***see*** Alzira
75 *X8* **Suez** ***Ar.*** As Suways, El Suweis. NE Egypt *29°59′N 32°33′E*
75 *W7* **Suez Canal** ***Ar.*** Qanāt as Suways. *canal* NE Egypt
Suez, Gulf of ***see*** Khalīj as Suways
11 *R17* **Suffield** Alberta, SW Canada *50°15′N 111°05′W*
21 *X7* **Suffolk** Virginia, NE USA *36°44′N 76°37′W*
97 *P20* **Suffolk** *cultural region* E England, United Kingdom
142 *J2* **Şūfīān** Āżarbāyjān-e Sharqī, N Iran *38°15′N 45°59′E*
31 *N12* **Sugar Creek** ≈ Illinois, N USA
30 *L13* **Sugar Creek** ≈ Illinois, N USA
31 *R3* **Sugar Island** *island* Michigan, N USA
25 *V11* **Sugar Land** Texas, SW USA *29°37′N 95°37′W*
19 *P6* **Sugarloaf Mountain** ▲ Maine, NE USA *45°01′N 70°18′W*
65 *G24* **Sugar Loaf Point** *headland* N Saint Helena *15°54′S 05°43′W*
136 *G16* **Suğla Gölü** ◎ SW Turkey
123 *T8* **Sugoy** ≈ E Russian Federation
158 *F7* **Sugun** Xinjiang Uygur Zizhiqu, W China *39°46′N 76°45′E*
147 *U11* **Sugut, Gora** ▲ SW Kyrgyzstan *39°52′N 73°36′E*
169 *V6* **Sugut, Sungai** ≈ East Malaysia
159 *O9* **Suhai Hu** ◎ C China
162 *K14* **Suhait** Nei Mongol Zizhiqu, N China *39°29′N 105°11′E*
141 *X7* **Şuḩār** ***var.*** Sohar. N Oman *24°20′N 56°43′E*
114 *N8* **Suha Reka** ≈ NE Bulgaria
113 *M17* **Suharekë** ***Serb.*** Suva Reka. S Kosovo *42°23′N 20°50′E*
162 *L6* **Sühbaatar** Selenge, N Mongolia *50°12′N 106°14′E*
163 *P8* **Sühbaatar** ***var.*** Haylaastay. Sühbaatar, E Mongolia *46°44′N 113°51′E*
163 *P9* **Sühbaatar** ◈ *province* E Mongolia
114 *J8* **Suhindol** ***var.*** Sukhindol. Veliko Turnovo, N Bulgaria *43°11′N 24°10′E*
101 *K17* **Suhl** Thüringen, C Germany *50°37′N 10°43′E*
108 *F7* **Suhr** Aargau, N Switzerland *47°23′N 08°05′E*
Sui'an ***see*** Zhangpu
Suicheng ***see*** Suixi
161 *O12* **Suichuan** ***var.*** Quanjiang. Jiangxi, S China *26°26′N 114°34′E*
Suid-Afrika ***see*** South Africa
160 *L4* **Suide** ***var.*** Mingzhou. Shaanxi, C China *37°30′N 110°10′E*
Suidwes-Afrika ***see*** Namibia
163 *Y9* **Suifenhe** Heilongjiang, NE China *44°22′N 131°12′E*
Suigen ***see*** Suwon
163 *W8* **Suihua** Heilongjiang, NE China *46°40′N 127°00′E*
Súilí, Loch ***see*** Swilly, Lough
161 *Q6* **Suining** Jiangsu, E China *33°54′N 117°58′E*
160 *I9* **Suining** Sichuan, C China *30°31′N 105°33′E*
103 *Q4* **Suippes** Marne, N France *49°08′N 04°31′E*
97 *E20* **Suir** ***Ir.*** An tSiúir. ≈ S Ireland
165 *J13* **Suita** Ōsaka, Honshū, SW Japan *34°39′N 135°27′E*
160 *L16* **Suixi** ***var.*** Suicheng. Guangdong, S China *21°23′N 110°14′E*
Sui Xian ***see*** Suizhou
163 *T13* **Suizhong** Liaoning, NE China *40°19′N 120°22′E*
161 *N8* **Suizhou** ***prev.*** Sui Xian. Hubei, C China *31°46′N 113°20′E*
149 *P17* **Sujāwal** Sind, SE Pakistan *24°36′N 68°06′E*
169 *O16* **Sukabumi** ***prev.*** Soekaboemi. Jawa, C Indonesia *06°55′S 106°56′E*
169 *Q12* **Sukadana, Teluk** *bay* Borneo, W Indonesia
165 *P11* **Sukagawa** Fukushima, Honshū, C Japan *37°16′N 140°20′E*
Sukarnapura ***see*** Jayapura
Sukarno, Puntjak ***see*** Jaya, Puncak
Sŭkh ***see*** Sokh
Sukhindol ***see*** Suhindol
126 *J5* **Sukhinichi** Kaluzhskaya Oblast', W Russian Federation *54°06′N 35°22′E*
Sukhne ***see*** As Sukhnah
129 *Q4* **Sukhona** ***var.*** Tot'ma. ≈ NW Russian Federation
167 *O8* **Sukhothai** ***var.*** Sukotai. Sukhothai, W Thailand *17°00′N 99°51′E*
Sukhumi ***see*** Sokhumi
Sukkertoppen ***see*** Maniitsoq
149 *Q13* **Sukkur** Sind, SE Pakistan *27°45′N 68°46′E*
Sukotai ***see*** Sukhothai
Sukra Bay ***see*** Şawqirah, Dawḩat
125 *V15* **Suksun** Permskiy Kray, NW Russian Federation *57°10′N 57°27′E*
165 *F15* **Sukumo** Kōchi, Shikoku, SW Japan *32°55′N 132°42′E*
94 *B12* **Sula** *island* S Norway
125 *Q5* **Sula** ≈ NW Russian Federation
117 *R5* **Sula** ≈ N Ukraine
42 *H6* **Sulaco, Río** ≈ NW Honduras
Sulaimaniya ***see*** As Sulaymānīyah
149 *S10* **Sulaimān Range** ▲▲ C Pakistan
127 *Q16* **Sulak** Respublika Dagestan, SW Russian Federation *43°19′N 47°28′E*
127 *Q16* **Sulak** ≈ SW Russian Federation
171 *Q13* **Sula, Kepulauan** *island group* C Indonesia
136 *I12* **Sulakyurt** ***var.*** Konur. Kırıkkale, N Turkey *40°10′N 33°42′E*
171 *P17* **Sulamu** Timor, S Indonesia *09°57′S 123°33′E*
96 *F5* **Sula Sgeir** *island* NW Scotland, United Kingdom
171 *N13* **Sulawesi** ***Eng.*** Celebes. *island* C Indonesia
171 *N13* **Sulawesi Barat** ***off.*** Provinsi Sulawesi Barat, ***var.*** Sulbar. ◈ *province* C Indonesia
Sulawesi Barat, Provinsi ***see*** Sulawesi Barat
Sulawesi, Laut ***see*** Celebes Sea
171 *N14* **Sulawesi Selatan** ***off.*** Propinsi Sulawesi Selatan, ***var.*** Sulsel, ***Eng.*** South Celebes, South Sulawesi. ◈ *province* C Indonesia
Sulawesi Selatan, Propinsi ***see*** Sulawesi Selatan
171 *P12* **Sulawesi Tengah** ***off.*** Propinsi Sulawesi Tengah, ***var.*** Sulteng, ***Eng.*** Central Celebes, Central Sulawesi. ◈ *province* N Indonesia
Sulawesi Tengah, Propinsi ***see*** Sulawesi Tengah
171 *O14* **Sulawesi Tenggara** ***off.*** Propinsi Sulawesi Tenggara, ***var.*** Sultenggara, ***Eng.*** South-East Celebes, South-East Sulawesi. ◈ *province* C Indonesia
Sulawesi Tenggara, Propinsi ***see*** Sulawesi Tenggara
171 *P11* **Sulawesi Utara** ***off.*** Propinsi Sulawesi Utara, ***var.*** Sulut, ***Eng.*** North Celebes, North Sulawesi. ◈ *province* N Indonesia
Sulawesi Utara, Propinsi ***see*** Sulawesi Utara
139 *T5* **Sulaymān Beg** At Ta'mím, N Iraq
Sulbar ***see*** Sulawesi Barat
95 *D15* **Suldalsvatnet** ◎ S Norway
Sulden ***see*** Solda
110 *E12* **Sulechów** ***Ger.*** Züllichau. Lubuskie, W Poland *52°05′N 15°37′E*
110 *E11* **Sulęcin** Lubuskie, W Poland *52°29′N 15°06′E*
77 *U14* **Suleja** Niger, C Nigeria *09°15′N 07°10′E*
111 *K14* **Sulejów** Łodzkie, S Poland *51°21′N 19°57′E*
96 *I5* **Sule Skerry** *island* N Scotland, United Kingdom
Suliag ***see*** Sawhāj
76 *J16* **Sulima** S Sierra Leone *06°59′N 11°34′W*
117 *O13* **Sulina** Tulcea, SE Romania *45°07′N 29°40′E*
117 *N13* **Sulina, Braţul** ≈ SE Romania
100 *H12* **Sulingen** Niedersachsen, NW Germany *52°40′N 08°48′E*
Sulisjielmmá ***see*** Sulitjelma
92 *H12* **Suliskongen** ▲ C Norway *67°10′N 16°16′E*
92 *H12* **Sulitjelma** ***Lapp.*** Sulisjielmmá. Nordland, C Norway *67°10′N 16°05′E*
56 *A9* **Sullana** Piura, NW Peru *04°54′S 80°42′W*
23 *N3* **Sulligent** Alabama, S USA *33°54′N 88°07′W*
30 *M14* **Sullivan** Illinois, N USA *39°36′N 88°36′W*
31 *N15* **Sullivan** Indiana, N USA *39°05′N 87°24′W*
27 *W5* **Sullivan** Missouri, C USA *38°12′N 91°09′W*
Sullivan Island ***see*** Lanbi Kyun
96 *M1* **Sullom Voe** NE Scotland, United Kingdom *60°24′N 01°09′W*
103 *O7* **Sully-sur-Loire** Loiret, C France *47°46′N 02°21′E*
Sulmo ***see*** Sulmona
107 *K15* **Sulmona** ***anc.*** Sulmo. Abruzzo, C Italy *42°03′N 13°56′E*
Sulo ***see*** Shule He
114 *M11* **Süloğlu** Edirne, NW Turkey *41°46′N 26°55′E*
22 *G9* **Sulphur** Louisiana, S USA *30°14′N 93°22′W*
27 *O12* **Sulphur** Oklahoma, C USA *34°31′N 96°58′W*
28 *K9* **Sulphur Creek** ≈ South Dakota, N USA
24 *M5* **Sulphur Draw** ≈ Texas, SW USA
25 *W5* **Sulphur River** ≈ Arkansas/Texas, SW USA
25 *V6* **Sulphur Springs** Texas, SW USA *33°09′N 95°36′W*
24 *M6* **Sulphur Springs Draw** ≈ Texas, SW USA
Sulsel ***see*** Sulawesi Selatan
14 *D8* **Sultan** Ontario, S Canada *47°34′N 82°45′W*
Sultānābād ***see*** Arāk
Sultan Alonto, Lake ***see*** Lanao, Lake
136 *G15* **Sultan Dağları** ▲▲ C Turkey
114 *N13* **Sultanköy** Tekirdağ, NW Turkey *41°01′N 27°58′E*
171 *Q7* **Sultan Kudarat** ***var.*** Nuling. Mindanao, S Philippines *07°20′N 124°16′E*
152 *M13* **Sultānpur** Uttar Pradesh, N India *26°15′N 82°04′E*
Sulteng ***see*** Sulawesi Tengah
Sultenggara ***see*** Sulawesi Tenggara
171 *O9* **Sulu Archipelago** *island group* SW Philippines
192 *F7* **Sulu Basin** *undersea feature* SE South China Sea *08°00′N 121°30′E*
Sülüktü ***see*** Sulyukta
Sulu, Laut ***see*** Sulu Sea
169 *X6* **Sulu Sea** ***var.*** Laut Sulu. *sea* SW Philippines
Sulut ***see*** Sulawesi Utara
145 *O15* **Sulutobe** ***Kaz.*** Sulūtöbe. Kzylorda, S Kazakhstan *44°31′N 66°17′E*
Sulūtöbe ***see*** Sulutobe
147 *Q11* **Sulyukta** ***Kir.*** Sülüktü. Batkenskaya Oblast', SW Kyrgyzstan *39°57′N 69°31′E*
Sulz ***see*** Sulz am Neckar
101 *G22* **Sulz am Neckar** ***var.*** Sulz. Baden-Württemberg, SW Germany *48°22′N 08°37′E*
101 *L20* **Sulzbach-Rosenberg** Bayern, SE Germany *49°30′N 11°43′E*
195 *N13* **Sulzberger Bay** *bay* Antarctica
81 *M14* **Sumalē** ***var.*** Somali. ◈ E Ethiopia
113 *F15* **Sumartin** Split-Dalmacija, S Croatia *43°17′N 16°52′E*
32 *H6* **Sumas** Washington, NW USA *49°00′N 122°15′W*
168 *J10* **Sumatera** ***Eng.*** Sumatra. *island* W Indonesia
168 *J12* **Sumatera Barat** ***off.*** Propinsi Sumatera Barat, ***var.*** Sumbar, ***Eng.*** West Sumatra. ◈ *province* W Indonesia
Sumatera Barat, Propinsi ***see*** Sumatera Barat
168 *L13* **Sumatera Selatan** ***off.*** Propinsi Sumatera Selatan, ***var.*** Sumsel, ***Eng.*** South Sumatra. ◈ *province* W Indonesia
Sumatera Selatan, Propinsi ***see*** Sumatera Selatan
168 *H10* **Sumatera Utara** ***off.*** Propinsi Sumatera Utara, ***var.*** Sumut, ***Eng.*** North Sumatra. ◈ *province* W Indonesia
Sumatera Utara, Propinsi ***see*** Sumatera Utara
Sumatra ***see*** Sumatera
Šumava ***see*** Bohemian Forest
Sumayl ***see*** Sêmêl
139 *U7* **Sumayr al Muḩammad** Diyālá, E Iraq *33°34′N 45°06′E*
171 *N17* **Sumba, Pulau** ***Eng.*** Sandalwood Island; ***prev.*** Soemba. *island* Nusa Tenggara, C Indonesia
146 *D12* **Sumbar** ≈ W Turkmenistan
Sumbar ***see*** Sumatera Barat
192 *E9* **Sumbawa** ***prev.*** Soembawa. *island* Nusa Tenggara, C Indonesia
170 *L16* **Sumbawabesar** Sumbawa, S Indonesia *08°30′S 117°25′E*
81 *F23* **Sumbawanga** Rukwa, W Tanzania *07°57′S 31°37′E*
82 *B12* **Sumbe** ***var.*** N'Gunza, ***Port.*** Novo Redondo. Kwanza Sul, W Angola *11°13′S 13°53′E*
96 *M3* **Sumburgh Head** *headland* NE Scotland, United Kingdom *59°51′N 01°16′W*
111 *H23* **Sümeg** Veszprém, W Hungary *47°01′N 17°13′E*
80 *C12* **Sumeih** Eastern Darfur, S Sudan *09°50′N 27°39′E*
169 *T16* **Sumenep** ***prev.*** Soemenep. Pulau Madura, C Indonesia *07°01′S 113°51′E*
Sumgait ***see*** Sumqayıtçay, Azerbaijan
Sumgait ***see*** Sumqayıt, Azerbaijan
165 *Y14* **Sumisu-jima** ***Eng.*** Smith Island. *island* SE Japan
Summēl ***see*** Sêmêl
31 *O5* **Summer Island** *island* Michigan, N USA
32 *H15* **Summer Lake** ◎ Oregon, NW USA
11 *N17* **Summerland** British Columbia, SW Canada *49°35′N 119°45′W*
13 *P14* **Summerside** Prince Edward Island, SE Canada *46°24′N 63°46′W*
21 *R5* **Summersville** West Virginia, NE USA *38°17′N 80°52′W*
21 *R5* **Summersville Lake** ⊠ West Virginia, NE USA
21 *S13* **Summerton** South Carolina, SE USA *33°36′N 80°21′W*
23 *R2* **Summerville** Georgia, SE USA *34°28′N 85°21′W*
21 *S14* **Summerville** South Carolina, SE USA *33°01′N 80°10′W*
39 *R10* **Summit** Alaska, USA *63°21′N 148°50′W*
35 *V6* **Summit Mountain** ▲ Nevada, W USA *39°23′N 116°25′W*
37 *R8* **Summit Peak** ▲ Colorado, C USA *37°21′N 106°42′W*
Summus Portus ***see*** Somport, Col du
29 *X12* **Sumner** Iowa, C USA *42°51′N 92°05′W*
22 *K3* **Sumner** Mississippi, S USA *33°58′N 90°22′W*
185 *H17* **Sumner, Lake** ◎ South Island, New Zealand
37 *U12* **Sumner, Lake** ⊠ New Mexico, SW USA
111 *G17* **Šumperk** ***Ger.*** Mährisch-Schönberg. Olomoucký Kraj, E Czech Republic *49°58′N 17°00′E*
42 *F7* **Sumpul, Río** ≈ El Salvador/Honduras
137 *Z11* **Sumqayıt** ***Rus.*** Sumgait. E Azerbaijan *40°33′N 49°41′E*
137 *Y11* **Sumqayıtçay** ***Rus.*** Sumgait. ≈ E Azerbaijan
147 *R9* **Sumsar** Dzhalal-Abadskaya Oblast', W Kyrgyzstan *41°12′N 71°16′E*
Sumsel ***see*** Sumatera Selatan
117 *S3* **Sums'ka Oblast'** ***var.*** Sumy, ***Rus.*** Sumskaya Oblast'. ◈ *province* NE Ukraine
Sumskaya Oblast' ***see*** Sums'ka Oblast'
124 *J8* **Sumskiy Posad** Respublika Kareliya, NW Russian Federation *64°12′N 35°22′E*
21 *S12* **Sumter** South Carolina, SE USA *33°54′N 80°22′W*
Sumut ***see*** Sumatera Utara
117 *T3* **Sumy** Sums'ka Oblast', NE Ukraine *50°54′N 34°49′E*
Sumy ***see*** Sums'ka Oblast'
159 *Q15* **Sumzom** Xizang Zizhiqu, W China *29°45′N 96°14′E*
125 *R15* **Suna** Kirovskaya Oblast', NW Russian Federation *57°53′N 50°04′E*
124 *I10* **Suna** ≈ NW Russian Federation
165 *S3* **Sunagawa** Hokkaidō, NE Japan *43°30′N 141°55′E*
153 *V13* **Sunamganj** Sylhet, NE Bangladesh *25°04′N 91°24′E*
163 *W14* **Sunan** ✈ (P'yŏngyang) SW North Korea *39°12′N 125°40′E*
Sunan/Sunan Yugurzu Zizhixian ***see*** Hongwansi
19 *N9* **Sunapee Lake** ◎ New Hampshire, NE USA
139 *P4* **Sunaysilah** *salt marsh* N Iraq
20 *M8* **Sunbright** Tennessee, S USA *36°12′N 84°39′W*
33 *R6* **Sunburst** Montana, NW USA *48°51′N 111°54′W*
183 *N12* **Sunbury** Victoria, SE Australia *37°36′S 114°45′E*
21 *X8* **Sunbury** North Carolina, SE USA *36°27′N 76°34′W*
18 *G14* **Sunbury** Pennsylvania, NE USA *40°51′N 76°47′W*
61 *A17* **Sunchales** Santa Fe, C Argentina *30°58′S 61°35′W*
163 *Y16* **Suncheon** ***prev.*** Sunch'ŏn. S South Korea *34°56′N 127°29′E*
Sunch'ŏn ***see*** Suncheon
36 *K13* **Sun City** Arizona, SW USA *33°36′N 112°16′W*
19 *O9* **Suncook** New Hampshire, NE USA *43°07′N 71°25′W*
161 *P5* **Suncun** ***prev.*** Xinwen. Shandong, E China *35°49′N 117°36′E*
Sunda Islands ***see*** Greater Sunda Islands
33 *Z12* **Sundance** Wyoming, C USA *44°24′N 104°22′W*
153 *T17* **Sundarbans** *wetland* Bangladesh/India
154 *M11* **Sundargarh** Odisha, E India *22°07′N 84°02′E*
129 *U15* **Sunda Shelf** *undersea feature* S South China Sea *05°00′N 107°00′E*
Sunda Trench ***see*** Java Trench
129 *U17* **Sunda Trough** *undersea feature* E Indian Ocean *08°50′S 109°30′E*
95 *O16* **Sundbyberg** Stockholm, C Sweden *59°22′N 17°58′E*
97 *M14* **Sunderland** ***var.*** Wearmouth. NE England, United Kingdom *54°55′N 01°23′W*
101 *F15* **Sundern** Nordrhein-Westfalen, W Germany *51°19′N 08°00′E*
136 *F12* **Sündiken Dağları** ▲▲ C Turkey
24 *M5* **Sundown** Texas, SW USA *33°27′N 102°29′W*
11 *P16* **Sundre** Alberta, SW Canada *51°49′N 114°46′W*
14 *H12* **Sundridge** Ontario, S Canada *45°45′N 79°25′W*
93 *H17* **Sundsvall** Västernorrland, C Sweden *62°22′N 17°20′E*
26 *H4* **Sunflower, Mount** ▲ Kansas, C USA *39°01′N 102°02′W*
Sunflower State ***see*** Kansas
169 *N14* **Sungaibuntu** Sumatera, SW Indonesia *04°04′S 105°37′E*
168 *K12* **Sungaidareh** Sumatera, W Indonesia *0°58′S 101°30′E*
167 *P17* **Sungai Kolok** ***var.*** Sungai Ko-Lok. Narathiwat, SW Thailand *06°02′N 101°58′E*
Sungai Ko-Lok ***see*** Sungai Kolok
168 *K12* **Sungaipenuh** ***prev.*** Soengaipenoeh. Sumatera, W Indonesia *02°S 101°28′E*
169 *P11* **Sungaipinyuh** Borneo, C Indonesia *0°16′N 109°06′E*
Sungari ***see*** Songhua Jiang
Sungaria ***see*** Dzungaria
Sungei Pahang ***see*** Pahang, Sungai
167 *O8* **Sung Men** Phrae, NW Thailand *17°59′N 100°07′E*
83 *M15* **Sungo** Tete, NW Mozambique *16°31′S 33°58′E*
Sungpu ***see*** Songpan
168 *M13* **Sungsang** Sumatera, W Indonesia *02°22′S 104°50′E*
114 *M9* **Sungurlare** Burgas, E Bulgaria *42°47′N 26°46′E*
136 *J12* **Sungurlu** Çorum, N Turkey *40°10′N 34°23′E*
112 *F9* **Sunja** Sisak-Moslavina, C Croatia *45°21′N 16°33′E*
153 *Q12* **Sun Koshī** ≈ E Nepal
94 *F9* **Sunndalen** *valley* S Norway
94 *F9* **Sunndalsøra** Møre og Romsdal, S Norway *62°39′N 08°37′E*
95 *K15* **Sunne** Värmland, C Sweden *59°52′N 13°05′E*
95 *O15* **Sunnersta** Uppsala, C Sweden *59°46′N 17°40′E*
94 *C11* **Sunnfjord** *physical region* S Norway
95 *C15* **Sunnhordland** *physical region* S Norway
94 *D10* **Sunnmøre** *physical region* S Norway
37 *N4* **Sunnyside** Utah, W USA *39°33′N 110°23′W*
32 *J10* **Sunnyside** Washington, NW USA *46°01′N 119°58′W*
35 *N9* **Sunnyvale** California, W USA *37°22′N 122°02′W*
30 *L8* **Sun Prairie** Wisconsin, N USA *43°12′N 89°12′W*
Sunqur ***see*** Sonqor
25 *N1* **Sunray** Texas, SW USA *36°01′N 101°49′W*
22 *I8* **Sunset** Louisiana, S USA *30°24′N 92°04′W*
25 *S5* **Sunset** Texas, SW USA *33°24′N 97°45′W*
Sunset State ***see*** Oregon
181 *Z10* **Sunshine Coast** *cultural region* Queensland, E Australia
Sunshine State ***see*** Florida
Sunshine State ***see*** New Mexico
Sunshine State ***see*** South Dakota
123 *O10* **Suntar** Respublika Sakha (Yakutiya), NE Russian Federation *62°10′N 117°34′E*
39 *R10* **Suntrana** Alaska, USA *63°51′N 148°51′W*
148 *J15* **Suntsar** Baluchistān, SW Pakistan *25°30′N 62°03′E*
163 *W15* **Sunwi-do** *island* SW North Korea
163 *W6* **Sunwu** Heilongjiang, NE China *49°29′N 127°15′E*
77 *O16* **Sunyani** W Ghana *07°22′N 02°18′W*
Suō ***see*** Su'ao
93 *M17* **Suolahti** Keski-Suomi, C Finland *62°32′N 25°51′E*
Suoločielgi ***see*** Saariselkä
Suomenlahti ***see*** Finland, Gulf of
Suomen Tasavalta/Suomi ***see*** Finland
93 *N14* **Suomussalmi** Kainuu, E Finland *64°54′N 29°05′E*
165 *E13* **Suō-nada** *sea* SW Japan
93 *M17* **Suonenjoki** Pohjois-Savo, C Finland *62°36′N 27°07′E*
167 *S13* **Suŏng** Kâmpóng Cham, C Cambodia *11°53′N 105°41′E*
124 *I10* **Suoyarvi** Respublika Kareliya, NW Russian Federation *62°02′N 32°24′E*
Supanburi ***see*** Suphan Buri
57 *D14* **Supe** Lima, W Peru *10°49′S 77°40′W*
15 *V7* **Supérieur, Lac** ◎ Québec, SE Canada
Supérieur, Lac ***see*** Superior, Lake
36 *M14* **Superior** Arizona, SW USA *33°17′N 111°06′W*
33 *O9* **Superior** Montana, NW USA *47°11′N 114°53′W*
29 *P17* **Superior** Nebraska, C USA *40°01′N 98°04′W*
30 *I3* **Superior** Wisconsin, N USA *46°42′N 92°04′W*
41 *S17* **Superior, Laguna** *lagoon* S Mexico
31 *N2* **Superior, Lake** ***Fr.*** Lac Supérieur. ◎ Canada/USA
36 *L13* **Superstition Mountains** ▲▲ Arizona, SW USA
113 *F14* **Supetar** ***It.*** San Pietro. Split-Dalmacija, S Croatia *43°22′N 16°34′E*
167 *O10* **Suphan Buri** ***var.*** Supanburi. Suphan Buri, W Thailand *14°29′N 100°10′E*
171 *V12* **Supiori, Pulau** *island* E Indonesia
188 *K2* **Supply Reef** *reef* N Northern Mariana Islands
195 *O7* **Support Force Glacier** *glacier* Antarctica
137 *R10* **Supsa** ***prev.*** Sup'sa. ≈ W Georgia
Sup'sa ***see*** Supsa
Sūq 'Abs ***see*** 'Abs
139 *W12* **Sūq ash Shuyūkh** Dhī Qār, SE Iraq *30°53′N 46°28′E*
138 *H4* **Şuqaylibīyah** Ḩamāh, W Syria *35°21′N 36°24′E*
161 *Q6* **Suqian** Jiangsu, E China *33°57′N 118°18′E*
Suqrah ***see*** Şawqirah
Suqrah Bay ***see*** Şawqirah, Dawḩat
141 *V16* **Suquṭrā** ***var.*** Sokotra, ***Eng.*** Socotra. *island* SE Yemen
141 *Z8* **Şūr** NE Oman *22°32′N 59°33′E*
Şūr ***see*** Soûr
127 *P5* **Sura** Penzenskaya Oblast', W Russian Federation *53°23′N 45°03′E*
127 *P4* **Sura** ≈ W Russian Federation
149 *N12* **Sūrāb** Baluchistān, SW Pakistan *28°28′N 66°15′E*
Surabaja ***see*** Surabaya
192 *E8* **Surabaya** ***prev.*** Surabaja, Soerabaja. Jawa, C Indonesia *07°14′S 112°45′E*
95 *N15* **Surahammar** Västmanland, C Sweden *59°43′N 16°13′E*
169 *Q16* **Surakarta** ***Eng.*** Solo; ***prev.*** Soerakarta. Jawa, S Indonesia *07°32′S 110°50′E*
Surakhany ***see*** Suraxanı
137 *S10* **Surami** C Georgia *41°59′N 43°36′E*
143 *X13* **Sūrān** Sīstān va Balūchestān, SE Iran *27°18′N 61°58′E*
111 *I21* **Šurany** ***Hung.*** Nagysurány. Nitriansky Kraj, SW Slovakia *48°05′N 18°10′E*
154 *D12* **Sūrat** Gujarāt, W India *21°10′N 72°54′E*
Suratdhani ***see*** Surat Thani
152 *G9* **Sūratgarh** Rājasthān, NW India *29°20′N 73°59′E*
167 *N14* **Surat Thani** ***var.*** Suratdhani. Surat Thani, SW Thailand *09°09′N 99°20′E*
119 *Q16* **Suraw** ***Rus.*** Surov. ≈ E Belarus
137 *Z11* **Suraxanı** ***Rus.*** Surakhany. E Azerbaijan *40°25′N 49°59′E*
141 *Y11* **Surayr** E Oman *19°56′N 57°47′E*
138 *K2* **Suraysāt** Ḩalab, N Syria *36°42′N 38°01′E*
118 *O12* **Surazh** Vitsyebskaya Voblasts', NE Belarus *55°25′N 30°44′E*
126 *H6* **Surazh** Bryanskaya Oblast', W Russian Federation *53°04′N 32°29′E*
191 *V17* **Sur, Cabo** *headland* Easter Island, Chile, E Pacific Ocean *27°11′S 109°26′W*
112 *L11* **Surčin** Serbia, N Serbia *44°48′N 20°19′E*
116 *H9* **Surduc** ***Hung.*** Szurduk. Sălaj, NW Romania *47°13′N 23°20′E*
113 *P16* **Surdulica** Serbia, SE Serbia *42°43′N 22°10′E*
99 *L24* **Sûre** ***var.*** Sauer. ≈ W Europe ***see also*** Sauer
Sûre ***see*** Sauer
154 *C10* **Surendranagar** Gujarāt, W India *22°44′N 71°43′E*
18 *K16* **Surf City** New Jersey, NE USA *39°21′N 74°24′W*
183 *V3* **Surfers Paradise** Queensland, E Australia *27°54′S 153°18′E*
21 *U13* **Surfside Beach** South Carolina, SE USA *33°36′N 78°58′W*
102 *J10* **Surgères** Charente-Maritime, W France *46°07′N 00°44′W*
122 *H10* **Surgut** Khanty-Mansiyskiy Avtonomnyy Okrug-Yugra, C Russian Federation *61°13′N 73°28′E*
122 *K10* **Surgutikha** Krasnoyarskiy Kray, N Russian Federation *64°44′N 87°13′E*
98 *M6* **Surhuisterveen** ***Fris.*** Surhusterfean. Fryslân, N Netherlands *53°11′N 06°10′E*
Surhusterfean ***see*** Surhuisterveen
105 *V5* **Súria** Cataluña, NE Spain *41°50′N 01°45′E*
143 *P10* **Sūrīān** Fārs, S Iran
155 *J15* **Suriāpet** Telangana, C India *17°10′N 79°42′E*
171 *Q6* **Surigao** Mindanao, S Philippines *09°43′N 125°31′E*
167 *R10* **Surin** Surin, E Thailand *14°53′N 103°29′E*
Surinam ***see*** Suriname
55 *U11* **Suriname** ***off.*** Republic of Suriname, ***var.*** Surinam; ***prev.*** Dutch Guiana, Netherlands Guiana. ◆ *republic* N South America
Suriname, Republic of ***see*** Suriname
Sūrīya/Sūrīyah, Al-Jumhūrīyah al-'Arabīyah as- ***see*** Syria
Surkhab, Darya-i- ***see*** Kahmard, Daryā-ye
Surkhandar'inskaya Oblast' ***see*** Surxondaryo Viloyati
Surkhandar'ya ***see*** Surxondaryo
Surkhet ***see*** Birendranagar
147 *R12* **Surkhob** ≈ C Tajikistan
137 *P11* **Sürmene** Trabzon, NE Turkey *40°56′N 40°03′E*
Surov ***see*** Suraw
127 *N11* **Surovikino** Volgogradskaya Oblast', SW Russian Federation *48°39′N 42°46′E*

◆ Country
● Country Capital
◇ Dependent Territory
○ Dependent Territory Capital
◈ Administrative Regions
✈ International Airport
▲ Mountain
▲▲ Mountain Range
⛰ Volcano
≈ River
◎ Lake
⊠ Reservoir

35 N11 **Sur, Point** *headland* California, W USA 36°18′N 121°54′W
187 N15 **Surprise, Île** *island* N New Caledonia
61 E22 **Sur, Punta** *headland* E Argentina 50°59′S 69°10′W
Surrentum *see* Sorrento
28 M3 **Surrey** North Dakota, N USA 48°13′N 101°05′W
97 O22 **Surrey** *cultural region* SE England, United Kingdom
21 X7 **Surry** Virginia, NE USA 37°08′N 81°34′W
108 F8 **Sursee** Luzern, W Switzerland 47°11′N 08°07′E
127 P6 **Sursk** Penzenskaya Oblast', W Russian Federation 53°06′N 45°46′E
127 P5 **Surskoye** Ul'yanovskaya Oblast', W Russian Federation 54°28′N 46°47′E
75 P8 **Surt** *var.* Sidra, Sirte. N Libya 31°13′N 16°35′E
95 I19 **Surte** Västra Götaland, S Sweden 57°49′N 12°01′E
75 Q8 **Surt, Khalīj** *Eng.* Gulf of Sidra, Gulf of Sirti, Sidra. *gulf* N Libya
92 I5 **Surtsey** *island* S Iceland
137 N17 **Suruç** Şanlıurfa, S Turkey 36°58′N 38°24′E
168 L13 **Surulangun** Sumatera, W Indonesia 02°35′S 102°47′E
147 P13 **Surxondaryo** *Rus.* Surkhandar'ya. ≈ Tajikistan/Uzbekistan
147 N13 **Surxondaryo Viloyati** *Rus.* Surkhandar'inskaya Oblast'. ◈ *province* S Uzbekistan
Süs *see* Susch
106 A8 **Susa** Piemonte, NE Italy 45°10′N 07°01′E
165 E12 **Susa** Yamaguchi, Honshū, SW Japan 34°35′N 131°34′E
Susa *see* Shūsh
113 E16 **Sušac** *It.* Cazza. *island* SW Croatia
Süsah *see* Sousse
164 G14 **Susaki** Kōchi, Shikoku, SW Japan 33°22′N 133°13′E
165 I15 **Susami** Wakayama, Honshū, SW Japan 33°32′N 135°32′E
142 K9 **Süsangerd** *var.* Susangird. Khūzestān, SW Iran 31°40′N 48°06′E
Susangird *see* Süsangerd
35 P4 **Susanville** California, W USA 40°25′N 120°39′W
108 J9 **Susch** *var.* Süs. Graubünden, SE Switzerland 46°45′N 10°04′E
137 N12 **Suşehri** Sivas, N Turkey 40°11′N 38°06′E
Susiana *see* Khūzestān
111 B18 **Sušice** *Ger.* Schüttenhofen. Plzeňský Kraj, W Czech Republic 49°14′N 13°32′E
39 R11 **Susitna** Alaska, USA 61°32′N 150°30′W
39 R11 **Susitna River** ≈ Alaska, USA
127 Q3 **Susloner** Respublika Mariy El, W Russian Federation 56°18′N 48°16′E
105 N14 **Suspiro del Moro, Puerto del** *pass* S Spain
18 H16 **Susquehanna River** ≈ New York/Pennsylvania, NE USA
13 O15 **Sussex** New Brunswick, SE Canada 45°43′N 65°32′W
18 J13 **Sussex** New Jersey, NE USA 41°12′N 74°34′W
21 W7 **Sussex** Virginia, NE USA 36°54′N 77°16′W
97 O23 **Sussex** *cultural region* S England, United Kingdom
183 S10 **Sussex Inlet** New South Wales, SE Australia 35°10′S 150°35′E
99 L17 **Susteren** Limburg, SE Netherlands 51°04′N 05°50′E
10 K12 **Sustut Peak** ▲ British Columbia, W Canada 56°25′N 126°34′W
123 S9 **Susuman** Magadanskaya Oblast', E Russian Federation 62°46′N 148°08′E
188 H6 **Susupe** ● (Northern Mariana Islands-judicial capital) Saipan, S Northern Mariana Islands
136 D12 **Susurluk** Balıkesir, NW Turkey 39°55′N 28°10′E
114 M13 **Susuzmüsellim** Tekirdağ, NW Turkey 41°04′N 27°03′E
136 F15 **Sütçüler** Isparta, SW Turkey 37°31′N 31°00′E
116 L13 **Şuţeşti** Brăila, SE Romania 45°13′N 27°27′E
83 F25 **Sutherland** Western Cape, SW South Africa 32°24′S 20°40′E
28 L15 **Sutherland** Nebraska, C USA 41°09′N 101°07′W
96 I7 **Sutherland** *cultural region* N Scotland, United Kingdom
185 B21 **Sutherland Falls** *waterfall* South Island, New Zealand
32 F14 **Sutherlin** Oregon, NW USA 43°23′N 123°18′W
149 V10 **Sutlej** ≈ India/Pakistan
Sutna *see* Satna
35 P7 **Sutter Creek** California, W USA 38°22′N 120°49′W
39 R11 **Sutton** Alaska, USA 61°42′N 148°53′W
29 Q16 **Sutton** Nebraska, C USA 40°36′N 97°52′W
21 R4 **Sutton** West Virginia, NE USA 38°41′N 80°43′W
12 F8 **Sutton** ≈ Ontario, C Canada
97 M19 **Sutton Coldfield** C England, United Kingdom 52°34′N 01°48′W
21 R4 **Sutton Lake** ▣ West Virginia, NE USA
15 P13 **Sutton, Monts** *hill range* Québec, SE Canada
12 F8 **Sutton Ridges** ▲▲ Ontario, C Canada
165 Q4 **Suttsu** Hokkaidō, NE Japan 42°46′N 140°12′E
39 P15 **Sutwik Island** *island* Alaska, USA
Süüj *see* Dashinchilen
118 H5 **Suure-Jaani** *Ger.* Gross-Sankt-Johannis. Viljandimaa, S Estonia 58°34′N 25°28′E
118 J7 **Suur Munamägi** *var.* Munamägi, *Ger.* Eier-Berg. ▲ SE Estonia 57°42′N 27°03′E
118 F5 **Suur Väin** *Ger.* Grosser Sund. *strait* W Estonia
147 U8 **Suusamyr** Chuyskaya Oblast', C Kyrgyzstan 42°07′N 73°55′E
187 X14 **Suva** ● (Fiji) Viti Levu, W Fiji 18°08′S 178°27′E
187 X15 **Suva** ✈ Viti Levu, C Fiji 18°01′S 178°30′E
113 N18 **Suva Gora** ▲▲ W FYR Macedonia
118 H11 **Suvainiškis** Panevėžys, NE Lithuania 56°09′N 25°15′E
Suvalkai/Suvalki *see* Suwałki
113 P15 **Suva Planina** ▲▲ SE Serbia
Suva Reka *see* Suharekë
126 K5 **Suvorov** Tul'skaya Oblast', W Russian Federation 54°08′N 36°33′E
117 N12 **Suvorove** Odes'ka Oblast', SW Ukraine 45°35′N 28°58′E
114 M8 **Suvorovo** Varna, E Bulgaria 43°19′N 27°26′E
Suvorovo *see* Ştefan Vodă
Suwaik *see* As Suwayq
Suwaira *see* Aş Şuwayrah
110 O7 **Suwałki** *Lith.* Suvalkai, *Rus.* Suvalki. Podlaskie, NE Poland 54°06′N 22°56′E
167 R10 **Suwannaphum** Roi Et, E Thailand 15°36′N 103°46′E
23 V8 **Suwannee River** ≈ Florida/Georgia, SE USA
Şuwār *see* Aş Şuwār
190 K14 **Suwarrow** *atoll* N Cook Islands
143 R16 **Suwaydān** *var.* Sweiham. Abū Ẓaby, E United Arab Emirates 24°30′N 55°19′E
Suwaydá/Suwaydā' *see* As Suwaydā'
Suwayqiyah, Hawr as *see* Shuwayjah, Hawr ash
Suways, Qanāt as *see* Suez Canal
Suweida *see* As Suwaydā'
Suweon *see* Suwon
163 X15 **Suwon** *var.* Suweon; *prev.* Suwŏn, *Jap.* Suigen. NW South Korea 37°17′N 127°03′E
Suwŏn *see* Suwon
Su Xian *see* Suzhou
143 R14 **Sūzā** Hormozgān, S Iran 26°50′N 56°05′E
Suzak *see* Sozak
165 K14 **Suzaka** Mie, Honshū, SW Japan 34°52′N 136°37′E
165 N12 **Suzaka** Nagano, Honshū, S Japan 36°38′N 138°20′E
126 M3 **Suzdal'** Vladimirskaya Oblast', W Russian Federation 56°27′N 40°29′E
161 P7 **Suzhou** *var.* Su Xian. Anhui, E China 33°38′N 117°02′E
161 R8 **Suzhou** *var.* Soochow, Su-chou, Suchow; *prev.* Wuhsien. Jiangsu, E China 31°23′N 120°34′E
Suzhou *see* Jiuquan
163 V12 **Suzi He** ≈ NE China
Suz, Mys *see* Soye, Mys
165 M10 **Suzu** Ishikawa, Honshū, SW Japan 37°24′N 137°12′E
165 M10 **Suzu-misaki** *headland* Honshū, SW Japan 37°31′N 137°19′E
Svågälv *see* Svågan
94 M10 **Svågan** *var.* Svågälv. ≈ C Sweden
Svalava/Svaljava *see* Svalyava
92 O2 **Svalbard** ◇ *constituent part of Norway* Arctic Ocean
92 J2 **Svalbarðseyri** Norðurland Eystra, N Iceland 65°43′N 18°03′W
95 K22 **Svalöv** Skåne, S Sweden 55°55′N 13°06′E
116 H7 **Svalyava** *Cz.* Svalava, Svaljava, *Hung.* Szolyva. Zakarpats'ka Oblast', W Ukraine 48°33′N 23°00′E
92 O2 **Svanbergfjellet** ▲ C Svalbard 78°40′N 18°10′E
95 M24 **Svaneke** Bornholm, E Denmark 55°07′N 15°08′E
95 L22 **Svängsta** Blekinge, S Sweden 56°16′N 14°46′E
95 J16 **Svanskog** Värmland, C Sweden 59°10′N 12°34′E
95 L16 **Svartå** Örebro, C Sweden 59°13′N 14°07′E
95 L15 **Svartälven** ≈ C Sweden
92 G12 **Svartisen** *glacier* C Norway
117 X6 **Svatove** *Rus.* Svatovo. Luhans'ka Oblast', E Ukraine 49°24′N 38°11′E
Svatovo *see* Svatove
Svatý Kříž nad Hronom *see* Žiar nad Hronom
167 Q11 **Svay Chék, Stœng** ≈ Cambodia/Thailand
167 S13 **Svay Riĕng** Svay Riĕng, S Cambodia 11°05′N 105°48′E
92 O3 **Sveagruva** Spitsbergen, W Svalbard 77°53′N 16°42′E
95 K23 **Svedala** Skåne, S Sweden 55°30′N 13°15′E
118 H12 **Svėdesai** Utena, NE Lithuania 55°42′N 25°22′E
93 G18 **Sveg** Jämtland, C Sweden 62°02′N 14°20′E
118 C12 **Švėkšna** Klaipėda, W Lithuania 55°31′N 21°37′E
94 C11 **Svelgen** Sogn Og Fjordane, S Norway 61°47′N 05°18′E
95 H15 **Svelvik** Vestfold, S Norway 59°37′N 10°24′E
118 I13 **Švenčionėliai** *Pol.* Nowo-Święciany. Vilnius, SE Lithuania 55°10′N 26°00′E
118 I13 **Švenčionys** *Pol.* Święciany. Vilnius, SE Lithuania 55°08′N 26°08′E
95 H24 **Svendborg** Syddtjylland, C Denmark 55°04′N 10°38′E
95 K19 **Svenljunga** Västra Götaland, S Sweden 57°30′N 13°05′E
92 P2 **Svenskøya** *island* E Svalbard
93 G17 **Svenstavik** Jämtland, C Sweden 62°40′N 14°24′E
95 G20 **Svenstrup** Nordjylland, N Denmark 56°58′N 09°52′E
118 H12 **Šventoji** ≈ C Lithuania
117 Z8 **Sverdlovs'k** *Rus.* Sverdlovsk; *prev.* Imeni Sverdlova Rudnik. Luhans'ka Oblast', E Ukraine 48°05′N 39°37′E
Sverdlovsk *see* Yekaterinburg
127 W2 **Sverdlovskaya Oblast'** ◈ *province* C Russian Federation
122 K6 **Sverdrupa, Ostrov** *island* N Russian Federation
Sverige *see* Sweden
113 D15 **Svetac** *prev.* Sveti Andrea, *It.* Sant'Andrea. *island* SW Croatia
Sveti Andrea *see* Svetac
Sveti Nikola *see* Sveti Nikole
113 O18 **Sveti Nikole** *prev.* Sveti Nikola. C FYR Macedonia 41°54′N 21°55′E
Sveti Vrach *see* Sandanski
123 T14 **Svetlaya** Primorskiy Kray, SE Russian Federation 46°33′N 138°20′E
126 B2 **Svetlogorsk** Kaliningradskaya Oblast', W Russian Federation 54°56′N 20°09′E
122 K9 **Svetlogorsk** Krasnoyarskiy Kray, N Russian Federation 66°51′N 88°29′E
Svetlogorsk *see* Svyetlahorsk
127 N14 **Svetlograd** Stavropol'skiy Kray, SW Russian Federation 45°20′N 42°53′E
Svetlovodsk *see* Svitlovods'k
119 A14 **Svetlyy** *Ger.* Zimmerbude. Kaliningradskaya Oblast', W Russian Federation 54°42′N 20°07′E
127 Y8 **Svetlyy** Orenburgskaya Oblast', W Russian Federation 50°34′N 60°42′E
127 P7 **Svetlyy** Saratovskaya Oblast', W Russian Federation 51°42′N 45°40′E
124 G11 **Svetogorsk** *Fin.* Enso. Leningradskaya Oblast', NW Russian Federation 61°06′N 28°52′E
Svetozarevo *see* Jagodina
111 B18 **Švihov** *Ger.* Schwihau. Plzeňský Kraj, W Czech Republic 49°31′N 13°18′E
112 E13 **Svilaja** ▲▲ SE Croatia
112 N12 **Svilajnac** Serbia, C Serbia 44°15′N 21°12′E
114 L11 **Svilengrad** *prev.* Mustafa-Pasha. Haskovo, S Bulgaria 41°45′N 26°14′E
Svinecea Mare, Munte *see* Svinecea Mare, Vârful
116 F13 **Svinecea Mare, Vârful** *var.* Munte Svinecea Mare. ▲ SW Romania 44°47′N 22°10′E
Svino *see* Svínoy
95 B18 **Svínoy** *Dan.* Svinø. *island* NE Faroe Islands
147 N14 **Svintsovyy Rudnik** *Turkm.* Swintsowyý Rudnik. Lebap Welaýaty, E Turkmenistan 37°54′N 66°25′E
118 I13 **Svir** *Rus.* Svir'. Minskaya Voblasts', NW Belarus 54°51′N 26°24′E
124 I12 **Svir'** *canal* NW Russian Federation
Svir', Ozero *see* Svir, Vozyera
119 I14 **Svir, Vozyera** *Rus.* Ozero Svir'. ◎ C Belarus
114 J7 **Svishtov** *prev.* Sistova. Veliko Tarnovo, N Bulgaria 43°37′N 25°20′E
119 F18 **Svislach** *Pol.* Świsłocz, *Rus.* Svisloch'. Hrodzyenskaya Voblasts', W Belarus 53°02′N 24°06′E
119 M17 **Svislach** *Rus.* Svisloch'. Mahilyowskaya Voblasts', E Belarus 53°26′N 28°59′E
119 L17 **Svislach** *Rus.* Svisloch'. ≈ E Belarus
Svisloch' *see* Svislach
111 F17 **Svitavy** *Ger.* Zwittau. Pardubický Kraj, C Czech Republic 49°45′N 16°27′E
117 S6 **Svitlovods'k** *Rus.* Svetlovodsk. Kirovohrads'ka Oblast', C Ukraine 49°05′N 33°15′E
Svizzera *see* Switzerland
123 Q13 **Svobodnyy** Amurskaya Oblast', SE Russian Federation 51°24′N 128°05′E
114 G9 **Svoge** Sofia, W Bulgaria 42°58′N 23°20′E
92 G11 **Svolvær** Nordland, C Norway 68°15′N 14°40′E
111 F18 **Svratka** *Ger.* Schwarzawa. ≈ SE Czech Republic
113 P14 **Svrljig** Serbia, E Serbia 43°26′N 22°07′E
197 U10 **Svyataya Anna Trough** *var.* Saint Anna Trough. *undersea feature* N Kara Sea
124 M4 **Svyatoy Nos, Mys** *headland* NW Russian Federation 68°07′N 39°49′E
119 N18 **Svyetlahorsk** *Rus.* Svetlogorsk. Homyel'skaya Voblasts', SE Belarus 52°38′N 29°46′E
Swabian Jura *see* Schwäbische Alb
97 P19 **Swaffham** E England, United Kingdom 52°39′N 00°40′E
23 V5 **Swainsboro** Georgia, SE USA 32°36′N 82°19′W
83 C19 **Swakop** ≈ W Namibia
83 C19 **Swakopmund** Erongo, W Namibia 22°40′S 14°34′E
97 M15 **Swale** ≈ N England, United Kingdom
Swallow Island *see* Nendö
99 M16 **Swalmen** Limburg, SE Netherlands 51°13′N 06°02′E
12 G8 **Swan** ≈ Ontario, C Canada
97 L24 **Swanage** S England, United Kingdom 50°37′N 01°59′W
182 M10 **Swan Hill** Victoria, SE Australia 35°23′S 143°37′E
11 P13 **Swan Hills** Alberta, W Canada 54°41′N 116°20′W
65 D24 **Swan Island** *island* C Falkland Islands
Swankalok *see* Sawankhalok
29 U10 **Swan Lake** ◎ Minnesota, N USA
21 Y10 **Swanquarter** North Carolina, SE USA 35°24′N 76°20′W
182 J9 **Swan Reach** South Australia 34°39′S 139°35′E
11 V15 **Swan River** Manitoba, S Canada 52°06′N 101°17′W
183 P17 **Swansea** Tasmania, SE Australia 42°09′S 148°03′E
97 J22 **Swansea** *Wel.* Abertawe. S Wales, United Kingdom 51°38′N 03°57′W
21 R13 **Swansea** South Carolina, SE USA 33°47′N 81°06′W
19 S7 **Swans Island** *island* Maine, NE USA
28 L17 **Swanson Lake** ▣ Nebraska, C USA
31 R11 **Swanton** Ohio, N USA 41°35′N 83°53′W
110 G11 **Swarzędz** Poznań, W Poland 52°24′N 17°05′E
Swatow *see* Shantou
83 L22 **Swaziland** *off.* Kingdom of Swaziland. ◆ *monarchy* S Africa
Swaziland, Kingdom of *see* Swaziland
93 G18 **Sweden** *off.* Kingdom of Sweden, *Swe.* Sverige. ◆ *monarchy* N Europe
Sweden, Kingdom of *see* Sweden
Swedru *see* Agona Swedru
25 V12 **Sweeny** Texas, SW USA 29°02′N 95°42′W
33 R6 **Sweetgrass** Montana, NW USA 48°58′N 111°58′W
32 G12 **Sweet Home** Oregon, NW USA 44°24′N 122°44′W
25 T12 **Sweet Home** Texas, SW USA 29°21′N 97°04′W
27 T4 **Sweet Springs** Missouri, C USA 38°57′N 93°24′W
20 M10 **Sweetwater** Tennessee, S USA 35°36′N 84°27′W
25 P7 **Sweetwater** Texas, SW USA 32°27′N 100°25′W
33 V15 **Sweetwater River** ≈ Wyoming, C USA
Sweiham *see* Suwaydān
83 F26 **Swellendam** Western Cape, SW South Africa 34°01′S 20°26′E
111 G15 **Świdnica** *Ger.* Schweidnitz. Wałbrzych, SW Poland 50°51′N 16°29′E
111 O14 **Świdnik** *Ger.* Streckenbach. Lubelskie, E Poland 51°14′N 22°41′E
110 F8 **Świdwin** *Ger.* Schivelbein. Zachodnio-pomorskie, NW Poland 53°47′N 15°44′E
111 F15 **Świebodzice** *Ger.* Freiburg in Schlesien, Swiebodzice. Wałbrzych, SW Poland 50°54′N 16°23′E
110 E11 **Świebodzin** *Ger.* Schwiebus. Lubuskie, W Poland 52°15′N 15°31′E
Święciany *see* Švenčionys
110 I9 **Świecie** *Ger.* Schwertberg. Kujawsko-pomorskie, C Poland 53°24′N 18°24′E
111 L15 **Świętokrzyskie** ◈ *province* S Poland
11 T16 **Swift Current** Saskatchewan, S Canada 50°17′N 107°49′W
98 K9 **Swifterbant** Flevoland, C Netherlands 52°36′N 05°33′E
183 Q12 **Swifts Creek** Victoria, SE Australia 37°17′S 147°41′E
96 E13 **Swilly, Lough** *Ir.* Loch Súilí. *inlet* N Ireland
97 M22 **Swindon** S England, United Kingdom 51°34′N 01°47′W
Swinemünde *see* Świnoujście
110 D8 **Świnoujście** *Ger.* Swinemünde. Zachodnio-pomorskie, NW Poland 53°54′N 14°13′E
Swintsowyý Rudnik *see* Svintsovyy Rudnik
Świsłocz *see* Svislach
Swiss Confederation *see* Switzerland
108 E9 **Switzerland** *off.* Swiss Confederation, *Fr.* La Suisse, *Ger.* Schweiz, *It.* Svizzera; *anc.* Helvetia. ◆ *federal republic* C Europe
97 F17 **Swords** *Ir.* Sord, Sórd Choluim Chille. Dublin, E Ireland 53°28′N 06°13′W
18 H13 **Swoyersville** Pennsylvania, NE USA 41°18′N 75°48′W
139 V3 **Syagwêz** *Ar.* Siyāh Gūz. As Sulaymānīyah, E Iraq 35°49′N 45°45′E
124 I10 **Syamozero, Ozero** ◎ NW Russian Federation
124 M13 **Syamzha** Vologodskaya Oblast', NW Russian Federation 60°02′N 41°09′E
118 N13 **Syanno** *Rus.* Senno. Vitsyebskaya Voblasts', NE Belarus 54°49′N 29°43′E
119 K16 **Syarhyeyevichy** *Rus.* Sergeyevichi. Minskaya Voblasts', C Belarus 53°30′N 27°45′E
124 I12 **Syas'stroy** Leningradskaya Oblast', NW Russian Federation 60°05′N 32°37′E
Sycaminum *see* Hefa
30 M10 **Sycamore** Illinois, N USA 41°59′N 88°41′W
126 J3 **Sychëvka** Smolenskaya Oblast', W Russian Federation 55°52′N 34°19′E
111 H14 **Syców** *Ger.* Gross Wartenberg. Dolnośląskie, SW Poland 51°18′N 17°42′E
95 F24 **Syddanmark** ◈ *county* SW Denmark
14 E17 **Sydenham** ≈ Ontario, S Canada
Sydenham Island *see* Nonouti
183 T9 **Sydney** *state capital* New South Wales, SE Australia 33°55′S 151°10′E
13 R14 **Sydney** Cape Breton Island, Nova Scotia, SE Canada 46°10′N 60°10′W
Sydney Island *see* Manra
13 R14 **Sydney Mines** Cape Breton Island, Nova Scotia, SE Canada 46°14′N 60°19′W
Syedpur *see* Saidpur
119 K18 **Syelishcha** *Rus.* Selishche. Minskaya Voblasts', C Belarus 53°01′N 27°25′E
119 J18 **Syemyezhava** *Rus.* Semezhevo. Minskaya Voblasts', C Belarus 52°58′N 27°00′E
Syene *see* Aswân
117 X6 **Syeverodonets'k** *Rus.* Severodonetsk. Luhans'ka Oblast', E Ukraine 48°59′N 38°28′E
161 T6 **Syjiao Shan** *island* SE China
100 H11 **Syke** Niedersachsen, NW Germany 52°55′N 08°49′E
94 D10 **Sykkylven** Møre og Romsdal, S Norway 62°23′N 06°35′E
Sykoúri *see* Sykoúrio
115 F15 **Sykoúrio** *var.* Sikouri, Sykoúri; *prev.* Sikoúrion. Thessalía, C Greece 39°46′N 22°35′E
125 R11 **Syktyvkar** *prev.* Ust'-Sysol'sk. Respublika Komi, NW Russian Federation 61°42′N 50°45′E
23 Q4 **Sylacauga** Alabama, S USA 33°10′N 86°15′W
Sylarna *see* Storsylen
153 V14 **Sylhet** Sylhet, NE Bangladesh 24°53′N 91°51′E
153 V13 **Sylhet** ◈ *division* NE Bangladesh
100 G6 **Sylt** *island* NW Germany
21 O10 **Sylva** North Carolina, SE USA 35°23′N 83°13′W
125 V15 **Sylva** ≈ NW Russian Federation
23 W5 **Sylvania** Georgia, SE USA 32°45′N 81°38′W
31 R11 **Sylvania** Ohio, N USA 41°43′N 83°42′W
11 Q15 **Sylvan Lake** Alberta, SW Canada 52°18′N 114°02′W
33 T13 **Sylvan Pass** *pass* Wyoming, C USA
23 T7 **Sylvester** Georgia, SE USA 31°31′N 83°50′W
25 P6 **Sylvester** Texas, SW USA 32°42′N 100°15′W
10 L11 **Sylvia, Mount** ▲ British Columbia, W Canada 58°03′N 124°26′W
122 K11 **Sym** ≈ C Russian Federation
115 N22 **Sými** *var.* Simi. *island* Dodekánisa, Greece, Aegean Sea
117 U8 **Synel'nykove** Dnipropetrovs'ka Oblast', E Ukraine 48°19′N 35°32′E
125 U6 **Synya** Respublika Komi, NW Russian Federation 65°21′N 58°01′E
117 P7 **Synyukha** *Rus.* Sinyukha. ≈ S Ukraine
195 V2 **Syowa** *Japanese research station* Antarctica 68°58′S 40°07′E
26 H6 **Syracuse** Kansas, C USA 38°00′N 101°43′W
29 S16 **Syracuse** Nebraska, C USA 40°39′N 96°11′W
18 H10 **Syracuse** New York, NE USA 43°03′N 76°09′W
Syracuse *see* Siracusa
Syrdar'inskaya Oblast' *see* Sirdaryo Viloyati
Syrdariya *see* Syr Darya
144 L14 **Syr Darya** *var.* Sai Hun, Sir Darya, Syrdarya, *Kaz.* Syrdariya, *Rus.* Syrdar'ya, *Uzb.* Sirdaryo; *anc.* Jaxartes. ≈ C Asia
Syrdarya *see* Syr Darya
138 J6 **Syria** *off.* Syrian Arab Republic, *var.* Siria, Syrie, *Ar.* Al-Jumhūrīyah al-'Arabīyah as-Sūrīyah, Sūrīya. ◆ *republic* SW Asia
Syrian Arab Republic *see* Syria
138 L9 **Syrian Desert** *Ar.* Al Hamad, Bādiyat ash Shām. *desert* SW Asia
Syrie *see* Syria
115 L22 **Sýrna** *var.* Sirna. *island* Kykládes, Greece, Aegean Sea
115 I20 **Sýros** *var.* Síros. *island* Kykládes, Greece, Aegean Sea
93 M18 **Sysmä** Päijät-Häme, S Finland 61°28′N 25°37′E
125 R12 **Sysola** ≈ NW Russian Federation
Syulemeshlii *see* Sredets
127 S2 **Syumsi** Udmurtskaya Respublika, NW Russian Federation 57°07′N 51°35′E
Syvash, Zaliv *see* Syvash, Zatoka
117 U12 **Syvash, Zatoka** *Rus.* Zaliv Syvash. *inlet* S Ukraine
127 Q6 **Syzran'** Samarskaya Oblast', W Russian Federation 53°10′N 48°23′E
Szabadka *see* Subotica
111 N21 **Szabolcs-Szatmár-Bereg** *off.* Szabolcs-Szatmár-Bereg Megye. ◈ *county* E Hungary
Szabolcs-Szatmár-Bereg Megye *see* Szabolcs-Szatmár-Bereg
110 G10 **Szamocin** *Ger.* Samotschin. Wielkopolskie, C Poland 53°02′N 17°04′E
116 H8 **Szamos** *var.* Someş, Somesul, *Ger.* Samosch, Somesch. ≈ Hungary/Romania
Szamosújvár *see* Gherla
110 G11 **Szamotuły** Poznań, W Poland 52°35′N 16°36′E
Szarkowszczyzna *see* Sharkawshchyna
111 M24 **Szarvas** Békés, SE Hungary 46°51′N 20°35′E
Szászmagyarós *see* Măieruş
Szászrégen *see* Reghin
Szászsebes *see* Sebeş
Szászváros *see* Orăştie
Szatmárnémeti *see* Satu Mare
Száva *see* Sava
111 P15 **Szczebrzeszyn** Lubelskie, E Poland 50°43′N 23°00′E
110 D9 **Szczecin** *Eng./Ger.* Stettin. Zachodnio-pomorskie, NW Poland 53°25′N 14°32′E
110 G8 **Szczecinek** *Ger.* Neustettin. Zachodnio-pomorskie, NW Poland 53°43′N 16°40′E
110 D8 **Szczeciński, Zalew** *var.* Stettiner Haff, *Ger.* Oderhaff. *bay* Germany/Poland
111 K15 **Szczekociny** Śląskie, S Poland 50°38′N 19°46′E
110 N8 **Szczuczyn** Podlaskie, NE Poland 53°34′N 22°17′E
Szczuczyn Nowogródzki *see* Shchuchyn
110 M8 **Szczytno** *Ger.* Ortelsburg. Warmińsko-Mazurskie, NE Poland 53°34′N 21°E
Szechuan/Szechwan *see* Sichuan
111 K21 **Szécsény** Nógrád, N Hungary 48°07′N 19°30′E
111 L25 **Szeged** *Ger.* Szegedin, *Rom.* Seghedin. Csongrád, SE Hungary 46°17′N 20°06′E
Szegedin *see* Szeged
111 N23 **Szeghalom** Békés, SE Hungary 47°01′N 21°09′E
Székelyhid *see* Săcueni
Székelykeresztúr *see* Cristuru Secuiesc
111 I23 **Székesfehérvár** *Ger.* Stuhlweissenberg; *anc.* Alba Regia. Fejér, W Hungary 47°13′N 18°24′E
Szeklerburg *see* Miercurea-Ciuc
Szekler Neumarkt *see* Târgu Secuiesc
111 J25 **Szekszárd** Tolna, S Hungary 46°21′N 18°41′E
Szempcz/Szenc *see* Senec
Szenice *see* Senica
Szentágota *see* Agnita
111 J22 **Szentendre** *Ger.* Sankt Andrä. Pest, N Hungary 47°40′N 19°02′E
111 L24 **Szentes** Csongrád, SE Hungary 46°40′N 20°17′E
111 F23 **Szentgotthárd** *Eng.* Saint Gotthard, *Ger.* Sankt Gotthard. Vas, W Hungary 46°57′N 16°18′E
Szentgyörgy *see* Đurđevac
Szenttamás *see* Srbobran
Széphely *see* Jebel
Szeping *see* Siping
Szered *see* Sereď
111 N21 **Szerencs** Borsod-Abaúj-Zemplén, NE Hungary 48°10′N 21°11′E
Szeret *see* Siret
Szeretfalva *see* Sărăţel
110 N7 **Szeska Góra** *var.* Szeskie Wygórza, *Ger.* Seesker Höhe. *hill* NE Poland
Szeskie Wygórza *see* Szeska Góra
111 H25 **Szigetvár** Baranya, SW Hungary 46°01′N 17°50′E
Szilágysomlyó *see* Şimleu Silvaniei
Szinna *see* Snina
Sziszek *see* Sisak
Szitás-Keresztúr *see* Cristuru Secuiesc
111 E15 **Szklarska Poręba** *Ger.* Schreiberhau. Dolnośląskie, SW Poland 50°50′N 15°30′E
Szkudy *see* Skuodas
Szlatina *see* Slatina
Szlavonia/Szlavonország *see* Slavonija
Szlovákia *see* Slovakia
Szluin *see* Slunj
111 L23 **Szolnok** Jász-Nagykun-Szolnok, C Hungary 47°11′N 20°12′E
Szolyva *see* Svalyava
111 G23 **Szombathely** *Ger.* Steinamanger; *anc.* Sabaria, Savaria. Vas, W Hungary 47°14′N 16°38′E
Szond/Szonta *see* Sonta
Szováta *see* Sovata
110 F13 **Szprotawa** *Ger.* Sprottau. Lubuskie, W Poland 51°33′N 15°32′E
Sztálinváros *see* Dunaújváros
Sztrazsó *see* Strážov
110 J8 **Sztum** *Ger.* Stuhm. Pomorskie, N Poland 53°54′N 19°01′E
110 H10 **Szubin** *Ger.* Schubin. Kujawsko-pomorskie, C Poland 53°04′N 17°49′E
Szucsava *see* Suceava
Szurduk *see* Surduc
111 M14 **Szydłowiec** *Ger.* Schlelau. Mazowieckie, C Poland 51°14′N 20°50′E

T

Taalintehdas *see* Dalsbruk
171 O4 **Taal, Lake** ◎ Luzon, NW Philippines
95 J23 **Taastrup** *var.* Tåstrup. Sjælland, E Denmark 55°39′N 12°19′E
111 I24 **Tab** Somogy, W Hungary 46°45′N 18°01′E
171 P4 **Tabaco** Luzon, N Philippines 13°22′N 123°42′E
186 G4 **Tabalo** Mussau Island, NE Papua New Guinea 01°22′S 149°37′E
104 K5 **Tábara** Castilla y León, N Spain 41°49′N 05°57′W
186 H5 **Tabar Islands** *island group* NE Papua New Guinea
Tabariya, Bahrat *see* Tiberias, Lake
143 S7 **Ţabas** *var.* Golshan. Yazd, C Iran 33°37′N 56°54′E
43 P15 **Tabasará, Serranía de** ▲▲ W Panama
41 U15 **Tabasco** ◈ *state* SE Mexico
Tabasco *see* Grijalva, Río
127 Q2 **Tabashino** Respublika Mariy El, W Russian Federation 57°00′N 47°47′E
58 B13 **Tabatinga** Amazonas, N Brazil 04°14′S 69°44′W
74 G9 **Tabelbala** W Algeria 29°22′N 03°01′W
11 Q17 **Taber** Alberta, SW Canada 49°48′N 112°09′W
171 V15 **Taberfane** Pulau Trangan, E Indonesia 06°14′S 134°08′E
95 L19 **Taberg** Jönköping, S Sweden 57°42′N 14°05′E
191 O3 **Tabiteuea** *prev.* Drummond Island. *atoll* Tungaru, W Kiribati
171 O5 **Tablas Island** *island* C Philippines
184 Q10 **Table Cape** *headland* North Island, New Zealand 39°07′S 178°00′E
13 S13 **Table Mountain** ▲ Newfoundland, Newfoundland and Labrador, E Canada 47°39′N 59°15′W
173 P17 **Table, Pointe de la** *headland* SE Réunion 21°19′S 55°49′E
27 S8 **Table Rock Lake** ▣ Arkansas/Missouri, C USA
36 K14 **Table Top** ▲ Arizona, SW USA 32°45′N 112°07′W
186 D8 **Tabletop, Mount** ▲ C Papua New Guinea 06°51′S 146°00′E
111 D18 **Tábor** Jihočeský Kraj, S Czech Republic 49°25′N 14°41′E
123 R7 **Tabor** Respublika Sakha (Yakutiya), NE Russian Federation 71°14′N 150°23′E
29 S15 **Tabor** Iowa, C USA 40°54′N 95°40′W
81 F21 **Tabora** Tabora, W Tanzania 05°04′S 32°49′E
81 F21 **Tabora** ◈ *region* C Tanzania
21 U12 **Tabor City** North Carolina, SE USA 34°09′N 78°52′W
147 Q10 **Taboshar** NW Tajikistan 40°37′N 69°33′E
76 L18 **Tabou** *var.* Tabu. S Ivory Coast 04°28′N 07°20′W
142 J2 **Tabrīz** *var.* Tebriz; *anc.* Tauris. Āžarbāyjān-e Sharqī, NW Iran 38°05′N 46°18′E
Tabu *see* Tabou
191 W1 **Tabuaeran** *prev.* Fanning Island. *atoll* Line Islands, E Kiribati
171 O2 **Tabuk** Luzon, N Philippines 17°26′N 121°25′E
140 J4 **Tabūk** Tabūk, NW Saudi Arabia 28°25′N 36°34′E
140 J5 **Tabūk** *off.* Mintaqat Tabūk. ◈ *province* NW Saudi Arabia
Tabūk, Mintaqat *see* Tabūk
187 Q13 **Tabwemasana, Mount** ▲ Espiritu Santo, W Vanuatu 15°22′S 166°44′E
95 O15 **Täby** Stockholm, C Sweden 59°29′N 18°04′E
41 N14 **Tacámbaro** Michoacán, SW Mexico 19°12′N 101°27′W
42 A5 **Tacaná, Volcán** ◬ Guatemala/Mexico 15°07′N 92°06′W
43 X16 **Tacarcuna, Cerro** ▲ SE Panama 08°08′N 77°15′W
Tachau *see* Tachov
158 J3 **Tacheng** *var.* Qoqek. Xinjiang Uygur Zizhiqu, NW China 46°45′N 82°55′E
54 H7 **Táchira** *off.* Estado Táchira. ◈ *state* W Venezuela
Táchira, Estado *see* Táchira
Tachoshui *see* Dazhoushui
111 A17 **Tachov** *Ger.* Tachau. Plzeňský Kraj, W Czech Republic 49°48′N 12°38′E
171 Q5 **Tacloban** *off.* Tacloban City. Leyte, C Philippines 11°15′N 125°E
Tacloban City *see* Tacloban
57 I19 **Tacna** Tacna, SE Peru 18°S 70°15′W
57 H18 **Tacna** *off.* Departamento de Tacna. ◈ *department* S Peru
Tacna, Departamento de *see* Tacna
32 H8 **Tacoma** Washington, NW USA 47°15′N 122°27′W
18 L11 **Taconic Range** ▲▲ NE USA
62 L6 **Taco Pozo** Formosa, N Argentina 25°35′S 63°15′W
57 M20 **Tacsara, Cordillera de** ▲▲ S Bolivia
61 F17 **Tacuarembó** *prev.* San Fructuoso. Tacuarembó, C Uruguay 31°42′S 56°W
61 E18 **Tacuarembó** ◈ *department* C Uruguay
61 F17 **Tacuarembó, Río** ≈ C Uruguay
83 I14 **Taculi** North Western, NW Zambia 14°17′S 26°51′E
171 Q8 **Tacurong** Mindanao, S Philippines 06°42′N 124°40′E
77 V8 **Tadek** ≈ NW Niger
74 J9 **Tademaït, Plateau du** *plateau* C Algeria
187 R17 **Tadine** Province des Îles Loyauté, E New Caledonia 21°33′S 167°54′E
80 M11 **Tadjoura, Golfe de** *Eng.* Gulf of Tajura. *inlet* E Djibouti
80 L11 **Tadjourah** E Djibouti 11°47′N 42°51′E
Tadmor/Tadmur *see* Tudmur
11 W10 **Tadoule Lake** ◎ Manitoba, C Canada
15 S8 **Tadoussac** Québec, SE Canada 48°09′N 69°43′W
155 H18 **Tādpatri** Andhra Pradesh, E India 14°55′N 77°59′E
Tadzhikabad *see* Tojikobod
Tadzhikistan *see* Tajikistan
163 Y14 **Taebaek-sanmaek** *prev.* T'aebaek-sanmaek. ▲▲ E South Korea
T'aebaek-sanmaek *see* Taebaek-sanmaek
Taechŏng-do *see* Daecheong-do
163 X13 **Taedong-gang** ≈ C North Korea
Taegu *see* Daegu
Taehan-haehyŏp *see* Korea Strait
Taehan Min'guk *see* South Korea
Taejŏn *see* Daejeon
193 Z13 **Tafahi** *island* N Tonga
105 Q4 **Tafalla** Navarra, N Spain 42°32′N 01°41′W
77 W7 **Tafassâsset, Ténéré du** *desert* N Niger
75 M12 **Tafassâsset, Oued** ≈ SE Algeria
55 U11 **Tafelberg** ▲ S Suriname 03°55′N 56°09′W
97 J21 **Taff** ≈ SE Wales, United Kingdom
Tafila/Ţafilah, Muḩāfaẓat at *see* Aţ Ţafilah
77 N15 **Tafiré** N Ivory Coast 09°04′N 05°10′W
142 M6 **Tafresh** Markazī, W Iran 34°40′N 50°00′E
143 Q9 **Taft** Yazd, C Iran 31°45′N 54°14′E
35 R13 **Taft** California, W USA 35°08′N 119°27′W
25 T14 **Taft** Texas, SW USA 27°58′N 97°24′W
143 W12 **Taftān, Kūh-e** ▲ SE Iran 28°38′N 61°06′E
35 R13 **Taft Heights** California, W USA 35°06′N 119°29′W
189 Y14 **Tafunsak** Kosrae, E Micronesia 05°21′N 162°58′E
192 G16 **Tāga** Savai'i, SW Samoa 13°46′S 172°31′W
149 O6 **Tagāb** Dāikondī, E Afghanistan 33°53′N 66°23′E
39 O8 **Tagagawik River** ≈ Alaska, USA
165 Q10 **Tagajō** *var.* Tagazyô. Miyagi, Honshū, C Japan 38°18′N 140°58′E
126 K12 **Taganrog** Rostovskaya Oblast', SW Russian Federation 47°10′N 38°55′E
126 K12 **Taganrog, Gulf of** *Rus.* Taganrogskiy Zaliv, *Ukr.* Tahanroz'ka Zatoka. *gulf* Russian Federation/Ukraine
Taganrogskiy Zaliv *see* Taganrog, Gulf of
76 J8 **Tagant** ◈ *region* C Mauritania
148 M14 **Tagas** Baluchistān, SW Pakistan 27°09′N 64°36′E
171 O4 **Tagaytay** Luzon, N Philippines 14°04′N 120°55′E
Tagazyô *see* Tagajō
171 P6 **Tagbilaran** *var.* Tagbilaran City. Bohol, C Philippines 09°41′N 123°54′E
Tagbilaran City *see* Tagbilaran
106 B10 **Taggia** Liguria, NW Italy 43°51′N 07°48′E
77 V9 **Taghouaji, Massif de** ▲ C Niger 17°13′N 08°37′E
107 J15 **Tagliacozzo** Lazio, C Italy 42°03′N 13°15′E
106 J7 **Tagliamento** ≈ NE Italy
149 N3 **Tagow Bāy** *var.* Bai. Sar-e Pul, N Afghanistan 35°41′N 66°01′E
146 H9 **Tagta** *var.* Tahta, *Rus.* Takhta. Daşoguz Welaýaty, N Turkmenistan 41°40′N 59°51′E
146 J16 **Tagtabazar** *var.* Takhtabazar. Mary Welaýaty, S Turkmenistan 35°57′N 62°49′E
59 L17 **Taguatinga** Tocantins, C Brazil 12°16′S 46°25′W
186 I10 **Tagula** Tagula Island, SE Papua New Guinea 11°21′S 153°11′E
186 I11 **Tagula Island** *prev.* Southeast Island, Sudest Island. *island* SE Papua New Guinea
171 Q7 **Tagum** Mindanao, S Philippines 07°33′N 125°51′E
54 C7 **Tagún, Cerro** *elevation* Colombia/Panama
105 P7 **Tagus** *Port.* Rio Tejo, *Sp.* Río Tajo. ≈ Portugal/Spain
64 M9 **Tagus Plain** *undersea feature* E Atlantic Ocean 37°30′N 12°00′W
191 S10 **Tahaa** *island* Îles Sous le Vent, W French Polynesia
191 U10 **Tahanea** *atoll* Îles Tuamotu, C French Polynesia

◆ Country
● Country Capital
◇ Dependent Territory
○ Dependent Territory Capital
◈ Administrative Regions
✈ International Airport
▲ Mountain
▲▲ Mountain Range
◬ Volcano
≈ River
◎ Lake
▣ Reservoir

Tahanroz'ka Zatoka see Taganrog, Gulf of
74 K12 **Tahat** ▲ SE Algeria 23°15′N 05°34′E
163 U4 **Tahe** Heilongjiang, NE China 52°21′N 124°42′E
Tahilt see Tsogt
191 T10 **Tahiti** island Îles du Vent, W French Polynesia
Tahiti, Archipel de see Société, Archipel de la
118 E4 **Tahkuna Nina** headland W Estonia 59°06′N 22°35′E
148 K12 **Tāhlāb** ≈ W Pakistan
148 K12 **Tāhlāb, Dasht-i** desert SW Pakistan
27 R10 **Tahlequah** Oklahoma, C USA 35°57′N 94°58′W
35 Q6 **Tahoe City** California, W USA 39°09′N 120°09′W
35 P6 **Tahoe, Lake** ◎ California/Nevada, W USA
25 N6 **Tahoka** Texas, SW USA 33°10′N 101°47′W
32 F8 **Taholah** Washington, NW USA 47°19′N 124°17′W
77 T11 **Tahoua** Tahoua, W Niger 14°53′N 05°18′E
77 T11 **Tahoua** ◈ department W Niger
31 P3 **Tahquamenon Falls** waterfall Michigan, N USA
31 P4 **Tahquamenon River** ≈ Michigan, N USA
139 V10 **Taḩrīr** Al Qādisīyah, S Iraq 31°58′N 45°34′E
10 K17 **Tahsis** Vancouver Island, British Columbia, SW Canada 49°42′N 126°31′W
75 W9 **Ṭahṭā** var. Ṭahta. C Egypt 26°47′N 31°31′E
Tahta see Tagta
136 L15 **Tahtalı Dağları** ▲▲ C Turkey
57 I14 **Tahuamanu, Río** ≈ Bolivia/Peru
56 F13 **Tahuanía, Río** ≈ E Peru
191 X7 **Tahuata** island Îles Marquises, NE French Polynesia
76 L17 **Taï** SW Ivory Coast 05°52′N 07°28′W
161 P5 **Tai'an** Shandong, E China 36°13′N 117°12′E
191 R8 **Taiarapu, Presqu'île de** peninsula Tahiti, W French Polynesia
Taibad see Tāybād
160 K7 **Taibai Shan** ▲ C China 33°57′N 107°31′E
161 T13 **Taibei** ● (Taiwan) N Taiwan 25°02′N 121°28′E
105 Q12 **Taibilla, Sierra de** ▲▲ S Spain
Taibus Qi see Baochang
Taichū see Taizhong
T'aichung see Taizhong
Taiden see Daejeon
161 T14 **Taidong** Jap. Taitō; prev. T'aitung. S Taiwan 22°43′N 121°10′E
185 E23 **Taieri** ≈ South Island, New Zealand
115 E21 **Taïgetos** ▲▲ S Greece
161 N4 **Taihang Shan** ▲▲ C China
184 M11 **Taihape** Manawatu-Wanganui, North Island, New Zealand 39°41′S 175°47′E
161 O7 **Taihe** Anhui, E China 33°14′N 115°35′E
161 O12 **Taihe** var. Chengjiang. Jiangxi, S China 26°47′N 114°52′E
Taihoku see Taibei
161 P9 **Taihu** Anhui, E China 30°22′N 116°20′E
161 R8 **Tai Hu** ◎ E China
159 O9 **Taikang** var. Dorbod, Dorbod Mongolzu Zizhixian. Heilongjiang, NE China 46°50′N 124°25′E
161 O6 **Taikang** Henan, C China 34°01′N 114°59′E
165 T5 **Taiki** Hokkaidō, NE Japan 42°29′N 143°15′E
166 L8 **Taikkyi** Yangon, SW Myanmar (Burma) 17°16′N 95°55′E
Taikyū see Daegu
163 U8 **Tailai** Heilongjiang, NE China 46°25′N 123°25′E
168 I12 **Taileleo** Pulau Siberut, W Indonesia 01°45′S 99°06′E
182 J10 **Tailem Bend** South Australia 35°20′S 139°34′E
96 I8 **Tain** N Scotland, United Kingdom 57°49′N 04°04′W
161 S14 **Tainan** prev. Dainan, T'ainan. S Taiwan 23°01′N 120°05′E
115 E22 **Taínaro, Akrotírio** cape S Greece
161 Q11 **Taining** var. Shancheng. Fujian, SE China 26°55′N 117°13′E
191 W7 **Taiohae** prev. Madisonville. Nuku Hiva, NE French Polynesia 08°55′S 140°04′W
Taipei see Taibei
T'aipei see Taibei
168 J7 **Taiping** Perak, Peninsular Malaysia 04°54′N 100°42′E
Taiping see Chongzuo
163 S8 **Taiping Ling** ▲ NE China 47°27′N 120°27′E
165 Q4 **Taisei** Hokkaidō, NE Japan 42°13′N 139°52′E
165 G12 **Taisha** Shimane, Honshū, SW Japan 35°23′N 132°40′E
Taishō-tō see Sekibi-sho
109 R4 **Taiskirchen** Oberösterreich, NW Austria 48°15′N 13°33′E
63 F20 **Taitao, Peninsula de** peninsula S Chile
81 J21 **Taita/Taveta** ◈ county S Kenya
Taitō see Taidong
T'aitung see Taidong
92 M13 **Taivalkoski** Pohjois-Pohjanmaa, E Finland 65°35′N 28°20′E
93 K19 **Taivassalo** Varsinais-Suomi, SW Finland 60°35′N 21°36′E
161 T14 **Taiwan** off. Republic of China, var. Formosa, Formo'sa. ◆ republic E Asia
192 F5 **Taiwan** var. Formosa. island E Asia
Taiwan see Taizhong
T'aiwan Haihsia/Taiwan Haixia see Taiwan Strait
Taiwan Shan see Chungyang Shanmo
161 R13 **Taiwan Strait** var. Formosa Strait, Chin. T'aiwan Haihsia, Taiwan Haixia. strait China/Taiwan
161 S12 **Taiwan Taoyuan** prev. Chiang Kai-shek. ✈ (T'aibei) N Taiwan 25°09′N 121°20′E
161 N4 **Taiyuan** var. T'ai-yuan, T'ai-yüan; prev. Yangku. province capital Shanxi, C China 37°48′N 112°33′E
T'ai-yuan/T'ai-yüan see Taiyuan
161 S13 **Taizhong** Jap. Taichū; prev. T'aichung. Taiwan, C Taiwan 24°09′N 120°40′E
161 R7 **Taizhou** Jiangsu, E China 32°36′N 119°52′E
161 S10 **Taizhou** var. Jiaojiang; prev. Haimen. Zhejiang, SE China 28°36′N 121°19′E
Taizhou see Linhai
141 O16 **Ta'izz** SW Yemen 13°36′N 44°04′E
141 O16 **Ta'izz** ✈ SW Yemen 13°40′N 44°10′E
75 P12 **Tajarhī** SW Libya 24°21′N 14°28′E
147 P13 **Tajikistan** off. Republic of Tajikistan, Rus. Tadzhikistan, Taj. Jumhurii Tojikiston; prev. Tajik S.S.R. ◆ republic C Asia
Tajikistan, Republic of see Tajikistan
Tajik S.S.R see Tajikistan
165 O11 **Tajima** Fukushima, Honshū, C Japan 37°10′N 139°46′E
Tajoe see Tayu
Tajo, Río see Tagus
42 B5 **Tajumulco, Volcán** ▲ W Guatemala 15°04′N 91°50′W
105 P7 **Tajuña** ≈ C Spain
Tajura, Gulf of see Tadjoura, Golfe de
167 O9 **Tak** var. Rahaeng. Tak, W Thailand 16°51′N 99°08′E
189 U4 **Taka Atoll** var. Tōke. atoll Ratak Chain, N Marshall Islands
165 P12 **Takahagi** Ibaraki, Honshū, S Japan 36°42′N 140°42′E
165 H13 **Takahashi** var. Takahasi. Okayama, Honshū, SW Japan 34°48′N 133°38′E
Takahasi see Takahashi
189 P12 **Takaieu Island** island E Micronesia
184 I13 **Takaka** Tasman, South Island, New Zealand 40°52′S 172°49′E
170 M14 **Takalar** Sulawesi, C Indonesia 05°28′S 119°24′E
165 H13 **Takamatsu** var. Takamatu. Kagawa, Shikoku, SW Japan 34°19′N 133°59′E
Takamatu see Takamatsu
165 D14 **Takamori** Kumamoto, Kyūshū, SW Japan 32°50′N 131°08′E
165 D16 **Takanabe** Miyazaki, Kyūshū, SW Japan 32°13′N 131°31′E
170 M16 **Takan, Gunung** ▲ Pulau Sumba, S Indonesia 08°52′S 117°32′E
165 Q7 **Takanosu** var. Kita-Akita. Akita, Honshū, C Japan 40°13′N 140°23′E
Takao see Gaoxiong
165 L11 **Takaoka** Toyama, Honshū, SW Japan 36°44′N 137°02′E
184 N12 **Takapau** Hawke's Bay, North Island, New Zealand 40°01′S 176°21′E
191 U9 **Takapoto** atoll Îles Tuamotu, C French Polynesia
184 L5 **Takapuna** Auckland, North Island, New Zealand 36°48′S 174°46′E
165 J3 **Takarazuka** Hyōgo, Honshū, SW Japan 34°49′N 135°21′E
191 U9 **Takaroa** atoll Îles Tuamotu, C French Polynesia
165 N12 **Takasaki** Gunma, Honshū, S Japan 36°20′N 139°00′E
164 L12 **Takayama** Gifu, Honshū, SW Japan 36°09′N 137°16′E
164 K12 **Takefu** var. Echizen, Takehu. Fukui, Honshū, SW Japan 35°55′N 136°11′E
Takehu see Takefu
164 C14 **Takeo** Saga, Kyūshū, SW Japan 33°13′N 130°00′E
Takeo see Takêv
164 C17 **Take-shima** island Nansei-shotō, SW Japan
142 M5 **Takestān** var. Takistan; prev. Siadehan. Qazvīn, N Iran 36°02′N 49°40′E
164 D14 **Taketa** Ōita, Kyūshū, SW Japan 33°00′N 131°23′E
167 R13 **Takêv** prev. Takeo. Takêv, S Cambodia 10°59′N 104°47′E
167 O10 **Tak Fah** Nakhon Sawan, C Thailand
139 T13 **Takhādīd** well S Iraq
149 R3 **Takhār** ◈ province NE Afghanistan
Takhiatash see Taxiatosh
Ta Khmau see Kândal
Takhta see Tagta
Takhtabazar see Tagtabazar
145 O8 **Takhtabrod** Severnyy Kazakhstan, N Kazakhstan 52°35′N 67°37′E
Takhtakupyr see Taxtako'pir
142 M8 **Takht-e Shāh, Kūh-e** ▲ C Iran
77 V12 **Takiéta** Zinder, S Niger 13°43′N 08°33′E
8 J8 **Takijuq Lake** ◎ Nunavut, NW Canada
165 S3 **Takikawa** Hokkaidō, NE Japan 43°35′N 141°54′E
165 U3 **Takinoue** Hokkaidō, NE Japan 44°10′N 143°09′E
Takistan see Tākestān
185 B23 **Takitimu Mountains** ▲▲ South Island, New Zealand
Takkaze see Tekezē
165 R7 **Takko** Aomori, Honshū, C Japan 40°19′N 141°11′E
10 L13 **Takla Lake** ◎ British Columbia, SW Canada
Takla Makan Desert see Taklimakan Shamo
158 H9 **Taklimakan Shamo** Eng. Takla Makan Desert. desert NW China
167 T12 **Takôk** Môndól Kiri, E Cambodia 12°37′N 106°30′E
39 P10 **Takotna** Alaska, USA 62°59′N 156°03′W
Takow see Gaoxiong
123 O12 **Taksimo** Respublika Buryatiya, S Russian Federation 56°18′N 114°53′E
164 C13 **Taku** Saga, Kyūshū, SW Japan 33°19′N 130°06′E
10 I10 **Taku** ≈ British Columbia, W Canada
166 M15 **Takua Pa** var. Ban Takua Pa. Phangnga, SW Thailand 08°55′N 98°20′E
77 W16 **Takum** Taraba, E Nigeria 07°16′N 10°00′E
191 V10 **Takume** atoll Îles Tuamotu, C French Polynesia
190 L16 **Takutea** island S Cook Islands
186 K6 **Takuu Islands** prev. Mortlock Group. island group NE Papua New Guinea
119 L18 **Tal'** Minskaya Voblasts', S Belarus 52°52′N 27°58′E
40 L13 **Tala** Jalisco, C Mexico 20°39′N 103°45′W
61 F19 **Tala** Canelones, S Uruguay 34°24′S 55°45′W
Talabriga see Aveiro, Portugal
Talabriga see Talavera de la Reina, Spain
119 N14 **Talachyn** Rus. Tolochin. Vitsyebskaya Voblasts', NE Belarus 54°25′N 29°42′E
149 U7 **Talagang** Punjab, E Pakistan 32°55′N 72°29′E
105 V11 **Talaiassa** ▲ Ibiza, Spain, W Mediterranean Sea 38°55′N 1°17′E
155 J23 **Talaimannar** Northern Province, NW Sri Lanka 09°05′N 79°43′E
117 R3 **Talalayivka** Chernihivs'ka Oblast', N Ukraine 50°51′N 33°09′E
43 O15 **Talamanca, Cordillera de** ▲▲ S Costa Rica
56 A9 **Talara** Piura, NW Peru 04°31′S 81°17′W
104 L11 **Talarrubias** Extremadura, W Spain 39°03′N 05°14′W
147 S8 **Talas** Talasskaya Oblast', NW Kyrgyzstan 42°29′N 72°21′E
147 S8 **Talas** ≈ NW Kyrgyzstan
186 G7 **Talasea** New Britain, E Papua New Guinea 05°20′S 150°01′E
Talas Oblasty see Talasskaya Oblast'
147 S8 **Talasskaya Oblast'** Kir. Talas Oblasty. ◈ province NW Kyrgyzstan
147 S8 **Talasskiy Alatau, Khrebet** ▲▲ Kazakhstan/Kyrgyzstan
77 U12 **Talata Mafara** Zamfara, NW Nigeria 12°33′N 06°01′E
171 R9 **Talaud, Kepulauan** island group E Indonesia
104 M9 **Talavera de la Reina** anc. Caesarobriga, Talabriga. Castilla-La Mancha, C Spain 39°58′N 04°50′W
104 J11 **Talavera la Real** Extremadura, W Spain 38°53′N 06°46′W
186 F7 **Talawe, Mount** ▲ New Britain, C Papua New Guinea 05°30′S 148°24′E
23 S5 **Talbotton** Georgia, SE USA 32°40′N 84°32′W
183 R7 **Talbragar River** ≈ New South Wales, SE Australia
62 G13 **Talca** Maule, C Chile 35°28′S 71°42′W
62 F13 **Talcahuano** Bío Bío, C Chile 36°43′S 73°07′W
154 N12 **Tālcher** Odisha, E India 20°57′N 85°13′E
25 W5 **Talco** Texas, SW USA 33°21′N 95°06′W
145 V14 **Taldykorgan** Kaz. Taldyqorghan; prev. Taldy-Kurgan. Taldykorgan, SE Kazakhstan 45°N 78°23′E
Taldy-Kurgan/Taldyqorghan see Taldykorgan
147 Y7 **Taldy-Suu** Issyk-Kul'skaya Oblast', E Kyrgyzstan 42°49′N 78°33′E
147 U10 **Taldy-Suu** Oshskaya Oblast', SW Kyrgyzstan 40°33′N 73°52′E
Tal-e Khosravī see Yāsūj
193 Y15 **Taleki Tonga** island Otu Tolu Group, C Tonga
193 Y15 **Taleki Vavu'u** island Otu Tolu Group, C Tonga
102 J13 **Talence** Gironde, SW France 44°49′N 00°35′W
145 U16 **Talgar** Kaz. Talghar. Almaty, SE Kazakhstan 43°17′N 77°15′E
Talghar see Talgar
171 Q12 **Taliabu, Pulau** island Kepulauan Sula, C Indonesia
115 L22 **Taliarós, Akrotírio** headland Astypálaia, Kykládes, Greece, Aegean Sea 36°31′N 26°18′E
Ta-lien see Dalian
27 Q12 **Talihina** Oklahoma, C USA 34°45′N 95°03′W
Talimardzhan see Tollimarjon
137 T12 **T'alin** Rus. Talin; prev. Verin T'alin. W Armenia 40°23′N 43°51′E
Talin see T'alin
81 E15 **Tali Post** Central Equatoria, S South Sudan 05°55′N 30°44′E
Taliq-an see Tāloqān
Taliş Dağları see Talish Mountains
142 L2 **Talish Mountains** Az. Taliş Dağları, Per. Kūhhā-ye Ṭavālesh, Rus. Talyshskiye Gory. ▲▲ Azerbaijan/Iran
170 M16 **Taliwang** Sumbawa, C Indonesia 08°45′S 116°55′E
119 L17 **Tal'ka** Minskaya Voblasts', C Belarus 53°22′N 28°21′E
39 R11 **Talkeetna** Alaska, USA 62°19′N 150°06′W
39 R11 **Talkeetna Mountains** ▲▲ Alaska, USA
Talkhof see Puurmani
92 H2 **Tálknafjördur** Vestfirdhir, W Iceland 65°38′N 23°51′W
139 Q3 **Tall 'Abṭah** Nīnawýa, N Iraq 35°52′N 42°40′E
138 M2 **Tall Abyaḍ** var. Tell Abiad. Ar Raqqah, N Syria 36°42′N 38°56′E
23 Q4 **Talladega** Alabama, S USA 33°26′N 86°06′W
139 Q2 **Tall 'Afar** Nīnawýa, N Iraq 36°22′N 42°27′E
23 S8 **Tallahassee** prev. Muskogean. state capital Florida, SE USA 30°26′N 84°17′W
22 L2 **Tallahatchie River** ≈ Mississippi, S USA
Tall al Abyaḍ see At Tall al Abyaḍ
139 W12 **Tall al Laḩm** Dhī Qār, S Iraq 30°46′N 46°22′E
183 P11 **Tallangatta** Victoria, SE Australia 36°15′S 147°13′E
23 R4 **Tallapoosa River** ≈ Alabama/Georgia, S USA
103 T13 **Tallard** Hautes-Alpes, SE France 44°30′N 06°04′E
139 Q3 **Tall ash Sha'īr** Nīnawýa, N Iraq 36°11′N 42°26′E
25 Q5 **Tallassee** Alabama, S USA 32°32′N 85°53′W
139 R4 **Tall 'Azbah** Nīnawýa, NW Iraq 35°47′N 43°13′E
138 I5 **Tall Bīsah** Ḩimş, W Syria 34°50′N 36°44′E
139 R3 **Tall Ḩassūnah** Al Anbār, N Iraq 36°05′N 43°10′E
139 Q2 **Tall Ḩuqnah** var. Tell Huqnah. Nīnawá, N Iraq 36°33′N 42°34′E
Tallin see Tallinn
118 G3 **Tallinn** Ger. Reval, Rus. Tallin; prev. Revel. ● (Estonia) Harjumaa, NW Estonia 59°26′N 24°42′E
118 H3 **Tallinn** ✈ Harjumaa, NW Estonia 59°23′N 24°52′E
138 H5 **Tall Kalakh** var. Tell Kalakh. Ḩimş, C Syria 34°40′N 36°18′E
139 R2 **Tall Kayf** Nīnawýa, NW Iraq 36°30′N 43°08′E
Tall Kūchak see Tall Kūshik
139 P2 **Tall Kūshik** var. Tall Kūchak. Al Ḩasakah, E Syria 36°48′N 42°01′E
31 U12 **Tallmadge** Ohio, N USA 41°06′N 81°26′W
22 J5 **Tallulah** Louisiana, S USA 32°22′N 91°12′W
139 Q2 **Tall 'Uwaynāt** Nīnawýa, NW Iraq 36°43′N 42°18′E
139 Q2 **Tall Ẓāhir** Nīnawá, N Iraq 36°51′N 42°29′E
122 J13 **Tal'menka** Altayskiy Kray, S Russian Federation 53°55′N 83°26′E
122 K8 **Talnakh** Krasnoyarskiy Kray, N Russian Federation 69°26′N 88°27′E
117 P7 **Tal'ne** Rus. Tal'noye. Cherkas'ka Oblast', C Ukraine 48°55′N 30°40′E
Tal'noye see Tal'ne
80 E12 **Talodi** Southern Kordofan, C Sudan 10°40′N 30°25′E
188 B16 **Talofofo** SE Guam 13°21′N 144°45′E
188 B16 **Talofofo Bay** bay SE Guam
26 L9 **Taloga** Oklahoma, C USA 36°02′N 98°58′W
123 T10 **Talon** Magadanskaya Oblast', E Russian Federation 59°47′N 148°46′E
14 H11 **Talon, Lake** ◎ Ontario, S Canada
149 R2 **Tāloqān** var. Taliq-an. Takhār, NE Afghanistan 36°44′N 69°33′E
126 M8 **Talovaya** Voronezhskaya Oblast', W Russian Federation 51°07′N 40°46′E
9 N6 **Taloyoak** prev. Spence Bay. Nunavut, N Canada 69°30′N 93°25′W
25 Q8 **Talpa** Texas, SW USA 31°46′N 99°42′W
40 K13 **Talpa de Allende** Jalisco, C Mexico 20°22′N 104°51′W
23 S9 **Talquin, Lake** ⊠ Florida, SE USA
Talsen see Talsi
Talshand see Chandmanĭ
118 E8 **Talsi** Ger. Talsen. NW Latvia 57°14′N 22°35′E
143 V11 **Tal Sīāh** Sīstān va Balūchestān, SE Iran 28°19′N 57°43′E
62 G6 **Taltal** Antofagasta, N Chile 25°22′S 70°27′W
8 K10 **Taltson** ≈ Northwest Territories, NW Canada
168 K11 **Taluk** Sumatera, W Indonesia 0°32′S 101°35′E
92 J8 **Talvik** Finnmark, N Norway 70°02′N 22°59′E
182 M7 **Talyawalka Creek** ≈ New South Wales, SE Australia
Talyshskiye Gory see Talish Mountains
29 W14 **Tama** Iowa, C USA 41°58′N 92°34′W
Tama Abu, Banjaran see Penambo, Banjaran
169 U9 **Tamabo, Banjaran** ▲▲ East Malaysia
190 B16 **Tamakautoga** SW Niue 19°05′S 169°55′W
127 N7 **Tamala** Penzenskaya Oblast', W Russian Federation 52°32′N 43°18′E
77 P15 **Tamale** C Ghana 09°21′N 00°54′W
191 P3 **Tamana** prev. Rotcher Island. atoll Tungaru, W Kiribati
74 K12 **Tamanrasset** var. Tamenghest. S Algeria 22°49′N 05°32′E
74 J13 **Tamanrasset** wadi Algeria/Mali
166 M2 **Tamanthi** Sagaing, N Myanmar (Burma) 25°17′N 95°18′E
97 I24 **Tamar** ≈ SW England, United Kingdom
Tamar see Tudmur
54 H9 **Támara** Casanare, C Colombia 05°51′N 72°09′W
54 F7 **Tamar, Alto de** ▲ C Colombia 07°25′N 74°28′W
173 X16 **Tamarin** E Mauritius 20°20′S 57°22′E
105 T5 **Tamarite de Litera** var. Tararite de Llitera. Aragón, NE Spain 41°52′N 00°25′E
111 I24 **Tamási** Tolna, S Hungary 46°39′N 18°17′E
Tamatave see Toamasina
41 O9 **Tamaulipas** ◈ state C Mexico
41 P10 **Tamaulipas, Sierra de** ▲▲ C Mexico
56 F12 **Tamaya, Río** ≈ E Peru
40 I9 **Tamazula** Durango, C Mexico 24°43′N 106°33′W
40 L14 **Tamazula** Jalisco, C Mexico 19°41′N 103°18′W
Tamazulapam see Tamazulapán
41 Q15 **Tamazulapán** var. Tamazulápam. Oaxaca, SE Mexico 17°41′N 97°33′W
41 P12 **Tamazunchale** San Luis Potosí, C Mexico 21°17′N 98°46′W
76 H11 **Tambacounda** SE Senegal 13°44′N 13°43′W
83 M16 **Tambara** Manica, C Mozambique 16°42′S 34°14′E
77 T13 **Tambawel** Sokoto, NW Nigeria 12°24′N 04°42′E
186 M9 **Tambea** Guadalcanal, C Solomon Islands 09°19′S 159°42′E
169 N10 **Tambelan, Kepulauan** island group W Indonesia
57 E15 **Tambo de Mora** Ica, W Peru 13°30′S 76°08′W
170 L16 **Tambora, Gunung** ▲ Sumbawa, S Indonesia 08°16′S 117°59′E
61 E17 **Tambores** Paysandú, W Uruguay 31°50′S 56°17′W
57 H14 **Tambo, Río** ≈ C Peru
56 F7 **Tamboryacu, Río** ≈ N Peru
126 M7 **Tambov** Tambovskaya Oblast', W Russian Federation 52°43′N 41°28′E
126 L6 **Tambovskaya Oblast'** ◈ province W Russian Federation
104 H3 **Tambre** ≈ NW Spain
169 V7 **Tambunan** Sabah, East Malaysia 05°40′N 116°22′E
81 C15 **Tambura** Western Equatoria, SW South Sudan 05°38′N 27°30′E
Tamchaket see Tâmchekket
76 J9 **Tâmchekket** var. Tamchaket. Hodh el Gharbi, S Mauritania 17°23′N 10°37′W
167 T7 **Tam Điệp** Ninh Bình, N Vietnam 20°09′N 105°54′E
Tamdybulak see Tomdibuloq
54 H8 **Tame** Arauca, C Colombia 06°27′N 71°42′W
104 H6 **Tâmega, Rio** Sp. Río Tâmega. ≈ Portugal/Spain
Támega, Río see Tâmega, Rio
115 H20 **Taméios, Akrotírio** headland Tziá, Kykládes, Greece, Aegean Sea 37°31′N 24°16′E
Tamenghest see Tamanrasset
77 V8 **Tamgak, Adrar** ▲ C Niger 19°10′N 08°39′E
76 I13 **Tamgue** ▲ NW Guinea 12°14′N 12°18′W
41 Q12 **Tamiahua** Veracruz-Llave, E Mexico 21°15′N 97°27′W
41 Q12 **Tamiahua, Laguna de** lagoon E Mexico
23 Y16 **Tamiami Canal** canal Florida, SE USA
188 F17 **Tamil Harbor** harbour Yap, W Micronesia
155 H21 **Tamil Nādu** prev. Madras. ◈ state SE India
99 H20 **Tamines** Namur, S Belgium 50°27′N 04°37′E
116 E12 **Tamiš** Ger. Temesch, Hung. Temes. ≈ Romania/Serbia
167 U10 **Tam Ky** Quang Nam-Đa Nẵng, C Vietnam 15°32′N 108°30′E
Tammerfors see Tampere
Tammisaari see Ekenäs
95 N14 **Tämnaren** ◎ C Sweden
191 Q7 **Tamotoe, Passe** passage Tahiti, W French Polynesia
23 V12 **Tampa** Florida, SE USA 27°57′N 82°27′W
23 V12 **Tampa** ✈ Florida, SE USA 27°57′N 82°29′W
23 V13 **Tampa Bay** bay Florida, SE USA
93 L18 **Tampere** Swe. Tammerfors. Pirkanmaa, W Finland 61°30′N 23°45′E
41 Q11 **Tampico** Tamaulipas, C Mexico 22°18′N 97°52′W
171 P14 **Tampo** Pulau Muna, C Indonesia 04°38′S 122°40′E
167 V11 **Tam Quan** Bình Định, C Vietnam 14°34′N 109°00′E
162 J13 **Tamsag Muchang** Nei Mongol Zizhiqu, N China 40°28′N 102°34′E
Tamsal see Tamsalu
118 I4 **Tamsalu** Ger. Tamsal. Lääne-Virumaa, NE Estonia 59°10′N 26°07′E
109 S8 **Tamsweg** Salzburg, SW Austria 47°08′N 13°49′E
188 C15 **Tamuning** NW Guam 13°29′N 144°47′E
183 T6 **Tamworth** New South Wales, SE Australia 31°07′S 150°54′E
97 M19 **Tamworth** C England, United Kingdom 52°39′N 01°40′W
81 K19 **Tana** Finn. Tenojoki, Lapp. Deatnu. ≈ SE Kenya see also Deatnu, Tenojoki
Tana see Deatnu/Tana
164 I15 **Tanabe** Wakayama, Honshū, SW Japan 33°43′N 135°22′E
92 L8 **Tana Bru** Finnmark, N Norway 70°11′N 28°06′E
39 T10 **Tanacross** Alaska, USA 63°30′N 143°21′W
92 L7 **Tanafjorden** Lapp. Deanuvuotna. fjord N Norway
38 G17 **Tanaga Island** island Aleutian Islands, Alaska, USA
38 G17 **Tanaga Volcano** ▲ Tanaga Island, Alaska, USA 51°53′N 178°08′W
107 M18 **Tanagro** ≈ S Italy
80 H11 **T'ana Hāyk'** var. Lake Tana. ◎ NW Ethiopia
168 H11 **Tanahbela, Pulau** island Kepulauan Batu, W Indonesia
171 H15 **Tanahjampea, Pulau** island W Indonesia
168 H11 **Tanahmasa, Pulau** island Kepulauan Batu, W Indonesia
Tanais see Don
152 L10 **Tanakpur** Uttarakhand, N India 29°04′N 80°06′E
Tana, Lake see T'ana Hāyk'
181 P5 **Tanami Desert** desert Northern Territory, N Australia
167 T14 **Tân An** Long An, S Vietnam 10°32′N 106°24′E
39 Q9 **Tanana** Alaska, USA 65°12′N 152°00′W
Tananarive see Antananarivo
39 Q9 **Tanana River** ≈ Alaska, USA
95 C16 **Tananger** Rogaland, S Norway 58°55′N 05°34′E
188 H5 **Tanapag** Saipan, S Northern Mariana Islands 15°14′S 145°45′E
188 H5 **Tanapag, Puetton** bay Saipan, S Northern Mariana Islands
81 J20 **Tana River** ◈ county SE Kenya
106 C9 **Tanaro** ≈ N Italy
163 Y12 **Tanch'ŏn** E North Korea 40°22′N 128°49′E
40 M14 **Tancítaro, Cerro** ▲ C Mexico 19°16′N 102°25′W
153 N12 **Tānda** Uttar Pradesh, N India 26°33′N 82°39′E
77 O15 **Tanda** E Ivory Coast 07°48′N 03°10′W
116 L14 **Ţăndărei** Ialomiţa, SE Romania 44°39′N 27°40′E
61 C23 **Tandil** Buenos Aires, E Argentina 37°18′S 59°10′W
78 H12 **Tandjilé** off. Région du Tandjilé. ◈ region SW Chad
Tandjilé, Région du see Tandjilé
Tandjoeng see Tanjung
Tandjoengkarang see Bandar Lampung
Tandjoengpandan see Tanjungpandan
Tandjoengpinang see Tanjungpinang
Tandjoengredeb see Tanjungredeb
149 Q16 **Tando Allāhyār** Sind, SE Pakistan 25°28′N 68°44′E
149 Q17 **Tando Bāgo** Sind, SE Pakistan 24°48′N 68°59′E
149 Q16 **Tando Muhammad Khān** Sind, SE Pakistan 25°07′N 68°35′E
182 L7 **Tandou Lake** seasonal lake New South Wales, SE Australia
94 L11 **Tandsjöborg** Gävleborg, C Sweden 61°40′N 14°40′E
155 H15 **Tāndūr** Telangana, C India 17°16′N 77°37′E
164 C17 **Tanega-shima** island Nansei-shotō, SW Japan
165 R7 **Taneichi** Iwate, Honshū, C Japan 40°23′N 141°42′E
Tanen Taunggyi see Tane Range
167 N8 **Tane Range** Bur. Tanen Taunggyi. ▲▲ W Thailand
111 P15 **Tanew** ≈ SE Poland
21 W2 **Taneytown** Maryland, NE USA 39°39′N 77°10′W
74 H12 **Tanezrouft** desert Algeria/Mali
138 L7 **Tanf, Jabal aţ** ▲ SE Syria 33°32′N 38°43′E
81 J21 **Tanga** Tanga, E Tanzania 05°07′S 39°05′E
81 I22 **Tanga** ◈ region E Tanzania
153 T14 **Tangail** Dhaka, C Bangladesh 24°15′N 89°55′E
186 I5 **Tanga Islands** island group NE Papua New Guinea
155 K26 **Tangalla** Southern Province, S Sri Lanka 06°02′N 80°47′E
Tanganyika and Zanzibar see Tanzania
68 J13 **Tanganyika, Lake** ◎ E Africa
56 E7 **Tangarana, Río** ≈ N Peru
191 V16 **Tangaroa, Maunga** ⛰ Easter Island, Chile, E Pacific Ocean
Tangdukou see Shaoyang
74 G5 **Tanger** var. Tangiers, Tangier, Fr./Ger. Tangerk, Sp. Tánger; anc. Tingis. NW Morocco 35°49′N 05°49′W
169 N15 **Tangerang** Jawa, C Indonesia 06°14′S 106°36′E
Tangerk see Tanger
100 M12 **Tangermünde** Sachsen-Anhalt, C Germany 52°35′N 11°57′E
159 O12 **Tanggulashan** var. Togton Heyan, Tuotuoheyan. Qinghai, C China 34°13′N 92°25′E
156 K10 **Tanggula Shan** var. Dangla, Tangla Range. ▲▲ W China
159 N13 **Tanggula Shan** ▲ W China 33°18′N 91°10′E
156 K10 **Tanggula Shankou** Tib. Dang La. pass W China
161 N7 **Tanghe** Henan, C China 32°40′N 112°53′E
149 T5 **Tāngi** Khyber Pakhtunkhwa, NW Pakistan 34°18′N 71°42′E
Tangier see Tanger
21 Y5 **Tangier Island** island Virginia, NE USA
Tangiers see Tanger
22 K8 **Tangipahoa River** ≈ Louisiana, S USA
Tangla Range see Tanggula Shan
156 I10 **Tangra Yumco** var. Tangro Tso. ◎ W China
Tangro Tso see Tangra Yumco
157 T7 **Tangshan** var. T'ang-shan. Hebei, E China 39°39′N 118°15′E
T'ang-shan see Tangshan
77 R14 **Tanguiéta** NW Benin 10°35′N 01°19′E
163 X7 **Tangwang He** ≈ NE China
163 X7 **Tangyuan** Heilongjiang, NE China 46°45′N 129°52′E
167 S14 **Tân Hiêp** var. Phung Hiêp. Cân Thơ, S Vietnam 09°50′N 105°48′E
92 M11 **Tanhua** Lappi, N Finland 67°31′N 27°30′E
171 U16 **Tanimbar, Kepulauan** island group Maluku, E Indonesia
Tanintharyi see Taninthayi
Tanintharyi see Taninthayi
167 N12 **Taninthayi** var. Tenasserim. Taninthayi, S Myanmar (Burma) 12°05′N 99°00′E
167 N11 **Taninthayi** var. Tenasserim; prev. Tanintharyi. ◈ region S Myanmar (Burma)
Tānjarō see Tancero
Tānjarū see Tancero
129 T15 **Tanjong Piai** headland Peninsular Malaysia
Tanjore see Thanjāvūr
169 U12 **Tanjung** prev. Tandjoeng. Borneo, C Indonesia 02°08′S 115°23′E
169 W9 **Tanjungbatu** Borneo, N Indonesia 02°19′N 118°03′E
Tanjungkarang/Tanjungkarang-Telukbetung see Bandar Lampung
169 N13 **Tanjungpandan** prev. Tandjoengpandan. Pulau Belitung, W Indonesia 02°44′S 107°36′E
168 M10 **Tanjungpinang** prev. Tandjoengpinang. Pulau Bintan, W Indonesia 0°55′N 104°28′E
169 V9 **Tanjungredeb** var. Tanjungredep; prev. Tandjoengredeb. Borneo, C Indonesia 02°09′N 117°29′E
Tanjungredep see Tanjungredeb
169 V9 **Tanjung Selor** Borneo, C Indonesia 2°50′N 117°22′E
149 S8 **Tānk** Khyber Pakhtunkhwa, NW Pakistan 32°14′N 70°29′E
187 S15 **Tanna** island S Vanuatu
93 F17 **Tännäs** Jämtland, C Sweden 62°27′N 12°40′E
Tannenhof see Krynica
108 K7 **Tannheim** Tirol, W Austria 47°30′N 10°32′E
Tannu-Tuva see Tyva, Respublika
171 Q12 **Tano** Pulau Taliabu, E Indonesia 01°51′S 124°55′E
77 O17 **Tano** ≈ S Ghana
152 D10 **Tanot** Rājasthān, NW India 27°44′N 70°17′E
77 V11 **Tanout** Zinder, C Niger 14°58′N 08°54′E
Tân Phu see Định Quan
41 P12 **Tanquián** San Luis Potosí, C Mexico 21°38′N 98°39′W
77 R13 **Tansarga** E Burkina Faso 11°51′N 01°51′E
167 T13 **Tan Son Nhat** ✈ (Hô Chi Minh) Tây Ninh, S Vietnam 10°52′N 106°38′E
75 V8 **Ţanţā** var. Tanta, Tantâ. N Egypt 30°42′N 31°00′E
74 D9 **Tan-Tan** SW Morocco 28°30′N 11°11′W
41 P12 **Tantoyuca** Veracruz-Llave, E Mexico 21°21′N 98°12′W
152 J12 **Tāntpur** Uttar Pradesh, N India 26°51′N 77°29′E
Tan-tung see Dandong
38 M12 **Tanunak** Alaska, USA 60°35′N 165°15′W
166 L5 **Ta-nyaung** Magway, W Myanmar (Burma) 20°49′N 94°40′E
167 S5 **Tân Yên** Tuyên Quang, N Vietnam 22°08′N 104°58′E
81 F22 **Tanzania** off. United Republic of Tanzania, Swa. Jamhuri ya Muungano wa Tanzania; prev. German East Africa, Tanganyika and Zanzibar. ◆ republic E Africa
Tanzania, Jamhuri ya Muungano wa see Tanzania
Tanzania, United Republic of see Tanzania
Tao'an see Taonan
163 T8 **Tao'er He** ≈ NE China
159 U11 **Tao He** ≈ C China
163 U9 **Taonan** var. Tao'an. Jilin, NE China 45°20′N 122°46′E
T'aon-an see Baicheng
Taongi see Bokaak Atoll
107 M23 **Taormina** anc. Tauromenium. Sicilia, Italy, C Mediterranean Sea 37°54′N 15°18′E
37 S9 **Taos** New Mexico, SW USA 36°24′N 105°35′W
Taoudenit see Taoudenni
77 O6 **Taoudenni** var. Taoudenit. Tombouctou, N Mali 22°46′N 03°54′W
74 G6 **Taounate** N Morocco 34°33′N 04°39′W
Taoyang see Lintao
161 S13 **Taoyuan** Jap. Tōen; prev. T'aoyüan. N Taiwan 25°00′N 121°15′E
118 I3 **Tapa** Ger. Taps. Lääne-Virumaa, NE Estonia 59°15′N 26°E
41 V17 **Tapachula** Chiapas, SE Mexico 14°53′N 92°18′W
Tapaiu see Gvardeysk
59 H14 **Tapajós, Rio** var. Tapajóz. ≈ NW Brazil
Tapajóz see Tapajós, Rio
61 C21 **Tapalqué** var. Tapalquén. Buenos Aires, E Argentina 36°21′S 60°01′W
Tapalquén see Tapalqué
Tapanahoni see Tapanahony Rivier
55 W11 **Tapanahony Rivier** var. Tapanahoni. ≈ E Suriname
41 T16 **Tapanatepec** var. San Pedro Tapanatepec. Oaxaca, SE Mexico 16°23′N 94°09′W
185 D23 **Tapanui** Otago, South Island, New Zealand 45°55′S 169°16′E
59 E14 **Tapauá** Amazonas, N Brazil 05°42′S 64°15′W
47 R7 **Tapauá, Rio** ≈ W Brazil
185 I14 **Tapawera** Tasman, South Island, New Zealand 41°24′S 172°50′E
61 I16 **Tapes** Rio Grande do Sul, S Brazil 30°40′S 51°25′W
76 K16 **Tapeta** C Liberia 06°36′N 08°52′W
154 H11 **Tāpi** prev. Tāpti. ≈ W India
104 J2 **Tapia de Casariego** Asturias, N Spain 43°34′N 06°56′W
56 F10 **Tapiche, Río** ≈ N Peru
167 N15 **Tapi, Mae Nam** var. Luang. ≈ SW Thailand
186 E8 **Tapini** Central, S Papua New Guinea 08°16′S 146°09′E
Tapirapecó, Serra see Tapirapecó, Sierra
55 N13 **Tapirapecó, Sierra** Port. Serra Tapirapecó. ▲▲ Brazil/Venezuela
77 R13 **Tapoa** ≈ Benin/Niger
188 H5 **Tapochau, Mount** ▲ Saipan, S Northern Mariana Islands
111 H24 **Tapolca** Veszprém, W Hungary 46°54′N 17°29′E
21 X5 **Tappahannock** Virginia, NE USA 37°55′N 76°54′W
31 U13 **Tappan Lake** ⊠ Ohio, N USA
165 Q6 **Tappi-zaki** headland Honshū, C Japan 41°15′N 140°19′E
Taps see Tapa
Tāpti see Tāpi
Tapuaemanu see Maiao
185 J16 **Tapuaenuku** ▲ South Island, New Zealand 42°00′S 173°39′E
171 N8 **Tapul Group** island group Sulu Archipelago, SW Philippines
58 E11 **Tapurmcuará** var. Tapuruquara. Amazonas, NW Brazil 0°17′S 65°00′W
Tapuruquara see Tapurmcuará
192 J17 **Taputapu, Cape** headland Tutuila, W American Samoa 14°20′S 170°51′W
141 W13 **Tāqah** S Oman 17°02′N 54°23′E
139 T3 **Taqtaq** Ar. Ţaqtaq. Arbil, N Iraq 35°54′N 44°36′E
Ţaqtaq see Taqtaq
61 J15 **Taquara** Rio Grande do Sul, S Brazil 29°36′S 50°46′W
59 H19 **Taquari, Rio** ≈ C Brazil
60 L8 **Taquaritinga** São Paulo, S Brazil 21°22′S 48°29′W
122 I11 **Tara** Omskaya Oblast', C Russian Federation 56°54′N 74°17′E
83 I16 **Tara** Southern, S Zambia 16°56′S 26°50′E
113 J15 **Tara** ≈ Montenegro
112 K13 **Tara** ≈ W Serbia
77 W15 **Taraba** ◈ state E Nigeria
77 X15 **Taraba** ≈ E Nigeria
75 O7 **Ţarābulus** var. Ţarābulus al Gharb, Eng. Tripoli. ● (Libya) NW Libya 32°54′N 13°11′E
75 O7 **Ţarābulus** ✈ NW Libya 32°37′N 13°07′E
Ţarābulus al Gharb see Ţarābulus
Ţarābulus/Ţarābulus ash Shām see Tripoli
105 O7 **Taracena** Castilla-La Mancha, C Spain 40°39′N 03°08′W
117 N12 **Taraclia** Rus. Tarakilya. S Moldova 45°55′N 28°40′E
139 V10 **Tarād al Kahf** Dhī Qār, SE Iraq 31°58′N 45°58′E
183 R10 **Tarago** New South Wales, SE Australia 35°04′S 149°40′E
162 J8 **Taragt** var. Hüremt. Övörhangay, C Mongolia 46°18′N 102°27′E
169 V8 **Tarakan** Borneo, C Indonesia 03°20′N 117°38′E

◆ Country ● Country Capital ◇ Dependent Territory ○ Dependent Territory Capital ◈ Administrative Regions ✈ International Airport ▲ Mountain ▲▲ Mountain Range ⛰ Volcano ≈ River ◎ Lake ⊠ Reservoir

169 V9 **Tarakan, Pulau** *island* N Indonesia
Tarakilya *see* Taraclia
165 P16 **Tarama-jima** *island* Sakishima-shotō, SW Japan
184 K10 **Taranaki** *off.* Taranaki Region. ◈ *region* North Island, New Zealand
184 K10 **Taranaki, Mount** *var.* Egmont. ⛰ North Island, New Zealand 39°16′S 174°04′E
Taranaki Region *see* Taranaki
105 O9 **Tarancón** Castilla-La Mancha, C Spain 40°01′N 03°01′W
188 M15 **Tarang Reef** *reef* C Micronesia
96 E7 **Taransay** *island* NW Scotland, United Kingdom
107 P18 **Taranto** *var.* Tarentum. Puglia, SE Italy 40°30′N 17°11′E
107 O19 **Taranto, Golfo di** *Eng.* Gulf of Taranto. *gulf* S Italy
Taranto, Gulf of *see* Taranto, Golfo di
62 G3 **Tarapacá** *off.* Región de Tarapacá. ◈ *region* N Chile
Tarapacá, Región de *see* Tarapacá
187 N9 **Tarapaina** Maramasike Island, N Solomon Islands 09°28′S 161°24′E
56 D10 **Tarapoto** San Martín, N Peru 06°31′S 76°23′W
138 M6 **Ţaraq an Na'jah** *hill range* E Syria
138 M6 **Ţaraq Sidāwī** *hill range* E Syria
103 Q11 **Tarare** Rhône, E France 45°54′N 04°26′E
Tararite de Llitera *see* Tamarite de Litera
184 M13 **Tararua Range** ▲▲ North Island, New Zealand
151 Q22 **Tarāsa Dwīp** *island* Nicobar Islands, India, NE Indian Ocean
103 Q15 **Tarascon** Bouches-du-Rhône, SE France 43°48′N 04°39′E
102 M17 **Tarascon-sur-Ariège** Ariège, S France 42°51′N 01°35′E
117 P6 **Tarashcha** Kyyivs'ka Oblast', N Ukraine 49°34′N 30°31′E
57 L18 **Tarata** Cochabamba, C Bolivia 17°35′S 66°04′W
57 I18 **Tarata** Tacna, SW Peru 17°30′S 70°00′W
190 H2 **Taratai** *atoll* Tungaru, W Kiribati
59 B15 **Tarauacá** Acre, W Brazil 08°06′S 70°45′W
59 B15 **Tarauacá, Rio** ≈ NW Brazil
191 Q8 **Taravao** Tahiti, W French Polynesia 17°44′S 149°19′W
191 R8 **Taravao, Baie de** *bay* Tahiti, W French Polynesia
191 Q8 **Taravao, Isthme de** *isthmus* Tahiti, W French Polynesia
103 X16 **Taravo** ≈ Corse, France, C Mediterranean Sea
190 J3 **Tarawa** ✈ Tarawa, W Kiribati 0°53′S 169°32′E
190 H2 **Tarawa** *atoll* Tungaru, W Kiribati
184 N10 **Tarawera** Hawke's Bay, North Island, New Zealand 39°03′S 176°34′E
184 N8 **Tarawera, Lake** ◎ North Island, New Zealand
184 N8 **Tarawera, Mount** ▲ North Island, New Zealand 38°13′S 176°29′E
105 S8 **Tarayuela** ▲ N Spain 40°28′N 00°22′W
145 R16 **Taraz** *prev.* Aulie Ata, Auliye-Ata, Dzhambul, Zhambyl. Zhambyl, S Kazakhstan 42°55′N 71°27′E
105 Q5 **Tarazona** Aragón, NE Spain 41°54′N 01°44′W
105 Q10 **Tarazona de la Mancha** Castilla-La Mancha, C Spain 39°16′N 01°55′W
145 X12 **Tarbagatay, Khrebet** ▲▲ China/Kazakhstan
96 J8 **Tarbat Ness** *headland* N Scotland, United Kingdom 57°51′N 03°48′W
149 U5 **Tarbela Reservoir** ⊠ N Pakistan
96 H12 **Tarbert** W Scotland, United Kingdom 55°52′N 05°26′W
96 F7 **Tarbert** NW Scotland, United Kingdom 57°54′N 06°48′W
102 K16 **Tarbes** *anc.* Bigorra. Hautes-Pyrénées, S France 43°14′N 00°04′E
21 W9 **Tarboro** North Carolina, SE USA 35°54′N 77°34′W
Tarca *see* Torysa
106 J6 **Tarcento** Friuli-Venezia Giulia, NE Italy 46°13′N 13°13′E
182 F5 **Tarcoola** South Australia 30°44′S 134°34′E
105 S5 **Tardienta** Aragón, NE Spain 41°58′N 00°31′W
102 L11 **Tardoire** ≈ W France
183 U7 **Taree** New South Wales, SE Australia 31°56′S 152°29′E
92 K12 **Tärendö** *Lapp.* Deargget. Norrbotten, N Sweden 67°10′N 22°40′E
Tarentum *see* Taranto
74 C9 **Tarfaya** SW Morocco 27°56′N 12°55′W
114 L8 **Targovishte** , Tărgovište; *prev.* Eski Dzhumaya. Targovishte, N Bulgaria 43°15′N 26°34′E
114 L8 **Targovishte** *var.* Tŭrgovishte. ◈ *province* N Bulgaria
116 J13 **Târgovişte** *prev.* Tîrgovişte. Dâmboviţa, S Romania 44°54′N 25°29′E
Tărgovište *see* Targovishte
116 M12 **Târgu Bujor** *prev.* Tîrgu Bujor. Galaţi, E Romania 45°52′N 27°55′E
116 H13 **Târgu Cărbuneşti** *prev.* Tîrgu. Gorj, SW Romania 44°57′N 23°32′E
116 L9 **Târgu Frumos** *prev.* Tîrgu Frumos. Iaşi, NE Romania 47°12′N 27°00′E
116 H13 **Targu Jiu** *prev.* Tîrgu Jiu. Gorj, W Romania 45°03′N 23°20′E
116 H9 **Târgu Lăpuş** *prev.* Tîrgu Lăpuş. Maramureş, N Romania 47°28′N 23°54′E
Târgul-Neamţ *see* Târgu-Neamţ
Târgul-Săcuiesc *see* Târgu Secuiesc
116 I10 **Târgu Mureş** *prev.* Oşorhei, Tîrgu Mures, *Ger.* Neumarkt, *Hung.* Marosvásárhely. Mureş, C Romania 46°33′N 24°36′E
116 K9 **Târgu-Neamţ** *var.* Târgul-Neamţ; *prev.* Tîrgu-Neamţ. Neamţ, NE Romania 47°12′N 26°25′E
116 K11 **Târgu Ocna** *Hung.* Aknavásár; *prev.* Tîrgu Ocna. Bacău, E Romania 46°17′N 26°37′E
116 K11 **Târgu Secuiesc** *Ger.* Neumarkt, Szekler Neumarkt, *Hung.* Kezdivásárhely; *prev.* Chezdi-Oşorheiu, Târgul-Săcuiesc, Tîrgu Secuiesc. Covasna, E Romania 46°00′N 26°08′E
145 X10 **Targyn** Vostochnyy Kazakhstan, E Kazakhstan 49°32′N 82°47′E
Tar Heel State *see* North Carolina
186 C7 **Tari** Hela, W Papua New Guinea 05°52′S 142°58′E
162 J6 **Tarialan** *var.* Badrah. Hövsgöl, N Mongolia 49°33′N 101°58′E
162 I7 **Tariat** *var.* Horgo. Arhangay, C Mongolia 48°06′N 99°52′E
143 P17 **Ţarīf** Abū Zaby, C United Arab Emirates 24°02′N 53°47′E
104 K16 **Tarifa** Andalucía, S Spain 36°01′N 05°36′W
84 C14 **Tarifa, Punta de** *headland* SW Spain 36°01′N 05°39′W
57 M21 **Tarija** Tarija, S Bolivia 21°33′S 64°42′W
57 M21 **Tarija** ◈ *department* S Bolivia
141 R14 **Tarīm** C Yemen 16°N 48°50′E
Tarim Basin *see* Tarim Pendi
81 G19 **Tarime** Mara, N Tanzania 01°20′S 34°24′E
129 S8 **Tarim He** ≈ NW China
159 H8 **Tarim Pendi** *Eng.* Tarim Basin. *basin* NW China
149 N7 **Tarīn Kōt** *var.* Terinkot; *prev.* Tarīn Kowt. Uruzgān, C Afghanistan 32°38′N 65°52′E
Tarīn Kowt *see* Tarīn Kōt
171 O12 **Taripa** Sulawesi, C Indonesia 01°51′S 120°46′E
117 Q12 **Tarkhankut, Mys** *headland* S Ukraine 45°20′N 32°32′E
27 Q1 **Tarkio** Missouri, C USA 40°25′N 95°24′W
122 J9 **Tarko-Sale** Yamalo-Nenetskiy Avtonomnyy Okrug, N Russian Federation 64°55′N 77°34′E
77 P17 **Tarkwa** S Ghana 05°16′N 01°59′W
171 O3 **Tarlac** Luzon, N Philippines 15°29′N 120°34′E
95 F22 **Tarm** Midtjylland, W Denmark 55°55′N 08°32′E
57 E14 **Tarma** Junín, C Peru 11°28′S 75°41′W
103 N15 **Tarn** ◈ *department* S France
102 M15 **Tarn** ≈ S France
111 L22 **Tarna** ≈ C Hungary
92 G13 **Tärnaby** Västerbotten, N Sweden 65°44′N 15°20′E
149 P8 **Tarnak Rūd** ≈ SE Afghanistan
116 J11 **Târnava Mare** *Ger.* Grosse Kokel, *Hung.* Nagy-Küküllő; *prev.* Tîrnava Mare. ≈ S Romania
116 I11 **Târnava Mică** *Ger.* Kleine Kokel, *Hung.* Kis-Küküllo; *prev.* Tîrnava Mică. ≈ C Romania
116 I11 **Târnăveni** *Ger.* Marteskirch, Martinskirch, *Hung.* Dicsőszentmárton; *prev.* Sinmartin, Tîrnăveni. Mureş, C Romania 46°20′N 24°17′E
102 L14 **Tarn-et-Garonne** ◈ *department* S France
111 P18 **Tarnica** ▲ SE Poland 49°05′N 22°43′E
111 N15 **Tarnobrzeg** Podkarpackie, SE Poland 50°35′N 21°40′E
125 N12 **Tarnogskiy Gorodok** Vologodskaya Oblast', NW Russian Federation 60°28′N 43°45′E
Tarnopol *see* Ternopil'
111 M16 **Tarnów** Małopolskie, S Poland 50°01′N 20°59′E
Tarnowice/Tarnowitz *see* Tarnowskie Góry
111 J16 **Tarnowskie Góry** *var.* Tarnowice, Tarnowskie Gory, *Ger.* Tarnowitz. Śląskie, S Poland 50°27′N 18°52′E
95 N14 **Tärnsjö** Västmanland, C Sweden 60°10′N 16°57′E
186 K7 **Taro** Choiseul, NW Solomon Islands 07°00′S 156°57′E
106 E9 **Taro** ≈ NW Italy
186 I6 **Taron** New Ireland, NE Papua New Guinea 04°22′S 153°04′E
74 E8 **Taroudannt** *var.* Taroudant. SW Morocco 30°31′N 08°50′W
Taroudant *see* Taroudannt
23 V12 **Tarpon, Lake** ◎ Florida, SE USA
23 V12 **Tarpon Springs** Florida, SE USA 28°09′N 82°45′W
107 G14 **Tarquinia** *anc.* Tarquinii, *hist.* Corneto. Lazio, C Italy 42°23′N 11°45′E
Tarquinii *see* Tarquinia
Tarraco *see* Tarragona
76 D10 **Tarrafal** Santiago, S Cape Verde 15°16′N 23°45′W
105 V6 **Tarragona** *anc.* Tarraco. Cataluña, E Spain 41°07′N 01°15′E
105 T7 **Tarragona** ◈ *province* Cataluña, NE Spain
183 O17 **Tarraleah** Tasmania, SE Australia 42°11′S 146°29′E
23 P3 **Tarrant City** Alabama, S USA 33°34′N 86°45′W
185 D21 **Tarras** Otago, South Island, New Zealand 44°48′S 169°25′E
Tarrasa *see* Terrassa
105 U5 **Tàrrega** *var.* Tarrega. Cataluña, NE Spain 41°39′N 01°09′E
21 W9 **Tar River** ≈ North Carolina, SE USA
Tarsatica *see* Rijeka
136 J17 **Tarsus** İçel, S Turkey 36°52′N 34°52′E
62 K4 **Tartagal** Salta, N Argentina 22°32′S 63°50′W
137 V12 **Tärtär** *Rus.* Terter. ≈ SW Azerbaijan
102 J15 **Tartas** Landes, SW France 43°52′N 00°45′W
Tartlau *see* Prejmer
Tartous/Tartouss *see* Ţarţūs
118 J5 **Tartu** *Ger.* Dorpat; *prev. Rus.* Yurev, Yury'ev. Tartumaa, SE Estonia 58°20′N 26°44′E
118 I5 **Tartumaa** *off.* Tartu Maakond. ◈ *province* E Estonia
Tartu Maakond *see* Tartumaa
138 H5 **Ţarţūs** *Fr.* Tartouss; *anc.* Tortosa. Ţarţūs, W Syria 34°55′N 35°52′E
138 H5 **Ţarţūs** *off.* Muḩāfaẓat Ţarţūs, *var.* Tartous, Tartus. ◈ *governorate* W Syria
Ţarţūs, Muḩāfaẓat *see* Ţarţūs
164 C16 **Tarumizu** Kagoshima, Kyūshū, SW Japan 31°30′N 130°40′E
126 K4 **Tarusa** Kaluzhskaya Oblast', W Russian Federation 54°45′N 37°10′E
117 N11 **Tarutyne** Odes'ka Oblast', SW Ukraine 46°11′N 29°09′E
162 I7 **Tarvagatyn Nuruu** ▲▲ N Mongolia
106 J6 **Tarvisio** Friuli-Venezia Giulia, NE Italy 46°31′N 13°33′E
Tarvisium *see* Treviso
57 O16 **Tarvo, Río** ≈ E Bolivia
14 G8 **Tarzwell** Ontario, S Canada 48°00′N 79°58′W
40 K5 **Tasajera, Sierra de la** ▲▲ N Mexico
145 S13 **Tasaral** Karaganda, C Kazakhstan 46°17′N 73°54′E
145 N15 **Tasboget** *Kaz.* Tasböget; *prev.* Tasbuget. Kzylorda, S Kazakhstan 44°46′N 65°38′E
Tasböget *see* Tasboget
Tasbuget *see* Tasboget
108 E11 **Tasch** Valais, SW Switzerland 46°04′N 07°43′E
Tasek Kenyir *see* Kenyir, Tasik
122 J14 **Tashanta** Respublika Altay, S Russian Federation 49°42′N 89°15′E
Tashauz *see* Daşoguz
Tashi Chho Dzong *see* Thimphu
153 U11 **Tashigang** E Bhutan 27°19′N 91°33′E
137 T11 **Tashir** *prev.* Kalinino. N Armenia 41°07′N 44°16′E
143 Q11 **Ţashk, Daryācheh-ye** ◎ C Iran
Tashkent *see* Toshkent
Tashkentskaya Oblast' *see* Toshkent Viloyati
Tashkepri *see* Daşköpri
Tash-Kömür *see* Tash-Kumyr
147 S9 **Tash-Kumyr** *Kir.* Tash-Kömür. Dzhalal-Abadskaya Oblast', W Kyrgyzstan 41°22′N 72°09′E
127 T7 **Tashla** Orenburgskaya Oblast', W Russian Federation 51°42′N 52°33′E
Tashqurghan *see* Khulm
122 J13 **Tashtagol** Kemerovskaya Oblast', S Russian Federation 52°49′N 88°00′E
95 H24 **Tåsinge** *island* C Denmark
12 M5 **Tasiujaq** Québec, E Canada 58°43′N 69°58′W
144 F8 **Taskala** *prev.* Kamenka. Zapadnyy Kazakhstan, NW Kazakhstan 51°06′N 51°16′E
77 W11 **Tasker** Zinder, C Niger 15°06′N 10°42′E
145 W12 **Taskesken** Vostochnyy Kazakhstan, E Kazakhstan 47°15′N 80°45′E
136 J10 **Taşköprü** Kastamonu, N Turkey 41°30′N 34°12′E
Taskuduk, Peski *see* Tosquduq Qumlari
186 G5 **Taskul** New Ireland, NE Papua New Guinea 02°34′S 150°25′E
137 S13 **Taşlıçay** Ağrı, E Turkey 39°37′N 43°23′E
185 H14 **Tasman** *off.* Tasman District. ◈ *unitary authority* South Island, New Zealand
192 J12 **Tasman Basin** *var.* East Australian Basin. *undersea feature* S Tasman Sea
185 I14 **Tasman Bay** *inlet* South Island, New Zealand
Tasman District *see* Tasman
192 I13 **Tasman Fracture Zone** *tectonic feature* S Indian Ocean
185 E19 **Tasman Glacier** *glacier* South Island, New Zealand
Tasman Group *see* Nukumanu Islands
183 N15 **Tasmania** *prev.* Van Diemen's Land. ◈ *state* SE Australia
183 Q16 **Tasmania** *island* SE Australia
185 H14 **Tasman Mountains** ▲▲ South Island, New Zealand
183 P17 **Tasman Peninsula** *peninsula* Tasmania, SE Australia
192 I11 **Tasman Plain** *undersea feature* W Tasman Sea
192 I12 **Tasman Plateau** *var.* South Tasmania Plateau. *undersea feature* SW Tasman Sea
192 I11 **Tasman Sea** *sea* SW Pacific Ocean
116 G9 **Tăşnad** *Ger.* Trestenberg, Trestendorf, *Hung.* Tasnád. Satu Mare, NW Romania 47°30′N 22°33′E
136 L11 **Taşova** Amasya, N Turkey 40°45′N 36°20′E
77 T10 **Tassara** Tahoua, W Niger 16°40′N 05°34′E
12 K4 **Tassialouc, Lac** ◎ Québec, C Canada
Tassili du Hoggar *see* Tassili Ta-n-Ahaggar
74 L11 **Tassili-n-Ajjer** *plateau* E Algeria
74 K14 **Tassili Ta-n-Ahaggar** *var.* Tassili du Hoggar, Tassili ta-n-Ahaggar. *plateau* S Algeria
Tassili ta-n-Ahaggar *see* Tassili Ta-n-Ahaggar
59 M15 **Tasso Fragoso** Maranhão, E Brazil 08°22′S 45°53′W
Tåstrup *see* Taastrup
145 O9 **Tasty-Taldy** Akmola, C Kazakhstan 50°47′N 66°31′E
143 W10 **Tāsūkī** Sīstān va Balūchestān, SE Iran
111 I22 **Tata** *Ger.* Totis. Komárom-Esztergom, NW Hungary 47°39′N 18°19′E
74 E8 **Tata** SW Morocco 29°38′N 08°04′W
Tataaihoa, Pointe *see* Vénus, Pointe
111 I22 **Tatabánya** Komárom-Esztergom, NW Hungary 47°33′N 18°23′E
191 X10 **Tatakoto** *atoll* Îles Tuamotu, E French Polynesia
75 N7 **Tataouine** *var.* Taţāwīn. SE Tunisia 32°48′N 10°27′E
55 O5 **Tataracual, Cerro** ▲ NE Venezuela 10°13′N 64°20′W
117 O12 **Tatarbunary** Odes'ka Oblast', SW Ukraine 45°50′N 29°37′E
119 M17 **Tatarka** Mahilyowskaya Voblasts', E Belarus 53°15′N 28°50′E
Tatar Pazardzhik *see* Pazardzhik
122 I12 **Tatarsk** Novosibirskaya Oblast', C Russian Federation 55°08′N 75°58′E
Tatarskaya ASSR *see* Tatarstan, Respublika
123 T13 **Tatarskiy Proliv** *Eng.* Tatar Strait. *strait* SE Russian Federation
127 R4 **Tatarstan, Respublika** *prev.* Tatarskaya ASSR. ◈ *autonomous republic* W Russian Federation
Tatar Strait *see* Tatarskiy Proliv
Taţāwīn *see* Tataouine
171 N12 **Tate** Sulawesi, N Indonesia 0°12′S 119°44′E
141 N11 **Tathlīth** 'Asīr, S Saudi Arabia 19°38′N 43°32′E
141 O11 **Tathlīth, Wādī** *dry watercourse* S Saudi Arabia
183 R11 **Tathra** New South Wales, SE Australia 36°46′S 149°58′E
127 P8 **Tatishchevo** Saratovskaya Oblast', W Russian Federation 51°43′N 45°35′E
39 S12 **Tatitlek** Alaska, USA 60°49′N 146°29′W
10 L15 **Tatla Lake** British Columbia, SW Canada 51°54′N 124°39′W
121 Q2 **Tatlısu** *Gk.* Akanthoú. N Cyprus 35°21′N 33°45′E
11 Z10 **Tatnam, Cape** *headland* Manitoba, C Canada 57°16′N 91°03′W
111 K18 **Tatra Mountains** *Ger.* Tatra, *Hung.* Tátra, *Pol./Slvk.* Tatry. ▲▲ Poland/Slovakia
Tatra/Tátra *see* Tatra Mountains
Tatry *see* Tatra Mountains
164 I13 **Tatsuno** *var.* Tatuno. Hyōgo, Honshū, SW Japan 34°54′N 134°30′E
Tatti *see* Tatty
145 S16 **Tatty** *prev.* Tatti. Zhambyl, S Kazakhstan 43°11′N 73°22′E
60 L10 **Tatuí** São Paulo, S Brazil 23°21′S 47°49′W
37 V14 **Tatum** New Mexico, SW USA 33°15′N 103°19′W
25 X7 **Tatum** Texas, SW USA 32°19′N 94°31′W
Ta-t'ung/Tatung *see* Datong
Tatuno *see* Tatsuno
137 R14 **Tatvan** Bitlis, SE Turkey 38°31′N 42°15′E
95 C16 **Tau** Rogaland, S Norway 59°04′N 05°55′E
192 L17 **Ta'ū** *var.* Tau. *island* Manua Islands, E American Samoa
193 W15 **Tau** *island* Tongatapu Group, N Tonga
59 O14 **Tauá** Ceará, E Brazil 06°04′S 40°26′W
60 N10 **Taubaté** São Paulo, S Brazil 23°S 45°36′W
101 I19 **Tauber** ≈ SW Germany
101 I19 **Tauberbischofsheim** Baden-Württemberg, C Germany 49°37′N 09°39′E
Tauchik *see* Taushyk
191 W10 **Tauere** *atoll* Îles Tuamotu, C French Polynesia
101 H17 **Taufstein** ▲ C Germany 50°31′N 09°18′E
190 I17 **Taukoka** *island* SE Cook Islands
145 T15 **Taukum, Peski** *desert* SE Kazakhstan
184 L10 **Taumarunui** Manawatu-Wanganui, North Island, New Zealand 38°52′S 175°14′E
59 A15 **Taumaturgo** Acre, W Brazil 08°54′S 72°48′W
27 X6 **Taum Sauk Mountain** ▲ Missouri, C USA 37°34′N 90°43′W
83 H22 **Taung** North-West, N South Africa 27°32′S 24°48′E
166 L6 **Taungdwingyi** Magway, C Myanmar (Burma) 20°01′N 95°20′E
166 M6 **Taunggyi** Shan State, C Myanmar (Burma) 20°47′N 97°00′E
166 M7 **Taungoo** Bago, C Myanmar (Burma) 18°57′N 96°26′E
166 L5 **Taungtha** Mandalay, C Myanmar (Burma) 21°16′N 95°25′E
Taungup *see* Toungup
149 S9 **Taunsa** Punjab, E Pakistan 30°43′N 70°41′E
97 K23 **Taunton** SW England, United Kingdom 51°01′N 03°06′W
19 O12 **Taunton** Massachusetts, NE USA 41°54′N 71°03′W
101 F18 **Taunus** ▲▲ W Germany
101 G18 **Taunusstein** Hessen, W Germany 50°09′N 08°09′E
184 N9 **Taupo** Waikato, North Island, New Zealand 38°42′S 176°05′E
184 M9 **Taupo, Lake** ◎ North Island, New Zealand
109 R8 **Taurach** *var.* Taurachbach. ≈ E Austria
Taurachbach *see* Taurach
118 D12 **Tauragė** *Ger.* Tauroggen. Tauragė, SW Lithuania 55°16′N 22°17′E
118 D13 **Tauragė** ◈ *province* Lithuania
54 G10 **Tauramena** Casanare, C Colombia 05°02′N 72°43′W
184 N7 **Tauranga** Bay of Plenty, North Island, New Zealand 37°42′S 176°09′E
15 O10 **Taureau, Réservoir** ⊠ Québec, SE Canada
107 N22 **Taurianova** Calabria, SW Italy 38°22′N 16°01′E
Tauris *see* Tabrīz
184 I2 **Tauroa Point** *headland* North Island, New Zealand 35°09′S 173°02′E
Tauroggen *see* Tauragė
Tauromenium *see* Taormina
Taurus Mountains *see* Toros Dağları
Taus *see* Domažlice
144 E14 **Taushyk** *Kaz.* Taūshyq; *prev.* Tauchik. Mangistau, SW Kazakhstan 44°17′N 51°22′E
Taūshyq *see* Taushyk
105 R5 **Tauste** Aragón, NE Spain 41°55′N 01°15′W
191 V16 **Tautara, Motu** *island* Easter Island, Chile, E Pacific Ocean
191 R8 **Tautira** Tahiti, W French Polynesia 17°45′S 149°10′W
Tauz *see* Tovuz
Ţavālesh, Kūhhā-ye *see* Talish Mountains
136 D15 **Tavas** Denizli, SW Turkey 37°33′N 29°04′E
Tavastehus *see* Hämeenlinna
Tavau *see* Davos
122 G10 **Tavda** Sverdlovskaya Oblast', C Russian Federation 58°01′N 65°07′E
122 G10 **Tavda** ≈ C Russian Federation
105 T11 **Tavernes de la Valldigna** Valenciana, E Spain 39°03′N 00°13′W
81 I20 **Taveta** Taita/Taveta, S Kenya 03°23′S 37°40′E
187 Y14 **Taveuni** *island* N Fiji
147 R13 **Tavildara** *Rus.* Tavil'dara, Tovil'-Dora. C Tajikistan 38°42′N 70°27′E
Tavin *see* Adaatsag
104 H14 **Tavira** Faro, S Portugal 37°07′N 07°39′W
97 I24 **Tavistock** SW England, United Kingdom 50°33′N 04°08′W
Tavoy *see* Dawei
Tavoy Island *see* Mali Kyun
115 E16 **Tavropoú, Technití Límni** ⊠ C Greece
136 E13 **Tavşanlı** Kütahya, NW Turkey 39°34′N 29°28′E
187 X14 **Tavua** Viti Levu, W Fiji 17°27′S 177°51′E
97 J23 **Taw** ≈ SW England, United Kingdom
185 L14 **Tawa** Wellington, North Island, New Zealand 41°10′S 174°50′E
25 V6 **Tawakoni, Lake** ⊠ Texas, SW USA
153 V11 **Tawang** Arunāchal Pradesh, NE India 27°34′N 91°54′E
169 R17 **Tawang, Teluk** *bay* Jawa, S Indonesia
31 R7 **Tawas Bay** ◎ Michigan, N USA
31 R7 **Tawas City** Michigan, N USA 44°16′N 83°33′W
169 V8 **Tawau** Sabah, East Malaysia 04°16′N 117°54′E
141 U10 **Ţawīl, Qalamat aţ** *well* SE Saudi Arabia
171 N9 **Tawitawi** *island* Tawitawi Group, SW Philippines
Ţawkar *see* Tokar
Tāwūq *see* Dāqūq
Tawzar *see* Tozeur
41 O15 **Taxco** *var.* Taxco de Alarcón. Guerrero, S Mexico 18°32′N 99°37′W
Taxco de Alarcón *see* Taxco
146 H8 **Taxiatosh** *Rus.* Takhiatash. Qoraqalpog'iston Respublikasi, W Uzbekistan 42°27′N 59°27′E
158 D9 **Taxkorgan** *var.* Taxkorgan Tajik Zizhixian. Xinjiang Uygur Zizhiqu, NW China 37°43′N 75°13′E
Taxkorgan Tajik Zizhixian *see* Taxkorgan
146 H7 **Taxtako'pir** *Rus.* Takhtakupyr. Qoraqalpog'iston Respublikasi, NW Uzbekistan 43°04′N 60°23′E
96 J10 **Tay** ≈ C Scotland, United Kingdom
143 V6 **Tāybād** *var.* Taibad, Tāyybād, Tayyebāt. Khorāsān-e Razavī, NE Iran 34°48′N 60°46′E
Taybert at Turkz *see* Ţayyibat at Turkī
124 J3 **Taybola** Murmanskaya Oblast', NW Russian Federation 68°30′N 33°18′E
81 M16 **Tayeeglow** Bakool, C Somalia 04°01′N 44°25′E
96 K11 **Tay, Firth of** *inlet* E Scotland, United Kingdom
122 J12 **Tayga** Kemerovskaya Oblast', S Russian Federation 56°02′N 85°26′E
Taygan *see* Delger
123 T9 **Taygonos, Mys** *headland* E Russian Federation 60°36′N 160°09′E
96 I11 **Tay, Loch** ◎ C Scotland, United Kingdom
11 N12 **Taylor** British Columbia, W Canada 56°09′N 120°43′W
29 O14 **Taylor** Nebraska, C USA 41°47′N 99°23′W
18 I13 **Taylor** Pennsylvania, NE USA 41°22′N 75°41′W
25 T10 **Taylor** Texas, SW USA 30°34′N 97°24′W
37 Q11 **Taylor, Mount** ▲ New Mexico, SW USA 35°14′N 107°36′W
37 R5 **Taylor Park Reservoir** ⊠ Colorado, C USA
37 R6 **Taylor River** ≈ Colorado, C USA
21 P11 **Taylors** South Carolina, SE USA 34°55′N 82°18′W
20 L5 **Taylorsville** Kentucky, S USA 38°01′N 85°21′W
21 R6 **Taylorsville** North Carolina, SE USA 35°56′N 81°10′W
30 L14 **Taylorville** Illinois, N USA 39°33′N 89°17′W
140 K5 **Taymā'** Tabūk, NW Saudi Arabia 27°39′N 38°32′E
122 M10 **Taymura** ≈ C Russian Federation
123 O7 **Taymylyr** Respublika Sakha (Yakutiya), NE Russian Federation 72°32′N 121°54′E
122 L7 **Taymyr, Ozero** ◎ N Russian Federation
122 M6 **Taymyr, Poluostrov** *peninsula* N Russian Federation
122 L8 **Taymyrskiy (Dolgano-Nenetskiy) Avtonomnyy Okrug** ◈ *autonomous district* Krasnoyarskiy Kray, N Russian Federation
167 S13 **Tây Ninh** Tây Ninh, S Vietnam 11°21′N 106°07′E
122 L12 **Tayshet** Irkutskaya Oblast', S Russian Federation 55°51′N 98°04′E
162 G8 **Tayshir** *var.* Tsagaan-Olom. Govĭ-Altay, C Mongolia 46°42′N 96°30′E
171 N5 **Taytay** Palawan, W Philippines 10°49′N 119°30′E
169 Q16 **Tayu** *prev.* Tajoe. Jawa, C Indonesia 06°32′S 111°02′E
Tāyybād/Tayyebāt *see* Tāybād
138 L5 **Ţayyibah** *var.* At Taybé. Ḩimş, C Syria 35°13′N 38°51′E
138 I4 **Ţayyibat at Turkī** *var.* Taybert at Turkz. Ḩamāh, W Syria 35°16′N 36°55′E
145 P7 **Tayynsha** *prev.* Krasnoarmeysk. Severnyy Kazakhstan, N Kazakhstan 53°52′N 69°51′E
122 J10 **Taz** ≈ N Russian Federation
74 G6 **Taza** NE Morocco 34°13′N 04°06′W
139 T4 **Tāza Khurmātū** Kirkūk, E Iraq 35°18′N 44°22′E
165 Q8 **Tazawa-ko** ◎ Honshū, C Japan
21 N8 **Tazewell** Tennessee, S USA 36°27′N 83°34′W
21 Q7 **Tazewell** Virginia, NE USA 37°08′N 81°34′W
75 S11 **Tāzirbū** SE Libya 25°43′N 21°16′E
39 S11 **Tazlina Lake** ◎ Alaska, USA
122 J8 **Tazovskiy** Yamalo-Nenetskiy Avtonomnyy Okrug, N Russian Federation 67°33′N 78°21′E
137 U10 **Tbilisi** *Eng.* Tiflis. ● (Georgia) SE Georgia 41°41′N 44°55′E
137 T10 **Tbilisi** ✈ S Georgia 41°43′N 44°49′E
79 E14 **Tchabal Mbabo** ▲ NW Cameroon 07°12′N 12°16′E
Tchad *see* Chad
Tchad, Lac *see* Chad, Lake
77 S15 **Tchaourou** E Benin 08°58′N 02°40′E
79 E20 **Tchibanga** Nyanga, S Gabon 02°49′S 11°00′E
Tchien *see* Zwedru
77 Z6 **Tchigaï, Plateau du** ▲▲ NE Niger
77 V9 **Tchighozérine** Agadez, C Niger 17°15′N 07°48′E
77 T10 **Tchin-Tabaradene** Tahoua, W Niger 15°57′N 05°49′E
78 G13 **Tcholliré** Nord, NE Cameroon 08°48′N 14°00′E
Tchongking *see* Chongqing
22 K4 **Tchula** Mississippi, S USA 33°10′N 90°13′W
110 I7 **Tczew** *Ger.* Dirschau. Pomorskie, N Poland 54°05′N 18°46′E
116 I10 **Teaca** *Ger.* Tekendorf, *Hung.* Teke; *prev. Ger.* Teckendorf. Bistriţa-Năsăud, N Romania 46°55′N 24°30′E
40 J11 **Teacapán** Sinaloa, C Mexico 22°33′N 105°44′W
190 A10 **Teafuafou** *island* Funafuti Atoll, C Tuvalu
25 U8 **Teague** Texas, SW USA 31°37′N 96°16′W
191 R9 **Teahupoo** Tahiti, W French Polynesia 17°51′S 149°15′W
190 H15 **Te Aiti Point** *headland* Rarotonga, S Cook Islands 21°11′S 59°47′W
65 D24 **Teal Inlet** East Falkland, Falkland Islands 51°34′S 58°25′W
185 B22 **Te Anau** Southland, South Island, New Zealand 45°25′S 167°45′E
185 B22 **Te Anau, Lake** ◎ South Island, New Zealand
41 U15 **Teapa** Tabasco, SE Mexico 17°36′N 92°57′W
184 Q7 **Te Araroa** Gisborne, North Island, New Zealand 37°37′S 178°21′E
184 M7 **Te Aroha** Waikato, North Island, New Zealand 37°32′S 175°58′E
Teate *see* Chieti
190 A9 **Te Ava Fuagea** *channel* Funafuti Atoll, SE Tuvalu
190 B8 **Te Ava I Te Lape** *channel* Funafuti Atoll, SE Tuvalu
190 B9 **Te Ava Pua Pua** *channel* Funafuti Atoll, SE Tuvalu
184 M8 **Te Awamutu** Waikato, North Island, New Zealand 38°00′S 177°18′E
171 X12 **Teba** Papua, E Indonesia 01°27′S 137°54′E
104 L15 **Teba** Andalucía, S Spain 36°59′N 04°54′W
126 M15 **Teberda** Karachayevo-Cherkesskaya Respublika, SW Russian Federation 43°28′N 41°45′E
74 M6 **Tébessa** NE Algeria 35°21′N 08°06′E
62 O7 **Tebicuary, Río** ≈ S Paraguay
168 L13 **Tebingtinggi** Sumatera, W Indonesia 03°33′S 103°00′E
168 I8 **Tebingtinggi** Sumatera, N Indonesia 03°20′N 99°08′E
Tebingtinggi, Pulau *see* Rantau, Pulau
Tebriz *see* Tabrīz
137 U9 **T'ebulos Mta** *Rus.* Gora Tebulosmta; *prev.* Tebulos Mta. ▲ Georgia/Russian Federation 42°33′N 45°21′E
Tebulos Mta *see* T'ebulos Mta
Tebulosmta, Gora *see* T'ebulos Mta
41 Q14 **Tecamachalco** Puebla, S Mexico 18°52′N 97°44′W
40 B1 **Tecate** Baja California Norte, NW Mexico 32°33′N 116°38′W
136 M13 **Tecer Dağları** ▲▲ C Turkey
103 O17 **Tech** ≈ S France
77 P16 **Techiman** W Ghana 07°35′N 01°56′W
117 N15 **Techirghiol** Constanţa, SE Romania 44°03′N 28°37′E
74 A12 **Techla** *var.* Techlé. SW Western Sahara 21°39′N 14°57′W
Techlé *see* Techla
63 H18 **Tecka, Sierra de** ▲▲ SW Argentina
Teckendorf *see* Teaca
40 K13 **Tecolotlán** Jalisco, SW Mexico 20°10′N 104°07′W
40 K14 **Tecomán** Colima, SW Mexico 18°53′N 103°54′W
35 V12 **Tecopa** California, W USA 35°51′N 116°14′W
40 G5 **Tecoripa** Sonora, NW Mexico 28°38′N 109°58′W
41 N16 **Tecpan** *var.* Tecpan de Galeana. Guerrero, S Mexico 17°12′N 100°39′W
Tecpan de Galeana *see* Tecpan
40 J11 **Tecuala** Nayarit, C Mexico 22°24′N 105°30′W
116 L12 **Tecuci** Galaţi, E Romania 45°50′N 27°27′E
31 R10 **Tecumseh** Michigan, N USA 42°00′N 83°57′W
29 S16 **Tecumseh** Nebraska, C USA 40°22′N 96°12′W
27 O11 **Tecumseh** Oklahoma, C USA 35°15′N 96°56′W
Tedzhen *see* Harīrūd/Tejen
Tedzhen *see* Tejen
146 H15 **Tedzhenstroy** *Turkm.* Tejenstroy. Ahal Welaýaty, S Turkmenistan 36°57′N 60°49′E
Teel *see* Öndör-Ulaan
97 L15 **Tees** ≈ N England, United Kingdom
14 E15 **Teeswater** Ontario, S Canada 44°00′N 81°17′W
190 A10 **Tefala** *island* Funafuti Atoll, C Tuvalu
58 D13 **Tefé** Amazonas, N Brazil 03°24′S 64°45′W
74 K11 **Tefedest** ▲▲ S Algeria
136 E16 **Tefenni** Burdur, SW Turkey 37°19′N 29°45′E
58 D13 **Tefé, Rio** ≈ NW Brazil
169 P16 **Tegal** Jawa, C Indonesia 06°52′S 109°07′E
100 O12 **Tegel** ✈ (Berlin) Berlin, NE Germany 52°33′N 13°16′E
99 M15 **Tegelen** Limburg, SE Netherlands 51°20′N 06°09′E
101 L24 **Tegernsee** ◎ SE Germany
107 M18 **Teggiano** Campania, S Italy 40°25′N 15°28′E
77 U14 **Tegina** Niger, C Nigeria 10°06′N 06°10′E
Tegucigalpa *see* Central District
Tegucigalpa *see* Toncontín
Tegucigalpa *see* Francisco Morazán
77 U9 **Teguidda-n-Tessoumt** Agadez, C Niger 17°27′N 06°40′E
64 Q11 **Teguise** Lanzarote, Islas Canarias, Spain, NE Atlantic Ocean 29°04′N 13°38′W
122 K12 **Tegul'det** Tomskaya Oblast', C Russian Federation 57°16′N 87°58′E
35 S13 **Tehachapi** California, W USA 35°07′N 118°27′W
35 S13 **Tehachapi Mountains** ▲▲ California, W USA
Tehama *see* Tihāmah
Teheran *see* Tehrān
77 O14 **Téhini** NE Ivory Coast 09°36′N 03°40′W
143 N5 **Tehrān** *var.* Teheran. ● (Iran) Tehrān, N Iran 35°44′N 51°27′E
143 N6 **Tehrān** *off.* Ostān-e Tehrān, *var.* Tehran. ◈ *province* N Iran
Tehrān, Ostān-e *see* Tehrān
Tehri *see* Tīkamgarh
Tehri *see* New Tehri
41 Q15 **Tehuacán** Puebla, S Mexico 18°29′N 97°24′W
41 S17 **Tehuantepec** *var.* Santo Domingo Tehuantepec. Oaxaca, SE Mexico 16°18′N 95°14′W
41 S17 **Tehuantepec, Golfo de** *var.* Gulf of Tehuantepec. *gulf* S Mexico
Tehuantepec, Gulf of *see* Tehuantepec, Golfo de
Tehuantepec, Isthmus of *see* Tehuantepec, Istmo de
41 T16 **Tehuantepec, Istmo de** *var.* Isthmus of Tehuantepec. *isthmus* SE Mexico
0 I16 **Tehuantepec Ridge** *undersea feature* E Pacific Ocean 13°30′N 98°00′W
41 S16 **Tehuantepec, Río** ≈ SE Mexico
191 W10 **Tehuata** *atoll* Îles Tuamotu, C French Polynesia
64 O11 **Teide, Pico del** ▲ Gran Canaria, Islas Canarias, Spain, NE Atlantic Ocean 28°16′N 16°39′W
97 I21 **Teifi** ≈ SW Wales, United Kingdom
80 B9 **Teiga Plateau** *plateau* W Sudan
97 J24 **Teignmouth** SW England, United Kingdom 50°34′N 03°29′W
Teisen *see* Jecheon
116 H1 **Teiuş** *Ger.* Dreikirchen, *Hung.* Tövis. Alba, C Romania 46°12′N 23°40′E
169 U17 **Tejakula** Bali, C Indonesia 08°09′S 115°19′E
146 H14 **Tejen** *Rus.* Tedzhen. Ahal Welaýaty, S Turkmenistan 37°24′N 60°29′E
146 I15 **Tejen** *Per.* Harīrūd, *Rus.* Tedzhen. ≈ Afghanistan/Iran *see also* Harīrūd
Tejen *see* Harīrūd
Tejenstroy *see* Tedzhenstroy
35 S14 **Tejon Pass** *pass* California, W USA
Tejo, Rio *see* Tagus
41 O14 **Tejupilco** *var.* Tejupilco de Hidalgo. México, S Mexico 18°55′N 100°10′W
Tejupilco de Hidalgo *see* Tejupilco
184 P7 **Te Kaha** Bay of Plenty, North Island, New Zealand 37°45′S 177°42′E
29 S14 **Tekamah** Nebraska, C USA 41°46′N 96°13′W
184 I1 **Te Kao** Northland, North Island, New Zealand 34°40′S 172°57′E
185 F20 **Tekapo** ≈ South Island, New Zealand
185 F19 **Tekapo, Lake** ◎ South Island, New Zealand
184 P9 **Te Karaka** Gisborne, North Island, New Zealand 38°30′S 177°52′E
184 L7 **Te Kauwhata** Waikato, North Island, New Zealand 37°22′S 175°07′E
41 X12 **Tekax** *var.* Tekax de Álvaro Obregón. Yucatán, SE Mexico 20°07′N 89°10′W
Tekax de Álvaro Obregón *see* Tekax
136 A14 **Teke Burnu** *headland* W Turkey 38°06′N 26°35′E
114 M12 **Teke Deresi** ≈ NW Turkey
146 D10 **Tekedzhik, Gory** *hill range* NW Turkmenistan
145 V14 **Tekeli** Almaty, SE Kazakhstan 44°50′N 78°47′E
145 R7 **Teke, Ozero** ◎ N Kazakhstan
158 I5 **Tekes** Xinjiang Uygur Zizhiqu, NW China 43°15′N 81°43′E
145 W16 **Tekes** Almaty, SE Kazakhstan 42°40′N 80°01′E

◆ Country
● Country Capital
◇ Dependent Territory
○ Dependent Territory Capital
◈ Administrative Regions
✈ International Airport
▲ Mountain
▲▲ Mountain Range
⛰ Volcano
≈ River
◎ Lake
⊠ Reservoir

Tekes see Tekes He
158 H5 **Tekes He** Rus. Tekes. China/Kazakhstan
Teke/Tekendorf see Teaca
80 I10 **Tekezē** var. Takkaze. Eritrea/Ethiopia
Tekhtin see Tsyakhtsin
136 C10 **Tekirdağ** It. Rodosto; anc. Bisanthe, Raidestos, Rhaedestus. Tekirdağ, NW Turkey 40°59′N 27°31′E
136 C10 **Tekirdağ** ◆ province NW Turkey
155 N14 **Tekkali** Andhra Pradesh, E India 18°37′N 84°15′E
115 K15 **Tekke Burnu** Turk. Ilyasbaba Burnu. headland NW Turkey 40°03′N 26°12′E
137 Q13 **Tekman** Erzurum, NE Turkey 39°39′N 41°31′E
32 M9 **Tekoa** Washington, NW USA 47°13′N 117°05′W
190 H16 **Te Kou** ▲ Rarotonga, S Cook Islands 21°14′S 159°46′W
Tekrit see Tikrit
171 P12 **Teku** Sulawesi, N Indonesia 0°46′S 123°25′E
184 L9 **Te Kuiti** Waikato, North Island, New Zealand 38°21′S 175°10′E
42 H4 **Tela** Atlántida, NW Honduras 15°46′N 87°25′W
138 F12 **Telalim** Southern, S Israel 30°58′N 34°47′E
Telanaipura see Jambi
155 I15 **Telangana** off. State of Telangana. ◆ state E India
Telangana, State of see Telangana
137 U10 **Telavi** prev. T'elavi. E Georgia 41°55′N 45°29′E
T'elavi see Telavi
138 F10 **Tel Aviv** ◆ district W Israel
Tel Aviv-Jaffa see Tel Aviv-Yafo
138 F10 **Tel Aviv-Yafo** var. Tel Aviv-Jaffa. Tel Aviv, C Israel 32°05′N 34°46′E
111 E18 **Telč** Ger. Teltsch. Vysočina, C Czech Republic 49°10′N 15°28′E
186 B6 **Telefomin** West Sepik, NW Papua New Guinea 05°08′S 141°31′E
10 J10 **Telegraph Creek** British Columbia, W Canada 57°56′N 131°10′W
190 B10 **Telele** island Funafuti Atoll, C Tuvalu
60 J11 **Telêmaco Borba** Paraná, S Brazil 24°20′S 50°44′W
95 E15 **Telemark** ◆ county S Norway
62 J13 **Telén** La Pampa, C Argentina 36°20′S 65°31′W
Teleneshty see Teleneşti
116 M9 **Teleneşti** Rus. Teleneshty. C Moldova 47°35′N 28°20′E
104 J4 **Teleno, El** ▲ NW Spain 42°19′N 06°21′W
116 I15 **Teleorman** ◆ county S Romania
116 I14 **Teleorman** S Romania
25 V5 **Telephone** Texas, SW USA 33°48′N 96°00′W
35 U11 **Telescope Peak** ▲ California, W USA 36°09′N 117°03′W
Teles Pirés see São Manuel, Rio
97 L19 **Telford** W England, United Kingdom 52°42′N 02°28′W
108 L7 **Telfs** Tirol, W Austria 47°19′N 11°05′E
42 J9 **Telica** León, NW Nicaragua 12°30′N 86°52′W
42 J6 **Telica, Río** C Honduras
76 I13 **Télimélé** W Guinea 10°45′N 13°02′W
43 O14 **Telire, Río** Costa Rica/Panama
114 J8 **Telish** prev. Azizie. Pleven, N Bulgaria 43°20′N 24°15′E
41 R16 **Telixtlahuaca** var. San Francisco Telixtlahuaca. Oaxaca, SE Mexico 17°18′N 96°54′W
10 K13 **Telkwa** British Columbia, SW Canada 54°39′N 126°51′W
25 P4 **Tell** Texas, SW USA 34°18′N 100°20′W
Tell Abiad see Tall Abyaḍ
Tell Abiad/Tell Abyad see At Tall al Abyaḍ
31 O16 **Tell City** Indiana, N USA 37°56′N 86°47′W
38 M9 **Teller** Alaska, USA 65°15′N 166°21′W
Tell Huqnah see Tall Ḥuqnah
155 F20 **Tellicherry** var. Thalashsheri, Thalassery. Kerala, SW India 11°44′N 75°29′E see also Thalassery
20 M10 **Tellico Plains** Tennessee, S USA 35°19′N 84°18′W
Tell Kalakh see Tall Kalakh
Tell Mardīkh see Ebla
54 E11 **Tello** Huila, C Colombia 03°06′N 75°08′W
Tell Shedadi see Ash Shadādah
37 Q7 **Telluride** Colorado, C USA 37°56′N 107°48′W
117 X9 **Tel'manove** Donets'ka Oblast', E Ukraine 47°24′N 38°03′E
Tel'man/Tel'mansk see Gubadag
162 H6 **Telmen** var. Övögdiy. Dzavhan, C Mongolia 48°38′N 97°39′E
162 H6 **Telmen Nuur** ◎ NW Mongolia
Teloekbetoeng see Bandar Lampung
41 O15 **Teloloapán** Guerrero, S Mexico 18°21′N 99°52′W
Telo Martius see Toulon
Telposiz, Gora see Telpoziz, Gora
125 V8 **Telpoziz, Gora** prev. Gora Telposiz. ▲ NW Russian Federation 63°52′N 59°15′E
Telschen see Telšiai
63 J17 **Telsen** Chubut, S Argentina 42°27′S 66°59′W
118 D11 **Telšiai** Ger. Telschen. Telšiai, NW Lithuania 55°59′N 22°21′E
118 D11 **Telšiai** ◆ province NW Lithuania
Teltsch see Telč
Telukbetung see Bandar Lampung
168 H10 **Telukdalam** Pulau Nias, W Indonesia 0°34′N 97°47′E
14 H9 **Temagami** Ontario, S Canada 47°03′N 79°47′W
14 G9 **Temagami, Lake** ◎ Ontario, S Canada
190 H16 **Te Manga** ▲ Rarotonga, S Cook Islands 21°13′S 159°45′W
191 W12 **Tematagi** prev. Tematangi. atoll Îles Tuamotu, S French Polynesia
Tematangi see Tematagi
41 X11 **Temax** Yucatán, SE Mexico 21°10′N 88°53′W
171 E14 **Tembagapura** Papua, E Indonesia 04°10′S 137°19′E
129 U5 **Tembenchi** N Russian Federation
55 P6 **Temblador** Monagas, NE Venezuela 08°59′N 62°44′W
105 N9 **Tembleque** Castilla-La Mancha, C Spain 39°41′N 03°30′W
Temboni see Mitemele, Río
35 U16 **Temecula** California, W USA 33°29′N 117°09′W
168 K7 **Temengor, Tasik** ◎ Peninsular Malaysia
112 L9 **Temerin** Vojvodina, N Serbia 45°25′N 19°54′E
Temeschburg/Temeschwar see Timişoara
Temes-Kubin see Kovin
Temes/Temesch see Tamiš
Temesvár/Temeswar see Timişoara
Teminaboean see Teminabuan
171 U12 **Teminabuan** prev. Teminaboean. Papua Barat, E Indonesia 01°30′S 131°59′E
145 P17 **Temirlan** prev. Temirlanovka. Yuzhnyy Kazakhstan, S Kazakhstan 42°36′N 69°17′E
Temirlanovka see Temirlan
145 R10 **Temirtau** prev. Samarkandski, Samarkandskoye. Karaganda, C Kazakhstan 50°05′N 72°55′E
14 H10 **Témiscaming** Québec, SE Canada 46°40′N 79°04′W
Témiscamingue, Lac see Timiskaming, Lake
15 T8 **Témiscouata, Lac** ◎ Québec, SE Canada
127 N5 **Temnikov** Respublika Mordoviya, W Russian Federation 54°39′N 43°09′E
191 Y13 **Temoe** island Îles Gambier, E French Polynesia
183 Q9 **Temora** New South Wales, SE Australia 34°28′S 147°33′E
40 H7 **Témoris** Chihuahua, W Mexico 27°16′N 108°15′W
40 I5 **Temósachic** Chihuahua, N Mexico 28°55′N 107°42′W
187 Q10 **Temotu** var. Temotu Province. ◆ province E Solomon Islands
Temotu Province see Temotu
36 L14 **Tempe** Arizona, SW USA 33°24′N 111°54′W
Tempelburg see Czaplinek
107 C17 **Tempio Pausania** Sardegna, Italy, C Mediterranean Sea 40°55′N 09°07′E
42 K12 **Tempisque, Río** NW Costa Rica
25 T9 **Temple** Texas, SW USA 31°06′N 97°22′W
100 O12 **Tempelhof** ✈ (Berlin) Berlin, NE Germany 52°28′N 13°24′E
97 D19 **Templemore** Ir. An Teampall Mór. Tipperary, C Ireland 52°48′N 07°50′W
100 O11 **Templin** Brandenburg, NE Germany 53°07′N 13°31′E
41 P12 **Tempoal** var. Tempoal de Sánchez. Veracruz-Llave, E Mexico 21°32′N 98°23′W
Tempoal de Sánchez see Tempoal
41 P13 **Tempoal, Río** C Mexico
83 E14 **Tempué** Moxico, C Angola 13°36′S 18°56′E
126 J14 **Temryuk** Krasnodarskiy Kray, SW Russian Federation 45°15′N 37°26′E
99 G17 **Temse** Oost-Vlaanderen, N Belgium 51°08′N 04°13′E
63 F15 **Temuco** Araucanía, C Chile 38°45′S 72°37′W
185 G20 **Temuka** Canterbury, South Island, New Zealand 44°14′S 171°17′E
189 P13 **Temwen Island** island E Micronesia
56 C6 **Tena** Napo, C Ecuador 01°00′S 77°48′W
41 W13 **Tenabo** Campeche, E Mexico 20°02′N 90°12′W
Tenaghau see Aola
25 X7 **Tenaha** Texas, SW USA 31°56′N 94°14′W
39 X13 **Tenake** Chichagof Island, Alaska, USA 57°46′N 135°13′W
155 K16 **Tenāli** Andhra Pradesh, E India 16°13′N 80°36′E
Tenan see Cheonan
41 O14 **Tenancingo** var. Tenencingo de Degollado. México, S Mexico 18°57′N 99°39′W
191 X12 **Tenararo** island Groupe Actéon, SE French Polynesia
Tenasserim see Tanintharyi
Tenasserim see Tanintharyi
98 O5 **Ten Boer** Groningen, NE Netherlands 53°16′N 06°42′E
97 I21 **Tenby** SW Wales, United Kingdom 51°41′N 04°43′W
80 K11 **Tendaho** Āfar, NE Ethiopia 11°39′N 40°59′E
103 V14 **Tende** Alpes Maritimes, SE France 44°04′N 07°34′E
151 Q20 **Ten Degree Channel** strait Andaman and Nicobar Islands, India, E Indian Ocean
80 F11 **Tendelti** White Nile, E Sudan 13°01′N 31°55′E
76 G8 **Te-n-Dghâmcha, Sebkhet** var. Sebkha de Ndrhamcha, Sebkra de Ndaghamcha. salt lake W Mauritania
165 P10 **Tendō** Yamagata, Honshū, C Japan 38°22′N 140°22′E
74 H7 **Tendrara** NE Morocco 33°06′N 01°58′W
117 Q11 **Tendrivs'ka Kosa** spit S Ukraine
117 Q11 **Tendrivs'ka Zatoka** gulf S Ukraine
Tenencingo de Degollado see Tenancingo
77 N11 **Ténenkou** Mopti, C Mali 14°28′N 04°55′W
77 W9 **Ténéré** physical region C Niger
77 W9 **Ténére, Erg du** desert C Niger
64 O11 **Tenerife** island Islas Canarias, Spain, NE Atlantic Ocean
74 J5 **Ténès** NW Algeria 36°35′N 01°18′E
170 M15 **Tengah, Kepulauan** island group C Indonesia
Tengcheng see Tengxian
169 V11 **Tenggarong** Borneo, C Indonesia 0°23′S 117°00′E
162 J15 **Tengger Shamo** desert N China
168 L8 **Tenggul, Pulau** island Peninsular Malaysia
Tengiz Köl see Teniz, Ozero
76 M14 **Tengréla** var. Tingréla. N Ivory Coast 10°26′N 06°20′W
160 M14 **Tengxian** var. Tengcheng, Tengxian, Teng Xian. Guangxi Zhuangzu Zizhiqu, S China 23°24′N 110°49′E
Teng Xian see Tengxian
Tengxian see Tengxian
194 H2 **Teniente Rodolfo Marsh** Chilean research station South Shetland Islands, Antarctica 61°57′S 58°23′W
32 G9 **Tenino** Washington, NW USA 46°51′N 122°51′W
145 P9 **Teniz, Ozero** Kaz. Tengiz Köl. salt lake C Kazakhstan
112 I9 **Tenja** Osijek-Baranja, E Croatia 45°30′N 18°45′E
188 B16 **Tenjo, Mount** ▲ W Guam
155 H23 **Tenkāsi** Tamil Nādu, SE India 08°58′N 77°22′E
79 N24 **Tenke** Katanga, SE Dem. Rep. Congo 10°34′S 26°12′E
Tenke see Tinca
123 Q7 **Tenkeli** Respublika Sakha (Yakutiya), NE Russian Federation 70°09′N 140°39′E
27 R10 **Tenkiller Ferry Lake** Oklahoma, C USA
77 Q13 **Tenkodogo** S Burkina Faso 11°54′N 00°19′W
181 Q5 **Tennant Creek** Northern Territory, C Australia 19°40′S 134°16′E
20 G9 **Tennessee** off. State of Tennessee, also known as The Volunteer State. ◆ state SE USA
37 R5 **Tennessee Pass** pass Colorado, C USA
20 H10 **Tennessee River** S USA
23 N2 **Tennessee Tombigbee Waterway** canal Alabama/Mississippi, S USA
99 K22 **Tenneville** Luxembourg, SE Belgium 50°05′N 05°31′E
92 M11 **Tenniöjoki** NE Finland
92 L9 **Tenojoki** Lapp. Deatnu, Nor. Tana. Finland/Norway see also Deatnu, Tana
Tenojoki to see Tana
169 U7 **Tenom** Sabah, East Malaysia 05°07′N 115°57′E
Tenos see Tínos
41 V15 **Tenosique** var. Tenosique de Pino Suárez. Tabasco, SE Mexico 17°30′N 91°24′W
Tenosique de Pino Suárez see Tenosique
22 I6 **Tensas River** Louisiana, S USA
23 O8 **Tensaw River** Alabama, S USA
74 E7 **Tensift** seasonal river W Morocco
171 O12 **Tentena** var. Tenteno. Sulawesi, C Indonesia 01°46′S 120°40′E
Tenteno see Tentena
183 U4 **Tenterfield** New South Wales, SE Australia 29°04′S 152°02′E
23 X16 **Ten Thousand Islands** island group Florida, SE USA
60 H10 **Teodoro Sampaio** São Paulo, S Brazil 22°30′S 52°13′W
59 N19 **Teófilo Otoni** var. Theophilo Ottoni. Minas Gerais, NE Brazil 17°52′S 41°31′W
116 K5 **Teofipol'** Khmel'nyts'ka Oblast', W Ukraine 50°00′N 26°22′E
Teohatu see Toahotu
41 P14 **Teotihuacán** ruins México, S Mexico
Teotilán see Teotitlán del Camino
41 Q15 **Teotitlán del Camino** var. Teotilán. Oaxaca, S Mexico 18°10′N 97°08′W
190 G12 **Tepa** Île Uvea, E Wallis and Futuna 13°19′S 176°09′W
191 P8 **Tepaee, Récif** reef Tahiti, W French Polynesia
40 L14 **Tepalcatepec** Michoacán, SW Mexico 19°11′N 102°50′W
190 A16 **Tepa Point** headland SW Niue 19°07′S 169°56′E
40 L13 **Tepatitlán** var. Tepatitlán de Morelos. Jalisco, SW Mexico 20°50′N 102°46′W
Tepatitlán de Morelos see Tepatitlán
40 J9 **Tepehuanes** var. Santa Catarina de Tepehuanes. Durango, C Mexico 25°22′N 105°42′W
Tepelena see Tepelenë
113 L22 **Tepelenë** var. Tepelena, It. Tepeleni. Gjirokastër, S Albania 40°18′N 20°00′E
Tepeleni see Tepelenë
40 K12 **Tepic** Nayarit, C Mexico 21°30′N 104°55′W
111 C15 **Teplice** Ger. Teplitz; prev. Teplice-Šanov, Teplitz-Schönau. Ústecký Kraj, NW Czech Republic 50°38′N 13°49′E
Teplice-Šanov/Teplitz/Teplitz-Schönau see Teplice
117 O7 **Teplyk** Vinnyts'ka Oblast', C Ukraine 48°40′N 29°46′E
123 R10 **Teplyy Klyuch** Respublika Sakha (Yakutiya), NE Russian Federation 62°46′N 137°01′E
40 E5 **Tepoca, Cabo** headland NW Mexico 29°19′N 112°24′W
191 W9 **Tepoto** island Îles du Désappointement, C French Polynesia
92 L11 **Tepsa** Lappi, N Finland 67°34′N 25°36′E
190 B8 **Tepuka** atoll Funafuti Atoll, C Tuvalu
184 N7 **Te Puke** Bay of Plenty, North Island, New Zealand 37°48′S 176°19′E
40 L13 **Tequila** Jalisco, SW Mexico 20°52′N 103°48′W
41 O13 **Tequisquiapan** Querétaro de Arteaga, C Mexico 20°34′N 99°52′W
77 Q12 **Téra** Tillabéri, W Niger 14°01′N 00°45′E
104 J5 **Tera** NW Spain
191 V1 **Teraina** prev. Washington Island. atoll Line Islands, E Kiribati
81 F15 **Terakeka** Central Equatoria, S South Sudan 05°26′N 31°45′E
107 J14 **Teramo** anc. Interamna. Abruzzi, C Italy 42°40′N 13°43′E
98 P7 **Ter Apel** Groningen, NE Netherlands 52°52′N 07°05′E
104 H11 **Tera, Ribeira de** S Portugal
185 K14 **Terawhiti, Cape** headland North Island, New Zealand 41°17′S 174°36′E
98 N12 **Terborg** Gelderland, E Netherlands 51°55′N 06°22′E
137 P13 **Tercan** Erzincan, NE Turkey 39°47′N 40°23′E
64 O2 **Terceira** ✈ Terceira, Azores, Portugal, NE Atlantic Ocean 38°43′N 27°13′W
64 O2 **Terceira** var. Ilha Terceira. island Azores, Portugal, NE Atlantic Ocean
Terceira, Ilha see Terceira
116 K6 **Terebovlya** Ternopil's'ka Oblast', W Ukraine 49°18′N 25°44′E
127 O15 **Terek** SW Russian Federation
Terekhovka see Tsyerakhowka
147 R9 **Terek-Say** Dzhalal-Abadskaya Oblast', W Kyrgyzstan 41°28′N 71°06′E
145 Z10 **Terekty** prev. Alekseevka, Alekseyevka. Vostochnyy Kazakhstan, E Kazakhstan 48°25′N 85°38′E
168 L7 **Terengganu** var. Trengganu. ◆ state Peninsular Malaysia
127 X7 **Terensay** Orenburgskaya Oblast', W Russian Federation 51°35′N 59°28′E
58 N13 **Teresina** var. Therezina. state capital Piauí, NE Brazil 05°09′S 42°46′W
60 P9 **Teresópolis** Rio de Janeiro, SE Brazil 22°25′S 42°59′W
110 P12 **Terespol** Lubelskie, E Poland 52°05′N 23°37′E
191 V16 **Terevaka, Maunga** ▲ Easter Island, Chile, E Pacific Ocean 27°05′S 109°23′W
103 P3 **Tergnier** Aisne, N France 49°39′N 03°18′E
43 O14 **Teribe, Río** NW Panama
124 K3 **Teriberka** Murmanskaya Oblast', NW Russian Federation 69°10′N 35°18′E
Terinkot see Tarīn Kōt
Terisaqqan see Tersakkan
24 K12 **Terlingua** Texas, SW USA 29°18′N 103°36′W
24 K11 **Terlingua Creek** Texas, SW USA
62 K7 **Termas de Río Hondo** Santiago del Estero, N Argentina 27°29′S 64°52′W
136 M11 **Terme** Samsun, N Turkey 41°12′N 37°00′E
Termez see Termiz
Termia see Kýthnos
107 J23 **Termini Imerese** anc. Thermae Himerenses. Sicilia, Italy, C Mediterranean Sea 37°59′N 13°42′E
41 V14 **Términos, Laguna de** lagoon SE Mexico
77 X10 **Termit-Kaoboul** Zinder, C Niger 15°34′N 11°31′E
147 O14 **Termiz** Rus. Termez. Surkhondaryo Viloyati, S Uzbekistan 37°17′N 67°12′E
107 L15 **Termoli** Molise, C Italy 42°00′N 14°58′E
Termonde see Dendermonde
98 P5 **Termunten** Groningen, NE Netherlands 53°18′N 07°02′E
171 R11 **Ternate** Pulau Ternate, E Indonesia 0°48′N 127°23′E
109 T5 **Ternberg** Oberösterreich, N Austria 47°57′N 14°22′E
99 E15 **Terneuzen** var. Neuzen. Zeeland, SW Netherlands 51°20′N 03°50′E
123 T14 **Terney** Primorskiy Kray, SE Russian Federation 45°03′N 136°43′E
107 I14 **Terni** anc. Interamna Nahars. Umbria, C Italy 42°34′N 12°38′E
109 X6 **Ternitz** Niederösterreich, E Austria 47°43′N 16°02′E
117 V7 **Ternivka** Dnipropetrovs'ka Oblast', E Ukraine 48°30′N 36°05′E
116 K6 **Ternopil'** Pol. Tarnopol, Rus. Ternopol'. Ternopil's'ka Oblast', W Ukraine 49°32′N 25°38′E
Ternopil' see Ternopil's'ka Oblast'
116 I6 **Ternopil's'ka Oblast'** var. Ternopil', Rus. Ternopol'skaya Oblast'. ◆ province NW Ukraine
Ternopol' see Ternopil'
Ternopol'skaya Oblast' see Ternopil's'ka Oblast'
123 U13 **Terpeniya, Mys** headland Ostrov Sakhalin, SE Russian Federation 48°37′N 144°40′E
Térraba, Río see Grande de Térraba, Río
10 J13 **Terrace** British Columbia, W Canada 54°31′N 128°32′W
12 D12 **Terrace Bay** Ontario, S Canada 48°47′N 87°06′W
107 I16 **Terracina** Lazio, C Italy 41°18′N 13°13′E
93 F14 **Terråk** Troms, N Norway 65°03′N 12°22′E
26 M13 **Terral** Oklahoma, C USA 33°55′N 97°54′W
107 B19 **Terralba** Sardegna, Italy, C Mediterranean Sea 39°47′N 08°35′E
Terranova di Sicilia see Gela
Terranova Pausania see Olbia
105 W5 **Terrassa** Cast. Tarrasa. Cataluña, E Spain 41°34′N 02°01′E
15 O12 **Terrebonne** Québec, SE Canada 45°42′N 73°37′W
22 J11 **Terrebonne Bay** bay Louisiana, SE USA
31 N14 **Terre Haute** Indiana, N USA 39°27′N 87°24′W
25 U6 **Terrell** Texas, SW USA 32°44′N 96°16′W
Terre Neuve see Newfoundland and Labrador
33 Q14 **Terreton** Idaho, NW USA 43°49′N 112°25′W
103 T7 **Territoire-de-Belfort** ◆ department E France
33 X9 **Terry** Montana, NW USA 46°46′N 105°16′W
28 J9 **Terry Peak** ▲ South Dakota, N USA 44°19′N 103°49′W
136 H14 **Tersakan Gölü** ◎ C Turkey
145 O10 **Tersakkan** Kaz. Terisaqqan. C Kazakhstan
98 J4 **Terschelling** Fris. Skylge. island Waddeneilanden, N Netherlands
78 H10 **Tersef** Hadjer-Lamis, C Chad 12°55′N 16°49′E
147 X8 **Terskey Ala-Too, Khrebet** ▲ Kazakhstan/Kyrgyzstan
Terter see Tärtär
105 R8 **Teruel** anc. Turba. Aragón, E Spain 40°21′N 01°06′W
105 R7 **Teruel** ◆ province Aragón, E Spain
114 M7 **Tervel** prev. Kurtbunar, Rom. Curtbunar. Dobrich, NE Bulgaria 43°45′N 27°25′E
93 M16 **Tervo** Pohjois-Savo, C Finland 62°57′N 26°48′E
92 L13 **Tervola** Lappi, NW Finland 66°04′N 24°49′E
Tervueren see Tervuren
99 H18 **Tervuren** var. Tervueren. Vlaams Brabant, C Belgium 50°48′N 04°28′E
162 G5 **Tes** var. Dzür. Dzavhan, W Mongolia 49°37′N 95°46′E
112 H11 **Tešanj** Federacija Bosne I Hercegovine, N Bosnia and Herzegovina 44°37′N 18°00′E
Teschen see Cieszyn
83 M19 **Tesenane** Inhambane, S Mozambique 22°48′S 34°02′E
80 I9 **Teseney** var. Tessenei. W Eritrea 15°05′N 36°42′E
39 P5 **Teshekpuk Lake** ◎ Alaska, USA
Teshevle see Tsyeshawlya
162 K6 **Teshig** Bulgan, N Mongolia 49°51′N 102°45′E
165 T2 **Teshio** Hokkaidō, NE Japan 44°49′N 141°46′E
165 T2 **Teshio-sanchi** ▲ Hokkaidō, NE Japan
14 L7 **Tessier, Lac** ◎ Québec, SE Canada
Tešín see Cieszyn
Tesiyn Gol see Tes-Khem
129 T7 **Tes-Khem** var. Tesiyn Gol. Mongolia/Russian Federation
112 H11 **Teslić** Republika Srpska, N Bosnia and Herzegovina 44°35′N 17°50′E
10 I9 **Teslin** Yukon, W Canada 60°12′N 132°44′W
10 I8 **Teslin** British Columbia/Yukon, W Canada
77 Q8 **Tessalit** Kidal, NE Mali 20°12′N 00°58′E
77 V12 **Tessaoua** Maradi, S Niger 13°46′N 07°55′E
99 J17 **Tessenderlo** Limburg, NE Belgium 51°05′N 05°04′E
Tessenei see Teseney
Tessin see Ticino
97 M23 **Test** S England, United Kingdom
Testama see Tõstamaa
55 P4 **Testigos, Islas los** island group N Venezuela
37 S10 **Tesuque** New Mexico, SW USA 35°45′N 105°55′W
103 O17 **Têt** var. Tet. S France
Tet see Têt
54 G5 **Tetas, Cerro de las** ▲ NW Venezuela 09°58′N 73°00′W
83 M15 **Tete** Tete, NW Mozambique 16°14′S 33°34′E
83 M15 **Tete** off. Província de Tete. ◆ province NW Mozambique
11 N15 **Tête Jaune Cache** British Columbia, SW Canada 52°52′N 119°22′W
184 O8 **Te Teko** Bay of Plenty, North Island, New Zealand 38°03′S 176°48′E
186 K9 **Tetepare** island New Georgia Islands, NW Solomon Islands
Tete, Província de see Tete
Teterev see Teteriv
116 M5 **Teteriv** Rus. Teterev. N Ukraine
100 M9 **Teterow** Mecklenburg-Vorpommern, NE Germany 53°47′N 12°34′E
114 I9 **Teteven** Lovech, N Bulgaria 42°54′N 24°18′E
191 T10 **Tetiaroa** atoll Îles du Vent, W French Polynesia
105 P14 **Tetica de Bacares** ▲ S Spain 37°15′N 02°24′W
Tetiyev see Tetiyiv
117 O6 **Tetiyiv** Rus. Tetiyev. Kyyivs'ka Oblast', N Ukraine 49°21′N 29°40′E
39 T10 **Tetlin** Alaska, USA 63°08′N 142°31′W
33 R8 **Teton River** Montana, NW USA
74 G5 **Tétouan** var. Tetouan, Tetuán. N Morocco 35°33′N 05°22′W
Tetova/Tetovë see Tetovo
114 L7 **Tetovo** Razgrad, N Bulgaria 43°49′N 26°21′E
113 N18 **Tetovo** Alb. Tetova, Tetovë, Turk. Kalkandelen. NW FYR Macedonia 42°01′N 20°58′E
115 E20 **Tetrázio** ▲ S Greece
Tetschen see Děčín
Tetuán see Tétouan
191 Q8 **Tetufera, Mont** ▲ Tahiti, W French Polynesia 17°40′S 149°26′W
127 R4 **Tetyushi** Respublika Tatarstan, W Russian Federation 54°55′N 48°46′E
108 I7 **Teufen** Ausser Rhoden, NE Switzerland 47°24′N 09°24′E
40 L12 **Teul** var. Teul de Gonzáles Ortega. Zacatecas, C Mexico 21°30′N 103°28′W
107 B21 **Teulada** Sardegna, Italy, C Mediterranean Sea 38°58′N 08°46′E
Teul de Gonzáles Ortega see Teul
11 X16 **Teulon** Manitoba, S Canada 50°20′N 97°14′W
42 I7 **Teupasenti** El Paraíso, S Honduras 14°13′N 86°42′W
165 S2 **Teuri-tō** island NE Japan
100 G13 **Teutoburger Wald** Eng. Teutoburg Forest. hill range NW Germany
Teutoburg Forest see Teutoburger Wald
93 K17 **Teuva** Swe. Östermark. Etelä-Pohjanmaa, W Finland 62°29′N 21°45′E
107 H15 **Tevere** Eng. Tiber. C Italy
Teverya see Tverya
96 K13 **Teviot** SE Scotland, United Kingdom
Tevli see Tewli
122 H11 **Tevriz** Omskaya Oblast', C Russian Federation 57°30′N 72°13′E
185 B24 **Te Waewae Bay** bay South Island, New Zealand
97 L21 **Tewkesbury** C England, United Kingdom 51°59′N 02°09′W
119 F19 **Tewli** Rus. Tevli. Brestskaya Voblasts', SW Belarus 52°20′N 24°15′E
159 U12 **Têwo** var. Dêngka; prev. Dêngkagoin. Gansu, C China 34°05′N 103°15′E
Tewulike see Hoxud
25 U12 **Texana, Lake** ◎ Texas, SW USA
27 S14 **Texarkana** Arkansas, C USA 33°26′N 94°02′W
25 X5 **Texarkana** Texas, SW USA 33°26′N 94°03′W
25 N9 **Texas** off. State of Texas, also known as Lone Star State. ◆ state S USA
25 W12 **Texas City** Texas, SW USA 29°23′N 94°55′W
41 P14 **Texcoco** México, C Mexico 19°32′N 98°52′W
98 I6 **Texel** island Waddeneilanden, NW Netherlands
26 H8 **Texhoma** Oklahoma, C USA 36°30′N 101°46′W
25 N1 **Texhoma** Texas, SW USA 36°30′N 101°46′W
37 W12 **Texico** New Mexico, SW USA 34°23′N 103°03′W
24 L1 **Texline** Texas, SW USA 36°22′N 103°01′W
41 P14 **Texmelucan** var. San Martín Texmelucan. Puebla, S Mexico 19°46′N 98°53′W
27 O13 **Texoma, Lake** Oklahoma/Texas, C USA
25 N9 **Texon** Texas, SW USA 31°13′N 101°42′W
83 J23 **Teyateyaneng** NW Lesotho 29°04′S 27°51′E
124 M16 **Teykovo** Ivanovskaya Oblast', W Russian Federation 56°49′N 40°31′E
124 M16 **Teza** W Russian Federation
41 Q13 **Teziutlán** Puebla, S Mexico 19°49′N 97°22′W
153 W12 **Tezpur** Assam, NE India 26°39′N 92°47′E
Thaa Atoll see Kolhumadulu
9 N10 **Tha-Anne** Nunavut, NE Canada
83 K23 **Thabana Ntlenyana** var. Thabantshonyana, Mount Ntlenyana. ▲ E Lesotho 29°26′S 29°16′E
Thabantshonyana see Thabana Ntlenyana
83 J23 **Thaba Putsoa** ▲ C Lesotho 29°48′S 27°46′E
167 Q8 **Tha Bo** Nong Khai, E Thailand 17°52′N 102°34′E
103 T12 **Thabor, Pic du** ▲ E France 45°07′N 06°34′E
Tha Chin see Samut Sakhon
166 M7 **Thagaya** Bago, C Myanmar (Burma) 19°19′N 96°16′E
Thai, Ao see Thailand, Gulf of
167 T6 **Thai Binh** Thai Binh, N Vietnam 20°27′N 106°20′E
167 S7 **Thai Hoa** var. Nghia Dan. Nghê An, N Vietnam 19°21′N 105°26′E
167 P9 **Thailand** off. Kingdom of Thailand, Th. Prathet Thai; prev. Siam. ◆ monarchy SE Asia
167 P13 **Thailand, Gulf of** var. Gulf of Siam, Th. Ao Thai, Vtn. Vinh Thai Lan. gulf SE Asia
Thailand, Kingdom of see Thailand
Thai Lan, Vinh see Thailand, Gulf of
167 T6 **Thai Nguyên** Bac Thai, N Vietnam 21°36′N 105°50′E
167 S8 **Thakhèk** var. Muang Khammouan. Khammouan, C Laos 17°25′N 104°51′E
153 S13 **Thakurgaon** Rajshahi, NW Bangladesh 26°05′N 88°34′E
149 S6 **Thal** Khyber Pakhtunkhwa, NW Pakistan 33°24′N 70°32′E
167 S11 **Thalabârivât** prev. Phumĭ Thalabârĭvăt. Stœng Trêng, N Cambodia 13°34′N 105°57′E
166 M15 **Thalang** Phuket, SW Thailand 08°00′N 98°21′E
Thalashsheri see Tellicherry/Thalassery
167 Q10 **Thalat Khae** Nakhon Ratchasima, C Thailand 15°15′N 102°24′E
109 Q5 **Thalgau** Salzburg, NW Austria 47°49′N 13°19′E
108 G7 **Thalwil** Zürich, NW Switzerland 47°17′N 08°35′E
83 I20 **Thamaga** Kweneng, SE Botswana 24°41′S 25°31′E
Thamarid see Thamarīt
141 V13 **Thamarīt** var. Thamarīd, Thumrayt. SW Oman 17°39′N 54°02′E
141 P16 **Thamar, Jabal** ▲ SW Yemen 13°46′N 45°32′E
184 M6 **Thames** Waikato, North Island, New Zealand 37°10′S 175°33′E
14 D17 **Thames** Ontario, S Canada
97 O22 **Thames** S England, United Kingdom
184 M6 **Thames, Firth of** gulf North Island, New Zealand
14 D17 **Thamesville** Ontario, S Canada 42°33′N 81°58′W
141 S13 **Thamūd** N Yemen 17°18′N 49°57′E
166 N9 **Thanbyuzayat** Mon State, S Myanmar (Burma) 15°58′N 97°44′E
166 K7 **Thandwe** var. Sandoway. Rakhine State, W Myanmar (Burma) 18°28′N 94°20′E
152 I9 **Thānesar** Haryāna, NW India 29°58′N 76°48′E
167 T7 **Thanh Hoa** Thanh Hoa, N Vietnam 19°49′N 105°48′E
Thanintari Taungdan see Bilauktaung Range
155 I21 **Thanjāvūr** prev. Tanjore. Tamil Nādu, SE India 10°46′N 79°09′E
Thanlwin see Salween
103 U7 **Thann** Haut-Rhin, NE France 47°51′N 07°04′E
167 O16 **Tha Nong Phrom** Phatthalung, SW Thailand 07°24′N 100°04′E
167 N13 **Thap Sakae** var. Thap Sakau. Prachuap Khiri Khan, SW Thailand 11°30′N 99°35′E
Thap Sakau see Thap Sakae
81 I19 **Tharaka-Nithi** ◆ county C Kenya
98 L10 **'t Harde** Gelderland, E Netherlands 52°25′N 05°53′E
152 D11 **Thar Desert** var. Great Indian Desert, Indian Desert. desert India/Pakistan
181 V10 **Thargomindah** Queensland, C Australia 28°00′S 143°47′E
150 D11 **Thar Pārkar** desert SE Pakistan
139 S7 **Tharthār al Furāt, Qanāt ath** canal C Iraq
139 R7 **Tharthār, Buḥayrat ath** ◎ C Iraq
139 R5 **Tharthār, Wādī ath** dry watercourse N Iraq
167 N13 **Tha Sae** Chumphon, SW Thailand
167 N15 **Tha Sala** Nakhon Si Thammarat, SW Thailand 08°43′N 99°54′E
114 I13 **Thásos** Thásos, E Greece 40°47′N 24°43′E
115 I14 **Thásos** island E Greece
37 N14 **Thatcher** Arizona, SW USA 32°47′N 109°46′W
167 T5 **Thât Khê** var. Tràng Dinh. Lang Sơn, N Vietnam 22°15′N 106°26′E
166 M8 **Thaton** Mon State, S Myanmar (Burma) 16°56′N 97°20′E
167 S9 **That Phanom** Nakhon Phanom, E Thailand 16°52′N 104°41′E
167 R10 **Tha Tum** Surin, E Thailand 15°18′N 103°39′E
103 P16 **Thau, Bassin de** var. Étang de Thau. ◎ S France
Thau, Étang de see Thau, Bassin de
166 L3 **Thaungdut** Sagaing, N Myanmar (Burma) 24°26′N 94°45′E
167 O8 **Thaungyin** Th. Mae Nam Moei. Myanmar (Burma)/Thailand
167 R8 **Tha Uthen** Nakhon Phanom, E Thailand 17°32′N 104°34′E
109 W2 **Thaya** var. Dyje. Austria/Czech Republic see also Dyje
Thaya see Dyje
27 V8 **Thayer** Missouri, C USA 36°31′N 91°34′W
166 L6 **Thayetmyo** Magway, C Myanmar (Burma) 19°20′N 95°10′E
33 S15 **Thayne** Wyoming, C USA 42°54′N 111°01′W
166 M5 **Thazi** Mandalay, C Myanmar (Burma) 20°50′N 96°04′E
Thebes see Thíva
44 L5 **The Carlton** var. Abraham Bay. Mayaguana, SE The Bahamas 22°21′N 72°56′W
45 O14 **The Crane** var. Crane. S Barbados 13°06′N 59°27′W
32 I11 **The Dalles** Oregon, NW USA 45°36′N 121°10′W
28 M14 **Thedford** Nebraska, C USA 41°59′N 100°32′W
The Flatts Village see Flatts Village
The Hague see 's-Gravenhage
Theiss see Tisa/Tisza
8 M9 **Thelon** Northwest Territories, N Canada
11 V15 **Theodore** Saskatchewan, S Canada 51°25′N 103°01′W
23 N8 **Theodore** Alabama, S USA 30°33′N 88°10′W
36 L13 **Theodore Roosevelt Lake** Arizona, SW USA
Theodosia see Feodosiya
Theophilo Ottoni see Teófilo Otoni
11 V13 **The Pas** Manitoba, C Canada 53°49′N 101°09′W
31 T14 **The Plains** Ohio, N USA 39°22′N 82°07′W
Thera see Santoríni
172 H17 **Thérèse, Île** island Inner Islands, NE Seychelles
Therezina see Teresina
115 L20 **Thérma** Ikaría, Dodekánisa, Greece, Aegean Sea 37°37′N 26°18′E
Thermae Himerenses see Termini Imerese
Thermae Pannonicae see Baden
Thermaic Gulf/Thermaicus Sinus see Thermaïkós Kólpos
121 Q8 **Thermaïkós Kólpos** Eng. Thermaic Gulf; anc. Thermaicus Sinus. gulf N Greece
Thermiá see Kýthnos
115 L17 **Thermís** Lésvos, E Greece 39°08′N 26°32′E
115 E18 **Thérmo** Dytikí Elláda, C Greece 38°32′N 21°42′E
33 V14 **Thermopolis** Wyoming, C USA 43°39′N 108°12′W
183 P10 **The Rock** New South Wales, SE Australia 35°18′S 147°07′E
195 O5 **Theron Mountains** ▲ Antarctica
The Sooner State see Oklahoma
115 G18 **Thespiés** Stereá Elláda, C Greece 38°18′N 23°08′E
115 E16 **Thessalía** Eng. Thessaly. ◆ region C Greece
14 C10 **Thessalon** Ontario, S Canada 46°15′N 83°34′W
115 G14 **Thessaloníki** Eng. Salonica, Salonika, SCr. Solun, Turk. Selânik. Kentrikí Makedonía, N Greece 40°38′N 22°58′E
115 G14 **Thessaloníki** ✈ Kentrikí Makedonía, N Greece 40°30′N 22°58′E
Thessaly see Thessalía
84 B12 **Theta Gap** undersea feature E Atlantic Ocean 12°40′W 43°30′N
97 P20 **Thetford** E England, United Kingdom 52°25′N 00°45′E
15 R11 **Thetford-Mines** Québec, SE Canada 46°07′N 71°16′W
113 K17 **Theth** var. Thethi. Shkodër, N Albania 42°25′N 19°45′E
Thethi see Theth
99 L20 **Theux** Liège, E Belgium 50°33′N 05°48′E
45 V9 **The Valley** ○ (Anguilla) E Anguilla 18°13′N 63°00′W
27 N10 **The Village** Oklahoma, C USA 35°33′N 97°33′W
The Volunteer State see Tennessee
25 W10 **The Woodlands** Texas, SW USA 30°09′N 95°27′E
Thiamis see Kalamás
Thian Shan see Tien Shan
Thibet see Xizang Zizhiqu
22 J9 **Thibodaux** Louisiana, S USA 29°48′N 90°49′W
29 S3 **Thief Lake** ◎ Minnesota, N USA
29 S3 **Thief River** Minnesota, N USA
29 S3 **Thief River Falls** Minnesota, N USA 48°07′N 96°10′W

Thièle see La Thielle
32 G14 **Thielsen, Mount** ▲ Oregon, NW USA 43°09′N 122°04′W
Thielt see Tielt
106 G7 **Thiene** Veneto, NE Italy 45°43′N 11°29′E
Thienen see Tienen
103 P11 **Thiers** Puy-de-Dôme, C France 45°51′N 03°33′E
76 F11 **Thiès** W Senegal 14°49′N 16°52′W
81 I19 **Thika** Kiambu, S Kenya 01°03′S 37°05′E
Thikombia see Cikobia
151 K18 **Thiladhunmathi Atoll** var. Tiladummati Atoll. atoll N Maldives
Thimbu see Thimphu
153 T11 **Thimphu** var. Thimbu; prev. Tashi Chho Dzong. ● (Bhutan) W Bhutan 27°28′N 89°37′E
92 H2 **Þingeyri** Vestfirðir, NW Iceland 65°52′N 23°28′W
92 I3 **Þingvellir** Suðurland, SW Iceland 64°15′N 21°06′W
187 Q17 **Thio** Province Sud, C New Caledonia 21°37′S 166°13′E
103 T4 **Thionville** Ger. Diedenhofen. Moselle, NE France 49°22′N 06°11′E
77 O12 **Thiou** NW Burkina Faso 13°42′N 02°34′W
115 K22 **Thíra** Santoríni, Kykládes, Greece, Aegean Sea 36°25′N 25°26′E
Thíra see Santoríni
115 J22 **Thirasía** island Kykládes, Greece, Aegean Sea
97 M16 **Thirsk** N England, United Kingdom 54°07′N 01°17′W
14 F12 **Thirty Thousand Islands** island group Ontario, S Canada
95 F20 **Thisted** Midtjylland, NW Denmark 56°58′N 08°42′E
Thistil Fjord see Þistilfjörður
92 L1 **Þistilfjörður** var. Thistil Fjord. fjord NE Iceland
182 G9 **Thistle Island** island South Australia
Thíthia see Cicia
Thiukhaoluang Phrahang see Luang Prabang Range
115 G18 **Thíva** Eng. Thebes; prev. Thívai. Stereá Elláda, C Greece 38°19′N 23°19′E
Thívai see Thíva
102 M12 **Thiviers** Dordogne, SW France 45°24′N 00°54′E
92 J4 **Þjórsá** ≈ C Iceland
9 N10 **Thlewiaza** ≈ Nunavut, NE Canada
8 L10 **Thoa** ≈ Northwest Territories, NW Canada
99 G14 **Tholen** Zeeland, SW Netherlands 51°31′N 04°13′E
99 F14 **Tholen** island SW Netherlands
26 L10 **Thomas** Oklahoma, C USA 35°44′N 98°45′W
21 T3 **Thomas** West Virginia, NE USA 39°09′N 79°28′W
27 U3 **Thomas Hill Reservoir** ▣ Missouri, C USA
23 S5 **Thomaston** Georgia, SE USA 32°53′N 84°19′W
19 R7 **Thomaston** Maine, NE USA 44°06′N 69°10′W
25 T12 **Thomaston** Texas, SW USA 28°56′N 97°07′W
23 O6 **Thomasville** Alabama, S USA 31°54′N 87°42′W
23 T8 **Thomasville** Georgia, SE USA 30°49′N 83°57′W
21 S9 **Thomasville** North Carolina, SE USA 35°52′N 80°04′W
35 N5 **Thomes Creek** ≈ California, W USA
11 W12 **Thompson** Manitoba, C Canada 55°45′N 97°54′W
29 R4 **Thompson** North Dakota, N USA 47°45′N 97°07′W
0 F8 **Thompson** ≈ Alberta/British Columbia, SW Canada
33 O8 **Thompson Falls** Montana, NW USA 47°36′N 115°20′W
29 Q10 **Thompson, Lake** ◎ South Dakota, N USA
34 M3 **Thompson Peak** ▲ California, W USA 41°00′N 123°01′W
27 S2 **Thompson River** ≈ Missouri, C USA
185 A22 **Thompson Sound** sound South Island, New Zealand
8 J5 **Thomsen** ≈ Banks Island, Northwest Territories, NW Canada
23 V4 **Thomson** Georgia, SE USA 33°28′N 82°30′W
103 T10 **Thonon-les-Bains** Haute-Savoie, E France 46°22′N 06°30′E
103 O15 **Thoré** var. Thore. ≈ S France
Thore see Thoré
37 P11 **Thoreau** New Mexico, SW USA 35°24′N 108°13′W
Thorenburg see Turda
92 J3 **Þórisvatn** ◎ C Iceland
92 P4 **Thor, Kapp** headland S Svalbard 76°25′N 25°01′E
92 I4 **Þorlákshöfn** Suðurland, SW Iceland 63°52′N 21°24′W
Thorn see Toruń
25 T10 **Thorndale** Texas, SW USA 30°36′N 97°12′W
14 H10 **Thorne** Ontario, S Canada 46°38′N 79°04′W
97 J14 **Thornhill** S Scotland, United Kingdom 55°13′N 03°46′W
25 U8 **Thornton** Texas, SW USA 31°24′N 96°34′W
Thornton Island see Millennium Island
14 H16 **Thorold** Ontario, S Canada 43°07′N 79°15′W
32 I9 **Thorp** Washington, NW USA 47°03′N 120°40′W
Thorshavn see Tórshavn
195 S3 **Thorshavnheiane** physical region Antarctica
92 L1 **Þórshöfn** Norðurland Eystra, NE Iceland 66°09′N 15°18′W
Thospitis see Van Gölü
167 S14 **Thốt Nốt** Cần Thơ, S Vietnam 10°17′N 105°31′E
102 K8 **Thouars** Deux-Sèvres, W France 46°59′N 00°13′W
153 X14 **Thoubal** Manipur, NE India 24°40′N 94°00′E
102 K9 **Thouet** ≈ W France
Thoune see Thun
18 H7 **Thousand Islands** island Canada/USA
35 S15 **Thousand Oaks** California, W USA 34°10′N 118°50′W
114 L12 **Thrace** cultural region SE Europe
114 J13 **Thracian Sea** Gk. Thrakikó Pélagos; anc. Thracium Mare. sea Greece/Turkey
Thracium Mare/Thrakikó Pélagos see Thracian Sea
Thrá Lí, Bá see Tralee Bay
33 R11 **Three Forks** Montana, NW USA 45°53′N 111°34′W
162 M8 **Three Gorges Dam** dam Hubei, C China
160 L9 **Three Gorges Reservoir** ▣ C China
11 Q16 **Three Hills** Alberta, SW Canada 51°43′N 113°15′W
183 N15 **Three Hummock Island** island Tasmania, SE Australia
184 H1 **Three Kings Islands** island group N New Zealand
175 P10 **Three Kings Rise** undersea feature W Pacific Ocean
77 O18 **Three Points, Cape** headland S Ghana 04°43′N 02°03′W
31 P10 **Three Rivers** Michigan, N USA 41°56′N 85°37′W
25 S13 **Three Rivers** Texas, SW USA 28°27′N 98°10′W
83 G24 **Three Sisters** Northern Cape, SW South Africa 31°51′S 23°04′E
32 H13 **Three Sisters** ▲ Oregon, NW USA 44°08′N 121°46′W
187 N10 **Three Sisters Islands** island group SE Solomon Islands
25 Q6 **Throckmorton** Texas, SW USA 33°11′N 99°12′W
180 M10 **Throssell, Lake** salt lake Western Australia
115 K25 **Thryptís** var. Thrýptis. ▲ Kríti, Greece, E Mediterranean Sea 35°06′N 25°51′E
167 U14 **Thuân Nam** prev. Ham Thuân Nam. Bình Thuận, S Vietnam 10°49′N 107°49′E
167 T13 **Thu Dâu Môt** var. Phu Cuông. Sông Be, S Vietnam 10°58′N 106°40′E
167 S6 **Thu Do** ✈ (Ha Nôi) Ha Nôi, N Vietnam 21°13′N 105°46′E
99 G21 **Thuin** Hainaut, S Belgium 50°21′N 04°18′E
149 Q12 **Thul** Sind, SE Pakistan 28°14′N 68°50′E
Thule see Qaanaaq
83 J18 **Thuli** var. Tuli. ≈ S Zimbabwe
Thumrayt see Thamarīt
108 D9 **Thun** Fr. Thoune. Bern, W Switzerland 46°46′N 07°38′E
12 C12 **Thunder Bay** Ontario, S Canada 48°27′N 89°12′W
30 M1 **Thunder Bay** lake bay S Canada
31 R6 **Thunder Bay** lake bay Michigan, N USA
31 R6 **Thunder Bay River** ≈ Michigan, N USA
27 N11 **Thunderbird, Lake** ▣ Oklahoma, C USA
28 L8 **Thunder Butte Creek** ≈ South Dakota, N USA
108 E9 **Thuner See** ◎ C Switzerland
167 N15 **Thung Song** var. Cha Mai. Nakhon Si Thammarat, SW Thailand 08°10′N 99°41′E
108 H7 **Thur** ≈ N Switzerland
108 G6 **Thurgau** Fr. Thurgovie. ◐ canton NE Switzerland
Thurgovie see Thurgau
Thuringe see Thüringen
108 J7 **Thüringen** Vorarlberg, W Austria 47°12′N 09°48′E
101 J17 **Thüringen** Eng. Thuringia, Fr. Thuringe. ◐ state C Germany
101 J17 **Thüringer Wald** Eng. Thuringian Forest. ▲ C Germany
Thuringia see Thüringen
Thuringian Forest see Thüringer Wald
97 D19 **Thurles** Ir. Durlas. S Ireland 52°41′N 07°49′W
21 W2 **Thurmont** Maryland, NE USA 39°36′N 77°22′W
Thurø see Thurø By
95 H24 **Thurø By** var. Thurø. Syddtjylland, C Denmark 55°03′N 10°43′E
14 M12 **Thurso** Québec, SE Canada 45°36′N 75°13′W
96 J6 **Thurso** N Scotland, United Kingdom 58°35′N 03°32′W
194 I10 **Thurston Island** island Antarctica
108 I9 **Thusis** Graubünden, S Switzerland 46°40′N 09°27′E
Thýamis see Kalamás
95 E21 **Thyborøn** var. Tyborøn. Midtjylland, W Denmark 56°40′N 08°12′E
195 U3 **Thyer Glacier** glacier Antarctica
115 L20 **Thýmaina** island Dodekánisa, Greece, Aegean Sea
83 N15 **Thyolo** var. Cholo. Southern, S Malawi 16°03′S 35°11′E
183 U6 **Tia** New South Wales, SE Australia 31°14′S 151°51′E
54 H5 **Tía Juana** Zulia, NW Venezuela 10°18′N 71°24′W
Tiancheng see Chongyang
160 J14 **Tiandong** var. Pingma. Guangxi Zhuangzu Zizhiqu, S China 23°37′N 107°06′E
161 O3 **Tianjin** var. Tientsin. Tianjin Shi, E China 39°13′N 117°06′E
Tianjin see Tianjin Shi
161 P3 **Tianjin Shi** var. Jin, Tianjin, T'ien-ching, Tientsin. ◐ municipality E China
159 S10 **Tianjun** var. Xinyuan. Qinghai, C China 37°16′N 99°03′E
160 J13 **Tianlin** var. Leli. Guangxi Zhuangzu Zizhiqu, S China 24°27′N 106°03′E
Tian Shan see Tien Shan
159 W11 **Tianshui** Gansu, C China 34°33′N 105°51′E
150 I7 **Tianshuihai** Xinjiang Uygur Zizhiqu, W China 35°17′N 79°30′E
161 S10 **Tiantai** Zhejiang, SE China 29°11′N 121°01′E
160 J14 **Tianyang** var. Tianzhou. Guangxi Zhuangzu Zizhiqu, S China 23°50′N 106°52′E
Tianzhou see Tianyang
159 U9 **Tianzhu** var. Huazangsi, Tianzhu Zangzu Zizhixian. Gansu, C China 37°01′N 103°04′E
Tianzhu Zangzu Zizhixian see Tianzhu
191 Q7 **Tiarei** Tahiti, W French Polynesia 17°32′S 149°20′W
74 J6 **Tiaret** var. Tihert. NW Algeria 35°20′N 01°20′E
77 N17 **Tiassalé** S Ivory Coast 05°54′N 04°50′W
192 I16 **Ti'avea** Upolu, SE Samoa 13°58′S 171°30′W
Tiba see Chiba
60 J11 **Tibagi** var. Tibají. Paraná, S Brazil 24°29′S 50°29′W
60 J10 **Tibagi, Rio** var. Rio Tibají. ≈ S Brazil
Tibají see Tibagi
Tibají, Rio see Tibagi, Rio
139 Q9 **Tibal, Wādī** dry watercourse S Iraq
54 G9 **Tibaná** Boyacá, C Colombia 05°19′N 73°25′W
79 F14 **Tibati** Adamaoua, N Cameroon 06°25′N 12°33′E
76 K15 **Tibé, Pic de** ▲ SE Guinea 08°39′N 08°58′W
Tiber see Tevere, Italy
Tiber see Tivoli, Italy
Tiberias see Tverya
138 G8 **Tiberias, Lake** var. Chinnereth, Sea of Bahr Tabariya, Sea of Galilee, Ar. Bahrat Tabariya, Heb. Yam Kinneret. ◎ N Israel
78 H5 **Tibesti** off. Région du Tibesti. ◐ region N Chad
78 H6 **Tibesti** var. Tibesti Massif, Ar. Tibistī. ▲▲ N Africa
Tibesti Massif see Tibesti
Tibesti, Région du see Tibesti
Tibet see Xizang Zizhiqu
Tibetan Autonomous Region see Xizang Zizhiqu
Tibet, Plateau of see Qingzang Gaoyuan
Tibistī see Tibesti
14 K7 **Tiblemont, Lac** ◎ Québec, SE Canada
139 X9 **Ţīb, Nahr aţ** ≈ S Iraq
Tibnī see At Tibnī
182 L4 **Tibooburra** New South Wales, SE Australia 29°24′S 142°01′E
95 L18 **Tibro** Västra Götaland, S Sweden 58°25′N 14°11′E
40 E5 **Tiburón, Isla** var. Isla del Tiburón. island NW Mexico
Tiburón, Isla del see Tiburón, Isla
23 W14 **Tice** Florida, SE USA 26°40′N 81°49′W
Tichau see Tychy
114 L8 **Ticha, Yazovir** ▣ NE Bulgaria
76 K9 **Tîchît** var. Tichitt. Tagant, C Mauritania 18°26′N 09°31′W
Tichitt see Tîchît
108 G11 **Ticino** Fr./Ger. Tessin. ◐ canton S Switzerland
106 D8 **Ticino** ≈ Italy/Switzerland
108 H11 **Ticino** Ger. Tessin. ≈ SW Switzerland
Ticinum see Pavia
41 X12 **Ticul** Yucatán, SE Mexico 20°22′N 89°30′W
95 K18 **Tidaholm** Västra Götaland, S Sweden 58°12′N 13°55′E
Tidjikdja see Tidjikja
76 J8 **Tidjikja** var. Tidjikdja; prev. Fort-Cappolani. Tagant, C Mauritania 18°31′N 11°24′W
Tidore see Soasiu
171 R11 **Tidore, Pulau** island E Indonesia
Tîdra, Île see Et Tîdra
77 N16 **Tiébissou** var. Tiebissou. C Ivory Coast 07°10′N 05°10′W
Tiebissou see Tiébissou
Tiefa see Diaobingshan
108 I9 **Tiefencastel** Graubünden, S Switzerland 46°40′N 09°33′E
Tiegenhof see Nowy Dwór Gdański
T'ieh-ling see Tieling
98 K13 **Tiel** Gelderland, C Netherlands 51°53′N 05°26′E
163 W7 **Tieli** Heilongjiang, NE China 46°57′N 128°01′E
163 V11 **Tieling** var. T'ieh-ling. Liaoning, NE China 42°19′N 123°52′E
152 L4 **Tielongtan** China/India 35°10′N 79°32′E
99 D17 **Tielt** var. Thielt. West-Vlaanderen, W Belgium 51°00′N 03°20′E
T'ien-ching see Tianjin Shi
99 I18 **Tienen** var. Thienen, Fr. Tirlemont. Vlaams Brabant, C Belgium 50°48′N 04°56′E
Tiên Giang, Sông see Mekong
147 X9 **Tien Shan** Chin. Thian Shan, Tian Shan, T'ien Shan, Rus. Tyan'-Shan'. ▲▲ C Asia
Tientsin see Tianjin
167 U6 **Tiên Yên** Quang Ninh, N Vietnam 21°19′N 107°24′E
95 O14 **Tierp** Uppsala, C Sweden 60°20′N 17°30′E
62 H7 **Tierra Amarilla** Atacama, N Chile 27°28′S 70°17′W
37 R9 **Tierra Amarilla** New Mexico, SW USA 36°42′N 106°31′W
41 R15 **Tierra Blanca** Veracruz-Llave, E Mexico 18°28′N 96°21′W
41 O16 **Tierra Colorada** Guerrero, S Mexico 17°10′N 99°30′W
63 J17 **Tierra Colorada, Bajo de la** basin SE Argentina
63 I25 **Tierra del Fuego** off. Provincia de la Tierra del Fuego. ◐ province S Argentina
63 J24 **Tierra del Fuego** island Argentina/Chile
Tierra del Fuego, Provincia de la see Tierra del Fuego
54 D7 **Tierralta** Córdoba, NW Colombia 08°10′N 76°04′W
104 K9 **Tiétar** ≈ W Spain
60 L10 **Tietê** São Paulo, S Brazil 23°04′S 47°41′W
60 J8 **Tietê, Rio** ≈ S Brazil
32 I9 **Tieton** Washington, NW USA 46°41′N 120°43′W
32 J6 **Tiffany Mountain** ▲ Washington, NW USA 48°40′N 119°55′W
31 S12 **Tiffin** Ohio, N USA 41°06′N 83°10′W
31 Q10 **Tiffin River** ≈ Ohio, N USA
Tiflis see Tbilisi
23 U7 **Tifton** Georgia, SE USA 31°27′N 83°31′W
171 R13 **Tifu** Pulau Buru, E Indonesia 03°46′S 126°36′E
38 L17 **Tigalda Island** island Aleutian Islands, Alaska, USA
115 I15 **Tigáni, Akrotírio** headland Límnos, E Greece 39°50′N 25°03′E
169 V6 **Tiga Tarok** Sabah, East Malaysia 06°57′N 117°07′E
117 O10 **Tighina** Rus. Bendery; prev. Bender. E Moldova 46°51′N 29°28′E
145 X9 **Tigiretskiy Khrebet** ▲ E Kazakhstan
79 F14 **Tignère** Adamaoua, N Cameroon 07°24′N 12°35′E
13 P14 **Tignish** Prince Edward Island, SE Canada 46°58′N 64°03′W
186 M10 **Tigoa** var. Tinggoa. Rennell, S Solomon Islands 11°39′S 160°13′E
Tigranocerta see Siirt
80 I11 **Tigray** ◐ federal region N Ethiopia
41 O11 **Tigre, Cerro del** ▲ C Mexico 23°06′N 99°13′W
56 F8 **Tigre, Río** ≈ N Peru
139 X10 **Tigris** Ar. Dijlah, Turk. Dicle. ≈ Iraq/Turkey
76 G9 **Tiguent** Trarza, SW Mauritania 17°15′N 16°00′W
74 M10 **Tiguentourine** E Algeria 27°59′N 09°16′E
77 V10 **Tiguidit, Falaise de** ridge C Niger
141 N13 **Tihāmah** var. Tehama. plain Saudi Arabia/Yemen
Tihert see Tiaret
Ti-hua/Tihwa see Ürümqi
41 Q13 **Tihuatlán** Veracruz-Llave, E Mexico 20°44′N 97°30′W
40 B1 **Tijuana** Baja California Norte, NW Mexico 32°32′N 117°01′W
42 E2 **Tikal** Petén, N Guatemala 17°11′N 89°36′W
154 I9 **Tīkamgarh** prev. Tehri. Madhya Pradesh, C India 24°44′N 78°50′E
158 L7 **Tikanlik** Xinjiang Uygur Zizhiqu, NW China 40°34′N 87°37′E
77 P12 **Tikaré** N Burkina Faso 13°16′N 01°39′W
39 O12 **Tikchik Lakes** lakes Alaska, USA
191 T9 **Tikehau** atoll Îles Tuamotu, C French Polynesia
191 V9 **Tikei** island Îles Tuamotu, C French Polynesia
126 L13 **Tikhoretsk** Krasnodarskiy Kray, SW Russian Federation 45°51′N 40°07′E
124 I13 **Tikhvin** Leningradskaya Oblast', NW Russian Federation 59°37′N 33°30′E
193 P9 **Tiki Basin** undersea feature S Pacific Ocean
76 K13 **Tikinsso** ≈ NE Guinea
Tikirarjuaq see Whale Cove
184 Q8 **Tikitiki** Gisborne, North Island, New Zealand 37°49′S 178°23′E
79 D16 **Tiko** Sud-Ouest, SW Cameroon 04°02′N 09°19′E
139 S6 **Tikrīt** var. Tekrit. Ṣalāḥ ad Dīn, N Iraq 34°36′N 43°42′E
124 I8 **Tiksha** Respublika Kareliya, NW Russian Federation 64°07′N 32°31′E
124 I6 **Tikshozero, Ozero** ◎ NW Russian Federation
123 P7 **Tiksi** Respublika Sakha (Yakutiya), NE Russian Federation 71°40′N 128°47′E
Tiladummati Atoll see Thiladhunmathi Atoll
42 A6 **Tilapa** San Marcos, SW Guatemala 14°31′N 92°11′W
42 L13 **Tilarán** Guanacaste, NW Costa Rica 10°28′N 84°57′W
99 J14 **Tilburg** Noord-Brabant, S Netherlands 51°34′N 05°05′E
14 D17 **Tilbury** Ontario, S Canada 42°15′N 82°26′W
182 K4 **Tilcha** South Australia 29°37′S 140°52′E
Tilcha Creek see Callabonna Creek
29 Q14 **Tilden** Nebraska, C USA 42°03′N 97°49′W
25 R13 **Tilden** Texas, SW USA 28°27′N 98°43′W
14 H10 **Tilden Lake** Ontario, S Canada 46°35′N 79°36′W
116 G9 **Tileagd** Hung. Mezőtelegd. Bihor, W Romania 47°03′N 22°11′E
145 P9 **Tilekey** prev. Ladyzhenka. Akmola, C Kazakhstan 50°58′N 68°44′E
77 Q8 **Tîlemsi, Vallée de** ≈ C Mali
123 V8 **Tilichiki** Krasnoyarskiy Kray, E Russian Federation 60°25′N 165°55′E
Tiligul see Tilihul
Tiligul'skiy Liman see Tilihul's'kyy Lyman
117 P9 **Tilihul** Rus. Tiligul. ≈ SW Ukraine
117 P10 **Tilihul's'kyy Lyman** Rus. Tiligul'skiy Liman. ◎ S Ukraine
Tilimsen see Tlemcen
Tilio Martius see Toulon
77 R11 **Tillabéri** var. Tillabéry. Tillabéri, W Niger 14°13′N 01°27′E
77 R11 **Tillabéri** ◐ department SW Niger
Tillabéry see Tillabéri
32 F11 **Tillamook** Oregon, NW USA 45°28′N 123°50′W
32 E11 **Tillamook Bay** inlet Oregon, NW USA
151 Q22 **Tillanchāng Dwīp** island Nicobar Islands, India, NE Indian Ocean
95 N15 **Tillberga** Västmanland, C Sweden 59°41′N 16°37′E
Tillenberg see Dyleň
21 S10 **Tillery, Lake** ▣ North Carolina, SE USA
77 T10 **Tillia** Tahoua, W Niger 16°13′N 04°51′E
23 N8 **Tillmans Corner** Alabama, S USA 30°35′N 88°10′W
14 F17 **Tillsonburg** Ontario, S Canada 42°53′N 80°44′W
115 N22 **Tílos** island Dodekánisa, Greece, Aegean Sea
183 N5 **Tilpa** New South Wales, SE Australia 30°56′S 144°24′E
Tilsit see Sovetsk
31 N13 **Tilton** Illinois, N USA 40°06′N 87°39′W
126 K7 **Tim** Kurskaya Oblast', W Russian Federation 51°39′N 37°11′E
54 D12 **Timaná** Huila, S Colombia 01°58′N 75°55′W
Timan Ridge see Timanskiy Kryazh
125 Q6 **Timanskiy Kryazh** Eng. Timan Ridge. ridge NW Russian Federation
185 G20 **Timaru** Canterbury, South Island, New Zealand 44°23′S 171°15′E
127 S6 **Timashevo** Samarskaya Oblast', W Russian Federation 53°22′N 51°13′E
126 K13 **Timashevsk** Krasnodarskiy Kray, SW Russian Federation 45°37′N 38°57′E
Timbaki/Timbákion see Tympáki
22 K10 **Timbalier Bay** bay Louisiana, S USA
22 K11 **Timbalier Island** island Louisiana, S USA
76 L10 **Timbedgha** var. Timbédra. Hodh ech Chargui, SE Mauritania 16°17′N 08°14′W
Timbédra see Timbedgha
32 G10 **Timber** Oregon, NW USA 45°42′N 123°19′W
181 O3 **Timber Creek** Northern Territory, N Australia 15°35′S 130°21′E
28 M8 **Timber Lake** South Dakota, N USA 45°25′N 101°01′W
54 D12 **Timbío** Cauca, SW Colombia 02°20′N 76°40′W
54 C12 **Timbiquí** Cauca, SW Colombia 02°43′N 77°45′W
83 O17 **Timbue, Ponta** headland C Mozambique 18°49′S 36°22′E
Timbuktu see Tombouctou
169 W8 **Timbun Mata, Pulau** island E Malaysia
77 P8 **Timétrine** var. Ti-n-Kâr. oasis C Mali
Timfi see Týmfi
Timfristos see Tymfristós
77 V9 **Timia** Agadez, C Niger 18°07′N 08°49′E
171 X14 **Timika** Papua, E Indonesia 04°39′S 137°15′E
74 I9 **Timimoun** C Algeria 29°18′N 00°21′E
Timiris, Cap see Timirist, Râs
76 F8 **Timirist, Râs** var. Cap Timiris. headland NW Mauritania 19°18′N 16°28′W
145 O7 **Timiryazevo** Severnyy Kazakhstan, N Kazakhstan 53°45′N 66°33′E
116 E11 **Timiş** ◐ county SW Romania
14 H9 **Timiskaming, Lake** Fr. Lac Témiscamingue. ◎ Ontario/Québec, SE Canada
116 E11 **Timişoara** Ger. Temeschwar, Temeswar, Hung. Temesvár; prev. Temeschburg. Timiş, W Romania 45°46′N 21°17′E
116 E11 **Timişoara** ✈ Timiş, SW Romania 45°50′N 21°21′E
Timkovichi see Tsimkavichy
77 U8 **Ti-m-Meghsoï** ≈ NW Niger
100 K8 **Timmerdorfer Strand** Schleswig-Holstein, N Germany 53°59′N 10°50′E
14 F7 **Timmins** Ontario, S Canada 48°09′N 80°01′W
21 S12 **Timmonsville** South Carolina, SE USA 34°07′N 79°56′W
30 K5 **Timms Hill** ▲ Wisconsin, N USA 45°27′N 90°12′W
112 P12 **Timok** ≈ E Serbia
58 N13 **Timon** Maranhão, E Brazil 05°08′S 42°52′W
171 Q17 **Timor Sea** sea E Indian Ocean
Timor Timur see East Timor
Timor Trench see Timor Trough
192 G8 **Timor Trough** var. Timor Trench. undersea feature NE Timor Sea
61 A21 **Timote** Buenos Aires, E Argentina 35°22′S 62°13′W
54 I6 **Timotes** Mérida, NW Venezuela 08°57′N 70°46′W
25 X8 **Timpson** Texas, SW USA 31°54′N 94°24′W
123 Q11 **Timpton** ≈ NE Russian Federation
93 H17 **Timrå** Västernorrland, C Sweden 62°29′N 17°20′E
20 J10 **Tims Ford Lake** ▣ Tennessee, S USA
168 L7 **Timur, Banjaran** ▲ Peninsular Malaysia
171 Q8 **Tinaca Point** headland Mindanao, S Philippines 05°35′N 125°18′E
54 K5 **Tinaco** Cojedes, N Venezuela 09°44′N 68°28′W
64 Q11 **Tinajo** Lanzarote, Islas Canarias, Spain, NE Atlantic Ocean 29°03′N 13°41′W
187 P10 **Tinakula** island Santa Cruz Islands, E Solomon Islands
54 K5 **Tinaquillo** Cojedes, N Venezuela 09°57′N 68°20′W
116 F10 **Tinca** Hung. Tenke. Bihor, W Romania 46°46′N 21°58′E
155 J20 **Tindivanam** Tamil Nādu, SE India 12°15′N 79°41′E
74 E9 **Tindouf** W Algeria 27°43′N 08°09′W
74 E9 **Tindouf, Sebkha de** salt lake W Algeria
104 J2 **Tineo** Asturias, N Spain 43°20′N 06°25′W
77 R9 **Ti-n-Essako** Kidal, E Mali 18°30′N 02°27′E
Tinggoa see Tigoa
183 T5 **Tingha** New South Wales, SE Australia 29°56′S 151°13′E
Tingis see Tanger
Tinglett see Tinglev
95 F24 **Tinglev** Ger. Tinglett. Syddanmark, SW Denmark 54°57′N 09°15′E
56 E12 **Tingo María** Huánuco, C Peru 09°10′S 76°00′W
Tingréla see Tengréla
158 K16 **Tingri** var. Xêgar. Xizang Zizhiqu, W China 28°40′N 87°04′E
95 M21 **Tingsryd** Kronoberg, S Sweden 56°30′N 15°E
95 P19 **Tingstäde** Gotland, SE Sweden 57°45′N 18°36′E
62 H12 **Tinguiririca, Volcán** ⛰ C Chile 34°52′S 70°24′W
94 F9 **Tingvoll** Møre og Romsdal, S Norway 62°55′N 08°13′E
188 K8 **Tinian** island S Northern Mariana Islands
Ti-n-Kâr see Timétrine
Tinnevelly see Tirunelveli
95 G15 **Tinnoset** Telemark, S Norway 59°43′N 09°03′E
95 F15 **Tinnsjå** prev. Tinnsjø, var. Tinnsjo. ◎ S Norway see also Tinnsjø
Tinnsjø see Tinnsjå
Tinnsjö see Tinnsjå
Tino see Chino
115 J20 **Tínos** Tínos, Kykládes, Greece, Aegean Sea 37°33′N 25°08′E
115 J20 **Tínos** anc. Tenos. island Kykládes, Greece, Aegean Sea
153 R14 **Tinpahar** Jhārkhand, NE India 25°00′N 87°43′E
153 X11 **Tinsukia** Assam, NE India 27°28′N 95°20′E
76 K10 **Tîntâne** Hodh el Gharbi, S Mauritania 16°25′N 10°08′W
62 L7 **Tintina** Santiago del Estero, N Argentina 27°00′S 62°45′W
182 K10 **Tintinara** South Australia 35°54′S 140°04′E
104 I14 **Tinto** ≈ SW Spain
77 S8 **Ti-n-Zaouâtene** Kidal, NE Mali 19°56′N 02°45′E
Tiobraid Árann see Tipperary
28 K3 **Tioga** North Dakota, N USA 48°24′N 102°56′W
18 G12 **Tioga** Pennsylvania, NE USA 41°54′N 77°07′W
25 T5 **Tioga** Texas, SW USA 33°28′N 96°55′W
35 Q8 **Tioga Pass** pass California, W USA
18 G12 **Tioga River** ≈ New York/Pennsylvania, NE USA
Tioman Island see Tioman, Pulau
168 M9 **Tioman, Pulau** var. Tioman Island. island Peninsular Malaysia
18 C12 **Tionesta** Pennsylvania, NE USA 41°31′N 79°30′W
18 D12 **Tionesta Creek** ≈ Pennsylvania, NE USA
168 J13 **Tiop** Pulau Pagai Selatan, W Indonesia 03°12′S 100°21′E
18 H11 **Tioughnioga River** ≈ New York, NE USA
74 J5 **Tipasa** var. Tipaza. N Algeria 36°35′N 02°27′E
Tipaza see Tipasa
42 J10 **Tipitapa** Managua, W Nicaragua 12°08′N 86°04′W
31 R13 **Tipp City** Ohio, N USA 39°57′N 84°10′W
31 O12 **Tippecanoe River** ≈ Indiana, N USA
97 D20 **Tipperary** Ir. Tiobraid Árann. S Ireland 52°29′N 08°10′W
97 D19 **Tipperary** Ir. Tiobraid Árann. cultural region S Ireland
35 R12 **Tipton** California, W USA 36°02′N 119°19′W
31 P13 **Tipton** Indiana, N USA 40°19′N 86°00′W
29 Y14 **Tipton** Iowa, C USA 41°46′N 91°07′W
27 U5 **Tipton** Missouri, C USA 38°39′N 92°46′W
36 I10 **Tipton, Mount** ▲ Arizona, SW USA 35°32′N 114°11′W
20 F8 **Tiptonville** Tennessee, S USA 36°21′N 89°30′W
12 E12 **Tip Top Mountain** ▲ Ontario, S Canada 48°18′N 86°06′W
155 G19 **Tiptūr** Karnātaka, W India 13°17′N 76°31′E
Tiquisate see Pueblo Nuevo Tiquisate
58 L13 **Tiracambu, Serra do** ▲ E Brazil
Tirana see Tiranë
113 K19 **Tirana Rinas** ✈ Durrës, W Albania 41°25′N 19°41′E
113 L20 **Tiranë** var. Tirana. ● (Albania) Tiranë, C Albania 41°20′N 19°50′E
113 K20 **Tiranë** ◐ district W Albania
140 I5 **Tirān, Jazīrat** island Egypt/Saudi Arabia
106 F6 **Tirano** Lombardia, N Italy 46°13′N 10°10′E
182 I2 **Tirari Desert** desert South Australia
117 O10 **Tiraspol** Rus. Tiraspol'. E Moldova 46°50′N 29°35′E
Tiraspol' see Tiraspol
184 M8 **Tirau** Waikato, North Island, New Zealand 37°59′S 175°44′E
136 C14 **Tire** İzmir, SW Turkey 38°04′N 27°45′E
137 O11 **Tirebolu** Giresun, N Turkey 41°01′N 38°49′E
96 F11 **Tiree** island W Scotland, United Kingdom
Tîrgovişte see Târgovişte
Tîrgu see Târgu Cărbuneşti
Tîrgu Bujor see Târgu Bujor
Tîrgu Frumos see Târgu Frumos
Tîrgu Jiu see Târgu Jiu
Tîrgu Lăpuş see Târgu Lăpuş
Tîrgu Mures see Târgu Mureş
Tîrgu-Neamţ see Târgu-Neamţ
Tîrgu Ocna see Târgu Ocna
Tîrgu Secuiesc see Târgu Secuiesc
149 T3 **Tirich Mīr** ▲ NW Pakistan 36°12′N 71°51′E
76 J5 **Tiris Zemmour** ◐ region N Mauritania
Tirlemont see Tienen
127 W5 **Tirlyanskiy** Respublika Bashkortostan, W Russian Federation 54°09′N 58°32′E
Tîrnava Mare see Târnava Mare
Tîrnava Mică see Târnava Mică
Tîrnăveni see Târnăveni
Tírnavos see Týrnavos
Tirnovo see Veliko Tarnovo
154 J11 **Tirodi** Madhya Pradesh, C India 21°40′N 79°44′E
108 K8 **Tirol** off. Land Tirol, var. Tyrol, It. Tirolo. ◐ state W Austria
Tirol, Land see Tirol
Tirolo see Tirol
Tirreno, Mare see Tyrrhenian Sea
107 B19 **Tirso** ≈ Sardegna, Italy, C Mediterranean Sea
95 H22 **Tirstrup** ✈ (Århus) Århus, C Denmark 56°17′N 10°36′E
155 I21 **Tiruchchirāppalli** prev. Trichinopoly. Tamil Nādu, SE India 10°50′N 78°43′E
155 H23 **Tirunelveli** var. Tinnevelly. Tamil Nādu, SE India 08°45′N 77°43′E
155 J19 **Tirupati** Andhra Pradesh, E India 13°39′N 79°25′E
155 I20 **Tiruppattūr** Tamil Nādu, SE India 12°28′N 78°31′E
155 H21 **Tiruppur** Tamil Nādu, SW India 11°05′N 77°20′E
Tiruvantapuram see Thiruvanathapuram/Trivandrum
155 I20 **Tiruvannāmalai** Tamil Nādu, SE India 12°13′N 79°07′E
112 L10 **Tisa** Ger. Theiss, Hung. Tisza, Rus. Tissa, Ukr. Tysa. ≈ SE Europe see also Tisza
Tisa see Tisza
Tischnowitz see Tišnov
11 U14 **Tisdale** Saskatchewan, S Canada 52°51′N 104°01′W
27 O13 **Tishomingo** Oklahoma, C USA 34°15′N 96°41′W
95 M17 **Tisnaren** ◎ S Sweden
111 F18 **Tišnov** Ger. Tischnowitz. Jihomoravský Kraj, SE Czech Republic 49°22′N 16°24′E
Tissa see Tisa/Tisza
74 J6 **Tissemsilt** N Algeria 35°37′N 01°48′E
153 S12 **Tista** ≈ NE India
112 L8 **Tisza** Ger. Theiss, Rom./Slvn./SCr. Tisa, Rus. Tissa, Ukr. Tysa. ≈ SE Europe see also Tisa
Tisza see Tisa
111 L23 **Tiszaföldvár** Jász-Nagykun-Szolnok, E Hungary 47°00′N 20°16′E
111 M22 **Tiszafüred** Jász-Nagykun-Szolnok, E Hungary 47°38′N 20°45′E
111 L23 **Tiszakécske** Bács-Kiskun, C Hungary 46°56′N 20°04′E
111 M21 **Tiszaújváros** prev. Leninváros. Borsod-Abaúj-Zemplén, NE Hungary 47°56′N 21°03′E
111 N21 **Tiszavasvári** Szabolcs-Szatmár-Bereg, NE Hungary 47°56′N 21°21′E
57 I17 **Titicaca, Lake** ◎ Bolivia/Peru
190 H17 **Titikaveka** Rarotonga, S Cook Islands 21°16′S 159°45′W
154 M13 **Titilāgarh** var. Titlagarh. Odisha, E India 20°18′N 83°09′E
168 K8 **Titiwangsa, Banjaran** ▲ Peninsular Malaysia
Titlagarh see Titilāgarh
Titograd see Podgorica
Titose see Chitose
Titova Mitrovica see Mitrovicë
Titovo Užice see Užice
113 M18 **Titov Vrv** ▲ NW FYR Macedonia 41°58′N 20°49′E
94 F7 **Titran** Sør-Trøndelag, S Norway 63°40′N 08°20′E
31 Q8 **Tittabawassee River** ≈ Michigan, N USA
116 J13 **Titu** Dâmboviţa, S Romania 44°40′N 25°32′E
79 M16 **Titule** Orientale, N Dem. Rep. Congo 03°20′N 25°23′E
23 X11 **Titusville** Florida, SE USA 28°37′N 80°50′W
18 C12 **Titusville** Pennsylvania, NE USA 41°36′N 79°39′W
76 G11 **Tivaouane** W Senegal 14°59′N 16°50′W
113 I17 **Tivat** SW Montenegro 42°25′N 18°43′E
14 E14 **Tiverton** Ontario, S Canada 44°15′N 81°31′W
97 J23 **Tiverton** SW England, United Kingdom 50°54′N 03°30′W
19 O12 **Tiverton** Rhode Island, NE USA 41°38′N 71°10′W
107 I15 **Tivoli** anc. Tiber. Lazio, C Italy 41°58′N 12°45′E
25 U13 **Tivoli** Texas, SW USA 28°26′N 96°54′W
141 Z8 **Ţīwī** NE Oman 22°43′N 59°20′E
41 Y11 **Tizimín** Yucatán, SE Mexico 21°10′N 88°09′W
74 K5 **Tizi Ouzou** var. Tizi-Ouzou. N Algeria 36°44′N 04°06′E
Tizi-Ouzou see Tizi Ouzou
74 D8 **Tiznit** SW Morocco 29°43′N 09°39′W
95 F23 **Tjæreborg** Syddtjylland, W Denmark 55°28′N 08°35′E
113 I14 **Tjentište** Republika Srpska, SE Bosnia and Herzegovina 43°23′N 18°42′E
98 L7 **Tjeukemeer** ◎ N Netherlands
Tjiamis see Ciamis
Tjiandjoer see Cianjur
Tjilatjap see Cilacap
Tjirebon see Cirebon
95 I18 **Tjörn** island S Sweden
92 O3 **Tjuvfjorden** fjord S Svalbard
Tkvarcheli see T'q'varcheli
40 L8 **Tlahualilo** Durango, N Mexico 26°06′N 103°25′W
41 P14 **Tlalnepantla** México, C Mexico 19°34′N 99°12′W
41 Q13 **Tlapacoyán** Veracruz-Llave, E Mexico 19°58′N 97°13′W
41 P16 **Tlapa de Comonfort** Guerrero, S Mexico 17°33′N 98°33′W
40 L13 **Tlaquepaque** Jalisco, C Mexico 20°36′N 103°19′W
Tlascala see Tlaxcala
41 P14 **Tlaxcala** var. Tlascala, Tlaxcala de Xicohténcatl. Tlaxcala, C Mexico 19°17′N 98°16′W
41 P14 **Tlaxcala** ◐ state S Mexico
Tlaxcala de Xicohténcatl see Tlaxcala
41 P14 **Tlaxco** var. Tlaxco de Morelos. Tlaxcala, S Mexico 19°38′N 98°06′W
Tlaxco de Morelos see Tlaxco
41 Q16 **Tlaxiaco** var. Santa María Asunción Tlaxiaco. Oaxaca, S Mexico 17°18′N 97°42′W
74 I6 **Tlemcen** var. Tilimsen, Tlemsen. NW Algeria 34°53′N 01°21′W
Tlemsen see Tlemcen
138 L4 **Tlété Ouâte Rharbi, Jebel** ▲ N Syria
116 J7 **Tlumach** Ivano-Frankivs'ka Oblast', W Ukraine 48°53′N 25°00′E
127 P17 **Tlyarata** Respublika Dagestan, SW Russian Federation 42°10′N 46°30′E
116 K10 **Toaca, Vârful** prev. Vîrful Toaca. ▲ NE Romania 46°58′N 25°55′E
Toaca, Vîrful see Toaca, Vârful
191 Q8 **Toahotu** prev. Teohatu. Tahiti, W French Polynesia
187 R13 **Toak** Ambrym, C Vanuatu 16°21′S 168°16′E
172 J4 **Toamasina** prev./Fr. Tamatave. Toamasina, E Madagascar 18°10′S 49°23′E
172 J4 **Toamasina** ◐ province E Madagascar

◆ Country
● Country Capital
◇ Dependent Territory
○ Dependent Territory Capital
◐ Administrative Regions
✈ International Airport
▲ Mountain
▲▲ Mountain Range
⛰ Volcano
≈ River
◎ Lake
▣ Reservoir

172 J4 **Toamasina** ✈ Toamasina, E Madagascar 18°10′S 49°23′E
21 X6 **Toano** Virginia, NE USA 37°22′N 76°46′W
191 U10 **Toau** *atoll* Îles Tuamotu, C French Polynesia
45 T6 **Toa Vaca, Embalse** ⊠ C Puerto Rico
62 K13 **Toay** La Pampa, C Argentina 36°43′S 64°22′W
159 R14 **Toba** Xizang Zizhiqu, W China 31°17′N 97°37′E
164 K14 **Toba** Mie, Honshū, SW Japan 34°29′N 136°51′E
168 I9 **Toba, Danau** ◎ Sumatera, W Indonesia
45 Y16 **Tobago** *island* NE Trinidad and Tobago
149 Q9 **Toba Kākar Range** ▲▲ NW Pakistan
105 Q12 **Tobarra** Castilla-La Mancha, C Spain 38°36′N 01°41′W
149 U9 **Toba Tek Singh** Punjab, E Pakistan 30°54′N 72°30′E
171 R11 **Tobelo** Pulau Halmahera, E Indonesia 01°45′N 127°59′E
14 E12 **Tobermory** Ontario, S Canada 45°15′N 81°39′W
96 G10 **Tobermory** W Scotland, United Kingdom 56°37′N 06°12′W
165 S4 **Tōbetsu** Hokkaidō, NE Japan 43°12′N 141°28′E
180 M6 **Tobin Lake** ◎ Western Australia
11 U14 **Tobin Lake** ◎ Saskatchewan, C Canada
35 T4 **Tobin, Mount** ▲ Nevada, W USA 40°25′N 117°28′W
165 O9 **Tobi-shima** *island* C Japan
169 N13 **Toboali** Pulau Bangka, W Indonesia 03°00′S 106°30′E
Tobol *see* Tobyl
122 H11 **Tobol'sk** Tyumenskaya Oblast', C Russian Federation 58°15′N 68°12′E
Tobruch/Tobruk *see* Ṭubruq
125 R3 **Tobseda** Nenetskiy Avtonomnyy Okrug, NW Russian Federation 68°37′N 52°24′E
144 M8 **Tobyl** *prev.* Tobol. Kustanay, N Kazakhstan 52°42′N 62°36′E
144 L8 **Tobyl** *prev.* Tobol. ≈ Kazakhstan/Russian Federation
125 Q6 **Tobysh** ≈ NW Russian Federation
54 F10 **Tocaima** Cundinamarca, C Colombia 04°30′N 74°38′W
59 K16 **Tocantins** *off.* Estado do Tocantins. ◈ *state* C Brazil
Tocantins, Estado do *see* Tocantins
59 K15 **Tocantins, Rio** ≈ N Brazil
23 T2 **Toccoa** Georgia, SE USA 34°34′N 83°19′W
165 O12 **Tochigi** *off.* Tochigi-ken, *var.* Totigi. ◈ *prefecture* Honshū, S Japan
Tochigi-ken *see* Tochigi
165 O11 **Tochio** *var.* Totio. Niigata, Honshū, C Japan 37°27′N 139°00′E
95 I15 **Töcksfors** Värmland, C Sweden 59°30′N 11°49′E
42 J5 **Tocoa** Colón, N Honduras 15°40′N 86°01′W
62 H4 **Tocopilla** Antofagasta, N Chile 22°06′S 70°08′W
62 I4 **Tocorpuri, Cerro de** ▲ Bolivia/Chile 22°26′S 67°53′W
183 O10 **Tocumwal** New South Wales, SE Australia 35°53′S 145°35′E
54 K4 **Tocuyo de la Costa** Falcón, NW Venezuela 11°04′N 68°23′W
152 H13 **Toda Rāisingh** Rājasthān, N India 26°02′N 75°35′E
106 H13 **Todi** Umbria, C Italy 42°47′N 12°25′E
108 G9 **Tödi** ▲ NE Switzerland 46°52′N 08°53′E
171 T12 **Todlo** Papua Barat, E Indonesia 0°46′S 130°50′E
165 S9 **Todoga-saki** *headland* Honshū, C Japan 39°33′N 142°02′E
59 P17 **Todos os Santos, Baía de** *bay* E Brazil
40 F10 **Todos Santos** Baja California Sur, NW Mexico 23°28′N 110°14′W
40 B2 **Todos Santos, Bahía de** *bay* NW Mexico
Toeban *see* Tuban
Toekang Besi Eilanden *see* Tukangbesi, Kepulauan
Toeloengagoeng *see* Tulungagung
Tōen *see* Taoyuan
185 D25 **Toetoes Bay** *bay* South Island, New Zealand
11 Q14 **Tofield** Alberta, SW Canada 53°22′N 112°39′W
10 K17 **Tofino** Vancouver Island, British Columbia, SW Canada 49°05′N 125°51′W
189 X17 **Tofol** Kosrae, E Micronesia
95 J20 **Tofta** Halland, S Sweden 57°10′N 12°19′E
95 H15 **Tofte** Buskerud, S Norway 59°31′N 10°33′E
95 F24 **Toftlund** Syddanmark, SW Denmark 55°12′N 09°04′E
193 X15 **Tofua** *island* Ha'apai Group, C Tonga
187 Q12 **Toga** *island* Torres Islands, N Vanuatu
80 N13 **Togdheer** *off.* Gobolka Togdheer. ◈ *region* NW Somalia
Togdheer, Gobolka *see* Togdheer
164 L11 **Togi** Ishikawa, Honshū, SW Japan 37°06′N 136°44′E
39 N13 **Togiak** Alaska, USA 59°03′N 160°31′W
171 O11 **Togian, Kepulauan** *island group* C Indonesia
77 Q15 **Togo** *off.* Togolese Republic; *prev.* French Togoland. ◆ *republic* W Africa
Togolese Republic *see* Togo
162 F8 **Tögrög** Govĭ-Altay, SW Mongolia 45°51′N 95°04′E
162 F8 **Tögrög** *var.* Hoolt. Övörhangay, C Mongolia 45°31′N 103°06′E
Tögrög *see* Manhan
159 N12 **Togton He** *var.* Tuotuo He. ≈ C China
Togton Heyan *see* Tanggulashan
Toguzak *see* Togyzak
144 L7 **Togyzak** *prev.* Toguzak. ≈ Kazakhstan/Russian Federation
37 P10 **Tohatchi** New Mexico, SW USA 35°51′N 108°45′W
191 O7 **Tohiea, Mont** ⛰ Moorea, W French Polynesia 17°33′S 149°48′W
137 N14 **Tohma Çayı** ≈ C Turkey
93 O17 **Tohmajärvi** Pohjois-Karjala, SE Finland 62°12′N 30°19′E
93 L16 **Toholampi** Keski-Pohjanmaa, W Finland 63°46′N 24°15′E
Töhöm *see* Mandah
23 X12 **Tohopekaliga, Lake** ◎ Florida, SE USA
164 M14 **Toi** Shizuoka, Honshū, S Japan 34°55′N 138°45′E
190 B15 **Toi** N Niue 18°57′S 169°51′W
93 L19 **Toijala** Pirkanmaa, SW Finland 61°09′N 23°51′E
171 P12 **Toima** Sulawesi, N Indonesia 0°48′S 122°21′E
164 D17 **Toi-misaki** Kyūshū, SW Japan
171 Q17 **Toineke** Timor, S Indonesia 10°06′S 124°22′E
Toirc, Inis *see* Inishturk
35 U6 **Toiyabe Range** ▲▲ Nevada, W USA
Tojikiston, Jumhurii *see* Tajikistan
147 R12 **Tojikobod** *Rus.* Tadzhikabad. C Tajikistan 39°08′N 70°54′E
164 G12 **Tōjō** Hiroshima, Honshū, SW Japan 34°54′N 133°15′E
39 T10 **Tok** Alaska, USA 63°20′N 142°59′W
164 K13 **Tōkai** Aichi, Honshū, SW Japan 35°01′N 136°51′E
111 N21 **Tokaj** Borsod-Abaúj-Zemplén, NE Hungary 48°08′N 21°25′E
165 N11 **Tōkamachi** Niigata, Honshū, C Japan 37°08′N 138°44′E
185 D25 **Tokanui** Southland, South Island, New Zealand 46°33′S 169°02′E
80 I7 **Tokar** *var.* Ṭawkar. Red Sea, NE Sudan 18°27′N 37°41′E
136 L12 **Tokat** Tokat, N Turkey 40°20′N 36°35′E
136 L12 **Tokat** ◈ *province* N Turkey
Tŏkchŏk-kundo *see* Deokjeok-gundo
Tōke *see* Taka Atoll
190 J9 **Tokelau** ◇ *NZ overseas territory* W Polynesia
Tőketerebes *see* Trebišov
Tokhtamyshbek *see* Tŭkhtamish
24 M6 **Tokio** Texas, SW USA 33°09′N 102°31′W
Tokio *see* Tōkyō
189 W11 **Toki Point** *point* NW Wake Island
Tokkuztara *see* Gongliu
117 V9 **Tokmak** *var.* Velykyy Tokmak. Zaporiz'ka Oblast', SE Ukraine 47°13′N 35°43′E
Tokmak *see* Tomok
184 Q8 **Tokomaru Bay** Gisborne, North Island, New Zealand 38°10′S 178°18′E
184 M8 **Tokoroa** Waikato, North Island, New Zealand 38°14′S 175°52′E
76 K14 **Tokounou** C Guinea 09°43′N 09°46′W
38 M12 **Toksook Bay** Alaska, USA 60°33′N 165°01′W
Toksu *see* Xinhe
Toksum *see* Toksun
158 L6 **Toksun** *var.* Toksum. Xinjiang Uygur Zizhiqu, NW China 42°47′N 88°38′E
147 T8 **Toktogul** Talasskaya Oblast', NW Kyrgyzstan 41°51′N 72°56′E
147 T9 **Toktogul'skoye Vodokhranilishche** ⊠ W Kyrgyzstan
Toktomush *see* Tŭkhtamish
193 Y14 **Toku** *island* Vava'u Group, N Tonga
165 U16 **Tokunoshima** Kagoshima, Tokuno-shima, SW Japan
165 U16 **Tokuno-shima** *island* Nansei-shotō, SW Japan
164 I14 **Tokushima** *var.* Tokusima. Tokushima, Shikoku, SW Japan 34°04′N 134°28′E
164 H14 **Tokushima** *off.* Tokushima-ken, *var.* Tokusima. ◈ *prefecture* Shikoku, SW Japan
Tokushima-ken *see* Tokushima
Tokusima *see* Tokushima
164 E13 **Tokuyama** *var.* Shūnan. Yamaguchi, Honshū, SW Japan 34°04′N 131°48′E
165 N13 **Tōkyō** *var.* Tokio. ● (Japan) Tōkyō, Honshū, S Japan 35°40′N 139°45′E
165 O13 **Tōkyō** *off.* Tōkyō-to. ◈ *capital district* Honshū, S Japan
Tōkyō-to *see* Tōkyō
145 T12 **Tokyrau** ≈ C Kazakhstan
149 O3 **Tokzār** *Pash.* Tukzār. Sar-e Pul, N Afghanistan 35°47′N 66°28′E
145 W13 **Tokzhaylau** *prev.* Dzerzhinskoye. Almaty, SE Kazakhstan 45°49′N 81°04′E
145 W13 **Tokzhaylau** *var.* Dzerzhinskoye. Taldykorgan, SE Kazakhstan 45°49′N 81°04′E
189 U12 **Tol** *atoll* Chuuk Islands, C Micronesia
184 Q9 **Tolaga Bay** Gisborne, North Island, New Zealand 38°22′S 178°17′E
172 I7 **Tôlañaro** *prev.* Faradofay, Fort-Dauphin. Toliara, SE Madagascar
162 D6 **Tolbo** Bayan-Ölgiy, W Mongolia 48°22′N 90°22′E
Tolbukhin *see* Dobrich
60 G11 **Toledo** Paraná, S Brazil 24°45′S 53°41′W
54 G8 **Toledo** Norte de Santander, N Colombia 07°16′N 72°28′W
105 N9 **Toledo** *anc.* Toletum. Castilla-La Mancha, C Spain 39°52′N 04°02′W
30 M14 **Toledo** Illinois, N USA 39°16′N 88°15′W
29 W13 **Toledo** Iowa, C USA 42°00′N 92°34′W
31 R11 **Toledo** Ohio, N USA 41°40′N 83°33′W
32 F12 **Toledo** Oregon, NW USA 44°37′N 123°56′W
32 G9 **Toledo** Washington, NW USA 46°27′N 122°49′W
42 F3 **Toledo** ◈ *district* S Belize
104 M9 **Toledo** ◈ *province* Castilla-La Mancha, C Spain
25 Y7 **Toledo Bend Reservoir** ⊠ Louisiana/Texas, SW USA
104 M10 **Toledo, Montes de** ▲▲ C Spain
106 J12 **Tolentino** Marche, C Italy 43°08′N 13°17′E
Toletum *see* Toledo
94 H10 **Tolga** Hedmark, S Norway 62°25′N 11°00′E
158 J3 **Toli** Xinjiang Uygur Zizhiqu, NW China 45°55′N 83°33′E
172 H7 **Toliara** *var.* Toliary; *prev.* Tuléar. Toliara, SW Madagascar 23°20′S 43°41′E
172 H7 **Toliara** ◈ *province* SW Madagascar
Toliary *see* Toliara
118 H11 **Toliejai** *prev.* Kamajai. Panevėžys, NE Lithuania 55°16′N 25°30′E
54 D11 **Tolima** *off.* Departamento del Tolima. ◈ *province* C Colombia
Tolima, Departamento del *see* Tolima
171 N11 **Tolitoli** Sulawesi, C Indonesia 01°05′N 120°50′E
95 K22 **Tollarp** Skåne, S Sweden 55°55′N 14°00′E
100 N9 **Tollense** ≈ NE Germany
100 N10 **Tollensesee** ◎ NE Germany
36 K13 **Tolleson** Arizona, SW USA 33°25′N 112°15′W
146 M13 **Tollimarjon** *Rus.* Talimardzhan. Qashqadaryo Viloyati, S Uzbekistan 38°22′N 65°31′E
Tolmein *see* Tolmin
106 J6 **Tolmezzo** Friuli-Venezia Giulia, NE Italy 46°27′N 13°01′E
109 S11 **Tolmin** *Ger.* Tolmein, *It.* Tolmino. W Slovenia 46°12′N 13°39′E
Tolmino *see* Tolmin
111 J25 **Tolna** *Ger.* Tolnau. Tolna, S Hungary 46°26′N 18°47′E
111 I24 **Tolna** *off.* Tolna Megye. ◈ *county* SW Hungary
Tolna Megye *see* Tolna
Tolnau *see* Tolna
79 I20 **Tolo** Bandundu, W Dem. Rep. Congo 02°57′S 18°35′E
Tolochin *see* Talachyn
190 D12 **Toloke** Île Futuna, W Wallis and Futuna
30 M13 **Tolono** Illinois, N USA 39°59′N 88°16′W
105 Q3 **Tolosa** País Vasco, N Spain 43°09′N 02°04′W
Tolosa *see* Toulouse
171 O13 **Tolo, Teluk** *bay* Sulawesi, C Indonesia
39 R9 **Tolovana River** ≈ Alaska, USA
123 U10 **Tolstoy, Mys** *headland* E Russian Federation 59°12′N 155°04′E
63 G15 **Toltén** Araucanía, C Chile 39°13′S 73°15′W
63 G15 **Toltén, Río** ≈ S Chile
54 E6 **Tolú** Sucre, NW Colombia 09°32′N 75°34′W
41 O14 **Toluca** *var.* Toluca de Lerdo. México, S Mexico 19°20′N 99°40′W
Toluca de Lerdo *see* Toluca
41 O14 **Toluca, Nevado de** ⛰ C Mexico 19°05′N 99°45′W
127 R6 **Tol'yatti** *prev.* Stavropol'. Samarskaya Oblast', W Russian Federation 53°32′N 49°27′E
77 O12 **Toma** NW Burkina Faso 12°46′N 02°51′W
30 K7 **Tomah** Wisconsin, N USA 43°59′N 90°31′W
30 L5 **Tomahawk** Wisconsin, N USA 45°27′N 89°40′W
117 T8 **Tomakivka** Dnipropetrovs'ka Oblast', E Ukraine 47°47′N 34°45′E
165 S4 **Tomakomai** Hokkaidō, NE Japan 42°38′N 141°32′E
165 S2 **Tomamae** Hokkaidō, NE Japan 44°18′N 141°38′E
104 G9 **Tomar** Santarém, W Portugal 39°36′N 08°25′W
123 T13 **Tomari** Ostrov Sakhalin, Sakhalinskaya Oblast', SE Russian Federation 47°47′N 142°09′E
115 C16 **Tómaros** ▲ W Greece 39°31′N 20°45′E
Tomaschow *see* Tomaszów Mazowiecki
Tomaschow *see* Tomaszów Lubelski
61 E16 **Tomás Gomensoro** Artigas, N Uruguay 30°28′S 57°28′W
117 N7 **Tomashpil'** Vinnyts'ka Oblast', C Ukraine 48°32′N 28°31′E
Tomaszów *see* Tomaszów Mazowiecki
111 P15 **Tomaszów Lubelski** *Ger.* Tomaschow. Lubelskie, E Poland 50°29′N 23°23′E
Tomaszów Mazowiecka *see* Tomaszów Mazowiecki
110 L13 **Tomaszów Mazowiecki** *var.* Tomaszów Mazowiecka; *prev.* Tomaszów, *Ger.* Tomaschow. Łódzkie, C Poland 51°33′N 20°E
40 J13 **Tomatlán** Jalisco, C Mexico 19°53′N 105°18′W
81 F15 **Tombe** Jonglei, E South Sudan 05°52′N 31°40′E
23 N4 **Tombigbee River** ≈ Alabama/Mississippi, S USA
82 A10 **Tomboco** Zaire Province, NW Angola 06°50′S 13°20′E
77 O10 **Tombouctou** *Eng.* Timbuktu. Tombouctou, N Mali 16°47′N 03°03′W
77 N9 **Tombouctou** ◈ *region* W Mali
37 N16 **Tombstone** Arizona, SW USA 31°42′N 110°04′W
83 A15 **Tombua** *Port.* Porto Alexandre. Namibe, SW Angola 15°49′S 11°53′E
83 J19 **Tom Burke** Limpopo, NE South Africa 23°07′S 28°01′E
146 L9 **Tomdibuloq** *Rus.* Tamdybulak. Navoiy Viloyati, N Uzbekistan 41°48′N 64°33′E
146 L9 **Tomditov-Togʻlari** ▲▲ N Uzbekistan
62 G13 **Tomé** Bío Bío, C Chile 36°38′S 72°57′W
58 L12 **Tomé-Açu** Pará, NE Brazil 02°25′S 48°09′W
95 L23 **Tomelilla** Skåne, S Sweden 55°33′N 14°00′E
105 O10 **Tomelloso** Castilla-La Mancha, C Spain 39°09′N 03°01′W
14 H10 **Tomiko Lake** ◎ Ontario, S Canada
77 N12 **Tominian** Ségou, C Mali 13°18′N 04°39′W
Tomini, Gulf of *see* Tomini, Teluk
171 N12 **Tomini, Teluk** *var.* Gulf of Tomini; *prev.* Teluk Gorontalo. *bay* Sulawesi, C Indonesia
165 Q11 **Tomioka** Fukushima, Honshū, C Japan 37°19′N 140°57′E
113 G14 **Tomislavgrad** Federacija Bosne I Hercegovine, SW Bosnia and Herzegovina 43°43′N 17°15′E
181 O9 **Tomkinson Ranges** ▲▲ South Australia/Western Australia
123 Q11 **Tommot** Respublika Sakha (Yakutiya), NE Russian Federation 58°57′N 126°24′E
171 Q11 **Tomohon** Sulawesi, N Indonesia 01°19′N 124°49′E
147 V7 **Tomok** *prev.* Tokmak. Chuyskaya Oblast', N Kyrgyzstan 42°50′N 75°18′E
54 K9 **Tomo, Río** ≈ E Colombia
113 L21 **Tomorrit, Mali i** ▲ S Albania 40°43′N 20°12′E
11 S17 **Tompkins** Saskatchewan, S Canada 50°03′N 108°49′W
20 K8 **Tompkinsville** Kentucky, S USA 36°43′N 85°41′W
171 N11 **Tompo** Sulawesi, N Indonesia 0°56′N 120°16′E
180 I8 **Tom Price** Western Australia 22°48′S 117°49′E
122 J12 **Tomsk** Tomskaya Oblast', C Russian Federation 56°30′N 85°05′E
122 I11 **Tomskaya Oblast'** ◈ *province* C Russian Federation
18 K16 **Toms River** New Jersey, NE USA 39°56′N 74°09′W
Tom Steed Lake *see* Tom Steed Reservoir
26 L12 **Tom Steed Reservoir** *var.* Tom Steed Lake. ⊠ Oklahoma, C USA
171 U13 **Tomu** Papua Barat, E Indonesia 02°07′S 133°01′E
158 H6 **Tomür Feng** *var.* Pobeda Peak, *Rus.* Pik Pobedy. ▲ China/Kyrgyzstan 42°02′N 80°07′E *see also* Pobedy, Pik
Tomür Feng *see* Pobedy, Pik
189 N13 **Tomworoahlang** Pohnpei, E Micronesia
41 U17 **Tonalá** Chiapas, SE Mexico 16°08′N 93°41′W
106 F6 **Tonale, Passo del** *pass* N Italy
164 I11 **Tonami** Toyama, Honshū, SW Japan 36°40′N 136°55′E
58 C12 **Tonantins** Amazonas, W Brazil 02°58′S 67°30′W
32 K6 **Tonasket** Washington, NW USA 48°41′N 119°27′W
55 Y9 **Tonate** *var.* Macouria. N French Guiana 05°00′N 52°28′W
18 D10 **Tonawanda** New York, NE USA 43°00′N 78°51′W
42 I7 **Toncontín** *prev.* Tegucigalpa. ● (Honduras) Francisco Morazán, SW Honduras 14°04′N 87°11′W
42 H7 **Toncontín** ✈ Central District, C Honduras 14°03′N 87°20′W
171 Q11 **Tondano** Sulawesi, C Indonesia 01°19′N 124°56′E
104 H7 **Tondela** Viseu, N Portugal 40°31′N 08°05′W
95 F24 **Tønder** *Ger.* Tondern. Syddanmark, SW Denmark 54°57′N 08°53′E
Tondern *see* Tønder
143 N4 **Tonekābon** *var.* Shahsawar, Tonkābon; *prev.* Shahsavār. Māzandarān, N Iran 36°40′N 51°25′E
Tonezh *see* Tonyezh
193 Y14 **Tonga** *off.* Kingdom of Tonga, *var.* Friendly Islands. ◆ *monarchy* SW Pacific Ocean
175 R9 **Tonga** *island group* SW Pacific Ocean
Tongaat *see* oThongathi
Tonga, Kingdom of *see* Tonga
161 Q13 **Tong'an** *var.* Datong, Tong an. Fujian, SE China 24°43′N 118°07′E
Tong an *see* Tong'an
27 Q4 **Tonganoxie** Kansas, C USA 39°06′N 95°05′W
39 Y13 **Tongass National Forest** *reserve* Alaska, USA
193 Y16 **Tongatapu** ✈ Tongatapu, S Tonga 21°10′S 175°10′W
193 Y16 **Tongatapu** *island* Tongatapu Group, S Tonga
193 Y16 **Tongatapu Group** *island group* S Tonga
175 S9 **Tonga Trench** *undersea feature* S Pacific Ocean
161 N8 **Tongbai Shan** ▲▲ C China
161 P8 **Tongcheng** Anhui, E China 31°16′N 117°00′E
160 L6 **Tongchuan** Shaanxi, C China 35°10′N 109°03′E
160 L12 **Tongdao** *var.* Tongdao Dongzu Zizhixian; *prev.* Shuangjiang. Hunan, S China 26°06′N 109°46′E
Tongdao Dongzu Zizhixian *see* Tongdao
159 T11 **Tongde** *var.* Gabasumdo. Qinghai, C China 35°13′N 100°39′E
99 K19 **Tongeren** *Fr.* Tongres. Limburg, NE Belgium 50°47′N 05°28′E
Tonghae *see* Donghae
160 G13 **Tonghai** *var.* Xiushan. Yunnan, SW China 24°06′N 102°45′E
163 X8 **Tonghe** Heilongjiang, NE China 46°00′N 128°45′E
163 W11 **Tonghua** Jilin, NE China 41°45′N 125°50′E
163 Z6 **Tongjiang** Heilongjiang, NE China 47°39′N 132°29′E
163 Y13 **Tongjosŏn-man** *prev.* Broughton Bay. *bay* E North Korea
163 V7 **Tongken He** ≈ NE China
Tongking, Gulf of *see* Tonkin, Gulf of
163 U10 **Tongliao** Nei Mongol Zizhiqu, N China 43°37′N 122°15′E
161 Q9 **Tongling** Anhui, E China 30°55′N 117°50′E
161 R9 **Tonglu** Zhejiang, SE China 29°50′N 119°38′E
187 R14 **Tongoa** *island* Shepherd Islands, S Vanuatu
62 G9 **Tongoy** Coquimbo, C Chile 30°16′S 71°31′W
160 L11 **Tongren** *var.* Rongwo. Guizhou, S China 27°44′N 109°10′E
159 T11 **Tongren** *var.* Rongwo. Qinghai, C China 35°31′N 101°58′E
Tongres *see* Tongeren
153 U11 **Tongsa** *var.* Tongsa Dzong. C Bhutan 27°33′N 90°30′E
Tongsa Dzong *see* Tongsa
Tongshan *see* Fuding, Fujian, China
Tongshan *see* Xuzhou, Jiangsu, China
Tongshi *see* Wuzhishan
159 P12 **Tongtian He** *var.* Zhi Qu. ≈ C China
96 I6 **Tongue** N Scotland, United Kingdom 58°30′N 04°25′W
44 H3 **Tongue of the Ocean** *strait* C The Bahamas
33 X10 **Tongue River** ≈ Montana, NW USA
33 W11 **Tongue River Resevoir** ⊠ Montana, NW USA
159 V11 **Tongwei** *var.* Pingxiang. Gansu, C China 35°09′N 105°15′E
159 W9 **Tongxin** Ningxia, N China 37°00′N 105°41′E
163 U9 **Tongyu** *var.* Kaitong. Jilin, NE China 44°49′N 123°08′E
160 J11 **Tongzi** *var.* Loushanguan. Guizhou, S China 28°08′N 106°49′E
162 F8 **Tonhil** *var.* Dzüyl. Govĭ-Altay, SW Mongolia 46°09′N 93°55′E
40 G5 **Tónichi** Sonora, NW Mexico 28°37′N 109°34′W
81 D14 **Tonj** Warap, W South Sudan 07°18′N 28°41′E
152 H13 **Tonk** Rājasthān, N India 26°10′N 75°50′E
Tonkābon *see* Tonekābon
27 N8 **Tonkawa** Oklahoma, C USA 36°40′N 97°18′W
167 T7 **Tonkin, Gulf of** *var.* Gulf of Tongking, *Chin.* Beibu Wan, *Vtn.* Vinh Băc Bô. *gulf* China/Vietnam
167 Q12 **Tônlé Sap** *Eng.* Great Lake. ◎ W Cambodia
102 L14 **Tonneins** Lot-et-Garonne, SW France 44°21′N 00°21′E
103 Q7 **Tonnerre** Yonne, C France 47°50′N 04°00′E
Tonoas *see* Dublon
35 U8 **Tonopah** Nevada, W USA 38°04′N 117°13′W
164 H13 **Tonoshō** Okayama, Shōdo-shima, SW Japan 34°29′N 134°10′E
43 S17 **Tonosí** Los Santos, S Panama 07°23′N 80°26′W
95 H16 **Tønsberg** Vestfold, S Norway 59°16′N 10°25′E
39 T11 **Tonsina** Alaska, USA 61°39′N 145°10′W
95 D17 **Tonstad** Vest-Agder, S Norway 58°40′N 06°42′E
193 X15 **Tonumea** *island* Nomuka Group, W Tonga
137 O11 **Tonya** Trabzon, NE Turkey 40°52′N 39°17′E
119 K20 **Tonyezh** *Rus.* Tonezh. Homyel'skaya Voblasts', SE Belarus 51°50′N 27°48′E
36 L3 **Tooele** Utah, W USA 40°32′N 112°18′W
122 L13 **Toora-Khem** Respublika Tyva, S Russian Federation 52°25′N 96°01′E
183 O5 **Toorale East** New South Wales, SE Australia 30°29′S 145°25′E
83 H25 **Toorberg** ▲ S South Africa 32°02′S 24°02′E
118 G5 **Tootsi** Pärnumaa, SW Estonia 58°34′N 24°43′E
183 U3 **Toowoomba** Queensland, E Australia 27°35′S 151°54′E
27 Q4 **Topeka** *state capital* Kansas, C USA 39°03′N 95°41′W
122 J12 **Topki** Kemerovskaya Oblast', S Russian Federation 55°12′N 85°40′E
111 M18 **Topľa** *Hung.* Toplya. ≈ NE Slovakia
Toplicza *see* Toplița
116 J10 **Toplița** *Ger.* Töplitz, *Hung.* Maroshevíz; *prev.* Toplița Română, *Hung.* Oláh-Toplicza, Toplicza. Harghita, C Romania 45°56′N 25°20′E
Toplița Română/Töplitz *see* Toplița
Toplya *see* Topľa
111 I20 **Topoľčany** *Hung.* Nagytapolcsány. Nitriansky Kraj, W Slovakia 48°33′N 18°10′E
40 G8 **Topolobampo** Sinaloa, C Mexico 25°36′N 109°04′W
116 I13 **Topoloveni** Argeș, S Romania 44°49′N 25°02′E
114 L11 **Topolovgrad** *prev.* Kavakli. Haskovo, S Bulgaria 42°06′N 26°20′E
Topolya *see* Bačka Topola
32 J10 **Toppenish** Washington, NW USA 46°22′N 120°18′W
181 P4 **Top Springs Roadhouse** Northern Territory, N Australia 16°37′S 131°49′E
189 U11 **Tora** *island* Chuuk, C Micronesia
Toraigh *see* Tory Island
189 U11 **Tora Island Pass** *passage* Chuuk Islands, C Micronesia
143 U5 **Torbat-e Ḩeydarīyeh** *var.* Turbat-i-Haidari. Khorāsān-e Raẓavī, NE Iran 35°18′N 59°12′E
143 V5 **Torbat-e Jām** *var.* Turbat-i-Jam. Khorāsān-e Raẓavī, NE Iran 35°16′N 60°36′E
39 Q11 **Torbert, Mount** ▲ Alaska, USA 61°30′N 152°15′W
31 P6 **Torch Lake** ◎ Michigan, N USA
Törcsvár *see* Bran
Torda *see* Turda
104 L6 **Tordesillas** Castilla y León, N Spain 41°30′N 05°00′W
92 K13 **Töre** Norrbotten, N Sweden 65°55′N 22°40′E
95 L17 **Töreboda** Västra Götaland, S Sweden 58°41′N 14°07′E
95 J21 **Torekov** Skåne, S Sweden 56°25′N 12°39′E
92 O3 **Torell Land** *physical region* SW Svalbard
105 U4 **Toretta de l'Orrí** *var.* Llorri; *prev.* Tossal de l'Orrí. ▲ NE Spain 42°24′N 01°15′E
144 M14 **Toretum** Kyzylorda, S Kazakhstan 45°38′N 63°20′E
117 Y8 **Torez** Donets'ka Oblast', SE Ukraine 48°00′N 38°38′E
101 N14 **Torgau** Sachsen, E Germany 51°34′N 13°01′E
145 R8 **Torgay** *Kaz.* Torghay; *prev.* Turgay. Akmola, W Kazakhstan 51°46′N 72°45′E
145 N10 **Torgay** *prev.* Turgay. ≈ C Kazakhstan
Torgay Üstirti *see* Turgayskaya Stolovaya Strana
Torghay *see* Torgay
95 N22 **Torhamn** Blekinge, S Sweden 56°04′N 15°49′E
99 C17 **Torhout** West-Vlaanderen, W Belgium 51°04′N 03°06′E
106 B8 **Torino** *Eng.* Turin. Piemonte, NW Italy 45°03′N 07°39′E
Tori-shima *see* Io-Tori-shima
81 F16 **Torit** Eastern Equatoria, S South Sudan 04°27′N 32°31′E
186 H6 **Toriu** New Britain, E Papua New Guinea 04°39′S 151°42′E
148 M4 **Torkestān, Selseleh-ye Band-e** *var.* Bandi-i Turkistan. ▲▲ NW Afghanistan
104 L7 **Tormes** ≈ W Spain
Tornacum *see* Tournai
Torneå *see* Tornio
92 K12 **Torneälven** *var.* Torniojoki, *Fin.* Tornionjoki. ≈ Finland/Sweden
92 I11 **Torneträsk** ◎ N Sweden
13 O4 **Torngat Mountains** ▲▲ Newfoundland and Labrador, NE Canada
24 H8 **Tornillo** Texas, SW USA 31°26′N 106°06′W
92 K13 **Tornio** *Swe.* Torneå. Lappi, NW Finland 65°50′N 24°18′E
Torniojoki/Tornionjoki *see* Torneälven
61 B23 **Tornquist** Buenos Aires, E Argentina 38°08′S 62°15′W
104 L6 **Toro** Castilla y León, N Spain 41°31′N 05°24′W
62 H9 **Toro, Cerro del** ▲ N Chile 29°10′S 69°43′W
77 R12 **Torodi** Tillabéri, SW Niger 13°05′N 01°46′E
Törökbecse *see* Novi Bečej
186 J7 **Torokina** Bougainville, NE Papua New Guinea 06°12′S 155°04′E
111 L23 **Törökszentmiklós** Jász-Nagykun-Szolnok, E Hungary 47°11′N 20°26′E
42 G7 **Torola, Río** ≈ El Salvador/Honduras
Toronaíos, Kólpos *see* Kassándras, Kólpos
14 H15 **Toronto** *province capital* Ontario, S Canada 43°42′N 79°25′W
31 V12 **Toronto** Ohio, N USA 40°27′N 80°36′W
Toronto *see* Lester B. Pearson
27 P6 **Toronto Lake** ⊠ Kansas, C USA
35 V16 **Toro Peak** ▲ California, W USA 33°31′N 116°25′W
124 H16 **Toropets** Tverskaya Oblast', W Russian Federation 56°29′N 31°37′E
81 G18 **Tororo** E Uganda 0°42′N 34°12′E
136 H16 **Toros Dağları** *Eng.* Taurus Mountains. ▲▲ S Turkey
183 N13 **Torquay** Victoria, SE Australia 38°21′S 144°18′E
97 J24 **Torquay** SW England, United Kingdom 50°28′N 03°30′W
104 M5 **Torquemada** Castilla y León, N Spain 42°02′N 04°17′W
35 S16 **Torrance** California, W USA 33°50′N 118°20′W
104 G12 **Torrão** Setúbal, S Portugal 38°18′N 08°13′W
104 H8 **Torre, Alto da** ▲ C Portugal 40°21′N 07°31′W
107 K18 **Torre Annunziata** Campania, S Italy 40°45′N 14°27′E
105 T8 **Torreblanca** Valenciana, E Spain 40°14′N 00°12′E
104 L15 **Torrecilla** ▲ S Spain 36°38′N 04°54′W
105 P4 **Torrecilla en Cameros** La Rioja, N Spain 42°18′N 02°33′W
105 N13 **Torredelcampo** Andalucía, S Spain 37°45′N 03°52′W
107 K17 **Torre del Greco** Campania, S Italy 40°46′N 14°22′E
104 I6 **Torre de Moncorvo** *var.* Moncorvo, Tôrre de Moncorvo. Bragança, N Portugal 41°10′N 07°03′W
104 J9 **Torrejoncillo** Extremadura, W Spain 39°54′N 06°28′W
105 O8 **Torrejón de Ardoz** Madrid, C Spain 40°27′N 03°29′W
105 N7 **Torrelaguna** Madrid, C Spain 40°50′N 03°33′W
105 N2 **Torrelavega** Cantabria, N Spain 43°21′N 04°03′W
107 M16 **Torremaggiore** Puglia, SE Italy 41°42′N 15°17′E
104 M15 **Torremolinos** Andalucía, S Spain 36°38′N 04°30′W
182 I6 **Torrens, Lake** *salt lake* South Australia
105 S10 **Torrent** *Cas.* Torrente, *var.* Torrent de l'Horta. Valenciana, E Spain 39°27′N 00°28′W
Torrent de l'Horta/Torrente *see* Torrent
40 L8 **Torreón** Coahuila, NE Mexico 25°47′N 103°21′W
105 R13 **Torre-Pacheco** Murcia, SE Spain 37°43′N 00°57′W
106 A8 **Torre Pellice** Piemonte, NE Italy 44°49′N 07°12′E
105 O13 **Torreperogil** Andalucía, S Spain 38°02′N 03°17′W
61 J15 **Torres** Rio Grande do Sul, S Brazil 29°20′S 49°43′W
Torrès, Îles *see* Torres Islands
187 Q11 **Torres Islands** *Fr.* Îles Torrès. *island group* N Vanuatu
104 G9 **Torres Novas** Santarém, C Portugal 39°28′N 08°32′W
181 V1 **Torres Strait** *strait* Australia/Papua New Guinea
104 F10 **Torres Vedras** Lisboa, C Portugal 39°05′N 09°15′W
105 S13 **Torrevieja** Valenciana, E Spain 37°59′N 00°40′W
186 B6 **Torricelli Mountains** ▲▲ NW Papua New Guinea
96 G8 **Torridon, Loch** *inlet* NW Scotland, United Kingdom
106 D9 **Torriglia** Liguria, NW Italy 44°31′N 09°08′E
104 M9 **Torrijos** Castilla-La Mancha, C Spain 39°59′N 04°18′W
18 L12 **Torrington** Connecticut, NE USA 41°48′N 73°07′W
33 Z15 **Torrington** Wyoming, C USA 42°04′N 104°10′W
Torröjen *see* Torrön
94 F16 **Torrön** *prev.* Torröjen. ◎ C Sweden
105 N15 **Torrox** Andalucía, S Spain 36°45′N 03°58′W
94 N13 **Torsåker** Gävleborg, C Sweden 60°31′N 16°30′E
95 N21 **Torsås** Kalmar, S Sweden 56°24′N 16°00′E
95 J14 **Torsby** Värmland, C Sweden 60°07′N 13°E
95 N16 **Torshälla** Södermanland, C Sweden 59°25′N 16°28′E
95 B19 **Tórshavn** *Dan.* Thorshavn. ○ Faroe Islands 62°02′N 06°47′W
Torshiz *see* Kāshmar
146 I9 **Toʻrtkokʻl** *var.* Türtkül, *Rus.* Turtkul'; *prev.* Petroaleksandrovsk. Qoraqalpogʻiston Respublikasi, W Uzbekistan 41°35′N 61°E
Tortoise Islands *see* Colón, Archipiélago de
45 T9 **Tortola** *island* C British Virgin Islands
106 D9 **Tortona** *anc.* Dertona. Piemonte, NW Italy 44°54′N 08°52′E
107 L23 **Tortorici** Sicilia, Italy, C Mediterranean Sea 38°02′N 14°49′E
105 U7 **Tortosa** *anc.* Dertosa. Cataluña, E Spain 40°49′N 00°31′E
Tortosa *see* Ṭarṭūs
105 U7 **Tortosa, Cap** *cape* E Spain
44 L8 **Tortue, Île de la** *var.* Tortuga Island. *island* N Haiti
55 Y10 **Tortue, Montagne** ▲▲ C French Guiana
Tortuga, Isla *see* La Tortuga, Isla
Tortuga Island *see* Tortue, Île de la
54 C11 **Tortugas, Golfo** *gulf* W Colombia
45 T5 **Tortuguero, Laguna** *lagoon* N Puerto Rico
137 Q12 **Tortum** Erzurum, NE Turkey 40°20′N 41°36′E
Torugart, Pereval *see* Turugart Shankou
137 O12 **Torul** Gümüşhane, NE Turkey 40°35′N 39°18′E
110 J10 **Toruń** *Ger.* Thorn. Toruń, Kujawsko-pomorskie, C Poland 53°02′N 18°36′E
95 K20 **Torup** Halland, S Sweden 56°57′N 13°04′E
118 I6 **Tõrva** *Ger.* Törwa. Valgamaa, S Estonia 58°00′N 25°54′E
Törwa *see* Tõrva
96 D13 **Tory Island** *Ir.* Toraigh. *island* NW Ireland
111 N19 **Torysa** *Hung.* Tarca. ≈ NE Slovakia
Törzburg *see* Bran
124 J16 **Torzhok** Tverskaya Oblast', W Russian Federation 57°04′N 34°55′E
164 F15 **Tosa-Shimizu** *var.* Tosasimizu. Kōchi, Shikoku, SW Japan 32°47′N 132°58′E
Tosasimizu *see* Tosa-Shimizu
164 G15 **Tosa-wan** *bay* SW Japan
83 H21 **Tosca** North-West, N South Africa 25°51′S 23°56′E
106 F12 **Toscana** *Eng.* Tuscany. ◈ *region* C Italy
107 E14 **Toscano, Archipelago** *Eng.* Tuscan Archipelago. *island group* C Italy
106 G10 **Tosco-Emiliano, Appennino** *Eng.* Tuscan-Emilian Mountains. ▲▲ C Italy
Tōsei *see* Dongshi
165 N15 **To-shima** *island* Izu-shotō, SE Japan
147 Q9 **Toshkent** *Eng./Rus.* Tashkent. ● Toshkent Viloyati, E Uzbekistan 41°19′N 69°17′E
147 Q9 **Toshkent** ✈ Toshkent Viloyati, E Uzbekistan 41°13′N 69°15′E
147 P9 **Toshkent Viloyati** *Rus.* Tashkentskaya Oblast'. ◈ *province* E Uzbekistan
124 H13 **Tosno** Leningradskaya Oblast', NW Russian Federation 59°34′N 30°48′E
159 Q10 **Toson Hu** ◎ C China
162 H6 **Tosontsengel** Dzavhan, NW Mongolia 48°42′N 98°14′E
162 J6 **Tosontsengel** *var.* Tsengel. Hövsgöl, N Mongolia 49°29′N 101°09′E
146 I8 **Tosquduq Qumlari** *var.* Goshquduq Qum, Taskuduk, Peski. *desert* W Uzbekistan
Tossal de l'Orrí *see* Toretta de l'Orrí
61 A15 **Tostado** Santa Fe, C Argentina 29°15′S 61°45′W
118 F6 **Tõstamaa** *Ger.* Testama. Pärnumaa, SW Estonia 58°20′N 23°59′E
100 I10 **Tostedt** Niedersachsen, NW Germany 53°16′N 09°42′E
136 J11 **Tosya** Kastamonu, N Turkey 41°02′N 34°02′E
95 F15 **Totak** ◎ S Norway
105 R13 **Totana** Murcia, SE Spain 37°45′N 01°30′W
94 H13 **Toten** *physical region* S Norway
83 G18 **Toteng** North-West, C Botswana 20°25′S 23°00′E
102 M3 **Tôtes** Seine-Maritime, N France 49°40′N 01°02′E
Totigi *see* Tochigi
Totio *see* Tochio
Totis *see* Tata
189 U13 **Totiw** *island* Chuuk, C Micronesia
125 N13 **Tot'ma** *var.* Totma. Vologodskaya Oblast', NW Russian Federation 59°58′N 42°42′E
Tot'ma *see* Sukhona
55 V9 **Totness** Coronie, N Suriname 05°53′N 56°19′W
42 C5 **Totonicapán** Totonicapán, W Guatemala 14°58′N 91°12′W
42 A2 **Totonicapán** *off.* Departamento de Totonicapán. ◈ *department* W Guatemala
Totonicapán, Departamento de *see* Totonicapán
61 B18 **Totoras** Santa Fe, C Argentina 32°35′S 61°11′W
187 Y15 **Totoya** *island* S Fiji
183 Q7 **Tottenham** New South Wales, SE Australia 32°16′S 147°23′E

◆ Country
● Country Capital
◇ Dependent Territory
○ Dependent Territory Capital
◈ Administrative Regions
✈ International Airport
▲ Mountain
▲▲ Mountain Range
⛰ Volcano
≈ River
◎ Lake
⊠ Reservoir

164 I12 **Tottori** Tottori, Honshū, SW Japan 35°29´N 134°14´E
164 H12 **Tottori** *off.* Tottori-ken. ◈ *prefecture* Honshū, SW Japan
Tottori-ken *see* Tottori
76 I6 **Touâjîl** Tiris Zemmour, N Mauritania 22°03´N 12°40´W
76 L15 **Touba** W Ivory Coast 08°17´N 07°41´W
76 G11 **Touba** W Senegal 14°55´N 15°53´W
74 E7 **Toubkal, Jbel** ▲ W Morocco 31°00´N 07°50´W
32 K10 **Touchet** Washington, NW USA 46°03´N 118°40´W
103 P7 **Toucy** Yonne, C France 47°45´N 03°18´E
77 O12 **Tougan** W Burkina Faso 13°06´N 03°03´W
74 L7 **Touggourt** NE Algeria 33°08´N 06°04´E
77 Q12 **Tougouri** N Burkina Faso 13°22´N 00°25´W
76 J13 **Tougué** NW Guinea 11°29´N 11°48´W
76 K12 **Toukoto** Kayes, W Mali 13°27´N 09°52´W
103 S5 **Toul** Meurthe-et-Moselle, NE France 48°41´N 05°54´E
76 L16 **Toulépleu** *var.* Toulobli. W Ivory Coast 06°37´N 08°27´W
Touliu *see* Douliu
15 U3 **Toulnustouc** ≈ Québec, SE Canada
Toulobli *see* Toulépleu
103 T16 **Toulon** *anc.* Telo Martius, Tilio Martius. Var, SE France 43°07´N 05°56´E
30 K12 **Toulon** Illinois, N USA 41°05´N 89°54´W
102 M15 **Toulouse** *anc.* Tolosa. Haute-Garonne, S France 43°37´N 01°25´E
102 M15 **Toulouse** ✈ Haute-Garonne, S France 43°38´N 01°19´E
77 N16 **Toumodi** C Ivory Coast 06°34´N 05°01´W
74 G9 **Tounassine, Hamada** *hill range* W Algeria
Toungoo *see* Taungoo
166 K7 **Toungup** *var.* Taungup. Rakhine State, W Myanmar (Burma) 18°50´N 94°14´E
102 L8 **Touraine** *cultural region* C France
Tourane *see* Đa Năng
103 P1 **Tourcoing** Nord, N France 50°44´N 03°10´E
104 F2 **Touriñán, Cabo** *headland* NW Spain 43°02´N 09°20´W
76 J6 **Tourîne** Tiris Zemmour, N Mauritania 22°23´N 11°50´W
102 J3 **Tourlaville** Manche, N France 49°38´N 01°34´W
99 D19 **Tournai** *var.* Tournay, *Dut.* Doornik; *anc.* Tornacum. Hainaut, SW Belgium 50°36´N 03°24´E
102 L16 **Tournay** Hautes-Pyrénées, S France 43°10´N 00°16´E
Tournay *see* Tournai
103 R12 **Tournon** Ardèche, E France 45°05´N 04°49´E
103 R9 **Tournus** Saône-et-Loire, C France 46°33´N 04°53´E
59 Q14 **Touros** Rio Grande do Norte, E Brazil 05°10´S 35°29´W
102 L8 **Tours** *anc.* Caesarodunum, Turoni. Indre-et-Loire, C France 47°22´N 00°40´E
183 Q17 **Tourville, Cape** *headland* Tasmania, SE Australia 42°09´S 148°20´E
44 M9 **Toussaint Louverture** ✈ E Haiti 18°38´N 72°13´W
162 L8 **Töv** ◈ *province* C Mongolia
54 H7 **Tovar** Mérida, NW Venezuela 08°22´N 71°50´W
126 L5 **Tovarkovskiy** Tul'skaya Oblast', W Russian Federation 53°41´N 38°18´E
Tovil'-Dora *see* Tavildara
Tövis *see* Teiuş
137 V11 **Tovuz** *Rus.* Tauz. W Azerbaijan 40°58´N 45°41´E
165 R7 **Towada** Aomori, Honshū, C Japan 40°35´N 141°13´E
184 K3 **Towai** Northland, North Island, New Zealand 35°29´S 174°06´E
18 H12 **Towanda** Pennsylvania, NE USA 41°45´N 76°25´W
29 W4 **Tower** Minnesota, N USA 47°48´N 92°16´W
171 N12 **Towera** Sulawesi, N Indonesia 0°29´S 120°01´E
Tower Island *see* Genovesa, Isla
180 M13 **Tower Peak** ▲ Western Australia 33°23´S 123°27´E
35 U11 **Towne Pass** *pass* California, W USA
29 N3 **Towner** North Dakota, N USA 48°20´N 100°27´W
33 R10 **Townsend** Montana, NW USA 46°19´N 111°31´W
181 X6 **Townsville** Queensland, NE Australia 19°24´S 146°53´E
Towoeti Meer *see* Towuti, Danau
148 K4 **Towraghoudī** Herāt, NW Afghanistan 35°13´N 62°19´E
21 X3 **Towson** Maryland, NE USA 39°25´N 76°36´W
171 O13 **Towuti, Danau** *Dut.* Towoeti Meer. ◎ Sulawesi, C Indonesia
Toxkan He *see* Ak-say
24 K9 **Toyah** Texas, SW USA 31°18´N 103°47´W
165 R4 **Tōya-ko** ◎ Hokkaidō, NE Japan
164 L11 **Toyama** Toyama, Honshū, SW Japan 36°41´N 137°13´E
164 L11 **Toyama** *off.* Toyama-ken. ◈ *prefecture* Honshū, SW Japan
Toyama-ken *see* Toyama
164 L11 **Toyama-wan** *bay* W Japan
164 H15 **Tōyo** Kōchi, Shikoku, SW Japan 33°22´N 134°18´E
Toyohara *see* Yuzhno-Sakhalinsk
164 L14 **Toyohashi** *var.* Toyohasi. Aichi, Honshū, SW Japan 34°46´N 137°22´E
Toyohasi *see* Toyohashi
164 L14 **Toyokawa** Aichi, Honshū, SW Japan 34°47´N 137°24´E
164 I14 **Toyooka** Hyōgo, Honshū, SW Japan 35°35´N 134°48´E
164 L13 **Toyota** Aichi, Honshū, SW Japan 35°04´N 137°09´E
165 T1 **Toyotomi** Hokkaidō, NE Japan 45°07´N 141°45´E
147 Q10 **To'ytepa** *Rus.* Toytepa. Toshkent Viloyati, E Uzbekistan 41°04´N 69°22´E
Toytepa *see* To'ytepa
74 M6 **Tozeur** *var.* Tawzar. W Tunisia 34°00´N 08°09´E
39 Q8 **Tozi, Mount** ▲ Alaska, USA 65°45´N 151°01´W
137 Q9 **T'q'varcheli** *Rus.* Tkvarcheli; *prev.* Tqvarch'eli. NW Georgia 42°51´N 41°42´E
Tqvarch'eli *see* T'q'varcheli
Trâblous *see* Tripoli
137 O11 **Trabzon** *Eng.* Trebizond; *anc.* Trapezus. Trabzon, NE Turkey 41°N 39°43´E
137 O11 **Trabzon** *Eng.* Trebizond. ◈ *province* NE Turkey
13 P13 **Tracadie** New Brunswick, SE Canada 47°32´N 64°57´W
Trachenberg *see* Żmigród
15 O11 **Tracy** Québec, SE Canada 45°59´N 73°07´W
35 O8 **Tracy** California, W USA 37°43´N 121°27´W
29 S10 **Tracy** Minnesota, N USA 44°14´N 95°37´W
20 K10 **Tracy City** Tennessee, S USA 35°15´N 85°44´W
106 D7 **Tradate** Lombardia, N Italy 45°43´N 08°57´E
84 F6 **Traena Bank** *undersea feature* E Norwegian Sea 66°15´N 09°45´E
29 W13 **Traer** Iowa, C USA 42°11´N 92°28´W
104 J16 **Trafalgar, Cabo de** *headland* SW Spain 36°10´N 06°03´W
Traiectum ad Mosam/ Traiectum Tungorum *see* Maastricht
Tráigh Mhór *see* Tramore
11 O17 **Trail** British Columbia, SW Canada 49°94´N 117°39´W
58 B11 **Traíra, Serra do** ▲▲ NW Brazil
109 V5 **Traisen** Niederösterreich, NE Austria 48°03´N 15°37´E
109 W4 **Traisen** ≈ NE Austria
109 X4 **Traiskirchen** Niederösterreich, NE Austria 48°01´N 16°18´E
Trajani Portus *see* Civitavecchia
Trajectum ad Rhenum *see* Utrecht
119 H14 **Trakai** *Ger.* Traken, *Pol.* Troki. Vilnius, SE Lithuania 54°39´N 24°58´E
Traken *see* Trakai
97 B20 **Tralee** *Ir.* Trá Lí. SW Ireland 52°16´N 09°42´W
97 A20 **Tralee Bay** *Ir.* Bá Thrá Lí. *bay* SW Ireland
Trá Lí *see* Tralee
Trälleborg *see* Trelleborg
Tralles Aydin *see* Aydın
61 J16 **Tramandaí** Rio Grande do Sul, S Brazil 30°01´S 50°11´W
108 C7 **Tramelan** Bern, W Switzerland 47°13´N 07°07´E
Trá Mhór *see* Tramore
97 E20 **Tramore** *Ir.* Tráigh Mhór, Trá Mhór. Waterford, S Ireland 52°10´N 07°10´W
114 F9 **Tran** *var.* Trŭn. Pernik, W Bulgaria 42°51´N 22°37´E
95 L18 **Tranås** Jönköping, S Sweden 58°03´N 15°00´E
62 J7 **Trancas** Tucumán, N Argentina 26°11´S 65°20´W
104 I7 **Trancoso** Guarda, N Portugal 40°46´N 07°21´W
95 H22 **Tranebjerg** Midtjylland, C Denmark 55°51´N 10°36´E
95 K19 **Tranemo** Västra Götaland, S Sweden 57°30´N 13°20´E
167 N16 **Trang** Trang, S Thailand 07°33´N 99°36´E
171 V15 **Trangan, Pulau** *island* Kepulauan Aru, E Indonesia
Trâng Dinh *see* Thât Khê
183 Q7 **Trangie** New South Wales, SE Australia 32°01´S 147°58´E
94 K12 **Trängslet** Dalarna, C Sweden 61°22´N 13°43´E
107 N16 **Trani** Puglia, SE Italy 41°17´N 16°25´E
61 F17 **Tranqueras** Rivera, NE Uruguay 31°13´S 55°45´W
63 G17 **Tranqui, Isla** *island* S Chile
39 V6 **Trans-Alaska pipeline** *oil pipeline* Alaska, USA
195 Q10 **Transantarctic Mountains** ▲▲ Antarctica
Transcarpathian Oblast *see* Zakarpats'ka Oblast'
122 E9 **Trans-Siberian Railway** *railway* Russian Federation
Transilvania *see* Transylvania
Transilvaniei, Alpi *see* Carpaţii Meridionalii
Transjordan *see* Jordan
172 L11 **Transkei Basin** *undersea feature* SW Indian Ocean 35°30´S 29°00´E
117 O10 **Transnistria** *cultural region* E Moldavia
81 H18 **Trans Nzoia** ◈ *county* W Kenya
Transsylvanische Alpen/ Transylvanian Alps *see* Carpaţii Meridionalii
94 K12 **Transtrand** Dalarna, C Sweden 61°06´N 13°19´E
116 G10 **Transylvania** *Eng.* Ardeal, Transilvania, *Ger.* Siebenbürgen, *Hung.* Erdély. *cultural region* NW Romania
167 S14 **Tra Ôn** Vinh Long, S Vietnam 09°58´N 105°58´E
107 H23 **Trapani** *anc.* Drepanum. Sicilia, Italy, C Mediterranean Sea 38°02´N 12°32´E
Trâpeăng Vêng *see* Kâmpóng Thum
Trapezus *see* Trabzon
114 L9 **Trapoklovo** Sliven, C Bulgaria 42°40´N 26°36´E
183 P13 **Traralgon** Victoria, SE Australia 38°15´S 146°36´E
76 H9 **Trarza** ◈ *region* SW Mauritania
Trasimenischersee *see* Trasimeno, Lago
106 H12 **Trasimeno, Lago** *Eng.* Lake of Perugia, *Ger.* Trasimenischersee. ◎ C Italy
95 J20 **Träslövsläge** Halland, S Sweden 57°02´N 12°18´E
Trás-os-Montes *see* Cucumbi
104 I6 **Trás-os-Montes e Alto Douro** *former province* N Portugal
167 Q12 **Trat** *var.* Bang Phra. Trat, S Thailand 12°16´N 102°30´E
Trá Tholl, Inis *see* Inishtrahull
Traù *see* Trogir
109 T4 **Traun** Oberösterreich, N Austria 48°14´N 14°15´E
109 S5 **Traun** ≈ N Austria
Traun, Lake *see* Traunsee
101 N23 **Traunreut** Bayern, SE Germany 47°58´N 12°36´E
109 S5 **Traunsee** *var.* Gmundner See, *Eng.* Lake Traun. ◎ N Austria
Trautenau *see* Trutnov
21 P11 **Travelers Rest** South Carolina, SE USA 34°58´N 82°26´W
182 L8 **Travellers Lake** *seasonal lake* New South Wales, SE Australia
31 P6 **Traverse City** Michigan, N USA 44°45´N 85°37´W
29 R7 **Traverse, Lake** ◎ Minnesota/South Dakota, N USA
185 I16 **Travers, Mount** ▲ South Island, New Zealand 42°01´S 172°46´E
11 P17 **Travers Reservoir** ⊠ Alberta, SW Canada
167 T14 **Tra Vinh** *var.* Phu Vinh. Tra Vinh, S Vietnam 09°57´N 106°20´E
25 S10 **Travis, Lake** ⊠ Texas, SW USA
112 H12 **Travnik** Federacija Bosne I Hercegovine, C Bosnia and Herzegovina 44°14´N 17°40´E
109 V11 **Trbovlje** *Ger.* Trifail. C Slovenia 46°10´N 15°03´E
23 V13 **Treasure Island** Florida, SE USA 27°46´N 82°46´W
Treasure State *see* Montana
186 I8 **Treasury Islands** *island group* NW Solomon Islands
106 D9 **Trebbia** *anc.* Trebia. ≈ NW Italy
100 N8 **Trebel** ≈ NE Germany
103 O16 **Trèbes** Aude, S France 43°12´N 02°26´E
Trebia *see* Trebbia
111 F18 **Třebíč** *Ger.* Trebitsch. Vysočina, C Czech Republic 49°13´N 15°52´E
113 I16 **Trebinje** Republika Srpska, S Bosnia and Herzegovina 42°42´N 18°19´E
Trebišnica *see* Trebišnjica
113 H16 **Trebišnjica** *var.* Trebišnica. ≈ S Bosnia and Herzegovina
111 N20 **Trebišov** *Hung.* Tőketerebes. Košický Kraj, E Slovakia 48°37´N 21°44´E
Trebitsch *see* Třebíč
Trebizond *see* Trabzon
Trebnitz *see* Trzebnica
109 V12 **Trebnje** SE Slovenia 45°54´N 15°01´E
111 D19 **Třeboň** *Ger.* Wittingau. Jihočeský Kraj, S Czech Republic 49°00´N 14°46´E
104 J15 **Trebujena** Andalucía, S Spain 36°52´N 06°11´W
100 I7 **Treene** ≈ N Germany
Tree Planters State *see* Nebraska
109 S9 **Treffen** Kärnten, S Austria 46°40´N 13°51´E
Trefynwy *see* Monmouth
102 G5 **Tréguier** Côtes d'Armor, NW France 48°50´N 03°12´W
61 G18 **Treinta y Tres** Treinta y Tres, E Uruguay 33°16´S 54°17´W
61 F18 **Treinta y Tres** ◈ *department* E Uruguay
122 F11 **Trëkhgornyy** Chelyabinskaya Oblast', C Russian Federation 54°42´N 58°25´E
114 F9 **Treklyanska Reka** ≈ W Bulgaria
102 K8 **Trélazé** Maine-et-Loire, NW France 47°27´N 00°28´W
63 K17 **Trelew** Chubut, SE Argentina 43°13´S 65°15´W
95 K23 **Trelleborg** *var.* Trälleborg. Skåne, S Sweden 55°22´N 13°10´E
113 P15 **Trem** ▲ SE Serbia 43°10´N 22°12´E
15 N11 **Tremblant, Mont** ▲ Québec, SE Canada 46°13´N 74°34´W
99 H17 **Tremelo** Vlaams Brabant, C Belgium 51°N 04°34´E
107 M15 **Tremiti, Isole** *island group* SE Italy
30 K12 **Tremont** Illinois, N USA 40°30´N 89°31´W
36 L1 **Tremonton** Utah, W USA 41°42´N 112°09´W
105 U4 **Tremp** Cataluña, NE Spain 42°09´N 00°53´E
30 J7 **Trempealeau** Wisconsin, N USA 44°00´N 91°25´W
15 P8 **Trenche** ≈ Québec, SE Canada
15 O7 **Trenche, Lac** ◎ Québec, SE Canada
111 I20 **Trenčiansky Kraj** ◈ *region* W Slovakia
111 I19 **Trenčín** *Ger.* Trentschin, *Hung.* Trencsén. Trenčiansky Kraj, W Slovakia 48°54´N 18°03´E
Trencsén *see* Trenčín
Trengganu *see* Terengganu
Trengganu, Kuala *see* Kuala Terengganu
61 A21 **Trenque Lauquen** Buenos Aires, E Argentina 36°01´S 62°47´W
14 J14 **Trent** ≈ Ontario, SE Canada
97 N18 **Trent** ≈ C England, United Kingdom
Trent *see* Trento
106 F5 **Trentino-Alto Adige** *Eng.* South Tyrol, *Ger.* Trentino-Südtirol; *prev.* Venezia Tridentina. ◈ *region* N Italy
Trentino-Südtirol *see* Trentino-Alto Adige
106 G6 **Trento** *Eng.* Trent, *Ger.* Trient; *anc.* Tridentum. Trentino-Alto Adige, N Italy 46°05´N 11°08´E
14 J15 **Trenton** Ontario, SE Canada 44°07´N 77°34´W
23 V10 **Trenton** Florida, SE USA 29°36´N 82°49´W
23 R1 **Trenton** Georgia, SE USA 34°52´N 85°27´W
31 S10 **Trenton** Michigan, N USA 42°08´N 83°10´W
27 S2 **Trenton** Missouri, C USA 40°04´N 93°37´W
28 M17 **Trenton** Nebraska, C USA 40°10´N 101°00´W
18 J15 **Trenton** *state capital* New Jersey, NE USA 40°13´N 74°45´W
21 W10 **Trenton** North Carolina, SE USA 35°03´N 77°20´W
20 G9 **Trenton** Tennessee, S USA 35°59´N 88°59´W
36 L1 **Trenton** Utah, W USA 41°53´N 111°57´W
Trentschin *see* Trenčín
Treptow an der Rega *see* Trzebiatów
61 C23 **Tres Arroyos** Buenos Aires, E Argentina 38°22´S 60°17´W
61 J15 **Três Cachoeiras** Rio Grande do Sul, S Brazil 29°21´S 49°48´W
106 E7 **Trescore Balneario** Lombardia, N Italy 45°43´N 09°52´E
41 V17 **Tres Cruces, Cerro** ▲ SE Mexico 15°28´N 92°27´W
57 K18 **Tres Cruces, Cordillera** ▲▲ W Bolivia
113 N18 **Treska** ≈ NW FYR Macedonia
113 I14 **Treskavica** ▲▲ SE Bosnia and Herzegovina
59 J20 **Três Lagoas** Mato Grosso do Sul, SW Brazil 20°46´S 51°43´W
40 H12 **Tres Marías, Islas** *island group* C Mexico
59 M19 **Três Marias, Represa** ⊠ SE Brazil
63 F20 **Tres Montes, Península** *headland* S Chile 46°49´S 75°29´W
105 O3 **Trespaderne** Castilla y León, N Spain 42°47´N 03°24´W
60 G13 **Três Passos** Rio Grande do Sul, S Brazil 27°33´S 53°55´W
61 A23 **Tres Picos, Cerro** ▲ E Argentina 38°10´S 61°54´W
63 G17 **Tres Picos, Cerro** ▲ SW Argentina 42°22´S 71°51´W
60 I12 **Três Pinheiros** Paraná, S Brazil 25°25´S 51°57´W
59 M21 **Três Pontas** Minas Gerais, SE Brazil 21°33´S 45°18´W
Tres Puntas, Cabo *see* Manabique, Punta
60 P9 **Três Rios** Rio de Janeiro, SE Brazil 22°06´S 43°15´W
Tres Tabernae *see* Saverne
Trestenberg/Trestendorf *see* Tăşnad
41 R15 **Tres Valles** Veracruz-Llave, SE Mexico 18°14´N 96°03´W
94 H12 **Tretten** Oppland, S Norway 61°19´N 10°19´E
Treuburg *see* Olecko
101 K21 **Treuchtlingen** Bayern, S Germany 48°57´N 10°55´E
100 N13 **Treuenbrietzen** Brandenburg, E Germany 52°06´N 12°52´E
63 H17 **Trevelín** Chubut, SW Argentina 43°02´S 71°27´W
Treves/Trèves *see* Trier
106 I13 **Trevi** Umbria, C Italy 42°52´N 12°46´E
106 E7 **Treviglio** Lombardia, N Italy 45°32´N 09°35´E
104 J4 **Trevinca, Peña** ▲ NW Spain 42°10´N 06°49´W
105 P3 **Treviño** Castilla y León, N Spain 42°45´N 02°41´W
106 I7 **Treviso** *anc.* Tarvisium. Veneto, NE Italy 45°40´N 12°15´E
97 G24 **Trevose Head** *headland* SW England, United Kingdom 50°33´N 05°03´W
Trg *see* Feldkirchen in Kärnten
183 P17 **Triabunna** Tasmania, SE Australia 42°33´S 147°55´E
21 W4 **Triangle** Virginia, NE USA 38°30´N 77°17´W
83 L18 **Triangle** Masvingo, SE Zimbabwe 20°58´S 31°28´E
115 L23 **Tría Nisía** *island* Kykládes, Greece, Aegean Sea
Triberg *see* Triberg im Schwarzwald
101 G23 **Triberg im Schwarzwald** *var.* Triberg. Baden-Württemberg, SW Germany 48°07´N 08°13´E
153 P11 **Tribhuvan** ✈ (Kathmandu) Central, C Nepal
54 C9 **Tribugá, Golfo de** *gulf* W Colombia
181 W4 **Tribulation, Cape** *headland* Queensland, NE Australia 16°14´S 145°48´E
108 M8 **Tribulaun** ▲ SW Austria 46°59´N 11°18´E
11 U17 **Tribune** Saskatchewan, S Canada 49°16´N 103°50´W
26 H5 **Tribune** Kansas, C USA 38°27´N 101°46´W
107 N18 **Tricarico** Basilicata, S Italy 40°37´N 16°09´E
107 Q19 **Tricase** Puglia, SE Italy 39°56´N 18°21´E
Trichinopoly *see* Tiruchchirāppalli
115 D18 **Trichonída, Límni** ◎ C Greece
155 G22 **Trichūr** *var.* Thrissur. Kerala, SW India 10°32´N 76°14´E *see also* Thrissur
Tricorno *see* Triglav
183 O8 **Trida** New South Wales, SE Australia 33°02´S 145°03´E
35 S1 **Trident Peak** ▲ Nevada, W USA 41°52´N 118°22´W
Tridentum/Trient *see* Trento
109 T6 **Trieben** Steiermark, SE Austria 47°29´N 14°30´E
101 D19 **Trier** *Eng.* Treves, *Fr.* Trèves; *anc.* Augusta Treverorum. Rheinland-Pfalz, SW Germany 49°45´N 06°39´E
106 K7 **Trieste** *Slvn.* Trst. Friuli-Venezia Giulia, NE Italy 45°39´N 13°45´E
Trieste, Golfo di/Triest, Golf von *see* Trieste, Gulf of
106 J8 **Trieste, Gulf of** *Cro.* Tršćanski Zaljev, *Ger.* Golf von Triest, *It.* Golfo di Trieste, *Slvn.* Tržaški Zaliv. *gulf* S Europe
109 W4 **Triesting** ≈ W Austria
Triêu Hai *see* Quang Tri
Trifail *see* Trbovlje
116 L9 **Trifeşti** Iaşi, NE Romania 47°30´N 27°31´E
109 S10 **Triglav** *It.* Tricorno. ▲ NW Slovenia 46°22´N 13°40´E
104 I14 **Trigueros** Andalucía, S Spain 37°24´N 06°50´W
115 E16 **Tríkala** *prev.* Trikkala. Thessalía, C Greece 39°33´N 21°46´E
115 E17 **Trikeriótis** ≈ C Greece
Trikkala *see* Tríkala
Trikomo/Tríkomon *see* İskele
97 F17 **Trim** *Ir.* Baile Átha Troim. Meath, E Ireland 53°34´N 06°47´W
108 E7 **Trimbach** Solothurn, NW Switzerland 47°22´N 07°53´E
109 Q5 **Trimmelkam** Oberösterreich, N Austria 48°02´N 12°55´E
29 U11 **Trimont** Minnesota, N USA 43°45´N 94°42´W
Trimontium *see* Plovdiv
Trinacria *see* Sicilia
155 K24 **Trincomalee** *var.* Trinkomali. Eastern Province, NE Sri Lanka 08°34´N 81°13´E
65 K16 **Trindade, Ilha da** *island* Brazil, W Atlantic Ocean
47 Y9 **Trindade Spur** *undersea feature* SW Atlantic Ocean 21°00´S 35°00´W
111 J17 **Třinec** *Ger.* Trzynietz. Moravskoslezský Kraj, E Czech Republic 49°41´N 18°39´E
57 M16 **Trinidad** El Beni, N Bolivia 14°52´S 64°54´W
54 H9 **Trinidad** Casanare, E Colombia 05°25´N 71°39´W
44 E6 **Trinidad** Sancti Spíritus, C Cuba 21°48´N 80°00´W
61 E19 **Trinidad** Flores, S Uruguay 33°35´S 56°54´W
37 U8 **Trinidad** Colorado, C USA 37°11´N 104°31´W
45 Y17 **Trinidad** *island* C Trinidad and Tobago
Trinidad *see* Jose Abad Santos
45 Y16 **Trinidad and Tobago** *off.* Republic of Trinidad and Tobago. ◆ *republic* SE West Indies
Trinidad and Tobago, Republic of *see* Trinidad and Tobago
63 F22 **Trinidad, Golfo** *gulf* S Chile
61 B24 **Trinidad, Isla** *island* E Argentina
107 N16 **Trinitapoli** Puglia, SE Italy 41°22´N 16°06´E
55 X10 **Trinité, Montagnes de la** ▲▲ C French Guiana
25 W9 **Trinity** Texas, SW USA 30°57´N 95°22´W
13 U12 **Trinity Bay** *inlet* Newfoundland, Newfoundland and Labrador, E Canada
39 P15 **Trinity Islands** *island group* Alaska, USA
35 N2 **Trinity Mountains** ▲▲ California, W USA
35 S4 **Trinity Peak** ▲ Nevada, W USA 40°13´N 118°43´W
35 S5 **Trinity Range** ▲▲ Nevada, W USA
35 N2 **Trinity River** ≈ California, W USA
25 V8 **Trinity River** ≈ Texas, SW USA
Trinkomali *see* Trincomalee
173 Y15 **Triolet** NW Mauritius 20°05´S 57°32´E
107 O20 **Trionto, Capo** *headland* S Italy 39°37´N 16°46´E
Tripití, Ákra *see* Trypití, Akrotírio
115 F20 **Trípoli** *prev.* Trípolis. Pelopónnisos, S Greece 37°31´N 22°22´E
138 G6 **Tripoli** *var.* Tarābulus, Ţarābulus ash Shām, Trâblous; *anc.* Tripolis. N Lebanon 34°30´N 35°42´E
29 X12 **Tripoli** Iowa, C USA 42°48´N 92°15´W
Tripoli *see* Ţarābulus
Trípolis *see* Trípoli, Greece
Tripolis *see* Tripoli, Lebanon
29 Q12 **Tripp** South Dakota, N USA 43°12´N 97°57´W
153 V15 **Tripura** *var.* Hill Tippera. ◈ *state* NE India
108 K8 **Trisanna** ≈ W Austria
100 H8 **Trischen** *island* NW Germany
Tristan da Cunha ◇ *dependency of Saint Helena* SE Atlantic Ocean
67 P15 **Tristan da Cunha** *island* SE Atlantic Ocean
Tristan da Cunha *see* Saint Helena, Ascension and Tristan da Cunha
65 L18 **Tristan da Cunha Fracture Zone** *tectonic feature* S Atlantic Ocean
167 S14 **Tri Tôn** An Giang, S Vietnam 10°26´N 105°01´E
Tri Tôn, Đao *see* Triton Island
167 W10 **Triton Island** *Chin.* Zhongjian Dao, *Viet.* Đao Tri Tôn. *island* S Paracel Islands
155 G24 **Trivandrum** *var.* Thiruvanathapuram, Tiruvantapuram. *state capital* Kerala, SW India 08°30´N 76°57´E
111 H20 **Trnava** *Ger.* Tyrnau, *Hung.* Nagyszombat. Trnavský Kraj, W Slovakia 48°22´N 17°36´E
111 H20 **Trnavský Kraj** ◈ *region* W Slovakia
Trnovo *see* Veliko Tarnovo
Trobriand Island *see* Kiriwina Island
Trobriand Islands *see* Kiriwina Islands
11 Q16 **Trochu** Alberta, SW Canada 51°48´N 113°12´W
109 U7 **Trofaiach** Steiermark, SE Austria 47°25´N 15°01´E
93 F14 **Trofors** Troms, N Norway 65°31´N 13°19´E
113 E14 **Trogir** *It.* Traù. Split-Dalmacija, S Croatia 43°32´N 16°13´E
112 F13 **Troglav** ▲ Bosnia and Herzegovina/Croatia 44°00´N 16°36´E
107 M16 **Troia** Puglia, SE Italy 41°21´N 15°19´E
107 K24 **Troina** Sicilia, Italy, C Mediterranean Sea 37°47´N 14°37´E
173 O16 **Trois-Bassins** W Réunion 21°05´S 55°18´E
101 E17 **Troisdorf** Nordrhein-Westfalen, W Germany 50°49´N 07°09´E
74 H5 **Trois Fourches, Cap des** *headland* NE Morocco 35°27´N 02°58´W
15 T8 **Trois-Pistoles** Québec, SE Canada 48°08´N 69°10´W
99 L21 **Trois-Ponts** Liège, E Belgium 50°22´N 05°52´E
15 P11 **Trois-Rivières** Québec, SE Canada 46°21´N 72°34´W
55 Y12 **Trois Sauts** S French Guiana 02°15´N 52°52´W
99 M22 **Troisvierges** Diekirch, N Luxembourg 50°07´N 06°00´E
122 F11 **Troitsk** Chelyabinskaya Oblast', S Russian Federation 54°04´N 61°31´E
125 T9 **Troitsko-Pechorsk** Respublika Komi, NW Russian Federation 62°39´N 56°06´E
127 V7 **Troitskoye** Orenburgskaya Oblast', W Russian Federation 52°23´N 56°24´E
Troki *see* Trakai
94 F9 **Trolla** ▲ S Norway 62°41´N 09°47´E
95 J18 **Trollhättan** Västra Götaland, S Sweden 58°17´N 12°20´E
94 G9 **Trollheimen** ▲▲ S Norway
94 E9 **Trolltindane** ▲ S Norway 62°30´N 07°43´E
58 H11 **Trombetas, Rio** ≈ NE Brazil
128 L16 **Tromelin, Île** *island* N Réunion
92 I9 **Troms** ◈ *county* N Norway
92 I9 **Tromsø** *Fin.* Tromssa. Troms, N Norway 69°39´N 19°01´E
84 F5 **Tromsøflaket** *undersea feature* W Barents Sea 18°30´E 71°30´N
Tromssa *see* Tromsø
94 H10 **Tron** ▲ S Norway 62°12´N 10°46´E
35 U12 **Trona** California, W USA 35°46´N 117°21´W
63 G16 **Tronador, Cerro** ▲ S Chile 41°12´S 71°51´W
94 H8 **Trondheim** *Ger.* Drontheim; *prev.* Nidaros, Trondhjem. Sør-Trøndelag, S Norway 63°25´N 10°24´E
94 H7 **Trondheimsfjorden** *fjord* S Norway
Trondhjem *see* Trondheim
107 J14 **Tronto** ≈ C Italy
121 P3 **Troódos** *var.* Troodos Mountains. ▲▲ C Cyprus
Troodos *see* Ólympos
Troodos Mountains *see* Troódos
96 I13 **Troon** W Scotland, United Kingdom 55°32´N 04°41´W
107 M22 **Tropea** Calabria, SW Italy 38°40´N 15°52´E
36 L7 **Tropic** Utah, W USA 37°37´N 112°04´W
64 L10 **Tropic Seamount** *var.* Banc du Tropique. *undersea feature* E Atlantic Ocean 23°50´N 20°40´W
Tropique, Banc du *see* Tropic Seamount
Tropoja *see* Tropojë
113 L17 **Tropojë** *var.* Tropoja. Kukës, N Albania 42°25´N 20°09´E
Troppau *see* Opava
95 O16 **Trosa** Södermanland, C Sweden 58°54´N 17°35´E
118 H12 **Troškūnai** Utena, E Lithuania 55°36´N 24°55´E
101 G23 **Trossingen** Baden-Württemberg, SW Germany 48°04´N 08°37´E
117 T4 **Trostyanets'** *Rus.* Trostyanets. Sums'ka Oblast', NE Ukraine 50°30´N 34°59´E
117 N7 **Trostyanets'** *Rus.* Trostyanets. Vinnyts'ka Oblast', C Ukraine 48°35´N 29°10´E
Trostyanets *see* Trostyanets'
116 L11 **Trotuş** ≈ E Romania
44 M8 **Trou-du-Nord** N Haiti 19°34´N 71°57´W
25 W7 **Troup** Texas, SW USA 32°08´N 95°07´W
8 I10 **Trout** ≈ Northwest Territories, NW Canada
33 N8 **Trout Creek** Montana, NW USA 47°51´N 115°40´W
32 H10 **Trout Lake** Washington, NW USA 45°59´N 121°33´W
12 B9 **Trout Lake** ◎ Ontario, S Canada
33 T12 **Trout Peak** ▲ Wyoming, C USA 44°36´N 109°33´W
102 L4 **Trouville** Calvados, N France 49°21´N 00°07´E
97 L22 **Trowbridge** S England, United Kingdom 51°20´N 02°13´W
23 Q6 **Troy** Alabama, S USA 31°48´N 85°58´W
27 Q3 **Troy** Kansas, C USA 39°45´N 95°06´W
27 W4 **Troy** Missouri, C USA 38°59´N 90°59´W
18 L10 **Troy** New York, NE USA 42°43´N 73°37´W
21 S10 **Troy** North Carolina, SE USA 35°23´N 79°54´W
31 R13 **Troy** Ohio, N USA 40°02´N 84°12´W
25 T9 **Troy** Texas, SW USA 31°12´N 97°18´W
114 I9 **Troyan** Lovech, N Bulgaria 42°52´N 24°42´E
114 I9 **Troyanski Prohod** *var.* Troyanski Prokhod. *pass* N Bulgaria
Troyanski Prokhod *see* Troyanski Prohod
145 N6 **Troyebratskiy** Severnyy Kazakhstan, N Kazakhstan 54°25´N 66°03´E
103 Q6 **Troyes** *anc.* Augustobona Tricassium. Aube, N France 48°18´N 04°04´E
117 X5 **Troyits'ke** Luhans'ka Oblast', E Ukraine 49°55´N 38°18´E
35 W7 **Troy Peak** ▲ Nevada, W USA 38°18´N 115°27´W
113 G15 **Trpanj** Dubrovnik-Neretva, S Croatia 43°00´N 17°18´E
Tršćanski Zaljev *see* Trieste, Gulf of
Trst *see* Trieste
113 N14 **Trstenik** Serbia, C Serbia 43°38´N 21°01´E
126 I6 **Trubchevsk** Bryanskaya Oblast', W Russian Federation 52°33´N 33°45´E
Trubchular *see* Orlyak
37 S10 **Truchas Peak** ▲ New Mexico, SW USA 35°57´N 105°38´W
143 P16 **Trucial Coast** *physical region* C United Arab Emirates
Trucial States *see* United Arab Emirates
35 Q6 **Truckee** California, W USA 39°18´N 120°10´W
35 R5 **Truckee River** ≈ Nevada, W USA
127 Q13 **Trudfront** Astrakhanskaya Oblast', SW Russian Federation 45°56´N 47°42´E
14 I9 **Truite, Lac à la** ◎ Québec, SE Canada
42 K4 **Trujillo** Colón, NE Honduras 15°59´N 85°54´W
56 C12 **Trujillo** La Libertad, NW Peru 08°04´S 79°02´W
104 K10 **Trujillo** Extremadura, W Spain 39°28´N 05°53´W
54 I6 **Trujillo** Trujillo, NW Venezuela 09°20´N 70°38´W
54 I6 **Trujillo** *off.* Estado Trujillo. ◈ *state* W Venezuela
Trujillo, Estado *see* Trujillo
Truk *see* Chuuk
Truk Islands *see* Chuuk Islands
29 U10 **Truman** Minnesota, N USA 43°49´N 94°26´W
27 X10 **Trumann** Arkansas, C USA 35°40´N 90°30´W
36 J9 **Trumbull, Mount** ▲ Arizona, SW USA 36°22´N 113°09´W
Trŭn *see* Tran
183 Q8 **Trundle** New South Wales, SE Australia 32°55´S 147°43´E
129 U13 **Trung Phân** *physical region* S Vietnam
Tr ng Sa L n, Đao *see* Spratly Island
Tr ng Sa, Quân Đao *see* Spratly Islands
Trupcilar *see* Orlyak
13 Q15 **Truro** Nova Scotia, SE Canada 45°24´N 63°18´W
97 H25 **Truro** SW England, United Kingdom 50°16´N 05°03´W
25 P5 **Truscott** Texas, SW USA 33°43´N 99°48´W
116 K9 **Truşeşti** Botoşani, NE Romania 47°45´N 27°01´E
116 H6 **Truskavets'** L'vivs'ka Oblast', W Ukraine 49°15´N 23°30´E
95 H22 **Trustrup** Midtjylland, C Denmark 56°20´N 10°46´E
10 M11 **Trutch** British Columbia, W Canada 57°42´N 123°00´W
37 Q14 **Truth Or Consequences** New Mexico, SW USA 33°07´N 107°15´W
111 F15 **Trutnov** *Ger.* Trautenau. Královéhradecký Kraj, N Czech Republic 50°34´N 15°55´E
103 P13 **Truyère** ≈ C France
114 K9 **Tryavna** Lovech, N Bulgaria 42°52´N 25°30´E
28 M14 **Tryon** Nebraska, C USA 41°33´N 100°57´W
115 J16 **Trypití, Akrotírio** *var.* Ákra Tripití. *headland* Ágios Efstrátios, E Greece 39°28´N 24°58´E
94 J12 **Trysil** Hedmark, S Norway 61°18´N 12°16´E
94 I11 **Trysilelva** ≈ S Norway
112 D10 **Tržac** Federacija Bosne I Hercegovine, NW Bosnia and Herzegovina 44°58´N 15°48´E
Tržaški Zaliv *see* Trieste, Gulf of
110 G10 **Trzcianka** *Ger.* Schönlanke. Pila, Wielkopolskie, C Poland 53°02´N 16°24´E
110 E7 **Trzebiatów** *Ger.* Treptow an der Rega. Zachodnio-pomorskie, NW Poland 54°04´N 15°14´E
111 G14 **Trzebnica** *Ger.* Trebnitz. Dolnośląskie, SW Poland 51°19´N 17°03´E
109 T10 **Tržič** *Ger.* Neumarktl. NW Slovenia 46°22´N 14°17´E
Trzynietz *see* Třinec
83 G21 **Tsabong** *var.* Tshabong. Kgalagadi, SW Botswana 26°03´S 22°27´E
162 G7 **Tsagaanchuluut** Dzavhan, C Mongolia 47°06´N 96°40´E
162 M8 **Tsagaandelger** *var.* Haraat. Dundgovĭ, C Mongolia 46°30´N 107°39´E
Tsagaanders *see* Bayantümen
162 G7 **Tsagaanhayrhan** *var.* Shiree. Dzavhan, W Mongolia 47°30´N 96°48´E
Tsagaannuur *see* Halhgol
Tsagaan-Olom *see* Tayshir
Tsagaan-Ovoo *see* Nariynteel
Tsagaantüngi *see* Altantsögts
162 H6 **Tsagaan-Uul** *var.* Sharga. Hövsgöl, N Mongolia 49°33´N 98°36´E
162 J5 **Tsagaan-Üür** *var.* Bulgan. Hövsgöl, N Mongolia 50°30´N 101°28´E
127 P12 **Tsagan Aman** Respublika Kalmykiya, SW Russian Federation 47°37´N 46°43´E
23 V11 **Tsala Apopka Lake** ◎ Florida, SE USA
Tsamkong *see* Zhanjiang
Tsangpo *see* Brahmaputra
Tsant *see* Deren
Tsao *see* Tsau
172 I4 **Tsaratanana** Mahajanga, C Madagascar 16°46´S 47°40´E
114 N10 **Tsarevo** *prev.* Michurin. Burgas, E Bulgaria 42°10´N 27°51´E
Tsarigrad *see* İstanbul
Tsaritsyn *see* Volgograd
114 K7 **Tsar Kaloyan** Ruse, N Bulgaria 43°36´N 26°14´E
Tsarskoye Selo *see* Pushkin
117 T7 **Tsarychanka** Dnipropetrovs'ka Oblast', E Ukraine 48°56´N 34°29´E
83 H21 **Tsatsu** Southern, S Botswana 25°21´S 24°45´E
83 G17 **Tsau** *var.* Tsao. North-West, NW Botswana 20°08´S 22°29´E
81 J20 **Tsavo** Taita/Taveta, S Kenya 02°59´S 38°28´E
83 E21 **Tsawisis** Karas, S Namibia 26°18´S 18°09´E
Tschakathurn *see* Čakovec
Tschaslau *see* Čáslav
Tschenstochau *see* Częstochowa
Tschernembl *see* Črnomelj
28 K6 **Tschida, Lake** ⊠ North Dakota, N USA
Tschorna *see* Mustvee
162 G8 **Tseel** Govĭ-Altay, SW Mongolia 45°45´N 95°54´E
138 G8 **Tsefat** *var.* Safed, *Ar.* Safad; *prev.* Zefat. Northern, N Israel 32°57´N 35°27´E
126 M13 **Tselina** Rostovskaya Oblast', SW Russian Federation 46°31´N 41°01´E
Tselinograd *see* Astana
Tselinogradskaya Oblast *see* Akmola
Tsengel *see* Tosontsengel
162 J8 **Tsenher** *var.* Altan-Ovoo. Arhangay, C Mongolia 47°24´N 101°51´E
Tsenher *see* Mönhhayrhan
163 N8 **Tsenhermandal** *var.* Modot. Hentiy, C Mongolia 47°45´N 109°03´E
Tsentral'nyye Nizmennyye Garagumy *see* Merkezi Garagumy

◆ Country
● Country Capital
◇ Dependent Territory
○ Dependent Territory Capital
◈ Administrative Regions
✈ International Airport
▲ Mountain
▲▲ Mountain Range
Volcano
≈ River
◎ Lake
⊠ Reservoir

83 E21 **Tses** Karas, S Namibia 25°58′S 18°08′E
Tseshevlya see Tsyeshawlya
162 E7 **Tsetseg** var. Tsetsegnuur. Hovd, W Mongolia 46°30′N 93°16′E
Tsetsegnuur see Tsetseg
Tsetsen Khan see Öndörhaan
162 J8 **Tsetserleg** Arhangay, C Mongolia 47°29′N 101°19′E
162 H6 **Tsetserleg** var. Halban. Hövsgöl, N Mongolia 49°30′N 97°33′E
162 J8 **Tsetserleg** var. Hujirt. Övörhangay, C Mongolia 46°50′N 102°38′E
77 R16 **Tsévié** S Togo 06°25′N 01°13′E
Tshabong see Tsabong
83 G20 **Tshane** Kgalagadi, SW Botswana 24°05′S 21°54′E
Tshangalele, Lac see Lufira, Lac de Retenue de la
83 H17 **Tshauxaba** Central, C Botswana 19°56′S 25°09′E
79 F21 **Tshela** Bas-Congo, W Dem. Rep. Congo 04°56′S 13°02′E
79 K22 **Tshibala** Kasai-Occidental, S Dem. Rep. Congo 06°53′S 22°01′E
79 J22 **Tshikapa** Kasai-Occidental, SW Dem. Rep. Congo 06°23′S 20°47′E
79 L22 **Tshilenge** Kasai Oriental , S Dem. Rep. Congo 06°17′S 23°48′E
79 L24 **Tshimbalanga** Katanga, S Dem. Rep. Congo 09°42′S 23°04′E
79 L22 **Tshimbulu** Kasai-Occidental, S Dem. Rep. Congo 06°27′S 22°54′E
Tshiumbe see Chiumbe
79 M21 **Tshofa** Kasai-Oriental, C Dem. Rep. Congo 05°13′S 25°13′E
79 K18 **Tshuapa** ≈ C Dem. Rep. Congo
114 G7 **Tsibritsa** ≈ NW Bulgaria
Tsien Tang see Fuchun Jiang
114 I12 **Tsigansko Gradishte** Gr. Giftokastro. ▲ Bulgaria/Greece 41°24′N 24°41′E
Tsihombe see Tsiombe
8 H7 **Tsiigehtchic** prev. Arctic Red River. Northwest Territories, NW Canada 67°24′N 133°40′W
125 Q7 **Tsil'ma** ≈ NW Russian Federation
119 J17 **Tsimkavichy** Rus. Timkovichi. Minskaya Voblasts', C Belarus 53°04′N 26°59′E
126 M11 **Tsimlyansk** Rostovskaya Oblast', SW Russian Federation 47°39′N 42°05′E
127 N11 **Tsimlyanskoye Vodokhranilishche** var. Tsimlyansk Vodoskhovshche, Eng. Tsimlyansk Reservoir. ⊠ SW Russian Federation
Tsimlyansk Reservoir see Tsimlyanskoye Vodokhranilishche
Tsimlyansk Vodoskhovshche see Tsimlyanskoye Vodokhranilishche
Tsinan see Jinan
Tsing Hai see Qinghai Hu, China
Tsinghai see Qinghai, China
Tsingtao/Tsingtau see Qingdao
Tsingyuan see Baoding
Tsinkiang see Quanzhou
Tsintao see Qingdao
83 D17 **Tsintsabis** Oshikoto, N Namibia 18°45′S 17°51′E
172 H8 **Tsiombe** var. Tsihombe. Toliara, S Madagascar
123 O13 **Tsipa** ≈ S Russian Federation
172 H5 **Tsiribihina** ≈ W Madagascar
172 I5 **Tsiroanomandidy** Antananarivo, C Madagascar 18°44′S 46°02′E
189 U13 **Tsis** island Chuuk, C Micronesia
Tsitsihar see Qiqihar
127 Q3 **Tsivil'sk** Chuvashskaya Respublika, W Russian Federation 55°51′N 47°30′E
137 T9 **Tskhinvali** prev. Staliniri, Ts'khinvali. C Georgia 42°12′N 43°58′E
119 J19 **Tsna** ≈ SW Belarus
124 I15 **Tsna** var. Zna. ≈ W Russian Federation
162 G9 **Tsogt** var. Tahilt. Govĭ-Altay, W Mongolia 45°20′N 96°42′E
162 K10 **Tsogt-Ovoo** var. Doloon. Ömnögovĭ, S Mongolia 44°28′N 105°22′E
162 L10 **Tsogttsetsiy** var. Baruunsuu. Ömnögovĭ, S Mongolia 43°46′N 105°28′E
114 M9 **Tsonevo, Yazovir** prev. Yazovir Georgi Traykov. ⊠ NE Bulgaria
Tsoohor see Hürmen
164 K14 **Tsu** var. Tu. Mie, Honshū, SW Japan 34°41′N 136°30′E
165 O10 **Tsubame** var. Tubame. Niigata, Honshū, C Japan 37°40′N 138°56′E
165 V3 **Tsubetsu** Hokkaidō, NE Japan 43°43′N 144°01′E
165 O13 **Tsuchiura** var. Tutiura. Ibaraki, Honshū, S Japan 36°05′N 140°11′E
165 Q6 **Tsugaru-kaikyō** strait N Japan
164 E14 **Tsukumi** var. Tukumi. Ōita, Kyūshū, SW Japan 33°00′N 131°51′E
Tsul-Ulaan see Bayannuur
Tsul-Ulaan see Bayanuur
83 D17 **Tsumeb** Oshikoto, N Namibia 19°13′S 17°42′E
83 F17 **Tsumkwe** Otjozondjupa, NE Namibia 19°37′S 20°30′E
164 D15 **Tsuno** Miyazaki, Kyūshū, SW Japan 32°43′N 131°32′E
164 D12 **Tsuno-shima** island SW Japan
164 K12 **Tsuruga** var. Turuga. Fukui, Honshū, SW Japan 35°38′N 136°01′E
164 H12 **Tsurugi-san** ▲ Shikoku, SW Japan 33°50′N 134°04′E
165 P9 **Tsuruoka** var. Turuoka. Yamagata, Honshū, C Japan 38°44′N 139°48′E
164 C12 **Tsushima** prev. Izuhara. Nagasaki, Tsushima, SW Japan 34°11′N 129°16′E
164 C12 **Tsushima** var. Tsushima-tō, Tusima. island group SW Japan
Tsushima-tō see Tsushima
164 H12 **Tsuyama** var. Tuyama. Okayama, Honshū, SW Japan 35°04′N 134°01′E
83 G19 **Tswaane** Ghanzi, W Botswana 22°21′S 21°52′E
119 N16 **Tsyakhtsin** Rus. Tekhtin. Mahilyowskaya Voblasts', E Belarus 53°51′N 29°44′E
119 P19 **Tsyerakhowka** Rus. Terekhovka. Homyel'skaya Voblasts', SE Belarus 52°13′N 31°24′E
119 I17 **Tsyeshawlya** prev. Cheshevlya, Tseshevlya, Rus. Teshevle. Brestskaya Voblasts', SW Belarus 53°14′N 25°49′E
Tsyurupinsk see Tsyurupyns'k
117 R10 **Tsyurupyns'k** Rus. Tsyurupinsk. Khersons'ka Oblast', S Ukraine 46°35′N 32°43′E
Tu see Tsu
186 C7 **Tua** ≈ C Papua New Guinea
Tuaim see Tuam
184 L6 **Tuakau** Waikato, North Island, New Zealand 37°16′S 174°56′E
97 C17 **Tuam** Ir. Tuaim. Galway, W Ireland 53°31′N 08°50′W
185 K14 **Tuamarina** Marlborough, South Island, New Zealand 41°27′S 174°00′E
Tuamotu, Archipel des see Tuamotu, Îles
193 Q9 **Tuamotu Fracture Zone** tectonic feature E Pacific Ocean
191 W9 **Tuamotu, Îles** var. Archipel des Tuamotu, Dangerous Archipelago, Tuamotu Islands. island group N French Polynesia
Tuamotu Islands see Tuamotu, Îles
175 X10 **Tuamotu Ridge** undersea feature C Pacific Ocean
167 R5 **Tuân Giao** Lai Châu, N Vietnam 21°34′N 103°24′E
171 O2 **Tuao** Luzon, N Philippines 17°42′N 121°25′E
190 B15 **Tuapa** NW Niue 18°57′S 169°59′W
43 N7 **Tuapi** Región Autónoma Atlántico Norte, NE Nicaragua 14°10′N 83°20′W
126 K15 **Tuapse** Krasnodarskiy Kray, SW Russian Federation 44°08′N 39°07′E
169 U6 **Tuaran** Sabah, East Malaysia 06°12′N 116°12′E
104 I6 **Tua, Rio** ≈ N Portugal
192 H15 **Tuasivi** Savai'i, C Samoa 13°38′S 172°08′W
185 B24 **Tuatapere** Southland, South Island, New Zealand 46°09′S 167°43′E
36 M9 **Tuba City** Arizona, SW USA 36°08′N 111°14′W
138 H11 **Ṭūbah, Qaṣr aṭ** castle 'Ammān, C Jordan
Tubame see Tsubame
169 R16 **Tuban** prev. Toeban. Jawa, C Indonesia 06°55′S 112°01′E
141 O16 **Tuban, Wādī** dry watercourse SW Yemen
61 K14 **Tubarão** Santa Catarina, S Brazil 28°29′S 49°00′W
98 O10 **Tubbergen** Overijssel, E Netherlands 52°25′N 06°46′E
Tubeke see Tubize
101 H22 **Tübingen** var. Tuebingen. Baden-Württemberg, SW Germany 48°32′N 09°04′E
127 W6 **Tubinskiy** Respublika Bashkortostan, W Russian Federation 52°48′N 58°18′E
99 G19 **Tubize** Dut. Tubeke. Walloon Brabant, C Belgium 50°43′N 04°14′E
76 J16 **Tubmanburg** NW Liberia 06°50′N 10°53′W
75 T7 **Ṭubruq** Eng. Tobruk, It. Tobruch. NE Libya 32°05′N 23°59′E
191 T13 **Tubuai** island Îles Australes, SW French Polynesia
Tubuai, Îles/Tubuai Islands see Australes, Îles
Tubuai-Manu see Maiao
40 F3 **Tubutama** Sonora, NW Mexico 30°51′N 111°31′W
54 K4 **Tucacas** Falcón, N Venezuela 10°50′N 68°22′W
59 P16 **Tucano** Bahia, E Brazil 10°52′S 38°48′W
57 P19 **Tucavaca, Río** ≈ E Bolivia
110 H8 **Tuchola** Kujawsko-pomorskie, C Poland 53°36′N 17°50′E
111 M17 **Tuchów** Małopolskie, S Poland 49°53′N 21°04′E
23 S3 **Tucker** Georgia, SE USA 33°53′N 84°10′W
27 W10 **Tuckerman** Arkansas, C USA 35°43′N 91°12′W
64 B12 **Tucker's Town** E Bermuda 32°20′N 64°42′W
Tuckum see Tukums
36 M15 **Tucson** Arizona, SW USA 32°14′N 111°01′W
62 J7 **Tucumán** off. Provincia de Tucumán. ✧ province N Argentina
Tucumán see San Miguel de Tucumán
Tucumán, Provincia de see Tucumán
37 V11 **Tucumcari** New Mexico, SW USA 35°10′N 103°43′W
58 H13 **Tucunaré** Pará, N Brazil 05°15′S 55°49′W
55 Q6 **Tucupita** Delta Amacuro, NE Venezuela 09°02′N 62°04′W
58 K13 **Tucuruí, Represa de** ⊠ NE Brazil
110 F9 **Tuczno** Zachodnio-pomorskie, NW Poland 53°12′N 16°08′E
Tuddo see Tudu
105 Q5 **Tudela** Basq. Tutera; anc. Tutela. Navarra, N Spain 42°04′N 01°37′W
104 M6 **Tudela de Duero** Castilla y León, N Spain 41°35′N 04°34′W
162 G6 **Tüdevtey** var. Oygon. Dzavhan, N Mongolia 48°57′N 96°33′E
138 K6 **Tudmur** var. Tadmur, Tamar, Gk. Palmyra, Bibl. Tadmor. Ḥimṣ, C Syria 34°36′N 38°15′E
118 J4 **Tudu** Ger. Tuddo. Lääne-Virumaa, NE Estonia 59°12′N 26°52′E
Tuebingen see Tübingen
122 J14 **Tuekta** Respublika Altay, S Russian Federation 50°51′N 85°52′E
104 I5 **Tuela, Rio** ≈ N Portugal
153 X12 **Tuensang** Nāgāland, NE India 26°16′N 94°45′E
136 L15 **Tufanbeyli** Adana, C Turkey 38°15′N 36°13′E
Tüffer see Laško
186 F9 **Tufi** Northern, S Papua New Guinea 09°08′S 149°20′S
193 O3 **Tufts Plain** undersea feature N Pacific Ocean
Tugalan see Kolkhozobod
67 V14 **Tugela** ≈ SE South Africa
21 P6 **Tug Fork** ≈ S USA
39 P15 **Tugidak Island** island Trinity Islands, Alaska, USA
171 O2 **Tuguegarao** Luzon, N Philippines 17°37′N 121°48′E
123 S12 **Tugur** Khabarovskiy Kray, SE Russian Federation 53°43′N 137°00′E
161 P4 **Tuhai He** ≈ E China
104 G4 **Tui** Galicia, NW Spain 42°02′N 08°37′W
77 O13 **Tui** var. Grand Balé. ≈ W Burkina Faso
57 J16 **Tuichi, Río** ≈ W Bolivia
64 Q11 **Tuineje** Fuerteventura, Islas Canarias, Spain, NE Atlantic Ocean 28°18′N 14°03′W
43 X16 **Tuira, Río** ≈ SE Panama
Tuisarkan see Tūysarkān
Tujiabu see Yongxiu
127 W5 **Tukan** Respublika Bashkortostan, W Russian Federation 53°58′N 57°29′E
171 P14 **Tukangbesi, Kepulauan** Dut. Toekang Besi Eilanden. island group C Indonesia
147 V13 **Tŭkhtamish** Rus. Toktomush; prev. Tokhtamyshbek. SE Tajikistan 37°51′N 74°41′E
184 O12 **Tukituki** ≈ North Island, New Zealand
Tu-k'ou see Panzhihua
121 P12 **Tūkrah** NE Libya 32°32′N 20°35′E
8 H6 **Tuktoyaktuk** Northwest Territories, NW Canada 69°27′N 133°W
168 I9 **Tuktuk** Pulau Samosir, W Indonesia 02°32′N 98°43′E
Tukumi see Tsukumi
118 E9 **Tukums** Ger. Tuckum. W Latvia 56°58′N 23°12′E
81 G24 **Tukuyu** prev. Neu-Langenburg. Mbeya, S Tanzania 09°14′S 33°39′E
Tukzār see Tokzār
41 O13 **Tula** var. Tula de Allende. Hidalgo, C Mexico 20°01′N 99°21′W
41 O11 **Tula** Tamaulipas, C Mexico 22°59′N 99°43′W
126 K5 **Tula** Tul'skaya Oblast', W Russian Federation 54°11′N 37°39′E
Tulach Mhór see Tullamore
Tula de Allende see Tula
186 M9 **Tulaghi** var. Tulagi. Florida Islands, C Solomon Islands 09°04′S 160°09′E
Tulagi see Tulaghi
159 N10 **Tulagt Ar Gol** ≈ W China
41 P13 **Tulancingo** Hidalgo, C Mexico 20°34′N 98°25′W
35 R11 **Tulare** California, W USA 36°12′N 119°21′W
29 P9 **Tulare** South Dakota, N USA 44°43′N 98°29′W
35 Q12 **Tulare Lake Bed** salt flat California, W USA
37 S14 **Tularosa** New Mexico, SW USA 33°04′N 106°01′W
37 P13 **Tularosa Mountains** ▲▲ New Mexico, SW USA
37 S15 **Tularosa Valley** basin New Mexico, SW USA
83 E25 **Tulbagh** Western Cape, SW South Africa 33°17′S 19°09′E
56 C5 **Tulcán** Carchi, N Ecuador 0°44′N 77°43′W
117 N13 **Tulcea** Tulcea, E Romania 45°11′N 28°49′E
117 N13 **Tulcea** ✧ county SE Romania
Tul'chin see Tul'chyn
117 N7 **Tul'chyn** Rus. Tul'chin. Vinnyts'ka Oblast', C Ukraine 48°40′N 28°49′E
Tuléar see Toliara
35 O1 **Tulelake** California, W USA 41°57′N 121°30′W
116 J10 **Tulgheș** Hung. Gyergyótölgyes. Harghita, C Romania 46°57′N 25°46′E
Tuli see Thuli
25 N4 **Tulia** Texas, SW USA 34°22′N 101°46′W
8 I9 **Tulita** prev. Fort Norman, Norman. Northwest Territories, NW Canada 64°55′N 125°25′W
20 J10 **Tullahoma** Tennessee, S USA 35°21′N 86°12′W
183 N12 **Tullamarine** ✈ (Melbourne) Victoria, SE Australia 37°40′S 144°46′E
183 Q7 **Tullamore** New South Wales, SE Australia 32°39′S 147°35′E
97 E18 **Tullamore** Ir. Tulach Mhór. Offaly, C Ireland 53°16′N 07°30′W
103 N12 **Tulle** anc. Tutela. Corrèze, C France 45°16′N 01°46′E
109 X3 **Tulln** var. Oberhollabrunn. Niederösterreich, NE Austria 48°20′N 16°02′E
109 W4 **Tulln** ≈ NE Austria
22 H6 **Tullos** Louisiana, S USA 31°48′N 92°19′W
97 F19 **Tullow** Ir. An Tullach. Carlow, SE Ireland 52°48′N 06°44′W
181 W5 **Tully** Queensland, NE Australia 18°03′S 145°56′E
124 J3 **Tuloma** ≈ NW Russian Federation
27 P9 **Tulsa** Oklahoma, C USA 36°09′N 96°W
153 N11 **Tulsipur** Mid Western, W Nepal 28°01′N 82°22′E
126 K6 **Tul'skaya Oblast'** ✧ province W Russian Federation
126 L14 **Tul'skiy** Respublika Adygeya, SW Russian Federation 44°26′N 40°12′E
186 E5 **Tulu** Manus Island, N Papua New Guinea 01°58′S 146°50′E
54 D10 **Tuluá** Valle del Cauca, W Colombia 04°01′N 76°16′W
116 M12 **Tulucești** Galați, E Romania 45°35′N 28°01′E
39 N12 **Tuluksak** Alaska, USA 61°06′N 160°57′W
41 Z12 **Tulum, Ruinas de** ruins Quintana Roo, SE Mexico
169 R17 **Tulungagung** prev. Toeloengagoeng. Jawa, C Indonesia 08°03′S 111°54′E
186 J6 **Tulun Islands** var. Kilinailau Islands; prev. Carteret Islands. island group NE Papua New Guinea
126 M4 **Tuma** Ryazanskaya Oblast', W Russian Federation 55°09′N 40°27′E
54 B12 **Tumaco** Nariño, SW Colombia 01°51′N 78°46′W
54 B12 **Tumaco, Bahía de** bay SW Colombia
Tuman-gang see Tumen
42 L8 **Tuma, Río** ≈ N Nicaragua
95 O16 **Tumba** Stockholm, C Sweden 59°12′N 17°49′E
Tumba, Lac see Ntomba, Lac
169 S12 **Tumbangsenamang** Borneo, C Indonesia 01°17′S 112°21′E
183 Q10 **Tumbarumba** New South Wales, SE Australia 35°47′S 148°03′E
56 A8 **Tumbes** Tumbes, NW Peru 03°33′S 80°27′W
56 A9 **Tumbes** off. Departamento de Tumbes. ✧ department NW Peru
Tumbes, Departamento de see Tumbes
19 P5 **Tumbledown Mountain** ▲ Maine, NE USA 45°27′N 70°28′W
11 N13 **Tumbler Ridge** British Columbia, W Canada 55°06′N 120°51′W
95 N16 **Tumbo** prev. Rekarne. Västmanland, C Sweden 59°25′N 16°04′E
167 Q12 **Tumbŏt, Phnum** ▲ W Cambodia 12°23′N 102°57′E
182 G9 **Tumby Bay** South Australia 34°22′S 136°05′E
163 Y10 **Tumen** Jilin, NE China 42°56′N 129°47′E
163 Y11 **Tumen** Chin. Tumen Jiang, Kor. Tuman-gang, Rus. Tumyn'tszyan. ≈ E Asia
Tumen Jiang see Tumen
55 Q8 **Tumeremo** Bolívar, E Venezuela 07°17′N 61°30′W
155 G19 **Tumkūr** Karnātaka, W India 13°20′N 77°06′E
96 I10 **Tummel** ≈ C Scotland, United Kingdom
188 B15 **Tumon Bay** bay W Guam
77 P14 **Tumu** NW Ghana 10°55′N 01°59′W
58 I10 **Tumuc-Humac Mountains** var. Serra Tumucumaque. ▲▲ N South America
Tumucumaque, Serra see Tumuc-Humac Mountains
183 Q10 **Tumut** New South Wales, SE Australia 35°20′S 148°14′E
158 F7 **Tumxuk** var. Urad Qianqi. Xinjiang Uygur Zizhiqu, NW China 78°40′N 39°54′E
Tumyn'tszyan see Tumen
Tūn see Ferdows
45 U14 **Tunapuna** Trinidad, Trinidad and Tobago 10°38′N 61°23′W
60 K11 **Tunas** Paraná, S Brazil 24°57′S 49°05′W
Tunbridge Wells see Royal Tunbridge Wells
114 L11 **Tunca Nehri** Bul. Tundzha. ≈ Bulgaria/Turkey see also Tundzha
Tunca Nehri see Tundzha
137 O14 **Tunceli** var. Kalan. Tunceli, E Turkey 39°07′N 39°34′E
137 O14 **Tunceli** ✧ province C Turkey
152 J12 **Tūndla** Uttar Pradesh, N India 27°13′N 78°14′E
81 I25 **Tunduru** Ruvuma, S Tanzania 11°08′S 37°21′E
114 L10 **Tundzha** Turk. Tunca Nehri. ≈ Bulgaria/Turkey see also Tunca Nehri
Tundzha see Tunca Nehri
162 I6 **Tünel** var. Bulag. Hövsgöl, N Mongolia 49°51′N 100°41′E
155 H17 **Tungabhadra** ≈ S India
155 F17 **Tungabhadra Reservoir** ◎ S India
191 P2 **Tungaru** prev. Gilbert Islands. island group W Kiribati
171 P7 **Tungawan** Mindanao, S Philippines 07°33′N 122°22′E
Tungdor see Mainling
T'ung-shan see Xuzhou
161 Q16 **Tungsha Tao** Chin. Dongsha Qundao, Eng. Pratas Island. island S Taiwan
Tungshih see Dongshi
8 H9 **Tungsten** Northwest Territories, W Canada 62°N 128°09′W
Tung-t'ing Hu see Dongting Hu
56 A13 **Tungurahua** ✧ province C Ecuador
95 F14 **Tunhovdfjorden** ◎ S Norway
22 K2 **Tunica** Mississippi, S USA 34°40′N 90°22′W
75 N5 **Tunis** var. Tūnis. ● (Tunisia) NE Tunisia 36°53′N 10°10′E
75 N5 **Tunis, Golfe de** Ar. Khalīj Tūnis. gulf NE Tunisia
75 N6 **Tunisia** off. Tunisian Republic, Ar. Al Jumhūrīyah at Tūnisīyah, Fr. République Tunisienne. ◆ republic N Africa
Tunisian Republic see Tunisia
Tunisienne, République see Tunisia
Tūnisīyah, Al Jumhūrīyah at see Tunisia
Tūnis, Khalīj see Tunis, Golfe de
54 G9 **Tunja** Boyacá, C Colombia 05°33′N 73°23′W
93 F14 **Tunnsjøen** Lapp. Dåtnejaevrie. ◎ C Norway
39 N12 **Tuntutuliak** Alaska, USA 60°21′N 162°40′W
197 P14 **Tunu** ✧ province E Greenland
147 U8 **Tunuk** Chuyskaya Oblast', C Kyrgyzstan 42°11′N 73°55′E
13 Q6 **Tunungayualok Island** island Newfoundland and Labrador, E Canada
62 H11 **Tunuyán** Mendoza, W Argentina 33°35′S 69°00′W
62 I11 **Tunuyán, Río** ≈ W Argentina
Tunxi see Huangshan
Tuodian see Shuangbai
Tuoji see Zhongba
Tuokezhake see Shufu
35 P9 **Tuolumne River** ≈ California, W USA
Tuong Buong see Tương Đương
167 R7 **Tương Đương** var. Tuong Buong. Nghê An, N Vietnam 19°15′N 104°30′E
160 I13 **Tuoniang Jiang** ≈ S China
Tuotiereke see Jeminay
Tuotuo He see Togton He
Tuotuoheyan see Tanggulashan
Tüp see Tyup
60 J9 **Tupã** São Paulo, S Brazil 21°57′S 50°28′W
191 S10 **Tupai** var. Motu Iti. atoll Îles Sous le Vent, W French Polynesia
61 G15 **Tupanciretã** Rio Grande do Sul, S Brazil 29°06′S 53°48′W
22 M2 **Tupelo** Mississippi, S USA 34°15′N 88°42′W
59 K18 **Tupiraçaba** Goiás, S Brazil 14°33′S 48°40′W
57 L21 **Tupiza** Potosí, S Bolivia 21°27′S 65°45′W
144 D14 **Tupkaragan, Mys** prev. Mys Tyub-Karagan. headland SW Kazakhstan 44°40′N 50°19′E
11 N13 **Tupper** British Columbia, W Canada 55°30′N 119°59′W
18 J8 **Tupper Lake** ◎ New York, NE USA
146 J10 **Tuproqqal'a** Khorazm Viloyati, W Uzbekistan 40°52′N 62°00′E
146 J10 **Tuproqqal'a** Rus. Turpakkla. Xorazm Viloyati, W Uzbekistan 40°52′N 62°00′E
62 H11 **Tupungato, Volcán** ▲ W Argentina 33°27′S 69°42′W
163 T9 **Tuquan** Nei Mongol Zizhiqu, N China 45°21′N 121°36′E
54 C13 **Túquerres** Nariño, SW Colombia 01°06′N 77°37′W
153 U13 **Tura** Meghālaya, NE India 25°33′N 90°14′E
122 M10 **Tura** Krasnoyarskiy Kray, N Russian Federation 64°20′N 100°17′E
122 G10 **Tura** ≈ C Russian Federation
140 M10 **Turabah** Makkah, W Saudi Arabia 22°00′N 42°00′E
55 O8 **Turagua, Cerro** ▲ C Venezuela 06°59′N 64°34′W
184 L12 **Turakina** Manawatu-Wanganui, North Island, New Zealand 40°03′S 175°13′E
185 K15 **Turakirae Head** headland North Island, New Zealand 41°26′S 174°54′E
186 B8 **Turama** ≈ S Papua New Guinea
122 K13 **Turan** Respublika Tyva, S Russian Federation 52°11′N 93°40′E
184 M10 **Turangi** Waikato, North Island, New Zealand 39°01′S 175°47′E
146 F11 **Turan Lowland** var. Turan Plain, Kaz. Turan Oypaty, Rus. Turanskaya Nizmennost', Turk. Turan Pesligi, Uzb. Turan Pasttekisligi. plain C Asia
Turan Oypaty/Turan Pesligi/Turan Plain/Turanskaya Nizmennost' see Turan Lowland
Turan Pasttekisligi see Turan Lowland
138 K7 **Ṭurāq al 'Ilab** hill range S Syria
119 K20 **Turaw** Rus. Turov. Homyel'skaya Voblasts', SE Belarus 52°04′N 27°44′E
140 L2 **Ṭurayf** Al Ḥudūd ash Shamālīyah, NW Saudi Arabia 31°43′N 38°40′E
Turba see Teruel
54 E5 **Turbaco** Bolívar, N Colombia 10°20′N 75°25′W
148 K15 **Turbat** Baluchistān, SW Pakistan 26°02′N 62°56′E
Turbat-i-Haidari see Torbat-e Ḥeydarīyeh
Turbat-i-Jam see Torbat-e Jām
54 D7 **Turbo** Antioquia, NW Colombia 08°06′N 76°44′W
Turčianský Svätý Martin see Martin
116 H10 **Turda** Ger. Thorenburg, Hung. Torda. Cluj, NW Romania 46°35′N 23°50′E
142 M7 **Ṭūreh** Markazī, W Iran
191 X12 **Tureia** atoll Îles Tuamotu, SE French Polynesia
110 I12 **Turek** Wielkopolskie, C Poland 52°01′N 18°30′E
93 L19 **Turenki** Kanta-Häme, SW Finland 60°55′N 24°38′E
Turfan see Turpan
Turgay see Torgay
Turgay see Torgay
144 M8 **Turgayskaya Stolovaya Strana** Kaz. Torgay Üstirti. plateau Kazakhstan/Russian Federation
Turgel see Türi
Tŭrgovishte see Targovishte
Tŭrgovishte see Targovishte
136 C14 **Turgutlu** Manisa, W Turkey 38°30′N 27°43′E
136 L12 **Turhal** Tokat, N Turkey 40°23′N 36°05′E
118 H4 **Türi** Ger. Turgel. Järvamaa, N Estonia 58°48′N 25°28′E
105 S9 **Tùria** ≈ E Spain
58 M12 **Turiaçu** Maranhão, E Brazil 01°40′S 45°22′W
Turin see Torino
116 J2 **Turiya** Pol. Turja, Rus. Tur'ya; prev. Tur"ya. ≈ NW Ukraine
116 I3 **Turiys'k** Volyns'ka Oblast', NW Ukraine 51°05′N 24°31′E
Turja see Turiya
116 H6 **Turka** L'vivs'ka Oblast', W Ukraine 49°07′N 23°01′E
81 H16 **Turkana** ✧ county Kenya
81 H16 **Turkana, Lake** var. Lake Rudolf. ◎ N Kenya
Turkestan see Turkistan
147 Q12 **Turkestan Range** Rus. Turkestanskiy Khrebet. ▲▲ C Asia
Turkestanskiy Khrebet see Turkestan Range
111 M23 **Túrkeve** Jász-Nagykun-Szolnok, E Hungary 47°06′N 20°42′E
25 O4 **Turkey** Texas, SW USA 34°23′N 100°54′W
136 H14 **Turkey** off. Republic of Turkey, Turk. Türkiye Cumhuriyeti. ◆ republic SW Asia
181 N4 **Turkey Creek** Western Australia 16°54′S 128°12′E
26 M9 **Turkey Creek** ≈ Oklahoma, C USA
37 T9 **Turkey Mountains** ▲▲ New Mexico, SW USA
Turkey, Republic of see Turkey
29 X11 **Turkey River** ≈ Iowa, C USA
127 N7 **Turki** Saratovskaya Oblast', W Russian Federation 52°00′N 43°16′E
121 O1 **Turkish Republic of Northern Cyprus** ◇ disputed territory Cyprus
145 P16 **Turkistan** prev. Turkestan. Yuzhnyy Kazakhstan, S Kazakhstan 43°18′N 68°18′E
Turkistan, Bandi-i see Torkestān, Selseleh-ye Band-e
Türkiye Cumhuriyeti see Turkey
146 K12 **Türkmenabat** prev. Rus. Chardzhev, Chardzhou, Chardzhui, Lenin-Turkmenski, Turkm. Chärjew. Lebap Welaýaty, E Turkmenistan 39°07′N 63°30′E
146 A11 **Türkmen Aylagy** Rus. Turkmenskiy Zaliv. lake gulf W Turkmenistan
Turkmenbashi see Türkmenbaşy
146 A10 **Türkmenbaşy** Rus. Turkmenbashi; prev. Krasnovodsk. Balkan Welaýaty, W Turkmenistan 40°N 53°04′E
146 A10 **Türkmenbasy Aylagy** prev. Rus. Krasnovodskiy Zaliv, Turkm. Krasnowodsk Aylagy. lake Gulf W Turkmenistan
146 J14 **Türkmengala** Rus. Turkmen-kala; prev. Turkmen-Kala. Mary Welaýaty, S Turkmenistan 37°25′N 62°19′E
146 G13 **Turkmenistan** ; prev. Turkmenskaya Soviet Socialist Republic. ◆ republic C Asia
Turkmen-kala/Turkmen-Kala see Türkmengala
Turkmenskaya Soviet Socialist Republic see Turkmenistan
Turkmenskiy Zaliv see Türkmen Aylagy
136 L16 **Türkoğlu** Kahramanmaraş, S Turkey 37°24′N 36°49′E
44 L6 **Turks and Caicos Islands** ◇ UK dependent territory N West Indies
64 G10 **Turks and Caicos Islands** UK dependant territory N West Indies
45 N6 **Turks Islands** island group SE Turks and Caicos Islands
93 K19 **Turku** Swe. Åbo. Varsinais-Suomi, SW Finland 60°27′N 22°17′E
81 H17 **Turkwel** seasonal river NW Kenya
27 P9 **Turley** Oklahoma, C USA 36°14′N 95°58′W
35 P9 **Turlock** California, W USA 37°29′N 120°52′W
118 I12 **Turmantas** Utena, NE Lithuania 55°41′N 26°27′E
Turmberg see Wieżyca
54 L5 **Turmero** Aragua, N Venezuela South America 10°14′N 66°40′W
184 N13 **Turnagain, Cape** headland North Island, New Zealand 40°30′S 176°36′E
Turnau see Turnov
42 H2 **Turneffe Islands** island group E Belize
18 M11 **Turners Falls** Massachusetts, NE USA 42°36′N 72°31′W
11 P16 **Turner Valley** Alberta, SW Canada 50°43′N 114°19′W
99 I16 **Turnhout** Antwerpen, N Belgium 51°19′N 04°57′E
109 V5 **Türnitz** Niederösterreich, E Austria 47°56′N 15°26′E
11 S12 **Turnor Lake** ◎ Saskatchewan, C Canada
111 E15 **Turnov** Ger. Turnau. Liberecký Kraj, N Czech Republic 50°36′N 15°10′E
Tŭrnovo see Veliko Tarnovo
116 I15 **Turnu Măgurele** var. Turnu-Măgurele. Teleorman, S Romania 43°44′N 24°53′E
Turnu Severin see Drobeta-Turnu Severin
Turócszentmárton see Martin
Turoni see Tours
Turov see Turaw
Turpakkla see Tuproqqal'a
158 M6 **Turpan** var. Turfan. Xinjiang Uygur Zizhiqu, NW China 42°55′N 89°06′E
Turpan Depression see Turpan Pendi
158 M6 **Turpan Pendi** Eng. Turpan Depression. depression NW China
158 M5 **Turpan Zhan** Xinjiang Uygur Zizhiqu, W China 43°10′N 89°06′E
Turpentine State see North Carolina
44 H8 **Turquino, Pico** ▲ E Cuba 19°54′N 76°55′W
27 Y10 **Turrell** Arkansas, C USA 35°22′N 90°13′W
43 N14 **Turrialba** Cartago, E Costa Rica 09°56′N 83°40′W
96 K8 **Turriff** NE Scotland, United Kingdom 57°32′N 02°28′W
139 V7 **Tursāq** Diyālá, E Iraq 33°27′N 45°47′E
Turshiz see Kāshmar
Tursunzade see Tursunzoda
147 P13 **Tursunzoda** Rus. Tursunzade; prev. Regar. W Tajikistan 38°30′N 68°10′E
Turt see Hanh
Tŭrtkŭl/Turtkul' see To'rtkok'l
29 O9 **Turtle Creek** ≈ South Dakota, N USA
30 K4 **Turtle Flambeau Flowage** ⊠ Wisconsin, N USA
11 S14 **Turtleford** Saskatchewan, S Canada 53°21′N 108°48′W
28 M4 **Turtle Lake** North Dakota, N USA 47°31′N 100°53′W
92 K12 **Turtola** Lappi, NW Finland 66°39′N 23°55′E
122 M10 **Turu** ≈ N Russian Federation
Turuga see Tsuruga
147 V10 **Turugart Pass** pass China/Kyrgyzstan
158 E7 **Turugart Shankou** var. Pereval Torugart. pass China/Kyrgyzstan
122 K9 **Turukhan** ≈ N Russian Federation
122 K9 **Turukhansk** Krasnoyarskiy Kray, N Russian Federation 65°50′N 87°48′E
139 N3 **Ṭurumbah** well NE Syria
Turuoka see Tsuruoka
Turush see Turysh
60 K7 **Turvo, Rio** ≈ S Brazil
Tur"ya see Turiya
144 H14 **Turysh** prev. Turush. Mangistau, SW Kazakhstan 45°24′N 56°02′E
23 O4 **Tuscaloosa** Alabama, S USA 33°13′N 87°34′W
23 O4 **Tuscaloosa, Lake** ◎ Alabama, S USA
Tuscan Archipelago see Toscano, Archipelago
Tuscan-Emilian Mountains see Tosco-Emiliano, Appennino
Tuscany see Toscana
35 V2 **Tuscarora** Nevada, W USA 41°16′N 116°13′W
18 F15 **Tuscarora Mountain** ridge Pennsylvania, NE USA
30 M14 **Tuscola** Illinois, N USA 39°46′N 88°19′W
25 P7 **Tuscola** Texas, SW USA 32°12′N 99°48′W
23 O2 **Tuscumbia** Alabama, S USA 34°43′N 87°42′W
92 O4 **Tusenøyane** island group S Svalbard
144 K13 **Tushchybas, Zaliv** prev. Zaliv Paskevicha. lake gulf SW Kazakhstan
Tusima see Tsushima
171 Y15 **Tusirah** Papua, E Indonesia 06°46′S 140°19′E
23 Q5 **Tuskegee** Alabama, S USA 32°25′N 85°41′W
94 E8 **Tustna** island S Norway
39 R12 **Tustumena Lake** ◎ Alaska, USA
110 K13 **Tuszyn** Łódzkie, C Poland 51°36′N 19°31′E
137 S13 **Tutak** Ağrı, E Turkey 39°34′N 42°48′E
185 C20 **Tutamoe Range** ▲▲ North Island, New Zealand
Tutasev see Tutayev
124 L15 **Tutayev** var. Tutasev. Yaroslavskaya Oblast', W Russian Federation 57°51′N 39°29′E
Tutela see Tulle, France
Tutela see Tudela, Spain
Tutera see Tudela
155 H23 **Tuticorin** Tamil Nādu, SE India 08°48′N 78°10′E
113 L15 **Tutin** Serbia, S Serbia 43°00′N 20°20′E
184 O10 **Tutira** Hawke's Bay, North Island, New Zealand 39°14′S 176°53′E
Tutiura see Tsuchiura
122 K10 **Tutonchny** Krasnoyarskiy Kray, N Russian Federation 64°12′N 93°52′E
114 L6 **Tutrakan** Silistra, NE Bulgaria 44°03′N 26°38′E
29 N5 **Tuttle** North Dakota, N USA 47°07′N 99°58′W
26 M11 **Tuttle** Oklahoma, C USA 35°17′N 97°48′W
27 O3 **Tuttle Creek Lake** ⊠ Kansas, C USA
101 H23 **Tuttlingen** Baden-Württemberg, S Germany 47°59′N 08°49′E
171 R16 **Tutuala** East Timor 08°23′S 127°12′E
192 K17 **Tutuila** island W American Samoa
83 I18 **Tutume** Central, E Botswana 20°26′S 27°02′E
39 N7 **Tututalak Mountain** ▲ Alaska, USA 67°51′N 161°27′W
22 K3 **Tutwiler** Mississippi, S USA 34°00′N 90°25′W
162 L8 **Tuul Gol** ≈ N Mongolia
93 O16 **Tuupovaara** Pohjois-Karjala, E Finland 62°30′N 30°40′E
Tuva see Tyva, Respublika
190 E7 **Tuvalu** prev. Ellice Islands. ◆ commonwealth republic SW Pacific Ocean
Tuvinskaya ASSR see Tyva, Respublika
163 O9 **Tuvshinshiree** var. Sergelen. Sühbaatar, E Mongolia 46°12′N 111°48′E
141 P9 **Ṭuwayq, Jabal** ▲▲ C Saudi Arabia
138 H13 **Ṭuwayyil ash Shihāq** desert S Jordan
11 U16 **Tuxford** Saskatchewan, S Canada 50°33′N 105°32′W
167 U12 **Tu Xoay** Đăc Lăc, S Vietnam 12°18′N 107°33′E
40 L14 **Tuxpan** Jalisco, C Mexico 19°33′N 103°21′W
40 J12 **Tuxpan** Nayarit, C Mexico 21°57′N 105°12′W
41 Q12 **Tuxpán** var. Tuxpán de Rodríguez Cano. Veracruz-Llave, E Mexico 20°58′N 97°23′W
Tuxpán de Rodríguez Cano see Tuxpán
41 R15 **Tuxtepec** var. San Juan Bautista Tuxtepec. Oaxaca, S Mexico 18°02′N 96°05′W
41 U16 **Tuxtla** var. Tuxtla Gutiérrez. Chiapas, SE Mexico 16°44′N 93°03′W
Tuxtla see San Andrés Tuxtla
Tuxtla Gutiérrez see Tuxtla
Tuyama see Tsuyama
167 T5 **Tuyên Quang** Tuyên Quang, N Vietnam 21°48′N 105°18′E
167 U13 **Tuy Hoa** Binh Thuân, S Vietnam 11°03′N 108°12′E
167 V12 **Tuy Hoa** Phu Yên, S Vietnam 13°02′N 109°15′E
127 U5 **Tuymazy** Respublika Bashkortostan, W Russian Federation 54°36′N 53°40′E
Tuy Phong see Liên Hương
142 L6 **Tūysarkān** var. Tuisarkan, Tuyserkān. Hamadān, W Iran 34°31′N 48°30′E
Tuyserkān see Tūysarkān
Tuyuk see Tuyyk
145 W16 **Tuyyk** Kaz. Tuyyq; prev. Tuyuk. Taldykorgan, SE Kazakhstan 43°07′N 79°24′E
Tuyyq see Tuyyk
136 I14 **Tuz Gölü** ◎ C Turkey
125 Q15 **Tuzha** Kirovskaya Oblast', NW Russian Federation 57°37′N 48°02′E
113 K17 **Tuzi** S Montenegro 42°22′N 19°21′E
139 T5 **Ṭūz Khurmātū** At Ta'mím, N Iraq 34°56′N 44°38′E
112 I11 **Tuzla** Federacija Bosne I Hercegovine, NE Bosnia and Herzegovina 44°33′N 18°40′E
117 N15 **Tuzla** Constanța, SE Romania 43°58′N 28°38′E

◆ Country
● Country Capital
◇ Dependent Territory
○ Dependent Territory Capital
✧ Administrative Regions
✈ International Airport
▲ Mountain
▲▲ Mountain Range
▲ Volcano
≈ River
◎ Lake
⊠ Reservoir

137 T12 **Tuzluca** Iğdır, E Turkey 40°02′N 43°39′E
95 J20 **Tvååker** Halland, S Sweden 57°04′N 12°25′E
95 F17 **Tvedestrand** Aust-Agder, S Norway 58°36′N 08°55′E
124 J16 **Tver'** *prev.* Kalinin. Tverskaya Oblast', W Russian Federation 56°53′N 35°52′E
126 I15 **Tverskaya Oblast'** ◆ *province* W Russian Federation
124 I15 **Tvertsa** W Russian Federation
138 G9 **Tverya** *var.* Tiberias; *prev.* Teverya. Northern, N Israel 32°48′N 35°32′E
95 F16 **Tvietsund** Telemark, S Norway 59°00′N 08°34′E
110 H13 **Twardogóra** *Ger.* Festenberg. Dolnośląskie, SW Poland 51°21′N 17°27′E
14 J14 **Tweed** Ontario, SE Canada 44°29′N 77°19′W
96 K13 **Tweed** England/Scotland, United Kingdom
98 O7 **Tweede-Exloërmond** Drenthe, NE Netherlands 52°55′N 06°55′E
183 V3 **Tweed Heads** New South Wales, SE Australia 28°10′S 153°32′E
98 M11 **Twello** Gelderland, E Netherlands 52°14′N 06°07′E
35 W15 **Twentynine Palms** California, W USA 34°08′N 116°03′W
25 P9 **Twin Buttes Reservoir** ▣ Texas, SW USA
33 O15 **Twin Falls** Idaho, NW USA 42°34′N 114°28′W
39 N13 **Twin Hills** Alaska, USA 59°06′N 160°21′W
11 O11 **Twin Lakes** Alberta, W Canada 57°47′N 117°30′W
33 O12 **Twin Peaks** ▲ Idaho, NW USA 44°37′N 114°24′W
185 I14 **Twins, The** ▲ South Island, New Zealand 41°14′S 172°38′E
29 S5 **Twin Valley** Minnesota, N USA 47°15′N 96°15′W
100 G11 **Twistringen** Niedersachsen, NW Germany 52°48′N 08°39′E
185 E20 **Twizel** Canterbury, South Island, New Zealand 44°04′S 171°12′E
29 X5 **Two Harbors** Minnesota, N USA 47°01′N 91°40′W
11 R14 **Two Hills** Alberta, SW Canada 53°40′N 111°43′W
31 N7 **Two Rivers** Wisconsin, N USA 44°10′N 87°33′W
116 H8 **Tyachiv** Zakarpats'ka Oblast', W Ukraine 48°02′N 23°35′E
Tyan'-Shan' *see* Tien Shan
166 L3 **Tyao** Myanmar (Burma)/India
117 R6 **Tyasmin** N Ukraine
23 X6 **Tybee Island** Georgia, SE USA 32°00′N 80°51′W
Tyborøn *see* Thyborøn
111 J16 **Tychy** *Ger.* Tichau. Śląskie, S Poland 50°12′N 19°01′E
111 O16 **Tyczyn** Podkarpackie, SE Poland 49°58′N 22°03′E
94 I8 **Tydal** Sør-Trøndelag, S Norway 63°01′N 11°36′E
115 H24 **Tyflós** Kríti, Greece, E Mediterranean Sea
21 S3 **Tygart Lake** ▣ West Virginia, NE USA
123 Q13 **Tygda** Amurskaya Oblast', SE Russian Federation 53°07′N 126°12′E
21 Q11 **Tyger River** South Carolina, SE USA
32 I11 **Tygh Valley** Oregon, NW USA 45°15′N 121°12′W
94 F12 **Tyin** ◎ S Norway
29 S10 **Tyler** Minnesota, N USA 44°16′N 96°07′W
25 W7 **Tyler** Texas, SW USA 32°21′N 95°18′W
25 W7 **Tyler, Lake** ▣ Texas, SW USA
22 K7 **Tylertown** Mississippi, S USA 31°07′N 90°08′W
117 P10 **Tylihuls'kyy Lyman** ◎ SW Ukraine
Tylos *see* Bahrain
115 C15 **Týmfi** *var.* Timfi. ▲ W Greece 39°58′N 20°51′E
115 E17 **Tymfristós** *var.* Timfristos. ▲ C Greece 38°57′N 21°49′E
115 J25 **Tympáki** *var.* Timbaki; *prev.* Timbákion. Kríti, Greece, E Mediterranean Sea 35°04′N 24°47′E
123 Q12 **Tynda** Amurskaya Oblast', SE Russian Federation 55°09′N 124°44′E
29 Q12 **Tyndall** South Dakota, N USA 42°57′N 97°52′W
97 L14 **Tyne** N England, United Kingdom
97 M14 **Tynemouth** NE England, United Kingdom 55°01′N 01°24′W
97 L14 **Tyneside** *cultural region* NE England, United Kingdom
94 H10 **Tynset** Hedmark, S Norway 61°45′N 10°49′E
39 Q12 **Tyonek** Alaska, USA 61°04′N 151°08′W
Tyôsi *see* Chōshi
Tyras *see* Dniester
Tyras *see* Bilhorod-Dnistrovs'kyy
Tyre *see* Soûr
95 G14 **Tyrifjorden** ◎ S Norway
95 K22 **Tyringe** Skåne, S Sweden 56°09′N 13°35′E
123 R13 **Tyrma** Khabarovskiy Kray, SE Russian Federation 50°00′N 132°04′E
Tyrnau *see* Trnava
115 F15 **Týrnavos** *var.* Tírnavos. Thessalía, C Greece 39°45′N 22°18′E
127 N16 **Tyrnyauz** Kabardino-Balkarskaya Respublika, SW Russian Federation 43°19′N 42°55′E
Tyrol *see* Tirol
18 E14 **Tyrone** Pennsylvania, NE USA 40°41′N 78°12′W
97 E15 **Tyrone** *cultural region* W Northern Ireland, United Kingdom
Tyros *see* Bahrain
182 M10 **Tyrrell, Lake** *salt lake* Victoria, SE Australia
84 H14 **Tyrrhenian Basin** *undersea feature* Tyrrhenian Sea, C Mediterranean Sea 39°30′N 13°00′E
120 L8 **Tyrrhenian Sea** *It.* Mare Tirreno. *sea* N Mediterranean Sea
94 J12 **Tyrsil** Hedmark, S Norway
Tysa *see* Tisa/Tisza
116 J7 **Tysmenytsya** Ivano-Frankivs'ka Oblast', W Ukraine 48°54′N 24°50′E
95 C14 **Tysnesøya** *island* S Norway
95 C14 **Tysse** Hordaland, S Norway 60°23′N 05°46′E
95 D14 **Tyssedal** Hordaland, S Norway 60°07′N 06°36′E
95 O17 **Tystberga** Södermanland, C Sweden 58°51′N 17°15′E
118 E12 **Tytuvėnai** Siaulai, C Lithuania 55°36′N 23°14′E
Tyub-Karagan, Mys *see* Tupkaragan, Mys
147 V8 **Tyugel'-Say** Narynskaya Oblast', C Kyrgyzstan 41°57′N 74°40′E
122 H11 **Tyukalinsk** Omskaya Oblast', C Russian Federation 55°56′N 72°02′E
127 V7 **Tyul'gan** Orenburgskaya Oblast', W Russian Federation 52°27′N 56°08′E
122 G11 **Tyumen'** Tyumenskaya Oblast', C Russian Federation 57°11′N 65°29′E
122 H11 **Tyumenskaya Oblast'** ◆ *province* C Russian Federation
147 Y7 **Tyup** *Kir.* Tüp. Issyk-Kul'skaya Oblast', NE Kyrgyzstan 42°44′N 78°18′E
122 L14 **Tyva, Respublika** *prev.* Tannu-Tuva, Tuva, Tuvinskaya ASSR. ◆ *autonomous republic* C Russian Federation
117 N7 **Tyvriv** Vinnyts'ka Oblast', C Ukraine 49°01′N 28°28′E
97 J21 **Tywi** S Wales, United Kingdom
97 I19 **Tywyn** W Wales, United Kingdom 52°35′N 04°06′W
83 K20 **Tzaneen** Limpopo, NE South Africa 23°50′S 30°09′E
Tzekung *see* Zigong
115 I20 **Tziá** *prev.* Kéa, Kéos; *anc.* Ceos. *island* Kykládes, Greece, Aegean Sea
41 X12 **Tzucacab** Yucatán, SE Mexico 20°04′N 89°03′W

U

82 B12 **Uaco Cungo** *var.* Waku Kungo, *Port.* Santa Comba. Kwanza Sul, C Angola 11°21′S 15°04′E
UAE *see* United Arab Emirates
191 X7 **Ua Huka** *island* Îles Marquises, NE French Polynesia
58 E10 **Uaiacás** Roraima, N Brazil 03°28′N 63°13′W
Uamba *see* Wamba
Uanle Uen *see* Wanlaweyn
191 W7 **Ua Pu** *island* Îles Marquises, NE French Polynesia
81 L17 **Uar Garas** *spring/well* SW Somalia 01°19′N 41°22′E
81 H18 **Uasin Gishu** ◆ *county* W Kenya
58 G12 **Uatumã, Rio** C Brazil
Ua Uíbh Fhailí *see* Offaly
58 C11 **Uaupés, Rio** *var.* Río Vaupés. Brazil/Colombia *see also* Vaupés, Río
Uaupés, Rio *see* Vaupés, Río
Uba *see* Oba
145 N6 **Ubagan** *Kaz.* Obagan. Kazakhstan/Russian Federation
186 G7 **Ubai** New Britain, E Papua New Guinea 05°38′S 150°45′E
79 J15 **Ubangi** *Fr.* Oubangui. C Africa
Ubangi-Shari *see* Central African Republic
119 L20 **Ubarts'** *Rus.,Ukr.* Ubort'. Belarus/Ukraine *see also* Ubort'
Ubarts' *see* Ubort'
54 F9 **Ubaté** Cundinamarca, C Colombia 05°20′N 73°50′W
60 N10 **Ubatuba** São Paulo, S Brazil 23°26′S 45°04′W
149 R12 **Ubauro** Sind, SE Pakistan 28°08′N 69°43′E
171 Q6 **Ubay** Bohol, C Philippines 10°02′N 124°29′E
103 U14 **Ubaye** SE France
Ubayid, Wadi al *see* Ubayyiḍ, Wādī al
139 N8 **Ubaylah** Al Anbār, W Iraq 33°06′N 40°13′E
139 O10 **Ubayyiḍ, Wādī al** *var.* Wadi al Ubayid. *dry watercourse* SW Iraq
98 L13 **Ubbergen** Gelderland, E Netherlands 51°49′N 05°54′E
164 E13 **Ube** Yamaguchi, Honshū, SW Japan 33°57′N 131°15′E
105 O13 **Úbeda** Andalucía, S Spain 38°01′N 03°22′W
109 V7 **Übelbach** *var.* Markt-Übelbach. Steiermark, SE Austria 47°13′N 15°15′E
59 L20 **Uberaba** Minas Gerais, SE Brazil 19°47′S 47°57′W
57 Q19 **Uberaba, Laguna** ◎ E Bolivia
59 K19 **Uberlândia** Minas Gerais, SE Brazil 18°17′S 48°17′W
101 H24 **Überlingen** Baden-Württemberg, S Germany 47°46′N 09°10′E
77 U16 **Ubiaja** Edo, S Nigeria 06°39′N 06°23′E
104 K3 **Ubiña, Peña** ▲ NW Spain 43°01′N 05°58′W
57 H17 **Ubinas, Volcán** S Peru 16°16′S 70°49′W
Ubol Rajadhani/Ubol Ratchathani *see* Ubon Ratchathani
167 P9 **Ubolratna Reservoir** ▣ C Thailand
167 S10 **Ubon Ratchathani** *var.* Muang Ubon, Ubol Rajadhani, Ubol Ratchathani, Udon Ratchathani. Ubon Ratchathani, E Thailand 15°15′N 104°50′E
116 M3 **Ubort'** *Bel.* Ubarts'. Belarus/Ukraine *see also* Ubarts'
Ubort' *see* Ubarts'
104 K15 **Ubrique** Andalucía, S Spain 36°42′N 05°27′W
Ubsu-Nur, Ozero *see* Uvs Nuur
79 M18 **Ubundu** Orientale, C Dem. Rep. Congo 0°24′S 25°30′E
146 J13 **Üçajy** *var.* Üchajy, *Rus.* Uch-Adzhi. Mary Welaýaty, C Turkmenistan 38°06′N 62°44′E
137 X11 **Ucar** *Rus.* Udzhary. C Azerbaijan 40°31′N 47°40′E
56 G13 **Ucayali** *off.* Departamento de Ucayali. ◆ *department* E Peru
Ucayali, Departamento de *see* Ucayali
56 F10 **Ucayali, Río** C Peru
Uccle *see* Ukkel
Uch-Adzhi/Üchajy *see* Üçajy
127 X4 **Uchaly** Respublika Bashkortostan, W Russian Federation 54°19′N 59°33′E
Ucharal *see* Usharal
164 C17 **Uchinoura** Kagoshima, Kyūshū, SW Japan 31°16′N 131°04′E
165 R5 **Uchiura-wan** *bay* NW Pacific Ocean
Uchkuduk *see* Uchquduq
147 S9 **Uchqo'rg'on** *Rus.* Uchkurghan. Namangan Viloyati, E Uzbekistan 41°06′N 72°04′E
146 K8 **Uchquduq** *Rus.* Uchkuduk. Navoiy Viloyati, N Uzbekistan 42°12′N 63°27′E
Uchsay *see* Uchsoy
146 G6 **Uchsoy** *Rus.* Uchsay. Qoraqalpog'iston Respublikasi, NW Uzbekistan 43°51′N 58°51′E
Uchtagan Gumy/Uchtagan, Peski *see* Uçtagan Gumy
123 R11 **Uchur** E Russian Federation
100 O10 **Uckermark** *cultural region* E Germany
10 K17 **Ucluelet** Vancouver Island, British Columbia, SW Canada 48°55′N 125°34′W
146 D10 **Uçtagan Gumy** *var.* Uchtagan Gumy, *Rus.* Uchtagan, Peski. *desert* NW Turkmenistan
122 M13 **Uda** S Russian Federation
123 R12 **Uda** E Russian Federation
123 N6 **Udachnyy** Respublika Sakha (Yakutiya), NE Russian Federation 66°27′N 112°18′E
155 G21 **Udagamandalam** *var.* Ooty, Udhagamandalam; *prev.* Ootacamund. Tamil Nādu, SW India 11°28′N 76°42′E
152 F14 **Udaipur** *prev.* Oodeypore. Rājasthān, N India 24°35′N 73°41′E
Udayadhani *see* Uthai Thani
143 N16 **'Udayd, Khawr al** *var.* Khor al Udeid. *inlet* Qatar/Saudi Arabia
112 D11 **Udbina** Lika-Senj, W Croatia 44°33′N 15°46′E
95 I18 **Uddevalla** Västra Götaland, S Sweden 58°20′N 11°56′E
Uddjaur *see* Uddjaure
92 H13 **Uddjaure** *var.* Uddjaur. ◎ N Sweden
Udeid, Khor al *see* 'Udayd, Khawr al
99 K14 **Uden** Noord-Brabant, SE Netherlands 51°40′N 05°37′E
Uden *see* Udenhout
99 J14 **Udenhout** *var.* Uden. Noord-Brabant, S Netherlands 51°37′N 05°09′E
155 H14 **Udgīr** Mahārāshtra, C India 18°23′N 77°06′E
Udhagamandalam *see* Udagamandalam
152 H6 **Udhampur** Jammu and Kashmir, NW India 32°55′N 75°07′E
139 X14 **'Udhaybah, 'Uqlat al** *well* S Iraq
106 J7 **Udine** *anc.* Utina. Friuli-Venezia Giulia, NE Italy 46°05′N 13°10′E
175 T14 **Udintsev Fracture Zone** *tectonic feature* S Pacific Ocean
Udipi *see* Udupi
Udmurtia *see* Udmurtskaya Respublika
127 S2 **Udmurtskaya Respublika** *Eng.* Udmurtia. ◆ *autonomous republic* NW Russian Federation
124 J15 **Udomlya** Tverskaya Oblast', W Russian Federation 57°53′N 34°59′E
Udon Ratchathani *see* Ubon Ratchathani
167 Q8 **Udon Thani** *var.* Ban Mak Khaeng, Udorndhani. Udon Thani, N Thailand 17°25′N 102°45′E
Udorndhani *see* Udon Thani
189 U12 **Udot** *atoll* Chuuk Islands, C Micronesia
123 S12 **Udskaya Guba** *bay* E Russian Federation
123 R12 **Udskoye** Khabarovskiy Kray, SE Russian Federation 54°32′N 134°26′E
155 E19 **Udupi** *var.* Udipi. Karnātaka, SW India 13°18′N 74°46′E
Udzhary *see* Ucar
100 O9 **Uecker** NE Germany
100 P9 **Ueckermünde** Mecklenburg-Vorpommern, NE Germany 53°43′N 14°03′E
164 M12 **Ueda** *var.* Uyeda. Nagano, Honshū, S Japan 36°27′N 138°13′E
79 L16 **Uele** *var.* Welle. NE Dem. Rep. Congo
123 W5 **Uelen** Chukotskiy Avtonomnyy Okrug, NE Russian Federation 66°01′N 169°52′W
Uele (upper course) *see* Kibali, Dem. Rep. Congo
Uele (upper course) *see* Uolo, Río, Equatorial Guinea/Gabon
100 J11 **Uelzen** Niedersachsen, N Germany 52°58′N 10°34′E
164 J14 **Ueno** Mie, Honshū, SW Japan 34°45′N 136°08′E
127 V4 **Ufa** Respublika Bashkortostan, W Russian Federation 54°46′N 56°02′E
127 V4 **Ufa** W Russian Federation
Ufra *see* Kenar
83 C18 **Ugab** C Namibia
118 D8 **Ugāle** NW Latvia 57°16′N 21°58′E
81 F17 **Uganda** *off.* Republic of Uganda. ◆ *republic* E Africa
Uganda, Republic of *see* Uganda
138 G4 **Ugarit** *Ar.* Ra's Shamrah. *site of ancient city* Al Lādhiqīyah, NW Syria
39 O14 **Ugashik** Alaska, USA 57°30′N 157°24′W
107 Q19 **Ugento** Puglia, SE Italy 39°53′N 18°09′E
105 O15 **Ugíjar** Andalucía, S Spain 36°58′N 03°03′W
103 T11 **Ugine** Savoie, E France 45°45′N 06°25′E
123 R13 **Uglegorsk** Amurskaya Oblast', S Russian Federation 51°40′N 128°05′E
Ugleural'sk *see* Ugleural'skiy
125 V13 **Ugleural'skiy** *prev.* Polovinka, Ugleural'sk. Permskiy Kray, NW Russian Federation 58°57′N 57°37′E
124 L15 **Uglich** Yaroslavskaya Oblast', W Russian Federation 57°33′N 38°23′E
126 I4 **Ugra** W Russian Federation
147 V9 **Ugyut** Narynskaya Oblast', C Kyrgyzstan 41°22′N 74°49′E
111 H19 **Uherské Hradiště** *Ger.* Ungarisch-Hradisch. Zlínský Kraj, E Czech Republic 49°05′N 17°26′E
111 H19 **Uherský Brod** *Ger.* Ungarisch-Brod. Zlínský Kraj, E Czech Republic 49°01′N 17°40′E
111 B17 **Úhlava** *Ger.* Angel. W Czech Republic
Uhorshchyna *see* Hungary
31 T13 **Uhrichsville** Ohio, N USA 40°23′N 81°21′W
Uhuru Peak *see* Kilimanjaro
96 G8 **Uig** N Scotland, United Kingdom 57°35′N 06°22′W
82 B10 **Uíge** *Port.* Carmona, Vila Marechal Carmona. Uíge, NW Angola 07°37′S 15°02′E
82 B10 **Uíge** ◆ *province* N Angola
193 Y15 **Uiha** *island* Ha'apai Group, C Tonga
189 U13 **Uijec** *island* Chuuk, C Micronesia
163 X14 **Uijeongbu** *Jap.* Giseifu; *prev.* Ŭijŏngbu. NW South Korea 37°42′N 127°02′E
Ŭijŏngbu *see* Uijeongbu
Uil *see* Oyyl
36 M3 **Uinta Mountains** ▲ Utah, W USA
83 C18 **Uis** Erongo, NW Namibia 21°08′S 14°49′E
83 I25 **Uitenhage** Eastern Cape, S South Africa 33°44′S 25°27′E
98 H9 **Uitgeest** Noord-Holland, W Netherlands 52°32′N 04°43′E
98 I11 **Uithoorn** Noord-Holland, C Netherlands 52°14′N 04°50′E
98 O4 **Uithuizen** Groningen, NE Netherlands 53°24′N 06°40′E
98 O4 **Uithuizermeeden** Groningen, NE Netherlands 53°25′N 06°43′E
189 R6 **Ujae Atoll** *var.* Wūjae. *atoll* Ralik Chain, W Marshall Islands
Ujain *see* Ujjain
111 I16 **Ujazd** Opolskie, S Poland 50°22′N 18°20′E
Új-Becse *see* Novi Bečej
Ujda *see* Oujda
189 N5 **Ujelang Atoll** *var.* Wujlān. *atoll* Ralik Chain, W Marshall Islands
111 N21 **Újfehértó** Szabolcs-Szatmár-Bereg, E Hungary 47°48′N 21°40′E
Újgradiska *see* Nova Gradiška
164 J13 **Uji** *var.* Uzi. Kyōto, Honshū, SW Japan 34°54′N 135°48′E
81 E21 **Ujiji** Kigoma, W Tanzania 04°55′S 29°39′E
154 G10 **Ujjain** *prev.* Ujain. Madhya Pradesh, C India 23°11′N 75°50′E
Újlak *see* Ilok
'Ujmān *see* 'Ajmān
Újmoldova *see* Moldova Nouă
Újszentanna *see* Sântana
Ujungpandang *see* Makassar
Ujung Salang *see* Phuket
Újvidék *see* Novi Sad
UK *see* United Kingdom
154 E11 **Ukāi Reservoir** ▣ W India
81 G19 **Ukara Island** *island* N Tanzania
'Ukash, Wādī *see* 'Akāsh, Wādī
81 F19 **Ukerewe Island** *island* N Tanzania
139 S9 **Ukhaydir** Al Anbār, C Iraq 32°28′N 43°36′E
153 X13 **Ukhrul** Manipur, NE India 25°08′N 94°24′E
125 S9 **Ukhta** Respublika Komi, NW Russian Federation 63°31′N 53°48′E
34 L6 **Ukiah** California, W USA 39°07′N 123°14′W
32 K12 **Ukiah** Oregon, NW USA 45°06′N 118°57′W
99 G18 **Ukkel** *Fr.* Uccle. Brussels, C Belgium 50°47′N 04°19′E
118 G13 **Ukmergė** *Pol.* Wiłkomierz. Vilnius, C Lithuania 55°16′N 24°46′E
Ukraina *see* Ukraine
116 L6 **Ukraine** *off.* Ukraine, *Rus.* Ukraina, *Ukr.* Ukrayina; *prev.* Ukrainian Soviet Socialist Republic, Ukrainskay S.S.R. ◆ *republic* SE Europe
Ukraine *see* Ukraine
Ukrainian Soviet Socialist Republic *see* Ukraine
Ukrainskay S.S.R/Ukrayina *see* Ukraine
82 B13 **Uku** Kwanza Sul, NW Angola 11°25′S 14°18′E
164 B13 **Uku-jima** *island* Gotō-rettō, SW Japan
83 F20 **Ukwi** Kgalagadi, SW Botswana 23°41′S 20°26′E
118 M13 **Ula** *Rus.* Ulla. Vitsyebskaya Voblasts', N Belarus 55°14′N 29°15′E
136 C16 **Ula** Muğla, SW Turkey 37°08′N 28°25′E
118 M13 **Ula** *Rus.* Ulla. N Belarus
162 L7 **Ulaanbaatar** *Eng.* Ulan Bator; *prev.* Urga. ● (Mongolia) Töv, C Mongolia 47°55′N 106°57′E
Ulaan-Ereg *see* Bayanmönh
162 E5 **Ulaangom** Uvs, NW Mongolia 49°56′N 92°06′E
162 D5 **Ulaanhus** *var.* Bilüü. Bayan-Ölgiy, W Mongolia 48°54′N 89°40′E
Ulaantolgoy *see* Möst
Ulaan-Uul *see* Öldziyt, Bayanhongor, Mongolia
Ulaan-Uul *see* Erdene, Dornogovĭ, Mongolia
162 M14 **Ulan** *var.* Otog Qi. Nei Mongol Zizhiqu, N China 39°05′N 107°58′E
159 R10 **Ulan** *var.* Xireg; *prev.* Xiligou. Qinghai, C China 36°59′N 98°21′E
Ulan Bator *see* Ulaanbaatar
162 L13 **Ulan Buh Shamo** *desert* N China
Ulanhad *see* Chifeng
163 T8 **Ulanhot** Nei Mongol Zizhiqu, N China 46°02′N 122°E
127 Q14 **Ulan Khol** Respublika Kalmykiya, SW Russian Federation 45°27′N 46°48′E
163 P13 **Ulan Qab** *var.* Jining. Nei Mongol Zizhiqu, N China 40°59′N 113°08′E
162 M13 **Ulansuhai Nur** ◎ N China
123 N14 **Ulan-Ude** *prev.* Verkhneudinsk. Respublika Buryatiya, S Russian Federation 51°55′N 107°40′E
159 N12 **Ulan Ul Hu** ◎ C China
187 N9 **Ulawa Island** *island* SE Solomon Islands
138 J7 **'Ulayyānīyah, Bi'r al** *var.* Al Hilbeh. *well* S Syria
123 S12 **Ul'banskiy Zaliv** *strait* E Russian Federation
Ulbo *see* Olib
113 J18 **Ulcinj** S Montenegro 41°56′N 19°14′E
Uldz *see* Norovlin
Uleåborg *see* Oulu
Uleälv *see* Oulujoki
95 G16 **Ulefoss** Telemark, S Norway 59°17′N 09°15′E
Uleträsk *see* Oulujärvi
113 L19 **Ulëz** *var.* Ulëza. Dibër, C Albania 41°42′N 19°52′E
Ulëza *see* Ulëz
95 F22 **Ulfborg** Midtjylland, W Denmark 56°16′N 08°21′E
98 N13 **Ulft** Gelderland, E Netherlands 51°53′N 06°23′E
162 G7 **Uliastay** *prev.* Jibhalanta. Dzavhan, W Mongolia 47°47′N 96°53′E
188 F8 **Ulimang** Babeldaob, N Palau
67 T10 **Ulindi** W Dem. Rep. Congo
188 H14 **Ulithi Atoll** *atoll* Caroline Islands, W Micronesia
112 N10 **Uljma** Vojvodina, NE Serbia 45°04′N 21°08′E
Ul'kayak *see* Ol'keyyek
145 Q7 **Ul'ken-Karaoy, Ozero** *prev.* Ul'ken-Karoy, Ozero. ◎ N Kazakhstan
Ul'ken-Karoy, Ozero *see* Ul'ken-Karaoy, Ozero
Ülkenözen *see* Karaozen
Ülkenqobda *see* Kobda
104 G3 **Ulla** NW Spain
Ulla *see* Ula
183 S10 **Ulladulla** New South Wales, SE Australia 35°21′S 150°25′E
153 T14 **Ullapara** Rajshahi, W Bangladesh 24°20′N 89°34′E
96 H7 **Ullapool** N Scotland, United Kingdom 57°54′N 05°10′W
95 J20 **Ullared** Halland, S Sweden 57°07′N 12°45′E
105 T7 **Ulldecona** Cataluña, NE Spain 40°36′N 00°27′E
92 I9 **Ullsfjorden** *fjord* N Norway
97 K15 **Ullswater** ◎ NW England, United Kingdom
101 I22 **Ulm** Baden-Württemberg, S Germany 48°24′N 09°59′E
33 R8 **Ulm** Montana, NW USA 47°27′N 111°32′W
183 V5 **Ulmarra** New South Wales, SE Australia 29°37′S 153°06′E
116 K13 **Ulmeni** Buzău, C Romania 45°08′N 26°43′E
116 K14 **Ulmeni** Călărași, S Romania 44°08′N 26°43′E
42 L7 **Ulmukhuás** Región Autónoma Atlántico Norte, NE Nicaragua 14°20′N 84°30′W
188 C8 **Ulong** *var.* Aulong. *island* Palau Islands, N Palau
83 N14 **Ulongué** *var.* Ulongwé. Tete, NW Mozambique 14°34′S 34°21′E
Ulongwé *see* Ulongué
95 K19 **Ulricehamn** Västra Götaland, S Sweden 57°47′N 13°25′E
98 N5 **Ulrum** Groningen, NE Netherlands 53°24′N 06°20′E
163 Z16 **Ulsan** *Jap.* Urusan. SE South Korea 35°33′N 129°19′E
94 D10 **Ulsteinvik** Møre og Romsdal, S Norway 62°21′N 05°53′E
97 D15 **Ulster** ◆ *province* Northern Ireland, United Kingdom/Ireland
171 Q10 **Ulu** Pulau Siau, N Indonesia 02°46′N 125°22′E
123 Q11 **Ulu** Respublika Sakha (Yakutiya), NE Russian Federation 60°18′N 127°27′E
42 H5 **Ulúa, Río** NW Honduras
136 D12 **Ulubat Gölü** ◎ NW Turkey
136 E12 **Uludağ** ▲ NW Turkey 40°08′N 29°13′E
158 D7 **Ulugqat** Xinjiang Uygur Zizhiqu, W China 39°45′N 74°10′E
136 J16 **Ulukışla** Niğde, S Turkey 37°33′N 34°29′E
189 O15 **Ulul** *island* Caroline Islands, C Micronesia
83 L22 **Ulundi** KwaZulu/Natal, E South Africa 28°18′S 31°26′E
158 M3 **Ulungur He** NW China
158 K2 **Ulungur Hu** ◎ NW China
181 P8 **Uluru** *var.* Ayers Rock. *monolith* Northern Territory, C Australia
97 K16 **Ulverston** NW England, United Kingdom 54°13′N 03°08′W
183 O16 **Ulverstone** Tasmania, SE Australia 41°09′S 146°10′E
94 D13 **Ulvik** Hordaland, S Norway 60°34′N 06°53′E
93 J18 **Ulvila** Satakunta, W Finland 61°26′N 21°55′E
117 O8 **Ul'yanovka** Kirovohrads'ka Oblast', C Ukraine 48°18′N 30°15′E
127 Q5 **Ul'yanovsk** *prev.* Simbirsk. Ul'yanovskaya Oblast', W Russian Federation 54°17′N 48°21′E
127 Q5 **Ul'yanovskaya Oblast'** ◆ *province* W Russian Federation
Ul'yanovskiy *see* Botakara
Ul'yanovskiy Kanal *see* Ul'yanow Kanali
146 M13 **Ul'yanow Kanali** *Rus.* Ul'yanovskiy Kanal. *canal* Turkmenistan/Uzbekistan
Ulyshylanshyq *see* Uly-Zhylanshyk
26 H6 **Ulysses** Kansas, C USA 37°36′N 101°23′W
145 O12 **Ulytau, Gory** ▲ C Kazakhstan
145 N11 **Uly-Zhylanshyk** *Kaz.* Ulyshylanshyq. C Kazakhstan
112 A9 **Umag** *It.* Umago. Istra, NW Croatia 45°25′N 13°32′E
Umago *see* Umag
41 W12 **Umán** Yucatán, SE Mexico 20°51′N 89°43′W
117 O7 **Uman'** *Rus.* Uman. Cherkas'ka Oblast', C Ukraine 48°45′N 30°10′E
189 V13 **Uman** *atoll* Chuuk Islands, C Micronesia
Uman *see* Uman'
Umanak/Umanaq *see* Uummannaq
'Umān, Khalīj *see* Oman, Gulf of
'Umān, Salṭanat *see* Oman
154 K10 **Umaria** Madhya Pradesh, C India 23°34′N 80°49′E
149 R16 **Umarkot** Sind, SE Pakistan 25°22′N 69°48′E
188 B17 **Umatac** SW Guam 13°17′N 144°40′E
188 A17 **Umatac Bay** *bay* SW Guam
139 S6 **Umayqah** Şalāḥ ad Dīn, C Iraq 34°32′N 43°45′E
124 J5 **Umba** Murmanskaya Oblast', NW Russian Federation 66°39′N 34°24′E
138 I8 **Umbāshī, Khirbat al** *ruins* As Suwaydā', S Syria
80 A12 **Umbelasha** W South Sudan
106 H12 **Umbertide** Umbria, C Italy 43°16′N 12°21′E
61 B17 **Umberto** *var.* Humberto. Santa Fe, C Argentina 30°52′S 61°19′W
186 E7 **Umboi Island** *var.* Rooke Island. *island* C Papua New Guinea
124 J4 **Umbozero, Ozero** ◎ NW Russian Federation
106 H13 **Umbria** ◆ *region* C Italy
Umbrian-Machigian Mountains *see* Umbro-Marchigiano, Appennino
106 I12 **Umbro-Marchigiano, Appennino** *Eng.* Umbrian-Machigian Mountains. ▲ C Italy
93 J16 **Umeå** Västerbotten, N Sweden 63°50′N 20°15′E
93 H14 **Umeälven** N Sweden
39 Q5 **Umiat** Alaska, USA 69°22′N 152°09′W
83 K23 **Umlazi** KwaZulu/Natal, E South Africa 29°58′S 30°50′E
139 X10 **Umm al Baqar, Hawr** *var.* Birkat ad Dawaymah. *spring* S Iraq
Umm al Fatūr *see* Umm al Fuṭūr
139 Q5 **Umm al Fuṭūr** *var.* Umm at Tūz, Umm al Fatūr. Şalāḥ ad Dīn, C Iraq 34°53′N 42°42′E
141 U12 **Umm al Ḥayt, Wādī** *var.* Wādī Amilḥayt. *seasonal river* SW Oman
Umm al Qaiwain *see* Umm al Qaywayn
143 R15 **Umm al Qaywayn** *var.* Umm al Qaiwain. Umm al Qaywayn, NE United Arab Emirates 25°43′N 55°55′E
138 J3 **Umm 'Āmūd** Ḥalab, N Syria 35°57′N 37°39′E
141 Y10 **Umm ar Ruṣāṣ** *var.* Umm Ruṣayṣ. W Oman 20°26′N 58°48′E
141 X9 **Umm as Samīn** *salt flat* C Oman
Umm at Tūz *see* Umm al Fuṭūr
141 V9 **Umm az Zumūl** *oasis* E Saudi Arabia
80 A9 **Umm Buru** Western Darfur, W Sudan 15°01′N 23°36′E
80 A12 **Umm Dafag** Southern Darfur, W Sudan 10°28′N 23°20′E
Umm Durmān *see* Omdurman
138 F9 **Umm el Fahm** Haifa, N Israel 32°30′N 35°06′E
80 F9 **Umm Inderab** Northern Kordofan, C Sudan 15°12′N 31°54′E
80 C10 **Umm Keddada** Northern Darfur, W Sudan 13°36′N 26°42′E
140 J7 **Umm Lajj** Tabūk, W Saudi Arabia 25°02′N 37°19′E
138 L10 **Umm Maḥfur** N Jordan
139 Y13 **Umm Qaṣr** Al Baṣrah, SE Iraq 30°02′N 47°55′E
Umm Ruṣayṣ *see* Umm ar Ruṣāṣ
80 F11 **Umm Ruwaba** *var.* Umm Ruwābah, Um Ruwāba. Northern Kordofan, C Sudan 12°54′N 31°13′E
Umm Ruwābah *see* Umm Ruwaba
143 N16 **Umm Sa'īd** *var.* Musay'īd. S Qatar 24°57′N 51°32′E
139 Y10 **Umm Sawān, Hawr** S Iraq
138 K10 **Umm Ṭuways, Wādī** *dry watercourse* N Jordan
38 J17 **Umnak Island** *island* Aleutian Islands, Alaska, USA
32 F13 **Umpqua River** Oregon, NW USA
82 D13 **Umpulo** Bié, C Angola 12°43′S 17°42′E
154 I12 **Umred** Mahārāshtra, C India 20°54′N 79°19′E
Um Ruwāba *see* Umm Ruwaba
Umtali *see* Mutare
Umtata *see* Mthatha
77 V17 **Umuahia** Abia, SW Nigeria 05°30′N 07°33′E
60 H10 **Umuarama** Paraná, S Brazil 23°45′S 53°20′W
Umvoma *see* Mvuma
83 K18 **Umzingwani** S Zimbabwe
112 D11 **Una** Bosnia and Herzegovina/Croatia
112 E12 **Unac** W Bosnia and Herzegovina
23 T6 **Unadilla** Georgia, SE USA 32°15′N 83°44′W
18 I10 **Unadilla River** New York, NE USA
59 L18 **Unaí** Minas Gerais, SE Brazil 16°24′S 46°49′W
39 N10 **Unalakleet** Alaska, USA 63°52′N 160°47′W
38 K17 **Unalaska Island** *island* Aleutian Islands, Alaska, USA
185 I16 **Una, Mount** ▲ South Island, New Zealand 42°12′S 172°34′E
82 N13 **Unango** Niassa, N Mozambique 12°45′S 35°28′E
Unao *see* Unnāo
92 L12 **Unari** Lappi, N Finland 67°07′N 25°37′E
141 O6 **'Unayzah** *var.* Anaiza. Al Qaşīm, C Saudi Arabia 26°03′N 44°00′E
138 L10 **'Unayzah, Jabal** ▲ Jordan/Saudi Arabia 32°09′N 39°11′E
Unci *see* Almería
57 K19 **Uncía** Potosí, C Bolivia 18°30′S 66°29′W
37 Q7 **Uncompahgre Peak** ▲ Colorado, C USA 38°04′N 107°27′W
37 P6 **Uncompahgre Plateau** *plain* Colorado, C USA
95 L17 **Unden** ◎ S Sweden
28 M4 **Underwood** North Dakota, N USA 47°25′N 101°09′W
171 T13 **Undur** Pulau Seram, E Indonesia 03°41′S 130°38′E
Undur Khan *see* Öndörhaan
126 H6 **Unecha** Bryanskaya Oblast', W Russian Federation 52°51′N 32°38′E
39 N16 **Unga** Unga Island, Alaska, USA 55°14′N 160°34′W
Ungaria *see* Hungary
183 P8 **Ungarie** New South Wales, SE Australia 33°39′S 146°54′E
Ungarisch-Brod *see* Uherský Brod
Ungarisches Erzgebirge *see* Slovenské rudohorie
Ungarisch-Hradisch *see* Uherské Hradiště
Ungarn *see* Hungary
12 M4 **Ungava Bay** *bay* Québec, E Canada
12 J2 **Ungava, Péninsule d'** *peninsula* Québec, SE Canada
Ungeny *see* Ungheni
116 M9 **Ungheni** *Rus.* Ungeny. W Moldova 47°13′N 27°48′E
Unguja *see* Zanzibar
146 G10 **Üngüz Angyrsyndaky Garagum** *Rus.* Zaunguzskiye Garagumy. *desert* N Turkmenistan
146 H11 **Unguz, Solonchakovyye Vpadiny** *salt marsh* C Turkmenistan
Ungvár *see* Uzhhorod
60 I12 **União da Vitória** Paraná, S Brazil 26°13′S 51°05′W
111 G17 **Uničov** *Ger.* Mährisch-Neustadt. Olomoucký Kraj, E Czech Republic 49°48′N 17°05′E
110 J12 **Uniejów** Łódzkie, C Poland 51°58′N 18°46′E
112 A11 **Unije** *island* W Croatia
38 L16 **Unimak Island** *island* Aleutian Islands, Alaska, USA
38 L16 **Unimak Pass** *strait* Aleutian Islands, Alaska, USA
62 J12 **Unión** San Luis, C Argentina 35°09′S 65°55′W
27 W5 **Union** Missouri, C USA 38°27′N 91°01′W
32 L12 **Union** Oregon, NW USA 45°12′N 117°51′W
21 Q11 **Union** South Carolina, SE USA 34°42′N 81°37′W
21 R6 **Union** West Virginia, NE USA 37°36′N 80°34′W
61 B25 **Unión, Bahía** *bay* E Argentina
31 Q13 **Union City** Indiana, N USA 40°12′N 84°50′W
31 Q10 **Union City** Michigan, N USA 42°03′N 85°06′W
18 C12 **Union City** Pennsylvania, NE USA 41°54′N 79°51′W
20 G8 **Union City** Tennessee, S USA 36°26′N 89°03′W
32 G14 **Union Creek** Oregon, NW USA 42°54′N 122°26′W
83 G25 **Uniondale** Western Cape, SW South Africa 33°40′S 23°07′E
40 K13 **Unión de Tula** Jalisco, SW Mexico 19°58′N 104°16′W
30 M9 **Union Grove** Wisconsin, N USA 42°39′N 88°03′W
45 Y15 **Union Island** *island* S Saint Vincent and the Grenadines
46 K5 **Union Reefs** *reef* SW Mexico
0 D7 **Union Seamount** *undersea feature* NE Pacific Ocean 49°35′N 132°45′W
23 Q6 **Union Springs** Alabama, S USA 32°08′N 85°43′W
20 H6 **Uniontown** Kentucky, S USA 37°46′N 87°55′W
18 C16 **Uniontown** Pennsylvania, NE USA 39°54′N 79°44′W
27 T1 **Unionville** Missouri, C USA 40°28′N 93°00′W
141 V8 **United Arab Emirates** *Ar.* Al Imārāt al 'Arabīyah al Muttaḥidah, *abbrev.* UAE; *prev.* Trucial States. ◆ *federation* SW Asia
United Arab Republic *see* Egypt
97 H14 **United Kingdom** *off.* United Kingdom of Great Britain and Northern Ireland, *abbrev.* UK. ◆ *monarchy* NW Europe
United Kingdom of Great Britain and Northern Ireland *see* United Kingdom
United Mexican States *see* Mexico
United Provinces *see* Uttar Pradesh
16 L10 **United States of America** *off.* United States of America, *var.* America, The States, *abbrev.* U.S., USA. ◆ *federal republic* North America
United States of America *see* United States of America
124 J10 **Unitsa** Respublika Kareliya, NW Russian Federation 62°31′N 34°31′E
11 S15 **Unity** Saskatchewan, S Canada 52°27′N 109°10′W
80 D13 **Unity** *var.* Wahda. ◆ *state* N South Sudan
105 Q8 **Universales, Montes** ▲ C Spain
27 X4 **University City** Missouri, C USA 38°40′N 90°19′W
187 Q13 **Unmet** Malekula, C Vanuatu 16°09′S 167°16′E
101 F15 **Unna** Nordrhein-Westfalen, W Germany 51°32′N 07°41′E
Unnan *see* Kisuki
152 L12 **Unnāo** *prev.* Unao. Uttar Pradesh, N India 26°32′N 80°30′E
187 R15 **Unpongkor** Erromango, S Vanuatu 18°48′S 169°01′E
Unruhstadt *see* Kargowa
96 M1 **Unst** *island* NE Scotland, United Kingdom
101 K16 **Unstrut** C Germany
Unterdrauburg *see* Dravograd
Unterlimbach *see* Lendava
101 L23 **Unterschleissheim** Bayern, SE Germany 48°16′N 11°34′E
101 H24 **Untersee** ◎ Germany/Switzerland
100 O10 **Unterueckersee** ◎ NE Germany

◆ Country
● Country Capital
◇ Dependent Territory
○ Dependent Territory Capital
◆ Administrative Regions
✈ International Airport
▲ Mountain
▲ Mountain Range
Volcano
River
◎ Lake
▣ Reservoir

55 N12 **Unturán, Sierra de** ▲▲ Brazil/Venezuela
159 N11 **Unuli Horog** Qinghai, W China 35°10′N 91°50′E
136 M11 **Ünye** Ordu, W Turkey 41°08′N 37°14′E
Unza see Unzha
125 O14 **Unzha** var. Unza. ≈ NW Russian Federation
79 E17 **Uolo, Río** var. Eyo (lower course), Mbini, Uele (upper course), Woleu; prev. Benito. ≈ Equatorial Guinea/Gabon
55 Q10 **Uonán** Bolívar, SE Venezuela 04°33′N 62°10′W
161 T12 **Uotsuri-shima** Chin. Diaoyu Dao. island (disputed) China/Japan/Taiwan
165 M11 **Uozu** Toyama, Honshū, SW Japan 36°50′N 137°25′E
42 L12 **Upala** Alajuela, NW Costa Rica 10°52′N 85°W
55 P7 **Upata** Bolívar, E Venezuela 08°02′N 62°25′W
79 M23 **Upemba, Lac** ◎ SE Dem. Rep. Congo
145 R11 **Upenskoye** prev. Uspenskiy. Karaganda, C Kazakhstan 48°45′N 72°46′E
197 O12 **Upernavik** var. Upernivik. Qaasuitsup, C Greenland 73°06′N 55°42′W
Upernivik see Upernavik
83 F22 **Upington** Northern Cape, W South Africa 28°28′S 21°14′E
Uplands see Ottawa
192 I16 **'Upolu** island SE Samoa
38 G11 **'Upolu Point** var. Upolu Point. headland Hawai'i, USA, C Pacific Ocean 20°15′N 155°51′W
Upper Austria see Oberösterreich
Upper Bann see Bann
14 M13 **Upper Canada Village** tourist site Ontario, SE Canada
18 I16 **Upper Darby** Pennsylvania, NE USA 39°57′N 75°15′W
28 L2 **Upper Des Lacs Lake** ⊠ North Dakota, N USA
185 L14 **Upper Hutt** Wellington, North Island, New Zealand 41°06′S 175°06′E
29 X11 **Upper Iowa River** ≈ Iowa, C USA
32 H15 **Upper Klamath Lake** ◎ Oregon, NW USA
34 M6 **Upper Lake** California, W USA 39°07′N 122°53′W
35 Q1 **Upper Lake** ◎ California, W USA
10 K9 **Upper Liard** Yukon, W Canada 60°01′N 128°59′W
97 E16 **Upper Lough Erne** ◎ SW Northern Ireland, United Kingdom
80 F12 **Upper Nile** ◈ state NE South Sudan
29 T3 **Upper Red Lake** ◎ Minnesota, N USA
31 S12 **Upper Sandusky** Ohio, N USA 40°49′N 83°16′W
Upper Volta see Burkina Faso
95 O15 **Upplands Väsby** var. Upplandsväsby. Stockholm, C Sweden 59°29′N 18°04′E
Upplandsväsby see Upplands Väsby
95 O15 **Uppsala** Uppsala, C Sweden 59°52′N 17°38′E
95 O14 **Uppsala** ◈ county C Sweden
38 J12 **Upright Cape** headland Saint Matthew Island, Alaska, USA 60°19′N 172°15′W
20 K6 **Upton** Kentucky, S USA 37°25′N 85°53′W
33 Y13 **Upton** Wyoming, C USA 44°06′N 104°37′W
141 N7 **'Uqlat aş Şuqūr** Al Qaşīm, W Saudi Arabia 25°51′N 42°13′E
Uqsuqtuuq see Gjoa Haven
Uqturpan see Wushi
54 C7 **Urabá, Golfo de** gulf NW Colombia
Uracas see Farallon de Pajaros
uradqianqi see Wulashan, N China
Uradar'ya see O'radaryo
Urad Qianqi see Xishanzui, N China
165 U5 **Urahoro** Hokkaidō, NE Japan 42°47′N 143°41′E
165 T5 **Urakawa** Hokkaidō, NE Japan 42°11′N 142°42′E
Ural see Zhayyk
183 T6 **Uralla** New South Wales, SE Australia 30°39′S 151°30′E
Ural Mountains see Ural'skiye Gory
144 F8 **Ural'sk** Kaz. Oral. Zapadnyy Kazakhstan, NW Kazakhstan 51°12′N 51°17′E
Ural'skaya Oblast' see Zapadnyy Kazakhstan
127 W5 **Ural'skiye Gory** var. Ural'skiy Khrebet, Eng. Ural Mountains. ▲▲ Kazakhstan/Russian Federation
Ural'skiy Khrebet see Ural'skiye Gory
138 I3 **Urām aş Şughrá** Ḩalab, N Syria 36°10′N 36°55′E
183 P10 **Urana** New South Wales, SE Australia 35°22′S 146°16′E
11 S10 **Uranium City** Saskatchewan, C Canada 59°30′N 108°46′W
58 F10 **Uraricoera** Roraima, N Brazil 03°26′N 60°54′W
47 S5 **Uraricoera, Rio** ≈ N Brazil
Ura-Tyube see Üroteppa
165 O13 **Urawa** var. Saitama. Saitama, Honshū, S Japan 35°52′N 139°40′E
122 H10 **Uray** Khanty-Mansiyskiy Avtonomnyy Okrug-Yugra, C Russian Federation 60°07′N 64°38′E
141 R7 **'Uray'irah** Ash Sharqīyah, E Saudi Arabia 25°59′N 48°52′E
30 M13 **Urbana** Illinois, N USA 40°06′N 88°12′W
31 R13 **Urbana** Ohio, N USA 40°04′N 83°46′W
29 V14 **Urbandale** Iowa, C USA 41°37′N 93°42′W
106 I11 **Urbania** Marche, C Italy 43°40′N 12°33′E
106 I11 **Urbino** Marche, C Italy 43°45′N 12°38′E
57 H16 **Urcos** Cusco, S Peru 13°40′S 71°38′W
105 N10 **Urda** Castilla-La Mancha, C Spain 39°25′N 03°43′W
Urda see Khan Ordasy
Urdgol see Chandmanī
105 O3 **Urduña** var. Orduña. País Vasco, N Spain 43°00′N 03°00′W
Urdunn see Jordan
Urdzhar see Urzhar
97 L16 **Ure** ≈ N England, United Kingdom
119 K18 **Urechcha** Rus. Urech'ye. Minskaya Voblasts', S Belarus 52°57′N 27°54′E
Urech'ye see Urechcha
127 P2 **Uren'** Nizhegorodskaya Oblast', W Russian Federation 57°30′N 45°48′E
122 J9 **Urengoy** Yamalo-Nenetskiy Avtonomnyy Okrug, N Russian Federation 65°52′N 78°42′E
184 K10 **Urenui** Taranaki, North Island, New Zealand 38°59′S 174°25′E
187 Q12 **Ureparapara** island Banks Islands, N Vanuatu
40 G5 **Ures** Sonora, NW Mexico 29°26′N 110°24′W
Urfa see Şanlıurfa
Urga see Ulaanbaatar
162 F6 **Urgamal** var. Hungiy. Dzavhan, W Mongolia 48°31′N 94°15′E
146 H9 **Urganch** Rus. Urgench; prev. Novo-Urgench. Xorazm Viloyati, W Uzbekistan 41°40′N 60°32′E
Urgench see Urganch
136 J14 **Ürgüp** Nevşehir, C Turkey 38°39′N 34°55′E
147 O12 **Urgut** Samarqand Viloyati, C Uzbekistan 39°26′N 67°15′E
158 K3 **Urho** Xinjiang Uygur Zizhiqu, W China 46°05′N 84°51′E
152 G5 **Uri** Jammu and Kashmir, NW India 34°05′N 74°03′E
108 G9 **Uri** ◈ canton C Switzerland
54 F11 **Uribe** Meta, C Colombia 03°01′N 74°33′W
54 H4 **Uribia** La Guajira, N Colombia 11°45′N 72°19′W
116 G12 **Uricani** Hung. Hobicaurikány. Hunedoara, SW Romania 45°18′N 23°03′E
57 M21 **Uriondo** Tarija, S Bolivia 21°43′S 64°40′W
40 I7 **Urique** Chihuahua, N Mexico 27°16′N 107°51′W
40 I7 **Urique, Río** ≈ N Mexico
56 E9 **Uritiyacu, Río** ≈ N Peru
Uritskiy see Sarykol'
98 K8 **Urk** Flevoland, N Netherlands 52°40′N 05°35′E
136 B14 **Urla** İzmir, W Turkey 38°19′N 26°47′E
116 K13 **Urlați** Prahova, SE Romania 44°59′N 26°15′E
127 V4 **Urman** Respublika Bashkortostan, W Russian Federation 54°53′N 56°52′E
147 P12 **Urmetan** W Tajikistan 39°27′N 68°13′E
Urmia see Orūmīyeh
Urmia, Lake see Orūmīyeh, Daryācheh-ye
Urmiyeh see Orūmīyeh
Uroševac see Ferizaj
147 P11 **Üroteppa** Rus. Ura-Tyube. NW Tajikistan 39°55′N 68°57′E
54 D8 **Urrao** Antioquia, W Colombia 06°16′N 76°10′W
Ursat'yevskaya see Xovos
Urt see Gurvantes
127 X7 **Urtazym** Orenburgskaya Oblast', W Russian Federation 52°12′N 58°48′E
59 K18 **Uruaçu** Goiás, C Brazil 14°38′S 49°06′W
40 M14 **Uruapan** var. Uruapan del Progreso. Michoacán, SW Mexico 19°26′N 102°04′W
Uruapan del Progreso see Uruapan
57 G15 **Urubamba, Cordillera** ▲▲ C Peru
57 G14 **Urubamba, Río** ≈ C Peru
58 G12 **Urucará** Amazonas, N Brazil 02°30′S 57°45′W
61 E16 **Uruguaiana** Rio Grande do Sul, S Brazil 29°45′S 57°05′W
Uruguai, Rio see Uruguay
61 E18 **Uruguay** off. Oriental Republic of Uruguay; prev. La Banda Oriental. ◆ republic E South America
61 E15 **Uruguay** var. Rio Uruguai, Río Uruguay. ≈ E South America
Uruguay, Oriental Republic of see Uruguay
Uruguay, Río see Uruguay
Urukthapel see Ngeruktabel
Urumchi see Ürümqi
Urumi Yeh see Orūmīyeh, Daryācheh-ye
158 L5 **Ürümqi** var. Tihwa, Urumchi, Urumqi, Urumtsi, Wu-lu-k'o-mu-shi, Wu-lu-mu-ch'i; prev. Ti-hua. Xinjiang Uygur Zizhiqu, NW China 43°52′N 87°31′E
Urumtsi see Ürümqi
Urundi see Burundi
183 V6 **Urunga** New South Wales, SE Australia 30°33′S 152°58′E
188 C15 **Uruno Point** headland NW Guam 13°37′N 144°50′E
123 U13 **Urup, Ostrov** island Kuril'skiye Ostrova, SE Russian Federation
141 P11 **'Uruq al Mawārid** desert S Saudi Arabia
Urusan see Ulsan
127 T5 **Urussu** Respublika Tatarstan, W Russian Federation 54°34′N 53°23′E
184 K10 **Uruti** Taranaki, North Island, New Zealand 38°57′S 174°32′E
57 K19 **Uru Uru, Lago** ◎ W Bolivia
55 P9 **Uruyén** Bolívar, SE Venezuela 05°40′N 62°26′W
149 O7 **Uruzgān** ; prev. Orūzgān. Uruzgān, C Afghanistan 32°58′N 66°39′E
149 N6 **Uruzgān** prev. Orūzgān. ◈ province C Afghanistan
165 T3 **Uryū-gawa** ≈ Hokkaidō, NE Japan
165 T2 **Uryū-ko** ◎ Hokkaidō, NE Japan
127 N8 **Uryupinsk** Volgogradskaya Oblast', SW Russian Federation 50°51′N 41°59′E
145 X12 **Urzhar** prev. Urdzhar. Vostochnyy Kazakhstan, E Kazakhstan 47°06′N 81°33′E
125 R16 **Urzhum** Kirovskaya Oblast', NW Russian Federation 57°09′N 49°56′E
116 K13 **Urziceni** Ialomița, SE Romania 44°43′N 26°39′E
164 E14 **Usa** Ōita, Kyūshū, SW Japan 33°31′N 131°22′E
125 T6 **Usa** ≈ NW Russian Federation
136 E14 **Uşak** prev. Ushak. Uşak, W Turkey 38°42′N 29°25′E
136 D14 **Uşak** var. Ushak. ◈ province W Turkey
83 C19 **Usakos** Erongo, W Namibia 22°01′S 15°32′E
81 J21 **Usambara Mountains** ▲▲ NE Tanzania
81 G23 **Usangu Flats** wetland SW Tanzania
65 D24 **Usborne, Mount** ▲ East Falkland, Falkland Islands 51°35′S 58°57′W
100 O8 **Usedom** island NE Germany
99 M24 **Useldange** Diekirch, C Luxembourg 49°47′N 05°59′E
119 L16 **Usha** ≈ C Belarus
Ushachi see Ushachy
118 L13 **Ushachy** Rus. Ushachi. Vitsyebskaya Voblasts', N Belarus 55°11′N 28°37′E
Ushak see Uşak
122 L4 **Ushakova, Ostrov** island Severnaya Zemlya, N Russian Federation
Ushant see Ouessant, Île d'
145 W13 **Usharal** prev. Ucharal. Almaty, E Kazakhstan 46°08′N 80°55′E
164 B15 **Ushibuka** var. Usibuka. Kumamoto, Shimo-jima, SW Japan 32°12′N 130°00′E
Ushi Point see Sabaneta, Puntan
145 V14 **Ushtobe** Kaz. Üshtöbe. Almaty, SE Kazakhstan 45°15′N 77°59′E
Üshtöbe see Ushtobe
63 I25 **Ushuaia** Tierra del Fuego, S Argentina 54°48′S 68°19′W
39 R10 **Usibelli** Alaska, USA 63°54′N 148°41′W
Usibuka see Ushibuka
186 D7 **Usino** Madang, N Papua New Guinea 05°40′S 145°31′E
125 U6 **Usinsk** Respublika Komi, NW Russian Federation 66°01′N 57°37′E
97 K22 **Usk** Wel. Wysg. ≈ SE Wales, United Kingdom
Uskočke Planine/Uskokengebirge see Gorjanci
Uskoplje see Gornji Vakuf
Üsküb/Üsküp see Skopje
114 M11 **Üsküpdere** Kırklareli, NW Turkey 41°41′N 27°21′E
126 L7 **Usman'** Lipetskaya Oblast', W Russian Federation 52°04′N 39°41′E
118 D8 **Usmas Ezers** ◎ NW Latvia
125 U13 **Usol'ye** Permskiy Kray, NW Russian Federation 59°27′N 56°33′E
41 T16 **Uspanapa, Río** ≈ SE Mexico
Uspenskiy see Upenskoye
103 O11 **Ussel** Corrèze, C France 45°33′N 02°18′E
163 Z6 **Ussuri** var. Usuri, Wusuri, Chin. Wusuli Jiang. ≈ China/Russian Federation
123 S15 **Ussuriysk** prev. Nikol'sk, Nikol'sk-Ussuriyskiy, Voroshilov. Primorskiy Kray, SE Russian Federation 43°48′N 131°59′E
136 J10 **Usta Burnu** headland N Turkey 41°58′N 34°30′E
149 P13 **Usta Muhammad** Baluchistān, SW Pakistan 28°07′N 68°00′E
123 V11 **Ust'-Bol'sheretsk** Kamchatskiy Kray, E Russian Federation 52°48′N 156°12′E
127 N9 **Ust'-Buzulukskaya** Volgogradskaya Oblast', SW Russian Federation 50°12′N 42°06′E
111 C16 **Ustecký Kraj** ◈ region NW Czech Republic
108 G7 **Uster** Zürich, NE Switzerland 47°21′N 08°49′E
107 I22 **Ustica, Isola d'** island S Italy
122 M11 **Ust'-Ilimsk** Irkutskaya Oblast', C Russian Federation 57°57′N 102°30′E
111 C15 **Ústí nad Labem** Ger. Aussig. Ústecký Kraj, NW Czech Republic 50°41′N 14°04′E
111 F17 **Ústí nad Orlicí** Ger. Wildenschwert. Pardubický Kraj, C Czech Republic 49°58′N 16°24′E
Ustinov see Izhevsk
113 J14 **Ustiprača** ◈ Republika Srpska, SE Bosnia and Herzegovina
122 H11 **Ust'-Ishim** Omskaya Oblast', C Russian Federation 57°42′N 70°58′E
110 G6 **Ustka** Ger. Stolpmünde. Pomorskie, N Poland 54°35′N 16°50′E
123 V9 **Ust'-Kamchatsk** Kamchatskiy Kray, E Russian Federation 56°14′N 162°28′E
145 X9 **Ust'-Kamenogorsk** Kaz. Öskemen. Vostochnyy Kazakhstan, E Kazakhstan 49°58′N 82°36′E
123 T10 **Ust'-Khayryuzovo** Krasnoyarskiy Kray, E Russian Federation 57°07′N 156°37′E
122 I14 **Ust'-Koksa** Respublika Altay, S Russian Federation 50°15′N 85°45′E
125 S11 **Ust'-Kulom** Respublika Komi, NW Russian Federation 61°42′N 53°42′E
123 Q8 **Ust'-Kuyga** Respublika Sakha (Yakutiya), NE Russian Federation 69°59′N 135°27′E
126 L14 **Ust'-Labinsk** Krasnodarskiy Kray, SW Russian Federation 44°40′N 40°46′E
123 R10 **Ust'-Maya** Respublika Sakha (Yakutiya), NE Russian Federation 60°27′N 134°28′E
123 R9 **Ust'-Nera** Respublika Sakha (Yakutiya), NE Russian Federation 64°28′N 143°01′E
123 P12 **Ust'-Nyukzha** Amurskaya Oblast', S Russian Federation 56°30′N 121°32′E
123 O7 **Ust'-Olenëk** Respublika Sakha (Yakutiya), NE Russian Federation 73°03′N 119°34′E
123 T9 **Ust'-Omchug** Magadanskaya Oblast', E Russian Federation 61°07′N 149°17′E
122 M13 **Ust'-Ordynskiy** Irkutskaya Oblast', S Russian Federation 52°50′N 104°42′E
125 N8 **Ust'-Pinega** Arkhangel'skaya Oblast', NW Russian Federation 64°09′N 41°55′E
122 K8 **Ust'-Port** Krasnoyarskiy Kray, N Russian Federation 69°42′N 84°25′E
114 L11 **Ustrem** prev. Vakav. Yambol, E Bulgaria 42°01′N 26°28′E
111 O18 **Ustrzyki Dolne** Podkarpackie, SE Poland 49°26′N 22°36′E
Ust'-Sysol'sk see Syktyvkar
125 R7 **Ust'-Tsil'ma** Respublika Komi, NW Russian Federation 65°25′N 52°09′E
Ust Urt see Ustyurt Plateau
125 O11 **Ust'ya** ≈ NW Russian Federation
124 K6 **Ust'ye Varzugi** Murmanskaya Oblast', NW Russian Federation 66°16′N 36°47′E
123 V10 **Ust'yevoye** prev. Kirovskiy. Kamchatskiy Kray, E Russian Federation 54°06′N 155°48′E
117 R8 **Ustynivka** Kirovohrads'ka Oblast', C Ukraine 47°58′N 32°32′E
144 H15 **Ustyurt Plateau** var. Ust Urt, Uzb. Ustyurt Platosi. plateau Kazakhstan/Uzbekistan
Ustyurt Platosi see Ustyurt Plateau
124 K14 **Ustyuzhna** Vologodskaya Oblast', NW Russian Federation 58°50′N 36°25′E
158 J4 **Usu** Xinjiang Uygur Zizhiqu, NW China 44°27′N 84°37′E
171 O13 **Usu** Sulawesi, C Indonesia 02°34′S 120°58′E
164 E14 **Usuki** Ōita, Kyūshū, SW Japan 33°07′N 131°48′E
42 G8 **Usulután** Usulután, SE El Salvador 13°20′N 88°26′W
42 B9 **Usulután** ◈ department SE El Salvador
41 W16 **Usumacinta, Río** ≈ Guatemala/Mexico
Usumbura see Bujumbura
Usuri see Ussuri
U.S./USA see United States of America
171 W14 **Uta** Papua, E Indonesia 04°28′S 136°03′E
36 K5 **Utah** off. State of Utah, also known as Beehive State, Mormon State. ◈ state W USA
36 L3 **Utah Lake** ◎ Utah, W USA
Utaidhani see Uthai Thani
93 M14 **Utajärvi** Pohjois-Pohjanmaa, C Finland 64°45′N 26°25′E
Utamboni see Mitemele, Río
Utaradit see Uttaradit
165 T3 **Utashinai** var. Utasinai. Hokkaidō, NE Japan 43°32′N 142°03′E
Utasinai see Utashinai
193 Y14 **'Uta Vava'u** island Vava'u Group, N Tonga
37 V9 **Ute Creek** ≈ New Mexico, SW USA
118 H12 **Utena** Utena, E Lithuania 55°30′N 25°34′E
118 H12 **Utena** ◈ province Lithuania
37 V10 **Ute Reservoir** ⊠ New Mexico, SW USA
167 O10 **Uthai Thani** var. Muang Uthai Thani, Udayadhani, Utaidhani. Uthai Thani, W Thailand 15°22′N 100°03′E
149 O15 **Uthal** Baluchistān, SW Pakistan 25°53′N 66°37′E
uThongathi see oThongathi
18 I10 **Utica** New York, NE USA 43°06′N 75°15′W
105 R10 **Utiel** Valenciana, E Spain 39°33′N 01°13′W
11 O13 **Utikuma Lake** ◎ Alberta, W Canada
42 I4 **Utila, Isla de** island Islas de la Bahía, N Honduras
Utina see Udine
59 O17 **Utinga** Bahia, E Brazil 12°05′S 41°07′W
Utirik see Utrik Atoll
95 M22 **Utlängan** island S Sweden
117 U11 **Utlyuts'kyy Lyman** bay S Ukraine
95 P16 **Utö** Stockholm, C Sweden 58°55′N 18°19′E
25 Q12 **Utopia** Texas, SW USA 29°30′N 99°31′W
98 J11 **Utrecht** Lat. Trajectum ad Rhenum. Utrecht, C Netherlands 52°06′N 05°07′E
83 K22 **Utrecht** KwaZulu/Natal, E South Africa 27°40′S 30°20′E
98 I11 **Utrecht** ◈ province C Netherlands
104 K14 **Utrera** Andalucía, S Spain 37°10′N 05°47′W
189 V4 **Utrik Atoll** var. Utirik, Utrōk, Utrōnk. atoll Ratak Chain, N Marshall Islands
Utrōk/Utrōnk see Utrik Atoll
95 B16 **Utsira** island SW Norway
92 L8 **Utsjoki** var. Ohcejohka. Lappi, N Finland 69°54′N 27°01′E
165 O12 **Utsunomiya** var. Utunomiya. Tochigi, Honshū, S Japan 36°36′N 139°53′E
127 P13 **Utta** Respublika Kalmykiya, SW Russian Federation 46°22′N 46°03′E
167 O8 **Uttaradit** var. Utaradit. Uttaradit, N Thailand 17°38′N 100°05′E
152 J9 **Uttarakhand** ◈ state N India
152 J8 **Uttarkāshi** Uttarakhand, N India 30°45′N 78°19′E
152 K11 **Uttar Pradesh** prev. United Provinces, United Provinces of Agra and Oudh. ◈ state N India
45 T5 **Utuado** C Puerto Rico 18°17′N 66°41′W
158 K3 **Utubulak** Xinjiang Uygur Zizhiqu, W China 46°50′N 86°15′E
39 N5 **Utukok River** ≈ Alaska, USA
Utunomiya see Utsunomiya
187 P10 **Utupua** island Santa Cruz Islands, E Solomon Islands
Utva see Shyngyrlau
189 Y15 **Utwe** Kosrae, E Micronesia
189 X15 **Utwe Harbor** harbour Kosrae, E Micronesia
Uubulan see Hayrhan
163 O8 **Uulbayan** var. Dzüünbulag. Sühbaatar, E Mongolia 46°30′N 112°22′E
118 G6 **Uulu** Pärnumaa, SW Estonia 58°15′N 24°32′E
197 N13 **Uummannaq** var. Umanak, Umanaq. ◈ Qaasuitsup, C Greenland
Uummannarsuaq see Nunap Isua
162 E4 **Üüreg Nuur** ◎ NW Mongolia
Uusikaarlepyy see Nykarleby
93 J19 **Uusikaupunki** Swe. Nystad. Varsinais-Suomi, SW Finland 60°48′N 21°25′E
93 M20 **Uusimaa** Swe. Nyland. ◈ region S Finland
127 S2 **Uva** Udmurtskaya Respublika, NW Russian Federation 56°41′N 52°15′E
155 K25 **Uva** ◈ province SE Sri Lanka
113 L14 **Uvac** ≈ W Serbia
25 Q12 **Uvalde** Texas, SW USA 29°14′N 99°49′W
119 O18 **Uvaravichy** Rus. Uvarovichi. Homyel'skaya Voblasts', SE Belarus 52°36′N 30°44′E
54 J11 **Uvá, Río** ≈ E Colombia
Uvarovichi see Uvaravichy
127 N7 **Uvarovo** Tambovskaya Oblast', W Russian Federation 51°58′N 42°13′E
122 H10 **Uvat** Tyumenskaya Oblast', C Russian Federation 59°11′N 68°37′E
190 G12 **Uvea, Île** island N Wallis and Futuna
81 E21 **Uvinza** Kigoma, W Tanzania 05°08′S 30°23′E
79 O20 **Uvira** Sud-Kivu, E Dem. Rep. Congo 03°24′S 29°05′E
162 E5 **Uvs** ◈ province NW Mongolia
162 F5 **Uvs Nuur** var. Ozero Ubsu-Nur. ◎ Mongolia/Russian Federation
164 F14 **Uwa** var. Seiyo. Ehime, Shikoku, SW Japan 33°22′N 132°29′E
164 F14 **Uwajima** var. Uwazima. Ehime, Shikoku, SW Japan 33°13′N 132°32′E
80 B5 **'Uwaynāt, Jabal al** var. Jebel Uweinat. ▲ Libya/Sudan 21°51′N 25°01′E
Uwazima see Uwajima
Uweinat, Jebel see 'Uwaynāt, Jabal al
14 H14 **Uxbridge** Ontario, S Canada 44°07′N 79°07′W
Uxellodunum see Issoudun
Uxin Qi see Dabqig, N China
41 X12 **Uxmal, Ruinas** ruins Yucatán, SE Mexico
129 Q5 **Uy** ≈ Kazakhstan/Russian Federation
144 K15 **Uyaly** Kzylorda, S Kazakhstan 44°22′N 61°16′E
123 R8 **Uyandina** ≈ NE Russian Federation
162 J8 **Uyanga** var. Ongi. Övörhangay, C Mongolia 46°30′N 102°18′E
Üydzen see Manlay
Uyeda see Ueda
122 K5 **Uyedineniya, Ostrov** island N Russian Federation
77 V17 **Uyo** Akwa Ibom, S Nigeria 05°00′N 07°57′E
162 D8 **Üyönch** Hovd, W Mongolia 46°04′N 92°05′E
Uyuk see Oyyk
141 V13 **'Uyūn** SW Oman 17°19′N 53°50′E
57 K20 **Uyuni** Potosí, W Bolivia 20°27′S 66°48′W
57 J20 **Uyuni, Salar de** wetland SW Bolivia
146 I9 **Uzbekistan** off. Republic of Uzbekistan. ◆ republic C Asia
Uzbekistan, Republic of see Uzbekistan
158 D8 **Uzbel Shankou** Rus. Pereval Kyzyl-Dzhiik. pass China/Tajikistan
146 B11 **Uzboý** prev. Rus. Imeni 26 Bakinskikh Komissarov, Turkm. 26 Baku Komissarlary Adyndaky. Balkan Welaýaty, W Turkmenistan 39°24′N 54°04′E
119 J17 **Uzda** Minskaya Voblasts', C Belarus 53°29′N 27°10′E
103 N12 **Uzerche** Corrèze, C France 45°24′N 01°35′E
103 R14 **Uzès** Gard, S France 44°00′N 04°25′E
147 T10 **Uzgen** Kir. Özgön. Oshskaya Oblast', SW Kyrgyzstan 40°42′N 73°17′E
117 O3 **Uzh** ≈ N Ukraine
Uzhgorod see Uzhhorod
116 G7 **Uzhhorod** Rus. Uzhgorod; prev. Ungvár. Zakarpats'ka Oblast', W Ukraine 48°36′N 22°19′E
Uzi see Uji
112 K13 **Užice** prev. Titovo Užice. Serbia, W Serbia 43°52′N 19°51′E
Uzin see Uzyn
126 L5 **Uzlovaya** Tul'skaya Oblast', W Russian Federation 54°01′N 38°15′E
108 H7 **Uznach** Sankt Gallen, NE Switzerland 47°12′N 09°00′E
Uzunagach see Uzynagash
136 B10 **Uzunköprü** Edirne, NW Turkey 41°18′N 26°40′E
118 D11 **Užventis** Šiaulai, C Lithuania 55°49′N 22°38′E
117 P5 **Uzyn** Rus. Uzin. Kyyivs'ka Oblast', N Ukraine 49°52′N 30°27′E
145 U16 **Uzynagash** prev. Uzunagach. Almaty, SE Kazakhstan 43°08′N 76°20′E
145 N7 **Uzynkol'** prev. Lenīn, Leninskoye. Kustanay, N Kazakhstan 54°05′N 65°23′E

V

Vääksy see Asikkala
83 H23 **Vaal** ≈ C South Africa
93 M14 **Vaala** Kainuu, C Finland 64°34′N 26°49′E
93 N19 **Vaalimaa** Etelä-Karjala, SE Finland 60°34′N 27°49′E
99 M19 **Vaals** Limburg, SE Netherlands 50°46′N 06°01′E
93 J16 **Vaasa** Swe. Vasa; prev. Nikolainkaupunki. Österbotten, W Finland 63°07′N 21°39′E
98 L10 **Vaassen** Gelderland, E Netherlands 52°18′N 05°59′E
118 G11 **Vabalninkas** Panevėžys, NE Lithuania 55°59′N 24°45′E
Vabkent see Vobkent
111 J22 **Vác** Ger. Waitzen. Pest, N Hungary 47°46′N 19°08′E
61 I14 **Vacaria** Rio Grande do Sul, S Brazil 28°31′S 50°52′W
35 N7 **Vacaville** California, W USA 38°21′N 121°59′W
103 R15 **Vaccarès, Étang de** ◎ SE France
114 I11 **Vacha** var. Vŭcha. ≈ SW Bulgaria
44 L10 **Vache, Île à** island SW Haiti
173 Y16 **Vacoas** W Mauritius 20°18′S 57°29′E
32 G10 **Vader** Washington, NW USA 46°23′N 122°58′W
94 D12 **Vadheim** Sogn Og Fjordane, S Norway 61°12′N 05°48′E
154 D11 **Vadodara** prev. Baroda. Gujarāt, W India 22°19′N 73°14′E
92 M8 **Vadsø** Fin. Vesisaari. Finnmark, N Norway 70°07′N 29°47′E
95 L17 **Vadstena** Östergötland, S Sweden 58°26′N 14°55′E
108 I8 **Vaduz** ● (Liechtenstein) W Liechtenstein 47°08′N 09°32′E
Vág see Váh
125 N12 **Vaga** ≈ NW Russian Federation
94 G11 **Vågåmo** Oppland, S Norway 61°52′N 09°06′E
112 D12 **Vaganski Vrh** ▲ W Croatia 44°24′N 15°32′E
95 A19 **Vágar** Dan. Vågø. island W Faroe Islands
Vágbeszterce see Považská Bystrica
95 L19 **Vaggeryd** Jönköping, S Sweden 57°30′N 14°10′E
137 T12 **Vagharshapat** var. Ejmiadzin, Ejmiatsin, Etchmiadzin, Rus. Echmiadzin. W Armenia 40°10′N 44°17′E
95 O16 **Vagnhärad** Södermanland, C Sweden 58°57′N 17°32′E
Vågø see Vágar
104 G7 **Vagos** Aveiro, N Portugal 40°33′N 08°42′W
Vágsellye see Sal'a
92 H10 **Vågsfjorden** fjord N Norway
94 C10 **Vågsøy** island S Norway
Vágújhely see Nové Mesto nad Váhom
111 I21 **Váh** Ger. Waag, Hung. Vág. ≈ W Slovakia
93 K16 **Vähäkyrö** Österbotten, W Finland 63°04′N 22°05′E
191 X11 **Vahitahi** atoll Îles Tuamotu, E French Polynesia
Váhtjer see Gällivare
190 C9 **Vaiaku** var. Funafuti. Funafuti Atoll, SE Tuvalu 08°31′N 179°11′E
Vaidei see Vulcan
22 L4 **Vaiden** Mississippi, S USA 33°19′N 89°42′W
155 I23 **Vaigai** ≈ SE India
191 V16 **Vaihu** Easter Island, Chile, E Pacific Ocean 27°10′S 109°22′W
118 I6 **Väike Emajõgi** ≈ S Estonia
118 I4 **Väike-Maarja** Ger. Klein-Marien. Lääne-Virumaa, NE Estonia 59°07′N 26°16′E
Väike-Salatsi see Mazsalaca
37 R4 **Vail** Colorado, C USA 39°36′N 106°20′W
193 V15 **Vaina** Tongatapu, S Tonga 21°12′S 175°10′W
118 E5 **Väinameri** prev. Muhu Väin, Ger. Moon-Sund. sea E Baltic Sea
93 N18 **Vainikkala** Etelä-Karjala, SE Finland 60°54′N 28°18′E
118 D10 **Vaiņode** SW Latvia 56°25′N 21°52′E
155 H23 **Vaippār** ≈ SE India
191 W11 **Vairaatea** atoll Îles Tuamotu, C French Polynesia
191 R8 **Vairao** Tahiti, W French Polynesia 17°48′S 149°17′W
103 R14 **Vaison-la-Romaine** Vaucluse, SE France 44°15′N 05°04′E
190 G11 **Vaitupu** Île Uvea, E Wallis and Futuna 13°14′S 176°09′W
190 F7 **Vaitupu** atoll C Tuvalu
Vajdahunyad see Hunedoara
Vajdej see Vulcan
78 K12 **Vakaga** ◈ prefecture NE Central African Republic
114 H10 **Vakarel** Sofia, W Bulgaria 42°35′N 23°40′E
Vakav see Ustrem
137 O11 **Vakfıkebir** Trabzon, NE Turkey 41°03′N 39°19′E
122 J10 **Vakh** ≈ C Russian Federation
Vakhon, Qatorkŭhi see Nicholas Range
147 P14 **Vakhsh** SW Tajikistan 37°46′N 68°48′E
147 Q12 **Vakhsh** ≈ SW Tajikistan
127 P1 **Vakhtan** Nizhegorodskaya Oblast', W Russian Federation 58°00′N 46°43′E
94 C13 **Vaksdal** Hordaland, S Norway 60°29′N 05°45′E
Valachia see Wallachia
108 D11 **Valais** Ger. Wallis. ◈ canton SW Switzerland
113 M21 **Valamarës, Mali i** ▲ SE Albania 40°48′N 20°31′E
127 S2 **Valamaz** Udmurtskaya Respublika, NW Russian Federation 57°36′N 52°07′E
113 Q19 **Valandovo** SE FYR Macedonia 41°20′N 22°33′E
111 I18 **Valašské Meziříčí** Ger. Wallachisch-Meseritsch, Pol. Wałeckie Międzyrzecze. Zlínský Kraj, E Czech Republic 49°29′N 17°57′E
115 I17 **Valáxa** island Vóreies Sporádes, Greece, Aegean Sea
95 K16 **Vålberg** Värmland, C Sweden 59°24′N 13°12′E
116 H12 **Vâlcea** prev. Vîlcea. ◈ county SW Romania
Vălčedrăm see Valchedram
114 G7 **Valchedram** var. Vălčedrăm, Vŭlchedrŭm. Montana, NW Bulgaria 43°42′N 23°25′E
63 J16 **Valcheta** Río Negro, E Argentina 40°42′S 66°08′W
114 N8 **Valchi Dol** var. Vŭlchidol; prev. Kurt-Dere. Varna, E Bulgaria 43°25′N 27°33′E
15 P12 **Valcourt** Québec, SE Canada 45°28′N 72°18′W
Valdai Hills see Valdayskaya Vozvyshennost'
104 M3 **Valdavia** ≈ N Spain
124 I15 **Valday** Novgorodskaya Oblast', W Russian Federation 57°57′N 33°20′E
124 I15 **Valdayskaya Vozvyshennost'** var. Valdai Hills. hill range W Russian Federation
104 L9 **Valdecañas, Embalse de** ⊠ W Spain
118 E8 **Valdemārpils** Ger. Sassmacken. NW Latvia 57°23′N 22°36′E
95 N18 **Valdemarsvik** Östergötland, S Sweden 58°13′N 16°35′E
105 N8 **Valdemoro** Madrid, C Spain 40°12′N 03°40′W
105 O11 **Valdepeñas** Castilla-La Mancha, C Spain 38°46′N 03°24′W
104 L5 **Valderaduey** ≈ NE Spain
104 L5 **Valderas** Castilla y León, N Spain 42°05′N 05°27′W
105 T7 **Valderrobres** var. Vall-de-roures. Aragón, NE Spain 40°53′N 00°08′E
63 K17 **Valdés, Península** peninsula SE Argentina
56 C5 **Valdéz** var. Limones. Esmeraldas, NW Ecuador 01°13′N 79°00′W
39 S11 **Valdez** Alaska, USA 61°08′N 146°21′W
Valdia see Weldiya
103 U11 **Val d'Isère** Savoie, E France 45°23′N 07°03′E
63 G15 **Valdivia** Los Ríos, C Chile 39°50′S 73°13′W
Valdivia Bank see Valdivia Seamount
65 P17 **Valdivia Seamount** var. Valdivia Bank. undersea feature E Atlantic Ocean 26°15′S 06°25′E
103 N4 **Val-d'Oise** ◈ department N France
14 J8 **Val-d'Or** Québec, SE Canada 48°06′N 77°42′W
23 U8 **Valdosta** Georgia, SE USA 30°49′N 83°16′W
94 G13 **Valdres** physical region S Norway
32 L13 **Vale** Oregon, NW USA 43°59′N 117°15′W
116 F9 **Valea lui Mihai** Hung. Érmihályfalva. Bihor, NW Romania 47°31′N 22°08′E
11 N15 **Valemount** British Columbia, SW Canada 52°46′N 119°17′W
59 O17 **Valença** Bahia, E Brazil 13°22′S 39°06′W
104 F4 **Valença do Minho** Viana do Castelo, N Portugal 42°02′N 08°38′W
59 N14 **Valença do Piauí** Piauí, E Brazil 06°26′S 41°46′W
103 N8 **Valençay** Indre, C France 47°10′N 01°31′E
103 R13 **Valence** anc. Valentia, Valentia Julia, Ventia. Drôme, E France 44°56′N 04°54′E
105 S10 **Valencia** Valenciana, E Spain 39°29′N 00°24′W
54 K5 **Valencia** Carabobo, N Venezuela 10°12′N 68°02′W
105 R10 **Valencia** Cat. València. ◈ province Valenciana, E Spain
105 S10 **Valencia** ✕ Valencia, E Spain
104 I10 **Valencia de Alcántara** Extremadura, W Spain 39°25′N 07°14′W
104 L4 **Valencia de Don Juan** Castilla y León, N Spain 42°17′N 05°31′W
105 U9 **Valencia, Golfo de** var. Gulf of Valencia. gulf E Spain
Valencia, Gulf of see Valencia, Golfo de
97 A21 **Valencia Island** Ir. Dairbhre. island SW Ireland
105 R10 **Valenciana** var. Valencia, Cat. València; anc. Valentia. ◈ autonomous community NE Spain
Valencia/València see Valenciana
103 P2 **Valenciennes** Nord, N France 50°21′N 03°32′E
116 K13 **Vălenii de Munte** Prahova, SE Romania 45°11′N 26°02′E
Valentia see Valence, France
Valentia see Valenciana
Valentia Julia see Valence
103 T8 **Valentigney** Doubs, E France 47°27′N 06°49′E
28 M12 **Valentine** Nebraska, C USA 42°53′N 100°31′W
24 J10 **Valentine** Texas, SW USA 30°35′N 104°30′W
Valentine State see Oregon
106 C8 **Valenza** Piemonte, NW Italy 45°01′N 08°37′E
94 I13 **Våler** Hedmark, S Norway 60°39′N 11°52′E
54 I6 **Valera** Trujillo, NW Venezuela 09°21′N 70°38′W
192 M11 **Valerie Guyot** S Pacific Ocean 33°00′S 164°00′W
Valetta see Valletta
118 I7 **Valga** Ger. Walk, Latv. Valka. Valgamaa, S Estonia 57°48′N 26°04′E
118 I7 **Valgamaa** var. Valga Maakond. ◈ province S Estonia
Valga Maakond see Valgamaa
43 Q15 **Valiente, Península** peninsula NW Panama
103 X16 **Valinco, Golfe de** gulf Corse, France, C Mediterranean Sea
112 L12 **Valjevo** Serbia, W Serbia 44°17′N 19°54′E
Valjok see Válljohka
118 I7 **Valka** Ger. Walk. N Latvia 57°48′N 26°01′E
Valka see Valga
93 L18 **Valkeakoski** Pirkanmaa, W Finland 61°17′N 24°05′E
93 M19 **Valkeala** Kymenlaakso, S Finland 60°55′N 26°49′E
99 L18 **Valkenburg** Limburg, SE Netherlands 50°52′N 05°50′E
99 K15 **Valkenswaard** Noord-Brabant, S Netherlands 51°21′N 05°29′E
119 G15 **Valkininkai** Alytus, S Lithuania 54°22′N 24°51′E
117 U5 **Valky** Kharkivs'ka Oblast', E Ukraine 49°51′N 35°40′E
41 Y12 **Valladolid** Yucatán, SE Mexico 20°39′N 88°13′W
104 M5 **Valladolid** Castilla y León, NW Spain 41°39′N 04°45′W
104 L5 **Valladolid** ◈ province Castilla y León, N Spain
103 U15 **Vallauris** Alpes-Maritimes, SE France 43°34′N 07°03′E
Vall-de-roures see Valderrobres
Vall D'Uxó see La Vall d'Uixó
95 E16 **Valle** Aust-Agder, S Norway 59°13′N 07°33′E
105 N2 **Valle** Cantabria, N Spain 43°14′N 04°16′W
42 H8 **Valle** ◈ department S Honduras
105 N8 **Vallecas** Madrid, C Spain 40°23′N 03°37′W
37 Q8 **Vallecito Reservoir** ⊠ Colorado, C USA
106 A7 **Valle d'Aosta** Fr. Vallée d'Aoste. ◈ region NW Italy
41 O14 **Valle de Bravo** México, S Mexico 19°19′N 100°08′W
55 N5 **Valle de Guanape** Anzoátegui, N Venezuela 09°54′N 65°41′W

◆ Country
● Country Capital
◇ Dependent Territory
○ Dependent Territory Capital
◈ Administrative Regions
✕ International Airport
▲ Mountain
▲▲ Mountain Range
Volcano
≈ River
◎ Lake
⊠ Reservoir

54 M6 **Valle de La Pascua** Guárico, N Venezuela *09°15′N 66°00′W*
54 B11 **Valle del Cauca** *off.* Departamento del Valle del Cauca. ◆ *province* W Colombia
Valle del Cauca, Departamento del *see* Valle del Cauca
41 N13 **Valle de Santiago** Guanajuato, C Mexico *20°25′N 101°15′W*
40 J7 **Valle de Zaragoza** Chihuahua, N Mexico *27°25′N 105°50′W*
54 G5 **Valledupar** Cesar, N Colombia *10°31′N 73°16′W*
Vallée d'Aoste *see* Valle d'Aosta
76 G10 **Vallée de Ferlo** ⇝ NW Senegal
57 M19 **Vallegrande** Santa Cruz, C Bolivia *18°30′S 64°06′W*
41 P8 **Valle Hermoso** Tamaulipas, C Mexico *25°39′N 97°49′W*
35 N8 **Vallejo** California, W USA *38°08′N 122°16′W*
62 G8 **Vallenar** Atacama, N Chile *28°35′S 70°44′W*
95 O15 **Vallentuna** Stockholm, C Sweden *59°32′N 18°05′E*
121 P16 **Valletta** *prev.* Valetta. ● (Malta) E Malta *35°54′N 14°31′E*
27 N6 **Valley Center** Kansas, C USA *37°49′N 97°22′W*
29 Q5 **Valley City** North Dakota, N USA *46°57′N 97°58′W*
32 I15 **Valley Falls** Oregon, NW USA *42°28′N 120°16′W*
Valleyfield *see* Salaberry-de-Valleyfield
21 S4 **Valley Head** West Virginia, NE USA *38°33′N 80°01′W*
25 T8 **Valley Mills** Texas, SW USA *31°36′N 97°27′W*
75 W10 **Valley of the Kings** *ancient monument* E Egypt
29 R11 **Valley Springs** South Dakota, N USA *43°34′N 96°28′W*
20 K5 **Valley Station** Kentucky, S USA *38°06′N 85°52′W*
11 O13 **Valleyview** Alberta, W Canada *55°02′N 117°17′W*
25 T5 **Valley View** Texas, SW USA *33°27′N 97°08′W*
61 C21 **Vallimanca, Arroyo** ⇝ E Argentina
92 L9 **Válljohka** *var.* Valjok. Finnmark, N Norway *69°40′N 25°52′E*
107 M19 **Vallo della Lucania** Campania, S Italy *40°13′N 15°15′E*
108 B9 **Vallorbe** Vaud, W Switzerland *46°43′N 06°21′E*
105 V6 **Valls** Cataluña, NE Spain *41°18′N 01°15′E*
94 N11 **Vallsta** Gävleborg, C Sweden *61°30′N 16°25′E*
94 N12 **Vallvik** Gävleborg, C Sweden *61°10′N 17°15′E*
11 T17 **Val Marie** Saskatchewan, S Canada *49°15′N 107°44′W*
118 H7 **Valmiera** *Est.* Volmari, *Ger.* Wolmar. N Latvia *57°34′N 25°26′E*
105 N3 **Valnera** ▲ N Spain *43°08′N 03°39′W*
102 J3 **Valognes** Manche, N France *49°31′N 01°28′W*
Valona *see* Vlorë
Valona Bay *see* Vlorës, Gjiri i
104 G6 **Valongo** *var.* Valongo de Gaia. Porto, N Portugal *41°11′N 08°30′W*
Valongo de Gaia *see* Valongo
104 M5 **Valoria la Buena** Castilla y León, N Spain *41°48′N 04°33′W*
119 J15 **Valozhyn** *Pol.* Wołożyn, *Rus.* Volozhin. Minskaya Voblasts', C Belarus *54°05′N 26°32′E*
104 I5 **Valpaços** Vila Real, N Portugal *41°36′N 07°17′W*
62 G11 **Valparaíso** Valparaíso, C Chile *33°05′S 71°18′W*
40 L11 **Valparaíso** Zacatecas, C Mexico *22°49′N 103°28′W*
23 P8 **Valparaiso** Florida, SE USA *30°30′N 86°28′W*
31 N11 **Valparaiso** Indiana, N USA *41°28′N 87°04′W*
62 G11 **Valparaíso** *off.* Región de Valparaíso. ◆ *region* C Chile
Valparaiso, Región de *see* Valparaíso
Valpo *see* Valpovo
112 I9 **Valpovo** *Hung.* Valpó. Osijek-Baranja, E Croatia *45°40′N 18°25′E*
103 R14 **Valréas** Vaucluse, SE France *44°22′N 05°00′E*
Vals *see* Vals-Platz
154 D12 **Valsād** *prev.* Bulsar. Gujarāt, W India *20°40′N 72°55′E*
Valsbaai *see* False Bay
171 T12 **Valse Pisang, Kepulauan** *island group* E Indonesia
108 H9 **Vals-Platz** *var.* Vals. Graubünden, S Switzerland *46°39′N 09°09′E*
171 X16 **Vals, Tanjung** *headland* Papua, SE Indonesia *08°26′S 137°35′E*
93 N15 **Valtimo** Pohjois-Karjala, E Finland *63°39′N 28°49′E*
115 D17 **Váltou** ▲ C Greece
127 O12 **Valuyevka** Rostovskaya Oblast', SW Russian Federation *46°48′N 43°49′E*
126 K9 **Valuyki** Belgorodskaya Oblast', W Russian Federation *50°11′N 38°07′E*
36 L2 **Val Verda** Utah, W USA *40°51′N 111°53′W*
64 N12 **Valverde** Hierro, Islas Canarias, Spain, NE Atlantic Ocean *27°48′N 17°55′W*
104 I13 **Valverde del Camino** Andalucía, S Spain *37°35′N 06°45′W*
95 G23 **Vamdrup** Syddanmark, C Denmark *55°26′N 09°18′E*
94 L12 **Våmhus** Dalarna, C Sweden *61°07′N 14°30′E*
93 K18 **Vammala** Pirkanmaa, SW Finland *61°20′N 22°55′E*
Vámosudvarhely *see* Odorheiu Secuiesc
137 S14 **Van** Van, E Turkey *38°30′N 43°23′E*
25 V7 **Van** Texas, SW USA *32°31′N 95°38′W*
137 T14 **Van** ◆ *province* E Turkey
137 T11 **Vanadzor** *prev.* Kirovakan. N Armenia *40°49′N 44°29′E*
25 U5 **Van Alstyne** Texas, SW USA *33°25′N 96°34′W*
33 W10 **Vananda** Montana, NW USA *46°22′N 106°58′W*
116 I11 **Vânători** *Hung.* Héjjasfalva; *prev.* Vînători. Mureş, C Romania *46°14′N 24°56′E*
191 W12 **Vanavana** *atoll* Îles Tuamotu, SE French Polynesia
Vana-Vändra *see* Vändra
122 M11 **Vanavara** Krasnoyarskiy Kray, C Russian Federation *60°19′N 102°19′E*
15 Q8 **Van Bruyssel** Québec, SE Canada *47°56′N 72°08′W*
27 R10 **Van Buren** Arkansas, C USA *35°28′N 94°25′W*
19 S1 **Van Buren** Maine, NE USA *47°07′N 67°57′W*
27 W7 **Van Buren** Missouri, C USA *37°00′N 91°00′W*
19 T5 **Vanceboro** Maine, NE USA *45°31′N 67°25′W*
21 W10 **Vanceboro** North Carolina, SE USA *35°16′N 77°06′W*
21 O4 **Vanceburg** Kentucky, S USA *38°36′N 83°40′W*
45 W10 **Vance W. Amory** ✈ Nevis, Saint Kitts and Nevis *17°08′N 62°36′W*
Vanch *see* Vanj
10 L17 **Vancouver** British Columbia, SW Canada *49°13′N 123°06′W*
32 G11 **Vancouver** Washington, NW USA *45°38′N 122°39′W*
10 L17 **Vancouver** ✈ British Columbia, SW Canada *49°03′N 123°00′W*
10 K16 **Vancouver Island** *island* British Columbia, SW Canada
Vanda *see* Vantaa
171 X13 **Van Daalen** ⇝ Papua, E Indonesia
30 L15 **Vandalia** Illinois, N USA *38°57′N 89°05′W*
27 V3 **Vandalia** Missouri, C USA *39°18′N 91°29′W*
31 R13 **Vandalia** Ohio, N USA *39°53′N 84°12′W*
25 U13 **Vanderbilt** Texas, SW USA *28°45′N 96°37′W*
31 Q10 **Vandercook Lake** Michigan, N USA *42°11′N 84°23′W*
10 L14 **Vanderhoof** British Columbia, SW Canada *53°54′N 124°00′W*
18 K8 **Vanderwhacker Mountain** ▲ New York, NE USA *43°54′N 74°06′W*
181 P1 **Van Diemen Gulf** *gulf* Northern Territory, N Australia
Van Diemen's Land *see* Tasmania
118 H5 **Vändra** *Ger.* Fennern; *prev.* Vana-Vändra. Pärnumaa, SW Estonia *58°39′N 25°00′E*
Vandsburg *see* Więcbork
34 L4 **Van Duzen River** ⇝ California, W USA
118 F13 **Vandžiogala** Kaunas, C Lithuania *55°07′N 23°55′E*
41 N10 **Vanegas** San Luis Potosí, C Mexico *23°53′N 100°55′W*
Vaner, Lake *see* Vänern
95 K17 **Vänern** *Eng.* Lake Vaner; *prev.* Lake Vener. ◎ S Sweden
95 J18 **Vänersborg** Västra Götaland, S Sweden *58°16′N 12°22′E*
94 F12 **Vang** Oppland, S Norway *61°07′N 08°34′E*
172 I7 **Vangaindrano** Fianarantsoa, SE Madagascar *23°21′S 47°35′E*
137 S14 **Van Gölü** *Eng.* Lake Van; *anc.* Thospitis. *salt lake* E Turkey
186 L9 **Vangunu** *island* New Georgia Islands, NW Solomon Islands
24 J9 **Van Horn** Texas, SW USA *31°03′N 104°51′W*
187 Q11 **Vanikolo** *var.* Vanikoro. *island* Santa Cruz Islands, E Solomon Islands
Vanikoro *see* Vanikolo
186 A5 **Vanimo** West Sepik, NW Papua New Guinea *02°40′S 141°17′E*
123 T13 **Vanino** Khabarovskiy Kray, SE Russian Federation *49°10′N 140°18′E*
155 G19 **Vānīvilāsa Sāgara** ◎ SW India
147 S13 **Vanj** *Rus.* Vanch. S Tajikistan *38°22′N 71°27′E*
116 G14 **Vânju Mare** *prev.* Vînju Mare. Mehedinţi, SW Romania *44°25′N 22°52′E*
15 N12 **Vankleek Hill** Ontario, SE Canada *45°32′N 74°39′W*
Van, Lake *see* Van Gölü
93 I16 **Vännäs** Västerbotten, N Sweden *63°54′N 19°43′E*
93 I15 **Vännäsby** Västerbotten, N Sweden *63°55′N 19°53′E*
102 H7 **Vannes** *anc.* Dariorigum. Morbihan, NW France *47°40′N 02°45′W*
92 I8 **Vannøya** *island* N Norway
103 T12 **Vanoise, Massif de la** ▲ E France
83 E24 **Vanrhynsdorp** Western Cape, SW South Africa *31°36′S 18°45′E*
21 P7 **Vansant** Virginia, NE USA *37°13′N 82°03′W*
94 L13 **Vansbro** Dalarna, C Sweden *60°32′N 14°15′E*
95 D18 **Vanse** Vest-Agder, S Norway *58°04′N 06°40′E*
9 P7 **Vansittart Island** *island* Nunavut, NE Canada
93 M20 **Vantaa** *Swe.* Vanda. Uusimaa, S Finland *60°18′N 25°01′E*
93 L19 **Vantaa** ✈ (Helsinki) Uusimaa, S Finland *60°18′N 25°01′E*
32 J9 **Vantage** Washington, NW USA *46°55′N 119°55′W*
187 Z14 **Vanua Balavu** *prev.* Vanua Mbalavu. *island* Lau Group, E Fiji
187 R12 **Vanua Lava** *island* Banks Islands, N Vanuatu
187 Y13 **Vanua Levu** *island* N Fiji
Vanua Mbalavu *see* Vanua Balavu
187 R12 **Vanuatu** *off.* Republic of Vanuatu; *prev.* New Hebrides. ◆ *republic* SW Pacific Ocean
175 P8 **Vanuatu** *island group* SW Pacific Ocean
Vanuatu, Republic of *see* Vanuatu
31 Q12 **Van Wert** Ohio, N USA *40°52′N 84°34′W*
187 Q17 **Vao** Province Sud, S New Caledonia *22°40′S 167°29′E*
Vapincum *see* Gap
117 N7 **Vapnyarka** Vinnyts'ka Oblast', C Ukraine *48°31′N 28°44′E*
103 T15 **Var** ◆ *department* SE France
103 U14 **Var** ⇝ SE France
95 J18 **Vara** Västra Götaland, S Sweden *58°16′N 12°57′E*
Varadinska Županija *see* Varaždin
118 J10 **Varakļāni** C Latvia *56°36′N 26°40′E*
106 C7 **Varallo** Piemonte, NE Italy *45°51′N 08°16′E*
143 O5 **Varāmīn** *var.* Veramin. Tehrān, N Iran *35°19′N 51°40′E*
153 N14 **Vārānasi** *prev.* Banaras, Benares, *hist.* Kasi. Uttar Pradesh, N India *25°20′N 83°E*
125 T3 **Varandey** Nenetskiy Avtonomnyy Okrug, NW Russian Federation *68°48′N 57°54′E*
92 M8 **Varangerbotn** *Lapp.* Vuonnabahta. Finnmark, N Norway *70°09′N 28°28′E*
92 M8 **Varangerfjorden** *Lapp.* Várjjatvuotna. *fjord* N Norway
92 M8 **Varangerhalvøya** *Lapp.* Várnjárga. *peninsula* N Norway
Varannó *see* Vranov nad Topl'ou
107 M15 **Varano, Lago di** ◎ SE Italy
118 J13 **Varapayeva** *Rus.* Voropayevo. Vitsyebskaya Voblasts', NW Belarus *55°09′N 27°13′E*
Varasd *see* Varaždin
112 E7 **Varaždin** *Ger.* Warasdin, *Hung.* Varasd. Varaždin, N Croatia *46°18′N 16°21′E*
112 E7 **Varaždin** *off.* Varadinska Županija. ◆ *province* N Croatia
106 C10 **Varazze** Liguria, NW Italy *44°21′N 08°35′E*
95 J20 **Varberg** Halland, S Sweden *57°06′N 12°15′E*
114 J11 **Varbitsa** *var.* Vŭrbitsa; *prev.* Filevo. Haskovo, S Bulgaria *42°02′N 25°25′E*
114 J12 **Varbitsa** ⇝ S Bulgaria
113 Q19 **Vardar** *Gk.* Axiós. ⇝ FYR Macedonia/Greece *see also* Axiós
Vardar *see* Axiós
95 F23 **Varde** Syddtjylland, W Denmark *55°38′N 08°31′E*
137 V12 **Vardenis** E Armenia *40°11′N 45°43′E*
92 N8 **Vardø** *Fin.* Vuoreija. Finnmark, N Norway *70°22′N 31°06′E*
115 E18 **Vardoúsia** ▲ C Greece
Vareia *see* Logroño
100 G10 **Varel** Niedersachsen, NW Germany *53°24′N 08°07′E*
119 G15 **Varėna** *Pol.* Orany. Alytus, S Lithuania *54°13′N 24°35′E*
15 O12 **Varennes** Québec, SE Canada *45°42′N 73°25′W*
103 P10 **Varennes-sur-Allier** Allier, C France *46°17′N 03°24′E*
112 I12 **Vareš** Federacija Bosne I Hercegovine, E Bosnia and Herzegovina *44°12′N 18°19′E*
106 D7 **Varese** Lombardia, N Italy *45°49′N 08°50′E*
116 J12 **Vârful Moldoveanu** *var.* Moldoveanul; *prev.* Virful Moldoveanu. ▲ C Romania *45°35′N 24°48′E*
Varganzi *see* Warganza
95 J18 **Vårgårda** Västra Götaland, S Sweden *58°00′N 12°49′E*
54 J4 **Vargas** *off.* Estado Vargas. ◆ *state* N Venezuela
95 J18 **Vargön** Västra Götaland, S Sweden *58°21′N 12°22′E*
95 C17 **Varhaug** Rogaland, S Norway *58°37′N 05°39′E*
Várjjatvuotna *see* Varangerfjorden
93 N17 **Varkaus** Pohjois-Savo, C Finland *62°20′N 27°50′E*
92 J2 **Varmahlíð** Norðurland Vestra, N Iceland *65°32′N 19°33′W*
95 J15 **Värmland** ◆ *county* C Sweden
95 K16 **Värmlandsnäs** *peninsula* S Sweden
114 N8 **Varna** *prev.* Stalin; *anc.* Odessus. Varna, E Bulgaria *43°14′N 27°56′E*
114 N8 **Varna** ◆ *province* E Bulgaria
114 N8 **Varna** ✈ Varna, E Bulgaria *43°16′N 27°52′E*
95 L20 **Värnamo** Jönköping, S Sweden *57°11′N 14°03′E*
114 N8 **Varnenski Zaliv** *prev.* Stalinski Zaliv. *bay* E Bulgaria
114 N8 **Varnensko Ezero** *estuary* E Bulgaria
118 D11 **Varniai** Telšiai, W Lithuania *55°45′N 22°22′E*
Várnjárga *see* Varangerhalvøya
Varnoús *see* Baba
111 D14 **Varnsdorf** *Ger.* Warnsdorf. Ústecký Kraj, NW Czech Republic *50°57′N 14°35′E*
111 I23 **Várpalota** Veszprém, W Hungary *47°12′N 18°08′E*
Varshava *see* Warszawa
114 G8 **Varshets** *var.* Vŭrshets. Montana, NW Bulgaria *43°14′N 23°20′E*
93 K20 **Varsinais-Suomi** *Swe.* Egentliga Finland. ◆ *region* W Finland
118 K6 **Värska** Põlvamaa, SE Estonia *57°58′N 27°37′E*
98 N12 **Varsseveld** Gelderland, E Netherlands *51°55′N 06°28′E*
115 D19 **Vartholomió** *prev.* Vartholomión. Dytikí Elláda, S Greece *37°52′N 21°12′E*
Vartholomión *see* Vartholomió
137 Q14 **Varto** Muş, E Turkey *39°10′N 41°28′E*
95 K18 **Vartofta** Västra Götaland, S Sweden *58°06′N 13°40′E*
93 O17 **Värtsilä** Pohjois-Karjala, E Finland *62°10′N 30°35′E*
Värtsilä *see* Vyartsilya
117 R4 **Varva** Chernihivs'ka Oblast', NE Ukraine *50°31′N 32°43′E*
59 H18 **Várzea Grande** Mato Grosso, SW Brazil *15°39′S 56°08′W*
106 D9 **Varzi** Lombardia, N Italy *44°51′N 09°13′E*
Varzimanor Ayni *see* Ayní
124 K5 **Varzuga** ⇝ NW Russian Federation
103 P8 **Varzy** Nièvre, C France *47°22′N 03°22′E*
111 G23 **Vas** *off.* Vas Megye. ◆ *county* W Hungary
Vasa *see* Vaasa
190 A9 **Vasafua** *island* Funafuti Atoll, C Tuvalu
111 O21 **Vásárosnamény** Szabolcs-Szatmár-Bereg, E Hungary *48°10′N 22°18′E*
104 H13 **Vascão, Ribeira de** ⇝ S Portugal
116 G10 **Vaşcău** *Hung.* Vaskoh. Bihor, NE Romania *46°28′N 22°30′E*
Vascongadas, Provincias *see* País Vasco
Vashess Bay *see* Vaskess Bay
125 O8 **Vashka** ⇝ NW Russian Federation
Vāsht *see* Khāsh
Vasilevichi *see* Vasilyevichy
115 G14 **Vasiliká** Kentrikí Makedonía, NE Greece *40°28′N 23°08′E*
115 C18 **Vasilikí** Lefkáda, Iónia Nisiá, Greece, C Mediterranean Sea *38°36′N 20°37′E*
115 K25 **Vasilikí** Kríti, Greece, E Mediterranean Sea *35°04′N 25°49′E*
119 G16 **Vasilishki** *Pol.* Wasiliszki. Hrodzyenskaya Voblasts', W Belarus *53°47′N 24°51′E*
Vasil Kolarov *see* Pamporovo
Vasil'kov *see* Vasyl'kiv
119 N19 **Vasilyevichy** *Rus.* Vasilevichi. Homyel'skaya Voblasts', SE Belarus *52°15′N 29°50′E*
191 Y3 **Vaskess Bay** *var.* Vashess Bay. *bay* Kiritimati, E Kiribati
Vaskoh *see* Vaşcău
Vaskohsziklás *see* Ştei
116 M10 **Vaslui** Vaslui, C Romania *46°38′N 27°44′E*
116 L11 **Vaslui** ◆ *county* NE Romania
Vas Megye *see* Vas
31 R8 **Vassar** Michigan, N USA *43°22′N 83°34′W*
95 E15 **Vassdalsegga** ▲ S Norway *59°47′N 07°07′E*
60 P9 **Vassouras** Rio de Janeiro, SE Brazil *22°24′S 43°38′W*
95 N15 **Västerås** Västmanland, C Sweden *59°37′N 16°33′E*
93 G15 **Västerbotten** ◆ *county* N Sweden
94 K12 **Västerdalälven** ⇝ C Sweden
95 O16 **Västerhaninge** Stockholm, C Sweden *59°07′N 18°06′E*
94 M10 **Västernorrland** ◆ *county* C Sweden
95 N19 **Västervik** Kalmar, S Sweden *57°44′N 16°40′E*
95 M15 **Västmanland** ◆ *county* C Sweden
107 L15 **Vasto** *anc.* Histonium. Abruzzo, C Italy *42°07′N 14°43′E*
95 J19 **Västra Götaland** ◆ *county* S Sweden
95 J16 **Västra Silen** ◎ S Sweden
111 G23 **Vasvár** *Ger.* Eisenburg. Vas, W Hungary *47°03′N 16°48′E*
117 U9 **Vasylivka** Zaporiz'ka Oblast', SE Ukraine *47°26′N 35°18′E*
117 O5 **Vasyl'kiv** *var.* Vasil'kov. Kyyivs'ka Oblast', N Ukraine *50°12′N 30°18′E*
117 V8 **Vasyl'kivka** Dnipropetrovs'ka Oblast', E Ukraine *48°12′N 36°00′E*
122 I11 **Vasyugan** ⇝ C Russian Federation
103 N8 **Vatan** Indre, C France *47°06′N 01°49′E*
Vaté *see* Efate
115 C18 **Vathy** *prev.* Itháki. Itháki, Iónia Nisiá, Greece, C Mediterranean Sea *38°22′N 20°43′E*
107 G15 **Vatican City** *off.* Vatican City. ◆ *papal state* S Europe
Vatican City *see* Vatican City
107 M22 **Vaticano, Capo** *headland* S Italy *38°37′N 15°49′E*
92 K3 **Vatnajökull** *glacier* SE Iceland
95 P15 **Vätö** Stockholm, C Sweden *59°48′N 18°55′E*
187 Z16 **Vatoa** *island* Lau Group, SE Fiji
172 J5 **Vatomandry** Toamasina, E Madagascar *19°20′S 48°58′E*
116 J9 **Vatra Dornei** *Ger.* Dorna Watra. Suceava, NE Romania *47°20′N 25°21′E*
116 J9 **Vatra Moldoviţei** Suceava, NE Romania *47°37′N 25°36′E*
Vatter, Lake *see* Vättern
95 L18 **Vättern** *Eng.* Lake Vatter; *prev.* Lake Vetter. ◎ S Sweden
187 X5 **Vatulele** *island* SW Fiji
117 P7 **Vatutine** Cherkas'ka Oblast', C Ukraine *49°01′N 31°04′E*
187 W15 **Vatu Vara** *island* Lau Group, E Fiji
103 R14 **Vaucluse** ◆ *department* SE France
103 S5 **Vaucouleurs** Meuse, NE France *48°37′N 05°38′E*
108 B9 **Vaud** *Ger.* Waadt. ◆ *canton* SW Switzerland
15 N12 **Vaudreuil** Québec, SE Canada *45°24′N 74°01′W*
37 T12 **Vaughn** New Mexico, SW USA *34°36′N 105°12′W*
54 I14 **Vaupés** *off.* Comisaría del Vaupés. ◆ *province* SE Colombia
Vaupés, Comisaría del *see* Vaupés
54 J13 **Vaupés, Río** *var.* Rio Uaupés. ⇝ Brazil/Colombia *see also* Uaupés, Rio
Vaupés, Río *see* Uaupés, Rio
103 Q15 **Vauvert** Gard, S France *43°42′N 04°16′E*
11 R17 **Vauxhall** Alberta, SW Canada *50°05′N 112°09′W*
99 K23 **Vaux-sur-Sûre** Luxembourg, SE Belgium *49°55′N 05°34′E*
172 J4 **Vavatenina** Toamasina, E Madagascar *17°25′S 49°11′E*
193 Y14 **Vava'u Group** *island group* N Tonga
76 M16 **Vavoua** W Ivory Coast *07°23′N 06°29′W*
127 S2 **Vavozh** Udmurtskaya Respublika, NW Russian Federation *56°48′N 51°53′E*
155 K23 **Vavuniya** Northern Province, N Sri Lanka *08°45′N 80°30′E*
119 G17 **Vawkavysk** *Pol.* Wołkowysk, *Rus.* Volkovysk. Hrodzyenskaya Voblasts', W Belarus *53°10′N 24°28′E*
119 F17 **Vawkavyskaye Wzvyshsha** *Rus.* Volkovyskiye Vysoty. *hill range* W Belarus
95 P15 **Vaxholm** Stockholm, C Sweden *59°25′N 18°21′E*
95 L21 **Växjö** *var.* Vexiö. Kronoberg, S Sweden *56°52′N 14°50′E*
125 T1 **Vaygach, Ostrov** *island* NW Russian Federation
137 V13 **Vayk'** *prev.* Azizbekov. SE Armenia *39°41′N 45°28′E*
Vazáš *see* Vittangi
125 P8 **Vazhgort** *prev.* Chasovo. Respublika Komi, NW Russian Federation *64°06′N 46°44′E*
45 V10 **V. C. Bird** ✈ (St. John's) Antigua, Antigua and Barbuda *17°07′N 61°49′W*
167 R13 **Veal Renh** *prev.* Phumĭ Veal Renh. Kâmpôt, SW Cambodia *10°43′N 103°49′E*
29 Q7 **Veblen** South Dakota, N USA *45°50′N 97°17′W*
98 N9 **Vecht** *Ger.* Vechte. ⇝ Germany/Netherlands *see also* Vechte
Vecht *see* Vechte
100 G12 **Vechta** Niedersachsen, NW Germany *52°44′N 08°16′E*
100 E12 **Vechte** *Dut.* Vecht. ⇝ Germany/Netherlands *see also* Vecht
Vechte *see* Vecht
118 I8 **Vecpiebalga** C Latvia *57°03′N 25°47′E*
118 G9 **Vecumnieki** C Latvia *56°36′N 24°30′E*
95 C16 **Vedavågen** Rogaland, S Norway *59°18′N 5°13′E*
Vedāvati *see* Hagari
95 J20 **Veddige** Halland, S Sweden *57°16′N 12°19′E*
116 J15 **Vedea** ⇝ S Romania
127 P16 **Vedeno** Chechenskaya Respublika, SW Russian Federation *42°57′N 46°02′E*
95 C16 **Vedvågen** Rogaland, S Norway *59°18′N 05°13′E*
98 O6 **Veendam** Groningen, NE Netherlands *53°05′N 06°53′E*
98 K12 **Veenendaal** Utrecht, C Netherlands *52°03′N 05°33′E*
99 E14 **Veere** Zeeland, SW Netherlands *51°33′N 03°40′E*
24 M2 **Vega** Texas, SW USA *35°14′N 102°26′W*
92 E13 **Vega** *island* C Norway
45 T5 **Vega Baja** C Puerto Rico *18°27′N 66°23′W*
38 D17 **Vega Point** *headland* Kiska Island, Alaska, USA *51°49′N 177°19′E*
95 F17 **Vegår** ◎ S Norway
99 K14 **Veghel** Noord-Brabant, S Netherlands *51°37′N 05°33′E*
Veglia *see* Krk
114 E13 **Vegorítida, Límni** *var.* Límni Vegorítis. ◎ N Greece
Vegorítis, Límni *see* Vegorítida, Límni
11 Q14 **Vegreville** Alberta, SW Canada *53°30′N 112°02′W*
95 K21 **Veinge** Halland, S Sweden *56°33′N 13°04′E*
61 B21 **Veinticinco de Mayo** *var.* 25 de Mayo. Buenos Aires, E Argentina *35°27′S 60°11′W*
63 I14 **Veinticinco de Mayo** La Pampa, C Argentina *37°45′S 67°40′W*
119 F15 **Veisiejai** Alytus, S Lithuania *54°06′N 23°42′E*
95 F23 **Vejen** Syddtjylland, W Denmark *55°29′N 09°13′E*
104 K16 **Vejer de la Frontera** Andalucía, S Spain *36°15′N 05°58′W*
95 G23 **Vejle** Syddanmark, C Denmark *55°43′N 09°33′E*
114 M7 **Vekilski** Shumen, NE Bulgaria *43°33′N 27°19′E*
54 G3 **Vela, Cabo de la** *headland* NE Colombia *12°14′N 72°13′W*
Vela Goa *see* Goa
113 F15 **Vela Luka** Dubrovnik-Neretva, S Croatia *42°57′N 16°43′E*
61 G19 **Velázquez** Rocha, E Uruguay *34°05′S 54°16′W*
101 E15 **Velbert** Nordrhein-Westfalen, W Germany *51°22′N 07°03′E*
109 S9 **Velden** Kärnten, S Austria *46°37′N 13°59′E*
Veldes *see* Bled
99 K15 **Veldhoven** Noord-Brabant, S Netherlands *51°24′N 05°24′E*
112 C11 **Velebit** ▲ C Croatia
114 N11 **Veleka** ⇝ SE Bulgaria
109 V10 **Velenje** *Ger.* Wöllan. N Slovenia *46°22′N 15°07′E*
190 E12 **Vele, Pointe** *headland* Île Futuna, S Wallis and Futuna
113 O18 **Veles** *Turk.* Köprülü. C FYR Macedonia *41°43′N 21°49′E*
113 M20 **Veleshta** SW FYR Macedonia *41°16′N 20°37′E*
115 F16 **Velestíno** *prev.* Velestínon. Thessalía, C Greece *39°23′N 22°45′E*
Velestínon *see* Velestíno
Velevshchina *see* Vyelyewshchyna
54 F9 **Vélez** Santander, C Colombia *06°02′N 73°43′W*
105 Q13 **Vélez Blanco** Andalucía, S Spain *37°43′N 02°07′W*
104 M17 **Vélez de la Gomera, Peñon de** *island group* S Spain
105 N15 **Vélez-Málaga** Andalucía, S Spain *36°47′N 04°06′W*
105 Q13 **Vélez Rubio** Andalucía, S Spain *37°39′N 02°04′W*
Velha Goa *see* Goa
Velho *see* Porto Velho
112 E8 **Velika Gorica** Zagreb, N Croatia *45°43′N 16°03′E*
112 C9 **Velika Kapela** ▲ NW Croatia
Velika Kikinda *see* Kikinda
112 D10 **Velika Kladuša** Federacija Bosne I Hercegovine, NW Bosnia and Herzegovina *45°10′N 15°48′E*
112 N11 **Velika Morava** *var.* Glavn'a Morava, Morava, *Ger.* Grosse Morava. ⇝ C Serbia
112 N12 **Velika Plana** Serbia, C Serbia *44°20′N 21°01′E*
109 U10 **Velika Raduha** ▲ N Slovenia *46°24′N 14°46′E*
123 V7 **Velikaya** ⇝ NE Russian Federation
124 F15 **Velikaya** ⇝ W Russian Federation
Velikaya Berestovitsa *see* Vyalikaya Byerastavitsa
Velikaya Lepetikha *see* Velyka Lepetykha
Veliki Bečkerek *see* Zrenjanin
112 P12 **Veliki Krš** *var.* Stol. ▲ E Serbia *44°10′N 22°09′E*
114 L8 **Veliki Preslav** *prev.* Preslav. Shumen, NE Bulgaria *43°09′N 26°50′E*
112 B9 **Veliki Risnjak** ▲ NW Croatia *45°30′N 14°31′E*
109 T13 **Veliki Snežnik** *Ger.* Schneeberg, *It.* Monte Nevoso. ▲ SW Slovenia *45°36′N 14°25′E*
112 J13 **Veliki Stolac** ▲ E Bosnia and Herzegovina *43°55′N 19°15′E*
Velikiy Bor *see* Vyaliki Bor
124 G16 **Velikiye Luki** Pskovskaya Oblast', W Russian Federation *56°20′N 30°27′E*
124 H14 **Velikiy Novgorod** *prev.* Novgorod. Novgorodskaya Oblast', W Russian Federation *58°32′N 31°15′E*
125 P12 **Velikiy Ustyug** Vologodskaya Oblast', NW Russian Federation *60°46′N 46°18′E*
112 N11 **Veliko Gradište** Serbia, NE Serbia *44°46′N 21°28′E*
155 I18 **Velikonda Range** ▲ SE India
114 K9 **Veliko Tarnovo** *prev.* Tirnovo, Trnovo, Tŭrnovo, *var.* Veliko Tărnovo, Veliko Tŭrnovo. Veliko Tarnovo, N Bulgaria *43°05′N 25°40′E*
114 K8 **Veliko Tarnovo** *var.* Veliko Tŭrnovo. ◆ *province* N Bulgaria
Veliko Tărnovo *see* Veliko Tarnovo
Veliko Tŭrnovo *see* Veliko Tarnovo
Veliko Tŭrnovo *see* Veliko Tarnovo
Velikovec *see* Völkermarkt
125 R5 **Velikovisochnoye** Nenetskiy Avtonomnyy Okrug, NW Russian Federation *67°13′N 52°00′E*
76 H12 **Vélingara** C Senegal *15°00′N 14°39′W*
76 H11 **Vélingara** S Senegal *13°12′N 14°05′W*
114 H11 **Velingrad** Pazardzhik, C Bulgaria *42°01′N 24°00′E*
126 H3 **Velizh** Smolenskaya Oblast', W Russian Federation *55°30′N 31°06′E*
111 F16 **Velká Deštná** *var.* Deštná, Grosskoppe, *Ger.* Deschnaer Koppe. ▲ NE Czech Republic *50°18′N 16°25′E*
111 F18 **Velké Meziříčí** *Ger.* Grossmeseritsch. Vysočina, C Czech Republic *49°22′N 16°02′E*
92 N1 **Velkomstpynten** *headland* NW Svalbard *79°51′N 11°37′E*
111 K21 **Veľký Krtíš** Banskobystrický Kraj, C Slovakia *48°13′N 19°21′E*
186 J8 **Vella Lavella** *var.* Mbilua. *island* New Georgia Islands, NW Solomon Islands
107 I15 **Velletri** Lazio, C Italy *41°41′N 12°47′E*
95 K23 **Vellinge** Skåne, S Sweden *55°29′N 13°00′E*
155 I19 **Vellore** Tamil Nādu, SE India *12°56′N 79°09′E*
Velobriga *see* Viana do Castelo
115 G21 **Velopoúla** *island* S Greece
98 M12 **Velp** Gelderland, SE Netherlands *52°00′N 05°59′E*
Velsen *see* Velsen-Noord
98 H9 **Velsen-Noord** *var.* Velsen. Noord-Holland, W Netherlands *52°27′N 04°40′E*
125 N12 **Vel'sk** *var.* Velsk. Arkhangel'skaya Oblast', NW Russian Federation *61°03′N 42°01′E*
Velsuna *see* Orvieto
98 K10 **Veluwemeer** *lake channel* C Netherlands
28 M3 **Velva** North Dakota, N USA *48°03′N 100°55′W*
Velvendos/Velvendós *see* Velventós
115 E14 **Velventós** *var.* Velvendos, Velvendós. Dytikí Makedonía, N Greece *40°15′N 22°04′E*
117 S5 **Velyka Bahachka** Poltavs'ka Oblast', C Ukraine *49°46′N 33°44′E*
117 S9 **Velyka Lepetykha** *Rus.* Velikaya Lepetikha. Khersons'ka Oblast', S Ukraine *47°09′N 33°59′E*
117 O10 **Velyka Mykhaylivka** Odes'ka Oblast', SW Ukraine *47°07′N 29°49′E*
117 W8 **Velyka Novosilka** Donets'ka Oblast', E Ukraine *47°49′N 36°49′E*
117 S9 **Velyka Oleksandrivka** Khersons'ka Oblast', S Ukraine *47°17′N 33°16′E*
117 T4 **Velyka Pysarivka** Sums'ka Oblast', NE Ukraine *50°25′N 35°28′E*
116 G6 **Velykyy Bereznyy** Zakarpats'ka Oblast', W Ukraine *48°54′N 22°27′E*
117 W4 **Velykyy Burluk** Kharkivs'ka Oblast', E Ukraine *50°04′N 37°25′E*
Velykyy Tokmak *see* Tokmak
173 P7 **Vema Fracture Zone** *tectonic feature* W Indian Ocean
65 P18 **Vema Seamount** *undersea feature* SW Indian Ocean *31°38′S 08°19′E*
93 F17 **Vemdalen** Jämtland, C Sweden *62°26′N 13°50′E*
95 N19 **Vena** Kalmar, S Sweden *57°31′N 16°00′E*
41 N11 **Venado** San Luis Potosí, C Mexico *22°56′N 101°05′W*
62 L11 **Venado Tuerto** Entre Ríos, E Argentina *33°45′S 61°56′W*
61 A19 **Venado Tuerto** Santa Fe, C Argentina *33°46′S 61°57′W*
107 K16 **Venafro** Molise, C Italy *41°28′N 14°03′E*
55 Q9 **Venamo, Cerro** ▲ E Venezuela *05°56′N 61°25′W*
106 B8 **Venaria** Piemonte, NW Italy *45°09′N 07°40′E*
103 U15 **Vence** Alpes-Maritimes, SE France *43°45′N 07°07′E*
104 H5 **Venda Nova** Vila Real, N Portugal *41°40′N 07°58′W*
104 G11 **Vendas Novas** Évora, S Portugal *38°41′N 08°27′W*
102 J9 **Vendée** ◆ *department* NW France
103 Q6 **Vendeuvre-sur-Barse** Aube, NE France *48°08′N 04°17′E*
102 M7 **Vendôme** Loir-et-Cher, C France *47°48′N 01°04′E*
Venedig *see* Venezia
Vener, Lake *see* Vänern
106 I8 **Veneta, Laguna** *lagoon* NE Italy
Venetia *see* Venezia
39 S7 **Venetie** Alaska, USA *67°00′N 146°25′W*
106 H8 **Veneto** *var.* Venezia Euganea. ◆ *region* NE Italy
114 M7 **Venets** Shumen, NE Bulgaria *43°33′N 26°56′E*
126 L5 **Venev** Tul'skaya Oblast', W Russian Federation *54°18′N 38°16′E*
106 I8 **Venezia** *Eng.* Venice, *Fr.* Venise, *Ger.* Venedig; *anc.* Venetia. Veneto, NE Italy *45°26′N 12°20′E*
Venezia Euganea *see* Veneto
Venezia, Golfo di *see* Venice, Gulf of
Venezia Tridentina *see* Trentino-Alto Adige
54 K8 **Venezuela** *off.* Republic of Venezuela; *prev.* Estados Unidos de Venezuela, United States of Venezuela. ◆ *republic* N South America
Venezuela, Cordillera de *see* Costa, Cordillera de la
Venezuela, Estados Unidos de *see* Venezuela
54 I4 **Venezuela, Golfo de** *Eng.* Gulf of Maracaibo, Gulf of Venezuela. *gulf* NW Venezuela
Venezuela, Gulf of *see* Venezuela, Golfo de
64 F11 **Venezuelan Basin** *undersea feature* E Caribbean Sea
Venezuela, Republic of *see* Venezuela
Venezuela, United States of *see* Venezuela
155 D16 **Vengurla** Mahārāshtra, W India *15°55′N 73°39′E*
39 O15 **Veniaminof, Mount** ▲ Alaska, USA *56°12′N 159°24′W*
23 V14 **Venice** Florida, SE USA *27°06′N 82°27′W*
22 L10 **Venice** Louisiana, S USA *29°15′N 89°20′W*
Venice *see* Venezia
106 J8 **Venice, Gulf of** *It.* Golfo di Venezia, *Slvn.* Beneški Zaliv. *gulf* N Adriatic Sea
Venise *see* Venezia
94 K13 **Venjan** Dalarna, C Sweden *60°58′N 13°55′E*
94 K13 **Venjanssjön** ◎ C Sweden
155 J18 **Venkatagiri** Andhra Pradesh, E India *14°00′N 79°39′E*
99 M15 **Venlo** *prev.* Venloo. Limburg, SE Netherlands *51°22′N 06°11′E*
Venloo *see* Venlo
95 E18 **Vennesla** Vest-Agder, S Norway *58°15′N 08°00′E*
107 M17 **Venosa** *anc.* Venusia. Basilicata, S Italy *40°57′N 15°49′E*
Venoste, Alpi *see* Ötztaler Alpen
Venraij *see* Venray
99 M14 **Venray** *var.* Venraij. Limburg, SE Netherlands *51°32′N 05°59′E*
118 C8 **Venta** *Ger.* Windau. ⇝ Latvia/Lithuania
Venta Belgarum *see* Winchester
40 G9 **Ventana, Punta Arena de la** *var.* Punta de la Ventana. *headland* NW Mexico *24°03′N 109°49′W*
Ventana, Punta de la *see* Ventana, Punta Arena de la
61 B23 **Ventana, Sierra de la** *hill range* E Argentina
Ventia *see* Valence
191 S11 **Vent, Îles du** *var.* Windward Islands. *island group* Archipel de la Société, W French Polynesia
191 R10 **Vent, Îles Sous le** *var.* Leeward Islands. *island group* Archipel de la Société, W French Polynesia
106 B11 **Ventimiglia** Liguria, NW Italy *43°47′N 07°37′E*
97 M24 **Ventnor** S England, United Kingdom *50°36′N 01°11′W*
18 J17 **Ventnor City** New Jersey, NE USA *39°19′N 74°27′W*
103 S14 **Ventoux, Mont** ▲ SE France *44°12′N 05°21′E*
118 C8 **Ventspils** *Ger.* Windau. NW Latvia *57°22′N 21°34′E*
54 M10 **Ventuari, Río** ⇝ S Venezuela
35 R15 **Ventura** California, W USA *34°15′N 119°18′W*
182 F8 **Venus Bay** South Australia *33°15′S 134°42′E*
Venusia *see* Venosa
191 P7 **Vénus, Pointe** *var.* Pointe Tataaihoa. *headland* Tahiti, W French Polynesia *17°28′S 149°29′W*
41 V16 **Venustiano Carranza** Chiapas, SE Mexico *16°21′N 92°33′W*
41 N7 **Venustiano Carranza, Presa** ▣ NE Mexico
61 B15 **Vera** Santa Fe, C Argentina *29°28′S 60°10′W*
105 Q14 **Vera** Andalucía, S Spain *37°15′N 01°51′W*
63 K18 **Vera, Bahía** *bay* E Argentina
41 R14 **Veracruz** *var.* Veracruz Llave. Veracruz-Llave, E Mexico *19°10′N 96°09′W*
Veracruz *see* Veracruz-Llave
41 Q13 **Veracruz-Llave** *var.* Veracruz. ◆ *state* E Mexico
Veracruz Llave *see* Veracruz
43 Q16 **Veraguas** *off.* Provincia de Veraguas. ◆ *province* W Panama
Veraguas, Provincia de *see* Veraguas
Veramin *see* Varāmīn
154 B12 **Verāval** Gujarāt, W India *20°54′N 70°22′E*
106 C6 **Verbania** Piemonte, NW Italy *45°56′N 08°34′E*
107 N20 **Verbicaro** Calabria, SW Italy *39°44′N 15°51′E*
108 D11 **Verbier** Valais, SW Switzerland *46°06′N 07°14′E*
Vercellae *see* Vercelli
106 C8 **Vercelli** *anc.* Vercellae. Piemonte, NW Italy *45°19′N 08°25′E*
103 S13 **Vercors** *physical region* E France
Verdal *see* Verdalsøra
93 E16 **Verdalsøra** *var.* Verdal. Nord-Trøndelag, C Norway *63°47′N 11°27′E*
Verde, Cabo *see* Cape Verde
44 J5 **Verde, Cape** *headland* Long Island, C The Bahamas *22°51′N 75°50′W*

◆ Country
● Country Capital
◇ Dependent Territory
○ Dependent Territory Capital
◆ Administrative Regions
✈ International Airport
▲ Mountain
▲ Mountain Range
Volcano
⇝ River
◎ Lake
▣ Reservoir

104 M2 **Verde, Costa** coastal region N Spain
Verde Grande, Río/Verde Grande y de Belem, Río see Verde, Río
100 H11 **Verden** Niedersachsen, NW Germany 52°55′N 09°14′E
57 P16 **Verde, Río** Bolivia/Brazil
59 J19 **Verde, Rio** SE Brazil
40 M12 **Verde, Río** var. Río Verde Grande, Río Verde Grande y de Belem. C Mexico
41 Q16 **Verde, Río** SE Mexico
36 L13 **Verde River** Arizona, SW USA
Verdhikoúsa/Verdhikoússa see Verdikoússa
27 Q8 **Verdigris River** Kansas/Oklahoma, C USA
115 E15 **Verdikoússa** var. Verdhikoúsa, Verdhikoússa. Thessalía, C Greece 39°47′N 21°59′E
103 S15 **Verdon** SE France
15 O12 **Verdun** Québec, SE Canada 45°27′N 73°36′W
103 S4 **Verdun** var. Verdun-sur-Meuse; anc. Verodunum. Meuse, NE France 49°09′N 05°25′E
Verdun-sur-Meuse see Verdun
83 J21 **Vereeniging** Gauteng, NE South Africa 26°41′S 27°56′E
Veremeyki see Vyeramyeyki
125 T14 **Vereshchagino** Permskiy Kray, NW Russian Federation 58°06′N 54°38′E
76 G14 **Verga, Cap** headland W Guinea 10°12′N 14°27′W
61 G18 **Vergara** Treinta y Tres, E Uruguay 32°58′S 53°54′W
108 G11 **Vergeletto** Ticino, S Switzerland 46°13′N 08°34′E
18 L8 **Vergennes** Vermont, NE USA 44°09′N 73°13′W
Veria see Véroia
104 I5 **Verín** Galicia, NW Spain 41°55′N 07°26′W
Verin T'alin see T'alin
118 K6 **Veriora** Põlvamaa, SE Estonia 57°57′N 27°23′E
117 T7 **Verkhivtseve** Dnipropetrovs'ka Oblast', E Ukraine 48°27′N 34°15′E
Verkhnedvinsk see Vyerkhnyadzvinsk
122 K10 **Verkhneimbatsk** Krasnoyarskiy Kray, N Russian Federation 63°06′N 88°03′E
124 I3 **Verkhnetulomskiy** Murmanskaya Oblast', NW Russian Federation 68°37′N 31°46′E
124 I3 **Verkhnetulomskoye Vodokhranilishche** NW Russian Federation
Verkhneudinsk see Ulan-Ude
123 P10 **Verkhnevilyuysk** Respublika Sakha (Yakutiya), NE Russian Federation 63°44′N 119°59′E
127 W5 **Verkhniy Avzyan** Respublika Bashkortostan, W Russian Federation 53°31′N 57°26′E
127 Q11 **Verkhniy Baskunchak** Astrakhanskaya Oblast', SW Russian Federation 48°14′N 46°43′E
127 W3 **Verkhniye Kigi** Respublika Bashkortostan, W Russian Federation 55°25′N 58°40′E
117 T9 **Verkhniy Rohachyk** Khersons'ka Oblast', S Ukraine 47°16′N 34°16′E
123 Q11 **Verkhnyaya Amga** Respublika Sakha (Yakutiya), NE Russian Federation 59°34′N 127°07′E
125 V6 **Verkhnyaya Inta** Respublika Komi, NW Russian Federation 65°55′N 60°07′E
125 O10 **Verkhnyaya Toyma** Arkhangel'skaya Oblast', NW Russian Federation 62°12′N 44°57′E
126 K6 **Verkhov'ye** Orlovskaya Oblast', W Russian Federation 52°49′N 37°20′E
116 I8 **Verkhovyna** Ivano-Frankivs'ka Oblast', W Ukraine 48°09′N 24°48′E
123 P8 **Verkhoyanskiy Khrebet** ▲▲ NE Russian Federation
117 T7 **Verkn'odniprovs'k** Dnipropetrovs'ka Oblast', E Ukraine 48°40′N 34°17′E
101 G14 **Verl** Nordrhein-Westfalen, NW Germany 51°52′N 08°30′E
92 N1 **Verlegenhuken** headland N Svalbard 80°03′N 16°15′E
82 A9 **Vermelha, Ponta** headland NW Angola 05°40′S 12°09′E
103 P7 **Vermenton** Yonne, C France 47°40′N 03°43′E
11 R14 **Vermilion** Alberta, SW Canada 53°21′N 110°52′W
31 T11 **Vermilion** Ohio, N USA 41°25′N 82°21′W
22 I10 **Vermilion Bay** bay Louisiana, S USA
29 V4 **Vermilion Lake** Minnesota, N USA
14 F9 **Vermilion River** Ontario, S Canada
30 L12 **Vermilion River** Illinois, N USA
29 R12 **Vermillion** South Dakota, N USA 42°46′N 96°55′W
29 R12 **Vermillion River** South Dakota, N USA
15 O9 **Vermillon, Rivière** Québec, SE Canada
115 E14 **Vérmio** ▲▲ N Greece
18 L8 **Vermont** off. State of Vermont, also known as Green Mountain State. ⬖ state NE USA
113 K16 **Vermosh** var. Vermoshi. Shkodër, N Albania 42°37′N 19°42′E
Vermoshi see Vermosh
37 O3 **Vernal** Utah, W USA 40°27′N 109°31′W
14 G11 **Verner** Ontario, S Canada 46°24′N 80°04′W
102 M5 **Verneuil-sur-Avre** Eure, N France 48°44′N 00°55′E
114 D13 **Vérno** ▲▲ N Greece
11 N17 **Vernon** British Columbia, SW Canada 50°17′N 119°19′W
102 M4 **Vernon** Eure, N France 49°04′N 01°28′E
23 N3 **Vernon** Alabama, S USA 33°45′N 88°06′W
31 P15 **Vernon** Indiana, N USA 38°59′N 85°39′W
25 Q4 **Vernon** Texas, SW USA 34°11′N 99°17′W
32 G10 **Vernonia** Oregon, NW USA 45°51′N 123°11′W
14 G12 **Vernon, Lake** Ontario, S Canada
22 G7 **Vernon Lake** Louisiana, S USA
23 Y13 **Vero Beach** Florida, SE USA 27°38′N 80°24′W
Verőcze see Virovitica
Verodunum see Verdun
115 E14 **Véroia** var. Veria, Vérroia, Turk. Karaferiye. Kentrikí Makedonía, N Greece 40°32′N 22°11′E
106 E8 **Verolanuova** Lombardia, N Italy 45°20′N 10°06′E
14 K14 **Verona** Ontario, SE Canada 44°30′N 76°42′W
106 G8 **Verona** Veneto, NE Italy 45°27′N 11°E
29 P6 **Verona** North Dakota, N USA 46°19′N 98°03′W
30 L9 **Verona** Wisconsin, N USA 42°59′N 89°33′W
61 E20 **Verónica** Buenos Aires, E Argentina 35°25′S 57°16′W
22 J9 **Verret, Lake** Louisiana, S USA
Vérroia see Véroia
103 N5 **Versailles** Yvelines, N France 48°48′N 02°08′E
31 P15 **Versailles** Indiana, N USA 39°04′N 85°16′W
20 M5 **Versailles** Kentucky, S USA 38°02′N 84°45′W
27 U5 **Versailles** Missouri, C USA 38°25′N 92°51′W
31 Q13 **Versailles** Ohio, N USA 40°13′N 84°28′W
Versecz see Vršac
108 A10 **Versoix** Genève, SW Switzerland 46°17′N 06°10′E
15 Z6 **Verte, Pointe** headland Québec, SE Canada 48°36′N 64°10′W
111 I22 **Vértes** ▲▲ NW Hungary
44 G6 **Vertientes** Camagüey, C Cuba 21°18′N 78°11′W
114 G13 **Vertískos** ▲▲ N Greece
102 I8 **Vertou** Loire-Atlantique, NW France 47°10′N 01°28′W
Verulamium see St Albans
99 L19 **Verviers** Liège, E Belgium 50°36′N 05°52′E
103 Y14 **Vescovato** Corse, France, C Mediterranean Sea 42°30′N 09°27′E
99 L20 **Vesdre** E Belgium
117 U10 **Vesele** Rus. Veseloye. Zaporiz'ka Oblast', S Ukraine 47°00′N 34°52′E
111 D18 **Veselí nad Lužnicí** var. Weseli an der Lainsitz, Ger. Frohenbruck. Jihočeský Kraj, S Czech Republic 49°11′N 14°40′E
114 M9 **Veselinovo** Shumen, NE Bulgaria 43°01′N 27°02′E
126 L12 **Veselovskoye Vodokhranilishche** SW Russian Federation
Veseloye see Vesele
117 Q9 **Veselynove** Mykolayivs'ka Oblast', S Ukraine 47°21′N 31°15′E
Veseya see Vyasyeya
126 M10 **Veshenskaya** Rostovskaya Oblast', SW Russian Federation 49°37′N 41°43′E
127 Q5 **Veshkayma** Ul'yanovskaya Oblast', W Russian Federation 54°04′N 47°06′E
Vesisaari see Vadsø
Vesontio see Besançon
103 T7 **Vesoul** anc. Vesulium, Vesulum. Haute-Saône, E France 47°37′N 06°09′E
95 J20 **Vessigebro** Halland, S Sweden 56°58′N 12°40′E
95 D17 **Vest-Agder** ⬖ county S Norway
23 P4 **Vestavia Hills** Alabama, S USA 33°27′N 86°47′W
84 F6 **Vesterålen** island NW Norway
92 G10 **Vesterålen** island group N Norway
87 V3 **Vestervig** Midtjylland, NW Denmark 56°46′N 08°20′E
92 H2 **Vestfirðir** ⬖ region NW Iceland
92 G11 **Vestfjorden** fjord C Norway
95 G16 **Vestfold** ⬖ county S Norway
Vestmanhavn see Vestmanna
95 B18 **Vestmanna** Dan. Vestmanhavn. Streymoy, N Faroe Islands 62°09′N 07°11′W
92 I4 **Vestmannaeyjar** Suðurland, S Iceland 63°26′N 20°14′W
94 E9 **Vestnes** Møre og Romsdal, S Norway 62°39′N 07°00′E
92 H3 **Vesturland** ⬖ region W Iceland
92 G11 **Vestvågøya** island C Norway
Vesulium/Vesulum see Vesoul
Vesuna see Périgueux
107 K17 **Vesuvio** Eng. Vesuvius. S Italy 40°48′N 14°29′E
Vesuvius see Vesuvio
124 K14 **Ves'yegonsk** Tverskaya Oblast', W Russian Federation 58°40′N 37°13′E
111 I23 **Veszprém** Ger. Veszprim. Veszprém, W Hungary 47°06′N 17°54′E
111 H23 **Veszprém** off. Veszprém Megye. ⬖ county W Hungary
Veszprém Megye see Veszprém
Veszprim see Veszprém
Vetka see Vyetka
95 M19 **Vetlanda** Jönköping, S Sweden 57°26′N 15°05′E
127 P1 **Vetluga** Nizhegorodskaya Oblast', W Russian Federation 57°51′N 45°45′E
125 P14 **Vetluga** NW Russian Federation
125 O14 **Vetluzhskiy** Kostromskaya Oblast', NW Russian Federation 58°21′N 45°25′E
127 P2 **Vetluzhskiy** Nizhegorodskaya Oblast', W Russian Federation 57°10′N 45°07′E
114 K7 **Vetovo** Ruse, N Bulgaria 43°42′N 26°16′E
107 H14 **Vetralla** Lazio, C Italy 42°18′N 12°03′E
Vetrino see Vyetryna
122 L7 **Vetrovaya, Gora** ▲ N Russian Federation 73°54′N 95°00′E
Vetter, Lake see Vättern
106 J13 **Vettore, Monte** ▲ C Italy 42°49′N 13°15′E
99 A17 **Veurne** var. Furnes. West-Vlaanderen, W Belgium 51°04′N 02°40′E
31 Q15 **Vevay** Indiana, N USA 38°45′N 85°08′W
108 C10 **Vevey** Ger. Vivis; anc. Vibiscum. Vaud, SW Switzerland 46°28′N 06°51′E
Vexiö see Växjö
103 S13 **Veynes** Hautes-Alpes, SE France 44°33′N 05°51′E
103 N11 **Vézère** W France
114 I9 **Vezhen** ▲ C Bulgaria 42°45′N 24°22′E
136 K11 **Vezirköprü** Samsun, N Turkey 41°09′N 35°27′E
57 J18 **Viacha** La Paz, W Bolivia 16°40′S 68°17′W
27 R10 **Vian** Oklahoma, C USA 35°30′N 94°56′W
Viana de Castelo see Viana do Castelo
104 H12 **Viana do Alentejo** Évora, S Portugal 38°20′N 08°00′W
104 I4 **Viana do Bolo** Galicia, NW Spain 42°10′N 07°06′W
104 G5 **Viana do Castelo** var. Viana de Castelo; anc. Velobriga. Viana do Castelo, NW Portugal 41°41′N 08°50′W
104 G5 **Viana do Castelo** var. Viana de Castelo. ⬖ district N Portugal
98 J12 **Vianen** Utrecht, C Netherlands 52°N 05°06′E
167 Q8 **Viangchan** Eng./Fr. Vientiane. ● (Laos) C Laos 17°58′N 102°38′E
167 P6 **Viangphoukha** var. Vieng Pou Kha. Louang Namtha, N Laos 20°41′N 101°03′E
104 K13 **Viar** SW Spain
106 E11 **Viareggio** Toscana, C Italy 43°52′N 10°15′E
103 O14 **Viaur** S France
Vibiscum see Vevey
95 G21 **Viborg** Midtjylland, NW Denmark 56°28′N 09°25′E
29 R12 **Viborg** South Dakota, N USA 43°10′N 97°04′W
107 N22 **Vibo Valentia** prev. Monteleone di Calabria; anc. Hipponium. Calabria, SW Italy 38°40′N 16°06′E
105 W5 **Vic** var. Vich; anc. Ausa, Vicus Ausonensis. Cataluña, NE Spain 41°56′N 02°16′E
102 K16 **Vic-en-Bigorre** Hautes-Pyrénées, S France 43°23′N 00°04′E
40 K10 **Vicente Guerrero** Durango, C Mexico 23°30′N 104°24′W
41 P10 **Vicente Guerrero, Presa** var. Presa de las Adjuntas. NE Mexico
Vicentia see Vicenza
106 G8 **Vicenza** anc. Vicentia. Veneto, NE Italy 45°32′N 11°31′E
Vich see Vic
54 J10 **Vichada** off. Comisaría del Vichada. ⬖ province E Colombia
Vichada, Comisaría del see Vichada
54 K10 **Vichada, Río** E Colombia
61 G17 **Vichadero** Rivera, NE Uruguay 31°45′S 54°41′W
Vichegda see Vychegda
124 M16 **Vichuga** Ivanovskaya Oblast', W Russian Federation 57°13′N 41°51′E
103 P10 **Vichy** Allier, C France 46°08′N 03°26′E
26 K9 **Vici** Oklahoma, C USA 36°09′N 99°18′W
95 I19 **Vickan** Halland, S Sweden 57°25′N 12°00′E
31 P10 **Vicksburg** Michigan, N USA 42°07′N 85°31′W
22 J5 **Vicksburg** Mississippi, S USA 32°21′N 90°52′W
103 O12 **Vic-sur-Cère** Cantal, C France 45°00′N 02°36′E
59 I21 **Víctor** Mato Grosso do Sul, SW Brazil 21°39′S 53°21′W
29 X14 **Victor** Iowa, C USA 41°45′N 92°18′W
182 I10 **Victor Harbor** South Australia 35°33′S 138°37′E
61 C18 **Victoria** Entre Ríos, E Argentina 32°40′S 60°10′W
10 L17 **Victoria** province capital Vancouver Island, British Columbia, SW Canada 48°25′N 123°22′W
45 R14 **Victoria** NW Grenada 12°12′N 61°42′W
42 H6 **Victoria** Yoro, NW Honduras 15°01′N 87°28′W
121 O15 **Victoria** var. Rabat. Gozo, NW Malta 36°02′N 14°14′E
116 I12 **Victoria** Ger. Viktoriastadt. Brașov, C Romania 45°44′N 24°41′E
172 H17 **Victoria** ● (Seychelles) Mahé, SW Seychelles 04°38′S 28°28′E
25 U13 **Victoria** Texas, SW USA 28°47′N 96°59′W
183 N12 **Victoria** ⬖ state SE Australia
174 K7 **Victoria** Western Australia
Victoria see Labuan, East Malaysia
Victoria see Masvingo, Zimbabwe
Victoria Bank see Vitória Seamount
11 Y15 **Victoria Beach** Manitoba, S Canada 50°40′N 96°30′W
Victoria de Durango see Durango
Victoria de las Tunas see Las Tunas
83 I16 **Victoria Falls** Matabeleland North, W Zimbabwe 17°55′S 25°51′E
83 I16 **Victoria Falls** waterfall Zambia/Zimbabwe
83 I16 **Victoria Falls** ✈ Matabeleland North, W Zimbabwe 18°03′S 25°48′E
Victoria Falls see Iguaçu, Salto do
63 F19 **Victoria, Isla** island Archipiélago de los Chonos, S Chile
8 K6 **Victoria Island** island Northwest Territories/Nunavut, NW Canada
182 L8 **Victoria, Lake** New South Wales, SE Australia
68 I12 **Victoria, Lake** var. Victoria Nyanza. E Africa
195 S13 **Victoria Land** physical region Antarctica
187 X14 **Victoria, Mount** ▲ Viti Levu, W Fiji 17°37′S 178°00′E
166 L5 **Victoria, Mount** ▲ W Myanmar (Burma) 21°13′N 93°53′E
186 E9 **Victoria, Mount** ▲ S Papua New Guinea 08°51′S 147°36′E
81 F17 **Victoria Nile** var. Somerset Nile. C Uganda
Victoria Nyanza see Victoria, Lake
42 G3 **Victoria Peak** ▲ SE Belize 16°50′N 88°38′W
185 H16 **Victoria Range** ▲▲ South Island, New Zealand
181 O3 **Victoria River** Northern Territory, N Australia
181 P3 **Victoria River Roadhouse** Northern Territory, N Australia 15°37′S 131°07′E
15 Q11 **Victoriaville** Québec, SE Canada 46°04′N 71°57′W
Victoria-Wes see Victoria West
83 G24 **Victoria West** Afr. Victoria-Wes. Northern Cape, W South Africa 31°25′S 23°08′E
62 J13 **Victorica** La Pampa, C Argentina 36°15′S 65°25′W
Victor, Mount see Dronning Fabiolafjella
35 U14 **Victorville** California, W USA 34°32′N 117°17′W
62 G9 **Vicuña** Coquimbo, N Chile 30°00′S 70°44′W
62 K11 **Vicuña Mackenna** Córdoba, C Argentina 33°53′S 64°25′W
Vicus Ausonensis see Vic
Vicus Elbii see Viterbo
33 X7 **Vida** Montana, NW USA 47°52′N 105°30′W
23 V6 **Vidalia** Georgia, SE USA 32°13′N 82°24′W
22 J7 **Vidalia** Louisiana, S USA 31°34′N 91°25′W
95 F22 **Videbæk** Midtjylland, C Denmark 56°08′N 08°38′E
60 I13 **Videira** Santa Catarina, S Brazil 27°00′S 51°08′W
116 J14 **Videle** Teleorman, S Romania 44°15′N 25°27′E
Videm-Krško see Krško
Vídeň see Wien
104 H12 **Vidigueira** Beja, S Portugal 38°12′N 07°48′W
114 J9 **Vidima** N Bulgaria
114 G7 **Vidin** anc. Bononia. Vidin, NW Bulgaria 44°00′N 22°52′E
114 F8 **Vidin** ⬖ province NW Bulgaria
154 H10 **Vidisha** Madhya Pradesh, C India 23°30′N 77°50′E
25 Y10 **Vidor** Texas, SW USA 30°07′N 94°01′W
95 L20 **Vidöstern** S Sweden
92 J13 **Vidsel** Norrbotten, N Sweden 65°49′N 20°31′E
118 H9 **Vidzemes Augstiene** ▲▲ C Latvia
118 J12 **Vidzy** Vitsyebskaya Voblasts', NW Belarus 55°24′N 26°38′E
63 L16 **Viedma** Río Negro, E Argentina 40°50′S 62°58′W
63 H22 **Viedma, Lago** S Argentina
45 O11 **Vieille Case** var. Itassi. N Dominica 15°36′N 61°24′W
104 M2 **Vieja, Peña** ▲ N Spain 43°09′N 04°47′W
24 J10 **Vieja, Sierra** ▲▲ Texas, SW USA
40 E4 **Viejo, Cerro** ▲ NW Mexico 30°16′N 112°18′W
56 B9 **Viejo, Cerro** ▲ N Peru 04°54′S 79°24′W
118 E10 **Viekšniai** Telšiai, NW Lithuania 56°14′N 22°33′E
105 U3 **Vielha** var. Viella. Cataluña, NE Spain 42°41′N 00°47′E
Viella see Vielha
99 L21 **Vielsalm** Luxembourg, E Belgium 50°17′N 05°55′E
Vieng Pou Kha see Viangphoukha
23 T6 **Vienna** Georgia, SE USA 32°05′N 83°48′W
30 L17 **Vienna** Illinois, N USA 37°24′N 88°55′W
27 V5 **Vienna** Missouri, C USA 38°12′N 91°59′W
21 Q3 **Vienna** West Virginia, NE USA 39°19′N 81°33′W
Vienna see Wien, Austria
Vienna see Vienne, France
103 R11 **Vienne** anc. Vienna. Isère, E France 45°32′N 04°53′E
102 L10 **Vienne** ⬖ department W France
102 L9 **Vienne** W France
Vientiane see Viangchan
Vientos, Paso de los see Windward Passage
45 V6 **Vieques** var. Isabel Segunda. E Puerto Rico 18°08′N 65°25′W
45 V6 **Vieques, Isla de** island E Puerto Rico
45 V6 **Vieques, Pasaje de** passage E Puerto Rico
45 V5 **Vieques, Sonda de** sound E Puerto Rico
Vierdörfer see Săcele
93 M15 **Vieremä** Pohjois-Savo, C Finland 63°42′N 27°02′E
99 M14 **Vierlingsbeek** Noord-Brabant, SE Netherlands 51°36′N 06°01′E
101 G20 **Viernheim** Hessen, W Germany 49°32′N 08°35′E
101 D15 **Viersen** Nordrhein-Westfalen, W Germany 51°16′N 06°24′E
108 G8 **Vierwaldstätter See** Eng. Lake of Lucerne. C Switzerland
103 N8 **Vierzon** Cher, C France 47°13′N 02°04′E
40 L8 **Viesca** Coahuila, NE Mexico 25°25′N 102°45′W
118 H10 **Viesīte** Ger. Eckengraf. S Latvia 56°21′N 25°30′E
107 N15 **Vieste** Puglia, SE Italy 41°52′N 16°11′E
167 T8 **Vietnam** off. Socialist Republic of Vietnam, Vtn. Công Hoa Xa Hôi Chu Nghia Viêt Nam. ◆ republic SE Asia
Vietnam, Socialist Republic of see Vietnam
167 S5 **Viêt Quang** Ha Giang, N Vietnam 22°24′N 104°48′E
Vietri see Viêt Tri
167 S6 **Viêt Tri** var. Vietri. Vinh Phu, N Vietnam 21°20′N 105°26′E
30 L4 **Vieux Desert, Lac** Michigan/Wisconsin, N USA
45 Y13 **Vieux Fort** S Saint Lucia 13°43′N 60°57′W
45 X6 **Vieux-Habitants** Basse Terre, SW Guadeloupe 16°04′N 61°45′W
119 G14 **Vievis** Vilnius, S Lithuania 54°46′N 24°51′E
171 N2 **Vigan** Luzon, N Philippines 17°34′N 102°21′E
106 D8 **Vigevano** Lombardia, N Italy 45°19′N 08°51′E
107 N18 **Viggiano** Basilicata, S Italy 40°21′N 15°54′E
58 L12 **Vigia** Pará, NE Brazil 0°50′S 48°07′W
41 Y12 **Vigía Chico** Quintana Roo, SE Mexico 19°49′N 87°31′W
45 T11 **Vigie** prev. George F L Charles. ✈ (Castries) NE Saint Lucia 14°01′N 60°59′W
102 K17 **Vignemale** var. Pic de Vignemale. ▲ France/Spain 42°48′N 00°06′W
Vignemale, Pic de see Vignemale
106 G10 **Vignola** Emilia-Romagna, C Italy 44°28′N 11°00′E
104 G4 **Vigo** Galicia, NW Spain 42°15′N 08°44′W
104 G4 **Vigo, Ría de** estuary NW Spain
94 D9 **Vigra** island S Norway
95 C17 **Vigrestad** Rogaland, S Norway 58°34′N 05°42′E
93 L15 **Vihanti** Pohjois-Pohjanmaa, C Finland 64°29′N 25°E
149 U10 **Vihāri** Punjab, E Pakistan 30°03′N 72°32′E
102 K8 **Vihiers** Maine-et-Loire, NW France 47°09′N 00°37′W
81 H18 **Vihiga** ⬖ county W Kenya
111 O19 **Vihorlat** ▲ E Slovakia 48°54′N 22°09′E
114 G11 **Vihren** var. Vikhren. ▲ SW Bulgaria 41°45′N 23°24′E
93 L19 **Vihti** Uusimaa, S Finland 60°25′N 24°16′E
Viipuri see Vyborg
93 M16 **Viitasaari** Keski-Suomi, C Finland 63°05′N 25°52′E
118 K3 **Viivikonna** Ida-Virumaa, NE Estonia 59°19′N 27°41′E
155 K16 **Vijayawāda** prev. Bezwada. Andhra Pradesh, SE India 16°34′N 80°40′E
Vijosa/Vijosë see Aóos, Albania/Greece
Vijosa/Vijosë see Vjosës, Lumi i, Albania/Greece
92 J4 **Vík** Suðurland, S Iceland 63°25′N 18°58′W
Vik see Vikøyri
94 L13 **Vika** Dalarna, C Sweden 60°55′N 14°30′E
92 L12 **Vikajärvi** Lappi, N Finland 66°37′N 26°10′E
94 L13 **Vikarbyn** Dalarna, C Sweden 60°57′N 15°00′E
95 J22 **Viken** Skåne, S Sweden 56°09′N 12°36′E
95 L17 **Viken** C Sweden
95 G15 **Vikersund** Buskerud, S Norway 59°58′N 09°59′E
Vikhren see Vihren
11 R15 **Viking** Alberta, SW Canada 53°07′N 111°50′W
84 E7 **Viking Bank** undersea feature N North Sea 60°35′N 02°35′E
95 M14 **Vikmanshyttan** Dalarna, C Sweden 60°19′N 15°55′E
94 D12 **Vikøyri** var. Vik. Sogn Og Fjordane, S Norway 61°04′N 06°34′E
93 H17 **Viksjö** Västernorrland, C Sweden 62°45′N 17°30′E
Viktoriastadt see Victoria
Vila see Port-Vila
Vila Arriaga see Bibala
Vila Artur de Paiva see Cubango
Vila Baleira see Porto Santo
Vila Bela da Santissima Trindade see Mato Grosso
58 B12 **Vila Bittencourt** Amazonas, NW Brazil 01°25′S 69°24′W
Vila da Ponte see Cubango
64 O2 **Vila da Praia da Vitória** Terceira, Azores, Portugal, NE Atlantic Ocean 38°44′N 27°04′W
Vila de Aljustrel see Cangamba
Vila de Almoster see Chiange
Vila de João Belo see Xai-Xai
Vila de Macia see Macia
Vila de Manhiça see Manhiça
Vila de Manica see Manica
Vila de Mocímboa da Praia see Mocímboa da Praia
83 N16 **Vila de Sena** var. Sena. Sofala, C Mozambique 17°25′S 34°59′E
104 F14 **Vila do Bispo** Faro, S Portugal 37°05′N 08°53′W
104 G6 **Vila do Conde** Porto, NW Portugal 41°21′N 08°45′W
Vila do Maio see Maio
64 P3 **Vila do Porto** Santa Maria, Azores, Portugal, NE Atlantic Ocean 36°57′N 25°10′W
83 K15 **Vila do Zumbo** prev. Vila do Zumbu, Zumbo. Tete, NW Mozambique 15°36′S 30°30′E
Vila do Zumbu see Vila do Zumbo
105 T8 **Vilafamés** prev. Villafamés. Valenciana, E Spain 40°07′N 00°03′W
104 I6 **Vila Flor** var. Vila Flôr. Bragança, N Portugal 41°18′N 07°09′W
105 V6 **Vilafranca del Penedès** var. Villafranca del Panadés. Cataluña, NE Spain 41°21′N 01°42′E
104 F10 **Vila Franca de Xira** var. Vilafranca de Xira. Lisboa, C Portugal 38°57′N 08°59′W
Vila Gago Coutinho see Lumbala N'Guimbo
104 G3 **Vilagarcía** var. Villagarcía de Arosa. Galicia, NW Spain 42°35′N 08°45′W
Vila General Machado see Camacupa
Vila Henrique de Carvalho see Saurimo
102 I7 **Vilaine** NW France
Vila João de Almeida see Chibia
118 K8 **Viļaka** Ger. Marienhausen. NE Latvia 57°12′N 27°43′E
104 I2 **Vilalba** Galicia, NW Spain 43°17′N 07°41′W
Vila Marechal Carmona see Uíge
Vila Mariano Machado see Ganda
172 G3 **Vilanandro, Tanjona** prev./Fr. Cap Saint-André. headland W Madagascar 16°10′S 44°27′E
Vilanculos see Vilankulo
118 J10 **Viļāni** E Latvia 56°33′N 26°55′E
83 N19 **Vilankulo** var. Vilanculos. Inhambane, E Mozambique 22°01′S 35°19′E
Vila Norton de Matos see Balombo
104 G6 **Vila Nova de Famalicão** var. Vila Nova de Famalicao. Braga, N Portugal 41°24′N 08°31′W
104 I6 **Vila Nova de Foz Côa** var. Vila Nova de Fozcôa. Guarda, N Portugal 41°05′N 07°09′W
Vila Nova de Fozcôa see Vila Nova de Foz Côa
104 F6 **Vila Nova de Gaia** Porto, NW Portugal 41°08′N 08°37′W
Vila Nova de Portimão see Portimão
105 V6 **Vilanova i La Geltru** Cataluña, NE Spain 41°15′N 01°42′E
Vila Pereira de Eça see Ondjiva
104 H6 **Vila Pouca de Aguiar** Vila Real, N Portugal 41°30′N 07°38′W
104 H6 **Vila Real** var. Vila Rial. Vila Real, N Portugal 41°17′N 07°45′W
105 T9 **Vila-real** var. Vila-real de los Infantes; prev. Villarreal. Valenciana, E Spain 39°56′N 00°08′W
104 H6 **Vila Real** ⬖ district N Portugal
Vila-real de los Infantes see Vila-real
104 H14 **Vila Real de Santo António** Faro, S Portugal 37°12′N 07°25′W
104 J7 **Vilar Formoso** Guarda, N Portugal 40°37′N 06°50′W
Vila Rial see Vila Real
59 J15 **Vila Rica** Mato Grosso, W Brazil 09°52′S 50°44′W
Vila Robert Williams see Caála
Vila Salazar see N'Dalatando
Vila Serpa Pinto see Menongue
Vila Teixeira da Silva see Bailundo
Vila Teixeira de Sousa see Luau
104 H9 **Vila Velha de Ródão** Castelo Branco, C Portugal 39°39′N 07°40′W
104 G5 **Vila Verde** Braga, N Portugal 41°39′N 08°27′W
104 H11 **Vila Viçosa** Évora, S Portugal 38°46′N 07°25′W
57 G15 **Vilcabamba, Cordillera de** ▲▲ C Peru
Vîlcea see Vâlcea
122 J4 **Vil'cheka, Zemlya** Eng. Wilczek Land. island Zemlya Frantsa-Iosifa, NW Russian Federation
95 F22 **Vildbjerg** Midtjylland, C Denmark 56°12′N 08°47′E
Vileyka see Vilyeyka
93 H15 **Vilhelmina** Västerbotten, N Sweden 64°38′N 16°40′E
59 F17 **Vilhena** Rondônia, W Brazil 12°40′S 60°08′W
115 G19 **Vília** Attikí, C Greece 38°09′N 23°21′E
119 I14 **Viliya** Lith. Neris. W Belarus
Viliya see Neris
118 H5 **Viljandi** Ger. Fellin. Viljandimaa, S Estonia 58°22′N 25°30′E
118 H5 **Viljandimaa** var. Viljandi Maakond. ⬖ province SW Estonia
Viljandi Maakond see Viljandimaa
119 E14 **Vilkaviškis** Pol. Wyłkowyszki. Marijampolė, SW Lithuania 54°39′N 23°03′E
118 F13 **Vilkija** Kaunas, C Lithuania 55°02′N 23°36′E
197 V9 **Vil'kitskogo, Proliv** strait N Russian Federation
Vilkovo see Vylkove
57 L21 **Villa Abecia** Chuquisaca, S Bolivia 21°00′S 65°18′W
41 N5 **Villa Acuña** var. Ciudad Acuña. Coahuila, NE Mexico 29°18′N 100°58′W
40 J4 **Villa Ahumada** Chihuahua, N Mexico 30°38′N 106°30′W
45 O9 **Villa Altagracia** C Dominican Republic 18°43′N 70°13′W
56 L13 **Villa Bella** El Beni, N Bolivia 10°21′S 65°25′W
104 J3 **Villablino** Castilla y León, N Spain 42°55′N 06°21′W
54 K6 **Villa Bruzual** Portuguesa, N Venezuela 09°20′N 69°06′W
105 O9 **Villacañas** Castilla-La Mancha, C Spain 39°38′N 03°20′W
105 O12 **Villacarrillo** Andalucía, S Spain 38°07′N 03°05′W
104 M7 **Villacastín** Castilla y León, N Spain 40°46′N 04°25′W
Villa Cecilia see Ciudad Madero
109 S9 **Villach** Slvn. Beljak. Kärnten, S Austria 46°36′N 13°49′E
107 B20 **Villacidro** Sardegna, Italy, C Mediterranean Sea 39°28′N 08°43′E
Villa Concepción see Concepción
104 L4 **Villada** Castilla y León, N Spain 42°15′N 04°59′W
40 M10 **Villa de Cos** Zacatecas, C Mexico 23°20′N 102°20′W
54 L5 **Villa de Cura** var. Cura. Aragua, N Venezuela 10°00′N 67°30′W
Villa del Nevoso see Ilirska Bistrica
Villa del Pilar see Pilar
104 M13 **Villa del Río** Andalucía, S Spain 37°59′N 04°17′W
Villa de Méndez see Méndez
42 H6 **Villa de San Antonio** Comayagua, W Honduras 14°24′N 87°37′W
105 N4 **Villadiego** Castilla y León, N Spain 42°31′N 04°01′W
Villafamés see Vilafamés
41 U16 **Villa Flores** Chiapas, SE Mexico 16°12′N 93°16′W
104 J3 **Villafranca del Bierzo** Castilla y León, N Spain 42°36′N 06°49′W
105 S8 **Villafranca del Cid** Valenciana, E Spain 40°25′N 00°15′W
104 J11 **Villafranca de los Barros** Extremadura, W Spain 38°34′N 06°20′W
105 N10 **Villafranca de los Caballeros** Castilla-La Mancha, C Spain 39°26′N 03°21′W
Villafranca del Panadés see Vilafranca del Penedès
106 F8 **Villafranca di Verona** Veneto, NE Italy 45°22′N 10°51′E
107 J23 **Villafrati** Sicilia, Italy, C Mediterranean Sea 37°53′N 13°30′E
Villagarcía de Arosa see Vilagarcía
41 O9 **Villagrán** Tamaulipas, C Mexico 24°29′N 99°30′W
61 C17 **Villaguay** Entre Ríos, E Argentina 31°55′S 59°01′W
62 O6 **Villa Hayes** Presidente Hayes, S Paraguay 25°05′S 57°25′W
41 U15 **Villahermosa** prev. San Juan Bautista. Tabasco, SE Mexico 17°56′N 92°50′W
105 O11 **Villahermosa** Castilla-La Mancha, C Spain 38°46′N 02°52′W
64 O11 **Villahermoso** Gomera, Islas Canarias, Spain, NE Atlantic Ocean 38°46′N 02°52′W
Villa Hidalgo see Hidalgo
105 T12 **Villajoyosa** Cat. La Vila Joíosa. Valenciana, E Spain 38°31′N 00°14′W
Villa Juárez see Juárez
Villalba see Collado Villalba
41 N8 **Villaldama** Nuevo León, NE Mexico 26°29′N 100°27′W
104 L5 **Villalón de Campos** Castilla y León, N Spain 42°05′N 05°03′W
61 A25 **Villalonga** Buenos Aires, E Argentina 39°55′S 62°35′W
104 L5 **Villalpando** Castilla y León, N Spain 41°51′N 05°25′W
40 K9 **Villa Madero** var. Francisco l. Madero. Durango, C Mexico 24°28′N 104°20′W
41 O9 **Villa Mainero** Tamaulipas, C Mexico 24°32′N 99°39′W
Villamaña see Villamañán
104 L4 **Villamañán** var. Villamaña. Castilla y León, N Spain 42°19′N 05°35′W
62 L10 **Villa María** Córdoba, C Argentina 32°23′S 63°15′W
61 C17 **Villa María Grande** Entre Ríos, E Argentina 31°39′S 59°54′W
57 K21 **Villa Martín** Potosí, SW Bolivia 20°46′S 67°45′W
104 K15 **Villamartin** Andalucía, S Spain 36°52′N 05°38′W
62 J8 **Villa Mazán** La Rioja, NW Argentina 28°43′S 66°25′W
62 J11 **Villa Mercedes** var. Mercedes. San Luis, C Argentina 33°40′S 65°25′W
Villamil see Puerto Villamil
Villa Nador see Nador
54 G5 **Villanueva** La Guajira, N Colombia 10°37′N 72°58′W
42 H5 **Villanueva** Cortés, NW Honduras 15°14′N 88°00′W
40 L11 **Villanueva** Zacatecas, C Mexico 22°24′N 102°53′W
42 I9 **Villa Nueva** Chinandega, NW Nicaragua 12°58′N 86°46′W
37 T11 **Villanueva** New Mexico, SW USA 35°18′N 105°20′W
104 M12 **Villanueva de Córdoba** Andalucía, S Spain 38°20′N 04°38′W
105 O12 **Villanueva del Arzobispo** Andalucía, S Spain 38°10′N 03°00′W
104 K11 **Villanueva de la Serena** Extremadura, W Spain 38°58′N 05°48′W
104 L5 **Villanueva del Campo** Castilla y León, N Spain 41°59′N 05°25′W
105 O11 **Villanueva de los Infantes** Castilla-La Mancha, C Spain 38°45′N 03°01′W
61 C14 **Villa Ocampo** Santa Fe, C Argentina 28°28′S 59°22′W
40 J8 **Villa Ocampo** Durango, C Mexico 26°29′N 105°30′W
40 J7 **Villa Orestes Pereyra** Durango, C Mexico 26°30′N 105°38′W
105 N3 **Villarcayo** Castilla y León, N Spain 42°56′N 03°34′W
104 L5 **Villardefrades** Castilla y León, N Spain 41°43′N 05°15′W
105 S9 **Villar del Arzobispo** Valenciana, E Spain 39°44′N 00°50′W
105 Q6 **Villaroya de la Sierra** Aragón, NE Spain 41°28′N 01°46′W
Villarreal see Vila-real
62 P6 **Villarrica** Guairá, SE Paraguay 25°45′S 56°28′W
63 G15 **Villarrica, Volcán** S Chile 39°28′S 71°57′W
105 P10 **Villarrobledo** Castilla-La Mancha, C Spain 39°16′N 02°36′W
105 N10 **Villarrubia de los Ojos** Castilla-La Mancha, C Spain 39°14′N 03°36′W
18 J17 **Villas** New Jersey, NE USA 39°01′N 74°54′W
105 O3 **Villasana de Mena** Castilla y León, N Spain 43°05′N 03°16′W
107 M23 **Villa San Giovanni** Calabria, S Italy 38°12′N 15°39′E
61 D18 **Villa San José** Entre Ríos, E Argentina 32°01′S 58°20′W
Villa Sanjurjo see Al-Hoceïma
105 P6 **Villasayas** Castilla y León, N Spain 41°19′N 02°36′W
107 C20 **Villasimius** Sardegna, Italy, C Mediterranean Sea 39°10′N 09°30′E
41 N6 **Villa Unión** Coahuila, NE Mexico 28°18′N 100°43′W
40 K10 **Villa Unión** Durango, C Mexico 23°59′N 104°01′W
40 J10 **Villa Unión** Sinaloa, C Mexico 23°10′N 106°12′W
62 K12 **Villa Valeria** Córdoba, C Argentina 34°21′S 64°56′W
105 N8 **Villaverde** Madrid, C Spain 40°21′N 03°43′W
54 F10 **Villavicencio** Meta, C Colombia 04°09′N 73°38′W
104 L2 **Villaviciosa** Asturias, N Spain 43°29′N 05°26′W
104 L12 **Villaviciosa de Córdoba** Andalucía, S Spain 38°04′N 05°00′W
57 L22 **Villazón** Potosí, S Bolivia 22°05′S 65°35′W
14 J8 **Villebon, Lac** Québec, SE Canada
Ville de Kinshasa see Kinshasa
102 J5 **Villedieu-les-Poêles** Manche, N France 48°51′N 01°12′W
Villefranche see Villefranche-sur-Saône

◆ Country
● Country Capital
◇ Dependent Territory
○ Dependent Territory Capital
⬖ Administrative Regions
✈ International Airport
▲ Mountain
▲▲ Mountain Range
Volcano
River
Lake
Reservoir

103 N16 **Villefranche-de-Lauragais** Haute-Garonne, S France 43°24′N 01°42′E
103 N14 **Villefranche-de-Rouergue** Aveyron, S France 44°21′N 02°02′E
103 R10 **Villefranche-sur-Saône** *var.* Villefranche. Rhône, E France 46°00′N 04°40′E
14 H9 **Ville-Marie** Québec, SE Canada 47°21′N 79°26′W
102 M15 **Villemur-sur-Tarn** Haute-Garonne, S France 43°50′N 01°32′E
105 S11 **Villena** Valenciana, E Spain 38°39′N 00°52′W
Villeneuve-d'Agen *see* Villeneuve-sur-Lot
102 L13 **Villeneuve-sur-Lot** *var.* Villeneuve-d'Agen, *hist.* Gajac. Lot-et-Garonne, SW France 44°24′N 00°43′E
103 P6 **Villeneuve-sur-Yonne** Yonne, C France 48°04′N 03°21′E
22 H8 **Ville Platte** Louisiana, S USA 30°41′N 92°16′W
103 R11 **Villeurbanne** Rhône, E France 45°46′N 04°54′E
101 G23 **Villingen-Schwenningen** Baden-Württemberg, S Germany 48°04′N 08°27′E
29 T15 **Villisca** Iowa, C USA 40°55′N 94°58′W
Villmanstrand *see* Lappeenranta
Vilna *see* Vilnius
119 H14 **Vilnius** *Pol.* Wilno, *Ger.* Wilna; *prev. Rus.* Vilna. ● (Lithuania) Vilnius, SE Lithuania 54°41′N 25°20′E
119 H14 **Vilnius** ✈ Vilnius, SE Lithuania 54°33′N 25°17′E
117 S7 **Vil'nohirs'k** Dnipropetrovs'ka Oblast', E Ukraine 48°31′N 34°01′E
117 U8 **Vil'nyans'k** Zaporiz'ka Oblast', SE Ukraine 47°56′N 35°22′E
93 L17 **Vilppula** Pirkanmaa, W Finland 62°02′N 24°30′E
101 M20 **Vils** ∿ SE Germany
118 C5 **Vilsandi** *island* W Estonia
117 P8 **Vil'shanka** *Rus.* Olshanka. Kirovohrads'ka Oblast', C Ukraine 48°12′N 30°54′E
101 O22 **Vilshofen** Bayern, SE Germany 48°36′N 13°10′E
155 J20 **Viluppuram** Tamil Nādu, SE India 12°54′N 79°40′E
113 I16 **Vilusi** W Montenegro 42°44′N 18°34′E
99 G18 **Vilvoorde** *Fr.* Vilvorde. Vlaams Brabant, C Belgium 50°56′N 04°25′E
Vilvorde *see* Vilvoorde
119 J14 **Vilyeyka** *Pol.* Wilejka, *Rus.* Vileyka. Minskaya Voblasts', NW Belarus 54°30′N 26°55′E
122 V11 **Vilyuchinsk** Kamchatskiy Kray, E Russian Federation 52°55′N 158°28′E
123 P10 **Vilyuy** ∿ NE Russian Federation
123 P10 **Vilyuysk** Respublika Sakha (Yakutiya), NE Russian Federation 63°42′N 121°20′E
123 N10 **Vilyuyskoye Vodokhranilishche** ▦ NE Russian Federation
104 G2 **Vimianzo** Galicia, NW Spain 43°06′N 09°03′W
95 M19 **Vimmerby** Kalmar, S Sweden 57°40′N 15°50′E
102 L5 **Vimoutiers** Orne, N France 48°56′N 00°10′E
93 L16 **Vimpeli** Etelä-Pohjanmaa, W Finland 63°10′N 23°50′E
79 G14 **Vina** ∿ Cameroon/Chad
62 G11 **Viña del Mar** Valparaíso, C Chile 33°02′S 71°35′W
19 R8 **Vinalhaven Island** *island* Maine, NE USA
105 T8 **Vinaròs** Valenciana, E Spain 40°29′N 00°28′E
Vînători *see* Vânători
31 N15 **Vincennes** Indiana, N USA 38°42′N 87°30′W
195 Y12 **Vincennes Bay** *bay* Antarctica
25 O7 **Vincent** Texas, SW USA 32°30′N 101°10′W
95 H24 **Vindeby** Syddtjylland, C Denmark 54°55′N 11°09′E
93 I15 **Vindeln** Västerbotten, N Sweden 64°11′N 19°45′E
95 F21 **Vinderup** Midtjylland, C Denmark 56°29′N 08°48′E
Vindhya Mountains *see* Vindhya Range
153 N14 **Vindhya Range** *var.* Vindhya Mountains. ▲▲ N India
Vindobona *see* Wien
20 K6 **Vine Grove** Kentucky, S USA 37°48′N 85°58′W
18 J17 **Vineland** New Jersey, NE USA 39°29′N 75°02′W
116 E11 **Vinga** Arad, W Romania 46°00′N 21°14′E
95 M16 **Vingåker** Södermanland, C Sweden 59°02′N 15°52′E
167 S8 **Vinh** Nghê An, N Vietnam 18°42′N 105°41′E
104 I5 **Vinhais** Bragança, N Portugal 41°50′N 07°00′W
Vinh Linh *see* Hô Xa
Vinh Loi *see* Bac Liêu
167 S14 **Vinh Long** *var.* Vinhlong. Vinh Long, S Vietnam 10°15′N 105°59′E
Vinhlong *see* Vinh Long
113 Q18 **Vinica** NE FYR Macedonia 41°53′N 22°30′E
109 V13 **Vinica** SE Slovenia 45°28′N 15°12′E
114 G8 **Vinishte** Montana, NW Bulgaria 43°30′N 23°04′E
27 Q8 **Vinita** Oklahoma, C USA 36°38′N 95°09′W
Vînju Mare *see* Vânju Mare
98 I11 **Vinkeveen** Utrecht, C Netherlands 52°13′N 04°55′E
116 L6 **Vin'kivtsi** Khmel'nyts'ka Oblast', W Ukraine 49°02′N 27°13′E
112 I10 **Vinkovci** *Ger.* Winkowitz, *Hung.* Vinkovcze. Vukovar-Srijem, E Croatia 45°18′N 18°45′E
Vinkovcze *see* Vinkovci
Vinnitsa *see* Vinnytsya
Vinnitskaya Oblast'/Vinnytsya *see* Vinnyts'ka Oblast'
116 M7 **Vinnyts'ka Oblast'** *var.* Vinnytsya, *Rus.* Vinnitskaya Oblast'. ◈ *province* C Ukraine
117 N6 **Vinnytsya** *Rus.* Vinnitsa. Vinnyts'ka Oblast', C Ukraine 49°14′N 28°30′E
117 N6 **Vinnytsya** ✈ Vinnyts'ka Oblast', N Ukraine 49°13′N 28°40′E
Vinogradov *see* Vynohradiv
194 L8 **Vinson Massif** ▲ Antarctica 78°45′S 85°19′W
94 G11 **Vinstra** Oppland, S Norway 61°36′N 09°45′E
116 K12 **Vintilă Vodă** Buzău, SE Romania 45°28′N 26°43′E
29 X13 **Vinton** Iowa, C USA 42°10′N 92°01′W
22 F9 **Vinton** Louisiana, S USA 30°10′N 93°37′W
155 J17 **Vinukonda** Andhra Pradesh, E India 16°03′N 79°41′E
Vioara *see* Ocnele Mari
83 E23 **Vioolsdrif** Northern Cape, SW South Africa 28°50′S 17°38′E
82 M13 **Viphya Mountains** ▲▲ C Malawi
171 Q4 **Virac** Catanduanes Island, N Philippines 13°39′N 124°17′E
124 K8 **Virandozero** Respublika Kareliya, NW Russian Federation 63°59′N 36°00′E
137 P16 **Viranşehir** Şanlıurfa, SE Turkey 37°13′N 39°32′E
154 D13 **Virār** Mahārāshtra, W India 19°30′N 72°48′E
11 W16 **Virden** Manitoba, S Canada 49°50′N 100°57′W
30 K14 **Virden** Illinois, N USA 39°30′N 89°46′W
Virdois *see* Virrat
102 J5 **Vire** Calvados, N France 48°50′N 00°53′W
102 J4 **Vire** ∿ N France
83 A15 **Virei** Namibe, SW Angola 15°43′S 12°54′E
Vîrful Moldoveanu *see* Vârful Moldoveanu
35 R5 **Virgina Peak** ▲ Nevada, W USA 39°46′N 119°26′W
45 U9 **Virgin Gorda** *island* C British Virgin Islands
83 I22 **Virginia** Free State, C South Africa 28°06′S 26°53′E
30 K13 **Virginia** Illinois, N USA 39°57′N 90°12′W
29 W4 **Virginia** Minnesota, N USA 47°31′N 92°32′W
21 T6 **Virginia** *off.* Commonwealth of Virginia, *also known as* Mother of Presidents, Mother of States, Old Dominion. ◈ *state* NE USA
21 Y7 **Virginia Beach** Virginia, NE USA 36°51′N 75°59′W
33 R11 **Virginia City** Montana, NW USA 45°17′N 111°54′W
35 Q6 **Virginia City** Nevada, W USA 39°19′N 119°39′W
14 H8 **Virginiatown** Ontario, S Canada 48°09′N 79°35′W
Virgin Islands *see* British Virgin Islands
45 T9 **Virgin Islands (US)** *var.* Virgin Islands of the United States; *prev.* Danish West Indies. ◇ *US unincorporated territory* E West Indies
Virgin Islands of the United States *see* Virgin Islands (US)
45 T9 **Virgin Passage** *passage* Puerto Rico/Virgin Islands (US)
35 Y10 **Virgin River** ∿ Nevada/Utah, W USA
Virihaur *see* Virihaure
92 H12 **Virihaure** *Lapp.* Virihávrre, *var.* Virihaur. ⊚ N Sweden
Virihávrre *see* Virihaure
167 T11 **Virôchey** Rôtânôkiri, NE Cambodia 13°59′N 106°49′E
93 N19 **Virolahti** Kymenlaakso, S Finland 60°33′N 27°37′E
30 J8 **Viroqua** Wisconsin, N USA 43°33′N 90°54′W
112 G8 **Virovitica** *Ger.* Virovititz, *Hung.* Verőcze; *prev. Ger.* Werowitz. Virovitica-Podravina, NE Croatia 45°49′N 17°25′E
112 G8 **Virovitica-Podravina** *off.* Virovitičko-Podravska Županija. ◈ *province* NE Croatia
Virovitičko-Podravska Županija *see* Virovitica-Podravina
Virovititz *see* Virovitica
113 J17 **Virpazar** S Montenegro 42°15′N 19°06′E
93 L17 **Virrat** *Swe.* Virdois. Pirkanmaa, W Finland 62°14′N 23°50′E
95 M20 **Virserum** Kalmar, S Sweden 57°17′N 15°18′E
99 K25 **Virton** Luxembourg, SE Belgium 49°34′N 05°32′E
118 F5 **Virtsu** *Ger.* Werder. Läänemaa, W Estonia 58°35′N 23°33′E
56 C12 **Virú** La Libertad, C Peru 08°24′S 78°40′W
Virudhunagar *see* Virudunagar
155 H23 **Virudunagar** *var.* Virudhunagar; *prev.* Virudupatti. Tamil Nādu, SE India 09°35′N 77°57′E
Virudupatti *see* Virudunagar
118 I3 **Viru-Jaagupi** *Ger.* Sankt-Jakobi. Lääne-Virumaa, NE Estonia 59°14′N 26°29′E
57 N19 **Viru-Viru** *var.* Santa Cruz. ✈ (Santa Cruz) Santa Cruz, C Bolivia 17°49′S 63°12′W
113 E15 **Vis** *It.* Lissa; *anc.* Issa. *island* S Croatia
Vis *see* Fish
118 I12 **Visaginas** *prev.* Sniečkus. Utena, E Lithuania 55°36′N 26°22′E
155 M15 **Visākhapatnam** *var.* Vishakhapatnam. Andhra Pradesh, SE India 17°45′N 83°19′E
35 R11 **Visalia** California, W USA 36°19′N 119°19′W
Vişău *see* Vişeu
95 P19 **Visby** *Ger.* Wisby. Gotland, SE Sweden 57°37′N 18°20′E
197 N9 **Viscount Melville Sound** *prev.* Melville Sound. *sound* Northwest Territories, N Canada
99 L19 **Visé** Liège, E Belgium 50°44′N 05°42′E
112 K13 **Višegrad** Republika Srpska, SE Bosnia and Herzegovina 43°46′N 19°18′E
58 L12 **Viseu** Pará, NE Brazil 01°10′S 46°09′W
104 H7 **Viseu** *prev.* Vizeu. Viseu, N Portugal 40°40′N 07°55′W
104 H7 **Viseu** *var.* Vizeu. ◈ *district* N Portugal
116 I8 **Vişeu** *Hung.* Visó; *prev.* Vişău. ∿ NW Romania
116 I8 **Vişeu de Sus** *var.* Vişeul de Sus, *Ger.* Oberwischau, *Hung.* Felsővisó. Maramureş, N Romania 47°43′N 23°24′E
Vişeul de Sus *see* Vişeu de Sus
Vishakhapatnam *see* Visākhapatnam
125 R10 **Vishera** ∿ NW Russian Federation
95 J19 **Viskafors** Västra Götaland, S Sweden 57°37′N 12°50′E
95 J20 **Viskan** ∿ S Sweden
95 L21 **Vislanda** Kronoberg, S Sweden 56°46′N 14°30′E
Vislinskiy Zaliv *see* Vistula Lagoon
Visó *see* Vişeu
112 H13 **Visoko** ◈ Federacija Bosne I Hercegovine, C Bosnia and Herzegovina
106 A9 **Viso, Monte** ▲ NW Italy 44°42′N 07°04′E
108 E10 **Visp** Valais, SW Switzerland 46°18′N 07°53′E
108 E10 **Vispa** ∿ S Switzerland
95 M21 **Vissefjärda** Kalmar, S Sweden 56°31′N 15°34′E
100 I11 **Visselhövede** Niedersachsen, NW Germany 52°58′N 09°36′E
95 G23 **Vissenbjerg** Syddtjylland, C Denmark 55°23′N 10°08′E
35 U17 **Vista** California, W USA 33°12′N 117°14′W
58 C11 **Vista Alegre** Amazonas, NW Brazil 01°23′N 68°13′W
114 J13 **Vistonída, Límni** ⊚ NE Greece
Vistula *see* Wisła
119 A14 **Vistula Lagoon** *Ger.* Frisches Haff, *Pol.* Zalew Wiślany, *Rus.* Vislinskiy Zaliv. *lagoon* Poland/Russian Federation
114 I8 **Vit** ∿ NW Bulgaria
Vitebsk *see* Vitsyebsk
Vitebskaya Oblast' *see* Vitsyebskaya Voblasts'
107 H14 **Viterbo** *anc.* Vicus Elbii. Lazio, C Italy 42°25′N 12°08′E
112 H12 **Vitez** Federacija Bosne I Hercegovine, C Bosnia and Herzegovina 44°08′N 17°47′E
167 S14 **Vi Thanh** Cân Thơ, S Vietnam 09°45′N 105°28′E
Viti *see* Fiji
186 E7 **Vitiaz Strait** *strait* NE Papua New Guinea
104 J7 **Vitigudino** Castilla y León, N Spain 41°00′N 06°26′W
175 Q9 **Viti Levu** *island* W Fiji
187 W15 **Viti Levu** *island* W Fiji
123 O11 **Vitim** ∿ C Russian Federation
123 O12 **Vitimskiy** Irkutskaya Oblast', C Russian Federation 58°12′N 113°10′E
109 V2 **Vitis** Niederösterreich, N Austria 48°45′N 15°09′E
59 O20 **Vitória** *state capital* Espírito Santo, SE Brazil 20°19′S 40°21′W
Vitoria *see* Vitoria-Gasteiz
Vitoria Bank *see* Vitória Seamount
59 N18 **Vitória da Conquista** Bahia, E Brazil 14°53′S 40°52′W
105 P3 **Vitoria-Gasteiz** *var.* Vitoria, *Eng.* Vittoria. País Vasco, N Spain 42°51′N 02°40′W
65 J16 **Vitória Seamount** *var.* Victoria Bank, Vitoria Bank. *undersea feature* C Atlantic Ocean 18°48′S 37°24′W
112 F13 **Vitorog** ▲ SW Bosnia and Herzegovina 44°06′N 17°03′E
102 J6 **Vitré** Ille-et-Vilaine, NW France 48°07′N 01°12′W
103 R5 **Vitry-le-François** Marne, N France 48°43′N 04°36′E
114 D13 **Vítsi** *var.* Vítsoi. ▲ N Greece 40°39′N 21°23′E
Vítsoi *see* Vítsi
118 N13 **Vitsyebsk** *Rus.* Vitebsk. Vitsyebskaya Voblasts', NE Belarus 55°11′N 30°10′E
118 K13 **Vitsyebskaya Voblasts'** *Rus.* Vitebskaya Oblast'. ◈ *province* N Belarus
92 J11 **Vittangi** *Lapp.* Vazáš. Norrbotten, N Sweden 67°40′N 21°39′E
103 R8 **Vitteaux** Côte d'Or, C France 47°24′N 04°31′E
103 S6 **Vittel** Vosges, NE France 48°13′N 05°57′E
95 N15 **Vittinge** Västmanland, C Sweden 59°52′N 17°04′E
107 K25 **Vittoria** Sicilia, Italy, C Mediterranean Sea 36°56′N 14°30′E
Vittoria *see* Vitoria-Gasteiz
106 I7 **Vittorio Veneto** Veneto, NE Italy 45°59′N 12°18′E
175 Q7 **Vityaz Trench** *undersea feature* W Pacific Ocean
108 G8 **Vitznau** Luzern, W Switzerland 47°01′N 08°28′E
104 I1 **Viveiro** Galicia, NW Spain 43°39′N 07°35′W
105 S9 **Viver** Valenciana, E Spain 39°55′N 00°36′W
103 Q13 **Viverais, Monts du** ▲▲ C France
122 L9 **Vivi** ∿ N Russian Federation
22 F4 **Vivian** Louisiana, S USA 32°52′N 93°59′W
29 N10 **Vivian** South Dakota, N USA 43°53′N 100°16′W
103 R13 **Viviers** Ardèche, E France 44°31′N 04°40′E
Vivis *see* Vevey
83 K19 **Vivo** Limpopo, NE South Africa 22°58′S 29°13′E
102 L10 **Vivonne** Vienne, W France 46°25′N 00°15′E
Vizakna *see* Ocna Sibiului
Vizcaya *see* Bizkaia
Vizcaya, Golfo de *see* Biscay, Bay of
136 C10 **Vize** Kırklareli, NW Turkey 41°34′N 27°45′E
122 K4 **Vize, Ostrov** *island* Severnaya Zemlya, N Russian Federation
Vizeu *see* Viseu
155 M15 **Vizianagaram** *var.* Vizianagram. Andhra Pradesh, E India 18°07′N 83°25′E
Vizianagram *see* Vizianagaram
103 S12 **Vizille** Isère, E France 45°05′N 05°46′E
125 R11 **Vizinga** Respublika Komi, NW Russian Federation 61°06′N 50°09′E
116 M13 **Viziru** Brăila, SE Romania 45°00′N 27°43′E
113 K21 **Vjosës, Lumi i** *var.* Vijosa, Vijosë, *Gk.* Aóos. ∿ Albania/Greece *see also* Aóos
Vjosës, Lumi i *see* Aóos
99 H18 **Vlaams Brabant** ◈ *province* C Belgium
99 G18 **Vlaanderen** *Eng.* Flanders, *Fr.* Flandre. ◈ Belgium/France
98 G12 **Vlaardingen** Zuid-Holland, SW Netherlands 51°55′N 04°21′E
116 F10 **Vlădeasa, Vârful** *prev.* Vîrful Vlădeasa. ▲ NW Romania 46°45′N 22°46′E
Vlădeasa, Vîrful *see* Vlădeasa, Vârful
113 P16 **Vladičin Han** Serbia, SE Serbia 42°44′N 22°04′E
127 O16 **Vladikavkaz** *prev.* Dzaudzhikau, Ordzhonikidze. Respublika Severnaya Osetiya, SW Russian Federation 42°58′N 44°41′E
126 M3 **Vladimir** Vladimirskaya Oblast', W Russian Federation 56°09′N 40°21′E
Vladimirets *see* Voldymyerets'
144 M7 **Vladimirovka** Kostanay, N Kazakhstan 53°30′N 64°02′E
Vladimirovka *see* Yuzhno-Sakhalinsk
126 L3 **Vladimirskaya Oblast'** ◈ *province* W Russian Federation
126 I3 **Vladimirskiy Tupik** Smolenskaya Oblast', W Russian Federation 55°45′N 33°25′E
Vladimir-Volynskiy *see* Volodymyr-Volyns'kyy
123 Q7 **Vladivostok** Primorskiy Kray, SE Russian Federation 43°09′N 131°53′E
117 U13 **Vladyslavivka** Avtonomna Respublika Krym, S Ukraine 45°09′N 35°25′E
98 P6 **Vlagtwedde** Groningen, NE Netherlands 53°02′N 07°07′E
Vlajna *see* Kukavica
112 J12 **Vlasenica** ◈ Republika Srpska, E Bosnia and Herzegovina
112 G12 **Vlašić** ▲ C Bosnia and Herzegovina 44°18′N 17°40′E
111 D17 **Vlašim** *Ger.* Wlaschim. Středočesky Kraj, C Czech Republic 49°42′N 14°54′E
113 P15 **Vlasotince** Serbia, SE Serbia 42°58′N 22°07′E
123 Q7 **Vlasovo** Respublika Sakha (Yakutiya), NE Russian Federation 70°41′N 134°49′E
98 I11 **Vleuten** Utrecht, C Netherlands 52°07′N 05°01′E
98 I5 **Vlieland** *Fris.* Flylân. *island* Waddeneilanden, N Netherlands
98 I5 **Vliestroom** *strait* NW Netherlands
99 J14 **Vlijmen** Noord-Brabant, S Netherlands 51°42′N 05°14′E
99 E15 **Vlissingen** *Eng.* Flushing, *Fr.* Flessingue. Zeeland, SW Netherlands 51°26′N 03°34′E
Vlodava *see* Włodawa
Vlonë/Vlora *see* Vlorë
113 K22 **Vlorë** *prev.* Vlonë, *It.* Valona, Vlora. Vlorë, SW Albania 40°28′N 19°31′E
113 K22 **Vlorë** ◈ *district* SW Albania
113 K22 **Vlorës, Gjiri i** *var.* Valona Bay. *bay* SW Albania
Vlotslavsk *see* Włocławek
111 C16 **Vltava** *Ger.* Moldau. ∿ W Czech Republic
126 K3 **Vnukovo** ✈ (Moskva) Gorod Moskva, W Russian Federation 55°30′N 36°52′E
146 L11 **Vobkent** *Rus.* Vabkent. Buxoro Viloyati, C Uzbekistan 40°01′N 64°25′E
25 Q9 **Voca** Texas, SW USA 30°58′N 99°09′W
109 R5 **Vöcklabruck** Oberösterreich, NW Austria 48°01′N 13°38′E
112 D13 **Vodice** Šibenik-Knin, S Croatia 43°46′N 15°46′E
124 K10 **Vodlozero, Ozero** ⊚ NW Russian Federation
112 A10 **Vodnjan** *It.* Dignano d'Istria. Istra, NW Croatia 44°57′N 13°51′E
125 S9 **Vodnyy** Respublika Komi, NW Russian Federation 63°31′N 53°21′E
145 V15 **Vodokhranilishche Kapshagay** *Kaz.* Qapshagay Böyeni; *prev.* Kapchagayskoye Vodokhranilishche. ▦ SE Kazakhstan
124 I6 **Vodokhranilishche, Kumskoye** ⊚ NW Russian Federation
95 G20 **Vodskov** Nordjylland, N Denmark 57°07′N 10°02′E
92 H4 **Vogar** Suðurnes, SW Iceland 63°58′N 22°20′W
Vogelkop *see* Doberai, Jazirah
77 X15 **Vogel Peak** *prev.* Dimlang. ▲ E Nigeria 08°16′N 11°44′E
101 H17 **Vogelsberg** ▲▲ C Germany
106 D8 **Voghera** Lombardia, N Italy 44°59′N 09°01′E
112 I13 **Vogošća** Federacija Bosne I Hercegovine, SE Bosnia and Herzegovina 43°55′N 18°20′E
101 M17 **Vogtland** *historical region* E Germany
125 V12 **Vogul'skiy Kamen', Gora** ▲ NW Russian Federation 60°10′N 58°41′E
187 P16 **Voh** Province Nord, C New Caledonia 20°57′S 164°41′E
Vohémar *see* Iharaña
172 H8 **Vohimena, Tanjona** *Fr.* Cap Sainte Marie. *headland* S Madagascar 25°20′S 45°06′E
172 J6 **Vohipeno** Fianarantsoa, SE Madagascar 22°21′S 47°51′E
118 H5 **Võhma** *Ger.* Wöchma. Viljandimaa, S Estonia 58°37′N 25°34′E
81 J20 **Voi** Taita/Taveta, S Kenya 03°23′S 38°35′E
76 K15 **Voinjama** N Liberia 08°25′N 09°42′W
103 S12 **Voiron** Isère, E France 45°22′N 05°35′E
109 V8 **Voitsberg** Steiermark, SE Austria 47°04′N 15°09′E
95 F24 **Vojens** *Ger.* Woyens. Syddanmark, SW Denmark 55°15′N 09°19′E
112 K9 **Vojvodina** *Ger.* Wojwodina. Vojvodina, N Serbia
15 S6 **Volant** ∿ Québec, SE Canada
Volaterrae *see* Volterra
43 P15 **Volcán** *var.* Hato del Volcán. Chiriquí, W Panama 08°45′N 82°38′W
Volcano Islands *see* Kazan-rettō
Volchansk *see* Vovchans'k
Volchya *see* Vovcha
94 D10 **Volda** Møre og Romsdal, S Norway 62°07′N 06°04′E
116 K3 **Voldymyerets'** *Russ.* Vladimirets. Rivnens'ka Oblast', NW Ukraine 51°24′N 25°52′E
98 J9 **Volendam** Noord-Holland, C Netherlands 52°30′N 05°04′E
124 L15 **Volga** Yaroslavskaya Oblast', W Russian Federation 57°56′N 38°23′E
29 R10 **Volga** South Dakota, N USA 44°19′N 96°55′W
122 C11 **Volga** ∿ NW Russian Federation
Volga-Baltic Waterway *see* Volgo-Baltiyskiy Kanal
Volga Uplands *see* Privolzhskaya Vozvyshennost'
124 L13 **Volgo-Baltiyskiy Kanal** *var.* Volga-Baltic Waterway. *canal* NW Russian Federation
126 M12 **Volgodonsk** Rostovskaya Oblast', SW Russian Federation 47°35′N 42°03′E
127 O10 **Volgograd** *prev.* Stalingrad, Tsaritsyn. Volgogradskaya Oblast', SW Russian Federation 48°42′N 44°29′E
127 N9 **Volgogradskaya Oblast'** ◈ *province* SW Russian Federation
127 P10 **Volgogradskoye Vodokhranilishche** ▦ SW Russian Federation
101 J19 **Volkach** Bayern, C Germany 49°51′N 10°15′E
109 U9 **Völkermarkt** *Slvn.* Velikovec. Kärnten, S Austria 46°40′N 14°38′E
124 I12 **Volkhov** Leningradskaya Oblast', NW Russian Federation 59°56′N 32°19′E
101 D20 **Völklingen** Saarland, SW Germany 49°15′N 06°51′E
Volkovysk *see* Vawkavysk
Volkovyskiye Vysoty *see* Vawkavyskaye Wzvyshsha
83 K22 **Volksrust** Mpumalanga, E South Africa 27°22′S 29°54′E
98 L8 **Vollenhove** Overijssel, N Netherlands 52°40′N 05°58′E
119 L16 **Volma** ∿ C Belarus
Volmari *see* Valmiera
117 W9 **Volnovakha** Donets'ka Oblast', SE Ukraine 47°36′N 37°32′E
116 K6 **Volochys'k** Khmel'nyts'ka Oblast', W Ukraine 49°32′N 26°14′E
117 O6 **Volodarka** Kyyivs'ka Oblast', N Ukraine 49°31′N 29°55′E
117 W9 **Volodars'ke** Donets'ka Oblast', E Ukraine 47°11′N 37°19′E
127 R13 **Volodarskiy** Astrakhanskaya Oblast', SW Russian Federation 46°23′N 48°39′E
Volodarskoye *see* Saumalkol'
117 N8 **Volodars'k-Volyns'kyy** Zhytomyrs'ka Oblast', N Ukraine 50°37′N 28°28′E
116 I3 **Volodymyr-Volyns'kyy** *Pol.* Włodzimierz, *Rus.* Vladimir-Volynskiy. Volyns'ka Oblast', NW Ukraine 50°51′N 24°19′E
124 L14 **Vologda** Vologodskaya Oblast', W Russian Federation 59°10′N 39°55′E
124 L12 **Vologodskaya Oblast'** ◈ *province* NW Russian Federation
126 K3 **Volokolamsk** Moskovskaya Oblast', W Russian Federation 56°03′N 35°57′E
126 K9 **Volokonovka** Belgorodskaya Oblast', W Russian Federation 50°30′N 37°54′E
115 G16 **Vólos** Thessalía, C Greece 39°21′N 22°58′E
124 M11 **Voloshka** Arkhangel'skaya Oblast', NW Russian Federation 61°19′N 40°06′E
Vološinovo *see* Novi Bečej
116 H7 **Volovets'** Zakarpats'ka Oblast', W Ukraine 48°42′N 23°12′E
Volozhin *see* Valozhyn
127 Q7 **Vol'sk** Saratovskaya Oblast', W Russian Federation 52°04′N 47°20′E
77 Q17 **Volta** ∿ SE Ghana
Volta Blanche *see* White Volta
77 P16 **Volta, Lake** ▦ SE Ghana
Volta Noire *see* Black Volta
60 O9 **Volta Redonda** Rio de Janeiro, SE Brazil 22°31′S 44°05′W
Volta Rouge *see* Red Volta
106 F12 **Volterra** *anc.* Volaterrae. Toscana, C Italy 43°23′N 10°52′E
107 K17 **Volturno** ∿ S Italy
113 I15 **Volujak** ▲▲ NW Montenegro
Volunteer Island *see* Starbuck Island
65 F24 **Volunteer Point** *headland* East Falkland, Falkland Islands 51°32′S 57°44′W
114 H13 **Vólvi, Límni** ⊚ N Greece
Volyn *see* Volyns'ka Oblast'
116 I3 **Volyns'ka Oblast'** *var.* Volyn, *Rus.* Volynskaya Oblast'. ◈ *province* NW Ukraine
Volynskaya Oblast' *see* Volyns'ka Oblast'
127 Q3 **Volzhsk** Respublika Mariy El, W Russian Federation 55°53′N 48°21′E
127 O10 **Volzhskiy** Volgogradskaya Oblast', SW Russian Federation 48°49′N 44°40′E
172 I7 **Vondrozo** Fianarantsoa, SE Madagascar 22°50′S 47°20′E
39 P10 **Von Frank Mountain** ▲ Alaska, USA 63°36′N 154°29′W
115 C17 **Vónitsa** Dytikí Elláda, W Greece 38°55′N 20°53′E
118 J6 **Võnnu** *Ger.* Wendau. Tartumaa, SE Estonia 58°17′N 27°06′E
98 G12 **Voorburg** Zuid-Holland, W Netherlands 52°04′N 04°22′E
98 H11 **Voorschoten** Zuid-Holland, W Netherlands 52°08′N 04°26′E
98 M11 **Voorst** Gelderland, E Netherlands 52°10′N 06°10′E
98 K11 **Voorthuizen** Gelderland, C Netherlands 52°12′N 05°36′E
92 L2 **Vopnafjördhur** *bay* E Iceland
92 L2 **Vopnafjörður** Austurland, E Iceland 65°45′N 14°51′W
Vora *see* Vorë
119 H15 **Voranava** *Pol.* Werenów, *Rus.* Voronovo. Hrodzyenskaya Voblasts', W Belarus 54°09′N 25°19′E
108 I8 **Vorarlberg** *off.* Land Vorarlberg. ◈ *state* W Austria
Vorarlberg, Land *see* Vorarlberg
109 X7 **Vorau** Steiermark, E Austria 47°22′N 15°55′E
98 N11 **Vorden** Gelderland, E Netherlands 52°07′N 06°18′E
108 H9 **Vorderrhein** ∿ SE Switzerland
95 I24 **Vordingborg** Sjælland, SE Denmark 55°01′N 11°55′E
92 J2 **Vorðufell** ▲ N Iceland 65°42′N 18°45′W
113 K19 **Vorë** *var.* Vora. Tiranë, W Albania 41°23′N 19°37′E
115 H17 **Vóreies Sporádes** *var.* Vóreioi Sporádes, Vórioi Sporádhes, *Eng.* Northern Sporades. *island group* E Greece
Vóreioi Sporádes *see* Vóreies Sporádes
115 J17 **Vóreion Aigaíon** *Eng.* Aegean North. ◈ *region* SE Greece
115 G18 **Vóreios Evvoïkós Kólpos** *var.* Voreiós Evvoïkós Kólpos. *gulf* E Greece
197 S16 **Voring Plateau** *undersea feature* N Norwegian Sea 67°00′N 04°00′E
Vórioi Sporádhes *see* Vóreies Sporádes
125 W4 **Vorkuta** Respublika Komi, NW Russian Federation 67°27′N 64°E
95 I14 **Vorma** ∿ S Norway
118 E4 **Vormsi** *var.* Vormsi Saar, *Ger.* Worms, *Swed.* Ormsö. *island* W Estonia
Vormsi Saar *see* Vormsi
127 N7 **Vorona** ∿ W Russian Federation
126 L7 **Voronezh** Voronezhskaya Oblast', W Russian Federation 51°40′N 39°13′E
126 L7 **Voronezh** ∿ W Russian Federation
126 K8 **Voronezhskaya Oblast'** ◈ *province* W Russian Federation
Voronovitsya *see* Voronovytsya
Voronovo *see* Voranava
117 N6 **Voronovytsya** *Rus.* Voronovitsya. Vinnyts'ka Oblast', C Ukraine 49°06′N 28°49′E
122 K7 **Vorontsovo** Krasnoyarskiy Kray, N Russian Federation 71°45′N 83°31′E
124 K3 **Voron'ya** ∿ NW Russian Federation
Voropayevo *see* Varapayeva
Voroshilov *see* Ussuriysk
Voroshilovgrad *see* Luhans'ka Oblast', Ukraine
Voroshilovgrad *see* Luhans'k, Ukraine
Voroshilovgradskaya Oblast' *see* Luhans'ka Oblast'
Voroshilovsk *see* Stavropol', Russian Federation
Voroshilovsk *see* Alchevs'k
137 V13 **Vorotan** *Az.* Bärguşad. ∿ Armenia/Azerbaijan
127 P3 **Vorotynets** Nizhegorodskaya Oblast', W Russian Federation 56°06′N 46°06′E
117 S3 **Vorozhba** Sums'ka Oblast', NE Ukraine 51°10′N 34°15′E
117 T5 **Vorskla** ∿ Russian Federation/Ukraine
99 I17 **Vorst** Antwerpen, N Belgium 51°06′N 05°01′E
83 G21 **Vorstershoop** North-West, N South Africa 25°46′S 22°57′E
118 H6 **Võrtsjärv** *Ger.* Wirz-See. ⊚ SE Estonia
118 J7 **Võru** *Ger.* Werro. Võrumaa, SE Estonia 57°51′N 27°01′E
147 R11 **Vorukh** N Tajikistan 39°51′N 70°34′E
118 I7 **Võrumaa** *off.* Võru Maakond. ◈ *province* SE Estonia
Võru Maakond *see* Võrumaa
83 G24 **Vosburg** Northern Cape, W South Africa 30°35′S 22°52′E
147 Q14 **Vose'** *Rus.* Vose; *prev.* Aral. SW Tajikistan 37°51′N 69°31′E
103 S6 **Vosges** ◈ *department* NE France
103 U6 **Vosges** ▲▲ NE France
124 K13 **Voskresenkoye** Vologodskaya Oblast', NW Russian Federation 59°25′N 37°56′E
126 L4 **Voskresensk** Moskovskaya Oblast', W Russian Federation 55°19′N 38°42′E
127 P2 **Voskresenskoye** Nizhegorodskaya Oblast', W Russian Federation 57°00′N 45°33′E
127 V6 **Voskresenskoye** Respublika Bashkortostan, W Russian Federation 53°07′N 56°07′E
94 D13 **Voss** Hordaland, S Norway 60°38′N 06°25′E
94 D13 **Voss** *physical region* S Norway
99 I16 **Vosselaar** Antwerpen, N Belgium 51°19′N 04°55′E
94 D13 **Vosso** ∿ S Norway
Vostochno-Kazakhstanskaya Oblast' *see* Vostochnyy Kazakhstan
123 S5 **Vostochno-Sibirskoye More** *Eng.* East Siberian Sea. *sea* Arctic Ocean
145 X10 **Vostochnyy Kazakhstan** *off.* Vostochno-Kazakhstanskaya Oblast', *var.* East Kazakhstan, *Kaz.* Shyghys Qazagastan Oblysy. ◈ *province* E Kazakhstan
122 L13 **Vostochnyy Sayan** *Eng.* Eastern Sayans, *Mong.* Dzüün Soyonï Nuruu. ▲▲ Mongolia/Russian Federation
Vostock Island *see* Vostok Island
195 U10 **Vostok** *Russian research station* Antarctica 77°18′S 105°32′E
191 X5 **Vostok Island** *var.* Vostock Island; *prev.* Stavers Island. *island* Line Islands, SE Kiribati
127 T2 **Votkinsk** Udmurtskaya Respublika, NW Russian Federation 57°04′N 54°00′E
125 U15 **Votkinskoye Vodokhranilishche** *var.* Votkinsk Reservoir. ▦ NW Russian Federation
Votkinsk Reservoir *see* Votkinskoye Vodokhranilishche
60 J7 **Votuporanga** São Paulo, S Brazil 20°26′S 49°53′W
104 H7 **Vouga, Rio** ∿ N Portugal
115 E14 **Voúrinos** ▲▲ N Greece
115 G24 **Voúxa, Akrotírio** *headland* Kríti, Greece, E Mediterranean Sea 35°37′N 23°34′E
103 R4 **Vouziers** Ardennes, N France 49°24′N 04°42′E
117 V7 **Vovcha** *Rus.* Volchya. ∿ E Ukraine
117 V4 **Vovchans'k** *Rus.* Volchansk. Kharkivs'ka Oblast', E Ukraine 50°19′N 36°55′E
103 N6 **Voves** Eure-et-Loir, C France 48°18′N 01°39′E
79 M14 **Vovodo** ∿ S Central African Republic
94 M12 **Voxna** Gävleborg, C Sweden 61°21′N 15°35′E
94 L11 **Voxnan** ∿ C Sweden
114 F7 **Voynishka Reka** ∿ NW Bulgaria
125 T9 **Voyvozh** Respublika Komi, NW Russian Federation 62°54′N 54°52′E
124 M12 **Vozhega** Vologodskaya Oblast', NW Russian Federation 60°27′N 40°11′E
124 L12 **Vozhe, Ozero** ⊚ NW Russian Federation
117 Q9 **Voznesens'k** *Rus.* Voznesensk. Mykolayivs'ka Oblast', S Ukraine 47°34′N 31°21′E
124 J12 **Voznesen'ye** Leningradskaya Oblast', NW Russian Federation 61°00′N 35°24′E
144 J14 **Vozrozhdeniya, Ostrov** *Uzb.* Wozrojdeniye Oroli. *island* Kazakhstan/Uzbekistan
95 G20 **Vrå** *var.* Vraa. Nordjylland, N Denmark 57°21′N 09°57′E
Vraa *see* Vrå
114 H9 **Vrachesh** Sofia, W Bulgaria 42°52′N 23°45′E
115 C19 **Vrachíonas** ▲ Zákynthos, Iónia Nisiá, Greece, C Mediterranean Sea 37°49′N 20°43′E
117 P8 **Vradiyivka** Mykolayivs'ka Oblast', S Ukraine 47°51′N 30°37′E
113 G14 **Vran** ▲ SW Bosnia and Herzegovina 43°35′N 17°30′E
116 K12 **Vrancea** ◈ *county* E Romania
147 T14 **Vrang** SE Tajikistan 37°03′N 72°26′E
123 T4 **Vrangelya, Ostrov** *Eng.* Wrangel Island. *island* NE Russian Federation
112 H13 **Vranica** ▲ C Bosnia and Herzegovina 43°57′N 17°43′E
113 O16 **Vranje** Serbia, SE Serbia 42°33′N 21°55′E
Vranov *see* Vranov nad Topl'ou
111 N19 **Vranov nad Topl'ou** *var.* Vranov, *Hung.* Varannó. Prešovský Kraj, E Slovakia 48°54′N 21°41′E
114 H8 **Vratsa** Vratsa, NW Bulgaria 43°13′N 23°34′E
114 H8 **Vratsa** ◈ *province* NW Bulgaria
114 F10 **Vrattsa** *prev.* Mirovo. Kyustendil, W Bulgaria 42°15′N 22°33′E
112 G11 **Vrbanja** ∿ NW Bosnia and Herzegovina
112 K9 **Vrbas** Vojvodina, NW Serbia 45°34′N 19°39′E
112 G13 **Vrbas** ∿ N Bosnia and Herzegovina
112 E8 **Vrbovec** Zagreb, N Croatia 45°53′N 16°24′E
112 C9 **Vrbovsko** Primorje-Gorski Kotar, NW Croatia 45°22′N 15°06′E
111 E15 **Vrchlabí** *Ger.* Hohenelbe. Královéhradecký Kraj, N Czech Republic 50°38′N 15°35′E
83 J22 **Vrede** Free State, E South Africa 27°25′S 29°10′E
100 E13 **Vreden** Nordrhein-Westfalen, NW Germany 52°01′N 06°50′E
83 E25 **Vredenburg** Western Cape, SW South Africa 32°55′S 18°00′E
99 I23 **Vresse-sur-Semois** Namur, SE Belgium 49°52′N 04°56′E
95 L16 **Vretstorp** Örebro, C Sweden 59°03′N 14°51′E
113 G15 **Vrgorac** *prev.* Vrhgorac. Split-Dalmacija, SE Croatia 43°10′N 17°24′E
Vrhgorac *see* Vrgorac
109 T12 **Vrhnika** *Ger.* Oberlaibach. W Slovenia 45°57′N 14°18′E
155 I21 **Vriddhāchalam** Tamil Nādu, SE India 11°33′N 79°18′E
98 N6 **Vries** Drenthe, NE Netherlands 53°04′N 06°34′E
98 O10 **Vriezenveen** Overijssel, E Netherlands 52°25′N 06°37′E
95 L20 **Vrigstad** Jönköping, S Sweden 57°19′N 14°30′E
108 H9 **Vrin** Graubünden, S Switzerland 46°40′N 09°06′E
112 E13 **Vrlika** Split-Dalmacija, S Croatia 43°54′N 16°24′E
113 M14 **Vrnjačka Banja** Serbia, C Serbia 43°36′N 20°55′E
Vrondádhes/Vrondados *see* Vrontádos
115 L18 **Vrontádos** *var.* Vrondados; *prev.* Vrondádhes. Chíos, E Greece 38°25′N 26°08′E
98 N9 **Vroomshoop** Overijssel, E Netherlands 52°28′N 06°35′E
112 N10 **Vršac** *Ger.* Werschetz, *Hung.* Versecz. Vojvodina, NE Serbia 45°08′N 21°18′E
112 M10 **Vršački Kanal** *canal* N Serbia
83 H21 **Vryburg** North-West, N South Africa 26°57′S 24°44′E
83 K22 **Vryheid** KwaZulu/Natal, E South Africa 27°45′S 30°48′E
111 I18 **Vsetín** *Ger.* Wsetin. Zlínský Kraj, E Czech Republic 49°21′N 17°57′E
111 J20 **Vtáčnik** *Hung.* Madaras, Ptacsnik; *prev.* Ptačnik. ▲ W Slovakia 48°38′N 18°38′E
Vuadil' *see* Wodil
Vŭcha *see* Vacha
Vučitrn *see* Vushtrri

◆ Country
● Country Capital
◇ Dependent Territory
○ Dependent Territory Capital
◈ Administrative Regions
✈ International Airport
▲ Mountain
▲▲ Mountain Range
⛰ Volcano
∿ River
⊚ Lake
▦ Reservoir

99 J14 **Vught** Noord-Brabant, S Netherlands 51°37′N 05°19′E
117 W8 **Vuhledar** Donets'ka Oblast', E Ukraine 47°48′N 37°11′E
112 I9 **Vuka** ≈ E Croatia
113 K17 **Vukël** *var.* Vukli. Shkodër, N Albania 42°29′N 19°39′E
Vukli *see* Vukël
112 J9 **Vukovar** *Hung.* Vukovár. Vukovar-Srijem, E Croatia 45°18′N 18°45′E
Vukovarsko-Srijemska Županija *see* Vukovar-Srijem
112 I10 **Vukovar-Srijem** *off.* Vukovarsko-Srijemska Županija. ◈ *province* E Croatia
125 U8 **Vuktyl** Respublika Komi, NW Russian Federation 63°49′N 57°07′E
11 Q17 **Vulcan** Alberta, SW Canada 50°27′N 113°12′W
116 G12 **Vulcan** *Ger.* Wulkan, *Hung.* Zsilyvajdevulkán; *prev.* Crivadia Vulcanului, Vaidei, *Hung.* Sily-Vajdej, Vajdej. Hunedoara, W Romania 45°22′N 23°16′E
116 M12 **Vulcănești** *Rus.* Vulkaneshty. S Moldova 45°41′N 28°25′E
107 L22 **Vulcano, Isola** *island* Isole Eolie, S Italy
Vŭlchedrŭm *see* Valchedram
Vŭlchidol *see* Valchi Dol
Vulkaneshty *see* Vulcănești
123 V11 **Vulkannyy** Kamchatskiy Kray, E Russian Federation 53°01′N 158°26′E
36 J13 **Vulture Mountains** ▲▲ Arizona, SW USA
167 T14 **Vung Tau** *prev. Fr.* Cape Saint Jacques, Cap Saint-Jacques. Ba Ria-Vung Tau, S Vietnam 10°21′N 107°04′E
187 X15 **Vunisea** Kadavu, SE Fiji 19°04′S 178°10′E
Vuohčču *see* Vuotso
93 N15 **Vuokatti** Kainuu, C Finland 64°08′N 28°16′E
93 M15 **Vuolijoki** Kainuu, C Finland 64°09′N 27°00′E
Vuolleriebme *see* Vuollerim
92 J13 **Vuollerim** *Lapp.* Vuolleriebme. Norrbotten, N Sweden 66°24′N 20°36′E
Vuonnabahta *see* Varangerbotn
Vuoreija *see* Vardø
92 L10 **Vuotso** *Lapp.* Vuohčču. Lappi, N Finland 68°04′N 27°05′E
Vŭrbitsa *see* Varbitsa
127 Q4 **Vurnary** Chuvashskaya Respublika, W Russian Federation 55°30′N 46°59′E
Vŭrshets *see* Varshets
Vusan *see* Busan
113 N16 **Vushtrri** *Serb.* Vučitrn. N Kosovo 42°49′N 21°00′E
119 F17 **Vyalikaya Byerastavitsa** *Pol.* Brzostowica Wielka, *Rus.* Bol'shaya Berëstovitsa; *prev.* Velikaya Berestovitsa. Hrodzyenskaya Voblasts', SW Belarus 53°12′N 24°03′E
119 N20 **Vyaliki Bor** *Rus.* Velikiy Bor. Homyel'skaya Voblasts', SE Belarus 52°02′N 29°56′E
119 J18 **Vyaliki Rozhan** *Rus.* Bol'shoy Rozhan. Minskaya Voblasts', S Belarus 52°46′N 27°07′E
124 H10 **Vyartsilya** *Fin.* Värtsilä. Respublika Kareliya, NW Russian Federation 62°07′N 30°43′E
119 K17 **Vyasyeya** *Rus.* Veseya. Minskaya Voblasts', C Belarus 53°04′N 27°41′E
125 R15 **Vyatka** ≈ NW Russian Federation
Vyatka *see* Kirov
125 S16 **Vyatskiye Polyany** Kirovskaya Oblast', NW Russian Federation 56°15′N 51°06′E
123 S14 **Vyazemskiy** Khabarovskiy Kray, SE Russian Federation 47°28′N 134°39′E
126 I4 **Vyaz'ma** Smolenskaya Oblast', W Russian Federation 55°09′N 34°20′E
127 N3 **Vyazniki** Vladimirskaya Oblast', W Russian Federation 56°15′N 42°06′E
127 O8 **Vyazovka** Volgogradskaya Oblast', SW Russian Federation 50°57′N 43°57′E
119 J14 **Vyazyn'** Minskaya Voblasts', NW Belarus 54°25′N 27°10′E
124 G11 **Vyborg** *Fin.* Viipuri. Leningradskaya Oblast', NW Russian Federation 60°44′N 28°47′E
125 P11 **Vychegda** *var.* Vichegda. ≈ NW Russian Federation
119 L14 **Vyelyewshchyna** *Rus.* Velevshchina. Vitsyebskaya Voblasts', N Belarus 54°44′N 28°35′E
119 P16 **Vyeramyeyki** *Rus.* Veremeyki. Mahilyowskaya Voblasts', E Belarus 53°46′N 31°17′E
118 K11 **Vyerkhnyadzvinsk** *Rus.* Verkhnedvinsk. Vitsyebskaya Voblasts', N Belarus 55°47′N 27°56′E
119 P18 **Vyetka** *Rus.* Vetka. Homyel'skaya Voblasts', SE Belarus 52°33′N 31°10′E
118 L12 **Vyetryna** *Rus.* Vetrino. Vitsyebskaya Voblasts', N Belarus 55°25′N 28°28′E
Vygonovskoye, Ozero *see* Vyhanashchanskaye, Vozyera
124 J9 **Vygozero, Ozero** ◎ NW Russian Federation
119 I18 **Vyhanashchanskaye, Vozyera** *prev.* Vozyera Vyhanawskaye, *Rus.* Ozero Vygonovskoye. ◎ SW Belarus
Vyhanawskaye, Vozyera *see* Vyhanashchanskaye, Vozyera
127 N4 **Vyksa** Nizhegorodskaya Oblast', W Russian Federation 55°21′N 42°10′E
117 O12 **Vylkove** *Rus.* Vilkovo. Odes'ka Oblast', SW Ukraine 45°24′N 29°37′E
125 R9 **Vym'** ≈ NW Russian Federation
116 H8 **Vynohradiv** *Cz.* Sevluš, *Hung.* Nagyszöllös, *Rus.* Vinogradov; *prev.* Sevlyush. Zakarpats'ka Oblast', W Ukraine 48°09′N 23°01′E
124 G13 **Vyritsa** Leningradskaya Oblast', NW Russian Federation 59°25′N 30°20′E
97 J19 **Vyrnwy** *Wel.* Afon Efyrnwy. ≈ E Wales, United Kingdom
145 X9 **Vyshe Ivanovskiy Belak, Gora** ▲ E Kazakhstan 50°16′N 83°46′E
117 P4 **Vyshhorod** Kyyivs'ka Oblast', N Ukraine 50°36′N 30°28′E
124 I15 **Vyshniy Volochek** Tverskaya Oblast', W Russian Federation 57°37′N 34°33′E
111 G18 **Vyškov** *Ger.* Wischau. Jihomoravský Kraj, SE Czech Republic 49°17′N 17°01′E
111 E18 **Vysočina** *prev.* Jihlavský Kraj. ◈ *region* N Czech Republic
119 E19 **Vysokaye** *Rus.* Vysokoye. Brestskaya Voblasts', SW Belarus 52°20′N 23°18′E
111 F17 **Vysoké Mýto** *Ger.* Hohenmauth. Pardubický Kraj, C Czech Republic 49°57′N 16°10′E
117 S9 **Vysokopillya** Khersons'ka Oblast', S Ukraine 47°28′N 33°30′E
126 K3 **Vysokovsk** Moskovskaya Oblast', W Russian Federation 56°12′N 36°42′E
Vysokoye *see* Vysokaye
124 K12 **Vytegra** Vologodskaya Oblast', NW Russian Federation 60°59′N 36°27′E
116 J8 **Vyzhnytsya** Chernivets'ka Oblast', W Ukraine 48°14′N 25°10′E

W

77 O14 **Wa** NW Ghana 10°07′N 02°28′W
Waadt *see* Vaud
Waag *see* Váh
Waagbistritz *see* Považská Bystrica
Waagneustadtl *see* Nové Mesto nad Váhom
81 M16 **Waajid** Gedo, SW Somalia 03°37′N 43°19′E
98 L13 **Waal** ≈ S Netherlands
187 O16 **Waala** Province Nord, W New Caledonia 19°46′S 163°41′E
99 I14 **Waalwijk** Noord-Brabant, S Netherlands 51°42′N 05°04′E
99 E16 **Waarschoot** Oost-Vlaanderen, NW Belgium 51°09′N 03°35′E
186 C7 **Wabag** Enga, W Papua New Guinea 05°28′S 143°40′E
15 N7 **Wabano** ≈ Québec, SE Canada
11 P11 **Wabasca** ≈ Alberta, SW Canada
31 P12 **Wabash** Indiana, N USA 40°47′N 85°48′W
29 X9 **Wabasha** Minnesota, N USA 44°22′N 92°01′W
31 N13 **Wabash River** ≈ N USA
14 C7 **Wabatongushi Lake** ◎ Ontario, S Canada
81 L15 **Wabē Gestro Wenz** ≈ SE Ethiopia
14 B9 **Wabos** Ontario, S Canada 46°48′N 84°06′W
11 W13 **Wabowden** Manitoba, C Canada 54°57′N 98°38′W
110 J9 **Wąbrzeźno** Kujawsko-pomorskie, C Poland 53°18′N 18°55′E
21 U12 **Waccamaw River** ≈ South Carolina, SE USA
23 U11 **Waccasassa Bay** *bay* Florida, SE USA
99 F16 **Wachtebeke** Oost-Vlaanderen, NW Belgium 51°10′N 03°52′E
25 T8 **Waco** Texas, SW USA 31°33′N 97°10′W
26 M3 **Waconda Lake** *var.* Great Elder Reservoir. ⊠ Kansas, C USA
Wadai *see* Ouaddaï
Wad Al-Hajarah *see* Guadalajara
164 I12 **Wadayama** Hyōgo, Honshū, SW Japan 35°19′N 134°51′E
80 D10 **Wad Banda** Western Kordofan, C Sudan 13°08′N 27°56′E
75 P9 **Waddān** NW Libya 29°10′N 16°08′E
98 J4 **Waddeneilanden** *Eng.* West Frisian Islands. *island group* N Netherlands
98 J6 **Waddenzee** *var.* Wadden Zee. *sea* SE North Sea
Wadden Zee *see* Waddenzee
10 L16 **Waddington, Mount** ▲ British Columbia, SW Canada 51°17′N 125°16′W
98 H12 **Waddinxveen** Zuid-Holland, C Netherlands 52°03′N 04°38′E
11 U15 **Wadena** Saskatchewan, S Canada 51°57′N 103°48′W
29 T6 **Wadena** Minnesota, N USA 46°27′N 95°07′W
108 G7 **Wädenswil** Zürich, N Switzerland 47°14′N 08°41′E
21 S11 **Wadesboro** North Carolina, SE USA 34°59′N 80°03′W
155 G16 **Wādi** Karnātaka, C India 17°00′N 76°58′E
138 G10 **Wādī as Sīr** *var.* Wadi es Sir. 'Ammān, NW Jordan 31°57′N 35°49′E
Wadi es Sir *see* Wādī as Sīr
78 J9 **Wadi Fira** *off.* Région du Wadi Fira; *prev.* Préfecture de Biltine. ◈ *region* E Chad
Wadi Fira, Région du *see* Wadi Fira
80 F5 **Wadi Halfa** *var.* Wādī Ḥalfā'. Northern, N Sudan 21°46′N 31°17′E
138 G13 **Wādī Mūsā** *var.* Petra. Ma'ān, S Jordan 30°19′N 35°29′E
23 V4 **Wadley** Georgia, SE USA 32°52′N 82°24′W
Wad Madanī *see* Wad Medani
80 G10 **Wad Medani** *var.* Wad Madanī. Gezira, C Sudan 14°24′N 33°30′E
80 F10 **Wad Nimr** White Nile, C Sudan 14°32′N 32°10′E
165 U16 **Wadomari** Kagoshima, Okinoerabu-jima, SW Japan 27°25′N 128°40′E
111 K17 **Wadowice** Małopolskie, S Poland 49°54′N 19°29′E
35 R5 **Wadsworth** Nevada, W USA 39°39′N 119°16′W
31 T12 **Wadsworth** Ohio, N USA 41°01′N 81°43′W
25 T11 **Waelder** Texas, SW USA 29°42′N 97°16′W
Waereghem *see* Waregem
163 U13 **Wafangdian** *var.* Fuxian, Fu Xian. Liaoning, NE China 39°36′N 122°00′E
171 R13 **Waflia** Pulau Buru, E Indonesia 03°10′S 126°05′E
Wagadugu *see* Ouagadougou
98 K12 **Wageningen** Gelderland, SE Netherlands 51°58′N 05°40′E
55 V9 **Wageningen** Nickerie, NW Suriname 05°44′N 56°45′W
9 O8 **Wager Bay** *inlet* Nunavut, N Canada
183 P10 **Wagga Wagga** New South Wales, SE Australia 35°11′S 147°22′E
180 J13 **Wagin** Western Australia 33°16′S 117°26′E
108 H8 **Wägitaler See** ◎ SW Switzerland
29 P12 **Wagner** South Dakota, N USA 43°04′N 98°17′W
27 Q9 **Wagoner** Oklahoma, C USA 35°58′N 92°23′W
37 U10 **Wagon Mound** New Mexico, SW USA 36°00′N 104°42′W
32 J14 **Wagontire** Oregon, NW USA 43°15′N 119°51′W
110 H10 **Wągrowiec** Wielkopolskie, C Poland 52°49′N 17°11′E
149 U6 **Wāh** Punjab, NE Pakistan 33°50′N 72°44′E
171 S13 **Wahai** Pulau Seram, E Indonesia 02°48′S 129°29′E
169 V10 **Wahau, Sungai** ≈ Borneo, C Indonesia
Wahaybah, Ramlat Al *see* Wahībah, Ramlat Āl
Wahda *see* Unity
38 D9 **Wahiawā** *var.* Wahiawa. O'ahu, Hawaii, USA, C Pacific Ocean 21°30′N 158°01′W
Wahībah, Ramlat Ahl *see* Wahībah, Ramlat Āl
141 Y9 **Wahībah, Ramlat Āl** *var.* Ramlat Ahl Wahībah, Ramlat Al Wahaybah, *Eng.* Wahībah Sands. *desert* N Oman
Wahībah Sands *see* Wahībah, Ramlat Āl
101 E16 **Wahn** ✈ (Köln) Nordrhein-Westfalen, W Germany 50°51′N 07°09′E
29 R15 **Wahoo** Nebraska, C USA 41°12′N 96°37′W
29 R6 **Wahpeton** North Dakota, N USA 46°16′N 96°36′W
Wahran *see* Oran
36 J6 **Wah Wah Mountains** ▲▲ Utah, W USA
38 D9 **Waialua** O'ahu, Hawaii, USA, C Pacific Ocean 21°34′N 158°07′W
38 D9 **Wai'anae** *var.* Waianae. O'ahu, Hawaii, USA, C Pacific Ocean 21°26′N 158°11′W
184 Q8 **Waiapu** ≈ North Island, New Zealand
185 I17 **Waiau** Canterbury, South Island, New Zealand 42°39′S 173°03′E
185 I17 **Waiau** ≈ South Island, New Zealand
185 B23 **Waiau** ≈ South Island, New Zealand
101 H21 **Waiblingen** Baden-Württemberg, S Germany 48°49′N 09°19′E
Waidhofen *see* Waidhofen an der Ybbs, Niederösterreich, Austria
Waidhofen *see* Waidhofen an der Thaya, Niederösterreich, Austria
109 V2 **Waidhofen an der Thaya** *var.* Waidhofen. Niederösterreich, NE Austria 48°49′N 15°17′E
109 U5 **Waidhofen an der Ybbs** *var.* Waidhofen. Niederösterreich, E Austria 47°58′N 14°47′E
171 T11 **Waigeo, Pulau** *island* Papua Barat, E Indonesia
184 L5 **Waiheke Island** *island* N New Zealand
184 M7 **Waihi** Waikato, North Island, New Zealand 37°22′S 175°51′E
185 C20 **Waihou** ≈ North Island, New Zealand
Waikaboebak *see* Waikabubak
171 N17 **Waikabubak** *prev.* Waikaboebak. Pulau Sumba, C Indonesia 09°40′S 119°25′E
185 D23 **Waikaia** ≈ South Island, New Zealand
185 D23 **Waikaka** Southland, South Island, New Zealand 45°55′S 168°59′E
184 L13 **Waikanae** Wellington, North Island, New Zealand 40°52′S 175°03′E
184 M7 **Waikare, Lake** ◎ North Island, New Zealand
184 O9 **Waikaremoana, Lake** ◎ North Island, New Zealand
185 I17 **Waikari** Canterbury, South Island, New Zealand 42°50′S 172°41′E
184 L8 **Waikato** *off.* Waikato Region. ◈ *region* North Island, New Zealand
184 M8 **Waikato** ≈ North Island, New Zealand
Waikato Region *see* Waikato
182 J9 **Waikerie** South Australia 34°12′S 139°57′E
185 F23 **Waikouaiti** Otago, South Island, New Zealand 45°36′S 170°39′E
38 H11 **Wailea** Hawaii, USA, C Pacific Ocean 19°53′N 155°07′W
38 F10 **Wailuku** Maui, Hawaii, USA, C Pacific Ocean 20°53′N 156°30′W
185 H18 **Waimakariri** ≈ South Island, New Zealand
38 D9 **Waimānalo Beach** *var.* Waimanalo Beach. O'ahu, Hawaii, USA, C Pacific Ocean 21°20′N 157°42′W
185 G15 **Waimangaroa** West Coast, South Island, New Zealand 41°41′S 171°49′E
185 G21 **Waimate** Canterbury, South Island, New Zealand 44°44′S 171°03′E
38 B8 **Waimea** Kaua'i, Hawaii, USA, C Pacific Ocean 21°57′N 159°40′W
38 G11 **Waimea** *var.* Kamuela. Hawaii, USA, C Pacific Ocean 20°02′N 155°20′W
38 D9 **Waimea** *var.* Maunawai. O'ahu, Hawaii, USA, C Pacific Ocean 21°39′N 158°04′W
99 M20 **Waimes** Liège, E Belgium 50°25′N 06°10′E
154 J11 **Wainganga** *var.* Wain River. ≈ C India
Waingapoe *see* Waingapu
171 N17 **Waingapu** *prev.* Waingapoe. Pulau Sumba, C Indonesia 09°40′S 120°16′E
55 S7 **Waini** ≈ N Guyana
55 S7 **Waini Point** *headland* NW Guyana 08°24′N 59°48′W
Wain River *see* Wainganga
11 R15 **Wainwright** Alberta, SW Canada 52°50′N 110°51′W
39 O5 **Wainwright** Alaska, USA 70°38′N 160°02′W
184 K4 **Waiotira** Northland, North Island, New Zealand 35°56′S 174°11′E
184 M11 **Waiouru** Manawatu-Wanganui, North Island, New Zealand 39°28′S 175°41′E
171 W14 **Waipa** Papua, E Indonesia 03°47′S 136°16′E
184 L8 **Waipa** ≈ North Island, New Zealand
184 P9 **Waipaoa** ≈ North Island, New Zealand
185 D25 **Waipapa Point** *headland* South Island, New Zealand 46°39′S 168°51′E
185 I18 **Waipara** Canterbury, South Island, New Zealand 43°04′S 172°45′E
184 N12 **Waipawa** Hawke's Bay, North Island, New Zealand 39°57′S 176°36′E
184 K4 **Waipu** Northland, North Island, New Zealand 35°58′S 174°25′E
184 N12 **Waipukurau** Hawke's Bay, North Island, New Zealand 40°01′S 176°34′E
171 U14 **Wair** Pulau Kai Besar, E Indonesia 05°16′S 133°09′E
Wairakai *see* Wairakei
184 N9 **Wairakei** *var.* Wairakai. Waikato, North Island, New Zealand 38°37′S 176°05′E
185 M14 **Wairarapa, Lake** ◎ North Island, New Zealand
185 J15 **Wairau** ≈ South Island, New Zealand
184 P10 **Wairoa** Hawke's Bay, North Island, New Zealand 39°03′S 177°26′E
184 P10 **Wairoa** ≈ North Island, New Zealand
184 J4 **Wairoa** ≈ North Island, New Zealand
184 N9 **Waitahanui** Waikato, North Island, New Zealand 38°48′S 176°04′E
184 M6 **Waitakaruru** Waikato, North Island, New Zealand 37°14′S 175°22′E
185 F21 **Waitaki** ≈ South Island, New Zealand
184 K10 **Waitara** Taranaki, North Island, New Zealand 39°01′S 174°14′E
184 M7 **Waitoa** Waikato, North Island, New Zealand 37°36′S 175°37′E
184 L8 **Waitomo Caves** Waikato, North Island, New Zealand 38°17′S 175°06′E
184 L11 **Waitotara** Taranaki, North Island, New Zealand 39°49′S 174°43′E
184 L11 **Waitotara** ≈ North Island, New Zealand
32 L10 **Waitsburg** Washington, NW USA 46°16′N 118°09′W
Waitzen *see* Vác
184 L6 **Waiuku** Auckland, North Island, New Zealand 37°15′S 174°45′E
164 L10 **Wajima** *var.* Wazima. Ishikawa, Honshū, SW Japan 37°23′N 136°53′E
81 K17 **Wajir** Wajir, NE Kenya 01°46′N 40°05′E
81 K18 **Wajir** ◈ *county* NE Kenya
79 J17 **Waka** Equateur, NW Dem. Rep. Congo 01°04′N 20°11′E
81 I14 **Waka** Southern Nationalities, S Ethiopia 07°12′N 37°19′E
14 D9 **Wakami Lake** ◎ Ontario, S Canada
164 I12 **Wakasa** Tottori, Honshū, SW Japan 35°18′N 134°25′E
164 J12 **Wakasa-wan** *bay* C Japan
185 C22 **Wakatipu, Lake** ◎ South Island, New Zealand
11 T15 **Wakaw** Saskatchewan, S Canada 52°40′N 105°45′W
164 I14 **Wakayama** Wakayama, Honshū, SW Japan 34°12′N 135°09′E
164 I15 **Wakayama** *off.* Wakayama-ken. ◈ *prefecture* Honshū, SW Japan
Wakayama-ken *see* Wakayama
26 K4 **Wa Keeney** Kansas, C USA 39°02′N 99°53′W
185 I14 **Wakefield** Tasman, South Island, New Zealand 41°24′S 173°03′E
97 M17 **Wakefield** N England, United Kingdom 53°42′N 01°29′W
27 O4 **Wakefield** Kansas, C USA 39°12′N 97°00′W
30 L4 **Wakefield** Michigan, N USA 46°27′N 89°55′W
21 U9 **Wake Forest** North Carolina, SE USA 35°58′N 78°30′W
Wakeham Bay *see* Kangiqsujuaq
189 Y11 **Wake Island** ◇ *US unincorporated territory* NW Pacific Ocean
189 Y12 **Wake Island** ✈ NW Pacific Ocean
189 Y12 **Wake Island** *atoll* NW Pacific Ocean
189 X12 **Wake Lagoon** *lagoon* Wake Island, NW Pacific Ocean
166 L8 **Wakema** Ayeyawady, SW Myanmar (Burma) 16°36′N 95°11′E
Wakhan *see* Khandūd
164 H14 **Waki** Tokushima, Shikoku, SW Japan 34°04′N 134°10′E
165 T1 **Wakkanai** Hokkaidō, NE Japan 45°25′N 141°39′E
83 K22 **Wakkerstroom** Mpumalanga, E South Africa 27°21′S 30°10′E
14 C10 **Wakomata Lake** ◎ Ontario, S Canada
183 N10 **Wakool** New South Wales, SE Australia 35°30′S 144°22′E
Wakra *see* Al Wakrah
Waku Kungo *see* Uaco Cungo
186 J7 **Wakunai** Bougainville, NE Papua New Guinea 05°52′S 155°10′E
Walachei/Walachia *see* Wallachia
155 K26 **Walawe Ganga** ≈ S Sri Lanka
111 F15 **Wałbrzych** *Ger.* Waldenburg, Waldenburg in Schlesien. Dolnośląskie, SW Poland 50°45′N 16°20′E
183 T6 **Walcha** New South Wales, SE Australia 31°01′S 151°38′E
101 K24 **Walchensee** ◎ SE Germany
99 D14 **Walcheren** *island* SW Netherlands
29 Z14 **Walcott** Iowa, C USA 41°34′N 90°46′W
33 W16 **Walcott** Wyoming, C USA 41°46′N 106°46′W
99 G21 **Walcourt** Namur, S Belgium 50°16′N 04°26′E
110 G9 **Wałcz** *Ger.* Deutsch Krone. Zachodnio-pomorskie, NW Poland 53°17′N 16°29′E
108 H7 **Wald** Zürich, N Switzerland 47°17′N 08°56′E
109 U3 **Waldaist** ≈ N Austria
180 I9 **Waldburg Range** ▲▲ Western Australia
37 R3 **Walden** Colorado, C USA 40°43′N 106°16′W
18 K13 **Walden** New York, NE USA 41°35′N 74°09′W
Waldenburg/Waldenburg in Schlesien *see* Wałbrzych
11 T15 **Waldheim** Saskatchewan, S Canada 52°38′N 106°35′W
Waldia *see* Weldiya
101 M23 **Waldkraiburg** Bayern, SE Germany 48°10′N 12°23′E
27 T14 **Waldo** Arkansas, C USA 33°21′N 93°18′W
23 V9 **Waldo** Florida, SE USA 29°47′N 82°07′W
19 R7 **Waldoboro** Maine, NE USA 44°06′N 69°22′W
21 W4 **Waldorf** Maryland, NE USA 38°36′N 76°54′W
32 F12 **Waldport** Oregon, NW USA 44°25′N 124°04′W
27 S11 **Waldron** Arkansas, C USA 34°54′N 94°09′W
195 Y13 **Waldron, Cape** *headland* Antarctica 66°08′S 116°00′E
101 F24 **Waldshut-Tiengen** Baden-Württemberg, S Germany 47°37′N 08°13′E
171 P12 **Walea, Selat** *strait* Sulawesi, C Indonesia
Wałeckie Międzyrzecze *see* Valašské Meziříčí
108 H8 **Walensee** ◎ NW Switzerland
38 L8 **Wales** Alaska, USA 65°36′N 168°03′W
97 J20 **Wales** *Wel.* Cymru. ◈ *national region* Wales, United Kingdom
9 O7 **Wales Island** *island* Nunavut, NE Canada
77 P14 **Walewale** N Ghana 10°21′N 00°48′W
99 M24 **Walferdange** Luxembourg, C Luxembourg 49°39′N 06°08′E
183 Q5 **Walgett** New South Wales, SE Australia 30°02′S 148°14′E
194 K10 **Walgreen Coast** *physical region* Antarctica
29 Q2 **Walhalla** North Dakota, N USA 48°55′N 97°55′W
21 O11 **Walhalla** South Carolina, SE USA 34°46′N 83°05′W
79 O19 **Walikale** Nord-Kivu, E Dem. Rep. Congo 01°29′S 28°05′E
Walk *see* Valga, Estonia
Walk *see* Valka, Latvia
29 U5 **Walker** Minnesota, N USA 47°06′N 94°35′W
15 V4 **Walker, Lac** ◎ Québec, SE Canada
35 S7 **Walker Lake** ◎ Nevada, W USA
35 R6 **Walker River** ≈ Nevada, W USA
28 K10 **Wall** South Dakota, N USA 43°58′N 102°12′W
173 U9 **Wallaby Plateau** *undersea feature* E Indian Ocean
33 N8 **Wallace** Idaho, NW USA 47°28′N 115°55′W
21 V11 **Wallace** North Carolina, SE USA 34°42′N 77°59′W
14 D17 **Wallaceburg** Ontario, S Canada 42°34′N 82°22′W
22 F5 **Wallace Lake** ⊠ Louisiana, S USA
11 P13 **Wallace Mountain** ▲ Alberta, W Canada 54°50′N 115°57′W
116 J14 **Wallachia** *var.* Walachia, *Ger.* Walachei, *Rom.* Valachia. *cultural region* S Romania
Wallachisch-Meseritsch *see* Valašské Meziříčí
183 U4 **Wallangarra** New South Wales, SE Australia 28°56′S 151°57′E
182 I8 **Wallaroo** South Australia 33°56′S 137°38′E
32 L10 **Walla Walla** Washington, NW USA 46°03′N 118°20′W
101 H19 **Walldürn** Baden-Württemberg, SW Germany 49°34′N 09°22′E
100 F12 **Wallenhorst** Niedersachsen, NW Germany 52°21′N 08°01′E
Wallenthal *see* Hațeg
109 S4 **Wallern** Oberösterreich, N Austria 48°13′N 13°58′E
Wallern *see* Wallern im Burgenland
109 Z5 **Wallern im Burgenland** *var.* Wallern. Burgenland, E Austria 47°44′N 16°57′E
18 M9 **Wallingford** Vermont, NE USA 43°27′N 72°56′W
25 V11 **Wallis** Texas, SW USA 29°38′N 96°05′W
Wallis *see* Valais
192 K9 **Wallis and Futuna** *Fr.* Territoire de Wallis et Futuna. ◇ *French overseas collectivity* C Pacific Ocean
108 G7 **Wallisellen** Zürich, N Switzerland 47°25′N 08°36′E
Wallis et Futuna, Territoire de *see* Wallis and Futuna
190 H11 **Wallis, Îles** *island group* N Wallis and Futuna
99 G20 **Wallonia** ◈ *cultural region* SW Belgium
31 Q5 **Walloon Lake** ◎ Michigan, N USA
32 K10 **Wallula** Washington, NW USA 46°03′N 118°54′W
32 K10 **Wallula, Lake** ◎ Washington, NW USA
21 S8 **Walnut Cove** North Carolina, SE USA 36°18′N 80°08′W
35 N8 **Walnut Creek** California, W USA 37°52′N 122°04′W
26 K5 **Walnut Creek** ≈ Kansas, C USA
27 W9 **Walnut Ridge** Arkansas, C USA 36°06′N 90°56′W
25 S7 **Walnut Springs** Texas, SW USA 32°05′N 97°42′W
182 L10 **Walpeup** Victoria, SE Australia 35°09′S 142°01′E
187 R17 **Walpole, Île** *island* SE New Caledonia
39 N13 **Walrus Islands** *island group* Alaska, USA
97 L19 **Walsall** C England, United Kingdom 52°35′N 01°58′W
37 T7 **Walsenburg** Colorado, C USA 37°37′N 104°46′W
11 S17 **Walsh** Alberta, SW Canada 49°58′N 110°03′W
37 W7 **Walsh** Colorado, C USA 37°20′N 102°17′W
100 I11 **Walsrode** Niedersachsen, NW Germany 52°52′N 09°36′E
Waltenberg *see* Zalău
21 R14 **Walterboro** South Carolina, SE USA 32°54′N 80°21′W
Walter F.George Lake *see* Walter F. George Reservoir
23 R6 **Walter F. George Reservoir** *var.* Walter F.George Lake. ⊠ Alabama/Georgia, SE USA
26 M12 **Walters** Oklahoma, C USA 34°22′N 98°18′W
101 J16 **Waltershausen** Thüringen, C Germany 50°53′N 10°33′E
173 N10 **Walters Shoal** *var.* Walters Shoals. *reef* S Madagascar
Walters Shoals *see* Walters Shoal
22 M3 **Walthall** Mississippi, S USA 33°36′N 89°16′W
20 M4 **Walton** Kentucky, S USA 38°52′N 84°36′W
18 J11 **Walton** New York, NE USA 42°10′N 75°07′W
79 O20 **Walungu** Sud-Kivu, E Dem. Rep. Congo 02°40′S 28°37′E
Walvisbaai *see* Walvis Bay
83 C19 **Walvis Bay** *Afr.* Walvisbaai. Erongo, NW Namibia 22°59′S 14°34′E
83 B19 **Walvis Bay** *bay* NW Namibia
Walvish Ridge *see* Walvis Ridge
65 O17 **Walvis Ridge** *var.* Walvish Ridge. *undersea feature* E Atlantic Ocean 28°00′S 03°00′E
171 X16 **Wamal** Papua, E Indonesia 08°00′S 139°06′E
171 U15 **Wamar, Pulau** *island* Kepulauan Aru, E Indonesia
79 O17 **Wamba** Orientale, NE Dem. Rep. Congo 02°10′N 27°59′E
77 V15 **Wamba** Nassarawa, C Nigeria 08°57′N 08°35′E
79 H22 **Wamba** *var.* Uamba. ≈ Angola/Dem. Rep. Congo
27 P4 **Wamego** Kansas, C USA 39°12′N 96°18′W
18 I10 **Wampsville** New York, NE USA 43°03′N 75°40′W
42 K6 **Wampú, Río** ≈ E Honduras
171 X16 **Wan** Papua, E Indonesia 08°15′S 138°00′E
Wan *see* Anhui
183 N4 **Wanaaring** New South Wales, SE Australia 29°42′S 144°07′E
185 D21 **Wanaka** Otago, South Island, New Zealand 44°42′S 169°09′E
185 D20 **Wanaka, Lake** ◎ South Island, New Zealand
171 W14 **Wanapiri** Papua, E Indonesia 04°21′S 135°52′E
14 F9 **Wanapitei** ≈ Ontario, S Canada
14 F10 **Wanapitei Lake** ◎ Ontario, S Canada
18 K14 **Wanaque** New Jersey, NE USA 41°02′N 74°17′W
171 U12 **Wanau** Papua Barat, E Indonesia 01°20′S 132°40′E
185 F22 **Wanbrow, Cape** *headland* South Island, New Zealand 45°07′S 170°59′E
Wancheng *see* Wanning
Wanchuan *see* Zhangjiakou
171 W13 **Wandai** *var.* Komeyo. Papua, E Indonesia 03°35′S 136°15′E
163 Z8 **Wanda Shan** ▲▲ NE China
197 R11 **Wandel Sea** *sea* Arctic Ocean
160 D13 **Wanding** *var.* Wandingzhen. Yunnan, SW China 24°01′N 98°00′E
Wandingzhen *see* Wanding
99 H20 **Wanfercée-Baulet** Hainaut, S Belgium 50°27′N 04°37′E
184 L12 **Wanganui** Manawatu-Wanganui, North Island, New Zealand 39°56′S 175°02′E
184 L11 **Wanganui** ≈ North Island, New Zealand
183 P11 **Wangaratta** Victoria, SE Australia 36°22′S 146°17′E
160 J8 **Wangcang** *var.* Donghe; *prev.* Fengjiaba, Hongjiang. Sichuan, C China 32°15′N 106°16′E
Wangda *see* Zogang
101 I24 **Wangen im Allgäu** Baden-Württemberg, S Germany 47°40′N 09°49′E
Wangerin *see* Węgorzyno
100 F9 **Wangerooge** *island* NW Germany
171 W13 **Wanggar** Papua, E Indonesia 03°22′S 135°15′E
160 J13 **Wangmo** *var.* Fuxing. Guizhou, S China 25°08′N 106°08′E
Wangolodougou *see* Ouangolodougou
161 S9 **Wangpan Yang** *sea* E China
163 Y10 **Wangqing** Jilin, NE China 43°19′N 129°42′E
167 P8 **Wang Saphung** Loei, C Thailand 17°18′N 101°45′E
167 O6 **Wan Hsa-la** Shan State, E Myanmar (Burma) 20°27′N 98°39′E
55 W9 **Wanica** ◈ *district* N Suriname
79 M18 **Wanie-Rukula** Orientale, C Dem. Rep. Congo 0°13′N 25°34′E
Wankie *see* Hwange
Wanki, Río *see* Coco, Río
81 N17 **Wanlaweyn** *var.* Wanle Weyn, *It.* Uanle Uen. Shabeellaha Hoose, SW Somalia 02°36′N 44°47′E
Wanle Weyn *see* Wanlaweyn
180 I12 **Wanneroo** Western Australia 31°40′S 115°35′E
160 L17 **Wanning** *var.* Wancheng. Hainan, S China 18°55′N 110°27′E
167 Q8 **Wanon Niwat** Sakon Nakhon, E Thailand 17°39′N 103°45′E
155 H16 **Wanparti** Telangana, C India 16°19′N 78°06′E
Wansen *see* Wiązów
160 L11 **Wanshan** Guizhou, S China 27°45′N 109°12′E
99 M14 **Wanssum** Limburg, SE Netherlands 51°31′N 06°04′E
184 N12 **Wanstead** Hawke's Bay, North Island, New Zealand 40°09′S 176°31′E
Wanxian *see* Wanzhou
188 F16 **Wanyaan** Yap, Micronesia
160 K8 **Wanyuan** Sichuan, C China 32°05′N 108°08′E
161 O11 **Wanzai** *var.* Kangle. Jiangxi, S China 28°06′N 114°27′E
99 J20 **Wanze** Liège, E Belgium 50°32′N 05°16′E
160 K9 **Wanzhou** *var.* Wanxian. Chongqing Shi, C China 30°48′N 108°21′E
31 R12 **Wapakoneta** Ohio, N USA 40°34′N 84°11′W
12 D7 **Wapaseese** ≈ Ontario, C Canada
32 I10 **Wapato** Washington, NW USA 46°27′N 120°25′W
29 Y15 **Wapello** Iowa, C USA 41°10′N 91°13′W
11 N13 **Wapiti** ≈ Alberta/British Columbia, SW Canada
27 X7 **Wappapello Lake** ⊠ Missouri, C USA
18 K13 **Wappingers Falls** New York, NE USA 41°36′N 73°54′W
29 X13 **Wapsipinicon River** ≈ Iowa, C USA
14 L9 **Wapus** ≈ Québec, SE Canada
160 H7 **Waqên** Sichuan, C China 33°05′N 102°34′E
21 Q7 **War** West Virginia, NE USA 37°18′N 81°39′W
Warab *see* Warrap
155 J15 **Warangal** Telangana, C India 18°N 79°35′E
Warasdin *see* Varaždin
183 O16 **Waratah** Tasmania, SE Australia 41°28′S 145°34′E
183 O14 **Waratah Bay** *bay* Victoria, SE Australia
101 H15 **Warburg** Nordrhein-Westfalen, W Germany 51°30′N 09°11′E
182 I1 **Warburton Creek** *seasonal river* South Australia
180 M9 **Warbuton** Western Australia 26°17′S 126°18′E
99 M20 **Warche** ≈ E Belgium
Wardag/Wardak *see* Wardak
149 P5 **Wardak** *prev.* Vardak, *Pash.* Wardag. ◈ *province* E Afghanistan
32 K9 **Warden** Washington, NW USA 46°58′N 119°02′W
154 I12 **Wardha** Mahārāshtra, W India 20°41′N 78°40′E
Wardija Point *see* Wardija, Ras il-
121 N15 **Wardija, Ras il-** *var.* Ras il- Wardija, Wardija Point. *headland* Gozo, NW Malta 36°03′N 14°11′E
Wardija, Ras il- *see* Wardija, Ras il-
139 P3 **Wardīyah** Nīnawá, N Iraq 36°18′N 41°45′E
185 E19 **Ward, Mount** ▲ South Island, New Zealand 43°49′S 169°54′E
10 L11 **Ware** British Columbia, W Canada 57°26′N 125°41′W
99 D18 **Waregem** *var.* Waereghem. West-Vlaanderen, W Belgium 50°53′N 03°26′E
99 J19 **Waremme** Liège, E Belgium 50°41′N 05°15′E
100 N10 **Waren** Mecklenburg-Vorpommern, NE Germany 53°32′N 12°42′E
171 W13 **Waren** Papua, E Indonesia 02°13′S 136°21′E
101 F14 **Warendorf** Nordrhein-Westfalen, W Germany 51°57′N 08°00′E
21 P12 **Ware Shoals** South Carolina, SE USA 34°24′N 82°15′W
98 N4 **Warffum** Groningen, NE Netherlands 53°22′N 06°34′E
81 O15 **Wargalo** Mudug, E Somalia 06°06′N 47°40′E
146 M12 **Warganza** *Rus.* Varganzi. Qashqadaryo Viloyati, S Uzbekistan 39°18′N 66°00′E
Wargla *see* Ouargla
183 T4 **Warialda** New South Wales, SE Australia 29°34′S 150°35′E
154 F13 **Wāri Godri** Mahārāshtra, C India 19°28′N 75°43′E
167 R10 **Warin Chamrap** Ubon Ratchathani, E Thailand 15°11′N 104°51′E
25 R11 **Waring** Texas, SW USA 29°56′N 98°48′W
39 O8 **Waring Mountains** ▲▲ Alaska, USA
110 M12 **Warka** Mazowieckie, E Poland 51°45′N 21°12′E
184 L5 **Warkworth** Auckland, North Island, New Zealand 36°23′S 174°42′E
171 U12 **Warmandi** Papua Barat, E Indonesia 0°21′S 132°38′E
83 E22 **Warmbad** Karas, S Namibia 28°29′S 18°41′E
98 H8 **Warmenhuizen** Noord-Holland, NW Netherlands 52°43′N 04°45′E
110 M8 **Warmińsko-Mazurskie** ◈ *province* C Poland
97 L22 **Warminster** S England, United Kingdom 51°13′N 02°12′W
18 I15 **Warminster** Pennsylvania, NE USA 40°11′N 75°04′W
35 V8 **Warm Springs** Nevada, W USA 38°10′N 116°21′W
32 H12 **Warm Springs** Oregon, NW USA 44°51′N 121°24′W
21 S5 **Warm Springs** Virginia, NE USA 38°03′N 79°48′W
100 M8 **Warnemünde** Mecklenburg-Vorpommern, NE Germany 54°10′N 12°03′E
27 Q10 **Warner** Oklahoma, C USA 35°29′N 95°18′W
35 Q2 **Warner Mountains** ▲▲ California, W USA
23 T5 **Warner Robins** Georgia, SE USA 32°38′N 83°38′W
57 N18 **Warnes** Santa Cruz, C Bolivia 17°30′S 63°11′W
100 M9 **Warnow** ≈ NE Germany
Warnsdorf *see* Varnsdorf
98 M11 **Warnsveld** Gelderland, E Netherlands 52°08′N 06°14′E
154 I13 **Warora** Mahārāshtra, C India 20°12′N 79°01′E
182 L11 **Warracknabeal** Victoria, SE Australia 36°17′S 142°26′E
183 O13 **Warragul** Victoria, SE Australia 38°11′S 145°55′E
80 D13 **Warrap** Warrap, W South Sudan 08°08′N 28°37′E
81 D14 **Warrap** *var.* Warab. ◈ *state* W South Sudan

◆ Country
● Country Capital
◇ Dependent Territory
○ Dependent Territory Capital
◈ Administrative Regions
✈ International Airport
▲ Mountain
▲▲ Mountain Range
⋀ Volcano
≈ River
◎ Lake
⊠ Reservoir

183 O4 **Warrego River** *seasonal river* New South Wales/ Queensland, E Australia
183 Q6 **Warren** New South Wales, SE Australia *31°41′S 147°51′E*
11 X16 **Warren** Manitoba, S Canada *50°05′N 97°33′W*
27 V14 **Warren** Arkansas, C USA *33°38′N 92°05′W*
31 S10 **Warren** Michigan, N USA *42°29′N 83°02′W*
29 R3 **Warren** Minnesota, N USA *48°12′N 96°46′W*
31 U11 **Warren** Ohio, N USA *41°14′N 80°49′W*
18 D12 **Warren** Pennsylvania, NE USA *41°52′N 79°09′W*
25 X10 **Warren** Texas, SW USA *30°33′N 94°24′W*
97 G16 **Warrenpoint** ***Ir.*** An Pointe. SE Northern Ireland, United Kingdom *54°07′N 06°16′W*
27 S4 **Warrensburg** Missouri, C USA *38°46′N 93°44′W*
83 H22 **Warrenton** Northern Cape, N South Africa *28°07′S 24°51′E*
23 U4 **Warrenton** Georgia, SE USA *33°24′N 82°39′W*
27 W4 **Warrenton** Missouri, C USA *38°48′N 91°08′W*
21 V8 **Warrenton** North Carolina, SE USA *36°24′N 78°11′W*
21 V4 **Warrenton** Virginia, NE USA *38°43′N 77°48′W*
77 U17 **Warri** Delta, S Nigeria *05°26′N 05°34′E*
97 L18 **Warrington** C England, United Kingdom *53°24′N 02°37′W*
23 O9 **Warrington** Florida, SE USA *30°22′N 87°16′W*
23 P3 **Warrior** Alabama, S USA *33°49′N 86°49′W*
182 L13 **Warrnambool** Victoria, SE Australia *38°23′S 142°30′E*
29 T2 **Warroad** Minnesota, N USA *48°55′N 95°18′W*
183 S6 **Warrumbungle Range** ▲▲ New South Wales, SE Australia
154 J12 **Wārsa** Mahārāshtra, C India *20°42′N 79°58′E*
31 P11 **Warsaw** Indiana, N USA *41°13′N 85°52′W*
20 L4 **Warsaw** Kentucky, S USA *38°47′N 84°55′W*
27 T5 **Warsaw** Missouri, C USA *38°14′N 93°23′W*
18 E10 **Warsaw** New York, NE USA *42°44′N 78°06′W*
21 V10 **Warsaw** North Carolina, SE USA *35°00′N 78°05′W*
21 X5 **Warsaw** Virginia, NE USA *37°57′N 76°46′W*
Warsaw/Warschau *see* Warszawa
81 N17 **Warshiikh** Shabeellaha Dhexe, C Somalia *02°22′N 45°52′E*
101 G15 **Warstein** Nordrhein-Westfalen, W Germany *51°27′N 08°21′E*
110 M11 **Warszawa** ***Eng.*** Warsaw, ***Ger.*** Warschau, ***Rus.*** Varshava. ● (Poland) Mazowieckie, C Poland *52°15′N 21°E*
110 J13 **Warta** Sieradz, C Poland *51°43′N 18°37′E*
110 D11 **Warta** ***Ger.*** Warthe. ≈ W Poland
Wartberg *see* Senec
20 M9 **Wartburg** Tennessee, S USA *36°08′N 84°37′W*
108 J7 **Warth** Vorarlberg, NW Austria *47°16′N 10°11′E*
Warthe *see* Warta
169 U12 **Waru** Borneo, C Indonesia *01°24′S 116°37′E*
171 T13 **Waru** Pulau Seram, E Indonesia *03°24′S 130°38′E*
139 N6 **Wa'r, Wādī al** *dry watercourse* E Syria
183 U3 **Warwick** Queensland, E Australia *28°12′S 152°E*
15 Q11 **Warwick** Québec, SE Canada *45°55′N 72°00′W*
97 M20 **Warwick** C England, United Kingdom *52°17′N 01°34′W*
18 K13 **Warwick** New York, NE USA *41°15′N 74°21′W*
29 P4 **Warwick** North Dakota, N USA *47°49′N 98°42′W*
19 O12 **Warwick** Rhode Island, NE USA *41°40′N 71°21′W*
97 L20 **Warwickshire** *cultural region* C England, United Kingdom
14 G14 **Wasaga Beach** Ontario, S Canada *44°30′N 80°00′W*
77 U13 **Wasagu** Kebbi, NW Nigeria *11°25′N 05°48′E*
36 M2 **Wasatch Range** ▲▲ W USA
35 R12 **Wasco** California, W USA *35°34′N 119°20′W*
29 V10 **Waseca** Minnesota, N USA *44°04′N 93°30′W*
14 H13 **Washago** Ontario, S Canada *44°46′N 78°48′W*
19 S2 **Washburn** Maine, NE USA *46°46′N 68°08′W*
28 M5 **Washburn** North Dakota, N USA *47°15′N 101°02′W*
30 K3 **Washburn** Wisconsin, N USA *46°41′N 90°53′W*
31 S14 **Washburn Hill** *hill* Ohio, N USA
154 H13 **Wāshīm** Mahārāshtra, C India *20°06′N 77°08′E*
97 M14 **Washington** NE England, United Kingdom *54°54′N 01°31′W*
23 U3 **Washington** Georgia, SE USA *33°44′N 82°44′W*
30 L12 **Washington** Illinois, N USA *40°42′N 89°24′W*
31 N15 **Washington** Indiana, N USA *38°40′N 87°10′W*
29 X15 **Washington** Iowa, C USA *41°18′N 91°41′W*
27 O3 **Washington** Kansas, C USA *39°49′N 97°03′W*
27 W5 **Washington** Missouri, C USA *38°31′N 91°01′W*
21 X9 **Washington** North Carolina, SE USA *35°33′N 77°04′W*
18 B15 **Washington** Pennsylvania, NE USA *40°11′N 80°16′W*
25 V10 **Washington** Texas, SW USA *30°18′N 96°08′W*
36 J8 **Washington** Utah, W USA *37°07′N 113°30′W*
21 V4 **Washington** Virginia, NE USA *38°43′N 78°11′W*
32 I9 **Washington** ***off.*** State of Washington, ***also known as*** Chinook State, Evergreen State. ◈ *state* NW USA
Washington *see* Washington Court House
31 S14 **Washington Court House** ***var.*** Washington. Ohio, NE USA *39°32′N 83°29′W*
21 W4 **Washington DC** ● (USA) District of Columbia, NE USA *38°54′N 77°02′W*
31 O5 **Washington Island** *island* Wisconsin, N USA
Washington Island *see* Teraina
19 O7 **Washington, Mount** ▲ New Hampshire, NE USA *44°16′N 71°18′W*
26 M11 **Washita River** ≈ Oklahoma/Texas, C USA
97 O18 **Wash, The** *inlet* E England, United Kingdom
32 L9 **Washtucna** Washington, NW USA *46°44′N 118°19′W*
Wasiliszki *see* Vasilishki
110 P9 **Wasilków** Podlaskie, NE Poland *53°12′N 23°15′E*
39 R11 **Wasilla** Alaska, USA *61°34′N 149°26′W*
139 V9 **Wāsiṭ** ***off.*** Muḥāfa at Wāsiṭ, ***var.*** Al Kūt. ◈ *governorate* E Iraq
Wāsiṭ, Muḥāfa at *see* Wāsiṭ
55 U9 **Wasjabo** Sipaliwini, NW Suriname *05°09′N 57°09′W*
12 I10 **Waskaganish** ***prev.*** Fort Rupert, Rupert House. Québec, SE Canada *51°30′N 79°45′W*
11 X11 **Waskaiowaka Lake** ◎ Manitoba, C Canada
11 T14 **Waskesiu Lake** Saskatchewan, C Canada *53°56′N 106°05′W*
25 X7 **Waskom** Texas, SW USA *32°28′N 94°03′W*
110 G13 **Wąsosz** Dolnośląskie, SW Poland *51°36′N 16°30′E*
42 M6 **Waspam** ***var.*** Waspán. Región Autónoma Atlántico Norte, NE Nicaragua *14°41′N 84°04′W*
Waspán *see* Waspam
165 T3 **Wassamu** Hokkaidō, NE Japan *44°01′N 142°25′E*
108 G9 **Wassen** Uri, C Switzerland *46°42′N 08°34′E*
98 G11 **Wassenaar** Zuid-Holland, W Netherlands *52°09′N 04°23′E*
99 N24 **Wasserbillig** Grevenmacher, E Luxembourg *49°43′N 06°30′E*
Wasserburg *see* Wasserburg am Inn
101 M23 **Wasserburg am Inn** ***var.*** Wasserburg. Bayern, SE Germany *48°02′N 12°12′E*
101 I17 **Wasserkuppe** ▲ C Germany *50°30′N 09°55′E*
103 R5 **Wassy** Haute-Marne, N France *48°32′N 04°54′E*
171 N14 **Watampone** ***var.*** Bone. Sulawesi, C Indonesia *04°33′S 120°20′E*
171 R13 **Watawa** Pulau Buru, E Indonesia *03°36′S 127°13′E*
Watenstedt-Salzgitter *see* Salzgitter
18 M13 **Waterbury** Connecticut, NE USA *41°33′N 73°01′W*
21 R11 **Wateree Lake** ⊡ South Carolina, SE USA
21 R12 **Wateree River** ≈ South Carolina, SE USA
97 E20 **Waterford** ***Ir.*** Port Láirge. Waterford, S Ireland *52°15′N 07°08′W*
31 S9 **Waterford** Michigan, N USA *42°42′N 83°24′W*
97 E20 **Waterford** ***Ir.*** Port Láirge. *cultural region* S Ireland
97 E21 **Waterford Harbour** ***Ir.*** Cuan Phort Láirge. *inlet* S Ireland
98 G12 **Wateringen** Zuid-Holland, W Netherlands *52°02′N 04°16′E*
99 G19 **Waterloo** Walloon Brabant, C Belgium *50°43′N 04°24′E*
14 F16 **Waterloo** Ontario, S Canada *43°28′N 80°32′W*
15 P12 **Waterloo** Québec, SE Canada *45°20′N 72°28′W*
30 K16 **Waterloo** Illinois, N USA *38°20′N 90°09′W*
29 X13 **Waterloo** Iowa, C USA *42°31′N 92°16′W*
18 G10 **Waterloo** New York, NE USA *42°54′N 76°51′W*
30 L4 **Watersmeet** Michigan, N USA *46°16′N 89°10′W*
23 V9 **Watertown** Florida, SE USA *30°11′N 82°36′W*
18 I8 **Watertown** New York, NE USA *43°37′N 75°56′W*
29 R9 **Watertown** South Dakota, N USA *44°54′N 97°07′W*
30 M8 **Watertown** Wisconsin, N USA *43°12′N 88°44′W*
22 L3 **Water Valley** Mississippi, S USA *34°09′N 89°37′W*
27 O3 **Waterville** Kansas, C USA *39°41′N 96°45′W*
17 V6 **Waterville** Maine, NE USA *44°34′N 69°41′W*
29 V10 **Waterville** Minnesota, N USA *44°13′N 93°34′W*
18 I10 **Waterville** New York, NE USA *42°55′N 75°18′W*
14 E16 **Watford** Ontario, S Canada *42°57′N 81°51′W*
97 N21 **Watford** E England, United Kingdom *51°39′N 00°24′W*
28 K4 **Watford City** North Dakota, N USA *47°48′N 103°16′W*
141 X12 **Wāṭif** S Oman *18°34′N 56°31′E*
18 G11 **Watkins Glen** New York, NE USA *42°23′N 76°53′W*
Watlings Island *see* San Salvador
171 U15 **Watnil** Pulau Kai Kecil, E Indonesia *05°45′S 132°39′E*
26 M10 **Watonga** Oklahoma, C USA *35°52′N 98°26′W*
11 T16 **Watrous** Saskatchewan, S Canada *51°40′N 105°29′W*
37 T10 **Watrous** New Mexico, SW USA *35°48′N 104°58′W*
79 P16 **Watsa** Orientale, NE Dem. Rep. Congo *03°00′N 29°31′E*
31 N12 **Watseka** Illinois, N USA *40°46′N 87°44′W*
79 J19 **Watsikengo** Equateur, C Dem. Rep. Congo *0°49′S 20°34′E*
182 C5 **Watson** South Australia *30°32′S 131°29′E*
11 U15 **Watson** Saskatchewan, S Canada *52°13′N 104°30′W*
195 O10 **Watson Escarpment** ▲▲ Antarctica
10 K9 **Watson Lake** Yukon, W Canada *60°05′N 128°47′W*
35 N10 **Watsonville** California, W USA *36°53′N 121°43′W*
167 Q8 **Wattay** ✈ (Viangchan) Viangchan, C Laos *18°03′N 102°36′E*
109 N7 **Wattens** Tirol, W Austria *47°18′N 11°37′E*
20 M9 **Watts Bar Lake** ⊡ Tennessee, S USA
108 H7 **Wattwil** Sankt Gallen, NE Switzerland *47°18′N 09°06′E*
171 T14 **Watubela, Kepulauan** *island group* E Indonesia
101 N24 **Watzmann** ▲ SE Germany *47°32′N 12°56′E*
186 E8 **Wau** Morobe, C Papua New Guinea *07°22′S 146°40′E*
81 D14 **Wau** ***var.*** Wâw. Western Bahr el Ghazal, W South Sudan *07°43′N 28°01′E*
29 Q8 **Waubay** South Dakota, N USA *45°19′N 97°18′W*
29 Q8 **Waubay Lake** ◎ South Dakota, N USA
183 U7 **Wauchope** New South Wales, SE Australia *31°30′S 152°46′E*
23 W13 **Wauchula** Florida, SE USA *27°33′N 81°48′W*
30 M10 **Wauconda** Illinois, N USA *42°15′N 88°08′W*
182 J7 **Waukaringa** South Australia *32°19′S 139°27′E*
31 N10 **Waukegan** Illinois, N USA *42°21′N 87°50′W*
30 M9 **Waukesha** Wisconsin, N USA *43°01′N 88°14′W*
29 X11 **Waukon** Iowa, C USA *43°16′N 91°28′W*
30 L8 **Waunakee** Wisconsin, N USA *43°13′N 89°28′W*
30 L7 **Waupaca** Wisconsin, N USA *44°23′N 89°04′W*
30 M8 **Waupun** Wisconsin, N USA *43°40′N 88°43′W*
26 M13 **Waurika** Oklahoma, C USA *34°10′N 98°00′W*
26 M12 **Waurika Lake** ⊡ Oklahoma, C USA
30 L6 **Wausau** Wisconsin, N USA *44°58′N 89°40′W*
31 R11 **Wauseon** Ohio, N USA *41°33′N 84°08′W*
30 L7 **Wautoma** Wisconsin, N USA *44°05′N 89°17′W*
30 M9 **Wauwatosa** Wisconsin, N USA *43°03′N 88°03′W*
22 L9 **Waveland** Mississippi, S USA *30°17′N 89°22′W*
97 Q20 **Waveney** ≈ E England, United Kingdom
184 L11 **Waverley** Taranaki, North Island, New Zealand *39°45′S 174°35′E*
29 W12 **Waverly** Iowa, C USA *42°43′N 92°28′W*
27 T4 **Waverly** Missouri, C USA *39°12′N 93°31′W*
29 R15 **Waverly** Nebraska, C USA *40°56′N 96°27′W*
18 G12 **Waverly** New York, NE USA *42°00′N 76°33′W*
20 H8 **Waverly** Tennessee, S USA *36°04′N 87°49′W*
21 W7 **Waverly** Virginia, NE USA *37°02′N 77°06′W*
99 H19 **Wavre** Walloon Brabant, C Belgium *50°43′N 04°37′E*
166 M8 **Waw** Bago, SW Myanmar (Burma) *17°26′N 96°40′E*
Wāw *see* Wau
14 B7 **Wawa** Ontario, S Canada *47°59′N 84°43′W*
77 T14 **Wawa** Niger, W Nigeria *09°52′N 04°33′E*
75 Q11 **Wāw al Kabīr** S Libya *25°21′N 16°41′E*
43 N7 **Wawa, Río** ***var.*** Rio Huahua. ≈ NE Nicaragua
186 B8 **Wawoi** ≈ SW Papua New Guinea
25 T7 **Waxahachie** Texas, SW USA *32°23′N 96°52′W*
158 L9 **Waxxari** Xinjiang Uygur Zizhiqu, NW China *38°43′N 87°11′E*
Wayaobu *see* Zichang
23 V7 **Waycross** Georgia, SE USA *31°13′N 82°21′W*
180 K10 **Way, Lake** ◎ Western Australia
31 P9 **Wayland** Michigan, N USA *42°40′N 85°38′W*
29 R13 **Wayne** Nebraska, C USA *42°13′N 97°01′W*
18 K14 **Wayne** New Jersey, NE USA *40°57′N 74°16′W*
21 P5 **Wayne** West Virginia, NE USA *38°14′N 82°27′W*
23 V4 **Waynesboro** Georgia, SE USA *33°04′N 82°01′W*
22 M7 **Waynesboro** Mississippi, S USA *31°40′N 88°39′W*
20 H10 **Waynesboro** Tennessee, S USA *35°20′N 87°49′W*
21 U5 **Waynesboro** Virginia, NE USA *38°04′N 78°54′W*
18 B16 **Waynesburg** Pennsylvania, NE USA *39°51′N 80°10′W*
27 U6 **Waynesville** Missouri, C USA *37°48′N 92°11′W*
21 O10 **Waynesville** North Carolina, SE USA *35°29′N 82°59′W*
26 L8 **Waynoka** Oklahoma, C USA *36°36′N 98°53′W*
Wazan *see* Ouazzane
Wazima *see* Wajima
149 V7 **Wazīrābād** Punjab, NE Pakistan *32°28′N 74°04′E*
Wazzan *see* Ouazzane
110 I8 **Wda** ***var.*** Czarna Woda, ***Ger.*** Schwarzwasser. ≈ N Poland
187 Q16 **Wé** Province des Îles Loyauté, E New Caledonia *20°55′S 167°15′E*
97 O23 **Weald, The** *lowlands* SE England, United Kingdom
186 A9 **Weam** Western, SW Papua New Guinea *08°33′S 141°10′E*
97 L15 **Wear** ≈ N England, United Kingdom
Wearmouth *see* Sunderland
26 L10 **Weatherford** Oklahoma, C USA *35°31′N 98°42′W*
25 S6 **Weatherford** Texas, SW USA *32°47′N 97°48′W*
34 M3 **Weaverville** California, W USA *40°42′N 122°57′W*
27 R7 **Webb City** Missouri, C USA *37°07′N 94°28′W*
192 G8 **Weber Basin** *undersea feature* S Ceram Sea
Webfoot State *see* Oregon
18 F9 **Webster** New York, NE USA *43°12′N 77°25′W*
29 Q8 **Webster** South Dakota, N USA *45°19′N 97°31′W*
29 V13 **Webster City** Iowa, C USA *42°28′N 93°49′W*
27 X5 **Webster Groves** Missouri, C USA *38°32′N 90°20′W*
21 S4 **Webster Springs** ***var.*** Addison. West Virginia, NE USA *38°29′N 80°52′W*
171 S11 **Weda, Teluk** *bay* Pulau Halmahera, E Indonesia
65 B25 **Weddell Island** ***var.*** Isla de San Jorge. *island* W Falkland Islands
65 K22 **Weddell Plain** *undersea feature* SW Atlantic Ocean *65°00′S 40°00′W*
65 K23 **Weddell Sea** *sea* SW Atlantic Ocean
65 B25 **Weddell Settlement** Weddell Island, W Falkland Islands *52°53′S 60°54′W*
182 M11 **Wedderburn** Victoria, SE Australia *36°26′S 143°37′E*
100 I9 **Wedel** Schleswig-Holstein, N Germany *53°35′N 09°42′E*
92 N3 **Wedel Jarlsberg Land** *physical region* SW Svalbard
100 I12 **Wedemark** Niedersachsen, NW Germany *52°33′N 09°43′E*
10 M17 **Wedge Mountain** ▲ British Columbia, SW Canada *50°10′N 122°43′W*
23 R4 **Wedowee** Alabama, S USA *33°16′N 85°28′W*
171 U15 **Weduar** Pulau Kai Besar, E Indonesia *05°55′S 132°51′E*
35 N2 **Weed** California, W USA *41°26′N 122°24′W*
15 Q12 **Weedon Centre** Québec, SE Canada *45°40′N 71°28′W*
18 E13 **Weedville** Pennsylvania, NE USA *41°15′N 78°28′W*
100 F10 **Weener** Niedersachsen, NW Germany *53°09′N 07°19′E*
29 S16 **Weeping Water** Nebraska, C USA *40°52′N 96°08′W*
99 L16 **Weert** Limburg, SE Netherlands *51°15′N 05°43′E*
98 I10 **Weesp** Noord-Holland, C Netherlands *52°18′N 05°03′E*
183 S5 **Wee Waa** New South Wales, SE Australia *30°16′S 149°27′E*
110 N7 **Węgorzewo** ***Ger.*** Angerburg. Warmińsko-Mazurskie, NE Poland *54°12′N 21°49′E*
110 E9 **Węgorzyno** ***Ger.*** Wangerin. Zachodnio-pomorskie, NW Poland *53°34′N 15°35′E*
110 N11 **Węgrów** ***Ger.*** Bingerau. Mazowieckie, C Poland *52°22′N 22°00′E*
98 N5 **Wehe-Den Hoorn** Groningen, NE Netherlands *53°20′N 06°29′E*
98 M12 **Wehl** Gelderland, E Netherlands *51°58′N 06°13′E*
Wehlau *see* Znamensk
168 F7 **Weh, Pulau** *island* NW Indonesia
Wei *see* Weifang
161 P1 **Weichang** ***prev.*** Zhuizishan. Hebei, E China *41°55′N 117°45′E*
Weichang *see* Weishan
Weichsel *see* Wisła
101 M16 **Weida** Thüringen, C Germany *50°46′N 12°05′E*
Weiden *see* Weiden in der Oberpfalz
101 M19 **Weiden in der Oberpfalz** ***var.*** Weiden. Bayern, SE Germany *49°40′N 12°10′E*
161 Q4 **Weifang** ***var.*** Wei, Wei-fang; ***prev.*** Weihsien. Shandong, E China *36°44′N 119°10′E*
161 S4 **Weihai** Shandong, E China *37°30′N 122°04′E*
160 K6 **Wei He** ≈ C China
Weihsien *see* Weifang
101 G17 **Weilburg** Hessen, W Germany *50°31′N 08°18′E*
101 K24 **Weilheim in Oberbayern** Bayern, SE Germany *47°50′N 11°09′E*
183 P4 **Weilmoringle** New South Wales, SE Australia *29°13′S 146°51′E*
101 L16 **Weimar** Thüringen, C Germany *50°59′N 11°20′E*
25 U11 **Weimar** Texas, SW USA *29°42′N 96°46′W*
160 L6 **Weinan** Shaanxi, C China *34°30′N 109°30′E*
108 H6 **Weinfelden** Thurgau, NE Switzerland *47°33′N 09°09′E*
101 I24 **Weingarten** Baden-Württemberg, S Germany *47°49′N 09°37′E*
101 G20 **Weinheim** Baden-Württemberg, SW Germany *49°33′N 08°40′E*
160 H11 **Weining** ***var.*** Caohai, Weining Yizu Huizu Miaozu Zizhixian. Guizhou, S China *26°51′N 104°16′E*
Weining Yizu Huizu Miaozu Zizhixian *see* Weining
181 V2 **Weipa** Queensland, NE Australia *12°43′S 142°01′E*
11 Y11 **Weir River** Manitoba, C Canada *56°44′N 94°06′W*
21 R1 **Weirton** West Virginia, NE USA *40°24′N 80°37′W*
32 M13 **Weiser** Idaho, NW USA *44°15′N 116°58′W*
160 F12 **Weishan** ***var.*** Weichang. Yunnan, SW China *25°22′N 100°19′E*
161 P6 **Weishan Hu** ◎ E China
101 M15 **Weisse Elster** ***Eng.*** White Elster. ≈ Czech Republic/ Germany
Weisse Körös/Weisse Kreisch *see* Crişul Alb
108 L7 **Weissenbach am Lech** Tirol, W Austria *47°27′N 10°39′E*
Weissenburg *see* Wissembourg, France
Weissenburg *see* Alba Iulia, Romania
101 K21 **Weissenburg in Bayern** Bayern, SE Germany *49°02′N 10°59′E*
101 M15 **Weissenfels** ***var.*** Weißenfels. Sachsen-Anhalt, C Germany *51°12′N 11°58′E*
109 R9 **Weissensee** ◎ S Austria
Weissenstein *see* Paide
108 E11 **Weisshorn** ***var.*** Flüela Wisshorn. ▲ SW Switzerland *46°06′N 07°43′E*
Weisskirchen *see* Bela Crkva
23 R3 **Weiss Lake** ⊡ Alabama, S USA
101 Q14 **Weisswasser** ***Lus.*** Běla Woda. Sachsen, E Germany *51°30′N 14°37′E*
99 M22 **Weiswampach** Diekirch, N Luxembourg *50°08′N 06°05′E*
109 U2 **Weitra** Niederösterreich, N Austria *48°41′N 14°54′E*
161 O4 **Weixian** ***var.*** Wei Xian. Hebei, E China *36°59′N 115°15′E*
Wei Xian *see* Weixian
159 V11 **Weiyuan** ***var.*** Qingyuan. Gansu, C China *35°07′N 104°12′E*
Weiyuan *see* Shuangjiang
160 F14 **Weiyuan Jiang** ≈ SW China
109 W7 **Weiz** Steiermark, SE Austria *47°13′N 15°38′E*
Weizhou *see* Wenchuan
160 K16 **Weizhou Dao** *island* S China
110 I6 **Wejherowo** Pomorskie, NW Poland *54°36′N 18°12′E*
27 Q8 **Welch** Oklahoma, C USA *36°52′N 95°06′W*
24 M6 **Welch** Texas, SW USA *32°52′N 102°06′W*
21 Q6 **Welch** West Virginia, NE USA *37°26′N 81°36′W*
45 O14 **Welchman Hall** C Barbados *13°10′N 59°34′W*
80 J11 **Weldiya** ***var.*** Waldia, ***It.*** Valdia. Āmara, N Ethiopia *11°45′N 39°39′E*
21 W8 **Weldon** North Carolina, SE USA *36°25′N 77°36′W*
25 V9 **Weldon** Texas, SW USA *31°00′N 95°33′W*
99 M19 **Welkenraedt** Liège, E Belgium *50°40′N 05°58′E*
193 O2 **Welker Seamount** *undersea feature* N Pacific Ocean *55°07′N 140°18′W*
83 I22 **Welkom** Free State, C South Africa *27°59′S 26°44′E*
14 H16 **Welland** Ontario, S Canada *45°59′N 79°14′W*
14 G16 **Welland** ≈ Ontario, S Canada
97 O19 **Welland** ≈ C England, United Kingdom
14 H17 **Welland Canal** *canal* Ontario, S Canada
155 K25 **Wellawaya** Uva Province, SE Sri Lanka *06°44′N 81°07′E*
Welle *see* Uele
181 T4 **Wellesley Islands** *island group* Queensland, N Australia
99 J22 **Wellin** Luxembourg, SE Belgium *50°06′N 05°05′E*
97 N20 **Wellingborough** C England, United Kingdom *52°19′N 00°42′W*
183 R7 **Wellington** New South Wales, SE Australia *32°33′S 148°59′E*
14 J15 **Wellington** Ontario, SE Canada *43°59′N 77°21′W*
185 L14 **Wellington** ● Wellington, North Island, New Zealand *41°17′S 174°47′E*
83 E26 **Wellington** Western Cape, SW South Africa *33°39′S 19°00′E*
37 T2 **Wellington** Colorado, C USA *40°42′N 105°00′W*
27 N7 **Wellington** Kansas, C USA *37°17′N 97°25′W*
35 R7 **Wellington** Nevada, W USA *38°45′N 119°22′W*
31 T11 **Wellington** Ohio, N USA *41°10′N 82°13′W*
25 P3 **Wellington** Texas, SW USA *34°52′N 100°13′W*
36 M4 **Wellington** Utah, W USA *39°31′N 110°45′W*
185 M14 **Wellington** ***off.*** Wellington Region. ◈ *region* (New Zealand) North Island, New Zealand
185 L14 **Wellington** ✈ Wellington, North Island, New Zealand *41°19′S 174°48′E*
Wellington *see* Wellington, Isla
63 F22 **Wellington, Isla** ***var.*** Wellington. *island* S Chile
183 P12 **Wellington, Lake** ◎ Victoria, SE Australia
Wellington Region *see* Wellington
29 X14 **Wellman** Iowa, C USA *41°27′N 91°50′W*
24 M6 **Wellman** Texas, SW USA *33°03′N 102°25′W*
97 K22 **Wells** SW England, United Kingdom *51°13′N 02°39′W*
29 V11 **Wells** Minnesota, N USA *43°45′N 93°43′W*
35 X2 **Wells** Nevada, W USA *41°07′N 114°58′W*
25 W8 **Wells** Texas, SW USA *31°28′N 94°54′W*
18 F12 **Wellsboro** Pennsylvania, NE USA *41°43′N 77°39′W*
21 R1 **Wellsburg** West Virginia, NE USA *40°15′N 80°37′W*
184 K4 **Wellsford** Auckland, North Island, New Zealand *36°17′S 174°30′E*
180 L9 **Wells, Lake** ◎ Western Australia
181 N4 **Wells, Mount** ▲ Western Australia *17°39′S 127°08′E*
97 P18 **Wells-next-the-Sea** E England, United Kingdom *52°58′N 00°48′E*
31 T15 **Wellston** Ohio, N USA *39°07′N 82°31′W*
27 O10 **Wellston** Oklahoma, C USA *35°41′N 97°03′W*
18 E11 **Wellsville** New York, NE USA *42°06′N 77°55′W*
31 V12 **Wellsville** Ohio, N USA *40°36′N 80°39′W*
36 L1 **Wellsville** Utah, W USA *41°38′N 111°55′W*
36 I14 **Wellton** Arizona, SW USA *32°40′N 114°09′W*
109 S4 **Wels** ***anc.*** Ovilava. Oberösterreich, N Austria *48°10′N 14°02′E*
99 K15 **Welschap** ✈ (Eindhoven) Noord-Brabant, S Netherlands *51°27′N 05°22′E*
100 P10 **Welse** ≈ NE Germany
22 H9 **Welsh** Louisiana, S USA *30°12′N 92°49′W*
97 K19 **Welshpool** ***Wel.*** Y Trallwng. E Wales, United Kingdom *52°38′N 03°06′W*
97 O21 **Welwyn Garden City** E England, United Kingdom *51°48′N 00°13′W*
79 K18 **Wema** Equateur, NW Dem. Rep. Congo *0°25′S 21°33′E*
81 G21 **Wembere** ≈ C Tanzania
11 N13 **Wembley** Alberta, W Canada *55°07′N 119°12′W*
12 I9 **Wemindji** ***prev.*** Nouveau-Comptoir, Paint Hills. Québec, C Canada *53°00′N 78°42′W*
99 G18 **Wemmel** Vlaams Brabant, C Belgium *50°54′N 04°18′E*
32 J8 **Wenatchee** Washington, NW USA *47°50′N 120°48′W*
160 M17 **Wenchang** Hainan, S China *19°34′N 110°46′E*
161 R11 **Wencheng** ***var.*** Daxue. Zhejiang, SE China *27°48′N 120°01′E*
77 P16 **Wenchi** W Ghana *07°45′N 02°02′W*
Wen-chou/Wenchow *see* Wenzhou
160 H8 **Wenchuan** ***var.*** Weizhou. Sichuan, C China *31°29′N 103°39′E*
Wendau *see* Võnnu
Wenden *see* Cēsis
161 S4 **Wendeng** Shandong, E China *37°10′N 122°00′E*
81 J14 **Wendo** Southern Nationalities, S Ethiopia *06°34′N 38°28′E*
36 J2 **Wendover** Utah, W USA *40°41′N 114°00′W*
14 D9 **Wenebegon** ≈ Ontario, S Canada
14 D8 **Wenebegon Lake** ◎ Ontario, S Canada
108 E9 **Wengen** Bern, W Switzerland *46°38′N 07°57′E*
161 O13 **Wengyuan** ***var.*** Longxian. Guangdong, S China *24°22′N 114°06′E*
189 P15 **Weno** ***prev.*** Moen. Chuuk, C Micronesia
189 V12 **Weno** ***prev.*** Moen. *atoll* Chuuk Islands, C Micronesia
158 N13 **Wenquan** Qinghai, C China *33°16′N 91°44′E*
159 H4 **Wenquan** ***var.*** Arixang, Bogeda'er. Xinjiang Uygur Zizhiqu, NW China *45°00′N 81°02′E*
Wenquan *see* Yingshan
160 H14 **Wenshan** ***var.*** Kaihua. Yunnan, SW China *23°22′N 104°21′E*
158 H6 **Wensu** Xinjiang Uygur Zizhiqu, W China *41°15′N 80°11′E*
182 L8 **Wentworth** New South Wales, SE Australia *34°04′S 141°53′E*
27 W4 **Wentzville** Missouri, C USA *38°48′N 90°51′W*
159 V12 **Wenxian** ***var.*** Wen Xian. Gansu, C China *32°57′N 104°42′E*
Wen Xian *see* Wenxian
161 S10 **Wenzhou** ***var.*** Wen-chou, Wenchow. Zhejiang, SE China *28°02′N 120°36′E*
34 L4 **Weott** California, W USA *40°19′N 123°57′W*
99 I20 **Wépion** Namur, SE Belgium *50°24′N 04°53′E*
100 O11 **Werbellinsee** ◎ NE Germany
99 L21 **Werbomont** Liège, E Belgium *50°22′N 05°43′E*
83 G20 **Werda** Kgalagadi, S Botswana *25°13′S 23°16′E*
81 N14 **Werdēr** Sumalē, E Ethiopia *06°59′N 45°20′E*
Werder *see* Virtsu
Werenów *see* Voranava
171 U13 **Weri** Papua Barat, E Indonesia *03°10′S 132°39′E*
98 I13 **Werkendam** Noord-Brabant, S Netherlands *51°48′N 04°54′E*
101 M20 **Wernberg-Köblitz** Bayern, SE Germany *49°31′N 12°10′E*
101 J18 **Werneck** Bayern, C Germany *50°00′N 10°06′E*
101 K14 **Wernigerode** Sachsen-Anhalt, C Germany *51°51′N 10°48′E*
Werowitz *see* Virovitica
101 J16 **Werra** ≈ C Germany
183 N12 **Werribee** Victoria, SE Australia *37°55′S 144°39′E*
183 T6 **Werris Creek** New South Wales, SE Australia *31°22′S 150°40′E*
Werro *see* Võru
Werschetz *see* Vršac
101 K23 **Wertach** ≈ S Germany
101 I19 **Wertheim** Baden-Württemberg, SW Germany *49°45′N 09°31′E*
98 J8 **Wervershoof** Noord-Holland, NW Netherlands *52°43′N 05°09′E*
Wervicq *see* Wervik
99 C18 **Wervik** ***var.*** Wervicq, Werwick. West-Vlaanderen, W Belgium *50°47′N 03°03′E*
Werwick *see* Wervik
101 D14 **Wesel** Nordrhein-Westfalen, W Germany *51°39′N 06°37′E*
Weseli an der Lainsitz *see* Veselí nad Lužnicí
Wesenberg *see* Rakvere
100 H12 **Weser** ≈ NW Germany
Wes-Kaap *see* Western Cape
25 S17 **Weslaco** Texas, SW USA *26°09′N 97°59′W*
14 J13 **Weslemkoon Lake** ◎ Ontario, SE Canada
181 R1 **Wessel Islands** *island group* Northern Territory, N Australia
29 P9 **Wessington** South Dakota, N USA *44°27′N 98°40′W*
29 P10 **Wessington Springs** South Dakota, N USA *44°02′N 98°33′W*
25 T8 **West** Texas, SW USA *31°48′N 97°05′W*
West *see* Ouest
30 M9 **West Allis** Wisconsin, N USA *43°01′N 88°00′W*
182 E8 **Westall, Point** *headland* South Australia *32°54′S 134°04′E*
194 M10 **West Antarctica** ***prev.*** Lesser Antarctica. *physical region* Antarctica
14 G11 **West Arm** Ontario, S Canada *46°16′N 80°25′W*
West Australian Basin *see* Wharton Basin
West Azerbaijan *see* Āžarbāyjān-e Gharbī
11 N17 **Westbank** British Columbia, SW Canada *49°50′N 119°37′W*
138 F10 **West Bank** *disputed region* SW Asia
14 E11 **West Bay** Manitoulin Island, Ontario, S Canada *45°48′N 82°09′W*
22 L11 **West Bay** *bay* Louisiana, S USA
30 M8 **West Bend** Wisconsin, N USA *43°26′N 88°13′W*
153 R16 **West Bengal** ◈ *state* NE India
West Borneo *see* Kalimantan Barat
29 Y14 **West Branch** Iowa, C USA *41°40′N 91°21′W*
31 R7 **West Branch** Michigan, N USA *44°16′N 84°14′W*
18 F13 **West Branch Susquehanna River** ≈ Pennsylvania, NE USA
97 L20 **West Bromwich** C England, United Kingdom *52°29′N 01°59′W*
19 P8 **Westbrook** Maine, NE USA *43°42′N 70°21′W*
29 T10 **Westbrook** Minnesota, N USA *44°02′N 95°26′W*
29 Y15 **West Burlington** Iowa, C USA *40°49′N 91°09′W*
96 L2 **West Burra** *island* NE Scotland, United Kingdom
30 J8 **Westby** Wisconsin, N USA *43°39′N 90°52′W*
44 L6 **West Caicos** *island* W Turks and Caicos Islands
185 A24 **West Cape** *headland* South Island, New Zealand *45°51′S 166°26′E*
174 L4 **West Caroline Basin** *undersea feature* SW Pacific Ocean *04°00′N 138°00′E*
18 I16 **West Chester** Pennsylvania, NE USA *39°56′N 75°35′W*
185 E18 **West Coast** ***off.*** West Coast Region. ◈ *region* South Island, New Zealand
West Coast Region *see* West Coast
25 V12 **West Columbia** Texas, SW USA *29°08′N 95°39′W*
29 W10 **West Concord** Minnesota, N USA *44°09′N 92°54′W*
29 V14 **West Des Moines** Iowa, C USA *41°33′N 93°42′W*
37 Q6 **West Elk Peak** ▲ Colorado, C USA *38°43′N 107°12′W*
44 F1 **West End** Grand Bahama Island, N The Bahamas *26°36′N 78°55′W*
44 F1 **West End Point** *headland* Grand Bahama Island, N The Bahamas *26°40′N 78°58′W*
98 O7 **Westerbork** Drenthe, NE Netherlands *52°49′N 06°36′E*
98 N3 **Westereems** *strait* Germany/ Netherlands
98 O9 **Westerhaar-Vriezenveensewijk** Overijssel, E Netherlands *52°28′N 06°38′E*
100 G6 **Westerland** Schleswig-Holstein, N Germany *54°54′N 08°19′E*
99 I17 **Westerlo** Antwerpen, N Belgium *51°05′N 04°55′E*
19 N13 **Westerly** Rhode Island, NE USA *41°22′N 71°45′W*
153 N11 **Western** ◈ *zone* C Nepal
186 A8 **Western** ***var.*** Fly River. ◈ *province* SW Papua New Guinea
186 K9 **Western** ◈ *province* S Solomon Islands
186 J8 **Western** ***off.*** Western Province. ◈ *province* NW Solomon Islands
155 J26 **Western** ◈ *province* SW Sri Lanka
83 G15 **Western** ◈ *province* SW Zambia
180 K8 **Western Australia** ◈ *state* W Australia
80 A13 **Western Bahr el Ghazal** ◈ *state* W South Sudan
Western Bug *see* Bug
83 F25 **Western Cape** ***off.*** Western Cape Province, ***Afr.*** Wes-Kaap. ◈ *province* SW South Africa
Western Cape Province *see* Western Cape
80 A11 **Western Darfur** ◈ *state* W Sudan
Western Desert *see* Şaḩrā' al Gharbīyah
118 G9 **Western Dvina** ***Bel.*** Dzvina, ***Ger.*** Düna, ***Latv.*** Daugava, ***Rus.*** Zapadnaya Dvina. ≈ W Europe
81 D15 **Western Equatoria** ◈ *state* SW South Sudan
155 E16 **Western Ghats** ▲▲ SW India
186 C7 **Western Highlands** ◈ *province* C Papua New Guinea
Western Isles *see* Outer Hebrides
80 D11 **Western Kordofan** ◈ *state* S Sudan
21 T3 **Westernport** Maryland, NE USA *39°29′N 79°03′W*
Western Province *see* Western
Western Punjab *see* Punjab
74 B10 **Western Sahara** ◇ *disputed territory* N Africa
Western Samoa *see* Samoa
Western Sayans *see* Zapadnyy Sayan
Western Scheldt *see* Westerschelde
Western Sierra Madre *see* Madre Occidental, Sierra
99 E15 **Westerschelde** ***Eng.*** Western Scheldt; ***prev.*** Honte. *inlet* S North Sea
31 S13 **Westerville** Ohio, N USA *40°07′N 82°55′W*
101 F17 **Westerwald** ▲▲ W Germany
65 C25 **West Falkland** ***var.*** Gran Malvina, Isla Gran Malvina. *island* W Falkland Islands
29 R5 **West Fargo** North Dakota, N USA *46°49′N 96°51′W*
188 M15 **West Fayu Atoll** *atoll* Caroline Islands, C Micronesia
18 C11 **Westfield** New York, NE USA *42°18′N 79°34′W*
30 L7 **Westfield** Wisconsin, N USA *43°56′N 89°31′W*
West Flanders *see* West-Vlaanderen
27 S10 **West Fork** Arkansas, C USA *35°55′N 94°11′W*
29 P16 **West Fork Big Blue River** ≈ Nebraska, C USA
29 U12 **West Fork Des Moines River** ≈ Iowa/Minnesota, C USA
25 S5 **West Fork Trinity River** ≈ Texas, SW USA
30 L16 **West Frankfort** Illinois, N USA *37°54′N 88°55′W*
98 I8 **West-Friesland** *physical region* NW Netherlands
West Frisian Islands *see* Waddeneilanden
19 T5 **West Grand Lake** ◎ Maine, NE USA
18 M12 **West Hartford** Connecticut, NE USA *41°44′N 72°45′W*
18 M13 **West Haven** Connecticut, NE USA *41°16′N 72°57′W*
27 X12 **West Helena** Arkansas, C USA *34°33′N 90°38′W*
28 M2 **Westhope** North Dakota, N USA *48°54′N 101°01′W*
195 Y8 **West Ice Shelf** *ice shelf* Antarctica
47 R2 **West Indies** *island group* SE North America
West Irian *see* Papua
West Java *see* Jawa Barat
36 L3 **West Jordan** Utah, W USA *40°37′N 111°55′W*
West Kalimantan *see* Kalimantan Barat

◆ Country
● Country Capital
◇ Dependent Territory
○ Dependent Territory Capital
◈ Administrative Regions
✈ International Airport
▲ Mountain
▲▲ Mountain Range
Volcano
≈ River
◎ Lake
⊡ Reservoir

99 D14 **Westkapelle** Zeeland, SW Netherlands 51°32′N 03°26′E
West Kazakhstan *see* Zapadnyy Kazakhstan
31 O13 **West Lafayette** Indiana, N USA 40°24′N 86°54′W
31 T13 **West Lafayette** Ohio, N USA 40°16′N 81°45′W
West Lake *see* Kagera
29 Y14 **West Liberty** Iowa, C USA 41°34′N 91°15′W
21 O5 **West Liberty** Kentucky, S USA 37°56′N 83°16′W
Westliche Morava *see* Zapadna Morava
10 J13 **Westlock** Alberta, SW Canada 54°12′N 113°50′W
14 E17 **West Lorne** Ontario, S Canada 42°36′N 81°35′W
96 J12 **West Lothian** *cultural region* S Scotland, United Kingdom
99 H16 **Westmalle** Antwerpen, N Belgium 51°18′N 04°40′E
192 G6 **West Mariana Basin** *var.* Perece Vela Basin. *undersea feature* W Pacific Ocean 15°00′N 137°00′E
97 E17 **Westmeath** *Ir.* An Iarmhí, Na H-Iarmhidhe. *cultural region* C Ireland
27 Y11 **West Memphis** Arkansas, C USA 35°09′N 90°11′W
21 W2 **Westminster** Maryland, NE USA 39°34′N 77°00′W
21 O11 **Westminster** South Carolina, SE USA 34°39′N 83°06′W
22 I5 **West Monroe** Louisiana, S USA 32°31′N 92°09′W
18 D15 **Westmont** Pennsylvania, NE USA 40°16′N 78°55′W
27 O3 **Westmoreland** Kansas, C USA 39°23′N 96°30′W
35 W17 **Westmorland** California, W USA 33°02′N 115°37′W
186 E6 **West New Britain** ◆ *province* E Papua New Guinea
West New Guinea *see* Papua
83 K18 **West Nicholson** Matabeleland South, S Zimbabwe 21°06′S 29°25′E
29 T14 **West Nishnabotna River** ≈ Iowa, C USA
175 P11 **West Norfolk Ridge** *undersea feature* W Pacific Ocean
25 P12 **West Nueces River** ≈ Texas, SW USA
West Nusa Tenggara *see* Nusa Tenggara Barat
29 T11 **West Okoboji Lake** ◎ Iowa, C USA
33 R16 **Weston** Idaho, NW USA 42°01′N 119°29′W
21 R4 **Weston** West Virginia, NE USA 39°03′N 80°28′W
97 J22 **Weston-super-Mare** SW England, United Kingdom 51°21′N 02°59′W
23 Z14 **West Palm Beach** Florida, SE USA 26°43′N 80°03′W
West Papua *see* Papua Barat
West Papua *see* Papua
188 E9 **West Passage** *passage* Babeldaob, N Palau
23 O9 **West Pensacola** Florida, SE USA 30°25′N 87°16′W
27 V8 **West Plains** Missouri, C USA 36°44′N 91°51′W
35 P7 **West Point** California, W USA 38°21′N 120°33′W
23 R5 **West Point** Georgia, SE USA 32°52′N 85°10′W
22 M3 **West Point** Mississippi, S USA 33°36′N 88°39′W
29 R14 **West Point** Nebraska, C USA 41°50′N 96°42′W
21 X6 **West Point** Virginia, NE USA 37°31′N 76°48′W
182 G10 **West Point** *headland* South Australia 35°01′S 135°58′E
65 B24 **Westpoint Island Settlement** Westpoint Island, NW Falkland Islands 51°21′S 60°41′W
23 R4 **West Point Lake** ⊠ Alabama/Georgia, SE USA
81 H18 **West Pokot** ◆ *county* W Kenya
97 B16 **Westport** *Ir.* Cathair na Mart. Mayo, W Ireland 53°48′N 09°32′W
185 G15 **Westport** West Coast, South Island, New Zealand 41°46′S 171°37′E
32 F10 **Westport** Oregon, NW USA 46°07′N 123°22′W
32 F9 **Westport** Washington, NW USA 46°53′N 124°06′W
31 S15 **West Portsmouth** Ohio, N USA 38°45′N 83°01′W
West Punjab *see* Punjab
11 V14 **Westray** Manitoba, C Canada 53°30′N 101°19′W
96 J4 **Westray** *island* NE Scotland, United Kingdom
14 F9 **Westree** Ontario, S Canada 47°25′N 81°32′W
97 L16 **West Riding** *cultural region* N England, United Kingdom
West River *see* Xi Jiang
30 J7 **West Salem** Wisconsin, N USA 43°54′N 91°04′W
65 H21 **West Scotia Ridge** *undersea feature* W Scotia Sea
186 B5 **West Sepik** *prev.* Sandaun. ◆ *province* NW Papua New Guinea
173 N4 **West Sheba Ridge** *undersea feature* W Indian Ocean 12°45′N 48°15′E
West Siberian Plain *see* Zapadno-Sibirskaya Ravnina
31 S11 **West Sister Island** *island* Ohio, N USA
West-Skylge *see* West-Terschelling
West Sumatra *see* Sumatera Barat
98 J5 **West-Terschelling** *Fris.* West-Skylge. Fryslân, N Netherlands 53°23′N 05°15′E
64 J7 **West Thulean Rise** *undersea feature* N Atlantic Ocean
29 X12 **West Union** Iowa, C USA 42°57′N 91°48′W
31 R15 **West Union** Ohio, N USA 38°47′N 83°33′W
21 R3 **West Union** West Virginia, NE USA 39°18′N 80°47′W
31 N13 **Westville** Illinois, N USA 40°02′N 87°38′W
21 R3 **West Virginia** *off.* State of West Virginia, *also known as* Mountain State. ◆ *state* NE USA
99 A17 **West-Vlaanderen** *Eng.* West Flanders. ◆ *province* W Belgium
35 R7 **West Walker River** ≈ California/Nevada, W USA
35 P4 **Westwood** California, W USA 40°18′N 121°02′W
183 P9 **West Wyalong** New South Wales, SE Australia 33°56′S 147°10′E
171 Q16 **Wetar, Pulau** *island* Kepulauan Damar, E Indonesia
Wetar, Selat *see* Wetar Strait
171 R16 **Wetar Strait** *var.* Selat Wetar. *strait* Nusa Tenggara, S Indonesia
11 Q15 **Wetaskiwin** Alberta, SW Canada 52°57′N 113°20′W
81 K21 **Wete** Pemba, E Tanzania 05°03′S 39°41′E
166 M4 **Wetlet** Sagaing, C Myanmar (Burma) 22°43′N 95°22′E
37 T6 **Wet Mountains** ▲▲ Colorado, C USA
101 E15 **Wetter** Nordrhein-Westfalen, W Germany 51°22′N 07°24′E
101 H17 **Wetter** ≈ W Germany
99 F17 **Wetteren** Oost-Vlaanderen, NW Belgium 51°06′N 03°59′E
108 F7 **Wettingen** Aargau, N Switzerland 47°28′N 08°20′E
27 P11 **Wetumka** Oklahoma, C USA 35°14′N 96°14′W
23 Q5 **Wetumpka** Alabama, S USA 32°32′N 86°12′W
108 G7 **Wetzikon** Zürich, N Switzerland 47°19′N 08°48′E
101 G17 **Wetzlar** Hessen, W Germany 50°33′N 08°30′E
99 C18 **Wevelgem** West-Vlaanderen, W Belgium 50°48′N 03°12′E
38 M6 **Wevok** *var.* Wewuk. Alaska, USA 68°52′N 166°05′W
23 R9 **Wewahitchka** Florida, SE USA 30°06′N 85°12′W
186 C6 **Wewak** East Sepik, NW Papua New Guinea 03°35′S 143°35′E
27 O11 **Wewoka** Oklahoma, C USA 35°09′N 96°30′W
Wewuk *see* Wevok
97 F20 **Wexford** *Ir.* Loch Garman. SE Ireland 52°21′N 06°31′W
97 F20 **Wexford** *Ir.* Loch Garman. *cultural region* SE Ireland
30 L7 **Weyauwega** Wisconsin, N USA 44°16′N 88°54′W
11 U17 **Weyburn** Saskatchewan, S Canada 49°39′N 103°51′W
Weyer *see* Weyer Markt
109 U5 **Weyer Markt** *var.* Weyer. Oberösterreich, N Austria 47°52′N 14°39′E
100 H11 **Weyhe** Niedersachsen, NW Germany 53°00′N 08°52′E
97 L24 **Weymouth** S England, United Kingdom 50°36′N 02°28′W
19 P11 **Weymouth** Massachusetts, NE USA 42°13′N 70°56′W
99 H18 **Wezembeek-Oppem** Vlaams Brabant, C Belgium 50°51′N 04°28′E
98 M9 **Wezep** Gelderland, E Netherlands 52°28′N 06°E
184 M9 **Whakamaru** Waikato, North Island, New Zealand 38°27′S 175°48′E
184 O8 **Whakatane** Bay of Plenty, North Island, New Zealand 37°58′S 177°E
184 O8 **Whakatane** ≈ North Island, New Zealand
9 O9 **Whale Cove** *var.* Tikirarjuaq. Nunavut, C Canada 62°14′N 92°10′W
96 M2 **Whalsay** *island* NE Scotland, United Kingdom
184 L11 **Whangaehu** ≈ North Island, New Zealand
184 M6 **Whangamata** Waikato, North Island, New Zealand 37°13′S 175°54′E
184 Q9 **Whangara** Gisborne, North Island, New Zealand 38°34′S 178°12′E
184 K3 **Whangarei** Northland, North Island, New Zealand 35°44′S 174°18′E
184 K3 **Whangaruru Harbour** *inlet* North Island, New Zealand
25 V12 **Wharton** Texas, SW USA 29°19′N 96°08′W
173 U8 **Wharton Basin** *var.* West Australian Basin. *undersea feature* E Indian Ocean
185 E18 **Whataroa** West Coast, South Island, New Zealand 43°17′S 170°20′E
8 K10 **Wha Ti** *prev.* Lac la Martre. Northwest Territories, W Canada 63°10′N 117°12′W
8 J9 **Wha Ti** Northwest Territories, W Canada 63°10′N 117°12′W
184 K6 **Whatipu** Auckland, North Island, New Zealand 37°17′S 174°44′E
33 Y16 **Wheatland** Wyoming, C USA 42°03′N 104°57′W
14 D18 **Wheatley** Ontario, S Canada 42°06′N 82°27′W
30 M10 **Wheaton** Illinois, N USA 41°52′N 88°06′W
29 R7 **Wheaton** Minnesota, N USA 45°48′N 96°30′W
37 T4 **Wheat Ridge** Colorado, C USA 39°44′N 105°06′W
25 P2 **Wheeler** Texas, SW USA 35°26′N 100°17′W
23 O2 **Wheeler Lake** ⊠ Alabama, S USA
35 Y6 **Wheeler Peak** ▲ Nevada, W USA 39°00′N 114°17′W
37 T9 **Wheeler Peak** ▲ New Mexico, SW USA 36°34′N 105°25′W
31 S15 **Wheelersburg** Ohio, N USA 38°43′N 82°51′W
21 R2 **Wheeling** West Virginia, NE USA 40°05′N 80°43′W
97 L16 **Whernside** ▲ N England, United Kingdom 54°13′N 02°27′W
182 F9 **Whidbey, Point** *headland* South Australia 34°36′S 135°08′E
180 I7 **Whim Creek** Western Australia 20°51′S 117°54′E
10 L17 **Whistler** British Columbia, SW Canada 50°07′N 122°57′W
21 W8 **Whitakers** North Carolina, SE USA 36°06′N 77°43′W
14 H15 **Whitby** Ontario, S Canada 43°52′N 78°56′W
97 N15 **Whitby** N England, United Kingdom 54°29′N 00°37′W
10 G6 **White** ≈ Yukon, W Canada
13 T11 **White Bay** *bay* Newfoundland, Newfoundland and Labrador, E Canada
20 I8 **White Bluff** Tennessee, S USA 36°06′N 87°13′W
28 J6 **White Butte** ▲ North Dakota, N USA 46°23′N 103°18′W
19 R5 **White Cap Mountain** ▲ Maine, NE USA 45°33′N 69°15′W
22 J9 **White Castle** Louisiana, S USA 30°10′N 91°09′W
182 M5 **White Cliffs** New South Wales, SE Australia 30°52′S 143°04′E
31 P8 **White Cloud** Michigan, N USA 43°34′N 85°47′W
11 P14 **Whitecourt** Alberta, SW Canada 54°10′N 115°38′W
25 O2 **White Deer** Texas, SW USA 35°26′N 101°10′W
White Elster *see* Weisse Elster
24 M5 **Whiteface** Texas, SW USA 33°36′N 102°36′W
18 K7 **Whiteface Mountain** ▲ New York, NE USA 44°22′N 73°54′W
29 W5 **Whiteface Reservoir** ⊠ Minnesota, N USA
33 O7 **Whitefish** Montana, NW USA 48°24′N 114°20′W
31 N9 **Whitefish Bay** Wisconsin, N USA 43°09′N 87°54′W
31 Q3 **Whitefish Bay** *lake bay* Canada/USA
14 E11 **Whitefish Falls** Ontario, S Canada 46°06′N 81°42′W
14 B7 **Whitefish Lake** ◎ Ontario, S Canada
29 U6 **Whitefish Lake** ◎ Minnesota, C USA
31 Q3 **Whitefish Point** *headland* Michigan, N USA 46°46′N 84°57′W
31 O4 **Whitefish River** ≈ Michigan, N USA
25 O4 **Whiteflat** Texas, SW USA 34°06′N 100°55′W
27 V12 **White Hall** Arkansas, C USA 34°18′N 92°05′W
30 K14 **White Hall** Illinois, N USA 39°26′N 90°24′W
31 O8 **Whitehall** Michigan, N USA 43°24′N 86°21′W
18 L9 **Whitehall** New York, NE USA 43°33′N 73°24′W
31 S13 **Whitehall** Ohio, N USA 39°58′N 82°53′W
30 J7 **Whitehall** Wisconsin, N USA 44°22′N 91°20′W
97 J15 **Whitehaven** NW England, United Kingdom 54°33′N 03°35′W
10 I8 **Whitehorse** *territory capital* Yukon, W Canada 60°41′N 135°08′W
184 O7 **White Island** *island* NE New Zealand
14 K13 **White Lake** ◎ Ontario, SE Canada
22 H10 **White Lake** ◎ Louisiana, S USA
186 G7 **Whiteman Range** ▲▲ New Britain, E Papua New Guinea
183 Q15 **Whitemark** Tasmania, SE Australia 40°10′S 148°01′E
35 S9 **White Mountains** ▲▲ California/Nevada, W USA
19 N7 **White Mountains** ▲▲ Maine/New Hampshire, NE USA
80 F11 **White Nile** ◆ *state* C Sudan
81 E14 **White Nile** *Ar.* Al Bahr al Abyad, An Nil al Abyad, Bahr el Jebel. ≈ NE South Sudan
67 U7 **White Nile** *var.* Bahr el Jebel. ≈ S Sudan
25 W5 **White Oak Creek** ≈ Texas, SW USA
10 H9 **White Pass** *pass* Canada/USA
32 I9 **White Pass** *pass* Washington, NW USA
21 O9 **White Pine** Tennessee, S USA 36°06′N 83°17′W
18 K14 **White Plains** New York, NE USA 41°01′N 73°45′W
37 N13 **Whiteriver** Arizona, SW USA 33°50′N 109°57′W
28 M11 **White River** South Dakota, N USA 43°34′N 100°45′W
27 W12 **White River** ≈ Arkansas, SE USA
37 P3 **White River** ≈ Colorado/Utah, C USA
31 N15 **White River** ≈ Indiana, N USA
31 O8 **White River** ≈ Michigan, N USA
28 K11 **White River** ≈ South Dakota, N USA
25 O5 **White River** ≈ Texas, SW USA
18 M8 **White River** ≈ Vermont, NE USA
25 O5 **White River Lake** ⊠ Texas, SW USA
32 H11 **White Salmon** Washington, NW USA 45°43′N 121°29′W
18 I10 **Whitesboro** New York, NE USA 43°07′N 75°17′W
25 T5 **Whitesboro** Texas, SW USA 33°39′N 96°54′W
21 O7 **Whitesburg** Kentucky, S USA 37°07′N 82°52′W
White Sea *see* Beloye More
White Sea-Baltic Canal/White Sea Canal *see* Belomorsko-Baltiyskiy Kanal
63 I25 **Whiteside, Canal** *channel* S Chile
33 S10 **White Sulphur Springs** Montana, NW USA 46°33′N 110°54′W
21 R6 **White Sulphur Springs** West Virginia, NE USA 37°48′N 80°18′W
20 J6 **Whitesville** Kentucky, S USA 37°40′N 86°48′W
32 I10 **White Swan** Washington, NW USA 46°22′N 120°46′W
21 U12 **Whiteville** North Carolina, SE USA 34°20′N 78°42′W
20 F10 **Whiteville** Tennessee, S USA 35°19′N 89°09′W
77 Q13 **White Volta** *var.* Nakambé, *Fr.* Volta Blanche. ≈ Burkina Faso/Ghana
30 M9 **Whitewater** Wisconsin, N USA 42°51′N 88°43′W
37 P14 **Whitewater Baldy** ▲ New Mexico, SW USA 33°19′N 108°38′W
23 X17 **Whitewater Bay** *bay* Florida, SE USA
31 Q14 **Whitewater River** ≈ Indiana/Ohio, N USA
11 V16 **Whitewood** Saskatchewan, S Canada 50°19′N 102°16′W
28 J9 **Whitewood** South Dakota, N USA 44°27′N 103°38′W
25 U5 **Whitewright** Texas, SW USA 33°30′N 96°23′W
97 I15 **Whithorn** S Scotland, United Kingdom 54°44′N 04°26′W
184 M6 **Whitianga** Waikato, North Island, New Zealand 36°50′S 175°42′E
19 N11 **Whitinsville** Massachusetts, NE USA 42°06′N 71°40′W
20 M8 **Whitley City** Kentucky, S USA 36°45′N 84°29′W
21 Q11 **Whitmire** South Carolina, SE USA 34°30′N 81°36′W
31 R10 **Whitmore Lake** Michigan, N USA 42°26′N 83°44′W
195 N9 **Whitmore Mountains** ▲▲ Antarctica
14 I12 **Whitney** Ontario, SE Canada 45°29′N 78°11′W
25 T8 **Whitney** Texas, SW USA 31°56′N 97°20′W
25 S8 **Whitney, Lake** ⊠ Texas, SW USA
35 S11 **Whitney, Mount** ▲ California, W USA 37°45′N 119°55′W
181 Y6 **Whitsunday Group** *island group* Queensland, E Australia
25 S6 **Whitt** Texas, SW USA 32°55′N 98°01′W
29 U12 **Whittemore** Iowa, C USA 43°03′N 94°25′W
39 R12 **Whittier** Alaska, USA 60°46′N 148°40′W
35 T15 **Whittier** California, W USA 33°58′N 118°01′W
83 I25 **Whittlesea** Eastern Cape, S South Africa 32°08′S 26°51′E
20 K10 **Whitwell** Tennessee, S USA 35°12′N 85°31′W
8 L10 **Wholdaia Lake** ◎ Northwest Territories, NW Canada
182 H7 **Whyalla** South Australia 33°04′S 137°34′E
Whydah *see* Ouidah
14 F13 **Wiarton** Ontario, S Canada 44°44′N 81°10′W
171 O13 **Wiau** Sulawesi, C Indonesia 03°08′S 121°22′E
111 H15 **Wiązów** *Ger.* Wansen. Dolnośląskie, SW Poland 50°49′N 17°13′E
33 Y8 **Wibaux** Montana, NW USA 46°57′N 104°11′W
27 N6 **Wichita** Kansas, C USA 37°42′N 97°20′W
25 R5 **Wichita Falls** Texas, SW USA 33°55′N 98°30′W
26 L11 **Wichita Mountains** ▲▲ Oklahoma, C USA
25 R5 **Wichita River** ≈ Texas, SW USA
96 K6 **Wick** N Scotland, United Kingdom 58°26′N 03°06′W
36 K13 **Wickenburg** Arizona, SW USA 33°57′N 112°41′W
24 L8 **Wickett** Texas, SW USA 31°34′N 103°00′W
180 I7 **Wickham** Western Australia 20°40′S 117°11′E
182 M14 **Wickham, Cape** *headland* Tasmania, SE Australia 39°36′S 143°55′E
20 G7 **Wickliffe** Kentucky, S USA 36°58′N 89°04′W
97 G19 **Wicklow** *Ir.* Cill Mhantáin. E Ireland 52°59′N 06°03′W
97 F19 **Wicklow** *Ir.* Cill Mhantáin. *cultural region* E Ireland
97 G19 **Wicklow Head** *Ir.* Ceann Chill Mhantáin. *headland* E Ireland 52°57′N 06°00′W
97 F18 **Wicklow Mountains** *Ir.* Sléibhte Chill Mhantáin. ▲▲ E Ireland
14 H10 **Wicksteed Lake** ◎ Ontario, S Canada
Wida *see* Ouidah
65 G15 **Wideawake Airfield** ✈ (Georgetown) SW Ascension Island
97 K18 **Widnes** NW England, United Kingdom 53°22′N 02°44′W
110 H9 **Więcbork** *Ger.* Vandsburg. Kujawsko-pomorskie, C Poland 53°21′N 17°31′E
101 E17 **Wied** ≈ W Germany
101 F16 **Wiehl** Nordrhein-Westfalen, W Germany 50°57′N 07°33′E
111 L17 **Wieliczka** Małopolskie, S Poland 50°N 20°02′E
110 H12 **Wielkopolskie** ◆ *province* SW Poland
111 J14 **Wieluń** Sieradz, C Poland 51°14′N 18°33′E
109 X4 **Wien** *Eng.* Vienna, *Hung.* Bécs, *Slvk.* Vídeň, *Slvn.* Dunaj; *anc.* Vindobona. ● (Austria) Wien, NE Austria 48°13′N 16°22′E
109 X4 **Wien** *off.* Land Wien, *Eng.* Vienna. ◆ *state* NE Austria
109 X5 **Wiener Neustadt** Niederösterreich, E Austria 47°49′N 16°08′E
Wien, Land *see* Wien
110 G7 **Wieprza** *Ger.* Wipper. ≈ NW Poland
98 O10 **Wierden** Overijssel, E Netherlands 52°22′N 06°35′E
98 I7 **Wieringerwerf** Noord-Holland, NW Netherlands 52°51′N 05°01′E
Wieruschow *see* Wieruszów
111 I14 **Wieruszów** *Ger.* Wieruschow. Łódzkie, C Poland 51°18′N 18°09′E
109 V9 **Wies** Steiermark, SE Austria 46°40′N 15°16′E
Wiesbachhorn *see* Grosses Wiesbachhorn
101 G18 **Wiesbaden** Hessen, W Germany 50°06′N 08°14′E
Wieselburg and Ungarisch-Altenburg/Wieselburg-Ungarisch-Altenburg *see* Mosonmagyaróvár
Wiesenhof *see* Ostrołęka
101 G20 **Wiesloch** Baden-Württemberg, SW Germany 49°18′N 08°42′E
100 F10 **Wiesmoor** Niedersachsen, NW Germany 53°22′N 07°46′E
110 I7 **Wieżyca** *Ger.* Turmberg. *hill* Pomorskie, N Poland
97 L17 **Wigan** NW England, United Kingdom 53°33′N 02°38′W
37 U3 **Wiggins** Colorado, C USA 40°11′N 104°03′W
22 M8 **Wiggins** Mississippi, S USA 30°50′N 89°09′W
Wigorna Ceaster *see* Worcester
97 I14 **Wigtown** S Scotland, United Kingdom 54°53′N 04°27′W
97 H14 **Wigtown** *cultural region* SW Scotland, United Kingdom
97 I15 **Wigtown Bay** *bay* SW Scotland, United Kingdom
98 L13 **Wijchen** Gelderland, SE Netherlands 51°48′N 05°44′E
92 N1 **Wijdefjorden** *fjord* NW Svalbard
98 M10 **Wijhe** Overijssel, E Netherlands 52°22′N 06°07′E
98 J12 **Wijk bij Duurstede** Utrecht, C Netherlands 51°58′N 05°21′E
98 J13 **Wijk en Aalburg** Noord-Brabant, S Netherlands 51°46′N 05°06′E
99 H16 **Wijnegem** Antwerpen, N Belgium 51°13′N 04°32′E
14 E11 **Wikwemikong** Manitoulin Island, Ontario, S Canada 45°46′N 81°43′W
108 H7 **Wil** Sankt Gallen, NE Switzerland 47°28′N 09°03′E
29 Q15 **Wilber** Nebraska, C USA 40°28′N 96°57′W
32 K8 **Wilbur** Washington, NW USA 47°45′N 118°42′W
27 Q11 **Wilburton** Oklahoma, C USA 34°57′N 95°20′W
182 M6 **Wilcannia** New South Wales, SE Australia 31°34′S 143°23′E
18 D12 **Wilcox** Pennsylvania, NE USA 41°34′N 78°40′W
Wilczek Land *see* Vil'cheka, Zemlya
109 U6 **Wildalpen** Steiermark, E Austria 47°40′N 14°54′E
31 O13 **Wildcat Creek** ≈ Indiana, N USA
108 L9 **Wilde Kreuzspitze** *It.* Picco di Croce. ▲ Austria/Italy 46°53′N 10°51′E
Wildenschwert *see* Ústí nad Orlicí
98 O6 **Wildervank** Groningen, NE Netherlands 53°04′N 06°52′E
100 G11 **Wildeshausen** Niedersachsen, NW Germany 52°54′N 08°26′E
108 D10 **Wildhorn** ▲ SW Switzerland 46°21′N 07°22′E
11 R17 **Wild Horse** Alberta, SW Canada 49°00′N 110°19′W
27 N12 **Wildhorse Creek** ≈ Oklahoma, C USA
28 L14 **Wild Horse Hill** ▲ Nebraska, C USA 41°52′N 101°56′W
109 W8 **Wildon** Steiermark, SE Austria 46°53′N 15°29′E
24 M2 **Wildorado** Texas, SW USA 35°12′N 102°10′W
29 R6 **Wild Rice River** ≈ Minnesota/North Dakota, N USA
Wilejka *see* Vilyeyka
195 Y9 **Wilhelm II Coast** *physical region* Antarctica
195 X9 **Wilhelm II Land** *physical region* Antarctica
55 U11 **Wilhelmina Gebergte** ▲▲ C Suriname
18 B13 **Wilhelm, Lake** ⊠ Pennsylvania, NE USA
92 O2 **Wilhelmøya** *island* C Svalbard
Wilhelm-Pieck-Stadt *see* Guben
109 W4 **Wilhelmsburg** Niederösterreich, E Austria 48°07′N 15°37′E
100 G10 **Wilhelmshaven** Niedersachsen, NW Germany 53°31′N 08°07′E
Wilia/Wilja *see* Neris
18 H13 **Wilkes Barre** Pennsylvania, NE USA 41°15′N 75°50′W
21 R9 **Wilkesboro** North Carolina, SE USA 36°08′N 81°09′W
195 W15 **Wilkes Coast** *physical region* Antarctica
189 W12 **Wilkes Island** *island* N Wake Island
195 X12 **Wilkes Land** *physical region* Antarctica
11 S15 **Wilkie** Saskatchewan, S Canada 52°27′N 108°42′W
194 I6 **Wilkins Ice Shelf** *ice shelf* Antarctica
182 D4 **Wilkinsons Lakes** *salt lake* South Australia
Wilkomierz *see* Ukmergė
182 K11 **Willalooka** South Australia 36°24′S 140°20′E
32 G11 **Willamette River** ≈ Oregon, NW USA
183 O8 **Willandra Billabong Creek** *seasonal river* New South Wales, SE Australia
32 F9 **Willapa Bay** *inlet* Washington, NW USA
27 T7 **Willard** Missouri, C USA 37°18′N 93°25′W
37 S12 **Willard** New Mexico, SW USA 34°36′N 106°01′W
31 S12 **Willard** Ohio, N USA 41°03′N 82°43′W
36 L1 **Willard** Utah, W USA 41°23′N 112°01′W
186 G6 **Willaumez Peninsula** *headland* New Britain, E Papua New Guinea 05°03′S 150°04′E
37 N15 **Willcox** Arizona, SW USA 32°13′N 109°49′W
37 N16 **Willcox Playa** *salt flat* Arizona, SW USA
99 G17 **Willebroek** Antwerpen, C Belgium 51°04′N 04°22′E
45 P16 **Willemstad** ○ Curaçao 12°07′N 68°54′W
99 G14 **Willemstad** Noord-Brabant, S Netherlands 51°40′N 04°27′E
11 S11 **William** ≈ Saskatchewan, C Canada
23 O6 **William "Bill" Dannelly Reservoir** ⊠ Alabama, S USA
182 G3 **William Creek** South Australia 28°55′S 136°23′E
181 T15 **William, Mount** ▲ South Australia
36 K11 **Williams** Arizona, SW USA 35°15′N 112°11′W
29 X14 **Williamsburg** Iowa, C USA 41°39′N 92°00′W
20 M8 **Williamsburg** Kentucky, S USA 36°44′N 84°10′W
31 R15 **Williamsburg** Ohio, N USA 39°00′N 84°02′W
21 X6 **Williamsburg** Virginia, NE USA 37°17′N 76°43′W
10 M15 **Williams Lake** British Columbia, SW Canada 52°08′N 122°09′W
21 P6 **Williamson** West Virginia, NE USA 37°42′N 82°16′W
31 N13 **Williamsport** Indiana, N USA 40°18′N 87°18′W
18 G13 **Williamsport** Pennsylvania, NE USA 41°16′N 77°03′W
21 W9 **Williamston** North Carolina, SE USA 35°53′N 77°05′W
21 P11 **Williamston** South Carolina, SE USA 34°37′N 82°28′W
20 M4 **Williamstown** Kentucky, S USA 38°39′N 84°32′W
18 L10 **Williamstown** Massachusetts, NE USA 42°41′N 73°11′W
18 J16 **Willingboro** New Jersey, NE USA 40°01′N 74°52′W
11 Q14 **Willingdon** Alberta, SW Canada 53°49′N 112°08′W
25 W10 **Willis** Texas, SW USA 30°25′N 95°28′W
108 F8 **Willisau** Luzern, W Switzerland 47°07′N 08°00′E
83 F24 **Williston** Northern Cape, W South Africa 31°20′S 20°52′E
23 V10 **Williston** Florida, SE USA 29°23′N 82°27′W
28 J3 **Williston** North Dakota, N USA 48°07′N 103°37′W
21 Q13 **Williston** South Carolina, SE USA 33°24′N 81°25′W
10 L12 **Williston Lake** ⊠ British Columbia, W Canada
34 L5 **Willits** California, W USA 39°24′N 123°22′W
29 T8 **Willmar** Minnesota, N USA 45°07′N 95°02′W
10 K11 **Will, Mount** ▲ British Columbia, W Canada 57°31′N 128°48′W
31 T11 **Willoughby** Ohio, N USA 41°38′N 81°24′W
11 U17 **Willow Bunch** Saskatchewan, S Canada 49°30′N 105°41′W
32 J11 **Willow Creek** ≈ Oregon, NW USA
39 R11 **Willow Lake** Alaska, USA 61°44′N 150°02′W
8 I9 **Willowlake** ≈ Northwest Territories, NW Canada
83 H25 **Willowmore** Eastern Cape, S South Africa 33°18′S 23°30′E
30 L5 **Willow Reservoir** ⊠ Wisconsin, N USA
35 N5 **Willows** California, W USA 39°28′N 122°12′W
27 V7 **Willow Springs** Missouri, C USA 36°59′N 91°58′W
182 I7 **Wilmington** South Australia 32°42′S 138°08′E
21 Y2 **Wilmington** Delaware, NE USA 39°45′N 75°33′W
21 V12 **Wilmington** North Carolina, SE USA 34°14′N 77°55′W
31 R14 **Wilmington** Ohio, N USA 39°27′N 83°49′W
20 M6 **Wilmore** Kentucky, S USA 37°51′N 84°39′W
29 R8 **Wilmot** South Dakota, N USA 45°24′N 96°51′W
Wilna/Wilno *see* Vilnius
101 G16 **Wilnsdorf** Nordrhein-Westfalen, W Germany 50°49′N 08°06′E
99 G16 **Wilrijk** Antwerpen, N Belgium 51°11′N 04°25′E
100 I10 **Wilseder Berg** *hill* NW Germany
67 Z12 **Wilshaw Ridge** *undersea feature* W Indian Ocean 17°30′S 56°30′E
21 V9 **Wilson** North Carolina, SE USA 35°43′N 77°56′W
25 N5 **Wilson** Texas, SW USA 33°21′N 101°44′W
182 A7 **Wilson Bluff** *headland* South Australia/Western Australia 31°41′S 129°01′E
35 Y7 **Wilson Creek Range** ▲▲ Nevada, W USA
23 O1 **Wilson Lake** ⊠ Alabama, S USA
26 M4 **Wilson Lake** ⊠ Kansas, SE USA
37 P7 **Wilson, Mount** ▲ Colorado, C USA 37°50′N 107°59′W
183 P13 **Wilsons Promontory** *peninsula* Victoria, SE Australia
29 Y14 **Wilton** Iowa, C USA 41°35′N 91°01′W
19 P7 **Wilton** Maine, NE USA 44°35′N 70°15′W
28 M5 **Wilton** North Dakota, N USA 47°09′N 100°46′W
97 L22 **Wiltshire** *cultural region* S England, United Kingdom
99 M23 **Wiltz** Diekirch, NW Luxembourg 49°58′N 05°56′E
180 K9 **Wiluna** Western Australia 26°34′S 120°14′E
99 M23 **Wilwerwiltz** Diekirch, NE Luxembourg 49°59′N 06°00′E
29 P5 **Wimbledon** North Dakota, N USA 47°08′N 98°25′W
42 K7 **Wina** *var.* Güina. Jinotega, N Nicaragua 14°00′N 85°14′W
31 O12 **Winamac** Indiana, N USA 41°03′N 86°37′W
81 G19 **Winam Gulf** *var.* Kavirondo Gulf. *gulf* SW Kenya
83 I22 **Winburg** Free State, C South Africa 28°31′S 27°01′E
19 N10 **Winchendon** Massachusetts, NE USA 42°41′N 72°01′W
14 M13 **Winchester** Ontario, SE Canada 45°07′N 75°19′W
97 M23 **Winchester** *hist.* Wintanceaster, *Lat.* Venta Belgarum. S England, United Kingdom 51°04′N 01°19′W
32 M10 **Winchester** Idaho, NW USA 46°13′N 116°35′W
30 J14 **Winchester** Illinois, N USA 39°38′N 90°28′W
31 Q13 **Winchester** Indiana, N USA 40°11′N 84°57′W
20 M5 **Winchester** Kentucky, S USA 38°00′N 84°10′W
18 M10 **Winchester** New Hampshire, NE USA 42°46′N 72°21′W
20 K10 **Winchester** Tennessee, S USA 35°11′N 86°06′W
21 V3 **Winchester** Virginia, NE USA 39°11′N 78°12′W
99 L22 **Wincrange** Diekirch, NW Luxembourg 50°03′N 05°55′E
10 I5 **Wind** ≈ Yukon, NW Canada
183 S8 **Windamere, Lake** ◎ New South Wales, SE Australia
Windau *see* Ventspils, Latvia
Windau *see* Venta, Latvia/Lithuania
18 D15 **Windber** Pennsylvania, NE USA 40°12′N 78°47′W
23 T3 **Winder** Georgia, SE USA 33°59′N 83°43′W
97 K15 **Windermere** NW England, United Kingdom 54°24′N 02°54′W
14 C7 **Windermere Lake** ◎ Ontario, S Canada
31 U11 **Windham** Ohio, N USA 41°14′N 81°03′W
83 D19 **Windhoek** *Ger.* Windhuk. ● (Namibia) Khomas, C Namibia 22°34′S 17°06′E
83 D20 **Windhoek** ✈ Khomas, C Namibia 22°31′S 17°04′E
Windhuk *see* Windhoek
15 O8 **Windigo** Québec, SE Canada 47°45′N 73°19′W
15 O8 **Windigo** ≈ Québec, SE Canada
Windischfeistritz *see* Slovenska Bistrica
109 T6 **Windischgarsten** Oberösterreich, W Austria 47°42′N 14°21′E
Windischgraz *see* Slovenj Gradec
37 T16 **Wind Mountain** ▲ New Mexico, SW USA 32°01′N 105°35′W
29 T10 **Windom** Minnesota, N USA 43°52′N 95°07′W
37 Q7 **Windom Peak** ▲ Colorado, C USA 37°37′N 107°35′W
181 U9 **Windorah** Queensland, C Australia 25°25′S 142°41′E
37 O10 **Window Rock** Arizona, SW USA 35°40′N 109°03′W
31 N9 **Wind Point** *headland* Wisconsin, N USA
33 U14 **Wind River** ≈ Wyoming, C USA
13 P15 **Windsor** Nova Scotia, SE Canada 45°00′N 64°09′W
14 C17 **Windsor** Ontario, S Canada 42°18′N 83°W
15 Q12 **Windsor** Québec, SE Canada 45°34′N 72°00′W
97 N22 **Windsor** S England, United Kingdom 51°29′N 00°39′W
37 T3 **Windsor** Colorado, C USA 40°28′N 104°54′W
18 M12 **Windsor** Connecticut, NE USA 41°51′N 72°38′W
27 T5 **Windsor** Missouri, C USA 38°31′N 93°31′W
21 X9 **Windsor** North Carolina, SE USA 36°00′N 76°57′W
18 M12 **Windsor Locks** Connecticut, NE USA 41°55′N 72°37′W
25 R5 **Windthorst** Texas, SW USA 33°34′N 98°26′W
45 Z14 **Windward Islands** *island group* E West Indies
Windward Islands *see* Barlavento, Ilhas de, Cape Verde
Windward Islands *see* Vent, Îles du, Archipel de la Société, French Polynesia
44 K8 **Windward Passage** *Sp.* Paso de los Vientos. *channel* Cuba/Haiti
55 T9 **Wineperu** C Guyana 06°10′N 58°34′W
23 O3 **Winfield** Alabama, S USA 33°55′N 87°49′W
29 Y15 **Winfield** Iowa, C USA 41°07′N 91°26′W
27 O7 **Winfield** Kansas, C USA 37°14′N 97°00′W
25 W6 **Winfield** Texas, SW USA 33°10′N 95°06′W
21 Q4 **Winfield** West Virginia, NE USA 38°30′N 81°54′W
29 N5 **Wing** North Dakota, N USA 47°06′N 100°16′W
183 U7 **Wingham** New South Wales, SE Australia 31°52′S 152°24′E
12 G16 **Wingham** Ontario, S Canada 43°54′N 81°19′W
33 T8 **Winifred** Montana, NW USA 47°33′N 109°26′W
12 E9 **Winisk Lake** ◎ Ontario, C Canada
24 L8 **Wink** Texas, SW USA 31°45′N 103°09′W
36 M14 **Winkelman** Arizona, SW USA 32°59′N 110°46′W
11 X17 **Winkler** Manitoba, S Canada 49°12′N 97°55′W
109 Q9 **Winklern** Tirol, W Austria 46°54′N 12°54′E
Winkowitz *see* Vinkovci
32 G9 **Winlock** Washington, NW USA 46°29′N 122°56′W
77 P17 **Winneba** SE Ghana 05°22′N 00°38′W
29 U11 **Winnebago** Minnesota, N USA 43°46′N 94°10′W
29 R13 **Winnebago** Nebraska, C USA 42°14′N 96°28′W
30 M7 **Winnebago, Lake** ◎ Wisconsin, N USA
30 M7 **Winneconne** Wisconsin, N USA 44°07′N 88°44′W
35 T3 **Winnemucca** Nevada, W USA 40°59′N 117°44′W
35 R4 **Winnemucca Lake** ◎ Nevada, W USA
101 H21 **Winnenden** Baden-Württemberg, SW Germany 48°52′N 09°22′E
29 N11 **Winner** South Dakota, N USA 43°22′N 99°51′W
33 U9 **Winnett** Montana, NW USA 47°00′N 108°18′W
19 I9 **Winneway** Québec, SE Canada 47°35′N 78°33′W
22 H6 **Winnfield** Louisiana, S USA 31°55′N 92°38′W
29 U4 **Winnibigoshish, Lake** ◎ Minnesota, N USA
25 X11 **Winnie** Texas, SW USA 29°49′N 94°22′W
11 Y16 **Winnipeg** *province capital* Manitoba, S Canada 49°53′N 97°10′W
11 X16 **Winnipeg** ✈ Manitoba, S Canada 49°56′N 97°16′W
0 J8 **Winnipeg** ≈ Manitoba, S Canada
11 X16 **Winnipeg Beach** Manitoba, S Canada 50°25′N 96°59′W
11 W14 **Winnipeg, Lake** ◎ Manitoba, C Canada
11 W15 **Winnipegosis** Manitoba, S Canada 51°36′N 99°59′W

◆ Country
● Country Capital
◇ Dependent Territory
○ Dependent Territory Capital
◆ Administrative Regions
✈ International Airport
▲ Mountain
▲▲ Mountain Range
Volcano
≈ River
◎ Lake
⊠ Reservoir

11 W15 **Winnipegosis, Lake** ◎ Manitoba, C Canada
19 O8 **Winnipesaukee, Lake** ◎ New Hampshire, NE USA
22 I6 **Winnsboro** Louisiana, S USA 32°09′N 91°43′W
21 R12 **Winnsboro** South Carolina, SE USA 34°22′N 81°05′W
25 W6 **Winnsboro** Texas, SW USA 33°01′N 95°16′W
29 X10 **Winona** Minnesota, N USA 44°03′N 91°37′W
22 L4 **Winona** Mississippi, S USA 33°30′N 89°42′W
27 W7 **Winona** Missouri, C USA 37°00′N 91°19′W
25 W7 **Winona** Texas, SW USA 32°29′N 95°10′W
18 M7 **Winooski River** ≈ Vermont, NE USA
98 P6 **Winschoten** Groningen, NE Netherlands 53°09′N 07°03′E
100 J10 **Winsen** Niedersachsen, N Germany 53°22′N 10°13′E
36 M11 **Winslow** Arizona, SW USA 35°01′N 110°42′W
19 Q7 **Winslow** Maine, NE USA 44°33′N 69°35′W
18 M12 **Winsted** Connecticut, NE USA 41°55′N 73°03′W
32 F14 **Winston** Oregon, NW USA 43°07′N 123°24′W
21 S9 **Winston Salem** North Carolina, SE USA 36°06′N 80°15′W
98 N5 **Winsum** Groningen, NE Netherlands 53°20′N 06°31′E
Wintanceaster *see* Winchester
23 W11 **Winter Garden** Florida, SE USA 28°34′N 81°35′W
10 J16 **Winter Harbour** Vancouver Island, British Columbia, SW Canada 50°28′N 128°03′W
23 W12 **Winter Haven** Florida, SE USA 28°01′N 81°43′W
23 X11 **Winter Park** Florida, SE USA 28°36′N 81°20′W
25 P8 **Winters** Texas, SW USA 31°57′N 99°57′W
29 U15 **Winterset** Iowa, C USA 41°19′N 94°00′W
98 O12 **Winterswijk** Gelderland, E Netherlands 51°58′N 06°44′E
108 G6 **Winterthur** Zürich, NE Switzerland 47°30′N 08°43′E
29 U9 **Winthrop** Minnesota, N USA 44°32′N 94°22′W
32 J7 **Winthrop** Washington, NW USA 48°28′N 120°13′W
181 V7 **Winton** Queensland, E Australia 22°22′S 143°04′E
185 C24 **Winton** Southland, South Island, New Zealand 46°08′S 168°20′E
21 X8 **Winton** North Carolina, SE USA 36°24′N 76°57′W
101 K15 **Wipper** ≈ C Germany
101 K14 **Wipper** ≈ C Germany
Wipper *see* Wieprza
182 G6 **Wirraminna** South Australia 31°10′S 136°13′E
182 F4 **Wirrida** South Australia 29°34′S 134°33′E
182 F7 **Wirrulla** South Australia 32°27′S 134°33′E
Wirsitz *see* Wyrzysk
Wirz-See *see* Võrtsjärv
97 O19 **Wisbech** E England, United Kingdom 52°39′N 00°08′E
Wisby *see* Visby
19 Q8 **Wiscasset** Maine, NE USA 44°01′N 69°41′W
Wischau *see* Vyškov
30 J5 **Wisconsin** *off.* State of Wisconsin, *also known as* Badger State. ◆ state N USA
30 L8 **Wisconsin Dells** Wisconsin, N USA 43°37′N 89°43′W
30 L8 **Wisconsin, Lake** ⊠ Wisconsin, N USA
30 L7 **Wisconsin Rapids** Wisconsin, N USA 44°24′N 89°50′W
30 L7 **Wisconsin River** ≈ Wisconsin, N USA
33 P11 **Wisdom** Montana, NW USA 45°36′N 113°27′W
21 P7 **Wise** Virginia, NE USA 37°00′N 82°36′W
39 Q7 **Wiseman** Alaska, USA 67°24′N 150°06′W
96 J12 **Wishaw** W Scotland, United Kingdom 55°47′N 03°56′W
29 O6 **Wishek** North Dakota, N USA 46°12′N 99°33′W
32 I11 **Wishram** Washington, NW USA 45°40′N 120°53′W
111 J17 **Wisła** Śląskie, S Poland 49°39′N 18°50′E
110 K11 **Wisła** *Eng.* Vistula, *Ger.* Weichsel. ≈ C Poland
Wiślany, Zalew *see* Vistula Lagoon
111 M16 **Wisłoka** ≈ SE Poland
100 L9 **Wismar** Mecklenburg-Vorpommern, N Germany 53°54′N 11°28′E
29 R14 **Wisner** Nebraska, C USA 41°59′N 96°54′W
103 V4 **Wissembourg** *var.* Weissenburg. Bas-Rhin, NE France 49°03′N 07°57′E
30 J6 **Wissota, Lake** ⊠ Wisconsin, N USA
97 O18 **Witham** ≈ E England, United Kingdom
97 O17 **Withernsea** E England, United Kingdom 53°46′N 00°01′W
37 Q13 **Withington, Mount** ▲ New Mexico, SW USA 33°52′N 107°29′W
23 U8 **Withlacoochee River** ≈ Florida/Georgia, SE USA
110 H11 **Witkowo** Wielkopolskie, C Poland 52°27′N 17°49′E
97 M21 **Witney** S England, United Kingdom 51°47′N 01°30′W
101 E15 **Witten** Nordrhein-Westfalen, W Germany 51°25′N 07°19′E
101 K14 **Wittenberg** Sachsen-Anhalt, E Germany 51°53′N 12°39′E
30 L6 **Wittenberg** Wisconsin, N USA 44°53′N 89°20′W
100 L11 **Wittenberge** Brandenburg, N Germany 52°59′N 11°45′E
103 U7 **Wittenheim** Haut-Rhin, NE France 47°49′N 07°19′E
180 I7 **Wittenoom** Western Australia 22°17′S 118°22′E
Wittingau *see* Třeboň
100 K12 **Wittingen** Niedersachsen, C Germany 52°42′N 10°43′E
101 E18 **Wittlich** Rheinland-Pfalz, SW Germany 49°59′N 06°54′E
100 F9 **Wittmund** Niedersachsen, NW Germany 53°34′N 07°46′E
100 M10 **Wittstock** Brandenburg, NE Germany 53°10′N 12°29′E
186 F6 **Witu Islands** *island group* E Papua New Guinea
110 O7 **Wiżajny** Podlaskie, NE Poland 54°22′N 22°51′E
55 W10 **W. J. van Blommesteinmeer** ⊠ E Suriname
110 L11 **Wkra** *Ger.* Soldau. ≈ C Poland
110 I6 **Władysławowo** Pomorskie, N Poland 54°48′N 18°25′E
Wlaschim *see* Vlašim
111 E14 **Wleń** *Ger.* Lähn. Dolnośląskie, SW Poland 51°00′N 15°39′E
110 J11 **Włocławek** *Ger./Rus.* Vlotslavsk. Kujawsko-pomorskie, C Poland 52°39′N 19°03′E
110 P13 **Włodawa** *Rus.* Vlodava. Lubelskie, SE Poland 51°33′N 23°31′E
Włodzimierz *see* Volodymyr-Volyns'kyy
111 K15 **Włoszczowa** Świętokrzyskie, C Poland 50°51′N 19°58′E
83 C19 **Wlotzkasbaken** Erongo, W Namibia 22°26′S 14°30′E
15 R12 **Woburn** Québec, SE Canada 45°22′N 70°52′W
19 O11 **Woburn** Massachusetts, NE USA 42°28′N 71°09′W
Wocheiner Feistritz *see* Bohinjska Bistrica
Wöchma *see* Võhma
147 S11 **Wodil** *var.* Vuadil'. Farg'ona Viloyati, E Uzbekistan 40°10′N 71°43′E
181 V14 **Wodonga** Victoria, SE Australia 36°11′S 146°55′E
111 I17 **Wodzisław Śląski** *Ger.* Loslau. Śląskie, S Poland 49°59′N 18°27′E
98 I11 **Woerden** Zuid-Holland, C Netherlands 52°06′N 04°54′E
98 I8 **Wognum** Noord-Holland, NW Netherlands 52°40′N 05°01′E
Wohlau *see* Wołów
108 F7 **Wohlen** Aargau, NW Switzerland 47°21′N 08°17′E
195 R2 **Wohlthat Massivet** *Eng.* Wohlthat Mountains. ▲▲ Antarctica
Wohlthat Mountains *see* Wohlthat Massivet
Wojerecy *see* Hoyerswerda
Wōjjā *see* Wotje Atoll
Wojwodina *see* Vojvodina
171 V15 **Wokam, Pulau** *island* Kepulauan Aru, E Indonesia
97 N22 **Woking** SE England, United Kingdom 51°20′N 00°34′W
Woldenberg Neumark *see* Dobiegniew
188 K15 **Woleai Atoll** *atoll* Caroline Islands, W Micronesia
Woleu *see* Uolo, Río
79 E17 **Woleu-Ntem** *off.* Province du Woleu-Ntem, *var.* Le Woleu-Ntem. ◆ province W Gabon
Woleu-Ntem, Province du *see* Woleu-Ntem
32 F15 **Wolf Creek** Oregon, NW USA 42°40′N 123°22′W
26 K9 **Wolf Creek** ≈ Oklahoma/Texas, SW USA
37 R7 **Wolf Creek Pass** *pass* Colorado, C USA
19 O9 **Wolfeboro** New Hampshire, NE USA 43°34′N 71°10′W
25 U5 **Wolfe City** Texas, SW USA 33°22′N 96°04′W
14 L15 **Wolfe Island** *island* Ontario, SE Canada
101 M14 **Wolfen** Sachsen-Anhalt, E Germany 51°40′N 12°16′E
100 J13 **Wolfenbüttel** Niedersachsen, C Germany 52°10′N 10°33′E
109 T4 **Wolfern** Oberösterreich, N Austria 48°06′N 14°16′E
109 Q6 **Wolfgangsee** *var.* Abersee, St Wolfgangsee. ◎ N Austria
29 P9 **Wolf Mountain** ▲ Alaska, USA 65°20′N 154°08′W
33 X7 **Wolf Point** Montana, NW USA 48°05′N 105°40′W
22 L8 **Wolf River** ≈ Mississippi, S USA
30 M7 **Wolf River** ≈ Wisconsin, N USA
109 U9 **Wolfsberg** Kärnten, SE Austria 46°50′N 14°50′E
100 K12 **Wolfsburg** Niedersachsen, N Germany 52°25′N 10°47′E
57 B17 **Wolf, Volcán** ▲ Galapagos Islands, Ecuador, E Pacific Ocean 0°01′N 91°22′W
100 O8 **Wolgast** Mecklenburg-Vorpommern, NE Germany 54°04′N 13°47′E
108 F8 **Wolhusen** Luzern, W Switzerland 47°04′N 08°06′E
110 D8 **Wolin** *Ger.* Wollin. Zachodnio-pomorskie, NW Poland 53°52′N 14°35′E
109 Y3 **Wolkersdorf** Niederösterreich, NE Austria 48°24′N 16°31′E
Wołkowysk *see* Vawkavysk
Wöllan *see* Velenje
8 J6 **Wollaston, Cape** *headland* Victoria Island, Northwest Territories, NW Canada 71°00′N 118°21′W
63 J25 **Wollaston, Isla** *island* S Chile
11 U11 **Wollaston Lake** Saskatchewan, C Canada 58°05′N 103°38′W
11 T10 **Wollaston Lake** ◎ Saskatchewan, C Canada
8 J6 **Wollaston Peninsula** *peninsula* Victoria Island, Northwest Territories/Nunavut NW Canada
Wollin *see* Wolin
183 S9 **Wollongong** New South Wales, SE Australia 34°25′S 150°52′E
Wolmar *see* Valmiera
100 L13 **Wolmirstedt** Sachsen-Anhalt, C Germany 52°15′N 11°37′E
110 M11 **Wołomin** Mazowieckie, C Poland 52°20′N 21°11′E
110 G13 **Wołów** *Ger.* Wohlau. Dolnośląskie, SW Poland 51°21′N 16°40′E
Wołożyn *see* Valozhyn
14 G11 **Wolseley Bay** Ontario, S Canada 46°05′N 80°16′W
29 P10 **Wolsey** South Dakota, N USA 44°22′N 98°28′W
110 F12 **Wolsztyn** Wielkopolskie, C Poland 52°07′N 16°07′E
98 M7 **Wolvega** *Fris.* Wolvegea. Fryslân, N Netherlands 52°53′N 06°E
Wolvegea *see* Wolvega
97 K19 **Wolverhampton** C England, United Kingdom 52°36′N 02°08′W
Wolverine State *see* Michigan
99 G18 **Wolvertem** Vlaams Brabant, C Belgium 50°55′N 04°19′E
99 H16 **Wommelgem** Antwerpen, N Belgium 51°12′N 04°32′E
186 D7 **Wonenara** *var.* Wonerara. Eastern Highlands, C Papua New Guinea 06°46′S 145°54′E
Wonerara *see* Wonenara
Wongalara Lake *see* Wongalaroo Lake
183 N6 **Wongalarroo Lake** *var.* Wongalara Lake. *seasonal lake* New South Wales, SE Australia
163 Y15 **Wonju** *Jap.* Genshū; *prev.* Wŏnju. N South Korea 37°21′N 127°57′E
Wŏnju *see* Wonju
10 M12 **Wonowon** British Columbia, W Canada 56°46′N 121°54′W
163 X13 **Wŏnsan** SE North Korea 39°11′N 127°21′E
183 O13 **Wonthaggi** Victoria, SE Australia 38°38′S 145°37′E
23 N2 **Woodall Mountain** ▲ Mississippi, S USA 34°47′N 88°14′W
23 W7 **Woodbine** Georgia, SE USA 30°58′N 81°43′W
29 S14 **Woodbine** Iowa, C USA 41°44′N 95°42′W
18 J17 **Woodbine** New Jersey, NE USA 39°12′N 74°47′W
21 W4 **Woodbridge** Virginia, NE USA 38°40′N 77°17′W
183 V4 **Woodburn** New South Wales, SE Australia 29°07′S 153°23′E
32 G11 **Woodburn** Oregon, NW USA 45°08′N 122°51′W
20 K9 **Woodbury** Tennessee, S USA 35°49′N 86°06′W
183 V5 **Wooded Bluff** *headland* New South Wales, SE Australia 29°24′S 153°22′E
183 V3 **Woodenbong** New South Wales, SE Australia 28°24′S 152°39′E
35 R11 **Woodlake** California, W USA 36°24′N 119°06′W
35 N7 **Woodland** California, W USA 38°41′N 121°46′W
19 T5 **Woodland** Maine, NE USA 45°10′N 67°25′W
32 G10 **Woodland** Washington, NW USA 45°54′N 122°44′W
37 T5 **Woodland Park** Colorado, C USA 38°59′N 105°03′W
186 I9 **Woodlark Island** *var.* Murua Island. *island* SE Papua New Guinea
Woodle Island *see* Kuria
11 T17 **Wood Mountain** ▲ Saskatchewan, S Canada
30 K15 **Wood River** Illinois, N USA 38°51′N 90°06′W
29 P16 **Wood River** Nebraska, C USA 40°48′N 98°33′W
39 R9 **Wood River** ≈ Alaska, USA
39 O13 **Wood River Lakes** *lakes* Alaska, USA
182 C1 **Woodroffe, Mount** ▲ South Australia 26°19′S 131°42′E
21 P11 **Woodruff** South Carolina, SE USA 34°44′N 82°02′W
30 K4 **Woodruff** Wisconsin, N USA 45°55′N 89°41′W
25 T14 **Woodsboro** Texas, SW USA 28°14′N 97°19′W
31 U13 **Woodsfield** Ohio, N USA 39°45′N 81°07′W
181 P4 **Woods, Lake** ◎ Northern Territory, N Australia
11 Z16 **Woods, Lake of the** *Fr.* Lac des Bois. ◎ Canada/USA
25 Q6 **Woodson** Texas, SW USA 33°00′N 99°01′W
13 N14 **Woodstock** New Brunswick, SE Canada 46°10′N 67°38′W
14 F16 **Woodstock** Ontario, S Canada 43°07′N 80°46′W
30 M10 **Woodstock** Illinois, N USA 42°18′N 88°27′W
18 M9 **Woodstock** Vermont, NE USA 43°37′N 72°33′W
21 U4 **Woodstock** Virginia, NE USA 38°55′N 78°31′W
19 N8 **Woodsville** New Hampshire, NE USA 44°08′N 72°02′W
184 M12 **Woodville** Manawatu-Wanganui, North Island, New Zealand 40°22′S 175°59′E
22 J7 **Woodville** Mississippi, S USA 31°06′N 91°18′W
25 X9 **Woodville** Texas, SW USA 30°47′N 94°26′W
26 K9 **Woodward** Oklahoma, C USA 36°26′N 99°25′W
29 O5 **Woodworth** North Dakota, N USA 47°06′N 99°19′W
171 W12 **Wool** Papua, E Indonesia 01°38′S 135°34′E
183 V5 **Woolgoolga** New South Wales, E Australia 30°04′S 153°09′E
182 H6 **Woomera** South Australia 31°12′S 136°52′E
19 O12 **Woonsocket** Rhode Island, NE USA 41°58′N 71°27′W
29 P10 **Woonsocket** South Dakota, N USA 44°03′N 98°16′W
31 T12 **Wooster** Ohio, N USA 40°48′N 81°56′W
80 L12 **Woqooyi Galbeed** *off.* Gobolka Woqooyi Galbeed. ◆ region NW Somalia
Woqooyi Galbeed, Gobolka *see* Woqooyi Galbeed
108 E8 **Worb** Bern, C Switzerland 46°54′N 07°36′E
83 F26 **Worcester** Western Cape, SW South Africa 33°41′S 19°22′E
97 L20 **Worcester** *hist.* Wigorna Ceaster. W England, United Kingdom 52°11′N 02°13′W
19 N11 **Worcester** Massachusetts, NE USA 42°17′N 71°48′W
97 L20 **Worcestershire** *cultural region* C England, United Kingdom
32 H16 **Worden** Oregon, NW USA 42°04′N 121°50′W
109 O6 **Wörgl** Tirol, W Austria 47°29′N 12°04′E
171 V15 **Workai, Pulau** *island* Kepulauan Aru, E Indonesia
97 J15 **Workington** NW England, United Kingdom 54°39′N 03°33′W
98 K7 **Workum** Fryslân, N Netherlands 52°58′N 05°25′E
33 V13 **Worland** Wyoming, C USA 44°01′N 107°57′W
see São João de Cortes
Wormatia *see* Worms
99 N25 **Wormeldange** Grevenmacher, E Luxembourg 49°37′N 06°25′E
98 I9 **Wormer** Noord-Holland, C Netherlands 52°30′N 04°50′E
101 G19 **Worms** *anc.* Augusta Vangionum, Borbetomagus, Wormatia. Rheinland-Pfalz, SW Germany 49°38′N 08°22′E
Worms *see* Vormsi
101 K21 **Wörnitz** ≈ S Germany
25 U8 **Wortham** Texas, SW USA 31°47′N 96°27′W
101 G21 **Wörth am Rhein** Rheinland-Pfalz, SW Germany 49°04′N 08°16′E
109 S9 **Wörther See** ◎ S Austria
97 O23 **Worthing** SE England, United Kingdom 50°48′N 00°23′W
29 S11 **Worthington** Minnesota, N USA 43°37′N 95°37′W
31 S13 **Worthington** Ohio, N USA 40°05′N 83°01′W
35 W8 **Worthington Peak** ▲ Nevada, W USA 37°57′N 115°32′W
171 Y13 **Wosi** Papua, E Indonesia 03°55′S 138°54′E
171 V13 **Wosimi** Papua Barat, E Indonesia 02°44′S 134°34′E
189 R5 **Wotho Atoll** *var.* Wōtto. *atoll* Ralik Chain, W Marshall Islands
189 V5 **Wotje Atoll** *var.* Wōjjā. *atoll* Ratak Chain, E Marshall Islands
Wotoe *see* Wotu
Wottawa *see* Otava
Wōtto *see* Wotho Atoll
171 O13 **Wotu** *prev.* Wotoe. Sulawesi, C Indonesia 02°34′S 120°46′E
98 K11 **Woudenberg** Utrecht, C Netherlands 52°05′N 05°25′E
98 I13 **Woudrichem** Noord-Brabant, S Netherlands 51°49′N 05°00′E
43 N8 **Wounta** *var.* Huaunta. Región Autónoma Atlántico Norte, NE Nicaragua 13°30′N 83°32′W
171 P14 **Wowoni, Pulau** *island* C Indonesia
81 J17 **Woyamdero Plain** *plain* E Kenya
Woyens *see* Vojens
Wozrojdeniye Oroli *see* Vozrozhdeniya, Ostrov
Wrangel Island *see* Vrangelya, Ostrov
39 Y13 **Wrangell** Wrangell Island, Alaska, USA 56°28′N 132°22′W
38 C15 **Wrangell, Cape** *headland* Attu Island, Alaska, USA 52°55′N 172°28′E
39 S11 **Wrangell, Mount** ▲ Alaska, USA 62°00′N 144°01′W
39 T11 **Wrangell Mountains** ▲▲ Alaska, USA
197 S7 **Wrangel Plain** *undersea feature* Arctic Ocean
96 H6 **Wrath, Cape** *headland* N Scotland, United Kingdom 58°37′N 05°01′W
37 W3 **Wray** Colorado, C USA 40°01′N 102°12′W
44 K13 **Wreck Point** *headland* C Jamaica 17°50′N 76°55′W
83 C23 **Wreck Point** *headland* W South Africa 28°52′S 16°17′E
23 V4 **Wrens** Georgia, SE USA 33°12′N 82°23′W
97 K18 **Wrexham** NE Wales, United Kingdom 53°03′N 03°W
27 R13 **Wright City** Oklahoma, C USA 34°03′N 95°00′W
194 J12 **Wright Island** *island* Antarctica
13 N9 **Wright, Mont** ▲ Québec, E Canada 52°36′N 67°40′W
25 X5 **Wright Patman Lake** ⊠ Texas, SW USA
36 M16 **Wrightson, Mount** ▲ Arizona, SW USA 31°42′N 110°51′W
23 U5 **Wrightsville** Georgia, SE USA 32°43′N 82°43′W
21 W12 **Wrightsville Beach** North Carolina, SE USA 34°12′N 77°48′W
35 T15 **Wrightwood** California, W USA 34°21′N 117°37′W
8 H9 **Wrigley** Northwest Territories, W Canada 63°16′N 123°39′W
111 G14 **Wrocław** *Eng./Ger.* Breslau. Dolnośląskie, SW Poland 51°07′N 17°01′E
110 F10 **Wronki** *Ger.* Fronicken. Wielkopolskie, C Poland 52°42′N 16°22′E
110 H11 **Września** Wielkopolskie, C Poland 52°19′N 17°34′E
110 F12 **Wschowa** Lubuskie, W Poland 51°49′N 16°15′E
Wsetin *see* Vsetín
161 O5 **Wu'an** Hebei, E China 36°45′N 114°12′E
180 I12 **Wubin** Western Australia 30°05′S 116°43′E
163 W9 **Wuchang** Heilongjiang, NE China 44°55′N 127°13′E
Wuchang *see* Wuhan
Wu-chou/Wuchow *see* Wuzhou
160 M16 **Wuchuan** *var.* Meilu. Guangdong, S China 21°28′N 110°49′E
160 K10 **Wuchuan** *var.* Duru, Gelaozu Miaozu Zhizhixian. Guizhou, S China 28°40′N 108°04′E
163 O13 **Wuchuan** Nei Mongol Zizhiqu, N China 41°04′N 111°28′E
163 V6 **Wudalianchi** *var.* Qingshan; *prev.* Dedu. Heilongjiang, NE China 48°40′N 126°06′E
159 O11 **Wudaoliang** Qinghai, C China 35°16′N 93°03′E
141 Q13 **Wuday'ah** *spring/well* S Saudi Arabia 17°03′N 47°06′E
77 V13 **Wudil** Kano, N Nigeria 11°46′N 08°49′E
160 G12 **Wuding** *var.* Jincheng. Yunnan, SW China 25°30′N 102°21′E
182 G8 **Wudinna** South Australia 33°06′S 135°30′E
Wudu *see* Longnan
160 L9 **Wufeng** Hubei, C China 30°09′N 110°31′E
161 O11 **Wugong Shan** ▲▲ S China
157 P7 **Wuhai** *var.* Haibowan. Nei Mongol Zizhiqu, N China 39°40′N 106°48′E
161 O9 **Wuhan** *var.* Han-kou, Han-k'ou, Hanyang, Wuchang, Wu-han; *prev.* Hankow. *province capital* Hubei, C China 30°35′N 114°19′E
Wu-han *see* Wuhan
161 Q7 **Wuhe** Anhui, E China 33°05′N 117°55′E
Wuhsien *see* Suzhou
Wuhsi/Wu-his *see* Wuxi
161 Q8 **Wuhu** *var.* Wu-na-mu. Anhui, E China 31°23′N 118°25′E
Wūjae *see* Ujae Atoll
160 K11 **Wu Jiang** ≈ C China
158 L5 **Wujiaqu** Xinjiang Uygur Zizhiqu, NW China 44°11′N 87°30′E
Wūjlān *see* Ujelang Atoll
77 W15 **Wukari** Taraba, E Nigeria 07°51′N 09°49′E
Wulan *see* Jingyuan
152 H4 **Wular Lake** ◎ NE India
162 M13 **Wulashan** Nei Mongol Zizhiqu, N China 40°43′N 108°45′E
160 H11 **Wulian Feng** ▲▲ SW China
160 F13 **Wuliang Shan** ▲▲ SW China
160 K11 **Wuling Shan** ▲▲ S China
109 Y5 **Wulka** ≈ E Austria
Wulkan *see* Vulcan
109 T3 **Wullowitz** Oberösterreich, N Austria 48°37′N 14°27′E
Wu-lu-k'o-mu-shi/Wu-lu-mu-ch'i *see* Ürümqi
79 D14 **Wum** Nord-Ouest, NE Cameroon 06°24′N 10°04′E
160 H12 **Wumeng Shan** ▲▲ SW China
160 K14 **Wuming** Guangxi Zhuangzu Zizhiqu, S China 23°12′N 108°11′E
100 G10 **Wümme** ≈ NW Germany
Wu-na-mu *see* Wuhu
171 X13 **Wunen** Papua, E Indonesia 03°40′S 138°31′E
12 D9 **Wunnummin Lake** ◎ Ontario, C Canada
80 D13 **Wun Rog** Warap, W South Sudan 09°00′N 28°20′E
101 M18 **Wunsiedel** Bayern, E Germany 50°02′N 12°00′E
100 I12 **Wunstorf** Niedersachsen, NW Germany 52°25′N 09°25′E
166 M3 **Wuntho** Sagaing, N Myanmar (Burma) 23°52′N 95°43′E
101 F15 **Wupper** ≈ W Germany
101 E15 **Wuppertal** *prev.* Barmen-Elberfeld. Nordrhein-Westfalen, W Germany 51°16′N 07°12′E
160 K5 **Wuqi** Shaanxi, C China 36°57′N 108°15′E
158 E7 **Wuqia** Xinjiang Uygur Zizhiqu, NW China 39°50′N 75°19′E
161 P4 **Wuqiao** *var.* Sangyuan. Hebei, E China 37°40′N 116°21′E
101 J23 **Würm** ≈ SE Germany
77 T12 **Wurno** Sokoto, NW Nigeria 13°15′N 05°24′E
101 I19 **Würzburg** Bayern, SW Germany 49°48′N 09°56′E
101 N15 **Wurzen** Sachsen, E Germany 51°21′N 12°48′E
160 L9 **Wu Shan** ▲▲ C China
158 G7 **Wushi** *var.* Uqturpan. Xinjiang Uygur Zizhiqu, NW China 41°07′N 79°09′E
Wusih *see* Wuxi
65 N18 **Wüst Seamount** *undersea feature* S Atlantic Ocean 32°00′S 00°06′E
Wusuli Jiang/Wusuri *see* Ussuri
161 N3 **Wutai Shan** *var.* Beitai Ding. ▲ C China 39°00′N 114°00′E
160 H10 **Wutongqiao** Sichuan, C China 29°21′N 103°48′E
159 P6 **Wutongwozi Quan** *spring* NW China
99 H15 **Wuustwezel** Antwerpen, N Belgium 51°24′N 04°34′E
186 B4 **Wuvulu Island** *island* NW Papua New Guinea
159 U9 **Wuwei** *var.* Liangzhou. Gansu, C China 37°58′N 102°40′E
161 R8 **Wuxi** *var.* Wuhsi, Wu-hsi, Wusih. Jiangsu, E China 31°35′N 120°19′E
Wuxing *see* Huzhou
160 L14 **Wuxuan** Guangxi Zhuangzu Zizhiqu, S China 23°40′N 109°41′E
Wuyang *see* Zhenyuan
160 K11 **Wuyang He** ≈ S China
163 X6 **Wuyiling** Heilongjiang, NE China 48°36′N 129°24′E
161 Q11 **Wuyishan** *prev.* Chong'an. Fujian, SE China 27°48′N 118°03′E
157 T12 **Wuyi Shan** ▲▲ SE China
162 M13 **Wuyuan** Nei Mongol Zizhiqu, N China 41°05′N 108°15′E
160 L17 **Wuzhishan** *prev.* Tongshi. Hainan, S China 18°37′N 109°34′E
160 L17 **Wuzhi Shan** ▲ S China 18°52′N 109°36′E
159 W8 **Wuzhong** Ningxia, N China 37°58′N 106°09′E
160 M14 **Wuzhou** *var.* Wu-chou, Wuchow. Guangxi Zhuangzu Zizhiqu, S China 23°30′N 111°21′E
18 H12 **Wyalusing** Pennsylvania, NE USA 41°40′N 76°13′W
182 M10 **Wycheproof** Victoria, SE Australia 36°06′S 143°13′E
97 K21 **Wye** *Wel.* Gwy. ≈ England/Wales, United Kingdom
Wylkowyszki *see* Vilkaviškis
97 P19 **Wymondham** E England, United Kingdom 52°29′N 01°10′E
29 R17 **Wymore** Nebraska, C USA 40°07′N 96°39′W
182 E5 **Wynbring** South Australia 30°34′S 133°27′E
181 N3 **Wyndham** Western Australia 15°28′S 128°08′E
29 R6 **Wyndmere** North Dakota, N USA 46°16′N 97°07′W
27 X11 **Wynne** Arkansas, C USA 35°14′N 90°48′W
27 N12 **Wynnewood** Oklahoma, C USA 34°39′N 97°09′W
183 O15 **Wynyard** Tasmania, SE Australia 40°57′S 145°33′E
11 U15 **Wynyard** Saskatchewan, S Canada 51°46′N 104°10′W
33 V11 **Wyola** Montana, NW USA 45°07′N 107°23′W
182 A4 **Wyola Lake** *salt lake* South Australia
31 P9 **Wyoming** Michigan, N USA 42°54′N 85°42′W
33 V14 **Wyoming** *off.* State of Wyoming, *also known as* Equality State. ◆ state C USA
33 S15 **Wyoming Range** ▲▲ Wyoming, C USA
183 T8 **Wyong** New South Wales, SE Australia 33°18′S 151°27′E
110 G9 **Wyrzysk** *Ger.* Wirsitz. Wielkopolskie, C Poland 53°09′N 17°15′E
Wysg *see* Usk
110 O10 **Wysokie Mazowieckie** Łomża, E Poland 52°54′N 22°34′E
110 M11 **Wyszków** *Ger.* Probstberg. Mazowieckie, NE Poland 52°36′N 21°28′E
110 L11 **Wyszogród** Mazowieckie, C Poland 52°24′N 20°14′E
21 R7 **Wytheville** Virginia, NE USA 36°57′N 81°07′W
111 L15 **Wyżyna Małopolska** *plateau*

X

80 Q12 **Xaafuun** *It.* Hafun. Bari, NE Somalia 10°25′N 51°17′E
80 Q12 **Xaafuun, Raas** *var.* Ras Hafun. *cape* NE Somalia
Xabia *see* Jávea
42 C4 **Xaclbal, Río** *var.* Xalbal. ≈ Guatemala/Mexico
137 Y10 **Xaçmaz** *Rus.* Khachmas. N Azerbaijan 41°26′N 48°47′E
80 O12 **Xadeed** *var.* Haded. *physical region* N Somalia
159 O14 **Xagquka** Xizang Zizhiqu, W China 31°47′N 92°46′E
Xai *see* Oudômxai
158 F10 **Xaidulla** Xinjiang Uygur Zizhiqu, W China 36°27′N 77°46′E
167 Q7 **Xaignabouli** *prev.* Muang Xaignabouri, *Fr.* Sayaboury. Xaignabouli, N Laos 19°16′N 101°43′E
167 R7 **Xai Lai Leng, Phou** ▲ Laos/Vietnam 19°13′N 104°09′E
158 L15 **Xainza** Xizang Zizhiqu, W China 30°54′N 88°36′E
158 L16 **Xaitongmoin** Xizang Zizhiqu, W China 29°27′N 88°13′E
83 F17 **Xaixai** *var.* Caecae. North-West, NW Botswana 19°52′S 21°04′E
83 M20 **Xai-Xai** *prev.* João Belo, Vila de João Belo. Gaza, S Mozambique 25°01′S 33°37′E
Xalbal *see* Xaclbal, Río
80 P13 **Xalin** Sool, N Somalia 09°16′N 49°00′E
146 H7 **Xalqobod** *Rus.* Khalkabad. Qoraqalpog'iston Respublikasi, W Uzbekistan 42°42′N 59°46′E
160 F11 **Xamgyi'nyilha** *var.* Jiantang; *prev.* Zhongdian. Yunnan, SW China 27°48′N 99°41′E
167 R6 **Xam Nua** *var.* Sam Neua. Houaphan, N Laos 20°24′N 104°03′E
82 D11 **Xá-Muteba** *Port.* Cinco de Outubro. Lunda Norte, NE Angola 09°34′S 17°50′E
Xangda *see* Nangqên
83 C16 **Xangongo** *Port.* Rocadas. Cunene, SW Angola 16°43′S 15°01′E
137 W12 **Xankändi** *Rus.* Khankendi; *prev.* Stepanakert. SW Azerbaijan 39°50′N 46°44′E
Xanlar *see* Göygöl
139 U5 **Xan Sûr** *Ar.* Khān Zūr. As Sulaymānīyah, E Iraq 35°03′N 45°08′E
114 J13 **Xánthi** Anatolikí Makedonía kai Thráki, NE Greece 41°09′N 24°54′E
60 H13 **Xanxerê** Santa Catarina, S Brazil 26°52′S 52°25′W
81 O15 **Xarardheere** Mudug, E Somalia 04°45′N 47°54′E
137 Z11 **Xärä Zirä Adasi** *Rus.* Ostrov Bulla. *Island* E Azerbaijan
162 K13 **Xar Burd** *prev.* Bayan Nuru. Nei Mongol Zizhiqu, N China 40°09′N 104°48′E
163 T11 **Xar Moron** ≈ NE China
163 T11 **Xar Moron** ≈ N China
Xarra *see* Xarrë
113 L23 **Xarrë** *var.* Xarra. Vlorë, S Albania 39°45′N 20°01′E
82 D12 **Xassengue** Lunda Sul, NW Angola 10°28′S 18°32′E
105 S11 **Xàtiva** *Cas.* Xátiva; *anc.* Setabis, *var.* Jativa. Valenciana, E Spain 39°N 00°32′W
Xauen *see* Chefchaouen
60 K10 **Xavantes, Represa de** *var.* Represa de Chavantes. ⊠ S Brazil
158 I7 **Xayar** Xinjiang Uygur Zizhiqu, W China 41°16′N 82°52′E
Xäzär Dänizi *see* Caspian Sea
167 S8 **Xé Bangfai** ≈ C Laos
167 T9 **Xé Banghiang** *var.* Bang Hieng. ≈ S Laos
Xêgar *see* Tingri
167 T10 **Xékong** *vaar.* Lamam. Xékong, S Laos 15°22′N 106°40′E
31 R14 **Xenia** Ohio, N USA 39°40′N 83°55′W
Xeres *see* Jerez de la Frontera
115 E15 **Xeriás** ≈ C Greece
115 G17 **Xeró** ▲ Évvoia, C Greece 38°52′N 23°18′E
139 S1 **Xērzok** *Ar.* Khayrūzuk, *var.* Kharwazawk. Arbil, E Iraq 36°58′N 44°19′E
Xhumo *see* Cum
161 N15 **Xiachuan Dao** *island* S China
Xiacun *see* Rushan
Xiaguan *see* Dali
159 U11 **Xiahe** *var.* Labrang. Gansu, C China 35°12′N 102°28′E
161 Q13 **Xiamen** *var.* Hsia-men; *prev.* Amoy. Fujian, SE China 24°28′N 118°05′E
160 L6 **Xi'an** *var.* Changan, Sian, Signan, Siking, Singan, Xian. *province capital* Shaanxi, C China 34°16′N 108°54′E
160 L10 **Xianfeng** *var.* Gaoleshan. Hubei, C China 29°45′N 109°10′E
Xiang *see* Hunan
161 N7 **Xiangcheng** Henan, C China 33°51′N 113°27′E
160 F10 **Xiangcheng** *var.* Sampê, *Tib.* Qagchêng. Sichuan, C China 28°52′N 99°45′E
160 M8 **Xiangfan** *var.* Xiangyang. Hubei, C China 32°07′N 112°00′E
Xianggang Tebie Xingzhengqu *see* Hong Kong
161 N10 **Xiang Jiang** ≈ S China
Xiangkhoang *see* Phônsaven
167 Q7 **Xiangkhoang, Plateau de** *var.* Plain of Jars. *plateau* N Laos
161 N11 **Xiangtan** *var.* Hsiang-t'an, Siangtan. Hunan, S China 27°53′N 112°55′E
161 N11 **Xiangxiang** Hunan, S China 27°50′N 112°31′E
Xiangyang *see* Xiangfan
161 S10 **Xianju** Zhejiang, SE China 28°53′N 120°41′E
Xianshui *see* Dawu
160 F8 **Xianshui He** ≈ C China
161 N9 **Xiantao** *var.* Mianyang. Hubei, C China 30°20′N 113°31′E
161 R10 **Xianxia Ling** ▲▲ SE China
160 K6 **Xianyang** Shaanxi, C China 34°26′N 108°40′E
158 L5 **Xiaocaohu** Xinjiang Uygur Zizhiqu, W China 45°44′N 90°07′E
161 O9 **Xiaogan** Hubei, C China 30°55′N 113°54′E
Xiaogang *see* Dongxiang
163 W6 **Xiao Hinggan Ling** *Eng.* Lesser Khingan Range. ▲▲ NE China
160 M6 **Xiao Shan** ▲▲ C China
160 M12 **Xiao Shui** ≈ S China
Xiaoxi *see* Pinghe
161 P6 **Xiaoxian** *var.* Longcheng, Xiao Xian. Anhui, E China 34°11′N 116°56′E
Xiao Xian *see* Xiaoxian
160 G11 **Xichang** Sichuan, C China 27°52′N 102°16′E
41 P11 **Xicoténcatl** Tamaulipas, C Mexico 22°59′N 98°54′W
Xieng Khouang *see* Phônsaven
Xieng Ngeun *see* Muong Xiang Ngeun
160 J11 **Xifeng** *var.* Yongjing. Guizhou, S China 27°15′N 106°44′E
Xifeng *see* Qingcheng
Xigang *see* Helan
158 L16 **Xigazê** *var.* Jih-k'a-tse, Shigatse, Xigaze. Xizang Zizhiqu, W China 29°18′N 88°50′E
159 W11 **Xihe** *var.* Hanyuan. Gansu, C China 34°00′N 105°24′E
160 I8 **Xi He** ≈ C China
Xihuachi *see* Heshui
159 W10 **Xiji** Ningxia, N China 36°02′N 105°33′E
160 M14 **Xi Jiang** *var.* Hsi Chiang, *Eng.* West River. ≈ S China
159 Q7 **Xijian Quan** *spring* NW China
160 K15 **Xijin Shuiku** ⊠ S China
Xilaganí *see* Xylaganí
Xiligou *see* Ulan
160 I13 **Xilin** *var.* Bada. Guangxi Zhuangzu Zizhiqu, S China 24°30′N 105°00′E
163 Q10 **Xilinhot** *var.* Silinhot. Nei Mongol Zizhiqu, N China 43°58′N 116°07′E
Xilinji *see* Mohe
Xilokastro *see* Xylókastro
Xin *see* Xinjiang Uygur Zizhiqu
Xin'an *see* Anlong
Xin'anjiang Shuiku *see* Qiandao Hu
Xin'anzhen *see* Xinyi
Xin Barag Youqi *see* Altan Emel
Xin Barag Zuoqi *see* Amgalang
163 W12 **Xinbin** *var.* Xinbin Manzu Zizhixian. Liaoning, NE China 41°44′N 125°02′E
Xinbin Manzu Zizhixian *see* Xinbin
161 O7 **Xincai** Henan, C China 32°47′N 114°58′E
Xincheng *see* Zhaojue
Xindu *see* Luhuo
161 O13 **Xinfeng** *var.* Jiading. Jiangxi, S China 25°23′N 114°48′E
161 O14 **Xinfengjiang Shuiku** ⊠ S China
Xing'an *see* Ankang
Xingba *see* Lhünzê
163 T13 **Xingcheng** Liaoning, NE China 40°38′N 120°47′E
Xingcheng *see* Xingning
82 E11 **Xinge** Lunda Norte, NE Angola 09°44′S 19°10′E
161 P12 **Xingguo** *var.* Lianjiang. Jiangxi, S China 26°25′N 115°22′E
159 S11 **Xinghai** *var.* Ziketan. Qinghai, C China 35°12′N 100°28′E
161 R7 **Xinghua** Jiangsu, E China 32°54′N 119°48′E
Xingkai Hu *see* Khanka, Lake
161 P13 **Xingning** *prev.* Xingcheng. Guangdong, S China 24°05′N 115°47′E
160 I13 **Xingren** Guizhou, S China 25°26′N 105°08′E
161 O4 **Xingtai** Hebei, E China 37°08′N 114°29′E
59 J14 **Xingu, Rio** ≈ C Brazil
159 P6 **Xingxingxia** Xinjiang Uygur Zizhiqu, NW China 41°48′N 95°01′E
160 I13 **Xingyi** Guizhou, S China 25°04′N 104°51′E
158 I6 **Xinhe** *var.* Toksu. Xinjiang Uygur Zizhiqu, NW China 41°32′N 82°39′E

◆ Country
● Country Capital
◇ Dependent Territory
○ Dependent Territory Capital
◆ Administrative Regions
✈ International Airport
▲ Mountain
▲▲ Mountain Range
▲ Volcano
≈ River
◎ Lake
⊠ Reservoir

163 Q10 **Xin Hot** Nei Mongol Zizhiqu, N China 43°58′N 114°59′E
Xinhua *see* Funing
163 T12 **Xinhui** *var.* Aohan Qi. Nei Mongol Zizhiqu, N China 42°12′N 119°57′E
159 T10 **Xining** *var.* Hsining, Hsi-ning, Sining. *province capital* Qinghai, C China 36°37′N 101°46′E
161 O4 **Xinji** *prev.* Shulu. Hebei, E China 37°55′N 115°14′E
161 P10 **Xinjian** Jiangxi, S China 28°37′N 115°46′E
Xinjiang *see* Xinjiang Uygur Zizhiqu
162 D8 **Xinjiang Uygur Zizhiqu** *var.* Sinkiang, Sinkiang Uighur Autonomous Region, Xin, Xinjiang. ◆ *autonomous region* NW China
160 H9 **Xinjin** *var.* Meixing, *Tib.* Zainlha. Sichuan, C China 30°27′N 103°46′E
Xinjin *see* Pulandian
Xinjing *see* Jingxi
163 U12 **Xinmin** Liaoning, NE China 41°58′N 122°51′E
160 M12 **Xinning** *var.* Jinshi. Hunan, S China 26°34′N 110°57′E
Xinning *see* Ningxian
Xinning *see* Fusui
Xinpu *see* Lianyungang
Xinshan *see* Anyuan
161 P5 **Xintai** Shandong, E China 35°54′N 117°44′E
Xinwen *see* Suncun
Xin Xian *see* Xinzhou
161 N6 **Xinxiang** Henan, C China 35°13′N 113°48′E
161 O8 **Xinyang** *var.* Hsin-yang, Sinyang. Henan, C China 32°09′N 114°04′E
161 Q6 **Xinyi** *var.* Xin'anzhen. Jiangsu, E China 34°17′N 118°14′E
161 Q6 **Xinyi He** E China
161 S14 **Xinying** *var.* Sinying, *Jap.* Shinei; *prev.* Hsinying. C Taiwan 23°12′N 120°15′E
161 O11 **Xinyu** Jiangxi, S China 27°51′N 115°00′E
158 I5 **Xinyuan** *var.* Künes. Xinjiang Uygur Zizhiqu, NW China 43°25′N 83°12′E
Xinyuan *see* Tianjun
162 M13 **Xinzhao Shan** ▲ N China 39°37′N 107°51′E
161 N3 **Xinzhou** *var.* Xin Xian. Shanxi, C China 38°24′N 112°43′E
Xinzhou *see* Longlin
161 S13 **Xinzhu** *var.* Hsinchu. N Taiwan 24°48′N 120°59′E
104 H4 **Xinzo de Limia** Galicia, NW Spain 42°05′N 07°45′W
Xions *see* Książ Wielkopolski
161 O7 **Xiping** Henan, C China 33°22′N 114°00′E
Xiping *see* Songyang
159 T11 **Xiqing Shan** ▲ C China
59 N16 **Xique-Xique** Bahia, E Brazil 10°47′S 42°44′W
Xireg *see* Ulan
115 E14 **Xirovoúni** ▲ N Greece 40°31′N 21°58′E
162 M13 **Xishanzui** *prev.* Urad Qianqi. Nei Mongol Zizhiqu, N China 40°43′N 108°41′E
Xisha Qundao *see* Paracel Islands
160 J11 **Xishui** *var.* Donghuang. Guizhou, S China 28°24′N 106°09′E
Xi Ujimqin Qi *see* Bayan Ul
160 K11 **Xiushan** *var.* Zhonghe. Chongqing Shi, C China 28°23′N 108°52′E
Xiushan *see* Tonghai
160 O10 **Xiu Shui** S China
Xiuyan *see* Qingjian
146 H9 **Xiva** *Rus.* Khiva, Khiwa. Xorazm Viloyati, W Uzbekistan 41°22′N 60°22′E
158 J16 **Xixabangma Feng** ▲ W China 28°25′N 85°47′E
160 M7 **Xixia** Henan, C China 33°30′N 111°25′E
Xixón *see* Gijón
Xixona *see* Jijona
Xizang *see* Xizang Zizhiqu
Xizang Gaoyuan *see* Qingzang Gaoyuan
160 E9 **Xizang Zizhiqu** *var.* Thibet, Tibetan Autonomous Region, Xizang, *Eng.* Tibet. ◆ *autonomous region* W China
163 U14 **Xizhong Dao** *island* N China
Xoi *see* Qüxü
146 H8 **Xo'jayli** *Rus.* Khodzheyli. Qoraqalpog'iston Respublikasi, W Uzbekistan 42°23′N 59°27′E
Xolotlán *see* Managua, Lago de
147 I9 **Xonqa** *var.* Khonqa, *Rus.* Khanka. Xorazm Viloyati, W Uzbekistan 41°31′N 60°39′E
146 H9 **Xorazm Viloyati** *Rus.* Khorezmskaya Oblast'. ◆ *province* W Uzbekistan
159 N9 **Xorkol** Xinjiang Uygur Zizhiqu, NW China 38°45′N 91°07′E
147 P11 **Xovos** *var.* Ursat'yevskaya, *Rus.* Khavast. Sirdaryo Viloyati, E Uzbekistan 40°14′N 68°46′E
41 X14 **Xpujil** Quintana Roo, E Mexico 18°30′N 89°24′W
161 Q8 **Xuancheng** *var.* Xuanzhou. Anhui, E China 30°57′N 118°53′E
Xuande Qundao *see* Amphitrite Group
167 T9 **Xuân Đuc** Quang Binh, C Vietnam 17°19′N 106°38′E
160 L9 **Xuan'en** *var.* Zhushan. Hubei, C China 30°03′N 109°26′E
160 K8 **Xuanhan** Sichuan, C China 31°25′N 107°41′E
161 O2 **Xuanhua** Hebei, E China 40°36′N 115°01′E
161 P4 **Xuanhui He** E China
167 T8 **Xuân Sơn** Quang Binh, C Vietnam 17°42′N 105°58′E
H12 **Xuanwei** Yunnan, China 26°08′N 104°04′E
Xuanzhou *see* Xuancheng
161 N7 **Xuchang** Henan, C China 34°03′N 113°48′E
Xucheng *see* Xuwen
137 X10 **Xudat** *Rus.* Khudat. NE Azerbaijan 41°37′N 48°39′E
81 M16 **Xuddur** *var.* Hudur, *It.* Oddur. Bakool, SW Somalia 04°07′N 43°47′E
80 O13 **Xudun** Sool, N Somalia 09°12′N 47°34′E
160 L11 **Xuefeng Shan** ▲▲ S China
161 S12 **Xue Shan** *prev.* Hsüeh Shan. ▲ N Taiwan
147 O13 **Xufar** Surkhondaryo Viloyati, S Uzbekistan 38°31′N 67°45′E
Xulun Hobot Qagan *see* Qagan Nur
42 F2 **Xunantunich** *ruins* Cayo, W Belize
163 W6 **Xun He** NE China
160 L7 **Xun He** C China
160 L14 **Xun Jiang** S China
163 W5 **Xunke** *var.* Bianjiang; *prev.* Qike. Heilongjiang, NE China 49°35′N 128°27′E
161 P13 **Xunwu** *var.* Changning. Jiangxi, S China 24°59′N 115°33′E
139 V4 **Xurmał** *Ar.* Khūrmāl, *var.* Khormal. As Sulaymānīyah, NE Iraq 35°19′N 46°06′E
161 O3 **Xushui** Hebei, E China 39°01′N 115°38′E
160 L16 **Xuwen** *var.* Xucheng. Guangdong, S China 20°21′N 110°09′E
160 I11 **Xuyong** *var.* Yongning. Sichuan, C China 28°17′N 105°21′E
161 P6 **Xuzhou** *var.* Hsu-chou, Suchow, Tongshan; *prev.* T'ung-shan. Jiangsu, E China 34°17′N 117°09′E
114 K13 **Xylagani** *var.* Xilagani. Anatolikí Makedonía kai Thráki, NE Greece 40°58′N 25°27′E
115 F19 **Xylókastro** *var.* Xilokastro. Pelopónnisos, S Greece 38°04′N 22°36′E

Y

160 H9 **Ya'an** *var.* Yaan. Sichuan, C China 30°N 102°57′E
182 L10 **Yaapeet** Victoria, SE Australia 35°48′S 142°03′E
79 D15 **Yabassi** Littoral, W Cameroon 04°30′N 09°59′E
81 J15 **Yabēlo** Oromīya, C Ethiopia 04°53′N 38°01′E
114 H9 **Yablanitsa** Lovech, N Bulgaria 43°01′N 24°06′E
43 N7 **Yablis** Región Autónoma Atlántico Norte, NE Nicaragua 14°08′N 83°44′W
123 O14 **Yablonovyy Khrebet** ▲▲ S Russian Federation
162 J14 **Yabrai Shan** ▲▲ NE China
45 U6 **Yabucoa** E Puerto Rico 18°02′N 65°53′W
160 J11 **Yachi He** S China
32 H10 **Yacolt** Washington, NW USA 45°49′N 122°22′W
54 M10 **Yacuaray** Amazonas, S Venezuela 04°12′N 66°30′W
57 M22 **Yacuiba** Tarija, S Bolivia 22°00′S 63°43′W
57 K16 **Yacuma, Río** C Bolivia
155 H16 **Yādgīr** Karnātaka, C India 16°46′N 77°09′E
21 R8 **Yadkin River** North Carolina, SE USA
21 R9 **Yadkinville** North Carolina, SE USA 36°07′N 80°40′W
127 P3 **Yadrin** Chuvashskaya Respublika, W Russian Federation 55°55′N 46°10′E
Yaegama-shotō *see* Yaeyama-shotō
Yaeme-saki *see* Paimi-saki
165 O16 **Yaeyama-shotō** *var.* Yaegama-shotō. *island group* SW Japan
75 O8 **Yafran** NW Libya 32°04′N 12°31′E
165 S2 **Yagashiri-tō** *island* NE Japan
65 H21 **Yaghan Basin** *undersea feature* SE Pacific Ocean
123 S9 **Yagodnoye** Magadanskaya Oblast', E Russian Federation 62°37′N 149°18′E
Yagotin *see* Yahotyn
78 G12 **Yagoua** Extrême-Nord, NE Cameroon 10°23′N 15°13′E
159 Q11 **Yagradagzê Shan** ▲ C China 35°06′N 95°41′E
Yaguachi *see* Yaguachi Nuevo
56 B7 **Yaguachi Nuevo** *var.* Yaguachi. Guayas, W Ecuador 02°06′S 79°43′W
Yaguarón, Río *see* Jaguarão, Rio
117 Q11 **Yahorlyts'kyy Lyman** *bay* S Ukraine
117 Q5 **Yahotyn** *Rus.* Yagotin. Kyyivs'ka Oblast', N Ukraine 50°15′N 31°48′E
40 L12 **Yahualica** Jalisco, SW Mexico 21°11′N 102°29′W
79 L17 **Yahuma** Orientale, N Dem. Rep. Congo 01°12′N 23°00′E
136 K15 **Yahyalı** Kayseri, C Turkey 38°08′N 35°23′E
167 N15 **Yai, Khao** ▲ SW Thailand 08°45′N 99°32′E
164 M14 **Yaizu** Shizuoka, Honshū, S Japan 34°52′N 138°20′E
160 G9 **Yajiang** *var.* Hekou, *Tib.* Nyagquka. Sichuan, C China 30°05′N 100°57′E
119 O14 **Yakawlyevichi** *Rus.* Yakovlevichi. Vitsyebskaya Voblasts', NE Belarus 54°20′N 30°31′E
163 S6 **Yakeshi** Nei Mongol Zizhiqu, N China 49°16′N 120°42′E
32 I9 **Yakima** Washington, NW USA 46°36′N 120°30′W
32 J10 **Yakima River** Washington, NW USA
Yakkabag *see* Yakkabog'
147 N12 **Yakkabog'** *Rus.* Yakkabag. Qashqadaryo Viloyati, S Uzbekistan 38°57′N 66°35′E
148 L12 **Yakmach** Baluchistān, SW Pakistan 28°48′N 63°48′E
77 O12 **Yako** W Burkina Faso 12°59′N 02°15′W
39 W13 **Yakobi Island** *island* Alexander Archipelago, Alaska, USA
79 K16 **Yakoma** Equateur, N Dem. Rep. Congo 04°04′N 22°23′E
114 H11 **Yakoruda** Blagoevgrad, SW Bulgaria 42°01′N 23°40′E
Yakovlevichi *see* Yakawlyevichi
127 T2 **Yakshur-Bod'ya** Udmurtskaya Respublika, NW Russian Federation 57°10′N 53°10′E
165 Q5 **Yakumo** Hokkaidō, NE Japan 42°18′N 140°15′E
164 B17 **Yaku-shima** *island* Nansei-shotō, SW Japan
39 V12 **Yakutat** Alaska, USA 59°33′N 139°44′W
39 U12 **Yakutat Bay** *inlet* Alaska, USA
Yakutia/Yakutiya/Yakutiya, Respublika *see* Sakha (Yakutiya), Respublika
123 Q10 **Yakutsk** Respublika Sakha (Yakutiya), NE Russian Federation 62°10′N 129°50′E
167 O17 **Yala** Yala, SW Thailand 06°32′N 101°19′E
182 D6 **Yalata** South Australia 31°30′S 131°53′E
31 S9 **Yale** Michigan, N USA 43°07′N 82°45′W
180 I11 **Yalgoo** Western Australia 28°23′S 116°43′E
114 O12 **Yalıköy** İstanbul, NW Turkey 41°29′N 28°19′E
79 L14 **Yalinga** Haute-Kotto, C Central African Republic 06°47′N 23°09′E
119 M17 **Yalizava** *Rus.* Yelizovo. Mahilyowskaya Voblasts', E Belarus 53°24′N 29°01′E
44 L13 **Yallahs Hill** ▲ E Jamaica 17°53′N 76°31′W
22 L3 **Yalobusha River** Mississippi, S USA
79 H15 **Yaloké** Ombella-Mpoko, W Central African Republic 05°15′N 17°12′E
160 E7 **Yalong Jiang** C China
136 E11 **Yalova** Yalova, NW Turkey 40°40′N 29°17′E
136 E11 **Yalova** ◆ *province* NW Turkey
Yaloveny *see* Ialoveni
Yalpug *see* Ialpug
Yalpug, Ozero *see* Yalpuh, Ozero
117 N12 **Yalpuh, Ozero** *Rus.* Ozero Yalpug. ◎ SW Ukraine
117 T14 **Yalta** Avtonomna Respublika Krym, S Ukraine 44°30′N 34°09′E
163 W12 **Yalu** *Chin.* Yalu Jiang, *Jap.* Oryokko, *Kor.* Amnok-kang. China/North Korea
Yalu Jiang *see* Yalu
136 F14 **Yalvaç** Isparta, SW Turkey 38°16′N 31°09′E
165 R9 **Yamada** Iwate, Honshū, C Japan 39°27′N 141°56′E
165 D14 **Yamaga** Kumamoto, Kyūshū, SW Japan 33°02′N 130°41′E
165 P10 **Yamagata** Yamagata, Honshū, C Japan 38°15′N 140°19′E
165 P9 **Yamagata** *off.* Yamagata-ken. ◆ *prefecture* Honshū, C Japan
Yamagata-ken *see* Yamagata
164 C16 **Yamagawa** Kagoshima, Kyūshū, SW Japan 31°12′N 130°37′E
164 E13 **Yamaguchi** *var.* Yamaguti. Yamaguchi, Honshū, SW Japan 34°11′N 131°26′E
164 E13 **Yamaguchi** *off.* Yamaguchi-ken, *var.* Yamaguti. ◆ *prefecture* Honshū, SW Japan
Yamaguchi-ken *see* Yamaguchi
Yamaguti *see* Yamaguchi
125 X5 **Yamalo-Nenetskiy Avtonomnyy Okrug** ◆ *autonomous district* N Russian Federation
122 J7 **Yamal, Poluostrov** *peninsula* N Russian Federation
165 N13 **Yamanashi** *off.* Yamanashi-ken, *var.* Yamanasi. ◆ *prefecture* Honshū, S Japan
Yamanashi-ken *see* Yamanashi
Yamanasi *see* Yamanashi
Yamanīyah, Al Jumhūrīyah al *see* Yemen
127 W5 **Yamantau** ▲ W Russian Federation 53°11′N 57°30′E
Yamasaki *see* Yamazaki
15 P12 **Yamaska** Québec, SE Canada
192 G4 **Yamato Ridge** *undersea feature* S Sea of Japan 39°20′N 135°00′E
164 I13 **Yamazaki** *var.* Yamasaki. Hyōgo, Honshū, SW Japan 35°00′N 134°31′E
183 V5 **Yamba** New South Wales, SE Australia 29°28′S 153°22′E
81 D16 **Yambio** *var.* Yambiyo. Western Equatoria, S South Sudan 04°34′N 28°21′E
Yambiyo *see* Yambio
114 L10 **Yambol** *Turk.* Yanboli. Yambol, E Bulgaria 42°29′N 26°30′E
114 M11 **Yambol** ◆ *province* E Bulgaria
79 M17 **Yambuya** Orientale, N Dem. Rep. Congo 01°22′N 24°21′E
171 T15 **Yamdena, Pulau** *prev.* Jamdena. *island* Kepulauan Tanimbar, E Indonesia
165 O14 **Yame** Fukuoka, Kyūshū, SW Japan 33°14′N 130°32′E
166 M6 **Yamethin** Mandalay, C Myanmar (Burma) 20°24′N 96°08′E
186 C6 **Yaminbot** East Sepik, NW Papua New Guinea 04°34′S 143°56′E
181 U9 **Yamma Yamma, Lake** ◎ Queensland, C Australia
76 M16 **Yamoussoukro** ● (Ivory Coast) C Ivory Coast 06°51′N 05°21′W
37 P3 **Yampa River** Colorado, C USA
117 S2 **Yampil'** Sums'ka Oblast', NE Ukraine 51°57′N 33°49′E
116 M8 **Yampil'** Vinnyts'ka Oblast', C Ukraine 48°15′N 28°18′E
123 T9 **Yamsk** Magadanskaya Oblast', E Russian Federation 59°33′N 154°04′E
152 J8 **Yamuna** *prev.* Jumna. N India
152 I9 **Yamunānagar** Haryāna, N India 30°07′N 77°17′E
Yamundá *see* Nhamundá, Rio
145 U8 **Yamyshevo** Pavlodar, NE Kazakhstan 51°49′N 77°28′E
159 N16 **Yamzho Yumco** ◎ W China
123 Q8 **Yana** NE Russian Federation
186 H9 **Yanaba Island** *island* SE Papua New Guinea
155 L16 **Yanam** *var.* Yanaon. Puducherry, E India 16°45′N 82°16′E
160 L5 **Yan'an** *var.* Yanan. Shaanxi, C China 36°35′N 109°27′E
Yanaon *see* Yanam
127 U3 **Yanaul** Respublika Bashkortostan, W Russian Federation 56°15′N 54°57′E
118 O12 **Yanavichy** *Rus.* Yanovichi. Vitsyebskaya Voblasts', NE Belarus 55°17′N 30°42′E
Yanboli *see* Yambol
140 K8 **Yanbu' al Baḩr** Al Madīnah, W Saudi Arabia 24°07′N 38°03′E
21 T8 **Yanceyville** North Carolina, SE USA 36°25′N 79°22′W
161 R7 **Yancheng** Jiangsu, E China 33°28′N 120°10′E
159 W8 **Yanchi** Ningxia, N China 37°49′N 107°24′E
160 L5 **Yanchuan** Shaanxi, C China 36°54′N 110°04′E
183 O10 **Yanco Creek** *seasonal river* New South Wales, SE Australia
183 O6 **Yanda Creek** *seasonal river* New South Wales, SE Australia
182 K4 **Yandama Creek** *seasonal river* New South Wales/South Australia
161 S11 **Yandang Shan** ▲▲ SE China
159 O6 **Yandun** Xinjiang Uygur Zizhiqu, W China 42°24′N 94°08′E
76 L13 **Yanfolila** Sikasso, SW Mali 11°08′N 08°12′W
79 M18 **Yangambi** Orientale, N Dem. Rep. Congo 0°46′N 24°24′E
158 M15 **Yangbajain** Xizang Zizhiqu, W China 30°05′N 90°35′E
Yangcheng *see* Yangshan, Guangdong, S China
Yangchow *see* Yangzhou
160 M15 **Yangchun** *var.* Chuncheng. Guangdong, S China 22°16′N 111°49′E
161 N2 **Yanggao** *var.* Longquan. Shanxi, C China 40°24′N 113°51′E
Yanggeta *see* Yaqeta
Yangiabad *see* Yangiobod
Yangibazar *see* Dzhany-Bazar, Kyrgyzstan
Yangi-Bazar *see* Kofarnihon, Tajikistan
Yangiklshak *see* Yangiqishloq
146 M13 **Yangi-Nishon** *Rus.* Yangi-Nishan. Qashqadaryo Viloyati, S Uzbekistan 38°37′N 65°39′E
147 Q9 **Yangiobod** *Rus.* Yangiabad. Toshkent Viloyati, E Uzbekistan 41°10′N 70°10′E
147 O10 **Yangiqishloq** *Rus.* Yangiklshak. Jizzax Viloyati, C Uzbekistan 40°27′N 67°06′E
147 P11 **Yangiyer** Sirdaryo Viloyati, E Uzbekistan 40°19′N 68°48′E
147 P9 **Yangiyo'l** *Rus.* Yangiyul'. Toshkent Viloyati, E Uzbekistan 41°12′N 69°05′E
Yangiyul' *see* Yangiyo'l
160 M15 **Yangjiang** Guangdong, S China 21°50′N 112°02′E
Yangku *see* Taiyuan
Yang-Nishan *see* Yangi-Nishon
166 L8 **Yangon** *Eng.* Rangoon. Yangon, S Myanmar (Burma) 16°50′N 96°11′E
166 M8 **Yangon** *Eng.* Rangoon. ◆ *region* SW Myanmar (Burma)
161 N4 **Yangquan** Shanxi, C China 37°52′N 113°29′E
161 N13 **Yangshan** *var.* Yangcheng. Guangdong, S China 24°32′N 112°36′E
167 U12 **Yang Sin, Chu** ▲ S Vietnam 12°23′N 108°25′E
Yangtze *see* Chang Jiang/Jinsha Jiang
Yangtze *see* Chang Jiang
Yangtze Kiang *see* Chang Jiang
161 R7 **Yangzhou** *var.* Yangchow. Jiangsu, E China 32°22′N 119°22′E
160 L5 **Yan He** C China
163 Y10 **Yanji** Jilin, NE China 42°54′N 129°31′E
Yanji *see* Longjing
Yanjing *see* Yanyuan
29 Q12 **Yankton** South Dakota, N USA 42°52′N 97°24′W
161 O12 **Yanling** *prev.* Lingxian, Ling Xian. Hunan, S China 26°32′N 113°48′E
Yannina *see* Ioánnina
123 Q7 **Yano-Indigirskaya Nizmennost'** *plain* NE Russian Federation
Yanovichi *see* Yanavichy
155 K24 **Yan Oya** N Sri Lanka
158 K6 **Yanqi** *var.* Yanqi Huizu Zizhixian. Xinjiang Uygur Zizhiqu, NW China 42°04′N 86°32′E
Yanqi Huizu Zizhixian *see* Yanqi
161 Q10 **Yanshan** *var.* Hekou. Jiangxi, S China 28°18′N 117°43′E
160 H14 **Yanshan** *var.* Jiangna. Yunnan, SW China 23°36′N 104°20′E
161 P2 **Yan Shan** ▲▲ E China
163 X8 **Yanshou** Heilongjiang, NE China 45°27′N 128°19′E
123 Q7 **Yanskiy Zaliv** *bay* N Russian Federation
183 O4 **Yantabulla** New South Wales, SE Australia 29°22′S 145°00′E
161 R4 **Yantai** *var.* Yan-t'ai; *prev.* Chefoo, Chih-fu. Shandong, E China 37°30′N 121°22′E
118 A13 **Yantarnyy** *Ger.* Palmnicken. Kaliningradskaya Oblast', W Russian Federation 54°53′N 19°59′E
114 J9 **Yantra** Gabrovo, N Bulgaria 42°58′N 25°19′E
114 K9 **Yantra** N Bulgaria
160 G11 **Yanyuan** *var.* Yanjing. Sichuan, C China 27°30′N 101°22′E
161 P5 **Yanzhou** Shandong, E China 35°35′N 116°53′E
79 E16 **Yaoundé** *var.* Yaunde. ● (Cameroon) Centre, S Cameroon 03°51′N 11°31′E
188 I14 **Yap** ◆ *state* W Micronesia
188 F16 **Yap** *island* Caroline Islands, W Micronesia
57 M18 **Yapacani, Rio** C Bolivia
171 W14 **Yapa Kopra** Papua, E Indonesia 04°18′S 135°05′E
Yapan *see* Yapen, Selat
Yapanskoye More *see* East Sea/Japan, Sea of
77 P15 **Yapei** N Ghana 09°10′N 01°08′W
12 M10 **Yapeitso, Mont** ▲ Québec, E Canada 52°18′N 70°24′W
171 W12 **Yapen, Pulau** *prev.* Japen. *island* E Indonesia
171 W12 **Yapen, Selat** *var.* Yapan. *strait* Papua, E Indonesia
61 E15 **Yapeyú** Corrientes, NE Argentina 29°28′S 56°50′W
136 I11 **Yapraklı** Çankırı, N Turkey 40°45′N 33°46′E
174 M3 **Yap Trench** *var.* Yap Trough. *undersea feature* SE Philippine Sea 08°30′N 138°00′E
Yap Trough *see* Yap Trench
Yapurá *see* Caquetá, Río, Brazil/Colombia
Yapurá *see* Japurá, Rio, Brazil/Colombia
197 I12 **Yaqaga** *island* N Fiji
197 H12 **Yaqeta** *prev.* Yanggeta. *island* Yasawa Group, NW Fiji
40 G6 **Yaqui** Sonora, NW Mexico 27°21′N 109°59′W
32 E12 **Yaquina Bay** *bay* Oregon, NW USA
40 G6 **Yaqui, Río** NW Mexico
54 K5 **Yaracuy** *off.* Estado Yaracuy. ◆ *state* NW Venezuela
Yaracuy, Estado *see* Yaracuy
146 E13 **Yarajy** *Rus.* Yaradzhi. Ahal Welaýaty, C Turkmenistan 38°12′N 57°40′E
Yaradzhi *see* Yarajy
125 Q15 **Yaransk** Kirovskaya Oblast', NW Russian Federation 57°18′N 47°52′E
136 F17 **Yardımcı Burnu** *headland* SW Turkey 36°10′N 30°25′E
97 Q19 **Yare** E England, United Kingdom
125 S9 **Yarega** Respublika Komi, NW Russian Federation 63°27′N 53°28′E
116 I7 **Yaremcha** Ivano-Frankivs'ka Oblast', W Ukraine 48°27′N 24°34′E
189 Q9 **Yaren (district)** ● (Nauru) SW Nauru 0°33′S 166°54′E
125 Q10 **Yarensk** Arkhangel'skaya Oblast', NW Russian Federation 62°09′N 49°03′E
155 F16 **Yargatti** Karnātaka, W India 16°07′N 75°11′E
164 M12 **Yariga-take** ▲ Honshū, S Japan 36°20′N 137°38′E
141 O15 **Yarīm** W Yemen 14°15′N 44°23′E
54 F14 **Yarí, Río** SW Colombia
54 K5 **Yaritagua** Yaracuy, N Venezuela 10°05′N 69°07′W
Yarkand *see* Yarkant He
Yarkant *see* Shache
158 E9 **Yarkant He** *var.* Yarkand. NW China
149 U3 **Yarkhūn** NW Pakistan
Yarlung Zangbo Jiang *see* Brahmaputra
116 L6 **Yarmolyntsi** Khmel'nyts'ka Oblast', W Ukraine 49°13′N 26°53′E
13 O16 **Yarmouth** Nova Scotia, SE Canada 43°53′N 66°09′W
Yarmouth *see* Great Yarmouth
Yaroslav *see* Jarosław
124 L15 **Yaroslavl'** Yaroslavskaya Oblast', W Russian Federation 57°38′N 39°53′E
124 K14 **Yaroslavskaya Oblast'** ◆ *province* W Russian Federation
123 N11 **Yaroslavskiy** Respublika Sakha (Yakutiya), NE Russian Federation 60°10′N 114°12′E
183 P13 **Yarram** Victoria, SE Australia 38°36′S 146°40′E
183 O11 **Yarrawonga** Victoria, SE Australia 36°04′S 145°58′E
182 L4 **Yarriarraburra Swamp** *wetland* New South Wales, SE Australia
122 I8 **Yar-Sale** Yamalo-Nenetskiy Avtonomnyy Okrug, N Russian Federation 66°52′N 70°42′E
122 K11 **Yartsevo** Krasnoyarskiy Kray, C Russian Federation 60°15′N 90°09′E
126 I4 **Yartsevo** Smolenskaya Oblast', W Russian Federation 55°03′N 32°46′E
54 E8 **Yarumal** Antioquia, NW Colombia 06°59′N 75°25′W
187 W14 **Yasawa Group** *island group* NW Fiji
Yasel'da *see* Yasyel'da
77 V12 **Yashi** Katsina, N Nigeria 12°21′N 07°56′E
77 S14 **Yashikera** Kwara, W Nigeria 09°40′N 03°19′E
147 T14 **Yashilkŭl** *Rus.* Ozero Yashil'kul'. ◎ SE Tajikistan
Yashil'kul', Ozero *see* Yashilkŭl
165 P9 **Yashima** Akita, Honshū, C Japan 39°10′N 140°10′E
127 P13 **Yashkul'** Respublika Kalmykiya, SW Russian Federation 46°09′N 45°22′E
146 F13 **Yashlyk** Ahal Welaýaty, C Turkmenistan 37°46′N 58°51′E
Yasinovataya *see* Yasynuvata
114 N10 **Yasna Polyana** Burgas, E Bulgaria 42°18′N 27°35′E
167 R10 **Yasothon** Yasothon, E Thailand 15°46′N 104°12′E
183 R10 **Yass** New South Wales, SE Australia 34°52′S 148°55′E
Yassy *see* Iaşi
164 H12 **Yasugi** Shimane, Honshū, SW Japan 35°25′N 133°12′E
143 N10 **Yāsūj** *var.* Yesuj; *prev.* Tal-e Khosravī. Kohkīlūyeh va Būyer AḤmad, C Iran 30°40′N 51°34′E
136 M11 **Yasun Burnu** *headland* N Turkey 41°07′N 37°40′E
119 I20 **Yasyel'da** *Rus.* Yasel'da. Brestskaya Voblasts', SW Belarus Europe
117 X8 **Yasynuvata** *Rus.* Yasinovataya. Donets'ka Oblast', SE Ukraine 48°05′N 37°57′E
136 C15 **Yatağan** Muğla, SW Turkey 37°22′N 28°08′E
165 Q7 **Yatate-tōge** *pass* Honshū, C Japan
187 Q17 **Yaté** Province Sud, S New Caledonia 22°10′S 166°56′E
27 P6 **Yates Center** Kansas, C USA 37°54′N 95°44′W
185 B21 **Yates Point** *headland* South Island, New Zealand 44°30′S 167°49′E
9 N9 **Yathkyed Lake** ◎ Nunavut, NE Canada
171 T16 **Yatoke** Pulau Babar, E Indonesia 07°51′S 129°49′E
79 M18 **Yatolema** Orientale, N Dem. Rep. Congo 0°25′N 24°35′E
164 C15 **Yatsushiro** *var.* Yatusiro. Kumamoto, Kyūshū, SW Japan 32°30′N 130°34′E
164 C15 **Yatsushiro-kai** *bay* SW Japan
138 F11 **Yatta** *var.* Yuta. S West Bank 31°29′N 35°10′E
81 J20 **Yatta Plateau** *plateau* SE Kenya
Yatusiro *see* Yatsushiro
57 F17 **Yauca, Río** SW Peru
45 S6 **Yauco** W Puerto Rico 18°02′N 66°51′W
Yaunde *see* Yaoundé
Yavan *see* Yovon
Yavarí *see* Javari, Rio
56 G9 **Yavari Mirim, Río** NE Peru
40 G7 **Yavaros** Sonora, NW Mexico 26°40′N 109°32′W
154 I13 **Yavatmāl** Mahārāshtra, C India 20°22′N 78°11′E
54 M9 **Yaví, Cerro** ▲ C Venezuela 05°43′N 65°51′W
43 W16 **Yaviza** Darién, SE Panama 08°09′N 77°41′W
138 F10 **Yavne** Central, W Israel 31°52′N 34°45′E
116 H5 **Yavoriv** *Pol.* Jaworów, *Rus.* Yavorov. L'vivs'ka Oblast', NW Ukraine 49°57′N 23°22′E
Yavorov *see* Yavoriv
164 F14 **Yawatahama** Ehime, Shikoku, SW Japan 33°27′N 132°24′E
Ya Xian *see* Sanya
136 L17 **Yayladağı** Hatay, S Turkey 35°55′N 36°00′E
125 V13 **Yayva** Permskiy Kray, NW Russian Federation 59°19′N 57°15′E
125 V12 **Yayva** NW Russian Federation
143 Q9 **Yazd** *var.* Yezd. Yazd, C Iran 31°55′N 54°22′E
143 Q8 **Yazd** *off.* Ostān-e Yazd, *var.* Yezd. ◆ *province* C Iran
Yazd, Ostān-e *see* Yazd
Yazgulemskiy Khrebet *see* Yazgulom, Qatorkŭhi
147 S13 **Yazgulom, Qatorkŭhi** *Rus.* Yazgulemskiy Khrebet. ▲▲ S Tajikistan
22 K5 **Yazoo City** Mississippi, S USA 32°51′N 90°24′W
22 K5 **Yazoo River** Mississippi, S USA
Yazovir Georgi Traykov *see* Tsonevo, Yazovir
127 Q5 **Yazykovka** Ul'yanovskaya Oblast', W Russian Federation 54°19′N 47°22′E
109 U4 **Ybbs** Niederösterreich, NE Austria 48°10′N 15°03′E
109 U4 **Ybbs** C Austria
95 G22 **Yding Skovhøj** *hill* C Denmark
115 G20 **Ýdra** *var.* Ídhra, Idra. Ýdra, S Greece 37°20′N 23°28′E
115 G21 **Ýdra** *var.* Ídhra. *island* Ýdra, S Greece
115 G20 **Ýdras, Kólpos** *strait* S Greece
167 N10 **Ye** Mon State, S Myanmar (Burma) 15°15′N 97°50′E
183 O12 **Yea** Victoria, SE Australia 37°15′S 145°27′E
167 P13 **Yeay Sên** *prev.* Phumĭ Yeay Sên. Kaôh Kŏng, SW Cambodia 11°09′N 103°09′E
Yebaishou *see* Jianping
78 I5 **Yebbi-Bou** Tibesti, N Chad 21°12′N 17°55′E
158 F9 **Yecheng** *var.* Kargilik. Xinjiang Uygur Zizhiqu, NW China 37°54′N 77°26′E
105 R11 **Yecla** Murcia, SE Spain 38°36′N 01°07′W
40 H6 **Yécora** Sonora, NW Mexico 28°23′N 108°56′W
Yedintsy *see* Edineţ
124 J13 **Yefimovskiy** Leningradskaya Oblast', NW Russian Federation 59°32′N 34°34′E
126 K6 **Yefremov** Tul'skaya Oblast', W Russian Federation 53°10′N 38°02′E
159 T11 **Yêgainnyin** *var.* Henan Mongolzu Zizhixian. Qinghai, C China 34°42′N 101°36′E
137 U12 **Yegbhegis** *Rus.* Yekhegis. C Armenia
137 U12 **Yeghegnadzor** C Armenia 39°45′N 45°20′E
145 T10 **Yegindybulak** *Kaz.* Egindibulaq. Karaganda, C Kazakhstan 49°45′N 75°45′E
126 L4 **Yegor'yevsk** Moskovskaya Oblast', W Russian Federation 55°29′N 39°03′E
Yehuda, Haré *see* Judaean Hills
81 E15 **Yei** S South Sudan
161 P8 **Yeji** *var.* Yejiaji. Anhui, E China 31°52′N 115°58′E
Yejiaji *see* Yeji
122 G10 **Yekaterinburg** *prev.* Sverdlovsk. Sverdlovskaya Oblast', C Russian Federation 56°52′N 60°35′E
Yekaterinodar *see* Krasnodar
Yekaterinoslav *see* Dnipropetrovs'k
123 R13 **Yekaterinoslavka** Amurskaya Oblast', SE Russian Federation 50°23′N 129°03′E
127 O7 **Yekaterinovka** Saratovskaya Oblast', W Russian Federation 52°01′N 44°11′E
76 K16 **Yekepa** NE Liberia 07°35′N 08°32′W
Yekhegis *see* Yegbhegis
145 T8 **Yekibastuz** *prev.* Ekibastuz. Pavlodar, NE Kazakhstan 51°42′N 75°22′E
127 T3 **Yelabuga** Respublika Tatarstan, W Russian Federation 55°46′N 52°07′E
Yela Island *see* Rossel Island
127 O8 **Yelan'** Volgogradskaya Oblast', SW Russian Federation 51°00′N 43°40′E
117 Q9 **Yelanets'** *Rus.* Yelanets. Mykolayivs'ka Oblast', S Ukraine 47°40′N 31°51′E
144 I9 **Yelek** *Kaz.* Elek; *prev.* Ilek. Kazakhstan/Russian Federation
126 L7 **Yelets** Lipetskaya Oblast', W Russian Federation 52°37′N 38°29′E
125 W4 **Yeletskiy** Respublika Komi, NW Russian Federation 67°03′N 64°05′E
76 J11 **Yélimané** Kayes, W Mali 15°06′N 10°43′W
Yelisavetpol *see* Gäncä
Yelizavetgrad *see* Kirovohrad
123 T12 **Yelizavety, Mys** *headland* SE Russian Federation 54°20′N 142°39′E
Yelizovo *see* Yalizava
127 S5 **Yelkhovka** Samarskaya Oblast', W Russian Federation 53°51′N 50°16′E
96 M1 **Yell** *island* NE Scotland, United Kingdom
155 E17 **Yellāpur** Karnātaka, W India 15°06′N 74°50′E
11 U17 **Yellow Grass** Saskatchewan, S Canada 49°51′N 104°09′W
Yellowhammer State *see* Alabama
11 O15 **Yellowhead Pass** *pass* Alberta/British Columbia, SW Canada
8 K10 **Yellowknife** *territory capital* Northwest Territories, W Canada 62°30′N 114°29′W
8 K9 **Yellowknife** Northwest Territories, NW Canada
23 P8 **Yellow River** Alabama/Florida, S USA
30 K7 **Yellow River** Wisconsin, N USA
30 I4 **Yellow River** Wisconsin, N USA
30 J6 **Yellow River** Wisconsin, N USA
Yellow River *see* Huang He
157 V8 **Yellow Sea** *Chin.* Huang Hai, *Kor.* Hwang-Hae. *sea* E Asia
33 S13 **Yellowstone Lake** ◎ Wyoming, C USA
33 T13 **Yellowstone National Park** *national park* Wyoming, NW USA
33 Y8 **Yellowstone River** Montana/Wyoming, NW USA
96 L1 **Yell Sound** *strait* N Scotland, United Kingdom
27 U9 **Yellville** Arkansas, C USA 36°12′N 92°41′W
122 K10 **Yeloguy** C Russian Federation
Yëloten *see* Ýolöten
119 M20 **Yel'sk** Homyel'skaya Voblasts', SE Belarus 51°48′N 29°09′E
77 T13 **Yelwa** Kebbi, W Nigeria 10°52′N 04°46′E
21 R15 **Yemassee** South Carolina, SE USA 32°41′N 80°51′W
141 O15 **Yemen** *off.* Republic of Yemen, *Ar.* Al Jumhuriyah al Yamaniyah, Al Yaman. ◆ *republic* SW Asia
Yemen, Republic of *see* Yemen
116 M4 **Yemil'chyne** Zhytomyrs'ka Oblast', N Ukraine 50°51′N 27°49′E
124 M10 **Yemtsa** Arkhangel'skaya Oblast', NW Russian Federation 63°04′N 40°18′E
124 M10 **Yemtsa** NW Russian Federation
125 R10 **Yemva** *prev.* Zheleznodorozhnyy. Respublika Komi, NW Russian Federation 62°38′N 50°59′E
77 U17 **Yenagoa** Bayelsa, S Nigeria 4°58′N 6°16′E
117 X7 **Yenakiyeve** *Rus.* Yenakiyevo; *prev.* Ordzhonikidze, Rykovo. Donets'ka Oblast', E Ukraine 48°13′N 38°13′E
Yenakiyevo *see* Yenakiyeve
166 L6 **Yenangyaung** Magway, W Myanmar (Burma) 20°28′N 94°54′E
167 S5 **Yên Bai** Yên Bai, N Vietnam 21°43′N 104°54′E
183 P9 **Yenda** New South Wales, SE Australia 34°16′S 146°15′E
77 Q14 **Yendi** NE Ghana 09°30′N 00°01′W
Yêndum *see* Zhag'yab
158 E8 **Yengisar** Xinjiang Uygur Zizhiqu, NW China 38°50′N 76°11′E
136 H11 **Yenice Çayı** *var.* Filyos Çayı. N Turkey
121 R1 **Yenierenköy** *var.* Yialousa, *Gk.* Agialoúsa. NE Cyprus 35°33′N 34°13′E
Yenipazar *see* Novi Pazar
136 E12 **Yenişehir** Bursa, NW Turkey 40°17′N 29°38′E
Yenisei Bay *see* Yeniseyskiy Zaliv
122 K12 **Yeniseysk** Krasnoyarskiy Kray, C Russian Federation 58°23′N 92°06′E
197 W10 **Yeniseyskiy Zaliv** *var.* Yenisei Bay. *bay* N Russian Federation
127 Q12 **Yenotayevka** Astrakhanskaya Oblast', SW Russian Federation 47°16′N 47°01′E
124 L4 **Yenozero, Ozero** ◎ NW Russian Federation
Yenping *see* Nanping
39 Q11 **Yentna River** Alaska, USA
180 M10 **Yeo, Lake** *salt lake* Western Australia
163 Z15 **Yeongcheon** *Jap.* Eisen; *prev.* Yŏngch'ŏn. SE South Korea 35°56′N 128°55′E

163 Y15 **Yeongju** *Jap.* Eishū; *prev.* Yŏngju. C South Korea 36°48′N 128°37′E
163 Y17 **Yeosu** *Jap.* Reisui; *prev.* Yŏsu. S South Korea 34°45′N 127°41′E
183 R7 **Yeoval** New South Wales, SE Australia 32°45′S 148°39′E
97 K23 **Yeovil** SW England, United Kingdom 50°57′N 02°39′W
40 H6 **Yepachic** Chihuahua, N Mexico 28°27′N 108°25′W
181 Y8 **Yeppoon** Queensland, E Australia 23°05′S 150°42′E
126 M5 **Yeraktur** Ryazanskaya Oblast', W Russian Federation 54°45′N 41°09′E
Yeraliyev *see* Kuryk
146 F12 **Yerbent** Ahal Welaýaty, C Turkmenistan 39°19′N 58°34′E
123 N11 **Yerbogachën** Irkutskaya Oblast', C Russian Federation 61°07′N 108°03′E
137 T12 **Yerevan** *Eng.* Erivan. ● (Armenia) C Armenia 40°12′N 44°31′E
137 U12 **Yerevan** ✈ C Armenia 40°07′N 44°34′E
145 R9 **Yereymentau** *var.* Jermentau, *Kaz.* Ereymentaū. Akmola, C Kazakhstan 51°38′N 73°10′E
145 R9 **Yereymentau, Gory** *prev.* Gory Yermentau. ▲▲ C Kazakhstan
127 O12 **Yergeni** *hill range* SW Russian Federation
Yeriho *see* Jericho
35 R6 **Yerington** Nevada, W USA 38°58′N 119°10′W
136 J13 **Yerköy** Yozgat, C Turkey 39°39′N 34°28′E
114 L13 **Yerlisu** Edirne, NW Turkey 40°45′N 26°38′E
Yermak *see* Aksu
Yermentau, Gory *see* Yereymentau, Gory
125 R5 **Yërmitsa** Respublika Komi, NW Russian Federation 66°57′N 52°15′E
35 V14 **Yermo** California, W USA 34°54′N 116°49′W
123 P13 **Yerofey Pavlovich** Amurskaya Oblast', SE Russian Federation 53°58′N 121°49′E
99 F15 **Yerseke** Zeeland, SW Netherlands 51°30′N 04°03′E
127 Q8 **Yershov** Saratovskaya Oblast', W Russian Federation 51°18′N 48°16′E
145 S7 **Yertis** *Kaz.* Ertis; *prev.* Irtyshsk. Pavlodar, NE Kazakhstan 53°21′N 75°27′E
129 R5 **Yertis** *var.* Irtish, *Kaz.* Ertis; *prev.* Irtysh. ≈ C Asia
125 P9 **Yërtom** Respublika Komi, NW Russian Federation 63°27′N 47°52′E
56 D13 **Yerupaja, Nevado** ▲ C Peru 10°23′S 76°58′W
Yerushalayim *see* Jerusalem
105 R4 **Yesa, Embalse de** ⊠ NE Spain
144 F11 **Yesbol** *prev.* Kulagino. Atyrau, W Kazakhstan 48°30′N 51°33′E
144 F9 **Yesensay** Zapadnyy Kazakhstan, NW Kazakhstan 49°59′N 51°19′E
144 F9 **Yesensay** Zapadnyy Kazakhstan, NW Kazakhstan 49°58′N 51°19′E
145 V15 **Yesik** *Kaz.* Esik; *prev.* Issyk. Almaty, SE Kazakhstan 42°23′N 77°25′E
145 O8 **Yesil'** *Kaz.* Esil. Akmola, C Kazakhstan 51°58′N 66°24′E
129 R6 **Yesil'** *Kaz.* Esil. ≈ Kazakhstan/Russian Federation
136 K15 **Yeşilhisar** Kayseri, C Turkey 38°22′N 35°08′E
136 L11 **Yeşilırmak** *var.* Iris. ≈ N Turkey
37 U12 **Yeso** New Mexico, SW USA 34°25′N 104°36′W
Yeso *see* Hokkaidō
127 N15 **Yessentuki** Stavropol'skiy Kray, SW Russian Federation 44°06′N 42°51′E
122 M9 **Yessey** Krasnoyarskiy Kray, N Russian Federation 68°18′N 101°49′E
105 P12 **Yeste** Castilla-La Mancha, C Spain 38°21′N 02°18′W
Yesuj *see* Yāsūj
183 T4 **Yetman** New South Wales, SE Australia 28°56′S 150°47′E
76 L4 **Yetti** *physical region* N Mauritania
166 M4 **Ye-u** Sagaing, C Myanmar (Burma) 22°49′N 95°26′E
102 H9 **Yeu, Île d'** *island* NW France
Yevlakh *see* Yevlax
137 W11 **Yevlax** *Rus.* Yevlakh. C Azerbaijan 40°36′N 47°10′E
117 S13 **Yevpatoriya** Avtonomna Respublika Krym, S Ukraine 45°12′N 33°23′E
Ye Xian *see* Laizhou
126 K12 **Yeya** ≈ SW Russian Federation
158 I10 **Yeyik** Xinjiang Uygur Zizhiqu, W China 36°44′N 83°14′E
126 K12 **Yeysk** Krasnodarskiy Kray, SW Russian Federation 46°41′N 38°15′E
Yezd *see* Yazd
Yezerishche *see* Yezyaryshcha
Yezhou *see* Jianshi
Yezo *see* Hokkaidō
118 N11 **Yezyaryshcha** *Rus.* Yezerishche. Vitsyebskaya Voblasts', NE Belarus 55°50′N 29°59′E
Yialí *see* Gyalí
Yialousa *see* Yenierenköy
163 V7 **Yi'an** Heilongjiang, NE China 47°52′N 125°13′E
Yiannitsá *see* Giannitsá
160 I10 **Yibin** Sichuan, C China 28°50′N 104°35′E
158 K13 **Yibug Caka** ◎ W China
160 M9 **Yichang** Hubei, C China 30°37′N 111°02′E
160 L5 **Yichuan** *var.* Danzhou. Shaanxi, C China 36°05′N 110°02′E
157 W3 **Yichun** Heilongjiang, NE China 47°41′N 129°10′E
161 O11 **Yichun** Jiangxi, S China 27°45′N 114°22′E
160 M9 **Yidu** *prev.* Zhicheng. Hubei, C China 30°21′N 111°27′E
Yidu *see* Qingzhou
188 C15 **Yigo** NE Guam 13°33′N 144°53′E
161 Q5 **Yi He** ≈ E China
163 X8 **Yilan** Heilongjiang, NE China 46°18′N 129°36′E
136 C9 **Yıldız Dağları** ▲▲ NW Turkey
136 L13 **Yıldızeli** Sivas, N Turkey 39°52′N 36°37′E
163 U4 **Yilehuli Shan** ▲▲ NE China
163 S7 **Yimin He** ≈ NE China
159 W8 **Yinchuan** *var.* Yinch'uan, Yin-ch'uan, Yinchwan. *province capital* Ningxia, N China 38°30′N 106°19′E
Yinchwan *see* Yinchuan
Yindu He *see* Indus
Yingcheng *see* Yingde
161 N14 **Yingde** *var.* Yingcheng. Guangdong, S China 24°08′N 113°21′E
163 U13 **Yingkou** *var.* Ying-k'ou, Yingkow; *prev.* Newchwang, Niuchwang. Liaoning, NE China 40°40′N 122°17′E
Yingkow *see* Yingkou
161 P9 **Yingshan** *var.* Wenquan. Hubei, C China 30°45′N 115°41′E
161 Q10 **Yingtan** Jiangxi, S China 28°17′N 117°03′E
Yin-hsien *see* Ningbo
158 H5 **Yining** *var.* I-ning, *Uigh.* Gulja, Kuldja. Xinjiang Uygur Zizhiqu, NW China 43°53′N 81°18′E
160 K11 **Yinjiang** *var.* Yinjiang Tujiazu Miaozu Zizhixian. Guizhou, S China 28°22′N 108°07′E
Yinjiang Tujiazu Miaozu Zizhixian *see* Yinjiang
166 L4 **Yinmabin** Sagaing, C Myanmar (Burma) 22°05′N 94°57′E
163 N13 **Yin Shan** ▲▲ N China
Yinshan *see* Guangshui
Yin-tu Ho *see* Indus
159 P15 **Yi'ong Zangbo** ≈ W China
Yioúra *see* Gyáros
81 J14 **Yirga 'Alem** *It.* Irgalem. Southern Nationalities, S Ethiopia 06°43′N 38°24′E
61 E19 **Yí, Río** ≈ C Uruguay
81 E14 **Yirol** Lakes, C South Sudan 06°34′N 30°33′E
163 S8 **Yirshi** *var.* Yirxie. Nei Mongol Zizhiqu, N China 47°16′N 119°51′E
Yirxie *see* Yirshi
Yishan *see* Guanyun
Yishi *see* Linyi
161 Q5 **Yishui** Shandong, E China 35°50′N 118°39′E
Yisrael/Yisra'el *see* Israel
Yíthion *see* Gýtheio
Yitiaoshan *see* Jingtai
163 W10 **Yitong** *var.* Yitong Manzu Zizhixian. Jilin, NE China 43°23′N 125°19′E
Yitong Manzu Zizhixian *see* Yitong
159 P5 **Yiwu** *var.* Aratürük. Xinjiang Uygur Zizhiqu, NW China 43°16′N 94°38′E
163 U12 **Yiwulü Shan** ▲▲ N China
163 T12 **Yixian** *var.* Yizhou. Liaoning, NE China 41°29′N 121°21′E
159 R15 **Yixing** Jiangsu, China 31°14′N 119°18′E
161 N10 **Yiyang** Hunan, S China 28°39′N 112°10′E
161 Q10 **Yiyang** Jiangxi, S China 28°21′N 117°23′E
161 N13 **Yizhang** Hunan, S China 25°24′N 112°51′E
Yizhou *see* Yixian
93 K19 **Yläne** Varsinais-Suomi, SW Finland 60°51′N 22°26′E
93 L14 **Yli-Ii** Pohjois-Pohjanmaa, C Finland 65°21′N 25°55′E
93 L14 **Ylikiiminki** Pohjois-Pohjanmaa, C Finland 65°00′N 26°10′E
92 N13 **Yli-Kitka** ◎ NE Finland
93 K17 **Ylistaro** Etelä-Pohjanmaa, W Finland 62°58′N 22°30′E
92 K13 **Ylitornio** Lappi, NW Finland 66°19′N 23°40′E
93 L15 **Ylivieska** Pohjois-Pohjanmaa, W Finland 64°05′N 24°30′E
93 L18 **Ylöjärvi** Pirkanmaa, W Finland 61°31′N 23°37′E
95 N17 **Yngaren** ◎ C Sweden
25 T12 **Yoakum** Texas, SW USA 29°17′N 97°09′W
77 X13 **Yobe** ◈ *state* NE Nigeria
165 R3 **Yobetsu-dake** ⛰ Hokkaidō, NE Japan 43°15′N 140°27′E
80 L11 **Yoboki** C Djibouti 11°30′N 42°04′E
22 M4 **Yockanookany River** ≈ Mississippi, S USA
22 L2 **Yocona River** ≈ Mississippi, S USA
171 Y15 **Yodom** Papua, E Indonesia 07°12′S 139°24′E
169 Q16 **Yogyakarta** *prev.* Djokjakarta, Jogjakarta, Jokyakarta. Jawa, C Indonesia 07°48′S 110°24′E
169 P17 **Yogyakarta** *off.* Daerah Istimewa Yogyakarta, *var.* Djokjakarta, Jogjakarta, Jokyakarta. ◈ *autonomous district* S Indonesia
Yogyakarta, Daerah Istimewa *see* Yogyakarta
165 Q3 **Yoichi** Hokkaidō, NE Japan 43°11′N 140°45′E
42 G6 **Yojoa, Lago de** ◎ NW Honduras
79 G16 **Yokadouma** Est, SE Cameroon 03°26′N 15°06′E
164 K13 **Yokkaichi** *var.* Yokkaiti. Mie, Honshū, SW Japan 34°58′N 136°38′E
Yokkaiti *see* Yokkaichi
79 E15 **Yoko** Centre, C Cameroon 05°29′N 12°19′E
165 V15 **Yokoate-jima** *island* Nansei-shotō, SW Japan
165 R6 **Yokohama** Aomori, Honshū, C Japan 41°04′N 141°14′E
165 O14 **Yokosuka** Kanagawa, Honshū, S Japan 35°18′N 139°39′E
164 G12 **Yokota** Shimane, Honshū, SW Japan 35°10′N 133°03′E
165 Q9 **Yokote** Akita, Honshū, C Japan 39°20′N 140°33′E
77 Y14 **Yola** Adamawa, E Nigeria 09°08′N 12°24′E
79 L19 **Yolombo** Equateur, C Dem. Rep. Congo 01°36′S 23°13′E
146 J14 **Ýolöten** *Rus.* Yëloten; *prev.* Iolotan'. Mary Welaýaty, S Turkmenistan 37°15′N 62°18′E
165 Y15 **Yome-jima** *island* Ogasawara-shotō, SE Japan
76 K16 **Yomou** SE Guinea 07°30′N 09°13′W
171 Y15 **Yomuka** Papua, E Indonesia 07°25′S 138°36′E
188 C16 **Yona** E Guam 13°24′N 144°46′E
164 H12 **Yonago** Tottori, Honshū, SW Japan 35°30′N 134°15′E
165 N16 **Yonaguni** Okinawa, SW Japan 24°29′N 123°00′E
165 N16 **Yonaguni-jima** *island* Nansei-shotō, SW Japan
165 T16 **Yonaha-dake** ▲ Okinawa, SW Japan 26°43′N 128°13′E
163 X14 **Yŏnan** SW North Korea 37°50′N 126°15′E
165 P10 **Yonezawa** Yamagata, Honshū, C Japan 37°56′N 140°06′E
161 Q12 **Yong'an** *var.* Yongan. Fujian, SE China 25°58′N 117°26′E
Yong'an *see* Fengjie
159 T9 **Yongchang** Gansu, N China 38°15′N 101°56′E
161 P7 **Yongcheng** Henan, C China 33°56′N 116°21′E
Yŏngch'ŏn *see* Yeongcheon
160 J10 **Yongchuan** Chongqing Shi, C China 29°27′N 105°56′E
159 U10 **Yongdeng** Gansu, C China 35°58′N 103°27′E
Yongding *see* Yongren
129 W9 **Yongding He** ≈ E China
161 P11 **Yongfeng** *var.* Enjiang. Jiangxi, S China 27°19′N 115°23′E
158 L5 **Yongfeng** *var.* Yongfengqu. Xinjiang Uygur Zizhiqu, W China 43°28′N 87°09′E
Yongfengqu *see* Yongfeng
160 L13 **Yongfu** Guangxi Zhuangzu Ziziqu, S China 24°57′N 109°59′E
163 X13 **Yŏnghŭng** E North Korea 39°31′N 127°14′E
159 U10 **Yongjing** *var.* Liujiaxia. Gansu, C China 36°00′N 103°30′E
Yongjing *see* Xifeng
Yŏngju *see* Yeongju
Yongle Qundao *see* Crescent Group
Yongning *see* Xuyong
160 E12 **Yongping** Yunnan, SW China 25°30′N 99°28′E
160 G12 **Yongren** *var.* Yongding. Yunnan, SW China 26°09′N 101°40′E
160 L10 **Yongshun** *var.* Lingxi. Hunan, S China 29°02′N 109°46′E
161 P10 **Yongxiu** *var.* Tujiabu. Jiangxi, S China 29°09′N 115°47′E
Yongzhou *see* Lingling
Yongzhou *see* Zhishan
18 K14 **Yonkers** New York, NE USA 40°56′N 73°51′W
103 Q7 **Yonne** ◈ *department* C France
103 P6 **Yonne** ≈ C France
54 H9 **Yopal** *var.* El Yopal. Casanare, C Colombia 05°20′N 72°19′W
158 E8 **Yopurga** *var.* Yukuriawat. Xinjiang Uygur Zizhiqu, NW China 39°13′N 76°44′E
147 S11 **Yordon** *var.* Iordan, *Rus.* Jardan. Farg'ona Viloyati, E Uzbekistan 39°59′N 71°44′E
180 J12 **York** Western Australia 31°55′S 116°52′E
97 M16 **York** *anc.* Eboracum, Eburacum. N England, United Kingdom 53°58′N 01°05′W
23 N5 **York** Alabama, S USA 32°29′N 88°18′W
29 Q15 **York** Nebraska, C USA 40°52′N 97°35′W
18 G16 **York** Pennsylvania, NE USA 39°57′N 76°44′W
21 R11 **York** South Carolina, SE USA 34°59′N 81°14′W
14 J13 **York** ≈ Ontario, SE Canada
15 X6 **York** ≈ Québec, SE Canada
181 V1 **York, Cape** *headland* Queensland, NE Australia 10°40′S 142°36′E
182 I9 **Yorke Peninsula** *peninsula* South Australia
182 I9 **Yorketown** South Australia 35°01′S 137°38′E
19 P9 **York Harbor** Maine, NE USA 43°10′N 70°37′W
York, Kap *see* Innaanganeq
21 X6 **York River** ≈ Virginia, NE USA
97 M16 **Yorkshire** *cultural region* N England, United Kingdom
97 L16 **Yorkshire Dales** *physical region* N England, United Kingdom
11 V16 **Yorkton** Saskatchewan, S Canada 51°12′N 102°29′W
25 T12 **Yorktown** Texas, SW USA 28°58′N 97°30′W
21 X6 **Yorktown** Virginia, NE USA 37°14′N 76°32′W
30 M11 **Yorkville** Illinois, N USA 41°38′N 88°27′W
42 I5 **Yoro** Yoro, C Honduras 15°08′N 87°10′W
42 H5 **Yoro** ◈ *department* N Honduras
165 T16 **Yoron-jima** *island* Nansei-shotō, SW Japan
77 N13 **Yorosso** Sikasso, S Mali 12°21′N 04°47′W
35 R8 **Yosemite National Park** *national park* California, W USA
127 Q3 **Yoshkar-Ola** Respublika Mariy El, W Russian Federation 56°38′N 47°54′E
Yösönbulag *see* Altay
162 K8 **Yösöndzüyl** *var.* Mönhbulag. Övörhangay, C Mongolia 46°48′N 103°25′E
171 Y16 **Yos Sudarso, Pulau** *var.* Pulau Dolak, Pulau Kolepom; *prev.* Jos Sudarso. *island* E Indonesia
Yŏsu *see* Yeosu
165 R4 **Yotei-zan** ▲ Hokkaidō, NE Japan 42°50′N 140°46′E
97 D21 **Youghal** *Ir.* Eochaill. Cork, S Ireland 51°57′N 07°50′W
97 D21 **Youghal Bay** *Ir.* Cuan Eochaille. *inlet* S Ireland
18 C15 **Youghiogheny River** ≈ Pennsylvania, NE USA
160 K14 **You Jiang** ≈ S China
183 Q9 **Young** New South Wales, SE Australia 34°19′S 148°20′E
11 T15 **Young** Saskatchewan, S Canada 51°44′N 105°44′W
61 E18 **Young** Río Negro, W Uruguay 32°44′S 57°36′W
182 G5 **Younghusband, Lake** *salt lake* South Australia
182 J10 **Younghusband Peninsula** *peninsula* South Australia
184 Q10 **Young Nicks Head** *headland* North Island, New Zealand 39°38′S 177°03′E
185 D20 **Young Range** ▲▲ South Island, New Zealand
191 Q15 **Young's Rock** *island* Pitcairn Island, Pitcairn Islands
11 R16 **Youngstown** Alberta, SW Canada 51°32′N 111°12′W
31 V12 **Youngstown** Ohio, N USA 41°06′N 80°39′W
159 N9 **Youshashan** Qinghai, C China 38°12′N 90°58′E
Youth, Isle of *see* Juventud, Isla de la
77 N11 **Youvarou** Mopti, C Mali 15°19′N 04°15′W
160 K10 **Youyang** *var.* Zhongduo. Chongqing Shi, C China 28°48′N 108°48′E
163 Y7 **Youyi** Heilongjiang, NE China 46°51′N 131°54′E
147 P13 **Yovon** *Rus.* Yavan. SW Tajikistan 38°19′N 69°02′E
136 J13 **Yozgat** Yozgat, C Turkey 39°49′N 34°48′E
136 K13 **Yozgat** ◈ *province* C Turkey
62 O6 **Ypacaraí** *var.* Ypacaray. Central, S Paraguay 25°23′S 57°16′W
Ypacaray *see* Ypacaraí
62 P5 **Ypané, Río** ≈ C Paraguay
Ypres *see* Ieper
114 I13 **Ypsário** *var.* Ipsario. ▲ Thásos, E Greece 40°43′N 24°39′E
31 R10 **Ypsilanti** Michigan, N USA 42°12′N 83°36′W
34 M1 **Yreka** California, W USA 41°43′N 122°39′W
Yrendagüé *see* General Eugenio A. Garay
144 L11 **Yrghyz** *prev.* Irgiz. Aktyubinsk, C Kazakhstan 48°36′N 61°14′E
186 G5 **Ysabel Channel** *channel* N Papua New Guinea
14 K8 **Yser, Lac** ◎ Québec, SE Canada
147 Y8 **Yshtyk** Issyk-Kul'skaya Oblast', E Kyrgyzstan 41°34′N 78°21′E
Yssel *see* IJssel
103 Q12 **Yssingeaux** Haute-Loire, C France 45°09′N 04°07′E
95 K23 **Ystad** Skåne, S Sweden 55°25′N 13°51′E
Ysyk-Köl *see* Issyk-Kul', Ozero
Ysyk-Köl *see* Balykchy
Ysyk-Köl Oblasty *see* Issyk-Kul'skaya Oblast'
96 L8 **Ythan** ≈ NE Scotland, United Kingdom
Y Trallwng *see* Welshpool
94 C13 **Ytre Arna** Hordaland, S Norway 60°28′N 05°25′E
94 B12 **Ytre Sula** *island* S Norway
93 G17 **Ytterhogdal** Jämtland, C Sweden 62°10′N 14°55′E
Yu *see* Henan
Yuan *see* Red River
Yuancheng *see* Heyuan, Guangdong, S China
Yuan Jiang *see* Red River
161 S13 **Yuanlin** *Jap.* Inrin; *prev.* Yüanlin. C Taiwan 23°57′N 120°33′E
161 N3 **Yuanping** Shanxi, C China 38°26′N 112°42′E
Yuanquan *see* Anxi
Yuanshan *see* Lianping
161 O11 **Yuan Shui** ≈ S China
35 O6 **Yuba City** California, W USA 39°07′N 121°40′W
35 O6 **Yuba River** ≈ California, W USA
80 H13 **Yubdo** Oromīya, C Ethiopia 09°05′N 35°28′E
41 X12 **Yucatán** ◈ *state* SE Mexico
47 O3 **Yucatan Basin** *var.* Yucatan Deep. *undersea feature* N Caribbean Sea 20°00′N 84°00′W
Yucatán, Canal de *see* Yucatan Channel
41 Y10 **Yucatan Channel** *Sp.* Canal de Yucatán. *channel* Cuba/Mexico
Yucatan Deep *see* Yucatan Basin
Yucatan Peninsula *see* Yucatán, Península de
41 X13 **Yucatán, Península de** *Eng.* Yucatan Peninsula. *peninsula* Guatemala/Mexico
36 I11 **Yucca** Arizona, SW USA 34°49′N 114°06′W
35 V15 **Yucca Valley** California, W USA 34°06′N 116°30′W
161 P4 **Yucheng** Shandong, E China 37°01′N 116°37′E
Yuci *see* Jinzhong
129 X5 **Yudoma** ≈ E Russian Federation
161 P12 **Yudu** *var.* Gongjiang. Jiangxi, C China 26°02′N 115°24′E
Yue *see* Guangdong
Yuecheng *see* Yuexi
160 M12 **Yuecheng Ling** ▲▲ S China
Yuegai *see* Qumarlêb
Yuegaitan *see* Qumarlêb
181 P7 **Yuendumu** Northern Territory, N Australia 22°19′S 131°51′E
Yue Shan, Tai *see* Lantau Island
160 H10 **Yuexi** *var.* Yuecheng. Sichuan, C China 28°50′N 102°36′E
161 N10 **Yueyang** Hunan, S China 29°24′N 113°08′E
125 U14 **Yug** Permskiy Kray, NW Russian Federation 57°49′N 56°08′E
125 P13 **Yug** ≈ NW Russian Federation
123 R10 **Yugorënok** Respublika Sakha (Yakutiya), NE Russian Federation 59°46′N 137°36′E
122 H9 **Yugorsk** Khanty-Mansiyskiy Avtonomnyy Okrug-Yugra, C Russian Federation 61°17′N 63°25′E
122 H7 **Yugorskiy Poluostrov** *peninsula* NW Russian Federation
Yugoslavia *see* Serbia
146 K14 **Yugo-Vostochnyye Garagumy** *prev.* Yugo-Vostochnyye Karakumy. *desert* E Turkmenistan
Yugo-Vostochnyye Karakumy *see* Yugo-Vostochnyye Garagumy
Yuhu *see* Eryuan
161 S10 **Yuhuan Dao** *island* SE China
160 L14 **Yu Jiang** ≈ S China
Yujin *see* Qianwei
159 P9 **Yuka** Qinghai, W China 38°03′N 94°45′E
123 S7 **Yukagirskoye Ploskogor'ye** *plateau* NE Russian Federation
159 P9 **Yuke He** ≈ C China
118 L11 **Yukhavichy** *Rus.* Yukhovichi. Vitsyebskaya Voblasts', N Belarus 56°02′N 28°39′E
126 J4 **Yukhnov** Kaluzhskaya Oblast', W Russian Federation 54°43′N 35°15′E
Yukhovichi *see* Yukhavichy
79 J20 **Yuki** *var.* Yuki Kengunda. Bandundu, W Dem. Rep. Congo 03°57′S 19°30′E
Yuki Kengunda *see* Yuki
26 M10 **Yukon** Oklahoma, C USA 35°30′N 97°45′W
10 I5 **Yukon** *prev.* Yukon Territory, *Fr.* Territoire du Yukon. ◈ *territory* NW Canada
0 F4 **Yukon** ≈ Canada/USA
39 S7 **Yukon Flats** *salt flat* Alaska, USA
Yukon, Territoire du *see* Yukon
Yukon Territory *see* Yukon
137 T16 **Yüksekova** Hakkâri, SE Turkey 37°35′N 44°17′E
123 N10 **Yukta** Krasnoyarskiy Kray, C Russian Federation 63°16′N 106°04′E
165 O13 **Yukuhashi** *var.* Yukuhasi. Fukuoka, Kyūshū, SW Japan 33°41′N 131°00′E
Yukuhasi *see* Yukuhashi
Yukuriawat *see* Yopurga
125 O9 **Yula** ≈ NW Russian Federation
181 P8 **Yulara** Northern Territory, N Australia 25°15′S 130°57′E
127 W6 **Yuldybayevo** Respublika Bashkortostan, W Russian Federation 52°22′N 57°55′E
23 W8 **Yulee** Florida, SE USA 30°37′N 81°36′W
158 K7 **Yuli** *var.* Lopnur. Xinjiang Uygur Zizhiqu, NW China 41°24′N 86°12′E
161 T14 **Yuli** *prev.* Yüli. C Taiwan 23°23′N 121°18′E
160 L15 **Yulin** Guangxi Zhuangzu Zizhiqu, S China 22°37′N 110°08′E
160 L4 **Yulin** Shaanxi, C China 38°14′N 109°48′E
161 T14 **Yuli Shan** *prev.* Yüli Shan. ▲ E Taiwan 23°23′N 121°14′E
160 F11 **Yulong Xueshan** ▲ SW China 27°09′N 100°10′E
36 H14 **Yuma** Arizona, SW USA 32°40′N 114°38′W
37 W3 **Yuma** Colorado, C USA 40°07′N 102°43′W
54 K5 **Yumare** Yaracuy, N Venezuela 10°37′N 68°41′W
63 G14 **Yumbel** Bío Bío, C Chile 37°05′S 72°40′W
79 N19 **Yumbi** Maniema, E Dem. Rep. Congo 01°14′S 26°14′E
159 Q7 **Yumendong** *prev.* Yumenzhen. Gansu, N China 40°19′N 97°12′E
159 R8 **Yumendong** *prev.* Laojunmiao. Gansu, N China 39°49′N 97°47′E
Yumenzhen *see* Yumendong
158 J3 **Yumin** *var.* Karabura. Xinjiang Uygur Zizhiqu, NW China 46°14′N 82°52′E
Yun *see* Yunnan
136 G14 **Yunak** Konya, W Turkey 38°50′N 31°42′E
45 O8 **Yuna, Río** ≈ E Dominican Republic
38 I17 **Yunaska Island** *island* Aleutian Islands, Alaska, USA
160 M6 **Yuncheng** Shanxi, C China 35°07′N 110°45′E
Yuncheng *see* Yunfu
161 N14 **Yunfu** *var.* Yuncheng. Guangdong, S China 22°56′N 112°02′E
57 L18 **Yungas** *physical region* E Bolivia
Yungki *see* Jilin
Yung-ning *see* Nanning
160 I12 **Yungui Gaoyuan** *plateau* SW China
Yunjinghong *see* Jinghong
160 M15 **Yunkai Dashan** ▲▲ S China
Yunki *see* Jilin
160 E11 **Yun Ling** ▲▲ SW China
Yunling *see* Yunxiao
161 N9 **Yunmeng** Hubei, C China 31°04′N 113°45′E
157 N14 **Yunnan** *var.* Yun, Yunnan Sheng, Yünnan, Yun-nan. ◈ *province* SW China
Yunnan *see* Kunming
Yunnan Sheng *see* Yunnan
Yünnan/Yun-nan *see* Yunnan
165 P15 **Yunomae** Kumamoto, Kyūshū, SW Japan 32°16′N 131°00′E
161 N8 **Yun Shui** ≈ C China
182 J7 **Yunta** South Australia 32°37′S 139°33′E
161 Q14 **Yunxiao** *var.* Yunling. Fujian, SE China 23°56′N 117°16′E
160 K9 **Yunyang** Sichuan, C China 31°03′N 109°43′E
Yunzhong *see* Huairen
193 S9 **Yupanqui Basin** *undersea feature* E Pacific Ocean
Yuping *see* Libo, Guizhou, China
Yuping *see* Pingbian, Yunnan, China
Yuratishki *see* Yuratsishki
119 I15 **Yuratsishki** *Pol.* Juraciszki, *Rus.* Yuratishki. Hrodzyenskaya Voblasts', W Belarus 54°02′N 25°56′E
Yurev *see* Tartu
122 J12 **Yurga** Kemerovskaya Oblast', S Russian Federation 55°42′N 84°59′E
Yurihonjō *see* Honjō
56 E10 **Yurimaguas** Loreto, N Peru 05°54′S 76°07′W
127 P3 **Yurino** Respublika Mariy El, W Russian Federation 56°19′N 46°15′E
41 N13 **Yuriria** Guanajuato, C Mexico 20°12′N 101°09′W
125 T13 **Yurla** Komi-Permyatskiy Okrug, NW Russian Federation 59°18′N 54°19′E
Yuruá, Río *see* Juruá, Rio
114 M13 **Yürük** Tekirdağ, NW Turkey 40°58′N 27°09′E
158 G10 **Yurungkax He** ≈ W China
125 Q14 **Yur'ya** *var.* Jarja. Kirovskaya Oblast', NW Russian Federation 59°01′N 49°22′E
Yury'ev *see* Tartu
125 N16 **Yur'yevets** Ivanovskaya Oblast', W Russian Federation 57°19′N 43°01′E
126 M3 **Yur'yev-Pol'skiy** Vladimirskaya Oblast', W Russian Federation 56°28′N 39°39′E
117 V7 **Yur"yivka** Dnipropetrovs'ka Oblast', E Ukraine 48°45′N 36°01′E
42 I7 **Yuscarán** El Paraíso, S Honduras 13°55′N 86°51′W
161 P12 **Yu Shan** ▲▲ S China
124 I7 **Yushkozero** Respublika Kareliya, NW Russian Federation 64°46′N 32°13′E
124 I7 **Yushkozerskoye Vodokhranilishche** *var.* Ozero Kujto. ◎ NW Russian Federation
169 W9 **Yushu** Jilin, China E Asia 44°48′N 126°55′E
159 R13 **Yushu** *var.* Gyêgu. Qinghai, C China 33°04′N 97°E
127 P12 **Yusta** Respublika Kalmykiya, SW Russian Federation 47°06′N 46°16′E
124 I10 **Yustozero** Respublika Kareliya, NW Russian Federation 62°44′N 33°31′E
137 Q11 **Yusufeli** Artvin, NE Turkey 40°50′N 41°31′E
164 F14 **Yusuhara** Kōchi, Shikoku, SW Japan 33°22′N 132°52′E
125 T14 **Yus'va** Permskiy Kray, NW Russian Federation 58°48′N 54°59′E
Yuta *see* Yatta
161 P2 **Yutian** Hebei, E China 39°52′N 117°44′E
158 H10 **Yutian** *var.* Keriya, Mugalla. Xinjiang Uygur Zizhiqu, NW China 36°49′N 81°31′E
62 K5 **Yuto** Jujuy, NW Argentina 23°35′S 64°28′W
62 P7 **Yuty** Caazapá, S Paraguay 26°31′S 56°20′W
160 G13 **Yuxi** Yunnan, SW China 24°22′N 102°28′E
161 O2 **Yuxian** *prev.* Yu Xian. Hebei, E China 39°50′N 114°33′E
Yu Xian *see* Yuxian
165 Q9 **Yuzawa** Akita, Honshū, C Japan 39°11′N 140°29′E
125 N16 **Yuzha** Ivanovskaya Oblast', W Russian Federation 56°34′N 42°00′E
Yuzhno-Alichurskiy Khrebet *see* Alichuri Janubí, Qatorkŭhi
Yuzhno-Kazakhstanskaya Oblast' *see* Yuzhnyy Kazakhstan
123 T13 **Yuzhno-Sakhalinsk** *Jap.* Toyohara; *prev.* Vladimirovka. Ostrov Sakhalin, Sakhalinskaya Oblast', SE Russian Federation 46°58′N 142°45′E
127 P14 **Yuzhno-Sukhokumsk** Respublika Dagestan, SW Russian Federation 44°43′N 45°32′E
145 Z10 **Yuzhnyy Altay, Khrebet** ▲▲ E Kazakhstan
Yuzhnyy Bug *see* Pivdennyy Buh
145 O15 **Yuzhnyy Kazakhstan** *off.* Yuzhno-Kazakhstanskaya Oblast', *Eng.* South Kazakhstan, *Kaz.* Ongtüstik Qazaqstan Oblysy; *prev.* Chimkentskaya Oblast'. ◈ *province* S Kazakhstan
123 U10 **Yuzhnyy, Mys** *headland* E Russian Federation 57°44′N 156°49′E
122 H6 **Yuzhnyy, Ostrov** *island* NW Russian Federation
127 W6 **Yuzhnyy Ural** *var.* Southern Urals. ▲▲ W Russian Federation
159 V10 **Yuzhong** Gansu, C China 35°52′N 104°09′E
Yuzhou *see* Chongqing
103 N5 **Yvelines** ◈ *department* N France
108 B9 **Yverdon** *var.* Yverdon-les-Bains, *Ger.* Iferten; *anc.* Eborodunum. Vaud, W Switzerland 46°47′N 06°38′E
Yverdon-les-Bains *see* Yverdon
102 M3 **Yvetot** Seine-Maritime, N France 49°37′N 00°48′E
Ýylanly *see* Gurb ansoltan Eje

Z

147 T12 **Zaalayskiy Khrebet** *Taj.* Qatorkŭhi Pasi Oloy. ▲▲ Kyrgyzstan/Tajikistan
Zaamin *see* Zomin
Zaandam *see* Zaanstad
98 I10 **Zaanstad** *prev.* Zaandam. Noord-Holland, C Netherlands 52°27′N 04°49′E
Zabadani *see* Az Zabdānī
Zabalatstsye *see* Zabalotstsye
112 L9 **Žabalj** *Ger.* Josefsdorf, *Hung.* Zsablya; *prev.* Józseffalva. Vojvodina, N Serbia 45°22′N 20°01′E
119 L18 **Zabalotstsye** *prev.* Zabalatstsye, *Rus.* Zabolot'ye. Homyel'skaya Voblasts', SE Belarus 52°40′N 28°34′E
Zāb aş Şaghīr, Nahraz *see* Little Zab
123 P14 **Zabaykal'sk** Zabaykal'skiy Kray, S Russian Federation 49°37′N 117°20′E
123 O12 **Zabaykal'skiy Kray** ◈ *province* S Russian Federation
Zāb-e Kūchek, Rūdkhāneh-ye *see* Little Zab
Zabeln *see* Sabile
Zabéré *see* Zabré
Zabern *see* Saverne
141 N16 **Zabīd** W Yemen 14°N 43°E
141 O16 **Zabīd, Wādī** *dry watercourse* SW Yemen
Żabinka *see* Zhabinka
Ząbkowice *see* Ząbkowice Śląskie
111 G15 **Ząbkowice Śląskie** *var.* Ząbkowice, *Ger.* Frankenstein, Frankenstein in Schlesien. Dolnośląskie, SW Poland 50°35′N 16°48′E
110 P10 **Zabłudów** Podlaskie, NE Poland 53°00′N 23°21′E
112 D8 **Zabok** Krapina-Zagorje, N Croatia 46°00′N 15°48′E
143 W9 **Zābol** *var.* Shahr-i-Zabul, Zabul; *prev.* Nasratabad. Sīstān va Balūchestān, E Iran 31°N 61°32′E
Zābol *see* Zābul
143 W13 **Zābolī** Sīstān va Balūchestān, SE Iran 27°09′N 61°32′E
Zabolot'ye *see* Zabalotstsye
77 Q13 **Zabré** *var.* Zabéré. S Burkina Faso 11°13′N 00°34′W
111 G17 **Zábřeh** *Ger.* Hohenstadt. Olomoucký Kraj, E Czech Republic 49°52′N 16°53′E
111 J16 **Zabrze** *Ger.* Hindenburg, Hindenburg in Oberschlesien. Śląskie, S Poland 50°18′N 18°47′E
149 O7 **Zābul** *prev.* Zābol. ◈ *province* SE Afghanistan
Zabul/Zābul *see* Zābol
42 E6 **Zacapa** Zacapa, E Guatemala 14°59′N 89°33′W
42 A3 **Zacapa** *off.* Departamento de Zacapa. ◈ *department* E Guatemala
Zacapa, Departamento de *see* Zacapa
40 M14 **Zacapú** Michoacán, SW Mexico 19°49′N 101°48′W
41 V14 **Zacatal** Campeche, SE Mexico 18°40′N 91°52′W
40 M11 **Zacatecas** Zacatecas, C Mexico 22°46′N 102°33′W
40 L10 **Zacatecas** ◈ *state* C Mexico
42 F8 **Zacatecoluca** La Paz, S El Salvador 13°29′N 88°51′W
41 P15 **Zacatepec** Morelos, S Mexico 18°40′N 99°11′W
41 Q13 **Zacatlán** Puebla, S Mexico 19°56′N 97°58′W
144 F8 **Zachagansk** *Kaz.* Zashaghan. Zapadnyy Kazakhstan, NW Kazakhstan 51°04′N 51°13′E
115 D20 **Zacháro** *var.* Zaharo, Zakháro. Dytikí Elláda, S Greece 37°29′N 21°40′E
22 J8 **Zachary** Louisiana, S USA 30°39′N 91°09′W
117 U6 **Zachepylivka** Kharkivs'ka Oblast', E Ukraine 49°13′N 35°15′E
Zachist'ye *see* Zachystsye
110 E9 **Zachodnio-pomorskie** ◈ *province* NW Poland
119 L14 **Zachystsye** *Rus.* Zachist'ye. Minskaya Voblasts', NW Belarus 54°24′N 28°45′E
40 L13 **Zacoalco** *var.* Zacoalco de Torres. Jalisco, SW Mexico 20°14′N 103°33′W
Zacoalco de Torres *see* Zacoalco
41 P13 **Zacualtipán** Hidalgo, C Mexico 20°39′N 98°42′W
112 C12 **Zadar** *It.* Zara; *anc.* Iader. Zadar, SW Croatia 44°07′N 15°15′E
112 C12 **Zadar** *off.* Zadarsko-Kninska Županija, Zadar-Knin. ◈ *province* SW Croatia
Zadar-Knin *see* Zadar
Zadarsko-Kninska Županija *see* Zadar
166 M14 **Zadetkyi Kyun** *var.* St.Matthew's Island. *island* Mergui Archipelago, S Myanmar (Burma)
67 Q9 **Zadié** *var.* Djadié. ≈ NE Gabon
159 Q13 **Zadoi** *var.* Qapugtang. Qinghai, C China 32°56′N 95°21′E
126 L7 **Zadonsk** Lipetskaya Oblast', W Russian Federation 52°25′N 38°55′E
Za'farâna *see* Za'farānah
75 X8 **Za'farānah** *var.* Za'farâna. E Egypt 29°06′N 32°34′E
149 W7 **Zafarwāl** Punjab, E Pakistan 32°20′N 74°53′E
121 Q1 **Zafer Burnu** *var.* Cape Andreas, Cape Apostolas Andreas, *Gk.* Akrotíri Apostólou Andréa. *cape* NE Cyprus
107 J23 **Zafferano, Capo** *headland* Sicilia, Italy, C Mediterranean Sea 38°06′N 13°31′E
114 M7 **Zafirovo** Silistra, NE Bulgaria 44°00′N 26°51′E
Záfora *see* Sofraná
104 J12 **Zafra** Extremadura, W Spain 38°25′N 06°27′W
110 E13 **Żagań** *var.* Zagań, Żegań, *Ger.* Sagan. Lubuskie, W Poland 51°37′N 15°20′E
118 F10 **Žagarė** *Pol.* Žagory. Šiaulai, N Lithuania 56°22′N 23°16′E
Zagazig *see* Az Zaqāzīq
74 M5 **Zaghouan** *var.* Zaghwān. NE Tunisia 36°26′N 10°05′E
Zaghwān *see* Zaghouan
115 G16 **Zagorá** Thessalía, C Greece 39°27′N 23°06′E
Zagorod'ye *see* Zaharoddzye
Žagory *see* Žagarė
Zágráb *see* Zagreb
112 E8 **Zagreb** *Ger.* Agram, *Hung.* Zágráb. ● (Croatia) Zagreb, N Croatia 45°48′N 15°58′E
112 E8 **Zagreb** *prev.* Grad Zagreb. ◈ *province* N Croatia

◆ Country
● Country Capital
◇ Dependent Territory
○ Dependent Territory Capital
◈ Administrative Regions
✈ International Airport
▲ Mountain
▲▲ Mountain Range
⛰ Volcano
≈ River
◎ Lake
⊠ Reservoir

142 L7 **Zāgros, Kūhhā-ye** *Eng.* Zagros Mountains. ▲▲ W Iran
Zagros Mountains *see* Zāgros, Kūhhā-ye
112 O12 **Žagubica** Serbia, E Serbia 44°13′N 21°47′E
Zagunao *see* Lixian
111 L22 **Zagyva** ≈ N Hungary
Zaharo *see* Zacháro
119 G19 **Zaharoddzye** *Rus.* Zagorod'ye. *physical region* SW Belarus
143 W11 **Zāhedān** *var.* Zahidan; *prev.* Duzdab. Sīstān va Balūchestān, SE Iran 29°31′N 60°51′E
Zahidan *see* Zāhedān
Zaḩlah *see* Zahlé
138 H7 **Zahlé** *var.* Zaḩlah. C Lebanon 33°51′N 35°54′E
146 J14 **Zähmet** *Rus.* Zakhmet. Mary Welaýaty, C Turkmenistan 37°48′N 62°33′E
111 O20 **Záhony** Szabolcs-Szatmár-Bereg, NE Hungary 48°26′N 22°11′E
141 N13 **Z̧ahrān** 'Asīr, S Saudi Arabia 17°48′N 43°28′E
139 R12 **Z̧ahrat al Baţn** *hill range* S Iraq
120 H11 **Zahrez Chergui** *var.* Zahrez Chergúi. *marsh* N Algeria
Zainlha *see* Xinjin
127 S4 **Zainsk** Respublika Tatarstan, W Russian Federation 55°12′N 52°01′E
82 A10 **Zaire** *prev.* Congo. ◈ *province* NW Angola
Zaire *see* Congo (river)
Zaire *see* Congo (Democratic Republic of)
112 P13 **Zaječar** Serbia, E Serbia 43°54′N 22°16′E
83 L18 **Zaka** Masvingo, E Zimbabwe 20°20′S 31°29′E
122 M14 **Zakamensk** Respublika Buryatiya, S Russian Federation 50°18′N 102°57′E
116 G7 **Zakarpats'ka Oblast'** *Eng.* Transcarpathian Oblast, *Rus.* Zakarpatskaya Oblast'. ◈ *province* W Ukraine
Zakarpatskaya Oblast' *see* Zakarpats'ka Oblast'
Zakataly *see* Zaqatala
Zakháro *see* Zacháro
Zakhidnyy Buh/Zakhodni Buh *see* Bug
Zakhmet *see* Zähmet
Zākhō *see* Zaxo
Zākhū *see* Zaxo
Zákinthos *see* Zákynthos
111 L18 **Zakopane** Małopolskie, S Poland 49°17′N 19°57′E
78 J12 **Zakouma** Salamat, S Chad 10°47′N 19°51′E
115 L25 **Zákros** Kríti, Greece, E Mediterranean Sea 35°06′N 26°12′E
115 C19 **Zákynthos** *var.* Zákinthos. Zákynthos, W Greece 37°47′N 20°54′E
115 C20 **Zákynthos** *var.* Zákinthos, *It.* Zante. *island* Iónia Nísoi, Greece, C Mediterranean Sea
115 C19 **Zakýnthou, Porthmós** *strait* SW Greece
111 G24 **Zala** *off.* Zala Megye. ◈ *county* W Hungary
111 G24 **Zala** ≈ W Hungary
138 M4 **Zalābiyah** Dayr az Zawr, C Syria 35°39′N 39°51′E
111 G24 **Zalaegerszeg** Zala, W Hungary 46°51′N 16°49′E
104 K11 **Zalamea de la Serena** Extremadura, W Spain 38°38′N 05°37′W
104 J13 **Zalamea la Real** Andalucía, S Spain 37°41′N 06°40′W
Zala Megye *see* Zala
163 U7 **Zalantun** *var.* Butha Qi. Nei Mongol Zizhiqu, N China 47°58′N 122°44′E
111 G23 **Zalaszentgrót** Zala, SW Hungary 46°57′N 17°05′E
Zalatna *see* Zlatna
116 G9 **Zalău** *Ger.* Waltenberg, *Hung.* Zilah; *prev. Ger.* Zillenmarkt. Sălaj, NW Romania 47°11′N 23°03′E
109 V10 **Žalec** *Ger.* Sachsenfeld. C Slovenia 46°15′N 15°08′E
110 K8 **Zalewo** *Ger.* Saalfeld. Warmińsko-Mazurskie, NE Poland 53°54′N 19°39′E
141 N9 **Z̧alim** Makkah, W Saudi Arabia 22°46′N 42°12′E
80 A11 **Zalingei** *var.* Zalinje. Central Darfur, W Sudan 12°51′N 23°29′E
Zalinje *see* Zalingei
116 K7 **Zalishchyky** Ternopil's'ka Oblast', W Ukraine 48°40′N 25°43′E
Zallah *see* Zillah
`Załni Pjašaci *see* Zlatni Pyasatsi
98 J13 **Zaltbommel** Gelderland, C Netherlands 51°49′N 05°15′E
124 H15 **Zaluch'ye** Novgorodskaya Oblast', NW Russian Federation 57°40′N 31°45′E
Zamak *see* Zamakh
141 Q14 **Zamakh** *var.* Zamak. N Yemen 16°26′N 47°35′E
136 K15 **Zamantı Irmağı** ≈ C Turkey
Zambesi/Zambeze *see* Zambezi
83 G14 **Zambezi** North Western, W Zambia 13°34′S 23°08′E
83 K15 **Zambezi** *var.* Zambesi, *Port.* Zambeze. ≈ S Africa
83 O15 **Zambézia** *off.* Província da Zambézia. ◈ *province* C Mozambique
Zambézia, Província da *see* Zambézia
83 I14 **Zambia** *off.* Republic of Zambia; *prev.* Northern Rhodesia. ◆ *republic* S Africa
Zambia, Republic of *see* Zambia
171 O8 **Zamboanga** *off.* Zamboanga City. Mindanao, S Philippines 06°56′N 122°03′E
Zamboanga City *see* Zamboanga
54 E5 **Zambrano** Bolívar, N Colombia 09°45′N 74°50′W
110 N10 **Zambrów** Łomża, E Poland 52°59′N 22°14′E
83 L14 **Zâmbuè** Tete, NW Mozambique 15°03′S 30°49′E
77 T13 **Zamfara** ≈ NW Nigeria
Zamkog *see* Zamtang
56 C9 **Zamora** Zamora Chinchipe, S Ecuador 04°04′S 78°52′W
104 K6 **Zamora** Castilla y León, NW Spain 41°30′N 05°45′W
104 K5 **Zamora** ◈ *province* Castilla y León, NW Spain
Zamora *see* Barinas
56 A13 **Zamora Chinchipe** ◈ *province* S Ecuador
40 M13 **Zamora de Hidalgo** Michoacán, SW Mexico 20°N 102°18′W
111 P15 **Zamość** *Rus.* Zamoste. Lubelskie, E Poland 50°44′N 23°16′E
Zamoste *see* Zamość
160 G7 **Zamtang** *var.* Zamkog; *prev.* Gamba. Sichuan, C China 32°19′N 100°55′E
75 O8 **Zamzam, Wādī** *dry watercourse* NW Libya
79 F20 **Zanaga** Lékoumou, S Congo 02°50′S 13°53′E
41 T16 **Zanatepec** Oaxaca, SE Mexico 16°28′N 94°24′W
105 P9 **Záncara** ≈ C Spain
Zancle *see* Messina
158 G14 **Zanda** Xizang Zizhiqu, W China 31°29′N 79°50′E
98 H10 **Zandvoort** Noord-Holland, W Netherlands 52°22′N 04°31′E
39 P8 **Zane Hills** *hill range* Alaska, USA
31 T13 **Zanesville** Ohio, N USA 39°55′N 82°02′W
Zanga *see* Hrazdan
Zangoza *see* Sangüesa
142 L4 **Zanjān** *var.* Zenjan, Zinjan. Zanjān, NW Iran 36°40′N 48°30′E
142 L4 **Zanjān** *off.* Ostān-e Zanjān, *var.* Zenjan, Zinjan. ◈ *province* NW Iran
Zanjān, Ostān-e *see* Zanjān
Zante *see* Zákynthos
81 J22 **Zanzibar** Zanzibar, E Tanzania 06°10′S 39°12′E
81 J22 **Zanzibar** ◈ *region* E Tanzania
81 J22 **Zanzibar** *Swa.* Unguja. *island* E Tanzania
81 J22 **Zanzibar Channel** *channel* E Tanzania
161 N8 **Zaoyang** Hubei, C China 32°10′N 112°45′E
165 P10 **Zaō-zan** ▲ Honshū, C Japan 38°06′N 140°27′E
124 J2 **Zaozërsk** Murmanskaya Oblast', NW Russian Federation 69°25′N 32°25′E
161 Q6 **Zaozhuang** Shandong, E China 34°53′N 117°38′E
28 L4 **Zap** North Dakota, N USA 47°18′N 101°55′W
112 L13 **Zapadna Morava** *Ger.* Westliche Morava. ≈ C Serbia
124 H16 **Zapadnaya Dvina** Tverskaya Oblast', W Russian Federation 56°17′N 32°03′E
Zapadnaya Dvina *see* Western Dvina
Zapadno-Kazakhstanskaya Oblast' *see* Zapadnyy Kazakhstan
122 I9 **Zapadno-Sibirskaya Ravnina** *Eng.* West Siberian Plain. *plain* C Russian Federation
Zapadnyy Bug *see* Bug
144 E9 **Zapadnyy Kazakhstan** *off.* Zapadno-Kazakhstanskaya Oblast', *Eng.* West Kazakhstan, *Kaz.* Batys Qazaqstan Oblysy; *prev.* Ural'skaya Oblast'. ◈ *province* NW Kazakhstan
122 K13 **Zapadnyy Sayan** *Eng.* Western Sayans. ▲▲ S Russian Federation
63 H15 **Zapala** Neuquén, W Argentina 38°54′S 70°06′W
62 I4 **Zapaleri, Cerro** *var.* Cerro Sapaleri. ▲ N Chile 22°51′S 67°10′W
25 Q16 **Zapata** Texas, SW USA 26°57′N 99°17′W
44 D5 **Zapata, Península de** *peninsula* W Cuba
61 G19 **Zapicán** Lavalleja, S Uruguay 33°31′S 54°55′W
65 J19 **Zapiola Ridge** *undersea feature* SW Atlantic Ocean
65 L19 **Zapiola Seamount** *undersea feature* S Atlantic Ocean 38°15′S 26°15′W
124 I2 **Zapolyarnyy** Murmanskaya Oblast', NW Russian Federation 69°24′N 30°53′E
117 U8 **Zaporizhzhya** *Rus.* Zaporozh'ye; *prev.* Aleksandrovsk. Zaporiz'ka Oblast', SE Ukraine 47°47′N 35°12′E
Zaporizhzhya *see* Zaporiz'ka Oblast'
117 U9 **Zaporiz'ka Oblast'** *var.* Zaporizhzhya, *Rus.* Zaporozhskaya Oblast'. ◈ *province* SE Ukraine
Zaporozhskaya Oblast' *see* Zaporiz'ka Oblast'
Zaporozh'ye *see* Zaporizhzhya
40 L14 **Zapotiltic** Jalisco, SW Mexico 19°40′N 103°29′W
158 G13 **Zapug** Xizang Zizhiqu, W China
137 V10 **Zaqatala** *Rus.* Zakataly. NW Azerbaijan 41°38′N 46°38′E
159 P13 **Zaqên** Qinghai, W China 33°23′N 94°31′E
159 Q13 **Za Qu** ≈ C China
136 M13 **Zara** Sivas, C Turkey 39°55′N 37°44′E
Zara *see* Zadar
Zarafshan *see* Zarafshon
147 P12 **Zarafshon** *Rus.* Zeravshan. W Tajikistan 39°12′N 68°36′E
146 L9 **Zarafshon** *Rus.* Zarafshan. Navoiy Viloyati, N Uzbekistan 41°33′N 64°09′E
Zarafshon *see* Zeravshan
147 O12 **Zarafshon, Qatorkŭhi** *Rus.* Zeravshanskiy Khrebet, *Uzb.* Zarafshon Tizmasi. ▲▲ Tajikistan/Uzbekistan
Zarafshon Tizmasi *see* Zarafshon, Qatorkŭhi
54 E7 **Zaragoza** Antioquia, N Colombia 07°30′N 74°52′W
40 I5 **Zaragoza** Chihuahua, N Mexico 29°36′N 107°41′W
41 N6 **Zaragoza** Coahuila, NE Mexico 28°31′N 100°54′W
41 O10 **Zaragoza** Nuevo León, NE Mexico 23°59′N 99°49′W
105 R5 **Zaragoza** *Eng.* Saragossa; *anc.* Caesaraugusta, Salduba. Aragón, NE Spain 41°39′N 00°54′W
105 R6 **Zaragoza** ◈ *province* Aragón, NE Spain
105 R5 **Zaragoza** ✈ Aragón, NE Spain 41°38′N 00°53′W
143 S10 **Zarand** Kermān, C Iran 30°50′N 56°35′E
148 J9 **Zaranj** Nīmrōz, SW Afghanistan 30°59′N 61°54′E
118 I11 **Zarasai** Utena, E Lithuania 55°44′N 26°17′E
62 N12 **Zárate** *prev.* General José F.Uriburu. Buenos Aires, E Argentina 34°06′S 59°03′W
105 Q2 **Zarautz** *var.* Zarauz. País Vasco, N Spain 43°17′N 02°10′W
Zarauz *see* Zarautz
Zaravecchia *see* Biograd na Moru
Zarāyīn *see* Zarēn
126 L4 **Zaraysk** Moskovskaya Oblast', W Russian Federation 54°48′N 38°54′E
55 N6 **Zaraza** Guárico, N Venezuela 09°23′N 65°20′W
Zarbdar *see* Zarbdor
147 P11 **Zarbdor** *Rus.* Zarbdar. Jizzax Viloyati, C Uzbekistan 40°04′N 68°10′E
142 M8 **Zard Kūh** ▲ SW Iran 32°19′N 50°03′E
124 I5 **Zarechensk** Murmanskaya Oblast', NW Russian Federation 66°39′N 31°27′E
127 P6 **Zarechnyy** Penzenskaya Oblast', W Russian Federation 53°12′N 45°12′E
Zareh Sharan *see* Sharan
39 Y14 **Zarembo Island** *island* Alexander Archipelago, Alaska, USA
139 V4 **Zarēn** *var.* Zarāyīn. As Sulaymānīyah, E Iraq 35°16′N 45°43′E
149 Q7 **Zarghūn Shahr** *var.* Katawaz. Paktīkā, SE Afghanistan 32°40′N 68°20′E
77 V13 **Zaria** Kaduna, C Nigeria 11°06′N 07°42′E
116 K2 **Zarichne** Rivnens'ka Oblast', NW Ukraine 51°49′N 26°09′E
122 J13 **Zarinsk** Altayskiy Kray, S Russian Federation 53°34′N 85°22′E
116 J12 **Zărneşti** *Hung.* Zernest. Braşov, C Romania 45°34′N 25°18′E
115 J25 **Zarós** Kríti, Greece, E Mediterranean Sea 35°08′N 24°54′E
100 O9 **Zarow** ≈ NE Germany
Zarqa *see* Az Zarqā'
Zarqa/Muḩāfazat az Zarqā' *see* Az Zarqā'
111 G20 **Záruby** ▲ W Slovakia 48°30′N 17°24′E
56 B8 **Zaruma** El Oro, SW Ecuador 03°46′S 79°38′W
110 E13 **Żary** *Ger.* Sorau, Sorau in der Niederlausitz. Lubuskie, W Poland 51°44′N 15°09′E
54 D10 **Zarzal** Valle del Cauca, W Colombia 04°24′N 76°01′W
42 I7 **Zarzalar, Cerro** ▲ S Honduras 14°15′N 86°49′W
Zashaghan *see* Zachagansk
152 I5 **Zāskār** ≈ NE India
152 I5 **Zāskār Range** ▲▲ NE India
Zaslavl' *see* Zaslawye
119 K15 **Zaslawye** *Rus.* Zaslavl'. Minskaya Voblasts', C Belarus 54°01′N 27°16′E
116 K7 **Zastavna** Chernivets'ka Oblast', W Ukraine 48°30′N 25°51′E
111 B16 **Žatec** *Ger.* Saaz. Ústecký Kraj, NW Czech Republic 50°20′N 13°35′E
Zaumgarten *see* Chrzanów
Zaunguzskiye Garagumy *see* Üngüz Angyrsyndaky Garagum
25 X9 **Zavalla** Texas, SW USA 31°09′N 94°25′W
99 H18 **Zaventem** Vlaams Brabant, C Belgium 50°53′N 04°28′E
99 H18 **Zaventem** ✈ (Brussel/Bruxelles) Vlaams Brabant, C Belgium 50°55′N 04°28′E
Zavertse *see* Zawiercie
114 L7 **Zavet** Razgrad, NE Bulgaria 43°46′N 26°40′E
127 O12 **Zavetnoye** Rostovskaya Oblast', SW Russian Federation 47°10′N 43°54′E
156 M3 **Zavhan Gol** ≈ W Mongolia
112 H12 **Zavidovići** Federacija Bosne I Hercegovine, N Bosnia and Herzegovina 44°26′N 18°07′E
123 R13 **Zavitinsk** Amurskaya Oblast', SE Russian Federation 50°23′N 128°57′E
Zawia *see* Az Zāwiyah
111 K15 **Zawiercie** *Rus.* Zavertse. Śląskie, S Poland 50°29′N 19°24′E
75 P11 **Zawīlah** *var.* Zuwaylah, *It.* Zueila. C Libya 26°10′N 15°07′E
138 I4 **Zāwīyah, Jabal az** ▲▲ NW Syria
139 Q1 **Zaxo** *Ar.* Zākhū, *var.* Zākhō. Dahūk, N Iraq 37°09′N 42°40′E
109 Y3 **Zaya** ≈ NE Austria
166 M8 **Zayatkyi** Bago, C Myanmar (Burma) 17°48′N 96°27′E
145 Y11 **Zaysan** Vostochnyy Kazakhstan, E Kazakhstan 47°30′N 84°55′E
Zaysan Köl *see* Zaysan, Ozero
145 Y11 **Zaysan, Ozero** *Kaz.* Zaysan Köl. ◎ E Kazakhstan
159 R16 **Zayü** *var.* Gyigang. Xizang Zizhiqu, W China 28°36′N 97°25′E
Zayyq *see* Zhayyk
44 F6 **Zaza** ≈ C Cuba
116 K5 **Zbarazh** Ternopil's'ka Oblast', W Ukraine 49°40′N 25°47′E
116 J5 **Zboriv** Ternopil's'ka Oblast', W Ukraine 49°40′N 25°07′E
111 F18 **Zbraslav** Jihomoravský Kraj, SE Czech Republic 49°13′N 16°19′E
116 K6 **Zbruch** ≈ W Ukraine
Žd'ár *see* Žd'ár nad Sázavou
111 F17 **Žd'ár nad Sázavou** *Ger.* Saar in Mähren; *prev.* Žd'ár. Vysočina, C Czech Republic 49°34′N 16°00′E
116 K4 **Zdolbuniv** *Pol.* Zdolbunów, *Rus.* Zdolbunov. Rivnens'ka Oblast', NW Ukraine 50°33′N 26°15′E
Zdolbunov/Zdolbunów *see* Zdolbuniv
110 J13 **Zduńska Wola** Sieradz, C Poland 51°37′N 18°57′E
117 O4 **Zdvizh** ≈ N Ukraine
Zdzięciół *see* Dzyatlava
111 I16 **Zdzieszowice** *Ger.* Odertal. Opolskie, SW Poland 50°24′N 18°06′E
Zealand *see* Sjælland
188 K6 **Zealandia Bank** *undersea feature* C Pacific Ocean
63 H20 **Zeballos, Monte** ▲ S Argentina 47°04′S 71°32′W
83 K20 **Zebediela** Limpopo, NE South Africa 24°16′S 29°21′E
113 L18 **Zebës, Mali i** *var.* Mali i Zebës. ▲ NE Albania 41°57′N 20°16′E
Zebës, Mali i *see* Zebës, Mali i
21 V9 **Zebulon** North Carolina, SE USA 35°49′N 78°19′W
112 K8 **Žednik** *Hung.* Bácsjózseffalva. Vojvodina, N Serbia 45°58′N 19°40′E
99 C15 **Zeebrugge** West-Vlaanderen, NW Belgium 51°20′N 03°13′E
183 N16 **Zeehan** Tasmania, SE Australia 41°54′S 145°19′E
99 L14 **Zeeland** Noord-Brabant, SE Netherlands 51°42′N 05°40′E
29 N7 **Zeeland** North Dakota, N USA 45°57′N 99°49′W
99 E14 **Zeeland** ◈ *province* SW Netherlands
83 I21 **Zeerust** North-West, N South Africa 25°33′S 26°06′E
98 K10 **Zeewolde** Flevoland, C Netherlands 52°20′N 05°32′E
Zefat *see* Tsefat
Żegań *see* Żagań
Zehden *see* Cedynia
100 O11 **Zehdenick** Brandenburg, NE Germany 52°58′N 13°19′E
Zē-i Bādīnān *see* Great Zab
Zeiden *see* Codlea
146 M14 **Zeidskoye Vodokhranilishche** ⊠ E Turkmenistan
Zē-i Kōya *see* Little Zab
181 P7 **Zeil, Mount** ▲ Northern Territory, C Australia 23°31′S 132°41′E
98 J12 **Zeist** Utrecht, C Netherlands 52°05′N 05°15′E
101 M16 **Zeitz** Sachsen-Anhalt, E Germany 51°03′N 12°08′E
159 T11 **Zêkog** *var.* Zequ; *prev.* Sonag. Qinghai, C China 35°03′N 101°30′E
Zelaya Norte *see* Atlántico Norte, Región Autónoma
Zelaya Sur *see* Atlántico Sur, Región Autónoma
99 F17 **Zele** Oost-Vlaanderen, NW Belgium 51°04′N 04°02′E
110 N12 **Żelechów** Lubelskie, E Poland 51°49′N 21°57′E
113 H14 **Zelena Glava** ▲ SE Bosnia and Herzegovina 43°32′N 17°55′E
113 P20 **Zelen Breg** ▲ S Macedonia 41°10′N 22°14′E
113 I14 **Zelengora** ▲▲ S Bosnia and Herzegovina
124 I5 **Zelenoborskiy** Murmanskaya Oblast', NW Russian Federation 66°52′N 32°25′E
127 R3 **Zelenodol'sk** Respublika Tatarstan, W Russian Federation 55°52′N 48°49′E
117 S9 **Zelenodols'k** Dnipropetrovs'ka Oblast', E Ukraine 47°38′N 33°41′E
122 J12 **Zelenogorsk** Krasnoyarskiy Kray, C Russian Federation 56°08′N 94°29′E
126 K3 **Zelenograd** Moskovskaya Oblast', W Russian Federation 56°02′N 37°08′E
118 B13 **Zelenogradsk** *Ger.* Cranz, Kranz. Kaliningradskaya Oblast', W Russian Federation 54°58′N 20°30′E
127 O15 **Zelenokumsk** Stavropol'skiy Kray, SW Russian Federation 44°22′N 43°48′E
165 X4 **Zelënyy, Ostrov** *var.* Shibotsu-jima. *island* NE Russian Federation
Železna Kapela *see* Eisenkappel
Železna Vrata *see* Demir Kapija
112 L11 **Železniki** Serbia, N Serbia 44°45′N 20°23′E
98 N12 **Zelhem** Gelderland, E Netherlands 52°00′N 06°21′E
113 N18 **Želino** NW FYR Macedonia 42°00′N 21°06′E
113 M14 **Željin** ▲▲ C Serbia
101 K17 **Zella-Mehlis** Thüringen, C Germany 50°40′N 10°40′E
109 P7 **Zell am See** *var.* Zell-am-See. Salzburg, S Austria 47°19′N 12°47′E
Zell-am-See *see* Zell am See
109 N7 **Zell am Ziller** Tirol, W Austria 47°13′N 11°52′E
Zelle *see* Celle
109 W2 **Zellerndorf** Niederösterreich, NE Austria 48°40′N 15°57′E
109 U7 **Zeltweg** Steiermark, S Austria 47°12′N 14°46′E
119 G17 **Zel'va** *Pol.* Zelwa. Hrodzyenskaya Voblasts', W Belarus 53°09′N 24°49′E
118 H13 **Želva** Vilnius, C Lithuania 55°13′N 25°07′E
Zelwa *see* Zel'va
99 E16 **Zelzate** *var.* Selzaete. Oost-Vlaanderen, NW Belgium 51°12′N 03°49′E
118 E11 **Žemaičių Aukštumas** *physical region* W Lithuania
118 C12 **Žemaičių Naumiestis** Klaipėda, SW Lithuania 55°22′N 21°39′E
119 L14 **Zembin** *var.* Zyembin. Minskaya Voblasts', C Belarus 54°22′N 28°13′E
127 N6 **Zemetchino** Penzenskaya Oblast', W Russian Federation 53°31′N 42°35′E
79 M15 **Zémio** Haut-Mbomou, E Central African Republic 05°04′N 25°07′E
41 R16 **Zempoaltepec, Cerro** ▲ SE Mexico 17°04′N 95°54′W
99 G17 **Zemst** Vlaams Brabant, C Belgium 50°59′N 04°28′E
112 L11 **Zemun** Serbia, N Serbia 44°52′N 20°25′E
Zendajan *see* Zindah Jān
Zendeh Jān *see* Zindah Jān
Zengg *see* Senj
112 H12 **Zenica** Federacija Bosne I Hercegovine, C Bosnia and Herzegovina 44°12′N 17°53′E
Zenjan *see* Zanjān
Zen'kov *see* Zin'kiv
Zenshū *see* Jeonju
Zenta *see* Senta
82 B11 **Zenza do Itombe** Kwanza Norte, NW Angola 09°22′S 14°10′E
112 H12 **Žepče** Federacija Bosne I Hercegovine, N Bosnia and Herzegovina 44°26′N 18°00′E
23 W12 **Zephyrhills** Florida, SE USA 28°13′N 82°10′W
158 F9 **Zepu** *var.* Poskam. Xinjiang Uygur Zizhiqu, NW China 38°10′N 77°18′E
Zequ *see* Zêkog
147 Q12 **Zeravshan** *Taj./Uzb.* Zarafshon. ≈ Tajikistan/Uzbekistan
Zeravshan *see* Zarafshon
Zeravshanskiy Khrebet *see* Zarafshon, Qatorkŭhi
101 M14 **Zerbst** Sachsen-Anhalt, E Germany 51°57′N 12°05′E
Zerenda *see* Zerendy
145 P8 **Zerendy** *prev.* Zerenda. Akmola, N Kazakhstan 52°56′N 69°09′E
110 H12 **Żerków** Wielkopolskie, C Poland 52°03′N 17°33′E
108 E11 **Zermatt** Valais, SW Switzerland 46°00′N 07°45′E
Zernest *see* Zărneşti
108 J9 **Zernez** Graubünden, SE Switzerland 46°42′N 10°06′E
126 L12 **Zernograd** Rostovskaya Oblast', SW Russian Federation 46°52′N 40°13′E
Zestafoni *see* Zest'aponi
137 S9 **Zest'aponi** *Rus.* Zestafoni; *prev.* Zestap'oni. C Georgia 42°09′N 43°00′E
Zestap'oni *see* Zest'aponi
98 H12 **Zestienhoven** ✈ (Rotterdam) Zuid-Holland, SW Netherlands 51°57′N 04°30′E
113 J16 **Zeta** ≈ C Montenegro
8 L6 **Zeta Lake** ◎ Victoria Island, Northwest Territories, N Canada
98 L12 **Zetten** Gelderland, SE Netherlands 51°55′N 05°43′E
101 M17 **Zeulenroda** Thüringen, C Germany 50°40′N 11°58′E
100 H10 **Zeven** Niedersachsen, NW Germany 53°17′N 09°16′E
98 M12 **Zevenaar** Gelderland, SE Netherlands 51°55′N 06°05′E
99 H14 **Zevenbergen** Noord-Brabant, S Netherlands 51°39′N 04°36′E
129 X6 **Zeya** ≈ SE Russian Federation
Zeya Reservoir *see* Zeyskoye Vodokhranilishche
143 T11 **Zeynalābād** Kermān, C Iran 29°56′N 57°29′E
123 R12 **Zeyskoye Vodokhranilishche** *Eng.* Zeya Reservoir. ⊠ SE Russian Federation
104 H8 **Zêzere, Rio** ≈ C Portugal
Zgerzh *see* Zgierz
138 H6 **Zgharta** N Lebanon 34°24′N 35°54′E
110 K12 **Zgierz** *Ger.* Neuhof, *Rus.* Zgerzh. Łódź, C Poland 51°55′N 19°20′E
111 E14 **Zgorzelec** *Ger.* Görlitz. Dolnośląskie, SW Poland 51°10′N 15°E
119 F19 **Zhabinka** *Pol.* Żabinka. Brestskaya Voblasts', SW Belarus 52°12′N 24°01′E
Zhaggo *see* Luhuo
159 R15 **Zhag'yab** *var.* Yêndum. Xizang Zizhiqu, W China 30°42′N 97°33′E
Zhailma *see* Zhayylma
145 N14 **Zhalagash** *prev.* Dzhalagash. Kzylorda, S Kazakhstan 45°06′N 64°40′E
145 V16 **Zhalanash** Almaty, SE Kazakhstan 43°04′N 78°08′E
145 S7 **Zhalauly, Ozero** ◎ NE Kazakhstan
145 W10 **Zhalgyztobe** *prev.* Zhangiztobe. Vostochnyy Kazakhstan, E Kazakhstan 49°15′N 81°16′E
144 E9 **Zhalpaktal** *Kaz.* Zhalpaqtal; *prev.* Furmanovo. Zapadnyy Kazakhstan, W Kazakhstan 49°43′N 49°28′E
Zhalpaqtal *see* Zhalpaktal
119 G16 **Zhaludok** *Rus.* Zheludok. Hrodzyenskaya Voblasts', W Belarus 53°36′N 24°59′E
Zhaman-Akkol', Ozero *see* Akkol', Ozero
145 Q14 **Zhambyl** *off.* Zhambylskaya Oblast', *Kaz.* Zhambyl Oblysy; *prev.* Dzhambulskaya Oblast'. ◈ *province* S Kazakhstan
Zhambyl *see* Taraz
Zhambyl Oblysy/Zhambylskaya Oblast' *see* Zhambyl
Zhamo *see* Bomi
145 S12 **Zhamshy** ≈ C Kazakhstan
144 M15 **Zhanadariya** *prev.* Zhanadar'ya. Kzylorda, S Kazakhstan 44°41′N 64°39′E
Zhanadar'ya *see* Zhanadariya
144 E10 **Zhanakazan** *prev.* Novaya Kazanka. Zapadnyy Kazakhstan, W Kazakhstan 48°57′N 49°34′E
145 O15 **Zhanakorgan** *Kaz.* Zhangaqorghan. Kzylorda, S Kazakhstan 43°57′N 67°14′E
159 N16 **Zhanang** *var.* Chatang. Xizang Zizhiqu, W China 29°15′N 91°20′E
145 T12 **Zhanaortalyk** Karaganda, C Kazakhstan 47°31′N 75°42′E
144 F15 **Zhanaozen** *Kaz.* Zhangaözen; *prev.* Novyy Uzen'. Mangistau, W Kazakhstan 43°22′N 52°50′E
145 Q16 **Zhanatas** Zhambyl, S Kazakhstan 43°36′N 69°43′E
Zhangaözen *see* Zhanaozen
Zhangaqazaly *see* Ayteke Bi
Zhangaqorghan *see* Zhanakorgan
161 O2 **Zhangbei** Hebei, E China 41°13′N 114°43′E
Zhang-chia-k'ou *see* Zhangjiakou
Zhangdian *see* Zibo
Zhanggu *see* Danba
163 X9 **Zhangguangcai Ling** ▲▲ NE China
161 S13 **Zhanghua** *Jap.* Shōka; *prev.* Changhua. C Taiwan 24°06′N 120°31′E
Zhangiztobe *see* Zhalgyztobe
159 W11 **Zhangjiachuan** Gansu, N China 34°55′N 106°26′E
160 L10 **Zhangjiajie** *var.* Dayong. Hunan, S China 29°10′N 110°22′E
161 O2 **Zhangjiakou** *var.* Changkiakow, Zhang-chia-k'ou, *Eng.* Kalgan; *prev.* Wanchuan. Hebei, E China 40°48′N 114°51′E
161 Q13 **Zhangping** Fujian, SE China 25°21′N 117°29′E
161 Q13 **Zhangpu** *var.* Sui'an. Fujian, SE China 24°08′N 117°36′E
163 U11 **Zhangwu** Liaoning, NE China 42°21′N 122°32′E
159 S8 **Zhangye** *var.* Ganzhou. Gansu, N China 38°58′N 100°30′E
161 Q13 **Zhangzhou** Fujian, SE China 24°31′N 117°40′E
163 W6 **Zhan He** ≈ NE China
Zhänibek *see* Dzhanibek
160 L16 **Zhanjiang** *var.* Chanchiang, Chan-chiang, *Cant.* Tsamkong, *Fr.* Fort-Bayard. Guangdong, S China 21°10′N 110°20′E
145 V14 **Zhansugirov** *prev.* Dzhansugurov. Almaty, SE Kazakhstan 45°23′N 79°29′E
163 V8 **Zhaodong** Heilongjiang, NE China 46°03′N 125°58′E
Zhaoge *see* Qixian
160 H11 **Zhaojue** *var.* Xincheng. Sichuan, C China 28°03′N 102°50′E
161 N14 **Zhaoqing** Guangdong, S China 23°08′N 112°26′E
Zhaoren *see* Changwu
158 H5 **Zhaosu** *var.* Mongolküre. Xinjiang Uygur Zizhiqu, NW China 43°09′N 81°07′E
160 H11 **Zhaotong** Yunnan, SW China 27°20′N 103°29′E
163 V9 **Zhaoyuan** Heilongjiang, NE China 45°30′N 125°05′E
163 V9 **Zhaozhou** Heilongjiang, NE China 45°42′N 125°11′E
145 X13 **Zharbulak** Vostochnyy Kazakhstan, E Kazakhstan 46°04′N 82°05′E
158 J15 **Zhari Namco** ◎ W China
144 I12 **Zharkamys** *Kaz.* Zharqamys. Aktyubinsk, W Kazakhstan 47°58′N 56°33′E
145 W15 **Zharkent** *prev.* Panfilov. Taldykorgan, SE Kazakhstan 44°10′N 80°01′E
124 H17 **Zharkovskiy** Tverskaya Oblast', W Russian Federation 55°51′N 32°19′E
145 W11 **Zharma** Vostochnyy Kazakhstan, E Kazakhstan 48°48′N 80°55′E
144 F14 **Zharmysh** Mangistau, SW Kazakhstan 44°12′N 52°27′E
Zharqamys *see* Zharkamys
118 L13 **Zhary** Vitsyebskaya Voblasts', N Belarus 55°05′N 28°40′E
Zhaslyk *see* Jasliq
158 J14 **Zhaxi Co** ◎ W China
127 X6 **Zhayyk** *Kaz.* Zayyq, *var.* Ural. ≈ Kazakhstan/Russian Federation
144 L9 **Zhayylma** *prev.* Zhailma. Kostanay, N Kazakhstan 51°34′N 61°39′E
Zhdanov *see* Beyläqan
Zhdanov *see* Mariupol'
Zhe *see* Zhejiang
161 R10 **Zhejiang** *var.* Che-chiang, Chekiang, Zhe, Zhejiang Sheng. ◈ *province* SE China
Zhejiang Sheng *see* Zhejiang
145 S7 **Zhelezinka** Pavlodar, N Kazakhstan 53°35′N 75°16′E
119 C14 **Zheleznodorozhnyy** *Ger.* Gerdauen. Kaliningradskaya Oblast', W Russian Federation 54°21′N 21°17′E
Zheleznodorozhnyy *see* Yemva
122 K12 **Zheleznogorsk** Krasnoyarskiy, C Russian Federation 56°20′N 93°36′E
126 J7 **Zheleznogorsk** Kurskaya Oblast', W Russian Federation 52°22′N 35°21′E
127 N15 **Zheleznovodsk** Stavropol'skiy Kray, SW Russian Federation 44°12′N 43°01′E
Zhëltyye Vody *see* Zhovti Vody
Zheludok *see* Zhaludok
144 H12 **Zhem** *prev.* Emba. ≈ W Kazakhstan
160 K7 **Zhenba** Shaanxi, C China 32°42′N 107°55′E
160 I13 **Zhenfeng** *var.* Mingu. Guizhou, S China 25°27′N 105°38′E
Zhengjiatun *see* Shuangliao
159 X10 **Zhengning** *var.* Shanhe. Gansu, N China 35°29′N 108°21′E
Zhengxiangbai Qi *see* Qagan Nur
161 N6 **Zhengzhou** *var.* Ch'eng-chou, Chengchow; *prev.* Chenghsien. *province capital* Henan, C China 34°45′N 113°38′E
161 R8 **Zhenjiang** *var.* Chenkiang. Jiangsu, E China 32°08′N 119°30′E
163 U9 **Zhenlai** Jilin, NE China 45°52′N 123°11′E
160 I11 **Zhenxiong** Yunnan, SW China 27°31′N 104°52′E
160 K11 **Zhenyuan** *var.* Wuyang. Guizhou, S China 27°07′N 108°33′E
161 R11 **Zherong** *var.* Shuangcheng. Fujian, SE China 27°16′N 119°54′E
145 U15 **Zhetigen** *prev.* Nikolayevka. Almaty, SE Kazakhstan 43°39′N 77°10′E
Zhetiqara *see* Zhitikara
144 F15 **Zhetybay** Mangistau, SW Kazakhstan 43°35′N 52°05′E
145 P17 **Zhetysay** *var.* Dzhetysay. Yuzhnyy Kazakhstan 40°45′N 68°18′E
145 W14 **Zhetysuskiy Alatau** *prev.* Dzhungarskiy Alatau. ▲▲ China/Kazakhstan
160 M11 **Zhexi Shuiku** ⊠ C China
145 O12 **Zhezdy** Karaganda, C Kazakhstan 48°06′N 67°01′E
145 O12 **Zhezkazgan** *Kaz.* Zhezqazghan; *prev.* Dzhezkazgan. Karaganda, C Kazakhstan 47°49′N 67°44′E
Zhezqazghan *see* Zhezkazgan
Zhicheng *see* Yidu
Zhidachov *see* Zhydachiv
159 Q12 **Zhidoi** *var.* Gyaijêpozhanggê. Qinghai, C China 33°55′N 95°39′E
122 M13 **Zhigalovo** Irkutskaya Oblast', S Russian Federation 54°47′N 105°00′E
127 R6 **Zhigulevsk** Samarskaya Oblast', W Russian Federation 53°24′N 49°30′E
118 D13 **Zhilino** *Ger.* Schillen. Kaliningradskaya Oblast', W Russian Federation 54°55′N 21°54′E
Zhiloy, Ostrov *see* Çilov Adası
127 O8 **Zhirnovsk** Volgogradskaya Oblast', SW Russian Federation 51°01′N 44°49′E
160 M12 **Zhishan** *prev.* Yongzhou. Hunan, S China 26°12′N 111°36′E
Zhishan *see* Lingling
144 L8 **Zhitikara** *Kaz.* Zhetiqara; *prev.* Džetygara. Kostanay, NW Kazakhstan 52°14′N 61°12′E
Zhitkovichi *see* Zhytkavichy
Zhitomir *see* Zhytomyr
Zhitomirskaya Oblast' *see* Zhytomyrs'ka Oblast'
126 J5 **Zhizdra** Kaluzhskaya Oblast', W Russian Federation 53°38′N 34°39′E
119 N18 **Zhlobin** Homyel'skaya Voblasts', SE Belarus 52°53′N 30°01′E
Zhmerinka *see* Zhmerynka
116 M7 **Zhmerynka** *Rus.* Zhmerinka. Vinnyts'ka Oblast', C Ukraine 49°00′N 28°02′E
149 R9 **Zhob** *var.* Fort Sandeman. Baluchistān, SW Pakistan 31°21′N 69°31′E
149 R8 **Zhob** ≈ C Pakistan
Zhodino *see* Zhodzina
119 L15 **Zhodzina** *Rus.* Zhodino. Minskaya Voblasts', C Belarus 54°06′N 28°21′E
123 Q5 **Zhokhova, Ostrov** *island* Novosibirskiye Ostrova, NE Russian Federation
Zholkev/Zholkva *see* Zhovkva
Zhondor *see* Jondor
158 I15 **Zhongba** *var.* Tuoji. Xizang Zizhiqu, W China 29°37′N 84°11′E
Zhongba *see* Jiangyou
Zhongdian *see* Xamgyi'nyilha
Zhongduo *see* Youyang
Zhonghe *see* Xiushan
Zhonghua Renmin Gongheguo *see* China
Zhongjian Dao *see* Triton Island
159 V9 **Zhongning** Ningxia, N China 37°26′N 105°40′E
Zhongping *see* Huize
161 N15 **Zhongshan** Guangdong, S China 22°30′N 113°20′E
195 X7 **Zhongshan** *Chinese research station* Antarctica 69°23′S 76°34′E
160 M6 **Zhongtiao Shan** ▲▲ C China
159 V9 **Zhongwei** Ningxia, N China 37°31′N 105°10′E
160 K9 **Zhongxian** *var.* Zhongzhou. Chongqing Shi, C China 30°16′N 108°03′E
161 N9 **Zhongxiang** Hubei, C China 31°12′N 112°35′E
Zhongzhou *see* Zhongxian
144 M14 **Zhosaly** *prev.* Dzhusaly. Kzylorda, SW Kazakhstan 45°29′N 64°04′E
161 O7 **Zhoukou** *var.* Zhoukouzhen. Henan, C China 33°32′N 114°40′E
Zhoukouzhen *see* Zhoukou
161 S9 **Zhoushan** Zhejiang, S China
Zhoushan Islands *see* Zhoushan Qundao
161 S9 **Zhoushan Qundao** *Eng.* Zhoushan Islands. *island group* SE China
116 I5 **Zhovkva** *Pol.* Żółkiew, *Rus.* Zholkev, Zholkva; *prev.* Nesterov. L'vivs'ka Oblast', NW Ukraine 50°04′N 24°E
117 S7 **Zhovti Vody** *Rus.* Zhëltyye Vody. Dnipropetrovs'ka Oblast', E Ukraine 48°24′N 33°30′E
117 Q10 **Zhovtneve** *Rus.* Zhovtnevoye. Mykolayivs'ka Oblast', S Ukraine 46°50′N 32°00′E
Zhovtnevoye *see* Zhovtneve
Zhi Qu *see* Tongtian He
114 K9 **Zhrebchevo, Yazovir** ⊠ C Bulgaria
163 V13 **Zhuanghe** Liaoning, NE China 39°42′N 123°00′E
159 W11 **Zhuanglang** *var.* Shuiluo; *prev.* Shuilocheng. Gansu, C China 35°06′N 106°21′E
145 P15 **Zhuantobe** *Kaz.* Zhūantōbe. Yuzhnyy Kazakhstan, S Kazakhstan 44°45′N 68°50′E
161 Q5 **Zhucheng** Shandong, E China 35°58′N 119°24′E
159 V12 **Zhugqu** Gansu, C China 33°51′N 104°14′E
161 N15 **Zhuhai** Guangdong, S China 22°16′N 113°30′E
Zhuizishan *see* Weichang
Zhuji *see* Shangqiu

◆ Country
● Country Capital
◇ Dependent Territory
○ Dependent Territory Capital
◈ Administrative Regions
✈ International Airport
▲ Mountain
▲▲ Mountain Range
⛰ Volcano
≈ River
◎ Lake
⊠ Reservoir